Scalar, Vector, and Matrix Mathematics

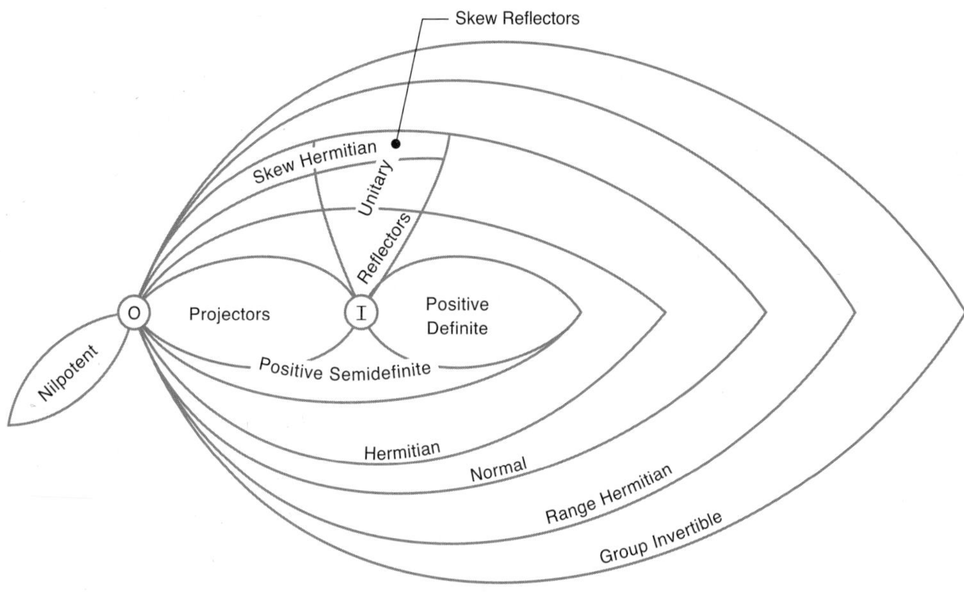
Skew Reflectors
Skew Hermitian
Unitary
Reflectors
O
I
Projectors
Positive
Definite
Nilpotent
Positive Semidefinite
Hermitian
Normal
Range Hermitian
Group Invertible

Scalar, Vector, and Matrix Mathematics

Theory, Facts, and Formulas

Revised and Expanded Edition

Dennis S. Bernstein

PRINCETON UNIVERSITY PRESS
PRINCETON AND OXFORD

Published by Princeton University Press,
41 William Street, Princeton, New Jersey 08540

In the United Kingdom: Princeton University Press,
6 Oxford Street, Woodstock, Oxfordshire, 0X20 1TW

Library of Congress Cataloging-in-Publication Data

Names: Bernstein, Dennis S., 1954– | Bernstein, Dennis S., 1954– Matrix mathematics.
Title: Scalar, vector, and matrix mathematics: theory, facts, and formulas / Dennis S. Bernstein.
Other titles: Matrix mathematics
Description: Revised and expanded edition. | Princeton: Princeton University Press, [2018] | "Revised and expanded edition of Matrix mathematics, retitled Scalar, vector, and matrix mathematics"–Preface. | Includes bibliographical references and index.
Identifiers: LCCN 2017009620 | ISBN 9780691151205 (hardcover: alk. paper) | ISBN 9780691176536 (pbk.)
Subjects: LCSH: Matrices. | Linear systems. | Vector analysis.
Classification: LCC QA188 .B475 2018 | DDC 512.9/434–dc23 LC record available at https://lccn.loc.gov/2017009620

British Library Cataloging-in-Publication Data is available

This book has been composed in Times New Roman and Helvetica.

The publisher would like to acknowledge the author of this volume for providing the camera-ready copy from which this book was printed.

Printed on acid-free paper. ∞

press.princeton.edu

Printed in the United States of America

10 9 8 7 6 5 4 3 2 1

To the memory of my parents,

Irma Shorrie (Hirshon) Bernstein and Milton Bernstein,

whose love and guidance are everlasting

To Susan, with love and gratitude

. . . vessels, unable to contain the great light flowing into them, shatter and break. . . . the remains of the broken vessels fall . . . into the lowest world, where they remain scattered and hidden

— D. W. Menzi and Z. Padeh,
The Tree of Life: Chayyim Vital's Introduction to the Kabbalah of Isaac Luria, Jason Aaronson, Northvale, 1999

Thor . . . placed the horn to his lips . . . He drank with all his might and kept drinking as long as ever he was able; when he paused to look, he could see that the level had sunk a little, . . . for the other end lay out in the ocean itself.

— P. A. Munch, *Norse Mythology*,
AMS Press, New York, 1970

Contents

Preface to the Revised and Expanded Edition

This third edition of *Matrix Mathematics*, retitled *Scalar, Vector, and Matrix Mathematics*, is the culmination of seven years of effort to expand the scope of the second edition of this work. In contrast to the first two editions, which were typeset in Computer Modern, this edition is typeset in New Times Roman, which, as a more compact font, accommodates more material per line. The resulting horizontal compression along with tighter line spacing facilitates one of the goals of this edition, which is to substantially expand the scope of the work to include more scalar and vector mathematics than was envisioned in the original "matrix" area. To this end, this edition includes extensive material on scalar inequalities, graphs, groups, geometry, combinatorics, number theory, finite sums and products, special functions, series, and integrals. As an indication of this augmented scope, the second edition of *Matrix Mathematics* cited 1540 references, whereas the present volume cites 3024.

After three editions and almost three decades of organizational and lexicographical labor, it is perhaps fitting to reflect on mathematics as an artistic and utilitarian endeavor. Mathematics is about the creation of concepts, ideas, and idealizations that are natural, attractive, interesting, and powerful. These attributes may or may not be motivated by physical applications, but what is natural, attractive, interesting, and powerful often turns out to be useful, and vice versa.

Mathematical ideas can be appreciated for either their inner beauty or their usefulness. In science and engineering, idealizations serve as approximations; applying and interpreting these approximations is an art guided by analysis, experience, and insight. The ability to think about real-world phenomena in terms of idealizations is essential to the ability of scientists and engineers to interpret and analyze data. The "gap" between idealizations and reality is the space inhabited by mathematically oriented scientists and engineers, who must reconcile the tangibility of data and reality with the ethereal nature of ideas and concepts.

Beyond approximation, mathematics is about characterization, classification, and connection. Characterization is the elucidation of properties possessed by an object such as a number or a function; classification is the construction of taxonomies and hierarchies of concepts and structures; and connections between concepts and constructions reveal deeper and hidden properties. Among the most surprising results are connections that link seemingly unrelated objects in elegant and unexpected ways. $e^{\pi J} = -1$ is the classic example, but is only one of many that can be found in this book. Each serendipitous connection reinforces the belief that mathematical objects have an independent existence that transcends the ink on the page and the thoughts in our heads.

Mathematics is also about composition and decomposition—putting things together and taking things apart. How can a matrix of one type be factored into matrices of another type? How can matrices of one type be combined to form a matrix of another type? Through its focus on conceptual atoms and their agglomeration into complex molecules, mathematics is the chemistry of ideas.

Mathematics strives for abstraction. However, the original motivation for this book was to minimize abstraction so that a user of mathematics could find a much-needed *result* in a sufficiently concrete form to facilitate its correct application. To achieve concreteness, this book avoids abstract structures such as fields and vector spaces. But abstractions are powerful. A result proved for topological/metric/normed/inner-product/symplectic spaces is valid for *all* such spaces, whether they are defined in terms of scalars, vectors, matrices, or functions in finite or (with suitable restrictions) infinite dimensions. Abstraction provides efficiency of effort, unity of thought, and depth of understanding. With a few exceptions, however, this book intentionally avoids abstraction in order to facilitate the accessibility of the material. But the price paid for this accessibility is to some extent a tunnel view of the larger picture of common ideas and structures. The perfect "handbook of mathematics" would embrace just enough abstraction to unify a huge body of results, and then systematically specialize those results to a multitude of accessible cases that mathematically oriented scientists and engineers might find useful. The perfect handbook remains to be written.

Why devote three decades to writing a book such as this one? The main goal is to provide a convenient resource for users of mathematics. This collection complements existing compendia, with coverage and organization that are unique and, hopefully, useful. An unexpected benefit of collecting and organizing this diverse material is the connections that are uncovered. In addition to connections, this collection reveals gaps in knowledge left for researchers to fill and explore. The inclusion of conjectures and problems is a reminder that much remains to be done. As in all worthwhile endeavors, we have finally reached the beginning.

Acknowledgments

A published review of the second edition of this book aptly suggested that a work of this scope would ideally be the labor of a team of authors. In fact, I have relied heavily on advice and input from many individuals, with 33 acknowledged in the first edition and 44 in the second. For this edition I am indebted to numerous individuals who answered my queries about their books and papers, provided feedback on portions of the manuscript, and contacted me with valuable suggestions. These include Shoshana Abramovich, Khaled Aljanaideh, Ovidiu Bagdasar, Oskar Baksalary, Ravindra Bapat, Sanjay Bhat, JC Bourin, Ryan Caverly, Naveen Crasta, Marco Cuturi, Anton de Ruiter, Ayhan Dil, Justin Edmondson, James Forbes, Daniel Franco, Ovidiu Furdui, Michael Gil, Chris Gilbreth, Ankit Goel, Wassim Haddad, Nicholas Higham, Jesse Hoagg, Matthew Holzel, Qing Hui, Jeffrey Humphery, Gidado-Yisa Immanuel, Fuad Kittaneh, Omran Kouba, Peter Larcombe, Minghua Lin, Zongli Lin, Florian Luca, Victor Moll, Robert Piziak, Olivier Ramare, Ranjan Roy, Sneha Sanjeevini, Meiyue Shao, Joseph Silverman, Alina Sintamarian, Wasin So, Valeriu Soltan, Yongge Tian, Götz Trenkler, Antai Xie, Doron Zeilberger, Fuzhen Zhang, Xuan Zhou, and Limin Zou. I am especially indebted to Oskar Baksalary, Omran Kouba, Minghua Lin, Götz Trenkler, and Yongge Tian for their substantial advice, encouragement, and assistance.

Finally, I take full responsibility for the inevitable errors. I encourage you to inform me of any that you may find, and I will post them on my website.

Dennis S. Bernstein
Ann Arbor, Michigan
dsbaero@umich.edu
November 2017

Preface to the Second Edition

This second edition of *Matrix Mathematics* represents a major expansion of the original work. While the total number of pages is increased 57% from 752 to 1181, the increase is actually greater since this edition is typeset in a smaller font to facilitate a manageable physical size.

The second edition expands on the first edition in several ways. For example, the new version includes material on graphs (developed within the framework of relations and partially ordered sets), as well as alternative partial orderings of matrices, such as rank subtractivity, star, and generalized Löwner. This edition also includes additional material on the Kronecker canonical form and matrix pencils; matrix representations of finite groups; zeros of multi-input, multi-output transfer functions; equalities and inequalities for real and complex numbers; bounds on the roots of polynomials; convex functions; and vector and matrix norms.

The additional material as well as works published subsequent to the first edition increased the number of cited works from 820 to 1540, an increase of 87%. To increase the utility of the bibliography, this edition uses the "back reference" feature of LATEX, which indicates where each reference is cited in the text. As in the first edition, the second edition includes an author index. The expansion of the first edition resulted in an increase in the size of the index from 108 pages to 161 pages.

The first edition included 57 problems, while the current edition has 74. These problems represent extensions or generalizations of known results, sometimes motivated by gaps in the literature.

In this edition, I have attempted to correct all errors that appeared in the first edition. As with the first edition, readers are encouraged to contact me about errors or omissions in the current edition, which I will periodically update on my home page.

Acknowledgments

I am grateful to many individuals who kindly provided advice and material for this edition. Some readers alerted me to errors, while others suggested additional material. In other cases I sought out researchers to help me understand the precise nature of interesting results. At the risk of omitting those who were helpful, I am pleased to acknowledge the following: Mark Balas, Jason Bernstein, Sanjay Bhat, Gerald Bourgeois, Adam Brzezinski, Francesco Bullo, Vijay Chellaboina, Naveena Crasta, Anthony D'Amato, Sever Dragomir, Bojana Drincic, Harry Dym, Matthew Fledderjohn, Haoyun Fu, Masatoshi Fujii, Takayumi Furuta, Steven Gillijns, Rishi Graham, Wassim Haddad, Nicholas Higham, Diederich Hinrichsen, Matthew Holzel, Qing Hui, Masatoshi Ito, Iman Izadi, Pierre Kabamba, Marthe Kassouf, Christopher King, Siddharth Kirtikar, Michael Margliot, Roy Mathias, Peter Mercer, Alex Olshevsky, Paul Otanez, Bela Palancz, Harish Palanthandalam-Madapusi, Fotios Paliogiannis, Isaiah Pantelis, Wei Ren, Ricardo Sanfelice, Mario Santillo, Amit Sanyal, Christoph Schmoeger, Demetrios Serakos, Wasin So, Robert Sullivan, Dogan Sumer, Yongge Tian, Götz Trenkler, Panagiotis Tsiotras, Takeaki Yamazaki, Jin Yan, Masahiro Yanagida, Vera Zeidan, Chenwei Zhang, Fuzhen Zhang, and Qing-Chang Zhong.

As with the first edition, I am especially indebted to my family, who endured four more years of my consistent absence to make this revision a reality. It is clear that any attempt to fully embrace the enormous body of mathematics known as matrix theory is a neverending task. After devoting more than two decades to this project of reassembling the scattered shards, I remain, like Thor, barely able to perceive a dent in the vast knowledge that resides in the hundreds of thousands of pages devoted to this fascinating and incredibly useful subject. Yet, it is my hope that this book will prove to be valuable to everyone who uses matrices, and will inspire interest in a mathematical construction whose secrets and mysteries have no bounds.

Dennis S. Bernstein
Ann Arbor, Michigan
dsbaero@umich.edu
March 2009

Preface to the First Edition

The idea for this book began with the realization that at the heart of the solution to many problems in science, mathematics, and engineering often lies a "matrix fact," that is, an identity, inequality, or property of matrices that is crucial to the solution of the problem. Although there are numerous excellent books on linear algebra and matrix theory, no one book contains all or even most of the vast number of matrix facts that appear throughout the scientific, mathematical, and engineering literature. This book is an attempt to organize many of these facts into a reference source for users of matrix theory in diverse applications areas.

Viewed as an extension of scalar mathematics, matrix mathematics provides the means to manipulate and analyze multidimensional quantities. Matrix mathematics thus provides powerful tools for a broad range of problems in science and engineering. For example, the matrix-based analysis of systems of ordinary differential equations accounts for interaction among all of the state variables. The discretization of partial differential equations by means of finite differences and finite elements yields linear algebraic or differential equations whose matrix structure reflects the nature of physical solutions [2553]. Multivariate probability theory and statistical analysis use matrix methods to represent probability distributions, to compute moments, and to perform linear regression for data analysis [1072, 1276, 1343, 1431, 1923, 2415]. The study of linear differential equations [1401, 1402, 1474] depends heavily on matrix analysis, while linear systems and control theory are matrix-intensive areas of engineering [5, 151, 298, 302, 688, 696, 782, 832, 834, 964, 1070, 1287, 1503, 1714, 1736, 1894, 2257, 2340, 2357, 2451, 2469, 2496, 2726, 2784, 2912, 2999]. In addition, matrices are widely used in rigid body dynamics [57, 1473, 1486, 1578, 1606, 1709, 1968, 2104, 2195, 2196, 2421, 2459, 2530, 2758], structural mechanics [1734, 2010, 2269], computational fluid dynamics [680, 1038, 2872], circuit theory [72], queuing and stochastic systems [1328, 1859, 2122], econometrics [884, 1924, 2300], geodesy [2556], game theory [488, 1802, 2542], computer graphics [122, 1062], computer vision [1912], optimization [558, 835, 1939], signal processing [1431, 2382, 2776], classical and quantum information theory [789, 1431, 2131, 2237], communications systems [1556, 1557], statistics [1221, 1343, 1924, 2300, 2403], statistical mechanics [34, 337, 338, 2789], demography [655, 1605], combinatorics, networks, and graph theory [275, 345, 418, 486, 503, 586, 588, 591, 677, 678, 745, 606, 805, 897, 932, 1040, 1067, 1183, 1256, 1323, 1431, 1700, 1861, 1884, 2338, 2809], optics [1163, 1359, 1593], dimensional analysis [1327, 2582], and number theory [1674].

In all applications involving matrices, computational techniques are essential for obtaining numerical solutions. The development of efficient and reliable algorithms for matrix computations is therefore an important area of research that has been extensively developed [201, 679, 869, 1196, 1389, 1391, 1467, 1526, 2533, 2534, 2536, 2538, 2694, 2785, 2874, 2878, 2882, 2958]. To facilitate the solution of matrix problems, entire computer packages have been developed using the language of matrices. However, this book is concerned with the analytical properties of matrices rather than their computational aspects.

This book encompasses a broad range of fundamental questions in matrix theory, which, in many cases can be viewed as extensions of related questions in scalar mathematics. A few such questions follow.

> What are the basic properties of matrices? How can matrices be characterized, classified, and quantified?
>
> How can a matrix be decomposed into simpler matrices? A matrix decomposition may involve addition, multiplication, and partition. Decomposing a matrix into its fundamental components provides insight into its algebraic and geometric properties. For example, the polar decomposition states that every square matrix can be written as the product of a rotation and a dilation analogous to the polar representation of a complex number.
>
> Given a pair of matrices having certain properties, what can be inferred about the sum, product, and concatenation of these matrices? In particular, if a matrix has a given property, to what extent does that property change or remain unchanged if the matrix is perturbed by another matrix of a certain type by means of addition, multiplication, or concatenation? For example, if a matrix is nonsingular, how large can an additive perturbation to that matrix be without the sum becoming singular?
>
> How can properties of a matrix be determined by means of simple operations? For example, how can the location of the eigenvalues of a matrix be estimated directly in terms of the entries of the matrix?
>
> To what extent do matrices satisfy the formal properties of the real numbers? For example, while $0 \le a \le b$ implies that $a^r \le b^r$ for real numbers a, b and a positive integer r, when does $0 \le A \le B$ imply $A^r \le B^r$ for positive-semidefinite matrices A and B and with the positive-semidefinite ordering?

Questions of these types have occupied matrix theorists for at least a century, with motivation from diverse applications. The existing scope and depth of knowledge are enormous. Taken together, this body of knowledge provides a powerful framework for developing and analyzing models for scientific and engineering applications.

This book is intended to be useful to at least four groups of readers. Since linear algebra is a standard course in the mathematical sciences and engineering, graduate students in these fields can use this book to expand the scope of their linear algebra text. For instructors, many of the facts can be used as exercises to augment standard material in matrix courses. For researchers in the mathematical sciences, including statistics, physics, and engineering, this book can be used as a general reference on matrix theory. Finally, for users of matrices in the applied sciences, this book will provide access to a large body of results in matrix theory. By collecting these results in a single source, it is my hope that this book will prove to be convenient and useful for a broad range of applications. The material in this book is thus intended to complement the large number of classical and modern texts and reference works on linear algebra and matrix theory [21, 838, 1071, 1139, 1140, 1184, 1228, 1429, 1580, 1762, 1909, 1943, 1952, 2053, 2139, 2169, 2263, 2338, 2445, 2553].

After a review of mathematical preliminaries in Chapter 1, fundamental properties of matrices are described in Chapter 2. Chapter 3 summarizes the major classes of matrices and various matrix transformations. In Chapter 4 we turn to polynomial and rational matrices whose basic properties are essential for understanding the structure of constant matrices. Chapter 5 is concerned with various decompositions of matrices including the Jordan, Schur, and singular value decompositions. Chapter 6 provides a brief treatment of generalized inverses, while Chapter 7 describes the Kronecker and Schur product operations. Chapter 8 is concerned with the properties of positive-semidefinite matrices. A detailed treatment of vector and matrix norms is given in Chapter 9, while formulas for matrix derivatives are given in Chapter 10. Next, Chapter 11 focuses on the matrix exponential and stability theory, which are central to the study of linear differential equations. In Chapter 12 we apply matrix theory to the analysis of linear systems, their state space realizations, and their transfer function representation. This chapter also includes a discussion of the matrix Riccati equation of control theory.

Each chapter provides a core of results with, in many cases, complete proofs. Sections at the end of each chapter provide a collection of Facts organized to correspond to the order of topics in the chapter. These Facts include corollaries and special cases of results presented in the chapter, as well as related results that go beyond the results of the chapter. In some cases the Facts include open problems, illuminating remarks, and hints regarding proofs. The Facts are intended to provide the reader with a useful reference collection of matrix results as well as a gateway to the matrix theory literature.

Acknowledgments

The writing of this book spanned more than a decade and a half, during which time numerous individuals contributed both directly and indirectly. I am grateful for the helpful comments of many people who contributed technical material and insightful suggestions, all of which greatly improved the presentation and content of the book. In addition, numerous individuals generously agreed to read sections or chapters of the book for clarity and accuracy. I wish to thank Jasim Ahmed, Suhail Akhtar, David Bayard, Sanjay Bhat, Tony Bloch, Peter Bullen, Steve Campbell, Agostino Capponi, Ramu Chandra, Jaganath Chandrasekhar, Nalin Chaturvedi, Vijay Chellaboina, Jie Chen, David Clements, Dan Davison, Dimitris Dimogianopoulos, Jiu Ding, D. Z. Djokovic, R. Scott Erwin, R. W. Farebrother, Danny Georgiev, Joseph Grcar, Wassim Haddad, Yoram Halevi, Jesse Hoagg, Roger Horn, David Hyland, Iman Izadi, Pierre Kabamba, Vikram Kapila, Fuad Kittaneh, Seth Lacy, Thomas Laffey, Cedric Langbort, Alan Laub, Alexander Leonessa, Kai-Yew Lum, Pertti Makila, Roy Mathias, N. Harris McClamroch, Boris Mordukhovich, Sergei Nersesov, JinHyoung Oh, Concetta Pilotto, Harish Palanthandalum-Madapusi, Michael Piovoso, Leiba Rodman, Phil Roe, Carsten Scherer, Wasin So, Andy Sparks, Edward Tate, Yongge Tian, Panagiotis Tsiotras, Feng Tyan, Ravi Venugopal, Jan Willems, Hong Wong, Vera Zeidan, Xingzhi Zhan, and Fuzhen Zhang for their assistance. Nevertheless, I take full responsibility for any remaining errors, and I encourage readers to alert me to any mistakes, corrections of which will be posted on the web. Solutions to the open problems are also welcome.

Portions of the manuscript were typed by Jill Straehla and Linda Smith at Harris Corporation, and by Debbie Laird, Kathy Stolaruk, and Suzanne Smith at the University of Michigan. John Rogosich of Techsetters, Inc., provided invaluable assistance with LATEX issues, and Jennifer Slater carefully copyedited the entire manuscript. I also thank JinHyoung Oh and Joshua Kang for writing C code to refine the index.

I especially thank Vickie Kearn of Princeton University Press for her wise guidance and constant encouragement. Vickie managed to address all of my concerns and anxieties, and helped me improve the manuscript in many ways.

Finally, I extend my greatest appreciation for the (uncountably) infinite patience of my family, who endured the days, weeks, months, and years that this project consumed. The writing of this book began with toddlers and ended with a teenager and a twenty-year old. We can all be thankful it is finally finished.

Dennis S. Bernstein
Ann Arbor, Michigan
dsbaero@umich.edu
January 2005

Special Symbols

General Notation

π	3.1415926535897932384626433832795028841971693993751058...
e	2.7182818284590452353602874713526624977572470936999595...
γ (Euler's constant)	0.5772156649015328606065120900824024310421593359399235...
G (Catalan's constant)	0.9159655941772190150546035149323841107741493742816721...
$\triangleq$	equals by definition
$\lim_{\varepsilon\uparrow 0}$	limit from the left
$\lim_{\varepsilon\downarrow 0}$	limit from the right
$\lim_{\varepsilon\to 0}$	limit
$0!$	1
$n!$	$n(n-1)\cdots(2)(1)$
$z^{\underline{k}}$	$z(z-1)\cdots(z-k+1)$
$z^{\overline{k}}$	$z(z+1)\cdots(z+k-1)$
$0!!$	1
$(-1)!!$	1
$n!!$	$n(n-2)(n-4)\cdots(2) = 2^{n/2}(n/2)!$ if n is even; $n(n-2)(n-4)\cdots(3)(1) = \dfrac{(n+1)!}{2^{(n+1)/2}[\frac{1}{2}(n+1)]!}$ if n is odd
$\binom{\alpha}{m}$	$\dfrac{\alpha(\alpha-1)\cdots(\alpha-m+1)}{m!}$ (p. 14)
$\binom{n}{m}$	$\dfrac{n!}{m!(n-m)!}$ (p. 15)
$\binom{n}{k_1,\ldots,k_l}$	$\dfrac{n!}{k_1!\cdots k_l!}$, where $\sum_{i=1}^{l} k_i = n$ (p. 15)
$\left[{n \atop k}\right]$	cycle number (p. 105, Fact 1.19.1)
$\left\{{n \atop k}\right\}$	subset number (p. 107, Fact 1.19.3)
$\lfloor a\rfloor$	largest integer less than or equal to (floor of) a
$\lceil a\rceil$	smallest integer greater than or equal to (ceiling of) a
$\delta_{i,j}$	1 if $i=j$, 0 if $i\neq j$ (Kronecker delta)
log	logarithm with base e

$\operatorname{sign}\alpha$	1 if $\alpha > 0$, -1 if $\alpha < 0$, 0 if $\alpha = 0$
d_n	derangement number (p. 104, Fact 1.18.2)
C_n	Catalan number (p. 104, Fact 1.18.4)
$\mathcal{B}_n$	Bell number (p. 112, Fact 1.19.6)
p_n	partition number (p. 113, Fact 1.20.1)
B_n	Bernoulli number (p. 977, Fact 13.1.6)
E_n	Euler number (p. 980, Fact 13.1.8)
$\zeta(n)$	$\sum_{i=1}^{\infty} \frac{1}{i^n}$, where $n \ge 2$ (p. 994, Fact 13.3.1)

Chapter 1

$\{\ \}$	set (p. 1)
$\in$	is an element of (p. 1)
$\notin$	is not an element of (p. 1)
$\varnothing$	empty set (p. 1)
$\cap$	intersection (p. 1)
$\cup$	union (p. 1)
$\mathcal{Y}\backslash\mathcal{X}$	complement of $\mathcal{X}$ relative to $\mathcal{Y}$ (p. 1)
$\mathcal{X}^{\sim}$	complement of $\mathcal{X}$ (p. 1)
$\ominus$	symmetric difference (p. 1)
$\subseteq$	is a subset of (p. 1)
$\subset$	is a proper subset of (p. 2)
$(x_1, \ldots, x_n)$	tuple or n-tuple (p. 2)
$\times_{i=1}^{n}\mathcal{X}_i$	$\mathcal{X}_1 \times \cdots \times \mathcal{X}_n$ (p. 2)
$\mathcal{X}^n$	$\times_{i=1}^{n}\mathcal{X}$ (p. 2)
$(x_i)_{i=1}^{\infty}$	sequence $(x_1, x_2, \ldots)$ (p. 2)
$\mathbb{N}$	nonnegative integers (p. 2)
$\mathbb{P}$	positive integers (p. 2)
$\mathbb{Z}$	integers (p. 2)
$\mathbb{Q}$	rational numbers (p. 2)
$\mathbb{R}$	real numbers (p. 2)
$\{\ \}_{\rm ms}$	multiset (p. 2)
$\operatorname{rev}(\mathcal{R})$	reversal of the relation $\mathcal{R}$ (p. 6)
$\mathcal{R}^{\sim}$	complement of the relation $\mathcal{R}$ (p. 6)
$\operatorname{ref}(\mathcal{R})$	reflexive hull of the relation $\mathcal{R}$ (p. 6)
$\operatorname{sym}(\mathcal{R})$	symmetric hull of the relation $\mathcal{R}$ (p. 6)
$\operatorname{trans}(\mathcal{R})$	transitive hull of the relation $\mathcal{R}$ (p. 6)
$\operatorname{equiv}(\mathcal{R})$	equivalence hull of the relation $\mathcal{R}$ (p. 6)
$x \equiv y$	(x, y) is an element of the equivalence relation $\mathcal{R}$ (p. 6)

$x \preceq y$	(x, y) is an element of the relation $\mathcal{R}$ that defines the partially ordered set $(\mathcal{X}, \mathcal{R})$ (p. 7)
$\mathrm{glb}(\mathcal{S})$	greatest lower bound of $\mathcal{S}$ (p. 7, Definition 1.3.9)
$\mathrm{lub}(\mathcal{S})$	least upper bound of $\mathcal{S}$ (p. 7, Definition 1.3.9)
$\inf(\mathcal{S})$	infimum of $\mathcal{S}$ (p. 8, Definition 1.3.9)
$\sup(\mathcal{S})$	supremum of $\mathcal{S}$ (p. 8, Definition 1.3.9)
$\mathrm{essglb}(\mathcal{S})$	essential greatest lower bound of $\mathcal{S}$ (p. 8, Definition 1.3.9)
$\mathrm{esslub}(\mathcal{S})$	essential least upper bound of $\mathcal{S}$ (p. 8, Definition 1.3.9)
$\mathrm{esslim}(\mathcal{S})$	essential limit of $\mathcal{S}$ (p. 9, Definition 1.3.20)
$\mathrm{rev}(\mathcal{G})$	reversal of the graph $\mathcal{G}$ (p. 10)
$\mathcal{G}^{\sim}$	complement of the graph $\mathcal{G}$ (p. 10)
$\mathrm{ref}(\mathcal{G})$	reflexive hull of the graph $\mathcal{G}$ (p. 10)
$\mathrm{sym}(\mathcal{G})$	symmetric hull of the graph $\mathcal{G}$ (p. 10)
$\mathrm{trans}(\mathcal{G})$	transitive hull of the graph $\mathcal{G}$ (p. 10)
$\mathrm{equiv}(\mathcal{G})$	equivalence hull of the graph $\mathcal{G}$ (p. 10)
$\mathrm{indeg}(x)$	indegree of the node x (p. 10)
$\mathrm{outdeg}(x)$	outdegree of the node x (p. 10)
$\deg(x)$	degree of the node x (p. 10)
$x \vert y$	x divides y (p. 12)
$x \nmid y$	x does not divide y (p. 12)
$\gcd\{n_1, \ldots, n_k\}$	greatest common divisor of $n_1, \ldots, n_k$ (p. 12)
$\mathrm{lcm}\{n_1, \ldots, n_k\}$	least common multiple of $n_1, \ldots, n_k$ (p. 12)
$\mathrm{rem}_k(n)$	remainder after dividing n by k (p. 12)
$n \overset{k}{\equiv} m$	n and m are congruent modulo k (p. 12)
H_n	$\sum_{i=1}^{n} \frac{1}{i}$ (p. 13)
$H_{n,k}$	$\sum_{i=1}^{n} \frac{1}{i^k}$ (p. 13)
$\mathbb{C}$	complex numbers (p. 13)
$\jmath$	$\sqrt{-1}$ (p. 13)
$\overline{z}$	complex conjugate of $z \in \mathbb{C}$ (p. 13)
$\mathrm{Re}\, z$	real part of $z \in \mathbb{C}$ (p. 13)
$\mathrm{Im}\, z$	imaginary part of $z \in \mathbb{C}$ (p. 13)
$\vert z \vert$	absolute value of $z \in \mathbb{C}$ (p. 13)
OLHP	open left half plane in $\mathbb{C}$ (p. 14)
CLHP	closed left half plane in $\mathbb{C}$ (p. 14)
ORHP	open right half plane in $\mathbb{C}$ (p. 14)
CRHP	closed right half plane in $\mathbb{C}$ (p. 14)
IA	imaginary numbers (p. 14)
OIUD	open inside unit disk in $\mathbb{C}$ (p. 14)
CIUD	closed inside unit disk in $\mathbb{C}$ (p. 14)

COUD	closed outside unit disk in $\mathbb{C}$ (p. 14)
OOUD	open outside unit disk in $\mathbb{C}$ (p. 14)
UC	unit circle in $\mathbb{C}$ (p. 14)
$\mathbb{F}$	$\mathbb{R}$ or $\mathbb{C}$ (p. 14)
$f\colon\ \mathcal{X} \mapsto \mathcal{Y}$	f is a function with domain $\mathcal{X}$ and codomain $\mathcal{Y}$ (p. 16)
Graph(f)	$\{(x, f(x))\colon\ x \in \mathcal{X}\}$ (p. 16)
$f \circ g$	composition of functions f and g (p. 16)
card	cardinality (p. 16)
f^{L}	left inverse of f (p. 17)
f^{R}	right inverse of f (p. 17)
f^{Inv}	inverse of f (p. 17)
f^{inv}	set-valued inverse of f (p. 17)
$(\sigma(1), \ldots, \sigma(n))$	permutation of $(1, \ldots, n)$ (p. 21)

Chapter 3

$\mathbb{R}^n$	$\mathbb{R}^{n\times 1}$ (real column vectors) (p. 277)
$\mathbb{C}^n$	$\mathbb{C}^{n\times 1}$ (complex column vectors) (p. 277)
$\mathbb{F}^n$	$\mathbb{R}^n$ or $\mathbb{C}^n$ (p. 277)
$x_{(i)}$	ith component of $x \in \mathbb{F}^n$ (p. 277)
$x \geq\geq y$	$x_{(i)} \geq y_{(i)}$ for all i ($x - y$ is nonnegative) (p. 277)
$x >> y$	$x_{(i)} > y_{(i)}$ for all i ($x - y$ is positive) (p. 277)
conv $\mathcal{S}$	convex hull of $\mathcal{S}$ (p. 279)
cone $\mathcal{S}$	conical hull of $\mathcal{S}$ (p. 279)
coco $\mathcal{S}$	convex conical hull of $\mathcal{S}$ (p. 279)
span $\mathcal{S}$	span of $\mathcal{S}$ (p. 279)
affin $\mathcal{S}$	affine hull of $\mathcal{S}$ (p. 279)
dim $\mathcal{S}$	dimension of $\mathcal{S}$ (p. 279)
$\mathbb{R}^{n\times m}$	$n \times m$ real matrices (p. 280)
$\mathbb{C}^{n\times m}$	$n \times m$ complex matrices (p. 280)
$\mathbb{F}^{n\times m}$	$\mathbb{R}^{n\times m}$ or $\mathbb{C}^{n\times m}$ (p. 280)
$\mathrm{row}_i(A)$	ith row of A (p. 280)
$\mathrm{col}_i(A)$	ith column of A (p. 280)
$A_{(i,j)}$	(i, j) entry of A (p. 280)
$A \overset{i}{\leftarrow} b$	matrix obtained from $A \in \mathbb{F}^{n\times m}$ by replacing $\mathrm{col}_i(A)$ with $b \in \mathbb{F}^n$ or $\mathrm{row}_i(A)$ with $b \in \mathbb{F}^{1\times m}$ (p. 280)
$\mathrm{d}_{\max}(A)$	largest diagonal entry of $A \in \mathbb{F}^{n\times m}$ having real diagonal entries (p. 280)
$\mathrm{d}_i(A)$	ith largest diagonal entry of $A \in \mathbb{F}^{n\times m}$ having real diagonal entries (p. 280)
$\mathrm{d}_{\min}(A)$	smallest diagonal entry of $A \in \mathbb{F}^{n\times m}$ having real diagonal entries (p. 280)

$\mathrm{d}(A)$	vector $\begin{bmatrix} \mathrm{d}_1(A) \\ \vdots \\ \mathrm{d}_{\min\{n,m\}} \end{bmatrix}$ of diagonal entries of $A \in \mathbb{F}^{n\times m}$ having real diagonal entries (p. 281)
$A_{(\mathcal{S}_1,\mathcal{S}_2)}$	submatrix of A formed by retaining the rows of A listed in $\mathcal{S}_1$ and the columns of A listed in $\mathcal{S}_2$ (p. 281)
$A_{(\mathcal{S})}$	$A_{(\mathcal{S},\mathcal{S})}$ (p. 281)
$A_{(\mathcal{S}_1,\cdot)}$	submatrix of A formed by retaining the rows of A listed in $\mathcal{S}_1$ (p. 281)
$A_{(\cdot,\mathcal{S}_2)}$	submatrix of A formed by retaining the columns of A listed in $\mathcal{S}_2$ (p. 281)
$A_{[\mathcal{S}_1,\mathcal{S}_2]}$	submatrix of A formed by deleting the rows of A listed in $\mathcal{S}_1$ and the columns of A listed in $\mathcal{S}_2$ (p. 281)
$A_{[\mathcal{S}]}$	$A_{[\mathcal{S},\mathcal{S}]}$ (p. 281)
$A_{[\mathcal{S}_1,\cdot]}$	submatrix of A obtained by deleting the rows of A listed in $\mathcal{S}_1$ (p. 281)
$A_{[\cdot,\mathcal{S}_2]}$	submatrix of A obtained by deleting the columns of A listed in $\mathcal{S}_2$ (p. 281)
$A_{[i,j]}$	$A_{[\{i\},\{j\}]}$ (p. 281)
$A \geq\geq B$	$A_{(i,j)} \geq B_{(i,j)}$ for all i, j ($A - B$ is nonnegative) (p. 281)
$A >> B$	$A_{(i,j)} > B_{(i,j)}$ for all i, j ($A - B$ is positive) (p. 281)
$A^{\mathrm{inv}}(\mathcal{S})$	image of the set $\mathcal{S}$ by the inverse of the map $f(x) = Ax$ (p. 282)
$[A, B]$	commutator $AB - BA$ (p. 283)
$\mathrm{ad}_A(X)$	adjoint operator $[A, X]$ (p. 283)
$x \times y$	cross product of vectors $x, y \in \mathbb{R}^3$ (p. 283)
$K(x)$	cross-product matrix for $x \in \mathbb{R}^3$ (p. 283)
$0_{n\times m}, 0$	$n \times m$ zero matrix (p. 283)
I_n, I	$n \times n$ identity matrix (p. 284)
$\hat{I}_n, \hat{I}$	$n \times n$ reverse permutation matrix $\begin{bmatrix} 0 & & 1 \\ & \iddots & \\ 1 & & 0 \end{bmatrix}$ (p. 284)
P_n	$n \times n$ cyclic permutation matrix (p. 284)
N_n, N	$n \times n$ standard nilpotent matrix (p. 284)
$e_{i,n}, e_i$	$\mathrm{col}_i(I_n)$ (p. 285)
$E_{i,j,n\times m}, E_{i,j}$	$e_{i,n}e_{j,m}^{\mathsf{T}}$ (p. 285)
$1_{n\times m}$	$n \times m$ ones matrix (p. 285)
$\lVert x\rVert_2$	Euclidean norm of $x \in \mathbb{F}^n$ (p. 286)
A^{T}	transpose of A (p. 286)
$\mathrm{tr}\, A$	trace of A (p. 287)
$\overline{C}$	complex conjugate of $C \in \mathbb{C}^{n\times m}$ (p. 287)
$\mathcal{S}^{\mathsf{T}}$	$\{A^{\mathsf{T}}\colon\ A \in \mathcal{S}\}$ or $\{A^{\mathsf{T}}\colon\ A \in \mathcal{S}\}_{\mathrm{ms}}$ (p. 287)
A^*	$\overline{A}^{\mathsf{T}}$ conjugate transpose of A (p. 288)
$\mathrm{Re}\, A$	real part of $A \in \mathbb{F}^{n\times m}$ (p. 288)
$\mathrm{Im}\, A$	imaginary part of $A \in \mathbb{F}^{n\times m}$ (p. 288)

$\overline{\mathcal{S}}$	$\{\overline{A}\colon\ A \in \mathcal{S}\}$ or $\{\overline{A}\colon\ A \in \mathcal{S}\}_{\mathrm{ms}}$ (p. 288)
$\mathcal{S}^*$	$\{A^*\colon\ A \in \mathcal{S}\}$ or $\{A^*\colon\ A \in \mathcal{S}\}_{\mathrm{ms}}$ (p. 288)
$A^{\hat{\mathsf{T}}}$	$\hat{I}A^{\mathsf{T}}\hat{I}$ reverse transpose of A (p. 288)
$A^{\hat{*}}$	$\hat{I}A^{*}\hat{I}$ reverse complex conjugate transpose of A (p. 288)
$\mathcal{S}^{\hat{\mathsf{T}}}$	$\{A^{\hat{\mathsf{T}}}\colon\ A \in \mathcal{S}\}$ or $\{A^{\hat{\mathsf{T}}}\colon\ A \in \mathcal{S}\}_{\mathrm{ms}}$ (p. 289)
$\mathcal{S}^{\hat{*}}$	$\{A^{\hat{*}}\colon\ A \in \mathcal{S}\}$ or $\{A^{\hat{*}}\colon\ A \in \mathcal{S}\}_{\mathrm{ms}}$ (p. 289)
$\lvert x\rvert$	absolute value of $x \in \mathbb{F}^n$ (p. 289)
$\lvert A\rvert$	absolute value of $A \in \mathbb{F}^{n\times n}$ (p. 289)
$\operatorname{sign} x$	sign of $x \in \mathbb{R}^n$ (p. 289)
$\operatorname{sign} A$	sign of $A \in \mathbb{R}^{n\times n}$ (p. 289)
$\mathcal{S}^{\perp}$	orthogonal complement of $\mathcal{S}$ (p. 289)
$\operatorname{polar} \mathcal{S}$	polar of $\mathcal{S}$ (p. 290)
$\operatorname{dcone} \mathcal{S}$	dual cone of $\mathcal{S}$ (p. 290)
$\mathcal{R}(A)$	range of A (p. 290)
$\mathcal{N}(A)$	null space of A (p. 291)
$\operatorname{rank} A$	rank of A (p. 292)
$\operatorname{def} A$	defect of A (p. 292)
A^{L}	left inverse of A (p. 294)
A^{R}	right inverse of A (p. 294)
A^{-1}	inverse of A (p. 297)
$A^{-\mathsf{T}}$	$(A^{\mathsf{T}})^{-1}$ (p. 298)
A^{-*}	$(A^{*})^{-1}$ (p. 298)
$\det A$	determinant of A (p. 299)
A^{A}	adjugate of A (p. 301)
$x^{\downarrow}$	vector with components of x in decreasing order (p. 305)
$x^{\uparrow}$	vector with components of x in increasing order (p. 305)
$x \overset{\mathrm{w}}{\prec} y$	y weakly majorizes x (p. 305)
$x \overset{\mathrm{s}}{\prec} y$	y strongly majorizes x (p. 305)
$x \overset{\mathrm{wlog}}{\prec} y$	y weakly log majorizes x (p. 305)
$x \overset{\mathrm{slog}}{\prec} y$	y strongly log majorizes x (p. 305)

Chapter 4

$\operatorname{diag}(a_1,\ldots,a_n)$	$\begin{bmatrix} a_1 & & 0 \\ & \ddots & \\ 0 & & a_n \end{bmatrix}$ (p. 365)
$\operatorname{diag}(x)$	$\operatorname{diag}(x_{(1)},\ldots,x_{(n)})$, where $x \in \mathbb{F}^n$ (p. 365)
$\operatorname{revdiag}(a_1,\ldots,a_n)$	$\begin{bmatrix} 0 & & a_1 \\ & ⋰ & \\ a_n & & 0 \end{bmatrix}$ (p. 365)

$\operatorname{revdiag}(x)$	$\operatorname{revdiag}(x_{(1)},\ldots,x_{(n)})$, where $x \in \mathbb{F}^n$ (p. 365)
$\operatorname{diag}(A_1,\ldots,A_k)$	block-diagonal matrix $\begin{bmatrix} A_1 & & 0 \\ & \ddots & \\ 0 & & A_k \end{bmatrix}$, where $A_i \in \mathbb{F}^{n_i \times m_i}$ (p. 365)
J_{2n}, J	$\begin{bmatrix} 0 & I_n \\ -I_n & 0 \end{bmatrix}$ (p. 367)
$\mathrm{gl}_{\mathbb{F}}(n), \mathrm{pl}_{\mathbb{C}}(n), \mathrm{sl}_{\mathbb{F}}(n)$, $\mathrm{u}(n), \mathrm{su}(n), \mathrm{so}(n)$, $\mathrm{symp}_{\mathbb{F}}(2n)$, $\mathrm{osymp}_{\mathbb{F}}(2n), \mathrm{aff}_{\mathbb{F}}(n)$, $\mathrm{se}_{\mathbb{F}}(n), \mathrm{trans}_{\mathbb{F}}(n)$	Lie algebras (p. 369)
$\mathcal{S}_1 \simeq \mathcal{S}_2$	the groups $\mathcal{S}_1$ and $\mathcal{S}_2$ are isomorphic (p. 372)
$\mathrm{GL}_{\mathbb{F}}(n), \mathrm{PL}_{\mathbb{F}}(n)$, $\mathrm{SL}_{\mathbb{F}}(n), \mathrm{U}(n), \mathrm{O}(n)$, $\mathrm{U}(n,m), \mathrm{O}(n,m)$, $\mathrm{SU}(n), \mathrm{SO}(n)$, $\mathrm{P}(n), \mathrm{A}(n), \mathrm{D}(n), \mathrm{C}(n)$, $\mathrm{Symp}_{\mathbb{F}}(2n)$, $\mathrm{OSymp}_{\mathbb{F}}(2n), \mathrm{Aff}_{\mathbb{F}}(n)$, $\mathrm{SE}_{\mathbb{F}}(n), \mathrm{Trans}_{\mathbb{F}}(n)$	groups (p. 372)
$A_{\perp}$	complementary idempotent matrix or projector $I - A$ of A (p. 374)
$\operatorname{ind} A$	index of A (p. 375)
$A \overset{\mathrm{rs}}{\leq} B$	rank subtractivity partial ordering (p. 430, Fact 4.30.3)
$A \overset{*}{\leq} B$	star partial ordering (p. 431, Fact 4.30.8)
$\mathbb{H}$	quaternions (p. 437, Fact 4.32.1)
$\mathrm{Sp}(n)$	symplectic group in $\mathbb{H}$ (p. 438, Fact 4.32.4)

Chapter 6

$\mathbb{F}[s]$	polynomials with coefficients in $\mathbb{F}$ (p. 499)
$\deg p$	degree of $p \in \mathbb{F}[s]$ (p. 499)
$\operatorname{mroots}(p)$	multiset of roots of $p \in \mathbb{F}[s]$ (p. 499)
$\operatorname{roots}(p)$	set of roots of $p \in \mathbb{F}[s]$ (p. 499)
$\operatorname{mult}_p(\lambda)$	multiplicity of λ as a root of $p \in \mathbb{F}[s]$ (p. 499)
$\delta(p)$	spread of p (p. 500)
$\rho_i(p)$	ith largest root modulus of p (p. 500)
$\rho_{\min}(p)$	minimum root modulus of p (p. 500)
$\rho_{\max}(p)$	root radius of p (p. 500)
$\alpha_i(p)$	ith largest root real part of p (p. 500)
$\alpha_{\min}(p)$	minimum root real part of p (p. 500)
$\alpha_{\max}(p)$	root real abscissa of p (p. 500)
$\beta_i(p)$	ith largest root imaginary part of p (p. 500)
$\beta_{\min}(p)$	minimum root imaginary part of p (p. 500)

$\beta_{\max}(p)$	root imaginary abscissa of p (p. 500)
$\mathbb{F}[s]^{n\times m}$	$n \times m$ matrices with entries in $\mathbb{F}[s]$ ($n \times m$ polynomial matrices with coefficients in $\mathbb{F}$) (p. 501)
rank P	rank of $P \in \mathbb{F}[s]^{n\times m}$ (p. 502)
Szeros(P)	set of Smith zeros of $P \in \mathbb{F}[s]^{n\times m}$ (p. 504)
mSzeros(P)	multiset of Smith zeros of $P \in \mathbb{F}[s]^{n\times m}$ (p. 504)
χ_A	characteristic polynomial of A (p. 506)
$\lambda_i(A)$	ith largest eigenvalue of $A \in \mathbb{F}^{n\times n}$ having real eigenvalues (p. 506)
$\lambda_{\max}(A)$	largest eigenvalue $\lambda_1(A)$ of $A \in \mathbb{F}^{n\times n}$ having real eigenvalues (p. 506)
$\lambda_{\min}(A)$	smallest eigenvalue $\lambda_n(A)$ of $A \in \mathbb{F}^{n\times n}$ having real eigenvalues (p. 506)
$\lambda(A)$	vector $\begin{bmatrix}\lambda_1(A)\\ \vdots \\ \lambda_n(A)\end{bmatrix}$ of eigenvalues of $A \in \mathbb{F}^{n\times n}$ having real eigenvalues (p. 506)
$\mathrm{amult}_A(\lambda)$	algebraic multiplicity of $\lambda \in \mathrm{spec}(A)$ (p. 506)
spec(A)	spectrum of A (p. 506)
mspec(A)	multispectrum of A (p. 506)
$\delta(A)$	spread of A (p. 510)
$\rho_i(A)$	ith largest spectral modulus of A (p. 510)
$\rho_{\max}(A)$	spectral radius of A (p. 510)
$\rho_{\min}(A)$	minimum spectral modulus of A (p. 510)
$\rho(A)$	vector $\begin{bmatrix}\rho_1(A)\\ \vdots \\ \rho_n(A)\end{bmatrix}$ of spectral moduli of $A \in \mathbb{F}^{n\times n}$ (p. 510)
$\mathrm{gmult}_A(\lambda)$	geometric multiplicity of $\lambda \in \mathrm{spec}(A)$ (p. 511)
$\alpha_i(A)$	ith largest spectral real part of A (p. 510)
$\alpha_{\max}(A)$	spectral abscissa of A (p. 510)
$\alpha_{\min}(A)$	minimum spectral real part of A (p. 510)
$\alpha(A)$	vector $\begin{bmatrix}\alpha_1(A)\\ \vdots \\ \alpha_n(A)\end{bmatrix}$ of spectral real parts of $A \in \mathbb{F}^{n\times n}$ (p. 510)
$\beta_i(A)$	ith largest spectral imaginary part of A (p. 510)
$\beta_{\max}(A)$	spectral imaginary abscissa of A (p. 511)
$\beta_{\min}(A)$	minimum spectral imaginary part of A (p. 510)
$\beta(A)$	vector $\begin{bmatrix}\beta_1(A)\\ \vdots \\ \beta_n(A)\end{bmatrix}$ of spectral imaginary parts of $A \in \mathbb{F}^{n\times n}$ (p. 511)
$\nu_-(A), \nu_0(A), \nu_+(A)$	number of eigenvalues of A counting algebraic multiplicity having negative, zero, and positive real part, respectively (p. 511)

$\operatorname{In} A$	inertia $\begin{bmatrix} \nu_-(A) \\ \nu_0(A) \\ \nu_+(A) \end{bmatrix}$ of A (p. 511)
$\operatorname{sig} A$	signature of A; that is, $\nu_+(A) - \nu_-(A)$ (p. 511)
μ_A	minimal polynomial of A (p. 512)
$\mathbb{F}(s)$	rational functions with coefficients in $\mathbb{F}$ (SISO rational transfer functions) (p. 513)
$\mathbb{F}(s)_{\rm prop}$	proper rational functions with coefficients in $\mathbb{F}$ (SISO proper rational transfer functions) (p. 513)
$\operatorname{reldeg} g$	relative degree of $g \in \mathbb{F}(s)_{\rm prop}$ (p. 513)
$\mathbb{F}(s)^{n\times m}$	$n \times m$ matrices with entries in $\mathbb{F}(s)$ (MIMO rational transfer functions) (p. 514)
$\mathbb{F}(s)^{n\times m}_{\rm prop}$	$n\times m$ matrices with entries in $\mathbb{F}(s)_{\rm prop}$ (MIMO proper rational transfer functions) (p. 514)
$\operatorname{reldeg} G$	relative degree of $G \in \mathbb{F}(s)^{n\times m}_{\rm prop}$ (p. 514)
$\operatorname{rank} G$	rank of $G \in \mathbb{F}(s)^{n\times m}$ (p. 514)
$\operatorname{poles}(G)$	set of poles of $G \in \mathbb{F}(s)^{n\times m}$ (p. 514)
$\operatorname{bzeros}(G)$	set of blocking zeros of $G \in \mathbb{F}(s)^{n\times m}$ (p. 514)
$\operatorname{Mcdeg} G$	McMillan degree of $G \in \mathbb{F}(s)^{n\times m}$ (p. 515)
$\operatorname{tzeros}(G)$	set of transmission zeros of $G \in \mathbb{F}(s)^{n\times m}$ (p. 515)
$\operatorname{mpoles}(G)$	multiset of poles of $G \in \mathbb{F}(s)^{n\times m}$ (p. 515)
$\operatorname{mtzeros}(G)$	multiset of transmission zeros of $G \in \mathbb{F}(s)^{n\times m}$ (p. 515)
$\operatorname{mbzeros}(G)$	multiset of blocking zeros of $G \in \mathbb{F}(s)^{n\times m}$ (p. 515)
$B(p,q)$	Bezout matrix of $p, q \in \mathbb{F}[s]$ (p. 519, Fact 6.8.8)
$H(g)$	Hankel matrix of $g \in \mathbb{F}(s)$ (p. 520, Fact 6.8.10)

Chapter 7

$C(p)$	companion matrix for monic polynomial p (p. 546)
$\mathcal{J}_l(q)$	$l \times l$ or $2l \times 2l$ real Jordan matrix (p. 549)
$\mathcal{H}_l(q)$	$l \times l$ or $2l \times 2l$ hypercompanion matrix (p. 550)
$\sigma_i(A)$	ith largest singular value of $A \in \mathbb{F}^{n\times m}$ (p. 555)
$\sigma(A)$	vector $\begin{bmatrix} \sigma_1(A) \\ \vdots \\ \sigma_{\min\{n,m\}}(A) \end{bmatrix}$ of singular values of $A \in \mathbb{F}^{n\times m}$ (p. 556)
$\operatorname{msval}(A)$	multiset $\{\sigma_1(A), \ldots, \sigma_{\min\{n,m\}}(A)\}_{\rm ms}$ of singular values of $A \in \mathbb{F}^{n\times m}$ (p. 556)
$\sigma_{\max}(A)$	largest singular value $\sigma_1(A)$ of $A \in \mathbb{F}^{n\times m}$ (p. 556)
$\sigma_{\min}(A)$	minimum singular value $\sigma_n(A)$ of a square matrix $A \in \mathbb{F}^{n\times n}$ (p. 556)
$\operatorname{ind}_A(\lambda)$	index of λ with respect to A (p. 558)
$P_{A,B}$	pencil of (A, B), where $A, B \in \mathbb{F}^{n\times n}$ (p. 563)
$\operatorname{spec}(A, B)$	generalized spectrum of (A, B), where $A, B \in \mathbb{F}^{n\times n}$ (p. 563)

$\mathrm{mspec}(A, B)$	generalized multispectrum of (A, B), where $A, B \in \mathbb{F}^{n\times n}$ (p. 563)
$\chi_{A,B}$	characteristic polynomial of (A, B), where $A, B \in \mathbb{F}^{n\times n}$ (p. 564)
$V(\lambda_1, \dots, \lambda_n)$	Vandermonde matrix (p. 613, Fact 7.18.3)
$\mathrm{circ}(a_0, \dots, a_{n-1})$	circulant matrix of $a_0, \dots, a_{n-1} \in \mathbb{F}$ (p. 614, Fact 7.18.13)

Chapter 8

A^+	(Moore-Penrose) generalized inverse of A (p. 621)
$D\|\mathcal{A}$	Schur complement of D with respect to $\mathcal{A}$ (p. 625)
A^{D}	Drazin generalized inverse of A (p. 625)
$A^{\#}$	group generalized inverse of A (p. 627)
$A \overset{\#}{\leq} B$	sharp partial ordering (p. 648, Fact 8.5.16)
$A \overset{\mathrm{c}}{\leq} B$	core partial ordering (p. 648, Fact 8.5.17)

Chapter 9

$\mathrm{vec}\, A$	vector formed by stacking columns of A (p. 681)
$\mathrm{vec}^{-1} A$	inverse vec operator (p. 681)
$\otimes$	Kronecker product (p. 681)
$P_{n,m}$	Kronecker permutation matrix (p. 683)
$\oplus$	Kronecker sum (p. 683)
$A \odot B$	Schur product of A and B (p. 685)
$A^{\odot\alpha}$	Schur power of A, $(A^{\odot\alpha})_{(i,j)} = (A_{(i,j)})^\alpha$ (p. 685)

Chapter 10

$\mathbf{H}^n$	$n \times n$ Hermitian matrices (p. 703)
$\mathbf{N}^n$	$n \times n$ positive-semidefinite matrices (p. 703)
$\mathbf{P}^n$	$n \times n$ positive-definite matrices (p. 703)
$A \geq B$	$A - B \in \mathbf{N}^n$ (p. 703)
$A > B$	$A - B \in \mathbf{P}^n$ (p. 703)
$\langle A \rangle$	$(A^*A)^{1/2}$ (p. 714)
$A\#B$	geometric mean of A and B (p. 743, Fact 10.11.68)
$A\#_\alpha B$	generalized geometric mean of A and B (p. 745, Fact 10.11.72)
$A{:}B$	parallel sum of A and B (p. 818, Fact 10.24.20)
$\mathrm{sh}(A, B)$	shorted operator (p. 819, Fact 10.24.21)

Chapter 11

$\\|x\\|_p$	Hölder norm $\left(\sum_{i=1}^{n} \|x_{(i)}\|^p\right)^{1/p}$ (p. 833)
$\\|A\\|_p$	Hölder norm $\left(\sum_{i,j=1}^{n,m} \|A_{(i,j)}\|^p\right)^{1/p}$ (p. 836)

$\|A\|_{\mathrm{F}}$	Frobenius norm $\sqrt{\operatorname{tr} A^*A}$ (p. 836)
$\|A\|_{p\|q}$	mixed Hölder norm $\left\|\left\|\begin{bmatrix} \|\mathrm{col}_1(A)\|_p \\ \vdots \\ \|\mathrm{col}_m(A)\|_p \end{bmatrix}\right\|\right\|_q$ (p. 836)
$\|A\|_{\sigma p}$	Schatten norm $\|\sigma(A)\|_p$ (p. 837)
$\|A\|_{q,p}$	Hölder-induced norm (p. 842)
$\|A\|_{p,q,\mathbb{F}}$	Hölder-induced norm over $\mathbb{F}$ (p. 842)
$\|A\|_{\mathrm{col}}$	column norm $\|A\|_{1,1} = \|A\|_{1\|\infty} = \max_{j\in\{1,\ldots,m\}}\|\mathrm{col}_j(A)\|_1$ (p. 844)
$\|A\|_{\mathrm{row}}$	row norm $\|A\|_{\infty,\infty} = \|A^{\mathsf{T}}\|_{1\|\infty} = \max_{i\in\{1,\ldots,n\}}\|\mathrm{row}_i(A)\|_1$ (p. 844)
$\ell(A)$	induced lower bound of A (p. 845)
$\ell_{q,p}(A)$	Hölder-induced lower bound of A (p. 846)
$\|\cdot\|_{\mathrm{D}}$	dual norm (p. 859, Fact 11.8.24)

Chapter 12

$\mathbb{B}_\varepsilon(x)$	open ball of radius ε centered at x (p. 913)
$\mathbb{S}_\varepsilon(x)$	sphere of radius ε centered at x (p. 913)
$\operatorname{int} \mathcal{S}$	interior of $\mathcal{S}$ (p. 913)
$\operatorname{int}_{\mathcal{S}'} \mathcal{S}$	interior of $\mathcal{S}$ relative to $\mathcal{S}'$ (p. 913)
$\operatorname{relint} \mathcal{S}$	interior of $\mathcal{S}$ relative to affin $\mathcal{S}$ (p. 913)
$\operatorname{cl} \mathcal{S}$	closure of $\mathcal{S}$ (p. 914)
$\lim \mathcal{S}$	set of limit points of $\mathcal{S}$ (p. 914)
$\operatorname{cl}_{\mathcal{S}'} \mathcal{S}$	closure of $\mathcal{S}$ relative to $\mathcal{S}'$ (p. 914)
$\operatorname{bd} \mathcal{S}$	boundary of $\mathcal{S}$ (p. 915)
$\operatorname{bd}_{\mathcal{S}'} \mathcal{S}$	boundary of $\mathcal{S}$ relative to $\mathcal{S}'$ (p. 915)
$\operatorname{relbd} \mathcal{S}$	boundary of $\mathcal{S}$ relative to affin $\mathcal{S}$ (p. 915)
$\operatorname{fcone} \mathcal{D}$	feasible cone of $\mathcal{D}$ (p. 924)
$\mathrm{D}_+f(x_0;\xi)$	one-sided directional derivative of f at x_0 in the direction ξ (p. 924)
$\frac{\partial f(x_0)}{\partial x_{(i)}}$	partial derivative of f with respect to $x_{(i)}$ at x_0 (p. 924)
$f'(x)$	derivative of f at x (p. 925)
$\frac{\mathrm{d}f(x_0)}{\mathrm{d}x_{(i)}}$	$f'(x_0)$ (p. 925)
$f^{(k)}(x)$	kth derivative of f at x (p. 926)
$\frac{\mathrm{d}^+f(x_0)}{\mathrm{d}x_{(i)}}$	right one-sided derivative (p. 926)
$\frac{\mathrm{d}^-f(x_0)}{\mathrm{d}x_{(i)}}$	left one-sided derivative (p. 926)
$\operatorname{Sign}(A)$	matrix sign of $A \in \mathbb{C}^{n\times n}$ (p. 932)

Chapter 14

e^A or $\exp(A)$	matrix exponential (p. 1179)
$\mathcal{L}$	Laplace transform (p. 1181)
$\mathcal{S}_{\rm s}(A)$	asymptotically stable subspace of A (p. 1199)
$\mathcal{S}_{\rm u}(A)$	unstable subspace of A (p. 1199)

Chapter 15

$\mathcal{U}(A,C)$	unobservable subspace of (A,C) (p. 1254)
$\mathcal{O}(A,C)$	$\begin{bmatrix} C \\ CA \\ CA^2 \\ \vdots \\ CA^{n-1} \end{bmatrix}$ (p. 1254)
$\mathcal{C}(A,B)$	controllable subspace of (A,B) (p. 1260)
$\mathcal{K}(A,B)$	$[B\ \ AB\ \ A^2B\ \ \cdots\ \ A^{n-1}B]$ (p. 1260)
$G \sim \left[\begin{array}{c\|c} A & B \\ \hline C & 0 \end{array}\right]$	state space realization of $G \in \mathbb{F}(s)^{l\times m}_{\rm prop}$ (p. 1270)
$\mathcal{H}_{i,j,k}(G)$	Markov block-Hankel matrix $\mathcal{O}_i(A,C)\mathcal{K}_j(A,B)$ (p. 1274)
$\mathcal{H}(G)$	Markov block-Hankel matrix $\mathcal{O}(A,C)\mathcal{K}(A,B)$ (p. 1275)
$G \overset{\rm min}{\sim} \left[\begin{array}{c\|c} A & B \\ \hline C & 0 \end{array}\right]$	minimal state space realization of $G \in \mathbb{F}(s)^{l\times m}_{\rm prop}$ (p. 1276)
$\mathcal{H}$	Hamiltonian $\begin{bmatrix} A & \Sigma \\ R_1 & -A^{\rm T} \end{bmatrix}$ (p. 1297)

Conventions, Notation, and Terminology

For convenience and clarity, this section summarizes conventions, notation, and terminology used in this book. Precise definitions are given in the main text.

Italic font is used to indicate that a word is being defined.

All definitions of words, phrases, and symbols are "if and only if" statements, although for brevity "only if" is omitted. The symbol $\triangleq$ means equal by definition, where $A \triangleq B$ means that the left-hand expression A is defined to be the right-hand expression B.

A mathematical object defined by means of a constructive procedure is *well-defined* if the construction produces a unique object.

A hypothesis established by the words "let" or "assume" is valid for all statements in the remainder of the paragraph, which may be a theorem, proposition, lemma, corollary, or fact. A hypothesis established by the word "if" is valid only within that sentence. This convention applies to both implications in an "if and only if" statement.

Every theorem, proposition, lemma, corollary, and fact consists of exactly one paragraph. A proof may consist of multiple paragraphs. The statements in a proof are valid only within the proof. No statement in a proof is accessed from outside the proof. The end of a proof is denoted by "□." An example may consist of multiple paragraphs. The end of an example is denoted by "◇."

Three types of existence statements are used. Existence is the statement "there exists $x \in \mathcal{X}$"; equivalently, "there exists at least one $x \in \mathcal{X}$." Existence with uniqueness is the statement "there exists a unique $x \in \mathcal{X}$"; equivalently, "there exists exactly one $x \in \mathcal{X}$." Existence with pre-uniqueness is the statement "there exists at most one $x \in \mathcal{X}$."

The phrases "for all," "for every," and "for each" are synonymous.

The words "always," "any," "provided," "some," "unless," "when," and "whenever" are not used for mathematical statements in this book.

Analogous statements are written in parallel using the following style: If n is (even, odd), then $n+1$ is (odd, even). If n is (even, odd), then so is $n+2$. n is (even, odd) if and only if $n+2$ is.

$\mathbb{N}$ denotes $\{0, 1, 2, 3, \ldots\}$, and $\mathbb{P}$ denotes $\{1, 2, 3, \ldots\}$, where "$\mathbb{N}$" and "$\mathbb{P}$" denote "nonnegative" and "positive," respectively. Traditionally, "$\mathbb{N}$" denotes the natural numbers $\{1, 2, 3, \ldots\}$.

Unless stated otherwise, the variables i, j, k, l, m, n denote integers. Hence, $k \geq 0$ denotes a nonnegative integer, $k \geq 1$ denotes a positive integer, and the limit $\lim_{k\to\infty} A^k$ is taken over positive integers.

The imaginary unit $\sqrt{-1}$ is denoted by dotless $\jmath$. The unit quaternions are denoted by $1, \hat{\imath}, \hat{\jmath}, \hat{k}$.

Unless stated otherwise, the letter s represents a complex scalar. The letter z may or may not represent a complex scalar.

A line over a variable has dual meaning, distinguishable from context. In particular, $\overline{z}$ is the complex conjugate of z, whereas $x^{\overline{n}}$ denotes the n-term rising factorial.

The inequalities $c \le a \le d$ and $c \le b \le d$ are written simultaneously as

$$c \le \left\{ \begin{array}{c} a \\ b \end{array} \right\} \le d.$$

The inequalities $a \le b \le c \le d \le e$ may be written as

$$\begin{aligned} a \le b &\le c \\ &\le d \le e. \end{aligned}$$

$xy + z$ denotes $(xy) + z$, $x/y + z$ denotes $(x/y) + z$, x^{y^z} denotes $x^{(y^z)}$, $\sin x + y$ denotes $(\sin x) + y$, $\sin xy + z$ denotes $(\sin xy) + z$, $\log x + y$ denotes $(\log x) + y$, $\log xy + z$ denotes $(\log xy) + z$, $\sum x_i y_i$ denotes $\sum (x_i y_i)$, and $\sum x_i + y$ denotes $(\sum x_i) + y$. x/yz is ambiguous and is not used. For clarity, we sometimes write $\log(x \sin y)$ for $\log x \sin y$, although the parentheses are superfluous.

$\sin(x+y)\sin(x-y)$ denotes $[\sin(x+y)]\sin(x-y)$; $\det(A+B)\det(A-B)$ denotes $[\det(A+B)]\det(A-B)$.

$\sum_{i\in\varnothing} \triangleq 0$, and $\prod_{i\in\varnothing} \triangleq 1$.

The prefix "non" means "not" in the words nonconstant, nonempty, nonintegral, nonnegative, nonreal, nonrepeated, nonsingular, nonsquare, nonunique, and nonzero. In some traditional usage, "non" may mean "not necessarily."

"Increasing" and "decreasing" indicate strict change for a change in the argument. The word "strict" is superfluous, and thus is omitted. Nonincreasing means nowhere increasing, while nondecreasing means nowhere decreasing.

A set can have a finite or infinite number of elements. A finite set has a finite number of elements.

A multiset can have repeated elements with at most finite multiplicity. Hence, the multisets $\{x\}_{\rm ms}$ and $\{x, x\}_{\rm ms}$ are different. The listed elements α, β, γ of the set $\{\alpha, \beta, \gamma\}$ need not be distinct. For example, $\{\alpha, \beta, \alpha\} = \{\alpha, \beta\}$. However, in a definition such as "Let $\operatorname{spec}(A) = \{\lambda_1, \ldots, \lambda_r\}$," the listed elements $\lambda_1, \ldots, \lambda_r$ are assumed to be distinct.

A set is a collection of distinct objects without a specified ordering. A multiset is a collection of possibly repeated objects without a specified ordering. A tuple can be viewed as an ordered multiset, which is a collection of possibly repeated objects with a specified ordering. Consequently, the order in which the elements of the set $\{x_1, \ldots, x_n\}$ and the elements of the multiset $\{x_1, \ldots, x_n\}_{\rm ms}$ are listed has no significance. The components of the n-tuple $(x_1, \ldots, x_n)$ are ordered.

$(x_i)_{i=1}^{\infty}$ denotes the sequence $(x_1, x_2, \ldots)$. A sequence can be viewed as a tuple with a countably infinite number of components, where the order of the components is specified and the components need not be distinct.

For clarity, square brackets alternate with parentheses. For example, $f[g(x)]$ denotes $f(g(x))$.

$\mathcal{S}_1 \subset \mathcal{S}_2$ means that $\mathcal{S}_1$ is a proper subset of $\mathcal{S}_2$, whereas $\mathcal{S}_1 \subseteq \mathcal{S}_2$ means that $\mathcal{S}_1$ is either a proper subset of $\mathcal{S}_2$ or is equal to $\mathcal{S}_2$. Hence, $\mathcal{S}_1 \subset \mathcal{S}_2$ is equivalent to both $\mathcal{S}_1 \subseteq \mathcal{S}_2$ and $\mathcal{S}_1 \ne \mathcal{S}_2$, while $\mathcal{S}_1 \subseteq \mathcal{S}_2$ is equivalent to either $\mathcal{S}_1 \subset \mathcal{S}_2$ or $\mathcal{S}_1 = \mathcal{S}_2$.

The word "graph" corresponds to what is commonly called a "simple directed graph," while "symmetric graph" corresponds to a "simple undirected graph."

The set-valued inverse of the function $f\colon \mathcal{X} \mapsto \mathcal{Y}$ is denoted by f^{inv}. If f is invertible, then its inverse is denoted by f^{Inv}. The traditional notation f^{-1} is not used in order to avoid ambiguity with $1/f$.

If $f\colon \mathcal{X} \mapsto \mathcal{Y}$ is one-to-one, then $\hat{f}\colon \mathcal{X} \mapsto f(\mathcal{X})$, where, for all $x \in \mathcal{X}$, $\hat{f}(x) \triangleq f(x)$, is onto and thus invertible. For convenience and with a slight abuse of notation, we write $f^{\mathrm{Inv}}\colon f(\mathcal{X}) \mapsto \mathcal{X}$. If f is not onto, then it does not have a unique left inverse. However, each left inverse restricted to $f(\mathcal{X})$ is an inverse of $\hat{f}$. Consequently, all left inverses of f are identical on $f(\mathcal{X})$. This observation explains why, if A is left invertible and $b \in \mathcal{R}(A)$, then, despite the fact that A has infinitely many left inverses, every left inverse of A yields the unique solution of $Ax = b$.

The matrix $A \in \mathbb{F}^{n\times m}$ may be square, tall, or wide depending on whether $m = n$, $m < n$, or $m > n$, respectively. If A is tall, then it is left invertible if and only if it has m linearly independent rows or columns; if A is wide, then it is right invertible if and only if it has n linearly independent rows or columns. It is helpful to keep this distinction in mind.

The inverse trigonometric functions are denoted by asin, acos, atan, acsc, asec, acot, while the inverse hyperbolic functions are denoted by asinh, acosh, atanh, acsch, asech, acoth . The inverse tangent function with two arguments is denoted by atan2.

For real arguments, the domain of asin and acos is $[-1, 1]$, the domain of acsc and asec is $(-\infty, -1]\cup [1, \infty)$, and the domain of atan and acot is $\mathbb{R}$. Furthermore, the range of asin is $[-\frac{\pi}{2}, \frac{\pi}{2}]$, the range of acos is $[0, \pi]$, the range of atan is $(-\frac{\pi}{2}, \frac{\pi}{2})$, the range of acsc is $[-\frac{\pi}{2}, 0) \cup (0, \frac{\pi}{2}]$, the range of asec is $[0, \frac{\pi}{2}) \cup (\frac{\pi}{2}, \pi]$, and the range of acot is $(-\frac{\pi}{2}, 0) \cup (0, \frac{\pi}{2}]$.

The principal argument of the nonzero complex number z is denoted by $\arg z \in (-\pi, \pi]$. Likewise, the principal logarithm of the nonzero complex number z is denoted by $\log z$. Traditional notation (not followed in this book) is to use “arg” and “log” to denote set-valued functions, and “Arg” and “Log” to denote principal functions. Set-valued inverse functions are denoted by f^{inv}. For example, $\sin^{\mathrm{inv}}(0) = \{i\pi\colon i \in \mathbb{Z}\}$.

For all $a < 0$ and $n \ge 1$, $(-1)^{1/n} = e^{(\pi/n)J}$. However, for all $a < 0$ and odd $n \ge 1$, $\sqrt[n]{a} \triangleq -\sqrt[n]{|a|}$. Hence, $\sqrt[3]{-1} = -1 \ne \frac{1}{2}(1 + \sqrt{3}J) = e^{(\pi/3)J} = (-1)^{1/3}$.

The *angle between two vectors* is an element of $[0, \pi]$. Therefore, by using acos, the inner product of two vectors can be used to compute the angle between two vectors.

$0! \triangleq 1$, $0/0 = (\sin 0)/0 = (1 - \cos 0)/0 = (\sinh 0)/0 \triangleq 1$, and $1/\infty \triangleq 0$.

For all $z \in \mathbb{C}$ and $k \in \mathbb{Z}$,

$$\binom{z}{k} \triangleq \begin{cases} \dfrac{z(z-1)\cdots(z-k+1)}{k!}, & k > 0, \\ 1, & k = 0, \\ 0, & k < 0. \end{cases}$$

In particular, if $n, k \in \mathbb{N}$, then

$$\binom{n}{k} = \begin{cases} \dfrac{n!}{(n-k)!k!}, & n \ge k \ge 0, \\ 0, & k > n \ge 0. \end{cases}$$

Hence,

$$\binom{n}{n} = \begin{cases} 1, & n \ge 0, \\ 0, & n < 0. \end{cases}$$

In particular, $\binom{0}{0} = 1$.

For all square matrices A, $A^0 \triangleq I$. In particular, $0^0_{n\times n} \triangleq I_n$. With this convention, we can write

$$\sum_{i=0}^{\infty} \alpha^i = \frac{1}{1-\alpha}$$

for all $\alpha \in (-1,1)$. Of course, $\lim_{x\downarrow 0} 0^x = 0$, $\lim_{x\uparrow 0} x^0 = \lim_{x\downarrow 0} x^0 = 1$, and $\lim_{x\uparrow 0} x^x = \lim_{x\downarrow 0} x^x = 1$.

The symbols ∞ and $-\infty$ are defined as limits of sequences of real numbers, and neither ∞ nor $-\infty$ is a real number. The set of *extended real numbers* is $\mathbb{R} \cup \{-\infty, \infty\}$. On the set of extended real numbers, we define $\infty + \infty = \infty$, $-\infty - \infty = -\infty$, $\infty\infty = \infty$, $(-\infty)\infty = \infty(-\infty) = -\infty$, and $(-\infty)(-\infty) = \infty$. Furthermore, for all real numbers α, $\alpha - \infty = -\infty$, $\alpha + \infty = \infty$, and $\alpha/\infty = 0$, whereas, for all nonzero real numbers α, $\alpha\infty = \operatorname{sign}(\alpha)\infty$. Hence, for all $\alpha \in (0,\infty]$, $\alpha\infty = \infty$, whereas, for all $\alpha \in [-\infty, 0)$, $\alpha\infty = -\infty$. The expressions 0∞, $\infty - \infty$, and ∞^∞ are not defined. Finally, for all $\alpha \in (0,\infty)$, we define $\alpha/0 = \infty$, whereas, for all $\alpha \in (-\infty, 0)$, we define $\alpha/0 = -\infty$. See [154, pp. 14, 15] and [2249, p. 44].

Let a and b be real numbers such that $a < b$. A *finite interval* is of the form (a,b), $[a,b)$, $(a,b]$, or $[a,b]$, whereas an *infinite interval* is of the form $(-\infty,a)$, $(-\infty,a]$, (a,∞), $[a,\infty)$, or $(-\infty,\infty)$. An *interval* is either a finite interval or an infinite interval. Neither the empty set nor a single point is an interval. An *extended infinite interval* includes either ∞ or $-\infty$. For example, $[-\infty,a) = \{-\infty\} \cup (-\infty,a)$, $[-\infty,a] = \{-\infty\} \cup (-\infty,a]$, $(a,\infty] = (a,\infty) \cup \{\infty\}$, $[a,\infty] = [a,\infty) \cup \{\infty\}$, and $[-\infty,\infty] = \{-\infty\} \cup (-\infty,\infty) \cup \{\infty\}$.

The symbol $\mathbb{F}$ denotes either $\mathbb{R}$ or $\mathbb{C}$ consistently in each theorem, proposition, lemma, corollary, and fact. For example, in Theorem 7.6.3, the three appearances of "$\mathbb{F}$" can be read as either all "$\mathbb{C}$" or all "$\mathbb{R}$."

The imaginary numbers are denoted by $\mathrm{I}\mathbb{A}$. Hence, 0 is both a real number and an imaginary number.

The notation $\operatorname{Re} A$ and $\operatorname{Im} A$ represents the real and imaginary parts of A, respectively. Some books use $\operatorname{Re} A$ and $\operatorname{Im} A$ to denote the Hermitian and skew-Hermitian matrices $\frac{1}{2}(A + A^*)$ and $\frac{1}{2}(A - A^*)$.

For the scalar ordering "$\le$," if $x \le y$, then $x \ne y$ if and only if $x < y$. For vector and matrix orderings, the conditions $x \le y$ and $x \ne y$ do not imply $x < y$.

Operations denoted by superscripts are applied before operations represented by preceding operators. For example, $\operatorname{tr}(A+B)^2$ means $\operatorname{tr}[(A+B)^2]$, and $\operatorname{cl} \mathcal{S}^\sim$ means $\operatorname{cl}(\mathcal{S}^\sim)$. This convention simplifies many expressions.

An element of $\mathbb{F}^n$ is a column vector, which is also a matrix with one column. The components of $x \in \mathbb{F}^n$ can be viewed as coordinates. In more general settings, "vector" typically refers to a coordinate-free object.

Sets have elements; vectors, tuples, and sequences have components; and matrices have entries. This terminology has no mathematical consequence.

All matrices have nonnegative integral dimensions. A matrix that has either zero rows or zero columns is an empty matrix.

The notation $x_{(i)}$ denotes the ith component of the vector x.

The entries of a submatrix $\hat{A}$ of a matrix A are the entries of A located in specified rows and columns of A. The submatrix $\hat{A}$ of A is a block of A if $\hat{A}$ is a submatrix of A whose entries are entries of adjacent rows and adjacent columns of A. Every matrix is both a submatrix and block of itself.

$A_{(i,j)}$ denotes the scalar (i, j) entry of A. $A_{i,j}$ or A_{ij} denotes a block or submatrix of A.

$A_{(\mathcal{S}_1,\mathcal{S}_2)}$ denotes the submatrix of A formed by retaining the rows of A listed in $\mathcal{S}_1$ and the columns of A listed in $\mathcal{S}_2$. $A_{(\mathcal{S})}$ denotes $A_{(\mathcal{S},\mathcal{S})}$.

$A_{[\mathcal{S}_1,\mathcal{S}_2]}$ denotes the submatrix of A formed by deleting the rows of A listed in $\mathcal{S}_1$ and the columns of A listed in $\mathcal{S}_2$. $A_{[\mathcal{S}]}$ denotes $A_{[\mathcal{S},\mathcal{S}]}$. $A_{[i,j]}$ denotes the submatrix of A obtained by deleting $\mathrm{row}_i(A)$ and $\mathrm{col}_j(A)$.

$A_{[\mathcal{S},\cdot]}$ denotes the submatrix of A obtained by deleting the rows of A listed in $\mathcal{S}$, and $A_{[\cdot,\mathcal{S}]}$ denotes the submatrix of A obtained by deleting the columns of A listed in $\mathcal{S}$. $A_{[i,\cdot]}$ denotes the submatrix of A obtained by deleting $\mathrm{row}_i(A)$, and $A_{[\cdot,j]}$ denotes the submatrix of A obtained by deleting $\mathrm{col}_j(A)$.

The determinant of a square submatrix is a subdeterminant. Some books use "minor." The determinant of a matrix is also a subdeterminant.

The dimension of the null space of a matrix is its defect. Some books use "nullity."

A block of a square matrix is diagonally located if the block is square and the diagonal entries of the block are also diagonal entries of the matrix; otherwise, the block is off-diagonally located. This terminology avoids confusion with a "diagonal block," which is a block that is also a square, diagonal submatrix.

For the partitioned matrix $\left[\begin{smallmatrix} A & B \\ C & D \end{smallmatrix}\right] \in \mathbb{F}^{(n+m)\times(k+l)}$, it can be inferred that $A \in \mathbb{F}^{n\times k}$ and similarly for B, C, and D.

The Schur product of matrices A and B is denoted by $A \odot B$. Matrix multiplication is given priority over Schur multiplication; that is, $A \odot BC$ means $A \odot (BC)$.

The adjugate of $A \in \mathbb{F}^{n\times n}$ is denoted by A^{A}. The traditional notation is $\mathrm{adj}\, A$, while the notation A^{A} is used in [2537]. If $A \in \mathbb{F}$ is a scalar, then $A^{\mathsf{A}} = 1$. In particular, $0^{\mathsf{A}}_{1\times 1} = 1$. However, for all $n \geq 2$, $0^{\mathsf{A}}_{n\times n} = 0_{n\times n}$.

If $\mathbb{F} = \mathbb{R}$, then $\overline{A}$ becomes A, A^* becomes A^{T}, "Hermitian" becomes "symmetric," "unitary" becomes "orthogonal," "unitarily" becomes "orthogonally," and "congruence" becomes "T-congruence." The square complex matrix A is symmetric if $A^{\mathsf{T}} = A$ and orthogonal if $A^{\mathsf{T}}A = I$.

The diagonal entries of $A \in \mathbb{F}^{n\times n}$, all of whose diagonal entries are real, are ordered as $\mathrm{d}_{\max}(A) = \mathrm{d}_1(A) \geq \mathrm{d}_2(A) \geq \cdots \geq \mathrm{d}_n(A) = \mathrm{d}_{\min}(A)$.

Every $n \times n$ matrix has n eigenvalues. Hence, eigenvalues are counted in accordance with their algebraic multiplicity. The phrase "distinct eigenvalues" ignores algebraic multiplicity.

The eigenvalues of $A \in \mathbb{F}^{n\times n}$, all of whose eigenvalues are real, are ordered as $\lambda_{\max}(A) = \lambda_1(A) \geq \lambda_2(A) \geq \cdots \geq \lambda_n(A) = \lambda_{\min}(A)$.

The inertia of $A \in \mathbb{F}^{n\times n}$ is written as

$$\mathrm{In}\, A \triangleq \begin{bmatrix} \nu_-(A) \\ \nu_0(A) \\ \nu_+(A) \end{bmatrix}.$$

Some books use the notation $(\nu(A), \delta(A), \pi(A))$.

For $A \in \mathbb{F}^{n\times n}$, $\mathrm{amult}_A(\lambda)$ is the number of copies of λ in the multispectrum of A, $\mathrm{gmult}_A(\lambda)$ is the number of Jordan blocks of A associated with λ, and $\mathrm{ind}_A(\lambda)$ is the size of the largest Jordan block of A associated with λ. The index of A, denoted by $\mathrm{ind}\, A = \mathrm{ind}_A(0)$, is the size of the largest Jordan block of A associated with the eigenvalue 0.

$A \in \mathbb{C}^{n\times n}$ is *semisimple* if the size of every Jordan block of A is 1. *Defective* means not semisimple.

$A \in \mathbb{C}^{n\times n}$ is *cyclic* if A has exactly one Jordan block associated with each distinct eigenvalue. *Derogatory* means not cyclic.

$A \in \mathbb{F}^{n\times n}$ is *diagonalizable over* $\mathbb{F}$ if it can be transformed into a diagonal matrix whose entries are in $\mathbb{F}$ by means of a similarity transformation whose entries are in $\mathbb{F}$. Therefore, $A \in \mathbb{C}^{n\times n}$ is diagonalizable over $\mathbb{C}$ if and only if it is semisimple. Furthermore, $A \in \mathbb{R}^{n\times n}$ is diagonalizable over $\mathbb{R}$ if and only if it is semisimple and all of its eigenvalues are real. The real matrix $\left[\begin{smallmatrix} 0 & 1 \\ -1 & 0 \end{smallmatrix}\right]$ is diagonalizable over $\mathbb{C}$, but not diagonalizable over $\mathbb{R}$.

$A \in \mathbb{F}^{n\times m}$ has exactly $\min\{n, m\}$ singular values, exactly $\operatorname{rank} A$ of which are positive.

The $\min\{n, m\}$ singular values of the matrix $A \in \mathbb{F}^{n\times m}$ are ordered as $\sigma_{\max}(A) \triangleq \sigma_1(A) \ge \sigma_2(A) \ge \cdots \ge \sigma_{\min\{n,m\}}(A) \ge 0$. If $n = m$, then $\sigma_{\min}(A) \triangleq \sigma_n(A)$. The notation $\sigma_{\min}(A)$ is defined only for square matrices.

By definition, positive-semidefinite and positive-definite matrices are Hermitian.

An idempotent matrix $A \in \mathbb{F}^{n\times n}$ satisfies $A^2 = A$, while a projector is a Hermitian, idempotent matrix. Some books use "projector" for idempotent and "orthogonal projector" for projector. A reflector is a Hermitian, involutory matrix. A projector is a normal matrix each of whose eigenvalues is 1 or 0, while a reflector is a normal matrix each of whose eigenvalues is 1 or -1.

An elementary matrix is a nonsingular matrix formed by adding an outer-product matrix to the identity matrix. An elementary reflector is a reflector exactly one of whose eigenvalues is -1. An elementary projector is a projector exactly one of whose eigenvalues is 0. Elementary reflectors are elementary matrices. However, elementary projectors are not elementary matrices since elementary projectors are singular.

A range-Hermitian matrix is a square matrix whose range is equal to the range of its complex conjugate transpose. These matrices are sometimes called EP matrices.

The polynomials 1 and $s^3 + 5s^2 - 4$ are monic. The zero polynomial is not monic.

The rank of the polynomial matrix P is the maximum rank of $P(s)$ over all $s \in \mathbb{C}$. This quantity is also called the normal rank. We denote this quantity by $\operatorname{rank} P$ as distinct from $\operatorname{rank} P(s)$, which denotes the rank of the matrix $P(s)$.

The rank of the rational transfer function G is the maximum rank of $G(s)$ over all $s \in \mathbb{C}$ excluding poles of the entries of G. This quantity is also called the normal rank. We denote this quantity by $\operatorname{rank} G$ as distinct from $\operatorname{rank} G(s)$, which denotes the rank of the matrix $G(s)$.

$\oplus$ denotes the Kronecker sum. Some books use $\oplus$ to denote a direct sum of matrices or subspaces.

$|A|$ represents the matrix obtained by replacing every entry of A by its absolute value.

$\langle A\rangle$ represents the matrix $(A^*A)^{1/2}$. Some books use $|A|$ to denote this matrix.

Statements about vector norms on $\mathbb{F}^{nm}$ apply to matrix norms on $\mathbb{F}^{n\times m}$, and vice versa.

The Hölder norm of $A \in \mathbb{F}^{n\times m}$ is denoted by $\|A\|_p$. The matrix norm induced by $\|\cdot\|_q$ on the domain and $\|\cdot\|_p$ on the codomain is denoted by $\|\cdot\|_{p,q}$.

The Schatten p norm of $A \in \mathbb{F}^{n\times m}$ is denoted by $\|A\|_{\sigma p}$, and the Frobenius norm of A is denoted by $\|A\|_{\mathrm{F}}$. Hence, $\|A\|_{\sigma\infty} = \|A\|_{2,2} = \sigma_{\max}(A)$, $\|A\|_{\sigma 2} = \|A\|_{\mathrm{F}}$, and $\|A\|_{\sigma 1} = \operatorname{tr}\langle A\rangle$.

Unitarily invariant norms are not necessarily normalized.

Terminology Relating to Inequalities

Let "$\leq$" be a partial ordering, let X be a set, and consider the inequality

$$f(x) \leq g(x) \text{ for all } x \in X. \tag{1}$$

Inequality (1) is *sharp* if there exists $x_0 \in X$ such that $f(x_0) = g(x_0)$.

The inequality

$$f(x) \leq f(y) \text{ for all } x \leq y \tag{2}$$

is a monotonicity result.

The inequality

$$f(x) \leq p(x) \leq g(x) \text{ for all } x \in X, \tag{3}$$

where p is not identically equal to either f or g on X, is an *interpolation* or *refinement* of (1). The inequality

$$g(x) \leq \alpha f(x) \text{ for all } x \in X, \tag{4}$$

where $\alpha > 1$, is a *reversal* of (1). The inequality

$$g(x) \leq f(x) + \alpha \text{ for all } x \in X, \tag{5}$$

where $\alpha > 0$, is a *reversal* of (1).

Defining $h(x) \triangleq g(x) - f(x)$, it follows that (1) is equivalent to

$$h(x) \geq 0 \text{ for all } x \in X. \tag{5}$$

Now, suppose that h has a global minimizer $x_0 \in X$. Then, (5) implies that

$$0 \leq h(x_0) = \min_{x \in X} h(x) \leq h(y) \text{ for all } y \in X. \tag{6}$$

Consequently, inequalities are often expressed equivalently in terms of minimizers, and vice versa.

Many inequalities are based on a function that is either monotonic or convex.

Scalar, Vector, and Matrix Mathematics

Chapter One
Sets, Logic, Numbers, Relations, Orderings, Graphs, and Functions

In this chapter we review basic terminology and results concerning sets, logic, numbers, relations, orderings, graphs, and functions. This material is used throughout the book.

1.1 Sets

A *set* $\{x, y, \ldots\}$ is a collection of elements. A set can include either a finite or infinite number of elements. The set $\mathcal{X}$ is *finite* if it has a finite number of elements; otherwise, $\mathcal{X}$ is *infinite*. The set $\mathcal{X}$ is *countably infinite* if $\mathcal{X}$ is infinite and its elements are in one-to-one correspondence with the positive integers. The set $\mathcal{X}$ is *countable* if it is either finite or countably infinite.

Let $\mathcal{X}$ be a set. Then,

$$x \in \mathcal{X} \tag{1.1.1}$$

means that x is an *element* of $\mathcal{X}$. If w is not an element of $\mathcal{X}$, then we write

$$w \notin \mathcal{X}. \tag{1.1.2}$$

No set can be an element of itself. Therefore, there does not exist a set that includes every set. The set with no elements, denoted by $\varnothing$, is the *empty set*. If $\mathcal{X} \neq \varnothing$, then $\mathcal{X}$ is *nonempty*.

Let $\mathcal{X}$ and $\mathcal{Y}$ be sets. The *intersection* of $\mathcal{X}$ and $\mathcal{Y}$ is the set of common elements of $\mathcal{X}$ and $\mathcal{Y}$, which is given by

$$\mathcal{X} \cap \mathcal{Y} \triangleq \{x\colon\ x \in \mathcal{X} \text{ and } x \in \mathcal{Y}\} = \{x \in \mathcal{X}\colon\ x \in \mathcal{Y}\} = \{x \in \mathcal{Y}\colon\ x \in \mathcal{X}\} = \mathcal{Y} \cap \mathcal{X}, \tag{1.1.3}$$

The *union* of $\mathcal{X}$ and $\mathcal{Y}$ is the set of elements in either $\mathcal{X}$ or $\mathcal{Y}$, which is the set

$$\mathcal{X} \cup \mathcal{Y} \triangleq \{x\colon\ x \in \mathcal{X} \text{ or } x \in \mathcal{Y}\} = \mathcal{Y} \cup \mathcal{X}. \tag{1.1.4}$$

The *complement* of $\mathcal{X}$ *relative* to $\mathcal{Y}$ is

$$\mathcal{Y}\backslash\mathcal{X} \triangleq \{x \in \mathcal{Y}\colon\ x \notin \mathcal{X}\}. \tag{1.1.5}$$

If $\mathcal{Y}$ is specified, then the *complement* of $\mathcal{X}$ is

$$\mathcal{X}^{\sim} \triangleq \mathcal{Y}\backslash\mathcal{X}. \tag{1.1.6}$$

The *symmetric difference* of $\mathcal{X}$ and $\mathcal{Y}$ is the set of elements that are in either $\mathcal{X}$ or $\mathcal{Y}$ but not both, which is given by

$$\mathcal{X} \ominus \mathcal{Y} \triangleq (\mathcal{X} \cup \mathcal{Y})\backslash(\mathcal{X} \cap \mathcal{Y}). \tag{1.1.7}$$

If $x \in \mathcal{X}$ implies that $x \in \mathcal{Y}$, then $\mathcal{X}$ is a *subset* of $\mathcal{Y}$ (equivalently, $\mathcal{Y}$ *contains* $\mathcal{X}$), which is written as

$$\mathcal{X} \subseteq \mathcal{Y}. \tag{1.1.8}$$

Equivalently,

$$\mathcal{Y} \supseteq \mathcal{X}. \tag{1.1.9}$$

Note that $\mathcal{X} \subseteq \mathcal{Y}$ if and only if $\mathcal{X}\backslash\mathcal{Y} = \varnothing$. Furthermore, $\mathcal{X} = \mathcal{Y}$ if and only if $\mathcal{X} \subseteq \mathcal{Y}$ and $\mathcal{Y} \subseteq \mathcal{X}$. If $\mathcal{X} \subseteq \mathcal{Y}$ and $\mathcal{X} \neq \mathcal{Y}$, then $\mathcal{X}$ is a *proper subset* of $\mathcal{Y}$ and we write $\mathcal{X} \subset \mathcal{Y}$. The sets $\mathcal{X}$ and $\mathcal{Y}$ are *disjoint* if $\mathcal{X} \cap \mathcal{Y} = \varnothing$. A *partition* of $\mathcal{X}$ is a set of pairwise-disjoint and nonempty subsets of $\mathcal{X}$ whose union is equal to $\mathcal{X}$.

The symbols $\mathbb{N}$, $\mathbb{P}$, $\mathbb{Z}$, $\mathbb{Q}$, and $\mathbb{R}$ denote the sets of nonnegative integers, positive integers, integers, rational numbers, and real numbers, respectively.

A set cannot have repeated elements. Therefore, $\{x, x\} = \{x\}$. A *multiset* is a finite collection of elements that allows for repetition. The multiset consisting of two copies of x is written as $\{x, x\}_{\rm ms}$. For example, the roots of the polynomial $p(x) = (x-1)^2$ are the elements of the multiset $\{1, 1\}_{\rm ms}$, while the prime factors of 72 are the elements of the multiset $\{2, 2, 2, 3, 3\}_{\rm ms}$.

The operations "$\cap$," "$\cup$," "$\backslash$," "$\ominus$," and "$\times$" and the relations "$\subset$" and "$\subseteq$" extend to multisets. For example,

$$\{x, x\}_{\rm ms} \cup \{x\}_{\rm ms} = \{x, x, x\}_{\rm ms}. \tag{1.1.10}$$

By ignoring repetitions, a multiset can be converted to a set, while a set can be viewed as a multiset with distinct elements.

The *Cartesian product* $\mathcal{X}_1 \times \cdots \times \mathcal{X}_n$ of sets $\mathcal{X}_1, \ldots, \mathcal{X}_n$ is the set consisting of *tuples* of the form $(x_1, \ldots, x_n)$, where, for all $i \in \{1, \ldots, n\}$, $x_i \in \mathcal{X}_i$. A tuple with n components is an *n-tuple*. The components of a tuple are ordered but need not be distinct. Therefore, a tuple can be viewed as an ordered multiset. We thus write

$$(x_1, \ldots, x_n) \in \times_{i=1}^{n} \mathcal{X}_i \triangleq \mathcal{X}_1 \times \cdots \times \mathcal{X}_n. \tag{1.1.11}$$

$\mathcal{X}^n$ denotes $\times_{i=1}^{n} \mathcal{X}$.

Definition 1.1.1. A *sequence* $(x_i)_{i=1}^{\infty} = (x_1, x_2, \ldots)$ is a tuple with a countably infinite number of components. Now, let $i_1 < i_2 < \cdots$. Then, $(x_{i_j})_{j=1}^{\infty}$ is a *subsequence* of $(x_i)_{i=1}^{\infty}$.

Let $\mathcal{X}$ be a set, and let $X \triangleq (x_i)_{i=1}^{\infty}$ be a sequence whose components are elements of $\mathcal{X}$; that is, $\{x_1, x_2, \ldots\} \subseteq \mathcal{X}$. For convenience, we write either $X \subseteq \mathcal{X}$ or $X \subset \mathcal{X}$, where X is viewed as a set and the multiplicity of the components of the sequence is ignored. For sequences $X, Y \subset \mathbb{F}^n$, define $X + Y \triangleq (x_i + y_i)_{i=1}^{\infty}$ and $X \odot Y \triangleq (x_i \odot y_i)_{i=1}^{\infty}$, where "$\odot$" denotes component-wise multiplication. In the case $n = 1$, we define $XY \triangleq (x_i y_i)_{i=1}^{\infty}$.

1.2 Logic

Every *statement* is either true or false, and no statement is both true and false. A *proof* is a collection of statements that verify that a statement is true. A *conjecture* is a statement that is believed to be true but whose proof is not known.

Let A and B be statements. The *not* of A is the statement (not A), the *and* of A and B is the statement (A and B), and the *or* of A and B is the statement (A or B). The statement (A or B) does not contradict the statement (A and B); hence, the word "or" is inclusive. The *exclusive or* of A and B is the statement (A xor B), which is [(A and not B) or (B and not A)]. Equivalently, (A xor B) is the statement [(A or B) and not(A and B)], that is, A or B, but not both. Note that (A and B) = (B and A), (A or B) = (B or A), and (A xor B) = (B xor A).

Let A, B, and C be statements. Then, the statements (A and B or C) and (A or B and C) are ambiguous. For clarity, we thus write, for example, [A and (B or C)] and [A or (B and C)]. In words, we write "A and either B or C" and "A or both B and C," respectively, where "either" and "both" signify parentheses. Furthermore,

$$(A \text{ and } B) \text{ or } C = (A \text{ and } C) \text{ or } (B \text{ and } C), \tag{1.2.1}$$

$$(A \text{ or } B) \text{ and } C = (A \text{ or } C) \text{ and } (B \text{ or } C). \tag{1.2.2}$$

Let A be a statement. To analyze statements involving logic operators, define truth$(A) = 1$ if A is true, and truth$(A) = 0$ if A is false. Then,

$$\text{truth}(\text{not } A) = \text{truth}(A) + 1, \tag{1.2.3}$$

where $0+0=0$, $1+0=0+1=1$, and $1+1=0$. Therefore, A is true if and only if (not A) is false, while A is false if and only if (not A) is true. Note that

$$\begin{aligned}\text{truth}[\text{not}(\text{not } A)] &= \text{truth}(\text{not } A) + 1\\ &= [\text{truth}(A) + 1] + 1\\ &= \text{truth}(A).\end{aligned}$$

Furthermore, note that truth(A) + truth$(A) = 0$ and truth(A) truth(A) = truth(A).

Let A and B be statements. Then,

$$\text{truth}(A \text{ and } B) = \text{truth}(A)\,\text{truth}(B), \tag{1.2.4}$$

$$\text{truth}(A \text{ or } B) = \text{truth}(A)\,\text{truth}(B) + \text{truth}(A) + \text{truth}(B), \tag{1.2.5}$$

$$\text{truth}(A \text{ xor } B) = \text{truth}(A) + \text{truth}(B). \tag{1.2.6}$$

Hence,

$$\text{truth}(A \text{ and } B) = \min\,\{\text{truth}(A), \text{truth}(B)\}, \tag{1.2.7}$$

$$\text{truth}(A \text{ or } B) = \max\,\{\text{truth}(A), \text{truth}(B)\}. \tag{1.2.8}$$

Consequently, truth(A and B) = truth(B and A), truth(A or B) = truth(B or A), and truth(A xor B) = truth(B xor A). Furthermore, truth(A and A) = truth(A or A) = truth(A), and truth(A xor A) = 0.

Let A and B be statements. The *implication* $(A \Longrightarrow B)$ is the statement [(not A) or B]. Therefore,

$$\text{truth}(A \Longrightarrow B) = \text{truth}(A)\,\text{truth}(B) + \text{truth}(A) + 1. \tag{1.2.9}$$

The implication $(A \Longrightarrow B)$ is read as either "if A, then B," "if A holds, then B holds," or "A implies B." The statement A is the *hypothesis*, while the statement B is the *conclusion*. If $(A \Longrightarrow B)$, then A is a *sufficient condition* for B, and B is a *necessary condition* for A. It follows from (1.2.9) that, if A and B are true, then $(A \Longrightarrow B)$ is true; if A is true and B is false, then $(A \Longrightarrow B)$ is false; and, if A is false, then $(A \Longrightarrow B)$ is true whether or not B is true. For example, both implications $[(2+2=5) \Longrightarrow (3+3=6)]$ and $[(2+2=5) \Longrightarrow (3+3=8)]$ are true. Finally, note that $[(A \Longrightarrow B)$ and $A]$ = A and B.

A *predicate* is a statement that depends on a variable. Let $\mathcal{X}$ be a set, let $x \in \mathcal{X}$, and let $A(x)$ be a predicate. There are two ways to use a predicate to create a statement. An *existential statement* has the form

$$\text{there exists } x \in \mathcal{X} \text{ such that } A(x) \text{ holds}, \tag{1.2.10}$$

whereas a *universal statement* has the form

$$\text{for all } x \in \mathcal{X}, A(x) \text{ holds}. \tag{1.2.11}$$

Note that

$$\text{truth}[\text{there exists } x \in \mathcal{X} \text{ such that } A(x) \text{ holds}] = \max_{x\in\mathcal{X}} \text{truth}[A(x)], \tag{1.2.12}$$

$$\text{truth}[\text{for all } x \in \mathcal{X}, A(x) \text{ holds}] = \min_{x\in\mathcal{X}} \text{truth}[A(x)]. \tag{1.2.13}$$

An *argument* is an implication whose hypothesis and conclusion are predicates that depend on the same variable. In particular, letting x denote a variable, and letting $A(x)$ and $B(x)$ be predicates, the implication $[A(x) \Longrightarrow B(x)]$ is an argument. For example, for each real number x, the implication $[(x = 1) \Longrightarrow (x + 1 = 2)]$ is an argument. Note that the variable x links the hypothesis and the conclusion, thereby making this implication useful for the purpose of *inference*. In particular, for all real numbers x, truth$[(x = 1) \Longrightarrow (x+1 = 2)] = 1$. The statements (for all x, $[A(x) \Longrightarrow B(x)]$ holds) and (there exists x such that $[A(x) \Longrightarrow B(x)]$ holds) are inferences.

Let A and B be statements. The *bidirectional implication* $(A \Longleftrightarrow B)$ is the statement $[(A \Longrightarrow B)$ and $(A \Longleftarrow B)]$, where $(A \Longleftarrow B)$ means $(B \Longrightarrow A)$. If $(A \Longleftrightarrow B)$, then A and B are *equivalent*. Furthermore,

$$\text{truth}(A \Longleftrightarrow B) = \text{truth}(A) + \text{truth}(B) + 1. \tag{1.2.14}$$

Therefore, A and B are equivalent if and only if either both A and B are true or both A and B are false.

Let A and B be statements, and assume that $(A \Longleftrightarrow B)$. Then, A holds *if and only if* B holds. The implication $A \Longrightarrow B$ (the "only if" part) is *necessity*, while $B \Longrightarrow A$ (the "if" part) is *sufficiency*.

Let A and B be statements. The *converse* of $(A \Longrightarrow B)$ is $(B \Longrightarrow A)$. Note that

$$\begin{aligned}(A \Longrightarrow B) &\Longleftrightarrow [(\text{not } A) \text{ or } B]\\ &\Longleftrightarrow [(\text{not } A) \text{ or not}(\text{not } B)]\\ &\Longleftrightarrow [\text{not}(\text{not } B) \text{ or not } A]\\ &\Longleftrightarrow (\text{not } B \Longrightarrow \text{not } A).\end{aligned}$$

Therefore, the statement $(A \Longrightarrow B)$ is equivalent to its *contrapositive* $[(\text{not } B) \Longrightarrow (\text{not } A)]$.

Let A, B, A', and B' be statements, and assume that $(A' \Longrightarrow A \Longrightarrow B \Longrightarrow B')$. Then, $(A' \Longrightarrow B')$ is a *corollary* of $(A \Longrightarrow B)$.

Let A, B, and A' be statements, and assume that $A \Longrightarrow B$. Then, $(A \Longrightarrow B)$ is a *strengthening* of $[(A \text{ and } A') \Longrightarrow B]$. If, in addition, $(A \Longrightarrow A')$, then the statement $[(A \text{ and } A') \Longrightarrow B]$ has a *redundant assumption*.

An *interpretation* is a feasible assignment of true or false to all statements that comprise a statement. For example, there are four interpretations of the statement (A and B), depending on whether A is assigned to be true or false and B is assigned to be true or false. Likewise, $[(x = 1)$ and $(x = 2)]$ has three interpretations, which depend on the value of x.

Let $A_1, A_2, \ldots$ be statements, and let B be a statement that depends on $A_1, A_2, \ldots$ Then, B is a *tautology* if B is true whether or not $A_1, A_2, \ldots$ are true. For example, let B denote the statement (A or not A). Then,

$$\text{truth}(A \text{ or not } A) = 1, \tag{1.2.15}$$

and thus the statement (A or not A) is true whether or not A is true. Hence, (A or not A) is a tautology. Likewise, $(A \Longrightarrow A)$ is a tautology. Furthermore, since

$$\text{truth}[(A \text{ and } B) \Longrightarrow A] = \text{truth}(A)^2\,\text{truth}(B) + \text{truth}(A)\,\text{truth}(B) + 1 = 1, \tag{1.2.16}$$

it follows that $[(A \text{ and } B) \Longrightarrow A]$ is a tautology. Likewise, truth$([A \text{ and not } A] \Longrightarrow B) = 1$, and thus $([A \text{ and not } A] \Longrightarrow B)$ is a tautology.

Let $A_1, A_2, \ldots$ be statements, and let B be a statement that depends on $A_1, A_2, \ldots$ Then, B is a *contradiction* if B is false whether or not $A_1, A_2, \ldots$ are true. For example, let B denote the statement (A and not A). Then,

$$\text{truth}(A \text{ and not } A) = 0, \tag{1.2.17}$$

and thus the statement (A and not A) is false whether or not A is true. Hence, (A and not A) is a contradiction.

Let A and B be statements. If the implication ($A \implies B$) is neither a tautology nor a contradiction, then truth($A \implies B$) depends on the truth of the statements that comprise A and B. For example, truth($A \implies$ not A) = truth(A) + 1, and thus the statement ($A \implies$ not A) is true if and only if A is false, and false if and only if A is true. Hence, ($A \implies$ not A) is neither a tautology nor a contradiction. A statement that is neither a tautology nor a contradiction is a *contingency*. For example, the implication [$A \implies$ (A and B)] is a contingency. Likewise, for each real number x, truth$[(x = 1) \implies (x = 2)]$ = truth$(x \neq 1)$, and thus the statement $[(x = 1) \implies (x = 2)]$ is a contingency.

An argument that is a contingency is a *theorem*, *proposition*, *corollary*, or *lemma*. A theorem is a significant result; a proposition is a theorem of less significance. The primary role of a lemma is to support the proof of a theorem or a proposition. A *corollary* is a consequence of a theorem or a proposition. A *fact* is either a theorem, proposition, lemma, or corollary.

In order to visualize logic operations on predicates, it is helpful to replace statements with sets and logic operations by set operations; the truth of a statement can then be visualized in terms of Venn diagrams. To do this, let $\mathcal{X}$ be a set, for all $x \in \mathcal{X}$, let $A(x)$ and $B(x)$ be predicates, and define $\mathcal{A} \triangleq \{x \in \mathcal{X}\colon \text{truth}[A(x)] = 1\}$ and $\mathcal{B} \triangleq \{x \in \mathcal{X}\colon \text{truth}[B(x)] = 1\}$. Then, the logic operations "and," "or," "xor," and "not" are equivalent to "$\cap$," "$\cup$," "$\ominus$," and "$^\sim$," respectively. For example, $\{x \in \mathcal{X}\colon \text{truth}[(\text{not } A(x)) \text{ and } B(x)] = 1\} = \mathcal{A}^\sim \cap \mathcal{B}$. Furthermore, since $[A(x) \implies B(x)]$ is equivalent to $[(\text{not } A(x)) \text{ or } B(x)]$, it follows that $\{x \in \mathcal{X}\colon \text{truth}[A(x) \implies B(x)] = 1\} = \mathcal{A}^\sim \cup \mathcal{B}$. Similarly, since $[A(x) \iff B(x)]$ is equivalent to $[(A(x) \text{ or not } B(x)) \text{ and } ([\text{not } A(x)] \text{ or } B(x))]$, it follows that $\{x \in \mathcal{X}\colon A(x) \iff B(x)\} = (\mathcal{A} \cup \mathcal{B}^\sim) \cap (\mathcal{A}^\sim \cup \mathcal{B}) = (\mathcal{A} \cap \mathcal{B}) \cup (\mathcal{A} \cup \mathcal{B})^\sim$.

Now, define $\mathcal{X}$, $A(x)$, $B(x)$, $\mathcal{A}$, and $\mathcal{B}$ as in the previous paragraph, and assume that, for all $x \in \mathcal{X}$, $A(x) \implies B(x)$. Therefore, $\mathcal{A}^\sim \cup \mathcal{B} = \{x \in \mathcal{X}\colon \text{truth}[(\text{not } A(x)) \text{ or } B(x)] = 1\} = \mathcal{X}$, and thus $\mathcal{A}\backslash\mathcal{B} = (\mathcal{A}^\sim \cup \mathcal{B})^\sim = \{x \in \mathcal{X}\colon \text{truth}[(\text{not } A(x)) \text{ or } B(x)] = 0\} = \varnothing$. Consequently, $\mathcal{A} \subseteq \mathcal{B}$. This means that the logic operator "$\implies$" is represented by "$\subseteq$." For example, for all $x \in \mathcal{X}$, let $C(x)$ be a predicate, and define $\mathcal{C} \triangleq \{x \in \mathcal{X}\colon \text{truth}[C(x)] = 1\}$. Then, for all $x \in \mathcal{X}$, truth$[(A(x) \text{ and } B(x)) \implies C(x)] = 1$ if and only if $\mathcal{A} \cap \mathcal{B} \subseteq \mathcal{C}$. Likewise, for all $x \in \mathcal{X}$,

$$\text{truth}([A(x) \text{ and } (B(x) \text{ or } C(x))] \iff [(A(x) \text{ and } B(x)) \text{ or } (A(x) \text{ and } C(x))]) = 1 \tag{1.2.18}$$

if and only if

$$\mathcal{A} \cap (\mathcal{B} \cup \mathcal{C}) = (\mathcal{A} \cap \mathcal{B}) \cup (\mathcal{A} \cap \mathcal{C}). \tag{1.2.19}$$

Note that (1.2.19) represents a tautology.

1.3 Relations and Orderings

Let $\mathcal{X}$, $\mathcal{X}_1$, and $\mathcal{X}_2$ be sets. A *relation* $\mathcal{R}$ on $(\mathcal{X}_1, \mathcal{X}_2)$ is a subset of $\mathcal{X}_1 \times \mathcal{X}_2$. A *relation* $\mathcal{R}$ on $\mathcal{X}$ is a subset of $\mathcal{X} \times \mathcal{X}$. Likewise, a *multirelation* $\mathcal{R}$ on $(\mathcal{X}_1, \mathcal{X}_2)$ is a multisubset of $\mathcal{X}_1 \times \mathcal{X}_2$, while a *multirelation* $\mathcal{R}$ on $\mathcal{X}$ is a multisubset of $\mathcal{X} \times \mathcal{X}$.

Let $\mathcal{X}$ be a set, and let $\mathcal{R}_1$ and $\mathcal{R}_2$ be relations on $\mathcal{X}$. Then, the sets $\mathcal{R}_1 \cap \mathcal{R}_2$, $\mathcal{R}_1\backslash\mathcal{R}_2$, and $\mathcal{R}_1 \cup \mathcal{R}_2$ are relations on $\mathcal{X}$. Furthermore, if $\mathcal{R}$ is a relation on $\mathcal{X}$ and $\mathcal{X}_0 \subseteq \mathcal{X}$, then we define the *restricted relation* $\mathcal{R}|_{\mathcal{X}_0} \triangleq \mathcal{R} \cap (\mathcal{X}_0 \times \mathcal{X}_0)$, which is a relation on $\mathcal{X}_0$.

Definition 1.3.1. Let $\mathcal{R}$ be a relation on the set $\mathcal{X}$. Then, the following terminology is defined:

i) $\mathcal{R}$ is *reflexive* if, for all $x \in \mathcal{X}$, it follows that $(x, x) \in \mathcal{R}$.

ii) $\mathcal{R}$ is *symmetric* if, for all $(x_1, x_2) \in \mathcal{R}$, it follows that $(x_2, x_1) \in \mathcal{R}$.

iii) $\mathcal{R}$ is *transitive* if, for all $(x_1, x_2) \in \mathcal{R}$ and $(x_2, x_3) \in \mathcal{R}$, it follows that $(x_1, x_3) \in \mathcal{R}$.

iv) $\mathcal{R}$ is an *equivalence relation* if $\mathcal{R}$ is reflexive, symmetric, and transitive.

Proposition 1.3.2. Let $\mathcal{R}_1$ and $\mathcal{R}_2$ be relations on the set $\mathcal{X}$. If $\mathcal{R}_1$ and $\mathcal{R}_2$ are (reflexive, symmetric) relations, then so are $\mathcal{R}_1 \cap \mathcal{R}_2$ and $\mathcal{R}_1 \cup \mathcal{R}_2$. If $\mathcal{R}_1$ and $\mathcal{R}_2$ are (transitive, equivalence) relations, then so is $\mathcal{R}_1 \cap \mathcal{R}_2$.

Definition 1.3.3. Let $\mathcal{R}$ be a relation on the set $\mathcal{X}$. Then, the following terminology is defined:

i) The *complement* $\mathcal{R}^\sim$ of $\mathcal{R}$ is the relation $\mathcal{R}^\sim \triangleq (\mathcal{X} \times \mathcal{X})\backslash\mathcal{R}$.

ii) The *support* $\operatorname{supp}(\mathcal{R})$ of $\mathcal{R}$ is the smallest subset $\mathcal{X}_0$ of $\mathcal{X}$ such that $\mathcal{R}$ is a relation on $\mathcal{X}_0$.

iii) The *reversal* $\operatorname{rev}(\mathcal{R})$ of $\mathcal{R}$ is the relation $\operatorname{rev}(\mathcal{R}) \triangleq \{(y,x)\colon (x,y) \in \mathcal{R}\}$.

iv) The *shortcut* $\operatorname{shortcut}(\mathcal{R})$ of $\mathcal{R}$ is the relation $\operatorname{shortcut}(\mathcal{R}) \triangleq \{(x,y) \in \mathcal{X} \times \mathcal{X}\colon x$ and y are distinct and there exist $k \ge 1$ and $x_1,\dots,x_k \in \mathcal{X}$ such that $(x,x_1),(x_1,x_2),\dots,(x_k,y) \in \mathcal{R}\}$.

v) The *reflexive hull* $\operatorname{ref}(\mathcal{R})$ of $\mathcal{R}$ is the smallest reflexive relation on $\mathcal{X}$ that contains $\mathcal{R}$.

vi) The *symmetric hull* $\operatorname{sym}(\mathcal{R})$ of $\mathcal{R}$ is the smallest symmetric relation on $\mathcal{X}$ that contains $\mathcal{R}$.

vii) The *transitive hull* $\operatorname{trans}(\mathcal{R})$ of $\mathcal{R}$ is the smallest transitive relation on $\mathcal{X}$ that contains $\mathcal{R}$.

viii) The *equivalence hull* $\operatorname{equiv}(\mathcal{R})$ of $\mathcal{R}$ is the smallest equivalence relation on $\mathcal{X}$ that contains $\mathcal{R}$.

Proposition 1.3.4. Let $\mathcal{R}$ be a relation on the set $\mathcal{X}$. Then, the following statements hold:

i) $\operatorname{ref}(\mathcal{R}) = \mathcal{R} \cup \{(x,x)\colon x \in \mathcal{X}\}$.

ii) $\operatorname{sym}(\mathcal{R}) = \mathcal{R} \cup \operatorname{rev}(\mathcal{R})$.

iii) $\operatorname{trans}(\mathcal{R}) = \mathcal{R} \cup \operatorname{shortcut}(\mathcal{R})$.

iv) If $\mathcal{R}$ is symmetric, then $\operatorname{trans}(\mathcal{R}) = \operatorname{sym}(\operatorname{trans}(\mathcal{R}))$.

v) $\operatorname{equiv}(\mathcal{R}) = \operatorname{trans}(\operatorname{sym}(\operatorname{ref}(\mathcal{R})))$.

Furthermore, the following statements hold:

vi) $\mathcal{R}$ is reflexive if and only if $\mathcal{R} = \operatorname{ref}(\mathcal{R})$.

vii) The following statements are equivalent:

a) $\mathcal{R}$ is symmetric.

b) $\mathcal{R} = \operatorname{sym}(\mathcal{R})$.

c) $\mathcal{R} = \operatorname{rev}(\mathcal{R})$.

viii) $\mathcal{R}$ is transitive if and only if $\mathcal{R} = \operatorname{trans}(\mathcal{R})$.

ix) $\mathcal{R}$ is an equivalence relation if and only if $\mathcal{R} = \operatorname{equiv}(\mathcal{R})$.

For an equivalence relation $\mathcal{R}$ on the set $\mathcal{X}$, $(x_1,x_2) \in \mathcal{R}$ is denoted by $x_1 \equiv x_2$. If $\mathcal{R}$ is an equivalence relation and $x \in \mathcal{X}$, then the subset $\mathcal{E}_x \triangleq \{y \in \mathcal{X}\colon\ y \equiv x\}$ of $\mathcal{X}$ is the *equivalence class of x induced by* $\mathcal{R}$.

Theorem 1.3.5. Let $\mathcal{R}$ be an equivalence relation on a set $\mathcal{X}$. Then, the set $\{\mathcal{E}_x\colon x \in \mathcal{X}\}$ of equivalence classes induced by $\mathcal{R}$ is a partition of $\mathcal{X}$.

Proof. Since $\mathcal{X} = \bigcup_{x\in\mathcal{X}} \mathcal{E}_x$, it suffices to show that, if $x,y \in \mathcal{X}$, then either $\mathcal{E}_x = \mathcal{E}_y$ or $\mathcal{E}_x \cap \mathcal{E}_y = \varnothing$. Hence, let $x,y \in \mathcal{X}$, and suppose that $\mathcal{E}_x$ and $\mathcal{E}_y$ are not disjoint so that there exists $z \in \mathcal{E}_x \cap \mathcal{E}_y$. Thus, $(x,z) \in \mathcal{R}$ and $(z,y) \in \mathcal{R}$. Now, let $w \in \mathcal{E}_x$. Then, $(w,x) \in \mathcal{R}$, $(x,z) \in \mathcal{R}$, and $(z,y) \in \mathcal{R}$ imply that $(w,y) \in \mathcal{R}$. Hence, $w \in \mathcal{E}_y$, which implies that $\mathcal{E}_x \subseteq \mathcal{E}_y$. By a similar argument, $\mathcal{E}_y \subseteq \mathcal{E}_x$. Consequently, $\mathcal{E}_x = \mathcal{E}_y$. □

The following result, which is the converse of Theorem 1.3.5, shows that a partition of a set $\mathcal{X}$ defines an equivalence relation on $\mathcal{X}$.

Theorem 1.3.6. Let $\mathcal{X}$ be a set, let $\mathcal{P}$ be a partition of $\mathcal{X}$, and define the relation $\mathcal{R}$ on $\mathcal{X}$ by $(x,y) \in \mathcal{R}$ if and only if x and y belong to the same element of $\mathcal{P}$. Then, $\mathcal{R}$ is an equivalence relation on $\mathcal{X}$.

Theorem 1.3.5 shows that every equivalence relation induces a partition, while Theorem 1.3.6

shows that every partition induces an equivalence relation.

Definition 1.3.7. Let $\mathcal{X}$ be a set, let $\mathcal{P}$ be a partition of $\mathcal{X}$, and let $X_0 \subseteq \mathcal{X}$. Then, X_0 is a *representative subset* of $\mathcal{X}$ *relative to* $\mathcal{P}$ if, for all $X \in \mathcal{P}$, exactly one element of X_0 is an element of X.

Definition 1.3.8. Let $\mathcal{R}$ be a relation on the set $\mathcal{X}$. Then, the following terminology is defined:

i) $\mathcal{R}$ is *antisymmetric* if $(x_1, x_2) \in \mathcal{R}$ and $(x_2, x_1) \in \mathcal{R}$ imply that $x_1 = x_2$.

ii) $\mathcal{R}$ is a *partial ordering* if $\mathcal{R}$ is reflexive, antisymmetric, and transitive.

iii) $(\mathcal{X}, \mathcal{R})$ is a *partially ordered set* if $\mathcal{R}$ is a partial ordering.

Let $(\mathcal{X}, \mathcal{R})$ be a partially ordered set. Then, $(x_1, x_2) \in \mathcal{R}$ is denoted by $x_1 \preceq x_2$. If $x_1 \preceq x_2$ and $x_2 \preceq x_1$, then, since $\mathcal{R}$ is antisymmetric, it follows that $x_1 = x_2$. Furthermore, if $x_1 \preceq x_2$ and $x_2 \preceq x_3$, then, since $\mathcal{R}$ is transitive, it follows that $x_1 \preceq x_3$.

Definition 1.3.9. Let $(\mathcal{X}, \mathcal{R})$ be a partially ordered set. Then, the following terminology is defined:

i) Let $\mathcal{S} \subseteq \mathcal{X}$. Then, $y \in \mathcal{X}$ is a *lower bound* for $\mathcal{S}$ if, for all $x \in \mathcal{S}$, it follows that $y \preceq x$.

ii) Let $\mathcal{S} \subseteq \mathcal{X}$. Then, $y \in \mathcal{X}$ is an *upper bound* for $\mathcal{S}$ if, for all $x \in \mathcal{S}$, it follows that $x \preceq y$.

The following result shows that every partially ordered set has at most one lower bound that is "greatest" and at most one upper bound that is "least."

Lemma 1.3.10. Let $(\mathcal{X}, \mathcal{R})$ be a partially ordered set, and let $\mathcal{S} \subseteq \mathcal{X}$. Then, there exists at most one lower bound $y \in \mathcal{X}$ for $\mathcal{S}$ such that every lower bound $x \in \mathcal{X}$ for $\mathcal{S}$ satisfies $x \preceq y$. Furthermore, there exists at most one upper bound $y \in \mathcal{X}$ for $\mathcal{S}$ such that every upper bound $x \in \mathcal{X}$ for $\mathcal{S}$ satisfies $y \preceq x$.

Proof. For $i = 1, 2$, let $y_i \in \mathcal{X}$ be such that y_i is a lower bound for $\mathcal{S}$ and, for all $x \in \mathcal{X}$, $x \preceq y_i$. Therefore, $y_1 \preceq y_2$ and $y_2 \preceq y_1$. Since "$\preceq$" is antisymmetric, it follows that $y_1 = y_2$. □

Definition 1.3.11. Let $(\mathcal{X}, \mathcal{R})$ be a partially ordered set. Then, the following terminology is defined:

i) Let $\mathcal{S} \subseteq \mathcal{X}$. Then, $y \in \mathcal{X}$ is the *greatest lower bound* for $\mathcal{S}$ if y is a lower bound for $\mathcal{S}$ and every lower bound $x \in \mathcal{X}$ for $\mathcal{S}$ satisfies $x \preceq y$. In this case, we write $y = \mathrm{glb}(\mathcal{S})$.

ii) Let $\mathcal{S} \subseteq \mathcal{X}$. Then, $y \in \mathcal{X}$ is the *least upper bound* for $\mathcal{S}$ if y is an upper bound for $\mathcal{S}$ and every upper bound $x \in \mathcal{X}$ for $\mathcal{S}$ satisfies $y \preceq x$. In this case, we write $y = \mathrm{lub}(\mathcal{S})$.

iii) $(\mathcal{X}, \preceq)$ is a *lattice* if, for all distinct $x, y \in \mathcal{X}$, the set $\{x, y\}$ has a least upper bound and a greatest lower bound.

iv) $(\mathcal{X}, \preceq)$ is a *complete lattice* on $\mathcal{X}$ if every subset $\mathcal{S}$ of $\mathcal{X}$ has a least upper bound and a greatest lower bound.

Example 1.3.12. Consider the partially ordered set $(\mathbb{P}, \preceq)$, where $m \preceq n$ indicates that n is an integer multiple of m. For example, $3 \preceq 21$, but it is not true that $2 \preceq 3$. Next, note that the greatest lower bound of a subset $\mathcal{S}$ of $\mathbb{P}$ is the greatest common divisor of the elements of $\mathcal{S}$. For example, $\mathrm{glb}\,\{9, 21\} = 3$. Likewise, the least upper bound of a subset $\mathcal{S}$ of $\mathbb{P}$ is the least common multiple of the elements of $\mathcal{S}$. For example, $\mathrm{lub}\,\{2, 3, 4\} = 12$. Therefore, $(\mathbb{P}, \preceq)$ is a lattice. Next, note that 1 is a lower bound for every subset of $\mathbb{P}$. Since every subset of $\mathbb{P}$ has a smallest element in the usual ordering, it follows that every subset of $\mathbb{P}$ has a greatest lower bound. In particular, $\mathrm{glb}(\mathbb{P}) = 1$. However, no subset of $\mathbb{P}$ that has an infinite number of elements has an upper bound. Therefore, $(\mathbb{P}, \preceq)$ is not a complete lattice. Now, consider $(\mathbb{N}, \preceq)$. Note that 1 is a lower bound for every subset of $\mathbb{N}$. Since every subset of $\mathbb{N}$ has a smallest element in the usual ordering, it follows that every subset of $\mathbb{N}$ has a greatest lower bound. In particular, $\mathrm{glb}(\mathbb{N}) = 1$. Furthermore, for all $m \in \mathbb{N}$, $0 = 0 \cdot m$, and thus 0 is an upper bound for every subset of $\mathbb{N}$. In particular, since 0 is the unique upper bound of $\mathbb{N}$, it follows that 0 is the least upper bound of $\mathbb{N}$. Hence, $(\mathbb{N}, \preceq)$ is a complete lattice. ◇

Proposition 1.3.13. Let $(\mathcal{X}, \preceq)$ be a lattice, and let $\mathcal{S}_1, \mathcal{S}_2 \subseteq \mathcal{X}$. Then,

$$\mathrm{glb}(\mathcal{S}_1 \cup \mathcal{S}_2) = \mathrm{glb}[\mathcal{S}_1 \cup \{\mathrm{glb}(\mathcal{S}_2)\}], \quad \mathrm{lub}(\mathcal{S}_1 \cup \mathcal{S}_2) = \mathrm{lub}[\mathcal{S}_1 \cup \{\mathrm{lub}(\mathcal{S}_2)\}]. \tag{1.3.1}$$

Definition 1.3.14. Let $(\mathcal{X}, \mathcal{R})$ be a partially ordered set. Then, $\mathcal{R}$ is a *total ordering* on $\mathcal{X}$ if, for all $x, y \in \mathcal{X}$, either $(x, y) \in \mathcal{R}$ or $(y, x) \in \mathcal{R}$.

Let $\mathcal{S} \subseteq \mathbb{R}$. Then, it is traditional to write $\inf \mathcal{S}$ and $\sup \mathcal{S}$ for $\mathrm{glb}(\mathcal{S})$ and $\mathrm{lub}(\mathcal{S})$, respectively, where "inf" and "sup" denote infimum and supremum, respectively. If $\mathcal{S} = \varnothing$, then we define $\inf \varnothing \triangleq \infty$ and $\sup \varnothing \triangleq -\infty$. Finally, if $\mathcal{S}$ has no lower bound, then we write $\inf \mathcal{S} = -\infty$, whereas, if $\mathcal{S}$ has no upper bound, then we write $\sup \mathcal{S} = \infty$.

The following result uses the fact that "$\subseteq$" is a partial ordering on every collection of sets.

Proposition 1.3.15. Let $\mathcal{S}$ be a collection of sets. Then,

$$\mathrm{glb}(\mathcal{S}) = \bigcap_{S \in \mathcal{S}} S, \quad \mathrm{lub}(\mathcal{S}) = \bigcup_{S \in \mathcal{S}} S. \tag{1.3.2}$$

Hence, for all $S \in \mathcal{S}$,

$$\mathrm{glb}(\mathcal{S}) \subseteq S \subseteq \mathrm{lub}(\mathcal{S}). \tag{1.3.3}$$

Let $\mathcal{S} \triangleq (S_i)_{i=1}^{\infty}$ be a sequence of sets. Then, by viewing $\mathcal{S}$ as the collection of sets $\{S_1, S_2, \ldots\}$, it follows that

$$\mathrm{glb}(\mathcal{S}) = \bigcap_{i=1}^{\infty} S_i, \quad \mathrm{lub}(\mathcal{S}) = \bigcup_{i=1}^{\infty} S_i. \tag{1.3.4}$$

Hence, for all $i \geq 1$,

$$\mathrm{glb}(\mathcal{S}) \subseteq S_i \subseteq \mathrm{lub}(\mathcal{S}). \tag{1.3.5}$$

Note that $\mathrm{glb}(\mathcal{S})$ and $\mathrm{lub}(\mathcal{S})$ are independent of the ordering of the sequence $\mathcal{S}$.

Proposition 1.3.16. Let $\mathcal{S}$ be a collection of sets, let A be a set, let $\mathcal{S}_0 \triangleq \{S \in \mathcal{S} \colon A \subseteq S\}$, and assume that $\mathcal{S}_0 \neq \varnothing$. Then, $A \subseteq \mathrm{glb}(\mathcal{S}_0)$. If, in addition, $\mathrm{glb}(\mathcal{S}_0) \in \mathcal{S}_0$, then $\mathrm{glb}(\mathcal{S}_0)$ is the smallest element of $\mathcal{S}$ that contains A in the sense that, if $S \in \mathcal{S}$ and $A \subseteq S$, then $\mathrm{glb}(\mathcal{S}_0) \subseteq S$.

Proposition 1.3.17. Let $\mathcal{S}$ be a collection of sets, let A be a set, and let $\mathcal{S}_0 \triangleq \{S \in \mathcal{S} \colon S \subseteq A\}$. Then, $\mathrm{lub}(\mathcal{S}_0) \subseteq A$. If, in addition, $\mathrm{lub}(\mathcal{S}_0) \in \mathcal{S}_0$, then $\mathrm{lub}(\mathcal{S}_0)$ is the largest element of $\mathcal{S}$ that is contained in A in the sense that, if $S \in \mathcal{S}$ and $S \subseteq A$, then $S \subseteq \mathrm{lub}(\mathcal{S}_0)$.

Definition 1.3.18. Let $\mathcal{S} \triangleq (S_i)_{i=1}^{\infty}$ be a sequence of sets. Then, the *essential greatest lower bound* of $\mathcal{S}$ is defined by

$$\mathrm{essglb}(\mathcal{S}) \triangleq \bigcup_{j=1}^{\infty} \bigcap_{i=j}^{\infty} S_i, \tag{1.3.6}$$

and the *essential least upper bound* of $\mathcal{S}$ is defined by

$$\mathrm{esslub}(\mathcal{S}) \triangleq \bigcap_{j=1}^{\infty} \bigcup_{i=j}^{\infty} S_i. \tag{1.3.7}$$

Let $\mathcal{S} \triangleq (S_i)_{i=1}^{\infty}$ be a sequence of sets. Then, the set $\mathrm{essglb}(\mathcal{S})$ consists of all elements of $\cup_{i=1}^{\infty} S_i$ that belong to all but finitely many of the sets in $\mathcal{S}$. Furthermore, the set $\mathrm{esslub}(\mathcal{S})$ consists of all elements of $\cup_{i-1}^{\infty} S_i$ that belong to infinitely many of the sets in $\mathcal{S}$. Therefore, $\mathrm{essglb}(\mathcal{S})$ and $\mathrm{esslub}(\mathcal{S})$ are independent of the ordering of the sequence $\mathcal{S}$, and

$$\mathrm{glb}(\mathcal{S}) \subseteq \mathrm{essglb}(\mathcal{S}) \subseteq \mathrm{esslub}(\mathcal{S}) \subseteq \mathrm{lub}(\mathcal{S}). \tag{1.3.8}$$

Note that $\mathrm{lub}(\mathcal{S})\backslash\mathrm{esslub}(\mathcal{S})$ is the set of elements of $\cup_{i=1}^{\infty} S_i$ that belong to at most finitely many of the sets in $\mathcal{S}$.

Example 1.3.19. Consider the sequence of sets given by

$$(\{1,4\},\{1,2\},\{1,2,3\},\{1,2\},\{1,2,3\},\{1,2\},\{1,2,3\},\ldots).$$

Then, (1.3.8) becomes $\{1\} \subseteq \{1,2\} \subseteq \{1,2,3\} \subseteq \{1,2,3,4\}$. ◇

Definition 1.3.20. Let $\mathcal{S} \triangleq (S_i)_{i=1}^{\infty}$ be a sequence of sets, and assume that $\mathrm{essglb}(\mathcal{S}) = \mathrm{esslub}(\mathcal{S})$. Then, the *essential limit* of $\mathcal{S}$ is defined by

$$\mathrm{esslim}(\mathcal{S}) \triangleq \mathrm{essglb}(\mathcal{S}) = \mathrm{esslub}(\mathcal{S}). \tag{1.3.9}$$

Let $\mathcal{S} \triangleq (S_i)_{i=1}^{\infty}$ be a sequence of sets. Then, $\mathcal{S}$ is *nonincreasing* if, for all $i \in \mathbb{P}$, $S_{i+1} \subseteq S_i$. Furthermore, $\mathcal{S}$ is *nondecreasing* if, for all $i \in \mathbb{P}$, $S_i \subseteq S_{i+1}$.

Proposition 1.3.21. Let $\mathcal{S} \triangleq (S_i)_{i=1}^{\infty}$ be a sequence of sets. If $\mathcal{S}$ is nonincreasing, then

$$\mathrm{esslim}(\mathcal{S}) = \mathrm{glb}(\mathcal{S}) = \mathrm{essglb}(\mathcal{S}) = \mathrm{esslub}(\mathcal{S}). \tag{1.3.10}$$

Furthermore, if $\mathcal{S}$ is nondecreasing, then

$$\mathrm{esslim}(\mathcal{S}) = \mathrm{essglb}(\mathcal{S}) = \mathrm{esslub}(\mathcal{S}) = \mathrm{lub}(\mathcal{S}). \tag{1.3.11}$$

Example 1.3.22. Consider the nonincreasing sequence of sets

$$(\mathbb{N}, \mathbb{N}\backslash\{1\}, \mathbb{N}\backslash\{1,2\}, \mathbb{N}\backslash\{1,2,3\},\ldots).$$

Then, (1.3.8) becomes $\{0\} = \{0\} = \{0\} \subseteq \mathbb{N}$. Now, consider the nondecreasing sequence of subsets of $\mathbb{R}$ given by

$$(\{1\},\{1,2\},\{1,2,3\},\{1,2,3,4\},\ldots).$$

Then, (1.3.8) becomes $\{1\} \subseteq \mathbb{P} = \mathbb{P} = \mathbb{P}$, where $\mathbb{P}$ is the set of positive integers. ◇

Let $\mathcal{S} \triangleq (S_i)_{i=1}^{\infty}$ be a sequence of sets. Then, the sequence $\hat{\mathcal{S}} \triangleq (\cap_{j=1}^{k}[\cup_{i=j}^{\infty} S_i])_{k=1}^{\infty} = (\cup_{i=k}^{\infty} S_i)_{k=1}^{\infty} = (\hat{S}_k)_{i=1}^{\infty}$ is nonincreasing. Hence,

$$\mathrm{esslub}(\mathcal{S}) = \mathrm{esslim}(\hat{\mathcal{S}}) = \mathrm{glb}(\hat{\mathcal{S}}) = \mathrm{essglb}(\hat{\mathcal{S}}) = \mathrm{esslub}(\hat{\mathcal{S}}). \tag{1.3.12}$$

Furthermore, the sequence $\tilde{\mathcal{S}} \triangleq (\cup_{j=1}^{k}[\cap_{i=j}^{\infty} S_i])_{k=1}^{\infty} = (\cap_{i=k}^{\infty} S_i)_{k=1}^{\infty} = (\tilde{S}_k)_{i=1}^{\infty}$ is nondecreasing. Hence,

$$\mathrm{essglb}(\mathcal{S}) = \mathrm{esslim}(\tilde{\mathcal{S}}) = \mathrm{essglb}(\tilde{\mathcal{S}}) = \mathrm{esslub}(\tilde{\mathcal{S}}) = \mathrm{lub}(\tilde{\mathcal{S}}). \tag{1.3.13}$$

1.4 Directed and Symmetric Graphs

Let $\mathcal{X}$ be a finite, nonempty set, and let $\mathcal{R}$ be a multirelation on $\mathcal{X}$. Then, the pair $\mathcal{G} = (\mathcal{X},\mathcal{R})$ is a *directed multigraph*. The elements of $\mathcal{X}$ are the *nodes* of $\mathcal{G}$, while the elements of $\mathcal{R}$ are the *directed edges* of $\mathcal{G}$. If $\mathcal{R}$ is a relation on $\mathcal{X}$, then $\mathcal{G} = (\mathcal{X},\mathcal{R})$ is a *directed graph*. We focus on directed graphs, which have distinct (that is, nonrepeated) directed edges.

The directed graph $\mathcal{G} = (\mathcal{X},\mathcal{R})$ can be visualized as a set of points in the plane representing the nodes in $\mathcal{X}$ connected by the directed edges in $\mathcal{R}$. Specifically, the directed edge $(x,y) \in \mathcal{R}$ from x to y can be visualized as a directed line segment or curve connecting node x to node y. The direction of a directed edge can be denoted by an arrowhead. A directed edge of the form (x,x) is a *self-directed edge*.

If the relation $\mathcal{R}$ is symmetric, then $\mathcal{G}$ is a *symmetric graph*. In this case, it is convenient to represent the pair of directed edges (x,y) and (y,x) in $\mathcal{R}$ by a single *edge* $\{x,y\}$, which is a subset of $\mathcal{X}$. For the self-directed edge (x,x), the corresponding edge is the single-element *self-edge* $\{x\}$. To illustrate these notions, consider a directed graph that represents a city with streets (directed edges) connecting intersections (nodes). Each directed edge represents a one-way street, while the presence of the one-way street (x, y) and its *reverse* (y, x) represents a two-way street. A symmetric relation is a street plan consisting entirely of two-way streets (that is, edges) and thus no one-way streets (directed edges), whereas an antisymmetric relation is a street plan consisting entirely of one-way streets (directed edges) and thus no two-way streets (edges).

Definition 1.4.1. Let $\mathcal{G} = (\mathcal{X}, \mathcal{R})$ be a directed graph. Then, the following terminology is defined:

i) If $x, y \in \mathcal{X}$ are distinct and $(x, y) \in \mathcal{R}$, then y is the *head* of (x, y) and x is the *tail* of (x, y).

ii) If $x, y \in \mathcal{X}$ are distinct and $(x, y) \in \mathcal{R}$, then x is a *parent* of y, and y is a *child* of x.

iii) If $x, y \in \mathcal{X}$ are distinct and either $(x, y) \in \mathcal{R}$ or $(y, x) \in \mathcal{R}$, then x and y are *adjacent*.

iv) If $x \in \mathcal{X}$ has no parent, then x is a *root*.

v) If $x \in \mathcal{X}$ has no child, then x is a *leaf*.

Definition 1.4.2. Let $\mathcal{G} = (\mathcal{X}, \mathcal{R})$ be a directed graph. Then, the following terminology is defined:

i) The *reversal* of $\mathcal{G}$ is the graph $\operatorname{rev}(\mathcal{G}) \triangleq (\mathcal{X}, \operatorname{rev}(\mathcal{R}))$.

ii) The *complement* of $\mathcal{G}$ is the graph $\mathcal{G}^{\sim} \triangleq (\mathcal{X}, \mathcal{R}^{\sim})$.

iii) The *reflexive hull* of $\mathcal{G}$ is the graph $\operatorname{ref}(\mathcal{G}) \triangleq (\mathcal{X}, \operatorname{ref}(\mathcal{R}))$.

iv) The *symmetric hull* of $\mathcal{G}$ is the graph $\operatorname{sym}(\mathcal{G}) \triangleq (\mathcal{X}, \operatorname{sym}(\mathcal{R}))$.

v) The *transitive hull* of $\mathcal{G}$ is the graph $\operatorname{trans}(\mathcal{G}) \triangleq (\mathcal{X}, \operatorname{trans}(\mathcal{R}))$.

vi) The *equivalence hull* of $\mathcal{G}$ is the graph $\operatorname{equiv}(\mathcal{G}) \triangleq (\mathcal{X}, \operatorname{equiv}(\mathcal{R}))$.

vii) $\mathcal{G}$ is *reflexive* if $\mathcal{R}$ is reflexive.

viii) $\mathcal{G}$ is *transitive* if $\mathcal{R}$ is transitive.

ix) $\mathcal{G}$ is an *equivalence graph* if $\mathcal{R}$ is an equivalence relation.

x) $\mathcal{G}$ is *antisymmetric* if $\mathcal{R}$ is antisymmetric.

xi) $\mathcal{G}$ is *partially ordered* if $\mathcal{R}$ is a partial ordering on $\mathcal{X}$.

xii) $\mathcal{G}$ is *totally ordered* if $\mathcal{R}$ is a total ordering on $\mathcal{X}$.

xiii) $\mathcal{G}$ is a *tournament* if $\mathcal{G}$ is antisymmetric and $\operatorname{sym}(\mathcal{R}) = \mathcal{X} \times \mathcal{X} \backslash \{(x, x)\colon x \in \mathcal{X}\}$.

Definition 1.4.3. Let $\mathcal{G} = (\mathcal{X}, \mathcal{R})$ be a directed graph. Then, the following terminology is defined:

i) The directed graph $\mathcal{G}' = (\mathcal{X}', \mathcal{R}')$ is a *directed subgraph* of $\mathcal{G}$ if $\mathcal{X}' \subseteq \mathcal{X}$ and $\mathcal{R}' \subseteq \mathcal{R}$.

ii) The directed subgraph $\mathcal{G}' = (\mathcal{X}', \mathcal{R}')$ of $\mathcal{G}$ is a *spanning directed subgraph* of $\mathcal{G}$ if $\operatorname{supp}(\mathcal{R}) = \operatorname{supp}(\mathcal{R}')$.

iii) If $\mathcal{X}_0 \subseteq \mathcal{X}$, then $\mathcal{G}|_{\mathcal{X}_0} \triangleq (\mathcal{X}_0, \mathcal{R}|_{\mathcal{X}_0})$.

iv) If $\mathcal{G}' = (\mathcal{X}', \mathcal{R}')$ is a directed graph, then $\mathcal{G} \cup \mathcal{G}' \triangleq (\mathcal{X} \cup \mathcal{X}', \mathcal{R} \cup \mathcal{R}')$ and $\mathcal{G} \cap \mathcal{G}' \triangleq (\mathcal{X} \cap \mathcal{X}', \mathcal{R} \cap \mathcal{R}')$.

v) For $x, y \in \mathcal{X}$, a *directed walk* in $\mathcal{G}$ from x to y is an n-tuple of directed edges of $\mathcal{G}$ of the form $((x, y)) \in \mathcal{R}$ for $n = 1$ and $((x, x_1), (x_1, x_2), \ldots, (x_{n-1}, y)) \in \mathcal{R}^n$ for all $n \geq 2$. The *length* of the directed walk is n. The nodes $x, x_1, \ldots, x_{n-1}, y$ are the *nodes* of the walk. Furthermore, if $n \geq 2$, then the nodes $x_1, \ldots, x_{n-1}$ are the *intermediate nodes* of the walk.

vi) For $x, y \in \mathcal{X}$, a *directed trail* in $\mathcal{G}$ from x to y is a directed walk in $\mathcal{G}$ from x to y whose directed edges are distinct.

vii) For $x, y \in \mathcal{X}$, a *directed path* in $\mathcal{G}$ from x to y is a directed trail in $\mathcal{G}$ from x to y whose intermediate nodes are distinct and do not include x and y.

viii) For $x \in \mathcal{X}$, a *directed cycle* in $\mathcal{G}$ at x is a directed path in $\mathcal{G}$ from x to x whose length is at least 2.

ix) $\mathcal{G}$ is *directionally acyclic* if $\mathcal{G}$ has no directed cycles.

x) If $\mathcal{G}$ has at least one directed cycle, then the *directed period* of $\mathcal{G}$ is the greatest common divisor of the lengths of the directed cycles of $\mathcal{G}$.

xi) $\mathcal{G}$ is *directionally aperiodic* if it has at least one directed cycle and the greatest common divisor of the lengths of the directed cycles in $\mathcal{G}$ is 1.

xii) A *directed Hamiltonian path* is a directed path whose nodes include all of the nodes of $\mathcal{X}$.

xiii) A *directed Hamiltonian cycle* is a directed cycle whose nodes include every node in $\mathcal{X}$.

xiv) $\mathcal{G}$ is a *directed tree* if $\mathcal{G}$ has exactly one root x and, for all $y \in \mathcal{X}$ such that $y \neq x$, y has exactly one parent.

xv) $\mathcal{G}$ is a *directed forest* if $\mathcal{G}$ is a union of disjoint directed trees.

xvi) $\mathcal{G}$ is a *directed chain* if $\mathcal{G}$ is a tree and has exactly one leaf.

xvii) $\mathcal{G}$ is *directionally connected* if, for all distinct $x, y \in \mathcal{X}$, there exist directed walks in $\mathcal{G}$ from x to y and from y to x.

xviii) $\mathcal{G}$ is *bipartite* if there exist nonempty, disjoint sets $\mathcal{X}_1$ and $\mathcal{X}_2$ such that $\mathcal{X} = \mathcal{X}_1 \cup \mathcal{X}_2$ and $\mathcal{R} \cap (\mathcal{X}_1 \times \mathcal{X}_1) = \mathcal{R} \cap (\mathcal{X}_2 \times \mathcal{X}_2) = \varnothing$.

xix) The *indegree* of $x \in \mathcal{X}$ is $\operatorname{indeg}(x) \triangleq \operatorname{card}\{y \in \mathcal{X}\colon y \text{ is a parent of } x\}$.

xx) The *outdegree* of $x \in \mathcal{X}$ is $\operatorname{outdeg}(x) \triangleq \operatorname{card}\{y \in \mathcal{X}\colon y \text{ is a child of } x\}$.

xxi) Let $\mathcal{X} = \mathcal{X}_1 \cup \mathcal{X}_2$, where $\mathcal{X}_1$ and $\mathcal{X}_2$ are nonempty and disjoint, and assume that $\mathcal{X} = \operatorname{supp}(\mathcal{G})$. Then, $(\mathcal{X}_1, \mathcal{X}_2)$ is a *directed cut* of $\mathcal{G}$ if, for all $x_1 \in \mathcal{X}_1$ and $x_2 \in \mathcal{X}_2$, there does not exist a directed walk from x_1 to x_2.

A self-directed edge is a directed path; however, a self-directed edge is not a directed cycle.

A directed Hamiltonian cycle is both a directed Hamiltonian path and a directed cycle, both of which are directed paths.

Definition 1.4.4. Let $\mathcal{G} = (\mathcal{X}, \mathcal{R})$ be a symmetric graph. Then, the following terminology is defined:

i) For $x, y \in \mathcal{X}$, a *walk* in $\mathcal{G}$ connecting x and y is an n-tuple of edges of $\mathcal{G}$ of the form $(\{x, y\}) \in \mathcal{E}$ for $n = 1$ and $(\{x, x_1\}, \{x_1, x_2\}, \ldots, \{x_{n-1}, y\}) \in \mathcal{E}^n$ for $n \geq 2$. The *length* of the walk is n. The nodes $x, x_1, \ldots, x_{n-1}, y$ are the *nodes* of the walk. Furthermore, if $n \geq 2$, then the nodes $x_1, \ldots, x_{n-1}$ are the *intermediate nodes* of the walk.

ii) For $x, y \in \mathcal{X}$, a *trail* in $\mathcal{G}$ connecting x and y is a walk in $\mathcal{G}$ connecting x to y whose edges are distinct.

iii) For $x, y \in \mathcal{X}$, a *path* in $\mathcal{G}$ connecting x and y is a trail in $\mathcal{G}$ connecting x and y whose intermediate nodes are distinct and do not include x and y.

iv) For $x \in \mathcal{X}$, a *cycle* in $\mathcal{G}$ at x is a path in $\mathcal{G}$ connecting x and x whose length is at least 3.

v) $\mathcal{G}$ is *acyclic* if $\mathcal{G}$ has no cycles.

vi) If $\mathcal{G}$ has at least one cycle, then the *period* of $\mathcal{G}$ is the greatest common divisor of the lengths of the cycles of $\mathcal{G}$.

vii) $\mathcal{G}$ is *aperiodic* if the period of $\mathcal{G}$ is 1.

viii) A *Hamiltonian path* is a path whose nodes include every node in $\mathcal{X}$.

ix) $\mathcal{G}$ is *Hamiltonian* if $\mathcal{G}$ has a *Hamiltonian cycle* $\mathcal{P}$, which is a cycle such that every node in $\mathcal{X}$ is a node of $\mathcal{P}$.

x) $\mathcal{G}$ is a *tree* if there exists a directed tree $\mathcal{G}' = (\mathcal{X}, \mathcal{R}')$ such that $\mathcal{G} = \operatorname{sym}(\mathcal{G}')$.

xi) $\mathcal{G}$ is a *forest* if $\mathcal{G}$ is a union of disjoint trees.

xii) $\mathcal{G}$ is a *chain* if there exists a directed chain $\mathcal{G}' = (\mathcal{X}, \mathcal{R}')$ such that $\mathcal{G} = \mathrm{sym}(\mathcal{G}')$.

xiii) $\mathcal{G}$ is *connected* if, for all distinct $x, y \in \mathcal{X}$, there exists a walk in $\mathcal{G}$ connecting x and y.

xiv) $\mathcal{G}$ is *bipartite* if there exist nonempty, disjoint sets $\mathcal{X}_1$ and $\mathcal{X}_2$ such that $\mathcal{X} = \mathcal{X}_1 \cup \mathcal{X}_2$ and $\{\{x, y\} \in \mathcal{R}\colon x \in \mathcal{X}_1 \text{ and } y \in \mathcal{X}_2\} = \emptyset$.

xv) The *degree* of $x \in \mathcal{X}$ is $\deg(x) \triangleq \mathrm{indeg}(x) = \mathrm{outdeg}(x)$.

A self-edge is a path; however, a self-edge is not a cycle.

A Hamiltonian cycle is both a Hamiltonian path and a cycle, both of which are paths.

Let $\mathcal{G} = (\mathcal{X}, \mathcal{R})$ be a directed graph, and let $w\colon \mathcal{X} \times \mathcal{X} \mapsto [0, \infty)$, where $w(x, y) > 0$ if $(x, y) \in \mathcal{R}$ and $w(x, y) = 0$ if $(x, y) \notin \mathcal{R}$. For each directed edge $(x, y) \in \mathcal{R}$, $w(x, y)$ is the *weight* associated with the directed edge (x, y), and the triple $\mathcal{G} = (\mathcal{X}, \mathcal{R}, w)$ is a *weighted directed graph*. The graph $\mathcal{G}' = (\mathcal{X}', \mathcal{R}', w')$ is a *weighted directed subgraph* of $\mathcal{G}$ if $\mathcal{X}' \subseteq \mathcal{X}$, $\mathcal{R}'$ is a relation on $\mathcal{X}'$, $\mathcal{R}' \subseteq \mathcal{R}$, and w' is the restriction of w to $\mathcal{R}'$. Finally, if $\mathcal{G}$ is symmetric, then w is *symmetric* if, for all $(x, y) \in \mathcal{R}$, $w(x, y) = w(y, x)$. In this case, w is defined on each edge $\{x, y\}$ of $\mathcal{G}$.

1.5 Numbers

Let x and y be real numbers. Then, x *divides* y if there exists an integer n such that $y = nx$, In this case, we write $x|y$. For example, $6|12$, $3|-9$, $\pi|-2\pi$, $3|0$, and $0|0$. The notation $x \nmid y$ means that x does not divide y.

Let $n_1, \ldots, n_k$ be integers, not all of which are zero. Then, the *greatest common divisor* of the set $\{n_1, \ldots, n_k\}$ is the positive integer defined by

$$\gcd\{n_1, \ldots, n_k\} \triangleq \max\{i \in \mathbb{P}\colon i \text{ divides } n_1, \ldots, n_k\}.$$

For example, $\gcd\{5, 10\} = 5$, and $\gcd\{0, 2\} = 2$. The set $\{n_1, \ldots, n_k\}$ is *coprime* if $\gcd\{n_1, \ldots, n_k\} = 1$. For example, $\gcd\{-3, -7\} = 1$, and thus $\{-3, -7\}$ is coprime.

Let $n_1, \ldots, n_k$ be nonzero integers. Then, the *least common multiple* of the set $\{n_1, \ldots, n_k\}$ is the positive integer defined by

$$\mathrm{lcm}\{n_1, \ldots, n_k\} \triangleq \min\{i \in \mathbb{P}\colon n_1, \ldots, n_k \text{ divide } i\}.$$

For example, $\mathrm{lcm}\{-3, -7\} = 21$, and $\mathrm{lcm}\{-2, 3\} = 6$.

Let m be a nonzero integer, and let n be an integer. Then, $m|n$ if and only if $\gcd\{m, n\} = |m|$.

Let n be an integer, and let k be a positive integer. Furthermore, let l be an integer, and let $r \in [0, k-1]$ be an integer satisfying $n = kl + r$. Then, we write

$$r = \mathrm{rem}_k(n). \tag{1.5.1}$$

where r is the *remainder* after dividing n by k. For example, $\mathrm{rem}_3(-11) = 1$ and $\mathrm{rem}_3(11) = 2$. Furthermore, $k|n$ if and only if $\mathrm{rem}_k(n) = 0$.

Proposition 1.5.1. Let m and n be integers, and let k be a positive integer. Then,

$$\mathrm{rem}_k(n - m) = \mathrm{rem}_k[\mathrm{rem}_k(n) - \mathrm{rem}_k(m)]. \tag{1.5.2}$$

Furthermore, $k|n - m$ if and only if $\mathrm{rem}_k(n) = \mathrm{rem}_k(m)$.

Definition 1.5.2. Let n and m be integers, and let k be a positive integer. Then, n and m are *congruent modulo* k if k divides $n - m$. In this case, we write

$$n \overset{k}{\equiv} m. \tag{1.5.3}$$

Proposition 1.5.1 implies that $n \overset{k}{\equiv} m$ if and only if the remainders of n and m after dividing by k

differ by a multiple of k. For example, $-1 \stackrel{3}{\equiv} 2 \stackrel{3}{\equiv} 8 \stackrel{3}{\equiv} 26 \stackrel{3}{\equiv} 29$.

Let n be an integer. Then, n is *even* if 2 divides n, whereas n is *odd* if 2 does not divide n. Now, assume that $n \ge 2$. Then, n is *prime* if, for all integers m such that $2 \le m < n$, m does not divide n. Note that 2 is prime, but 1 is not prime. Letting p_n denote the nth prime, it follows that

$$(p_i)_{i=1}^{25} = (2,3,5,7,11,13,17,19,23,29,31,37,41,43,47,53,59,61,67,71,73,79,83,89,97).$$

The nth *harmonic number* is denoted by

$$H_n \triangleq \sum_{i=1}^{n} \frac{1}{i}. \tag{1.5.4}$$

Then,

$$(H_i)_{i=0}^{12} = \left(0, 1, \frac{3}{2}, \frac{11}{6}, \frac{25}{12}, \frac{137}{60}, \frac{49}{20}, \frac{363}{140}, \frac{761}{280}, \frac{7129}{2520}, \frac{7381}{2520}, \frac{83711}{27720}, \frac{86021}{27720}\right).$$

For all $\alpha \in \mathbb{R}$, the nth *generalized harmonic number of order* α is denoted by

$$H_{n,\alpha} \triangleq \sum_{i=1}^{n} \frac{1}{i^\alpha}. \tag{1.5.5}$$

Define $H_0 \triangleq H_{0,\alpha} \triangleq 0$. Then,

$$(H_{i,2})_{i=0}^{10} = \left(0, 1, \frac{5}{4}, \frac{49}{36}, \frac{205}{144}, \frac{5269}{3600}, \frac{5369}{3600}, \frac{266681}{176400}, \frac{1077749}{705600}, \frac{9778141}{6350400}, \frac{1968329}{1270080}\right).$$

The symbol $\mathbb{C}$ denotes the set of complex numbers. The elements of $\mathbb{R}$ and $\mathbb{C}$ are *scalars*. Define

$$\jmath \triangleq \sqrt{-1}. \tag{1.5.6}$$

Let $z \in \mathbb{C}$. Then, $z = x + y\jmath$, where $x, y \in \mathbb{R}$. Define the *complex conjugate* $\overline{z}$ of z by

$$\overline{z} \triangleq x - y\jmath \tag{1.5.7}$$

and the real part $\operatorname{Re} z$ of z and the imaginary part $\operatorname{Im} z$ of z by

$$\operatorname{Re} z \triangleq \tfrac{1}{2}(z + \overline{z}) = x, \quad \operatorname{Im} z \triangleq \tfrac{1}{2\jmath}(z - \overline{z}) = \tfrac{1}{2}(\overline{z} - z)\jmath = y. \tag{1.5.8}$$

Furthermore, the *absolute value* $|z|$ of z is defined by

$$|z| \triangleq \sqrt{x^2 + y^2}. \tag{1.5.9}$$

Finally, the *argument* $\arg z \in (-\pi, \pi]$ of z is defined by

$$\arg z \triangleq \begin{cases} 0, & y = x = 0, \\ \operatorname{atan} \frac{y}{x}, & x > 0, \\ -\frac{\pi}{2}, & y < 0, x = 0, \\ \frac{\pi}{2}, & y > 0, x = 0, \\ -\pi + \operatorname{atan} \frac{y}{x}, & y < 0, x < 0, \\ \pi + \operatorname{atan} \frac{y}{x}, & y \ge 0, x < 0, \end{cases} \tag{1.5.10}$$

where atan: $\mathbb{R} \mapsto (-\frac{\pi}{2}, \frac{\pi}{2})$.

Let z be a complex number. Then,

$$z = |z| e^{(\arg z)\jmath}. \tag{1.5.11}$$

z is a nonnegative number if and only if $\arg z = 0$, and z is a negative number if and only if $\arg z = -\pi$. If z is not a nonnegative number, then $\arg z \in (-\pi, 0) \cup (0, \pi]$ is the angle from the positive real axis to the line segment connecting z to the origin in the complex plane, where clockwise angles are negative and confined to the set $(-\pi, 0)$, and counterclockwise angles are positive and confined to the set $(0, \pi]$. Furthermore, if z is nonzero, then

$$\arg \frac{1}{z} = \begin{cases} -\arg z, & \arg z \in (-\pi, \pi), \\ \pi, & \arg z = \pi. \end{cases} \tag{1.5.12}$$

Let z_1 and z_2 be nonzero complex numbers. Then, there exists $k \in \{-1, 0, 1\}$ such that

$$\arg z_1 z_2 = \arg z_1 + \arg z_2 + 2k\pi. \tag{1.5.13}$$

Hence, $2\pi | \arg z_1 z_2 - \arg z_1 - \arg z_2$. For example,

$$\arg(-1)(-1) = \arg 1 = 0 = \pi + \pi - 2\pi = \arg -1 + \arg -1 - 2\pi,$$
$$\arg(1)(-1) = \arg -1 = \pi = 0 + \pi = \arg 1 + \arg -1,$$
$$\arg(-\jmath)(-\jmath) = \arg -1 = \pi = -\pi/2 - \pi/2 + 2\pi = \arg -\jmath + \arg -\jmath + 2\pi.$$

The *closed left half plane* (CLHP), *open left half plane* (OLHP), *closed right half plane* (CRHP), and *open right half plane* (ORHP) are the subsets of $\mathbb{C}$ defined by

$$\text{OLHP} \triangleq \{x \in \mathbb{C}\colon \ \operatorname{Re} x < 0\}, \quad \text{ORHP} \triangleq \{x \in \mathbb{C}\colon \ \operatorname{Re} x > 0\}, \tag{1.5.14}$$

$$\text{CLHP} \triangleq \{x \in \mathbb{C}\colon \ \operatorname{Re} x \le 0\}, \quad \text{CRHP} \triangleq \{x \in \mathbb{C}\colon \ \operatorname{Re} x \ge 0\}. \tag{1.5.15}$$

The imaginary numbers are represented by IA. Note that 0 is a real number, an imaginary number, and a complex number.

Next, we define the *open inside unit disk* (OIUD) and the *closed inside unit disk* (CIUD) by

$$\text{OIUD} \triangleq \{x \in \mathbb{C}\colon \ |x| < 1\}, \quad \text{CIUD} \triangleq \{x \in \mathbb{C}\colon \ |x| \le 1\}. \tag{1.5.16}$$

The complements of the open inside unit disk and the closed inside unit disk are given, respectively, by the *closed outside unit disk* (COUD) and the *open outside unit disk*, which are defined by

$$\text{COUD} \triangleq \{x \in \mathbb{C}\colon \ |x| \ge 1\}, \quad \text{OOUD} \triangleq \{x \in \mathbb{C}\colon \ |x| > 1\}. \tag{1.5.17}$$

The unit circle in $\mathbb{C}$ is denoted by UC.

Since $\mathbb{R}$ is a proper subset of $\mathbb{C}$, we state many results for $\mathbb{C}$. In other cases, we treat $\mathbb{R}$ and $\mathbb{C}$ separately. To do this efficiently, we use the symbol $\mathbb{F}$ to consistently denote either $\mathbb{R}$ or $\mathbb{C}$.

Let $n \in \mathbb{N}$. Then,

$$n! \triangleq \begin{cases} n(n-1)\cdots(2)(1), & n \ge 1, \\ 1, & n = 0. \end{cases} \tag{1.5.18}$$

Then,

$$(i!)_{i=0}^{12} = (1, 1, 2, 6, 24, 120, 720, 5040, 40320, 362880, 3628800, 39916800, 479001600).$$

Let $z \in \mathbb{C}$ and $k \in \mathbb{Z}$. Then,

$$\binom{z}{k} \triangleq \begin{cases} \dfrac{z(z-1)\cdots(z-k+1)}{k!}, & k > 0, \\ 1, & k = 0, \\ 0, & k < 0. \end{cases} \tag{1.5.19}$$

In particular, if $n, k \in \mathbb{N}$, then

$$\binom{n}{k} = \begin{cases} \dfrac{n!}{(n-k)!k!}, & n \ge k \ge 0, \\ 0, & k > n \ge 0. \end{cases} \tag{1.5.20}$$

Hence,

$$\binom{n}{n} = \begin{cases} 1, & n \ge 0, \\ 0, & n < 0. \end{cases} \tag{1.5.21}$$

For example,

$$\binom{-1}{-1} = 0, \quad \binom{-1}{1} = -1, \quad \binom{1}{-1} = 0, \quad \binom{-1}{0} = 1, \quad \binom{0}{0} = 1,$$

$$\binom{-1}{3} = -1, \quad \binom{-\frac{1}{2}}{3} = \frac{-5}{16}, \quad \binom{0}{3} = 0, \quad \binom{\frac{1}{2}}{3} = \frac{1}{16}, \quad \binom{1}{3} = 0.$$

Note that, for all $n \ge k \ge 1$, $\binom{n}{k}$ is the number of k-element subsets of $\{1, \ldots, n\}$.

Let $z, w \in \mathbb{C}$, and assume that $z \notin -\mathbb{P}$, $w \notin -\mathbb{P}$, and $z - w \notin -\mathbb{P}$. Then,

$$\binom{z}{w} \triangleq \frac{\Gamma(z+1)}{\Gamma(w+1)\Gamma(z-w+1)}. \tag{1.5.22}$$

For $k_1, \ldots, k_l \in \mathbb{N}$, where $\sum_{i=1}^{l} k_i = n$, we define the *multinomial coefficient*

$$\binom{n}{k_1, \ldots, k_l} \triangleq \frac{n!}{k_1! \cdots k_l!}. \tag{1.5.23}$$

Note that, if $1 \le m \le n$, then

$$\binom{n}{m} = \binom{n}{m, n-m}.$$

For $z \in \mathbb{C}$ and $k \in \mathbb{N}$, we define the *falling factorial*

$$z^{\underline{k}} \triangleq \begin{cases} z(z-1)\cdots(z-k+1), & k \ge 0, \\ 1, & k = 0. \end{cases} \tag{1.5.24}$$

In particular, if $n \in \mathbb{N}$, then $n^{\underline{n}} = n!$. Hence, if $z \in \mathbb{C}$ and $k \in \mathbb{Z}$, then

$$\binom{z}{k} \triangleq \begin{cases} \dfrac{z^{\underline{k}}}{k!}, & k \ge 0, \\ 0, & k < 0. \end{cases} \tag{1.5.25}$$

Furthermore, for all $z \in \mathbb{C}$ and $k \in \mathbb{N}$, we define the *rising factorial*

$$z^{\overline{k}} \triangleq \begin{cases} z(z+1)\cdots(z+k-1), & k \ge 1, \\ 1, & k = 0. \end{cases} \tag{1.5.26}$$

In particular, if $n \in \mathbb{N}$, then $1^{\overline{n}} = n!$. Finally, if $z \in \mathbb{C}$ and $k \in \mathbb{N}$, then

$$z^{\underline{k}} = (z-k+1)^{\overline{k}}, \quad z^{\overline{k}} = (z+k-1)^{\underline{k}}, \quad z^{\underline{k}} = (-1)^k(-z)^{\overline{k}}. \tag{1.5.27}$$

The *double factorial* is defined by

$$n!! \triangleq \begin{cases} n(n-2)(n-4)\cdots(2) = 2^{n/2}(n/2)!, & n \text{ even}, \\ n(n-2)(n-4)\cdots(3)(1) = \dfrac{(n+1)!}{2^{(n+1)/2}[\frac{1}{2}(n+1)]!}, & n \text{ odd}. \end{cases} \tag{1.5.28}$$

By convention, $(-1)!! = 0!! = 1$. Finally, if $n \ge 1$, then $(2n)!!(2n-1)!! = (2n)!$ and $(2n+1)!!(2n)!! = (2n+1)!$.

1.6 Functions and Their Inverses

Let $\mathcal{X}$ and $\mathcal{Y}$ be nonempty sets. Then, a *function* f that maps $\mathcal{X}$ into $\mathcal{Y}$ is a rule $f\colon\ \mathcal{X} \mapsto \mathcal{Y}$ that assigns a unique element $f(x)$ (the *image* of x) of $\mathcal{Y}$ to each element x of $\mathcal{X}$. Equivalently, a function $f\colon\ \mathcal{X} \mapsto \mathcal{Y}$ can be viewed as a subset $\mathcal{F}$ of $\mathcal{X} \times \mathcal{Y}$ such that, for each $x \in \mathcal{X}$, there exists a unique $y \in \mathcal{Y}$ such that $(x, y) \in \mathcal{F}$. In this case,

$$\mathcal{F} = \operatorname{Graph}(f) \triangleq \{(x, f(x))\colon\ x \in \mathcal{X}\}. \tag{1.6.1}$$

The set $\mathcal{X}$ is the *domain* of f, while the set $\mathcal{Y}$ is the *codomain* of f. For $\mathcal{X}_1 \subseteq \mathcal{X}$, it is convenient to define

$$f(\mathcal{X}_1) \triangleq \{f(x)\colon\ x \in \mathcal{X}_1\}. \tag{1.6.2}$$

The *range* of f is the set $\mathcal{R}(f) \triangleq f(\mathcal{X})$. The function f is *one-to-one* if, for all $x_1, x_2 \in \mathcal{X}$ such that $f(x_1) = f(x_2)$, it follows that $x_1 = x_2$. The function f is *onto* if $\mathcal{R}(f) = \mathcal{Y}$. The function $I_{\mathcal{X}}\colon\ \mathcal{X} \mapsto \mathcal{X}$ defined by $I_{\mathcal{X}}(x) \triangleq x$ for all $x \in \mathcal{X}$ is the *identity mapping* on $\mathcal{X}$. Finally, if $\mathcal{S} \subseteq \mathcal{X}$, $f_{\mathcal{S}}\colon\ \mathcal{S} \mapsto \mathcal{Y}$, and, for all $x \in \hat{\mathcal{X}}$, $f_{\mathcal{S}}(x) = f(x)$, then $f_{\mathcal{S}}$ is the *restriction* of f to $\mathcal{S}$.

Note that the subset $\mathcal{F}$ of $\mathcal{X} \times \mathcal{Y}$ can be viewed as a relation on $(\mathcal{X}, \mathcal{Y})$. Consequently, a function can be viewed as a special case of a relation.

Let $\mathcal{X}$ be a set, and let $\hat{\mathcal{X}}$ be a partition of $\mathcal{X}$. Furthermore, let $f\colon\ \hat{\mathcal{X}} \mapsto \mathcal{X}$, where, for all $\mathcal{S} \in \hat{\mathcal{X}}$, it follows that $f(\mathcal{S}) \in \mathcal{S}$. Then, f is a *canonical mapping*, and $f(\mathcal{S})$ is a *canonical form*. That is, for each element $\mathcal{S} \subseteq \mathcal{X}$ in the partition $\hat{\mathcal{X}}$ of $\mathcal{X}$, the function f assigns an element of $\mathcal{S}$ to the set $\mathcal{S}$. For example, let $\mathcal{S} \triangleq \{1, 2, 3, 4\}$, $\hat{\mathcal{X}} \triangleq \{\{1, 3\}, \{2, 4\}\}$, $f(\{1, 3\}) = 1$, and $f(\{2, 4\}) = 2$.

Let $\mathcal{X}$ and $\mathcal{Y}$ be sets. If $f\colon\ \mathcal{X} \mapsto \mathcal{Y}$ is one-to-one and onto, then $\mathcal{X}$ and $\mathcal{Y}$ have the same *cardinality*, which is written as $\operatorname{card}(\mathcal{X}) = \operatorname{card}(\mathcal{Y})$. Consequently, if $\mathcal{X}$ is finite, then $\operatorname{card}(\mathcal{X})$ is the number of elements of $\mathcal{X}$. If $f\colon\ \mathcal{X} \mapsto \mathcal{Y}$ is one-to-one, then $\operatorname{card}(\mathcal{X}) \le \operatorname{card}(\mathcal{Y})$. If every function $f\colon\ \mathcal{X} \mapsto \mathcal{Y}$ that is one-to-one is not onto, then $\operatorname{card}(\mathcal{X}) < \operatorname{card}(\mathcal{Y})$. If $\operatorname{card}(\mathcal{X}) = \operatorname{card}(\mathbb{P})$, then $\mathcal{X}$ is *countable*. Note that $\operatorname{card}(\mathbb{N}) = \operatorname{card}(\mathbb{P}) = \operatorname{card}(\mathbb{Z}) = \operatorname{card}(\mathbb{Q}) < \operatorname{card}([0, 1]) = \operatorname{card}(\mathbb{R}) = \operatorname{card}(\mathbb{R}^2)$.

Let $\mathcal{X}$ be a finite multiset. Then, $\operatorname{card}(\mathcal{X})$ is the number of elements in $\mathcal{X}$. Cardinality is not defined for infinite multisets.

Let $\mathcal{X}$ be a set, and let $f\colon\ \mathcal{X} \mapsto \mathcal{X}$. Then, f is a function on $\mathcal{X}$. The element $x \in \mathcal{X}$ is a *fixed point* of f if $f(x) = x$.

Let $\mathcal{X}$, $\mathcal{Y}$, and $\mathcal{Z}$ be sets, let $f\colon\ \mathcal{X} \mapsto \mathcal{Y}$, and let $g\colon\ f(\mathcal{X}) \mapsto \mathcal{Z}$. Then, the *composition* of g and f is the function $g \circ f\colon\ \mathcal{X} \mapsto \mathcal{Z}$ defined by $(g \circ f)(x) \triangleq g[f(x)]$. The following result shows that function composition is associative.

Proposition 1.6.1. Let $\mathcal{X}$, $\mathcal{Y}$, $\mathcal{Z}$, and $\mathcal{W}$ be sets, and let $f\colon\ \mathcal{X} \mapsto \mathcal{Y}$, $g\colon\ \mathcal{Y} \mapsto \mathcal{Z}$, $h\colon\ \mathcal{Z} \mapsto \mathcal{W}$. Then,

$$h \circ (g \circ f) = (h \circ g) \circ f. \tag{1.6.3}$$

Hence, we write $h \circ g \circ f$ for $h \circ (g \circ f)$ and $(h \circ g) \circ f$.

Proposition 1.6.2. Let $\mathcal{X}$, $\mathcal{Y}$, and $\mathcal{Z}$ be sets, and let $f\colon\ \mathcal{X} \mapsto \mathcal{Y}$ and $g\colon\ \mathcal{Y} \mapsto \mathcal{Z}$. Then, the following statements hold:

i) If $g \circ f$ is onto, then g is onto.

ii) If $g \circ f$ is one-to-one, then f is one-to-one.

Proof. To prove *i)*, note that $\mathcal{Z} = g(f(\mathcal{X})) \subseteq g(\mathcal{Y}) \subseteq \mathcal{Z}$. Hence, $g(\mathcal{Y}) = \mathcal{Z}$. To prove *ii)*, suppose that f is not one-to-one. Then, there exist distinct $x_1, x_2 \in \mathcal{X}$ such that $f(x_1) = f(x_2)$. Therefore, $g(f(x_1)) = g(f(x_2))$, and thus $g \circ f$ is not one-to-one. □

Let $f\colon\ \mathcal{X} \mapsto \mathcal{Y}$. Then, f is *left invertible* if there exists a function $f^{\mathsf{L}}\colon\ \mathcal{Y} \mapsto \mathcal{X}$ (a *left inverse* of f) such that $f^{\mathsf{L}} \circ f = I_{\mathcal{X}}$, whereas f is *right invertible* if there exists a function $f^{\mathsf{R}}\colon\ \mathcal{Y} \mapsto \mathcal{X}$ (a *right inverse* of f) such that $f \circ f^{\mathsf{R}} = I_{\mathcal{Y}}$. In addition, the function $f\colon\ \mathcal{X} \mapsto \mathcal{Y}$ is *invertible* if there exists a function $f^{\mathsf{Inv}}\colon\ \mathcal{Y} \mapsto \mathcal{X}$ (the *inverse* of f) such that $f^{\mathsf{Inv}} \circ f = I_{\mathcal{X}}$ and $f \circ f^{\mathsf{Inv}} = I_{\mathcal{Y}}$; that is, f^{Inv} is both a left inverse of f and a right inverse of f.

Let $f\colon\ \mathcal{X} \mapsto \mathcal{Y}$, and let $\tilde{\mathcal{X}}$ denote the set of subsets of $\mathcal{X}$. Then, for all $y \in \mathcal{Y}$, the *set-valued inverse* $f^{\mathsf{inv}}\colon\ \mathcal{Y} \mapsto \tilde{\mathcal{X}}$ is defined by $f^{\mathsf{inv}}(y) \triangleq \{x \in \mathcal{X}\colon\ f(x) = y\}$. If f is one-to-one, then, for all $y \in \mathcal{R}(f)$, the set $f^{\mathsf{inv}}(y)$ has a single element, and thus $f^{\mathsf{inv}}\colon\ \mathcal{R}(f) \mapsto \mathcal{X}$ is a function. If f is invertible, then, for all $y \in \mathcal{Y}$, $f^{\mathsf{inv}}(y) = \{f^{\mathsf{Inv}}(y)\}$. The *inverse image* $f^{\mathsf{inv}}(\mathcal{S})$ of $\mathcal{S} \subseteq \mathcal{Y}$ is the set

$$f^{\mathsf{inv}}(\mathcal{S}) \triangleq \bigcup_{y\in\mathcal{S}} f^{\mathsf{inv}}(y) = \{x \in \mathcal{X}\colon\ f(x) \in \mathcal{S}\}. \tag{1.6.4}$$

Note that $f^{\mathsf{inv}}(\mathcal{S})$ is defined whether or not f is invertible. In fact, $f^{\mathsf{inv}}(\mathcal{Y}) = f^{\mathsf{inv}}[f(\mathcal{X})] = \mathcal{X}$ and $f[f^{\mathsf{inv}}(\mathcal{Y})] = f(\mathcal{X})$.

Proposition 1.6.3. Let $\mathcal{X}$ and $\mathcal{Y}$ be sets, let $f\colon\ \mathcal{X} \mapsto \mathcal{Y}$, and let $g\colon\ \mathcal{Y} \mapsto \mathcal{X}$. Then, the following statements are equivalent:

i) f is a left inverse of g.

ii) g is a right inverse of f.

Proposition 1.6.4. Let $\mathcal{X}$ and $\mathcal{Y}$ be sets, let $f\colon\ \mathcal{X} \mapsto \mathcal{Y}$, and assume that f is invertible. Then, f has a unique inverse. Now, let $g\colon\ \mathcal{Y} \mapsto \mathcal{X}$. Then, the following statements are equivalent:

i) g is the inverse of f.

ii) f is the inverse of g.

Theorem 1.6.5. Let $\mathcal{X}$ and $\mathcal{Y}$ be sets, and let $f\colon\ \mathcal{X} \mapsto \mathcal{Y}$. Then, the following statements hold:

i) f is left invertible if and only if f is one-to-one.

ii) f is right invertible if and only if f is onto.

Furthermore, the following statements are equivalent:

iii) f is invertible.

iv) f has a unique inverse.

v) f is one-to-one and onto.

vi) f is left invertible and right invertible.

vii) f has a unique right inverse.

viii) f has a one-to-one left inverse.

ix) f has an onto right inverse.

If, in addition, $\operatorname{card}(\mathcal{X}) \geq 2$, then the following statement is equivalent to *iii)*–*ix)*:

x) f has a unique left inverse.

Proof. To prove *i)*, suppose that f is left invertible with left inverse $g\colon\ \mathcal{Y} \mapsto \mathcal{X}$. Furthermore, suppose that $x_1, x_2 \in \mathcal{X}$ satisfy $f(x_1) = f(x_2)$. Then, $x_1 = g[f(x_1)] = g[f(x_2)] = x_2$, which shows that f is one-to-one. Conversely, suppose that f is one-to-one so that, for all $y \in \mathcal{R}(f)$, there exists a unique $x \in \mathcal{X}$ such that $f(x) = y$. Hence, define the function $g\colon\ \mathcal{Y} \mapsto \mathcal{X}$ by $g(y) \triangleq x$ for all

$y = f(x) \in \mathcal{R}(f)$ and by $g(y)$ arbitrary for all $y \in \mathcal{Y}\backslash\mathcal{R}(f)$. Consequently, $g[f(x)] = x$ for all $x \in \mathcal{X}$, which shows that g is a left inverse of f.

To prove *ii*), suppose that f is right invertible with right inverse $g\colon\ \mathcal{Y} \mapsto \mathcal{X}$. Then, for all $y \in \mathcal{Y}$, it follows that $f[g(y)] = y$, which shows that f is onto. Conversely, suppose that f is onto so that, for all $y \in \mathcal{Y}$, there exists at least one $x \in \mathcal{X}$ such that $f(x) = y$. Selecting one such x arbitrarily, define $g\colon\ \mathcal{Y} \mapsto \mathcal{X}$ by $g(y) \triangleq x$. Consequently, $f[g(y)] = y$ for all $y \in \mathcal{Y}$, which shows that g is a right inverse of f. □

Let $f\colon \mathcal{X} \mapsto \mathcal{Y}$, and assume that f is one-to-one. Then, the function $\hat{f}\colon \mathcal{X} \mapsto \mathcal{R}(f)$ defined by $\hat{f}(x) \triangleq f(x)$ is one-to-one and onto and thus invertible. For convenience, we write $f^{\mathsf{Inv}}\colon \mathcal{R}(f) \mapsto \mathcal{X}$.

The sine and cosine functions $\sin\colon \mathbb{R} \mapsto [-1, 1]$ and $\cos\colon \mathbb{R} \mapsto [-1, 1]$ can be defined in an elementary way in terms of ratios of sides of triangles. The additional trigonometric functions $\tan\colon \mathbb{R}\backslash\pi(\frac{1}{2} + \mathbb{Z}) \mapsto \mathbb{R}$, $\csc\colon \mathbb{R}\backslash\pi\mathbb{Z} \mapsto \mathbb{R}$, $\sec\colon \mathbb{R}\backslash\pi(\frac{1}{2} + \mathbb{Z}) \mapsto \mathbb{R}$, and $\cot\colon \mathbb{R}\backslash\pi\mathbb{Z} \mapsto \mathbb{R}$ are defined by

$$\tan x \triangleq \frac{\sin x}{\cos x}, \quad \csc x \triangleq \frac{1}{\sin x}, \quad \sec x \triangleq \frac{1}{\cos x}, \quad \cot x \triangleq \frac{\cos x}{\sin x}. \tag{1.6.5}$$

The exponential function $\exp\colon \mathbb{R} \mapsto (0, \infty)$ is defined by

$$\exp(x) \triangleq e^x, \tag{1.6.6}$$

where $e \triangleq \lim_{x\to\infty}(1 + 1/x)^x \approx 2.71828\ldots$. The exponential function can be extended to complex arguments as follows. For all $x \in \mathbb{R}$, the power series for "exp" is given by

$$\exp(x) = \sum_{i=0}^{\infty} \frac{x^i}{i!}. \tag{1.6.7}$$

Hence, for all $y \in \mathbb{R}$, we define

$$\exp(y\jmath) = e^{y\jmath} \triangleq \sum_{i=0}^{\infty} \frac{(y\jmath)^i}{i!} = \sum_{i=0}^{\infty}(-1)^i \frac{y^{2i}}{(2i)!} + \sum_{i=0}^{\infty}(-1)^{2i+1} \frac{y^{2i+1}}{(2i+1)!}\jmath = \cos y + (\sin y)\jmath. \tag{1.6.8}$$

Thus, for all $y \in \mathbb{R}$,

$$\sin y = \frac{1}{2\jmath}(e^{y\jmath} - e^{-y\jmath}), \quad \cos y = \tfrac{1}{2}(e^{y\jmath} + e^{-y\jmath}). \tag{1.6.9}$$

Now, let $z = x + y\jmath$, where $x, y \in \mathbb{R}$. Then, $\exp\colon \mathbb{C} \mapsto \mathbb{C}\backslash\{0\}$ is defined by

$$\exp(z) = \exp(x + y\jmath) \triangleq e^{x+y\jmath} = e^x e^{y\jmath} = e^x[\cos x + (\sin x)\jmath]. \tag{1.6.10}$$

In particular, $e^{\pi\jmath} = -1$.

The six trigonometric functions can now be extended to complex arguments. In particular, by replacing $y \in \mathbb{R}$ in (1.6.9) by $z \in \mathbb{C}$, we define $\sin\colon \mathbb{C} \mapsto \mathbb{C}$ and $\cos\colon \mathbb{C} \mapsto \mathbb{C}$ by

$$\sin z \triangleq \frac{1}{2\jmath}(e^{z\jmath} - e^{-z\jmath}), \quad \cos z \triangleq \tfrac{1}{2}(e^{z\jmath} + e^{-z\jmath}). \tag{1.6.11}$$

Hence,

$$e^{z\jmath} = \cos z + (\sin z)\jmath, \quad e^{-z\jmath} = \cos z - (\sin z)\jmath. \tag{1.6.12}$$

Likewise, $\tan\colon \mathbb{C}\backslash\pi(\frac{1}{2} + \mathbb{Z}) \mapsto \mathbb{R}$, $\csc\colon \mathbb{C}\backslash\pi\mathbb{Z} \mapsto \mathbb{R}$, $\sec\colon \mathbb{C}\backslash\pi(\frac{1}{2} + \mathbb{Z}) \mapsto \mathbb{R}$, and $\cot\colon \mathbb{C}\backslash\pi\mathbb{Z} \mapsto \mathbb{R}$ are defined by

$$\tan z \triangleq \frac{\sin z}{\cos z}, \quad \csc z \triangleq \frac{1}{\sin z}, \quad \sec z \triangleq \frac{1}{\cos z}, \quad \cot z \triangleq \frac{\cos z}{\sin z}. \tag{1.6.13}$$

Let $f\colon \mathcal{X} \mapsto \mathcal{Y}$. If f is not one-to-one, then f is not invertible. This is the case, for example, for a periodic function such as $\sin\colon \mathbb{R} \mapsto [-1, 1]$, respectively. In particular, $\sin^{\mathsf{inv}}(1) = \{(4k+1)\pi/2\colon k \in \mathbb{Z}\}$. However, it is convenient to define a *principal inverse* asin of sin by choosing an element of the set $\sin^{\mathsf{inv}}(y)$ for each $y \in [-1, 1]$. Although this choice can be made arbitrarily, it is traditional to define

$$\text{asin}\colon [-1, 1] \mapsto [-\tfrac{\pi}{2}, \tfrac{\pi}{2}]. \tag{1.6.14}$$

Similarly,

$$\text{acos}\colon [-1, 1] \mapsto [0, \pi], \quad \text{atan}\colon \mathbb{R} \mapsto (-\tfrac{\pi}{2}, \tfrac{\pi}{2}), \tag{1.6.15}$$

$$\text{acsc}\colon (-\infty, -1] \cup [1, \infty) \mapsto [-\tfrac{\pi}{2}, 0) \cup (0, \tfrac{\pi}{2}], \tag{1.6.16}$$

$$\text{asec}\colon (-\infty, -1] \cup [1, \infty) \mapsto [0, \tfrac{\pi}{2}) \cup (\tfrac{\pi}{2}, \pi], \tag{1.6.17}$$

$$\text{acot}\colon \mathbb{R} \mapsto (-\tfrac{\pi}{2}, 0) \cup (0, \tfrac{\pi}{2}]. \tag{1.6.18}$$

An analogous situation arises for the exponential function $f(z) = e^z$, which is not one-to-one and thus requires a principal inverse in the form of a logarithm defined on $\mathbb{C}\backslash\{0\}$. Let w be a nonzero complex number, and, for all $i \in \mathbb{Z}$, define

$$z_i \stackrel{\triangle}{=} \log|w| + (\arg w + 2i\pi)\jmath. \tag{1.6.19}$$

Then, for all $i \in \mathbb{Z}$,

$$e^{z_i} = |w|e^{(\arg w)\jmath}e^{2i\pi\jmath} = |w|e^{(\arg w)\jmath} = w. \tag{1.6.20}$$

Consequently, $f^{\mathsf{inv}}(w) = \{z_i\colon i \in \mathbb{Z}\}$. For example, $f^{\mathsf{inv}}(1) = \{2i\pi\jmath\colon i \in \mathbb{Z}\}$, and $f^{\mathsf{inv}}(-1) = \{(2i+1)\pi\jmath\colon i \in \mathbb{Z}\}$. The *principal logarithm* $\log w$ of w is defined by choosing z_0, which yields

$$\log w \stackrel{\triangle}{=} z_0 = \log|w| + (\arg w)\jmath. \tag{1.6.21}$$

Therefore,

$$\log\colon \mathbb{C}\backslash\{0\} \mapsto \{z\colon\ \text{Re}\, z \neq 0 \text{ and } -\pi < \text{Im}\, z \le \pi\}. \tag{1.6.22}$$

Hence,

$$\text{Re}\log w = \log|w|, \quad \text{Im}\log w = \arg w. \tag{1.6.23}$$

Let w_1 and w_2 be nonzero complex numbers. Then, with $f\colon \mathbb{C} \mapsto \mathbb{C}\backslash\{0\}$ given by (1.6.10),

$$f^{\mathsf{inv}}(w_1w_2) = f^{\mathsf{inv}}(w_1) + f^{\mathsf{inv}}(w_2). \tag{1.6.24}$$

However,

$$\log w_1w_2 = \log w_1 + \log w_2 \tag{1.6.25}$$

if and only if

$$\arg w_1w_2 = \arg w_1 + \arg w_2. \tag{1.6.26}$$

For example,

$$\arg\left(\tfrac{\sqrt{2}}{2} + \tfrac{\sqrt{2}}{2}\jmath\right)^2 = \arg\jmath = \frac{\pi}{2} = \frac{\pi}{4} + \frac{\pi}{4} = \arg\left(\tfrac{\sqrt{2}}{2} + \tfrac{\sqrt{2}}{2}\jmath\right) + \arg\left(\tfrac{\sqrt{2}}{2} + \tfrac{\sqrt{2}}{2}\jmath\right),$$

and thus

$$\log\left(\tfrac{\sqrt{2}}{2} + \tfrac{\sqrt{2}}{2}\jmath\right)^2 = \log\left(\tfrac{\sqrt{2}}{2} + \tfrac{\sqrt{2}}{2}\jmath\right) + \log\left(\tfrac{\sqrt{2}}{2} + \tfrac{\sqrt{2}}{2}\jmath\right).$$

However,

$$\arg(-1)^2 = \arg 1 = 0 \neq 2\pi = \pi + \pi = \arg(-1) + \arg(-1),$$

and thus

$$\log(-1)^2 = \log 1 = 0 \neq 2\pi\jmath = \pi\jmath + \pi\jmath = \log(-1) + \log(-1).$$

Therefore, there exist nonzero complex numbers w_1 and w_2 such that the principal logarithm does not satisfy (1.6.25).

Let w be a nonzero complex number. Then,

$$w = e^{\log w}. \tag{1.6.27}$$

Now, let z be a complex number. Then,

$$\log e^z = z - \left(\operatorname{round} \frac{\operatorname{Im} z}{2\pi}\right) 2\pi\jmath, \tag{1.6.28}$$

where, for all $x \in \mathbb{R}$, round(x) denotes the closest integer to x except in the case where $2x$ is an integer, in which case round$(x) = \lfloor x \rfloor$. Therefore, $\log e^z = z$ if and only if $\operatorname{Im} z \in (-\pi, \pi]$.

An analogous situation arises for nth roots. Consider $f\colon \mathbb{R} \mapsto [0,\infty)$ defined by $f(x) = x^2$. Then, for all $y \in [0,\infty)$, it follows that $f^{\mathrm{inv}}(y) = \{-\sqrt{y}, \sqrt{y}\}$, where $\sqrt{y}$ represents the nonnegative square root of $y \geq 0$. For complex-valued extensions, let $n \geq 1$, and define $f\colon \mathbb{C} \mapsto \mathbb{C}$ by $f(z) = z^n$. Let w be a nonzero complex number. If z satisfies $z^n = w$, then $\log z^n = \log w = \log|w| + (\arg w)\jmath$, where "log" is the principal log. Furthermore, z satisfies $z^n = w$ if and only if there exists an integer i such that $n \log z = \log|w| + (\arg w + 2i\pi)\jmath$. Therefore, for all $i \in \mathbb{Z}$, define

$$z_i \triangleq e^{\frac{1}{n}[\log|w| + (\arg w + 2i\pi)\jmath]}, \tag{1.6.29}$$

which satisfies

$$z_i^n = w. \tag{1.6.30}$$

Note that, for all $i \in \mathbb{Z}$, $z_{n+i} = z_i$. Therefore, for all $i \in \{0, \ldots, n-1\}$, define the n distinct numbers

$$z_i \triangleq \sqrt[n]{|w|} e^{\frac{\arg w}{n}\jmath} e^{\frac{2i\pi}{n}\jmath}, \tag{1.6.31}$$

where $\sqrt[n]{|w|}$ is the nonnegative nth root of $|w|$. Consequently, $f^{\mathrm{inv}}(w) = \{z_0, \ldots, z_{n-1}\}$. The *principal* *nth root* $w^{1/n}$ *of* w is defined by choosing z_0, which yields

$$w^{1/n} \triangleq z_0 = \sqrt[n]{|w|} e^{\frac{\arg w}{n}\jmath}. \tag{1.6.32}$$

In particular, if w is a positive number, then $w^{1/n} = \sqrt[n]{w}$, which is the positive nth root of w. However, for an odd integer n and a negative number a, a notational conflict arises between the principal nth root of a and the negative nth root of a. For example, $(-1)^{1/3} = e^{(\pi/3)\jmath}$, whereas, for all odd integers n, it is traditional to interpret $\sqrt[n]{-1}$ as -1. In other words, for all $a < 0$ and odd $n \geq 1$, $\sqrt[n]{a} \triangleq -\sqrt[n]{|a|}$, and thus

$$a^{1/n} = \sqrt[n]{|a|} e^{(\pi/n)\jmath} = \sqrt[n]{a} e^{[(1/n - 1)\pi]\jmath}. \tag{1.6.33}$$

Let z and α be complex numbers, and assume that z is not zero. As an extension of the functions $f(z) = z^n$ and $f(z) = z^{1/n}$, define

$$z^\alpha \triangleq e^{\alpha \log z}, \tag{1.6.34}$$

where $\log z$ is the principal logarithm of z. For example,

$$\frac{1}{\jmath^{2\jmath}} = e^{-2\jmath \log \jmath} = e^{-2\jmath(\pi/2)\jmath} = e^\pi.$$

Next, let z_1 and z_2 be complex numbers, and let α be a real number. Then, $(z_1 z_2)^\alpha = z_1^\alpha z_2^\alpha$. Now, let α be a complex number. Then, $\alpha^{z_1}\alpha^{z_2} = \alpha^{z_1+z_2}$. However, $(z_1 z_2)^\alpha$ and $z_1^\alpha z_2^\alpha$ are not necessarily equal. For example, $(-1)^{\jmath}(-1)^{\jmath} = e^{-\pi}e^{-\pi} = e^{-2\pi} \neq 1 = 1^{\jmath} = [(-1)(-1)]^{\jmath}$. However,

$$(z_1z_2)^{\alpha} = z_1^{\alpha}z_2^{\alpha}e^{2n\pi\alpha\jmath}, \tag{1.6.35}$$

where

$$n = \begin{cases} 1, & -2\pi < \arg z_1 + \arg z_2 \leq -\pi, \\ 0, & -\pi < \arg z_1 + \arg z_2 \leq \pi, \\ -1, & \pi < \arg z_1 + \arg z_2 \leq 2\pi. \end{cases} \tag{1.6.36}$$

Finally,

$$(\alpha^{z_1})^{z_2} = \alpha^{z_1z_2}e^{2n\pi z_2\jmath}, \tag{1.6.37}$$

where

$$n = \left\lfloor \frac{1}{2} - \frac{(\operatorname{Im} z_1)\log|\alpha| + (\operatorname{Re} z_1)\arg\alpha}{2\pi} \right\rfloor. \tag{1.6.38}$$

For example, setting $\alpha = -1$, $z_1 = -1$, and $z_2 = \frac{1}{2}$ yields $n = 1$, and thus $\jmath = (-1)^{1/2} = [(-1)^{-1}]^{1/2} = (-1)^{-1/2}e^{n\pi\jmath} = (1/\jmath)(-1) = \jmath$. Furthermore,

$$(e^{z_1})^{z_2} = e^{z_1z_2}e^{2n\pi z_2\jmath}, \tag{1.6.39}$$

where $n = \left\lfloor \frac{1}{2} - \frac{\operatorname{Im} z_1}{2\pi} \right\rfloor$. See [2216, pp. 108–114] and [2249, pp. 91, 114–119].

Finally, let z, α, and β be complex numbers. Then, $(z^{\alpha})^{\beta}$, $(z^{\beta})^{\alpha}$, and $z^{\alpha\beta}$ may be different as can be seen from the example $z = \frac{1}{2}\jmath$, $\alpha = 2 - \jmath$, and $\beta = -3 - \jmath$, where $(z^{\alpha})^{\beta} \approx 0.03 + 0.04\jmath$, $(z^{\beta})^{\alpha} \approx 9104 + 10961\jmath$, and $z^{\alpha\beta} \approx 17 + 20\jmath$. A similar situation can occur in the case where z, α, and β are real. For example, if $z = -1$, $\alpha = 1/2$, and $\beta = 2$, then $(z^{\alpha})^{\beta} = z^{\alpha\beta} = -1 \neq 1 = (z^{\beta})^{\alpha}$. As a final example, let $z = e$, $\alpha = 2\pi i\jmath$, where $i \geq 1$, and $\beta = \pi$. Then, $(z^{\beta})^{\alpha} = (e^{\pi})^{2\pi i\jmath} = e^{2\pi i\jmath\log e^{\pi}} = e^{2\pi^2 i\jmath} = z^{\alpha\beta} = \cos 2\pi^2 i + \jmath\sin 2\pi^2 i$ and $(z^{\alpha})^{\beta} = (e^{2\pi i\jmath})^{\pi} = 1^{\pi} = e^{\pi\log 1} = e^{\pi 0} = 1$. Since, for all $i \geq 1$, $\cos 2\pi^2 i + \jmath\sin 2\pi^2 i \neq 1$, it follows that $(z^{\beta})^{\alpha} = z^{\alpha\beta} \neq (z^{\alpha})^{\beta}$. See [2107, pp. 166, 167].

Definition 1.6.6. Let $\mathcal{I} \subset \mathbb{R}$ be a finite or infinite interval, and let $f\colon \mathcal{I} \mapsto \mathbb{R}$. Then, f is *convex* if, for all $\alpha \in [0, 1]$ and $x, y \in \mathcal{I}$,

$$f[\alpha x + (1-\alpha)y] \leq \alpha f(x) + (1-\alpha)f(y). \tag{1.6.40}$$

Furthermore, f is *strictly convex* if, for all $\alpha \in (0, 1)$ and distinct $x, y \in \mathcal{I}$,

$$f[\alpha x + (1-\alpha)y] < \alpha f(x) + (1-\alpha)f(y). \tag{1.6.41}$$

Finally, f is (*concave, strictly convex*) if $-f$ is (convex, strictly convex).

A more general definition of a convex function is given by Definition 10.6.14.

Let $\mathcal{X}$ be a set, and let $\sigma\colon \mathcal{X} \times \cdots \times \mathcal{X} \mapsto \mathcal{X} \times \cdots \times \mathcal{X}$, where each Cartesian product has n factors. Then, σ is a *permutation* if, for all $(x_1, \ldots, x_n) \in \mathcal{X} \times \cdots \times \mathcal{X}$, the tuples $(x_1, \ldots, x_n)$ and $\sigma[(x_1, \ldots, x_n)]$ have the same components with the same multiplicity but possibly in a different order. For convenience, we write $(\sigma(x_1), \ldots, \sigma(x_n))$ for $\sigma[(x_1, \ldots, x_n)]$. In particular, we write $(\sigma(1), \ldots, \sigma(n))$ for $\sigma[(1, \ldots, n)]$. The permutation σ is a *transposition* if $(\sigma(x_1), \ldots, \sigma(x_n))$ and $(x_1, \ldots, x_n)$ differ by exactly two distinct interchanged components. Finally, let $\operatorname{sign}(\sigma)$ denote -1 raised to the smallest number of transpositions needed to transform $(\sigma(1), \ldots, \sigma(n))$ to $(1, \ldots, n)$. Note that, if σ_1 and σ_2 are permutations of $(1, \ldots, n)$, then $\operatorname{sign}(\sigma_1 \circ \sigma_2) = \operatorname{sign}(\sigma_1)\operatorname{sign}(\sigma_2)$.

1.7 Facts on Logic

Fact 1.7.1. Let A and B be statements. Then, the following statements hold:

i) $[A \text{ and } (A \Longrightarrow B)] \Longrightarrow B$.

ii) not(A and B) $\Longleftrightarrow$ [(not A) or not B].

iii) not(A or B) $\Longleftrightarrow$ [(not A) and not B].

iv) (A or B) $\Longleftrightarrow$ [(not A) $\Longrightarrow B$] $\Longleftrightarrow$ [(A and B) xor (A xor B)].

v) ($A \Longrightarrow B$) $\Longleftrightarrow$ [(not A) or B] $\Longleftrightarrow$ not(A and not B)] $\Longleftrightarrow$ [(A and B) xor not A].

vi) not(A and B) $\Longleftrightarrow$ ($A \Longrightarrow$ not B) $\Longleftrightarrow$ ($B \Longrightarrow$ not A).

vii) [A and not B] $\Longleftrightarrow$ [not($A \Longrightarrow B$)].

Remark: Each statement is a tautology. **Remark:** *ii*) and *iii*) are *De Morgan's laws*. See [493, p. 24]. See Fact 1.8.1.

Fact 1.7.2. Let A and B be statements. Then, the following statements are equivalent:

i) $A \Longleftrightarrow B$.

ii) (A or not B) and not(A and not B).

iii) (A or not B) and [(not A) or B].

iv) (A and B) or [(not A) and not B].

v) not(A xor B).

Remark: The equivalence of each pair of statements is a tautology.

Fact 1.7.3. Let A, B, and C be statements. Then,

$$[(A \Longrightarrow B) \text{ and } (B \Longrightarrow C)] \Longrightarrow (A \Longrightarrow C).$$

Fact 1.7.4. Let A, B, and C be statements. Then, the following statements are equivalent:

i) $A \Longrightarrow$ (B or C).

ii) [A and (not B)] $\Longrightarrow C$.

Remark: The statement that *i*) and *ii*) are equivalent is a tautology.

Fact 1.7.5. Let A, B, and C be statements. Then, the following statements are equivalent:

i) (A and B) $\Longrightarrow C$.

ii) [B and (not C)] $\Longrightarrow$ (not A).

iii) [A and (not C)] $\Longrightarrow$ (not B).

Source: To prove *i*) $\Longrightarrow$ *ii*), note that [(A and B) or (not B)] $\Longrightarrow$ [C or (not B)], that is, [A or (not B)] $\Longrightarrow$ [C or (not B)], and thus $A \Longrightarrow$ [C or (not B)]. Hence, [B and (not C)] $\Longrightarrow$ (not A). Conversely, to prove *ii*) $\Longrightarrow$ *i*), note that [(B and (not C)) or (not B)] $\Longrightarrow$ [(not A) or (not B)], that is, [(not C) or (not B)] $\Longrightarrow$ [(not A) or (not B)], and thus (not C) $\Longrightarrow$ [(not A) or (not B)]. Hence, (A and B) $\Longrightarrow C$.

Fact 1.7.6. Let $\mathcal{X}$ and $\mathcal{Y}$ be sets, and let Z be a statement that depends on elements of $\mathcal{X}$ and $\mathcal{Y}$. Then, the following statements are equivalent:

i) Not[for all $x \in \mathcal{X}$, Z holds].

ii) There exists $x \in \mathcal{X}$ such that Z does not hold.

Furthermore, the following statements are equivalent:

iii) Not[there exists $y \in \mathcal{Y}$ such that Z holds].

iv) For all $y \in \mathcal{Y}$, Z does not hold.

Finally, the following statements are equivalent:

v) Not[for all $x \in \mathcal{X}$, there exists $y \in \mathcal{Y}$ such that Z holds].

vi) There exists $x \in \mathcal{X}$ such that, for all $y \in \mathcal{Y}$, Z does not hold.

1.8 Facts on Sets

Fact 1.8.1. Let $\mathcal{A}$ and $\mathcal{B}$ be subsets of a set $\mathcal{X}$. Then, the following statements hold:

i) $\mathcal{A} \cap \mathcal{A} = \mathcal{A} \cup \mathcal{A} = \mathcal{A}$.

ii) $\mathcal{A}\backslash\mathcal{B} = \mathcal{A} \cap \mathcal{B}^\sim$.

iii) $(\mathcal{A} \cup \mathcal{B})^\sim = \mathcal{A}^\sim \cap \mathcal{B}^\sim$.

iv) $(\mathcal{A} \cap \mathcal{B})^\sim = \mathcal{A}^\sim \cup \mathcal{B}^\sim$.

v) $(\mathcal{A}\backslash\mathcal{B}) \cup (\mathcal{A} \cap \mathcal{B}) = \mathcal{A}$.

vi) $\mathcal{A}\backslash(\mathcal{A} \cap \mathcal{B}) = \mathcal{A} \cap \mathcal{B}^\sim$.

vii) $\mathcal{A} \cap (\mathcal{A}^\sim \cup \mathcal{B}) = \mathcal{A} \cap \mathcal{B}$.

viii) $(\mathcal{A} \cup \mathcal{B}) \cap (\mathcal{A} \cup \mathcal{B}^\sim) = \mathcal{A}$.

ix) $[\mathcal{A}\backslash(\mathcal{A} \cap \mathcal{B})] \cup \mathcal{B} = \mathcal{A} \cup \mathcal{B}$.

x) $(\mathcal{A} \cup \mathcal{B}) \cap (\mathcal{A}^\sim \cup \mathcal{B}) \cap (\mathcal{A} \cup \mathcal{B}^\sim) = \mathcal{A} \cap \mathcal{B}$.

xi) $(\mathcal{A}^\sim \cup \mathcal{B}) \cap (\mathcal{A} \cup \mathcal{B}^\sim) = (\mathcal{A} \cap \mathcal{B}) \cup (\mathcal{A}^\sim \cap \mathcal{B}^\sim) = [(\mathcal{A} \cup \mathcal{B})\backslash(\mathcal{A} \cap \mathcal{B})]^\sim = [(\mathcal{A} \cap \mathcal{B}^\sim) \cup (\mathcal{A}^\sim \cap \mathcal{B})]^\sim$.

Remark: *iii)* and *iv)* are De Morgan's laws. See Fact 1.7.1.

Fact 1.8.2. Let $\mathcal{A}$, $\mathcal{B}$, and $\mathcal{C}$ be subsets of a set $\mathcal{X}$. Then, the following statements hold:

i) $\mathcal{A} \cap (\mathcal{B} \cup \mathcal{C}) = (\mathcal{A} \cap \mathcal{B}) \cup (\mathcal{A} \cap \mathcal{C})$.

ii) $\mathcal{A} \cup (\mathcal{B} \cap \mathcal{C}) = (\mathcal{A} \cup \mathcal{B}) \cap (\mathcal{A} \cup \mathcal{C})$.

iii) $(\mathcal{A}\backslash\mathcal{B})\backslash\mathcal{C} = \mathcal{A}\backslash(\mathcal{B} \cup \mathcal{C})$.

iv) $(\mathcal{A} \cap \mathcal{B})\backslash\mathcal{C} = (\mathcal{A}\backslash\mathcal{C}) \cap (\mathcal{B}\backslash\mathcal{C})$.

v) $(\mathcal{A} \cap \mathcal{B})\backslash(\mathcal{C} \cap \mathcal{B}) = (\mathcal{A}\backslash\mathcal{C}) \cap \mathcal{B}$.

vi) $(\mathcal{A} \cup \mathcal{B})\backslash\mathcal{C} = (\mathcal{A}\backslash\mathcal{C}) \cup (\mathcal{B}\backslash\mathcal{C}) = [\mathcal{A}\backslash(\mathcal{B} \cup \mathcal{C})] \cup (\mathcal{B}\backslash\mathcal{C})$.

vii) $(\mathcal{A} \cup \mathcal{B})\backslash(\mathcal{C} \cap \mathcal{B}) = (\mathcal{A}\backslash\mathcal{B}) \cup (\mathcal{B}\backslash\mathcal{C})$.

viii) $\mathcal{A}\backslash(\mathcal{B} \cup \mathcal{C}) = (\mathcal{A}\backslash\mathcal{B}) \cap \mathcal{A}\backslash\mathcal{B})$.

ix) $\mathcal{A}\backslash(\mathcal{B} \cap \mathcal{C}) = (\mathcal{A}\backslash\mathcal{B}) \cup \mathcal{A}\backslash\mathcal{B})$.

Fact 1.8.3. Let $\mathcal{A}$, $\mathcal{B}$, and $\mathcal{C}$ be subsets of a set $\mathcal{X}$. Then, the following statements hold:

i) $\mathcal{A} \ominus \varnothing = \varnothing \ominus \mathcal{A} = \mathcal{A}$, $\mathcal{A} \ominus \mathcal{A} = \varnothing$.

ii) $\mathcal{A} \ominus \mathcal{B} = \mathcal{B} \ominus \mathcal{A}$.

iii) $\mathcal{A} \ominus \mathcal{B} = (\mathcal{A} \cap \mathcal{B}^\sim) \cup (\mathcal{B} \cap \mathcal{A}^\sim) = (\mathcal{A}\backslash\mathcal{B}) \cup (\mathcal{B}\backslash\mathcal{A}) = (\mathcal{A} \cup \mathcal{B})\backslash(\mathcal{A} \cap \mathcal{B})$.

iv) $\mathcal{A} \ominus \mathcal{B} = \{x \in \mathcal{X}\colon (x \in \mathcal{A}) \text{ xor } (x \in \mathcal{B})\}$.

v) $\mathcal{A} \ominus \mathcal{B} = \varnothing$ if and only if $\mathcal{A} = \mathcal{B}$.

vi) $\mathcal{A} \ominus (\mathcal{B} \ominus \mathcal{C}) = (\mathcal{A} \ominus \mathcal{B}) \ominus \mathcal{C}$.

vii) $(\mathcal{A} \ominus \mathcal{B}) \ominus (\mathcal{B} \ominus \mathcal{C}) = \mathcal{A} \ominus \mathcal{C}$.

viii) $\mathcal{A} \cap (\mathcal{B} \ominus \mathcal{C}) = (\mathcal{A} \cap \mathcal{B}) \ominus (\mathcal{A} \cap \mathcal{C})$.

If, in addition, $\mathcal{A}$ and $\mathcal{B}$ are finite, then

$$\operatorname{card}(\mathcal{A} \ominus \mathcal{B}) = \operatorname{card}(\mathcal{A}) + \operatorname{card}(\mathcal{B}) - 2\operatorname{card}(\mathcal{A} \cap \mathcal{B}).$$

Fact 1.8.4. Let $\mathcal{A}$, $\mathcal{B}$, and $\mathcal{C}$ be finite sets. Then,

$$\operatorname{card}(\mathcal{A} \times \mathcal{B}) = \operatorname{card}(\mathcal{A})\operatorname{card}(\mathcal{B}),$$

$$\operatorname{card}(\mathcal{A} \cup \mathcal{B}) = \operatorname{card}(\mathcal{A}) + \operatorname{card}(\mathcal{B}) - \operatorname{card}(\mathcal{A} \cap \mathcal{B}),$$

$$\begin{aligned}\operatorname{card}(\mathcal{A} \cup \mathcal{B} \cup \mathcal{C}) = {} & \operatorname{card}(\mathcal{A}) + \operatorname{card}(\mathcal{B}) + \operatorname{card}(\mathcal{C}) - \operatorname{card}(\mathcal{A} \cap \mathcal{B}) - \operatorname{card}(\mathcal{A} \cap \mathcal{C}) - \operatorname{card}(\mathcal{B} \cap \mathcal{C})\\ & + \operatorname{card}(\mathcal{A} \cap \mathcal{B} \cap \mathcal{C}).\end{aligned}$$

Remark: The second and third equalities are versions of the *inclusion-exclusion principle*. See [411, p. 82], [1372, p. 67], and [2520, pp. 64–67]. **Remark:** The inclusion-exclusion principle holds for multisets $\mathcal{A}$ and $\mathcal{B}$ with "$\mathcal{A} \cup \mathcal{B}$" defined as the smallest multiset that contains both $\mathcal{A}$ and $\mathcal{B}$. For example, $\text{card}(\{1,1,2,2\}) = \text{card}(\{1,1,2\} \cup \{1,2,2\}) = \text{card}(\{1,1,2\}) + \text{card}(\{1,2,2\}) - \text{card}(\{1,2\})$; that is, $4 = 3 + 3 - 2$. See [2879].

Fact 1.8.5. Define $\mathcal{A} \triangleq \{x_1,\dots,x_1,\dots,x_n,\dots,x_n\}_{\text{ms}}$, where, for all $i \in \{1,\dots,n\}$, k_i is the number of repetitions of x_i. Then, the number of multisubsets of $\mathcal{A}$ is $\prod_{i=1}^n (k_i+1)$. **Source:** [2460].

Fact 1.8.6. Let $\mathcal{A}, \mathcal{B} \subseteq \mathbb{R}$. Then, the following statements hold:

i) $\sup(-\mathcal{A}) = -\inf A$.

ii) $\inf(-\mathcal{A}) = -\sup A$.

iii) $\sup(\mathcal{A}+\mathcal{B}) = \sup\mathcal{A} + \sup\mathcal{B}$.

iv) $\sup(\mathcal{A}-\mathcal{B}) = \sup\mathcal{A} - \inf\mathcal{B}$.

v) $\inf(\mathcal{A}+\mathcal{B}) = \inf\mathcal{A} + \inf\mathcal{B}$.

vi) $\inf(\mathcal{A}-\mathcal{B}) = \inf\mathcal{A} - \sup\mathcal{B}$.

vii) $\sup(\mathcal{A}\cup\mathcal{B}) = \max\{\sup\mathcal{A}, \sup\mathcal{B}\}$.

viii) $\inf(\mathcal{A}\cup\mathcal{B}) = \min\{\inf\mathcal{A}, \inf\mathcal{B}\}$.

ix) If $0 \notin \mathcal{A}$, then

$$\sup\left\{\frac{1}{x}\colon x \in \mathcal{A}\right\} = \max\left\{\frac{1}{\inf[\mathcal{A}\cap(-\infty,0)]}, \frac{1}{\inf[\mathcal{A}\cap(0,\infty)]}\right\}.$$

x) $\sup\{xy\colon x\in\mathcal{A}, y\in\mathcal{B}\} = \max\{(\inf\mathcal{A})\inf\mathcal{B}, (\inf\mathcal{A})\sup\mathcal{B}, (\sup\mathcal{A})\inf\mathcal{B}, (\sup\mathcal{A})\sup\mathcal{B}\}$.

Source: [1566, p. 3].

Fact 1.8.7. Let $S_1,\dots,S_m$ be finite sets, and let $n \triangleq \sum_{i=1}^m \text{card}(S_i)$. Then,

$$\left\lceil \frac{n}{m} \right\rceil \le \max_{i\in\{1,\dots,m\}} \text{card}(S_i).$$

In particular, if $m < n$, then there exists $i \in \{1,\dots,m\}$ such that $\text{card}(S_i) \ge 2$. **Remark:** This is the *pigeonhole principle.*

Fact 1.8.8. Let $S_1,\dots,S_m$ be sets, assume that, for all $i \in \{1,\dots,m\}$, $\text{card}(S_i) = n$, and assume that, for all distinct $i,j \in \{1,\dots,m\}$, $\text{card}(S_i \cap S_j) \le k$. Then,

$$\frac{n^2 m}{n+(m-1)k} \le \text{card}\left(\cup_{i=1}^m S_i\right).$$

Source: [1561, p. 23].

Fact 1.8.9. Let X be a set, let $n \triangleq \text{card}(X)$, let $S_1,\dots,S_m \subseteq X$, and assume that, for all distinct $i,j \in \{1,\dots,m\}$, $S_i\backslash S_j$ and $S_j\backslash S_i$ are nonempty. Then, $m \le \binom{n}{\lfloor n/2\rfloor}$. **Source:** [1992, p. 57]. **Remark:** This is a *Sperner lemma.*

Fact 1.8.10. Let X be a set, let $n \triangleq \text{card}(X)$, let $S_1,\dots,S_m \subseteq X$, let $k \le n/2$, assume that, for all $i \in \{1,\dots,m\}$, $\text{card}(S_i) = k$, and, for all distinct $i,j \in \{1,\dots,m\}$, $S_i \cap S_j$ is nonempty. Then, $m \le \binom{n-1}{k-1}$. **Source:** [1992, p. 57]. **Remark:** This is the *Erdös-Ko-Rado theorem.*

Fact 1.8.11. Let X be a set, let $n \triangleq \text{card}(X)$, let $S_1,\dots,S_m \subseteq X$, assume that, for all $i \in \{1,\dots,m\}$, $\text{card}(S_i)$ is odd, and, for all distinct $i,j \in \{1,\dots,m\}$, $\text{card}(S_i \cap S_j)$ is even. Then, $m \le n$. **Source:** [1992, p. 57]. **Remark:** This is the *oddtown theorem.*

Fact 1.8.12. Let X be a set, let $n \triangleq \text{card}(X)$, let $S_1,\dots,S_m \subseteq X$, let $p \ge 2$ be prime, and assume that, for all $i \in \{1,\dots,m\}$, $\text{card}(S_i) = 2p-1$, and, for all distinct $i,j \in \{1,\dots,m\}$, $\text{card}(S_i \cap S_j) \ne p-1$. Then, $m \le \sum_{i=1}^{p-1}\binom{n}{i}$. **Source:** [1992, p. 58]. **Remark:** Excluding intersections of cardinality $p-1$ restricts the number of possible subsets of X.

Fact 1.8.13. Let X be a set, let $S_1, \ldots, S_m, T_1, \ldots, T_m \subseteq X$, let $k \ge 1$ and $l \ge 1$, and assume that, for all $i \in \{1, \ldots, m\}$, $\operatorname{card}(S_i) = k$, $\operatorname{card}(T_i) = l$, and $S_i \cap T_i = \varnothing$, and, for all $i, j \in \{1, \ldots, m\}$ such that $i < j$, $S_i \cap T_j \ne \varnothing$. Then, $m \le \binom{k+l}{l}$. **Source:** [1992, pp. 171–173].

Fact 1.8.14. Let S be a set, and let $\mathcal{S}$ denote the set of all subsets of S. Then, "$\subset$" and "$\subseteq$" are transitive relations on $\mathcal{S}$, and "$\subseteq$" is a partial ordering on $\mathcal{S}$.

Fact 1.8.15. Define the relation $\mathcal{R}$ on $\mathbb{R} \times \mathbb{R}$ by

$$\mathcal{R} \triangleq \{((x_1, y_1), (x_2, y_2)) \in (\mathbb{R} \times \mathbb{R}) \times (\mathbb{R} \times \mathbb{R})\colon\ x_1 \le x_2 \text{ and } y_1 \le y_2\}.$$

Then, $\mathcal{R}$ is a partial ordering.

Fact 1.8.16. Define the relation $\mathcal{L}$ on $\mathbb{R} \times \mathbb{R}$ by

$$\mathcal{L} \triangleq \{((x_1, y_1), (x_2, y_2)) \in (\mathbb{R} \times \mathbb{R}) \times (\mathbb{R} \times \mathbb{R})\colon\ x_1 \le x_2 \text{ and, if } x_1 = x_2, \text{ then } y_1 \le y_2\}.$$

Then, $\mathcal{L}$ is a total ordering on $\mathbb{R} \times \mathbb{R}$.

Remark: Denoting this total ordering by "$\preceq$," note that $(1,4) \preceq (2,3)$ and $(1,4) \preceq (1,5)$. **Remark:** This ordering is the *lexicographic ordering* or *dictionary ordering*, where "book" $\preceq$ "box". Note that the ordering of words in a dictionary is reflexive, antisymmetric, and transitive, and that every pair of words can be ordered. **Related:** Fact 3.11.23.

Fact 1.8.17. Let $n \ge 1$ and $x_1, \ldots, x_{n^2+1} \in \mathbb{R}$. Then, at least one of the following statements holds:

i) There exist $1 \le i_1 \le \cdots \le i_{n+1} \le n^2 + 1$ such that $x_{i_1} \le \cdots \le x_{i_{n+1}}$.

ii) There exist $1 \le i_1 \le \cdots \le i_{n+1} \le n^2 + 1$ such that $x_{i_1} \ge \cdots \ge x_{i_{n+1}}$.

Source: [2294, p. 53] and [2526]. **Remark:** This is the *Erdös-Szekeres theorem.*

1.9 Facts on Graphs

Fact 1.9.1. Let $\mathcal{G} = (\mathcal{X}, \mathcal{R})$ be a directed graph. Then, the following statements hold:

i) $\mathcal{R}$ is the graph of a function on $\mathcal{X}$ if and only if every node in $\mathcal{X}$ has exactly one child.

Furthermore, the following statements are equivalent:

ii) $\mathcal{R}$ is the graph of a one-to-one function on $\mathcal{X}$.

iii) $\mathcal{R}$ is the graph of an onto function on $\mathcal{X}$.

iv) $\mathcal{R}$ is the graph of a one-to-one and onto function on $\mathcal{X}$.

v) Every node in $\mathcal{X}$ has exactly one child and not more than one parent.

vi) Every node in $\mathcal{X}$ has exactly one child and at least one parent.

vii) Every node in $\mathcal{X}$ has exactly one child and exactly one parent.

Related: Fact 1.10.1.

Fact 1.9.2. Let $\mathcal{G} = (\mathcal{X}, \mathcal{R})$ be a directed graph, and assume that $\mathcal{R}$ is the graph of a function $f\colon \mathcal{X} \mapsto \mathcal{X}$. Then, either f is the identity function or $\mathcal{G}$ has a directed cycle.

Fact 1.9.3. Let $\mathcal{G} = (\mathcal{X}, \mathcal{R})$ be a directed graph, and assume that $\mathcal{G}$ has a directed Hamiltonian cycle. Then, $\mathcal{G}$ has no roots and no leaves.

Fact 1.9.4. Let $\mathcal{G} = (\mathcal{X}, \mathcal{R})$ be a directed graph. Then, $\mathcal{G}$ has either a root or a directed cycle.

Fact 1.9.5. Let $\mathcal{G} = (\mathcal{X}, \mathcal{R})$ be a directed graph. If $\mathcal{G}$ is a directed tree, then it is not transitive.

Fact 1.9.6. Let $\mathcal{G} = (\mathcal{X}, \mathcal{R})$ be a directed graph, and assume that $\mathcal{G}$ is directionally acyclic. Furthermore, for all $x, y \in \mathcal{X}$, let "$x \le y$" denote the existence of directional path from x to y. Then, "$\le$" is a partial ordering on $\mathcal{X}$. **Remark:** This result provides the foundation for the *Hasse diagram*, which illustrates the structure of a partially ordered set. See [2405, 2734].

Fact 1.9.7. Let $\mathcal{G} = (\mathcal{X}, \mathcal{R})$ be a directed graph. If $\mathcal{G}$ is a directed forest, then $\mathcal{G}$ is directionally acyclic.

Fact 1.9.8. Let $\mathcal{G} = (\mathcal{X}, \mathcal{R})$ be a symmetric graph, and let $n = \text{card}(\mathcal{X})$. Then, the following statements are equivalent:

i) $\mathcal{G}$ is a forest.

ii) $\mathcal{G}$ is acyclic.

iii) No pair of nodes in $\mathcal{X}$ is connected by more than one path.

Furthermore, the following statements are equivalent:

iv) $\mathcal{G}$ is a tree.

v) $\mathcal{G}$ is a connected forest.

vi) $\mathcal{G}$ is connected and has no cycles.

vii) $\mathcal{G}$ is connected and has $n-1$ edges.

viii) $\mathcal{G}$ has no cycles and has $n-1$ edges.

ix) Every pair of nodes in $\mathcal{X}$ is connected by exactly one path.

Fact 1.9.9. Let $\mathcal{G} = (\mathcal{X}, \mathcal{R})$ be a tournament. Then, $\mathcal{G}$ has a directed Hamiltonian path. If, in addition, $\mathcal{G}$ is directionally connected, then $\mathcal{G}$ has a directed Hamiltonian cycle. **Remark:** The second statement is *Camion's theorem.* See [276, p. 16]. **Remark:** The directed edges in a tournament distinguish winners and losers in a contest where every player (that is, node) encounters every other player exactly once.

Fact 1.9.10. Let $\mathcal{G} = (\mathcal{X}, \mathcal{R})$ be a symmetric graph without self-edges, where $\mathcal{X} \subset \mathbb{R}^2$, assume that $v \triangleq \text{card}(\mathcal{X}) \ge 3$, assume that $\mathcal{G}$ is connected, and assume that the edges in $\mathcal{R}$ can be represented by line segments that lie in the same plane and that pairwise either are disjoint or intersect at a node. Furthermore, let e denote the number of edges of $\mathcal{G}$, and let f denote the number of disjoint regions in $\mathbb{R}^2$ whose boundaries are the edges of $\mathcal{G}$. Then,

$$f + v - e = 2, \quad \frac{3}{2}f \le e \le 3v - 6, \quad f \le 2v - 4.$$

If, in addition, $\mathcal{G}$ has no triangles, then $e \le 2v - 4$. **Source:** [754, pp. 162–166] and [2735, pp. 97–116]. **Remark:** The equality gives the *Euler characteristic* for a planar graph. A related result for the surfaces of a convex polyhedron is given by Fact 5.4.8. See [2307].

1.10 Facts on Functions

Fact 1.10.1. Let $\mathcal{X}$ and $\mathcal{Y}$ be finite sets, and let $f\colon\ \mathcal{X} \mapsto \mathcal{Y}$. Then, the following statements hold:

i) If $\text{card}(\mathcal{X}) < \text{card}(\mathcal{Y})$, then f is not onto.

ii) If $\text{card}(\mathcal{Y}) < \text{card}(\mathcal{X})$, then f is not one-to-one.

iii) If f is one-to-one and onto, then $\text{card}(\mathcal{X}) = \text{card}(\mathcal{Y})$.

Now, assume that $\text{card}(\mathcal{X}) = \text{card}(\mathcal{Y})$. Then, the following statements are equivalent:

iv) f is one-to-one.

v) f is onto.

vi) $\text{card}[f(\mathcal{X})] = \text{card}(\mathcal{X})$.

Related: Fact 1.9.1.

Fact 1.10.2. Let $f\colon\ \mathcal{X} \mapsto \mathcal{Y}$ be invertible. Then, f^{Inv} is invertible, and $(f^{\text{Inv}})^{\text{Inv}} = f$.

Fact 1.10.3. Let $f\colon\ \mathcal{X} \mapsto \mathcal{Y}$. Then, for all $A, B \subseteq \mathcal{X}$, the following statements hold:

i) $A \subseteq f^{\text{inv}}[f(A)] \subseteq \mathcal{X}$.

ii) $f^{\text{inv}}[f(\mathcal{X})] = \mathcal{X} = f^{\text{inv}}(\mathcal{Y})$.

iii) If $A \subseteq B$, then $f(A) \subseteq f(B)$.

iv) $f(A \cap B) \subseteq f(A) \cap f(B)$.

v) $f(A \cup B) = f(A) \cup f(B)$.

vi) $f(A) \backslash f(B) \subseteq f(A \backslash B)$.

Furthermore, the following statements are equivalent:

vii) f is one-to-one.

viii) For all $A \subseteq \mathcal{X}$, $f^{\mathsf{inv}}[f(A)] = A$.

ix) For all $A, B \subseteq \mathcal{X}$, $f(A \cap B) = f(A) \cap f(B)$.

x) For all disjoint $A, B \subseteq \mathcal{X}$, $f(A)$ and $f(B)$ are disjoint.

xi) For all $A, B \subseteq \mathcal{X}$, $f(A) \backslash f(B) = f(A \backslash B)$.

Source: [154, pp. 44, 45] and [643, p. 64]. **Remark:** To show that equality does not necessarily hold in *iv)*, let $f(x) = x^2$, $A = [-2, 1]$, and $B = [-1, 2]$. Then, $f(A \cap B) = [0, 1] \subset [0, 4] = f(A) \cap f(B)$. **Related:** Fact 3.12.7.

Fact 1.10.4. Let $f\colon\ \mathcal{X} \mapsto \mathcal{Y}$. Then, for all $A, B \subseteq \mathcal{Y}$, the following statements hold:

i) $f[f^{\mathsf{inv}}(A)] = A \cap f(\mathcal{X}) \subseteq A$.

ii) $f[f^{\mathsf{inv}}(\mathcal{Y})] = f(\mathcal{X})$.

iii) If $A \subseteq B$, then $f^{\mathsf{inv}}(A) \subseteq f^{\mathsf{inv}}(B)$.

iv) $f^{\mathsf{inv}}(A \cap B) = f^{\mathsf{inv}}(A) \cap f^{\mathsf{inv}}(B)$.

v) $f^{\mathsf{inv}}(A \cup B) = f^{\mathsf{inv}}(A) \cup f^{\mathsf{inv}}(B)$.

vi) $f^{\mathsf{inv}}(A) \backslash f^{\mathsf{inv}}(B) = f^{\mathsf{inv}}(A \backslash B)$.

In addition, the following statements are equivalent:

vii) f is onto.

viii) For all $A \subseteq \mathcal{Y}$, $f[f^{\mathsf{inv}}(A)] = A$.

Source: [154, pp. 44, 45] and [643, p. 64]. **Related:** Fact 3.12.8.

Fact 1.10.5. Let $f\colon\ \mathcal{X} \mapsto \mathcal{Y}$. Then, the following statements hold:

i) If f is invertible, then, for all $y \in \mathcal{Y}$, $f^{\mathsf{inv}}(y) = \{f^{\mathsf{Inv}}(y)\}$.

ii) Assume that f is left invertible, and define $\hat{f}\colon\ \mathcal{X} \mapsto \mathcal{R}(f)$, where, for all $x \in \mathcal{X}$, $\hat{f}(x) \triangleq f(x)$. Then, $\hat{f}$ is invertible, and, for all $y \in \mathcal{R}(f)$, $f^{\mathsf{inv}}(y) = \{\hat{f}^{\mathsf{Inv}}(y)\}$.

iii) If f is left invertible and f^{L} is a left inverse of f, then, for all $y \in \mathcal{R}(f)$, $f^{\mathsf{inv}}(y) = \{f^{\mathsf{L}}(y)\}$.

iv) If f is right invertible and f^{R} is a right inverse of f, then, for all $y \in \mathcal{Y}$, $f^{\mathsf{R}}(y) \in f^{\mathsf{inv}}(y)$.

Related: Fact 3.18.8.

Fact 1.10.6. Let $g\colon\ \mathcal{X} \mapsto \mathcal{Y}$ and $f\colon\ \mathcal{Y} \mapsto \mathcal{Z}$. Then, the following statements hold:

i) If $A \subseteq \mathcal{Z}$, then $(f \circ g)^{\mathsf{inv}}(A) = g^{\mathsf{inv}}[f^{\mathsf{inv}}(A)]$.

ii) $f \circ g$ is one-to-one if and only if g is one-to-one and the restriction $\hat{f}\colon\ g(\mathcal{X}) \mapsto \mathcal{Z}$ of f is one-to-one. If these conditions hold and g^{L} and $\hat{f}^{\mathsf{L}}$ are left inverses of g and $\hat{f}$, respectively, then $g^{\mathsf{L}} \circ \hat{f}^{\mathsf{L}}$ is a left inverse of $f \circ g$.

iii) $f \circ g$ is onto if and only if the restriction $\hat{f}\colon\ g(\mathcal{X}) \mapsto \mathcal{Z}$ of f is onto. Let $\hat{g}\colon\ \mathcal{X} \mapsto g(\mathcal{X})$, where, for all $x \in \mathcal{X}$, $\hat{g}(x) = g(x)$. If these conditions hold and $\hat{g}^{\mathsf{R}}$ and $\hat{f}^{\mathsf{R}}$ are right inverses of $\hat{g}$ and $\hat{f}$, respectively, then $\hat{g}^{\mathsf{R}} \circ \hat{f}^{\mathsf{R}}$ is a right inverse of $f \circ g$.

iv) $f \circ g$ is invertible if and only if g is one-to-one and the restriction $\hat{f}\colon\ g(\mathcal{X}) \mapsto \mathcal{Z}$ of f is one-to-one and onto. If these conditions hold, g^{L} is a left inverse of g, and $\hat{f}^{\mathsf{Inv}}$ is the inverse of $\hat{f}$, then $(f \circ g)^{\mathsf{Inv}} = g^{\mathsf{L}} \circ \hat{f}^{\mathsf{Inv}}$.

Remark: A matrix version of this result is given by Fact 3.18.9 and Fact 3.18.10.

Fact 1.10.7. Let $f\colon\ \mathcal{X} \mapsto \mathcal{Y}$, let $g\colon\ \mathcal{Y} \mapsto \mathcal{X}$, and assume that f and g are one-to-one. Then, there exists $h\colon\ \mathcal{X} \mapsto \mathcal{Y}$ such that h is one-to-one and onto. **Source:** [968, pp. 311, 312] and [2092, pp. 16, 17]. **Remark:** This is the *Schroeder-Bernstein theorem.*

Fact 1.10.8. Let $\mathcal{X}$ and $\mathcal{Y}$ be sets, let $f\colon \mathcal{X} \mapsto \mathcal{Y}$, and, for $i \in \{1,2\}$, let $g_i\colon \mathcal{R}(f) \mapsto \mathbb{F}^n$ and $\alpha_i \in \mathbb{F}$. Then, $(\alpha_1 g_1 + \alpha_2 g_2) \circ f = \alpha_1(g_1 \circ f) + \alpha_2(g_2 \circ f)$. **Remark:** The composition operator $\mathcal{C}(g,f) \triangleq g \circ f$ is linear in its first argument.

1.11 Facts on Integers

Fact 1.11.1. Let $n,m \ge 0$ and $k,l \ge 2$. Then, $\prod_{i=1}^k (n+i) \ne m^l$. **Source:** [997]. **Remark:** A product of consecutive integers cannot be a power of an integer.

Fact 1.11.2. Let n be an integer. Then, $n(n+1)(n+2)(n+3)+1 = (n^2+3n+1)^2$. Hence, $n(n+1)(n+2)(n+3)+1$ is a square. **Example:** $5(6)(7)(8)+1 = 41^2$. **Related:** Fact 2.1.2.

Fact 1.11.3. Let x be a real number, and assume that $x+\frac{1}{2}$ is not an integer. Then, the integer closest to x is $\lfloor x+\frac{1}{2}\rfloor$.

Fact 1.11.4. Let w, x, y, and z be real numbers, and let n and m be integers. Then, the following statements hold:

i) If $w|x$ and $y|z$, then $wy|xz$.

ii) If $x|y$ and $x|z$, then $x^2|yz$.

iii) If $x|y$, then $x|ny$.

iv) If $x|y$ and $y|z$, then $x|z$.

v) If $x|y$ and $x|z$, then $x|my+nz$.

Fact 1.11.5. Let n and m be integers, at least one of which is nonzero. Then, the following statements hold:

i) Assume that m is positive. Then, there exist unique integers q and $r \in [0, m-1]$ such that $n = qm + r$. In particular, $q = \lfloor n/m \rfloor$ and $r = \operatorname{rem}_m(n) = n - qm = n - m\lfloor n/m\rfloor \in [0, m-1]$.

ii) If m is positive, then $\lceil n/m \rceil = \lfloor (n+m-1)/m \rfloor$.

iii) If $n|m$, then $\gcd\{n,m\} = |n|$.

iv) If k is prime and $k|mn$, then either $k|m$ or $k|n$.

v) $\gcd\{n/\gcd\{n,m\}, m/\gcd\{n,m\}\} = 1$.

vi) If both n and m are prime and $m \ne n$, then n and m are coprime.

vii) If $n > 0$ and $m > 0$, then $1 \le \gcd\{n,m\} \le \min\{n, m, |n-m|\}$.

viii) $(\operatorname{lcm}\{n,m\})\gcd\{n,m\} = |nm|$.

ix) n and m are coprime if and only if $\operatorname{lcm}\{n,m\} = |nm|$.

x) There exist integers k, l such that $\gcd\{n,m\} = kn + lm$.

Now, assume that n and m are coprime, and let k be an integer. Then, the following statements hold:

xi) $\gcd\{n-m, n+m, nm\} = 1$.

xii) $\gcd\{n^k - m^k, n^k + m^k\} \le 2$.

xiii) $\gcd\{(n-m)^k, (n+m)^k\} \le 2^k$.

xiv) $\gcd\{n^2 - nm + m^2, n+m\} \le 3$.

xv) $\gcd\{nk, m\} = \gcd\{k, m\}$.

Finally, let $n_1, \dots, n_k$ and $m_1, \dots, m_l$ be integers. Then, the following statement holds:

xvi) $\gcd\{n_1m_1, n_1m_2, \dots, n_km_l\} = (\gcd\{n_1, \dots, n_k\})\gcd\{m_1, \dots, m_l\}$.

Source: [2380, p. 12]. *x)*–*xiv)* are given in [1757, pp. 86, 89, 105]; *xv)* is given in [1241, p. 123]. **Example:** $\gcd\{221, 754\} = 13 = -17(221) + 5(754)$. See [1757, pp. 86, 87]. **Remark:** The first set in *xvi)* contains kl products. **Remark:** *x)* is the *GCD identity*. See [79, p. 17].

Fact 1.11.6. Let $l, m, n \ge 1$. Then, the following statements hold:

i) $\gcd\{l,m,n\} = \gcd\{\gcd\{l,m\}, \gcd\{m,n\}, \gcd\{n,l\}\}$.

ii) $lmn = (\gcd\{lm, mn, nl\})\,\mathrm{lcm}\{l,m,n\}$.

iii) $\gcd\{l, \mathrm{lcm}\{m,n\}\} = \mathrm{lcm}\{\gcd\{l,m\}, \gcd\{l,n\}\}$.

iv) $\mathrm{lcm}\{l, \gcd\{m,n\}\} = \gcd\{\mathrm{lcm}\{l,m\}, \mathrm{lcm}\{l,n\}\}$.

v) $\gcd\{\mathrm{lcm}\{l,m\}, \mathrm{lcm}\{m,n\}, \mathrm{lcm}\{n,l\}\} = \mathrm{lcm}\{\gcd\{l,m\}, \gcd\{m,n\}, \gcd\{n,l\}\}$.

vi) $lmn \gcd\{l,m,n\} = (\mathrm{lcm}\{l,m,n\})(\gcd\{l,m\})(\gcd\{m,n\}) \gcd\{n,l\}$.

vii) $\gcd\{l,m\} = \gcd\{l+m, \mathrm{lcm}\{l,m\}\}$.

viii)

$$\frac{(\gcd\{l,m,n\})^2}{\gcd\{l,m\}\gcd\{m,n\}\gcd\{n,l\}} = \frac{(\mathrm{lcm}\{l,m,n\})^2}{\mathrm{lcm}\{l,m\}\,\mathrm{lcm}\{m,n\}\,\mathrm{lcm}\{n,l\}}.$$

Source: [1757, p. 105]. *i)* is given in [289, pp. 25, 144]; *viii)* is given in [1158, p. 310].

Fact 1.11.7. Let $n \ge 1$. Then, $\gcd\{n^2+1, (n+1)^2+1\} \in \{1,5\}$. Furthermore, $\gcd\{n^2+1, (n+1)^2+1\} = 5$ if and only if $n \overset{5}{\equiv} 2$. **Source:** [289, pp. 31, 165].

Fact 1.11.8. Let $k_1, \dots, k_n$ be positive integers, and assume that $k_1 < \cdots < k_n$. Then,

$$\sum_{i=1}^{n-1} \frac{1}{\mathrm{lcm}\{k_i, k_{i+1}\}} \le 1 - \frac{1}{2^{n-1}}.$$

Source: [2380, p. 12].

Fact 1.11.9. Let m and n be integers. Then, the following statements are equivalent:

i) Either both m and n are even or both m and n are odd.

ii) $n \overset{2}{\equiv} m$.

Furthermore, the following statements are equivalent:

iii) $m|n$.

iv) $n \overset{|m|}{\equiv} 0$.

v) $n \overset{|m|}{\equiv} m$.

Fact 1.11.10. Let $k \ge 1$, and let m, n, p, q be integers. Then, the following statements hold:

i) If $n = m$, then $n \overset{k}{\equiv} m$.

ii) $n \overset{k}{\equiv} n$.

Furthermore, the following statements are equivalent:

iii) $k|(n-m)$.

iv) $n \overset{k}{\equiv} m$.

v) $m \overset{k}{\equiv} n$.

vi) $-n \overset{k}{\equiv} -m$.

vii) $n - m \overset{k}{\equiv} 0$.

Furthermore, the following statement holds:

viii) If $n \overset{k}{\equiv} m$ and $m \overset{k}{\equiv} p$, then $n \overset{k}{\equiv} p$.

Next, if $p \overset{k}{\equiv} q$ and $n \overset{k}{\equiv} m$, then the following statements hold:

ix) $n + p \overset{k}{\equiv} m + q$.

x) $n - p \overset{k}{\equiv} m - q$.

xi) $np \stackrel{k}{\equiv} mq$.

Finally, the following statements hold:

xii) If $n \stackrel{k}{\equiv} m$, and p is a positive integer, then $pn \stackrel{k}{\equiv} pm$.

xiii) If $n \stackrel{k}{\equiv} m$, and p is a positive integer, then $n^p \stackrel{k}{\equiv} m^p$.

xiv) If $pn \stackrel{k}{\equiv} pm$, then $n \stackrel{k/\gcd\{k,p\}}{\equiv} m$.

xv) If $pn \stackrel{k}{\equiv} pm$ and $\gcd\{k,p\} = 1$, then $n \stackrel{k}{\equiv} m$.

xvi) $k!|\prod_{i=0}^{k-1}(n+i)$. For example, $11(12)(13) = 6(286)$ and $(22)(23)\cdots(28) = 5040(1184040)$.

xvii) If $n \stackrel{k}{\equiv} n_0$ and $m \stackrel{k}{\equiv} m_0$, then $nm \stackrel{k}{\equiv} \text{rem}_k(n_0 m_0)$.

Source: *xiv)* is given in [2763, pp. 30, 31]. **Remark:** "$\stackrel{k}{\equiv}$" is an equivalence relation on $\mathbb{Z}$, which partitions $\mathbb{Z}$ into *residue classes*.

Fact 1.11.11. Let $n \geq 1$, and let m be the sum of the decimal digits of n. Then, the following statements hold:

i) $3|n$ if and only if $3|m$.

ii) $n \stackrel{9}{\equiv} m$.

Source: [2763, pp. 31, 32].

Fact 1.11.12. Let n be a positive integer. Then, the following statements hold:

i) $n^2 \stackrel{3}{\equiv} 0$ if and only if $n \stackrel{3}{\equiv} 0$.

ii) $n^2 \stackrel{3}{\equiv} 1$ if and only if either $n \stackrel{3}{\equiv} 1$ or $n \stackrel{3}{\equiv} 2$.

Source: [2114]. **Example:** $3 \stackrel{3}{\equiv} 6 \stackrel{3}{\equiv} 9 \stackrel{3}{\equiv} 12 \stackrel{3}{\equiv} 15 \stackrel{3}{\equiv} 0$, $9 \stackrel{3}{\equiv} 36 \stackrel{3}{\equiv} 81 \stackrel{3}{\equiv} 144 \stackrel{3}{\equiv} 225 \stackrel{3}{\equiv} 0$, $1 \stackrel{3}{\equiv} 4 \stackrel{3}{\equiv} 7 \stackrel{3}{\equiv} 10 \stackrel{3}{\equiv} 13 \stackrel{3}{\equiv} 1$, $2 \stackrel{3}{\equiv} 5 \stackrel{3}{\equiv} 8 \stackrel{3}{\equiv} 11 \stackrel{3}{\equiv} 14 \stackrel{3}{\equiv} 2$, and $1 \stackrel{3}{\equiv} 4 \stackrel{3}{\equiv} 16 \stackrel{3}{\equiv} 25 \stackrel{3}{\equiv} 49 \stackrel{3}{\equiv} 64 \stackrel{3}{\equiv} 100 \stackrel{3}{\equiv} 121 \stackrel{3}{\equiv} 169 \stackrel{3}{\equiv} 196 \stackrel{3}{\equiv} 1$.

Fact 1.11.13. Let $k, l, m, n \geq 1$. Then, the following statements hold:

i) If $m \leq n$ is prime, then m does not divide $n!+1$. Hence, there exists a prime $k \in [n+1, n!+1]$ such that $k|n!+1$.

ii) None of the integers $n!+2, n!+3, \ldots, n!+n$ are prime.

iii) Assume that $n \geq 2$ is not prime, and let k be the smallest prime such that $k|n$. Then, $k \leq \sqrt{n}$. If, in addition, $\sqrt[3]{n} < k$, then n/k is prime.

iv) If n is prime, then $(2^{n-1}-1)/n$ is an integer.

v) If $n \geq 3$ is odd, then $n^2 \stackrel{8}{\equiv} 1$.

vi) If n is prime and $n \geq 5$, then either $n \stackrel{6}{\equiv} 1$ or $n \stackrel{6}{\equiv} 5$.

vii) If $n \stackrel{8}{\equiv} 7$, then n is not the sum of three squares of integers.

viii) If $n \stackrel{9}{\equiv} 4$, then n is not the sum of three cubes of integers.

ix) The last digit of n^2 is neither 2, 3, 7, nor 8.

x) Neither 3 nor 5 divides $(n+1)^3 - n^3$.

xi) If $n \geq 2$, then $n^4 + 4^n$ is not prime.

xii) $3|n(n^2-3n+8)$, $6|n^3+5n$, $8|(n-1)(n^3-5n^2+18n-8)$.

xiii) $9|4^n+15n-1$, $30|n^5-n$, $120|n^5-5n^3+4n$.

xiv) 121 does not divide n^2+3n+5.

xv) $3^{n+1}|2^{3^n}+1$.

xvi) 2^n does not divide $n!$.

xvii) If $m \le n$, then $m!|n^{\underline{m}}$.

xviii) $\gcd\{2^m-1, 2^n-1\} = 2^{\gcd\{m,n\}}-1$. Hence, $n|m$ if and only if $2^n-1|2^m-1$.

xix) If n and 6 are coprime, then $24|n^2-1$.

xx) If n is even, then $n^2-1|2^{n!}-1$.

xxi) If $6|k+l+m$, then $6|k^3+l^3+m^3$.

xxii) If $n \ge 4$ and $m \ge 4$ are prime, then $24|n^2-m^2$.

xxiii) If n is not prime, then 2^n-1 is not prime. Furthermore, if $n \in \{2,3,5,7,13,17,19,31,61, 89,107,127,521,607,1279\}$, then 2^n-1 is prime.

xxiv) If $n \ge 1$ and 2^n+1 is prime, then there exists $k \ge 0$ such that $n = 2^k$. If $k \in \{0,1,2,3,4\}$ then $2^{2^k}+1$ is prime. If $5 \le k \le 32$, then $2^{2^k}+1$ is not prime.

xxv) If $n \ge 3$ is odd, then $3 \nmid 2^n-1$.

xxvi) If $n \ge 4$ is even, then $3 \nmid 2^n+1$.

xxvii) If 4^n+2^n+1 is prime, then there exists a positive integer m such that $n = 3^m$.

xxviii) If $n \ge 5$ is prime, then there exists a positive integer m such that $n = \sqrt{24m+1}$.

xxix) $\frac{n^5}{5}+\frac{n^4}{2}+\frac{n^3}{3}-\frac{n}{30}$ is a positive integer.

xxx) If $n \ge 12$, then $\sqrt{n^2-19n+89}$ is not an integer.

xxxi) If $n = k^2+l^2+m^2$, then there exist positive integers p,q,r such that $n^2 = p^2+q^2+r^2$.

xxxii) There exist infinitely many multisets of integers $\{x,y,z,p,q,r\}_{\rm ms}$ such that $x^2+y^2+z^2 = p^3+q^3+r^3$.

xxxiii) If m and n are coprime, then $\{m+in\colon i \ge 1\}$ contains an infinite number of primes.

xxxiv) If n is prime, then $(k+l)^n \overset{n}{\equiv} k^n+l^n$.

xxxv) If $kl = mn$, then $k+l+m+n$ is not prime.

xxxvi) $2(k^4-l^4) \ne m^2$.

xxxvii) If k,l,m,n are nonnegative, $\{k,l\} \ne \{m,n\}$, and $k^2+l^2 = m^2+n^2$, then k^2+l^2 is not prime.

xxxviii) If m and n are prime and $m < n$, then $mn|\binom{n+m}{m}-\binom{n}{m}-1$.

xxxix) If $n \ge 3$ is prime, then $([(n-1)/2]!)^2 \overset{n}{\equiv} (-1)^{(n+1)/2}$.

xl) $10|1+8^n-3^n-6^n$.

xli) $5|1+2^n+3^n+4^n$ if and only if $n/4$ is not an integer.

xlii) If $k \ge 3$ and $l \ge 5$ are consecutive primes, then $k+l$ is the product of at least three primes.

xliii) $\sqrt{n^4+2n^3+2n^2+2n+1}$ is not an integer.

Source: [107, pp. 595–598], [993, pp. 118, 131–137, 208]. *iii*) is given in [2763, pp. 13, 19]; *viii*) and *ix*) are given in [2763, pp. 31, 33]; *xi*) is given in [993, p. 120]; *xx*) is given in [1158, p. 266]; *xxix*) is given in [1757, p. 64]; *xxxiii*) is given in [2529, chapter 8]; *xxxiv*) is given in [2068, p. 68]; *xxxv*)–*xxxvii*) are given in [107, pp. 595–599]; *xlviii*) is given in [108, pp. 51, 294, 295]; *xxxix*) is given in [14]; *xl*)–*xliii*) are given in [289, pp. 7, 11, 32, 36, 72, 73, 82, 167, 178]. **Remark:** *vi*) implies that, if n is prime and $n \ge 5$, then there exists a positive integer k such that either $n = 6k-1$ or $n = 6k+1$. For example, $23 \overset{6}{\equiv} 5$ and $31 \overset{6}{\equiv} 1$. For $k = 20$, neither $n = 6k-1 = 119 = 7(17)$ nor $n = 6k+1 = 121 = 11^2$ is prime. **Remark:** *i*) and *xxiii*) imply that there are an infinite number of primes. **Remark:** *xxxiv*) is *Dirichlet's theorem*. **Remark:** The prime numbers 2^n-1 listed in *xxiii*) are *Mersenne primes*. It is unknown whether or not there exist infinitely many Mersenne primes. **Remark:** The prime numbers 2^n+1 listed in *xxiv*), namely, 3, 5, 17, 257, 65537, are *Fermat primes*. These are the only known Fermat primes. **Example:** In *xxxvii*), $1^2+7^2=5^2+5^2=50$, $1^2+8^2=4^2+7^2=65$, and $0^2+10^2=6^2+8^2=100$ are not prime. **Example:** In *xlviii*), $2(3)|\binom{5}{2}-\binom{3}{2}-1$; that is, $6|6$; $3(5)|\binom{8}{3}-\binom{5}{3}-1$, that is, $15|45$; and $11(13)|\binom{24}{11}-\binom{13}{11}-1$, that is, $143|2496065$.

Fact 1.11.14. Let $n\geq 1$. Then, there exist n consecutive positive integers whose sum of squares is prime if and only if $n\in\{2,3,6\}$. **Source:** [109, pp. 74, 75]. **Example:** $1^2+2^2=5$, $2^2+3^2+4^2=29$, and $2^2+3^2+4^2+5^2+6^2+7^2=139$.

Fact 1.11.15. Let $n\geq 2$ be prime, and let $k\geq 1$. Then, $n|k^n-k$. Equivalently, $k^n\overset{n}{\equiv}k$. **Source:** [411, p. 115], [947], [993, p. 119]. **Remark:** This is *Fermat's little theorem.* **Remark:** An equivalent statement is the following: Let n be prime, let k be a positive integer, and assume that n and k are coprime. Then, $k^{n-1}\overset{n}{\equiv}1$. See [2763, p. 42]. **Example:** $4^7-4=7(2340)$ and $13^3-13=3(728)$. **Remark:** $341|2^{341}-2$, but $341=11(31)$ is not prime. See [993, p. 120].

Fact 1.11.16. Let $n\geq 2$ be prime, and let k and l be positive integers. Then, $(k+l)^n\overset{n}{\equiv}k^n+l^n$. **Source:** [149].

Fact 1.11.17. Let $n\geq 2$. Then, n is prime if and only if $\sum_{i=1}^{n-1}i^{n-1}\overset{n}{\equiv}n-1$. **Remark:** Necessity follows from Fermat's little theorem given by Fact 1.11.15. Sufficiency is a conjecture. **Example:** $1^6+2^6+3^6+4^6+5^6+6^6=67171=7(9595)+6\overset{7}{\equiv}6$.

Fact 1.11.18. Let n be prime, and let $k\geq 1$. Then,

$$\sum_{i=1}^{n}i^k\overset{p}{\equiv}\begin{cases}-1, & n-1|k,\\ 0, & n-1\nmid k.\end{cases}$$

Source: [1918]. **Example:** Let $n=3$ and $k=2$. Then, $1^2+2^2+3^2=14\overset{3}{\equiv}-1$.

Fact 1.11.19. Let $n\geq 5$ be prime. Then,

$$\sum_{i=0}^{n-1}\binom{2i}{i}\overset{n}{\equiv}\begin{cases}1, & n\overset{3}{\equiv}1,\\ -1, & n\overset{3}{\equiv}2.\end{cases}$$

Source: [149].

Fact 1.11.20. Let $n\geq 2$. Then, n is prime if and only if $n|(n-1)!+1$. **Remark:** This is *Wilson's theorem.* **Remark:** $n|(n-1)!+1$ is equivalent to $(n-1)!\overset{n}{\equiv}-1$. **Example:** $4!+1=5(5)$ and $12!+1=13(36846277)$.

Fact 1.11.21. Let $n\geq 3$. Then, n is prime if and only if $\prod_{i=1}^{n-1}(2^i-1)\overset{2^n-1}{\equiv}n$. **Remark:** This is *Vantieghem's theorem.* **Example:** $4!+1=5(5)$. **Source:** [1132].

Fact 1.11.22. Let $p\geq 2$ be prime and let $1\leq n\leq p$. Then, $p|(n-1)!(p-n)!+(-1)^{n+1}$. **Source:** [2068, p. 67]. **Remark:** This is an extension of Wilson's theorem given by Fact 1.11.20. **Example:** $4!6!+1=11(1571)$ and $13!9!-1=23(98246143821913)$.

Fact 1.11.23. Let $m,n\geq 1$. Then, $(m^2-n^2)^2+(2mn)^2=(m^2+n^2)^2$. **Remark:** This result characterizes all *Pythagorean triples* within an integer multiple. **Example:** If $m=2$ and $n=1$, then $3^2+4^2=5^2$; if $m=3$ and $n=2$, then $5^2+12^2=13^2$; if $m=4$ and $n=1$, then $8^2+15^2=17^2$; if $m=4$ and $n=3$, then $7^2+24^2=25^2$.

Fact 1.11.24. Let $n\geq 1$. Then, there exist $k\geq 1$ and $\delta_1,\dots,\delta_k\in\{-1,1\}$ such that $n=\sum_{i=1}^{k}\delta_i i^2$. **Source:** [289, pp. 33, 171] and [1158, p. 9]. **Example:** $7=1-4-9+16+25-36$, $12=-1+4+9$, and $18=1-4-9+16+25-36-49-64+81-100+121$.

Fact 1.11.25. Let n be a positive integer. Then, the number of 4-tuples of integers (j,k,l,m) such that $j^2+k^2+l^2+m^2=n$ is equal to 8 times the sum of the distinct divisors of n that are not divisible by 4. **Source:** Fact 13.5.5 and [117, 2970]. **Remark:** This is *Jacobi's four-square theorem.* **Example:** The distinct divisors of 4 that are not divisible by 4 are 1 and 2. Accordingly, the number of ways of writing 4 as a sum of squares of the components of a 4-tuple of integers is 24. Two of these are $0^2+0^2+0^2+2^2$ and $1^2+(-1)^2+1^2+1^2$.

Fact 1.11.26. Let $n \ge 0$. Then, the following statements hold:

i) There exist nonnegative integers $m_1, \dots, m_4$ such that $n = \sum_{i=1}^{4} m_i^2$.

ii) There exist nonnegative integers $m_1, \dots, m_9$ such that $n = \sum_{i=1}^{9} m_i^3$.

iii) There exist nonnegative integers $m_1, \dots, m_{19}$ such that $n = \sum_{i=1}^{19} m_i^4$.

iv) There exist nonnegative integers $m_1, \dots, m_{37}$ such that $n = \sum_{i=1}^{37} m_i^5$.

v) There exist nonnegative integers $m_1, \dots, m_{73}$ such that $n = \sum_{i=1}^{73} m_i^6$.

Source: [1260, pp. 372, 373]. **Remark:** These are solutions of *Waring's problem.* The first result is *Lagrange's four-square theorem.* For example, $3 = 0^2+1^2+1^2+1^2$ and $310 = 1^2+2^2+4^2+17^2$.

Fact 1.11.27. Let $n \ge 0$. Then, the following statements hold:

i) There exist nonnegative integers $m_1, \dots, m_4$ such that $n = m_1^2+m_2^2+m_3^2+m_4^2$.

ii) There exist nonnegative integers $m_1, \dots, m_4$ such that $n = m_1^2+m_2^2+2m_3^2+2m_4^2$.

iii) There exist nonnegative integers $m_1, \dots, m_4$ such that $n = m_1^2+2m_2^2+4m_3^2+14m_4^2$.

iv) $\operatorname{rem}(n,4) \ne 3$ if and only if there exist nonnegative integers $m_1, \dots, m_4$ such that $n = m_1^2+m_2^2+4m_3^2+4m_4^2$.

v) If $n \ge 2$, then there exist nonnegative integers $m_1, \dots, m_4$ such that $n = 2m_1^2+3m_2^2+4m_3^2+5m_4^2$.

vi) Let k_1, k_2, k_3, k_4 be positive integers, and assume that, for all $k \in \{1,2,3,5,6,7,10,14,15\}$, there exist nonnegative integers $m_1, \dots, m_4$ such that $k = k_1m_1^2+k_2m_2^2+k_3m_3^2+k_4m_4^2$. Then, there exist nonnegative integers $m_1, \dots, m_4$ such that $n = k_1m_1^2+k_2m_2^2+k_3m_3^2+k_4m_4^2$.

Remark: *i)–iii)* are universal positive integer-matrix quaternary quadratic forms. There are 54 such forms. See [2316, pp. 123–125] and [2886]. **Related:** Fact 10.18.27.

Fact 1.11.28. Let i, j, k, l be odd positive integers. Then, there exist even nonnegative integers q, r, s, t such that $q^2+r^2+s^2+t^2 = i^2+j^2+k^2+l^2$. If, in addition, i, j, k, l are distinct, then so are q, r, s, t. **Example:** $1^2+3^2+5^2+7^2 = 0^2+2^2+4^2+8^2$. **Source:** [2116]. **Related:** Fact 2.4.8.

Fact 1.11.29. Let $n \ge 1$, let $d_1, \dots, d_l$ be the distinct positive divisors of n, and, for all $i \in \{1, \dots, l\}$, let a_i denote the number of distinct positive divisors of d_i. Then,

$$\sum_{i=1}^{l} a_i^3 = \left(\sum_{i=1}^{l} a_i\right)^2.$$

Source: [2380, p. 64]. **Remark:** This is *Liouville's theorem.* **Related:** Fact 1.12.1. **Example:** Let $n = 8$ so that $d_1 = 1$, $d_2 = 2$, $d_3 = 4$, $d_4 = 8$, $a_1 = 1$, $a_2 = 2$, $a_3 = 3$, and $a_4 = 4$. Then, $1^3+2^3+3^3+4^3 = (1+2+3+4)^2$. Let $n = 15$ so that $d_1 = 1$, $d_2 = 3$, $d_3 = 5$, $d_4 = 15$, $a_1 = 1$, $a_2 = 2$, $a_3 = 2$, and $a_4 = 4$. Then, $1^3+2^3+2^3+4^3 = (1+2+2+4)^2$.

Fact 1.11.30. The following statements hold:

i) $1^2+7^2 = 5^2+5^2 = 50$, $1^2+8^2 = 4^2+7^2 = 65$, $2^2+9^2 = 6^2+7^2 = 85$, $2^2+11^2 = 5^2+10^2 = 125$.

ii) $5^2+14^2 = 10^2+11^2 = 221$, $4^2+19^2 = 11^2+16^2 = 377$, $7^2+24^2 = 15^2+20^2 = 25^2 = 625$.

iii) $1^2+18^2 = 6^2+17^2 = 10^2+15^2 = 325$, $20^2+107^2 = 43^2+100^2 = 68^2+85^2 = 11849$.

iv) $15^2+70^2 = 30^2+65^2 = 34^2+63^2 = 47^2+54^2 = 5125$, $10^2+11^2+12^2 = 13^2+14^2 = 365$.

v) $25^2+60^2 = 33^2+56^2 = 16^2+63^2 = 39^2+52^2 = 65^2 = 4225$, $1+3+3^2+3^3+3^4 = 11^2$.

vi) $7^2+74^2 = 14^2+73^2 = 22^2+71^2 = 25^2+70^2 = 41^2+62^2 = 50^2+55^2 = 5525$.

vii) $5^2 + 17^2 + 18^2 = 9^2 + 14^2 + 19^2 = 638$, $21^2 + 22^2 + 23^2 + 24^2 = 25^2 + 26^2 + 27^2 = 2030$.

viii) $36^2 + 37^2 + 38^2 + 39^3 + 40^2 = 41^2 + 42^2 + 43^2 + 44^2 = 7230$.

ix) $55^2 + 56^2 + 57^2 + 58^2 + 59^2 + 60^2 = 61^2 + 62^2 + 63^2 + 64^2 + 65^2 = 19855$.

x) $297^2 = (88 + 209)^2 = 88209$, $7777^2 = (6048 + 1729)^2 = 60481729$.

xi) $3^3 + 4^3 + 5^3 = 6^3 = 216$, $58^3 + 59^3 + 69^3 = 90^3 = 729000$, $1^3 + 12^3 = 9^3 + 10^3 = 1729$.

xii) $10^3 + 27^3 = 19^3 + 24^3 = 20683$, $4^3 + 48^3 = 36^3 + 40^3 = 110656$, $1 + 18 + 18^2 = 7^3$.

xiii) $167^3 + 436^3 = 228^3 + 423^3 = 255^3 + 414^3 = 87539319$, $11^3 + 12^3 + 13^3 + 14^3 = 20^3 = 8000$.

xiv) $31^3 + 33^3 + 35^3 + 37^3 + 39^3 + 41^3 = 66^3 = 287496$, $2^4 + 2^4 + 3^4 + 4^4 + 4^4 = 5^4$.

xv) $59^4 + 158^4 = 133^4 + 134^4 = 635318657$, $30^4 + 120^4 + 272^4 + 315^4 = 353^4 = 15527402881$.

xvi) $240^4 + 340^4 + 430^4 + 599^4 = 651^4 = 179607287601$.

xvii) $27^5 + 84^5 + 110^5 + 133^5 = 144^5 = 61917364224$, $1 + 7 + 7^2 + 7^3 = 20^2$.

xviii) $1^6 - 2^6 + 3^6 = 3(6 + 6^3) = 2^2 + 3^2 + 5^2 + 7^2 + 11^2 + 13^2 + 17^2 = 666$.

xix) $95800^4 + 217519^4 + 414560^4 = 422481^4 = 31858749840007945920321$.

xx) $3^6 + 19^6 + 22^6 = 10^6 + 15^6 + 23^6 = 160426514$, $13^2 + 7^3 = 2^9$, $2^7 + 17^3 = 71^2$.

xxi) $10^7 + 14^7 + 123^7 + 149^7 = 15^7 + 90^7 + 129^7 + 146^7 = 2056364173794800$.

xxii) $81^8 + 539^8 + 966^8 = 158^8 + 310^8 + 481^8 + 725^8 + 954^8 = 765381793634649192581218$.

xxiii) $42^9 + 99^9 + 179^9 + 475^9 + 542^9 + 574^9 + 625^9 + 668^9 + 822^9 + 851^9 = 917^9$ $= 458483827502199203411828597$.

xxiv) $62^{10} + 115^{10} + 172^{10} + 245^{10} + 295^{10} + 533^{10} + 689^{10} + 927^{10} + 1011^{10} + 1234^{10} + 1603^{10}$ $+ 1684^{10} = 1772^{10} = 303518810756415395921574821458201$.

xxv) For all $i \in \{1, 2, 3\}$, $1^i + 21^i + 36^i + 56^i = 2^i + 18^i + 39^i + 55^i$.

xxvi) For all $i \in \{1, 3, 9\}$, $1^i + 13^i + 13^i + 14^i + 18^i + 23^i = 5^i + 9^i + 10^i + 15^i + 21^i + 22^i$.

xxvii) For all $i \in \{-1, 1\}$, $4^i + 10^i + 12^i = 5^i + 6^i + 15^i$, $6^i + 14^i + 14^i = 7^i + 9^i + 18^i$, and $3^i + 40^i = 4^i + 15^i + 24^i = 5^i + 8^i + 30^i$.

xxviii) For all $i \in \{-2, -1, 1, 2\}$, $(-230)^i + (-92)^i + 23^i + 46^i = (-220)^i + (-110)^i + 22^i + 55^i$.

xxix) For all $i \in \{1, 2, 6\}$, $83^i + 211^i + (-300)^i = (-124)^i + (-185)^i + 303^i$, and $43^i + 371^i + (-372)^i = 140^i + 307^i + (-405)^i$.

xxx) For all $i \in \{1, 3, 5\}$, $(-51)^i + (-33)^i + (-24)^i + 7^i + 13^i + 38^i + 50^i = (-134)^i + (-75)^i + (-66)^i + 8^i + 47^i + 87^i + 133^i = 0$.

xxxi) For all $i \in \{1, 2, 3, 9\}$, $(-621)^i + 51^i + 253^i + 412^i + 600^i = (-624)^i + 187^i + 100^i + 429^i + 603^i$.

xxxii) For all $i \in \{1, 3, 5, 7\}$, $(-98)^i + (-82)^i + (-58)^i + (-34)^i + 13^i + 16^i + 69^i + 75^i + 99^i = (-169)^i + (-161)^i + (-119)^i + (-63)^i + 8^i + 50^i + 132^i + 148^i + 174^i = 0$.

xxxiii) For all $i \in \{1, 2, 3, 4, 5\}$, $(-461)^i + (-233)^i + (-199)^i + 465^i + 237^i + 203^i = (-435)^i + (-343)^i + 1^i + 3^i + 347^i + 439^i$.

xxxiv) $13! = 112296^2 - 79896^2 = 6227020800$.

Source: [564, 651, 731, 981, 1321, 2000, 2232] and [2259, pp. 48, 49]. **Remark:** *xvii)* and *xix)* are counterexamples to Euler's conjecture, which states that, for all $n \geq 4$, the nth power of a positive integer cannot be decomposed into the sum of $n-1$ or fewer nth powers of integers. Euler's conjecture is true in the case $n = 3$; that is, the cube of a positive integer cannot be the sum of the cubes of two positive integers. This case is given by Fact 1.11.39.

Fact 1.11.31. Let i, j, k, l be positive integers. Then, there exist positive integers m, n, r, s such that $\{m, n\} \neq \{r, s\}$ and

$$(i^2 + j^2)(k^2 + l^2) = m^2 + n^2 = r^2 + s^2.$$

In particular, $m = |ik - jl|$, $n = jk + il$, $r = ik + jl$, and $s = |il - jk|$. **Source:** Fact 2.4.7 and [2107, pp. 25, 26]. **Example:** $(2^2 + 3^2)(4^2 + 5^2) = 533 = 7^2 + 22^2 = 23^2 + 2^2$.

Fact 1.11.32. Let $k, m, n \geq 1$, assume that $k > m + n$, let $x_1, \ldots, x_m, y_1, \ldots, y_n$ be integers, and assume that $\sum_{i=1}^m x_i^k = \sum_{i=1}^n y_i^k$. Then, $m = n$ and $x^{\downarrow} = y^{\downarrow}$. **Remark:** This is *Euler's extended conjecture*. See [982, 1740].

Fact 1.11.33. Let $n \geq 0$. Then, there exist $k, l \geq 0$ such that $n = k^2 + l^2$ if and only if n does not have a prime factor of the form $4k + 3$ raised to an odd exponent. **Source:** [20, Chapter 4] and [2450, p. 378]. **Remark:** $29 = 2^2 + 5^2$, but neither 27, 71, nor 243 is the sum of two squares.

Fact 1.11.34. Let $n \geq 0$. Then, there exist $k, l, m \geq 0$ such that $n = k^2 + l^2 + m^2$ if and only if there do not exist $i, j \geq 0$ such that $n = 4^i(8j + 7)$. Hence, if $k, l \geq 1$, $k \overset{8}{\equiv} 3$, and $l \overset{8}{\equiv} 5$, then $kl \overset{8}{\equiv} 7$, and thus kl is not the sum of three squares. **Source:** [1258, p. 38] and [2316, p. 59]. **Remark:** $14 = 1^2 + 2^2 + 3^2$, but $15 = 4^0(8 \cdot 1 + 7)$ is not the sum of three squares.

Fact 1.11.35. Let $n \geq 0$. Then, there exist positive integers k, l, m such that $k < l < m$ and $n = k^2 + l^2 - m^2$. **Source:** [2380, pp. 56, 57]. **Example:** $0 = 3^2 + 4^2 - 5^2$, $1 = 4^2 + 7^2 - 8^2$, and $2 = 5^2 + 11^2 - 12^2$.

Fact 1.11.36. Let $l, m, n \geq 1$. Then, there exist integers j, k such that $j^2 + k^2 = (l^2 + m^2)^n$. **Source:** [1757, p. 115]. **Example:** $(2^2 + 3^2)^3 = 2197 = 9^2 + 46^2$.

Fact 1.11.37. Let $n \geq 1$. Then, the following statements are equivalent:

i) There exist $k, l \geq 1$ such that $n = k^3 + l^3$.

ii) There exists a divisor m of n such that $\sqrt[3]{n} \leq m \leq 2^{2/3}\sqrt[3]{n}$, $3|m^2 - n/m$, and $\sqrt{\frac{4n}{3m} - \frac{m^2}{3}}$ is an integer.

Furthermore, the following statements are equivalent:

iii) There exist $k, l \geq 1$ such that $n = k^3 - l^3$.

iv) There exists a divisor m of n such that $1 \leq m \leq \sqrt[3]{n}$, $3|m^2 - \frac{n}{m}$, and $\sqrt{\frac{4n}{3m} - \frac{m^2}{3}}$ is an integer.

Source: [580]. **Example:** $91 = 3^3 + 4^3$ and $m = 7$.

Fact 1.11.38. Let $n \geq 2$. Then, H_n is not an integer. **Source:** [1757, p. 105].

Fact 1.11.39. Let $k, l, m \geq 1$ and $n \geq 3$. Then, $k^n + l^n \neq m^n$. **Remark:** This is *Fermat's last theorem*. **Credit:** A. Wiles.

Fact 1.11.40. Let $n \geq 2$ be prime, and assume that $n \overset{4}{\equiv} 1$. Then, there exist $k, l \geq 1$ such that $n = k^2 + l^2$. **Source:** [116, p. 41] and [2963]. **Credit:** P. de Fermat. **Example:** $29 = 4 + 25$ and $89 = 25 + 64$.

Fact 1.11.41. Let $k, l, m, n \geq 2$, and assume that $k^l - m^n = 1$. Then, $k = 3$, $l = 2$, $m = 2$, and $n = 3$. **Remark:** This is *Catalan's conjecture*. **Credit:** P. Mihăilescu.

Fact 1.11.42. Let $n \geq 1$. Then, there exists a prime $m \in (n, 2n]$. If, in addition, $n \geq 2898242$, then there exists a prime $m \in (n, n + n/(111 \log^2 n))$. **Source:** [20, Chapter 2] and [202, 2736]. **Remark:** The first statement is *Bertrand's postulate*.

Fact 1.11.43. Let $n \geq 20$, and, for all $i \geq 1$, let p_i denote the ith prime. Then,

$$n(\log n + \log\log n - \tfrac{3}{2}) < p_n < n(\log n + \log\log n - \tfrac{1}{2}).$$

Source: [1350, p. 183] and [2342].

Fact 1.11.44. Let $n \geq 1$, and, for all $i \geq 1$, let p_i denote the ith prime. Then, $\prod p_i \leq 4^n$, where the product is taken over all i such that $p_i \leq n$. **Source:** [2068, p. 90].

Fact 1.11.45. For all $i \geq 1$, let p_i denote the ith prime. Then, for all $k \geq 4$, $p_{k+1}^2 < \prod_{i=1}^k p_i$. **Remark:** This is *Bonse's inequality*. **Remark:** $121 < 210$ and $169 < 2310$.

Fact 1.11.46. Let $n \geq 4$ be even. Then, there exist primes k and l such that $n = k + l$. **Remark:**

This is the *Goldbach conjecture*. **Example:** $44 = 13 + 31$ and $100 = 17 + 83$. **Remark:** An incomplete proof is given in [594].

Fact 1.11.47. Let $n \geq 1$, and let d_n denote the sum of all positive integers (not counting multiplicity) that divide n. Then, $d_n \leq H_n + e^{H_n} \log H_n$. **Remark:** This result is equivalent to the *Riemann hypothesis*. See [524, p. 48] and [1724]. Equivalent statements are given by Fact 13.3.1. **Remark:** Let $r_n \triangleq d_n/(H_n + e^{H_n} \log H_n)$. Then, $r_{12} \approx .98864$, $r_{120} \approx .98344$, $r_{360} \approx .97111$, and $r_{2520} \approx .97831$.

Fact 1.11.48. Let $n \geq 1$, let $\{a_1, \ldots, a_n\} \cup \{b_1, \ldots, b_n\} = \{1, \ldots, 2n\}$, and assume that $a_1 < \cdots < a_n$ and $b_n < \cdots < b_1$. Then, $\sum_{i=1}^n |a_i - b_i| = n^2$. **Source:** [2380, p. 66].

Fact 1.11.49. If $n \geq 1$, then there exist finitely many multisets $\{k_1, \ldots, k_n\}_{\rm ms}$ of positive integers such that $\sum_{i=1}^n \frac{1}{k_i} = 1$. Now, define $S_1 \triangleq 2$ and, for all $n \geq 2$, define $S_n \triangleq 1 + \prod_{i=1}^{n-1} S_i$. In particular, $(S_i)_{i=1}^6 = (2, 3, 7, 43, 1807, 3263443)$. If $n \geq 2$ and the positive integers $k_1, \ldots, k_n$ satisfy $\sum_{i=1}^n \frac{1}{k_i} = 1$, then $\max\{k_1, \ldots, k_n\} \leq S_{n-1} - 1$. **Source:** [2336, p. 288] and [2494].

Fact 1.11.50. Let $n \geq 1$. Then,

$$\frac{4}{4n+1} = \frac{1}{n} - \frac{1}{n(4n+1)}, \qquad \frac{4}{4n-1} = \frac{1}{n} + \frac{1}{n(4n-1)}.$$

If n is odd, then

$$\frac{4}{n} = \frac{2}{n-1} + \frac{2}{n+1} - \frac{4}{n(n^2-1)}.$$

If $n \stackrel{3}{\equiv} 2$, then

$$\frac{4}{n} = \frac{1}{n} + \frac{3}{n+1} + \frac{3}{n(n+1)}.$$

Source: [131]. **Remark:** These equalities concern *Egyptian fractions* and are associated with the Erdös-Straus conjecture. See Fact 1.11.51.

Fact 1.11.51. Let $n \geq 2$. Then, there exist $k, l, m \geq 1$ such that $4/n = 1/k + 1/l + 1/m$. **Example:** $4/5 = 1/2 + 1/4 + 1/20 = 1/2 + 1/5 + 1/10$. **Remark:** This is the *Erdös-Straus conjecture*. **Related:** Fact 1.11.50.

Fact 1.11.52. Let $n \geq 1$. Then, $\lfloor \sqrt{n} + \sqrt{n+1} \rfloor = \lfloor \sqrt{4n+2} \rfloor$. **Source:** [289, pp. 19, 119].

1.12 Facts on Finite Sums

Fact 1.12.1. Let $n, k \geq 1$. Then,

$$\sum_{i=1}^n i^k = \frac{1}{k+1} \sum_{i=0}^k B_i \binom{k+1}{i} (n+1)^{k+1-i} = \frac{1}{k+1} \left[\left(\sum_{i=0}^{k+1} B_{k+1-i} \binom{k+1}{i} (n+1)^i \right) - B_{k+1} \right].$$

In particular,

$$\sum_{i=1}^n i = \binom{n+1}{2} = \tfrac{1}{2} n(n+1) = \tfrac{1}{2} n^2 + \tfrac{1}{2} n,$$

$$\sum_{i=1}^n i^2 = \frac{1}{4} \binom{2n+2}{3} = \binom{n+1}{2} + 2\binom{n+1}{3} = \tfrac{1}{6} n(n+1)(2n+1) = \tfrac{1}{3} n^3 + \tfrac{1}{2} n^2 + \tfrac{1}{6} n,$$

$$\sum_{i=1}^n i^3 = \left(\sum_{i=1}^n i \right)^2 = \binom{n+1}{2}^2 = \tfrac{1}{4} n^2 (n+1)^2 = \tfrac{1}{4} n^4 + \tfrac{1}{2} n^3 + \tfrac{1}{4} n^2,$$

$$\sum_{i=1}^n i^4 = \tfrac{1}{30} n(n+1)(2n+1)(3n^2+3n-1) = \tfrac{1}{5} n^5 + \tfrac{1}{2} n^4 + \tfrac{1}{3} n^3 - \tfrac{1}{30} n,$$

$$\sum_{i=1}^{n} i^5 = \tfrac{1}{12}n^2(n+1)^2(2n^2+2n-1) = \tfrac{1}{6}n^6 + \tfrac{1}{2}n^5 + \tfrac{5}{12}n^4 - \tfrac{1}{12}n^2,$$

$$\sum_{i=1}^{n} i^6 = \tfrac{1}{42}n(n+1)(2n+1)(3n^4+6n^3-3n+1) = \tfrac{1}{7}n^7 + \tfrac{1}{2}n^6 + \tfrac{1}{2}n^5 - \tfrac{1}{6}n^3 + \tfrac{1}{42}n,$$

$$\sum_{i=1}^{n} i^7 = \tfrac{1}{24}n^2(n+1)^2(3n^4+6n^3-4n+2),$$

$$\sum_{i=1}^{n} i^8 = \tfrac{1}{90}n(n+1)(2n+1)(5n^6+15n^5+5n^4-15n^3-n^2+9n-3),$$

$$\sum_{i=1}^{n} i(i+1) = \tfrac{1}{3}n(n+1)(n+2), \quad \sum_{i=1}^{n} i(i+1)(i+2) = \tfrac{1}{4}n(n+1)(n+2)(n+3),$$

$$\sum_{i=1}^{n} i(i+1)^2 = \tfrac{1}{12}n(n+1)(n+2)(3n+5), \quad \sum_{i=1}^{n} i(i+1)^3 = \tfrac{1}{60}n(n+1)(n+2)(12n^2+39n+29),$$

$$\sum_{i=0}^{n-1}(2i+1) = n^2, \quad \sum_{i=0}^{n-1}(2i+1)^2 = \tfrac{1}{3}n(4n^2-1), \quad \sum_{i=0}^{n-1}(2i+1)^3 = n^2(2n^2-1),$$

$$\sum_{i=0}^{n-1}(2i+1)^4 = \tfrac{1}{15}n(48n^4-40n^2+7), \quad \sum_{i=0}^{n-1}(2i+1)^5 = \tfrac{1}{3}n^2(16n^4-20n^2+7),$$

$$\sum_{i=0}^{n-1}(2i+1)^6 = \tfrac{1}{21}n(4n^2-1)(48n^4-72n^2+31), \quad \sum_{i=0}^{n-1}(2i+1)^7 = \tfrac{1}{3}n^2(48n^6-112n^4+98n^2-31).$$

Now, let $k \ge 1$ and $n \ge 1$, and define $p_k(n) \triangleq \sum_{i=1}^{n} i^k$. Then, the following statements hold:

i) $p_k(n)$ is a polynomial whose degree is $k+1$ and whose leading coefficient is $1/(k+1)$.

ii) The coefficient of n in $p_k(n)$ is $(-1)^k B_k$.

iii) $p_k(1) = 1$.

iv) For all $z \in \mathbb{C}$, $p_k'(z) = kp_{k-1}(z) + (-1)^k B_k$.

v) p_2 divides p_{2k}, and p_3 divides p_{2k+1}.

vi) $p_1(n)$ divides $p_{2k+1}(n)$.

vii) $\sum_{i=1}^{k} \binom{k+1}{i} p_i(n) = (n+1)^{k+1} - n - 1$.

Source: The first equality is the *Bernoulli formula*, where B_i is the ith Bernoulli number. See Fact 13.1.6. See [771, pp. 153–155], [1217, pp. 2, 3], [1219, pp. 283, 284], and [2689]. *i)*–*iv)* are given in [2915]; *v)* is given in [771, p. 155]; *vi)* is given in [1919]; and *vii)* is given in [2504, p. 135]. **Remark:** *v)* is a statement about polynomials, whereas *vi)* is a statement about integers. **Remark:** A matrix approach to sums of powers of integers is given in [959]. The expressions involving binomial coefficients are given in [410] and [411, pp. 109–112]. See also [1917]. **Related:** Fact 1.11.29, Fact 1.12.2, Fact 2.11.27, and [1371, p. 11].

Fact 1.12.2. Let $n \ge 1$, let $k \ge 0$, and define $\sigma_k \triangleq \sum_{i=1}^{n} i^k$. Then,

$$\sigma_1 = \tfrac{1}{2}(n+\tfrac{1}{2})^2 - \tfrac{1}{8}, \quad \sigma_2 = \tfrac{1}{3}(n+\tfrac{1}{2})^3 - \tfrac{1}{12}(n+\tfrac{1}{2}), \quad 2\sigma_1^4 = \sigma_5 + \sigma_7,$$

$$\sigma_3 = \sigma_1^2, \quad \sigma_4 = (\tfrac{6}{5}\sigma_1 - \tfrac{1}{5})\sigma_2, \quad \sigma_5 = \tfrac{4}{3}\sigma_1^3 - \tfrac{1}{3}\sigma_1^2, \quad \sigma_6 = (\tfrac{12}{7}\sigma_1^2 - \tfrac{6}{7}\sigma_1 + \tfrac{1}{7})\sigma_2,$$

$$\sigma_7 = 2\sigma_1^4 - \tfrac{4}{3}\sigma_1^3 + \tfrac{1}{3}\sigma_1^2, \quad \sigma_1^3 = \tfrac{1}{4}\sigma_3 + \tfrac{3}{4}\sigma_5, \quad \sigma_1^5 = \tfrac{1}{16}\sigma_5 + \tfrac{5}{8}\sigma_7 + \tfrac{5}{16}\sigma_9,$$

$$8\sigma_1^3 + \sigma_1^2 - 9\sigma_2^2 = 0, \quad 81\sigma_2^4 - 18\sigma_2^2\sigma_3 + \sigma_3^2 - 64\sigma_3^3 = 0, \quad 16\sigma_3^3 - \sigma_3^2 - 6\sigma_3\sigma_5 - 9\sigma_5^2 = 0.$$

Furthermore,

$$\sum_{i=0}^{k} \binom{k+1}{i} \sigma_i = (n+1)^{k+1} - 1.$$

Next, define the polynomial

$$F_k(s) \triangleq \frac{1}{k+1} \sum_{i=0}^{k} B_i \binom{k+1}{i} (s+1)^{k+1-i}.$$

Then,

$$F_3(s) = s^2,\ F_4(s) = \tfrac{6}{5}s - \tfrac{1}{5},\ F_5(s) = \tfrac{4}{3}s^3 - \tfrac{1}{3}s^2,\ F_6(s) = \tfrac{12}{7}s^2 - \tfrac{6}{7}s + \tfrac{1}{7},\ F_7(s) = 2s^4 - \tfrac{4}{3}s^3 + \tfrac{1}{3}s^2.$$

If $k \geq 3$ is odd, then $\sigma_k = F_k(\sigma_1)$ and $\deg F_k = \frac{1}{2}(k+1)$. If $k \geq 2$ is even, then $\sigma_k = \sigma_2 F_k(\sigma_1)$ and $\deg F_k = \frac{1}{2}(k-2)$. **Source:** [334]. **Remark:** F_k is a *Faulhaber polynomial.* Generating functions are given in [334]. **Remark:** B_i is the ith Bernoulli number. See Fact 13.1.6. **Related:** Fact 1.12.1.

Fact 1.12.3. For all $n \geq 0$, define the nth *triangular number* by $T_n \triangleq \frac{1}{2}n(n+1)$. Then, the following statements hold:

i) $(T_i)_{i=0}^{20} = (0, 1, 3, 6, 10, 15, 21, 28, 36, 45, 55, 66, 78, 91, 105, 120, 136, 153, 171, 190, 210)$.

ii) If $n \geq 1$, then $T_n = \sum_{i=1}^n i = \binom{n+1}{2} = \frac{1}{2}n(n+1) = \sqrt{\sum_{i=1}^n i^3}$.

iii) $2T_nT_{n+1} = T_{n^2+2n}$, $T_nT_{n+2} = 2T_{(n^2+3n)/2}$, and $T_n \overset{2}{\equiv} \lfloor (n+1)/2 \rfloor$.

iv) If $n \geq 1$, then $8T_n + 1 = (2n+1)^2$ and $T_n + T_{n+1} = (n+1)^2$.

v) If $n \geq 2$, then $8T_n + 1 = T_{n+1} + 6T_n + T_{n-1}$.

vi) If $n \geq 1$, then $T_n^2 = \sum_{i=1}^n i^3$, $T_{2n^2-1} = \sum_{i=1}^n (2i-1)^3$, and $\sum_{i=1}^{2n-1} (-1)^{i+1} T_i = n^2$.

vii) If $n \geq 1$, then $9T_n + 1 = T_{3n+1}$, $T_{(n+1)^2} = T_n^2 + T_{n+1}^2 = T_{n^2} + T_{n+1}^2 - T_{n-1}^2$, and $\sum_{i=0}^n 9^i = T_{(3^{n+1}-1)/2}$.

viii) If $n \geq 1$, then $\sum_{i=1}^n T_i = \frac{n}{3}T_{n+1} = \frac{1}{3}(n+2)T_n = \frac{1}{6}n(n+1)(n+2)$ and $2T_n^2 = T_{n^2} + n^3$.

ix) If $n \geq 2$, then $T_n^2 = T_n + T_{n-1}T_{n+1}$.

x) If $n \geq 2$, then none of $\sqrt[3]{T_n}$, $\sqrt[4]{T_n}$, $\sqrt[5]{T_n}$ are integers.

xi) If $n \geq 1$, then the last digit of T_n is not an element of $\{2, 4, 7, 9\}$.

xii) If $n \geq 1$ and T_n is prime, then $n = 2$ and $T_n = 3$.

xiii) If $n, m \geq 1$, then $T_{m+n} = T_m + T_n + mn$, $T_{mn} = T_mT_n + T_{m-1}T_{n-1}$, $T_{mn-1} = T_{m-1}T_n + T_mT_{n-1}$.

xiv) For all $n \geq 1$, let τ_n be the nth positive integer such that $T_{\tau_n - 1}$ is square. Then, $(\tau_i)_{i=1}^{12} = (1, 2, 9, 50, 289, 1682, 9801, 57122, 332929, 1940450, 11309769, 65918162)$.

xv) There are infinitely many square triangular numbers.

xvi) For all $n \geq 1$, $\tau_{n+2} = 6\tau_{n+1} - \tau_n - 2$.

xvii) For all $n \geq 1$, $\tau_n = \frac{1}{2}[(3 - 2\sqrt{2})^{n-1} + 1]^2/(3 - 2\sqrt{2})^{n-1}$.

xviii) For all $n \geq 1$, $\sqrt{\tau_n} - \sqrt{\tau_n - 1} = (\sqrt{2} - 1)^{n-1}$.

xix) For all $n \geq 2$,

$$T_{\tau_n - 1} = \left[\sum_{i=0}^{\lfloor (n-1)/2 \rfloor} 2^i \binom{n-1}{2i}\right]^2 \left[\sum_{i=0}^{\lfloor (n-2)/2 \rfloor} 2^i \binom{n-1}{2i+1}\right]^2.$$

In particular, $(T_{\tau_i - 1})_{i=1}^{9} = (0, 1, 36, 1225, 41616, 1413721, 48024900, 1631432881)$.

xx) For all $n \geq 1$, $T_{\tau_{n+2}} = (6\sqrt{T_{\tau_{n+1}}} - \sqrt{T_{\tau_n}})^2$.

xxi) $(\sqrt{T_{\tau_i - 1}})_{i=1}^{12} = (0, 1, 6, 35, 204, 1189, 6930, 40391, 235416, 1372105, 7997214, 46611179)$.

xxii) For all $n \geq 1$, $T_{T_{n+2}-1} = 34T_{T_{n+1}-1} - T_{T_n-1} + 2$.

xxiii) For all $n \geq 1$, $T_{T_n-1} = \frac{1}{32}[(1+\sqrt{2})^{2n-2} - (1-\sqrt{2})^{2n-2}]^2$.

xxiv) n is a triangular number and a Fibonacci number if and only if $n \in \{1, 3, 21, 55\}$.

xxv) Every nonnegative integer is the sum of three triangular numbers.

xxvi) Every triangular number except T_1 and T_3 is the sum of three positive triangular numbers.

xxvii) $T_{132}^2 + T_{143}^2 = T_{164}^2$.

xxviii) If $n \geq 3$, then $\prod_{i=1}^n T_i < T_{n!}$.

xxix) If $n \geq 0$, then $\sum_{i=0}^n \binom{n}{i} T_i = 2^{n-2}(T_{n+1} - 1)$.

xxx) If $n \geq 1$, then $T_{n^2+n-1} + T_{n^2+3n+1} = (n+1)^4$.

Source: [132, 2193, 2733]. *xiii)* and *xxviii)* are given in [177]; *xviii)* is given in [1497]; *xix)* is given in [108, pp. 55, 312, 313]; *xx)* is given in [1598]; *xxv)* is given in [1258, p. 25] and [2569]; *xxix)* is given in [1197]. *xxx)* is given in [1958]. **Remark:** T_n is given by $P_2(n)$ in Fact 1.12.6. See Fact 1.20.1. **Related:** Fact 13.4.6.

Fact 1.12.4. For all $n \in \mathbb{N}$, define the nth *pentagonal number* by $P_n \triangleq \frac{1}{2}n(3n-1)$. Then, the following statements hold:

i) $(P_i)_{i=0}^{18} = (0, 1, 5, 12, 22, 35, 51, 70, 92, 117, 145, 176, 210, 247, 287, 330, 376, 425, 477)$.

ii) If $n \geq 1$, then $\frac{1}{n}\sum_{i=1}^n P_i = T_n$ and $3P_n = T_{3n-1}$.

iii) If n is a pentagonal number, then $\frac{1}{6}(\sqrt{24n+1} + 1) = n$.

iv) Let $n \geq 1$. Then, the following statements are equivalent:

a) n is a pentagonal number.

b) $24n + 1$ is a square, and $\sqrt{24n+1} \overset{6}{\equiv} 5$.

c) $\frac{1}{6}(\sqrt{24n+1} + 1)$ is an integer.

v) Every nonnegative integer is the sum of five pentagonal numbers.

vi) If $n \geq 1$ and $n \notin \{9, 21, 31, 43, 55, 89\}$, then n is the sum of four pentagonal numbers.

Finally, for all $n \geq 1$, define the nth *dual pentagonal number* by $P'_n \triangleq \frac{1}{2}n(3n+1)$. Then, the following statements hold:

vii) For all $n \geq 1$, $P_n < P'_n < P_{n+1}$.

viii) $(P'_i)_{i=1}^{18} = (2, 7, 15, 26, 40, 57, 77, 100, 126, 155, 187, 222, 260, 301, 345, 392, 442, 495)$.

Source: [2569, 2862]. **Remark:** For all $n \geq 1$, $P'_n = P_{-n}$. **Remark:** See [1083, pp. 45–47]. **Related:** Fact 13.4.7.

Fact 1.12.5. For all $n \geq 0$, define the nth *generalized pentagonal number* by

$$g_n \triangleq \begin{cases} \frac{1}{8}(n+1)(3n+1), & n \text{ odd}, \\ \frac{1}{8}n(3n+2), & n \text{ even}. \end{cases}$$

Then, the following statements hold:

i) $(g_i)_{i=0}^{21} = (0, 1, 2, 5, 7, 12, 15, 22, 26, 35, 40, 51, 57, 70, 77, 92, 100, 117, 126, 145, 155, 176)$.

ii) For all $n \geq 1$, $g_{2n-1} = \frac{1}{2}n(3n-1)$ and $g_{2n} = \frac{1}{2}n(3n+1)$.

iii) $(g_i)_{i=-\infty}^{\infty} = (\frac{1}{2}i(3i-1))_{i=-\infty}^{\infty}$.

iv) For all $n \geq 0$,

$$g_n \triangleq \begin{cases} P_{(n+1)/2}, & n \text{ odd}, \\ P_{-n/2} = P'_{n/2}, & n \text{ even}. \end{cases}$$

v) $(g_0, g_1, g_2, g_3, g_4, \ldots) = (P_0, P_1, P_1', P_2, P_2', \ldots)$, where P_n is the nth pentagonal number and P_n' is the nth dual pentagonal number.

vi) Every nonnegative integer is the sum of three generalized pentagonal numbers.

Source: [136, 1282, 2569]. **Related:** Fact 1.12.4 and Fact 13.10.27.

Fact 1.12.6. Let $n, k \geq 1$, and define $P_k(n) \triangleq \operatorname{card}\{(i_1, \ldots, i_k) : 1 \leq i_1 \leq \cdots \leq i_k \leq n\}$. Then,

$$P_k(n) = \binom{n+k-1}{k}.$$

In particular,

$$P_2(n) = \sum_{i=1}^{n} i = \binom{n+1}{2} = \frac{n(n+1)}{2}, \quad P_2(n) = \binom{n+2}{3} = \sum_{i=1}^{n}\sum_{j=1}^{i} j = \frac{n(n+1)(n+2)}{6},$$

$$P_3(n) = \binom{n+3}{4} = \sum_{i=1}^{n}\sum_{j=1}^{i}\sum_{l=1}^{j} l = \frac{n(n+1)(n+2)(n+3)}{24}.$$

Remark: $P_2(n)$, $P_3(n)$, and $P_4(n)$ are the *triangular*, *tetrahedral*, and *pentatopic* numbers. **Remark:** $P_k(n)$ is the number of k-element multisubsets of $\{1, \ldots, n\}$; that is, $P_k(n) = \binom{n}{k}_{\mathrm{r}}$. See Fact 1.16.16. **Related:** Fact 1.12.3.

Fact 1.12.7. Let $n \geq 0$ and $k \geq 3$, and define the *(n, k) polygonal number* $p_k(n) \triangleq (k-2)\binom{n}{2} + n$. Then, the following statements hold:

i) $p_k(n) = \frac{1}{2}n[(k-2)n + 4 - k]$.

ii) $p_3(n) = \frac{1}{2}n(n+1)$ is the nth triangular number.

iii) $p_4(n) = n^2$.

iv) $p_5(n) = \frac{1}{2}n(3n-1)$ is the nth pentagonal number.

v) $p_k(n) = \left.\dfrac{\mathrm{d}^n}{\mathrm{d}x^n} \dfrac{x[(k-3)x+1]}{n!(1-x)^3}\right|_{x=0}$.

vi) Let $m \geq 0$. Then, there exist nonnegative integers $n_1, \ldots, n_k$ such that $m = \sum_{i=1}^{k} p_k(n_i)$.

Source: [115, 1282, 2569, 2863]. **Credit:** The last statement is due to A. L. Cauchy.

Fact 1.12.8. Let $n \geq 1$. Then,

$$\sum_{i,j=1}^{n} |i-j| = \frac{1}{3}n(n^2-1), \quad \sum_{i,j=1}^{n} (i-j)^2 = \frac{1}{6}n^2(n^2-1).$$

Now, let $k \geq 1$, and define $\sigma_k \triangleq \sum_{i=1}^{n} i^k$. Then,

$$\sum_{i,j=1}^{n} |i^k - j^k| = 4\sigma_{k+1} - 2(n+1)\sigma_k, \quad \sum_{i,j=1}^{n} |(i-j)(i^k - j^k)| = 2n\sigma_{k+1} - n(n+1)\sigma_k.$$

Source: [394]. **Related:** Fact 2.11.10.

Fact 1.12.9. Let $n \geq 1$. Then,

$$\exp \sum_{i,j=1}^{n} \left|\log \frac{i}{j}\right| = \frac{\prod_{i=1}^{n} i^{4i}}{(n!)^{2n+2}}, \quad \exp \sum_{i,j=1}^{n} \left|(i-j)\log \frac{i}{j}\right| = \left(\frac{\prod_{i=1}^{n} i^{2i}}{(n!)^{n+1}}\right)^n.$$

Source: [394]. **Related:** Fact 2.11.10.

Fact 1.12.10. Let $1 \le k \le n$, let $r \in \mathbb{R}$, and define

$$S_{k,r} \triangleq \sum \prod_{j=1}^{k} i_j^r,$$

where the sum is taken over all k-tuples $(i_1, \ldots, i_k)$ such that $1 \le i_1 < \cdots < i_k \le n$. Then,

$$S_{1,1} = \tfrac{1}{2}n(n+1), \quad S_{2,1} = \frac{1}{24}n(n^2-1)(3n+2), \quad S_{3,1} = \frac{1}{48}(n-2)(n-1)n^2(n+1)^2,$$

$$S_{1,2} = \frac{1}{6}n(n+1)(2n+1), \quad S_{2,2} = \frac{1}{432}n(n+1)(2n+1)(10n^3-3n^2-13n+6).$$

Furthermore, for all $r \in \mathbb{R}$,

$$S_{3,r} = \frac{1}{6}(2S_{1,3r} - 3S_{1,r}S_{1,2r} + S_{1,r}^2).$$

Source: [366].

Fact 1.12.11. Let $k \ge 1$ and $n \ge 1$. If $n^2 \le k \le (n+1)^2 - 1$, then $\lfloor \sqrt{k} \rfloor = n$. If $n^3 \le k \le (n+1)^3 - 1$, then $\lfloor \sqrt[3]{k} \rfloor = n$. Now, assume that $n \ge 2$. Then,

$$\sum_{i=1}^{n^2-1} \lfloor \sqrt{i} \rfloor = \frac{1}{6}n(n-1)(4n+1), \quad \sum_{i=1}^{n^3-1} \lfloor \sqrt[3]{i} \rfloor = \frac{1}{4}(n-1)n^2(3n+1).$$

Source: [289, pp. 39, 187].

Fact 1.12.12. Let $n \ge 1$. Then,

$$\sum_{i=1}^{n} \left\lfloor \frac{i}{2} \right\rfloor = \frac{n^2}{4} + \frac{(-1)^n - 1}{8}, \quad \sum_{i=1}^{\lfloor n/2 \rfloor} \left\lfloor \frac{i}{2} \right\rfloor = \left\lfloor \frac{n}{4} \right\rfloor \left\lfloor \frac{n+2}{4} \right\rfloor.$$

Source: [335].

Fact 1.12.13. Let $n \ge 3$ and $m \ge 1$, assume that n is prime, and assume that $n \nmid m$. Then,

$$\sum_{i=1}^{n-1} \left\lfloor \frac{im}{n} \right\rfloor = \tfrac{1}{2}(n-1)(m-1), \quad \sum_{i=1}^{n-1} \left\lfloor \frac{(-1)^i i^2 m}{n} \right\rfloor = \tfrac{1}{2}(n-1)(m-1),$$

$$\sum_{i=1}^{n-1} \left\lfloor \frac{i^3 m}{n} \right\rfloor = \tfrac{1}{4}(n-1)(n^2m - nm - 2), \quad \sum_{i=1}^{n-1} \left\lfloor \frac{(-1)^i i^4 m}{n} \right\rfloor = \tfrac{1}{2}(n-1)[m(n^2-n-1)-1].$$

In particular,

$$\sum_{i=1}^{n-1} \left\lfloor \frac{i}{n} \right\rfloor = 0, \quad \sum_{i=1}^{n-1} \left\lfloor \frac{(-1)^i i^2}{n} \right\rfloor = 0, \quad \sum_{i=1}^{n-1} \left\lfloor \frac{i^3}{n} \right\rfloor = \tfrac{1}{4}(n-2)(n^2-1), \quad \sum_{i=1}^{n-1} \left\lfloor \frac{(-1)^i i^4}{n} \right\rfloor = \tfrac{1}{2}(n-2)(n^2-1).$$

If n is odd, then

$$\sum_{i=1}^{n-1} (-1)^i \left\lfloor \frac{i^2}{n} \right\rfloor = \tfrac{1}{2}(n-1).$$

If n is even and $\frac{n}{2} \overset{4}{\equiv} 1$, then

$$\sum_{i=1}^{n-1} (-1)^i \left\lfloor \frac{i^2}{n} \right\rfloor = 1 - \frac{n}{2}.$$

Finally, if n is an odd prime, then

$$\sum_{i=1}^{n-1}\left\lfloor\frac{i^n}{n^2}\right\rfloor = \frac{1}{n^2}\left(\sum_{i=1}^{n-1} i^n\right) - \tfrac{1}{2}(n-1).$$

Source: [107, pp. 428–432] and [1673].

Fact 1.12.14. Let $1 \le m \le n$. Then,

$$\sum\prod_{j=1}^{m} i_j = \frac{n}{(2m-1)!}\prod_{i=1}^{m-1}(n^2 - i^2),$$

where the sum is taken over all m-tuples $(i_1, \ldots, i_m)$ of positive integers such that $\sum_{j=1}^{m} i_j = n$. In particular,

$$\sum ij = \frac{1}{6}n(n^2-1),$$

where the sum is taken over all ordered pairs (i, j) of positive integers such that $i + j = n$. **Source:** [771, pp. 33, 85].

Fact 1.12.15. Let $1 \le m < n$. Then,

$$\sum_{i=m+1}^{n} i\prod_{j=1}^{m}(i^2 - j^2) = \frac{(n+m+1)!}{2(m+1)(n-m-1)!}.$$

Source: [108, pp. 31, 188].

Fact 1.12.16. Let $1 \le k \le n$. Then,

$$\sum \operatorname{card}(\cap_{i=1}^{k}\mathcal{S}_i) = 2^{k(n-1)}n,$$

where the sum is taken over all k-tuples $(\mathcal{S}_1, \ldots, \mathcal{S}_k)$ of subsets of $\{1, \ldots, n\}$. In particular,

$$\sum \operatorname{card}(\mathcal{S}_1 \cap \mathcal{S}_2) = 4^{n-1}n,$$

where the sum is taken over all ordered pairs $(\mathcal{S}_1, \mathcal{S}_2)$ of subsets of $\{1, \ldots, n\}$. **Source:** [771, pp. 33, 34].

Fact 1.12.17. Let $n \ge 2$. Then,

$$\operatorname{card}(\{(i, j)\colon i, j \in \{1, \ldots, n\} \text{ and } i < j\}) = \binom{n}{2}.$$

Fact 1.12.18. Let $n \ge 1$. Then,

$$\sum_{i=1}^{n}(2i-1) = n^2, \quad \sum_{i=1}^{2n} i = \sum_{i=1}^{n}(4i-1) = (2n+1)n, \quad \sum_{i=1}^{2n-1} i = \sum_{i=1}^{n}(4i-3) = (2n-1)n.$$

Fact 1.12.19. Let $m, n \ge 1$. Then,

$$\sum_{i=1}^{n}(mi-1) = \tfrac{1}{2}mn(n+1) - n, \quad \sum_{i=1}^{n}(mi-1)^2 = \tfrac{1}{6}m^2n(n+1)(2n+1) - mn(n+1) + n,$$

$$\sum_{i=1}^{n}(mi-1)^3 = \tfrac{1}{4}m^3n^2(n+1)^2 - \tfrac{1}{2}m^2n(n+1)(2n+1) + \tfrac{3}{2}mn(n+1) - n.$$

In particular,

$$\sum_{i=1}^{n}(2i-1) = n^2, \quad \sum_{i=1}^{n}(3i-1) = \tfrac{3}{2}n^2 + \tfrac{1}{2}n, \quad \sum_{i=1}^{n}(2i-1)^2 = \tfrac{4}{3}n^3 - \tfrac{1}{3}n,$$

$$\sum_{i=1}^{n}(3i-1)^2 = 3n^3 - \tfrac{3}{2}n^2 - \tfrac{1}{2}n, \qquad \sum_{i=1}^{n}(2i-1)^3 = 2n^4 - n^2.$$

Source: [1217, pp. 2, 3] and [1524, p. 37].

Fact 1.12.20. Let $m \ge n \ge 1$. Then,

$$\sum_{i,j=1}^{m,n} \min\{i,j\} = \frac{1}{6}n(n+1)(3m-n+1), \qquad \sum_{i,j=1}^{m,n} \max\{i,j\} = \frac{1}{6}n(n^2-1) + \frac{1}{2}mn(m+1).$$

Source: [771, p. 168].

Fact 1.12.21. Let $n \ge 1$. Then,

$$\sum_{i=1}^{n} 2^i i = 2^{n+1}(n-1) + 2, \qquad \sum_{i=1}^{n} 2^i i^2 = 2^{n+1}(n^2-2n+3) - 6,$$

$$\sum_{i=1}^{n} 2^i i^3 = 2^{n+1}(n^3 - 3n^2 + 9n - 13) + 26.$$

If $n \ge 2$, then

$$\sum_{i=1}^{n-1} 2^{i-1}(n-i) = 2^n - n - 1.$$

Source: [2228, pp. 95, 97].

Fact 1.12.22. Let $n \ge 1$, let x be a complex number, and assume that $x \ne 1$. Then,

$$\sum_{i=0}^{n} x^i = \frac{1-x^{n+1}}{1-x}, \qquad \sum_{i=1}^{n} x^i = \frac{x-x^{n+1}}{1-x}, \qquad \sum_{i=0}^{n-1}(n-i)x^i = \frac{x^{n+1}-(n+1)x+n}{(x-1)^2},$$

$$\sum_{i=1}^{n} ix^i = \frac{[nx^{n+1}-(n+1)x^n+1]x}{(x-1)^2} = \frac{(nx^n \quad \sum_{i=0}^{n-1} x^i)x}{x-1},$$

$$\begin{aligned}
\sum_{i=1}^{n} i^2 x^i &= \frac{([n(x-1)-1]^2 + x)x^{n+1} - x^2 - x}{(x-1)^3} \\
&= \frac{[n^2(x-1)^2 - 2n(x-1) + x + 1]x^{n+1} - x^2 - x}{(x-1)^3} \\
&= \frac{[n^2x^{n+1} - (n^2+2n-1)x^n + 2\sum_{i=1}^{n-1} x^i + 1]x}{(x-1)^2} \\
&= \frac{[n^2x^n - \sum_{i=0}^{n-1}(2i+1)x^i]x}{x-1}.
\end{aligned}$$

In particular,

$$\sum_{i=1}^{n} \frac{i}{2^i} = \frac{2^{n+1}-n-2}{2^n}, \qquad \sum_{i=1}^{n} \frac{i^2}{2^i} = 6 - \frac{n^2+4n+6}{2^n}.$$

Source: [289, pp. 22, 132], [1757, pp. 54, 55], [1937], and [2228, pp. 95, 97]. **Related:** Fact 13.5.34.

Fact 1.12.23. Let $n \geq 1$. Then,

$$\sum_{i=1}^{n}(-1)^{i+1}i = \begin{cases} -\frac{1}{2}n, & n \text{ even}, \\ \frac{1}{2}(n+1), & n \text{ odd}, \end{cases} \qquad \sum_{i=1}^{n}(-1)^{i+1}i^2 = \begin{cases} -\frac{1}{2}n(n+1), & n \text{ even}, \\ \frac{1}{2}n(n+1), & n \text{ odd}. \end{cases}$$

Now, let $m \geq 1$. Then,

$$\sum_{i=1}^{n}(-1)^{i+1}(mi-1) = \tfrac{1}{4}(-1)^n[2-m(2n+1)] + \tfrac{1}{4}(m-2),$$

$$\sum_{i=1}^{n}(-1)^{i+1}(mi-1)^2 = \tfrac{1}{2}(-1)^{n+1}[m^2n(n+1) - m(2n+1)+1] + \tfrac{1}{2}(1-m).$$

Source: [867] and [1217, pp. 2, 3].

Fact 1.12.24. Let $n \geq 2$. Then,

$$\sum_{i=n+1}^{2n-1}\frac{1}{i^2} = 4\sum_{i=1}^{n-1}(-1)^{n-1-i}\frac{\left(\frac{i}{n^2-i^2}\right)^2}{\binom{2n}{n-i}}.$$

Source: [42]. **Example:** $\frac{1}{16}+\frac{1}{25} = \frac{8}{75} - \frac{1}{240} = \frac{41}{400}$.

Fact 1.12.25. Let $n, m, k \geq 1$. Then,

$$\sum_{i=1}^{n}\frac{1}{[m+k(i-1)](m+ki)} = \frac{n}{m(kn+m)}, \qquad \sum_{i=0}^{n}\frac{1}{(ki+m)(ki+m+k)} = \frac{n+1}{m(kn+m+k)}.$$

In particular,

$$\sum_{i=1}^{n}\frac{1}{i(i+1)} = \frac{n}{n+1}, \qquad \sum_{i=1}^{n}\frac{1}{4i^2-1} = \frac{n}{2n+1}, \qquad \sum_{i=1}^{n}\frac{1}{(i+1)(i+2)} = \frac{n}{2n+4},$$

$$\sum_{i=1}^{n}\frac{1}{(i+2)(i+3)} = \frac{n}{3n+9}, \qquad \sum_{i=1}^{n}\frac{1}{(3i+1)(3i-2)} = \frac{n}{3n+1}, \qquad \sum_{i=1}^{n}\frac{1}{(5i+2)(5i-3)} = \frac{n}{10n+4}.$$

Source: [1217, p. 3]. **Related:** Fact 13.5.27.

Fact 1.12.26. Let $n, k \geq 1$. Then,

$$\sum_{i=0}^{n}\frac{1}{(i+k)(i+k+1)} = \frac{n+1}{k(n+k+1)},$$

$$\sum_{i=0}^{n}\frac{1}{(i+k)(i+k+2)} = \frac{(n+1)[(2k+1)n+2(k+1)^2]}{2k(k+1)(n+k+1)(n+k+2)}.$$

Related: Fact 13.5.68.

Fact 1.12.27. Let $n \geq 1$. Then,

$$\sum_{i=1}^{n}\frac{1}{i(2i+1)} = 2 - 2\sum_{i=1}^{2n+1}(-1)^{i+1}\frac{1}{i}, \qquad \sum_{i=1}^{n}\frac{i}{4i^4+1} = \frac{1}{4} - \frac{1}{4(2n^2+2n+1)},$$

$$\sum_{i=1}^{n}\frac{2i^2-1}{4i^4+1} = \frac{1}{2} - \frac{2n+1}{2(2n^2+2n+1)}, \qquad \sum_{i=1}^{n}\frac{i}{i^4+4} = \frac{3}{8} - \frac{2n^2+2n+3}{4(n^2+1)(n^2+2n+2)},$$

$$\sum_{i=1}^{n}\frac{i}{\prod_{j=0}^{i}(2j+1)} = \frac{1}{2} - \frac{1}{2}\frac{1}{\prod_{i=0}^{n}(2j+1)}, \qquad \sum_{i=1}^{n}\frac{1}{i(i+1)(i+2)} = \frac{n^2+3n}{4(n+1)(n+2)},$$

$$\sum_{i=1}^{n}\frac{1}{i(i+1)(2i+1)}=1+4\sum_{i=3}^{2n+1}(-1)^i\frac{1}{i}+\frac{1}{n+1},\quad \sum_{i=1}^{n}\frac{2^i(i^2-2i-1)}{i^2(i+1)^2}=\frac{2^{n+1}}{(n+1)^2}-2,$$

$$\sum_{i=1}^{n}\frac{3i^2+3i+1}{i^3(i+1)^3}=\frac{-1}{(n+1)^3}+1,\quad \sum_{i=1}^{n}\frac{6i+3}{4i^4+8i^3+8i^2+4i+3}=\frac{n^2+2n}{2n^2+4n+3},$$

$$\sum_{i=1}^{n}\frac{i^2+3i+3}{i^4+2i^3-3i^2-4i+2}=-\frac{2n^2+5n}{n^2+2n-1},\quad \sum_{i=1}^{n}\frac{4^ii^2}{(i+1)(i+2)}=\frac{2}{3}+\frac{4^{n+1}(n-1)}{3(n+2)},$$

$$\sum_{i=1}^{n}\frac{2^i(i^3-3i^2-3i-1)}{i^3(i+1)^3}=\frac{2^n}{n^3},\quad \sum_{i=0}^{n}\frac{i^3+6i^2+11i+5}{(i+3)!}=\frac{5}{2}-\frac{n^2+6n+10}{(n+3)!}.$$

If $n\ge 2$, then

$$\sum_{i=2}^{n}\frac{1}{i^2-1}=\frac{3n^2-n-2}{4n^2+4n}=\frac{3}{4}-\frac{2n+1}{2n(n+1)}.$$

If $n\ge 3$, then

$$\sum_{i=1,\,i\ne 2}^{n}\frac{1}{i^2-4}=\frac{3}{16}-\frac{1}{4(n-1)}-\frac{1}{4n}-\frac{1}{4(n+1)}-\frac{1}{4(n+2)}=\frac{3}{16}-\frac{2n^3+3n^2-n-1}{2(n-1)n(n+1)(n+2)}.$$

Source: [1217, pp. 2, 3]. The first equality is given in [506, p. 119]. The second equality is given in [112, p. 41]. The third equality is given in [506, p. 235]. The fourth equality is given in [506, p. 122]. The fifth equality is given in [1757, p. 171]. The sixth equality is given in [506, p. 118]. The penultimate equality is given in [1217, p. 3]. **Related:** Fact 13.5.29 and Fact 13.5.71.

Fact 1.12.28. Let $n\ge 1$. Then,

$$\sum_{i=1}^{n}\frac{1}{i^2(i+1)^2}=2\sum_{i=1}^{n}\frac{1}{i^2}-3+\frac{1}{(n+1)^2}+\frac{2}{n+1},$$

$$\sum_{i=1}^{n}\frac{(-1)^{i+1}}{i^2(i+1)^2}=3-4\sum_{i=1}^{n}\frac{(-1)^{i+1}}{i}+\frac{(-1)^{n+1}}{(n+1)^2}+\frac{(-1)^{n+1}2}{n+1}.$$

Source: [108, pp. 34, 203–205]. **Related:** Fact 13.5.103.

Fact 1.12.29. Let $n\ge 1$. Then,

$$\sum_{i=1}^{n}\frac{\prod_{j=1}^{n-1}(4i^4+j^4)}{i^2\prod_{j=1,j\ne i}^{n}(i^4-j^4)}=\frac{1}{2n^2}\binom{2n}{n}.$$

Source: [42]. **Remark:** For $n=1$, both products are set to 1.

Fact 1.12.30. Let $n\ge 1$. Then,

$$\sqrt{1+\frac{1}{n^2}+\frac{1}{(n+1)^2}}=1+\frac{1}{n}-\frac{1}{n+1},$$

$$\sum_{i=1}^{n}\sqrt{1+\frac{1}{i^2}+\frac{1}{(i+1)^2}}=\sum_{i=1}^{n}\frac{i^2+i+1}{i(i+1)}=\frac{n(n+2)}{n+1},\quad \sum_{i=1}^{n}\frac{i^2+i-1}{i(i+1)}=\frac{n^2}{n+1}.$$

Source: [384].

Fact 1.12.31. Let $n \geq 1$. Then,

$$\sum_{i=1}^{n} \frac{1}{\sqrt{i}+\sqrt{i+1}} = \sqrt{n+1}-1.$$

Source: [1158, p. 121].

Fact 1.12.32. Let $n \geq 1$. Then,

$$\sum_{i=1}^{n} \frac{1}{\sqrt{1+(1+1/i)^2}+\sqrt{1+(1-1/i)^2}} = \frac{1}{4}(\sqrt{(n+1)^2+n^2}-1).$$

Source: [107, pp. 3, 70, 71].

Fact 1.12.33. Let $n \geq 1$. Then,

$$\sum_{i=1}^{n} \frac{1}{(\sqrt{i}+\sqrt{i+1})(\sqrt[4]{i}+\sqrt[4]{i+1})} = \sqrt[4]{n+1}-1.$$

Source: [107, pp. 4, 73].

Fact 1.12.34. Let $n \geq 1$. Then,

$$\sum_{1\le i\le n+1} \frac{1}{i} + \sum_{1\le i<j\le n+1} \frac{1}{ij} + \cdots + \sum \frac{1}{\prod_{j=1}^{n} i_j} = n+1-\frac{1}{(n+1)!},$$

where the last sum is taken over all n-tuples $(i_1,\ldots,i_n)$ such that $1 \le i_1 < \cdots < i_n \le n+1$. Furthermore,

$$\sum_{i=1}^{n}\sum \frac{1}{\prod_{j=1}^{i} k_j} = n,$$

where the last sum is taken over all i-tuples $(k_1,\ldots,k_i)$ such that $1 \le k_1 < \cdots < k_i \le n$. Now, let $n \geq 2$. Then,

$$\sum_{i=1}^{n-1}(-1)^{i+1}\sum \frac{1}{\prod_{j=1}^{i} k_j} = \frac{n-1}{n}, \quad \sum_{i=1}^{n-1}(-1)^{i+1}\sum \frac{1}{\prod_{j=1}^{i} k_j^2} = \frac{n-1}{2n},$$

$$\sum_{i=1}^{n-1}(-1)^{i+1}\sum \frac{2^i}{\prod_{j=1}^{i}(k_j^3+1)} = \frac{(n-1)(n+2)}{3n(n+1)},$$

where the second sum in each equality is taken over all i-tuples $(k_1,\ldots,k_i)$ such that $2 \le k_1 < \cdots < k_i \le n$. **Source:** [398, 894].

Fact 1.12.35. Let $n \geq 1$. Then,

$$\frac{2}{3}n^{3/2} < \left(\frac{2n}{3}+\frac{1}{8}-\frac{1}{8\sqrt{n+1}}\right)\sqrt{n+1} < \sum_{i=1}^{n}\sqrt{i} < \left(\frac{2n}{3}+\frac{1}{6}-\frac{1}{6\sqrt{n+1}}\right)\sqrt{n+1} < \frac{2}{3}n^{3/2}+\frac{1}{2}\sqrt{n}.$$

Source: [2020]. **Remark:** It is conjectured in [2020] that

$$\left\lfloor \frac{1}{n}\sum_{i=1}^{n}\sqrt{i} \right\rfloor = \left\lfloor \left(\frac{2}{3}+\frac{1}{6n}\right)\sqrt{n+1} \right\rfloor.$$

Fact 1.12.36. Let $n \geq 1$. Then,

$$\sum_{i=1}^{n-1} i^2 < \frac{n^3}{3} < \sum_{i=1}^{n} i^2, \quad \sum_{i=1}^{n-1} i^3 < \frac{n^4}{4} < \sum_{i=1}^{n} i^3.$$

Fact 1.12.37. Let $n \geq 1$ and $p \geq 1$. Then,

$$n\left(\frac{n+1}{2}\right)^p \leq \sum_{i=1}^{n} i^p.$$

Source: [806, p. 103].

Fact 1.12.38. Let $n \geq 1$. Then,

$$\frac{2}{3} < \sum_{i=n}^{2n} \frac{1}{i}, \quad 1 < \sum_{i=n+1}^{3n+1} \frac{1}{i}, \quad \frac{1}{2} < \sum_{i=3n+1}^{5n+1} \frac{1}{i} < \frac{2}{3}.$$

Source: [1566, p. 9].

Fact 1.12.39. Let $n > m \geq 1$. Then, $\sum_{i=m}^{n} \frac{1}{i}$ is not an integer. **Source:** [1350, p. 24].

Fact 1.12.40. Let $n \geq 1$ and $p \in (0, \infty)$. Then,

$$\sum_{i=1}^{n} \left(\frac{1}{i}\right)^{1/p} < \frac{p}{p-1} n^{1-1/p}, \quad \sum_{i=1}^{n} \frac{1}{\sqrt{i}} < 2\sqrt{n}.$$

Source: [1757, p. 63] and [2294, p. 282].

Fact 1.12.41. Let $n \geq 1$. Then,

$$\frac{n}{\sqrt{n^2+n}} < \sum_{i=1}^{n} \frac{1}{\sqrt{n^2+i}} < \frac{n}{\sqrt{n^2+1}}.$$

Source: [1757, p. 278].

Fact 1.12.42. Let $n \geq 1$ and $r > 0$. Then,

$$\frac{n}{n+1} \leq \left[\frac{(n+1)\sum_{i=1}^{n} i^r}{n\sum_{i=1}^{n+1} i^r}\right]^{1/r} \leq \frac{\sqrt[n]{n!}}{\sqrt[n+1]{(n+1)!}} \leq \sqrt{\frac{n}{n+1}}.$$

Source: [6, 413, 2274]. **Remark:** The first and second inequality are *Alzer's inequality* and *Martins's inequality*, respectively. **Related:** Fact 1.13.13.

Fact 1.12.43. Let $n \geq 1$, let p be a real number, and define the sequences

$$\mathcal{S}_1(p) \triangleq \left(\frac{\sum_{i=1}^{n} i^p}{n(n!)^{p/n}}\right)_{n=1}^{\infty}, \quad \mathcal{S}_2(p) \triangleq \left(\frac{n(n+1)^p}{\sum_{i=1}^{n} i^p}\right)_{n=1}^{\infty}.$$

Then, the following statements hold:

i) If $p > 0$, then $\mathcal{S}_1(p)$ is increasing.

ii) If $p \in (0, 1)$, then $\mathcal{S}_2(p)$ is increasing.

iii) If $p \in (-\infty, 0) \cup (1, \infty)$, then $\mathcal{S}_2(p)$ is decreasing.

Source: [413].

Fact 1.12.44. Let $n \geq 1$. Then,

$$\sum_{i=1}^{2n} (-1)^{i+1} \frac{1}{i} = H_{2n} - H_n.$$

Source: [1158, p. 10] and [2380, pp. 21, 22]. **Remark:** This is *Catalan's identity*. **Example:** $1 - \frac{1}{2} + \frac{1}{3} - \frac{1}{4} = \frac{1}{3} + \frac{1}{4}$.

Fact 1.12.45. Let $n \geq 1$. Then,

$$\sum_{i=1}^{n} H_i = (n+1)(H_{n+1} - 1) = (n+1)H_n - n, \quad \sum_{i=1}^{n} iH_i = \frac{1}{4}n(n+1)(2H_{n+1} - 1),$$

$$\sum_{i=1}^{n} i^2 H_i = \frac{1}{36}n(n+1)[6(2n+1)H_{n+1} - 4n - 5],$$

$$\sum_{i=1}^{n} i^3 H_i = \frac{1}{48}n(n+1)[12n(n+1)H_{n+1} - 3n^2 - 7n - 2], \quad \sum_{i=1}^{n} \frac{H_i}{i} = \sum_{i=1}^{n}\sum_{j=1}^{i} \frac{1}{ij} = \frac{1}{2}(H_n^2 + H_{n,2}).$$

If, in addition, $k \geq 0$, then

$$(H_{n+k} - H_k)^2 + H_{n+k,2} - H_{k,2} = \left(\sum_{i=1}^{n} \frac{1}{i+k}\right)^2 + \sum_{i=1}^{n} \frac{1}{(i+k)^2} = \sum_{i=1}^{n}\sum_{j=1}^{i} \frac{2}{(i+k)(j+k)}.$$

Furthermore,

$$\sum_{i=1}^{n} (-1)^i H_i = \begin{cases} \frac{1}{2}H_{n/2}, & n \text{ even},\\ \frac{1}{2}H_{(n+1)/2} - H_{n+1}, & n \text{ odd}. \end{cases}$$

Source: [315], [411, p. 91], [1219, pp. 279, 280], and [2482].

Fact 1.12.46. Let $n \geq 1$. Then,

$$\log n + \frac{1}{2n} + \frac{1}{2} < H_n < \log n + \frac{1}{n} + \frac{29}{50}.$$

Now, let $n \geq 2$. Then,

$$\log n + \frac{1}{n} < H_n < \log n + 1, \quad \frac{1}{n} < H_n - \log n < 1,$$

$$n(\sqrt[n]{n+1} - 1) < H_n \leq n - \frac{n-1}{\sqrt[n-1]{n}} < n - \frac{n}{\sqrt[n]{n+1}} + \frac{n}{n+1} < 1 + n\left(1 - \frac{1}{\sqrt[n]{n}}\right).$$

Source: [1350, p. 47], [1371, pp. 158, 161], [1566, p. 9], [1757, p. 250], and [2294, p. 10]. In the last string, the ordering of the third and fifth terms follows from the arithmetic-mean–geometric mean inequality. **Remark:** The second inequality in the last string is strict for all $n \geq 3$. **Related:** Fact 12.18.33.

Fact 1.12.47. Let $n \geq 2$. Then,

$$\tfrac{1}{2}(\lfloor \log_2 n \rfloor + 1) < H_n \leq \lfloor \log_2 n \rfloor + 1.$$

Equivalently,

$$\frac{1}{2}\left(\left\lfloor \frac{\log n}{\log 2} \right\rfloor + 1\right) < H_n \leq \left\lfloor \frac{\log n}{\log 2} \right\rfloor + 1.$$

Source: [1219, p. 276].

Fact 1.12.48. Let $n \geq 1$. Then,

$$\sum_{i=1}^{n} \frac{1}{iH_i^2} < 1.85, \quad \sum_{i=1}^{n} \frac{1}{iH_i^3} < 1.34.$$

Source: [806, p. 183].

Fact 1.12.49. Let $n \geq 1$. Then,

$$\frac{n(3n+5)}{2(n+1)^2} \leq H_{n,2} \leq 2 - \frac{1}{n}.$$

Source: [399].

Fact 1.12.50. Let $n \geq 5$ be prime. Then, the following statements hold:

i) Let $H_{n-1} = N_{n-1}/D_{n-1}$, where N_{n-1} and D_{n-1} are positive coprime integers. Then, $n^2 | N_{n-1}$.

ii) Let $H_{n-1,2} = N_{n-1,2}/D_{n-1,2}$, where $N_{n-1,2}$ and $D_{n-1,2}$ are positive coprime integers. Then, $n|N_{n-1,2}$.

Source: [145] and [2068, pp. 29, 305]. **Example:** $H_{18} = 14274301/4084080 = (19)^2 39541/4084080$ and $H_{10,2} = 1968329/1270080 = (11)178939/1270080$.

Fact 1.12.51. Le $n \geq 1$. Then,

$$\exp \frac{2n}{2n+1} \leq \exp \frac{en^{n+1}}{(n+1)^{n+1}} \leq \left(1+\frac{1}{n}\right)^n \leq \exp \sqrt{\frac{n}{n+1}},$$

$$\exp \frac{2}{n+2} \leq \exp \frac{e}{(n+1)^{(n+1)/n}} \leq (n+1)^{1/n} \leq \exp \frac{1}{\sqrt{n+1}},$$

$$\sum_{i=1}^{n} \log^2\left(1+\frac{1}{i}\right) \leq \frac{n}{n+1}, \quad \prod_{i=1}^{n} \log(i+1) \leq \sqrt{\frac{n!}{n+1}}, \quad (n+1)! \leq \exp \sum_{i=1}^{n} \frac{i}{\sqrt{i+1}}.$$

Source: [370].

1.13 Facts on Factorials

Fact 1.13.1. Let n and m be positive integers such that $m < n$. Then, $n^{\underline{m}}$ is the number of m-tuples whose components are distinct elements of $\{1, \ldots, n\}$. **Remark:** $n^{\underline{m}}$ is the number of permutations of m distinct elements chosen from a set of n elements.

Fact 1.13.2. Let $n \geq 3$, and assume that n is prime. Then,

$$(n-1)!! \overset{n}{\equiv} \prod_{i=1}^{n-1} i! \overset{n}{\equiv} (-1)^{(n-1)/2} \prod_{i=1}^{n-1} i^i.$$

Source: [14].

Fact 1.13.3. Let $n \geq 1$. Then,

$$\sum_{i=1}^{n} i(i!) = (n+1)! - 1, \quad \sum_{i=1}^{n} (i^2+1)i! = n(n+1)!,$$

$$\sum_{i=0}^{n} \frac{1}{i!} = \frac{\lfloor en! \rfloor}{n!}, \quad \sum_{i=1}^{n} \frac{i}{(i+1)!} = 1 - \frac{1}{(n+1)!}, \quad \sum_{i=1}^{n} \frac{i^2+i-1}{(i+2)!} = \frac{1}{2} - \frac{n+1}{(n+2)!},$$

$$\sum_{i=1}^{n} \frac{i^2+3i+1}{(i+2)!} = \frac{3}{2} - \frac{n+3}{(n+2)!}, \quad \sum_{i=1}^{n} \frac{(4i+1)i!}{(2i+1)!} = 1 - \frac{n!}{(2n+1)!},$$

$$\sum_{i=1}^{n} \frac{i^3+6i^2+11i+5}{(i+3)!} = \frac{5}{3} - \frac{n^2+6n+10}{(n+3)!} = \frac{5}{3} - \frac{1}{(n+1)!} - \frac{1}{(n+2)!} - \frac{1}{(n+3)!}.$$

Source: The second equality is given in [1158, p. 123]. The third equality is given in [621, pp. 33, 34] and [1345]. The fourth equality is given in [2380, pp. 4, 5]. The seventh equality is given in [2228, p. 78]. The last inequality is given in [1566, pp. 28, 165].

Fact 1.13.4. Let $n \geq 1$ and $k \geq 1$. Then,

$$\sum_{i=1}^{n} i!(i^2+ki+1) = (n+1)!(n+k) - k.$$

In particular,

$$\sum_{i=1}^{n} i!(i^2+i+1) = (n+1)!(n+1) - 1, \quad \sum_{i=1}^{n} i!(i+1)^2 = (n+1)!(n+2) - 2.$$

Source: [112, p. 39] and [411, p. 92].

Fact 1.13.5. Let $n \geq 2$ and $k \in \{1, \ldots, n\}$. Then,

$$\prod_{i=1,\, i\neq k}^{n} (k-i) = (-1)^{n-k}(n-k)!(k-1)!.$$

Source: [1672].

Fact 1.13.6. Let $n \geq 1$. Then,

$$\sum_{i=0}^{n}(1+i-\sqrt{i})\sqrt{i!} = (n+1)\sqrt{n!}.$$

Source: [2068, p. 8].

Fact 1.13.7. Let $n \geq 2$. Then,

$$\sum_{i=1}^{n} \frac{1}{(i-1)!}\sum_{j=0}^{n-i}(-1)^j\frac{1}{j!} = 1, \quad \sum_{i=1}^{n} \frac{i}{(i-1)!}\sum_{j=0}^{n-i}(-1)^j\frac{1}{j!} = 2.$$

Source: [1158, p. 313].

Fact 1.13.8. Let $n \geq 2$. Then,

$$\sum_{i=1}^{n}(-1)^i\frac{1}{i!}\sum \frac{1}{\prod_{j=1}^{i} k_j} = \sum_{i=1}^{n}(-1)^i\frac{1}{i}\sum \frac{1}{\prod_{j=1}^{i} k_j!} = 0,$$

where the second and fourth sums are taken over all i-tuples $(k_1, \ldots, k_i)$ of positive integers such that $\sum_{j=1}^{i} k_j = n$. **Source:** [1082].

Fact 1.13.9. Let $n \geq 1$. Then,

$$(n+1)! \leq \sqrt[n]{\prod_{i=1}^{n}(2i)!}.$$

Source: [1757, p. 63].

Fact 1.13.10. Let $n \geq 2$. Then, $2^{H_n} \leq \sqrt[n]{n!} < \frac{1}{2}(n+1) < e^{H_n}$. **Source:** [109, pp. 172–174].

Fact 1.13.11. Let $n \geq 1$. Then,

$$\sqrt[n]{n!} \leq \prod p^{1/(p-1)},$$

where the product is taken over all primes p that divide n. **Source:** [1757, p. 169]. **Remark:** This implies there are infinitely many primes.

Fact 1.13.12. If $n \geq 2$, then

$$(n-1)! < \frac{n^n}{e^{n-1}} < n! \leq \frac{(n+1)^n}{2^n}, \quad n! \leq \left[\frac{(n+1)(2n+1)}{6}\right]^{n/2}.$$

If $n \geq 3$, then

$$n! < 2^{n(n-1)/2}.$$

If $n \geq 6$, then

$$\left(\frac{n}{3}\right)^n < n! < \left(\frac{n}{2}\right)^n.$$

Source: [289, pp. 13, 89, 90], [517, p. 210], [1371, p. 137], and [2294, p. 346].

Fact 1.13.13. Let $n \ge 1$. Then,

$$\sqrt[n]{n!} < \sqrt[n+1]{(n+1)!}, \quad \frac{n}{\sqrt[n]{n!}} < \frac{n+1}{\sqrt[n+1]{(n+1)!}}, \quad \frac{n+1}{\sqrt[n]{n!}} < \frac{n+2}{\sqrt[n+1]{(n+1)!}}.$$

Source: [413]. **Remark:** The first inequality is the *Minc-Sathre inequality*. The second inequality is given by Fact 1.12.42.

Fact 1.13.14. Let $n \ge 1$. Then,

$$\left(\frac{n}{e}\right)^n \sqrt{2\pi\left(n+\frac{1}{6}\right)} < n! \le \left(\frac{n}{e}\right)^n \sqrt{2\pi\left(n+\frac{e^2}{2\pi}-1\right)},$$

$$n^{n/2} \le n! \le \left(\frac{n+1}{2}\right)^n, \qquad \sqrt{2n\pi}\left(\frac{n}{e}\right)^n < n! < \sqrt{2n\pi}\left(\frac{n}{e}\right)^n e^{1/(12n)},$$

$$\frac{\sqrt{2\pi e}}{e^{(3-\sqrt{3})/3}}\left(\frac{n+(3-\sqrt{3})/3}{e}\right)^{n+1/2} < n! < \frac{\sqrt{2\pi e}}{e^{(3+\sqrt{3})/3}}\left(\frac{n+(3+\sqrt{3})/3}{e}\right)^{n+1/2}.$$

Now, let $n \ge 3$. Then, $n^{n/2} < n!$ and

$$\begin{aligned} 2\left(\frac{n}{e}\right)^n &< e\left(\frac{n}{e}\right)^n < \sqrt{2n\pi}\left(\frac{n}{e}\right)^n < \frac{n+\frac{13}{12}}{n+1}\sqrt{2n\pi}\left(\frac{n}{e}\right)^n \\ &< n! < \frac{n-\frac{23}{12}}{n-2}\sqrt{2n\pi}\left(\frac{n}{e}\right)^n < \sqrt{\frac{n}{n-1}}\sqrt{2n\pi}\left(\frac{n}{e}\right)^n < \left(\frac{n+1}{2}\right)^n < e\left(\frac{n}{2}\right)^n. \end{aligned}$$

Therefore,

$$\begin{aligned} \frac{2}{\sqrt[n]{e}} &< \frac{2n}{n+1} < \sqrt[2n]{\frac{n-1}{n}}\frac{e}{\sqrt[2n]{2n\pi}} < \sqrt[n]{\frac{n-2}{n-\frac{23}{12}}}\frac{e}{\sqrt[2n]{2n\pi}} \\ &< \frac{n}{\sqrt[n]{n!}} < \sqrt[n]{\frac{n+1}{n+\frac{13}{12}}}\frac{e}{\sqrt[2n]{2n\pi}} < \frac{e}{\sqrt[2n]{2n\pi}} < \frac{e}{\sqrt[n]{e}} < \frac{e}{\sqrt[n]{2}}. \end{aligned}$$

Finally,

$$\sqrt[2n]{2n\pi}\frac{n}{\sqrt[n]{n!}} < \sqrt[n]{\frac{n+\frac{13}{12}}{n+1}}\sqrt[2n]{2n\pi}\frac{n}{\sqrt[n]{n!}} < e < \sqrt[n]{\frac{n-\frac{23}{12}}{n-2}}\sqrt[2n]{2n\pi}\frac{n}{\sqrt[n]{n!}} < \sqrt[2n]{\frac{n}{n-1}}\sqrt[2n]{2n\pi}\frac{n}{\sqrt[n]{n!}}.$$

Now, let $6 \le n \le 9$. Then,

$$n^{n/2} \le 2\left(\frac{n}{e}\right)^n < e\left(\frac{n}{e}\right)^n < \sqrt{2n\pi}\left(\frac{n}{e}\right)^n < n! < \left(\frac{n}{2}\right)^n.$$

Finally, let $n \ge 10$. Then,

$$n^{n/2} < \left(\frac{n}{3}\right)^n < 2\left(\frac{n}{e}\right)^n < e\left(\frac{n}{e}\right)^n < \sqrt{2n\pi}\left(\frac{n}{e}\right)^n < n! < \left(\frac{n}{2}\right)^n.$$

Source: [516, p. 197], [325], [1158, p. 10], [1415], [1566, p. 10], [2373], and [2586, pp. 35–37]. **Remark:** $\sqrt{2n\pi}\left(\frac{n}{e}\right)^n < n!$ is Stirling's formula. See Fact 12.18.58. **Remark:** $\sqrt{2\pi} < e < 2\sqrt{\pi}$ and $e \approx (\pi^4 + \pi^5)^{1/6}$. See [650]. **Remark:** $0.16666 \approx 1/6 < e^2/(2\pi) - 1 \approx 0.17600$. **Related:** Fact 12.18.58.

Fact 1.13.15. Let $n \ge 1$. Then,

$$\sum_{i=0}^{n}(i+1)i!! = (n+1)!! + (n+2)!! - 2, \quad \sum_{i=0}^{n}(-1)^i(i+1)i!! = (-1)^n[(n+2)!! - (n+1)!!],$$

$$\sum_{i=0}^{n}\frac{i}{(i+1)!!}=2-\frac{1}{(n+1)!!}-\frac{1}{n!!},\quad \sum_{i=0}^{n}\frac{1}{(2i)!!(2n-2i)!!}=\frac{1}{n!},$$

$$\sum_{i=0}^{n}\binom{n}{i}(2i-1)!!(2n-2i-1)!!=(2n)!!,\quad \sum_{i=0}^{n}\frac{(2i-1)!!(2n-2i-1)!!}{(2i)!!(2n-2i)!!}=1,$$

$$\prod_{i=1}^{n}\frac{2i}{2i-1}=\frac{\sqrt{\pi}\Gamma(n+1)}{\Gamma(n+\frac{1}{2})}=\frac{(2n)!!}{(2n-1)!!}=\frac{[(2n)!!]^2}{(2n)!}=\frac{4^n}{\binom{2n}{n}}.$$

If, in addition, $n\ge 5$, then

$$\sqrt{2n}<\sqrt{2n+1}<\sqrt{3n+1}<\sqrt{\frac{2n(2n+1)\pi}{4n+1}}<\sqrt{(n+\tfrac{1}{4})\pi}<\frac{(2n)!!}{(2n-1)!!}$$
$$<\frac{\sqrt{\pi}(2n+1)}{\sqrt{4n+3}}<\sqrt{\frac{(4n+3)(2n+1)\pi}{8n+8}}<\sqrt{(n+\tfrac{4}{\pi}-1)\pi}<\sqrt{(n+\tfrac{1}{2})\pi}<2\sqrt{n}.$$

Source: [650, 692, 1516], [1675, pp. 49, 52], and [2380, p. 51]. **Remark:** This result yields the Wallis product given by Fact 13.10.3. **Related:** Fact 12.18.59 and Fact 13.10.11.

Fact 1.13.16. Let $n,m\ge 1$. Then, $m!n!(m+n)!|(2m)!(2n)!$. **Source:** [771, p. 79].

Fact 1.13.17. Let $n,m\ge 1$. Then, $(n!)^m|(mn)!$. Now, define $m \triangleq \max\{k^l\colon k \text{ is prime}, l\ge 1, \text{ and } k^l\le n\}$. Then, $(n!)^{m+1}|(mn)!$. **Source:** [2083].

1.14 Facts on Finite Products

Fact 1.14.1. If $n\ge 1$, then

$$\prod_{i=1}^{n}\left(1+\frac{1}{i}\right)=n+1,\quad \prod_{i=1}^{n}\left(1+\frac{1}{i^2}\right)=\frac{\sinh(\pi)\Gamma(n+1-\jmath)\Gamma(n+1+\jmath)}{\pi(n!)^2},$$

$$\prod_{i=1}^{n}\left(1+\frac{1}{i^3}\right)=\frac{(n+1)\cosh(\sqrt{3}\pi/2)\Gamma(n+1/2-\sqrt{3}\jmath/2)\Gamma(n+1/2+\sqrt{3}\jmath/2)}{\pi(n!)^2},$$

$$\prod_{i=1}^{n}\frac{1}{4i^2-1}=\frac{(n!)(n+1)!2^{2n+1}}{(2n)!(2n+2)!}.$$

If $n\ge 2$, then

$$\prod_{i=2}^{n}\left(1-\frac{1}{i}\right)=\frac{1}{n},\quad \prod_{i=2}^{n}\left(1-\frac{1}{i^2}\right)=\frac{n+1}{2n},\quad \prod_{i=2}^{n}\left(1-\frac{1}{i^4}\right)=\frac{(n+1)\prod_{i=1}^{n}(1+i^2)}{4n^3[(n-1)!]^2},$$

$$\prod_{i=2}^{n}\frac{i^2}{i^2-1}=\frac{2n}{n+1},\quad \prod_{i=2}^{n}\frac{i^3-1}{i^3+1}=\frac{2(n^2+n+1)}{3n(n+1)}.$$

Source: [112, p. 39]. **Related:** Fact 13.10.11.

1.15 Facts on Numbers

Fact 1.15.1. Let $n\ge 1$, let $q_1,\dots,q_n$ and $p_1,\dots,p_n$ be positive rational numbers, and assume that $q_1^{p_1}$ is an irrational number. Then, $\sum_{i=1}^{n} q_i^{p_i}$ is an irrational number. **Source:** [1352, p. 129] and [2207]. **Related:** Fact 6.8.2.

Fact 1.15.2. Let a be a nonzero rational number. Then, e^a is irrational. **Source:** [20, Chapter 7]. **Remark:** If $x\in(0,\infty)$ is transcendental, then, for all $n\ge 1$, x^n is irrational.

Fact 1.15.3. There exist positive irrational numbers a and b such that a^b is rational.

Source: Note that $\sqrt{2}$ is irrational, and define $\alpha \triangleq \sqrt{2}^{\sqrt{2}}$. Then, $\alpha^{\sqrt{2}} = 2$. Suppose that α is irrational. Then, the result holds with $a = \alpha$ and $b = \sqrt{2}$. Alternatively, suppose that α is rational. Then, the result holds with $a = b = \sqrt{2}$. **Remark:** This proof does not depend on knowing whether or not $\sqrt{2}^{\sqrt{2}}$ is irrational. In fact, $\sqrt{2}^{\sqrt{2}}$ and e^{π} are irrational.

Fact 1.15.4.

$$\sqrt{3+2\sqrt{2}} = 1+\sqrt{2}, \quad \sqrt{5+2\sqrt{6}} = \sqrt{2}+\sqrt{3}, \quad \sqrt{3\sqrt{2}-4} = \sqrt[4]{2}(\sqrt{2}-1),$$

$$\sqrt{19-4\sqrt{21}} = 2\sqrt{3}-\sqrt{7}, \quad \sqrt{21-4\sqrt{17}} = \sqrt{17}-2, \quad \sqrt{25-4\sqrt{21}} = \sqrt{21}-2,$$

$$\sqrt[3]{2+\sqrt{5}} = \tfrac{1}{2}(1+\sqrt{5}), \quad \sqrt[3]{\sqrt{5}-2} = \tfrac{1}{2}(\sqrt{5}-1),$$

$$\sqrt[3]{2+\sqrt{5}} + \sqrt[3]{\sqrt{5}-2} = \sqrt{5}, \quad \sqrt[3]{2+\sqrt{5}} - \sqrt[3]{2-\sqrt{5}} = 1,$$

$$\sqrt[3]{2+\tfrac{10\sqrt{3}}{9}} + \sqrt[3]{2-\tfrac{10\sqrt{3}}{9}} = 2, \quad \sqrt[3]{16+12\sqrt[3]{7}+9\sqrt[3]{49}} = 2\sqrt[3]{3\sqrt[3]{7}-5} + \sqrt[3]{2+3\sqrt[3]{49}},$$

$$(19+17\sqrt[3]{2}+22\sqrt[3]{4})(5\sqrt[3]{2}-\sqrt[3]{4}-3) = 129, \quad \sqrt[3]{\sqrt[3]{2}-1} = \sqrt[3]{1/9}-\sqrt[3]{2/9}+\sqrt[3]{4/9},$$

$$\sqrt{\sqrt[3]{5}-\sqrt[3]{4}} = \tfrac{1}{3}(\sqrt[3]{2}+\sqrt[3]{20}-\sqrt[3]{25}), \quad \sqrt{\sqrt[3]{28}-\sqrt[3]{27}} = \tfrac{1}{3}(\sqrt[3]{98}-\sqrt[3]{28}-1),$$

$$\sqrt[4]{\frac{3+2\sqrt[4]{5}}{3-2\sqrt[4]{5}}} = \frac{\sqrt[4]{5}+1}{\sqrt[4]{5}-1}, \quad \sqrt[5]{1+\sqrt[5]{2}+\sqrt[5]{8}}\,\sqrt[10]{5} = \sqrt{1+\sqrt[5]{4}},$$

$$\sqrt[3]{\sqrt[5]{32/5}-\sqrt[5]{27/5}} = \sqrt[5]{1/25}+\sqrt[5]{3/25}-\sqrt[5]{9/25}, \quad \sqrt[6]{7\sqrt[3]{20}-19} = \sqrt[3]{5/3}-\sqrt[3]{2/3}.$$

Source: [416, 1414, 1739, 2480, 2903]. **Remark:** $\sqrt[3]{2-\sqrt{5}} = \tfrac{1}{2}(1-\sqrt{5})$.

Fact 1.15.5. Let $n \ge 1$. Then,

$$\sqrt[3]{3n-1+n\sqrt{8n-3}} + \sqrt[3]{3n-1-n\sqrt{8n-3}} = 1.$$

In particular,

$$\sqrt[3]{2+\sqrt{5}} + \sqrt[3]{2-\sqrt{5}} = \sqrt[3]{5+2\sqrt{13}} + \sqrt[3]{5-2\sqrt{13}} = \sqrt[3]{8+3\sqrt{21}} + \sqrt[3]{8-3\sqrt{21}} = 1.$$

Source: [2480]. **Remark:** For $a < 0$, $\sqrt[3]{a} = -\sqrt[3]{|a|}$.

Fact 1.15.6. Let $n \ge 1$, and define $\alpha \triangleq \tfrac{1}{2}(\sqrt{5}+1)$. Then,

$$\underbrace{\sqrt{-2+\sqrt{2+\cdots+\sqrt{2+\sqrt{5}}}}}_{n+1 \text{ square roots}} = \alpha^{2^{-n}} - \alpha^{-2^{-n}}, \quad \underbrace{\sqrt{2+\sqrt{2+\cdots+\sqrt{2+\sqrt{5}}}}}_{n+1 \text{ square roots}} = \alpha^{2^{-n}} + \alpha^{-2^{-n}}.$$

In particular,

$$\sqrt{\alpha} - \frac{1}{\sqrt{\alpha}} = \sqrt{\sqrt{5}-2}, \quad \sqrt{\alpha} + \frac{1}{\sqrt{\alpha}} = \sqrt{2+\sqrt{5}}.$$

Source: [2144].

Fact 1.15.7. Define

$$\pi_{-12} \triangleq \frac{223}{71}, \quad \pi_{-11} \triangleq \sqrt[3]{31}, \quad \pi_{-10} \triangleq \frac{1}{3}\sqrt{120-18\sqrt{3}}, \quad \pi_{-9} \triangleq \frac{7^7}{4^9}, \quad \pi_{-8} \triangleq \frac{52163}{16604},$$

$$\pi_{-7} \triangleq \frac{20^3+47^3}{30^3} - 1, \quad \pi_{-6} \triangleq \frac{689}{396\log\frac{689}{396}}, \quad \pi_{-5} \triangleq \frac{66\sqrt{2}}{33\sqrt{29}-148}, \quad \pi_{-4} \triangleq \sqrt[4]{\frac{2143}{22}},$$

$$\pi_{-3} \triangleq \frac{3\log 5280}{\sqrt{67}}, \quad \pi_{-2} \triangleq \frac{3\log(640320)}{\sqrt{163}}, \quad \pi_{-1} \triangleq \frac{\log(640320^3+743)}{\sqrt{163}},$$

$$\pi_1 \triangleq \frac{\log(640320^3+744)}{\sqrt{163}}, \quad \pi_2 \triangleq \sqrt[4]{\frac{35444733}{363875}}, \quad \pi_3 \triangleq \frac{63(17+15\sqrt{5})}{25(7+15\sqrt{5})}, \quad \pi_4 \triangleq \frac{104348}{33215},$$

$$\pi_5 \triangleq \sqrt[5]{\frac{77729}{254}}, \quad \pi_6 \triangleq \frac{99^2}{2206\sqrt{2}}, \quad \pi_7 \triangleq \frac{99}{80}\left(\frac{7}{7-3\sqrt{2}}\right), \quad \pi_8 \triangleq \frac{355}{113}, \quad \pi_9 \triangleq \log_5 157,$$

$$\pi_{10} \triangleq \frac{7}{3}\left(1+\frac{\sqrt{3}}{5}\right), \quad \pi_{11} \triangleq \sqrt{7+\sqrt{6+\sqrt{5}}}, \quad \pi_{12} \triangleq \frac{9}{5}+\frac{3}{\sqrt{5}}, \quad \pi_{13} \triangleq \frac{19\sqrt{7}}{16},$$

$$\pi_{14} \triangleq \frac{22}{7}, \quad \pi_{15} \triangleq \sqrt{2}+\sqrt{3}.$$

Then, $\pi_{-12} < \cdots < \pi_{-1} < \pi < \pi_1 < \cdots < \pi_{15}$. **Source:** [516, 650, 1096], [1317, pp. xxxiv, 34, 35], [1350, p. 96], and [2202].

Fact 1.15.8. Let x be a nonzero rational number. Then, $\tan x$ is an irrational number. **Source:** [1352, pp. 104–107]. **Remark:** Since $\tan \pi/4 = 1$, it follows that π is an irrational number.

Fact 1.15.9. Let $x < 0$ and $p \in \mathbb{R}$. Then, the following statements hold:

i) x^p is real if and only if p is an integer.

ii) x^p is positive if and only if p is an even integer.

Remark: $(-1)^{1/3} = \frac{1}{2} + \frac{\sqrt{3}}{2}\jmath$, and $\sqrt[3]{-1} = -1$.

Fact 1.15.10. Let $x \in [-11/4, -5/4] \cup [5/4, 11/4]$. Then,

$$[e^{(x-1/4)\pi\jmath}]^{x+1/4} + [e^{(x+1/4)\pi\jmath}]^{x-1/4} = 0.$$

1.16 Facts on Binomial Coefficients

Fact 1.16.1. Let n and k be positive integers. Then, the following statements hold:

i) $\binom{n}{k}$ is an integer.

ii) $\binom{2n}{n}$ is an even integer.

iii) $\frac{1}{n+1}\binom{2n}{n}$, $\frac{3}{n}\binom{2n}{n-3}$, $\frac{3}{n}\binom{3n}{n+1}$, and $\frac{4}{(3n+1)(3n+2)}\binom{3n+2}{n}$ are integers.

iv) If $n \ge 3$, then $\frac{(3n)!}{n!(n+1)!(n+2)!}$ is an integer.

v) $\frac{(nk)!}{n!(k!)^n}$, $\frac{(2n)!(2k)!}{n!k!(n+k)!}$, $\frac{\gcd\{n,k\}(n+k-1)!}{n!k!}$, and $\frac{\left[\binom{2n}{n}\binom{2k}{k}\right]^2}{\binom{n+k}{k}}$ are integers.

vi) If $1 \le k \le n$, then $\frac{\gcd\{n,k\}}{n}\binom{n}{k}$ and $\frac{\gcd\{n+1,k\}}{n-k+1}\binom{n}{k}$ are integers. If, in addition, $\gcd\{n,k\} = 1$, then $n|\binom{n}{k}$.

vii) Let $n \ge 2$ be prime, and assume that $k < n < 2k$. Then, $n|(2k)!$ and $n|\binom{2k}{k}$. In addition, $\frac{(k)!}{n}$ is not an integer.

viii) If n is prime, then $\binom{2n}{n} \overset{n}{\equiv} 2$.

ix) Let $p \ge 2$ be prime, and assume that $\max\{k, n-k\} < p \le n$. Then, $p|\binom{n}{k}$. In particular, if $k \le p-1$, then $p|\binom{p}{k}$.

x) If $\gcd\{n,k\} = \gcd\{n-1,k\} = 1$, then $\frac{1}{2}n(n-1)|\binom{n}{k}$.

xi) If $2k \le n$, then $\binom{n}{k}$ has a prime factor $p_1 \ge k+1$ and a prime factor $p_2 \le \max\{n/k, n/2\}$.

xii) If k is prime, then $\binom{n}{k} \overset{k}{\equiv} \lfloor \frac{n}{k} \rfloor$.

xiii) If n is prime and $k \le n-2$, then $\binom{n-1}{k} \overset{n}{\equiv} (-1)^k$.

xiv) If n is prime and $k \le n-3$, then $\binom{n-2}{k} \overset{n}{\equiv} (-1)^k(k+1)$.

xv) If n is prime and $k \le n-4$, then $\binom{n-3}{k} \overset{n}{\equiv} (-1)^k\binom{k+2}{2}$.

xvi) Assume that $n \ge 3$. Then, n is prime if and only if, for all $i \in \{1,\dots,n-1\}$, $\binom{n-1}{i} \overset{n}{\equiv} (-1)^i$.

xvii) If $n \ne 3$ is prime, then $n^2 | \sum_{i=1}^{n^2-1} \binom{2i}{i}$.

xviii) If $n \ge 5$ is prime and $k \ge l \ge 1$, then $\binom{kn-1}{n-1} \overset{n^3}{\equiv} 1$, $\binom{kn}{ln} \overset{n^3}{\equiv} \binom{k}{l}$, and $\binom{kn}{n} \overset{n^3}{\equiv} k$.

xix) There exist integers $0 \le m_1 < \cdots < m_k$ such that $n = \sum_{i=1}^{k} \binom{m_i}{i}$.

xx) If $n \ge 5$ is prime, then $\binom{n^2}{n} \overset{n^5}{\equiv} n$.

xxi) If $n \ge 5$ is prime and $k, l, m \ge 1$, then $\binom{ln^k}{mn^k} \overset{n^{3k}}{\equiv} \binom{ln^{k-1}}{mn^{k-1}}$.

xxii) $2(2n+1)\binom{2n}{n}\Big|\binom{6n}{3n}\binom{3n}{n}$, $\binom{2k}{k}\Big|\binom{4n+2k+2}{2n+k+1}\binom{2n+k+1}{2k}\binom{2n-k+1}{n}$, $\binom{2k}{k}\Big|C_{n+k}(2n+1)\binom{2n}{n}\binom{n+k+1}{2k}$.

Source: [1675, pp. 9–11, 15, 18, 21, 23, 25, 44, 45, 65]. *i)* is given in [1158, p. 296]; *xii)–xv)* are given in [771, pp. 78, 79]; *xvi)* is given in [108, pp. 21, 142]; *xvii)* is given in [617]; *xviii)* is given in [2466]; *xix)* is given in [771, p. 75]; *xx)* and *xxi)* are given in [1083, pp. 37–39]; *xxii)* is given in [2571]. **Example:** To illustrate *xix)*, let $k=3$ and note that $1 = \binom{0}{1}+\binom{1}{2}+\binom{3}{3}$, $2 = \binom{0}{1}+\binom{2}{2}+\binom{3}{3}$, $3 = \binom{1}{1}+\binom{2}{2}+\binom{3}{3}$, $4 = \binom{0}{1}+\binom{1}{2}+\binom{4}{3}$, $5 = \binom{0}{1}+\binom{2}{2}+\binom{4}{3}$, and $6 = \binom{1}{1}+\binom{2}{2}+\binom{4}{3}$.

Fact 1.16.2. Let $n \ge 1$. Then, the following statements hold:

i) $2^{n-1} \le \operatorname{lcm}\{1,2,\dots,n\} \le 3^n$.

ii) $\sqrt{n}2^{n-2} \le n\binom{n-1}{\lfloor (n-1)/2 \rfloor} \le \operatorname{lcm}\{1,2,\dots,n\} \le 3^n$.

iii) As $n \to \infty$, $\log \operatorname{lcm}\{1,2,\dots,n\} \sim n$.

iv) $\operatorname{lcm}\left\{\binom{n}{0},\binom{n}{1},\dots,\binom{n}{n}\right\} = \dfrac{\operatorname{lcm}\{1,2,\dots,n+1\}}{n+1}$.

Source: [1020]. **Remark:** If $n \ge 4$, then $2^{n-1} \le \sqrt{n}2^{n-2}$.

Fact 1.16.3. Let $n \ge 5$, and assume that n is prime. Then,

$$4^{n-1} \overset{n^3}{\equiv} \pm\binom{n-1}{\frac{1}{2}(n-1)}.$$

Source: [13]. **Remark:** For each $n \ge 5$, the congruence holds for either "+" or "−." **Credit:** F. Morley. **Example:** $256 = 4^4 \overset{125}{\equiv} 6 = \binom{4}{2}$ and $4096 = 4^6 \overset{343}{\equiv} -20 = -\binom{6}{3}$.

Fact 1.16.4. Let $n \ge 1$ and $k \ge 1$. Then,

$$n^k = \sum_{i=1}^{k} \alpha_{i,k}\binom{n}{i},$$

where, for all $i \in \{1,\dots,k\}$,

$$\alpha_{i,k} = \sum_{j=0}^{i} (-1)^{i-j}\binom{i}{j}j^k.$$

In particular,

$$n = \binom{n}{1}, \quad n^2 = 2\binom{n}{2} + \binom{n}{1}, \quad n^3 = 6\binom{n}{3} + 6\binom{n}{2} + \binom{n}{1}, \quad n^4 = 24\binom{n}{4} + 36\binom{n}{3} + 14\binom{n}{2} + \binom{n}{1}.$$

Source: [1407].

Fact 1.16.5. Let $n \ge 1$. Then,

$$\binom{n}{2} = \tfrac{1}{2}(n^2 - n), \quad \binom{n+1}{2} = \binom{n}{2} + n = \tfrac{1}{2}(n^2 + n), \quad \binom{n+1}{2} + \binom{n}{2} = n^2,$$

$$\binom{2n+2}{n+1} = \frac{4n+2}{n+1}\binom{2n}{n}, \quad \binom{\binom{n}{2}}{2} = 3\binom{n+1}{4},$$

$$\sum_{i=1}^{n} \tfrac{1}{2}i(i+1) = \sum_{i=1}^{n}\binom{i+1}{2} = \binom{n+2}{3} = \tfrac{1}{6}n(n+1)(n+2),$$

$$\sum_{i=1}^{n}(2i-1)^2 = \binom{2n+1}{3} = \tfrac{1}{3}n(2n-1)(2n+1),$$

$$\sum_{i=1}^{n} i^2 = \frac{1}{4}\binom{2n+2}{3} = 2\binom{n+1}{3} + \binom{n+1}{2} = \tfrac{1}{6}n(n+1)(2n+1),$$

$$\sum_{i=1}^{n} i^3 = \left(\sum_{i=1}^{n} i\right)^2 = \binom{n+1}{2}^2 = \tfrac{1}{4}n^2(n+1)^2 = \tfrac{1}{4}n^4 + \tfrac{1}{2}n^3 + \tfrac{1}{4}n^2,$$

$$\sum_{i=1}^{n} i^4 = \tfrac{1}{20}(3n^2 + 3n - 1)\binom{2n+2}{3} = \tfrac{1}{30}n(n+1)(2n+1)(3n^2+3n-1),$$

$$\sum_{i=1}^{n} i^5 = \binom{n+1}{2} + 30\binom{n+2}{4} + 120\binom{n+3}{6}.$$

Source: [289, pp. 17, 110] and [334, 866].

Fact 1.16.6. Let $n \ge 0$. Then,

$$\prod_{i=0}^{n}\binom{n}{i} = \frac{(n!)^{n+1}}{\prod_{i=0}^{n}(i!)^2}.$$

Source: [1410].

Fact 1.16.7. Let $x_0, x_1, \ldots, x_n, y_0, y_1, \ldots, y_n$ be complex numbers. Then, for all $k \in \{0, 1, \ldots, n\}$,

$$y_k = \sum_{i=0}^{k}\binom{k}{i}x_i$$

if and only if, for all $k \in \{0, 1, \ldots, n\}$,

$$x_k = \sum_{i=0}^{k}(-1)^{k-i}\binom{k}{i}y_i.$$

Furthermore, for all $k \in \{0, 1, \ldots, n\}$,

$$y_k = \sum_{i=0}^{k}(-1)^{i+1}\binom{k}{i}x_i$$

if and only if, for all $k \in \{0, 1, \dots, n\}$,

$$x_k = \sum_{i=0}^{k} (-1)^{i+1} \binom{k}{i} y_i.$$

Source: Each equality is a *binomial transform*. See [554].

Fact 1.16.8. The following statements hold:

i) Let $n \geq 0$. Then,

$$\binom{0}{n} = \operatorname{truth}(n = 0), \quad \binom{n}{0} = \binom{n}{n} = 1, \quad \binom{n}{1} = n + \operatorname{truth}(n = 0).$$

ii) Let $k, n \geq 0$. Then,

$$\binom{n}{k} = \binom{n}{n-k}, \quad \binom{n+k}{n} = \binom{n+k}{k}, \quad (n-k)\binom{n}{k} = n\binom{n-1}{k}, \quad \binom{n}{k} = \frac{n+1-k}{n+1}\binom{n+1}{k},$$

$$k\binom{n}{k} = n\binom{n-1}{k-1} = (n+1-k)\binom{n}{k-1}, \quad k(k-1)\binom{n}{k} = n(n-1)\binom{n-2}{k-2}, \quad \binom{kn}{k} = n\binom{kn-1}{k-1},$$

$$\binom{n+k}{2} = \binom{n}{2} + \binom{k}{2} + nk, \quad \binom{n}{k} = \binom{n-1}{k} + \binom{n-1}{k-1} = (-1)^k\binom{k-n-1}{k}.$$

iii) Let $n \geq 1$. Then,

$$\binom{2n}{n} = 2\binom{2n-1}{n} = \frac{n+1}{2n+1}\binom{2n+1}{n} = \frac{n+1}{2(2n+1)}\binom{2n+2}{n+1} = \frac{(2n)!}{(n!)^2},$$

$$\binom{2n}{n} = (n+1)\binom{2n+1}{n+1} - 2(n+1)\binom{2n}{n+1} = \sum_{i=0}^{n} \frac{(n!)^2}{(i!)^2[(n-i)!]^2}, \quad \binom{2n}{n}^2 = \sum_{i=0}^{n} \frac{(2n)!}{(i!)^2[(n-i)!]^2}.$$

iv) Let $0 \leq k \leq n$. Then,

$$\binom{-n}{k} = \frac{-n(-n-1)\cdots(-n-k+1)}{k!} = (-1)^k\binom{n+k-1}{k}.$$

For example, $\binom{-1}{n} = (-1)^n$.

v) Let $k \geq 0$ and $n \geq 1$. Then,

$$\binom{n+k}{k+1} = \frac{1}{k!}\sum_{i=0}^{n-1} \frac{(k+i)!}{i!}.$$

vi) Let $n \geq k \geq 0$, and let m be prime. Then,

$$\binom{mn}{mk} \overset{m}{\equiv} \binom{n}{k}, \quad \binom{mn}{mk} \overset{m^2}{\equiv} \binom{n}{k}.$$

If, in addition, $m \geq 5$, then

$$\binom{mn}{mk} \overset{m^3}{\equiv} \binom{n}{k}.$$

vii) Let m be prime, and let n, k, q, r be nonnegative integers such that $q < m$ and $r < m$. Then,

$$\binom{mn+q}{mk+r} \overset{m}{\equiv} \binom{n}{k}\binom{q}{r}.$$

viii) Let $0 \le k \le n$. Then,

$$\binom{n-\frac{1}{2}}{k} = \frac{\binom{2n}{n}\binom{n}{k}}{4^k\binom{2n-2k}{n-k}} = \frac{(2n-1)!!}{(2n-2k-1)!!(2k)!!},$$

$$\binom{n+\frac{1}{2}}{n-k} = \frac{2n+1}{2k+1}\frac{\binom{2n}{n}\binom{n}{k}}{4^{n-k}\binom{2k}{k}} = \frac{(2n+1)!!}{(2n-2k)!!(2k+1)!!}.$$

ix) Let $1 \le k \le n$. Then,

$$\frac{1}{n+1}\binom{n+1}{k}\binom{n+1}{k+1} = \binom{n}{k}^2 - \binom{n}{k-1}\binom{n}{k+1},$$

$$\frac{n+2}{n+1}\binom{n+1}{k}\binom{n+1}{k+1} = \binom{n}{k-1}\binom{n}{k} + \binom{n}{k}\binom{n}{k+1} + 2\binom{n}{k}^2,$$

$$\frac{n}{n+1}\binom{n+1}{k}\binom{n+1}{k+1}\binom{k+1}{2} = \frac{n}{n+1}\binom{n+1}{k-1}\binom{n+1}{k}\binom{n-k+2}{2} = \binom{n}{k-1}\binom{n}{k}\binom{n+1}{2}.$$

x) Let $1 \le k \le n$. Then,

$$\binom{n}{k}^3 + \binom{n}{k+1}^3 + 3\binom{n}{k}\binom{n}{k+1}\binom{n+1}{k+1} = \binom{n+1}{k+1}^3,$$

$$\frac{1}{\binom{n+1}{k}^3} + \frac{1}{\binom{n+1}{k+1}^3} + \frac{3(n+2)}{(n+1)\binom{n}{k}\binom{n+1}{k}\binom{n+1}{k+1}} = \frac{(n+2)^3}{(n+1)^3\binom{n}{k}^3}.$$

Source: [366], [411, pp. 123, 124], [740], [1083, p. 31], [1206], [1219, p. 174], [1371, p. 10], and [1675, pp. 5, 43]. *viii*) is given in [1241, p. 274]. **Example:** $252 = \binom{10}{5} \overset{125}{\equiv} \binom{2}{1} = 2$.

Fact 1.16.9. The following statements hold:

i) Let $x \in \mathbb{C}$. Then,

$$\binom{x}{0} = 1, \quad \binom{x}{1} = x.$$

ii) Let $k \ge 1$ and $x \in \mathbb{C}$. Then,

$$\binom{x}{k} = \frac{x}{k}\binom{x-1}{k-1} = \frac{x+1-k}{k}\binom{x}{k-1}.$$

iii) Let $k \ge 2$ and $x \in \mathbb{C}$. Then,

$$\binom{x}{k} = \frac{x(x-1)}{k(k-1)}\binom{x-2}{k-2}.$$

iv) Let $k \ge 0$ and $x \in \mathbb{C}$. Then,

$$(x-k)\binom{x}{k} = x\binom{x-1}{k}.$$

v) Let $0 \le k \le n$ and $x \in \mathbb{C}$. Then,

$$\binom{x}{k} = \binom{x-1}{k} + \binom{x-1}{k-1}.$$

vi) Let $k \ge 0$ and $x \in \mathbb{C}$. Then,

$$\binom{x}{k} = (-1)^k\binom{k-x-1}{k}.$$

vii) Let $n, k \geq 0$ and $x \in \mathbb{C}$. Then,

$$\binom{n}{k}\binom{x+n}{n} = \binom{x+n}{n-k}\binom{x+k}{k}.$$

Fact 1.16.10. The following statements hold:

i) Let $x, y, z \in \mathbb{C}$ and $n \geq 0$. Then,

$$(x+y)^n = \sum_{i=0}^{n} \binom{n}{i} x(x-iz)^{i-1}(y+iz)^{n-i}.$$

In particular,

$$(x+y)^n = \sum_{i=0}^{n} \binom{n}{i} x^i y^{n-i} = \sum_{i=0}^{n} \binom{n}{i} x^{n-i} y^i.$$

ii) Let $0 \leq k \leq m \leq n$. Then,

$$\sum_{i=m}^{n} \binom{i}{k} = \binom{n+1}{k+1} - \binom{m}{k+1}.$$

In particular,

$$\sum_{i=k}^{n} \binom{i}{k} = \binom{n+1}{k+1} = \frac{n+1}{k+1}\binom{n}{k}.$$

iii) Let $n \geq k \geq 0$ and $m \geq 0$. Then,

$$\sum_{i=k}^{n} \binom{m+i}{i} = \binom{n+m+1}{n} - \binom{k+m}{k-1}.$$

In particular,

$$\sum_{i=0}^{n} \binom{m+i}{i} = \sum_{i=0}^{n} \binom{m+i}{m} = \binom{n+m+1}{n} = \binom{n+m+1}{m+1},$$

$$\sum_{i=0}^{n} \binom{n+i}{i} = \sum_{i=0}^{n} \binom{n+i}{n} = \binom{2n+1}{n}.$$

iv) Let $n, m \geq 1$. Then,

$$\sum_{i=1}^{m} \binom{n+m-i}{n} = \binom{n+m}{n+1}.$$

v) Let $n \geq 0$. Then,

$$\sum_{i=0}^{n} \frac{1}{2^i}\binom{n+i}{i} = 2^n, \quad \sum_{i=0}^{n} \frac{i}{2^i}\binom{n+i}{i} = \frac{n+1}{2^{n+1}}\left[2^{2n+1} - \binom{2n+2}{n+1}\right].$$

vi) Let $n \geq 0$. Then,

$$\sum_{i=0}^{n} \binom{n}{i} = 2^n.$$

Let $n \geq 3$ be odd. Then,

$$\sum_{i=1}^{\lfloor n/2 \rfloor} \binom{n}{i} = 2^{n-1} - 1.$$

Let $n \geq 2$ be even. Then,

$$\sum_{i=1}^{n/2} \binom{n}{i} = \frac{1}{2}\binom{n}{n/2} + 2^{n-1} - 1.$$

vii) Let $n, k \geq 1$. Then,

$$\sum_{i=0}^{\lfloor n/k \rfloor} \binom{n}{ki} = \frac{2^n}{k} \sum_{i=1}^{k} \cos^n \frac{i\pi}{k} \cos \frac{ni\pi}{k}.$$

In particular,

$$\sum_{i=0}^{\lfloor n/2 \rfloor} \binom{n}{2i} = \sum_{i=0}^{\lfloor (n-1)/2 \rfloor} \binom{n}{2i+1} = 2^{n-1}.$$

viii) Let $n \geq 0$ and $x \in \mathbb{C}$. Then,

$$\sum_{i=0}^{n} (2i - x)\binom{x}{i} = (n - x)\binom{x}{n} = -(n+1)\binom{x}{n+1}.$$

ix) Let $n \geq 1$. Then,

$$\sum_{i=1}^{n} i\binom{n}{i} = 2^{n-1}n.$$

x) Let $n, k \geq 1$, and define

$$S_{n,k} \triangleq \sum_{i=1}^{n} i^k \binom{n}{i}.$$

Then,

$$S_{n,k+1} = n(S_{n,k} - S_{n-1,k}).$$

xi) Let $n \geq 1$. Then,

$$\sum_{i=1}^{n} i^2 \binom{n}{i} = 2^{n-2}n(n+1), \quad \sum_{i=1}^{n} i^3 \binom{n}{i} = 2^{n-3}n^2(n+3),$$

$$\sum_{i=1}^{n} i^4 \binom{n}{i} = 2^{n-4}n(n+1)(n^2+5n-2).$$

xii) Let $n \geq 0$. Then,

$$\sum_{i=0}^{n} \frac{1}{i+1}\binom{n}{i} = \frac{2^{n+1}-1}{n+1}.$$

xiii) Let $n \geq 0$. Then,

$$\sum_{i=0}^{n} \frac{(i+1)!}{(n+1)^{i+1}}\binom{n}{i} = 1.$$

xiv) Let $n \geq 0$. Then,

$$\sum_{i=1}^{\lfloor (n+1)/2 \rfloor} \frac{1}{2i}\binom{n+1}{2i} = \sum_{i=0}^{n} \frac{2^i - 1}{i+1}.$$

xv) Let $n \geq 0$ and $x \in \mathbb{C}$. Then,

$$\sum_{i=0}^{n} \frac{x^{i+1}}{i+1}\binom{n}{i} = \frac{(x+1)^{n+1}-1}{n+1}.$$

xvi) Let $n \geq 0$. Then,

$$\sum_{i=0}^{n} \binom{2n+1}{i} = \sum_{i=0}^{n} \binom{2n+1}{2i} = \sum_{i=0}^{2n} \binom{2n}{i} = 4^n.$$

xvii) Let $n \geq 0$. Then,

$$\sum_{i=0}^{n}\binom{2n}{n-i} = \sum_{i=0}^{n}\binom{2n}{i} = \frac{1}{2}\left[4^n + \binom{2n}{n}\right], \quad \sum_{i=0}^{\lfloor n/2\rfloor}\binom{2n}{n-2i} = \frac{1}{2}\binom{2n}{n} + 4^{n-1},$$

$$\sum_{i=0}^{\lfloor n/3\rfloor}\binom{2n}{n-3i} = \frac{1}{2}\binom{2n}{n} + \frac{1}{3}(2^{2n-1}+1), \quad \sum_{i=0}^{\lfloor n/4\rfloor}\binom{2n}{n-4i} = \frac{1}{2}\binom{2n}{n} + 2^{2n-3} + 2^{n-2},$$

$$\sum_{i=0}^{\lfloor n/5\rfloor}\binom{2n}{n-5i} = \frac{1}{2}\binom{2n}{n} + \frac{1}{5(2^n)}[2^{3n-1} + (3+\sqrt{5})^n + (3-\sqrt{5})^n],$$

$$\sum_{i=0}^{\lfloor n/6\rfloor}\binom{2n}{n-6i} = \frac{1}{2}\binom{2n}{n} + \frac{1}{6}(2^{2n-1} + 3^n + 1), \quad \sum_{i=1}^{n}\binom{2n-1}{n-i} = 4^{n-1}.$$

If $n \geq 2$, then

$$\sum_{i=1}^{\lfloor n/3\rfloor}\binom{2n-3}{n-3i} = \frac{1}{3}(4^{n-2}-1).$$

If $n \geq 3$, then

$$\sum_{i=1}^{\lfloor n/5\rfloor}\binom{2n-5}{n-5i} = \frac{1}{5}4^{n-3} - \frac{1}{5(2^{2n-5})}[(\sqrt{5}+1)^{2n-5} - (\sqrt{5}-1)^{2n-5}].$$

xviii) Let $n \geq 0$. Then,

$$\sum_{i=0}^{n} i\binom{2n}{i} = \tfrac{1}{2}4^n n.$$

xix) Let $n > 1$. Then,

$$\sum_{i=0}^{n} i^2\binom{2n}{i} = 4^{n-1}n(2n+1) - \tfrac{1}{2}n^2\binom{2n}{n}.$$

xx) Let $n > 1$. Then,

$$\sum_{i=0}^{n-1}(n-i)^2\binom{2n}{i} = 4^{n-1}n.$$

xxi) Let $n \geq 0$. Then,

$$\sum_{i=0}^{n} i\binom{2n+1}{2i} = (2n+1)2^{2n-2}.$$

xxii) Let $n \geq 0$. Then,

$$\sum_{i=0}^{n} i^2\binom{2n+1}{2i} = (n+1)(2n+1)2^{2n-3}.$$

xxiii) Let $n \geq 1$ and $k \geq 1$. Then,

$$\sum_{i=0}^{n} k^i\binom{n}{i} = (k+1)^n, \quad \sum_{i=0}^{n} k^i i\binom{n}{i} = kn(k+1)^{n-1},$$

$$\sum_{i=0}^{n} k^i i^2\binom{n}{i} = kn(kn+1)(k+1)^{n-2}, \quad \sum_{i=0}^{n} k^i i^3\binom{n}{i} = kn(k^2n^2 + 3kn + 1 - k)(k+1)^{n-3},$$

$$\sum_{i=0}^{n} k^i i^4\binom{n}{i} = kn[k^3n^3 + 6k^2n^2 + (7k-4k^2)n + (k-2)^2 - 3](k+1)^{n-4}.$$

In particular,

$$\sum_{i=0}^{n} 2^i \binom{n}{i} = 3^n, \quad \sum_{i=0}^{n} 2^i i \binom{n}{i} = 2n3^{n-1}, \quad \sum_{i=0}^{n} 2^i i^2 \binom{n}{i} = 2n(2n+1)3^{n-2},$$

$$\sum_{i=0}^{n} 2^i i^3 \binom{n}{i} = 2n(4n^2+6n-1)3^{n-3}, \quad \sum_{i=0}^{n} 2^i i^4 \binom{n}{i} = 2n(8n^3+24n^2-2n-3)3^{n-4},$$

$$\sum_{i=0}^{n} 3^i \binom{n}{i} = 4^n, \quad \sum_{i=0}^{n} 3^i i \binom{n}{i} = 3n4^{n-1}, \quad \sum_{i=0}^{n} 3^i i^2 \binom{n}{i} = 3n(3n+1)4^{n-2},$$

$$\sum_{i=0}^{n} 3^i i^3 \binom{n}{i} = 3n(9n^2+9n-2)4^{n-3}, \quad \sum_{i=0}^{n} 3^i i^4 \binom{n}{i} = 3n(27n^3+54n^2-15n-2)4^{n-4},$$

$$\sum_{i=0}^{n} 4^i \binom{n}{i} = 5^n, \quad \sum_{i=0}^{n} 4^i i \binom{n}{i} = 4n5^{n-1}, \quad \sum_{i=0}^{n} 4^i i^2 \binom{n}{i} = 4n(4n+1)5^{n-2},$$

$$\sum_{i=0}^{n} 4^i i^3 \binom{n}{i} = 4n(16n^2+12n-3)5^{n-3}, \quad \sum_{i=0}^{n} 4^i i^4 \binom{n}{i} = 4n(64^3+96n^2-36n+1)5^{n-4}.$$

xxiv) Let $n \ge 1$. Then,

$$\sum_{i=0}^{\lfloor n/2 \rfloor} 4^i \binom{n}{2i} = \tfrac{1}{2}[3^n + (-1)^n], \quad \sum_{i=0}^{\lfloor (n-1)/2 \rfloor} 5^{2i+1} \binom{n}{2i+1} = \tfrac{1}{2}[6^n - (-4)^n].$$

xxv) Let $n \ge 1$, and let F_n denote the nth Fibonacci number. Then,

$$\sum_{i=1}^{\lfloor (n+1)/2 \rfloor} \binom{n-i}{i-1} = F_k, \quad \sum_{i=0}^{\lfloor n/3 \rfloor} 2^{n-3i} \binom{n-i}{2i} = F_{2n} + 1.$$

xxvi) Let $n \ge 1$, and let L_n denote the nth Lucas number. Then,

$$\sum_{i=0}^{\lfloor n/2 \rfloor} \frac{n}{n-i} \binom{n-i}{i} = L_n, \quad \sum_{i=0}^{\lfloor n/2 \rfloor} \frac{2^i n}{n-i} \binom{n-i}{i} = 2^n + (-1)^n,$$

$$\sum_{i=0}^{\lfloor n/2 \rfloor} 2^i \binom{n-i}{i} = \frac{1}{3}[2^{n+1} + (-1)^n], \quad \sum_{i=0}^{\lfloor n/2 \rfloor} \frac{2^i i}{n-i} \binom{n-i}{i} = \frac{1}{3}[2^n + (-1)^n 2].$$

xxvii) Let $n \ge 1$, and let F_n and L_n denote the nth Fibonacci number and nth Lucas number, respectively. Then,

$$\sum_{i=1}^{\lfloor (n-1)/2 \rfloor} i \binom{n-i-1}{i} = \frac{1}{10}[(5n-4)F_n - nL_n],$$

$$\sum_{i=1}^{\lfloor (n-1)/2 \rfloor} i^2 \binom{n-i-1}{i} = \frac{1}{50}[(15n^2-20n+4)F_n - (5n^2-6n)L_n].$$

xxviii) Let $n \ge 1$, and let P_n denote the nth Pell number. Then,

$$\sum_{i=0}^{\lfloor n/4 \rfloor} 2^{n+1-4i} \binom{n-2i}{2i} = P_{n+1} + n + 1.$$

xxix) Let $n \ge k \ge 0$. Then,

$$\sum_{i=k}^{n} \binom{n}{i} = \sum_{i=0}^{n-k} 2^i \binom{n-i-1}{k-1}.$$

xxx) Let $n \ge 1$. Then,

$$S_n \triangleq \sum_{i=0}^{n} \binom{n}{i}^{-1} = \frac{n+1}{2^n} \sum_{i=0}^{\lfloor n/2 \rfloor} \frac{1}{2i+1} \binom{n+1}{2i+1} = \frac{n+1}{2^n} \sum_{i=0}^{n} \frac{2^i}{i+1},$$

$$\sum_{i=0}^{n} i \binom{n}{i}^{-1} = \frac{n}{2} S_n = \frac{n(n+1)}{2^{n+1}} \sum_{i=0}^{n} \frac{2^i}{i+1}, \quad \sum_{i=0}^{n} i^2 \binom{n}{i}^{-1} = \frac{(n+1)(n-2)}{4} S_n + \frac{(n+1)^2}{2},$$

$$\sum_{i=0}^{n} i^3 \binom{n}{i}^{-1} = \frac{n(n^2-3n-6)}{8} S_n + \frac{3n(n+1)^2}{4},$$

$$\sum_{i=0}^{n} i^4 \binom{n}{i}^{-1} = \frac{(n+1)(n^3-7n^2-2n+16)}{16} S_n + \frac{(7n-8)(n+1)^3}{8}.$$

Furthermore, for all $n \ge 2$,

$$S_n = \frac{n+1}{2n} S_{n-1} + 1.$$

xxxi) Let $n \ge 1$ and $z \in \mathbb{C}$, and assume that $\operatorname{Re} z \ne -1$. Then,

$$\sum_{i=0}^{n} \binom{n}{i}^{-1} z^i = (n+1) \left(\frac{z}{z+1} \right)^{n+1} \sum_{i=1}^{n+1} \frac{1}{i} \frac{(1+z^i)(1+z)^{i-1}}{z^i}.$$

In particular,

$$\sum_{i=0}^{n} \binom{n}{i}^{-1} = \frac{n+1}{2^{n+1}} \sum_{i=1}^{n+1} \frac{2^i}{i} = \frac{n+1}{2^n} \sum_{i=0}^{n} \frac{2^i}{i+1}.$$

xxxii) Let $n \ge 2$ and $m \ge 0$. Then,

$$\sum_{i=n}^{n+m} \binom{i}{n}^{-1} = \frac{n}{n-1} \left[1 - \binom{n+m}{n-1}^{-1} \right].$$

xxxiii) Let $n \ge 0$ and $x, y \in \mathbb{C}$, assume that $x + y \ne 0$, and define

$$S_n \triangleq \sum_{i=0}^{n} \binom{n}{i}^{-1} x^i y^{n-i}.$$

Then,

$$S_n = x^n + (n+1) \left(\frac{xy}{x+y} \right)^n \sum_{i=0}^{n-1} \frac{[(i+1)y^{i+2} + yx^{i+1}](x+y)^i}{(xy)^{i+1}(i+1)(i+2)}.$$

Furthermore, for all $n \ge 2$,

$$S_n = \frac{(n+1)xy}{n(x+y)} S_{n-1} + \frac{x^{n+1} + y^{n+1}}{x+y}.$$

xxxiv) Let $n \ge 0$, and define

$$S_n \triangleq \sum_{i=0}^{n} \binom{n}{i}^{-2}.$$

Then,

$$S_n = 1 + \frac{(n+1)^2(n+1)!}{4^n(\frac{1}{2})^{\overline{n+2}}} \sum_{i=0}^{n-1} \frac{4^i(\frac{1}{2})^{\overline{i+2}}(3i^2 + 12i^2 + 18i + 10)}{(i+1)^2(i+2)^3(i+1)!}.$$

Furthermore, for all $n \ge 2$,

$$S_n = \frac{(n+1)^3}{2n^2(2n+3)} S_{n-1} + \frac{3n+3}{2n+3}.$$

xxxv) Let $n \ge 1$ and $k \ge 1$. Then,

$$\sum_{i=0}^{n} \binom{n}{i}^{-k} = (n+1)^k \sum_{i=0}^{n} \left[\sum_{j=0}^{i} (-1)^i \frac{1}{n+1+j-i} \binom{i}{j} \right]^k.$$

In particular,

$$\sum_{i=0}^{n} \binom{n}{i}^{-1} = (n+1) \sum_{i=0}^{n} \frac{1}{2^i(n+1-i)} = (n+1) \sum_{i=0}^{n} \sum_{j=0}^{i} \frac{(-1)^j}{n+1+j-i} \binom{i}{j},$$

$$\sum_{i=0}^{n} \binom{n}{i}^{-2} = (n+1)^2 \sum_{i=0}^{n} \frac{2}{n+1-i} \sum_{j=0}^{i} \frac{(-1)^j}{n+2+j} \binom{i}{j} = (n+1)^2 \sum_{i=0}^{n} \left[\sum_{j=0}^{i} \frac{(-1)^j}{n+1+j-i} \binom{i}{j} \right]^2.$$

Furthermore,

$$\lim_{n\to\infty} \sum_{i=0}^{n} \binom{n}{i}^{-k} = 2.$$

xxxvi) Let $n \ge 1$. Then,

$$\sum_{i=0}^{n} \binom{2n}{2i}^{-1} = \frac{2(2n+1)}{2^{2n+2}} \sum_{i=0}^{2n+1} \frac{2^i}{i+1}.$$

xxxvii) Let $n, m \ge 0$ and $1 \le k \le m$. Then,

$$\sum_{i=0}^{n} \frac{\binom{n}{i}}{\binom{n+m}{k+i}} = \frac{n+m+1}{(m+1)\binom{m}{k}}.$$

xxxviii) Let $n \ge 1$. Then,

$$\sum_{i=n}^{n^2-n+1} \frac{\binom{(n-1)^2}{i-n}}{i\binom{n^2}{i}} = \frac{1}{n\binom{2n-1}{n}}.$$

xxxix) Let $n \ge 1$. Then,

$$\sum_{i=1}^{n} \frac{1}{i^2\binom{2i}{i}} = \frac{2}{3}H_{n,2} - \frac{1}{3} \sum_{i,j=1}^{n} \frac{(i-1)!(j-1)!}{(i+j)!}.$$

xl) Let $n \ge 0$. Then,

$$\sum_{i=0}^{n} \frac{4^i}{\binom{2i}{i}} = \frac{4^{n+1}(2n+1)}{3\binom{2n+2}{n+1}} + \frac{1}{3}.$$

xli) Let $n \ge 1$. Then,

$$\sum_{i=1}^{n} \frac{4^i}{i\binom{2i}{i}} = \frac{2(4^n)}{\binom{2n}{n}} - 2.$$

xlii) Let $n \geq 2$. Then,

$$\sum_{i=1}^{n-1} \frac{4^i}{i(2i+1)\binom{2i}{i}} = 2 - \frac{4^n}{n\binom{2n}{n}}.$$

xliii) Let $n \geq 3$. Then,

$$\sum_{i=2}^{n-1} \frac{4^i}{i(i-1)\binom{2i}{i}} = 4 - \frac{4^n(2n-1)}{n(n-1)\binom{2n}{n}}.$$

xliv) Let $n \geq 0$ and $x \in \mathbb{C}$. Then,

$$\sum_{i=0}^{\lfloor n/2 \rfloor} \binom{n}{2i} x^{n-2i} = \tfrac{1}{2}(x+1)^n + \tfrac{1}{2}(x-1)^n,$$

$$2x \sum_{i=0}^{n} \binom{n}{i} x^{2\lfloor i/2 \rfloor} = (1+x)^{n+1} - (1-x)^{n+1}.$$

xlv) Let $n \geq 0$ and $x \in \mathbb{C}$. Then,

$$x \sum_{i=0}^{n} \binom{n+i}{2i} \left(\frac{x^2-1}{4} \right)^{n-i} = [\tfrac{1}{2}(x+1)]^{2n+1} + [\tfrac{1}{2}(x-1)]^{2n+1}.$$

xlvi) Let $n \geq 1$ and $x \in \mathbb{C}$. Then,

$$\sum_{i=1}^{n} \frac{1}{i} \binom{n+i-1}{2i-1} (x-1)^{2i} x^{n-i} = \frac{1}{n}(x^n - 1)^2.$$

xlvii) Let $n \geq 1$. Then,

$$\sum_{i=1}^{n} H_i \binom{n}{i} = 2^n \left(H_n \quad \sum_{i=1}^{n} \frac{1}{2^i i} \right).$$

xlviii) Let $n \geq k \geq 1$. Then,

$$\sum_{i=k}^{n} H_i \binom{i}{k} = \binom{n+1}{k+1} \left(H_{n+1} - \frac{1}{k+1} \right).$$

xlix) Let $n \geq 1$. Then,

$$\sum_{i=1}^{n} H_i \binom{n}{i}^2 = (2H_n - H_{2n}) \binom{2n}{n}.$$

l) Let $n \geq k \geq 1$. Then,

$$\sum_{i=k}^{n} \frac{1}{n+1-i} \binom{i}{k} = \binom{n+1}{k} (H_{n+1} - H_k).$$

li) Let $n \geq k \geq 1$. Then,

$$\sum_{i=0}^{k-1} \sum_{j=k}^{n} (-1)^{i+j-1} \frac{1}{j-i} \binom{n}{i} \binom{n}{j} = \sum_{i=0}^{k-1} \binom{n}{i}^2 (H_{n-i} - H_i),$$

$$\sum_{i=1}^{n} i \binom{n}{i}^2 (H_i - H_{n-i}) = \sum_{i=1}^{n} (2i-n) \binom{n}{i}^2 H_i = \binom{2n-1}{n}.$$

lii) Let $n \geq 1$. Then,

$$\sum_{i=1}^{n} \frac{i}{4^i} \binom{2i}{i} = \frac{n(n+1)(n+2)}{6(4^n)} C_{n+1} = \frac{(2n+2)!}{6(4^n)(n+1)!(n-1)!}.$$

liii) Let $n \geq 0$ and $m \geq 1$. Then,

$$\sum_{i=0}^{n} \binom{n+1}{i} \sum_{j=1}^{m} j^i = (m+1)^{n+1} - 1.$$

liv) Let p be prime and $n \geq 1$. Then,

$$\sum \binom{n}{i} \overset{p}{\equiv} 0,$$

where the sum is taken over all $i \in \{1, \ldots, n-1\}$ such that $(p-1)|i$. Equivalently,

$$\sum_{i=1}^{\left\lfloor \frac{n-1}{p-1} \right\rfloor} \binom{n}{(p-1)i} \overset{p}{\equiv} 0.$$

lv) Let $n \geq 2$. Then, $\sum_{i=1}^{n-1} \binom{n}{i}$ is even.

lvi) Let $n \geq 1$. Then,

$$\sum_{i=1}^{n-1} i^{n-i}(n-i)^i \binom{n}{i} = (-1)^n n! \left(1 + \sum_{i=1}^{n-2} (-1)^i \frac{n^i}{i!}\right).$$

lvii) Let p be prime and $n \in [p, 2p-2]$. Then, $p | \binom{n}{p-1}$.

lviii) Let $n \geq 0$ and $x, y \in \mathbb{C}$. Then,

$$(x+y)^{\underline{n}} = \sum_{i=0}^{n} \binom{n}{i} x^{\underline{n-i}} y^{\underline{i}}.$$

lix) Let $n, m \geq 0$ and $x \in \mathbb{C}$. Then,

$$(1-x)^{n+1} \sum_{i=0}^{m} \binom{n+i}{i} x^i + x^{m+1} \sum_{i=0}^{n} \binom{m+i}{i} (1-x)^i = 1.$$

lx) Let $n \geq 0$ and $x \in \mathbb{C}$. Then,

$$\sum_{i=0}^{n} \binom{n}{i}^2 x^i = \sum_{i=1}^{n} \binom{n}{i} \binom{2n-i}{n} (x-1)^i.$$

lxi) Let $x_0, x_1, \ldots, x_n \in \mathbb{C}$. Then,

$$\sum_{i=0}^{n} x_i \binom{n}{i} = \sum_{i=0}^{n-1} (x_i + x_{i+1}) \binom{n-1}{i}.$$

lxii) Let $n \geq 1$. Then,

$$\sum_{i=1,\, i \text{ odd}}^{n} \binom{n}{i} = \sum_{i=0,\, i \text{ even}}^{n} \binom{n}{i} = \frac{1}{2} \sum_{i=0}^{n} \binom{n}{i} = 2^{n-1}.$$

lxiii) Let $n \geq 1$ and let $z \in \mathbb{C}$, where $|z| < 1/4$. Then,

$$\sum_{i=0}^{\lfloor n/2 \rfloor} \binom{n-i}{i} z^i = \frac{1}{\sqrt{1+4z}} \left[\left(\frac{1+\sqrt{1+4z}}{2} \right)^{n+1} - \left(\frac{1+\sqrt{1-4z}}{2} \right)^{n+1} \right],$$

$$\sum_{i=0}^{\lfloor n/2 \rfloor} \frac{n}{n-i} \binom{n-i}{i} z^i = \left(\frac{1+\sqrt{1+4z}}{2} \right)^n + \left(\frac{1+\sqrt{1+4z}}{2} \right)^n.$$

lxiv) Let $n \geq 1$ and let $z \in \mathbb{C}$, where $|z| > 2$. Then,

$$\sum_{i=0}^{\lfloor n/2 \rfloor} (-1)^i \binom{n-i}{i} z^{n-2i} = \frac{1}{\sqrt{z^2-4}} \left[\left(\frac{z+\sqrt{z^2-4}}{2} \right)^{n+1} - \left(\frac{z-\sqrt{z^2-4}}{2} \right)^{n+1} \right].$$

lxv) Let $n \geq 2$. Then,

$$\sum_{i=0}^{\lfloor (n-1)/2 \rfloor} \binom{n}{i}(n-2i) = n\binom{n-1}{\lfloor n/2 \rfloor}.$$

lxvi) Let $n \geq 0$, and assume that n is even. Then,

$$\sum_{i=0}^{2n} \binom{2n}{i}|n-i| = n\binom{2n}{n}.$$

lxvii) Let $n \geq 1$. Then,

$$\sum_{i=1}^{n} \frac{i}{i+2}\binom{n}{i} = 2^n - \frac{2(2^{n+1}n+1)}{(n+1)(n+2)}, \quad \sum_{i=1}^{n} \frac{i}{i+3}\binom{n}{i} = 2^n - \frac{6(2^n n^2 + 2^n n + 2^{n+1} - 1)}{(n+1)(n+2)(n+3)},$$

$$\sum_{i=1}^{n} \frac{i}{i+4}\binom{n}{i} = 2^n - \frac{8(2^n n^3 + 3(2^n)n^2 + 2^{n+3}n + 3)}{(n+1)(n+2)(n+3)(n+4)}.$$

lxviii) Let $n \geq 0$. Then,

$$\sum_{i=0}^{n} 2^{n-i}\binom{n+i}{2i} = \frac{1}{3}(2^{2n+1}+1).$$

lxix) Let $n \geq 0$ and $k \geq 0$, and define

$$S_k(n) \triangleq \sum_{i=0}^{n} i^k \binom{n+i}{i}.$$

Then,

$$S_0(n) = \binom{2n+1}{n}, \quad S_1(n) = \frac{n(n+1)}{n+2}\binom{2n+1}{n}, \quad S_2(n) = \frac{n(n+1)^3}{(n+2)(n+3)}\binom{2n+1}{n},$$

$$S_3(n) = \frac{n^2(n+1)^2(n^2+4n+5)}{(n+2)(n+3)(n+4)}\binom{2n+1}{n}, \quad S_4(n) = \frac{n(n+1)^3(n^4+7n^3+17n^2+9n-4)}{(n+2)(n+3)(n+4)(n+5)}\binom{2n+1}{n},$$

$$S_{k+1}(n) = (n+1)\left[S_k(n+1) - S_k(n) - (n+1)^k \binom{2n+2}{n+1} \right].$$

Furthermore, as $n \to \infty$, $S_k(n) \sim 2n^k\binom{2n}{n}$.

lxx) Let $n \geq 2$. Then,

$$\sum_{i=1}^{n-1} \frac{in^{n-i}}{i+1}\binom{n}{i} = \frac{n(n^n-1)}{n+1}.$$

Source: [411, pp. 64–68, 78], [1207], [1225, pp. 1, 2], and [1371, pp. 2–10]. *ii)* and *iii)* are given in [2240]; *iv)* is given in [570, p. 159]; *v)* is given in [993, pp. 96, 97, 207], [1219, p. 167], and [1749]; *vi)* follows from *i)* with $x = y = 1$; *vii)* is given in [411, pp. 65, 81]; *viii)* is given in [2228, pp. 95, 97]; *ix)* is given in [411, p. 66]; *x)* is given in [1197]; *xi)* is given in [993, pp. 95, 96], [1197], and [1372, p. 62]; *xii)* is given in [1524, pp. 35, 36]; *xiii)* is given in [771, p. 173]; *xiv)* is given in [2577]; *xv)* is given in [1217, p. 5]; *xvi)* is given in [1371, p. 3], [511, p. 16], and [740]; *xvii)* is given in [1217, p. 12] and [2015]; *xviii)* and *xix)* are given in [1217, p. 12]; *xx)* is given in

[740]; *xxi*) and *xxii*) are given in [511, p. 16]; *xxiii*) is given in [504, p. 77] and [1372, p. 55]; *xxiv*) is given in [504, p. 78]; *xxv*) is given in [597], [993, p. 205], and [1158, p. 301]; *xxvi*) is given in [1622, 1677]; *xxvii*) is given in [1152]; *xxviii*) is given in [598]; *xxix*) is given in [771, p. 72]; *xxx*) is given in [517, p. 55] and [2235, 2321, 2510, 2577]; *xxxi*) is given in [771, p. 294]; *xxxii*) is given in [506, pp. 136, 137]; *xxxiii*) is given in [148]; *xxxiv*) is given in [148]; *xxxv*) is given in [148, 1944]; *xxxvi*) is given in [1944]; *xxxvii*) is given in [2728]; *xxxviii*) is given in [2518]; *xxxix*) is given in [2859]; *xl*)–*xliii*) are given in [2509]; *xliv*) is given in [411, p. 113]; *xlv*) and *xlvi*) are given in [2880, p. 174]; *xlvii*) is given in [554]; *xlviii*) is given in [411, p. 92] and [1219, pp. 279, 280]; *xlviv*) is given in [739]; *l*) is given in [411, p. 92]; *li*) is given in [2490]; *lii*) is given in [1675, p. 57]; *liii*)–*lv*) and *lvii*) are given in [1918]; *lvi*) is given in [2013, p. 179]; *lviii*) is given in [1158, p. 298] and [1197]; *lix*) is given in [1670]; *lx*) is given in [771, p. 168]; *lxi*) and *lxii*) are given in [1197]; *lxiii*) is given in [1219]; *lxiv*) is given in [2880, pp. 131, 132]; *lxv*) and *lxvi*) are given in [2744]; *lxvii*) is given in [1886]; *lxviii*) is given in [2880, p. 133]; *lxix*) is given in [1750, 2198]; *lxx*) is given in [385]. **Remark:** *i*) is the *Abel identity*; the special case $z = 0$ is the *binomial identity*. See [771, pp. 128–130] and [1274]. An extension is given in [2]. *liii*) is *Pascal's identity*. See [1918].

Fact 1.16.11. The following statements hold:

i) Let $0 \le k \le n$. Then,

$$\sum_{i=0}^{k}(-1)^i\binom{n}{i} = (-1)^k\binom{n-1}{k}.$$

ii) Let $n \ge 0$. Then,

$$\sum_{i=0}^{n}(-1)^i\binom{n}{i} = 0.$$

iii) Let $n \ge 1$. Then,

$$\sum_{i=0}^{n}(-1)^i\binom{2n}{i} = (-1)^n\frac{1}{2}\binom{2n}{n}, \quad \sum_{i=0}^{n}(-1)^i\binom{2n}{n-i} = \frac{1}{2}\binom{2n}{n},$$

$$\sum_{i=0}^{\lfloor n/2\rfloor}(-1)^i\binom{2n}{n-2i} = \frac{1}{2}\binom{2n}{n} + 2^{n-1}, \quad \sum_{i=0}^{\lfloor n/3\rfloor}(-1)^i\binom{2n}{n-3i} = \frac{1}{2}\binom{2n}{n} + 3^{n-1},$$

$$\sum_{i=0}^{\lfloor n/4\rfloor}(-1)^i\binom{2n}{n-4i} = \frac{1}{2}\binom{2n}{n} + \frac{1}{4}[(2+\sqrt{2})^n + (2-\sqrt{2})^n],$$

$$\sum_{i=0}^{\lfloor n/5\rfloor}(-1)^i\binom{2n}{n-5i} = \frac{1}{2}\binom{2n}{n} + \frac{1}{5(2^n)}[(5+\sqrt{5})^n + (5-\sqrt{5})^n].$$

iv) Let $n \ge 2$. Then,

$$\sum_{i=1}^{n}(-1)^i i\binom{n}{i} = 0.$$

v) Let $n \ge 0$. Then,

$$\sum_{i=0}^{n}(-1)^i 2^i\binom{n}{i} = (-1)^n.$$

vi) Let $n \ge 1$ and $1 \le k \le n-1$. Then,

$$\sum_{i=1}^{n}(-1)^i i^{2k}\binom{2n}{n+i} = 0.$$

Furthermore,

$$\sum_{i=1}^{n}(-1)^i\binom{2n}{n+i} = -\frac{1}{2}\binom{2n}{n}, \quad \sum_{i=1}^{n}(-1)^i i^{2n}\binom{2n}{n+i} = (-1)^n\tfrac{1}{2}(2n)!.$$

vii) Let $n > k \ge 1$. Then,

$$\sum_{i=1}^{n}(-1)^i i^k\binom{n}{i} = 0.$$

In particular, if $n \ge 3$, then

$$\sum_{i=1}^{n}(-1)^i i^2\binom{n}{i} = 0.$$

viii) Let $n \ge 1$. Then,

$$\sum_{i=1}^{n}(-1)^i i^n\binom{n}{i} = (-1)^n n!.$$

ix) Let $n \ge 1$ and $x \in \mathbb{C}$. Then,

$$\sum_{i=0}^{n}(-1)^i(x+i)^n\binom{n}{i} = (-1)^n n!.$$

x) Let $n > k \ge 0$ and $x \in \mathbb{C}$. Then,

$$\sum_{i=0}^{n}(-1)^i(x+i)^k\binom{n}{i} = 0.$$

xi) Let $n \ge 0$. Then,

$$\sum_{i=0}^{\lfloor n/2\rfloor}(-1)^i\binom{n}{2i} = 2^{n/2}\cos\frac{n\pi}{4}, \quad \sum_{i=0}^{\lfloor n/3\rfloor}(-1)^i\binom{n}{3i} = (2)3^{n/2-1}\cos\frac{n\pi}{6}.$$

xii) Let $n \ge 0$. Then,

$$\sum_{i=0}^{\lfloor (n-1)/2\rfloor}(-1)^i\binom{n}{2i+1} = 2^{n/2}\sin\frac{n\pi}{4}.$$

xiii) Let $n \ge 0$. Then,

$$\sum_{i=0}^{\lfloor n/2\rfloor}(-1)^i\binom{n-i}{i} = \begin{cases} -1, & n \overset{6}{\equiv} 3 \text{ or } n \overset{6}{\equiv} 4,\\ 0, & n \overset{6}{\equiv} 2 \text{ or } n \overset{6}{\equiv} 5,\\ 1, & n \overset{6}{\equiv} 0 \text{ or } n \overset{6}{\equiv} 1.\end{cases}$$

xiv) Let $n \ge 0$. Then,

$$\sum_{i=0}^{\lfloor n/2\rfloor}(-1)^i 2^{n-2i}\binom{n-i}{i} = n+1, \quad \sum_{i=0}^{\lfloor n/3\rfloor}(-1)^i 2^{n-3i}\binom{n-2i}{2i} = F_{n+3} - 1.$$

xv) Let $n \ge 0$. Then,

$$\sum_{i=0}^{n}(-1)^i 4^i\binom{n+i}{2i} = (-1)^n(2n+1).$$

xvi) Let $n \geq 0$. Then,

$$\sum_{i=0}^{4n+1}(-1)^{i(i+1)/2}\binom{4n+1}{i} = 0, \quad \sum_{i=0}^{4n+3}(-1)^{i(i+1)/2}\binom{4n+3}{i} = (-4)^{n+1},$$

$$\sum_{i=0}^{2n+2}(-1)^{i(i+1)/2}\binom{2n+2}{i} = (-1)^{(n+1)(n+2)/2}2^{n+1}.$$

xvii) Let $n, k \geq 0$. Then,

$$\sum_{i=0}^{n}(-1)^i(n-i)^k\binom{n}{i} = \begin{cases} 0, & k < n, \\ n!, & k = n, \\ \frac{1}{2}n(n+1)!, & k = n+1. \end{cases}$$

xviii) Let $n \geq 1$. Then,

$$\sum_{i=0}^{\lfloor n/2 \rfloor}(-1)^i\frac{n}{n-i}\binom{n-i}{i} = \begin{cases} -2, & n \stackrel{6}{\equiv} 3, \\ -1, & n \stackrel{6}{\equiv} 2 \text{ or } n \stackrel{6}{\equiv} 4, \\ 1, & n \stackrel{6}{\equiv} 1 \text{ or } n \stackrel{6}{\equiv} 5, \\ 2, & n \stackrel{6}{\equiv} 0. \end{cases}$$

xix) Let $k > n \geq 0$. Then,

$$\sum_{i=0}^{n}(-1)^{n-i}\frac{1}{k-i}\binom{n}{i} = \frac{1}{(k-n)\binom{k}{n}}.$$

xx) Let $n \geq 1$. Then,

$$\sum_{i=0}^{\lfloor n/2 \rfloor}(-1)^i2^{n-2i}\frac{n}{n-i}\binom{n-i}{i} = 2.$$

xxi) Let $n \geq 1$ and $x, y \in \mathbb{C}$. Then,

$$\sum_{i=1}^{n}(-1)^{i-1}\binom{n}{i}\frac{1}{i}(x+iy)^n = x^nH_n + nx^{n-1}y.$$

xxii) Let $n \geq 1$. Then,

$$\sum_{i=1}^{n}(-1)^{i-1}\binom{n}{i}\frac{1}{i}(n-i)^n = n^n(H_n - 1).$$

xxiii) Let $n \geq 1$ and $1 \leq k \leq n$. Then,

$$\sum_{i=1}^{n}(-1)^{i+1}\frac{1}{i^k}\binom{n}{i} = \sum\prod_{j=1}^{k}\frac{1}{i_j},$$

where the sum is taken over all k-tuples $(i_1, \ldots, i_k)$ of positive integers such that $1 \leq i_1 \leq \cdots \leq i_k \leq n$. In particular,

$$\sum_{i=1}^{n}(-1)^{i+1}\frac{1}{i}\binom{n}{i} = H_n.$$

xxiv) Let $n \geq 0$, $0 \leq k \leq n$, and $x \in \mathbb{C}$, and assume that $-x \notin \{0, \ldots, n\}$. Then,

$$\sum_{i=0}^{n}(-1)^i\frac{i^k}{i+x}\binom{n}{i} = (-1)^k\frac{n!x^k}{\overline{x^{n+1}}}, \quad \sum_{i=0}^{n}(-1)^i\frac{1}{i+x}\binom{n}{i} = \frac{n!}{\overline{x^{n+1}}} = \frac{1}{x}\prod_{i=1}^{n}\frac{i}{x+i} = \frac{1}{x\binom{x+n}{n}} = \frac{1}{(n+1)\binom{x+n}{n+1}},$$

$$\sum_{i=0}^{n}(-1)^i\frac{x}{i+x}\binom{n}{i} = \frac{1}{\binom{n+x}{n}}, \quad \sum_{i=0}^{n}(-1)^i\binom{n}{i}\left(\frac{x}{x+i}\right)^2 = \left(\prod_{i=1}^{n}\frac{i}{x+i}\right)\left(1+\sum_{i=1}^{n}\frac{x}{x+i}\right).$$

If $n \ge 2$, then

$$\sum_{i=1}^{n}(-1)^{n-i+1}\frac{i^{n-2}}{x+i}\binom{n-1}{i-1} = \frac{(n-1)!x^{n-2}}{\prod_{i=1}^{n}(x+i)}.$$

xxv) Let $n \ge 1$. Then,

$$\sum_{i=0}^{n}(-1)^i\frac{1}{i+1}\binom{n}{i} = \frac{1}{n+1}, \quad \sum_{i=1}^{n}(-1)^{i+1}\frac{1}{i+1}\binom{n}{i} = \frac{n}{n+1}.$$

xxvi) Let $n \ge 0$. Then,

$$\sum_{i=0}^{n}(-1)^i\frac{1}{2i+1}\binom{n}{i} = \frac{4^n}{(2n+1)\binom{2n}{n}} = \frac{(2n)!!}{(2n+1)!!} = \frac{4^n(n!)^2}{(2n+1)!}.$$

xxvii) Let $n, m \ge 1$. Then,

$$\sum_{i=0}^{n}(-1)^i\frac{1}{m+i+1}\binom{n}{i} = \frac{1}{(n+m+1)\binom{n+m}{m}} = \frac{n!m!}{(n+m+1)!}.$$

xxviii) Let $n \ge 0$. Then,

$$\sum_{i=0}^{n}(-1)^{n+i}\frac{1}{2n-2i+1}\binom{2n+1}{i} = \frac{2^{4n}}{(n+1)\binom{2n+1}{n}}.$$

xxix) Let $n \ge 1$. Then,

$$\sum_{i=1}^{n}(-1)^{i-1}\frac{1}{i}\binom{n}{i} = H_n, \quad \sum_{i-1}^{n}(-1)^{i-1}\frac{1}{i+1}\binom{n}{i} = \frac{n}{n+1}, \quad \sum_{i-1}^{n}(-1)^{i-1}\frac{H_i}{i+1}\binom{n}{i} = \frac{H_n}{n+1},$$

$$\sum_{i=1}^{n}(-1)^{i-1}\frac{1}{i+2}\binom{n}{i} = \frac{n(n+3)}{2(n+1)(n+2)}, \quad \sum_{i=1}^{n}(-1)^{i-1}\frac{1}{i+3}\binom{n}{i} = \frac{n(n^2+6n+11)}{3(n+1)(n+2)(n+3)}.$$

xxx) Let $n \ge 0$ and $m \ge 1$. Then,

$$\sum_{i=0}^{n}(-1)^i\frac{1}{(i+m)^2}\binom{n}{i} = \frac{(H_{n+m}-H_{m-1})}{m\binom{n+m}{n}}.$$

In particular,

$$\sum_{i=0}^{n}(-1)^i\frac{1}{(i+1)^2}\binom{n}{i} = \frac{H_{n+1}}{n+1}, \quad \sum_{i=0}^{n}(-1)^i\frac{1}{(i+2)^2}\binom{n}{i} = \frac{H_{n+2}-1}{(n+1)(n+2)},$$

$$\sum_{i=0}^{n}(-1)^i\frac{1}{(i+3)^2}\binom{n}{i} = \frac{2H_{n+3}-3}{(n+1)(n+2)(n+3)}.$$

Furthermore,

$$\sum_{i=0}^{n}(-1)^i\frac{1}{(i+m)^3}\binom{n}{i} = \frac{(H_{n+m}-H_{m-1})^2+H_{n+m,2}-H_{m-1,2}}{2m\binom{n+m}{n}}, \quad \sum_{i=0}^{n}(-1)^i\frac{1}{(i+m)^4}\binom{n}{i}$$

$$= \frac{(H_{n+m}-H_{m-1})^3+3(H_{n+m}-H_{m-1})(H_{n+m,2}-H_{m-1,2})+2(H_{n+m,3}-H_{m-1,3})}{6m\binom{n+m}{n}}.$$

xxxi) Let $n \geq 1$, $k \in \{0, 1, \ldots, n\}$, and $x, y \in \mathbb{C}$. Then,

$$\sum_{i=0}^{n} (-1)^i \binom{n}{i} (ix + y)^k = \begin{cases} 0, & 0 \leq k \leq n-1, \\ n!(-x)^n, & k = n. \end{cases}$$

xxxii) Let $n \geq 1$ and $k \in \{1, \ldots, n\}$, and assume that n and k are odd. Then,

$$\sum_{i=0}^{(n-1)/2} (-1)^i \binom{n}{i} (n - 2i)^k = \begin{cases} 0, & k < n, \\ 2^{n-1} n!, & k = n. \end{cases}$$

xxxiii) Let $n \geq 2$ and $k \in \{2, \ldots, n\}$, and assume that n and k are even. Then,

$$\sum_{i=0}^{(n-2)/2} (-1)^i \binom{n}{i} (n - 2i)^k = \begin{cases} 0, & k < n, \\ 2^{n-1} n!, & k = n. \end{cases}$$

xxxiv) Let $k, n \geq 1$ and $x \in \mathbb{C}$. Then,

$$\sum_{i=0}^{n} (-1)^i (x - ki)^n \binom{n}{i} = k^n n!.$$

xxxv) Let $n \geq 1$, let $p \in \mathbb{F}[s]$, assume that $\deg p \leq n$, and let $\alpha \in \mathbb{C}$ be the coefficient of s^n in $p(s)$. Then,

$$\sum_{i=1}^{n} (-1)^i \binom{n}{i} p(i) = \alpha (-1)^n n!.$$

If, in particular, $\deg p < n$, then,

$$\sum_{i=1}^{n} (-1)^i \binom{n}{i} p(i) = 0.$$

xxxvi) Let $n \geq 1$ and $k \in \{0, 1, \ldots, n\}$. Then,

$$\sum_{i=0}^{n} (-1)^i i^k \binom{n}{i} = \begin{cases} 0, & k < n, \\ (-1)^n n!, & k = n. \end{cases}$$

xxxvii) Let $n, k \geq 1$, and assume that n is prime. Then,

$$\sum_{i=1}^{n-1} (-1)^i i^k \binom{n-1}{i} \overset{n}{\equiv} \begin{cases} -1, & n - 1 | k, \\ 0, & \text{otherwise}. \end{cases}$$

xxxviii) Let $n \geq 1$ and $x \in \mathbb{C}$. Then,

$$\sum_{i=1}^{n} H_i \binom{n}{i} x^{n-i} = H_n (x+1)^n - \sum_{i=1}^{n} \frac{1}{i} x^i (x+1)^{n-i}.$$

In particular,

$$\sum_{i=1}^{n} (-1)^{i+1} H_i \binom{n}{i} = \frac{1}{n}, \quad \sum_{i=1}^{n} H_i \binom{n}{i} = 2^n H_n - \sum_{i=1}^{n} \frac{2^{n-i}}{i},$$

$$\sum_{i=1}^{n} (-1)^i 2^{n-i} H_i \binom{n}{i} = H_n - \sum_{i=1}^{n} \frac{2^i}{i}, \quad \sum_{i=1}^{n} (-1)^{n-i} 2^i H_i \binom{n}{i} = H_n - \sum_{i=1}^{n} (-1)^i \frac{1}{i}.$$

xxxix) Let $n \ge 1$ and $1 \le k \le \lfloor n/2 \rfloor$. Then,

$$\sum_{i=0}^{\lfloor n/2 \rfloor} (-1)^i \binom{n-i}{i} C_{n-k-1} = 0.$$

xl) Let $n \ge 1$ and $k \in \{0, 1, \ldots, n\}$. Then,

$$\sum_{i=k}^{n} (-1)^i \binom{n}{i}\binom{i}{k} = \begin{cases} 0, & k < n, \\ (-1)^n, & k = n. \end{cases}$$

xli) Let $n \ge 1$ and $k \in \{0, 1, \ldots, n\}$. Then,

$$\sum_{i=0}^{n} (-1)^i \binom{n}{i} (n-i)^k = \begin{cases} 0, & k < n, \\ n!, & k = n. \end{cases}$$

xlii) Let $n, m \ge 1$ and $k \in \{0, 1, \ldots, n-1\}$. Then,

$$\sum_{i=0}^{n} (-1)^i \binom{n}{i} (mi)^k = 0.$$

xliii) Let $n \ge 0$. Then,

$$\sum_{i=0}^{n} (-1)^i \frac{1}{\binom{n}{i}} = \begin{cases} 0, & n \text{ odd}, \\ \dfrac{2n+2}{n+2}, & n \text{ even}, \end{cases} \qquad \sum_{i=0}^{2n} (-1)^i \frac{1}{\binom{2n}{i}} = \frac{2n+1}{n+1}.$$

xliv) Let $n \ge 1$. Then,

$$\sum_{i=1}^{n} (-1)^{i+1} \frac{1}{\binom{2n}{i}} = \frac{1}{2(n+1)} + \frac{(-1)^{i+1}}{2\binom{2n}{n}}, \qquad \sum_{i=1}^{2n-1} (-1)^{i+1} \frac{i}{\binom{2n}{i}} = \frac{n}{n+1}.$$

xlv) Let $n \ge 0$. Then,

$$\sum_{i=0}^{2n} (-1)^i \frac{\binom{2n}{i}}{\binom{4n}{2i}} = \frac{4n+1}{2n+1}, \qquad \sum_{i=0}^{2n} (-1)^i \frac{\binom{4n}{2i}}{\binom{2n}{i}} = \frac{1}{1-2n}, \qquad \sum_{i=0}^{n} (-1)^i \frac{4^i \binom{n}{i}}{\binom{2i}{i}} = \frac{1}{1-2n}.$$

xlvi) Let $n \ge 1$. Then,

$$\sum_{i=1}^{n} (-1)^i \frac{4^i \binom{n}{i}}{i \binom{2i}{i}} = H_n - 2H_{2n}, \qquad \sum_{i=1}^{n} (-1)^i \frac{4^i \binom{n-1}{i-1}}{i^2 \binom{2i}{i}} = \frac{1}{n}(H_n - 2H_{2n}).$$

xlvii) Let $n \ge 0$. Then,

$$\sum_{i=0}^{n} (-1)^i \binom{n}{i} (H_i - 2H_{2i}) = \frac{4^n}{n\binom{2n}{n}}.$$

xlviii) Let $n \ge 1$ and $k \ge 2$. Then,

$$\sum_{i=0}^{n} (-1)^i \frac{\binom{kn}{i}}{\binom{2kn}{2i}} = \begin{cases} 0, & kn \text{ odd}, \\ \dfrac{2kn+1}{2(kn+1)}, & kn \text{ even}, \end{cases} \qquad \sum_{i=0}^{n} (-1)^i \frac{i\binom{kn}{i}}{\binom{2kn}{2i}} = \begin{cases} -\dfrac{(kn+1)(2kn+1)}{2(kn+2)}, & kn \text{ odd}, \\ \dfrac{kn(2kn+1)}{2(kn+1)}, & kn \text{ even}. \end{cases}$$

xlix) Let $n, k \ge 1$. Then,

$$\sum_{i=0}^{n} (-1)^i \frac{1}{\binom{n+k}{i+k}} = \frac{n+k+1}{n+k+2} \left[\binom{n+k+1}{k}^{-1} + (-1)^n \right],$$

$$\sum_{i=0}^{n}(-1)^i\frac{i}{\binom{n+k}{i+k}} = (-1)^n(n+1)\frac{n+k+1}{n+k+2} - \frac{n+k+1}{n+k+3}\left[\binom{n+k+2}{k+1}^{-1} + (-1)^n\right].$$

l) Let $n \geq 1$. Then,

$$\sum_{i=0}^{n}(-1)^i\sum_{j=1}^{n}\frac{1}{i+j}\binom{n}{i} = \frac{1}{n}\left[1-\binom{2n}{n}^{-1}\right].$$

li) Let $n \geq 1$. Then,

$$\sum_{i=0}^{n}(-1)^i\binom{n}{i}\frac{1}{(i+k)^2} = \frac{(k-1)!}{(n+1)^{\overline{k}}}(H_{n+k}-H_{k-1}), \qquad \frac{(k-1)!}{(n+1)^{\overline{k}}} = \sum_{i=1}^{k}(-1)^{i-1}\frac{1}{n+i}\binom{k-1}{i-1}.$$

In particular,

$$\sum_{i=1}^{n}(-1)^i\binom{n}{i}\frac{1}{(i+1)^2} = \frac{H_{n+1}}{n+1}, \qquad \sum_{i=1}^{n}(-1)^i\binom{n}{i}\frac{1}{(i+2)^2} = \frac{H_{n+2}-1}{(n+1)(n+2)},$$

$$\sum_{i=1}^{n}(-1)^i\binom{n}{i}\frac{1}{(i+3)^2} = \frac{2H_{n+3}-3}{(n+1)(n+2)(n+3)}.$$

lii) Let $n, k \geq 1$. Then,

$$\sum_{i=1}^{n}(-1)^i\binom{n}{i}\frac{H_i}{\binom{k+i}{i}} = \frac{k}{n+k}(H_{k-1}-H_{n+k-1}), \qquad \sum_{i=1}^{n}(-1)^i\binom{n}{i}\frac{iH_i}{\binom{k+i}{i}} = \frac{nk(H_{n+k-2}-H_{k-1}-1)}{(n+k)(n+k-1)},$$

$$\sum_{i=1}^{n}(-1)^i\binom{n}{i}\frac{i^2H_i}{\binom{k+i}{i}} = \frac{nk[2n-k+(k-n)(H_{n+k-3}-H_{k-1})}{(n+k)(n+k-1)(n+k-2)}.$$

liii) Let $n, k \geq 1$. Then,

$$\sum_{i=1}^{n}(-1)^i\binom{n}{i}\frac{H_{k+i}-H_k}{\binom{k+i}{i}} = \frac{kH_{n+k}}{n+k}, \qquad \sum_{i=1}^{n}(-1)^i\binom{n}{i}\frac{i(H_{k+i}-H_i)}{\binom{k+i}{i}} = \frac{nk(1-H_{n+k})}{(n+k)(n+k-1)},$$

$$\sum_{i=1}^{n}(-1)^i\binom{n}{i}\frac{i(H_{k+i}-H_k)}{\binom{k+i}{i}} = \frac{n(n^2-n-k^2)}{(n+k)^2(n+k-1)^2},$$

$$\sum_{i=1}^{n}(-1)^i\binom{n}{i}\frac{i^2(H_{k+i}-H_i)}{\binom{k+i}{i}} = \frac{nk(1+k-2n)+nk(n-k)H_{n+k}}{(n+k)(n+k-1)(n+k-2)}.$$

liv) Let $n \geq 1$. Then,

$$\sum_{i=1}^{n}(-1)^{i-1}\frac{\binom{n}{i}}{i^2\binom{n+i}{n}} = \sum_{i=1}^{n}(-1)^{i-1}\frac{\binom{2n}{n+i}}{i^2\binom{2n}{n}} = \tfrac{1}{2}H_{n,2}.$$

lv) Let $n \geq 1$. Then,

$$\sum_{i=1}^{n}(-1)^{i-1}\frac{1}{i}\binom{n}{i}\sum_{j=1}^{i}\frac{H_j}{j} = 3\sum_{i=1}^{n}(-1)^{i-1}\frac{(n!)^2\binom{n}{i}}{i^3\prod_{j=1}^{n}(i^2+ij+j^2)} = H_{n,3}.$$

lvi) Let $n \geq 1$ and $k \geq 1$. Then,

$$\sum_{i=1}^{n}(-1)^{i-1}\frac{\binom{n}{i}}{i^k\prod_{j=1}^{n}\left(\sum_{l=0}^{k-1}(j/i)^l\right)} = \tfrac{1}{k}H_{n,k}.$$

lvii) Let $n \geq 0$. Then,

$$\sum_{i=0}^{n}(-1)^i \frac{1}{1-2i}\binom{n}{i} = \frac{4^n}{\binom{2n}{n}}.$$

lviii) Let $n \geq 1$. Then,

$$\sum_{i=1}^{n}(-1)^{i-1} \frac{i}{(2i-1)^2}\binom{n}{i} = \frac{4^n}{2\binom{2n}{n}} \sum_{i=1}^{n} \frac{1}{2i-1}.$$

lix) Let $n \geq 1$. Then,

$$\sum_{i=1}^{n}(-1)^{i-1} \frac{n+i}{(2i-1)^3}\binom{2n}{n+i} = \frac{16^n}{4\binom{2n}{n}} \sum_{i=1}^{n} \frac{1}{(2i-1)^2}.$$

lx) Let $n \geq 1$ and $z \in \mathbb{C}$. Then,

$$\sum_{i=1}^{n}(-1)^{i+1}\binom{n}{i}(1-z^i) = (1-z)^n.$$

lxi) Let $m \geq 0$, $k \geq 1$, and $n \geq km+1$. Then,

$$\sum_{i=1}^{n-m}(-1)^{i+1}\binom{n}{i}\binom{n-i}{m}^k = \binom{n}{m}^k.$$

lxii) Let $n \geq 1$. If $0 \leq k \leq n-1$, then

$$\sum_{i=k+1}^{n}(-1)^{k+i-1} \frac{1}{i-k}\binom{n}{i} = \binom{n}{k}(H_n - H_k).$$

If $1 \leq k \leq n$, then

$$\sum_{i=0}^{k-1}(-1)^{k+i-1} \frac{1}{k-i}\binom{n}{i} = \binom{n}{k}(H_n - H_{n-k}), \quad \sum_{i=0}^{k-1}\sum_{j=k}^{n}(-1)^{i+j-1} \frac{1}{j-i}\binom{n}{i}\binom{n}{j} = \sum_{i=0}^{k-1}\binom{n}{i}^2 (H_{n-i} - H_i).$$

lxiii) Let $n \geq 1$. Then,

$$\sum_{i=1}^{n}(-1)^i\binom{n}{i}(H_i^2 - 3H_{i,2}) = \frac{2}{n}(H_n + H_{n-1}),$$

$$\sum_{i=1}^{n}(-1)^i\binom{n}{i}(H_i^3 - 9H_iH_{i,2} + 14H_{i,3}) = -\frac{3}{n}[(H_n + H_{n-1})^2 + H_{n,2} - H_{n-1,2}].$$

Source: *i)* and *ii)* are given in [1158, p. 295]; *iii)* is given in [1675, p. 84] and [2015]; *iv)*, *vii)*–*x)* are given in [1524, pp. 35, 36] and [1217, pp. 4, 5]; *v)* is given in [2228, p. 133]; *vi)* is given in [1675, pp. 40–42]; the first equality *xi)* and *xii)* are given in [1371, p. 75]; the second equality in *xi)* is given in [2880, pp. 54, 55]; *xiii)*, *xvii)*, and *xviii)* are given in [411, pp. 85, 86, 89, 90]; *xiv)* is given in [596] and [771, p. 168]; *xv)* is given in [1757, pp. 183–185]; *xvi)* is given in [2241]; *xix)* is given in [511, pp. 48–52]; *xx)* follows from Fact 2.2.3; *xxi)* and *xxii)* are given in [1219, pp. 280–282]; *xxiii)* is given in [554, 900], [1217, p. 5], and [1757, pp. 160, 161]; *xxiv)* is given in [737, 1104, 2226] and [2228, p. 31]; *xxvi)* is given in [511, p. 166]; *xxvii)* is given in [2508, 2577]; *xxviii)* is given in [756]; *xxix)* is given in [2513, p. 254]; *xxx)* is given in [739, 1752] and [1757, p. 163]; *xxxi)*–*xxxiii)* are given in [1592]; *xxxiv)* is given in [125]; *xxxv)* is given in [1034, 1197, 1592]; *xxxvi)* is given in [1034]; *xxxvii)* is given in [155, p. 275]; *xxxviii)* is given in [556, 2100]; *xxxix)* is given in [1675, p. 131]; *xl)*–*xliii)* are given in [1034]; *xliv)* is given in [2511]; *xlv)* is given in [2509, 2728]; *xlvi)* and *xlvii)* are given in [2509]; *l)* is given in [2577]; *xlix)* is given in [2577, 2728]; *l)* is given in [1675, p.

38]; *li)*–*liii)* are given in [739]; *liv)* is given in [1672]; *lv)* is given in [1273, 1672]; *lvi)* is given in [1672]; *lvii)* is given in [2509]; *lviii)* and *lix)* are given in [1672]; *lx)* is given in [887]; *lxi)* is given in [3]; *lxii)* is given in [2492] and [2513, p. 254]; *lxiii)* is given in [2830]. **Related:** Fact 14.2.32.

Fact 1.16.12. The following statements hold:

i) Let $n \ge 0$ and $l \ge 1$, and define $\omega \triangleq e^{(2\pi/l)J}$. Then,

$$\sum_{i=0}^{\lfloor n/l \rfloor} \binom{n}{li} = \frac{1}{l} \sum_{i=0}^{l-1} (1+\omega^i)^n.$$

ii) Let $n \ge 0$. Then,

$$\sum_{i=0}^{\lfloor n/2 \rfloor} \binom{n}{2i} = 2^{n-1}, \quad \sum_{i=0}^{\lfloor n/3 \rfloor} \binom{n}{3i} = \frac{1}{3}\left(2^n + 2\cos\frac{n\pi}{3}\right), \quad \sum_{i=0}^{\lfloor n/4 \rfloor} \binom{n}{4i} = \frac{1}{2}\left(2^{n-1} + 2^{n/2}\cos\frac{n\pi}{4}\right).$$

iii) Let $k, l, n \ge 0$, assume that $k < l$, and define $\omega \triangleq e^{(2\pi/l)J}$. Then,

$$\sum_{i=0}^{\lfloor (n-k)/l \rfloor} \binom{n}{li+k} = \frac{1}{l} \sum_{i=0}^{l-1} \omega^{-ik}(1+\omega^i)^n.$$

iv) Let $n \ge 0$. Then,

$$\sum_{i=0}^{\lfloor (n-1)/2 \rfloor} \binom{n}{2i+1} = 2^{n-1}, \quad \sum_{i=0}^{\lfloor (n-1)/3 \rfloor} \binom{n}{3i+1} = \frac{1}{3}\left(2^n + 2\cos\frac{(n-2)\pi}{3}\right),$$

$$\sum_{i=0}^{\lfloor (n-2)/3 \rfloor} \binom{n}{3i+2} = \frac{1}{3}\left(2^n + 2\cos\frac{(n-4)\pi}{3}\right), \quad \sum_{i=0}^{\lfloor (n-1)/4 \rfloor} \binom{n}{4i+1} = \frac{1}{2}\left(2^{n-1} + 2^{n/2}\sin\frac{n\pi}{4}\right),$$

$$\sum_{i=0}^{\lfloor (n-2)/4 \rfloor} \binom{n}{4i+2} = \frac{1}{2}\left(2^{n-1} - 2^{n/2}\cos\frac{n\pi}{4}\right), \quad \sum_{i=0}^{\lfloor (n-3)/4 \rfloor} \binom{n}{4i+3} = \frac{1}{2}\left(2^{n-1} - 2^{n/2}\sin\frac{n\pi}{4}\right).$$

v) Let $n \ge 1$. Then,

$$\sum_{i=0}^{\lfloor n/3 \rfloor} \frac{2^i n}{n-i}\binom{n-i}{2i} = 2^{n-1} + \cos\frac{n\pi}{2}.$$

vi) Let $n, k \ge 1$. Then,

$$\sum_{\substack{0 \le i \le n \\ k|2i-n}} \binom{n}{i} = \frac{2^n}{k} \sum_{i=0}^{k-1} \cos^n \frac{2i\pi}{k}.$$

vii) Let $n, k \ge 1$, and assume that k divides n. Then,

$$\sum_{i=0}^{n} \binom{n}{ki} = \frac{2^n}{k} \sum_{i=0}^{k-1} (-1)^{ni/k} \cos^n \frac{i\pi}{k}.$$

Source: [408], [1524, p. 35], [1702], and [2228, p. 104]. **Remark:** If $l = 1$, then $\omega = 1$ and *i)* yields *v)* of Fact 1.16.10. If $l = 2$, then $\omega = -1$ and *i)* yields the first equality in *ii)*. If $l = 3$, then $\omega = \frac{1}{2}(-1 + \sqrt{3}J)$ and *i)* yields the second equality in *ii)*. For example,

$$\binom{8}{0} + \binom{8}{3} + \binom{8}{6} = 1 + 56 + 28 = \tfrac{1}{3}(2^8 - 1) = 85.$$

If $l = 4$, then $\omega = \jmath$, and *i*) yields the third equality in *ii*). For example,

$$\binom{13}{0} + \binom{13}{4} + \binom{13}{8} + \binom{13}{12} = 1 + 715 + 1287 + 13 = \tfrac{1}{2}(2^{12} - 2^6) = 2016.$$

Remark: The second equality in *ii*) can be written as

$$\sum_{i=0}^{\lfloor n/3 \rfloor} \binom{n}{3i} = \tfrac{1}{3}(2^n + m),$$

where $m = 2, 1, -1, -2, -1, 1$ correspond to $n \overset{6}{\equiv} 0, 1, 2, 3, 4, 5$, respectively. Likewise,

$$\sum_{i=0}^{\lfloor n/4 \rfloor} \binom{n}{4i} = \tfrac{1}{4}(2^n + m2^{\lceil n/2 \rceil}),$$

where $m = 3, 1, 0, -1, -2, -1, 0, 1$ correspond to $n \overset{8}{\equiv} 0, 1, 2, 3, 4, 5, 6, 7$, respectively. See [412].

Fact 1.16.13. The following statements hold:

i) Let $k, m \geq 0$ and $x \in \mathbb{C}$. Then,

$$\binom{x}{k}\binom{k}{m} = \binom{x}{m}\binom{x-m}{k-m}.$$

ii) Let $k, m \geq 0$ and $x \in \mathbb{C}$. Then,

$$\binom{x}{k}\binom{x-k}{m} = \binom{x}{m}\binom{x-m}{k}.$$

iii) Let $k, m \geq 0$ and $x \in \mathbb{C}$. Then,

$$\binom{k}{m}\binom{x-m}{k} = \binom{x-m}{k-m}\binom{x-k}{m}.$$

iv) Let $k, l, m, n \geq 0$. Then,

$$\sum_{i=l-m}^{n-k} \binom{n}{k+i}\binom{m}{l-i} = \binom{n+m}{k+l}, \qquad \sum_{i=\max\{-k,-l\}}^{\min\{n-k,m-l\}} \binom{n}{k+i}\binom{m}{l+i} = \binom{n+m}{n-k+l}.$$

In particular,

$$\sum_{i=1}^{n} \binom{n}{i}\binom{n}{i-1} = \binom{2n}{n+1}.$$

v) Let $m, n \geq 0$ and $0 \leq k \leq m$. Then,

$$\sum_{i=0}^{\min\{n,k\}} \binom{n}{i}\binom{m}{k-i} = \binom{n+m}{k}.$$

vi) Let $k, m, n \geq 0$. If $k \leq m$, then

$$\sum_{i=0}^{\min\{n,k\}} i\binom{n}{i}\binom{m}{k-i} = n\binom{n+m-1}{k-1}.$$

If $k \leq m-1$ and $n \leq m$, then

$$\sum_{i=\max\{1,n+k-m\}}^{\max\{n,k\}} i\binom{n}{i}\binom{m-n}{k-i} = \frac{kn}{m-k}\binom{m-1}{k}.$$

vii) Let $n, k \ge 0$ and $x \in \mathbb{C}$. Then,

$$\sum_{i=0}^{n} \binom{n}{i}\binom{x}{k+i} = \binom{n+x}{n+k}.$$

viii) Let $n, m \ge 0$. Then,

$$\sum_{i=0}^{\min\{n,m\}} \binom{n}{i}\binom{m}{i} = \binom{n+m}{n}.$$

In particular,

$$\sum_{i=0}^{n} \binom{n}{i}^2 = \binom{2n}{n}, \quad \sum_{i=0}^{n} \binom{n}{i}\binom{2n}{i} = \binom{3n}{n}.$$

ix) Let $n \ge 0$. Then,

$$\sum_{i=0}^{n} \binom{n}{i}\binom{2n}{n-i} = \binom{3n}{n}.$$

x) Let $n \ge 0$ and $x, y \in \mathbb{C}$. Then,

$$\sum_{i=0}^{n} \binom{x}{i}\binom{y}{n-i} = \binom{x+y}{n}.$$

xi) Let $l, m \ge 0$, and assume that $n \ge k \ge 0$. Then,

$$\sum_{i=0}^{l} \binom{l-i}{m}\binom{k+i}{n} = \binom{l+k+1}{m+n+1}.$$

xii) Let $0 \le k \le n/2$. Then,

$$\sum_{i=k}^{n-k} \binom{i}{k}\binom{n-i}{k} = \binom{n+1}{2k+1}.$$

xiii) Let $n, m, k \ge 0$. Then,

$$\sum_{i=0}^{k} \binom{n+k-i}{n}\binom{m+i}{m} = \binom{n+m+k+1}{k}.$$

xiv) Let $0 \le k \le n$. Then,

$$\sum_{i=0}^{n-k} \binom{n}{i}\binom{n}{i+k} = \sum_{i=k}^{n} \binom{n}{i}\binom{n}{i-k} = \binom{2n}{n+k} = \frac{(2n)!}{(n-k)!(n+k)!}.$$

xv) Let $n \ge 0$ and $1 \le l \le k$. Then,

$$\sum_{i=l}^{n+l-k} \binom{i-1}{l-1}\binom{n-i}{k-l} = \binom{n}{k}.$$

xvi) Let $0 \le k \le n$. Then,

$$\sum_{i=0}^{k} \binom{n}{i}\binom{n-i}{k-i} = 2^k\binom{n}{k}.$$

xvii) Let $0 \le k \le n$. Then,

$$\sum_{i=k}^{n} \binom{n}{i}\binom{i}{k} = 2^{n-k}\binom{n}{k}.$$

xviii) Let $n \geq 1$ and $0 \leq k \leq \lfloor n/2 \rfloor$. Then,

$$\sum_{i=k}^{\lfloor n/2 \rfloor} \binom{n}{2i}\binom{i}{k} = \frac{2^{n-2k-1}n}{n-k}\binom{n}{k}.$$

xix) Let $0 \leq k \leq n-1$. Then,

$$\sum_{i=\lceil k/2 \rceil}^{\lfloor n/2 \rfloor} \binom{n}{2i}\binom{2i}{k} = 2^{n-k-1}\binom{n}{k}.$$

xx) Let $n, k \geq 1$. If $k \leq n$, then

$$\sum_{i=0}^{k} 2^{2i}\binom{n}{k-i}\binom{n-k+i}{2i} = \binom{2n}{2k}.$$

If $2k+1 \leq n$, then

$$\sum_{i=0}^{k} 2^{2i+1}\binom{n}{k-i}\binom{n-k+i}{2i+1} = \binom{2n}{2k+1}.$$

xxi) Let $n \geq 1$ and $x \in \mathbb{C}$. Then,

$$\sum_{i=0}^{\lfloor x/2 \rfloor} 2^{2i+1}\binom{x-2i}{n-i}\binom{x+1}{2i+1} = \binom{2x+2}{2n+1}.$$

xxii) Let $n \geq 1$ and $x \in \mathbb{C}$. Then,

$$\sum_{i=1}^{n} i\binom{n}{i}\binom{x}{i} = n\binom{n+x-1}{n}.$$

xxiii) Let $n \geq 1$. Then,

$$\sum_{i=1}^{n} i\binom{n}{i}^2 = \frac{n}{2}\binom{2n}{n} = n\binom{2n-1}{n-1} = (2n-1)\binom{2n-2}{n-1}.$$

xxiv) Let $n \geq 1$. Then,

$$\sum_{i=1}^{n} i^2\binom{n}{i}^2 = \frac{n^3}{2n-1}\binom{2n-1}{n-1}.$$

xxv) Let $n \geq 0$. Then,

$$\sum_{i=0}^{n} \frac{1}{4^{2i}(i+1)}\binom{2i}{i}^2 = \frac{(2n+1)^2}{4^{2n}(n+1)}\binom{2n}{n}^2.$$

xxvi) Let $n \geq 0$. Then,

$$\sum_{i=0}^{n} \frac{(4i-1)}{4^{2i}(2i-1)^2}\binom{2i}{i}^2 = \frac{(4n-1)}{4^{2n}(4n-1)}\binom{2n}{n}^2.$$

xxvii) Let $n \geq 1$. Then,

$$\sum_{i=0}^{n} \frac{\binom{n}{i}^2}{(2i+1)\binom{2n}{2i}} = \frac{2^{4n}(n!)^4}{(2n+1)!(2n)!}.$$

xxviii) Let $m, n \geq 1$. Then,

$$\sum_{i=0}^{\min\{m,n\}} 4^i \frac{\binom{m}{i}\binom{n}{i}}{\binom{2i}{i}} = \frac{\binom{2m+2n}{2m}}{\binom{m+n}{m}}.$$

In particular,

$$\sum_{i=0}^{n} 4^i \frac{\binom{n}{i}^2}{\binom{2i}{i}} = \frac{\binom{4n}{2n}}{\binom{2n}{n}}.$$

xxix) Let $n, k \geq 1$. Then,

$$\sum_{i=0}^{n} \binom{2i}{i}\binom{2n-2i}{n-i} = \sum_{i=0}^{n-m} \binom{2i+k}{i}\binom{2n-2i-k}{n-i} = 4^n, \quad \sum_{i=1}^{n} i^2 \binom{2i}{i}\binom{2n-2i}{n-i} = \frac{4^{n-1}}{2} n(3n+1),$$

$$\sum_{i=0}^{n} \binom{4i}{2i}\binom{4n-4i}{2n-2i} = 2^{4n-1} + 2^{2n-1}\binom{2n}{n}.$$

xxx) Let $n, k \geq 1$. Then,

$$\sum_{i=1}^{n} i\binom{n}{i}\binom{k}{k-i} = n\binom{n+k-1}{k-1},$$

$$\sum_{i=1}^{\min\{n,k\}} \binom{2n}{n-i}\binom{2k}{k-i} = \frac{1}{2}\left[\binom{2n+2k}{n+k} - \binom{2n}{n}\binom{2k}{k}\right],$$

$$\sum_{i=1}^{\min\{n,k\}} i\binom{2n}{n-i}\binom{2k}{k-i} = \frac{nk}{2(n+k)}\binom{2n}{n}\binom{2k}{k},$$

$$\sum_{i=0}^{\min\{n,k\}} (2i+1)\binom{2n+1}{n-i}\binom{2k+1}{k-i} = \frac{(2n+1)(2k+1)}{n+k+1}\binom{2n}{n}\binom{2k}{k}.$$

In particular,

$$\sum_{i=1}^{n} \binom{2n}{n-i}^2 = \frac{1}{2}\left[\binom{4n}{2n} - \binom{2n}{n}^2\right].$$

xxxi) Let $n \geq 0$. Then,

$$\sum_{i=0}^{n} \frac{1}{2i+1}\binom{2i}{i}\binom{2n-2i}{n-i} = \frac{16^n}{(2n+1)\binom{2n}{n}},$$

$$\sum_{i=0}^{n} \frac{1}{(2i+1)(2n-2i+1)}\binom{2i}{i}\binom{2n-2i}{n-i} = \frac{16^n}{(n+1)(2n+1)\binom{2n}{n}}.$$

xxxii) Let $n \geq 1$ and $0 \leq k \leq n$. Then,

$$\sum_{i=k}^{n} \frac{1}{i+1}\binom{2i}{i}\binom{2n-2i}{n-i} = \frac{n-k+1}{2(n+1)}\binom{2k}{k}\binom{2n+2-2k}{n+1-k}.$$

In particular,

$$\sum_{i=0}^{n} \frac{1}{i+1}\binom{2i}{i}\binom{2n-2i}{n-i} = \binom{2n+1}{n}.$$

xxxiii) Let $n \geq 0$. Then,

$$\sum_{i=0}^{\lfloor n/2 \rfloor} 2^{n-2i}\binom{2i}{i}\binom{n}{2i} = \sum_{i=0}^{\lfloor n/2 \rfloor} 2^{n-2i}\binom{n}{i}\binom{n-i}{i} = \binom{2n}{n}.$$

xxxiv) Let $n \geq 1$. Then,

$$\sum_{i=0}^{\lfloor n/2 \rfloor} \frac{1}{4^i}\binom{2i}{i}\binom{n}{2i} = \frac{1}{2^{n-1}}\binom{2n-1}{n-1}.$$

xxxv) Let $n \geq 0$ and $x, y, z \in \mathbb{C}$. Then,

$$\sum_{i=0}^{n} \frac{x}{x+iz}\binom{x+iz}{i}\binom{y-iz}{n-i} = \binom{x+y}{n}.$$

xxxvi) $n \geq k \geq 0$. Then,

$$\sum_{i=0}^{n-1} \frac{1}{i+1}\binom{2i}{i}\binom{n+k-2i-1}{n-i-1} = \binom{n+k}{n-1}.$$

xxxvii) $n \geq 1$. Then,

$$\sum_{i=1}^{n} \frac{n+1}{i(n-i+1)}\binom{2i-2}{i-1}\binom{2n-2i}{n-i} = \binom{2n}{n}.$$

xxxviii) Let $n \geq 1$. Then,

$$\sum_{i=0}^{n-1}\sum_{j=i+1}^{n+1} \binom{n}{i}\binom{n+1}{j} = 4^n - 1.$$

xxxix) Let $n \geq 1$. Then,

$$\sum_{i=0}^{n} 2^{n-i}\binom{n}{i}\binom{i}{\lfloor i/2 \rfloor} = \binom{2n+1}{n}.$$

xl) Let $n \geq 1$ and $1 \leq k \leq (n-1)/2$. Then,

$$\sum_{i=k}^{(n-1)/2} \binom{n}{2i+1}\binom{i}{k} = 2^{n-2k-1}\binom{n-k-1}{k}.$$

xli) Let $n \geq 1$. Then,

$$\sum_{i=0}^{n} 2^i\binom{2n-2i}{n-i}\binom{n+i}{n} = \frac{2^n}{n!}\prod_{i=1}^{n}(4i-1).$$

xlii) Let $n \geq 1$ and $k \geq 2$. Then,

$$\sum_{i=0}^{n} \frac{1}{kn-ki+1}\binom{kn-ki+1}{i}\binom{ki}{n-i} = \frac{1}{kn+1}\binom{kn+1}{n}.$$

In particular,

$$\sum_{i=0}^{n} \frac{1}{2n-2i+1}\binom{2n-2i+1}{i}\binom{2i}{n-i} = C_n.$$

xliii) Let $n \geq 1$. Then,

$$\sum_{i=0}^{n} \binom{n}{i}^2\binom{3n+i}{2n} = \binom{3n}{n}^2.$$

xliv) Let $n \geq k \geq 1$. Then,

$$\sum_{i=0}^{k} \binom{k}{i}^2\binom{n+2k-i}{2k} = \binom{n+k}{n}^2, \quad \sum_{i=0}^{n} \binom{n}{i}^2\binom{3n-i}{2n} = \binom{2n}{n}^2.$$

xlv) Let $n \geq 1$. Then,

$$\sum_{i=0}^{n} \binom{n}{i}\sum_{j=0}^{i}\binom{i}{j}^3 = \sum_{i=0}^{n}\binom{n}{i}^2\binom{2i}{i}.$$

xlvi) Let $n \geq 1$ and $0 \leq k \leq n-1$. Then,

$$\sum_{i=0}^{k}\binom{2n}{2i+1}\binom{n-i-1}{k-i} = 2^{2k+1}\binom{n+k}{2k+1}, \quad \sum_{i=0}^{k}\binom{2n-1}{2i}\binom{n-i-1}{k-i} = 4^k\binom{n+k-1}{2k},$$

$$\sum_{i=0}^{k}\binom{2n}{2i}\binom{n-i}{k-i} = \frac{n4^k}{2k}\binom{n+k-1}{2k-1}, \quad \sum_{i=0}^{k}\binom{m+i}{i}\binom{n-i}{k-i} = \binom{n+m+1}{k},$$

$$\sum_{i=0}^{k}\binom{2n-1}{2i+1}\binom{n-i-1}{k-i} = \frac{(2n-1)4^k}{2k+1}\binom{n+k-1}{2k}.$$

xlvii) Let $n \geq 1$ and $x \in \mathbb{C}$. Then,

$$\sum_{i=0}^{n}(-1)^{n-i}\binom{n}{i}\binom{n+i}{i}(1+x)^i = \sum_{i=0}^{n}\binom{n}{i}\binom{n+i}{i}x^i,$$

$$\sum_{i=0}^{n}\binom{n}{i}\binom{n+i}{i}(x-\tfrac{1}{2})^i = (-1)^n\sum_{i=0}^{n}\binom{n}{i}\binom{n+i}{i}(-x-\tfrac{1}{2})^i.$$

xlviii) Let $k, m, n \geq 0$. Then,

$$\sum_{i=0}^{m-k+l}\binom{m-k+l}{i}\binom{n+k-l}{n-i}\binom{k+i}{m+n} = \binom{k}{m}\binom{l}{n}.$$

xlix) Let $n \geq 0$. Then,

$$\sum\binom{i+j}{i}\binom{j+k}{j}\binom{k+i}{k} = \sum_{i=0}^{n}\binom{2i}{i}, \quad \sum\binom{2i}{i}\binom{2j}{j}\binom{2k}{k} = (2n+1)\binom{2n}{n},$$

where both sums are taken over all 3-tuples (i, j, k) of nonnegative integers such that $i+j+k=n$.

l) Let $n \geq 0$. Then,

$$\sum_{i=0}^{n}\binom{n}{i}\binom{i}{\lfloor i/2\rfloor}\binom{n-i}{\lfloor (n-i)/2\rfloor} = \binom{n}{\lfloor n/2\rfloor}\binom{n+1}{\lfloor (n+1)/2\rfloor}.$$

li) Let $n \geq 0$, and define

$$S(n) \triangleq \sum_{i=0}^{n}\binom{n}{i}^3 = \sum_{i=0}^{n}\binom{n}{i}^2\binom{2i}{n}.$$

Then, for all $n \geq 2$,

$$S(n) = \frac{7n^2-7n+2}{n^2}S(n-1) + \frac{8(n-1)^2}{n^2}S(n-2).$$

In particular, $S(0) = 1$, $S(1) = 2$, $S(2) = 10$, $S(3) = 56$, $S(4) = 346$, and $S(5) = 2252$.

lii) Let $n \geq 0$. Then,

$$\sum_{i=0}^{n}\left[\sum_{j=0}^{i}\binom{n}{j}\right]^3 = n2^{3n-1} - 2^{3n} - 3n2^{n-2}\binom{2n}{n}.$$

liii) Let $n \geq 0$, and define

$$S(n) \triangleq \sum_{i=0}^{n}\binom{n}{i}^2\binom{n+i}{i}^2.$$

Then,

$$S(n) = \frac{34n^3 - 51n^2 + 27n - 5}{n^3} S(n-1) - \frac{(n-1)^3}{n^3} S(n-2).$$

In particular, $S(0) = 1$, $S(1) = 5$, $S(2) = 73$, $S(3) = 1445$, $S(4) = 33001$, and $S(5) = 819005$.

liv) Let $n \ge 1$. Then,

$$S(n) \triangleq \sum_{i=0}^{n} \binom{n}{i}^4 = \sum_{i,j=0}^{n} (-1)^{n+i+j} \binom{n}{i}\binom{n}{j}\binom{n+i}{i}\binom{n+j}{j}\binom{2n-i-j}{n}.$$

Furthermore, for all $n \ge 2$,

$$S(n) = \frac{2(2n-1)(3n^2-3n+1)}{n^3} S(n-1) + \frac{(4n-3)(4n-4)(4n-5)}{n^3} S(n-2).$$

In particular, $S(0) = 1$, $S(1) = 2$, $S(2) = 18$, $S(3) = 164$, $S(4) = 1810$, and $S(5) = 21252$.

lv) Let $n \ge k \ge 0$ and $z \in \mathbb{C}$.

$$\sum_{i=0}^{k} \binom{z}{i}\binom{-z}{n-i} = \frac{n-k}{n}\binom{z-1}{k}\binom{-z}{n-k}.$$

In particular,

$$\sum_{i=0}^{n} \binom{z}{i}\binom{-z}{n-i} = 0.$$

lvi) Let $n > k \ge 0$ and $z \in \mathbb{C}$.

$$\sum_{i=0}^{n} \binom{z}{i}\binom{1-z}{n-i} = \frac{(n-1)(1-z)-k}{n(n-1)}\binom{z-1}{k}\binom{-z}{n-k-1}.$$

lvii) Let $n \ge 1$. Then,

$$\sum_{i=0}^{n} \binom{n}{i}^2 \binom{n+i}{i}^2 = \sum_{i=0}^{n}\sum_{j=0}^{i} \binom{n}{i}\binom{n+i}{i}\binom{i}{j}^3.$$

lviii) Let $n \ge 1$ and $z \in \mathbb{C}$. Then,

$$\sum_{i=1}^{n} \binom{n+i}{2i}\binom{2i}{i}\binom{2i}{i+1} z^{i-1}(z+1)^{i+1} = n(n+1)\sum_{i=0}^{n}\left[\binom{n+i}{2i} C_i z^i\right]^2.$$

lix) Let $n \ge 1$. Then,

$$\sum_{i=1}^{n} H_i \binom{n}{i}^2 = H_n\binom{2n}{n} - \sum_{i=1}^{n}\frac{1}{i}\binom{2n-i}{n-i}, \quad \sum_{i=1}^{n} H_i\binom{n}{i}\binom{2n}{i} = H_n\binom{3n}{n} - \sum_{i=1}^{n}\frac{1}{i}\binom{3n-i}{n-i}.$$

lx) Let $n \ge 0$ and $z_1, \ldots, z_m \in \mathbb{C}$. Then,

$$\sum \prod_{j=1}^{m} \binom{i_j + z_j}{i_j} = \binom{n+m-1+\sum_{i=1}^{m} z_i}{n},$$

where the sum is taken over all multisets $\{i_1, \ldots, i_m\}_{\mathrm{ms}}$ of nonnegative integers such that $\sum_{j=1}^{m} i_j = n$.

lxi) Let $n \geq 1$ and $k \geq 1$. Then,

$$\sum (-1)^{\sum_{j=1}^{n} i_j} \binom{\sum_{j=1}^{n} i_j}{i_1, \ldots, i_n} \prod_{j=1}^{n} \binom{n}{j}^{i_j} = \binom{n+k-1}{k},$$

where the sum is over all n-tuples $(i_1, \ldots, i_n)$ of nonnegative integers such that $\sum_{j=1}^{n} j i_j = n$.

lxii) Let $n \geq 1$. Then,

$$\sum_{i=0}^{n} \sum_{j=0}^{n} \binom{i+j}{i}^2 \binom{4n-2i-2j}{2n-2i} = (2n+1)\binom{2n}{n}^2.$$

lxiii) Let $n \geq 1$. Then,

$$\sum_{i=0}^{n} \sum_{j=0}^{n} \binom{i+j}{i} \binom{n-i}{j} \binom{n-j}{n-i-j} = \sum_{i=0}^{n} \binom{2i}{i}.$$

lxiv) Let $m, n \geq 1$. Then,

$$\sum_{i=0}^{m} \sum_{j=0}^{n} \binom{i+j}{i} \binom{m-i+j}{j} \binom{n-j+i}{i} \binom{m+n-i-j}{m-i} = \frac{(m+n+1)!}{m!n!} \sum_{i=0}^{\min\{m,n\}} \frac{1}{2i+1} \binom{m}{i} \binom{n}{i}.$$

lxv) Let $n \geq 0$. Then,

$$\sum_{i=0}^{n} \frac{1}{4^i} \binom{2i}{i} \binom{2n-i}{n} = \sum_{i=0}^{n} \frac{1}{4^i} \binom{2i}{i} \binom{2n+1}{2i} = \frac{1}{4^n} \binom{4n+1}{2n}.$$

lxvi) Let $n \geq 1$. Then,

$$\sum \prod_{j=1}^{m} \binom{2i_j}{i_j} = \frac{4^n \Gamma(n+m/2)}{\Gamma(m/2)},$$

where the sum is taken over all m-tuples $(i_1, \ldots, i_m)$ of nonnegative integers such that $\sum_{j=1}^{m} i_j = n$.

lxvii) Let $n \geq 0$ and $z_1, \ldots, z_m \in \mathbb{C}$. Then,

$$\sum \prod_{j=1}^{m} \binom{z_j}{i_j} = \binom{\sum_{i=1}^{m} z_i}{n},$$

where the sum is taken over all m-tuples $(i_1, \ldots, i_m)$ of nonnegative integers such that $\sum_{j=1}^{m} i_j = n$.

lxviii) Let $n, m \geq 1$. Then,

$$\sum_{i,j=-n}^{n} \binom{2n}{n+i} \binom{2n}{n+j} |i^2 - j^2| = 2n^2 \binom{2n}{n}^2.$$

lxix) Let $n \geq 1$. Then,

$$\sum_{i=1}^{2n} 4^i \binom{\frac{1}{2}}{i} \binom{-\frac{1}{2}}{i} \binom{-2i}{2n-i} = \frac{4n+1}{16^n} \binom{2n}{n}^2.$$

lxx) Let $n \geq 0$, $0 \leq k \leq 2n+1$, and $x \in \mathbb{C}$, and assume that $-x \notin \{0, 1, \ldots, n\}$. Then,

$$\sum_{i=0}^{n} \binom{n}{i}^2 \left[\frac{(-i)^k}{(x+i)^2} + \frac{(-i)^{k-1}}{x+i} [k - 2i(H_i - H_{n-i})] \right] = \left(\frac{n!}{x^{\overline{n+1}}} \right)^2 x^k,$$

$$\sum_{i=0}^{n} \binom{n}{i}^2 \left[\frac{1}{(x+i)^2} + \frac{2}{x+i} (H_i - H_{n-i}) \right] = \left(\frac{n!}{x^{\overline{n+1}}} \right)^2,$$

$$\sum_{i=0}^{n}\binom{n}{i}^2\binom{n+i}{i}\left[\frac{1}{(x+i)^2}+\frac{1}{x+i}(3H_i-2H_{n-i}-H_{n+i})\right]=\frac{n!(1-x)^{\overline{n}}}{(x^{\overline{n+1}})^2},$$

$$\sum_{i=0}^{n}\binom{n}{i}^2\binom{n+i}{i}^2\left[\frac{1}{(x+i)^2}+\frac{2}{x+i}(2H_i-H_{n-i}-H_{n+i})\right]=\left(\frac{(1-x)^{\overline{n}}}{x^{\overline{n+1}}}\right)^2,$$

$$\sum_{i=0}^{n}\binom{n}{i}^2(H_i-H_{n-i})=0,\quad \sum_{i=0}^{n}\binom{n}{i}^2\binom{n+i}{i}(3H_i-2H_{n-i}-H_{n+i})=0,$$

$$\sum_{i=0}^{n}\binom{n}{i}^2\binom{n+i}{i}^2(2H_i-H_{n-i}-H_{n+i})=0,\quad \sum_{i=0}^{n}(2i-n)\binom{n}{i}^2(H_i-H_{n-i})=\binom{2n}{n}.$$

lxxi) Let $n\ge 0$ and $x\in\mathbb{C}$, and assume that $-x\notin\{0,1,\ldots,n\}$. Then,

$$\frac{1}{x}+\sum_{i=1}^{n}\binom{n}{i}^2\binom{n+i}{i}^2\left[\frac{-i}{(x+i)^2}+\frac{1+2iH_{n+i}+2iH_{n-i}-4H_i}{x+i}\right]=x\left[\frac{(1-x)^{\overline{n}}}{x^{\overline{n+1}}}\right]^2,$$

$$1+\sum_{i=1}^{n}\binom{n}{i}^2\binom{n+i}{i}^2\left[\frac{i^2}{(x+i)^2}-\frac{2i^2}{x+i}\left(\frac{1}{i}+H_{n+i}+H_{n-i}-2H_i\right)\right]=\left[\frac{(1-x)^{\overline{n}}}{(1+x)^{\overline{n}}}\right]^2.$$

Furthermore,

$$\sum_{i=1}^{n}\binom{n}{i}^2\binom{n+i}{i}^2(1+2iH_{n+i}+2iH_{n-i}-4H_i)=0,$$

$$\sum_{i=1}^{n}i^2\binom{n}{i}^2\binom{n+i}{i}^2\left(\frac{1}{i}+H_{n+i}+H_{n-i}-2H_i\right)=n(n+1).$$

lxxii) Let $n\ge 1$. Then,

$$\sum_{i=1}^{n}\binom{n}{i}^2[2H_i+(n-2i)(2H_i^2+H_{i,2})]=-\frac{1}{n},$$

$$\sum_{i=0}^{n}\binom{n}{i}^3[1+3(n-2i)H_i]=(-1)^n,\quad \sum_{i=1}^{n}\binom{n}{i}^3[2H_i+(n-2i)(3H_i^2+H_{i,2})]=(-1)^n 2H_n,$$

$$\sum_{i=1}^{n}\binom{n}{i}^4[1+4(n-2i)H_i]=(-1)^n\binom{2n}{n},\quad \sum_{i=0}^{n}\binom{n}{i}\binom{2n}{i}\binom{2n}{n+i}[1+(n-2i)(2H_i+H_{n+i}]=(-1)^n.$$

lxxiii) Let $n\ge 0$ and $0\le k\le 2n+1$. Then,

$$\sum_{i=0}^{n}i^{k-1}\binom{n}{i}^2[k-2i(H_i-H_{n-i})]=\begin{cases}0, & 0\le k\le 2n,\\ (n!)^2, & k=2n+1.\end{cases}$$

lxxiv) Let $n\ge 1$. Then,

$$\sum_{i=0}^{n}\frac{1}{(2i+1)(2n-2i-1)}\binom{2i}{i}\binom{2(n-i-1)}{n-i-1}=\frac{16^n}{8n^2\binom{2n}{n}}.$$

lxxv) Let $n, m, k_1, \ldots, k_m$ be positive integers, and define $k \triangleq \sum_{i=1}^{m}k_i$. Then,

$$\sum\prod_{j=1}^{m}\binom{i_j+k_j-1}{i_j}=\binom{n+k-1}{n},$$

where the sum is taken over all $\binom{n+m-1}{n}$ m-tuples $(i_1, \ldots, i_m)$ of nonnegative integers such that $\sum_{j=1}^{m} i_j = n$.

Source: *i*), *iii*), *iv*), *xi*), and *lxvii*) are given in [1219, pp. 167, 169, 171, 172]; *ii*) is given in [504, p. 75]; *v*), *xii*), *xv*), *xvii*), *xix*), *xxiii*), and *xxiv*) are given in [411, pp. 64–68, 78]; *vi*) is given in [1197, 2134]; *vii*) is given in [1225, p. 2] and [2228, p. 31]; *viii*) is given in [411, p. 78] and [2228, p. 130]; *ix*) is given in [2880, p. 138]; *x*) is given in [2799]; *xiii*) is given in [785]; *xiv*) is given in [1371, p. 9]; *xvi*) is given in [1372, p. 62]; *xviii*) is given in [2434]; *xx*) is given in [2402]; *xxi*) is given in [2228, p. 31]; *xxii*) is given in [2228, p. 138]; *xxiv*) follows from [1757, p. 163]; *xxv*) and *xxvi*) are given in [2228, pp. 95, 96]; *xxvii*) is given in [2243]; *xxviii*) is given in [666]; *xxix*) is given in [599, 671, 1154, 2799] and [1219, p. 187]; *xxx*) is given in [411, p. 79] and [666, 2255]; *xxxi*) is given in [1675, p. 84] and [2509]; *xxxii*) is given in [1675, p. 140]; *xxxiii*) is given in [69], [411, p. 78], and [1675, p. 97]; *xxxiv*) is given in [2228, p. 113]; *xxxv*) is given in [738], [1219, p. 201], [1274], and [2228, p. 142]; *xxxvi*) and *xxxvii*) are given in [1372, p. 66]; *xxxviii*) is given in [1757, pp. 161, 162]; *xxxix*)–*xl*) are given in [1158, pp. 300, 303]; *xli*) is given in [511, p. 156]; *xlii*) is given in [618]; *xliv*) is given in [771, p. 173]; *xlv*) is given in [616]; *xlvi*) is given in [1405, 2353] and [1675, p. 96]; *xlvii*) is given in [1210]; *xlix*) is given in [1166] and [2228, p. 22]; *l*) is given in [1158, p. 304]; *li*) is given in [771, p. 90], [2068, p. 171], and [2558]; *lii*) is given in [614] *liii*) is given in [116, pp. 399, 400] and [2068, p. 171]; *liv*) is given in [771, p. 90], [2068, p. 171], and [2228, p. 33]; *lv*) and *lvi*) are given in [771, p. 169]; *lvii*) is given in [2558]; *lviii*) is given in [2570]; *lix*) is given in [2100]; *lx*) is given in [2]; *lxi*) is given in [2016]; *lxii*)–*lxiv*) are given in [711]; *lxv*) is given in [67]; *lxvi*)–*lxvii*) are given in [2799]; *lxviii*) is given in [2744]; *lxix*) is given in [1164]; *lxx*) and *lxxi*) are given in [736, 737]; *lxxii*) is given in [2830]; *lxxiii*) is given in [737]; *lxxiv*) is given in [700]; *lxxv*) is given in [1587]. **Remark:** *v*) is *Vandermonde's convolution*. *xxxv*) is *Rothe's identity*; see [738, 1274]. $S(n)$ in *l*) is the nth *Franel number*. See Fact 13.2.8.

Fact 1.16.14. The following statements hold:

i) Let $n, m \geq 0$. Then,

$$\sum_{i=0}^{\min\{n,m\}} (-1)^i \binom{n+m}{n+i}\binom{m+n}{m+i} = \binom{n+m}{n} = \binom{m+n}{m}.$$

ii) Let $n, m \geq 0$. Then,

$$\sum_{i=-m}^{m} (-1)^i \binom{2n}{n-i}\binom{2m}{m-i} = \sum_{i=-n}^{n} (-1)^i \binom{2n}{n-i}\binom{2m}{m-i} = \frac{\binom{2n}{n}\binom{2m}{m}}{\binom{n+m}{n}} = \frac{\binom{2n}{n}\binom{2m}{m}}{\binom{n+m}{m}} = \frac{(2n)!(2m)!}{n!m!(n+m)!}.$$

iii) Let $0 \leq n \leq k \leq m$. Then,

$$\sum_{i=0}^{n} (-1)^i \binom{n}{i}\binom{m+i}{k} = (-1)^n \binom{m}{k-n}.$$

In particular,

$$\sum_{i=0}^{k} (-1)^i \binom{k}{i}\binom{m+i}{k} = (-1)^k.$$

iv) Let $0 \leq k \leq n$. Then,

$$\sum_{i=k}^{n} (-1)^i \binom{n}{i}\binom{i}{k} = \begin{cases} 0, & k < n, \\ (-1)^n, & k = n. \end{cases}$$

v) Let $n, m, k \geq 0$. Then,

$$\sum_{i=0}^{n}(-1)^i\binom{n}{i}\binom{m+i}{k} = \begin{cases} 0, & k < n, \\ (-1)^n, & k = n, \\ (-1)^n\dfrac{m^{\underline{k-n}}}{(k-n)!}, & k > n. \end{cases}$$

vi) Let $n, m, k, l \geq 0$. Then,

$$\sum_{i=\max\{-m,n-k\}}^{l-m}(-1)^i\binom{l}{m+i}\binom{k+i}{n} = (-1)^{l+m}\binom{k-m}{n-l}.$$

vii) Let $n, m, k, l \geq 0$. Then,

$$\sum_{i=n}^{\min\{l-m,k+n\}}(-1)^i\binom{l-i}{m}\binom{k}{i-n} = (-1)^{l+m}\binom{k-m-1}{l-m-n}.$$

viii) Let $n \geq k \geq 1$. Then,

$$\sum_{i=0}^{k}(-1)^{i+1}\binom{n}{i}\binom{n-i}{k-i} = 0.$$

ix) Let $n \geq 0$. Then,

$$\sum_{i=0}^{n}(-1)^i\frac{1}{2^i}\binom{2i}{i}\binom{n}{i} = \begin{cases} 0, & n \text{ even}, \\ \dfrac{1}{2^n}\dbinom{n}{n/2}, & n \text{ odd}. \end{cases}$$

x) Let $n \geq 0$ and $x \in \mathbb{C}$. Then,

$$\sum_{i=0}^{n}(-1)^i\binom{n}{i}\binom{i+x}{i} = (-1)^n\binom{x}{n}.$$

xi) Let $n \geq 1$.

$$\sum_{i=0}^{n}(-1)^i\frac{1}{i+1}\binom{2i}{i}\binom{n+i}{2i} = 0.$$

xii) Let $n \geq k \geq 1$. Then,

$$\sum_{i=0}^{n-k}(-1)^i\frac{1}{i+1}\binom{2i}{i}\binom{n+i}{k+2i} = \binom{n-1}{k-1}.$$

xiii) Let $n, k \geq 0$. Then,

$$\sum_{i=0}^{2n}(-1)^i\binom{2n}{i}\binom{2k}{k-n+i} = (-1)^n\frac{\binom{2n}{n}\binom{2k}{k}}{\binom{n+k}{n}}.$$

In particular,

$$\sum_{i=0}^{2n}(-1)^i\binom{2n}{i}^2 = (-1)^n\binom{2n}{n}.$$

xiv) Let $n \geq 0$. Then,

$$\sum_{i=0}^{\lfloor n/2\rfloor}(-1)^i\binom{n}{i}\binom{2n-2i}{n} = \sum_{i=0}^{\lfloor n/2\rfloor}(-1)^i\binom{n-i}{i}\binom{2n-2i}{n-i} = 2^n,$$

$$\sum_{i=0}^{\lfloor n/2\rfloor}(-1)^i\binom{n+1}{i}\binom{2n-2i}{n}=n+1,\quad \sum_{i=0}^{n}(-1)^i\binom{2i}{i}\binom{2n-2i}{n-i}=\begin{cases}2^n\binom{n}{n/2}, & n \text{ even},\\ 0, & n \text{ odd}.\end{cases}$$

xv) Let $n,k\ge 1$. Then,

$$\sum_{i=0}^{n}(-1)^i\binom{n}{i}\binom{(n-i)k}{n+1}=nk^{n-1}\binom{k}{2}.$$

xvi) Let $n\ge 0$. Then,

$$\sum_{i=0}^{2n+1}(-1)^i\binom{2n+1}{i}^2=0.$$

xvii) Let $n,k\ge 0$. Then,

$$\sum_{i=0}^{n}(-1)^i\binom{n+k+1}{k+i+l}\binom{k+i}{k}=1.$$

xviii) Let $n,m\ge 0$. Then,

$$\sum_{i=0}^{n}(-1)^i\binom{n}{i}\binom{2n-i}{m-i}=\binom{n}{m}.$$

xix) Let $n\ge 1$ and $1\le m\le 2n$. Then,

$$\sum_{i=0}^{\min\{n,2n-m\}}(-4)^i\binom{n}{i}\binom{2n-2i}{m-i}=(-1)^m\binom{2n}{m}.$$

xx) Let $n\ge 1$ and $1\le m\le n$. Then,

$$\sum_{i=m}^{n}(-1)^i4^{n-i}\binom{n}{i}\binom{2i}{i-m}=\binom{2n}{n-m}.$$

In particular,

$$\sum_{i=0}^{n}(-1)^i4^{n-i}\binom{n}{i}\binom{2i}{i}=\binom{2n}{n}.$$

xxi) Let $n,k\ge 0$. Then,

$$\sum_{i=0}^{n}(-1)^{i+k}2^{n-2i}\binom{n-i}{i}\binom{i}{k}=\binom{n+1}{2k+1}.$$

xxii) Let $n\ge 0$ and $k\ge n+1$. Then,

$$\sum_{i=0}^{n}(-1)^i\frac{1}{i+1}\binom{k}{i}\binom{k-1-i}{n-i}=\frac{1}{k+1}\left[\binom{k}{n+1}+(-1)^n\right].$$

xxiii) Let $n\ge k\ge 0$. Then,

$$\sum_{j=k}^{n}\sum_{i=0}^{n-j}(-1)^i\binom{i+j}{j}\binom{j}{k}=1.$$

In particular,

$$\sum_{j=0}^{n}\sum_{i=0}^{n-j}(-1)^i\binom{i+j}{j}=\sum_{j=1}^{n}\sum_{i=0}^{n-j}(-1)^i j\binom{i+j}{j}=1.$$

xxiv) Let $n\ge 1$. Then,

$$\sum_{i=0}^{n}(-1)^i\frac{4^i\binom{n}{i}}{\binom{2i}{i}}=\frac{1}{1-2n}.$$

xxv) Let $n \geq 1$. Then,

$$\sum_{i=0}^{n}(-1)^i\frac{\binom{n}{i}^2}{\binom{2n}{i}} = \frac{1}{\binom{2n}{n}}.$$

xxvi) Let $n \geq 0$ and $x \in \mathbb{C}$, and, if $n \geq 1$, assume that $-x \notin \{1,\ldots,n\}$. Then,

$$\sum_{i=0}^{n}(-1)^i\frac{\binom{n}{i}}{\binom{x+i}{i}} = \frac{x}{x+n}.$$

xxvii) Let $n \geq 1$. Then,

$$\sum_{i=0}^{n}(-1)^i\frac{\binom{n}{i}}{\binom{n+i}{i}} = \sum_{i=1}^{n}(-1)^{i+1}\frac{\binom{n}{i}}{\binom{n+i}{i}} = \frac{1}{2}.$$

xxviii) Let $n \geq 1$. Then,

$$(-1)^n\sum_{i=0}^{2n}(-1)^i\binom{2n}{i}^3 = \sum_{i=-n}^{n}(-1)^i\binom{2n}{n+i}^3 = \frac{(3n)!}{(n!)^3}.$$

xxix) Let $n \geq 0$. Then,

$$\sum_{i=0}^{2n}(-1)^i\binom{2n}{i}\binom{2i}{i}\binom{4n-2i}{2n-i} = \binom{2n}{n}^2.$$

xxx) Let $k,m,n \geq 0$. Then,

$$\sum_{i=-k}^{k}(-1)^i\binom{2k}{k+i}\binom{2m}{m+i}\binom{2n}{n+i} = \frac{(k+m+n)!(2k)!(2m)!(2n)!}{(k+m)!(m+n)!(n+k)!k!m!n!}.$$

xxxi) Let $k,l,m,n,p \geq 0$. Then,

$$\sum_{i,j=0}^{\min\{k,n\}}(-1)^{i+j}\binom{i+j}{j+l}\binom{k}{i}\binom{n}{j}\binom{p+n-i-j}{m-i} = (-1)^l\binom{n+k}{n+l}\binom{p-k}{m-n-l}.$$

xxxii) Let $n \geq 1$ and $x,y \in \mathbb{C}$. Then,

$$\sum_{i=0}^{n}(-1)^i\binom{x+i}{i}\binom{y+n}{n-i} = \binom{y-x+n-1}{n}, \quad \sum_{i=0}^{n}(-1)^i\frac{\binom{n}{i}\binom{x+i}{i}}{\binom{y+i}{i}} = \frac{(y-x)^{\overline{n}}}{(y+1)^{\overline{n}}}.$$

xxxiii) Let $n,k \geq 1$ and $x \in \mathbb{C}$, where $x \notin \{-n,\ldots,-1,0\}$. Then,

$$\sum_{i=0}^{n}(-1)^i\frac{\binom{n}{i}}{\binom{x+i}{i}}\sum\prod_{j=1}^{k}\frac{1}{x+i_j} = \frac{x}{(x+n)^{k+1}},$$

where the second sum is taken over all k-tuples of integers $(i_1,\ldots,i_k)$ such that $0 \leq i_1 \leq \cdots \leq i_k \leq i$. In particular,

$$\sum_{i=1}^{n}(-1)^{i+1}\binom{n}{i}\sum\prod_{j=1}^{k}\frac{1}{i_j} = \frac{1}{n^k}, \quad \sum_{i=1}^{n}(-1)^{i+1}\binom{n}{i}\sum_{j=1}^{i}\sum_{l=1}^{j}\frac{1}{jl} = \frac{1}{n^2},$$

$$\sum_{i=0}^{n}(-1)^i\frac{\binom{n}{i}}{\binom{x+i}{i}}\sum_{j=1}^{i}\sum_{l=1}^{j}\frac{1}{(x+j)(x+l)} = \frac{x}{(x+n)^3},$$

$$\sum_{i=1}^{n}(-1)^{i+1}\frac{\binom{n}{i}}{\binom{x+i}{i}}\sum_{j=1}^{i}\sum_{l=1}^{j}\frac{1}{(x+j)(x+l)}=\frac{n}{(x+n)^3},$$

$$\sum_{i=1}^{n}(-1)^{i+1}\binom{n}{i}\left(\sum_{j=1}^{i}\frac{1}{j^3}+\sum_{1\le j<k\le i}\frac{1}{jk(j+k)}+\sum_{1\le j<k<l\le i}\frac{1}{jkl}\right)=\frac{1}{n^3}.$$

xxxiv) For all $n\ge 1$ and $k\ge 0$,

$$\sum_{i=1}^{n}(-1)^{i+1}\frac{1}{i^2}\binom{n}{i}=\frac{1}{2}(H_n^2+H_{n,2})=\sum_{i=1}^{n}\sum_{j=1}^{i}\frac{1}{ij},$$

$$\sum_{i=0}^{n}(-1)^{i}\frac{\binom{n}{i}}{\binom{i+k}{k}}\sum_{j=1}^{i}\frac{1}{j+k}=-\frac{n}{(n+k)^2},\quad \sum_{i=0}^{n}(-1)^{i}H_i\binom{n}{i}=-\frac{1}{n},$$

$$\sum_{i=1}^{n}(-1)^{i}\frac{\binom{n}{i}}{\binom{i+k}{k}}\sum_{j=1}^{i}\sum_{l=1}^{j}\frac{1}{(j+k)(l+k)}=-\frac{n}{(n+k)^3},$$

$$\sum_{i=1}^{n}(-1)^{i}\binom{n}{i}\sum_{j=1}^{i}(-1)^{j+1}\frac{1}{j^2}\binom{i}{j}=-\frac{1}{2}\sum_{i=1}^{n}(-1)^{i}(H_i^2+H_{i,2})\binom{n}{i}=\frac{1}{n^2},$$

$$\sum_{i=1}^{n}(-1)^{i}\frac{\binom{n}{i}}{\binom{i+k}{k}}\left[\left(\sum_{j=1}^{i}\frac{1}{j+k}\right)^3+3\left(\sum_{j=1}^{i}\frac{1}{j+k}\right)\sum_{j=1}^{i}\frac{1}{(j+k)^2}+2\sum_{j=1}^{i}\frac{1}{(j+k)^3}\right]=-\frac{6n}{(n+k)^4},$$

$$\sum_{i=1}^{n}(-1)^{i}(H_i^3+3H_iH_{i,2}+2H_{i,3})\binom{n}{i}=-\frac{6}{n^3},$$

$$\sum_{i=1}^{n}(-1)^{i}(H_i^4+6H_i^2H_{i,2}+3H_{i,2}^2+8H_iH_{i,3}+6H_{i,4})\binom{n}{i}=-\frac{24}{n^4},$$

$$\sum_{i=1}^{n}(-1)^{i}H_i^3\binom{n}{i}=\frac{1}{2n}\left(5H_{n-1,2}+\frac{4}{n}H_{n-1}-H_{n-1}^2-\frac{2}{n^2}\right),$$

$$\sum_{i=1}^{n}(-1)^{i}H_iH_{i,2}\binom{n}{i}=\frac{1}{2n}\left(H_{n-1}^2-H_{n-1,2}-\frac{2}{n^2}\right),$$

$$\sum_{i=1}^{n}(-1)^{i}H_{i,3}\binom{n}{i}=-\frac{1}{2n}\left(H_{n-1,2}+H_{n-1}^2+\frac{2}{n}H_{n-1}+\frac{2}{n^2}\right).$$

If $n+k\ge 2$, then

$$\sum_{i=1}^{n}(-1)^{i}i\frac{\binom{n}{i}}{\binom{i+k}{k}}\sum_{j=1}^{i}\frac{1}{j+k}=\frac{n(n^2-n-k^2)}{(n+k)^2(n+k-1)^2},$$

$$\sum_{i=1}^{n}(-1)^{i}i\frac{\binom{n}{i}}{\binom{i+k}{k}}\left[\left(\sum_{j=1}^{i}\frac{1}{j+k}\right)^2+\sum_{j=1}^{i}\frac{1}{(j+k)^2}\right]=\frac{2n[2n^3+3n^2(k-1)-n(3k-1)-k^3]}{(n+k)^3(n+k-1)^3}.$$

If $n\ge 2$, then

$$\sum_{i=1}^{n}(-1)^{i}iH_i\binom{n}{i}=\frac{1}{n-1},\quad \sum_{i=1}^{n}(-1)^{i}i(H_i^2+H_{i,2})\binom{n}{i}=\frac{2(2n-1)}{n(n-1)^2},$$

$$\sum_{i=1}^{n}(-1)^i i(H_i^3 + 3H_iH_{i,2} + 2H_{i,3})\binom{n}{i} = \frac{6(3n^2 - 3n + 1)}{n^2(n-1)^3}.$$

xxxv) Let $n \geq 0$, $m \in \mathbb{Z}$, and $x \in \mathbb{C}$. Then,

$$\sum_{i=0}^{2n}(-1)^i\binom{2n}{i}\binom{x+i}{2n+m}\binom{x+2n-i}{2n+m} = (-1)^n\binom{2n}{n}\binom{x+n}{2n+m}\frac{\binom{x+n}{n+m}}{\binom{x+n}{n}}.$$

In particular,

$$\sum_{i=0}^{2n}(-1)^i\binom{2n}{i}\binom{x+i}{2n}\binom{x+2n-i}{2n} = (-1)^n\binom{x}{n}\binom{x+n}{n},$$

$$\sum_{i=0}^{2n}(-1)^i\binom{2n}{i}\binom{x+i}{2n+1}\binom{x+2n-i}{2n+1} = (-1)^n\frac{x}{n+1}\binom{2n}{n}\binom{x+n}{2n+1}.$$

xxxvi) Let $n, m, k \geq 0$. Then,

$$\sum_{i=0}^{\min\{n,m,k\}}(-1)^i\binom{n+m}{n+i}\binom{m+k}{m+i}\binom{k+n}{k+i} = \frac{(n+m+k)!}{(n!)(m!)(k!)}.$$

xxxvii) Let $n \geq 0$ and $x \in \mathbb{C}$, and assume that $-x \notin \{0, \ldots, n\}$. Then,

$$\sum_{i=0}^{n}(-1)^i\frac{1}{x+i}\binom{n}{i}\binom{n+i}{i} = \frac{(1-x)^{\overline{n}}}{x^{\overline{n+1}}},$$

$$\sum_{i=0}^{n}(-1)^i\binom{n}{i}^3\left(\frac{1}{(x+i)^3} + \frac{3}{(x+i)^2}(H_i - H_{n-i}) + \frac{3}{2(x+i)}[3(H_i - H_{n-i})^2 + H_{i,2} + H_{n-i,2}]\right) = \left(\frac{n!}{x^{\overline{n+1}}}\right)^3.$$

Furthermore,

$$\sum_{i=0}^{n}(-1)^i[3(H_i - H_{n-i})^2 + H_{i,2} + H_{n-i,2}]\binom{n}{i}^3 = 0.$$

xxxviii) Let $n \geq 0$ and $x \in \mathbb{C}$, and assume that $-x \notin \{0, \pm 1 \ldots, \pm(n+1)\}$. Then,

$$\sum_{i=1}^{n}(-1)^i\frac{1}{(n+1)(x-i)}\frac{\binom{2n}{n+i}}{\binom{n+i}{n+1}} + \sum_{i=0}^{n}\left[\frac{1}{(x+i)^2} + \frac{H_i + H_{n+i} - 2H_{n-i}}{x+i}\right]\binom{n}{i}\binom{2n}{n+i} = \frac{n!(2n)!}{(x^{\overline{n+1}})^2(1-x)^{\overline{n}}}.$$

Furthermore,

$$\sum_{i=1}^{n}(-1)^i\frac{1}{n+1}\frac{\binom{2n}{n+i}}{\binom{n+i}{n+1}} = \sum_{i=0}^{n}(2H_{n-i} - H_i - H_{n+i})\binom{n}{i}\binom{2n}{n+i}.$$

xxxix) Let $n \geq 1$ and $x, y \in \mathbb{C}$. Then,

$$\sum_{i=0}^{n}(-1)^i\binom{n}{i}\frac{\binom{y+i}{i}}{\binom{x+i}{i}} = \frac{\binom{x-y+n-1}{n}}{\binom{x+n}{n}}.$$

xl) Let $n \geq 1$. Then,

$$\sum_{i=1}^{n}(-1)^i\binom{n}{i}\binom{n+i}{i}H_i = (-1)^n 2H_n, \quad \sum_{i=1}^{n}(-1)^i\binom{n}{i}\binom{n+i}{i}(H_i^2 + H_{i,2}) = (-1)^n 4H_n^2,$$

$$\sum_{i=1}^{n}(-1)^i\binom{n}{i}\binom{n+i}{i}(H_i^3+3H_iH_{i,2}+2H_{i,3}) = (-1)^n4(2H_n^3+H_{n,3}),$$

$$\sum_{i=1}^{n}(-1)^i\binom{n}{i}\binom{n+i}{i}(H_i^4+6H_i^2H_{i,2}+8H_iH_{i,3}+3H_{i,2}^2+6H_{i,4}) = (-1)^n16H_n(H_n^3+2H_{n,3}),$$

$$\sum_{i=1}^{n}(-1)^i\binom{n}{i}\binom{2n+i}{i}H_i = (-1)^n\binom{2n}{n}H_{2n},$$

$$\sum_{i=1}^{n}(-1)^i\binom{n}{i}\binom{2n+i}{i}(H_i^2+H_{i,2}) = (-1)^n\binom{2n}{n}(2H_{n,2}+H_{2n}^2-H_{2n,2}),$$

$$\sum_{i=1}^{n}(-1)^i\binom{n}{i}\binom{2n+i}{i}(H_i^3+3H_iH_{i,2}+2H_{i,3}) = (-1)^n\binom{2n}{n}(H_{2n}^3+2H_{2n,3}+3H_{2n}(2H_{n,2}-H_{2n,2}).$$

Source: *i*), *vi*), *vii*), *xxxi*), and *xxxvi*) are given in [1219, pp. 167, 169, 171, 172, 187]; *ii*) is given in [1754]; *iii*) is given in [1225, p. 2]; *iv*) is given in [411, p. 84]; *v*) is a corrected version of a result given in [1369, p. 26]; *viii*) is given in [1372, p. 62]; *ix*) is given in [2228, pp. 63, 114]; *x*) is given in [2228, p. 133]; *xi*) is given in [2228, p. 115]; *xii*) is given in [116, p. 108]; *xiii*) is given in [1217, p. 5] and [2014]; *xiv*) is given in [107, pp. 45, 247, 248] and [2507]; *xv*) is given in [790]; *xvi*) is given in [1217, p. 5]; *xvii*) is given in [511, p. 65]; *xviii*) is given in [1369, p. 26]; *xviv*) is given in [1675, p. 72]; *xx*) is given in [2006]; *xxi*) is given in [314]; *xxii*) is given in [116, pp. 107, 108]; *xxiii*) is given in [1519]; *xxiv*), *xxvi*), *xxviii*), and *xxix*) are given in [2228, pp. 22, 32, 44–47]; *xxv*) is given in [1675, p. 85]; *xxvii*) is given in [1761]; *xxx*) is given in [116, pp. 108, 109]; *xxxii*) is given in [739]; *xxxiii*) is given in [739, 887, 895, 2515, 2516]; *xxxiv*) is given in [736, 2482, 2830]; *xxxv*) is given in [1211]; *xxxvii*) and *xxxviii*) are given in [736, 737]; *xxxix*) and *xl*) are given in [2830]. **Remark:** If $n > k$, then both terms in *iv*) are zero. *ii*) is *Dixon's identity*. See [2228, p. 43]. *xxviii*) is a special case. **Remark:** *xxxv*) is *Vosmansky's identity*. **Remark:** Additional equalities for products of binomial coefficients are given in [2312, pp. 141–146].

Fact 1.16.15. The following statements hold:

i) If $n \ge 2$, then

$$2^n < \frac{4^n}{n+1} < \frac{4^n}{2\sqrt{n}} < \binom{2n}{n} < \frac{4^n}{\sqrt{(n+1/4)\pi}} < \left\{\begin{array}{c}\dfrac{4^n}{\sqrt{n\pi}}\\[2mm] \dfrac{4^n}{\sqrt{3n+1}}\end{array}\right\} < \frac{4^n}{\sqrt{2n+1}} < \frac{4^n}{\sqrt{2n}} < \frac{4^n\log 3}{\log(2n+3)} < 4^n.$$

ii) If $n \ge 3$, then

$$2^n < \frac{4^n}{n+1} < \frac{4^n}{2\sqrt{n}} < \frac{4^n}{\sqrt{n\pi}}\left(1-\frac{1}{4n}\right) < \binom{2n}{n} < \frac{4^n}{\sqrt{(n+1/4)\pi}} < \frac{4^n}{\sqrt{n\pi}}.$$

iii) If $n \ge 1$, then

$$\binom{2n+1}{n} < 4^n.$$

If, in addition, $n \ge 4$, then

$$\binom{2n+2}{n+1} < 4^n.$$

iv) If $n \ge 11$, then

$$\frac{2^{2n+1}}{n+1} < \binom{2n}{n}.$$

v) If $n \geq 1$, then

$$\prod_{\substack{n+1 \leq i \leq 2n, \\ i \text{ prime}}} i \leq \binom{2n}{n}.$$

vi) If $n \geq 2$ and $1 \leq k \leq n-1$, then

$$\binom{n}{k-1}\binom{n}{k+1} \leq \binom{n}{k}^2.$$

vii) If $n \geq 1$ and $0 \leq k \leq n-1$, then

$$\binom{n-1}{k}\binom{n+1}{k} \leq \binom{n}{k}^2.$$

viii) If $1 \leq k \leq n$, then

$$\left(\frac{n}{k}\right)^k \leq \binom{n}{k} \leq \min\left\{\frac{n^k}{k!}, 2^n\right\}.$$

ix) If $1 \leq k < n/2$, then

$$\binom{n}{k} \leq \binom{n}{k+1}.$$

x) If $0 \leq k \leq n$, then

$$(n+1)^k\binom{n}{k} \leq n^k\binom{n+1}{k}.$$

xi) If $1 \leq k \leq n-1$, then

$$\sum_{i=1}^{k} i(i+1)\binom{2n}{k-i} < \frac{2^{2n-2}k(k+1)}{n}.$$

xii) If $1 \leq k < n$, then

$$n^k \leq (k+1)^{k-1}\binom{n}{k} \leq k^{k/2}(k+1)^{(k-1)/2}\binom{n}{k}.$$

xiii) If $n \geq 2$, then

$$\prod_{i=0}^{n}\binom{n}{i} \leq \left(\frac{2^n-2}{n-1}\right)^{n-1}.$$

xiv) If $n \geq 1$, then

$$\sum_{i=1}^{n}\sqrt{\binom{n}{i}} \leq \sqrt{n(2^n-1)}.$$

xv) If $n \geq 2$, and $1 \leq k \leq n-1$, then

$$\frac{1}{n+1} < \binom{n}{k}\left(\frac{k}{n}\right)^k\left(1-\frac{k}{n}\right)^{n-k}.$$

xvi) If $n \geq 2$, $k \geq 0$, and $2k+1 < n$, then

$$\sum_{i=0}^{k}\binom{n}{i} < \frac{2^n(k+1)}{n+1}.$$

xvii) If $n \geq 2$, $k \geq 1$, and $n < 2k+1$, then

$$\frac{2^n(k+1)}{n+1} < \sum_{i=0}^{k}\binom{n}{i}.$$

xviii) If $n \ge 1$, then

$$\left|\sum_{i=1}^{n}(-1)^i\binom{n}{i}\binom{2n}{i}\right| \le (2\sqrt{2})^n.$$

xix) If $n \ge 1$, then

$$\left|\sum_{i=0}^{n}(-1)^i\binom{n}{i}\binom{3n}{i}\right| \le 4^n.$$

xx) If $n \ge 1$, then

$$\sum_{i=1}^{n}\frac{1}{i!}\binom{n-1}{i-1} < \frac{2^n}{n}.$$

xxi) If $n \ge 1$, then

$$\frac{4}{n+1} \le \sum_{i=0}^{n}\frac{\binom{n}{i}^2}{\binom{2n-1}{i}^2}.$$

Source: *i)* is given in [594], [517, p. 210], [1371, p. 137], and [1675, pp. 45–50]; *ii)* is given in [1413]; *iii)* is given in [1241, p. 156]; *iv)* is given in [2620]; *v)* is given in [217, p. 287]; *vi)* and *vii)* are given in [504, pp. 76, 80]; *viii* and *ix)* is given in [2968]; *x)* is given in [1371, p. 111]; *xi)* is given in [740]; *xii)* is given in [986]; *xiii)* is given in [1566, p. 14] and [1757, p. 253]; *xiv)* is given in [1566, p. 14]; *xv)*–*xvii)* are given in [1885]; *xviii)* and *xix)* are given in [217, pp. 157, 160]; *xx)* is given in [2620]; *xxi)* is given in [894]. **Remark:** If $k = n \ge 1$, then $\sum_{i=0}^{k}\binom{n}{i} = \frac{2^n(k+1)}{n+1} = 2^n$, while, if $n \ge 1$ is odd and $2k+1 = n$, then $\sum_{i=0}^{k}\binom{n}{i} = \frac{2^n(k+1)}{n+1} = 4^n$. See Fact 1.16.10.

Fact 1.16.16. Let $n \ge k \ge 1$, and let $\binom{n}{k}_{\rm r}$ denote the number of k-element multisubsets of $\{1,\dots,n\}$. Then,

$$\binom{n}{k}_{\rm r} = \binom{n+k-1}{k}.$$

Furthermore, for all $z \in \mathbb{C}$,

$$\binom{z}{k}_{\rm r} = \binom{z+k-1}{k} = \frac{z^{\overline{k}}}{k!}, \quad \binom{z}{k}_{\rm r} = \binom{z}{k-1}_{\rm r} + \binom{z-1}{k}_{\rm r},$$

$$\binom{z}{k}_{\rm r} = \frac{z}{k}\binom{z+1}{k-1}_{\rm r} = \frac{z+k-1}{k}\binom{z}{k-1}_{\rm r}, \quad \binom{z+1}{k}_{\rm r} = \sum_{i=0}^{k}\binom{z}{i}_{\rm r}.$$

Source: [771, pp. 15–17]. **Remark:** $\binom{n}{k}_{\rm r}$ is the *binomial coefficient with repetition.* **Related:** Fact 1.12.6 and Fact 13.4.2.

Fact 1.16.17. Let $n \ge 1$. Then,

$$n^n = \sum\binom{n}{k_1,\dots,k_{n-1}} = \sum\frac{n!}{k_1!\cdots k_{n-1}!},$$

where the sum is taken over all $n-1$-tuples $(k_1,\dots,k_{n-1})$ such that $0 \le k_1 \le 1$, $0 \le k_1 + k_2 \le 2$, $\dots, 0 \le k_1 + \cdots + k_{n-1} \le n-1$. **Source:** [409]. **Example:** $3^3 = \binom{3}{0,0} + \binom{3}{0,1} + \binom{3}{0,2} + \binom{3}{1,0} + \binom{3}{1,1} = 6+6+3+6+6 = 27$.

Fact 1.16.18. Let $n, m \ge 1$, and assume that $m \le n$. Then,

$$\binom{2n}{m} = \sum_{i=0}^{\lfloor m/2\rfloor} 2^{m-2i}\binom{n}{i, m-2i, n-m+i}.$$

Source: [2566].

1.17 Facts on Fibonacci, Lucas, and Pell Numbers

Fact 1.17.1. Define $F_1 \triangleq F_2 \triangleq 1$ and, for all $k \in \mathbb{Z}$, define F_k by $F_{k+2} = F_{k+1} + F_k$. Then,

$$(F_i)_{i=-5}^{18} = (5, -3, 2, -1, 1, 0, 1, 1, 2, 3, 5, 8, 13, 21, 34, 55, 89, 144, 233, 377, 610, 987, 1597, 2584).$$

Furthermore, for all $k, l \in \mathbb{Z}$, the following statements hold:

i) If $k \geq 2$, then F_k is the number of tuples each of whose components is either 1 or 2 and the sum of whose components is $n - 1$.

ii) If $k \geq 3$, then F_k is the number of subsets of $\{1, \ldots, k - 2\}$ that do not contain a pair of consecutive integers.

iii) $F_{-k} = (-1)^{k+1} F_k$.

iv) If $3|k$, then F_k is even.

v) If $4|k$, then $3|F_k$; if $5|k$, then $5|F_k$; if $6|k$, then $8|F_k$.

vi) If $k \geq 3$ is prime, then $k|F_{2k} - F_k$.

vii) $\gcd\{F_k, F_l\} = F_{\gcd\{k,l\}}$.

viii) If $k|l$, then $F_k|F_l$. Hence, $F_k|F_{lk}$.

ix) $\gcd\{F_k, F_{k+1}\} = \gcd\{F_k, F_{k+2}\} = 1$.

x) If $k \geq 4$, then $F_k + 1$ is not prime.

xi) If $n \geq 1$, then there exists a unique set $\{i_1, \ldots, i_m\}$ of integers $2 \leq i_1 < \cdots < i_m$ such that, for all $j \in \{1, \ldots, m - 1\}$, $i_j + 1 < i_{j+1}$ and such that $n = \sum_{j=1}^m F_{i_j}$.

xii) For all $k \in \mathbb{Z}$,

$$F_{2k+2} = 2F_{2k} + F_{2k-1} = F_{k+1}^2 + 2F_kF_{k+1} = F_{k+2}^2 - F_k^2 = F_kF_{k+1} + F_{k+1}F_{k+2},$$
$$F_{2k+3} = F_kF_{k+2} + F_{k+1}F_{k+3} = F_{k+2}F_{k+3} - F_kF_{k+1}, \quad F_kF_{k+2} = F_{k+1}^2 + (-1)^{k+1},$$
$$F_{k+5}F_{k+2} = F_{2k+5} + F_{k+3}F_k, \quad F_{k+4} + F_k = 3F_{k+2}, \quad F_{3k+3} = F_{k+2}^3 + F_{k+1}^3 - F_k^3,$$
$$F_{3k+6} = 4F_{3k+3} + F_{3k}, \quad F_k^2 = F_{k+3}F_{k-3} + 4(-1)^{3-k}, \quad F_{2k+1} = F_{k+1}^2 + F_k^2,$$
$$F_k^2 + F_{k+3}^2 = 2(F_{k+2}^2 + F_{k+1}^2), \quad F_{2k}^2 + 1 = F_{2k-1}F_{2k+1}, \quad F_{2k}^2 = F_{2k-2}F_{2k+2} + 1,$$
$$F_{2k+1}^2 = F_{2k}F_{2k+2} + 1, \quad F_{k+2}^2 + F_k^2 = F_{k+1}^2 + F_kF_{k+3} = 3F_{k+1}^2 + 2(-1)^{k+1},$$
$$F_{2k+3}^2 + 1 = F_{2k+1}F_{2k+5}, \quad F_{k+1}F_{k+2} - F_kF_{k+3} = (-1)^k,$$
$$F_{k+2}^2 - F_{k+1}^2 = F_kF_{k+3}, \quad F_{k+1}^2 - F_kF_{k+1} - F_k^2 = (-1)^k, \quad 2F_{2k+1} = 5F_k^2 + F_{2k} + 2(-1)^k,$$
$$F_{2k+2} = F_kF_{k+2} + F_{k+2}^2 + (-1)^k, \quad F_k^2 + F_{k+4}^2 = F_{k+1}^2 + 4F_{k+2}^2 + F_{k+3}^2,$$
$$F_{2k+1}^2 = (2F_kF_{k+1})^2 + (F_{k+1}^2 - F_k^2)^2, \quad F_{2k+3}^2 = (F_kF_{k+3})^2 + (2F_{k+1}F_{k+2})^2,$$
$$F_{k+3}^2 = 2F_{k+2}^2 + 2F_{k+1}^2 - F_k^2 = 4F_{k+2}F_{k+1} + F_k^2, \quad F_{k+4}^2 = (F_{k+2} + F_{k+3})^2 = (2F_k + 3F_{k+1})^2,$$
$$5F_k^3 = F_{3k} + 3(-1)^{k+1}F_k, \quad 25F_k^5 = F_{5k} + 5(-1)^{k+1}F_{3k} + 10F_k,$$
$$F_{k+1}^3 = F_{k+3}F_k^2 + (-1)^kF_{k+2}, \quad F_{k+4}^3 = 3F_{k+3}^3 + 6F_{k+2}^3 - 3F_{k+1}^3 - F_k^3,$$
$$F_{k+2}^3 = F_{3k+3} - F_{k+1}^3 + F_k^3 = F_{k+1}^3 + F_k^3 + 3F_kF_{k+1}F_{k+2} = F_kF_{k+3}^2 + (-1)^kF_{k+1},$$
$$F_kF_{k+1}F_{k+2} = F_{k+1}^3 + (-1)^{k+1}F_{k+1}, \quad F_{k+1}F_{k+2}F_{k+6} = F_{k+3}^3 + (-1)^kF_k,$$
$$F_kF_{k+4}F_{k+5} = F_{k+3}^3 + (-1)^{k+1}F_{k+6}, \quad F_{3k+9} = F_{k+2}F_{k+5}^2 + F_{k+1}F_{k+4}^2 - F_kF_{k+3}^2,$$
$$F_{k+2}^4 = F_kF_{k+1}F_{k+3}F_{k+4} + 1, \quad (F_k^2 + F_{k+1}^2 + F_{k+2}^2)^2 = 2(F_k^4 + F_{k+1}^4 + F_{k+2}^4),$$
$$2F_{4k+6} = F_{k+3}^4 + 2F_{k+2}^4 - 2F_{k+1}^4 - F_k^4, \quad F_kF_{k+4}^3 = F_{k+3}^4 + (-1)^{k+1}(F_{k+2}F_{k+6} + 2F_{k+3}^2),$$
$$6F_{4k+4} = F_{k+3}^4 + 3F_{k+2}^4 - 3F_k^4 - F_{k-1}^4, \quad F_{k+5}^4 = 5F_{k+4}^4 + 15F_{k+3}^4 - 15F_{k+2}^4 - 5F_{k+1}^4 + F_k^4,$$

$$6F_{2k+3}^2 = F_k^4 + 4F_{k+1}^4 + 4F_{k+2}^4 + F_{k+3}^4, \quad 10F_{2k+3}^3 = F_k^6 + 8F_{k+1}^6 + 8F_{k+2}^6 + F_{k+3}^6,$$
$$6F_{2k+4}^2 = F_k^4 - 4F_{k+1}^4 - 10F_{k+2}^4 - F_{k+3}^4 + F_{k+4}^4,$$
$$56F_{2k+6}^2 = F_k^4 - 6F_{k+2}^4 - 6F_{k+4}^4 + F_{k+6}^4 - 20,$$
$$216F_{2k+8}^2 = -F_k^4 + 81F_{k+2}^4 - 520F_{k+4}^4 + 81F_{k+6}^4 - F_{k+8}^4,$$
$$1224F_{2k+9}^2 = F_k^4 + 19F_{k+3}^4 + 19F_{k+6}^4 + F_{k+9}^4 + 480,$$
$$20304F_{2k+12}^2 = F_k^4 - 46F_{k+4}^4 - 46F_{k+8}^4 + F_{k+12}^4 - 8100,$$
$$3264F_{2k+12}^2 = F_k^4 - 256F_{k+3}^4 - 4930F_{k+6}^4 - 256F_{k+9}^4 + F_{k+12}^4,$$
$$(F_kF_{k+1})^2 + (F_kF_{k+2})^2 + (F_{k+1}F_{k+2})^2 = (F_k^2 + F_{k+1}F_{k+2})^2, \quad F_{k+1}^2F_{k+3}^2 - F_k^2F_{k+4}^2 = 4(-1)^kF_{k+2}^2,$$
$$F_{k+1}^5 = F_{k+1} + F_k^3F_{k+2}F_{k+3} + (-1)^kF_kF_{k+2}F_{k+3}, \quad 6F_{5k+10} = F_{k+4}^5 + 3F_{k+3}^5 - 6F_{k+2}^5 - 3F_{k+1}^5 + F_k^5,$$
$$(F_k^2 + F_{k+1}^2)(F_{k+2}^2 + F_{k+3}^2) = F_{2k+3}^2 + 1, \quad (F_k^2 + F_{k+2}^2)(F_{k+4}^2 + F_{k+6}^2) = F_{2k+6}^2 + [2F_{k+3}^2 - 5(-1)^k]^2,$$
$$\sum_{i=0}^{5} F_{k+i} = 4F_{k+4}, \quad \sum_{i=0}^{9} F_{k+i} = 11F_{k+6}.$$

xiii) For all $k \ge 1$,

$$\operatorname{atan}\frac{1}{F_{2k+1}} + \operatorname{atan}\frac{1}{F_{2k+2}} = \operatorname{atan}\frac{1}{F_{2k}}.$$

xiv) For all $k \ge 0$,

$$\sum_{i=0}^{\lfloor k/2 \rfloor} \binom{k-i}{i} = (-1)^k \sum_{j=0}^{k} \sum_{i=0}^{\lfloor (k-j)/2 \rfloor} (-2)^j \binom{k-i}{i}\binom{k-2i}{j} = F_{k+1},$$

$$\sum_{i=0}^{\lfloor k/2 \rfloor} 5^i \binom{k+1}{2i+1} = 2^kF_{k+1}, \quad \sum_{i=0}^{k} \binom{k+i}{2i} = F_{2k+1}, \quad \sum_{i=0}^{k} \binom{k+i+1}{2i+1} = F_{2k+2}.$$

xv) For all $k \ge 1$,

$$\sum_{i=1}^{k} F_i = F_{k+2} - 1, \quad \sum_{i=1}^{k} F_{2i-1} = F_{2k}, \quad \sum_{i=1}^{k} F_{2i} = F_{2k+1} - 1, \quad \sum_{i=1}^{k} F_{4i-2} = F_{2k}^2,$$

$$\sum_{i=1}^{k} F_{4i-1} = F_{2k}F_{2k+1}, \quad \sum_{i=1}^{k} (-1)^{i+1}F_{i+1} = (-1)^{k-1}F_k, \quad \sum_{i=1}^{k} iF_i = kF_{k+2} - F_{k+3} + 2,$$

$$5\sum_{i=1}^{k} F_{2i-1}^2 = F_{4k} + 2k, \quad \sum_{i=1}^{k} F_i^2 = F_kF_{k+1}, \quad \sum_{i=1}^{2k-1} F_iF_{i+1} = F_{2k}^2, \quad \sum_{i=1}^{2k} F_iF_{i+1} = F_{2k+1}^2 - 1,$$

$$\sum_{i=1}^{k} F_iF_{i+1} = F_{k+1}^2 - \tfrac{1}{2}[1 + (-1)^k], \quad \sum_{i=1}^{k} F_iF_{3i} = F_kF_{k+1}F_{2k+1}, \quad \sum_{i=1}^{k-1} F_iF_{k-i} = \frac{1}{5}[2kF_{k+1} - (k+1)F_k,$$

$$\sum_{i=1}^{k} F_i^2F_{i+1} = \frac{1}{2}F_kF_{k+1}F_{k+2}, \quad \sum_{i=1}^{k} (F_iF_{i+1})^3 = \frac{1}{4}(F_kF_{k+1}F_{k+2})^2, \quad \sum_{i=1}^{k} \frac{1}{F_{2^i}} = 2 - \frac{F_{2^k-1}}{F_{2^k}},$$

$$\sum_{i=1}^{k} \binom{k}{i} F_i = F_{2k}, \quad \sum_{i=0}^{k} \binom{k}{i} F_{i+1} = F_{2k+1}, \quad \sum_{i=1}^{k} \binom{k+1}{i+1} F_i = F_{2k+1} - 1,$$

$$\sum_{i=1}^{k} 2^i \binom{k}{i} F_i = \sum_{i=1}^{k} 2^{k-i} \binom{k}{i} F_{k-i} = F_{3k}, \quad \sum_{i=0}^{k} 2^i \binom{k}{i} F_{i+1} = F_{3k+1}, \quad \sum_{i=0}^{k} 2^i \binom{k}{i} F_{i+2} = F_{3k+2},$$

$$\sum_{i=1}^{2k} \frac{1}{2^i}\binom{2k}{i}F_{2i} = \left(\frac{5}{4}\right)^k F_{2k}, \quad \sum_{i=0}^{2k+1} \frac{1}{2^i}\binom{2k+1}{i}F_{2i} = \frac{1}{2}\left(\frac{5}{4}\right)^k L_{2k+1},$$

$$\sum_{i=1}^{k} 2^i\binom{k}{i}F_{3i} = F_{6k}, \quad \sum_{i=1}^{2k} \frac{1}{3^i}\binom{2k}{i}F_{3i} = \left(\frac{20}{9}\right)^k F_{2k}, \quad \sum_{i=0}^{2k+1} \frac{1}{3^i}\binom{2k+1}{i}F_{3i} = \frac{2}{3}\left(\frac{20}{9}\right)^k L_{2k+1},$$

$$\sum_{i=1}^{2k} 2^{2k-i}\binom{2k}{i}F_i = 5^k F_{2k}, \quad \sum_{i=1}^{2k+1} 2^{2k+1-i}\binom{2k}{i}F_i = 5^k L_{2k+1}, \quad \sum_{i=1}^{k}\left(\frac{3}{2}\right)^i\binom{k}{i}F_i = \frac{1}{2^k}F_{4k},$$

$$\sum_{i=1}^{k}\binom{2k}{2i}F_{4i} = \tfrac{1}{2}(5^k+1)F_{2k}, \quad \sum_{i=1}^{k}\binom{2k+1}{2i}F_{4i} = \tfrac{1}{2}(5^k L_{2k+1} - F_{2k+1}),$$

$$\sum_{i=0}^{k-1}\binom{2k}{2i+1}F_{4i+2} = \tfrac{1}{2}(5^k-1)F_{2k}, \quad \sum_{i=0}^{k}\binom{2k+1}{2i+1}F_{4i+2} = \tfrac{1}{2}(5^{k+1/2}L_{2k+1} + F_{2k+1}),$$

$$\sum_{i=1}^{k} 4^i\binom{k}{i}F_i = F_{3k}, \quad \sum_{i=1}^{2k}(-1)^{2k-i}4^i\binom{2k}{i}F_{2i} = 5^k F_{6k}, \quad \sum_{i=1}^{2k+1}(-1)^{2k+1-i}4^i\binom{2k+1}{i}F_{2i} = 5^k L_{6k+3},$$

$$\sum_{i=1}^{2k}\binom{2k}{i}F_{2i} = 5^k F_{2k}, \quad \sum_{i=0}^{k}\binom{2k}{2i}F_{2i} = \frac{2}{5}\sum_{i=0}^{4}(\sin\tfrac{2i\pi}{5})(\sin\tfrac{4i\pi}{5})(1+2\cos\tfrac{i\pi}{5})^k = \tfrac{1}{2}(F_{2k} - F_k),$$

$$\sum_{i=0}^{k}\binom{2k}{2i}F_{2i-1} = \tfrac{1}{2}(F_{2k-1} + F_{k+1}), \quad \sum_{i=0}^{k}\binom{2k+1}{2i+1}F_{2i+1} = \sum_{i=0}^{k}F_{2i-1}F_{k-i} = \tfrac{1}{2}(F_{2k} + F_k),$$

$$\sum_{i=0}^{k}\binom{2k+1}{2i+1}F_{2i} = \sum_{i=0}^{k-1}F_{2i}F_{k-i-1} = \tfrac{1}{2}(F_{2k-1} - F_{k+1}),$$

$$\sum_{i=0}^{k}\binom{k}{i}F_{3i} = 2^k F_{2k}, \quad \sum_{i=0}^{k}\binom{2k}{2i}F_{4i} = \tfrac{1}{2}F_{2k}(5^k+1), \quad \sum_{i=0}^{k}\binom{2k+1}{2i}F_{4i} = \tfrac{1}{2}(5^k L_k - F_k),$$

$$\sum_{i=1}^{k}(-1)^{i+1}\binom{k}{i}F_i = \sum_{i=1}^{k}(-1)^{k-i}\binom{k}{i}F_{2i} = F_k, \quad \sum_{i=1}^{k}(-1)^{k-i}2^i\binom{k}{i}F_{2i} = F_{3k},$$

$$\sum_{i=1}^{k}(-1)^{k-i}\binom{k}{i}F_{3i} = 2^k F_k, \quad \sum_{i=1}^{k}(-1)^{k-i}\binom{k}{i}F_{6i} = 4^k F_{3k},$$

$$\sum_{i=1}^{2k}(-1)^i 2^{i-1}\binom{2k}{i}F_i = 0, \quad \sum_{i=1}^{2k+1}\binom{2k+1}{i}F_i^2 = 5^k F_{2k+1},$$

$$\sum_{i=0}^{\lfloor k/2\rfloor}(-1)^i\frac{k}{k-i}\binom{k-i}{i}F_{2k-3i} = F_k, \quad \sum_{i=0}^{\lfloor k/2\rfloor}(-1)^i\frac{k}{k-i}\binom{k-i}{i}L_{2k-3i} = L_k + 2,$$

$$\sum_{i=1}^{k}\binom{2k}{2i}F_{6i} = 2^{2k-1}(F_{4k} + F_{2k}), \quad \sum_{i=1}^{k}\binom{2k+1}{2i}F_{6i} = 4^k(F_{4k+2} - F_{2k+1}),$$

$$\sum_{i=0}^{k}\binom{2k}{2i+1}F_{6i+3} = 2^{2k-1}(F_{4k} - F_{2k}), \quad \sum_{i=0}^{k}\binom{2k+1}{2i+1}F_{6i+3} = 4^k(F_{4k+2} + F_{2k+1}),$$

$$2F_{k+2} + 2\sum_{1\le i<j\le k}F_iF_j = F_{2k+1} + F_kF_{k+1} + 1, \quad F_k^5 + F_{k+1}^5 + \frac{5}{7}\left(\frac{F_{k+2}^7 - F_{k+1}^7 - F_k^7}{F_{k+2}^2 - F_kF_{k+1}}\right) = F_{k+2}^5,$$

$$\sum_{j=1}^{k}\sum_{i=1}^{j}\sum_{h=1}^{i}\sum_{g=1}^{h}F_g^4=\frac{1}{100}[4F_{k+2}^4+(k+2)^4-5(k+2)^2].$$

xvi) For all $k\geq 2$,

$$\sum_{i=1}^{k}F_{4i-6}=F_{2k-3}F_{2k-1}-1,\qquad \sum_{i=1}^{k}F_{4i-7}=F_{2k-3}F_{2k-2}.$$

xvii) For all $n,k\geq 1$,

$$5^nF_k^{2n+1}=\sum_{i=0}^{n}(-1)^{i(k+1)}\binom{2n+1}{i}F_{[2(n-i)+1]k},$$

$$5^nF_k^{2n}=\sum_{i=0}^{n-1}(-1)^{i(k+1)}\binom{2n}{i}L_{2(n-i)k}+(-1)^{n(k+1)}\binom{2n}{n}.$$

xviii) For all $n\geq 4$,

$$\left\lfloor\left(\sum_{i=n}^{2n}\frac{1}{F_i}\right)^{-1}\right\rfloor=F_{n-2}.$$

xix) For all $n\geq 1$ and $k\geq 3$,

$$\left\lfloor\left(\sum_{i=2n}^{2kn}\frac{1}{F_i}\right)^{-1}\right\rfloor=F_{2n-2},\qquad \left\lfloor\left(\sum_{i=2n+1}^{k(2n+1)}\frac{1}{F_i}\right)^{-1}\right\rfloor=F_{2n-1}-1.$$

xx) For all $n\geq 1$ and $k\geq 2$,

$$\left\lfloor\left(\sum_{i=2n}^{2kn}\frac{1}{F_i^2}\right)^{-1}\right\rfloor=F_{2n}F_{2n-1}-1,\qquad \left\lfloor\left(\sum_{i=2n-1}^{k(2n-1)}\frac{1}{F_i^2}\right)^{-1}\right\rfloor=F_{2n-1}F_{2n-2}.$$

xxi) For all $k,l\in\mathbb{Z}$,

$$F_k=F_{l+1}F_{k-l}+F_lF_{k-l-1},\qquad F_{k+l+1}=F_{k+1}F_{l+1}+F_kF_l,$$
$$(-1)^kF_{k-l}=F_{l+1}F_k-F_lF_{k+1},\qquad F_{k+l+2}=F_{k+1}F_l+F_{k+2}F_{l+1}=F_{k+2}F_{l+2}-F_kF_l,$$
$$F_{k+l+1}^2=F_{k+1}^2F_{l+1}^2+\tfrac{1}{3}F_kF_l(F_{k+3}F_{l+3}+2F_{k-1}F_{l-1}),$$
$$F_{k+l+1}^2+F_{k-l}^2=F_{2k+1}F_{2l+1},\qquad F_k^2+F_{k+2l+1}^2=F_{2l+1}F_{2k+2l+1}^2,$$
$$F_{k+2l}^2-F_k^2=F_{2l}F_{2k+2l},\qquad F_k^2=F_{k-l}F_{k+l}+(-1)^{k+l}F_l^2,$$
$$F_{k+l+1}F_{k+l-1}=F_{k+l}^2+(F_{k+1}F_{k-1}-F_k)^2(F_{l+1}F_{l-1}-F_l^2),$$
$$F_{k+l+2}=F_kF_{l+1}+F_{k+1}F_l+F_{k+1}F_{l+1},\qquad F_{k+2}F_{l+2}=F_kF_l+F_kF_{l+1}+F_{k+1}F_l+F_{k+1}F_{l+1}.$$

xxii) For all $k,l\in\mathbb{Z}$,

$$\sum_{i=1}^{2k}2^{2k-i}\binom{2k}{i}F_{i+l}=5^kF_{2k+l},\qquad \sum_{i=1}^{2k+1}2^{2k+1-i}\binom{2k}{i}F_{i+l}=5^kL_{2k+1+l}.$$

xxiii) For all $k\geq 1$ and $l\geq 1$,

$$\sum_{i=0}^{k}\binom{k}{i}F_{3i+l}=2^kF_{2k+l}.$$

xxiv) For all $k\geq 1$ and $l\geq 2$,

$$\sum_{i=1}^{k}(-1)^{k-i}\binom{k}{i}\frac{F_{il}}{F_{l-1}^i}=\left(\frac{F_l}{F_{l-1}}\right)^kF_k.$$

xxv) For all $k \ge 1$ and $l \ge 3$,

$$\sum_{i=1}^{k} \binom{k}{i} \left(\frac{F_l}{F_{l-1}} \right)^i F_i = \frac{F_{kl}}{F_{l-1}^k}.$$

xxvi) For all $k \ge 1$ and $l \in \{0, \ldots, k\}$, $F_{2k+1-l}^2 + F_l^2 = F_{2k+1}F_{2k-2l+1}$.

xxvii) For all $k \in \mathbb{Z}$ and $l \in \{0, 1, 2, 3\}$, $F_{3k+1}F_{k+l+1}^3 + F_{3k+2}F_{k+l}^3 = F_{l-2k-1}^3 + F_{3k+1}F_{3k+2}F_{3l}$.

xxviii) For all $k \in \mathbb{Z}$ and $l \ge 1$,

$$\sum_{i=1}^{l} F_{k+i} = F_{k+l+2} - F_{k+2}.$$

xxix) For all $k \ge 0$ and $l \in \mathbb{Z}$,

$$\sum_{i=0}^{k} \binom{k}{i} F_{i+l} = F_{2k+l}.$$

xxx) For all $k, l, m \in \mathbb{Z}$,

$$F_{k+l}F_{k+m} = F_k F_{k+l+m} + (-1)^k F_l F_m,$$

$$F_{k+l+m+3} + F_k F_l F_m = F_{k+2}F_{l+2}F_{m+2} + F_{k+1}F_{l+1}F_{m+1}.$$

xxxi) For all $n \ge 1$,

$$\prod_{i=1}^{n} \left(1 + 4\sin^2 \frac{2i\pi}{n} \right) = [1 + F_n - 2F_{n+1} + (-1)^n]^2,$$

$$F_{2n+1} = \prod_{i=1}^{n} \left(5 - 4\sin^2 \frac{2i\pi}{2n+1} \right), \quad F_{4n+2} = \prod_{i=1}^{n} \left(5 + 4\sin^2 \frac{i\pi}{2n+1} \right).$$

xxxii) For all $n \ge 1$ and $z \in \mathbb{C}$,

$$\prod_{i=0}^{n-1} \left[3 + 2\cos\left(\frac{2i\pi}{n} - z \right) \right] = 5F_n^2 + 4(-1)^n \sin^2 \frac{nz}{2}.$$

xxxiii) For all $n \ge 2$,

$$F_n = \frac{2^{n-1}}{n} \sqrt{\prod_{i=1}^{n-1} \left[1 - \left(\cos \frac{i\pi}{n} \right) \cos \frac{3i\pi}{n} \right]}.$$

xxxiv) For all $n \ge 4$,

$$F_n = \prod_{i=1}^{\lfloor (n-1)/2 \rfloor} \left(1 + 4\sin^2 \frac{i\pi}{n} \right) = \prod_{i=1}^{\lfloor (n-1)/2 \rfloor} \left(1 + 4\cos^2 \frac{i\pi}{n} \right) = \prod_{i=1}^{\lfloor (n-1)/2 \rfloor} \left(3 + 2\cos \frac{2i\pi}{n} \right).$$

xxxv) For all $k \ge 1$,

$$\sum_{0 \le j \le i \le k} \binom{k}{i-j} \binom{k-i}{j} = F_{2k-1}, \quad \sum_{i,j=1}^{k} \binom{k-i}{j-1} \binom{k-j}{i-1} = F_{2k},$$

$$\sum_{i,j=0}^{k} \binom{k+i}{2j} \binom{k+j}{2i} = F_{4k-1}, \quad \sum_{i,j=0}^{k} \binom{k+i}{2j-1} \binom{k+j}{2i} = F_{4k}, \quad \sum_{i,j=0}^{k} \binom{k+i}{2j+1} \binom{k+j}{2i+1} = F_{4k-3}.$$

xxxvi) Let $n \ge 0$. Then, there exists $k \ge 0$ such that $n = F_k$ if and only if either $\sqrt{5n^2 - 4}$ or $\sqrt{5n^2 + 4}$ is an integer.

xxxvii) Let $n \ge 1$, and define $\mathcal{A}_n \triangleq \{(i_1, \ldots, i_k) \in \times_{i=1}^{k} \{1, 2\} \colon k \ge 1 \text{ and } \sum_{j=1}^{k} i_j = n\}$. Then, $\operatorname{card}(\mathcal{A}_n) = F_{n+1}$.

xxxviii) For all $n \ge 0$, $9F_n^2 \le F_{n+3}^2$.

xxxix) For all $n \notin \{-2, -1\}$,

$$\left(1 + \frac{F_n}{F_{n+1}} - \frac{F_n}{F_{n+2}}\right)^2 = 1 + \left(\frac{F_n}{F_{n+1}}\right)^2 + \left(\frac{F_n}{F_{n+2}}\right)^2.$$

Source: [107, pp. 63, 330, 331], [109, pp. 186, 187, 239–241], [399], [411, pp. 10–12, 70, 78, 125, 126, 144], [483, 663, 665, 666, 687, 755, 885, 890, 892], [993, p. 206], [1008], [1083, p. 63], [1148], [1158, p. 297], [1241, pp. 10, 11, 12, 38, 39, 56, 57, 61, 109, 115, 116, 117, 121], [1236, 1237, 1238, 1239, 1240, 1585, 1603, 1612], [1674, pp. 6–8, 239–241, 362, 363], [1675, pp. 78, 79], [1741], [1757, pp. 72, 175], [1928, 2004, 2006, 2008, 2005], [2068, pp. 110–113], [2155, 2148, 2157, 2165, 2454, 2598], [2764, pp. 37, 70–72, 182, 183], [2810, 2864, 2865]. **Remark:** F_n is the nth *Fibonacci number*. **Remark:** Concerning *ii)*, $\binom{n-k-1}{k}$ is the number of k-element subsets of $\{1,\dots,n-2\}$ that do not contain a pair of consecutive integers. See [109, p. 187]. **Remark:** *xi)* is *Zeckendorf's theorem*. **Related:** Fact 13.9.2. The generating function is given by Fact 13.9.3.

Fact 1.17.2. Define $L_1 \triangleq 1$, $L_2 \triangleq 3$ and, for all $k \in \mathbb{Z}$, define L_k by $L_{k+2} = L_{k+1} + L_k$. Then,

$$(L_i)_{i=-5}^{16} = (-11, 7, -4, 3, -1, 2, 1, 3, 4, 7, 11, 18, 29, 47, 76, 123, 199, 322, 521, 843, 1364, 2207).$$

Then, for all $k \in \mathbb{Z}$, the following statements hold:

i) $L_{-k} = (-1)^k L_k$.

ii) L_{3k} is even.

iii) If k is prime, then $L_k \overset{k}{\equiv} 1$.

iv) 5 divides $L_{k+1} - 3L_k$, $3^{k-1} - L_k$, $kL_{k+1} + 2F_k$, and $T_k L_{k+1} + (k+1)F_k$.

v) 10 divides $4kL_{k+1} - 2F_k$.

vi) $\gcd\{L_k, L_{k+1}\} = \gcd\{L_k, L_{k+2}\} = 1$.

For all $k \in \mathbb{Z}$,

$$L_k^2 = L_{2k} + 2(-1)^k, \quad L_{k+1}^2 = L_k L_{k+2} + 5(-1)^{k+1}, \quad L_{2k}^2 = L_{4k} + 2,$$
$$L_{k+1}^2 - L_{k+1}L_k - L_k^2 = 5(-1)^{k+1}, \quad L_k^2 + L_{k+1}^2 = L_{2k} + L_{2k+2},$$
$$L_{3k} = L_k[L_{2k} + (-1)^{k+1}], \quad (L_k^2 + L_{k+1}^2 + L_{k+2}^2)^2 = 2(L_k^4 + L_{k+1}^4 + L_{k+2}^4),$$
$$L_{k+3}^2 = 2L_{k+2}^2 + 2L_{k+1}^2 - L_k^2, \quad L_{k+4}^3 = 3L_{k+3}^3 + 6L_{k+2}^3 - 3L_{k+1}^3 - L_k^3,$$
$$L_{k+2}^3 + L_{k+1}^3 - L_k^3 = 5L_{3k+3}, \quad L_{k+5}^4 = 5L_{k+4}^4 + 15L_{k+3}^4 - 15L_{k+2}^4 - 5L_{k+1}^4 + L_k^4,$$
$$2L_{k+1} = L_k + 5F_k, \quad 2L_{k+2} = 3L_k + 5F_k, \quad L_{k+1} = F_k + F_{k+2}, \quad L_{k+2} = 3F_{k+1} + F_k,$$
$$5F_{k+1} = L_k + L_{k+2}, \quad 2F_{k+1} = F_k + L_k, \quad 2F_{k+2} = 3F_k + L_k, \quad F_{k+4} = F_k + L_{k+2}, \quad F_{2k+4} = F_{2k} + L_{2k+2},$$
$$2L_k = L_k^2 + 5F_k^2, \quad L_{2k} = 5F_k^2 + 2(-1)^k, \quad 5F_{2k+1} = L_k^2 + L_{k+1}^2, \quad F_{2k+1} + F_k L_k = F_{k+1}L_{k+1},$$
$$F_{2k} = F_k L_k, \quad F_{k+1}L_k = F_{2k+1} + (-1)^k, \quad 2L_{2k} = L_k^2 + 5F_k^2, \quad L_k^2 = 5F_k^2 + 4(-1)^k,$$
$$L_{k+3}^2 = F_k F_{k+4} + F_{k+4}F_{k+5} + F_{k+2}^2, \quad 5F_{2k+3}F_{2k-3} = L_{4k} + 18, \quad F_k \overset{2}{\equiv} L_k,$$
$$L_{k+1}^2 + F_{k+1}^2 = 2F_{k+2}^2 + 2F_k^2, \quad 25F_{k+1}^2 + L_{k+1}^2 = 2L_{k+2}^2 + 2L_k^2,$$
$$L_{2k+7} = F_{k+4}^2 + L_{k+3}^2 - F_k F_{k+4} + F_{2k} + (-1)^k,$$
$$F_k^4 + L_{k+2}^4 + F_{k+4}^4 = 9(F_{k+1}^4 + F_{k+2}^4 + F_{k+3}^4), \quad L_k^4 + L_{k+4}^4 + 625F_{k+2}^4 = 9(L_{k+1}^4 + L_{k+2}^4 + L_{k+3}^4),$$
$$2F_k^2 L_{k+6}^2 = -17F_{k+1}^4 + 57F_{k+2}^4 + 402F_{k+3}^4 + 113F_{k+4}^4 - 25F_{k+5}^4,$$
$$50L_k^2 F_{k+6}^2 = -17L_{k+1}^4 + 57L_{k+2}^4 + 402L_{k+3}^4 + 113L_{k+4}^4 - 25L_{k+5}^4,$$

$$F_k^3 + F_{k+1}^3 + 3F_kF_{k+1}F_{k+2} = F_{k+2}^3, \quad L_k^3 + L_{k+1}^3 + 3L_kL_{k+1}L_{k+2} = L_{k+2}^3,$$

$$F_{k+1} = \frac{1}{2}\left[F_k + (-1)^{\min\{k,0\}}\sqrt{5F_k^2 + 4(-1)^k}\right], \quad L_{k+1} = \frac{1}{2}\left[L_k + (-1)^{\min\{k+1,0\}}\sqrt{5(L_k^2 - 4(-1)^k)}\right].$$

For all $k \geq 1$,

$$\sum_{i=0}^{\lfloor k/2 \rfloor} \frac{k}{k-i}\binom{k-i}{i} = L_k,$$

$$\sum_{i=0}^{k} L_i = L_{k+2} - 1, \quad \sum_{i=1}^{k} L_{2i-1} = L_{2i} - 2, \quad \sum_{i=0}^{k} L_{2i} = L_{2k+1} + 1, \quad \sum_{i=0}^{k} L_i^2 = L_kL_{k+1} + 2,$$

$$\sum_{i=1}^{k} L_{2i-1}^2 = F_{4k} - 2k, \quad \sum_{i=1}^{k} iL_i = kL_{k+2} - L_{k+3} + 4, \quad \sum_{i=0}^{k} 2^iL_i = 2^{n+1}F_{k+1},$$

$$\sum_{i=0}^{k} 3^iL_i + \sum_{i=0}^{k+1} 3^{i-1}F_i = 3^{k+1}F_{k+1}, \quad \sum_{i=0}^{k}\binom{k}{i}L_i = L_{2k}, \quad \sum_{i=0}^{2k}(-1)^i2^{i-1}\binom{2k}{i}L_i = 5^i,$$

$$\sum_{i=0}^{k}\binom{k}{i}L_iL_{k-i} = 2^kL_k + 2, \quad 5\sum_{i=0}^{k}\binom{k}{i}F_iF_{k-i} = 2^kL_k - 2, \quad \sum_{i=0}^{k}\binom{k}{i}F_iL_{k-i} = 2^kF_k,$$

$$\sum_{i=0}^{k}(-1)^iL_{k-2i} = 2F_{k+1}, \quad \sum_{i=0}^{k}(-1)^{k-i}\binom{k}{i}L_i = (-1)^kL_k, \quad \sum_{i=0}^{k}(-1)^{k-i}\binom{k}{i}L_{2i} = L_k,$$

$$\sum_{i=0}^{k-1}(-1)^i\binom{2k}{i}L_{2k-2i} = 1 + (-1)^{n-1}\binom{2k}{k}, \quad \sum_{i=0}^{k}(-1)^i\binom{2k+1}{i}L_{2k+1-2i} = 1,$$

$$\sum_{i=0}^{2k}\binom{2k}{i}L_{2i} = \sum_{i=0}^{2k}\binom{2k}{i}L_i^2 = 5^kL_{2k}, \quad \sum_{i=0}^{2k+1}\binom{2k+1}{i}L_{2i} = \sum_{i=0}^{2k+1}\binom{2k+1}{i}L_i^2 = 5^{k+1}F_{2k+1},$$

$$\sum_{i=0}^{2k+1}\binom{2k+1}{i}F_{2i} = 5^kL_{2k+1}, \quad \sum_{i=0}^{2k}\binom{2k}{i}F_i^2 = 5^{k-1}L_{2k}, \quad \sum_{i=0}^{k}F_iL_{k-i} = (k+1)F_k, \quad \sum_{i=0}^{\lfloor k/2 \rfloor}5^i\binom{k}{2i} = 2^{k-1}L_k,$$

$$L_{2k} = 2 + \prod_{i=0}^{k-1}\left(1 + 4\sin^2\frac{i\pi}{k}\right) = 2\prod_{i=1}^{k}\left(\frac{1}{4} + \frac{5}{4}\tan^2\frac{(2i-1)\pi}{4k}\right), \quad L_{2k+1} = \prod_{i=1}^{k}\left(\frac{1}{4} + \frac{5}{4}\tan^2\frac{2i\pi}{2k+1}\right),$$

$$\sum_{i=1}^{n}\frac{\tan^2\frac{2i\pi}{2k+1}}{1 + 5\tan^2\frac{2i\pi}{2k+1}} = \frac{(2k+1)F_{2k}}{4L_{2k+1}}, \quad \sum_{i=1}^{n}\frac{\tan^2\frac{(2i-1)\pi}{4k}}{1 + 5\tan^2\frac{(2i-1)\pi}{4k}} = \frac{kF_{2k-1}}{2L_{2k}},$$

$$1 + 10\sum_{i=1}^{\lfloor (n-1)/2 \rfloor}\frac{\cos^2\frac{i\pi}{k}}{3 + 2\cos\frac{2i\pi}{k}} = \frac{kL_{k-1}}{2F_k}.$$

For all $k, l \in \mathbb{Z}$,

$$L_{k+l} + (-1)^lL_{k-l} = L_kL_l, \quad L_{k+l} = 5F_kF_l + (-1)^lL_{k-l}, \quad L_{2k}L_{2l} = L_{k+l}^2 + 5F_{k-l}^2,$$

$$5F_kF_l = L_{k+l} + (-1)^{l+1}L_{k-l}, \quad 5F_k^2 = L_{2k} + 2(-1)^{k+1},$$

$$2L_{k+l} = L_kL_l + 5F_kF_l, \quad 2(-1)^lL_{k+l} = L_kL_l - 5F_kF_l, \quad 2(-1)^lL_{k-l} = L_kL_l - 5F_kF_l,$$

$$2F_{k+l} = F_kL_l + F_lL_k, \quad F_{k+l} + (-1)^lF_{k-l} = F_kL_l, \quad F_{k+l} + (-1)^lF_{k-l} = F_kL_l,$$

$$F_kL_l = L_kF_l + 2(-1)^lF_{k-l}, \quad L_{k+l+1} = F_{k+1}L_{l+1} + F_kL_l, \quad L_{k+l+1}^2 + L_{k-l}^2 = 5F_{2k+1}F_{2l+1},$$

$$F_k^4 + [(-1)^{l+1}L_{2l} + 1](F_{k+l}^4 + F_{k+2l}^4) + F_{k+3l}^4 = F_lL_{2l}F_{3l}F_{2k+3l}^2 + 10(-1)^lF_{l-1}F_l^4F_{l+1},$$

$$(-1)^{l+1}F_k^6 + (L_{4l}+1)[F_{k+l}^6 + (-1)^{l+1}F_{k+2l}^6] + F_{k+3l}^6 = F_lF_{3l}F_{5l}F_{2k+3l}^3 + 15(-1)^lF_{l-1}F_l^4F_{l+1}F_{3l}F_{2k+3l},$$

$$(-1)^{l+1}F_lF_k^4 + (-1)^lL_l^3F_{2l}(F_{k+l}^4 + F_{k+3l}^4) - [L_{4l} + 2(-1)^lL_{2l} + 4]F_{3l}F_{k+2l}^4 + (-1)^{l+1}F_lF_{k+4l}^4 = 3F_{2l}^2F_{3l}F_{2k+4l}^2.$$

For all $k, l, m \in \mathbb{Z}$,

$$L_{m+k}L_{m+l-1} = L_{m-1}L_{m+l+k} + (-1)^{m-1}F_{k+1}(L_l - 2L_{l+1}).$$

If p is prime, then

$$\sum_{i=0}^{\lfloor (p-1)/2 \rfloor} (-1)^i \binom{2i}{i} \overset{p}{\equiv} F_p, \quad \sum_{i=0}^{\lfloor (p-1)/2 \rfloor} 5^i \binom{p}{2i} \overset{p}{\equiv} L_p, \quad \sum_{i=0}^{\lfloor (p-1)/2 \rfloor} (-1)^i (i+1)^{p-2} \binom{2i}{i} \overset{p}{\equiv} L_{p-1}.$$

For all $n \geq 1$,

$$\sum_{i=1}^{n} F_{n-i}H_i = F_{2n}H_n - \sum_{i=1}^{n} \frac{F_{2n-i}}{i}, \quad \sum_{i=1}^{n} 2^{n-i}F_{n-i}H_i = F_{3n}H_n - \sum_{i=1}^{n} \frac{2^iF_{3n-2i}}{i},$$

$$\sum_{i=1}^{n} L_{n-i}H_i = L_{2n}H_n - \sum_{i=1}^{n} \frac{L_{2n-i}}{i}, \quad \sum_{i=1}^{n} 2^{n-i}L_{n-i}H_i = L_{3n}H_n - \sum_{i=1}^{n} \frac{2^iL_{3n-2i}}{i},$$

$$\sum_{i=1}^{n} \binom{n+i+1}{n-i} H_i = \sum_{i=1}^{n} \frac{1}{i}(L_{2i} - 2)F_{2n-2i+2}, \quad \sum_{i=1}^{n} \binom{n+i}{n-i} \frac{2n+1}{2i+1} H_i = \sum_{i=1}^{n} \frac{1}{i}(L_{2i} - 2)L_{2n-2i+1},$$

$$\sum_{i=1}^{2n} F_{2i}^3 = \frac{1}{4}F_{2n-1}F_{2n}^2L_{2n+1}^2L_{2n+2}, \quad \sum_{i=1}^{2n+1} F_{2i}^3 = \frac{1}{4}L_{2n}L_{2n+1}^2F_{2n+2}^2F_{2n+3},$$

$$\sum_{i=1}^{n} F_{2i}^3 = \frac{1}{4}(F_{2n+1} - 1)^2(F_{2n+1} + 2), \quad \prod_{i=0}^{n-1}[1 + e^{(2\pi i/n)j} - e^{(4\pi i/n)j}] = (-1)^{n+1}L_n + (-1)^n + 1,$$

$$\operatorname{atan} \frac{2}{L_{2n-1}} = 2 \operatorname{atan} \frac{1}{L_{2n}} + \operatorname{atan} \frac{2}{L_{2n+1}},$$

$$\operatorname{atan} \frac{2}{L_{2n-1}} = \operatorname{atan} \frac{1}{F_{2n}} + \operatorname{atan} \frac{1}{L_{2n}}, \quad \operatorname{atan} \frac{2}{L_{2n+1}} = \operatorname{atan} \frac{1}{F_{2n}} - \operatorname{atan} \frac{1}{L_{2n}},$$

$$\operatorname{atan} \frac{1}{F_{2n-1}} = \operatorname{atan} \frac{1}{L_{2n-2}} + \operatorname{atan} \frac{1}{L_{2n}}, \quad \operatorname{atan} \frac{2}{\sqrt{5}F_{2n}} = \operatorname{atan} \frac{\sqrt{5}}{L_{2n+1}} + \operatorname{atan} \frac{1}{\sqrt{5}F_{2n+1}},$$

$$\operatorname{atan} \frac{F_{2n}}{F_{2n+1}} = \sum_{i=1}^{2n} \operatorname{atan} \frac{1}{L_{2i}} = \operatorname{atan} 1 - \frac{1}{2} \operatorname{atan} \frac{1}{2} - \frac{1}{2} \operatorname{atan} \frac{2}{L_{4n+1}},$$

$$\operatorname{atan} \frac{F_{2n-1}}{F_{2n}} = \operatorname{atan} 2 - \sum_{i=1}^{2n-1} \operatorname{atan} \frac{1}{L_{2i}} = \operatorname{atan} 1 - \frac{1}{2} \operatorname{atan} \frac{1}{2} + \frac{1}{2} \operatorname{atan} \frac{2}{L_{4n-1}},$$

$$[5F_k^2 + (-1)^k 4]^{2/3} L_k^{2/3} = 5^{1/3}[L_k^2 + (-1)^{k+1}4]^{2/3} F_k^{2/3} + (-1)^k 4,$$

$$\sum_{i=1}^{n} \frac{2}{F_{i+3}} \leq \log F_{n+2}, \quad \sum_{i=1}^{n} \frac{2}{L_{i+3}} \leq \log L_{n+2}.$$

For all $n, k \geq 1$,

$$L_k^{2n+1} = \sum_{i=0}^{n} (-1)^{ik} \binom{2n+1}{i} L_{[2(n-i)+1]k}, \quad L_k^{2n} = \sum_{i=0}^{n-1} (-1)^{ik} \binom{2n}{i} L_{2(n-i)k} + (-1)^{nk} \binom{2n}{n}.$$

For all $k \geq 1$ and $n \geq 1$,

$$[\tfrac{1}{2}(L_k \pm \sqrt{5}F_k)]^n = \tfrac{1}{2}(L_{nk} \pm \sqrt{5}F_{nk}), \quad F_{2^k n} = F_n \prod_{i=1}^{k} L_{2^{k-i}n}, \quad \sum_{i=0}^{k} (-1)^{in} \frac{1}{L_{(i+1)n}L_{in}} = \frac{F_{(k+1)n}}{2F_n L_{(k+1)n}}.$$

Source: [12], [107, pp. 36, 195], [109, p. 187], [366, 370], [411, pp. 10–12, 78, 125, 126, 144], [483, 665, 687, 878, 1239], [1241, pp. 97–99, 108, 110], [1397, 1603], [1674, pp. 6–8, 239–241, 362, 363], [1675, pp. 78, 79], [1677, 1793, 2576, 2004, 2007], [2068, p. 112], [2100, 2149, 2575, 2598], [2764, pp. 70–72, 182, 183], and [2806]. **Remark:** L_n is the nth *Lucas number*. **Remark:** The generating function is given by Fact 13.9.3. **Remark:** F_n and L_n are analogous to the sine and cosine functions, respectively. See [1793].

Fact 1.17.3. Define $P_1 \triangleq 1$, $P_2 \triangleq 2$ and, for all $k \in \mathbb{Z}$, define P_k by $P_{k+2} = 2P_{k+1} + P_k$. Then,

$$(P_i)_{i=-5}^{14} = (29, -12, 5, -2, 1, 0, 1, 2, 5, 12, 29, 70, 169, 408, 985, 2378, 5741, 13860, 33461, 80782).$$

For all $k \in \mathbb{Z}$,

$$P_k = \frac{\sqrt{2}}{4}[(1+\sqrt{2})^k - (1-\sqrt{2})^k], \quad P_{k+1}P_{k-1} = P_k^2 + (-1)^k,$$

$$P_{2k+1} = P_k^2 + P_{k+1}^2, \quad 5P_{6k+3} = P_{3k}^2 + P_{3k+3}^2,$$

$$P_{k+3}^2 = 5P_{k+2}^2 + 5P_{k+1}^2 - P_k^2, \quad P_{k+4}^3 = 12P_{k+3}^3 + 30P_{k+2}^3 - 12P_{k+1}^3 - P_k^3,$$

$$P_{k+5}^4 = 29P_{k+4}^4 + 174P_{k+3}^4 - 174P_{k+2}^4 - 29P_{k+1}^4 + P_k^4,$$

$$(2P_kP_{k+1})^2 + (P_{k+1}^2 - P_k^2)^2 = (P_{k+1}^2 + P_k^2)^2.$$

For all $k \geq 1$,

$$P_k = \sum_{i=0}^{\lfloor (k-1)/2 \rfloor} 2^i \binom{k}{2i+1}, \quad P_{k+1} = \sum_{0 \le i \le j \le k} \binom{k-i}{j}\binom{j}{i},$$

$$\sum_{i=1}^{4k+1} P_i = \left[\sum_{i=1}^{n} 2^i \binom{2n+1}{2i}\right]^2 = (P_{2k} + P_{2k+1})^2, \quad \sum_{i=0}^{\lfloor (k-1)/4 \rfloor} \frac{1}{16^i}\binom{k-1-2i}{2i} = \frac{1}{2^k}(P_k + k),$$

$$2^{2-k} \sum_{i=0}^{\lfloor (k-3)/4 \rfloor} \binom{4k-2}{2k-8i-5} = 2^{k-2}(4^{k-1}+1) - P_{2k-1}, \quad \sum_{i=0}^{n} 5^i 2^{n-i}\binom{n}{i} P_i = P_{3n}.$$

Source: [687, 893, 1155, 1612, 1926, 2419]. **Remark:** P_n is the nth *Pell number*.

1.18 Facts on Arrangement, Derangement, and Catalan Numbers

Fact 1.18.1. For all $n \geq 1$, let a_n denote the nth *arrangement number*, which is the number of k-tuples whose components are distinct elements of $\{1, \ldots, n\}$ and where $0 \leq k \leq n$. Define $a_0 = 1$. Then,

$$(a_i)_{i=0}^{12} = (1, 2, 5, 16, 65, 326, 1957, 13700, 109601, 986410, 108505112, 1302061345, 16926797486).$$

For all $n \geq 1$,

$$a_n = \sum_{i=0}^{n} n^{\underline{i}} = n! \sum_{i=0}^{n} \frac{1}{i!} = \lfloor n!e \rfloor, \quad a_{n+1} = na_{n-1} + 1.$$

Remark: a_n is the nth *arrangement number*. See [771, p. 75]. **Remark:** The five arrangements of $\{1, 2\}$ are $\varnothing$, (1), (2), (1, 2), and (2, 1). **Remark:** The generating function is given by Fact 13.4.9. **Related:** Fact 14.8.1.

Fact 1.18.2. For all $n \ge 1$, let d_n denote the number of permutations of $(1,\dots,n)$ that leave no component unchanged. Define $d_0 = 1$. Then,

$$(d_i)_{i=0}^{13} = (1,0,1,2,9,44,265,1854,14833,133496,1334961,14684570,176214841,2290792932).$$

For all $n \ge 1$,

$$d_n = n!\sum_{i=0}^{n}(-1)^i\frac{1}{i!} = \sum_{i=0}^{n}(-1)^i\binom{n}{i}(n-i)! = \left\lfloor\frac{n!}{e}+\frac{1}{2}\right\rfloor = \left\lfloor\frac{n!}{e}+\frac{1}{n}\right\rfloor = \int_0^\infty e^{-x}(x-1)^n\,\mathrm{d}x,$$

and, for all $n \ge 1$,

$$d_{n+1} = n(d_n + d_{n-1}) = (n+1)d_n + (-1)^{n+1}, \quad n! = \sum_{i=0}^{n}\binom{n}{i}d_{n-i}, \quad d_n \overset{n}{\equiv} (-1)^n.$$

Finally,

$$\lim_{n\to\infty}\frac{d_n}{n!} = \frac{1}{e}.$$

Source: $d_n \overset{n}{\equiv} (-1)^n$ is given in [14]. **Remark:** d_n is the nth *derangement number*. See [621, pp. 57, 58] and [1345]. **Remark:** The permutation $(1,2,3,4,5) \mapsto (3,1,2,4,5)$ is not a derangement, but $(1,2,3,4,5) \mapsto (3,1,2,5,4)$ is a derangement. Each derangement is represented by a permutation matrix whose diagonal entries are zero. **Remark:** The generating function is given by Fact 13.4.9. **Related:** Fact 1.18.3.

Fact 1.18.3. For all $n_1,\dots,n_k \ge 1$, let $D_{n_1,\dots,n_k}$ denote the number of permutations of $(1,\dots,1,2,\dots,2,\dots,n,\dots,n)$ that leave no component unchanged, where i appears n_i times. Then,

$$D_{n_1,\dots,n_k} = (-1)^{\sum_{i=1}^k n_i}\int_0^\infty e^{-x}\prod_{i=1}^{n} L_{n_i}(x)\,\mathrm{d}x.$$

Source: [1003]. **Remark:** $D_{n_1,\dots,n_k}$ is a *generalized derangement number*, where the nth derangement number is $d_n = D_{1,\dots,1}$. See Fact 1.18.2. L_n is the nth Laguerre polynomial. See Fact 13.2.9.

Fact 1.18.4. Let $n \ge 0$, and let C_n denote the number of ways that n factors can be grouped for multiplication. Then,

$$(C_i)_{i=0}^{15} = (1,1,2,5,14,42,132,429,1430,4862,16796,58786,208012,742900,2674440,9694845).$$

For all $n \ge 1$,

$$C_n = \binom{2n}{n} - \binom{2n}{n+1} = 2\binom{2n}{n} - \binom{2n+1}{n} = 4\binom{2n-1}{n} - \binom{2n+1}{n} = \binom{2n+1}{n+1} - 2\binom{2n}{n+1},$$

$$C_n = \frac{1}{n}\binom{2n}{n-1} = \frac{1}{n}\binom{2n}{n+1} = \frac{1}{n+1}\binom{2n}{n} = \frac{1}{2n+1}\binom{2n+1}{n},$$

$$C_n = \frac{2^n(2n-1)!!}{n!} = \prod_{i=2}^{n}\frac{n+i}{i} = \frac{4^n\Gamma(n+\frac{1}{2})}{\sqrt{\pi}\,\Gamma(n+2)},$$

$$C_n = \frac{1}{n+1}\sum_{i=0}^{n}\binom{n}{i}^2 = \frac{1}{n}\sum_{i=1}^{n}\binom{n}{i}\binom{n}{i-1} = \sum_{i=0}^{\lfloor n/2\rfloor}\left[\binom{n}{i} - \binom{n}{i-1}\right]^2 = \sum_{i=0}^{\lfloor n/2\rfloor}\left[\frac{n+1-2i}{n+1}\binom{n+1}{i}\right]^2,$$

$$C_{n+1} = \frac{4n+2}{n+2}C_n - 2C_n + \frac{2}{n}\binom{2n}{n-2} = \sum_{i=0}^{\lfloor n/2\rfloor}\binom{n}{2i}2^{n-2i}C_i, \quad \sum_{i=0}^{n}\binom{2n-2i}{n-i}C_i = \binom{2n+1}{n},$$

$$C_{n+1} = \sum_{i=0}^{n} C_iC_{n-i} = \frac{n+3}{2n}\sum_{i=1}^{n} C_iC_{n+1-i} = \frac{1}{4^{n+1}}\sum_{i=0}^{n+1} C_{2i}C_{2n+2-2i},$$

$$\sum_{i=1}^{n} iC_iC_{n-i} = \frac{n}{2}C_{n+1}, \quad \sum_{i=1}^{n} i^2C_iC_{n-i} = \frac{n^2+2n+2}{2}C_{n+1} - 4^n,$$

$$\sum_{i=1}^{n} i^3C_iC_{n-i} = \frac{n}{2}[(n^2+3n+3)C_{n+1} - 3(4^n)], \quad C_{2n+1} = \sum_{i=1}^{n+1}\left[\frac{2i}{n+1+i}\binom{2n+1}{n+1-i}\right]^2,$$

$$\sum_{i=0}^{n-1} \frac{i+1}{2i+1}C_iC_{n-i+1} = \frac{1}{2(2n+1)}\left[(n+1)C_n + \frac{2^{4n-1}}{n(n+1)C_n}\right].$$

For all $n \ge 2$,

$$\sum_{i=1}^{\lfloor n/2 \rfloor}(-1)^i\binom{n-i}{i}C_{n-1-i} = 0.$$

Furthermore, C_n is odd if and only if there exists $k \ge 1$ such that $n = 2^k - 1$. In addition, C_n is prime if and only if $n = 3$. Finally,

$$\lim_{n\to\infty} \frac{C_{n+1}}{C_n} = 4.$$

Source: [124, 666, 766, 1151, 1153], [1158, p. 299], [1675, pp. 112, 123, 127, 129, 329, 330], [1743, 1755], [2068, pp. 184, 186], and [2217]. **Remark:** C_n is the nth *Catalan number*. See Fact 13.4.2 and Fact 13.9.1. **Remark:** The generating function is given by Fact 13.4.9. **Remark:** Additional interpretations of the Catalan numbers are given in [2521].

1.19 Facts on Cycle, Subset, Eulerian, Bell, and Ordered Bell Numbers

Fact 1.19.1. For $n \ge k \ge 1$, let $\begin{bmatrix} n \\ k \end{bmatrix}$ denote the number of permutations of $(1,\ldots,n)$ that have exactly k cycles. Furthermore, define $\begin{bmatrix} 0 \\ 0 \end{bmatrix} \triangleq 1$, and, for all $k \ge 1$, define $\begin{bmatrix} k \\ 0 \end{bmatrix} \triangleq 0$. Then, the following statements hold:

i) Let $n \ge 1$. Then,

$$\begin{bmatrix} n \\ 1 \end{bmatrix} = (n-1)!, \quad \begin{bmatrix} n \\ 2 \end{bmatrix} = (n-1)!H_{n-1},$$

$$\begin{bmatrix} n \\ 3 \end{bmatrix} = \frac{(n-1)!}{2}(H_{n-1}^2 - H_{n-1,2}), \quad \begin{bmatrix} n \\ 4 \end{bmatrix} = \frac{(n-1)!}{3!}(H_{n-1}^3 - 3H_{n-1}H_{n-1,2} + 2H_{n-1,3}).$$

ii) Let $n \ge 0$. Then, $\begin{bmatrix} n \\ n \end{bmatrix} = 1$ and $\sum_{i=0}^{n}\begin{bmatrix} n \\ i \end{bmatrix} = n!$.

iii) Let $n \ge 1$. Then, $\begin{bmatrix} n \\ n-1 \end{bmatrix} = \binom{n}{2}$.

iv) Let $n \ge 2$. Then, $\begin{bmatrix} n \\ n-2 \end{bmatrix} = 2\binom{n}{3} + 3\binom{n}{4} = \frac{3n-1}{4}\binom{n}{3}$.

v) Let $n \ge 3$. Then, $\begin{bmatrix} n \\ n-3 \end{bmatrix} = 6\binom{n}{4} + 20\binom{n}{5} + 15\binom{n}{6} = \binom{n}{2}\binom{n}{4}$.

vi) Let $n \ge 4$. Then, $\begin{bmatrix} n \\ n-4 \end{bmatrix} = \frac{1}{48}(15n^3 - 30n^2 + 5n + 2)$.

vii) Let $n > k \geq 1$. Then, $\left[\begin{matrix} n \\ n-k \end{matrix}\right] = (-1)^k \binom{n-1}{k} \frac{\mathrm{d}^k}{\mathrm{d}z^k} \frac{z^n}{(e^z-1)^n}\bigg|_{z=0}$.

viii) Let $n \geq k \geq 1$. Then,

$$\left[\begin{matrix} n \\ n-k \end{matrix}\right] = (-1)^k \frac{1}{(n-k-1)!} \sum (-1)^\kappa \frac{(n+\kappa-1)!}{\prod_{j=1}^k i_j![(j+1)!]^{i_j}},$$

where $\kappa \triangleq \sum_{j=1}^k i_j$ and the sum is taken over all k-tuples $(i_1, \ldots, i_k)$ of nonnegative integers such that $\sum_{j=1}^k ji_j = k$.

ix) Let $n \geq k \geq 1$. Then,

$$\sum \frac{1}{\prod_{j=1}^k i_j} = \frac{k!}{n!} \left[\begin{matrix} n \\ k \end{matrix}\right],$$

where the sum is taken over all k-tuples $(i_1, \ldots, i_k)$ of positive integers such that $\sum_{j=1}^k i_j = n$.

Source: [411, pp. 93–96], [1219, pp. 257–267], and [3024, p. 139]. *viii*) is given in [1930], and *ix*) is given in [771, p. 172]. **Remark:** The permutation $(1,2,3,4,5) \mapsto (3,1,2,5,4)$ has two cycles, while the permutation $(1,2,3,4,5) \mapsto (3,1,2,4,5)$ has three cycles. Each cycle is represented by a diagonally located block in the canonical form of a permutation matrix given by Fact 7.18.14. **Remark:** $\left[\begin{matrix} n \\ k \end{matrix}\right]$ is a *cycle number*, which is related to the *Stirling number of the first kind* $s(n,k) = (-1)^{n-k} \left[\begin{matrix} n \\ k \end{matrix}\right]$. See [411, pp. 103–107]. **Remark:** *vii*) relates the cycle number $\left[\begin{matrix} n \\ n-k \end{matrix}\right]$ to the coefficients of the power series for $z^n/(e^z-1)^n$. See [1930]. In particular,

$$\frac{\mathrm{d}}{\mathrm{d}z} \frac{z^n}{(e^z-1)^n}\bigg|_{z=0} = -\frac{n}{2}, \quad \frac{\mathrm{d}^2}{\mathrm{d}z^2} \frac{z^n}{(e^z-1)^n}\bigg|_{z=0} = \frac{1}{12}n(3n-1), \quad \frac{\mathrm{d}^3}{\mathrm{d}z^3} \frac{z^n}{(e^z-1)^n}\bigg|_{z=0} = -\frac{1}{8}n^2(n-1),$$

$$\frac{\mathrm{d}^4}{\mathrm{d}z^4} \frac{z^n}{(e^z-1)^n}\bigg|_{z=0} = \frac{1}{240}n(15n^3 - 30n^2 + 5n + 2), \quad \frac{\mathrm{d}^5}{\mathrm{d}z^5} \frac{z^n}{(e^z-1)^n}\bigg|_{z=0} = -\frac{1}{96}n^2(3n^3 - 10n^2 + 5n + 2),$$

$$\frac{\mathrm{d}^6}{\mathrm{d}z^6} \frac{z^n}{(e^z-1)^n}\bigg|_{z=0} = \frac{1}{4032}n(63n^5 - 315n^4 + 315n^3 + 91n^2 - 42n - 16).$$

For the case $k = 3$, note that, for all $n \geq 4$, $8\binom{n}{2}\binom{n}{4} = n^2(n-1)\binom{n-1}{3}$. **Example:** $\left[\begin{matrix} 4 \\ 2 \end{matrix}\right] = 11$, $\left[\begin{matrix} 4 \\ 3 \end{matrix}\right] = \binom{4}{2} = \frac{3!}{2}(H_3^2 - H_{3,2}) = 6$, $\left[\begin{matrix} 5 \\ 2 \end{matrix}\right] = 50$, $\left[\begin{matrix} 5 \\ 3 \end{matrix}\right] = 35$, $\left[\begin{matrix} 5 \\ 4 \end{matrix}\right] = 10$, $\left[\begin{matrix} 6 \\ 2 \end{matrix}\right] = 274$, $\left[\begin{matrix} 6 \\ 3 \end{matrix}\right] = 225$, $\left[\begin{matrix} 6 \\ 4 \end{matrix}\right] = 85$, $\left[\begin{matrix} 6 \\ 5 \end{matrix}\right] = 15$. To illustrate *ix*), note that

$$\frac{1}{3\cdot 1} + \frac{1}{1\cdot 3} + \frac{1}{2\cdot 2} = \frac{2!}{4!}\left[\begin{matrix} 4 \\ 2 \end{matrix}\right] = \frac{11}{12}, \quad \frac{1}{1\cdot 1\cdot 2} + \frac{1}{1\cdot 2\cdot 1} + \frac{1}{2\cdot 1\cdot 1} = \frac{3!}{4!}\left[\begin{matrix} 4 \\ 3 \end{matrix}\right] = \frac{3}{2}.$$

Related: Fact 13.4.4 and Fact 13.5.40.

Fact 1.19.2. The following statements hold:

i) Let $n \geq 1$ and $x \in \mathbb{C}$. Then,

$$\sum_{i=1}^n \left[\begin{matrix} n \\ i \end{matrix}\right] x^i = x^{\overline{n}}.$$

In particular,

$$\sum_{i=1}^n 2^i \left[\begin{matrix} n \\ i \end{matrix}\right] = (n+1)!.$$

ii) Let $n \geq 1$ and $x \in \mathbb{C}$. Then,

$$\sum_{i=1}^{n}(-1)^{n-i}\begin{bmatrix} n \\ i \end{bmatrix} x^i = x^{\underline{n}}.$$

iii) Let $n \geq 2$. Then,

$$\sum_{i=0}^{n}(-1)^{i}\begin{bmatrix} n \\ i \end{bmatrix} = 0.$$

iv) Let $n \geq k \geq 1$. Then,

$$\begin{bmatrix} n \\ k \end{bmatrix} = (n-1)\begin{bmatrix} n-1 \\ k \end{bmatrix} + \begin{bmatrix} n-1 \\ k-1 \end{bmatrix}.$$

v) Let $k, n \geq 0$. Then,

$$\sum_{i=k}^{n}\begin{bmatrix} n \\ i \end{bmatrix}\binom{i}{k} = \begin{bmatrix} n+1 \\ k+1 \end{bmatrix}.$$

vi) Let $n \geq 1$. Then,

$$\sum_{i=1}^{n} i\begin{bmatrix} n \\ i \end{bmatrix} = \begin{bmatrix} n+1 \\ 2 \end{bmatrix} = n!H_n.$$

vii) Let $n, k \geq 0$. Then,

$$\sum_{i=0}^{k}(n+i)\begin{bmatrix} n+i \\ i \end{bmatrix} = \begin{bmatrix} n+k+1 \\ k \end{bmatrix}.$$

viii) Let $n, k \geq 0$. Then,

$$\sum_{i=k}^{n}\frac{n!}{i!}\begin{bmatrix} i \\ k \end{bmatrix} = \sum_{i=k}^{n} n^{\underline{n-i}}\begin{bmatrix} i \\ k \end{bmatrix} = \begin{bmatrix} n+1 \\ k+1 \end{bmatrix}.$$

ix) Let $n, k, l \geq 0$. Then,

$$\sum_{i=0}^{n}\binom{n}{i}\begin{bmatrix} i \\ l \end{bmatrix}\begin{bmatrix} n-i \\ k \end{bmatrix} = \begin{bmatrix} n \\ l+k \end{bmatrix}\binom{l+k}{l}.$$

x) Let $n \geq k \geq 0$. Then,

$$\sum_{i=k}^{n}(-1)^{k-i}\binom{i}{k}\begin{bmatrix} n+1 \\ i+1 \end{bmatrix} = \begin{bmatrix} n \\ k \end{bmatrix}.$$

xi) Let $n \geq 1$. Then,

$$\sum_{i=0}^{n}(-1)^{i}\binom{2n}{i}\begin{bmatrix} 2n-i \\ n-i \end{bmatrix} = \prod_{i=1}^{n}(2i-1).$$

Source: [411, pp. 93–96], [1594], and [3024, p. 139]. **Example:**

$$\begin{bmatrix} 3 \\ 1 \end{bmatrix} z - \begin{bmatrix} 3 \\ 2 \end{bmatrix} z^2 + \begin{bmatrix} 3 \\ 3 \end{bmatrix} z^3 = z - 3z^2 + z^3 = z(z-1)(z-2) - z^{\underline{3}}.$$

Fact 1.19.3. For $n \geq k \geq 1$, let $\left\{ \begin{matrix} n \\ k \end{matrix} \right\}$ denote the number of partitions of a set of n elements into k subsets. Furthermore, define $\left\{ \begin{matrix} 0 \\ 0 \end{matrix} \right\} \triangleq 1$ and, for all $n, k \geq 1$, define $\left\{ \begin{matrix} n \\ 0 \end{matrix} \right\} \triangleq 0$ and $\left\{ \begin{matrix} 0 \\ k \end{matrix} \right\} \triangleq 0$. Then, the following statements hold:

i) Let $n \geq k \geq 1$. Then,

$$\left\{ \begin{matrix} n \\ k \end{matrix} \right\} = \frac{1}{k!}\sum_{i=0}^{k-1}(-1)^i\binom{k}{i}(k-i)^n = \frac{1}{k!}\sum_{i=1}^{k}(-1)^{k-i}\binom{k}{i}i^n = \sum_{i=1}^{k}(-1)^{k-i}\frac{i^{n-1}}{(i-1)!(k-i)!}$$

$$= \frac{n!}{k!} \sum \frac{1}{i_1! \cdots i_k!} = \sum \frac{n!}{(1!)^{i_1} i_1! (2!)^{i_2} i_2! \cdots (n!)^{i_n} i_n!},$$

where the penultimate sum is taken over all k-tuples $(i_1, \ldots, i_k)$ of positive integers whose sum is n, and the last sum is taken over all n-tuples $(i_1, \ldots, i_n)$ of nonnegative integers whose sum is k and satisfy $\sum_{j=1}^{n} j i_j = n$. In particular, if $n \geq 1$, then

$$\left\{ {n \atop 1} \right\} = \left\{ {n \atop n} \right\} = 1, \quad \left\{ {n \atop 2} \right\} = 2^{n-1} - 1, \quad \left\{ {n \atop n-1} \right\} = \binom{n}{2},$$

$$\left\{ {n \atop n-2} \right\} = \binom{n}{3} + 3\binom{n}{4} = \frac{1}{4}\binom{n}{3}(3n-5), \quad \left\{ {n \atop n-3} \right\} = \binom{n}{4} + 10\binom{n}{5} + 15\binom{n}{6} = \frac{1}{2}\binom{n}{4}(n^2 - 5n + 6).$$

ii) Let $n \geq k \geq 0$. Then,

$$\left\{ {n \atop k} \right\} \leq \left[{n \atop k} \right], \quad k^{n-k} \leq \left\{ {n \atop k} \right\} \leq \binom{n-1}{k-1} k^{n-k}.$$

iii) Let $n \geq k \geq 1$. Then,

$$\left\{ {n \atop n-k} \right\} = \sum \frac{n!}{(n-k-\kappa)! \prod_{j=1}^{k} i_j! [(j+1)!]^{i_j}},$$

where $\kappa \triangleq \sum_{j=1}^{k} i_j$ and the sum is taken over all k-tuples $(i_1, \ldots, i_k)$ of nonnegative integers such that $\sum_{j=1}^{k} j i_j = k$.

iv) Let $n, k, m \geq 1$. Then,

$$\sum x_{i_j}^n = k \sum_{i=1}^{n} i! \left\{ {n \atop i} \right\} \binom{k+m-1}{m-i},$$

where the sum is over all k-tuples $(i_1, \ldots, i_k)$ of nonnegative integers such that $\sum_{j=1}^{k} i_j = m$.

v) Let $n \geq 2$. Then,

$$\sum_{i=1}^{n} (-1)^i (n-1)! \left\{ {n \atop i} \right\} = 0.$$

Source: [411, p. 103] and [3024, p. 140]. *i)* is given in [34, p. 159], [1082], and [1552] (see (2.1)); *ii)* is given in [1219, p. 260] and [771, p. 292]; *iii)* is given in [1930]; *iv)* is given in [771, pp. 172, 173]; *v)* is given in [2105]. **Remark:** $\left\{ {n \atop k} \right\}$ is a *subset number*, which is also called a *Stirling number of the second kind* denoted by $S(n, k)$. The curly braces are reminiscent of set notation. See [411, pp. 103–107], [1219, pp. 257–267], and [1649]. **Example:** $\left\{ {3 \atop 2} \right\} = 3$, $\left\{ {4 \atop 2} \right\} = 7$, $\left\{ {5 \atop 2} \right\} = 15$, $\left\{ {6 \atop 2} \right\} = 31$, $\left\{ {4 \atop 3} \right\} = 6$, $\left\{ {5 \atop 3} \right\} = 25$, $\left\{ {6 \atop 3} \right\} = 90$, $\left\{ {6 \atop 4} \right\} = 65$. **Related:** Fact 13.4.4.

Fact 1.19.4. The following statements hold:

i) Let $n \geq 1$ and $x \in \mathbb{C}$. Then,

$$\sum_{i=1}^{n} \left\{ {n \atop i} \right\} x^{\underline{i}} = x^n.$$

ii) Let $n \geq 1$ and $x \in \mathbb{C}$. Then,

$$\sum_{i=1}^{n} (-1)^{n-i} \left\{ {n \atop i} \right\} x^{\overline{i}} = x^n.$$

iii) Let $n \ge 1$ and $x \in \mathbb{C}$. Then,

$$x^{\overline{n}} = \sum_{i=1}^{n} \begin{bmatrix} n \\ i \end{bmatrix} x^i = \sum_{i=1}^{n} \sum_{j=i}^{n} \begin{bmatrix} n \\ j \end{bmatrix} \begin{Bmatrix} j \\ i \end{Bmatrix} x^{\underline{i}} = \sum_{i=1}^{n} \binom{n}{i} \frac{(n-1)!}{(i-1)!} x^{\underline{i}}.$$

iv) Let $n \ge 1$ and $x \in \mathbb{C}$. Then,

$$x^{\underline{n}} = \sum_{i=1}^{n} (-1)^{n-i} \begin{bmatrix} n \\ i \end{bmatrix} x^i = \sum_{i=1}^{n} \sum_{j=1}^{i} (-1)^{n-j} \begin{bmatrix} n \\ i \end{bmatrix} \begin{Bmatrix} i \\ j \end{Bmatrix} x^{\overline{j}}.$$

v) Let $k \ge 1$ and $n \ge k+1$. Then,

$$\begin{Bmatrix} n \\ k \end{Bmatrix} = k \begin{Bmatrix} n-1 \\ k \end{Bmatrix} + \begin{Bmatrix} n-1 \\ k-1 \end{Bmatrix}.$$

vi) Let $n, k \ge 0$. Then,

$$\sum_{i=1}^{\min\{k,n\}} i! \begin{Bmatrix} k \\ i \end{Bmatrix} \binom{n}{i} = n^k.$$

vii) Let $n, k \ge 0$. Then,

$$\sum_{i=1}^{\min\{k,n\}} i! \begin{Bmatrix} k \\ i \end{Bmatrix} \binom{n+1}{i+1} = \sum_{i=1}^{n} i^k.$$

viii) Let $n \ge 1$. Then,

$$\sum_{i=1}^{n} (-1)^i (i-1)! \begin{Bmatrix} n \\ i \end{Bmatrix} = 0.$$

ix) Let $n \ge k \ge 1$. Then,

$$\sum_{i=1}^{k} (-1)^{k-i} i^n \binom{k}{i} = k! \begin{Bmatrix} n \\ k \end{Bmatrix}.$$

x) Let $n, k \ge 0$. Then,

$$\sum_{i=k}^{n} \binom{n}{i} \begin{Bmatrix} i \\ k \end{Bmatrix} = \begin{Bmatrix} n+1 \\ k+1 \end{Bmatrix}.$$

xi) Let $n, k \ge 0$. Then,

$$\sum_{i=1}^{k} i \begin{Bmatrix} n+i \\ i \end{Bmatrix} = \begin{Bmatrix} n+k+1 \\ k \end{Bmatrix}.$$

xii) Let $n, k \ge 0$. Then,

$$\sum_{i=0}^{n} \begin{Bmatrix} i \\ k \end{Bmatrix} (k+1)^{n-i} = \begin{Bmatrix} n+1 \\ k+1 \end{Bmatrix}.$$

xiii) Let $n \ge k \ge 0$. Then,

$$\sum_{i=k}^{n} (-1)^{n-i} \binom{n}{i} \begin{Bmatrix} i+1 \\ k+1 \end{Bmatrix} = \begin{Bmatrix} n \\ k \end{Bmatrix}.$$

xiv) Let $n, k \ge 0$. Then,

$$\sum_{i=1}^{k} (-1)^i \begin{Bmatrix} k \\ i \end{Bmatrix} i^n = (-1)^k k! \begin{Bmatrix} n \\ k \end{Bmatrix}.$$

xv) Let $n \ge 0$. Then,

$$\sum_{i=1}^{n} (-1)^i \frac{i!}{i+1} \begin{Bmatrix} n \\ i \end{Bmatrix} = B_n.$$

xvi) Let $n \geq k \geq 1$. Then,

$$\sum_{i=k}^{n} \begin{bmatrix} n \\ i \end{bmatrix} \begin{Bmatrix} i \\ k \end{Bmatrix} = \binom{n}{k} \frac{(n-1)!}{(k-1)!}.$$

xvii) Let $n \geq k \geq 1$. Then,

$$\sum_{i=k}^{n} (-1)^{n-i} \begin{bmatrix} n \\ i \end{bmatrix} \begin{Bmatrix} i \\ k \end{Bmatrix} = \begin{cases} 1, & n = k, \\ 0, & n \neq k. \end{cases}$$

xviii) Let $n \geq k \geq 1$. Then,

$$\sum_{i=k}^{n} (-1)^{n-i} \begin{Bmatrix} n \\ i \end{Bmatrix} \begin{bmatrix} i \\ k \end{bmatrix} = \begin{cases} 1, & n = k, \\ 0, & n \neq k. \end{cases}$$

xix) Let $n, k \geq 0$. Then,

$$\sum_{i=k}^{n} (-1)^{k-i} \begin{Bmatrix} n+1 \\ i+1 \end{Bmatrix} \begin{bmatrix} i \\ k \end{bmatrix} = \binom{n}{k}.$$

xx) Let $n \geq k \geq 0$. Then,

$$\sum_{i=k}^{n} \begin{bmatrix} n+1 \\ i+1 \end{bmatrix} \begin{Bmatrix} i \\ k \end{Bmatrix} = \frac{n!}{m!}.$$

xxi) Let $n \geq k \geq 0$. Then,

$$\sum_{i=k}^{n} (-1)^{k-i} \begin{bmatrix} n+1 \\ i+1 \end{bmatrix} \begin{Bmatrix} i \\ k \end{Bmatrix} = n^{\underline{n-k}}.$$

xxii) Let $n, k \geq 0$, and assume that $k \leq n + 1$. Then,

$$\sum_{i=0}^{n} (-1)^{i+1-k} \frac{1}{i+1} \begin{bmatrix} i+1 \\ k \end{bmatrix} \begin{Bmatrix} n \\ i \end{Bmatrix} = \frac{1}{n+1} \binom{n+1}{k} B_{n+1-k}.$$

xxiii) Let $n, k \geq 0$. Then,

$$\sum_{i=1}^{k} (-1)^{i+1} i \begin{bmatrix} n+1 \\ n+1-i \end{bmatrix} \begin{Bmatrix} n+k-i \\ n \end{Bmatrix} = \sum_{i=1}^{n} i^k.$$

xxiv) Let $n, k, l \geq 0$. Then,

$$\sum_{i=0}^{n} \binom{n}{i} \begin{Bmatrix} i \\ l \end{Bmatrix} \begin{Bmatrix} n-i \\ k \end{Bmatrix} = \binom{l+k}{l} \begin{Bmatrix} n \\ l+k \end{Bmatrix}.$$

xxv) Let $n \geq k \geq 0$. Then,

$$\sum_{i=0}^{n} \binom{k-n}{k+i} \binom{k+n}{n+i} \begin{bmatrix} k+i \\ i \end{bmatrix} = \begin{Bmatrix} n \\ n-k \end{Bmatrix}.$$

xxvi) Let $n \geq k \geq 0$. Then,

$$\sum_{i=0}^{n} \binom{k-n}{k+i} \binom{k+n}{n+i} \begin{Bmatrix} k+i \\ i \end{Bmatrix} = \begin{bmatrix} n \\ n-k \end{bmatrix}.$$

xxvii) Let $n \geq k \geq 0$. Then,

$$\begin{bmatrix} n \\ k \end{bmatrix} = \sum_{i=0}^{n-k} (-1)^i \binom{n-1+i}{n-k+i} \binom{2n-k}{n-k-i} \begin{Bmatrix} n-k-i \\ i \end{Bmatrix},$$

where $\kappa \triangleq \sum_{j=1}^{k} i_j$ and the sum is taken over all k-tuples $(i_1, \ldots, i_k)$ such that, for all $j \in \{1, \ldots, k\}$, $0 \leq i_j \leq k$ and such that $\sum_{j=1}^{k} j i_j = k$.

Source: [411, pp. 103, 106, 107], [555, 809], [993, p. 95], [1082, 1197], [1219, pp. 264, 265, 289], and [2018, 2312]. **Remark:** In *xv*), B_n is the nth Bernoulli number. See Fact 13.1.6. **Remark:** The coefficient $L(n,k) \triangleq \binom{n}{k}\binom{n-1}{k-1}(n-k)! = \binom{n}{k}\frac{(n-1)!}{(k-1)!}$ in *iii*) is a *Lah number*. See [809]. **Example:**

$$1^2+2^2+3^2 = \left\{\begin{matrix}2\\1\end{matrix}\right\}\binom{4}{2}1! + \left\{\begin{matrix}2\\2\end{matrix}\right\}\binom{4}{3}2! = 1(6)(1)+1(4)(2) = 14,$$

$$\left\{\begin{matrix}3\\1\end{matrix}\right\}z + \left\{\begin{matrix}3\\2\end{matrix}\right\}z(z-1) + \left\{\begin{matrix}3\\3\end{matrix}\right\}z(z-1)(z-2) = z+3z(z-1)+z(z-1)(z-2) = z^3.$$

Fact 1.19.5. Let $n \ge 1$, let $0 \le k \le n-1$, and let $\left\langle\begin{matrix}n\\k\end{matrix}\right\rangle$ denote the number of permutations of $(1,\dots,n)$ in which exactly k components are larger than the previous component. Then, the following statements hold:

i) Let $n \ge 1$ and $0 \le k \le n-1$. Then,

$$\left\langle\begin{matrix}n\\k\end{matrix}\right\rangle = \sum_{i=0}^{k}(-1)^i\binom{n+1}{i}(k+1-i)^n,$$

$$\sum_{i=0}^{n-1}\left\langle\begin{matrix}n\\i\end{matrix}\right\rangle = n!, \quad \sum_{i=0}^{n-1}(-1)^i\left\langle\begin{matrix}n\\i\end{matrix}\right\rangle = \frac{2^{n+1}(2^{n+1}-1)B_{n+1}}{n+1}.$$

ii) Let $n \ge 1$. Then,

$$\left\langle\begin{matrix}n\\0\end{matrix}\right\rangle = \left\langle\begin{matrix}n\\n-1\end{matrix}\right\rangle = 1, \quad \left\langle\begin{matrix}n\\1\end{matrix}\right\rangle = 2^n-n-1, \quad \left\langle\begin{matrix}n\\2\end{matrix}\right\rangle = 3^n-(n+1)2^n+\binom{n+1}{2}, \quad \left\langle\begin{matrix}n\\n\end{matrix}\right\rangle \triangleq 0,$$

$$\left\langle\begin{matrix}2\\1\end{matrix}\right\rangle = 1, \quad \left\langle\begin{matrix}3\\1\end{matrix}\right\rangle = 4, \quad \left\langle\begin{matrix}4\\1\end{matrix}\right\rangle = \left\langle\begin{matrix}4\\2\end{matrix}\right\rangle = 11, \quad \left\langle\begin{matrix}5\\1\end{matrix}\right\rangle = \left\langle\begin{matrix}5\\3\end{matrix}\right\rangle = 26, \quad \left\langle\begin{matrix}5\\2\end{matrix}\right\rangle = 66.$$

iii) Let $n \ge 1$ and $0 \le k \le n-1$. Then,

$$\left\langle\begin{matrix}n\\k\end{matrix}\right\rangle = \left\langle\begin{matrix}n\\n-1-k\end{matrix}\right\rangle, \quad \left\langle\begin{matrix}n\\k\end{matrix}\right\rangle = (k+1)\left\langle\begin{matrix}n-1\\k\end{matrix}\right\rangle + (n-k)\left\langle\begin{matrix}n-1\\k-1\end{matrix}\right\rangle,$$

where $\left\langle\begin{matrix}0\\0\end{matrix}\right\rangle \triangleq 1$ and $\left\langle\begin{matrix}0\\k\end{matrix}\right\rangle \triangleq 0$.

iv) Let $n \ge 1$ and $x \in \mathbb{C}$. Then,

$$x^n = \sum_{i=0}^{n-1}\left\langle\begin{matrix}n\\i\end{matrix}\right\rangle\binom{x+i}{n}.$$

In particular,

$$x^2 = \binom{x}{2}+\binom{x+1}{2}, \quad x^3 = \binom{x}{3}+4\binom{x+1}{3}+\binom{x+2}{3}, \quad x^4 = \binom{x}{4}+11\binom{x+1}{4}+11\binom{x+2}{4}+\binom{x+3}{4}.$$

v) Let $k \ge 1$ and $1 \le n \le k$, Then,

$$k^n = \sum_{i=0}^{n-1}\left\langle\begin{matrix}n\\i\end{matrix}\right\rangle\binom{k+i}{n} = \sum_{i=0}^{n-1}\left\langle\begin{matrix}n\\i\end{matrix}\right\rangle\binom{k+n-i-1}{n}.$$

vi) If $k,n \ge 1$, then

$$\sum_{i=1}^{k} i^n = \sum_{i=0}^{n-1}\left\langle\begin{matrix}n\\i\end{matrix}\right\rangle\binom{k+i+1}{n+1}.$$

vii) If $k,n \geq 1$, then

$$\left\{ {n \atop k} \right\} = \frac{1}{k!}\sum_{i=n-k}^{n}\left\{ {n \atop i} \right\}\binom{i}{n-k}, \quad \left\langle {n \atop k} \right\rangle = \sum_{i=1}^{n-k}(-1)^{n-k-i}i!\left\{ {n \atop i} \right\}\binom{n-i}{k}.$$

viii) If $n \geq 2$, then

$$\sum_{i=0}^{n-1}(-1)^i\frac{\left\langle {n \atop i} \right\rangle}{\binom{n-1}{i}} = 0, \quad \sum_{i=0}^{n-1}(-1)^i\frac{\left\langle {n \atop i} \right\rangle}{\binom{n}{i}} = (n+1)B_n.$$

Source: [1219, pp. 267–269], [1652, pp. 35–39], and [2225]. **Remark:** $\left\langle {n \atop m} \right\rangle$ is an *Eulerian number*. An alternative definition is used in [349]. **Remark:** *iv)* is *Worpitzky's identity*. **Related:** Fact 13.1.2.

Fact 1.19.6. Let $\mathcal{B}_n$ denote the number of partitions of $\{1,\ldots,n\}$, and define $\mathcal{B}_0 \triangleq 1$. Then, the following statements hold:

i) Let $n \geq 1$. Then,

$$\mathcal{B}_n = \sum_{i=1}^{n}\left\{ {n \atop i} \right\} = \sum_{i=0}^{n}\frac{1}{i!}\sum_{j=0}^{i}(-1)^{i-j}\binom{i}{j}j^n = 1 + \left\lfloor \frac{1}{e}\sum_{i=1}^{2n}\frac{i^n}{i!} \right\rfloor.$$

ii) $(\mathcal{B}_i)_{i=0}^{13} = (1,1,2,5,15,52,203,877,4140,21147,115975,678570,4213597,27644437)$.

iii) If p is prime, then $\mathcal{B}_{n+p} \overset{p}{\equiv} \mathcal{B}_n + \mathcal{B}_{n+1}$.

iv) If $n,m \geq 1$, then

$$\mathcal{B}_{n+m} = \sum_{i=0}^{n}\sum_{j=1}^{m}\left\{ {m \atop j} \right\}\binom{n}{i}j^{n-i}\mathcal{B}_i, \quad \mathcal{B}_{n+1} = \sum_{i=0}^{n}\binom{n}{i}\mathcal{B}_i.$$

v) Let $k \geq n \geq 1$. Then,

$$\mathcal{B}_n = \sum_{i=1}^{k}\frac{i^n}{i!}\sum_{j=0}^{k-i}(-1)^j\frac{1}{j!} = \sum_{i=1}^{n}\frac{i^n}{i!}\sum_{j=0}^{n-i}(-1)^j\frac{1}{j!}.$$

Remark: $\mathcal{B}_n$ is the nth *Bell number*. See [34, p. 160], [571, p. 623], and [2506]. **Related:** Fact 13.1.4.

Fact 1.19.7. For all $n \geq 1$, let $\mathcal{O}_n$ denote the number of possible orderings of the multiset $\{i_1,\ldots,i_n\}_{\rm ms}$ of real numbers, and define $\mathcal{O}_0 \triangleq 1$. Then, the following statements hold:

i) Let $n \geq 1$. Then,

$$\mathcal{O}_n = \sum_{i=1}^{n}i!\left\{ {n \atop i} \right\}.$$

ii) $(\mathcal{O}_i)_{i=0}^{11} = (1,1,3,13,75,541,4683,47293,545835,7087261,102247563,1622632573)$.

iii) Let $n \geq 1$. Then,

$$\mathcal{O}_n = \sum_{i=0}^{n-1}\binom{n}{i}\mathcal{O}_i = (-1)^{n-1} + 2\sum_{i=0}^{n}(-1)^{n-i}\binom{n}{i}\mathcal{O}_i = \sum_{i=0}^{n}\sum_{j=0}^{i}(-1)^{i-j}\binom{i}{j}j^n = \sum_{i=0}^{\infty}\frac{i^n}{2^{i+1}} = \sum_{i=0}^{n-1}2^i\left\langle {n \atop i} \right\rangle.$$

iv) Let $n \geq 1$. Then,

$$\sum_{i=1}^{n} \mathcal{O}_{i-1}\mathcal{O}_{n-i}\binom{n}{i} = \frac{n}{2}\mathcal{O}_{n-1} + \frac{1}{2}\sum_{i=1}^{n} i!H_i \left\{ {n \atop i} \right\},$$

$$\sum_{i=1}^{n} \mathcal{O}_{i-1}\mathcal{O}_{n-i}\binom{n}{i} = \frac{n+1}{2}\mathcal{O}_{n-1} - \frac{1}{4}\delta_{n,1} + \frac{1}{4}\sum_{i=1}^{n} i!H_i \left\{ {n+1 \atop i+1} \right\}.$$

Source: [899, 898, 2100]. **Remark:** $\mathcal{O}_n$ is the nth *ordered Bell number*. **Remark:** For $n = 3$, the 13 possible orderings of $\{x, y, z\}_{\rm ms}$ are $x = y = z$, $x < y = z$, $y = z < x$, $y < x = z$, $x = z < y$, $z < x = y$, $x = y < z$, $x < y < z$, $x < z < y$, $y < x < z$, $y < z < x$, $z < x < y$, $z < y < x$. **Remark:** The generating function is given by Fact 13.1.5.

1.20 Facts on Partition Numbers, the Totient Function, and Divisor Sums

Fact 1.20.1. For all $n \geq 1$, let p_n denote the number of partitions of the n-element multiset $\{1, \ldots, 1\}_{\rm ms}$. Equivalently, for all $n \geq 1$, let p_n denote the number of ways of representing n as a sum of one or more positive integers. Define $p_0 \triangleq 1$. Furthermore, for all $n, k \geq 1$, let $p_{n,k}$ denote the number of ways of representing n as a sum of k positive integers, and, for all $n, k, l \geq 1$, let $p_{n,k,l}$ denote the number of ways of representing n as a sum of k positive integers the largest of which is l. Then, the following statements hold:

i) For all $n \geq 1$, $p_n = \operatorname{card}\{(k_1, \ldots, k_n) \in \mathbb{N}^n\colon k_1 \leq \cdots \leq k_n \text{ and } \sum_{i=1}^n k_i = n\}$.

ii) For all $n \geq 1$, $p_n = \operatorname{card}\{(k_1, \ldots, k_n) \in \mathbb{N}^n\colon \sum_{i=1}^n ik_i = n\}$.

iii) $(p_i)_{i=0}^{20} = (1, 1, 2, 3, 5, 7, 11, 15, 22, 30, 42, 56, 77, 101, 135, 176, 231, 297, 385, 490, 627)$.

iv) For all $n \in \{1, 2\}$, $p_n = \mathcal{B}_n$. For all $n \geq 3$, $p_n < \mathcal{B}_n$.

v) For all $n \in \{1, 2, 3, 4\}$, $p_n = F_{n+1}$. For all $n > 5$, $p_n < F_{n+1}$.

vi) For all $n, k \geq 1$, $p_{n,k}$ is the number of ways of representing n as a sum of positive integers, the largest of which is k.

vii) Let $n \geq 1$. Then, $\sum_{i=1}^n p_{n,i} = p_n$.

viii) Let $n \geq 1$. Then,

$$p_{n,1} = p_{n,n-1} = p_{n,n} = 1, \quad p_{n,2} = \tfrac{1}{4}[2n + 3 + (-1)^n], \quad p_{n,3} = \left\lfloor \tfrac{1}{12}(n+3)^2 + \tfrac{1}{2} \right\rfloor,$$

$$p_{n,4} = \left\lfloor \tfrac{1}{144}(n+5)(n^2 + n + 22 + 18\lfloor n/2 \rfloor) + \tfrac{1}{2} \right\rfloor,$$

$$p_{n,5} = \left\lfloor \tfrac{1}{2880}(n+8)(n^3 + 22n^2 + 44n + 248 + 180\lfloor n/2 \rfloor) + \tfrac{1}{2} \right\rfloor,$$

$$\sum_{i=1}^{2} p_{n,i} = 1 + \left\lfloor \frac{n}{2} \right\rfloor, \quad \sum_{i=1}^{3} p_{n,i} = 1 + \left\lfloor \frac{n^2 + 6n}{12} \right\rfloor.$$

ix) Let $k \geq 1$. Then, as $n \to \infty$,

$$\sum_{i=1}^{k} p_{n,i} \sim \frac{n^{k\ 1}}{k!(k-1)!}.$$

x) Let $n \geq k \geq 1$. Then,

$$\sum_{i=1}^{n} p_{n,i,k} = \sum_{i=1}^{n} p_{n,k,i} = p_{n,k}.$$

In particular,

$$\sum_{i=1}^{n} p_{n,i,1} = \sum_{i=1}^{n} p_{n,1,i} = \sum_{i=1}^{n} p_{n,i,n-1} = \sum_{i=1}^{n} p_{n,n-1,i} = \sum_{i=1}^{n} p_{n,i,n} = \sum_{i=1}^{n} p_{n,n,i} = 1.$$

xi) Let $n \geq k \geq 1$. Then, the number of ways of representing n as a sum of k or fewer positive integers is

$$P_{n,k} \triangleq \sum_{i,j=1}^{n,k} p_{n,i,j} = \sum_{i,j=1}^{n,k} p_{n,j,i} = \sum_{i=1}^{k} p_{n,i}.$$

Furthermore,

$$P_{n,k+1} = P_{n-k-1,k+1} + P_{n,k}.$$

xii) Let $n, m \geq 1$. Then, the number of ways of representing all of the positive integers less than or equal to mn as the sum of n or fewer positive integers, the largest of which is less than or equal to m, is

$$\sum_{i,j,k=1}^{mn,m,n} p_{i,j,k} = \binom{n+m}{m} - 1.$$

xiii) For all $n \geq 1$, $5|p_{5n+4}$, $7|p_{7n+5}$, $11|p_{11n+6}$, and $13|p_{17303n+237}$.

xiv) For each prime m, there exists $n \geq 1$ such that $m|p_n$.

xv) For all $n \geq 1$,

$$p_n = \frac{1}{n}\sum_{i=1}^{n} s_i p_{n-i},$$

where s_i is the sum of the divisors of i.

xvi) For all $n \geq 1$,

$$p_n = \sum (-1)^{\lfloor (i-1)/2 \rfloor} p_{n-g_i} = \sum (-1)^i (p_{n-P_i} + p_{n-P_i'}) = \sum_{i=1}^{n-1} e_i p_{n-i},$$

where the first sum is taken over all $i \geq 1$ such that $g_i \leq n$, the second sum is taken over all $i \geq 1$ such that $P_i \leq n$ and $P_i' \leq n$, and, for all $k \geq 0$,

$$e_k \triangleq \begin{cases} 1, & k = 0, \\ (-1)^i, & k \in \{g_{2i-1}, g_{2i}\} = \{\tfrac{1}{2}i(3i-1), \tfrac{1}{2}i(3i+1)\}, \\ 0, & \text{otherwise.} \end{cases}$$

xvii) For all $n \geq 2$, let $(n_{\rm e}, n_{\rm o})$ denote the number of partitions of $\{1, \ldots, n\}$ into an (even, odd) number of subsets. Then, $n_{\rm e} - n_{\rm o} = e_n$.

xviii) Let $n \geq 1$, define $\mathcal{P}_n \triangleq \{(k_1, \ldots, k_n) \in \mathbb{N}^n \colon k_1 \leq \cdots \leq k_n \text{ and } \sum_{i=1}^n k_i = n\}$, and let $i \in \{1, \ldots, n\}$. Then,

$$\sum \sum_{j=1}^{n} \operatorname{truth}(k_j = i) = \sum \sum_{j=1}^{n} \operatorname{card}\left\{ j \in \{1, \ldots, n\} \colon \sum_{l=1}^{n} \operatorname{truth}(k_l = j) \geq i \right\},$$

where the first sum on both sides is taken over all $(k_1, \ldots, k_n) \in \mathcal{P}_n$. (Example: The number 1 appears 12 times over all 7 elements of $\mathcal{P}_5$, while the number of times that an integer appears 1 or more times in each element of $\mathcal{P}_n$ summed over all elements of $\mathcal{P}_5$ is 12. Likewise, the number 2 appears 4 times over all the 7 elements of $\mathcal{P}_5$, while the number of times that an integer appears 2 or more times in each element of $\mathcal{P}_n$ summed over all 7 elements of $\mathcal{P}_5$ is 4.)

Remark: p_n is the nth *partition number*. See [114, p. 70], [116, pp. 553–576], [118, pp. 58–61, 125], [411, p. 79], [571, pp. 624–627], [771, Chapter II], [1404], [2175], [3024, p. 138]. *xvii)* is given in [2374]. **Remark:** To illustrate $p_5 = 7$, note that $5 = 1+1+1+1+1 = 2+1+1+1 = 2+2+1 = 3+1+1 = 4+1 = 3+2$. **Remark:** P_i is the ith pentagonal number, P_i' is the i dual pentagonal number, and g_i is the ith generalized pentagonal number. See Fact 1.12.5 and Fact 13.10.27. For example, $15 = p_7 = p_6 + p_5 - p_2 - p_0 = 11 + 7 - 2 - 1 = 15$. **Remark:** *xii*) implies that the number of ways of representing positive integers less than or equal to mn as the sum of less than or equal to n positive integers, the largest of which is less than or equal to m, is equal to the number of ways of representing positive integers less than or equal to mn as the sum of less than or equal to m positive integers, the largest of which is less than or equal to n. **Remark:** $(e_i)_{i=0}^{6} = (1,-1,-1,0,0,1,0)$. **Remark:** The generating function for $(p_i)_{i=1}^{\infty}$ is given by Fact 13.10.25. **Remark:** *xviii*) is *Elder's theorem*. The case $i = 1$ is *Stanley's theorem*. See [1175]. **Remark:** For all $n \geq 1$, the number of tuples of positive integers whose components sum to n is 2^{n-1}. **Credit:** *xiii*) is due to S. A. Ramanujan and A. O. L. Atkin. **Related:** Fact 4.13.32, Fact 13.1.3, and Fact 13.10.25.

Fact 1.20.2. For all $n \geq 1$, let s_n denote the sum of the distinct positive divisors of n, and define $(p_i)_{i=0}^{\infty}$ and $(e_i)_{i=1}^{\infty}$ as in Fact 1.20.1. Then, the following statements hold:

i) If n is prime and $k \geq 1$, then $s_{n^k} = \frac{n^{k+1}-1}{n-1}$.

ii) If $n \geq 1$ and $m \geq 1$ are coprime, then $s_{nm} = s_n s_m$.

iii) Let $n \geq 2$, and let $m_1, \dots, m_l$ be distinct primes and $k_1, \dots, k_l$ be positive integers such that $n = \prod_{i=1}^{l} m_i^{k_i}$. Then, $s_n = \prod_{i=1}^{l} s_{m_i^{k_i}}$.

iv) Let $n \geq 1$. Then, $s_n = -ne_n - \sum_{i=0}^{n-1} e_{n-i}s_i = -ne_n - \sum_{i=1}^{n} e_n s_{n-i}$, where $s_0 \triangleq n$.

v) Let $n \geq 1$. Then, $s_n = -\sum_{i=1}^{n} ie_i p_{n-i}$.

Source: [2175]. **Remark:** The generating function for $(s_i)_{i=1}^{\infty}$ is given by Fact 13.4.2. **Remark:** To illustrate *iv*), note that $12 = s_6 = -6e_6 - e_6s_0 - e_5s_1 - e_4s_2 - e_3s_3 - e_2s_4 - e_1s_5 = -6(0) - 0(12) - (1)1 - 0(3) - 0(4) - (-1)(7) - (-1)6 = 12$. **Remark:** To illustrate *v*), note that $12 = s_6 = -[e_1p_5 + 2e_2p_4 + 3e_3p_3 + 4e_4p_2 + 5e_5p_1 + 6e_6p_0] = -[1(-1)7 + 2(-1)5 + 3(0)3 + 4(0)2 + 5(1)1 + 6(0)1] = 12$.

Fact 1.20.3. Let $n \geq 1$, let $\tau(n)$ be the number of positive divisors of n, and let $\sigma(n)$ be the sum of the positive divisors of n, where $\sqrt{n}$ is counted and summed twice if $\sqrt{n}$ is an integer. Then, $\sqrt{n} \leq \sigma(n)/\tau(n)$. **Source:** [1566, p. 17]. **Example:** $\sqrt{4} \leq (1+2+2+4)/4 = 9/4$ and $\sqrt{20} \leq (1+2+4+5+10+20)/6 = 7$. **Related:** Fact 12.18.21.

Fact 1.20.4. Let $k \geq 1$, and let $\phi(k) \triangleq \operatorname{card}\{i \in \{1,\dots,k\}\colon \gcd\{k,i\} = 1\}$. Then, the following statements hold:

i) $(\phi(i))_{i=1}^{28} = (1,1,2,2,4,2,6,4,6,4,10,4,12,6,8,8,16,6,18,8,12,10,22,8,20,12,18,12)$.

ii) Let $n \geq 2$ be prime. Then, $\phi(n) = n - 1$.

iii) Let m be an integer, let $n \geq 1$, and assume that m and n are coprime. Then, $m^{\phi(n)} \overset{n}{\equiv} 1$.

iv) Let $n \geq 1$, let $n_1, \dots, n_l$ be distinct primes, let $i_1, \dots, i_l \geq 1$, and assume that $n = \prod_{j=1}^{l} n_j^{i_j}$. Then,

$$\phi(n) = n\prod_{i=1}^{l}\left(1 - \frac{1}{n_i}\right).$$

If, in addition, $i_1 = \cdots = i_l = 1$, then $\phi(n) = \prod_{i=1}^{l}(n_i - 1)$.

v) Let $n \geq 1$ and $m \geq 1$, and assume that n and m are coprime. Then, $\phi(nm) = \phi(n)\phi(m)$.

vi) Let $n \geq 1$. Then, $\sum \phi(i) = n$, where the sum is taken over all $i \geq 1$ that divide n.

vii) If $n \geq 2$, then $\sqrt{n/2} \leq \phi(n)$. If, in addition, $n \geq 3$ and $n \neq 6$, then $\sqrt{n} \leq \phi(n)$.

viii) Let c_n denote the nth positive number such that n and $\phi(n)$ are coprime. Then, $(c_i)_{i=1}^{27} = (1,2,3,5,7,11,13,15,17,19,23,29,31,33,35,37,41,43,47,51,53,59,61,65,67,69,71)$.

ix) Let $n \geq 3$. Then, n and $\phi(n)$ are coprime if and only if n has distinct prime factors and, for all distinct prime factors k and l of n such that $k < l$, $k \nmid l - 1$.

Source: [411, pp. 116, 117]. *vii*) is given in [1599]. **Remark:** ϕ is the *totient function*. See [155, pp. 25–28]. **Remark:** *iii*) is *Euler's theorem*. See [1757, p. 148]. **Example:** For *iii*), note that, for $m = 4$ and $n = 3$, $4^2 - 1 = 3 \cdot 5$. Furthermore, for $m = 7$ and $n = 5$, $7^4 - 1 = 5 \cdot 480$. **Example:** For *iv*), note that $\phi(23) = 22 = 23(1 - 1/23)$, $6 = \phi(9) = 9(1 - 1/3)$, and $40 = \phi(55) = 55(1 - 1/5)(1 - 1/11)$. **Example:** For *v*), note that $\phi(35) = 24 = 4 \cdot 6 = \phi(5)\phi(7)$ and $\phi(68) = 32 = 2 \cdot 16 = \phi(4)\phi(17)$. **Example:** For *vi*), note that $20 = \phi(1) + \phi(2) + \phi(4) + \phi(5) + \phi(10) + \phi(20) = 1 + 1 + 2 + 4 + 4 + 8$ and $23 = \phi(1) + \phi(23) = 1 + 22$. **Remark:** c_n is the nth *cyclic number*. See Fact 4.31.17. **Remark:** The first ten cyclic numbers that are not prime are $1, 15, 33, 35, 51, 65, 69, 77, 85, 87$. **Related:** Fact 12.18.21.

1.21 Facts on Convex Functions

Fact 1.21.1. let $a, b \in \mathbb{R}$, assume that $a < b$, and let $f\colon (a,b) \mapsto \mathbb{R}$. Then, the following statements are equivalent:

i) f is convex.

ii) For all $x_1, x, x_2 \in (a,b)$ such that $x_1 < x < x_2$,

$$\frac{f(x) - f(x_1)}{x - x_1} \le \frac{f(x_2) - f(x)}{x_2 - x}.$$

iii) For all $x_1, x, x_2 \in (a,b)$ such that $x_1 < x < x_2$,

$$\frac{f(x) - f(x_1)}{x - x_1} \le \frac{f(x_2) - f(x_1)}{x_2 - x_1} \le \frac{f(x_2) - f(x)}{x_2 - x}.$$

iv) For all $x_1, y_1, x_2, y_2 \in (a,b)$ such that $x_1 \ne y_1$, $x_2 \ne y_2$, $x_1 \le x_2$, and $y_1 \le y_2$,

$$\frac{f(x_1) - f(y_1)}{x_1 - y_1} \le \frac{f(x_2) - f(y_2)}{x_2 - y_2}.$$

Furthermore, the following statements are equivalent:

v) f is strictly convex.

vi) For all $x_1, x, x_2 \in (a,b)$ such that $x_1 < x < x_2$,

$$\frac{f(x) - f(x_1)}{x - x_1} < \frac{f(x_2) - f(x)}{x_2 - x}.$$

vii) For all $x_1, y_1, x_2, y_2 \in (a,b)$ such that $x_1 \ne y_1$, $x_2 \ne y_2$, $x_1 < x_2$, and $y_1 < y_2$,

$$\frac{f(x_1) - f(y_1)}{x_1 - y_1} < \frac{f(x_2) - f(y_2)}{x_2 - y_2}.$$

Source: [1260, p. 4] and [1569].

Fact 1.21.2. Let a and b be nonnegative numbers such that $a < b$, let $f\colon [a,b] \mapsto \mathbb{R}$, assume that f is convex, and let $x_1, \dots, x_n$ be positive numbers such that $\sum_{i=1}^n x_i \le b - a$. Then,

$$\sum_{i=1}^n f(a + x_i) \le f\left(a + \sum_{i=1}^n x_i\right) + (n-1)f(a).$$

Source: [1569].

Fact 1.21.3. Let a, b, c be nonnegative numbers such that $0 \le a \le b \le c$, let $f\colon [0,c] \mapsto \mathbb{R}$, and assume that f is convex. Then,

$$f(c - b + a) + f(b) \le f(c) + f(a).$$

Source: Fact 1.21.2 and [1569].

Fact 1.21.4. Let $\mathcal{I}$ be a finite or infinite interval, and let $f\colon \mathcal{I} \mapsto \mathbb{R}$. Then, in each case below, f is convex:

i) $\mathcal{I} = (0, \infty)$, $f(x) = -\log x$.

ii) $\mathcal{I} = (0, \infty)$, $f(x) = x \log x$.

iii) $\mathcal{I} = (0, \infty)$, $f(x) = x^p$, where $p < 0$.

iv) $\mathcal{I} = [0, \infty)$, $f(x) = -x^p$, where $p \in (0, 1)$.

v) $\mathcal{I} = [0, \infty)$, $f(x) = x^p$, where $p \in (1, \infty)$.

vi) $\mathcal{I} = [0, \infty)$, $f(x) = (1 + x^p)^{1/p}$, where $p \in (1, \infty)$.

vii) $\mathcal{I} = \mathbb{R}$, $f(x) = \frac{a^x - b^x}{c^x - d^x}$, where $0 < d < c < b < a$ and $f(0) \triangleq (\log a/b)/\log c/d$.

viii) $\mathcal{I} = \mathbb{R}$, $f(x) = \log \frac{a^x - b^x}{c^x - d^x}$, $0 < d < c < b < a$, $ad \ge bc$, and $f(0) \triangleq \log[(\log a/b)/(\log c/d)]$.

ix) $\mathcal{I} = \mathbb{R}$, $f(x) = \log \frac{c^x - d^x}{a^x - b^x}$, $0 < d < c < b < a$, $ad < bc$, and $f(0) \triangleq \log[(\log c/d)/(\log a/b)]$.

x) $\mathcal{I} = (0, \infty)$, $f(x) = \log \Gamma(x)\Gamma(1/x)$.

Source: *vii)* and *viii)* are given in [517, p. 39]; *x)* is given in [1517].

Fact 1.21.5. Let $\mathcal{I} \subseteq (0, \infty)$ be a finite or infinite interval, let $f\colon \mathcal{I} \mapsto \mathbb{R}$, and define $g\colon \mathcal{I} \mapsto \mathbb{R}$ by $g(x) = xf(1/x)$. Then, f is (convex, strictly convex) if and only if g is (convex, strictly convex). **Source:** [2128, p. 13].

Fact 1.21.6. Let $f\colon \mathbb{R} \mapsto \mathbb{R}$, assume that f is convex, and assume that there exists $\alpha \in \mathbb{R}$ such that, for all $x \in \mathbb{R}$, $f(x) \le \alpha$. Then, f is constant. **Source:** [2128, p. 35].

Fact 1.21.7. Let $\mathcal{I} \subseteq \mathbb{R}$ be a finite or infinite interval, let $f\colon \mathcal{I} \mapsto \mathbb{R}$, and assume that f is continuous. Then, the following statements are equivalent:

i) f is convex.

ii) For all $n \in \mathbb{P}$, $x_1, \ldots, x_n \in \mathcal{I}$, and $\alpha_1, \ldots, \alpha_n \in [0, 1]$ such that $\sum_{i=1}^n \alpha_i = 1$, it follows that

$$f\left(\sum_{i=1}^n \alpha_i x_i\right) \le \sum_{i=1}^n \alpha_i f(x_i).$$

Remark: This is *Jensen's inequality.* **Remark:** Setting $f(x) = x^p$ yields Fact 2.11.131, whereas setting $f(x) = \log x$ for all $x \in (0, \infty)$ yields the arithmetic-mean–geometric-mean inequality given by Fact 2.11.81. **Related:** Fact 12.13.19.

Fact 1.21.8. Let $[a, b] \subset \mathbb{R}$, let $f\colon [a, b] \mapsto \mathbb{R}$ be convex, and let $x, y \in [a, b]$. Then,

$$\tfrac{1}{2}[f(x) + f(y)] - f[\tfrac{1}{2}(x + y)] \le \tfrac{1}{2}[f(a) + f(b)] - f[\tfrac{1}{2}(a + b)].$$

Remark: This is *Niculescu's inequality.* See [209, p. 13].

Fact 1.21.9. Let $\mathcal{I} \subseteq \mathbb{R}$ be a finite or infinite interval, let $f\colon \mathcal{I} \mapsto \mathbb{R}$. Then, the following statements are equivalent:

i) f is convex.

ii) f is continuous, and, for all $x, y, z \in \mathcal{I}$,

$$\tfrac{2}{3}(f[\tfrac{1}{2}(x + y)] + f[\tfrac{1}{2}(y + z)] + f[\tfrac{1}{2}(z + x)]) \le \tfrac{1}{3}[f(x) + f(y) + f(z)] + f[\tfrac{1}{3}(x + y + z)].$$

Remark: This is *Popoviciu's inequality.* See [2128, p. 12]. **Remark:** For a scalar argument and $f(x) = |x|$, this result implies Hlawka's inequality given by Fact 11.8.3. See Fact 2.21.8 and [2130]. **Problem:** Extend this result so that it yields Hlawka's inequality for vector arguments.

Fact 1.21.10. Let $[a, b] \subset \mathbb{R}$, let $f\colon [a, b] \mapsto \mathbb{R}$, and assume that f is convex. Then,

$$f[\tfrac{1}{2}(a + b)] \le \frac{1}{b - a}\int_a^b f(x)\,\mathrm{d}x \le \tfrac{1}{2}[f(a) + f(b)].$$

Source: [2128, pp. 50–53], [2362, 2364]. **Remark:** This is the *Hermite-Hadamard inequality.*

Fact 1.21.11. Let $[a,b] \subset \mathbb{R}$, let $f\colon [a,b] \mapsto \mathbb{R}$, assume that f is concave, and let p and q be positive numbers such that $p < q$. Then,

$$\left(\frac{q+1}{b-a}\int_a^b [f(x)]^q \,\mathrm{d}x\right)^{1/q} \le \left(\frac{p+1}{b-a}\int_a^b [f(x)]^p \,\mathrm{d}x\right)^{1/p}.$$

Source: [2213, p. 216].

1.22 Notes

Some of the preliminary material in this chapter can be found in [2112]. A related treatment of mathematical preliminaries is given in [2314]. An extensive introduction to logic and mathematical fundamentals is given in [493]. In [493], the notation "$A \to B$" denotes an implication, which is called a *disjunction*, while "$A \Longrightarrow B$" denotes a tautology.

The "truth" operator is represented by square brackets in [1219]. See [1649].

Multisets are discussed in [492, 1220, 2460, 2879].

Partially ordered sets are considered in [2405, 2734]. Lattices are discussed in [493]. For a pair of elements x, y, $\mathrm{glb}(\{x,y\})$ is alternatively written as $x \wedge y$, where "$\wedge$" is the *meet* operator. Similarly, $\mathrm{lub}(\{x,y\})$ is alternatively written as $x \vee y$, where "$\vee$" is the *join* operator.

A directed graph is also called a *digraph*. A directionally connected graph is traditionally called *strongly connected* [2873, p. 56].

Alternative terminology for "one-to-one" and "onto" is *injective* and *surjective*, respectively, while a function that is injective and surjective is *bijective*.

Subtle aspects of compositions of complex functions are discussed in [498].

Chapter Two
Equalities and Inequalities

2.1 Facts on Equalities and Inequalities in One Variable

Fact 2.1.1. Let x be a complex number. Then,

$$(2x)^2 + (2x+1)^2 + (2x+2)^2 + (6x^2+6x+2)^2 = (6x^2+6x+3)^2,$$
$$(2x+1)^2 + [2x(x+1)]^2 = [2x(x+1)+1]^2, \quad (4x)^2 + (4x^2-1)^2 = (4x^2+1)^2,$$
$$(x^3+1)^3 + (2x^3-1)^3 + (x^4-2x)^3 = (x^4+x)^3,$$
$$(3x^2)^3 + (6x^2-3x+1)^3 + [3x(3x^2-2x+1)-1]^3 = [3x(3x^2-2x+1)]^3,$$
$$(3x^2)^3 + (6x^2+3x+1)^3 + [3x(3x^2+2x+1)]^3 = [3x(3x^2+2x+1)+1]^3.$$

Source: [289, pp. 30, 36, 37, 161, 181].

Fact 2.1.2. Let x be a complex number. Then,

$$x(x+1)(x+2) + x + 1 = (x+1)^3, \quad x(x+1)(x+2)(x+3) + 1 = (x^2+3x+1)^2,$$
$$x(x+1)(x+2)(x+3)(x+4) + x(5x^2+30x+56) + 32 = (x+2)^5.$$

Example: $J(J+1)(J+2)(J+3) = -10$. **Related:** Fact 1.11.2.

Fact 2.1.3. Let x be a complex number. Then,

$$x(x+2)^3 = (2x+1)^3 + (x+1)(x-1)^3, \quad 4x^4+1 = (2x^2+2x+1)(2x^2-2x+1),$$
$$(x^2+x+1)^4 = x^4 + (x+1)^4 + (x^4+2x^3+3x^2+2x)^2,$$
$$(x^3+6x^2+10x+4)(x^2+4x+3)x+1 = [(x^3+6x^2+10x+4)-(x^2+4x+3)]^2 = (x^3+5x^2+6x+1)^2,$$
$$\begin{aligned}(x+1)^5 &+ (x-1)^5 + (x+18)^5 + (x-18)^5 + (x+19)^5 + (x-19)^5 \\ &= (x+7)^5 + (x-7)^5 + (x+14)^5 + (x-14)^5 + (x+21)^5 + (x-21)^5 \\ &= 6x^5 + 13720x^3 + 2352980x.\end{aligned}$$

Source: [27], [108, p. 290], [1651, p. 392], and [1935].

Fact 2.1.4. Let $x \in \mathbb{C}$ and $n \ge 1$. Then,

$$x^{4n} + x^{2n} + 1 = (x^{2n} - x^n + 1)(x^{2n} + x^n + 1).$$

In particular,

$$x^4 + x^2 + 1 = (x^2 - x + 1)(x^2 + x + 1), \quad x^8 + x^4 + 1 = (x^4 - x^2 + 1)(x^4 + x^2 + 1),$$
$$x^{12} + x^6 + 1 = (x^6 - x^3 + 1)(x^6 + x^3 + 1), \quad x^{16} + x^8 + 1 = (x^8 - x^4 + 1)(x^8 + x^4 + 1).$$

Furthermore,

$$x^{5n} + x^n - 1 = (x^{3n} + x^{2n} - 1)(x^{2n} - x^n + 1).$$

In particular,

$$x^5 + x - 1 = (x^3 + x^2 - 1)(x^2 - x + 1), \quad x^{10} + x^2 - 1 = (x^6 + x^4 - 1)(x^4 - x^2 + 1),$$

$$x^{15} + x^3 - 1 = (x^9 + x^6 - 1)(x^6 - x^3 + 1).$$

Furthermore,

$$x^{5n} + x^n + 1 = (x^{3n} - x^{2n} + 1)(x^{2n} + x^n + 1).$$

In particular,

$$x^5 + x + 1 = (x^3 - x^2 + 1)(x^2 + x + 1), \quad x^{10} + x^2 + 1 = (x^6 - x^4 + 1)(x^4 + x^2 + 1),$$
$$x^{15} + x^3 + 1 = (x^9 - x^6 + 1)(x^6 + x^3 + 1).$$

Finally,

$$x^{2^{n+1}} + x^{2^n} + 1 = (x^2 + x + 1)\prod_{i=1}^{n}(1 + x^{2^i} - x^{2^{i-1}}).$$

In particular,

$$x^4 + x^2 + 1 = (x^2 + x + 1)(x^2 - x + 1), \quad x^8 + x^4 + 1 = (x^2 + x + 1)(x^4 - x^2 + 1)(x^2 - x + 1),$$
$$x^{16} + x^8 + 1 = (x^2 + x + 1)(x^8 - x^4 + 1)(x^4 - x^2 + 1)(x^2 - x + 1).$$

Source: [398].

Fact 2.1.5. Let x be a complex number. Then,

$$x^{10} - x^5 + 1 = (x^2 - x + 1)(x^8 + x^7 - x^5 - x^4 - x^3 + x + 1),$$
$$x^{10} + x^5 + 1 = (x^2 + x + 1)(x^8 - x^7 + x^5 - x^4 + x^3 - x + 1),$$
$$x^{14} - x^7 + 1 = (x^2 - x + 1)(x^{12} + x^{11} - x^9 - x^8 + x^6 - x^4 - x^3 + x + 1),$$
$$x^{14} + x^7 + 1 = (x^2 + x + 1)(x^{12} - x^{11} + x^9 - x^8 + x^6 - x^4 + x^3 - x + 1).$$

Consequently, if $x \neq 0$ and $x + 1/x = \pm 1$, then $x^5 + 1/x^5 = \pm 1$ and $x^7 + 1/x^7 = \pm 1$.

Fact 2.1.6. Let a, b, c, x be complex numbers, and assume that $a \neq 0$. Then,

$$ax^2 + bx + c = a\left(x + \frac{b}{2a}\right)^2 + \frac{4ac - b^2}{4a}.$$

Related: This is *completing the square*.

Fact 2.1.7. Let x be a complex number. Then,

$$x^3 + 1 = (x + 1)(x^2 - x + 1), \quad x^5 + 1 = (x + 1)(x^4 - x^3 + x^2 - x + 1),$$
$$x^7 + 1 = (x + 1)(x^6 - x^5 + x^4 - x^3 + x^2 - x + 1),$$
$$x^2 - x + 1 = (x - e^{(\pi/3)J})(x + e^{(2\pi/3)J}),$$
$$x^4 - x^3 + x^2 - x + 1 = (x - e^{(\pi/5)J})(x + e^{(2\pi/5)J})(x - e^{(3\pi/5)J})(x + e^{(4\pi/5)J}),$$
$$x^6 - x^5 + x^4 - x^3 + x^2 - x + 1 = (x - e^{(\pi/7)J})(x + e^{(2\pi/7)J})(x - e^{(3\pi/7)J})(x + e^{(4\pi/7)J})(x - e^{(5\pi/7)J})(x + e^{(6\pi/7)J}).$$

Source: [732, p. 252].

Fact 2.1.8. Let x be a complex number. Then,

$$x^3 - 1 = (x - 1)(x^2 + x + 1), \quad x^5 - 1 = (x - 1)(x^4 + x^3 + x^2 + x + 1),$$
$$x^7 - 1 = (x - 1)(x^6 + x^5 + x^4 + x^3 + x^2 + x + 1),$$
$$x^2 + x + 1 = (x + e^{(\pi/3)J})(x - e^{(2\pi/3)J}),$$
$$x^4 + x^3 + x^2 + x + 1 = (x + e^{(\pi/5)J})(x - e^{(2\pi/5)J})(x + e^{(3\pi/5)J})(x - e^{(4\pi/5)J}),$$
$$x^6 + x^5 + x^4 + x^3 + x^2 + x + 1 = (x + e^{(\pi/7)J})(x - e^{(2\pi/7)J})(x + e^{(3\pi/7)J})(x - e^{(4\pi/7)J})(x + e^{(5\pi/7)J})(x - e^{(6\pi/7)J}).$$

Fact 2.1.9. Let x be a complex number such that division by zero does not occur in the expressions below. Then,

$$\left(1+\frac{1}{x}-\frac{1}{x+1}\right)^2 = 1+\frac{1}{x^2}+\frac{1}{(x+1)^2}, \quad \left(1-x-\frac{x}{x-1}\right)^2 = 1+x^2+\left(\frac{x}{x-1}\right)^2,$$

$$\left(1+x-\frac{x}{x+1}\right)^2 = 1+x^2+\left(\frac{x}{x+1}\right)^2, \ \left(1+\frac{1}{x(x+1)}-\frac{1}{x^2+x+1}\right)^2 = 1+\frac{1}{x^2(x+1)^2}+\frac{1}{(x^2+x+1)^2},$$

$$\left(1+\frac{1}{x}+\frac{1}{x+1}-\frac{2(x+1)}{x^2+3x+1}\right)^2 = 1+\frac{1}{x^2}+\frac{1}{(x+1)^2}+\frac{4(x+1)^2}{(x^2+3x+1)^2}.$$

Source: [399]. **Related:** Fact 2.2.11.

Fact 2.1.10. Let $x \in \mathbb{C}$ and $n \ge 1$. Then,

$$(x^2+x+1)^n = \sum_{i=0}^{2n}\sum_{j=0}^{n}\binom{n}{j}\binom{n-j}{i-2j}x^i.$$

Source: [2068, pp. 96, 97]. **Remark:** The coefficient of x^i is a *trinomial coefficient*. **Related:** Fact 2.3.1.

Fact 2.1.11. Let $x \in \mathbb{C}$ and $n \ge 1$. Then,

$$\left(\sum_{i=0}^{n} x^i\right)^2 = x^n + \left(\sum_{i=0}^{n-1} x^i\right)\left(\sum_{i=0}^{n+1} x^i\right).$$

Source: [993, pp. 256, 264].

Fact 2.1.12. Let $x \in \mathbb{C}$ and $n \ge 1$. Then,

$$\sum_{i=0}^{n}(-1)^{n+i}\frac{(n+i)!}{(n-i)!(i!)^2}(x+1)^i = \sum_{i=0}^{n}\frac{(n+i)!}{(n-i)!(i!)^2}x^i.$$

Furthermore, for all $x \in \mathbb{C}$, define

$$f(x) \triangleq \sum_{i=0}^{n}\frac{(n+i)!}{(n-i)!(i!)^2}(x-\tfrac{1}{2})^i.$$

Then, for all $x \in \mathbb{C}$, $f(-x) = (-1)^n f(x)$. **Source:** [2456]. **Example:** If $n = 1$, then $f(x) = 2x$. If $n = 2$, then $f(x) = 6x^2 - \frac{1}{2}$.

Fact 2.1.13. Let $m \ge 1$, $n \ge 1$, and $x \in \mathbb{C}$. Then,

$$(1-x^{m/n})\sum_{i=0}^{n-1} x^{i/n} = (1-x)\sum_{i=0}^{m-1} x^{i/n}.$$

Source: [570, p. 250].

Fact 2.1.14. Let $n \ge 1$ and $m \ge 2$, and let x be a complex number such that division by zero does not occur in the expressions below. Then,

$$\frac{x^n}{(1-x^n)^2} = \sum_{i=0}^{m-2}\frac{4^i x^{2^i n}}{(1+x^{2^i n})^2} + \frac{4^{m-1}x^{2^{m-1}n}}{(1-x^{2^{m-1}n})^2},$$

$$\frac{x^n}{(1-x^n)^2} = \frac{x^n}{(1+x^n)^2} + \frac{4x^{2n}}{(1-x^{2n})^2},$$

$$\frac{x^n}{(1-x^n)^2} = \frac{x^n}{(1+x^n)^2} + \frac{4x^{2n}}{(1+x^{2n})^2} + \frac{16x^{4n}}{(1-x^{4n})^2}.$$

Source: [2150]. **Related:** Fact 13.5.2.

Fact 2.1.15. Let $n \ge 1$, for all $x \in \mathbb{C}$, define $f_n(x) \triangleq \prod_{i=1}^n (x^i - 1)$, and let $m \ge 1$. Then, there exists a polynomial $p_{n,m}$ with integer coefficients such that $f_{n+m} = p_{n,m} f_n f_m$. **Source:** [1158, pp. 295, 296]. **Example:** $p_{4,2}(x) = x^8 + x^7 + 2x^6 + 2x^5 + 3x^4 + 2x^3 + 2x^2 + x + 1$.

Fact 2.1.16. Let $n \ge 1$, let $x \in \mathbb{C}$, and assume that, for all $i \in \{1, \dots, n\}$, $x^i \ne 1$. Then,

$$\sum \prod_{i=1}^n \frac{1}{j_i! i^{j_i} (1 - x^i)^{j_i}} = \prod_{i=1}^n \frac{1}{1 - x^i},$$

where the sum is taken over all n-tuples $(j_1, \dots, j_n)$ of nonnegative integers such that $\sum_{i=1}^n i j_i = n$. **Source:** [2590]. **Example:** $\frac{1}{2(1-x)^2} + \frac{1}{2(1-x^2)} = \frac{1}{(1-x)(1-x^2)}$.

Fact 2.1.17. Let $n \ge 1$, and, for all $x \in \mathbb{C}$ such that, for all $i \in \{1, \dots, n\}$, $x^i \ne 1$, define $f_0(x) \triangleq 1$ and

$$f_n(x) \triangleq (-1)^n \prod_{i=1}^n \frac{x^i + 1}{x^i - 1}.$$

Then, for all $x \in \mathbb{C}$ such that, for all $i \in \{1, \dots, n\}$, $x^i \notin \{-1, 1\}$,

$$\sum_{i=-n}^n (-1)^i \frac{4x^i}{(x^i + 1)^2} f_n^2(x) f_{n+i}(x) f_{n-i}(x) = 1, \quad \sum_{i=0}^n (-1)^i \frac{2x^{i(n+1)} f_i(x)}{(x^i + 1) f_n(x)} = \sum_{i=-n}^n (-1)^i x^{i^2}.$$

Source: [117, 2970]. **Related:** Fact 13.5.5.

Fact 2.1.18. Let $m, n \ge 2$, let $x \in \mathbb{C}$, and assume that, for all $i \in \{1, \dots, \min\{m, n\}\}$, $x^{2i} \ne 1$. Then,

$$\sum_{i=1}^{\min\{m,n\}} x^{(m-i)^2 + (n-i)^2} \prod_{j=1}^{i-1} \frac{(1 - x^{2m-2j})(1 - x^{2n-2j})}{1 - x^{2j}} = x^{(m-n)^2}.$$

In particular,

$$\sum_{i=1}^n x^{2(n-i)^2} \prod_{j=1}^{i-1} \frac{(1 - x^{2n-2j})^2}{1 - x^{2j}} = 1.$$

Source: [666].

Fact 2.1.19. Let n be a positive integer, and, for all $x \in \mathbb{C}$, define $g_0(x) \triangleq 1$ and

$$g_n(x) \triangleq \prod_{i=1}^n (1 - x^i).$$

Then, for all $x \in \mathbb{C}$ such that, for all $i \in \{0, \dots, 2n\}$, $x^i \ne 1$,

$$\sum_{i=0}^n \frac{x^{i^2}}{g_i(x) g_{n-i}(x)} = \sum_{i=-n}^n \frac{(-1)^i x^{(5i^2 - i)/2}}{g_{n-i}(x) g_{n+i}(x)}.$$

Source: [2970]. **Related:** Fact 13.5.6.

Fact 2.1.20. Let $m \ge 1$, $n \ge 1$, and $x \in \mathbb{C}$. If $a, b \in \mathbb{C}$, $a \ne b$, $x + a \ne 0$, and $x + b \ne 0$, then

$$\begin{aligned} \frac{1}{(x+a)^m (x+b)^n} &= \sum_{i=1}^m (-1)^{m-i} \binom{m+n-i-1}{m-i} \frac{1}{(b-a)^{m+n-i} (x+a)^i} \\ &\quad + \sum_{i=1}^n (-1)^{n-i} \binom{m+n-i-1}{n-i} \frac{1}{(a-b)^{m+n-i} (x+b)^i}. \end{aligned}$$

If $x \notin \{-1, 1\}$, then

$$\frac{1}{(1+x)^m(1-x)^n} = \sum_{i=1}^{m} \frac{1}{2^{m+n-i}}\binom{m+n-i-1}{m-i}\frac{1}{(1+x)^i} + \sum_{i=1}^{n} \frac{1}{2^{m+n-i}}\binom{m+n-i-1}{n-i}\frac{1}{(1-x)^i},$$

$$\frac{1}{(1-x^2)^{n+1}} = \sum_{i=1}^{n+1} \frac{1}{2^{2n+2-i}}\binom{2n+1-i}{n}\left[\frac{1}{(1+x)^i} + \frac{1}{(1-x)^i}\right].$$

Source: [704].

Fact 2.1.21. Let x and α be real numbers, and assume that $x \ge -1$. Then, the following statements hold:

i) If $\alpha \le 0$, then

$$1 + \alpha x \le (1+x)^\alpha.$$

Equality holds if and only if either $x = 0$ or $\alpha = 0$.

ii) If $\alpha \in [0, 1]$, then

$$(1+x)^\alpha \le 1 + \alpha x.$$

Equality holds if and only if either $x = 0$, $\alpha = 0$, or $\alpha = 1$.

iii) If $\alpha \ge 1$, then

$$1 + \alpha x \le (1+x)^\alpha.$$

Equality holds if and only if either $x = 0$ or $\alpha = 1$.

Source: [77], [604, p. 4], and [2061, p. 65]. Alternatively, use Fact 2.15.3. See [2901]. **Remark:** This is *Bernoulli's inequality*. An equivalent version is given by Fact 2.1.23. An extension to n variables is given in Fact 2.11.11. **Remark:** The proof of *i)* and *iii)* in [77] is based on the fact that, for all $x \ge -1$, the function $f(x) \triangleq \frac{(1+x)^\alpha - 1}{x}$ for $x \ne 0$ and $f(0) \triangleq \alpha$, is increasing.

Fact 2.1.22. Let $x \in (0, 1)$ and $k \ge 1$. Then,

$$(1-x)^k < \frac{1}{1+kx}.$$

Source: Use *i)* of Fact 2.1.21 with x replaced by $-x$ and $\alpha = -k$. See [1371, p. 137].

Fact 2.1.23. Let $x \ge 0$ and $\alpha \in \mathbb{R}$. If $\alpha \in [0, 1]$, then

$$\alpha + x^\alpha \le 1 + \alpha x,$$

whereas, if either $\alpha \le 0$ or $\alpha \ge 1$, then

$$1 + \alpha x \le \alpha + x^\alpha.$$

Source: Set $y = x + 1$ in Fact 2.1.21. Alternatively, in the case where $\alpha \in [0, 1]$, set $y = 1$ in the right-hand inequality in Fact 2.2.53. For $\alpha \ge 1$, note that $f(x) \triangleq \alpha + x^\alpha - 1 - \alpha x$ satisfies $f(1) = 0$, $f'(1) = 0$, and, for all $x \ge 0$, $f''(x) = \alpha(\alpha-1)x^{\alpha-2} > 0$. **Remark:** This result is equivalent to Bernoulli's inequality. See Fact 2.1.21. **Remark:** For $\alpha \in [0, 1]$, a matrix version is given by Fact 10.10.47. **Problem:** Compare the second inequality to Fact 2.2.54 with $y = 1$.

Fact 2.1.24. Let x and y be positive numbers. If $x, y \in (0, 1]$, then

$$\left(1 + \frac{1}{x}\right)^y \le 1 + \frac{y}{x}.$$

Equality holds if and only if either $y = 0$ or $x = y = 1$. If $x \in (0, 1)$, then

$$\left(1 + \frac{1}{x}\right)^x < 2.$$

If $x > 1$ and $y \in [1, x]$, then

$$1 + \frac{y}{x} \le \left(1 + \frac{1}{x}\right)^y < 1 + \frac{y}{x} + \frac{y^2}{x^2}.$$

The left-hand inequality is an equality if and only if $y = 1$. Furthermore, if $x > 1$, then

$$2 < \left(1 + \frac{1}{x}\right)^x < 3.$$

Source: Fact 2.1.21 and [1371, p. 137].

Fact 2.1.25. Let $a > 0$ satisfy $\sqrt{a}^{\sqrt{a+1}} = \sqrt{a+1}^{\sqrt{a}}$. Then, $a \approx 6.91156$. Furthermore, if $x \in (0, a)$, then $\sqrt{x}^{\sqrt{x+1}} < \sqrt{x+1}^{\sqrt{x}}$, whereas, if $x \in (a, \infty)$, then $\sqrt{x+1}^{\sqrt{x}} < \sqrt{x}^{\sqrt{x+1}}$. **Source:** [289, pp. 3, 54].

Fact 2.1.26. Let $x \in [0, \infty)$ and $1 \le k \le l \le n$. Then,

$$\sqrt[n]{1 + x} \le \sqrt[l]{1 + \frac{l}{n}x} \le \sqrt[k]{1 + \frac{k}{n}x} \le 1 + \frac{1}{n}x.$$

Source: [356].

Fact 2.1.27. Let x and α be real numbers, assume that either $\alpha \le 0$ or $\alpha \ge 1$, and assume that $x \in [0, 1]$. Then,

$$(1 + x)^\alpha \le 1 + (2^\alpha - 1)x.$$

Equality holds if and only if either $\alpha = 0$, $\alpha = 1$, $x = 0$, or $x = 1$. **Source:** [77].

Fact 2.1.28. Let $x \in \mathbb{R}$ and $n \ge 1$. Then,

$$x[nx^{n+1} - (n + 1)x^n + 1] = (x - 1)^2 \sum_{i=1}^n ix^i.$$

Now, assume that x is nonnegative. Then,

$$(n + 1)x^n \le nx^{n+1} + 1, \quad (n + 1)x^{n/(n+1)} \le nx + 1.$$

Furthermore, each inequality is an equality if and only if $x = 1$. **Source:** [658]. **Remark:** For $x > 0$, setting $n = 3$ and replacing x by $1/x$ yields $4x \le x^4 + 3$. **Related:** Fact 2.1.29 and Fact 2.2.39.

Fact 2.1.29. Let $x \in \mathbb{R}$ and $n \ge 2$. Then,

$$n(1 - x^{n-1}) - (n - 1)(1 - x^n) = (x - 1)\left[(n - 1)x^{n-1} - \sum_{i=0}^{n-2} x^i\right].$$

Now, assume that x is nonnegative. Then,

$$\frac{1 - x^n}{n} \le \frac{1 - x^{n-1}}{n - 1}.$$

Furthermore, equality holds if and only if $x = 1$. **Source:** [2332]. **Related:** This result follows from Fact 2.1.28 by replacing n with $n - 1$.

Fact 2.1.30. Let $x \in [-1, 1]$ and $n \ge 1$. Then, $(1 + x)^n + (1 - x)^n \le 2^n$. **Source:** [1158, p. 146].

Fact 2.1.31. Let a, b, c, x be real numbers, and assume that $a^2 + c^2 \le 4b$. Then,

$$x^4 + ax^3 + bx^2 + cx + 1 \ge 0.$$

Source: [112, p. 35].

Fact 2.1.32. Let $x \ge 0$ and $n \ge 2$. Then, $nx \le x^n + n - 1$. **Source:** [2527, p. 34].

Fact 2.1.33. Let x be a nonnegative number. Then,

$$8x < x^4 + 9,$$
$$3x^2 \le x^3 + 4,$$
$$27x^3 < (x+1)^5,$$
$$3x^5 \le x^{11} + x^4 + 1,$$
$$2x^3 + x^2 \le 2x^4 + 1,$$
$$x^9 + x < x^{12} + x^4 + 1,$$
$$4x^2 \le x^4 + x^3 + x + 1,$$
$$8x^2 < x^4 + x^3 + 4x + 4,$$
$$3(2x^{27} + x^6) \le (2x^{15} + 1)^2.$$

Source: [806, p. 5], [1033], and [1371, pp. 117, 123, 152, 153, 155].

Fact 2.1.34. Let $n \ge 1$, $1 \le k \le n$, and $x \ge 0$. Then,

$$x^k + x^{n-k} \le x^n + 1, \quad 2\sum_{i=1}^n x^i \le n(x^{n+1} + 1), \quad (2n+1)x^n \le \sum_{i=0}^{2n} x^i = \frac{1 - x^{2n+1}}{1 - x},$$

$$(n+1)\left(\frac{x+1}{2}\right)^n \le \sum_{i=0}^n x^i \le (n+1)\frac{x^n + 1}{2}, \quad \sum_{i=0}^{n-1} x^i \le n\left(\frac{x^n + 1}{2}\right)^{1-1/n}.$$

Source: [393] and [1566, p. 10]. **Related:** Fact 2.2.38.

Fact 2.1.35. Let $n \ge 1$ and $x > 0$. Then,

$$\frac{(n+1)(x+1)}{2n} < \frac{x^n + \cdots + x^2 + x + 1}{x^{n-1} + \cdots + x^2 + x + 1}, \quad \frac{n+1}{n} \le \frac{x^{2n} + \cdots + x^4 + x^2 + 1}{x^{2n-1} + \cdots + x^3 + x}.$$

If $n \ge 2$, then

$$\frac{(n+1)x}{n-1} \le \frac{x^n + \cdots + x^2 + x + 1}{x^{n-2} + \cdots + x^2 + x + 1}.$$

Source: [378]. **Remark:** For $n = 3$, the last inequality implies that, for all $x \ge 0$, $x^2 + x \le x^3 + 1$.

Fact 2.1.36. Let $n \ge 1$ and $x \in (0, 1)$. Then,

$$\sum_{i=1}^n \left(\frac{1}{i}\sum_{j=0}^{i-1} x^i\right)^2 \le 4(\log 2)\sum_{i=0}^{n-1} x^{2i}.$$

Source: [2527, p. 177]. **Related:** Fact 2.11.132.

Fact 2.1.37. Let x be a positive number. Then,

$$1 + \tfrac{1}{2}x - \tfrac{1}{8}x^2 < \sqrt{1+x} < 1 + \tfrac{1}{2}x - \tfrac{1}{8}x^2 + \tfrac{1}{16}x^3.$$

Source: [1567, p. 55].

Fact 2.1.38. Let x be a positive number. Then,

$$\frac{x(x+1)^2}{2(x^3+1)} \le \frac{x(x^2+1)}{x^3+1} \le \frac{2x}{x+1} \le \sqrt{x} \le \frac{x+1}{2} \le \sqrt{\frac{x^2+1}{2}} \le \sqrt[4]{\frac{x^4+1}{2}} \le \sqrt{x^2 - x + 1} \le \frac{x^3+1}{x^2+1}.$$

Source: [379].

Fact 2.1.39. Let $a, b \in (0, \infty)$, assume that $a < b$, and let $x \in [a, b]$. Then,

$$\frac{x}{ab} + \frac{1}{x} \le \frac{a+b}{ab}.$$

Source: [2991, p. 249].

Fact 2.1.40. Let $x \in (0, 1)$. Then,

$$4 \le \frac{1}{x} + \frac{1}{1-x}, \quad \frac{25}{2} \le \left(x + \frac{1}{x}\right)^2 + \left(1 - x + \frac{1}{1-x}\right)^2.$$

Source: [51].

Fact 2.1.41. Let $x \in (0, 1)$. Then,

$$\frac{1}{2-x} < x^x < x^2 - x + 1.$$

Source: [1371, p. 164].

Fact 2.1.42. Let $x \in (1, \infty)$. Then,

$$\frac{3}{x} < \frac{1}{x-1} + \frac{1}{x} + \frac{1}{x+1}.$$

Source: [2527, p. 99].

Fact 2.1.43. Let $x, p \in [1, \infty)$. Then,

$$x^{1/p}(x-1) < px(x^{1/p} - 1).$$

Equality holds if and only if either $p = 1$ or $x = 1$. **Source:** [1124, p. 194].

Fact 2.1.44. If $p \in [\sqrt{2}, 2)$, then, for all $x \in (0, 1)$,

$$\left[\frac{1 - x^p}{p(1-x)}\right]^2 \le \tfrac{1}{2}(1 + x^{p-1}).$$

Furthermore, if $p \in (1, \sqrt{2})$, then there exists $x \in (0, 1)$ such that

$$\tfrac{1}{2}(1 + x^{p-1}) < \left[\frac{1 - x^p}{p(1-x)}\right]^2.$$

Source: [463].

Fact 2.1.45. Let $x, p \in [1, \infty)$. Then,

$$(p-1)^{p-1}(x^p - 1)^p \le p^p(x-1)(x^p - x)^{p-1}x^{p-1}.$$

Equality holds if and only if either $p = 1$ or $x = 1$. **Source:** [1124, p. 194].

Fact 2.1.46. Let $x \in [1, \infty)$, and let $p, q \in (1, \infty)$ satisfy $1/p + 1/q = 1$. Then,

$$px^{1/q} \le 1 + (p-1)x.$$

Equality holds if and only if $x = 1$. **Source:** [1124, p. 194].

Fact 2.1.47. Let $x \in [1, \infty)$, and let $p, q \in (1, \infty)$ satisfy $1/p + 1/q = 1$. Then,

$$x - 1 \le p^{1/p}q^{1/q}(x^{1/p} - 1)^{1/p}(x^{1/q} - 1)^{1/q}x^{2/(pq)}.$$

Equality holds if and only if $x = 1$. **Source:** [1124, p. 195].

Fact 2.1.48. Let $x \in [0, 1]$ and $p, q \in (0, \infty)$. Then,

$$x^q(1 - x^p) \le \frac{p}{q}\left(\frac{q}{p+q}\right)^{1+q/p}.$$

Equality holds if and only if $x = [q/(p+q)]^{1+q/p}$. **Source:** [2348].

Fact 2.1.49. Let $x \in [0, \infty)$. Then,

$$\frac{2x^2}{x^3+1} + \frac{x^2}{x^6+1} \le \frac{3}{2}.$$

Source: [2601].

Fact 2.1.50. Let $x \in (0, \infty)$. Then, $x^{1/x} < 1 + 1/\sqrt{x}$. **Source:** [2412].

Fact 2.1.51. Let x and α be nonnegative numbers. Then,

$$x^\alpha \le \left(\frac{\alpha}{e}\right)^\alpha e^x.$$

Source: [2039]. **Related:** Fact 2.11.144.

Fact 2.1.52. Let x be a real number, and let $p, q \in (1, \infty)$ satisfy $1/p + 1/q = 1$. Then,

$$\tfrac{1}{p}e^{px} + \tfrac{1}{q}e^{-qx} \le e^{p^2q^2x^2/8}.$$

Source: [1757, p. 260].

Fact 2.1.53. Let $x \in (0, 1)$. Then,

$$e^{2x} < \frac{1+x}{1-x}.$$

Source: [968, p. 106] and [1757, p. 271].

Fact 2.1.54. Let x be a real number. Then,

$$2 - e^{-x} \le 1 + x \le e^x.$$

Fact 2.1.55. Let x be a positive number. Then,

$$1 + (1+x)\log(1+x) < 1 + x + \frac{x^2}{2} < e^x.$$

Source: [1757, p. 271].

Fact 2.1.56. Let x be a nonnegative number, and let p and q be real numbers such that $0 < p \le q$. Then,

$$e^x\left(1 + \frac{1}{p}\right)^{-x} \le \left(1 + \frac{x}{p}\right)^p \le \left(1 + \frac{x}{q}\right)^q \le e^x.$$

Furthermore, if $p < q$, then equality holds if and only if $x = 0$. Finally,

$$\lim_{q\to\infty}\left(1 + \frac{x}{q}\right)^q = e^x.$$

Source: [604, pp. 7, 8]. **Remark:** As $q \to \infty$, $(1 + 1/q)^q = e + O(1/q)$, whereas $(1 + 1/q)^q[1 + 1/(2q)] = e + O(1/q^2)$. See [1648].

Fact 2.1.57. Let x be a positive number. Then,

$$e^{1-\frac{1}{2x}} < \left(1 + \frac{1}{x}\right)^x < \frac{2x+2}{2x+1}\left(1 + \frac{1}{x}\right)^x < e^{1/[6(2x+1)^2]}\frac{2x+2}{2x+1}\left(1 + \frac{1}{x}\right)^x$$

$$< e < e^{-1/[3(2x+1)^2]}\left(1 + \frac{1}{x}\right)^{x+1/2} < \left(1 + \frac{1}{x}\right)^{x+1/2} < \left(\frac{2x+1}{2x}\right)\left(1 + \frac{1}{x}\right)^x < \left(1 + \frac{1}{x}\right)^{x+1},$$

$$\left(1 + \frac{1}{x}\right)^x\left(1 + \frac{6}{12x+5}\right) < e < \left(1 + \frac{1}{x}\right)^{x+1}\left(1 - \frac{6}{12x+7}\right), \quad e^{\frac{2x}{2x+1}} \le \left(1 + \frac{1}{x}\right)^x \le e^{\frac{2x+1}{2x+2}},$$

$$\left.\begin{matrix} e^{1-\frac{1}{2x}} \\ e^{\frac{1}{2x}-\frac{1}{3x^2}}\left(1+\frac{1}{x}\right)^x \end{matrix}\right\} < e^{-1/[12x(x+1)]}\left(1+\frac{1}{x}\right)^{x+1/2} < e^{1/[6(2x+1)^2]}\frac{2x+2}{2x+1}\left(1+\frac{1}{x}\right)^x < e,$$

$$\left(1+\frac{1}{x}\right)^x < e\left[1-\frac{1}{2(x+1)}-\frac{1}{24(x+1)^2}-\frac{1}{48(x+1)^3}\right], \quad \frac{e}{2x+2} < e-\left(1+\frac{1}{x}\right)^x < \frac{e}{2x+1}.$$

If $x \ge 2/3$, then

$$\left(1+\frac{1}{x}\right)^x < e^{1-\frac{1}{2x}+\frac{1}{3x^2}} < e.$$

Source: [369, 1214, 1609], [1757, p. 278], and [2373, 2948, 3012].

Fact 2.1.58. Let $x > 4/5$. Then,

$$\left(1+\frac{1}{x+\frac{1}{5}}\right)^{1/2} < \left(1+\frac{2}{3x+1}\right)^{3/4} < \left(1+\frac{1}{\frac{5}{4}x+\frac{1}{3}}\right)^{5/8} < \frac{e}{\left(1+\frac{1}{x}\right)^x} < \left(1+\frac{1}{x+\frac{1}{6}}\right)^{1/2}.$$

Source: [1863].

Fact 2.1.59. Let $x \ge 1$. Then,

$$e\left(1-\frac{7}{14x+12}\right) < \left(1+\frac{1}{x}\right)^x < e\left(1-\frac{6}{12x+11}\right).$$

Source: [327, 3012].

Fact 2.1.60. Let $n \ge 1$. Then,

$$\left(1+\frac{1}{n}\right)^{n+1/\log 2-1} < e < \left(1+\frac{1}{n}\right)^{n+1/2}, \quad 1+\frac{e/2-1}{n} < \frac{e}{\left(1+\frac{1}{n}\right)^n} < 1+\frac{1}{2n},$$

$$\frac{e\log 2}{2n\log(1+1/n)} < \frac{e}{(1+1/n)^n} \le \frac{1}{n\log(1+1/n)}, \quad e\left(1+\frac{4/e-1}{n}\right) < \left(1+\frac{1}{n}\right)^{n+1} < e\left(1+\frac{1}{2n}\right),$$

$$e^{1-\frac{n}{2[n-1+(2-2\log 2)^{-1/2}]^2}} < \left(1+\frac{1}{n}\right)^n < e^{1-\frac{n}{2(n+1/3)^2}}, \quad e^{1+\frac{1}{2(n+1/3)}} < \left(1+\frac{1}{n}\right)^{n+1} < e^{1+\frac{1}{2[n-1+1/(4\log 2-2)]}}.$$

Source: [327].

Fact 2.1.61. Let $x > 0$ and $\alpha \in [0,1]$. Then,

$$x^\alpha + x^{-\alpha} \le x + x^{-1}.$$

Source: [1982].

Fact 2.1.62. Let $x \in [0,1]$. If $\alpha \in [0,1]$, then

$$(1-x)^\alpha + (1+x)^\alpha \le 2.$$

If $\alpha \in [1,\infty)$, then

$$2 \le (1-x)^\alpha + (1+x)^\alpha.$$

Source: [2227].

Fact 2.1.63. Let $x \in [0,1)$ and $n \ge 2$. Then,

$$2\sin\frac{\operatorname{asin} x}{n} \le \sqrt[n]{1+x} - \sqrt[n]{1-x}.$$

Source: [2227].

2.2 Facts on Equalities and Inequalities in Two Variables

Fact 2.2.1. Let x and y be complex numbers. Then,

$$x^2 - y^2 = (x-y)(x+y),$$
$$(x+y)^2 - (x-y)^2 = 4xy,$$
$$x^3 - y^3 = (x-y)(x^2+xy+y^2),$$
$$x^3 + y^3 = (x+y)^3 - 3xy(x+y),$$
$$x^4 - y^4 = (x-y)(x+y)(x^2+y^2),$$
$$x^4 + (x+y)^4 + y^4 = 2(x^2+xy+y^2)^2,$$
$$4(x^3+y^3) = (x+y)^3 + 3(x+y)(x-y)^2,$$
$$(x+y)^4 = 2(x^2+y^2)(x+y)^2 - (x^2-y^2)^2,$$
$$(6x^3+y^3)^3 = (6x^3-y^3)^3 + 2(y^3)^3 + (6x^2y)^3,$$
$$x^4 + 4y^4 = (x^2+2xy+2y^2)(x^2-2xy+2y^2),$$
$$x^4 + x^2y^2 + y^4 = (x^2+xy+y^2)(x^2-xy+y^2),$$
$$6(x^2+y^2)^2 = (x-y)^4 + 4x^4 + 4y^4 + (x+y)^4,$$
$$x^5 - y^5 = (x-y)(x^4+x^3y+x^2y^2+xy^3+y^4),$$
$$x^5 + y^5 = (x+y)^5 - 5xy(x+y)(x^2+xy+y^2),$$
$$10(x^2+y^2)^3 = (x-y)^6 + 8x^6 + 8y^6 + (x+y)^6,$$
$$x^7 + y^7 = (x+y)^7 - 7xy(x+y)(x^2+xy+y^2)^2,$$
$$x^9 - y^9 = (x-y)(x^2+xy+y^2)(x^6+x^3y^3+y^6),$$
$$(2x^4+y^4)^4 = (2x^4-y^4)^4 + 4(2x^3y)^4 + (2xy^3)^4,$$
$$(4x^4+y^4)^4 = (4x^4-y^4)^4 + 2(4x^3y)^4 + 2(2xy^3)^4,$$
$$xy + (1+x)(1+y)(x+y) = (1+x+y)(xy+x+y),$$
$$x^6 - y^6 = (x-y)(x+y)(x^2+xy+y^2)(x^2-xy+y^2),$$
$$(x+y)(x^4+y^4) = (x^2+y^2)(x^3+y^3) + xy(x+y)(x-y)^2,$$
$$(x^2-y^2)^4 + (x^2+2xy)^4 + (2xy+y^2)^4 = 2(x^2+xy+y^2)^4,$$
$$x^7 - y^7 = (x-y)(x^6+x^5y+x^4y^2+x^3y^3+x^2y^4+xy^5+y^6),$$
$$(x^2+7xy-9y^2)^3 + (2x^2-4xy+12y^2)^3 = (2x^2+10y^2)^3 + (x^2-9xy-y^2)^3,$$
$$x^{12} - y^{12} = (x-y)(x+y)(x^2+y^2)(x^2-xy+y^2)(x^2+xy+y^2)(x^4-x^2y^2+y^4),$$
$$x^{11} + y^{11} = (x+y)^{11} - 11xy(x+y)(x^2+xy+y^2)[x^2y^2(x+y)^2 + (x^2+xy+y^2)^3],$$
$$27xy(x+y)(x^2+xy+y^2)^3 = [x^3-y^3+3xy(2x+y)]^3 + [y^3-x^3+3xy(2y+x)]^3,$$
$$x^{10} - y^{10} = (x-y)(x+y)(x^4-x^3y+x^2y^2-xy^3+y^4)(x^4+x^3y+x^2y^2+xy^3+y^4),$$
$$x^{11} - y^{11} = (x-y)(x^{10}+x^9y+x^8y^2+x^7y^3+x^6y^4+x^5y^5+x^4y^6+x^3y^7+x^2y^8+x,$$
$$(3x^2+5xy-5y^2)^3 + (4x^2-4xy+6y^2)^3 + (5x^2-5xy-3y^2)^3 = (6x^2-4xy+4y^2)^3,$$
$$(9x^2-11xy+y^2)^3 + (15x^2-5xy-y^2)^3 + (12x^2-4xy+2y^2)^3 = (18x^2-8xy+2y^2)^3,$$
$$(3x^2+xy-7y^2)^3 + (4x^2+12xy+14y^2)^3 + (5x^2+15xy+7y^2)^3 = (6x^2+16xy+14y^2)^3,$$
$$(49x^2+12xy-y^2)^3 + (42x^2+4xy+2y^2)^3 + (15x^2+20xy+y^2)^3 = (58x^2+12xy+2y^2)^3,$$
$$(76x^2-6xy-y^2)^3 + (72x^2-16xy+2y^2)^3 + (84x^2-26xy+y^2)^3 = (112x^2-24xy+2y^2)^3,$$

$$(9x^2+45xy-135y^2)^3+(10x^2-20xy+172y^2)^3=(12x^2+12xy+138y^2)^3+(x^2-83xy+y^2)^3,$$

$$[1-(x-3y)(x^2+3y^2)]^3+[(x+3y)(x^2+3y^2)-1]^3=[x+3y-(x^2+3y^2)^2]^3+[(x^2+3y^2)^2-(x-3y)]^3,$$

$$(x^2-2xy+21y^2)^4+(3x^2-4xy+41y^2)^4+(5x^2+6xy-71y^2)^4$$
$$=(3x^2+6xy-69y^2)^4+(5x^2+4xy-49y^2)^4+(x^2-22xy+y^2)^4,$$

$$(x^7+x^5y^2-2x^3y^4+3x^2y^5+xy^6)^4+(x^6y-3x^5y^2-2x^4y^3+x^2y^5+y^7)^4$$
$$=(x^7+x^5y^2-2x^3y^4-3x^2y^5+xy^6)^4+(x^6y+3x^5y^2-2x^4y^3+x^2y^5+y^7)^4.$$

Remark: The eighth and 28th equalities are given in [732, p. 57]. The tenth equality, which can be written as

$$x^4+4y^4=[(x+y)^2+y^2][(x-y)^2+y^2],$$

is the *identity of Sophie Germain*. See [993, p. 121]. The ninth-to-last equality is due to S. A. Ramanujan. See [1318, pp. 259, 260] and [1321]. The sum-of-cubes equalities are given in [776, 1308, 1309, 1321].

Fact 2.2.2. Let $n\ge 1$, and let x and y be complex numbers. Then,

$$x^n-y^n=(x-y)\sum_{i=0}^{n-1}x^{n-1-i}y^i=\prod_{i=1}^{n}(x-e^{2i\pi j/n}y),\quad x^{2^n}-y^{2^n}=(x-y)\prod_{i=1}^{n-1}(x^{2^i}+y^{2^i}),$$

$$x^{2n}-y^{2n}=(x-y)\sum_{i=0}^{2n-1}x^{2n-1-i}y^i=\prod_{i=1}^{2n}(x-e^{i\pi j/n}y)=(x^2-y^2)\prod_{i=1}^{n-1}[x^2-2(\cos\tfrac{i\pi}{n})xy+y^2],$$

$$x^{2n+1}-y^{2n+1}=(x-y)\sum_{i=0}^{2n}x^{2n-i}y^i=\prod_{i=1}^{2n+1}(x-e^{2i\pi j/(2n+1)}y)=(x-y)\prod_{i=1}^{n}[x^2-2(\cos\tfrac{2i\pi}{2n+1})xy+y^2].$$

If n is odd, then

$$x^{2n}-y^{2n}=(x^n-y^n)(x^n+y^n)=(x^2-y^2)\left(\sum_{i=0}^{n-1}x^{n-1-i}y^i\right)\left(\sum_{i=0}^{n-1}(-1)^ix^{n-1-i}y^i\right).$$

In particular,

$$x^3-y^3=(x-y)(x^2+xy+y^2),\quad x^4-y^4=(x-y)(x^3+x^2y+xy^2+y^3)=(x^2-y^2)(x^2+y^2),$$

$$x^5-y^5=(x-y)(x^4+x^3y+x^2y^2+xy^3+y^4)$$
$$=(x-y)[x^2-\tfrac{1}{2}(\sqrt{5}-1)xy+y^2][x^2+\tfrac{1}{2}(\sqrt{5}+1)xy+y^2],$$

$$x^6-y^6=(x-y)(x^5+x^4y+x^3y^2+x^2y^3+xy^4+y^5)=(x^2-y^2)(x^2-xy+y^2)(x^2+xy+y^2),$$

$$x^7-y^7=(x-y)(x^6+x^5y+x^4y^2+x^3y^3+x^2y^4+xy^5+y^6)$$
$$=(x-y)[x^2-2(\cos\tfrac{2\pi}{7})xy+y^2][x^2-2(\cos\tfrac{4\pi}{7})xy+y^2][x^2-2(\cos\tfrac{6\pi}{7})xy+y^2].$$

Source: [665], [771, pp. 155, 156], [2173], and [2504, p. 6]. **Related:** Fact 2.2.3.

Fact 2.2.3. Let $n\ge 1$, and let x and y be complex numbers. Then,

$$x^n+y^n=(x+y)^n+\sum_{i=1}^{\lfloor n/2\rfloor}(-1)^i\left[\binom{n-i}{i}+\binom{n-i-1}{i-1}\right](xy)^i(x+y)^{n-2i}$$
$$=\sum_{i=0}^{\lfloor n/2\rfloor}(-1)^i\left[\binom{n-i}{i}+\binom{n-i-1}{i-1}\right](xy)^i(x+y)^{n-2i}$$

$$= \sum_{i=0}^{\lfloor n/2\rfloor}(-1)^i\frac{n}{n-i}\binom{n-i}{i}(xy)^i(x+y)^{n-2i} = \prod_{i=1}^{n}(x - e^{(2i+1)\pi\jmath/n}y),$$

$$x^{2n}+y^{2n} = \sum_{i=0}^{n}(-1)^i\frac{2n}{2n-i}\binom{2n-i}{i}(xy)^i(x+y)^{2n-2i} = \prod_{i=1}^{n}\left[x^2+2xy\cos\frac{(2i-1)\pi}{2n}+y^2\right],$$

$$x^{2n+1}+y^{2n+1} = (x+y)\sum_{i=0}^{2n}(-1)^ix^{2n-i}y^i = (x+y)\prod_{i=1}^{n}\left[x^2+2xy\cos\frac{2i\pi}{2n+1}+y^2\right].$$

In particular,

$$x^2+y^2 = (x+y)^2-2xy, \quad x^3+y^3 = (x+y)^3-3xy(x+y) = (x+y)(x^2-xy+y^2),$$

$$x^4+y^4 = (x+y)^4-4xy(x+y)^2+2(xy)^2 = (x^2+\sqrt{2}xy+y^2)(x^2-\sqrt{2}xy+y^2),$$

$$x^5+y^5 = (x+y)^5-5xy(x+y)^3+5(xy)^2(x+y) = (x+y)(x^4-x^3y+x^2y^2-xy^3+y^4)$$
$$= (x-y)[x^2-\tfrac{1}{2}(\sqrt{5}-1)xy+y^2][x^2+\tfrac{1}{2}(\sqrt{5}+1)xy+y^2],$$

$$x^6+y^6 = (x^2+y^2)(x^2+\sqrt{3}xy+y^2)(x^2-\sqrt{3}xy+y^2),$$

$$x^7+y^7 = (x+y)(x^6-x^5y+x^4y^2-x^3y^3+x^2y^4-xy^5+y^6)$$
$$= (x+y)^7-7xy(x+y)^5+14(xy)^2(x+y)^3-7(xy)^3(x+y)$$
$$= (x+y)^7-7xy(x+y)(x^2+xy+y^2)^2$$
$$= (x+y)^7-7xy(x+y)(x-\omega y)^2(x-\omega^2y)^2$$
$$= (x+y)[x^2+2(\cos\tfrac{2\pi}{7})xy+y^2][x^2+2(\cos\tfrac{4\pi}{7})xy+y^2][x^2+2(\cos\tfrac{6\pi}{7})xy+y^2],$$

where $\omega \triangleq (-1)^{1/3} = \frac{1}{2}(-1+\sqrt{3}\jmath)$. **Source:** [289, pp. 25, 143, 144], [771, pp. 155, 156], [1677], and [2504, p. 6]. **Related:** Fact 1.16.11 and Fact 2.2.2.

Fact 2.2.4. Let $n \ge 1$, and let x and y be complex numbers. Then,

$$(x+y)^n = \sum_{i=0}^{n}\binom{n}{i}x^{n-i}y^i, \quad (x+y)^{2n} = \sum_{i=1}^{n}\binom{2n-i-1}{n-1}(x^i+y^i)(x+y)^i(xy)^{n-i}.$$

Source: [771, pp. 155, 156].

Fact 2.2.5. Let x and y be complex numbers. Then,

$$x^2-xy+y^2 = (x+e^{(2\pi/3)\jmath}y)(x+e^{-(2\pi/3)\jmath}y), \quad x^2+xy+y^2 = (x-e^{(2\pi/3)\jmath}y)(x-e^{-(2\pi/3)\jmath}y),$$

$$x^2-\sqrt{2}xy+y^2 = (x-e^{(\pi/4)\jmath}y)(x-e^{-(\pi/4)\jmath}y), \quad x^2+\sqrt{2}xy+y^2 = (x+e^{(\pi/4)\jmath}y)(x+e^{-(\pi/4)\jmath}y),$$

$$x^4+y^4 = (x+e^{(\pi/4)\jmath}y)(x+e^{-(\pi/4)\jmath}y)(x+e^{(3\pi/4)\jmath}y)(x+e^{-(3\pi/4)\jmath}y).$$

Source: [732, pp. 136, 137].

Fact 2.2.6. Let $n \ge 1$, and let x and y be complex numbers. Then,

$$x^{2n}-x^ny^n+y^{2n} = \prod_{\substack{1\le i<3n\\ \gcd\{i,6\}=1}}\left(x^2-2xy\cos\frac{i\pi}{3n}+y^2\right).$$

In particular,

$$x^4-x^2y^2+y^4 = (x^2-\sqrt{3}xy+y^2)(x^2+\sqrt{3}xy+y^2).$$

Source: [2904].

Fact 2.2.7. Let $n \geq 1$, let x and y be complex numbers, and let α be a real number. Then,

$$x^{2n} - 2x^n y^n \cos n\alpha + y^{2n} = \prod_{i=0}^{n-1} \left[x^2 - 2xy\cos\left(\alpha + \frac{2i\pi}{n}\right) + y^2\right] = \prod_{i=1}^{n} \left[x^2 - 2xy\cos\left(\alpha + \frac{2i\pi}{n}\right) + y^2\right].$$

In particular,

$$(x^n - y^n)^2 = \prod_{i=1}^{n}\left(x^2 - 2xy\cos\frac{2i\pi}{n} + y^2\right), \quad (x^n + y^n)^2 = \prod_{i=1}^{n}\left(x^2 - 2xy\cos\frac{(2i+1)\pi}{n} + y^2\right).$$

Source: [1217, p. 41] and [2249, p. 21].

Fact 2.2.8. Let x, y be complex numbers, assume that $xy \neq 1$, define $z \triangleq (x+y)/(xy-1)$, and assume that $x, y, z \neq \pm\sqrt{3}/3$. Then,

$$\frac{(3x - x^3)(3y - y^3)(3z - z^3)}{(3x^2 - 1)(3y^2 - 1)(3z^2 - 1)} = \frac{3x - x^3}{3x^2 - 1} + \frac{3y - y^3}{3y^2 - 1} + \frac{3z - z^3}{3z^2 - 1}.$$

Source: [1158, p. 241].

Fact 2.2.9. Let x and y be complex numbers. Then,

$$\frac{x^5 + y^5 - (x+y)^5}{5} = \left(\frac{x^3 + y^3 - (x+y)^3}{3}\right)\left(\frac{x^2 + y^2 + (x+y)^2}{2}\right),$$

$$\frac{x^7 + y^7 - (x+y)^7}{7} = \left(\frac{x^5 + y^5 - (x+y)^5}{5}\right)\left(\frac{x^2 + y^2 + (x+y)^2}{2}\right).$$

Source: [1158, p. 48] and [1757, pp. 35, 138].

Fact 2.2.10. Let x and y be complex numbers, and, for all $n \geq 0$, define

$$K_n \triangleq (x+y+1)^n + (-xy - x - y)^n - (-xy - y - 1)^n - (xy + x + 1)^n + (xy - 1)^n - (y - x)^n.$$

Then,

$$K_0 = K_1 = K_2 = K_4 = 0, \quad 5K_3K_8 = 8K_5K_6, \quad 15K_6K_7 = 7K_3K_{10},$$

$$45K_8^2 = 64K_6K_{10}, \quad 21K_5^2 = 25K_3K_7,$$

$$245K_3K_{11} + 330K_7^2 = 539K_5K_9, \quad 300K_6K_{14} + 308K_{10}^2 = 525K_8K_{12},$$

$$1260K_3K_9 = 35K_3^4 + 945K_6^2 + 972K_5K_7.$$

Now, assume that $x+y+1$, $xy+x+y$, $xy+y+1$, $xy+x+1$, $xy-1$, and $-x+y$ are nonzero, and, for all $n \leq -1$, define K_n as above. Then,

$$K_{-2}K_3^2 + 3K_{-1}^2K_6 = 0, \quad 5K_6K_{-1}^4 = 12K_{-2}K_{-1}K_5 + 5K_{-2}^2K_6.$$

Source: [664]. **Related:** In Fact 2.4.15, set $b = ax$, $c = ay$, and $d = axy$.

Fact 2.2.11. Let x and y be complex numbers such that $x + y \neq -1$. Then,

$$\left(1 + x + y - \frac{xy + x + y}{x + y + 1}\right)^2 = 1 + x^2 + y^2 + \left(\frac{xy + x + y}{x + y + 1}\right)^2.$$

Source: [399]. **Related:** Fact 2.1.9.

Fact 2.2.12. Let x and y be real numbers. Then,

$$4xy \leq 4xy + (x-y)^2 = (x+y)^2 < 2(x^2 + y^2).$$

Hence,

$$2xy \leq \tfrac{1}{2}(x+y)^2 \leq x^2 + y^2.$$

If, in addition, α is a positive real number, then

$$2xy \le \frac{x^2}{\alpha} + \alpha y^2.$$

Finally, if x and y are positive, then

$$2 \le \frac{(x+y)^2}{2xy} \le \frac{x}{y} + \frac{y}{x}.$$

Fact 2.2.13. Let α be a real number. Then, the following statements are equivalent:

i) For all positive numbers x, y, $\alpha xy \le x^2 + y^2$.

ii) $\alpha \le 2$.

Fact 2.2.14. Let α be a real number. Then, following statements are equivalent:

i) For all real numbers x, y, $0 \le x^2 + \alpha xy + y^2$.

ii) $\alpha \in [-2, 2]$.

Fact 2.2.15. Let x and y be positive numbers. Then,

$$\left\{\begin{matrix} \dfrac{xy(x+y)^2}{2(x^3+y^3)} \\ \min\{x,y\} \end{matrix}\right\} \le \frac{xy(x^2+y^2)}{x^3+y^3} \le \frac{2}{\frac{1}{x}+\frac{1}{y}} \le \sqrt{xy} \le \frac{x+y}{2} \le \sqrt{\frac{x^2+y^2}{2}}$$

$$\le \left\{\begin{matrix} \frac{\sqrt{2}}{2}(x+y) \\ \sqrt[3]{\dfrac{x^3+y^3}{2}} \le \left\{\begin{matrix} \dfrac{x^2+y^2}{x+y} \\ \sqrt[4]{\dfrac{x^4+y^4}{2}} \end{matrix}\right\} \le \sqrt{x^2-xy+y^2} \le \dfrac{x^3+y^3}{x^2+y^2} \le \max\{x,y\} \end{matrix}\right\} \le x+y.$$

Hence,

$$\frac{4}{x+y} \le \frac{2}{\sqrt{xy}} \le \frac{1}{x} + \frac{1}{y}, \quad [\tfrac{1}{2}(x+y)]^3 \le \tfrac{1}{2}(x^3+y^3) \le \left(\frac{x^2+y^2}{x+y}\right)^3.$$

Source: [47, pp. 3, 11], [109, pp. 96, 97], and [379]. **Remark:** $\sqrt{xy} \le \frac{x+y}{2}$ is the *arithmetic-mean–geometric-mean inequality*. **Remark:** $\frac{2}{\frac{1}{x}+\frac{1}{y}}$ is the *harmonic mean* of x and y. **Remark:** $\frac{x^2+y^2}{x+y}$ is the *contraharmonic mean* of x and y. **Related:** Fact 2.2.38, Fact 2.3.40, and Fact 10.11.18.

Fact 2.2.16. Let x and y be positive numbers, and assume that $0 < x \le y$. Then,

$$\frac{(x-y)^2}{8y} \le \frac{(x-y)^2}{4(x+y)} \le \tfrac{1}{2}(x+y) - \sqrt{xy} \le \frac{(x-y)^2}{8x}.$$

Equality holds if and only if $x = y$. **Source:** [288, p. 231] and [993, p. 183].

Fact 2.2.17. Let x and y be distinct positive numbers. Then,

$$\frac{1}{(x-y)^2} + \frac{1}{x^2} + \frac{1}{y^2} = \frac{4}{xy} + \frac{(x^2+y^2-3xy)^2}{x^2y^2(x-y)^2}.$$

Hence,

$$\frac{4}{xy} \le \frac{1}{(x-y)^2} + \frac{1}{x^2} + \frac{1}{y^2}.$$

Source: [1938, p. 204].

Fact 2.2.18. Let x and y be real numbers. Then,

$$x+y \le \sqrt{(x^2+1)(y^2+1)}, \quad \tfrac{1}{2}(x^2y^2+1) \le \sqrt{\tfrac{1}{4}(x^4+1)(y^4+1)} \le (x^2-x+1)(y^2-y+1).$$

If, in addition, x and y are positive numbers, then

$$\tfrac{1}{2}(x^2y^2+1) \le \sqrt{\tfrac{1}{4}(x^4+1)(y^4+1)} \le (x^2-x+1)(y^2-y+1) \le \frac{(x^3+1)^2(y^3+1)^2}{(x^2+1)^2(y^2+1)^2}.$$

Source: [379].

Fact 2.2.19. Let x and y be positive numbers. Then,

$$\frac{9}{2x^2+y^2} + \frac{9}{x^2+2y^2} \le \frac{1}{x^2} + \frac{1}{y^2} + \frac{4}{xy}, \quad \frac{32(x^2+y^2)}{(x+y)^4} \le \frac{1}{x^2} + \frac{1}{y^2} + \frac{4}{x^2+y^2}.$$

Source: The first inequality follows from Problem 257 of [806, p. 210] in the case of a rectangle. The second inequality is given in [752, p. 1].

Fact 2.2.20. Let x and y be positive numbers, and let $\alpha \ge 1$. Then,

$$\frac{2}{1+\alpha} \le \frac{x}{x+\alpha y} + \frac{y}{y+\alpha x}.$$

Related: This is a special case of Fact 2.11.80.

Fact 2.2.21. Let x and y be real numbers, and let $\alpha \in [0,1]$. Then,

$$\sqrt{\alpha}x + \sqrt{1-\alpha}y \le \sqrt{x^2+y^2}.$$

Equality holds if and only if exactly one of the following statements holds:

i) $x = y = 0$.

ii) $x = 0$, $y > 0$, and $\alpha = 0$.

iii) $x > 0$, $y = 0$, and $\alpha = 1$.

iv) $x > 0$, $y > 0$, and $\alpha = x^2/(x^2+y^2)$.

Fact 2.2.22. Let x and y be real numbers. Then,

$$3x^4 + y^4 \le 4x^3y,$$
$$8xy \le x^4 + y^4 + 8,$$
$$x^3y + y^3x \le x^4 + y^4,$$
$$3x^2y^2 \le x^4y^2 + x^2y^4 + 1,$$
$$4xy(x-y)^2 \le (x^2-y^2)^2,$$
$$2x + 2xy \le x^2y^2 + x^2 + 2,$$
$$\sqrt{3}x^2 \le x^4 + x^2y^2 + xy + 1,$$
$$3(x+y-1) \le x^2 + xy + y^2,$$
$$3(x+y+xy) \le (x+y+1)^2,$$
$$2|(x+y)(1-xy)| \le (1+x^2)(1+y^2).$$

If $xy(x+y) \ge 0$, then

$$(x^2+y^2)(x^3+y^3) \le (x+y)(x^4+y^4).$$

Source: [289, pp. 19, 118], [806, p. 184], [993, pp. 183, 185, 258], [1371, p. 117], [1938, p. 9], and [2264, p. 244].

Fact 2.2.23. Let x and y be real numbers. Then,

$$[x^2 + y^2 + (x + y)^2]^2 = 2[x^4 + y^4 + (x + y)^4].$$

Therefore,

$$\tfrac{1}{2}(x^2 + y^2)^2 \le x^4 + y^4 + (x + y)^4, \quad x^4 + y^4 \le \tfrac{1}{2}[x^2 + y^2 + (x + y)^2]^2.$$

Now, assume that x and y are nonnegative. Then,

$$2(x^2 + y^2)^2 \le x^4 + y^4 + (x + y)^4, \quad x^4 + y^4 \le \tfrac{1}{4}[x^2 + y^2 + (x + y)^2]^2.$$

Remark: This is *Candido's identity*. See [46].

Fact 2.2.24. Let x and y be real numbers, at least one of which is not zero, and let $\alpha \ge -4/3$. Then,

$$\frac{\sqrt{x^2 + \alpha xy + y^2}}{x^2 - xy + y^2} \le \frac{\sqrt{2(\alpha + 2)}}{\sqrt{x^2 + y^2}}.$$

In particular,

$$\frac{|x + y|}{x^2 - xy + y^2} \le \frac{2\sqrt{2}}{\sqrt{x^2 + y^2}}, \quad \frac{\sqrt{x^2 + xy + y^2}}{x^2 - xy + y^2} \le \frac{\sqrt{6}}{\sqrt{x^2 + y^2}},$$

$$\frac{\sqrt{x^2 + 3xy + y^2}}{x^2 - xy + y^2} \le \frac{\sqrt{10}}{\sqrt{x^2 + y^2}}, \quad \frac{\sqrt{x^2 + 4xy + y^2}}{x^2 - xy + y^2} \le \frac{2\sqrt{3}}{\sqrt{x^2 + y^2}}, \quad \frac{\sqrt{x^2 + 6xy + y^2}}{x^2 - xy + y^2} \le \frac{4}{\sqrt{x^2 + y^2}}.$$

Source: [420].

Fact 2.2.25. Let x and y be real numbers. Then,

$$-\frac{1}{2} \le \frac{(x + y)(1 - xy)}{(1 + x^2)(1 + y^2)} \le \frac{1}{2}.$$

Source: [112, p. 31].

Fact 2.2.26. Let p, q be positive numbers such that $\frac{1}{p} + \frac{1}{q} = 1$, and let $x, y \in (0, 1)$. Then,

$$\frac{pq}{1 - xy} \le \frac{q}{1 - x^p} + \frac{p}{1 - y^q}, \quad \frac{xy}{(1 - xy)^2} \le \frac{x^p}{p(1 - x^p)^2} + \frac{y^q}{q(1 - y^q)^2}.$$

Source: [2088].

Fact 2.2.27. Let x and y be real numbers. Then,

$$54x^2y^2(x + y)^2 \le [x^2 + y^2 + (x + y)^2]^3.$$

Equivalently,

$$[x^2y^2(x + y)^2]^{1/3} \le \tfrac{1}{\sqrt[3]{2}}\tfrac{1}{3}[x^2 + y^2 + (x + y)^2].$$

Remark: For $x^2, y^2, (x+y)^2$, this result interpolates the arithmetic-mean–geometric-mean inequality given by Fact 2.11.81 due to the factor $1/\sqrt[3]{2} \approx .794 < 1$. **Remark:** This is used in Fact 6.10.7.

Fact 2.2.28. Let x and y be real numbers, and let $p \in [1, \infty)$. Then,

$$\tfrac{p-1}{2^{p+1}}(x - y)^2 + \tfrac{1}{4}(x + y)^2 \le [\tfrac{1}{2}(|x|^p + |y|^p)]^{2/p}.$$

Source: [1146, p. 148].

Fact 2.2.29. Let x and y be complex numbers. If $p \in [1, 2]$, then

$$[|x|^2 + (p - 1)|y|^2]^{1/2} \le [\tfrac{1}{2}(|x + y|^p + |x - y|^p)]^{1/p}.$$

If $p \in [2, \infty]$, then

$$[\tfrac{1}{2}(|x + y|^p + |x - y|^p)]^{1/p} \le [|x|^2 + (p - 1)|y|^2]^{1/2}.$$

Source: Fact 11.10.64.

Fact 2.2.30. Let x and y be positive numbers. If $p \in \{-1, 0, 1\} \cup [2, \infty)$, then

$$[\tfrac{1}{2}(x^p + y^p)]^{1/p} \le \frac{(1+p)(x-y)^2 + 8xy}{4(x+y)}.$$

If $p \in (-\infty, -1] \cup [1, 3/2]$, then

$$\frac{(1+p)(x-y)^2 + 8xy}{4(x+y)} \le [\tfrac{1}{2}(x^p + y^p)]^{1/p}.$$

If either $x = y$ or $p \in \{-1, 1\}$, then equality holds in both inequalities. **Source:** [107, pp. 414, 415]. **Remark:** If $p = 0$, then $[\frac{1}{2}(x^p + y^p)]^{1/p}$ is interpreted as $\sqrt{xy}$.

Fact 2.2.31. Let x and y be positive numbers, and let p and q be real numbers such that $0 \le p \le q$. Then,

$$\frac{x^p + y^p}{(xy)^{p/2}} \le \frac{x^q + y^q}{(xy)^{q/2}}.$$

Related: Fact 10.9.13.

Fact 2.2.32. Let x and y be real numbers, let p and q be real numbers, and assume that $1 \le p \le q$. Then,

$$\left[\frac{1}{2}\left(\left|x + \frac{y}{\sqrt{q-1}}\right|^q + \left|x - \frac{y}{\sqrt{q-1}}\right|^q\right)\right]^{1/q} \le \left[\frac{1}{2}\left(\left|x + \frac{y}{\sqrt{p-1}}\right|^p + \left|x - \frac{y}{\sqrt{p-1}}\right|^p\right)\right]^{1/p}.$$

Source: [1146, p. 206]. **Remark:** This is a scalar version of Bonami's inequality. See Fact 11.7.15.

Fact 2.2.33. Let x and y be nonnegative numbers. Then,

$$\begin{gathered}
9xy^2 \le 3x^3 + 7y^3,\\
27x^2y \le 4(x+y)^3,\\
2x^2y^2 \le x^3y + y^3x,\\
6xy^2 \le x^3 + y^6 + 8,\\
x^2y + y^2x \le x^3 + y^3,\\
x^3y + y^3x \le x^4 + y^4,\\
x^4y + y^4x \le x^5 + y^5,\\
5x^6y^6 \le 2x^{15} + 3y^{10},\\
(x+y)^3 \le 4(x^3 + y^3),\\
x^3y^2 + y^3x^2 \le x^5 + y^5,\\
8xy(x^2 + y^2) \le (x+y)^4,\\
(x^2 + y^2)^3 \le 2(x^3 + y^3)^2,\\
4x^2y \le x^4 + x^3y^2 + y^2 + x,\\
4x^2y \le x^4 + x^3y + y^2 + xy,\\
12xy \le 4x^2y + 4y^2x + 4x + y,\\
(x+y)^2(x^2 + y^2) \le 4(x^4 + y^4),\\
(x+y)^3(x^3 + y^3) \le 8(x^6 + y^6),\\
(x+y)^4 \le 8x^2y^2 + 2(x^2 + y^2)^2,\\
6x^2y^2 \le x^4 + 2x^3y + 2y^3x + y^4,
\end{gathered}$$

$$9xy \le (x^2+x+1)(y^2+y+1),$$
$$2xy(x^2+y^2) \le (x+y)(x^3+y^3),$$
$$(x^3+y^3)(x^5+y^5) \le 2(x^8+y^8),$$
$$(x+y)(xy^2+yx^2) \le (x^2+y^2)^2,$$
$$(x+y)(x^2+xy+y^2) \le 3(x^3+y^3),$$
$$(x^2+y^2)(x^3+y^3) \le (x+y)(x^4+y^4),$$
$$4(x^2y+y^2x) \le 2(x^2+y^2)^2+x^2+y^2,$$
$$(x^2+y^2)(x^3+y^3)(x^4+y^4) \le 4(x^9+y^9),$$
$$256x^4y^4+1 \le (x+1)^2(y+1)^3(xy+1)^4,$$
$$16(x^2+xy+y^2)^3 \le 27(x+y)^2(x^2+y^2)^2,$$
$$2(x^2y+y^2x+x^2y^2) \le 2(x^4+y^4)+x^2+y^2,$$
$$x^2y+2xy+2xy^2 \le x^3y^2+x^2+x+y^3+y^2,$$
$$(x^2+y^2)^2 \le (x+y)(x^3+y^3) \le \tfrac{1}{2}[x^4+y^4+(x+y)^4],$$
$$x^2y^3+x^2y^2+xy^3+xy+y^2+y \le x^3y^3+3xy^2+y^3+1.$$

Source: Fact 2.11.25, [108, pp. 5, 7, 75, 76, 81, 82], [806, pp. 45, 116, 184, 188, 210, 241, 262], [993, p. 183], [1033], [1371, pp. 117, 120, 123, 124, 150, 153, 155], [1938, p. 176], and [2380, p. 74]. **Remark:** $x^2y^3+x^2y^2+xy^3+xy+y^2+y \le x^3y^3+3xy^2+y^3+1$ is equivalent to $(xy+y-1)(xy-y+1)(-xy+y+1) \le xy^2$. Replacing x and y by x/y and y/z, respectively, yields the 3-variable inequality $(x+y-z)(x-y+z)(-x+y+z) \le xyz$, which is given by Fact 2.3.60. **Remark:** The penultimate string interpolates the third inequality in Fact 2.2.23.

Fact 2.2.34. Let x and y be nonnegative numbers, let $n \ge 1$, and let $0 \le k \le \lfloor n/2 \rfloor$. Then,

$$x^{n-k}y^k + x^k y^{n-k} \le x^n + y^n.$$

Source: [176]. **Related:** Fact 2.2.35.

Fact 2.2.35. Let x, y, p, q be nonnegative numbers. Then,

$$x^p y^q + x^q y^p \le x^{p+q} + y^{p+q}.$$

Equality holds if and only if either $pq = 0$ or $x = y$. **Source:** [1371, p. 96]. **Related:** Fact 2.2.34.

Fact 2.2.36. Let p, q, r, u, v, w be positive numbers, assume that $[p\ q\ r] \overset{\mathrm{s}}{\prec} [u\ v\ w]$, and let x and y be nonnegative numbers. Then,

$$(x^p+y^p)(x^q+y^q)(x^r+y^r) \le (x^u+y^u)(x^v+y^v)(x^w+y^w).$$

In particular,

$$(x^2+y^2)(x^4+y^4)^2 \le (x+y)(x^3+y^3)(x^6+y^6).$$

Related: Fact 2.2.62.

Fact 2.2.37. Let x and y be nonnegative numbers. Then,

$$\sqrt[3]{x}+\sqrt[3]{y} \le \sqrt[3]{4}\sqrt[3]{x+y}, \quad x\sqrt{y^2+1}+y\sqrt{x^2+1} < x^2+y^2+1,$$
$$\sqrt{xy}+\sqrt{\tfrac{1}{2}(x^2+y^2)} \le x+y, \quad \sqrt{x^2+xy}+\sqrt{y^2+yx} \le \sqrt{2}(x+y),$$
$$2\left(\sqrt{x^4+x^3y+xy^3}+\sqrt{y^4+y^3x+yx^3}\right) \le \sqrt{2x^4+y^4}+\sqrt{2y^4+x^4}+x\sqrt{x^2+2y^2}+y\sqrt{y^2+2x^2}.$$

Source: [377] and [806, pp. 6, 76].

Fact 2.2.38. Let x and y be positive numbers, and let $n \geq 1$. Then,

$$\left(\frac{x+y}{2}\right)^n \leq \frac{1}{n+1}\sum_{i=0}^{n} x^i y^{n-i} \leq \frac{x^n + y^n}{2} \leq \frac{1}{2}(x+y)^n, \quad 2^{n+1} \leq \left(1+\frac{x}{y}\right)^n + \left(1+\frac{y}{x}\right)^n.$$

If $\alpha \in [0, 1]$, then

$$\tfrac{1}{2}(x^\alpha + y^\alpha) \leq [\tfrac{1}{2}(x+y)]^\alpha.$$

If $\alpha \in [1, \infty)$, then

$$[\tfrac{1}{2}(x+y)]^\alpha \leq \tfrac{1}{2}(x^\alpha + y^\alpha).$$

Source: [393], [419, pp. 47–49], [1938, p. 26], and Fact 2.2.58 with $p = 1 \leq q$. **Related:** Fact 2.1.34 and Fact 2.2.15.

Fact 2.2.39. Let x and y be nonnegative numbers, and let r be a positive number. Then,

$$xy \leq \frac{x^{r+1}}{r+1} + \frac{ry^{1+1/r}}{r+1}.$$

If, in addition, $n \geq 1$, then

$$(n+1)(xy^n)^{1/(n+1)} \leq x + ny.$$

In particular,

$$4xy \leq (x+y)^2, \quad 27xy^2 \leq (x+2y)^3, \quad 256xy^3 \leq (x+3y)^4.$$

Equality holds if and only if $x = y$. **Source:** [1757, p. 252]. The case $x = 1$ is given in [658]. **Remark:** The inequality holds with n replaced by $\alpha > 0$. **Related:** Fact 2.1.28, Fact 2.2.39, and Fact 12.13.16.

Fact 2.2.40. Let x and y be nonnegative numbers, and let $n \geq 1$. Then,

$$nx^{n-1}y \leq (n-1)x^n + y^n.$$

In particular,

$$2xy \leq x^2 + y^2, \quad 3x^2y \leq 2x^3 + y^3, \quad 4x^3y \leq 3x^4 + y^4.$$

Equality holds if and only if either $n = 1$ or $x = y$. **Source:** Use the arithmetic-mean–geometric-mean inequality. See [1757, p. 62].

Fact 2.2.41. Let x and y be nonnegative numbers, and let $n \geq 1$. Then,

$$n(x-y)(xy)^{(n-1)/2} < x^n - y^n.$$

Source: [2527, p. 32].

Fact 2.2.42. Let x and y be positive numbers such that $x < y$, and let $n \geq 1$. Then,

$$(n+1)(y-x)x^n < y^{n+1} - x^{n+1} < (n+1)(y-x)y^n.$$

Source: [1757, p. 248].

Fact 2.2.43. Let $[a, b] \subset \mathbb{R}$ and $x, y \in [a, b]$. Then,

$$|x| + |y| - |x+y| \leq |a| + |b| - |a+b|.$$

Source: Fact 1.21.8.

Fact 2.2.44. Let $x, y \in \mathbb{R}$. Then,

$$\min\{x, y\} = \tfrac{1}{2}(x + y - |x-y|), \quad \max\{x, y\} = \tfrac{1}{2}(x + y + |x-y|),$$

$$\min\{|x+y|, |x-y|\} = \big||x| - |y|\big| = |x+y| + |x-y| - |x| - |y|, \quad \max\{|x+y|, |x-y|\} = |x| + |y|.$$

Source: [107, p. 363] and [1932]. **Related:** Fact 2.2.45 and Fact 11.7.7.

Fact 2.2.45. Let x and y be nonnegative numbers, and let $\alpha \in [0, 1]$. Then,

$$x + y \le |x - y| + 2x^{\alpha} y^{1-\alpha}.$$

Source: Fact 2.2.44 and [1426]. **Remark:** This is a scalar version of the Powers-Stormer inequality. See Fact 10.14.32.

Fact 2.2.46. Let $x, y \in [0, \sqrt{2}]$. Then,

$$\frac{1}{\sqrt{1+x^2}} + \frac{1}{\sqrt{1+y^2}} \le \frac{2}{\sqrt{1+xy}}.$$

Source: $f\colon (-\infty, \log \sqrt{2}] \mapsto \mathbb{R}$ defined by $f(x) \triangleq -(1+e^{2x})^{-1/2}$ is convex. See [1938, p. 31].

Fact 2.2.47. Let x and y be real numbers such that neither x, y, nor xy is equal to 1. Then,

$$1 \le \frac{1}{(1-x)^2} + \frac{1}{(1-y)^2} + \frac{x^2y^2}{(1-xy)^2}.$$

If, in addition, x and y are positive, then the inequality is strict. **Source:** [2136]. **Related:** Fact 2.11.147.

Fact 2.2.48. Let $[a, b] \subset (0, \infty)$ and $x, y \in [a, b]$. Then,

$$\sqrt{\frac{x}{y}} + \sqrt{\frac{y}{x}} \le \sqrt{\frac{a}{b}} + \sqrt{\frac{b}{a}}.$$

Source: Fact 1.21.8.

Fact 2.2.49. Let x and y be positive numbers. Then,

$$\frac{2}{xy+2} \le \frac{1}{2x+1} + \frac{1}{2y+1}, \quad \frac{1}{xy+1} \le \frac{1}{(x+1)^2} + \frac{1}{(y+1)^2},$$

$$\frac{3}{2} \le \frac{x}{x^2y^3+y^2} + \frac{y}{y^2x^3+x^2} + \frac{x^3y^3}{x+y}, \quad \sqrt{3(x^2+y^2+1)} \le xy^2 + \frac{1}{x} + \frac{x}{y},$$

$$\frac{1}{\sqrt{x+3y}} + \frac{1}{\sqrt{y+3x}} \le \frac{2}{\sqrt{x}+\sqrt{y}} \le \frac{1}{2\sqrt{x}} + \frac{1}{2\sqrt{y}}, \quad \sqrt{\frac{x+2y}{x^2+2y^2}} + \sqrt{\frac{y+2x}{y^2+2x^2}} \le \sqrt{\frac{8}{x+y}},$$

$$\frac{x}{\sqrt{x^2+y^2}} + \frac{y}{\sqrt{9x^2+y^2}} + \frac{2xy}{\sqrt{x^2+y^2}\sqrt{9x^2+y^2}} \le \frac{3}{2},$$

$$(1+\sqrt{5})(1-xy) \le \sqrt{x^2+1} + \sqrt{y^2+1} + \sqrt{(x-1)^2+(y-1)^2}.$$

If $xy \le \frac{1}{2}(3-\sqrt{5})$, then

$$\frac{1}{x^2+x+1} + \frac{1}{y^2+y+1} \le \frac{2}{xy+\sqrt{xy}+1}.$$

If $xy \ge 9/16$, then

$$\frac{2}{xy+\sqrt{xy}+1} \le \frac{1}{x^2+x+1} + \frac{1}{y^2+y+1}.$$

Source: [752, pp. 2–4] and [1569].

Fact 2.2.50. Let x and y be nonnegative numbers, and let $p, q \in (1, \infty)$ satisfy $1/p + 1/q = 1$. Then,

$$xy \le \frac{x^p}{p} + \frac{y^q}{q}.$$

Equality holds if and only if $x^p = y^q$. **Source:** [935, p. 12] and [936, p. 10]. **Remark:** This is *Young's inequality*. See Fact 2.2.53, Fact 10.11.73, Fact 10.14.8, Fact 10.14.33, Fact 10.14.34, and Fact 10.11.38. **Remark:** $1/p + 1/q = 1$ is equivalent to $(p-1)(q-1) = 1$.

Fact 2.2.51. Let $x, y \in (1, \infty)$, and assume that $\frac{1}{x} + \frac{1}{y} = 1$. Then,

$$\frac{1}{x^{x-1}} + \frac{1}{y^{y-1}} \le 1 \le \frac{1}{x^{y-1}} + \frac{1}{y^{x-1}}.$$

If, in addition, $x \in (1, 2]$, then $x^{1/x} \le y^{1/y}$. **Source:** [1036]. **Remark:** x and y are conjugate exponents.

Fact 2.2.52. Let x, y, and p be positive numbers. Then, the following statements hold:

i) If $p \in [0, 1]$, then $x^p y^{1-p} \le px + (1-p)y$.

ii) If $p \in (0, 1]$, then

$$\frac{1}{p}x^p + \frac{p-1}{p}y^{p/(p-1)} \le xy \le px^{1/p} + (1-p)y^{1/(1-p)}.$$

Equality holds in the first inequality if and only if $x^{1-p} = y^p$. Equality holds in the second inequality if and only if $x^{p-1} = y$.

iii) If $p \ge 1$, then $px + (1-p)y \le x^p y^{1-p}$.

iv) If $p > 1$, then

$$px^{1/p} + (1-p)y^{1/(1-p)} \le xy \le \frac{1}{p}x^p + \frac{p-1}{p}y^{p/(p-1)}.$$

Equality holds in the first inequality if and only if $x^{1-p} = y^p$. Equality holds in the second inequality if and only if $x^{p-1} = y$.

Source: [1941]. **Remark:** The second inequality in *iv)* is Young's inequality.

Fact 2.2.53. Let x and y be positive numbers, let $\alpha \in [0, 1]$, and define $\delta \triangleq \min\{\alpha, 1-\alpha\}$ and $\rho \triangleq \max\{\alpha, 1-\alpha\}$. Then,

$$[\alpha x^{-1} + (1-\alpha)y^{-1}]^{-1} \le x^\alpha y^{1-\alpha} \le x^\alpha y^{1-\alpha} + \delta(\sqrt{x} - \sqrt{y})^2 \le \alpha x + (1-\alpha)y,$$
$$x^\alpha y^{1-\alpha} + x^{1-\alpha}y^\alpha + 2\delta(\sqrt{x} - \sqrt{y})^2 \le x + y,$$
$$(x^\alpha y^{1-\alpha} + x^{1-\alpha}y^\alpha)^2 + 2\delta(x-y)^2 \le (x+y)^2,$$
$$(x^\alpha y^{1-\alpha})^2 + \delta(x-y)^2 \le \alpha x^2 + (1-\alpha)y^2,$$
$$(x^\alpha y^{1-\alpha})^2 + \delta^2(x-y)^2 \le [\alpha x + (1-\alpha)y]^2,$$
$$\alpha x^2 + (1-\alpha)y^2 \le (x^\alpha y^{1-\alpha})^2 + \rho(x-y)^2,$$
$$[\alpha x + (1-\alpha)y]^2 \le (x^\alpha y^{1-\alpha})^2 + \rho^2(x-y)^2,$$
$$(x+y)^2 \le (x^\alpha y^{1-\alpha} + x^{1-\alpha}y^\alpha)^2 + 2\rho(x-y)^2.$$

Source: [1109, 1356, 1418, 1640]. **Remark:** $x^\alpha y^{1-\alpha} \le \alpha x + (1-\alpha)y$ follows from the concavity of the logarithm function. This is *Young's inequality*. **Related:** Fact 2.2.50, Fact 2.2.54, Fact 2.2.55, Fact 10.11.38, Fact 10.11.73, Fact 10.14.8, Fact 10.14.33, and Fact 10.14.34.

Fact 2.2.54. Let x and y be positive numbers, and let $\alpha \in [0, 1]$. Then,

$$x^\alpha y^{1-\alpha} \le \alpha x + (1-\alpha)y \le S(x/y)x^\alpha y^{1-\alpha},$$

where S is Specht's ratio given by Fact 12.17.5. In particular,

$$\sqrt{xy} \le \tfrac{1}{2}(x+y) \le S(x/y)\sqrt{xy}.$$

Furthermore,

$$x^\alpha y^{1-\alpha} \le \alpha x + (1-\alpha)y \le x^\alpha y^{1-\alpha} + \varepsilon \log S(x/y),$$

where, if $x \ne y$, then $\varepsilon \triangleq (y-x)/(\log y - \log x)$ and, if $x = y$, then $\varepsilon \triangleq x$. **Source:** [1109, 1640]. **Remark:** This is the *reverse Young inequality*. See Fact 2.2.53. The case $\alpha = 1/2$ is the *reverse arithmetic-mean–geometric mean inequality*. See Fact 2.11.95. **Related:** Fact 2.2.55, Fact 2.2.66, Fact 2.11.95, Fact 15.15.23, and Fact 12.17.5. **Credit:** M. Tominaga. See [1091].

Fact 2.2.55. Let x and y be positive numbers, and let $\alpha \in [0,1]$. Then,

$$x^\alpha y^{1-\alpha} \le \alpha x + (1-\alpha)y \le x^\alpha y^{1-\alpha} e^{\alpha(1-\alpha)(x-y)^2/(\min\{x,y\})^2},$$

$$x^\alpha y^{1-\alpha} \le \alpha x + (1-\alpha)y \le x^\alpha y^{1-\alpha} + \alpha(1-\alpha)(\max\{x,y\})(\log x - \log y)^2.$$

Source: [1114]. **Related:** Fact 2.2.54, Fact 2.2.66, Fact 2.11.95, and Fact 15.15.23.

Fact 2.2.56. Let x and y be real numbers, and let $\alpha \in [0,1]$. Then,

$$[\alpha e^{-x} + (1-\alpha)e^{-y}]^{-1} \le e^{\alpha x + (1-\alpha)y} \le \alpha e^x + (1-\alpha)e^y.$$

Source: Replace x and y by e^x and e^y, respectively, in Fact 2.2.53. **Remark:** The right-hand inequality follows from the convexity of the exponential function.

Fact 2.2.57. Let x and y be real numbers, and assume that $x \ne y$. Then,

$$e^{(x+y)/2} < \frac{e^x - e^y}{x-y} < \tfrac{1}{2}(e^x + e^y).$$

Source: [45]. **Related:** Fact 2.2.63.

Fact 2.2.58. Let x and y be positive numbers, and let p and q be nonzero real numbers such that $p \le q$. Then,

$$\left(\frac{x^p + y^p}{2}\right)^{1/p} \le \left(\frac{x^q + y^q}{2}\right)^{1/q}.$$

Equality holds if and only if either $p = q$ or $x = y$. Furthermore,

$$\min\{x,y\} = \lim_{p\to-\infty}\left(\frac{x^p+y^p}{2}\right)^{1/p} \le \sqrt{xy} = \lim_{p\to 0}\left(\frac{x^p+y^p}{2}\right)^{1/p} \le \lim_{p\to\infty}\left(\frac{x^p+y^p}{2}\right)^{1/p} = \max\{x,y\}.$$

Hence, if $p \le -1 \le q \le 0 \le r \le 1 \le s$, then

$$\begin{aligned}\min\{x,y\} &\le \left(\frac{x^p+y^p}{2}\right)^{1/p} \le \frac{2}{\frac{1}{x}+\frac{1}{y}} \le \left(\frac{x^q+y^q}{2}\right)^{1/q} \le \sqrt{xy}\\ &\le \left(\frac{x^r+y^r}{2}\right)^{1/r} \le \tfrac{1}{2}(x+y) \le \left(\frac{x^s+y^s}{2}\right)^{1/s} \le \max\{x,y\},\end{aligned}$$

where each inequality is an equality if and only if $x = y$. **Source:** [413], [1597, pp. 63–65], and [1855]. **Remark:** This is the *power-mean inequality*. The arithmetic-mean–geometric-mean inequality $\sqrt{xy} \le \frac{1}{2}(x+y)$ is given by Fact 2.11.81. **Related:** Fact 2.2.38 and Fact 2.11.89.

Fact 2.2.59. Let x and y be positive numbers. If $p \in [0,1]$, then

$$(x+y)^p \le x^p + y^p \le 2^{1-p}(x+y)^p,$$

$$2^{p-1}(x^p+y^p) \le (x+y)^p \le x^p + y^p.$$

If $p \in [1,\infty)$, then

$$2^{1-p}(x+y)^p \le x^p + y^p \le (x+y)^p,$$

$$x^p + y^p \le (x+y)^p \le 2^{p-1}(x^p+y^p).$$

If $p \in [1, \infty)$ and $\alpha > 0$, then

$$x^p + y^p \le (x+y)^p \le (1+\alpha)^{p-1}x^p + (1+\alpha^{-1})^{p-1}y^p.$$

Equality holds in the second inequality if and only if $y = \alpha x$. **Source:** [1423, 1629]. The second inequality in the last string is given in [2428, p. 7]. **Related:** Fact 2.2.58, Fact 2.11.130, Fact 2.11.131, Fact 10.14.37, and Fact 11.8.21.

Fact 2.2.60. Let x, y, a, b be positive numbers, and let $p \in (1, \infty)$. Then,

$$(x^{1/p} + y^{1/p})^p \le [a^{1/(p-1)} + b^{1/(p-1)}]^{p-1}\left(\frac{x}{a} + \frac{y}{b}\right).$$

Furthermore, equality holds if and only if $ay^{(p-1)/p} = bx^{(p-1)/p}$. **Source:** [1260, pp. 142, 143].

Fact 2.2.61. Let p and q be real numbers such that $0 < p \le q \le 1$, and define $f\colon [0, \infty) \mapsto \mathbb{R}$ by

$$f(t) \triangleq \begin{cases} \dfrac{p(t^q - 1)}{q(t^p - 1)}, & t \ne 1,\\ 1, & t = 1. \end{cases}$$

Then, f is nondecreasing. **Source:** [1314].

Fact 2.2.62. Let x and y be positive numbers, and let p and q be nonzero real numbers such that $p \le q$. Then,

$$\frac{x^p + y^p}{x^{p-1} + y^{p-1}} \le \frac{x^q + y^q}{x^{q-1} + y^{q-1}}.$$

Equality holds if and only if either $x = y$ or $p = q$. **Source:** Fact 2.2.36 and [209, p. 23]. **Remark:** The quantity $\dfrac{x^p + y^p}{x^{p-1} + y^{p-1}}$ is the *Lehmer mean.*

Fact 2.2.63. Let x and y be distinct positive numbers, and define

$$H(x,y) \triangleq \frac{2xy}{x+y}, \quad G(x,y) \triangleq \sqrt{xy}, \quad L(x,y) \triangleq \frac{y-x}{\log y - \log x} = \int_0^1 x^t y^{1-t}\, dt,$$

$$P(x,y) \triangleq \frac{x-y}{4\,\mathrm{atan}\,\sqrt{x/y} - \pi}, \quad I(x,y) \triangleq \frac{1}{e}\left(\frac{x^x}{y^y}\right)^{1/(x-y)}, \quad A(x,y) \triangleq \tfrac{1}{2}(x+y),$$

$$N(x,y) \triangleq \frac{x-y}{2\,\mathrm{asinh}\,\frac{x-y}{x+y}}, \quad T(x,y) \triangleq \frac{x-y}{2\,\mathrm{atan}\,\frac{x-y}{x+y}}, \quad Q(x,y) \triangleq \sqrt{\tfrac{1}{2}(x^2+y^2)}, \quad C(x,y) \triangleq \frac{x^2+y^2}{x+y}.$$

Let H denote $H(x,y)$ and likewise for $G, L, P, I, A, N, T, Q, C$. Then,

$$\min\{x,y\} < H < G < L < P < I < A < N < T < Q < C < \max\{x,y\},$$

$$G + \frac{(x-y)^2(x+3y)(y+3x)}{8(x+y)(x^2+6xy+y^2)} \le A \le \frac{1}{4}L\left(1 + \sqrt{1 + \frac{8A^2}{G^2}}\right), \quad \frac{3AG}{2A+G} \le \sqrt[3]{G^2A} \le L,$$

$$2A < \pi P \quad N < \tfrac{1}{3}Q + \tfrac{2}{3}A, \quad \tfrac{\pi}{4\log(1+\sqrt{2})}T < N < \tfrac{1}{\log(1+\sqrt{2})}A, \quad A = L(1 + \log I/G),$$

$$\sqrt{2T^2 - Q^2}\,Q < NQ < T^2, \quad H(T,A) < N < L(A,Q), \quad H(N,Q) < T,$$

$$NP < A^2, \quad \sqrt[3]{A^2Q} < N < \tfrac{1}{3}(2A+Q), \quad \sqrt{AT} < N < \sqrt{A^2+T^2},$$

$$\sqrt[3]{A^2G} < P < \tfrac{1}{3}(2A+G), \quad \tfrac{2}{\pi}A + \tfrac{\pi-2}{\pi}H < P < \tfrac{1}{6}(5A+H), \quad A + L \le G + \sqrt{2A^2 - G^2},$$

$$2(A^2 - G^2) + AL \le \sqrt{8A^4 - 8A^2G^2 + G^4}, \quad (I-G)^4 + \frac{(I^2 - G^2)L}{A-L} \le \sqrt{2(I^4 + G^4)},$$

$$G < \sqrt{GA} < \sqrt[3]{G^2A} < \sqrt[3]{\tfrac{1}{4}(G+A)^2G} < L < \left\{ \begin{matrix} \tfrac{1}{3}(2G+A) < \tfrac{1}{3}(G+2A) \\ \sqrt{LA} < \tfrac{1}{2}(L+A) \end{matrix} \right\} < I < A,$$

$$G < G(\sqrt{x}, \sqrt{y})A(\sqrt{x}, \sqrt{y}) < L(x,y) < A(\sqrt{x}, \sqrt{y})^2 < A.$$

If $n \ge 1$, then

$$n\sqrt[3]{\tfrac{1}{2}x^ny^n(x^n+y^n)} \le L\sum_{i=1}^n x^{n-i}y^{i-1},$$

$$(xy)^{1/2^{n+1}}\prod_{i=1}^n A(x^{1/2^i}, y^{1/2^i}) < L < A(x^{1/2^n}, y^{1/2^n})\prod_{i=1}^n A(x^{1/2^i}, y^{1/2^i}).$$

If $\alpha \ge 2/9$, then $H^\alpha Q^{1-\alpha} < N$. If $\alpha \ge 1/3$, then $G^\alpha Q^{1-\alpha} < N$. If $\alpha \ge 5/12$, then $H^\alpha C^{1-\alpha} < N$. If $\alpha \ge 5/9$, then $G^\alpha C^{1-\alpha} < N$. If $\alpha \le 0$, then $N < H^\alpha Q^{1-\alpha}$, $N < G^\alpha Q^{1-\alpha}$, $N < H^\alpha C^{1-\alpha}$, and $N < G^\alpha C^{1-\alpha}$. If $x, y \in (0, 1/2]$, then

$$\frac{G(x,y)}{G(1-x,1-y)} < \frac{L(x,y)}{L(1-x,1-y)} < \frac{P(x,y)}{P(1-x,1-y)} < \frac{A(x,y)}{A(1-x,1-y)} < \frac{N(x,y)}{N(1-x,1-y)} < \frac{T(x,y)}{T(1-x,1-y)},$$

$$\frac{1}{A(1-x,1-y)} - \frac{1}{A(x,y)} < \frac{1}{N(1-x,1-y)} - \frac{1}{N(x,y)} < \frac{1}{T(1-x,1-y)} - \frac{1}{T(x,y)},$$

$$A(x,y)A(1-x,1-y) < N(x,y)N(1-x,1-y) < T(x,y)T(1-x,1-y).$$

For all $p \in \mathbb{R}$, define $M_p(x,y) \triangleq \left(\frac{x^p+y^p}{2}\right)^{1/p}$. Then,

$$M_{-1/3} < \tfrac{1}{3}(2G+H), \quad M_{-2/3} < \tfrac{1}{3}(G+2H),$$

$$M_{1/3} \le \tfrac{1}{3}(2G+A) \le M_{1/2}, \quad M_{(\log 2)/\log 3} \le \tfrac{1}{3}(G+2A) \le M_{2/3}.$$

If $1/3 \le p < 1 < q$, then

$$L < M_p < A < M_q.$$

If $p \in (0, \frac{1}{2}) \cup (1, \infty)$, then $M_p < pA + (1-p)G$. If $p \in (\frac{1}{2}, 1)$, then $pA + (1-p)G < M_p$. If $p \in (-\infty, 6/5)$, then $G^p + 2A^p < 3I^p$. If $p \in (1.33, \infty)$, then $3I^p < G^p + 2A^p$. **Source:** [374, 375, 635, 741, 742], [1371, p. 106] and [1855, 2361, 2368, 2551, 2568]. $3L < 2G + A$ is *Polya's inequality*. See [2128, p. 53]. $G + 2A < 3I$ is due to J. Sandor. See [209, p. 24]. The last four statements are given in [1687]. **Remark:** $(G, L, P, I, A, N, T, Q, C)$ is the (geometric, logarithmic, first Seiffert, identric, arithmetic, Newman-Sandor, second Seiffert, quadratic, contraharmonic) mean. **Remark:** These inequalities refine the arithmetic-mean–geometric-mean inequality given by Fact 2.11.81. **Remark:** L is the *logarithmic mean*, and I is the *identric mean*. See [2551]. **Remark:** $\min\{x,y\} < L < \max\{x,y\}$ is *Napier's inequality*. **Remark:** The logarithmic mean is extended to matrices in Fact 10.11.69. **Remark:** Inequalities for these and additional means are given in [2953]. **Related:** Fact 2.2.56, Fact 2.2.64, Fact 2.11.112, Fact 12.18.56, Fact 14.2.17, and Fact 14.2.19.

Fact 2.2.64. Let x and y be positive numbers. If $\alpha \in [0, 1]$, then

$$\sqrt{xy} \le \tfrac{1}{2}(x^{1-\alpha}y^\alpha + x^\alpha y^{1-\alpha}) \le \tfrac{1}{2}(x+y).$$

Furthermore, if $x \ne y$ and $\alpha \in [\frac{1}{2}(1 - 1/\sqrt{3}), \frac{1}{2}(1 + 1/\sqrt{3})]$, then

$$\sqrt{xy} \le \tfrac{1}{2}(x^{1-\alpha}y^\alpha + x^\alpha y^{1-\alpha}) \le \frac{y-x}{\log y - \log x} \le \tfrac{1}{2}(x+y).$$

Source: [949]. **Remark:** The first string refines the arithmetic-mean–geometric-mean inequality given by Fact 2.11.81. The center term is the *Heinz mean.* Monotonicity is considered in Fact 2.12.34, while matrix extensions are given by Fact 11.10.82. **Related:** Fact 2.2.63.

Fact 2.2.65. Let x and y be positive numbers. Then,

$$\sqrt{xy} \le \left(1 + \frac{1}{8}\log^2\frac{x}{y}\right)\sqrt{xy} \le \frac{1}{2}(x+y).$$

Source: [3019]. **Related:** Fact 2.2.63, Fact 2.2.64, and Fact 10.11.69.

Fact 2.2.66. Let x and y be positive numbers. Then,

$$\sqrt{xy} \le S(\sqrt{x/y})\sqrt{xy} \le \frac{1}{2}(x+y) \le S(x/y)\sqrt{xy},$$

where S is Specht's ratio given by Fact 12.17.5. **Source:** [1110, 3019] and Fact 2.11.95. **Related:** Fact 2.2.54, Fact 2.2.63, Fact 2.2.64, Fact 2.11.95, and Fact 15.15.23.

Fact 2.2.67. Let x and y be distinct positive numbers, and define

$$L \triangleq \frac{y-x}{\log y - \log x}, \quad H_p \triangleq \left(\frac{x^p + (xy)^{p/2} + y^p}{3}\right)^{1/p}, \quad M_p \triangleq \left(\frac{x^p + y^p}{2}\right)^{1/p}.$$

If p, q are positive numbers such that $p < q$, then

$$M_p < M_q, \quad H_p < H_q.$$

Now, let p, q, r be positive numbers such that $0.5283 \approx (\log 3)/(3\log 2) \le p \le 3q/2$ and $1/3 < r < [(\log 2)/\log 3]p \approx 0.6309p$. Then,

$$L < H_{1/2} < M_{1/3} < M_r < H_p < M_q.$$

In particular, if $r \le (\log 2)/\log 3 \approx 0.6309$ and $q \ge 2/3 \approx 0.6667$, then

$$\left(\frac{x^r + y^r}{2}\right)^{1/r} < \frac{x + \sqrt{xy} + y}{3} < \left(\frac{x^q + y^q}{2}\right)^{1/q}.$$

Finally, if $1/2 \le p \le 3q/2$, then

$$\frac{y-x}{\log y - \log x} < \left(\frac{x^p + (xy)^{p/2} + y^p}{3}\right)^{1/p} < \left(\frac{x^q + y^q}{2}\right)^{1/q}.$$

Source: [605, p. 350] and [1263, 1530]. **Remark:** The center term is the *Heron mean.*

Fact 2.2.68. Let $x, y \in [0, 1]$. Then, $1 - x^y \le (1-x)(1+x)^{y-1}$. **Source:** [376].

Fact 2.2.69. Let $x, y \in (0, 1]$. Then,

$$\frac{x}{x + y - xy} \le x^y.$$

Source: [806, p. 37].

Fact 2.2.70. Let x and y be positive numbers. Then, $1 < x^y + y^x$. **Source:** [993, p. 184], [1158, p. 45], [1567, p. 75], [2294, pp. 219, 220]. **Related:** Fact 2.11.32.

Fact 2.2.71. Let x and y be positive numbers. Then,

$$x^y y^x \le (xy)^{\frac{x+y}{2}} \le \left(\frac{x+y}{2}\right)^{x+y} \le x^x y^y.$$

Source: [109, pp. 227, 228], [644, p. 107], and [2294, pp. 267, 270, 271]. **Related:** Fact 2.11.116 and Fact 2.11.51.

Fact 2.2.72. Let x and y be positive numbers. Then, $x^{2y} + y^{2x} \leq x^{2x} + y^{2y}$. **Source:** [752, p. 241].

Fact 2.2.73. Let x and y be positive numbers, let $\alpha \in [0, e)$, and assume that at least one of the following conditions holds: $1 \leq y^2 \leq x; ey/2 \leq 1 \leq x; y \leq e \leq x; 2/e \leq y \leq 1 \leq x \leq e; 1 \leq y \leq x \leq e$. Then,

$$x^{\alpha y} + y^{\alpha x} \leq 2\sqrt{x^{\alpha x} y^{\alpha y}}.$$

Source: [783, 1979, 1980]. **Related:** Fact 2.3.55.

Fact 2.2.74. Let x and y be positive numbers. Then,

$$\left(\frac{x}{y}\right)^y \leq \left(\frac{x+1}{y+1}\right)^{y+1}.$$

Equality holds if and only if $x = y$. **Source:** [1757, p. 267].

Fact 2.2.75. Let x and y be real numbers. If either $0 < x < y < 1$ or $1 < x < y$, then

$$\frac{y^x}{x^y} < \frac{y}{x}, \quad \frac{y^y}{x^x} < \left(\frac{y}{x}\right)^{xy}.$$

If $0 < x < 1 < y$, then both inequalities are reversed. If either $0 < x < 1 < y$ or $0 < x < y < e$, then

$$1 < \left(\frac{y \log x}{x \log y}\right)\left(\frac{y^x - 1}{x^y - 1}\right) < \frac{y^x}{x^y}.$$

If $e < x < y$, then both inequalities are reversed. **Source:** [2277].

Fact 2.2.76. Let α be a positive number, let $y \in (0, \alpha]$, and let x be a nonnegative number. Then, $\alpha x + y \leq y x^{\alpha/y} + \alpha$. **Source:** [806, p. 124]. **Related:** Fact 2.3.38.

Fact 2.2.77. Let x and y be real numbers, let p and q be odd positive integers, assume that $p \leq q$, and define $\alpha \triangleq p/q$. Then,

$$|x^\alpha - y^\alpha| \leq 2^{1-\alpha}|x - y|^\alpha.$$

If $p = 1$, $k \geq 1$, and $q = 2k + 1$, then

$$|x - y|^{2k+1} \leq 2^{2k}|x^{2k+1} - y^{2k+1}|.$$

Now, assume that x and y are nonnegative. If $r \geq 1$, then

$$|x - y|^r \leq |x^r - y^r|.$$

Source: [1419, 1483]. **Remark:** Matrix versions of these results are given in [1419]. Applications to nonlinear control appear in [1483, 2281]. **Problem:** Unify the first and third inequalities.

Fact 2.2.78. Let $x > 0$. Then, the following statements hold:

i) If $y \in [0, 1]$, then $x(x + y)^{y-1} \leq \frac{\Gamma(x+y)}{\Gamma(x)} \leq x^y$.

ii) If $y \in [1, 2]$, then $x^y \leq \frac{\Gamma(x+y)}{\Gamma(x)} \leq x(x + 1)^{y-1}$.

iii) If $y \geq 2$, then $x(x + 1)^{y-1} \leq \frac{\Gamma(x+y)}{\Gamma(x)} \leq x(x + y)^{y-1}$.

Source: [1516].

Fact 2.2.79. Let $x > 0$, let $y \geq \frac{1}{6}\left(2 + \sqrt[3]{35 - 3\sqrt{129}} + \sqrt[3]{35 + 3\sqrt{129}}\right) \approx 1.17965$, and let $n \geq 1$. Then,

$$\frac{1}{x + y - 1} + \left(\frac{ny + 1}{y^2}\right)\log x \leq \frac{x^n}{y}.$$

Source: [2745]. **Remark:** The lower bound for y is the unique positive root of $4s^3 - 4s^2 - 1$.

2.3 Facts on Equalities and Inequalities in Three Variables

Fact 2.3.1. Let x, y, z be complex numbers, and let $n \geq 1$. Then,

$$(x+y+z)^n = \sum \frac{n!}{i!j!k!} x^i y^j z^k,$$

where the sum is taken over all $\frac{1}{2}(n+2)(n+1)$ triples (i, j, k) of nonnegative integers such that $i+j+k=n$. **Related:** Fact 2.1.10 and Fact 2.11.8.

Fact 2.3.2. Let x, y, z be complex numbers. Then,

$$x(y-z) + y(z-x) + z(x-y) = 0,$$
$$x^2(y-z) + y^2(z-x) + z^2(x-y) = (x-y)(y-z)(z-x),$$
$$x^3(y-z) + y^3(z-x) + z^3(x-y) = (x-y)(y-z)(z-x)(x+y+z),$$
$$x^4(y-z) + y^4(z-x) + z^4(x-y) = (x-y)(y-z)(z-x)(x^2+y^2+z^2+xy+yz+zx),$$
$$x^5(y-z) + y^5(z-x) + z^5(x-y)$$
$$= (x-y)(y-z)(z-x)(x^3+y^3+z^3+x^2y+y^2z+z^2x+xy^2+yz^2+zx^2+xyz),$$
$$x(y^2-z^2) + y(z^2-x^2) + z(x^2-y^2) = (x-y)(y-z)(z-x),$$
$$x^2(y^2-z^2) + y^2(z^2-x^2) + z^2(x^2-y^2) = 0,$$
$$x^3(y^2-z^2) + y^3(z^2-x^2) + z^3(x^2-y^2) = -(x-y)(y-z)(z-x)(xy+yz+zx),$$
$$x(y^3-z^3) + y(z^3-x^3) + z(x^3-y^3) = (x-y)(y-z)(z-x)(x+y+z),$$
$$x^2(y^3-z^3) + y^2(z^3-x^3) + z^2(x^3-y^3) = (x-y)(y-z)(z-x)(xy+yz+zx),$$
$$x^3(y^3-z^3) + y^3(z^3-x^3) + z^3(x^3-y^3) = 0,$$
$$x^4(y^3-z^3) + y^4(z^3-x^3) + z^4(x^3-y^3)$$
$$= (x-y)(y-z)(z-x)(x^2y^2+y^2z^2+z^2x^2+x^2yz+y^2zx+z^2xy).$$

Source: [289, pp. 14, 95].

Fact 2.3.3. Let x, y, z be complex numbers. Then,

$$(x-z)^2 = 2(x-y)^2 + 2(y-z)^2,$$
$$(x+z)^2 = (x-2y+z)^2 + 4(xy+yz) + 4y^2,$$
$$x^2+y^2+z^2 = (x+y+z)^2 - 2(xy+yz+zx),$$
$$(x-z)^2 + (z-y)^2 = \tfrac{1}{2}(x-y)^2 + 2[z - \tfrac{1}{2}(x+y)]^2,$$
$$(x^2+y^2+z^2)^2 = (x^2+y^2-z^2)^2 + (2xz)^2 + (2yz)^2,$$
$$x^3+y^3+z^3 = (x+y+z)^3 - 3(x+y)(y+z)(z+x),$$
$$3(x-y)(y-z)(z-x) = (x-y)^3 + (y-z)^3 + (z-x)^3,$$
$$x^2y^2+y^2z^2+z^2x^2 = (xy+yz+zx)^2 - 2(x+y+z)xyz,$$
$$(x+y)(y+z)(z+x) = (x+y+z)(xy+yz+zx) - xyz,$$
$$(x+1)(y+1)(z+1) = x+y+z+xy+yz+zx+xyz+1,$$
$$(x+y)^2 + (y+z)^2 + (z+x)^2 = x^2+y^2+z^2+(x+y+z)^2,$$
$$3(x^2+y^2+z^2) = (x-y)^2 + (y-z)^2 + (z-x)^2 + (x+y+z)^2,$$
$$x^3+y^3+z^3 = (x+y+z)(x^2+y^2+z^2-xy-yz-zx) + 3xyz,$$
$$x^3+y^3+z^3 = (x+y+z)^3 - 3(x+y+z)(xy+yz+zx) + 3xyz,$$

$$24xyz = (x+y+z)^3 - (x-y+z)^3 - (x+y-z)^3 + (x-y-z)^3,$$

$$2(x^2+y^2+z^2) = (x-y)^2 + (y-z)^2 + (z-x)^2 + 2(xy+yz+zx),$$

$$(xy+yz+zx-1)^2 = (x^2+1)(y^2+1)(z^2+1) + (x+y+z-xyz)^2,$$

$$9(x^2+y^2+z^2) = (x-2y-2z)^2 + (y-2x-2z)^2 + (z-2x-2y)^2,$$

$$2(x^3+y^3+z^3) = (x+y+z)[(x-y)^2+(y-z)^2+(z-x)^2] + 6xyz,$$

$$xy^3+yz^3+zx^3 = x^3y+y^3z+z^3x + (x+y+z)(x-y)(y-z)(z-x),$$

$$(x+y+z)(xy+yz+zx) = x(y^2+z^2) + y(z^2+x^2) + z(x^2+y^2) + 3xyz,$$

$$3(x^2y+y^2z+z^2x) = (x-y)^3 + (y-z)^3 + (z-x)^3 + 3(xy^2+yz^2+zx^2),$$

$$4(x^2+y^2+z^2) = (x+y-z)^2 + (y+z-x)^2 + (z+x-y)^2 + (x+y+z)^2,$$

$$(x+y+z)^2 + xy+yz+zx = (x+y)(y+z) + (y+z)(z+x) + (z+x)(x+y),$$

$$xy(x^2-y^2) + yz(y^2-z^2) + zx(z^2-x^2) + (x+y+z)(x-y)(y-z)(z-x) = 0,$$

$$(x+y)^5 + (x-y)^5 + (x+z)^5 + (x-z)^5 = 2x[2x^4 + 10x^2(y^2+z^2) + 5(y^4+z^4)],$$

$$x^3y^3+y^3z^3+z^3x^3 = (xy+yz+zx)^3 - 3(x+y+z)(xy+yz+zx)xyz + 3x^2y^2z^2,$$

$$x^5+y^5+z^5 = (x+y+z)^5 - 5(x+y)(y+z)(z+x)(x^2+y^2+z^2+xy+yz+zx),$$

$$2(x+y+z) + xy+yz+zx+3 = (x+1)(y+1) + (y+1)(z+1) + (z+1)(x+1),$$

$$(x-y)^2 + (y-z)^2 + (z-x)^2 = 2[(x-y)(x-z) + (y-z)(y-x) + (z-x)(z-y)],$$

$$2(x^2y^2+y^2z^2+z^2x^2) = x^4+y^4+z^4 + (x+y+z)(x+y-z)(y+z-x)(z+x-y),$$

$$(x+y-z)(y+z-x)(z+x-y) = 4(x+y+z)(xy+yz+zx) - (x+y+z)^3 - 8xyz,$$

$$3(xy^3+yz^3+zx^3) = (x+y+z)[(x-y)^3+(y-z)^3+(z-x)^3] + 3(x^3y+y^3z+z^3x),$$

$$(9x^3+y^3+z^3)^3 = (9x^3+y^3-z^3)^3 + (9x^3-y^3+z^3)^3 + (-9x^3+y^3+z^3)^3 + (6xyz)^3,$$

$$(x^2-y^2)^3 + (y^2-z^2)^3 + (z^2-x^2)^3 = (x+y)(y+z)(z+x)[(x-y)^3+(y-z)^3+(z-x)^3],$$

$$(x^2y-y^2z)^2 + (y^2z-z^2x)^2 + (z^2x-x^2y)^2 + 2xyz(xy^2+yz^2+zx^2) = 2x^4y^2 + y^4z^2 + z^4x^2,$$

$$(x-y)^6 + (y-z)^6 + (z-x)^6 = 3(x-y)^2(y-z)^2(z-x)^2 + 2(x^2+y^2+z^2-xy-yz-zx)^3,$$

$$3(x^3+y^3+z^3) = (2x+y)(x-y)^2 + (2y+z)(y-z)^2 + (2z+x)(z-x)^2 + 3(x^2y+y^2z+z^2x),$$

$$x^4+y^4+z^4 = (x+y+z)^4 - 4(x+y+z)^2(xy+yz+zx) + 2(xy+yz+zx)^2 + 4xyz(x+y+z),$$

$$2(x^4+y^4+z^4) = (x+y)^2(x-y)^2 + (y+z)^2(y-z)^2 + (z+x)^2(z-x)^2 + 2(x^2y^2+y^2z^2+z^2x^2),$$

$$x^7+y^7+z^7 = (x+y+z)^7 - 7(x+y)(y+z)(z+x)[(x^2+y^2+z^2+xy+yz+zx)^2 + xyz(x+y+z)],$$

$$x(y^3+z^3) + y(z^3+x^3) + z(x^3+y^3) = (x+y+z)^2(xy+yz+zx) - 2(xy+yz+zx)^2 - xyz(x+y+z),$$

$$\frac{x^5(y-z)+y^5(z-x)+z^5(x-y)}{(x-y)(y-z)(z-x)} + x^3+y^3+z^3 + xy^2+yz^2+zx^2+x^2y+y^2z+z^2x+xyz = 0,$$

$$\frac{(x-y)^8+(y-z)^8+(z-x)^8}{(x-y)^2+(y-z)^2+(z-x)^2} = 4(x-y)^2(y-z)^2(z-x)^2 + (x^2+y^2+z^2-xy-yz-zx)^3,$$

$$\begin{aligned}&(x-y)(y-z)(z-x)(x+y+z)^2\\ &\qquad = (x-y)(x^2+y^2-z^2)z^2 + (y-z)(y^2+z^2-x^2)x^2 + (z-x)(z^2+x^2-y^2)y^2,\end{aligned}$$

$$\begin{aligned}&[x^3y+y^3z+z^3x - xyz(x+y+z)][xy^3+yz^3+zx^3 - xyz(x+y+z)]\\ &\qquad = [x^2y^2+y^2z^2+z^2x^2 - xyz(x+y+z)]^2 + xyz(x+y+z)(x^2+y^2+z^2-xy-yz-zx)^2,\end{aligned}$$

$$(x^4+y^4+z^4-x^2y^2-y^2z^2-z^2x^2)[x^2y^2+y^2z^2+z^2x^2-xyz(x+y+z)]$$
$$=[x^3y+y^3z+z^3x-xyz(x+y+z)]^2+[xy^3+yz^3+zx^3-xyz(x+y+z)]^2$$
$$-[x^3y+y^3z+z^3x-xyz(x+y+z)][xy^3+yz^3+zx^3-xyz(x+y+z)],$$

$$[(x^3-y^3-z^3)z]^3+[(2x^3+3x^2y+3xy^2+y^3)z+z^4]^3$$
$$=[x^4+2x^3y+3x^2y^2+2xy^3+y^4+(2x+y)z^3]^3+[(x-y)z^3-(x^4+2x^3y+3x^2y^2+2xy^3+y^4)]^3,$$

$$6(x^5+y^5+z^5)-5(x^3+y^3+z^3)(x^2+y^2+z^2)$$
$$=(x+y+z)^2(x^3-2x^2y-2xy^2+y^3-2x^2z+6xyz-2y^2z-2xz^2-2yz^2+z^3),$$

$$2(x^7+y^7+z^7)-7xyz(x^4+y^4+z^4)$$
$$=(x+y+z)[2(x^6+y^6+z^6)+2(x^4y^2+y^4z^2+z^4x^2)+2(x^2y^4+y^2z^4+x^4z^2)$$
$$+x^3y^2z+y^3z^2x+z^3x^2y+x^2y^3z+y^2z^3x+z^2x^3y-2x^2y^2z^2-2(x^3y^3+y^3z^3+z^3x^3)$$
$$-2(x^5y+y^5z+z^5x+xy^5+yz^5+zx^5)-3(x^4yz+y^4zx+z^4xy)],$$

$$x^{11}+y^{11}+z^{11}=(x+y+z)^{11}$$
$$-11(x+y)(y+z)(z+x)(\alpha^6\beta-2\alpha^4\beta^2-2\alpha^3\beta\gamma+\alpha^2\beta^3+\alpha^2\gamma^2+5\alpha\beta^2\gamma+\beta\gamma^2+\beta^4),$$

where $\alpha \triangleq x+y+z$, $\beta \triangleq x^2+y^2+z^2+xy+yz+zx$, and $\gamma \triangleq xyz$. Now, assume that x, y, z are distinct. Then,

$$\frac{x^2(y^2z^2+1)}{(x-y)(x-z)}+\frac{y^2(z^2x^2+1)}{(y-z)(y-x)}+\frac{z^2(x^2y^2+1)}{(z-x)(z-y)}=1.$$

Source: [108, pp. 30, 184, 185], [288, pp. 242, 402], [289, pp. 11, 32, 83, 167, 168], [663], [728, 729], [732, p. 161], [750], [806, pp. 137, 156, 164], [1226], [1860, pp. 14, 79, 152], [2000, 2569, 2927].

Fact 2.3.4. Let x, y, z be complex numbers, and define $u \triangleq x+y+z$, $v \triangleq (x+y+z)/2$, and $w \triangleq (x+y+z)/3$. Then,

$$(xu+yz)(yu+zx)(zu+xy)=(x+y)^2(y+z)^2(z+x)^2, \quad (v-x)^3+(v-y)^3+(v-z)^3+3xyz=v^3,$$
$$(w-x)^4+(w-y)^4+(w-z)^4=2(w-x)^2(w-y)^2+2(w-y)^2(w-z)^2+2(w-z)^2(w-x)^2.$$

Source: [732, p. 71].

Fact 2.3.5. Let x, y, z be complex numbers, let $n \ge 0$, define $s_n \triangleq x^n+y^n+z^n$, and define $e_1 \triangleq x+y+z$, $e_2 \triangleq xy+yz+zx$, and $e_3 = xyz$. Then,

$$s_2=e_1s_1-2e_2, \quad s_3=e_1s_2-e_2s_1+3e_3, \quad s_{n+3}=e_1s_{n+2}-e_2s_{n+1}+e_3s_n.$$

Source: [732, p. 71].

Fact 2.3.6. Let x, y, z be complex numbers, let $n \ge 0$, and define $s_n \triangleq (x-y)^n+(y-z)^n+(z-x)^n$. Then,

$$s_{n+3}=(x^2+y^2+z^2-xy-yz-zx)s_{n+1}+(x-y)(y-z)(z-x)s_n.$$

Now, define $u \triangleq (x-y)(y-z)+(y-z)(z-x)+(z-x)(x-y)$ and $v \triangleq (x-y)(y-z)(z-x)$. Then,

$$s_2=-2u, \quad s_3=3v, \quad s_4=2u^2, \quad s_5=-5uv, \quad s_6=3v^2-2u^3, \quad s_7=7u^2v, \quad 25s_3s_7=25s_5^2.$$

Source: [732, p. 71].

Fact 2.3.7. Let x, y, z be complex numbers, assume that $x^2 \ne yz$ and $y^2 \ne zx$, and define $u \triangleq (xy-z^2)/(yz-x^2)$ and $v \triangleq (xy-z^2)/(zx-y^2)$. Then,

$$z(u^2-v)=xu(1-uv), \quad z(v^2-u)=yv(1-uv).$$

Source: [732, p. 161].

Fact 2.3.8. Let x, y, z be distinct complex numbers. Then,

$$\frac{1}{(x-y)^2}+\frac{1}{(y-z)^2}+\frac{1}{(z-x)^2}=\left(\frac{1}{x-y}+\frac{1}{y-z}+\frac{1}{z-x}\right)^2.$$

Source: [289, pp. 23, 133].

Fact 2.3.9. Let x, y, z be complex numbers, and assume that $x+y$, $y+z$, and $z+x$ are nonzero. Then,

$$\frac{x-y}{x+y}+\frac{y-z}{y+z}+\frac{z-x}{z+x}=-\frac{(x-y)(y-z)(z-x)}{(x+y)(y+z)(z+x)},$$

$$\left(\frac{x-y}{x+y}\right)^3+\left(\frac{y-z}{y+z}\right)^3+\left(\frac{z-x}{z+x}\right)^3=\frac{24xyz(x-y)(y-z)(z-x)}{(x+y)^2(y+z)^2(z+x)^2}-\left[\frac{(x-y)(y-z)(z-x)}{(x+y)(y+z)(z+x)}\right]^3,$$

$$\left(\frac{x-y}{x+y}\right)^3+\left(\frac{y-z}{y+z}\right)^3+\left(\frac{z-x}{z+x}\right)^3$$
$$=\frac{(x-y)(y-z)(z-x)}{(x+y)(y+z)(z+x)}\left[3-2\left(\frac{xz-y^2}{(x+y)(y+z)}\right)^2-2\left(\frac{yx-z^2}{(y+z)(z+x)}\right)^2-2\left(\frac{zy-x^2}{(z+x)(x+y)}\right)^2\right].$$

Source: [380].

Fact 2.3.10. Let x, y, z be complex numbers, define $\alpha \triangleq xy+yz+zx$, and assume that $x^2+\alpha$, $y^2+\alpha$, and $z^2+\alpha$ are nonzero. Then,

$$\frac{\alpha^2-x^2yz}{x^2+\alpha}+\frac{\alpha^2-y^2zx}{y^2+\alpha}+\frac{\alpha^2-z^2xy}{z^2+\alpha}=2\alpha.$$

Source: [293].

Fact 2.3.11. Let x, y, z be nonnegative numbers. Then,

$$4(x^2y^2+y^2z^2+z^2x^2)=\left(\sqrt{x^2+y^2}+\sqrt{y^2+z^2}+\sqrt{z^2+x^2}\right)\left(-\sqrt{x^2+y^2}+\sqrt{y^2+z^2}+\sqrt{z^2+x^2}\right)$$
$$\cdot\left(\sqrt{x^2+y^2}-\sqrt{y^2+z^2}+\sqrt{z^2+x^2}\right)\left(\sqrt{x^2+y^2}+\sqrt{y^2+z^2}-\sqrt{z^2+x^2}\right).$$

Source: [174]. **Remark:** This equality is equivalent to de Gua's theorem, which states that, for a tetrahedron with three mutually perpendicular edges incident at P, the square of the area of the face opposite to P equals the sum of the squares of the areas of the remaining faces. See [48, p. 193].

Fact 2.3.12. Let x, y, z be real numbers. Then, the following statements are equivalent:

i) x, y, and z are nonnegative.

ii) $x+y+z$, $xy+yz+zx$, and xyz are nonnegative.

Source: [1158, p. 51].

Fact 2.3.13. Let x, y, z be real numbers. Then, $(x+y)z \le \frac{1}{2}(x^2+y^2)+z^2$. **Source:** [288, p. 230].

Fact 2.3.14. Let x, y, z be real numbers. Then,

$$\max\{xy, yz, zx\} \le \max\{\tfrac{1}{2}(x^2+y^2), \tfrac{1}{2}(y^2+z^2), \tfrac{1}{2}(z^2+x^2)\} \le \tfrac{1}{2}(x^2+y^2+z^2).$$

Fact 2.3.15. Let x, y, z be real numbers. Then,

$$\left.\begin{matrix} xy+yz+zx \le \left\{\begin{matrix} xy+yz+zx+\frac{3}{4}(x-y)^2 \\ \frac{1}{3}(x+y+z)^2 \end{matrix}\right\} \\ x\sqrt{y^2+z^2}+y\sqrt{x^2+z^2} \end{matrix}\right\} \le x^2+y^2+z^2,$$

$$xy + x + y \le x^2 + y^2 + 1, \quad (\tfrac{1}{2}x + \tfrac{1}{3}y + \tfrac{1}{6}z)^2 \le \tfrac{1}{2}x^2 + \tfrac{1}{3}y^2 + \tfrac{1}{6}z^2,$$

$$0 \le \frac{y^2 - x^2}{2x^2 + 1} + \frac{z^2 - y^2}{2y^2 + 1} + \frac{x^2 - z^2}{2z^2 + 1}.$$

If, in addition, x, y, z are distinct, then

$$2 \le \frac{x^2}{(y - z)^2} + \frac{y^2}{(z - x)^2} + \frac{z^2}{(x - y)^2}.$$

Source: [112, p. 35], [748, p. 5], [993, p. 162], [1158, p. 42], [1371, p. 129], and [1938, p. 8].

Fact 2.3.16. Let x, y, z be real numbers. Then,

$$2xyz \le x^2 + y^2z^2,$$
$$\sqrt{8}xyz \le x^4 + y^4 + z^2,$$
$$4(xy + yz) \le (x + y + z)^2,$$
$$4(xy + yz) \le (x + z)^2 + 4y^2,$$
$$x^3y + y^3z + z^3x \le x^4 + y^4 + z^4,$$
$$2(xy + yz - zx) \le x^2 + y^2 + z^2,$$
$$2(x + y + z) \le x^2 + y^2 + z^2 + 3,$$
$$3xyz(x + y + z) \le (xy + yz + zx)^2,$$
$$3(x^3y + y^3z + z^3x) \le (x^2 + y^2 + z^2)^2,$$
$$0 \le x(x + y)^3 + y(y + z)^3 + z(z + x)^3,$$
$$3(x^2y^2 + y^2z^2 + z^2x^2) \le (x^2 + y^2 + z^2)^2,$$
$$0 \le 3(x^4 + y^4 + z^4) + 4(x^3y + y^3z + z^3x),$$
$$2x(xy^2 + z + 1) \le x^4 + y^4 + z^4 + 2x^2 + 1,$$
$$xyz(xy^2 + yz^2 + zx^2) \le x^4y^2 + y^4z^2 + z^4x^2,$$
$$(xy + yz + zx - 1)^2 \le (x^2 + 1)(y^2 + 1)(z^2 + 1),$$
$$\tfrac{4}{7}(x^4 + y^4 + z^4) \le (x + y)^4 + (y + z)^4 + (z + x)^4,$$
$$32(x^2y + y^2z + z^2x) \le 17(x^3 + y^3 + z^3) + 45xyz,$$
$$xyz(x + y + z) \le x^2y^2 + y^2z^2 + z^2x^2 \le x^4 + y^4 + z^4,$$
$$2(x^3y + y^3z + z^3x) \le xy^3 + yz^3 + zx^3 + x^4 + y^4 + z^4,$$
$$x^2y^2z^2 + 1 \le 3(x^2 - x + 1)(y^2 - y + 1)(z^2 - z + 1) + xyz,$$
$$0 \le (x - y)(2x + y)^3 + (y - z)(2y + z)^3 + (z - x)(2z + x)^3,$$
$$0 \le (x - y)(x + 2z)^3 + (y - z)(y + 2x)^3 + (z - x)(z + 2y)^3,$$
$$(xy + yz + zx)^3 \le (x^2 + xy + y^2)(y^2 + yz + z^2)(z^2 + zx + x^2),$$
$$x^3y^3 + y^3z^3 + z^3x^3 \le 3(x^2 - xy + y^2)(y^2 - yz + z^2)(z^2 - zx + x^2),$$
$$3(xy + yz + zx)(x^2 + y^2 + z^2) \le 8(x^2y^2 + y^2z^2 + z^2x^2) + x^4 + y^4 + z^4,$$
$$xy(x^2 + y^2) + yz(y^2 + z^2) + zx(z^2 + x^2) \le xyz(x + y + z) + x^4 + y^4 + z^4,$$
$$xyz(x + y + z) + 2\sqrt{2}(x^3y + y^3z + z^3x) \le x^4 + y^4 + z^4 + xy^3 + yz^3 + zx^3,$$
$$x^2(y - z)^2 + y^2(z - x)^2 + z^2(x - y)^2 + 3(x^3y + y^3z + z^3x) \le 3(x^4 + y^4 + z^4,$$
$$3x^2y^2z^2 + x^3y^2z + y^3z^2x + z^3x^2y \le x^4y^2 + y^4z^2 + z^4x^2 + x^4yz + y^4zx + z^4xy,$$

$$3(x^2y+y^2z+z^2x)(xy^2+yz^2+zx^2) \le (x^2+xy+y^2)(y^2+yz+z^2)(z^2+zx+x^2),$$

$$[x^3(y+z)+y^3(z+x)+z^3(x+y)-2(x^2yz+y^2zx+z^2xy)]^2$$
$$\le 4(x^4+y^4+z^4-x^2y^2-y^2z^2-z^2x^2)(x^2y^2+y^2z^2+z^2x^2-x^2yz-y^2zx-z^2xy),$$

$$3(x^3y+y^3z+z^3x-x^2yz-y^2zx-z^2xy)^2$$
$$\le 4(x^4+y^4+z^4-x^2y^2-y^2z^2-z^2x^2)(x^2y^2+y^2z^2+z^2x^2-x^2yz-y^2zx-z^2xy).$$

If, in addition, x, y, z are not all zero, then

$$-1 \le \frac{xy+yz-zx}{x^2+y^2+z^2} \le \frac{1}{2}.$$

Source: [98, 99], [107, pp. 24, 145, 146, 484], [108, pp. 55, 318], [109, pp. 98, 99], [748, pp. 110–112], [750, 751], [752, pp. 40, 42, 43], [806, pp. 3, 6, 149, 161, 216], [1158, p. 240], [1371, p. 117], [1860, p. 95], [1938, p. 13], [2088], and [2380, p. 71]. The fifth inequality follows from $(x-2y+z)^2 \ge 0$. The sixth inequality follows from Fact 2.11.35. The 16th inequality follows from $(x^2y-y^2z)^2+(y^2z-z^2x)^2+(z^2x-x^2y)^2 \ge 0$. The 17th inequality follows from $(x+y+z-xyz)^2 \ge 0$.
Remark: The third to last inequality is due to Schur. See [751], Fact 2.3.28, Fact 2.3.67, and Fact 2.3.70.

Fact 2.3.17. Let x, y, z be real numbers, and let $\alpha \in [-2, 1]$. Then,

$$\alpha(xy+yz+zx) \le x^2+y^2+z^2.$$

Source: [1615].

Fact 2.3.18. Let x, y, z be real numbers, and assume that $z \le y \le x$. Then,

$$(x-y)^2+(y-z)^2 \le (x-z)^2.$$

Source: [806, p. 162].

Fact 2.3.19. Let p, q, r be real numbers. Then, the following statements are equivalent:

i) $p^2+pq+q^2 \le 3(r+1)$.

ii) For all $x, y, z \in \mathbb{R}$,

$$p(x^3y+y^3z+z^3x)+q(xy^3+yz^3+zx^3)$$
$$\le (p+q-r-1)xyz(x+y+z)+r(x^2y^2+y^2z^2+z^2x^2)+x^4+y^4+z^4.$$

In particular, for all $x, y, z \in \mathbb{R}$,

$$x^3y+y^3z+z^3x \le x^4+y^4+z^4+2xyz(x+y+z),$$
$$\tfrac{2}{3}(xy+yz+zx)^2 \le x^4+y^4+z^4+x^3y+y^3z+z^3x,$$
$$6(x^3y+y^3z+z^3x) \le 3(x^4+y^4+z^4)+(xy+yz+zx)^2,$$
$$\sqrt{3}(x^3y+y^3z+z^3x) \le x^4+y^4+z^4+(\sqrt{3}-1)xyz(x+y+z),$$
$$(1+\sqrt{3})xyz(x+y+z) \le x^4+y^4+z^4+\sqrt{3}(x^3y+y^3z+z^3x),$$
$$\sqrt{3}(x^3y+y^3z+z^3x-xy^3-yz^3-zx^3)+xyz(x+y+z) \le x^4+y^4+z^4,$$
$$\tfrac{2}{3}[x^2y^2+y^2z^2+z^2x^2-xyz(x+y+z)]+x^3y+y^3z+z^3x \le x^4+y^4+z^4,$$
$$3[xy(x^2-y^2+z^2)+yz(y^2-z^2+x^2)+zx(z^2-x^2+y^2)] \le (x^2+y^2+z^2)^2,$$
$$6[x^3y+y^3z+z^3x+xyz(x+y+z)] \le x^4+y^4+z^4+11(x^2y^2+y^2z^2+z^2x^2),$$
$$\sqrt{6}[xy(x-z)^2+yz(y-x)^2+zx(z-y)^2] \le (x^2-yz)^2+(y^2-zx)^2+(z^2-xy)^2,$$

$$(1+\sqrt{3})(x^3y+y^3z+z^3x-x^2y^2-y^2z^2-z^2x^2)+xy^3+yz^3+zx^3\le x^4+y^4+z^4,$$
$$(\sqrt{3}-1)(x^2y^2+y^2z^2+z^2x^2-x^3y-y^3z-z^3x)+xy^3+yz^3+zx^3\le x^4+y^4+z^4,$$
$$\tfrac{1}{2}(3+\sqrt{21})(x^3y+y^3z+z^3x-x^2y^2-y^2z^2-z^2x^2)+xyz(x+y+z)\le x^4+y^4+z^4,$$
$$\tfrac{1}{2}(\sqrt{21}-3)(x^2y^2+y^2z^2+z^2x^2-x^3y-y^3z-z^3x)+xyz(x+y+z)\le x^4+y^4+z^4,$$
$$\sqrt{6}(x^3y+y^3z+z^3x-xy^3-yz^3-zx^3)\le(x^2+y^2+z^2)(x^2+y^2+z^2-xy-yz-zx),$$
$$\begin{aligned}&[x^3y+y^3z+z^3x-xyz(x+y+z)][xy^3+yz^3+zx^3-xyz(x+y+z)]\\&\quad\le\tfrac{1}{4}([x^3y+y^3z+z^3x-xyz(x+y+z)]+[xy^3+yz^3+zx^3-xyz(x+y+z)])^2\\&\quad\le\tfrac{1}{2}([x^3y+y^3z+z^3x-xyz(x+y+z)]^2+[xy^3+yz^3+zx^3-xyz(x+y+z)]^2)\\&\quad\le(x^4+y^4+z^4-x^2y^2-y^2z^2-z^2x^2)[x^2y^2+y^2z^2+z^2x^2-xyz(x+y+z)].\end{aligned}$$

If, in addition, x, y, z are nonnegative, then

$$2\sqrt{2}(x^3y+y^3z+z^3x-xy^3-yz^3-zx^3)+xyz(x+y+z)\le x^4+y^4+z^4,$$
$$[x^2y^2+y^2z^2+z^2x^2-xyz(x+y+z)]^2\le[x^3y+y^3z+z^3x-xyz(x+y+z)][xy^3+yz^3+zx^3-xyz(x+y+z)].$$

Source: [98, 750]. The last statement follows from Fact 2.3.20.

Fact 2.3.20. Let p and r be real numbers. Then, the following statements are equivalent:

i) $p^2-1\le r$.

ii) For all $x, y, z \in \mathbb{R}$,

$$\begin{aligned}&p[xy(x^2+y^2)+yz(y^2+z^2)+zx(z^2+x^2)]\\&\qquad\le(2p-r-1)xyz(x+y+z)+r(x^2y^2+y^2z^2+z^2x^2)+x^4+y^4+z^4.\end{aligned}$$

Furthermore, the following statements are equivalent:

iii) $(p-1)\max\{2, p+1\}\le r$.

iv) For all nonnegative numbers x, y, z, the inequality in *ii)* holds.

Source: [750].

Fact 2.3.21. Let x, y, z, r be real numbers. Then,

$$3r(x^3y+y^3z+z^3x)\le 3r(1-r)xyz(x+y+z)+(3r^2-1)(x^2y^2+y^2z^2+z^2x^2)+x^4+y^4+z^4.$$

Source: [748, p. 67].

Fact 2.3.22. Let x, y, z, p be real numbers. Then,

$$(x-y)(x-py)(x-z)(x-pz)+(y-z)(y-pz)(y-x)(y-px)+(z-x)(z-px)(z-y)(z-py)\ge 0.$$

Source: [750, 751].

Fact 2.3.23. Let x, y, z be real numbers. Then,

$$\begin{aligned}(x^2+y^2+z^2)^3&=(x^3+y^3+z^3)^2+(x^2y+y^2z+z^2x)^2+(xy^2+yz^2+zx^2)^2\\&\qquad+(x^2y^2+y^2z^2+z^2x^2)[(x-y)^2+(y-z)^2+(z-x)^2].\end{aligned}$$

Therefore,

$$(x^3+y^3+z^3)^2\le(x^2+y^2+z^2)^3,\quad \sqrt[3]{|x^3+y^3+z^3|}\le\sqrt{x^2+y^2+z^2}.$$

Furthermore,

$$(x^3+y^3+z^3)^2\le(x^2+y^2+z^2)(x^4+y^4+z^4)\le(x^2+y^2+z^2)^3.$$

Source: [1860, pp. 14, 76]. The first inequality in the last string is given in [1158, p. 147].

Fact 2.3.24. Let $n \ge 1$, and, for all real numbers x, y, z define $f_n(x, y, z) \triangleq x(x+y)^n + y(y+z)^n + z(z+x)^n$. Then, the following statements are equivalent:

i) n is odd and, for all $x, y, z \in \mathbb{R}$, $f_n(x, y, z) \ge 0$.

ii) $n \in \{1, 3, 5\}$.

Furthermore, for all real numbers x, y, z, define $g_n(x, y, z) \triangleq x^n(x + y) + y^n(y + z) + z^n(z + x)$, and assume that n is odd. Then, for all $x, y, z \in \mathbb{R}$, $g_n(x, y, z) \ge 0$. **Source:** [2293].

Fact 2.3.25. Let x, y, z be real numbers. Then,

$$0 \le (x-y)\sqrt[3]{2x+y} + (y-z)\sqrt[3]{2y+z} + (z-x)\sqrt[3]{2z+x},$$

$$0 \le (x-y)\sqrt[3]{x+2z} + (y-z)\sqrt[3]{y+2x} + (z-x)\sqrt[3]{z+2y},$$

$$\frac{|x-z|}{\sqrt{1+x^2}\sqrt{1+z^2}} \le \frac{|x-y|}{\sqrt{1+x^2}\sqrt{1+y^2}} + \frac{|y-z|}{\sqrt{1+y^2}\sqrt{1+z^2}},$$

$$\frac{3\sqrt{2}}{2} \le \sqrt{x^2+(y-1)^2} + \sqrt{y^2+(z-1)^2} + \sqrt{z^2+(x-1)^2},$$

$$(x+y)(y+z)(z+x) \le 4xyz + \sqrt{2(x^2+y^2)(y^2+z^2)(z^2+x^2)},$$

$$(x+1)(y+1)(z+1) \le 2(xyz+1) + \sqrt{2(x^2+1)(y^2+1)(z^2+1)},$$

$$\frac{1}{2} \le \frac{(x+y-z)^2}{(x+y)^2+2z^2} + \frac{(y+z-x)^2}{(y+z)^2+2x^2} + \frac{(z+x-y)^2}{(z+x)^2+2y^2}.$$

Now, assume that x, y, z are distinct. Then,

$$2 \le \left(\frac{x}{y-z}\right)^2 + \left(\frac{y}{z-x}\right)^2 + \left(\frac{z}{x-y}\right)^2, \quad \frac{6(xy+yz+zx)}{x^2+y^2+z^2} \le \left(\frac{x}{y-z}\right)^2 + \left(\frac{y}{z-x}\right)^2 + \left(\frac{z}{x-y}\right)^2 + 1,$$

$$\frac{3}{2} \le \frac{x^2y^2+1}{(x-y)^2} + \frac{y^2z^2+1}{(y-z)^2} + \frac{z^2x^2+1}{(z-x)^2}.$$

Source: [748, pp. 12, 110], [752, pp. 28, 39, 40, 42], [806, p. 206], [993, p. 184], and [1158, p. 240].

Fact 2.3.26. Let $f\colon [0, \infty) \mapsto [0, \infty)$ represent either of the functions $f(x) = \log(x + 1)$, $f(x) = x/(x + 1)$, or $f(x) = x^\alpha$, where $\alpha \in (0, 1]$. Then, for all real numbers x, y, z,

$$f(|x+y|) + f(|y+z|) + f(|z+x|) \le f(|x|) + f(|y|) + f(|z|) + f(|x+y+z|).$$

Furthermore, for all real numbers x, y, z,

$$e^{-|x|} + e^{-|y|} + e^{-|z|} + e^{-|x+y+z|} \le 1 + e^{-|x+y|} + e^{-|y+z|} + e^{-|z+x|}.$$

Source: [2305]. **Remark:** Each function f is a *Bernstein function*. See [2391].

Fact 2.3.27. Let a, x, y, z be real numbers such that $a > 0$ and $z < y < x$. Then,

$$a^x(y-z) + a^y(z-x) + a^z(x-y) \ge 0.$$

Source: Use the fact that $f(x) = a^x$ is convex. See [1860, pp. 19, 111].

Fact 2.3.28. Let x, y, z be real numbers, let α be a real number, and assume that at least one of the following statements holds:

i) α is a positive even integer.

ii) x, y, z are nonnegative, and either $xyz > 0$ or $\alpha \ge 0$.

Then,

$$x^\alpha(x-y)(x-z)+y^\alpha(y-z)(y-x)+z^\alpha(z-x)(z-y)\geq 0.$$

Equality holds if and only if either $x=y=z$ or two of the numbers x,y,z are equal and the third number is zero. **Source:** Case *i*) is given in [140]. Case *ii*) is given in [806, p. 121], where α can be an arbitrary real number, whereas $\alpha>0$ is assumed in [140]. **Remark:** This is *Schur's inequality.* **Remark:** Setting $\alpha=0$ yields $xy+yz+zx\leq x^2+y^2+z^2$. See Fact 2.3.36. Setting $\alpha=1$ yields

$$4(x+y+z)(xy+yz+zx)\leq(x+y+z)^3+9xyz,$$
$$x(y^2+z^2)+y(z^2+x^2)+z(x^2+y^2)\leq 3xyz+x^3+y^3+z^3,$$
$$0\leq(x-y)^2(x+y-z)+(y-z)^2(y+z-x)+(z-x)^2(z+x-y).$$

Equivalently, in the case where x,y,z are nonnegative,

$$2(xy+yz+zx)\leq x^2+y^2+z^2+\frac{9xyz}{x+y+z},\qquad 2\leq\frac{x}{y+z}+\frac{y}{z+x}+\frac{z}{x+y}+\frac{4xyz}{(x+y)(y+z)(z+x)}.$$

See [752, p. 329] and Fact 2.3.60. **Remark:** If x,y,z are nonnegative, $xyz=0$, and $\alpha<0$, then, defining $1/0\triangleq\infty$, the inequality has the form $\infty\geq 0$. With this convention, the second condition in *ii*) can be deleted. **Related:** Fact 2.3.16, Fact 2.3.67, and Fact 2.3.70.

Fact 2.3.29. Let w,x,y,z be nonnegative numbers. Then,

$$\begin{aligned}&(w^6-3w^4+2w^3-3w^2+1)xyz(x+y+z)\\&\quad+w^2(3-w^4)(x^3y+y^3z+z^3x)+(3w^4-1)(xy^3+yz^3+zx^3)\leq 2w^3(x^4+y^4+z^4).\end{aligned}$$

Source: [98].

Fact 2.3.30. Let x,y,z,a,b,c be nonnegative numbers, and assume that either $a\leq b\leq c$ or $c\leq b\leq a$. Then,

$$a(x-y)(x-z)+b(y-z)(y-x)+c(z-x)(z-y)\geq 0.$$

Source: [806, p. 123]. **Remark:** This is a consequence of Schur's inequality. See Fact 2.3.28.

Fact 2.3.31. For all $r\in\mathbb{R}$ and $x,y,z\in(0,\infty)$, define

$$\alpha_r\triangleq x^r(x-y)(x-z)+y^r(y-z)(y-x)+z^r(z-x)(z-y).$$

Now, let $p,r\in\mathbb{R}$, assume that $pr\geq 0$, and let $x,y,z\in(0,\infty)$. Then, $\alpha_p\alpha_r\leq\alpha_0\alpha_{p+r}$. **Source:** [748, p. 68].

Fact 2.3.32. Let $x,y,z\in[0,\infty)$, let $r_0\approx 1.5583$ be the unique positive solution of $(1+s)^{1+s}=(3s)^s$, and let $r\in[0,r_0]$. Then,

$$\frac{1}{3}(x^ry+y^rz+z^rx)\leq\left(\frac{x+y+z}{3}\right)^{r+1}.$$

Source: [748, p. 113].

Fact 2.3.33. Let $x,y,z\in[0,\infty)$. If $r\in[-1,2]$, then

$$0\leq x^2(x-y)(x-ry)+y^2(y-z)(y-rz)+z^2(z-x)(z-rx).$$

If $r\in[-2,2]$, then

$$0\leq x(x-y)(x^2-ry^2)+y(y-z)(y^2-rz^2)+z(z-x)(z^2-rx^2).$$

Source: [748, p. 109].

Fact 2.3.34. Let $x,y,z\in(-1,\infty)$. Then,

$$2\leq\frac{x^2+1}{z^2+y+1}+\frac{y^2+1}{x^2+z+1}+\frac{z^2+1}{y^2+x+1}.$$

Source: [748, p. 10].

Fact 2.3.35. Let $x, y, z \in [0, 1]$. Then,

$$1 \le (1-x)(1-y)(1-z) + x + y + z, \quad x(1-y^2) + y(1-z^2) + z(1-x^2) \le \tfrac{5}{4}.$$

Source: [752, p. 39] and [112, p. 37]. **Remark:** The first inequality is equivalent to $xyz \le xy + yz + zx$.

Fact 2.3.36. Let x, y, z be nonnegative numbers. Then,

$$\begin{gathered}
3xyz \le x^2y + y^2x + z^3,\\
3xyz \le x^2y + y^2z + z^2x,\\
8xyz \le (xy+1)(yz+1)(zx+1),\\
6(x^2 + 2y^2 + 3z^2) \le (x + 2y + 3z)^2,\\
5xy + 3yz + 7zx \le 6x^2 + 4y^2 + 5z^2,\\
81xyz \le (x+1)(x+y)(y+z)(z+16),\\
5(x+y+z) \le 2(x^2+y^2+z^2) + xyz + 8,\\
9(xy+yz+zx) \le (x^2+2)(y^2+2)(z^2+2),\\
(x+1)(y+1)(z+1) \le 2xyz + x^2 + y^2 + z^2 + 3,\\
6(xy+yz+zx) \le (x+y+z)(xy+yz+zx+3),\\
5(x+y+z+1)^2 \le 16(x^2+1)(y^2+1)(z^2+1),\\
8(x^2y^2z^2 + xyz + 1)^2 \le 9(x^4+1)(y^4+1)(z^4+1),\\
xy + yz + zx + x + y + z \le x^2 + y^2 + z^2 + xyz + 2,\\
8(xy+yz+zx) \le x^3 + y^3 + z^3 + 4(x+y+z) + 9xyz,\\
(x+y+z)^3 \le (x^5 - x^2 + 3)(y^5 - y^2 + 3)(z^5 - z^2 + 3),\\
xy + yz + zx \le (\sqrt{xy} + \sqrt{yz} + \sqrt{zx})^2 \le 3(xy+yz+zx),\\
x^2y^2z^2 + xyz + 1 \le 3(x^2 - x + 1)(y^2 - y + 1)(z^2 - z + 1),\\
2(xy+yz+zx) \le x^2 + y^2 + z^2 + \sqrt{xyz}(\sqrt{x} + \sqrt{y} + \sqrt{z}),\\
(x+1)(y+1)(z+1)(xyz+1) \le 2(x^2+1)(y^2+1)(z^2+1),\\
2(xy+yz+zx) \le x^2 + y^2 + z^2 + 3(xyz)^{2/3} \le x^2 + y^2 + z^2 + 2xyz + 1,\\
3(x+y)^2(y+z)^2(z+x)^2(x^2+y^2+z^2) \le 8(x^2+y^2)(y^2+z^2)(z^2+x^2)(x+y+z)^2,\\
3xyz \le x(xy - y + 1)(yz - z + 1) + y(yz - z + 1)(zx - x + 1) + z(zx - x + 1)(xy - y + 1),\\
\frac{3}{xyz+1} \le \frac{1}{x(y+1)} + \frac{1}{y(z+1)} + \frac{1}{z(x+1)}, \quad \frac{2(xy+yz+zx+1)^2}{(x+y)(y+z)(z+x)} \le x + y + z + \frac{1}{xyz},
\end{gathered}$$

$$\begin{aligned}
x\sqrt{yz} + y\sqrt{zx} + z\sqrt{xy} &\le xy + yz + zx \le \tfrac{1}{3}(x+y+z)^2\\
&\le x^2 + y^2 + z^2 \le (x+y+z)^2 \le 6^3\left(\frac{x^2}{3^3} + \frac{y^2}{4^3} + \frac{z^2}{5^3}\right).
\end{aligned}$$

Source: [107, pp. 475, 477, 478], [748, pp. 372, 373, 381], [752, pp. 27, 28], [806, pp. 12, 39, 92, 101, 112, 113, 196, 206, 212, 213], [1371, pp. 117, 126], [1938, pp. 8, 32, 111], and [2380, p. 101]. **Remark:** $xy + yz + zx \le x^2 + y^2 + z^2$ follows from Fact 2.11.50 as well as from Fact 2.3.28 with $\alpha = 0$. **Remark:** $\frac{1}{3}(x+y+z)^2 \le x^2 + y^2 + z^2$ relates the mean to the quadratic mean. See Fact

2.11.89. **Remark:** Note that

$$1 \le \frac{x^2+y^2+z^2}{xy+yz+zx}, \quad 1 \le \frac{(x+y+z)^2}{x^2+y^2+z^2} \le 3, \quad 3 \le \frac{(x+y+z)^2}{xy+yz+zx},$$

which can be contrasted with the corresponding bounds for the sides of a triangle given by Fact 5.2.25.

Fact 2.3.37. Let x, y, z be nonnegative numbers, and let r be a real number. If $r \in [1,3]$, then

$$x^r y^{4-r} + y^r z^{4-r} + z^r x^{4-r} \le \tfrac{1}{3}(x^2+y^2+z^2)^2.$$

If $x+y+z \ge 3$ and $r = \frac{1}{2}(x+y+z-1)$, then

$$x^{1+r}y^r + y^{1+r}z^r + z^{1+r}x^r \le r^r(r+1)^{r+1}.$$

If $x+y+z \ge 3$ and $r = x+y+z-1$, then

$$x^r y + y^r z + z^r x \le r^r.$$

Source: [748, pp. 68, 69].

Fact 2.3.38. Let x, y, z be nonnegative numbers, and let $\alpha \in (0,3]$. Then,

$$2(xy+yz+zx) + \alpha \le x^2+y^2+z^2 + \alpha(xyz)^{2/\alpha} + 3.$$

In particular,

$$2(xy+yz+zx) \le x^2+y^2+z^2 + 3(xyz)^{2/3},$$
$$2(xy+yz+zx) \le x^2+y^2+z^2 + 2xyz + 1,$$
$$2(xy+yz+zx) \le x^2+y^2+z^2 + (xyz)^2 + 2.$$

If, in addition, $xyz \le 1$, then

$$2(xy+yz+zx) \le x^2+y^2+z^2 + 3.$$

Source: [107, p. 475] and [806, p. 124]. **Related:** Fact 2.2.76.

Fact 2.3.39. Let x, y, z be real numbers. Then,

$$\left|\sqrt{x^2+y^2} - \sqrt{y^2+z^2}\right| \le |x-z|.$$

Source: [1566, p. 13].

Fact 2.3.40. Let x, y, z be nonnegative numbers. Then,

$$\sqrt{x+z} \le \sqrt{x+y} + \sqrt{y+z},$$
$$\sqrt{x^2+yz} + \sqrt{y^2+zx} + \sqrt{z^2+xy} \le \frac{3}{2}(x+y+z),$$
$$\sqrt{4x^2+yz} + \sqrt{4y^2+zx} + \sqrt{4z^2+xy} \le \tfrac{5}{2}(x+y+z),$$
$$\sqrt{x^2+2yz} + \sqrt{y^2+2zx} + \sqrt{z^2+2xy} \le \sqrt{3}(x+y+z),$$
$$0 \le (x^2-yz)\sqrt{y+z} + (y^2-zx)\sqrt{z+x} + (z^2-xy)\sqrt{x+y},$$
$$2(xy+yz+zx) \le x\sqrt{x^2+3yz} + y\sqrt{y^2+3zx} + z\sqrt{z^2+3xy},$$
$$4\sqrt{2(x+yz)(y+zx)(z+xy)} \le (xy+yz+zx+1)(x+y+z+1),$$
$$\sqrt{2}(x+y+z) \le \sqrt{x^2+y^2} + \sqrt{y^2+z^2} + \sqrt{z^2+x^2} \le 2(x+y+z),$$

$$3\sqrt{xy+yz+zx} \le \sqrt{x^2+xy+y^2} + \sqrt{y^2+yz+z^2} + \sqrt{z^2+zx+x^2},$$

$$0 \le (x^2-yz)\sqrt{x^2+4yz} + (y^2-zx)\sqrt{y^2+4zx} + (z^2-xy)\sqrt{z^2+4xy},$$

$$\sqrt{3(x^3y^3+y^3z^3+z^3x^3)} \le xy(x+y-z) + yz(y+z-x) + zx(z+x-y),$$

$$xy\sqrt{2(x^2+y^2)} + yz\sqrt{2(y^2+z^2)} + zx\sqrt{2(z^2+x^2)} \le x^3+y^3+z^3+3xyz,$$

$$\sqrt{(x+y)(y+z)(z+x)} \le \sqrt{x+y+z}\Big[\sqrt{x(y+z)} + \sqrt{y(z+x)} + \sqrt{z(x+y)}\Big],$$

$$\sqrt{x+y} + \sqrt{y+z} + \sqrt{z+x} \le \sqrt{6(x+y+z)} \le \sqrt{x^2+1} + \sqrt{y^2+1} + \sqrt{z^2+1},$$

$$\sqrt[3]{x^3+3(x+z)y^2+2xyz} + \sqrt[3]{y^3+3(y+x)z^2+2yzx} + \sqrt[3]{z^3+3(z+y)x^2+2zxy} \le \sqrt[3]{9}(x+y+z),$$

$$\begin{aligned} x^2+y^2+z^2 \le{}& \sqrt{(x^2-xy+y^2)(y^2-yz+z^2)} + \sqrt{(y^2-yz+z^2)(z^2-zx+x^2)} \\ &+ \sqrt{(z^2-zx+x^2)(x^2-xy+y^2)}, \end{aligned}$$

$$\begin{aligned} x+y+z+\sqrt{3(xy+yz+zx)} &\le \sqrt{(x+y)(y+z)} + \sqrt{(y+z)(z+x)} + \sqrt{(z+x)(x+y)} \\ &\le \tfrac{1}{3}[5(x+y+z) + \sqrt{3(xy+yz+zx)}]. \end{aligned}$$

If $r \in [0, \sqrt{2}]$, then

$$(1+r)\sqrt{x^3y+y^3z+z^3x} \le \sqrt{x^4+y^4+z^4} + r\sqrt{x^2y^2+y^2z^2+z^2x^2}.$$

If $r \ge 2$, then

$$\sqrt{x^2+rxy+y^2} + \sqrt{y^2+ryz+z^2} + \sqrt{z^2+rzx+x^2} \le \sqrt{4(x^2+y^2+z^2)+(3r+2)(xy+yz+zx)}.$$

If $a, b, c > 0$, then

$$\sqrt{3(xy+yz+zx)} \le \frac{a}{b+c}(x+y) + \frac{b}{c+a}(y+z) + \frac{c}{a+b}(z+x).$$

Source: [47, p. 3], [377], [748, pp. 6, 109, 375], [752, pp. 43, 243], [806, pp. 45, 89, 104, 201], and [1956].

Fact 2.3.41. Let x, y, z be positive numbers. Then,

$$3\sqrt{2} \le \left\{ \begin{array}{c} \sqrt{\dfrac{16(x+y+z)^3}{3(x+y)(y+z)(z+x)}} \\ \sqrt{\dfrac{6(x+y+z)}{\sqrt[3]{xyz}}} \end{array} \right\} \le \sqrt{\frac{x+y}{z}} + \sqrt{\frac{y+z}{x}} + \sqrt{\frac{z+x}{y}},$$

$$\frac{3\sqrt{3}}{2\sqrt{x+y+z}} \le \frac{\sqrt{x}}{y+z} + \frac{\sqrt{y}}{z+x} + \frac{\sqrt{z}}{x+y}, \quad 2 < \sqrt{\frac{x}{y+z}} + \sqrt{\frac{y}{z+x}} + \sqrt{\frac{z}{x+y}},$$

$$\sqrt{\frac{2x}{x+y}} + \sqrt{\frac{2y}{y+z}} + \sqrt{\frac{2z}{z+x}} \le 3, \quad \sqrt{\frac{x}{4x+5y}} + \sqrt{\frac{y}{4y+5z}} + \sqrt{\frac{z}{4z+5x}} \le 1,$$

$$3 \le \sqrt{\frac{x^2+yz}{(z+x)y}} + \sqrt{\frac{y^2+zx}{(x+y)z}} + \sqrt{\frac{z^2+xy}{(y+z)x}}, \quad 0 \le \frac{x^2-yz}{\sqrt{x^2+yz}} + \frac{y^2-zx}{\sqrt{y^2+zx}} + \frac{z^2-xy}{\sqrt{z^2+xy}},$$

$$2 \le \sqrt{\frac{x(y+z)}{x^2+yz}}+\sqrt{\frac{y(z+x)}{y^2+zx}}+\sqrt{\frac{z(x+y)}{z^2+xy}}, \quad 1 \le \sqrt{\frac{x^3}{x^3+(y+z)^3}}+\sqrt{\frac{y^3}{y^3+(z+x)^3}}+\sqrt{\frac{z^3}{z^3+(x+y)^3}},$$

$$\frac{3\sqrt{2}}{2} \le \frac{\sqrt{x^2+yz}}{y+z}+\frac{\sqrt{y^2+zx}}{z+x}+\frac{\sqrt{z^2+xy}}{x+y}, \quad 15 \le \sqrt{1+\frac{48x}{y+z}}+\sqrt{1+\frac{48y}{z+x}}+\sqrt{1+\frac{48z}{x+y}},$$

$$\frac{3}{\sqrt{2xyz}} \le \frac{1}{x\sqrt{x+y}}+\frac{1}{y\sqrt{y+z}}+\frac{1}{z\sqrt{z+x}}, \quad x+y+z \le \frac{x\sqrt{x^2+3yz}}{y+z}+\frac{y\sqrt{y^2+3zx}}{z+x}+\frac{z\sqrt{z^2+3xy}}{x+y},$$

$$x+y+z \le \frac{\sqrt{2}x^2}{\sqrt{x^2+y^2}}+\frac{\sqrt{2}y^2}{\sqrt{y^2+z^2}}+\frac{\sqrt{2}z^2}{\sqrt{z^2+x^2}}, \quad \frac{x}{\sqrt{x^2+2yz}}+\frac{y}{\sqrt{y^2+2zx}}+\frac{z}{\sqrt{z^2+2xy}} \le \frac{x+y+z}{\sqrt{xy+yz+zx}},$$

$$3 \le \left(\frac{2x}{y+z}\right)^{3/5}+\left(\frac{2y}{z+x}\right)^{3/5}+\left(\frac{2z}{x+y}\right)^{3/5}, \quad \frac{x}{2x^2+yz}+\frac{y}{2y^2+zx}+\frac{z}{2z^2+xy} \le \sqrt{\frac{x^{-1}+y^{-1}+z^{-1}}{x+y+z}},$$

$$2+\frac{1}{\sqrt[3]{2}} \le \sqrt[3]{\frac{x^2+yz}{y^2+z^2}}+\sqrt[3]{\frac{y^2+zx}{z^2+x^2}}+\sqrt[3]{\frac{z^2+xy}{x^2+y^2}}, \quad 3\sqrt{\frac{xy+yz+zx}{x^2+y^2+z^2}} \le \sqrt{\frac{2x}{x+y}}+\sqrt{\frac{2y}{y+z}}+\sqrt{\frac{2z}{z+x}},$$

$$\frac{9(x^3+y^3+z^3)}{2(x+y+z)^4} \le \frac{\sqrt{xy}}{xy+z^2}+\frac{\sqrt{yz}}{yz+x^2}+\frac{\sqrt{zx}}{zx+y^2}, \quad \sqrt{3xyz(x+y+z)} \le \frac{xy(x+z)}{y+z}+\frac{yz(y+x)}{z+x}+\frac{zx(z+y)}{x+y},$$

$$\sqrt{3(x^2+y^2+z^2)} \le \frac{x^2+y^2}{x+y}+\frac{y^2+z^2}{y+z}+\frac{z^2+x^2}{z+x}, \quad \frac{2}{\sqrt[3]{(x+y)(y+z)(z+x)}} \le \frac{1}{2x+y}+\frac{1}{2y+z}+\frac{1}{2z+x},$$

$$2\sqrt{3} \le \sqrt{x^2+y^2+z^2}+\sqrt{\frac{1}{x^2}+\frac{1}{y^2}+\frac{1}{z^2}}, \quad \frac{2\sqrt{2}}{\sqrt{xy+yz+zx}} \le \frac{1}{\sqrt{x^2+yz}}+\frac{1}{\sqrt{y^2+zx}}+\frac{1}{\sqrt{z^2+xy}},$$

$$\sqrt{3xyz(x+y+z)} \le \frac{xy(x+z)}{y+z}+\frac{yz(y+x)}{z+x}+\frac{zx(z+y)}{x+y},$$

$$2(\sqrt{xy}+\sqrt{yz}+\sqrt{zx}) \le \frac{x^2}{y}+\frac{y^2}{z}+\frac{z^2}{x}+\frac{y^2}{x}+\frac{z^2}{y}+\frac{x^2}{z},$$

$$\sqrt{\frac{x^3+y^3+z^3}{x^2+y^2+z^2}} \le \sqrt{\frac{x^3}{x^2+8y^2}}+\sqrt{\frac{y^3}{y^2+8z^2}}+\sqrt{\frac{z^3}{z^2+8x^2}},$$

$$0 \le \frac{x^2-yz}{\sqrt{2x^2+y^2+z^2}}+\frac{y^2-zx}{\sqrt{2y^2+z^2+x^2}}+\frac{z^2-xy}{\sqrt{2z^2+x^2+y^2}},$$

$$0 \le \frac{x^2-yz}{\sqrt{8x^2+(y+z)^2}}+\frac{y^2-zx}{\sqrt{8y^2+(z+x)^2}}+\frac{z^2-xy}{\sqrt{8z^2+(x+y)^2}},$$

$$2 \le \sqrt{\frac{x(x+y+z)}{(x+y)(x+z)}}+\sqrt{\frac{y(x+y+z)}{(y+z)(y+x)}}+\sqrt{\frac{z(x+y+z)}{(z+x)(z+y)}},$$

$$\sqrt{\frac{x^2(y^2+z^2)}{x^2+yz}}+\sqrt{\frac{y^2(z^2+x^2)}{y^2+zx}}+\sqrt{\frac{z^2(x^2+y^2)}{z^2+xy}} \le x+y+z,$$

$$\sqrt{10-\frac{3\sqrt[3]{xyz}}{x+y+z}} \le \sqrt[4]{\frac{x^2+yz}{x(y+z)}}+\sqrt[4]{\frac{y^2+zx}{y(z+x)}}+\sqrt[4]{\frac{z^2+xy}{z(x+y)}},$$

$$2\sqrt{(x^2+y^2+z^2)\left(\frac{x}{y}+\frac{y}{z}+\frac{z}{x}\right)} \le x+y+z+\frac{x^2}{y}+\frac{y^2}{z}+\frac{z^2}{x},$$

$$\sqrt{\frac{x^2}{y^2+(z+x)^2}}+\sqrt{\frac{y^2}{z^2+(x+y)^2}}+\sqrt{\frac{z^2}{x^2+(y+z)^2}} \le \frac{3\sqrt{5}}{5},$$

$$\frac{9}{2} \le \frac{\sqrt{4x(y+z)+yz}}{y+z}+\frac{\sqrt{4y(z+x)+zx}}{z+x}+\frac{\sqrt{4z(x+y)+xy}}{x+y},$$

$$1 \le \sqrt{\frac{x^3}{x^3+7xyz+y^3}}+\sqrt{\frac{y^3}{y^3+7xyz+z^3}}+\sqrt{\frac{z^3}{z^3+7xyz+x^3}},$$

$$\sqrt{3xyz(x^3+y^3+z^3)} \le \frac{xy(x^3+y^3)}{x^2+y^2}+\frac{yz(y^3+z^3)}{y^2+z^2}+\frac{zx(z^3+x^3)}{z^2+x^2},$$

$$x\sqrt{\frac{y}{z+x}}+y\sqrt{\frac{z}{x+y}}+z\sqrt{\frac{x}{y+z}} \le \frac{3\sqrt{3}}{4}\sqrt{\frac{(x+y)(y+z)(z+x)}{x+y+z}},$$

$$2 \le \sqrt{\frac{2x(y+z)}{(2y+z)(y+2z)}}+\sqrt{\frac{2y(z+x)}{(2z+x)(z+2x)}}+\sqrt{\frac{2z(x+y)}{(2x+y)(x+2y)}},$$

$$\frac{3}{2}\left[\sqrt{\frac{(x+y+z)(x^2+y^2+z^2)}{xyz}}-1\right] \le \frac{x^2+y^2}{(x+y)z}+\frac{y^2+z^2}{(y+z)x}+\frac{z^2+x^2}{(z+x)y},$$

$$\frac{9}{4} \le \frac{xy}{(x+y)^2}+\frac{yz}{(y+z)^2}+\frac{zx}{(z+x)^2}+\sqrt{\frac{x}{y+z}}+\sqrt{\frac{y}{z+x}}+\sqrt{\frac{z}{x+y}},$$

$$\frac{x}{x+\sqrt{(x+y)(x+z)}}+\frac{y}{y+\sqrt{(y+z)(y+x)}}+\frac{z}{z+\sqrt{(z+x)(z+y)}} \le 1,$$

$$\frac{2}{1+\sqrt{2}} \le \frac{x}{x+\sqrt{2(y^2+z^2)}}+\frac{y}{y+\sqrt{2(z^2+x^2)}}+\frac{z}{z+\sqrt{2(x^2+y^2)}} \le 1,$$

$$\frac{6(xy+yz+zx)}{(x+y+z)\sqrt{(x+y)(y+z)(z+x)}} \le \frac{\sqrt{x^3+y^3}}{x^2+y^2}+\frac{\sqrt{y^3+z^3}}{y^2+z^2}+\frac{\sqrt{z^3+x^3}}{z^2+x^2},$$

$$1+\sqrt{1+\sqrt{(x^2+y^2+z^2)\left(\frac{1}{x^2}+\frac{1}{y^2}+\frac{1}{z^2}\right)}} \le \sqrt{(x+y+z)\left(\frac{1}{y}+\frac{1}{z}+\frac{1}{x}\right)},$$

$$4 \le \frac{2(xy+yz+zx)}{x^2+y^2+z^2}+\sqrt{\frac{x^2-xy+y^2}{xy+z^2}}+\sqrt{\frac{y^2-yz+z^2}{yz+x^2}}+\sqrt{\frac{z^2-zx+x^2}{zx+y^2}},$$

$$2\left(\sqrt{\frac{x}{y+z}}+\sqrt{\frac{y}{z+x}}+\sqrt{\frac{z}{x+y}}\right) \le \sqrt{\frac{y+z}{2x}}+\sqrt{\frac{z+x}{2y}}+\sqrt{\frac{x+y}{2z}}+3\sqrt{2}-3,$$

$$\frac{x}{\sqrt{3(x^2+xy+yz)}}+\frac{y}{\sqrt{3(y^2+yz+zx)}}+\frac{z}{\sqrt{3(z^2+zx+xy)}} \le \frac{x}{x+2y}+\frac{y}{y+2z}+\frac{z}{z+2x}.$$

If $r \in \mathbb{R}$, then

$$\sqrt{\frac{8(x+y+z)(x^r y^r+y^r z^r+z^r x^r)}{(x+y)(y+z)(z+x)}} \le \frac{x^r+y^r}{x+y}+\frac{y^r+z^r}{y+z}+\frac{z^r+x^r}{z+x}.$$

If $n \geq 2$, then

$$\sqrt{\frac{3(x^{n-1}+y^{n-1}+z^{n-1})(x^n+y^n+z^n)}{x+y+z}} \leq \frac{x^n+y^n}{x+y}+\frac{y^n+z^n}{y+z}+\frac{z^n+x^n}{z+x}.$$

Source: [107, pp. 45, 58, 59, 157, 251, 252, 310, 314, 489, 490, 484–486, 586, 590, 591], [108, pp. 27, 171, 172, 356], [748, pp. 6, 8, 12, 151, 343, 345, 372, 375, 376], [752, pp. 31–36, 41, 42, 43, 242–245, 310–312, 318], [806, pp. 13, 91, 104, 185, 193, 202, 207], [960], [1771], [1956], [2294, pp. 271, 272], and [2527, p. 187].

Fact 2.3.42. Let x, y, z be positive numbers, and let $p \leq 1/2$. Then,

$$\left(\frac{x}{x+y}\right)^p+\left(\frac{y}{y+z}\right)^p+\left(\frac{z}{z+x}\right)^p \leq \frac{3}{2^p}.$$

In particular,

$$\sqrt{\frac{x}{x+y}}+\sqrt{\frac{y}{y+z}}+\sqrt{\frac{z}{z+x}} \leq \frac{3\sqrt{2}}{2}.$$

Source: [2850].

Fact 2.3.43. Let x, y, z be positive numbers. Then,

$$\sqrt[4]{\frac{x^2}{x^2-xy+y^2}}+\sqrt[4]{\frac{y^2}{y^2-yz+z^2}}+\sqrt[4]{\frac{z^2}{z^2-zx+x^2}} \leq \sqrt{\frac{2x}{x+y}}+\sqrt{\frac{2y}{y+z}}+\sqrt{\frac{2z}{z+x}} \leq 3.$$

Source: [162].

Fact 2.3.44. Let x, y, z be positive numbers, and let $p \geq (\log 3)/(\log 2) - 1 \approx 0.585$. Then,

$$\frac{3}{2^p} \leq \left(\frac{x}{y+z}\right)^p+\left(\frac{y}{z+x}\right)^p+\left(\frac{z}{y+z}\right)^p.$$

Source: [748, p. 151].

Fact 2.3.45. Let x, y, z be positive numbers, and let $p \geq 1$. Then,

$$\frac{9}{2\cdot 3^p} \leq \frac{x^p}{y+z}+\frac{y^p}{z+x}+\frac{z^p}{y+z}.$$

Source: [2527, p. 131].

Fact 2.3.46. Let x, y, and z be positive numbers, and let $\alpha \geq 8$. Then,

$$\frac{3}{\sqrt{1+\alpha}} \leq \frac{x}{\sqrt{x^2+\alpha yz}}+\frac{y}{\sqrt{y^2+\alpha zx}}+\frac{z}{\sqrt{z^2+\alpha xy}}.$$

In particular,

$$1 \leq \frac{x}{\sqrt{x^2+8yz}}+\frac{y}{\sqrt{y^2+8zx}}+\frac{z}{\sqrt{z^2+8xy}}.$$

Source: [108, pp. 442, 446]. **Related:** This result is a special case of Fact 2.11.80.

Fact 2.3.47. Let $x, y, z \in (0, 1)$. Then,

$$\sqrt{xyz}+\sqrt{(1-x)(1-y)(1-z)} < 1.$$

Source: [1158, p. 239] and [1938, p. 42].

Fact 2.3.48. Let $x, y, z \in [0, 1]$. Then,

$$\frac{3}{2} \leq (1-x+xy)^2+(1-y+yz)^2+(1-z+zx)^2.$$

Source: [752, p. 241].

Fact 2.3.49. Let x, y, z be nonnegative numbers. Then,

$$\sqrt[3]{xyz} \le \tfrac{1}{3}(\sqrt{xy} + \sqrt{yz} + \sqrt{zx}) \le \left\{ \begin{array}{c} \sqrt{\tfrac{1}{3}(xy + yz + zx)} \\ \tfrac{1}{6}(x + y + z) + \tfrac{1}{2}\sqrt[3]{xyz} \end{array} \right\} \le \tfrac{1}{3}(x + y + z).$$

Source: The first inequality is given by Fact 2.11.97; the second upper inequality is given in [2527, p. 179]; the second lower inequality is given in [1507, 2129, 2925]. **Remark:** See [50].

Fact 2.3.50. Let x, y, z be nonnegative numbers, let α, β, γ be real numbers, and assume that $\alpha + \beta + \gamma = \pi$. Then,

$$2(\cos\alpha)\sqrt{xy} + 2(\cos\beta)\sqrt{yz} + 2(\cos\gamma)\sqrt{zx} \le x + y + z.$$

Source: [1893].

Fact 2.3.51. Let x, y, z be nonnegative numbers. Then,

$$\sqrt[3]{xyz} \le \tfrac{1}{3}(x + y + z) \le \sqrt[3]{xyz} + \max\{(\sqrt{x} - \sqrt{y})^2, (\sqrt{y} - \sqrt{z})^2, (\sqrt{z} - \sqrt{x})^2\}.$$

Source: [107, p. 478] and [1158, p. 146].

Fact 2.3.52. Let x, y be nonnegative numbers, and let z be a positive number. Then,

$$x + y \le z^y x + z^{-x} y.$$

Source: [1371, p. 163].

Fact 2.3.53. Let x, y, z be nonnegative numbers. Then,

$$\min\{x^y y^z z^x, x^z y^x z^y\} \le \left\{ \begin{array}{c} x^y y^z z^x \\ x^z y^x z^y \\ (xyz)^{(x+y+z)/3} \end{array} \right\} \le \left(\frac{x + y + z}{3}\right)^{x+y+z} \le x^x y^y z^z.$$

Source: [644, p. 107], [806, p. 115], and [2294, p. 267].

Fact 2.3.54. Let x, y, z be positive numbers. Then,

$$\frac{3}{4} \le \frac{x^{y+z}}{(y + z)^2} + \frac{y^{z+x}}{(z + x)^2} + \frac{z^{z+y}}{(x + y)^2}.$$

Source: [752, p. 245].

Fact 2.3.55. Let x, y, z be positive numbers, assume that $x \le y \le z$, and let $\alpha \in [0, e)$. Then,

$$x^{\alpha y} + y^{\alpha z} + z^{\alpha x} \le x^{\alpha x} + y^{\alpha y} + z^{\alpha z}.$$

Source: [783]. **Related:** Fact 2.2.73.

Fact 2.3.56. Let $x, y, z \in (0, 1)$, and assume that $x + y + z = 1$. Then,

$$\frac{3}{2} \le \frac{x}{1 - x} + \frac{y}{1 - y} + \frac{z}{1 - z}.$$

Source: Use Nesbitt's inequality. See Fact 2.3.60 and [2088].

Fact 2.3.57. Let $x, y, z \in (-1, 1)$. Then,

$$\frac{1}{1 - xy} + \frac{1}{1 - yz} + \frac{1}{1 - zx} \le \frac{1}{1 - x^2} + \frac{1}{1 - y^2} + \frac{1}{1 - z^2},$$

$$\frac{1}{1 - x^2yz} + \frac{1}{1 - y^2zx} + \frac{1}{1 - z^2xy} \le \frac{1}{1 - x^4} + \frac{1}{1 - y^4} + \frac{1}{1 - z^4},$$

$$\frac{32}{1 - x^2y} + \frac{32}{1 - y^2z} + \frac{32}{1 - z^2x} \le \frac{17}{1 - x^3} + \frac{17}{1 - y^3} + \frac{17}{1 - z^3} + \frac{45}{1 - xyz},$$

$$\frac{x^2yz}{(1-x^2yz)^2}+\frac{y^2zx}{(1-y^2zx)^2}+\frac{z^2xy}{(1-z^2xy)^2}\le\left(\frac{x^2}{1-x^4}\right)^2+\left(\frac{y^2}{1-y^4}\right)^2+\left(\frac{z^2}{1-z^4}\right)^2.$$

Source: [2088].

Fact 2.3.58. Let x, y, z be distinct positive numbers. Then,

$$\sqrt{3}\le\left|\frac{x+y}{x-y}+\frac{y+z}{y-z}+\frac{z+x}{z-x}\right|,\quad \frac{4}{xy+yz+zx}\le\frac{1}{(x-y)^2}+\frac{1}{(y-z)^2}+\frac{1}{(z-x)^2}.$$

Source: [1938, p. 113].

Fact 2.3.59. Let n be a positive integer, let k be a nonnegative integer, and let x, y, z be positive numbers. Then,

$$x^k+y^k+z^k\le\frac{x^{n+k}}{y^n}+\frac{y^{n+k}}{z^n}+\frac{z^{n+k}}{x^n}\le\frac{x^{n+1+k}}{y^{n+1}}+\frac{y^{n+1+k}}{z^{n+1}}+\frac{z^{n+1+k}}{x^{n+1}}.$$

In particular,

$$3\le\frac{x}{y}+\frac{y}{z}+\frac{z}{x}\le\frac{x^2}{y^2}+\frac{y^2}{z^2}+\frac{z^2}{x^2}\le\frac{x^3}{y^3}+\frac{y^3}{z^3}+\frac{z^3}{x^3},$$

$$x+y+z\le\frac{x^2}{y}+\frac{y^2}{z}+\frac{z^2}{x}\le\frac{x^3}{y^2}+\frac{y^3}{z^2}+\frac{z^3}{x^2}\le\frac{x^4}{y^3}+\frac{y^4}{z^3}+\frac{z^4}{x^3},$$

$$x^2+y^2+z^2\le\frac{x^3}{y}+\frac{y^3}{z}+\frac{z^3}{x}\le\frac{x^4}{y^2}+\frac{y^4}{z^2}+\frac{z^4}{x^2}\le\frac{x^5}{y^3}+\frac{y^5}{z^3}+\frac{z^5}{x^3}.$$

Source: [806, pp. 131, 184, 193, 194], Fact 2.11.67, and Fact 2.11.70.

Fact 2.3.60. Let x, y, z be nonnegative numbers. Then,

$$\begin{aligned}
&6(x+y-z)(y+z-x)(z+x-y)\\
&\quad\le 2[x^2(z+y-x)+y^2(z+x-y)+z^2(x+y-z)]\\
&\quad= 2[x(y^2+z^2)+y(z^2+x^2)+z(x^2+y^2)-(x^3+y^3+z^3)]\\
&\quad\le 2(x+y+z)(xy+yz+zx)-\tfrac{4}{9}(x+y+z)^3-\tfrac{4}{9}[(x+y+z)^2-3(xy+yz+zx)]^{3/2}\\
&\quad\le 6xyz\\
&\quad\le\left\{\begin{array}{c}2(x+y+z)(xy+yz+zx)-\tfrac{4}{9}(x+y+z)^3\\ \qquad+\tfrac{4}{9}[(x+y+z)^2-3(xy+yz+zx)]^{3/2}\le\tfrac{2\sqrt{3}}{3}(xy+yz+zx)^{3/2}\\ \dfrac{3}{2}xyz\left(1+\dfrac{x+y+z}{\sqrt[3]{xyz}}\right)\end{array}\right\}\\
&\quad\le\tfrac{2}{3}(x+y+z)(xy+yz+zx)\\
&\quad\le\left\{\begin{array}{c}\tfrac{3}{4}(x+y)(y+z)(z+x)\le\left\{\begin{array}{c}x(y^2+z^2)+y(z^2+x^2)+z(x^2+y^2)\\ \tfrac{2}{9}(x+y+z)^3\end{array}\right\}\\ 3xyz+\tfrac{1}{3}(x+y+z)(x^2+y^2+z^2)\end{array}\right\}\\
&\quad\le\left\{\begin{array}{c}\tfrac{2}{5}(x^3+y^3+z^3)+\tfrac{4}{5}[x(y^2+z^2)+y(z^2+x^2)+z(x^2+y^2)]\le\tfrac{2}{3}(x+y+z)(x^2+y^2+z^2)\\ 3xyz+x^3+y^3+z^3\end{array}\right\}\\
&\quad\le 2xyz+\tfrac{4}{3}(x^3+y^3+z^3)\le\tfrac{3}{2}(xyz+x^3+y^3+z^3)\le 2(x^3+y^3+z^3),
\end{aligned}$$

$$6xyz\le\left\{\begin{array}{c}2(xy^2+yz^2+zx^2)\\ \tfrac{3}{8}(x+1)(y+1)(x+z)(y+z)\\ x(y^2+z^2)+y(z^2+x^2)+z(x^2+y^2)\end{array}\right\}\le x^2y^2+y^2z^2+z^2x^2+x^2+y^2+z^2,$$

$$6xyz \le \left\{\begin{matrix} 2(xy^2+yz^2+zx^2) \\ x(y^2+z^2)+y(z^2+x^2)+z(x^2+y^2) \\ 6xyz+2\sqrt{9+6\sqrt{3}}|(x-y)(y-z)(z-x)| \\ 6xyz+\frac{1}{2}\max\{(x+y-2z)^3,(y+z-2x)^3,(z+x-2y)^3\} \end{matrix}\right\} \le 2(x^3+y^3+z^3),$$

$$6xyz \le \tfrac{3}{4}(x+y)(y+z)(z+x) \le \tfrac{4}{9}(x+y+z)^3 - [x(y^2+z^2)+y(z^2+x^2)+z(x^2+y^2)]$$
$$\le 3xyz + x^3+y^3+z^3,$$

$$6xyz \le \tfrac{3}{2}(x^2y^2z^2+xy+yz+zx),$$
$$6xyz+3(x^2+y^2)z \le \tfrac{4}{9}(x+y+z)^3,$$
$$6xyz+12xy \le 6x^2+y^2(z+2)(2z+3),$$
$$108xyz \le (x+2y)(x+5y)(3x+2y+z),$$
$$27(xyz+xy^2+yz^2+zx^2) \le 4(x+y+z)^3,$$
$$100xyz \le (x+y+z)[(x+y)^2+(x+y+4z)^2],$$
$$(x+y)^3+(y+z)^3+(z+x)^3 \le 8(x^3+y^3+z^3),$$
$$4(x+y+z)(xy+yz+zx) \le (x+y+z)^3+9xyz,$$
$$7(x+y+z)(xy+yz+zx) \le 2(x+y+z)^3+9xyz,$$
$$x^2y+y^2z+z^2x \le \tfrac{\sqrt[3]{4}}{3}(x^3+y^3+z^3)+(3-\sqrt[3]{4})xyz,$$
$$9xyz \le 3(xy^2+yz^2+zx^2) \le 2(x^3+y^3+z^3)+2(x^2y+y^2z+z^2x),$$
$$x^3+y^3+z^3 \le (x^2+y^2+z^2)^{3/2}+3(1-\sqrt{3})xyz \le (x^2+y^2+z^2)^{3/2},$$
$$3xyz+15(xy^2+yz^2+zx^2) \le 12(x^2y+y^2z+z^2x)+4(x^3+y^3+z^3),$$
$$6(x+y+z)(x^2y+y^2z+z^2x) \le x^4+y^4+z^4+17(x^2y^2+y^2z^2+z^2x^2),$$
$$6xyz \le \tfrac{2}{3}(x+y+z)(xy+yz+zx) \le (x^2+x+1)(y^2+y+1)(z^2+z+1),$$
$$6xyz \le 2(xy^2+yz^2+zx^2) \le \tfrac{2}{5}(x^3+y^3+z^3)+\tfrac{4}{5}[x(y^2+z^2)+y(z^2+x^2)+z(x^2+y^2)],$$
$$\left(\sqrt{4\sqrt{2}+13/4}+\tfrac{1}{2}\right)(xy^2+yz^2+zx^2) \le \left(\sqrt{4\sqrt{2}+13/4}-\tfrac{1}{2}\right)(x^2y+y^2z+z^2x)+x^3+y^3+z^3,$$
$$3\log 2 \le \log\frac{(x+y)(y+z)(z+x)}{xyz} \le (x+y+z)\left(\frac{1}{x}+\frac{1}{y}+\frac{1}{z}\right)+\log 8-9,$$
$$\frac{9}{x+y+z} \le \frac{2}{x+y}+\frac{2}{y+z}+\frac{2}{z+x} \le \frac{1}{x}+\frac{1}{y}+\frac{1}{z}$$
$$\le \frac{3}{2(x+y+z)}\left(\frac{x}{y}+\frac{y}{z}+\frac{z}{x}+\frac{y}{x}+\frac{z}{y}+\frac{x}{z}\right)$$
$$\le \frac{3}{x+y+z}\max\left\{\frac{x}{y}+\frac{y}{z}+\frac{z}{x},\frac{y}{x}+\frac{z}{y}+\frac{x}{z}\right\},$$

$$3 \le \frac{x+y}{x+z}+\frac{y+z}{y+x}+\frac{z+x}{z+y} \le \frac{x}{y}+\frac{y}{z}+\frac{z}{x}, \quad 1 \le \frac{x}{x+2y}+\frac{y}{y+2z}+\frac{z}{z+2x} < 2,$$
$$\frac{1}{2} \le \frac{x}{2x+y+z}+\frac{y}{2y+z+x}+\frac{z}{2z+x+y} \le \frac{3}{4}, \quad \frac{1}{2} \le \frac{x}{x+2y+3z}+\frac{y}{y+2z+3x}+\frac{z}{z+2x+3y},$$
$$\frac{x}{4x+4y+z}+\frac{y}{4y+4z+x}+\frac{z}{4z+4x+y} \le \frac{1}{3},$$

$$3 \le \frac{2x+y}{2x+z} + \frac{2y+z}{2y+x} + \frac{2z+x}{2z+y}, \quad 2 \le \frac{4xyz}{(x+y)(y+z)(z+x)} + \frac{x}{y+z} + \frac{y}{z+x} + \frac{z}{x+y},$$

$$\frac{1}{2(x^2+y^2+z^2)} + \frac{1}{xy+yz+zx} \le \frac{1}{x+y+z}\left(\frac{1}{x+y} + \frac{1}{y+z} + \frac{1}{z+x}\right),$$

$$\frac{x^2+yz}{(x+y)(x+z)} + \frac{y^2+zx}{(y+z)(y+x)} + \frac{z^2+xy}{(z+x)(z+y)} \le \frac{x}{y+z} + \frac{y}{z+x} + \frac{z}{x+y},$$

$$\begin{aligned} &9(x+y+z)(xy+yz+zx) + 3(xy+yz+zx) \\ &\qquad \le 27xyz + 2(x+y+z)^3 + 2(x+y+z)^2 + [(x+y+z)^2 - 3(xy+yz+zx)]^2. \end{aligned}$$

If $x, y, z \in [\frac{1}{2}, 2]$, then

$$5\left(\frac{y}{x} + \frac{z}{y} + \frac{x}{z}\right) + 9 \le 8\left(\frac{x}{y} + \frac{y}{z} + \frac{z}{x}\right).$$

Source: Fact 2.3.61, [107, pp. 10, 24, 92, 93, 148, 149], [108, pp. 45, 264, 265, 446], [176, 287], [748, pp. 5, 7, 111], [752, pp. 29, 30], [806, pp. 138, 181, 188, 198, 204, 214], [993, pp. 166, 169, 179, 182], [1158, pp. 36, 45], [1371, pp. 117, 120, 152], [1757, pp. 247, 257], [1772], [1938, pp. 11, 34], [1962, 2002, 2381], and [2527, p. 204]. $3xyz \le xy^2+yz^2+zx^2$ is equivalent to $3 \le \frac{x}{y}+\frac{y}{z}+\frac{z}{x}$, which follows from Fact 2.3.66 and Fact 2.11.50. $x(y^2+z^2)+y(z^2+x^2)+z(x^2+y^2) \le 3xyz+x^3+y^3+z^3$ is given in [1938, p. 49]. This inequality follows from Schur's inequality given by Fact 2.3.28. See [1569]. $\frac{2}{9}(x+y+z)^3 \le \frac{3}{2}(xyz+x^3+y^3+z^3)$ is a slight improvement of the inequality $\frac{1}{7}(x+y+z)^3 \le xyz + x^3 + y^3 + z^3$ given in [1938, p. 48]. $6xyz \le \frac{4}{9}(x+y+z)^3 - 3(x^2+y^2)z$ follows from Fact 2.11.104. $\frac{2}{3}(x+y+z)(x^2+y^2+z^2) \le 2xyz + \frac{4}{3}(x^3+y^3+z^3)$ follows from Suranyi's inequality given by Fact 2.11.22. **Remark:** $x^3+y^3+z^3-3xyz = \frac{1}{2}(x+y+z)[(x-y)^2+(y-z)^2+(z-x)]^2$ yields $6xyz \le 2(x^3+y^3+z^3)$. **Remark:** For $x,y,z > 0$, $9xyz \le (x+y+z)(xy+yz+zx)$ is given by Fact 2.11.43. **Remark:** For $x,y,z > 0$, $3xyz \le xy^2+yz^2+zx^2$ is given by Fact 2.11.48. **Remark:** $x(y^2+z^2)+y(z^2+x^2)+z(x^2+y^2) \le 3xyz+x^3+y^3+z^3$ is a special case of Schur's inequality. See Fact 2.3.28. **Remark:** $x(y^2+z^2)+y(z^2+x^2)+z(x^2+y^2) \le 2(x^3+y^3+z^3)$ can be written as

$$\frac{9}{x+y+z} \le \frac{2}{x+y} + \frac{2}{y+z} + \frac{2}{z+x}.$$

Remark: For $x,y,z > 0$, $x(y^2+z^2)+y(z^2+x^2)+z(x^2+y^2) \le 2(x^3+y^3+z^3)$ is equivalent to

$$\frac{3}{2} \le \frac{x}{y+z} + \frac{y}{z+x} + \frac{z}{x+y},$$

which is Nesbitt's inequality. See [1757, p. 267]. Nesbitt's inequality is interpolated by

$$\begin{aligned} \frac{3}{2} &\le \frac{(x+y+z)^2}{2(xy+yz+zx)} + \frac{1}{2}\left(\frac{\sqrt[3]{xyz}}{\frac{1}{3}(x+y+z)} - 1\right) \\ &\le \left\{ \begin{matrix} \dfrac{(x+y+z)^2}{2(xy+yz+zx)} \\ \dfrac{x}{y+z} + \dfrac{y}{z+x} + \dfrac{z}{x+y} + \dfrac{1}{2}\left(\dfrac{\sqrt[3]{xyz}}{\frac{1}{3}(x+y+z)} - 1\right) \end{matrix} \right\} \le \frac{x}{y+z} + \frac{y}{z+x} + \frac{z}{x+y}, \end{aligned}$$

$$\begin{aligned} \frac{3}{2} &\le \min\left\{\frac{x}{y+z} + \frac{2(y+z)}{2x+y+z}, \frac{y}{z+x} + \frac{2(z+x)}{2y+z+x}, \frac{z}{x+y} + \frac{2(x+y)}{2z+x+y}\right\} \\ &\le \max\left\{\frac{x}{y+z} + \frac{2(y+z)}{2x+y+z}, \frac{y}{z+x} + \frac{2(z+x)}{2y+z+x}, \frac{z}{x+y} + \frac{2(x+y)}{2z+x+y}\right\} \end{aligned}$$

$$\leq \frac{x}{y+z} + \frac{y}{z+x} + \frac{z}{x+y}.$$

See [397, 1770]. An upper bound for $\frac{x}{y+z} + \frac{y}{z+x} + \frac{z}{x+y}$ is given by Fact 2.3.61. In the case where x, y, z are the sides of a triangle, an upper bound for $\frac{x}{y+z} + \frac{y}{z+x} + \frac{z}{x+y}$ is given by Fact 5.2.25. A generalization is given by Fact 2.11.53. **Remark:**

$$2[x(y^2+z^2) + y(z^2+x^2) + z(x^2+y^2) - (x^3+y^3+z^3)] \leq 6xyz$$

is equivalent to

$$4(x+y+z)(xy+yz+zx) \leq (x+y+z)^3 + 9xyz,$$

which in turn is equivalent to

$$x(x-y)(x-z) + y(y-z)(y-x) + z(z-x)(z-y) \geq 0,$$

which is Schur's inequality. See Fact 2.3.28. An equivalent form is given by $2(xy+yz+zx) - (x^2+y^2+z^2) \leq 9xyz(x+y+z)$. **Remark:** $3xyz + 3(xy^2+yz^2+zx^2) \leq 2(x^3+y^3+z^3) + 2(x^2y+y^2z+z^2x)$ can be written as

$$1 \leq \frac{x}{y+2z} + \frac{y}{z+2x} + \frac{z}{x+2y}.$$

See [806, p. 40] and [1938, p. 104]. **Remark:** $6xyz \leq x(y^2+z^2) + y(z^2+x^2) + z(x^2+y^2)$ follows from Fact 2.11.78. **Remark:** The left-hand inequality in the penultimate string is $6xyz \leq x(y^2+z^2)+y(z^2+x^2)+z(x^2+y^2)$. In the case where x, y, z represent the sides of a triangle, this string is interpolated by Fact 5.2.25. **Remark:** The bound $\frac{4}{27}(x+y+z)^3 \leq x^3+y^3+z^3+xyz$ is sharp and thus tighter than the bound given in [806, p. 21], where 4/27 is replaced by 1/7.

Fact 2.3.61. Let x, y, z be positive numbers. Then,

$$\frac{3}{2} \leq \frac{x}{y+z} + \frac{y}{z+x} + \frac{z}{x+y} \leq \frac{1}{4}\left(\frac{x+y}{z} + \frac{y+z}{x} + \frac{z+x}{y}\right) \leq \frac{x+y}{z} + \frac{y+z}{x} + \frac{z+x}{y} - \frac{9}{2}.$$

Source: The second inequality is given in [1938, p. 32]. This inequality is equivalent to

$$6x^2y^2z^2 + 2xyz(x^3+y^3+z^3) \leq 2(x^3y^3+y^3z^3+z^3x^3) + x^4(y^2+z^2) + y^4(z^2+x^2) + z^4(x^2+y^2),$$

which can be written as

$$2(x^3y^3+y^3z^3+z^3x^3) - 6x^2y^2z^2 + x^4(y^2+z^2-2yz) + y^4(z^2+x^2-2zx) + z^4(x^2+y^2-2xy) \geq 0.$$

Remark: The left-most inequality is Nesbitt's inequality. See Fact 2.3.60. **Remark:** The third inequality follows from $\frac{3}{2} \leq \frac{1}{4}\left(\frac{x+y}{z} + \frac{y+z}{x} + \frac{z+x}{y}\right)$. **Remark:** The third term interpolates

$$\frac{3}{2} \leq \frac{x}{y+z} + \frac{y}{z+x} + \frac{z}{x+y} \leq \frac{x+y}{z} + \frac{y+z}{x} + \frac{z+x}{y} - \frac{9}{2},$$

which is given in [209, pp. 33, 34]. This string is equivalent to

$$6 \leq \frac{9}{2} + \frac{x}{y+z} + \frac{y}{z+x} + \frac{z}{x+y} \leq \frac{x+y}{z} + \frac{y+z}{x} + \frac{z+x}{y},$$

which in turn is equivalent to

$$6xyz \leq \frac{9}{2}xyz + \frac{x^2yz}{y+z} + \frac{xy^2z}{z+x} + \frac{xyz^2}{x+y} \leq x(y^2+z^2) + y(z^2+x^2) + z(x^2+y^2).$$

The left-hand inequality and the second inequality yield

$$6xyz \leq 4\left(\frac{x^2yz}{y+z} + \frac{xy^2z}{z+x} + \frac{xyz^2}{x+y}\right) \leq x(y^2+z^2) + y(z^2+x^2) + z(x^2+y^2),$$

which interpolates $6xyz \le x(y^2+z^2)+y(z^2+x^2)+z(x^2+y^2)$ in Fact 2.3.60. This interpolating term is not comparable with the three interpolating terms in Fact 2.3.60.

Fact 2.3.62. Let x, y, z be positive numbers. Then,

$$6 \le \left\{ \begin{array}{c} \dfrac{9}{2} + \dfrac{x}{y+z} + \dfrac{y}{z+x} + \dfrac{z}{x+y} \\ 6 + 11\left(1 - \sqrt{\dfrac{xy+yz+zx}{x^2+y^2+z^2}}\right) \end{array} \right\} \le \frac{x+y}{z} + \frac{y+z}{x} + \frac{z+x}{y}.$$

Hence,

$$17 \le \frac{x+y}{z} + \frac{y+z}{x} + \frac{z+x}{y} + 11\sqrt{\frac{xy+yz+zx}{x^2+y^2+z^2}}.$$

Source: [107, pp. 483, 484].

Fact 2.3.63. Let x, y, z, p be positive numbers. Then,

$$\frac{3}{1+p} \le \frac{x}{py+z} + \frac{y}{pz+x} + \frac{z}{px+y}.$$

Source: [2850]. **Remark:** Setting $p = 1$ yields Nesbitt's inequality. See Fact 2.3.60.

Fact 2.3.64. Let x, y, z be positive numbers, and let $r \ge 1$. Then,

$$\tfrac{3}{2}[\tfrac{1}{3}(x+y+z)]^{r-1} \le \frac{x^r}{y+z} + \frac{y^r}{z+x} + \frac{z^r}{x+y}.$$

Source: [2850]. **Remark:** Setting $r = 1$ yields Nesbitt's inequality. See Fact 2.3.60.

Fact 2.3.65. Let x, y, z be positive numbers, let p and q be nonnegative numbers, and assume that $p + q > 0$. Then,

$$\frac{3}{p+q} \le \frac{(x+y+z)^2}{(p+q)(xy+yz+zx)} \le \frac{x}{py+qz} + \frac{y}{pz+qx} + \frac{z}{px+qy}.$$

Source: [2743]. **Remark:** Setting $p = q = 1$ yields Nesbitt's inequality. See Fact 2.3.60.

Fact 2.3.66. Let x, y, z be nonnegative numbers. Then,

$$\left. \begin{array}{r} xyz(x+y+z) \\ 2xyz|x+y-z| \\ 2xyz|x-y+z| \\ 2xyz|-x+y+z| \end{array} \right\} \le \left\{ \begin{array}{c} x^2y^2+y^2z^2+z^2x^2 \\ 3xyz(x+y+z) \end{array} \right\} \le (xy+yz+zx)^2 \le 3(x^2y^2+y^2z^2+z^2x^2)$$

$$\le \tfrac{3}{2}(x^3y+y^3z+z^3x+xy^3+yz^3+zx^3) \le (x^2+y^2+z^2)^2$$

$$\le (x+y+z)(x^3+y^3+z^3) \le \left\{ \begin{array}{c} 3(x^4+y^4+z^4) \\ (x+y+z)^4 \end{array} \right\} \le 27(x^4+y^4+z^4),$$

$$x^2y^2+y^2z^2+z^2x^2 \le \tfrac{1}{2}[x^4+y^4+z^4+xyz(x+y+z)] \le x^4+y^4+z^4 \le (x^2+y^2+z^2)^2,$$

$$\left. \begin{array}{r} 2xyz|x+y-z| \\ 2xyz|x-y+z| \\ 2xyz|-x+y+z| \end{array} \right\} \le 3(x^3y+y^3z+z^3x) \le (x^2+y^2+z^2)^2,$$

$$xyz(x+y+z) \le \left\{ \begin{array}{c} x^3y+y^3z+z^3x \\ xy^3+yz^3+zx^3 \end{array} \right\} \le x^4+y^4+z^4,$$

$$xyz(x+y+z) \le \tfrac{1}{4}(x+y)^2(x+z)^2, \quad x[x^3+(y+z)^3)] \le (x^2+y^2+z^2)^2,$$

$$\tfrac{4}{27}(x+y+z)^4 \le x^4+y^4+z^4+3(x^2y^2+y^2z^2+z^2x^2),$$
$$6(x^2y^2+y^2z^2+z^2x^2) \le x^4+y^4+z^4+5(x^3y+y^3z+z^3x),$$
$$|(x^3y+y^3z+z^3x-(xy^3+yz^3+zx^3)| \le \tfrac{9\sqrt{2}}{32}(x^2+y^2+z^2)^2$$
$$4(x+y+z)^2(xy+yz+zx) \le (x+y+z)^4+3(xy+yz+zx)^2,$$
$$0 \le (x-y)(3x+2y)^3+(y-z)(3y+2z)^3+(z-x)(3z+2x)^3,$$
$$\tfrac{4\sqrt[4]{3}}{3}(x^3y+y^3z+z^3x) \le (\tfrac{4\sqrt[4]{3}}{3}-1)xyz(x+y+z)+x^4+y^4+z^4,$$
$$4(xy+yz+zx)^2 \le (x+y+z)^2(xy+yz+zx)+3xyz(x+y+z),$$
$$x^2(y+z)^2+y^2(z+x)^2+z^2(x+y)^2 \le \tfrac{4}{3}(x+y+z)(x^3+y^3+z^3),$$
$$x^2y^2+y^2z^2+z^2x^2+2|x^3y+y^3z+z^3x-xy^3-yz^3-zx^3| \le x^4+y^4+z^4,$$
$$\sqrt{6}[xy(z-x)^2+yz(x-y)^2+zx(y-z)^2] \le (x^2-yz)^2+(y^2-zx)^2+(z^2-xy)^2,$$
$$5(x+y+z)^2(xy+yz+zx) \le (x+y+z)^4+4(xy+yz+zx)^2+6xyz(x+y+z),$$
$$\frac{3(x+y+z)}{3+x+y+z} \le \frac{x}{y+1}+\frac{y}{z+1}+\frac{z}{x+1}.$$

If $x, y, z \in [1/\sqrt{4+3\sqrt{2}}, \sqrt{4+3\sqrt{2}}]$, then

$$(x+y+z)^4 \le 9(x^2+y^2+z^2)(xy+yz+zx).$$

Now, assume that x, y, z are positive numbers. Then,

$$3 \le \frac{x^2+1}{y+z}+\frac{y^2+1}{z+x}+\frac{z^2+1}{x+y}, \quad 1 \le \frac{x}{x+2y}+\frac{y}{y+2z}+\frac{z}{z+2x},$$
$$0 \le \frac{z^2-y^2}{x+y}+\frac{x^2-z^2}{y+z}+\frac{y^2-x^2}{z+x}, \quad 0 \le \frac{x^2-yz}{x+y}+\frac{y^2-zx}{y+z}+\frac{z^2-xy}{z+x},$$
$$\frac{1}{x+y}+\frac{1}{y+z}+\frac{1}{z+x} \le \frac{1}{2x}+\frac{1}{2y}+\frac{1}{2z}, \quad 3 \le \frac{x+y}{z+x}+\frac{z+x}{y+z}+\frac{y+z}{x+y} \le \frac{x}{y}+\frac{y}{z}+\frac{z}{x},$$
$$\frac{3}{2}(x+y+z) \le \frac{x^2+2yz}{y+z}+\frac{y^2+2zx}{z+x}+\frac{z^2+2xy}{x+y}, \quad \frac{4xy}{2x+y+z}+\frac{4yz}{2y+z+x}+\frac{4zx}{2z+x+y} \le x+y+z,$$
$$0 \le \frac{x^2-yz}{3x+y+z}+\frac{y^2-zx}{3y+z+x}+\frac{x^2-xy}{3z+x+y},$$
$$\frac{9xy}{3x+4y+2z}+\frac{9yz}{3y+4z+2x}+\frac{9zx}{3z+4x+2y} \le x+y+z,$$
$$\frac{2xy}{x+y}+\frac{2yz}{y+z}+\frac{2zx}{z+x} \le x+y+z \le \begin{cases} \dfrac{x^2+y^2}{x+y}+\dfrac{y^2+z^2}{y+z}+\dfrac{z^2+x^2}{x+y} \\[2ex] \dfrac{x^2+y^2}{2z}+\dfrac{y^2+z^2}{2x}+\dfrac{z^2+x^2}{2y}, \end{cases}$$
$$\frac{4xy}{x+y+2z}+\frac{4yz}{y+z+2x}+\frac{4zx}{z+x+2y} \le x+y+z \le \frac{2x^2}{y+z}+\frac{2y^2}{z+x}+\frac{2z^2}{x+y}.$$

Source: [108, pp. 11, 100], [287], [748, pp. 111, 112], [752, p. 29], [806, pp. 11, 40, 41, 64, 65, 138, 186, 252, 324], [993, pp. 170, 180], [1158, pp. 172, 173], [1371, pp. 106, 122, 147, 149], [1569], [1757, pp. 247, 257], [1938, pp. 12, 40, 41, 49, 112], Fact 2.11.15, and Fact

2.11.98. **Remark:** $3xyz(x+y+z) \le \frac{3}{2}(x^3y+y^3z+z^3x+xy^3+yz^3+zx^3)$ is given in [1371, p. 147]. **Remark:** $2xyz(x+y-z) \le x^2y^2+y^2z^2+z^2x^2$ follows from $(xy-yz-zx)^2$, and thus is valid for all real x,y,z. See [993, p. 194]. **Remark:** $3xyz(x+y+z) \le (xy+yz+zx)^2$ follows from Newton's inequality. See Fact 2.11.35. **Remark:** $2xyz(z+x-y) \le x^2y^2+y^2z^2+z^2x^2$ is equivalent to $\sqrt{x^2+xz+z^2} \le \sqrt{x^2-xy+y^2}+\sqrt{y^2-yz+z^2}$. See [47, p. 17], [993, p. 184], [1158, p. 36], and [1938, p. 52].

Fact 2.3.67. Let xyz be nonnegative numbers, let α and β be real numbers, and assume that, if $\beta \le 0$, then $\alpha \ge 2\beta$, whereas, if $\beta \ge 0$, then $\alpha \ge \beta^2+2\beta$. Then,

$$(1+\beta)[xy(x^2+y^2)+yz(y^2+z^2)+zx(z^2+x^2)] \le (1+2\beta-\alpha)xyz(x+y+z) \le \alpha(x^2y^2+y^2z^2+z^2x^2)+x^4+y^4+z^4.$$

Source: [751]. **Remark:** Setting $\alpha = \beta = 0$ yields Schur's inequality, which is given by Fact 2.3.16, Fact 2.3.28, and Fact 2.3.70, and which holds for real x,y,z.

Fact 2.3.68. Let x,y,z be nonnegative numbers. Then,

$$x^4(y+z)+y^4(z+x)+z^4(x+y) \le \tfrac{1}{12}(x+y+z)^5,$$
$$xyz(xy+yz+zx) \le xy^4+yz^4+zx^4 \le x^5+y^5+z^5,$$
$$xyz(x+y+z)^2 \le \tfrac{3}{8}[xy(x+y)^3+yz(y+z)^3+zx(z+x)^3],$$
$$(x^2y^2+y^2z^2+z^2x^2)(x+y+z) \le (x^2+y^2+z^2)(x^3+y^3+z^3),$$
$$2xyz(x+y+z)^2+3xyz(xy+yz+zx) \le (x+y+z)(xy+yz+zx)^2,$$
$$3xyz(x^2+y^2+z^2) \le (x^2+y^2+z^2)(x^3+y^3+z^3) \le 3(x^5+y^5+z^5),$$
$$33xyz(x^2+y^2+z^2) \le 2(x^2+y^2+z^2)(x^3+y^3+z^3)+9xyz(x+y+z)^2,$$
$$x^2y^3+y^2z^3+z^2x^3+x^3y^2+y^3z^2+z^3x^2 \le xy^4+yz^4+zx^4+x^4y+y^4z+z^4x,$$
$$6xyz[(y+z)(z+x)+(z+x)(x+y)+(x+y)(y+z)] \le (x+y)(y+z)(z+x)(x+y+z)^2,$$
$$x^2(z+x)(x+y)(y+z-x)+y^2(x+y)(y+z)(z+x-y)+z^2(y+z)(z+x)(x+y-z)$$
$$\le \tfrac{1}{2}(x+y)(y+z)(z+x)(xy+yz+zx).$$

Now, assume that x,y,z are positive. Then,

$$\frac{x+y}{y+z}+\frac{y+z}{x+y}+1 \le \frac{x}{y}+\frac{y}{z}+\frac{z}{x}, \quad \frac{(x-z)^2}{xy+yz+zx}+3 \le \frac{x}{y}+\frac{y}{z}+\frac{z}{x},$$

$$\frac{xy+yz+zx}{x^2+y^2+z^2} \le \frac{xyz}{x^3+y^3+z^3}+\frac{2}{3}, \quad 2 \le \frac{x^2+y^2+z^2}{xy+yz+zx}+\frac{8xyz}{(x+y)(y+z)(z+x)},$$

$$4 \le \frac{x+y}{y+z}+\frac{y+z}{z+x}+\frac{z+x}{x+y}+\frac{3(xy+yz+zx)}{(x+y+z)^2}, \quad x^2+y^2+z^2 \le \frac{x^3+xyz}{y+z}+\frac{y^3+xyz}{z+x}+\frac{z^3+xyz}{x+y},$$

$$\frac{13}{6} \le \frac{x}{y+z}+\frac{y}{z+x}+\frac{z}{x+y}+\frac{2(xy+yz+zx)}{3(x^2+y^2+z^2)}, \quad \frac{2}{3}(x^2+y^2+z^2) \le \frac{x^3+3y^3}{5x+y}+\frac{y^3+3z^3}{5y+z}+\frac{z^3+3x^3}{5z+x},$$

$$\frac{6(x^2+y^2+z^2)}{x+y+z} \le x+y+z+\frac{x^2}{y}+\frac{y^2}{z}+\frac{z^2}{x}, \quad \frac{1}{2}(x+y+z)^2 \le \frac{x^3+2xyz}{y+z}+\frac{y^3+2xyz}{z+x}+\frac{z^3+2xyz}{x+y},$$

$$\frac{6(x-z)^2}{(x+y+z)^2}+3 \le \frac{x}{y}+\frac{y}{z}+\frac{z}{x}, \quad \frac{27(x-z)^2}{2(x+y+z)^2}+9 \le (x+y+z)\left(\frac{1}{x}+\frac{1}{y}+\frac{1}{z}\right),$$

$$\frac{7(x-z)^2}{16(xy+yz+zx)}+\frac{3}{2} \le \frac{x}{y+z}+\frac{y}{z+x}+\frac{z}{x+y}, \quad \frac{17}{2} \le \frac{x}{y}+\frac{y}{z}+\frac{z}{x}+\frac{7(xy+yz+zx)}{x^2+y^2+z^2},$$

$$2(xy+yz+zx) \le \frac{x^3+3xyz}{y+z} + \frac{y^3+3xyz}{z+x} + \frac{z^3+3xyz}{x+y},$$

$$\frac{5}{x+y+z} \le \frac{1}{x+y} + \frac{1}{y+z} + \frac{1}{z+x} + \frac{3xyz}{2(xy+yz+zx)^2},$$

$$\frac{x+y}{x+y+2z} + \frac{y+z}{y+z+2x} + \frac{z+x}{z+x+2y} + \frac{2(xy+yz+zx)}{3(x^2+y^2+z^2)} \le \frac{13}{6},$$

$$\frac{x^3-y^3}{x+y} + \frac{y^3-z^3}{y+z} + \frac{z^3-x^3}{z+x} \le \frac{1}{8}[(x-y)^2+(y-z)^2+(z-x)^2],$$

$$\frac{x^2+y^2+z^2-xy-yz-zx}{x+y+z} \le \frac{2x^2-xy-y^2}{x+y} + \frac{2y^2-yz-z^2}{y+z} + \frac{2z^2-zx-x^2}{z+x}.$$

If x, y, z are positive and distinct, then

$$\frac{4}{xy+yz+zx} \le \frac{1}{(x-y)^2} + \frac{1}{(y-z)^2} + \frac{1}{(z-x)^2}.$$

Source: [107, pp. 9, 51, 88, 274, 578, 579], [108, pp. 25, 162], [109, pp. 72, 113], [287], [748, pp. 13, 216, 377, 380], [752, pp. 34, 242, 244, 317], [806, pp. 63, 138, 157, 165, 186, 212], [1006, 1731], [1938, pp. 11, 49], and [2527, p. 184].

Fact 2.3.69. Let x, y, z be nonnegative numbers. Then,

$$\begin{aligned}
0 &\le 27x^2y^2z^2 \le 3xyz(x+y+z)(xy+yz+zx) \le \tfrac{27}{8}xyz(x+y)(y+z)(z+x) \\
&\le \left\{ \begin{matrix} \tfrac{27}{64}xyz(2x+y+z)(2y+z+x)(2z+x+y) \le xyz(x+y+z)^3 \\ (xy+yz+zx)^3 \end{matrix} \right\} \le \tfrac{27}{64}(x+y)^2(y+z)^2(z+x)^2 \\
&\le \left\{ \begin{matrix} \tfrac{9}{16}[x^2(x+y)(y+z)^2(z+x) + y^2(x+y)(y+z)(z+x)^2 + z^2(x+y)^2(y+z)(z+x)] \\ (x^2+y^2+z^2)(xy+yz+zx)^2 \end{matrix} \right\} \\
&\le \tfrac{1}{27}(x+y+z)^6 \le \tfrac{3}{16}(x+y+z)[x(x+y)^2(x+z)^2 + y(y+z)^2(y+x)^2 + z(z+x)^2(z+y)^2] \\
&\le \tfrac{9}{64}[(x+y)^6 + (y+z)^6 + (z+x)^6] \le 9(x^6+y^6+z^6),
\end{aligned}$$

$$\begin{aligned}
&\tfrac{9}{16}[x^2(x+y)(y+z)^2(z+x) + y^2(x+y)(y+z)(z+x)^2 + z^2(x+y)^2(y+z)(z+x)] \\
&\quad \le (xy+yz+zx)^3 \le (x^2+xy+y^2)(y^2+yz+z^2)(z^2+zx+x^2) \\
&\quad \le 3(x^2+y^2+z^2)(x^2y^2+y^2z^2+z^2x^2) \le \tfrac{27}{8}(x^2+y^2)(y^2+z^2)(z^2+x^2) \\
&\quad \le (x^2+y^2+z^2)^3 \le 3(x^3+y^3+z^3)^2,
\end{aligned}$$

$$3x^2y^2z^2 \le \left\{ \begin{matrix} x^3yz^2 + y^3zx^2 + z^3xy^2 \\ xy^3z^2 + yz^3x^2 + zx^3y^2 \end{matrix} \right\} \le \left\{ \begin{matrix} x^2y^4 + y^2z^4 + z^2x^4 \\ x^4y^2 + y^4z^2 + z^4x^2 \end{matrix} \right\},$$

$$\begin{aligned}
12x^2y^2z^2 &\le 6x^2y^2z^2 + 2xyz(x^3+y^3+z^3) \\
&\le 2(x^3y^3+y^3z^3+z^3x^3) + x^4(y^2+z^2) + y^4(z^2+x^2) + z^4(x^2+y^2),
\end{aligned}$$

$$\begin{aligned}
&\tfrac{3}{4}(x+y)(y+z)(z+x)(x+y+z)(x^2+y^2+z^2) \\
&\qquad \le \left\{ \begin{matrix} (x+y+z)(x^2+y^2+z^2)[x(y^2+z^2) + y(z^2+x^2) + z(x^2+y^2)] \\ \tfrac{2}{9}(x+y+z)(x^2+y^2+z^2)(x+y+z)^3 \end{matrix} \right\} \\
&\qquad \le 9x^2y^2z^2 + (x+y+z)(x^2+y^2+z^2)(x^3+y^3+z^3+2xyz),
\end{aligned}$$

$$432xy^2z^3 \le (x+y+z)^6,$$
$$(x^4+y^4+z^4)(xy+yz+zx) \le (x^3+y^3+z^3)^2,$$
$$9(x^2+yz)(y^2+zx)(z^2+xy) \le 8(x^3+y^3+z^3)^2,$$
$$3xyz(x^3+y^3+z^3) \le (xy+yz+zx)(x^4+y^4+z^4),$$
$$x^2y^2z^2+xyz+1 \le 3(x^2-x+1)(y^2-y+1)(z^2-z+1),$$
$$(xy+yz+zx)^3 \le (x^2+xy+yz)(y^2+yz+zx)(z^2+zx+xy),$$
$$4xyz(xy^2+yz^2+zx^2) \le (x^2+yz)(y^2+zx)(z^2+xy)+4x^2y^2z^2,$$
$$2xyz(xy^2+yz^2+zx^2)+2x^2y^2z^2 \le (x^2+yz)(y^2+zx)(z^2+xy),$$
$$xyz(x^3+y^3+z^3) \le 3(x^2-xy+y^2)(y^2-yz+z^2)(z^2-zx+x^2),$$
$$9x^2y^2z^2 \le \tfrac{1}{3}(xy+yz+zx)^3 \le (x^2y+y^2z+z^2x)(xy^2+yz^2+zx^2),$$
$$(x^3+y^3+z^3)^2+36x^2y^2z^2 \le (x^2+y^2+z^2)^3+6xyz(x^3+y^3+z^3),$$
$$3(x^3y^3+y^3z^3+z^3x^3) \le [xy(x+y-z)+yz(y+z-x)+zx(z+x-y)]^2,$$
$$(x+y+z)^2(xy+yz+zx)^2 \le 3(x^2+xy+y^2)(y^2+yz+z^2)(z^2+zx+x^2),$$
$$xyz(x+y+z)^3 \le 9(x^4yz+y^4zx+z^4xy) \le \tfrac{9}{64}[(x+y)^6+(y+z)^6+(z+x)^6],$$
$$6x^2y^2z^2 \le 2(x^3y^3+y^3z^3+z^3x^3) \le \tfrac{1}{2}[(x^3+y^3+z^3)^2+3x^2y^2z^2] \le x^6+y^6+z^6+3x^2y^2z^2,$$
$$[27xyz+(x+y+z)(2x^2+2y^2+2z^2-5xy-5yz-5zx)]^2 \le 4(x^2+y^2+z^2-xy-yz-zx)^3,$$
$$(x^2+xy+y^2)(y^2+yz+z^2)(z^2+zx+x^2) \le (x^3+y^3+z^3)(xy+yz+zx)(x+y+z) \le (x^2+y^2+z^2)^3.$$

If $\alpha \in [3/7, 7/3]$, then

$$(\alpha+1)^6(xy+yz+zx)^3 \le 27(\alpha x+y)^2(\alpha y+z)^2(\alpha z+x)^2.$$

In particular,

$$64(xy+yz+zx)^3 \le 27(x+y)^2(y+z)^2(z+x)^2, \quad 27(xy+yz+zx)^3 \le (2x+y)^2(2y+z)^2(2z+x)^2.$$

Now, assume that x, y, z are positive. Then,

$$\frac{9}{4} \le \frac{x^2+2yz}{(y+z)^2}+\frac{y^2+2zx}{(z+x)^2}+\frac{z^2+2xy}{(x+y)^2}, \quad 10 \le \frac{x^2+16yz}{y^2+z^2}+\frac{y^2+16zx}{z^2+x^2}+\frac{z^2+16xy}{x^2+y^2},$$

$$0 \le \frac{xy-2yz+zx}{y^2-yz+z^2}+\frac{yz-2zx+xy}{z^2-zx+x^2}+\frac{zx-2xy+yz}{x^2-xy+y^2}, \quad \frac{9}{2} \le \frac{2x^2+yz}{y^2+z^2}+\frac{2y^2+zx}{z^2+x^2}+\frac{2z^2+xy}{x^2+y^2},$$

$$2 \le \frac{x^2+yz}{y^2+yz+z^2}+\frac{y^2+zx}{z^2+zx+x^2}+\frac{z^2+xy}{x^2+xy+y^2}, \quad 3 \le \frac{2x^2-yz}{y^2-yz+z^2}+\frac{2y^2-zx}{z^2-zx+x^2}+\frac{2z^2-xy}{x^2-xy+y^2},$$

$$4 \le \frac{xy+4yz+zx}{y^2+z^2}+\frac{yz+4zx+xy}{z^2+x^2}+\frac{zx+4xy+yz}{x^2+y^2},$$

$$1 \le \frac{x^2}{2y^2-yz+2z^2}+\frac{y^2}{2z^2-zx+2x^2}+\frac{z^2}{2x^2-xy+2y^2},$$

$$0 \le \frac{x^2-yz}{2y^2-3yz+2z^2}+\frac{y^2-zx}{2z^2-3zx+2x^2}+\frac{z^2-xy}{2x^2-3xy+2y^2},$$

$$\frac{x^2}{(2x+y)(2x+z)}+\frac{y^2}{(2y+z)(2y+x)}+\frac{z^2}{(2z+x)(2z+y)} \le \frac{1}{3},$$

$$5 \le \frac{x}{y} + \frac{y}{z} + \frac{z}{x} + \frac{6xyz}{x^2y + y^2z + z^2x}, \quad \frac{xy}{x^2 + 2y^2} + \frac{yz}{y^2 + 2z^2} + \frac{zx}{z^2 + 2x^2} \le 1,$$

$$\frac{(x+y)^2}{x^2 + y^2 + 2z^2} \le \frac{x^2}{x^2 + z^2} + \frac{y^2}{y^2 + z^2}, \quad 3 \le \frac{x+y}{y+z} + \frac{y+z}{z+x} + \frac{z+x}{x+y} \le \frac{x}{y} + \frac{y}{z} + \frac{z}{x},$$

$$\frac{6xyz}{x^2y + y^2z + z^2x} \le \frac{3xyz}{xy^2 + yz^2 + zx^2} + 1, \quad x + y + z \le \frac{x^2 + yz}{y+z} + \frac{y^2 + zx}{z+x} + \frac{z^2 + xy}{x+y},$$

$$6 \le \frac{(x+y)^2}{z^2 + xy} + \frac{(y+z)^2}{x^2 + yz} + \frac{(z+x)^2}{y^2 + zx}, \quad \frac{8}{(x+y+z)^2} \le \frac{1}{2x^2 + yz} + \frac{1}{2y^2 + zx} + \frac{1}{2z^2 + xy},$$

$$2 \le \frac{x(y+z)}{x^2 + yz} + \frac{y(z+x)}{y^2 + zx} + \frac{z(x+y)}{z^2 + xy}, \quad \frac{1}{2x^2 + yz} + \frac{1}{2y^2 + zx} + \frac{1}{2z^2 + zx} \le \frac{1}{9}\left(\frac{1}{x} + \frac{1}{y} + \frac{1}{z}\right)^2,$$

$$\frac{3(x^3 + y^3 + z^3)}{2(x^2 + y^2 + z^2)} \le \frac{x^2}{y+z} + \frac{y^2}{z+x} + \frac{z^2}{x+y}, \quad 0 \le \frac{x^2 - yz}{2x^2 + y^2 + z^2} + \frac{y^2 - zx}{2y^2 + z^2 + x^2} + \frac{z^2 - xy}{2z^2 + x^2 + y^2},$$

$$4 \le \frac{x^2 + y^2 + z^2}{xy + yz + zx} + \frac{3(x^3y + y^3z + z^3x)}{x^2y^2 + y^2z^2 + z^2x^2}, \quad \frac{3x^2 - yz}{2x^2 + y^2 + z^2} + \frac{3y^2 - zx}{2y^2 + z^2 + x^2} + \frac{3z^2 - xy}{2z^2 + x^2 + y^2} \le \frac{3}{2},$$

$$\frac{12}{(x+y+z)^2} \le \frac{1}{x^2 + yz} + \frac{1}{y^2 + zx} + \frac{1}{z^2 + xy}, \quad 1 \le \frac{x^2}{x^2 + xy + y^2} + \frac{y^2}{y^2 + yz + z^2} + \frac{z^2}{z^2 + zx + x^2},$$

$$2(x+y+z) \le \frac{xy}{z} + \frac{yz}{x} + \frac{zx}{y} + \frac{xyz}{xy + yz + zx}, \quad \frac{27}{(x+y+z)^2} \le \frac{2}{(x+y)y} + \frac{2}{(y+z)z} + \frac{2}{(z+x)x},$$

$$\frac{2}{xy + yz + zx} \le \frac{1}{x^2 + 2yz} + \frac{1}{y^2 + 2zx} + \frac{1}{z^2 + 2xy}, \quad \frac{3}{xy + yz + zx} \le \frac{1}{x^2 + yz} + \frac{1}{y^2 + zx} + \frac{1}{z^2 + xy},$$

$$\frac{6}{x+y+z} \le \frac{x+y}{2z^2 + xy} + \frac{y+z}{2x^2 + yz} + \frac{z+x}{2y^2 + zx}, \quad \frac{yz}{3x^2 + y^2 + z^2} + \frac{zx}{3y^2 + z^2 + x^2} + \frac{xy}{3z^2 + x^2 + y^2} \le \frac{3}{5},$$

$$2 \le \frac{x(y+z)}{y^2 + yz + z^2} + \frac{y(z+x)}{z^2 + zx + x^2} + \frac{z(x+y)}{x^2 + xy + y^2}, \quad \frac{3}{2} \le \frac{xy - yz + zx}{y^2 + z^2} + \frac{yz + xy}{z^2 - zx + xy} + \frac{zx - xy + yz}{x^2 + y^2},$$

$$\frac{9}{4(xy + yz + zx)} \le \frac{1}{(x+y)^2} + \frac{1}{(y+z)^2} + \frac{1}{(z+x)^2}, \quad \frac{9}{4(x+y+z)} \le \frac{z}{(x+y)^2} + \frac{x}{(y+z)^2} + \frac{y}{(z+x)^2},$$

$$\frac{10}{(x+y+z)^2} \le \frac{1}{x^2 + y^2} + \frac{1}{y^2 + z^2} + \frac{1}{z^2 + x^2}, \quad 6 \le \frac{xy + 4yz + zx}{x^2 + yz} + \frac{yz + 4zx + xy}{y^2 + zx} + \frac{zx + 4xy + yz}{z^2 + xy},$$

$$\frac{1}{xy + yz + zx} \le \frac{1}{(x+2y)^2} + \frac{1}{(y+2z)^2} + \frac{1}{(z+2x)^2},$$

$$\frac{4x^2 - y^2 - z^2}{x(y+z)} + \frac{4y^2 - z^2 - x^2}{y(z+x)} + \frac{4z^2 - x^2 - y^2}{z(x+y)} \le 3,$$

$$\frac{x}{y+z} + \frac{y}{z+x} + \frac{z}{x+y} \le \frac{x^2}{y^2 + z^2} + \frac{y^2}{z^2 + x^2} + \frac{z^2}{x^2 + y^2},$$

$$\frac{x}{x^2 + yz} + \frac{y}{y^2 + zx} + \frac{z}{z^2 + xy} \le \frac{1}{x+y} + \frac{1}{y+z} + \frac{1}{z+x},$$

$$\frac{1}{(x+y+z)^2} \le \frac{1}{22x^2 + 5yz} + \frac{1}{22y^2 + 5zx} + \frac{1}{22z^2 + 5xy},$$

$$\frac{(2x+y)^2}{4x^2 + y^2 + 4z^2} + \frac{(2y+z)^2}{4y^2 + z^2 + 4x^2} + \frac{(2z+x)^2}{4z^2 + x^2 + 4y^2} \le 3,$$

$$0 \le \frac{x^2 - yz}{4x^2 + 4y^2 + z^2} + \frac{y^2 - zx}{4y^2 + 4z^2 + x^2} + \frac{z^2 - xy}{4z^2 + 4x^2 + y^2},$$

$$\frac{9xyz}{2(x+y+z)} \le \frac{xy^2}{x+y} + \frac{yz^2}{y+z} + \frac{zx^2}{z+x} \le \tfrac{1}{2}(x^2 + y^2 + z^2),$$

$$2(xy + yz + zx) \le \frac{x^2(y+z)^2}{y^2 + z^2} + \frac{y^2(z+x)^2}{z^2 + x^2} + \frac{z^2(x+y)^2}{x^2 + y^2},$$

$$\frac{1}{3} \le \frac{x^2}{3x^2 + (y+z)^2} + \frac{y^2}{3y^2 + (z+x)^2} + \frac{z^2}{3z^2 + (x+y)^2} \le \frac{1}{2},$$

$$\frac{9}{(x+y+z)^2} \le \frac{1}{x^2 + xy + y^2} + \frac{1}{y^2 + yz + z^2} + \frac{1}{z^2 + zx + x^2},$$

$$\frac{4}{5}\left(\frac{1}{x+y} + \frac{1}{y+z} + \frac{1}{z+x}\right) \le \frac{x}{y^2 + z^2} + \frac{y}{z^2 + x^2} + \frac{z}{x^2 + y^2},$$

$$\frac{3}{xy + yz + zx} \le \frac{1}{x^2 - xy + y^2} + \frac{1}{y^2 - yz + z^2} + \frac{1}{z^2 - zx + x^2},$$

$$1 \le \frac{y^2}{(x+y)^2} + \frac{z^2}{(y+z)^2} + \frac{x^2}{(z+x)^2} + \frac{2xyz}{(x+y)(y+z)(z+x)},$$

$$\frac{x^2}{(2x+y)(2x+z)} + \frac{y^2}{(2y+z)(2y+x)} + \frac{z^2}{(2z+x)(2z+y)} \le \frac{1}{3},$$

$$\frac{6}{x^2 + y^2 + z^2 + xy + yz + zx} \le \frac{1}{2x^2 + yz} + \frac{1}{2y^2 + zx} + \frac{1}{2z^2 + xy},$$

$$\frac{3}{2} \le \frac{3x(y+z) - 2yz}{(2x+y+z)(y+z)} + \frac{3y(z+x) - 2zx}{(2y+z+x)(z+x)} + \frac{3z(x+y) - 2xy}{(2z+x+y)(x+y)},$$

$$\frac{1}{x^2 + y^2 + z^2} \le \frac{1}{5(x^2 + y^2) - xy} + \frac{1}{5(y^2 + z^2) - yz} + \frac{1}{5(z^2 + x^2) - zx},$$

$$1 \le \frac{x^2 + y^2}{2(x^2 + y^2) + (x+y)z} + \frac{y^2 + z^2}{2(y^2 + z^2) + (y+z)x} + \frac{z^2 + x^2}{2(z^2 + x^2) + (z+x)y}.$$

If $r \le \frac{5}{2}$, then

$$\frac{3(1+r)}{4} \le \frac{x^2 + ryz}{(y+z)^2} + \frac{y^2 + rzx}{(z+x)^2} + \frac{z^2 + rxy}{(x+y)^2}.$$

In particular,

$$-\frac{3}{4} \le \frac{x^2 - 2yz}{(y+z)^2} + \frac{y^2 - 2zx}{(z+x)^2} + \frac{z^2 - 2xy}{(x+y)^2}, \quad 0 \le \frac{x^2 - yz}{(y+z)^2} + \frac{y^2 - zx}{(z+x)^2} + \frac{z^2 - xy}{(x+y)^2},$$

$$\frac{3}{4} \le \left(\frac{x}{y+z}\right)^2 + \left(\frac{y}{z+x}\right)^2 + \left(\frac{z}{x+y}\right)^2, \quad \frac{3}{2} \le \frac{x^2 + yz}{(y+z)^2} + \frac{y^2 + zx}{(z+x)^2} + \frac{z^2 + xy}{(x+y)^2}.$$

If $r > -2$, then

$$\frac{3(3+2r)}{2+r} \le \frac{2x^2 + (1+2r)yz}{y^2 + ryz + z^2} + \frac{2y^2 + (1+2r)zx}{z^2 + \alpha zx + x^2} + \frac{2z^2 + (1+2r)xy}{x^2 + rxy + y^2}.$$

In particular,

$$\frac{9}{2} \le \frac{2x^2 + yz}{y^2 + z^2} + \frac{2y^2 + zx}{z^2 + x^2} + \frac{2z^2 + xy}{x^2 + y^2}, \quad \frac{21}{4} \le \frac{2x^2 + 5yz}{(y+z)^2} + \frac{2y^2 + 5zx}{(z+x)^2} + \frac{2z^2 + 5xy}{(x+y)^2},$$

$$5 \le \frac{2x^2+3yz}{y^2+yz+z^2} + \frac{2y^2+3zx}{z^2+zx+x^2} + \frac{2z^2+3xy}{x^2+xy+y^2}, \quad 3 \le \frac{2x^2-yz}{y^2-yz+z^2} + \frac{2y^2-zx}{z^2-zx+x^2} + \frac{2z^2-xy}{x^2-xy+y^2},$$

$$1 \le \frac{x^2}{2y^2-yz+2z^2} + \frac{y^2}{2z^2-zx+2x^2} + \frac{z^2}{2x^2-xy+2y^2},$$

$$0 \le \frac{x^2-yz}{2y^2-3yz+2z^2} + \frac{y^2-zx}{2z^2-3zx+2x^2} + \frac{z^2-xy}{2x^2-3xy+2y^2},$$

$$\frac{9}{7(x^2+y^2+z^2)} \le \frac{1}{4y^2-yz+4z^2} + \frac{1}{4z^2-zx+4x^2} + \frac{1}{4x^2-xy+4y^2}.$$

If $r \ge 0$, then

$$\frac{rx+y}{ry+z} + \frac{ry+z}{rz+x} + \frac{rz+x}{rx+y} \le \frac{x}{y} + \frac{y}{z} + \frac{z}{x}.$$

If $r \in [0, \frac{5}{2}]$, then

$$\frac{3(r+1)}{x^2+y^2+z^2+r(xy+yz+zx)} \le \frac{1}{x^2+xy+y^2} + \frac{1}{y^2+yz+z^2} + \frac{1}{z^2+zx+x^2}.$$

If $r \in [\frac{2}{3}, 3+\sqrt{7}]$, then

$$\frac{r+2}{r(xy+yz+zx)} \le \frac{1}{rx^2+yz} + \frac{1}{ry^2+zx} + \frac{1}{rz^2+xy}.$$

If $r \ge 3+\sqrt{7}$, then

$$\frac{9}{(r+1)(xy+yz+zx)} \le \frac{1}{rx^2+yz} + \frac{1}{ry^2+zx} + \frac{1}{rz^2+xy}.$$

Source: [107, pp. 30, 54, 168, 289, 290], [108, pp. 12, 104, 353], [109, pp. 179, 180, 388, 389], [135], [287], [289, pp. 20, 120, 121], [371], [603, p. 244], [725, p. 114], [748, pp. 6, 9, 11, 12, 217, 218, 301, 302, 318, 342, 344, 375, 377, 378, 381], [749], [752, pp. 27, 29, 31, 35, 37–40, 242], [806, pp. 15, 100, 101, 111, 114, 127, 138, 139, 186, 188, 189, 191, 196, 201, 204, 211, 212, 213, 214], [961], [993, pp. 179, 182], [1025], [1371, pp. 105, 134, 149, 150, 155, 169], [1507], [1757, pp. 247, 252, 257], [1903], [1938, pp. 31, 107, 110, 111], [2128, p. 14], [2380, p. 72], [2791]. **Remark:** The inequalities $2(x^3y^3+y^3z^3+z^3x^3) \le \frac{1}{2}[(x^3+y^3+z^3)^2+3(x^2y^2z^2] \le x^6+y^6+z^6+3x^2y^2z^2$ hold for all real numbers. The left-hand inequality is given in [1860, pp. 20, 114, 115]. **Related:** Fact 2.3.61, Fact 2.11.15, Fact 2.11.25, and Fact 5.2.25. **Credit:** $(xy+yz+zx)^2(xyz^2+x^2yz+xy^2z) \le 3(y^2z^2+z^2x^2+x^2y^2)^2$ is due to M. Klamkin. See Fact 5.2.7 and [2791].

Fact 2.3.70. Let x, y, z be nonnegative numbers. Then,

$$x^2y^2z^2(xy+yz+zx) \le x^8+y^8+z^8,$$
$$(xyz+1)^3 \le (x^3+1)(y^3+1)(z^3+1),$$
$$x^3y^3z^3(x+y+z) \le x^5y^5+y^5z^5+z^5x^5,$$
$$x^2y^2z^2(x+y+z) \le x^5yz+y^5zx+z^5xy,$$
$$x^2y^2z^2(x^3+y^3+z^3) \le x^3y^6+y^3z^6+z^3x^6,$$
$$(x^5+y^5+z^5)(x+y+z)^3 \le 9(x^4+y^4+z^4)^2,$$
$$(x^3+y^3+z^3)(x^5+y^5+z^5) \le 3(x^8+y^8+z^8),$$
$$xyz(x+y+z)(x^3+y^3+z^3) \le (xy+yz+zx)(x^5+y^5+z^5),$$
$$(xy+yz+zx)^2(xyz^2+yzx^2+zxy^2) \le 3(x^2y^2+y^2z^2+z^2x^2)^2,$$
$$(x^3+y^3+z^3+xyz)(x+y-z)(y+z-x)(z+x-y) \le 4x^2y^2z^2,$$

$$x^2y^2z^2(x+y+z)^2 \le 3x^2y^2z^2(x^2+y^2+z^2) \le (x^2y^2+y^2z^2+z^2x^2)^2,$$
$$(x^3y^3+y^3z^3+z^3x^3)(x+y)(y+z)(z+x) \le 3(x^3+y^3)(y^3+z^3)(z^3+x^3),$$
$$[2(xy+yz+zx)-(x^2+y^2+z^2)][2(x^3y^3+y^3z^3+z^3x^3)-(x^6+y^6+z^6)]$$
$$\le [2(x^2y^2+y^2z^2+z^2x^2)-(x^4+y^4+z^4)]^2.$$

Now, assume that x, y, z are positive. Then,

$$6 \le \frac{3x^3+xyz}{y^3+z^3} + \frac{3y^3+yzx}{z^3+x^3} + \frac{3z^3+zxy}{x^3+y^3}, \quad \frac{xy^2}{x^3+y^3} + \frac{yz^2}{y^3+z^3} + \frac{zx^2}{z^3+x^3} \le \frac{3}{2},$$

$$\frac{(xy^2+yz^2+zx^2)^5}{(x^3+y^3+z^3)^4} \le \frac{x^5}{y^2} + \frac{y^5}{z^2} + \frac{z^5}{x^2}, \quad 1 \le \frac{x^3}{x^3+y^3+xyz} + \frac{y^3}{y^3+z^3+xyz} + \frac{z^3}{z^3+x^3+xyz},$$

$$x+y+z \le \frac{2x^3}{y^2+z^2} + \frac{2y^3}{z^2+x^2} + \frac{2z^3}{x^2+y^2}, \quad \frac{3}{2} \le \frac{x^3y}{z^2(y^2+zx)} + \frac{y^3z}{x^2(z^2+xy)} + \frac{z^3x}{y^2(x^2+yz)},$$

$$\frac{x^2yz}{x^2+yz} + \frac{y^2zx}{y^2+zx} + \frac{z^2xy}{z^2+xy} \le \tfrac{1}{2}(x^2+y^2+z^2), \quad \frac{(x+y+z)^6}{81(x^2+y^2+z^2)^2} \le \frac{x^3}{x+2y} + \frac{y^3}{y+2z} + \frac{z^3}{z+2x},$$

$$0 \le \frac{x^2y^2(y-z)}{x+y} + \frac{y^2z^2(z-x)}{y+z} + \frac{z^2x^2(x-y)}{z+x}, \quad \frac{3(x^3+y^3+z^3)}{2(x^2+y^2+z^2)} \le \frac{x^3+xyz}{(y+z)^2} + \frac{y^3+yzx}{(z+x)^2} + \frac{z^3+zxy}{(x+y)^2},$$

$$\frac{9}{4(xy+yz+zx)} \le \frac{x}{y(y+z)^2} + \frac{y}{z(z+x)^2} + \frac{z}{x(x+y)^2}, \quad \frac{3}{2} \le \frac{x^3y}{z^2(y^2+xz)} + \frac{y^3z}{x^2(z^2+yx)} + \frac{z^3x}{y^2(x^2+yz)},$$

$$x+y+z \le \frac{2x^4}{x^3+y^3} + \frac{2y^4}{y^3+z^3} + \frac{2z^4}{z^3+x^3}, \quad x+y+z \le \frac{x^3+3xyz}{(y+z)^2} + \frac{y^3+3xyz}{(z+x)^2} + \frac{z^3+3xyz}{(x+y)^2},$$

$$3(x^6+y^6+z^6)^2 \le \left(\frac{x^5}{y} + \frac{y^5}{z} + \frac{z^5}{x}\right)^3, \quad 1 \le \frac{x^2}{x^2+xy+y^2} + \frac{y^2}{y^2+yz+z^2} + \frac{z^2}{z^2+zx+x^2},$$

$$\left(\frac{x^2+y^2+z^2}{xy+yz+zx}\right)^2 \le \frac{x^3+y^3+z^3}{4xyz} + \frac{1}{4}, \quad \frac{3}{2} \le \frac{x^3+3xyz}{(y+z)^3} + \frac{y^3+3xyz}{(z+x)^3} + \frac{z^3+3xyz}{(x+y)^3},$$

$$\frac{x+y+z}{x^2+y^2+z^2} \le \frac{x}{2x^2+yz} + \frac{y}{2y^2+zx} + \frac{z}{2z^2+xy},$$

$$1 \le \frac{x^3}{x^3+y^3+xyz} + \frac{y^3}{y^3+z^3+xyz} + \frac{z^3}{z^3+x^3+xyz},$$

$$\frac{xy+yz+zx}{x^2+y^2+z^2} + 1 \le \frac{x(y+z)}{x^2+2yz} + \frac{y(z+x)}{y^2+2zx} + \frac{z(x+y)}{z^2+2xy},$$

$$\frac{9(x+y+z)^2}{xy+yz+zx} \le \left(\frac{2x}{y}+1\right)^2 + \left(\frac{2y}{z}+1\right)^2 + \left(\frac{2z}{x}+1\right)^2,$$

$$\frac{xy}{z^2+xy} + \frac{yz}{x^2+yz} + \frac{zx}{y^2+zx} \le \frac{x}{y+z} + \frac{y}{z+x} + \frac{z}{x+y},$$

$$\frac{9}{2(x^3+y^3+z^3)} \le \frac{1}{xy(x+y)} + \frac{1}{yz(y+z)} + \frac{1}{zx(z+x)},$$

$$\frac{1}{x^3+y^3+xyz} + \frac{1}{y^3+z^3+xyz} + \frac{1}{z^3+x^3+xyz} \le \frac{1}{xyz},$$

$$\frac{x}{y+z} + \frac{y}{z+x} + \frac{z}{x+y} \le \frac{x^2}{y^2+z^2} + \frac{y^2}{z^2+x^2} + \frac{z^2}{x^2+y^2},$$

$$\frac{x}{x^2+yz}+\frac{y}{y^2+zx}+\frac{z}{z^2+xy}\le\frac{1}{x+y}+\frac{1}{y+z}+\frac{1}{z+x},$$

$$x+y+z\le\frac{x^3}{y^2-yz+z^2}+\frac{y^3}{z^2-zx+x^2}+\frac{z^3}{x^2-xy+y^2},$$

$$\frac{2x}{3x^2+yz}+\frac{2y}{3y^2+zx}+\frac{2z}{3z^2+xy}\le\frac{1}{x+y}+\frac{1}{y+z}+\frac{1}{z+x},$$

$$\frac{x^3+y^3+z^3}{x+y+z}\le\frac{x^4}{x^2+xy+y^2}+\frac{y^4}{y^2+yz+z^2}+\frac{z^4}{z^2+zx+x^2},$$

$$\frac{x^2}{x^2+xy+yz}+\frac{y^2}{y^2+yz+zx}+\frac{z^2}{z^2+zx+xy}\le\frac{x^2+y^2+z^2}{xy+yz+zx},$$

$$\frac{9}{(x+y+z)^2}\le\frac{1}{x^2+xy+yz}+\frac{1}{y^2+yz+zx}+\frac{1}{z^2+zx+xy},$$

$$\frac{xy}{y^2+yz+z^2}+\frac{yz}{z^2+zx+x^2}+\frac{zx}{x^2+xy+y^2}\le\frac{x^2+y^2+z^2}{xy+yz+zx},$$

$$\frac{xy}{x^2+yz+zx}+\frac{yz}{y^2+zx+xy}+\frac{zx}{z^2+xy+yz}\le\frac{x^2+y^2+z^2}{xy+yz+zx},$$

$$\frac{x^3+y^3+z^3}{x^2+y^2+z^2}\le\frac{x^4}{x^3+xyz+y^3}+\frac{y^4}{y^3+xyz+z^3}+\frac{z^4}{z^3+xyz+x^3},$$

$$\frac{9(xy+yz+zx)}{(x+y+z)^3}\le\frac{x}{x^2+xy+yz}+\frac{y}{y^2+yz+zx}+\frac{z}{z^2+zx+xy},$$

$$\frac{1}{xy+yz+zx}+\frac{2}{x^2+y^2+z^2}\le\frac{1}{2x^2+yz}+\frac{1}{2y^2+zx}+\frac{1}{2z^2+xy},$$

$$6xyz\le\frac{4xyz(xy+yz+zx)}{x^2+y^2+z^2}+\frac{xy^4+yz^4+zx^4+x^4y+y^4z+z^4x}{xy+yz+zx},$$

$$\frac{12}{3x+y}+\frac{12}{3y+z}+\frac{12}{3z+x}\le\frac{1}{x}+\frac{1}{y}+\frac{1}{z}+\frac{4}{x+y}+\frac{4}{y+z}+\frac{4}{z+x},$$

$$\frac{3(x+1)(y+1)(z+1)}{xyz+1}\le 3+x+y+z+\frac{1}{x}+\frac{1}{y}+\frac{1}{z}+\frac{x}{y}+\frac{y}{z}+\frac{z}{x},$$

$$\frac{3}{2}\left(\sqrt{\frac{(x+y+z)(x^2+y^2+z^2)}{xyz}}-1\right)\le\frac{y^2+z^2}{x(y+z)}+\frac{z^2+x^2}{y(z+x)}+\frac{x^2+y^2}{z(x+y)},$$

$$1\le\frac{x^3+4xyz}{x^3+(y+z)^3+6xyz}+\frac{y^3+4xyz}{y^3+(z+x)^3+6xyz}+\frac{z^3+4xyz}{z^3+(x+y)^3+6xyz},$$

$$\frac{[(x+y+z)(x^2+y^2+z^2)-(x^3+y^3+z^3)]^4}{(x^3+y^3+z^3)^3}\le\frac{(x+y)^4}{z}+\frac{(y+z)^4}{x}+\frac{(z+x)^4}{y},$$

$$\frac{xy}{x^2+xy+yz}+\frac{yz}{y^2+yz+zx}+\frac{zx}{z^2+zx+xy}\le\frac{x}{2x+z}+\frac{y}{2y+x}+\frac{z}{2z+y}\le 1,$$

$$\frac{1}{x^2+yz}+\frac{1}{y^2+zx}+\frac{1}{z^2+xy}\le\frac{(x+y+z)^2}{3(xy+yz+zx)}\left(\frac{1}{x^2+y^2}+\frac{1}{y^2+z^2}+\frac{1}{z^2+x^2}\right),$$

$$\frac{x^3}{(2x^2+y^2)(2x^2+z^2)}+\frac{y^3}{(2y^2+z^2)(2y^2+x^2)}+\frac{z^3}{(2z^2+x^2)(2z^2+y^2)}\le\frac{1}{x+y+z},$$

$$\frac{4xyz(x^2-y^2)^2}{(x+y+z)(x+y)(y+z)(z+x)} \le x(x-y)(x-z)+y(y-z)(y-x)+z(z-x)(z-y),$$

$$\left(\frac{2}{x+y}\right)^5+\left(\frac{6}{3x+y+2z}\right)^5+\left(\frac{6}{3x+3y+z}\right)^5 \le \frac{1}{x^5}+\frac{1}{y^5}+\frac{1}{z^5},$$

$$\frac{2}{x+y}+\frac{2}{y+z}+\frac{2}{z+x}+\frac{8}{3x+y}+\frac{8}{3y+z}+\frac{8}{3z+x} \le \frac{1}{x}+\frac{1}{y}+\frac{1}{z}+\frac{8}{x+3y}+\frac{8}{y+3z}+\frac{8}{z+3x},$$

$$\tfrac{2}{3}xyz\left(\frac{1}{x^2+yz}+\frac{1}{y^2+zx}+\frac{1}{z^2+xy}\right) \le \tfrac{1}{3}(x+y+z) \le \frac{x^2+y^2+z^2}{x+y+z}$$
$$\le \frac{x^3}{x^2+xy+y^2}+\frac{y^3}{y^2+yz+z^2}+\frac{z^3}{z^2+zx+x^2}.$$

If $r > -2$, then

$$4r+10 \le \frac{x^2+4(r+2)^2yz}{y^2+ryz+z^2}+\frac{y^2+4(r+2)^2zx}{z^2+rzx+x^2}+\frac{z^2+4(r+2)^2xy}{x^2+rxy+y^2},$$

$$\frac{3(2r+3)}{r+2} \le \frac{2x^2+(2r+1)yz}{y^2+ryz+z^2}+\frac{2y^2+(2r+1)zx}{z^2+rzx+x^2}+\frac{2z^2+(2r+1)xy}{x^2+rxy+y^2},$$

$$\frac{3(r+1)}{r+2} \le \frac{xy+(r-1)yz+zx}{y^2+ryz+z^2}+\frac{yz+(r-1)zx+xy}{z^2+rzx+x^2}+\frac{zx+(r-1)xy+yz}{x^2+rxy+y^2},$$

$$r+4 \le \frac{xy+(r+2)^2yz+zx}{y^2+ryz+z^2}+\frac{yz+(r+2)^2zx+xy}{z^2+rzx+x^2}+\frac{zx+(r+2)^2xy+yz}{x^2+rxy+y^2}.$$

If $r > -2$ and $p \le (2r+1)/2$, then

$$\frac{3(p+2)}{r+2} \le \frac{xy+pyz}{y^2+ryz+z^2}+\frac{yz+pzx}{z^2+rzx+x^2}+\frac{zx+pxy}{x^2+rxy+y^2}.$$

If $r > -2$ and $p \in [(2r+1)/2, 4(r+2)^2]$, then

$$\frac{p}{r+2}+2 \le \frac{xy+pyz}{y^2+ryz+z^2}+\frac{yz+pzx}{z^2+rzx+x^2}+\frac{zx+pxy}{x^2+rxy+y^2}.$$

If $r > -2$ and $p \ge 1$, then

$$4pr+12p^2-2 \le \frac{xy+4p(r+2p)^2yz}{y^2+ryz+z^2}+\frac{yz+4p(r+2p)^2zx}{z^2+rzx+x^2}+\frac{zx+4p(r+2p)^2xy}{x^2+rxy+y^2}.$$

If $r > -2$ and $p \le r-1$, then

$$\frac{3(p+2)}{r+2} \le \frac{xy+pyz+zx}{y^2+ryz+z^2}+\frac{yz+pzx+xy}{z^2+rzx+x^2}+\frac{zx+pxy+yz}{x^2+rxy+y^2}.$$

If $r > -2$ and $p \in [r-1, (r+2)^2]$, then

$$\frac{p}{r+2}+2 \le \frac{xy+pyz+zx}{y^2+ryz+z^2}+\frac{yz+pzx+xy}{z^2+rzx+x^2}+\frac{zx+pxy+yz}{x^2+rxy+y^2}.$$

If $r > -2$ and $p \ge (r+2)^2$, then

$$2\sqrt{p}-r \le \frac{xy+pyz+zx}{y^2+ryz+z^2}+\frac{yz+pzx+xy}{z^2+rzx+x^2}+\frac{zx+pxy+yz}{x^2+rxy+y^2}.$$

If $n \ge 0$, then

$$\tfrac{1}{2}(x^n+y^n+z^n) \le \frac{x^{n+1}}{y+z}+\frac{y^{n+1}}{z+x}+\frac{z^{n+1}}{x+y}, \qquad \left(\frac{x+2y}{3}\right)^n+\left(\frac{y+2z}{3}\right)^n+\left(\frac{z+2x}{3}\right)^n \le x^n+y^n+z^n,$$

$$(xyz)^{n+2}(x^n + y^n + z^n) \le (xy)^{2n+3} + (yz)^{2n+3} + (zx)^{2n+3}.$$

If $n \ge 2$, then

$$\sqrt{\frac{3(x^{n-1} + y^{n-1} + z^{n-1})(x^n + y^n + z^n)}{x + y + z}} \le \frac{x^n + y^n}{x + y} + \frac{y^n + z^n}{y + z} + \frac{z^n + x^n}{z + x}.$$

If $n \ge 3$, then

$$x^{n-1}y + y^{n-1}z + z^{n-1}x + xy^{n-1} + yz^{n-1} + zx^{n-1} \le x^n + y^n + z^n + xyz(x^{n-3} + y^{n-3} + z^{n-3}).$$

If $n \ge 1$, then

$$0 \le \frac{2x^n - y^n - z^n}{y^2 - yz + z^2} + \frac{2y^n - z^n - x^n}{z^2 - zx + x^2} + \frac{2z^n - x^n - y^n}{x^2 - xy + y^2},$$

$$\sqrt[n]{\tfrac{1}{3}(x^n + y^n + z^n)}\left(\frac{x^n}{y + z} + \frac{y^n}{z + x} + \frac{z^n}{x + y}\right) \le \frac{x^{n+1}}{y + z} + \frac{y^{n+1}}{z + x} + \frac{z^{n+1}}{x + y}.$$

Source: [55, 99], [107, pp. 12, 15, 54, 99, 100, 108, 109, 290, 487, 488], [108, pp. 15, 55, 115, 116, 318, 354–356], [109, p. 156], [288, p. 229], [318, 751], [748, pp. 12, 13, 111, 319, 373, 376, 382], [752, pp. 29, 30, 32, 33, 35–38, 42, 244], [806, pp. 45, 116, 129, 157, 187, 198, 199, 213, 215], [993, p. 184], [1371, pp. 108, 117, 134, 150], [1569], [1860, pp. 19, 112, 113, 134, 228], [1772], [1938, pp. 12, 43, 49, 107, 108], [2527, p. 204], [2601], and [2791]. **Remark:** Setting $n = 0$ in the third to last inequality yields Nesbitt's inequality. See Fact 2.3.60. **Remark:** The last inequality is due to Schur. See [99], Fact 2.3.16, Fact 2.3.28, and Fact 2.3.67.

Fact 2.3.71. Let x, y, z be positive numbers. Then,

$$\frac{3}{3x + y} + \frac{3}{3y + z} + \frac{3}{3z + x} \le \frac{1}{x + y} + \frac{1}{y + z} + \frac{1}{z + x} + \frac{1}{4}\left(\frac{1}{x} + \frac{1}{y} + \frac{1}{z}\right),$$

$$\frac{2}{3x + y} + \frac{2}{3y + z} + \frac{2}{3z + x} \le \frac{1}{x + 3y} + \frac{1}{y + 3z} + \frac{1}{z + 3x} + \frac{1}{4}\left(\frac{1}{x} + \frac{1}{y} + \frac{1}{z}\right),$$

$$\frac{3}{x + y + z} + \frac{15}{x + 2y} + \frac{15}{y + 2z} + \frac{15}{z + 2x} \le \frac{12}{2x + y} + \frac{12}{2y + z} + \frac{12}{2z + x} + \frac{4}{3}\left(\frac{1}{x} + \frac{1}{y} + \frac{1}{z}\right).$$

If $r \in [1, 3]$, then

$$\frac{3}{rx + (4 - r)y} + \frac{3}{ry + (4 - r)z} + \frac{3}{rz + (4 - r)x} \le \frac{1}{x + y} + \frac{1}{y + z} + \frac{1}{z + x} + \frac{1}{4}\left(\frac{1}{x} + \frac{1}{y} + \frac{1}{z}\right).$$

If $r \in [-1, 2]$, then

$$\frac{r + 1}{3x + y} + \frac{r + 1}{3y + z} + \frac{r + 1}{3z + x} \le \frac{r}{2}\left(\frac{1}{x + y} + \frac{1}{y + z} + \frac{1}{z + x}\right) + \frac{1}{4}\left(\frac{1}{x} + \frac{1}{y} + \frac{1}{z}\right).$$

Source: [381].

2.4 Facts on Equalities and Inequalities in Four Variables

Fact 2.4.1. Let a, b be positive numbers, let x, y be real numbers, and assume that $x/a \le y/b$. Then,

$$\frac{x}{a} \le \frac{x + y}{a + b} \le \frac{y}{b}.$$

Source: [47, p. 32]. **Remark:** The center term is the *mediant* of x/a and y/b. See Fact 2.12.6.

Fact 2.4.2. Let a, b, x, y be positive numbers, and assume that $a \le x$ and $b \le y$. Then,

$$\frac{x + y}{a + y} \le \frac{x + b}{a + b}.$$

Fact 2.4.3. Let a, b be positive numbers, and let x, y be real numbers. Then,

$$\frac{(x+y)^2}{a+b} \le \frac{x^2}{a} + \frac{y^2}{b}.$$

Source: Fact 2.12.19. **Related:** Fact 2.6.7.

Fact 2.4.4. Let x_1, x_2, x_3, x_4 be complex numbers. Then,

$$6(x_1^2 + x_2^2 + x_3^2 + x_4^2) = \sum_{1\le i<j\le 4} (x_i + x_j)^2 + \sum_{1\le i<j\le 4} (x_i - x_j)^2,$$

$$6x_1(x_1^2 + x_2^2 + x_3^2 + x_4^2) = \sum_{2\le i\le 4} (x_1 + x_i)^3 + \sum_{2\le i\le 4} (x_1 - x_i)^3,$$

$$6(x_1^2 + x_2^2 + x_3^2 + x_4^2)^2 = \sum_{1\le i<j\le 4} (x_i + x_j)^4 + \sum_{1\le i<j\le 4} (x_i - x_j)^4.$$

Remark: The second equality is *Maillet's identity*, while the second and third equalities are *Liouville's identity*. See [309, p. 14] and [2232].

Fact 2.4.5. Let w, x, y, z be complex numbers. Then,

$$(w^2 + x^2 + y^2 + z^2)^2 = (w^2 + x^2 - y^2 - z^2)^2 + 4(wy + xz)^2 + 4(wz - xy)^2.$$

Source: [2316, p. 58]. **Remark:** This is *Lebesgue's identity*.

Fact 2.4.6. Let w, x, y, z be complex numbers. Then,

$$(x-y)(w-z)^2 + (y-z)(w-x)^2 + (z-x)(w-y)^2 + (x-y)(y-z)(z-x) = 0,$$

$$z^2[(3wxy - 2x^3 - w^2z)^2 + 4(wy - x^2)^3] = w^2[(3xyz - 2y^3 - z^2w)^2 + 4(xz - y^2)^3],$$

$$2(w^4+x^4+y^4+z^4)+8wxyz = (w^2+x^2+y^2+z^2)^2+(w+x-y-z)(w-x-y+z)(w-x+y-z)(w+x+y+z).$$

Now, let $s \triangleq w + x + y + z$. Then,

$$16(s-w)(s-x)(s-y)(s-z) + (w^2 - x^2 - y^2 + z^2)^2 = 4(wz + xy)^2,$$

$$s(s-w-z)(s-x-z)(s-y-z) + wxyz = (s-w)(s-x)(s-y)(s-z).$$

Source: [732, pp. 57, 71] and [289, pp. 27, 154].

Fact 2.4.7. Let a, w, x, y, z be complex numbers. Then,

$$(w^2 - ax^2)(y^2 - az^2) = (wy + axz)^2 - a(wz + xy)^2 = (wy - axz)^2 - a(wz - xy)^2.$$

In particular,

$$(w^2 + x^2)(y^2 + z^2) = (wz + xy)^2 + (wy - xz)^2 = (wy + xz)^2 + (wz - xy)^2.$$

If, in addition, w, x, y, z are real, then

$$\max\{(wz + xy)^2, (wy - xz)^2, (wy + xz)^2, (wz - xy)^2\} \le (w^2 + x^2)(y^2 + z^2).$$

Remark: The first equality is *Brahmagupta's identity*. See [1016]. **Remark:** The case $a = -1$ is *Diophantus's identity*, which is a special case of Lagrange's identity given by Fact 2.12.13. This equality is a statement of the fact that $|w + x\jmath|^2|y + z\jmath|^2 = |(w + x\jmath)(y + z\jmath)|^2$. See [773, p. 77], [1171], [1258, pp. 13, 14], [2107, pp. 25, 26], [2380, p. 6], and [2527, p. 47]. **Related:** Fact 1.11.31.

Fact 2.4.8. Let w, x, y, z be complex numbers. Then,

$$(4w + 1)^2 + (4x + 1)^2 + (4y + 1)^2 + (4z + 1)^2$$
$$= 4(w + x + y + z + 1)^2 + 4(w + x - y - z)^2 + 4(w - x + y - z)^2 + 4(-w + x + y - z)^2.$$

Source: [2116]. **Related:** Fact 1.11.28.

Fact 2.4.9. Let w, x, y, z be complex numbers. Then,

$$(w+x+y+z)^3 = w^3+x^3+y^3+z^3+3[w^2(x+y+z)+w(x+y+z)^2+(x+y)(x+z)(y+z)].$$

Fact 2.4.10. Let w, x, y, z be distinct, nonzero complex numbers, and let a, b, c be complex numbers. Then,

$$\frac{(w+a)(w+b)(w+c)}{w(w-x)(w-y)(w-z)} + \frac{(x+a)(x+b)(x+c)}{x(x-y)(x-z)(x-w)} + \frac{(y+a)(y+b)(y+c)}{y(y-z)(y-x)(y-x)} + \frac{(y+a)(y+b)(y+c)}{y(y-z)(y-x)(y-x)} = -\frac{abc}{wxyz}.$$

Source: [732, p. 161].

Fact 2.4.11. Let w, x, y, z be complex numbers such that division by zero does not occur in the expressions below. Then,

$$\frac{(wx-yz)(w^2-x^2+y^2-z^2)+(wy-xz)(w^2+x^2-y^2-z^2)}{(w^2-x^2+y^2-z^2)(w^2+x^2-y^2-z^2)+4(wx-yz)(wy-xz)} = \frac{(w+z)(x+y)}{(w+z)^2+(x+y)^2}.$$

Source: [732, p. 161].

Fact 2.4.12. Let w, x, y, z be complex numbers. Then,

$$(w^2-x^2)^2+(y^2-z^2)^2+2(wx-yz)^2+4wxyz = w^4+x^4+y^4+z^4.$$

If w, x, y, z are real, then

$$4wxyz \le w^4+x^4+y^4+z^4.$$

Furthermore, if w, x, y, z are nonnegative, then

$$(w^2-x^2)^2+(y^2-z^2)^2+2(wx-yz)^2 \le w^4+x^4+y^4+z^4.$$

Remark: This result yields the arithmetic-mean–geometric-mean inequality for four variables. See [288, pp. 226, 367].

Fact 2.4.13. Let w, x, y, z be complex numbers. Then,

$$(w^2+wx+x^2)[wy^3+xz^3+(w+x)(y+z)^3] = (y^2+yz+z^2)[yw^3+zx^3+(y+z)(w+x)^3].$$

Fact 2.4.14. Let x_1, x_2, x_3, x_4 be complex numbers. Then,

$$60(x_1^2+x_2^2+x_3^2+x_4^2)^3 = \sum_{1\le i<j<k\le 4}^{4} (x_i \pm x_j \pm x_k)^6 + 2\sum_{1\le i<j\le 4}^{4} (x_i \pm x_j)^6 + 36\sum_{i=1} x_i^6,$$

where the sums are taken over all choices of pluses and minuses as indicated by $\pm$. **Remark:** This is *Fleck's identity*. See [309, p. 14].

Fact 2.4.15. Let w, x, y, z be complex numbers, assume that $wz = xy$, and, for all $k \ge 1$, define

$$R_k \triangleq (w+x+y)^k + (x+y+z)^k + (w-z)^k - (w+y+z)^k - (w+x+z)^k - (x-y)^k,$$
$$H_k \triangleq (w+x+y)^k - (x+y+z)^k - (w-z)^k + (w+y+z)^k - (w+x+z)^k + (x-y)^k.$$

Then,

$$64R_6R_{10} = 45R_8^2, \quad 25H_3H_7 = 21H_5^2.$$

Source: [664, 708, 1403]. **Remark:** The first equality is *Ramanujan's 6-10-8 identity*. The second equality is *Hirschhorn's 3-7-5 identity*. **Related:** Fact 2.2.10.

Fact 2.4.16. Let w, x, y, z be real numbers. Then,

$$\left.\begin{matrix} wx+xy+yz+zw \\ 2(wy+xz) \end{matrix}\right\} \le w^2+x^2+y^2+z^2,$$

$$4wxyz \le w^2x^2 + x^2y^2 + y^2w^2 + z^4,$$
$$(wxyz+1)^4 \le (w^4+1)(x^4+1)(y^4+1)(z^4+1),$$
$$0 \le (w-y)^2(x-z)^2 + 4(w-x)(x-y)(y-z)(z-w),$$
$$2(w^2y^2 - wxyz + x^2z^2) \le 3(w^2 - wx + x^2)(y^2 - yz + z^2),$$
$$12(wx+xy+yz) \le 6(w^2+x^2+y^2+z^2) + (w+x+y+z)^2,$$
$$(5wy + wz + xy + 3xz)^2 \le (5w^2 + 2wx + 3x^2)(5y^2 + 2yz + 3z^2),$$
$$8wxyz(wx+wy+wz+xy+xz+yz) \le 3(wxy+xyz+yzw+zwx)^2,$$
$$8(wx+xy+yz+zw+wy+xz) \le 3(w+x+y+z)^2 \le 12(w^2+x^2+y^2+z^2),$$
$$9(w+x+y+z)(wxy+xyz+yzw+zwx) \le 4(wx+wy+wz+xy+xz+yz)^2.$$

The fifth string is an equality if and only if either $w/x = y/z = \frac{1}{2}(3+\sqrt{5})$ or $w/x = y/z = \frac{1}{2}(3-\sqrt{5})$. **Source:** [108, p. 450], [289, pp. 31, 162, 163], [309, p. 78], [748, p. 10], [752, pp. 161, 162], [1158, pp. 171, 561, 562], [1371, p. 134], and [2527, p. 14]. **Related:** Fact 2.11.19.

Fact 2.4.17. Let w, x, y, z be real numbers. Then,

$$3w(x^3y + y^3z + z^3x) + 3w(w-1)xyz(x+y+z) \le (3w^2-1)(x^2y^2+y^2z^2+z^2x^2) + x^4+y^4+z^4,$$
$$w(x^3y+y^3z+z^3x+xy^3+xz^3+zx^3) \le \tfrac{1}{2}w(w-2)xyz(x+y+z)+\tfrac{1}{4}(w^2+8)(x^2y^2+y^2z^2+z^2x^2)+x^4+y^4+z^4.$$

Source: [98]. The second inequality is equivalent to $[w(xy+yz+zx) - 2(x^2+y^2+z^2)]^2 \ge 0$.

Fact 2.4.18. Let $w, x, y, z \in [0,1]$. Then,

$$1 \le (1-w)(1-x)(1-y)(1-z) + w + x + y + z.$$

Source: [112, p. 37].

Fact 2.4.19. Let a_1, a_2, a_3, a_4 be nonnegative numbers. Then,

$$\sqrt[4]{a_1a_2a_3a_4} \le \tfrac{1}{4}(a_1+a_2+a_3+a_4) \le \sqrt[4]{a_1a_2a_3a_4} + \tfrac{3}{2}\max_{i,j\in\{1,2,3,4\}}(\sqrt{a_i} - \sqrt{a_j})^2.$$

Source: [888].

Fact 2.4.20. Let w, x, y, z be nonnegative numbers. Then,

$$\sqrt{wx} + \sqrt{yz} \le \sqrt{(w+y)(x+z)}, \quad \sqrt[3]{wy} + \sqrt[3]{xz} \le \sqrt[3]{(w+x+y)(w+x+z)},$$
$$\sqrt{(w+y)^2 + (x+z)^2} \le \sqrt{w^2+x^2} + \sqrt{y^2+z^2},$$
$$6\sqrt[4]{wxyz} \le \sqrt{(w+x)(y+z)} + \sqrt{(w+y)(x+z)} + \sqrt{(w+z)(x+y)},$$
$$16wxyz(w+1)(x+1)(y+1)(z+1) \le (w+x)(x+y)(y+z)(z+w)(\sqrt[4]{wxyz}+1)^4,$$
$$\sqrt{w^2 + x(2y+z)} + \sqrt{x^2 + y(2z+w)} + \sqrt{y^2 + z(2w+x)} + \sqrt{z^2 + w(2x+y)} \le 2(w+x+y+z).$$

Source: The first inequality follows from Fact 2.2.12. The second inequality is given in [752, p. 161]. The third inequality is a special case of Minkowski's inequality given by Fact 2.12.51. The fourth inequality is given in [1371, p. 120]. The second last inequality is given in [752, p. 163]. The last inequality is given in [377].

Fact 2.4.21. Let w, x, y, z, α be nonnegative numbers. Then,

$$(\alpha+1)^2 \le \left(\frac{\alpha w}{x+y} + 1\right)\left(\frac{\alpha x}{y+z} + 1\right)\left(\frac{\alpha y}{z+w} + 1\right)\left(\frac{\alpha z}{w+x} + 1\right).$$

Source: [748, p. 69].

Fact 2.4.22. Let w and x be positive numbers, and let y and z be nonnegative numbers. Then,

$$4(z-y)\left(\frac{1}{x+z}-\frac{1}{w+y}\right)\le(w-x)\left(\frac{1}{x}-\frac{1}{w}\right).$$

Source: [355].

Fact 2.4.23. Let w, x, y, z be nonnegative numbers. Then,

$$8wxyz \le (wx+yz)(w+x)(y+z),$$
$$4wxyz \le wxy^2+xyz^2+yzw^2+zwx^2,$$
$$4(wx+xy+yz+zw)\le(w+x+y+z)^2,$$
$$16(wxy+xyz+yzw+zwx)\le(w+x+y+z)^3,$$
$$(wxyz+1)^3\le(w^3+1)(x^3+1)(y^3+1)(z^3+1),$$
$$4wxyz\le wx(y^2+z^2)+yz(w^2+x^2)=(yx+wz)(wy+xz),$$
$$81wxyz\le(w^2+w+1)(x^2+x+1)(y^2+y+1)(z^2+z+1),$$
$$wx+xy+yz+zw+wy+xz\le 1+w^2+x^2+y^2+z^2+wxyz,$$
$$(wxyz+1)^2\le 4(w^2-w+1)(x^2-x+1)(y^2-y+1)(z^2-z+1),$$
$$(w+x+y+z)(w^3+x^3+y^3+z^3)\le 3(w^4+x^4+y^4+z^4)+4wxyz,$$
$$(w+x+y+z)^3\le 4(w^3+x^3+y^3+z^3)+15(wxy+xyz+yzx+zxy),$$
$$27(w+x+y+z)(wxy+xyz+yzw+zwx)\le(w+x+y+z)^4+176wxyz,$$
$$8(w^4+x^4+y^4+z^4)(wx+wy+wz+xy+xz+yz)\le 3(w^2+x^2+y^2+z^2)^3,$$
$$64(w^2+x^2+y^2)(x^2+y^2+z^2)(y^2+z^2+w^2)(z^2+w^2+x^2)\le(w+x+y+z)^8,$$
$$108(wxy+xyz+yzw+zwx)^2\le 8(wx+wy+wz+xy+xz+yz)^3\le 27(w^2+x^2+y^2+z^2)^3,$$
$$(w^3+x^3+y^3+z^3)(wx+wy+wz+xy+xz+yz)\le\tfrac{1}{2}(w^2+x^2+y^2+z^2)^{5/2}+2wxyz(w+x+y+z),$$
$$\begin{aligned}256wxyz&\le 16(w+x+y+z)(wxy+xyz+yzw+zwx)\\&\le 32\sqrt[3]{2}(wxy+xyz+yzw+zwx)^{4/3}\le\tfrac{64}{9}(wx+xy+yz+zw+wy+xz)^2\\&\le(w+x+y+z)^4\le 16(w+x+y+z)(w^3+x^3+y^3+z^3),\end{aligned}$$
$$\begin{aligned}(wx+wy+wz+xy+xz+yz)^2&\le 6(w^2x^2+w^2y^2+w^2z^2+x^2y^2+x^2z^2+y^2z^2)\\&\le 6(w^4+x^4+y^4+z^4+2wxyz),\end{aligned}$$
$$\begin{aligned}\tfrac{1}{2}(w+x+y+z)(w^3+x^3+y^3+z^3)&+(wx+wy+wz+xy+xz+yz)(w^2+x^2+y^2+z^2)\\&\le(w+x+y+z)(w^2+x^2+y^2+z^2)^{3/2},\end{aligned}$$
$$\begin{aligned}w^3x^3y^3+x^3y^3z^3+y^3z^3w^3+z^3w^3x^3&\le(wxy+xyz+yzw+zwx)^3\\&\le 16(w^3x^3y^3+x^3y^3z^3+y^3z^3w^3+z^3w^3x^3),\end{aligned}$$
$$\frac{1}{2}(wxyz+1)\le\frac{(w^3+1)(x^3+1)(y^3+1)(z^3+1)}{(w^2+1)(x^2+1)(y^2+1)(z^2+1)},$$
$$\frac{1}{w+x+y+1}+\frac{1}{w+x+1+z}+\frac{1}{w+1+y+z}+\frac{1}{1+x+y+z}\le 1.$$

Now, assume that w, x, y, z are positive. Then,

$$\frac{64}{w+x+y+z} \le \frac{1}{w}+\frac{1}{x}+\frac{4}{y}+\frac{16}{z}, \quad 0 \le \frac{w-x}{x+y}+\frac{x-y}{y+z}+\frac{y-z}{z+w}+\frac{z-w}{w+x},$$

$$1 \le \left(\frac{w}{w+x}\right)^2+\left(\frac{x}{x+y}\right)^2+\left(\frac{y}{y+z}\right)^2+\left(\frac{z}{z+w}\right)^2,$$

$$1 \le \frac{y}{w+3x}+\frac{z}{x+3y}+\frac{w}{y+3z}+\frac{x}{z+3w}, \quad \frac{1}{8} \le \left(\frac{w}{w+x}\right)^5+\left(\frac{x}{x+y}\right)^5+\left(\frac{y}{y+z}\right)^5+\left(\frac{z}{z+w}\right)^5,$$

$$\frac{4}{wy+xz} \le \frac{1}{w^2+wx}+\frac{1}{x^2+xy}+\frac{1}{y^2+yz}+\frac{1}{z^2+zx},$$

$$w+x+y+z \le \frac{2w^3}{w^2+x^2}+\frac{2x^3}{x^2+y^2}+\frac{2y^3}{y^2+z^2}+\frac{2z^3}{z^2+w^2},$$

$$0 \le \frac{w-x}{w+2x+y}+\frac{x-y}{x+2y+z}+\frac{y-z}{y+2z+w}+\frac{z-x}{z+2w+x},$$

$$\frac{1}{wx}+\frac{1}{xy}+\frac{1}{yz}+\frac{1}{zx}+\frac{1}{wy}+\frac{1}{wz} \le \frac{3}{8}\left(\frac{1}{w}+\frac{1}{x}+\frac{1}{y}+\frac{1}{z}\right)^2,$$

$$\frac{16}{1+8\sqrt{wxyz}} \le \frac{1}{w(x+1)}+\frac{1}{x(y+1)}+\frac{1}{y(z+1)}+\frac{1}{z(w+1)},$$

$$\frac{2}{3} \le \frac{w}{x+2y+3z}+\frac{x}{y+2z+3w}+\frac{y}{z+2w+3x}+\frac{z}{w+2x+3y},$$

$$3(wxy+xyz+yzw+zwx) \le w^3+x^3+y^3+z^3+\frac{32wxyz}{w+x+y+z},$$

$$\frac{4}{9} \le \left(\frac{w}{w+x+y}\right)^2+\left(\frac{x}{x+y+z}\right)^2+\left(\frac{y}{y+z+w}\right)^2+\left(\frac{z}{z+w+x}\right)^2,$$

$$\frac{16}{3(w+x+y+z)} \le \frac{1}{w+x+y}+\frac{1}{x+y+z}+\frac{1}{y+z+w}+\frac{1}{z+w+x},$$

$$0 \le \frac{(w-x)(w-y)}{w+x+y}+\frac{(x-y)(x-z)}{x+y+z}+\frac{(y-z)(y-w)}{y+z+w}+\frac{(z-w)(z-x)}{z+w+x},$$

$$w+x+y+z \le \frac{w^2+x^2+y^2}{w+x+y}+\frac{x^2+y^2+z^2}{x+y+z}+\frac{y^2+z^2+x^2}{y+z+x}+\frac{z^2+x^2+y^2}{z+x+y},$$

$$\frac{12}{(w+x+y+z)^2} \le \frac{1}{w^2+x^2+y^2}+\frac{1}{x^2+y^2+z^2}+\frac{1}{y^2+z^2+w^2}+\frac{1}{z^2+w^2+x^2},$$

$$w^2+x^2+y^2+z^2 \le \frac{w^3+x^3+y^3}{w+x+y}+\frac{x^3+y^3+z^3}{x+y+z}+\frac{y^3+z^3+w^3}{y+z+w}+\frac{z^3+w^3+x^3}{z+w+x},$$

$$\frac{1}{(w+x)(y+z)}+\frac{1}{(w+y)(x+z)}+\frac{1}{(w+z)(x+y)}$$
$$\le \frac{9}{16}\left[\frac{1}{w(x+y+z)}+\frac{1}{x(y+z+w)}+\frac{1}{y(z+w+x)}+\frac{1}{z(w+x+y)}\right].$$

If $wxyz = 1$, then

$$(w+x+y+z)^2 \le (w^2+1)(x^2+1)(y^2+1)(z^2+1).$$

Source: [107, pp. 139, 140], [108, pp. 20, 134], [109, pp. 347–349], [289, pp. 15, 169], [748, pp. 6, 8–10, 373, 374], [752, pp. 161–163, 246], [806, pp. 39, 118, 131, 190, 194, 197, 198, 204], [814], [993, p. 179], [1158, p. 34], [1371, pp. 120, 123, 124, 134, 144, 161], [1591, 2620, 2619, 2621]. **Remark:** $(w+x+y+z)^3 \le 16(w^3+x^3+y^3+z^3)$ is given by Fact 2.11.15. **Remark:** $16wxyz \le (w+x+y+z)(wxy+xyz+yzw+zwx)$ is given by Fact 2.11.43. **Remark:** $4wxyz \le w^2xy+xyz^2+y^2zw+zwx^2$ follows from Fact 2.11.48 with $n = 2$. **Remark:** $4wxyz \le wx^2z+xy^2w+yz^2x+zw^2y$ is given by Fact 2.11.48. **Remark:** $w^2x^2+w^2y^2+w^2z^2+x^2y^2+x^2z^2+y^2z^2 \le w^4+x^4+y^4+z^4+2wxyz$ is *Turkevici's inequality*. See [107, p. 458]. **Related:** Fact 2.11.35, Fact 2.11.96, and Fact 2.11.98.

2.5 Facts on Equalities and Inequalities in Five Variables

Fact 2.5.1. Let v, w, x, y, z be positive numbers. Then,

$$\frac{3}{v+w} \le \frac{x}{vy+wz} + \frac{y}{vz+wx} + \frac{z}{vx+wy},$$

$$0 \le \frac{v-w}{w+x} + \frac{w-x}{x+y} + \frac{x-y}{y+z} + \frac{y-z}{z+v} + \frac{z-v}{v+w},$$

$$\frac{3}{(v+w)^2} \le \frac{x^2}{(vy+wz)^2} + \frac{y^2}{(vz+wx)^2} + \frac{z^2}{(vx+wy)^2},$$

$$\frac{243}{32} \le \left(\frac{v+w+x}{w+x}\right)\left(\frac{w+x+y}{x+y}\right)\left(\frac{x+y+z}{y+z}\right)\left(\frac{y+z+v}{z+v}\right)\left(\frac{z+v+w}{v+w}\right),$$

$$0 \le \frac{v-w}{v+2w+x} + \frac{w-x}{w+2x+y} + \frac{x-y}{x+2y+z} + \frac{y-z}{y+2z+v} + \frac{z-v}{z+2v+w},$$

$$\sqrt{v^2+2(w+x)y} + \sqrt{w^2+2(x+y)z} + \sqrt{x^2+2(y+z)w} + \sqrt{y^2+2(z+w)v} + \sqrt{z^2+2(v+w)x}$$
$$\le \sqrt{5}(v+w+x+y+z).$$

Source: [320, 377], [752, pp. 246, 320], and [1938, p. 38].

Fact 2.5.2. Let $v, w, x, y, z \in [0, 1]$. Then,

$$\frac{5}{1+2\sqrt[5]{vwxyz}} \le \frac{1}{v+w+1} + \frac{1}{w+x+1} + \frac{1}{x+y+1} + \frac{1}{y+z+1} + \frac{1}{z+v+1}.$$

Source: [752, p. 320].

Fact 2.5.3. Let v, w, x, y, z be nonnegative numbers, and assume that $v \ne w \ne x \ne y \ne z \ne v$. Then,

$$3 \le \frac{v}{|w-x|} + \frac{w}{|x-y|} + \frac{x}{|y-z|} + \frac{y}{|z-v|} + \frac{z}{|v-w|}.$$

Source: [752, p. 182].

Fact 2.5.4. Let v, w, x, y, z be positive numbers. Then,

$$\sqrt[5]{vwxyz} + \frac{(v-w)^2+(w-x)^2+(x-y)^2+(y-z)^2+(z-v)^2}{20(v+w+x+y+z)} \le \frac{1}{5}(v+w+x+y+z).$$

Source: [752, p. 182].

Fact 2.5.5. Let x_1, x_2, x_3, x_4, x_5 be nonnegative numbers, and define $x_6 \triangleq x_1$ and $x_7 \triangleq x_2$. Then,

$$25\sum_{i=1}^{5} x_ix_{i+1}x_{i+2} \le \left(\sum_{i=1}^{5} x_i\right)^3, \quad 125\sum_{i=1}^{5} x_ix_{i+1}x_{i+2}x_{i+3} \le \left(\sum_{i=1}^{5} x_i\right)^4, \quad 3125x_1x_2x_3x_4x_5 \le \left(\sum_{i=1}^{5} x_i\right)^5.$$

Source: Fact 2.11.16. **Related:** Fact 2.6.17.

Fact 2.5.6. Let v, w, x, y, z be real numbers. Then,

$$(vwy^2 + v^2yz + wx)^2 \le (v^2y^2 + w^2)[(vy^2 - x)^2 + (vz + 2wy)^2].$$

Source: Use the Cauchy-Schwarz inequality given by Fact 2.12.17 or use the fact that $(v^2y^3+2w^2y-vxy + vwz)^2 \ge 0$. **Remark:** Consider a particle moving in a plane, and define $v = r$, $w = \dot{r}$, $x = \ddot{r}$, $y = \dot{\theta}$, and $z = \ddot{\theta}$, and let L and R denote the left-hand side and right-hand side of the inequality, respectively. Then, $R - L = V^6/\rho^2$, where V is the particle speed and ρ is the radius of curvature of its path.

Fact 2.5.7. Let v, w, x, y, z be real numbers, and assume that $z \le y \le x \le w \le v$. Then,

$$8(vx + wy + xz) \le (v + w + x + y + z)^2.$$

Source: [748, p. 10].

2.6 Facts on Equalities and Inequalities in Six Variables

Fact 2.6.1. Let u, v, w, x, y, z be complex numbers. Then,

$$x^6 + y^6 + z^6 + u^6 + v^6 + w^6 - 6xyzuvw = \tfrac{1}{2}(x^2 + y^2 + z^2)[(x^2 - y^2)^2 + (y^2 - z^2)^2 + (z^2 - x^2)^2] + \tfrac{1}{2}(u^2 + v^2 + w^2)[(u^2 - v^2)^2 + (v^2 - w^2)^2 + (w^2 - u^2)^2] + 3(xyz - uvw)^2.$$

Remark: This equality yields the arithmetic-mean–geometric-mean inequality for six variables. See [288, p. 226].

Fact 2.6.2. Let u, v, w, x, y, z be complex numbers. Then,

$$(u^3 + v^3 + w^3 - 3uvw)(x^3 + y^3 + z^3 - 3xyz) = (ux + vy + wz)^3 + (uy + vz + wx)^3 + (uz + vx + wy)^3 - 3(ux + vy + wz)(uy + vz + wx)(uz + vx + wy).$$

Source: [1016].

Fact 2.6.3. Let a, b, c, x, y, z be real numbers. Then,

$$3(bcx + cay + abz)^2 \le 4(a^2 + x^2)(b^2 + y^2)(c^2 + z^2).$$

Source: [748, p. 10] and [752, p. 182].

Fact 2.6.4. Let a, b, c, x, y, z be real numbers, let $\alpha \in [-1, 2]$, and assume that $ax + by + cz \ne 0$ and $(\alpha + 2)(a^2 + b^2 + c^2) \ne 2\alpha(ab + bc + ca)$. Then,

$$\frac{(2 - \alpha)(\alpha + 1)}{(\alpha + 2)(a^2 + b^2 + c^2) - 2\alpha(ab + bc + ca)} \le \frac{x^2 + y^2 + z^2 + \alpha(xy + yz + zx)}{(ax + by + cz)^2}.$$

Source: [1615].

Fact 2.6.5. Let a, b, c, x, y, z be real numbers. Then,

$$(2az + 2acx - by)^2 = 4[a^2(z - cx)^2 + b(z - cx)(bx - ay) + c(bx - ay)^2] + (b^2 - 4a^2c)(y^2 - 4xz).$$

Fact 2.6.6. Let x and y be nonnegative numbers, and let a, b, c, d be real numbers. Then,

$$\max\{ax + by, cx + dy\} \le (\max\{a, c\})x + (\max\{b, d\})y \le (a + c)x + (b + d)y.$$

Fact 2.6.7. Let a, b, c be positive numbers, and let x, y, z be real numbers. Then,

$$\frac{(x + y + z)^2}{a + b + c} \le \frac{(x + y)^2}{a + b} + \frac{z^2}{c} \le \frac{x^2}{a} + \frac{y^2}{b} + \frac{z^2}{c}.$$

Source: [1938, p. 35]. **Related:** Fact 2.4.3 and Fact 2.12.19.

Fact 2.6.8. Let a, b, c be positive numbers, and let x, y, z be real numbers. Then,

$$ax(y+z)+by(z+x)+cz(x+y) \le \frac{(x+y+z)^2(a+b)(b+c)(c+a)}{4(ab+bc+ca)},$$

$$\frac{xy}{(b+c)(c+a)}+\frac{yz}{(c+a)(a+b)}+\frac{zx}{(a+b)(b+c)} \le \frac{(x+y+z)^2}{4(ab+bc+ca)}.$$

Source: [108, pp. 460, 461].

Fact 2.6.9. Let a, b, c be nonnegative numbers, and let x, y, z be real numbers. Then,

$$4(ab+bc+ca)(xy+yz+zx) \le [x(b+c)+y(c+a)+z(a+b)]^2.$$

Now, assume that x, y, z are nonnegative. Then,

$$2\sqrt{(ab+bc+ca)(xy+yz+zx)} \le a(y+z)+b(z+x)+c(x+y),$$

$$ax+by+cz+2\sqrt{(ab+bc+ca)(xy+yz+zx)} \le (a+b+c)(x+y+z).$$

Furthermore, if x, y, z are positive, then

$$3\sqrt{ab+bc+ca} \le \frac{x(b+c)}{y+z}+\frac{y(c+a)}{z+x}+\frac{z(a+b)}{x+y}.$$

Source: [108, pp. 351–353, 461] and [806, p. 32, 33, 202]. **Credit:** The last inequality is due to W. Janous.

Fact 2.6.10. Let a, b, c, x, y, z be nonnegative numbers. Then,

$$(ayz+bzx+cxy-xyz)^2 \le (a^2+x^2)(b^2+y^2)(c^2+z^2).$$

Source: [108, pp. 48, 277, 278].

Fact 2.6.11. Let a, b, c, x, y, z be positive numbers. Then,

$$\frac{9}{(a+b)z+(b+c)x+(c+a)y} \le \frac{2}{(a+b)(x+y)}+\frac{2}{(b+c)(y+z)}+\frac{2}{(c+a)(z+x)}.$$

Source: [752, p. 184].

Fact 2.6.12. Let a, b, c, d, x, y be positive numbers, and assume that $\frac{c}{d} < \frac{a}{b}$ and $x < y$. Then,

$$\frac{ax+c}{bx+d} < \frac{ay+c}{by+d}.$$

Source: [1096].

Fact 2.6.13. Let a, b, c, x, y, z be positive numbers. Then,

$$\frac{ax}{a+x}+\frac{by}{b+y}+\frac{cz}{c+z} \le \frac{(a+b+c)(x+y+z)}{a+b+c+x+y+z}.$$

Source: [806, p. 183].

Fact 2.6.14. Let a, b, c, x, y, z be positive numbers. Then,

$$\frac{(x+y+z)^3}{3(a+b+c)} \le \frac{x^3}{a}+\frac{y^3}{b}+\frac{z^3}{c}.$$

Source: [806, p. 100].

Fact 2.6.15. Let a, b, c, x, y, z be positive numbers. Then, the following statements are equivalent:

i) The following statements hold:

a) $a + b + c = x + y + z$.

b) $a^2 + b^2 + c^2 = x^2 + y^2 + z^2$.

c) $\max\{a, b, c\} \le \max\{x, y, z\}$.

ii) The following statements hold:

a) For all $p \in [0, 1] \cup [2, \infty)$, $a^p + b^p + c^p \le x^p + y^p + z^p$.

b) For all $p \in [1, 2]$, $x^p + y^p + z^p \le a^p + b^p + c^p$.

Now, suppose that *i)* and *ii)* hold, and let $p \in [0, \infty)$. Then, the following statements are equivalent:

iii) $a^p + b^p + c^p = x^p + y^p + z^p$.

iv) Either $p \in \{0, 1, 2\}$ or $\{a, b, c\}_{\rm ms} = \{x, y, z\}_{\rm ms}$.

Source: [414].

Fact 2.6.16. Let $a, b, c, x, y, z > 0$. Then,

$$\frac{3(a+b+c)}{2} \le \frac{ax+by+cz}{x+y} + \frac{ay+bz+cx}{y+z} + \frac{az+bx+cy}{z+x},$$

$$\frac{3(a+b+c)^2}{4} \le \frac{(ax+by+cz)(ay+bz+cx)}{(x+y)(y+z)} + \frac{(ay+bz+cx)(z+x+4y)}{(y+z)(z+x)} + \frac{(az+bx+cy)(ax+by+cz)}{(z+x)(x+y)},$$

$$\frac{(a+b+c)^3}{8} \le \frac{(ax+by+cz)(ay+bz+cx)(az+bx+cy)}{(x+y)(y+z)(z+x)}.$$

Source: [1772].

Fact 2.6.17. Let $x_1, x_2, x_3, x_4, x_5, x_6$ be nonnegative numbers, and define $x_7 \triangleq x_1$ and $x_8 \triangleq x_2$. Then,

$$216\sum_{i=1}^{6} x_i x_{i+1} x_{i+2} x_{i+3} \le \left(\sum_{i=1}^{6} x_i\right)^4, \quad 1296\sum_{i=1}^{6} x_i x_{i+1} x_{i+2} x_{i+3} x_{i+4} \le \left(\sum_{i=1}^{6} x_i\right)^5,$$

$$46656 x_1 x_2 x_3 x_4 x_5 x_6 \le \left(\sum_{i=1}^{6} x_i\right)^6.$$

Source: Fact 2.11.16. The first inequality is given in [108, p. 462]. **Related:** Fact 2.5.5.

Fact 2.6.18. Let $x_1, x_2, x_3, x_4, x_5, x_6 \in [\sqrt{3}/3, \sqrt{3}]$. Then,

$$0 \le \frac{x_1 - x_2}{x_2 + x_3} + \frac{x_2 - x_3}{x_3 + x_4} + \frac{x_3 - x_4}{x_4 + x_5} + \frac{x_4 - x_5}{x_5 + x_6} + \frac{x_5 - x_6}{x_6 + x_5} + \frac{x_6 - x_1}{x_1 + x_2}.$$

Source: [748, p. 379].

2.7 Facts on Equalities and Inequalities in Seven Variables

Fact 2.7.1. Let x, y, z be positive numbers, let p, q, r, s be nonnegative numbers, and assume that $r \le s$ and $p \le q$. Then,

$$\frac{(x^s + y^s + z^s)(x^q + y^q + z^q)}{(x^r + y^r + z^r)(x^p + y^p + z^p)} \le \frac{x^{s+q} + y^{s+q} + z^{s+q}}{x^{r+p} + y^{r+p} + z^{r+p}}.$$

Source: [53].

2.8 Facts on Equalities and Inequalities in Eight Variables

Fact 2.8.1. Let $x_1, x_2, x_3, x_4, y_1, y_2, y_3, y_4$ be complex numbers. Then,

$$(x_1^2 + x_2^2 + x_3^2 + x_4^2)(y_1^2 + y_2^2 + y_3^2 + y_4^2) = (x_1y_1 - x_2y_2 - x_3y_3 - x_4y_4)^2 + (x_1y_2 + x_2y_1 + x_3y_4 - x_4y_3)^2 + (x_1y_3 - x_2y_4 + x_3y_1 + x_4y_2)^2 + (x_1y_4 + x_2y_3 - x_3y_2 + x_4y_1)^2.$$

Hence,

$$\left.\begin{array}{r} (x_1y_1 - x_2y_2 - x_3y_3 - x_4y_4)^2 + (x_1y_2 + x_2y_1 + x_3y_4 - x_4y_3)^2 \\ + (x_1y_3 - x_2y_4 + x_3y_1 + x_4y_2)^2 \\ (x_1y_1 - x_2y_2 - x_3y_3 - x_4y_4)^2 + (x_1y_2 + x_2y_1 + x_3y_4 - x_4y_3)^2 \\ + (x_1y_4 + x_2y_3 - x_3y_2 + x_4y_1)^2 \\ (x_1y_1 - x_2y_2 - x_3y_3 - x_4y_4)^2 + (x_1y_3 - x_2y_4 + x_3y_1 + x_4y_2)^2 \\ + (x_1y_4 + x_2y_3 - x_3y_2 + x_4y_1)^2 \\ (x_1y_2 + x_2y_1 + x_3y_4 - x_4y_3)^2 + (x_1y_3 - x_2y_4 + x_3y_1 + x_4y_2)^2 \\ + (x_1y_4 + x_2y_3 - x_3y_2 + x_4y_1)^2 \end{array}\right\} \le (x_1^2 + x_2^2 + x_3^2 + x_4^2)(y_1^2 + y_2^2 + y_3^2 + y_4^2).$$

Credit: L. Euler. See [2228, p. 8]. **Remark:** Replacing x_2, x_3, x_4 by $-x_2, -x_3, -x_4$ yields

$$(x_1^2 + x_2^2 + x_3^2 + x_4^2)(y_1^2 + y_2^2 + y_3^2 + y_4^2) = (x_1y_1 + x_2y_2 + x_3y_3 + x_4y_4)^2 + (x_1y_2 - x_2y_1 - x_3y_4 + x_4y_3)^2 + (x_1y_3 + x_2y_4 - x_3y_1 - x_4y_2)^2 + (x_1y_4 - x_2y_3 + x_3y_2 - x_4y_1)^2.$$

Remark: This equality represents a relationship between a pair of quaternions. An analogous equality holds for two sets of eight variables representing a pair of octonions. See [773, p. 77].

Fact 2.8.2. Let a, b, c, x, y, z, p, q be positive numbers. Then,

$$\frac{18}{(p+1)(q+1)[(b+c)x + (c+a)y + (a+b)z]} \le \frac{1}{(a+pb)(x+qy)} + \frac{1}{(b+pc)(y+qz)} + \frac{1}{(c+pa)(z+qx)}.$$

Source: [752, p. 320].

2.9 Facts on Equalities and Inequalities in Nine Variables

Fact 2.9.1. Let $x_1, x_2, x_3, y_1, y_2, y_3, z_1, z_2, z_3$ be nonnegative numbers. Then,

$$(x_1y_1z_1 + x_2y_2z_2 + x_3y_3z_3)^3 \le (x_1^3 + x_2^3 + x_3^3)(y_1^3 + y_2^3 + y_3^3)(z_1^3 + z_2^3 + z_3^3).$$

Source: [806, p. 98].

2.10 Facts on Equalities and Inequalities in Sixteen Variables

Fact 2.10.1. Let $x_1, \ldots, x_8, y_1, \ldots, y_n$ be complex numbers. Then,

$$\begin{aligned}\left(\sum_{i=1}^{8} x_i^2\right)\sum_{i=1}^{8} y_i^2 = {} & (x_1y_1 - x_2y_2 - x_3y_3 - x_4y_4 - x_5y_5 - x_6y_6 - x_7y_7 - x_8y_8)^2 \\ & + (x_1y_2 + x_2y_1 + x_3y_4 - x_4y_3 + x_5y_6 - x_6y_5 - x_7y_8 + x_8y_7)^2 \\ & + (x_1y_3 - x_2y_4 + x_3y_1 + x_4y_2 + x_5y_7 + x_6y_8 - x_7y_5 - x_8y_6)^2 \\ & + (x_1y_4 + x_2y_3 - x_3y_2 + x_4y_1 + x_5y_8 - x_6y_7 + x_7y_6 - x_8y_5)^2 \\ & + (x_1y_5 - x_2y_6 - x_3y_7 - x_4y_8 + x_5y_1 + x_6y_2 + x_7y_3 + x_8y_4)^2\end{aligned}$$

$$+ (x_1y_6 + x_2y_5 - x_3y_8 + x_4y_7 - x_5y_2 + x_6y_1 - x_7y_4 + x_8y_3)^2$$
$$+ (x_1y_7 + x_2y_8 + x_3y_5 - x_4y_6 - x_5y_3 + x_6y_4 + x_7y_1 - x_8y_2)^2$$
$$+ (x_1y_8 - x_2y_7 + x_3y_6 + x_4y_5 - x_5y_4 - x_6y_3 + x_7y_2 + x_8y_1)^2.$$

Source: [127] and [2316, p. 94]. **Remark:** This is *Degen's eight-square identity.*

2.11 Facts on Equalities and Inequalities in n Variables

Fact 2.11.1. Let $z_1, \ldots, z_n$ be complex numbers. Then,

$$\sum_{i=1}^{n}\sum_{j=1}^{i} z_iz_j = \frac{1}{2}\left[\left(\sum_{i=1}^{n} z_i\right)^2 + \sum_{i=1}^{n} z_i^2\right].$$

Source: [1219, p. 37].

Fact 2.11.2. Let $z_1, \ldots, z_n \in \mathbb{C}$, let $\alpha_1, \ldots, \alpha_n \in [0,1]$, and assume that $\sum_{i=1}^{n} \alpha_i = 1$. Then,

$$\sum_{1\le i<j\le n} \alpha_i\alpha_j(z_i - z_j)^2 = \sum_{j=1}^{n} \alpha_j\left(z_j - \sum_{i=1}^{n}\alpha_iz_i\right)^2 = \sum_{i=1}^{n}\alpha_iz_i^2 - \left(\sum_{i=1}^{n}\alpha_iz_i\right)^2.$$

Source: [1115].

Fact 2.11.3. Let $n \ge 2$, and let $z_1, \ldots, z_n$ be distinct complex numbers. Then,

$$\sum_{i=1}^{n}\prod_{j\in\{1,\ldots,n\},\, j\ne i}\frac{z_i + z_j}{z_i - z_j} = \begin{cases} 0, & n \text{ even},\\ 1, & n \text{ odd}.\end{cases}$$

Source: [1042].

Fact 2.11.4. Let $n \ge 2$, let $z_1, \ldots, z_n$ be distinct complex numbers, and let $0 \le k \le n$. Then,

$$\sum_{i=1}^{n}\frac{z_i^k}{\prod_{j=1, j\ne i}^{n}(z_i - z_j)} = \begin{cases} 0, & 0 \le k < n-1,\\ 1, & k = n-1,\\ \sum_{i=1}^{n} z_i, & k = n.\end{cases}$$

In particular,

$$\frac{1}{z_1 - z_2} + \frac{1}{z_2 - z_1} = 0, \quad \frac{z_1}{z_1 - z_2} + \frac{z_2}{z_2 - z_1} = 1, \quad \frac{z_1^2}{z_1 - z_2} + \frac{z_2^2}{z_2 - z_1} = z_1 + z_2,$$
$$\frac{1}{(z_1 - z_2)(z_1 - z_3)} + \frac{1}{(z_2 - z_1)(z_2 - z_3)} + \frac{1}{(z_3 - z_1)(z_3 - z_2)} = 0,$$
$$\frac{z_1}{(z_1 - z_2)(z_1 - z_3)} + \frac{z_2}{(z_2 - z_1)(z_2 - z_3)} + \frac{z_3}{(z_3 - z_1)(z_3 - z_2)} = 0,$$
$$\frac{z_1^2}{(z_1 - z_2)(z_1 - z_3)} + \frac{z_2^2}{(z_2 - z_1)(z_2 - z_3)} + \frac{z_3^2}{(z_3 - z_1)(z_3 - z_2)} = 1,$$
$$\frac{z_1^3}{(z_1 - z_2)(z_1 - z_3)} + \frac{z_2^3}{(z_2 - z_1)(z_2 - z_3)} + \frac{z_3^3}{(z_3 - z_1)(z_3 - z_2)} = z_1 + z_2 + z_3.$$

Source: [1650, pp. 36–38]. **Related:** Fact 2.11.5. The first equality is a special case of Abel's theorem given by Fact 12.16.6.

Fact 2.11.5. Let $n \ge 2$, let $z_1, \ldots, z_n$ be distinct complex numbers, and let $k \ge 0$. Then,

$$\sum_{i=1}^{n}\frac{z_i^{n+k}}{\prod_{j=1, j\ne i}^{n}(z_i - z_j)} = \sum\prod_{j=1}^{n} z_j^{i_j},$$

where the second sum is taken over all n-tuples $(i_1, \ldots, i_n)$ of nonnegative integers whose sum is $k+1$. In particular,

$$\frac{z_1^5}{(z_1-z_2)(z_1-z_3)(z_1-z_4)} + \frac{z_2^5}{(z_2-z_1)(z_2-z_3)(z_2-z_4)} + \frac{z_3^5}{(z_3-z_1)(z_3-z_2)(z_3-z_4)}$$
$$+ \frac{z_4^5}{(z_4-z_1)(z_4-z_2)(z_4-z_3)} = z_1^2+z_2^2+z_3^2+z_4^2+z_1z_2+z_1z_3+z_1z_4+z_2z_3+z_2z_4+z_3z_4.$$

Source: [763]. **Remark:** Setting $k=0$ yields the third case in Fact 2.11.4.

Fact 2.11.6. Let $n \ge 2$, and let $z_1, \ldots, z_n$ be distinct, nonzero complex numbers. Then,

$$\sum_{i=1}^n \frac{1}{z_i} = (-1)^{n-1}\left(\prod_{i=1}^n z_i\right)\sum_{i=1}^n \frac{1}{z_i^2 \prod_{j=1, j\ne i}^n (z_i - z_j)}.$$

Source: [1672]. **Related:** Fact 2.11.4 and Fact 2.11.5.

Fact 2.11.7. Let $z, z_1, \ldots, z_n \in \mathbb{C}$, and assume that $-z, z_1, \ldots, z_n$ are distinct. Then,

$$\prod_{i=1}^n \frac{1}{z+z_i} = \sum_{i=1}^n \frac{1}{z+z_i}\prod_{\substack{j=1\\ j\ne i}}^n \frac{1}{z_j - z_i}.$$

Fact 2.11.8. Let $z_1, \ldots, z_n$ be complex numbers, and let $k \ge 1$. Then,

$$\left(\sum_{i=1}^n z_i\right)^k = \sum \binom{k!}{i_1, \ldots, i_n} z_1^{i_1}\cdots z_n^{i_n},$$

where the sum is taken over all $\binom{n+k-1}{n}$ n-tuples $(i_1, \ldots, i_n)$ of nonnegative integers such that $\sum_{j=1}^n i_j = k$. **Remark:** This is the *multinomial theorem.* **Remark:** Probabilistic interpretations are discussed in [1587]. **Related:** Fact 2.3.1.

Fact 2.11.9. Let $n \ge 2$, and let $x_1 \ldots, x_n$ be complex numbers. Then,

$$\sum_{(i_1,\ldots,i_n)\in\{1,2\}^n} (-1)^{\sum_{j=1}^n i_j}\left(\sum_{j=1}^n (-1)^{i_j} x_{i_j}\right)^n = 2^{n-1} n! \prod_{i=1}^n x_i.$$

In particular, let w, x, y, z be complex numbers. Then,

$$(x+y)^2 - (x-y)^2 = 4xy, \quad (x+y+z)^3 - (x-y+z)^3 - (x+y-z)^3 + (x-y-z)^3 = 24xyz,$$
$$(w+x+y+z)^4 - (w-x+y+z)^4 - (w+x-y+z)^4 - (w+x+y-z)^4$$
$$+ (w-x-y+z)^4 + (w-x+y-z)^4 + (w+x-y-z)^4 - (w-x-y-z)^4 = 192wxyz.$$

Remark: This is *Boutin's identity*. See [2232].

Fact 2.11.10. Let $x_1 \ldots, x_n$ be real numbers, and assume that $x_1 \le \cdots \le x_n$. Then,

$$\sum_{i,j=1}^n |(i-j)(x_i - x_j)| = \frac{n}{2}\sum_{i,j=1}^n |x_i - x_j| = n\sum_{i=1}^n (2i-1-n)x_i.$$

Source: [394]. **Related:** Fact 1.12.8.

Fact 2.11.11. Let $x_1, \ldots, x_n \in [-1, \infty)$, and let $\alpha_1, \ldots, \alpha_n$ be real numbers. Then, the following statements hold:

i) Assume that either $\alpha_1,\ldots,\alpha_n \in (-\infty,0]$ or $\alpha_1,\ldots,\alpha_n \in [1,\infty)$, and assume that either $x_1,\ldots,x_n \in [-1,0]$ or $x_1,\ldots,x_n \in [0,\infty)$. Then,

$$1+\sum_{i=1}^{n}\alpha_i x_i \le \prod_{i=1}^{n}(1+x_i)^{\alpha_i}.$$

ii) Assume that $\alpha_1,\ldots,\alpha_n \in [0,1]$ and $\sum_{i=1}^{n}\alpha_i \le 1$. Then,

$$\prod_{i=1}^{n}(1+x_i)^{\alpha_i} \le 1+\sum_{i=1}^{n}\alpha_i x_i.$$

Source: [2443]. **Remark:** This is a multivariable extension of Bernoulli's inequality given by Fact 2.1.21. **Related:** Fact 2.11.12.

Fact 2.11.12. Let $x_1,\ldots,x_n$ be nonnegative numbers, and assume that either $x_1,\ldots,x_n \in [0,1]$ or $x_1,\ldots,x_n \in [1,\infty)$. Then,

$$1+\sum_{i=1}^{n}x_i \le n+\prod_{i=1}^{n}x_i.$$

Source: [112, p. 37], [806, p. 27], and Fact 2.11.11.

Fact 2.11.13. Let $x_1,\ldots,x_n \in [0,1]$, and define $x_{n+1} \triangleq x_1$. Then,

$$\sum_{i=1}^{n}x_i \le \left\lfloor \frac{n}{2}\right\rfloor + \sum_{i=1}^{n}x_i x_{i+1}.$$

Source: [112, p. 37].

Fact 2.11.14. Let $x_1,\ldots,x_n$ be nonnegative numbers. Then,

$$\sum_{i=1}^{n}ix_i \le \binom{n}{2}+\sum_{i=1}^{n}x_i^i.$$

Source: [806, p. 197].

Fact 2.11.15. Let $x_1,\ldots,x_n$ be nonnegative numbers, and let $k \ge 1$. Then,

$$\sum_{i=1}^{n}x_i^k \le \left(\sum_{i=1}^{n}x_i\right)^k \le n^{k-1}\sum_{i=1}^{n}x_i^k.$$

Equality holds in the second inequality if and only if $x_1=\cdots=x_n$. **Remark:** The case $n=4$, $k=3$ is given by $(w+x+y+z)^3 \le 16(w^3+x^3+y^3+z^3)$ of Fact 2.4.23.

Fact 2.11.16. Let $n \ge 2$, let $x_1,\ldots,x_n$ be nonnegative numbers, let k be a positive integer such that $n-2 \le k \le n$, and define $x_{n+1} \triangleq x_1,\ldots,x_{n+k-1} \triangleq x_{k-1}$. Then,

$$n^{k-1}\sum_{i=1}^{n}\prod_{j=i}^{i+k-1}x_j \le \left(\sum_{i=1}^{n}x_i\right)^k.$$

If $k=n=2$, then $4x_1x_2 \le (x_1+x_2)^2$. If $n=3$ and $k=2,3$, then

$$3(x_1x_2+x_2x_3+x_3x_1) \le (x_1+x_2+x_3)^2, \quad 27x_1x_2x_3 \le (x_1+x_2+x_3)^3.$$

If $n=4$ and $k=2,3,4$, then

$$4(x_1x_2+x_2x_3+x_3x_4+x_4x_1) \le (x_1+x_2+x_3+x_4)^2,$$
$$16(x_1x_2x_3+x_2x_3x_4+x_3x_4x_1+x_4x_1x_2) \le (x_1+x_2+x_3+x_4)^3,$$
$$264x_1x_2x_3x_4 \le (x_1+x_2+x_3+x_4)^4.$$

If $k = n$, then

$$n^n \prod_{i=i}^{n} x_i \le \left(\sum_{i=1}^{n} x_i\right)^n.$$

Source: The case $n = 4$ and $k = 2$ is given in [1371, p. 144]. The case $n = 6$ and $k = 4$ is given in [108, pp. 462, 463]. See Fact 2.11.61. **Related:** The case $n = 5$ and $k = 3, 4, 5$ is given by Fact 2.5.5. The case $n = 6$ and $k = 4, 5, 6$ is given by Fact 2.6.17. **Conjecture:** Let $n \ge 4$, let $x_1, \ldots, x_n$ be nonnegative numbers, let $1 \le k \le n - 3$, and define $x_{n+1} \triangleq x_1, \ldots, x_{n+k-1} \triangleq x_{k-1}$. Then,

$$k^k \sum_{i=1}^{n} \prod_{j=i}^{i+k-1} x_j \le \left(\sum_{i=1}^{n} x_i\right)^k.$$

If $n = 6$ and $k = 3$, then

$$27 \sum_{i=1}^{6} x_i x_{i+1} x_{i+2} \le \left(\sum_{i=1}^{6} x_i\right)^3.$$

The case $n \ge 4$ and $k = 2$ is given in [1371, p. 144].

Fact 2.11.17. Let $x_1, \ldots, x_n$ be nonnegative numbers. Then,

$$\left(\sum_{i=1}^{n} x_i\right)^2 \le n \sum_{i=1}^{n} x_i^2.$$

Equality holds if and only if $x_1 = \cdots = x_n$. **Remark:** This result is equivalent to *i*) of Fact 11.9.23 with $m = 1$.

Fact 2.11.18. Let $x_1, \ldots, x_n$ be real numbers. Then,

$$\left(\sum_{i=1}^{n} x_i\right)^2 \le (n-1)\left(2x_1x_2 + \sum_{i=1}^{n} x_i^2\right).$$

Source: [1566, p. 12].

Fact 2.11.19. Let $x_1, \ldots, x_n$ be real numbers, let $1 \le k \le n-1$, and define $x_{n+1} \triangleq x_1, \ldots, x_{n+k} \triangleq x_k$. Then,

$$\sum_{i=1}^{n} x_i x_{i+k} \le \sum_{i=1}^{n} x_i^2.$$

In particular, let w, x, y, z be real numbers. Then,

$$\left.\begin{matrix} wx + xy + yz + zw \\ 2(wy + xz) \end{matrix}\right\} \le w^2 + x^2 + y^2 + z^2.$$

Source: [176]. **Related:** Fact 2.4.16.

Fact 2.11.20. Let $x_0, x_1, \ldots, x_{n+1}$ be real numbers. Then, the following statements hold:

i) If $x_1 = 0$, then

$$4\left(\sin^2 \frac{\pi}{2(2n-1)}\right) \sum_{i=2}^{n} x_i^2 \le \sum_{i=1}^{n-1} (x_{i+1} - x_i)^2 \le 4\left(\cos^2 \frac{\pi}{2n-1}\right) \sum_{i=2}^{n} x_i^2.$$

ii) If $x_0 = x_{n+1} = 0$, then

$$4\left(\sin^2 \frac{\pi}{2(n+1)}\right) \sum_{i=1}^{n} x_i^2 \le \sum_{i=0}^{n} (x_{i+1} - x_i)^2.$$

iii) If $x_{n+1} = x_1$ and $\sum_{i=1}^n x_i = 0$, then

$$4\left(\sin^2 \frac{\pi}{n}\right)\sum_{i=1}^n x_i^2 \le \sum_{i=1}^n (x_{i+1} - x_i)^2.$$

iv) If $x_n = 0$ and $x_{n+1} = x_1$, then

$$4\left(\sin^2 \frac{\pi}{2n}\right)\sum_{i=1}^n x_i^2 \le \sum_{i=1}^n (x_{i+1} - x_i)^2.$$

v) If $\sum_{i=1}^n x_i = 0$, then

$$4\left(\sin^2 \frac{\pi}{2n}\right)\sum_{i=1}^n x_i^2 \le \sum_{i=1}^{n-1} (x_{i+1} - x_i)^2.$$

vi) If n is even and $x_{n+1} = x_1$, then

$$4\left(\sin^2 \frac{\pi}{n}\right)\sum_{i=1}^n x_i^2 + n\left(\sin \frac{\pi}{n}\right)\left(\sin \frac{2\pi}{n} - \sin \frac{\pi}{n}\right)(x_{n/2} - x_n)^2 \le \sum_{i=1}^n (x_{i+1} - x_i)^2.$$

vii) If $\sum_{i=1}^n x_i = 0$, then

$$4\left(\sin^2 \frac{\pi}{2n}\right)\sum_{i=1}^n x_i^2 + 2n\left(\sin \frac{\pi}{n}\right)\left(\sin \frac{\pi}{n} - \sin \frac{\pi}{2n}\right)(x_1 + x_n)^2 \le \sum_{i=1}^{n-1} (x_{i+1} - x_i)^2.$$

viii) If $x_0 = x_{n+1} = 0$, then

$$4\left(\sin^2 \frac{\pi}{2(n+1)}\right)\sum_{i=1}^n x_i^2 \le \sum_{i=0}^n (x_{i+1} - x_i)^2 \le 4\left(\cos^2 \frac{\pi}{2(n+1)}\right)\sum_{i=1}^n x_i^2.$$

ix) If $a \ge 0$ and $x_0 = x_{n+1} = 0$, then

$$\begin{aligned}4\left(\sin^2 \frac{\pi}{2(n+1)}\right)\sum_{i=1}^n (i+a)^2 x_i^2 &\le \sum_{i=0}^n (i+a)(i+a+1)(x_{i+1} - x_i)^2\\ &\le 4\left(\cos^2 \frac{\pi}{2(n+1)}\right)\sum_{i=1}^n (i+a)^2 x_i^2.\end{aligned}$$

x) If $x_0 = x_1$, $x_{n+1} = x_n$, and $\sum_{i=1}^n x_i = 0$, then

$$16\left(\sin^4 \frac{\pi}{2n}\right)\sum_{i=1}^n x_i^2 \le \sum_{i=0}^{n-1} (x_{i+2} - 2x_{i+1} + x_i)^2.$$

xi) If $x_0 = x_1$ and $x_{n+1} = x_n$, then

$$\sum_{i=0}^{n-1} (x_{i+2} - 2x_{i+1} + x_i)^2 \le 16\left(\cos^4 \frac{\pi}{2n}\right)\sum_{i=1}^n x_i^2.$$

Source: [2045]. **Remark:** These are inequalities of *Wirtinger's type*.

Fact 2.11.21. Let $x_1, \ldots, x_n$ be nonnegative numbers, and let k be a positive integer. Then,

$$\sum_{i=1}^n x_i^k \le \left(\sum_{i=1}^n x_i\right)\left(\sum_{i=1}^n x_i^{k-1}\right) \le n\sum_{i=1}^n x_i^k.$$

Source: [1757, pp. 257, 258].

Fact 2.11.22. Let $x_1, \ldots, x_n$ be nonnegative numbers. Then,

$$\sum_{i=1}^n x_i^n \le \left(\sum_{i=1}^n x_i\right)\left(\sum_{i=1}^n x_i^{n-1}\right) \le (n-1)\sum_{i=1}^n x_i^n + n\prod_{i=1}^n x_i \le n\sum_{i=1}^n x_i^n.$$

Source: [806, pp. 35, 36]. **Remark:** This result interpolates the right-hand inequality in Fact 2.11.21 in the case $k = n$. **Remark:** The second inequality is *Suranyi's inequality*.

Fact 2.11.23. Let $x_1, \ldots, x_n$ be nonnegative numbers, and let $\alpha \in [0, 1]$. Then,

$$\left(\sum_{i=1}^n \sqrt{x_i}\right)^2 \le \left(\sum_{i=1}^n x_i^\alpha\right)\left(\sum_{i=1}^n x_i^{1-\alpha}\right).$$

Equality holds if and only if either $\alpha = 1/2$ or $x_1 = \cdots = x_n$. **Source:** [2991, p. 250].

Fact 2.11.24. Let $x_1, \ldots, x_n$ be nonnegative numbers. Then,

$$\left(\sum_{i=1}^n x_i^3\right)^2 \le \left(\sum_{i=1}^n x_i^2\right)^3 \le n\left(\sum_{i=1}^n x_i^3\right)^2.$$

Source: Set $p = 2$ and $q = 3$ in Fact 2.11.90 and square all terms.

Fact 2.11.25. Let $x_1, \ldots, x_n$ be positive numbers. Then,

$$\left(\sum_{i=1}^n x_i\right)\left(\sum_{i=1}^n \frac{1}{x_i}\right) \le \left(\sum_{i=1}^n x_i^2\right)\left(\sum_{i=1}^n \frac{1}{x_i^2}\right), \quad \frac{n}{2}\left(\sum_{i=1}^n x_i^2\right) \le \left(\sum_{i=1}^n x_i^3\right)\left(\sum_{i=1}^n \frac{1}{x_i}\right),$$

$$\left(\sum_{i=1}^n x_i\right)\left(\sum_{i=1}^n x_i^3\right) \le \left(\sum_{i=1}^n x_i^5\right)\left(\sum_{i=1}^n \frac{1}{x_i}\right).$$

Source: [1371, p. 150] and [1809, 2750].

Fact 2.11.26. Let $x_1, \ldots, x_n$ be positive numbers. Then,

$$\left(\prod_{i=1}^n x_i\right)\left(\sum_{i=1}^n x_i\right)\left(\sum_{i=1}^n \frac{1}{x_i}\right) \le \left(\sum_{i=1}^n x_i^n\right) + n(n-1)\prod_{i=1}^n x_i, \quad \left(\sum_{i=1}^n x_i\right)\left(\sum_{i=1}^n x_i^{n-1}\right) \le (n-1)\left(\sum_{i=1}^n x_i^n\right) + n\prod_{i=1}^n x_i,$$

$$\left(\sum_{i=1}^n x_i\right)\left[\left(\sum_{i=1}^n x_i^n\right) - \prod_{i=1}^n x_i\right] \le (n-1)\sum_{i=1}^n x_i^{n+1}, \quad 2 \le \left[\left(\sum_{i=1}^n x_i\right) - n\right]\left[\left(\sum_{i=1}^n \frac{1}{x_i}\right) - n\right] + \prod_{i=1}^n x_i + \prod_{i=1}^n \frac{1}{x_i}.$$

Source: [748, pp. 219, 220].

Fact 2.11.27. Let $x_1, \ldots, x_n$ be positive numbers, assume that $x_1 \le 1$, and assume that, for all $i \in \{1, \ldots, n-1\}$, $x_i \le x_{i+1} \le x_i + 1$. Then,

$$\sum_{i-1}^n x_i^3 \le \left(\sum_{i-1}^n x_i\right)^2.$$

Source: [993, p. 183]. **Remark:** Equality holds in the case where $x_i = i$, as shown in Fact 1.12.1.

Fact 2.11.28. Let $x_1, \ldots, x_n$ be nonnegative numbers. Then,

$$\sum_{1 \le i < j \le n} x_i x_j (x_i^2 + x_j^2) \le \frac{1}{8}\left(\sum_{i=1}^n x_i\right)^4.$$

Source: [752, p. 215] and [1938, p. 106].

Fact 2.11.29. Let $x_1, \ldots, x_n$ be positive numbers. Then,

$$\left(\prod_{i=1}^n x_i\right)\sum_{i=1}^n x_i \le \sum_{i=1}^n x_i^{n+1}.$$

Source: [993, p. 170].

Fact 2.11.30. Let $x_1, \ldots, x_n$ be real numbers, and define $x_{n+1} \triangleq x_1$. Then,

$$0 \le \sum_{i=1}^n (x_i - x_{i+1})(3x_i + x_{i+1})^3.$$

Now, let $r \in [0, (\sqrt{3} - 1)/2]$. Then,

$$(r+1)\sum_{i=1}^n x_i^3 x_{i+1} \le \sum_{i=1}^n (x_i^4 + r x_i x_{i+1}^3).$$

Source: [748, p. 110].

Fact 2.11.31. Let $x_1, \ldots, x_n$ be nonnegative numbers, and define $x_{n+1} \triangleq x_1$. Then,

$$\frac{3}{2}\sum_{i=1}^n x_i^3 x_{i+1} \le \sum_{i=1}^n (x_i^4 + \frac{1}{2} x_i x_{i+1}^3).$$

Now, let $r \in [1/(\sqrt[3]{4} - 1), \infty)$. Then,

$$0 \le \sum_{i=1}^n (x_i - x_{i+1})(r x_i + x_{i+1})^3.$$

Source: [748, p. 110].

Fact 2.11.32. Let $x_1, \ldots, x_n$ be positive numbers, and define $x_{n+1} \triangleq x_1$. Then,

$$1 + (n-2)\min\{x_1^{x_2}, \ldots, x_n^{x_{n+1}}\} < \sum_{i=1}^n x_i^{x_{i+1}}.$$

Source: [2294, p. 286]. **Related:** Fact 2.2.70.

Fact 2.11.33. Let $x_1, \ldots, x_n$ be positive numbers, define $x_{n+1} \triangleq x_1$, and let $m \ge 1$. Then,

$$\sum_{i=1}^n (x_i^2 - x_i x_{i+1} + x_{i+1}^2)^m \le 3^m \sum_{i=1}^n x_i^{2m}.$$

Source: [383].

Fact 2.11.34. Let $z_1, \ldots, z_n$ be complex numbers, define $E_0 \triangleq 1$, and, for all $k \in \{1, \ldots, n\}$, define

$$E_k \triangleq \sum_{i_1 < \cdots < i_k} \prod_{j=1}^k z_{i_j}.$$

Furthermore, for each positive integer k define $\mu_k \triangleq \sum_{i=1}^n z_i^k$. Then, for all $k \in \{1, \ldots, n\}$,

$$kE_k = \sum_{i=1}^k (-1)^{i-1} E_{k-i}\mu_i.$$

In particular,

$$E_1 = \mu_1, \quad 2E_2 = E_1\mu_1 - \mu_2, \quad 3E_3 = E_2\mu_2 - E_1\mu_2 + \mu_3.$$

Furthermore,

$$E_1 = \mu_1, \quad E_2 = \tfrac{1}{2}(\mu_1^2 - \mu_2), \quad E_3 = \tfrac{1}{6}(\mu_1^3 - 3\mu_1\mu_2 + 2\mu_3),$$
$$\mu_1 = E_1, \quad \mu_2 = E_1^2 - 2E_2, \quad \mu_3 = E_1^3 - 3E_1E_2 + 3E_3.$$

Remark: These equalities are *Newton's identities*. An application to roots of polynomials is given by Fact 6.8.4. **Remark:** E_k is the kth *elementary symmetric polynomial*. **Related:** Fact 2.11.35.

Fact 2.11.35. Let $n \ge 2$, let $x_1, \ldots, x_n$ be real numbers, and, for all $k \in \{1, \ldots, n\}$, define

$$S_k \triangleq \binom{n}{k}^{-1} \sum_{i_1 < \cdots < i_k} \prod_{j=1}^{k} x_{i_j}.$$

Then, for all $k \in \{2, \ldots, n-1\}$,

$$S_{k-1}S_{k+1} \le S_k^2.$$

Now, assume that $x_1, \ldots, x_n$ are nonnegative. Then,

$$S_n^{1/n} \le \cdots \le S_2^{1/2} \le S_1.$$

Remark: See [806, pp. 117, 118]. **Remark:** The first inequality is *Newton's inequality*. The case $n = 3$, $k = 2$ is given by Fact 2.3.16 for real numbers and Fact 2.3.66 for nonnegative numbers. The cases $n = 4$ and $k = 2, 3$ are given by Fact 2.4.16. The second inequality is *Maclaurin's inequality*, which interpolates the arithmetic-mean–geometric-mean inequality. See [356, 658]. For $n = 3$, this result is given by Fact 2.3.60. For $n = 4$, this result is given by Fact 2.4.23. **Remark:** S_k is the kth *elementary symmetric mean*. **Related:** Fact 2.11.34 and Fact 2.11.96.

Fact 2.11.36. Let $x_1, \ldots, x_n$ be positive numbers, and assume that $\sum_{i=1}^n x_i = 1$. Then,

$$\sum_{i=1}^{n} |x_i - \tfrac{1}{n}| \le \begin{cases} 2 - 2/n, & n \le 4, \\ 1 + 1/\sqrt{n}, & n \ge 4. \end{cases}$$

Source: The bound $2 - 2/n$ follows from *vii*) of Fact 2.15.28 with $y_1 = \cdots = y_n = 1/n$.

Fact 2.11.37. Let $x_1, \ldots, x_n$ be real numbers, and define

$$\mu \triangleq \frac{1}{n}\sum_{j=1}^{n} x_j, \quad \sigma \triangleq \sqrt{\frac{1}{n}\sum_{j=1}^{n}(x_j - \mu)^2} = \sqrt{\left(\frac{1}{n}\sum_{j=1}^{n} x_j^2\right) - \mu^2},$$

$\alpha \triangleq \min\{x_1, \ldots, x_n\}$, and $\beta \triangleq \max\{x_1, \ldots, x_n\}$. Then, for all $i \in \{1, \ldots, n\}$,

$$|x_i - \mu| \le \sqrt{n-1}\,\sigma.$$

Equality holds if and only if all of the elements of $\{x_1, \ldots, x_n\}_{\mathrm{ms}}\backslash\{x_i\}$ are equal. In addition,

$$\frac{\sigma}{\sqrt{n-1}} \le \beta - \mu \le \sqrt{n-1}\,\sigma.$$

Equality holds in either the left-hand inequality or the right-hand inequality if and only if all of the elements of $\{x_1, \ldots, x_n\}_{\mathrm{ms}}\backslash\{\max\{x_1, \ldots, x_n\}\}$ are equal. Finally,

$$\frac{\sigma}{\sqrt{n-1}} \le \mu - \alpha \le \sqrt{n-1}\,\sigma.$$

Equality holds in either the left-hand inequality or the right-hand inequality if and only if all of the elements of $\{x_1, \ldots, x_n\}_{\mathrm{ms}}\backslash\{a\}$ are equal. **Source:** The first result is the *Laguerre-Samuelson inequality*. See [1205, 1482, 1528, 2132, 2330, 2720]. The lower bounds in the second and third strings are given in [2905]. See also [2330]. **Remark:** A vector extension of the Laguerre-Samuelson inequality is given by Fact 10.10.40. An application to eigenvalue bounds is given by Fact 7.12.48.

Fact 2.11.38. Let $n \geq 2$, let $x_1, \ldots, x_n$ be real numbers, define $\mu, \sigma, \alpha, \beta$ as in Fact 2.11.37, and define

$$\delta \triangleq \beta - \alpha, \quad \mu_3 \triangleq \frac{1}{n}\sum_{i=1}^{n}(x_i - \mu)^3.$$

Then,

$$\frac{\delta^2}{2n} \leq \sigma^2 \leq \sigma^2 + \left(\frac{\mu_3}{2\sigma^2}\right)^2 \leq \frac{\delta^2}{4}, \quad \sigma^2 \leq (\beta - \mu)(\mu - \alpha).$$

Next, assume that $x_1, \ldots, x_n$ are positive numbers, and define $h \triangleq \left(\frac{1}{n}\sum_{i=1}^{n}\frac{1}{x_i}\right)^{-1}$. Then,

$$\sigma^2 \leq \frac{1}{4}\beta(\beta - h), \quad \frac{\sigma}{\mu} \leq \sqrt{\frac{\beta - h}{4h}}.$$

If $\alpha < h$, then

$$\frac{\alpha(\mu - h)(\mu - \alpha)}{h - \alpha} \leq \sigma^2.$$

If $h < \beta$, then

$$\sigma^2 \leq \frac{\beta(\mu - h)(\beta - \mu)}{\beta - h}.$$

Source: [2433]. **Remark:** $(\mu, \sigma, \delta, \mu_3, h)$ is the (mean, standard deviation, third central moment, harmonic mean). **Remark:** σ/μ is the *coefficient of dispersion.*

Fact 2.11.39. Let $x_1, \ldots, x_n$ be real numbers, and let α, δ, p be positive numbers. If $p \geq 1$, then

$$\left(\frac{\alpha}{\alpha + n}\right)^{p-1}\delta^p \leq \left|\delta - \sum_{i=1}^{n}x_i\right|^p + \alpha^{p-1}\sum_{i=1}^{n}|x_i|^p.$$

In particular,

$$\frac{\alpha\delta^2}{\alpha + n} \leq \left(\delta - \sum_{i=1}^{n}x_i\right)^2 + \alpha\sum_{i=1}^{n}x_i^2.$$

Furthermore, if $p \leq 1$, $x_1, \ldots, x_n$ are nonnegative, and $\sum_{i=1}^{n} x_i \leq \delta$, then

$$\left|\delta - \sum_{i=1}^{n}x_i\right|^p + \alpha^{p-1}\sum_{i=1}^{n}|x_i|^p \leq \left(\frac{\alpha}{\alpha + n}\right)^{p-1}\delta^p.$$

For all $p > 0$, equality holds if and only if $x_1 = \cdots = x_n = \delta/(\alpha + n)$. **Source:** [2583]. **Remark:** This is *Wang's inequality*. The special case $p = 2$ is *Hua's inequality*. **Related:** Extensions are given by Fact 11.8.18 and Fact 11.8.19.

Fact 2.11.40. Let $x_1, \ldots, x_n$ be real numbers. Then,

$$\left(\sum_{i,j=1}^{n}|x_i - x_j|\right)^2 \leq \frac{2}{3}(n^2 - 1)\sum_{i,j=1}^{n}(x_i - x_j)^2.$$

Source: [752, p. 215] and [1938, pp. 109, 183, 184].

Fact 2.11.41. Let $x_1, \ldots, x_n$ be nonnegative numbers. Then,

$$\left(\frac{4}{3}\right)^n\sqrt{\prod_{i=1}^{n}(x_i + x_{i+1} + 1)} \leq \prod_{i=1}^{n}\left(\frac{2x_i + 4}{3}\right)^{2x_i+1}.$$

Source: [752, p. 216].

Fact 2.11.42. Let $x_1, \dots, x_n$ be positive numbers, and define $\alpha \triangleq \min\{x_1, \dots, x_n\}$ and $\beta \triangleq \max\{x_1, \dots, x_n\}$. Then, for all $i, j \in \{1, \dots, n\}$,

$$\frac{(x_i + x_j)^2}{x_i x_j} \le \frac{(\alpha + \beta)^2}{\alpha\beta}.$$

Source: [2991, p. 251].

Fact 2.11.43. Let $x_1, \dots, x_n$ be positive numbers. Then,

$$n^2 \le \sum_{i=1}^n \frac{1}{x_i} \sum_{i=1}^n x_i.$$

Consequently, if $\alpha > 0$ and $\sum_{i=1}^n x_i \le \alpha$, then $\frac{n^2}{\alpha} \le \sum_{i=1}^n \frac{1}{x_i}$. **Source:** Fact 2.11.89, [1371, p. 130], and [1566, p. 10]. **Remark:** $n = 3$ yields $9xyz \le (x + y + z)(xy + yz + zx)$ of Fact 2.3.60. **Remark:** $n = 4$ yields $16wxyz \le (w + x + y + z)(wxy + xyz + yzw + zwx)$ of Fact 2.4.23.

Fact 2.11.44. Let $n \ge 2$, let $x_1, \dots, x_n$ be positive numbers, and define $s \triangleq \sum_{i=1}^n x_i$. Then,

$$\frac{n}{n-1} \le \frac{1}{n}\sum_{i=1}^n \frac{s}{s - x_i} \le \sum_{i=1}^n \frac{x_i}{s - x_i}, \quad n(n-1) \le \sum_{i=1}^n \frac{s - x_i}{x_i},$$

$$n(n+1) \le \sum_{i=1}^n \frac{s + x_i}{x_i}, \quad \frac{1}{2} \le \sum_{i=1}^n \frac{x_i}{s + x_i} \le \frac{n}{n+1},$$

$$\frac{n(n+1)}{n-1} \le \sum_{i=1}^n \frac{s + x_i}{s - x_i}, \quad \frac{n(n-1)}{n+1} \le \sum_{i=1}^n \frac{s - x_i}{s + x_i} \le n - 1.$$

Source: [1566, p. 10] and [2991, p. 348].

Fact 2.11.45. Let $x_1, \dots, x_n$ be positive numbers. Then,

$$\sum_{i=1}^n \frac{i}{\sum_{j=1}^i x_j} \le 2\sum_{i=1}^n \frac{1}{x_i}.$$

Source: [506, pp. 177, 178] and [3010]. **Remark:** This is *Knopp's inequality*. **Related:** Fact 13.5.9.

Fact 2.11.46. Let $x_1, \dots, x_n$ be positive numbers. Then,

$$\sum_{i=1}^n \frac{x_i}{(1 + \sum_{j=1}^i x_j)^2} \le \frac{\sum_{i=1}^n x_i}{1 + \sum_{i=1}^n x_i}.$$

Source: [109, pp. 78, 79].

Fact 2.11.47. Let $x_1, \dots, x_n$ and $a_1, \dots, a_n$ be positive numbers. Then,

$$\sum_{i=1}^n \frac{a_i x_i}{1 + \sum_{j=1}^i x_j^2} < \sqrt{\sum_{i=1}^n a_i^2}.$$

In particular,

$$\sum_{i=1}^n \frac{x_i}{1 + \sum_{j=1}^i x_j^2} < \sqrt{n}, \quad \sum_{i=1}^n \sqrt{\binom{n}{i}} \frac{x_i}{1 + \sum_{j=1}^i x_j^2} < \sqrt{2^n - 1}.$$

Source: [894] and [1158, pp. 187, 188].

Fact 2.11.48. Let $x_1, \ldots, x_n$ be positive numbers, and define $x_{n+1} \triangleq x_1$. Then,

$$n \le \sum_{i=1}^{n} \frac{x_i}{x_{i+1}} \le \sum_{i=1}^{n} \left(\frac{x_{i+1}}{x_i} \right)^n.$$

Source: The left-hand inequality follows from Fact 2.11.50. The right-hand inequality is given in [1757, pp. 251, 252]. **Remark:** $n = 3$ yields $3xyz \le xy^2 + yz^2 + zx^2$ of Fact 2.3.60. **Remark:** $n = 4$ yields $4wxyz \le wx^2z + xy^2w + yz^2x + zw^2y$ of Fact 2.4.23.

Fact 2.11.49. Let $n \ge 2$, let $x_1, \ldots, x_n$ be positive numbers, assume that $x_n \le \cdots \le x_1$, and let $x_{n+1} \triangleq x_1$. Then,

$$\sum_{i=1}^{n} \frac{x_i}{x_{i+1}} \le \sum_{i=1}^{n} \frac{x_{i+1}}{x_i}.$$

Source: [993, p. 186].

Fact 2.11.50. Let $x_1, \ldots, x_n$ be positive numbers, and let $\{j_1, \ldots, j_n\} = \{1, \ldots, n\}$. Then,

$$n \le \sum_{i=1}^{n} \frac{x_{j_i}}{x_i}, \quad \sum_{i=1}^{n} x_i x_{j_i} \le \sum_{i=1}^{n} x_i^2.$$

Source: [1938, pp. 11, 13]. **Related:** Fact 2.11.48.

Fact 2.11.51. Let $x_1, \ldots, x_n$ be positive numbers, and let $\{j_1, \ldots, j_n\} = \{1, \ldots, n\}$. Then,

$$\prod_{i=1}^{n} x_{j_i}^{x_i} \le \prod_{i=1}^{n} x_i^{x_i}, \quad \prod_{i=1}^{n} x_i^{1/x_i} \le \prod_{i=1}^{n} x_{j_i}^{1/x_i}.$$

Source: [2294, pp. 270, 271]. **Related:** Fact 2.2.71.

Fact 2.11.52. Let $n \ge 2$, let $x_1, \ldots, x_n$ be positive numbers, and define $x_{n+1} \triangleq x_1$ and $x_{n+2} \triangleq x_2$. Then,

$$\frac{n}{4} \le \sum_{i=1}^{n} \frac{x_i}{x_{i+1} + x_{i+2}}.$$

Source: [1566, p. 16]. **Related:** Fact 2.11.53.

Fact 2.11.53. Let $n \in \{3, 4, 5, 6, 7, 8, 9, 10, 11, 12, 13, 15, 17, 19, 21, 23\}$, let $x_1, \ldots, x_n$ be positive numbers, and define $x_{n+1} \triangleq x_1$ and $x_{n+2} \triangleq x_2$. Then,

$$\frac{n}{2} \le \sum_{i=1}^{n} \frac{x_i}{x_{i+1} + x_{i+2}}.$$

Source: [1052]. **Remark:** This is *Shapiro's inequality*. For $n = 3$, this result yields Nesbitt's inequality given by Fact 2.3.60. Equality holds for $n = 2$. **Remark:** For integers $n \ge 3$ that are not listed, the best lower bound is approximately $0.9891n/2$.

Fact 2.11.54. Let $x_1, \ldots, x_n$ be positive numbers, and define $x_{n+1} \triangleq x_1$ and $x_{n+2} \triangleq x_2$. Then,

$$2 \le \sum_{i=1}^{n} \frac{x_{i+1}}{x_i + x_{i+2}}.$$

Source: [993, p. 185].

Fact 2.11.55. Let $x_1, \ldots, x_n$ be positive numbers, and define $x_{n+1} \triangleq x_1$ and $x_{n+2} \triangleq x_2$. Then,

$$\sum_{i=1}^{n} \frac{x_i}{(n-2)x_i + x_{i+1} + x_{i+2}} \le 1.$$

Source: [752, p. 322].

Fact 2.11.56. Let $x_1, \ldots, x_n$ be positive numbers, and define $x_{n+1} \triangleq x_1$ and $x_{n+2} \triangleq x_2$. Then,

$$0 \le \sum_{i=1}^{n} \frac{x_i - x_{i+2}}{x_{i+1} + x_{i+2}}.$$

Source: [806, p. 197].

Fact 2.11.57. Let $n \ge 2$, let $x_1, \ldots, x_n$ be positive numbers, and let k satisfy $1 \le k \le n-1$. Then,

$$\frac{kn}{n-k} \le \frac{x_1 + x_2 + \cdots + x_k}{x_{k+1} + \cdots + x_n} + \frac{x_2 + x_3 + \cdots + x_{k+1}}{x_{k+2} + \cdots + x_n + x_1} + \cdots + \frac{x_n + x_1 + \cdots + x_{k-1}}{x_k + \cdots + x_{n-1}}.$$

In particular,

$$\frac{n}{n-1} \le \frac{x_1}{x_2 + \cdots + x_n} + \frac{x_2}{x_3 + \cdots + x_n + x_1} + \cdots + \frac{x_n}{x_1 + \cdots + x_{n-1}},$$

$$2 \le \frac{x_1}{x_2} + \frac{x_2}{x_1}, \quad \frac{3}{2} \le \frac{x_1}{x_2 + x_3} + \frac{x_2}{x_3 + x_1} + \frac{x_3}{x_1 + x_2},$$

$$\frac{4}{3} \le \frac{x_1}{x_2 + x_3 + x_4} + \frac{x_2}{x_3 + x_4 + x_1} + + \frac{x_3}{x_4 + x_1 + x_2} + \frac{x_4}{x_1 + x_2 + x_3}.$$

Furthermore,

$$\frac{n^2}{n-1} \le \frac{\sum_{i=1}^n x_i}{x_2 + \cdots + x_n} + \frac{\sum_{i=1}^n x_i}{x_3 + \cdots + x_n + x_1} + \cdots + \frac{\sum_{i=1}^n x_i}{x_1 + \cdots + x_{n-1}},$$

$$\frac{n^2}{2n-1} \le \frac{\sum_{i=1}^n x_i}{2\sum_{i=1}^n x_i - x_1} + \frac{\sum_{i=1}^n x_i}{2\sum_{i=1}^n x_i - x_2} + \cdots + \frac{\sum_{i=1}^n x_i}{2\sum_{i=1}^n x_i - x_n}.$$

Source: [210]. The second and penultimate inequalities are given in [1938, p. 19]. The last inequality is given in [2527, p. 131]. **Remark:** The first inequality is due to D. S. Mitrinovic, the second inequality is due to M. Peixoto, and the fourth inequality is Nesbitt's inequality. See Fact 2.3.60, Fact 2.3.61, Fact 2.11.53, and Fact 5.2.25. **Related:** Fact 2.11.60 and Fact 2.11.61.

Fact 2.11.58. Let $n \ge 2$, let $x_1, \ldots, x_n$ be positive numbers, and let $p \ge 1$. Then,

$$\frac{n}{(n-1)^p} \le n^{1-p}\left(\sum_{i=1}^{n} \frac{x_i}{\sum_{j=1, j\ne i}^n x_j}\right)^p \le \sum_{i=1}^{n}\left(\frac{x_i}{\sum_{j=1, j\ne i}^n x_j}\right)^p.$$

Source: [2850].

Fact 2.11.59. Let $n \ge 2$, let $x_1, \ldots, x_n$ be positive numbers, let $\lambda \ge 1$, and let $r \ge s > 0$.

$$\frac{n}{(n-1)^\lambda}\left(\frac{1}{n}\sum_{i=1}^{n} x_i^s\right)^{\lambda(r/s-1)} \le \sum_{i=1}^{n}\left(\frac{x_i^r}{\sum_{j=1, j\ne i} x_i^s}\right)^\lambda.$$

Source: [2850]. **Related:** Fact 2.3.64.

Fact 2.11.60. Let $n \ge 2$, let $x_1, \ldots, x_n$ be positive numbers, let $p_1, \ldots, p_n \in [0, 1]$, assume that $\sum_{i=1}^n p_i = 1$, and let $\alpha \in (-\infty, 0] \cup [1, \infty)$. Then,

$$\frac{\left(\sum_{i=1}^n p_i x_i\right)^\alpha}{\sum_{i=1}^n p_i \sum_{j=1, j\ne i}^n x_j} \le \sum_{i=1}^{n} \frac{p_i x_i^\alpha}{\sum_{j=1, j\ne i}^n x_j}.$$

In particular,

$$\frac{\sum_{i=1}^n p_ix_i}{\sum_{i=1}^n p_i\sum_{j=1,j\neq i}^n x_j}\le\sum_{i=1}^n\frac{p_ix_i}{\sum_{j=1,j\neq i}^n x_j},\quad \frac{n^{2-\alpha}}{n-1}\left(\sum_{i=1}^n x_i\right)^{\alpha-1}\le\frac{n^{2-\alpha}\sum_{i=1}^n x_i}{\sum_{i=1}^n\sum_{j=1,j\neq i}^n x_j}\le\sum_{i=1}^n\frac{x_i}{\sum_{j=1,j\neq i}^n x_j},$$

$$\frac{n}{n-1}=\frac{n\sum_{i=1}^n x_i}{\sum_{i=1}^n\sum_{j=1,j\neq i}^n x_j}\le\sum_{i=1}^n\frac{x_i}{\sum_{j=1,j\neq i}^n x_j},$$

$$\frac{1}{\sum_{i=1}^n p_i\sum_{j=1,j\neq i}^n x_j}\le\sum_{i=1}^n\frac{p_i}{\sum_{j=1,j\neq i}^n x_j},\quad \frac{n^2}{\sum_{i=1}^n\sum_{j=1,j\neq i}^n x_j}\le\sum_{i=1}^n\frac{1}{\sum_{j=1,j\neq i}^n x_j}.$$

Source: [397]. **Related:** Fact 2.11.57 and Fact 2.11.61.

Fact 2.11.61. Let $n\ge 4$, let $x_1,\dots,x_n$ be positive numbers, and let $r\ge\frac{1}{2}$. Then,

$$\frac{4^r}{n^{r-1}(n-1)}(x_1x_2+x_2x_3+\cdots+x_nx_1)^r$$
$$\le\frac{x_1^{2r+1}}{x_2+\cdots+x_n}+\frac{x_2^{2r+1}}{x_3+\cdots+x_n+x_1}+\cdots+\frac{x_n^{2r+1}}{x_1+\cdots+x_{n-1}}.$$

Furthermore, define $x_{n+1}\triangleq x_1$. Then,

$$\frac{4^r}{n^{2r-1}(n-1)}\left(\sum_{i=1}^n x_ix_{i+1}\right)^r\le\frac{n^{1-2r}}{n-1}\left(\sum_{i=1}^n x_i\right)^{2r}\le\frac{n^{1-r}}{n-1}\left(\sum_{i=1}^n x_i^2\right)^r\le\frac{1}{n-1}\sum_{i=1}^n x_i^{2r}$$
$$\le\frac{n}{(n-1)\sum_{i=1}^n x_i}\sum_{i=1}^n x_i^{2r+1}\le\sum_{i=1}^n\frac{x_i^{2r+1}}{\sum_{j=1,j\neq i}^n x_j}.$$

Source: [1860, pp. 18, 102–104]. **Related:** Fact 2.11.16, Fact 2.11.57, and Fact 2.11.60.

Fact 2.11.62. Let $x_1,\dots,x_n$ be positive numbers, and define $x_{n+1}\triangleq x_1$. Then,

$$\sum_{i=1}^n\frac{x_ix_{i+1}}{x_i+x_{i+1}}\le\frac{1}{2}\sum_{i=1}^n x_i\le\sum_{i=1}^n\frac{x_i^2}{x_i+x_{i+1}}.$$

Source: [993, p. 185].

Fact 2.11.63. Let $n\ge 2$, and let $x_1,\dots,x_n$ be real numbers. Then,

$$\sum_{1\le i<j\le n}x_ix_j\le\frac{n-1}{2}\sum_{i=1}^n x_i^2.$$

Source: [2527, p. 189].

Fact 2.11.64. Let $n\ge 2$, and let $x_1,\dots,x_n$ be positive numbers. Then,

$$\sum_{1\le i<j\le n}\frac{x_ix_j}{x_i+x_j}\le\frac{n-1}{4}\sum_{i=1}^n x_i.$$

Equality holds if and only if $x_1=\cdots=x_n$. **Source:** [219].

Fact 2.11.65. Let $n\ge 2$, and let $x_1,\dots,x_n$ be positive numbers. Then,

$$\frac{n^2(n-1)}{4\sum_{i=1}^n x_i}\le\sum_{1\le i<j\le n}\frac{1}{x_i+x_j}.$$

Equality holds if and only if $x_1=\cdots=x_n$. **Source:** [792].

Fact 2.11.66. Let $n \ge 3$, and let $x_1, \dots, x_n$ be positive numbers. Then,

$$\sum_{1 \le i < j < k \le n} \frac{x_i x_j x_k}{(x_i + x_j + x_k)^2} \le \frac{(n-1)(n-2)}{54} \sum_{i=1}^{n} x_i.$$

Equality holds if and only if $x_1 = \cdots = x_n$. **Source:** [219]. **Credit:** C. Delrome.

Fact 2.11.67. Let $x_1, \dots, x_n$ be positive numbers, and let $\{j_1, \dots, j_n\} = \{1, \dots, n\}$. Then,

$$\sum_{i=1}^{n} x_i \le \sum_{i=1}^{n} \frac{x_i^2}{x_{j_i}}.$$

In particular,

$$\sum_{i=1}^{n} x_i \le \sum_{i=1}^{n} \frac{x_i^2}{x_{i+1}},$$

where $x_{n+1} \triangleq x_1$. **Source:** [806, p. 184].

Fact 2.11.68. Let $x_1, \dots, x_n$ be positive numbers. Then,

$$\frac{n^3+1}{(n^2+1)^2} \left(\sum_{i=1}^{n} \frac{1}{x_i} + \frac{1}{\sum_{i=1}^{n} x_i} \right)^2 \le \sum_{i=1}^{n} \frac{1}{x_i^2} + \frac{1}{\left(\sum_{i=1}^{n} x_i\right)^2}.$$

Source: [109, pp. 264, 265].

Fact 2.11.69. Let $x_1, \dots, x_n$ be positive numbers, assume that $\sum_{i=1}^{n} x_i = 1$, and let $p > 0$. Then,

$$\frac{(n^2+1)^p}{n^{p-1}} \le \sum_{i=1}^{n} \left(x_i + \frac{1}{x_i} \right)^p.$$

In particular,

$$n^3 + 2n + \frac{1}{n} \le \sum_{i=1}^{n} \left(x_i + \frac{1}{x_i} \right)^2, \quad n^4 + 3n^2 + 2n + 3 + \frac{1}{n^2} \le \sum_{i=1}^{n} \left(x_i + \frac{1}{x_i} \right)^3.$$

Source: [2527, p. 14].

Fact 2.11.70. Let $x_1, \dots, x_n$ be positive numbers, and define $x_{n+1} \triangleq x_1$. Then,

$$\sum_{i=1}^{n} x_i^2 \le \sum_{i=1}^{n} \frac{x_i^3}{x_{i+1}}.$$

Source: [806, p. 131] and [1569].

Fact 2.11.71. Let $x_1, \dots, x_n$ be positive numbers, and define $x_{n+1} \triangleq x_1$. Then,

$$\frac{n^2}{2 \sum_{i=1}^{n} x_i x_{i+1}} \le \sum_{i=1}^{n} \frac{1}{x_i^2 + x_i x_{i+1}}.$$

Source: [752, p. 321].

Fact 2.11.72. Let $x_1, \dots, x_n$ be positive numbers, and define $x_{n+1} \triangleq x_1$. Then,

$$\prod_{i=1}^{n} (x_i + 1) \le \prod_{i=1}^{n} \left(\frac{x_i^2}{x_{i+1}} + 1 \right).$$

Source: [806, p. 214].

Fact 2.11.73. Let $x_1, \ldots, x_n$ be positive numbers, and define $x_{n+1} \triangleq x_1$. Then,

$$\frac{1}{3}\sum_{i=1}^n x_i \le \sum_{i=1}^n \frac{x_i^3}{x_i^2 + x_i x_{i+1} + x_{i+1}^2} = \sum_{i=1}^n \frac{x_{i+1}^3}{x_i^2 + x_i x_{i+1} + x_{i+1}^2}.$$

Source: [2380, pp. 73, 74].

Fact 2.11.74. Let $x_1, \ldots, x_n$ be positive numbers, and define $x_0 \triangleq x_n$ and $x_{n+1} \triangleq x_1$. Then,

$$\sum_{i=1}^n \frac{x_i}{x_{i-1} + 2x_i + x_{i+1}} \le \frac{n}{4}.$$

Source: [752, p. 213].

Fact 2.11.75. Let $x_1, \ldots, x_n$ be nonnegative numbers. Then,

$$\frac{\sum_{i=1}^n x_i}{\prod_{i=1}^n (x_i^2 + 1)} \le \frac{(2n-1)^{n-1/2}}{2^n n^{n-1}}, \quad \frac{\left(\sum_{i=1}^n x_i\right)^2}{\prod_{i=1}^n (x_i^2 + 1)} \le \frac{(n-1)^{n-1}}{n^{n-2}}.$$

Source: [748, p. 372] and [752, p. 214].

Fact 2.11.76. Let $x_1, \ldots, x_n$ be nonzero real numbers, and define $x_{n+1} \triangleq x_1$. Then,

$$\frac{\left(\sum_{i=1}^n x_i\right)^2}{9\sum_{i=1}^n x_i^4} \le \sum_{i=1}^n \left(\frac{x_i}{x_i^2 + x_i x_{i+1} + x_{i+1}^2}\right)^2.$$

Source: [891].

Fact 2.11.77. Let $x_1, \ldots, x_n$ be positive numbers, and define $x_{n+1} \triangleq x_1$. Then,

$$\frac{n}{2^{n-1}} \le \sum_{i=1}^n \frac{x_i^n + x_{i+1}^n}{(x_i + x_{i+1})^n}.$$

Source: [1729].

Fact 2.11.78. Let $n \ge 2$, let $x_1, \ldots, x_n$ be positive numbers, for all $i \in \{1, \ldots, n-1\}$, define $x_{n+i} \triangleq x_i$, and let $p > 0$. Then,

$$n(n-1)^p \le \sum_{i=1}^n \left(\frac{x_{i+1} + \cdots + x_{i+n-1}}{x_i}\right)^p.$$

In particular,

$$6 \le \frac{x_2 + x_3}{x_1} + \frac{x_3 + x_1}{x_2} + \frac{x_1 + x_2}{x_3}, \quad 3\sqrt{2} \le \sqrt{\frac{x_2 + x_3}{x_1}} + \sqrt{\frac{x_3 + x_1}{x_2}} + \sqrt{\frac{x_1 + x_2}{x_3}}.$$

Equality holds if and only if $x_1 = x_2 = \cdots = x_n$. **Source:** This result follows from two applications of the arithmetic-mean–geometric-mean inequality. See [1860, pp. 7, 56]. The case $p = 1$ is given in [1969, p. 72] and [1971, p. 103]. **Remark:** $n = 3$ yields $6xyz \le x(y^2 + z^2) + y(z^2 + x^2) + z(x^2 + y^2)$ of Fact 2.3.60.

Fact 2.11.79. Let $x_1, \ldots, x_n$ be nonnegative numbers. Then,

$$\sqrt{n}\sqrt{\sum_{i=1}^n \left(\frac{1}{i}\sum_{j=1}^i x_i\right)^2} \le \sum_{i=1}^n \sqrt{\frac{1}{i}\sum_{j=1}^i x_i^2}.$$

Source: [752, p. 323].

Fact 2.11.80. Let $n \geq 2$, let α be a positive number, assume that $\alpha \geq n^{n-1} - 1$, and let $x_1, \ldots, x_n$ be positive numbers. Then,

$$\frac{n}{\sqrt[n-1]{1+\alpha}} \leq \sum_{i=1}^{n} \frac{x_i}{\sqrt[n-1]{x_i^{n-1} + \alpha \prod_{j\in\{1,\ldots,n\}, j\neq i} x_j}}.$$

Source: [2378].

Fact 2.11.81. Let $x_1, \ldots, x_n$ be nonnegative numbers. Then,

$$\left(\prod_{i=1}^{n} x_i\right)^{1/n} \leq \frac{1}{n}\sum_{i=1}^{n} x_i.$$

Equality holds if and only if $x_1 = \cdots = x_n$. **Source:** [605, 698]. **Remark:** This is the arithmetic-mean–geometric-mean inequality in n variables. **Remark:** Bounds for the difference between these quantities are given in [60, 648, 2746]. **Remark:** An extension to complex numbers is given in Fact 2.21.18.

Fact 2.11.82. Let $n \geq 2$, and let $x_1, \ldots, x_n$ be nonnegative numbers. Then,

$$x_n\left(\frac{1}{n-1}\sum_{i=1}^{n-1} x_i\right)^{n-1} \leq \left(\frac{1}{n}\sum_{i=1}^{n} x_i\right)^{n}.$$

Remark: This is *Akerberg's refinement* of the arithmetic-mean–geometric-mean inequality. **Source:** [2527, p. 34].

Fact 2.11.83. Let $n \geq 2$, and define $p \in \mathbb{R}[s]$ and $q\colon (0,\infty) \mapsto (1,\infty)$ by

$$p(x) \triangleq \frac{1}{n!}\prod_{i=0}^{n-2}\left(\frac{x+i}{n-i}\right)^{n-i-1}, \quad q(x) \triangleq \prod_{i=1}^{n-1}\left(1 + \frac{n-i}{x+i-1}\right).$$

Furthermore, let $x_1, \ldots, x_n$ be distinct positive numbers, and let $\lambda > 0$ satisfy

$$p(\lambda) = \frac{\left(\prod_{i=1}^{n} x_i\right)^{n-1}}{\prod(x_i - x_j)^2},$$

where the product in the denominator is taken over all $i, j \in \{1, \ldots, n\}$ such that $i < j$. Then,

$$\sqrt[n]{\prod_{i=1}^{n} x_i} < \sqrt[n]{q(\lambda)}\sqrt[n]{\prod_{i=1}^{n} x_i} \leq \frac{1}{n}\sum_{i=1}^{n} x_i.$$

Source: [116, pp. 419–421]. **Remark:** This is *Siegel's inequality.*

Fact 2.11.84. Let $x_1, \ldots, x_n$ be nonnegative numbers, define $A \triangleq \frac{1}{n}\sum_{i=1}^{n} x_i$ and $G \triangleq \left(\prod_{i=1}^{n} x_i\right)^{1/n}$, assume that $\varepsilon \triangleq (A-G)/A < 1$, and let r_1 and r_2 be the unique solutions of $xe^{1-x} = (1-\varepsilon)^n$ satisfying $0 < r_1 \leq 1 \leq r_2$. Then, for all $i \in \{1, \ldots, n\}$, $Ar_1 \leq x_i \leq Ar_2$. **Source:** [2527, p. 35].

Fact 2.11.85. Let $x_1, \ldots, x_n$ be positive numbers. Then,

$$1 + \log\left(\prod_{i=1}^{n} x_i\right)^{1/n} \leq \left(\prod_{i=1}^{n} x_i\right)^{1/n} \leq \frac{1}{n}\sum_{i=1}^{n} x_i.$$

Source: For all $x > 0$, $1 + \log x \leq x$.

Fact 2.11.86. Let $x_1, \ldots, x_n$ be positive numbers, let r be a real number, and define

$$M_r \triangleq \begin{cases} \left(\prod_{i=1}^n x_i\right)^{1/n}, & r = 0, \\ \left(\frac{1}{n}\sum_{i=1}^n x_i^r\right)^{1/r}, & r \neq 0. \end{cases}$$

Then, $M_0 = \lim_{r\to 0} M_r$. Now, let p and q be real numbers such that $p \le q$. Then, $M_p \le M_q$. Therefore,

$$\lim_{r\to-\infty} M_r = \min\{x_1, \ldots, x_n\} \le M_{-1} \le M_0 \le M_1 \le \lim_{r\to\infty} M_r = \max\{x_1, \ldots, x_n\}.$$

Finally, $p < q$ and at least two of the numbers $x_1, \ldots, x_n$ are distinct if and only if $M_p < M_q$. **Source:** [603, p. 210] and [1952, p. 105]. To verify the limit for M_0, take the log and use l'Hôpital's rule given by Fact 12.17.13. **Remark:** This is the *power-mean inequality*. $M_0 \le M_1$ is the arithmetic-mean–geometric-mean inequality given by Fact 2.11.81. $M_{-1} \le M_0$ is the harmonic-mean–geometric-mean inequality. **Related:** Fact 2.2.58, Fact 2.11.87, Fact 2.11.89, Fact 2.11.135, and Fact 10.13.8.

Fact 2.11.87. Let $x_1, \ldots, x_n$ be positive numbers, let $\alpha_1, \ldots, \alpha_n$ be nonnegative numbers, and assume that $\sum_{i=1}^n \alpha_i = 1$. Then,

$$\frac{1}{\sum_{i=1}^n \frac{\alpha_i}{x_i}} \le \prod_{i=1}^n x_i^{\alpha_i} \le \sum_{i=1}^n \alpha_i x_i.$$

Now, let r be a real number, define

$$M_r \triangleq \begin{cases} \left(\prod_{i=1}^n x_i^{\alpha_i}\right)^{1/n}, & r = 0, \\ \left(\sum_{i=1}^n \alpha_i x_i^r\right)^{1/r}, & r \neq 0, \end{cases}$$

and let p and q be real numbers such that $p \le q$. Then, $M_p \le M_q$. If, in addition, $\alpha_1, \ldots, \alpha_n$ are positive, then

$$\lim_{r\to-\infty} M_r = \min\{x_1, \ldots, x_n\} \le \lim_{r\to 0} M_r = M_0 \le \lim_{r\to\infty} M_r = \max\{x_1, \ldots, x_n\}.$$

Equality holds if and only if $x_1 = x_2 = \cdots = x_n$. **Source:** Since $f(x) = -\log x$ is convex, it follows that

$$\log \prod_{i=1}^n x_i^{\alpha_i} = \sum_{i=1}^n \alpha_i \log x_i \le \log \sum_{i=1}^n \alpha_i x_i.$$

To prove the last statement, define $f\colon\ [0,\infty)^n \mapsto [0,\infty)$ by $f(\mu_1, \ldots, \mu_n) \triangleq \sum_{i=1}^n \alpha_i\mu_i - \prod_{i=1}^n \mu_i^{\alpha_i}$. Note that $f(\mu, \ldots, \mu) = 0$ for all $\mu \ge 0$. If $x_1, \ldots, x_n$ minimizes f, then $\partial f/\partial\mu_i(x_1, \ldots, x_n) = 0$ for all $i \in \{1, \ldots, n\}$, which implies that $x_1 = x_2 = \cdots = x_n$. **Source:** [1074] and [2128, p. 11]. **Remark:** This is the *weighted arithmetic-mean–geometric-mean* inequality. Setting $\alpha_1 = \cdots = \alpha_n = 1/n$ yields Fact 2.11.81. **Remark:** The second inequality generalizes Young's inequality. See Fact 2.2.50 and Fact 2.2.53.

Fact 2.11.88. Let $x_1, \ldots, x_n$ be positive numbers, let $\alpha_1, \ldots, \alpha_n \in [1,\infty)$, and let p and q be real numbers. If either $p < q < 0$ or $0 < p < q$, then

$$\left(\sum_{i=1}^{n} \alpha_i x_i^q\right)^{1/q} \le \left(\sum_{i=1}^{n} \alpha_i x_i^p\right)^{1/p}.$$

If $p < 0 < q$, then

$$\left(\sum_{i=1}^{n} \alpha_i x_i^p\right)^{1/p} \le \left(\sum_{i=1}^{n} \alpha_i x_i^q\right)^{1/q}.$$

Source: [1621].

Fact 2.11.89. Let $x_1, \ldots, x_n$ be positive numbers, and let p and q be nonzero real numbers such that $p \le q$. Then,

$$\left(\frac{1}{n}\sum_{i=1}^{n} x_i^p\right)^{1/p} \le \left(\frac{1}{n}\sum_{i=1}^{n} x_i^q\right)^{1/q}.$$

Furthermore,

$$\min\{x_1, \ldots, x_n\} \le \frac{n}{\frac{1}{x_1} + \cdots + \frac{1}{x_n}} \le \sqrt[n]{x_1 \cdots x_n} \le \frac{x_1 + \cdots + x_n}{n} \le \sqrt{\frac{1}{n}\sum_{i=1}^{n} x_i^2} \le \max\{x_1, \ldots, x_n\}.$$

Equality holds in each inequality if and only if $x_1 = x_2 = \cdots = x_n$. **Source:** [2060, pp. 28–30]. **Remark:** The lower bound for the geometric mean is the harmonic mean, while the second and third terms are the harmonic-mean–geometric-mean inequality. See Fact 2.11.135. **Remark:** The upper bound for the arithmetic mean is the *quadratic mean*. See [1283]. **Related:** Fact 2.2.58 and Fact 2.11.86.

Fact 2.11.90. Let $x_1, \ldots, x_n$ be nonnegative numbers, and let $p, q \in [1, \infty)$, where $p \le q$. Then,

$$\left(\sum_{i=1}^{n} x_i^q\right)^{1/q} \le \left(\sum_{i=1}^{n} x_i^p\right)^{1/p} \le n^{1/p-1/q}\left(\sum_{i=1}^{n} x_i^q\right)^{1/q}.$$

Equivalently,

$$\sum_{i=1}^{n} x_i^q \le \left(\sum_{i=1}^{n} x_i^p\right)^{q/p} \le n^{q/p-1}\sum_{i=1}^{n} x_i^q.$$

Furthermore, the first inequality is strict if and only if $p < q$ and at least two of the numbers $x_1, \ldots, x_n$ are positive. **Source:** Fact 11.8.7. **Remark:** This is the *power-sum inequality*. See [603, p. 213]. This result implies that the Hölder norm is a monotonic function of the exponent. **Remark:** Setting $p = 1$ and $q = k$ yields Fact 2.11.15. **Remark:** The power-mean inequality is given by Fact 2.11.86. **Related:** Fact 11.8.7.

Fact 2.11.91. Let $x_1, \ldots, x_n$ be nonnegative numbers, and let $p, q \in (0, 1]$, where $p \le q$. Then,

$$\left(\sum_{i=1}^{n} x_i^q\right)^{1/q} \le \left(\sum_{i=1}^{n} x_i^p\right)^{1/p}.$$

Furthermore, this inequality is strict if and only if $p < q$ and at least two of the numbers $x_1, \ldots, x_n$ are positive. **Remark:** This is the *reverse power-sum inequality*. **Related:** Fact 2.11.90 and Fact 11.8.21.

Fact 2.11.92. Let $0 \le a < b$, and let $x_1, \ldots, x_n \in [a, b]$. Then,

$$\left(\prod_{i=1}^{n} x_i\right)^{1/n} \le \frac{1}{n}\sum_{i=1}^{n} x_i \le \left(\prod_{i=1}^{n} x_i\right)^{1/n} + \frac{n-1}{n}(\sqrt{b} - \sqrt{a})^2 \le \left(\prod_{i=1}^{n} x_i\right)^{1/n} + \frac{1}{n}\sum_{1 \le i < j \le n} |x_i - x_j|.$$

Source: [748, p. 373] and [993, p. 186].

Fact 2.11.93. Let $x_1,\ldots,x_n$ be nonnegative numbers. Then,

$$(n-1)\sqrt[n]{\prod_{i=1}^n x_i}+\sqrt{\frac{1}{n}\sum_{i=1}^n x_i^2}\le\sum_{i=1}^n x_i.$$

Source: [748, p. 382].

Fact 2.11.94. Let $0<a<b$, and let $x_1,\ldots,x_n\in[a,b]$. Then,

$$\frac{1}{n}\sum_{i=1}^n x_i\le\left[\frac{1}{2}\left(\sqrt{\frac{a}{b}}+\sqrt{\frac{b}{a}}\right)\right]^{2-2/n}\left(\prod_{i=1}^n x_i\right)^{1/n}.$$

Source: [752, p. 217].

Fact 2.11.95. Let $0<a<b$, and let $x_1,\ldots,x_n\in[a,b]$. Then,

$$\left(\prod_{i=1}^n x_i\right)^{1/n}\le\frac{1}{n}\sum_{i=1}^n x_i\le S(b/a)\left(\prod_{i=1}^n x_i\right)^{1/n},$$

where S is Specht's ratio given by Fact 12.17.5. **Remark:** The right-hand inequality is a *reverse arithmetic-mean–geometric mean inequality*. See [1087, 1092, 2501, 2938]. **Related:** Fact 2.2.54, Fact 2.2.66, and Fact 15.15.23. **Credit:** W. Specht.

Fact 2.11.96. Let $x_1,\ldots,x_n$ be positive numbers, and let k and l satisfy $1\le k\le l\le n$. Then,

$$\left(\prod_{i=1}^n x_i\right)^{1/n}\le\left(\binom{n}{l}^{-1}\sum_{i_1<\cdots<i_l}\prod_{j=1}^l x_{i_j}\right)^{1/l}\le\left(\binom{n}{k}^{-1}\sum_{i_1<\cdots<i_k}\prod_{j=1}^k x_{i_j}\right)^{1/k}\le\frac{1}{n}\sum_{i=1}^n x_i.$$

Source: Fact 2.11.35 and [1172, pp. 17, 18]. **Remark:** This result shows that the kth elementary symmetric function is Schur-concave. See Definition 3.10.3 and [1146, p. 102, Ex. 7.11].

Fact 2.11.97. Let $x_1,\ldots,x_n$ be nonnegative numbers, and let k and l satisfy $1\le k\le l\le n$. Then,

$$\left(\prod_{i=1}^n x_i\right)^{1/n}\le\binom{n}{l}^{-1}\sum_{i_1<\cdots<i_l}\prod_{j=1}^l x_{i_j}^{1/l}\le\binom{n}{k}^{-1}\sum_{i_1<\cdots<i_k}\prod_{j=1}^k x_{i_j}^{1/k}\le\frac{1}{n}\sum_{i=1}^n x_i.$$

Source: [1146, p. 23] and [1591]. **Remark:** The case $n=3$, $k=2$ is given by Fact 2.3.49.

Fact 2.11.98. Let $x_1,\ldots,x_n$ be nonnegative numbers, and let k be a positive integer such that $1\le k\le n$. Then,

$$\left(\sum_{i_1<\cdots<i_k}\prod_{j=1}^k x_{i_j}\right)^k\le\binom{n}{k}^{k-1}\sum_{i_1<\cdots<i_k}\prod_{j=1}^k x_{i_j}^k.$$

Remark: Equality holds in the cases $k=1$ and $k=n$. The case $n=3$, $k=2$ is given by Fact 2.3.66. The cases $n=4$, $k=3$ and $n=4$, $k=2$ are given by Fact 2.4.23.

Fact 2.11.99. Let $x_1,\ldots,x_n$ be nonnegative numbers, and let k satisfy $1\le k\le n$. Then,

$$\left(\prod_{i=1}^n x_i\right)^{1/n}\le\binom{n}{k}^{-1}\sum_{i_1<\cdots<i_k}\prod_{j=1}^k x_{i_j}^{1/k}\le\left(\binom{n}{k}^{-1}\sum_{i_1<\cdots<i_k}\prod_{j=1}^k x_{i_j}\right)^{1/k}\le\frac{1}{n}\sum_{i=1}^n x_i.$$

Source: Use Fact 2.11.98 to combine the first, third, and fourth terms of Fact 2.11.96 with the first, third, and fourth terms of Fact 2.11.97.

Fact 2.11.100. Let $x_1,\ldots,x_n$ be nonnegative numbers. Then,

$$\left(\prod_{i=1}^n x_i\right)^{1/n} \le \frac{2}{n^2-n}\sum_{1\le i<j\le n}\sqrt{x_ix_j}.$$

Source: [2527, p. 189].

Fact 2.11.101. Let $x_1,\ldots,x_n$ be nonnegative numbers, let $\alpha_1,\ldots,\alpha_n$ be nonnegative numbers, and assume that $\sum_{i=1}^n \alpha_i = 1$. Then,

$$\left(\prod_{i=1}^n x_i\right)^{1/n} \le \frac{1}{n!}\sum\prod_{j=1}^n x_{i_j}^{\alpha_j} \le \frac{1}{n}\sum_{i=1}^n x_i,$$

where the sum is taken over all $n!$ permutations $(i_1,\ldots,i_n)$ of $(1,\ldots,n)$. **Source:** [1146, p. 100].

Fact 2.11.102. Let $x_1,\ldots,x_n,\beta$ be nonnegative numbers, and let $\alpha_1,\ldots,\alpha_n$ be positive numbers. Then,

$$\left(\prod_{i=1}^n x_i^{\alpha_i}\right)^{1/\sum_{i=1}^n\alpha_i} \le \left[\prod_{i=1}^n (x_i+\beta)^{\alpha_i}\right]^{1/\sum_{i=1}^n\alpha_i} - \beta \le \frac{1}{\sum_{i=1}^n\alpha_i}\sum_{i=1}^n \alpha_ix_i.$$

Source: [2042].

Fact 2.11.103. Let $x_1,\ldots,x_n$ be real numbers. Then,

$$\left(\prod_{i=1}^n x_i\right)^{1/n} \le \log\left(1+\sqrt[n]{\prod_{i=1}^n (e^{x_i}-1)}\right) \le \frac{1}{n}\sum_{i=1}^n x_i.$$

Source: [2294, p. 283].

Fact 2.11.104. Let $n\ge 2$, and let $x_1,\ldots,x_n$ be nonnegative numbers. Then,

$$x_n\left(\frac{1}{n-1}\sum_{i=1}^{n-1} x_i\right)^{n-1} \le \left(\frac{1}{n}\sum_{i=1}^n x_i\right)^n.$$

Source: [1757, p. 77]. **Remark:** The case $n=3$ is given by Fact 2.3.60.

Fact 2.11.105. Let $x_1,\ldots,x_n$ be nonnegative numbers. Then,

$$\left(\prod_{i=1}^n x_i\right)^{(n-1)/n}\sqrt{\frac{1}{n}\sum_{i=1}^n x_i^2} \le \left(\frac{1}{n}\sum_{i=1}^n x_i\right)^n.$$

Now, assume that $x_1,\ldots,x_n$ are positive. Then,

$$(n+1)\left(\prod_{i=1}^n x_i\right)^{1/n} \le \sum_{i=1}^n x_i + \frac{n}{\sum_{i=1}^n 1/x_i}.$$

Source: [108, pp. 22, 23, 142, 143, 149]. **Remark:** The case $n=3$ is given by Fact 2.3.60.

Fact 2.11.106. Let $x_1,\ldots,x_n$ be nonnegative numbers. Then,

$$\sqrt[n]{\prod_{i=1}^n x_i} = \min\ \frac{1}{n}\sum_{i=1}^n \alpha_ix_i,$$

where the minimum is taken over all positive numbers $\alpha_1,\ldots,\alpha_n$ such that $\prod_{i=1}^n\alpha_i = 1$. **Source:** [2527, p. 133].

Fact 2.11.107. Let $x_0, x_1, \ldots, x_n$ be real numbers. Then,

$$\sum_{i=1}^{n} \frac{x_i}{i} \le \sqrt{2\sum_{i=1}^{n} x_i^2}, \quad \sum_{i=1}^{n} \frac{x_i}{\sqrt{n+i}} \le \sqrt{(\log 2)\sum_{i=1}^{n} x_i^2}, \quad \sum_{i=0}^{n} \binom{n}{i} x_i \le \sqrt{\binom{2n}{n}\sum_{i=0}^{n} x_i^2}.$$

Source: [2527, pp. 14, 15].

Fact 2.11.108. Let $x_1, \ldots, x_n$ be real numbers. Then,

$$\left(\sum_{i=1}^{n} \frac{x_i}{i}\right)^2 \le \sum_{i,j=1}^{n} \frac{x_i x_j}{i+j-1}.$$

Source: [2294, pp. 282, 369]. **Related:** Fact 2.11.142.

Fact 2.11.109. Let $x_1, \ldots, x_n$ be real numbers. Then,

$$\sum_{i,j=1}^{n} \frac{x_i x_j}{i+j} \le \pi \sum_{i=1}^{n} x_i^2.$$

Source: Fact 2.12.25.

Fact 2.11.110. Let $x_1, \ldots, x_n$ be nonnegative numbers, and let $k \ge 2$ be an integer. Then,

$$\frac{1}{k}\sum_{i=1}^{n}\left(\frac{1}{i}\sum_{j=1}^{i} x_j\right)^k \le \frac{1}{k-1}\sum_{i=1}^{n} x_i \left(\frac{1}{i}\sum_{j=1}^{i} x_j\right)^{k-1}.$$

Source: [1566, p. 15].

Fact 2.11.111. Let $n \ge 2$, and let $x_1, \ldots, x_n$ be nonnegative numbers. Then,

$$\prod (x_i - x_j)^2 \le \left(\prod_{i=2}^{n} i^i (i-1)^{i-1}\right)\left(\frac{1}{n(n-1)}\sum_{i=1}^{n} x_i\right)^{n(n-1)},$$

where the product is taken over all $i, j \in \{1, \ldots, n\}$ such that $i < j$. In particular, if x, y, z are positive numbers, then

$$6\sqrt{3}xy|x-y| \le (x+y)^3, \quad 108(x-y)^2(y-z)^2(z-x)^2 \le (x+y+z)^6.$$

Source: [116, p. 423]. **Credit:** I. Schur.

Fact 2.11.112. Let $x_1, \ldots, x_n$ be positive numbers. Then,

$$\left(\prod_{i=1}^{n} x_i\right)^{1/n} < \frac{1}{n}\left(\frac{x_2 - x_1}{\log x_2 - \log x_1} + \frac{x_3 - x_2}{\log x_3 - \log x_2} + \cdots + \frac{x_1 - x_n}{\log x_1 - \log x_n}\right) < \frac{1}{n}\sum_{i=1}^{n} x_i.$$

Source: [209, p. 44]. **Remark:** This result extends Fact 2.2.63 to n variables. See also [2930]. **Credit:** M. Bencze.

Fact 2.11.113. Let $x_1, \ldots, x_n \in [a, b]$, where $a > 0$. Then,

$$\frac{a}{2n^2}\sum_{i<j} (\log x_i - \log x_j)^2 \le \frac{1}{n}\sum_{i=1}^{n} x_i - \left(\prod_{i=1}^{n} x_i\right)^{1/n} \le \frac{b}{2n^2}\sum_{i<j} (\log x_i - \log x_j)^2.$$

Source: [2128, p. 86] and [2129].

Fact 2.11.114. Let $x_1, \ldots, x_n \in (0, 1/2]$, and define

$$A \triangleq \frac{1}{n}\sum_{i=1}^{n} x_i, \quad G \triangleq \prod_{i=1}^{n} x_i^{1/n}, \quad H \triangleq \frac{n}{\sum_{i=1}^{n} \frac{1}{x_i}},$$

$$A' \triangleq \frac{1}{n}\sum_{i=1}^{n}(1 - x_i), \quad G' \triangleq \prod_{i=1}^{n}(1 - x_i)^{1/n}, \quad H' \triangleq \frac{n}{\sum_{i=1}^{n} \frac{1}{1 - x_i}}.$$

Then, the following statements hold:

i) $A'/G' \le A/G$. Equality holds if and only if $x_1 = \cdots = x_n$.

ii) $A' - G' \le A - G$. Equality holds if and only if $x_1 = \cdots = x_n$.

iii) $A^n - G^n \le A'^n - G'^n$. Equality holds for $n = 1$ and $n = 2$; for all $n \ge 3$, equality holds if and only if $x_1 = \cdots = x_n$.

iv) $G'/H' \le G/H$.

Source: [2331] and Fact 3.25.6. *iv)* is given in [2365]. **Credit:** *i)* is due to K. Fan. See [1559].

Fact 2.11.115. Let $x_1, \ldots, x_n$ be positive numbers, and, for all $k \in \{1, \ldots, n\}$, define

$$A_k \triangleq \frac{1}{k}\sum_{i=1}^{k} x_i, \quad G_k \triangleq \prod_{i=1}^{k} x_i^{1/k}.$$

Then,

$$1 = \frac{A_1}{G_1} \le \left(\frac{A_2}{G_2}\right)^2 \le \cdots \le \left(\frac{A_n}{G_n}\right)^n, \quad 0 = A_1 - G_1 \le 2(A_2 - G_2) \le \cdots \le n(A_n - G_n).$$

Source: [658] and [2128, p. 13]. **Credit:** The first result is due to T. Popoviciu. The second result is due to R. Rado.

Fact 2.11.116. Let $x_1, \ldots, x_n$ be positive numbers. Then,

$$\left(\prod_{i=1}^{n} x_i\right)^{\frac{1}{n}\sum_{i=1}^{n} x_i} \le \prod_{i=1}^{n} x_i^{x_i}.$$

Source: [2294, p. 267]. **Related:** Fact 2.2.71.

Fact 2.11.117. Let $x_1, \ldots, x_n \in (0, 1]$. Then,

$$\sum_{i=1}^{n}\prod_{j=1}^{n} x_j^{x_i} \le n\prod_{i=1}^{n} x_i^{x_i}.$$

Source: [783].

Fact 2.11.118. Let $x_1, \ldots, x_n$ be positive numbers, let $\alpha_1, \ldots, \alpha_n$ be nonnegative numbers, and assume that $\sum_{i=1}^{n} \alpha_i = 1$. Then,

$$e^{2\beta}\prod_{i=1}^{n} x_i^{\alpha_i} \le \sum_{i=1}^{n} \alpha_i x_i,$$

where

$$\beta \triangleq 1 - \frac{\sum_{i=1}^{n} \alpha_i \sqrt{x_i}}{\sqrt{\sum_{i=1}^{n} \alpha_i x_i}}.$$

Source: [32]. **Remark:** This is a refinement of the weighted arithmetic-mean–geometric-mean inequality. See Fact 2.11.87. **Remark:** The convexity of $f(x) = x^2$ implies that $\beta \ge 0$.

Fact 2.11.119. Let $x_1, \ldots, x_n$ be nonnegative numbers. Then,

$$1 + \left(\prod_{i=1}^n x_i\right)^{1/n} \le \left[\prod_{i=1}^n (1 + x_i)\right]^{1/n}.$$

Equality holds if and only if $x_1 = \cdots = x_n$. **Source:** Use Fact 2.11.102 or the convexity of $f\colon \mathbb{R} \mapsto \mathbb{R}$, where $f(x) \triangleq \log(1 + e^x)$. See [116, p. 410] and [517, p. 210]. **Remark:** This inequality is used to prove Corollary 10.4.15.

Fact 2.11.120. Let $x_1, \ldots, x_n$ be real numbers, and let α be a real number such that, for all $i \in \{1, \ldots, n\}$, $x_i < \alpha$. Then,

$$\frac{n}{\alpha - \frac{1}{n}\sum_{i=1}^n x_i} \le \sum_{i=1}^n \frac{1}{\alpha - x_i}.$$

Furthermore, if $x_1, \ldots, x_n \in (0, 1)$, then

$$\frac{n\sum_{i=1}^n x_i}{n - \sum_{i=1}^n x_i} \le \sum_{i=1}^n \frac{x_i}{1 - x_i}.$$

Source: [1566, pp. 14, 15].

Fact 2.11.121. Let $x_1, \ldots, x_n \in (0, 1]$. Then,

$$\sum_{i=1}^n \frac{1}{1 + x_i} \le \frac{n\sum_{i=1}^n x_i}{\sum_{i=1}^n x_i + n\prod_{i=1}^n x_i}.$$

Source: [1566, p. 16].

Fact 2.11.122. Let $x_1, \ldots, x_n \in \mathbb{R}$, where either $x_1, \ldots, x_n \in (-1, 0)$ or $x_1, \ldots, x_n > 0$. Then,

$$1 + \sum_{i=1}^n x_i < \prod_{i=1}^n (1 + x_i).$$

Source: [2060, p. 35].

Fact 2.11.123. Let $x_1, \ldots, x_n \in (0, 1)$, and assume that $\sum_{i=1}^n x_i < 1$. Then,

$$1 + \sum_{i=1}^n x_i < \prod_{i=1}^n (1 + x_i) < \frac{1}{1 - \sum_{i=1}^n x_i}, \quad 1 - \sum_{i=1}^n x_i < \prod_{i=1}^n (1 - x_i) < \frac{1}{1 + \sum_{i=1}^n x_i}.$$

Source: [1566, p. 15]. **Remark:** These are the *Weierstrass inequalities*.

Fact 2.11.124. Let $n \ge 2$ and $x_1, \ldots, x_n \in (0, 1)$. Then,

$$1 < \prod_{i=1}^n (1 - x_i) + \sum_{i=1}^n x_i < \prod_{i=1}^n (1 - x_i) + \prod_{i=1}^n (1 + x_i) - 1.$$

Source: [2227].

Fact 2.11.125. Let $x_1, \ldots, x_n \in [1, \infty)$. Then,

$$\frac{n}{1 + \sqrt[n]{x_1 \cdots x_n}} \le \sum_{i=1}^n \frac{1}{1 + x_i}.$$

Source: [1158, p. 9] and [1559].

Fact 2.11.126. Let $n \ge 2$, let $x_1, \ldots, x_n \in [-1, 1]$, and assume that $x_1, \ldots, x_n$ are distinct. Then,

$$\frac{1}{8} n \log n \le \max_{k\in\{2,\ldots,n\}} \sum_{i=1}^{k-1} \frac{1}{|x_k - x_i|}.$$

Now, assume that $x_1 < \cdots < x_n$. Then,

$$\frac{1}{8} n^2 \log n \le \sum_{1\le i<j\le n} \frac{1}{x_j - x_i}.$$

Source: [2527, p. 134].

Fact 2.11.127. Let $x_1, \ldots, x_n \in [-1, 1]$, and assume that $x_1, \ldots, x_n$ are distinct. Then,

$$2^{n-2} \le \sum_{i=1}^{n} \frac{1}{\prod_{j\ne i} |x_j - x_i|}.$$

Source: [1158, p. 61].

Fact 2.11.128. Let $x_1, \ldots, x_n$ be positive numbers, and define $\alpha \triangleq \sum_{i=1}^n x_i$. Then,

$$\prod_{i=1}^{n} x_i \le \prod_{i=1}^{n} \frac{\alpha - x_i}{n-1} \le \prod_{i=1}^{n} \frac{\alpha + x_i}{n+1}.$$

Source: [1566, p. 16].

Fact 2.11.129. Let $x_1, \ldots, x_n$ be nonnegative numbers, and let $\alpha \ge 2$. Then,

$$\left(1 + \frac{1}{n}\sum_{i=1}^{n} x_i\right)^{\alpha} \le \frac{1}{n}\sum_{i=1}^{n}(1 + x_i)^{\alpha}.$$

Source: Apply Jensen's inequality Fact 1.21.7 to $f(x) = (1 + x)^{\alpha}$. See [382].

Fact 2.11.130. Let $x_1, \ldots, x_n$ be nonnegative numbers. If $p \in [0, 1]$, then

$$n^{p-1}\sum_{i=1}^{n} x_i^p \le \left(\sum_{i=1}^{n} x_i\right)^p \le \sum_{i=1}^{n} x_i^p.$$

If $p \in [1, \infty)$, then

$$\sum_{i=1}^{n} x_i^p \le \left(\sum_{i=1}^{n} x_i\right)^p \le n^{p-1}\sum_{i=1}^{n} x_i^p.$$

Source: [1423, 1629]. **Remark:** With $x \triangleq [x_1 \ \cdots \ x_n]^{\mathsf{T}}$, the last string can be written as $\|x\|_p \le \|p\|_1 \le n^{1-1/p}\|x\|_p$. **Remark:** Note that, if $p > 0$, then

$$\sum_{i=1}^{n} x_i^p \le n^{1-\min\{p,1\}}\left(\sum_{i=1}^{n} x_i\right)^p.$$

Related: Fact 2.2.59, Fact 2.11.90, and Fact 11.10.59.

Fact 2.11.131. Let $x_1, \ldots, x_n$ be positive numbers, and let $\alpha_1, \ldots, \alpha_n \in [0,1]$ satisfy $\sum_{i=1}^n \alpha_i = 1$. If either $p \le 0$ or $p \ge 1$, then

$$\left(\sum_{i=1}^{n} \alpha_i x_i\right)^p \le \sum_{i=1}^{n} \alpha_i x_i^p, \quad \left(\frac{1}{n}\sum_{i=1}^{n} x_i\right)^p \le \frac{1}{n}\sum_{i=1}^{n} x_i^p.$$

Furthermore, if $p \in [0, 1]$, then

$$\sum_{i=1}^{n} \alpha_i x_i^p \le \left(\sum_{i=1}^{n} \alpha_i x_i\right)^p.$$

Equality in both cases holds if and only if either $p = 0$ or $p = 1$ or $x_1 = \cdots = x_n$. **Remark:** The special case $\alpha_1 = \cdots = \alpha_n = 1/n$ is given in [644, p. 105]. **Remark:** This is a consequence of Jensen's inequality given by Fact 1.21.7. Fact 2.2.59 is a special case of the third inequality.

Fact 2.11.132. Let $x_1, \ldots, x_n$ be nonnegative numbers, and let $p > 1$. Then,

$$\sum_{i=1}^{n} \left(\frac{1}{i} \sum_{j=1}^{i} x_j \right)^p \le \left(\frac{p}{p-1} \right)^p \sum_{i=1}^{n} x_i^p.$$

In particular,

$$\sum_{i=1}^{n} \left(\frac{1}{i} \sum_{j=1}^{i} x_j \right)^2 \le 4 \sum_{i=1}^{n} x_i^2.$$

Source: [1708]. **Remark:** This is the *discrete Hardy inequality*. See [753, 1708] and [2527, p. 169]. **Related:** Fact 2.1.36 and Fact 12.13.17.

Fact 2.11.133. Let $x_1, \ldots, x_n$ be nonnegative numbers, and let $p > 1$. Then,

$$\sum_{i=1}^{n} \left(\sum_{j=i}^{n} \frac{x_j}{j} \right)^p \le p^p \sum_{i=1}^{n} x_i^p.$$

Source: [1708]. **Remark:** This is the *Copson inequality*.

Fact 2.11.134. Let $x_1, \ldots, x_n$ be positive numbers, assume that $x_1 = \min\{x_1, \ldots, x_n\}$ and $x_n = \max\{x_1, \ldots, x_n\}$, let $\alpha_1, \ldots, \alpha_n$ be nonnegative numbers, and assume that $\sum_{i=1}^{n} \alpha_i = 1$. Then,

$$1 \le \left(\sum_{i=1}^{n} \alpha_i x_i \right) \left(\sum_{i=1}^{n} \frac{\alpha_i}{x_i} \right) \le \frac{(x_1 + x_n)^2}{4 x_1 x_n}.$$

Source: [47, p. 122] and [1566, p. 17]. **Remark:** This is the *Kantorovich inequality*. See Fact 10.18.8 and [1876]. This result is used in [1267, pp. 369, 370] to estimate the convergence rate of a steepest descent algorithm. **Related:** Fact 2.11.135.

Fact 2.11.135. Let $x_1, \ldots, x_n$ be positive numbers, and define $\alpha \triangleq \min_{i \in \{1,\ldots,n\}} x_i$ and $\beta \triangleq \max_{i \in \{1,\ldots,n\}} x_i$. Then,

$$1 \le \left(\frac{1}{n} \sum_{i=1}^{n} x_i \right) \left(\frac{1}{n} \sum_{i=1}^{n} \frac{1}{x_i} \right) \le \frac{(\alpha + \beta)^2}{4 \alpha \beta}.$$

Source: Fact 2.11.134, Fact 2.12.44, [935, p. 94], and [936, p. 119]. **Remark:** The left-hand inequality is the arithmetic-mean–harmonic-mean inequality. See Fact 2.11.89. The right-hand inequality is *Schweitzer's inequality*. See [2819, 2843] for historical details. **Remark:** A matrix extension is given by Fact 10.11.53.

Fact 2.11.136. Let $x_1, \ldots, x_n$ be positive numbers, and let p and q be positive numbers. Then,

$$\left(\frac{1}{n} \sum_{i=1}^{n} x_i^p \right) \left(\frac{1}{n} \sum_{i=1}^{n} x_i^q \right) \le \frac{1}{n} \sum_{i=1}^{n} x_i^{p+q}.$$

In particular, if $p \in [0, 1]$, Then,

$$\left(\frac{1}{n} \sum_{i=1}^{n} x_i^p \right) \left(\frac{1}{n} \sum_{i=1}^{n} x_i^{1-p} \right) \le \frac{1}{n} \sum_{i=1}^{n} x_i.$$

Source: [2827]. **Remark:** These inequalities are interpolated in [2827].

Fact 2.11.137. Let $x_1,\dots,x_n$ be positive numbers, let p,q,r,s be nonnegative numbers, and assume that $0\le p<q<r<s$ and $p+s=q+r$. Then,

$$\left(\frac{1}{n}\sum_{i=1}^n x_i^q\right)\left(\frac{1}{n}\sum_{i=1}^n x_i^r\right)\le\left(\frac{1}{n}\sum_{i=1}^n x_i^p\right)\left(\frac{1}{n}\sum_{i=1}^n x_i^s\right).$$

Related: $s=0$ yields Fact 2.11.136.

Fact 2.11.138. Let $x_1,\dots,x_n$ be positive numbers. Then,

$$\frac{1}{n}\sum_{k=1}^n\left(\prod_{i=1}^k x_i\right)^{1/k}\le\left[\prod_{k=1}^n\left(\frac{1}{k}\sum_{i=1}^k x_i\right)\right]^{1/k}.$$

Equality holds if and only if $x_1=\cdots=x_n$. **Remark:** This result can be expressed as

$$\frac{1}{n}(z_1+\cdots+z_n)\le\sqrt[n]{y_1\cdots y_n},$$

where $z_k\triangleq\sqrt[k]{x_1\cdots x_k}\le y_k\triangleq\frac{1}{k}(x_1+\cdots+x_k)$. **Remark:** This is the *mixed arithmetic-geometric mean inequality*. **Credit:** T. S. Nanjundiah. See [753, 1993].

Fact 2.11.139. Let $n\ge 2$, and let $x_1,\dots,x_n$ be nonnegative numbers. Then,

$$\sqrt[n]{\prod_{i=1}^n x_i}\le\frac{e}{2n^2}\sum_{i=1}^n(2i-1)x_i.$$

Source: [2527, p. 129]. **Related:** Fact 2.11.140.

Fact 2.11.140. Let $n\ge 2$, and let $x_1,\dots,x_n$ be positive numbers. Then,

$$\sum_{j=1}^n\left(\prod_{i=1}^j x_i\right)^{1/k}\le\frac{n}{\sqrt[n]{n!}}\sum_{i=1}^n x_i\le e^{(n-1)/n}\sum_{i=1}^n x_i\le e\sum_{i=1}^n x_i.$$

Equality holds in all of these inequalities if and only if $x_1=\cdots=x_n=0$. **Remark:** $n/\sqrt[n]{n!}<e^{(n-1)/n}$, which is equivalent to $e(n/e)^n<n!$, follows from Fact 1.13.14. **Remark:** The first and last terms are *Carleman's inequality*. See [753], [1146, p. 22], and [2294, pp. 45–51]. **Related:** Fact 2.12.11.

Fact 2.11.141. Let $x_1,\dots,x_n$ be nonnegative numbers, not all of which are zero. Then,

$$\left(\sum_{i=1}^n x_i\right)^4<(2\,\mathrm{atan}\,n)^2\left(\sum_{i=1}^n x_i^2\right)\sum_{i=1}^n i^2x_i^2<\pi^2\left(\sum_{i=1}^n x_i^2\right)\sum_{i=1}^n i^2x_i^2.$$

Furthermore,

$$\left(\sum_{i=1}^n x_i\right)^2<\frac{\pi^2}{6}\sum_{i=1}^n i^2x_i^2.$$

Source: [310] and [1758, p. 18]. **Remark:** The first and third terms in the first inequality are a finite version of the *Carlson inequality*. The last inequality follows from the Cauchy-Schwarz inequality. See [993, pp. 175, 176].

Fact 2.11.142. Let $x_1,\dots,x_n$ be nonnegative numbers. Then,

$$\left(\sum_{i=1}^n\frac{x_i}{i}\right)^4\le 2\pi^2\left(\sum_{i,j=1}^n\frac{x_ix_j}{i+j}\right)\sum_{i,j=1}^n\frac{x_ix_j}{(i+j)^3}.$$

Source: [501]. **Remark:** This is a discrete version of the integral version of Carlson's inequality

$$\left(\int_0^\infty f(x)\,\mathrm{d}x\right)^4 \le \pi^2\left(\int_0^\infty f^2(x)\,\mathrm{d}x\right)\int_0^\infty x^2f^2(x)\,\mathrm{d}x,$$

where $f(x) \triangleq \sum_{i=1}^n x_ie^{-ix}$. **Related:** Fact 2.11.108.

Fact 2.11.143. Let $x_1,\dots,x_n,\alpha,\beta$ be positive numbers, let p and q be real numbers, and assume that at least one of the following statements holds:

i) $p \in (-\infty, 1]\backslash\{0\}$ and $(n-1)\alpha \le \beta$.

ii) $p \ge 1$ and $(n^p - 1)\alpha \le \beta$.

Then,

$$\frac{n}{(\alpha+\beta)^{1/p}} \le \sum_{i=1}^n \left(\frac{x_i^q}{\alpha x_i^q + \beta\prod_{k=1}^n x_k^{q/n}}\right)^{1/p}.$$

Source: [2923].

Fact 2.11.144. Let $x_1,\dots,x_n$ be nonnegative numbers, and let $k \ge 1$. Then,

$$\frac{e^k}{k^k}\left(\sum_{i=1}^n x_i\right)^k \le \prod_{i=1}^n e^{x_i}, \quad \frac{e^2}{4}\sum_{i=1}^n x_i^2 \le \frac{e^2}{4}\left(\sum_{i=1}^n x_i\right)^2 \le \prod_{i=1}^n e^{x_i}.$$

Source: [2039, 2275]. **Related:** Fact 2.1.51

Fact 2.11.145. Let $x_1,\dots,x_n$ be positive numbers, and define $f\colon [0,\infty) \mapsto (0,\infty)$ by $f(\alpha) \triangleq \sum_{i=1}^n(x_i^\alpha + x_i^{-\alpha})$. Then, f is nondecreasing. **Source:** [1809].

Fact 2.11.146. Let $x_1,\dots,x_n$ be real numbers, assume that either $x_1,\dots,x_n \in [0,1]$ or $x_1,\dots, x_n \in [1,\infty)$, and let $\alpha \in (0,1)$. Then,

$$\prod_{i=1}^n [\alpha + (1-\alpha)x_i] \le \alpha + (1-\alpha)\prod_{i=1}^n x_i.$$

Source: [830].

Fact 2.11.147. Let $n \ge 2$, let $x_1,\dots,x_n$ be real numbers such that, for all $i \in \{1,\dots,n\}$, $x_i \ne 1$, and such that $\prod_{i=1}^n x_i = 1$, and define $S_n \triangleq \sum_{i=1}^n 1/(1-x_i)^2$. Then, the following statements hold:

i) If $n = 2$, then $S_n \ge \frac{1}{2}$. Equality holds if and only if $x_1 = x_2 = -1$.

ii) If $n = 3$, then $S_n \ge 1$. Equality holds if and only if $x_1 + x_2 + x_3 = 3$.

iii) If $n = 4$, then $S_n \ge 1$. Equality holds if and only if $x_1 = x_2 = x_3 = x_4 = -1$.

iv) If either $n \ge 5$ or $x_1,\dots,x_n$ are positive, then $S_n > 1$.

Source: [2136]. **Related:** Fact 2.2.47.

Fact 2.11.148. Let $x_1,\dots,x_n$ be positive numbers, and define $x_{n+1} \triangleq x_1$ and $x_{n+2} \triangleq x_2$. Then,

$$\sum_{i=1}^n \frac{x_{i+2}(x_ix_{i+1}+1)(x_{i+2}^2+1)}{x_{i+2}^3+1} \le 2\sum_{i=1}^n \frac{(x_i^3+1)(x_{i+1}^3+1)(x_{i+2}^3+1)}{(x_i^2+1)(x_{i+1}^2+1)(x_{i+2}^2+1)},$$

$$\sum_{i=1}^n \frac{x_ix_{i+1}(x_{i+2}+1)(x_i^2x_{i+1}^2+1)}{x_i^3x_{i+1}^3+1} \le 2\sum_{i=1}^n \frac{(x_i^3+1)(x_{i+1}^3+1)(x_{i+2}^3+1)}{(x_i^2+1)(x_{i+1}^2+1)(x_{i+2}^2+1)}.$$

Source: [379].

2.12 Facts on Equalities and Inequalities in $2n$ Variables

Fact 2.12.1. Let $x_1, \ldots, x_n$ be real numbers, and let $y_1, \ldots, y_n$ be nonnegative numbers. Then,

$$\sum_{1\le i,j\le n} x_i x_j \min\{y_i, y_j\} \ge 0.$$

Source: [107, pp. 361, 362].

Fact 2.12.2. Let $x_1, \ldots, x_n, y_1, \ldots, y_n$ be real numbers. Then,

$$2\sum_{1\le i,j\le n} \min\{x_i, y_j\} \le \sum_{1\le i,j\le n} \min\{x_i, x_j\} + \sum_{1\le i,j\le n} \min\{y_i, y_j\}.$$

Source: [107, p. 365].

Fact 2.12.3. Let $x_1, \ldots, x_n, y_1, \ldots, y_n$ be nonnegative numbers. Then,

$$\left(\sum_{1\le i,j\le n} \min\{x_i, y_j\}\right)^2 \le \left(\sum_{1\le i,j\le n} \min\{x_i, x_j\}\right) \sum_{1\le i,j\le n} \min\{y_i, y_j\}.$$

Source: [107, p. 365].

Fact 2.12.4. Let $x_1, \ldots, x_n, y_1, \ldots, y_n$ be complex numbers, and let $\alpha \in (0, 2]$. Then,

$$\sum_{1\le i<j\le n} (|x_i - x_j|^\alpha + |y_i - y_j|^\alpha) \le \sum_{1\le i,j\le n} |x_i - y_j|^\alpha.$$

Source: [107, pp. 364, 365] and [1217, p. 1060].

Fact 2.12.5. Let $x_1, \ldots, x_n, y_1, \ldots, y_n$ be nonnegative numbers. Then,

$$\sum_{1\le i,j\le n} \min\{x_i x_j, y_i y_j\} \le \sum_{1\le i,j\le n} \min\{x_i y_j, x_j y_i\}.$$

Source: [107, pp. 361, 366, 367].

Fact 2.12.6. Let $x_1, \ldots, x_n$ be real numbers, and let $y_1, \ldots, y_n$ be positive numbers Then,

$$\min\left\{\frac{x_1}{y_1}, \ldots, \frac{x_n}{y_n}\right\} \le \frac{\sum_{i=1}^n x_i}{\sum_{i=1}^n y_i} \le \max\left\{\frac{x_1}{y_1}, \ldots, \frac{x_n}{y_n}\right\}.$$

Source: [47, p. 39] and [2527, p. 82]. **Related:** Fact 2.4.1.

Fact 2.12.7. Let $x_1, \ldots, x_n$ and $y_1, \ldots, y_n$ be real numbers, and assume that either $x_1 \le \cdots \le x_n$ and $y_1 \le \cdots \le y_n$ or $x_n \le \cdots \le x_1$ and $y_n \le \cdots \le y_1$. Then,

$$\sum_{i=1}^n x_i y_{n-i+1} \le \frac{1}{n}\left(\sum_{i=1}^n x_i\right)\left(\sum_{i=1}^n y_i\right) \le \sum_{i=1}^n x_i y_i.$$

Furthermore, each inequality is an equality if and only if either $x_1 = \cdots = x_n$ or $y_1 = \cdots = y_n$. Now, for all $i \in \{1, \ldots, n\}$, let $\alpha_i \ge 0$. Then,

$$\left(\sum_{i=1}^n \alpha_i x_i\right)\left(\sum_{i=1}^n \alpha_i y_i\right) \le \left(\sum_{i=1}^n \alpha_i\right) \sum_{i=1}^n \alpha_i x_i y_i.$$

Source: [154, p. 27], [1316, p. 43], [1371, pp. 148, 149], and [2060, p. 36]. **Remark:** This is *Chebyshev's inequality*. **Related:** Fact 10.25.69.

Fact 2.12.8. Let $x_1,\ldots,x_n$ and $y_1,\ldots,y_n$ be real numbers, let $\{j_1,\ldots,j_n\} = \{k_1,\ldots,k_n\} = \{1,\ldots,n\}$, and assume that $x_{j_n} \le \cdots \le x_{j_1}$ and $y_{k_n} \le \cdots \le y_{k_1}$. Then,

$$\sum_{i=1}^n x_{j_i} y_{k_{n-i+1}} \le \sum_{i=1}^n x_i y_i \le \sum_{i=1}^n x_{j_i} y_{k_i}.$$

If $x_1,\ldots,x_n$ are nonnegative, then, for all $l \in \{1,\ldots,n\}$,

$$\sum_{i=1}^l x_{j_{n+i-l}} y_{k_{n-i+l}} \le \sum_{i=1}^l x_i y_i.$$

If $x_1,\ldots,x_n$ and $y_1,\ldots,y_n$ are nonnegative, then, for all $l \in \{1,\ldots,n\}$,

$$\max \sum_{i=1}^l x_{j_{\sigma(i)}} y_{k_{\sigma(n-i+1)}} \le \max \sum_{i=1}^l x_{\sigma(i)} y_{\sigma(i)} \le \sum_{i=1}^l x_{j_i} y_{k_i},$$

where both maxima are taken over all permutations σ of $\{1,\ldots,n\}$. **Source:** [507, p. 127], [844, 1664], [1969, p. 141], and [1971, p. 207]. **Remark:** The first string of inequalities is the *Hardy-Littlewood rearrangement inequality*. **Related:** The last string of inequalities is given more succinctly by *iv*) of Fact 3.25.3. **Related:** Fact 10.22.25.

Fact 2.12.9. Let $x_1,\ldots,x_n$ and $y_1,\ldots,y_n$ be real numbers, let $\{j_1,\ldots,j_n\} = \{k_1,\ldots,k_n\} = \{1,\ldots,n\}$, and assume that $x_{j_n} \le \cdots \le x_{j_1}$ and $y_{k_n} \le \cdots \le y_{k_1}$. Then,

$$\sum_{i=1}^n (x_{j_i} - y_{k_i})^2 \le \sum_{i=1}^n (x_i - y_i)^2.$$

If, in addition, $x_1,\ldots,x_n$ are distinct and $y_1,\ldots,y_n$ are distinct, then the inequality is strict. **Source:** [993, p. 180], [1158, p. 285], and [1938, p. 14]. **Related:** This result is equivalent to the second inequality in Fact 2.12.8.

Fact 2.12.10. Let $x_1,\ldots,x_n$ and $y_1,\ldots,y_n$ be positive numbers, and let p, q be positive numbers such that, for all $i \in \{1,\ldots,n\}$, $q \le x_i/y_i \le p$. Furthermore, let $\{j_1,\ldots,j_n\} = \{k_1,\ldots,k_n\} = \{1,\ldots,n\}$, and assume that $x_{j_n} \le \cdots \le x_{j_1}$ and $y_{k_n} \le \cdots \le y_{k_1}$. Then,

$$\sum_{i=1}^n x_{j_i} y_{k_i} \le \frac{p+q}{2\sqrt{pq}} \sum_{i=1}^n x_i y_i.$$

Source: [542] and [2991, p. 348]. **Remark:** This is a reverse rearrangement inequality. **Remark:** Equality holds for $x_1 = 2$, $x_2 = 1$, $y_1 = 1/2$, $y_2 = 2$, $q = 1$, and $p = 4$. Consequently, if $q = \min_{i\in\{1,\ldots,n\}} x_i/y_i$ and $p = \max_{i\in\{1,\ldots,n\}} x_i/y_i$, then the coefficient $\frac{p+q}{2\sqrt{pq}}$ is the best possible.

Fact 2.12.11. Let $n \ge 2$, let $x_1,\ldots,x_n,y_1,\ldots,y_n$ be positive numbers, and, for all $k \in \{1,\ldots,n\}$, define $s_k \triangleq \sum_{i=1}^k y_i$. Then,

$$\sum_{k=1}^n y_k \left(\prod_{i=1}^k x_i^{y_i}\right)^{1/s_k} \le e \sum_{i=1}^n y_i x_i.$$

Remark: This is Polya's generalization of Carleman's inequality given by Fact 2.11.140. See [327].

Fact 2.12.12. Let $x_1,\ldots,x_n$ and $y_1,\ldots,y_n$ be nonnegative numbers, let $\{j_1,\ldots,j_n\} = \{k_1,\ldots, k_n\} = \{1,\ldots,n\}$, and assume that $x_{j_n} \le \cdots \le x_{j_1}$ and $y_{k_n} \le \cdots \le y_{k_1}$. Then,

$$\prod_{i=1}^n (x_{j_i} + y_{k_i}) \le \prod_{i=1}^n (x_i + y_i) \le \prod_{i=1}^n (x_{j_i} + y_{k_{n-i+1}}).$$

Fact 2.12.13. Let $x_1,\ldots,x_n$ and $y_1,\ldots,y_n$ be complex numbers. Then,

$$\sum_{i=1}^n x_i \sum_{i=1}^n y_i = n\sum_{i=1}^n x_iy_i - \sum_{i<j}(x_j - x_i)(y_j - y_i),$$

$$\sum_{i=1}^n x_i^2 \sum_{i=1}^n y_i^2 = \left(\sum_{i=1}^n x_iy_i\right)^2 + \sum_{i<j}(x_iy_j - x_jy_i)^2,$$

$$\sum_{i=1}^n |x_i|^2 \sum_{i=1}^n |y_i|^2 = \left|\sum_{i=1}^n x_iy_i\right|^2 + \sum_{i<j}|\overline{x}_iy_j - \overline{x}_jy_i|^2.$$

Source: The first equality is given in [1219, p. 38]. The second equality is *Lagrange's identity*. For the complex case, see [935, p. 6], [936, p. 3], and [1171]. For the real case, see [698, 2703]. Both versions follow from Cauchy's identity given by Fact 2.14.1. This result also follows from Fact 2.12.19. See [2252]. **Remark:** The second equality yields Aczel's inequality. See Fact 2.12.37. **Remark:** Fact 3.15.37 gives a matrix extension of Lagrange's identity. **Related:** The case $n = 2$ is given by Fact 2.4.7.

Fact 2.12.14. Let $x_1,\ldots,x_n$ and $y_1,\ldots,y_n$ be complex numbers. Then,

$$\sum_{i=1}^n x_iy_i = \sum_{i=1}^{n-1}\left(\sum_{j=1}^i y_j\right)(x_i - x_{i+1}) + x_n\sum_{i=1}^n y_i,$$

$$\sum_{i=1}^{n-1} x_iy_i = \sum_{i=1}^{n-1}\left(\sum_{j=1}^i y_j\right)(x_i - x_{i+1}) + x_n\sum_{i=1}^{n-1} y_i.$$

Source: [506, p. 55], [752, p. 325], [2294, p. 63], and [2418, p. 277]. **Remark:** This is *Abel's identity*.

Fact 2.12.15. Let $y_1,\ldots,y_n$ be real numbers, define $a \triangleq \min\{y_1, y_1+y_2,\ldots,y_1+y_2+\cdots+y_n\}$ and $b \triangleq \max\{y_1, y_1+y_2,\ldots,y_1+y_2+\cdots+y_n\}$, let $x_1,\ldots,x_n$ be nonnegative numbers, and assume that $x_1 \ge x_2 \ge \cdots \ge x_n \ge 0$. Then,

$$ax_1 \le \sum_{i=1}^n x_iy_i \le bx_1.$$

Source: [2060, pp. 32, 33] and [2294, p. 30]. **Remark:** This is *Abel's inequality*.

Fact 2.12.16. Let $y_1,\ldots,y_n$ be complex numbers, define $b \triangleq \max\{|y_1|, |y_1+y_2|,\ldots,|y_1+y_2+\cdots+y_n|\}$, let $x_1,\ldots,x_n$ be nonnegative numbers, and assume that $x_1 \ge x_2 \ge \cdots \ge x_n \ge 0$. Then,

$$\left|\sum_{i=1}^n x_iy_i\right| \le bx_1.$$

Source: [2527, p. 208]. **Remark:** This is a complex version of Abel's inequality. See Fact 2.12.15.

Fact 2.12.17. Let $x_1,\ldots,x_n$ and $y_1,\ldots,y_n$ be real numbers. Then,

$$\left|\sum_{i=1}^n x_iy_i\right| \le \left(\sum_{i=1}^n x_i^2\right)^{1/2}\left(\sum_{i=1}^n y_i^2\right)^{1/2}.$$

Equality holds if and only if there exists $\alpha \in \mathbb{R}$ such that either $[x_1 \ \cdots \ x_n] = \alpha[y_1 \ \cdots \ y_n]$ or $[y_1 \ \cdots \ y_n] = \alpha[x_1 \ \cdots \ x_n]$. **Remark:** This is the *Cauchy-Schwarz inequality*. **Remark:** Let $\beta > 0$.

Then, Fact 2.2.12 and the Cauchy-Schwarz inequality imply

$$\sum_{i=1}^n x_iy_i \le \left(\sum_{i=1}^n x_i^2\right)^{1/2}\left(\sum_{i=1}^n y_i^2\right)^{1/2} \le \frac{1}{2\beta}\sum_{i=1}^n x_i^2 + \frac{\beta}{2}\sum_{i=1}^n y_i^2.$$

Equality holds in the first inequality if and only if there exists $\alpha \ge 0$ such that either $[x_1 \ \cdots \ x_n] = \alpha[y_1 \ \cdots \ y_n]$ or $[y_1 \ \cdots \ y_n] = \alpha[x_1 \ \cdots \ x_n]$.

Fact 2.12.18. Let $x_1, \dots, x_n$ and $y_1, \dots, y_n$ be real numbers. Then,

$$\left(\sum_{i=1}^n x_i\right)\sum_{i=1}^n y_i \le \frac{n}{2}\sum_{i=1}^n x_iy_i + \frac{n}{2}\sqrt{\left(\sum_{i=1}^n x_i^2\right)\sum_{i=1}^n y_i^2}.$$

Source: [748, p. 372].

Fact 2.12.19. Let $x_1, \dots, x_n$ be real numbers, and let $y_1, \dots, y_n$ be positive numbers. Then,

$$\frac{\left(\sum_{i=1}^n x_i\right)^2}{\sum_{i=1}^n y_i} \le \frac{\left(\sum_{i=1}^n x_i\right)^2}{\sum_{i=1}^n y_i} + \max_{1\le i<j\le n} \frac{(y_ix_j - y_jx_i)^2}{y_iy_j(y_i+y_j)} \le \sum_{i=1}^n \frac{x_i^2}{y_i}.$$

Now, assume that $x_1, \dots, x_n$ are positive. Then,

$$\frac{\left(\sum_{i=1}^n x_i\right)^2}{\sum_{i=1}^n x_iy_i} \le \sum_{i=1}^n \frac{x_i}{y_i}, \quad \frac{1}{\sum_{i=1}^n x_i}\left(\sum_{i=1}^n \frac{x_i}{y_i}\right)^2 \le \sum_{i=1}^n \frac{x_i}{y_i^2}.$$

Source: [318], [1938, p. 35], and [2252]. **Remark:** The first and third terms in the first string are *Bergstrom's inequality*. **Related:** Fact 2.4.3, Fact 2.6.7, Fact 2.12.22, and Fact 2.21.20.

Fact 2.12.20. Let $x_1, \dots, x_n, y_1, \dots, y_n$ be positive numbers, and assume that $\sum_{i=1}^n x_i = \sum_{i=1}^n y_i$. Then,

$$\sum_{i=1}^n x_i \le \sum_{i=1}^n \frac{2x_i^2}{x_i+y_i} \le \sum_{i=1}^n \frac{x_i^2}{y_i}.$$

Source: [1938, p. 12]. **Related:** Fact 2.12.19.

Fact 2.12.21. Let $x_1, \dots, x_n$ and $y_1, \dots, y_n$ be positive numbers, and let $p \in [0,\infty)$ and $r \in [1,\infty)$. Then,

$$\frac{\left(\sum_{i=1}^n x_iy_i^{r-1}\right)^{p+r}}{\left(\sum_{i=1}^n y_i^r\right)^{p+r-1}} \le \sum_{i=1}^n \frac{x_i^{p+r}}{y_i^p}.$$

In particular,

$$\frac{\left(\sum_{i=1}^n x_i\right)^{p+1}}{\left(\sum_{i=1}^n y_i\right)^p} \le \sum_{i=1}^n \frac{x_i^{p+1}}{y_i^p}.$$

Source: [318, 1957]. **Remark:** The case $r = 1$ is *Radon's inequality*. Setting $p = 1$ in the last inequality yields Bergstrom's inequality. See Fact 2.12.19.

Fact 2.12.22. Let $x_1, \dots, x_n$ be complex numbers, and let $y_1, \dots, y_n$ be nonzero real numbers such $\sum_{i=1}^n y_i \ne 0$. Then,

$$\sum_{i=1}^n \frac{|x_i|^2}{y_i} = \frac{\left|\sum_{i=1}^n x_i\right|^2}{\sum_{i=1}^n y_i} + \frac{1}{\sum_{i=1}^n y_i}\sum_{1\le i<j\le n} \frac{|y_ix_j - y_jx_i|^2}{y_iy_j}.$$

Source: [2252]. **Related:** Fact 2.12.13, Fact 2.12.19, and Fact 2.21.8.

Fact 2.12.23. Let $x_1, \ldots, x_n$ and $y_1, \ldots, y_n$ be nonnegative numbers, and let $\alpha \in [0, 1]$. Then,

$$\sum_{i=1}^{n} x_i^{\alpha} y_i^{1-\alpha} \le \left(\sum_{i=1}^{n} x_i\right)^{\alpha} \left(\sum_{i=1}^{n} y_i\right)^{1-\alpha}.$$

Now, let $p, q \in [1, \infty]$ satisfy $1/p + 1/q = 1$. Then, equivalently,

$$\sum_{i=1}^{n} x_i y_i \le \left(\sum_{i=1}^{n} x_i^p\right)^{1/p} \left(\sum_{i=1}^{n} y_i^q\right)^{1/q}.$$

Equality holds if and only if there exists $\alpha \ge 0$ such that either $[x_1^p \ \cdots \ x_n^p] = \alpha[y_1^q \ \cdots \ y_n^q]$ or $[y_1^q \ \cdots \ y_n^q] = \alpha[x_1^p \ \cdots \ x_n^p]$. **Remark:** This is *Hölder's inequality*. **Remark:** Note the relationship between the *conjugate parameters* p, q and the *barycentric coordinates* $\alpha, 1 - \alpha$. See Fact 10.25.68. **Related:** Fact 11.8.28 and Fact 11.10.38.

Fact 2.12.24. Let $x_1, \ldots, x_n$ and $y_1, \ldots, y_n$ be complex numbers, let p, q, r be positive numbers, and assume that $1/p + 1/q = 1/r$. If $p \in (0, 1)$, $q < 0$, and $r = 1$, then

$$\left(\sum_{i=1}^{n} |x_i|^p\right)^{1/p} \left(\sum_{i=1}^{n} |y_i|^q\right)^{1/q} \le \sum_{i=1}^{n} |x_i y_i|.$$

Furthermore, if $p, q, r > 0$, then

$$\left(\sum_{i=1}^{n} |x_i y_i|^r\right)^{1/r} \le \left(\sum_{i=1}^{n} |x_i|^p\right)^{1/p} \left(\sum_{i=1}^{n} |y_i|^q\right)^{1/q}.$$

Source: [2128, p. 19]. **Remark:** This is the *Rogers-Hölder inequality*. **Remark:** Extensions with negative values of p, q, and r are considered in [2128, p. 19]. **Related:** Proposition 11.1.6.

Fact 2.12.25. Let $x_1, \ldots, x_n$ and $y_1, \ldots, y_n$ be nonnegative numbers, and let $p, q \in (1, \infty)$ satisfy $1/p + 1/q = 1$. Then,

$$\sum_{i,j=1}^{n} \frac{x_i y_j}{i + j} \le \frac{\pi}{\sin(\pi/p)} \left(\sum_{i=1}^{n} x_i^p\right)^{1/p} \left(\sum_{i=1}^{n} y_i^q\right)^{1/q}.$$

In particular,

$$\sum_{i,j=1}^{n} \frac{x_i y_j}{i + j} \le \pi \left(\sum_{i=1}^{n} x_i^2\right)^{1/2} \left(\sum_{i=1}^{n} y_i^2\right)^{1/2}.$$

Source: [515], [1146, p. 66], and [1708]. **Remark:** This is the *Hardy-Hilbert inequality* for finite sequences. The constant is sharp for sums over infinite sequences, but is conservative for finite sequences. **Remark:** Fact 2.12.23 implies that

$$\sum_{i,j-1}^{n} x_i y_j \le n \left(\sum_{i=1}^{n} x_i^p\right)^{1/p} \left(\sum_{i=1}^{n} y_i^q\right)^{1/q}.$$

Fact 2.12.26. Let $x_1, \ldots, x_n$ and $y_1, \ldots, y_n$ be nonnegative numbers, let $p, q \in (1, \infty)$ satisfy $1 \le 1/p + 1/q < 2$, and define $\lambda \in (0, 1]$ by $\lambda = 2 - 1/p - 1/q$. Then,

$$\sum_{i,j=1}^{n} \frac{x_i y_j}{(i + j)^{\lambda}} \le \left[\pi \csc\left(\frac{\pi(q - 1)}{\lambda q}\right)\right]^{\lambda} \left(\sum_{i=1}^{n} x_i^p\right)^{1/p} \left(\sum_{i=1}^{n} y_i^q\right)^{1/q}.$$

Source: [515] and [1051, p. 216]. **Related:** Fact 2.12.25.

Fact 2.12.27. Let $x_1,\dots,x_n$ and $y_1,\dots,y_n$ be nonnegative numbers, and let $p,q\in(1,\infty)$ satisfy $1/p+1/q=1$. Then,

$$\sum_{i,j=1}^{n}\frac{x_iy_j}{\max\{i,j\}}\le pq\left(\sum_{i=1}^{n}x_i^p\right)^{1/p}\left(\sum_{i=1}^{n}y_i^q\right)^{1/q}.$$

Furthermore,

$$\sum_{i,j=2}^{n}\frac{x_iy_j}{\log ij}\le\frac{\pi}{\sin(\pi/p)}\left(\sum_{i=2}^{n}i^{p-1}x_i^p\right)^{1/p}\left(\sum_{i=2}^{n}i^{q-1}y_i^q\right)^{1/q}.$$

In particular,

$$\sum_{i,j=2}^{n}\frac{x_iy_j}{\log ij}\le\pi\left(\sum_{i=2}^{n}ix_i^2\right)^{1/2}\left(\sum_{i=2}^{n}iy_i^2\right)^{1/2}.$$

Source: [204, 2944]. **Remark:** Related inequalities are given in [2945].

Fact 2.12.28. Let $x_1,\dots,x_n$ and $y_1,\dots,y_n$ be nonnegative numbers. Then,

$$4\left(\sum_{1\le i<j\le n}x_ix_j\right)\left(\sum_{1\le i<j\le n}y_iy_j\right)\le\left(\sum_{i=1}^{n}x_i\sum_{j=1,j\ne i}^{n}y_j\right)^2.$$

Source: [752, p. 246].

Fact 2.12.29. Let $x_1,\dots,x_n$ and $y_1,\dots,y_n$ be positive numbers. Then,

$$\left(\sum_{i=1}^{n}x_iy_i\right)^2\le\sum_{i=1}^{n}(x_i^2+y_i^2)\sum_{i=1}^{n}\frac{x_i^2y_i^2}{x_i^2+y_i^2}\le\sum_{i=1}^{n}x_i^2\sum_{i=1}^{n}y_i^2.$$

Source: [935, p. 37], [936, p. 51], and [1957, 2808]. **Remark:** This interpolation of the Cauchy-Schwarz inequality is *Milne's inequality*. **Related:** Fact 2.12.30.

Fact 2.12.30. Let $x_1,\dots,x_n$ and $y_1,\dots,y_n$ be positive numbers, and let $p\in[-1,\infty)$. Then,

$$\sum_{i=1}^{n}x_iy_i(x_i+y_i)^p\le\left(\sum_{i=1}^{n}x_i\right)\left(\sum_{i=1}^{n}y_i\right)\left[\sum_{i=1}^{n}(x_i+y_i)\right]^p.$$

Source: [440]. **Related:** This result generalizes the second inequality in Fact 2.12.29.

Fact 2.12.31. Let $x_1,\dots,x_n$ and $y_1,\dots,y_n$ be nonnegative numbers, and assume that, for all $i\in\{1,\dots,n\}$, $x_i+y_i>0$. Then,

$$\sum_{i=1}^{n}\frac{x_iy_i}{x_i+y_i}\sum_{i=1}^{n}(x_i+y_i)\le\sum_{i=1}^{n}x_i\sum_{i=1}^{n}y_i.$$

Source: [935, p. 36] and [936, p. 42]. **Remark:** For $x,y>0$, define the harmonic mean $H(x,y)$ of x and y by $H(x,y)\triangleq\frac{2}{\frac{1}{x}+\frac{1}{y}}$. Then, this result is equivalent to

$$\sum_{i=1}^{n}H(x_i,y_i)\le H\left(\sum_{i=1}^{n}x_i,\sum_{i=1}^{n}y_i\right).$$

See [935, p. 37] and [936, p. 43]. The factor of 2 appearing on the right-hand side in [935, 936] is not needed. **Remark:** Letting α,β be positive numbers and defining the arithmetic mean $A(\alpha,\beta)\triangleq\frac{1}{2}(\alpha+\beta)$, it follows from [2843] that

$$\frac{(\alpha+\beta)^2}{4\alpha\beta}=\frac{A(\alpha,\beta)}{H(\alpha,\beta)}.$$

Related: Fact 2.12.29.

Fact 2.12.32. Let $x_1,\dots,x_n$ be complex numbers, at least one of which is not zero, and let $y_1,\dots,y_n$ be distinct positive numbers. Then,

$$\sum_{i,j=1}^{n}\frac{x_i\overline{x_j}}{y_i+y_j}>0.$$

Source: Fact 10.9.10 and [2527, p. 162].

Fact 2.12.33. Let $x_1,\dots,x_n$ and $y_1,\dots,y_n$ be nonnegative numbers, and let $\alpha\in[0,1]$. Then,

$$\left(\sum_{i=1}^{n}x_iy_i\right)^2\le\sum_{i=1}^{n}x_i^{1+\alpha}y_i^{1-\alpha}\sum_{i=1}^{n}x_i^{1-\alpha}y_i^{1+\alpha}\le\sum_{i=1}^{n}x_i^2\sum_{i=1}^{n}y_i^2.$$

Source: [619], [935, p. 43], [936, p. 51], and [2808]. **Remark:** This interpolation of the Cauchy-Schwarz inequality is *Callebaut's inequality*. **Related:** Fact 2.12.34.

Fact 2.12.34. Let $x_1,\dots,x_n$ and $y_1,\dots,y_n$ be nonnegative numbers, let $\alpha,\beta\in\mathbb{R}$, and assume that either $0\le\beta\le\alpha\le\frac{1}{2}$ or $\frac{1}{2}\le\alpha\le\beta\le1$. Then,

$$\sum_{i=1}^{n}x_i^{1-\alpha}y_i^{\alpha}\sum_{i=1}^{n}x_i^{\alpha}y_i^{1-\alpha}\le\sum_{i=1}^{n}x_i^{1-\beta}y_i^{\beta}\sum_{i=1}^{n}x_i^{\beta}y_i^{1-\beta}.$$

Furthermore, if x and y are nonnegative numbers, then

$$x^{1-\alpha}y^{\alpha}+x^{\alpha}y^{1-\alpha}\le x^{1-\beta}y^{\beta}+x^{\beta}y^{1-\beta}.$$

Source: [619]. **Remark:** This result implies Fact 2.12.33.

Fact 2.12.35. Let $x_1,\dots,x_n$ and $y_1,\dots,y_n$ be real numbers. Then,

$$\begin{aligned}\left(\sum_{i=1}^{n}x_iy_i\right)^2&\le\left(\sum_{i=1}^{n}\min\{x_i^2,y_i^2\}\right)\left(\sum_{i=1}^{n}\max\{x_i^2,y_i^2\}\right)\\&\le\left(\sum_{i=1}^{n}\min\{x_i^2,y_i^2\}\right)\left(\sum_{i=1}^{n}\max\{x_i^2,y_i^2\}\right)+\left(\sum_{\{i:\ y_i^2<x_i^2\}}(x_i^2-y_i^2)\right)\left(\sum_{\{i:\ x_i^2<y_i^2\}}(y_i^2-x_i^2)\right)\\&=\sum_{i=1}^{n}x_i^2\sum_{i=1}^{n}y_i^2.\end{aligned}$$

Source: [2929]. **Remark:** This is an interpolation of the Cauchy-Schwarz inequality.

Fact 2.12.36. Let $x_1,\dots,x_{2n}$ and $y_1,\dots,y_{2n}$ be real numbers. Then,

$$\left(\sum_{i=1}^{2n}x_iy_i\right)^2\le\left(\sum_{i=1}^{2n}x_iy_i\right)^2+\left[\sum_{i=1}^{n}(x_iy_{n+i}-x_{n+i}y_i)\right]^2\le\sum_{i=1}^{2n}x_i^2\sum_{i=1}^{2n}y_i^2.$$

Source: [935, p. 41] and [936, p. 49]. **Remark:** This interpolation of the Cauchy-Schwarz inequality is *McLaughlin's inequality*.

Fact 2.12.37. Let $x_1,\dots,x_n$ and $y_1,\dots,y_n$ be positive numbers, and assume that either $\sum_{i=2}^{n}x_i^2<x_1^2$ or $\sum_{i=2}^{n}y_i^2<y_1^2$. Then,

$$\left(x_1^2-\sum_{i=2}^{n}x_i^2\right)\left(y_1^2-\sum_{i=2}^{n}y_i^2\right)\le\left(x_1y_1-\sum_{i=2}^{n}x_iy_i\right)^2.$$

If, in addition, $\sum_{i=2}^n x_i^2 < x_1^2$ and $\sum_{i=2}^n y_i^2 < y_1^2$. Then,

$$\left(x_1^2 - \sum_{i=2}^n x_i^2\right)^{1/2} \left(y_1^2 - \sum_{i=2}^n y_i^2\right)^{1/2} \le x_1y_1 - \sum_{i=2}^n x_iy_i.$$

Source: Use Fact 2.12.13 with x_i, y_i replaced by x_iJ, y_iJ for all $i \in \{2,\dots,n\}$. See [47, p. 107]. **Remark:** This is *Aczel's inequality.* See [603, p. 16]. **Related:** Fact 2.12.38, Fact 2.12.39, and Fact 11.8.3.

Fact 2.12.38. Let $x_1,\dots,x_n$ and $y_1,\dots,y_n$ be positive numbers, let p and q be nonzero real numbers, assume that $1/p + 1/q = 1$, and assume that $\sum_{i=2}^n x_i^p < x_1^p$ and $\sum_{i=2}^n y_i^q < y_1^q$. If $p > 1$, then

$$\left(x_1^p - \sum_{i=2}^n x_i^p\right)^{1/p} \left(y_1^q - \sum_{i=2}^n y_i^q\right)^{1/q} \le x_1y_1 - \sum_{i=2}^n x_iy_i.$$

If $p < 1$, then

$$x_1y_1 - \sum_{i=2}^n x_iy_i \le \left(x_1^p - \sum_{i=2}^n x_i^p\right)^{1/p} \left(y_1^q - \sum_{i=2}^n y_i^q\right)^{1/q}.$$

Remark: The case $p > 1$ is Popoviciu's inequality, which generalizes Aczel's inequality. See [2622]. **Related:** Fact 2.12.37 and Fact 2.12.39.

Fact 2.12.39. Let $x_1,\dots,x_n$ and $y_1,\dots,y_n$ be positive numbers, let $p, q \in (0,\infty)$, assume that $1/p + 1/q = 1$, and assume that $\sum_{i=2}^n x_i^p < x_1^p$ and $\sum_{i=2}^n x_i^q < x_1^q$. Then,

$$\left(x_1^p - \sum_{i=2}^n x_i^p\right)^{1/p} \left(y_1^q - \sum_{i=2}^n y_i^q\right)^{1/q} \le x_1y_1 - \sum_{i=2}^n x_iy_i - \frac{x_1y_1}{\max\{p,q\}}\left[\sum_{i=2}^n \left(\frac{x_i^p}{x_1^p} - \frac{y_i^q}{y_1^q}\right)\right]^2.$$

Equality holds if and only if, for all $i,j \in \{1,\dots,n\}$, $x_i^p/y_i^q = x_j^p/y_j^q$. **Source:** [2921]. **Remark:** Omitting the last term on the right-hand side yields *Popoviciu's inequality.* This inequality extends and unifies Aczel's inequality and Popoviciu's inequality. **Related:** Fact 2.12.37 and Fact 2.12.38.

Fact 2.12.40. Let $x_1,\dots,x_n$ and $y_1,\dots,y_n$ be nonnegative numbers, let $p \in [1,\infty)$, and assume that $\sum_{i=2}^n x_i^p \le x_1^p$ and $\sum_{i=2}^n y_i^p \le y_1^p$. Then,

$$\left(x_1^p - \sum_{i=2}^n x_i^p\right)^{1/p} + \left(y_1^p - \sum_{i=2}^n y_i^p\right)^{1/p} \le \left[(x_1 + y_1)^p - \sum_{i=2}^n (x_i + y_i)^p\right]^{1/p}.$$

Remark: This is *Bellman's inequality.* See [3000].

Fact 2.12.41. Let $x_1,\dots,x_n$ and $y_1,\dots,y_n$ be complex numbers, and let $\alpha \in [0,1]$. Then,

$$\left(\sum_{i=1}^n \operatorname{Re}(x_i\overline{y_i}) + \alpha \sum_{i,j=1,i\ne j}^n \operatorname{Re}(x_i\overline{y_j})\right)^2 \le \left(\sum_{i=1}^n |x_i|^2 + 2\alpha \sum_{i,j=1,i<j}^n \operatorname{Re}(x_i\overline{x_j})\right)\left(\sum_{i=1}^n |y_i|^2 + 2\alpha \sum_{i,j=1,i<j}^n \operatorname{Re}(y_i\overline{y_j})\right).$$

If, in particular, $x_1,\dots,x_n$ and $y_1,\dots,y_n$ are real, then

$$\left(\sum_{i=1}^n x_iy_i + \alpha \sum_{i,j=1,i\ne j}^n x_iy_j\right)^2 \le \left(\sum_{i=1}^n x_i^2 + 2\alpha \sum_{1\le i<j\le n} x_ix_j\right)\left(\sum_{i=1}^n y_i^2 + 2\alpha \sum_{1\le i<j\le n} y_iy_j\right).$$

Source: [603, p. 47], [660, pp. 33–35], and [2061, pp. 85, 86]. **Remark:** This is *Wagner's inequality.*

Fact 2.12.42. Let $x_1,\dots,x_n$ and $y_1,\dots,y_n$ be nonnegative numbers. Then,

$$\left(\sum_{i=1}^n x_i^2\right)^{1/2}\left(\sum_{i=1}^n y_i^2\right)^{1/2}+2\left(\sum_{1\le i<j\le n} x_ix_j\right)^{1/2}\left(\sum_{1\le i<j\le n} y_iy_j\right)^{1/2}\le\sum_{i=1}^n x_i\sum_{i=1}^n y_i.$$

Source: [368].

Fact 2.12.43. Let $x_1,\dots,x_n$ be real numbers, and let $z_1,\dots,z_n$ be complex numbers. Then,

$$\left|\sum_{i=1}^n x_iz_i\right|^2\le\frac{1}{2}\sum_{i=1}^n x_i^2\left(\sum_{i=1}^n |z_i|^2+\left|\sum_{i=1}^n z_i^2\right|\right)\le\sum_{i=1}^n x_i^2\sum_{i=1}^n |z_i|^2.$$

Source: [935, p. 40] and [936, p. 48]. **Remark:** Conditions for equality in the left-hand inequality are given in [935, p. 40] and [936, p. 48]. **Remark:** This interpolation of the Cauchy-Schwarz inequality is *De Bruijn's inequality.*

Fact 2.12.44. Let $x_1,\dots,x_n$ and $y_1,\dots,y_n$ be positive numbers, and define $\alpha\triangleq\min_{i\in\{1,\dots,n\}}x_i/y_i$ and $\beta\triangleq\max_{i\in\{1,\dots,n\}}x_i/y_i$. Then,

$$\left(\sum_{i=1}^n x_iy_i\right)^2\le\sum_{i=1}^n x_i^2\sum_{i=1}^n y_i^2\le\frac{(\alpha+\beta)^2}{4\alpha\beta}\left(\sum_{i=1}^n x_iy_i\right)^2.$$

Equivalently, let $m_x\triangleq\min_{i\in\{1,\dots,n\}}x_i$, $m_y\triangleq\min_{i\in\{1,\dots,n\}}y_i$, $M_x\triangleq\max_{i\in\{1,\dots,n\}}x_i$, and $M_y\triangleq\max_{i\in\{1,\dots,n\}}y_i$. Then,

$$\left(\sum_{i=1}^n x_iy_i\right)^2\le\sum_{i=1}^n x_i^2\sum_{i=1}^n y_i^2\le\frac{(m_xm_y+M_xM_y)^2}{4m_xm_yM_xM_y}\left(\sum_{i=1}^n x_iy_i\right)^2.$$

Source: [935, p. 73] and [936, p. 92]. **Remark:** This reversal of the Cauchy-Schwarz inequality is the *Polya-Szego inequality.*

Fact 2.12.45. Let $x_1,\dots,x_n$ and $y_1,\dots,y_n$ be nonnegative numbers, and define $m_x\triangleq\min_{i\in\{1,\dots,n\}}x_i$, $m_y\triangleq\min_{i\in\{1,\dots,n\}}y_i$, $M_x\triangleq\max_{i\in\{1,\dots,n\}}x_i$, and $M_y\triangleq\max_{i\in\{1,\dots,n\}}y_i$. Then,

$$\left(\sum_{i=1}^n x_iy_i\right)^2\le\sum_{i=1}^n x_i^2\sum_{i=1}^n y_i^2\le\left(\sum_{i=1}^n x_iy_i\right)^2+\frac{n^2}{3}(M_xM_y-m_xm_y)^2.$$

Source: [1510]. **Remark:** This reversal of the Cauchy-Schwarz inequality is *Ozeki's inequality.*

Fact 2.12.46. Let a,b,c,d be real numbers, assume that $a\le b$ and $c\le d$, let $x_1,\dots,x_n\in[a,b]$, and let $y_1,\dots,y_n\in[c,d]$. Then,

$$\left|\sum_{i=1}^n x_iy_i-\frac{1}{n}\sum_{i=1}^n x_i\sum_{i=1}^n y_i\right|\le(b-a)(d-c)\left\lfloor\frac{n}{2}\right\rfloor\left(1-\frac{1}{n}\left\lfloor\frac{n}{2}\right\rfloor\right).$$

Source: [946] or Theorem A of [1991]. **Related:** Fact 2.15.27.

Fact 2.12.47. Let $x_1,\dots,x_n$ and $y_1,\dots,y_n$ be positive numbers, let $a\triangleq\min_{i\in\{1,\dots,n\}}x_i$, $A\triangleq\max_{i\in\{1,\dots,n\}}x_i$, $b\triangleq\min_{i\in\{1,\dots,n\}}y_i$, and $B\triangleq\max_{i\in\{1,\dots,n\}}y_i$, let p,q be positive numbers, and assume that $1/p+1/q=1$. Then,

$$\sum_{i=1}^n x_iy_i\le\left(\sum_{i=1}^n x_i^p\right)^{1/p}\left(\sum_{i=1}^n y_i^q\right)^{1/q}\le\gamma\sum_{i=1}^n x_iy_i,$$

where

$$\gamma\triangleq\frac{A^pB^q-a^pb^q}{[p(AbB^q-aBb^q)]^{1/p}[q(aBA^p-Aba^p)]^{1/q}}.$$

Source: [2819]. **Remark:** The right-hand inequality, which is a reversal of Hölder's inequality, is the *Diaz-Goldman-Metcalf inequality*. **Remark:** Setting $p = q = 1/2$ yields Fact 2.12.44. **Remark:** The case where $1/p + 1/q = 1/r$ is discussed in [2819].

Fact 2.12.48. Let $x_1, \dots, x_n$ and $y_1, \dots, y_n$ be positive numbers. Then,

$$\frac{\sum_{i=1}^n (x_i^2 + y_i^2)}{\sum_{i=1}^n (x_i + y_i)} \le \sum_{i=1}^n \frac{x_i^2 + y_i^2}{x_i + y_i}.$$

Source: [1957].

Fact 2.12.49. Let $x_1, \dots, x_n$ and $y_1, \dots, y_n$ be positive numbers, and let $p \in [1, 2]$. Then,

$$\frac{\sum_{i=1}^n (x_i + y_i)^p}{\sum_{i=1}^n (x_i + y_i)^{p-1}} \le \frac{\sum_{i=1}^n x_i^p}{\sum_{i=1}^n x_i^{p-1}} + \frac{\sum_{i=1}^n y_i^p}{\sum_{i=1}^n y_i^{p-1}}.$$

Source: [1957, 2787]. **Credit:** E. F. Beckenbach.

Fact 2.12.50. Let $x_1, \dots, x_n$ and $y_1, \dots, y_n$ be positive numbers. Then,

$$4n^2 \le \sum_{i=1}^n \frac{1}{x_i y_i} \sum_{i=1}^n (x_i + y_i)^2.$$

Source: [1938, p. 12].

Fact 2.12.51. Let $x_1, \dots, x_n$ and $y_1, \dots, y_n$ be nonnegative numbers. If $p \in (0, 1]$, then

$$\left(\sum_{i=1}^n x_i^p\right)^{1/p} + \left(\sum_{i=1}^n y_i^p\right)^{1/p} \le \left[\sum_{i=1}^n (x_i + y_i)^p\right]^{1/p}.$$

If $p \ge 1$, then

$$\left[\sum_{i=1}^n (x_i + y_i)^p\right]^{1/p} \le \left(\sum_{i=1}^n x_i^p\right)^{1/p} + \left(\sum_{i=1}^n y_i^p\right)^{1/p},$$

$$\left[\left(\sum_{i=1}^n x_i\right)^p + \left(\sum_{i=1}^n y_i\right)^p\right]^{1/p} \le \sum_{i=1}^n (x_i^p + y_i^p)^{1/p}.$$

In particular,

$$\sqrt{\sum_{i=1}^n (x_i + y_i)^2} \le \sqrt{\sum_{i=1}^n x_i^2} + \sqrt{\sum_{i=1}^n y_i^2}, \quad \sqrt{\left(\sum_{i=1}^n x_i\right)^2 + \left(\sum_{i=1}^n y_i\right)^2} \le \sum_{i=1}^n \sqrt{x_i^2 + y_i^2}.$$

Equality holds in the second string if and only if either $p = 1$ or there exists $\alpha \ge 0$ such that either $[x_1 \ \cdots \ x_n] = \alpha[y_1 \ \cdots \ y_n]$ or $[y_1 \ \cdots \ y_n] = \alpha[x_1 \ \cdots \ x_n]$. **Source:** [581] and [806, pp. 31, 32, 99]. **Remark:** The second string is *Minkowski's inequality*.

Fact 2.12.52. Let $x_1, \dots, x_n, y_1, \dots, y_n \in (-1, 1)$, and let m be a positive integer. Then,

$$\left[\sum_{i=1}^n \frac{1}{(1 - x_i y_i)^m}\right]^2 \le \left[\sum_{i=1}^n \frac{1}{(1 - x_i^2)^m}\right]\left[\sum_{i=1}^n \frac{1}{(1 - y_i^2)^m}\right].$$

Source: [935, p. 19] and [936, p. 19].

Fact 2.12.53. Let $x_1, \dots, x_n$ and $y_1, \dots, y_n$ be nonnegative numbers, and assume that $\alpha \triangleq$

$\sum_{i=1}^n x_i = \sum_{i=1}^n y_i > 0$. Then,

$$\frac{1}{2}\left[\prod_{i=1}^n(\alpha - x_i) + \prod_{i=1}^n(\alpha - y_i)\right] \le \prod_{i=1}^n \left[\alpha - \tfrac{1}{2}(x_i + y_i)\right].$$

Source: [928]. **Related:** Fact 10.13.12.

Fact 2.12.54. Let $x_1, \dots, x_n$ and $y_1, \dots, y_n$ be nonnegative numbers, let $\alpha_1, \dots, \alpha_n$ be nonnegative numbers, and assume that $\sum_{i=1}^n \alpha_i = 1$. Then,

$$\prod_{i=1}^n x_i^{\alpha_i} + \prod_{i=1}^n y_i^{\alpha_i} \le \prod_{i=1}^n (x_i + y_i)^{\alpha_i}, \qquad \sqrt[n]{\prod_{i=1}^n x_i} + \sqrt[n]{\prod_{i=1}^n y_i} \le \sqrt[n]{\prod_{i=1}^n (x_i + y_i)}.$$

Source: [806, p. 100], [1158, p. 41], [1567, p. 64], and [1957]. **Remark:** The geometric mean is superadditive. See [47, p. 40], [2294, p. 270], and [2527, p. 100].

Fact 2.12.55. Let $x_1, \dots, x_n$ and $y_1, \dots, y_n$ be nonnegative numbers, and assume that $\sum_{i=1}^n x_i > 0$ and $\sum_{i=1}^n y_i > 0$. Then,

$$\left(\frac{\sum_{i=1}^n x_i}{\sum_{i=1}^n y_i}\right)^{\sum_{i=1}^n x_i} \prod_{i=1}^n y_i^{x_i} \le \prod_{i=1}^n x_i^{x_i}.$$

Equality holds if and only if there exists $\alpha > 0$ such that, for all $i \in \{1, \dots, n\}$, $x_i = \alpha y_i$. If $\sum_{i=1}^n x_i = \sum_{i=1}^n y_i$, then

$$\prod_{i=1}^n y_i^{x_i} \le \prod_{i=1}^n x_i^{x_i}.$$

Finally,

$$\left(\frac{1}{n}\sum_{i=1}^n x_i\right)^{\sum_{i=1}^n x_i} \le \prod_{i=1}^n x_i^{x_i}.$$

Source: Fact 2.12.55, [279, 2373], and [2418, p. 276].

Fact 2.12.56. Let $x_1, \dots, x_n$ and $y_1, \dots, y_n$ be nonnegative numbers, and assume that $\sum_{i=1}^n x_i$ is nonzero. Then,

$$\prod_{i=1}^n y_i^{x_i} \le \left(\frac{\sum_{i=1}^n x_i y_i}{\sum_{i=1}^n x_i}\right)^{\sum_{i=1}^n x_i}.$$

Source: [658]. **Remark:** This is the *Rogers inequality*.

Fact 2.12.57. Let $x_1, \dots, x_n$ and $y_1, \dots, y_n$ be nonnegative numbers, and assume that, for all $i \in \{1, \dots, n\}$, $x_i \le y_i$. If $\alpha \in (0, 1]$, then

$$\left[\left(\sum_{i=1}^n y_i\right)^\alpha - \left(\sum_{i=1}^n x_i\right)^\alpha\right]^{1/\alpha} \le \sum_{i=1}^n (y_i^\alpha - x_i^\alpha)^{1/\alpha}.$$

If $\alpha \in [1, \infty)$, then

$$\sum_{i=1}^n (y_i^\alpha - x_i^\alpha)^{1/\alpha} \le \left[\left(\sum_{i=1}^n y_i\right)^\alpha - \left(\sum_{i=1}^n x_i\right)^\alpha\right]^{1/\alpha}.$$

Source: [390]. **Credit:** M. Chirita.

Fact 2.12.58. Let $x_0 \in [0, \infty)$, $y_0 \in [0, \infty)$, $x \in \mathbb{C}^n$, and $x \in \mathbb{C}^n$, and assume that $x_0 \le \|x\|_2$ and $y_0 \le \|y\|_2$. Then,

$$\sqrt{(x_0^2 - x^*x)(y_0^2 - y^*y)} \le x_0 y_0 - \operatorname{Re} x^* y.$$

Source: [2527, p. 63]. **Remark:** This is the *light cone inequality*.

2.13 Facts on Equalities and Inequalities in $3n$ Variables

Fact 2.13.1. Let $x_1,\ldots,x_n,y_1,\ldots,y_n,z_1,\ldots,z_n$ be positive numbers, let $p,q,r\in[1,\infty)$, and assume that $1/p+1/q+1/r=1$. Then,

$$\sum_{i=1}^n x_iy_iz_i \le \left(\sum_{i=1}^n x_i^p\right)^{1/p}\left(\sum_{i=1}^n y_i^q\right)^{1/q}\left(\sum_{i=1}^n z_i^r\right)^{1/r}.$$

In particular,

$$\left(\sum_{i=1}^n x_iy_iz_i\right)^2 \le \left(\sum_{i=1}^n x_i^2\right)\left(\sum_{i=1}^n y_i^2\right)\left(\sum_{i=1}^n z_i^2\right),\quad \left(\sum_{i=1}^n x_iy_iz_i\right)^3 \le \left(\sum_{i=1}^n x_i^3\right)\left(\sum_{i=1}^n y_i^3\right)\left(\sum_{i=1}^n z_i^3\right),$$

$$\left(\sum_{i=1}^n x_iy_iz_i\right)^4 \le \left(\sum_{i=1}^n x_i^2\right)^2\left(\sum_{i=1}^n y_i^4\right)\left(\sum_{i=1}^n z_i^4\right),\quad \left(\sum_{i=1}^n x_iy_iz_i\right)^{12} \le \left(\sum_{i=1}^n x_i^2\right)^6\left(\sum_{i=1}^n y_i^3\right)^4\left(\sum_{i=1}^n z_i^6\right)^2.$$

Source: [154, p. 27], [1938, p. 32], [2380, p. 70], and [2527, p. 13]. **Remark:** The inequality for $p=2$ and $q=r=4$ is valid for all real $x_1,\ldots,x_n,y_1,\ldots,y_n,z_1,\ldots,z_n$. The inequality for $p=2$, $q=3$, and $r=6$ is valid for all real $x_1,\ldots,x_n,z_1,\ldots,z_n$ and nonnegative $y_1,\ldots,y_n$.

Fact 2.13.2. Let $x_1,\ldots,x_n,y_1,\ldots,y_n,z_1,\ldots,z_n$ be complex numbers. Then,

$$\left|\sum_{i=1}^n x_i\overline{z_i}\sum_{i=1}^n z_i\overline{y_i}\right| \le \frac{1}{2}\left(\sqrt{\sum_{i=1}^n |x_i|^2\sum_{i=1}^n |y_i|^2}+\left|\sum_{i=1}^n x_i\overline{y_i}\right|\right)\sum_{i=1}^n |z_i|^2.$$

Source: [1090, 1908]. **Remark:** This extension of the Cauchy-Schwarz inequality is *Buzano's inequality*. **Related:** *vii)* of Fact 11.8.5.

Fact 2.13.3. Let $x_1,\ldots,x_n,y_1,\ldots,y_n,z_1,\ldots,z_n$ be nonnegative numbers. Then,

$$\prod_{i=1}^n(x_i+y_i)+\prod_{i=1}^n(y_i+z_i)+\prod_{i=1}^n(z_i+x_i)\le\prod_{i=1}^n(x_i+y_i+z_i)+\prod_{i=1}^n x_i+\prod_{i=1}^n y_i+\prod_{i=1}^n z_i.$$

Source: [1835]. **Related:** Fact 10.16.35.

2.14 Facts on Equalities and Inequalities in $4n$ Variables

Fact 2.14.1. Let $w_1,\ldots,w_n,x_1,\ldots,x_n,y_1,\ldots,y_n,z_1,\ldots,z_n$ be complex numbers. Then,

$$\left(\sum_{i=1}^n w_iy_i\right)\left(\sum_{i=1}^n x_iz_i\right)=\left(\sum_{i=1}^n w_iz_i\right)\left(\sum_{i=1}^n x_iy_i\right)+\sum_{i<j}(w_ix_j-w_jx_i)(y_iz_j-y_jz_i).$$

Source: [2238, p. 519] and [2527, p. 49]. **Remark:** This is *Cauchy's identity*, which generalizes Lagrange's identity given by Fact 2.12.13.

2.15 Facts on Equalities and Inequalities for the Logarithm Function

Fact 2.15.1. Let $x,y\in(0,\infty)$. Then, $x^{\log y}=y^{\log x}$.

Fact 2.15.2. Let $x\in(0,\infty)$. Then,

$$\frac{x-1}{x}\le\log x\le x-1.$$

Equality holds if and only if $x=1$. Now, let $a\in(0,\infty)$. Then,

$$\frac{ax-1}{ax}-\log a\le\log x\le ax-\log a-1.$$

In particular, $(x-e)/x\le\log x\le x/e$. **Source:** [946] and Fact 15.15.27.

Fact 2.15.3. Let x be a positive number, and let p and q be real numbers such that $0 < p \le q$. Then,

$$\log x \le \frac{x^p - 1}{p} \le \frac{x^q - 1}{q} \le x^q \log x.$$

In particular,

$$\log x \le 2(\sqrt{x} - 1) \le x - 1.$$

Equality holds in the second inequality if and only if either $p = q$ or $x = 1$. Finally,

$$\lim_{p \downarrow 0} \frac{x^p - 1}{p} = \log x.$$

Source: [77], [604, p. 8], and [2901]. **Related:** Proposition 10.6.4 and Fact 10.15.1.

Fact 2.15.4. For all $x \in (-1, 0)$,

$$\log(x+1) < x - \tfrac{1}{2}x^2 + \tfrac{1}{3}x^3 - \tfrac{1}{4}x^4 < x - \tfrac{1}{2}x^2 + \tfrac{1}{3}x^3 < x - \tfrac{1}{2}x^2 < x.$$

For all $x > 0$,

$$\left.\begin{matrix} x - \tfrac{1}{2}x^2 \\ x - \tfrac{1}{2}x^2 + \tfrac{1}{3}x^3 - \tfrac{1}{4}x^4 \end{matrix}\right\} < \log(x+1) < \left\{\begin{matrix} x \\ x - \tfrac{1}{2}x^2 + \tfrac{1}{3}x^3. \end{matrix}\right.$$

Source: [806, p. 136] and [1567, p. 55].

Fact 2.15.5. Let $x \in (0, \infty)$. Then,

$$|\log x| \le \frac{|x-1|}{\sqrt{x}}, \quad \log(x+1) \le \frac{x}{\sqrt{x+1}}.$$

Equality in the first inequality holds if and only if $x = 1$. **Source:** [2294, p. 309].

Fact 2.15.6.

$$\lim_{x \to 1} \frac{\log x}{x^2 - 1} = \frac{1}{2}.$$

Furthermore, for all $x > 0$,

$$\frac{1}{x^2+1} \le \frac{\log x}{x^2 - 1} \le \frac{1}{2x},$$

where $\dfrac{\log x}{x^2-1} \triangleq \dfrac{1}{2}$ for $x = 1$. Finally, for all $x \in (0, 1)$,

$$\frac{x^2-1}{2x} < \log x < \frac{x^2-1}{x^2+1},$$

whereas, for all $x > 1$,

$$\frac{x^2-1}{x^2+1} < \log x < \frac{x^2-1}{2x}.$$

Fact 2.15.7. If $x \in (0, 1]$, then

$$\frac{x-1}{x} \le \frac{x^2-1}{2x} \le \frac{x-1}{\sqrt{x}} \le \frac{(x-1)(1+x^{1/3})}{x + x^{1/3}} \le \log x \le \frac{2(x-1)}{x+1} \le \frac{x^2-1}{x^2+1} \le x - 1.$$

If $x \ge 1$, then

$$\frac{x-1}{x} \le \frac{x^2-1}{x^2+1} \le \frac{2(x-1)}{x+1} \le \log x \le \frac{(x-1)(1+x^{1/3})}{x + x^{1/3}} \le \frac{x-1}{\sqrt{x}} \le \frac{x^2-1}{2x} \le x - 1.$$

If $x > 0$, then

$$\frac{2|x-1|}{x+1} \le |\log x| \le \frac{|x-1|(1+x^{1/3})}{x + x^{1/3}} \le \frac{|x-1|}{\sqrt{x}}.$$

Equality holds in each inequality if and only if $x = 1$. **Source:** [604, p. 8], [946], and [1300].

Fact 2.15.8. Let $x \in (-1, 1)$. Then,

$$\frac{|x|}{|x|+1} \le |\log(x+1)| \le \frac{|x|(|x|+1)}{|x+1|}.$$

Source: [2294, p. 309].

Fact 2.15.9. Let $x \ge 1$. Then,

$$\frac{3(x^2-1)}{x^2+4x+1} \le \log x \le \frac{(x^3-1)(x+1)}{3x(x^2+1)}.$$

Source: [374].

Fact 2.15.10. Let $x \ge 1$. Then,

$$\log x \le (x-1)\sqrt[3]{\frac{2}{x(x+1)}}, \quad \log x \le \frac{x-1}{2(x+1)}\left(1 + \sqrt{\frac{2x^2+5x+2}{x}}\right).$$

Source: [374].

Fact 2.15.11. Let $x \ge 1$. Then,

$$(x-1)^2 + \frac{x^2-1}{\log x} \le \sqrt{2(x^4+1)}.$$

Source: [375].

Fact 2.15.12. Let $x > 1$. Then,

$$\frac{x-1}{\log x} < \left(\frac{x^{1/2}+x^{1/4}+1}{3}\right)^2 < \left(\frac{x^{1/3}+1}{2}\right)^3.$$

Source: [1530].

Fact 2.15.13. Let $x \in (0, 1)$. Then,

$$0 < \log\frac{1+x}{1-x} - 2x < \frac{2x^3}{3(1-x^2)}.$$

Source: [2294, p. 414].

Fact 2.15.14. Let $x \in (0, \infty)$. Then,

$$\log\left(1+\frac{1}{x}\right) < \frac{1}{x+1} + \frac{1}{x(2x+1)}, \quad 1 < \left(x+\frac{1}{2}\right)\log\left(1+\frac{1}{x}\right) < 1 + \frac{1}{12x} - \frac{1}{12(x+1)}.$$

Source: [1415].

Fact 2.15.15. Let $x > 0$. Then,

$$\frac{x^2}{e^x - 1} < \log(x+1).$$

Source: [2294, p. 308].

Fact 2.15.16. Let x be a positive number. Then,

$$x - 1 \le \frac{3(x-1)^2}{2(x+2)} + x - 1 \le x\log x.$$

Equality holds in each inequality if and only if $x = 1$. **Source:** [523, p. 63].

Fact 2.15.17. Let $x \in [0, 1)$. Then,

$$(e-1)\log\frac{e-1}{e-e^x} \le \frac{x}{1-x}.$$

Source: [2294, p. 176].

Fact 2.15.18. Let $\alpha > 1$. If $x \in [0, 1)$, then

$$\log\frac{1}{1-x^\alpha} \le \left(\log\frac{1}{1-x}\right)^\alpha.$$

If $x \in [0, \infty)$, then

$$\log\frac{1}{1-(1-e^{-x})^\alpha} \le x^\alpha.$$

Source: [2294, p. 175].

Fact 2.15.19. Let $n \ge 1$. Then,

$$\frac{1}{3}n(n+1)(n+2) < \sum_{i=1}^n \frac{1}{\log^2(1+1/i)} \le \frac{1}{4}n + \frac{1}{3}n(n+1)(n+2).$$

Source: [386].

Fact 2.15.20. Let $x \in (-1, \infty)$ and $y \in \mathbb{R}$. Then,

$$xy + x + y + 1 \le (x+1)\log(x+1) + e^y.$$

Equality holds if and only if $y = 1 + \log x$. **Source:** Fact 12.13.16. **Related:** Fact 2.2.39.

Fact 2.15.21. Let x and y be positive numbers. Then,

$$(x+y)\log[\tfrac{1}{2}(x+y)] \le x\log x + y\log y.$$

Source: Use the fact that $f(x) = x\log x$ is convex on $(0, \infty)$. See [1567, p. 62].

Fact 2.15.22. Let $x, y \in (0, 1]$. Then, $|x\log x - y\log y| \le |x-y|^{1-1/e}$. **Source:** [2550].

Fact 2.15.23. Let a and b be positive numbers, assume that $a < b$, and let $x \ge 0$. Then,

$$\frac{x}{\frac{1}{2}(a+b)[x+\frac{1}{2}(a+b)]} \le \frac{1}{b-a}\log\frac{b(x+a)}{a(x+b)} \le \frac{x}{\sqrt{ab}(x+\sqrt{ab})}.$$

Source: [1965].

Fact 2.15.24. Let $x_1, \dots, x_n$ be positive numbers, and assume that $\sum_{i=1}^n x_i = 1$. Then, the following statements hold:

i) $0 \le \sum_{i=1}^n x_i \log\frac{1}{x_i} \le \log n$.

ii) $\sum_{i=1}^n x_i \log\frac{1}{x_i} = 0$ if and only if $n = 1$.

iii)

$$\frac{1}{2}\left(\sum_{i=1}^n |x_i - \tfrac{1}{n}|\right)^2 \le \frac{3}{2}\sum_{i=1}^n \frac{(x_i - \frac{1}{n})^2}{x_i + \frac{2}{n}} \le \log n - \sum_{i=1}^n x_i\log\frac{1}{x_i} = \log\left(n\prod_{i=1}^n x_i^{x_i}\right).$$

iv) $\sum_{i=1}^n x_i \log\frac{1}{x_i} = \log n$ if and only if $x_1 = \cdots = x_n = 1/n$.

Source: [2061, Chapter XXIII]. *iii)* is given in [523, p. 63]. **Remark:** $\sum_{i=1}^n x_i \log\frac{1}{x_i}$ is the *entropy*.

Fact 2.15.25. Let $x_1, \dots, x_n$ be positive numbers, and assume that $\sum_{i=1}^n x_i = 1$. Then,

$$\log\frac{1}{\sum_{i=1}^n x_i^2} \le \sum_{i=1}^n x_i\log\frac{1}{x_i} \le \log n.$$

Furthermore, each inequality is an equality if and only if $x_1 = \cdots = x_n = 1/n$. **Source:** Theorem 8 of [1991] and [2991, p. 348].

Fact 2.15.26. Let $x_1, \ldots, x_n$ be positive numbers, and assume that $\sum_{i=1}^n x_i = 1$. Then,

$$0 \le \log n - \sum_{i=1}^n x_i \log \frac{1}{x_i} \le \frac{1}{2}\sum_{i,j=1}^n (x_i - x_j)^2 = \left(n\sum_{i=1}^n x_i^2\right) - 1 \le \begin{cases} \frac{1}{2}(n^2-n)\max_{i,j\in\{1,\ldots,n\}}(x_i-x_j)^2 \\ \left(\sum_{i=1}^n x_i^3\right)^{1/2}\left[\left(\sum_{i=1}^n \frac{1}{x_i}\right) - n^2\right]^{1/2}. \end{cases}$$

Equality holds in the first and second inequalities if and only if $x_1 = \cdots = x_n = 1/n$. **Source:** [944, 945]. **Remark:** Fact 2.11.135 implies that $n^2 \le \sum_{i=1}^n \frac{1}{x_i}$.

Fact 2.15.27. Let $x_1, \ldots, x_n$ be positive numbers, assume that $\sum_{i=1}^n x_i = 1$, and define $a \triangleq \min_{i\in\{1,\ldots,n\}} x_i$ and $b \triangleq \max_{i\in\{1,\ldots,n\}} x_i$. Then,

$$0 \le \log n - \sum_{i=1}^n x_i \log\frac{1}{x_i} \le \frac{1}{n}\left\lfloor\frac{n^2}{4}\right\rfloor(b-a)\log\frac{b}{a} \le \frac{1}{n}\left\lfloor\frac{n^2}{4}\right\rfloor\frac{(b-a)^2}{\sqrt{ab}}.$$

Equality holds in each inequality if and only if $x_1 = \cdots = x_n = 1/n$. **Source:** [946]. **Related:** Fact 2.12.46 and Fact 3.25.7.

Fact 2.15.28. Let $x_1, \ldots, x_n, y_1, \ldots, y_n$ be positive numbers. Then, the following statements hold:

i)

$$\left(\sum_{i=1}^n x_i\right)\log\frac{\sum_{j=1}^n x_j}{\sum_{j=1}^n y_j} \le \sum_{i=1}^n x_i \log\frac{x_i}{y_i}.$$

ii) Assume that $\sum_{i=1}^n x_i = 1$. Then,

$$\sum_{i=1}^n x_i \log\frac{1}{x_i} \le \sum_{i=1}^n x_i\log\frac{1}{y_i} + \log\sum_{i=1}^n y_i.$$

iii) Assume that $\sum_{i=1}^n y_i < \sum_{i=1}^n x_i$. Then,

$$0 < \sum_{i=1}^n x_i\log\frac{x_i}{y_i},$$

or, equivalently,

$$\sum_{i=1}^n x_i\log\frac{1}{x_i} < \sum_{i=1}^n x_i\log\frac{1}{y_i}.$$

iv) Assume that $\sum_{i=1}^n y_i \le \sum_{i=1}^n x_i$. Then,

$$0 \le \sum_{i=1}^n x_i\log\frac{x_i}{y_i},$$

or, equivalently,

$$\sum_{i=1}^n x_i\log\frac{1}{x_i} \le \sum_{i=1}^n x_i\log\frac{1}{y_i}.$$

Equality holds if and only if, for all $i \in \{1, \ldots, n\}$, $x_i = y_i$.

v) Assume that $\sum_{i=1}^n y_i \le \sum_{i=1}^n x_i$, and assume that, for all $i \in \{1, \ldots, n\}$, $x_i \le 1$ and $y_i \le 1$. Then,

$$0 \le \sum_{i=1}^n x_i\log\frac{x_i}{y_i} \le \frac{1}{2}\sum_{i=1}^n x_i(x_i - y_i)^2.$$

vi) Assume that $\sum_{i=1}^n x_i = \sum_{i=1}^n y_i$. Then,

$$\sum_{i=1}^n \frac{y_i(y_i - x_i)^2}{y_i^2 + (\max\{x_i, y_i\})^2} \le \sum_{i=1}^n x_i \log \frac{x_i}{y_i} \le \sum_{i=1}^n \frac{y_i(y_i - x_i)^2}{y_i^2 + (\min\{x_i, y_i\})^2}.$$

vii) Assume that $\sum_{i=1}^n x_i = \sum_{i=1}^n y_i = 1$. Then,

$$\frac{1}{2}\sum_{i=1}^n |x_i - y_i| \le -\log\left(1 - \frac{1}{2}\sum_{i=1}^n |x_i - y_i|\right) \le \sum_{i=1}^n x_i \log \frac{x_i}{y_i} + \sum_{i=1}^n x_i \log \frac{1}{x_i} = \sum_{i=1}^n x_i \log \frac{1}{y_i}.$$

viii) Assume that $\sum_{i=1}^n x_i = \sum_{i=1}^n y_i = 1$. Then,

$$\frac{1}{2}\left(\sum_{i=1}^n |x_i - y_i|\right)^2 \le \frac{3}{2}\sum_{i=1}^n \frac{(x_i - y_i)^2}{x_i + 2y_i} \le \sum_{i=1}^n x_i \log \frac{x_i}{y_i}.$$

Source: *i)* is given in [789, p. 31]; *ii)* follows from *i)*, see Theorem 1 in [1991]; *iii)* follows from *i)*; *iv)* follows from *i)* and is given in [2418, p. 276]; *v)* is given in [2418, p. 276]; *vi)* is given in [1300]; *vii)* is given by Theorem 7 of [1991]; *viii)* is given in [523, p. 63]. **Related:** Fact 3.25.7.

2.16 Facts on Equalities for Trigonometric Functions

Fact 2.16.1. The following statements hold:

i) $\sin 0 = 0, \quad \cos 0 = 1.$

ii) $\sin \frac{\pi}{80} = \frac{1}{4}\sqrt{8 - 2\sqrt{8 + 2\sqrt{8 + 2\sqrt{10 + 2\sqrt{5}}}}}, \quad \cos \frac{\pi}{80} = \frac{1}{4}\sqrt{8 + 2\sqrt{8 + 2\sqrt{8 + 2\sqrt{10 + 2\sqrt{5}}}}}.$

iii) $\sin \frac{\pi}{60} = \frac{1}{16}[\sqrt{2}(\sqrt{3} + 1)(\sqrt{5} - 1) - 2(\sqrt{3} - 1)\sqrt{5 + \sqrt{5}}].$

iv) $\sin \frac{\pi}{40} = \frac{1}{4}\sqrt{8 - 2\sqrt{8 + 2\sqrt{10 + 2\sqrt{5}}}}, \quad \cos \frac{\pi}{40} = \frac{1}{4}\sqrt{8 + 2\sqrt{8 + 2\sqrt{10 + 2\sqrt{5}}}}.$

v) $\sin \frac{\pi}{32} = \frac{1}{2}\sqrt{2 - \sqrt{2 + \sqrt{2 + \sqrt{2}}}}, \quad \cos \frac{\pi}{32} = \frac{1}{2}\sqrt{2 + \sqrt{2 + \sqrt{2 + \sqrt{2}}}}.$

vi) $\sin \frac{\pi}{30} = \frac{1}{4}\sqrt{9 - \sqrt{5} - \sqrt{30 + 6\sqrt{5}}} = \frac{1}{8}(-1 - \sqrt{5} + \sqrt{30 - 6\sqrt{5}}), \quad \cos \frac{\pi}{30} = \frac{1}{4}\sqrt{7 + \sqrt{5} + \sqrt{30 + 6\sqrt{5}}}.$

vii) $\sin \frac{\pi}{20} = \frac{1}{4}\sqrt{8 - 2\sqrt{10 + 2\sqrt{5}}}, \quad \cos \frac{\pi}{20} = \frac{1}{4}\sqrt{8 + 2\sqrt{10 + 2\sqrt{5}}}, \quad \tan \frac{\pi}{20} = 1 + \sqrt{5} - \sqrt{5 + 2\sqrt{5}}.$

viii) $\cos \frac{2\pi}{17} = \frac{1}{16}\left(-1 + \sqrt{17} + \sqrt{34 - 2\sqrt{17}} + 2\sqrt{17 + 3\sqrt{17} - \sqrt{34 - 2\sqrt{17}} - 2\sqrt{34 + 2\sqrt{17}}}\right).$

ix) $\sin \frac{\pi}{16} = \frac{1}{2}\sqrt{2 - \sqrt{2 + \sqrt{2}}}, \quad \cos \frac{\pi}{16} = \frac{1}{2}\sqrt{2 + \sqrt{2 + \sqrt{2}}}, \quad \tan \frac{\pi}{16} = \sqrt{4 + 2\sqrt{2}} - 1 - \sqrt{2}.$

x) $\sin \frac{\pi}{15} = \frac{1}{4}\sqrt{7 - \sqrt{5} - \sqrt{30 - 6\sqrt{5}}}, \quad \cos \frac{\pi}{15} = \frac{1}{4}\sqrt{9 + \sqrt{5} + \sqrt{30 - 6\sqrt{5}}} = \frac{1}{8}(-1 + \sqrt{5} + \sqrt{30 + 6\sqrt{5}}).$

xi) $\tan \frac{\pi}{15} = \sqrt{23 - 10\sqrt{5} - 2\sqrt{255 - 114\sqrt{5}}}.$

xii) $\sin \frac{\pi}{12} = \frac{1}{4}(\sqrt{6} - \sqrt{2}), \quad \cos \frac{\pi}{12} = \frac{1}{4}(\sqrt{6} + \sqrt{2}), \quad \tan \frac{\pi}{12} = 2 - \sqrt{3}.$

xiii) $\sin \frac{\pi}{10} = \frac{1}{4}(\sqrt{5} - 1), \quad \cos \frac{\pi}{10} = \frac{1}{4}\sqrt{10 + 2\sqrt{5}}, \quad \tan \frac{\pi}{10} = \sqrt{1 - \frac{2\sqrt{5}}{5}}.$

xiv) $\sin \frac{\pi}{8} = \frac{1}{2}\sqrt{2 - \sqrt{2}}, \quad \cos \frac{\pi}{8} = \frac{1}{2}\sqrt{2 + \sqrt{2}}, \quad \cos \frac{\pi}{8} = \sqrt{2} - 1.$

xv) $\sin \frac{\pi}{6} = \frac{1}{2}, \quad \cos \frac{\pi}{6} = \frac{\sqrt{3}}{2}, \quad \tan \frac{\pi}{6} = \frac{\sqrt{3}}{3}.$

xvi) $\sin \frac{3\pi}{16} = \frac{1}{2}\sqrt{2 - \sqrt{2 - \sqrt{2}}}, \quad \cos \frac{3\pi}{16} = \frac{1}{2}\sqrt{2 + \sqrt{2 - \sqrt{2}}}, \quad \tan \frac{3\pi}{16} = 1 - \sqrt{2} + \sqrt{4 - 2\sqrt{2}}.$

xvii) $\sin \frac{\pi}{5} = \frac{1}{4}\sqrt{10 - 2\sqrt{5}}, \quad \cos \frac{\pi}{5} = \frac{1}{4}(\sqrt{5} + 1), \quad \tan \frac{\pi}{5} = \sqrt{5 - 2\sqrt{5}}.$

xviii) $\sin\frac{\pi}{4} = \cos\frac{\pi}{4} = \frac{\sqrt{2}}{2}, \quad \tan\frac{\pi}{4} = 1.$

xix) $\sin\frac{5\pi}{16} = \frac{1}{2}\sqrt{2+\sqrt{2-\sqrt{2}}}, \quad \cos\frac{5\pi}{16} = \frac{1}{2}\sqrt{2-\sqrt{2-\sqrt{2}}}, \quad \tan\frac{5\pi}{16} = \sqrt{2}-1+\sqrt{4-2\sqrt{2}}.$

xx) $\sin\frac{\pi}{3} = \frac{\sqrt{3}}{2}, \quad \cos\frac{\pi}{3} = \frac{1}{2}, \quad \tan\frac{\pi}{3} = \sqrt{3}.$

xxi) $\sin\frac{3\pi}{8} = \frac{1}{2}\sqrt{2+\sqrt{2}}, \quad \cos\frac{3\pi}{8} = \frac{1}{2}\sqrt{2-\sqrt{2}}, \quad \tan\frac{3\pi}{8} = 1+\sqrt{2}.$

xxii) $\sin\frac{2\pi}{5} = \frac{1}{4}\sqrt{10+2\sqrt{5}}, \quad \cos\frac{2\pi}{5} = \frac{1}{4}(\sqrt{5}-1), \quad \tan\frac{2\pi}{5} = \sqrt{5+2\sqrt{5}}, \quad 1+2\cos\frac{2\pi}{5} = 2\cos\frac{\pi}{5}.$

xxiii) $\sin\frac{5\pi}{12} = \frac{1}{4}(\sqrt{2}+\sqrt{6}) = \frac{1}{2}\sqrt{2+\sqrt{3}}, \ \cos\frac{5\pi}{12} = \frac{1}{4}(\sqrt{6}-\sqrt{2}) = \frac{1}{2}\sqrt{2-\sqrt{3}}, \ \tan\frac{5\pi}{12} = 2+\sqrt{3}.$

xxiv) $\sin\frac{7\pi}{16} = \frac{1}{2}\sqrt{2+\sqrt{2+\sqrt{2}}}, \quad \cos\frac{7\pi}{16} = \frac{1}{2}\sqrt{2-\sqrt{2+\sqrt{2}}}, \quad \tan\frac{7\pi}{16} = 1+\sqrt{2}+\sqrt{4+2\sqrt{2}}.$

xxv) $\sin\frac{\pi}{2} = 1, \quad \cos\frac{\pi}{2} = 0, \quad \sin\pi = 0, \quad \cos\pi = -1, \quad \sin\frac{3\pi}{2} = -1, \quad \cos\frac{3\pi}{2} = 0.$

xxvi) $\sin\frac{\pi}{2^{n+1}} = \frac{1}{2}\underbrace{\sqrt{2-\sqrt{2+\cdots+\sqrt{2}}}}_{n\text{ roots}}, \quad \cos\frac{\pi}{2^{n+1}} = \frac{1}{2}\underbrace{\sqrt{2+\sqrt{2+\cdots+\sqrt{2}}}}_{n\text{ roots}}.$

xxvii) Let $n \ge 2$, let $\alpha_1,\ldots,\alpha_{n-1} \in \{-1,1\}$, and let $\alpha_n \in [-2,2]$. Then,

$$\sin\left(\pi\sum_{i=1}^{n}\frac{\alpha_1\alpha_2\cdots\alpha_i}{2^{i+1}}\right) = \frac{\alpha_1}{2}\sqrt{2+\alpha_2\sqrt{2+\alpha_3\sqrt{2+\cdots+\alpha_{n-1}\sqrt{2+2\sin\frac{\alpha_n\pi}{4}}}}}.$$

In particular,

$$\sin\left[\left(\frac{\alpha_1}{4}+\frac{\alpha_1\alpha_2}{8}\right)\pi\right] = \frac{\alpha_1}{2}\sqrt{2+2\sin\frac{\alpha_2\pi}{4}}.$$

xxviii) Let $n \ge 2$, and let $\alpha_1,\ldots,\alpha_n \in \{-1,1\}$. Then,

$$\sin\left(\pi\sum_{i=1}^{n}\frac{\alpha_1\alpha_2\cdots\alpha_i}{2^{i+1}}\right) = \frac{\alpha_1}{2}\sqrt{2+\alpha_2\sqrt{2+\alpha_3\sqrt{2+\cdots+\alpha_{n-1}\sqrt{2+\alpha_n\sqrt{2}}}}}.$$

xxix) $\cos\frac{\pi}{7} - \cos\frac{2\pi}{7} + \cos\frac{3\pi}{7} = \frac{1}{2}.$

xxx) $\sin^2\frac{\pi}{7}+\sin^2\frac{2\pi}{7}+\sin^2\frac{3\pi}{7} = \frac{7}{4}, \ \cos^2\frac{\pi}{7}+\cos^2\frac{2\pi}{7}+\cos^2\frac{3\pi}{7} = \frac{5}{4}, \ \tan^2\frac{\pi}{7}+\tan^2\frac{2\pi}{7}+\tan^2\frac{3\pi}{7} = 21.$

xxxi) $\csc^2\frac{\pi}{7}+\csc^2\frac{2\pi}{7}+\csc^2\frac{3\pi}{7} = 8, \ \sec^2\frac{\pi}{7}+\sec^2\frac{2\pi}{7}+\sec^2\frac{3\pi}{7} = 24, \ \cot^2\frac{\pi}{7}+\cot^2\frac{2\pi}{7}+\cot^2\frac{3\pi}{7} = 5.$

xxxii) $\tan\frac{\pi}{11}+4\sin\frac{3\pi}{11} = \tan\frac{3\pi}{11}+4\sin\frac{2\pi}{11} = \tan\frac{4\pi}{11}+4\sin\frac{\pi}{11} = \sqrt{11}.$

xxxiii) $\tan\frac{3\pi}{7}-4\sin\frac{\pi}{7} = 4\sin\frac{2\pi}{7}-\tan\frac{\pi}{7} = 4\sin\frac{3\pi}{7}-\tan\frac{2\pi}{7} = \sqrt{7}.$

xxxiv) Let $\theta = \frac{2\pi}{7}$. Then,

$$\sqrt[3]{\cos\theta}+\sqrt[3]{\cos 2\theta}+\sqrt[3]{\cos 4\theta} = \sqrt[3]{\tfrac{1}{2}(5-3\sqrt[3]{7})}, \quad \sqrt[3]{\sec\theta}+\sqrt[3]{\sec 2\theta}+\sqrt[3]{\sec 4\theta} = \sqrt[3]{8-6\sqrt[3]{7}},$$

$$\sqrt[3]{\frac{\cos\theta}{\cos 2\theta}}+\sqrt[3]{\frac{\cos 2\theta}{\cos 4\theta}}+\sqrt[3]{\frac{\cos 4\theta}{\cos\theta}} = -\sqrt[3]{7}, \quad \sqrt[3]{\frac{\cos\theta}{\cos 4\theta}}+\sqrt[3]{\frac{\cos 4\theta}{\cos 2\theta}}+\sqrt[3]{\frac{\cos 2\theta}{\cos\theta}} = 0,$$

$$\frac{\sqrt[3]{\cos\theta}}{\cos 4\theta}+\frac{\sqrt[3]{\cos 4\theta}}{\cos 2\theta}+\frac{\sqrt[3]{\cos 2\theta}}{\cos\theta} = \sqrt[3]{4(26-6\sqrt[3]{7}-3\sqrt[3]{49})}.$$

Source: *iii*) is given in [1695]; *viii*) is due to C. F. Gauss and is given in [1358] and [2880, p. 68]; *xx*) is given in [1566, p. 53]; *xxi*) is given in [2425]; *xxii*), which is a special case of *xxi*), is given in [977] and [1566, p. 53]; *xxiii*) is given in [108, p. 426]; *xxvi*) and *xxvii*) are given in [2288]; *xxviii*) is given in [2903]. **Credit:** The first equality in *xxvii*) is due to S. A. Ramanujan.

Fact 2.16.2. Let z be a complex number such that division by zero does not occur in the expressions below. Then, the following statements hold:

i) $e^0 = e^{2\pi J} = 1, \quad e^{-\pi J} = e^{\pi J} = -1, \quad e^{-(\pi/2)J} = -J, \quad e^{(\pi/2)J} = J.$

ii) $\sin(-z) = -\sin z, \quad \cos(-z) = \cos z, \quad \sin(z \pm 2\pi) = \sin z, \quad \cos(z \pm 2\pi) = \cos z.$

iii) $\sin(z \pm \pi) = -\sin z, \quad \cos(z \pm \pi) = -\cos z, \quad \sin(\pi - z) = \sin z, \quad \cos(\pi - z) = -\cos z.$

iv) $\sin(z \pm \frac{\pi}{2}) = \pm\cos z, \quad \cos(z \pm \frac{\pi}{2}) = \mp\sin z, \quad \sin(\frac{\pi}{2} - z) = \cos z, \quad \cos(\frac{\pi}{2} - z) = \sin z.$

v) $|\sin(\frac{z}{2} + \frac{\pi}{4})| = |\cos(\frac{\pi}{4} - \frac{z}{2})| = \frac{1}{2}\sqrt{|2 + 2\sin z|}, \quad |\sin(\frac{\pi}{4} - \frac{z}{2})| = |\cos(\frac{z}{2} + \frac{\pi}{4})| = \frac{1}{2}\sqrt{|2 - 2\sin z|}.$

vi) $\csc(-z) = -\csc z, \quad \cot(-z) = -\cot z.$

vii) $\csc(z \pm 2\pi) = \csc z, \quad \csc(z \pm \pi) = -\csc z, \quad \cot(z \pm 2\pi) = \cot z, \quad \cot(z \pm \pi) = \cot z.$

viii) $\tan(z \pm \frac{\pi}{2}) = -\cot z, \quad \tan(\frac{\pi}{2} - z) = \cot z, \quad \sec(z \pm \frac{\pi}{2}) = \mp\csc z.$

ix) $\tan(-z) = -\tan z, \quad \sec(-z) = \sec z.$

x) $\tan(z \pm 2\pi) = \tan z, \quad \sec(z \pm 2\pi) = \sec z, \quad \tan(z \pm \pi) = \tan z.$

xi) $\sec(z \pm \pi) = -\sec z, \quad \tan(\pi - z) = -\tan z.$

xii) $\csc(z \pm \frac{\pi}{2}) = \pm\sec z, \quad \cot(z \pm \frac{\pi}{2}) = -\tan z, \quad \cot(\frac{\pi}{2} - z) = \tan z.$

xiii) $\tan(\frac{z}{2} + \frac{\pi}{4}) + \tan(\frac{z}{2} - \frac{\pi}{4}) = 2\tan z, \quad \tan(\frac{z}{2} + \frac{\pi}{4}) - \tan(\frac{z}{2} - \frac{\pi}{4}) = 2\sec z.$

xiv) $\tan(\frac{z}{2} + \frac{\pi}{4}) = \tan z + \sec z = \dfrac{1 + \sin z}{\cos z} = \dfrac{\cos z}{1 - \sin z}.$

xv) $|\tan(\frac{z}{2} + \frac{\pi}{4})| = |\cot(\frac{\pi}{4} - \frac{z}{2})| = \sqrt{\frac{|1+\sin z|}{|1-\sin z|}}, \quad |\tan(\frac{\pi}{4} - \frac{z}{2})| = |\cot(\frac{z}{2} + \frac{\pi}{4})| = \sqrt{\frac{|1-\sin z|}{|1+\sin z|}}.$

xvi) $\tan(z + \frac{\pi}{4}) = \dfrac{1 + \tan z}{1 - \tan z}, \quad \tan(\frac{\pi}{4} - z) = \dfrac{1 - \tan z}{1 + \tan z}.$

Source: For real z, *v)* follows from *xxvii)* of Fact 2.16.1 with $\alpha_1 = 1$ and $\alpha_2 = 4z/\pi$.

Fact 2.16.3. Let z be a complex number such that division by zero does not occur in the expressions below, and let n be an integer. Then, the following statements hold:

i) $(\cos z + J\sin z)^n = \cos nz + J\sin nz.$

ii) $\sin^2 \frac{z}{2} = \frac{1}{2}(1 - \cos z), \quad \cos^2 \frac{z}{2} = \frac{1}{2}(1 + \cos z).$

iii) $\tan \frac{z}{2} = \dfrac{\sin z}{1 + \cos z} = \dfrac{1 - \cos z}{\sin z} = \csc z - \cot z, \quad \tan^2 \frac{z}{2} = \dfrac{1 - \cos z}{1 + \cos z}.$

iv) $\dfrac{1 - \sin z}{\cos z} = \dfrac{\cos z}{1 + \sin z} = \dfrac{1 - \tan \frac{z}{2}}{1 + \tan \frac{z}{2}}, \quad \dfrac{\sin z}{1 + \cos z} + \dfrac{1 + \cos z}{\sin z} = 2\csc z.$

v) $\dfrac{\cos z}{1 + \sin z} = \dfrac{1 - \sin z - \cos z}{-1 - \sin z + \cos z}.$

vi) $\sin z \pm \cos z = \sqrt{2}\sin(z \pm \frac{\pi}{4}) = \sqrt{2}\cos(z \mp \frac{\pi}{4}), \quad \sin z + \cos z = \dfrac{\sec z + \csc z}{\tan z + \cot z}.$

vii) $\dfrac{\csc z - \sec z}{\csc z + \sec z} = \dfrac{1 - \tan z}{1 + \tan z} = \dfrac{1 - \sin 2z}{\cos 2z}, \quad \sec z = 1 + (\tan z)\tan \frac{z}{2}.$

viii) $\sin z = 2(\sin \frac{z}{2})\cos \frac{z}{2} = \dfrac{2\tan \frac{z}{2}}{1 + \tan^2 \frac{z}{2}} = 1 - (\sin \frac{z}{2} - \cos \frac{z}{2})^2.$

ix) $\cos z = 2\cos^2 \frac{z}{2} - 1 = 1 - 2\sin^2 \frac{z}{2} = \cos^2 \frac{z}{2} - \sin^2 \frac{z}{2} = \dfrac{1 - \tan^2 \frac{z}{2}}{1 + \tan^2 \frac{z}{2}}.$

x) $\tan z + \sec z = \dfrac{1}{\sec z - \tan z} = \dfrac{-1 + \tan z + \sec z}{1 + \tan z - \sec z}, \quad \tan z + \cot z = 2\csc 2z = (\csc z)\sec z.$

xi) $\tan z = \dfrac{2\tan \frac{z}{2}}{1 - \tan^2 \frac{z}{2}} = \dfrac{2\cot \frac{z}{2}}{\csc^2 \frac{z}{2} - 2} = \dfrac{\sin 2z}{1 + \cos 2z} = \dfrac{1 - \cos 2z}{\sin 2z} = \dfrac{1 + \sin 2z - \cos 2z}{1 + \sin 2z + \cos 2z}.$

xii) $\dfrac{1-\tan z}{1+\tan z} = \dfrac{1-\sin 2z+\cos 2z}{1+\sin 2z+\cos 2z}.$

xiii) $\cot z = \tan z + 2\cot 2z = \csc 2z + \cot 2z = \dfrac{\cot^2\frac{z}{2}-1}{2\cot\frac{z}{2}} = \dfrac{1+\cos 2z}{\sin 2z}.$

xiv) $\cot z = \frac{1}{2}(\cot\frac{z}{2} - \tan\frac{z}{2}) = \left(1+\frac{2}{e^{2zJ}-1}\right)J.$

xv) $\sin 2z = 2(\sin z)\cos z = 2(\cot z)\sin^2 z, \quad (\csc^2 z)\sec^2 z = \csc^2 z + \sec^2 z.$

xvi) $\sin 2z = \dfrac{2\tan z}{1+\tan^2 z} = 2 - 2\dfrac{\sin^3 z+\cos^3 z}{\sin z+\cos z} = 2\dfrac{\sin^3 z-\cos^3 z}{\sin z-\cos z} - 2.$

xvii) $\cos 2z = 2\cos^2 z - 1 = 1 - 2\sin^2 z = \cos^2 z - \sin^2 z = (\cos^2 z)(1-\tan^2 z) = \cos^4 z - \sin^4 z.$

xviii) $\cos 2z = (\cot z)\sin 2z - 1 = (\sin^2 z)(\cot^2 z - 1) = \dfrac{1-\tan^2 z}{1+\tan^2 z} = (1-\sin 2z)\dfrac{1+\tan z}{1-\tan z}.$

xix) $\tan 2z = \dfrac{2\tan z}{1-\tan^2 z} = \dfrac{2\cot z}{\cot^2 z-1} = \dfrac{2}{\cot z-\tan z} = \dfrac{\sin z+\sin 3z}{\cos z+\cos 3z}.$

xx) $\cot 2z = \dfrac{\cot^2 z-1}{2\cot z} = \frac{1}{2}(\cot z - \tan z) = \frac{1}{2}[\cot z + \cot(z+\frac{1}{2}\pi)].$

xxi) $\sin 3z = 3\sin z - 4\sin^3 z = (\sin z)(1+2\cos 2z).$

xxii) $\sin 3z = (\sin^3 z)(3\cot^2 z - 1) = (\cos^3 z)(3\tan z - \tan^3 z).$

xxiii) $\cos 3z = -3\cos z + 4\cos^3 z = \cos z - 4(\cos z)\sin^2 z = (\sin^3 z)(-3\cot z + \cot^3 z) = (\cos^3 z)(1 - 3\tan^2 z).$

xxiv) $\tan 3z = \dfrac{3\tan z-\tan^3 z}{1-3\tan^2 z} = \dfrac{\sin 2z+\sin 4z}{\cos 2z+\cos 4z}, \quad \dfrac{\sin 3z}{\sin z} - \dfrac{\cos 3z}{\cos z} = 2.$

xxv) $\sin 4z = 2(\sin 2z)(1-2\sin^2 z) = 2(\sin 2z)(2\cos^2 z - 1).$

xxvi) $\sin 4z = (\sin^4 z)(-4\cot z + 4\cot^3 z) = (\cos^4 z)(4\tan z - 4\tan^3 z).$

xxvii) $\cos 4z = 1 - 8\cos^2 z + 8\cos^4 z = (\sin^4 z)(1 - 6\cot^2 z + \cot^4 z) = (\cos^4 z)(1 - 6\tan^2 z + \tan^4 z).$

xxviii) $\tan 4z = \dfrac{2\tan 2z}{1-\tan^2 2z} = \dfrac{\sin 8z}{1+\cos 8z} = \dfrac{4\tan z - 4\tan^3 z}{1-6\tan^2 z+\tan^4 z}.$

xxix) $\sin 5z = 5\sin z - 20\sin^3 z + 16\sin^5 z = 5(\sin z)\cos^4 z - 10(\sin^3 z)\cos^2 z + \sin^5 z.$

xxx) $\sin 5z = (\sin^5 z)(1 - 10\cot^2 z + 5\cot^4 z) = (\cos^5 z)(5\tan z - 10\tan^3 z + \tan^5 z).$

xxxi) $\cos 5z = 5\cos z - 20\cos^3 z + 16\cos^5 z = \cos^5 z - 10(\cos^3 z)\sin^2 z + 5(\cos z)\sin^4 z.$

xxxii) $\cos 5z = (\cos^5 z)(1 - 10\tan^2 z + 5\tan^4 z) = (\sin^5 z)(\cot^5 z - 10\cot^3 z + 5\cot z).$

xxxiii) $\tan 5z = \dfrac{5\tan z - 10\tan^3 z + \tan^5 z}{1 - 10\tan^2 z + 5\tan^4 z}.$

xxxiv) $\sin 6z = (\cos z)(6\sin z - 32\sin^3 z + 32\sin^5 z).$

xxxv) $\sin 6z = (\sin^6 z)(6\cot z - 20\cot^3 z + 6\cot^5 z) = (\cos^6 z)(6\tan z - 20\tan^3 z + 6\tan^5 z).$

xxxvi) $\cos 6z = (\cos^6 z)(1 - 15\tan^2 z + 15\tan^4 z - \tan^6 z) = -1 + 18\cos^2 z - 48\cos^4 z + 32\cos^6 z.$

xxxvii) $\cos 6z = 2(\cos^2 3z) - 1 = 1 - 2(\sin^2 3z) = (\sin^6 z)(-1 + 15\cot^2 z - 15\cot^4 z + \cot^6 z).$

xxxviii) $\tan 6z = \dfrac{\sin 3z + \sin 5z + \sin 7z + \sin 9z}{\cos 3z + \cos 5z + \cos 7z + \cos 9z}.$

xxxix) $\sin 7z = 7\sin z - 56\sin^3 z + 112\sin^5 z - 64\sin^7 z.$

xl) $\cos 7z = -7\cos z + 56\cos^3 z - 112\cos^5 z + 64\cos^7 z.$

xli) $\cos(n-\frac{1}{2})z - \cos(n+\frac{1}{2})z = 2(\sin\frac{z}{2})\sin nz.$

xlii) $\sin nz = 2(\sin[(n-1)z])\cos z - \sin(n-2)z, \quad \cos nz = 2(\cos[(n-1)z])\cos z - \cos(n-2)z.$

xliii) $\tan nz = \dfrac{\tan[(n-1)z] + \tan z}{1 - (\tan[(n-1)z])\tan z} = \dfrac{\sum_{i=0}^{\lfloor (n-1)/2 \rfloor} (-1)^i \binom{n}{2i+1} \tan^{2i+1} z}{\sum_{i=0}^{\lfloor n/2 \rfloor} (-1)^i \binom{n}{2i} \tan^{2i} z}.$

xliv) $\sin^2 \frac{z}{2} = \frac{1}{2}(1 - \cos z) = \frac{1}{2}(\tan \frac{z}{2})\sin z, \quad \cos^2 \frac{z}{2} = \frac{1}{2}(1 + \cos z) = \dfrac{\cos z}{1 - \tan^2 \frac{z}{2}}.$

xlv) $\sin^2 z = \frac{1}{2}(1 - \cos 2z) = \frac{1}{2}(\tan z)\sin 2z, \quad \cos^2 z = \frac{1}{2}(1 + \cos 2z) = \dfrac{\cos 2z}{1 - \tan^2 z}.$

xlvi) $(\sin z)\cos z = \frac{1}{2}\sin 2z, \quad \sin^2 z + \cos^2 z = 1.$

xlvii) $\tan^2 z = \dfrac{1 - \cos 2z}{1 + \cos 2z} = 1 - 2(\cot 2z)\tan z = 2(\csc 2z)\tan z - 1.$

xlviii) $\sec^2 z = 1 + \tan^2 z, \quad \csc^2 z = 1 + \cot^2 z, \quad \csc^2 z + \sec^2 z = 4\csc^2 2z.$

xlix) $\sin^3 z = \frac{1}{4}(3\sin z - \sin 3z) = \dfrac{\sin 3z}{3\cot^2 z - 1} = \dfrac{\cos 3z}{\cot^3 z - 3\cot z}.$

l) $\sin^3 z = (\cos z)(1 + \cos z)(\tan z - \sin z).$

li) $\cos^3 z = \frac{1}{4}(3\cos z + \cos 3z) = \dfrac{\cos 3z}{1 - 3\tan^2 z} = \dfrac{\sin 3z}{3\tan z - \tan^3 z}.$

lii) $\tan^3 z = \dfrac{3\sin z - \sin 3z}{3\cos z + \cos 3z}.$

liii) $(\sin^2 z)\cos z = \frac{1}{4}(\cos z - \cos 3z), \quad (\sin z)\cos^2 z = \frac{1}{4}(\sin z + \sin 3z).$

liv) $\sin^4 z = \sin^2 z - \frac{1}{4}\sin^2(2z).$

lv) $\sin^4 z = \frac{1}{8}(3 - 4\cos 2z + \cos 4z) = \dfrac{\sin 4z}{4\cot^3 z - 4\cot z} = \dfrac{\cos 4z}{\cot^4 z - 6\cot^2 z + 1}.$

lvi) $\cos^4 z = \frac{1}{8}(3 + 4\cos 2z + \cos 4z) = \dfrac{\sin 4z}{4\tan z - 4\tan^3 z} = \dfrac{\cos 4z}{1 - 6\tan^2 z + \tan^4 z}.$

lvii) $\tan^4 z = \dfrac{3 - 4\cos 2z + \cos 4z}{3 + 4\cos 2z + \cos 4z}.$

lviii) $(\sin^3 z)\cos z = \frac{1}{8}(2\sin 2z - \sin 4z), \quad (\sin^2 z)\cos^2 z = \frac{1}{8}(1 - \cos 4z).$

lix) $(\sin z)\cos^3 z = \frac{1}{8}(2\sin 2z + \sin 4z).$

lx) $\sin^5 z = \frac{1}{16}(10\sin z - 5\sin 3z + \sin 5z).$

lxi) $\sin^5 z = \dfrac{\sin 5z}{5\cot^4 z - 10\cot^2 z + 1} = \dfrac{\cos 5z}{\cot^5 z - 10\cot^3 z + 5\cot z}.$

lxii) $\cos^5 z = \frac{1}{16}(10\cos z + 5\cos 3z + \cos 5z).$

lxiii) $\cos^5 z = \dfrac{\sin 5z}{5\tan z - 10\tan^3 z + \tan^5 z} = \dfrac{\cos 5z}{1 - 10\tan^2 z + 5\tan^4 z}.$

lxiv) $(\sin^4 z)\cos z = \frac{1}{16}(2\cos z - 3\cos 3z + \cos 5z).$

lxv) $(\sin^3 z)\cos^2 z = \frac{1}{16}(2\sin z + \sin 3z - \sin 5z).$

lxvi) $(\sin^2 z)\cos^3 z = \frac{1}{16}(2\cos z - \cos 3z - \cos 5z).$

lxvii) $(\sin z)\cos^4 z = \frac{1}{16}(2\sin z + 3\sin 3z + \sin 5z).$

lxviii) $\sin^6 z = \frac{1}{32}(10 - 15\cos 2z + 6\cos 4z - \cos 6z).$

lxix) $\sin^6 z = \dfrac{\sin 6z}{6\cot^5 z - 20\cot^3 z + 6\cot z} = \dfrac{\cos 6z}{\cot^6 z - 15\cot^4 z + 15\cot^2 z - 1}.$

lxx) $\cos^6 z = \frac{1}{32}(10 + 15\cos 2z + 6\cos 4z + \cos 6z).$

lxxi) $\cos^6 z = \dfrac{\sin 6z}{6\tan z - 20\tan^3 z + 6\tan^5 z}.$

lxxii) $8(\sin^6 z + \cos^6 z) = 3\cos 4z + 5, \quad 64(\sin^8 z + \cos^8 z) = \cos 8z + 28\cos 4z + 35.$

lxxiii) $(\sin z)(\sin 2z)(\sin 4z)\sin 8z = -\frac{1}{8}(\cos z - \cos 3z - \cos 5z + \cos 7z - \cos 9z + \cos 11z + \cos 13z - \cos 15z).$

lxxiv) $$\sin 2nz = (\sin z)(\cos z)\sum_{i=0}^{n-1}(-1)^i 2^{2i+1}\binom{n+i}{2i+1}\sin^{2i} z.$$

lxxv) $$\cos 2nz = 1 + n\sum_{i=1}^{n}(-1)^i \frac{2^{2i-1}}{i}\binom{n+i-1}{2i-1}\sin^{2i} z.$$

lxxvi) $$\sin(2n-1)z = (2n-1)(\sin z)\sum_{i=0}^{n-1}(-1)^i \frac{4^i}{2i+1}\binom{n+i-1}{2i}\sin^{2i} z.$$

lxxvii) $$\cos(2n-1)z = (\cos z)\sum_{i=0}^{n-1}(-1)^i 4^i\binom{n+i-1}{2i}\sin^{2i} z.$$

lxxviii) If z is real, then $[(\csc z - \sin z)^2]^{1/3} + [(\sec z - \cos z)^2]^{1/3} = [(\csc z - \sin z)^2(\sec z - \cos z)^2]^{-1/3}.$

lxxix) $$\cot nz + \cot(n+1)z = \frac{\sin(2n+1)z}{(\sin nz)\sin(n+1)z}, \quad \cot nz - \cot(n+1)z = \frac{\sin z}{(\sin nz)\sin(n+1)z}.$$

lxxx) If z is real and $0 < nz < \frac{\pi}{2}$, then

$$\tan nz + \tan(n-1)z = \frac{\sin(2n-1)z}{(\cos nz)\cos(n-1)z}, \quad \tan nz - \tan(n-1)z = \frac{\sin z}{(\cos nz)\cos(n-1)z}.$$

Source: [203, Ch. 10], [1083, p. 31], [1757, p. 118], and [1524, pp. 114–116]. *lxxiii)–lxxvi)* are given in [2353]; *lxxvii)* is given in [1397].

Fact 2.16.4. Let x and y be complex numbers such that division by zero does not occur in the expressions below. Then, the following statements hold:

i) $\sin(x+y) = (\sin x)(\cos y) + (\cos x)\sin y, \quad \sin(x-y) = (\sin x)(\cos y) - (\cos x)\sin y.$

ii) $\cos(x+y) = (\cos x)(\cos y) - (\sin x)\sin y, \quad \cos(x-y) = (\cos x)(\cos y) + (\sin x)\sin y.$

iii) $(\sin x)\sin y = \frac{1}{2}[\cos(x-y) - \cos(x+y)], \quad (\sin x)\cos y = \frac{1}{2}[\sin(x-y) + \sin(x+y)].$

iv) $(\cos x)\cos y = \frac{1}{2}[\cos(x-y) + \cos(x+y)].$

v) $$(\tan x)\tan y = \frac{\tan x + \tan y}{\cot x + \cot y}.$$

vi) $\sin^2 x - \sin^2 y = [\sin(x+y)]\sin(x-y), \quad \sin^2 x - \cos^2 y = -[\cos(x+y)]\cos(y-x).$

vii) $\cos^2 x - \sin^2 y = [\cos(x+y)]\cos(x-y), \quad \cos^2 x - \cos^2 y = [\sin(x+y)]\sin(y-x).$

viii) $\sin x + \sin y = 2[\sin\frac{1}{2}(x+y)]\cos\frac{1}{2}(x-y), \quad \sin x - \sin y = 2[\sin\frac{1}{2}(x-y)]\cos\frac{1}{2}(x+y).$

ix) $\cos x + \cos y = 2[\cos\frac{1}{2}(x+y)]\cos\frac{1}{2}(x-y), \quad \cos x - \cos y = 2[\sin\frac{1}{2}(x+y)]\sin\frac{1}{2}(y-x).$

x) $$\tan(x+y) = \frac{\tan x + \tan y}{1 - (\tan x)\tan y}, \quad \tan(x-y) = \frac{\tan x - \tan y}{1 + (\tan x)\tan y}.$$

xi) $$\tan x + \tan y = \frac{\sin(x+y)}{(\cos x)\cos y}, \quad \tan x - \tan y = \frac{\sin(x-y)}{(\cos x)\cos y}.$$

xii) $$\csc(x+y) = \frac{2(\sin x)\cos y - 2(\sin y)\cos x}{\cos 2y - \cos 2x}, \quad \sec(x+y) = \frac{2(\sin x)\sin y + 2(\cos x)\cos y}{\cos 2x + \cos 2y}.$$

xiii) $$\cot(x+y) = \frac{(\cot x)\cot y - 1}{\cot x + \cot y}, \quad \cot(x-y) = \frac{(\cot x)\cot y + 1}{\cot y - \cot x}.$$

xiv) $$\cot x + \cot y = \frac{\sin(x+y)}{(\sin x)\sin y}, \quad \cot x - \cot y = \frac{\sin(y-x)}{(\sin x)\sin y}.$$

xv) $\dfrac{\sin x - \sin y}{\sin x + \sin y} = \dfrac{\tan \frac{1}{2}(x-y)}{\tan \frac{1}{2}(x+y)}, \quad \dfrac{\sin x + \sin y}{\cos x + \cos y} = \tan \frac{1}{2}(x+y).$

xvi) $\dfrac{\sin x - \sin y}{\cos x + \cos y} = \tan \frac{1}{2}(x-y), \quad \dfrac{\sin x + \sin y}{\cos x - \cos y} = \cot \frac{1}{2}(y-x).$

xvii) $\dfrac{\cos x - \cos y}{\cos x + \cos y} = [\tan \frac{1}{2}(y-x)] \tan \frac{1}{2}(x+y).$

xviii) $\sin \frac{x+y}{2} = \frac{1}{2}(\sin x + \sin y) + 2(\sin \frac{x+y}{2}) \sin^2 \frac{x-y}{4}.$

xix) $\cos \frac{x+y}{2} = \frac{1}{2}(\cos x + \cos y) + 2(\cos \frac{x+y}{2}) \sin^2 \frac{x-y}{4}.$

xx) $1 + 2(\sin \frac{x+y}{2}) \cos \frac{x-y}{2} = \sin x + \sin y + \cos(x+y) + 2\sin^2 \frac{x+y}{2}.$

Source: [391, 367], [806, p. 80], [1523, pp. 681–684], and [1524, pp. 127–132].

Fact 2.16.5. Let $\omega, t, \phi, \psi, a, b$ be real numbers. Then,

$$ae^{\phi J} + be^{\psi J} = \sqrt{a^2 + b^2 + 2ab\cos(\phi - \psi)}e^{\operatorname{atan2}(a\sin\phi + b\sin\psi, a\cos\phi + b\cos\psi)J},$$

$$a\cos(\omega t + \phi) + b\cos(\omega t + \psi)$$
$$= \sqrt{a^2 + b^2 + 2ab\cos(\phi - \psi)}\cos[\omega t + \operatorname{atan2}(a\sin\phi + b\sin\psi, a\cos\phi + b\cos\psi],$$
$$a\cos(\omega t) + b\sin(\omega t) = a\cos(\omega t) - b\cos(\omega t + \frac{\pi}{2}) =$$
$$= \sqrt{a^2 + b^2}\cos[\omega t + \operatorname{atan2}(-b, a)] = \sqrt{a^2 + b^2}\sin[\omega t + \operatorname{atan2}(a, b)].$$

Source: [2685]. **Remark:** These equalities concern phasor addition. **Related:** Fact 2.19.3.

Fact 2.16.6. Let x, y, and z be complex numbers. Then, the following statements hold:

i) $\sin(x+y+z) = (\sin x)(\cos y)\cos z + (\sin y)(\cos z)\cos x + (\sin z)(\cos x)\cos y - (\sin x)(\sin y)\sin z.$

ii) $\cos(x+y+z) = (\cos x)(\cos y)\cos z - (\cos x)(\sin y)\sin z - (\cos y)(\sin z)\sin x - (\cos z)(\sin x)\sin y.$

iii) $\sin(x+y+z) + \cos(x+y+z) = 2(\cos x)(\cos y)\cos z - 2(\sin x)(\sin y)\sin z$
$- (\cos x - \sin x)(\cos y - \sin y)(\cos z - \sin z).$

iv) $\sin x + \sin y + \sin z = \sin(x+y+z) + 4(\sin \frac{x+y}{2})(\sin \frac{y+z}{2}) \sin \frac{z+x}{2}.$

v) $\cos x + \cos y + \cos z + \cos(x+y+z) = 4(\cos \frac{x+y}{2})(\cos \frac{y+z}{2}) \cos \frac{z+x}{2}.$

vi) $\sin(x+y+z) = \dfrac{\tan x + \tan y + \tan z - (\tan x)(\tan y)\tan z}{(\sec x)(\sec y)\sec z}.$

vii) $\cos(x+y+z) = \dfrac{1 - (\tan x)\tan y - (\tan y)\tan z - (\tan z)\tan x}{(\sec x)(\sec y)\sec z}.$

viii) $\tan(x+y+z) = \dfrac{\tan x + \tan y + \tan z - (\tan x)(\tan y)\tan z}{1 - (\tan x)\tan y - (\tan y)\tan z - (\tan z)\tan x}.$

ix) $\cos^2 x + \cos^2 y + \cos^2 z + 2(\cos x)(\cos y)\cos z$
$= 1 + 4[\cos \frac{1}{2}(x+y+z)][\cos \frac{1}{2}(-x+y+z)][\cos \frac{1}{2}(x-y+z)] \cos \frac{1}{2}(x+y-z).$

x) $\cos^2 x + \cos^2 y + \cos^2 z + 4[\sin \frac{1}{2}(x+y+z)][\sin \frac{1}{2}(-x+y+z)][\sin \frac{1}{2}(x-y+z)] \sin \frac{1}{2}(x+y-z)$
$= 1 + 2(\cos x)(\cos y)\cos z.$

xi) $\cos 2x + \cos 2y + \cos 2z + 4(\cos x)(\cos y)\cos z + 1$
$= 8[\cos \frac{1}{2}(x+y+z)][\cos \frac{1}{2}(-x+y+z)][\cos \frac{1}{2}(x-y+z)] \cos \frac{1}{2}(x+y-z).$

xii) $\sin 2x + \sin 2y + \sin 2z = 4(\sin x)(\sin y)\sin z + 2\sin(x+y+z)$
$+ 8[\cos \frac{1}{2}(x+y+z)][\sin \frac{1}{2}(-x+y+z)][\sin \frac{1}{2}(x-y+z)] \sin \frac{1}{2}(x+y-z).$

xiii) $4(\cos \frac{x}{2})(\cos \frac{y}{2}) \cos \frac{z}{2} = \cos \frac{1}{2}(x+y+z) + \cos \frac{1}{2}(-x+y+z) + \cos \frac{1}{2}(x-y+z) + \cos \frac{1}{2}(x+y-z).$

xiv) $4(\sin \frac{x}{2})(\sin \frac{y}{2}) \sin \frac{z}{2} + \sin \frac{1}{2}(x+y+z) = \sin \frac{1}{2}(-x+y+z) + \sin \frac{1}{2}(x-y+z) + \sin \frac{1}{2}(x+y-z).$

Source: [107, p. 117], [644, pp. 165, 183], and [1294, 1319].

Fact 2.16.7. Let x, y be real numbers such that $\cos x \neq \cos y$ and $\sin y \neq 0$, and let $n \geq 2$. Then,

$$\frac{\cos nx - \cos ny}{\cos x - \cos y} = \frac{2\sum_{i=1}^{n-1}(\sin iy)\cos(n-i)x}{\sin y} + \frac{\sin ny}{\sin y}.$$

In particular,

$$\frac{\cos 2x - \cos 2y}{\cos x - \cos y} = 2\cos x + \frac{\sin 2y}{\sin y}, \quad \frac{\cos 3x - \cos 3y}{\cos x - \cos y} = \frac{2[(\sin y)\cos 2x + (\sin 2y)\cos x]}{\sin y} + \frac{\sin 3y}{\sin y}.$$

Source: [877].

Fact 2.16.8. Let $n, m \geq 0$ and $z \in \mathbb{C}$. Then,

$$[\sin(n+1)z]\sin(m+1)z = \sum_{i=0}^{\min\{n,m\}}(\sin z)\sin(n+m+1-2i)z.$$

Source: [116, p. 317].

Fact 2.16.9. Let z be a complex number such that division by zero does not occur in the expressions below, and let $n \geq 1$. Then, the following statements hold:

$$\sum_{i=1}^{n}\sin iz = \frac{1}{2}\cot\frac{z}{2} - \frac{\cos(n+\frac{1}{2})z}{2\sin\frac{z}{2}}, \quad \sum_{i=1}^{n}\cos iz = \frac{\sin(n+\frac{1}{2})z}{2\sin\frac{z}{2}} - \frac{1}{2},$$

$$\sum_{i=1}^{n}\sin(2i-1)z = \frac{\sin^2 nz}{\sin z} = \frac{1-\cos 2nz}{2\sin z}, \quad \sum_{i=1}^{n}\cos(2i-1)z = \frac{\sin 2nz}{2\sin z},$$

$$\sum_{i=1}^{n}(-1)^i\sin(2i-1)z = (-1)^n\frac{\sin 2nz}{2\cos z}, \quad \sum_{i=1}^{n}(-1)^i\cos\tfrac{1}{2}(2i-1)z = \frac{(-1)^n\cos nz - 1}{2\cos\frac{1}{2}z},$$

$$\sum_{i=1}^{n}(-1)^i\cos iz = (-1)^n\frac{\cos(n+\frac{1}{2})z}{2\cos\frac{1}{2}z} - \frac{1}{2}, \quad \sum_{i=0}^{2n-1}\tan\left(z + \frac{i\pi}{2n}\right) = -2n\cot 2nz,$$

$$\sum_{i=1}^{n}i\sin iz = \frac{\sin(n+1)z}{4\sin^2\frac{1}{2}z} - \frac{(n+1)\cos(n+\frac{1}{2})z}{2\sin\frac{1}{2}z}, \quad \sum_{i=1}^{n}i\cos iz = \frac{(n+1)\sin(n+\frac{1}{2})z}{2\sin\frac{1}{2}z} - \frac{1-\cos(n+1)z}{4\sin^2\frac{1}{2}z},$$

$$1 + 2\sum_{i=1}^{n}\left(1 - \frac{i}{n+1}\right)\cos iz = \frac{1}{n+1}\left[\frac{\sin\frac{1}{2}(n+1)z}{\sin\frac{1}{2}z}\right]^2,$$

$$\sum_{i=1}^{n}\sin^2 iz = \frac{n}{2} - \frac{[\cos(n+1)z]\sin nz}{2\sin z}, \quad \sum_{i=1}^{n}\cos^2 iz = \frac{n}{2} + \frac{[\cos(n+1)z]\sin nz}{2\sin z},$$

$$\sum_{i=1}^{n}\sin^2(2i-1)z = \frac{n}{2} - \frac{\sin 4nz}{4\sin 2z}, \quad \sum_{i=1}^{n}\cos^2(2i-1)z = \frac{n}{2} + \frac{\sin 4nz}{4\sin 2z},$$

$$\sum_{i=1}^{n}3^{i-1}\sin^3\frac{z}{3^i} = \frac{3^n}{4}\sin\frac{z}{3^n} - \frac{1}{4}\sin z, \quad \sum_{i=1}^{n}4^i\sin^4\frac{z}{2^i} = 4^n\sin^2\frac{z}{2^n} - \sin^2 z,$$

$$\sum_{i=1}^{n}\binom{n}{i}\sin 2iz = 2^n(\cos^n z)\sin nz, \quad \sum_{i=0}^{n}\binom{n}{i}\cos 2iz = 2^n(\cos^n z)\cos nz,$$

$$\sum_{i=1}^{n}(-1)^{n+i}2^i\binom{n}{i}(\cos^i z)\sin iz = \sin 2nz, \quad \sum_{i=0}^{n}(-1)^{n+i}2^i\binom{n}{i}(\cos^i z)\cos iz = \cos 2nz,$$

$$\sum_{i=0}^{n}(-1)^i\frac{\cos^3 3^i z}{3^i} = \frac{1}{4}\left[\left(\frac{-1}{3}\right)^n \cos 3^{n+1}z + 3\cos z\right],$$

$$\sin^n z = \frac{1}{2^n}\sum_{i=0}^{n}(-1)^i\binom{n}{i}\sin(n-2i)z,\quad \cos^n z = \frac{1}{2^{n-1}}\sum_{i=0}^{n-1}\binom{n-1}{i}\cos(n-2i)z = \frac{1}{2^n}\sum_{i=0}^{n}\binom{n}{i}\cos(n-2i)z,$$

$$\sin^{2n} z = \frac{1}{2^{2n-1}}\left[\frac{1}{2}\binom{2n}{n} + \sum_{i=0}^{n-1}(-1)^{n-i}\binom{2n}{i}\cos 2(n-i)z\right],$$

$$\cos^{2n} z = \frac{1}{2^{2n-1}}\left[\frac{1}{2}\binom{2n}{n} + \sum_{i=0}^{n-1}\binom{2n}{i}\cos 2(n-i)z\right],$$

$$\sin^{2n+1} z = \frac{1}{2^{2n}}\sum_{i=0}^{n}(-1)^{n-i}\binom{2n+1}{i}\sin(2n-2i+1)z,\quad \cos^{2n+1} z = \frac{1}{2^{2n}}\sum_{i=0}^{n}\binom{2n+1}{i}\cos(2n-2i+1)z,$$

$$\sum_{i=0}^{n}\frac{1}{2^i}\tan\frac{z}{2^i} = \frac{1}{2^n}\cot\frac{z}{2^n} - 2\cot 2z,\quad \sum_{i=1}^{n}\frac{1}{2^i}\tan\frac{z}{2^i} = \frac{1}{2^n}\cot\frac{z}{2^n} - \cot z,$$

$$\sum_{i=0}^{n}\frac{1}{4^i}\tan^2\frac{z}{2^i} = \frac{4^{n+1}-1}{3(2^{2n-1})} + 4\cot^2 2z - \frac{1}{4^n}\cot^2\frac{z}{2^n},\quad \sum_{i=0}^{\lfloor n/2\rfloor}(-1)^i\binom{n}{2i}\tan^{2i} z = \frac{\cos nz}{\cos^n z},$$

$$\sum_{i=1}^{n}\frac{1}{4^i}\sec^2\frac{z}{2^i} = \csc^2 z - \frac{1}{4^n}\csc^2\frac{z}{2^n},\quad \sum_{i=1}^{n}\csc 2^i z = \cot z - \cot 2^n z,$$

$$\sum_{i=1}^{n}\csc\frac{z}{2^{i-1}} = \cot\frac{z}{2^n} - \cot z,\quad \sum_{i-0}^{n}\frac{1}{(\cos iz)\cos(i+1)z} = \frac{\tan(n+1)z}{\sin z},$$

$$\sum_{i=1}^{n}\frac{\sin z}{\cos\frac{2\pi i}{n} - \cos z} = n\cot\frac{nz}{2}.$$

Source: [112, p. 28], [116, pp. 371, 602], [178, pp. 1–3], [512], [526, p. 165], [1024], [1158, p. 237], [1217, p. 39], [1757, pp. 119, 132, 166, 169, 173], and [1793, 1889, 2961]. **Remark:** The fifth equality corrects a misprint in [1757, p. 169]. **Remark:** Setting $n = 1$ in the eighth equality yields

$$\sin z = \frac{\sin 2z}{4\sin^2\frac{1}{2}z} - \frac{\cos\frac{3}{2}z}{\sin\frac{1}{2}z}.$$

Related: Fact 2.17.12 and Fact 13.6.19.

Fact 2.16.10. If $n \ge 1$, then

$$\sum_{i=1}^{n}\sin\frac{2i\pi}{n} = 0,\quad \sum_{i=1}^{n}\cos\frac{2i\pi}{n} = \operatorname{truth}(n=1),$$

$$\sum_{i=1}^{2n+1} e^{\frac{2i\pi}{2n+1}J} = 0,\quad \sum_{i=1}^{n}\cos\frac{2i\pi}{2n+1} = -\frac{1}{2},\quad \sum_{i=1}^{2n}\cos\frac{2i\pi}{2n+1} = -1,\quad \sum_{i=1}^{2n+1}\cos\frac{2i\pi}{2n+1} = 0,$$

$$\sum_{i=1}^{n}(-1)^{i+1}\cos\frac{i\pi}{2n+1} = \frac{1}{2},\quad \sum_{i=1}^{n}(-1)^{i+1}\sin\frac{(2i-1)\pi}{4n+2} = (-1)^{n+1}\frac{1}{2},\quad \sum_{i=1}^{n}\sin\frac{i\pi}{n+1} = \cot\frac{\pi}{2n+2},$$

$$\sum_{i=1}^{n}\cos\frac{i\pi}{n+1} = \sum_{i=1}^{n}\cot\frac{i\pi}{n+1} = 0,\quad \sum_{i=1}^{2n}\sec\frac{i\pi}{2n+1} = \sum_{i=1}^{2n}\tan\frac{i\pi}{2n+1} = 0,\quad \sum_{i=1}^{2n}\sec\frac{2i\pi}{4n+1} = 2n,$$

$$\sum_{i=1}^{2n+1}\sec\frac{2i\pi}{4n+3}=-2(n+1),\quad \sum_{i=0}^{n}(-1)^i\tan\frac{(2i+1)\pi}{4n+4}=(-1)^n(n+1),$$

$$\sum_{i=1}^{n}\tan^2\frac{i\pi}{2n+1}=2n^2+n,\quad \sum_{i=1}^{n}\csc^2\frac{i\pi}{n+1}=\frac{1}{3}(n^2+2n),\quad \sum_{i=1}^{n}\cot^2\frac{i\pi}{n+1}=\frac{1}{3}(n^2-n),$$

$$\sum_{i=1}^{2n}\sec\frac{2i\pi}{4n+1}=2n,\quad \sum_{i=1}^{2n-1}\sec\frac{2i\pi}{4n-1}=-2n-2,$$

$$\sum_{i=1}^{n-1}\sec^2\frac{i\pi}{2n}=\frac{2}{3}(n^2-1),\quad \sum_{i=1}^{n-1}\sec^4\frac{i\pi}{2n}=\frac{4}{45}(2n^4+5n^2-7),$$

$$\sum_{i=1}^{n}\sin^2\frac{i\pi}{2n+2}=\sum_{i=1}^{n}\cos^2\frac{i\pi}{2n+2}=\frac{n}{2},\quad \sum_{i=1}^{n}\sec^4\frac{i\pi}{2n+2}=\frac{4}{45}(n^2+2n)(2n^2+4n+9),$$

$$\sum_{i=0}^{n}\csc^2\frac{(2i+1)\pi}{2n+2}=(n+1)^2,\quad \sum_{i=1}^{n}\cot^4\frac{i\pi}{n+1}=\frac{1}{45}(n^2-n)(n^2+5n-9),$$

$$\sum_{i=1}^{n}\csc^2\frac{i\pi}{2n+2}=\sum_{i=1}^{n}\sec^2\frac{i\pi}{2n+2}=\frac{2}{3}(n^2+2n),\quad \sum_{i=1}^{n}\tan^2\frac{i\pi}{2n+2}=\sum_{i=1}^{n}\cot^2\frac{i\pi}{2n+2}=\frac{1}{3}(2n^2+n),$$

$$\sum_{i=1}^{n}\sin^2\frac{i\pi}{2n+1}=\sum_{i=1}^{n}\sin^2\frac{2i\pi}{2n+1}=\frac{1}{4}(2n+1),\quad \sum_{i=1}^{n}\cos^2\frac{i\pi}{2n+1}=\sum_{i=1}^{n}\cos^2\frac{2i\pi}{2n+1}=\frac{1}{4}(2n-1),$$

$$\sum_{i=1}^{n}\tan^2\frac{i\pi}{2n+1}=\sum_{i=1}^{n}\tan^2\frac{2i\pi}{2n+1}=2n^2+n,\quad \sum_{i=1}^{n}\csc^2\frac{i\pi}{2n+1}=\sum_{i=1}^{n}\csc^2\frac{2i\pi}{2n+1}=\frac{2}{3}(n^2+n),$$

$$\sum_{i=1}^{n}\sec^2\frac{i\pi}{2n+1}=\sum_{i=1}^{n}\sec^2\frac{2i\pi}{2n+1}=2(n^2+n),\quad \sum_{i=1}^{n}\cot^2\frac{i\pi}{2n+1}=\sum_{i=1}^{n}\cot^2\frac{2i\pi}{2n+1}=\frac{1}{3}(2n^2-n),$$

$$\sum_{i=1}^{n}\sin^2\frac{(2i-1)\pi}{4n+2}=\frac{1}{4}(2n-1),\quad \sum_{i=1}^{n}\cos^2\frac{(2i-1)\pi}{4n+2}=\frac{1}{4}(2n+1),\quad \sum_{i=0}^{n-1}\tan^2\frac{(2i+1)\pi}{4n+2}=\frac{1}{3}(2n^2-n),$$

$$\sum_{i=1}^{n}\csc^2\frac{(2i-1)\pi}{4n+2}=2(n^2+n),\quad \sum_{i=1}^{n}\sec^2\frac{(2i-1)\pi}{4n+2}=\frac{2}{3}(n^2+n),\quad \sum_{i=1}^{n}\cot^2\frac{(2i-1)\pi}{4n+2}=2n^2+n,$$

$$\sum_{i=1}^{n}\sin^2\frac{(2i-1)\pi}{4n}=\frac{1}{2}n,\quad \sum_{i=1}^{n}\tan^2\frac{(2i-1)\pi}{4n}=\sum_{i=1}^{n}\cot^2\frac{(2i-1)\pi}{4n}=2n^2-n,$$

$$\sum_{i=1}^{n}\sin\frac{2i^2\pi}{n+1}=\frac{\sqrt{n+1}}{2}\left(1+\cos\frac{(n+1)\pi}{2}-\sin\frac{(n+1)\pi}{2}\right),$$

$$\sum_{i=0}^{n}\cos\frac{2i^2\pi}{n+1}=\frac{\sqrt{n+1}}{2}\left(1+\cos\frac{(n+1)\pi}{2}+\sin\frac{(n+1)\pi}{2}\right).$$

If $n\ge 3$, $m\in\mathbb{Z}$, and $n|2m$, then

$$\sum_{i=1}^{n-1}\sin^2\frac{2mi\pi}{n}=0,\quad \sum_{i=1}^{n-1}\cos^2\frac{2mi\pi}{n}=n,\quad \sum_{i=1}^{n}\left(\sin\frac{2mi\pi}{n}\right)\cos\frac{2mi\pi}{n}=0.$$

If $n \ge 3$, $m \in \mathbb{Z}$, and $n \nmid 2m$, then

$$\sum_{i=1}^{n-1} \sin^2 \frac{2mi\pi}{n} = \sum_{i=1}^{n-1} \cos^2 \frac{2mi\pi}{n} = \frac{n}{2}, \quad \sum_{i=1}^{n-1} \left(\sin \frac{2mi\pi}{n}\right) \cos \frac{2mi\pi}{n} = 0.$$

If $n \ge 1$ and $a, b \in \mathbb{R}$, then

$$\sum_{i=1}^{n} \sin(a + bi) = \frac{\sin nb/2}{\sin b/2} \sin[a + \tfrac{1}{2}(n+1)b], \quad \sum_{i=1}^{n} \cos(a + bi) = \frac{\sin nb/2}{\sin b/2} \cos[a + \tfrac{1}{2}(n+1)b].$$

If $n \ge 2$ and $1 \le m \le n - 1$, then

$$\sum_{i=1}^{\lfloor n/2 \rfloor} \cos^{2m} \frac{(2i-1)\pi}{2n} = \frac{n}{2^{2m+1}} \binom{2m}{m}, \quad \sum_{i=1}^{\lfloor n/2 \rfloor} \cos^{2n} \frac{(2i-1)\pi}{2n} = \frac{n}{2^{2n+1}} \binom{2n}{n} - \frac{n}{4^n}.$$

If $n \ge 3$ and $1 \le m \le n - 1$, then

$$\sum_{i=1}^{\lfloor (n-1)/2 \rfloor} \cos^{2m} \frac{i\pi}{n} = \frac{n}{2^{2m+1}} \binom{2m}{m} - \frac{1}{2}, \quad \sum_{i=1}^{\lfloor (n-1)/2 \rfloor} \cos^{2n} \frac{i\pi}{n} = \frac{n}{2^{2n+1}} \binom{2n}{n} + \frac{n}{4^n} - \frac{1}{2},$$

$$\sum_{i=1}^{\lfloor (n-1)/2 \rfloor} (-1)^i \cos^n \frac{i\pi}{n} = \frac{n}{2^n} - \frac{1}{2}.$$

If $n \ge 3$, $1 \le m \le n - 1$, and $n + m$ is even, then

$$\sum_{i=1}^{\lfloor (n-1)/2 \rfloor} (-1)^i \cos^m \frac{i\pi}{n} = -\frac{1}{2}.$$

If $n \ge 0$, then

$$\sum_{i=0}^{2^n-1} (-1)^i \cot \frac{(2i+1)\pi}{2^{n+2}} = 2^n, \quad \sum_{i=0}^{2^n-1} \csc^2 \frac{(2i+1)\pi}{2^{n+2}} = 2^{2n+1},$$

$$\sum_{i=0}^{n} (-1)^i \frac{\cos^3 \frac{\pi}{3^{n-i}}}{3^i} = \frac{3}{4} \left[\left(\frac{-1}{3} \right)^{n+1} + \cos \frac{\pi}{3^n} \right].$$

If $n \ge 1$ and $m \ge 1$, then

$$\sum \binom{m}{k_1, \ldots, k_n} \cos \left[\left(\sum_{i=1}^{n} ik_i \right) \frac{2\pi}{n} \right] = 0,$$

where the sum is taken over all n-tuples $(k_1, \ldots, k_n)$ of positive integers whose sum is m. If $n, m \ge 3$ are odd and coprime, then

$$n \sum_{i=1}^{m-1} \frac{\tan \frac{ni\pi}{m}}{\tan \frac{2i\pi}{m}} + m \sum_{i=1}^{n-1} \frac{\tan \frac{mi\pi}{n}}{\tan \frac{2i\pi}{n}} = -\tfrac{1}{2}(n - m)^2.$$

If $1 \le m < n$, then

$$\sum_{i=1}^{n-1} \left(\sin \frac{2im\pi}{n} \right) \cot \frac{i\pi}{n} = n - 2m.$$

If $n \ge 1$, then

$$\sum \left(\cot^2 \frac{i\pi}{2n+2} \right) \csc^2 \frac{i\pi}{2n+2} = \frac{1}{6}(n+1)^2(n^2 + 2n),$$

where the sum is taken over all odd integers $i \in \{1,\ldots,n\}$. If $n \ge 2$, then

$$\sum |e^{2i\pi\jmath/n} - e^{2j\pi\jmath/n}|^{-2} = \frac{1}{12}(n^3 - n),$$

where the sum is taken over all subsets $\{i, j\}$ of $\{0,\ldots,n-1\}$ such that $i \ne j$. If $n \ge 1$, then

$$\sum_{i=1}^{n}(e^{2i\pi\jmath/(n+1)} - 1)^{-1} = -\frac{n}{2}, \quad \sum_{i=1}^{n}(e^{2i\pi\jmath/(n+1)} - 1)^{-2} = -\frac{1}{12}(n^2 - 4n), \quad \sum_{i=1}^{n}(e^{2i\pi\jmath/(n+1)} - 1)^{-3} = \frac{1}{8}(n^2 - 2n).$$

If $n \ge 1$ and $x \in (0,1)$, then

$$\sum_{i=0}^{n} \cot^2 \frac{(i+x)\pi}{n+1} = (n+1)^2 \csc^2 \pi x - n - 1.$$

If $1 \le m \le n$, then

$$\sum \prod_{i=1}^{m} \cot^2 \frac{j_i\pi}{2n+1} = \frac{(2n)!}{(2m+1)!(2n-2m)!},$$

where the sum is taken over all m-tuples $(j_1,\ldots,j_m)$ such that $1 \le j_1 \le \cdots \le j_m \le n$. **Source:** [10], [107, pp. 65, 340, 341], [108, pp. 59, 336, 337, 404–409], [109, pp. 279–281], [425], [516, pp. 204, 205], [1158, p. 237], [1217, pp. 37, 39], and [1645, 1689, 1694, 2015, 2182, 2353]. **Related:** Fact 10.9.6.

Fact 2.16.11. Let $n \ge 1$. Then,

$$\sum_{i=0}^{n-1} \sin^{2n} \frac{i\pi}{n} = \frac{n}{4^n}\left[\binom{2n}{n} + (-1)^n 2\right], \quad \sum_{i=0}^{n-1} \sin^{2n} \frac{(2i+1)\pi}{2n} = \frac{n}{4^n}\left[\binom{2n}{n} + (-1)^{n+1} 2\right],$$

$$\sum_{i=0}^{n-1} \cos^{2n} \frac{i\pi}{n} = \frac{n}{4^n}\left[\binom{2n}{n} + 2\right], \quad \sum_{i=1}^{n} \cos^{2n} \frac{(2i-1)\pi}{2n} = \frac{n}{4^n}\left[\binom{2n}{n} - 2\right], \quad \sum_{i=0}^{n-1} (-1)^i \cos^n \frac{i\pi}{n} = \frac{n}{2^{n-1}}.$$

If $1 \le m \le n$, then

$$\sum_{i=1}^{n} (-1)^i \cos^{2m} \frac{i\pi}{2n+2} = -\frac{1}{2}.$$

If n is even, then

$$\sum_{i=0}^{n-1} (-1)^i \sin^n \frac{i\pi}{n} = (-1)^{n/2} \frac{n}{2^{n-1}}.$$

If n is odd, then

$$\sum_{i=0}^{n-1} (-1)^i \sin^n \frac{(2i+1)\pi}{2n} = (-1)^{(n-1)/2} \frac{n}{2^{n-1}}.$$

If $1 \le m < n$, then

$$\sum_{i=0}^{n-1} \sin^{2m} \frac{i\pi}{n} = \sum_{i=0}^{n-1} \sin^{2m} \frac{(2i+1)\pi}{2n} = \frac{n}{4^m}\binom{2m}{m},$$

$$\sum_{i=0}^{n-1} (-1)^i \cos^m \frac{i\pi}{n} = \begin{cases} 0, & n+m \text{ is even,} \\ 1, & n+m \text{ is odd.} \end{cases}$$

If $1 \le m < n$ and either m is even or both n and m are odd, then

$$\sum_{i=0}^{n-1} (-1)^i \sin^m \frac{i\pi}{n} = 0.$$

If $1 \le m < n$ and either m is odd or both n and m are even, then

$$\sum_{i=0}^{n-1}(-1)^i \sin^m \frac{(2i+1)\pi}{2n} = 0.$$

Source: [807, 2017].

Fact 2.16.12. Let x be a complex number such that division by zero does not occur in the expressions below, and let $n \ge 1$. Then,

$$\sum_{i=0}^{n-1} \frac{1}{x^2 - 2x\cos\frac{2i\pi}{n} + 1} = \frac{n(x^n+1)}{(x^2-1)(x^n-1)}.$$

In particular,

$$\sum_{i=0}^{n-1} \frac{1}{1 + 8\sin^2\frac{i\pi}{n}} = \sum_{i=0}^{n-1} \frac{1}{5 - 4\cos\frac{2i\pi}{n}} = \frac{n(2^n+1)}{3(2^n-1)}.$$

Source: [107, pp. 39, 218–220], [108, pp. 401–409], and [699].

Fact 2.16.13. Let $n \ge 1$, let $x \in \mathbb{R}$, and assume that $|x| > 1$. Then,

$$\sum_{i=0}^{n-1} \frac{1}{x - \cos\frac{2i\pi}{n}} = \frac{2n(x+\sqrt{x^2-1})[(x+\sqrt{x^2-1})^n+1]}{[(x+\sqrt{x^2-1})^2-1][(x+\sqrt{x^2-1})^n-1]}.$$

In particular,

$$\sum_{i=0}^{n-1} \frac{1}{1 + 2\sin^2\frac{i\pi}{n}} = \frac{n[(2+\sqrt{3})^n+1]}{\sqrt{3}[(2+\sqrt{3})^n-1]}.$$

Source: Replace x by $x+\sqrt{x^2-1}$ in Fact 2.16.12. See [108, p. 409].

Fact 2.16.14. Let $n \ge 2$ and $m \in \{1,\dots,n-1\}$. Then,

$$\sum_{i=1}^{n-1}\left(\cot\frac{i\pi}{n}\right)\sin\frac{2mi\pi}{n} = n - 2m, \quad \sum_{i=1}^{n-1}\left(\csc^2\frac{i\pi}{n}\right)\cos\frac{2mi\pi}{n} = \frac{1}{3}(n^2-1) + 2m^2 - 2mn,$$

$$\sum_{i=1}^{n-1}\left(\csc^4\frac{i\pi}{n}\right)\cos\frac{2mi\pi}{n} = \frac{1}{45}(n^2-1)(n^2+11) - \frac{2}{3}m(n-m)(nm-m^2+2), \quad \sum_{i=1}^{n-1}\csc^2\frac{i\pi}{n} = \frac{1}{3}(n^2-1),$$

$$\sum_{i=1}^{n-1}\csc^4\frac{i\pi}{n} = \frac{1}{45}(n^2-1)(n^2+11), \quad \sum_{i=1}^{n-1}\csc^6\frac{i\pi}{n} = \frac{1}{945}(n^2-1)(2n^4+23n^2+191),$$

$$\sum_{i=1}^{n-1}\csc^8\frac{i\pi}{n} = \frac{1}{14175}(n^2-1)(n^2+11)(3n^4+10n^2+227).$$

Source: [1217, p. 1090].

Fact 2.16.15. Let $n \ge 2$. Then,

$$\prod_{i=1}^{n-1}\sin\frac{i\pi}{n} = \frac{n}{2^{n-1}}, \quad \prod_{i=1}^{n-1}\cos\frac{i\pi}{n} = \frac{\sin\frac{n\pi}{2}}{2^{n-1}}, \quad \prod_{i=1}^{2n}\tan\frac{i\pi}{2n+1} = \frac{n}{\sin\frac{(2n+1)\pi}{2}},$$

$$\prod_{i=1}^{n-1}\sin\frac{i\pi}{2n} = \prod_{i=1}^{n-1}\cos\frac{i\pi}{2n} = \frac{\sqrt{n}}{2^{n-1}}, \quad \prod_{i=1}^{n-1}\tan\frac{i\pi}{2n} = 1, \quad \prod_{i=1}^{2n-1}\sin\frac{i\pi}{4n} = \prod_{i=1}^{2n-1}\cos\frac{i\pi}{4n} = \sqrt{\frac{8n}{16^n}},$$

$$\prod_{i=1}^{n}\sin\frac{i\pi}{2n+1} = \frac{\sqrt{2n+1}}{2^n}, \quad \prod_{i=1}^{n}\cos\frac{i\pi}{2n+1} = \frac{1}{2^n}, \quad \prod_{i=1}^{n-1}\tan\frac{i\pi}{2n+1} = \sqrt{2n+1},$$

$$\prod_{i=1}^{n}\sin\frac{(2i-1)\pi}{4n}=\prod_{i=1}^{n}\cos\frac{(2i-1)\pi}{4n}=\frac{\sqrt{2}}{2^n},\quad \prod_{i=1}^{n}\tan\frac{(2i-1)\pi}{4n}=1,$$

$$\prod_{i=1}^{n}\cos\frac{2^i\pi}{2^n-1}=\prod_{i=0}^{n-1}\cos\frac{2^i\pi}{2^n+1}=\frac{1}{2^n},\quad \prod_{i=0}^{n-1}\cos\frac{2^i\pi}{2^n-1}=\prod_{i=1}^{n}\cos\frac{2^i\pi}{2^n+1}=-\frac{1}{2^n},$$

$$\prod_{i=1}^{n}\frac{1}{1-\tan^2 2^{-i}}=\frac{\tan 1}{2^n\tan 2^{-n}},\quad \prod_{i=1}^{n}\left(1-\tan^2\frac{2^i\pi}{2^n+1}\right)=-2^n,$$

$$\prod_{i=1}^{n}\sin\frac{(2i-1)\pi}{4n+2}=\frac{1}{2^n},\quad \prod_{i=1}^{n}\cos\frac{(2i-1)\pi}{4n+2}=\frac{\sqrt{2n+1}}{2^n},\quad \prod_{i=1}^{n}\tan\frac{(2i-1)\pi}{4n+2}=\frac{1}{\sqrt{2n+1}},$$

$$\prod_{i=1}^{\lfloor (n-1)/2\rfloor}\left(5+4\cos\frac{2i\pi}{n}\right)=\frac{1}{3}[2^n-(-1)^n],$$

$$\prod_{i=1}^{n}\cos\frac{2i\pi}{2n+1}=\begin{cases}1/2^n, & \text{rem}_4(n)\in\{0,3\},\\ -1/2^n, & \text{rem}_4(n)\in\{1,2\},\end{cases}\quad \prod_{i=1}^{n}\tan\frac{2i\pi}{2n+1}=\begin{cases}\sqrt{2n+1}, & \text{rem}_4(n)\in\{0,3\},\\ -\sqrt{2n+1}, & \text{rem}_4(n)\in\{1,2\},\end{cases}$$

$$\prod_{i=1}^{n}\left(1-\tan^2\frac{2i\pi}{2n+1}\right)=\begin{cases}2^n, & \text{rem}_4(n)\in\{0,3\},\\ -2^n, & \text{rem}_4(n)\in\{1,2\},\end{cases}$$

$$\prod_{i=1}^{n}\left(1-3\tan^2\frac{2i\pi}{2n+1}\right)=\begin{cases}4^n, & \text{rem}_3(n)\in\{0,2\},\\ -2^{2n+1}, & n\overset{3}{\equiv}1,\end{cases}$$

$$\prod_{i=1}^{n}\left(1-\tan^2\frac{(2i-1)\pi}{4n}\right)=\begin{cases}0, & n \text{ odd},\\ (-1)^{n/2}2^n, & n \text{ even},\end{cases}$$

$$\prod_{i=1}^{n}\left(1-3\tan^2\frac{(2i-1)\pi}{4n}\right)=\begin{cases}-2^{2n-1}, & \text{rem}_3(n)\in\{1,2\},\\ 4^n, & 3|n.\end{cases}$$

Source: [112, p. 29], [665], [1158, pp. 51, 244], [1757, p. 132], [2133, 2137, 2352, 2353, 2410]. **Related:** Fact 10.9.6.

Fact 2.16.16. Let $x, y \in \mathbb{C}$ and $n \ge 1$. Then,

$$\prod_{i=0}^{n-1}\left[\cos x-\cos\left(y+\frac{2i\pi}{n}\right)\right]=\frac{\cos nx-\cos ny}{2^{n-1}},\quad \prod_{i=0}^{n-1}\left[\cosh x-\cos\left(y+\frac{2i\pi}{n}\right)\right]=\frac{\cosh nx-\cos ny}{2^{n-1}}.$$

Source: [1217, p. 41] and [2249, p. 21].

Fact 2.16.17. Let $z \in \mathbb{C}$ and $n \ge 1$. Then,

$$2^n(\sin z)\prod_{i=0}^{n-1}\cos 2^i z=\sin 2^n z.$$

In particular,

$$\left(\cos\frac{\pi}{5}\right)\left(\cos\frac{2\pi}{5}\right)\cos\frac{3\pi}{5}=\frac{1}{16}(1-\sqrt{5}),\quad \left(\cos\frac{\pi}{5}\right)\left(\cos\frac{2\pi}{5}\right)\cos\frac{4\pi}{5}=-\frac{1}{16}(1+\sqrt{5}),$$

$$\left(\cos\frac{\pi}{7}\right)\left(\cos\frac{2\pi}{7}\right)\cos\frac{3\pi}{7} = \frac{1}{8}, \quad \left(\cos\frac{\pi}{7}\right)\left(\cos\frac{2\pi}{7}\right)\cos\frac{4\pi}{7} = -\frac{1}{8},$$

$$\left(\cos\frac{\pi}{9}\right)\left(\cos\frac{2\pi}{9}\right)\cos\frac{4\pi}{9} = \frac{1}{8}, \quad \left(\cos\frac{\pi}{10}\right)\left(\cos\frac{\pi}{5}\right)\cos\frac{2\pi}{5} = \frac{\sqrt{2}}{16}\sqrt{5+\sqrt{5}}.$$

Source: [76, 357, 442, 1793, 2080]. **Remark:** The case $n = 3$ and $z = \frac{\pi}{9}$ is *Morrie's law.*

Fact 2.16.18. Let $z \in \mathbb{C}$ and $n \ge 1$. Then,

$$\prod_{i=0}^{n-1} \sin\left(z + \frac{i\pi}{n}\right) = \frac{\sin nz}{2^{n-1}}, \quad \prod_{i=1}^{n} \sin\left(z + \frac{(2i-1)\pi}{2n}\right) = \frac{\cos nz}{2^{n-1}}.$$

Consequently,

$$\prod_{i=1}^{n} \sin\frac{(2i-1)\pi}{2n} = \frac{1}{2^{n-1}}.$$

Furthermore,

$$\prod_{i=0}^{n-1} \sin\left(z + \frac{2i\pi}{n}\right) = \begin{cases} (-1)^{(n-1)/2}\dfrac{\sin nz}{2^{n-1}}, & n \text{ odd},\\[2mm] (-1)^{n/2}\dfrac{1-\cos nz}{2^{n-1}}, & n \text{ even},\end{cases}$$

$$\prod_{i=0}^{n-1} \cos\left(z + \frac{2i\pi}{n}\right) = \begin{cases} \dfrac{\cos nz}{2^{n-1}}, & n \text{ odd},\\[2mm] \dfrac{(-1)^{n/2}-\cos nz}{2^{n-1}}, & n \text{ even}.\end{cases}$$

Source: [1217, p. 41] and [2346, p. 118].

Fact 2.16.19. Let $z \in \mathbb{C}$ and $n \ge 1$. Then,

$$\prod_{i=1}^{n-1}\left[z^2 - 2z\cos\frac{i\pi}{n} + 1\right] = \frac{z^{2n}-1}{z^2-1}, \quad \prod_{i=0}^{n-1}\left[z^2 - 2z\cos\frac{(2i+1)\pi}{2n} + 1\right] = z^{2n}+1,$$

$$\prod_{i=1}^{n}\left[z^2 - 2z\cos\frac{2i\pi}{2n+1} + 1\right] = \frac{z^{2n+1}-1}{z-1}, \quad \prod_{i=1}^{n}\left[z^2 + 2z\cos\frac{2i\pi}{2n+1} + 1\right] = \frac{z^{2n+1}+1}{z+1},$$

$$4^n z\prod_{i=1}^{n-1}\left[1 + (z^2-1)\cos^2\frac{i\pi}{2n}\right] = (z+1)^{2n} - (z-1)^{2n},$$

$$2^{2n+1} z\prod_{i=1}^{n}\left[1 + (z^2-1)\cos^2\frac{i\pi}{2n+1}\right] = (z+1)^{2n+1} + (z-1)^{2n+1},$$

$$2\prod_{i=1}^{n}\left(1 + z^2\tan^2\frac{2i\pi}{2n+1}\right) = (z+1)^{2n+1} - (z-1)^{2n+1}.$$

Source: [665] and [1217, p. 41].

Fact 2.16.20. Let $a_1, \ldots, a_n \in \mathbb{C}$, and assume that, for all $i, j \in \{1, \ldots, n\}$, $(a_i - a_j)/\pi$ is not an integer. Furthermore, let $z \in \mathbb{C}$, and, for all $i \in \{1, \ldots, n\}$, define $\alpha_i \triangleq \prod_{j=1, j\ne i}^{n} \cot(a_i - a_j)$. Then,

$$\prod_{i=1}^{n} \cot(z - a_i) = \cos\frac{n\pi}{2} + \sum_{i=1}^{n} \alpha_i \cot(z - a_i).$$

Source: [1553]. **Credit:** C. Hermite.

Fact 2.16.21. Let $a_1,\dots,a_n,b_1,\dots,b_n \in \mathbb{R}$, and assume that $a_1,\dots,a_n$ are distinct. Then,

$$\sum_{i=1}^{n} \frac{\prod_{j=1}^{n}\sin(a_i-b_j)}{\prod_{j=1,j\neq i}^{n}\sin(a_i-a_j)} = \sin\sum_{i=1}^{n}(a_i-b_i).$$

If, in addition, $b_1,\dots,b_n$ are distinct, then

$$\sum_{i=1}^{n} \frac{\prod_{j=1}^{n}\sin(b_i-a_j)}{\prod_{j=1,j\neq i}^{n}\sin(a_i-a_j)} + \sum_{i=1}^{n} \frac{\prod_{j=1}^{n}\sin(a_i-b_j)}{\prod_{j=1,j\neq i}^{n}\sin(b_i-b_j)} = 0.$$

Consequently, if $a,b,c,f,g,h \in \mathbb{R}$ and $a+b+c+f+g+h=0$, then

$$\det\begin{bmatrix} \sin(a+f)\sin(b+f)\sin(c+f) & \cos f & \sin f \\ \sin(a+g)\sin(b+g)\sin(c+g) & \cos g & \sin g \\ \sin(a+h)\sin(b+h)\sin(c+h) & \cos h & \sin h \end{bmatrix} = 0.$$

Source: [1553]. **Credit:** The last equality is due to A. Cayley.

Fact 2.16.22. Let $z \in \mathbb{C}$ be such that division by zero does not occur in the expressions below. Then,

$$\frac{\mathrm{d}}{\mathrm{d}z}\sin z = \cos z, \quad \frac{\mathrm{d}}{\mathrm{d}z}\cos z = -\sin z, \quad \frac{\mathrm{d}}{\mathrm{d}z}\tan z = \sec^2 z,$$

$$\frac{\mathrm{d}}{\mathrm{d}z}\csc z = -(\csc z)\cot z, \quad \frac{\mathrm{d}}{\mathrm{d}z}\sec z = (\sec z)\tan z, \quad \frac{\mathrm{d}}{\mathrm{d}z}\cot z = -\csc^2 z.$$

Fact 2.16.23. Let $n \ge 1$, and let $z \in \mathbb{C}$ be such that all terms below are defined. Then,

$$\left[\prod_{i=1}^{n}\left(\frac{\mathrm{d}^2}{\mathrm{d}z^2}+(2i-1)^2\right)\right]\sec z = (2n)!\sec^{2n+1} z, \quad \left[\prod_{i=1}^{n}\left(\frac{\mathrm{d}^2}{\mathrm{d}z^2}+4i^2\right)\right]\sec^2 z = (2n+1)!\sec^{2n+2} z,$$

$$\frac{\mathrm{d}^2}{\mathrm{d}z^2}\csc^n z + n^2\csc^n z = n(n+1)\csc^{n+2} z.$$

Source: [1149].

2.17 Facts on Inequalities for Trigonometric Functions

Fact 2.17.1. Let x be a real number. Then, the following statements hold:

i) If $x \in [0, \frac{\pi}{2}]$, then

$$\tfrac{2}{\pi}x \le \left\{\begin{matrix} x - \frac{2(\pi-2)}{\pi^2}x^2 \\ \frac{2}{\pi}x + \frac{x}{\pi^3}(\pi^2-4x^2) \end{matrix}\right\} \le \tfrac{1}{\pi}x(3-\tfrac{4}{\pi^2}x^2) + \tfrac{\pi-3}{\pi^3}x(2x-\pi)^2 \le \sin x,$$

$$\begin{aligned}\tfrac{2}{\pi}x &\le \left\{\begin{matrix} \frac{2}{\pi}x + \frac{8}{\pi^4}x^3(\pi-2x) \le \frac{2}{\pi}x + \frac{2}{\pi^4}x^2(\pi^2-4x^2) \le \frac{2}{\pi}x + \frac{4}{\pi^3}x^2(\pi-2x) \\ \frac{2}{\pi}x + \frac{1}{12\pi}x(\pi^2-4x^2) \end{matrix}\right\} \\ &\le \tfrac{1}{\pi}x(3-\tfrac{4}{\pi^2}x^2) \le \tfrac{1}{\pi}x(3-\tfrac{4}{\pi^2}x^2) + \tfrac{\pi-3}{\pi^3}x(2x-\pi)^2 \le \sin x,\end{aligned}$$

$$x\cos x \le \left\{\begin{array}{c}\left.\begin{array}{r}\frac{x}{\sqrt{x^2+1}}\\ \frac{x}{(1-2/\pi)x+1}\end{array}\right\} \le \dfrac{x}{\sqrt{(1-4/\pi^2)x^2+1}}\\ \left\{\begin{array}{c}\frac{1}{2}x(1+\cos x) \le \frac{2}{\pi}x + \frac{\pi-2}{\pi}x\cos x\\ x\sqrt[3]{\cos x} \le x\cos^{4/3}\frac{x}{2}\end{array}\right\}\end{array}\right\} \le \sin x \le \left\{\begin{array}{l}1 - \frac{1}{4\pi}(\pi-2x)^2 \le 1\\ \frac{4}{\pi}x - \frac{4}{\pi^2}x^2\\ x - \frac{4(\pi-2)}{\pi^3}x^3 \le x - \frac{1}{3\pi}x^3\\ \frac{2}{\pi}x + \frac{\pi-2}{\pi^3}x(\pi^2-4x^2)\\ x \le x + \frac{1}{3}x^3 \le \tan x \le x\sec x\\ \frac{1}{3}x(\cos^2\frac{x}{2} + 2\cos\frac{x}{2})\\ \quad \le \frac{1}{3}x(2+\cos x)\\ \quad \le \frac{1}{2}x\left(\frac{x}{\sin x} + \cos x\right).\end{array}\right.$$

ii) If $x \in [0, \frac{\pi}{2}]$, then

$$\left.\begin{array}{r}1 - \frac{2}{\pi}x \le 1 - \frac{2}{\pi}x + \frac{\pi-2}{\pi^2}x(\pi-2x)\\ \dfrac{\pi^2-4x^2}{\pi^2+4x^2}\end{array}\right\} \le \cos x \le 1 - \frac{2}{\pi}x + \frac{2}{\pi^2}x(\pi-2x) \le 1 - \frac{1}{\pi}x^2,$$

$$\frac{3x^2}{8} \le 1 - (\cos x)\sec\frac{x}{2} \le \frac{4x^2}{\pi^2}, \quad \frac{4(2-\sqrt{2})x^2}{\pi^2} \le 2 - (\sin x)\csc\frac{x}{2} \le \frac{x^2}{4}.$$

iii) If $x \in [0, \pi]$, then

$$\cos x \le 1 - \frac{2}{\pi^2}x^2, \quad \frac{\pi^2 x - x^3}{\pi^2 + x^2} \le \sin x \le \frac{x(\pi^2 - x^2)}{\sqrt{\pi^4 + 3x^4}},$$

$$\frac{1}{\pi}x(\pi - x) \le \sin x \le \frac{4}{\pi^2}x(\pi - x), \quad \frac{x(\pi - x)}{\pi + x} \le \sin x \le \left(3 - \frac{x}{\pi}\right)\frac{x(\pi - x)}{\pi + x}.$$

iv) If $x \in [-4, 4]$, then

$$\cos x \le \frac{\sin x}{x} \le 1.$$

v) If $x \in (0, \frac{\pi}{2})$, then

$$\cos x < \left\{\begin{array}{c}\dfrac{\cos x}{1 - x^2/3} < \sqrt[3]{\cos x}\\ \cos^2\frac{x}{2}\end{array}\right\} \le \cos\frac{\sqrt{3}x}{3} < \cos^{4/3}\frac{x}{2} = \left(\frac{1+\cos x}{2}\right)^{2/3}$$

$$< \frac{\sin x}{x} < \left\{\begin{array}{c}\cos^{5/3}\dfrac{\sqrt{5}x}{5} < \cos^3\dfrac{x}{3} < \dfrac{2+\cos x}{3}\\ \cos\left[\frac{2}{\pi}\left(\operatorname{acos}\frac{2}{\pi}\right)x\right]\end{array}\right\} < \cos\frac{x}{2}.$$

vi) If $x \in [-\frac{\pi}{2}, \frac{\pi}{2}]$ and $p \in [0, 3]$, then

$$\cos x \le \left(\frac{\sin x}{x}\right)^p \le 1.$$

vii) If $x \in \mathbb{R}$, then

$$1 - \frac{1}{2}x^2 \le \cos x \le 1 - \frac{1}{2}x^2 + \frac{1}{24}x^4.$$

viii) If $x \in [0, \infty)$, then

$$\left.\begin{array}{r}x - \frac{1}{6}x^3\\ \dfrac{x(\pi^2 - x^2)}{\pi^2 + x^2}\end{array}\right\} \le \sin x \le x - \frac{1}{6}x^3 + \frac{1}{120}x^5.$$

ix) If $x \in (0, \frac{\pi}{2})$, then

$$\frac{28/\pi + 6\cos x}{14 + \cos x} < \frac{\sin x}{x} < \frac{9 + 6\cos x}{14 + \cos x}, \quad 4 + \frac{1}{10}x^4 + \frac{1}{210}x^6 < \cos x + \frac{3x}{\sin x}.$$

x) If $x \geq \sqrt{3}$, then $1 + x\cos\frac{\pi}{x} < (x+1)\cos\frac{\pi}{x+1}$.

xi) If $x \in [0, \frac{\pi}{2})$,

$$\left.\begin{matrix} \dfrac{4x}{\pi(\pi - 2x)} < \dfrac{8x}{\pi^2 - 4x^2} \\ \left(\dfrac{\pi^2 x}{\pi^2 - 4x^2}\right)^{\pi^2/12} \end{matrix}\right\} \leq \tan x \leq \frac{\pi^2 x}{\pi^2 - 4x^2}, \quad \frac{\pi^2}{\pi^2 - 4x^2} < \sec x < \left\{\begin{matrix} \dfrac{4\pi}{\pi^2 - 4x^2} \\ \left(\dfrac{\pi^2}{\pi^2 - 4x^2}\right)^{\pi^2/8} \end{matrix}\right. .$$

xii) If $x \in (0, \frac{\pi}{2})$, then

$$x + \frac{1}{3}x^3 + \frac{2}{15}x^4 \tan x < \tan x < x + \frac{1}{3}x^3 + \frac{16}{\pi^4}x^4 \tan x, \quad 2 \leq \tan x + \cot x,$$

$$x + \frac{1}{3}x^2 \tan x < \tan x < x + \frac{1}{3}x^{9/5} \tan^{6/5} x,$$

$$3 < \frac{1}{60}x^3 \sin x + 3 < \frac{2x}{\sin x} + \frac{x}{\tan x} < \left\{\begin{matrix} \dfrac{\sin x}{x} + \dfrac{4\tan x/2}{x} \\ 3 + \dfrac{8\pi - 24}{\pi^3}x^3 \sin x \end{matrix}\right\} < \frac{2\cos x + 1}{(\cos x)^{2/3}}$$

$$< \frac{3}{20}x^3 \tan x + 3 < \frac{2\sin x}{x} + \frac{\tan x}{x} < \frac{16}{\pi^4}x^3 \tan x + 3,$$

$$\frac{1}{40}x^3 \sin x + 3 < \frac{\sin x}{x} + \frac{4\tan x/2}{x} < \frac{80 - 24\pi}{\pi^4}x^3 \sin x + 3,$$

$$2 < \left\{\begin{matrix} \dfrac{2}{45}x^3 \sin x + 2 \leq \dfrac{2}{45}x^4 + 2 \\ \dfrac{x}{\sin x} + \left(\dfrac{x}{2\tan x/2}\right)^2 < \dfrac{\sin x}{x} + \left(\dfrac{2\tan x/2}{x}\right)^2 \end{matrix}\right\} < \left(\frac{x}{\sin x}\right)^2 + \frac{x}{\tan x}$$

$$\leq \left(\frac{2}{\pi} - \frac{16}{\pi^3}\right)x^3 \sin x + 2 \leq \frac{16}{\pi^4}x^3 \tan x + 2 < \left(\frac{\sin x}{x}\right)^2 + \frac{\tan x}{x} \leq \frac{8}{45}x^3 \tan x + 2,$$

$$2 \leq 2\sqrt{\frac{\sin^3 x}{x^3 \cos x}} \leq \frac{16}{\pi^4}x^3 \tan x + 2 \leq \left(\frac{\sin x}{x}\right)^2 + \frac{\tan x}{x}.$$

xiii) If $x \in (0, \frac{\pi}{2})$ and $p > 0$, then

$$2 \leq \left\{\begin{matrix} \left(\dfrac{x}{\sin x}\right)^{2p} + \left(\dfrac{x}{\tan x}\right)^p \\ 2\left(\dfrac{\sin^3 x}{x^3 \cos x}\right)^{p/2} \end{matrix}\right\} \leq \left(\frac{\sin x}{x}\right)^{2p} + \left(\frac{\tan x}{x}\right)^p,$$

$$3\,\mathrm{truth}(p \geq 1) < 2\left(\frac{x}{\sin x}\right)^p + \left(\frac{x}{\tan x}\right)^p < 2\left(\frac{\sin x}{x}\right)^p + \left(\frac{\tan x}{x}\right)^p,$$

$$2\,\mathrm{truth}(p \geq 3/20) < \left(\frac{x}{\sin x}\right)^p + \left(\frac{x/2}{\tan x/2}\right)^{2p} < \left(\frac{\sin x}{x}\right)^p + \left(\frac{\tan x/2}{x/2}\right)^{2p},$$

$$3\text{truth}(p \geq 1/5) < \left(\frac{x}{\sin x}\right)^p + 2\left(\frac{x/2}{\tan x/2}\right)^p.$$

xiv) If $x \in (0, \frac{\pi}{2})$, then

$$3 < \frac{2x}{\sin x} + \frac{x}{\tan x} < \frac{\sin x}{x} + \frac{2\tan x/2}{x/2} < \frac{2\sin x}{x} + \frac{\tan x}{x}.$$

xv) If $x \in (0, \frac{\pi}{2})$, then $x^2 < \log(1+\tan^2 x) = 2\log\sec x < (\sin x)\tan x$.

xvi) If $x \in (0, \frac{\pi}{2})$, then

$$\frac{\log(1-\sin x)}{\log\cos x} < \frac{x+2}{x}.$$

xvii) If $x \in (-\frac{\pi}{2}, \frac{\pi}{2})$, then

$$2x^2 \leq (\sin x)\log\frac{1+\sin x}{1-\sin x}.$$

xviii) If $x \in [0, \pi]$ and $n \geq 0$, then $|\sin nx| \leq n\sin x$.

xix) If $x \in (0, 1.364)$, then $\sin^2 x < \sin x^2$.

xx) If $x \in [0, \frac{\pi}{2}]$, then

$$-0.00163 < \cos x - \frac{\pi^2 - 4x^2}{\pi^2 + x^2} < 0.00135.$$

xxi) If $\alpha > 0$ and $x \in [0, \frac{\pi}{2})$, then

$$\tan\left[\left(\frac{2}{\pi}\right)^\alpha x^{\alpha+1}\right] \leq \left(\frac{2}{\pi}\right)^\alpha \tan^{\alpha+1} x.$$

xxii) If $x \in [0, \frac{\pi}{2})$, then

$$\tan\frac{2x^2}{\pi} \leq \frac{2}{\pi}\tan^2 x.$$

xxiii) If $n \geq 0$ and $x \in (0, \frac{\pi}{2})$, then $2^{n+1} \leq \sec^{2n} x + \csc^{2n} x$.

xxiv) If $x \in (0, \frac{\pi}{2}]$, then $\cot x < \frac{1}{x} < \csc x < x + \cot x$.

xxv) If $x \in \mathbb{R}$, then

$$|\sin\cos x| \leq |\cos x| \leq \cos\sin x,$$

$$\tfrac{\sqrt{2}}{2}(\csc\tfrac{\sqrt{2}}{2})|\sin\cos x + \sin\sin x| \leq |\sin x + \cos x| \leq \tfrac{\sqrt{2}}{2}(\sec\tfrac{\sqrt{2}}{2})(\cos\cos x + \cos\sin x).$$

xxvi) If $x \in (0, 1)$, then

$$\frac{2x}{1-x^2} < \frac{1}{x} - \pi\cot\pi x < \frac{\pi^2 x}{3(1-x^2)}.$$

xxvii) If $x \in (-1, 0) \cup (0, 1)$, then

$$\log\frac{\pi x}{\sin\pi x} < \frac{\pi^2 x^2}{6(1-x^2)}, \quad \log\sec\tfrac{1}{2}\pi x < \frac{\pi^2 x^2}{8(1-x^2)}, \quad \log\frac{2\tan\frac{1}{2}\pi x}{\pi x} < \frac{\pi^2 x^2}{12(1-x^2)}.$$

Source: *i)*, *ii)*, *iv)*, and *x)* are given in [603, pp. 250, 251]; for *i)*, see also [415, 693, 1348], [806, p. 136], [1567, p. 75], [1812, 2021, 2278]; the second string in *ii)* is given in [2371]; *iii)* is given in [1348], [1567, p. 72], [2278], and [2294, p. 309]; *v)* is given in [2952]; *vi)* is given in [3004]; *vii)* is given in [2278]; *viii)* is given in [1567, p. 68]; *ix)* is given in [694, 2087]; *xi)* is given in [603, pp. 250, 251] and [690, 694]; *xii)* is given in [77, 691, 1270, 1478, 1679, 2087, 2115, 2881]. See also [604, p. 9], [1757, pp. 270–271], and [3003, 3004]; *xiii)* is given in [1567, p. 71], [1757, p. 266], [1860, pp. 22, 147], and [2121]; *xv)* is given in [2121]; *xv)* is given in [603, p. 250] and [2294, p. 309]; *xvi)* is given in [1860, pp. 39, 235, 236]; *xvii)* is given in [2294, p. 337]; *xviii)* is given in [1757, p. 66]; *xix)* is given in [1757, p. 271]; *xx)* is given in [2446]; *xxi)* and *xxii)* are given in [2294, p. 176]; *xxiii*) is given in [1158, p. 234]; *xxiv*) follows from $\sin x < x < \tan x$ in *i*) and [1975]; *xxv*) is given in [61]; *xxvi*) and *xxvii*) are given in [1270]. **Remark:** $2/\pi \le (\sin x)/x$ in *i*) is *Jordan's inequality*. See [47, p. 15], [1812, 2278]. **Remark:** *xiii*) is *Huygens's inequality*. See [2278]. **Remark:** $3\sin x < (2+\cos x)x$ in *i*) is the *Cusa-Huygens* inequality. See [1567, p. 71], [1757, p. 266], and [2369]. **Remark:** $2 < \left(\frac{\sin x}{x}\right)^2 + \frac{\tan x}{x}$ in *xiii*) with $p = 1$ is *Wilker's inequality*. See [3003, 3004]. **Remark:** See [2279] for additional inequalities. **Credit:** The first string in *xxv*) is due to R. J. Webster. See [61].

Fact 2.17.2. The following statements hold:

i) If $x \in \mathbb{R}$, then

$$\frac{1-x^2}{1+x^2} \le \frac{\sin \pi x}{\pi x}.$$

ii) If $|x| \ge 1$, then

$$\frac{1-x^2}{1+x^2} + \frac{(1-x)^2}{x(1+x^2)} \le \frac{\sin \pi x}{\pi x}.$$

iii) If $x \in (0,1)$, then

$$\frac{(1-x^2)(4-x^2)(9-x^2)}{x^6-2x^4+13x^2+36} \le \frac{\sin \pi x}{\pi x} \le \frac{1-x^2}{\sqrt{1+3x^4}}.$$

iv) Let a and b be positive numbers, and assume that $a < b$. Then,

$$1-\frac{a}{b} \le \sup_{x>0}\left|\frac{\sin ax}{ax} - \frac{\sin bx}{bx}\right| \le \frac{7}{5}\left(1-\frac{a}{b}\right).$$

Source: [1812]. *iv*) is a sharper version of [2294, pp. 228, 229].

Fact 2.17.3. Let $x \in (0, \frac{\pi}{2})$. Then,

$$\begin{aligned}\frac{(\sin x)(\cos x)(\sin x+\cos x)}{2[1-(\sin x)\cos x]} &\le \frac{(\sin x)\cos x}{(\sin x+\cos x)[1-(\sin x)\cos x]} \le \frac{2(\sin x)\cos x}{\sin x+\cos x} \le \sqrt{(\sin x)\cos x}\\ &\le \tfrac{1}{2}(\sin x+\cos x) \le \frac{\sqrt{2}}{2} \le \sqrt[4]{\tfrac{1}{2}-(\sin^2 x)\cos^2 x}\\ &\le \sqrt{1-(\sin x)\cos x} \le (\sin x+\cos x)[1-(\sin x)\cos x].\end{aligned}$$

Source: [379].

Fact 2.17.4. Let w, x, y, z be real numbers. Then, the following statements hold:

i) If $0 \le x < y < \pi/2$, then

$$\frac{\tan x}{\tan y} \le \frac{x}{y} \le \frac{\sin x}{\sin y} \le \frac{\pi}{2}\left(\frac{x}{y}\right).$$

ii) If either $x, y \in [0,1]$ or both $x, y \in [1, \frac{\pi}{2}]$ and $xy < \frac{\pi}{2}$, then $(\tan x)\tan y \le (\tan 1)\tan xy$.

iii) If $x, y \in [0,1]$, then $(\operatorname{asin} x)\operatorname{asin} y \le \frac{\pi}{2}\operatorname{asin} xy$.

iv) If $y \in (0, \frac{\pi}{2}]$ and $x \in [0, y]$, then

$$\left(\frac{\sin y}{y}\right)x \le \sin x \le \sin\left[y\left(\frac{x}{y}\right)^{y\cot y}\right].$$

v) If $x, y \in [0,\pi]$ are distinct, then

$$\tfrac{1}{2}[\sin x + \sin y] < \frac{\cos x - \cos y}{y-x} < \sin\tfrac{1}{2}(x+y).$$

vi) If $0 \le x < y < \pi/2$, then

$$\frac{1}{\cos^2 x} < \frac{\tan x - \tan y}{x - y} < \frac{1}{\cos^2 y}.$$

vii) If $x, y \in [0, \frac{\pi}{2})$, then

$$\sqrt{(\cos x)\cos y} \le \tfrac{1}{2}(\cos x + \cos y) \le \cos \tfrac{1}{2}(x+y),$$

$$\left.\begin{array}{c}\sqrt{(\tan x)\tan y}\\ \tan\frac{1}{2}(x+y)\end{array}\right\} \le \tfrac{1}{2}(\tan x + \tan y).$$

viii) If $x, y \in [0, \frac{\pi}{4})$, then

$$\sqrt{(\tan x)\tan y} \le \tan\tfrac{1}{2}(x+y) \le \tfrac{1}{2}(\tan x + \tan y).$$

ix) If $x, y \in (0, \frac{\pi}{2})$, then

$$\sec(x-y) \le \frac{\sin^3 x}{\sin y} + \frac{\cos^3 x}{\cos y}.$$

x) If $\sqrt{x^2+y^2} \le \frac{\pi}{2}$, then

$$\cos\sqrt{\frac{x^2+y^2}{2}} \le \frac{\sin x + \sin y}{x+y} \le \frac{\sin\sqrt{\frac{x^2+y^2}{2}}}{\sqrt{\frac{x^2+y^2}{2}}} \le \frac{1}{2}\left(\frac{\sin x}{x} + \frac{\sin y}{y}\right) \le \frac{1}{2}\left(1 + \frac{\sin\sqrt{x^2+y^2}}{\sqrt{x^2+y^2}}\right).$$

If $\sqrt{x^2+y^2} \le \frac{\sqrt{2}\pi}{2}$, then

$$\cos\sqrt{\frac{x^2+y^2}{2}} \le \left|\frac{\cos x + \cos y}{x+y}\right|\sqrt{\frac{x^2+y^2}{2}}.$$

xi) If $x \in [0,\pi]$, $y, z \in [0, \frac{\pi}{2}]$, and $x \le y+z$, then $\sin x \le \sin y + \sin z$.

xii) If $x, y, z \in \mathbb{R}$, then

$$|(\sin^2 x)\cos y + (\sin^2 y)\cos x| \le 1, \quad |(\sin x)\cos y + (\sin y)\cos z + (\sin z)\cos x| \le \tfrac{3}{2},$$

$$|(\sin^2 x)\cos y + (\sin^2 y)\cos z + (\sin^2 z)\cos x| \le \tfrac{5}{4},$$

$$(\sin^2 x)\cos^2 y + (\sin^2 y)\cos^2 z + (\sin^2 z)\cos^2 x \le 1.$$

xiii) If $w, x, y, z \in [0, \pi]$, then

$$(\sin x)\sin y \le \sin^2 \tfrac{x+y}{2},$$

$$(\sin x)(\sin y)\sin z \le \sin^3 \tfrac{x+y+z}{3},$$

$$(\sin w)(\sin x)(\sin y)\sin z \le \sin^4 \tfrac{w+x+y+z}{4}.$$

xiv) If $x, y \in [0,1]$, then

$$\sin\left(\frac{x}{1+y^2} + \frac{y}{1+x^2}\right) \le x\sqrt{1+y^2} + y\sqrt{1+x^2},$$

$$\sqrt{(1-x^2)(1-y^2)} \le xy + \cos\left(\frac{x}{1+y^2} + \frac{y}{1+x^2}\right),$$

$$\frac{2}{\pi}\left(x\sqrt{1-y^2} + y\sqrt{1-x^2}\right) \le x\sqrt{\frac{1-y^2}{1+x^2}} + y\sqrt{\frac{1-x^2}{1+y^2}} \le \sin(x+y) \le \frac{x}{\sqrt{1+y^2}} + \frac{y}{\sqrt{1+x^2}}$$

$$\le \frac{1}{\pi}\left(x\sqrt{\pi^2-4y^2} + y\sqrt{\pi^2-4x^2}\right) \le \frac{x+y}{\sqrt{(1-x^2)(1-y^2)}}.$$

xv) If $x, y, z \in [0,1]$, then

$$\sqrt{(1-x^2)(1-y^2)(1-z^2)} \le xy\sqrt{1-z^2}+yz\sqrt{1-x^2}+zx\sqrt{1-y^2}+\cos\left(\tfrac{x}{1+x^2}+\tfrac{y}{1+y^2}+\tfrac{z}{1+z^2}\right),$$

$$\frac{2}{\pi}\left(x\sqrt{(1-y^2)(1-z^2)}+y\sqrt{(1-z^2)(1-x^2)}+z\sqrt{(1-x^2)(1-y^2)}\right) \le \sin(x+y+z)$$
$$\le \frac{1}{\pi^2}\left(x\sqrt{(\pi^2-4y^2)(\pi^2-4z^2)}+y\sqrt{(\pi^2-4z^2)(\pi^2-4x^2)}+z\sqrt{(\pi^2-4x^2)(\pi^2-4y^2)}\right)$$
$$\le \frac{x+y+z-3xyz}{\sqrt{(1-x^2)(1-y^2)(1-z^2)}}.$$

Source: The first two inequalities in *i*) are given in [47, p. 127] and [1757, p. 267]; the second and third inequalities in *i*) as well as *ii*) and *iii*) are given in [603, pp. 250, 251]; *iv*) is given in [2128, p. 26]; *v*) is a consequence of the Hermite-Hadamard inequality given by Fact 1.21.9, see [2128, p. 51]; *vi*) follows from the mean value theorem and monotonicity of the cosine function, see [1757, p. 264]; *vii*) is given in [1559]; *viii*) is given in [952]; *ix*) is given in [1158, p. 36]; *x*) is given in [1975]; *xi*) is given in [1829]; *xii*) is given in [2823]. *xiii*) is given in [1938, p. 31]; *xiv*) and *xv*) are given in [389].

Fact 2.17.5. Let $n \ge 1$. Then,

$$\frac{n}{2} \le \sum_{i=0}^{n} |\cos i|.$$

Source: [2090].

Fact 2.17.6. Let $n \ge 1$. Then,

$$0 < \frac{4^n}{\pi\binom{2n}{n}} - \sum_{i=1}^{n}\cos^{2n+1}\frac{i\pi}{2n+1} < 1.$$

Source: [2019].

Fact 2.17.7. Let $n \ge 1$. Then,

$$\sum_{i=1}^{n}\sec^2\frac{(n-i)\pi}{4n} \le \frac{4n}{\pi} \le \sum_{i=1}^{n}\sec^2\frac{i\pi}{4n}.$$

Source: [886].

Fact 2.17.8. Let $x_1,\dots,x_n \in (0,\pi)$, and define $x \triangleq \frac{1}{n}\sum_{i=1}^n x_i$. Then,

$$\prod_{i=1}^{n}\frac{\sin x_i}{x_i} \le \left(\frac{\sin x}{x}\right)^n.$$

Source: [2294, p. 286].

Fact 2.17.9. If $x_1,\dots,x_n \in [0,\pi]$, then

$$\frac{1}{n}\sum_{i=1}^{n}\sin x_i \le \sin\left(\frac{1}{n}\sum_{i=1}^{n}x_i\right).$$

If $x_1,\dots,x_n \in [0,\frac{\pi}{2}]$, then

$$\frac{1}{n}\sum_{i=1}^{n}\cos x_i \le \cos\left(\frac{1}{n}\sum_{i=1}^{n}x_i\right).$$

Source: [391, p. 286].

Fact 2.17.10. Let $x_1, \ldots, x_n \in [-\frac{\pi}{6}, \frac{\pi}{6}]$, and define $x_{n+1} \triangleq x_1$. Then,

$$\sum_{i=1}^{n} \cos(2x_i - x_{i+1}) \le \sum_{i=1}^{n} \cos x_i.$$

Source: [1569].

Fact 2.17.11. Let $n \ge 2$, and let $x_1, \ldots, x_n$ be real numbers such that $0 < x_1 < \cdots < x_n < \pi/2$. Then,

$$\tan x_1 < \frac{\sum_{i=1}^{n} \sin x_i}{\sum_{i=1}^{n} \cos x_i} < \tan x_n.$$

Source: [1566, p. 14].

Fact 2.17.12. Let $x \in \mathbb{R}$ and $n \ge 1$. If $x \in [0, \frac{2\pi}{3}]$, then

$$\sum_{i=0}^{n} (n-i+1)(n-i+2)(i+1)\sin(i+1)x \ge 0.$$

If $x \in (0, \pi)$, then

$$0 < \sum_{i=1}^{n} \frac{\sin ix}{i} < \pi - x, \quad \frac{\sin(n-1)x}{(n-1)\sin x} - \frac{\sin(n+1)x}{(n+1)\sin x} < \frac{4n}{n^2-1}\left(1 - \frac{\sin nx}{n \sin x}\right),$$

$$\sum_{i=0}^{n} \frac{3^{\overline{n-i}}}{(n-i)!}(i+\tfrac{1}{2})\sin(i+\tfrac{1}{2})x > 0, \quad \sum_{i=0}^{n} \frac{3^{\overline{n-i}}}{(n-i)!}(i+2)\sin(i+1)x > 0,$$

$$\sum_{i=0}^{n} \frac{4^{\overline{n-i}}}{(n-i)!}(i+1)\sin(i+1)x > 0, \quad -1 < \sum_{i=1}^{n} \frac{\cos ix}{i} \le \begin{cases} -\log(\sin\frac{x}{2}) + \frac{1}{2}(\pi - 1 - x) \\ -\log(2\sin\frac{x}{2}) + \frac{1}{2}. \end{cases}$$

If $x \in [0, \pi]$, then

$$\sum_{i=1}^{n} \binom{n-i+k}{k} \sin ix \ge 0.$$

In particular,

$$\sum_{i=1}^{n} (n+1-i)\sin ix \ge 0, \quad \sum_{i=1}^{n} (n+1-i)(n+2-i)\sin ix \ge 0.$$

If $x \in [0, \pi]$, then

$$-\frac{9}{200} \le \sum_{i=1}^{n} \frac{\sin ix}{i+1}, \quad -\frac{1}{2} \le \sum_{i=1}^{n} \frac{\cos ix}{i+1}, \quad -\frac{1}{2} \le \sum_{i=1}^{n} \frac{\sin ix + \cos ix}{i+1}.$$

If $n \ge 2$ and $x \in [0, \pi]$, then

$$-\frac{5}{6} \le \sum_{i=1}^{n} \frac{\cos ix}{i}, \quad -\frac{41}{96} \le \sum_{i=1}^{n} \frac{\cos ix}{i+1}.$$

If $x \in [0, \pi]$ and $p \in [0, 9/2]$, then

$$-\frac{1}{1+p} \le \sum_{i=1}^{n} \frac{\cos ix}{i+p}.$$

If $x \in [0, 2\pi]$, then

$$\sum_{i=0}^{n} \sin(i+\tfrac{1}{2})x = \frac{1-\cos(n+1)x}{2\sin\frac{1}{2}x} = \frac{\sin^2\frac{1}{2}(n+1)x}{\sin\frac{1}{2}x} \ge 0,$$

$$1+2\sum_{i=1}^{n}\left(1-\frac{i}{n+1}\right)\cos ix=\frac{1}{n+1}\left(\frac{\sin\frac{1}{2}(n+1)x}{\sin\frac{1}{2}x}\right)^2\ge 0,$$

$$1+2\sum_{i=1}^{n}\frac{(n!)^2}{(n-i)!(n+i)!}\cos ix=\frac{(n!)^2}{(2n)!}(2\cos\tfrac{1}{2}x)^{2n}\ge 0.$$

Source: [63, 64, 65], [116, pp. 371, 372, 382], [178, pp. 1–6], and [1691]. **Related:** Fact 2.16.9.

2.18 Facts on Equalities and Inequalities for Inverse Trigonometric Functions

Fact 2.18.1. The following statements hold:

i) asin: $[-1,1]\mapsto[-\frac{\pi}{2},\frac{\pi}{2}]$ is given by

$$\operatorname{asin}x=\frac{\pi}{2}+\jmath\log(x+\sqrt{1-x^2}\jmath).$$

ii) asin: $(-\infty,-1]\cup[1,\infty)\mapsto[-\frac{\pi}{2},\frac{\pi}{2}]+\mathrm{IA}$ is given by

$$\operatorname{asin}x=\frac{\pi}{2}+\jmath\log(x-\sqrt{x^2-1})=(\operatorname{sign}x)\left(\frac{\pi}{2}-\jmath\log(\sqrt{x^2-1}+|x|)\right).$$

iii) asin: $\mathbb{C}\mapsto[-\frac{\pi}{2},\frac{\pi}{2}]+\mathrm{IA}$ is given by

$$\operatorname{asin}z=\frac{\pi}{2}+\jmath\log(z+\sqrt{1-z^2}\jmath)=-\jmath\log(\sqrt{1-z^2}+z\jmath).$$

iv) acos: $[-1,1]\mapsto[0,\pi]$ is given by

$$\operatorname{acos}x=\frac{\pi}{2}+\jmath\log(\sqrt{1-x^2}+x\jmath)]=-\jmath\log(x+\sqrt{1-x^2}\jmath).$$

v) acos: $(-\infty,-1]\cup[1,\infty)\mapsto\mathbb{C}$ is given by

$$\operatorname{acos}x=\frac{\pi}{2}+\jmath\log(x\jmath+\sqrt{x^2-1}\jmath).$$

vi) acos: $\mathbb{C}\mapsto\mathbb{C}$ is given by

$$\operatorname{acos}z=\frac{\pi}{2}+\jmath\log(\sqrt{1-z^2}+z\jmath)]=-\jmath\log(z+\sqrt{1-z^2}\jmath).$$

vii) atan: $\mathbb{R}\mapsto(-\frac{\pi}{2},\frac{\pi}{2})$ is given by

$$\operatorname{atan}x=\frac{\jmath}{2}[\log(1-x\jmath)-\log(1+x\jmath)].$$

viii) atan: $\mathbb{C}\backslash\{-\jmath,\jmath\}\mapsto(-\frac{\pi}{2},\frac{\pi}{2})+\mathrm{IA}$ is given by

$$\operatorname{atan}z=\frac{\jmath}{2}[\log(1-z\jmath)-\log(1+z\jmath)].$$

ix) acsc: $(-\infty,-1]\cup[1,\infty)\mapsto[-\frac{\pi}{2},0)\cup(0,\frac{\pi}{2}]$ is given by

$$\operatorname{acsc}x=-\jmath\log\left(\sqrt{1-\frac{1}{x^2}}+\frac{1}{x}\jmath\right).$$

x) acsc: $\mathbb{C}\backslash\{0\}\mapsto[-\frac{\pi}{2},\frac{\pi}{2}]\backslash\{0\}+\mathrm{IA}$ is given by

$$\operatorname{acsc}z=-\jmath\log\left(\sqrt{1-\frac{1}{z^2}}+\frac{1}{z}\jmath\right).$$

xi) asec: $(-\infty,-1]\cup[1,\infty)\mapsto[0,\frac{\pi}{2})\cup(\frac{\pi}{2},\pi]$ is given by

$$\operatorname{asec} x = \frac{\pi}{2} + \jmath\log\left(\sqrt{1-\frac{1}{x^2}}+\frac{1}{x}\jmath\right).$$

xii) asec: $\mathbb{C}\backslash\{0\}\mapsto[0,\pi]\backslash\{\frac{\pi}{2}\}+\mathrm{IA}$ is given by

$$\operatorname{asec} z = \frac{\pi}{2} + \jmath\log\left(\sqrt{1-\frac{1}{z^2}}+\frac{1}{z}\jmath\right).$$

xiii) acot: $\mathbb{R}\mapsto(-\frac{\pi}{2},0)\cup(0,\frac{\pi}{2}]$ is given by

$$\operatorname{acot} x = \begin{cases}\frac{\pi}{2}, & x=0,\\ \frac{\jmath}{2}[\log(-1-x\jmath)-\log(-1+x\jmath)+\log x\jmath-\log -x\jmath], & x\neq 0.\end{cases}$$

xiv) acot: $\mathbb{C}\backslash\{-\jmath,\jmath\}\mapsto((-\frac{\pi}{2},0)\cup(0,\frac{\pi}{2}])+\mathrm{IA}$ is given by

$$\operatorname{acot} z = \begin{cases}\frac{\pi}{2}, & z=0,\\ \frac{\jmath}{2}[\log(-1-z\jmath)-\log(-1+z\jmath)+\log z\jmath-\log -z\jmath], & z\neq 0.\end{cases}$$

Source: To prove *i)*, note that

$$\begin{aligned}\sin[\tfrac{\pi}{2}+\jmath\log(x+\sqrt{1-x^2}\jmath)] &= \cos[\jmath\log(x+\sqrt{1-x^2}\jmath)] = \cosh\log(x+\sqrt{1-x^2}\jmath)\\ &= \frac{1}{2}\left(e^{\log(x+\sqrt{1-x^2}\jmath)}+e^{-\log(x+\sqrt{1-x^2}\jmath)}\right)\\ &= \frac{1}{2}\left(x+\sqrt{1-x^2}\jmath+\frac{1}{x+\sqrt{1-x^2}\jmath}\right) = x.\end{aligned}$$

Remark: The logarithm and square root are the principal inverses. **Remark:** The range of acot and its value at zero are consistent with Matlab and Mathematica. The expression for acot for nonzero $z\in\jmath(-1,1)$ is consistent with Mathematica but differs from Matlab by the sign of the real part. The expression for atan for nonzero $z\in\jmath[(-\infty,-1)\cup(1,\infty)]$ is consistent with Mathematica but differs from Matlab by the sign of the real part.

Fact 2.18.2. Let z be a complex number. Then, the following statements hold:

i) $\operatorname{asin}(-z) = -\operatorname{asin} z$, $\operatorname{acos}(-z) = \pi - \operatorname{acos} z$.

ii) $\operatorname{acsc}(-z) = -\operatorname{acsc} z$, $\operatorname{asec}(-z) = \pi - \operatorname{asec} z$.

iii) $\operatorname{asin} z + \operatorname{acos} z = \frac{\pi}{2}$, $\operatorname{acsc} z + \operatorname{asec} z = \frac{\pi}{2}$.

iv) If $z\neq 0$, then $\operatorname{acsc} z = \operatorname{asin}\frac{1}{z}$ and $\operatorname{asec} z = \operatorname{acos}\frac{1}{z}$.

v) If $z\notin\{-\jmath,\jmath\}$, then $\operatorname{atan}(-z) = -\operatorname{atan} z$.

vi) If $z\notin\{-\jmath,\jmath,0\}$, then $\operatorname{acot}(-z) = -\operatorname{acot} z$ and $\operatorname{acot} z = \operatorname{atan}\frac{1}{z}$.

If $\operatorname{Re} z\neq 0$, then the following statement holds:

vii) $\operatorname{atan} z + \operatorname{atan}\frac{1}{z} = \operatorname{atan} z + \operatorname{acot} z = \operatorname{acot} z + \operatorname{acot}\frac{1}{z} = (\operatorname{sign}\operatorname{Re} z)\frac{\pi}{2}$.

Fact 2.18.3. Let x be a real number. Then, the following statements hold:

i) $\operatorname{asin} x + \operatorname{acos} x = \frac{\pi}{2}$.

ii) $\operatorname{acos}^2 x - \operatorname{asin}^2 x + \pi\operatorname{asin} x = \frac{\pi^2}{4}$.

iii) $\operatorname{atan} 0 = 0$ and $\operatorname{acot} 0 = \frac{\pi}{2}$.

iv) If $x\neq 0$, then $\operatorname{atan} x + \operatorname{acot} x = (\operatorname{sign} x)\frac{\pi}{2}$.

v) If $|x|\geq 1$, then $\operatorname{acsc} x = \operatorname{asin}\frac{1}{x}$ and $\operatorname{asec} x = \operatorname{acos}\frac{1}{x}$.

vi) If $x \neq 0$, then $\operatorname{acot} x = \operatorname{atan} \frac{1}{x}$.

vii) If $x \in [-1, 0) \cup (0, 1]$, then $\operatorname{asin} x + \operatorname{asec} \frac{1}{x} = \frac{\pi}{2}$.

viii) If $x \neq 0$, then $\operatorname{atan} x + \operatorname{atan} \frac{1}{x} = (\operatorname{sign} x)\frac{\pi}{2}$.

ix) If $x \in [0, 1]$, then $\operatorname{asin} x = \operatorname{acos} \sqrt{1 - x^2} = \operatorname{atan} \dfrac{x}{\sqrt{1 - x^2}}$.

x) If $x \in [0, 1]$, then $\operatorname{acos} x = \operatorname{asin} \sqrt{1 - x^2} = \frac{1}{2} \operatorname{acos}(2x^2 - 1)$.

xi) If $|x| < 1$, then $\operatorname{atan} \dfrac{1 - x}{\sqrt{1 - x^2}} + \dfrac{1}{2} \operatorname{asin} x = \dfrac{\pi}{4}$.

xii) If $x \in [-1, 0]$, then $\operatorname{asin} x + \operatorname{acos} \sqrt{1 - x^2} = 0$ and $\operatorname{acos} x + \operatorname{asin} \sqrt{1 - x^2} = \pi$.

xiii) If $x \in (0, 1]$, then $\operatorname{asin} x = \operatorname{acot} \dfrac{\sqrt{1 - x^2}}{x}$ and $\operatorname{acos} x = \operatorname{atan} \dfrac{\sqrt{1 - x^2}}{x}$.

xiv) If $x \in [-1, 0)$, then $\operatorname{asin} x + \pi = \operatorname{acot} \dfrac{\sqrt{1 - x^2}}{x}$ and $\operatorname{acos} x = \operatorname{atan} \dfrac{\sqrt{1 - x^2}}{x} + \pi$.

xv) If $x \in [-1, 0]$, then $\operatorname{acos}(2x^2 - 1) + 2 \operatorname{acos} x = 2\pi$.

xvi) If $|x| < 1$, then $\operatorname{acos} x = \operatorname{acot} \dfrac{x}{\sqrt{1 - x^2}} = 2 \operatorname{atan} \sqrt{\dfrac{1 - x}{1 + x}}$.

xvii) $\sin \operatorname{atan} x = \dfrac{x}{\sqrt{x^2 + 1}}$ and $\cos \operatorname{atan} x = \dfrac{1}{\sqrt{x^2 + 1}}$.

xviii) $\operatorname{atan} x = \operatorname{asin} \dfrac{x}{\sqrt{x^2 + 1}} = (\operatorname{sign} x) \operatorname{acos} \dfrac{1}{\sqrt{x^2 + 1}} = \dfrac{\pi}{2} - \operatorname{acos} \dfrac{x}{\sqrt{x^2 + 1}}$.

xix) $\sin \operatorname{acot} x = \dfrac{\operatorname{sign} x}{\sqrt{x^2 + 1}}$ and $\cos \operatorname{acot} x = \dfrac{|x|}{\sqrt{x^2 + 1}}$.

xx) $\operatorname{acot} x = (\operatorname{sign} x) \operatorname{asin} \dfrac{1}{\sqrt{x^2 + 1}} = \operatorname{acos} \dfrac{|x|}{\sqrt{x^2 + 1}} = \operatorname{acos} \dfrac{x}{\sqrt{x^2 + 1}} - \operatorname{truth}(x < 0)\pi$.

xxi) $\operatorname{atan} x = 2 \operatorname{atan} \dfrac{x}{1 + \sqrt{1 + x^2}}$.

xxii) If $x \neq 0$, then $\operatorname{atan} \frac{1}{x} = 2 \operatorname{atan} \frac{1}{2x} - \operatorname{atan} \frac{1}{4x^3 + 3x}$.

xxiii) If $x \geq 0$, then $\operatorname{atan} x = \dfrac{1}{2} \operatorname{asin} \dfrac{x^2 - 1}{x^2 + 1} + \dfrac{\pi}{4} = \operatorname{acos} \dfrac{1}{\sqrt{1 + x^2}}$ and $\operatorname{acot} x = \operatorname{asin} \dfrac{1}{\sqrt{1 + x^2}}$.

xxiv) If $x \leq 0$, then $\operatorname{atan} x + \operatorname{acos} \dfrac{1}{\sqrt{1 + x^2}} = 0$ and $\operatorname{acot} x + \operatorname{asin} \dfrac{1}{\sqrt{1 + x^2}} = \pi$.

xxv) If $|x| \leq 1$, then $\operatorname{asin} x = 2 \operatorname{atan} \dfrac{x}{1 + \sqrt{1 - x^2}}$ and $2 \operatorname{atan} x = \operatorname{asin} \dfrac{2x}{x^2 + 1}$.

xxvi) If $|x| \geq 1$, then $2 \operatorname{atan} x + \operatorname{asin} \dfrac{2x}{x^2 + 1} = (\operatorname{sign} x)\pi$.

xxvii) If $|x| < 1$, then $2 \operatorname{atan} x = \operatorname{atan} \dfrac{2x}{1 - x^2}$.

xxviii) If $|x| > 1$, then $2 \operatorname{atan} x = \operatorname{atan} \dfrac{2x}{1 - x^2} + (\operatorname{sign} x)\pi$.

xxix) If $x \in (-1, 1]$, then $\operatorname{acos} x = 2 \operatorname{atan} \dfrac{\sqrt{1 - x^2}}{1 + x}$.

xxx) $\operatorname{atan} x = 2 \operatorname{atan}(x - \sqrt{x^2 + 1}) + \frac{\pi}{2}$ and $2 \operatorname{atan}(x + \sqrt{x^2 + 1}) = \operatorname{atan} x + \frac{\pi}{2}$.

xxxi) If $x < -1$, then $\operatorname{atan} x + \operatorname{atan} \dfrac{1 - x}{1 + x} = -\dfrac{3\pi}{4}$.

xxxii) If $x > -1$, then $\operatorname{atan} x + \operatorname{atan} \dfrac{1-x}{1+x} = \dfrac{\pi}{4}$.

xxxiii) $\operatorname{acos} \dfrac{1-x^2}{1+x^2} = 2(\operatorname{sign} x) \operatorname{atan} x$.

xxxiv) If $|x| \le \sqrt{2}/2$, then $2 \operatorname{acos} x + \operatorname{asin}(2x\sqrt{1-x^2}) = \pi$ and $2 \operatorname{asin} x = \operatorname{asin}(2x\sqrt{1-x^2})$.

xxxv) If $|x| \ge \sqrt{2}/2$, then $2 \operatorname{asin} x + \operatorname{asin}(2x\sqrt{1-x^2}) = (\operatorname{sign} x)\pi$.

xxxvi) If $|x| \le 1$, then $\operatorname{asin} \dfrac{x+3}{\sqrt{12+4x^2}} = \operatorname{asin} \dfrac{x-3}{\sqrt{12+4x^2}} + \dfrac{2\pi}{3}$.

xxxvii) If $x \ne 0$, then $\operatorname{atan} \dfrac{1}{x} = 2 \operatorname{atan} \dfrac{1}{2x} - \operatorname{atan} \dfrac{1}{4x^3+3x}$.

xxxviii) If $|x| < 1$, then $\operatorname{atan} \frac{\sqrt{5}x}{1-x^2} = \operatorname{atan}\left(\frac{1+\sqrt{5}}{2}\right)x - \operatorname{atan}\left(\frac{1-\sqrt{5}}{2}\right)x$.

xxxix) If $|x| < 1 + \sqrt{2}$, then $4 \operatorname{atan} x = \operatorname{atan} \frac{4x(1-x^2)}{x^4-6x^2+1}$.

xl) If $x > 0$, then $2 \operatorname{atan} x = \frac{\pi}{2} + \operatorname{atan} \frac{1}{2}(x - \frac{1}{x})$.

xli) If $x > 0$, then $2 \operatorname{asin} \dfrac{1}{\sqrt{x+2}} = \operatorname{asin} \dfrac{2\sqrt{x+1}}{x+2} = \operatorname{asec}(1 + \frac{2}{x})$.

xlii) $(\operatorname{sign} x) \operatorname{atan} \sinh x + \operatorname{asin} \operatorname{sech} x = \frac{\pi}{2}$.

xliii) $e^{2(\operatorname{acot} x)J} = \dfrac{xJ-1}{xJ+1} = \dfrac{x^2-1+2xJ}{x^2+1}$.

Remark: See [650, 1514, 1619, 2458] and [1217, pp. 57–59]. **Remark:** Some of these results can be extended to complex numbers.

Fact 2.18.4. The following equalities hold:

i) $\operatorname{atan} \frac{\sqrt{3}}{3} = \frac{\pi}{6}$, $\operatorname{atan} 1 = \frac{\pi}{4}$, $\operatorname{atan} \sqrt{3} = \frac{\pi}{3}$.

ii) $\operatorname{atan}(2-\sqrt{3}) = \frac{\pi}{12}$, $\operatorname{atan}(2+\sqrt{3}) = \frac{5\pi}{12}$.

iii) $\operatorname{atan} \sqrt{5-2\sqrt{5}} = \frac{\pi}{5}$, $\operatorname{atan} \sqrt{5+2\sqrt{5}} = \frac{2\pi}{5}$.

iv) $\operatorname{atan} \frac{3}{4} = 2 \operatorname{atan} \frac{1}{3}$, $2 \operatorname{atan} \frac{1}{2}(\sqrt{5}-1) = \operatorname{atan} 2$.

v) $\operatorname{asin} \frac{3}{5} + \operatorname{asin} \frac{4}{5} = \frac{\pi}{2}$.

vi) $\operatorname{atan} \frac{3}{5} + \operatorname{atan} \frac{1}{4} = \frac{\pi}{4}$.

vii) $\operatorname{atan} \frac{1}{2} + \operatorname{atan} 2 = \frac{\pi}{2}$.

viii) $\operatorname{atan} \frac{3}{4} + \operatorname{atan} \frac{4}{3} = \frac{\pi}{2}$.

ix) $\operatorname{atan} \frac{1}{2} + \operatorname{atan} \frac{1}{3} = \frac{\pi}{4}$.

x) $\operatorname{atan} \frac{3}{4} + \operatorname{atan} \frac{1}{7} = \frac{\pi}{4}$.

xi) $2 \operatorname{atan} \frac{1}{2} - \operatorname{atan} \frac{1}{7} = \frac{\pi}{4}$.

xii) $3 \operatorname{atan} \frac{1}{4} + \operatorname{atan} \frac{5}{99} = \frac{\pi}{4}$.

xiii) $4 \operatorname{atan} \frac{1}{5} - \operatorname{atan} \frac{1}{239} = \frac{\pi}{4}$.

xiv) $\operatorname{atan} \sqrt{7} + \operatorname{acot} \sqrt{7} = \frac{\pi}{2}$.

xv) $5 \operatorname{atan} \frac{1}{7} + 2 \operatorname{atan} \frac{3}{79} = \frac{\pi}{4}$.

xvi) $\operatorname{atan} \frac{1}{2} = \operatorname{atan} \frac{1}{3} + \operatorname{atan} \frac{1}{7}$.

xvii) $\operatorname{atan} \frac{1}{3} = \operatorname{atan} \frac{1}{7} + \operatorname{atan} \frac{2}{11}$.

xviii) $5 \operatorname{atan} \frac{29}{278} + 7 \operatorname{atan} \frac{3}{79} = \frac{\pi}{4}$.

xix) $\operatorname{atan}\frac{2}{11} = \operatorname{atan}\frac{1}{7} + \operatorname{atan}\frac{3}{79}$.

xx) $\operatorname{atan}1 + \operatorname{atan}2 + \operatorname{atan}3 = \pi$.

xxi) $\operatorname{atan}\frac{1}{2} + \operatorname{atan}\frac{1}{5} + \operatorname{atan}\frac{1}{8} = \frac{\pi}{4}$.

xxii) $2\operatorname{atan}\frac{1}{5} + \operatorname{atan}\frac{1}{7} + 2\operatorname{atan}\frac{1}{8} = \frac{\pi}{4}$.

xxiii) $3\operatorname{atan}\frac{1}{4} + \operatorname{atan}\frac{1}{20} + \operatorname{atan}\frac{1}{1985} = \frac{\pi}{4}$.

xxiv) $6\operatorname{atan}\frac{1}{8} + 2\operatorname{atan}\frac{1}{57} + \operatorname{atan}\frac{1}{239} = \frac{\pi}{4}$.

xxv) $8\operatorname{atan}\frac{1}{10} - \operatorname{atan}\frac{1}{239} - 4\operatorname{atan}\frac{1}{515} = \frac{\pi}{4}$.

xxvi) $12\operatorname{atan}\frac{1}{18} + 8\operatorname{atan}\frac{1}{57} - 5\operatorname{atan}\frac{1}{239} = \frac{\pi}{4}$.

xxvii) $\operatorname{atan}\frac{1}{2} + \operatorname{atan}\frac{1}{3} + 4\operatorname{atan}\frac{1}{5} - \operatorname{atan}\frac{1}{239} = \frac{\pi}{2}$.

xxviii) $\operatorname{atan}1 + \operatorname{atan}3 + \operatorname{atan}5 + \operatorname{atan}7 + \operatorname{atan}8 = 2\pi$.

xxix) $22\operatorname{atan}\frac{1}{28} + 2\operatorname{atan}\frac{1}{443} - 5\operatorname{atan}\frac{1}{1393} - 10\operatorname{atan}\frac{1}{11018} = \frac{\pi}{4}$.

xxx) $44\operatorname{atan}\frac{1}{57} + 7\operatorname{atan}\frac{1}{239} - 12\operatorname{atan}\frac{1}{682} + 24\operatorname{atan}\frac{1}{12943} = \frac{\pi}{4}$.

xxxi) $48\operatorname{atan}\frac{1}{49} + 128\operatorname{atan}\frac{1}{57} - 20\operatorname{atan}\frac{1}{239} + 48\operatorname{atan}\frac{1}{110443} = \pi$.

xxxii) $176\operatorname{atan}\frac{1}{57} + 28\operatorname{atan}\frac{1}{239} - 48\operatorname{atan}\frac{1}{682} + 96\operatorname{atan}\frac{1}{12943} = \pi$.

xxxiii) $\operatorname{atan}2 + \operatorname{atan}3 + \operatorname{atan}4 + \operatorname{atan}5 + \operatorname{atan}7 + \operatorname{atan}8 + \operatorname{atan}13 = 3\pi$.

xxxiv) $\operatorname{atan}1 + \operatorname{atan}5 + \operatorname{atan}7 + \operatorname{atan}8 + \operatorname{atan}12 + \operatorname{atan}13 + \operatorname{atan}17 + \operatorname{atan}18 + \operatorname{atan}21 = 4\pi$.

Source: *xxviii*) and the last two equalities are given in [1180].

Fact 2.18.5. Let $n \ge 1$. Then,

$$\sum_{i=1}^{n}\operatorname{atan}\frac{1}{2i^2} = \operatorname{atan}\frac{n}{n+1}, \quad \sum_{i=1}^{n}(-1)^i\operatorname{atan}\frac{2}{i^2} = \operatorname{acot}(n+1) - \operatorname{acot}n - \frac{\pi}{4},$$

$$\sum_{i=0}^{n}\operatorname{atan}\frac{1}{i^2+i+1} = \frac{\pi}{2} - \operatorname{atan}\frac{1}{n+1} = \operatorname{atan}(n+1).$$

Source: [112, pp. 27, 28], [1311, p. 277], and [2180]. **Related:** Fact 13.6.13.

Fact 2.18.6. The following statements hold:

i) Let $x, y \in [-1, 1]$. Then,

$$\operatorname{asin}x + \operatorname{asin}y = \begin{cases} \operatorname{acos}(\sqrt{(1-y^2)(1-x^2)} - xy), & x \ge 0 \text{ and } y > 0,\\ \operatorname{acos}(\sqrt{(1-y^2)(1-x^2)} - xy), & xy < 0 \text{ and } x+y > 0,\\ -\operatorname{acos}(\sqrt{(1-y^2)(1-x^2)} - xy), & x < 0 \text{ and } y < 0,\\ -\operatorname{acos}(\sqrt{(1-y^2)(1-x^2)} - xy), & xy < 0 \text{ and } x+y < 0. \end{cases}$$

ii) Let $x, y \in [-1, 1]$. Then,

$$\operatorname{asin}x + \operatorname{asin}y = \begin{cases} \operatorname{asin}(x\sqrt{1-y^2} + y\sqrt{1-x^2}), & x^2+y^2 \le 1 \text{ or } xy \le 0,\\ -\operatorname{asin}(x\sqrt{1-y^2} + y\sqrt{1-x^2}) + \pi, & x^2+y^2 > 1, x > 0, \text{ and } y > 0,\\ -\operatorname{asin}(x\sqrt{1-y^2} + y\sqrt{1-x^2}) - \pi, & x^2+y^2 > 1, x < 0, \text{ and } y < 0. \end{cases}$$

iii) Let $x, y \in [-1, 1]$. Then,

$$\operatorname{asin} x + \operatorname{asin} y = \begin{cases} \operatorname{atan} \dfrac{x\sqrt{1-y^2} + y\sqrt{1-x^2}}{\sqrt{(1-x^2)(1-y^2)} - xy} + (\operatorname{sign} x)\pi, & x^2 + y^2 > 1 \text{ and } xy > 0, \\ \operatorname{atan} \dfrac{x\sqrt{1-y^2} + y\sqrt{1-x^2}}{\sqrt{(1-x^2)(1-y^2)} - xy}, & \text{else.} \end{cases}$$

iv) Let $x, y \in [-1, 1]$. Then,

$$\operatorname{acos} x + \operatorname{acos} y = \begin{cases} -\operatorname{acos}[xy - \sqrt{(1-y^2)(1-x^2)}] + 2\pi, & x + y < 0, \\ \operatorname{acos}[xy - \sqrt{(1-y^2)(1-x^2)}], & x + y > 0, \end{cases}$$

$$\operatorname{acos} x - \operatorname{acos} y = \operatorname{sign}(y - x) \operatorname{acos}[xy + \sqrt{(1-y^2)(1-x^2)}].$$

v) Let $x, y \in \mathbb{R}$. Then,

$$\operatorname{atan} x + \operatorname{atan} y = \begin{cases} \operatorname{atan} \dfrac{x+y}{1-xy}, & xy < 1, \\ \operatorname{atan} \dfrac{x+y}{1-xy} + (\operatorname{sign} x)\pi, & xy > 1. \end{cases}$$

vi) Let $x, y \in \mathbb{R}$, and assume that $x \neq 0$, $y \neq 0$, and $x + y \neq 0$. Then,

$$\operatorname{acot} x + \operatorname{acot} y = \begin{cases} \operatorname{acot} \dfrac{xy-1}{x+y}, & xy < 0 \text{ or } xy \geq 1, \\ \operatorname{acot} \dfrac{xy-1}{x+y} + (\operatorname{sign} x)\pi, & 0 < xy < 1. \end{cases}$$

vii) Let p, q, r be positive numbers, and assume that $qr = 1 + p^2$. Then,

$$\operatorname{atan} \frac{1}{p} = \operatorname{atan} \frac{1}{p+q} + \operatorname{atan} \frac{1}{p+r}.$$

In particular, $\operatorname{atan} \frac{1}{3} = \operatorname{atan} \frac{1}{5} + \operatorname{atan} \frac{1}{8}$ and $\operatorname{atan} \frac{1}{70} = \operatorname{atan} \frac{1}{99} + \operatorname{atan} \frac{1}{239}$.

viii) Let $x, y \in \mathbb{C}$, and assume that $x \neq 0$, $x + y \neq 0$, and $x^2 + xy + 1 \neq 0$. Then,

$$\operatorname{atan} \frac{1}{x+y} = \operatorname{atan} \frac{1}{x} - \operatorname{atan} \frac{y}{x^2 + xy + 1}.$$

ix) Let $x, y \in \mathbb{C}$, and assume that $x \neq 0$, $x - y \neq 0$, and $x^2 - xy + 1 \neq 0$. Then,

$$\operatorname{atan} \frac{1}{x-y} = \operatorname{atan} \frac{1}{x} + \operatorname{atan} \frac{y}{x^2 - xy + 1}.$$

x) Let $x, y \in \mathbb{R}$, and assume that $x^2 < y^2$. Then,

$$\operatorname{atan} \frac{x+y}{\sqrt{y^2 - x^2}} - \operatorname{atan} \frac{x}{\sqrt{y^2 - x^2}} = \frac{1}{2} \operatorname{acos} \frac{x}{y}.$$

Remark: See [649], [1217, pp. 57, 58], and [2013, p. 166].

Fact 2.18.7. Let x and y be real numbers, and define the four-quadrant inverse tangent function atan2 by

$$\operatorname{atan2}(y, x) \triangleq \arg(x + \jmath y).$$

If $x \neq 0$, then

$$\operatorname{atan} \frac{y}{x} = \operatorname{atan2}[(\operatorname{sign} x)y, |x|].$$

Furthermore,

$$\operatorname{atan2}(y,x)=\begin{cases}0, & y=x=0,\\ \tan^{-1}\frac{y}{x}, & x>0,\\ -\frac{\pi}{2}, & y<0, x=0,\\ \frac{\pi}{2}, & y>0, x=0,\\ -\pi+\tan^{-1}\frac{y}{x}, & y<0, x<0,\\ \pi+\tan^{-1}\frac{y}{x}, & y\ge 0, x<0.\end{cases}$$

Equivalently,

$$\operatorname{atan2}(y,x)=\begin{cases}0, & y=x=0,\\ 2\operatorname{atan}\frac{y}{\sqrt{x^2+y^2}+x}, & \sqrt{x^2+y^2}+x>0,\\ \pi, & y=0, x<0.\end{cases}$$

Finally, if x_1, y_1, x_2, y_2 are real numbers, then there exists $k\in\{-2,0,2\}$ such that

$$\operatorname{atan2}(y_1,x_1)-\operatorname{atan2}(y_2,x_2)=\operatorname{atan2}(y_1x_2-y_2x_1, y_1y_2+x_1x_2)+k\pi.$$

Remark: The range of atan is $(-\frac{\pi}{2},\frac{\pi}{2})$, whereas the range of atan2 is $(-\pi,\pi]$.

Fact 2.18.8. The following statements hold:

i) If $x\in[0,1)$, then

$$\frac{2}{\pi}\left(\sin\frac{\pi x}{2}\right)\operatorname{asin}x\le x^2\le(\sin x)\operatorname{asin}x,$$

$$\left.\begin{array}{r}\dfrac{3x}{2+\sqrt{1-x^2}}\le\dfrac{6(\sqrt{1+x}-\sqrt{1-x})}{4+\sqrt{1+x}+\sqrt{1-x}}\\ \dfrac{(\pi^2/4)x}{\frac{\pi}{2}+\sqrt{1-x^2}}\\ x\end{array}\right\}\le\operatorname{asin}x\le\frac{\pi x}{2+(\pi-2)\sqrt{1-x^2}}\le\left\{\begin{array}{l}\dfrac{\pi x}{2+\sqrt{1-x^2}}\\ \dfrac{x}{1-x^2}.\end{array}\right.$$

ii) If $x\in(0,1)$, then $\operatorname{acos}x<\frac{\sqrt{1-x^2}}{x}$.

iii) If $x>0$, then

$$\left.\begin{array}{r}\dfrac{x}{x^2+1}\le\dfrac{\log(x^2+1)}{x}\\ \dfrac{\pi x}{2x+\pi}\\ x-\frac{1}{3}x^3\end{array}\right\}<\frac{3x}{1+2\sqrt{x^2+1}}<\operatorname{atan}x<\frac{\pi x}{\pi-2+2\sqrt{x^2+1}}$$

$$<\left\{\begin{array}{l}\dfrac{\pi x}{1+2\sqrt{x^2+1}}\\ \dfrac{(\pi+2)x}{2+\pi\sqrt{x^2+1}}<\dfrac{2x}{1+\sqrt{x^2+1}}<x.\end{array}\right.$$

iv) If $x\ge 0$, then

$$\left|\frac{\pi}{4}-\operatorname{atan}x\right|\le\frac{\pi|x-1|}{4\sqrt{x^2+1}}.$$

v) If $x\in(0,\frac{\pi}{2})$, then $x^2<(\tan x)\operatorname{atan}x$.

vi) If $0 < x < y < \frac{\pi}{2}$, then

$$(\tan x)\,\text{atan}\,\frac{x\tan y}{y} < \frac{x^2\tan y}{y}.$$

Source: *i)* is given in [1757, p. 271], [2192], [2294, p. 307], and [3005]; *ii)* is given in [1975]; *iii)* is given in [1567, p. 75], [2192], [2294, p. 309], and [2372, 2280]; *iv)* and *v)* are given in [2294, pp. 308, 309]; *vi)* is given in [2370].

Fact 2.18.9. Let $z \in \mathbb{C}$ be such that all terms below are defined. Then,

$$\frac{\mathrm{d}}{\mathrm{d}z}\,\text{asin}\,z = \frac{1}{\sqrt{1-z^2}}, \quad \frac{\mathrm{d}}{\mathrm{d}z}\,\text{acos}\,z = -\frac{1}{\sqrt{1-z^2}}, \quad \frac{\mathrm{d}}{\mathrm{d}z}\,\text{atan}\,z = \frac{1}{z^2+1},$$

$$\frac{\mathrm{d}}{\mathrm{d}z}\,\text{acsc}\,z = -\frac{1}{z^2\sqrt{1-\frac{1}{z^2}}}, \quad \frac{\mathrm{d}}{\mathrm{d}z}\,\text{asec}\,z = \frac{1}{z^2\sqrt{1-\frac{1}{z^2}}}, \quad \frac{\mathrm{d}}{\mathrm{d}z}\,\text{acot}\,z = -\frac{1}{z^2+1}.$$

2.19 Facts on Equalities and Inequalities for Hyperbolic Functions

Fact 2.19.1. Let $z = x + y\jmath$, where x and y are real numbers, be such that division by zero does not occur in the expressions below. Then, the following statements hold:

i) $\sin z = (\sin x)\cosh y + (\cos x)(\sinh y)\jmath$, $\cos z = (\cos x)\cosh y - (\sin x)(\sinh y)\jmath$.

ii) $\tan z = \dfrac{\sin 2x + (\sinh 2y)\jmath}{\cos 2x + \cosh 2y}$, $\cot z = \dfrac{\sin 2x - (\sinh 2y)\jmath}{\cosh 2y - \cos 2x}$.

iii) $|\sin z| = \sqrt{\sin^2 x + \sinh^2 y} = \sqrt{\frac{1}{2}(\cosh 2y - \cos 2x)}$.

iv) $|\cos z| = \sqrt{\cos^2 x + \sinh^2 y} = \sqrt{\frac{1}{2}(\cosh 2y + \cos 2x)}$, $|\tan z| = \sqrt{\dfrac{\cosh 2y - \cos 2x}{\cosh 2y + \cos 2x}}$.

v) $\sinh z = (\sinh x)\cos y + (\cosh x)(\sin y)\jmath$, $\cosh z = (\cosh x)\cos y - (\sinh x)(\sin y)\jmath$.

vi) $|\sinh z| = \sqrt{\sin^2 y + \sinh^2 x} = \sqrt{\frac{1}{2}(\cosh 2x - \cos 2y)}$.

vii) $|\cosh z| = \sqrt{\cos^2 y + \sinh^2 x} = \sqrt{\frac{1}{2}(\cosh 2x + \cos 2y)}$.

viii) $\tanh z = \dfrac{\sinh 2x + (\sin 2y)\jmath}{\cosh 2x + \cos 2y}$, $\coth z = \dfrac{\sinh 2x - (\sin 2y)\jmath}{\cosh 2x - \cos 2y}$.

ix) $\sinh z = \frac{1}{2}(e^z - e^{-z})$, $\cosh z = \frac{1}{2}(e^z + e^{-z})$, $\tanh z = \dfrac{\sinh z}{\cosh z} = \dfrac{e^z - e^{-z}}{e^z + e^{-z}}$.

x) $\text{csch}\,z = \dfrac{1}{\sinh z} = \dfrac{2}{e^z - e^{-z}}$, $\text{sech}\,z = \dfrac{1}{\cosh z} = \dfrac{2}{e^z + e^{-z}}$, $\coth z = \dfrac{1}{\tanh z} = \dfrac{e^z + e^{-z}}{e^z - e^{-z}}$.

xi) $\sin z\jmath = (\sinh z)\jmath$, $\cos z\jmath = \cosh z$, $\tan z\jmath = (\tanh z)\jmath$.

xii) $\cosh z + \sinh z = e^z$, $\cosh z - \sinh z = e^{-z}$.

xiii) $\csc z\jmath = -(\text{csch}\,z)\jmath$, $\sec z\jmath = \text{sech}\,z$, $\cot z\jmath = -(\coth z)\jmath$.

xiv) $\sinh z\jmath = (\sin z)\jmath$, $\cosh z\jmath = \cos z$, $\tanh z\jmath = (\tan z)\jmath$.

xv) $\text{csch}\,z\jmath = -(\csc z)\jmath$, $\text{sech}\,z\jmath = \sec z$, $\coth z\jmath = -(\cot z)\jmath$.

xvi) $\sinh\log z = \frac{z^2-1}{2z}$, $\cosh\log z = \frac{z^2+1}{2z}$, $\tanh\log z = \frac{z^2-1}{z^2+1}$.

xvii) $\sinh\log\frac{1}{2}(1+\sqrt{5}) = \frac{1}{2}$, $\cosh\log\frac{1}{2}(1-\sqrt{3}\jmath) = \frac{1}{2}$, $\tanh\log\sqrt{3} = \frac{1}{2}$, $\coth\log(1+\sqrt{2}) = \sqrt{2}$.

xviii) $\cosh^2 z - \sinh^2 z = 1$, $\text{sech}^2 z = 1 - \tanh^2 z$, $\text{csch}^2 z = \coth^2 z - 1$.

xix) $\sinh^2\frac{z}{2} = \frac{1}{2}(\cosh z - 1)$, $\cosh^2\frac{z}{2} = \frac{1}{2}(\cosh z + 1)$.

xx) $\tanh\frac{z}{2} = \dfrac{\cosh z - 1}{\sinh z} = \dfrac{\sinh z}{\cosh z + 1}$, $\tanh^2\frac{z}{2} = \dfrac{\cosh z - 1}{\cosh z + 1}$.

xxi) $\sinh 2z = 2(\sinh z)\cosh z = \dfrac{2\tanh z}{1-\tanh^2 z}$, $\coth 2z + \operatorname{csch} 2z = \coth z$.

xxii) $\cosh 2z = 2\cosh^2 z - 1 = 2\sinh^2 z + 1 = \cosh^2 z + \sinh^2 z$.

xxiii) $\tanh 2z = \dfrac{2\tanh z}{1+\tanh^2 z}$, $\sinh 3z = 3\sinh z + 4\sinh^3 z$, $\cosh 3z = -3\cosh z + 4\cosh^3 z$.

xxiv) $\sinh 4z = 4(\sinh^3 z)\cosh z + 4(\cosh^3 z)\sinh z$, $\cosh 4z = \cosh^4 z + 6(\sinh^2 z)\cosh^2 z + \sinh^4 z$.

xxv) If z is real, then $\sqrt[3]{(\operatorname{csch} z + \sinh z)^2} - \sqrt[3]{(\operatorname{sech} z - \cosh z)^2} = \frac{1}{\sqrt[3]{(\operatorname{csch} z + \sinh z)^2(\operatorname{sech} z - \cosh z)^2}}$.

xxvi) Let $n \in \mathbb{Z}$. Then,

$$(\cosh z + \sinh z)^n = \cosh nz + \sinh nz, \quad (\cosh z - \sinh z)^n = \cosh nz - \sinh nz,$$
$$\cosh 2^{n+1}z - \cos 2^{n+1}z = 2(\sin^2 2^n z + \sinh^2 2^n z).$$

xxvii) $2\,\mathrm{atan}\,\frac{\tanh z}{\tan z} = \mathrm{atan}\,\frac{\tanh 2z}{\tan 2z} + \mathrm{atan}\,\frac{\sinh 2z}{\sin 2z} + \varepsilon\pi$, where $\varepsilon \in \{-1, 0, 1\}$. If z is real, then $\varepsilon = 0$.

Remark: See [1524, pp. 117–119] and [2346, p. 118]. *xxv)* is given in [1397]. *xxvi)* and *xxvii)* are given in [512].

Fact 2.19.2. Let x and y be complex numbers such that division by zero does not occur in the expressions below. Then, the following statements hold:

i) $\sinh(x+y) = (\sinh x)(\cosh y) + (\cosh x)\sinh y$, $\sinh(x-y) = (\sinh x)(\cosh y) - (\cosh x)\sinh y$.

ii) $\cosh(x+y) = (\cosh x)(\cosh y) + (\sinh x)\sinh y$, $\cosh(x-y) = (\cosh x)(\cosh y) - (\sinh x)\sinh y$.

iii) $\tanh(x+y) = \dfrac{\tanh x + \tanh y}{1 + (\tanh x)\tanh y}$, $\tanh(x-y) = \dfrac{\tanh x - \tanh y}{1 - (\tanh x)\tanh y}$.

iv) $\coth(x+y) = \dfrac{1 + (\coth x)\coth y}{\coth x + \coth y}$, $\coth(x-y) = \dfrac{1 - (\coth x)\coth y}{\coth x - \coth y}$.

v) $\sinh x + \sinh y = 2\left(\sinh \frac{x+y}{2}\right)\cosh \frac{x-y}{2}$, $\sinh x - \sinh y = 2\left(\cosh \frac{x+y}{2}\right)\sinh \frac{x-y}{2}$.

vi) $\cosh x + \cosh y = 2\left(\cosh \frac{x+y}{2}\right)\cosh \frac{x-y}{2}$, $\cosh x - \cosh y = 2\left(\sinh \frac{x+y}{2}\right)\sinh \frac{x-y}{2}$.

vii) $\tanh x + \tanh y = \dfrac{\sinh(x+y)}{(\cosh x)\cosh y}$, $\tanh x - \tanh y = \dfrac{\sinh(x-y)}{(\cosh x)\cosh y}$.

viii) $\coth x + \coth y = \dfrac{\sinh(x+y)}{(\sinh x)\sinh y}$, $\coth y - \coth x = \dfrac{\sinh(x-y)}{(\sinh x)\sinh y}$.

ix) $2(\sinh x)\sinh y = \cosh(x+y) - \cosh(x-y)$, $2(\cosh x)\cosh y = \cosh(x+y) + \cosh(x-y)$.

x) $2(\sinh x)\cosh y = \sinh(x+y) + \sinh(x-y)$.

xi) $\sinh^2 x - \sinh^2 y = \cosh^2 x - \cosh^2 y = \sinh(x+y)\sinh(x-y) = \frac{1}{2}(\cosh 2x - \cosh 2y)$.

xii) $\sinh^2 x + \cosh^2 y = \cosh(x+y)\cosh(x-y) = \frac{1}{2}(\cosh 2x + \cosh 2y)$.

xiii) $\coth\left(\dfrac{1}{2}\log\dfrac{x}{y}\right) = \dfrac{x+y}{x-y}$.

Fact 2.19.3. Let ω, t, a, b be real numbers. Then,

$$a\cosh\omega t + b\sinh\omega t = \begin{cases} ae^{\omega t}, & a = b,\\ ae^{-\omega t}, & a = -b,\\ \sqrt{|a^2-b^2|}\cosh(\frac{1}{2}\log\left|\frac{a+b}{a-b}\right| + \omega t), & |b| < a,\\ -\sqrt{|a^2-b^2|}\cosh(\frac{1}{2}\log\left|\frac{a+b}{a-b}\right| + \omega t), & |b| < -a,\\ \sqrt{|a^2-b^2|}\sinh(\frac{1}{2}\log\left|\frac{a+b}{a-b}\right| + \omega t), & |a| < b,\\ -\sqrt{|a^2-b^2|}\sinh(\frac{1}{2}\log\left|\frac{a+b}{a-b}\right| + \omega t), & |a| < -b, \end{cases}$$

Source: [2685]. **Related:** Fact 2.16.5.

Fact 2.19.4. The following statements hold:

i) If $x \in (0, \frac{\pi}{2})$, then

$$2 < \left(\frac{\sinh x}{x}\right)^2 + \frac{\sin x}{x}, \qquad \frac{\sinh x}{x} < \frac{x}{\sin x} < \left(\frac{\sinh x}{x}\right)^2.$$

ii) If $x \in \mathbb{R}$ is nonzero, then

$$3 < \frac{2x}{\sinh x} + \frac{x}{\tanh x} < \frac{\sinh x}{x} + \frac{4\tanh x/2}{x},$$

$$\left\{\begin{array}{c} 3 + \frac{3}{20}x^4 - \frac{3}{56}x^6 \\ \dfrac{2\cosh x + 1}{(\cosh x)^{2/3}} \end{array}\right\} < \frac{2\sinh x}{x} + \frac{\tanh x}{x} < 3 + \frac{3}{20}x^3 \sinh x,$$

$$2 < \frac{x}{\sinh x} + \left(\frac{x}{2\tanh x/2}\right)^2 < \frac{\sinh x}{x} + \left(\frac{2\tanh x/2}{x}\right)^2,$$

$$\begin{aligned} 1 &\le (\cosh x)^{1/3} \le (\cosh x/2)^{4/3} \le \frac{\sinh x}{x} \\ &\le \tfrac{1}{3}(\cosh^2 x/2 + 2\cosh x/2) \le \tfrac{1}{3}(2 + \cosh x) \le \cosh x \le \left(\frac{\sinh x}{x}\right)^3. \end{aligned}$$

iii) If $x \in \mathbb{R}$ is nonzero and $p > 0$, then

$$3\operatorname{truth}(p \ge 1) < 2\left(\frac{x}{\sinh x}\right)^p + \left(\frac{x}{\tanh x}\right)^p < 2\left(\frac{\sinh x}{x}\right)^p + \left(\frac{\tanh x}{x}\right)^p,$$

$$2\operatorname{truth}(p \ge 1) < \left(\frac{x}{\sinh x}\right)^p + \left(\frac{x}{2\tanh x/2}\right)^{2p} < \left(\frac{\sinh x}{x}\right)^p + \left(\frac{2\tanh x/2}{x}\right)^{2p},$$

$$3\operatorname{truth}(p \ge 1) < \left(\frac{x}{\sinh x}\right)^p + 2\left(\frac{x}{2\tanh x/2}\right)^p < \left(\frac{\sinh x}{x}\right)^p + 2\left(\frac{2\tanh x/2}{x}\right)^p.$$

iv) If $x \in [-2.676, 2.676]$, then

$$\left(\frac{\sinh x}{x}\right)^2 \le \cosh x.$$

v) If $x > 0$, then

$$2 \le \tfrac{8}{45}x^3 \tanh x + 2 \le \left(\frac{\sinh x}{x}\right)^2 + \frac{\tanh x}{x}.$$

vi) If $x > 0$, then

$$\left.\begin{array}{c} \dfrac{\sinh x}{\sqrt{\sinh^2 x + \cosh^2 x}} \\ (\sin x)\cos x \end{array}\right\} < \tanh x < \left\{\begin{array}{l} x < \sinh x < \frac{1}{2}\sinh 2x \\ 1. \end{array}\right.$$

vii) Let $x > 0$, let $p > 0$, and let $q \ge p + 1$. Then,

$$\frac{p\cosh x}{2} + \frac{(2-p)\sinh x}{2x} < \left(\frac{\sinh x}{x}\right)^q.$$

In particular, if $q \ge 3$, then

$$\cosh x < \left(\frac{\sinh x}{x}\right)^q.$$

viii) If $x \geq 0$, then $\operatorname{atan} x \leq \frac{\pi}{2} \tanh x$.

ix) If $x \geq 0$ and $y \geq 0$,

$$\sinh x + \sinh y \leq \sinh(x+y) \leq \tfrac{1}{2}(\sinh 2x + \sinh 2y),$$

$$(\sinh x) \sinh xy \leq xy \sinh(x + xy), \quad \cosh \tfrac{1}{2}(x+y) \leq \sqrt{(\cosh x)\cosh y},$$

$$\tfrac{1}{2}(\tanh x + \tanh y) \leq \tanh \tfrac{1}{2}(x+y), \quad |x-y|\sqrt{(\sinh x)\sinh y} \leq |\cosh x - \cosh y|.$$

x) If x and y are real numbers and $z = x + y\jmath$, then

$$|\sinh y| \leq \left\{ \begin{matrix} |\sin z| \\ |\cos z| \end{matrix} \right\} \leq \cosh y \leq \cosh |z|,$$

$$|\sin z| \leq \sinh |z| \leq \cosh |z|, \quad \operatorname{csch} |z| \leq |\csc z| \leq \operatorname{csch} |y|.$$

Source: *i*) and *ii*) are given in [1478, 2121, 2369]; *iii*) is given in [603, p. 131], [1300], [1379, p. 71], [2121]; *iv*) are given in [2278]; *v*) are given in [3004]; *vi*) is given in [1567, p. 74] and [2346, p. 125]; *vii*) is given in [3006]; *viii*) is given in [2346, p. 125]; *ix*) is given in [1379, p. 71], [1559], and [2346, p. 125]; *x*) is given in [2346, p. 116]. **Remark:** $(\sinh 0)/0 = 1$.

Fact 2.19.5. The following statements hold:

i) $f\colon [0,\infty) \mapsto [0,\infty)$ defined by $f(x) = \sinh x$ is increasing.

ii) $f\colon [0,\infty) \mapsto [1,\infty)$ defined by $f(x) = \cosh x$ is increasing.

iii) $f\colon [0,\infty) \mapsto [1,\infty)$ defined by $f(x) = (\sinh x)/x$ is increasing.

iv) Define $f\colon (0,\infty) \mapsto (0,\infty)$ by $f(x) = (\cosh x)/x$. Then, there exists a unique $\alpha \in (0,\infty)$ such that $f'(\alpha) = 0$. Furthermore, $\alpha = \coth \alpha \approx 1.1996786$.

Now, let $\alpha > 0$ satisfy $\alpha = \coth \alpha$, and define $\beta \triangleq (\cosh \alpha)/\alpha \approx 1.5088$. Then, the following statements hold:

v) For all $x \in (0,\infty)$, $\beta \leq (\cosh x)/x$.

vi) $f\colon (0,\alpha) \mapsto (\beta,\infty)$ defined by $f(x) = (\cosh x)/x$ is decreasing.

vii) $f\colon (\alpha,\infty) \mapsto (\beta,\infty)$ defined by $f(x) = (\cosh x)/x$ is increasing.

Remark: $(\sinh 0)/0 = 1$.

Fact 2.19.6. Let $z \in \mathbb{C}$ be such that all terms below are defined. Then,

$$\frac{\mathrm{d}}{\mathrm{d}z} \sinh z = \cosh z, \quad \frac{\mathrm{d}}{\mathrm{d}z} \cosh z = \sinh z, \quad \frac{\mathrm{d}}{\mathrm{d}z} \tanh z = \operatorname{sech}^2 z,$$

$$\frac{\mathrm{d}}{\mathrm{d}z} \operatorname{csch} z = -(\operatorname{csch} z)\coth z, \quad \frac{\mathrm{d}}{\mathrm{d}z} \operatorname{sech} z = -(\operatorname{sech} z)\tanh z, \quad \frac{\mathrm{d}}{\mathrm{d}z} \coth z = -\operatorname{csch}^2 z.$$

2.20 Facts on Equalities and Inequalities for Inverse Hyperbolic Functions

Fact 2.20.1. The following statements hold:

i) $\operatorname{asinh}\colon \mathbb{R} \mapsto \mathbb{R}$ is given by

$$\operatorname{asinh} x = \log(x + \sqrt{x^2+1}).$$

ii) $\operatorname{asinh}\colon \mathbb{C} \mapsto \mathbb{R} + \jmath[-\frac{\pi}{2}, \frac{\pi}{2}]$ is given by

$$\operatorname{asinh} z = \log(z + \sqrt{z^2+1}).$$

iii) $\operatorname{acosh}\colon [1,\infty) \mapsto [0,\infty)$ is given by

$$\operatorname{acosh} x = \log(x + \sqrt{x^2-1}).$$

iv) acosh: $\mathbb{C} \mapsto \mathbb{R} + \jmath[0,\pi]$ is given by

$$\operatorname{acosh} z = \log(z + \sqrt{z-1}\,\sqrt{z+1}).$$

v) atanh: $(-1,1) \mapsto (-\infty,\infty)$ is given by

$$\operatorname{atanh} x = \frac{1}{2}\log\frac{1+x}{1-x}.$$

vi) atanh: $\mathbb{C}\backslash\{-1,1\} \mapsto \mathbb{R} + \jmath(-\frac{\pi}{2},\frac{\pi}{2})$ is given by

$$\operatorname{atanh} z = \tfrac{1}{2}[\log(1+z) - \log(1-z)].$$

vii) acsch: $\mathbb{R}\backslash\{0\} \mapsto \mathbb{R}\backslash\{0\}$ is given by

$$\operatorname{acsch} x = \log\left(\sqrt{1+\frac{1}{x^2}} + \frac{1}{x}\right).$$

viii) acsch: $\mathbb{C}\backslash\{0\} \mapsto \mathbb{R} + \jmath(-\frac{\pi}{2},\frac{\pi}{2}]$ is given by

$$\operatorname{acsch} z = \log\left(\sqrt{1+\frac{1}{z^2}} + \frac{1}{z}\right).$$

ix) asech: $(0,1) \mapsto \mathbb{R}$ is given by

$$\operatorname{asech} x = \log\left(\sqrt{\frac{1}{x^2}-1} + \frac{1}{x}\right).$$

x) asech: $\mathbb{C}\backslash\{0\} \mapsto \mathbb{R} + \jmath(-\pi,\pi]$ is given by

$$\operatorname{asech} z = \log\left(\sqrt{\frac{1}{z}-1}\,\sqrt{\frac{1}{z}+1} + \frac{1}{z}\right).$$

xi) acoth: $(-\infty,-1]\cup[1,\infty) \mapsto \mathbb{R}\backslash\{0\}$ is given by

$$\operatorname{acoth} x = \frac{1}{2}\log\frac{x+1}{x-1}.$$

xii) acoth: $\mathbb{C}\backslash\{0\} \mapsto \mathbb{R} + \jmath(-\frac{\pi}{2},\frac{\pi}{2}]$ is given by

$$\operatorname{acoth} z = \frac{1}{2}\left(\log\frac{z+1}{z} - \log\frac{z-1}{z}\right).$$

Fact 2.20.2. If $x \in (-1,0)\cup(0,1)$, then

$$1 < \left(\frac{\operatorname{asin} x}{x}\right)\frac{\operatorname{asinh} x}{x}, \quad 1 < \left(\frac{\operatorname{atan} x}{x}\right)\frac{\operatorname{atanh} x}{x},$$

$$2 < \frac{\operatorname{asin} x}{x} + \frac{\operatorname{asinh} x}{x}, \quad 2 < \frac{\operatorname{atan} x}{x} + \frac{\operatorname{atanh} x}{x},$$

$$\left(\frac{\operatorname{asinh} x}{x}\right)^2 < \frac{\operatorname{atan} x}{x} < \frac{\operatorname{asinh} x}{x} < 1 < \frac{\operatorname{asin} x}{x} < \left(\frac{\operatorname{asin} x}{x}\right)^2 < \frac{\operatorname{atanh} x}{x}.$$

If $x \in (0,\infty)$, then

$$\frac{x-x^3}{\sqrt{x^2+1}} < \operatorname{asinh} x, \quad x^2 < (\sinh x)\operatorname{asinh} x,$$

$$(\operatorname{asinh} x)\operatorname{atan} x < \frac{x^2}{\sqrt{x^2+1}}, \quad \frac{\operatorname{asinh} x}{\sqrt{x^2+1}} + \operatorname{atan} x < \frac{x^3+2x}{x^2+1}.$$

If $x \in (0,1)$, then

$$x^2 < (\tanh x)\,\mathrm{atanh}\, x.$$

If $0 < x < y$, then

$$y(\sinh x)\,\mathrm{asinh}\,\frac{x \sinh y}{y} < x^2 \sinh y.$$

If $0 < x < y < 1$, then

$$\frac{x^2 y}{\mathrm{atanh}\, y} < (\mathrm{atanh}\, x) \tanh \frac{xy}{\mathrm{atanh}\, y}.$$

Source: [2120, 2370, 2953].

Fact 2.20.3. Let $z \in \mathbb{C}$ be such that all terms below are defined. Then,

$$\frac{\mathrm{d}}{\mathrm{d}z}\,\mathrm{asinh}\, z = \frac{1}{\sqrt{z^2+1}}, \quad \frac{\mathrm{d}}{\mathrm{d}z}\,\mathrm{acosh}\, z = \frac{1}{\sqrt{z-1}\,\sqrt{z+1}}, \quad \frac{\mathrm{d}}{\mathrm{d}z}\,\mathrm{atanh}\, z = \frac{1}{1-z^2},$$

$$\frac{\mathrm{d}}{\mathrm{d}z}\,\mathrm{acsch}\, z = -\frac{1}{z^2\sqrt{1+1/z^2}}, \quad \frac{\mathrm{d}}{\mathrm{d}z}\,\mathrm{asech}\, z = -\frac{1}{z\sqrt{1-z^2}}, \quad \frac{\mathrm{d}}{\mathrm{d}z}\,\mathrm{acoth}\, z = \frac{1}{1-z^2}.$$

2.21 Facts on Equalities and Inequalities in Complex Variables

Fact 2.21.1. Let a, b, c be complex numbers, and assume that $a \neq 0$. Then, $z \in \mathbb{C}$ satisfies

$$az^2 + bz + c = 0$$

if and only if

$$z = \frac{y-b}{2a},$$

where

$$y \triangleq \pm\tfrac{\sqrt{2}}{2}\big(\sqrt{|\Delta| + \mathrm{Re}\,\Delta} + \mathrm{sign}(\mathrm{Im}\,\Delta)\sqrt{|\Delta| - \mathrm{Re}\,\Delta}\,\jmath\big), \quad \Delta \triangleq b^2 - 4ac.$$

If, in addition, a, b, c are real, then $z \in \mathbb{C}$ satisfies

$$az^2 + bz + c = 0$$

if and only if

$$z = \frac{-b \pm \sqrt{b^2 - 4ac}}{2a}.$$

Source: [110, pp. 15, 16].

Fact 2.21.2. Let $a, b, c \in \mathbb{C}$, and define $p \triangleq b - a^2/3$ and $q \triangleq c + 2a^3/27 - ab/3$. Then, the following statements hold:

i) There exist unique $z_1, z_2, z_3 \in \mathbb{C}$ such that, for all $z \in \mathbb{C}$, $z^3 + az^2 + bz + c = (z - z_1)(z - z_2)(z - z_3)$.

ii) Let $z, y \in \mathbb{C}$, and assume that $y = z + a/3$. Then, $z^3 + az^2 + bz + c = 0$ if and only if $y^3 + py + q = 0$.

iii) Let $y, u \in \mathbb{C}$, and assume that $u^2 - yu - p/3 = 0$. Then, $y^3 + py + q = 0$ if and only if $u^6 + qu^3 - p^3/27 = 0$. If these conditions are satisfied, then the following statements hold:

a) If $p = 0$, then $u \in \{0, y\}$ and at least three of the six roots of $u^6 + qu^3 - p^3/27$ are zero.

b) If $p \neq 0$ and a, b, c are real, then the multiplicity of each root of $u^6 + qu^3 - p^3/27$ is at least two.

iv) Assume that $u \in \mathbb{C}$ is nonzero and satisfies $u^6 + qu^3 - p^3/27 = 0$. Then, $y \triangleq u - p/(3u)$ satisfies $y^3 + py + q = 0$, and $z \triangleq y - a/3$ satisfies $z^3 + az^2 + bz + c = 0$.

Now, assume that a, b, c are real. Then, the following statements hold:

v) $y^3 + py + q = 0$ has exactly one real root if and only if $27q^2 + 4p^3 > 0$.

vi) $y^3 + py + q = 0$ has three real roots if and only if $27q^2 + 4p^3 < 0$.

Source: [1485, pp. 673–676]. **Remark:** This is *Cardano's formula*. **Remark:** In the case where a, b, c are not real and p is nonzero, the six roots of $u^6 + qu^3 - p^3/27$ may be distinct. However, computing $y = u - p/(3u)$ yields at most three distinct roots of $y^3 + py + q$. **Related:** Fact 2.21.2, Fact 6.10.7, Fact 15.18.4, Fact 15.21.27, and Fact 15.21.29.

Fact 2.21.3. Let z be a complex number with complex conjugate $\overline{z}$, real part $\operatorname{Re} z$, and imaginary part $\operatorname{Im} z$. Then, the following statements hold:

i) $\overline{\overline{z}} = z$, $0 \le |z| = |-z| = |\overline{z}|$, $\operatorname{Re} jz = -\operatorname{Im} z$, $\operatorname{Im} jz = \operatorname{Re} z$.

ii) $-|z| \le \operatorname{Re} z \le |\operatorname{Re} z| \le |z|$, $-|z| \le \operatorname{Im} z \le |\operatorname{Im} z| \le |z|$.

iii) $\operatorname{Re} z = |\operatorname{Re} z| = |z|$ if and only if $\operatorname{Re} z \ge 0$ and $\operatorname{Im} z = 0$.

iv) $\operatorname{Im} z = |\operatorname{Im} z| = |z|$ if and only if $\operatorname{Im} z \ge 0$ and $\operatorname{Re} z = 0$.

v) If $z \ne 0$, then $\overline{z^{-1}} = \overline{z}^{-1}$, $z^{-1} = \overline{z}/|z|^2$, and $|z^{-1}| = 1/|z|$.

vi) If $|z| = 1$, then $z^{-1} = \overline{z}$.

vii) If $z \ne 0$, then $\operatorname{Re} z^{-1} = (\operatorname{Re} z)/|z|^2$.

viii) $\operatorname{Re} z \ne 0$ if and only if $\operatorname{Re} z^{-1} \ne 0$.

ix) If $\operatorname{Re} z \ne 0$, then $|z| = \sqrt{(\operatorname{Re} z)/(\operatorname{Re} z^{-1})}$.

x) $|z^2| = |z|^2 = |\overline{z}^2| = |\overline{z}|^2 = z\overline{z} = (\operatorname{Re} z)^2 + (\operatorname{Im} z)^2 = (\operatorname{Re} \overline{z})^2 + (\operatorname{Im} \overline{z})^2$.

xi) $z^2 \in [0, \infty)$ if and only if $\operatorname{Im} z = 0$.

xii) $z^2 \in (-\infty, 0]$ if and only if $\operatorname{Re} z = 0$.

xiii) $z^2 + \overline{z}^2 + 4(\operatorname{Im} z)^2 = 2|z|^2$, $z^2 + \overline{z}^2 + 2|z|^2 = 4(\operatorname{Re} z)^2$, $z^2 + \overline{z}^2 + 2(\operatorname{Im} z)^2 = 2(\operatorname{Re} z)^2$.

xiv) $z^2 + \overline{z}^2 \le \left\{ \begin{matrix} |z^2 + \overline{z}^2| \\ 2(\operatorname{Re} z)^2 \end{matrix} \right\} \le 2|z|^2$.

xv) If at least two of the quantities $z^2 + \overline{z}^2$, $2(\operatorname{Re} z)^2$, and $2|z|^2$ are equal, then $\operatorname{Im} z = 0$.

xvi) Either $\operatorname{Re} z = 0$ or $\operatorname{Im} z = 0$ if and only if $\overline{z}^2 = z^2$.

xvii) Both $\operatorname{Re} z = 0$ and $\operatorname{Im} z = 0$ if and only if $z = 0$.

xviii) Let n be a positive integer. Then,

$$\sum_{i=0}^{n-1} z^i = 1 + z + \cdots + z^{n-1}.$$

If $z \ne 1$, then

$$\frac{1 - z^n}{1 - z} = \sum_{i=0}^{n-1} z^i = 1 + z + \cdots + z^{n-1}.$$

Furthermore,

$$\lim_{z \to 1} \frac{1 - z^n}{1 - z} = n.$$

xix) Let $n, k, m \ge 1$, and define

$$S \triangleq \sum_{i=0}^{n-1} e^{(m\pi i/k)j}.$$

If $2k \nmid m$, then

$$S = \frac{1 - e^{(mn\pi/k)j}}{1 - e^{(m\pi/k)j}}.$$

Furthermore,

$$S = \sum_{i=0}^{n-1} \cos \frac{m\pi i}{k} = \begin{cases} n, & 2k|m, \\ 1, & 2k\nmid m \text{ and } 2k|m(n-1), \\ 0, & 2k\nmid m \text{ and } 2k|mn. \end{cases}$$

In all other cases, $\operatorname{Im} S \neq 0$.

xx) Let $n, m \geq 1$, and define

$$S \triangleq \sum_{i=0}^{n-1} e^{(2m\pi i/n)J}.$$

Then,

$$S = \begin{cases} n, & n|m, \\ 0, & n\nmid m. \end{cases}$$

xxi) Let $n \geq 1$, and let k be an integer. Then,

$$\sum_{i=k}^{k+n-1} e^{(2\pi i/n)J} = \begin{cases} 1, & n = 1, \\ 0, & n \geq 2. \end{cases}$$

Source: *xx)* is given in [672]; *xx)* implies *xxi)*. **Remark:** A matrix version of *i)* is given in [2613].

Fact 2.21.4. Let $n \geq 1$, and let z be a complex number. Then,

$$\prod_{i=0}^{n-1}[z - e^{(2\pi i/n)J}] = \prod_{i=1}^{n}[z - e^{(2\pi i/n)J}] = z^n - 1, \quad \prod_{i=0}^{n-1}[1 - e^{(2\pi i/n)J}z] = \prod_{i=1}^{n}[1 - e^{(2\pi i/n)J}z] = 1 - z^n.$$

Now, define

$$p_n(z) \triangleq \prod [z - e^{(2\pi i/n)J}],$$

where the product is taken over all $i \in \{1, \ldots, n\}$ such that $\gcd\{i, n\} = 1$. Then, p_n is monic, all of its coefficients are integers, and

$$z^n - 1 = \prod p_i(z),$$

where the product is taken over all $i \in \{1, \ldots, n\}$ such that $i|n$. In particular,

$$p_1(z) = z - 1, \quad p_2(z) = z + 1, \quad z_3(z) = z^2 + z + 1, \quad p_4(z) = z^2 + 1,$$
$$p_5(z) = z^4 + z^3 + z^2 + z + 1, \quad p_6(z) = z^2 - z + 1, \quad p_7(z) = z^6 + z^5 + z^4 + z^3 + z^2 + z + 1,$$
$$p_8(z) = z^4 + 1, \quad p_9(z) = z^6 + z^3 + 1, \quad p_{10}(z) = z^4 - z^3 + z^2 - z + 1,$$
$$p_{11}(z) = z^{10}+z^9+z^8+z^7+z^6+z^5+z^4+z^3+z^2+z+1, \quad p_{12}(z) = z^4-z^2+1, \quad p_{13}(z) = z^{12}+z^{11}+p_{11}(z),$$
$$z^2 - 1 = p_1(z)p_2(z), \quad z^3 - 1 = p_1(z)p_3(z), \quad z^4 - 1 = p_1(z)p_4(z), \quad z^5 - 1 = p_1(z)p_5(z),$$
$$z^6 - 1 = p_1(z)p_2(z)p_3(z)p_6(z), \quad z^7 - 1 = p_1(z)p_7(z), \quad z^8 - 1 = p_1(z)p_2(z)p_4(z)p_8(z),$$
$$z^9 - 1 = p_1(z)p_3(z)p_9(z), \quad z^{10} - 1 = p_1(z)p_2(z)p_5(z), \quad z^{11} - 1 = p_1(z)p_{11}(z),$$
$$z^{12} - 1 = p_1(z)p_2(z)p_3(z)p_4(z)p_6(z)p_{12}(z), \quad z^{13} - 1 = p_1(z)p_{13}(z).$$

Furthermore, the following statements hold:

i) If n is prime, then $p_n(z) = \sum_{i=0}^{n} z^i$.

ii) If n is odd, then $p_{2n}(z) = p_n(-z)$.

iii) If m and n are distinct primes, then all of the coefficients of p_{mn} are either -1, 0, or 1.

iv) Let $n \geq 2$, and define $m \triangleq \deg p_n$. Then, $p_n(z) = z^m p_n(1/z)$.

v) If m and n are distinct primes, then $p_n(z^m) = p_{mn}(z)p_n(z)$.

vi) If m and n are distinct primes, then $(z^{mn} - 1)p_{mn}(z) = p_m(z^n)p_n(z^m)(z - 1)$.

Source: [126, 577]. **Remark:** p_n is the nth *cyclotomic polynomial.* **Remark:** p_{105} is the first cyclotomic polynomial that has a coefficient that is neither 1, −1, nor 0. In particular, the coefficient of z^7 in p_{105} is 2.

Fact 2.21.5. Let $n, m \geq 1$, and let z be a complex number. Then,

$$\prod_{i=1}^{n}\prod_{j=1}^{m}(z - e^{\frac{2\pi i}{n}J}e^{\frac{2\pi j}{m}J}) = (z^{\mathrm{lcm}\,\{n,m\}} - 1)^{\mathrm{gcd}\,\{n,m\}}.$$

Source: [107, pp. 60, 319, 321].

Fact 2.21.6. Let z_1 and z_2 be complex numbers. If $n \geq 0$, then

$$\prod_{i=0}^{2n}(z_1 + e^{[2i\pi/(2n+1)]J}z_2) = z_1^{2n+1} + z_2^{2n+1}.$$

If $n \geq 1$, then

$$\prod_{i=0}^{2n-1}(z_1 + e^{(2i\pi/n)J}z_2) = (z_1^n - (-1)^n z_2^n)^2.$$

Source: [109, pp. 298–300].

Fact 2.21.7. Let z be a complex number, and assume that $|z| \leq 1$. Then,

$$\tfrac{4}{5}|z| \leq (\sin 1)|z| \leq |\sin z| \leq \tfrac{1}{2}(e - \tfrac{1}{e})|z| \leq \tfrac{6}{5}|z|, \quad \tfrac{1}{2}|z| \leq (\cos 1) \leq |\cos z| \leq \tfrac{1}{2}(e + \tfrac{1}{e}) \leq \tfrac{8}{5},$$
$$\tfrac{3}{4}|z| \leq \tfrac{e^2-1}{e^2+1}|z| \leq |\tan z| \leq (\tan 1)|z| \leq \tfrac{8}{5}|z|, \quad \tfrac{3}{5}|z| \leq (1 - \tfrac{1}{e})|z| \leq |e^z - 1| \leq (e-1)|z| \leq \tfrac{7}{4}|z|.$$

Source: [2346, p. 116] and [2994].

Fact 2.21.8. Let z_1 and z_2 be complex numbers, and let α be a real number. Then, the following statements hold:

i) $z_1z_2 = 0$ if and only if either $z_1 = 0$ or $z_2 = 0$.

ii) $\overline{z_1z_2} = \overline{z_1}\,\overline{z_2}$.

iii) $|z_1z_2| = |z_1||z_2|$.

iv) If $z_2 \neq 0$, then $|z_1/z_2| = |z_1|/|z_2|$.

v) $\big||z_1| - |z_2|\big| \leq |z_1 + z_2| \leq |z_1| + |z_2|$.

vi) $|z_1 + z_2| = |z_1| + |z_2|$ if and only if $\mathrm{Re}(z_1\overline{z_2}) = |z_1||z_2|$.

vii) $|z_1 + z_2| = |z_1| + |z_2|$ if and only if there exists $\alpha \geq 0$ such that either $z_1 = \alpha z_2$ or $z_2 = \alpha z_1$.

viii) $\big||z_1| - |z_2|\big| \leq |z_1 - z_2|$.

ix) $\big||z_1| - |z_2|\big| - |z_1 - z_2|$ if and only if there exists $\alpha \geq 0$ such that either $z_1 = \alpha z_2$ or $z_2 = \alpha z_1$.

x) $|1 + \overline{z_1}z_2|^2 = (1 - |z_1|^2)(1 - |z_2|^2) + |z_1 + z_2|^2 = (1 + |z_1|^2)(1 + |z_2|^2) - |z_1 - z_2|^2$.

xi) $|z_1 - z_2|^2 \leq (1 + |z_1|^2)(1 + |z_2|^2)$.

xii) If z_1 and z_2 are nonzero, then $\frac{1}{2}\Big|z_1 - z_2 + \Big|\frac{z_2}{z_1}\Big|z_1 - \Big|\frac{z_1}{z_2}\Big|z_2\Big| = \frac{1}{2}(|z_1| + |z_2|)\Big|\frac{z_1}{|z_1|} - \frac{z_2}{|z_2|}\Big| \leq |z_1 - z_2|$.

xiii) $2\,\mathrm{Re}(z_1z_2) \leq 2|\mathrm{Re}(z_1z_2)| \leq 2|z_1z_2| \leq |z_1|^2 + |z_2|^2$.

xiv) $2|z_1z_2| = |z_1|^2 + |z_2|^2$ if and only if $|z_1| = |z_2|$.

xv) $2\,\mathrm{Re}(z_1z_2) = |z_1|^2 + |z_2|^2$ if and only if $z_1 = \overline{z}_2$.

xvi) $|z_1 + z_2|^2 = |z_1|^2 + |z_2|^2 + 2\,\mathrm{Re}\,z_1\overline{z}_2$.

xvii) $|z_1 + z_2|^2 = |z_1|^2 + |z_2|^2$ if and only if $\mathrm{Re}\,z_1\overline{z}_2 = 0$.

xviii) If $\alpha \neq 0$, then $\frac{1}{\alpha}(\alpha z_1 + z_2)^2 + (z_1 - z_2)^2 = (1+\alpha)z_1^2 + (1+\frac{1}{\alpha})z_2^2$.

xix) $(z_1 + z_2)^2 + (z_1 - z_2)^2 = 2z_1^2 + 2z_2^2$.

xx) If $\alpha \neq 0$, then $\frac{1}{\alpha}|\alpha z_1 + z_2|^2 + |z_1 - z_2|^2 = (1+\alpha)|z_1|^2 + (1+\frac{1}{\alpha})|z_2|^2$.

xxi) $|z_1 + z_2|^2 + |z_1 - z_2|^2 = 2|z_1|^2 + 2|z_2|^2$.

xxii) Let a_1 and a_2 be real numbers, and assume that a_1, a_2, and $a_1 + a_2$ are nonzero. Then,

$$\frac{|z_1|^2}{a_1} + \frac{|z_2|^2}{a_2} = \frac{|z_1 + z_2|^2}{a_1 + a_2} + \frac{|a_1 z_2 - a_2 z_1|^2}{a_1 a_2 (a_1 + a_2)}.$$

xxiii) $(z_1 + z_2)^2 + (z_1 + z_2 \jmath)^2 \jmath = (z_1 - z_2)^2 + (z_1 - z_2 \jmath)^2 \jmath$.

xxiv) $4\overline{z_1} z_2 = |z_1 + z_2|^2 - |z_1 - z_2|^2 + (|z_1 - z_2 \jmath|^2 - |z_1 + z_2 \jmath|^2)\jmath$.

xxv) Let $n \geq 3$. Then,

$$\overline{z_1} z_2 = \frac{1}{n}\sum_{i=0}^{n-1} e^{-(2i\pi/n)\jmath}|z_1 + e^{(2i\pi/n)\jmath} z_2|^2.$$

xxvi) $2\overline{z_1} z_2 = |z_1 + z_2|^2 - |z_1|^2 - |z_2|^2 + (|z_1|^2 + |z_2|^2 - |z_1 + z_2 \jmath|^2)\jmath$.

xxvii) If $z_1 z_2 \neq 0$ and $\arg z_1 + \arg z_2 \in (-\pi, \pi]$, then $\arg z_1 z_2 = \arg z_1 + \arg z_2$. In particular, if $z_1 z_2 \neq 0$, $\operatorname{Re} z_1 > 0$, and $\operatorname{Re} z_2 > 0$, then $\arg z_1 z_2 = \arg z_1 + \arg z_2$.

xxviii) $|z_1| + |z_2| \leq |1 + z_1| + |1 + z_2| + |1 + z_1 z_2|$.

xxix) $(|z_1 - z_2| - |z_1 z_2 - 1|)^2 \leq |(z_1^2 - 1)(z_2^2 - 1)| \leq (|z_1 - z_2| + |z_1 z_2 - 1|)^2$.

xxx) $\dfrac{|z_1 + z_2|}{1 + |z_1 + z_2|} \leq \dfrac{|z_1|}{1 + |z_1|} + \dfrac{|z_2|}{1 + |z_2|}$.

xxxi) If z_1 and z_2 are nonzero, then

$$\left|\frac{z_1}{|z_1|} - |z_1| z_2\right| = \left|\frac{z_2}{|z_2|} - |z_2| z_1\right|.$$

xxxii) If $p \in [1, 2]$, then

$$|z_1 + z_2|^p \leq 2^{p-1}(|z_1|^p + |z_2|^p) \leq |z_1 + z_2|^p + |z_1 - z_2|^p \leq 2(|z_1|^p + |z_2|^p).$$

xxxiii) If $p \geq 2$, then

$$2(|z_1|^p + |z_2|^p) \leq |z_1 + z_2|^p + |z_1 - z_2|^p \leq 2^{p-1}(|z_1|^p + |z_2|^p).$$

xxxiv) If $p \in (1, 2]$, $q \geq 2$, and $1/p + 1/q = 1$, then

$$|z_1 + z_2|^q + |z_1 - z_2|^q \leq 2(|z_1|^p + |z_2|^p)^{q-1}.$$

xxxv) If $p \geq 2$, $q \in (1, 2]$, and $1/p + 1/q = 1$, then

$$2(|z_1|^p + |z_2|^p)^{q-1} \leq |z_1 + z_2|^q + |z_1 - z_2|^q.$$

xxxvi) If $p, q > 1$ and $1/p + 1/q = 1$, then

$$|z_1 + z_2|^2 \leq p|z_1|^2 + q|z_2|^2.$$

Equality holds if and only if $z_2 = (p-1)z_1$.

xxxvii) Let $n \geq 1$. If $z_1 \neq z_2$, then

$$\frac{z_1^n - z_2^n}{z_1 - z_2} = z_1^{n-1} + z_2 z_1^{n-2} + \cdots + z_2^{n-1}, \quad \lim_{z_2 \to z_1} \frac{z_1^n - z_2^n}{z_1 - z_2} = n z_1^{n-1}.$$

xxxviii) Let $p \in [0, 1]$. Then,

$$||z_1|^p - |z_2|^p| \leq ||z_1| - |z_2||^p \leq |z_1 - z_2|^p.$$

xxxix) Let $p \in [1, \infty)$. Then,

$$||z_1| - |z_2||^p \le \begin{cases} ||z_1|^p - |z_2|^p| \\ |z_1 - z_2|^p. \end{cases}$$

xl) Let $n \ge 1$. Then,

$$||z_1| - |z_2||^n \le \begin{cases} ||z_1|^n - |z_2|^n| \le |z_1^n - z_2^n| \\ |z_1 - z_2|^n. \end{cases}$$

xli) If $\alpha \in [0, 1]$, then

$$|z_1 - z_2|^2 + |\alpha z_1 + z_2|^2 \le (1 + \alpha)|z_1|^2 + (1 + \tfrac{1}{\alpha})|z_2|^2.$$

xlii) If either $\alpha < 0$ or $\alpha \ge 1$, then

$$(1 + \alpha)|z_1|^2 + (1 + \tfrac{1}{\alpha})|z_2|^2 \le |z_1 - z_2|^2 + |\alpha z_1 + z_2|^2.$$

xliii) If $r \ge 1$, then

$$|e^{z_1} - e^{z_2}| \le |z_1 - z_2|[\tfrac{1}{2}(|e^{z_1}|^r + |e^{z_2}|^r)]^{1/r}.$$

xliv) If $x \in \mathbb{R}$, then $|e^{xȷ} - 1| = 2|\sin \frac{x}{2}|$ and $|e^{xȷ} - 1 - xȷ| \le \frac{1}{2}x^2$.

xlv) If $p, q \in (1, \infty)$ satisfy $1/p + 1/q = 1$, then

$$|z_1 z_2| \le \frac{|z_1|^p}{p} + \frac{|z_2|^q}{q}.$$

Equality holds if and only if $|z_1|^p = |z_2|^q$.

Remark: Matrix versions of *iii), v), vii)–ix)* are given in [2613]; *x)* is given in [110, p. 19] and [2933]; *xii)* is the *Dunkl-Williams inequality*, see [935, p. 43], [936, p. 52], and *ii)* of Fact 11.8.3; *xvii)* is the *Pythagorean theorem*; *xx)* is the *generalized parallelogram law*, see [1095]; *xxi)* is the *parallelogram law*, see [978] and Fact 11.8.3; *xxii)* is given in [2060, p. 315] and [2252] and follows from Fact 2.12.22; *xxiv)* is the *polarization identity*, see [828, p. 54], [2112, p. 276], and Fact 11.8.3; *xxv)* is given in [828, p. 54]; *xxvi)* is given in [2238, p. 261]; *xxvii)* is given in [582, p. 23]; *xxviii)* is given in [2682]; *xxix)* is given in [365]; *xxx)* is given in [993, p. 183]; *xxxii)–xxxv)* are due to J. A. Clarkson; see [1419], [2061, p. 536], [2294, p. 253], Fact 11.8.3, and Fact 11.10.54; *xxxvi)* is given in [1984]; *xli)* and *xlii)* are given in [1095]; *xliii)* is given in [942]; *xliv)* is given in [968, p. 274]. **Remark:** The absolute value $|z| = |x + yȷ|$, where x and y are real, is equal to the Euclidean norm $\|[\begin{smallmatrix} x \\ y \end{smallmatrix}]\|_2$. Hence, results involving the Euclidean norm on $\mathbb{R}^2$ can be recast in terms of complex numbers. **Remark:** *xxxvi)* is *Bohr's inequality*. Extensions are given in Fact 2.21.23 and Fact 10.11.83. **Remark:** The lower bounds for $|z_1 - z_2|$ given by *viii)* and *xii)* cannot be ordered. **Remark:** *xlv)* is Young's inequality. See Fact 2.2.50.

Fact 2.21.9. Let z_1 and z_2 be nonzero complex numbers. Then,

$$\frac{|z_1 - z_2| - \big||z_1| - |z_2|\big|}{\min\{|z_1|, |z_2|\}} \le \left|\frac{z_1}{|z_1|} - \frac{z_2}{|z_2|}\right| \le \left\{\begin{array}{c} \dfrac{|z_1 - z_2| + \big||z_1| - |z_2|\big|}{\max\{|z_1|, |z_2|\}} \\[2ex] \dfrac{2|z_1 - z_2|}{|z_1| + |z_2|} \end{array}\right\}$$

$$\le \left\{\begin{array}{c} \dfrac{2|z_1 - z_2|}{\max\{|z_1|, |z_2|\}} \\[2ex] \dfrac{2(|z_1 - z_2| + \big||z_1| - |z_2|\big|)}{|z_1| + |z_2|} \end{array}\right\} \le \frac{4|z_1 - z_2|}{|z_1| + |z_2|}.$$

Source: Fact 11.7.11 and Fact 11.8.3. **Remark:** $\left|\frac{z_1}{|z_1|} - \frac{z_2}{|z_2|}\right| \le \frac{2|z_1 - z_2|}{|z_1|+|z_2|}$ is given by *xii)* in Fact 2.21.8.

Fact 2.21.10. Let $w, z \in \mathbb{C}$, and assume that $|w| < 1$ and $|z| < 1$. Then,

$$|(1-w)^{-1} - (1-z)^{-1}| \le \begin{cases} \frac{1}{2}|w-z|[(1-\frac{1}{2}|w+z|)^{-2} + \frac{1}{2}(1-|w|)^{-2} + \frac{1}{2}(1-|z|)^{-2}], \\ |w-z|(1-|w|)^{-1}(1-|z|)^{-1}, & |w| \ne |z|, \\ |w-z|(1-|w|)^{-2}, & |w| = |z|, \end{cases}$$

$$|\log(1-w)^{-1} - \log(1-z)^{-1}| \le \begin{cases} \frac{1}{2}|w-z|[(1-\frac{1}{2}|w+z|)^{-1} + \frac{1}{2}(1-|w|)^{-1} + \frac{1}{2}(1-|z|)^{-1}], \\ \dfrac{|w-z|}{|w|-|z|}[\log(1-|w|)^{-1} - \log(1-|z|)^{-1}], & |w| \ne |z|, \\ |w-z|(1-|w|)^{-1}, & |w| = |z|. \end{cases}$$

Source: [942].

Fact 2.21.11. Let z_1, z_2, z_3 be complex numbers. Then, the following statements hold:

i) $(z_1 - z_3)^2 + (z_3 - z_2)^2 = \frac{1}{2}(z_1 - z_2)^2 + 2[z_3 - \frac{1}{2}(z_1 + z_2)]^2$.

ii) $(z_1 + z_2)^2 + (z_2 + z_3)^2 + (z_3 + z_1)^2 = z_1^2 + z_2^2 + z_3^2 + (z_1 + z_2 + z_3)^2$.

iii) $(z_1 - z_2)^2 + (z_2 - z_3)^2 + (z_3 - z_1)^2 + (z_1 + z_2 + z_3)^2 = 3(z_1^2 + z_2^2 + z_3^2)$.

iv) $(z_1 + z_2 - z_3)^2 + (z_2 + z_3 - z_1)^2 + (z_3 + z_1 - z_2)^2 + (z_1 + z_2 + z_3)^2 = 4(z_1^2 + z_2^2 + z_3^2)$.

v) $|z_1 - z_3|^2 + |z_3 - z_2|^2 = \frac{1}{2}|z_1 - z_2|^2 + 2|z_3 - \frac{1}{2}(z_1 + z_2)|^2$.

vi) $|z_1 + z_2|^2 + |z_2 + z_3|^2 + |z_3 + z_1|^2 = |z_1|^2 + |z_2|^2 + |z_3|^2 + |z_1 + z_2 + z_3|^2$.

vii) $|z_1 - z_2|^2 + |z_2 - z_3|^2 + |z_3 - z_1|^2 + |z_1 + z_2 + z_3|^2 = 3(|z_1|^2 + |z_2|^2 + |z_3|^2)$.

viii) $|z_1 + z_2 - z_3|^2 + |z_2 + z_3 - z_1|^2 + |z_3 + z_1 - z_2|^2 + |z_1 + z_2 + z_3|^2 = 4(|z_1|^2 + |z_2|^2 + |z_3|^2)$.

ix) $|z_1 + z_2| + |z_2 + z_3| + |z_3 + z_1| \le |z_1| + |z_2| + |z_3| + |z_1 + z_2 + z_3|$.

x) $|z_1| + |z_2| + |z_3| \le |z_1 + z_2 - z_3| + |z_2 + z_3 - z_1| + |z_3 + z_1 - z_2|$.

xi) If z_1, z_2, z_3 are nonzero and $z_1^7 + z_2^7 + z_3^7 = z_1 + z_2 + z_3 = 0$, then $|z_1| = |z_2| = |z_3|$.

Source: The first four equalities are given by Fact 2.3.2; *v)* is the *Appolonius identity*, see [2238, p. 260]; *vi)* is given in [110, p. 19]; *vii)* is given in [2238, p. 244]; *viii)* is given in [110, p. 19] and [978]; *ix)* is *Hlawka's inequality*, see Fact 1.21.9 and Fact 11.8.3; *x)* is given in [993, p. 181]; *xi)* is given in [110, pp. 186, 187]. **Remark:** If z_1, z_2, z_3 are positive numbers that represent the lengths of the sides of a triangle, then equality holds in *x)*. See Fact 5.2.14.

Fact 2.21.12. Let z_1, z_2, z_3, z_4 be complex numbers. Then,

$$(z_1 - z_3)(z_2 - z_4) = (z_1 - z_2)(z_3 - z_4) + (z_1 - z_4)(z_2 - z_3).$$

Consequently,

$$|z_1 - z_3|\,|z_2 - z_4| \le |z_1 - z_2|\,|z_3 - z_4| + |z_1 - z_4|\,|z_2 - z_3|.$$

Source: [2238, p. 473]. **Remark:** This is *Ptolemy's inequality*.

Fact 2.21.13. Let $n \ge 1$, define $p \in \mathbb{R}[s]$ by $p(s) = s^n + 1$, define $\{\lambda_1, \ldots, \lambda_n\} \triangleq \text{roots}(p)$, and let $a \in (0, \infty)$. Then,

$$\frac{1}{n}\sum_{i=1}^{n} \frac{1}{|\lambda_i - a|^2} = \frac{1}{(a^n + 1)^2}\sum_{i=0}^{n-1} a^{2i}.$$

Source: [2547].

Fact 2.21.14. Let $z_1, \ldots, z_n \in \mathbb{C}$, let $\alpha_1, \ldots, \alpha_n$ be nonzero real numbers, and assume that

$\sum_{i=1}^n \alpha_i = 1$. Then,

$$\sum_{i=1}^n \frac{1}{\alpha_i}|z_i|^2 - \left|\sum_{i=1}^n z_i\right|^2 = \sum_{1\le i\le j\le n} \frac{\alpha_i}{\alpha_j}\left|\frac{\alpha_j}{\alpha_i}z_i - z_j\right|^2.$$

Source: [1095, 2989]. **Related:** Fact 10.11.85 and Fact 11.8.4.

Fact 2.21.15. Let $z, z_1, \dots, z_n$ be complex numbers. Then,

$$\frac{1}{n}\sum_{i=1}^n |z - z_i|^2 = \left|z - \frac{1}{n}\sum_{i=1}^n z_i\right|^2 + \frac{1}{n^2}\sum_{1\le i<j\le n}|z_i - z_j|^2, \quad \frac{1}{n}\sum_{1\le i<j\le n}|z_i - z_j|^2 \le \sum_{i=1}^n |z_i|^2.$$

Source: [110, p. 146] and [2252].

Fact 2.21.16. Let $n \ge 3$, and let $z_1, \dots, z_n$ be distinct complex numbers. Then, the following statements are equivalent:

i) In the ordering listed, $z_1, \dots, z_n$ lie on the same line or circle in the complex plane.

ii)

$$\frac{|z_n - z_2|}{|z_2 - z_1||z_n - z_1|} = \sum_{i=2}^{n-1} \frac{|z_{i+1} - z_i|}{|z_{i+1} - z_1||z_i - z_1|}.$$

Fact 2.21.17. Let $z_1, \dots, z_n$ be complex numbers. Then,

$$\left(\sum_{i=1}^n |\mathrm{Re}\, z_i| - \left|\sum_{i=1}^n \mathrm{Re}\, z_i\right|\right)^2 \le \left(\sum_{i=1}^n |z_i|\right)^2 - \left|\sum_{i=1}^n z_i\right|^2.$$

Source: [2548].

Fact 2.21.18. Let $z_1, \dots, z_n$ be complex numbers, and assume that there exists $\phi \in [0, \frac{\pi}{2}]$ such that, for all $i \in \{1, \dots, n\}$, $|\arg z_i| \le \phi$. Then,

$$(\cos\phi)\left|\prod_{i=1}^n z_i\right|^{1/n} \le \frac{1}{n}\left|\sum_{i=1}^n z_i\right|.$$

Remark: This is an extension of the arithmetic-mean–geometric-mean inequality to complex numbers. **Source:** [2527, p. 36].

Fact 2.21.19. Let $z_1, \dots, z_n$ be complex numbers, and assume that $r \triangleq \max\{|z_1|, \dots, |z_n|\} < 1$. Then, there exists $z \in \mathbb{C}$ such that $|z| \le r$ and $(z+1)^n = \prod_{i=1}^n (z_i + 1)$. **Source:** [2527, p. 103].

Fact 2.21.20. Let $n \ge 2$, let $z_1, \dots, z_n \in \mathbb{C}$, let $a_1, \dots, a_n$ be nonzero real numbers, and assume that $\sum_{i=1}^n a_i \ne 0$. Then,

$$\sum_{i=1}^n \frac{|z_i|^2}{a_i} = \frac{\left|\sum_{i=1}^n z_i\right|^2}{\sum_{i=1}^n a_i} + \frac{1}{\sum_{i=1}^n a_i}\sum_{1\le i<j\le n}\frac{|a_i z_j - a_j z_i|^2}{a_i a_j}.$$

If, in addition, $a_1, \dots, a_n$ are positive numbers, then

$$\frac{1}{\sum_{i=1}^n a_i}\sum_{1\le i<j\le n}\frac{|a_i z_j - a_j z_i|^2}{a_i a_j} \le \sum_{i=1}^n \frac{|z_i|^2}{a_i}.$$

Source: [2252]. **Related:** Fact 2.12.19. The case $n = 2$ is given by Fact 2.21.8.

Fact 2.21.21. Let $z_1, \dots, z_n$ be complex numbers and, for all $i \in \{1, \dots, n\}$, let $z_i = r_i e^{\phi_i J}$, where $r_i \ge 0$ and $\phi_i \in \mathbb{R}$. Furthermore, assume that there exist $\theta_1, \theta_2 \in \mathbb{R}$ such that $0 < \theta_2 - \theta_1 < \pi$ and such that, for all $i \in \{1,\ldots,n\}$, $\theta_1 \le \phi_i \le \theta_2$. Then,

$$[\cos \tfrac{1}{2}(\theta_2-\theta_1)]\sum_{i=1}^n |z_i| \le \left|\sum_{i=1}^n z_i\right|.$$

Source: [938]. **Credit:** M. Petrovich.

Fact 2.21.22. Let $z_1,\ldots,z_n$ be complex numbers. Then,

$$\sum_{i=1}^n |z_i| \le \pi \max \left|\sum_{i\in\mathcal{S}} z_i\right|,$$

where the maximum is taken over all subsets $\mathcal{S}$ of $\{1,\ldots,n\}$. **Source:** [2527, p. 224].

Fact 2.21.23. Let $z_1,\ldots,z_n$ be complex numbers, let $p_1,\ldots,p_n$ be positive numbers, and let $r>1$. Then,

$$\left|\sum_{i=1}^n z_i\right|^r \le \left(\sum_{i=1}^n p_i^{\frac{1}{1-r}}\right)^{r-1} \sum_{i=1}^n p_i|z_i|^r.$$

Source: [1984]. **Related:** The special case $n=r=2$ is Bohr's inequality. See Fact 2.21.8. A matrix version of this result is given in Fact 11.10.56.

Fact 2.21.24. Let $n,m\ge 1$, and let $z_1,\ldots,z_m$ be complex numbers. Then,

$$\prod \prod_{j=1}^m z_i^{i_j} = \left(\prod_{i=1}^m z_i\right)^{\binom{m+n-1}{m}},$$

where the first product is taken over all m-tuples $(i_1,\ldots,i_m)$ of nonnegative integers such that $\sum_{j=1}^m i_j = n$. **Source:** [771, p. 35].

Fact 2.21.25. Let $n\ge 3$, let $z_1,\ldots,z_n$ be complex numbers, and, for all $k\in\{1,\ldots,n\}$, define $S_k \triangleq \sum |z_{i_1}+\cdots+z_{i_k}|$, where the sum is taken over all k-tuples $(i_1,\ldots,i_k)$ such that $1\le i_1<\cdots< i_k\le n$. Then, for all $k\in\{2,\ldots,n-1\}$,

$$S_k \le \binom{n-2}{k-1}S_1 + \binom{n-2}{k-2}S_n.$$

In addition,

$$\sum_{i=2}^{n-1} S_i \le (2^{n-2}-1)(S_1+S_n).$$

Source: [2470]. **Example:** If $n=3$, then $|z_1+z_2|+|z_2+z_3|+|z_3+z_1| \le |z_1|+|z_2|+|z_3|+|z_1+z_2+z_3|$. If $n=4$, then, with $z \triangleq z_1+z_2+z_3+z_4$,

$$|z_1+z_2|+|z_1+z_3|+|z_1+z_4|+|z_2+z_3|+|z_2+z_4|+|z_3+z_4| \le 2(|z_1|+|z_2|+|z_3|+|z_4|)+|z|,$$
$$|z_1+z_2+z_3|+|z_1+z_2+z_4|+|z_1+z_3+z_4|+|z_2+z_3+z_4| \le |z_1|+|z_2|+|z_3|+|z_4|+2|z|.$$

Remark: These inequalities concern the diagonals of a polygon. **Related:** Fact 11.8.8.

Fact 2.21.26. Let $z_1,\ldots,z_n,w_1,\ldots,w_n$ be complex numbers. Then,

$$\sum_{i,j=1}^n |z_i-z_j|^2 + \sum_{i,j=1}^n |w_i-w_j|^2 + 2\left|\sum_{i=1}^n (z_i-w_i)\right|^2 = 2\sum_{i,j=1}^n |z_i-w_j|^2.$$

Equivalently,

$$\sum_{i,j=1,i<j}^n |z_i-z_j|^2 + \sum_{i,j=1,i<j}^n |w_i-w_j|^2 + \left|\sum_{i=1}^n (z_i-w_i)\right|^2 = \sum_{i,j=1}^n |z_i-w_j|^2.$$

Source: [2093]. **Remark:** This is a generalized parallelogram law. Setting $z_1 = -z_2 = z$, $z_3 = -z_4 = w$, and $w_1 = w_2 = w_3 = w_4 = 0$ yields the parallelogram law $|z+w|^2 + |z-w|^2 = 2|z|^2 + 2|w|^2$ given by *xix*) of Fact 2.21.8. **Related:** Fact 11.8.17.

Fact 2.21.27. Let z be a complex number. Then, the following statements hold:

i) $0 < |e^z| \le e^{|z|}$.

ii) $|e^z| = e^{|z|}$ if and only if $\operatorname{Im} z = 0$ and $\operatorname{Re} z \ge 0$.

iii) $|e^z| = 1$ if and only if $\operatorname{Re} z = 0$.

iv) $\big||e^z| - 1\big| \le |e^z - 1| \le e^{|z|} - 1$.

v) $e^z = e^{\operatorname{Re} z}[\cos \operatorname{Im} z + (\sin \operatorname{Im} z)J]$.

vi) $\operatorname{Re} e^z = 0$ if and only if $\operatorname{Im} z$ is an odd integer multiple of $\pm\frac{\pi}{2}$.

vii) $\operatorname{Im} e^z = 0$ if and only if $\operatorname{Im} z$ is an integer multiple of $\pm\pi$.

viii) If z is nonzero, then $|z^J| < e^{\pi}$.

ix) $J^J = e^{-\pi/2}$.

x) If θ is a real number and $\theta/(2\pi)$ is not an integer, then $\lim_{k\to\infty} \frac{1}{k}\sum_{i=1}^k e^{i\theta J} = 0$.

Furthermore, let θ_1 and θ_2 be real numbers. Then, the following statements hold:

xi) $|e^{\theta_1 J} - e^{\theta_2 J}| = 2|\sin \frac{1}{2}(\theta_1 - \theta_2)|$, $|e^{\theta_1 J} + e^{\theta_2 J}| = 2|\cos \frac{1}{2}(\theta_1 - \theta_2)|$.

xii) $|e^{\theta_1 J} - e^{\theta_2 J}| \le |\theta_1 - \theta_2|$, and $|e^{\theta_1 J} - e^{\theta_2 J}| = |\theta_1 - \theta_2|$ if and only if $\theta_1 = \theta_2$.

xiii) $|e^{e^{\theta_1 J}} - e^{e^{\theta_2 J}}| \le 2e|\sin \frac{1}{2}(\theta_1 - \theta_2)|$.

xiv) $|\sinh e^{\theta_1 J} - \sinh e^{\theta_2 J}| \le \frac{e^2+1}{e}|\sin \frac{1}{2}(\theta_1 - \theta_2)| \le \frac{e^2+1}{2e}|\theta_1 - \theta_2|$.

Finally, let r_1 and r_2 be nonnegative numbers, at least one of which is positive. Then, the following statement holds:

xv) $|e^{\theta_1 J} - e^{\theta_2 J}| \le \dfrac{2|r_1 e^{\theta_1 J} - r_2 e^{\theta_2 J}|}{r_1 + r_2}$.

Source: *ix*) is discussed in [1351, p. 48]; *x*) is given in [970, p. 503]; *xi*)–*xiv*) are given in [942]; *xv*) is given in [1391, p. 218]. **Remark:** A matrix version of *x*) is given by Fact 15.17.15.

Fact 2.21.28. Let $p, q \in (1, \infty)$, assume that $1/p + 1/q = 1$, and let x and y be complex numbers such that $|x| < 1$ and $|y| < 1$. Then,

$$\frac{1}{|1 - xy|^2} \le \frac{1}{p(1-|x|^p)(1-|y|^p)} + \frac{1}{q(1-|x|^q)(1-|y|^q)},$$

$$\frac{1}{|1 - x|y|^{q-1}||1 - x|y|^{p-1}|} \le \frac{1}{p(1-|x|^p)(1-|y|^q)} + \frac{1}{q(1-|x|^q)(1-|y|^p)},$$

$$|e^{xy}|^2 \le \frac{1}{p}e^{|x|^p + |y|^p} + \frac{1}{q}e^{|x|^q + |y|^q}, \quad |e^{x|y|^{q-1} + x|y|^{p-1}}| \le \frac{1}{p}e^{|x|^p + |y|^q} + \frac{1}{q}e^{|x|^q + |y|^p},$$

$$|\log(1 - xy)|^2 \le \frac{1}{p}\log(1-|x|^p)\log(1-|y|^p) + \frac{1}{q}\log(1-|x|^q)\log(1-|y|^q),$$

$$|\log(1 - x|y|^{q-1})\log(1 - x|y|^{p-1})| \le \frac{1}{p}\log(1-|x|^p)\log(1-|y|^q) + \frac{1}{q}\log(1-|x|^q)\log(1-|y|^p).$$

Source: [1496].

Fact 2.21.29. Let z be a complex number. Define $\mathcal{D} \triangleq \{z \in \mathbb{C}\colon z \ne 0 \text{ and } |z - 1| \le 1\}$. Then, the following statements hold:

i) For all $z \in \mathcal{D}$, $\log z$ is given by the convergent series

$$\log z = \sum_{i=1}^{\infty} \frac{(-1)^{i+1}}{i}(z-1)^i.$$

ii) If $z \in \mathbb{C}$ and $-\pi < \operatorname{Im} z \le \pi$, then $\log e^z = z$.

iii) If $z_1, z_2 \in \mathcal{D}$, then $\log z_1 z_2 = \log z_1 + \log z_2$.

iv) If $|z| < 1$, then

$$|\log(1+z)| \le -\log(1-|z|), \quad \frac{|z|}{1+|z|} \le |\log(1+z)| \le \frac{|z|(1+|z|)}{|1+z|}.$$

Remark: Let $z = re^{\theta j} \in \mathbb{C}$ satisfy $|z-1| < 1$. Then, $-\frac{\pi}{2} < \theta < \frac{\pi}{2}$. Furthermore, $\log z = \log r + \theta j$, and thus $-\frac{\pi}{2} < \operatorname{Im} \log z < \frac{\pi}{2}$. Consequently, the series gives the principal log of z.

Fact 2.21.30. For all $z \in$ OIUD, define

$$f(z_1, z_2) \triangleq \operatorname{atanh}\left|\frac{z_1 - z_2}{1 - z_1\overline{z_2}}\right|.$$

Then, the following statements hold:

i) For all $z_1, z_2 \in$ OIUD, $f(z_1, z_2) \ge 0$.

ii) Let $z_1, z_2 \in$ OIUD. Then, $f(z_1, z_2) = 0$ if and only if $z_1 = z_2$.

iii) For all $z_1, z_2, z_3 \in$ OIUD, $f(z_1, z_3) \le f(z_1, z_2) + f(z_2, z_3)$.

Remark: f is a *Poincaré metric*. See [137]. **Related:** Fact 2.21.31.

Fact 2.21.31. Let OUHP $\triangleq \{z \in \mathbb{C}\colon \operatorname{Im} z > 0\}$, and, for all $z_1, z_2 \in$ OUHP, define

$$f(z_1, z_2) \triangleq 2\operatorname{atanh}\frac{|z_1 - z_2|}{|z_1 - \overline{z_2}|} = \log\frac{|z_1 - \overline{z_2}| + |z_1 - z_2|}{|z_1 - \overline{z_2}| - |z_1 - z_2|}.$$

Then, the following statements hold:

i) For all $z_1, z_2 \in$ OUHP, $f(z_1, z_2) \ge 0$.

ii) Let $z_1, z_2 \in$ OUHP. Then, $f(z_1, z_2) = 0$ if and only if $z_1 = z_2$.

iii) For all $z_1, z_2, z_3 \in$ OUHP, $f(z_1, z_3) \le f(z_1, z_2) + f(z_2, z_3)$.

Remark: f is a *Poincaré metric*. See [137]. **Related:** Fact 2.21.30.

2.22 Notes

Reference works on inequalities include [340, 603, 604, 605, 764, 1316, 1952, 1969, 1971, 2061]. Texts on complex variables include [217, 582, 645, 1136, 1369, 1472, 1697, 2113, 2191]. A collection of identities involving binomial coefficients is given in [1208].

Chapter Three
Basic Matrix Properties

In this chapter we provide a detailed treatment of the basic properties of matrices, such as range, null space, rank, and invertibility. We also consider properties of convex sets, cones, and subspaces.

3.1 Vectors

The set $\mathbb{F}^n$ consists of *vectors* x of the form

$$x = \begin{bmatrix} x_{(1)} \\ \vdots \\ x_{(n)} \end{bmatrix}, \tag{3.1.1}$$

where $x_{(1)}, \ldots, x_{(n)} \in \mathbb{F}$ are the *components* of x, and $\mathbb{F}$ represents either $\mathbb{R}$ or $\mathbb{C}$. Hence, the elements of $\mathbb{F}^n$ are *column vectors*. Since $\mathbb{F}^1 = \mathbb{F}$, it follows that every scalar is also a vector. If $x \in \mathbb{R}^n$ and every component of x is nonnegative, then x is *nonnegative*, and, if every component of x is positive, then x is *positive*.

If $\alpha \in \mathbb{F}$ and $x \in \mathbb{F}^n$, then $\alpha x \in \mathbb{F}^n$ is given by

$$\alpha x = \begin{bmatrix} \alpha x_{(1)} \\ \vdots \\ \alpha x_{(n)} \end{bmatrix}. \tag{3.1.2}$$

If $x, y \in \mathbb{F}^n$, then x and y are *linearly dependent* if there exists $\alpha \in \mathbb{F}$ such that either $x = \alpha y$ or $y = \alpha x$. Furthermore, vectors add component by component; that is, if $x, y \in \mathbb{F}^n$, then

$$x + y = \begin{bmatrix} x_{(1)} + y_{(1)} \\ \vdots \\ x_{(n)} + y_{(n)} \end{bmatrix}. \tag{3.1.3}$$

Thus, if $\alpha, \beta \in \mathbb{F}$, then the *linear combination* $\alpha x + \beta y$ is given by

$$\alpha x + \beta y = \begin{bmatrix} \alpha x_{(1)} + \beta y_{(1)} \\ \vdots \\ \alpha x_{(n)} + \beta y_{(n)} \end{bmatrix}. \tag{3.1.4}$$

If $x \in \mathbb{R}^n$ and x is nonnegative, then we write $x \geq\geq 0$, and, if x is positive, then we write $x >> 0$. If $x, y \in \mathbb{R}^n$, then $x \geq\geq y$ means that $x - y \geq\geq 0$, and $x >> y$ means that $x - y >> 0$.

Let $\mathcal{S} \subseteq \mathbb{F}^n$. For $\alpha \in \mathbb{F}$, define $\alpha\mathcal{S} \triangleq \{\alpha x\colon\ x \in \mathcal{S}\}$. We write $-\mathcal{S}$ for $(-1)\mathcal{S}$. The set $\mathcal{S}$ is *symmetric* if $\mathcal{S} = -\mathcal{S}$; that is, $x \in \mathcal{S}$ if and only if $-x \in \mathcal{S}$.

For $\mathcal{S}_1, \mathcal{S}_2 \subseteq \mathbb{F}^n$, define the *Minkowski sum*

$$\mathcal{S}_1 + \mathcal{S}_2 \triangleq \{x + y\colon\ x \in \mathcal{S}_1 \text{ and } y \in \mathcal{S}_2\}. \tag{3.1.5}$$

If $\mathcal{S}_1$ and $\mathcal{S}_2$ are multisets, then

$$\mathcal{S}_1 + \mathcal{S}_2 \triangleq \{x + y\colon\ x \in \mathcal{S}_1 \text{ and } y \in \mathcal{S}_2\}_{\text{ms}}. \tag{3.1.6}$$

Note that $\mathcal{S}_1 + \mathcal{S}_2 = \mathcal{S}_2 + \mathcal{S}_1$ and, if $\mathcal{S}_3 \subseteq \mathbb{F}^n$, then $(\mathcal{S}_1 + \mathcal{S}_2) + \mathcal{S}_3 = \mathcal{S}_1 + (\mathcal{S}_2 + \mathcal{S}_3)$. Furthermore, for all $y \in \mathbb{F}^n$ and $\mathcal{S} \subseteq \mathbb{F}^n$, define $y + \mathcal{S} \triangleq \{y\} + \mathcal{S} = \{y + x\colon\ x \in \mathcal{S}\}$. Note that, for all $\alpha, \beta \in \mathbb{F}$, $(\alpha + \beta)\mathcal{S} \subseteq \alpha\mathcal{S} + \beta\mathcal{S}$. Finally, $\mathcal{S} + \{0\} = \{0\} + \mathcal{S} = \mathcal{S}$ and $\mathcal{S} + \varnothing = \varnothing + \mathcal{S} = \varnothing$.

Let $\mathcal{S}_1 \subseteq \mathbb{F}^n$ and $\mathcal{S}_2 \subseteq \mathbb{F}^m$. Then, the Cartesian product $\mathcal{S}_1 \times \mathcal{S}_2 \subseteq \mathbb{F}^n \times \mathbb{F}^m$ is the subset of $\mathbb{F}^{n+m}$ given by

$$\begin{bmatrix} \mathcal{S}_1 \\ \mathcal{S}_2 \end{bmatrix} \triangleq \begin{bmatrix} I_n \\ 0_{m\times n} \end{bmatrix} \mathcal{S}_1 + \begin{bmatrix} 0_{n\times m} \\ I_m \end{bmatrix} \mathcal{S}_2 \tag{3.1.7}$$

and likewise for multisets.

Let $\mathcal{S} \subseteq \mathbb{F}^n$. Then, $\mathcal{S}$ is a *cone* if, for all $x \in \mathcal{S}$ and $\alpha > 0$, the vector αx is an element of $\mathcal{S}$. Now, assume that $\mathcal{S}$ is a cone. Then, $\mathcal{S}$ is *pointed* if $0 \in \mathcal{S}$, while $\mathcal{S}$ is *blunt* if $0 \notin \mathcal{S}$. Furthermore, $\mathcal{S}$ is a *one-sided cone* if $x, -x \in \mathcal{S}$ implies that $x = 0$. Hence, $\mathcal{S}$ is one-sided if and only if $\mathcal{S} \cap -\mathcal{S} \subseteq \{0\}$. The empty set is a cone.

For $x, y \in \mathbb{F}^n$ and $\alpha \in [0, 1]$, the vector $\alpha x + (1 - \alpha)y$ is a *convex combination* of x and y with *barycentric coordinates* α and $1 - \alpha$. The set $\mathcal{S} \subseteq \mathbb{F}^n$ is a *convex set* if, for all $x, y \in \mathcal{S}$, every convex combination of x and y is an element of $\mathcal{S}$. The empty set is a convex set. Furthermore, $\mathcal{S}$ is a *convex cone* if it is a convex set and a cone. The empty set is a convex cone.

For $x, y \in \mathbb{F}^n$ and $\alpha \in \mathbb{F}$, the vector $\alpha x + (1 - \alpha)y$ is an *affine combination* of x and y. The set $\mathcal{S} \subseteq \mathbb{F}^n$ is an *affine subspace* of $\mathbb{F}^n$ if, for all $x, y \in \mathcal{S}$, every affine combination of x and y is an element of $\mathcal{S}$. The empty set is an affine subspace.

For $x, y \in \mathbb{F}^n$ and $\alpha, \beta \in \mathbb{F}$, the vector $\alpha x + \beta y$ is a *linear combination* of x and y. The set $\mathcal{S} \subseteq \mathbb{F}^n$ is a *subspace* of $\mathbb{F}^n$ if, for all $x, y \in \mathcal{S}$, every linear combination of x and y is an element of $\mathcal{S}$. The empty set is a subspace.

Note that $\mathbb{R}$ is a convex cone and a subspace of $\mathbb{R}^2$. However, $\mathbb{R}$ is not a subspace of $\mathbb{C}$.

The following result shows how affine subspaces and subspaces are related.

Proposition 3.1.1. Let $\mathcal{S} \subseteq \mathbb{F}^n$. Then, the following statements hold:

i) $\mathcal{S}$ is an affine subspace if and only if there exist $z, x_1, \ldots, x_r \in \mathbb{F}^n$ such that

$$\mathcal{S} = z + \left\{ \sum_{i=1}^{r} \alpha_i x_i\colon\ \alpha_1, \ldots, \alpha_r \in \mathbb{F} \right\}. \tag{3.1.8}$$

ii) $\mathcal{S}$ is a subspace if and only if there exist $x_1, \ldots, x_r \in \mathbb{F}^n$ such that

$$\mathcal{S} = \left\{ \sum_{i=1}^{r} \alpha_i x_i\colon\ \alpha_1, \ldots, \alpha_r \in \mathbb{F} \right\}. \tag{3.1.9}$$

iii) $\mathcal{S}$ is an affine subspace if and only if there exists $z \in \mathbb{F}^n$ such that $\mathcal{S} + z$ is a subspace.

iv) $\mathcal{S}$ is a subspace if and only if there exists $z \in \mathbb{F}^n$ such that $\mathcal{S} + z$ is an affine subspace.

Proof. The third statement is given in [1267, Theorem 4.4]. □

The affine subspaces $\mathcal{S}_1, \mathcal{S}_2 \subseteq \mathbb{F}^n$ are *parallel* if there exists $z \in \mathbb{F}^n$ such that $\mathcal{S}_1 + z = \mathcal{S}_2$. If $\mathcal{S}$ is an affine subspace, then there exists a unique subspace parallel to $\mathcal{S}$.

Proposition 1.3.15 implies that the greatest lower bound of a collection of sets is given by their intersection.

Proposition 3.1.2. Let **S** be a collection of subsets of $\mathbb{F}^n$, each of which is a (cone, convex set, convex cone, affine subspace, subspace). Then, the intersection glb(**S**) of all elements of **S** is a (cone, convex set, convex cone, affine subspace, subspace).

Proof. The case of convex sets is given in [2487, p. 75]. □

Let $\mathcal{S} \subseteq \mathbb{F}^n$, and let **S** denote the collection of all (cones, convex sets, convex cones, affine subspaces, subspaces) that contain $\mathcal{S}$. Then, Proposition 1.3.16 implies that the intersection $\mathcal{S}_0 = \text{glb}(\mathbf{S})$ of all elements of $\mathbf{S}$ is the smallest (cone, convex set, convex cone, affine subspace, subspace) that contains $\mathcal{S}$ in the sense that, if $\mathcal{S}_1$ is a (cone, convex set, convex cone, affine subspace, subspace) that contains $\mathcal{S}$, then $\mathcal{S}_0 \subseteq \mathcal{S}_1$.

Let $\mathcal{S} \subseteq \mathbb{F}^n$. Then, the *conical hull* of $\mathcal{S}$, denoted by cone $\mathcal{S}$, is the smallest cone in $\mathbb{F}^n$ containing $\mathcal{S}$. The *convex hull* of $\mathcal{S}$, denoted by conv $\mathcal{S}$, is the smallest convex set containing $\mathcal{S}$. The *convex conical hull* of $\mathcal{S}$, denoted by coco $\mathcal{S}$, is the smallest convex cone in $\mathbb{F}^n$ containing $\mathcal{S}$. The *affine hull* of $\mathcal{S}$, denoted by affin $\mathcal{S}$, is the smallest affine subspace in $\mathbb{F}^n$ containing $\mathcal{S}$, while the *span* of $\mathcal{S}$, denoted by span $\mathcal{S}$, is the smallest subspace in $\mathbb{F}^n$ containing $\mathcal{S}$. If $x, y \in \mathbb{R}^3$ are distinct, then affin $\{x, y\}$ is the line passing through x and y, while, if $x, y, z \in \mathbb{R}^3$ are not contained in a single line, then affin $\{x, y, z\}$ is the plane passing through x, y, and z. Note that cone $\varnothing$ = conv $\varnothing$ = coco $\varnothing$ = affin $\varnothing$ = span $\varnothing$ = $\varnothing$.

Let $x_1, \ldots, x_r \in \mathbb{F}^n$. Then, $x_1, \ldots, x_r$ are *linearly independent* if $\alpha_1, \ldots, \alpha_r \in \mathbb{F}$ and

$$\sum_{i=1}^{r} \alpha_i x_i = 0 \tag{3.1.10}$$

imply that $\alpha_1 = \alpha_2 = \cdots = \alpha_r = 0$. Note that $x_1, \ldots, x_r$ are linearly independent if and only if $\overline{x_1}, \ldots, \overline{x_r}$ are linearly independent. If $x_1, \ldots, x_r$ are not linearly independent, then $x_1, \ldots, x_r$ are *linearly dependent*. Note that $0_{n\times 1}$ is linearly dependent.

Let $\mathcal{S} \subseteq \mathbb{F}^n$, assume that $\mathcal{S}$ is nonempty, and assume that $\mathcal{S}$ is a subspace. If $\mathcal{S}$ is not equal to $\{0_{n\times 1}\}$, then there exist $x_1, \ldots, x_r \in \mathbb{F}^n$ such that $x_1, \ldots, x_r$ are linearly independent over $\mathbb{F}$ and such that span $\{x_1, \ldots, x_r\} = \mathcal{S}$. The set $\{x_1, \ldots, x_r\}$ is a *basis* for $\mathcal{S}$. The positive integer r is the *dimension* $\dim \mathcal{S}$ of $\mathcal{S}$. For $n \geq 0$, we define $\dim \{0_{n\times 1}\} = 0$. The subspace $\mathcal{S}$ is a *hyperplane* if $\dim \mathcal{S} = n - 1$. We define $\dim \varnothing \triangleq -\infty$.

Let $\mathcal{S} \subseteq \mathbb{F}^n$. If $\mathcal{S}$ is an affine subspace, then the *dimension* $\dim \mathcal{S}$ of $\mathcal{S}$ is the dimension of the subspace parallel to affin $\mathcal{S}$. If, however, $\mathcal{S}$ is not an affine subspace, then $\dim \mathcal{S}$ is defined to be the dimension of the affine hull of $\mathcal{S}$; that is,

$$\dim \mathcal{S} \triangleq \dim \operatorname{affin} \mathcal{S}. \tag{3.1.11}$$

The affine subspace $\mathcal{S}$ is an *affine hyperplane* if $\dim \mathcal{S} = n - 1$.

The following result is the *subspace dimension theorem.*

Theorem 3.1.3. Let $\mathcal{S}_1, \mathcal{S}_2 \subseteq \mathbb{F}^n$ be subspaces. Then,

$$\dim(\mathcal{S}_1 + \mathcal{S}_2) + \dim(\mathcal{S}_1 \cap \mathcal{S}_2) = \dim \mathcal{S}_1 + \dim \mathcal{S}_2. \tag{3.1.12}$$

Proof. See [1305, p. 227]. □

For the next result, note that "$\subset$" indicates proper inclusion.

Proposition 3.1.4. Let $\mathcal{S}_1, \mathcal{S}_2 \subseteq \mathbb{F}^n$ be affine subspaces such that $\mathcal{S}_1 \subseteq \mathcal{S}_2$. Then, $\mathcal{S}_1 \subset \mathcal{S}_2$ if and only if $\dim \mathcal{S}_1 < \dim \mathcal{S}_2$. Equivalently, $\mathcal{S}_1 = \mathcal{S}_2$ if and only if $\dim \mathcal{S}_1 = \dim \mathcal{S}_2$.

Corollary 3.1.5. Let $\mathcal{S}_1, \mathcal{S}_2 \subseteq \mathbb{F}^n$ be subspaces such that $\mathcal{S}_1 \subseteq \mathcal{S}_2$. Then, $\mathcal{S}_1 \subset \mathcal{S}_2$ if and only if $\dim \mathcal{S}_1 < \dim \mathcal{S}_2$. Equivalently, $\mathcal{S}_1 = \mathcal{S}_2$ if and only if $\dim \mathcal{S}_1 = \dim \mathcal{S}_2$.

Let $\mathcal{S}_1, \mathcal{S}_2 \subseteq \mathbb{F}^n$ be subspaces. Then, $\mathcal{S}_1$ and $\mathcal{S}_2$ are *complementary subspaces* if $\mathcal{S}_1 + \mathcal{S}_2 = \mathbb{F}^n$ and $\mathcal{S}_1 \cap \mathcal{S}_2 = \{0\}$. In this case, we say that $\mathcal{S}_1$ is complementary to $\mathcal{S}_2$, and vice versa.

Corollary 3.1.6. Let $\mathcal{S}_1, \mathcal{S}_2 \subseteq \mathbb{F}^n$ be subspaces, and consider the following statements:

i) $\dim(\mathcal{S}_1 + \mathcal{S}_2) = n$.

ii) $\mathcal{S}_1 \cap \mathcal{S}_2 = \{0\}$.

iii) $\dim \mathcal{S}_1 + \dim \mathcal{S}_2 = n$.

iv) $\mathcal{S}_1$ and $\mathcal{S}_2$ are complementary subspaces.

Then,

$$[i), ii)] \iff [i), iii)] \iff [ii), iii)] \iff [i), ii), iii)] \iff [iv)].$$

The following result shows that cones can be used to induce relations on $\mathbb{F}^n$.

Proposition 3.1.7. Let $\mathcal{S} \subseteq \mathbb{F}^n$ be a cone and, for $x, y \in \mathbb{F}^n$, let $x \preceq y$ denote the relation $y - x \in \mathcal{S}$. Then, the following statements hold:

i) "$\preceq$" is reflexive if and only if $\mathcal{S}$ is a pointed cone.

ii) "$\preceq$" is antisymmetric if and only if $\mathcal{S}$ is a one-sided cone.

iii) "$\preceq$" is symmetric if and only if $\mathcal{S}$ is a symmetric cone.

iv) "$\preceq$" is transitive if and only if $\mathcal{S}$ is a convex cone.

Proof. The proofs of *i)*, *ii)*, and *iii)* are immediate. To prove necessity in *iv)*, let $x, y \in \mathcal{S}$ so that, for all $\alpha \in [0, 1]$, $0 \preceq \alpha x \preceq \alpha x + (1-\alpha)y$. Since "$\preceq$" is transitive, it follows that $\alpha x + (1-\alpha)y \in \mathcal{S}$ for all $\alpha \in [0, 1]$, and thus $\mathcal{S}$ is a convex set. Conversely, suppose that $\mathcal{S}$ is a convex cone, and assume that $x \preceq y$ and $y \preceq z$. Then, $y - x \in \mathcal{S}$ and $z - y \in \mathcal{S}$ imply that $z - x = 2\left[\frac{1}{2}(y - x) + \frac{1}{2}(z - y)\right] \in \mathcal{S}$. Hence, $x \preceq z$, and thus "$\preceq$" is transitive. □

3.2 Matrices

The vectors $x_1, \ldots, x_m \in \mathbb{F}^n$ placed side by side form the *matrix*

$$A \triangleq [x_1 \ \cdots \ x_m], \tag{3.2.1}$$

which has *n rows* and *m columns*. The components of the vectors $x_1, \ldots, x_m$ are the *entries* of A. We write $A \in \mathbb{F}^{n\times m}$ and say that A has *size* $n \times m$. Since $\mathbb{F}^n = \mathbb{F}^{n\times 1}$, it follows that every vector is also a matrix. Note that $\mathbb{F}^{1\times 1} = \mathbb{F}^1 = \mathbb{F}$. If $n = m$, then A is *square* and has size n. The ith row of A and the jth column of A are denoted by $\mathrm{row}_i(A)$ and $\mathrm{col}_j(A)$, respectively. Hence,

$$A = \begin{bmatrix} \mathrm{row}_1(A) \\ \vdots \\ \mathrm{row}_n(A) \end{bmatrix} = [\mathrm{col}_1(A) \ \cdots \ \mathrm{col}_m(A)]. \tag{3.2.2}$$

The entry $x_{j(i)}$ of A in both the ith row of A and the jth column of A is denoted by $A_{(i,j)}$. Therefore, $x \in \mathbb{F}^n$ can be written as

$$x = \begin{bmatrix} x_{(1)} \\ \vdots \\ x_{(n)} \end{bmatrix} = \begin{bmatrix} x_{(1,1)} \\ \vdots \\ x_{(n,1)} \end{bmatrix}. \tag{3.2.3}$$

Let $A \in \mathbb{F}^{n\times m}$. For $b \in \mathbb{F}^n$, the matrix obtained from A by replacing $\mathrm{col}_i(A)$ with b is denoted by

$$A \overset{i}{\leftarrow} b. \tag{3.2.4}$$

Likewise, for $b \in \mathbb{F}^{1\times m}$, (3.2.4) denotes the matrix obtained from A by replacing $\mathrm{row}_i(A)$ with b.

Let $A \in \mathbb{F}^{n\times m}$, and let $l \triangleq \min\{n, m\}$. Then, the entries $A_{(i,i)}$ for all $i \in \{1, \ldots, l\}$ and $A_{(i,j)}$ for all $i \neq j$ are the *diagonal entries* and *off-diagonal entries* of A, respectively. Moreover, for all $i \in \{1, \ldots, l-1\}$, the entries $A_{(i,i+1)}$ and $A_{(i+1,i)}$ are the *superdiagonal entries* and *subdiagonal entries* of A, respectively. In addition, the entries $A_{(i,l+1-i)}$ for all $i \in \{1, \ldots, l\}$ are the *reverse-diagonal entries* of A. If the diagonal entries $A_{(1,1)}, \ldots, A_{(l,l)}$ of A are real, then the diagonal entries of A are labeled as

$$\mathrm{d}_l(A) \le \cdots \le \mathrm{d}_1(A), \tag{3.2.5}$$

and we define

$$\mathrm{d}_{\max}(A) \triangleq \mathrm{d}_1(A), \quad \mathrm{d}_{\min}(A) \triangleq \mathrm{d}_l(A). \tag{3.2.6}$$

With this notation, the vector of diagonal entries of A is defined by

$$\mathrm{d}(A) \triangleq \begin{bmatrix} \mathrm{d}_1(A) \\ \vdots \\ \mathrm{d}_l(A) \end{bmatrix}. \tag{3.2.7}$$

Partitioned matrices are of the form

$$\begin{bmatrix} A_{11} & \cdots & A_{1l} \\ \vdots & \cdot\vdots\cdot & \vdots \\ A_{k1} & \cdots & A_{kl} \end{bmatrix}, \tag{3.2.8}$$

where, for all $i \in \{1,\dots,k\}$ and $j \in \{1,\dots,l\}$, the *block* A_{ij} of A is a matrix of size $n_i \times m_j$. If $n_i = m_j$ and the diagonal entries of A_{ij} lie on the diagonal of A, then the square matrix A_{ij} is a *diagonally located block*; otherwise, A_{ij} is an *off-diagonally located block*.

Let $A \in \mathbb{F}^{n\times m}$. Then, a *submatrix* of A is formed by deleting rows and columns of A. In particular, A is a submatrix of A, as is the matrix obtained by deleting rows and columns of A. Alternatively, a submatrix can be specified in terms of the rows and columns that are retained. If A is a partitioned matrix, then every block of A is a submatrix of A. A block of A is thus a submatrix of A whose entries are entries of adjacent rows and adjacent columns of A. If like-numbered rows and columns of A are retained, then the resulting square submatrix of A is a *principal submatrix* of A. Every diagonally located block is a principal submatrix. Finally, if rows and columns $1,\dots,j$ of A are retained, then the resulting $j \times j$ submatrix of A is a *leading principal submatrix* of A.

Let $A \in \mathbb{F}^{n\times m}$, and let $\mathcal{S}_1$ and $\mathcal{S}_2$ be subsets of $\{1,\dots,n\}$ and $\{1,\dots,m\}$, respectively. Then, $A_{(\mathcal{S}_1,\mathcal{S}_2)}$ is the $\mathrm{card}(\mathcal{S}_1) \times \mathrm{card}(\mathcal{S}_2)$ submatrix of A formed by retaining the rows of A listed in $\mathcal{S}_1$ and the columns of A listed in $\mathcal{S}_2$. Hence, $A_{(i,j)}$ denotes $A_{(\{i\},\{j\})}$. If $\mathcal{S} \subseteq \{1,\dots,\min\{n,m\}\}$, then we define $A_{(\mathcal{S})} \triangleq A_{(\mathcal{S},\mathcal{S})}$, which is a principal submatrix of A. Furthermore, $A_{(\mathcal{S}_1,\cdot)}$ denotes the submatrix of A formed by retaining the rows listed in $\mathcal{S}_1$, and $A_{(\cdot,\mathcal{S}_2)}$ denotes the submatrix of A formed by retaining the columns listed in $\mathcal{S}_2$. Likewise, $A_{[\mathcal{S}_1,\mathcal{S}_2]}$ is the $[n - \mathrm{card}(\mathcal{S}_1)] \times [m - \mathrm{card}(\mathcal{S}_2)]$ submatrix of A formed by deleting the rows of A listed in $\mathcal{S}_1$ and the columns of A listed in $\mathcal{S}_2$. In particular, $A_{[i,j]}$ denotes the $(n-1)\times(m-1)$ submatrix of A formed by deleting the ith row and jth column of A. Furthermore, $A_{[\mathcal{S}_1,\cdot]}$ denotes the submatrix of A formed by deleting the rows listed in $\mathcal{S}_1$, and $A_{[\cdot,\mathcal{S}_2]}$ denotes the submatrix of A formed by deleting the columns listed in $\mathcal{S}_2$. If $\mathcal{S} \subseteq \{1,\dots,\min\{n,m\}\}$, then we define $A_{[\mathcal{S}]} \triangleq A_{[\mathcal{S},\mathcal{S}]}$.

Let $A, B \in \mathbb{F}^{n\times m}$. Then, A and B add entry by entry; that is, for all $i \in \{1,\dots,n\}$ and $j \in \{1,\dots,m\}$, $(A+B)_{(i,j)} = A_{(i,j)} + B_{(i,j)}$. Furthermore, for all $i \in \{1,\dots,n\}$ and $j \in \{1,\dots,m\}$, it follows that, for all $\alpha \in \mathbb{F}$, $(\alpha A)_{(i,j)} = \alpha A_{(i,j)}$. Hence, for all $\alpha,\beta \in \mathbb{F}$, $(\alpha A + \beta B)_{(i,j)} = \alpha A_{(i,j)} + \beta B_{(i,j)}$. If $A, B \in \mathbb{F}^{n\times m}$, then A and B are *linearly dependent* if there exists $\alpha \in \mathbb{F}$ such that either $A = \alpha B$ or $B = \alpha A$.

Let $A \in \mathbb{R}^{n\times m}$. If every entry of A is nonnegative, then A is *nonnegative*, which is written as $A \geq\geq 0$. If every entry of A is positive, then A is *positive*, which is written as $A >> 0$. If $A, B \in \mathbb{R}^{n\times m}$, then $A \geq\geq B$ means that $A - B \geq\geq 0$, while $A >> B$ means that $A - B >> 0$.

Let $z \in \mathbb{F}^{1\times n}$ and $y \in \mathbb{F}^n = \mathbb{F}^{n\times 1}$. Then, the scalar $zy \in \mathbb{F}$ is defined by

$$zy \triangleq \sum_{i=1}^{n} z_{(1,i)} y_{(i)}. \tag{3.2.9}$$

Now, let $A \in \mathbb{F}^{n\times m}$ and $x \in \mathbb{F}^m$. Then, the matrix-vector product Ax is defined by

$$Ax \triangleq \begin{bmatrix} \mathrm{row}_1(A)x \\ \vdots \\ \mathrm{row}_n(A)x \end{bmatrix}. \tag{3.2.10}$$

It can be seen that Ax is a linear combination of the columns of A, that is,

$$Ax = \sum_{i=1}^{m} x_{(i)}\mathrm{col}_i(A). \tag{3.2.11}$$

Let $A \in \mathbb{F}^{n\times m}$. Then, A can be associated with the function $f\colon\ \mathbb{F}^m \mapsto \mathbb{F}^n$ defined by $f(x) \triangleq Ax$ for all $x \in \mathbb{F}^m$. For all $\alpha, \beta \in \mathbb{F}$ and $x, y \in \mathbb{F}^m$, it follows that

$$f(\alpha x + \beta y) = \alpha f(x) + \beta f(y) = \alpha Ax + \beta Ay. \tag{3.2.12}$$

Therefore, f is *linear*. For all $\mathcal{S} \subseteq \mathbb{F}^m$, we define

$$A\mathcal{S} \triangleq f(\mathcal{S}) = \{Ax\colon\ x \in \mathcal{S}\}, \tag{3.2.13}$$

and, for all $\mathcal{S} \subseteq \mathbb{F}^n$, we define

$$A^{\mathsf{inv}}(\mathcal{S}) \triangleq f^{\mathsf{inv}}(\mathcal{S}) = \{x \in \mathbb{F}^m\colon\ Ax \in \mathcal{S}\}. \tag{3.2.14}$$

Note that $A\varnothing = A^{\mathsf{inv}}(\varnothing) = \varnothing$. Now, let $b \in \mathbb{F}^n$. Then, the function $f\colon\ \mathbb{F}^m \mapsto \mathbb{F}^n$ defined by

$$f(x) \triangleq Ax + b \tag{3.2.15}$$

is *affine*.

Theorem 3.2.1. Let $A \in \mathbb{F}^{n\times m}$ and $B \in \mathbb{F}^{m\times l}$, and define $f\colon\ \mathbb{F}^m \mapsto \mathbb{F}^n$ and $g\colon\ \mathbb{F}^l \mapsto \mathbb{F}^m$ by $f(x) \triangleq Ax$ and $g(y) \triangleq By$. Furthermore, define the composition $h \triangleq f \circ g\colon\ \mathbb{F}^l \mapsto \mathbb{F}^n$. Then, for all $y \in \mathbb{R}^l$,

$$h(y) = f[g(y)] = A(By) = (AB)y, \tag{3.2.16}$$

where, for all $i \in \{1,\dots,n\}$ and $j \in \{1,\dots,l\}$, $AB \in \mathbb{F}^{n\times l}$ is defined by

$$(AB)_{(i,j)} \triangleq \sum_{k=1}^{m} A_{(i,k)}B_{(k,j)}. \tag{3.2.17}$$

Hence, we write ABy for $(AB)y$ and $A(By)$.

Let $A \in \mathbb{F}^{n\times m}$ and $B \in \mathbb{F}^{m\times l}$. Then, $AB \in \mathbb{F}^{n\times l}$ is the *product* of A and B. The matrices A and B are *conformable*, and the product (3.2.17) defines *matrix multiplication*.

Let $A \in \mathbb{F}^{n\times m}$ and $B \in \mathbb{F}^{m\times l}$. Then, AB can be written as

$$AB = [A\mathrm{col}_1(B)\ \cdots\ A\mathrm{col}_l(B)] = \begin{bmatrix} \mathrm{row}_1(A)B \\ \vdots \\ \mathrm{row}_n(A)B \end{bmatrix}. \tag{3.2.18}$$

Thus, for all $i \in \{1,\dots,n\}$ and $j \in \{1,\dots,l\}$,

$$(AB)_{(i,j)} = \mathrm{row}_i(A)\mathrm{col}_j(B), \tag{3.2.19}$$

$$\mathrm{col}_j(AB) = A\mathrm{col}_j(B), \tag{3.2.20}$$

$$\mathrm{row}_i(AB) = \mathrm{row}_i(A)B. \tag{3.2.21}$$

For conformable matrices A, B, C, the associative and distributive equalities

$$(AB)C = A(BC), \tag{3.2.22}$$

$$A(B+C) = AB + AC, \tag{3.2.23}$$

$$(A+B)C = AC + BC \tag{3.2.24}$$

are valid. Hence, we write ABC for $(AB)C$ and $A(BC)$. Note that (3.2.22) is a special case of (1.6.3).

For $\mathcal{S}_1 \subseteq \mathbb{F}^{n\times m}$ and $\mathcal{S}_2 \subseteq \mathbb{F}^{m\times l}$, define the *Minkowski product*

$$\mathcal{S}_1\mathcal{S}_2 \triangleq \{AB\colon\ A \in \mathcal{S}_1 \text{ and } B \in \mathcal{S}_2\}. \tag{3.2.25}$$

If $\mathcal{S}_1$ and $\mathcal{S}_2$ are multisets, then

$$\mathcal{S}_1\mathcal{S}_2 \triangleq \{AB\colon\ A \in \mathcal{S}_1 \text{ and } B \in \mathcal{S}_2\}_{\text{ms}}. \tag{3.2.26}$$

In particular, for $A \in \mathbb{F}^{n\times m}$ and $\mathcal{S} \subseteq \mathbb{F}^{m\times l}$, $A\mathcal{S} = \{A\}\mathcal{S} = \{AB\colon\ B \in \mathcal{S}\}$, which generalizes (3.2.13).

Let $A, B \in \mathbb{F}^{n\times n}$. Then, the *commutator* $[A, B] \in \mathbb{F}^{n\times n}$ of A and B is the matrix

$$[A, B] \triangleq AB - BA. \tag{3.2.27}$$

The *adjoint operator* $\text{ad}_A\colon\ \mathbb{F}^{n\times n} \mapsto \mathbb{F}^{n\times n}$ is defined by

$$\text{ad}_A(X) \triangleq [A, X]. \tag{3.2.28}$$

Let $x, y \in \mathbb{R}^3$. Then, the *cross product* $x \times y \in \mathbb{R}^3$ of x and y is defined by

$$x \times y \triangleq \begin{bmatrix} x_{(2)}y_{(3)} - x_{(3)}y_{(2)} \\ x_{(3)}y_{(1)} - x_{(1)}y_{(3)} \\ x_{(1)}y_{(2)} - x_{(2)}y_{(1)} \end{bmatrix}. \tag{3.2.29}$$

Furthermore, the 3×3 *cross-product matrix* is defined by

$$K(x) \triangleq \begin{bmatrix} 0 & -x_{(3)} & x_{(2)} \\ x_{(3)} & 0 & -x_{(1)} \\ -x_{(2)} & x_{(1)} & 0 \end{bmatrix}. \tag{3.2.30}$$

Note that

$$x \times y = K(x)y. \tag{3.2.31}$$

Multiplication of partitioned matrices is analogous to matrix multiplication with scalar entries. For example, for matrices with conformable blocks,

$$[A\ \ B]\begin{bmatrix} C \\ D \end{bmatrix} = AC + BD, \quad [A\ \ B]C = \begin{bmatrix} AC \\ BC \end{bmatrix}, \tag{3.2.32}$$

$$\begin{bmatrix} A \\ B \end{bmatrix}[C\ \ D] = \begin{bmatrix} AC & AD \\ BC & BD \end{bmatrix}, \quad \begin{bmatrix} A & B \\ C & D \end{bmatrix}\begin{bmatrix} E & F \\ G & H \end{bmatrix} = \begin{bmatrix} AE + BG & AF + BH \\ CE + DG & CF + DH \end{bmatrix}. \tag{3.2.33}$$

The $n \times m$ *zero matrix*, all of whose entries are zero, is written as $0_{n\times m}$. If the dimensions are unambiguous, then we write just 0. Let $x \in \mathbb{F}^m$ and $A \in \mathbb{F}^{n\times m}$. Then, the zero matrix satisfies

$$0_{k\times m}x = 0_{k\times 1}, \quad A0_{m\times l} = 0_{n\times l}, \quad 0_{k\times n}A = 0_{k\times m}. \tag{3.2.34}$$

Another special matrix is the *empty matrix*. For $n \in \mathbb{N}$, the $0 \times n$ empty matrix, which is written as $0_{0\times n}$, has zero rows and n columns, while the $n \times 0$ empty matrix, which is written as $0_{n\times 0}$, has n rows and zero columns. For $A \in \mathbb{F}^{n\times m}$, where $n, m \in \mathbb{N}$, the empty matrix satisfies the multiplication rules

$$0_{0\times n}A = 0_{0\times m}, \quad A0_{m\times 0} = 0_{n\times 0}. \tag{3.2.35}$$

Although empty matrices have no entries, it is useful to define the product

$$0_{n\times 0}0_{0\times m} \triangleq 0_{n\times m}. \tag{3.2.36}$$

Also, we define

$$I_0 \triangleq \hat{I}_0 \triangleq 0_{0\times 0}. \tag{3.2.37}$$

For $n, m \in \mathbb{N}$, we define $\mathbb{F}^{0\times m} \triangleq \{0_{0\times m}\}$, $\mathbb{F}^{n\times 0} \triangleq \{0_{n\times 0}\}$, and $\mathbb{F}^0 \triangleq \mathbb{F}^{0\times 1}$. Note that

$$\begin{bmatrix} 0_{n\times 0} & 0_{n\times m} \\ 0_{0\times 0} & 0_{0\times m} \end{bmatrix} = 0_{n\times m}. \tag{3.2.38}$$

The empty matrix is analogous to 0 for real numbers and $\varnothing$ for sets.

The $n \times n$ *identity matrix*, which has 1's on the diagonal and 0's elsewhere, is denoted by I_n or just I. Let $x \in \mathbb{F}^n$ and $A \in \mathbb{F}^{n\times m}$. Then, the identity matrix satisfies

$$I_n x = x, \quad AI_m = I_n A = A. \tag{3.2.39}$$

Let $A \in \mathbb{F}^{n\times n}$. Then, $A^2 \triangleq AA$ and, for all $k \ge 1$, $A^k \triangleq AA^{k-1}$. We use the convention $A^0 \triangleq I$. In particular, $0^0_{n\times n} = I_n$.

The $n\times n$ *reverse permutation matrix*, which has 1's on the reverse diagonal and 0's elsewhere, is denoted by $\hat{I}_n$ or just $\hat{I}$. In particular, $\hat{I}_1 \triangleq 1$. Multiplication of $x \in \mathbb{F}^n$ by $\hat{I}_n$ reverses the components of x. Likewise, left multiplication of $A \in \mathbb{F}^{n\times m}$ by $\hat{I}_n$ reverses the rows of A, while right multiplication of A by $\hat{I}_m$ reverses the columns of A. Note that $\hat{I}_n^2 = I_n$.

The $n \times n$ *cyclic permutation matrix* P_n is defined by $P_1 \triangleq 1$ and, for all $n \ge 2$,

$$P_n \triangleq \begin{bmatrix} 0 & 1 & 0 & \cdots & 0 & 0 \\ 0 & 0 & 1 & \ddots & 0 & 0 \\ 0 & 0 & 0 & \ddots & 0 & 0 \\ \vdots & \ddots & \ddots & \ddots & \ddots & \vdots \\ 0 & 0 & 0 & \ddots & 0 & 1 \\ 1 & 0 & 0 & \cdots & 0 & 0 \end{bmatrix}.$$

Note that $P_1 = \hat{I}_1$, $P_2 = \hat{I}_2$, and, for all $n \ge 1$, $P_n^n = I_n$.

The $n \times n$ *standard nilpotent matrix* N_n, or just N, is defined by $N_0 \triangleq 0_{0\times 0}$, $N_1 \triangleq 0$, and, for all $n \ge 2$,

$$N_n \triangleq \begin{bmatrix} 0 & 1 & 0 & \cdots & 0 & 0 \\ 0 & 0 & 1 & \ddots & 0 & 0 \\ 0 & 0 & 0 & \ddots & 0 & 0 \\ \vdots & \ddots & \ddots & \ddots & \ddots & \vdots \\ 0 & 0 & 0 & \ddots & 0 & 1 \\ 0 & 0 & 0 & \cdots & 0 & 0 \end{bmatrix}.$$

Note that, for all $n \ge 0$, $N_n^n = 0$.

3.3 Transpose and Inner Product

A fundamental vector and matrix operation is the transpose. If $x \in \mathbb{F}^n$, then the *transpose* x^{T} of x is defined to be the row vector

$$x^{\mathsf{T}} \triangleq [x_{(1)} \ \cdots \ x_{(n)}] \in \mathbb{F}^{1\times n}. \tag{3.3.1}$$

Similarly, if $x = [x_{(1,1)} \ \cdots \ x_{(1,n)}] \in \mathbb{F}^{1\times n}$, then

$$x^{\mathsf{T}} = \begin{bmatrix} x_{(1,1)} \\ \vdots \\ x_{(1,n)} \end{bmatrix} \in \mathbb{F}^{n\times 1}. \tag{3.3.2}$$

Let $x, y \in \mathbb{F}^n$. Then, $x^{\mathsf{T}}y \in \mathbb{F}$ is a scalar, and

$$x^{\mathsf{T}}y = y^{\mathsf{T}}x = \sum_{i=1}^{n} x_{(i)}y_{(i)}. \tag{3.3.3}$$

Note that

$$x^{\mathsf{T}}x = \sum_{i=1}^{n} x_{(i)}^2. \tag{3.3.4}$$

The vector $e_{i,n} \in \mathbb{R}^n$, or just e_i, has 1 as its ith component and 0's elsewhere. Thus,

$$e_{i,n} = \operatorname{col}_i(I_n). \tag{3.3.5}$$

Let $A \in \mathbb{F}^{n\times m}$. Then, $e_i^{\mathsf{T}}A = \operatorname{row}_i(A)$ and $Ae_i = \operatorname{col}_i(A)$. Furthermore, the (i, j) entry of A can be written as

$$A_{(i,j)} = e_i^{\mathsf{T}}Ae_j. \tag{3.3.6}$$

The $n \times m$ matrix $E_{i,j,n\times m} \in \mathbb{R}^{n\times m}$, or just $E_{i,j}$, has 1 as its (i, j) entry and 0's elsewhere. Thus,

$$E_{i,j,n\times m} = e_{i,n}e_{j,m}^{\mathsf{T}}. \tag{3.3.7}$$

Note that $E_{i,1,n\times 1} = e_{i,n}$ and

$$I_n = E_{1,1} + \cdots + E_{n,n} = \sum_{i=1}^{n} e_i e_i^{\mathsf{T}}. \tag{3.3.8}$$

Finally, the $n \times m$ *ones matrix*, all of whose entries are 1, is written as $1_{n\times m}$ or just 1. Thus,

$$1_{n\times m} = \sum_{i,j=1}^{n,m} E_{i,j,n\times m}. \tag{3.3.9}$$

Note that

$$1_{n\times 1} = \sum_{i=1}^{n} e_{i,n} = \begin{bmatrix} 1 \\ \vdots \\ 1 \end{bmatrix}, \qquad 1_{n\times m} = 1_{n\times 1}1_{1\times m}. \tag{3.3.10}$$

Lemma 3.3.1. Let $x \in \mathbb{R}$. Then, $x^{\mathsf{T}}x = 0$ if and only if $x = 0$.

Let $x, y \in \mathbb{R}^n$. Then, $x^{\mathsf{T}}y \in \mathbb{R}$ is the *inner product* of x and y. Furthermore, x and y are *mutually orthogonal* if $x^{\mathsf{T}}y = 0$. If x and y are nonzero, then the *angle* $\theta \in [0, \pi]$ between x and y is defined by

$$\theta \triangleq \operatorname{acos} \frac{x^{\mathsf{T}}y}{\sqrt{x^{\mathsf{T}}xy^{\mathsf{T}}y}}. \tag{3.3.11}$$

Note that x and y are mutually orthogonal if and only if $\theta = \frac{\pi}{2}$.

Let $x \in \mathbb{C}^n$. Then, $x = y + \jmath z$, where $y, z \in \mathbb{R}^n$. Therefore, the transpose x^{T} of x is given by

$$x^{\mathsf{T}} = y^{\mathsf{T}} + \jmath z^{\mathsf{T}}. \tag{3.3.12}$$

The *complex conjugate* $\overline{x}$ of x is defined by

$$\overline{x} \triangleq y - \jmath z, \tag{3.3.13}$$

while the *complex conjugate transpose* x^* of x is defined by

$$x^* \triangleq \overline{x}^{\mathsf{T}} = y^{\mathsf{T}} - \jmath z^{\mathsf{T}}. \tag{3.3.14}$$

The vectors y and z are the *real* and *imaginary* parts $\operatorname{Re} x$ and $\operatorname{Im} x$ of x, respectively, which are defined by

$$\operatorname{Re} x \triangleq \tfrac{1}{2}(x + \overline{x}) = y, \quad \operatorname{Im} x \triangleq \tfrac{1}{2\jmath}(x - \overline{x}) = z. \tag{3.3.15}$$

Note that

$$x^*x = \sum_{i=1}^{n} \overline{x}_{(i)}x_{(i)} = \sum_{i=1}^{n} |x_{(i)}|^2 = \sum_{i=1}^{n} \left(y_{(i)}^2 + z_{(i)}^2\right). \tag{3.3.16}$$

If $w, x \in \mathbb{C}^n$, then $w^{\mathsf{T}}x = x^{\mathsf{T}}w$.

Let $x \in \mathbb{F}^n$. Then, the *Euclidean norm* of x is defined by

$$\|x\|_2 \triangleq \left(\sum_{i=1}^{n} |x_{(i)}|^2\right)^{1/2} = \sqrt{x^*x}. \tag{3.3.17}$$

If $x \in \mathbb{R}^n$, then

$$\|x\|_2 = \left(\sum_{i=1}^{n} x_{(i)}^2\right)^{1/2} = \sqrt{x^{\mathsf{T}}x}. \tag{3.3.18}$$

Lemma 3.3.2. Let $x \in \mathbb{C}^n$. Then, $x^*x = 0$ if and only if $x = 0$.

Let $x, y \in \mathbb{C}^n$. Then, $x^*y \in \mathbb{C}$ is the *inner product* of x and y, which is given by

$$x^*y = \sum_{i=1}^{n} \overline{x}_{(i)}y_{(i)}. \tag{3.3.19}$$

Furthermore, x and y are *mutually orthogonal* if $x^*y = 0$. If x and y are nonzero, then the *angle* $\theta \in [0, \pi]$ between x and y is defined by

$$\theta \triangleq \operatorname{acos} \frac{\operatorname{Re} x^*y}{\sqrt{x^*xy^*y}}. \tag{3.3.20}$$

Note that x and y are mutually orthogonal if and only if $\theta = \frac{\pi}{2}$. It follows from the Cauchy-Schwarz inequality given by Corollary 11.1.7 that the arguments of acos in (3.3.11) and (3.3.20) are elements of the interval $[-1, 1]$. Furthermore, $\theta \in \mathcal{R}(\operatorname{acos}) = [0, \pi]$.

Let $A \in \mathbb{F}^{n \times m}$. Then, the *transpose* $A^{\mathsf{T}} \in \mathbb{F}^{m \times n}$ of A is defined by

$$A^{\mathsf{T}} \triangleq [[\operatorname{row}_1(A)]^{\mathsf{T}} \ \cdots \ [\operatorname{row}_n(A)]^{\mathsf{T}}] = \begin{bmatrix} [\operatorname{col}_1(A)]^{\mathsf{T}} \\ \vdots \\ [\operatorname{col}_m(A)]^{\mathsf{T}} \end{bmatrix}. \tag{3.3.21}$$

Note that $(A^{\mathsf{T}})^{\mathsf{T}} = A$. Furthermore, for all $i \in \{1, \ldots, n\}$, $\operatorname{col}_i(A^{\mathsf{T}}) = [\operatorname{row}_i(A)]^{\mathsf{T}}$; for all $j \in \{1, \ldots, m\}$, $\operatorname{row}_j(A^{\mathsf{T}}) = [\operatorname{col}_j(A)]^{\mathsf{T}}$; and, for all $i \in \{1, \ldots, n\}$ and $j \in \{1, \ldots, m\}$, $(A^{\mathsf{T}})_{(j,i)} = A_{(i,j)}$. If $B \in \mathbb{F}^{m \times l}$, then

$$(AB)^{\mathsf{T}} = B^{\mathsf{T}}A^{\mathsf{T}}. \tag{3.3.22}$$

In particular, if $x \in \mathbb{F}^m$, then

$$(Ax)^{\mathsf{T}} = x^{\mathsf{T}}A^{\mathsf{T}}, \tag{3.3.23}$$

and if, in addition, $y \in \mathbb{F}^n$, then $y^{\mathsf{T}}Ax$ is a scalar and

$$y^{\mathsf{T}}Ax = (y^{\mathsf{T}}Ax)^{\mathsf{T}} = x^{\mathsf{T}}A^{\mathsf{T}}y. \tag{3.3.24}$$

If $B \in \mathbb{F}^{n\times m}$, then, for all $\alpha, \beta \in \mathbb{F}$,

$$(\alpha A + \beta B)^{\mathsf{T}} = \alpha A^{\mathsf{T}} + \beta B^{\mathsf{T}}. \tag{3.3.25}$$

Let $\mathcal{S} \subseteq \mathbb{C}^{n\times m}$. Then,

$$\mathcal{S}^{\mathsf{T}} \triangleq \{A^{\mathsf{T}}\colon\ A \in \mathcal{S}\}, \tag{3.3.26}$$

and likewise for multisets.

Let $x \in \mathbb{F}^n$ and $y \in \mathbb{F}^m$. Then, the matrix $xy^{\mathsf{T}} \in \mathbb{F}^{n\times m}$ is the *outer product* of x and y. The outer product xy^{T} is nonzero if and only if both x and y are nonzero.

Let $A \in \mathbb{F}^{n\times m}$ and $B \in \mathbb{F}^{m\times l}$. Then,

$$AB = \sum_{i=1}^{m} \operatorname{col}_i(A)\operatorname{row}_i B. \tag{3.3.27}$$

Therefore, the product of two matrices can be written as the sum of outer-product matrices.

The *trace* of a square matrix $A \in \mathbb{F}^{n\times n}$, denoted by $\operatorname{tr} A$, is defined to be the sum of its diagonal entries; that is,

$$\operatorname{tr} A \triangleq \sum_{i=1}^{n} A_{(i,i)}. \tag{3.3.28}$$

Note that

$$\operatorname{tr} A = \operatorname{tr} A^{\mathsf{T}}. \tag{3.3.29}$$

Let $A \in \mathbb{F}^{n\times m}$ and $B \in \mathbb{F}^{m\times n}$. Then, AB and BA are square,

$$\operatorname{tr} AB = \operatorname{tr} BA = \operatorname{tr} A^{\mathsf{T}}B^{\mathsf{T}} = \operatorname{tr} B^{\mathsf{T}}A^{\mathsf{T}} = \sum_{i,j=1}^{n,m} A_{(i,j)}B_{(j,i)} = \sum_{i=1}^{m} \operatorname{row}_i(A)\operatorname{col}_i B, \tag{3.3.30}$$

and

$$\operatorname{tr} AA^{\mathsf{T}} = \operatorname{tr} A^{\mathsf{T}}A = \sum_{i,j=1}^{n,m} A_{(i,j)}^2. \tag{3.3.31}$$

Furthermore, if $n = m$, then, for all $\alpha, \beta \in \mathbb{F}$,

$$\operatorname{tr}(\alpha A + \beta B) = \alpha \operatorname{tr} A + \beta \operatorname{tr} B. \tag{3.3.32}$$

Lemma 3.3.3. Let $A \in \mathbb{R}^{n\times m}$. Then, $\operatorname{tr} A^{\mathsf{T}}A = 0$ if and only if $A = 0$.

Let $A, B \in \mathbb{R}^{n\times m}$. Then, the *inner product* of A and B is $\operatorname{tr} A^{\mathsf{T}}B$. Furthermore, A and B are *mutually orthogonal* if $\operatorname{tr} A^{\mathsf{T}}B = 0$.

Let $C \in \mathbb{C}^{n\times m}$. Then, $C = A + \jmath B$, where $A, B \in \mathbb{R}^{n\times m}$. Then, the transpose C^{T} of C is given by

$$C^{\mathsf{T}} = A^{\mathsf{T}} + \jmath B^{\mathsf{T}}. \tag{3.3.33}$$

The *complex conjugate* $\overline{C}$ of C is

$$\overline{C} \triangleq A - \jmath B, \tag{3.3.34}$$

while the *complex conjugate transpose* C^* of C is

$$C^* \triangleq \overline{C}^{\mathsf{T}} = A^{\mathsf{T}} - \jmath B^{\mathsf{T}}. \tag{3.3.35}$$

Note that $\overline{C} = C$ if and only if $B = 0$, and that

$$(C^{\mathsf{T}})^{\mathsf{T}} = \overline{\overline{C}} = (C^*)^* = C. \tag{3.3.36}$$

The matrices A and B are the real and imaginary parts $\operatorname{Re} C$ and $\operatorname{Im} C$ of C, respectively, which are denoted by

$$\operatorname{Re} C \triangleq \tfrac{1}{2}(C + \overline{C}) = A, \quad \operatorname{Im} C \triangleq \tfrac{1}{2\jmath}(C - \overline{C}) = B. \tag{3.3.37}$$

If C is square, then

$$\operatorname{tr} C = \operatorname{tr} A + (\operatorname{tr} B)\jmath = \operatorname{tr} C^{\mathsf{T}} = \overline{\operatorname{tr} \overline{C}} = \overline{\operatorname{tr} C^*}. \tag{3.3.38}$$

Let $\mathcal{S} \subseteq \mathbb{C}^{n\times m}$. Then,

$$\overline{\mathcal{S}} \triangleq \{\overline{A}\colon\ A \in \mathcal{S}\}, \quad \mathcal{S}^* \triangleq \{A^*\colon\ A \in \mathcal{S}\}, \tag{3.3.39}$$

and likewise for multisets.

Let $A \in \mathbb{F}^{n\times n}$. Then, for all $k \in \mathbb{N}$,

$$A^{k\mathsf{T}} \triangleq (A^k)^{\mathsf{T}} = (A^{\mathsf{T}})^k, \quad \overline{A^k} = \overline{A}^k, \quad A^{k*} \triangleq (A^k)^* = (A^*)^k. \tag{3.3.40}$$

Lemma 3.3.4. Let $A \in \mathbb{C}^{n\times m}$. Then, $\operatorname{tr} A^*A = 0$ if and only if $A = 0$.

Let $A, B \in \mathbb{C}^{n\times m}$. Then, the *inner product* of A and B is $\operatorname{tr} A^*B$. Furthermore, A and B are *mutually orthogonal* if $\operatorname{tr} A^*B = 0$.

If $A, B \in \mathbb{C}^{n\times m}$, then, for all $\alpha, \beta \in \mathbb{C}$,

$$(\alpha A + \beta B)^* = \overline{\alpha}A^* + \overline{\beta}B^*, \tag{3.3.41}$$

while, if $A \in \mathbb{C}^{n\times m}$ and $B \in \mathbb{C}^{m\times l}$, then

$$\overline{AB} = \overline{A}\,\overline{B}, \quad (AB)^* = B^*A^*. \tag{3.3.42}$$

In particular, if $A \in \mathbb{C}^{n\times m}$ and $x \in \mathbb{C}^m$, then

$$(Ax)^* = x^*A^*, \tag{3.3.43}$$

while, if, in addition, $y \in \mathbb{C}^n$, then

$$y^*Ax = (y^*Ax)^{\mathsf{T}} = x^{\mathsf{T}}A^{\mathsf{T}}\overline{y} \tag{3.3.44}$$

and

$$(y^*Ax)^* = (\overline{y^*Ax})^{\mathsf{T}} = (y^{\mathsf{T}}\overline{Ax})^{\mathsf{T}} = x^*A^*y. \tag{3.3.45}$$

For $A \in \mathbb{F}^{n\times m}$, define the *reverse transpose* of A by

$$A^{\hat{\mathsf{T}}} \triangleq \hat{I}_m A^{\mathsf{T}} \hat{I}_n \tag{3.3.46}$$

and the *reverse complex conjugate transpose* of A by

$$A^{\hat{*}} \triangleq \hat{I}_m A^* \hat{I}_n. \tag{3.3.47}$$

For example,

$$\begin{bmatrix} 1 & 2 & 3 \\ 4 & 5 & 6 \end{bmatrix}^{\hat{\mathsf{T}}} = \begin{bmatrix} 6 & 3 \\ 5 & 2 \\ 4 & 1 \end{bmatrix}. \tag{3.3.48}$$

Furthermore,

$$(A^*)^{\hat{*}} = (A^{\hat{*}})^* = (A^{\mathsf{T}})^{\hat{\mathsf{T}}} = (A^{\hat{\mathsf{T}}})^{\mathsf{T}} = \hat{I}_n A \hat{I}_m, \quad (A^{\hat{*}})^{\hat{*}} = (A^{\hat{\mathsf{T}}})^{\hat{\mathsf{T}}} = A. \tag{3.3.49}$$

Finally, if $B \in \mathbb{F}^{m\times l}$, then

$$(AB)^{\hat{*}} = B^{\hat{*}}A^{\hat{*}}, \quad (AB)^{\hat{\mathsf{T}}} = B^{\hat{\mathsf{T}}}A^{\hat{\mathsf{T}}}. \tag{3.3.50}$$

Let $\mathcal{S} \subseteq \mathbb{C}^{n\times m}$. Then,

$$\mathcal{S}^{\hat{\mathsf{T}}} \triangleq \{A^{\hat{\mathsf{T}}}\colon\ A \in \mathcal{S}\}, \quad \mathcal{S}^{\hat{*}} \triangleq \{A^{\hat{*}}\colon\ A \in \mathcal{S}\}, \tag{3.3.51}$$

and likewise for multisets.

For $x \in \mathbb{F}^m$ and $A \in \mathbb{F}^{n\times m}$, every component of x can be replaced by its absolute value to obtain $|x| \in \mathbb{R}^m$, where, for all $i \in \{1,\dots,n\}$,

$$|x|_{(i)} \triangleq |x_{(i)}|, \tag{3.3.52}$$

and every entry of A can be replaced by its absolute value to obtain $|A| \in \mathbb{R}^{n\times m}$, where, for all $i \in \{1,\dots,n\}$ and $j \in \{1,\dots,m\}$,

$$|A|_{(i,j)} \triangleq |A_{(i,j)}|. \tag{3.3.53}$$

Note that

$$|Ax| \leq\leq |A||x|, \tag{3.3.54}$$

and, if $B \in \mathbb{F}^{m\times l}$, then

$$|AB| \leq\leq |A||B|. \tag{3.3.55}$$

For $x \in \mathbb{R}^n$ and $A \in \mathbb{R}^{n\times m}$, every component of x can be replaced by its sign to obtain $\operatorname{sign} x \in \mathbb{R}^n$, where, for all $i \in \{1,\dots,n\}$,

$$(\operatorname{sign} x)_{(i)} \triangleq \operatorname{sign} x_{(i)}, \tag{3.3.56}$$

and every entry of A can be replaced by its sign to obtain $\operatorname{sign} A \in \mathbb{R}^{n\times m}$, where, for all $i \in \{1,\dots,n\}$ and $j \in \{1,\dots,m\}$,

$$(\operatorname{sign} A)_{(i,j)} \triangleq \operatorname{sign} A_{(i,j)}. \tag{3.3.57}$$

Let $\mathcal{S}_1, \mathcal{S}_2 \subseteq \mathbb{F}^n$ be sets. Then, $\mathcal{S}_1$ and $\mathcal{S}_2$ are *mutually orthogonal* if $x^*y = 0$ for all $x \in \mathcal{S}_1$ and $y \in \mathcal{S}_2$.

Let $\mathcal{S} \subseteq \mathbb{F}^n$ be nonempty. Then, the *orthogonal complement* $\mathcal{S}^\perp$ of $\mathcal{S}$ is defined by

$$\mathcal{S}^\perp \triangleq \{x \in \mathbb{F}^n\colon\ x^*y = 0 \text{ for all } y \in \mathcal{S}\}. \tag{3.3.58}$$

The orthogonal complement $\mathcal{S}^\perp$ of $\mathcal{S}$ is a subspace. Furthermore, $\mathcal{S}$ and $\mathcal{S}^\perp$ are mutually orthogonal.

Proposition 3.3.5. Let $\mathcal{S}_1, \mathcal{S}_2 \subseteq \mathbb{F}^n$ be mutually orthogonal sets. Then,

$$\mathcal{S}_1 \cap \mathcal{S}_2 \subseteq \{0\}. \tag{3.3.59}$$

Furthermore, $0 \in \mathcal{S}_1 \cap \mathcal{S}_2$ if and only if

$$\mathcal{S}_1 \cap \mathcal{S}_2 = \{0\}. \tag{3.3.60}$$

Proposition 3.3.5 shows that, if $\mathcal{S}_1$ and $\mathcal{S}_2$ are mutually orthogonal subspaces, then (3.3.60) holds.

Let $\mathcal{S}_1, \mathcal{S}_2 \subseteq \mathbb{F}^n$ be subspaces. Then, $\mathcal{S}_1$ and $\mathcal{S}_2$ are *orthogonally complementary* if $\mathcal{S}_1$ and $\mathcal{S}_2$ are complementary and mutually orthogonal.

Proposition 3.3.6. Let $\mathcal{S}_1, \mathcal{S}_2 \subseteq \mathbb{F}^n$ be subspaces. Then, $\mathcal{S}_1$ and $\mathcal{S}_2$ are orthogonally complementary if and only if $\mathcal{S}_1 = \mathcal{S}_2^\perp$. If these conditions hold, then $\dim \mathcal{S}_1 + \dim \mathcal{S}_2 = n$.

Corollary 3.3.7. Let $\mathcal{S} \subseteq \mathbb{F}^n$ be a subspace. Then, $\mathcal{S}$ and $\mathcal{S}^\perp$ are orthogonally complementary.

3.4 Geometrically Defined Sets

Let $\mathcal{S} \subseteq \mathbb{F}^n$. Then, $\mathcal{S}$ is a hyperplane if and only if there exists a nonzero vector $y \in \mathbb{F}^n$ such that $\mathcal{S} = \{y\}^\perp$. Furthermore, $\mathcal{S}$ is an affine hyperplane if and only if there exists $z \in \mathbb{F}^n$ such that $\mathcal{S} + z$ is a hyperplane.

Let $\mathcal{S} \subseteq \mathbb{F}^n$. Then, the *dual cone* of $\mathcal{S}$ is defined by

$$\operatorname{dcone} \mathcal{S} \triangleq \{x \in \mathbb{F}^n\colon\ \operatorname{Re} x^*y \le 0 \text{ for all } y \in \mathcal{S}\}. \tag{3.4.1}$$

Note that $\operatorname{dcone} \mathcal{S}$ is a pointed convex cone. Furthermore,

$$\operatorname{dcone} \mathcal{S} = \operatorname{dcone} \operatorname{conv} \mathcal{S} = \operatorname{dcone} \operatorname{cone} \mathcal{S} = \operatorname{dcone} \operatorname{coco} \mathcal{S}. \tag{3.4.2}$$

The set $\{x \in \mathbb{F}^n\colon\ \operatorname{Re} x^*y \le 0\} = \operatorname{dcone} \{y\}$ is a *closed half space*, while the set $\{x \in \mathbb{F}^n\colon\ \operatorname{Re} x^*y < 0\}$ is an *open half space*. Furthermore, $\mathcal{S}$ is an *affine (closed, open) half space* if there exists $z \in \mathbb{F}^n$ such that $\mathcal{S} + z$ is a (closed, open) half space.

Let $\mathcal{S} \subseteq \mathbb{F}^n$. Then, $\mathcal{S}$ is a *polytope* if $\mathcal{S}$ is the union of a finite number of sets, each of which is the intersection of a finite number of closed half spaces. If $\mathbb{F} = \mathbb{R}$ and $\dim \mathcal{S} = n = 3$, then $\mathcal{S}$ is a *polyhedron*, whereas, if $\mathbb{F} = \mathbb{R}$ and $\dim \mathcal{S} = n = 2$, then $\mathcal{S}$ is a *polygon*. Note that every polytope, and thus every polyhedron and polygon, is a closed set (see Definition 12.1.6) that is not necessarily convex, connected, or bounded (see Definition 12.1.13). If $\mathcal{S} \subset \mathbb{F}^n$, $\operatorname{card} \mathcal{S} = n + 1$, and $\dim \mathcal{S} = n$, then the convex polytope $\operatorname{conv} \mathcal{S}$ is a *simplex*. $\mathcal{S}$ is a *polyhedral cone* if $\mathcal{S}$ is a cone and a polytope. $\mathcal{S}$ is a *zonotope* if there exist $z, x_1, \ldots, x_m \in \mathbb{F}^n$ such that

$$\mathcal{S} = z + \left\{ \sum_{i=1}^m \alpha_i x_i\colon 0 \le \alpha_i \le 1 \text{ for all } i \in \{1, \ldots, m\} \right\}.$$

$\mathcal{S}$ is a zonotope if and only if $\mathcal{S}$ is the Minkowski sum of a finite number of line segments. $\mathcal{S}$ is a *parallelotope* if $\mathcal{S}$ is a zonotope and $\dim \mathcal{S} = m = n$. If $\mathbb{F} = \mathbb{R}$ and $\dim \mathcal{S} = n = 3$, then the parallelotope $\mathcal{S}$ is a *parallelepiped*, while, if $\mathbb{F} = \mathbb{R}$ and $\dim \mathcal{S} = n = 2$, then the parallelotope $\mathcal{S}$ is a *parallelogram*.

Let $\mathcal{S} \subseteq \mathbb{F}^n$. Then,

$$\operatorname{polar} \mathcal{S} \triangleq \{x \in \mathbb{F}^n\colon\ \operatorname{Re} x^*y \le 1 \text{ for all } y \in \mathcal{S}\} \tag{3.4.3}$$

is the *polar* of $\mathcal{S}$. Note that $\operatorname{polar} \mathcal{S}$ is a convex set and

$$\operatorname{polar} \mathcal{S} = \operatorname{polar} \operatorname{conv} \mathcal{S}. \tag{3.4.4}$$

3.5 Range and Null Space

Two key features of a matrix $A \in \mathbb{F}^{n\times m}$ are its range and null space, denoted by $\mathcal{R}(A)$ and $\mathcal{N}(A)$, respectively. The *range* of A is defined by

$$\mathcal{R}(A) \triangleq \{Ax\colon\ x \in \mathbb{F}^m\} = A\mathbb{F}^m. \tag{3.5.1}$$

Note that, for all nonnegative n and m, it follows that $\mathcal{R}(0_{n\times 0}) = \{0_{n\times 1}\}$ and $\mathcal{R}(0_{0\times m}) = \{0_{0\times 1}\}$. Letting α_i denote $x_{(i)}$, it can be seen that

$$\mathcal{R}(A) = \left\{ \sum_{i=1}^m \alpha_i \operatorname{col}_i(A)\colon\ \alpha_1, \ldots, \alpha_m \in \mathbb{F} \right\}, \tag{3.5.2}$$

which shows that $\mathcal{R}(A)$ is a subspace of $\mathbb{F}^n$. It follows from Fact 3.11.5 that

$$\mathcal{R}(A) = \operatorname{span} \{\operatorname{col}_1(A), \ldots, \operatorname{col}_m(A)\}. \tag{3.5.3}$$

The *null space* of $A \in \mathbb{F}^{n\times m}$ is defined by

$$\mathcal{N}(A) \triangleq \{x \in \mathbb{F}^m\colon\ Ax = 0\}. \tag{3.5.4}$$

Note that $\mathcal{N}(0_{n\times 0}) = \mathbb{F}^0 = \{0_{0\times 1}\}$ and $\mathcal{N}(0_{0\times m}) = \mathbb{F}^m$. Equivalently,

$$\mathcal{N}(A) = \left\{x \in \mathbb{F}^m\colon\ x^{\mathsf{T}}[\mathrm{row}_i(A)]^{\mathsf{T}} = 0 \text{ for all } i \in \{1,\dots,n\}\right\} \tag{3.5.5}$$

$$= \left\{[\mathrm{row}_1(A)]^{\mathsf{T}},\dots,[\mathrm{row}_n(A)]^{\mathsf{T}}\right\}^{\perp}, \tag{3.5.6}$$

which shows that $\mathcal{N}(A)$ is a subspace of $\mathbb{F}^m$. Note that, if $\alpha \in \mathbb{F}$ is nonzero, then $\mathcal{R}(\alpha A) = \mathcal{R}(A)$ and $\mathcal{N}(\alpha A) = \mathcal{N}(A)$. Finally, if $\mathbb{F} = \mathbb{C}$, then $\mathcal{R}(A)$ and $\mathcal{R}(\overline{A})$ are not necessarily identical. For example, let $A \triangleq \left[\begin{smallmatrix} j \\ 1 \end{smallmatrix}\right]$.

Let $A \in \mathbb{F}^{n\times n}$, and let $\mathcal{S} \subseteq \mathbb{F}^n$ be a subspace. Then, $\mathcal{S}$ is an *invariant subspace* of A if $A\mathcal{S} \subseteq \mathcal{S}$. Note that $A\mathcal{R}(A) \subseteq A\mathbb{F}^n = \mathcal{R}(A)$ and $A\mathcal{N}(A) = \{0_n\} \subseteq \mathcal{N}(A)$. Hence, $\mathcal{R}(A)$ and $\mathcal{N}(A)$ are invariant subspaces of A.

If $A \in \mathbb{F}^{n\times m}$ and $B \in \mathbb{F}^{m\times l}$, then

$$\mathcal{R}(AB) = A\mathcal{R}(B). \tag{3.5.7}$$

Lemma 3.5.1. Let $A \in \mathbb{F}^{n\times m}$, $B \in \mathbb{F}^{m\times l}$, and $C \in \mathbb{F}^{k\times n}$. Then,

$$\mathcal{R}(AB) \subseteq \mathcal{R}(A), \tag{3.5.8}$$

$$\mathcal{N}(A) \subseteq \mathcal{N}(CA). \tag{3.5.9}$$

Proof. Since $\mathcal{R}(B) \subseteq \mathbb{F}^m$, it follows that $\mathcal{R}(AB) = A\mathcal{R}(B) \subseteq A\mathbb{F}^m = \mathcal{R}(A)$. Furthermore, $y \in \mathcal{N}(A)$ implies that $Ay = 0$, and thus $CAy = 0$. □

Corollary 3.5.2. Let $A \in \mathbb{F}^{n\times n}$, and let $k \ge 1$. Then,

$$\mathcal{R}(A^k) \subseteq \mathcal{R}(A), \tag{3.5.10}$$

$$\mathcal{N}(A) \subseteq \mathcal{N}(A^k). \tag{3.5.11}$$

Although $\mathcal{R}(AB) \subseteq \mathcal{R}(A)$ for arbitrary conformable matrices A, B, we now show that equality holds in the special case $B = A^*$. This result, along with others, is the subject of the following basic theorem.

Theorem 3.5.3. Let $A \in \mathbb{F}^{n\times m}$. Then, the following statements hold:

i) $\mathcal{R}(A)^{\perp} = \mathcal{N}(A^*)$.

ii) $\mathcal{R}(A) = \mathcal{R}(AA^*)$.

iii) $\mathcal{N}(A) = \mathcal{N}(A^*A)$.

Proof. To prove *i*), we first show that $\mathcal{R}(A)^{\perp} \subseteq \mathcal{N}(A^*)$. Let $x \in \mathcal{R}(A)^{\perp}$. Then, $x^*z = 0$ for all $z \in \mathcal{R}(A)$. Hence, $x^*Ay = 0$ for all $y \in \mathbb{R}^m$. Equivalently, $y^*A^*x = 0$ for all $y \in \mathbb{R}^m$. Letting $y = A^*x$, it follows that $x^*AA^*x = 0$. Now, Lemma 3.3.2 implies that $A^*x = 0$. Thus, $x \in \mathcal{N}(A^*)$. Conversely, let us show that $\mathcal{N}(A^*) \subseteq \mathcal{R}(A)^{\perp}$. Letting $x \in \mathcal{N}(A^*)$, it follows that $A^*x = 0$, and, hence, $y^*A^*x = 0$ for all $y \in \mathbb{R}^m$. Equivalently, $x^*Ay = 0$ for all $y \in \mathbb{R}^m$. Hence, $x^*z = 0$ for all $z \in \mathcal{R}(A)$. Thus, $x \in \mathcal{R}(A)^{\perp}$, which proves *i*).

To prove *ii*), note that Lemma 3.5.1 with $B = A^*$ implies that $\mathcal{R}(AA^*) \subseteq \mathcal{R}(A)$. To show that $\mathcal{R}(A) \subseteq \mathcal{R}(AA^*)$, let $x \in \mathcal{R}(A)$, and suppose that $x \notin \mathcal{R}(AA^*)$. Then, it follows from Proposition 3.3.6 that $x = x_1 + x_2$, where $x_1 \in \mathcal{R}(AA^*)$ and $x_2 \in \mathcal{R}(AA^*)^{\perp}$ with $x_2 \ne 0$. Thus, $x_2^*AA^*y = 0$ for all $y \in \mathbb{R}^n$, and setting $y = x_2$ yields $x_2^*AA^*x_2 = 0$. Hence, Lemma 3.3.2 implies that $A^*x_2 = 0$, so that, by *i*), $x_2 \in \mathcal{N}(A^*) = \mathcal{R}(A)^{\perp}$. Since $x \in \mathcal{R}(A)$, it follows that $0 = x_2^*x = x_2^*x_1 + x_2^*x_2$. However, $x_2^*x_1 = 0$ so that $x_2^*x_2 = 0$ and $x_2 = 0$, which is a contradiction. This proves *ii*).

To prove *iii*), note that *ii*) with A replaced by A^* implies that $\mathcal{R}(A^*A)^\perp = \mathcal{R}(A^*)^\perp$. Furthermore, replacing A by A^* in *i*) yields $\mathcal{R}(A^*)^\perp = \mathcal{N}(A)$. Hence, $\mathcal{N}(A) = \mathcal{R}(A^*A)^\perp$. Now, *i*) with A replaced by A^*A implies that $\mathcal{R}(A^*A)^\perp = \mathcal{N}(A^*A)$. Hence, $\mathcal{N}(A) = \mathcal{N}(A^*A)$, which proves *iii*). □

i) of Theorem 3.5.3 can be written equivalently as

$$\mathcal{N}(A)^\perp = \mathcal{R}(A^*), \tag{3.5.12}$$

$$\mathcal{N}(A) = \mathcal{R}(A^*)^\perp, \tag{3.5.13}$$

$$\mathcal{N}(A^*)^\perp = \mathcal{R}(A), \tag{3.5.14}$$

while replacing A by A^* in *ii*) and *iii*) of Theorem 3.5.3 yields

$$\mathcal{R}(A^*) = \mathcal{R}(A^*A), \tag{3.5.15}$$

$$\mathcal{N}(A^*) = \mathcal{N}(AA^*). \tag{3.5.16}$$

Using *ii*) of Theorem 3.5.3 and (3.5.15), it follows that

$$\mathcal{R}(AA^*A) = A\mathcal{R}(A^*A) = A\mathcal{R}(A^*) = \mathcal{R}(AA^*) = \mathcal{R}(A). \tag{3.5.17}$$

Letting $A \triangleq [1 \ \ \jmath]$ shows that $\mathcal{R}(A)$ and $\mathcal{R}(AA^\mathsf{T})$ may be different.

3.6 Rank and Defect

Let $A \in \mathbb{F}^{n\times m}$. Then, the *rank* of A is defined by

$$\operatorname{rank} A \triangleq \dim \mathcal{R}(A). \tag{3.6.1}$$

It can be seen that the rank of A is equal to the number of linearly independent columns of A over $\mathbb{F}$. For example, if $\mathbb{F} = \mathbb{C}$, then $\operatorname{rank}\,[1 \ \ \jmath] = 1$, and, if either $\mathbb{F} = \mathbb{R}$ or $\mathbb{F} = \mathbb{C}$, then $\operatorname{rank}\,[1 \ \ 1] = 1$. If all of the entries of A are real, then $\operatorname{rank} A$ is the same whether $\mathbb{F}$ in (3.5.2) is chosen to be either $\mathbb{R}$ or $\mathbb{C}$. Furthermore, $\operatorname{rank} A = \operatorname{rank} \overline{A}$, $\operatorname{rank} A^\mathsf{T} = \operatorname{rank} A^*$, $\operatorname{rank} A \le m$, and $\operatorname{rank} A^\mathsf{T} \le n$. If $\operatorname{rank} A = m$, then A has *full column rank*, and, if $\operatorname{rank} A^\mathsf{T} = n$, then A has *full row rank*. If A has either full column rank or full row rank, then A has *full rank*; otherwise, A is *rank deficient*. For all nonnegative n and m, $\operatorname{rank}(0_{n\times m}) = \operatorname{rank}(0_{n\times 0}) = \operatorname{rank}(0_{0\times m}) = 0$. Finally, the *defect* of A is

$$\operatorname{def} A \triangleq \dim \mathcal{N}(A). \tag{3.6.2}$$

The following result follows from Theorem 3.5.3.

Corollary 3.6.1. Let $A \in \mathbb{F}^{n\times m}$. Then, the following statements hold:

i) $\operatorname{rank} A^* + \operatorname{def} A = m$.

ii) $\operatorname{rank} A = \operatorname{rank} AA^*$.

iii) $\operatorname{def} A = \operatorname{def} A^*A$.

Proof. (3.5.12) and Proposition 3.1.6 imply that $\operatorname{rank} A^* = \dim \mathcal{R}(A^*) = \dim \mathcal{N}(A)^\perp = m - \dim \mathcal{N}(A) = m - \operatorname{def} A$, which proves *i*). *ii*) and *iii*) follow from *ii*) and *iii*) of Theorem 3.5.3. □

Replacing A by A^* in Corollary 3.6.1 yields

$$\operatorname{rank} A + \operatorname{def} A^* = n, \tag{3.6.3}$$

$$\operatorname{rank} A^* = \operatorname{rank} A^*A, \tag{3.6.4}$$

$$\operatorname{def} A^* = \operatorname{def} AA^*. \tag{3.6.5}$$

Furthermore, note that

$$\operatorname{def} A = \operatorname{def} \overline{A}, \quad \operatorname{def} A^\mathsf{T} = \operatorname{def} A^*. \tag{3.6.6}$$

Lemma 3.6.2. Let $A \in \mathbb{F}^{n\times m}$ and $B \in \mathbb{F}^{m\times l}$. Then,

$$\operatorname{rank} AB \le \min\{\operatorname{rank} A, \operatorname{rank} B\}. \tag{3.6.7}$$

Proof. Since, by Lemma 3.5.1, $\mathcal{R}(AB) \subseteq \mathcal{R}(A)$, it follows that $\operatorname{rank} AB \le \operatorname{rank} A$. Next, suppose that $\operatorname{rank} B < \operatorname{rank} AB$. Let $\{y_1, \dots, y_r\} \subset \mathbb{F}^n$ be a basis for $\mathcal{R}(AB)$, where $r \triangleq \operatorname{rank} AB$, and, since $y_i \in A\mathcal{R}(B)$ for all $i \in \{1, \dots, r\}$, let $x_i \in \mathcal{R}(B)$ be such that $y_i = Ax_i$ for all $i \in \{1, \dots, r\}$. Since $\operatorname{rank} B < r$, it follows that $x_1, \dots, x_r$ are linearly dependent. Hence, there exist $\alpha_1, \dots, \alpha_r \in \mathbb{F}$, not all zero, such that $\sum_{i=1}^r \alpha_i x_i = 0$, which implies that $\sum_{i=1}^r \alpha_i A x_i = \sum_{i=1}^r \alpha_i y_i = 0$. Thus, $y_1, \dots, y_r$ are linearly dependent, which is a contradiction. □

Corollary 3.6.3. Let $A \in \mathbb{F}^{n \times m}$. Then,

$$\operatorname{rank} A = \operatorname{rank} A^*, \tag{3.6.8}$$

$$\operatorname{def} A = \operatorname{def} A^* + m - n. \tag{3.6.9}$$

Therefore,

$$\operatorname{rank} A = \operatorname{rank} A^*A. \tag{3.6.10}$$

If, in addition, $n = m$, then

$$\operatorname{def} A = \operatorname{def} A^*. \tag{3.6.11}$$

Proof. It follows from (3.6.7) with $B = A^*$ that $\operatorname{rank} AA^* \le \operatorname{rank} A^*$. Furthermore, *ii*) of Corollary 3.6.1 implies that $\operatorname{rank} A = \operatorname{rank} AA^*$. Hence, $\operatorname{rank} A \le \operatorname{rank} A^*$. Interchanging A and A^* and repeating this argument yields $\operatorname{rank} A^* \le \operatorname{rank} A$. Hence, $\operatorname{rank} A = \operatorname{rank} A^*$. Next, *i*) of Corollary 3.6.1, (3.6.8), and (3.6.3) imply that $\operatorname{def} A = m - \operatorname{rank} A^* = m - \operatorname{rank} A = m - (n - \operatorname{def} A^*)$, which proves (3.6.9). □

Corollary 3.6.4. Let $A \in \mathbb{F}^{n \times m}$. Then,

$$\operatorname{rank} A \le \min\{m, n\}. \tag{3.6.12}$$

Proof. By definition, $\operatorname{rank} A \le m$, while it follows from (3.6.8) that $\operatorname{rank} A = \operatorname{rank} A^* \le n$. □

The *dimension theorem* is given by (3.6.13) in the following result.

Corollary 3.6.5. Let $A \in \mathbb{F}^{n \times m}$. Then,

$$\operatorname{rank} A + \operatorname{def} A = m, \tag{3.6.13}$$

$$\operatorname{rank} A = \operatorname{rank} A^*A. \tag{3.6.14}$$

Proof. (3.6.13) follows from *i*) of Corollary 3.6.1 and (3.6.8), while (3.6.14) follows from (3.6.4) and (3.6.8). □

Corollary 3.6.6. Let $A \in \mathbb{F}^{n \times n}$ and $l \ge k \ge 1$. Then,

$$\operatorname{rank} A^l \le \operatorname{rank} A^k, \tag{3.6.15}$$

$$\operatorname{def} A^k \le \operatorname{def} A^l. \tag{3.6.16}$$

Proposition 3.6.7. Let $A \in \mathbb{F}^{n \times n}$. If $\operatorname{rank} A^2 = \operatorname{rank} A$, then, for all $k \ge 1$, $\operatorname{rank} A^k = \operatorname{rank} A$. Equivalently, if $\operatorname{def} A^2 = \operatorname{def} A$, then, for all $k \ge 1$, $\operatorname{def} A^k = \operatorname{def} A$.

Proof. Since $\operatorname{rank} A^2 = \operatorname{rank} A$ and $\mathcal{R}(A^2) \subset \mathcal{R}(A)$, it follows from Lemma 3.1.5 that $\mathcal{R}(A^2) = \mathcal{R}(A)$. Hence, $\mathcal{R}(A^3) = A\mathcal{R}(A^2) = A\mathcal{R}(A) = \mathcal{R}(A^2)$. Thus, $\operatorname{rank} A^3 = \operatorname{rank} A$. For all $k \ge 1$, similar arguments yield $\operatorname{rank} A^k = \operatorname{rank} A$. □

The following results follow from the subspace dimension theorem (3.1.12) and the dimension theorem (3.6.13).

Corollary 3.6.8. Let $A \in \mathbb{F}^{n \times n}$. Then,

$$\dim[\mathcal{R}(A) + \mathcal{N}(A)] + \dim[\mathcal{R}(A) \cap \mathcal{N}(A)] = n. \tag{3.6.17}$$

Corollary 3.6.9. Let $A \in \mathbb{F}^{n \times m}$ and $B \in \mathbb{F}^{m \times l}$. Then,

$$\dim[\mathcal{N}(A) + \mathcal{R}(B)] + \dim[\mathcal{N}(A) \cap \mathcal{R}(B)] = \operatorname{def} A + \operatorname{rank} B. \tag{3.6.18}$$

The following result is *Sylvester's rank formula.*

Proposition 3.6.10. Let $A \in \mathbb{F}^{n\times m}$ and $B \in \mathbb{F}^{m\times l}$. Then,

$$\operatorname{rank} AB + \dim[\mathcal{N}(A) \cap \mathcal{R}(B)] = \operatorname{rank} B. \tag{3.6.19}$$

Proof. Using Proposition 8.1.7 and Fact 8.9.15 it follows that

$$\begin{aligned}\dim[\mathcal{N}(A) + \mathcal{R}(B)] &= \dim[\mathcal{R}(I - A^{+}A) + \mathcal{R}(B)] = \dim \mathcal{R}([B \quad I - A^{+}A]) \\ &= \operatorname{rank}\,[B \quad I - A^{+}A] = m + \operatorname{rank} AB - \operatorname{rank} A.\end{aligned}$$

Therefore, using (3.6.18) it follows that

$$\begin{aligned}\operatorname{rank} AB + \dim[\mathcal{N}(A) \cap \mathcal{R}(B)] &= \operatorname{rank} AB + \operatorname{def} A + \operatorname{rank} B - \dim[\mathcal{N}(A) + \mathcal{R}(B)] \\ &= \operatorname{rank} AB + \operatorname{def} A + \operatorname{rank} B - (m + \operatorname{rank} AB - \operatorname{rank} A) \\ &= \operatorname{def} A + \operatorname{rank} B - m + \operatorname{rank} A \\ &= \operatorname{rank} B.\end{aligned}$$

□

The next result is *Sylvester's inequality.*

Proposition 3.6.11. Let $A \in \mathbb{F}^{n\times m}$ and $B \in \mathbb{F}^{m\times l}$. Then,

$$\operatorname{rank} A + \operatorname{rank} B \le m + \operatorname{rank} AB. \tag{3.6.20}$$

Furthermore, equality holds if and only if $\mathcal{N}(A) \subseteq \mathcal{R}(B)$.

Proof. It follows from (3.6.13) and (3.6.19) that

$$\operatorname{rank} B - \operatorname{rank} AB = \dim[\mathcal{N}(A) \cap \mathcal{R}(B)] \le \dim \mathcal{N}(A) = \operatorname{def} A = m - \operatorname{rank} A,$$

which implies (3.6.20). Alternatively, using (3.6.7) to obtain the second inequality below, it follows that

$$\begin{aligned}\operatorname{rank} A + \operatorname{rank} B &= \operatorname{rank}\begin{bmatrix} 0 & A \\ B & 0 \end{bmatrix} \le \operatorname{rank}\begin{bmatrix} 0 & A \\ B & I \end{bmatrix} = \operatorname{rank}\begin{bmatrix} -I & A \\ 0 & I \end{bmatrix}\begin{bmatrix} AB & 0 \\ B & I \end{bmatrix} \\ &= \operatorname{rank}\begin{bmatrix} AB & 0 \\ B & I \end{bmatrix} \le \operatorname{rank}\,[AB \quad 0] + \operatorname{rank}\,[B \quad I] = \operatorname{rank} AB + m.\end{aligned}$$

□

Combining (3.6.7) with (3.6.20) yields lower and upper bounds for $\operatorname{rank} AB$.

Corollary 3.6.12. Let $A \in \mathbb{F}^{n\times m}$ and $B \in \mathbb{F}^{m\times l}$. Then,

$$\max\{0, \operatorname{rank} A + \operatorname{rank} B - m\} \le \operatorname{rank} AB \le \min\{\operatorname{rank} A, \operatorname{rank} B\} \le \min\{n, m, l\}. \tag{3.6.21}$$

Corollary 3.6.13. Let $A \in \mathbb{F}^{n\times m}$ and $B \in \mathbb{F}^{m\times l}$. Then, the following statements hold:

i) If $\operatorname{rank} AB = n$, then $\operatorname{rank} A = n$.

ii) If $\operatorname{rank} AB = l$, then $\operatorname{rank} B = l$.

iii) $\operatorname{rank} AB = m$ if and only if $\operatorname{rank} A = \operatorname{rank} B = m$.

3.7 Invertibility

Let $A \in \mathbb{F}^{n\times m}$. Then, A is *left invertible* if there exists $A^{\mathsf{L}} \in \mathbb{F}^{m\times n}$ such that $A^{\mathsf{L}}A = I_m$, while A is *right invertible* if there exists $A^{\mathsf{R}} \in \mathbb{F}^{m\times n}$ such that $AA^{\mathsf{R}} = I_n$. These definitions are consistent with the definitions of left and right invertibility given in Chapter 1 applied to the function $f\colon\ \mathbb{F}^m \mapsto \mathbb{F}^n$ given by $f(x) = Ax$. Note that A^{L} (if it exists) and A^* are the same size, and likewise for A^{R}.

Theorem 3.7.1. Let $A \in \mathbb{F}^{n\times m}$. Then, the following statements are equivalent:

i) A is left invertible.

ii) A is one-to-one.

iii) $\operatorname{def} A = 0$.

iv) $\operatorname{rank} A = m$.

v) A has full column rank.

The following statements are equivalent:

vi) A is right invertible.

vii) A is onto.

viii) $\operatorname{def} A = m - n$.

ix) $\operatorname{rank} A = n$.

x) A has full row rank.

Proposition 3.7.2. Let $A \in \mathbb{F}^{n\times m}$. Then, the following statements are equivalent:

i) A has a unique left inverse.

ii) A has a unique right inverse.

iii) $\operatorname{rank} A = n = m$.

Proof. To prove *i*) $\Longrightarrow$ *iii*), suppose that $\operatorname{rank} A = m < n$ so that A is left invertible but nonsquare. Then, it follows from the dimension theorem Corollary 3.6.5 that $\operatorname{def} A^{\mathsf{T}} = n - m > 0$. Hence, there exist infinitely many matrices $A^{\mathsf{L}} \in \mathbb{F}^{m\times n}$ such that $A^{\mathsf{L}}A = I_m$. Conversely, suppose that $B \in \mathbb{F}^{n\times n}$ and $C \in \mathbb{F}^{n\times n}$ are left inverses of A. Then, $(B - C)A = 0$, and it follows from Sylvester's inequality Proposition 3.6.11 that $B = C$. □

The following result shows that the rank and defect of a matrix are not affected by either left multiplication by a left invertible matrix or right multiplication by a right invertible matrix. This result is an extension of Lemma 3.5.1.

Proposition 3.7.3. Let $A \in \mathbb{F}^{n\times m}$, let $B \in \mathbb{F}^{m\times l}$ be right invertible, and let $C \in \mathbb{F}^{k\times n}$ be left invertible. Then,

$$\mathcal{R}(A) = \mathcal{R}(AB), \tag{3.7.1}$$

$$\mathcal{N}(A) = \mathcal{N}(CA), \tag{3.7.2}$$

$$\operatorname{rank} A = \operatorname{rank} CA = \operatorname{rank} AB, \tag{3.7.3}$$

$$\operatorname{def} A = \operatorname{def} CA = \operatorname{def} AB + m - l. \tag{3.7.4}$$

Proof. Let C^{L} be a left inverse of C. Using both inequalities in (3.6.21) and the fact that $\operatorname{rank} A \le n$, it follows that

$$\operatorname{rank} A = \operatorname{rank} A + \operatorname{rank} C^{\mathsf{L}}C - n \le \operatorname{rank} C^{\mathsf{L}}CA \le \operatorname{rank} CA \le \operatorname{rank} A,$$

which implies that $\operatorname{rank} A = \operatorname{rank} CA$. Finally, (3.6.13) and (3.7.3) imply that

$$\operatorname{def} A = m - \operatorname{rank} A = m - \operatorname{rank} AB = m - (l - \operatorname{def} AB) = \operatorname{def} AB + m - l.$$ □

Proposition 3.7.4. Let $A \in \mathbb{F}^{n\times m}$, assume that A is left invertible, let $B \in \mathbb{F}^{m\times n}$ and $C \in \mathbb{F}^{m\times n}$ be left inverses of A, and let $y \in \mathcal{R}(A)$. Then, $By = Cy$.

Proof. Let $x \in \mathbb{F}^m$ satisfy $y = Ax$. Then, $By = BAx = x = CAx = Cy$. □

As shown in Proposition 3.7.2, left and right inverses of nonsquare matrices are not unique. For example, the matrix $A = \begin{bmatrix} 0 \\ 1 \end{bmatrix}$ is left invertible and has left inverses $[0\ \ 1]$ and $[1\ \ 1]$. In spite of this nonuniqueness, however, left inverses are useful for solving equations of the form $Ax = b$, where $A \in \mathbb{F}^{n\times m}$, $x \in \mathbb{F}^m$, and $b \in \mathbb{F}^n$. If A is left invertible, it follows from $Ax = b$ that $x = A^{\mathsf{L}}Ax = A^{\mathsf{L}}b$, where $A^{\mathsf{L}} \in \mathbb{R}^{m\times n}$ is a left inverse of A. However, it is necessary to determine beforehand whether or not there exists x satisfying $Ax = b$. For example, if $A = \begin{bmatrix} 0 \\ 1 \end{bmatrix}$ and $b = \begin{bmatrix} 1 \\ 0 \end{bmatrix}$, then A is left invertible but there does not exist a vector x satisfying $Ax = b$. The following result addresses the various possibilities that can arise. A consequence of Proposition 3.7.4 is the fact that, if there exists a solution of $Ax = b$ and A is left invertible, then $Ax = b$ has a unique solution in the case where A does not have a unique left inverse, which holds if and only if $m < n$.

The notation $[A\ \ b]$ denotes the $n \times (m+1)$ partitioned matrix formed from $A \in \mathbb{F}^{n\times m}$ and $b \in \mathbb{F}^n$. Note that $\operatorname{rank} A \le \operatorname{rank}[A\ \ b] \le m+1$, while $\operatorname{rank} A = \operatorname{rank}[A\ \ b]$ is equivalent to $b \in \mathcal{R}(A)$.

Theorem 3.7.5. Let $A \in \mathbb{F}^{n\times m}$ and $b \in \mathbb{F}^n$. Then, the following statements hold:

i) The following statements are equivalent:

a) $Ax = b$ has no solution.

b) $\operatorname{rank} A < \operatorname{rank}[A\ \ b]$.

ii) The following statements are equivalent:

a) $Ax = b$ has at least one solution.

b) $\operatorname{rank} A = \operatorname{rank}[A\ \ b]$.

c) $b \in \mathcal{R}(A)$.

If these statements hold and $\hat{x} \in \mathbb{F}^m$ satisfies $A\hat{x} = b$, then the set of solutions of $Ax = b$ is given by $\hat{x} + \mathcal{N}(A)$.

iii) The following statements are equivalent:

a) $Ax = b$ has a unique solution.

b) $\operatorname{rank} A = \operatorname{rank}[A\ \ b] = m$.

c) $Ax = b$ has at least one solution, and A is left invertible.

iv) The following statements are equivalent:

a) $Ax = b$ has infinitely many solutions.

b) $\operatorname{rank} A = \operatorname{rank}[A\ \ b] < m$.

c) $Ax = b$ has at least one solution, and A is not left invertible.

v) Assume that A is left invertible. Then, the following statements hold:

a) $Ax = b$ has at most one solution.

b) $Ax = b$ has no solution if and only if $\operatorname{rank}[A\ \ b] = m+1$.

c) $Ax = b$ has a unique solution if and only if $\operatorname{rank}[A\ \ b] = m$. If these conditions hold and $A^{\mathsf{L}} \in \mathbb{F}^{m\times n}$ is a left inverse of A, then the unique solution of $Ax = b$ is given by $x = A^{\mathsf{L}}b$.

vi) Assume that A is right invertible. Then, the following statements hold:

a) $Ax = b$ has at least one solution.

b) Let $A^{\mathsf{R}} \in \mathbb{F}^{m\times n}$ be a right inverse of A. Then, $x = A^{\mathsf{R}}b$ is a solution of $Ax = b$.

c) $Ax = b$ has a unique solution if and only if $n = m$. If these conditions hold and $A^{\mathsf{R}} \in \mathbb{F}^{m\times n}$ is a right inverse of A, then $A^{\mathsf{R}} = A^{-1}$.

Proof. To prove *i)*, note that $\operatorname{rank} A < \operatorname{rank}[A\ \ b]$ is equivalent to the statement that b is not a linear combination of columns of A; equivalently, $Ax = b$ does not have a solution $x \in \mathbb{F}^m$.

To prove *ii)*, note that the equivalence of *a)* and *b)* follows from *i)*. To prove *b)* $\Longrightarrow$ *c)*, note that, if $b \notin \mathcal{R}(A)$, then $[A\ \ b]$ has more linearly independent columns than A, and thus $\operatorname{rank} A < \operatorname{rank}[A\ \ b]$. To prove *c)* $\Longrightarrow$ *b)*, let $x \in \mathbb{F}^n$ satisfy $Ax = b$. Then, $\operatorname{rank} A \le \operatorname{rank}[A\ \ b] = \operatorname{rank}[A\ \ Ax] = \operatorname{rank} A[I\ \ x] \le \operatorname{rank} A$. Hence, $\operatorname{rank} A = \operatorname{rank}[A\ \ b]$. Finally, assume that $b \in \mathcal{R}(A)$, and let $\hat{x} \in \mathbb{F}^n$ satisfy $A\hat{x} = b$. Then, $x \in \mathbb{F}^n$ satisfies $Ax = b$ if and only if $A(x - \hat{x}) = 0$, which is equivalent to $x \in \hat{x} + \mathcal{N}(A)$.

To prove *a)* $\Longrightarrow$ *b)* of *iii)*, note that it follows from *ii)* that $\mathcal{N}(A) = \{0\}$. Theorem 3.7.1 thus implies that $\operatorname{rank} A = m$, which, along with *ii)*, implies *b)*. Conversely, it follows from *b)* that

$\operatorname{rank} A = m$, and thus it follows from the last statement of *ii*) that $Ax = b$ has a unique solution. Finally, *c*) is a restatement of *b*).

iv) and *v*) are consequences of *i*)–*iii*).

To prove *vi*), note that, since $\operatorname{rank} A = n$, it follows that $\operatorname{rank} A = \operatorname{rank}[A \ \ b]$, and thus it follows from *ii*) that $Ax = b$ has at least one solution. In fact, $AA^{\mathrm{R}}b = b$. The last statement follows from Proposition 3.7.2. □

If A is right invertible, then it does not follow that, for every solution x of $Ax = b$, there exists a right inverse A^{R} of A such that $x = A^{\mathrm{R}}b$. For example, let $A = [1 \ \ 0]$ and $b = 0$. Then, $x = \begin{bmatrix} 0 \\ 1 \end{bmatrix}$ satisfies $Ax = b$, but $A^{\mathrm{R}}b = 0$ for every right inverse of A. See Fact 8.3.16.

The set of solutions of $Ax = b$ is further characterized by Proposition 8.1.8, while connections to least squares solutions are discussed in Fact 11.17.10.

Let $A \in \mathbb{F}^{n \times m}$. Proposition 3.7.2 considers the uniqueness of left and right inverses of A, but does not consider the case where a matrix is both a left inverse and a right inverse of A. Consequently, we say that A is *nonsingular* if there exists $B \in \mathbb{F}^{m \times n}$, the *inverse* of A, such that $BA = I_m$ and $AB = I_n$; that is, B is both a left and right inverse of A.

Proposition 3.7.6. Let $A \in \mathbb{F}^{n \times m}$. Then, the following statements are equivalent:

i) A is nonsingular.

ii) $\operatorname{rank} A = n = m$.

If these statements hold, then A has a unique inverse.

Proof. If A is nonsingular, then, since B is both left and right invertible, it follows from Theorem 3.7.1 that $\operatorname{rank} A = m$ and $\operatorname{rank} A = n$. Hence, *ii*) holds. Conversely, it follows from Theorem 3.7.1 that A has both a left inverse B and a right inverse C. Then, $B = BI_n = BAC = I_nC = C$. Hence, B is also a right inverse of A. Thus, A is nonsingular. In fact, the same argument shows that A has a unique inverse. □

The following result can be viewed as a specialization of Theorem 1.6.5 to the function $f\colon \mathbb{F}^n \mapsto \mathbb{F}^n$, where $f(x) = Ax$.

Corollary 3.7.7. Let $A \in \mathbb{F}^{n \times n}$. Then, the following statements are equivalent:

i) A is nonsingular.

ii) A has a unique inverse.

iii) A is one-to-one.

iv) A is onto.

v) A is left invertible.

vi) A is right invertible.

vii) A has a unique left inverse.

viii) A has a unique right inverse.

ix) $\operatorname{rank} A = n$.

x) $\operatorname{def} A = 0$.

Let $A \in \mathbb{F}^{n \times n}$ be nonsingular. Then, the inverse of A, denoted by A^{-1}, is a unique $n \times n$ matrix with entries in $\mathbb{F}$. If A is not nonsingular, then A is *singular*.

The following results follow from Theorem 3.7.5 in the case $n = m$.

Corollary 3.7.8. Let $A \in \mathbb{F}^{n \times n}$ and $b \in \mathbb{F}^n$. Then, the following statements hold:

i) A is nonsingular if and only if there exists a unique vector $x \in \mathbb{F}^n$ satisfying $Ax = b$. If these conditions hold, then $x = A^{-1}b$.

ii) A is singular and $\operatorname{rank} A = \operatorname{rank}[A \ \ b]$ if and only if there exist infinitely many $x \in \mathbb{R}^n$

satisfying $Ax = b$. Assume that these conditions hold, and let $\hat{x} \in \mathbb{F}^n$ satisfy $A\hat{x} = b$. Then, the set of solutions of $Ax = b$ is given by $\hat{x} + \mathcal{N}(A)$.

Corollary 3.7.9. Let $A \in \mathbb{F}^{n\times n}$. Then, the following statements are equivalent:

i) A is nonsingular.

ii) For all $b \in \mathbb{F}^n$, there exists a unique solution $x \in \mathbb{F}^m$ of $Ax = b$.

iii) For all $b \in \mathbb{F}^n$, there exists a solution $x \in \mathbb{F}^m$ of $Ax = b$.

iv) There exists $b \in \mathbb{F}^n$ such that $Ax = b$ has a unique solution $x \in \mathbb{F}^m$.

Proposition 3.7.10. Let $\mathcal{S} \subseteq \mathbb{F}^m$, let $\hat{\mathcal{S}} \subseteq \mathbb{F}^n$, and let $A \in \mathbb{F}^{n\times m}$. Then,

$$\mathcal{S} \subseteq A^{\mathsf{inv}}(A\mathcal{S}), \tag{3.7.5}$$

$$AA^{\mathsf{inv}}(\hat{\mathcal{S}}) = \hat{\mathcal{S}} \cap \mathcal{R}(A) \subseteq \hat{\mathcal{S}}. \tag{3.7.6}$$

Furthermore, if A is left invertible and A^{L} is a left inverse of A, then

$$\mathcal{S} = A^{\mathsf{inv}}(A\mathcal{S}), \tag{3.7.7}$$

$$A^{\mathsf{inv}}(\hat{\mathcal{S}}) = A^{\mathsf{L}}[\hat{\mathcal{S}} \cap \mathcal{R}(A)] \subseteq A^{\mathsf{L}}\hat{\mathcal{S}}. \tag{3.7.8}$$

In addition, if A is right invertible and A^{R} is a right inverse of A, then

$$AA^{\mathsf{inv}}(\hat{\mathcal{S}}) = \hat{\mathcal{S}}, \tag{3.7.9}$$

$$A^{\mathsf{R}}\hat{\mathcal{S}} \subseteq A^{\mathsf{inv}}(\hat{\mathcal{S}}). \tag{3.7.10}$$

Finally, if A is nonsingular, then

$$A^{\mathsf{inv}}(\mathcal{S}) = A^{-1}\mathcal{S}. \tag{3.7.11}$$

Proof. The inclusions (3.7.5) and (3.7.6) follow from Fact 1.10.3 and Fact 1.10.4, respectively. To prove (3.7.8), multiply (3.7.6) by A^{L}. To prove (3.7.7), replace $\hat{\mathcal{S}}$ in (3.7.8) with $A\mathcal{S}$ and use (3.7.5). To prove (3.7.10), replace $\mathcal{S}$ in (3.7.5) with $A^{\mathsf{R}}\hat{\mathcal{S}}$. To prove (3.7.9), multiply (3.7.10) by A and use (3.7.6). □

A more complete characterization of $A^{\mathsf{inv}}(\mathcal{S})$ is given by Proposition 8.1.10.

Proposition 3.7.11. Let $A \in \mathbb{F}^{n\times n}$. Then, the following statements are equivalent:

i) A is nonsingular.

ii) $\overline{A}$ is nonsingular.

iii) A^{T} is nonsingular.

iv) A^* is nonsingular.

If these statements hold, then

$$(\overline{A})^{-1} = \overline{A^{-1}}, \tag{3.7.12}$$

$$(A^{\mathsf{T}})^{-1} = (A^{-1})^{\mathsf{T}}, \tag{3.7.13}$$

$$(A^*)^{-1} = (A^{-1})^*. \tag{3.7.14}$$

Proof. Since $AA^{-1} = I$, it follows that $(A^{-1})^*A^* = I$. Hence, $(A^{-1})^* = (A^*)^{-1}$. □

We thus use $A^{-\mathsf{T}}$ to denote $(A^{\mathsf{T}})^{-1}$ and $(A^{-1})^{\mathsf{T}}$ and A^{-*} to denote $(A^*)^{-1}$ and $(A^{-1})^*$.

Proposition 3.7.12. Let $A, B \in \mathbb{F}^{n\times n}$ be nonsingular. Then,

$$(AB)^{-1} = B^{-1}A^{-1}, \tag{3.7.15}$$

$$(AB)^{-\mathsf{T}} = A^{-\mathsf{T}}B^{-\mathsf{T}}, \tag{3.7.16}$$

$$(AB)^{-*} = A^{-*}B^{-*}. \tag{3.7.17}$$

Proof. Note that $ABB^{-1}A^{-1} = AIA^{-1} = I$, which shows that $B^{-1}A^{-1}$ is the inverse of AB. Similarly, $(AB)^*A^{-*}B^{-*} = B^*A^*A^{-*}B^{-*} = B^*IB^{-*} = I$, which shows that $A^{-*}B^{-*}$ is the inverse of $(AB)^*$. □

For a nonsingular matrix $A \in \mathbb{F}^{n\times n}$ and $r \in \mathbb{Z}$ we write

$$A^{-r} \triangleq (A^r)^{-1} = (A^{-1})^r, \tag{3.7.18}$$

$$A^{-r\mathsf{T}} \triangleq (A^r)^{-\mathsf{T}} = (A^{-\mathsf{T}})^r = (A^{-r})^{\mathsf{T}} = (A^{\mathsf{T}})^{-r}, \tag{3.7.19}$$

$$A^{-r*} \triangleq (A^r)^{-*} = (A^{-*})^r = (A^{-r})^* = (A^*)^{-r}. \tag{3.7.20}$$

For example, $A^{-2*} = (A^{-*})^2$.

3.8 The Determinant

One of the most useful quantities associated with a square matrix is its determinant. In this section we develop some basic results pertaining to the determinant of a matrix.

The *determinant* of $A \in \mathbb{F}^{n\times n}$ is defined by

$$\det A \triangleq \sum \operatorname{sign}(\sigma) \prod_{i=1}^{n} A_{(i,\sigma(i))}, \tag{3.8.1}$$

where the sum is taken over all $n!$ permutations σ of the column indices $(1,\ldots,n)$, and where

$$\operatorname{sign}(\sigma) \triangleq \prod_{1\le i<j\le n} \frac{\sigma(j)-\sigma(i)}{j-i}. \tag{3.8.2}$$

It can be seen that $\operatorname{sign}(\sigma) = (-1)^k$, where k is the smallest number of transpositions needed to transform $(\sigma(1),\ldots,\sigma(n))$ to $(1,\ldots,n)$. The following result is an immediate consequence of this definition.

Proposition 3.8.1. Let $A \in \mathbb{F}^{n\times n}$. Then,

$$\det A^{\mathsf{T}} = \det A, \tag{3.8.3}$$

$$\det \overline{A} = \overline{\det A}, \tag{3.8.4}$$

$$\det A^* = \overline{\det A}, \tag{3.8.5}$$

and, for all $\alpha \in \mathbb{F}$,

$$\det \alpha A = \alpha^n \det A. \tag{3.8.6}$$

If, in addition, $B \in \mathbb{F}^{m\times n}$ and $C \in \mathbb{F}^{m\times m}$, then

$$\det \begin{bmatrix} A & 0 \\ B & C \end{bmatrix} = (\det A)(\det C). \tag{3.8.7}$$

The following observations are immediate consequences of the definition of the determinant.

Proposition 3.8.2. Let $A, B \in \mathbb{F}^{n\times n}$. Then, the following statements hold:

i) If every off-diagonal entry of A is zero, then

$$\det A = \prod_{i=1}^{n} A_{(i,i)}. \tag{3.8.8}$$

In particular, $\det I_n = 1$.

ii) If A has a row or column consisting entirely of 0's, then $\det A = 0$.

iii) If A has two identical rows or two identical columns, then $\det A = 0$.

iv) If $x \in \mathbb{F}^n$ and $i \in \{1,\dots,n\}$, then

$$\det(A + xe_i^\mathsf{T}) = \det A + \det(A \overset{i}{\leftarrow} x). \tag{3.8.9}$$

v) If $x \in \mathbb{F}^{1\times n}$ and $i \in \{1,\dots,n\}$, then

$$\det(A + e_i x) = \det A + \det(A \overset{i}{\leftarrow} x). \tag{3.8.10}$$

vi) If B is equal to A except that, for some $i \in \{1,\dots,n\}$ and $\alpha \in \mathbb{F}$, either $\mathrm{col}_i(B) = \alpha\mathrm{col}_i(A)$ or $\mathrm{row}_i(B) = \alpha\mathrm{row}_i(A)$, then $\det B = \alpha \det A$.

vii) If B is formed from A by interchanging two rows or two columns of A, then $\det B = -\det A$.

viii) If B is formed from A by adding a multiple of a (row, column) of A to another (row, column) of A, then $\det B = \det A$.

vi)–*viii*) correspond, respectively, to multiplying the matrix A on the left or right by matrices of the form

$$I_n + (\alpha - 1)E_{i,i} = \begin{bmatrix} I_{i-1} & 0 & 0 \\ 0 & \alpha & 0 \\ 0 & 0 & I_{n-i} \end{bmatrix}, \tag{3.8.11}$$

$$I_n + E_{i,j} + E_{j,i} - E_{i,i} - E_{j,j} = \begin{bmatrix} I_{i-1} & 0 & 0 & 0 & 0 \\ 0 & 0 & 0 & 1 & 0 \\ 0 & 0 & I_{j-i-1} & 0 & 0 \\ 0 & 1 & 0 & 0 & 0 \\ 0 & 0 & 0 & 0 & I_{n-j} \end{bmatrix}, \tag{3.8.12}$$

where $i \neq j$, and

$$I_n + \beta E_{i,j} = \begin{bmatrix} I_{i-1} & 0 & 0 & 0 & 0 \\ 0 & 1 & 0 & \beta & 0 \\ 0 & 0 & I_{j-i-1} & 0 & 0 \\ 0 & 0 & 0 & 1 & 0 \\ 0 & 0 & 0 & 0 & I_{n-j} \end{bmatrix}, \tag{3.8.13}$$

where $\beta \in \mathbb{F}$ and $i \neq j$. The matrices in (3.8.12) and (3.8.13) illustrate the case $i < j$. Since $I+(\alpha-1)E_{i,i} = I+(\alpha-1)e_ie_i^\mathsf{T}$, $I+E_{i,j}+E_{j,i}-E_{i,i}-E_{j,j} = I-(e_i-e_j)(e_i-e_j)^\mathsf{T}$, and $I+\beta E_{i,j} = I+\beta e_ie_j^\mathsf{T}$, it follows that all of these matrices are of the form $I - xy^\mathsf{T}$. In terms of Definition 4.1.1, (3.8.11) is an elementary matrix if and only if $\alpha \neq 0$, (3.8.12) is an elementary matrix, and (3.8.13) is an elementary matrix if and only if either $i \neq j$ or $\beta \neq -1$.

Proposition 3.8.3. Let $A, B \in \mathbb{F}^{n\times n}$. Then,

$$\det AB = \det BA = (\det A)(\det B). \tag{3.8.14}$$

Proof. First note the equality

$$\begin{bmatrix} A & 0 \\ I & B \end{bmatrix} = \begin{bmatrix} I & A \\ 0 & I \end{bmatrix}\begin{bmatrix} -AB & 0 \\ 0 & I \end{bmatrix}\begin{bmatrix} I & 0 \\ B & I \end{bmatrix}\begin{bmatrix} 0 & I \\ I & 0 \end{bmatrix}.$$

The first and third matrices on the right-hand side of this equality add multiples of rows and columns of $\begin{bmatrix} -AB & 0 \\ 0 & I \end{bmatrix}$ to other rows and columns of $\begin{bmatrix} -AB & 0 \\ 0 & I \end{bmatrix}$. These operations do not affect the determinant of $\begin{bmatrix} -AB & 0 \\ 0 & I \end{bmatrix}$. In addition, the fourth matrix on the right-hand side of this equality interchanges n pairs of columns of $\begin{bmatrix} 0 & A \\ B & I \end{bmatrix}$. Using (3.8.6), (3.8.7), and the fact that every interchange of a pair of columns of $\begin{bmatrix} 0 & A \\ B & I \end{bmatrix}$ multiplies the determinant by -1, it thus follows that $(\det A)(\det B) = \det \begin{bmatrix} A & 0 \\ I & B \end{bmatrix} = (-1)^n \det \begin{bmatrix} -AB & 0 \\ 0 & I \end{bmatrix} = (-1)^n \det(-AB) = \det AB$. □

Corollary 3.8.4. Let $A \in \mathbb{F}^{n \times n}$ be nonsingular. Then, $\det A \neq 0$ and

$$\det A^{-1} = (\det A)^{-1}. \tag{3.8.15}$$

Proof. Since $AA^{-1} = I_n$, it follows that $\det AA^{-1} = (\det A)\det A^{-1} = 1$. Hence, $\det A \neq 0$. In addition, $\det A^{-1} = 1/\det A$. □

Let $A \in \mathbb{F}^{n \times m}$. The determinant of a square submatrix of A is a *subdeterminant* of A. By convention, the determinant of A is a subdeterminant of A. The determinant of a $j \times j$ (principal, leading principal) submatrix of A is a $j \times j$ *(principal, leading principal) subdeterminant* of A.

Let $A \in \mathbb{F}^{n \times n}$. Then, the *cofactor* of $A_{(i,j)}$ is the $(n-1) \times (n-1)$ submatrix $A_{[i,j]}$ of A obtained by deleting the ith row and jth column of A. The following result provides a cofactor expansion of $\det A$.

Proposition 3.8.5. Let $A \in \mathbb{F}^{n \times n}$. Then, for all $i \in \{1, \ldots, n\}$,

$$\sum_{k=1}^{n} (-1)^{i+k} A_{(i,k)} \det A_{[i,k]} = \det A. \tag{3.8.16}$$

Furthermore, for all $i, j \in \{1, \ldots, n\}$ such that $j \neq i$,

$$\sum_{k=1}^{n} (-1)^{i+k} A_{(j,k)} \det A_{[i,k]} = 0. \tag{3.8.17}$$

Proof. Equality (3.8.16) is an equivalent recursive form of the definition $\det A$, while the right-hand side of (3.8.17) is equal to $\det B$, where B is obtained from A by replacing $\mathrm{row}_i(A)$ by $\mathrm{row}_j(A)$. Hence, $\det B = 0$. □

Let $A \in \mathbb{F}^{n \times n}$, where $n \geq 2$. To simplify (3.8.16) and (3.8.17) it is useful to define the *adjugate* of A, denoted by $A^{\mathsf{A}} \in \mathbb{F}^{n \times n}$, where, for all $i, j \in \{1, \ldots, n\}$,

$$(A^{\mathsf{A}})_{(i,j)} \triangleq (-1)^{i+j} \det A_{[j,i]} = \det(A \stackrel{i}{\leftarrow} e_j). \tag{3.8.18}$$

Then, (3.8.16) implies that, for all $i \in \{1, \ldots, n\}$,

$$\sum_{k=1}^{n} A_{(i,k)} (A^{\mathsf{A}})_{(k,i)} = (AA^{\mathsf{A}})_{(i,i)} = (A^{\mathsf{A}}A)_{(i,i)} = \det A, \tag{3.8.19}$$

while (3.8.17) implies that, for all $i, j \in \{1, \ldots, n\}$ such that $j \neq i$,

$$\sum_{k=1}^{n} A_{(i,k)} (A^{\mathsf{A}})_{(k,j)} = (AA^{\mathsf{A}})_{(i,j)} = (A^{\mathsf{A}}A)_{(i,j)} = 0. \tag{3.8.20}$$

Thus,

$$AA^{\mathsf{A}} = A^{\mathsf{A}}A = (\det A)I. \tag{3.8.21}$$

Consequently, if $\det A \neq 0$, then

$$A^{-1} = \frac{1}{\det A} A^{\mathsf{A}}, \tag{3.8.22}$$

whereas, if $\det A = 0$, then

$$AA^{\mathsf{A}} = A^{\mathsf{A}}A = 0. \tag{3.8.23}$$

For a scalar $A \in \mathbb{F}$, we define $A^{\mathsf{A}} \triangleq 1$. In particular, $0_{1\times 1}^{\mathsf{A}} = 1$.

The following result provides the converse of Corollary 3.8.4 by using (3.8.22) to construct A^{-1} in terms of $(n-1)\times(n-1)$ subdeterminants of A.

Corollary 3.8.6. Let $A \in \mathbb{F}^{n\times n}$. Then, A is nonsingular if and only if $\det A \neq 0$. If these conditions hold, then, for all $i, j \in \{1,\dots,n\}$, the (i, j) entry of A^{-1} is given by

$$(A^{-1})_{(i,j)} = (-1)^{i+j}\frac{\det A_{[j,i]}}{\det A}. \tag{3.8.24}$$

Finally, the following result characterizes the rank of a matrix in terms of the nonsingularity of its submatrices.

Proposition 3.8.7. Let $A \in \mathbb{F}^{n\times m}$. Then, $\operatorname{rank} A$ is the largest size of all nonsingular submatrices of A.

3.9 Partitioned Matrices

Partitioned matrices were used to state or prove several results in this chapter including Proposition 3.6.11, Theorem 3.7.5, Proposition 3.8.1, and Proposition 3.8.3. In this section we give several useful equalities for partitioned matrices.

Proposition 3.9.1. Let $A_{ij} \in \mathbb{F}^{n_i\times m_j}$ for all $i \in \{1,\dots,k\}$ and $j \in \{1,\dots,l\}$. Then,

$$\begin{bmatrix} A_{11} & \cdots & A_{1l} \\ \vdots & \iddots & \vdots \\ A_{k1} & \cdots & A_{kl} \end{bmatrix}^{\mathsf{T}} = \begin{bmatrix} A_{11}^{\mathsf{T}} & \cdots & A_{k1}^{\mathsf{T}} \\ \vdots & \iddots & \vdots \\ A_{1l}^{\mathsf{T}} & \cdots & A_{kl}^{\mathsf{T}} \end{bmatrix}, \tag{3.9.1}$$

$$\begin{bmatrix} A_{11} & \cdots & A_{1l} \\ \vdots & \iddots & \vdots \\ A_{k1} & \cdots & A_{kl} \end{bmatrix}^{*} = \begin{bmatrix} A_{11}^{*} & \cdots & A_{k1}^{*} \\ \vdots & \iddots & \vdots \\ A_{1l}^{*} & \cdots & A_{kl}^{*} \end{bmatrix}. \tag{3.9.2}$$

If, in addition, $k = l$ and $n_i = m_i$ for all $i \in \{1,\dots,m\}$, then

$$\operatorname{tr}\begin{bmatrix} A_{11} & \cdots & A_{1k} \\ \vdots & \iddots & \vdots \\ A_{k1} & \cdots & A_{kk} \end{bmatrix} = \sum_{i=1}^{k} \operatorname{tr} A_{ii}, \quad \det\begin{bmatrix} A_{11} & A_{12} & \cdots & A_{1k} \\ 0 & A_{22} & \cdots & A_{2k} \\ \vdots & \ddots & \ddots & \vdots \\ 0 & 0 & \cdots & A_{kk} \end{bmatrix} = \prod_{i=1}^{k} \det A_{ii}. \tag{3.9.3}$$

Lemma 3.9.2. Let $B \in \mathbb{F}^{n\times m}$ and $C \in \mathbb{F}^{m\times n}$. Then,

$$\begin{bmatrix} I & B \\ 0 & I \end{bmatrix}^{-1} = \begin{bmatrix} I & -B \\ 0 & I \end{bmatrix}, \quad \begin{bmatrix} I & 0 \\ C & I \end{bmatrix}^{-1} = \begin{bmatrix} I & 0 \\ -C & I \end{bmatrix}. \tag{3.9.4}$$

Let $A \in \mathbb{F}^{n\times n}$ and $D \in \mathbb{F}^{m\times m}$ be nonsingular. Then,

$$\begin{bmatrix} A & 0 \\ 0 & D \end{bmatrix}^{-1} = \begin{bmatrix} A^{-1} & 0 \\ 0 & D^{-1} \end{bmatrix}. \tag{3.9.5}$$

Proposition 3.9.3. Let $A \in \mathbb{F}^{n\times n}$, $B \in \mathbb{F}^{n\times m}$, $C \in \mathbb{F}^{l\times n}$, and $D \in \mathbb{F}^{l\times m}$, and assume that A is nonsingular. Then,

$$\begin{bmatrix} A & B \\ C & D \end{bmatrix} = \begin{bmatrix} I & 0 \\ CA^{-1} & I \end{bmatrix}\begin{bmatrix} A & 0 \\ 0 & D - CA^{-1}B \end{bmatrix}\begin{bmatrix} I & A^{-1}B \\ 0 & I \end{bmatrix}, \tag{3.9.6}$$

$$\operatorname{rank}\begin{bmatrix} A & B \\ C & D \end{bmatrix} = n + \operatorname{rank}(D - CA^{-1}B). \tag{3.9.7}$$

If, furthermore, $l = m$, then

$$\det\begin{bmatrix} A & B \\ C & D \end{bmatrix} = (\det A)\det(D - CA^{-1}B). \tag{3.9.8}$$

Proposition 3.9.4. Let $A \in \mathbb{F}^{n\times m}$, $B \in \mathbb{F}^{n\times l}$, $C \in \mathbb{F}^{l\times m}$, and $D \in \mathbb{F}^{l\times l}$, and assume that D is nonsingular. Then,

$$\begin{bmatrix} A & B \\ C & D \end{bmatrix} = \begin{bmatrix} I & BD^{-1} \\ 0 & I \end{bmatrix}\begin{bmatrix} A - BD^{-1}C & 0 \\ 0 & D \end{bmatrix}\begin{bmatrix} I & 0 \\ D^{-1}C & I \end{bmatrix}, \tag{3.9.9}$$

$$\operatorname{rank}\begin{bmatrix} A & B \\ C & D \end{bmatrix} = l + \operatorname{rank}(A - BD^{-1}C). \tag{3.9.10}$$

If, furthermore, $n = m$, then

$$\det\begin{bmatrix} A & B \\ C & D \end{bmatrix} = (\det D)\det(A - BD^{-1}C). \tag{3.9.11}$$

Corollary 3.9.5. Let $A \in \mathbb{F}^{n\times m}$ and $B \in \mathbb{F}^{m\times n}$. Then,

$$\begin{bmatrix} I_n & A \\ B & I_m \end{bmatrix} = \begin{bmatrix} I_n & 0 \\ B & I_m \end{bmatrix}\begin{bmatrix} I_n & 0 \\ 0 & I_m - BA \end{bmatrix}\begin{bmatrix} I_n & A \\ 0 & I_m \end{bmatrix} \tag{3.9.12}$$

$$= \begin{bmatrix} I_n & A \\ 0 & I_m \end{bmatrix}\begin{bmatrix} I_n - AB & 0 \\ 0 & I_m \end{bmatrix}\begin{bmatrix} I_n & 0 \\ B & I_m \end{bmatrix}. \tag{3.9.13}$$

Hence,

$$\operatorname{rank}\begin{bmatrix} I_n & A \\ B & I_m \end{bmatrix} = n + \operatorname{rank}(I_m - BA) = m + \operatorname{rank}(I_n - AB), \tag{3.9.14}$$

$$\det\begin{bmatrix} I_n & A \\ B & I_m \end{bmatrix} = \det(I_m - BA) = \det(I_n - AB). \tag{3.9.15}$$

Hence, $I_n + AB$ is nonsingular if and only if $I_m + BA$ is nonsingular.

Lemma 3.9.6. Let $A \in \mathbb{F}^{n\times n}$, $B \in \mathbb{F}^{n\times m}$, $C \in \mathbb{F}^{m\times n}$, and $D \in \mathbb{F}^{m\times m}$, and assume that A and D are nonsingular. Then,

$$(\det A)\det(D - CA^{-1}B) = (\det D)\det(A - BD^{-1}C). \tag{3.9.16}$$

Furthermore, $D - CA^{-1}B$ is nonsingular if and only if $A - BD^{-1}C$ is nonsingular.

Proposition 3.9.7. Let $A \in \mathbb{F}^{n\times n}$, $B \in \mathbb{F}^{n\times m}$, $C \in \mathbb{F}^{m\times n}$, and $D \in \mathbb{F}^{m\times m}$. If A and $D - CA^{-1}B$ are nonsingular, then

$$\begin{bmatrix} A & B \\ C & D \end{bmatrix}^{-1} = \begin{bmatrix} A^{-1} + A^{-1}B(D - CA^{-1}B)^{-1}CA^{-1} & -A^{-1}B(D - CA^{-1}B)^{-1} \\ -(D - CA^{-1}B)^{-1}CA^{-1} & (D - CA^{-1}B)^{-1} \end{bmatrix}. \tag{3.9.17}$$

If D and $A - BD^{-1}C$ are nonsingular, then

$$\begin{bmatrix} A & B \\ C & D \end{bmatrix}^{-1} = \begin{bmatrix} (A - BD^{-1}C)^{-1} & -(A - BD^{-1}C)^{-1}BD^{-1} \\ -D^{-1}C(A - BD^{-1}C)^{-1} & D^{-1} + D^{-1}C(A - BD^{-1}C)^{-1}BD^{-1} \end{bmatrix}. \tag{3.9.18}$$

If A, D, and $D - CA^{-1}B$ are nonsingular, then $A - BD^{-1}C$ is nonsingular, and

$$\begin{bmatrix} A & B \\ C & D \end{bmatrix}^{-1} = \begin{bmatrix} (A - BD^{-1}C)^{-1} & -(A - BD^{-1}C)^{-1}BD^{-1} \\ -(D - CA^{-1}B)^{-1}CA^{-1} & (D - CA^{-1}B)^{-1} \end{bmatrix}. \tag{3.9.19}$$

The following result is the *matrix inversion lemma*. A special case is the Sherman-Morrison-Woodbury formula given by Fact 3.21.3.

Corollary 3.9.8. Let $A \in \mathbb{F}^{n\times n}$, $B \in \mathbb{F}^{n\times m}$, $C \in \mathbb{F}^{m\times n}$, and $D \in \mathbb{F}^{m\times m}$. If A, $D - CA^{-1}B$, and D are nonsingular, then $A - BD^{-1}C$ is nonsingular,

$$(A - BD^{-1}C)^{-1} = A^{-1} + A^{-1}B(D - CA^{-1}B)^{-1}CA^{-1}, \tag{3.9.20}$$

$$C(A - BD^{-1}C)^{-1}A = D(D - CA^{-1}B)^{-1}C. \tag{3.9.21}$$

If A and $I - CA^{-1}B$ are nonsingular, then $A - BC$ is nonsingular and

$$(A - BC)^{-1} = A^{-1} + A^{-1}B(I - CA^{-1}B)^{-1}CA^{-1}. \tag{3.9.22}$$

If $D - CB$, and D are nonsingular, then $I - BD^{-1}C$ is nonsingular and

$$(I - BD^{-1}C)^{-1} = I + B(D - CB)^{-1}C. \tag{3.9.23}$$

If $I - CB$ is nonsingular, then $I - BC$ is nonsingular and

$$(I - BC)^{-1} = I + B(I - CB)^{-1}C. \tag{3.9.24}$$

Corollary 3.9.9. Let $A, B, C, D \in \mathbb{F}^{n\times n}$. If A, B, $C - DB^{-1}A$, and $D - CA^{-1}B$ are nonsingular, then

$$\begin{bmatrix} A & B \\ C & D \end{bmatrix}^{-1} = \begin{bmatrix} A^{-1} - (C - DB^{-1}A)^{-1}CA^{-1} & (C - DB^{-1}A)^{-1} \\ -(D - CA^{-1}B)^{-1}CA^{-1} & (D - CA^{-1}B)^{-1} \end{bmatrix}. \tag{3.9.25}$$

If A, C, $B - AC^{-1}D$, and $D - CA^{-1}B$ are nonsingular, then

$$\begin{bmatrix} A & B \\ C & D \end{bmatrix}^{-1} = \begin{bmatrix} A^{-1} - A^{-1}B(B - AC^{-1}D)^{-1} & -A^{-1}B(D - CA^{-1}B)^{-1} \\ (B - AC^{-1}D)^{-1} & (D - CA^{-1}B)^{-1} \end{bmatrix}. \tag{3.9.26}$$

If A, B, C, $B - AC^{-1}D$, and $D - CA^{-1}B$ are nonsingular, then $C - DB^{-1}A$ is nonsingular and

$$\begin{bmatrix} A & B \\ C & D \end{bmatrix}^{-1} = \begin{bmatrix} A^{-1} - A^{-1}B(B - AC^{-1}D)^{-1} & (C - DB^{-1}A)^{-1} \\ (B - AC^{-1}D)^{-1} & (D - CA^{-1}B)^{-1} \end{bmatrix}. \tag{3.9.27}$$

If B, D, $A - BD^{-1}C$, and $C - DB^{-1}A$ are nonsingular, then

$$\begin{bmatrix} A & B \\ C & D \end{bmatrix}^{-1} = \begin{bmatrix} (A - BD^{-1}C)^{-1} & (C - DB^{-1}A)^{-1} \\ -D^{-1}C(A - BD^{-1}C)^{-1} & D^{-1} - D^{-1}C(C - DB^{-1}A)^{-1} \end{bmatrix}. \tag{3.9.28}$$

If C, D, $A - BD^{-1}C$, and $B - AC^{-1}D$ are nonsingular, then

$$\begin{bmatrix} A & B \\ C & D \end{bmatrix}^{-1} = \begin{bmatrix} (A - BD^{-1}C)^{-1} & -(A - BD^{-1}C)^{-1}BD^{-1} \\ (B - AC^{-1}D)^{-1} & D^{-1} - (B - AC^{-1}D)^{-1}BD^{-1} \end{bmatrix}. \tag{3.9.29}$$

If B, C, D, $A - BD^{-1}C$, and $C - DB^{-1}A$ are nonsingular, then $B - AC^{-1}D$ is nonsingular and

$$\begin{bmatrix} A & B \\ C & D \end{bmatrix}^{-1} = \begin{bmatrix} (A - BD^{-1}C)^{-1} & (C - DB^{-1}A) \\ (B - AC^{-1}D)^{-1} & D^{-1} - D^{-1}C(C - DB^{-1}A)^{-1} \end{bmatrix}. \tag{3.9.30}$$

Finally, if A, B, C, D, $A - BD^{-1}C$, and $B - AC^{-1}D$ are nonsingular, then $C - DB^{-1}A$ and $D - CA^{-1}B$ are nonsingular and

$$\begin{bmatrix} A & B \\ C & D \end{bmatrix}^{-1} = \begin{bmatrix} (A - BD^{-1}C)^{-1} & (C - DB^{-1}A)^{-1} \\ (B - AC^{-1}D)^{-1} & (D - CA^{-1}B)^{-1} \end{bmatrix}. \tag{3.9.31}$$

Corollary 3.9.10. Let $A, B \in \mathbb{F}^{n\times n}$, and assume that A and $I - A^{-1}B$ are nonsingular. Then, $A - B$ is nonsingular, and

$$(A - B)^{-1} = A^{-1} + A^{-1}B(I - A^{-1}B)^{-1}A^{-1}. \tag{3.9.32}$$

If, in addition, B is nonsingular, then

$$(A - B)^{-1} = A^{-1} + A^{-1}(B^{-1} - A^{-1})^{-1}A^{-1}. \tag{3.9.33}$$

3.10 Majorization

Let $x \in \mathbb{R}^n$. Then, the components of $x^{\downarrow} \in \mathbb{R}^n$ are the same as the components of x and satisfy $(x^{\downarrow})_{(n)} \le \cdots \le (x^{\downarrow})_{(1)}$. An analogous definition is used for $x \in \mathbb{R}^{1\times n}$. Furthermore, the components of $x^{\uparrow} \in \mathbb{R}^n$ are the same as the components of x and satisfy $(x^{\uparrow})_{(1)} \le \cdots \le (x^{\uparrow})_{(n)}$. An analogous definition is used for $x \in \mathbb{R}^{1\times n}$.

Definition 3.10.1. Let $x, y \in \mathbb{R}^n$ or $x, y \in \mathbb{R}^{1\times n}$. Then, the following terminology is defined:

i) y *weakly majorizes* x ($x \overset{\mathrm{w}}{\prec} y$) if, for all $k \in \{1, \ldots, n\}$,

$$\sum_{i=1}^{k}(x^{\downarrow})_{(i)} \le \sum_{i=1}^{k}(y^{\downarrow})_{(i)}. \tag{3.10.1}$$

ii) y *strongly majorizes* ($x \overset{\mathrm{s}}{\prec} y$) x if y weakly majorizes x and

$$\sum_{i=1}^{n}(x^{\downarrow})_{(i)} = \sum_{i=1}^{n}(y^{\downarrow})_{(i)}. \tag{3.10.2}$$

Now, assume that x and y are nonnegative. Then, the following terminology is defined:

iii) y *weakly log majorizes* x ($x \overset{\mathrm{wlog}}{\prec} y$) if, for all $k \in \{1, \ldots, n\}$,

$$\prod_{i=1}^{k}(x^{\downarrow})_{(i)} \le \prod_{i=1}^{k}(y^{\downarrow})_{(i)}. \tag{3.10.3}$$

iv) y *strongly log majorizes* x ($x \overset{\mathrm{slog}}{\prec} y$) if y weakly log majorizes x and

$$\prod_{i=1}^{n}(x^{\downarrow})_{(i)} = \prod_{i=1}^{n}(y^{\downarrow})_{(i)}. \tag{3.10.4}$$

An equivalent formulation of (3.10.1) is the condition

$$\max\sum_{j=1}^{k} x_{(i_j)} \le \sum_{i=1}^{k}(y^{\downarrow})_{(i)}. \tag{3.10.5}$$

where the maximum is taken over all $1 \le i_1 < \cdots < i_k \le n$. Therefore, (3.10.1) implies that

$$\sum_{i=1}^{k} x_{(i)} \le \sum_{i=1}^{k}(y^{\downarrow})_{(i)}, \tag{3.10.6}$$

but the converse is not true. If y strongly majorizes x, then y weakly majorizes x, and, if y strongly log majorizes x, then y weakly log majorizes x. Fact 3.25.15 states that, if y weakly log majorizes x, then y weakly majorizes x. Furthermore, Fact 3.25.3 states that y strongly log majorizes x and y strongly majorizes x if and only if x and y have the same components.

Note that, if x weakly majorizes y and y weakly majorizes x, then x and y have the same components but are not necessarily equal. Therefore, "$\overset{\mathrm{w}}{<}$" is not an antisymmetric relation and thus is not a partial ordering.

Proposition 3.10.2. Weak majorization "$\overset{\mathrm{w}}{<}$" and strong majorization "$\overset{\mathrm{s}}{<}$" are reflexive and transitive relations on $\mathbb{R}^n$. Furthermore, weak log majorization "$\overset{\mathrm{wlog}}{<}$" and strong log majorization "$\overset{\mathrm{slog}}{<}$" are reflexive and transitive relations on $[0,\infty)^n$.

Definition 3.10.3. Let $\mathcal{S} \subseteq \mathbb{R}^n$, and let $f\colon \mathcal{S} \mapsto \mathbb{R}$. Then, f is *Schur-convex* if, for all $x, y \in \mathcal{S}$ such that $x \overset{\mathrm{s}}{<} y$, it follows that $f(x) \le f(y)$. Furthermore, f is *Schur-concave* if $-f$ is Schur-convex.

3.11 Facts on One Set

Fact 3.11.1. Let $\alpha \in \mathbb{F}$, and let $\mathcal{S} \subseteq \mathbb{F}^n$ be a (cone, pointed cone, convex set, polytope, convex cone, polyhedral cone, affine subspace, subspace). Then, so is $\alpha\mathcal{S}$.

Fact 3.11.2. Let $\mathcal{S} \subseteq \mathbb{F}^n$, let $\alpha, \beta \in \mathbb{F}$, and assume that at least one of the following statements holds:

i) $\operatorname{card}(\mathcal{S}) \le 1$.

ii) $\mathcal{S}$ is a convex set, and α and β are nonnegative numbers.

iii) $\mathcal{S}$ is an affine subspace, and either $\alpha = \beta = 0$ or $\alpha + \beta \ne 0$.

Then, $(\alpha+\beta)\mathcal{S} = \alpha\mathcal{S} + \beta\mathcal{S}$. **Source:** For *ii)*, let $x, y \in \mathcal{S}$ and let α and β be nonnegative numbers such that $\alpha + \beta > 0$. Then, $\alpha x + \beta y \in \alpha\mathcal{S} + \beta\mathcal{S}$. Define $\theta \triangleq \alpha/(\alpha+\beta) \in (0,1)$. Then, $\theta x + (1-\theta)y \in \mathcal{S}$, and thus $\alpha x + \beta y = (\alpha+\beta)[\theta x + (1-\theta)y] \in (\alpha+\beta)\mathcal{S}$. See [2606, p. 6]. For *iii)*, let $x, y \in \mathcal{S}$ and let $\alpha, \beta \in \mathbb{F}$ satisfy $\alpha + \beta \ne 0$. Then, $\alpha x + \beta y \in \alpha\mathcal{S} + \beta\mathcal{S}$. Define $\theta \triangleq \alpha/(\alpha+\beta) \in \mathbb{F}$. Then, $\theta x + (1-\theta)y \in \mathcal{S}$, and thus $\alpha x + \beta y = (\alpha+\beta)[\theta x + (1-\theta)y] \in (\alpha+\beta)\mathcal{S}$.

Fact 3.11.3. Let $\mathcal{S} \subseteq \mathbb{F}^n$. Then, the following statements are equivalent:

i) $\mathcal{S}$ is convex.

ii) For all $\alpha, \beta > 0$, $\alpha\mathcal{S} + \beta\mathcal{S} = (\alpha+\beta)\mathcal{S}$.

iii) For all $k \ge 2$ and $\alpha_1, \ldots, \alpha_k \in [0,\infty)$ such that $\sum_{i=1}^k \alpha_i = 1$, $\sum_{i=1}^k \alpha_i\mathcal{S} = \mathcal{S}$.

Source: [1267, p. 107] and [2487, p. 73].

Fact 3.11.4. Let $\mathcal{S} \subseteq \mathbb{F}^n$. Then, the following statements hold:

i) $\mathcal{S} = \operatorname{cone}\mathcal{S}$ if and only if $\mathcal{S}$ is a cone.

ii) $\mathcal{S} = \operatorname{conv}\mathcal{S}$ if and only if $\mathcal{S}$ is a convex set.

iii) $\mathcal{S} = \operatorname{coco}\mathcal{S}$ if and only if $\mathcal{S}$ is a convex cone.

iv) $\mathcal{S} = \operatorname{affin}\mathcal{S}$ if and only if $\mathcal{S}$ is an affine subspace.

v) $\mathcal{S} = \operatorname{span}\mathcal{S}$ if and only if $\mathcal{S}$ is a subspace.

Fact 3.11.5. Let $\mathcal{S} \subseteq \mathbb{R}^n$ be nonempty. Then,

$$\operatorname{cone}\mathcal{S} = \{\alpha x\colon\ x \in \mathcal{S} \text{ and } \alpha > 0\},$$

$$\begin{aligned}\operatorname{conv}\mathcal{S} &= \bigcup_{k\in\mathbb{P}} \left\{\sum_{i=1}^k \alpha_i x_i\colon\ \alpha_i > 0,\ x_i \in \mathcal{S},\ i = 1,\ldots,k, \text{ and } \sum_{i=1}^k \alpha_i = 1\right\}\\ &= \bigcup_{1\le k\le n+1} \left\{\sum_{i=1}^k \alpha_i x_i\colon\ \alpha_i > 0,\ x_i \in \mathcal{S},\ i = 1,\ldots,k, \text{ and } \sum_{i=1}^k \alpha_i = 1\right\},\end{aligned}$$

$$\operatorname{coco} \mathcal{S} = \bigcup_{k\in\mathbb{P}} \left\{\sum_{i=1}^{k} \alpha_i x_i \colon \ \alpha_i > 0, \ x_i \in \mathcal{S}, \ i = 1,\dots,k\right\}$$
$$= \bigcup_{1\le k\le n} \left\{\sum_{i=1}^{k} \alpha_i x_i \colon \ \alpha_i > 0, \ x_i \in \mathcal{S}, \ i = 1,\dots,k\right\},$$

$$\operatorname{affin} \mathcal{S} = \bigcup_{k\in\mathbb{P}} \left\{\sum_{i=1}^{k} \alpha_i x_i \colon \ \alpha_i \in \mathbb{R}, \ x_i \in \mathcal{S}, \ i = 1,\dots,k, \text{ and } \sum_{i=1}^{k} \alpha_i = 1\right\}$$
$$= \bigcup_{1\le k\le n+1} \left\{\sum_{i=1}^{k} \alpha_i x_i \colon \ \alpha_i \in \mathbb{R}, \ x_i \in \mathcal{S}, \ i = 1,\dots,k, \text{ and } \sum_{i=1}^{k} \alpha_i = 1\right\},$$

$$\operatorname{span} \mathcal{S} = \bigcup_{k\in\mathbb{P}} \left\{\sum_{i=1}^{k} \alpha_i x_i \colon \ \alpha_i \in \mathbb{R} \text{ and } x_i \in \mathcal{S}, \ i = 1,\dots,k\right\}$$
$$= \bigcup_{1\le k\le n} \left\{\sum_{i=1}^{k} \alpha_i x_i \colon \ \alpha_i \in \mathbb{R} \text{ and } x_i \in \mathcal{S}, \ i = 1,\dots,k\right\}.$$

Source: The second expression for $\operatorname{conv} \mathcal{S}$ is *Caratheodory's theorem.* See [309, p. 10], [1260, p. 43], and [1769, Theorem 2.23]. The first expression for $\operatorname{conv} \mathcal{S}$ is given in [1769, Theorem 2.15], while the second expression for $\operatorname{conv} \mathcal{S}$ is given in [1769, Theorem 2.23]. The second expression for $\operatorname{coco} \mathcal{S}$ is given in [1267, Theorem 4.21].

Fact 3.11.6. Let $\mathcal{S} \subseteq \mathbb{C}^n$ be nonempty. Then,

$$\operatorname{cone} \mathcal{S} = \{\alpha x \colon \ x \in \mathcal{S} \text{ and } \alpha > 0\},$$

$$\operatorname{conv} \mathcal{S} = \bigcup_{k\in\mathbb{P}} \left\{\sum_{i=1}^{k} \alpha_i x_i \colon \ \alpha_i > 0, \ x_i \in \mathcal{S}, \ i = 1,\dots,k, \text{ and } \sum_{i=1}^{k} \alpha_i = 1\right\}$$
$$= \bigcup_{1\le k\le 2n+1} \left\{\sum_{i=1}^{k} \alpha_i x_i \colon \ \alpha_i > 0, \ x_i \in \mathcal{S}, \ i = 1,\dots,k, \text{ and } \sum_{i=1}^{k} \alpha_i = 1\right\},$$

$$\operatorname{coco} \mathcal{S} = \bigcup_{k\in\mathbb{P}} \left\{\sum_{i=1}^{k} \alpha_i x_i \colon \ \alpha_i > 0, \ x_i \in \mathcal{S}, \ i = 1,\dots,k\right\}$$
$$= \bigcup_{1\le k\le 2n} \left\{\sum_{i=1}^{k} \alpha_i x_i \colon \ \alpha_i > 0, \ x_i \in \mathcal{S}, \ i = 1,\dots,k\right\},$$

$$\operatorname{affin} \mathcal{S} = \bigcup_{k\in\mathbb{P}} \left\{\sum_{i=1}^{k} \alpha_i x_i \colon \ \alpha_i \in \mathbb{C}, \ x_i \in \mathcal{S}, \ i = 1,\dots,k, \text{ and } \sum_{i=1}^{k} \alpha_i = 1\right\}$$
$$= \bigcup_{1\le k\le n+1} \left\{\sum_{i=1}^{k} \alpha_i x_i \colon \ \alpha_i \in \mathbb{C}, \ x_i \in \mathcal{S}, \ i = 1,\dots,k, \text{ and } \sum_{i=1}^{k} \alpha_i = 1\right\},$$

$$\operatorname{span} \mathcal{S} = \bigcup_{k\in\mathbb{P}} \left\{\sum_{i=1}^{k} \alpha_i x_i \colon \ \alpha_i \in \mathbb{C} \text{ and } x_i \in \mathcal{S}, \ i = 1,\dots,k\right\}$$

$$= \bigcup_{1\le k\le n} \left\{ \sum_{i=1}^{k} \alpha_i x_i\colon\ \alpha_i \in \mathbb{C} \text{ and } x_i \in \mathcal{S},\ i = 1,\dots,k \right\}.$$

Fact 3.11.7. Let $\mathcal{S} \subset \mathbb{F}^n$. Then, the following statements hold:

i) If $\mathcal{S}$ is a closed half space, then $\mathcal{S}$ is a pointed, polyhedral, convex cone that is not one-sided.

ii) If $\mathcal{S}$ is an open half space, then $\mathcal{S}$ is a blunt, one-sided, convex cone.

iii) If $\mathcal{S}$ is a convex cone, then $\mathcal{S}$ is contained in a closed half space.

iv) If $\mathcal{S}$ is an open convex cone, then $\mathcal{S}$ is a blunt, one-sided cone and is contained in an open half space.

Fact 3.11.8. Let $\mathcal{S} \subseteq \mathbb{F}^n$. Then, the following statements hold:

i) $\operatorname{coco} \mathcal{S} = \operatorname{conv} \operatorname{cone} \mathcal{S} = \operatorname{cone} \operatorname{conv} \mathcal{S}$.

ii) $\operatorname{coco}(\mathcal{S} \cup -\mathcal{S}) = \operatorname{affin}(\mathcal{S} \cup -\mathcal{S}) = \operatorname{affin}(\mathcal{S} \cup \{0\}) = \operatorname{span} \mathcal{S} = \mathcal{S}^{\perp\perp}$.

iii) $\mathcal{S} \subseteq \operatorname{conv} \mathcal{S} \subseteq \left\{ \begin{matrix} \operatorname{affin} \mathcal{S} \\ \operatorname{coco} \mathcal{S} \end{matrix} \right\} \subseteq \operatorname{span} \mathcal{S}$.

iv) $\mathcal{S} \subseteq \operatorname{cone} \mathcal{S} \subseteq \operatorname{coco} \mathcal{S} \subseteq \operatorname{span} \mathcal{S}$.

Remark: See [362, p. 52]. “Pointed” in [362] means one-sided.

Fact 3.11.9. Let $\mathcal{S} \subseteq \mathbb{R}^n$, and assume that $\mathcal{S}$ is a convex cone. Then,

$$\operatorname{affin} \mathcal{S} = \operatorname{span} \mathcal{S} = \mathcal{S} - \mathcal{S} = \operatorname{conv}[\mathcal{S} \cup (-\mathcal{S})].$$

Hence, $\mathcal{S} - \mathcal{S}$ is a subspace. **Source:** [2487, p. 152]. **Remark:** $\mathbb{R}$ is not a subspace of $\mathbb{C}$, and thus the result does not hold for $\mathcal{S} \subseteq \mathbb{C}^n$. **Related:** Fact 3.11.12.

Fact 3.11.10. Let $\mathcal{S} \subseteq \mathbb{F}^n$, and consider the following statements:

i) $\mathcal{S}$ is a cone.

ii) $\mathcal{S}$ is a convex set.

iii) $\mathcal{S}$ is a convex cone.

iv) $\mathcal{S}$ is an affine subspace.

v) $\mathcal{S}$ is a subspace.

vi) $0 \in \mathcal{S}$.

vii) $0 \in \operatorname{affin} \mathcal{S}$.

viii) $\operatorname{affin} \mathcal{S}$ is a subspace.

ix) $\operatorname{affin} \mathcal{S} = \operatorname{span} \mathcal{S}$.

x) $\dim \mathcal{S} = n$.

xi) $\operatorname{affin} \mathcal{S} = \mathbb{F}^n$.

xii) $\operatorname{span} \mathcal{S} = \mathbb{F}^n$.

Then,

$$v) \Longleftrightarrow \{i), iv)\} \Longleftrightarrow \{iii), iv)\} \Longleftrightarrow \{iv), vi)\} \Longrightarrow iii) \Longleftrightarrow \{i), ii)\},$$

$$\left. \begin{matrix} x) \Longleftrightarrow xi) \Longleftrightarrow xii) \\ vi) \end{matrix} \right\} \Longrightarrow vii) \Longleftrightarrow viii) \Longleftrightarrow ix).$$

Fact 3.11.11. Let $\mathcal{S} \subseteq \mathbb{F}^n$, assume that $\mathcal{S}$ is an affine subspace, and let $z \in \mathbb{F}^n$. Then, $\mathcal{S} + z$ is an affine subspace. If, in addition, $z \in \mathcal{S}$, then $\mathcal{S} - z$ is a subspace. **Source:** Fact 3.11.10.

Fact 3.11.12. Let $\mathcal{S} \subseteq \mathbb{F}^n$, and assume that $\mathcal{S}$ is an affine subspace. Then, there exist a unique vector $x \in \mathbb{F}^n$ and a unique subspace $\mathcal{S}_0 \subseteq \mathbb{F}^n$ such that $\mathcal{S} = x + \mathcal{S}_0$. **Source:** [2487, p. 12]. **Remark:** $\mathcal{S} - \mathcal{S} = \{y - z\colon\ y, z \in \mathcal{S}\} = \mathcal{S}_0$ is a subspace. **Related:** Fact 3.11.9.

Fact 3.11.13. Let $\mathcal{S} \subset \mathbb{F}^n$. Then, the following statements hold:

i) $\mathcal{S}$ is an affine hyperplane if and only if there exist a nonzero vector $y \in \mathbb{F}^n$ and $\alpha \in \mathbb{R}$ such that $\mathcal{S} = \{x \in \mathbb{F}^n\colon\ \operatorname{Re} x^*y = \alpha\}$.

ii) $\mathcal{S}$ is an affine closed half space if and only if there exist a nonzero vector $y \in \mathbb{F}^n$ and $\alpha \in \mathbb{R}$ such that $\mathcal{S} = \{x \in \mathbb{F}^n\colon\ \operatorname{Re} x^*y \le \alpha\}$.

iii) $\mathcal{S}$ is an affine open half space if and only if there exist a nonzero vector $y \in \mathbb{F}^n$ and $\alpha \in \mathbb{R}$ such that $\mathcal{S} = \{x \in \mathbb{F}^n\colon\ \operatorname{Re} x^*y < \alpha\}$.

Source: Let $z \in \mathbb{F}^n$ satisfy $z^*y = \alpha$. Then, $\{x\colon\ x^*y = \alpha\} = \{y\}^\perp + z$.

Fact 3.11.14. Let $x_1, \ldots, x_k \in \mathbb{F}^n$. Then,

$$\operatorname{affin}\{x_1, \ldots, x_k\} = x_1 + \operatorname{span}\{x_2 - x_1, \ldots, x_k - x_1\}.$$

Related: Fact 12.11.13.

Fact 3.11.15. Let $A \in \mathbb{F}^{n\times m}$, and let $\mathcal{S} \subseteq \mathbb{F}^m$. Then,

$$\operatorname{cone} A\mathcal{S} = A\operatorname{cone}\mathcal{S}, \quad \operatorname{conv} A\mathcal{S} = A\operatorname{conv}\mathcal{S}, \quad \operatorname{coco} A\mathcal{S} = A\operatorname{coco}\mathcal{S},$$
$$\operatorname{affin} A\mathcal{S} = A\operatorname{affin}\mathcal{S}, \quad \operatorname{span} A\mathcal{S} = A\operatorname{span}\mathcal{S}.$$

Hence, if $\mathcal{S}$ is a (cone, convex set, polytope, convex cone, polyhedral cone, affine subspace, subspace), then so is $A\mathcal{S}$. Now, assume that A is left invertible, and let $A^{\mathsf{L}} \in \mathbb{F}^{m\times n}$ be a left inverse of A. Then,

$$\operatorname{cone}\mathcal{S} = A^{\mathsf{L}}\operatorname{cone} A\mathcal{S}, \quad \operatorname{conv}\mathcal{S} = A^{\mathsf{L}}\operatorname{conv} A\mathcal{S}, \quad \operatorname{coco}\mathcal{S} = A^{\mathsf{L}}\operatorname{coco} A\mathcal{S},$$
$$\operatorname{affin}\mathcal{S} = A^{\mathsf{L}}\operatorname{affin} A\mathcal{S}, \quad \operatorname{span}\mathcal{S} = A^{\mathsf{L}}\operatorname{span} A\mathcal{S}.$$

Hence, if $A\mathcal{S}$ is a (cone, convex set, polytope, convex cone, polyhedral cone, affine subspace, subspace), then so is $\mathcal{S}$. **Related:** Fact 3.11.18.

Fact 3.11.16. Let $A \in \mathbb{F}^{n\times m}$, let $\mathcal{S} \subseteq \mathbb{F}^n$, and assume that $\mathcal{S}$ is a (cone, convex set, convex cone, affine subspace, subspace). Then, so is $A^{\mathsf{inv}}(\mathcal{S})$.

Fact 3.11.17. Let $A \in \mathbb{F}^{n\times m}$ and $\mathcal{S} \subseteq \mathbb{F}^n$. Then,

$$\operatorname{conv} A^{\mathsf{inv}}(\mathcal{S}) = A^{\mathsf{inv}}(\operatorname{conv}[\mathcal{S} \cap \mathcal{R}(A)]) \subseteq A^{\mathsf{inv}}(\operatorname{conv}\mathcal{S}).$$

Source: [2487, pp. 126–128]. **Related:** Fact 12.11.22.

Fact 3.11.18. Let $A \in \mathbb{F}^{n\times m}$, let $X \triangleq \{\operatorname{col}_1(A), \ldots, \operatorname{col}_m(A)\} \subset \mathbb{F}^m$, and define

$$\Theta^m \triangleq \{\theta \in [0,\infty)^m\colon\ \sum_{i=1}^m \theta_{(i)} = 1\}, \quad \Gamma^m \triangleq \{\alpha e_{i,m}\colon \alpha \in [0,\infty), i \in \{1, \ldots, m\}\},$$
$$\Phi^m \triangleq \{\alpha \in \mathbb{F}^m\colon\ \sum_{i=1}^m \alpha_{(i)} = 1\}.$$

Then, the following statements hold:

i) $\Gamma^m = \operatorname{cone}\{e_{1,m}, \ldots, e_{m,m}\}$ is a cone.

ii) $\Theta^m = \operatorname{conv}\{e_{1,m}, \ldots, e_{m,m}\}$ is a convex polytope.

iii) $[0,\infty)^m = \operatorname{coco}\{e_{1,m}, \ldots, e_{m,m}\}$ is a polyhedral cone.

iv) $\Phi^m = \operatorname{affin}\{e_{1,m}, \ldots, e_{m,m}\} = \mathcal{N}(1_{1\times m}) + e_{1,m}$ is an affine subspace.

v) $\mathbb{F}^m = \operatorname{span}\{e_{1,m}, \ldots, e_{m,m}\}$ is a subspace.

vi) $\operatorname{cone} X = A\Gamma^m$.

vii) $\operatorname{conv} X = A\Theta^m$.

viii) $\operatorname{coco} X = A[0,\infty)^m$.

ix) $\operatorname{affin} X = A\Phi^m$.

x) $\operatorname{span} X = A\mathbb{F}^m = \mathcal{R}(A)$.

Related: Fact 3.11.15.

Fact 3.11.19. Let $\mathcal{S} \subseteq \mathbb{F}^n$. Then, the following statements hold:

i) $\mathcal{S}$ is a convex polytope if and only if $\mathcal{S}$ is the intersection of a finite number of closed half spaces.

ii) $\mathcal{S}$ is a bounded, convex polytope if and only if $\mathcal{S}$ is the convex hull of a finite number of points.

iii) $\mathcal{S}$ is a polyhedral cone if and only if $\mathcal{S}$ is the convex conical hull of a finite number of points that includes 0.

iv) Let $z, x_1, \ldots, x_m \in \mathbb{F}^n$, and let $\mathcal{S}$ be the zonotope

$$\mathcal{S} = z + \left\{ \sum_{i=1}^{m} \alpha_i x_i \colon 0 \le \alpha_i \le 1 \text{ for all } i \in \{1, \ldots, m\} \right\}.$$

Then, $\mathcal{S}$ is convex. In particular,

$$\mathcal{S} = z + \operatorname{conv}\left(\{0\} \cup \bigcup_{k=1}^{m} \left\{ \sum_{j=1}^{k} x_{i_j} \colon 1 \le i_1 < \cdots < i_k \le m \right\} \right).$$

Remark: The number of vertices of the zonotope in *iv*) does not exceed 2^m. **Related:** Fact 5.4.3 and Fact 5.4.4.

Fact 3.11.20. Let $\mathcal{S} \subseteq \mathbb{F}^n$ be a subspace. Then, for all $m \ge \dim \mathcal{S}$, there exists $A \in \mathbb{F}^{n\times m}$ such that $\mathcal{S} = \mathcal{R}(A)$.

Fact 3.11.21. Let $A \in \mathbb{F}^{n\times n}$, let $\mathcal{S} \subseteq \mathbb{F}^n$, assume that $\mathcal{S}$ is a subspace, let $k \triangleq \dim \mathcal{S}$, let $S \in \mathbb{F}^{n\times k}$, and assume that $\mathcal{R}(S) = \mathcal{S}$. Then, $\mathcal{S}$ is an invariant subspace of A if and only if there exists $M \in \mathbb{F}^{k\times k}$ such that $AS = SM$. **Source:** Use Fact 7.14.1 with $B = I$. To prove sufficiency, note that $\mathcal{S} \subseteq A\mathcal{S} + \mathcal{S}$. It then follows from $\dim \mathcal{S} = \dim(A\mathcal{S} + \mathcal{S})$ and Corollary 3.1.5 that $\mathcal{S} = A\mathcal{S} + \mathcal{S}$. Finally, Fact 3.12.12 implies that $A\mathcal{S} \subseteq \mathcal{S}$. See [1762, pp. 89, 90].

Fact 3.11.22. Let $\mathcal{S} \subseteq \mathbb{F}^n$. Then, $\mathcal{S}^\perp = (\operatorname{span} \mathcal{S})^\perp$. Now, assume that $\mathcal{S}$ is a subspace. Then, $\mathcal{S}$ and $\mathcal{S}^\perp$ are orthogonally complementary.

Fact 3.11.23. Define the convex pointed cone $\mathcal{S} \subset \mathbb{R}^2$ by

$$\mathcal{S} \triangleq \{(x_1, x_2) \in [0, \infty) \times \mathbb{R} \colon \text{ if } x_1 = 0, \text{ then } x_2 \ge 0\} = ([0, \infty) \times \mathbb{R}) \backslash [\{0\} \times (-\infty, 0)].$$

Furthermore, for all $x, y \in \mathbb{R}^2$, define $x \overset{\mathrm{d}}{\le} y$ if and only if $y - x \in \mathcal{S}$. Then, "$\overset{\mathrm{d}}{\le}$" is a total ordering on $\mathbb{R}^2$. **Remark:** "$\overset{\mathrm{d}}{\le}$" is the lexicographic (dictionary) ordering. See Fact 1.8.16 and [309, p. 161].

3.12 Facts on Two or More Sets

Fact 3.12.1. Let $\mathcal{S}_1, \mathcal{S}_2 \subseteq \mathbb{F}^n$, and assume that $\mathcal{S}_1 \subseteq \mathcal{S}_2$. Then,

$$\operatorname{cone} \mathcal{S}_1 \subseteq \operatorname{cone} \mathcal{S}_2, \quad \operatorname{conv} \mathcal{S}_1 \subseteq \operatorname{conv} \mathcal{S}_2, \quad \operatorname{coco} \mathcal{S}_1 \subseteq \operatorname{coco} \mathcal{S}_2,$$
$$\operatorname{affin} \mathcal{S}_1 \subseteq \operatorname{affin} \mathcal{S}_2, \quad \operatorname{span} \mathcal{S}_1 \subseteq \operatorname{span} \mathcal{S}_2, \quad \dim \mathcal{S}_1 \le \dim \mathcal{S}_2.$$

Furthermore, $\dim \mathcal{S}_1 = \dim \mathcal{S}_2$ if and only if $\operatorname{affin} \mathcal{S}_1 = \operatorname{affin} \mathcal{S}_2$. **Remark:** The last statement follows from Proposition 3.1.4.

Fact 3.12.2. Let $\mathcal{S}_1, \mathcal{S}_2 \subseteq \mathbb{F}^n$. Then,

$$\operatorname{cone}(\mathcal{S}_1 + \mathcal{S}_2) \subseteq \operatorname{cone} \mathcal{S}_1 + \operatorname{cone} \mathcal{S}_2, \quad \operatorname{conv}(\mathcal{S}_1 + \mathcal{S}_2) = \operatorname{conv} \mathcal{S}_1 + \operatorname{conv} \mathcal{S}_2,$$
$$\operatorname{coco}(\mathcal{S}_1 + \mathcal{S}_2) = \operatorname{coco} \mathcal{S}_1 + \operatorname{coco} \mathcal{S}_2, \quad \operatorname{affin}(\mathcal{S}_1 + \mathcal{S}_2) = \operatorname{affin} \mathcal{S}_1 + \operatorname{affin} \mathcal{S}_2,$$

$$\text{span}(\mathcal{S}_1 + \mathcal{S}_2) \subseteq \text{span}\,\mathcal{S}_1 + \text{span}\,\mathcal{S}_2.$$

Source: [1267, p. 138] and [2398, pp. 2, 3]. To prove $\text{conv}\,\mathcal{S}_1 + \text{conv}\,\mathcal{S}_2 \subseteq \text{conv}(\mathcal{S}_1 + \mathcal{S}_2)$, let $x = \sum_{i=1}^k \theta_i x_i \in \text{conv}\,\mathcal{S}_1$ and $y = \sum_{i=1}^k \phi_i y_i \in \text{conv}\,\mathcal{S}_2$, where $\theta_1, \ldots, \theta_k, \phi_1, \ldots, \phi_k \geq 0$ and $\sum_{i=1}^k \theta_i = \sum_{i=1}^k \phi_i = 1$. Then, $x + y = \sum_{i,j=1}^k \theta_i \phi_j (x_i + y_j) \in \text{conv}(\mathcal{S}_1 + \mathcal{S}_2)$.

Fact 3.12.3. Let $\mathcal{S}_1, \mathcal{S}_2 \subseteq \mathbb{F}^m$, and let $A \in \mathbb{F}^{n\times m}$. Then, $A(\mathcal{S}_1 + \mathcal{S}_2) = A\mathcal{S}_1 + A\mathcal{S}_2$.

Fact 3.12.4. Let $\mathcal{S}_1, \mathcal{S}_2 \subseteq \mathbb{F}^n$ be (cones, pointed cones, convex sets, polytopes, convex polytopes, convex cones, polyhedral cones, affine subspaces, subspaces), and let $\alpha, \beta \in \mathbb{F}$. Then, so are $\alpha\mathcal{S}_1 \cap \beta\mathcal{S}_2$ and $\alpha\mathcal{S}_1 + \beta\mathcal{S}_2$. **Source:** Fact 3.14.7. See [1267, p. 90] for the case where $\mathcal{S}_1$ and $\mathcal{S}_2$ are convex.

Fact 3.12.5. Let $\mathcal{S}_1, \ldots, \mathcal{S}_k \subseteq \mathbb{F}^n$, and let $\alpha_1, \ldots, \alpha_k \in \mathbb{R}$. Then,

$$\text{conv} \sum_{i=1}^k \alpha_i \mathcal{S}_i = \sum_{i=1}^k \alpha_i \,\text{conv}\,\mathcal{S}_i.$$

Source: [2487, p. 124]. **Related:** Fact 12.12.12.

Fact 3.12.6. Let $\mathcal{S}_1, \mathcal{S}_2 \subseteq \mathbb{F}^n$. Then, $(\mathcal{S}^\sim - \mathcal{S}_2)^\sim = \{x \in \mathbb{F}^n \colon x + \mathcal{S}_2 \subseteq \mathcal{S}_1\}$. **Remark:** These expressions define the *Minkowski difference*.

Fact 3.12.7. Let $A \in \mathbb{F}^{n\times m}$. Then, for all $\mathcal{S}_1, \mathcal{S}_2 \subseteq \mathbb{F}^m$, the following statements hold:

i) $\mathcal{S}_1 \subseteq A^{\text{inv}}(A\mathcal{S}_1)$.

ii) $A\mathcal{S}_1 = AA^{\text{inv}}(A\mathcal{S}_1)$.

iii) If $\mathcal{S}_1 \subseteq \mathcal{S}_2$, then $A\mathcal{S}_1 \subseteq A\mathcal{S}_2$.

iv) $A(\mathcal{S}_1 \cap \mathcal{S}_2) \subseteq A\mathcal{S}_1 \cap A\mathcal{S}_2$.

v) $A(\mathcal{S}_1 \cup \mathcal{S}_2) = A\mathcal{S}_1 \cup A\mathcal{S}_2$.

vi) $A(\mathcal{S}_1 + \mathcal{S}_2) = A\mathcal{S}_1 + A\mathcal{S}_2$.

vii) $(A\mathcal{S}_1)\backslash(A\mathcal{S}_2) \subseteq A(\mathcal{S}_1\backslash\mathcal{S}_2)$.

Furthermore, the following statements are equivalent:

viii) A is left invertible.

ix) For all $\mathcal{S} \subseteq \mathbb{F}^m$, $A^{\text{inv}}(A\mathcal{S}) = \mathcal{S}$.

iv) For all $\mathcal{S}_1, \mathcal{S}_2 \subseteq \mathbb{F}^m$, $A(\mathcal{S}_1 \cap \mathcal{S}_2) = A\mathcal{S}_1 \cap A\mathcal{S}_2$.

x) For all disjoint $\mathcal{S}_1, \mathcal{S}_2 \subseteq \mathbb{F}^m$, $A\mathcal{S}_1$ and $A\mathcal{S}_2$ are disjoint.

xi) For all $\mathcal{S}_1, \mathcal{S}_2 \subseteq \mathbb{F}^m$, $(A\mathcal{S}_1)\backslash(A\mathcal{S}_2) = A(\mathcal{S}_1\backslash\mathcal{S}_2)$.

Source: Fact 1.10.3 and [688, p. 12].

Fact 3.12.8. Let $A \in \mathbb{F}^{n\times m}$. Then, for all $\mathcal{S}_1, \mathcal{S}_2 \subseteq \mathbb{F}^n$, the following statements hold:

i) $AA^{\text{inv}}(\mathcal{S}_1) \subseteq \mathcal{S}_1$.

ii) If $\mathcal{S}_1 \subseteq \mathcal{S}_2$, then $A^{\text{inv}}(\mathcal{S}_1) \subseteq A^{\text{inv}}(\mathcal{S}_2)$.

iii) $A^{\text{inv}}(\mathcal{S}_1 \cap \mathcal{S}_2) = A^{\text{inv}}(\mathcal{S}_1) \cap A^{\text{inv}}(\mathcal{S}_2)$.

iv) $A^{\text{inv}}(\mathcal{S}_1 \cup \mathcal{S}_2) = A^{\text{inv}}(\mathcal{S}_1) \cup A^{\text{inv}}(\mathcal{S}_2)$.

v) $A^{\text{inv}}(\mathcal{S}_1) + A^{\text{inv}}(\mathcal{S}_2) \subseteq A^{\text{inv}}(\mathcal{S}_1 + \mathcal{S}_2)$.

vi) $(A^{\text{inv}}\mathcal{S}_1)\backslash(A^{\text{inv}}\mathcal{S}_2) = A^{\text{inv}}(\mathcal{S}_1\backslash\mathcal{S}_2)$.

In addition, the following statements are equivalent:

vii) A is right invertible.

viii) For all $\mathcal{S} \subseteq \mathbb{F}^n$, $AA^{\text{inv}}(\mathcal{S}) = \mathcal{S}$.

Source: Fact 1.10.4 and [688, p. 12].

Fact 3.12.9. Let $A_1 \in \mathbb{F}^{n\times m}$, $A_2 \in \mathbb{F}^{n\times l}$, $\mathcal{S}_1 \subseteq \mathbb{F}^m$, and $\mathcal{S}_2 \subseteq \mathbb{F}^l$. Then,

$$[A_1 \ \ A_2]\begin{bmatrix}\mathcal{S}_1\\ \mathcal{S}_2\end{bmatrix} = A_1\mathcal{S}_1 + A_2\mathcal{S}_2.$$

Fact 3.12.10. Let $A \in \mathbb{F}^{n\times m}$, and let $\mathcal{S}_1 \subseteq \mathbb{F}^m$ and $\mathcal{S}_2 \subseteq \mathbb{F}^n$ be subspaces. Then, the following statements are equivalent:

i) $A\mathcal{S}_1 \subseteq \mathcal{S}_2$.

ii) $A^*\mathcal{S}_2^\perp \subseteq \mathcal{S}_1^\perp$.

Source: It follows from *i)* and Fact 3.18.19 that $\mathcal{S}_2^\perp \subseteq (A\mathcal{S}_1)^\perp = A^{*\mathrm{inv}}(\mathcal{S}_1^\perp)$. It now follows from Proposition 3.7.10 that $A^*\mathcal{S}_2^\perp \subseteq A^*A^{*\mathrm{inv}}(\mathcal{S}_1^\perp) \subseteq \mathcal{S}_1^\perp$. See [688, p. 12].

Fact 3.12.11. Let $\mathcal{S}_1, \mathcal{S}_2 \subseteq \mathbb{F}^n$. Then,

$$\left.\begin{array}{r}(\operatorname{span}\mathcal{S}_1) \cup \operatorname{span}\mathcal{S}_2\\ \operatorname{span}(\mathcal{S}_1 + \mathcal{S}_2)\end{array}\right\} \subseteq \operatorname{span}(\mathcal{S}_1 \cup \mathcal{S}_2) = (\operatorname{span}\mathcal{S}_1) + \operatorname{span}\mathcal{S}_2,$$

$$\operatorname{span}(\mathcal{S}_1 \cap \mathcal{S}_2) \subseteq (\operatorname{span}\mathcal{S}_1) \cap \operatorname{span}\mathcal{S}_2.$$

If, in addition, $0 \in \mathcal{S}_1 \cap \mathcal{S}_2$, then $\operatorname{span}(\mathcal{S}_1 + \mathcal{S}_2) = \operatorname{span}(\mathcal{S}_1 \cup \mathcal{S}_2) = (\operatorname{span}\mathcal{S}_1) + \operatorname{span}\mathcal{S}_2$. **Source:** [2238, p. 532], [2418, p. 11], and [2487, p. 4].

Fact 3.12.12. Let $\mathcal{S}_1, \mathcal{S}_2 \subseteq \mathbb{F}^n$. Then, the following statements hold:

i) If $0 \in \mathcal{S}_2$ and $\mathcal{S}_1 + \mathcal{S}_2 \subseteq \mathcal{S}_2$, then $\mathcal{S}_1 \subseteq \mathcal{S}_2$.

ii) If $\mathcal{S}_2$ is a convex cone and $\mathcal{S}_1 \subseteq \mathcal{S}_2$, then $\mathcal{S}_1 + \mathcal{S}_2 \subseteq \mathcal{S}_2$.

Now, assume that $\mathcal{S}_2$ is a subspace. Then, the following statements are equivalent:

iii) $\mathcal{S}_1 \subseteq \mathcal{S}_2$.

iv) $\mathcal{S}_1 + \mathcal{S}_2 \subseteq \mathcal{S}_2$.

v) $\mathcal{S}_1 + \mathcal{S}_2 = \mathcal{S}_2$.

Source: To prove *i)*, note that $\mathcal{S}_1 = \mathcal{S}_1 + \{0\} \subseteq \mathcal{S}_1 + \mathcal{S}_2 \subseteq \mathcal{S}_2$. To prove *ii)*, note that $\mathcal{S}_1 + \mathcal{S}_2 \subseteq \mathcal{S}_2 + \mathcal{S}_2 = \mathcal{S}_2$. To prove *iv)* $\Longrightarrow$ *v)*, let $x \in \mathcal{S}_1$ and $z \in \mathcal{S}_2$. Then, $x \in \mathcal{S}_2$, and thus $z - x \in \mathcal{S}_2$. Finally, $z = x + (z - x) \in \mathcal{S}_1 + \mathcal{S}_2$.

Fact 3.12.13. Let $\mathcal{S}_1, \mathcal{S}_2 \subseteq \mathbb{F}^n$ be cones. Then, the following statements hold:

i) $\mathcal{S}_1 \cup \mathcal{S}_2$ is a cone.

ii) If either $\mathcal{S}_1$ or $\mathcal{S}_2$ is a pointed cone, then $\mathcal{S}_1 \cup \mathcal{S}_2$ is a pointed cone.

iii) If $\mathcal{S}_1$ and $\mathcal{S}_2$ are convex cones, then $\mathcal{S}_1 + \mathcal{S}_2 \subseteq \operatorname{conv}(\mathcal{S}_1 \cup \mathcal{S}_2)$.

iv) If $\mathcal{S}_1$ and $\mathcal{S}_2$ are pointed convex cones, then $\mathcal{S}_1 + \mathcal{S}_2 = \operatorname{conv}(\mathcal{S}_1 \cup \mathcal{S}_2)$.

Source: [1267, p. 107].

Fact 3.12.14. Let $\mathcal{S}_1, \mathcal{S}_2 \subseteq \mathbb{F}^n$, and assume that $\mathcal{S}_1 \subseteq \mathcal{S}_2$. Then, $\mathcal{S}_2^\perp \subseteq \mathcal{S}_1^\perp$.

Fact 3.12.15. Let $\mathcal{S}_1, \mathcal{S}_2 \subseteq \mathbb{F}^n$. Then, $\mathcal{S}_1^\perp \cap \mathcal{S}_2^\perp \subseteq (\mathcal{S}_1 + \mathcal{S}_2)^\perp$.

Fact 3.12.16. Let $\mathcal{S}_1, \mathcal{S}_2 \subseteq \mathbb{F}^n$ be subspaces. Then, the following statements hold:

i) $\mathcal{S}_1 \cup \mathcal{S}_2 \subseteq \operatorname{span}(\mathcal{S}_1 \cup \mathcal{S}_2) = \mathcal{S}_1 + \mathcal{S}_2$.

ii) If $\mathcal{S}_3 \subseteq \mathbb{F}^n$ is a subspace and $\mathcal{S}_1 \cup \mathcal{S}_2 \subseteq \mathcal{S}_3$, then $\mathcal{S}_1 + \mathcal{S}_2 \subseteq \mathcal{S}_3$.

iii) $\mathcal{S}_1 \cup \mathcal{S}_2$ is a subspace if and only if either $\mathcal{S}_1 \subseteq \mathcal{S}_2$ or $\mathcal{S}_2 \subseteq \mathcal{S}_1$. If these conditions hold, then

$$\mathcal{S}_1 \cup \mathcal{S}_2 = \mathcal{S}_1 + \mathcal{S}_2 = \begin{cases}\mathcal{S}_1, & \mathcal{S}_2 \subseteq \mathcal{S}_1,\\ \mathcal{S}_2, & \mathcal{S}_1 \subseteq \mathcal{S}_2.\end{cases}$$

iv) $\mathcal{S}_1 + \mathcal{S}_2$ is the intersection of all subspaces containing $\mathcal{S}_1 \cup \mathcal{S}_2$ and thus is the smallest subspace containing $\mathcal{S}_1 \cup \mathcal{S}_2$.

v) The following statements are equivalent:

a) $\mathcal{S}_1 \subseteq \mathcal{S}_2$.

b) $\mathcal{S}_2^\perp \subseteq \mathcal{S}_1^\perp$.

c) $\mathcal{S}_1 + \mathcal{S}_2 = \mathcal{S}_2$.

d) $\mathcal{S}_1 \cap \mathcal{S}_2 = \mathcal{S}_1$.

e) $\mathcal{S}_1$ and $\mathcal{S}_2^\perp$ are mutually orthogonal.

f) For each subspace $\mathcal{S} \subseteq \mathbb{F}^n$, $\mathcal{S}_1 + (\mathcal{S}_2 \cap \mathcal{S}) = \mathcal{S}_2 \cap (\mathcal{S}_1 + \mathcal{S})$.

g) There exists a subspace $\mathcal{S} \subseteq \mathbb{F}^n$ such that $\mathcal{S}_1 + (\mathcal{S}_2 \cap \mathcal{S}) = \mathcal{S}_2 \cap (\mathcal{S}_1 + \mathcal{S})$.

vi) $\mathcal{S}_1 \subset \mathcal{S}_2$ if and only if $\mathcal{S}_2^\perp \subset \mathcal{S}_1^\perp$.

vii) $\mathcal{S}_1 = (\mathcal{S}_1 \cap \mathcal{S}_2) + [\mathcal{S}_1 \cap (\mathcal{S}_1 \cap \mathcal{S}_2)^\perp]$.

Source: [2403, p. 81].

Fact 3.12.17. Let $\mathcal{S}_1, \mathcal{S}_2 \subseteq \mathbb{F}^n$ be subspaces. Then, the following statements hold:

i) $(\mathcal{S}_1 \cap \mathcal{S}_2)^\perp = \mathcal{S}_1^\perp + \mathcal{S}_2^\perp$.

ii) $(\mathcal{S}_1 + \mathcal{S}_2)^\perp = \mathcal{S}_1^\perp \cap \mathcal{S}_2^\perp$.

iii) $\mathcal{S}_1 = \mathcal{S}_2$ if and only if $\mathcal{S}_1 \cap (\mathcal{S}_1^\perp + \mathcal{S}_2^\perp) = \mathcal{S}_2 \cap (\mathcal{S}_1^\perp + \mathcal{S}_2^\perp)$.

iv) $\mathcal{S}_1 \cap (\mathcal{S}_1^\perp + \mathcal{S}_2^\perp)$ and $\mathcal{S}_2 + (\mathcal{S}_1^\perp \cap \mathcal{S}_2^\perp)$ are complementary subspaces.

v) $\mathcal{S}_1 + \mathcal{S}_2 = \mathcal{S}_1 + [\mathcal{S}_1^\perp \cap (\mathcal{S}_1 + \mathcal{S}_2)]$.

vi) The following statements are equivalent:

a) $\mathcal{S}_1$ and $\mathcal{S}_2$ are complementary subspaces.

b) $\mathcal{S}_1^\perp$ and $\mathcal{S}_2^\perp$ are complementary subspaces.

c) $\mathcal{S}_1 + \mathcal{S}_2 = \mathcal{S}_1^\perp + \mathcal{S}_2^\perp$.

d) $\mathcal{S}_1 \cap \mathcal{S}_2 = \mathcal{S}_1^\perp \cap \mathcal{S}_2^\perp$.

vii) The following statements are equivalent:

a) $\mathcal{S}_1 \cap \mathcal{S}_2^\perp = \mathcal{S}_1^\perp \cap \mathcal{S}_2 = \{0\}$.

b) $\mathcal{S}_1 + \mathcal{S}_2^\perp = \mathcal{S}_1^\perp + \mathcal{S}_2$.

c) $\mathcal{S}_1 \cap \mathcal{S}_2^\perp = \mathcal{S}_1^\perp \cap \mathcal{S}_2$.

Source: [255]. **Related:** Fact 4.15.4.

Fact 3.12.18. Let $\mathcal{S}_1 \subseteq \mathbb{F}^n$ and $\mathcal{S}_2 \subseteq \mathbb{F}^m$ be (cones, pointed cones, convex sets, polytopes, convex cones, polyhedral cones, affine subspaces, subspaces). Then, so is $\begin{bmatrix}\mathcal{S}_1\\\mathcal{S}_2\end{bmatrix}$. Furthermore,

$$\dim \begin{bmatrix}\mathcal{S}_1\\\mathcal{S}_2\end{bmatrix} = \dim \mathcal{S}_1 + \dim \mathcal{S}_2.$$

Fact 3.12.19. Let $\mathcal{S}_1, \mathcal{S}_2 \subseteq \mathbb{F}^n$ be subspaces. Then,

$$\begin{aligned}\dim \mathcal{S}_1 + \dim \mathcal{S}_2 - n &\le \dim(\mathcal{S}_1 \cap \mathcal{S}_2) \le \min\{\dim \mathcal{S}_1, \dim \mathcal{S}_2\}\\ &\le \left\{\begin{matrix}\dim \mathcal{S}_1\\ \dim \mathcal{S}_2\end{matrix}\right\} \le \max\{\dim \mathcal{S}_1, \dim \mathcal{S}_2\}\\ &\le \dim(\mathcal{S}_1 + \mathcal{S}_2) = \dim \mathcal{S}_1 + \dim \mathcal{S}_2 - \dim(\mathcal{S}_1 \cap \mathcal{S}_2)\\ &\le \min\{\dim \mathcal{S}_1 + \dim \mathcal{S}_2, n\}.\end{aligned}$$

Furthermore, the following statements hold:

i) If $\dim(\mathcal{S}_1 + \mathcal{S}_2) = \dim(\mathcal{S}_1 \cap \mathcal{S}_2) + 1$, then either $\mathcal{S}_1 \subseteq \mathcal{S}_2$ or $\mathcal{S}_2 \subseteq \mathcal{S}_1$.

ii) If $\mathcal{S}_1 \cap \mathcal{S}_2 = \{0\}$, then $\dim \mathcal{S}_1 + \dim \mathcal{S}_2 = \dim(\mathcal{S}_1 + \mathcal{S}_2) \le n$.

iii) $\dim \mathcal{S}_1 + \dim(\mathcal{S}_1^\perp \cap \mathcal{S}_2) = \dim \mathcal{S}_2 + \dim(\mathcal{S}_1 \cap \mathcal{S}_2^\perp)$.

iv) $\dim[\mathcal{S}_1 \cap (\mathcal{S}_1^\perp + \mathcal{S}_2)] = \dim[\mathcal{S}_2 \cap (\mathcal{S}_1 + \mathcal{S}_2^\perp)]$.

v) $\dim(\mathcal{S}_1 + \mathcal{S}_2) = \dim(\mathcal{S}_1 \cap \mathcal{S}_2) + \dim[(\mathcal{S}_1 + \mathcal{S}_2) \cap (\mathcal{S}_1^\perp + \mathcal{S}_2^\perp)]$.

vi) If $\operatorname{affin} \mathcal{S}_1 \cap \operatorname{affin} \mathcal{S}_2$ is nonempty, then $\dim(\mathcal{S}_1 \cup \mathcal{S}_2) = \dim(\mathcal{S}_1 + \mathcal{S}_2)$.

vii) If $\operatorname{affin} \mathcal{S}_1 \cap \operatorname{affin} \mathcal{S}_2$ is empty, then $\dim(\mathcal{S}_1 \cup \mathcal{S}_2) = \dim(\mathcal{S}_1 + \mathcal{S}_2) + 1$.

Source: Theorem 3.1.3. To prove the first inequality, note that $\dim \mathcal{S}_1 + \dim \mathcal{S}_2 = \dim(\mathcal{S}_1 + \mathcal{S}_2) + \dim(\mathcal{S}_1 \cap \mathcal{S}_2) \le n + \dim(\mathcal{S}_1 \cap \mathcal{S}_2)$. *i)* is given in [2991, p. 7]; *iii)*–*v)* are given in [255]; and *vi)* and *vii)* are given in [2487, p. 45]. **Related:** Fact 4.18.1.

Fact 3.12.20. Let $\mathcal{S}_1, \mathcal{S}_2 \subseteq \mathbb{F}^n$ be subspaces, and define $f(\mathcal{S}_1, \mathcal{S}_2) \triangleq \dim(\mathcal{S}_1 + \mathcal{S}_2) - \dim(\mathcal{S}_1 \cap \mathcal{S}_2)$. Then, the following statements hold:

i) If $\mathcal{S}_1 \ne \mathcal{S}_2$, then $f(\mathcal{S}_1, \mathcal{S}_2) > 0$.

ii) $f(\mathcal{S}_1, \mathcal{S}_2) = 0$ if and only if $\mathcal{S}_1 = \mathcal{S}_2$.

iii) $f(\mathcal{S}_1, \mathcal{S}_2) = f(\mathcal{S}_2, \mathcal{S}_1)$.

iv) Let $\mathcal{S}_3 \subseteq \mathbb{F}^n$ be a subspace. Then, $f(\mathcal{S}_1, \mathcal{S}_3) \le f(\mathcal{S}_1, \mathcal{S}_2) + f(\mathcal{S}_2, \mathcal{S}_3)$.

v) $f(\mathcal{S}_1, \mathcal{S}_2) = f[\mathcal{S}_1 \cap (\mathcal{S}_1 \cap \mathcal{S}_2)^\perp, \mathcal{S}_2 \cap (\mathcal{S}_1 \cap \mathcal{S}_2)^\perp]$.

Source: [249, 253]. **Remark:** f is a metric on the vector space of subspaces of $\mathbb{F}^n$.

Fact 3.12.21. Let $\mathcal{S}_1, \mathcal{S}_2 \subseteq \mathbb{F}^n$ be nonzero subspaces, and define $\theta \in [0, \frac{\pi}{2}]$ by

$$\cos\theta = \max\{|x^*y| \colon (x, y) \in \mathcal{S}_1 \times \mathcal{S}_2 \text{ and } x^*x = y^*y = 1\}.$$

Then,

$$\cos\theta = \max\{|x^*y| \colon (x, y) \in \mathcal{S}_1^\perp \times \mathcal{S}_2^\perp \text{ and } x^*x = y^*y = 1\}.$$

Furthermore, $\theta = 0$ if and only if $\mathcal{S}_1 \cap \mathcal{S}_2 = \{0\}$, and $\theta = \frac{\pi}{2}$ if and only if $\mathcal{S}_1 = \mathcal{S}_2^\perp$. **Remark:** θ is a *principal angle*. See [1134, 1504]. **Related:** Fact 7.10.29, Fact 7.12.42, and Fact 7.13.27.

Fact 3.12.22. Let $\mathcal{S}_1, \mathcal{S}_2 \subseteq \mathbb{F}^n$ be subspaces, assume that $\mathcal{S}_1$ and $\mathcal{S}_2$ are complementary, assume that $\dim \mathcal{S}_2 \ge 1$, and let $\mathcal{S}_3 \subseteq \mathbb{F}^n$. Then, the following statements hold:

i) If $\mathcal{S}_3 \subset \mathcal{S}_2$, then $\mathcal{S}_1 + \mathcal{S}_3 \subset \mathbb{F}^n$.

ii) If $\mathcal{S}_3$ is a subspace and $\mathcal{S}_1 + \mathcal{S}_3 = \mathbb{F}^n$, then $\mathcal{S}_2 \subseteq \mathcal{S}_3$.

Fact 3.12.23. Let $\mathcal{S}_1, \mathcal{S}_2 \subseteq \mathbb{F}^n$ be subspaces, and assume that $\dim \mathcal{S}_2 < \dim \mathcal{S}_1$. Then, there exists a subspace $\mathcal{S}_3 \subseteq \mathbb{F}^n$ such that the following statements hold:

i) $\mathcal{S}_3 \subseteq \mathcal{S}_1$.

ii) $\mathcal{S}_2$ and $\mathcal{S}_3$ are mutually orthogonal.

iii) $\dim \mathcal{S}_1 \le \dim \mathcal{S}_2 + \dim \mathcal{S}_3$.

Source: [2991, p. 34].

Fact 3.12.24. Let $\mathcal{S}_1, \mathcal{S}_2, \mathcal{S}_3 \subseteq \mathbb{F}^n$ be subspaces. Then,

$$\begin{aligned}\dim(\mathcal{S}_1 + \mathcal{S}_2 + \mathcal{S}_3) + \max\{\dim(\mathcal{S}_1 \cap \mathcal{S}_2), \dim(\mathcal{S}_1 \cap \mathcal{S}_3), \dim(\mathcal{S}_2 \cap \mathcal{S}_3)\} &\le \dim \mathcal{S}_1 + \dim \mathcal{S}_2 + \dim \mathcal{S}_3\\ &\le \dim(\mathcal{S}_1 \cap \mathcal{S}_2 \cap \mathcal{S}_3) + 2n.\end{aligned}$$

Source: [860, p. 124], [2238, p. 127], and [2991, p. 267]. **Remark:** Setting $\mathcal{S}_3 = \{0\}$ yields a weaker version of Theorem 3.1.3.

Fact 3.12.25. Let $\mathcal{S}_1, \mathcal{S}_2, \mathcal{S}_3 \subseteq \mathbb{F}^n$ be subspaces. Then,

$$\mathcal{S}_1 + (\mathcal{S}_2 \cap \mathcal{S}_3) \subseteq (\mathcal{S}_1 + \mathcal{S}_2) \cap (\mathcal{S}_1 + \mathcal{S}_3) = \mathcal{S}_1 + (\mathcal{S}_1 + \mathcal{S}_2) \cap \mathcal{S}_3,$$

$$\mathcal{S}_1 \cap (\mathcal{S}_2 + \mathcal{S}_3) \supseteq (\mathcal{S}_1 \cap \mathcal{S}_2) + (\mathcal{S}_1 \cap \mathcal{S}_3).$$

If, in addition, $\mathcal{S}_2 \subseteq \mathcal{S}_1$, then
$$\mathcal{S}_1 \cap (\mathcal{S}_2 + \mathcal{S}_3) = \mathcal{S}_2 + (\mathcal{S}_1 \cap \mathcal{S}_3).$$
Source: [688, pp. 11, 12] and [2991, p. 7].

Fact 3.12.26. Let $\mathcal{S}_1, \ldots, \mathcal{S}_k \subseteq \mathbb{F}^n$ be convex sets, and let $\alpha_1, \ldots, \alpha_k \in \mathbb{F}$. Then, $\sum_{i=1}^k \alpha_i \mathcal{S}_i$ is convex. **Source:** [2487, pp. 75, 76].

Fact 3.12.27. Let $\mathcal{S}_1, \ldots, \mathcal{S}_k \subseteq \mathbb{F}^n$ be subspaces having the same dimension. Then, there exists a subspace $\hat{\mathcal{S}} \subseteq \mathbb{F}^n$ such that, for all $i \in \{1, \ldots, k\}$, $\hat{\mathcal{S}}$ and $\mathcal{S}_i$ are complementary. **Source:** [1304, pp. 78, 79, 259, 260].

3.13 Facts on Range, Null Space, Rank, and Defect

Fact 3.13.1. Let $x_1, \ldots, x_r \in \mathbb{F}^n$. Then, $\dim \cap_{i=1}^r \{x_i\}^\perp = n - \operatorname{rank}\,[x_1 \ \cdots \ x_r]$. **Remark:** This result determines the dimension of an intersection of hyperplanes in terms of the number of linearly independent vectors that define the hyperplanes. **Related:** Fact 4.17.9.

Fact 3.13.2. Let $A \in \mathbb{F}^{n \times n}$. Then, $\mathcal{N}(A) \subseteq \mathcal{R}(I - A)$ and $\mathcal{N}(I - A) \subseteq \mathcal{R}(A)$. **Related:** Fact 4.15.5.

Fact 3.13.3. Let $A \in \mathbb{F}^{n \times m}$, and let $S \in \mathbb{F}^{m \times l}$. Then, $S\mathcal{N}(AS) \subseteq \mathcal{N}(A)$. If, in addition, S is right invertible, then $S\mathcal{N}(AS) = \mathcal{N}(A)$. **Source:** $S\mathcal{N}(AS) = S\{x \in \mathbb{F}^l\colon ASx = 0\} = \{Sx \in \mathbb{F}^l\colon ASx = 0\} \subseteq \{y \in \mathbb{F}^l\colon Ay = 0\} = \mathcal{N}(A)$. Now, assume that S is right invertible. Then, S is onto, and thus $S\mathcal{N}(AS) = S\{x \in \mathbb{F}^l\colon ASx = 0\} = \{Sx \in \mathbb{F}^l\colon ASx = 0\} = \{y \in \mathbb{F}^l\colon Ay = 0\} = \mathcal{N}(A)$. **Related:** Proposition 3.7.3.

Fact 3.13.4. Let $\mathcal{S} \subseteq \mathbb{F}^m$, assume that $\mathcal{S}$ is an affine subspace, and let $A \in \mathbb{F}^{n \times m}$. Then, the following statements hold:

i) $\operatorname{rank} A + \dim \mathcal{S} - m \le \dim A\mathcal{S} \le \min\{\operatorname{rank} A, \dim \mathcal{S}\}$.

ii) $\dim(A\mathcal{S}) + \dim[\mathcal{N}(A) \cap \mathcal{S}] = \dim \mathcal{S}$.

iii) $\dim A\mathcal{S} \le \dim \mathcal{S}$.

iv) If A is left invertible, then $\dim A\mathcal{S} = \dim \mathcal{S}$.

Source: For *ii*), see [2314, p. 413]. For *iv*), note that $\dim A\mathcal{S} \le \dim \mathcal{S} = \dim A^{\mathsf{L}}A\mathcal{S} \le \dim A\mathcal{S}$. **Remark:** The proof of Proposition 3.6.10 uses the Moore-Penrose generalized inverse defined in Chapter 8. An alternative proof that avoids this technique is given in [2238, p. 111]. Similarly, define $\mathcal{S} \triangleq \mathcal{R}(B)$. Then, it follows from *ii*) that $\dim \mathcal{R}(B) = \dim[A\mathcal{R}(B)] + \dim[\mathcal{N}(A) \cap \mathcal{R}(B)] = \operatorname{rank} AB + \dim[\mathcal{R}(B) \cap \mathcal{N}(A)]$. **Related:** Fact 3.11.15 and Fact 12.11.28.

Fact 3.13.5. Let $A \in \mathbb{F}^{n \times m}$, let $B, C \in \mathbb{F}^{m \times p}$, and assume that $A^*AB = A^*AC$. Then, $AB = AC$. **Source:** For all $i \in \{1, \ldots, p\}$, $\operatorname{col}_i(B) - \operatorname{col}_i(C) \in \mathcal{N}(A^*A) = \mathcal{N}(A)$.

Fact 3.13.6. Let $A \in \mathbb{F}^{n \times m}$, let $B, C \in \mathbb{F}^{n \times p}$, and assume that $AA^*B = AA^*C$. Then, $A^*B = A^*C$. **Source:** Fact 3.13.5.

Fact 3.13.7. Let $A \in \mathbb{F}^{n \times m}$, let $B, C \in \mathbb{F}^{p \times n}$, and assume that $BAA^* = CAA^*$. Then, $BA = CA$.

Fact 3.13.8. Let $A \in \mathbb{F}^{n \times m}$, let $B, C \in \mathbb{F}^{p \times m}$, and assume that $BA^*A = CA^*A$. Then, $BA^* = CA^*$.

Fact 3.13.9. Let $A \in \mathbb{F}^{n \times m}$, let $x, y \in \mathcal{R}(A^*)$, and assume that $Ax = Ay$. Then, $x = y$. **Source:** Fact 3.13.6.

Fact 3.13.10. Let $A \in \mathbb{F}^{n \times m}$ and $B \in \mathbb{F}^{1 \times m}$. Then, $\mathcal{N}(A) \subseteq \mathcal{N}(B)$ if and only if there exists $y \in \mathbb{F}^n$ such that $B = y^*A$.

Fact 3.13.11. Let $A \in \mathbb{F}^{n \times m}$ and $b \in \mathbb{F}^n$. Then, there exists $x \in \mathbb{F}^n$ satisfying $Ax = b$ if and only if, for all $y \in \mathcal{N}(A^*)$, $b^*y = 0$. **Source:** Assume that $A^*y = 0$ implies that $b^*y = 0$. Then, $\mathcal{N}(A^*) \subseteq \mathcal{N}(b^*)$. Hence, $b \in \mathcal{R}(b) \subseteq \mathcal{R}(A)$.

Fact 3.13.12. Let $A \in \mathbb{F}^{n \times m}$ and $B \in \mathbb{F}^{l \times m}$. Then, $\mathcal{N}(B) \subseteq \mathcal{N}(A)$ if and only if there exists $C \in \mathbb{F}^{n \times l}$ such that $A = CB$. Now, let $A \in \mathbb{F}^{n \times m}$ and $B \in \mathbb{F}^{n \times l}$. Then, $\mathcal{R}(A) \subseteq \mathcal{R}(B)$ if and only if there exists $C \in \mathbb{F}^{l \times m}$ such that $A = BC$.

Fact 3.13.13. Let $A, B \in \mathbb{F}^{n\times m}$, and let $C \in \mathbb{F}^{m\times l}$ be right invertible. Then, $\mathcal{R}(A) \subseteq \mathcal{R}(B)$ if and only if $\mathcal{R}(AC) \subseteq \mathcal{R}(BC)$. Furthermore, $\mathcal{R}(A) = \mathcal{R}(B)$ if and only if $\mathcal{R}(AC) = \mathcal{R}(BC)$. **Source:** Since C is right invertible, it follows that $\mathcal{R}(A) = \mathcal{R}(AC)$.

Fact 3.13.14. Let $A, B \in \mathbb{F}^{n\times n}$, and assume that there exists $\alpha \in \mathbb{F}$ such that $\alpha A+B$ is nonsingular. Then, $\mathcal{N}(A) \cap \mathcal{N}(B) = \{0\}$. **Remark:** The converse is false. Let $A \triangleq \begin{bmatrix} 1 & 0 \\ 2 & 0 \end{bmatrix}$ and $B \triangleq \begin{bmatrix} 0 & 1 \\ 0 & 2 \end{bmatrix}$.

Fact 3.13.15. Let $A, B \in \mathbb{F}^{n\times m}$. Then, $\mathcal{R}(A + B) \subseteq \mathcal{R}(A) + \mathcal{R}(B)$. Furthermore, the following statements are equivalent:

i) $\mathcal{R}(A) \subseteq \mathcal{R}(A + B)$.

ii) $\mathcal{R}(B) \subseteq \mathcal{R}(A + B)$.

iii) $\mathcal{R}(A + B) = \mathcal{R}(A) + \mathcal{R}(B)$.

Source: To prove *i)* $\Longrightarrow$ *ii)*, $\mathcal{R}(A) \subseteq \mathcal{R}(A + B)$ implies that $\mathcal{R}(A + B) = \mathcal{R}([A\ \ A + B]) = \mathcal{R}([A\ \ B]) = \mathcal{R}([B\ \ A + B])$. Hence, $\mathcal{R}(B) \subseteq \mathcal{R}(A + B)$. See [2991, p. 56].

Fact 3.13.16. Let $A, B \in \mathbb{F}^{n\times m}$. Then, $\mathcal{N}(A) \cap \mathcal{N}(B) \subseteq \mathcal{N}(A + B)$. Furthermore, the following statements are equivalent:

i) $\mathcal{N}(A + B) \subseteq \mathcal{N}(A)$.

ii) $\mathcal{N}(A + B) \subseteq \mathcal{N}(B)$.

iii) $\mathcal{N}(A + B) = \mathcal{N}(A) \cap \mathcal{N}(B)$.

Source: Fact 3.13.15.

Fact 3.13.17. Let $A, B \in \mathbb{F}^{n\times m}$, and let $\alpha \in \mathbb{F}$ be nonzero. Then,

$$\mathcal{N}(A) \cap \mathcal{N}(B) = \mathcal{N}(A) \cap \mathcal{N}(A + \alpha B) = \mathcal{N}(\alpha A + B) \cap \mathcal{N}(B).$$

Related: Fact 3.14.10.

Fact 3.13.18. Let $x \in \mathbb{F}^n$ and $y \in \mathbb{F}^m$. If either $x = 0$ or $y \neq 0$, then $\mathcal{R}(xy^{\mathsf{T}}) = \mathcal{R}(x) = \operatorname{span}\{x\}$. Furthermore, if either $x \neq 0$ or $y = 0$, then $\mathcal{N}(xy^{\mathsf{T}}) = \mathcal{N}(y^{\mathsf{T}}) = \{\overline{y}\}^{\perp}$.

Fact 3.13.19. Let $A \in \mathbb{F}^{n\times m}$ and $B \in \mathbb{F}^{m\times l}$. Then, $\operatorname{rank} AB = \operatorname{rank} A$ if and only if $\mathcal{R}(AB) = \mathcal{R}(A)$. **Source:** If $\mathcal{R}(AB) \subset \mathcal{R}(A)$, then Lemma 3.1.5 implies that $\operatorname{rank} AB < \operatorname{rank} A$.

Fact 3.13.20. Let $A \in \mathbb{F}^{n\times m}$, $B \in \mathbb{F}^{m\times l}$, and $C \in \mathbb{F}^{l\times k}$, and assume that $\operatorname{rank} AB = \operatorname{rank} B$. Then, $\operatorname{rank} ABC = \operatorname{rank} BC$. **Source:** $\operatorname{rank} B^{\mathsf{T}}A^{\mathsf{T}} = \operatorname{rank} B^{\mathsf{T}}$ implies that $\mathcal{R}(C^{\mathsf{T}}B^{\mathsf{T}}A^{\mathsf{T}}) = \mathcal{R}(C^{\mathsf{T}}B^{\mathsf{T}})$.

Fact 3.13.21. Let $A \in \mathbb{F}^{n\times m}$ and $B \in \mathbb{F}^{m\times l}$. Then, the following statements hold:

i) $\operatorname{rank} AB + \dim[\mathcal{N}(A) \cap \mathcal{R}(B)] = \operatorname{rank} B$.

ii) $\operatorname{rank} AB + \dim[\mathcal{N}(B^*) \cap \mathcal{R}(A^*)] = \operatorname{rank} A$.

iii) $\operatorname{rank} AB + m = \operatorname{rank} A + \dim[\mathcal{N}(A) + \mathcal{R}(B)]$.

iv) $\operatorname{rank} AB + \operatorname{def} A = \dim[\mathcal{N}(A) + \mathcal{R}(B)]$.

v) $\operatorname{rank} AB + \operatorname{def} A + \dim[\mathcal{N}(B^*) \cap \mathcal{R}(A^*)] = m$.

vi) $\operatorname{rank} AB + \operatorname{def} A = \dim[\mathcal{N}(A) + \mathcal{R}(B)]$.

vii) $\operatorname{rank} AB + \operatorname{def} B + \dim[\mathcal{N}(A) \cap \mathcal{R}(B)] = l$.

viii) $\operatorname{def} AB + \operatorname{rank} B = \dim[\mathcal{N}(A) \cap \mathcal{R}(B)] + l$.

ix) $\operatorname{def} AB = \operatorname{def} B + \dim[\mathcal{N}(A) \cap \mathcal{R}(B)]$.

x) $\operatorname{def} AB + \operatorname{rank} A = \dim[\mathcal{N}(B^*) \cap \mathcal{R}(A^*)] + l$.

xi) $\operatorname{def} AB + m = \operatorname{def} A + \dim[\mathcal{N}(B^*) \cap \mathcal{R}(A^*)] + l$.

xii) $\operatorname{def} AB + \operatorname{rank} A + \dim[\mathcal{N}(A) + \mathcal{R}(B)] = l + m$.

xiii) $\operatorname{def} AB + \dim[\mathcal{N}(A) + \mathcal{R}(B)] = \operatorname{def} A + l$.

Remark: *i)* is Sylvester's rank formula given by Proposition 3.6.10. **Related:** Fact 8.9.2, Fact 8.9.3, and Fact 8.9.4.

Fact 3.13.22. Let $A \in \mathbb{F}^{n\times m}$ and $B \in \mathbb{F}^{m\times l}$. Then,

$$\max\{\operatorname{def} A + l - m, \operatorname{def} B\} \le \operatorname{def} AB \le \operatorname{def} A + \operatorname{def} B.$$

If, in addition, $m = l$, then

$$\max\{\operatorname{def} A, \operatorname{def} B\} \le \operatorname{def} AB.$$

Remark: The first inequality is *Sylvester's law of nullity.*

Fact 3.13.23. Let $A \in \mathbb{F}^{n\times m}$, $B \in \mathbb{F}^{n\times l}$, and $k \ge 0$. Then, there exists $X \in \mathbb{F}^{m\times l}$ such that $AX = B$ and $\operatorname{rank} X = k$ if and only if $\operatorname{rank} B \le k \le \min\{m + \operatorname{rank} B - \operatorname{rank} A, l\}$. **Source:** [2760].

Fact 3.13.24. The following statements hold:

i) Let $A \in \mathbb{F}^{n\times m}$. Then, $\operatorname{rank} A = 0$ if and only if $A = 0$.

ii) For all $\alpha \in \mathbb{F}$ and $A \in \mathbb{F}^{n\times m}$, $\operatorname{rank} \alpha A = (\operatorname{sign}|\alpha|) \operatorname{rank} A$.

iii) For all $A, B \in \mathbb{F}^{n\times m}$, $\operatorname{rank}(A + B) \le \operatorname{rank} A + \operatorname{rank} B$

Remark: Compare these statements to the properties of a matrix norm given by Definition 11.2.1.

Fact 3.13.25. Let $n, m, k \in \mathbb{P}$. Then, $\operatorname{rank} 1_{n\times m} = 1$ and $1_{n\times n}^k = n^{k-1}1_{n\times n}$.

Fact 3.13.26. Let $A \in \mathbb{F}^{n\times m}$. Then, $\operatorname{rank} A = 1$ if and only if there exist $x \in \mathbb{F}^n$ and $y \in \mathbb{F}^m$ such that $x \ne 0$, $y \ne 0$, and $A = xy^{\mathsf{T}}$. If these statements hold, then $\operatorname{tr} A = y^{\mathsf{T}}x$. **Related:** Fact 6.10.5.

Fact 3.13.27. Let $A \in \mathbb{F}^{n\times n}$, $k \ge 1$, and $l \in \mathbb{N}$. Then, the following statements hold:

i) $\mathcal{R}[(AA^*)^k] = \mathcal{R}[(AA^*)^l A]$.

ii) $\mathcal{N}[(A^*A)^k] = \mathcal{N}[A(A^*A)^l]$.

iii) $\operatorname{rank}(AA^*)^k = \operatorname{rank}(AA^*)^l A$.

iv) $\operatorname{def}(A^*A)^k = \operatorname{def} A(A^*A)^l$.

Fact 3.13.28. Let $A \in \mathbb{F}^{n\times n}$. Then,

$$\begin{aligned}
\mathcal{R}(I - A) &= \mathcal{R}(A - A^2) + \mathcal{R}(I - A^2), & \mathcal{N}(I - A) &= \mathcal{N}(A - A^2) \cap \mathcal{N}(I - A^2),\\
\mathcal{R}(I + A) &= \mathcal{R}(A + A^2) + \mathcal{R}(I - A^2), & \mathcal{N}(I + A) &= \mathcal{N}(A + A^2) \cap \mathcal{N}(I - A^2),\\
\mathcal{R}(A) &= \mathcal{R}(A^2) + \mathcal{R}(A - A^3), & \mathcal{N}(A) &= \mathcal{N}(A^2) \cap \mathcal{N}(A - A^3),\\
\mathcal{R}(A) &= \mathcal{R}(A - A^2) + \mathcal{R}(A + A^2), & \mathcal{N}(A) &= \mathcal{N}(A - A^2) \cap \mathcal{N}(A + A^2).
\end{aligned}$$

Source: [257].

Fact 3.13.29. Let $A \in \mathbb{F}^{n\times n}$. Then,

$$\begin{aligned}
\operatorname{rank}(I - A^2) &= \operatorname{rank}(I + A) + \operatorname{rank}(I - A) - n,\\
\operatorname{rank}(A - A^2) &= \operatorname{rank} A + \operatorname{rank}(I - A) - n,\\
\operatorname{rank}(A + A^2) &= \operatorname{rank} A + \operatorname{rank}(I + A) - n,\\
\operatorname{rank}(A - A^3) &= \operatorname{rank} A + \operatorname{rank}(I - A^2) - n,\\
\operatorname{rank}(A - A^3) + \operatorname{rank}(I - A) &= \operatorname{rank}(A - A^2) + \operatorname{rank}(I - A^2),\\
\operatorname{rank}(A - A^3) + \operatorname{rank}(I + A) &= \operatorname{rank}(A + A^2) + \operatorname{rank}(I - A^2),\\
\operatorname{rank} A + \operatorname{rank}(A - A^3) &= \operatorname{rank}(A + A^2) + \operatorname{rank}(A - A^2),\\
\operatorname{rank} A + \operatorname{rank}(A^2 - A^4) &= \operatorname{rank} A^2 + \operatorname{rank}(A - A^3),\\
\operatorname{rank} A + \operatorname{rank}(A^2 - A^5) &= \operatorname{rank} A^2 + \operatorname{rank}(A - A^4).
\end{aligned}$$

Source: [257, 2659]. **Related:** Fact 4.9.2 and Fact 4.21.3.

Fact 3.13.30. Let $A \in \mathbb{F}^{n\times n}$. Then, the following statements hold:

i) $2\operatorname{rank} A \le \operatorname{rank} A^2 + n$.

ii) $\operatorname{rank} A = \operatorname{rank} A^2 + n$ if and only if $\mathcal{N}(A) \subseteq \mathcal{R}(A)$.

iii) $2\operatorname{rank} A^2 \le \operatorname{rank} A + \operatorname{rank} A^3$.

iv) If $k \ge 1$, then $k\operatorname{rank} A \le \operatorname{rank} A^k + (k-1)n$.

v) If $l \ge k \ge 0$, then $\operatorname{rank}(A^k - A^{k+l}) = \operatorname{rank} A^k + \operatorname{rank}(I - A^l) - n$.

vi) If $k \ge 0$, then $\operatorname{rank}(I - A^{k+1}) = \operatorname{rank}(I - A) + \left(\operatorname{rank} \sum_{i=0}^{k} A^i\right) - n$.

vii) If $k \ge 0$ and $l \ge 2$, then $\operatorname{rank}(A^k - A^{kl}) = \operatorname{rank} A^k + \operatorname{rank}(I - A^{k(l-1)}) - n$.

Furthermore, consider the following statements:

viii) There exists $k \ge 2$ such that $A^k = A$.

ix) $2\operatorname{rank} A^2 = \operatorname{rank} A + \operatorname{rank} A^3$.

x) There exist $X, Y \in \mathbb{F}^{n\times n}$ such that $A = A^2X + YA^2$.

Then, *viii*) $\Longrightarrow$ *ix*) $\Longleftrightarrow$ *x*). **Source:** [257] and [860, p. 126].

Fact 3.13.31. Let $x, y \in \mathbb{F}^n$. Then, the following statements hold:

i) $\mathcal{R}(xy^{\mathsf{T}} + yx^{\mathsf{T}}) \subseteq \mathcal{R}([x\ \ y])$.

ii) $\{x\}^\perp \cap \{y\}^\perp \subseteq \mathcal{N}(xy^{\mathsf{T}} + yx^{\mathsf{T}})$.

iii) $\operatorname{rank}(xy^{\mathsf{T}} + yx^{\mathsf{T}}) \le 2$.

Furthermore, the following statements are equivalent:

iv) Either x or y is zero.

v) $xy^{\mathsf{T}} + yx^{\mathsf{T}} = 0$.

vi) $\operatorname{rank}(xy^{\mathsf{T}} + yx^{\mathsf{T}}) = 0$.

vii) $\operatorname{def}(xy^{\mathsf{T}} + yx^{\mathsf{T}}) = n$.

In addition, the following statements are equivalent:

viii) There exists $\alpha \in \mathbb{F}$ such that $x = \alpha y \ne 0$.

ix) $\operatorname{rank}(xy^{\mathsf{T}} + yx^{\mathsf{T}}) = 1$.

x) $\operatorname{def}(xy^{\mathsf{T}} + yx^{\mathsf{T}}) = n - 1$.

Moreover, the following statements are equivalent:

xi) x and y are linearly independent.

xii) $\operatorname{rank}(xy^{\mathsf{T}} + yx^{\mathsf{T}}) = 2$.

xiii) $\operatorname{def}(xy^{\mathsf{T}} + yx^{\mathsf{T}}) = n - 2$.

Finally, the following statements are equivalent:

xiv) x and y are nonzero.

xv) x or y is nonzero and $\mathcal{R}(xy^{\mathsf{T}} + yx^{\mathsf{T}}) = \mathcal{R}([x\ \ y])$.

xvi) x or y is nonzero and $\operatorname{rank}(xy^{\mathsf{T}} + yx^{\mathsf{T}}) = \operatorname{rank}([x\ \ y])$.

xvii) x or y is nonzero and $\{x\}^\perp \cap \{y\}^\perp = \mathcal{N}(xy^{\mathsf{T}} + yx^{\mathsf{T}})$.

xviii) x or y is nonzero and $\dim(\{x\}^\perp \cap \{y\}^\perp) = \operatorname{def}(xy^{\mathsf{T}} + yx^{\mathsf{T}})$.

xix) $1 \le \operatorname{rank}(xy^{\mathsf{T}} + yx^{\mathsf{T}}) \le 2$.

Remark: $xy^{\mathsf{T}} + yx^{\mathsf{T}}$ is a *doublet*. See [835, pp. 539, 540].

Fact 3.13.32. Let $A \in \mathbb{F}^{n\times m}$, $x \in \mathbb{F}^n$, and $y \in \mathbb{F}^m$. Then,

$$(\operatorname{rank} A) - 1 \le \operatorname{rank}(A + xy^*) \le (\operatorname{rank} A) + 1.$$

Related: Fact 8.4.10.

Fact 3.13.33. Let $A \triangleq \begin{bmatrix} 1 & 0 \\ 0 & 0 \end{bmatrix}$ and $B \triangleq \begin{bmatrix} 0 & 1 \\ 0 & 0 \end{bmatrix}$. Then, $\operatorname{rank} AB = 1$ and $\operatorname{rank} BA = 0$. **Related:** Fact 4.10.32.

Fact 3.13.34. Let $A \in \mathbb{F}^{n\times m}$ and $B \in \mathbb{F}^{n\times l}$. Then,

$$\mathcal{R}(AA^* + BB^*) = \mathcal{R}(A) + \mathcal{R}(B), \quad \mathcal{N}(AA^* + BB^*) = \mathcal{N}(A^*) \cap \mathcal{N}(B^*).$$

Source: [2239].

Fact 3.13.35. Let $A \in \mathbb{F}^{n\times m}$ and $B \in \mathbb{F}^{m\times l}$. Then, $\operatorname{rank} AB = \operatorname{rank} A^*AB = \operatorname{rank} ABB^*$. **Source:** [2418, p. 37].

Fact 3.13.36. Let $A, B \in \mathbb{F}^{n\times m}$. Then,

$$|\operatorname{rank} A - \operatorname{rank} B| \le \left\{\begin{matrix}\operatorname{rank}(A+B)\\ \operatorname{rank}(A-B)\end{matrix}\right\} \le \operatorname{rank} A + \operatorname{rank} B.$$

If, in addition, $\operatorname{rank} B \le k$, then

$$(\operatorname{rank} A) - k \le \left\{\begin{matrix}\operatorname{rank}(A+B)\\ \operatorname{rank}(A-B)\end{matrix}\right\} \le (\operatorname{rank} A) + k.$$

Fact 3.13.37. Let $A \in \mathbb{F}^{n\times m}$ and $B \in \mathbb{F}^{n\times l}$. Then, $A^*B = 0$ if and only if $\mathcal{R}(A)$ and $\mathcal{R}(B)$ are mutually orthogonal. If these statements hold, then $\mathcal{R}(A) \cap \mathcal{R}(B) = \{0\}$. **Source:** Let $x = Az \in \mathcal{R}(A)$ and $y = Bw \in \mathcal{R}(B)$. Then, $x^*y = z^*A^*Bw = 0$. Hence, $\mathcal{R}(A)$ and $\mathcal{R}(B)$ are mutually orthogonal subspaces. Now, Proposition 3.3.5 implies that $\mathcal{R}(A) \cap \mathcal{R}(B) = \{0\}$. See [2991, p. 34]. **Related:** Fact 8.4.8.

Fact 3.13.38. Let $A, B \in \mathbb{F}^{n\times m}$, and assume that $A^*B = 0$ and $BA^* = 0$. Then,

$$\operatorname{rank}(A+B) = \operatorname{rank} A + \operatorname{rank} B.$$

Source: Since $A^*B = 0$, it follows from Fact 3.13.37 that $\mathcal{R}(A) \cap \mathcal{R}(B) = \{0\}$. Likewise, $BA^* = 0$ implies that $\mathcal{R}(A^*) \cap \mathcal{R}(B^*) = \{0\}$. The result now follows from Fact 3.14.11. **Source:** Fact 8.4.8 and [762, 1377]. **Remark:** The converse is false. Let $A = \left[\begin{smallmatrix}1&0\\0&0\end{smallmatrix}\right]$ and $B = \left[\begin{smallmatrix}0&1\\0&1\end{smallmatrix}\right]$. Fact 3.14.11 gives necessary and sufficient conditions for rank to be additive.

Fact 3.13.39. Let $A, B \in \mathbb{F}^{n\times n}$, assume that A is Hermitian, and assume that $[A, B] = 0$. Then,

$$\operatorname{rank}(A+B) = \operatorname{rank} A + \operatorname{rank} B.$$

Source: Fact 3.13.38.

Fact 3.13.40. Let $A, B \in \mathbb{F}^{n\times m}$. Then,

$$\begin{aligned}\operatorname{rank} A + \operatorname{rank} B &= \operatorname{rank}(A+B) + \dim\left[\mathcal{R}\left(\left[\begin{smallmatrix}A\\B\end{smallmatrix}\right]\right) \cap \mathcal{N}([I_n \ \ I_n])\right] + \dim[\mathcal{R}(A^*) \cap \mathcal{R}(B^*)]\\ &= \operatorname{rank}(A+B) + \dim\left[\mathcal{R}\left(\left[\begin{smallmatrix}A\\B\end{smallmatrix}\right]\right) \cap \mathcal{N}([I_n \ \ I_n])\right] + \dim([\mathcal{N}(A) + \mathcal{N}(B)]^\perp)\\ &= \operatorname{rank}(A+B) + \dim\left[\mathcal{R}\left(\left[\begin{smallmatrix}A\\B\end{smallmatrix}\right]\right) \cap \mathcal{N}([I_n \ \ I_n])\right] + m - \dim[\mathcal{N}(A) + \mathcal{N}(B)].\end{aligned}$$

Remark: See [2238, pp. 114, 115] and [2991, p. 53]. **Problem:** Use this result to prove Fact 3.13.38.

Fact 3.13.41. Let $A, B \in \mathbb{F}^{n\times n}$. Then,

$$\mathcal{R}(A - ABA) = \mathcal{R}(A) \cap \mathcal{R}(I - AB), \quad \mathcal{N}(A - ABA) = \mathcal{N}(A) + \mathcal{N}(I - BA),$$
$$\mathcal{R}(I - A^2) = \mathcal{R}(I - A) \cap \mathcal{R}(I + A), \quad \mathcal{N}(I - A^2) = \mathcal{N}(I - A) + \mathcal{N}(I + A).$$

Source: [257].

Fact 3.13.42. Let $A, B \in \mathbb{F}^{n\times n}$. Then,

$$\operatorname{rank}(AB - I) \le \operatorname{rank}(A - I) + \operatorname{rank}(B - I),$$
$$\operatorname{rank}(AB - I) + \operatorname{rank} B = \operatorname{rank}(B - BAB) + n,$$
$$\operatorname{rank}(I - BA) + \operatorname{rank} A = \operatorname{rank}(A - ABA) + n.$$

Source: $\operatorname{rank}(AB-I) = \operatorname{rank}[A(B-I)+A-I] \le \operatorname{rank}[A(B-I)]+\operatorname{rank}(A-I) \le \operatorname{rank}(A-I)+\operatorname{rank}(B-I)$. See [2991, p. 55].

Fact 3.13.43. Let $A \in \mathbb{F}^{n\times m}$ and $B \in \mathbb{F}^{m\times n}$. Then,

$$\operatorname{rank}(A - ABA) + m = \operatorname{rank} A + \operatorname{rank}(I_m - BA),$$
$$\operatorname{rank}(A - ABA) + n = \operatorname{rank} A + \operatorname{rank}(I_n - AB),$$
$$\operatorname{rank}(A - ABA) - \operatorname{rank}(B - BAB) = \operatorname{rank} A - \operatorname{rank} B.$$

Source: [257].

Fact 3.13.44. Let $A_1, \dots, A_k \in \mathbb{F}^{n\times m}$, and define $A \triangleq \sum_{i=1}^k A_i$. Then, the following statements are equivalent:

i) $\operatorname{rank} A = \sum_{i=1}^k \operatorname{rank} A_i$.

ii) For all $i \in \{1, \dots, k\}$, $\operatorname{rank}(A - A_i) = \operatorname{rank} A - \operatorname{rank} A_i$.

Source: [231, 1334].

Fact 3.13.45. Let $a, b \in \mathbb{F}$ be nonzero, let $n, m \ge 2$, and define $A \in \mathbb{F}^{n\times m}$ by

$$A \triangleq \begin{bmatrix} a & a+b & \cdots & a+(m-1)b \\ a+mb & a+(m+1)b & \cdots & a+(2m-1)b \\ \vdots & \vdots & \iddots & \vdots \\ a+(n-1)mb & a+[(n-1)m+1]b & \cdots & a+(nm-1)b \end{bmatrix}.$$

Then, $\operatorname{rank} A = 2$. **Source:** [1777]. **Remark:** As stated in [1777], "Given a homogeneous recurrence sequence, the rank of the associated recurrence matrix is bounded above by the order r of the recurrence. For inhomogeneous sequences, the upper bound on matrix rank is $r + 1$."

3.14 Facts on the Range, Rank, Null Space, and Defect of Partitioned Matrices

Fact 3.14.1. Let $A \in \mathbb{F}^{n\times m}$ and $B \in \mathbb{F}^{k\times l}$, and assume that B is a submatrix of A. Then,

$$\operatorname{rank} B \le \operatorname{rank} A \le \operatorname{rank} B + n + m - k - l.$$

If, in particular, $B = 0$, then $\operatorname{rank} A \le n + m - k - l$. **Source:** [284]. **Related:** Fact 3.13.36 and Fact 3.16.4.

Fact 3.14.2. Let $A \in \mathbb{F}^{n\times m}$ and $B \in \mathbb{F}^{n\times l}$. Then,

$$\mathcal{R}([A\ \ B]) = \mathcal{R}([B\ \ A]), \quad \operatorname{rank}[A\ \ B] = \operatorname{rank}[B\ \ A].$$

Fact 3.14.3. Let $A \in \mathbb{F}^{n\times m}$ and $B \in \mathbb{F}^{n\times l}$. Then, $\operatorname{rank} A^*B = \operatorname{rank}[A\ \ B]$ if and only if $\mathcal{R}(A) = \mathcal{R}(B)$. **Source:** [252].

Fact 3.14.4. Let $A \in \mathbb{F}^{n\times m}$, let $B, C \in \mathbb{F}^{n\times l}$, and assume that $\mathcal{R}(B) = \mathcal{R}(C)$. Then,

$$\mathcal{R}([A\ \ B]) = \mathcal{R}([A\ \ C]), \quad \operatorname{rank}[A\ \ B] = \operatorname{rank}[A\ \ C].$$

Fact 3.14.5. Let $A \in \mathbb{F}^{n\times m}$ and $B \in \mathbb{F}^{l\times m}$. Then,

$$\mathcal{N}\left(\begin{bmatrix} A \\ B \end{bmatrix}\right) = \mathcal{N}\left(\begin{bmatrix} B \\ A \end{bmatrix}\right), \quad \operatorname{def}\begin{bmatrix} A \\ B \end{bmatrix} = \operatorname{def}\begin{bmatrix} B \\ A \end{bmatrix}.$$

Fact 3.14.6. Let $A \in \mathbb{F}^{n\times m}$, let $B, C \in \mathbb{F}^{l\times m}$, and assume that $\mathcal{N}(B) = \mathcal{N}(C)$. Then,

$$\mathcal{N}\left(\begin{bmatrix} A \\ B \end{bmatrix}\right) = \mathcal{N}\left(\begin{bmatrix} A \\ C \end{bmatrix}\right), \quad \operatorname{def}\begin{bmatrix} A \\ B \end{bmatrix} = \operatorname{def}\begin{bmatrix} A \\ C \end{bmatrix}.$$

Fact 3.14.7. Let $A \in \mathbb{F}^{n\times m}$ and $B \in \mathbb{F}^{n\times l}$, and let $x, y \in \mathbb{F}^n$. Then, $[\mathcal{R}(A) + x] \cap [\mathcal{R}(B) + y]$ is nonempty if and only if $x - y \in \mathcal{R}([A\ \ B])$. Now, assume that these conditions hold. Then,

$$[\mathcal{R}(A) + x] \cap [\mathcal{R}(B) + y] = [-A\ \ 0_{n\times l}][A\ \ B]^{\mathrm{inv}}(x - y) + x = [0_{n\times m}\ \ B][A\ \ B]^{\mathrm{inv}}(x - y) + y.$$

Finally, assume that $[A\ \ B]$ has full column rank, and let $w \in \mathbb{F}^{m+l}$ satisfy $x - y = [A\ \ B]w$. Then,

$$[\mathcal{R}(A) + x] \cap [\mathcal{R}(B) + y] = \{[-A\ \ 0_{n\times l}]w + x\} = \{[0_{n\times m}\ \ B]w + y\}.$$

Remark: The intersection of two affine subspaces is an affine subspace. See Fact 3.12.4.

Fact 3.14.8. Let $A \in \mathbb{F}^{n\times m}$ and $B \in \mathbb{F}^{n\times l}$. Then,

$$\mathcal{R}(A) + \mathcal{R}(B) = \mathcal{R}([A\ \ B]) = \mathcal{R}(AA^* + BB^*) = \operatorname{span}[\mathcal{R}(A) \cup \mathcal{R}(B)].$$

Consequently,

$$\dim[\mathcal{R}(A) + \mathcal{R}(B)] = \operatorname{rank}\,[A\ \ B] = \operatorname{rank}(AA^* + BB^*).$$

Furthermore, the following statements are equivalent:

i) $\operatorname{rank}\,[A\ \ B] = n$.

ii) $\operatorname{def}\begin{bmatrix} A^* \\ B^* \end{bmatrix} = 0$.

iii) $\mathcal{N}(A^*) \cap \mathcal{N}(B^*) = \{0\}$.

Source: $\mathcal{R}(A) + \mathcal{R}(B) = \operatorname{span}[\mathcal{R}(A) \cup \mathcal{R}(B)]$ follows from Fact 3.12.16.

Fact 3.14.9. Let $A \in \mathbb{F}^{n\times m}$ and $B \in \mathbb{F}^{l\times m}$. Then,

$$\dim[\mathcal{R}(A^*) + \mathcal{R}(B^*)] = \operatorname{rank}\begin{bmatrix} A \\ B \end{bmatrix}.$$

Source: Fact 3.14.8.

Fact 3.14.10. Let $A \in \mathbb{F}^{n\times m}$ and $B \in \mathbb{F}^{l\times m}$. Then,

$$\mathcal{N}(A) \cap \mathcal{N}(B) = \mathcal{N}\left(\begin{bmatrix} A \\ B \end{bmatrix}\right), \quad \dim[\mathcal{N}(A) \cap \mathcal{N}(B)] = \operatorname{def}\begin{bmatrix} A \\ B \end{bmatrix}.$$

Furthermore, the following statements are equivalent:

i) $\operatorname{rank}\begin{bmatrix} A \\ B \end{bmatrix} = m$.

ii) $\operatorname{def}\begin{bmatrix} A \\ B \end{bmatrix} = 0$.

iii) $\mathcal{N}(A) \cap \mathcal{N}(B) = \{0\}$.

Related: Fact 3.13.17.

Fact 3.14.11. Let $A, B \in \mathbb{F}^{n\times m}$. Then, the following statements are equivalent:

i) $\operatorname{rank}(A + B) = \operatorname{rank} A + \operatorname{rank} B$.

ii) $\operatorname{rank}\,[A\ \ B] = \operatorname{rank}\begin{bmatrix} A \\ B \end{bmatrix} = \operatorname{rank} A + \operatorname{rank} B$.

iii) $\dim[\mathcal{R}(A) \cap \mathcal{R}(B)] = \dim[\mathcal{R}(A^*) \cap \mathcal{R}(B^*)] = 0$.

iv) $\mathcal{R}(A) \cap \mathcal{R}(B) = \{0\}$ and $\mathcal{R}(A^*) \cap \mathcal{R}(B^*) = \{0\}$.

v) There exists $C \in \mathbb{F}^{m\times n}$ such that $ACA = A$, $CB = 0$, and $BC = 0$.

Source: [615, 762, 1966, 2666]. **Remark:** Equivalent statements are given by Fact 8.4.31 assuming that $A + B$ is nonsingular. **Remark:** Equivalent statements for $\operatorname{rank}\,[A\ \ B] = \operatorname{rank}\left[\begin{smallmatrix} A \\ B \end{smallmatrix}\right]$ are given by Fact 8.4.7. **Related:** Fact 3.13.38, Fact 3.13.40, and Fact 3.14.18.

Fact 3.14.12. Let $A \in \mathbb{F}^{n\times m}$ and $B \in \mathbb{F}^{n\times l}$. Then, $\mathcal{R}(A) = \mathcal{R}(B)$ if and only if $\operatorname{rank} A = \operatorname{rank} B = \operatorname{rank}\,[A\ \ B]$.

Fact 3.14.13. Let $A \in \mathbb{F}^{n\times m}$, $B \in \mathbb{F}^{k\times m}$, $C \in \mathbb{F}^{m\times l}$, and $D \in \mathbb{F}^{m\times p}$, and assume that

$$\operatorname{rank}\begin{bmatrix} A \\ B \end{bmatrix} = \operatorname{rank} A, \quad \operatorname{rank}\,[C\ \ D] = \operatorname{rank} C.$$

Then,

$$\operatorname{rank}\begin{bmatrix} A \\ B \end{bmatrix}[C\ \ D] = \operatorname{rank} AC.$$

Source: *i*) of Fact 3.13.21.

Fact 3.14.14. Let $A \in \mathbb{F}^{n\times m}$ and $B \in \mathbb{F}^{n\times l}$. Then,

$$\mathcal{R}(A) \cap \mathcal{R}(B) = [A\ \ 0_{n\times l}]\mathcal{N}([A\ \ B]) \cap [0_{n\times m}\ \ B]\mathcal{N}([A\ \ B]),$$
$$\dim([A\ \ 0_{n\times l}]\mathcal{N}([A\ \ B])) = \dim([0_{n\times m}\ \ B]\mathcal{N}([A\ \ B])) = \operatorname{def}\,[A\ \ B] - \operatorname{def} A - \operatorname{def} B.$$

Source: Let $z \in \mathcal{R}(A) \cap \mathcal{R}(B)$. Then, there exist $x \in \mathbb{R}^m$ and $y \in \mathbb{F}^l$ such that $z = Ax = -By$. Therefore, $[A\ \ B]\left[\begin{smallmatrix} x \\ y \end{smallmatrix}\right] = 0$, that is, $\left[\begin{smallmatrix} x \\ y \end{smallmatrix}\right] \in \mathcal{N}([A\ \ B])$, and thus $z \in [A\ \ 0_{n\times l}]\mathcal{N}([A\ \ B])$. Likewise, $z \in [0_{n\times m}\ \ B]\mathcal{N}([A\ \ B])$. The reverse inclusion is immediate. To prove the second equality, note that *ii*) of Fact 3.13.4 implies that

$$\begin{aligned}\dim([A\ \ 0_{n\times l}]\mathcal{N}([A\ \ B])) &= \dim \mathcal{N}([A\ \ B]) - \dim[\mathcal{N}([A\ \ 0_{n\times l}]) \cap \mathcal{N}([A\ \ B])] \\ &= \dim \mathcal{N}([A\ \ B]) - \dim \begin{bmatrix} \mathcal{N}(A) \\ \mathcal{N}(B) \end{bmatrix} = \operatorname{def}\,[A\ \ B] - (\operatorname{def} A + \operatorname{def} B).\end{aligned}$$

Related: Fact 8.9.1.

Fact 3.14.15. Let $A \in \mathbb{F}^{n\times m}$ and $B \in \mathbb{F}^{n\times l}$. Then,

$$\begin{aligned}\operatorname{rank} A + \operatorname{rank} B - n &\le \dim[\mathcal{R}(A) \cap \mathcal{R}(B)] \le \min\{\operatorname{rank} A, \operatorname{rank} B\} \le \left\{\begin{matrix} \operatorname{rank} A \\ \operatorname{rank} B \end{matrix}\right\} \\ &\le \max\{\operatorname{rank} A, \operatorname{rank} B\} \le \operatorname{rank}\,[A\ \ B] \\ &= \operatorname{rank} A + \operatorname{rank} B - \dim[\mathcal{R}(A) \cap \mathcal{R}(B)] \le \min\{\operatorname{rank} A + \operatorname{rank} B, n\},\end{aligned}$$

$$\begin{aligned}\max\{\operatorname{def} A + \operatorname{def} B, m + l - n\} &\le \operatorname{def} A + \operatorname{def} B + \dim[\mathcal{R}(A) \cap \mathcal{R}(B)] = \operatorname{def}\,[A\ \ B] \\ &\le \min\{l + \operatorname{def} A, m + \operatorname{def} B\} \le \left\{\begin{matrix} l + \operatorname{def} A \\ m + \operatorname{def} B \end{matrix}\right\} \\ &\le \max\{l + \operatorname{def} A, m + \operatorname{def} B\} \le m + l - \dim[\mathcal{R}(A) \cap \mathcal{R}(B)] \\ &\le \operatorname{def} A + \operatorname{def} B + n.\end{aligned}$$

Consequently, the following statements are equivalent:

i) $\operatorname{rank}\,[A\ \ B] = \operatorname{rank} A + \operatorname{rank} B$.

ii) $\operatorname{def}\,[A\ \ B] = \operatorname{def} A + \operatorname{def} B$.

iii) $\mathcal{R}(A) \cap \mathcal{R}(B) = \{0\}$.

If, in addition, $A^*B = 0$, then

$$\operatorname{rank}\,[A\ \ B] = \operatorname{rank} A + \operatorname{rank} B, \quad \operatorname{def}\,[A\ \ B] = \operatorname{def} A + \operatorname{def} B.$$

Source: Theorem 3.1.3, Fact 3.12.19, and Fact 3.14.8. For the case $A^*B = 0$, note that

$$\operatorname{rank}\,[A\ \ B] = \operatorname{rank}\begin{bmatrix} A^* \\ B^* \end{bmatrix}[A\ \ B] = \begin{bmatrix} A^*A & 0 \\ 0 & B^*B \end{bmatrix} = \operatorname{rank} A^*A + \operatorname{rank} B^*B = \operatorname{rank} A + \operatorname{rank} B.$$

Remark: Note that $\operatorname{rank}\,[A\ \ B] + \dim[\mathcal{R}(A)\cap\mathcal{R}(B)] = \operatorname{rank} A + \operatorname{rank} B$. Using Fact 3.13.37, $A^*B = 0$ implies that $\mathcal{R}(A)\cap\mathcal{R}(B) = \{0\}$, and thus $\operatorname{rank}\,[A\ \ B] = \operatorname{rank} A + \operatorname{rank} B$. **Related:** Fact 8.9.7.

Fact 3.14.16. Let $A \in \mathbb{F}^{n\times m}$ and $B \in \mathbb{F}^{l\times m}$. Then,

$$\operatorname{rank}\begin{bmatrix} A \\ B \end{bmatrix} + \dim[\mathcal{R}(A^*)\cap\mathcal{R}(B^*)] = \operatorname{rank} A + \operatorname{rank} B.$$

Source: Fact 3.14.15.

Fact 3.14.17. Let $A \in \mathbb{F}^{n\times m}$ and $B \in \mathbb{F}^{l\times m}$. Then,

$$\max\{\operatorname{rank} A, \operatorname{rank} B\} \le \operatorname{rank}\begin{bmatrix} A \\ B \end{bmatrix} = \operatorname{rank} A + \operatorname{rank} B - \dim[\mathcal{R}(A^*)\cap\mathcal{R}(B^*)] \le \operatorname{rank} A + \operatorname{rank} B,$$

$$\operatorname{def} A - \operatorname{rank} B \le \operatorname{def} A - \operatorname{rank} B + \dim[\mathcal{R}(A^*)\cap\mathcal{R}(B^*)] = \operatorname{def}\begin{bmatrix} A \\ B \end{bmatrix} \le \min\{\operatorname{def} A, \operatorname{def} B\}.$$

If, in addition, $AB^* = 0$, then

$$\operatorname{rank}\begin{bmatrix} A \\ B \end{bmatrix} = \operatorname{rank} A + \operatorname{rank} B, \quad \operatorname{def}\begin{bmatrix} A \\ B \end{bmatrix} = \operatorname{def} A - \operatorname{rank} B.$$

Source: Fact 3.12.19 and Fact 3.14.15. **Related:** Fact 8.9.7.

Fact 3.14.18. Let $A, B \in \mathbb{F}^{n\times m}$. Then,

$$\left.\begin{matrix} \max\{\operatorname{rank} A, \operatorname{rank} B\} \\ \operatorname{rank}(A+B) \end{matrix}\right\} \le \left\{\begin{matrix} \operatorname{rank}\,[A\ \ B] \\ \operatorname{rank}\begin{bmatrix} A \\ B \end{bmatrix} \end{matrix}\right\} \le \operatorname{rank} A + \operatorname{rank} B,$$

$$\begin{aligned}\operatorname{rank}\,[A\ \ B] + \operatorname{rank}\begin{bmatrix} A \\ B \end{bmatrix} &\le \operatorname{rank} A + \operatorname{rank} B + \operatorname{rank}(A+B)\\ &\le \left\{\begin{matrix} \operatorname{rank} A + \operatorname{rank} B + \operatorname{rank}\,[A\ \ B] \\ \operatorname{rank} A + \operatorname{rank} B + \operatorname{rank}\begin{bmatrix} A \\ B \end{bmatrix} \end{matrix}\right\} \le 2(\operatorname{rank} A + \operatorname{rank} B),\end{aligned}$$

$$\operatorname{def} A - \operatorname{rank} B \le \left\{\begin{matrix} \operatorname{def}\,[A\ \ B] - m \\ \operatorname{def}\begin{bmatrix} A \\ B \end{bmatrix} \end{matrix}\right\} \le \left\{\begin{matrix} \min\{\operatorname{def} A, \operatorname{def} B\} \\ \operatorname{def}(A+B). \end{matrix}\right.$$

Source: $\operatorname{rank}(A+B) = \operatorname{rank}\,[A\ \ B]\left[\begin{smallmatrix} I \\ I \end{smallmatrix}\right] \le \operatorname{rank}\,[A\ \ B]$, and $\operatorname{rank}(A+B) = \operatorname{rank}\,[I\ \ I]\left[\begin{smallmatrix} A \\ B \end{smallmatrix}\right] \le \operatorname{rank}\left[\begin{smallmatrix} A \\ B \end{smallmatrix}\right]$.

Fact 3.14.19. Let $A \in \mathbb{F}^{n\times m}$, $B \in \mathbb{F}^{l\times k}$, and $C \in \mathbb{F}^{l\times m}$. Then,

$$\operatorname{rank} A + \operatorname{rank} B = \operatorname{rank}\begin{bmatrix} A & 0 \\ 0 & B \end{bmatrix} \le \operatorname{rank}\begin{bmatrix} A & 0 \\ C & B \end{bmatrix},$$

$$\operatorname{rank} A + \operatorname{rank} B = \operatorname{rank}\begin{bmatrix} 0 & A \\ B & 0 \end{bmatrix} \le \operatorname{rank}\begin{bmatrix} 0 & A \\ B & C \end{bmatrix}.$$

Finally, let $D \in \mathbb{F}^{k\times m}$ and $E \in \mathbb{F}^{l\times n}$. Then,

$$\operatorname{rank} A + \operatorname{rank} B = \operatorname{rank}\begin{bmatrix} A & 0 \\ BD + EA & B \end{bmatrix} = \operatorname{rank}\begin{bmatrix} 0 & A \\ B & BD + EA \end{bmatrix}.$$

Fact 3.14.20. Let $A \in \mathbb{F}^{n\times m}$, $B \in \mathbb{F}^{m\times l}$, and $C \in \mathbb{F}^{l\times k}$. Then,

$$\operatorname{rank} AB + \operatorname{rank} BC \le \operatorname{rank}\begin{bmatrix} 0 & AB \\ BC & B \end{bmatrix} = \operatorname{rank} B + \operatorname{rank} ABC,$$

$$\operatorname{rank} A + \operatorname{rank} B + \operatorname{rank} C - m - l \le \operatorname{rank} AB + \operatorname{rank} BC - \operatorname{rank} B \le \operatorname{rank} ABC.$$

Furthermore, the following statements are equivalent:

i) $\operatorname{rank}\begin{bmatrix} 0 & AB \\ BC & B \end{bmatrix} = \operatorname{rank} AB + \operatorname{rank} BC.$

ii) $\operatorname{rank} AB + \operatorname{rank} BC - \operatorname{rank} B = \operatorname{rank} ABC.$

iii) There exist $X \in \mathbb{F}^{k\times l}$ and $Y \in \mathbb{F}^{m\times n}$ such that $BCX + YAB = B$.

Remark: This is the *Frobenius inequality*. **Source:** Use $\left[\begin{smallmatrix} 0 & AB \\ BC & B \end{smallmatrix}\right] = \left[\begin{smallmatrix} I & A \\ 0 & I \end{smallmatrix}\right]\left[\begin{smallmatrix} -ABC & 0 \\ 0 & B \end{smallmatrix}\right]\left[\begin{smallmatrix} I & 0 \\ C & I \end{smallmatrix}\right]$ and Fact 3.14.19. The last statement follows from Fact 7.11.26. See [2674, 2675]. **Related:** Fact 8.9.16 for the case of equality.

Fact 3.14.21. Let $A, B \in \mathbb{F}^{n\times m}$. Then,

$$\operatorname{rank}\,[A\ \ B] + \operatorname{rank}\begin{bmatrix} A \\ B \end{bmatrix} \le \operatorname{rank}\begin{bmatrix} 0 & A & B \\ A & A & 0 \\ B & 0 & B \end{bmatrix} = \operatorname{rank} A + \operatorname{rank} B + \operatorname{rank}(A+B).$$

Source: Use the Frobenius inequality with $A \triangleq C^{\mathsf{T}} \triangleq [I\ \ I]$ and B replaced by $\left[\begin{smallmatrix} A & 0 \\ 0 & B \end{smallmatrix}\right]$.

Fact 3.14.22. Let $A \in \mathbb{F}^{n\times m}$, $B \in \mathbb{F}^{n\times l}$, and $C \in \mathbb{F}^{n\times k}$. Then,

$$\begin{aligned}\operatorname{rank}\,[A\ \ B\ \ C] &\le \operatorname{rank}\,[A\ \ B] + \operatorname{rank}\,[B\ \ C] - \operatorname{rank} B \\ &\le \operatorname{rank}\,[A\ \ B] + \operatorname{rank} C \\ &\le \operatorname{rank} A + \operatorname{rank} B + \operatorname{rank} C.\end{aligned}$$

Source: [1896].

Fact 3.14.23. Let $A \in \mathbb{F}^{n\times m}$ and $B \in \mathbb{F}^{m\times n}$. Then,

$$\begin{aligned}\begin{bmatrix} I_n & I_n - AB \\ B & 0 \end{bmatrix} &= \begin{bmatrix} I_n & A \\ 0 & I_m \end{bmatrix}\begin{bmatrix} 0 & I_n - AB \\ B & 0 \end{bmatrix}\begin{bmatrix} I_n & 0 \\ I_n & I_n \end{bmatrix} \\ &= \begin{bmatrix} I_n & 0 \\ B & I_m \end{bmatrix}\begin{bmatrix} I_n & 0 \\ 0 & BAB - B \end{bmatrix}\begin{bmatrix} I_n & I_n - AB \\ 0 & I_m \end{bmatrix},\end{aligned}$$

$$\operatorname{rank}\begin{bmatrix} I_n & I_n - AB \\ B & 0 \end{bmatrix} = \operatorname{rank} B + \operatorname{rank}(I_n - AB) = n + \operatorname{rank}(BAB - B).$$

Related: Fact 3.17.7.

Fact 3.14.24. Let $A \in \mathbb{F}^{n\times m}$ and $B \in \mathbb{F}^{m\times n}$. Then,

$$\begin{aligned}\begin{bmatrix} A & AB \\ BA & B \end{bmatrix} &= \begin{bmatrix} I_n & 0 \\ B & I_m \end{bmatrix}\begin{bmatrix} A & 0 \\ 0 & B - BAB \end{bmatrix}\begin{bmatrix} I_m & B \\ 0 & I_n \end{bmatrix} \\ &= \begin{bmatrix} I_n & A \\ 0 & I_m \end{bmatrix}\begin{bmatrix} A - ABA & 0 \\ 0 & B \end{bmatrix}\begin{bmatrix} I_m & 0 \\ A & I_n \end{bmatrix},\end{aligned}$$

$$\operatorname{rank}\begin{bmatrix} A & AB \\ BA & B \end{bmatrix} = \operatorname{rank} A + \operatorname{rank}(B - BAB) = \operatorname{rank} B + \operatorname{rank}(A - ABA).$$

Related: Fact 3.17.9.

Fact 3.14.25. Let $\mathcal{A} \triangleq \left[\begin{smallmatrix} A & B \\ C & D \end{smallmatrix}\right] \in \mathbb{F}^{(n+m)\times(n+m)}$, and consider the following statements:

i) $\operatorname{rank}\,[A\ \ B] = \operatorname{rank}\left[\begin{smallmatrix} A \\ C \end{smallmatrix}\right] = \operatorname{rank} \mathcal{A} = n.$

ii) A is nonsingular.

iii) $\operatorname{rank}\,[A\ \ B] = \operatorname{rank}\left[\begin{smallmatrix} A \\ C \end{smallmatrix}\right] = n \le \operatorname{rank}\mathcal{A}$.

Then, *i*) $\Longrightarrow$ *ii*) $\Longrightarrow$ *iii*). **Source:** [1451, p. 20].

Fact 3.14.26. Let $\left[\begin{smallmatrix} A & B \\ C & D \end{smallmatrix}\right] \in \mathbb{F}^{(n+k)\times(m+l)}$, and define $r \triangleq \operatorname{rank}\left[\begin{smallmatrix} A & B \\ C & D \end{smallmatrix}\right]$. Then,

$$\operatorname{rank}\begin{bmatrix} A \\ C \end{bmatrix} + \operatorname{rank}\begin{bmatrix} B \\ D \end{bmatrix} + \operatorname{rank}\,[A\ \ B] + \operatorname{rank}\,[C\ \ D] - \operatorname{rank}A - \operatorname{rank}B - \operatorname{rank}C - \operatorname{rank}D$$
$$\le \operatorname{rank}\begin{bmatrix} A & B \\ C & D \end{bmatrix} \le \operatorname{rank}A + \operatorname{rank}B + \operatorname{rank}C + \operatorname{rank}D,$$

$$\max\{0, r-k-l\} \le \operatorname{rank}A \le \min\{r,n,m\}, \quad \max\{0, r-m-k\} \le \operatorname{rank}B \le \min\{r,n,l\},$$
$$\max\{0, r-n-l\} \le \operatorname{rank}C \le \min\{r,m,k\}, \quad \max\{0, r-n-m\} \le \operatorname{rank}D \le \min\{r,k,l\}.$$

Source: [2238, p. 117] and [2638, 2664].

Fact 3.14.27. Let $\left[\begin{smallmatrix} A & B \\ C & D \end{smallmatrix}\right] \in \mathbb{F}^{(n_1+n_2)\times(m_1+m_2)}$, assume that $\left[\begin{smallmatrix} A & B \\ C & D \end{smallmatrix}\right]$ is nonsingular, and define $\left[\begin{smallmatrix} E & F \\ G & H \end{smallmatrix}\right] \in \mathbb{F}^{(m_1+m_2)\times(n_1+n_2)}$ by

$$\begin{bmatrix} E & F \\ G & H \end{bmatrix} \triangleq \begin{bmatrix} A & B \\ C & D \end{bmatrix}^{-1}.$$

Then, $\operatorname{def}A = \operatorname{def}H$, $\operatorname{def}B = \operatorname{def}F$, $\operatorname{def}C = \operatorname{def}G$, and $\operatorname{def}D = \operatorname{def}E$. **Source:** [2557, 2779] and [2780, p. 38]. **Remark:** The sizes of the matrix blocks differ from the sizes in Fact 3.17.32. **Remark:** This is the *nullity theorem*. See [2557] and Fact 4.24.2. **Remark:** A and H are complementary. The matrices $U \in \mathbb{F}^{n\times m}$ and $V \in \mathbb{F}^{m\times n}$ are *complementary submatrices* if the row numbers not used to create U are the column numbers used to create V, and the column numbers not used to create U are the row numbers used to create V. See Fact 3.14.28.

Fact 3.14.28. Let $A \in \mathbb{F}^{n\times n}$, assume that A is nonsingular, and let $\mathcal{S}_1, \mathcal{S}_2 \subseteq \{1,\ldots,n\}$. Then,

$$\operatorname{rank}\,(A^{-1})_{(\mathcal{S}_1,\mathcal{S}_2)} = \operatorname{rank}A_{(\mathcal{S}_2^\sim,\mathcal{S}_1^\sim)} + \operatorname{card}(\mathcal{S}_1) + \operatorname{card}(\mathcal{S}_2) - n,$$
$$\operatorname{def}\,(A^{-1})_{(\mathcal{S}_1,\mathcal{S}_2)} = \operatorname{def}A_{(\mathcal{S}_2^\sim,\mathcal{S}_1^\sim)}.$$

Source: [1451, p. 19] and [2780, p. 40]. **Remark:** The submatrices $(A^{-1})_{(\mathcal{S}_1,\mathcal{S}_2)}$ and $A_{(\mathcal{S}_2^\sim,\mathcal{S}_1^\sim)}$ are complementary. **Related:** Fact 3.14.27, Fact 3.14.29, and Fact 3.17.33.

Fact 3.14.29. Let $A \in \mathbb{F}^{n\times n}$, assume that A is nonsingular, and let $\mathcal{S} \subseteq \{1,\ldots,n\}$. Then,

$$\operatorname{rank}\,(A^{-1})_{(\mathcal{S},\mathcal{S}^\sim)} = \operatorname{rank}A_{(\mathcal{S},\mathcal{S}^\sim)}.$$

Source: Apply Fact 3.14.28 with $\mathcal{S}_2 = \mathcal{S}_1^\sim$.

Fact 3.14.30. Let $A_1,\ldots,A_k \in \mathbb{F}^{n\times n}$. Then,

$$\operatorname{rank}\sum_{i=1}^{k} A_i \le \sum_{i=1}^{k} \operatorname{rank}A_i.$$

Now, assume that $\sum_{i=1}^{k} A_i$ is nonsingular. Then,

$$\operatorname{rank}\begin{bmatrix} A_1 & A_2 & A_3 & \cdots & A_n & 0 & \cdots & 0 \\ 0 & A_1 & A_2 & A_3 & \cdots & A_n & \cdots & 0 \\ \vdots & \ddots & \ddots & \ddots & \ddots & \ddots & \ddots & \vdots \\ 0 & 0 & \cdots & A_1 & A_2 & A_3 & \cdots & A_n \end{bmatrix} = kn.$$

Source: [2238, p. 107].

3.15 Facts on the Inner Product, Outer Product, Trace, and Matrix Powers

Fact 3.15.1. Let $x, y, z \in \mathbb{F}^n$, assume that $x^*x = y^*y = z^*z = 1$, and let $p \geq 2$. Then,

$$\sqrt[p]{1 - |x^*y|^2} \leq \sqrt[p]{1 - |x^*z|^2} + \sqrt[p]{1 - |z^*y|^2}, \quad \sqrt[p]{1 - |\operatorname{Re} x^*y|^2} \leq \sqrt[p]{1 - |\operatorname{Re} x^*z|^2} + \sqrt[p]{1 - |\operatorname{Re} z^*y|^2}.$$

In particular,

$$\sqrt{1 - |x^*y|^2} \leq \sqrt{1 - |x^*z|^2} + \sqrt{1 - |z^*y|^2}, \quad \sqrt{1 - |\operatorname{Re} x^*y|^2} \leq \sqrt{1 - |\operatorname{Re} x^*z|^2} + \sqrt{1 - |\operatorname{Re} z^*y|^2}.$$

Equality holds in each inequality if and only if there exists $\alpha \in \mathbb{F}$ such that either $z = \alpha x$ or $z = \alpha y$. **Source:** [1829], [2983, p. 155], and [2991, pp. 195–197]. **Related:** Fact 4.13.27, Fact 5.1.5, and Fact 11.8.9.

Fact 3.15.2. Let $x, y \in \mathbb{F}^n$. Then, $x^*x = y^*y$ and $\operatorname{Im} x^*y = 0$ if and only if $(x - y)^*(x + y) = 0$.

Fact 3.15.3. Let $x, y \in \mathbb{R}^n$. Then, $xx^\mathsf{T} = yy^\mathsf{T}$ if and only if either $x = y$ or $x = -y$.

Fact 3.15.4. Let $x, y \in \mathbb{R}^n$. Then, $xy^\mathsf{T} = yx^\mathsf{T}$ if and only if x and y are linearly dependent.

Fact 3.15.5. Let $x, y \in \mathbb{R}^n$. Then, $xy^\mathsf{T} = -yx^\mathsf{T}$ if and only if either $x = 0$ or $y = 0$. **Source:** If $x_{(i)} \neq 0$ and $y_{(j)} \neq 0$, then $x_{(j)} = y_{(i)} = 0$ and $0 \neq x_{(i)}y_{(j)} \neq x_{(j)}y_{(i)} = 0$.

Fact 3.15.6. Let $x, y \in \mathbb{R}^n$. Then, $yx^\mathsf{T} + xy^\mathsf{T} = y^\mathsf{T}yxx^\mathsf{T}$ if and only if either $x = 0$ or $y = \frac{1}{2}y^\mathsf{T}yx$.

Fact 3.15.7. Let $x, y \in \mathbb{F}^n$. Then, $(xy^*)^r = (y^*x)^{r-1}xy^*$.

Fact 3.15.8. Let $A \in \mathbb{C}^{n \times n}$, and let $x, y \in \mathbb{C}^n$. Then,

$$\begin{aligned} 4y^*Ax &= \sum_{i=0}^{3} \jmath^i (x + \jmath^i y)^* A (x + \jmath^i y) \\ &= (x + y)^*A(x + y) - (x - y)^*A(x - y) + [(x + \jmath y)^*A(x + \jmath y) - (x - \jmath y)^*A(x - \jmath y)]\jmath. \end{aligned}$$

Source: [2238, p. 261] and [2979, p. 3]. **Remark:** $A = I_n$ yields the polarization identity in Fact 11.8.3.

Fact 3.15.9. Let $x_1, \ldots, x_k \in \mathbb{F}^n$ and $y_1, \ldots, y_k \in \mathbb{F}^m$. Then, the following statements are equivalent:

i) $x_1, \ldots, x_k$ are linearly independent, and $y_1, \ldots, y_k$ are linearly independent.

ii) $\operatorname{rank}\left(\sum_{i=1}^k x_i y_i^\mathsf{T}\right) = k$.

Source: [835, p. 537].

Fact 3.15.10. Let $A, B, C \in \mathbb{R}^{2 \times 2}$. Then,

$$\operatorname{tr}(ABC + ACB) + (\operatorname{tr} A)(\operatorname{tr} B)\operatorname{tr} C = (\operatorname{tr} A)\operatorname{tr} BC + (\operatorname{tr} B)\operatorname{tr} AC + (\operatorname{tr} C)\operatorname{tr} AB.$$

Source: [591, p. 330]. **Related:** Fact 6.9.21.

Fact 3.15.11. Let $A \in \mathbb{F}^{n \times m}$ and $B \in \mathbb{F}^{l \times k}$. Then, $AE_{i,j,m \times l}B = \operatorname{col}_i(A)\operatorname{row}_j(B)$.

Fact 3.15.12. Let $A \in \mathbb{F}^{n \times m}$, $B \in \mathbb{F}^{m \times l}$, and $C \in \mathbb{F}^{l \times n}$. Then, $\operatorname{tr} ABC = \sum_{i=1}^n \operatorname{row}_i(A)B\operatorname{col}_i(C)$.

Fact 3.15.13. Let $A \in \mathbb{F}^{n \times m}$. Then, the following statements are equivalent:

i) $A = 0$.

ii) $Ax = 0$ for all $x \in \mathbb{F}^m$.

iii) $\operatorname{tr} AA^* = 0$.

Fact 3.15.14. Let $A \in \mathbb{F}^{n \times n}$ and $k \geq 1$. Then, $\operatorname{Re} \operatorname{tr} A^{2k} \leq \operatorname{tr} A^k A^{k*} \leq \operatorname{tr}(AA^*)^k$. **Remark:** To prove the left-hand inequality, consider $\operatorname{tr}(A^k - A^{k*})(A^{k*} - A^k)$. For the right-hand inequality in the case $k = 2$, consider $\operatorname{tr}(AA^* - A^*A)^2$.

Fact 3.15.15. Let $A \in \mathbb{F}^{n \times n}$. Then, $\operatorname{tr} A^k = 0$ for all $k \in \{1, \ldots, n\}$ if and only if $A^n = 0$. **Source:** For sufficiency, Fact 6.10.12 implies that $\operatorname{spec}(A) = \{0\}$, and thus the Jordan form of A is a block-diagonal matrix each of whose diagonally located blocks is a standard nilpotent matrix. For necessity, see [2983, p. 112].

Fact 3.15.16. Let $A \in \mathbb{F}^{n\times n}$, and assume that $\operatorname{tr} A = 0$. If $A^2 = A$, then $A = 0$. If $A^k = A$, where $k \geq 4$ and $2 \leq n < p$, where p is the smallest prime divisor of $k-1$, then $A = 0$. **Source:** [770].

Fact 3.15.17. Let $A, B \in \mathbb{F}^{2\times 2}$, and define $\operatorname{str} A \triangleq A_{(1,1)} - A_{(2,2)}$. Then,

$$(\operatorname{tr} AB)^2 = \operatorname{tr} A^2B^2 + \operatorname{tr} ABA^{\mathsf{A}}B^{\mathsf{A}},$$

$$\operatorname{str} AB + \operatorname{str} BA = (\operatorname{tr} A)\operatorname{str} B + (\operatorname{tr} B)\operatorname{str} A.$$

Source: [1611]. **Related:** Fact 4.29.2.

Fact 3.15.18. Let $A, B \in \mathbb{F}^{n\times m}$. Then,

$$\operatorname{Re}\operatorname{tr} A^*B \leq |\operatorname{Re}\operatorname{tr} A^*B| \leq |\operatorname{tr} A^*B| \leq \sqrt{\operatorname{tr}(AA^*)\operatorname{tr}(BB^*)} \leq \tfrac{1}{2}\operatorname{tr}(AA^* + BB^*).$$

Source: [2991, p. 32]. **Related:** Fact 3.15.19.

Fact 3.15.19. Let $A, B \in \mathbb{F}^{n\times n}$. Then,

$$\left.\begin{aligned} \operatorname{Re}\operatorname{tr} A^*B \leq |\operatorname{Re}\operatorname{tr} A^*B| \leq |\operatorname{tr} A^*B| \\ \operatorname{Re}\operatorname{tr} AB \leq |\operatorname{Re}\operatorname{tr} AB| \leq |\operatorname{tr} AB| \end{aligned}\right\} \leq \sqrt{\operatorname{tr}(AA^*)\operatorname{tr}(BB^*)} \leq \tfrac{1}{2}\operatorname{tr}(AA^* + BB^*).$$

Source: Fact 3.15.18, Fact 10.14.22, [1477], and [2991, p. 32].

Fact 3.15.20. Let $A, B \in \mathbb{R}^{n\times n}$, and assume that $\operatorname{tr}(AA^{\mathsf{T}} + BB^{\mathsf{T}}) = \operatorname{tr}(AB + A^{\mathsf{T}}B^{\mathsf{T}})$. Then, $A = B^{\mathsf{T}}$. **Source:** [1158, p. 62].

Fact 3.15.21. Let $A, B \in \mathbb{F}^{n\times n}$, and let $k \geq 0$. Then, $\operatorname{tr}(AB)^k = \operatorname{tr}(BA)^k$.

Fact 3.15.22. Let $A, B \in \mathbb{F}^{n\times n}$, assume that $AB = 0$, and let $k \geq 0$. Then, $\operatorname{tr}(A+B)^k = \operatorname{tr} A^k + \operatorname{tr} B^k$.

Fact 3.15.23. Let $A \in \mathbb{F}^{n\times n}$, and assume that $\operatorname{tr} A = 0$. Then, the following statements hold:

i) If $n = 2$, then $A^2 = \frac{1}{2}(\operatorname{tr} A^2)I_2 = -(\det A)I_2$ and $\operatorname{tr} A^2 = -2\det A$.

ii) If $n = 3$, then $\operatorname{tr} A^3 = 3\det A$.

iii) If $n = 4$, then $\operatorname{tr} A^4 = \frac{1}{2}(\operatorname{tr} A^2)^2 - 4\det A$.

Source: Fact 6.9.1, Fact 6.9.2, and Fact 6.9.3. **Remark:** These results apply to commutators $A = [B, C]$.

Fact 3.15.24. Let $A \in \mathbb{R}^{n\times n}$, $x, y \in \mathbb{R}^n$, and $k \geq 1$. Then,

$$(A + xy^{\mathsf{T}})^k = A^k + B\hat{I}_kC^{\mathsf{T}},$$

where $B \triangleq [x \;\; Ax \;\; \cdots \;\; A^{k-1}x]$ and $C \triangleq [y \quad (A^{\mathsf{T}} + yx^{\mathsf{T}})y \quad \cdots \quad (A^{\mathsf{T}} + yx^{\mathsf{T}})^{k-1}y]$. **Source:** [437].

Fact 3.15.25. Let $A, B \in \mathbb{F}^{n\times n}$. Then, the following statements hold:

i) $AB + BA = \frac{1}{2}[(A + B)^2 - (A - B)^2]$.

ii) $(A + B)(A - B) = A^2 - B^2 - [A, B]$.

iii) $(A - B)(A + B) = A^2 - B^2 + [A, B]$.

iv) $A^2 - B^2 = \frac{1}{2}[(A + B)(A - B) + (A - B)(A + B)]$.

v) $A^*A - B^*B = \frac{1}{2}[(A + B)^*(A - B) + (A - B)^*(A + B)]$.

Fact 3.15.26. Let $A, B \in \mathbb{F}^{n\times n}$ and $k \geq 1$. Then,

$$A^k - B^k = \sum_{i=0}^{k-1} A^i(A - B)B^{k-1-i} = \sum_{i=1}^{k} A^{k-i}(A - B)B^{i-1}.$$

Fact 3.15.27. Let $A, B \in \mathbb{F}^{n\times n}$ and $k \ge 1$. Then,

$$A^k + B^k = \frac{1}{k}\sum_{i=0}^{k-1}(A + e^{(2i\pi/k)J}B)^k.$$

Source: [454, p. 27].

Fact 3.15.28. Let $A, B \in \mathbb{F}^{n\times n}$, and assume that A, $B^{-1}-A$, and $A^{-1}+(B^{-1}-A)^{-1}$ are nonsingular. Then,

$$ABA = A - [A^{-1} + (B^{-1} - A)^{-1}]^{-1}.$$

Source: [769, p. 345]. **Remark:** This is *Hua's identity.*

Fact 3.15.29. Let $A, B \in \mathbb{F}^{n\times m}$, let $C \in \mathbb{F}^{m\times n}$, and assume that $A+B = ACB$. Then, $A+B = BCA$. **Source:** [2937].

Fact 3.15.30. Let $A \in \mathbb{F}^{n\times n}$, $B \in \mathbb{F}^{n\times m}$, and $C \in \mathbb{F}^{m\times m}$, and let $k \ge 1$. Then,

$$\begin{bmatrix} A & B \\ 0 & C \end{bmatrix}^k = \begin{bmatrix} A^k & \sum_{i=1}^k A^{k-i}BC^{i-1} \\ 0 & C^k \end{bmatrix}.$$

Fact 3.15.31. Let $A, B \in \mathbb{F}^{n\times n}$, and define $\mathcal{A} \triangleq \left[\begin{smallmatrix} A & A \\ A & A \end{smallmatrix}\right]$ and $\mathcal{B} \triangleq \left[\begin{smallmatrix} B & -B \\ -B & B \end{smallmatrix}\right]$. Then, $\mathcal{AB} = \mathcal{BA} = 0$.

Fact 3.15.32. Let $A, B \in \mathbb{C}^{2\times 2}$, assume that $A_{(2,2)} = B_{(2,2)} = 0$ and $A_{(1,2)}A_{(2,1)} = B_{(1,2)}B_{(2,1)}$, and let $n \ge 1$. Then, $(A^n)_{(1,2)}(A^n)_{(2,1)} = (B^n)_{(1,2)}(B^n)_{(2,1)}$. **Source:** [1751].

Fact 3.15.33. Let $A \in \mathbb{C}^{2\times 2}$, assume that A is nonsingular and $\operatorname{tr} A \ne 0$, and let $n \ge 1$. Then, $A_{(1,2)}(A^n)_{(2,1)} = (A^n)_{(1,2)}A_{(2,1)}$. **Source:** [1747].

Fact 3.15.34. Let $A \in \mathbb{R}^{2\times 2}$, and define $a \triangleq \frac{1}{2}\operatorname{tr} A$. Then, the following are equivalent:

i) There exists $B \in \mathbb{R}^{2\times 2}$ such that $A = B^2$.

ii) $\det A \ge 0$ and either $A + \sqrt{\det A}I = 0$ or $\operatorname{tr} A + 2\sqrt{\det A} > 0$.

B is a *square root* of A. If these statements hold, then the following statements hold:

iii) A has finitely many square roots if and only if $A \ne aI_2$. If these statements hold, then A has either exactly zero, two, or four square roots.

iv) A has exactly zero square roots if and only if $A \ne aI_2$ and either $\det A < 0$ or both $\det A \ge 0$ and $\operatorname{tr} A + 2\sqrt{\det A} \le 0$.

v) A has exactly four square roots if and only if $A \ne aI_2$, $\det A > 0$, and $\operatorname{tr} A - 2\sqrt{\det A} > 0$. In this case, the four square roots of A are $\pm(\operatorname{tr} A + 2\sqrt{\det A})^{-1/2}(A + \sqrt{\det A}I_2)$ and $\pm(\operatorname{tr} A - 2\sqrt{\det A})^{-1/2}(A - \sqrt{\det A}I_2)$.

vi) If A has exactly two square roots, then the two square roots of A are $\pm(\operatorname{tr} A + 2\sqrt{\det A})^{-1/2}$ $(A + \sqrt{\det A}I_2)$.

vii) A has infinitely many square roots if and only if $A = aI_2$. If these statements hold, then the set $\mathcal{B}$ of square roots of A is given by

$$\mathcal{B} = \begin{cases} \mathcal{B}_1, & a < 0, \\ \mathcal{B}_1 \cup \mathcal{B}_2, & a = 0, \\ \mathcal{B}_1 \cup \mathcal{B}_3 \cup \mathcal{B}_4 \cup \mathcal{B}_5, & a > 0, \end{cases}$$

where

$$\mathcal{B}_1 \triangleq \left\{\begin{bmatrix} \alpha & \beta \\ \frac{a-\alpha^2}{\beta} & -\alpha \end{bmatrix} : \alpha, \beta \in \mathbb{R},\ \beta \ne 0\right\}, \quad \mathcal{B}_2 \triangleq \left\{\begin{bmatrix} 0 & 0 \\ \alpha & 0 \end{bmatrix} : \alpha \in \mathbb{R}\right\},$$

$$\mathcal{B}_3 \triangleq \begin{bmatrix} \sqrt{a} & 0 \\ 0 & -\sqrt{a} \end{bmatrix} + \mathcal{B}_2, \quad \mathcal{B}_4 \triangleq \begin{bmatrix} -\sqrt{a} & 0 \\ 0 & \sqrt{a} \end{bmatrix} + \mathcal{B}_2 \quad \mathcal{B}_5 \triangleq \{\sqrt{a}I_2, -\sqrt{a}I_2\}.$$

Source: [123]. **Remark:** $A + \sqrt{\det A}I = 0$ implies $\operatorname{tr} A + 2\sqrt{\det A} = 0$. **Related:** Fact 10.10.8.

Fact 3.15.35. Two cube roots of I_2 are given by

$$\begin{bmatrix} -\frac{1}{2} & \frac{\sqrt{3}}{2} \\ \frac{-\sqrt{3}}{2} & -\frac{1}{2} \end{bmatrix}^3 = \begin{bmatrix} -1 & -1 \\ 1 & 0 \end{bmatrix}^3 = I_2.$$

Fact 3.15.36. Let $A, B, C, D \in \mathbb{R}^{n\times n}$, assume that A, B, C, D are nonnegative, and assume that $0 \leq\leq B \leq\leq A$ and $0 \leq\leq D \leq\leq C$. Then, $0 \leq\leq BD \leq\leq AC$.

Fact 3.15.37. Let $A, B \in \mathbb{F}^{n\times m}$. Then,

$$AA^{\mathsf{T}}BB^{\mathsf{T}} = (AB^{\mathsf{T}})^2 + \sum[(\operatorname{col}_i A)^{\mathsf{T}}\operatorname{col}_j B - (\operatorname{col}_j A)^{\mathsf{T}}\operatorname{col}_i B][\operatorname{col}_i A(\operatorname{col}_j B)^{\mathsf{T}} - \operatorname{col}_j A(\operatorname{col}_i B)^{\mathsf{T}}],$$

where the sum is taken over all $i, j \in \{1, \dots, m\}$ such that $i < j$. **Source:** [2703]. **Remark:** This is a matrix extension of Lagrange's identity. See Fact 2.12.13. **Related:** Fact 3.15.38, Fact 4.13.33, and Fact 8.3.3.

Fact 3.15.38. Let $A, B \in \mathbb{R}^{n\times n}$. Then, the following statements are equivalent:

i) $AA^{\mathsf{T}}BB^{\mathsf{T}} = (AB^{\mathsf{T}})^2$.

ii) $\operatorname{tr} AA^{\mathsf{T}}BB^{\mathsf{T}} = \operatorname{tr}(AB^{\mathsf{T}})^2$.

iii) $A^{\mathsf{T}}B = B^{\mathsf{T}}A$.

Source: [2703]. **Related:** Fact 3.15.37.

Fact 3.15.39. Let $A \in \mathbb{R}^{n\times m}$. Then,

$$n\sum_{i=1}^{n}\left(\sum_{j=1}^{m}A_{(i,j)}\right)^2 + m\sum_{j=1}^{m}\left(\sum_{i=1}^{n}A_{(i,j)}\right)^2 \leq \left(\sum_{i=1}^{n}\sum_{j=1}^{m}A_{(i,j)}\right)^2 + nm\sum_{i=1}^{n}\sum_{j=1}^{m}A_{(i,j)}^2.$$

Furthermore, equality holds if and only if there exist $x \in \mathbb{R}^n$ and $y \in \mathbb{R}^m$ such that $A = x1_{1\times m} + 1_{n\times 1}y^{\mathsf{T}}$. **Source:** [1032]. **Remark:** This is an extension of the *Cauchy-Khinchin inequality*. **Credit:** E. R. van Dam.

3.16 Facts on the Determinant

Fact 3.16.1. Let $n \geq 1$. Then, $\det \hat{I}_n = (-1)^{\lfloor n/2 \rfloor} = (-1)^{n(n-1)/2} = (-1)^{n-1}\det \hat{I}_{n-1}$. Consequently, $\hat{I}_n$ is an (even, odd) permutation matrix if and only if $\frac{1}{2}n(n-1)$ is (even, odd). **Source:** Since $\hat{I}_n$ is a permutation matrix, its determinant reflects whether it permutes the components of a vector in an odd or even manner, which reflects the parity of pairwise component swaps that it performs on a vector. The total number of swaps performed by a permutation matrix is given by the sum over all rows of the number of 1's in subsequent rows that are to the left of the 1 in each row. For $\hat{I}_n$, this number is $(n-1) + (n-2) + \cdots + 2 + 1 = n(n-1)/2$. See [1560, pp. 29–32]. **Related:** Fact 4.13.18 and Fact 4.31.14.

Fact 3.16.2. Let $n \geq 1$. Then, $\det P_n = (-1)^{n-1}$. Consequently, P_n is an (even, odd) permutation matrix if and only if n is (odd, even). **Related:** Fact 4.13.18 and Fact 7.18.13.

Fact 3.16.3. Let α be a complex number. Then, $\det(I_n + \alpha 1_{n\times n}) = 1 + \alpha n$.

Fact 3.16.4. Let $A \in \mathbb{F}^{n\times n}$, and assume that A has a zero submatrix of size $r \times s$, where $n + 1 \leq r + s$. Then, $\det A = 0$. **Source:** [281, p. 14] and [2991, p. 158]. **Related:** Fact 3.14.1.

Fact 3.16.5. Let $A \in \mathbb{R}^{n\times n}$. Then, $\det(I + A^2) \geq 0$. **Source:** [112, p. 55] and [1158, p. 68]. **Related:** Fact 4.10.15 and Fact 4.13.20.

Fact 3.16.6. Let $A \in \mathbb{F}^{n\times n}$ and $\mathcal{S} \subseteq \{1, \dots, n\}$, and assume that $A_{(\mathcal{S})}$ is nonsingular. Then,

$$\det A = (\det A_{(\mathcal{S})})\det[A_{[\mathcal{S}]} - A_{(\mathcal{S}^{\sim},\mathcal{S})}(A_{(\mathcal{S})})^{-1}A_{(\mathcal{S},\mathcal{S}^{\sim})}].$$

Remark: This result generalizes (3.9.8).

Fact 3.16.7. Let $A \in \mathbb{F}^{n\times m}$ and $B \in \mathbb{F}^{m\times n}$, and assume that $m < n$. Then, $\det AB = 0$.

Fact 3.16.8. Let $A \in \mathbb{F}^{n\times m}$, $B \in \mathbb{F}^{m\times l}$, $k \le \min\{n,m,l\}$, $\mathcal{S}_1 \subseteq \{1,\dots,n\}$, and $\mathcal{S}_2 \subseteq \{1,\dots,l\}$, and assume that $\operatorname{card}(\mathcal{S}_1) = \operatorname{card}(\mathcal{S}_2) = k$. Then,

$$\det (AB)_{(\mathcal{S}_1,\mathcal{S}_2)} = \sum \det A_{(\mathcal{S}_1,\mathcal{S})} \det B_{(\mathcal{S},\mathcal{S}_2)},$$

where the sum is taken over all subsets $\mathcal{S}$ of $\{1,\dots,m\}$ having k elements. **Source:** [1882, pp. 116, 117] and [2991, p. 123]. **Remark:** This is the *Binet-Cauchy formula.* **Remark:** This result is equivalent to *ix*) of Fact 9.5.18. The case $k = n = l$ is given by Fact 3.16.9. **Related:** Fact 3.16.9 and Fact 6.9.37.

Fact 3.16.9. Let $A \in \mathbb{F}^{n\times m}$ and $B \in \mathbb{F}^{m\times n}$, and assume that $n \le m$. Then,

$$\det AB = \sum \det A_{(\cdot,\mathcal{S})} \det B_{(\mathcal{S},\cdot)},$$

where the sum is taken over all subsets $\mathcal{S}$ of $\{1,\dots,m\}$ having n elements. **Source:** [970, p. 102]. **Remark:** $\det AB$ is equal to the sum of all $\binom{m}{n}$ products of pairs of subdeterminants of A and B formed by choosing n columns of A and the corresponding n rows of B. **Remark:** Determinantal and minor equalities are given in [592, 1773]. **Related:** Fact 3.17.8. This is a special case of the Binet-Cauchy formula given by Fact 3.16.8 and Fact 9.5.18. The special case $n = m$ is given by Proposition 3.8.3.

Fact 3.16.10. Let $A \in \mathbb{F}^{n\times m}$, define $r \triangleq \operatorname{rank} A$, let $\mathcal{S}_1, \mathcal{S}_2 \subseteq \{1,\dots,n\}$, let $\mathcal{S}_3, \mathcal{S}_4 \subseteq \{1,\dots,m\}$, and assume that $\operatorname{card}(\mathcal{S}_1) = \operatorname{card}(\mathcal{S}_2) = \operatorname{card}(\mathcal{S}_3) = \operatorname{card}(\mathcal{S}_4) = r$. Then,

$$\det A_{(\mathcal{S}_1,\mathcal{S}_3)} \det A_{(\mathcal{S}_2,\mathcal{S}_4)} = \det A_{(\mathcal{S}_1,\mathcal{S}_4)} \det A_{(\mathcal{S}_2,\mathcal{S}_3)}.$$

Source: [1451, p. 20].

Fact 3.16.11. Let $A \in \mathbb{F}^{n\times n}$, let S be the $n \times n$ matrix whose (i, j) entry is $(-1)^{i+j}$, let $\mathcal{S}_1, \mathcal{S}_2 \subseteq \{1,\dots,n\}$, assume that $\operatorname{card}(\mathcal{S}_1) = \operatorname{card}(\mathcal{S}_2)$, and let l be the sum of the elements in the multiset $\mathcal{S}_1 \cup \mathcal{S}_2$. Then,

$$\det (S \odot A)_{(\mathcal{S}_1,\mathcal{S}_2)} = (-1)^l \det A_{(\mathcal{S}_1,\mathcal{S}_2)}.$$

Related: Fact 3.17.35.

Fact 3.16.12. Let $A \in \mathbb{F}^{n\times n}$, assume that A is nonsingular, and let $b \in \mathbb{F}^n$. Then, the solution $x \in \mathbb{F}^n$ of $Ax = b$ is given by

$$x = \begin{bmatrix} \dfrac{\det(A \overset{1}{\leftarrow} b)}{\det A} \\ \vdots \\ \dfrac{\det(A \overset{n}{\leftarrow} b)}{\det A} \end{bmatrix}.$$

Source: Note that $A(I \overset{i}{\leftarrow} x) = A \overset{i}{\leftarrow} b$. Since $\det(I \overset{i}{\leftarrow} x) = x_{(i)}$, it follows that $(\det A)x_{(i)} = \det(A \overset{i}{\leftarrow} b)$. **Remark:** This is *Cramer's rule.* **Remark:** See Fact 3.16.13 for nonsquare extensions. **Related:** Fact 3.19.8.

Fact 3.16.13. Let $A \in \mathbb{F}^{n\times m}$, assume that A is right invertible, and let $b \in \mathbb{F}^n$. Then, a solution $x \in \mathbb{F}^m$ of $Ax = b$ is given, for all $i \in \{1,\dots,m\}$, by

$$x_{(i)} = \frac{\det[(A \overset{i}{\leftarrow} b)A^*] - \det[(A \overset{i}{\leftarrow} 0)A^*]}{\det(AA^*)}.$$

Source: [1726]. **Remark:** This result extends Cramer's rule. See Fact 3.16.12. Extensions to generalized inverses are given in [417, 1529, 1717] and [2821, Chapter 3].

Fact 3.16.14. Let $A, B \in \mathbb{R}^{n\times n}$, and assume that A and B are symmetric. Then, the following statements are equivalent:

i) $AB = 0$.

ii) For all $\alpha, \beta \in \mathbb{R}$, $\det(I - \alpha A - \beta B) = [\det(I - \alpha A)]\det(I - \beta B)$.

Source: [259, 1795]. **Remark:** This is the *Craig-Sakamoto theorem.* **Related:** Fact 4.18.8.

Fact 3.16.15. Let $A \in \mathbb{F}^{n\times n}$, and assume that either $A_{(i,j)} = 0$ for all i, j such that $i + j < n + 1$ or $A_{(i,j)} = 0$ for all i, j such that $i + j > n + 1$. Then,

$$\det A = (-1)^{\lfloor n/2 \rfloor} \prod_{i=1}^{n} A_{(i,n+1-i)}.$$

Remark: A is either *lower reverse triangular* or *upper reverse triangular.*

Fact 3.16.16. Let $a_1, \dots, a_n \in \mathbb{F}$. Then,

$$\det \begin{bmatrix} 1 + a_1 & a_2 & \cdots & a_n \\ a_1 & 1 + a_2 & \cdots & a_n \\ \vdots & \vdots & \ddots & \vdots \\ a_1 & a_2 & \cdots & 1 + a_n \end{bmatrix} = 1 + \sum_{i=1}^{n} a_i.$$

Fact 3.16.17. Let $a_1, \dots, a_n \in \mathbb{F}$ be nonzero. Then,

$$\det \begin{bmatrix} \frac{1+a_1}{a_1} & 1 & \cdots & 1 \\ 1 & \frac{1+a_2}{a_2} & \cdots & 1 \\ \vdots & \vdots & \ddots & \vdots \\ 1 & 1 & \cdots & \frac{1+a_n}{a_n} \end{bmatrix} = \frac{1 + \sum_{i=1}^{n} a_i}{\prod_{i=1}^{n} a_i}.$$

Source: Multiply the matrix in Fact 3.16.16 on the right by $\operatorname{diag}(1/a_1, \dots, 1/a_n)$.

Fact 3.16.18. Let $a, b, c_1, \dots, c_n \in \mathbb{F}$, define $A \in \mathbb{F}^{n\times n}$ by

$$A \triangleq \begin{bmatrix} c_1 & a & a & \cdots & a \\ b & c_2 & a & \cdots & a \\ b & b & c_3 & \ddots & a \\ \vdots & \vdots & \ddots & \ddots & \vdots \\ b & b & b & \cdots & c_n \end{bmatrix},$$

define $p(x) \triangleq (c_1 - x)(c_2 - x) \cdots (c_n - x)$, and, for all $i \in \{1, \dots, n\}$, define $p_i(x) \triangleq p(x)/(c_i - x)$. Then,

$$\det A = \begin{cases} \dfrac{bp(a) - ap(b)}{b - a}, & b \neq a, \\ a \sum_{i=1}^{n-1} p_i(a) + c_n p_n(a) = p(a) - ap'(a), & b = a. \end{cases}$$

Source: [1757, pp. 65, 66] and [2980, p. 10].

Fact 3.16.19. Let $a, b \in \mathbb{F}$, and define $A, B \in \mathbb{F}^{n\times n}$ by

$$A \triangleq (a-b)I_n + b1_{n\times n} = \begin{bmatrix} a & b & b & \cdots & b \\ b & a & b & \cdots & b \\ b & b & a & \ddots & b \\ \vdots & \vdots & \ddots & \ddots & \vdots \\ b & b & b & \cdots & a \end{bmatrix},$$

$$B \triangleq aI_n + b1_{n\times n} = \begin{bmatrix} a+b & b & b & \cdots & b \\ b & a+b & b & \cdots & b \\ b & b & a+b & \ddots & b \\ \vdots & \vdots & \ddots & \ddots & \vdots \\ b & b & b & \cdots & a+b \end{bmatrix}.$$

Then, $\det A = (a-b)^{n-1}[a + b(n-1)]$ and, if $\det A \neq 0$, then

$$A^{-1} = \frac{1}{a-b}I_n + \frac{b}{(b-a)[a+b(n-1)]}1_{n\times n}.$$

Furthermore, $\det B = a^{n-1}(a+nb)$, and, if $\det B \neq 0$, then

$$B^{-1} = \frac{1}{a}\left(I_n - \frac{b}{a+nb}1_{n\times n}\right).$$

Remark: $aI_n + b1_{n\times n}$ arises in combinatorics. See [589, 591]. **Related:** Fact 3.17.30, Fact 6.10.21, and Fact 10.10.39.

Fact 3.16.20. Let $n \geq 2$. Then,

$$\det[\operatorname{diag}(2,\ldots,n) + 1_{(n-1)\times(n-1)}] = n!H_n.$$

Source: [1350, p. 132]. **Remark:** H_n is the nth harmonic number.

Fact 3.16.21. Let $a_1,\ldots,a_n \in \mathbb{F}$, and define $A \in \mathbb{F}^{n\times n}$ by

$$A \triangleq \sum_{i=1}^{n} \operatorname{diag}[0_{(i-1)\times(i-1)}, a_i 1_{(n+1-i)\times(n+1-i)}].$$

Then, $\det A = \prod_{i=1}^{n} a_i$. **Example:** $\det\begin{bmatrix} a_1 & a_1 & a_1 \\ a_1 & a_1+a_2 & a_1+a_2 \\ a_1 & a_1+a_2 & a_1+a_2+a_3 \end{bmatrix} = a_1a_2a_3$.

Fact 3.16.22. Let $a_1,\ldots,a_n \in \mathbb{F}$, define $s \triangleq \sum_{i=1}^{n} a_i$, and define

$$A \triangleq \begin{bmatrix} -a_1 & s-a_1 & s-a_1 & \cdots & s-a_1 \\ s-a_2 & -a_2 & s-a_2 & \cdots & s-a_2 \\ s-a_3 & s-a_3 & -a_3 & \ddots & s-a_3 \\ \vdots & \vdots & \ddots & \ddots & \vdots \\ s-a_n & s-a_n & s-a_n & \cdots & -a_n \end{bmatrix}.$$

Then, $\det A = (-1)^{n-1}(n-2)s^n$. In particular, for all $a, b, c, d \in \mathbb{F}$,

$$\det\begin{bmatrix} -a & b+c & b+c \\ a+c & -b & a+c \\ a+b & a+b & -c \end{bmatrix} = (a+b+c)^3,$$

$$\det\begin{bmatrix} -a & b+c+d & b+c+d & b+c+d \\ a+c+d & -b & a+c+d & a+c+d \\ a+b+d & a+b+d & -c & a+b+d \\ a+b+c & a+b+c & a+b+c & -d \end{bmatrix} = -2(a+b+c+d)^4.$$

Source: [1641] and [1860, pp. 34, 219].

Fact 3.16.23. Let $x, y, z \in \mathbb{C}$. Then,

$$\det\begin{bmatrix} 0 & 1 & 1 & 1 \\ 1 & 0 & x^2 & y^2 \\ 1 & x^2 & 0 & z^2 \\ 1 & y^2 & z^2 & 0 \end{bmatrix} = (x+y+z)(x-y-z)(x-y+z)(x+y-z).$$

Related: Fact 5.4.7.

Fact 3.16.24. Let $A \in \mathbb{F}^{n\times n}$, and define $\gamma \triangleq \max_{i,j\in\{1,\ldots,n\}} |A_{(i,j)}|$. Then, $|\det A| \le \gamma^n n^{n/2}$. **Source:** This result is a consequence of the arithmetic-mean–geometric-mean inequality Fact 2.11.81 and Schur's inequality Fact 10.21.10. See [970, p. 200]. **Related:** Fact 10.15.10.

Fact 3.16.25. Let $A \in \mathbb{R}^{n\times n}$, and, for all $i \in \{1,\ldots,n\}$, let α_i denote the sum of the positive components in $\operatorname{row}_i(A)$ and let β_i denote the sum of the positive components in $\operatorname{row}_i(-A)$. Then,

$$|\det A| \le \prod_{i=1}^{n} \max\{\alpha_i, \beta_i\} - \prod_{i=1}^{n} \min\{\alpha_i, \beta_i\}.$$

Source: [1543]. **Credit:** This is an extension of a result due to A. Schinzel.

Fact 3.16.26. For $i \in \{1,2,3,4\}$, let $A_i, B_i \in \mathbb{F}^{2\times 2}$, where $\det A_i = \det B_i = 1$. Furthermore, let $\mathcal{A}, \mathcal{B}, \mathcal{C}, \mathcal{D} \in \mathbb{F}^{4\times 4}$, where, for all $i, j \in \{1,2,3,4\}$,

$$\mathcal{A}_{(i,j)} \triangleq \operatorname{tr} A_iA_j, \quad \mathcal{B}_{(i,j)} \triangleq \operatorname{tr} B_iB_j, \quad \mathcal{C}_{(i,j)} \triangleq \operatorname{tr} A_iB_j, \quad \mathcal{D}_{(i,j)} \triangleq \operatorname{tr} A_iB_j^{-1}.$$

Then,

$$\det\mathcal{C} + \det\mathcal{D} = 0, \quad \det\mathcal{A}\mathcal{B} = (\det\mathcal{C})^2.$$

Credit: W. Magnus. See [1488].

Fact 3.16.27. Let $\mathcal{I} \subseteq \mathbb{R}$ be a finite or infinite interval, and let $f\colon \mathcal{I} \mapsto \mathbb{R}$. Then, the following statements are equivalent:

i) f is convex.

ii) For all distinct $x, y, z \in \mathcal{I}$,

$$\frac{\det\begin{bmatrix} 1 & x & f(x) \\ 1 & y & f(y) \\ 1 & z & f(z) \end{bmatrix}}{\det\begin{bmatrix} 1 & x & x^2 \\ 1 & y & y^2 \\ 1 & z & z^2 \end{bmatrix}} \ge 0.$$

iii) For all $x, y, z \in \mathcal{I}$ such that $x < y < z$,

$$\det\begin{bmatrix} 1 & x & f(x) \\ 1 & y & f(y) \\ 1 & z & f(z) \end{bmatrix} \ge 0.$$

Source: [2128, p. 21].

Fact 3.16.28. Let $A \in \mathbb{R}^{n\times n}$, where, for all $i, j \in \{1,\ldots,n\}$, $A_{(i,j)}$ is defined below. Then, the following statements hold:

i) If $A_{(i,j)} \triangleq (i+j-2)!$, then $\det A = [1!2!\cdots(n-1)!]^2$.

ii) If $A_{(i,j)} \triangleq 1/\min\{i,j\}$, then $\det A = (-1)^{n-1}/[n!(n-1)!]$.

iii) If $A_{(i,j)} \triangleq \binom{i+j-2}{i-1}$, then $\det A = 1$.

iv) If $A_{(i,j)} \triangleq \binom{2i-2}{j-1}$, then $\det A = 2^{\binom{n}{2}}$.

v) If $A_{(i,j)} \triangleq \binom{2n}{n+i-j}$, then $\det A = \prod_{i,j,k=1}^{n} \frac{i+j+k-1}{i+j+k-2}$.

vi) Let $k \ge 1$. If $A_{(i,j)} \triangleq C_{k+i+j-2n-2}$, then $\det A = \prod_{i+j\le k-2n-1} \frac{i+j+2(n-1)}{i+j}$.

vii) Let $k,m \ge 1$. If $A_{(i,j)} \triangleq \binom{ki+m-2}{j-1}$, then $\det A = k^{\binom{n}{2}}$.

Source: [68], [166, p. 62], and [1699]. *ii)* is given in [146]. **Remark:** In *i)* and *vi)*, A is a Hankel matrix. **Remark:** In *vii)*, $\det A$ is independent of m. **Related:** Fact 7.18.6 and Fact 10.9.8.

Fact 3.16.29. Let $A \in \mathbb{R}^{(n+1)\times n}$, where, for all $i,j \in \{1,\ldots,n\}$,

$$A_{(i,j)} \triangleq \begin{cases} \binom{i}{j-1}, & i \ge j, \\ 0, & i < j. \end{cases}$$

Then, for all $i \in \{1,\ldots,n+1\}$,

$$\det A_{[i,\cdot]} = \begin{cases} (-1)^{n+i+1}\frac{(n+1)!}{(n+1-i)!}\binom{n}{n-i}B_{n+1-i}, & i \in \{1,\ldots,n\}, \\ n!, & i = n+1. \end{cases}$$

Source: [1150]. **Remark:** B_i is the ith Bernoulli number. See Fact 13.1.6. **Example:** Let $n = 4$. Then, $\det A_{[3,\cdot]} = (-1)^8(5!/2!)\binom{4}{1}B_2 = 60(4)(1/6) = 40$.

3.17 Facts on the Determinant of Partitioned Matrices

Fact 3.17.1. Let $A \in \mathbb{F}^{n\times n}$, let $k \in \{1,\ldots,n-1\}$, let A_0 be the $k\times k$ leading principal submatrix of A, and let $B \in \mathbb{F}^{(n-k)\times(n-k)}$, where, for all $i,j \in \{1,\ldots,n-k\}$, $B_{(i,j)} \triangleq \det A_{(\{1,\ldots,k,k+i\},\{1,\ldots,k,k+j\})}$. Then,

$$\det B = (\det A_0)^{n-k-1}\det A.$$

If, in addition, A_0 is nonsingular, then

$$\det A = \frac{\det B}{(\det A_0)^{n-k-1}}.$$

Remark: If $k = n-1$, then $\det B = \det A$. **Remark:** This is *Sylvester's identity.*

Fact 3.17.2. Let $A \in \mathbb{F}^{n\times n}$, $x,y \in \mathbb{F}^n$, and $a \in \mathbb{F}$. Then,

$$\det\begin{bmatrix} A & x \\ y^{\mathsf{T}} & a \end{bmatrix} = a(\det A) - y^{\mathsf{T}}A^{\mathsf{A}}x = (a+1)\det A - \det(A + xy^{\mathsf{T}}).$$

Hence,

$$\det\begin{bmatrix} A & x \\ y^{\mathsf{T}} & a \end{bmatrix} = \begin{cases} (\det A)(a - y^{\mathsf{T}}A^{-1}x), & \det A \ne 0, \\ a\det(A - a^{-1}xy^{\mathsf{T}}), & a \ne 0, \\ -y^{\mathsf{T}}A^{\mathsf{A}}x, & a = 0 \text{ or } \det A = 0. \end{cases}$$

In particular,

$$\det\begin{bmatrix} A & Ax \\ y^{\mathsf{T}}A & y^{\mathsf{T}}Ax \end{bmatrix} = 0.$$

Furthermore,

$$\det(A + xy^{\mathsf{T}}) = \det A + y^{\mathsf{T}}A^{\mathsf{A}}x = -\det\begin{bmatrix} A & x \\ y^{\mathsf{T}} & -1 \end{bmatrix}.$$

If, in addition, A is nonsingular, then $\det(A + xy^{\mathsf{T}}) = (\det A)(1 + y^{\mathsf{T}}A^{-1}x)$. **Related:** Fact 3.17.3, Fact 3.21.4, and Fact 3.21.5.

Fact 3.17.3. Let $A \in \mathbb{F}^{n\times n}$, $b \in \mathbb{F}^n$, and $a \in \mathbb{F}$. Then,

$$\det\begin{bmatrix} A & b \\ b^* & a \end{bmatrix} = a(\det A) - b^*A^{\mathsf{A}}b.$$

In particular,

$$\det\begin{bmatrix} A & b \\ b^* & a \end{bmatrix} = \begin{cases} (\det A)(a - b^*A^{-1}b), & \det A \neq 0, \\ a\det(A - a^{-1}bb^*), & a \neq 0, \\ -b^*A^{\mathsf{A}}b, & a = 0. \end{cases}$$

Remark: This is a special case of Fact 3.17.2 with $x = b$ and $y = \overline{b}$. **Related:** Fact 10.18.5.

Fact 3.17.4. Let $A \in \mathbb{F}^{n\times n}$. Then,

$$\operatorname{rank}\begin{bmatrix} A & A \\ A & A \end{bmatrix} = \operatorname{rank}\begin{bmatrix} A & -A \\ -A & A \end{bmatrix} = \operatorname{rank} A, \quad \operatorname{rank}\begin{bmatrix} A & A \\ -A & A \end{bmatrix} = 2\operatorname{rank} A,$$

$$\det\begin{bmatrix} A & A \\ A & A \end{bmatrix} = \det\begin{bmatrix} A & -A \\ -A & A \end{bmatrix} = 0, \quad \det\begin{bmatrix} A & A \\ -A & A \end{bmatrix} = 2^n(\det A)^2.$$

Related: Fact 3.17.5.

Fact 3.17.5. Let $a, b, c, d \in \mathbb{F}$, let $A \in \mathbb{F}^{n\times n}$, and define $\mathcal{A} \triangleq \left[\begin{smallmatrix} aA & bA \\ cA & dA \end{smallmatrix}\right]$. Then,

$$\operatorname{rank}\mathcal{A} = \left(\operatorname{rank}\begin{bmatrix} a & b \\ c & d \end{bmatrix}\right)\operatorname{rank} A, \quad \det\mathcal{A} = (ad - bc)^n(\det A)^2.$$

Source: Proposition 9.1.11 and Fact 9.4.20. **Related:** Fact 3.17.4.

Fact 3.17.6. $\det\left[\begin{smallmatrix} 0 & I_n \\ I_m & 0 \end{smallmatrix}\right] = (-1)^{nm}$. In particular, $\det\left[\begin{smallmatrix} 0 & I_n \\ I_n & 0 \end{smallmatrix}\right] = (-1)^n$.

Fact 3.17.7. Let $A, B, C \in \mathbb{F}^{n\times n}$. Then,

$$\det\begin{bmatrix} 0 & A \\ B & C \end{bmatrix} = \det\begin{bmatrix} C & A \\ B & 0 \end{bmatrix} = (-1)^n \det AB.$$

Furthermore,

$$\det\begin{bmatrix} 0 & I - AB \\ B & C \end{bmatrix} = \det\begin{bmatrix} C & I - AB \\ B & 0 \end{bmatrix} = \det(BAB - B).$$

Source: Fact 3.17.6. **Related:** Fact 3.14.23.

Fact 3.17.8. Let $A \in \mathbb{F}^{n\times m}$, $C \in \mathbb{F}^{m\times m}$, and $B \in \mathbb{F}^{m\times n}$, and assume that C is nonsingular. Then,

$$\det\begin{bmatrix} A & 0 \\ C & B \end{bmatrix} = (-1)^{(m+1)n}(\det C)\det AC^{-1}B.$$

In particular,

$$\det\begin{bmatrix} A & 0 \\ I_m & B \end{bmatrix} = (-1)^{(m+1)n}\det AB, \quad \det\begin{bmatrix} A & 0 \\ -I_m & B \end{bmatrix} = (-1)^{(n+1)m}\det AB.$$

Source: [970]. **Related:** Fact 3.16.9.

Fact 3.17.9. Let $A, B \in \mathbb{F}^{n\times n}$. Then,

$$\begin{bmatrix} AB & A \\ 0 & I \end{bmatrix} = \begin{bmatrix} A & 0 \\ I & B \end{bmatrix}\begin{bmatrix} 0 & I \\ I & 0 \end{bmatrix}\begin{bmatrix} -I & 0 \\ B & I \end{bmatrix}, \quad \begin{bmatrix} AB & 0 \\ B & I \end{bmatrix} = \begin{bmatrix} -I & A \\ 0 & I \end{bmatrix}\begin{bmatrix} A & 0 \\ I & B \end{bmatrix}\begin{bmatrix} 0 & I \\ I & 0 \end{bmatrix},$$

$$\begin{bmatrix} I & I \\ A & B \end{bmatrix} = \begin{bmatrix} I & 0 \\ A & I \end{bmatrix}\begin{bmatrix} I & 0 \\ 0 & B-A \end{bmatrix}\begin{bmatrix} I & I \\ 0 & I \end{bmatrix}, \quad \det\begin{bmatrix} I & I \\ A & B \end{bmatrix} = \det(B-A),$$

$$\begin{bmatrix} I & A \\ B & I \end{bmatrix} = \begin{bmatrix} I & A \\ 0 & I \end{bmatrix}\begin{bmatrix} I-AB & 0 \\ 0 & I \end{bmatrix}\begin{bmatrix} I & 0 \\ B & I \end{bmatrix}, \quad \det\begin{bmatrix} I & A \\ B & I \end{bmatrix} = \det(I-BA),$$

$$\begin{bmatrix} A+B & B \\ B & B \end{bmatrix} = \begin{bmatrix} I & I \\ 0 & I \end{bmatrix}\begin{bmatrix} A & 0 \\ 0 & B \end{bmatrix}\begin{bmatrix} I & 0 \\ I & I \end{bmatrix}, \quad \det\begin{bmatrix} A+B & B \\ B & B \end{bmatrix} = (\det A)\det B,$$

$$\begin{bmatrix} A & AB \\ BA & B \end{bmatrix} = \begin{bmatrix} I & A \\ 0 & I \end{bmatrix}\begin{bmatrix} A-ABA & 0 \\ BA-B^2A & B \end{bmatrix}\begin{bmatrix} I & 0 \\ BA & I \end{bmatrix}, \quad \det\begin{bmatrix} A & AB \\ BA & B \end{bmatrix} = (\det B)\det(A-ABA),$$

$$\begin{bmatrix} A & AB \\ BA & B \end{bmatrix} = \begin{bmatrix} I & 0 \\ BA & I \end{bmatrix}\begin{bmatrix} A & 0 \\ BA-BA^2 & B-BAB \end{bmatrix}\begin{bmatrix} I & B \\ 0 & I \end{bmatrix}, \quad \det\begin{bmatrix} A & AB \\ BA & B \end{bmatrix} = (\det A)\det(B-BAB),$$

$$\begin{bmatrix} A & B \\ B & A \end{bmatrix} = \frac{1}{2}\begin{bmatrix} I & I \\ I & -I \end{bmatrix}\begin{bmatrix} A+B & 0 \\ 0 & A-B \end{bmatrix}\begin{bmatrix} I & I \\ I & -I \end{bmatrix}, \quad \det\begin{bmatrix} A & B \\ B & A \end{bmatrix} = \det(A+B)\det(A-B).$$

Fact 3.17.10. Let $A, B \in \mathbb{F}^{n\times n}$. Then,

$$\det\begin{bmatrix} I-A^*A & I-A^*B \\ I-B^*A & I-B^*B \end{bmatrix} = (-1)^n|\det(A-B)|^2.$$

Source: Consider

$$\begin{bmatrix} I-A^*A & I-A^*B \\ I-B^*A & I-B^*B \end{bmatrix} = \begin{bmatrix} I & A^* \\ I & B^* \end{bmatrix}\begin{bmatrix} I & I \\ -A & -B \end{bmatrix}.$$

See [2991, p. 233]. **Related:** Fact 3.20.16 and Fact 10.16.28.

Fact 3.17.11. Let A, B, C, D be conformable matrices with entries in $\mathbb{F}$. Then,

$$\begin{bmatrix} A & AB \\ C & D \end{bmatrix} = \begin{bmatrix} I & 0 \\ C & I \end{bmatrix}\begin{bmatrix} A & 0 \\ C-CA & D-CB \end{bmatrix}\begin{bmatrix} I & B \\ 0 & I \end{bmatrix}, \quad \det\begin{bmatrix} A & AB \\ C & D \end{bmatrix} = (\det A)\det(D-CB),$$

$$\begin{bmatrix} A & B \\ CA & D \end{bmatrix} = \begin{bmatrix} I & 0 \\ C & I \end{bmatrix}\begin{bmatrix} A & B-AB \\ 0 & D-CB \end{bmatrix}\begin{bmatrix} I & B \\ 0 & I \end{bmatrix}, \quad \det\begin{bmatrix} A & B \\ CA & D \end{bmatrix} = (\det A)\det(D-CB),$$

$$\begin{bmatrix} A & BD \\ C & D \end{bmatrix} = \begin{bmatrix} I & B \\ 0 & I \end{bmatrix}\begin{bmatrix} A-BC & 0 \\ C-DC & D \end{bmatrix}\begin{bmatrix} I & 0 \\ C & I \end{bmatrix}, \quad \det\begin{bmatrix} A & BD \\ C & D \end{bmatrix} = \det(A-BC)\det D,$$

$$\begin{bmatrix} A & B \\ DC & D \end{bmatrix} = \begin{bmatrix} I & B \\ 0 & I \end{bmatrix}\begin{bmatrix} A-BC & B-BD \\ 0 & D \end{bmatrix}\begin{bmatrix} I & 0 \\ C & I \end{bmatrix}, \quad \det\begin{bmatrix} A & B \\ DC & D \end{bmatrix} = \det(A-BC)\det D.$$

Related: Fact 8.9.34.

Fact 3.17.12. Let $A, B, C, D \in \mathbb{F}^{n\times n}$, and assume that $\operatorname{rank}\left[\begin{smallmatrix} A & B \\ C & D \end{smallmatrix}\right] = n$. Then,

$$\det\begin{bmatrix} \det A & \det B \\ \det C & \det D \end{bmatrix} = 0.$$

Fact 3.17.13. Let $A, B, C, D \in \mathbb{F}^{n\times n}$. Then,

$$\det\begin{bmatrix} A & B \\ C & D \end{bmatrix} = \begin{cases} (\det A)\det(D-CA^{-1}B), & \det A \ne 0, \\ (-1)^n(\det B)\det(C-DB^{-1}A), & \det B \ne 0, \\ (-1)^n(\det C)\det(B-AC^{-1}D), & \det C \ne 0, \\ (\det D)\det(A-BD^{-1}C), & \det D \ne 0. \end{cases}$$

Fact 3.17.14. Let $A, B, C, D \in \mathbb{F}^{n\times n}$. Then,

$$\det\begin{bmatrix} A & B \\ C & D \end{bmatrix} = \begin{cases} \det(DA - CB), & AB = BA, \\ \det(AD - CB), & AC = CA, \\ \det(AD - BC), & DC = CD, \\ \det(DA - BC), & DB = BD. \end{cases}$$

Source: If A is nonsingular and $AB = BA$, then

$$\det\begin{bmatrix} A & B \\ C & D \end{bmatrix} = (\det A)\det(D - CA^{-1}B) = \det(DA - CA^{-1}BA) = \det(DA - CB).$$

Alternatively, note that

$$\begin{bmatrix} A & B \\ C & D \end{bmatrix} = \begin{bmatrix} A & 0 \\ C & DA - CB \end{bmatrix}\begin{bmatrix} I & BA^{-1} \\ 0 & A^{-1} \end{bmatrix}.$$

If A is singular, then replace A with $A + \varepsilon I$ and use continuity. **Remark:** These are *Schur's formulas.* See [302, p. 11]. **Problem:** Prove this result in the case where A is singular without invoking continuity.

Fact 3.17.15. Let $A, B, C, D \in \mathbb{F}^{n\times n}$. Then,

$$\det\begin{bmatrix} A & B \\ C & D \end{bmatrix} = \begin{cases} \det(AD^{\mathsf{T}} - B^{\mathsf{T}}C^{\mathsf{T}}), & AB = BA^{\mathsf{T}}, \\ \det(AD^{\mathsf{T}} - BC), & DC = CD^{\mathsf{T}}, \\ \det(A^{\mathsf{T}}D - CB), & A^{\mathsf{T}}C = CA, \\ \det(A^{\mathsf{T}}D - C^{\mathsf{T}}B^{\mathsf{T}}), & D^{\mathsf{T}}B = BD. \end{cases}$$

Source: Define the nonsingular matrix $A_\varepsilon \triangleq A + \varepsilon I$, which satisfies $A_\varepsilon B = BA_\varepsilon^{\mathsf{T}}$. Then,

$$\det\begin{bmatrix} A_\varepsilon & B \\ C & D \end{bmatrix} = (\det A_\varepsilon)\det(D - CA_\varepsilon^{-1}B) = \det(DA_\varepsilon^{\mathsf{T}} - CA_\varepsilon^{-1}BA_\varepsilon^{\mathsf{T}}) = \det(DA_\varepsilon^{\mathsf{T}} - CB).$$

Fact 3.17.16. Let $A, B, C, D \in \mathbb{F}^{n\times n}$. Then,

$$\det\begin{bmatrix} A & B \\ C & D \end{bmatrix} = \begin{cases} (-1)^{\operatorname{rank} C}\det(A^{\mathsf{T}}D + C^{\mathsf{T}}B), & A^{\mathsf{T}}C = -C^{\mathsf{T}}A, \\ (-1)^{n+\operatorname{rank} A}\det(A^{\mathsf{T}}D + C^{\mathsf{T}}B), & A^{\mathsf{T}}C = -C^{\mathsf{T}}A, \\ (-1)^{\operatorname{rank} B}\det(A^{\mathsf{T}}D + C^{\mathsf{T}}B), & B^{\mathsf{T}}D = -D^{\mathsf{T}}B, \\ (-1)^{n+\operatorname{rank} D}\det(A^{\mathsf{T}}D + C^{\mathsf{T}}B), & B^{\mathsf{T}}D = -D^{\mathsf{T}}B, \\ (-1)^{\operatorname{rank} B}\det(AD^{\mathsf{T}} + BC^{\mathsf{T}}), & AB^{\mathsf{T}} = -BA^{\mathsf{T}}, \\ (-1)^{n+\operatorname{rank} A}\det(AD^{\mathsf{T}} + BC^{\mathsf{T}}), & AB^{\mathsf{T}} = -BA^{\mathsf{T}}, \\ (-1)^{\operatorname{rank} C}\det(AD^{\mathsf{T}} + BC^{\mathsf{T}}), & CD^{\mathsf{T}} = -DC^{\mathsf{T}}, \\ (-1)^{n+\operatorname{rank} D}\det(AD^{\mathsf{T}} + BC^{\mathsf{T}}), & CD^{\mathsf{T}} = -DC^{\mathsf{T}}. \end{cases}$$

Source: [1949, 2839]. **Remark:** If $A^{\mathsf{T}}C = -C^{\mathsf{T}}A$ and $\operatorname{rank} A + \operatorname{rank} C + n$ is odd, then $\left[\begin{smallmatrix} A & B \\ C & D \end{smallmatrix}\right]$ is singular. **Credit:** D. Callan. See [2839].

Fact 3.17.17. Let $A, B, C, D \in \mathbb{F}^{n\times n}$. Then,

$$\det\begin{bmatrix} A & B \\ C & D \end{bmatrix} = \begin{cases} \det(AD^{\mathsf{T}} - BC^{\mathsf{T}}), & AB^{\mathsf{T}} = BA^{\mathsf{T}}, \\ \det(AD^{\mathsf{T}} - BC^{\mathsf{T}}), & DC^{\mathsf{T}} = CD^{\mathsf{T}}, \\ \det(A^{\mathsf{T}}D - C^{\mathsf{T}}B), & A^{\mathsf{T}}C = C^{\mathsf{T}}A, \\ \det(A^{\mathsf{T}}D - C^{\mathsf{T}}B), & D^{\mathsf{T}}B = B^{\mathsf{T}}D. \end{cases}$$

Source: [1949].

Fact 3.17.18. Let $A, B, C, D \in \mathbb{F}^{n\times n}$, and assume that A, B, C, D are nonsingular. Then,

$$\det\begin{bmatrix} A^{-1} & B^{-1} \\ C^{-1} & D^{-1} \end{bmatrix} = \frac{(-1)^n}{\det ABCD}\det\begin{bmatrix} A & C \\ B & D \end{bmatrix} = \frac{1}{\det ABCD}\det\begin{bmatrix} B & D \\ A & C \end{bmatrix} = \frac{\det(B - DC^{-1}A)}{\det ABD}.$$

Source: [2991, p. 232].

Fact 3.17.19. Let $A \in \mathbb{F}^{n\times m}$, $B \in \mathbb{F}^{n\times l}$, $C \in \mathbb{F}^{k\times m}$, and $D \in \mathbb{F}^{k\times l}$, and assume that $n + k = m + l$. If $AC^{\mathsf{T}} + BD^{\mathsf{T}} = 0$, then

$$\det\begin{bmatrix} A & B \\ C & D \end{bmatrix}^2 = \det(AA^{\mathsf{T}} + BB^{\mathsf{T}})\det(CC^{\mathsf{T}} + DD^{\mathsf{T}}).$$

Alternatively, if $A^{\mathsf{T}}B + C^{\mathsf{T}}D = 0$, then

$$\det\begin{bmatrix} A & B \\ C & D \end{bmatrix}^2 = \det(A^{\mathsf{T}}A + C^{\mathsf{T}}C)\det(B^{\mathsf{T}}B + D^{\mathsf{T}}D).$$

Source: Consider $\left[\begin{smallmatrix} A & B \\ C & D \end{smallmatrix}\right]\left[\begin{smallmatrix} A & B \\ C & D \end{smallmatrix}\right]^{\mathsf{T}}$ and $\left[\begin{smallmatrix} A & B \\ C & D \end{smallmatrix}\right]^{\mathsf{T}}\left[\begin{smallmatrix} A & B \\ C & D \end{smallmatrix}\right]$.

Fact 3.17.20. Let $A \in \mathbb{F}^{n\times m}$, $B \in \mathbb{F}^{n\times m}$, $C \in \mathbb{F}^{k\times m}$, and $D \in \mathbb{F}^{k\times m}$, and assume that $n + k = 2m$. If $AD^{\mathsf{T}} + BC^{\mathsf{T}} = 0$, then

$$\det\begin{bmatrix} A & B \\ C & D \end{bmatrix}^2 = (-1)^m\det(AB^{\mathsf{T}} + BA^{\mathsf{T}})\det(CD^{\mathsf{T}} + DC^{\mathsf{T}}).$$

Alternatively, if either $AB^{\mathsf{T}} + BA^{\mathsf{T}} = 0$ or $CD^{\mathsf{T}} + DC^{\mathsf{T}} = 0$, then

$$\det\begin{bmatrix} A & B \\ C & D \end{bmatrix}^2 = (-1)^{m+n}\det(AD^{\mathsf{T}} + BC^{\mathsf{T}})^2.$$

Source: Consider $\left[\begin{smallmatrix} A & B \\ C & D \end{smallmatrix}\right]\left[\begin{smallmatrix} B^{\mathsf{T}} & D^{\mathsf{T}} \\ A^{\mathsf{T}} & C^{\mathsf{T}} \end{smallmatrix}\right]$ and $\left[\begin{smallmatrix} A & B \\ C & D \end{smallmatrix}\right]\left[\begin{smallmatrix} D^{\mathsf{T}} & B^{\mathsf{T}} \\ C^{\mathsf{T}} & A^{\mathsf{T}} \end{smallmatrix}\right]$. See [2839].

Fact 3.17.21. Let $A \in \mathbb{F}^{n\times m}$, $B \in \mathbb{F}^{n\times l}$, $C \in \mathbb{F}^{n\times m}$, and $D \in \mathbb{F}^{n\times l}$, and assume that $m + l = 2n$. If $A^{\mathsf{T}}D + C^{\mathsf{T}}B = 0$, then

$$\det\begin{bmatrix} A & B \\ C & D \end{bmatrix}^2 = (-1)^n\det(C^{\mathsf{T}}A + A^{\mathsf{T}}C)\det(D^{\mathsf{T}}B + B^{\mathsf{T}}D).$$

Alternatively, if either $B^{\mathsf{T}}D + D^{\mathsf{T}}B = 0$ or $A^{\mathsf{T}}C + C^{\mathsf{T}}A = 0$, then

$$\det\begin{bmatrix} A & B \\ C & D \end{bmatrix}^2 = (-1)^{n+m}\det(A^{\mathsf{T}}D + C^{\mathsf{T}}B)^2.$$

Source: Consider $\left[\begin{smallmatrix} C^{\mathsf{T}} & A^{\mathsf{T}} \\ D^{\mathsf{T}} & B^{\mathsf{T}} \end{smallmatrix}\right]\left[\begin{smallmatrix} A & B \\ C & D \end{smallmatrix}\right]$ and $\left[\begin{smallmatrix} D^{\mathsf{T}} & B^{\mathsf{T}} \\ C^{\mathsf{T}} & A^{\mathsf{T}} \end{smallmatrix}\right]\left[\begin{smallmatrix} A & B \\ C & D \end{smallmatrix}\right]$.

Fact 3.17.22. Let $A \in \mathbb{F}^{n\times n}$, $B \in \mathbb{F}^{n\times k}$, $C \in \mathbb{F}^{k\times n}$, and $D \in \mathbb{F}^{k\times k}$. If either $AB + BD = 0$ or $CA + DC = 0$, then

$$\det\begin{bmatrix} A & B \\ C & D \end{bmatrix}^2 = \det(A^2 + BC)\det(CB + D^2).$$

Alternatively, if either $A^2 + BC = 0$ or $CB + D^2 = 0$, then

$$\det\begin{bmatrix} A & B \\ C & D \end{bmatrix}^2 = (-1)^{nk}\det(AB + BD)\det(CA + DC).$$

Source: Consider $\left[\begin{smallmatrix} A & B \\ C & D \end{smallmatrix}\right]^2$ and $\left[\begin{smallmatrix} A & B \\ C & D \end{smallmatrix}\right]\left[\begin{smallmatrix} B & A \\ D & C \end{smallmatrix}\right]$.

Fact 3.17.23. Let $A \in \mathbb{F}^{n\times m}$, $B \in \mathbb{F}^{n\times n}$, $C \in \mathbb{F}^{m\times m}$, and $D \in \mathbb{F}^{m\times n}$. If either $AD + B^2 = 0$ or $C^2 + DA = 0$, then

$$\det\begin{bmatrix} A & B \\ C & D \end{bmatrix}^2 = (-1)^{nm}\det(AC + BA)\det(CD + DB).$$

Alternatively, if either $AC + BA = 0$ or $CD + DB = 0$, then

$$\det\begin{bmatrix} A & B \\ C & D \end{bmatrix}^2 = \det(AD + B^2)\det(C^2 + DA).$$

Source: Consider $\left[\begin{smallmatrix} A & B \\ C & D \end{smallmatrix}\right]\left[\begin{smallmatrix} C & D \\ A & B \end{smallmatrix}\right]$ and $\left[\begin{smallmatrix} A & B \\ C & D \end{smallmatrix}\right]\left[\begin{smallmatrix} D & C \\ B & A \end{smallmatrix}\right]$.

Fact 3.17.24. Let $A \in \mathbb{F}^{n\times m}$, $B \in \mathbb{F}^{n\times l}$, $C \in \mathbb{F}^{k\times m}$, and $D \in \mathbb{F}^{k\times l}$, and assume that $n + k = m + l$. If $AC^* + BD^* = 0$, then

$$\left|\det\begin{bmatrix} A & B \\ C & D \end{bmatrix}\right|^2 = \det(AA^* + BB^*)\det(CC^* + DD^*).$$

Alternatively, if $A^*B + C^*D = 0$, then

$$\left|\det\begin{bmatrix} A & B \\ C & D \end{bmatrix}\right|^2 = \det(A^*A + C^*C)\det(B^*B + D^*D).$$

Source: Consider $\left[\begin{smallmatrix} A & B \\ C & D \end{smallmatrix}\right]\left[\begin{smallmatrix} A & B \\ C & D \end{smallmatrix}\right]^*$ and $\left[\begin{smallmatrix} A & B \\ C & D \end{smallmatrix}\right]^*\left[\begin{smallmatrix} A & B \\ C & D \end{smallmatrix}\right]$. **Related:** Fact 10.16.30.

Fact 3.17.25. Let $A \in \mathbb{F}^{n\times m}$, $B \in \mathbb{F}^{n\times m}$, $C \in \mathbb{F}^{k\times m}$, and $D \in \mathbb{F}^{k\times m}$, and assume that $n + k = 2m$. If $AD^* + BC^* = 0$, then

$$\left|\det\begin{bmatrix} A & B \\ C & D \end{bmatrix}\right|^2 = (-1)^m\det(AB^* + BA^*)\det(CD^* + DC^*).$$

Alternatively, if either $AB^* + BA^* = 0$ or $CD^* + DC^* = 0$, then

$$\left|\det\begin{bmatrix} A & B \\ C & D \end{bmatrix}\right|^2 = (-1)^{m+n}|\det(AD^* + BC^*)|^2.$$

Source: Consider $\left[\begin{smallmatrix} A & B \\ C & D \end{smallmatrix}\right]\left[\begin{smallmatrix} B^* & D^* \\ A^* & C^* \end{smallmatrix}\right]$ and $\left[\begin{smallmatrix} A & B \\ C & D \end{smallmatrix}\right]\left[\begin{smallmatrix} D^* & B^* \\ C^* & A^* \end{smallmatrix}\right]$. **Remark:** If $m^2 + nk$ is odd, then $\left[\begin{smallmatrix} A & B \\ C & D \end{smallmatrix}\right]$ is singular.

Fact 3.17.26. Let $A \in \mathbb{F}^{n\times m}$, $B \in \mathbb{F}^{n\times l}$, $C \in \mathbb{F}^{n\times m}$, and $D \in \mathbb{F}^{n\times l}$, and assume that $m + l = 2n$. If $A^*D + C^*B = 0$, then

$$\left|\det\begin{bmatrix} A & B \\ C & D \end{bmatrix}\right|^2 = (-1)^m\det(C^*A + A^*C)\det(D^*B + B^*D).$$

Alternatively, if either $D^*B + B^*D = 0$ or $C^*A + A^*C = 0$, then

$$\left|\det\begin{bmatrix} A & B \\ C & D \end{bmatrix}\right|^2 = (-1)^{n+m}|\det(A^*D + C^*B)|^2.$$

Source: Consider $\begin{bmatrix} C^* & A^* \\ D^* & B^* \end{bmatrix}\begin{bmatrix} A & B \\ C & D \end{bmatrix}$ and $\begin{bmatrix} D^* & B^* \\ C^* & A^* \end{bmatrix}\begin{bmatrix} A & B \\ C & D \end{bmatrix}$. **Remark:** If $n + m$ is odd, then $\begin{bmatrix} A & B \\ C & D \end{bmatrix}$ is singular.

Fact 3.17.27. Let $A \in \mathbb{F}^{n\times m}$ and $B \in \mathbb{F}^{n\times l}$. Then,

$$\det\begin{bmatrix} I_m & A^* \\ A & AA^* + BB^* \end{bmatrix} = \det BB^*.$$

Source: [2991, p. 49].

Fact 3.17.28. Let $A \in \mathbb{F}^{n\times m}$ and $B \in \mathbb{F}^{n\times l}$. Then,

$$\det\begin{bmatrix} A^*A & A^*B \\ B^*A & B^*B \end{bmatrix} = \begin{cases} \det(A^*A)\det[B^*B - B^*A(A^*A)^{-1}A^*B], & \operatorname{rank} A = m, \\ \det(B^*B)\det[A^*A - A^*B(B^*B)^{-1}B^*A], & \operatorname{rank} B = l, \\ 0, & n < m + l. \end{cases}$$

If, in addition, $m + l = n$, then

$$\det\begin{bmatrix} A^*A & A^*B \\ B^*A & B^*B \end{bmatrix} = \det(AA^* + BB^*).$$

Related: Fact 8.9.36.

Fact 3.17.29. Let $A \in \mathbb{F}^{n\times n}$, $B \in \mathbb{F}^{n\times m}$, $C \in \mathbb{F}^{m\times n}$, and $D \in \mathbb{F}^{m\times m}$, and define $\mathcal{A} \triangleq \begin{bmatrix} A & B \\ C & D \end{bmatrix}$. If A is singular, then the following statements are equivalent:

i) $\mathcal{A}$ is nonsingular.

ii) $\mathcal{N}(A) \cap \mathcal{N}(C) = \{0\}$, $\mathcal{N}(D) \cap \mathcal{N}(B) = \{0\}$, and $\mathcal{R}\left(\begin{bmatrix} A \\ C \end{bmatrix}\right) \cap \mathcal{R}\left(\begin{bmatrix} B \\ D \end{bmatrix}\right) = \{0\}$.

Furthermore, if A is nonsingular, then the following statements are equivalent:

iii) $\mathcal{A}$ is nonsingular.

iv) $D - CA^{-1}B$ is nonsingular.

Source: [212]. **Problem:** Assume that A is nonsingular. Does *i)* $\Longleftrightarrow$ *ii)* hold?

Fact 3.17.30. Let $A, B \in \mathbb{F}^{n\times n}$, and define $\mathcal{A} \in \mathbb{F}^{kn\times kn}$ by

$$\mathcal{A} \triangleq \begin{bmatrix} A & B & B & \cdots & B \\ B & A & B & \cdots & B \\ B & B & A & \ddots & B \\ \vdots & \vdots & \ddots & \ddots & \vdots \\ B & B & B & \cdots & A \end{bmatrix}.$$

Then,

$$\det \mathcal{A} = \det[A + (k-1)B][\det(A - B)]^{k-1}.$$

If $k = 2$, then

$$\det\begin{bmatrix} A & B \\ B & A \end{bmatrix} = \det(A + B)(A - B) = \det(A^2 - B^2 - [A, B]).$$

Source: [1203]. For $k = 2$, the result follows from Fact 6.10.31. **Related:** Fact 3.16.19, Fact 3.22.6, and Fact 6.10.31.

Fact 3.17.31. Let $A \in \mathbb{F}^{n\times n}$, $B \in \mathbb{F}^{n\times m}$, $C \in \mathbb{F}^{m\times n}$, and $D \in \mathbb{F}^{m\times m}$, and define $M \triangleq \left[\begin{smallmatrix} A & B \\ C & D \end{smallmatrix}\right] \in \mathbb{F}^{(n+m)\times(n+m)}$. Furthermore, let $\left[\begin{smallmatrix} A' & B' \\ C' & D' \end{smallmatrix}\right] \triangleq M^{\mathsf{A}}$, where $A' \in \mathbb{F}^{n\times n}$ and $D' \in \mathbb{F}^{m\times m}$. Then,

$$\det D' = (\det M)^{m-1}\det A, \quad \det A' = (\det M)^{n-1}\det D.$$

Source: [2418, p. 297]. **Related:** Fact 3.17.32.

Fact 3.17.32. Let $A \in \mathbb{F}^{n\times n}$, $B \in \mathbb{F}^{n\times m}$, $C \in \mathbb{F}^{m\times n}$, and $D \in \mathbb{F}^{m\times m}$, define $M \triangleq \left[\begin{smallmatrix} A & B \\ C & D \end{smallmatrix}\right] \in \mathbb{F}^{(n+m)\times(n+m)}$, and assume that M is nonsingular. Furthermore, let $\left[\begin{smallmatrix} A' & B' \\ C' & D' \end{smallmatrix}\right] \triangleq M^{-1}$, where $A' \in \mathbb{F}^{n\times n}$ and $D' \in \mathbb{F}^{m\times m}$. Then,

$$\det D' = \frac{\det A}{\det M}, \quad \det A' = \frac{\det D}{\det M}.$$

Hence, A is nonsingular if and only if D' is nonsingular, and D is nonsingular if and only if A' is nonsingular. **Source:** Use $M\left[\begin{smallmatrix} I & B' \\ 0 & D' \end{smallmatrix}\right] = \left[\begin{smallmatrix} A & 0 \\ C & I \end{smallmatrix}\right]$. See [2426]. **Related:** Fact 3.14.27, Fact 3.17.31, and Fact 4.13.21. This is a special case of Fact 3.17.34.

Fact 3.17.33. Let $A \in \mathbb{F}^{n\times n}$, assume that A is nonsingular, let $\mathcal{S}_1, \mathcal{S}_2 \subseteq \{1, \ldots, n\}$, and assume that $\operatorname{card}(\mathcal{S}_1) = \operatorname{card}(\mathcal{S}_2)$. Then,

$$|\det (A^{-1})_{(\mathcal{S}_1,\mathcal{S}_2)}| = \frac{|\det A_{(\mathcal{S}_2^\sim,\mathcal{S}_1^\sim)}|}{|\det A|}.$$

Source: [2780, p. 39] or use Fact 3.17.34. **Remark:** For $\operatorname{card}(\mathcal{S}_1) = \operatorname{card}(\mathcal{S}_2) = 1$, this result yields the absolute value of (3.8.24). **Related:** Fact 3.14.28.

Fact 3.17.34. Let $A \in \mathbb{F}^{n\times n}$, let $k \le n$, let $\mathcal{R}, \mathcal{C} \subseteq \{1, \ldots, n\}$, where $\operatorname{card}(\mathcal{R}) = \operatorname{card}(\mathcal{C}) = k$, and let l be the sum of the elements of $\mathcal{R} \cup \mathcal{C}$. Then,

$$\det (A^{\mathsf{A}})_{[\mathcal{R},\mathcal{C}]} = (-1)^l (\det A)^{n-k-1} \det A_{(\mathcal{C},\mathcal{R})}.$$

If, in addition, A is nonsingular, then

$$\det (A^{-1})_{[\mathcal{R},\mathcal{C}]} = (-1)^l \frac{\det A_{(\mathcal{C},\mathcal{R})}}{\det A}.$$

Source: [1448, p. 21]. **Remark:** This is *Jacobi's identity*. **Related:** Fact 3.17.32 and Fact 3.19.4.

Fact 3.17.35. Let $A \in \mathbb{F}^{n\times n}$, let $k \le n$, let $\mathcal{R}, \mathcal{C} \subseteq \{1, \ldots, n\}$, where $\operatorname{card}(\mathcal{R}) = \operatorname{card}(\mathcal{C}) = k$, and let $M \in \mathbb{F}^{n\times n}$, where, for all $i, j \in \{1, \ldots, n\}$, $M_{(i,j)} \triangleq \det A_{[i,j]}$. Then,

$$\det M_{(\mathcal{R},\mathcal{C})} = (\det A)^{k-1} \det A_{[\mathcal{R},\mathcal{C}]}.$$

In particular, $(\det A)^{n-1} = \det A^{\mathsf{A}}$ and

$$(\det A) \det A_{[\{r_1,r_2\},\{c_1,c_2\}]} = (\det A_{[r_1,c_1]}) \det A_{[r_2,c_2]} - (\det A_{[r_1,c_2]}) \det A_{[r_2,c_1]}.$$

Source: [23, 129, 1699, 2940]. **Remark:** The case $k = 2$ is the *Desnanot-Jacobi identity*, which is used for *Dodgson condensation*. See [129]. **Remark:** The second equality uses $\det 0_{0\times 0} = 1$. **Remark:** Let S be the $n \times n$ matrix whose (i, j) entry is $(-1)^{i+j}$, and let l be the sum of the $2k$ elements of the multiset $\mathcal{R} \cup \mathcal{C}$. Then, $M = (S \odot A^{\mathsf{A}})^{\mathsf{T}}$, and Fact 3.16.11 implies that

$$\det(S \odot A^{\mathsf{A}})_{(\mathcal{R},\mathcal{C})} = (-1)^l \det (A^{\mathsf{A}})_{(\mathcal{R},\mathcal{C})},$$

which can be used to show that this result implies Fact 3.17.34.

Fact 3.17.36. Let $x_1, x_2, x_3, x_4 \in \mathbb{F}^2$. Then,

$$\det([x_1\ x_2][x_3\ x_4]) - \det([x_1\ x_3][x_2\ x_4]) + \det([x_1\ x_4][x_2\ x_3]) = 0.$$

Related: Fact 3.17.37.

Fact 3.17.37. Let $n \ge 2$, let $1 \le m < n$, let $A \in \mathbb{F}^{n\times(n-m)}$, and let $x_1,\ldots,x_{2m} \in \mathbb{F}^n$. Then,

$$\sum(-1)^{\sum_{j=1}^m i_j}\det([A\ x_{i_1}\ \cdots\ x_{i_m}][A\ x_{i_{m+1}}\ \cdots\ x_{i_{2m}}]) = 0,$$

where the sum is taken over all permutations $(i_1,\ldots,i_{2m})$ of $(1,\ldots,2m)$ such that $i_1 < \cdots < i_m$ and $i_{m+1} < \cdots < i_{2m}$. In particular, let $n > 2$, $m = 2$, $A \in \mathbb{F}^{n\times(n-2)}$, and $x_1, x_2, x_3, x_4 \in \mathbb{F}^n$. Then,

$$\det([A\ x_1\ x_2][A\ x_3\ x_4]) - \det([A\ x_1\ x_3][A\ x_2\ x_4]) + \det([A\ x_1\ x_4][A\ x_2\ x_3]) = 0.$$

Source: [2940]. **Remark:** This is a *Plücker relation.* **Related:** Fact 3.17.36.

Fact 3.17.38. Let $A_1, A_2, B_1, B_2 \in \mathbb{F}^{n\times m}$, and define $\mathcal{A} \triangleq \begin{bmatrix} A_1 & A_2 \\ A_2 & A_1 \end{bmatrix}$ and $\mathcal{B} \triangleq \begin{bmatrix} B_1 & B_2 \\ B_2 & B_1 \end{bmatrix}$. Then,

$$\operatorname{rank}\begin{bmatrix} \mathcal{A} & \mathcal{B} \\ \mathcal{B} & \mathcal{A} \end{bmatrix} = \sum_{i=1}^4 \operatorname{rank} C_i,$$

where $C_1 \triangleq A_1+A_2+B_1+B_2$, $C_2 \triangleq A_1+A_2-B_1-B_2$, $C_3 \triangleq A_1-A_2+B_1-B_2$, and $C_4 \triangleq A_1-A_2-B_1+B_2$. If, in addition, $n = m$, then

$$\det\begin{bmatrix} \mathcal{A} & \mathcal{B} \\ \mathcal{B} & \mathcal{A} \end{bmatrix} = \prod_{i=1}^4 \det C_i.$$

Source: [2671]. **Related:** Fact 4.32.8.

3.18 Facts on Left and Right Inverses

Fact 3.18.1. Let $A \in \mathbb{F}^{n\times m}$. Then, the following statements are equivalent:

i) A is left invertible.

ii) $\overline{A}$ is left invertible.

iii) A^{T} is right invertible.

iv) A^* is right invertible.

Now, assume that A is left invertible, and let $A^{\mathsf{L}} \in \mathbb{F}^{m\times n}$ be a left inverse of A. Then, the following statements hold:

v) $\overline{A^{\mathsf{L}}} \in \mathbb{F}^{m\times n}$ is a left inverse of $\overline{A}$.

vi) $A^{\mathsf{LT}} \in \mathbb{F}^{n\times m}$ is a right inverse of A^{T}.

vii) $A^{\mathsf{L}*} \in \mathbb{F}^{n\times m}$ is a right inverse of A^*.

viii) A^{L} is right invertible.

ix) A is a right inverse of A^{L}.

x) $\mathcal{R}(A) = \mathcal{R}(AA^{\mathsf{L}})$.

xi) $\operatorname{rank} A = \operatorname{rank} A^{\mathsf{L}} = \operatorname{rank} AA^{\mathsf{L}} = m$.

xii) $\mathcal{N}(A^{\mathsf{L}}) = \mathcal{N}(AA^{\mathsf{L}})$.

xiii) $\operatorname{def} A^{\mathsf{L}} = \operatorname{def} AA^{\mathsf{L}} = n - m$.

xiv) $\operatorname{def} A = 0$.

Source: $\mathcal{R}(AA^{\mathsf{L}}) \subseteq \mathcal{R}(A) = \mathcal{R}(AA^{\mathsf{L}}A) \subseteq \mathcal{R}(AA^{\mathsf{L}})$. **Related:** Fact 3.18.19.

Fact 3.18.2. Let $A \in \mathbb{F}^{n\times m}$. Then, the following statements are equivalent:

i) A is right invertible.

ii) $\overline{A}$ is right invertible.

iii) A^{T} is left invertible.

iv) A^* is left invertible.

Now, assume that A is right invertible, and let A^{R} be a right inverse of A. Then, the following statements hold:

v) $\overline{A^{\mathsf{R}}} \in \mathbb{F}^{m\times n}$ is a right inverse of $\overline{A}$.

vi) $A^{\mathsf{RT}} \in \mathbb{F}^{n\times m}$ is a left inverse of A^{T}.

vii) $A^{\mathsf{R}*} \in \mathbb{F}^{n\times m}$ is a left inverse of A^*.

viii) A^{R} is left invertible.

ix) A is a left inverse of A^{R}.

x) $\mathcal{R}(A^{\mathsf{R}}) = \mathcal{R}(A^{\mathsf{R}}A)$.

xi) $\operatorname{rank} A = \operatorname{rank} A^{\mathsf{R}} = \operatorname{rank} A^{\mathsf{R}}A = n$.

xii) $\mathcal{N}(A) = \mathcal{N}(A^{\mathsf{R}}A)$.

xiii) $\operatorname{def} A = \operatorname{def} A^{\mathsf{R}}A = m - n$.

xiv) $\operatorname{def} A^{\mathsf{R}} = 0$.

Fact 3.18.3. Let $A \in \mathbb{F}^{n\times m}$, assume that A is left invertible, and let A^{L} be a left inverse of A. Then, $B \in \mathbb{F}^{m\times n}$ is a left inverse of A if and only if there exists $S \in \mathbb{F}^{m\times n}$ such that $B = A^{\mathsf{L}} + S$ and $SA = 0$. **Source:** For necessity, let $S = B - A^{\mathsf{L}}$. **Related:** Fact 8.3.14 and [2238, p. 150].

Fact 3.18.4. Let $A \in \mathbb{F}^{n\times m}$, assume that A is right invertible, and let A^{R} be a right inverse of A. Then, $B \in \mathbb{F}^{m\times n}$ is a right inverse of A if and only if there exists $S \in \mathbb{F}^{m\times n}$ such that $B = A^{\mathsf{R}} + S$ and $AS = 0$. **Source:** For necessity, let $S = B - A^{\mathsf{R}}$. **Related:** Fact 8.3.15.

Fact 3.18.5. Let $A \in \mathbb{F}^{n\times m}$. If $\operatorname{rank} A = m$, then $(A^*A)^{-1}A^*$ is a left inverse of A. If $\operatorname{rank} A = n$, then $A^*(AA^*)^{-1}$ is a right inverse of A. **Related:** Fact 4.10.25, Fact 4.10.26, and Fact 4.17.8.

Fact 3.18.6. Let $A \in \mathbb{F}^{n\times m}$, and assume that A is left invertible. Then, $A^{\mathsf{L}} \in \mathbb{F}^{m\times n}$ is a left inverse of A if and only if there exists $B \in \mathbb{F}^{m\times n}$ such that BA is nonsingular and $A^{\mathsf{L}} = (BA)^{-1}B$. **Source:** For necessity, let $B = A^{\mathsf{L}}$.

Fact 3.18.7. Let $A \in \mathbb{F}^{n\times m}$, and assume that A is right invertible. Then, $A^{\mathsf{R}} \in \mathbb{F}^{m\times n}$ is a right inverse of A if and only if there exists $B \in \mathbb{F}^{m\times n}$ such that AB is nonsingular and $A^{\mathsf{R}} = B(AB)^{-1}$. **Source:** For necessity, let $B = A^{\mathsf{R}}$.

Fact 3.18.8. Let $A \in \mathbb{F}^{n\times m}$ and $b \in \mathbb{F}^n$, and define $f\colon \mathbb{F}^m \mapsto \mathbb{F}^n$ by $f(x) = Ax + b$. Then, the following statements hold:

i) f is invertible if and only if A is nonsingular. Now, assume that these conditions hold. Then, for all $y \in \mathbb{F}^n$, $f^{\mathsf{Inv}}(y) = A^{-1}(y - b)$.

ii) f is left invertible if and only if A is left invertible. Now, assume that these conditions hold, and let A^{L} be a left inverse of A. Then, for all $y \in \mathbb{F}^n$, $f^{\mathsf{inv}}(y) = \{A^{\mathsf{L}}(y - b)\}$.

iii) f is right invertible if and only if A is right invertible. Now, assume that these conditions hold, and let A^{R} be a right inverse of A. Then, for all $y \in \mathbb{F}^n$, $A^{\mathsf{R}}(y - b) \in f^{\mathsf{inv}}(y)$.

Related: Fact 1.10.5 and Proposition 3.7.10.

Fact 3.18.9. Let $A \in \mathbb{F}^{n\times m}$ and $B \in \mathbb{F}^{m\times l}$, and assume that A and B are left invertible. Then, AB is left invertible. If, in addition, A^{L} is a left inverse of A and B^{L} is a left inverse of B, then $B^{\mathsf{L}}A^{\mathsf{L}}$ is a left inverse of AB. **Source:** Fact 1.10.6, Corollary 3.6.12, and Proposition 3.7.3. **Remark:** If A and B have full column rank, then so does AB. The example $A = \begin{bmatrix} 1 & 0 \\ 0 & 0 \end{bmatrix}$ and $B = \begin{bmatrix} 1 \\ 0 \end{bmatrix}$ shows that the converse is not true. **Related:** Fact 8.4.20 and Fact 9.4.34. Fact 3.18.13 provides necessary and sufficient conditions for AB to be left invertible.

Fact 3.18.10. Let $A \in \mathbb{F}^{n\times m}$ and $B \in \mathbb{F}^{m\times l}$, and assume that A and B are right invertible. Then, AB is right invertible. If, in addition, A^{R} is a right inverse of A and B^{R} is a right inverse of B, then $B^{\mathsf{R}}A^{\mathsf{R}}$ is a right inverse of AB. **Remark:** If A and B have full row rank, then so does AB. **Related:** Fact 8.4.21 and Fact 9.4.35. Fact 3.18.14 provides necessary and sufficient conditions for AB to be right invertible.

Fact 3.18.11. Let $A \in \mathbb{F}^{n\times m}$ and $B \in \mathbb{F}^{m\times l}$, and assume that AB is left invertible. Then, B is left invertible. Now, let $(AB)^{\mathsf{L}}$ be a left inverse of AB, and define $B^{\mathsf{L}} \triangleq (AB)^{\mathsf{L}}A$. Then, B^{L} is a left inverse of B. Finally, assume that A is right invertible, and let A^{R} be a right inverse of A. Then, $(AB)^{\mathsf{L}} = B^{\mathsf{L}}A^{\mathsf{R}}$.

Fact 3.18.12. Let $A \in \mathbb{F}^{n\times m}$ and $B \in \mathbb{F}^{m\times l}$, and assume that AB is right invertible. Then, A is right invertible. Now, let $(AB)^{\mathsf{R}}$ be a right inverse of AB, and define $B^{\mathsf{R}} \triangleq B(AB)^{\mathsf{R}}$. Then, A^{R} is a right inverse of A. Finally, assume that B is left invertible, and let B^{L} be a left inverse of B. Then, $(AB)^{\mathsf{R}} = A^{\mathsf{R}}B^{\mathsf{L}}$.

Fact 3.18.13. Let $A \in \mathbb{F}^{n\times m}$ and $B \in \mathbb{F}^{m\times l}$. Then, the following statements are equivalent:

i) AB is left invertible.

ii) B is left invertible, and $\mathcal{N}(A) \cap \mathcal{R}(B) = \{0\}$.

iii) B and $\begin{bmatrix} A \\ I_m - B(B^*B)^{-1}B^* \end{bmatrix}$ are left invertible.

iv) $\operatorname{rank} B = l$ and $\operatorname{rank} \begin{bmatrix} A \\ I_m - B(B^*B)^{-1}B^* \end{bmatrix} = m$.

Source: The equivalence of *ii)* and *iii)* follows from Fact 8.9.4. **Remark:** If A and B are left invertible, then $\left[\begin{smallmatrix} A \\ I_m-B(B^*B)^{-1}B^* \end{smallmatrix}\right]$ is left invertible. See Fact 3.18.9.

Fact 3.18.14. Let $A \in \mathbb{F}^{n\times m}$ and $B \in \mathbb{F}^{m\times l}$. Then, the following statements are equivalent:

i) AB is right invertible.

ii) A is right invertible and $\mathcal{N}(A) + \mathcal{R}(B) = \mathbb{F}^m$.

iii) A and $[I_m - A^*(AA^*)^{-1}A \;\; B]$ are right invertible.

iv) $\operatorname{rank} A = n$ and $\operatorname{rank}\, [I_m - A^*(AA^*)^{-1}A \;\; B] = m$.

Source: Fact 3.18.13 and Fact 8.9.4. **Remark:** If A and B are right invertible, then $[I_m-A^*(AA^*)^{-1}A \;\; B]$ is right invertible. See Fact 3.18.10.

Fact 3.18.15. Let $A \in \mathbb{F}^{n\times m}$, assume that A is left invertible, let $S \in \mathbb{F}^{m\times m}$, assume that S is nonsingular, and let $(AS)^{\mathsf{L}} \in \mathbb{F}^{m\times n}$ be a left inverse of AS. Then, there exists a left inverse A^{L} of A such that $(AS)^{\mathsf{L}} = S^{-1}A^{\mathsf{L}}$. **Source:** It follows from $(AS)^{\mathsf{L}}AS = I$ that $NA = I$, where $N \triangleq S(AS)^{\mathsf{L}}$ is a left inverse of A. Hence, $(AS)^{\mathsf{L}} = S^{-1}N$.

Fact 3.18.16. Let $A \in \mathbb{F}^{n\times m}$ and $B \in \mathbb{F}^{m\times l}$, assume that A and B are left invertible, and let $(AB)^{\mathsf{L}}$ be a left inverse of AB. Then, there exist a left inverse A^{L} of A and a left inverse B^{L} of B such that $(AB)^{\mathsf{L}} = B^{\mathsf{L}}A^{\mathsf{L}}$. **Source:** Let $S_1 \in \mathbb{F}^{n\times n}$ and $S_2 \in \mathbb{F}^{m\times m}$ be nonsingular matrices such that $A = S_1\left[\begin{smallmatrix} I_m \\ 0 \end{smallmatrix}\right]S_2$, let $(S_2B)^{\mathsf{L}}$ be a left inverse of S_2B, and define $B^{\mathsf{L}} \triangleq (S_2B)^{\mathsf{L}}S_2$, which is a left inverse of B. Then, there exists $C \in \mathbb{F}^{l\times(n-m)}$ such that

$$(AB)^{\mathsf{L}} = \left(S_1\begin{bmatrix} I_m \\ 0 \end{bmatrix}S_2B\right)^{\mathsf{L}} = \left(S_1\begin{bmatrix} S_2B \\ 0 \end{bmatrix}\right)^{\mathsf{L}} = \begin{bmatrix} S_2B \\ 0 \end{bmatrix}^{\mathsf{L}} S_1^{-1} = [(S_2B)^{\mathsf{L}} \;\; C]S_1^{-1}$$
$$= [B^{\mathsf{L}}S_2^{-1} \;\; C]S_1^{-1} = B^{\mathsf{L}}S_2^{-1}[I \;\; S_2BC]S_1^{-1} = B^{\mathsf{L}}A^{\mathsf{L}},$$

where $A^{\mathsf{L}} \triangleq S_2^{-1}[I \;\; S_2BC]S_1^{-1}$ is a left inverse of A. **Problem:** Extend this result to the composition $f \circ g$ of left-invertible functions f and g.

Fact 3.18.17. Let $A \in \mathbb{F}^{n\times m}$, assume that A is right invertible, let $S \in \mathbb{F}^{n\times n}$, assume that S is nonsingular, and let $(SA)^{\mathsf{R}} \in \mathbb{F}^{n\times m}$ be a right inverse of AS. Then, there exists a right inverse A^{R} of A such that $(SA)^{\mathsf{R}} = A^{\mathsf{R}}S^{-1}$. **Related:** Fact 3.18.15.

Fact 3.18.18. Let $A \in \mathbb{F}^{n\times m}$ and $B \in \mathbb{F}^{m\times l}$, assume that A and B are right invertible, and let $(AB)^{\mathsf{R}}$ be a right inverse of AB. Then, there exist a right inverse A^{R} of A and a right inverse B^{R} of B such that $(AB)^{\mathsf{R}} = B^{\mathsf{R}}A^{\mathsf{R}}$. **Related:** Fact 3.18.16.

Fact 3.18.19. Let $\mathcal{S} \subseteq \mathbb{F}^m$, let $\hat{\mathcal{S}} \subseteq \mathbb{F}^n$, and let $A \in \mathbb{F}^{n\times m}$. Then,

$$(A\mathcal{S})^\perp = A^{*\mathsf{inv}}(\mathcal{S}^\perp), \quad A\mathcal{S}^\perp = [A^{*\mathsf{inv}}(\mathcal{S})]^\perp, \quad (A^*\hat{\mathcal{S}})^\perp = A^{\mathsf{inv}}(\hat{\mathcal{S}}^\perp), \quad A^*\hat{\mathcal{S}}^\perp = [A^{\mathsf{inv}}(\hat{\mathcal{S}})]^\perp.$$

Furthermore, the following statements hold:

i) If A is left invertible and A^{L} is a left inverse of A, then $A\mathcal{S}^\perp \subseteq (A^{\mathsf{L}*}\mathcal{S})^\perp$.

ii) If A is right invertible and A^{R} is a right inverse of A, then $(A\mathcal{S})^\perp \subseteq A^{\mathsf{R}*}\mathcal{S}^\perp$.

iii) If $n = m$ and A is nonsingular, then $(A\mathcal{S})^\perp = A^{-*}\mathcal{S}^\perp$.

Source: Note that $(A\mathcal{S})^\perp = \{y \in \mathbb{F}^n\colon y^*Ax = 0 \text{ for all } x \in \mathcal{S}\} = \{y \in \mathbb{F}^n\colon (A^*y)^*x = 0 \text{ for all } x \in \mathcal{S}\} = \{y \in \mathbb{F}^n\colon A^*y \in \mathcal{S}^\perp\} = A^{*\mathsf{inv}}(\mathcal{S}^\perp)$. The third equality is given in [688, p. 12] and [2726, p. 19]. *i)* follows from the first equality, (3.7.10), and Fact 3.18.1. *ii)* follows from the second equality, (3.7.8), and Fact 3.18.2. *iii)* follows from (3.7.11).

Fact 3.18.20. For all $i \in \{1,\dots,k\}$, let $A_{ij} \in \mathbb{F}^{n_i\times m_j}$, define $A \in \mathbb{F}^{n\times m}$, where $n \triangleq \sum_{i=1}^k n_i$ and $m \triangleq \sum_{i=1}^k m_i$, by

$$A \triangleq \begin{bmatrix} A_{11} & \cdots & A_{1k} \\ \vdots & \because & \vdots \\ A_{k1} & \cdots & A_{kk} \end{bmatrix},$$

assume that A is (upper block triangular, lower block triangular), and, for all $i \in \{1,\dots,k\}$, assume that A_{ii} is (left invertible, right invertible). Then, A is (left invertible, right invertible) and has (an upper block-triangular left inverse, a lower block-triangular right inverse).

Fact 3.18.21. Let $A \in \mathbb{F}^{n\times m}$, $B \in \mathbb{F}^{k\times m}$, and $C \in \mathbb{F}^{k\times l}$, assume that A and C are left invertible, let $A^{\mathsf{L}} \in \mathbb{F}^{m\times n}$ be a left inverse of A, let $C^{\mathsf{L}} \in \mathbb{F}^{l\times k}$ be a left inverse of C, define $\mathcal{A} \triangleq \left[\begin{smallmatrix} A & 0 \\ B & C \end{smallmatrix}\right] \in \mathbb{F}^{(n+k)\times(m+l)}$, and $\mathcal{B} \triangleq \left[\begin{smallmatrix} A^{\mathsf{L}} & 0 \\ -C^{\mathsf{L}}BA^{\mathsf{L}} & C^{\mathsf{L}} \end{smallmatrix}\right]$. Then, $\mathcal{B}$ is a left inverse of $\mathcal{A}$. **Related:** This result provides an explicit left inverse for a special case of Fact 3.18.20.

3.19 Facts on the Adjugate

Fact 3.19.1. Let $A \in \mathbb{F}^{n\times n}$. Then, the following statements hold:

i) $(\overline{A})^{\mathsf{A}} = \overline{A^{\mathsf{A}}}$.

ii) $A^{\mathsf{AT}} \triangleq (A^{\mathsf{T}})^{\mathsf{A}} = (A^{\mathsf{A}})^{\mathsf{T}}$.

iii) $A^{\mathsf{A}*} \triangleq (A^*)^{\mathsf{A}} = (A^{\mathsf{A}})^*$.

iv) If $\alpha \in \mathbb{F}$, then $(\alpha A)^{\mathsf{A}} = \alpha^{n-1}A^{\mathsf{A}}$.

v) $\det A^{\mathsf{A}} = (\det A)^{n-1}$.

vi) $(A^{\mathsf{A}})^{\mathsf{A}} = (\det A)^{n-2}A$.

vii) $\det (A^{\mathsf{A}})^{\mathsf{A}} = (\det A)^{(n-1)^2}$.

viii) $\operatorname{tr} A^{\mathsf{A}} = \sum_{i=1}^n \det A_{[i,i]}$.

ix) If $k \ge 0$, then $A^{k\mathsf{A}} \triangleq (A^{\mathsf{A}})^k = (A^k)^{\mathsf{A}}$, $A^{k\mathsf{AT}} \triangleq (A^{\mathsf{AT}})^k = (A^k)^{\mathsf{AT}}$, and $A^{k\mathsf{A}*} \triangleq (A^{\mathsf{A}*})^k = (A^k)^{\mathsf{A}*}$.

x) A is nonsingular if and only if A^{A} is nonsingular.

Now, assume that A is nonsingular. Then, the following statements hold:

xi) $\operatorname{tr} A^{\mathsf{A}} = (\det A) \operatorname{tr} A^{-1}$.

xii) $A^{-\mathsf{A}} \triangleq (A^{-1})^{\mathsf{A}} = (A^{\mathsf{A}})^{-1}$.

xiii) $\operatorname{tr} A^{-\mathsf{A}} = (\operatorname{tr} A)/\det A$.

xiv) $A^{-\mathsf{AT}} \triangleq (A^{-1})^{\mathsf{AT}} = (A^{\mathsf{AT}})^{-1}$.

xv) If $k \in \mathbb{Z}$, then $A^{k\mathsf{A}} \triangleq (A^{\mathsf{A}})^k = (A^k)^{\mathsf{A}}$, $A^{k\mathsf{AT}} \triangleq (A^{\mathsf{AT}})^k = (A^k)^{\mathsf{AT}}$, and $A^{k\mathsf{A}*} \triangleq (A^{\mathsf{A}*})^k = (A^k)^{\mathsf{A}*}$.

Source: [1394]. *viii)* follows from (6.4.20). **Remark:** With $0/0 \triangleq 1$ and using $0_{1\times1}^{\mathsf{A}} = 1$ in *vi)*, all of these results hold in the case $n = 1$. **Related:** Fact 6.9.1, Fact 6.9.2, and Fact 6.10.13.

Fact 3.19.2. Let $A \in \mathbb{F}^{n\times n}$, and assume that A is singular. Then, $\mathcal{R}(A) \subseteq \mathcal{N}(A^{\mathsf{A}})$. Hence, $\operatorname{rank} A \le \operatorname{def} A^{\mathsf{A}}$ and $\operatorname{rank} A + \operatorname{rank} A^{\mathsf{A}} \le n$. Furthermore, $\mathcal{R}(A) = \mathcal{N}(A^{\mathsf{A}})$ if and only if $\operatorname{rank} A = n - 1$.

Fact 3.19.3. Let $A \in \mathbb{F}^{n\times n}$. Then, the following statements hold:

i) $\operatorname{rank} A^{\mathsf{A}} = n$ if and only if $\operatorname{rank} A = n$.

ii) $\operatorname{rank} A^{\mathsf{A}} = 1$ if and only if $\operatorname{rank} A = n - 1$.

iii) $A^{\mathsf{A}} = 0$ if and only if $\operatorname{rank} A \le n - 2$.

iv) $N_n^{\mathsf{A}} = (-1)^{n+1} N_n^{n-1}$.

Source: [2263, p. 12] and [2980, p. 18]. **Remark:** Fact 8.3.21 provides an expression for A^{A} in the case where $\operatorname{rank} A^{\mathsf{A}} = 1$. **Related:** Fact 6.10.13.

Fact 3.19.4. Let $A \in \mathbb{F}^{n\times n}$ and $k \ge 1$. Then,

$$\det\,(A^{\mathsf{A}})_{(\{1,\ldots,k\})} = (\det A)^{k-1} \det A_{(\{k+1,\ldots,n\})}.$$

Source: [2263, p. 12]. **Related:** This is a special case of Fact 3.17.34.

Fact 3.19.5. Let $A \in \mathbb{F}^{n\times n}$. Then, the following statements are equivalent:

i) $(A^{\mathsf{A}})^2 = 0$.

ii) For all $k \in \{1, \ldots, n\}$, $\operatorname{tr}\,(A^{\mathsf{A}})^k = 0$.

Source: [1898]. **Problem:** For the case $\operatorname{rank} A \le n - 1$, determine conditions on A under which A^{A} is semisimple. See Fact 6.10.13.

Fact 3.19.6. Let $A \in \mathbb{F}^{n\times n}$ and $\mathcal{A} \in \mathbb{F}^{(n-1)\times(n-1)}$, where, for all $i, j \in \{1, \ldots, n-1\}$, $\mathcal{A}_{(i,j)} \triangleq (e_i - e_{i+1})^{\mathsf{T}} A (e_j - e_{j+1})$. Then,

$$\det(A + 1_{n\times n}) - \det A = 1_{1\times n} A^{\mathsf{A}} 1_{n\times 1} = \sum_{i=1}^{n} \det(A \overset{i}{\leftarrow} 1_{n\times 1}) = \det \mathcal{A}.$$

Source: [484]. **Related:** Fact 3.17.2, Fact 3.19.8, and Fact 12.16.19.

Fact 3.19.7. Let $n \ge 2$, and let $A \in \mathbb{F}^{n\times n}$. Then,

$$[(A^{\mathsf{A}})_{[i,\cdot]} + (A_{[i,i]})^{\mathsf{A}} A_{[i,\cdot]}] e_{i,n} = 0.$$

If, in addition, A and $A_{[i,i]}$ are nonsingular, then

$$[(\det A)(A^{-1})_{[i,\cdot]} + (\det A_{[i,i]})(A_{[i,i]})^{-1} A_{[i,\cdot]}] e_{i,n} = 0.$$

Source: [40].

Fact 3.19.8. Let $A \in \mathbb{F}^{n\times n}$ and $b \in \mathbb{F}^n$. Then, for all $i \in \{1, \ldots, n\}$,

$$(A^{\mathsf{A}} b)_{(i)} = \det(A \overset{i}{\leftarrow} b).$$

Now, let $B \in \mathbb{F}^{n\times m}$. Then, for all $i \in \{1, \ldots, n\}$ and $j \in \{1, \ldots, m\}$,

$$(A^{\mathsf{A}} B)_{(i,j)} = \det[A \overset{i}{\leftarrow} \operatorname{col}_j(B)].$$

In particular,

$$(A^{\mathsf{A}})_{(i,j)} = \det(A \overset{i}{\leftarrow} e_j).$$

Remark: See Fact 12.16.19 and [1451, p. 24]. The first equality implies Cramer's rule. See Fact 3.16.12.

Fact 3.19.9. Let $A, B \in \mathbb{F}^{n\times n}$. Then, the following statements hold:

i) $(AB)^{\mathsf{A}} = B^{\mathsf{A}} A^{\mathsf{A}}$.

ii) $A(A + B)^{\mathsf{A}} B = B(A + B)^{\mathsf{A}} A$.

iii) If B is nonsingular, then $(BAB^{-1})^{\mathsf{A}} = BA^{\mathsf{A}}B^{-1}$.

iv) If $AB = BA$, then $A^{\mathsf{A}}B = BA^{\mathsf{A}}$, $AB^{\mathsf{A}} = B^{\mathsf{A}}A$, and $A^{\mathsf{A}}B^{\mathsf{A}} = B^{\mathsf{A}}A^{\mathsf{A}}$.

Source: [1451, p. 23], [2263, p. 11], and [2980, p. 18]. **Related:** Fact 4.10.9.

Fact 3.19.10. Let $A \in \mathbb{F}^{n\times n}$, $B \in \mathbb{F}^{n\times m}$, and $C \in \mathbb{F}^{m\times n}$, where $B = [B_1 \ \cdots \ B_m]$ and $C = \begin{bmatrix} C_1 \\ \vdots \\ C_m \end{bmatrix}$. Then,

$$\det\begin{bmatrix} A & B \\ C & 0 \end{bmatrix} = \begin{cases} -C_1A^{\mathsf{A}}B_1, & m = 1, \\ -C_2[A^{\mathsf{A}} - (A + B_1C_1)^{\mathsf{A}}]B_2, & m = 2, \\ -C_3[A^{\mathsf{A}} - (A + B_1C_1)^{\mathsf{A}} - (A + B_2C_2)^{\mathsf{A}} + (A + B_1C_1 + B_2C_2)^{\mathsf{A}}]B_3, & m = 3. \end{cases}$$

Furthermore, define $f_1(A) \triangleq A^{\mathsf{A}}$ and, for all $i \in \{1, \dots, m-1\}$, define $f_{i+1}(A) \triangleq f_i(A) - f_i(A + B_iC_i)$. Then,

$$\det\begin{bmatrix} A & B \\ C & 0 \end{bmatrix} = -C_m f_m(A) B_m.$$

Fact 3.19.11. Let $A \in \mathbb{F}^{n\times n}$, $B \in \mathbb{F}^{n\times m}$, $C \in \mathbb{F}^{p\times n}$, and $K \in \mathbb{F}^{m\times p}$, and assume that either $m = 1$ or $p = 1$. Then,

$$CA^{\mathsf{A}}B = C(A + BKC)^{\mathsf{A}}B.$$

Source: Note that

$$\begin{bmatrix} A + BKC & B \\ C & 0 \end{bmatrix} = \begin{bmatrix} A & B \\ C & 0 \end{bmatrix}\begin{bmatrix} I & 0 \\ KC & I \end{bmatrix}.$$

In the case $m = p = 1$, the result follows from Fact 3.19.10. Now, assume that $p = 2$ and $m = 1$, and let $C = \begin{bmatrix} C_1 \\ C_2 \end{bmatrix}$. Then,

$$C_1A^{\mathsf{A}}B = C_1(A + BK_1C_1)^{\mathsf{A}}B, \quad C_2A^{\mathsf{A}}B = C_2(A + BK_2C_2)^{\mathsf{A}}B.$$

Therefore,

$$C_1(A + BK_2C_2)^{\mathsf{A}}B = C_1(A + BK_2C_2 + BK_1C_1)^{\mathsf{A}}B = C_1(A + BKC)^{\mathsf{A}}B,$$
$$C_2(A + BK_1C_1)^{\mathsf{A}}B = C_2(A + BK_1C_1 + BK_2C_2)^{\mathsf{A}}B = C_2(A + BKC)^{\mathsf{A}}B.$$

Furthermore, since $\det\begin{bmatrix} A & [B \ B] \\ C & 0 \end{bmatrix} = 0$, it follows from Fact 3.19.10 that

$$C_2A^{\mathsf{A}}B = C_2(A+BK_1C_1)^{\mathsf{A}}B = C_2(A+BK_2C_2)^{\mathsf{A}}B, \quad C_1A^{\mathsf{A}}B = C_1(A+BK_2C_2)^{\mathsf{A}}B = C_1(A+BK_1C_1)^{\mathsf{A}}B.$$

Hence,

$$C_2(A + BK_2C_2)^{\mathsf{A}}B = C_2(A + BKC)^{\mathsf{A}}B, \quad C_1(A + BK_1C_1)^{\mathsf{A}}B = C_1(A + BKC)^{\mathsf{A}}B,$$

and thus

$$C_2A^{\mathsf{A}}B = C_2(A + BKC)^{\mathsf{A}}B, \quad C_1A^{\mathsf{A}}B = C_1(A + BKC)^{\mathsf{A}}B,$$

which implies $CA^{\mathsf{A}}B = C(A + BKC)^{\mathsf{A}}B$. **Remark:** If $m = p \geq 1$, then

$$\begin{bmatrix} A + BKC & B \\ C & 0 \end{bmatrix}^{\mathsf{A}} = \begin{bmatrix} I & 0 \\ -KC & I \end{bmatrix}\begin{bmatrix} A & B \\ C & 0 \end{bmatrix}^{\mathsf{A}},$$

and thus

$$\left(\det\begin{bmatrix} A + BKC & B \\ C & 0 \end{bmatrix}\right)^{n+p-1} = \left(\det\begin{bmatrix} A & B \\ C & 0 \end{bmatrix}\right)^{n+p-1}.$$

Remark: If $m \geq 2$ and $p \geq 2$, then the result does not hold. **Related:** Proposition 16.10.10 and Fact 16.24.15. In Proposition 16.10.11, $C = I$.

Fact 3.19.12. Let $A_1 \in \mathbb{F}^{n\times n}$, $A_2 \in \mathbb{F}^{m\times m}$, $B_1 \in \mathbb{F}^{n\times 1}$, $B_2 \in \mathbb{F}^{m\times 1}$, $C_1 \in \mathbb{F}^{1\times n}$, $C_2 \in \mathbb{F}^{1\times m}$, and define

$$\mathcal{A} \triangleq \begin{bmatrix} A_1 & B_1C_2 \\ B_2C_1 & A_2 \end{bmatrix}.$$

Then,

$$\det\mathcal{A} = (\det A_1)\det A_2 - C_1A_1^{\mathrm{A}}B_1C_2A_2^{\mathrm{A}}B_2.$$

If, in addition, A_1 and A_2 are nonsingular, then

$$\det\mathcal{A} = (\det A_1)(\det A_2)(1 - C_1A_1^{-1}B_1C_2A_2^{-1}B_2).$$

Credit: K. Aljanaideh.

3.20 Facts on the Inverse

Fact 3.20.1. Let $A, B, C, D \in \mathbb{F}^{n\times n}$ and $ABCD = I$. Then, $ABCD = DABC = CDAB = BCDA$.

Fact 3.20.2. Let $A = \left[\begin{smallmatrix} a & b \\ c & d \end{smallmatrix}\right] \in \mathbb{F}^{2\times 2}$, where $ad - bc \neq 0$. Then,

$$A^{-1} = (ad - bc)^{-1}\begin{bmatrix} d & -b \\ -c & a \end{bmatrix}.$$

Furthermore, if $A = \left[\begin{smallmatrix} a & b & c \\ d & e & f \\ g & h & i \end{smallmatrix}\right] \in \mathbb{F}^{3\times 3}$ and $\beta = a(ei - fh) - b(di - fg) + c(dh - eg) \neq 0$, then

$$A^{-1} = \beta^{-1}\begin{bmatrix} ei - fh & -(bi - ch) & bf - ce \\ -(di - fg) & ai - cg & -(af - cd) \\ dh - eg & -(ah - bg) & ae - bd \end{bmatrix}.$$

Fact 3.20.3. Let $A \in \mathbb{F}^{n\times n}$, and assume that $I + A$ is nonsingular. Then,

$$(I + A)^{-1} = I - A(I + A)^{-1} = I - (I + A)^{-1}A, \quad A(I + A)^{-1} = (I + A)^{-1}A.$$

Fact 3.20.4. Let $A \in \mathbb{F}^{n\times n}$, and assume that A and $I + A$ are nonsingular. Then,

$$(I + A)^{-1} + (I + A^{-1})^{-1} = (I + A)^{-1} + (I + A)^{-1}A = I.$$

Fact 3.20.5. Let $A, B \in \mathbb{F}^{n\times n}$, and assume that B is nonsingular. Then, $A = B[I + B^{-1}(A - B)]$.

Fact 3.20.6. Let $A \in \mathbb{F}^{n\times m}$ and $B \in \mathbb{F}^{m\times n}$. Then, $I + AB$ is nonsingular if and only if $I + BA$ is nonsingular. Now, assume that these conditions hold. Then,

$$(I_n + AB)^{-1}A = A(I_m + BA)^{-1}, \quad (I_m + BA)^{-1}B = B(I_n + AB)^{-1},$$
$$(I_n + AB)^{-1} = I_n - A(I_m + BA)^{-1}B, \quad (I_m + BA)^{-1} = I_m - B(I_n + AB)^{-1}A,$$
$$(I_n + AB)^{-1} = I_n - (I_n + AB)^{-1}AB = I_n - A(I_m + BA)^{-1}B,$$
$$(I_m + BA)^{-1} = I_m - (I_m + BA)^{-1}BA = I_m - B(I_n + AB)^{-1}A.$$

Remark: The first equality is the *push-through identity*. **Remark:** Fact 10.11.15.

Fact 3.20.7. Let $A \in \mathbb{F}^{n\times m}$. Then,

$$(I_n + AA^*)^{-1} = I_n - A(I_m + A^*A)^{-1}A^*, \quad (I_m + A^*A)^{-1} = I_m - A^*(I_n + AA^*)^{-1}A,$$
$$(I_m + A^*A)^{-1}A^*A = A^*A(I_m + A^*A)^{-1} = I_m - (I_m + A^*A)^{-1}.$$

Fact 3.20.8. Let $A, B \in \mathbb{F}^{n\times n}$, and assume that $A + B$ is nonsingular. Then,

$$A(A + B)^{-1}B = B(A + B)^{-1}A = A - A(A + B)^{-1}A = B - B(A + B)^{-1}B.$$

Now, assume that A is nonsingular. Then,

$$(A + B)^{-1} = A^{-1} - (I + A^{-1}B)^{-1}A^{-1}BA^{-1} = A^{-1} - A^{-1}(I + BA^{-1})^{-1}BA^{-1}$$

$$= A^{-1} - A^{-1}B(I + A^{-1}B)^{-1}A^{-1} = A^{-1} - A^{-1}BA^{-1}(I + BA^{-1})^{-1}.$$

Fact 3.20.9. Let $A, B \in \mathbb{F}^{n\times n}$, and assume that A and B are nonsingular. Then,

$$A + B = A(A^{-1} + B^{-1})B, \quad A^{-1} + B^{-1} = A^{-1}(A + B)B^{-1}, \quad \operatorname{rank}(A + B) = \operatorname{rank}(A^{-1} + B^{-1}).$$

In particular, $A^{-1} + B^{-1}$ is nonsingular if and only if $A + B$ is nonsingular. In this case,

$$(A + B)^{-1} = B^{-1} - B^{-1}(A^{-1} + B^{-1})^{-1}B^{-1} = A^{-1} - A^{-1}(A^{-1} + B^{-1})^{-1}A^{-1},$$

$$(A^{-1} + B^{-1})^{-1} = A(A + B)^{-1}B = B(A + B)^{-1}A = A - A(A + B)^{-1}A = B - B(A + B)^{-1}B.$$

Fact 3.20.10. Let $A, B \in \mathbb{F}^{n\times n}$, and assume that A and B are nonsingular. Then,

$$A - B = A(B^{-1} - A^{-1})B, \quad A^{-1} - B^{-1} = A^{-1}(B - A)B^{-1}, \quad \operatorname{rank}(A - B) = \operatorname{rank}(A^{-1} - B^{-1}).$$

In particular, $A^{-1} - B^{-1}$ is nonsingular if and only if $A - B$ is nonsingular. In this case,

$$(A - B)^{-1} = B^{-1}(B^{-1} - A^{-1})^{-1}B^{-1} - B^{-1} = A^{-1}(A^{-1} - B^{-1})^{-1}A^{-1} - A^{-1},$$

$$(A^{-1} - B^{-1})^{-1} = A(B - A)^{-1}B = B(B - A)^{-1}A = A + A(B - A)^{-1}A = B(B - A)^{-1}B - B.$$

Source: Fact 8.9.7 implies that $\operatorname{rank}\left[\begin{smallmatrix} A & I \\ I & B^{-1} \end{smallmatrix}\right] = n + \operatorname{rank}(A - B) = n + \operatorname{rank}(B^{-1} - A^{-1})$. See [2142].

Fact 3.20.11. Let $A, B \in \mathbb{F}^{n\times n}$, and assume A and $A + B$ are nonsingular. Then, for all $k \geq 0$,

$$\begin{aligned}(A + B)^{-1} &= \sum_{i=0}^{k} A^{-1}(-BA^{-1})^i + (-A^{-1}B)^{k+1}(A + B)^{-1} \\ &= \sum_{i=0}^{k} A^{-1}(-BA^{-1})^i + A^{-1}(-BA^{-1})^{k+1}(I + BA^{-1})^{-1}.\end{aligned}$$

Fact 3.20.12. Let $A, B \in \mathbb{F}^{n\times n}$, let $\alpha, \beta \in \mathbb{F}$, and assume that $\alpha I - A$ and $\beta I - A$ are nonsingular. Then,

$$(\alpha I - A)^{-1} - (\beta I - A)^{-1} = (\beta - \alpha)(\alpha I - A)^{-1}(\beta I - A)^{-1}.$$

Fact 3.20.13. Let $A, B \in \mathbb{F}^{n\times n}$ and $\alpha \in \mathbb{F}$, and assume that A, B, $\alpha A^{-1} + (1 - \alpha)B^{-1}$, and $\alpha B + (1 - \alpha)A$ are nonsingular. Then,

$$\alpha A + (1 - \alpha)B - [\alpha A^{-1} + (1 - \alpha)B^{-1}]^{-1} = \alpha(1 - \alpha)(A - B)[\alpha B + (1 - \alpha)A]^{-1}(A - B).$$

Related: *iv*) of Proposition 10.6.17.

Fact 3.20.14. Let $A \in \mathbb{F}^{n\times n}$, assume that A is nonsingular, let $B \in \mathbb{F}^{n\times m}$, let $C \in \mathbb{F}^{m\times n}$, and assume that $A + BC$ and $I + CA^{-1}B$ are nonsingular. Then,

$$(A + BC)^{-1}B = A^{-1}B(I + CA^{-1}B)^{-1}.$$

In particular, if $A + BB^*$ and $I + B^*A^{-1}B$ are nonsingular, then

$$(A + BB^*)^{-1}B = A^{-1}B(I + B^*A^{-1}B)^{-1}.$$

Fact 3.20.15. Let $A \in \mathbb{F}^{n\times n}$, $B \in \mathbb{F}^{n\times m}$, $C \in \mathbb{F}^{l\times n}$, and $D \in \mathbb{F}^{m\times l}$, and assume that A and $A + BDC$ are nonsingular. Then,

$$\begin{aligned}(A + BDC)^{-1} &= A^{-1} - (I_n + A^{-1}BDC)^{-1}A^{-1}BDCA^{-1} = A^{-1} - A^{-1}(I_n + BDCA^{-1})^{-1}BDCA^{-1} \\ &= A^{-1} - A^{-1}B(I_m + DCA^{-1}B)^{-1}DCA^{-1} = A^{-1} - A^{-1}BD(I_l + CA^{-1}BD)^{-1}CA^{-1} \\ &= A^{-1} - A^{-1}BDC(I_n + A^{-1}BDC)^{-1}A^{-1} = A^{-1} - A^{-1}BDCA^{-1}(I_n + BDCA^{-1})^{-1}.\end{aligned}$$

Source: [1366]. **Remark:** Since D is not necessarily either square or nonsingular, the third equality generalizes the matrix inversion lemma given by Corollary 3.9.8 in the form

$$(A + BDC)^{-1} = A^{-1} - A^{-1}B(D^{-1} + CA^{-1}B)^{-1}CA^{-1}.$$

Fact 3.20.16. Let $A, B \in \mathbb{F}^{n\times m}$, let $C, D \in \mathbb{F}^{n\times m}$, and assume that $I + DB$ is nonsingular. Then,

$$I + AC = (A + B)(I + DB)^{-1}(C + D) + (I - AD)(I + BD)^{-1}(I - BC).$$

Source: Compare blocks after inverting both sides of

$$\begin{bmatrix} I + BD & I - BC \\ I - AD & I + AC \end{bmatrix} = \begin{bmatrix} I & -B \\ I & A \end{bmatrix}\begin{bmatrix} I & I \\ -D & C \end{bmatrix}.$$

See [1851, 2184, 2933] and [2991, p. 233]. **Remark:** This result generalizes Hua's matrix equality. See Fact 10.12.52. **Related:** Fact 3.17.10, Fact 3.20.17, and Fact 3.20.19.

Fact 3.20.17. Let $A, B, C, D \in \mathbb{F}^{n\times m}$, and assume that $I + B^*D$ is nonsingular. Then,

$$I + A^*C = (A - B)^*(I + DB^*)^{-1}(C - D) + (I + A^*D)(I + B^*D)^{-1}(I + B^*C).$$

Remark: This is equivalent to Fact 3.20.16 with A and B replaced by A^* and B^* and with B and D replaced by $-B$ and $-D$.

Fact 3.20.18. Let $A, B, C, D \in \mathbb{F}^{n\times m}$, let $R, S \in \mathbb{F}^{n\times n}$, and assume $I + B^*D$ is nonsingular. Then,

$$R^*S + A^*C = (A - BR)^*(I + DB^*)^{-1}(C - DS) + (R^* + A^*D)(I + B^*D)^{-1}(S + B^*C).$$

Source: [1851]. **Remark:** This result generalizes Fact 3.20.17.

Fact 3.20.19. Let $A, B, C \in \mathbb{F}^{n\times m}$. Then,

$$I + AC^* = (A + B)(I + B^*B)^{-1}(B + C)^* + (I - AB^*)(I + BB^*)^{-1}(I - BC^*).$$

Source: Set $D = B^*$ and replace C with C^* in Fact 3.20.16.

Fact 3.20.20. Let $A \in \mathbb{F}^{n\times n}$, assume that A is either upper triangular or lower triangular, let $D = I \odot A$ denote the diagonal part of A, and assume that D is nonsingular. Then,

$$A^{-1} = \sum_{i=0}^{n}(I - D^{-1}A)^iD^{-1}.$$

Fact 3.20.21. Let $A \in \mathbb{F}^{n\times n}$, assume that A is nonsingular, and define $A_0 \triangleq I_n$. Furthermore, for all $k \in \{1,\dots,n\}$, let $\alpha_k = \frac{1}{k}\operatorname{tr} AA_{k-1}$, and, for all $k \in \{1,\dots,n-1\}$, let $A_k = AA_{k-1} - \alpha_kI$. Then,

$$A^{-1} = \frac{1}{\alpha_n}A_{n-1}.$$

Source: [2979, p. 198]. **Credit:** J. S. Frame. See [353, p. 99].

Fact 3.20.22. Let $A \in \mathbb{F}^{n\times n}$, assume that A is nonsingular, and define $(B_i)_{i=1}^{\infty}$ by

$$B_{i+1} \triangleq 2B_i - B_iAB_i,$$

where $B_0 \in \mathbb{F}^{n\times n}$ satisfies $\rho_{\max}(I - B_0A) < 1$. Then, $B_i \to A^{-1}$ as $i \to \infty$. **Source:** [300, p. 167]. **Remark:** This sequence is given by a Newton-Raphson algorithm. **Related:** Fact 8.3.38 for the case where A is either singular or nonsquare.

Fact 3.20.23. Let $A \in \mathbb{F}^{n\times n}$, and assume that A is nonsingular. Then, $A + A^{-*}$ is nonsingular. **Source:** Note that $AA^* + I$ is positive definite.

3.21 Facts on Bordered Matrices

Fact 3.21.1. Let $x, y \in \mathbb{F}^n$. Then,

$$(I + xy^\mathsf{T})^\mathsf{A} = (1 + y^\mathsf{T}x)I - xy^\mathsf{T}, \quad \det(I + xy^\mathsf{T}) = \det(I + yx^\mathsf{T}) = 1 + x^\mathsf{T}y = 1 + y^\mathsf{T}x.$$

If, in addition, $x^\mathsf{T}y \neq -1$, then

$$(I + xy^\mathsf{T})^{-1} = I - (1 + x^\mathsf{T}y)^{-1}xy^\mathsf{T}.$$

Fact 3.21.2. Let $A \in \mathbb{F}^{n\times n}$ and $x, y \in \mathbb{F}^n$. Then,

$$y^\mathsf{T}(A + xy^\mathsf{T})^\mathsf{A} = y^\mathsf{T}A^\mathsf{A}, \quad A(A + xy^\mathsf{T})^\mathsf{A} = (A - xy^\mathsf{T})A^\mathsf{A} + y^\mathsf{T}A^\mathsf{A}xI.$$

If, in addition, A is singular, then

$$\det(A + xy^\mathsf{T}) = y^\mathsf{T}A^\mathsf{A}x = y^\mathsf{T}(A + xy^\mathsf{T})^\mathsf{A}x, \quad y^\mathsf{T}A^\mathsf{A}xA^\mathsf{A} = A^\mathsf{A}xy^\mathsf{T}A^\mathsf{A}.$$

Source: Use Fact 3.21.3 and the last equality in Fact 3.21.5.

Fact 3.21.3. Let $A \in \mathbb{F}^{n\times n}$, assume that A is nonsingular, and let $x, y \in \mathbb{F}^n$. Then,

$$\det(A + xy^\mathsf{T}) = (1 + y^\mathsf{T}A^{-1}x)\det A, \quad (A + xy^\mathsf{T})^\mathsf{A} = A^\mathsf{A} + \frac{1}{\det A}(y^\mathsf{T}A^\mathsf{A}xA^\mathsf{A} - A^\mathsf{A}xy^\mathsf{T}A^\mathsf{A}).$$

Furthermore, the following statements are equivalent:

i) $\det(A + xy^\mathsf{T}) \neq 0$.

ii) $y^\mathsf{T}A^{-1}x \neq -1$.

iii) $\begin{bmatrix} A & x \\ y^\mathsf{T} & -1 \end{bmatrix}$ is nonsingular.

If these statements hold, then

$$(A + xy^\mathsf{T})^{-1} = A^{-1} - \frac{1}{1 + y^\mathsf{T}A^{-1}x}A^{-1}xy^\mathsf{T}A^{-1}.$$

Remark: The last equality is the *Sherman-Morrison-Woodbury formula*, which is a special case of the matrix inversion lemma given by Corollary 3.9.8. **Related:** Fact 3.17.2 and Fact 3.21.4. **Problem:** Obtain expressions for $(A + xy^\mathsf{T})^\mathsf{A}$ in the cases *i)* $\operatorname{rank}(A + xy^\mathsf{T}) = n - 1$, where $n - 2 \le \operatorname{rank} A \le n - 1$, and *ii)* $\operatorname{rank}(A + xy^\mathsf{T}) = n$, where A is singular.

Fact 3.21.4. Let $A \in \mathbb{F}^{n\times n}$, $x, y \in \mathbb{F}^n$, and $a \in \mathbb{F}$. Then,

$$\begin{bmatrix} A & x \\ y^\mathsf{T} & a \end{bmatrix} = \begin{cases} \begin{bmatrix} I & 0 \\ y^\mathsf{T}A^{-1} & 1 \end{bmatrix}\begin{bmatrix} A & 0 \\ 0 & a - y^\mathsf{T}A^{-1}x \end{bmatrix}\begin{bmatrix} I & A^{-1}x \\ 0 & 1 \end{bmatrix}, & \det A \neq 0, \\[2ex] \begin{bmatrix} I & 0 \\ y^\mathsf{T} & 1 \end{bmatrix}\begin{bmatrix} A & 0 \\ y^\mathsf{T} - y^\mathsf{T}A & a - y^\mathsf{T}A^{-1}x \end{bmatrix}\begin{bmatrix} I & A^{-1}x \\ 0 & 1 \end{bmatrix}, & \det A \neq 0, \\[2ex] \begin{bmatrix} I & a^{-1}x \\ 0 & 1 \end{bmatrix}\begin{bmatrix} A - a^{-1}xy^\mathsf{T} & 0 \\ 0 & a \end{bmatrix}\begin{bmatrix} I & 0 \\ a^{-1}y^\mathsf{T} & 1 \end{bmatrix}, & a \neq 0. \end{cases}$$

Remark: The second factorization follows from Fact 8.9.34 in the case where A is nonsingular. Fact 8.9.34 provides a factorization in the case where A is singular and $a = 0$, but with the additional assumption that $x \in \mathcal{R}(A)$.

Fact 3.21.5. Let $A \in \mathbb{F}^{n\times n}$, let $x, y \in \mathbb{F}^n$, and let $a \in \mathbb{F}$. Then,

$$\begin{bmatrix} A & x \\ y^{\mathsf{T}} & a \end{bmatrix}^{\mathsf{A}} = \begin{bmatrix} (a+1)A^{\mathsf{A}} - (A + xy^{\mathsf{T}})^{\mathsf{A}} & -A^{\mathsf{A}}x \\ -y^{\mathsf{T}}A^{\mathsf{A}} & \det A \end{bmatrix}.$$

Now, assume that $\left[\begin{smallmatrix} A & x \\ y^{\mathsf{T}} & a \end{smallmatrix}\right]$ is nonsingular. Then,

$$\begin{bmatrix} A & x \\ y^{\mathsf{T}} & a \end{bmatrix}^{-1} = \begin{cases} \dfrac{1}{a - y^{\mathsf{T}}A^{-1}x} \begin{bmatrix} (a - y^{\mathsf{T}}A^{-1}x)A^{-1} + A^{-1}xy^{\mathsf{T}}A^{-1} & -A^{-1}x \\ -y^{\mathsf{T}}A^{-1} & 1 \end{bmatrix}, & \det A \neq 0, \\[2ex] \dfrac{1}{a\det(A - a^{-1}xy^{\mathsf{T}})} \begin{bmatrix} (a+1)A^{\mathsf{A}} - (A + xy^{\mathsf{T}})^{\mathsf{A}} & -A^{\mathsf{A}}x \\ -y^{\mathsf{T}}A^{\mathsf{A}} & \det A \end{bmatrix}, & a \neq 0, \\[2ex] \dfrac{1}{y^{\mathsf{T}}A^{\mathsf{A}}x} \begin{bmatrix} (A + xy^{\mathsf{T}})^{\mathsf{A}} - A^{\mathsf{A}} & A^{\mathsf{A}}x \\ y^{\mathsf{T}}A^{\mathsf{A}} & -\det A \end{bmatrix}, & a = 0. \end{cases}$$

Source: Fact 3.17.2 and [991, 1394].

3.22 Facts on the Inverse of Partitioned Matrices

Fact 3.22.1. Let $A \in \mathbb{F}^{n\times n}$, $B \in \mathbb{F}^{n\times m}$, $C \in \mathbb{F}^{m\times n}$, and $D \in \mathbb{F}^{m\times m}$, and assume that A and D are nonsingular. Then,

$$\begin{bmatrix} A & B \\ 0 & D \end{bmatrix}^{-1} = \begin{bmatrix} A^{-1} & -A^{-1}BD^{-1} \\ 0 & D^{-1} \end{bmatrix}, \quad \begin{bmatrix} A & 0 \\ C & D \end{bmatrix}^{-1} = \begin{bmatrix} A^{-1} & 0 \\ -D^{-1}CA^{-1} & D^{-1} \end{bmatrix}.$$

Fact 3.22.2. Let $A \in \mathbb{F}^{n\times n}$, $B \in \mathbb{F}^{m\times m}$, and $C \in \mathbb{F}^{m\times n}$. Then,

$$\det\begin{bmatrix} 0 & A \\ B & C \end{bmatrix} = \det\begin{bmatrix} C & B \\ A & 0 \end{bmatrix} = (-1)^{nm}(\det A)(\det B).$$

If, in addition, A and B are nonsingular, then

$$\begin{bmatrix} 0 & A \\ B & C \end{bmatrix}^{-1} = \begin{bmatrix} -B^{-1}CA^{-1} & B^{-1} \\ A^{-1} & 0 \end{bmatrix}, \quad \begin{bmatrix} C & B \\ A & 0 \end{bmatrix}^{-1} = \begin{bmatrix} 0 & A^{-1} \\ B^{-1} & -B^{-1}CA^{-1} \end{bmatrix}.$$

Fact 3.22.3. Let $A \in \mathbb{F}^{n\times n}$, $B \in \mathbb{F}^{n\times m}$, and $C \in \mathbb{F}^{m\times m}$, and assume that C is nonsingular. Then,

$$\begin{bmatrix} A & B \\ B^{\mathsf{T}} & C \end{bmatrix} = \begin{bmatrix} A - BC^{-1}B^{\mathsf{T}} & B \\ 0 & C \end{bmatrix} \begin{bmatrix} I & 0 \\ C^{-1}B^{\mathsf{T}} & I \end{bmatrix}.$$

If, in addition, $A - BC^{-1}B^{\mathsf{T}}$ is nonsingular, then $\left[\begin{smallmatrix} A & B \\ B^{\mathsf{T}} & C \end{smallmatrix}\right]$ is nonsingular and

$$\begin{bmatrix} A & B \\ B^{\mathsf{T}} & C \end{bmatrix}^{-1} = \begin{bmatrix} (A - BC^{-1}B^{\mathsf{T}})^{-1} & -(A - BC^{-1}B^{\mathsf{T}})^{-1}BC^{-1} \\ -C^{-1}B^{\mathsf{T}}(A - BC^{-1}B^{\mathsf{T}})^{-1} & C^{-1}B^{\mathsf{T}}(A - BC^{-1}B^{\mathsf{T}})^{-1}BC^{-1} + C^{-1} \end{bmatrix}.$$

Fact 3.22.4. Let $A, B \in \mathbb{F}^{n\times n}$. Then,

$$\det\begin{bmatrix} I & A \\ B & I \end{bmatrix} = \det(I - AB) = \det(I - BA).$$

If $\det(I - BA) \neq 0$, then

$$\begin{bmatrix} I & A \\ B & I \end{bmatrix}^{-1} = \begin{bmatrix} I + A(I - BA)^{-1}B & -A(I - BA)^{-1} \\ -(I - BA)^{-1}B & (I - BA)^{-1} \end{bmatrix} = \begin{bmatrix} (I - AB)^{-1} & -(I - AB)^{-1}A \\ -B(I - AB)^{-1} & I + B(I - AB)^{-1}A \end{bmatrix}.$$

Fact 3.22.5. Let $A, B \in \mathbb{F}^{n\times m}$. Then,

$$\begin{bmatrix} A & B \\ B & A \end{bmatrix} = \frac{1}{2}\begin{bmatrix} I_n & I_n \\ I_n & -I_n \end{bmatrix}\begin{bmatrix} A + B & 0 \\ 0 & A - B \end{bmatrix}\begin{bmatrix} I_m & I_m \\ I_m & -I_m \end{bmatrix}.$$

Therefore,

$$\operatorname{rank}\begin{bmatrix} A & B \\ B & A \end{bmatrix} = \operatorname{rank}(A + B) + \operatorname{rank}(A - B).$$

Now, assume that $n = m$. Then,

$$\det\begin{bmatrix} A & B \\ B & A \end{bmatrix} = \det[(A + B)(A - B)] = \det(A^2 - B^2 - [A, B]).$$

Hence, $\left[\begin{smallmatrix} A & B \\ B & A \end{smallmatrix}\right]$ is nonsingular if and only if $A + B$ and $A - B$ are nonsingular. If these conditions hold, then

$$\begin{bmatrix} A & B \\ B & A \end{bmatrix}^{-1} = \frac{1}{2}\begin{bmatrix} (A + B)^{-1} + (A - B)^{-1} & (A + B)^{-1} - (A - B)^{-1} \\ (A + B)^{-1} - (A - B)^{-1} & (A + B)^{-1} + (A - B)^{-1} \end{bmatrix},$$

$$(A + B)^{-1} = \frac{1}{2}[I_n \;\; I_n]\begin{bmatrix} A & B \\ B & A \end{bmatrix}^{-1}\begin{bmatrix} I_n \\ I_n \end{bmatrix}, \quad (A - B)^{-1} = \frac{1}{2}[I_n \;\; -I_n]\begin{bmatrix} A & B \\ B & A \end{bmatrix}^{-1}\begin{bmatrix} I_n \\ -I_n \end{bmatrix}.$$

Related: Fact 8.9.32.

Fact 3.22.6. Let $A, B \in \mathbb{F}^{n\times m}$. Then,

$$\begin{bmatrix} A & B & B \\ B & A & B \\ B & B & A \end{bmatrix} = S_n\begin{bmatrix} A + 2B & 0 & 0 \\ 0 & A - B & 0 \\ 0 & 0 & A - B \end{bmatrix}S_m^{\mathsf{T}},$$

where

$$S_n \triangleq \begin{bmatrix} \frac{\sqrt{3}}{3}I_n & \frac{\sqrt{2}}{2}I_n & -\frac{\sqrt{6}}{6}I_n \\ \frac{\sqrt{3}}{3}I_n & 0 & \frac{\sqrt{6}}{3}I_n \\ \frac{\sqrt{3}}{3}I_n & -\frac{\sqrt{2}}{2}I_n & -\frac{\sqrt{6}}{6}I_n \end{bmatrix}.$$

Therefore,

$$\operatorname{rank}\begin{bmatrix} A & B & B \\ B & A & B \\ B & B & A \end{bmatrix} = \operatorname{rank}(A + 2B) + 2\operatorname{rank}(A - B).$$

Now, assume that $n = m$. Then, $S_n^{\mathsf{T}} = S_n^{-1}$, and

$$\det\begin{bmatrix} A & B & B \\ B & A & B \\ B & B & A \end{bmatrix} = \det(A + 2B)[\det(A - B)]^2.$$

Hence, $\left[\begin{smallmatrix} A & B & B \\ B & A & B \\ B & B & A \end{smallmatrix}\right]$ is nonsingular if and only if $A + 2B$ and $A - B$ are nonsingular. If these conditions hold, then

$$\begin{bmatrix} A & B & B \\ B & A & B \\ B & B & A \end{bmatrix}^{-1} = S_n\begin{bmatrix} (A + 2B)^{-1} & 0 & 0 \\ 0 & (A - B)^{-1} & 0 \\ 0 & 0 & (A - B)^{-1} \end{bmatrix}S_n^{\mathsf{T}}.$$

Source: [2667]. **Related:** Fact 3.17.30.

Fact 3.22.7. Let $A_1,\ldots,A_k \in \mathbb{F}^{n\times n}$, and assume that the $kn \times kn$ partitioned matrix below is nonsingular. Then, $A_1 + \cdots + A_k$ is nonsingular, and

$$(A_1 + \cdots + A_k)^{-1} = \frac{1}{k}[I_n \ \cdots \ I_n]\begin{bmatrix} A_1 & A_2 & \cdots & A_k \\ A_k & A_1 & \cdots & A_{k-1} \\ \vdots & \vdots & \ddots & \vdots \\ A_2 & A_3 & \cdots & A_1 \end{bmatrix}^{-1}\begin{bmatrix} I_n \\ \vdots \\ I_n \end{bmatrix}.$$

Source: [2629]. **Remark:** This matrix is *block circulant.* See Fact 8.9.33 and Fact 8.12.5.

Fact 3.22.8. Let $\mathcal{A} \triangleq \left[\begin{smallmatrix} A & B \\ 0_{m\times m} & C \end{smallmatrix}\right]$, where $A \in \mathbb{F}^{n\times m}$, $B \in \mathbb{F}^{n\times n}$, and $C \in \mathbb{F}^{m\times n}$, and assume that CA is nonsingular. Furthermore, define $P \triangleq A(CA)^{-1}C$ and $P_\perp \triangleq I - P$. Then, $\mathcal{A}$ is nonsingular if and only if $P + P_\perp BP_\perp$ is nonsingular. If these conditions hold, then

$$\mathcal{A}^{-1} = \begin{bmatrix} (CA)^{-1}(C - CBD) & -(CA)^{-1}CB(A - DBA)(CA)^{-1} \\ D & (A - DBA)(CA)^{-1} \end{bmatrix},$$

where $D \triangleq (P + P_\perp BP_\perp)^{-1}P_\perp$. **Source:** [1325].

Fact 3.22.9. Let $A \in \mathbb{F}^{n\times m}$ and $B \in \mathbb{F}^{n\times(n-m)}$, and assume that $[A \ \ B]$ is nonsingular and $A^*B = 0$. Then,

$$[A \ \ B]^{-1} = \begin{bmatrix} (A^*A)^{-1}A^* \\ (B^*B)^{-1}B^* \end{bmatrix}.$$

Related: Fact 8.9.21. **Problem:** Find an expression for $[A \ \ B]^{-1}$ without assuming $A^*B = 0$.

Fact 3.22.10. Let $A \in \mathbb{F}^{n\times m}$, $B \in \mathbb{F}^{n\times l}$, and $C \in \mathbb{F}^{m\times l}$. Then,

$$\begin{bmatrix} I_n & A & B \\ 0 & I_m & C \\ 0 & 0 & I_l \end{bmatrix}^{-1} = \begin{bmatrix} I_n & -A & AC - B \\ 0 & I_m & -C \\ 0 & 0 & I_l \end{bmatrix}.$$

Fact 3.22.11. Let $A \in \mathbb{F}^{n\times n}$, and assume that A is nonsingular. Then, $X = A^{-1}$ is the unique matrix satisfying

$$\operatorname{rank}\begin{bmatrix} A & I \\ I & X \end{bmatrix} = \operatorname{rank} A.$$

Source: [1043]. **Related:** Fact 8.3.34 and Fact 8.10.11.

3.23 Facts on Commutators

Fact 3.23.1. Let $A, B \in \mathbb{F}^{n\times n}$. Then,

$$\operatorname{tr}[A,B]^2 = 2[\operatorname{tr}(AB)^2 - \operatorname{tr}A^2B^2], \quad \operatorname{tr}[A,B]^3 = 3\operatorname{tr}(A^2B^2AB - B^2A^2BA) = -3\operatorname{tr}AB^2A[A,B].$$

If, in addition, $n = 3$, then $\operatorname{tr}[A,B]^3 = 3\det[A,B]$.

Fact 3.23.2. Let $A, B \in \mathbb{F}^{n\times n}$, assume that $[A,B] = 0$, and let $k, l \in \mathbb{N}$. Then, $[A^k, B^l] = 0$.

Fact 3.23.3. Let $A, B, C \in \mathbb{F}^{n\times n}$. Then, the following statements hold:

i) $[A,A] = 0$.

ii) $[A,B] = [-A,-B] = -[B,A]$.

iii) $[A,B+C] = [A,B] + [A,C]$.

iv) $[\alpha A,B] = [A,\alpha B] = \alpha[A,B]$ for all $\alpha \in \mathbb{F}$.

v) $[A,[B,C]] + [B,[C,A]] + [C,[A,B]] = 0$.

vi) $[A,B]^\mathsf{T} = [B^\mathsf{T}, A^\mathsf{T}] = -[A^\mathsf{T}, B^\mathsf{T}]$.

vii) $\operatorname{tr}[A,B] = 0$.

viii) $\operatorname{tr} A^k[A,B] = \operatorname{tr} B^k[A,B] = 0$ for all $k \ge 1$.

ix) $[[A,B], B-A] = [[B,A], A-B]$.

x) $[A,[A,B]] = -[A,[B,A]]$.

Remark: *v)* is the *Jacobi identity*. **Related:** Fact 3.23.4.

Fact 3.23.4. Let $A, B, C \in \mathbb{F}^{n\times n}$. Then, the following statements hold:

i) $\operatorname{ad}_A(A) = 0$.

ii) For all $\alpha \in \mathbb{F}$, $\operatorname{ad}_A(\alpha B) = \alpha \operatorname{ad}_A(\alpha B)$.

iii) $\operatorname{ad}_A(B+C) = \operatorname{ad}_A(B) + \operatorname{ad}_A(C)$.

iv) $\operatorname{ad}_A(BC) = [\operatorname{ad}_A(B)]C + B\operatorname{ad}_A(C)$.

v) $\operatorname{ad}_{[A,B]} = [\operatorname{ad}_A, \operatorname{ad}_B]$.

vi) $\operatorname{ad}_{[A,B]}(C) = [\operatorname{ad}_A, \operatorname{ad}_B](C) = \operatorname{ad}_A[\operatorname{ad}_B(C)] - \operatorname{ad}_B[\operatorname{ad}_A(C)]$.

Remark: *vi)* is equivalent to the Jacobi identity given by Fact 3.23.3.

Fact 3.23.5. Let $A \in \mathbb{F}^{n\times n}$ and, for all $X \in \mathbb{F}^{n\times n}$, define

$$\operatorname{ad}_A^k(X) \triangleq \begin{cases} \operatorname{ad}_A(X), & k=1,\\ \operatorname{ad}_A^{k-1}[\operatorname{ad}_A(X)], & k\ge 2.\end{cases}$$

Then, for all $X \in \mathbb{F}^{n\times n}$ and $k \ge 1$,

$$\operatorname{ad}_A^2(X) = [A,[A,X]] - [[A,X],A], \quad \operatorname{ad}_A^k(X) = \sum_{i=0}^{k}(-1)^{k-i}\binom{k}{i}A^iXA^{k-i}.$$

Source: For the last equality, see [2263, pp. 176, 207]. **Remark:** The proof of Proposition 15.4.12 is based on $g(e^{t\operatorname{ad}_A}e^{t\operatorname{ad}_B})$, where $g(z) \triangleq (\log z)/(z-1)$. See [2379, p. 35]. **Related:** Fact 15.15.5.

Fact 3.23.6. Let $A, B \in \mathbb{F}^{n\times n}$, and assume that $[A,B] = A$. Then, A is singular. **Source:** If A is nonsingular, then $\operatorname{tr} B = \operatorname{tr} ABA^{-1} = \operatorname{tr} B + n$, which is false.

Fact 3.23.7. Let $A, B \in \mathbb{C}^{n\times n}$, and assume that there exist nonzero $a, b \in \mathbb{C}$ such that $AB = aA + bB$. Then, $AB = BA$. **Source:** [1158, p. 72].

Fact 3.23.8. Let $A, B \in \mathbb{R}^{n\times n}$, and assume that $AB = BA$. Then, there exists $C \in \mathbb{R}^{n\times n}$ such that $A^2 + B^2 = C^2$. **Source:** [913]. **Remark:** This result does not hold for complex matrices.

Fact 3.23.9. Let $A \in \mathbb{F}^{n\times n}$. Then,

$$n \le \dim\{X \in \mathbb{F}^{n\times n}\colon AX = XA\}, \quad \dim\{[A,X]\colon X \in \mathbb{F}^{n\times n}\} \le n^2 - n.$$

Source: [860, pp. 125, 142, 493, 537]. **Remark:** The first set is the *centralizer* (also called the *commutant*) of A. See Fact 9.5.3. **Remark:** These quantities are the defect and rank, respectively, of the operator $f\colon \mathbb{F}^{n\times n} \mapsto \mathbb{F}^{n\times n}$ defined by $f(X) \triangleq AX - XA$. See Fact 9.5.3. **Related:** Fact 7.16.8 and Fact 7.16.9.

Fact 3.23.10. Let $A \in \mathbb{F}^{n\times n}$. Then, the following statements are equivalent:

i) There exists $\alpha \in \mathbb{F}$ such that $A = \alpha I$.

ii) For all $X \in \mathbb{F}^{n\times n}$, $AX = XA$.

Source: To prove sufficiency, note that $A^\mathsf{T} \oplus -A = 0$. Hence, $\{0\} = \operatorname{spec}(A^\mathsf{T} \oplus -A) = \{\lambda - \mu\colon \lambda, \mu \in \operatorname{spec}(A)\}$. Therefore, $\operatorname{spec}(A) = \{\alpha\}$, and thus $A = \alpha I + N$, where N is nilpotent. Consequently, for all $X \in \mathbb{F}^{n\times n}$, $NX = XN$. Setting $X = N^*$, it follows that N is normal. Hence, $N = 0$. **Remark:** This result determines the center subgroup of $\mathrm{GL}(n)$.

Fact 3.23.11. Define $\mathcal{S} \subseteq \mathbb{F}^{n\times n}$ by $\mathcal{S} \triangleq \{[X,Y]\colon X,Y \in \mathbb{F}^{n\times n}\}$. Then, $\mathcal{S}$ is a subspace. Furthermore,

$$\mathcal{S} = \{Z \in \mathbb{F}^{n\times n}\colon \operatorname{tr} Z = 0\}, \quad \dim \mathcal{S} = n^2 - 1.$$

Consequently, if $Z \in \mathbb{F}^{n\times n}$ and $\operatorname{tr} Z = 0$, then there exist $X, Y \in \mathbb{F}^{n\times n}$ such that $Z = [X,Y]$. **Source:** [860, pp. 125, 493]. Alternatively, note $\operatorname{tr}\colon \mathbb{F}^{n^2} \mapsto \mathbb{F}$ is onto, and use Corollary 3.6.5.

Fact 3.23.12. Let $A, B, C, D \in \mathbb{F}^{n\times n}$. Then, there exist $E, F \in \mathbb{F}^{n\times n}$ such that

$$[E,F] = [A,B] + [C,D].$$

Source: Fact 3.23.11. **Problem:** Construct E and F.

3.24 Facts on Complex Matrices

Fact 3.24.1. Let $a, b \in \mathbb{R}$. Then, $\left[\begin{smallmatrix} a & b \\ -b & a \end{smallmatrix}\right]$ is a representation of the complex number $a + \jmath b$ that preserves addition, multiplication, and inversion of complex numbers. In particular, if $a^2 + b^2 \neq 0$, then

$$\begin{bmatrix} a & b \\ -b & a \end{bmatrix}^{-1} = \begin{bmatrix} \dfrac{a}{a^2+b^2} & \dfrac{-b}{a^2+b^2} \\ \dfrac{b}{a^2+b^2} & \dfrac{a}{a^2+b^2} \end{bmatrix}, \quad (a + b\jmath)^{-1} = \frac{a}{a^2+b^2} - \jmath\frac{b}{a^2+b^2}.$$

Remark: $\left[\begin{smallmatrix} a & b \\ -b & a \end{smallmatrix}\right]$ is a *rotation-dilation*. See Fact 4.32.6.

Fact 3.24.2. Let $\nu, \omega \in \mathbb{R}$. Then,

$$\begin{aligned}
\begin{bmatrix} \nu & \omega \\ -\omega & \nu \end{bmatrix} &= \frac{1}{\sqrt{2}}\begin{bmatrix} 1 & 1 \\ \jmath & -\jmath \end{bmatrix}\begin{bmatrix} \nu+\omega\jmath & 0 \\ 0 & \nu-\omega\jmath \end{bmatrix}\frac{1}{\sqrt{2}}\begin{bmatrix} 1 & 1 \\ \jmath & -\jmath \end{bmatrix}^* \\
&= \frac{1}{\sqrt{2}}\begin{bmatrix} 1 & \jmath \\ \jmath & 1 \end{bmatrix}\begin{bmatrix} \nu+\omega\jmath & 0 \\ 0 & \nu-\omega\jmath \end{bmatrix}\frac{1}{\sqrt{2}}\begin{bmatrix} 1 & \jmath \\ \jmath & 1 \end{bmatrix}^* \\
&= \frac{1}{\sqrt{2}}\begin{bmatrix} 1 & -\jmath \\ \jmath & -1 \end{bmatrix}\begin{bmatrix} \nu+\omega\jmath & 0 \\ 0 & \nu-\omega\jmath \end{bmatrix}\frac{1}{\sqrt{2}}\begin{bmatrix} 1 & -\jmath \\ \jmath & -1 \end{bmatrix},
\end{aligned}$$

$$\begin{bmatrix} \nu & \omega \\ -\omega & \nu \end{bmatrix}^{-1} = \frac{1}{\nu^2+\omega^2}\begin{bmatrix} \nu & -\omega \\ \omega & \nu \end{bmatrix}.$$

Remark: All three transformations are unitary. The third transformation is also Hermitian. **Related:** Fact 3.24.1.

Fact 3.24.3. Let $A, B \in \mathbb{R}^{n\times m}$. Then, the real matrices $\left[\begin{smallmatrix} A & B \\ -B & A \end{smallmatrix}\right]$ and $\left[\begin{smallmatrix} A & -B \\ B & A \end{smallmatrix}\right]$ are representations of the complex matrices $A + \jmath B$ and $\overline{A + \jmath B}$, respectively. Furthermore, the real matrices $\left[\begin{smallmatrix} A^\mathsf{T} & B^\mathsf{T} \\ -B^\mathsf{T} & A^\mathsf{T} \end{smallmatrix}\right]$ and $\left[\begin{smallmatrix} A^\mathsf{T} & -B^\mathsf{T} \\ B^\mathsf{T} & A^\mathsf{T} \end{smallmatrix}\right]$ are representations of the complex matrices $(A + \jmath B)^\mathsf{T}$ and $(A + \jmath B)^*$, respectively.

Fact 3.24.4. Let $A, B \in \mathbb{R}^{n\times m}$ and $C, D \in \mathbb{R}^{m\times l}$. Then, for all $\alpha, \beta \in \mathbb{R}$, the real matrices $\left[\begin{smallmatrix} A & B \\ -B & A \end{smallmatrix}\right]$, $\left[\begin{smallmatrix} C & D \\ -D & C \end{smallmatrix}\right]$, and

$$\begin{bmatrix} \alpha A + \beta C & \alpha B + \beta D \\ -(\alpha B + \beta D) & \alpha A + \beta C \end{bmatrix} = \alpha\begin{bmatrix} A & B \\ -B & A \end{bmatrix} + \beta\begin{bmatrix} C & D \\ -D & C \end{bmatrix}$$

are representations of the complex matrices $A + \jmath B$, $C + \jmath D$, and $\alpha(A + \jmath B) + \beta(C + \jmath D)$, respectively.

Fact 3.24.5. Let $A, B \in \mathbb{R}^{n\times m}$ and $C, D \in \mathbb{R}^{m\times l}$. Then, the real matrices $\left[\begin{smallmatrix} A & B \\ -B & A \end{smallmatrix}\right]$, $\left[\begin{smallmatrix} C & D \\ -D & C \end{smallmatrix}\right]$, and

$$\begin{bmatrix} AC - BD & AD + BC \\ -(AD + BC) & AC - BD \end{bmatrix} = \begin{bmatrix} A & B \\ -B & A \end{bmatrix}\begin{bmatrix} C & D \\ -D & C \end{bmatrix}$$

are representations of the complex matrices $A + \jmath B$, $C + \jmath D$, and $(A + \jmath B)(C + \jmath D)$, respectively.

Fact 3.24.6. Let $A, B \in \mathbb{C}^{n\times m}$. Then,

$$\begin{bmatrix} A & B \\ -B & A \end{bmatrix} = \frac{1}{2}\begin{bmatrix} I & I \\ \jmath I & -\jmath I \end{bmatrix}\begin{bmatrix} A+\jmath B & 0 \\ 0 & A-\jmath B \end{bmatrix}\begin{bmatrix} I & -\jmath I \\ I & \jmath I \end{bmatrix}$$

$$= \frac{1}{2}\begin{bmatrix} I & \jmath I \\ -\jmath I & -I \end{bmatrix}\begin{bmatrix} A-\jmath B & 0 \\ 0 & A+\jmath B \end{bmatrix}\begin{bmatrix} I & \jmath I \\ -\jmath I & -I \end{bmatrix}$$

$$= \begin{bmatrix} I & 0 \\ \jmath I & I \end{bmatrix}\begin{bmatrix} A+\jmath B & B \\ 0 & A-\jmath B \end{bmatrix}\begin{bmatrix} I & 0 \\ -\jmath I & I \end{bmatrix}.$$

Consequently,

$$\begin{bmatrix} A+\jmath B & 0 \\ 0 & A-\jmath B \end{bmatrix} = \frac{1}{2}\begin{bmatrix} I & -\jmath I \\ I & \jmath I \end{bmatrix}\begin{bmatrix} A & B \\ -B & A \end{bmatrix}\begin{bmatrix} I & I \\ \jmath I & -\jmath I \end{bmatrix},$$

and thus

$$A+\jmath B = \frac{1}{2}[I \;\; -\jmath I]\begin{bmatrix} A & B \\ -B & A \end{bmatrix}\begin{bmatrix} I \\ \jmath I \end{bmatrix}, \quad A-\jmath B = \frac{1}{2}[I \;\; \jmath I]\begin{bmatrix} A & B \\ -B & A \end{bmatrix}\begin{bmatrix} I \\ -\jmath I \end{bmatrix}.$$

Furthermore,

$$\operatorname{rank}(A+\jmath B) + \operatorname{rank}(A-\jmath B) = \operatorname{rank}\begin{bmatrix} A & B \\ -B & A \end{bmatrix}.$$

Finally, if A and B are real, then

$$\operatorname{rank}(A+\jmath B) = \operatorname{rank}(A-\jmath B) = \frac{1}{2}\operatorname{rank}\begin{bmatrix} A & B \\ -B & A \end{bmatrix}.$$

Source: Fact 6.10.32 and [181, 2628].

Fact 3.24.7. Let $A, B \in \mathbb{C}^{n\times n}$. Then, the following statements hold:

i) $\det\begin{bmatrix} A & B \\ -B & A \end{bmatrix} = \det(A+\jmath B)\det(A-\jmath B) = \det[A^2+B^2 \pm \jmath(AB-BA)]$.

ii) If $AB = BA$, then $\det\begin{bmatrix} A & B \\ -B & A \end{bmatrix} = \det(A^2+B^2)$.

iii) $\left[\begin{smallmatrix} A & B \\ -B & A \end{smallmatrix}\right]$ is nonsingular if and only if $A+\jmath B$ and $A-\jmath B$ are nonsingular. If these conditions hold, then

$$(A+\jmath B)^{-1} = \frac{1}{2}[I \;\; -\jmath I]\begin{bmatrix} A & B \\ -B & A \end{bmatrix}^{-1}\begin{bmatrix} I \\ \jmath I \end{bmatrix}, \quad (A-\jmath B)^{-1} = \frac{1}{2}[I \;\; \jmath I]\begin{bmatrix} A & B \\ -B & A \end{bmatrix}^{-1}\begin{bmatrix} I \\ -\jmath I \end{bmatrix}.$$

iv) Assume that A is nonsingular. Then,

$$\det\begin{bmatrix} A & B \\ -B & A \end{bmatrix} = \det(A^2 + ABA^{-1}B).$$

Furthermore, $A+\jmath B$ and $A-\jmath B$ are nonsingular if and only if $A+BA^{-1}B$ is nonsingular. If these conditions hold, then

$$\begin{bmatrix} A & B \\ -B & A \end{bmatrix}^{-1} = \begin{bmatrix} (A+BA^{-1}B)^{-1} & -A^{-1}B(A+BA^{-1}B)^{-1} \\ A^{-1}B(A+BA^{-1}B)^{-1} & (A+BA^{-1}B)^{-1} \end{bmatrix},$$

$$(A+\jmath B)^{-1} = (A+BA^{-1}B)^{-1} - \jmath A^{-1}B(A+BA^{-1}B)^{-1},$$

$$(A-\jmath B)^{-1} = (A+BA^{-1}B)^{-1} + \jmath A^{-1}B(A+BA^{-1}B)^{-1}.$$

v) Assume that B is nonsingular. Then,

$$\det\begin{bmatrix} A & B \\ -B & A \end{bmatrix} = \det(B^2 + BAB^{-1}A).$$

Furthermore, $A + \jmath B$ and $A - \jmath B$ are nonsingular if and only if $B + AB^{-1}A$ is nonsingular. If these conditions hold, then

$$\begin{bmatrix} A & B \\ -B & A \end{bmatrix}^{-1} = \begin{bmatrix} B^{-1}A(B + AB^{-1}A)^{-1} & -(B + AB^{-1}A)^{-1} \\ (B + AB^{-1}A)^{-1} & B^{-1}A(B + AB^{-1}A)^{-1} \end{bmatrix},$$

$$(A + \jmath B)^{-1} = B^{-1}A(B + AB^{-1}A)^{-1} - \jmath(B + AB^{-1}A)^{-1},$$
$$(A - \jmath B)^{-1} = B^{-1}A(B + AB^{-1}A)^{-1} + \jmath(B + AB^{-1}A)^{-1}.$$

vi) Assume that A and B are nonsingular. Then,

$$\det\begin{bmatrix} A & B \\ -B & A \end{bmatrix} = \det(A^2 + ABA^{-1}B) = \det(B^2 + BAB^{-1}A).$$

Furthermore, the following statements are equivalent:

a) $A + \jmath B$ and $A - \jmath B$ are nonsingular.

b) $A + BA^{-1}B$ is nonsingular.

c) $B + AB^{-1}A$ is nonsingular.

Now, assume that *a)*–*c)* hold. Then,

$$\begin{bmatrix} A & B \\ -B & A \end{bmatrix}^{-1} = \begin{bmatrix} (A + BA^{-1}B)^{-1} & -(B + AB^{-1}A)^{-1} \\ (B + AB^{-1}A)^{-1} & (A + BA^{-1}B)^{-1} \end{bmatrix},$$

$$(A + \jmath B)^{-1} = (A + BA^{-1}B)^{-1} - \jmath(B + AB^{-1}A)^{-1},$$
$$(A - \jmath B)^{-1} = (A + BA^{-1}B)^{-1} + \jmath(B + AB^{-1}A)^{-1}.$$

vii) If A and B are real, then $\det\begin{bmatrix} A & B \\ -B & A \end{bmatrix} = |\det(A + \jmath B)|^2 \geq 0.$

viii) Assume that A and B are real. Then, the following conditions are equivalent:

a) $\left[\begin{smallmatrix} A & B \\ -B & A \end{smallmatrix}\right]$ is nonsingular.

b) $A + \jmath B$ is nonsingular.

c) $A - \jmath B$ is nonsingular.

Source: If A is nonsingular, then

$$\begin{bmatrix} A & B \\ -B & A \end{bmatrix} = \begin{bmatrix} A & 0 \\ 0 & A \end{bmatrix}\begin{bmatrix} I & A^{-1}B \\ -A^{-1}B & I \end{bmatrix}, \quad \det\begin{bmatrix} I & A^{-1}B \\ -A^{-1}B & I \end{bmatrix} = \det[I + (A^{-1}B)^2].$$

Remark: See [2629]. **Related:** Fact 4.10.30, Fact 4.13.12, Fact 8.9.32, and Fact 10.16.6.

Fact 3.24.8. Let $A, B \in \mathbb{F}^{n\times n}$. Then,

$$\det\begin{bmatrix} A & B \\ -\overline{B} & \overline{A} \end{bmatrix} \geq 0.$$

If, in addition, A is nonsingular, then

$$\det\begin{bmatrix} A & B \\ -\overline{B} & \overline{A} \end{bmatrix} = |\det A|^2\det(I + \overline{A^{-1}B}A^{-1}B).$$

Source: [2982] and [2991, p. 106]. **Remark:** Fact 10.15.15 implies that $\det(I + \overline{A^{-1}BA^{-1}B}) \ge 0$. **Related:** Fact 6.10.33.

Fact 3.24.9. Let $A, B \in \mathbb{C}^{n\times n}$. Then,

$$(A + \jmath AB)(A - \jmath BA) = (A - \jmath AB)(A + \jmath BA) = A^2 + AB^2A,$$
$$(A + \jmath BA)(A - \jmath AB) = (A - \jmath BA)(A + \jmath AB) = A^2 + BA^2B - \jmath(BA^2 - A^2B).$$

Source: [1845].

Fact 3.24.10. Let $A, B \in \mathbb{R}^{n\times n}$, and define $C \in \mathbb{R}^{2n\times 2n}$ by $C \triangleq \begin{bmatrix} C_{11} & C_{12} & \cdots \\ C_{21} & \cdots & \\ \vdots & & \end{bmatrix}$, where, for all $i, j \in \{1, \dots, n\}$, $C_{ij} \triangleq \begin{bmatrix} A_{(i,j)} & B_{(i,j)} \\ -B_{(i,j)} & A_{(i,j)} \end{bmatrix} \in \mathbb{R}^{2\times 2}$. Then, $\det C = |\det(A + \jmath B)|^2$. **Source:** [568] and note that

$$C = A \otimes I_2 + B \otimes J_2 = P_{2,n}(I_2 \otimes A + J_2 \otimes B)P_{2,n} = P_{2,n}\begin{bmatrix} A & B \\ -B & A \end{bmatrix}P_{2,n}.$$

3.25 Facts on Majorization

Fact 3.25.1. Let $x, y \in \mathbb{R}^n$. Then, the following statements are equivalent:

i) $x \overset{\mathrm{w}}{\prec} y$.

ii) There exists $z \in \mathbb{R}^n$ such that $x \le\le z$ and $z \overset{\mathrm{s}}{\prec} y$.

iii) There exists $z \in \mathbb{R}^n$ such that $x \overset{\mathrm{s}}{\prec} z$ and $z \le\le y$.

Source: [2991, p. 328].

Fact 3.25.2. Let $x, y \in \mathbb{R}^n$. Then, the following statements hold:

i) $x \overset{\mathrm{w}}{\prec} |x|$.

ii) $|x + y| \overset{\mathrm{w}}{\prec} |x|^{\downarrow} + |y|^{\downarrow}$.

iii) $x \overset{\mathrm{s}}{\prec} y$ if and only if $-x \overset{\mathrm{s}}{\prec} -y$.

iv) $x \overset{\mathrm{s}}{\prec} y$ if and only if $x \overset{\mathrm{w}}{\prec} y$ and $-x \overset{\mathrm{w}}{\prec} -y$.

v) $x - y \overset{\mathrm{s}}{\prec} x^{\downarrow} - y^{\uparrow}$.

vi) $x^{\downarrow} + y^{\uparrow} \overset{\mathrm{s}}{\prec} x + y \overset{\mathrm{s}}{\prec} x^{\downarrow} + y^{\downarrow}$.

vii) $x^{\downarrow\mathsf{T}}y^{\uparrow} \le x^{\mathsf{T}}y \le x^{\downarrow\mathsf{T}}y^{\downarrow}$.

viii) If $x \overset{\mathrm{w}}{\prec} y$, then $(x^{\downarrow})_{(1)} \le (y^{\downarrow})_{(1)}$.

ix) If $x \overset{\mathrm{s}}{\prec} y$, then $(y^{\downarrow})_{(n)} \le (x^{\downarrow})_{(n)} \le (x^{\downarrow})_{(1)} \le (y^{\downarrow})_{(1)}$.

x) $(x^{\downarrow} + y^{\downarrow})^{\downarrow} = x^{\downarrow} + y^{\downarrow}$.

xi) If x and y are nonnegative, then $(x^{\downarrow} \odot y^{\downarrow})^{\downarrow} - x^{\downarrow} \odot y^{\downarrow}$.

xii) $x \overset{\mathrm{w}}{\prec} y$ if and only if, for all $z \in [0, \infty)^n$, $x^{\downarrow\mathsf{T}}z^{\downarrow} \le y^{\downarrow\mathsf{T}}z^{\downarrow}$.

xiii) $x \overset{\mathrm{s}}{\prec} y$ if and only if, for all $z \in \mathbb{R}^n$, $x^{\downarrow\mathsf{T}}z^{\downarrow} \le y^{\downarrow\mathsf{T}}z^{\downarrow}$.

xiv) If $x \overset{\mathrm{s}}{\prec} y$ and $z \in \mathbb{R}^n$, then $z^{\downarrow\mathsf{T}}y^{\uparrow} \le z^{\downarrow\mathsf{T}}x \le z^{\downarrow\mathsf{T}}x^{\downarrow} \le z^{\downarrow\mathsf{T}}y^{\downarrow}$.

xv) If $u, v \in \mathbb{R}^m$, $x \overset{\mathrm{w}}{\prec} y$, and $u \overset{\mathrm{w}}{\prec} v$, then $\left[\begin{smallmatrix} x \\ u \end{smallmatrix}\right] \overset{\mathrm{w}}{\prec} \left[\begin{smallmatrix} y \\ v \end{smallmatrix}\right]$.

xvi) If $u, v \in \mathbb{R}^m$, $x \overset{\mathrm{s}}{\prec} y$, and $u \overset{\mathrm{s}}{\prec} v$, then $\left[\begin{smallmatrix} x \\ u \end{smallmatrix}\right] \overset{\mathrm{s}}{\prec} \left[\begin{smallmatrix} y \\ v \end{smallmatrix}\right]$.

xvii) If $z \in \mathbb{R}^m$ and $x \overset{\mathrm{s}}{\prec} y$, then $\left[\begin{smallmatrix} x \\ z \end{smallmatrix}\right] \overset{\mathrm{s}}{\prec} \left[\begin{smallmatrix} y \\ z \end{smallmatrix}\right]$.

xviii) If $z \in \mathbb{R}^m$ and $\left[\begin{smallmatrix} x \\ z \end{smallmatrix}\right] \overset{\mathrm{s}}{\prec} \left[\begin{smallmatrix} y \\ z \end{smallmatrix}\right]$, then $x \overset{\mathrm{w}}{\prec} y$.

xix) If $u, v \in \mathbb{R}^n$, $x \overset{\rm w}{\prec} y$, and $u \overset{\rm w}{\prec} v$, then $x + u \overset{\rm w}{\prec} y^{\downarrow} + v^{\downarrow}$.

xx) If $u, v \in \mathbb{R}^n$, $x \overset{\rm s}{\prec} y$, and $u \overset{\rm s}{\prec} v$, then $x + u \overset{\rm s}{\prec} y^{\downarrow} + v^{\downarrow}$.

xxi) If $z \in \mathbb{R}^n$ and $x \overset{\rm w}{\prec} \frac{1}{2}(y+z)$, then $\left[\begin{smallmatrix} x \\ x \end{smallmatrix}\right] \overset{\rm w}{\prec} \left[\begin{smallmatrix} y \\ z \end{smallmatrix}\right]$.

xxii) If $z \in \mathbb{R}^n$ and $x \overset{\rm s}{\prec} \frac{1}{2}(y+z)$, then $\left[\begin{smallmatrix} x \\ x \end{smallmatrix}\right] \overset{\rm s}{\prec} \left[\begin{smallmatrix} y \\ z \end{smallmatrix}\right]$.

xxiii) If $z \in \mathbb{R}^n$, $x, y, z \geq\geq 0$, and $x \overset{\rm w}{\prec} (y \odot z)^{\odot 1/2}$, then $\left[\begin{smallmatrix} x \\ x \end{smallmatrix}\right] \overset{\rm w}{\prec} \left[\begin{smallmatrix} y \\ z \end{smallmatrix}\right]$.

xxiv) If $z \in \mathbb{R}^n$, $x, y, z \geq\geq 0$, and $x \overset{\rm w}{\prec} y \odot z$, then $\left[\begin{smallmatrix} x \\ x \end{smallmatrix}\right] \overset{\rm w}{\prec} \left[\begin{smallmatrix} y\odot y \\ z\odot z \end{smallmatrix}\right]$.

xxv) If $x \overset{\rm w}{\prec} y$, $z \in \mathbb{R}^n$, and $z \geq\geq 0$, then $x \odot z \overset{\rm w}{\prec} y^{\downarrow} \odot z^{\downarrow}$.

xxvi) If $u, v \in \mathbb{R}^n$, $u, v \geq\geq 0$, $x \overset{\rm w}{\prec} u$, and $y \overset{\rm w}{\prec} v$, then $x \odot y \overset{\rm w}{\prec} u^{\downarrow} \odot v^{\downarrow}$.

xxvii) If $z \in \mathbb{R}^n$ and $x \overset{\rm s}{\prec} y^{\downarrow} - z^{\uparrow}$, then $x^{\downarrow} + z^{\downarrow} \overset{\rm s}{\prec} y^{\downarrow} + z^{\downarrow} - z^{\uparrow}$.

Source: [1969, p. 95], [1971, p. 136], [2750], and [2991, pp. 327–333].

Fact 3.25.3. Let $x, y \in \mathbb{R}^n$. Then, the following statement holds:

i) $x^{\downarrow} + y^{\uparrow} \overset{\rm s}{\prec} x + y \overset{\rm s}{\prec} x^{\downarrow} + y^{\downarrow}$.

Now, assume that x and y are nonnegative. Then, the following statements hold:

ii) If $x \overset{\rm slog}{\prec} y$ and $x \overset{\rm s}{\prec} y$, then $x^{\downarrow} = y^{\downarrow}$.

iii) If $x \overset{\rm wlog}{\prec} y$, then $x \overset{\rm w}{\prec} y$.

iv) $x^{\downarrow} \odot y^{\uparrow} \overset{\rm w}{\prec} x \odot y \overset{\rm w}{\prec} x^{\downarrow} \odot y^{\downarrow}$.

Source: [1969, p. 117], [1971, p. 168, 224], [2750] and [2991, pp. 345, 348]. **Related:** Fact 2.12.8.

Fact 3.25.4. Let $x \in \mathbb{R}^n$ be nonnegative, and assume that $\sum_{i=1}^n x_{(i)} = 1$. Then, $\frac{1}{n}1_{n\times 1} \overset{\rm s}{\prec} x \overset{\rm s}{\prec} e_{1,n}$.
Source: [1969, p. 7] and [1971, p. 9].

Fact 3.25.5. Let $a < b$, let $f\colon (a,b)^n \mapsto \mathbb{R}$, and assume that f is C^1. Then, f is Schur-convex if and only if f is symmetric and, for all $x \in (a,b)^n$,

$$(x_{(1)} - x_{(2)})\left(\frac{\partial f(x)}{\partial x_{(1)}} - \frac{\partial f(x)}{\partial x_{(2)}}\right) \geq 0.$$

Furthermore, the following functions are Schur-convex:

i) $f\colon \mathbb{R}^n \mapsto \mathbb{R}$, where $f(x) = \max_{i\in\{1,\ldots,n\}} |x_{(i)}|$.

ii) $f\colon \mathbb{R}^n \mapsto \mathbb{R}$, where $f(x) = \sum_{i=1}^n |x_{(i)}|^p$ and $p \geq 1$.

iii) $f\colon \mathbb{R}^n \mapsto \mathbb{R}$, where $f(x) = \left(\sum_{i=1}^n |x_{(i)}|^p\right)^{1/p}$ and $p \geq 1$.

iv) $f\colon \mathbb{R}^n \mapsto \mathbb{R}$, where $f(x) = \sum_{i=1}^k [(|x|^{\downarrow})_{(i)}]^p$, $k \leq n$, and $p \geq 1$.

v) $f\colon \mathbb{R}^n \mapsto \mathbb{R}$, where $f(x) = \left(\sum_{i=1}^k [(|x|^{\downarrow})_{(i)}]^p\right)^{1/p}$, $k \leq n$, and $p \geq 1$.

vi) $f\colon [0,\infty)^n \mapsto \mathbb{R}$, where $f(x) = \sum_{i=1}^n \frac{1}{x_{(i)}}$.

vii) $f\colon [0,\infty)^n \mapsto \mathbb{R}$, where $f(x) = \prod_{i=1}^n x_i - \prod_{i=1}^n (x_i+1)$.

Source: [1835], [1969, p. 57], [1971, p. 84], and [2991, pp. 347, 376]. **Remark:** f is symmetric means that $f(Ax) = f(x)$ for all $x \in (a,b)^n$ and every permutation matrix $A \in \mathbb{R}^{n\times n}$. **Remark:** See [1557].

Fact 3.25.6. Let $x, y \in \mathbb{R}^n$ be nonnegative, assume that $x = x^{\downarrow}$, $y = y^{\downarrow}$, and $x \overset{\rm s}{\prec} y$, and let

$p_1, \dots, p_n$ be nonnegative numbers. Then,

$$\sum \prod_{j=1}^{n} p_{i_j}^{x_{(j)}} \le \frac{1}{n!} \sum \prod_{j=1}^{n} p_{i_j}^{y_{(j)}},$$

where both sums are taken over all $n!$ permutations $\{i_1, \dots, i_n\}$ of $\{1, \dots, n\}$. **Source:** [1146, p. 99], [1569], [1969, p. 87], [1971, p. 125], [1938, pp. 44–47], and [2264, pp. 81–83]. **Remark:** This is *Muirhead's theorem*, which is based on a function that is Schur-convex. **Remark:** Let $x = [5\ \ 0]$ and $y = [3\ \ 2]$. Then, x strongly majorizes y. Therefore, for all nonnegative a, b, it follows that $a^3b^2 + b^3a^2 \le a^5 + b^5$. As another example, let $x = [2\ \ 2\ \ 0]$ and $y = [2\ \ 1\ \ 1]$. Then, x strongly majorizes y. Therefore, for all nonnegative a, b, c, it follows that $a^2bc + b^2ca + c^2ab \le a^2b^2 + b^2c^2 + c^2a^2$. See [1938, p. 44]. **Remark:** Let $x = ne_{1,n}^{\mathsf{T}}$ and $y = 1_{1\times n}$. Then, $y \overset{\mathrm{s}}{\prec} x$. Therefore, for all nonnegative $z_1, \dots, z_n$, it follows that $\prod_{i=1}^{n} z_i \le \frac{1}{n} \sum_{i=1}^{n} z_i^n$, which is the arithmetic-mean–geometric-mean inequality. See Fact 2.11.114.

Fact 3.25.7. Let $x, y \in \mathbb{R}^n$ be positive, and assume that $x \overset{\mathrm{s}}{\prec} y$ and $\sum_{i=1}^{n} x_{(i)} = 1$. Then,

$$\sum_{i=1}^{n} y_i \log \tfrac{1}{y_{(i)}} \le \sum_{i=1}^{n} x_i \log \tfrac{1}{x_{(i)}} \le \log n.$$

Source: [1146, p. 102], [1969, pp. 71, 405], and [1971, pp. 101, 556]. **Remark:** For $x_{(1)}, x_{(2)} > 0$, note that $(x_{(1)} - x_{(2)}) \log(x_{(1)}/x_{(2)}) \ge 0$. Hence, Fact 3.25.5 implies that the entropy function is Schur-concave. **Related:** Entropy bounds are given in Fact 2.15.27.

Fact 3.25.8. Let $x, y \in \mathbb{R}^n$, assume that $x \overset{\mathrm{s}}{\prec} y$, let $f\colon \mathbb{R} \mapsto \mathbb{R}$, and assume that f is convex. Then,

$$[f(x_{(1)})\ \cdots\ f(x_{(n)})] \overset{\mathrm{w}}{\prec} [f(y_{(1)})\ \cdots\ f(y_{(n)})].$$

Source: [449, p. 42], [1450, p. 173], [1969, p. 116], [1971, p. 167], and [2991, p. 342].

Fact 3.25.9. Let $x, y \in \mathbb{R}^n$ be nonnegative, assume that $x \overset{\mathrm{slog}}{\prec} y$, let $f\colon [0, \infty) \mapsto \mathbb{R}$, and assume that $g\colon \mathbb{R} \mapsto \mathbb{R}$ defined by $g(z) \triangleq f(e^z)$ is convex. Then,

$$[f(x_{(1)})\ \cdots\ f(x_{(n)})] \overset{\mathrm{w}}{\prec} [f(y_{(1)})\ \cdots\ f(y_{(n)})].$$

Source: Fact 3.25.8.

Fact 3.25.10. Let $x, y \in \mathbb{R}^n$, assume that $x \overset{\mathrm{w}}{\prec} y$, let $f\colon \mathbb{R} \mapsto \mathbb{R}$, and assume that f is convex and nondecreasing. Then,

$$[f(x_{(1)})\ \cdots\ f(x_{(n)})] \overset{\mathrm{w}}{\prec} [f(y_{(1)})\ \cdots\ f(y_{(n)})].$$

Source: [449, p. 42], [1450, p. 173], [1969, p. 116], [1971, p. 167], and [2991, p. 342]. **Related:** Fact 3.25.12.

Fact 3.25.11. Let $x, y \in \mathbb{R}^n$, assume that $x \overset{\mathrm{w}}{\prec} y$, and define $e^x \triangleq [e^{x_{(1)}}\ \cdots\ e^{x_{(n)}}]^{\mathsf{T}}$. Then, $e^x \overset{\mathrm{slog}}{\prec} e^y$. **Source:** [2750].

Fact 3.25.12. Let $x, y \in \mathbb{R}^n$ be nonnegative, assume that $x \overset{\mathrm{s}}{\prec} y$, and let $r \ge 1$. Then,

$$[x_{(1)}^r\ \cdots\ x_{(n)}^r] \overset{\mathrm{w}}{\prec} [y_{(1)}^r\ \cdots\ y_{(n)}^r].$$

Source: Fact 3.25.10. **Remark:** The majorization can be written as $x^{\odot r} \overset{\mathrm{w}}{\prec} y^{\odot r}$.

Fact 3.25.13. Let $x, y \in \mathbb{R}^n$ be positive, assume that $x \overset{\mathrm{wlog}}{\prec} y$, let $f\colon [0, \infty) \mapsto \mathbb{R}$, and assume that $g\colon \mathbb{R} \mapsto \mathbb{R}$ defined by $g(z) \triangleq f(e^z)$ is convex and nondecreasing. Then,

$$[f(x_{(1)})\ \cdots\ f(x_{(n)})] \overset{\mathrm{w}}{\prec} [f(y_{(1)})\ \cdots\ f(y_{(n)})].$$

Source: Fact 3.25.10 and [449, p. 42].

Fact 3.25.14. Let $x, y \in \mathbb{R}^n$ be nonnegative, and assume that $x \overset{\mathrm{s}}{\prec} y$. Then, $\prod_{i=1}^n y_{(i)} \le \prod_{i=1}^n x_{(i)}$. **Source:** [2991, p. 347]. **Related:** Fact 10.16.1.

Fact 3.25.15. Let $x, y \in \mathbb{R}^n$ be nonnegative, and assume that $x \overset{\mathrm{wlog}}{\prec} y$. Then, $x \overset{\mathrm{w}}{\prec} y$. **Source:** Use Fact 3.25.13 with $f(t) = t$ and the fact that $g(z) = e^z$ is convex and increasing. See [2977, p. 19] and [2991, p. 345]. **Example:** Let $x = [4 \;\; 3]$ and $y = [7 \;\; 2]$. Then, $4 \le 7$ and $4 \cdot 3 \le 7 \cdot 2$. Hence, y weakly log majorizes x. Furthermore, $4 \le 7$ and $4+3 \le 7+2$. Thus, y weakly majorizes x. **Remark:** The converse is false. Let $x = [2 \;\; 1]$ and $y = [3 \;\; \frac{1}{2}]$. Then, $2 \le 3$ and $2 + 1 \le 3 + \frac{1}{2}$. Thus, y weakly majorizes x. However, $2 \le 3$ and $2 \cdot 1 > 3 \cdot \frac{1}{2}$. Therefore, y does not weakly log majorize x.

Fact 3.25.16. Let $x, y \in \mathbb{R}^n$, be nonnegative, assume that $x = x^{\downarrow}$ and $y = y^{\downarrow}$, assume that $x \overset{\mathrm{w}}{\prec} y$, and let $p \in [1, \infty)$. Then, for all $k \in \{1, \ldots, n\}$,

$$\left(\sum_{i=1}^k x_{(i)}^p\right)^{1/p} \le \left(\sum_{i=1}^k y_{(i)}^p\right)^{1/p}.$$

Source: Fact 3.25.10, [1969, p. 96], and [1971, p. 138]. **Remark:** $\phi(x) \triangleq (\sum_{i=1}^k x_{(i)}^p)^{1/p}$ is a symmetric gauge function. See Fact 11.9.59.

Fact 3.25.17. Let $x, y \in \mathbb{R}^n$. Then, the following statements are equivalent:

i) $x \overset{\mathrm{s}}{\prec} y$.

ii) For all convex functions $f\colon \mathbb{R} \mapsto \mathbb{R}$, it follows that $\sum_{i=1}^n f(x_i) \le \sum_{i=1}^n f(y_i)$.

Furthermore, the following statements are equivalent:

iii) $x \overset{\mathrm{w}}{\prec} y$.

iv) For all nondecreasing functions $f\colon \mathbb{R} \mapsto \mathbb{R}$, it follows that $\sum_{i=1}^n f(x_i) \le \sum_{i=1}^n f(y_i)$.

Source: [1569], [1969, p. 108], and [1971, p. 156]. **Remark:** This is *Karamata's inequality*. See [1569].

3.26 Notes

The theory of determinants is discussed in [2097, 2098, 2752]. A graph-theoretic interpretation is given in [588, Chapter 4]. Applications to physics are described in [2788, 2789]. Contributors to the development of this subject are highlighted in [1222]. The empty matrix is discussed in [847, 2118], [2314, pp. 462–464], and [2544, p. 3]. Recent versions of Matlab follow the properties of the empty matrix given in this chapter [1384, pp. 305, 306]. Convexity is the subject of [421, 523, 558, 980, 1769, 2213, 2319, 2544, 2767, 2846]. Convex optimization theory is developed in [362, 558]. In [523] the dual cone is called the *polar cone*.

The development of rank properties is based on [1966]. Theorem 3.7.5 is based on [2139]. The term "subdeterminant" is used in [2221] and is equivalent to *minor*. The notation A^{A} for adjugate is used in [2537]. Numerous papers on basic topics in matrix theory and linear algebra are collected in [641, 642]. A geometric interpretation of $\mathcal{N}(A)$, $\mathcal{R}(A)$, $\mathcal{N}(A^*)$, and $\mathcal{R}(A^{\mathsf{T}})$ is given in [2554]. Some reflections on matrix theory are given in [2593, 2618]. Applications of the matrix inversion lemma are discussed in [1291]. Some historical notes on the determinant and inverse of partitioned matrices as well as the matrix inversion lemma are given in [1366].

Combinatorial proofs of several matrix theorems are given in [2969].

A detailed treatment of 2×2 matrices is given in [2253].

Chapter Four
Matrix Classes and Transformations

This chapter presents definitions of various types of matrices as well as transformations for analyzing matrices.

4.1 Types of Matrices

In this section we categorize various types of matrices based on their algebraic and patterned properties. The following definition introduces various types of square matrices. Note that, if $\mathbb{F} = \mathbb{R}$, then $A^* = A^{\mathsf{T}}$, whereas, if $\mathbb{F} = \mathbb{C}$, then A^* and A^{T} may be different.

Definition 4.1.1. For $A \in \mathbb{F}^{n\times n}$ define the following types of matrices:

i) A is *group invertible* if $\mathcal{R}(A) = \mathcal{R}(A^2)$.

ii) A is *involutory* if $A^2 = I$.

iii) A is *skew involutory* if $A^2 = -I$.

iv) A is *idempotent* if $A^2 = A$.

v) A is *skew idempotent* if $A^2 = -A$.

vi) A is *tripotent* if $A^3 = A$.

vii) A is *nilpotent* if there exists $k \in \mathbb{P}$ such that $A^k = 0$.

viii) A is *unipotent* if $A - I$ is nilpotent.

ix) A is *range Hermitian* if $\mathcal{R}(A) = \mathcal{R}(A^*)$.

x) A is *range symmetric* if $\mathcal{R}(A) = \mathcal{R}(A^{\mathsf{T}})$.

xi) A is *range disjoint* if $\mathcal{R}(A) \cap \mathcal{R}(A^*) = \{0\}$.

xii) A is *range spanning* if $\mathcal{R}(A) + \mathcal{R}(A^*) = \mathbb{F}^n$.

xiii) A is *Hermitian* if $A = A^*$.

xiv) A is *symmetric* if $A = A^{\mathsf{T}}$.

xv) A is *skew Hermitian* if $A = -A^*$.

xvi) A is *skew symmetric* if $A = -A^{\mathsf{T}}$.

xvii) A is *normal* if $AA^* = A^*A$.

xviii) A is *positive semidefinite* ($A \geq 0$) if A is Hermitian and $x^*Ax \geq 0$ for all $x \in \mathbb{F}^n$.

xix) A is *negative semidefinite* ($A \leq 0$) if $-A$ is positive semidefinite.

xx) A is *positive definite* ($A > 0$) if A is Hermitian and $x^*Ax > 0$ for all $x \in \mathbb{F}^n$ such that $x \neq 0$.

xxi) A is *negative definite* ($A < 0$) if $-A$ is positive definite.

xxii) A is *semidissipative* if $A + A^*$ is negative semidefinite.

xxiii) A is *dissipative* if $A + A^*$ is negative definite.

xxiv) A is *unitary* if $A^*A = I$.

xxv) A is *shifted unitary* if $A + A^* = 2A^*A$; equivalently, $2A - I$ is unitary.

xxvi) A is *orthogonal* if $A^{\mathsf{T}}A = I$.

xxvii) A is *shifted orthogonal* if $A + A^{\mathsf{T}} = 2A^{\mathsf{T}}A$; equivalently, $2A - I$ is orthogonal.

xxviii) A is a *projector* if A is Hermitian and idempotent.

xxix) A is a *generalized projector* if $A^2 = A^*$.

xxx) A is a *partial isometry* if AA^* is a projector.

xxxi) A is a *reflector* if A is Hermitian and unitary.

xxxii) A is a *skew reflector* if A is skew Hermitian and unitary.

xxxiii) A is an *elementary projector* if there exists a nonzero vector $x \in \mathbb{F}^n$ such that $A = I - (x^*x)^{-1}xx^*$.

xxxiv) A is an *elementary reflector* if there exists a nonzero vector $x \in \mathbb{F}^n$ such that $A = I - 2(x^*x)^{-1}xx^*$.

xxxv) A is an *elementary matrix* if there exist $x, y \in \mathbb{F}^n$ such that $A = I - xy^\mathsf{T}$ and $x^\mathsf{T}y \neq 1$.

xxxvi) A is *reverse Hermitian* if $A = A^{\hat{*}}$.

xxxvii) A is *reverse symmetric* if $A = A^{\hat{\mathsf{T}}}$.

xxxviii) A is a *permutation matrix* if each row of A and each column of A possesses one 1 and zeros otherwise. A is an *(even, odd) permutation matrix* if A is a permutation matrix and $(\det A = 1, \det A = -1)$. A is a *transposition matrix* if it is a permutation matrix and A has exactly two off-diagonal entries that are nonzero. The *cyclic permutation matrix* P_n is defined by

$$P_n \triangleq \begin{bmatrix} 0_{(n-1)\times 1} & I_{n-1} \\ 1 & 0_{1\times(n-1)} \end{bmatrix} = \begin{bmatrix} 0 & 1 & 0 & \cdots & 0 & 0 \\ 0 & 0 & 1 & \ddots & 0 & 0 \\ 0 & 0 & 0 & \ddots & 0 & 0 \\ \vdots & \ddots & \ddots & \ddots & \ddots & \vdots \\ 0 & 0 & 0 & \ddots & 0 & 1 \\ 1 & 0 & 0 & \cdots & 0 & 0 \end{bmatrix}, \tag{4.1.1}$$

where $P_1 \triangleq 1$.

xxxix) A is *reducible* if either $A = 0_{1\times 1}$ or both $n \geq 2$ and there exist $k \geq 1$ and a permutation matrix $S \in \mathbb{R}^{n\times n}$ such that $SAS^\mathsf{T} = \begin{bmatrix} B & C \\ 0_{k\times(n-k)} & D \end{bmatrix}$, where $B \in \mathbb{F}^{(n-k)\times(n-k)}$, $C \in \mathbb{F}^{(n-k)\times k}$, and $D \in \mathbb{F}^{k\times k}$.

xl) A is *irreducible* if A is not reducible.

xli) A is (*totally nonnegative, totally positive*) if every subdeterminant of A is (nonnegative, positive).

Let $A \in \mathbb{F}^{n\times n}$ be Hermitian. Then, the function $f\colon\ \mathbb{F}^n \mapsto \mathbb{R}$ defined by

$$f(x) \triangleq x^*Ax \tag{4.1.2}$$

is a *quadratic form*.

The following definition concerns matrices that are not necessarily square.

Definition 4.1.2. For $A \in \mathbb{F}^{n\times m}$ define the following types of matrices:

i) A is *semicontractive* if $I_n - AA^*$ is positive semidefinite.

ii) A is *contractive* if $I_n - AA^*$ is positive definite.

iii) A is *left inner* if $A^*A = I_m$.

iv) A is *right inner* if $AA^* = I_n$.

v) A is *centrohermitian* if $A = \hat{I}_n\overline{A}\hat{I}_m$.

vi) A is *centrosymmetric* if $A = \hat{I}_n A \hat{I}_m$.

vii) A is an *outer-product matrix* if there exist $x \in \mathbb{F}^n$ and $y \in \mathbb{F}^m$ such that $A = xy^{\mathsf{T}}$.

The following definition introduces several types of structured and patterned matrices.

Definition 4.1.3. For $A \in \mathbb{F}^{n\times m}$ define the following types of matrices:

i) A is *diagonal* if $A_{(i,j)} = 0$ for all $i \neq j$. If $n = m$, then A is diagonal if and only if

$$A = \operatorname{diag}(A_{(1,1)}, \ldots, A_{(n,n)}) \triangleq \begin{bmatrix} A_{(1,1)} & 0 & \cdots & 0 \\ 0 & A_{(2,2)} & \ddots & 0 \\ \vdots & \ddots & \ddots & 0 \\ 0 & 0 & \cdots & A_{(n,n)} \end{bmatrix}. \tag{4.1.3}$$

ii) A is *tridiagonal* if $A_{(i,j)} = 0$ for all $|i - j| > 1$.

iii) A is *reverse diagonal* if $A_{(i,j)} = 0$ for all $i + j \neq \min\{n, m\} + 1$. If $n = m$, then A is reverse diagonal if and only if

$$A = \operatorname{revdiag}(A_{(1,n)}, \ldots, A_{(n,1)}) \triangleq \begin{bmatrix} 0 & \cdots & 0 & A_{(1,n)} \\ 0 & \iddots & A_{(2,n-1)} & 0 \\ \vdots & \iddots & \iddots & \vdots \\ A_{(n,1)} & 0 & \cdots & 0 \end{bmatrix}. \tag{4.1.4}$$

iv) A is (*upper triangular, strictly upper triangular*) if $A_{(i,j)} = 0$ for all $(i > j, i \geq j)$.

v) A is (*lower triangular, strictly lower triangular*) if $A_{(i,j)} = 0$ for all $(i < j, i \leq j)$.

vi) A is (*upper bidiagonal, lower bidiagonal*) if A is tridiagonal and (upper triangular, lower triangular).

vii) A is (*upper Hessenberg, lower Hessenberg*) if $A_{(i,j)} = 0$ for all $(i > j + 1, i < j + 1)$.

viii) A is *Toeplitz* if $A_{(i,j)} = A_{(k,l)}$ for all $k - i = l - j$; that is,

$$A = \begin{bmatrix} a & b & c & \cdots \\ d & a & b & \ddots \\ e & d & a & \ddots \\ \vdots & \ddots & \ddots & \ddots \end{bmatrix}.$$

ix) A is *Hankel* if $A_{(i,j)} = A_{(k,l)}$ for all $i + j = k + l$; that is,

$$A = \begin{bmatrix} a & b & c & \cdots \\ b & c & d & \iddots \\ c & d & e & \iddots \\ \vdots & \iddots & \iddots & \iddots \end{bmatrix}.$$

The following definition introduces several types of partitioned matrices whose blocks are structured or patterned.

Definition 4.1.4. For $A \in \mathbb{F}^{n\times m}$ define the following types of matrices:

i) A is *block diagonal* if

$$A = \operatorname{diag}(A_1, \ldots, A_k) \triangleq \begin{bmatrix} A_1 & & 0 \\ & \ddots & \\ 0 & & A_k \end{bmatrix},$$

where $A_i \in \mathbb{F}^{n_i \times m_i}$ for all $i \in \{1, \ldots, k\}$.

ii) A is *reverse block diagonal* if

$$A = \operatorname{revdiag}(A_1, \ldots, A_k) \triangleq \begin{bmatrix} 0 & & A_1 \\ & \cdot^{\cdot^{\cdot}} & \\ A_k & & 0 \end{bmatrix},$$

where $A_i \in \mathbb{F}^{n_i \times m_i}$ for all $i \in \{1, \ldots, k\}$.

iii) A is *upper block triangular* if

$$A = \begin{bmatrix} A_{11} & A_{12} & \cdots & A_{1k} \\ 0 & A_{22} & \cdots & A_{2k} \\ \vdots & \ddots & \ddots & \vdots \\ 0 & 0 & \cdots & A_{kk} \end{bmatrix},$$

where $A_{ij} \in \mathbb{F}^{n_i \times n_j}$ for all $i, j \in \{1, \ldots, k\}$ such that $i \le j$.

iv) A is *lower block triangular* if

$$A = \begin{bmatrix} A_{11} & 0 & \cdots & 0 \\ A_{21} & A_{22} & \ddots & 0 \\ \vdots & \vdots & \ddots & \vdots \\ A_{k1} & A_{k2} & \cdots & A_{kk} \end{bmatrix},$$

where $A_{ij} \in \mathbb{F}^{n_i \times n_j}$ for all $i, j \in \{1, \ldots, k\}$ such that $j \le i$.

v) A is *block Toeplitz* if

$$A = \begin{bmatrix} A_1 & A_2 & A_3 & \cdots \\ A_4 & A_1 & A_2 & \ddots \\ A_5 & A_4 & A_1 & \ddots \\ \vdots & \ddots & \ddots & \ddots \end{bmatrix},$$

where $A_i \in \mathbb{F}^{n_i \times m_i}$ for all $i \in \{1, \ldots, k\}$.

vi) A is *block Hankel* if

$$A = \begin{bmatrix} A_1 & A_2 & A_3 & \cdots \\ A_2 & A_3 & A_4 & \cdot^{\cdot^{\cdot}} \\ A_3 & A_4 & A_5 & \cdot^{\cdot^{\cdot}} \\ \vdots & \cdot^{\cdot^{\cdot}} & \cdot^{\cdot^{\cdot}} & \cdot^{\cdot^{\cdot}} \end{bmatrix},$$

where $A_i \in \mathbb{F}^{n_i \times m_i}$ for all $i \in \{1, \ldots, k\}$.

For $x \in \mathbb{F}^n$, define

$$\operatorname{diag}(x) \triangleq \operatorname{diag}(x_{(1)}, \ldots, x_{(n)}), \tag{4.1.5}$$

$$\operatorname{revdiag}(x) \triangleq \operatorname{revdiag}(x_{(1)}, \ldots, x_{(n)}). \tag{4.1.6}$$

Definition 4.1.5. For $A \in \mathbb{R}^{n\times m}$ define the following types of matrices:

i) A is *nonnegative* ($A \geq\geq 0$) if $A_{(i,j)} \geq 0$ for all $i \in \{1,\dots,n\}$ and $j \in \{1,\dots,m\}$.

ii) A is *row stochastic* if A is nonnegative and $A1_{m\times 1} = 1_{n\times 1}$.

iii) A is *column stochastic* if A is nonnegative and $1_{1\times n}A = 1_{1\times m}$.

iv) A is *doubly stochastic* if A is both row stochastic and column stochastic.

v) A is *positive* ($A >> 0$) if $A_{(i,j)} > 0$ for all $i \in \{1,\dots,n\}$ and $j \in \{1,\dots,m\}$.

Now, assume that $n = m$. Then, define the following types of matrices:

vi) A is *almost nonnegative* if $A_{(i,j)} \geq 0$ for all $i, j \in \{1,\dots,n\}$ such that $i \neq j$.

vii) A is a *Z-matrix* if $-A$ is almost nonnegative.

Define the *unit imaginary matrix* $J_{2n} \in \mathbb{R}^{2n\times 2n}$ (or just J) by

$$J_{2n} \triangleq \begin{bmatrix} 0 & I_n \\ -I_n & 0 \end{bmatrix}. \tag{4.1.7}$$

In particular,

$$J_2 = \begin{bmatrix} 0 & 1 \\ -1 & 0 \end{bmatrix}. \tag{4.1.8}$$

Note that J_{2n} is skew symmetric and orthogonal; that is,

$$J_{2n}^{\mathrm{T}} = -J_{2n} = J_{2n}^{-1}. \tag{4.1.9}$$

Hence, J_{2n} is skew involutory, and J_{2n} is a skew reflector.

The following definition introduces two types of matrices of even size that are defined in terms of J. Note that $\mathbb{F}$ can represent either $\mathbb{R}$ or $\mathbb{C}$.

Definition 4.1.6. For $A \in \mathbb{F}^{2n\times 2n}$ define the following types of matrices:

i) A is *Hamiltonian* if $J^{-1}A^{\mathrm{T}}J = -A$.

ii) A is *symplectic* if A is nonsingular and $J^{-1}A^{\mathrm{T}}J = A^{-1}$.

Proposition 4.1.7. Let $A \in \mathbb{F}^{n\times n}$. Then, the following statements hold:

i) If A is either Hermitian, skew Hermitian, or unitary, then A is normal.

ii) If A is either nonsingular or normal, then A is range Hermitian.

iii) If A is either range Hermitian, idempotent, or tripotent, then A is group invertible.

iv) If A is a reflector, then A is tripotent.

v) If A is a permutation matrix, then A is orthogonal.

Proof. *i)* is immediate. To prove *ii)*, note that, if A is nonsingular, then $\mathcal{R}(A) = \mathcal{R}(A^*) = \mathbb{F}^n$, and thus A is range Hermitian. If A is normal, then Theorem 3.5.3 implies that $\mathcal{R}(A) = \mathcal{R}(AA^*) = \mathcal{R}(A^*A) = \mathcal{R}(A^*)$, which proves that A is range Hermitian. To prove *iii)*, note that, if A is range Hermitian, then $\mathcal{R}(A) = \mathcal{R}(AA^*) = A\mathcal{R}(A^*) = A\mathcal{R}(A) = \mathcal{R}(A^2)$, while, if A is idempotent, then $\mathcal{R}(A) = \mathcal{R}(A^2)$. If A is tripotent, then $\mathcal{R}(A) = \mathcal{R}(A^3) = A^2\mathcal{R}(A) \subseteq \mathcal{R}(A^2) = A\mathcal{R}(A) \subseteq \mathcal{R}(A)$. Hence, $\mathcal{R}(A) = \mathcal{R}(A^2)$. □

Proposition 4.1.8. Let $\mathcal{A} \in \mathbb{F}^{2n\times 2n}$. Then, $\mathcal{A}$ is Hamiltonian if and only if there exist $A, B, C \in \mathbb{F}^{n\times n}$ such that B and C are symmetric and

$$\mathcal{A} = \begin{bmatrix} A & B \\ C & -A^{\mathrm{T}} \end{bmatrix}. \tag{4.1.10}$$

4.2 Matrices Related to Graphs

Definition 4.2.1. Let $\mathcal{G} = (\mathcal{X}, \mathcal{R})$ be a directed graph, where $\mathcal{X} = \{x_1,\dots,x_n\}$. Then, the following terminology is defined:

i) Let $\{a_1, \ldots, a_m\}$ denote the set of directed edges in $\mathcal{R}$ that are not self-directed. Then, the *incidence matrix* $B \in \mathbb{R}^{n\times m}$ of $\mathcal{G}$ is given by $B_{(i,j)} = 1$ for all $i \in \{1, \ldots, n\}$ and $j \in \{1, \ldots, m\}$ such that x_i is the tail of a_j, $B_{(i,j)} = -1$ for all $i \in \{1, \ldots, n\}$ and $j \in \{1, \ldots, m\}$ such that x_i is the head of a_j, and $B_{(i,j)} = 0$ otherwise.

ii) The *adjacency matrix* $A \in \mathbb{R}^{n\times n}$ of $\mathcal{G}$ is given by $A_{(i,j)} = 1$ for all $i, j \in \{1, \ldots, n\}$ such that $(x_j, x_i) \in \mathcal{R}$ and $A_{(i,j)} = 0$ for all $i, j \in \{1, \ldots, n\}$ such that $(x_j, x_i) \notin \mathcal{R}$.

iii) The *inbound Laplacian matrix* $L_{\text{in}} \in \mathbb{R}^{n\times n}$ of $\mathcal{G}$ is given by $L_{\text{in}(i,i)} = \sum_{j=1, j\neq i}^{n} A_{(i,j)}$ for all $i \in \{1, \ldots, n\}$, and $L_{\text{in}(i,j)} = -A_{(i,j)}$ for all distinct $i, j \in \{1, \ldots, n\}$.

iv) The *outbound Laplacian matrix* $L_{\text{out}} \in \mathbb{R}^{n\times n}$ of $\mathcal{G}$ is given by $L_{\text{out}(i,i)} = \sum_{j=1, j\neq i}^{n} A_{(j,i)}$ for all $i \in \{1, \ldots, n\}$, and $L_{\text{out}(i,j)} = -A_{(i,j)}$ for all distinct $i, j \in \{1, \ldots, n\}$.

v) The *indegree matrix* $D_{\text{in}} \in \mathbb{R}^{n\times n}$ is the diagonal matrix such that $D_{\text{in}(i,i)} = \text{indeg}(x_i)$ for all $i \in \{1, \ldots, n\}$.

vi) The *outdegree matrix* $D_{\text{out}} \in \mathbb{R}^{n\times n}$ is the diagonal matrix such that $D_{\text{out}(i,i)} = \text{outdeg}(x_i)$ for all $i \in \{1, \ldots, n\}$.

vii) If $\mathcal{G}$ is symmetric, then the *Laplacian matrix* of $\mathcal{G}$ is given by $L \triangleq L_{\text{in}} = L_{\text{out}}$.

viii) If $\mathcal{G}$ is symmetric, then the *degree matrix* $D \in \mathbb{R}^{n\times n}$ of $\mathcal{G}$ is given by $D \triangleq D_{\text{in}} = D_{\text{out}}$.

ix) If $\mathcal{G} = (\mathcal{X}, \mathcal{R}, w)$ is a weighted directed graph, then the *adjacency matrix* $A \in \mathbb{R}^{n\times n}$ of $\mathcal{G}$ is given by $A_{(i,j)} = w(x_j, x_i)$ for all $i, j \in \{1, \ldots, n\}$ such that $(x_j, x_i) \in \mathcal{R}$, and $A_{(i,j)} = 0$ for all $i, j \in \{1, \ldots, n\}$ such that $(x_j, x_i) \notin \mathcal{R}$.

Note that the adjacency matrix is nonnegative, while the inbound Laplacian, outbound Laplacian, and Laplacian matrices are Z-matrices. Furthermore, note that the inbound Laplacian, outbound Laplacian, and Laplacian matrices are unaffected by the presence of self-directed edges. However, the indegree and outdegree matrices account for self-directed edges. For the directed edge (x_k, x_l), the ith column of the incidence matrix B is given by $\text{col}_i(B) = e_l - e_k$. Finally, if $\mathcal{G}$ is a symmetric graph, then A and L are symmetric.

Theorem 4.2.2. Let $\mathcal{G} = (\mathcal{X}, \mathcal{R})$ be a directed graph, where $\mathcal{X} = \{x_1, \ldots, x_n\}$, and let L_{in}, L_{out}, D_{in}, D_{out}, and A denote the inbound Laplacian, outbound Laplacian, indegree, outdegree, and adjacency matrices of $\mathcal{G}$, respectively. Then,

$$L_{\text{in}} = D_{\text{in}} - A, \tag{4.2.1}$$

$$L_{\text{out}} = D_{\text{out}} - A. \tag{4.2.2}$$

Theorem 4.2.3. Let $\mathcal{G} = (\mathcal{X}, \mathcal{R})$ be a symmetric graph, where $\mathcal{X} = \{x_1, \ldots, x_n\}$, and let A, L, D, and B denote the adjacency, Laplacian, degree, and incidence matrices of $\mathcal{G}$, respectively. Then,

$$L = D - A. \tag{4.2.3}$$

Now, assume that $\mathcal{G}$ has no self-edges. Then,

$$L = \tfrac{1}{2}BB^{\mathsf{T}}. \tag{4.2.4}$$

Definition 4.2.4. Let $M \in \mathbb{F}^{n\times n}$ and $\mathcal{X} = \{x_1, \ldots, x_n\}$. Then, the *directed graph of M* is $\mathcal{G}(M) \triangleq (\mathcal{X}, \mathcal{R})$, where, for all $i, j \in \{1, \ldots, n\}$, $(x_j, x_i) \in \mathcal{R}$ if and only if $M_{(i,j)} \neq 0$.

Proposition 4.2.5. Let $M \in \mathbb{F}^{n\times n}$. Then, the adjacency matrix A of $\mathcal{G}(M)$ satisfies

$$A = \text{sign}\,|M|. \tag{4.2.5}$$

4.3 Lie Algebras

In the following definition, note that α and β are assumed to be real.

Definition 4.3.1. Let $\mathcal{S} \subseteq \mathbb{F}^{n\times n}$. Then, $\mathcal{S}$ is a *Lie algebra* if the following statements hold:

i) If $A, B \in \mathcal{S}$ and $\alpha, \beta \in \mathbb{R}$, then $\alpha A + \beta B \in \mathcal{S}$.

ii) If $A, B \in \mathcal{S}$, then $[A, B] \in \mathcal{S}$.

If $\mathbb{F} = \mathbb{R}$, then *i*) is equivalent to the statement that $\mathcal{S}$ is a subspace. However, if $\mathbb{F} = \mathbb{C}$ and $\mathcal{S}$ contains matrices that are not real, then $\mathcal{S}$ is not a subspace.

Proposition 4.3.2. The following sets are Lie algebras:

i) $\mathrm{gl}_{\mathbb{F}}(n) \triangleq \mathbb{F}^{n\times n}$.

ii) $\mathrm{pl}_{\mathbb{C}}(n) \triangleq \{A \in \mathbb{C}^{n\times n}\colon\ \operatorname{tr} A \in \mathbb{R}\}$.

iii) $\mathrm{sl}_{\mathbb{F}}(n) \triangleq \{A \in \mathbb{F}^{n\times n}\colon\ \operatorname{tr} A = 0\}$.

iv) $\mathrm{u}(n) \triangleq \{A \in \mathbb{C}^{n\times n}\colon\ A \text{ is skew Hermitian}\}$.

v) $\mathrm{su}(n) \triangleq \{A \in \mathbb{C}^{n\times n}\colon\ A \text{ is skew Hermitian and } \operatorname{tr} A = 0\}$.

vi) $\mathrm{so}(n) \triangleq \{A \in \mathbb{R}^{n\times n}\colon\ A \text{ is skew symmetric}\}$.

vii) $\mathrm{su}(n,m) \triangleq \{A \in \mathbb{C}^{(n+m)\times(n+m)}\colon\ \operatorname{diag}(I_n, -I_m)A^*\operatorname{diag}(I_n, -I_m) = -A \text{ and } \operatorname{tr} A = 0\}$.

viii) $\mathrm{so}(n,m) \triangleq \{A \in \mathbb{R}^{(n+m)\times(n+m)}\colon\ \operatorname{diag}(I_n, -I_m)A^{\mathsf{T}}\operatorname{diag}(I_n, -I_m) = -A\}$.

ix) $\mathrm{symp}_{\mathbb{F}}(2n) \triangleq \{A \in \mathbb{F}^{2n\times 2n}\colon\ A \text{ is Hamiltonian}\}$.

x) $\mathrm{osymp}_{\mathbb{C}}(2n) \triangleq \mathrm{su}(2n) \cap \mathrm{symp}_{\mathbb{C}}(2n)$.

xi) $\mathrm{osymp}_{\mathbb{R}}(2n) \triangleq \mathrm{so}(2n) \cap \mathrm{symp}_{\mathbb{R}}(2n)$.

xii) $\mathrm{aff}_{\mathbb{F}}(n) \triangleq \left\{\begin{bmatrix} A & b \\ 0 & 0 \end{bmatrix}\colon\ A \in \mathrm{gl}_{\mathbb{F}}(n),\ b \in \mathbb{F}^n\right\}$.

xiii) $\mathrm{se}_{\mathbb{C}}(n) \triangleq \left\{\begin{bmatrix} A & b \\ 0 & 0 \end{bmatrix}\colon\ A \in \mathrm{su}(n),\ b \in \mathbb{C}^n\right\}$.

xiv) $\mathrm{se}_{\mathbb{R}}(n) \triangleq \left\{\begin{bmatrix} A & b \\ 0 & 0 \end{bmatrix}\colon\ A \in \mathrm{so}(n),\ b \in \mathbb{R}^n\right\}$.

xv) $\mathrm{trans}_{\mathbb{F}}(n) \triangleq \left\{\begin{bmatrix} 0 & b \\ 0 & 0 \end{bmatrix}\colon\ b \in \mathbb{F}^n\right\}$.

4.4 Abstract Groups

Definition 4.4.1. Let $\mathcal{S}$ be a nonempty set, and let $\phi\colon \mathcal{S}\times\mathcal{S} \mapsto \mathcal{S}$. Then, $(\mathcal{S}, \phi)$ is a *group* if the following statements hold:

i) For all $x, y, z \in \mathcal{S}$, $\phi[x, \phi(y, z)] = \phi[\phi(x, y), z]$.

ii) There exists $\iota \in \mathcal{S}$ such that, for all $x \in \mathcal{S}$, $\phi(\iota, x) = \phi(x, \iota) = x$.

iii) For all $x \in \mathcal{S}$, there exists $y \in \mathcal{S}$ such that $\phi(x, y) = \phi(y, x) = \iota$.

Now, assume that $(\mathcal{S}, \phi)$ is a group. Then, the following terminology is defined:

iv) $(\mathcal{S}, \phi)$ is a *finite group* if $\operatorname{card}(\mathcal{S})$ is finite. Otherwise, $(\mathcal{S}, \phi)$ is an *infinite group*.

v) $(\mathcal{S}, \phi)$ is an *Abelian group* if, for all $x, y \in \mathcal{S}$, $\phi(x, y) = \phi(y, x)$.

vi) $(\mathcal{S}, \phi)$ is a *cyclic group* if there exists $x \in \mathcal{S}$ such that $\mathcal{S} = \{\phi^{(i)}(x)\colon i \in \mathbb{Z}\}$, where $\phi^{(0)}(x) \triangleq \iota$, $\phi^{(1)}(x) \triangleq x$, and, for all $i \ge 1$, $\phi^{(i+1)}(x) \triangleq \phi[\phi^{(i)}(x), x]$ and $\phi^{(-i)}(x)$ satisfies $\phi[\phi^{(-i)}(x), \phi^{(i)}(x)] = \iota$. In this case, x is a *generator* of $(\mathcal{S}, \phi)$.

Now, let $\mathcal{X} \subseteq \mathcal{S}$ be nonempty, let $\phi_{\mathcal{X}}$ denote the restriction of ϕ to $\mathcal{X}\times\mathcal{X}$, and assume that $\phi_{\mathcal{X}}\colon \mathcal{X}\times\mathcal{X} \mapsto \mathcal{X}$. Then, the following terminology is defined:

vii) $(\mathcal{X}, \phi_{\mathcal{X}})$ is a *subgroup* of $(\mathcal{S}, \phi)$ if $(\mathcal{X}, \phi_{\mathcal{X}})$ is a group.

viii) $(\mathcal{X}, \phi_{\mathcal{X}})$ is a *nontrivial subgroup* of $(\mathcal{S}, \phi)$ if $(\mathcal{X}, \phi_{\mathcal{X}})$ is a subgroup of $(\mathcal{S}, \phi)$ and $\mathcal{X} \ne \{\iota\}$.

ix) $(\mathcal{X}, \phi_{\mathcal{X}})$ is a *proper subgroup* of $(\mathcal{S}, \phi)$ if $(\mathcal{X}, \phi_{\mathcal{X}})$ is a subgroup of $(\mathcal{S}, \phi)$ and $\mathcal{X} \subset \mathcal{S}$.

x) $(\mathcal{X}, \phi_{\mathcal{X}})$ is a *minimal subgroup* of $(\mathcal{S}, \phi)$ if $(\mathcal{X}, \phi_{\mathcal{X}})$ is a nontrivial subgroup of $(\mathcal{S}, \phi)$ and there does not exist a nontrivial subgroup of $(\mathcal{S}, \phi)$ that is also a proper subgroup of $(\mathcal{X}, \phi_{\mathcal{X}})$.

xi) $(\mathcal{X}, \phi_{\mathcal{X}})$ is a *maximal subgroup* of $(\mathcal{S}, \phi)$ if $(\mathcal{X}, \phi_{\mathcal{X}})$ is a proper subgroup of $(\mathcal{S}, \phi)$ and there does not exist a proper subgroup of $(\mathcal{S}, \phi)$ that contains $(\mathcal{X}, \phi_{\mathcal{X}})$ as a proper subgroup.

xii) $(\mathcal{X}, \phi_{\mathcal{X}})$ is a *normal subgroup* of $(\mathcal{S}, \phi)$ if $(\mathcal{X}, \phi_{\mathcal{X}})$ is a subgroup of $(\mathcal{S}, \phi)$ and, for all $x \in \mathcal{S}$, $\phi(x, \mathcal{X}) = \phi(\mathcal{X}, x)$.

xiii) $(\mathcal{S}, \phi)$ is a *simple group* if $\operatorname{card}(\mathcal{S}) \geq 2$ and the only normal subgroups of $(\mathcal{S}, \phi)$ are $(\{\iota\}, \phi_{\{\iota\}})$ and $(\mathcal{S}, \phi)$.

Proposition 4.4.2. Let $(\mathcal{S}, \phi)$ be a group, and let $\mathcal{X} \subseteq \mathcal{S}$ be nonempty. Then, the following statements hold:

i) The following statements are equivalent:

a) $(\mathcal{X}, \phi_{\mathcal{X}})$ is a subgroup of $(\mathcal{S}, \phi)$.

b) $\phi_{\mathcal{X}}(\mathcal{X} \times \mathcal{X}) \subseteq \mathcal{X}$, and $\phi^{(-1)}(\mathcal{X}) \subseteq \mathcal{X}$.

c) For all $x, y \in \mathcal{X}$, $\phi[\phi^{(-1)}(x), y] \in \mathcal{X}$.

ii) If $\mathcal{X}$ is finite and $\phi_{\mathcal{X}}(\mathcal{X} \times \mathcal{X}) \subseteq \mathcal{X}$, then $(\mathcal{X}, \phi_{\mathcal{X}})$ is a subgroup of $(\mathcal{S}, \phi)$.

Proof. See [1713, pp. 4, 5] and [2336, p. 24]. □

The *identity element* ι of the group $(\mathcal{S}, \phi)$ is unique, and it is contained in every subgroup $(\mathcal{X}, \phi_{\mathcal{X}})$ of $(\mathcal{S}, \phi)$. Furthermore, for all $x \in \mathcal{S}$, there exists a unique $y \in \mathcal{S}$ such that $\phi(x, y) = \phi(y, x) = \iota$. In particular, $y = \phi^{(-1)}(x)$. Note that $\phi^{(-1)}(\mathcal{S}) = \{\phi^{(-1)}(x) \colon x \in \mathcal{S}\} = \mathcal{S}$. Every subgroup of an Abelian group is Abelian and normal.

Proposition 4.4.3. Let $(\mathcal{S}, \phi)$ be a group, and let $x \in \mathcal{S}$. Then, exactly one of the following statements holds:

i) For all distinct positive integers k and l, $\phi^{(k)}(x) \neq \phi^{(l)}(x)$.

ii) There exists a positive integer ℓ such that $\iota, \phi(x), \phi^{(2)}(x), \ldots, \phi^{(\ell-1)}(x)$ of $\mathcal{S}$ are distinct and $\phi^{(\ell)}(x) = \iota$.

Let $(\mathcal{S}, \phi)$ be a group, and let $x \in \mathcal{S}$. If *i)* holds, then $\mathcal{S}$ is infinite. By Definition 4.4.1, $(\mathcal{S}, \phi)$ is cyclic if and only if there exists $x \in \mathcal{S}$ such that $\mathcal{S} = \{\phi^{(i)}(x) \colon i \in \mathbb{Z}\}$. The positive integer ℓ given by *ii)* of Proposition 4.4.3 is the *order* of x. In other words, the order ℓ of x is the smallest positive integer such that $\phi^{(\ell)}(x) = \iota$. If *i)* holds, then the order of x is infinite. If $\mathcal{S}$ is finite, then *ii)* holds and $\ell \leq \operatorname{card}(\mathcal{S})$. If there exists $x \in \mathcal{S}$ and a positive integer ℓ such that $\mathcal{S} = \{\iota, \phi(x), \phi^{(2)}(x), \ldots, \phi^{(\ell-1)}(x)\}$, then $\mathcal{S}$ is a finite cyclic group. Consequently, the cyclic group $(\mathcal{S}, \phi)$ with generator x is finite if and only if there exists $n \geq 1$ such that $\phi^{(n)}(x) = \iota$.

As an example, let $\mathcal{S} \triangleq \mathbb{Z}_2 \triangleq \{0, 1\}$, and define $\phi(0, 0) \triangleq \phi(1, 1) \triangleq 0$ and $\phi(0, 1) \triangleq \phi(1, 0) \triangleq 1$. The identity element ι of $(\mathbb{Z}_2, \phi)$ is 0, and $(\mathbb{Z}_2, \phi)$ is Abelian. As another example, let $\mathcal{S} \triangleq \mathbb{Z}_6 \triangleq \{0, 1, 2, 3, 4, 5\}$ and, for all $k, l \in \mathcal{S}$, define $\phi(k, l) \triangleq \operatorname{rem}_6(k + l)$. Then, $\phi(4, 0) = \phi(0, 4) = 4$, and $\phi(4, 5) = \phi(5, 4) = 3$. The identity element ι of $(\mathbb{Z}_6, \phi)$ is 0, and $(\mathbb{Z}_6, \phi)$ is Abelian.

The following result is *Lagrange's theorem*.

Theorem 4.4.4. Let $(\mathcal{S}, \phi)$ be a finite group, and let $(\mathcal{X}, \phi_{\mathcal{X}})$ be a subgroup of $(\mathcal{S}, \phi)$. Then,

$$\operatorname{card}(\mathcal{S}) = \operatorname{card}(\mathcal{X}) \operatorname{card}(\{\phi(x, \mathcal{X}) \colon x \in \mathcal{S}\}). \tag{4.4.1}$$

Proof: See [79, Chapter 24], [966, pp. 89, 90], and [1560, pp. 77, 95]. □

Theorem 4.4.5. Let $(\mathcal{S}, \phi)$ be a group, and let $(\mathcal{X}, \phi_{\mathcal{X}})$ be a subgroup of $(\mathcal{S}, \phi)$. Then, $\{\phi(x, \mathcal{X}) \colon x \in \mathcal{S}\}$ is a partition of $\mathcal{S}$. Next, define $\tilde{\mathcal{S}} \triangleq \{\phi(x, \mathcal{X}) \colon x \in \mathcal{S}\}$, and, for all $A, B \in \tilde{\mathcal{S}}$, define $\tilde{\phi}(A, B) \triangleq \{\phi(x, y) \colon x \in A \text{ and } y \in B\}$. Then, the following statements are equivalent:

i) For all $(A, B) \in \tilde{\mathcal{S}} \times \tilde{\mathcal{S}}$, $\tilde{\phi}(A, B) \in \tilde{\mathcal{S}}$.

ii) $(\mathcal{X}, \phi)$ is a normal subgroup of $(\mathcal{S}, \phi)$.

If these statements hold and $\mathcal{S}$ is finite, then $\operatorname{card}(\mathcal{S}) = \operatorname{card}(\tilde{\mathcal{S}})\operatorname{card}(\mathcal{X})$.

Proof. See [966, pp. 76–82]. □

If *i)* and *ii)* of Theorem 4.4.5 hold, then the group $(\tilde{\mathcal{S}}, \tilde{\phi})$ is a *quotient group* of $(\mathcal{S}, \phi)$. This group is denoted by $(\mathcal{S}/\mathcal{X}, \tilde{\phi})$.

Definition 4.4.6. Let $(\mathcal{S}_1, \phi_1)$ and $(\mathcal{S}_2, \phi_2)$ be groups, and let $\Phi\colon \mathcal{S}_1 \mapsto \mathcal{S}_2$. Then, Φ is a *homomorphism* if, for all $x, y \in \mathcal{S}_1$, $\Phi(\phi_1(x,y)) = \phi_2(\Phi(x), \Phi(y))$. In addition, Φ is an *isomorphism* and $(\mathcal{S}_1, \phi_1)$ and $(\mathcal{S}_2, \phi_2)$ are *isomorphic* if Φ is one-to-one and onto. In this case, $\mathcal{S}_1 \simeq \mathcal{S}_2$.

4.5 Addition Groups

Note that $(\mathbb{F}^{n\times m}, \phi)$, where $\phi(A,B) \triangleq A+B$, is an Abelian group. An addition group is a subgroup of $(\mathbb{F}^{n\times m}, \phi)$.

Definition 4.5.1. Let $\mathcal{S} \subseteq \mathbb{F}^{n\times m}$. Then, $\mathcal{S}$ is an *addition group* if the following statements hold:

i) If $A \in \mathcal{S}$, then $-A \in \mathcal{S}$.

ii) If $A, B \in \mathcal{S}$, then $A + B \in \mathcal{S}$.

Now, assume that $\mathcal{S}$ is an addition group, and let $\mathcal{X} \subseteq \mathcal{S}$. Then, $\mathcal{X}$ is an *addition subgroup* of $\mathcal{S}$ if $\mathcal{X}$ is an addition group.

If $\mathcal{S} \subseteq \mathbb{F}^{n\times m}$ is an addition group, then $\mathcal{S} = -\mathcal{S}$ and $0_{n\times m} \in \mathcal{S}$.

Definition 4.5.2. Let $\mathcal{S} \subseteq \mathbb{F}^{n\times m}$, assume that $\mathcal{S}$ is an addition group, and let $\mathcal{X}$ be an addition subgroup of $\mathcal{S}$. Then, $\tilde{\mathcal{S}} \triangleq \{x + \mathcal{X}\colon\ x \in \mathcal{S}\}$ is a *quotient addition group* of $\mathcal{S}$.

Definition 4.5.3. Let $\mathcal{S}_1 \subseteq \mathbb{F}^{n\times m}$ and $\mathcal{S}_2 \subseteq \mathbb{F}^{l\times k}$ be addition groups, and let $\Phi\colon \mathcal{S}_1 \mapsto \mathcal{S}_2$. Then, Φ is a *homomorphism* if, for all $A, B \in \mathcal{S}_1$, $\Phi(A + B) = \Phi(A) + \Phi(B)$. In addition, Φ is an *isomorphism* and $\mathcal{S}_1$ and $\mathcal{S}_2$ are *isomorphic* if Φ is one-to-one and onto. In this case, $\mathcal{S}_1 \simeq \mathcal{S}_2$.

As an example, let $\mathcal{S} \triangleq \mathbb{Z} \subset \mathbb{R}^{1\times 1}$, and define the even integers $\mathbb{E} \triangleq 2\mathbb{Z}$ and the odd integers $\mathbb{O} \triangleq \mathbb{Z}\backslash\mathbb{E}$. Then, $\mathbb{E}$ is an addition subgroup of $\mathbb{Z}$. Furthermore, $-\mathbb{E} = \mathbb{E}$, $-\mathbb{O} = \mathbb{O}$, $\mathbb{E} + \mathbb{E} = \mathbb{E}$, $\mathbb{O} + \mathbb{O} = \mathbb{E}$, and $\mathbb{E} + \mathbb{O} = \mathbb{O}$. Therefore, $\tilde{\mathcal{S}} = \{\mathbb{E}, \mathbb{O}\}$ is a quotient addition group of $\mathbb{Z}$. Finally, $\mathbb{Z}_2 \simeq \mathbb{Z}/\mathbb{E}$, where $\mathbb{Z}_2$ is defined in the previous section.

As another example, let $\mathcal{S} \triangleq \mathbb{Z}$, and define the addition subgroup $\mathcal{X}_0 \triangleq \{\dots, -18, -12, -6, 0, 6, 12, 18, \dots\} = 6\mathbb{Z}$ of $\mathbb{Z}$. Consequently, $\tilde{\mathcal{S}} = \{\mathcal{X}_0, \dots, \mathcal{X}_5\} = \mathbb{Z}/6\mathbb{Z}$ is a quotient addition group of $\mathbb{Z}$, where, for all $i \in \{0, \dots, 5\}$, $\mathcal{X}_i \triangleq i + 6\mathbb{Z}$. For example, $\mathcal{X}_2 = \{\dots, -16, -10, -4, 2, 8, 14, 20, \dots\}$. Hence, for all $i \in \{0, \dots, 5\}$, $k \in \mathcal{X}_i$ if and only if $\operatorname{rem}_6(k) = i$, and $k, l \in \mathbb{Z}$ are in the same element of $\tilde{\mathcal{S}}$ if and only if $k \overset{6}{\equiv} l$. Furthermore, for all $i, j \in \{0, \dots, 6\}$, $\mathcal{X}_i + \mathcal{X}_j = \mathcal{X}_{\operatorname{rem}_6(i+j)}$. Finally, $\mathbb{Z}_6 \simeq \mathbb{Z}/6\mathbb{Z}$.

As a final example, let $n \ge 2$, and let $\mathcal{S} \subset \mathbb{F}^n$ be a subspace. Then, $\mathbb{F}^n$ is an addition group, and $\mathcal{S}$ is an addition subgroup of $\mathbb{F}^n$. Furthermore, $\tilde{\mathcal{S}} \triangleq \{x + \mathcal{S}\colon\ x \in \mathbb{F}^n\}$ is a quotient addition subgroup of $\mathbb{F}^n$.

4.6 Multiplication Groups

Let $\mathrm{GL}_{\mathbb{F}}(n)$ denote the set of $n \times n$ nonsingular matrices with entries in $\mathbb{F}$. Then, $(\mathrm{GL}_{\mathbb{F}}(n), \phi)$, where $\phi(A, B) \triangleq AB$, is a group. A multiplication group is a subgroup of $(\mathrm{GL}_{\mathbb{F}}(n), \phi)$.

Definition 4.6.1. Let $\mathcal{S} \subset \mathbb{F}^{n\times n}$. Then, $\mathcal{S}$ is a *multiplication group* if the following statements hold:

i) If $A \in \mathcal{S}$, then A is nonsingular.

ii) If $A \in \mathcal{S}$, then $A^{-1} \in \mathcal{S}$.

iii) If $A, B \in \mathcal{S}$, then $AB \in \mathcal{S}$.

Now, assume that $\mathcal{S}$ is a multiplication group. Then, $\mathcal{S}$ is an *Abelian multiplication group* if, for all $A, B \in \mathcal{S}$, $[A, B] = 0$. Now, let $\mathcal{X} \subseteq \mathcal{S}$. Then, $\mathcal{X}$ is a *multiplication subgroup* of $\mathcal{S}$ if $\mathcal{X}$ is a multiplication group. Furthermore, $\mathcal{X}$ is a *normal multiplication subgroup* of $\mathcal{S}$ if $\mathcal{X}$ is a multiplication subgroup of $\mathcal{S}$ and, for all $A \in \mathcal{S}$, $A\mathcal{X}A^{-1} = \mathcal{X}$.

If $\mathcal{S} \subset \mathbb{F}^{n\times n}$ is a multiplication group, then $\mathcal{S} = \mathcal{S}^{-1}$, and $I_n \in \mathcal{S}$.

Definition 4.6.2. Let $\mathcal{S} \subset \mathbb{F}^{n\times n}$, assume that $\mathcal{S}$ is a multiplication group, and let $\mathcal{X}$ be a normal multiplication subgroup of $\mathcal{S}$. Then, $\tilde{\mathcal{S}} \triangleq \{x\mathcal{X}\colon\ x \in \mathcal{S}\}$ is a *quotient multiplication group* of $\mathcal{S}$.

As an example, let $\mathcal{S}$ denote the nonzero real numbers. Then, $(0,\infty)$ is a normal multiplication subgroup of $\mathcal{S}$, and the corresponding quotient multiplication group of $\mathcal{S}$ is $\tilde{\mathcal{S}} = \{(-\infty,0),(0,\infty)\}$.

Proposition 4.6.3. Let $\mathcal{S} \subset \mathbb{F}^{n\times n}$ be a multiplication group, and let $A \in \mathcal{S}$. Then, exactly one of the following statements holds:

i) For all distinct positive integers k and l, $A^k \neq A^l$.

ii) There exists a positive integer ℓ such that $I_n, A, A^2, \ldots, A^{\ell-1}$ are distinct and $A^\ell = I_n$.

Let $\mathcal{S} \subset \mathbb{F}^{n\times n}$ be a multiplication group, and let $A \in \mathcal{S}$. If there exists $A \in \mathcal{S}$ such that $\mathcal{S} = \{A^k\colon\ k \in \mathbb{Z}\}$, then $\mathcal{S}$ is a *cyclic multiplication group*. If *i)* holds, then $\mathcal{S}$ is infinite. The positive integer ℓ given by *ii)* of Proposition 4.6.3 is the *order* of A. In other words, the order ℓ of A is the smallest positive integer such that $A^\ell = I_n$. If *i)* holds, then the order of A is infinite. If $\mathcal{S}$ is finite, then *ii)* holds and $\ell \le \operatorname{card}(\mathcal{S})$. If there exists $A \in \mathbb{F}^{n\times n}$ and a positive integer ℓ such that $\mathcal{S} = \{I_n, A, A^2, \ldots, A^{\ell-1}\}$, then $\mathcal{S}$ is a *finite cyclic multiplication group*.

Definition 4.6.4. Let $\mathcal{S}_1 \subset \mathbb{F}^{n\times n}$ and $\mathcal{S}_2 \subset \mathbb{F}^{m\times m}$ be multiplication groups, and let $\Phi\colon\ \mathcal{S}_1 \mapsto \mathcal{S}_2$. Then, Φ is a *homomorphism* if, for all $A, B \in \mathcal{S}_1$, $\Phi(AB) = \Phi(A)\Phi(B)$. Furthermore, $\mathcal{S}_1$ and $\mathcal{S}_2$ are *isomorphic* and Φ is an *isomorphism* if Φ is one-to-one and onto. In this case, $\mathcal{S}_1 \simeq \mathcal{S}_2$.

Proposition 4.6.5. Let $\mathcal{S}_1 \subset \mathbb{F}^{n\times n}$ and $\mathcal{S}_2 \subset \mathbb{F}^{m\times m}$ be multiplication groups, and assume that $\mathcal{S}_1$ and $\mathcal{S}_2$ are isomorphic with isomorphism $\Phi\colon\ \mathcal{S}_1 \mapsto \mathcal{S}_2$. Then, $\Phi(I_n) = I_m$, and, for all $A \in \mathcal{S}_1$, $\Phi(A^{-1}) = [\Phi(A)]^{-1}$.

The following result lists multiplication groups that arise in physics and engineering. For example, $\mathrm{O}(1,3)$ is the *Lorentz group*, see [2379, p. 16] and [2423, p. 126]. The special orthogonal group $\mathrm{SO}(n)$ consists of the real $n\times n$ orthogonal matrices whose determinant is 1. In particular, each matrix in $\mathrm{SO}(2)$ and $\mathrm{SO}(3)$ is a *rotation matrix*. Furthermore, $\mathrm{P}(n)$, $\mathrm{A}(n)$, $\mathrm{D}(n)$, and $\mathrm{C}(n)$ are the $n\times n$ *permutation group*, *alternating group*, *dihedral group*, and *cyclic group*, respectively.

Proposition 4.6.6. The following sets are multiplication groups:

i) $\mathrm{GL}_{\mathbb{F}}(n) \triangleq \{A \in \mathbb{F}^{n\times n}\colon\ \det A \neq 0\}$.

ii) $\mathrm{PL}_{\mathbb{F}}(n) \triangleq \{A \in \mathbb{F}^{n\times n}\colon\ \det A > 0\}$.

iii) $\mathrm{SL}_{\mathbb{F}}(n) \triangleq \{A \in \mathbb{F}^{n\times n}\colon\ \det A = 1\}$.

iv) $\mathrm{U}(n) \triangleq \{A \in \mathbb{C}^{n\times n}\colon\ A \text{ is unitary}\}$.

v) $\mathrm{O}(n) \triangleq \{A \in \mathbb{R}^{n\times n}\colon\ A \text{ is orthogonal}\}$.

vi) $\mathrm{SU}(n) \triangleq \{A \in \mathrm{U}(n)\colon\ \det A = 1\}$.

vii) $\mathrm{SO}(n) \triangleq \{A \in \mathrm{O}(n)\colon\ \det A = 1\}$.

viii) $\mathrm{P}(n) \triangleq \{A \in \mathbb{R}^{n\times n}\colon\ A \text{ is a permutation matrix}\}$.

ix) $\mathrm{A}(n) \triangleq \{A \in \mathrm{P}(n)\colon\ A \text{ is an even permutation matrix}\}$.

x) $\mathrm{D}(2) \triangleq \{I_2, -I_2, \hat{I}_2, -\hat{I}_2\}$.

xi) $\mathrm{D}(n) \triangleq \{I_n, P_n, P_n^2, \ldots, P_n^{n-1}, \hat{I}_n, \hat{I}_nP_n, \hat{I}_nP_n^2, \ldots, \hat{I}_nP_n^{n-1}\}$, where $n \ge 3$.

xii) $\mathrm{C}(n) \triangleq \{I_n, P_n, P_n^2, \ldots, P_n^{n-1}\}$.

xiii) $\mathrm{U}(n,m) \triangleq \{A \in \mathbb{C}^{(n+m)\times(n+m)}\colon\ A^*\operatorname{diag}(I_n,-I_m)A = \operatorname{diag}(I_n,-I_m)\}$.

xiv) $\mathrm{O}(n,m) \triangleq \{A \in \mathbb{R}^{(n+m)\times(n+m)}\colon\ A^{\mathsf{T}}\operatorname{diag}(I_n,-I_m)A = \operatorname{diag}(I_n,-I_m)\}$.

xv) $\mathrm{SU}(n,m) \triangleq \{A \in \mathrm{U}(n,m)\colon\ \det A = 1\}$.

xvi) $\mathrm{SO}(n,m) \triangleq \{A \in \mathrm{O}(n,m)\colon\ \det A = 1\}$.

xvii) $\mathrm{Symp}_{\mathbb{F}}(2n) \triangleq \{A \in \mathbb{F}^{2n\times 2n}\colon\ A \text{ is symplectic}\}$.

xviii) $\mathrm{OSymp}_{\mathbb{C}}(2n) \triangleq \mathrm{U}(2n) \cap \mathrm{Symp}_{\mathbb{C}}(2n)$.

xix) $\mathrm{OSymp}_{\mathbb{R}}(2n) \triangleq \mathrm{O}(2n) \cap \mathrm{Symp}_{\mathbb{R}}(2n)$.

xx) $\mathrm{Aff}_{\mathbb{F}}(n) \triangleq \left\{\left[\begin{smallmatrix} A & b\\ 0 & 1\end{smallmatrix}\right]\colon\ A \in \mathrm{GL}_{\mathbb{F}}(n),\ b \in \mathbb{F}^n\right\}$.

xxi) $\mathrm{SE}_{\mathbb{C}}(n) \triangleq \left\{\left[\begin{smallmatrix} A & b\\ 0 & 1\end{smallmatrix}\right]\colon\ A \in \mathrm{SU}(n),\ b \in \mathbb{C}^n\right\}$.

xxii) $\mathrm{SE}_{\mathbb{R}}(n) \triangleq \left\{\left[\begin{smallmatrix} A & b\\ 0 & 1\end{smallmatrix}\right]\colon\ A \in \mathrm{SO}(n),\ b \in \mathbb{R}^n\right\}$.

xxiii) $\mathrm{Trans}_{\mathbb{F}}(n) \triangleq \left\{\left[\begin{smallmatrix} I & b\\ 0 & 1\end{smallmatrix}\right]\colon\ b \in \mathbb{F}^n\right\}$.

4.7 Matrix Transformations

The following results use groups to define equivalence relations.

Proposition 4.7.1. Let $\mathcal{S}_1 \subset \mathbb{F}^{n\times n}$ and $\mathcal{S}_2 \subset \mathbb{F}^{m\times m}$ be multiplication groups, and let $\mathcal{M} \subseteq \mathbb{F}^{n\times m}$. Then, the subset of $\mathcal{M}\times\mathcal{M}$ defined by

$$\mathcal{R} \triangleq \{(A,B) \in \mathcal{M}\times\mathcal{M}\colon\ \text{there exist } S_1 \in \mathcal{S}_1 \text{ and } S_2 \in \mathcal{S}_2 \text{ such that } A = S_1BS_2\}$$

is an equivalence relation on $\mathcal{M}$.

Proposition 4.7.2. Let $\mathcal{S} \subset \mathbb{F}^{n\times n}$ be a multiplication group, and let $\mathcal{M} \subseteq \mathbb{F}^{n\times n}$. Then, the following subsets of $\mathcal{M}\times\mathcal{M}$ are equivalence relations:

i) $\mathcal{R} \triangleq \{(A,B) \in \mathcal{M}\times\mathcal{M}\colon\ \text{there exists } S \in \mathcal{S} \text{ such that } A = SBS^{-1}\}$.

ii) $\mathcal{R} \triangleq \{(A,B) \in \mathcal{M}\times\mathcal{M}\colon\ \text{there exists } S \in \mathcal{S} \text{ such that } A = SBS^{*}\}$.

iii) $\mathcal{R} \triangleq \{(A,B) \in \mathcal{M}\times\mathcal{M}\colon\ \text{there exists } S \in \mathcal{S} \text{ such that } A = SBS^{\mathsf{T}}\}$.

If, in addition, $\mathcal{S}$ is an Abelian multiplication group, then the following subset of $\mathcal{M}\times\mathcal{M}$ is an equivalence relation:

iv) $\mathcal{R} \triangleq \{(A,B) \in \mathcal{M}\times\mathcal{M}\colon\ \text{there exists } S \in \mathcal{S} \text{ such that } A = SBS\}$.

Various transformations can be employed for analyzing matrices. Propositions 4.7.1 and 4.7.2 imply that these transformations define equivalence relations.

Definition 4.7.3. Let $A, B \in \mathbb{F}^{n\times m}$. Then, the following terminology is defined:

i) A and B are *left equivalent* if there exists a nonsingular matrix $S_1 \in \mathbb{F}^{n\times n}$ such that $A = S_1B$.

ii) A and B are *right equivalent* if there exists a nonsingular matrix $S_2 \in \mathbb{F}^{m\times m}$ such that $A = BS_2$.

iii) A and B are *biequivalent* if there exist nonsingular matrices $S_1 \in \mathbb{F}^{n\times n}$ and $S_2 \in \mathbb{F}^{m\times m}$ such that $A = S_1BS_2$.

iv) A and B are *unitarily left equivalent* if there exists a unitary matrix $S_1 \in \mathbb{F}^{n\times n}$ such that $A = S_1B$.

v) A and B are *unitarily right equivalent* if there exists a unitary matrix $S_2 \in \mathbb{F}^{m\times m}$ such that $A = BS_2$.

vi) A and B are *unitarily biequivalent* if there exist unitary matrices $S_1 \in \mathbb{F}^{n\times n}$ and $S_2 \in \mathbb{F}^{m\times m}$ such that $A = S_1BS_2$.

Definition 4.7.4. Let $A, B \in \mathbb{F}^{n\times n}$. Then, the following terminology is defined:

i) A and B are *similar* if there exists a nonsingular matrix $S \in \mathbb{F}^{n\times n}$ such that $A = SBS^{-1}$.

ii) A and B are *congruent* if there exists a nonsingular matrix $S \in \mathbb{F}^{n\times n}$ such that $A = SBS^{*}$.

iii) A and B are *T-congruent* if there exists a nonsingular matrix $S \in \mathbb{F}^{n\times n}$ such that $A = SBS^{\mathsf{T}}$.

iv) A and B are *unitarily similar* if there exists a unitary matrix $S \in \mathbb{F}^{n\times n}$ such that $A = SBS^{*} = SBS^{-1}$.

The transformations that appear in Definition 4.7.3 and Definition 4.7.4 are *left equivalence, right equivalence, biequivalence, unitary left equivalence, unitary right equivalence, unitary biequivalence, similarity, congruence, T-congruence,* and *unitary similarity* transformations, respectively. The following results summarize some matrix properties that are preserved under left equivalence, right equivalence, biequivalence, similarity, congruence, and unitary similarity.

Proposition 4.7.5. Let $A, B \in \mathbb{F}^{n\times n}$. If the matrices A and B are similar, then the following statements hold:

i) A and B are biequivalent.

ii) $\operatorname{tr} A = \operatorname{tr} B$.

iii) $\det A = \det B$.

iv) A^k and B^k are similar for all $k \geq 1$.

v) A^{k*} and B^{k*} are similar for all $k \geq 1$.

vi) A is nonsingular if and only if B is; in this case, A^{-k} and B^{-k} are similar for all $k \geq 1$.

vii) A is (group invertible, involutory, skew involutory, idempotent, tripotent, nilpotent) if and only if B is.

If A and B are congruent, then the following statements hold:

viii) A and B are biequivalent.

ix) A^* and B^* are congruent.

x) A is nonsingular if and only if B is; in this case, A^{-1} and B^{-1} are congruent.

xi) A is (range Hermitian, Hermitian, skew Hermitian, positive semidefinite, positive definite) if and only if B is.

If A and B are unitarily similar, then the following statements hold:

xii) A and B are similar.

xiii) A and B are congruent.

xiv) A is (range Hermitian, group invertible, normal, Hermitian, skew Hermitian, positive semidefinite, positive definite, unitary, involutory, skew involutory, idempotent, tripotent, nilpotent) if and only if B is.

4.8 Projectors, Idempotent Matrices, and Subspaces

The following result shows that each subspace is associated with a unique projector.

Proposition 4.8.1. Let $\mathcal{S} \subseteq \mathbb{F}^n$ be a subspace. Then, there exists a unique projector $A \in \mathbb{F}^{n\times n}$ such that $\mathcal{S} = \mathcal{R}(A)$. Furthermore, $x \in \mathcal{S}$ if and only if $x = Ax$.

Proof. See [2036, p. 386] and Fact 4.18.2. □

For a subspace $\mathcal{S} \subseteq \mathbb{F}^n$, the projector $A \in \mathbb{F}^{n\times n}$ given by Proposition 4.8.1 is the *projector onto* $\mathcal{S}$. If, in addition, $\mathcal{S}' \subseteq \mathbb{F}^n$, then $A\mathcal{S}'$ is the *projection* of $\mathcal{S}'$ *into* $\mathcal{S}$.

Let $A \in \mathbb{F}^{n\times n}$ be a projector. Then, the *complementary projector* $A_\perp$ is the projector defined by

$$A_\perp \triangleq I - A. \tag{4.8.1}$$

Proposition 4.8.2. Let $\mathcal{S} \subseteq \mathbb{F}^n$ be a subspace, and let $A \in \mathbb{F}^{n\times n}$ be the projector onto $\mathcal{S}$. Then, $A_\perp$ is the projector onto $\mathcal{S}^\perp$. Furthermore,

$$\mathcal{R}(A)^\perp = \mathcal{N}(A) = \mathcal{R}(A_\perp) = \mathcal{S}^\perp. \tag{4.8.2}$$

The following result shows that each pair of complementary subspaces is associated with a unique idempotent matrix.

Proposition 4.8.3. Let $\mathcal{S}_1, \mathcal{S}_2 \subseteq \mathbb{F}^n$ be complementary subspaces; that is, $\mathcal{S}_1 + \mathcal{S}_2 = \mathbb{F}^n$ and

$\mathcal{S}_1 \cap \mathcal{S}_2 = \{0\}$. Then, there exists a unique idempotent matrix $A \in \mathbb{F}^{n \times n}$ such that $\mathcal{R}(A) = \mathcal{S}_1$ and $\mathcal{N}(A) = \mathcal{S}_2$.

Proof. See [423, p. 118] and [2036, p. 386]. □

For complementary subspaces $\mathcal{S}_1, \mathcal{S}_2 \subseteq \mathbb{F}^n$, the unique idempotent matrix $A \in \mathbb{F}^{n \times n}$ given by Proposition 4.8.3 is the *idempotent matrix onto* $\mathcal{S}_1 = \mathcal{R}(A)$ *along* $\mathcal{S}_2 = \mathcal{N}(A)$.

For an idempotent matrix $A \in \mathbb{F}^{n \times n}$, the *complementary idempotent matrix* $A_\perp$ defined by (4.8.1) is also idempotent.

Proposition 4.8.4. Let $\mathcal{S}_1, \mathcal{S}_2 \subseteq \mathbb{F}^n$ be complementary subspaces, and let $A \in \mathbb{F}^{n \times n}$ be the idempotent matrix onto $\mathcal{S}_1 = \mathcal{R}(A)$ along $\mathcal{S}_2 = \mathcal{N}(A)$. Then, $\mathcal{R}(A_\perp) = \mathcal{S}_2$ and $\mathcal{N}(A_\perp) = \mathcal{S}_1$; that is, the complementary idempotent matrix $A_\perp$ is the idempotent matrix onto $\mathcal{S}_2$ along $\mathcal{S}_1$.

Since, by Proposition 4.8.1 each subspace is associated with a unique projector, it follows that each pair of complementary subspaces is associated with a unique pair of projectors.

Definition 4.8.5. Let $A, B \in \mathbb{F}^{n \times n}$, and assume that A and B are projectors. Then, A and B are *complementary projectors* if $\mathcal{R}(A)$ and $\mathcal{R}(B)$ are complementary subspaces.

If A is a projector, then $\mathcal{R}(A)$ and $\mathcal{R}(A_\perp)$ are orthogonally complementary subspaces. Consequently, $\mathcal{R}(A)$ and $\mathcal{R}(A_\perp)$ are complementary subspace, and thus A and $A_\perp$ are complementary projectors. The following result characterizes complementary pairs of projectors.

Proposition 4.8.6. Let $A, B \in \mathbb{F}^{n \times n}$, and assume that A and B are projectors. Then, the following statements are equivalent:

i) A and B are complementary projectors.

ii) $\operatorname{rank}\,[A\ \ B] = \operatorname{rank} A + \operatorname{rank} B = n$.

Proposition 4.8.3 implies that every pair of complementary projectors can be associated with a unique idempotent matrix. In particular, for complementary projectors $A, B \in \mathbb{F}^{n \times n}$, Fact 8.8.14 provides an expression for the unique idempotent matrix onto $\mathcal{R}(A)$ along $\mathcal{R}(B)$. Conversely, for an idempotent matrix $\mathcal{A} \in \mathbb{F}^{n \times n}$, the unique complementary projectors $A, B \in \mathbb{F}^{n \times n}$ such that $\mathcal{R}(A) = \mathcal{R}(\mathcal{A})$ and $\mathcal{R}(B) = \mathcal{N}(\mathcal{A})$ are given by Fact 8.8.11.

Definition 4.8.7. The *index of* A, denoted by $\operatorname{ind} A$, is the smallest nonnegative integer k such that

$$\mathcal{R}(A^k) = \mathcal{R}(A^{k+1}). \tag{4.8.3}$$

Proposition 4.8.8. Let $A \in \mathbb{F}^{n \times n}$. Then, A is nonsingular if and only if $\operatorname{ind} A = 0$. Furthermore, A is group invertible if and only if $\operatorname{ind} A \le 1$.

Note that $\operatorname{ind} 0_{n \times n} = 1$.

Proposition 4.8.9. Let $A \in \mathbb{F}^{n \times n}$, and let $k \ge 1$. Then, $\operatorname{ind} A \le k$ if and only if $\mathcal{R}(A^k)$ and $\mathcal{N}(A^k)$ are complementary subspaces.

The following corollary of Proposition 4.8.9 shows that the null space and range of a group-invertible matrix are complementary subspaces. Note that every idempotent matrix is group invertible.

Corollary 4.8.10. Let $A \in \mathbb{F}^{n \times n}$. Then, A is group invertible if and only if $\mathcal{R}(A)$ and $\mathcal{N}(A)$ are complementary subspaces.

Fact 4.9.6 states that the range and null space of a range-Hermitian matrix are orthogonally complementary subspaces. Furthermore, Proposition 4.1.7 states that every range-Hermitian matrix is group invertible.

For a group-invertible matrix $A \in \mathbb{F}^{n \times n}$, the following result shows how to construct the idempotent matrix onto $\mathcal{R}(A)$ along $\mathcal{N}(A)$. This construction is based on the full-rank factorization given by Proposition 7.6.6.

Proposition 4.8.11. Let $A \in \mathbb{F}^{n \times n}$, and let $r \triangleq \operatorname{rank} A$. Then, A is group invertible if and only if there exist $B \in \mathbb{F}^{n\times r}$ and $C \in \mathbb{F}^{r\times n}$ such that $A = BC$ and $\operatorname{rank} B = \operatorname{rank} C = \operatorname{rank} CB = r$. If these conditions hold, then $P \triangleq B(CB)^{-1}C$ is the idempotent matrix onto $\mathcal{R}(A)$ along $\mathcal{N}(A)$. If, in addition, A is range Hermitian, then P is the projector onto $\mathcal{R}(A)$.

Proof. See [2036, p. 634]. □

An alternative expression for the idempotent matrix onto $\mathcal{R}(A)$ along $\mathcal{N}(A)$ is given by Proposition 8.2.3.

4.9 Facts on Elementary, Group-Invertible, Range-Hermitian, Range-Disjoint, and Range-Spanning Matrices

Fact 4.9.1. Let $A \in \mathbb{F}^{n\times m}$. Then, the following statements are equivalent:

i) A is an elementary matrix.

ii) There exist $x, y \in \mathbb{F}^n$ such that $A = I - xy^*$ and $x^*y \neq 1$.

iii) There exist $x, y \in \mathbb{F}^n$ such that $A = I - xy^{\mathsf{T}}$ and $x^{\mathsf{T}}y \neq 1$.

Fact 4.9.2. Let $A \in \mathbb{F}^{n\times n}$. Then, the following statements are equivalent:

i) A is group invertible.

ii) A^* is group invertible.

iii) A^{T} is group invertible.

iv) $\overline{A}$ is group invertible.

v) $\mathcal{R}(A) = \mathcal{R}(A^2)$.

vi) $\mathcal{N}(A) = \mathcal{N}(A^2)$.

vii) $\mathcal{N}(A) \cap \mathcal{R}(A) = \{0\}$.

viii) $\mathcal{N}(A) + \mathcal{R}(A) = \mathbb{F}^n$.

ix) $\mathcal{R}(A)$ and $\mathcal{N}(A)$ are complementary subspaces.

x) A and A^2 are left equivalent.

xi) A and A^2 are right equivalent.

xii) $\operatorname{ind} A \leq 1$.

xiii) $\operatorname{rank} A = \operatorname{rank} A^2$.

xiv) $\operatorname{rank}(A - A^4) = \operatorname{rank}(A^2 - A^5)$.

xv) $\operatorname{def} A = \operatorname{def} A^2$.

xvi) $\operatorname{def} A = \operatorname{amult}_A(0)$.

Related: Fact 3.13.29, Corollary 4.8.10, Proposition 4.8.11, and Corollary 7.7.9.

Fact 4.9.3. Let $A \in \mathbb{F}^{n\times n}$. Then, $\operatorname{ind} A \leq k$ if and only if A^k is group invertible.

Fact 4.9.4. Let $A \in \mathbb{F}^{n\times n}$. Then, the following statements hold:

i) A is range disjoint if and only if $\mathcal{N}(A) + \mathcal{N}(A^*) = \mathbb{F}^n$.

ii) A is range spanning if and only if $\mathcal{N}(A) \cap \mathcal{N}(A^*) = \{0\}$.

iii) A is range Hermitian and range disjoint if and only if $A = 0$.

iv) A is range Hermitian and range spanning if and only if A is nonsingular.

v) A is range disjoint and range spanning if and only if $\mathcal{R}(A)$ and $\mathcal{R}(A^*)$ are complementary subspaces.

vi) If A is range disjoint, then so is A^2.

vii) If A^2 is range spanning, then so is A.

viii) A is (range disjoint, range spanning) if and only if A^* is.

Source: [250].

Fact 4.9.5. Let $A \in \mathbb{F}^{n\times n}$, and assume that $AA^* + A^*A = A + A^*$. Then, A is range Hermitian. **Source:** [2719].

Fact 4.9.6. Let $A \in \mathbb{F}^{n\times n}$. Then, the following statements are equivalent:

i) A is range Hermitian.

ii) A^* is range Hermitian.

iii) $\mathcal{R}(A) = \mathcal{R}(A^*)$.

iv) $\mathcal{R}(A) \subseteq \mathcal{R}(A^*)$.

v) $\mathcal{R}(A^*) \subseteq \mathcal{R}(A)$.

vi) $\mathcal{N}(A) = \mathcal{N}(A^*)$.

vii) A and A^* are right equivalent.

viii) $\mathcal{R}(A)^\perp = \mathcal{N}(A)$.

ix) $\mathcal{N}(A)^\perp = \mathcal{R}(A)$.

x) $\mathcal{R}(A)$ and $\mathcal{N}(A)$ are orthogonally complementary subspaces.

xi) $\operatorname{rank} A = \operatorname{rank} [A \ \ A^*]$.

xii) $\mathcal{R}(A^2) = \mathcal{R}(A^*)$.

xiii) A is group invertible, and A^2 is range Hermitian.

Source: [257, 719, 2658]. **Remark:** Using Fact 4.18.2, Proposition 4.8.2, and Proposition 8.1.7, *vi)* is equivalent to $A^+A = I - (I - A^+A) = AA^+$. See Fact 8.5.2, Fact 8.5.3, and Fact 8.5.8. **Related:** Fact 8.10.24.

Fact 4.9.7. Let $A, B \in \mathbb{F}^{n\times n}$, and assume that A and B are range Hermitian. Then, $\mathcal{R}(A) \subseteq \mathcal{R}(B)$ if and only if $\mathcal{N}(B) \subseteq \mathcal{N}(A)$.

Fact 4.9.8. Let $A, B \in \mathbb{F}^{n\times n}$, and assume that A and B are range Hermitian. Then, $\operatorname{rank} AB = \operatorname{rank} BA$. **Source:** [268].

Fact 4.9.9. Let $A, B \in \mathbb{F}^{n\times n}$, and assume that A and B are range Hermitian. Then, the following statements are equivalent:

i) AB is range Hermitian.

ii) $\mathcal{R}(AB) = \mathcal{R}(A) \cap \mathcal{R}(B)$ and $\mathcal{N}(AB) = \mathcal{N}(A) + \mathcal{N}(B)$.

Source: [916].

Fact 4.9.10. Let $A, B \in \mathbb{F}^{n\times n}$, and assume that A, B, AB, and BA are range Hermitian. Then, $\mathcal{R}(AB) = \mathcal{R}(BA)$. **Source:** [845].

4.10 Facts on Normal, Hermitian, and Skew-Hermitian Matrices

Fact 4.10.1. Let $A \in \mathbb{F}^{n\times m}$. Then, AA^T and $A^\mathsf{T}A$ are symmetric.

Fact 4.10.2. Let $\alpha \in \mathbb{R}$ and $A \in \mathbb{R}^{n\times n}$. Then, the matrix equation $\alpha A + A^\mathsf{T} = 0$ has a nonzero solution A if and only if either $\alpha = 1$ or $\alpha = -1$.

Fact 4.10.3. Let $A \in \mathbb{F}^{n\times n}$, assume that A is Hermitian, and let $k \geq 1$. Then, $\mathcal{R}(A) = \mathcal{R}(A^k)$ and $\mathcal{N}(A) = \mathcal{N}(A^k)$.

Fact 4.10.4. Let $A \in \mathbb{R}^{n\times n}$. Then, the following statements hold:

i) The following statements are equivalent:

a) For all $x \in \mathbb{R}^n$, $x^\mathsf{T}Ax = 0$.

b) For all $x \in \mathbb{C}^n$, x^*Ax is imaginary.

c) A is skew symmetric.

ii) A is symmetric and, for all $x \in \mathbb{R}^n$, $x^\mathsf{T}Ax = 0$ if and only if $A = 0$.

iii) $x^*Ax = 0$ for all $x \in \mathbb{C}^n$ if and only if $A = 0$.

iv) x^*Ax is real for all $x \in \mathbb{C}^n$ if and only if A is symmetric.

Remark: Let $x = \begin{bmatrix} 1-\jmath \\ 1+\jmath \end{bmatrix}$ and $A = \begin{bmatrix} 0 & 1 \\ -1 & 0 \end{bmatrix}$. Then, $x^*Ax = 4\jmath$. Hence, "is imaginary" cannot be replaced by "= 0" in *b)* of *i)*. **Related:** Fact 4.10.5.

Fact 4.10.5. Let $A \in \mathbb{C}^{n\times n}$. Then, the following statements hold:

i) $x^*Ax = 0$ for all $x \in \mathbb{C}^n$ if and only if $A = 0$.

ii) x^*Ax is imaginary for all $x \in \mathbb{C}^n$ if and only if A is skew Hermitian.

iii) x^*Ax is real for all $x \in \mathbb{C}^n$ if and only if A is Hermitian.

iv) $x^*Ax \geq 0$ for all $x \in \mathbb{C}^n$ if and only if A is positive semidefinite.

v) $\mathrm{Re}(x^*Ax) \geq 0$ for all $x \in \mathbb{C}^n$ if and only if $A + A^*$ is positive semidefinite.

vi) $x^*Ax > 0$ for all nonzero $x \in \mathbb{C}^n$ if and only if A is positive definite.

Remark: For all $x \in \mathbb{R}^2$, $A = \begin{bmatrix} 1 & 1 \\ -1 & 0 \end{bmatrix}$ satisfies $x^{\mathsf{T}}Ax \geq 0$ but is not symmetric and thus is not positive semidefinite. Likewise, for all nonzero $x \in \mathbb{R}^2$, $A = \begin{bmatrix} 1 & 1 \\ -1 & 1 \end{bmatrix}$ satisfies $x^{\mathsf{T}}Ax > 0$ but is not symmetric and thus is not positive definite. Hence, $\mathbb{C}$ cannot be replaced by $\mathbb{R}$ in *ii)* and *iv)*.

Fact 4.10.6. Let $A \in \mathbb{R}^{n\times n}$. Then, the following statements are equivalent:

i) A is positive definite.

ii) A is symmetric and, for all nonzero $x \in \mathbb{R}^n$, $x^{\mathsf{T}}Ax > 0$.

iii) A is symmetric and, for all nonzero $x \in \mathbb{C}^n$, $x^*Ax > 0$.

Fact 4.10.7. Let $A \in \mathbb{F}^{n\times n}$, and assume that A is block diagonal. Then, A is (normal, Hermitian, skew Hermitian) if and only if every diagonally located block has the same property.

Fact 4.10.8. Let $A \in \mathbb{C}^{n\times n}$. Then, the following statements hold:

i) A is Hermitian if and only if $\jmath A$ is skew Hermitian.

ii) A is skew Hermitian if and only if $\jmath A$ is Hermitian.

iii) A is Hermitian if and only if $\mathrm{Re}\, A$ is symmetric and $\mathrm{Im}\, A$ is skew symmetric.

iv) A is skew Hermitian if and only if $\mathrm{Re}\, A$ is skew symmetric and $\mathrm{Im}\, A$ is symmetric.

v) If A is positive semidefinite, then $\mathrm{Re}\, A$ is positive semidefinite and $\mathrm{Im}\, A$ is skew symmetric.

vi) If A is positive definite, then $\mathrm{Re}\, A$ is positive definite and $\mathrm{Im}\, A$ is skew symmetric.

vii) A is symmetric if and only if $\begin{bmatrix} 0 & A \\ A & 0 \end{bmatrix}$ is symmetric.

viii) A is Hermitian if and only if $\begin{bmatrix} 0 & A \\ A & 0 \end{bmatrix}$ is Hermitian.

ix) A is symmetric if and only if $\begin{bmatrix} 0 & A \\ -A & 0 \end{bmatrix}$ is skew symmetric.

x) A is Hermitian if and only if $\begin{bmatrix} 0 & A \\ -A & 0 \end{bmatrix}$ is skew Hermitian.

Remark: *x)* is a real analogue of *i)* since $\begin{bmatrix} 0 & A \\ -A & 0 \end{bmatrix} = J_2 \otimes A$, and J_2 is a real representation of $\jmath$.

Fact 4.10.9. Let $A \in \mathbb{F}^{n\times n}$. Then, the following statements hold:

i) If A is (nonsingular, range Hermitian, normal, Hermitian, skew Hermitian, unitary, positive semidefinite, positive definite, diagonal, diagonalizable over $\mathbb{F}$, nilpotent), then so are $\overline{A}$, A^{T}, A^*, and A^{A}.

ii) Assume that A is nonsingular. If A is (range Hermitian, normal, Hermitian, skew Hermitian, unitary, positive definite, diagonal, diagonalizable over $\mathbb{F}$), then so is A^{-1}.

iii) If A is skew Hermitian and n is odd, then A^{A} is Hermitian.

iv) If A is skew Hermitian and n is even, then A^{A} is skew Hermitian.

v) If A is diagonal, then, for all $i \in \{1, \ldots, n\}$, $(A^{\mathsf{A}})_{(i,i)} = \prod_{j=1, j\neq i}^{n} A_{(j,j)}$.

Source: Fact 3.19.9. **Related:** Fact 6.10.13.

Fact 4.10.10. Let $A \in \mathbb{F}^{n\times n}$, assume that A is Hermitian, define $r \triangleq \operatorname{rank} A$, let $\lambda_1, \dots, \lambda_r$ denote the nonzero eigenvalues of A, let $A_1 \in \mathbb{F}^{r\times r}$ be a principal submatrix of A, and assume that $\det A_1 \neq 0$. Then,

$$\operatorname{sign}\det A_1 = \operatorname{sign}\prod_{i=1}^{r}\lambda_i.$$

Source: [2991, p. 259].

Fact 4.10.11. Let $A \in \mathbb{F}^{n\times n}$, assume that n is even and A is skew Hermitian, and let $x \in \mathbb{F}^n$ and $\alpha \in \mathbb{F}$. Then, $\det(A+\alpha xx^*) = \det A$. **Source:** Fact 3.17.2 and Fact 4.10.9 imply that $\det(A+\alpha xx^*) = \det A + \alpha x^*A^{\mathsf{A}}x = \det A$.

Fact 4.10.12. Let $A \in \mathbb{F}^{n\times n}$. Then, the following statements are equivalent:

i) A is normal.

ii) $[A, AA^*] = 0$.

iii) $[A, A^*A] = 0$.

iv) $[A, A + A^*] = 0$.

v) $[A, A - A^*] = 0$.

vi) $[A + A^*, A - A^*] = 0$.

vii) $[A, [A, A^*]] = 0$.

viii) $\operatorname{tr}(AA^*)^2 = \operatorname{tr} A^2A^{2*}$.

ix) $(AA^*)^2 = A^2A^{2*}$.

x) There exists $k \geq 1$ such that $\operatorname{tr}(AA^*)^k = \operatorname{tr} A^kA^{k*}$.

xi) There exist $k, l \in \mathbb{P}$ such that $\operatorname{tr}(AA^*)^{kl} = \operatorname{tr}(A^kA^{k*})^l$.

xii) A is range Hermitian, and $AA^*A^2 = A^2A^*A$.

xiii) $AA^* - A^*A$ is positive semidefinite.

xiv) $[\frac{1}{2}(A + A^*)]^2 + [\frac{1}{2j}(A - A^*)]^2 = AA^*$.

xv) $[\frac{1}{2}(A + A^*)]^2 + [\frac{1}{2j}(A - A^*)]^2 = A^*A$.

xvi) There exists a unitary matrix $S \in \mathbb{F}^{n\times n}$ such that $A^* = AS$.

xvii) There exists a unitary matrix $S \in \mathbb{F}^{n\times n}$ such that $A^* = SA$.

xviii) For all $p \in \mathbb{F}[s]$, $p(A)$ is normal.

xix) There exist $\mu_1, \dots, \mu_r \in \mathbb{C}$ and projectors $A_1, \dots, A_r \in \mathbb{F}^{n\times n}$ such that, for all distinct $i, j \in \{1, \dots, r\}$, $A_iA_j = 0$ and such that $\sum_{i=1}^{r} A_i = I$ and $A = \sum_{i=1}^{n} \mu_iA_i$.

xx) If $\mathcal{S} \subseteq \mathbb{F}^n$ is a subspace and $A\mathcal{S} \subseteq \mathcal{S}$, then $A\mathcal{S}^\perp \subseteq \mathcal{S}^\perp$.

Source: [240, 719, 987, 990, 1242, 2476], [2238, pp. 345, 346], and [2991, pp. 294, 295]. **Related:** Fact 4.13.6, Fact 7.15.16, Fact 7.17.5, Fact 8.6.1, Fact 8.10.17, Fact 10.10.31, Fact 10.13.17, Fact 10.21.10, Fact 15.16.4, and Fact 15.16.12.

Fact 4.10.13. Let $A \in \mathbb{F}^{n\times n}$. Then, the following statements are equivalent:

i) A is Hermitian.

ii) $A^2 = A^*A$.

iii) $A^2 = AA^*$.

iv) $A^{*2} = A^*A$.

v) $A^{*2} = AA^*$.

vi) There exists $\alpha \in \mathbb{F}$ such that $A^2 = \alpha A^*A + (1 - \alpha)AA^*$.

vii) There exists $\alpha \in \mathbb{F}$ such that $A^{*2} = \alpha A^*A + (1 - \alpha)AA^*$.

viii) $\operatorname{tr} A^2 = \operatorname{tr} A^*A$.

ix) $\operatorname{tr} A^2 = \operatorname{tr} AA^*$.

x) $\operatorname{tr} A^{*2} = \operatorname{tr} A^*A$.

xi) $\operatorname{tr} A^{*2} = \operatorname{tr} AA^*$.

If, in addition, $\mathbb{F} = \mathbb{R}$, then the following statement is equivalent to *i)*–*xi)*:

xii) There exist $\alpha, \beta \in \mathbb{R}$ such that $\alpha A^2 + (1-\alpha)A^{\mathsf{T}2} = \beta A^{\mathsf{T}}A + (1-\beta)AA^{\mathsf{T}}$.

Source: To prove *viii)* $\Longrightarrow$ *i)*, use the Schur decomposition Theorem 7.5.1 to replace A with $D+S$, where D is diagonal and S is strictly upper triangular. Then, $\operatorname{tr} D^*D + \operatorname{tr} S^*S = \operatorname{tr} D^2 \le \operatorname{tr} D^*D$. Hence, $S = 0$, and thus $\operatorname{tr} D^*D = \operatorname{tr} D^2$, which implies that D is real. See [240, 1718] and [2991, pp. 254, 255]. **Remark:** Fact 11.13.2 states that, for all $A \in \mathbb{F}^{n\times n}$, $|\operatorname{tr} A^2| \le \operatorname{tr} A^*A$. **Related:** Fact 4.17.4.

Fact 4.10.14. Let $A \in \mathbb{F}^{n\times n}$, let $\alpha, \beta \in \mathbb{F}$, and assume that $\alpha \ne 0$. Then, the following statements are equivalent:

i) A is normal.

ii) $\alpha A + \beta I$ is normal.

Now, assume, in addition, that $\alpha, \beta \in \mathbb{R}$. Then, the following statements are equivalent:

iii) A is Hermitian.

iv) $\alpha A + \beta I$ is Hermitian.

Remark: The function $f(A) = \alpha A + \beta I$ is an *affine mapping*.

Fact 4.10.15. Let $A \in \mathbb{R}^{n\times n}$, and assume that A is skew symmetric. Then, the following statements hold:

i) $\det A \ge 0$, and $-A^2$ is positive semidefinite.

ii) If n is odd, then $\det A = 0$.

iii) If α is a real number, then $\det(I + \alpha A^2) \ge 0$.

iv) If $\alpha > 0$, then $\det(\alpha I + A) > 0$.

Source: *iv)* is given in [1158, p. 69]. **Related:** Fact 3.16.5 and Fact 4.13.20.

Fact 4.10.16. Let $A \in \mathbb{F}^{n\times n}$, and assume that A is skew Hermitian. If n is even, then $\det A \ge 0$. If n is odd, then $\det A$ is imaginary. **Source:** The first statement follows from Proposition 7.7.21.

Fact 4.10.17. Let $x, y \in \mathbb{F}^n$, and define $A \triangleq [x\ \ y]$. Then, $xy^* - yx^* = AJ_2A^*$. Furthermore, $xy^* - yx^*$ is skew Hermitian, and $\operatorname{rank}(xy^* - yx^*) \in \{0, 2\}$.

Fact 4.10.18. Let $x, y \in \mathbb{F}^n$. Then, the following statements hold:

i) xy^{T} is idempotent if and only if either $xy^{\mathsf{T}} = 0$ or $x^{\mathsf{T}}y = 1$.

ii) xy^{T} is Hermitian if and only if there exists $\alpha \in \mathbb{R}$ such that either $y = \alpha\overline{x}$ or $x = \alpha\overline{y}$.

Fact 4.10.19. Let $x, y \in \mathbb{F}^n$, and define $A \triangleq I - xy^{\mathsf{T}}$. Then, the following statements hold:

i) $\det A = 1 - x^{\mathsf{T}}y$.

ii) A is nonsingular if and only if $x^{\mathsf{T}}y \ne 1$.

iii) A is nonsingular if and only if A is elementary.

iv) $\operatorname{rank} A = n - 1$ if and only if $x^{\mathsf{T}}y = 1$.

v) A is Hermitian if and only if there exists $\alpha \in \mathbb{R}$ such that either $y = \alpha\overline{x}$ or $x = \alpha\overline{y}$.

vi) A is positive semidefinite if and only if A is Hermitian and $x^{\mathsf{T}}y \le 1$.

vii) A is positive definite if and only if A is Hermitian and $x^{\mathsf{T}}y < 1$.

viii) A is idempotent if and only if either $xy^{\mathsf{T}} = 0$ or $x^{\mathsf{T}}y = 1$.

ix) A is orthogonal if and only if either $x = 0$ or $y = \frac{1}{2}y^{\mathsf{T}}yx$.

x) A is involutory if and only if $x^{\mathsf{T}}y = 2$.

xi) A is a projector if and only if either $y = 0$ or $x = x^*xy$.

xii) A is a reflector if and only if either $y = 0$ or $2x = x^*xy$.

xiii) A is an elementary projector if and only if $x \neq 0$ and $y = (x^*x)^{-1}x$.

xiv) A is an elementary reflector if and only if $x \neq 0$ and $y = 2(x^*x)^{-1}x$.

Related: Fact 4.17.11.

Fact 4.10.20. Let $x, y \in \mathbb{F}^n$ satisfy $x^\mathsf{T}y \neq 1$. Then, $I - xy^\mathsf{T}$ is nonsingular, and

$$(I - xy^\mathsf{T})^{-1} = I - \frac{1}{x^\mathsf{T}y - 1}xy^\mathsf{T}.$$

Remark: The inverse of an elementary matrix is an elementary matrix.

Fact 4.10.21. Let $A \in \mathbb{F}^{n\times n}$, and assume that A is Hermitian. Then, $\det A$ is real.

Fact 4.10.22. Let $A \in \mathbb{F}^{n\times n}$, and assume that A is Hermitian. Then,

$$(\operatorname{tr} A)^2 \le (\operatorname{rank} A)\operatorname{tr} A^2.$$

Furthermore, equality holds if and only if there exists $\alpha \in \mathbb{R}$ such that $A^2 = \alpha A$. **Related:** Fact 7.12.13 and Fact 11.15.14.

Fact 4.10.23. Let $A \in \mathbb{R}^{n\times n}$, and assume that A is skew symmetric. Then, $\operatorname{tr} A = 0$. If, in addition, $B \in \mathbb{R}^{n\times n}$ is symmetric, then $\operatorname{tr} AB = 0$.

Fact 4.10.24. Let $A \in \mathbb{F}^{n\times n}$, and assume that A is skew Hermitian. Then, $\operatorname{Re}\operatorname{tr} A = 0$. If, in addition, $B \in \mathbb{F}^{n\times n}$ is Hermitian, then $\operatorname{Re}\operatorname{tr} AB = 0$.

Fact 4.10.25. Let $A \in \mathbb{F}^{n\times m}$. Then, A^*A is positive semidefinite. Furthermore, A^*A is positive definite if and only if A is left invertible. If these conditions hold, then $A^\mathsf{L} \in \mathbb{F}^{m\times n}$ defined by $A^\mathsf{L} \triangleq (A^*A)^{-1}A^*$ is a left inverse of A. **Related:** Fact 3.18.5, Fact 4.10.26, and Fact 4.17.8.

Fact 4.10.26. Let $A \in \mathbb{F}^{n\times m}$. Then, AA^* is positive semidefinite. Furthermore, AA^* is positive definite if and only if A is right invertible. If these conditions hold, then $A^\mathsf{R} \in \mathbb{F}^{m\times n}$ defined by $A^\mathsf{R} \triangleq A^*(AA^*)^{-1}$ is a right inverse of A. **Related:** Fact 3.18.5, Fact 4.10.25, and Fact 4.17.8.

Fact 4.10.27. Let $A \in \mathbb{F}^{n\times m}$. Then, A^*A, AA^*, and $\left[\begin{smallmatrix} 0 & A^* \\ A & 0 \end{smallmatrix}\right]$ are Hermitian, and $\left[\begin{smallmatrix} 0 & A^* \\ -A & 0 \end{smallmatrix}\right]$ is skew Hermitian.

Fact 4.10.28. Let $A \in \mathbb{F}^{n\times n}$. Then, $A + A^*$, $\jmath(A - A^*)$, and $\frac{1}{2\jmath}(A - A^*)$ are Hermitian, and $A - A^*$ is skew Hermitian. Furthermore,

$$A = \tfrac{1}{2}(A + A^*) + \tfrac{1}{2}(A - A^*) = \tfrac{1}{2}(A + A^*) + \jmath[\tfrac{1}{2\jmath}(A - A^*)],$$

$$[\tfrac{1}{2}(A + A^*)]^2 + [\tfrac{1}{2\jmath}(A - A^*)]^2 = \tfrac{1}{2}(AA^* + A^*A), \quad 2[A, A^*] = [A - A^*, A + A^*].$$

Related: Fact 7.20.2 and Fact 7.20.3.

Fact 4.10.29. Let $A, B \in \mathbb{F}^{n\times n}$, assume that A and B are Hermitian, and assume that $A + B$ is nonsingular. Then, $A(A + B)^{-1}B$ is Hermitian. **Source:** If A and B are nonsingular, then $A(A + B)^{-1}B = B(A + B)^{-1}A = A^{-1} + B^{-1}$. In the case where either A or B is singular, use a continuity argument. Alternatively, define $C \triangleq A + B$. Then, $A(A + B)^{-1}B = AC^{-1}(C - A) = A - AC^{-1}A = A - (C - B)C^{-1}A = BC^{-1}A = [A(A + B)^{-1}B]^*$.

Fact 4.10.30. Let $A, B \in \mathbb{F}^{n\times n}$, assume that A and B are Hermitian, and assume that either A or B is either positive definite or negative definite. Then, $A + \jmath B$, $A - \jmath B$, $A + BA^{-1}B$, $B + AB^{-1}A$, and $\left[\begin{smallmatrix} A & B \\ -B & A \end{smallmatrix}\right]$ are nonsingular. **Source:** Consider the case where B is either positive definite or negative definite. Let $x \in \mathbb{F}^n$ be nonzero, and assume that $(A + \jmath B)x = 0$. Hence, $x^*(A + \jmath B)x = 0$, and thus $-\jmath = x^*Ax/x^*Bx \in \mathbb{R}$, which is a contradiction. The remaining results follow from Fact 3.24.7.

Fact 4.10.31. Let $A, B \in \mathbb{F}^{n\times n}$, and assume that A and B are Hermitian. Then,

$$\operatorname{rank} AB = \operatorname{rank} BA, \quad \mathcal{R}(A) + \mathcal{R}(B) = \mathcal{R}([A\ \ B]) = \mathcal{R}(A^2 + B^2) = \operatorname{span}[\mathcal{R}(A) \cup \mathcal{R}(B)],$$

$$\dim[\mathcal{R}(A) + \mathcal{R}(B)] = \operatorname{rank}\,[A\ \ B] = \operatorname{rank}(A^2 + B^2) = \operatorname{rank}\begin{bmatrix} A \\ B \end{bmatrix}.$$

Furthermore, the following statements are equivalent:

i) $\operatorname{rank}\,[A\ \ B] = n$.

ii) $\operatorname{def}\begin{bmatrix} A \\ B \end{bmatrix} = 0$.

iii) $\mathcal{N}(A) \cap \mathcal{N}(B) = \{0\}$.

Source: Fact 3.14.8 and Fact 3.14.9.

Fact 4.10.32. Let $A, B \in \mathbb{C}^{n\times n}$, assume that A is either Hermitian or skew Hermitian, and assume that B is either Hermitian or skew Hermitian. Then, $\operatorname{rank} AB = \operatorname{rank} BA$. **Source:** AB and $(AB)^* = BA$ have the same singular values. See Fact 7.12.22. **Related:** Fact 3.13.33.

Fact 4.10.33. Let $A, B \in \mathbb{R}^{3\times 3}$, and assume that A and B are skew symmetric. Then,

$$\operatorname{tr} AB^3 = \tfrac{1}{2}(\operatorname{tr} AB)(\operatorname{tr} B^2), \quad \operatorname{tr} A^3B^3 = \tfrac{1}{4}(\operatorname{tr} A^2)(\operatorname{tr} AB)(\operatorname{tr} B^2) + \tfrac{1}{3}(\operatorname{tr} A^3)(\operatorname{tr} B^3).$$

Source: [181].

Fact 4.10.34. Let $A \in \mathbb{F}^{n\times n}$ and $k \ge 1$. Then, the following statements hold:

i) If A is (normal, Hermitian, unitary, involutory, positive semidefinite, positive definite, idempotent, nilpotent), then so is A^k.

ii) If A is (skew Hermitian, skew involutory), then so is A^{2k+1}.

iii) If A is Hermitian, then A^{2k} is positive semidefinite.

iv) If A is tripotent, then so is A^{3k}.

Fact 4.10.35. Let $a, b, c, d, e, f \in \mathbb{R}$, and define the skew-symmetric matrix $A \in \mathbb{R}^{4\times 4}$ given by

$$A \triangleq \begin{bmatrix} 0 & a & b & c \\ -a & 0 & d & e \\ -b & -d & 0 & f \\ -c & -e & -f & 0 \end{bmatrix}.$$

Then,

$$\det A = (af - be + cd)^2.$$

Source: [2418, p. 63]. **Related:** Fact 6.8.16 and Fact 6.10.8.

Fact 4.10.36. Let $A \in \mathbb{R}^{n\times n}$, and assume that A is skew symmetric, where every entry of A above the diagonal is 1. If n is odd, then $\operatorname{rank} A = n - 1$. If n is even, then $\det A = 1$.

Fact 4.10.37. Let $A \in \mathbb{R}^{2n\times 2n}$, and assume that A is skew symmetric. Then, there exists a nonsingular matrix $S \in \mathbb{R}^{2n\times 2n}$ such that $S^{\mathsf{T}}AS = J_{2n}$. **Source:** [218, p. 231].

Fact 4.10.38. Let $A \in \mathbb{F}^{n\times n}$. Then, the following statements are equivalent:

i) A is reverse Hermitian.

ii) $\hat{I}A$ is Hermitian.

iii) $A\hat{I}$ is Hermitian.

Furthermore, the following statements are equivalent:

iv) A is reverse symmetric.

v) $\hat{I}A$ is symmetric.

vi) $A\hat{I}$ is symmetric.

Fact 4.10.39. Let $A \in \mathbb{F}^{n\times m}$, assume that A is nonzero, define $r \triangleq \operatorname{rank} A$, and let $x \in \mathbb{F}^m$. Then, $Ax = 0$ if and only if there exist $y_1, \ldots, y_r \in \mathbb{F}^n$ and skew-Hermitian matrices $S_1, \ldots, S_r \in \mathbb{F}^{m\times m}$ such that $A = \sum_{i=1}^r y_i x^* S_i$. **Source:** [2261]. **Related:** Fact 4.11.4.

Fact 4.10.40. Let $A \in \mathbb{R}^{n\times n}$ be skew symmetric, let $\alpha_1, \ldots, \alpha_k$ be nonnegative numbers, and define $\beta \triangleq \prod_{i=1}^k \alpha_i$. Then,

$$[\det(A + \beta^{1/k} I)]^k \le \prod_{i=1}^k \det(A + \alpha_i I).$$

Now, assume that either $\alpha_1, \ldots, \alpha_k \in [0, 1]$ or $\alpha_1, \ldots, \alpha_k \in [1, \infty)$. Then,

$$\prod_{i=1}^k \det(A + \alpha_i I) \le [\det(A + I)]^{k-1} \det(A + \beta I).$$

Source: [1137].

4.11 Facts on Linear Interpolation

Fact 4.11.1. Let $y \in \mathbb{F}^n$ and $x \in \mathbb{F}^m$. Then, there exists $A \in \mathbb{F}^{n\times m}$ such that $y = Ax$ if and only if either $y = 0$ or $x \ne 0$. If $y = 0$, then one such matrix is $A = 0$. If $x \ne 0$, then one such matrix is

$$A = (x^*x)^{-1} yx^*.$$

Finally, if $x \ne 0$, then $A \in \mathbb{F}^{n\times m}$ satisfies $y = Ax$ if and only if there exists $B \in \mathbb{F}^{n\times m}$ such that

$$A = (x^*x)^{-1} yx^* + B(x^*xI_m - xx^*).$$

Source: [2704]. **Remark:** This is a linear interpolation problem. See [1549, 2233].

Fact 4.11.2. Let $x, y \in \mathbb{F}^n$, and assume that $x \ne 0$. Then, there exists a Hermitian matrix $A \in \mathbb{F}^{n\times n}$ such that $y = Ax$ if and only if x^*y is real. One such matrix is

$$A = (x^*x)^{-1}[yx^* + xy^* - x^*yI].$$

Now, assume that x and y are real. Then, $A = (x^{\mathsf{T}}x)^{-1}[yx^{\mathsf{T}} + xy^{\mathsf{T}} - x^{\mathsf{T}}yI]$ satisfies

$$\sigma_{\max}(A) = \frac{\|y\|_2}{\|x\|_2} = \min\{\sigma_{\max}(B)\colon B \in \mathbb{R}^{n\times n} \text{ is symmetric and } y = Bx\}.$$

Source: The last statement is given in [2473].

Fact 4.11.3. Let $x, y \in \mathbb{F}^n$, and assume that $x \ne 0$. Then, there exists a positive-definite matrix $A \in \mathbb{F}^{n\times n}$ such that $y = Ax$ if and only if x^*y is real and positive. One such matrix is

$$A = I + (x^*y)^{-1} yy^* - (x^*x)^{-1} xx^*.$$

Source: To show that A is positive definite, note that the elementary projector $I - (x^*x)^{-1}xx^*$ is positive semidefinite and $\operatorname{rank}[I - (x^*x)^{-1}xx^*] = n - 1$. Since $(x^*y)^{-1}yy^*$ is positive semidefinite, it follows that $\mathcal{N}(A) \subseteq \mathcal{N}[I - (x^*x)^{-1}xx^*]$. Next, since $x^*y > 0$, it follows that $y^*x \ne 0$ and $y \ne 0$, and thus $x \notin \mathcal{N}(A)$. Consequently, $\mathcal{N}(A) \subset \mathcal{N}[I - (x^*x)^{-1}xx^*]$ (note proper inclusion), and thus $\operatorname{def} A < 1$. Hence, A is nonsingular.

Fact 4.11.4. Let $x, y \in \mathbb{F}^n$, where $x \ne 0$. Then, the following statements are equivalent:

i) $x^*y = 0$.

ii) There exists a skew-Hermitian matrix $A \in \mathbb{F}^{n\times n}$ such that $y = Ax$.

iii) $\operatorname{Re} x^*y = 0$.

If *i)* holds, then one such matrix satisfying *ii)* is $A = (x^*x)^{-1}(yx^* - xy^*)$. Now, assume that $\mathbb{F} = \mathbb{R}$. Then, the following statements are equivalent:

iv) $x^{\mathsf{T}}y = 0$.

v) There exists a skew-symmetric matrix $A \in \mathbb{R}^{n\times n}$ such that $y = Ax$.

If these statements hold, then one such matrix satisfying *v)* is $A = (x^{\mathsf{T}}x)^{-1}(yx^{\mathsf{T}} - xy^{\mathsf{T}})$. **Source:** [1872]. **Related:** Fact 4.10.39.

Fact 4.11.5. Let $x, y \in \mathbb{F}^n$. Then, there exists a unitary matrix $A \in \mathbb{F}^{n\times n}$ such that $Ax = y$ if and only if $x^*x = y^*y$. Now, let $\mathbb{F} = \mathbb{R}$. Then, one such matrix is given by a product of n plane rotations given by Fact 7.17.18. Another matrix is given by the product of elementary reflectors given by Fact 7.17.17. For $n = 3$, one such matrix is given by Fact 4.14.9, while another is given by the exponential of a skew-symmetric matrix given by Fact 15.12.7. **Related:** Fact 11.17.17. **Problem:** Construct A in the case where $\mathbb{F} = \mathbb{C}$.

Fact 4.11.6. Let $x, y \in \mathbb{R}^n$, where $x_{(1)} \geq \cdots \geq x_{(n)}$ and $y_{(1)} \geq \cdots \geq y_{(n)}$. Then, the following statements are equivalent:

i) $x \overset{\mathrm{s}}{\prec} y$.

ii) $x \in \operatorname{conv}\{Ay\colon A \in \mathrm{P}(n)\}$.

iii) There exists a doubly stochastic matrix $A \in \mathbb{R}^{n\times n}$ such that $y = Ax$.

Source: [449, p. 33], [1448, p. 197], [1969, p. 22], and [1971, p. 33]. **Remark:** The equivalence of *i)* and *ii)* is due to R. Rado. See [1969, p. 113] and [1971, p. 162]. The equivalence of *i)* and *iii)* is the *Hardy-Littlewood-Polya theorem.* **Related:** Fact 4.13.1 and Fact 10.21.11.

4.12 Facts on the Cross Product

Fact 4.12.1. Let $x, y, z, v, w \in \mathbb{R}^3$, and define the cross-product matrix $K(x) \in \mathbb{R}^{3\times 3}$ by

$$K(x) \triangleq \begin{bmatrix} 0 & -x_{(3)} & x_{(2)} \\ x_{(3)} & 0 & -x_{(1)} \\ -x_{(2)} & x_{(1)} & 0 \end{bmatrix}.$$

Then, the following statements hold:

i) $x \times x = K(x)x = 0$, $x^\mathsf{T}K(x) = 0$, $K^\mathsf{T}(x) = -K(x)$, $K^2(x) = xx^\mathsf{T} - (x^\mathsf{T}x)I$.

ii) $\operatorname{tr} K^\mathsf{T}(x)K(x) = -\operatorname{tr} K^2(x) = 2x^\mathsf{T}x$, $K^3(x) = -(x^\mathsf{T}x)K(x)$.

iii) $[I - K(x)]^{-1} = I + \frac{1}{1+x^\mathsf{T}x}[K(x) + K^2(x)]$.

iv) $[I + \frac{1}{2}K(x)][I - \frac{1}{2}K(x)]^{-1} = I + \frac{4}{4+x^\mathsf{T}x}[K(x) + \frac{1}{2}K^2(x)]$.

v) $[I - K(x)][I + K(x)]^{-1} = \frac{1}{1+x^\mathsf{T}x}[(1 - x^\mathsf{T}x)I + 2xx^\mathsf{T} - 2K(x)]$.

vi) Define $H(x) \triangleq \frac{1}{2}[\frac{1}{2}(1 - x^\mathsf{T}x)I + xx^\mathsf{T} + K(x)]$. Then, $H(x)H^\mathsf{T}(x) = \frac{1}{16}(1 + x^\mathsf{T}x)^2 I$.

vii) For all $\alpha, \beta \in \mathbb{R}$, $K(\alpha x + \beta y) = \alpha K(x) + \beta K(y)$.

viii) $x \times y = -(y \times x) = K(x)y = -K(y)x = K^\mathsf{T}(y)x$.

ix) If $x \times y \neq 0$, then $\mathcal{N}[(x \times y)^\mathsf{T}] = \{x \times y\}^\perp = \mathcal{R}([x\ \ y])$.

x) $K(x \times y) = K[K(x)y] = [K(x), K(y)]$.

xi) $K(x \times y) = yx^\mathsf{T} - xy^\mathsf{T} = [x\ \ y]\begin{bmatrix} -y^\mathsf{T} \\ x^\mathsf{T} \end{bmatrix} = -[x\ \ y]J_2[x\ \ y]^\mathsf{T}$.

xii) $(x \times y) \times x = [(x^\mathsf{T}x)I - xx^\mathsf{T}]y$.

xiii) $K[(x \times y) \times x] = (x^\mathsf{T}x)K(y) - (x^\mathsf{T}y)K(x)$.

xiv) $(x \times y)^\mathsf{T}(x \times y) = \det[x\ \ y\ \ x \times y]$.

xv) $(x \times y)^\mathsf{T}z = x^\mathsf{T}(y \times z) = \det[x\ \ y\ \ z]$.

xvi) $x \times (y \times z) = K(x)K(y)z = -K(x)K(z)y = (x^\mathsf{T}z)y - (x^\mathsf{T}y)z$.

xvii) $(x \times y) \times z = -K(z)K(x)y = K(z)K(y)x = (x^\mathsf{T}z)y - (y^\mathsf{T}z)x$.

xviii) $x \times (y \times z) + y \times (z \times x) + z \times (x \times y) = 0$.

xix) $K[(x \times y) \times z] = (x^\mathsf{T}z)K(y) - (y^\mathsf{T}z)K(x)$.

xx) $K[x \times (y \times z)] = (x^\mathsf{T}z)K(y) - (x^\mathsf{T}y)K(z)$.

xxi) $(x \times y)^\mathsf{T}(x \times y) = x^\mathsf{T}xy^\mathsf{T}y - (x^\mathsf{T}y)^2$.

xxii) $K(x)K(y) = [K(y)K(x)]^{\mathsf{T}} = yx^{\mathsf{T}} - (x^{\mathsf{T}}y)I.$

xxiii) $\operatorname{tr} K(x)K(y) = -2x^{\mathsf{T}}y.$

xxiv) $K(x)K(y)K(x) = -(x^{\mathsf{T}}y)K(x).$

xxv) $K^2(x)K(y) + K(y)K^2(x) = -(x^{\mathsf{T}}x)K(y) - (x^{\mathsf{T}}y)K(x).$

xxvi) $K^2(x)K^2(y) - K^2(y)K^2(x) = -(x^{\mathsf{T}}y)K(x\times y).$

xxvii) $K(z)K(y)K(x) - K(y)K(z)K(x) = zx^{\mathsf{T}}K(y) - yx^{\mathsf{T}}K(z).$

xxviii) $K(x)K(y)K(z) + K(z)K(y)K(x) + (x^{\mathsf{T}}y)K(z) = K(x)K(z)K(y) + K(y)K(z)K(x) + (x^{\mathsf{T}}z)K(y).$

xxix) $\|x\times y\|_2 = \|x\|_2\|x\|_2 \sin\theta$, where θ is the angle between x and y.

xxx) $2xx^{\mathsf{T}}K(y) + (x^{\mathsf{T}}y)K(x) = (x\times y)x^{\mathsf{T}} + x(x\times y)^{\mathsf{T}} + (x^{\mathsf{T}}x)K(y).$

xxxi) If $\|x\|_2 = \|y\|_2 = \|z\|_2 = 1$, then $1 + 2(x^{\mathsf{T}}y)(y^{\mathsf{T}}z)(z^{\mathsf{T}}x) = [x^{\mathsf{T}}(y\times z)]^2 + (x^{\mathsf{T}}y)^2 + (y^{\mathsf{T}}z)^2 + (z^{\mathsf{T}}x)^2.$

xxxii) $K(x)K(z)[(x^{\mathsf{T}}w)y - (x^{\mathsf{T}}y)w] = K(x)K(w)(x^{\mathsf{T}}z)y.$

xxxiii) $xz^{\mathsf{T}}K(y) + yx^{\mathsf{T}}K(z) + zy^{\mathsf{T}}K(x) = -(\det[x\;\;y\;\;z])I.$

xxxiv) $(x\times y)^{\mathsf{T}}(z\times w) = x^{\mathsf{T}}zy^{\mathsf{T}}w - x^{\mathsf{T}}wy^{\mathsf{T}}z = \det\begin{bmatrix} x^{\mathsf{T}}z & x^{\mathsf{T}}w \\ y^{\mathsf{T}}z & y^{\mathsf{T}}w \end{bmatrix}.$

xxxv) $(x\times y)\times(z\times w) = [x^{\mathsf{T}}(y\times w)]z - [x^{\mathsf{T}}(y\times z)]w = [x^{\mathsf{T}}(z\times w)]y - [y^{\mathsf{T}}(z\times w)]x.$

xxxvi) $(x\times y)\times(x\times z) = [(x\times y)^{\mathsf{T}}z]x = [x^{\mathsf{T}}(y\times z)]x.$

xxxvii) $x\times[y\times(z\times w)] = (y^{\mathsf{T}}w)(x\times z) - (y^{\mathsf{T}}z)(x\times w).$

xxxviii) $x\times[y\times(y\times x)] = y\times[x\times(y\times x)] = (y^{\mathsf{T}}x)(x\times y).$

xxxix) Let $A\in\mathbb{R}^{3\times 3}$. Then,

$$(\operatorname{tr} A)K(x) = K(Ax) + A^{\mathsf{T}}K(x) + K(x)A, \quad K(Ax)A = A^{\mathsf{AT}}K(x),$$
$$A^{\mathsf{T}}K(x)A = K(A^{\mathsf{A}}x), \quad Ax\times Ay = A^{\mathsf{AT}}(x\times y),$$
$$A^{\mathsf{T}}K(Ax)A = (\det A)K(x), \quad A^{\mathsf{T}}(Ax\times Ay) = (\det A)(x\times y),$$
$$K(Ax\times Ay) = AK(x\times y)A^{\mathsf{T}} = K[A^{\mathsf{AT}}(x\times y)].$$

xl) Let $A\in\mathbb{R}^{3\times 3}$, and assume that A is orthogonal. Then,

$$K(Ax) = (\det A)AK(x)A^{\mathsf{T}}, \quad Ax\times Ay = (\det A)A(x\times y).$$

xli) Let $A\in\mathbb{R}^{3\times 3}$, and assume that A is orthogonal and $\det A = 1$. Then,

$$K(Ax) = AK(x)A^{\mathsf{T}}, \quad Ax\times Ay = A(x\times y).$$

xlii) $K(x)^{\mathsf{A}} = xx^{\mathsf{T}}$, $[x\;\;y\;\;z]^{\mathsf{A}} = [y\times z\;\;z\times x\;\;x\times y]^{\mathsf{T}}.$

xliii) $\det\begin{bmatrix} K(x) & y \\ -y^{\mathsf{T}} & 0 \end{bmatrix} = (x^{\mathsf{T}}y)^2.$

xliv) $\begin{bmatrix} K(x) & y \\ -y^{\mathsf{T}} & 0 \end{bmatrix}^{\mathsf{A}} = -(x^{\mathsf{T}}y)\begin{bmatrix} K(y) & x \\ -x^{\mathsf{T}} & 0 \end{bmatrix}.$

xlv) If $x^{\mathsf{T}}y \neq 0$, then

$$\begin{bmatrix} K(x) & y \\ -y^{\mathsf{T}} & 0 \end{bmatrix}^{-1} = \frac{-1}{x^{\mathsf{T}}y}\begin{bmatrix} K(y) & x \\ -x^{\mathsf{T}} & 0 \end{bmatrix}.$$

xlvi) Let α and β be real numbers, and assume that either $\alpha\neq 0$ or $\beta\|x\|_2^2 \neq 1$. Then,

$$[I + \alpha K(x) + \beta K^2(x)]^{-1} = I - \frac{\alpha}{\alpha^2\|x\|_2^2 + (\beta\|x\|_2^2 - 1)^2}K(x) + \frac{\alpha^2 + \beta^2\|x\|_2^2 - \beta}{\alpha^2\|x\|_2^2 + (\beta\|x\|_2^2 - 1)^2}K^2(x).$$

xlvii) $[(x\times y)^{\mathsf T}v]z\times w+[(y\times z)^{\mathsf T}v]x\times w+[(z\times x)^{\mathsf T}v]y\times w=[(x\times y)^{\mathsf T}z]v\times w$.

xlviii) Let $\mathcal S$ be a triangle with vertices $x,y,z\in\mathbb R^3$ and sides given by the vectors $a\triangleq x-y$, $b\triangleq y-z$, and $c\triangleq z-x$. Then, the vectors $a'\triangleq a\times(b\times c)$, $b'\triangleq b\times(c\times a)$, and $c'\triangleq c\times(a\times b)$ are the sides of a triangle $\mathcal S'$ that is similar to $\mathcal S$.

xlix) Let $x,y,z\in\mathbb R^3$ be linearly independent, and define $w\triangleq x\times y+y\times z+z\times x$. Then, w is perpendicular to the plane passing through the points x,y,z.

l) $K(x)=\left(\sqrt{\frac{\|x\|_2}{2}}I+\frac{1}{\sqrt{2\|x\|_2}}K(x)-\frac{1}{\sqrt{2\|x\|_2^3}}xx^{\mathsf T}\right)^2=\left(\jmath\sqrt{\frac{\|x\|_2}{2}}I-\frac{\jmath}{\sqrt{2\|x\|_2}}K(x)-\frac{\jmath}{\sqrt{2\|x\|_2^3}}xx^{\mathsf T}\right)^2$.

li) $\|x\times y\|_2=\|K(x)y\|_2=\|K(y)x\|_2=[\|x\|_2^2\|y\|_2^2-(x^{\mathsf T}y)^2]^{1/2}$.

lii) $\|x\times(y-z)\|_2^2+\|y\times(z-x)\|_2^2+\|z\times(x-y)\|_2^2=\|x\times y\|_2^2+\|y\times z\|_2^2+\|z\times x\|_2^2+\|x\times y+y\times z+z\times x\|_2^2$.

liii) $\|x\times y\|_2^2I=\|x\|_2^2yy^{\mathsf T}+\|y\|_2^2xx^{\mathsf T}-(x^{\mathsf T}y)(xy^{\mathsf T}+yx^{\mathsf T})+(x\times y)(x\times y)^{\mathsf T}$.

liv) Let $n\ge1$. Then, $K^{2n}(x)=(-1)^n(x^{\mathsf T}x)^{n-1}[(x^{\mathsf T}x)I-xx^{\mathsf T}]$.

lv) Let $n\ge0$. Then, $K^{2n+1}(x)=(-1)^n(x^{\mathsf T}x)^nK(x)$.

lvi) Let $n\ge0$. Then, $\sigma_{\max}[K^n(x)]=\|x\|_2^n$.

lvii) Let $A\in\mathbb R^{3\times3}$, and assume that A is positive definite. Then,

$$\|K(x)Ax\|_2\le[\lambda_{\max}^2(A)-\lambda_{\min}^2(A)]^{1/2}\|x\|_2^2.$$

lviii) $\operatorname{mspec}[K(x)]=\{0,\jmath\|x\|_2,-\jmath\|x\|_2\}_{\rm ms}$, and x is an eigenvector of $K(x)$ associated with 0.

lix) Assume that $x\ne0$, $x^{\mathsf T}y=0$, and $\|y\|_2=\sqrt2/2$. Then,

$$K(x)=S\begin{bmatrix}0&0&0\\0&\jmath\|x\|_2&0\\0&0&-\jmath\|x\|_2\end{bmatrix}S^*,$$

where

$$S\triangleq\left[\frac{1}{\|x\|_2}x\quad y-\frac{\jmath}{\|x\|_2}K(x)y\quad y+\frac{\jmath}{\|x\|_2}K(x)y\right].$$

Furthermore, S is a unitary matrix.

lx) $\sigma[K(x)]=[\|x\|_2\ \ \|x\|_2\ \ 0]^{\mathsf T}$.

lxi) Assume that $x\ne0$, $\|y\|_2=1$, and $x^{\mathsf T}y=0$. Then,

$$K(x)=U\begin{bmatrix}\|x\|_2&0&0\\0&\|x\|_2&0\\0&0&0\end{bmatrix}V^{\mathsf T},$$

where

$$U\triangleq\left[\frac{1}{\|x\|_2}K(x)y\quad -y\quad \frac{1}{\|x\|_2}x\right],\quad V\triangleq\left[y\quad \frac{1}{\|x\|_2}K(x)y\quad \frac{1}{\|x\|_2}x\right].$$

Furthermore, U and V are orthogonal matrices.

lxii) Let $A\in\mathbb R^{3\times3}$, and assume that A is symmetric. Then, $\operatorname{tr}AK(x)=0$.

lxiii) Let $A\in\mathbb R^{3\times3}$, and assume that $A-A^{\mathsf T}=K(y)$. Then, $\operatorname{tr}AK(x)=-x^{\mathsf T}y$.

lxiv) $\|x\times y\|_2^2=\frac12\sum_{j=1}^n\sum_{i=1}^n(x_{(i)}y_{(j)}-x_{(j)}y_{(i)})^2$.

Source: *iii)*, *iv)*, and *xxiv)*–*xxvi)* are given in [1506, p. 363]; *v)* is given in [1961]; *vi)* is given in [2740]; *xxi)* is equivalent to $\sin^2\theta+\cos^2\theta=1$; *xxvii)* and *xxviii)* are given in [1158, p. 203]; *xxix)* arises from quaternion multiplication $|q_3|=|q_2||q_1|$, see Fact 4.14.8; *xxxi)* is due to N. Crasta; *xxxii)* follows from [1200, 1.10-7, p. 58]; *xxxiii)* is given in [2724]; *xxxv)* implies *xxxvi)*; *xxxix)* is given in [1961]; *xlii)* is given in [1961, 2701]; *xliii)*–*xlv)* are given in [2723], see [908, 1028, 1506, 2181, 2435, 2596, 2711]; *xlvii)* is given in [1509]; *xlviii)* and *xlix)* are given in [1158, p. 203]; *l)* is given in [2698]; *lii)* is given in [1860, pp. 33, 216, 217]; *lxiv)* is given in [968, p. 113]. **Remark:** Cross products of complex vectors are considered in [1253]. **Remark:** *xviii*) is the Jacobi identity, see Fact 3.23.3. **Remark:** For $\theta \in (0, \pi)$, *xxx*) gives twice the area of the triangle with vertices 0, x, and y. See Fact 5.2.6. **Credit:** The Schur decomposition in *lix*) and the singular value decomposition in *lxi*) are due to A. H. J. de Ruiter. **Related:** Fact 6.9.18, Fact 8.9.18, Fact 15.12.6, and Fact 15.12.11.

Fact 4.12.2. Let $x_1, \ldots, x_{n-1}, y \in \mathbb{R}^n$, define $M \triangleq [x_1 \ \cdots \ x_{n-1}] \in \mathbb{R}^{n\times(n-1)}$, for all $i \in \{1, \ldots, n\}$, define $\alpha_i \triangleq \det M_{[i,\cdot]}$, and define

$$x_1 \times \cdots \times x_{n-1} \triangleq \begin{bmatrix} \alpha_1 \\ -\alpha_2 \\ \alpha_3 \\ \vdots \\ (-1)^{n+1}\alpha_n \end{bmatrix} \in \mathbb{R}^n.$$

Then,

$$(x_1 \times \cdots \times x_{n-1})^{\mathsf{T}} y = \det [y \ x_1 \ x_2 \ \cdots \ x_{n-1}].$$

In addition, the following statements hold:

i) For all $i \in \{1, \ldots, n-1\}$, $(x_1 \times \cdots \times x_{n-1})^{\mathsf{T}} x_i = 0$.

ii) $x_1 \times \cdots \times x_{n-1} = 0$ if and only if $x_1, \ldots, x_{n-1}$ are linearly independent.

iii) $\det [x_1 \times \cdots \times x_{n-1} \ \ x_1 \ x_2 \ \cdots \ x_{n-1}] = \|x_1 \times \cdots \times x_{n-1}\|_2^2$.

iv) $\|x_1 \times \cdots \times x_{n-1}\|_2$ is the $(n-1)$-dimensional volume of the parallellotope generated by $x_1, \ldots, x_{n-1}$.

Source: [774, 908]. **Remark:** An extension of the cross product to higher dimensions is given by the exterior product in Clifford algebras. See Fact 11.8.13 and [784, 908, 925, 1178, 1268, 1375, 1376, 1759, 1892, 1977, 2265]. A cross product on $\mathbb{R}^7$ is defined in [1035, pp. 297–299]. **Remark:** For $n = 3$, this definition coincides with the usual cross product.

Fact 4.12.3. Let $A \in \mathbb{R}^{3\times3}$, assume that A is orthogonal, let $B \in \mathbb{C}^{3\times3}$, and assume that B is symmetric. Then,

$$\sum_{i=1}^{3} (Ae_i) \times (BAe_i) = 0.$$

Source: For $i = 1, 2, 3$, multiply by $e_i^{\mathsf{T}} A^{\mathsf{T}}$.

Fact 4.12.4. Let $\alpha_1, \alpha_2, \alpha_3$ be distinct positive numbers, let $A \in \mathbb{R}^{3\times3}$, assume that A is orthogonal, and assume that

$$\sum_{i=1}^{3} \alpha_i e_i \times Ae_i = 0.$$

Then, $A \in \{I, \operatorname{diag}(1,-1,-1), \operatorname{diag}(-1,1,-1), \operatorname{diag}(-1,-1,1)\}$. **Remark:** This result characterizes equilibria for a dynamical system on SO(3). See [681].

4.13 Facts on Inner, Unitary, and Shifted-Unitary Matrices

Fact 4.13.1. Let $A \in \mathbb{R}^{n\times n}$. Then, the following statements are equivalent:

i) A is a doubly stochastic matrix.

ii) $A \in \operatorname{conv} \mathrm{P}(n)$.

If these statements hold, then A is a convex combination of not more than $n^2 - 2n + 2$ permutation matrices. **Source:** [1448, p. 527]. **Related:** Fact 4.11.6 and Fact 6.11.11. **Credit:** G. Birkhoff.

Fact 4.13.2. Let $\mathcal{S} \subseteq \mathbb{F}^n$, assume that $\mathcal{S}$ is a subspace, let $A \in \mathbb{F}^{n\times n}$, and assume that A is unitary. Then, $(A\mathcal{S})^{\perp} = A\mathcal{S}^{\perp}$. **Source:** Fact 3.18.19.

Fact 4.13.3. Let $\mathcal{S}_1, \mathcal{S}_2 \subseteq \mathbb{F}^n$, assume that $\mathcal{S}_1$ and $\mathcal{S}_2$ are subspaces, and assume that $\dim \mathcal{S}_1 \le \dim \mathcal{S}_2$. Then, there exists a unitary matrix $A \in \mathbb{F}^{n\times n}$ such that $A\mathcal{S}_1 \subseteq \mathcal{S}_2$.

Fact 4.13.4. Let $\mathcal{S}_1, \mathcal{S}_2 \subseteq \mathbb{F}^n$, assume that $\mathcal{S}_1$ and $\mathcal{S}_2$ are subspaces, and assume that $\dim \mathcal{S}_1 + \dim \mathcal{S}_2 \le n$. Then, there exists a unitary matrix $A \in \mathbb{F}^{n\times n}$ such that $A\mathcal{S}_1 \subseteq \mathcal{S}_2^\perp$. **Source:** Fact 4.13.3.

Fact 4.13.5. Let $A \in \mathbb{F}^{n\times n}$, and assume that A is unitary. Then, the following statements hold:

i) $A = A^{-*}$.

ii) $A^\mathsf{T} = \overline{A}^{-1} = \overline{A}^{*}$.

iii) $\overline{A} = A^{-\mathsf{T}} = \overline{A}^{-*}$.

iv) $A^* = A^{-1}$.

Fact 4.13.6. Let $A \in \mathbb{F}^{n\times n}$, and assume that A is nonsingular. Then, the following statements are equivalent:

i) A is normal.

ii) $A = A^*AA^{-*}$.

iii) $A^{-1}A^*$ is unitary.

iv) $[A, A^*] = 0$.

v) $[A, A^{-*}] = 0$.

vi) $[A^{-1}, A^{-*}] = 0$.

vii) $[A, A^{-1}A^*] = 0$.

Source: [1242]. **Related:** Fact 4.10.12, Fact 7.17.5, Fact 8.6.1, and Fact 8.10.17.

Fact 4.13.7. Let $A \in \mathbb{F}^{n\times m}$. If A is (left inner, right inner), then A is (left invertible, right invertible) and A^* is a (left inverse, right inverse) of A.

Fact 4.13.8. Let $A \in \mathbb{F}^{n\times m}$. If A is (left inner, right inner), then (AA^*, A^*A) is idempotent.

Fact 4.13.9. Let $x, y \in \mathbb{F}^n$, let $A \in \mathbb{F}^{n\times n}$, and assume that A is unitary. Then, $x^*y = 0$ if and only if $(Ax)^*Ay = 0$.

Fact 4.13.10. Let $A \in \mathbb{F}^{n\times n}$, and assume that A is block diagonal. Then, A is (unitary, shifted unitary) if and only if every diagonally located block has the same property.

Fact 4.13.11. Let $A \in \mathbb{F}^{n\times n}$, and assume that A is unitary. Then, $\frac{1}{\sqrt{2}}\left[\begin{smallmatrix} A & -A \\ A & A \end{smallmatrix}\right]$ is unitary.

Fact 4.13.12. Let $A, B \in \mathbb{R}^{n\times n}$. Then, $A + \jmath B$ is (Hermitian, skew Hermitian, unitary) if and only if $\left[\begin{smallmatrix} A & B \\ -B & A \end{smallmatrix}\right]$ is (symmetric, skew symmetric, orthogonal). **Related:** Fact 3.24.7.

Fact 4.13.13. Let $A \in \mathbb{F}^{n\times n}$, and assume that A is unitary. Then,

$$|\mathrm{Re}\,\mathrm{tr}\,A| \le n, \quad |\mathrm{Im}\,\mathrm{tr}\,A| \le n, \quad |\mathrm{tr}\,A| \le n.$$

Remark: The third inequality does not follow from the first two inequalities.

Fact 4.13.14. Let $A \in \mathbb{R}^{n\times n}$, and assume that A is orthogonal. Then, $-1_{n\times n} \le\le A \le\le 1_{n\times n}$; that is, for all $i, j \in \{1, \ldots, n\}$, $|A_{(i,j)}| \le 1$. Hence, $|\mathrm{tr}\,A| \le n$. Furthermore, the following statements are equivalent:

i) $A = I$.

ii) $I \odot A = I$.

iii) $\mathrm{tr}\,A = n$.

Finally, if n is odd and $\det A = 1$, then $2 - n \le \mathrm{tr}\,A \le n$. **Related:** Fact 4.13.15.

Fact 4.13.15. Let $A \in \mathbb{R}^{n\times n}$, assume that A is orthogonal, let $B \in \mathbb{R}^{n\times n}$, and assume that B is diagonal and positive definite. Then,

$$-B1_{n\times n} \le\le BA \le\le B1_{n\times n}, \quad -\mathrm{tr}\,B \le \mathrm{tr}\,BA \le \mathrm{tr}\,B.$$

Furthermore, the following statements are equivalent:

i) $BA = B$.

ii) $I \odot (BA) = B$.

iii) $\operatorname{tr} BA = \operatorname{tr} B$.

Related: Fact 4.13.14.

Fact 4.13.16. Let $x \in \mathbb{C}^n$, where $n \geq 2$. Then, the following statements are equivalent:

i) There exists a unitary matrix $A \in \mathbb{C}^{n\times n}$ such that $x = \begin{bmatrix} A_{(1,1)} \\ \vdots \\ A_{(n,n)} \end{bmatrix}$.

ii) For all $j \in \{1,\dots,n\}$, $|x_{(j)}| \leq 1$ and $2(1 - |x_{(j)}|) + \sum_{i=1}^n |x_{(i)}| \leq n$.

Source: [2731]. **Remark:** This result is equivalent to the Schur-Horn theorem given by Fact 10.21.14. **Remark:** The inequalities in *ii)* define a polytope.

Fact 4.13.17. Let $A \in \mathbb{C}^{n\times n}$, and assume that A is unitary. Then, $|\det A| = 1$.

Fact 4.13.18. Let $A \in \mathbb{R}^{n\times n}$, and assume that A is orthogonal. Then, either $\det A = 1$ or $\det A = -1$. Now, assume that A is a permutation matrix. Then, the following statements hold:

i) A is either an even permutation matrix or an odd permutation matrix.

ii) If A is a transposition matrix, then A is odd.

iii) Let σ be a permutation of $(1,\dots,n)$, and assume that $[\sigma(1) \ \cdots \ \sigma(n)]^\mathsf{T} = A[1 \ \cdots \ n]^\mathsf{T}$. Then,

$$\det A = \prod_{1\leq i<j\leq n} \frac{\sigma(j) - \sigma(i)}{j - i}.$$

Related: Fact 3.16.1 and Fact 4.31.14.

Fact 4.13.19. Let $A, B \in \mathrm{SO}(3)$. Then, $\det(A + B) \geq 0$. **Source:** [2064].

Fact 4.13.20. Let $A \in \mathbb{F}^{n\times n}$, and assume that A is unitary. Then, $|\det(I + A)| \leq 2^n$. If, in addition, A is real, then $0 \leq \det(I + A) \leq 2^n$. **Related:** Fact 3.16.5 and Fact 4.10.15.

Fact 4.13.21. Let $M \triangleq \left[\begin{smallmatrix} A & B \\ C & D \end{smallmatrix}\right] \in \mathbb{F}^{(n+m)\times(n+m)}$, and assume that M is unitary. Then,

$$\det A = (\det M)\overline{\det D}, \quad |\det A| = |\det D|.$$

Source: Let $\left[\begin{smallmatrix} \hat{A} & \hat{B} \\ \hat{C} & \hat{D} \end{smallmatrix}\right] \triangleq M^{-1}$, and take the determinant of $M\left[\begin{smallmatrix} I & \hat{B} \\ 0 & \hat{D} \end{smallmatrix}\right] = \left[\begin{smallmatrix} A & 0 \\ C & I \end{smallmatrix}\right]$. See [24, 2426]. **Related:** Fact 3.17.7, Fact 3.17.32, Fact 4.13.22, and Fact 11.16.13.

Fact 4.13.22. Let $A \in \mathbb{F}^{n\times n}$, assume that A is unitary, and let $i \in \{1,\dots,n\}$. Then,

$$\det A_{[i,i]} = \overline{A_{(i,i)}} \det A, \quad |\det A_{[i,i]}| = |A_{(i,i)}|.$$

Source: Use $(A^\mathsf{A})_{(i,i)} = (\det A)(A^*)_{(i,i)}$. See Fact 4.13.21 and [2991, pp. 43, 172].

Fact 4.13.23. Let $A \in \mathbb{F}^{n\times n}$, assume that A is unitary, and let $x \in \mathbb{F}^n$ satisfy $x^*x = 1$ and $Ax = -x$. Then, the following statements hold:

i) $\det(A + I) = 0$.

ii) $A + 2xx^*$ is unitary.

iii) $A = (A + 2xx^*)(I - 2xx^*) = (I - 2xx^*)(A + 2xx^*)$.

iv) $\det(A + 2xx^*) = -\det A$.

Fact 4.13.24. The following statements hold:

i) If $A \in \mathbb{F}^{n\times n}$ is Hermitian, then $A + \jmath I$ is nonsingular, $B \triangleq (\jmath I - A)(\jmath I + A)^{-1}$ is unitary, and $I + B = 2\jmath(\jmath I + A)^{-1}$.

ii) If $B \in \mathbb{F}^{n\times n}$ is unitary and $\lambda \in \mathbb{C}$ is such that $|\lambda| = 1$ and $I + \lambda B$ is nonsingular, then

$A \triangleq \jmath(I - \lambda B)(I + \lambda B)^{-1}$ is Hermitian and $\jmath I + A = 2\jmath(I + \lambda B)^{-1}$.

iii) If $A \in \mathbb{F}^{n\times n}$ is Hermitian, then there exists a unique unitary matrix $B \in \mathbb{F}^{n\times n}$ such that $I + B$ is nonsingular and $A = \jmath(I - B)(I + B)^{-1}$. In fact, $B = (\jmath I - A)(\jmath I + A)^{-1}$.

iv) If $B \in \mathbb{F}^{n\times n}$ is unitary and $\lambda \in \mathbb{C}$ is such that $|\lambda| = 1$ and $I + \lambda B$ is nonsingular, then there exists a unique Hermitian matrix $A \in \mathbb{F}^{n\times n}$ such that $\lambda B = (\jmath I - A)(\jmath I + A)^{-1}$. In fact, $A = \jmath(I - \lambda B)(I + \lambda B)^{-1}$.

v) If A is nonsingular and skew Hermitian, then $A^3 + I$ is nonsingular and $(A^3 - I)(A^3 + I)^{-1}$ is unitary.

Source: [1084, pp. 168, 169] and [2991, p. 258]. **Remark:** The linear fractional transformation $f(s) = (\jmath - s)/(\jmath + s)$ maps the closed upper half plane in $\mathbb{C}$ onto the closed inside unit disk in $\mathbb{C}$, and the real line in $\mathbb{C}$ onto the unit circle in $\mathbb{C}$. **Remark:** $\mathcal{C}(A) \triangleq (A - I)(A + I)^{-1} = I - 2(A + I)^{-1}$ is the *Cayley transform* of A. **Related:** Fact 4.14.9, Fact 4.13.25, Fact 4.13.26, Fact 4.28.12, Fact 10.10.35, and Fact 15.22.10.

Fact 4.13.25. The following statements hold:

i) If $A \in \mathbb{F}^{n\times n}$ is skew Hermitian, then $I + A$ is nonsingular, $B \triangleq (I - A)(I + A)^{-1}$ is unitary, and $I + B = 2(I + A)^{-1}$. If, in addition, $\operatorname{mspec}(A) = \overline{\operatorname{mspec}(A)}$, then $\det B = 1$.

ii) If $B \in \mathbb{F}^{n\times n}$ is unitary and $\lambda \in \mathbb{C}$ is such that $|\lambda| = 1$ and $I + \lambda B$ is nonsingular, then $A \triangleq (I + \lambda B)^{-1}(I - \lambda B)$ is skew Hermitian and $I + A = 2(I + \lambda B)^{-1}$.

iii) If $A \in \mathbb{F}^{n\times n}$ is skew Hermitian, then there exists a unique unitary matrix $B \in \mathbb{F}^{n\times n}$ such that $I + B$ is nonsingular and $A = (I + B)^{-1}(I - B)$. In fact, $B = (I - A)(I + A)^{-1}$.

iv) If B is unitary and $\lambda \in \mathbb{C}$ is such that $|\lambda| = 1$ and $I + \lambda B$ is nonsingular, then there exists a unique skew-Hermitian matrix $A \in \mathbb{F}^{n\times n}$ such that $B = \overline{\lambda}(I - A)(I + A)^{-1}$. In fact, $A = (I + \lambda B)^{-1}(I - \lambda B)$.

Source: [1084, p. 184] and [1450, p. 440].

Fact 4.13.26. The following statements hold:

i) If $A \in \mathbb{R}^{n\times n}$ is skew symmetric, then $I+A$ is nonsingular, $B \triangleq (I-A)(I+A)^{-1}$ is orthogonal, $I + B = 2(I + A)^{-1}$, and $\det B = 1$.

ii) If $B \in \mathbb{R}^{n\times n}$ is orthogonal, $C \in \mathbb{R}^{n\times n}$ is diagonal with diagonally located entries ± 1, and $I+CB$ is nonsingular, then $A \triangleq (I+CB)^{-1}(I-CB)$ is skew symmetric, $I+A = 2(I+CB)^{-1}$, and $\det CB = 1$.

iii) If $A \in \mathbb{R}^{n\times n}$ is skew symmetric, then there exists a unique orthogonal matrix $B \in \mathbb{R}^{n\times n}$ such that $I + B$ is nonsingular and $A = (I + B)^{-1}(I - B)$. In fact, $B = (I - A)(I + A)^{-1}$.

iv) If $B \in \mathbb{R}^{n\times n}$ is orthogonal and $C \in \mathbb{R}^{n\times n}$ is diagonal with diagonally located entries ± 1, then there exists a unique skew-symmetric matrix $A \in \mathbb{R}^{n\times n}$ such that $CB = (I - A)(I + A)^{-1}$. In fact, $A = (I + CB)^{-1}(I - CB)$.

Remark: The Cayley transform is a one-to-one and onto map from the set of skew-symmetric matrices to the set of orthogonal matrices whose spectrum does not include -1. **Credit:** The last statement is due to P. L. Hsu. See [2263, p. 101].

Fact 4.13.27. Let $A, B \in \mathbb{F}^{n\times n}$, and assume that A and B are unitary. Then,

$$\sqrt{1 - \left|\tfrac{1}{n}\operatorname{tr} AB\right|^2} \le \sqrt{1 - \left|\tfrac{1}{n}\operatorname{tr} A\right|^2} + \sqrt{1 - \left|\tfrac{1}{n}\operatorname{tr} B\right|^2}.$$

Source: [2816] and [2991, p. 197]. **Related:** Fact 3.15.1 and Fact 4.13.28.

Fact 4.13.28. Let $A, B \in \mathbb{F}^{n\times n}$, and assume that A and B are unitary. Then,

$$\left|\tfrac{1}{n}\operatorname{tr} A\right|^2 + \left|\tfrac{1}{n}\operatorname{tr} B\right|^2 + \left|\tfrac{1}{n}\operatorname{tr} AB\right|^2 \le 1 + 2\left|\tfrac{1}{n}\operatorname{tr} A\right|\left|\tfrac{1}{n}\operatorname{tr} B\right|\left|\tfrac{1}{n}\operatorname{tr} AB\right|^2.$$

Source: [1897]. **Related:** Fact 4.13.27.

Fact 4.13.29. If $A \in \mathbb{F}^{n\times n}$ is shifted unitary, then $B \triangleq 2A - I$ is unitary. Conversely, If $B \in \mathbb{F}^{n\times n}$ is unitary, then $A \triangleq \frac{1}{2}(B + I)$ is shifted unitary. **Remark:** The affine mapping $f(A) \triangleq 2A - I$ from the shifted-unitary matrices to the unitary matrices is one-to-one and onto. See Fact 4.19.1 and Fact 4.20.3. **Related:** Fact 4.10.14 and Fact 4.17.14.

Fact 4.13.30. If $A \in \mathbb{F}^{n\times n}$ is shifted unitary, then A is normal. Furthermore, the following statements are equivalent:

i) A is shifted unitary.

ii) $A + A^* = 2A^*A$.

iii) $A + A^* = 2AA^*$.

Source: By Fact 4.13.29 there exists a unitary matrix B such that $A = \frac{1}{2}(B + I)$. Since B is normal, it follows from Fact 4.10.14 that A is normal.

Fact 4.13.31. The matrices

$$\frac{1}{\sqrt{2}}\begin{bmatrix} I_n & I_n \\ -I_n & I_n \end{bmatrix}, \quad \frac{1}{\sqrt{2}}\begin{bmatrix} I_n & I_n \\ \hat{I}_n & -\hat{I}_n \end{bmatrix}, \quad \frac{1}{\sqrt{2}}\begin{bmatrix} I_n & -\hat{I}_n \\ \hat{I}_n & I_n \end{bmatrix}, \quad \frac{1}{\sqrt{2}}\begin{bmatrix} I_n & 0 & -\hat{I}_n \\ 0 & \sqrt{2} & 0 \\ \hat{I}_n & 0 & I_n \end{bmatrix}$$

are orthogonal, and the matrix $\frac{1}{\sqrt{2}}\begin{bmatrix} I & I \\ \jmath I & -\jmath I \end{bmatrix}$ is unitary. **Related:** Fact 7.10.23, Fact 7.10.24, and Fact 7.10.25.

Fact 4.13.32. Define $A \in \mathbb{R}^{5\times 5}$ by

$$A \triangleq \frac{1}{5}\begin{bmatrix} 2 & 1 & 0 & 4 & -2 \\ 2 & -2 & 4 & 0 & 1 \\ 2 & 0 & -2 & 1 & 4 \\ -3 & 2 & 2 & 2 & 2 \\ 2 & 4 & 1 & -2 & 0 \end{bmatrix}.$$

Then,

$$A^{\mathsf{T}}A = A^5 = I_5, \quad \operatorname{spec}(A) = \{1, \cos\tfrac{2\pi}{5} \pm \sin\tfrac{2\pi}{5}\jmath, -\cos\tfrac{\pi}{5} \pm \sin\tfrac{\pi}{5}\jmath\}, \quad A1_{5\times 1} = 1_{5\times 1}.$$

Source: [1404]. **Remark:** A is used to prove the partition congruence $5|p_{5n+4}$. See Fact 1.20.1.

Fact 4.13.33. Let $A, B \in \mathrm{O}(n)$. Then, the following statements are equivalent:

i) AB is involutory.

ii) $\operatorname{tr}(AB)^2 = n$.

iii) AB is symmetric.

Source: [2703]. **Related:** Fact 3.15.37.

4.14 Facts on Rotation Matrices

Fact 4.14.1. Let $\theta \in \mathbb{R}$, and define the orthogonal matrix

$$A(\theta) \triangleq \begin{bmatrix} \cos\theta & \sin\theta \\ -\sin\theta & \cos\theta \end{bmatrix}.$$

Now, let $\theta_1, \theta_2 \in \mathbb{R}$. Then,

$$A(\theta_1)A(\theta_2) = A(\theta_1 + \theta_2),$$
$$\cos(\theta_1 + \theta_2) = (\cos\theta_1)\cos\theta_2 - (\sin\theta_1)\sin\theta_2,$$
$$\sin(\theta_1 + \theta_2) = (\cos\theta_1)\sin\theta_2 + (\sin\theta_1)\cos\theta_2.$$

Furthermore, $\mathrm{SO}(2) = \{A(\theta)\colon\ \theta \in \mathbb{R}\}$ and

$$\mathrm{O}(2)\backslash\mathrm{SO}(2) = \left\{\begin{bmatrix} \cos\theta & \sin\theta \\ \sin\theta & -\cos\theta \end{bmatrix}:\ \theta \in \mathbb{R}\right\}.$$

Remark: See Proposition 4.6.6 and Fact 15.12.3.

Fact 4.14.2. Let $A \in \mathbb{R}^{3\times3}$. Then, $A \in \mathrm{O}(3)\backslash\mathrm{SO}(3)$ if and only if $-A \in \mathrm{SO}(3)$.

Fact 4.14.3. Let $A \in \mathbb{F}^{3\times3}$. Then, the following statements are equivalent:

i) $A \in \mathrm{SU}(3)$.

ii) $\|\mathrm{col}_1(A)\| = \|\mathrm{col}_2(A)\| = 1$, $[\mathrm{col}_1(A)]^*\mathrm{col}_2(A) = 0$, and $\mathrm{col}_3(A) = \overline{\mathrm{col}_1(A)} \times \overline{\mathrm{col}_2(A)}$.

iii) $\|\mathrm{col}_2(A)\| = \|\mathrm{col}_3(A)\| = 1$, $[\mathrm{col}_2(A)]^*\mathrm{col}_3(A) = 0$, and $\mathrm{col}_1(A) = \overline{\mathrm{col}_2(A)} \times \overline{\mathrm{col}_3(A)}$.

iv) $\|\mathrm{col}_3(A)\| = \|\mathrm{col}_1(A)\| = 1$, $[\mathrm{col}_3(A)]^*\mathrm{col}_1(A) = 0$, and $\mathrm{col}_2(A) = \overline{\mathrm{col}_3(A)} \times \overline{\mathrm{col}_1(A)}$.

Credit: A. H. J. de Ruiter.

Fact 4.14.4. Let $A \in \mathrm{SO}(3)$. Then,

$$\begin{gathered}
(\operatorname{tr} A)^2 = \operatorname{tr} A^2 + 2\operatorname{tr} A,\\
(\operatorname{tr} A)^3 + 3 = \operatorname{tr} A^3 + 3\operatorname{tr} A^2 + 6\operatorname{tr} A,\\
(\operatorname{tr} A)^3 + 2\operatorname{tr} A^3 = 3(\operatorname{tr} A)(\operatorname{tr} A^2) + 6,\\
(\operatorname{tr} A)^3 = (\operatorname{tr} A)(\operatorname{tr} A^2) + 2\operatorname{tr} A^2 + 4\operatorname{tr} A,\\
(\operatorname{tr} A)^4 + 12 = \operatorname{tr} A^4 + 4\operatorname{tr} A^3 + 10\operatorname{tr} A^2 + 16\operatorname{tr} A,\\
(\operatorname{tr} A)^5 + 45 = \operatorname{tr} A^5 + 5\operatorname{tr} A^4 + 15\operatorname{tr} A^3 + 30\operatorname{tr} A^2 + 45\operatorname{tr} A,\\
(\operatorname{tr} A)^6 + 153 = \operatorname{tr} A^6 + 6\operatorname{tr} A^5 + 21\operatorname{tr} A^4 + 50\operatorname{tr} A^3 + 90\operatorname{tr} A^2 + 126\operatorname{tr} A.
\end{gathered}$$

Remark: The first equality can be written as

$$\operatorname{tr} A + A_{(1,2)}A_{(2,1)} + A_{(1,3)}A_{(3,1)} + A_{(2,3)}A_{(3,2)} = A_{(1,1)}A_{(2,2)} + A_{(2,2)}A_{(3,3)} + A_{(3,3)}A_{(1,1)}.$$

Remark: These equalities hold for all $n \times n$ matrices A that have exactly three nonzero eigenvalues of the form 1, λ, and $1/\lambda$.

Fact 4.14.5. Let $A \in \mathbb{R}^{3\times3}$, and let $z \triangleq \begin{bmatrix} b \\ c \\ d \end{bmatrix}$, where $b^2 + c^2 + d^2 = 1$. Then, $A \in \mathrm{SO}(3)$, and A rotates every vector in $\mathbb{R}^3$ by the angle π about z if and only if

$$A = \begin{bmatrix} 2b^2 - 1 & 2bc & 2bd \\ 2bc & 2c^2 - 1 & 2cd \\ 2bd & 2cd & 2d^2 - 1 \end{bmatrix} = 2zz^{\mathrm{T}} - I_3 = I_3 + 2K^2(z).$$

Source: This formula follows from the last expression for A in Fact 4.14.6 with $\theta = \pi$. See [796, p. 30]. **Remark:** A is a reflector. **Remark:** z is uniquely determined up to a sign.

Fact 4.14.6. Let $A \in \mathbb{R}^{3\times3}$. Then, $A \in \mathrm{SO}(3)$ if and only if there exist real numbers a, b, c, d such that $a^2 + b^2 + c^2 + d^2 = 1$, $a \in (-1, 1]$, and

$$A = \begin{bmatrix} a^2 + b^2 - c^2 - d^2 & 2(bc - ad) & 2(ac + bd) \\ 2(ad + bc) & a^2 - b^2 + c^2 - d^2 & 2(cd - ab) \\ 2(bd - ac) & 2(ab + cd) & a^2 - b^2 - c^2 + d^2 \end{bmatrix}.$$

Assume that these conditions hold. Then,

$$a = \pm\tfrac{1}{2}\sqrt{1 + \operatorname{tr} A}.$$

If, in addition, $a \neq 0$, then b, c, and d are given by

$$b = \frac{A_{(3,2)} - A_{(2,3)}}{4a}, \quad c = \frac{A_{(1,3)} - A_{(3,1)}}{4a}, \quad d = \frac{A_{(2,1)} - A_{(1,2)}}{4a}.$$

Furthermore, the following statements are equivalent: *i)* $a = 1$; *ii)* $A = I_3$; *iii)* $b = c = d = 0$. Now, in the case where $a \neq 1$, define $v \triangleq [b\ \ c\ \ d]^{\mathsf{T}}$. If $a \neq 1$, then A represents a rotation about the unit-length vector $z \triangleq (\csc \frac{\theta}{2})v$ through an angle $\theta \in (0, 2\pi)$, which satisfies $a = \cos \frac{\theta}{2}$, and where the direction of rotation about z is determined by the right-hand rule. If $a = 1$, then define $\theta \triangleq 0$. Therefore, for all $a \in (-1, 1]$, $\theta = 2\operatorname{acos} a$. If $a \in [0, 1]$, then

$$\theta = 2\operatorname{acos}(\tfrac{1}{2}\sqrt{1 + \operatorname{tr} A}) = \operatorname{acos}(\tfrac{1}{2}[(\operatorname{tr} A) - 1]),$$

whereas, if $a \in (-1, 0]$, then

$$\theta = 2\operatorname{acos}(-\tfrac{1}{2}\sqrt{1 + \operatorname{tr} A}) = \pi + \operatorname{acos}(\tfrac{1}{2}[1 - \operatorname{tr} A]).$$

In particular, $a = 1$ if and only if $\theta = 0$, and $a = 0$ if and only if $\theta = \pi$. Furthermore,

$$\begin{aligned} A &= (2a^2 - 1)I_3 + 2aK(v) + 2vv^{\mathsf{T}} \\ &= (\cos\theta)I_3 + (\sin\theta)K(z) + (1 - \cos\theta)zz^{\mathsf{T}} \\ &= I_3 + (\sin\theta)K(z) + (1 - \cos\theta)K^2(z), \\ A - A^{\mathsf{T}} &= 4aK(v) = 2(\sin\theta)K(z). \end{aligned}$$

If $\theta \neq \pi$, then

$$A = I_3 + \frac{2}{\alpha^2 + 1}[K(\alpha z) + K^2(\alpha z)],$$

where $\alpha \triangleq \frac{\sin\theta}{1+\cos\theta} = \tan\frac{\theta}{2}$. If $\theta = 0$, then $v = z = 0$, whereas, if $\theta = \pi$, then $K^2(z) = \frac{1}{2}(A - I)$. Conversely, let $\theta \in \mathbb{R}$ be nonzero, let $z \in \mathbb{R}^3$, assume that $z^{\mathsf{T}}z = 1$, and define

$$B = \begin{bmatrix} z_{(1)}^2 + (z_{(2)}^2 + z_{(3)}^2)\cos\theta & z_{(1)}z_{(2)}(1-\cos\theta) - z_{(3)}\sin\theta & z_{(1)}z_{(3)}(1-\cos\theta) + z_{(2)}\sin\theta \\ z_{(1)}z_{(2)}(1-\cos\theta) + z_{(3)}\sin\theta & z_{(2)}^2 + (z_{(1)}^2 + z_{(3)}^2)\cos\theta & z_{(2)}z_{(3)}(1-\cos\theta) - z_{(1)}\sin\theta \\ z_{(1)}z_{(3)}(1-\cos\theta) - z_{(2)}\sin\theta & z_{(2)}z_{(3)}(1-\cos\theta) + z_{(1)}\sin\theta & z_{(3)}^2 + (z_{(1)}^2 + z_{(2)}^2)\cos\theta \end{bmatrix}.$$

Then, B represents a rotation about the unit-length vector z through the angle θ, where the direction of rotation is determined by the right-hand rule. Finally, define

$$\begin{bmatrix} a \\ b \\ c \\ d \end{bmatrix} \triangleq \begin{bmatrix} \cos\frac{\theta}{2} \\ (\sin\frac{\theta}{2})z \end{bmatrix}$$

and A as above in terms of a, b, c, d. Then, $A = B$. **Source:** [1035, p. 162], [1178, p. 22], [2421, p. 19], [2716, 2722], and use Fact 4.14.9. **Remark:** The quadruples (a, b, c, d) are *Euler parameters*, and $[a\ \ b\ \ c\ \ d]^{\mathsf{T}}$ is an element of the sphere S^3 in $\mathbb{R}^4$. The Euler parameter $(-1, 0, 0, 0)$ and point $[-1\ \ 0\ \ 0\ \ 0]^{\mathsf{T}} \in \mathrm{S}^3$ can be viewed as representing a rotation through the angle $\theta = 2\pi$, which is equivalent to the case $\theta = 0$. Each element of S^3 can be represented by a unit quaternion in Sp(1), thus giving S^3 a group structure. See Fact 4.31.8. **Remark:** A is unchanged if a, b, c, d are replaced by $-a, -b, -c, -d$. Replacing a by $-a$ in A but keeping b, c, d unchanged yields A^{T}. **Remark:** The entries of A are *direction cosines*. See [308, pp. 384–387] and Fact 4.32.1. **Remark:** For each rotation matrix A, there exist exactly two distinct Euler parameters (a, b, c, d) that parameterize A. Therefore, the Euler parameters, which parameterize the unit sphere S^3 in $\mathbb{R}^4$, provide a *double cover* of SO(3). See [1967, p. 304] and Fact 4.32.1. **Remark:** Sp(1) $\simeq$ SU(2) is a double cover of

SO(3) (see Fact 4.32.4), Sp(1)×Sp(1) ≃ SU(2)×SU(2) is a double cover of SO(4), Sp(2) is a double cover of SO(5), and SU(4) is a double cover of SO(6). For each n, SO(n) is double covered by the *spin group* Spin(n). See [803, p. 141], [2588, p. 130], and [2889, pp. 42–47]. Sp(2) is defined in Fact 4.32.4. **Remark:** Rotation matrices in $\mathbb{R}^{2\times2}$ are discussed in [2449]. **Related:** Fact 10.10.30 and Fact 15.16.11. **Credit:** O. Rodrigues. See [58].

Fact 4.14.7. Let $n \ge 4$, let $A \in \mathrm{SO}(n)$, and let $\{\lambda \in \mathrm{mspec}(A)\colon \lambda \ne 1\}_{\mathrm{ms}} = \{e^{\theta_1 J}, e^{-\theta_1 J}, \ldots, e^{\theta_m J}, e^{-\theta_m J}\}_{\mathrm{ms}}$, where $\theta_1, \ldots, \theta_m \in (0, \pi]$. Then, there exist skew-symmetric matrices $A_1, \ldots, A_m \in \mathbb{R}^{n\times n}$ such that, for all $i, j \in \{1, \ldots, m\}$, $A_iA_j = A_jA_i$, for all $i \in \{1, \ldots, m\}$, $A_i^3 = -A_i$, and

$$A = I + \sum_{i=1}^{m} [(\sin\theta_i)A_i + (1 - \cos\theta_i)A_i^2].$$

Source: [1135, 2326].

Fact 4.14.8. Let $\theta_1, \theta_2 \in \mathbb{R}$, let $z_1, z_2 \in \mathbb{R}^3$, assume that $z_1^\mathsf{T}z_1 = z_2^\mathsf{T}z_2 = 1$, and, for $i = 1, 2$, let $A_i \in \mathbb{R}^{3\times3}$ be the rotation matrix that represents the rotation about the unit-length vector z_i through the angle θ_i, where the direction of rotation is determined by the right-hand rule. Then, $A_3 \triangleq A_2A_1$ represents the rotation about the unit-length vector z_3 through the angle θ_3, where the direction of rotation is determined by the right-hand rule, and where θ_3 and z_3 are given by

$$\cos\tfrac{\theta_3}{2} = (\cos\tfrac{\theta_2}{2})\cos\tfrac{\theta_1}{2} - (\sin\tfrac{\theta_2}{2})\sin\tfrac{\theta_1}{2} z_2^\mathsf{T}z_1,$$

$$\begin{aligned} z_3 &= (\csc\tfrac{\theta_3}{2})[(\sin\tfrac{\theta_2}{2})(\cos\tfrac{\theta_1}{2})z_2 + (\cos\tfrac{\theta_2}{2})(\sin\tfrac{\theta_1}{2})z_1 + (\sin\tfrac{\theta_2}{2})(\sin\tfrac{\theta_1}{2})(z_2 \times z_1)] \\ &= \frac{\cot\frac{\theta_3}{2}}{1 - z_2^\mathsf{T}z_1(\tan\frac{\theta_2}{2})\tan\frac{\theta_1}{2}}[(\tan\tfrac{\theta_2}{2})z_2 + (\tan\tfrac{\theta_1}{2})z_1 + (\tan\tfrac{\theta_2}{2})(\tan\tfrac{\theta_1}{2})(z_2 \times z_1)]. \end{aligned}$$

Source: [58], [1178, pp. 22–24], and [2722]. **Remark:** These expressions are *Rodrigues's formulas*, which follow from the quaternion multiplication formula given in Fact 4.32.1. In particular, for $i = 1, 2$, define $q_i \triangleq a_i + b_i\hat{\imath} + c_i\hat{\jmath} + d_i\hat{k}$ and $v_i \triangleq [b_i\ \ c_i\ \ d_i]^\mathsf{T}$. Then, $q_3 \triangleq q_2q_1 = a_3 + b_3\hat{\imath} + c_3\hat{\jmath} + d_3\hat{k}$ is given by

$$\begin{bmatrix} a_3 \\ b_3 \\ c_3 \\ d_3 \end{bmatrix} = \begin{bmatrix} \cos\frac{\theta_3}{2} \\ (\sin\frac{\theta_3}{2})z_3 \end{bmatrix} = \begin{bmatrix} a_1a_2 - v_2^\mathsf{T}v_1 \\ a_1v_2 + a_2v_1 + v_2 \times v_1 \end{bmatrix},$$

where

$$\begin{bmatrix} a_2 \\ b_2 \\ c_2 \\ d_2 \end{bmatrix} = \begin{bmatrix} \cos\frac{\theta_2}{2} \\ (\sin\frac{\theta_2}{2})z_2 \end{bmatrix} = \begin{bmatrix} a_2 \\ v_2 \end{bmatrix}, \quad \begin{bmatrix} a_1 \\ b_1 \\ c_1 \\ d_1 \end{bmatrix} = \begin{bmatrix} \cos\frac{\theta_1}{2} \\ (\sin\frac{\theta_1}{2})z_1 \end{bmatrix} = \begin{bmatrix} a_1 \\ v_1 \end{bmatrix}.$$

Fact 4.14.9. Let $z, w \in \mathbb{R}^3$, assume that $\|z\|_2 = \|w\|_2 = 1$, let $\theta, \phi \in \mathbb{R}$, and define $A_z(\theta) \in \mathbb{R}^{3\times3}$ by

$$A_z(\theta) \triangleq (\cos\theta)I + (\sin\theta)K(z) + (1 - \cos\theta)zz^\mathsf{T}.$$

Then, the following statements hold:

i) $A_z(\theta) = I$ if and only if θ/π is an even integer.

ii) $A_z(\theta) = I + (\sin\theta)K(z) + (1 - \cos\theta)K^2(z)$.

iii) $A_z(\theta)$ is a rotation matrix.

iv) Both $A_z(\theta) \ne I$ and $A_z^2(\theta) = I$ if and only if θ/π is an odd integer. If these conditions hold, then $A_z(\theta) = -I + 2zz^\mathsf{T}$.

v) $A_{-z}(2\pi-\theta) = A_z(\theta)$, $A_z^{-1}(\theta) = A_z^{\mathsf{T}}(\theta) = A_z(-\theta) = A_{-z}(\theta)$.

vi) Let $x, y \in \mathbb{R}^3$, assume that $\|x\|_2 = \|y\|_2 \neq 0$, and let $\theta \in [0,\pi]$ denote the angle between x and y. Furthermore, if $\theta \in (0,\pi)$, then let $z \in \mathbb{R}^3$ be given by $z = \frac{1}{\|x\times y\|_2} x \times y$, whereas, if $\theta \in \{0,\pi\}$, then let $z \in \{x\}^\perp$ satisfy $\|z\|_2 = 1$. Then, $y = A_z(\theta)x$.

vii) Let $x \in \mathbb{R}^3$. Then, $A_z(\theta)x = x$ if and only if either θ/π is an even integer or $z \times x = 0$.

viii) If $\cos\frac{\theta}{2} \neq 0$, then $A_z(\theta) = [I + (\tan\frac{\theta}{2})K(z)][I - (\tan\frac{\theta}{2})K(z)]^{-1}$.

ix) If $\cos\frac{\theta}{2} \neq 0$, then $[I + A_z(\theta)]^{-1}[I - A_z(\theta)] = -(\tan\frac{\theta}{2})K(z)$.

x) $[A_z(\theta), A_w(\phi)] = (\sin\theta)(\sin\phi)K(z\times w) + (\sin\theta)(1-\cos\phi)[K(z), ww^{\mathsf{T}}]$
$+ (\sin\phi)(1-\cos\theta)[zz^{\mathsf{T}}, K(w)] + (1-\cos\theta)(1-\cos\phi)w^{\mathsf{T}}z(zw^{\mathsf{T}} - wz^{\mathsf{T}})$.

xi) Assume that $\cos\frac{\theta}{2} \neq 0$ and $\cos\frac{\phi}{2} \neq 0$, define $A \triangleq A_w(\phi)A_z(\theta)$ and $\psi \triangleq \operatorname{acos}\frac{1}{2}[(\operatorname{tr} A) - 1]$, assume that $\cos\frac{\psi}{2} \neq 0$, and let $v \in \mathbb{R}^3$ satisfy $\|v\|_2 = 1$ and $K(v) = \frac{1}{2\sqrt{1+\operatorname{tr} A}}(A - A^{\mathsf{T}})$. Then,

$$A = [I + (\tan\tfrac{\psi}{2})K(v)][I - (\tan\tfrac{\psi}{2})K(v)]^{-1},$$

$$\begin{aligned}(\tan\tfrac{\psi}{2})K(v) &= [I + (\tan\tfrac{\theta}{2})K(z)][I + (\tan\tfrac{\phi}{2})K(w)(\tan\tfrac{\theta}{2})K(z)]^{-1}\\ &\quad\cdot [(\tan\tfrac{\phi}{2})K(w) + (\tan\tfrac{\theta}{2})K(z)][I + (\tan\tfrac{\theta}{2})K(z)]^{-1}.\end{aligned}$$

xii) $[A_z(\theta), A_w(\phi)] = 0$ if and only if at least one of the following statements holds:

a) $z \times w = 0$.

b) Either $A_z(\theta) = I$ or $A_w(\phi) = I$.

c) $A_z^2(\theta) = I$, $A_w^2(\phi) = I$, and $w^{\mathsf{T}}z = 0$.

xiii) $\operatorname{mspec}[A_z(\theta)] = \{1, e^{\theta J}, e^{-\theta J}\}_{\mathrm{ms}}$, and z is an eigenvector of $A_z(\theta)$ associated with 1.

Source: *viii*), *ix*), and *xi*) are given in [2053, pp. 244, 245]; *xii*) is due to S. Bhat. **Remark:** If x and y in *vi*) are linearly independent, then $\theta = \operatorname{acos}\frac{x^{\mathsf{T}}y}{\|x\|_2\|y\|_2}$. Furthermore, *xxix*) of Fact 4.12.1 implies that $\sin\theta = \frac{\|x\times y\|_2}{\|x\|_2\|y\|_2}$. **Remark:** In the notation of *vi*), $A_z(\theta)$ can be written as

$$\begin{aligned}A_z(\theta) &= (\cos\theta)I + \frac{1}{\|x\|_2^2}(yx^{\mathsf{T}} - xy^{\mathsf{T}}) + \frac{1-\cos\theta}{\|x\times y\|_2^2}(x\times y)(x\times y)^{\mathsf{T}}\\
&= \frac{x^{\mathsf{T}}y}{x^{\mathsf{T}}x}I + \frac{1}{x^{\mathsf{T}}x}(yx^{\mathsf{T}} - xy^{\mathsf{T}}) + \frac{1-\cos\theta}{(x^{\mathsf{T}}x\sin\theta)^2}(x\times y)(x\times y)^{\mathsf{T}}\\
&= \frac{x^{\mathsf{T}}y}{x^{\mathsf{T}}x}I + \frac{1}{x^{\mathsf{T}}x}(yx^{\mathsf{T}} - xy^{\mathsf{T}}) + \frac{\tan\frac{\theta}{2}}{(x^{\mathsf{T}}x)^2\sin\theta}(x\times y)(x\times y)^{\mathsf{T}}\\
&= \frac{x^{\mathsf{T}}y}{x^{\mathsf{T}}x}I + \frac{1}{x^{\mathsf{T}}x}(yx^{\mathsf{T}} - xy^{\mathsf{T}}) + \frac{1}{(x^{\mathsf{T}}x)^2(1+\cos\theta)}(x\times y)(x\times y)^{\mathsf{T}}\\
&= \frac{x^{\mathsf{T}}y}{x^{\mathsf{T}}x}I + \frac{1}{x^{\mathsf{T}}x}(yx^{\mathsf{T}} - xy^{\mathsf{T}}) + \frac{1}{x^{\mathsf{T}}x(x^{\mathsf{T}}x + x^{\mathsf{T}}y)}(x\times y)(x\times y)^{\mathsf{T}}.\end{aligned}$$

Consequently,

$$\begin{aligned}A_z(\theta)x &= (\cos\theta)x + \frac{1}{\|x\|_2^2}(x^{\mathsf{T}}xy - y^{\mathsf{T}}xx) + \frac{1-\cos\theta}{\|x\times y\|_2^2}(x\times y)(x\times y)^{\mathsf{T}}x\\
&= \frac{x^{\mathsf{T}}y}{\|x\|_2^2}x + \frac{1}{\|x\|_2^2}(x^{\mathsf{T}}xy - y^{\mathsf{T}}xx) = y.\end{aligned}$$

Furthermore, $B_z(\theta) \triangleq -(\tan \frac{1}{2}\theta)K(z)$ can be written as

$$B_z(\theta) = \frac{\|x\|_2\|y\|_2 - x^{\mathsf{T}}y}{\|x \times y\|^2}(xy^{\mathsf{T}} - yx^{\mathsf{T}}).$$

These expressions satisfy $A_z(\theta) + B_z(\theta) + A_z(\theta)B_z(\theta) = I$. **Remark:** In *vi*), $A_z(\theta)$ represents a right-hand rule rotation of the nonzero vector x through the angle θ around z to yield y, which has the same length as x. The cases $x = y$ and $x = -y$ correspond, respectively, to $\theta = 0$ and $\theta = \pi$; in these cases, the axis of rotation z is not unique. **Remark:** Extensions of the Cayley transform are discussed in [2741]. **Remark:** *vi*) is a linear interpolation result. See Fact 4.11.5, Fact 15.12.7, and [285, 1549]. **Related:** Fact 15.12.6.

Fact 4.14.10. Let $x, y, z \in \mathbb{R}^2$. If x is rotated according to the right-hand rule through an angle $\theta \in \mathbb{R}$ about y, then the final point $\hat{x} \in \mathbb{R}^2$ has the coordinates

$$\hat{x} = \begin{bmatrix} \cos\theta & -\sin\theta \\ \sin\theta & \cos\theta \end{bmatrix} x + \begin{bmatrix} y_{(1)}(1-\cos\theta) + y_{(2)}\sin\theta \\ y_{(2)}(1-\cos\theta) - y_{(1)}\sin\theta \end{bmatrix}.$$

If x is reflected across the line passing through 0 and z in the direction of the line passing through 0 and y, then the final point $\hat{x} \in \mathbb{R}^2$ has the coordinates

$$\hat{x} = \begin{bmatrix} y_{(1)}^2 - y_{(2)}^2 & 2y_{(1)}y_{(2)} \\ 2y_{(1)}y_{(2)} & y_{(2)}^2 - y_{(1)}^2 \end{bmatrix} x + \begin{bmatrix} -z_{(1)}(y_{(1)}^2 - y_{(2)}^2 - 1) - 2z_{(2)}y_{(1)}y_{(2)} \\ -z_{(2)}(y_{(1)}^2 - y_{(2)}^2 - 1) - 2z_{(1)}y_{(1)}y_{(2)} \end{bmatrix}.$$

Remark: These *affine planar transformations* are used in computer graphics. See [122, 1062, 2258]. **Related:** Fact 4.14.9 and Fact 4.14.11.

Fact 4.14.11. Let $x, y \in \mathbb{R}^3$, and assume that $y^{\mathsf{T}}y = 1$. If x is rotated according to the right-hand rule through an angle $\theta \in \mathbb{R}$ about the line passing through 0 and y, then the final point $\hat{x} \in \mathbb{R}^3$ has the coordinates $\hat{x} = x + (\sin\theta)(y \times x) + (1 - \cos\theta)[y \times (y \times x)]$. **Source:** [44]. **Related:** Fact 4.14.9 and Fact 4.14.10.

4.15 Facts on One Idempotent Matrix

Fact 4.15.1. Let $A \in \mathbb{F}^{n\times n}$, assume that A is idempotent, and let $x \in \mathbb{F}^n$. Then, $x \in \mathcal{R}(A)$ if and only if $Ax = x$. **Related:** Fact 4.17.11.

Fact 4.15.2. Let $A \in \mathbb{F}^{n\times n}$. Then, A is idempotent if and only if A is semisimple and there exists a positive integer k such that $A^{k+1} = A^k$.

Fact 4.15.3. Let $A \in \mathbb{F}^{n\times m}$ and $B \in \mathbb{F}^{l\times n}$. Then, the following statements are equivalent:

i) $\mathcal{R}(A)$ and $\mathcal{N}(B)$ are complementary subspaces.

ii) $\operatorname{rank} A = \operatorname{rank} B = \operatorname{rank} BA$.

Source: [659]. **Related:** Fact 8.7.9.

Fact 4.15.4. Let $\mathcal{S}_1, \mathcal{S}_2 \subseteq \mathbb{F}^n$ be complementary subspaces, and let $A \in \mathbb{F}^{n\times n}$. Then, the following statements are equivalent:

i) A is the idempotent matrix onto $\mathcal{S}_1$ along $\mathcal{S}_2$.

ii) $A_\perp$ is the idempotent matrix onto $\mathcal{S}_2$ along $\mathcal{S}_1$.

iii) A^* is the idempotent matrix onto $\mathcal{S}_2^\perp$ along $\mathcal{S}_1^\perp$.

iv) $A_\perp^*$ is the idempotent matrix onto $\mathcal{S}_1^\perp$ along $\mathcal{S}_2^\perp$.

Related: Fact 4.15.5.

Fact 4.15.5. Let $A \in \mathbb{F}^{n\times n}$. Then, the following statements are equivalent:

i) A is idempotent.

ii) $\mathcal{R}(A) \subseteq \mathcal{N}(A_\perp)$.

iii) $\mathcal{R}(A_\perp) \subseteq \mathcal{N}(A)$.

iv) $\mathcal{R}(A) = \mathcal{N}(A_\perp)$.

v) $\mathcal{R}(A_\perp) = \mathcal{N}(A)$.

vi) $\mathcal{R}(A)$ and $\mathcal{R}(A_\perp)$ are complementary subspaces.

vii) For all $x \in \mathcal{R}(A)$, $Ax = x$.

If these statements hold, then the following statements hold:

viii) A is the idempotent matrix onto $\mathcal{R}(A)$ along $\mathcal{N}(A)$.

ix) $A_\perp$ is the idempotent matrix onto $\mathcal{N}(A)$ along $\mathcal{R}(A)$.

x) A^* is the idempotent matrix onto $\mathcal{N}(A)^\perp$ along $\mathcal{R}(A)^\perp$.

xi) $A_\perp^*$ is the idempotent matrix onto $\mathcal{R}(A)^\perp$ along $\mathcal{N}(A)^\perp$.

Source: [257] and [1343, p. 146]. **Related:** Fact 3.13.2 and Fact 7.13.28.

Fact 4.15.6. Let $A \in \mathbb{F}^{n\times n}$, and assume that A is idempotent. Then, $\mathcal{R}(I - AA^*) = \mathcal{R}(2I - A - A^*)$. **Source:** [2635].

Fact 4.15.7. Let $A \in \mathbb{F}^{n\times n}$. Then, the following statements are equivalent:

i) A is skew idempotent.

ii) $-A$ is idempotent.

iii) $\operatorname{rank} A = -\operatorname{tr} A$, and $\operatorname{rank}(A + I) = n + \operatorname{tr} A$.

Source: [2715].

Fact 4.15.8. Let $A \in \mathbb{F}^{n\times n}$. Then, A is idempotent and $\operatorname{rank} A = 1$ if and only if there exist $x, y \in \mathbb{F}^n$ such that $y^\mathsf{T}x = 1$ and $A = xy^\mathsf{T}$.

Fact 4.15.9. Let $A \in \mathbb{F}^{n\times n}$, and assume that A is idempotent. Then, A^T, $\overline{A}$, and A^* are idempotent.

Fact 4.15.10. Let $A \in \mathbb{F}^{n\times n}$, and assume that A is idempotent and skew Hermitian. Then, $A = 0$.

Fact 4.15.11. Let $A \in \mathbb{F}^{n\times n}$. Then, the following statements are equivalent:

i) A is idempotent.

ii) $\operatorname{rank} A + \operatorname{rank}(I - A) = n$.

iii) $\operatorname{rank} A = \operatorname{tr} A$, and $\operatorname{rank}(I - A) = \operatorname{tr}(A - I)$.

iv) $I + A$ is nonsingular, and $\operatorname{rank} A + \operatorname{rank}(I - A^2) = n$.

v) A is tripotent, and $I - A$ is tripotent.

vi) A is tripotent, and $I + A$ is nonsingular.

vii) A is tripotent, and $\operatorname{rank} A = \operatorname{tr} A$.

If these statements hold, then $(I + A)^{-1} = I - \frac{1}{2}A$ and $\det(I + A) = 2^{\operatorname{tr} A}$. **Source:** [257] and [2991, p. 130].

Fact 4.15.12. Let $A \in \mathbb{F}^{n\times m}$. If $A^\mathsf{L} \in \mathbb{F}^{m\times n}$ is a left inverse of A, then AA^L is idempotent and $\operatorname{rank} A^\mathsf{L} = \operatorname{rank} A$. Furthermore, if $A^\mathsf{R} \in \mathbb{F}^{m\times n}$ is a right inverse of A, then $A^\mathsf{R}A$ is idempotent and $\operatorname{rank} A^\mathsf{R} = \operatorname{rank} A$.

Fact 4.15.13. Let $A \in \mathbb{F}^{n\times n}$, and assume that A is nonsingular and idempotent. Then, $A = I_n$.

Fact 4.15.14. Let $A \in \mathbb{F}^{n\times n}$, and assume that A is idempotent. Then, so is $A_\perp \triangleq I - A$, and, furthermore, $AA_\perp = A_\perp A = 0$.

Fact 4.15.15. Let $A \in \mathbb{F}^{n\times n}$, assume that A is idempotent, and let $k \geq 1$. Then, $(A+I)^k = 2^kA + A_\perp$. **Source:** [2238].

Fact 4.15.16. Let $A \in \mathbb{F}^{n\times n}$, and assume that A is idempotent. Then, the following statements hold:

i) $\operatorname{rank}(A - A^*) = 2\operatorname{rank}\,[A \;\; A^*] - 2\operatorname{rank} A$.

ii) $\operatorname{rank}(I - A - A^*) = \operatorname{rank}(I + A - A^*) = n$.

iii) $\operatorname{rank}(A + A^*) = \operatorname{rank}(AA^* + A^*A) = \operatorname{rank}\,[A\ \ A^*]$.

iv) $\mathcal{R}(A) \subseteq \mathcal{R}(A + A^*)$ and $\mathcal{R}(A^*) \subseteq \mathcal{R}(A + A^*)$.

v) $\operatorname{rank}\,[A, A^*] = \operatorname{rank}(A - A^*)$.

vi) $\operatorname{rank}(I - AA^*) = \operatorname{rank}(2I - A - A^*)$.

Source: [2672].

Fact 4.15.17. Let $A \in \mathbb{F}^{n\times n}$, $B \in \mathbb{F}^{n\times m}$, and $C \in \mathbb{F}^{l\times n}$, and assume that A is idempotent, $\operatorname{rank}\,[C^*\ \ B] = n$, and $CB = 0$. Then, $\operatorname{rank} CAB = \operatorname{rank} CA + \operatorname{rank} AB - \operatorname{rank} A$. **Source:** [2674]. **Related:** Fact 4.15.22.

Fact 4.15.18. Let $A \in \mathbb{F}^{n\times m}$ and $B \in \mathbb{F}^{m\times n}$, and assume that AB is nonsingular. Then, $B(AB)^{-1}A$ is idempotent.

Fact 4.15.19. Let $A \in \mathbb{F}^{n\times n}$, let $r \triangleq \operatorname{rank} A$, and let $B \in \mathbb{F}^{n\times r}$ and $C \in \mathbb{F}^{r\times n}$ satisfy $A = BC$. Then, A is idempotent if and only if $CB = I$. **Source:** [2821, p. 16]. **Remark:** $A = BC$ is a full-rank factorization. See Proposition 7.6.6.

Fact 4.15.20. Let $A, B \in \mathbb{R}^{n\times n}$. Then, the following statements hold:

i) Assume that $A^3 = -A$ and $B = I + A + A^2$. Then, $B^4 = I$, $B^{-1} = I - A + A^2$, $B^3 - B^2 + B - I = 0$, $A = \frac{1}{2}(B - B^3)$, and $I + A^2$ is idempotent.

ii) Assume that $B^3 - B^2 + B - I = 0$ and $A = \frac{1}{2}(B - B^3)$. Then, $A^3 = -A$ and $B = I + A + A^2$.

iii) Assume that $B^4 = I$ and $A = \frac{1}{2}(B - B^{-1})$. Then, $A^3 = -A$, and $\frac{1}{4}(I + B + B^2 + B^3)$ is idempotent.

Remark: The geometric meaning of these results is discussed in [1028, pp. 153, 212–214, 242].

Fact 4.15.21. Let $A \in \mathbb{F}^{n\times n}$ and $\alpha \in \mathbb{F}$, where $\alpha \neq 0$. Then, the matrices

$$\begin{bmatrix} A & A^* \\ A^* & A \end{bmatrix}, \quad \begin{bmatrix} A & \alpha^{-1}A \\ \alpha(I - A) & I - A \end{bmatrix}, \quad \begin{bmatrix} A & \alpha^{-1}A \\ -\alpha A & -A \end{bmatrix}$$

are, respectively, normal, idempotent, and nilpotent.

Fact 4.15.22. Let $A \triangleq \begin{bmatrix} A_{11} & A_{12} \\ A_{21} & A_{22} \end{bmatrix} \in \mathbb{F}^{(n+m)\times(n+m)}$, and assume that A is idempotent. Then,

$$\operatorname{rank} A = \operatorname{rank}\begin{bmatrix} A_{12} \\ A_{22} \end{bmatrix} + \operatorname{rank}\,[A_{11}\ \ A_{12}] - \operatorname{rank} A_{12} = \operatorname{rank}\begin{bmatrix} A_{11} \\ A_{21} \end{bmatrix} + \operatorname{rank}\,[A_{21}\ \ A_{22}] - \operatorname{rank} A_{21}.$$

Source: [2674] and Fact 4.15.17. **Related:** Fact 4.17.13 and Fact 8.9.14.

4.16 Facts on Two or More Idempotent Matrices

Fact 4.16.1. Let $A, B \in \mathbb{F}^{n\times n}$, and assume that $AB = A$ and $BA = B$. Then, A and B are idempotent. **Source:** [2418, p. 169].

Fact 4.16.2. Let $A, B \in \mathbb{F}^{n\times n}$, and assume that $A^2 = A = AB$. Then, $B^2 = B = BA$ if and only if $\operatorname{rank} A = \operatorname{rank} B$. **Source:** [312].

Fact 4.16.3. Let $A, B \in \mathbb{F}^{n\times n}$, and assume that A and B are idempotent. Then, the following statements hold:

i) $AB = B$ if and only if $\mathcal{R}(B) \subseteq \mathcal{R}(A)$.

ii) $BA = A$ if and only if $\mathcal{N}(B) \subseteq \mathcal{N}(A)$.

Furthermore, the following statements are equivalent:

iii) $\mathcal{R}(A) \subseteq \mathcal{R}(B)$ and $\mathcal{N}(A) \subseteq \mathcal{N}(B)$.

iv) $\mathcal{R}(A) = \mathcal{R}(B)$ and $\mathcal{N}(A) = \mathcal{N}(B)$.

v) $A = B$.

Source: [778].

Fact 4.16.4. If $A \in \mathbb{F}^{n\times m}$ and $B \in \mathbb{F}^{n\times(n-m)}$, assume that $[A\ B]$ is nonsingular, and define

$$P \triangleq [A\ \ 0][A\ \ B]^{-1}, \quad Q \triangleq [0\ \ B][A\ \ B]^{-1}.$$

Then, the following statements hold:

i) P and Q are idempotent.

ii) $P + Q = I_n$.

iii) $PQ = 0$.

iv) $P[A\ \ 0] = [A\ \ 0]$.

v) $Q[0\ \ B] = [0\ \ B]$.

vi) $\mathcal{R}(P) = \mathcal{R}(A)$ and $\mathcal{N}(P) = \mathcal{R}(B)$.

vii) $\mathcal{R}(Q) = \mathcal{R}(B)$ and $\mathcal{N}(Q) = \mathcal{R}(A)$.

viii) $A^*B = 0$ if and only if P and Q are projectors. If these conditions hold, then $P = A(A^*A)^{-1}A^*$ and $Q = B(B^*B)^{-1}B^*$.

ix) $\mathcal{R}(A)$ and $\mathcal{R}(B)$ are complementary subspaces.

x) P is the idempotent matrix onto $\mathcal{R}(A)$ along $\mathcal{R}(B)$.

xi) Q is the idempotent matrix onto $\mathcal{R}(B)$ along $\mathcal{R}(A)$.

Source: [2996] and [2997, pp. 74, 75]. **Related:** Fact 4.18.14, Fact 4.18.19, Fact 8.8.14, and Fact 8.8.15.

Fact 4.16.5. Let $A, B \in \mathbb{F}^{n\times n}$, and assume that A and B are idempotent. Then,

$$\begin{aligned}
\operatorname{rank}(A+B) &= \operatorname{rank} A + \operatorname{rank}(A_\perp B A_\perp) = n - \dim[\mathcal{N}(A_\perp B) \cap \mathcal{N}(A)] \\
&= \operatorname{rank}\begin{bmatrix} 0 & A & B \\ A & 0 & 0 \\ B & 0 & 2B \end{bmatrix} - \operatorname{rank} A - \operatorname{rank} B = \operatorname{rank}\begin{bmatrix} A & B \\ B & 0 \end{bmatrix} - \operatorname{rank} B \\
&= \operatorname{rank}\begin{bmatrix} B & A \\ A & 0 \end{bmatrix} - \operatorname{rank} A = \operatorname{rank}(B_\perp A B_\perp) + \operatorname{rank} B = \operatorname{rank}(A_\perp B A_\perp) + \operatorname{rank} A \\
&= \operatorname{rank}(A - AB - BA + BAB) + \operatorname{rank} B = \operatorname{rank}(B - AB - BA + ABA) + \operatorname{rank} A \\
&= \operatorname{rank}(A + A_\perp B) = \operatorname{rank}(A + BA_\perp) = \operatorname{rank}(B + B_\perp A) = \operatorname{rank}(B + AB_\perp) \\
&= \operatorname{rank}(I - A_\perp B_\perp) = \operatorname{rank}(I - B_\perp A_\perp) = \operatorname{rank}\,[AB_\perp\ \ B] = \operatorname{rank}\,[BA_\perp\ \ A] \\
&= \operatorname{rank}\begin{bmatrix} B_\perp A \\ B \end{bmatrix} = \operatorname{rank}\begin{bmatrix} A_\perp B \\ A \end{bmatrix} = \operatorname{rank} A + \operatorname{rank} B - n + \operatorname{rank}\begin{bmatrix} A_\perp & A_\perp B_\perp \\ B_\perp A_\perp & B_\perp \end{bmatrix}.
\end{aligned}$$

Furthermore, the following statements hold:

i) If $\alpha, \beta \in \mathbb{F}$ are nonzero, $\gamma \in \mathbb{F}$, and $\gamma \neq \alpha + \beta$, then $\operatorname{rank}(A + B) = \operatorname{rank}(\alpha A + \beta B - \gamma AB)$.

ii) If $\alpha, \beta \in \mathbb{F}$ are nonzero and $\alpha + \beta \neq 0$, then $\operatorname{rank}(A + B) = \operatorname{rank}(\alpha A + \beta B)$.

iii) If $AB = 0$, then $\operatorname{rank}(A + B) = \operatorname{rank} BA_\perp + \operatorname{rank} A = \operatorname{rank} B_\perp A + \operatorname{rank} B$.

iv) If $BA = 0$, then $\operatorname{rank}(A + B) = \operatorname{rank} AB_\perp + \operatorname{rank} B = \operatorname{rank} A_\perp B + \operatorname{rank} A$.

v) If $AB = BA$, then $\operatorname{rank}(A + B) = \operatorname{rank}(A - AB) + \operatorname{rank} B = \operatorname{rank}(B - AB) + \operatorname{rank} A$.

vi) $A + B$ is idempotent if and only if $AB = BA = 0$. If these conditions hold, then $A + B$ is the idempotent matrix onto $\mathcal{R}(A) + \mathcal{R}(B)$ along $\mathcal{N}(A) \cap \mathcal{N}(B)$.

vii) $A + B = I$ if and only if $AB = BA = 0$ and $\operatorname{rank}(A + B) = n$.

viii) If either $AB = 0$ or $BA = 0$, then $\operatorname{rank}(A + B) = \operatorname{rank} A + \operatorname{rank} B$.

Finally, the following statements are equivalent:

ix) $A + B$ is nonsingular.

x) $\mathcal{R}(A) \cap \mathcal{R}(BA_\perp) = \mathcal{N}(A) \cap \mathcal{N}(B) = \{0\}$.

xi) $\mathcal{R}(AB_\perp) \cap \mathcal{R}(BA_\perp) = \mathcal{N}(A) \cap \mathcal{N}(B) = \{0\}$.

xii) $\mathcal{R}(A) \cap \mathcal{R}(BA_\perp) = \mathcal{N}(A) \cap \mathcal{N}(A_\perp B) = \{0\}$.

xiii) $\mathcal{R}(A) + \mathcal{R}(BA_\perp) = \mathcal{N}(A) + \mathcal{N}(A_\perp B) = \mathbb{F}^n$.

xiv) $\mathcal{R}(A)$ and $\mathcal{R}(BA_\perp)$ are complementary subspaces, and $\mathcal{N}(A)$ and $\mathcal{N}(A_\perp B)$ are complementary subspaces.

xv) $\operatorname{rank}\,[A\ \ B] = n$ and $\mathcal{R}\left(\begin{bmatrix} A \\ B \end{bmatrix}\right) \cap \mathcal{R}\left(\begin{bmatrix} B \\ 0 \end{bmatrix}\right) = \{0\}$.

xvi) There exist $\alpha, \beta, \gamma \in \mathbb{F}$ such that $\alpha \neq 0, \beta \neq 0, \gamma \neq \alpha + \beta$, and $\alpha A + \beta B - \gamma AB$ is nonsingular.

xvii) For all $\alpha, \beta, \gamma \in \mathbb{F}$ such that $\alpha \neq 0, \beta \neq 0$, and $\gamma \neq \alpha + \beta$, $\alpha A + \beta B - \gamma AB$ is nonsingular.

xviii) There exist nonzero $\alpha, \beta \in \mathbb{F}$ such that $\alpha + \beta \neq 0$ and $\alpha A + \beta B$ is nonsingular.

xix) For all nonzero $\alpha, \beta \in \mathbb{F}$ such that $\alpha + \beta \neq 0$, $\alpha A + \beta B$ is nonsingular.

Source: [221, 1251], [1275, p. 18], and [1659, 1661, 1662, 2672, 2673, 2676, 3023]. To prove necessity in *vi*), note that $(A + B)^2 = A + B$ implies $AB + BA = 0$, which implies $AB + ABA = ABA + BA = 0$. Hence, $AB - BA = 0$, and thus $AB = 0$. See [1305, p. 250] and [1343, p. 435]. **Related:** Fact 8.8.4.

Fact 4.16.6. Let $A, B \in \mathbb{F}^{n \times n}$, assume that A and B are idempotent, and let $\alpha \in \mathbb{F}$ be nonzero. Then,

$$\operatorname{rank}(A - B) = \operatorname{rank}\begin{bmatrix} 0 & A & B \\ A & 0 & 0 \\ B & 0 & 0 \end{bmatrix} - \operatorname{rank} A - \operatorname{rank} B = \operatorname{rank}\begin{bmatrix} A \\ B \end{bmatrix} + \operatorname{rank}\,[A\ \ B] - \operatorname{rank} A - \operatorname{rank} B$$

$$\begin{aligned}
&= \operatorname{rank}(A - AB) + \operatorname{rank}(AB - B) = \operatorname{rank}(A - BA) + \operatorname{rank}(BA - B) \\
&= \operatorname{rank}[A - AB + \alpha(AB - B)] = \operatorname{rank}(A + B - 2AB) \\
&= n - \dim[\mathcal{N}(A) \cap \mathcal{N}(B)] - \dim[\mathcal{R}(A) \cap \mathcal{R}(B)] = \operatorname{rank}(AB_\perp) + \operatorname{rank}(A_\perp B) \\
&\le \operatorname{rank}(A + B) \le \operatorname{rank} A + \operatorname{rank} B.
\end{aligned}$$

Furthermore, the following statements hold:

i) If either $AB = 0$ or $BA = 0$, then $\operatorname{rank}(A - B) = \operatorname{rank}(A + B) = \operatorname{rank} A + \operatorname{rank} B$.

ii) If $AB = 0$, then $\operatorname{rank}(A - BA) + \operatorname{rank}(BA - B) = \operatorname{rank} A + \operatorname{rank} B$.

iii) If $BA = 0$, then $\operatorname{rank}(A - AB) + \operatorname{rank}(AB - B) = \operatorname{rank} A + \operatorname{rank} B$.

iv) If $\alpha, \beta \in \mathbb{F}$ are nonzero, then $\operatorname{rank}(A - B) = \operatorname{rank}[\alpha A + \beta B - (\alpha + \beta)AB]$.

Finally, the following statements are equivalent:

v) $\operatorname{rank}(A - B) = \operatorname{rank} A - \operatorname{rank} B$.

vi) $ABA = B$.

vii) $\mathcal{R}(B) \subseteq \mathcal{R}(A)$ and $\mathcal{R}(B^*) \subseteq \mathcal{R}(A^*)$.

Source: [1251, 1662, 2657, 2672, 2673, 2676, 3023]. $\operatorname{rank}(A - B) \le \operatorname{rank}(A + B)$ follows from Fact 3.14.19 and the block 3×3 expressions in this result and in Fact 4.16.5. To prove *i*) in the case $AB = 0$, note that $\operatorname{rank} A + \operatorname{rank} B = \operatorname{rank}(A - B)$, which yields $\operatorname{rank}(A - B) \le \operatorname{rank}(A + B) \le \operatorname{rank} A + \operatorname{rank} B = \operatorname{rank}(A - B)$. **Related:** Fact 8.8.4.

Fact 4.16.7. Let $A \in \mathbb{F}^{n \times n}$, $B \in \mathbb{F}^{m \times m}$, and $C \in \mathbb{F}^{n \times m}$, and assume that A and B are idempotent.

Then,

$$\operatorname{rank}(AC - CB) = \operatorname{rank}\begin{bmatrix} AC \\ B \end{bmatrix} + \operatorname{rank}\,[A \;\; CB] - \operatorname{rank} A - \operatorname{rank} B$$
$$= \operatorname{rank}(AC - ACB) + \operatorname{rank}(ACB - CB).$$

Furthermore, the following statements hold:

i) If $ACB = 0$, then $\operatorname{rank}(AC - CB) = \operatorname{rank} AC + \operatorname{rank} CB$.

ii) $AC = CB$ if and only if $\mathcal{R}(CB) \subseteq \mathcal{R}(A)$ and $\mathcal{R}[(AC)^*] \subseteq \mathcal{R}(B^*)$.

iii) Assume that $n = m$. Then, $AC = CA$ if and only if $\mathcal{R}(CA) \subseteq \mathcal{R}(A)$ and $\mathcal{R}[(AC)^*] \subseteq \mathcal{R}(A^*)$.

Source: [2628, 2657, 2672].

Fact 4.16.8. Let $A, B \in \mathbb{F}^{n\times n}$, and assume that A and B are idempotent. Then, the following statements are equivalent:

i) $A - B$ is idempotent.

ii) $\operatorname{rank}(A_\perp + B) + \operatorname{rank}(A - B) = n$.

iii) $\operatorname{rank}(A - B) = \operatorname{rank} A - \operatorname{rank} B$.

iv) $\mathcal{R}(B) \subseteq \mathcal{R}(A)$ and $\mathcal{R}(B^*) \subseteq \mathcal{R}(A^*)$.

v) $ABA = B$.

vi) $AB = BA = B$.

If these statements hold, then $A - B$ is the idempotent matrix onto $\mathcal{R}(A) \cap \mathcal{N}(B)$ along $\mathcal{N}(A) + \mathcal{R}(B)$.

Source: [1275, p. 19] and [2672, 2675]. **Credit:** R. E. Hartwig and G. P. H. Styan.

Fact 4.16.9. Let $A, B \in \mathbb{F}^{n\times n}$, and assume that A and B are idempotent. Then, the following statements are equivalent:

i) $\mathcal{R}(A) \cap \mathcal{R}(B) = \{0\}$.

ii) $I - AB$ is nonsingular.

iii) $I - BA$ is nonsingular.

Furthermore, the following statements are equivalent:

iv) $A - B$ is nonsingular.

v) $\operatorname{rank}\left[\begin{smallmatrix} A \\ B \end{smallmatrix}\right] = \operatorname{rank}\,[A \;\; B] = \operatorname{rank} A + \operatorname{rank} B = n$.

vi) $I - AB$ is nonsingular, and there exist nonzero $\alpha, \beta \in \mathbb{F}$ such that $\alpha A + \beta B$ is nonsingular.

vii) $I - AB$ is nonsingular, and, for all nonzero $\alpha, \beta \in \mathbb{F}$, $\alpha A + \beta B$ is nonsingular.

viii) There exist nonzero $\alpha, \beta \in \mathbb{F}$ such that $\alpha A + \beta B - (\alpha + \beta)AB$ is nonsingular.

ix) For all nonzero $\alpha, \beta \in \mathbb{F}$, $\alpha A + \beta B - (\alpha + \beta)AB$ is nonsingular.

x) $I - AB$ and $A + A_\perp B$ are nonsingular.

xi) $I - BA$ and $A + BA_\perp$ are nonsingular.

xii) $I - AB$ and $A + B$ are nonsingular.

xiii) $\mathcal{R}(A)$ and $\mathcal{R}(B)$ are complementary subspaces, and $\mathcal{R}(A^*)$ and $\mathcal{R}(B^*)$ are complementary subspaces.

xiv) $\mathcal{R}(A)$ and $\mathcal{R}(B)$ are complementary subspaces, and $\mathcal{N}(A)$ and $\mathcal{N}(B)$ are complementary subspaces.

xv) $\mathcal{R}(A) \cap \mathcal{R}(B) = \mathcal{N}(A) \cap \mathcal{N}(B) = \{0\}$.

xvi) $\mathcal{R}(AB_\perp)$ and $\mathcal{R}(A_\perp B)$ are complementary subspaces.

xvii) $\mathcal{N}(AB_\perp)$ and $\mathcal{N}(A_\perp B)$ are complementary subspaces.

If *iv*)–*xvii*) hold, then the following statements hold:

xviii) $A(A-B)^{-1}A = A$, $A(A-B)^{-1}B = 0$, and $B(B-A)^{-1}B = B$.

xix) $A(A-B)^{-1} = (I-BA)^{-1}(I-B) = A(A+BA_\perp)^{-1}$ is the idempotent matrix onto $\mathcal{R}(A)$ along $\mathcal{R}(B)$.

xx) $A^*(A^*-B^*)^{-1}$ is the idempotent matrix onto $\mathcal{R}(A^*)$ along $\mathcal{R}(B^*)$.

Source: [221, 248, 1251, 1660, 1662, 2673]. **Related:** Fact 4.18.19.

Fact 4.16.10. Let $A, B \in \mathbb{F}^{n\times n}$, assume that A and B are idempotent, and assume that $A-B$ is nonsingular. Then, $A+B$ is nonsingular. Now, define $F, G \in \mathbb{F}^{n\times n}$ by

$$F \triangleq A(A-B)^{-1} = (A-B)^{-1}B_\perp, \quad G \triangleq (A-B)^{-1}A = A_\perp(A-B)^{-1}.$$

Then, F and G are idempotent. In particular, F is the idempotent matrix onto $\mathcal{R}(A)$ along $\mathcal{R}(B)$, G is the idempotent matrix onto $\mathcal{N}(B)$ along $\mathcal{N}(A)$, and G^* is the idempotent matrix onto $\mathcal{R}(A^*)$ along $\mathcal{R}(B^*)$. Furthermore,

$$F_\perp A = A_\perp F = FB = B_\perp F_\perp = GA_\perp = AG_\perp = BG = G_\perp B_\perp = 0,$$
$$(A-B)^{-1} = F - G_\perp = G - F_\perp = (A+B)^{-1}(A-B)(A+B)^{-1}, \quad (I-AB)^{-1} = FG + F_\perp,$$
$$(A+B)^{-1} = I - G_\perp F - GF_\perp = (2G-I)(F-G_\perp) = (A-B)^{-1}(A+B)(A-B)^{-1}.$$

Now, let $\alpha, \beta \in \mathbb{F}$ be nonzero. Then, $\alpha A + \beta B - (\alpha+\beta)AB$ is nonsingular, and

$$[\alpha A + \beta B - (\alpha+\beta)AB]^{-1} = \frac{1}{\beta}F_\perp + \frac{1}{\alpha}G.$$

If, in addition, $\gamma \in \mathbb{F}$ and $\gamma \neq \alpha+\beta$, then $\alpha A + \beta B - \gamma AB$ is nonsingular, and

$$(\alpha A + \beta B - \gamma AB)^{-1} = \frac{1}{\beta}F_\perp + \frac{\gamma-\alpha}{\alpha\beta}G + \frac{\alpha+\beta-\gamma}{\alpha\beta}GF.$$

Source: [873, 1662, 3023]. **Remark:** See [1662] for an explicit expression for $(A+B)^{-1}$ in the case where $A-B$ is singular. **Related:** Proposition 4.8.3 and Fact 8.11.14.

Fact 4.16.11. Let $A, B \in \mathbb{F}^{n\times n}$, assume that A and B are idempotent, and assume that $AB = BA$. Then, the following statements are equivalent:

i) $A-B$ is nonsingular.

ii) $(A-B)^2 = I$.

iii) $A+B = I$.

Source: [1251].

Fact 4.16.12. Let $A, B \in \mathbb{F}^{n\times n}$, and assume that A and B are idempotent. Then,

$$\begin{aligned}\operatorname{rank}(I-A-B) &= \operatorname{rank} AB + \operatorname{rank} BA - \operatorname{rank} A - \operatorname{rank} B + n\\ &= \operatorname{rank}(I-A-B+AB) + \operatorname{rank} AB\\ &= \operatorname{rank}(I-A-B+BA) + \operatorname{rank} BA.\end{aligned}$$

Furthermore, the following statements hold:

i) $A+B = I$ if and only if $AB = BA = 0$ and $\operatorname{rank}(A+B) = \operatorname{rank} A + \operatorname{rank} B = n$.

ii) $I-A-B$ is nonsingular if and only if $\operatorname{rank} AB = \operatorname{rank} BA = \operatorname{rank} A = \operatorname{rank} B$.

iii) $\operatorname{rank}(I+A-B) = \operatorname{rank} BAB - \operatorname{rank} B + n$.

iv) $\operatorname{rank}(2I-A-B) = \operatorname{rank}(B-BAB) - \operatorname{rank} B + n = \operatorname{rank}(A-ABA) - \operatorname{rank} A + n$.

v) $\operatorname{rank}(2I-A-B) = \operatorname{rank}(I-AB)$.

vi) $I+A-B$ is nonsingular if and only if $\operatorname{rank} BAB = \operatorname{rank} B$.

vii) $2I-A-B$ is nonsingular if and only if $\operatorname{rank}(A-ABA) = \operatorname{rank} A$.

viii) If $\alpha \in \mathbb{C}$ and $\alpha \neq 1$, then $\operatorname{rank}(I - A - B + \alpha AB) = \operatorname{rank}(I - A - B)$.

Source: [1251, 2672].

Fact 4.16.13. Let $A, B \in \mathbb{F}^{n\times n}$, and assume that A and B are idempotent. Then, the following statements hold:

i) $\mathcal{R}(A) = \mathcal{R}(B)$ if and only if $AB = B$ and $BA = A$.

ii) $\mathcal{R}(A) \cap \mathcal{R}(B) \subseteq \mathcal{R}(AB) = \mathcal{R}(A) \cap [\mathcal{N}(A) + \mathcal{R}(B)]$.

iii) $\mathcal{R}(A) \subseteq \mathcal{R}(B)$ if and only if $\mathcal{R}(B) = \mathcal{R}(A) + \mathcal{R}(B - AB)$ and $\mathcal{R}(A) \cap \mathcal{R}(B - AB) = \{0\}$.

iv) $\mathcal{N}(B) + [\mathcal{N}(A) \cap \mathcal{R}(B)] \subseteq \mathcal{N}(AB) = \mathcal{N}(B) + [\mathcal{N}(A) \cap \mathcal{R}(B)] \subseteq \mathcal{R}(I - AB) \subseteq \mathcal{N}(A) + \mathcal{N}(B)$.

v) If $AB = BA$, then AB is the idempotent matrix onto $\mathcal{R}(A) \cap \mathcal{R}(B)$ along $\mathcal{N}(A) + \mathcal{N}(B)$.

vi) If $B_\perp A_\perp = 0$, then AB is idempotent, $\mathcal{R}(AB) = \mathcal{R}(A) \cap \mathcal{R}(B)$, $\mathcal{N}(AB) = \mathcal{N}(A) + \mathcal{N}(B)$, and $\mathcal{N}(A) \cap \mathcal{N}(B) = \{0\}$.

vii) $\mathcal{R}(AB) = \mathcal{R}(BA)$ if and only if $ABA = BA$ and $BAB = AB$.

viii) $\mathcal{N}(AB) = \mathcal{N}(BA)$ if and only if $ABA = AB$ and $BAB = BA$.

ix) $AB = 0$ if and only if $\mathcal{R}(B) \subseteq \mathcal{R}(B - AB)$.

x) $AB = B$ if and only if $\mathcal{R}(B) \subseteq \mathcal{R}(AB)$.

xi) $\mathcal{R}([A, B]) = [\mathcal{R}(A) + \mathcal{R}(B)] \cap [\mathcal{R}(A) + \mathcal{N}(B)] \cap [\mathcal{N}(A) + \mathcal{R}(B)] \cap [\mathcal{N}(A) + \mathcal{N}(B)]$.

xii) $\mathcal{R}([A, B]) = \mathcal{R}(A - B) \cap \mathcal{R}(A_\perp - B)$.

xiii) $\mathcal{N}([A, B]) = [\mathcal{R}(A) \cap \mathcal{R}(B)] + [\mathcal{R}(A) \cap \mathcal{N}(B)] + [\mathcal{N}(A) \cap \mathcal{R}(B)] + [\mathcal{N}(A) \cap \mathcal{N}(B)]$.

xiv) $\mathcal{N}([A, B]) = \mathcal{N}(A - B) + \mathcal{N}(I - A - B)$.

xv) $\mathcal{N}(A - B) \cap \mathcal{N}(I - A - B) = \{0\}$.

The following statements are equivalent:

xvi) $\mathcal{R}(A) = \mathcal{R}(AB) + [\mathcal{R}(A) \cap \mathcal{N}(B)]$ and $\mathcal{R}(AB) \cap [\mathcal{R}(A) \cap \mathcal{N}(B)] = \{0\}$.

xvii) $\operatorname{rank} AB = \operatorname{rank} BA$ and $\mathcal{R}(AB) \cap \mathcal{N}(B) = \{0\}$.

The following statements are equivalent:

xviii) $\mathcal{N}(A) = \mathcal{N}(BA) \cap [\mathcal{N}(A) + \mathcal{R}(B)]$ and $\mathcal{R}(AB) \cap [\mathcal{R}(A) \cap \mathcal{R}(B)] = \{0\}$.

xix) $\operatorname{rank} AB = \operatorname{rank} BA$ and $\mathcal{N}(BA) + \mathcal{R}(B) = \mathbb{F}^n$.

The following statements are equivalent:

xx) $AB = BA$.

xxi) $\operatorname{rank} AB = \operatorname{rank} BA$, and AB is the idempotent matrix onto $\mathcal{R}(A) \cap \mathcal{R}(B)$ along $\mathcal{N}(A) + \mathcal{N}(B)$.

xxii) $\operatorname{rank} AB = \operatorname{rank} BA$, and $A + B - AB$ is the idempotent matrix onto $\mathcal{R}(A) + \mathcal{R}(B)$ along $\mathcal{N}(A) \cap \mathcal{N}(B)$.

The following statements are equivalent:

xxiii) AB is idempotent.

xxiv) $\mathcal{R}(AB) \subseteq \mathcal{N}(A - AB)$.

xxv) $\mathcal{R}(AB) \cap \mathcal{R}(AB_\perp A) = \{0\}$.

xxvi) $\mathcal{R}(AB) \subseteq \mathcal{R}(B) + [\mathcal{N}(A) \cap \mathcal{N}(B)]$.

xxvii) $\mathcal{R}(AB) = \mathcal{R}(A) \cap (\mathcal{R}(B) + [\mathcal{N}(A) \cap \mathcal{N}(B)])$.

xxviii) $\mathcal{N}(B) + [\mathcal{N}(A) \cap \mathcal{R}(B)] = \mathcal{R}(I - AB)$.

xxix) $\operatorname{rank} AB + \operatorname{rank} AB_\perp A = \operatorname{rank} A$.

xxx) $\operatorname{tr} A = \operatorname{rank} B$ and $\operatorname{tr} AB + \operatorname{rank} AB_\perp A = \operatorname{rank} A$.

The following statements are equivalent:

xxxi) $\mathcal{R}(AB) = \mathcal{R}(A) \cap \mathcal{R}(B)$.

xxxii) rank AB + rank$(A - BA)$ = rank A.

The following statements are equivalent:

xxxiii) $\mathcal{N}(AB) = \mathcal{N}(A) + \mathcal{N}(B)$.

xxxiv) rank AB + rank$(A - AB)$ = rank A.

The following statements are equivalent:

xxxv) AB is the idempotent matrix onto $\mathcal{R}(A) \cap \mathcal{R}(B)$ along $\mathcal{R}(A) + \mathcal{N}(B)$.

xxxvi) $\mathcal{R}(AB) = \mathcal{R}(A) \cap \mathcal{R}(B)$ and $\mathcal{N}(AB) = \mathcal{N}(A) + \mathcal{N}(B)$.

The following statements are equivalent:

xxxvii) AB is the idempotent matrix onto $\mathcal{R}(A)$ along $\mathcal{R}(B)$.

xxxviii) $\mathcal{R}(A) \subseteq \mathcal{N}(A - AB)$ and $\mathcal{N}(A) \cap \mathcal{R}(B) = \{0\}$.

xxxix) $\mathcal{R}(B - AB) \subseteq \mathcal{N}(B)$ and $\mathcal{N}(A) + \mathcal{R}(B) = \mathbb{F}^n$.

The following statements are equivalent:

xl) rank AB = rank A + rank $B - n$.

xli) $\mathcal{N}(A) \subseteq \mathcal{R}(B)$.

xlii) $A + B - AB = I$.

xliii) $\mathcal{R}(BA) + \mathcal{N}(A) = \mathcal{R}(B)$.

Finally, assume that AB is idempotent. Then, the following statements are equivalent:

xliv) $\mathcal{R}(AB) = \mathcal{N}(A - AB)$.

xlv) $\mathcal{N}(A - AB) \subseteq \mathcal{R}(A)$.

xlvi) $\mathcal{N}(A - AB) = \mathcal{N}(I - AB)$.

xlvii) rank$(A - AB)$ = rank$(I - AB)$.

Source: [256], [1133, p. 53], [1250], [1275, p. 19], and [2640, 2870]. **Related:** Fact 7.13.29.

Fact 4.16.14. Let $A, B \in \mathbb{F}^{n\times n}$, and assume that A and B are idempotent. Then, the following statements hold:

i) $\mathcal{R}(A - B) = [\mathcal{R}(A) + \mathcal{R}(B)] \cap [\mathcal{N}(A) + \mathcal{N}(B)]$.

ii) $\mathcal{N}(A - B) = [\mathcal{R}(A) \cap \mathcal{R}(B)] + [\mathcal{N}(A) \cap \mathcal{N}(B)] = \mathcal{N}(A - BA) \cap \mathcal{N}(B - BA)$.

iii) $\mathcal{R}(I - A - B) = [\mathcal{R}(A) + \mathcal{N}(B)] \cap [\mathcal{N}(A) + \mathcal{R}(B)]$.

iv) $\mathcal{N}(I - A - B) = [\mathcal{R}(A) \cap \mathcal{N}(B)] + [\mathcal{N}(A) \cap \mathcal{R}(B)]$.

v) $\mathcal{N}(I - A - B) = \mathcal{N}(AB) \cap \mathcal{N}(A_\perp B_\perp)$.

The following statements are equivalent:

vi) $A + B$ is idempotent.

vii) A is the idempotent matrix onto $\mathcal{R}(A) \cap \mathcal{N}(B)$ along $\mathcal{N}(A) + \mathcal{R}(B)$.

viii) $\mathcal{R}(B) \subseteq \mathcal{R}(B - AB)$ and $\mathcal{N}(B - BA) \subseteq \mathcal{N}(B)$.

The following statements are equivalent:

ix) $A - B$ is idempotent.

x) B is the idempotent matrix onto $\mathcal{R}(A) \cap \mathcal{R}(B)$ along $\mathcal{N}(A) + \mathcal{N}(B)$.

xi) $\mathcal{R}(B) \subseteq (AB)$ and $\mathcal{N}(BA) \subseteq \mathcal{N}(B)$.

The following statements are equivalent:

xii) $\mathcal{R}(A + B) = \mathcal{R}(A) + \mathcal{R}(B)$.

xiii) $\mathcal{R}(A) + \mathcal{R}(A_\perp BA_\perp) = \mathcal{R}(A) + \mathcal{R}(B)$.

The following statements are equivalent:

xiv) $\mathcal{N}(A + B) = \mathcal{N}(A) \cap \mathcal{N}(B)$.

xv) $\mathcal{N}(AB+BA) = [\mathcal{R}(A)\cap\mathcal{N}(B)] + [\mathcal{N}(A)\cap\mathcal{R}(B)] + [\mathcal{N}(A)\cap\mathcal{N}(B)]$.

xvi) $\operatorname{rank} A_\perp BA_\perp = \operatorname{rank}(B-BA)$.

xvii) $\mathcal{N}(A_\perp BA_\perp) = \mathcal{N}(B-BA)$.

xviii) $\mathcal{R}(A)\cap\mathcal{R}(B-BA) = \{0\}$.

The following statements are equivalent:

xix) $\mathcal{N}(I-AB) = \mathcal{R}(A)\cap\mathcal{R}(B)$.

xx) $\operatorname{rank} AB_\perp A = \operatorname{rank}(A-BA)$.

xxi) $\mathcal{N}(AB_\perp A) = \mathcal{N}(A-BA)$.

xxii) $\mathcal{N}(A)\cap\mathcal{R}(A-BA) = \{0\}$.

The following statements are equivalent:

xxiii) $\mathcal{R}(A)+\mathcal{R}(B-BA) = \mathcal{R}(A)+\mathcal{R}(B)$.

xxiv) $\mathcal{R}(A_\perp BA_\perp) = \mathcal{R}(B-AB)$.

xxv) $\operatorname{rank} A_\perp BA_\perp = \operatorname{rank}(B-AB)$.

The following statements are equivalent:

xxvi) $\mathcal{R}(A)\cap\mathcal{N}(A-AB) = \mathcal{R}(A)\cap\mathcal{R}(B)$.

xxvii) $\mathcal{N}(AB_\perp A) = \mathcal{N}(A-BA)$.

xxviii) $\operatorname{rank} AB_\perp A = \operatorname{rank}(A-BA)$.

Source: [256].

Fact 4.16.15. Let $A,B\in\mathbb{F}^{n\times n}$, and assume that A and B are idempotent. Then, the following statements hold:

i) $(A-B)^2+(A_\perp-B)^2 = I$.

ii) $[A,B] = [B,A_\perp] = [B_\perp,A] = [A_\perp,B_\perp]$.

iii) $A-B = AB_\perp - A_\perp B$.

iv) $AB_\perp + BA_\perp = AB_\perp A + A_\perp BA_\perp$.

v) $A[A,B] = [A,B]A_\perp$.

vi) $B[A,B] = [A,B]B_\perp$.

Source: [2138].

Fact 4.16.16. Let $A,B\in\mathbb{F}^{n\times n}$, and assume that A and B are idempotent. Then,

$$\begin{aligned}
\operatorname{rank}\,[A,B] &= \operatorname{rank}(A-B)+\operatorname{rank}(I-A-B)-n\\
&= \operatorname{rank}(A-B)+\operatorname{rank}AB+\operatorname{rank}BA-\operatorname{rank}A-\operatorname{rank}B\\
&= \operatorname{rank}\begin{bmatrix}A\\B\end{bmatrix}+\operatorname{rank}\,[A\;\;B]+\operatorname{rank}AB+\operatorname{rank}BA-2\operatorname{rank}A-2\operatorname{rank}B\\
&= \operatorname{rank}(A-AB)+\operatorname{rank}(AB-B)+\operatorname{rank}AB+\operatorname{rank}BA-\operatorname{rank}A-\operatorname{rank}B\\
&= \operatorname{rank}(AB-ABA)+\operatorname{rank}(BA-ABA)\\
&= \operatorname{rank}\begin{bmatrix}AB\\A\end{bmatrix}+\operatorname{rank}\,[A\;\;BA]-2\operatorname{rank}A\\
&= \operatorname{rank}\begin{bmatrix}AB\\BA\end{bmatrix}+\operatorname{rank}\,[AB\;\;BA]-\operatorname{rank}AB-\operatorname{rank}BA.
\end{aligned}$$

Source: [1662, 2657, 2672].

Fact 4.16.17. Let $A,B\in\mathbb{F}^{n\times n}$, and assume that A and B are idempotent. Then, following statements are equivalent:

i) $AB = BA$.

ii) $[A, B] = 0$.

iii) $\mathcal{R}(AB) = \mathcal{R}(BA)$ and $\mathcal{R}[(AB)^*] = \mathcal{R}[(BA)^*]$.

iv) $\operatorname{rank}(A - B) + \operatorname{rank}(I - A - B) = n$.

v) $\operatorname{rank}(A - B) = \operatorname{rank} A + \operatorname{rank} B - \operatorname{rank} AB - \operatorname{rank} BA$.

vi) $\operatorname{rank}(A - AB) = \operatorname{rank} A - \operatorname{rank} AB$ and $\operatorname{rank}(B - AB) = \operatorname{rank} B - \operatorname{rank} AB$.

The following statements are equivalent:

vii) $\operatorname{rank}\,[A, B] = \operatorname{rank}(A - B)$.

viii) $\operatorname{rank} AB = \operatorname{rank} BA = \operatorname{rank} A = \operatorname{rank} B$.

ix) $I - A - B$ is nonsingular.

The following statements are equivalent:

x) $[A, B]$ is nonsingular.

xi) $A - B$ and $I - A - B$ are nonsingular.

xii) $\operatorname{rank} AB = \operatorname{rank} BA = \operatorname{rank} A = \operatorname{rank} B$, $\mathcal{R}(A)$ and $\mathcal{R}(B)$ are complementary subspaces, and $\mathcal{R}(A^*)$ and $\mathcal{R}(B^*)$ are complementary subspaces.

Source: [2672].

Fact 4.16.18. Let $A, B \in \mathbb{F}^{n\times n}$, and assume that A and B are idempotent. Then,

$$\begin{aligned}\operatorname{rank}(AB + BA) &= \operatorname{rank}(A + B) + \operatorname{rank}(I - A - B) - n\\ &= \operatorname{rank}(A + B) + \operatorname{rank} AB + \operatorname{rank} BA - \operatorname{rank} A - \operatorname{rank} B\\ &= \operatorname{rank}(A - AB - BA + BAB) + \operatorname{rank} AB + \operatorname{rank} BA - \operatorname{rank} A\\ &= \operatorname{rank}\,[A, B] + \operatorname{rank}(A + B) - \operatorname{rank}(A - B).\end{aligned}$$

Furthermore, the following statements hold:

i) $AB + BA = 0$ if and only if $AB = BA = 0$.

ii) If $\alpha, \beta \in \mathbb{F}$ are nonzero and $\alpha + \beta \neq 0$, then $\operatorname{rank}(AB + BA) = \operatorname{rank}(\alpha AB + \beta BA)$.

iii) $\max\{\operatorname{rank} AB, \operatorname{rank} BA\} \le \operatorname{rank}(AB + BA)$.

iv) $\operatorname{rank}\,[A, B] \le \operatorname{rank}(AB + BA)$.

The following statements are equivalent:

v) $\operatorname{rank}(AB + BA) = \operatorname{rank}(A + B)$.

vi) $\operatorname{rank}\,[A, B] = \operatorname{rank}(A - B)$.

vii) $\operatorname{rank}(I - A - B) = n$.

viii) $\operatorname{rank} A = \operatorname{rank} B = \operatorname{rank} AB = \operatorname{rank} BA$.

Finally, the following statements are equivalent:

ix) $AB + BA$ is nonsingular.

x) $A + B$ and $I - A - B$ are nonsingular.

xi) $\mathcal{R}(B - BA) \cap \mathcal{R}(A) = \mathcal{N}(B - AB) \cap \mathcal{N}(A) = \{0\}$, $\mathcal{R}(A)$ and $\mathcal{N}(B)$ are complementary subspaces, and $\mathcal{N}(A)$ and $\mathcal{R}(B)$ are complementary subspaces.

Source: [221, 1251, 1662, 2642, 2657, 2672].

Fact 4.16.19. Let $A, B \in \mathbb{F}^{n\times n}$, and assume that A and B are idempotent. Then,

$$\operatorname{rank}[AB - (AB)^2] = \operatorname{rank}(2I - A - B) + \operatorname{rank} AB - n,$$

$$\operatorname{rank}[(A - B)^2 - (A - B)] = \operatorname{rank}(I - A + B) + \operatorname{rank}(A - B) - n$$

$$= \operatorname{rank} ABA + \operatorname{rank}(A - B) - \operatorname{rank} A.$$

Source: [1251, 2672].

Fact 4.16.20. Let $A, B \in \mathbb{F}^{n\times n}$, assume that A and B are idempotent, and assume that $A - B$ is group invertible. Then, $AB_{\perp}$ and $B_{\perp}A$ are group invertible. **Source:** [873].

Fact 4.16.21. Let $A_1, \ldots, A_r \in \mathbb{F}^{n\times n}$, assume that $A_1, \ldots, A_r$ are idempotent, and assume that $\sum_{i=1}^{r} A_i = I$. Then, for all distinct $i, j \in \{1, \ldots, r\}$, $A_iA_j = 0$. **Source:** [2991, p. 132]. **Related:** Fact 4.18.24.

4.17 Facts on One Projector

Fact 4.17.1. Let $A \in \mathbb{F}^{n\times n}$, and assume that A is Hermitian. Then, the following statements are equivalent:

i) A is a projector.

ii) $\operatorname{rank} A = \operatorname{tr} A = \operatorname{tr} A^2$.

Source: [2418, p. 55]. **Related:** Fact 4.17.3 and Fact 4.17.4.

Fact 4.17.2. Let $A \in \mathbb{F}^{n\times n}$, and assume that A is a projector. Then, $\mathcal{R}(A_{\perp}) = \mathcal{R}(A)^{\perp} = \mathcal{N}(A)$.

Fact 4.17.3. Let $A \in \mathbb{F}^{n\times n}$. Then, the following statements are equivalent:

i) A is a projector.

ii) A is idempotent and Hermitian.

iii) A is idempotent and normal.

iv) A is idempotent and range Hermitian.

v) A is idempotent, and $\mathcal{R}(A)$ and $\mathcal{N}(A)$ are mutually orthogonal.

vi) A is idempotent, and $\mathcal{R}(A)^{\perp} = \mathcal{N}(A)$.

vii) A is idempotent, and, for all $x \in \mathbb{F}^n$, $x^*Ax \ge 0$.

viii) A is idempotent, and, for all $x \in \mathbb{F}^n$, $x^*Ax \le x^*x$.

ix) A is idempotent, and, for all $x \in \mathbb{F}^n$, $\|Ax\|_2 \le \|x\|_2$.

x) A is idempotent, and $\operatorname{rank} A + \operatorname{rank}(I - A^*A) = n$.

xi) A is idempotent and $AA^*A = A$.

xii) A is idempotent, and $AA^* + A^*A = A + A^*$.

xiii) A is tripotent, range Hermitian, and $I - A$ is tripotent.

xiv) A is tripotent, range Hermitian, and $I + A$ is nonsingular.

xv) A is tripotent, range Hermitian, and $\operatorname{rank} A = \operatorname{tr} A$.

xvi) A is tripotent and positive semidefinite.

xvii) $\mathcal{R}(A^*) \subseteq \mathcal{N}(I - A)$.

Source: [257], [2238, p. 308], [2263, p. 105], [2705, 2725], and [2991, pp. 131, 322]. **Related:** Fact 4.17.1 and Fact 4.17.4.

Fact 4.17.4. Let $A \in \mathbb{F}^{n\times n}$. Then, the following statements are equivalent:

i) A is a projector.

ii) $A = AA^*$.

iii) $A = A^*A$.

iv) A is normal, and $A^3 = A^2$.

v) A and A^*A are idempotent.

vi) A and $\frac{1}{2}(A + A^*)$ are idempotent.

vii) A is tripotent, and $A^2 = A^*$.

viii) $AA^* = A^*AA^*$.

ix) $3\operatorname{tr} A^*A + \operatorname{tr} A^2A^{2*} = 2\operatorname{Re}\operatorname{tr}(A^2 + A^2A^*)$.

x) A is range Hermitian, and $\operatorname{rank} A + \operatorname{tr} A^*A = \operatorname{tr}(A + A^*)$.

Source: *ix)* is given in [2238, p. 307] and [2700]; *x)* is given in [234]. **Remark:** The matrix $A = \begin{bmatrix} 1/2 & 1/2 \\ 0 & 0 \end{bmatrix}$ satisfies $\operatorname{tr} A = \operatorname{tr} A^*A$ but is not a projector. See Fact 4.10.13. **Related:** Fact 4.17.1, Fact 4.17.3, and Fact 8.7.6.

Fact 4.17.5. Let $A \in \mathbb{F}^{n\times n}$, and assume that A is a projector. Then, A is positive semidefinite.

Fact 4.17.6. Let $n \ge 2$, let $A \in \mathbb{F}^{n\times n}$, and assume that A is a nonzero projector. Then,

$$|\operatorname{tr} \hat{I}A| \le \min\{\operatorname{rank} A, n + \tfrac{1}{2}[1 - (-1)^n] - \operatorname{rank} A\} < \sqrt{n \operatorname{rank} A}.$$

Source: [254, p. 55].

Fact 4.17.7. Let $A \in \mathbb{F}^{n\times n}$, assume that A is a projector, and let $x \in \mathbb{F}^n$. Then, $x \in \mathcal{R}(A)$ if and only if $x = Ax$.

Fact 4.17.8. Let $A \in \mathbb{F}^{n\times m}$. If $\operatorname{rank} A = m$, then $B \triangleq A(A^*A)^{-1}A^*$ is a projector and $\operatorname{rank} B = m$. If $\operatorname{rank} A = n$, then $B \triangleq A^*(AA^*)^{-1}A$ is a projector and $\operatorname{rank} B = n$. **Related:** Fact 3.18.5, Fact 4.10.25, and Fact 4.10.26.

Fact 4.17.9. Let $x \in \mathbb{F}^n$, assume that x is nonzero, and define the elementary projector $A \triangleq I - (x^*x)^{-1}xx^*$. Then, the following statements hold:

i) $\mathcal{R}(A) = \{x\}^\perp$.

ii) $\operatorname{rank} A = n - 1$.

iii) $\mathcal{N}(A) = \operatorname{span}\{x\}$.

iv) $\operatorname{def} A = 1$.

v) $2A - I$ is the elementary reflector $I - 2(x^*x)^{-1}xx^*$.

Remark: If $y \in \mathbb{F}^n$, then Ay is the *projection* of y into $\{x\}^\perp$. **Related:** Fact 3.13.1.

Fact 4.17.10. Let $n \ge 2$, let $\mathcal{S} \subset \mathbb{F}^n$, and assume that $\mathcal{S}$ is a hyperplane. Then, there exists a unique elementary projector $A \in \mathbb{F}^{n\times n}$ such that $\mathcal{R}(A) = \mathcal{S}$ and $\mathcal{N}(A) = \mathcal{S}^\perp$. Furthermore, if $x \in \mathbb{F}^n$ is nonzero and $\mathcal{S} \triangleq \{x\}^\perp$, then $A = I - (x^*x)^{-1}xx^*$.

Fact 4.17.11. Let $A \in \mathbb{F}^{n\times n}$. Then, A is a projector and $\operatorname{rank} A = n - 1$ if and only if there exists a nonzero vector $x \in \mathcal{N}(A)$ such that $A = I - (x^*x)^{-1}xx^*$. Now, assume that these conditions hold. Then, for all $y \in \mathbb{F}^n$,

$$y^*y - y^*Ay = \frac{|y^*x|^2}{x^*x}.$$

Furthermore, for all $y \in \mathbb{F}^n$, the following statements are equivalent:

i) $y^*x = 0$.

ii) $y^*Ay = y^*y$.

iii) $Ay = y$.

iv) $y \in \mathcal{R}(A)$.

Related: Fact 4.10.19, Fact 4.15.1, and Fact 4.17.12.

Fact 4.17.12. Let $A \in \mathbb{F}^{n\times n}$, assume that A is a projector, and let $x \in \mathbb{F}^n$. Then, $x^*Ax \le x^*x$. Furthermore, the following statements are equivalent:

i) $x^*Ax = x^*x$.

ii) $\|Ax\|_2 = \|x\|_2$.

iii) $Ax = x$.

iv) $x \in \mathcal{R}(A)$.

Related: Fact 4.15.1 and Fact 4.17.11.

Fact 4.17.13. Let $A \triangleq \begin{bmatrix} A_{11} & A_{12} \\ A_{12}^* & A_{22} \end{bmatrix} \in \mathbb{F}^{(n+m)\times(n+m)}$, and assume that A is a projector. Then,

$$\operatorname{rank} A = \operatorname{rank} A_{11} + \operatorname{rank} A_{22} - \operatorname{rank} A_{12}.$$

Source: [2675] and Fact 4.15.22. **Related:** Fact 8.9.14.

Fact 4.17.14. Let $A \in \mathbb{F}^{n\times n}$, and assume that A satisfies two out of the three properties (Hermitian, shifted unitary, idempotent). Then, A satisfies the remaining property. Furthermore, A satisfies all three properties if and only if A is a projector. **Source:** If A is idempotent and shifted unitary, then $(2A-I)^{-1} = 2A - I = (2A^* - I)^{-1}$. Hence, A is Hermitian. **Related:** Fact 4.13.30, Fact 4.19.2, and Fact 4.19.6.

4.18 Facts on Two or More Projectors

Fact 4.18.1. Let $A, B \in \mathbb{F}^{n\times n}$, assume that A and B are projectors, and define $\mathcal{S}_1 \triangleq \mathcal{R}(A)$ and $\mathcal{S}_2 \triangleq \mathcal{R}(B)$. Then, the following statements are equivalent:

i) AB is a projector.

ii) $\mathcal{S}_1 \cap \mathcal{S}_2 = \mathcal{S}_1 \cap (\mathcal{S}_1^\perp + \mathcal{S}_2)$.

iii) $\mathcal{S}_1 \cap (\mathcal{S}_1^\perp + \mathcal{S}_2) = \mathcal{S}_2 \cap (\mathcal{S}_2^\perp + \mathcal{S}_1)$.

The following statements are equivalent:

iv) $A + B$ is a projector.

v) $\mathcal{S}_1 \cap (\mathcal{S}_1^\perp + \mathcal{S}_2) = \{0\}$.

vi) $\mathcal{S}_2 \subseteq \mathcal{S}_1^\perp$.

The following statements are equivalent:

vii) $A - B$ is a projector.

viii) $\mathcal{S}_2 \cap (\mathcal{S}_1^\perp + \mathcal{S}_2^\perp) = \{0\}$.

ix) $\mathcal{S}_1^\perp \cap (\mathcal{S}_1 + \mathcal{S}_2) = \{0\}$.

x) $\mathcal{S}_2 \subseteq \mathcal{S}_1$.

The following statement holds:

xi) $\operatorname{rank}(A - B) = \dim(\mathcal{S}_1 + \mathcal{S}_2) - \dim(\mathcal{S}_1 \cap \mathcal{S}_2) = \dim[(\mathcal{S}_1 + \mathcal{S}_2) \cap (\mathcal{S}_1^\perp + \mathcal{S}_2^\perp)]$.

Source: [255]. **Related:** Fact 3.12.19 and Fact 7.13.27.

Fact 4.18.2. Let $A, B \in \mathbb{F}^{n\times n}$, and assume that A and B are projectors. Then, the following statements are equivalent:

i) $\mathcal{R}(A) = \mathcal{R}(B)$.

ii) $A = B$.

Related: Proposition 4.8.1.

Fact 4.18.3. Let $A \in \mathbb{F}^{n\times m}$, let $B \in \mathbb{F}^{n\times n}$, and assume that B is a projector. Then, the following statements are equivalent:

i) $\mathcal{R}(A) = \mathcal{R}(BA)$.

ii) $A = BA$.

Source: To prove *i)* $\Longrightarrow$ *ii)*, note that $0 = \mathcal{R}(B_\perp BA) = B_\perp\mathcal{R}(BA) = B_\perp\mathcal{R}(A) = \mathcal{R}(B_\perp A)$. Hence, $B_\perp A = 0$. Consequently, $BA = (B + B_\perp)A = A$. **Related:** Fact 8.8.3.

Fact 4.18.4. Let $A, B \in \mathbb{F}^{n\times n}$, and assume that A and B are projectors. Then, the following statements are equivalent:

i) $A < B$.

ii) $A = 0$ and $B = I$.

Fact 4.18.5. Let $A, B \in \mathbb{F}^{n\times n}$, and assume that A and B are projectors. Then, the following statements are equivalent:

i) $A \le B$.

ii) $\mathcal{R}(A) \subseteq \mathcal{R}(B)$.

iii) $\mathcal{R}(A) = \mathcal{R}(BA)$.

iv) $AB = BA$ and $\mathcal{R}(A) = \mathcal{R}(AB)$.

v) $A = AB$.

vi) $A = BA$.

vii) $A = ABA$.

viii) $B - A$ is idempotent.

ix) $B - A$ is a projector.

x) $A + B_\perp A$ is idempotent.

xi) $A + B_\perp A$ is a projector.

xii) $B_\perp \le A_\perp$.

xiii) $\mathcal{R}(B_\perp) \subseteq \mathcal{R}(A_\perp)$.

xiv) $B_\perp = B_\perp A_\perp$.

xv) $B_\perp = A_\perp B_\perp$.

xvi) $B_\perp = B_\perp A_\perp B_\perp$.

xvii) $A_\perp - B_\perp$ is a projector.

xviii) $B_\perp + AB_\perp$ is idempotent.

xix) For all $x \in \mathbb{F}^n$, $\|Ax\|_2 \le \|Bx\|_2$.

If these statements hold, then $AB = BA$. **Source:** [2418, pp. 24, 169]. **Related:** Fact 4.18.7 and Fact 10.11.9.

Fact 4.18.6. Let $A, B \in \mathbb{F}^{n\times n}$, and assume that A and B are projectors. Then,

$$\operatorname{tr}(AB)^2 \le \operatorname{tr} AB \le \min\{\operatorname{tr} A, \operatorname{tr} B, \operatorname{rank} AB\}.$$

In addition, the following statements are equivalent:

i) Either $A - B$ is a projector or $B - A$ is a projector.

ii) $\operatorname{tr} AB = \min\{\operatorname{tr} A, \operatorname{tr} B\}$.

Source: [234, 252] and [2942, p. 48]. **Related:** Fact 4.18.7.

Fact 4.18.7. Let $A, B \in \mathbb{F}^{n\times n}$, and assume that A and B are projectors. Then, the following statements are equivalent:

i) $AB = BA$.

ii) AB is a projector.

iii) AB is idempotent.

iv) AB is Hermitian.

v) AB is normal.

vi) AB is range Hermitian.

vii) $[AB(AB)^*, (AB)^*AB] = 0$.

viii) $AB = ABA$.

ix) $AB = BAB$.

x) $AB = (AB)^2$.

xi) There exist distinct tuples with alternating components A and B whose products are equal.

xii) $A + A_{\perp}B$ is a projector.

xiii) $A + A_{\perp}B$ is idempotent.

xiv) $A + A_{\perp}B$ is Hermitian.

xv) $A + B_{\perp}A$ is Hermitian.

xvi) AB is the projector onto $\mathcal{R}(A) \cap \mathcal{R}(B)$.

xvii) $A + A_{\perp}B$ is the projector onto $\mathcal{R}(A) + \mathcal{R}(B)$.

xviii) $\operatorname{tr}(AB)^2 = \operatorname{tr} AB$.

xix) $\mathcal{R}(AB) = \mathcal{R}(A) \cap \mathcal{R}(B)$.

xx) $\mathcal{R}(A) \cap \mathcal{R}(B)$ and $\mathcal{R}(A) \cap \mathcal{N}(B)$ are complementary subspaces.

xxi) $\mathcal{R}(AB) \subseteq \mathcal{R}(B)$.

xxii) $\mathcal{R}(BA) \subseteq \mathcal{R}(A)$.

xxiii) $\mathcal{R}(AB) = \mathcal{R}(BA)$.

xxiv) $\mathcal{N}(AB) = \mathcal{N}(BA)$.

xxv) $\mathcal{R}(AB)$ and $\mathcal{N}(AB)$ are complementary subspaces.

xxvi) $\mathcal{R}(A - B) \cap \mathcal{R}(AB) = \{0\}$.

xxvii) $\operatorname{rank}[A \;\; B] + \operatorname{rank} AB = \operatorname{rank} A + \operatorname{rank} B$.

xxviii) $\operatorname{rank}(A + B) + \operatorname{rank} AB = \operatorname{rank} A + \operatorname{rank} B$.

xxix) $\operatorname{rank}(I - AB) + \operatorname{rank} AB = n$.

xxx) $\operatorname{rank}(A - B) + \operatorname{rank} AB = \operatorname{rank}(A + B)$.

xxxi) $\operatorname{rank} AB = \operatorname{rank}(AB + BA)$.

xxxii) $\operatorname{tr}(AB)^2 = \operatorname{rank} AB$.

xxxiii) $\operatorname{tr} AB \le \operatorname{rank} AB$.

xxxiv) $\operatorname{tr} AB = \operatorname{rank} AB$.

xxxv) $\operatorname{tr} AB = \operatorname{rank} ABA$.

xxxvi) $\frac{1}{2}\operatorname{tr}(AB + BA) = \operatorname{rank}(AB + BA)$.

xxxvii) $\frac{1}{2}\operatorname{tr}(AB + BA)^2 = \operatorname{tr}(AB + BA)$.

xxxviii) ABA is a projector.

xxxix) $\operatorname{rank} ABA = \operatorname{tr} ABA$.

xl) $\operatorname{rank} ABA = \operatorname{tr}(ABA)^2$.

xli) $\operatorname{tr} ABA = \operatorname{tr}(ABA)^2$.

xlii) $AB - BA$ is a projector.

xliii) $\operatorname{rank}(AB - BA) = \operatorname{tr}(AB - BA)$.

xliv) $\operatorname{rank}(AB - BA) + \operatorname{tr}(AB - BA)^2 = 0$.

xlv) $A + B - AB$ is a projector.

xlvi) $\operatorname{rank}(A + B - AB) = \operatorname{tr}(A + B - AB)$.

Source: [234, 242, 252, 261], [1124, pp. 42–44], [2238, p. 308], and [2657, 2706, 2869]. **Remark:** To illustrate *xi*), consider (A, B, A, B, A) and (B, A, B). Then, $ABABA = BAB$ implies that $AB = BA$. See [235]. **Related:** Fact 4.18.5, Fact 7.13.5, Fact 8.4.32, and Fact 8.8.5.

Fact 4.18.8. Let $A, B \in \mathbb{F}^{n\times n}$, and assume that A and B are projectors. Then, the following statements are equivalent:

i) $AB = 0$.

ii) For all $\alpha,\beta \in \mathbb{F}$, $\mathcal{R}(I - \alpha A - \beta B) = \mathcal{R}[(I - \alpha A)(I - \beta B)]$.

iii) For all $\alpha,\beta \in \mathbb{F}$, $\operatorname{rank}(I - \alpha A - \beta B) = \operatorname{rank}\,(I - \alpha A)(I - \beta B)$.

iv) For all $\alpha,\beta \in \mathbb{F}$, $\operatorname{tr}(I - \alpha A - \beta B) = \operatorname{tr}\,(I - \alpha A)(I - \beta B)$.

Source: [259]. **Related:** Fact 3.16.14.

Fact 4.18.9. Let $A, B \in \mathbb{F}^{n\times n}$, and assume that A and B are projectors. Then, the following statements are equivalent:

i) $A + B$ is a projector.

ii) $AB = 0$.

iii) $BA = 0$.

iv) $AB = BA = 0$.

v) $AB + BA = 0$.

vi) $(A + B)^2 = A + B$.

vii) $\operatorname{tr}\,(A + B)^2 = \operatorname{tr}(A + B)$.

viii) $\mathcal{R}(A) \subseteq \mathcal{R}(B)^\perp$.

ix) $\mathcal{R}(B) \subseteq \mathcal{R}(A)^\perp$.

x) $\mathcal{R}(A)$ and $\mathcal{R}(B)$ are mutually orthogonal.

xi) AB is a projector, and $\operatorname{rank}(A + B) = \operatorname{tr}(A + B)$.

xii) AB is a projector, and $\operatorname{rank}(A + B) = \operatorname{tr}\,(A + B)^2$.

xiii) AB is a projector, and $\operatorname{rank}(I - A - B) = \operatorname{tr}(I - A - B)$.

Source: [261], [1124, pp. 42–44], and [2657]. **Remark:** See [244, 1134].

Fact 4.18.10. Let $A, B \in \mathbb{F}^{n\times n}$, and assume that A and B are projectors. Then, the following statements are equivalent:

i) $A - B$ is a projector.

ii) $\operatorname{tr}(A - B) = \operatorname{tr}\,(A - B)^2$.

iii) AB is a projector, and $\operatorname{rank}(A - B) = \operatorname{tr}(A - B)$.

Source: [261].

Fact 4.18.11. Let $A, B \in \mathbb{F}^{n\times n}$, and assume that A and B are projectors. Then, the following statements hold:

i) $\mathcal{R}(A + B) = \mathcal{R}(A) + \mathcal{R}(B) = \mathcal{R}([A\;\; B]) = \operatorname{span}[\mathcal{R}(A) \cup \mathcal{R}(B)]$.

ii) $\mathcal{R}(A + B) = \mathcal{R}(A + A_\perp B) = \mathcal{R}(A) + \mathcal{R}(A_\perp B)$.

iii) $\mathcal{R}(A + B) = \mathcal{R}(A - B) + \mathcal{R}(AB + BA) = \mathcal{R}(I - A_\perp B_\perp)$.

iv) $\mathcal{R}(A + B) = \mathcal{R}(A - B) + \mathcal{R}(AB)$.

v) $\mathcal{R}(A + B) = \mathcal{R}(B) + \mathcal{R}(B_\perp A B_\perp)$.

vi) $\mathcal{R}(A + B) = \mathcal{R}(A) + \mathcal{R}(B - AB - BA + ABA)$.

vii) $\mathcal{N}(A + B) = \mathcal{N}(A) \cap \mathcal{N}(B)$.

viii) $\mathcal{R}(A - B) + \mathcal{R}(A_\perp - B) = \mathbb{F}^n$.

ix) $\mathcal{R}(A - B) \cap \mathcal{R}(A_\perp - B) = \mathcal{R}([A, B]) \subseteq \mathcal{R}(A + B)$.

x) $\mathcal{R}(A - B) = [\mathcal{R}(A) + \mathcal{R}(B)] \cap [\mathcal{N}(A) + \mathcal{N}(B)]$.

xi) $\mathcal{R}(AB) = \mathcal{R}(A) \cap [\mathcal{R}(A) \cap \mathcal{N}(B)]^\perp$.

xii) $\mathcal{N}(AB) = \mathcal{N}(B) + [\mathcal{N}(A) \cap \mathcal{N}(B)]$.

xiii) $\mathcal{R}(ABA_\perp) = \mathcal{R}(A) \cap [\mathcal{R}(A) \cap \mathcal{R}(B)]^\perp \cap [\mathcal{R}(A) \cap \mathcal{N}(B)]^\perp$.

xiv) $\mathcal{N}(ABA_\perp) = \mathcal{R}(A) + [\mathcal{N}(A) \cap \mathcal{R}(B)] + [\mathcal{N}(A) \cap \mathcal{N}(B)]$.

xv) $\mathcal{R}(I - AB) = \mathcal{N}(A) + \mathcal{N}(B)$.

xvi) $\mathcal{N}(I - AB) = \mathcal{R}(A) \cap \mathcal{R}(B)$.

xvii) $\mathcal{R}([A, B]) = \mathcal{R}(ABA_\perp) + \mathcal{R}(A_\perp BA)$.

xviii) $\mathcal{R}(AB + BA) = [\mathcal{R}(A) + \mathcal{R}(B)] \cap [\mathcal{R}(A) + \mathcal{N}(B)] \cap [\mathcal{N}(A) + \mathcal{R}(B)]$.

xix) $\mathcal{N}(AB + BA) = [\mathcal{R}(A) \cap \mathcal{N}(B)] + [\mathcal{N}(A) \cap \mathcal{R}(B)] + [\mathcal{N}(A) \cap \mathcal{N}(B)]$.

Furthermore, the following statements are equivalent:

xx) $\mathcal{R}(A + B) = \mathbb{F}^n$.

xxi) $\mathcal{R}(B_\perp) = \mathcal{R}(B_\perp AB_\perp)$.

Source: [242, 256, 1247, 2239, 2712].

Fact 4.18.12. Let $A, B \in \mathbb{F}^{n\times n}$, and assume that A and B are projectors. Then,

$$\mathcal{R}(A) \cap \mathcal{R}(B) = \mathcal{N}\left(\begin{bmatrix} A_\perp \\ B_\perp \end{bmatrix}\right).$$

Furthermore, the following statements are equivalent:

i) $\mathcal{R}(A) \cap \mathcal{R}(B) = \{0\}$.

ii) $\mathcal{R}(A - B) = \mathcal{R}(A) + \mathcal{R}(B)$.

iii) $\operatorname{rank}(A - B) = \operatorname{rank} A + \operatorname{rank} B$.

iv) $\mathcal{R}(AB_\perp) + \mathcal{R}(A_\perp B) = \mathcal{R}(A) + \mathcal{R}(B)$.

v) $\operatorname{rank} AB_\perp + \operatorname{rank} A_\perp B = \operatorname{rank} A + \operatorname{rank} B$.

Source: Fact 8.9.2 and [242]. **Related:** Fact 8.8.17.

Fact 4.18.13. Let $A, B \in \mathbb{F}^{n\times n}$, and assume that A and B are projectors. Then, $A + A_\perp B$ is range Hermitian. **Source:** [258].

Fact 4.18.14. Let $A, B \in \mathbb{F}^{n\times n}$, assume that A and B are projectors, assume that $\operatorname{rank} A + \operatorname{rank} B = \operatorname{rank}(A + B) = n$, and define $P \triangleq A(A + B)^{-1}$. Then, the following statements hold:

i) P is the idempotent matrix onto $\mathcal{R}(A)$ along $\mathcal{R}(B)$.

ii) $A(A + B)^{-1}B = 0$.

iii) $A(A + B)^{-1}A = A$.

Related: Fact 4.16.4.

Fact 4.18.15. Let $A, B \in \mathbb{F}^{n\times n}$, where A and B are projectors. Then, AB is group invertible. **Source:** $\mathcal{N}(BA) \subseteq \mathcal{N}(BABA) \subseteq \mathcal{N}(ABABA) = \mathcal{N}(ABAABA) = \mathcal{N}(ABA) = \mathcal{N}(ABBA) = \mathcal{N}(BA)$. **Remark:** See [2869]. **Remark:** Fact 10.11.23 shows that AB is semisimple.

Fact 4.18.16. Let $A, B \in \mathbb{F}^{n\times n}$, where A and B are projectors. Then, $[A, B]$ is skew Hermitian, and $\operatorname{rank}\,[A, B]$ is even. **Source:** [406].

Fact 4.18.17. Let $A, B \in \mathbb{F}^{n\times n}$, assume that A and B are projectors, and assume that $\operatorname{rank} A = \operatorname{rank} B$. Then, there exists a reflector $S \in \mathbb{F}^{n\times n}$ such that $A = SBS$. If, in addition, $A + B - I$ is nonsingular, then one such reflector is given by $S = \langle A + B - I\rangle(A + B - I)^{-1}$. **Source:** [726]. **Remark:** $\langle\cdot\rangle$ is defined in Chapter 10.

Fact 4.18.18. Let $A, B \in \mathbb{F}^{n\times n}$, assume that A and B are projectors, and let $\alpha, \beta \in \mathbb{R}$, where $\alpha\beta \neq 0$ and $\alpha + \beta \neq 0$. Then,

$$\operatorname{rank}(A + B) = \operatorname{rank}\,[A\ \ B] = \operatorname{rank} A + \operatorname{rank} B - n + \operatorname{rank}(A_\perp + B_\perp)$$

$$= \operatorname{rank} A + \operatorname{rank}(B - AB) = \operatorname{rank} B + \operatorname{rank}(A - BA) = \operatorname{rank}(\alpha A + \beta B),$$

$$\mathcal{R}(\alpha A + \beta B) = \mathcal{R}([A \;\; B]), \quad \operatorname{sig}(A - B) = \operatorname{rank} A - \operatorname{rank} B,$$

$$\operatorname{rank}(A - B) = 2 \operatorname{rank} [A \;\; B] - \operatorname{rank} A - \operatorname{rank} B = \operatorname{rank}(A - AB) + \operatorname{rank}(AB - B),$$

$$\operatorname{rank}(I - A - B) = 2 \operatorname{rank} AB + n - \operatorname{rank} A - \operatorname{rank} B = \operatorname{rank} AB + \operatorname{rank} A_\perp B_\perp,$$

$$\operatorname{rank} [A \;\; B] = \operatorname{rank} A + \operatorname{rank} B - n + \operatorname{rank} [A_\perp \;\; B_\perp],$$

$$\begin{aligned} \operatorname{rank} [A, B] &= 2(\operatorname{rank} [AB \;\; BA] - \operatorname{rank} AB) = 2 \operatorname{rank} [AB - (AB)^2] \\ &= 2 \operatorname{rank}(AB - ABA) = 2 \operatorname{rank}(AB - BAB) \\ &= \operatorname{rank}(A - B) + \operatorname{rank}(A + B - I) - n \\ &= 2(\operatorname{rank} [A \;\; B] + \operatorname{rank} AB - \operatorname{rank} A - \operatorname{rank} B), \end{aligned}$$

$$[A, B] = (A - B)(A + B - I) = (A + B - I)(B - A),$$

$$AB + BA = (A + B)(A + B - I) = (A + B - I)(A + B),$$

$$\operatorname{rank}(I - AB) = \operatorname{rank}(I - BA) = \operatorname{rank} [A \;\; B] - \operatorname{rank} A - \operatorname{rank} B + n,$$

$$\mathcal{N}(I - AB) = \mathcal{N}(I - BA) = \mathcal{R}(A) \cap \mathcal{R}(B),$$

$$\begin{aligned} \operatorname{rank} [AB \;\; BA] &= \operatorname{rank}(AB + BA) = \operatorname{rank}[AB - (AB)^2] + \operatorname{rank} AB \\ &= \operatorname{rank}(A + B) + \operatorname{rank}(A + B - I) - n \\ &= \operatorname{rank} [A \;\; B] + 2 \operatorname{rank} AB - \operatorname{rank} A - \operatorname{rank} B, \end{aligned}$$

$$\begin{aligned} \operatorname{rank}(AB - ABA) &= \operatorname{rank} AB + \operatorname{rank} [A \;\; B] - \operatorname{rank} A - \operatorname{rank} B \\ &= \operatorname{rank} AB + \operatorname{rank}(B - AB) - \operatorname{rank} B \\ &= \operatorname{rank} BA + \operatorname{rank}(A - AB) - \operatorname{rank} A, \end{aligned}$$

$$\begin{aligned} \operatorname{rank} [AB - (AB)^2] &= \operatorname{rank}(I - AB) + \operatorname{rank} AB - n = \operatorname{rank} [A_\perp B_\perp - (A_\perp B_\perp)^2] \\ &= \operatorname{rank} [A \;\; BA] - \operatorname{rank} A = \operatorname{rank} [B \;\; AB] - \operatorname{rank} B = \operatorname{rank}[(AB)^+ - BA] \\ &= \operatorname{rank} [A \;\; B] + \operatorname{rank} AB - \operatorname{rank} A - \operatorname{rank} B \\ &= \operatorname{rank}[2A(A + B)^+ B - AB], \end{aligned}$$

$$\operatorname{rank} [A \;\; BA] = \operatorname{rank} [A \;\; B] + \operatorname{rank} AB - \operatorname{rank} B.$$

Hence, $2A(A + B)^+B = AB$ if and only if $(AB)^2 = AB$. Furthermore, $AB = BA$ if and only if

$$\operatorname{rank}(A + B) + \operatorname{rank} AB = \operatorname{rank} A + \operatorname{rank} B.$$

Finally, if $AB = BA$, then

$$\operatorname{rank}(A - B) + 2 \operatorname{rank} AB = \operatorname{rank} A + \operatorname{rank} B.$$

Source: [406, 2657, 2659, 2660, 2672, 2673, 2676].

Fact 4.18.19. Let $A, B \in \mathbb{F}^{n \times n}$, and assume that A and B are projectors. Then, the following statements are equivalent:

i) $\mathcal{R}(A) + \mathcal{R}(B) = \mathbb{F}^n$.

ii) $A + B - AB$ is nonsingular.

iii) $B + A - BA$ is nonsingular.

Furthermore, the following statements are equivalent:

iv) $A - B$ is nonsingular.

v) $\operatorname{rank} [A \;\; B] = \operatorname{rank} A + \operatorname{rank} B = n$.

vi) $\mathcal{R}(A)$ and $\mathcal{R}(B)$ are complementary subspaces.

If these statements hold, then the following statements hold:

vii) $I - AB$ and $I - BA$ are nonsingular.

viii) $A + B - AB$ and $B + A - BA$ are nonsingular.

ix) The idempotent matrix $M \in \mathbb{F}^{n\times n}$ onto $\mathcal{R}(A)$ along $\mathcal{R}(B)$ is given by

$$M = (I - AB)^{-1}A(I - AB) = A(I - BA)^{-1}(I - AB) = (I - BA)^{-1}(I - B) = A(A + B - BA)^{-1}.$$

x) M satisfies $M + M^* = (A - B)^{-1} + I$, and thus $(A - B)^{-1} = M + M^* - I = M - M_\perp^*$.

Source: Fact 7.13.27 and [15, 248, 601, 1134, 1232, 1251, 1504, 2296]. *ix)* follows from Fact 7.13.28. **Remark:** Fact 8.8.14 provides an alternative expression for M involving the Moore-Penrose generalized inverse. **Related:** Fact 4.16.4, Fact 4.16.9, and Fact 8.8.15.

Fact 4.18.20. Let $A, B \in \mathbb{F}^{n\times n}$, assume that A and B are projectors, assume that $A \neq 0$, $B \neq 0$, and $A \neq B$, and let $\alpha, \beta \in \mathbb{F}$, where α and β are nonzero. Then, $\alpha A + \beta B$ is a projector if and only if exactly one of the following statements holds:

i) $\alpha = \beta = 1$ and $AB = BA = 0$.

ii) $\alpha = -\beta = 1$ and $AB = BA = B$.

iii) $-\alpha = \beta = 1$ and $AB = BA = A$.

iv) $\alpha + \beta = 1$, $AB \neq BA$, and $(A - B)^2 = 0$.

Source: [220, 251]. **Related:** Fact 8.5.13 and Fact 8.8.8.

Fact 4.18.21. Let $A, B \in \mathbb{F}^{n\times n}$, and assume that A and B are projectors. Then,

$$\operatorname{tr} AB \le \operatorname{rank} AB \le \min\{\operatorname{tr} A, \operatorname{tr} B\}.$$

Furthermore, the first and third terms are equal if and only if either $A - B$ is a projector or $B - A$ is a projector. In addition, for all $k \ge 1$,

$$\dim[\mathcal{R}(A) \cap \mathcal{R}(B)] \le \operatorname{tr}(AB)^k \le \operatorname{rank}(AB)^k.$$

Source: [252].

Fact 4.18.22. Let $A, B \in \mathbb{F}^{n\times n}$, and assume that A and B are projectors. Then,

$$\operatorname{rank}\begin{bmatrix} A & B \\ B & A \end{bmatrix} = 3\operatorname{rank}[A \;\; B] - \operatorname{rank} A - \operatorname{rank} B.$$

Furthermore, the following statements are equivalent:

i) $\left[\begin{smallmatrix} A & B \\ B & A \end{smallmatrix}\right]$ is nonsingular.

ii) $A + B$ and $A - B$ are nonsingular.

iii) $3\operatorname{rank}[A \;\; B] = \operatorname{rank} A + \operatorname{rank} B + 2n$.

Source: [2667]. **Related:** Fact 7.9.25.

Fact 4.18.23. Let $A, B \in \mathbb{F}^{n\times n}$, and assume that A and B are projectors. Then,

$$\operatorname{rank}\begin{bmatrix} A+B & AB & & & \\ AB & A+B & \ddots & & \\ & \ddots & \ddots & \ddots & \\ & & \ddots & A+B & AB \\ & & & AB & A+B \end{bmatrix} = l\operatorname{rank}(A+B),$$

where the size of the matrix is $ln \times ln$. **Source:** [2676].

Fact 4.18.24. Let $A_1, \ldots, A_r \in \mathbb{F}^{n\times n}$, assume that $A_1, \ldots, A_r$ are Hermitian, and consider the following statements:

i) $A_1, \ldots, A_r$ are projectors.

ii) $\sum_{i=1}^r A_i$ is a projector.

iii) For all distinct $i, j \in \{1, \ldots, r\}$, it follows that $A_iA_j = 0$.

iv) $\operatorname{rank} \sum_{i=1}^r A_i = \sum_{i=1}^r \operatorname{rank} A_i$.

Then, if at least one of the pairs of statements [*i*),*ii*)], [*i*),*iii*)], [*ii*),*iii*)], [*ii*),*iv*)] holds, then *i*)–*iv*) hold. In particular, if $A_1, \ldots, A_r$ are projectors and $\sum_{i=1}^r A_i = I$, then, for all distinct $i, j \in \{1, \ldots, r\}$, $A_iA_j = 0$. **Source:** [2403, pp. 400–402]. **Remark:** The last result is *Cochran's theorem.* A stronger version is given by Fact 4.16.21. **Problem:** Extend this result to the case where $A_1, \ldots, A_r$ are idempotent but not necessarily projectors.

4.19 Facts on Reflectors

Fact 4.19.1. If $A \in \mathbb{F}^{n\times n}$ is a projector, then $B \triangleq 2A - I$ is a reflector. Conversely, if $B \in \mathbb{F}^{n\times n}$ is a reflector, then $A \triangleq \frac{1}{2}(B + I)$ is a projector. **Remark:** The affine mapping $f(A) \triangleq 2A - I$ from the projectors to the reflectors is one-to-one and onto. **Related:** Fact 4.13.29 and Fact 4.20.3.

Fact 4.19.2. Let $A \in \mathbb{F}^{n\times n}$, and assume that A satisfies two out of the three properties (Hermitian, unitary, involutory). Then, A also satisfies the remaining property. Furthermore, A satisfies all three properties if and only if A is a reflector. **Related:** Fact 4.17.14 and Fact 4.19.6. **Remark:** Properties of reflectors are discussed in [2084].

Fact 4.19.3. Let $x \in \mathbb{F}^n$ be nonzero, and define the elementary reflector $A \triangleq I - 2(x^*x)^{-1}xx^*$. Then, the following statements hold:

i) $\det A = -1$.

ii) If $y \in \mathbb{F}^n$, then Ay is the reflection of y across $\{x\}^\perp$.

iii) $Ax = -x$.

iv) $\frac{1}{2}(A + I)$ is the elementary projector $I - (x^*x)^{-1}xx^*$.

Fact 4.19.4. Let $x, y \in \mathbb{F}^n$. Then, there exists a unique elementary reflector $A \in \mathbb{F}^{n\times n}$ such that $Ax = y$ if and only if x^*y is real and $x^*x = y^*y$. If, in addition, $x \neq y$, then A is given by

$$A = I - 2[(x - y)^*(x - y)]^{-1}(x - y)(x - y)^*.$$

Remark: This is the *reflection theorem.* See [1184, pp. 16–18] and [2314, p. 357]. See Fact 4.11.5.

Fact 4.19.5. Let $n > 1$, let $\mathcal{S} \subset \mathbb{F}^n$, and assume that $\mathcal{S}$ is a hyperplane. Then, there exists a unique elementary reflector $A \in \mathbb{F}^{n\times n}$ such that, for all $y = y_1 + y_2 \in \mathbb{F}^n$, where $y_1 \in \mathcal{S}$ and $y_2 = \mathcal{S}^\perp$, it follows that $Ay = y_1 - y_2$. Furthermore, if $\mathcal{S} = \{x\}^\perp$, then $A = I - 2(x^*x)^{-1}xx^*$.

Fact 4.19.6. Let $A \in \mathbb{F}^{n\times n}$, and assume that A satisfies two out of the three properties (skew Hermitian, unitary, skew involutory). Then, A also satisfies the remaining property. Furthermore, these matrices are the skew reflectors. **Remark:** Properties of skew reflectors are discussed in [2084]. **Related:** Fact 4.17.14, Fact 4.19.2, and Fact 4.19.7.

Fact 4.19.7. Let $A \in \mathbb{C}^{n\times n}$. Then, A is a reflector if and only if $\jmath A$ is a skew reflector. **Remark:** The mapping $f(A) \triangleq \jmath A$ relates Fact 4.19.2 to Fact 4.19.6. **Problem:** Assuming A is real and n is even, determine a real transformation between the reflectors and the skew reflectors.

Fact 4.19.8. Let $A \in \mathbb{F}^{n\times n}$. Then, the following statements are equivalent:

i) A is a reflector.

ii) $A = AA^* + A^* - I$.

iii) $A = \frac{1}{2}(A + I)(A^* + I) - I$.

4.20 Facts on Involutory Matrices

Fact 4.20.1. Let $A \in \mathbb{F}^{n\times n}$, and assume that A is involutory. Then, $\operatorname{rank}(A+I)+\operatorname{rank}(A-I) = n$.

Fact 4.20.2. Let $A \in \mathbb{F}^{n\times n}$, and assume that A is involutory. Then, either $\det A = 1$ or $\det A = -1$.

Fact 4.20.3. The following statements hold:

i) If $A \in \mathbb{F}^{n\times n}$ is idempotent, then $B_1 \triangleq 2A - I$ and $B_2 \triangleq I - 2A$ are involutory.

ii) If $B \in \mathbb{F}^{n\times n}$ is involutory, then $A_1 \triangleq \frac{1}{2}(I + B)$ and $A_2 \triangleq \frac{1}{2}(I - B)$ are idempotent.

Remark: The affine mappings $f_1(A) \triangleq 2A-I$ and $f_2(A) \triangleq I-2A$ from the idempotent matrices to the involutory matrices are one-to-one and onto with inverses $f_1^{\mathrm{inv}}(B) = \frac{1}{2}(I + B)$ and $f_2^{\mathrm{inv}}(B) = \frac{1}{2}(I - B)$ from the involutory matrices to the idempotent matrices. See Fact 4.13.29 and Fact 4.19.1.

Fact 4.20.4. Let $A \in \mathbb{F}^{n\times n}$. Then, A is involutory if and only if $(A + I)(A - I) = 0$.

Fact 4.20.5. Let $n \ge 1$. Then, $\hat{I}_n$ is involutory.

Fact 4.20.6. Let $A \in \mathbb{F}^{n\times n}$, and assume that A is involutory. Then, $A + A^*$ is nonsingular, and

$$\begin{aligned}\operatorname{rank}(A - A^*) = \operatorname{rank}\,[A, A^*] &= 2\operatorname{rank}\,[I + A \;\; I + A^*] - 2\operatorname{rank}(I + A)\\ &= 2\operatorname{rank}\,[I - A \;\; I - A^*] - 2\operatorname{rank}(I - A).\end{aligned}$$

Source: [2672].

Fact 4.20.7. Let $n \ge 1$. Then, $P_n\hat{I}_nP_n = \hat{I}_n$. Consequently, $P_n\hat{I}_n$ is involutory. **Remark:** This equality helps to define the generators for the dihedral group $\mathrm{D}(n)$. See [1560, p. 169].

Fact 4.20.8. Let $A, B \in \mathbb{F}^{n\times n}$, and assume that A and B are involutory. Then,

$$\mathcal{R}([A,B]) = \mathcal{R}(A - B) \cap \mathcal{R}(A + B), \quad \mathcal{R}(A - B) + \mathcal{R}(A + B) = \mathbb{F}^n,$$
$$\mathcal{N}([A,B]) = \mathcal{N}(A - B) + \mathcal{N}(A + B), \quad \mathcal{N}(A - B) \cap \mathcal{N}(A + B) = \{0\}$$

Source: [2640].

Fact 4.20.9. Let $A, B \in \mathbb{F}^{n\times n}$, and assume that A and B are involutory. Then,

$$\begin{aligned}\operatorname{rank}(A + B) &= \operatorname{rank}\begin{bmatrix} I + A\\ I - B\end{bmatrix} + \operatorname{rank}\,[I + A \;\; I - B] - \operatorname{rank}(I + A) - \operatorname{rank}(I - B)\\ &= \operatorname{rank}[(I + A)(I + B)] + \operatorname{rank}[(I - A)(I - B)]\\ &= \operatorname{rank}[(I + A)(I + B)] + \operatorname{rank}[(I + B)(I + A)] - \operatorname{rank}(I + A) - \operatorname{rank}(I + B) + n,\\ \operatorname{rank}(A - B) &= \operatorname{rank}[(I + A)(I - B)] + \operatorname{rank}[(I - B)(I + A)] - \operatorname{rank}(I + A) - \operatorname{rank}(I - B) + n,\end{aligned}$$
$$\operatorname{rank}\,[A, B] = \operatorname{rank}(A + B) + \operatorname{rank}(A - B) - n.$$

Consequently, $AB = BA$ if and only if $\operatorname{rank}(A + B) + \operatorname{rank}(A - B) = n$. **Source:** [2672].

Fact 4.20.10. Let $A \in \mathbb{F}^{n\times m}$ and $B \in \mathbb{F}^{m\times n}$, and define

$$C \triangleq \begin{bmatrix} I - BA & B\\ 2A - ABA & AB - I\end{bmatrix}.$$

Then, C is involutory. **Source:** [2036, p. 113].

Fact 4.20.11. Let $A \in \mathbb{R}^{n\times n}$, and assume that A is skew involutory. Then, n is even.

4.21 Facts on Tripotent Matrices

Fact 4.21.1. Let $A \in \mathbb{F}^{n\times n}$, and assume that A is tripotent. Then, A^2 is idempotent. **Remark:** The converse is false. Let $A = \left[\begin{smallmatrix} 0 & 1\\ 0 & 0\end{smallmatrix}\right]$.

Fact 4.21.2. Let $A \in \mathbb{F}^{n\times n}$. Then, A is nonsingular and tripotent if and only if A is involutory.

Fact 4.21.3. Let $A \in \mathbb{F}^{n\times n}$. Then, the following statements are equivalent:

i) A is tripotent.

ii) $\mathcal{R}(A)$ and $\mathcal{R}(I - A^2)$ are complementary subspaces.

iii) $\mathcal{R}(A) \subseteq \mathcal{N}(I - A^2)$.

iv) $\mathcal{N}(A) \cap \mathcal{N}(I - A^2) = \{0\}$.

v) $\mathcal{R}(I - A^2) \subseteq \mathcal{N}(A)$.

vi) For all $x \in \mathcal{R}(A)$, $A^2x = x$.

vii) $\operatorname{rank} A = \operatorname{rank}(A + A^2) + \operatorname{rank}(A - A^2)$.

viii) $\operatorname{rank} A + \operatorname{rank}(I - A^2) = n$.

ix) $\operatorname{rank} A + \operatorname{rank}(I - A) + \operatorname{rank}(I + A) = 2n$.

x) A is group invertible, and A^2 is idempotent.

Source: [257, 585, 2675].

Fact 4.21.4. Let $A \in \mathbb{F}^{n\times n}$, and assume that A is tripotent. Then, the following statements are equivalent:

i) A is Hermitian.

ii) A is normal.

iii) A is range Hermitian and a partial isometry.

Source: [257].

Fact 4.21.5. Let $A \in \mathbb{R}^{n\times n}$ be tripotent. Then, $\operatorname{rank} A = \operatorname{rank} A^2 = \operatorname{tr} A^2$.

Fact 4.21.6. Let $A, B \in \mathbb{F}^{n\times n}$, and assume that A and B are tripotent. Then, the following statements hold:

i) If $AB = BA$, then AB is tripotent.

ii) If A and B are Hermitian, then the following statements are equivalent:

a) AB is tripotent and Hermitian.

b) $AB = BA$.

iii) If A and B are Hermitian and tripotent, then the following statements are equivalent:

a) $(AB)^+ = B^+A^+$.

b) A^2B^2 is Hermitian.

c) $(A^2B^2)^2 = B^2A^2$.

iv) If A and B are Hermitian and tripotent, then the following statements are equivalent:

a) $A \overset{*}{\le} B$.

b) $ABA = A$ and $BAB = A$.

c) $ABA = A$ and $A^2 \overset{*}{\le} B^2$.

Source: [257].

Fact 4.21.7. If $A, B \in \mathbb{F}^{n\times n}$ are idempotent and $AB = 0$, then $A + BA_\perp$ is idempotent and $C \triangleq A - B$ is tripotent. Conversely, if $C \in \mathbb{F}^{n\times n}$ is tripotent, then $A \triangleq \frac{1}{2}(C^2 + C)$ and $B \triangleq \frac{1}{2}(C^2 - C)$ are idempotent and satisfy $C = A - B$ and $AB = BA = 0$. **Source:** [1998, p. 215].

4.22 Facts on Nilpotent Matrices

Fact 4.22.1. Let $A \in \mathbb{F}^{n\times n}$. Then, the following statements are equivalent:

i) $\mathcal{R}(A) = \mathcal{N}(A)$.

ii) A is similar to a block-diagonal matrix each of whose diagonal blocks is N_2.

Source: To prove *i)* $\Longrightarrow$ *ii)*, let $S \in \mathbb{F}^{n\times n}$ transform A into its Jordan form. Then, it follows from Fact 3.13.3 that $\mathcal{R}(SAS^{-1}) = S\mathcal{R}(AS^{-1}) = S\mathcal{R}(A) = S\mathcal{N}(A) = S\mathcal{N}(AS^{-1}S) = \mathcal{N}(AS^{-1}) = \mathcal{N}(SAS^{-1})$. The only Jordan block J that satisfies $\mathcal{R}(J) = \mathcal{N}(J)$ is $J = N_2$. Using $\mathcal{R}(N_2) = \mathcal{N}(N_2)$ and reversing these steps yields the converse result. **Remark:** The fact that n is even follows from $\operatorname{rank} A + \operatorname{def} A = n$ and $\operatorname{rank} A = \operatorname{def} A$. **Related:** Fact 4.22.2 and Fact 4.22.3.

Fact 4.22.2. Let $A \in \mathbb{F}^{n\times n}$. Then, the following statements are equivalent:

i) $\mathcal{N}(A) \subseteq \mathcal{R}(A)$.

ii) A is similar to a block-diagonal matrix each of whose diagonal blocks is either nonsingular or N_2.

Related: Fact 4.22.1 and Fact 4.22.3.

Fact 4.22.3. Let $A \in \mathbb{F}^{n\times n}$. Then, the following statements are equivalent:

i) $\mathcal{R}(A) \subseteq \mathcal{N}(A)$.

ii) A is similar to a block-diagonal matrix each of whose diagonal blocks is either zero or N_2.

Related: Fact 4.22.1 and Fact 4.22.2.

Fact 4.22.4. Let $n \in \mathbb{P}$ and $k \in \{0,\dots,n\}$. Then, $\operatorname{rank} N_n^k = n - k$.

Fact 4.22.5. Let $A \in \mathbb{R}^{n\times n}$. Then, the following statements hold:

i) $\operatorname{rank} A^k$ is a nonincreasing function of $k \ge 1$.

ii) If there exists $k \in \{1,\dots,n\}$ such that $\operatorname{rank} A^{k+1} = \operatorname{rank} A^k$, then $\operatorname{rank} A^l = \operatorname{rank} A^k$ for all $l \ge k$.

iii) If A is nilpotent and $A^l \ne 0$, then $\operatorname{rank} A^{k+1} < \operatorname{rank} A^k$ for all $k \in \{1,\dots,l\}$.

Fact 4.22.6. Let $A \in \mathbb{F}^{n\times n}$. Then, A is nilpotent if and only if, for all $k \in \{1,\dots,n\}$, $\operatorname{tr} A^k = 0$. **Source:** [2263, p. 103] or use Fact 6.8.4 with $p = \chi_A$ and $\mu_1 = \cdots = \mu_n = 0$.

Fact 4.22.7. Let $\lambda \in \mathbb{F}$ and $n, k \in \mathbb{P}$. Then,

$$(\lambda I_n + N_n)^k = \begin{cases} \lambda^k I_n + \binom{k}{1}\lambda^{k-1}N_n + \cdots + \binom{k}{k}N_n^k, & k < n-1,\\[2mm] \lambda^k I_n + \binom{k}{1}\lambda^{k-1}N_n + \cdots + \binom{k}{n-1}\lambda^{k-n+1}N_n^{n-1}, & k \ge n-1.\end{cases}$$

Equivalently, for all $k \ge n-1$,

$$\begin{bmatrix} \lambda & 1 & \cdots & 0 & 0\\ 0 & \lambda & \ddots & 0 & 0\\ \vdots & \ddots & \ddots & \ddots & \vdots\\ 0 & 0 & \ddots & \lambda & 1\\ 0 & 0 & \cdots & 0 & \lambda \end{bmatrix}^k = \begin{bmatrix} \lambda^k & \binom{k}{1}\lambda^{k-1} & \cdots & \binom{k}{n-2}\lambda^{k-n+1} & \binom{k}{n-1}\lambda^{k-n+1}\\ 0 & \lambda^k & \ddots & \binom{k}{n-3}\lambda^{k-n+2} & \binom{k}{n-2}\lambda^{k-n+2}\\ \vdots & \ddots & \ddots & \ddots & \vdots\\ 0 & 0 & \ddots & \lambda^k & \binom{k}{1}\lambda^{k-1}\\ 0 & 0 & \cdots & 0 & \lambda^k \end{bmatrix}.$$

Fact 4.22.8. Let $A \in \mathbb{R}^{n\times n}$, assume that A is nilpotent, and let $k \ge 1$ satisfy $A^k = 0$. Then,

$$\det(I - A) = 1, \quad (I - A)^{-1} = \sum_{i=0}^{k-1} A^i.$$

Fact 4.22.9. Let $A \in \mathbb{R}^{n\times n}$, and assume that $A^3 = 0$. Then, $\frac{1}{2}A^2 + A + I$ is nonsingular, and

$$(\tfrac{1}{2}A^2 + A + I)^{-1} = \tfrac{1}{2}A^2 - A + I.$$

Fact 4.22.10. Let $A \in \mathbb{F}^{n\times n}$, and assume that A is nilpotent. Then, there exist idempotent matrices $B, C \in \mathbb{F}^{n\times n}$ such that $A = [B, C]$. **Source:** [951]. **Remark:** A necessary and sufficient condition for a matrix to be a commutator of a pair of idempotents is given in [951]. **Related:** Fact 11.11.2 in the case of projectors.

Fact 4.22.11. Let $A, B \in \mathbb{F}^{n\times n}$, assume that B is nilpotent, and assume that $AB = BA$. Then, $\det(A + B) = \det A$. **Source:** Assuming A is nonsingular, apply Fact 4.22.8 with A replaced by $-A^{-1}B$. Then, use continuity to remove the assumption that A is nonsingular. Alternatively, use Fact 7.19.5.

Fact 4.22.12. Let $A, B \in \mathbb{R}^{n\times n}$, assume that A and B are nilpotent, and assume that $AB = BA$. Then, $A + B$ is nilpotent. **Source:** If $k, l \geq 1$ and $A^k = B^l = 0$, then $(A + B)^{k+l} = 0$.

Fact 4.22.13. Let $A, B \in \mathbb{F}^{n\times n}$, and assume that A and B are either both upper triangular or both lower triangular. Then, $[A, B]$ is nilpotent. **Source:** [1065, 1066]. **Related:** Fact 7.19.7.

Fact 4.22.14. Let $A, B \in \mathbb{F}^{n\times n}$, and assume that there exist $k \in \mathbb{P}$ and nonzero $\alpha \in \mathbb{R}$ such that $[A^k, B] = \alpha A$. Then, A is nilpotent. **Source:** For all $l \in \mathbb{N}$, $A^{k+l}B - A^lBA^k = \alpha A^{l+1}$, and thus $\operatorname{tr} A^{l+1} = 0$. The result now follows from Fact 4.22.6. See [2340]. **Remark:** If $[A, B] = A$, then A is nilpotent.

Fact 4.22.15. Let $A, B \in \mathbb{F}^{n\times n}$. Then, the following statements hold:

i) If $[A^{\mathsf{A}}, B] = A$, then A is nilpotent.

ii) $[A^{\mathsf{A}}, B] = A^{\mathsf{A}}$, then $(A^{\mathsf{A}})^2 = 0$.

iii) If either $[A, [A^{\mathsf{A}}, B^{\mathsf{A}}]] = 0$ or $[A^{\mathsf{A}}, [A^{\mathsf{A}}, B^{\mathsf{A}}]] = 0$, then $([A, B]^{\mathsf{A}})^2 = 0$.

Source: [1899].

4.23 Facts on Hankel and Toeplitz Matrices

Fact 4.23.1. Let $A \in \mathbb{F}^{n\times m}$. Then, the following statements hold:

i) If A is Toeplitz, then $\hat{I}A$ and $A\hat{I}$ are Hankel.

ii) If A is Hankel, then $\hat{I}A$ and $A\hat{I}$ are Toeplitz.

iii) A is Toeplitz if and only if $\hat{I}A\hat{I}$ is Toeplitz.

iv) A is Hankel if and only if $\hat{I}A\hat{I}$ is Hankel.

Fact 4.23.2. Let $A \in \mathbb{C}^{n\times n}$. Then, the following statements hold:

i) If A is Hankel, then A is symmetric.

ii) If A is Hermitian and symmetric, then A is real.

iii) If A is Hankel and Hermitian, then A is symmetric and real.

Fact 4.23.3. Let $A \in \mathbb{F}^{n\times n}$, and assume that A is a partitioned matrix all of whose blocks are $k\times k$ (circulant, Hankel, Toeplitz) matrices. Then, A is similar to a block-(circulant, Hankel, Toeplitz) matrix. **Source:** [296].

Fact 4.23.4. For all $i, j \in \{1, \ldots, n\}$, define $A \in \mathbb{R}^{n\times n}$ by $A_{(i,j)} \triangleq 1/(i + j - 1)$. Then, A is Hankel, positive definite, and

$$\det A = \frac{[1!2!\cdots(n-1)!]^4}{1!2!\cdots(2n-1)!}.$$

Furthermore, for all $i, j \in \{1, \ldots, n\}$, A^{-1} has integer entries given by

$$(A^{-1})_{(i,j)} = (-1)^{i+j}(i + j - 1)\binom{n+i-1}{n-j}\binom{n+j-1}{n-i}\binom{i+j-2}{i-1}^2.$$

Finally, as $n \to \infty$, $\det A \sim 4^{-n^2}$. **Remark:** A is the *Hilbert matrix*, which is a Cauchy matrix. See [724] and [1389, p. 513]. **Related:** Fact 2.2.63, Fact 4.27.5, Fact 4.27.6, Fact 10.9.5, Fact 10.9.8, and Fact 16.22.18.

Fact 4.23.5. Let $A \in \mathbb{F}^{n\times n}$, and assume that A is Toeplitz. Then, A is reverse symmetric.

Fact 4.23.6. Let $A \in \mathbb{F}^{n\times n}$. Then, A is Toeplitz if and only if there exist $a_0,\dots,a_n \in \mathbb{F}$ and $b_1,\dots,b_n \in \mathbb{F}$ such that

$$A = \sum_{i=1}^{n} b_i N_n^{i\mathsf{T}} + \sum_{i=0}^{n} a_i N_n^i.$$

Fact 4.23.7. Let $A \in \mathbb{F}^{n\times n}$, let $k \ge 1$, and assume that A is (lower triangular, strictly lower triangular, upper triangular, strictly upper triangular). Then, so is A^k. If, in addition, A is Toeplitz, then so is A^k. **Remark:** If A is Toeplitz, then A^2 is not necessarily Toeplitz. **Related:** Fact 15.14.1.

Fact 4.23.8. Let $n \ge 2$ and $m \ge 2$, and define $A \in \mathbb{F}^{n\times m}$ by

$$A \triangleq \begin{bmatrix} \sin\theta & \sin 2\theta & \sin 3\theta & \cdots & \sin m\theta \\ \sin 2\theta & \sin 3\theta & \iddots & \iddots & \iddots \\ \sin 3\theta & \iddots & \iddots & \iddots & \iddots \\ \vdots & \iddots & \iddots & \iddots & \iddots \\ \sin n\theta & \iddots & \iddots & \iddots & \sin\,(m+n-1)\theta \end{bmatrix}.$$

Then, $\operatorname{rank} A = 2$. **Source:** Proposition 16.9.13.

Fact 4.23.9. Let $A \in \mathbb{F}^{n\times n}$ be the tridiagonal, Toeplitz matrix

$$A \triangleq \begin{bmatrix} b & c & 0 & \cdots & 0 & 0 \\ a & b & c & \cdots & 0 & 0 \\ 0 & a & b & \ddots & 0 & 0 \\ \vdots & \vdots & \ddots & \ddots & \ddots & \vdots \\ 0 & 0 & 0 & \ddots & b & c \\ 0 & 0 & 0 & \cdots & a & b \end{bmatrix},$$

and define $\alpha \triangleq \frac{1}{2}(b + \sqrt{b^2 - 4ac})$ and $\beta \triangleq \frac{1}{2}(b - \sqrt{b^2 - 4ac})$. Then,

$$\det A = \begin{cases} b^n, & ac = 0, \\ (n+1)(b/2)^n, & b^2 = 4ac, \\ (\alpha^{n+1} - \beta^{n+1})/(\alpha - \beta), & b^2 \ne 4ac. \end{cases}$$

Source: [2983, pp. 101, 102]. **Remark:** The square root is the principal square root. **Related:** Fact 4.24.3 and Fact 7.12.46.

Fact 4.23.10. Let $A \in \mathbb{R}^{n\times n}$ be the Toeplitz matrix

$$A \triangleq \begin{bmatrix} 1 & a & a^2 & \cdots & a^{n-2} & a^{n-1} \\ b & 1 & a & \cdots & a^{n-1} & a^{n-2} \\ b^2 & b & 1 & \ddots & a^{n-2} & a^{n-3} \\ \vdots & \vdots & \ddots & \ddots & \ddots & \vdots \\ b^{n-2} & b^{n-3} & b^{n-4} & \ddots & 1 & a \\ b^{n-1} & b^{n-2} & b^{n-3} & \cdots & b & 1 \end{bmatrix}.$$

Then, A is nonsingular if and only if $ab \neq 1$. If these conditions hold, then

$$A^{-1} = \begin{bmatrix} c & -ac & 0 & \cdots & 0 & 0 \\ -bc & (ab+1)c & -ac & \cdots & 0 & 0 \\ 0 & -bc & (ab+1)c & \ddots & 0 & 0 \\ \vdots & \vdots & \ddots & \ddots & \ddots & \vdots \\ 0 & 0 & 0 & \ddots & (ab+1)c & -ac \\ 0 & 0 & 0 & \cdots & -bc & c \end{bmatrix},$$

where $c \triangleq (1-ab)^{-1}$. Now, assume that $a = b$. Then, A is nonsingular if and only if $|a| < 1$. If these conditions hold, then A is positive definite. **Source:** [2403, pp. 348, 349].

Fact 4.23.11. Let $A \in \mathbb{F}^{n\times m}$, $B \in \mathbb{F}^{l\times m}$, $C \in \mathbb{F}^{n\times p}$, $D \in \mathbb{F}^{l\times p}$, and $E \in \mathbb{F}^{l\times q}$, and assume that A has full column rank and $\mathcal{R}(A) \cap \mathcal{R}(C) = \{0\}$. Then,

$$\mathcal{R}\left(\begin{bmatrix} A \\ B \end{bmatrix}\right) \cap \left(\begin{bmatrix} C & 0 \\ D & E \end{bmatrix}\right) = \{0\}.$$

Fact 4.23.12. Let $r \geq 3$, for all $i \in \{1,\dots,r\}$, let $A_i \in \mathbb{F}^{n\times m}$, and define the block-Toeplitz matrix

$$\begin{bmatrix} A_1 & 0 & \cdots & 0 & 0 \\ A_2 & A_1 & \ddots & 0 & 0 \\ \vdots & \ddots & \ddots & \ddots & \vdots \\ A_{r-1} & A_{r-2} & \ddots & A_1 & 0 \\ A_r & A_{r-1} & \cdots & A_2 & A_1 \end{bmatrix} = \begin{bmatrix} C_r & C_{r-1} & \cdots & C_2 & C_1 \end{bmatrix},$$

where, for all $i \in \{1,\dots,r\}$, $C_i \in \mathbb{F}^{rn\times m}$. Furthermore, let $l \in \{2,\dots,r-1\}$, and assume that C_l has full column rank and $\mathcal{R}(C_l) \cap \mathcal{R}([C_{l-1} \ \cdots \ C_1]) = \{0\}$. Then, $[C_r \ \cdots \ C_l]$ has full column rank, and $\mathcal{R}([C_r \ \cdots \ C_l]) \cap \mathcal{R}([C_{l-1} \ \cdots \ C_1]) = \{0\}$. **Credit:** A. Ansari.

4.24 Facts on Tridiagonal Matrices

Fact 4.24.1. Let $A \in \mathbb{F}^{n\times n}$. Then, the following statements hold:

i) If A is upper triangular, then A is upper Hessenberg and upper bidiagonal.

ii) If A is lower triangular, then A is lower Hessenberg and lower bidiagonal.

iii) The following statements are equivalent:

- *a)* A is diagonal.
- *b)* A is upper bidiagonal and lower bidiagonal.
- *c)* A is upper triangular and lower triangular.

iv) The following statements are equivalent:

- *a)* A is upper bidiagonal.
- *b)* A is tridiagonal and upper triangular.
- *c)* A is lower Hessenberg and upper triangular.

v) The following statements are equivalent:

- *a)* A is lower bidiagonal.
- *b)* A is tridiagonal and lower triangular.

c) A is upper Hessenberg and lower triangular.

vi) A is tridiagonal if and only if A is upper Hessenberg and lower Hessenberg.

Fact 4.24.2. Let $A \in \mathbb{F}^{n\times n}$, assume that A is nonsingular, and let $l \in \{0,\dots,n\}$ and $k \in \{1,\dots,n\}$. Then, the following statements are equivalent:

i) Every submatrix B of A whose entries are located above the lth superdiagonal of A satisfies $\operatorname{rank} B \le k-1$.

ii) Every submatrix C of A^{-1} whose entries are located above the lth subdiagonal of A^{-1} satisfies $\operatorname{rank} C \le l+k-1$.

Specifically, the following statements hold:

iii) A is lower triangular if and only if A^{-1} is lower triangular.

iv) A is upper triangular if and only if A^{-1} is upper triangular.

v) A is diagonal if and only if A^{-1} is diagonal.

vi) A is lower Hessenberg if and only if every submatrix C of A^{-1} whose entries are located either on or above above the diagonal of A^{-1} satisfies $\operatorname{rank} C \le 1$.

vii) A is upper Hessenberg if and only if every submatrix C of A^{-1} whose entries are located either on or below the superdiagonal of A^{-1} satisfies $\operatorname{rank} C \le 1$.

viii) A is tridiagonal if and only if every submatrix C of A^{-1} whose entries are located either on or above the diagonal of A^{-1} satisfies $\operatorname{rank} C \le 1$ and every submatrix C of A^{-1} whose entries are located either on or below the diagonal of A^{-1} satisfies $\operatorname{rank} C \le 1$.

ix) Every submatrix B of A whose entries are located above the diagonal of A satisfies $\operatorname{rank} B \le 1$ if and only if every submatrix C of A^{-1} above the diagonal of A^{-1} satisfies $\operatorname{rank} C \le 1$.

Source: [2557]. **Remark:** The 0th subdiagonal and the 0th superdiagonal are the diagonal. **Remark:** *iii)* corresponds to $l=0$ and $k=1$, *iv)* corresponds to $l=0$ and $k=1$ applied to A and A^{T}, *v)* corresponds to $l=1$ and $k=1$, and *vi)* corresponds to $l=1$ and $k=1$ applied to A and A^{T}. **Related:** Fact 3.14.27. **Remark:** Extensions to generalized inverses are considered in [280, 2317].

Fact 4.24.3. Let $A \in \mathbb{F}^{n\times n}$ be the tridiagonal, Toeplitz matrix

$$A \triangleq \begin{bmatrix} a+b & ab & 0 & \cdots & 0 & 0 \\ 1 & a+b & ab & \cdots & 0 & 0 \\ 0 & 1 & a+b & \ddots & 0 & 0 \\ \vdots & \vdots & \ddots & \ddots & \ddots & \vdots \\ 0 & 0 & 0 & \ddots & a+b & ab \\ 0 & 0 & 0 & \cdots & 1 & a+b \end{bmatrix}.$$

Then,

$$\det A = \begin{cases} (n+1)a^n, & a=b, \\ \dfrac{a^{n+1}-b^{n+1}}{a-b}, & a\ne b. \end{cases}$$

Source: [1674, pp. 401, 621].

Fact 4.24.4. Let $A \in \mathbb{R}^{n\times n}$, assume that A is tridiagonal with positive diagonal entries, and assume that, for all $i \in \{2,\dots,n\}$,

$$A_{(i,i-1)}A_{(i-1,i)} < \tfrac{1}{4}A_{(i,i)}A_{(i-1,i-1)}\sec^2\tfrac{\pi}{n+1}.$$

Then, $\det A > 0$. If, in addition, A is symmetric, then A is positive definite. **Source:** [1542]. **Related:** Fact 4.24.5 and Fact 10.9.22.

Fact 4.24.5. Let $A \in \mathbb{R}^{n\times n}$, assume that A is tridiagonal and symmetric with positive diagonal entries, define $\alpha \triangleq \min\{A_{(1,1)}, A_{(3,3)}, \ldots\}$ and $\beta \triangleq \min\{A_{(2,2)}, A_{(4,4)}, \ldots\}$, and assume that

$$\frac{8\alpha\beta}{n(n-1)} < 4\sum_{i=1}^{n-1} A_{(i,i+1)}^2 + \sum [A_{(i,i)} - A_{(j,j)}]^2,$$

where the second sum is taken over all $i, j \in \{1, \ldots, n\}$ such that $i < j$. Then, A is positive definite. **Source:** [720]. **Remark:** Necessary and sufficient conditions for A to be positive definite are given in [71].

Fact 4.24.6. Let $A \in \mathbb{R}^{n\times n}$, assume that A is tridiagonal, assume that every entry of the superdiagonal and subdiagonal of A is nonzero, assume that every leading principal subdeterminant of A and every trailing principal subdeterminant of A is nonzero. Then, every entry of A^{-1} is nonzero. **Source:** [1427]. **Remark:** Expressions for the inverse of symmetric tridiagonal matrices are given in [2033, 2565].

Fact 4.24.7. Let $A \in \mathbb{F}^{n\times n}$, assume that A is nonsingular, and assume that $A_{(2,2)}, \ldots, A_{(n-1,n-1)}$ are nonzero. Then, A^{-1} is tridiagonal if and only if, for all $i, j \in \{1, \ldots, n\}$ such that $|i - j| \geq 2$, and for all k satisfying $\min\{i, j\} < k < \max\{i, j\}$, it follows that $A_{(i,j)} = A_{(i,k)}A_{(k,j)}/A_{(k,k)}$. **Source:** [303].

Fact 4.24.8. Let $A \in \mathbb{R}^{n\times n}$ be the symmetric, tridiagonal matrix

$$A \triangleq \begin{bmatrix} 2 & -1 & 0 & \cdots & 0 & 0 \\ -1 & 2 & -1 & \cdots & 0 & 0 \\ 0 & -1 & 2 & \ddots & 0 & 0 \\ \vdots & \vdots & \ddots & \ddots & \ddots & \vdots \\ 0 & 0 & 0 & \ddots & 2 & -1 \\ 0 & 0 & 0 & \cdots & -1 & 1 \end{bmatrix}.$$

Then,

$$A^{-1} = \begin{bmatrix} 1 & 1 & 1 & \cdots & 1 & 1 \\ 1 & 2 & 2 & \cdots & 2 & 2 \\ 1 & 2 & 3 & \cdots & 3 & 3 \\ \vdots & \vdots & \vdots & .\vdots. & \vdots & \vdots \\ 1 & 2 & 3 & \cdots & n-1 & n-1 \\ 1 & 2 & 3 & \cdots & n-1 & n \end{bmatrix}.$$

Source: [2418, p. 182], where the (n, n) entry of A is incorrect. **Related:** Fact 4.24.6.

4.25 Facts on Triangular, Hessenberg, and Irreducible Matrices

Fact 4.25.1. Let $A \in \mathbb{F}^{n\times n}$, and assume that A is either upper triangular or lower triangular. Then, $\det A = \prod_{i=1}^n A_{(i,i)}$. **Related:** Fact 6.10.14.

Fact 4.25.2. Let $A, B \in \mathbb{F}^n$, and assume that A and B are (upper triangular, lower triangular). Then, AB is (upper triangular, lower triangular). If, in addition, either A or B is (strictly upper triangular, strictly lower triangular), then AB is (strictly upper triangular, strictly lower triangular). **Related:** Fact 4.31.11.

Fact 4.25.3. Let $n \geq 1$, let $A \in \mathbb{R}^{n\times n}$, for all $i, j \in \{1, \ldots, n\}$, define $A_{(i,j)} \triangleq \binom{i-1}{j-1}$, and define $M \triangleq A^{-1}$. Then, for all $i, j \in \{1, \ldots, n\}$, $M_{(i,j)} = (-1)^{i+j}\binom{i-1}{j-1}$. **Source:** [771, pp. 143, 144].

Example: $\begin{bmatrix} 1&0&0\\1&1&0\\1&2&1 \end{bmatrix}^{-1} = \begin{bmatrix} 1&0&0\\-1&1&0\\1&-2&1 \end{bmatrix}.$

Fact 4.25.4. Let $k \ge 1$. Then,

$$F_{k+1} = \det\begin{bmatrix} 1 & \jmath & 0 & \cdots & 0 & 0\\ \jmath & 1 & \jmath & \cdots & 0 & 0\\ 0 & \jmath & 1 & \ddots & 0 & 0\\ \vdots & \vdots & \ddots & \ddots & \ddots & \vdots\\ 0 & 0 & 0 & \ddots & 1 & \jmath\\ 0 & 0 & 0 & \cdots & \jmath & 1 \end{bmatrix} = \det\begin{bmatrix} 1 & 1 & 0 & \cdots & 0 & 0\\ -1 & 1 & 1 & \cdots & 0 & 0\\ 0 & -1 & 1 & \ddots & 0 & 0\\ \vdots & \vdots & \ddots & \ddots & \ddots & \vdots\\ 0 & 0 & 0 & \ddots & 1 & 1\\ 0 & 0 & 0 & \cdots & -1 & 1 \end{bmatrix},$$

$$F_{k+1} = \det\begin{bmatrix} 1 & 1 & 1 & \cdots & 1 & 1\\ -1 & 0 & 1 & \cdots & 1 & 1\\ 0 & -1 & 1 & \ddots & 1 & 1\\ \vdots & \vdots & \ddots & \ddots & \ddots & \vdots\\ 0 & 0 & 0 & \ddots & 0 & 1\\ 0 & 0 & 0 & \cdots & -1 & 0 \end{bmatrix} = \det\begin{bmatrix} 1 & 0 & 1 & \cdots & \mathrm{rem}_2(n+1) & \mathrm{rem}_2(n)\\ -1 & 1 & 0 & \cdots & \mathrm{rem}_2(n) & \mathrm{rem}_2(n+1)\\ 0 & -1 & 1 & \ddots & \mathrm{rem}_2(n+1) & \mathrm{rem}_2(n)\\ \vdots & \vdots & \ddots & \ddots & \ddots & \vdots\\ 0 & 0 & 0 & \ddots & 1 & 0\\ 0 & 0 & 0 & \cdots & -1 & 1 \end{bmatrix},$$

$$F_{2k} = \det\begin{bmatrix} 1 & 2 & 3 & \cdots & n-1 & n\\ -1 & 1 & 2 & \cdots & n-2 & n-1\\ 0 & -1 & 1 & \ddots & n-3 & n-2\\ \vdots & \vdots & \ddots & \ddots & \ddots & \vdots\\ 0 & 0 & 0 & \ddots & 1 & 2\\ 0 & 0 & 0 & \cdots & -1 & 1 \end{bmatrix}, \quad F_{2k+1} = \det\begin{bmatrix} 2 & 1 & 1 & \cdots & 1 & 1\\ -1 & 2 & 1 & \cdots & 1 & 1\\ 0 & -1 & 2 & \ddots & 1 & 1\\ \vdots & \vdots & \ddots & \ddots & \ddots & \vdots\\ 0 & 0 & 0 & \ddots & 2 & 1\\ 0 & 0 & 0 & \cdots & -1 & 2 \end{bmatrix},$$

$$F_{2k+2} = \det\begin{bmatrix} 3 & -1 & 0 & \cdots & 0 & 0\\ -1 & 3 & -1 & \cdots & 0 & 0\\ 0 & -1 & 3 & \ddots & 0 & 0\\ \vdots & \vdots & \ddots & \ddots & \ddots & \vdots\\ 0 & 0 & 0 & \ddots & 3 & -1\\ 0 & 0 & 0 & \cdots & -1 & 3 \end{bmatrix}, \quad L_{k+1} = \det\begin{bmatrix} 3 & \jmath & 0 & \cdots & 0 & 0\\ \jmath & 1 & \jmath & \cdots & 0 & 0\\ 0 & \jmath & 1 & \ddots & 0 & 0\\ \vdots & \vdots & \ddots & \ddots & \ddots & \vdots\\ 0 & 0 & 0 & \ddots & 1 & \jmath\\ 0 & 0 & 0 & \cdots & \jmath & 1 \end{bmatrix},$$

where all matrices are of size $k \times k$. **Source:** [611, 1519, 1520] and [2300, p. 515]. See Fact 6.9.8.

Fact 4.25.5. Let $n \ge 1$. Then,

$$\det\begin{bmatrix} 1 & \frac{1}{2!} & \frac{1}{3!} & \cdots & \frac{1}{(n-1)!} & \frac{1}{n!}\\ 1 & 1 & \frac{1}{2!} & \cdots & \frac{1}{(n-2)!} & \frac{1}{(n-1)!}\\ 0 & 1 & 1 & \ddots & \frac{1}{(n-3)!} & \frac{1}{(n-2)!}\\ \vdots & \vdots & \ddots & \ddots & \ddots & \vdots\\ 0 & 0 & 0 & \ddots & 1 & \frac{1}{2!}\\ 0 & 0 & 0 & \cdots & 1 & 1 \end{bmatrix} = \frac{1}{n!}.$$

Source: [407].

Fact 4.25.6. Let $a_0, a_1, \ldots, a_n \in \mathbb{F}$, and define $A \in \mathbb{F}^{n\times n}$ by

$$A \triangleq \begin{bmatrix} a_1 & a_2 & a_3 & \cdots & a_{n-1} & a_n \\ a_0 & a_1 & a_2 & \cdots & a_{n-2} & a_{n-1} \\ 0 & a_0 & a_1 & \ddots & a_{n-3} & a_{n-2} \\ \vdots & \vdots & \ddots & \ddots & \ddots & \vdots \\ 0 & 0 & 0 & \ddots & a_1 & a_2 \\ 0 & 0 & 0 & \cdots & a_0 & a_1 \end{bmatrix}.$$

Then,

$$\det A = \sum \binom{\sum_{i=1}^n k_i}{k_1, \ldots, k_n} (-a_0)^{n-\sum_{i=1}^n k_i} \prod_{i=1}^n a_i^{k_i},$$

where the sum is taken over all $(k_1, \ldots, k_n)$ such that $\sum_{i=1}^n ik_i = n$. **Source:** [2016]. **Remark:** This is *Trudi's formula.* **Remark:** The number of terms in the sum is the number p_n of partitions of $\{1, \ldots, 1\}_{\mathrm{ms}}$. See Fact 1.20.1. **Remark:** A is Toeplitz and upper Hessenberg.

Fact 4.25.7. Let $A \in \mathbb{F}^{n\times n}$, and assume that there exists $i \in \{1, \ldots, n\}$ such that either $\mathrm{row}_i(A) = 0$ or $\mathrm{col}_i(A) = 0$. Then, A is reducible.

Fact 4.25.8. Let $A \in \mathbb{F}^{n\times n}$, and assume that A is reducible. Then, A has at least $n-1$ entries that are equal to zero.

Fact 4.25.9. Let $A \in \mathbb{F}^{n\times n}$. Then, A is reducible if and only if $|A|$ is reducible. Furthermore, A is irreducible if and only if $|A|$ is irreducible.

4.26 Facts on Matrices Related to Graphs

Fact 4.26.1. Let $\mathcal{G} = (\mathcal{X}, \mathcal{R})$ be a directed graph, assume that $\mathcal{G}$ has n nodes and m directed edges, let $B \in \mathbb{R}^{n\times m}$ denote the incidence matrix of $\mathcal{G}$, and assume that $\mathrm{sym}(\mathcal{G})$ is connected. Then, $\mathcal{N}(B^{\mathsf{T}}) = \mathrm{span}\,\{1_{n\times 1}\}$, and $\mathrm{rank}\, B = n - 1 \le m$. **Source:** [281, p. 12].

Fact 4.26.2. Let $\mathcal{G} = (\{x_1, \ldots, x_n\}, \mathcal{R})$ be a directed graph without self-directed edges, assume that $\mathcal{G}$ is antisymmetric, let $A \in \mathbb{R}^{n\times n}$ denote the adjacency matrix of $\mathcal{G}$, let $L_{\mathrm{in}} \in \mathbb{R}^{n\times n}$ and $L_{\mathrm{out}} \in \mathbb{R}^{n\times n}$ denote the inbound and outbound Laplacians of $\mathcal{G}$, respectively, let D_{in} and D_{out} be the indegree and outdegree matrices of $\mathcal{G}$, respectively, and let A_{sym}, D_{sym}, and L_{sym} denote the adjacency, degree, and Laplacian matrices, respectively, of $\mathrm{sym}(\mathcal{G})$. Then,

$$D_{\mathrm{sym}} = D_{\mathrm{in}} + D_{\mathrm{out}}, \quad A_{\mathrm{sym}} = A + A^{\mathsf{T}},$$
$$L_{\mathrm{sym}} = L_{\mathrm{in}} + L_{\mathrm{out}}^{\mathsf{T}} = L_{\mathrm{in}}^{\mathsf{T}} + L_{\mathrm{out}} = D_{\mathrm{sym}} - A_{\mathrm{sym}}.$$

Fact 4.26.3. Let $\mathcal{G} = (\{x_1, \ldots, x_n\}, \mathcal{R})$ be a directed graph, let $A \in \mathbb{R}^{n\times n}$ be the adjacency matrix of $\mathcal{G}$, and let D_{out} be the outdegree matrix of $\mathcal{G}$. Then, $F \triangleq (I + D_{\mathrm{out}})^{-1}(I + A)$ is row stochastic. If, in addition, every node has a self-arc, then $F_0 \triangleq D_{\mathrm{out}}^{-1}A$ is row stochastic.

Fact 4.26.4. Let $\mathcal{G} = (\mathcal{X}, \mathcal{R})$ be a directed graph, define $n \triangleq \mathrm{card}(\mathcal{X})$, let $a_1, \ldots, a_m$ denote the directed edges of $\mathcal{G}$ that are not self-directed edges, let B denote the incidence matrix of $\mathcal{G}$, let $k \in \{1, \ldots, \min\{n, m\}\}$, and let $1 \le i_1 < \cdots < i_k \le m$. Then, the following statements are equivalent:

i) $\mathrm{rank}\, B_{(\{1,\ldots,n\},\{i_1,\ldots,i_k\})} = k$.

ii) $\mathrm{sym}(\mathcal{X}, \{a_{i_1}, \ldots, a_{i_k}\})$ is acyclic.

Source: [281, p. 13].

Fact 4.26.5. Let $\mathcal{G} = (\mathcal{X}, \mathcal{R})$ be a directed graph, let $n \triangleq \mathrm{card}(\mathcal{X})$, let B denote the incidence matrix of $\mathcal{G}$, and let A denote a square submatrix of B. Then, $\det A \in \{-1, 0, 1\}$. If, in addition, $\mathcal{G}$ is a tree and $A \in \mathbb{R}^{(n-1)\times(n-1)}$, then $|\det A| = 1$. **Source:** [281, p. 13].

4.27 Facts on Dissipative, Contractive, Cauchy, and Centrosymmetric Matrices

Fact 4.27.1. Let $A \in \mathbb{F}^{n\times n}$, and assume that A is dissipative. Then, A is nonsingular. **Source:** Assume that A is singular, and let $x \in \mathcal{N}(A)$. Then, $x^*(A+A^*)x = 0$. **Remark:** If $A+A^*$ is nonsingular, then A is not necessarily nonsingular. For example, $A = \left[\begin{smallmatrix} 0 & 1 \\ 0 & 0 \end{smallmatrix}\right]$.

Fact 4.27.2. Let $A \in \mathbb{F}^{n\times m}$. Then, A is (semicontractive, contractive) if and only if A^* is.

Fact 4.27.3. Let $A, B \in \mathbb{F}^{n\times n}$, and assume that A and B are semicontractive. Then, AB is semicontractive. If, in addition, B is contractive, then AB is contractive. **Source:** It follows from $A^*A \le I$ that $(AB)^*AB = B^*A^*AB \le B^*B \le I$. See [2991, p. 230].

Fact 4.27.4. Let $A \in \mathbb{F}^{n\times m}$. Then, A is semicontractive if and only if there exist unitary matrices $A_1, \dots, A_k$ and positive numbers $\alpha_1, \dots, \alpha_k$ such that $\sum_{i=1}^k \alpha_i = 1$ and $A = \sum_{i=1}^k \alpha_i A_i$. **Source:** [2991, pp. 189, 190].

Fact 4.27.5. Let $a_1, \dots, a_n, b_1, \dots, b_n \in \mathbb{R}$, assume that, for all $i, j \in \{1, \dots, n\}$, $a_i + b_j \ne 0$, and, for all $i, j \in \{1, \dots, n\}$, define $A \in \mathbb{R}^{n\times n}$ by $A_{(i,j)} \triangleq 1/(a_i + b_j)$. Then,

$$\det A = \frac{\displaystyle\prod_{1\le i<j\le n} (a_j - a_i)(b_j - b_i)}{\displaystyle\prod_{1\le i,j\le n} (a_i + b_j)}.$$

Now, assume that $a_1, \dots, a_n$ are distinct and $b_1, \dots, b_n$ are distinct. Then, A is nonsingular and

$$(A^{-1})_{(i,j)} = \frac{\displaystyle\prod_{1\le k\le n} (a_j + b_k)(a_k + b_i)}{\displaystyle(a_j + b_i)\prod_{\substack{1\le k\le n\\ k\ne j}} (a_j - a_k)\prod_{\substack{1\le k\le n\\ k\ne i}} (b_i - b_k)}.$$

Furthermore,

$$1_{1\times n} A^{-1} 1_{n\times 1} = \sum_{i=1}^n (a_i + b_i).$$

Source: [451, 1042], [1389, p. 515], and [2390]. **Remark:** A is a *Cauchy matrix*. **Related:** Fact 4.23.4, Fact 4.27.6, Fact 10.9.20, and Fact 16.22.18.

Fact 4.27.6. Let $x_1, \dots, x_n$ be distinct positive numbers, let $y_1, \dots, y_n$ be distinct positive numbers, and let $A \in \mathbb{R}^{n\times n}$, where, for all $i, j \in \{1, \dots, n\}$, $A_{(i,j)} \triangleq 1/(x_i + y_j)$. Then, A is nonsingular. **Source:** [1716]. **Related:** Fact 4.23.4, Fact 4.27.5, and Fact 16.22.18.

Fact 4.27.7. Let $A \in \mathbb{F}^{n\times m}$. Then, A is centrosymmetric if and only if $A^{\mathsf{T}} = A^{\hat{\mathsf{T}}}$. Furthermore, A is centrohermitian if and only if $A^* = A^{\hat{*}}$.

Fact 4.27.8. Let $A \in \mathbb{F}^{n\times m}$ and $B \in \mathbb{F}^{m\times l}$. If A and B are both (centrohermitian, centrosymmetric), then so is AB. **Source:** [1393]. **Remark:** See [1776, 2844].

4.28 Facts on Hamiltonian and Symplectic Matrices

Fact 4.28.1. Let $A \in \mathbb{F}^{2n\times 2n}$. Then, A is Hamiltonian if and only if $JA = (JA)^{\mathsf{T}}$. Furthermore, A is symplectic if and only if $A^{\mathsf{T}}JA = J$.

Fact 4.28.2. Assume that $n \in \mathbb{P}$ is even, let $A \in \mathbb{F}^{n\times n}$, and assume that A is Hamiltonian and symplectic. Then, A is skew involutory. **Related:** Fact 4.28.3.

Fact 4.28.3. The following statements hold:

i) I_{2n} is orthogonal, shifted orthogonal, a projector, a reflector, and symplectic.

ii) J_{2n} is skew symmetric, orthogonal, skew involutory, a skew reflector, symplectic, and Hamiltonian.

iii) $\hat{I}_{2n}$ is symmetric, orthogonal, involutory, shifted orthogonal, a projector, a reflector, and Hamiltonian.

Related: Fact 4.28.2 and Fact 7.10.24.

Fact 4.28.4. Let $A \in \mathbb{F}^{2n\times 2n}$, assume that A is Hamiltonian, and let $S \in \mathbb{F}^{2n\times 2n}$ be symplectic. Then, SAS^{-1} is Hamiltonian.

Fact 4.28.5. Let $A \in \mathbb{F}^{2n\times 2n}$, and assume that A is Hamiltonian and nonsingular. Then, A^{-1} is Hamiltonian.

Fact 4.28.6. Let $\mathcal{A} \in \mathbb{F}^{2n\times 2n}$. Then, $\mathcal{A}$ is Hamiltonian if and only if there exist $A, B, C \in \mathbb{F}^{n\times n}$ such that B and C are symmetric and

$$\mathcal{A} = \begin{bmatrix} A & B \\ C & -A^{\mathsf{T}} \end{bmatrix}.$$

Related: Fact 6.9.33.

Fact 4.28.7. Let $A \in \mathbb{F}^{2n\times 2n}$, and assume that A is Hamiltonian. Then, $\operatorname{tr} A = 0$.

Fact 4.28.8. Let $\mathcal{A} \in \mathbb{F}^{2n\times 2n}$. Then, $\mathcal{A}$ is skew symmetric and Hamiltonian if and only if there exist a skew-symmetric matrix $A \in \mathbb{F}^{n\times n}$ and a symmetric matrix $B \in \mathbb{F}^{n\times n}$ such that

$$\mathcal{A} = \begin{bmatrix} A & B \\ -B & A \end{bmatrix}.$$

Fact 4.28.9. Let $\mathcal{A} \triangleq \left[\begin{smallmatrix} A & B \\ C & D \end{smallmatrix}\right] \in \mathbb{F}^{2n\times 2n}$, where $A, B, C, D \in \mathbb{F}^{n\times n}$. Then, $\mathcal{A}$ is symplectic if and only if $A^{\mathsf{T}}C$ and $B^{\mathsf{T}}D$ are symmetric and $A^{\mathsf{T}}D - C^{\mathsf{T}}B = I$.

Fact 4.28.10. Let $A \in \mathbb{F}^{2n\times 2n}$, and assume that A is symplectic. Then, $\det A = 1$. **Source:** Using Fact 3.17.17 and Fact 4.28.9 it follows that $\det \mathcal{A} = \det(A^{\mathsf{T}}D - C^{\mathsf{T}}B) = \det I = 1$. See also [218, p. 27], [923], [1299, p. 8], and [2423, p. 128].

Fact 4.28.11. Let $A \in \mathbb{F}^{2\times 2}$. Then, A is symplectic if and only if $\det A = 1$. Hence, $\mathrm{SL}_{\mathbb{F}}(2) = \mathrm{Symp}_{\mathbb{F}}(2)$.

Fact 4.28.12. The following statements hold:

i) If $A \in \mathbb{F}^{2n\times 2n}$ is Hamiltonian and $A+I$ is nonsingular, then $B \triangleq (A-I)(A+I)^{-1}$ is symplectic, $I - B$ is nonsingular, and $(I - B)^{-1} = \frac{1}{2}(A + I)$.

ii) If $B \in \mathbb{F}^{2n\times 2n}$ is symplectic and $I-B$ is nonsingular, then $A = (I+B)(I-B)^{-1}$ is Hamiltonian, $A + I$ is nonsingular, and $(A + I)^{-1} = \frac{1}{2}(I - B)$.

iii) If $A \in \mathbb{F}^{2n\times 2n}$ is Hamiltonian, then there exists a unique symplectic matrix $B \in \mathbb{F}^{2n\times 2n}$ such that $I - B$ is nonsingular and $A = (I + B)(I - B)^{-1}$. In fact, $B = (A - I)(A + I)^{-1}$.

iv) If $B \in \mathbb{F}^{2n\times 2n}$ is symplectic and $I - B$ is nonsingular, then there exists a unique Hamiltonian matrix $A \in \mathbb{F}^{2n\times 2n}$ such that $B = (A - I)(A + I)^{-1}$. In fact, $A = (I + B)(I - B)^{-1}$.

Related: Fact 4.13.24, Fact 4.13.25, and Fact 4.13.26.

Fact 4.28.13. Let $\mathcal{A} \in \mathbb{R}^{2n\times 2n}$. Then, $\mathcal{A} \in \mathrm{osymp}_{\mathbb{R}}(2n)$ if and only if there exist $A, B \in \mathbb{R}^{n\times n}$ such that A is skew symmetric, B is symmetric, and $\mathcal{A} = \left[\begin{smallmatrix} A & B \\ -B & A \end{smallmatrix}\right]$. **Source:** [865]. **Remark:** $\mathrm{OSymp}_{\mathbb{R}}(2n)$ is the *orthosymplectic group*.

4.29 Facts on Commutators

Fact 4.29.1. Let $A, B \in \mathbb{F}^{n\times n}$, and assume that $A + B = AB$. Then, $AB = BA$. **Source:** [112, p. 54] and [2991, p. 15].

Fact 4.29.2. Let $A, B \in \mathbb{F}^{2\times 2}$, and define $\operatorname{str} A \triangleq A_{(1,1)} - A_{(2,2)}$. Then,

$$[A,B]^2 = \tfrac{1}{2}(\operatorname{tr}[A,B]^2)I_2, \quad [A,B]^2 = -(\det[A,B])I_2, \quad \operatorname{tr}[A,B]^2 + 2\det[A,B] = 0,$$
$$\det[A,B] = \operatorname{tr} A^2B^2 - \operatorname{tr}(AB)^2 = 2(\det A)\det B - \operatorname{tr} ABA^{\mathsf{A}}B^{\mathsf{A}},$$
$$\det[A,B] = 4\det AB - (\det A)(\operatorname{tr} B)^2 - (\det B)(\operatorname{tr} A)^2 + (\operatorname{tr} A)(\operatorname{tr} B)\operatorname{tr} AB - (\operatorname{tr} AB)^2,$$
$$\det[A,B] = (\operatorname{str} AB)\operatorname{str} BA - (\operatorname{str} A)(\operatorname{str} B)\operatorname{tr} AB - (\det A)(\operatorname{str} B)^2 - (\det B)(\operatorname{str} A)^2.$$

If, in addition, $\operatorname{tr} A = \operatorname{tr} B = 0$, then

$$\det[A,B] = 4\det AB - (\operatorname{tr} AB)^2.$$

Source: Fact 6.9.1, [1065, 1066], [1158, p. 444], and [1611]. **Related:** Fact 3.15.17.

Fact 4.29.3. Let $A, B \in \mathbb{F}^{3\times 3}$. Then, $\operatorname{tr}[A,B]^3 = 3\det[A,B]$. **Source:** [1158, p. 84]. **Related:** Fact 6.9.2.

Fact 4.29.4. Let $A, B \in \mathbb{F}^{4\times 4}$. Then, $\operatorname{tr}[A,B]^4 = \frac{1}{2}(\operatorname{tr}[A,B]^2)^2 - 4\det[A,B]$. **Related:** Fact 6.9.3.

Fact 4.29.5. Let $A, B \in \mathbb{F}^{n\times n}$. Then, $\operatorname{tr}(A[A,B]) = 0$.

Fact 4.29.6. Let $A, B \in \mathbb{F}^{n\times n}$. If either A and B are Hermitian or A and B are skew Hermitian, then $[A,B]$ is skew Hermitian. Furthermore, if either A is Hermitian and B is skew Hermitian or vice versa, then $[A,B]$ is Hermitian.

Fact 4.29.7. Let $A \in \mathbb{F}^{n\times n}$. Then, the following statements are equivalent:

i) $\operatorname{tr} A = 0$.

ii) There exist $B, C \in \mathbb{F}^{n\times n}$ such that B is Hermitian, $\operatorname{tr} C = 0$, and $A = [B,C]$.

iii) There exist $B, C \in \mathbb{F}^{n\times n}$ such that $A = [B,C]$.

Source: [1130] and Fact 7.10.19. Assuming $I \odot A = 0$, let $B \triangleq \operatorname{diag}(1,\dots,n)$, $I \odot C \triangleq 0$, and, for all $i \neq j$, $C_{(i,j)} \triangleq A_{(i,j)}/(i-j)$. See [2980, p. 110]. See also [2263, p. 172].

Fact 4.29.8. Let $A \in \mathbb{F}^{n\times n}$. Then, the following statements are equivalent:

i) A is Hermitian, and $\operatorname{tr} A = 0$.

ii) There exists a nonsingular matrix $B \in \mathbb{F}^{n\times n}$ such that $A = [B, B^*]$.

iii) There exist a Hermitian matrix $B \in \mathbb{F}^{n\times n}$ and a skew-Hermitian matrix $C \in \mathbb{F}^{n\times n}$ such that $A = [B,C]$.

iv) There exist a skew-Hermitian matrix $B \in \mathbb{F}^{n\times n}$ and a Hermitian matrix $C \in \mathbb{F}^{n\times n}$ such that $A = [B,C]$.

Source: [1130, 2608].

Fact 4.29.9. Let $A, B \in \mathbb{F}^{n\times n}$, assume that A and B are Hermitian, and assume that $[A - B, [A,B]] = 0$. Then, $[A,B] = 0$. **Source:** [2991, p. 258].

Fact 4.29.10. Let $A \in \mathbb{F}^{n\times n}$. Then, the following statements are equivalent:

i) A is skew Hermitian, and $\operatorname{tr} A = 0$.

ii) There exists a nonsingular matrix $B \in \mathbb{F}^{n\times n}$ such that $A = [\jmath B, B^*]$.

iii) There exist Hermitian matrices $B, C \in \mathbb{F}^{n\times n}$ such that $A = [B,C]$.

Source: [1130] and Fact 4.29.8.

Fact 4.29.11. Let $A \in \mathbb{F}^{n\times n}$, and assume that A is skew symmetric. Then, there exist symmetric matrices $B, C \in \mathbb{F}^{n\times n}$ such that $A = [B,C]$. **Source:** Fact 7.17.26 and [2263, pp. 83, 89]. **Remark:** "Symmetric" is correct for $\mathbb{F} = \mathbb{C}$.

Fact 4.29.12. Let $A \in \mathbb{F}^{n\times n}$. Then, there exist $B, C \in \mathbb{F}^{n\times n}$ such that B is normal, C is Hermitian, and $A = B + [C,B]$. **Source:** [956].

4.30 Facts on Partial Orderings

Fact 4.30.1. Let $A, B \in \mathbb{F}^{n\times m}$. Then, the following statements are equivalent:

i) $\operatorname{rank}(B - A) = \operatorname{rank} B - \operatorname{rank} A$.

ii) There exists $M \in \mathbb{F}^{m\times n}$ such that $A = BMB = BMBMB$.

iii) $\operatorname{rank} \left[\begin{smallmatrix} B & A \\ A & A \end{smallmatrix}\right] = \operatorname{rank} B$.

iv) $\mathcal{R}(B - A) \cap \mathcal{R}(A) = \{0\}$ and $\mathcal{R}(B^* - A^*) \cap \mathcal{R}(A^*) = \{0\}$.

v) $\mathcal{R}(A) \subseteq \mathcal{R}(B)$, $\mathcal{R}(A^*) \subseteq \mathcal{R}(B^*)$, and there exists $M \in \mathbb{F}^{m\times n}$ such that $A = AMA$ and $B = BMB$.

Source: This result corrects and extends results given in [762]. **Credit:** Y. Tian.

Fact 4.30.2. Let $A, B, C \in \mathbb{F}^{n\times m}$, and assume that

$$\operatorname{rank}(B - A) = \operatorname{rank} B - \operatorname{rank} A, \quad \operatorname{rank}(C - B) = \operatorname{rank} C - \operatorname{rank} B.$$

Then,

$$\operatorname{rank}(C - A) = \operatorname{rank} C - \operatorname{rank} A.$$

Source: $\operatorname{rank}(C - A) \le \operatorname{rank}(C - B) + \operatorname{rank}(B - A) = \operatorname{rank} C - \operatorname{rank} A$. Furthermore, $\operatorname{rank} C \le \operatorname{rank}(C - A) + \operatorname{rank} A$, and thus $\operatorname{rank}(C - A) \ge \operatorname{rank} C - \operatorname{rank} A$. Alternatively, use Fact 4.30.1. **Remark:** See [1333].

Fact 4.30.3. Let $A, B \in \mathbb{F}^{n\times m}$, and define $A \overset{\text{rs}}{\le} B$ if and only if $\operatorname{rank}(B - A) = \operatorname{rank} B - \operatorname{rank} A$. Then, "$\overset{\text{rs}}{\le}$" is a partial ordering on $\mathbb{F}^{n\times m}$. **Source:** Fact 4.30.2. **Remark:** "$\overset{\text{rs}}{\le}$" is the *rank subtractivity partial ordering*. **Related:** Fact 10.23.4.

Fact 4.30.4. Let $A, B \in \mathbb{F}^{n\times m}$, let $C \in \mathbb{F}^{l\times n}$, and consider the following statements:

i) $A \overset{\text{rs}}{\le} B$.

ii) $CA \overset{\text{rs}}{\le} CB$.

iii) $\dim[\mathcal{R}(B - A) \cap \mathcal{N}(C)] + \dim[\mathcal{R}(A) \cap \mathcal{N}(C)] = \dim[\mathcal{R}(B) \cap \mathcal{N}(C)]$.

If two of the statements *i)*, *ii)*, *iii)* hold, then the remaining statement holds. **Source:** [1880].

Fact 4.30.5. Let $A, B \in \mathbb{F}^{n\times m}$, let $C \in \mathbb{F}^{m\times l}$, and consider the following statements:

i) $A \overset{\text{rs}}{\le} B$.

ii) $AC \overset{\text{rs}}{\le} BC$.

iii) $\dim[\mathcal{R}(B^* - A^*) \cap \mathcal{N}(C^*)] + \dim[\mathcal{R}(A^*) \cap \mathcal{N}(C^*)] = \dim[\mathcal{R}(B^*) \cap \mathcal{N}(C^*)]$.

If two of the statements *i)*, *ii)*, *iii)* hold, then the remaining statement holds. **Source:** [1880].

Fact 4.30.6. Let $A, B \in \mathbb{F}^{n\times m}$. Then, the following statements are equivalent:

i) $A^*A = A^*B$ and $B^*B = B^*A$.

ii) $AA^* = BA^*$ and $BB^* = AB^*$.

iii) $A = B$.

Source: Consider $(A - B)^*(A - B)$. See [1339]. **Related:** Fact 4.30.7 and Fact 4.30.8.

Fact 4.30.7. Let $A, B, C \in \mathbb{F}^{n\times m}$, and assume that the following statements hold:

i) $A^*A = A^*B$.

ii) $AA^* = BA^*$.

iii) $B^*B = B^*C$.

iv) $BB^* = CB^*$.

Then, the following statements hold:

v) $A^*A = A^*C$.

vi) $AA^* = CA^*$.

Source: [1339]. **Related:** Fact 4.30.8.

Fact 4.30.8. Let $A, B \in \mathbb{F}^{n\times m}$, and let $A \overset{*}{\le} B$ denote the following statements:

i) $A^*A = A^*B$.

ii) $AA^* = BA^*$.

Then, "$\overset{*}{\le}$" is a partial ordering on $\mathbb{F}^{n\times m}$. Now, assume that $A \overset{*}{\le} B$. Then, the following statements hold:

iii) AB^* and B^*A are Hermitian.

iv) For all $k \ge 1$, $(AA^*)^k \overset{*}{\le} (BB^*)^k$, $(A^*A)^k \overset{*}{\le} (B^*B)^k$, $(AA^*)^kA \overset{*}{\le} (BB^*)^kB$, $(A^*A)^kA^* \overset{*}{\le} (B^*B)^kB$.

Source: Fact 4.30.6 and Fact 4.30.7. **Remark:** "$\overset{*}{\le}$" is the *star partial ordering*. See [230, 1339, 2058] and [2059, Chapter 5]. **Related:** Fact 4.30.7 and Fact 10.23.6.

Fact 4.30.9. Let $A, B \in \mathbb{F}^{n\times n}$, and assume that $A \overset{*}{\le} B$ and $AB = BA$. Then, $A^2 \overset{*}{\le} B^2$. **Source:** [223]. **Related:** Fact 10.11.47.

Fact 4.30.10. Let $A, C \in \mathbb{F}^{n\times m}$ and $B, D \in \mathbb{F}^{n\times l}$, and assume that $A \overset{*}{\le} C$, $B \overset{*}{\le} D$, and $\mathcal{R}(A) = \mathcal{R}(B)$. Then, $[A\ \ B] \overset{*}{\le} [C\ \ D]$. **Source:** [1880].

Fact 4.30.11. Let $A, C \in \mathbb{F}^{n\times m}$ and $B, D \in \mathbb{F}^{n\times l}$, and assume that $[A\ \ B] \overset{*}{\le} [C\ \ D]$. Then, $A \overset{*}{\le} C$ if and only if $B \overset{*}{\le} D$. **Source:** [1880].

Fact 4.30.12. Let $A, B \in \mathbb{F}^{n\times m}$ and $C \in \mathbb{F}^{n\times l}$. Then, the following statements hold:

i) If $A \overset{*}{\le} B$ and $\mathcal{R}(A) \subseteq \mathcal{R}(B)$, then $[A\ \ C] \overset{*}{\le} [B\ \ C]$ and $[C\ \ A] \overset{*}{\le} [C\ \ B]$.

ii) If either $[A\ \ C] \overset{*}{\le} [B\ \ C]$ or $[C\ \ A] \overset{*}{\le} [C\ \ B]$, then $A \overset{*}{\le} B$.

Source: [1880].

Fact 4.30.13. Let $A, C \in \mathbb{F}^{n\times m}$ and $B, D \in \mathbb{F}^{l\times m}$, and assume that $A \overset{*}{\le} C$, $B \overset{*}{\le} D$, and $\mathcal{N}(A) = \mathcal{N}(B)$. Then, $\left[\begin{smallmatrix} A \\ B \end{smallmatrix}\right] \overset{*}{\le} \left[\begin{smallmatrix} C \\ D \end{smallmatrix}\right]$. **Source:** [1880].

Fact 4.30.14. Let $A, C \in \mathbb{F}^{n\times m}$ and $B, D \in \mathbb{F}^{l\times m}$, and assume that $\left[\begin{smallmatrix} A \\ B \end{smallmatrix}\right] \overset{*}{\le} \left[\begin{smallmatrix} C \\ D \end{smallmatrix}\right]$. Then, $A \overset{*}{\le} B$ if and only if $B \overset{*}{\le} D$. **Source:** [1880].

Fact 4.30.15. Let $A, B \in \mathbb{F}^{n\times m}$ and $C \in \mathbb{F}^{l\times m}$. Then, the following statements hold:

i) If $A \overset{*}{\le} B$ and $\mathcal{R}(A) \subseteq \mathcal{R}(B)$, then $\left[\begin{smallmatrix} A \\ C \end{smallmatrix}\right] \overset{*}{\le} \left[\begin{smallmatrix} B \\ C \end{smallmatrix}\right]$ and $\left[\begin{smallmatrix} C \\ A \end{smallmatrix}\right] \overset{*}{\le} \left[\begin{smallmatrix} C \\ B \end{smallmatrix}\right]$.

ii) If either $\left[\begin{smallmatrix} A \\ C \end{smallmatrix}\right] \overset{*}{\le} \left[\begin{smallmatrix} B \\ C \end{smallmatrix}\right]$ or $\left[\begin{smallmatrix} C \\ A \end{smallmatrix}\right] \overset{*}{\le} \left[\begin{smallmatrix} C \\ B \end{smallmatrix}\right]$, then $A \overset{*}{\le} B$.

Source: [1880].

Fact 4.30.16. Let $A, B, C, D \in \mathbb{F}^{n\times m}$, and assume that $A \overset{*}{\le} C$, $B \overset{*}{\le} D$, $\mathcal{R}(A) = \mathcal{R}(B)$, and $\mathcal{N}(A) = \mathcal{N}(B)$. Then, $A + B \overset{*}{\le} C + D$. **Source:** [1880].

Fact 4.30.17. Let $A, B \in \mathbb{F}^{n\times n}$. If A is normal, B is Hermitian, and $A \overset{\rm rs}{\le} B$, then A is Hermitian. Now, assume that A and B are normal. Then, the following statements are equivalent:

i) $A \overset{*}{\le} B$.

ii) $A \overset{\rm rs}{\le} B$ and $AB = BA$.

If either B or $B - A$ is Hermitian, then the following statement is equivalent to *i*) and *ii*):

iii) $A \overset{\rm rs}{\le} B$ and $A^2 = B^2$.

Source: [2024].

4.31 Facts on Groups

Fact 4.31.1. Let $\mathcal{S}$ be a set, and let $\phi\colon \mathcal{S}\times\mathcal{S}\mapsto\mathcal{S}$. Then, $(\mathcal{S},\phi)$ is a group if and only if the following statements hold:

i) For all $x,y,z\in\mathcal{S}$, $\phi[x,\phi(y,z)] = \phi[\phi(x,y),z]$.

ii) There exists $\iota\in\mathcal{S}$ such that the following statements hold:

a) For all $x\in\mathcal{S}$, $\phi(\iota,x) = x$.

b) For all $x\in\mathcal{S}$, there exists $y\in\mathcal{S}$ such that $\phi(y,x) = \iota$.

Source: [2336, pp. 13, 14].

Fact 4.31.2. Let $(\mathcal{S},\phi)$ be a group. Then, the following statements hold:

i) Assume that $(\mathcal{S},\phi)$ is cyclic and finite, and define $\ell\triangleq\operatorname{card}(\mathcal{S})$. Then, $(\mathcal{S},\phi)\simeq\mathbb{Z}_\ell\triangleq(\{0,1,\ldots,\ell-1\},\psi)$, where $\psi(n,m)\triangleq\operatorname{rem}_\ell(n+m)$.

ii) If $(\mathcal{S},\phi)$ is cyclic and infinite, then $(\mathcal{S},\phi)\simeq(\mathbb{Z},\psi)$, where, for all $m,n\in\mathbb{Z}$, $\psi(m,n)\triangleq m+n$.

iii) If $(\mathcal{S},\phi)$ is cyclic, then $(\mathcal{S},\phi)$ is Abelian.

iv) Assume that $n\triangleq\operatorname{card}(\mathcal{X})$ is finite. Then, $(\mathcal{S},\phi)$ is cyclic if and only if, for all $m\ge 1$ such that $m|n$, there exists at most one cyclic subgroup of $(\mathcal{S},\phi)$ with m elements.

v) If $(\mathcal{S},\phi)$ is cyclic and $(\mathcal{X},\phi_{\mathcal{X}})$ is a subgroup of $(\mathcal{S},\phi)$, then $(\mathcal{X},\phi_{\mathcal{X}})$ is cyclic.

vi) If, for all $x\in\mathcal{S}$, $\phi(x,x)=\iota$, then $(\mathcal{S},\phi)$ is Abelian.

vii) The following statements are equivalent:

a) $\operatorname{card}(\mathcal{S})$ is prime.

b) $(\mathcal{S},\phi)$ is simple and cyclic.

c) $(\mathcal{S},\phi)$ is simple and Abelian.

viii) Let $x\in\mathcal{S}$, and define $\mathcal{S}_x\triangleq\{\phi^{(i)}(x)\colon i\in\mathbb{Z}\}$. Then, $(\mathcal{S}_x,\phi_{\mathcal{S}_x})$ is a subgroup of $(\mathcal{S},\phi)$.

ix) Define $\mathcal{S}_{\rm c}\triangleq\{x\in\mathcal{S}\colon \text{for all } y\in\mathcal{S}, \phi(x,y)=\phi(y,x)\}$. Then, $(\mathcal{S}_{\rm c},\phi_{\mathcal{S}_{\rm c}})$ is a normal, Abelian subgroup of $(\mathcal{S},\phi)$. Furthermore, $\mathcal{S}_{\rm c}=\mathcal{S}$ if and only if $(\mathcal{S},\phi)$ is Abelian.

x) Let $\mathcal{X}_1\subseteq\mathcal{X}_2\subseteq\mathcal{S}$, and assume that $(\mathcal{X}_1,\phi_{\mathcal{X}_1})$ and $(\mathcal{X}_2,\phi_{\mathcal{X}_2})$ are subgroups of $(\mathcal{S},\phi)$. Then, $(\mathcal{X}_1,\phi_{\mathcal{X}_1})$ is a subgroup of $(\mathcal{X}_2,\phi_{\mathcal{X}_2})$.

xi) Let $\mathcal{X}_1\subseteq\mathcal{S}$ and $\mathcal{X}_2\subseteq\mathcal{S}$, and assume that $(\mathcal{X}_1,\phi_{\mathcal{X}_1})$ is a subgroup of $(\mathcal{X}_2,\phi_{\mathcal{X}_2})$ and $(\mathcal{X}_2,\phi_{\mathcal{X}_2})$ is a subgroup of $(\mathcal{S},\phi)$. Then, $(\mathcal{X}_1,\phi_{\mathcal{X}_1})$ is a subgroup of $(\mathcal{S},\phi)$.

xii) Let $\mathcal{X}\subseteq\mathcal{S}$, and assume that $(\mathcal{X},\phi_{\mathcal{X}})$ is a subgroup of $(\mathcal{S},\phi)$. Then, $(\mathcal{X},\phi_{\mathcal{X}})$ is a normal subgroup of $(\mathcal{S},\phi)$ if and only if, for all $x\in\mathcal{S}$ and $y\in\mathcal{X}$, $\phi[\phi^{(-1)}(x),\phi(y,x)]\in\mathcal{X}$.

xiii) Let $\mathcal{X}_1\subseteq\mathcal{S}$ and $\mathcal{X}_2\subseteq\mathcal{S}$, and assume that $(\mathcal{X}_1,\phi_{\mathcal{X}_1})$ is a subgroup of $(\mathcal{X}_2,\phi_{\mathcal{X}_2})$, $(\mathcal{X}_2,\phi_{\mathcal{X}_2})$ is a subgroup of $(\mathcal{S},\phi)$, and $(\mathcal{X}_1,\phi_{\mathcal{X}_1})$ is a normal subgroup of $(\mathcal{S},\phi)$. Then, $(\mathcal{X}_1,\phi_{\mathcal{X}_1})$ is a normal subgroup of $(\mathcal{X}_2,\phi_{\mathcal{X}_2})$.

xiv) Let $\mathcal{X}_1\subseteq\mathcal{S}$ and $\mathcal{X}_2\subseteq\mathcal{S}$, assume that $(\mathcal{X}_1,\phi_{\mathcal{X}_1})$ and $(\mathcal{X}_2,\phi_{\mathcal{X}_2})$ are subgroups of $(\mathcal{S},\phi)$, and assume that either $(\mathcal{X}_1,\phi_{\mathcal{X}_1})$ or $(\mathcal{X}_2,\phi_{\mathcal{X}_2})$ is a normal subgroup of $(\mathcal{S},\phi)$. Then, $\phi(\mathcal{X}_1,\mathcal{X}_2)=\phi(\mathcal{X}_2,\mathcal{X}_1)$. If, in addition, both $(\mathcal{X}_1,\phi_{\mathcal{X}_1})$ and $(\mathcal{X}_2,\phi_{\mathcal{X}_2})$ are normal subgroups of $(\mathcal{S},\phi)$, then $(\phi(\mathcal{X}_1,\mathcal{X}_2),\phi_{\phi(\mathcal{X}_1,\mathcal{X}_2)})$ is a normal subgroup of $(\mathcal{S},\phi)$.

Source: [2336, pp. 24–33, 193]. *iv)* is given in [2345, p. 28]. **Remark:** $(\mathcal{S}_{\rm c},\phi_{\mathcal{S}_{\rm c}})$ is the *center subgroup*. **Remark:** If $(\mathcal{X}_1,\phi_{\mathcal{X}_1})$ is a normal subgroup of $(\mathcal{X}_2,\phi_{\mathcal{X}_2})$ and $(\mathcal{X}_2,\phi_{\mathcal{X}_2})$ is a normal subgroup of $(\mathcal{S},\phi)$, then it does not necessarily follow that $(\mathcal{X}_1,\phi_{\mathcal{X}_1})$ is a normal subgroup of $(\mathcal{S},\phi)$.

Fact 4.31.3. Let $(\mathcal{S},\phi)$ be a group. Then, the following statements hold:

i) Let $(\mathcal{X}_1,\phi_{\mathcal{X}_1})$ and $(\mathcal{X}_2,\phi_{\mathcal{X}_2})$ be subgroups of $(\mathcal{S},\phi)$, and define $\mathcal{X}\triangleq\mathcal{X}_1\cap\mathcal{X}_2$. Then, $(\mathcal{X},\phi_{\mathcal{X}})$ is a subgroup of $(\mathcal{S},\phi)$.

ii) Let $\mathcal{X}\subseteq\mathcal{S}$, and define $\mathcal{X}_0\triangleq\bigcap\hat{\mathcal{X}}$, where the intersection is taken over all $\hat{\mathcal{X}}\subseteq\mathcal{S}$ such that

$\mathcal{X} \subset \hat{\mathcal{X}}$ and $(\hat{\mathcal{X}}, \phi_{\hat{\mathcal{X}}})$ is a subgroup of $(\mathcal{S}, \phi)$. Then, $(\mathcal{X}_0, \phi_{\mathcal{X}_0})$ is a subgroup of $(\mathcal{S}, \phi)$.

iii) Let $\mathcal{X} \subseteq \mathcal{S}$, and define the subgroup $(\mathcal{X}_0, \phi_{\mathcal{X}_0})$ of $(\mathcal{S}, \phi)$ by *ii)*. Then, $\mathcal{X}_0 = \bigcup_{i=1}^{\infty} \mathcal{X}_i$, where $\mathcal{X}_1 \triangleq \mathcal{X} \cup \phi^{(-1)}(\mathcal{X})$ and, for all $i \geq 1$, $\mathcal{X}_{i+1} \triangleq \phi(\mathcal{X}_i, \mathcal{X}_1) \cup \phi(\mathcal{X}_1, \mathcal{X}_i)$.

iv) The set of subgroups of $(\mathcal{S}, \phi)$ is a lattice, where the greatest lower bound of the subgroups $(\mathcal{X}_1, \phi_{\mathcal{X}_1})$ and $(\mathcal{X}_2, \phi_{\mathcal{X}_2})$ of $(\mathcal{S}, \phi)$ is the subgroup $(\mathcal{X}, \phi_{\mathcal{X}})$ of $(\mathcal{S}, \phi)$ given by *i)* and the least upper bound of the subgroups $(\mathcal{X}_1, \phi_{\mathcal{X}_1})$ and $(\mathcal{X}_2, \phi_{\mathcal{X}_2})$ is the subgroup $(\mathcal{X}_0, \phi_{\mathcal{X}_0})$ of $(\mathcal{S}, \phi)$ given by *ii)* and *iii)* with $\mathcal{X} \triangleq \mathcal{X}_1 \cup \mathcal{X}_2$.

Source: [2336, pp. 25–32]. **Remark:** The subgroup $(\mathcal{X}_0, \phi_{\mathcal{X}_0})$ given by *ii)* and *iii)* is the *subgroup generated by* $\mathcal{X}$.

Fact 4.31.4. Let $l, m, n, p \geq 1$. Then, the following statements hold:

i) $\mathbb{C}^{n\times m}$, $\mathbb{R}^{n\times m}$, $\mathbb{Q}^{n\times m}$, $\mathbb{Z}^{n\times m}$, $\mathbb{Q}^{n\times m} + \jmath\mathbb{Q}^{n\times m}$, $\mathbb{Z}^{n\times m} + \jmath\mathbb{Z}^{n\times m}$, $\mathbb{Z}^{n\times m} + \jmath\mathbb{Q}^{n\times m}$, and $\mathbb{Q}^{n\times m} + \jmath\mathbb{Z}^{n\times m}$ are addition groups.

ii) Let $\mathcal{S} \subseteq \mathbb{C}^{n\times m}$, and assume that $\mathcal{S}$ is a subspace. Then, $\mathcal{S}$ is an addition group.

iii) Let $\mathcal{S} \subseteq \mathbb{C}^{n\times m}$ be an addition group. Then, $\mathcal{S}^{\mathsf{T}}$, $\overline{\mathcal{S}}$, and $\mathcal{S}^*$ are addition groups.

iv) Let $\mathcal{S} \subseteq \mathbb{C}^{n\times m}$ be an addition group, and let $A \in \mathbb{C}^{p\times n}$ and $B \in \mathbb{C}^{m\times l}$. Then, $A\mathcal{S}B$ is an addition group.

v) Let $\mathcal{S}_1 \subseteq \mathbb{C}^{n\times m}$ and $\mathcal{S}_2 \subseteq \mathbb{C}^{n\times m}$ be addition groups. Then, $\mathcal{S}_1 + \mathcal{S}_2$ is an addition group.

Remark: If $s \in \mathbb{C}$, then $s\mathbb{Z} = \{\ldots, -2s, -s, 0, s, 2s, \ldots\}$.

Fact 4.31.5. Let $\mathcal{S} \subset \mathbb{F}^{n\times n}$, assume that $\mathcal{S}$ is a multiplication group, and let $A \in \mathcal{S}$. Then, the following statements hold:

i) If $\mathcal{S}$ is finite, then $A^{\operatorname{card}(\mathcal{S})} = I_n$.

ii) If $A^k = I_n$, then the order of A divides k.

iii) If $\mathcal{S}$ is finite, then the order of A divides $\operatorname{card}(\mathcal{S})$.

iv) If $\operatorname{card}(\mathcal{S})$ is prime, then $\mathcal{S} = \{I_n, A, A^2, \ldots, A^{\operatorname{card}(\mathcal{S})-1}\}$ and $\mathcal{S}$ is a cyclic group.

v) If $\mathcal{S}$ is finite and m is a positive integer that is relatively prime to $\operatorname{card}(\mathcal{S})$, then there exists a unique matrix $B \in \mathcal{S}$ such that $B^m = A$.

Source: [1757, p. 147]. **Remark:** The order ℓ of A is the smallest positive integer ℓ such that $A^\ell = I_n$. **Example:** Let $A \triangleq e^{(2\pi/3)\jmath}$ and $\mathcal{S} = \{1, A, A^2\}$. Then, $\operatorname{card}(\mathcal{S}) = 3$, $A^3 = 1$, and the order of A is 3. **Example:** Let $A \triangleq \begin{bmatrix} 1 & 0 \\ 0 & -1 \end{bmatrix}$ and $\mathcal{S} = \{I_2, -I_2, A, -A\}$. Then, $\operatorname{card}(\mathcal{S}) = 4$, $A^4 = I_2$, and the order of A is 2. However, $\mathcal{S} \neq \{I, A, A^2, A^3\} = \{I_2, A\}$.

Fact 4.31.6. The following sets are multiplication groups:

i) $\{1\} = \mathrm{SO}(1) = \mathrm{SU}(1)$.

ii) $\{-1, 1\} = \mathrm{O}(1)$.

iii) $\{z \in \mathbb{C}\colon |z| = 1\} = \mathrm{U}(1)$.

iv) $\mathbb{C}\backslash\{0\}, \mathbb{R}\backslash\{0\}, (0, \infty), \mathbb{Q}\backslash\{0\}, (0, \infty) \cap \mathbb{Q}$.

Now, let $p \geq 2$ be prime. Then, $(\mathbb{Z}/p\mathbb{Z})^* \triangleq (\{1, \ldots, p-1\}, \psi)$, where $\psi(m, n) \triangleq \operatorname{rem}_p(mn)$, is a finite Abelian group. **Source:** The last statement is given in [2336, p. 18]. **Remark:** $(\mathbb{Z}/7\mathbb{Z})^* = (\{\psi^{(i)}(3)\colon i \in \mathbb{Z}\}, \psi)$, where $\psi(m, n) \triangleq \operatorname{rem}_7(mn)$. See [2336, p. 26].

Fact 4.31.7. Let $\mathcal{S} \subset \mathbb{F}^{n\times n}$, and assume that $\mathcal{S}$ is a multiplication group. Then, $\mathcal{S}^{\mathsf{T}}$, $\overline{\mathcal{S}}$, and $\mathcal{S}^*$ are multiplication groups. Now, let $m \geq 1$. Then, $\cup_{i=0}^{m-1} e^{(2i\pi/m)\jmath}\mathcal{S}$ is a multiplication group. In particular, $\mathcal{S} \cup (-\mathcal{S})$ and $\mathcal{S} \cup (-\mathcal{S}) \cup (\jmath\mathcal{S}) \cup (-\jmath\mathcal{S})$ are multiplication groups.

Fact 4.31.8. Let $n \geq 0$, and define $\mathrm{S}^n \triangleq \{x \in \mathbb{R}^{n+1}\colon x^{\mathsf{T}}x = 1\}$, which is the unit sphere in $\mathbb{R}^{n+1}$. Then, the following statements hold:

i) $\mathrm{S}^0 = \{-1, 1\} = \mathrm{O}(1)$.

ii) $\mathrm{U}(1) = \{e^{J\theta}\colon \theta \in [0, 2\pi)\} \simeq \mathrm{SO}(2)$.

iii) $\mathrm{S}^1 = \{[\cos\theta \;\; \sin\theta]^\mathsf{T} \in \mathbb{R}^2\colon \theta \in (-\pi, \pi]\} = \{[\mathrm{Re}\, z \;\; \mathrm{Im}\, z]^\mathsf{T}\colon z \in \mathrm{U}(1)\}$.

iv) $\mathrm{SU}(2) = \left\{\left[\begin{smallmatrix} z & w \\ -\bar{w} & \bar{z}\end{smallmatrix}\right] \in \mathbb{C}^{2\times 2}\colon z, w \in \mathbb{C} \text{ and } |z|^2 + |w|^2 = 1\right\} \simeq \mathrm{Sp}(1)$.

v) $\mathrm{S}^3 = \{[\mathrm{Re}\, z \;\; \mathrm{Im}\, z \;\; \mathrm{Re}\, w \;\; \mathrm{Im}\, w]^\mathsf{T} \in \mathbb{R}^4\colon z, w \in \mathbb{C} \text{ and } |z|^2 + |w|^2 = 1\}$.

Source: [2588, p. 40]. **Remark:** $\mathrm{Sp}(1) \subset \mathbb{H}^{1\times 1}$ is the group of unit quaternions. See Fact 4.32.1 and Fact 4.32.4. **Remark:** A group operation can be defined on S^n if and only if $n \in \{0, 1, 3\}$. See [2588, p. 40].

Fact 4.31.9. Let $n \ge 1$. Then, $\mathrm{U}(n) \simeq \mathrm{O}(2n) \cap \mathrm{Symp}_\mathbb{R}(2n)$. In particular, $\mathrm{U}(1) \simeq \mathrm{O}(2) \cap \mathrm{Symp}_\mathbb{R}(2) = \mathrm{SO}(2)$. **Source:** [206].

Fact 4.31.10. The following subsets of $\mathbb{F}^{n\times n}$ are Lie algebras:

i) $\mathrm{ut}(n) \triangleq \{A \in \mathrm{gl}_\mathbb{F}(n)\colon\; A \text{ is upper triangular}\}$.

ii) $\mathrm{sut}(n) \triangleq \{A \in \mathrm{gl}_\mathbb{F}(n)\colon\; A \text{ is strictly upper triangular}\}$.

iii) $\{0_{n\times n}\}$.

Fact 4.31.11. The following subsets of $\mathbb{F}^{n\times n}$ are multiplication groups:

i) $\mathrm{UT}(n) \triangleq \{A \in \mathrm{GL}_\mathbb{F}(n)\colon\; A \text{ is upper triangular}\}$.

ii) $\mathrm{UT}_+(n) \triangleq \{A \in \mathrm{UT}(n)\colon\; A_{(i,i)} > 0 \text{ for all } i \in \{1, \ldots, n\}\}$.

iii) $\mathrm{UT}_{\pm 1}(n) \triangleq \{A \in \mathrm{UT}(n)\colon\; A_{(i,i)} = \pm 1 \text{ for all } i \in \{1, \ldots, n\}\}$.

iv) $\mathrm{SUT}(n) \triangleq \{A \in \mathrm{UT}(n)\colon\; A_{(i,i)} = 1 \text{ for all } i \in \{1, \ldots, n\}\}$.

v) $\{I_n\}$.

Remark: The matrices in $\mathrm{SUT}(n)$ are unipotent. See Fact 7.17.10. **Remark:** $\mathrm{SUT}(3)$ with $\mathbb{F} = \mathbb{R}$ is the *Heisenberg group*. **Related:** Fact 4.25.2.

Fact 4.31.12. Let $n \ge 1$. Then, $(\mathrm{SL}(n), \mathrm{SO}(n), \mathrm{SU}(n), \mathrm{SUT}(n))$ is a normal subgroup of $(\mathrm{GL}(n), \mathrm{O}(n), \mathrm{U}(n), \mathrm{UT}(n)$. **Source:** [2336, p. 53].

Fact 4.31.13. Let $P \in \mathbb{F}^{n\times n}$, and define $\mathcal{S} \triangleq \{A \in \mathrm{GL}_\mathbb{F}(n)\colon\; A^\mathsf{T}PA = P\}$. Then, $\mathcal{S}$ is a multiplication group. If, in addition, P is nonsingular and skew symmetric, then every matrix $A \in \mathcal{S}$ satisfies $\det A = 1$. **Source:** [765]. **Remark:** If $\mathbb{F} = \mathbb{R}$, n is even, and $P = J_n$, then $\mathcal{S} = \mathrm{Symp}_\mathbb{R}(n)$. **Remark:** Necessary and sufficient conditions are given in [765] under which every matrix $A \in \mathcal{S}$ satisfies $\det A = 1$.

Fact 4.31.14. Let $A \in \mathrm{P}(n)$. Then, there exist transposition matrices $T_1, \ldots, T_k \in \mathbb{R}^{n\times n}$ such that $A = T_1 \cdots T_k$. Furthermore, the following statements hold:

i) $\det A = (-1)^k$.

ii) A is an even permutation matrix if and only if k is even.

iii) A is an odd permutation matrix if and only if k is odd.

Remark: Every permutation of n objects can be realized as a finite sequence of transpositions. See [966, pp. 106, 107] and [1061, p. 82]. **Example:** $P_3 = \left[\begin{smallmatrix} 0&1&0\\0&0&1\\1&0&0 \end{smallmatrix}\right] = \left[\begin{smallmatrix} 0&0&1\\0&1&0\\1&0&0 \end{smallmatrix}\right]\left[\begin{smallmatrix} 1&0&0\\0&0&1\\0&1&0 \end{smallmatrix}\right] = \left[\begin{smallmatrix} 1&0&0\\0&0&1\\0&1&0 \end{smallmatrix}\right]\left[\begin{smallmatrix} 0&1&0\\1&0&0\\0&0&1 \end{smallmatrix}\right]$, which represents a 3-cycle. **Remark:** As the above example shows, factorization in terms of transpositions is not unique. However, Fact 7.18.14 shows that every permutation matrix corresponds to a unique collection of disjoint cycles. **Related:** Fact 3.16.1 and Fact 4.13.18.

Fact 4.31.15. For all $n \ge 2$, the following statements hold:

i) $\mathrm{card}[\mathrm{P}(n)] = n!$.

ii) $\mathrm{card}[\mathrm{A}(n)] = \frac{1}{2}n!$.

iii) $\mathrm{card}[\mathrm{D}(n)] = 2n$.

iv) $\mathrm{card}[\mathrm{C}(n)] = n$.

In addition, the following statements hold:

v) $\mathrm{A}(2) \subset \mathrm{C}(2) = \mathrm{P}(2) \subset \mathrm{D}(2)$.

vi) $\mathrm{A}(3) = \mathrm{C}(3) \subset \mathrm{P}(3) = \mathrm{D}(3)$.

vii) If $n \geq 4$, then

$$\left.\begin{array}{r}\mathrm{C}(n) \subset \mathrm{D}(n)\\ \mathrm{A}(n)\end{array}\right\} \subset \mathrm{P}(n) \subset \mathrm{O}(n), \quad \mathrm{A}(n) \subset \mathrm{SO}(n) \subset \mathrm{O}(n).$$

viii) If $n \geq 5$ and n is odd, then

$$\mathrm{C}(n) \subset \left\{\begin{array}{c}\mathrm{D}(n)\\ \mathrm{A}(n)\end{array}\right\} \subset \mathrm{P}(n) \subset \mathrm{O}(n).$$

ix) If $n \geq 4$ and $n(n-1)/2$ is even, then $\mathrm{C}(n) \subset \mathrm{D}(n) \subset \mathrm{A}(n) \subset \mathrm{P}(n) \subset \mathrm{O}(n)$.

Source: Fact 3.16.1 and Fact 3.16.2.

Fact 4.31.16. Let k be a positive integer, define $R_k \in \mathbb{R}^{2\times 2}$ by

$$R_k \triangleq \begin{bmatrix} \cos\frac{2\pi}{k} & \sin\frac{2\pi}{k}\\ -\sin\frac{2\pi}{k} & \cos\frac{2\pi}{k}\end{bmatrix},$$

and note that $R_1 = I_2$, $R_2 = -I_2$, and $R_k^k = I_2$. Furthermore, define

$$\mathrm{O}_k(2) \triangleq \{I, R_k, \ldots, R_k^{k-1}, \hat{I}_2, \hat{I}_2R_k, \ldots, \hat{I}_2R_k^{k-1}\}, \quad \mathrm{SO}_k(2) \triangleq \{I, R_k, \ldots, R_k^{k-1}\},$$
$$\mathrm{SU}_k(1) \triangleq \{1, e^{2\pi \jmath/k}, e^{4\pi \jmath/k}, \ldots, e^{2(k-1)\pi \jmath/k}\}.$$

Then, the following statements hold:

i) $\mathrm{C}(1) = \mathrm{SU}_1(1) = \{1\} \simeq \mathrm{SO}_1(2) = \{I_2\}$.

ii) $\mathrm{O}_1(2) = \mathrm{P}(2) = \mathrm{C}(2) = \{I_2, \hat{I}_2\} \simeq \mathrm{SO}_2(2) = \{I_2, -\hat{I}_2\} \simeq \{I_2, -I_2\} \simeq \mathrm{SU}_2(1) = \{1, -1\}$.

iii) $\mathrm{O}_2(2) = \mathrm{D}(2) = \{I_2, -I_2, \hat{I}_2, -\hat{I}_2\}$.

iv) $\mathrm{C}(3) = \{I_3, P_3, P_3^2\} \simeq \mathrm{SO}_3(2) = \{I_2, R_3, R_3^2\} \simeq \mathrm{SU}_3(1) = \{1, -\frac{1}{2} + \frac{\sqrt{3}}{2}\jmath, -\frac{1}{2} - \frac{\sqrt{3}}{2}\jmath\}$.

v) $\mathrm{P}(3) = \{I_3, \hat{I}_3, P_3, P_3\hat{I}_3, \hat{I}_3P_3, \hat{I}_3P_3\hat{I}_3\}$.

vi) $\mathrm{P}(3) \simeq \left\{\begin{bmatrix}1&0\\0&1\end{bmatrix}, \begin{bmatrix}0&1\\-1&-1\end{bmatrix}, \begin{bmatrix}-1&-1\\1&0\end{bmatrix}, \begin{bmatrix}0&1\\1&0\end{bmatrix}, \begin{bmatrix}-1&-1\\0&1\end{bmatrix}, \begin{bmatrix}1&0\\-1&-1\end{bmatrix}\right\}$.

vii) $$\mathrm{D}(3) \simeq \left\{I_3, \begin{bmatrix}1&0&0\\0&-1&0\\0&0&-1\end{bmatrix}, \begin{bmatrix}-\frac{1}{2}&\frac{\sqrt{3}}{2}&0\\-\frac{\sqrt{3}}{2}&-\frac{1}{2}&0\\0&0&1\end{bmatrix}, \begin{bmatrix}-\frac{1}{2}&-\frac{\sqrt{3}}{2}&0\\\frac{\sqrt{3}}{2}&-\frac{1}{2}&0\\0&0&1\end{bmatrix}, \begin{bmatrix}-\frac{1}{2}&\frac{\sqrt{3}}{2}&0\\\frac{\sqrt{3}}{2}&\frac{1}{2}&0\\0&0&-1\end{bmatrix}, \begin{bmatrix}-\frac{1}{2}&-\frac{\sqrt{3}}{2}&0\\-\frac{\sqrt{3}}{2}&\frac{1}{2}&0\\0&0&-1\end{bmatrix}\right\}.$$

viii) $\mathrm{C}(4) = \{I_4, P_4, P_4^2, P_4^3\} \simeq \mathrm{SO}_4(2) = \{I_2, R_4, R_4^2, R_4^3\} \simeq \mathrm{SU}_4(1) = \{1, -1, \jmath, -\jmath\}$.

ix) $\mathrm{A}(4) \simeq \{I_3, D_1, D_2, D_3, P_3, D_1P_3, D_2P_3, D_3P_3, P_3^2, D_1P_3^2, D_2P_3^2, D_3P_3^2\}$, where $D_1 \triangleq \mathrm{diag}(1, -1, -1)$, $D_2 \triangleq \mathrm{diag}(-1, 1, -1)\}$, and $D_3 \triangleq \mathrm{diag}(-1, -1, 1)$.

x) $\mathrm{P}(4) \simeq \{A \in \mathrm{SO}(3)\colon |A| \in \mathrm{P}(3)\}$.

xi) $\mathrm{C}(6) \simeq \left\{I_5, \begin{bmatrix}P_2&0\\0&I_3\end{bmatrix}, \begin{bmatrix}I_2&0\\0&P_3\end{bmatrix}, \begin{bmatrix}I_2&0\\0&P_3^2\end{bmatrix}, \begin{bmatrix}P_2&0\\0&P_3\end{bmatrix}, \begin{bmatrix}P_2&0\\0&P_3^2\end{bmatrix}\right\}$.

xii) For all $k \geq 2$, $\mathrm{O}_k(2) \simeq \mathrm{D}(k)$ and $\mathrm{card}[\mathrm{O}_k(2)] = 2k$.

xiii) For all $k \geq 2$, $\mathrm{C}(k) \simeq \mathrm{SO}_k(2) \simeq \mathrm{SU}_k(1)$ and $\mathrm{card}[\mathrm{SO}_k(2)] = k$.

Remark: The multiplication groups $\mathrm{P}(k)$, $\mathrm{A}(k)$, $\mathrm{D}(k)$, and $\mathrm{C}(k)$ are isomorphic to *symmetry groups*, which are abstract groups consisting of transformations that map a set onto itself. Specifically, $\mathrm{P}(k)$, $\mathrm{A}(k)$, $\mathrm{D}(k) \simeq \mathrm{O}_k(2)$, and $\mathrm{C}(k) \simeq \mathrm{SO}_k(2)$ are isomorphic to the *symmetric group* S_k, the *alternating group* A_k, the *dihedral group* D_k, and the *cyclic group* C_k, respectively. The elements of $\mathrm{S}_n \simeq \mathrm{P}(n)$ permute n-tuples arbitrarily, while the elements of $\mathrm{A}_n \simeq \mathrm{A}(n)$ permute n-tuples evenly. See Fact 7.18.14 for the decomposition of a permutation matrix in terms of cyclic permutation matrices.

The elements of $\mathrm{SO}_k(2)$ perform counterclockwise rotations of planar figures by the angle $2\pi/k$ about a line perpendicular to the plane and passing through 0, while the elements of $\mathrm{O}_k(2)$ perform the rotations of $\mathrm{SO}_k(2)$ and reflect planar figures across the line $y = x$. See [966, pp. 41, 845]. Matrix representations of groups are discussed in [707, 1098, 1257, 1306, 1436, 2424]. **Remark:** Every finite subgroup of O(2) is isomorphic to either D_k or C_k for some k. Furthermore, every finite subgroup of SO(3) is isomorphic to either D_k for some k, C_k for some k, A_4, S_4, or A_5. The symmetry group D_3 is isomorphic to the group of bijective transformations of an equilateral triangle, namely, three planar rotations and three 180-degree out-of-plane rotations about the medians. The symmetry groups A_4, S_4, and A_5 are isomorphic to the group of bijective transformations of regular solids. Specifically, A_4 is represented by the *tetrahedral group*, which consists of $4!/2 = 12$ rotation matrices that map a regular tetrahedron onto itself; S_4 is represented by the *octahedral group*, which consists of $4! = 24$ rotation matrices that map either an octahedron or a cube onto itself; and A_5 is represented by the *icosahedral group*, which consists of $5!/2 = 60$ rotation matrices that map either a regular icosahedron or a regular dodecahedron onto itself. See [170, p. 184], [773, p. 32], [1199, pp. 176–193], [1259, pp. 9–23], [2345, p. 69], [2424, pp. 35–43], [2588, pp. 45–47], and [2589, Chapter 7]. **Remark:** The dihedral group D_2 is the *Klein four group*. **Remark:** *viii*) is given in [1199, p. 180]. **Remark:** For all $k \ge 3$, the permutation group S_k is not Abelian. The alternating group A_3 is Abelian, whereas, for all $k \ge 4$, A_k is not Abelian. For all $k \ge 5$, A_k is simple; see [1560, p. 145] and [2336, pp. 50, 51]. This result is related to the classical result of H. Abel and E. Galois that there exist polynomials of degree 5 and greater whose roots cannot be expressed in terms of radicals involving the coefficients. Two such polynomials are $p(x) = x^5 - x - 1$ and $p(x) = x^5 - 16x + 2$. See [170, p. 574], [173, Chapter 13], [966, pp. 32, 625–639], [1069, pp. 488–494], [1290, Theorem 34], [1618], and [2264, pp. 199–203]. Quintic polynomials that can be solved in terms of radicals are discussed in [965]. **Remark:** The 24 elements of the octahedral group representing either S_4 or P(4) are given in *ix*) by the 3×3 signed permutation matrices with determinant 1, where a *signed permutation matrix* has exactly one nonzero entry, which is either 1 or -1, in each row and column. **Remark:** *x*) shows that $\mathrm{C}_6 \simeq \mathrm{C}_2 \times \mathrm{C}_3$. See [1560, p. 169]. **Remark:** The converse of Lagrange's theorem is not true. Eleven proofs are given in [566] of the fact that A_4, which has 12 elements, has no subgroup with 6 elements. In particular, it can be shown that none of the $\binom{11}{5} = 462$ subsets of A_4 consisting of the identity and 5 nonidentity elements is closed under composition. See also [565].

Fact 4.31.17. The following statements hold:

i) There exists exactly one isomorphically distinct group consisting of one element.

ii) Let $n \ge 1$. Then, there exists exactly one isomorphically distinct group consisting of n elements if and only if n is a cyclic number. This group is C_n.

iii) The cyclic group C_2 is isomorphic to the permutation group S_2 and the multiplication groups P(2), C(2), $\mathrm{O}_1(2)$, $\mathrm{SO}_2(2)$, and $\mathrm{SU}_2(1) = \{1, -1\}$.

iv) The cyclic group C_3 is isomorphic to the alternating group A_3 and the multiplication groups A(3), C(3), $\mathrm{SO}_3(2)$, and $\mathrm{SU}_3(1)$.

v) There exist exactly two isomorphically distinct groups consisting of four elements, namely, the cyclic group C_4 and the dihedral group D_2. C_4 is isomorphic to the multiplication groups C(4), $\mathrm{SO}_4(2)$, and $\mathrm{SU}_4(1) = \{1, -1, \jmath, -\jmath\}$. D_2 is isomorphic to the multiplication group $\mathrm{O}_2(2)$.

vi) The cyclic group C_5 is isomorphic to the multiplication groups C(5), $\mathrm{SO}_5(2)$, and $\mathrm{SU}_5(1)$.

vii) There exist exactly two isomorphically distinct groups consisting of six elements, namely, the cyclic group C_6 and the dihedral group D_3. C_6 is isomorphic to the multiplication groups C(6), $\mathrm{SO}_6(2)$, and $\mathrm{SU}_6(1)$. D_3 is isomorphic to the multiplication groups P(3) and $\mathrm{O}_3(2)$. D_3 is the smallest group that is not Abelian.

viii) The cyclic group C_7 is isomorphic to the multiplication groups $C(7)$, $SO_7(2)$, and $SU_7(1)$.

ix) There exist exactly five isomorphically distinct groups containing eight elements, namely, C_8, $D_2 \times C_2$, $C_4 \times C_2$, D_4, and the quaternion multiplication group $\{\pm 1, \pm\hat{\imath}, \pm\hat{\jmath}, \pm\hat{k}\}$. Furthermore, C_8 is isomorphic to $C(8)$, $SO_8(2)$, and $SU_8(1)$; D_4 is isomorphic to $O_4(2)$; and $\{\pm 1, \pm\hat{\imath}, \pm\hat{\jmath}, \pm\hat{k}\}$ is isomorphic to the multiplication group given by *v*) of Fact 4.32.6.

Source: [1178, pp. 4–7] and [1560, pp. 168–172]. *ii*) is given in [2337, p. 7]. **Remark:** Cyclic numbers are defined in Fact 1.20.4. **Remark:** $SU_k(1)$ is defined in Fact 4.31.16. **Remark:** The Euler totient function is defined in Fact 1.20.4. **Remark:** There are 267 isomorphically distinct groups consisting of 64 elements, 2328 isomorphically distinct groups consisting of 128 elements, 56092 isomorphically distinct groups consisting of 264 elements, and 10494213 isomorphically distinct groups consisting of 512 elements. See [1560, p. 168] and [2336, p. 294]. There are 73 nonprime positive integers n between 1 to 520 inclusive for which there is exactly one isomorphically distinct group consisting of n elements. These 73 values of n are cyclic numbers that are not prime. See Fact 1.20.4. **Remark:** There are 173 isomorphically distinct simple groups whose cardinality is less than 1000. Five of these isomorphically distinct groups are not Abelian. **Remark:** The collection of finite simple groups, that is, groups that contain no normal subgroups other than the identity subgroup and the group itself, consists of 18 countably infinite sets of finite groups (one of which is the set of cyclic groups with a prime number of elements, while another is the set of alternating groups with either 5 or more elements) along with 26 finite groups called *sporadic groups* [2890, Chapter 5]. The largest sporadic group is the *monster group* M, which includes 19 sporadic groups as proper subgroups and has $2^{46} \cdot 3^{20} \cdot 5^9 \cdot 7^6 \cdot 11^2 \cdot 13^3 \cdot 17 \cdot 19 \cdot 23 \cdot 29 \cdot 31 \cdot 41 \cdot 47 \cdot 59 \cdot 71 = 808{,}017{,}424{,}794{,}512{,}875{,}886{,}459{,}904{,}961{,}710{,}757{,}005{,}754{,}368{,}000{,}000{,}000 \simeq 8 \times 10^{53}$ elements. None of the sporadic groups are Abelian. **Remark:** The *Rubik's cube group* has $43{,}252{,}003{,}274{,}489{,}856{,}000 = 2^{27} \cdot 3^{14} \cdot 5^3 \cdot 7^2 \cdot 11$ elements. This group is not simple and not Abelian, and the largest order of its elements is 1260. Rubik's cube can be solved from an arbitrary orientation in either 20 or fewer moves.

4.32 Facts on Quaternions

Fact 4.32.1. Let $\hat{\imath}, \hat{\jmath}, \hat{k}$ satisfy

$$\hat{\imath}^2 = \hat{\jmath}^2 = \hat{k}^2 = -1, \quad \hat{\imath}\hat{\jmath} = \hat{k} = -\hat{\jmath}\hat{\imath}, \quad \hat{\jmath}\hat{k} = \hat{\imath} = -\hat{k}\hat{\jmath}, \quad \hat{k}\hat{\imath} = \hat{\jmath} = -\hat{\imath}\hat{k},$$

and define

$$\mathbb{H} \triangleq \{a + b\hat{\imath} + c\hat{\jmath} + d\hat{k}\colon\ a, b, c, d \in \mathbb{R}\}.$$

Furthermore, for all $a, b, c, d \in \mathbb{R}$, define $q \triangleq a + b\hat{\imath} + c\hat{\jmath} + d\hat{k}$, $\overline{q} \triangleq a - b\hat{\imath} - c\hat{\jmath} - d\hat{k}$, and $|q| \triangleq \sqrt{q\overline{q}} = \sqrt{a^2 + b^2 + c^2 + d^2} = |\overline{q}|$. Then,

$$qI_4 = U\mathcal{Q}(q)U,$$

where

$$\mathcal{Q}(q) \triangleq \begin{bmatrix} a & -b & -c & -d \\ b & a & -d & c \\ c & d & a & -b \\ d & -c & b & a \end{bmatrix}, \quad U \triangleq \frac{1}{2}\begin{bmatrix} 1 & \hat{\imath} & \hat{\jmath} & k \\ -\hat{\imath} & 1 & \hat{k} & -\hat{\jmath} \\ -\hat{\jmath} & -\hat{k} & 1 & \hat{\imath} \\ -\hat{k} & \hat{\jmath} & -\hat{\imath} & 1 \end{bmatrix},$$

and U satisfies $U^2 = I_4$. In addition,

$$\det \mathcal{Q}(q) = (a^2 + b^2 + c^2 + d^2)^2.$$

Furthermore, if $|q| = 1$, then $\begin{bmatrix} a & -b & -c & -d \\ b & a & -d & c \\ c & d & a & -b \\ d & -c & b & a \end{bmatrix}$ is orthogonal. Next, for $i = 1,2$, let $a_i, b_i, c_i, d_i \in \mathbb{R}$ and define $q_i \triangleq a_i + b_i\hat{\imath} + c_i\hat{\jmath} + d_i\hat{k}$ and $v_i \triangleq [b_i \;\; c_i \;\; d_i]^\mathsf{T}$, and define $q_3 \triangleq q_2q_1 = a_3 + b_3\hat{\imath} + c_3\hat{\jmath} + d_3\hat{k}$. Then,

$$\mathcal{Q}(q_3) = \mathcal{Q}(q_2)\mathcal{Q}(q_1), \quad \overline{q_3} = \overline{q_2}\,\overline{q_1}, \quad |q_3| = |q_2q_1| = |q_1q_2| = |q_1\overline{q_2}| = |\overline{q_1}q_2| = |\overline{q_1}\,\overline{q_2}| = |q_1||q_2|,$$

$$\begin{bmatrix} a_3 \\ b_3 \\ c_3 \\ d_3 \end{bmatrix} = \mathcal{Q}(q_2)\begin{bmatrix} a_1 \\ b_1 \\ c_1 \\ d_1 \end{bmatrix} = \begin{bmatrix} a_2a_1 - v_2^\mathsf{T}v_1 \\ a_1v_2 + a_2v_1 + v_2 \times v_1 \end{bmatrix}.$$

Remark: q is a *quaternion*. See [1035, pp. 287–294]. Note the analogy between $\hat{\imath}, \hat{\jmath}, \hat{k}$ and the unit vectors in $\mathbb{R}^3$ under cross-product multiplication. See [218, p. 119]. **Remark:** The group Sp(1) of unit-length quaternions is isomorphic to SU(2). See [803, p. 30], [2588, p. 40], Fact 4.28.11, and Fact 4.32.4. **Remark:** The unit-length quaternions, which are called *Euler parameters*, comprise the unit sphere $\mathrm{S}^3 \subset \mathbb{R}^4$ and provide a double cover of SO(3) as shown by Fact 4.14.6. See [57], [308, p. 380], and [773, 1709, 2448]. **Remark:** An equivalent formulation of quaternion multiplication is given by Rodrigues's formulas. See Fact 4.14.8. **Remark:** Determinants and inverses of matrices with quaternion entries are discussed in [182, 2076], [2588, p. 31], and [2981]. Solutions of quaternion equations are given in [2323, 2753]. Calculus with quaternions is developed in [2076]. **Remark:** The *Clifford algebras* include the *quaternion algebra* $\mathbb{H}$ and the *octonion algebra* $\mathbb{O}$, which involves the *Cayley numbers*. See [1035, pp. 295–300]. These ideas form the basis for *geometric algebra*. See [2505, p. 100] and [207, 773, 784, 808, 909, 925, 926, 1035, 1268, 1277, 1315, 1374, 1375, 1376, 1392, 1656, 1759, 1892, 2077, 2263, 2421, 2580, 2588, 2626].

Fact 4.32.2. Let $a, b, c, d \in \mathbb{R}$, and let $q \triangleq a + b\hat{\imath} + c\hat{\jmath} + d\hat{k} \in \mathbb{H}$. Then, $q = a + b\hat{\imath} + (c + d\hat{\imath})\hat{\jmath}$. **Remark:** $\mathbb{H}$ denotes the quaternion algebra. For all $q \in \mathbb{H}$, there exist $z, w \in \mathbb{C}$ such that $q = z + w\hat{\jmath}$, where we interpret $\mathbb{C}$ as $\{a + b\hat{\imath}\colon a, b \in \mathbb{R}\}$. This observation is analogous to the fact that, for all $z \in \mathbb{C}$, there exist $a, b \in \mathbb{R}$ such that $z = a + b\jmath$, where $\jmath \triangleq \sqrt{-1}$. See [2588, p. 10].

Fact 4.32.3. The following sets are groups:

i) $\mathrm{Q} \triangleq \{\pm 1, \pm\hat{\imath}, \pm\hat{\jmath}, \pm\hat{k}\}$.

ii) $\mathrm{GL}_{\mathbb{H}}(1) \triangleq \mathbb{H}\backslash\{0\} = \{a + b\hat{\imath} + c\hat{\jmath} + d\hat{k}\colon a, b, c, d \in \mathbb{R} \text{ and } a^2 + b^2 + c^2 + d^2 > 0\}$.

iii) $\mathrm{Sp}(1) \triangleq \{a + b\hat{\imath} + c\hat{\jmath} + d\hat{k}\colon a, b, c, d \in \mathbb{R} \text{ and } a^2 + b^2 + c^2 + d^2 = 1\}$.

iv) $\mathrm{Q}_{\mathbb{R}} \triangleq \left\{\pm I_4, \pm\begin{bmatrix} 0 & -1 & 0 & 0 \\ 1 & 0 & 0 & 0 \\ 0 & 0 & 0 & -1 \\ 0 & 0 & 1 & 0 \end{bmatrix}, \pm\begin{bmatrix} 0 & 0 & -1 & 0 \\ 0 & 0 & 0 & 1 \\ 1 & 0 & 0 & 0 \\ 0 & -1 & 0 & 0 \end{bmatrix}, \pm\begin{bmatrix} 0 & 0 & 0 & -1 \\ 0 & 0 & -1 & 0 \\ 0 & 1 & 0 & 0 \\ 1 & 0 & 0 & 0 \end{bmatrix}\right\}$.

v) $\mathrm{GL}_{\mathbb{H},\mathbb{R}}(1) \triangleq \left\{\begin{bmatrix} a & -b & -c & -d \\ b & a & -d & c \\ c & d & a & -b \\ d & -c & b & a \end{bmatrix}\colon a^2 + b^2 + c^2 + d^2 > 0\right\}$.

vi) $\mathrm{GL}'_{\mathbb{H},\mathbb{R}}(1) \triangleq \left\{\begin{bmatrix} a & -b & -c & -d \\ b & a & -d & c \\ c & d & a & -b \\ d & -c & b & a \end{bmatrix}\colon a^2 + b^2 + c^2 + d^2 = 1\right\}$.

Furthermore, $\mathrm{Q} \simeq \mathrm{Q}_{\mathbb{R}}$, $\mathrm{GL}_{\mathbb{H}}(1) \simeq \mathrm{GL}_{\mathbb{H},\mathbb{R}}(1)$, $\mathrm{Sp}(1) \simeq \mathrm{GL}'_{\mathbb{H},\mathbb{R}}(1)$, and $\mathrm{GL}'_{\mathbb{H},\mathbb{R}}(1) \subset \mathrm{SO}(4) \cap \mathrm{Symp}_{\mathbb{R}}(4)$. **Remark:** $J_4 \in \mathrm{Symp}_{\mathbb{R}}(4) \cap \mathrm{SO}(4)$ but is not an element of $\mathrm{GL}'_{\mathbb{H},\mathbb{R}}(1)$. **Related:** Fact 4.32.1.

Fact 4.32.4. Define $\mathrm{Sp}(n) \triangleq \{A \in \mathbb{H}^{n\times n}\colon A^*A = I\}$, where $\mathbb{H}$ is the quaternion algebra, $A^* \triangleq \overline{A}^\mathsf{T}$, and, for $q = a + b\hat{\imath} + c\hat{\jmath} + d\hat{k} \in \mathbb{H}$, $\overline{q} \triangleq a - b\hat{\imath} - c\hat{\jmath} - d\hat{k}$. Then, the groups $\mathrm{Sp}(n)$ and $\mathrm{U}(2n) \cap \mathrm{Symp}_{\mathbb{C}}(2n)$ are isomorphic. In particular, Sp(1) and $\mathrm{U}(2) \cap \mathrm{Symp}_{\mathbb{C}}(2) = \mathrm{SU}(2)$ are isomorphic. **Source:** [206]. **Remark:** $\mathrm{U}(n) \simeq \mathrm{O}(2n) \cap \mathrm{Symp}_{\mathbb{R}}(2n)$. **Related:** Fact 4.32.3.

Fact 4.32.5. Let n be a positive integer. Then, $\mathrm{SO}(2n) \cap \mathrm{Symp}_{\mathbb{R}}(2n)$ is a multiplication group whose Lie algebra is $\mathrm{so}(2n) \cap \mathrm{symp}_{\mathbb{R}}(2n)$. Now, let $A \in \mathbb{R}^{2n\times 2n}$. Then, the following statements are equivalent:

i) $A \in \mathrm{SO}(2n) \cap \mathrm{Symp}_{\mathbb{R}}(2n)$.

ii) $A \in \mathrm{Symp}_{\mathbb{R}}(2n)$ and $AJ_{2n} = J_{2n}A$.

Furthermore, the following statements are equivalent:

iii) $A \in \mathrm{so}(2n) \cap \mathrm{symp}_{\mathbb{R}}(2n)$.

iv) $A \in \mathrm{symp}_{\mathbb{R}}(2n)$ and $AJ_{2n} = J_{2n}A$.

Source: [445].

Fact 4.32.6. Define $Q_0, Q_1, Q_2, Q_3 \in \mathbb{C}^{2\times 2}$ by

$$Q_0 \triangleq I_2,\quad Q_1 \triangleq \begin{bmatrix} 0 & -1 \\ 1 & 0 \end{bmatrix},\quad Q_2 \triangleq \begin{bmatrix} -\jmath & 0 \\ 0 & \jmath \end{bmatrix},\quad Q_3 \triangleq \begin{bmatrix} 0 & -\jmath \\ -\jmath & 0 \end{bmatrix}.$$

Then, the following statements hold:

i) $Q_0^* = Q_0$ and $Q_i^* = -Q_i$ for all $i \in \{1,2,3\}$.

ii) $Q_0^2 = Q_0$ and $Q_i^2 = -Q_0$ for all $i \in \{1,2,3\}$.

iii) $Q_iQ_j = -Q_jQ_i$ for all $1 \le i < j \le 3$.

iv) $Q_1Q_2 = Q_3$, $Q_2Q_3 = Q_1$, and $Q_3Q_1 = Q_2$.

v) $\{\pm Q_0, \pm Q_1, \pm Q_2, \pm Q_3\}$ is a group.

For $\beta \triangleq [\beta_0\ \beta_1\ \beta_2\ \beta_3]^{\mathsf{T}} \in \mathbb{R}^4$ define

$$Q(\beta) \triangleq \sum_{i=0}^{3} \beta_i Q_i = \begin{bmatrix} \beta_0 + \beta_1 \jmath & -(\beta_2 + \beta_3 \jmath) \\ \beta_2 - \beta_3 \jmath & \beta_0 - \beta_1 \jmath \end{bmatrix}.$$

Then,

$$Q(\beta)Q^*(\beta) = \beta^{\mathsf{T}}\beta I_2, \qquad \det Q(\beta) = \beta^{\mathsf{T}}\beta.$$

Hence, if $\beta^{\mathsf{T}}\beta = 1$, then $Q(\beta)$ is unitary. Furthermore, the complex matrices Q_0, Q_1, Q_2, Q_3, and $Q(\beta)$ have the real representations

$$\mathcal{Q}_0 = I_4, \qquad \mathcal{Q}_1 = \begin{bmatrix} -J_2 & 0 \\ 0 & -J_2 \end{bmatrix},$$

$$\mathcal{Q}_2 = \begin{bmatrix} 0 & 0 & -1 & 0 \\ 0 & 0 & 0 & 1 \\ 1 & 0 & 0 & 0 \\ 0 & -1 & 0 & 0 \end{bmatrix},\quad \mathcal{Q}_3 = \begin{bmatrix} 0 & 0 & 0 & -1 \\ 0 & 0 & -1 & 0 \\ 0 & 1 & 0 & 0 \\ 1 & 0 & 0 & 0 \end{bmatrix},\quad \mathcal{Q}(\beta) = \begin{bmatrix} \beta_0 & -\beta_1 & -\beta_2 & -\beta_3 \\ \beta_1 & \beta_0 & -\beta_3 & \beta_2 \\ \beta_2 & \beta_3 & \beta_0 & -\beta_1 \\ \beta_3 & -\beta_2 & \beta_1 & \beta_0 \end{bmatrix}.$$

Hence,

$$\mathcal{Q}(\beta)\mathcal{Q}^{\mathsf{T}}(\beta) = \beta^{\mathsf{T}}\beta I_4, \quad \det \mathcal{Q}(\beta) = (\beta^{\mathsf{T}}\beta)^2.$$

Remark: Q_0, Q_1, Q_2, Q_3 represent the quaternions $1, \hat{\imath}, \hat{\jmath}, \hat{k}$. See Fact 4.32.1. An alternative representation is given by the *Pauli spin matrices* $\sigma_0 = I_2, \sigma_1 = \jmath Q_3, \sigma_2 = \jmath Q_1, \sigma_3 = \jmath Q_2$. See [1315, pp. 143–144], [1555]. **Remark:** For applications of quaternions, see [57, 1277, 1315, 1709]. **Remark:** $\mathcal{Q}(\beta)$ has the form $\left[\begin{smallmatrix} A & B \\ -B & A \end{smallmatrix}\right]$, where A and $\hat{I}B$ are rotation-dilations. See Fact 3.24.1.

Fact 4.32.7. Let $A, B, C, D \in \mathbb{R}^{n\times m}$, define $\hat{\imath}, \hat{\jmath}, \hat{k}$ as in Fact 4.32.1, and let $Q \triangleq A + \hat{\imath}B + \hat{\jmath}C + \hat{k}D$. Then,

$$\mathrm{diag}(Q, Q) = U_n^* \begin{bmatrix} A + \hat{\imath}B & -C - \hat{\imath}D \\ C - \hat{\imath}D & A - \hat{\imath}B \end{bmatrix} U_m, \quad U_n \triangleq \frac{1}{\sqrt{2}} \begin{bmatrix} I_n & -\hat{\imath}I_n \\ -\hat{\jmath}I_n & \hat{k}I_n \end{bmatrix}.$$

Furthermore, $U_nU_n^* = I_{2n}$. **Source:** [2670, 2671]. **Remark:** In the case $n = m$, this equality uses a similarity transformation to construct a complex representation of quaternions. **Remark:** The complex conjugate U_n^* is constructed as in Fact 4.32.7.

Fact 4.32.8. Let $A, B, C, D \in \mathbb{R}^{n\times n}$, define $\hat{\imath}, \hat{\jmath}, \hat{k}$ as in Fact 4.32.1, and let $Q \triangleq A + \hat{\imath}B + \hat{\jmath}C + \hat{k}D$. Then,

$$\operatorname{diag}(Q, Q, Q, Q) = U_n \begin{bmatrix} A & -B & -C & -D \\ B & A & -D & C \\ C & D & A & -B \\ D & -C & B & A \end{bmatrix} U_n, \quad U_n \triangleq \frac{1}{2} \begin{bmatrix} I_n & \hat{\imath}I_n & \hat{\jmath}I_n & \hat{k}I_n \\ -\hat{\imath}I_n & I_n & \hat{k}I_n & -\hat{\jmath}I_n \\ -\hat{\jmath}I_n & -\hat{k}I_n & I_n & \hat{\imath}I_n \\ -\hat{k}I_n & \hat{\jmath}I_n & -\hat{\imath}I_n & I_n \end{bmatrix}.$$

Furthermore, $U_n^* = U_n$ and $U_n^2 = I_{4n}$. **Source:** [182, 568, 1017, 1254, 2670, 2671, 2981]. **Remark:** In the case $n = m$, this equality uses a similarity transformation to construct a real representation of quaternions. See Fact 3.17.38. **Remark:** The complex conjugate U_n^* is constructed by replacing $\hat{\imath}, \hat{\jmath}, \hat{k}$ in U_n^{T} by $-\hat{\imath}, -\hat{\jmath}, -\hat{k}$.

Fact 4.32.9. Let $A \in \mathbb{C}^{2\times 2}$. Then, A is unitary if and only if there exist $\theta \in \mathbb{R}$ and $\beta \in \mathbb{R}^4$ such that $A = e^{\theta J}Q(\beta)$, where $Q(\beta)$ is defined in Fact 4.32.6. **Source:** [2314, p. 228].

4.33 Notes

In the literature on generalized inverses, range-Hermitian matrices are traditionally called *EP matrices*. The name "EP" originated in [2408, p. 130], where "P" stands for "principal," and "E" apparently stands for "equal." However, since AA^+ is the projector onto $\mathcal{R}(A)$ and A^+A is the projector onto $\mathcal{R}(A^*)$, it is widely believed (see [405]) that "EP" stands for "equal projectors."

Elementary reflectors are also called either *Householder matrices* or *Householder reflections*.

An alternative term for irreducible is *indecomposable*, see [1952, p. 147].

Diagonal, bidiagonal, and tridiagonal matrices are examples of *sparse matrices*, which have a high percentage of zero entries.

Left equivalence, right equivalence, and biequivalence are treated in [2314]. Each of the groups defined in Proposition 4.6.6 is a *Lie group*; see Definition 15.6.1. Elementary treatments of Lie algebras and Lie groups are given in [170, 172, 218, 803, 998, 1023, 1176, 1177, 1471, 2210, 2343, 2421], while an advanced treatment appears in [2781]. Some additional groups of matrices are given in [1916]. Applications of group theory are discussed in [1560].

Almost nonnegative matrices are called *ML-matrices* in [2418, p. 208], *essentially nonnegative matrices* in [423, 434, 1288], and *Metzler matrices* in [2451, p. 402].

The terminology "idempotent" and "projector" is not standardized in the literature. Some writers use either "projector," "oblique projector," or "projection" [1133] for idempotent, and either "orthogonal projector" or "orthoprojector" for projector.

Matrices with set-valued entries are discussed in [1170]. Matrices with entries having physical dimensions are discussed in [1327, 2187].

Graphs with specialized nodes are used to analyze the structure of molecules in [1518, 2730]. Connections between groups are illustrated in [646] as directed graphs with specialized edges in the form of *Cayley diagrams*.

Chapter Five
Geometry

5.1 Facts on Angles, Lines, and Planes

Fact 5.1.1. Let $\mathcal{X} \triangleq \{x_1, \ldots, x_n\}$ be a set of points in $\mathbb{R}^2$, and assume that no three points in $\mathcal{X}$ lie in a single line. Furthermore, let $\mathcal{L} \triangleq \{L_1, \ldots, L_{n(n-1)/2}\}$ be the set of lines passing through all pairs of points in $\mathcal{X}$, and assume that no pair of lines in $\mathcal{L}$ is parallel and no three of the lines in $\mathcal{L}$ intersect at a point that is not in $\mathcal{X}$. Let $\mathcal{P}$ denote the set of polygons whose boundaries are subsets of $\cup_{i=1}^{n(n-1)/2} L_i$ and whose interiors are disjoint from $\cup_{i=1}^{n(n-1)/2} L_i$. Then, the lines in $\mathcal{L}$ intersect at exactly $\frac{1}{8}n(n-1)(n-2)(n-3)$ points that are not in $\mathcal{X}$, and the number of polygons in $\mathcal{P}$ is $\frac{1}{8}(n-1)(n^3-5n^2+18n-8)$, of which $n(n-1)$ are not bounded. **Source:** [771, p. 72].

Fact 5.1.2. Let $\mathcal{L} \triangleq \{L_1, \ldots, L_n\}$ be a set of lines in $\mathbb{R}^2$, assume that no pair of lines in $\mathcal{L}$ is parallel, and assume that the intersection of each triple of lines in $\mathcal{L}$ is empty. Let $\mathcal{P}$ denote the set of polygons whose boundaries are subsets of $\cup_{i=1}^{n} L_i$ and whose interiors are disjoint from $\cup_{i=1}^{n} L_i$. Then, the number of polygons in $\mathcal{P}$ is $\frac{1}{2}(n^2+n+2)$, of which $\frac{1}{2}(n-1)(n-2)$ are bounded. **Source:** [771, p. 72].

Fact 5.1.3. Let $\mathcal{P} \triangleq \{P_1, \ldots, P_n\}$ be a set of planes in $\mathbb{R}^3$, assume that no pair of planes in $\mathcal{P}$ is parallel, and assume that the intersection of each triple of planes in $\mathcal{P}$ is a single point. Let $\mathcal{H}$ denote the set of polyhedra in $\mathbb{R}^3$ whose boundaries are subsets of $\cup_{i=1}^{n} P_i$ and whose interiors are disjoint from $\cup_{i=1}^{n} P_i$. Then, the number of polyhedra in $\mathcal{H}$ is $\sum_{i=0}^{3}\binom{n}{i} = \frac{1}{6}(n^3+5n+6)$, of which $\binom{n-1}{3} = \frac{1}{3}(n-1)(n-2)(n-3)$ are bounded. **Source:** [771, p. 72]. **Remark:** Extensions to hyperplanes in $\mathbb{R}^n$ are discussed in [771, p. 72].

Fact 5.1.4. The points $x, y, z \in \mathbb{R}^2$ lie on one line if and only if

$$\det\begin{bmatrix} x & y & z \\ 1 & 1 & 1 \end{bmatrix} = 0.$$

Fact 5.1.5. Let $x, y, z \in \mathbb{R}^n$, assume that $\|x\|_2 = \|y\|_2 = \|z\|_2 = 1$, and let $\phi_{x,y} \in (-\pi, \pi]$ denote the angle between x and y. Then,

$$\phi_{x,z} \le \phi_{x,y} + \phi_{y,z}, \qquad \sin\phi_{x,z} \le \sin\phi_{x,y} + \sin\phi_{y,z}.$$

Furthermore, equality holds in each inequality if and only if x, y, z lie in a single plane. **Source:** [2991, pp. 31, 198]. **Related:** Fact 3.15.1 and Fact 11.8.9.

Fact 5.1.6. The points $w, x, y, z \in \mathbb{R}^3$ lie in one plane if and only if

$$\det\begin{bmatrix} w & x & y & z \\ 1 & 1 & 1 & 1 \end{bmatrix} = 0.$$

Fact 5.1.7. Let $x_1, \ldots, x_n \in \mathbb{R}^n$. Then,

$$\operatorname{rank}\begin{bmatrix} 1 & \cdots & 1 \\ x_1 & \cdots & x_n \end{bmatrix} = \operatorname{rank}\begin{bmatrix} 1 & 0 & \cdots & 0 \\ x_1 & x_2 - x_1 & \cdots & x_n - x_1 \end{bmatrix}.$$

Hence,

$$\operatorname{rank}\begin{bmatrix} 1 & \cdots & 1 \\ x_1 & \cdots & x_n \end{bmatrix} = n$$

if and only if

$$\operatorname{rank}\,[x_2 - x_1 \ \ \cdots \ \ x_n - x_1] = n - 1.$$

If these conditions hold, then

$$\operatorname{affin}\{x_1, \ldots, x_n\} = x_1 + \operatorname{span}\{x_2 - x_1, \ldots, x_n - x_1\},$$

and thus $\operatorname{affin}\{x_1, \ldots, x_n\}$ is an affine hyperplane. Finally,

$$\operatorname{affin}\{x_1, \ldots, x_n\} = \{x \in \mathbb{R}^n\colon \det\begin{bmatrix} 1 & 1 & \cdots & 1 \\ x & x_1 & \cdots & x_n \end{bmatrix} = 0\}.$$

Source: [2418, p. 31]. **Related:** Fact 5.1.8.

Fact 5.1.8. Let $x_1, \ldots, x_{n+1} \in \mathbb{R}^n$. Then, the following statements are equivalent:

i) $\operatorname{conv}\{x_1, \ldots, x_{n+1}\}$ is a simplex.

ii) $\operatorname{conv}\{x_1, \ldots, x_{n+1}\}$ has nonempty interior.

iii) $\operatorname{affin}\{x_1, \ldots, x_{n+1}\} = \mathbb{R}^n$.

iv) $\operatorname{span}\{x_2 - x_1, \ldots, x_{n+1} - x_1\} = \mathbb{R}^n$.

v) $\begin{bmatrix} 1 & \cdots & 1 \\ x_1 & \cdots & x_{n+1} \end{bmatrix}$ is nonsingular.

Source: The equivalence of *i*) and *ii*) follows from Fact 12.11.20. The equivalence of *i*) and *iv*) follows from Fact 3.11.14. Finally, the equivalence of *iv*) and *v*) follows from

$$\begin{bmatrix} 1 & \cdots & 1 \\ x_1 & \cdots & x_{n+1} \end{bmatrix} = \begin{bmatrix} 1 & 0 & \cdots & 0 \\ x_1 & x_2 - x_1 & \cdots & x_{n+1} - x_1 \end{bmatrix} \begin{bmatrix} 1 & 1 & 1 & \cdots & 1 \\ 0 & 1 & 0 & \cdots & 0 \\ 0 & 0 & 1 & \cdots & 0 \\ \vdots & \vdots & \ddots & \ddots & \vdots \\ 0 & 0 & \cdots & \cdots & 1 \end{bmatrix}.$$

Related: Fact 5.1.7 and Fact 12.11.13.

Fact 5.1.9. Let z_1, z_2, z be complex numbers, and assume that $z_1 \neq z_2$. Then, the following statements are equivalent:

i) z lies on the line passing through z_1 and z_2.

ii) $\dfrac{z - z_1}{z_2 - z_1}$ is real.

iii) $\det\begin{bmatrix} z - z_1 & \overline{z} - \overline{z_1} \\ z_2 - z_1 & \overline{z_2} - \overline{z_1} \end{bmatrix} = 0.$

iv) $\det\begin{bmatrix} z & \overline{z} & 1 \\ z_1 & \overline{z_1} & 1 \\ z_2 & \overline{z_2} & 1 \end{bmatrix} = 0.$

Furthermore, the following statements are equivalent:

v) z lies on the line segment connecting z_1 and z_2 excluding the endpoints.

vi) $\dfrac{z - z_1}{z_2 - z}$ is a positive number.

Source: [110, pp. 54–56].

5.2 Facts on Triangles

Fact 5.2.1. Let z_1, z_2, z_3 be distinct complex numbers. Then, the following statements are equivalent:

i) z_1, z_2, z_3 are the vertices of an equilateral triangle.

ii) $|z_1 - z_2| = |z_2 - z_3| = |z_3 - z_1|$.

iii) $z_1^2 + z_2^2 + z_3^2 = z_1z_2 + z_2z_3 + z_3z_1$.

iv) $\dfrac{z_3 - z_1}{z_3 - z_2} = \dfrac{z_3 - z_2}{z_1 - z_2}$.

Source: [110, pp. 70, 71] and [1757, p. 316].

Fact 5.2.2. Let $n \geq 2$, let $x_1, \ldots, x_n$ be positive numbers, assume that $x_1 = \max\{x_1, \ldots, x_n\}$, and assume that $x_1 \leq \sum_{i=2}^n x_i$. Then, there exist complex numbers $z_1, \ldots, z_n$ such that $\sum_{i=1}^n z_i = 0$ and, for all $i \in \{1, \ldots, n\}$, $|z_i| = x_i$. **Source:** [2979, pp. 189, 190].

Fact 5.2.3. Let $\mathcal{S} \subset \mathbb{R}^2$ denote the triangle with vertices $\left[\begin{smallmatrix}0\\0\end{smallmatrix}\right], \left[\begin{smallmatrix}x_1\\y_1\end{smallmatrix}\right], \left[\begin{smallmatrix}x_2\\y_2\end{smallmatrix}\right] \in \mathbb{R}^2$. Then,

$$\operatorname{area}(\mathcal{S}) = \frac{1}{2}\left|\det\begin{bmatrix} x_1 & x_2 \\ y_1 & y_2 \end{bmatrix}\right|.$$

Fact 5.2.4. Let $\mathcal{S} \subset \mathbb{R}^2$ denote the triangle with vertices $\left[\begin{smallmatrix}x_1\\y_1\end{smallmatrix}\right], \left[\begin{smallmatrix}x_2\\y_2\end{smallmatrix}\right], \left[\begin{smallmatrix}x_3\\y_3\end{smallmatrix}\right] \in \mathbb{R}^2$. Then,

$$\operatorname{area}(\mathcal{S}) = \frac{1}{2}\left|\det\begin{bmatrix} 1 & 1 & 1 \\ x_1 & x_2 & x_3 \\ y_1 & y_2 & y_3 \end{bmatrix}\right|.$$

Source: [2418, p. 32].

Fact 5.2.5. Let z_1, z_2, z_3 be complex numbers. Then, the area of the triangle $\mathcal{S}$ formed by z_1, z_2, z_3 is given by

$$\operatorname{area}(\mathcal{S}) = \frac{1}{4}\left|\det\begin{bmatrix} z_1 & \overline{z_1} & 1 \\ z_2 & \overline{z_2} & 1 \\ z_3 & \overline{z_3} & 1 \end{bmatrix}\right|.$$

Source: [110, p. 79].

Fact 5.2.6. Let $\mathcal{S} \subset \mathbb{R}^3$ denote the triangle with vertices $x, y, z \in \mathbb{R}^3$. Then,

$$\operatorname{area}(\mathcal{S}) = \tfrac{1}{2}\sqrt{[(y - x) \times (z - x)]^\mathsf{T}[(y - x) \times (z - x)]} = \tfrac{1}{2}\|(y - x) \times (z - x)\|_2.$$

Furthermore,

$$\operatorname{area}(\mathcal{S}) = \tfrac{1}{2}\|y - x\|_2\|z - x\|_2 \sin\theta,$$

where $\theta \in (0, \pi)$ is the angle between $y - x$ and $z - x$. Now, assume that $x = 0$. Then,

$$\operatorname{area}(\mathcal{S}) = \tfrac{1}{2}\sqrt{[y \times z]^\mathsf{T}[y \times z]} = \tfrac{1}{2}\|y \times z\|_2.$$

Furthermore,

$$\operatorname{area}(\mathcal{S}) = \tfrac{1}{2}\|y\|_2\|z\|_2 \sin\theta,$$

where $\theta \in (0, \pi)$ is the angle between y and z. **Remark:** The connection between the norm of the cross product of two vectors and the angle between the vectors is given by *xxviii)* of Fact 4.12.1.

Fact 5.2.7. Let $\mathcal{S} \subset \mathbb{R}^2$ denote a triangle whose sides a, b, c have lengths a, b, c, let A, B, C denote the vertices opposite a, b, c with radian measure A, B, C, respectively, define the semiperimeter $s \triangleq \frac{1}{2}(a + b + c)$, let r denote the *inner radius* of $\mathcal{S}$, that is, the radius of the *incircle*, which is the largest inscribed circle, and let R denote the *outer radius* of $\mathcal{S}$, that is, the radius of the *circumcircle*, which is the smallest circumscribed circle. The triangle $\mathcal{S}$ is *acute* if all of its angles are less than

$\frac{\pi}{2}$; *right* if one of its angles is $\frac{\pi}{2}$; *obtuse* if one of its angles is greater than $\frac{\pi}{2}$; *equilateral* if all of its sides are equal (equivalently, all of its angles are equal); and *isosceles* if two of its sides are equal (equivalently, two of its angles are equal). Then, the following statements hold:

i) $A + B + C = \pi.$

ii) $a^2 + b^2 = c^2 + 2ab\cos C.$

iii) $a^2 = (b+c)^2\sin^2\frac{A}{2} + (b-c)^2\cos^2\frac{A}{2}.$

iv) $a = b\cos C + c\cos B, \quad ab\cos C + bc\cos A + ca\cos B = \frac{1}{2}(a^2+b^2+c^2).$

v) $\dfrac{\sin A}{a} = \dfrac{\sin B}{b} = \dfrac{\sin C}{c} = \dfrac{1}{2R}.$

vi) $\dfrac{a+b}{a-b} = \dfrac{\tan\frac{A+B}{2}}{\tan\frac{A-B}{2}} = \dfrac{\cot\frac{C}{2}}{\tan\frac{A-B}{2}}.$

vii) $\dfrac{a-b}{c} = \dfrac{\sin\frac{A-B}{2}}{\cos\frac{C}{2}} = \dfrac{\sin A - \sin B}{\sin C} = \dfrac{2\cos\frac{A+B}{2}\sin\frac{A-B}{2}}{2\sin\frac{C}{2}\cos\frac{C}{2}} = \dfrac{\tan\frac{A}{2} - \tan\frac{B}{2}}{\tan\frac{A}{2} + \tan\frac{B}{2}}.$

viii) $\dfrac{a+b}{c} = \dfrac{\cos\frac{A-B}{2}}{\sin\frac{C}{2}}, \quad \dfrac{a^2-b^2}{c} = a\cos B - b\cos A, \quad \dfrac{a^2-b^2}{c^2} = \dfrac{\sin(A-B)}{\sin(A+B)}.$

ix) $\cot A + \cot B = \frac{c}{b}\csc A, \quad a\cos A + b\cos B = b\cos(A-B).$

x) $a\cos A + b\cos B + c\cos C = 2a(\sin B)\sin C.$

xi) $\sin\frac{A}{2} = \sqrt{\dfrac{(s-b)(s-c)}{bc}}, \quad \cos\frac{A}{2} = \sqrt{\dfrac{s(s-a)}{bc}}.$

xii) $\tan\frac{A}{2} = \sqrt{\dfrac{(s-b)(s-c)}{s(s-a)}} = \dfrac{r}{s-a}.$

xiii) $\sin A = \dfrac{2}{bc}\sqrt{s(s-a)(s-b)(s-c)}, \quad \cos A = \dfrac{b^2+c^2-a^2}{2bc}.$

xiv) $\tan A = \dfrac{4\sqrt{s(s-a)(s-b)(s-c)}}{b^2+c^2-a^2}.$

xv) $\operatorname{area}(\mathbb{S}) = \frac{1}{2}ab\sin C = \dfrac{c^2(\sin A)\sin B}{2\sin C} = \frac{1}{2}(a^2-b^2)\dfrac{(\sin A)\sin B}{\sin(A-B)} = \dfrac{a^2-(b-c)^2}{4\tan\frac{A}{2}}.$

xvi) $\operatorname{area}(\mathbb{S}) = rs = \dfrac{abc}{4R} = \frac{1}{4}(a+b+c)^2(\tan\frac{A}{2})(\tan\frac{B}{2})\tan\frac{C}{2} = \frac{1}{2}\sqrt[3]{(abc)^2(\sin A)(\sin B)\sin C}.$

xvii) $\operatorname{area}(\mathbb{S}) = \sqrt{s(s-a)(s-b)(s-c)} = \dfrac{(s-a)(s-b)(s-c)}{r}.$

xviii) $\operatorname{area}(\mathbb{S}) = \frac{1}{4}\sqrt{(a+b+c)(a+b-c)(b+c-a)(c+a-b)}.$

xix) $\operatorname{area}(\mathbb{S}) = \frac{1}{4}\sqrt{(a^2+b^2+c^2)^2 - 2(a^4+b^4+c^4)}.$

xx) $\operatorname{area}(\mathbb{S}) = \frac{1}{4}\sqrt{2(a^2b^2+b^2c^2+c^2a^2) - (a^4+b^4+c^4)}.$

xxi) $\operatorname{area}(\mathbb{S}) = \dfrac{ab+bc+ca}{2(\csc A + \csc B + \csc C)} = \dfrac{2abc}{a+b+c}(\cos\frac{A}{2})(\cos\frac{B}{2})\cos\frac{C}{2}.$

xxii) Let $S \triangleq \frac{1}{2}(\sin A + \sin B + \sin C)$. Then,

$$\operatorname{area}(\mathbb{S}) = 4R^2\sqrt{S(S-\sin A)(S-\sin B)(S-\sin C)}.$$

xxiii) $\sin A$, $\sin B$, and $\sin C$ are the lengths of the sides of a triangle whose area is $\operatorname{area}(\mathbb{S})/(4R^2)$.

xxiv) If $A \neq \frac{\pi}{2}$, then $\operatorname{area}(\mathbb{S}) = \frac{1}{4}(\tan A)(b^2+c^2-a^2).$

xxv) $r = \dfrac{c(\sin\frac{A}{2})\sin\frac{B}{2}}{\cos\frac{C}{2}} = \dfrac{ab\sin C}{2s} = (s-a)\tan\frac{A}{2} = s(\tan\frac{A}{2})(\tan\frac{B}{2})\tan\frac{C}{2}.$

xxvi) $r = \dfrac{\mathrm{area}(\mathcal{S})}{s} = \sqrt{\dfrac{(s-a)(s-b)(s-c)}{s}} = \dfrac{(a+b-c)(b+c-a)(c+a-b)R}{2abc}.$

xxvii) $R = \dfrac{a}{2\sin A} = \dfrac{a}{4(\sin\frac{A}{2})\cos\frac{A}{2}} = \dfrac{abc}{4rs} = \dfrac{abc}{4\,\mathrm{area}(\mathcal{S})} = \dfrac{abc}{4\sqrt{s(s-a)(s-b)(s-c)}}.$

xxviii) $\dfrac{R}{r} = \dfrac{2abc}{(a+b-c)(b+c-a)(c+a-b)} = \dfrac{1}{\cos A + \cos B + \cos C - 1}.$

xxix) $a = r\cot\frac{B}{2} + r\cot\frac{C}{2} = r\cot\frac{B}{2} + r\tan\frac{A+B}{2}.$

xxx) a, b, c are the roots of the cubic equation

$$x^3 - 2sx^2 + (s^2 + r^2 + 4Rr)x - 4Rrs = 0.$$

That is,

$$a + b + c = 2s, \quad ab + bc + ca = s^2 + r^2 + 4rR, \quad abc = 4Rrs.$$

xxxi) a, b, c satisfy

$$ab + bc + ca = 4Rr + r^2 + s^2,$$

$$a^2 + b^2 + c^2 = 2(s^2 - r^2 - 4Rr),$$

$$a^3 + b^3 + c^3 = 2s(s^2 - 3r^2 - 6Rr),$$

$$(a+b)(b+c)(c+a) = 2s(s^2 + r^2 + 2Rr),$$

$$a^2b^2c + b^2c^2a + c^2a^2b = 4Rrs(s^2 + 4Rr + r^2),$$

$$a^2b^2 + b^2c^2 + c^2a^2 = (s^2 + 4Rr + r^2)^2 - 16Rrs^2,$$

$$a^4 + b^4 + c^4 = 2s^4 - 4(4Rr + 3r^2)s^2 + 2(4Rr + r^2)^2,$$

$$a(b-c)^2 + b(c-a)^2 + c(a-b)^2 = 2s(s^2 - 14Rr + r^2),$$

$$a^2b + b^2c + c^2a + ab^2 + bc^2 + ca^2 = 2s(s^2 - 2Rr + r^2),$$

$$a^4b + b^4c + c^4a + ab^4 + bc^4 + ca^4 = 2s(s^4 - 3r^4 - 14Rr^3 - 8R^2r^2 - 2r^2s^2 - 6Rrs^2),$$

$$a^3b^2 + b^3c^2 + c^3a^2 + a^2b^3 + b^2c^3 + c^2a^3 = 2s(s^4 + r^4 + 6Rr^3 + 8R^2r^2 + 2r^2s^2 - 10Rrs^2),$$

$$\frac{1}{a} + \frac{1}{b} + \frac{1}{c} = \frac{s^2 + r^2 + 4Rr}{4Rrs}, \quad \frac{1}{ab} + \frac{1}{bc} + \frac{1}{ca} = \frac{1}{2Rr},$$

$$\frac{(a+b)^2}{ab} + \frac{(b+c)^2}{bc} + \frac{(c+a)^2}{ca} = \frac{s^2 + 10Rr + r^2}{2Rr},$$

$$\frac{1}{a^2} + \frac{1}{b^2} + \frac{1}{c^2} = \left(\frac{s^2 + 4Rr + r^2}{4Rrs}\right)^2 - \frac{1}{Rr}, \quad \frac{a+b}{c} + \frac{b+c}{a} + \frac{c+a}{b} = \frac{s^2 + r^2 - 2Rr}{2Rr},$$

$$\frac{1}{s-a} + \frac{1}{s-b} + \frac{1}{s-c} = \frac{4R+r}{rs}, \quad \frac{a}{s-a} + \frac{b}{s-b} + \frac{c}{s-c} = \frac{4R-2r}{r},$$

$$\frac{a^2}{s-a} + \frac{b^2}{s-b} + \frac{c^2}{s-c} = \frac{4s(R-r)}{r},$$

$$\frac{a}{(s-b)(s-c)} + \frac{b}{(s-c)(s-a)} + \frac{c}{(s-a)(s-b)} = \frac{2(4R+r)}{rs},$$

$$\frac{a^2}{(s-b)(s-c)} + \frac{b^2}{(s-c)(s-a)} + \frac{c^2}{(s-a)(s-b)} = \frac{4(R+r)}{r}.$$

xxxii) $\mathcal{S}$ is equilateral if and only if $a^2 + b^2 + c^2 = ab + bc + ca$. If these conditions hold, then $s = \frac{3}{2}a$, $r = \frac{\sqrt{3}}{6}a$, $R = \frac{\sqrt{3}}{3}a$, and $\text{area}(\mathcal{S}) = \frac{\sqrt{3}}{4}a^2 = 3\sqrt{3}r^2 = \frac{3\sqrt{3}}{4}R^2$.

Source: *ii)* is the law of cosines; *v)* is the law of sines; *vi)* is the law of tangents; *vii)* is Mollweide's formula; *viii)* is Newton's formula; *xvii)* is *Heron's formula*, see [3024, pp. 318–320, 512–515]; *iii)* is given in [757]; *ix)* is given in [757, 759]; the last expression in *xv)* is given in [2527, p. 102]; the last expression in *xvi)* is given in [1559]; *xxii)* is given in [47, p. 124]; *xxii)* follows from *v)*, see [2056]; *xxviii)* is given in [1938, p. 74]; *xxix)* is given in [215]; *xxxi)* is given in [1026, 1533] and [2062, pp. 52–54]. **Remark:** The role of triangle geometry in mathematics is discussed in [840]. Geometric constructions are described in [534, 1443].

Fact 5.2.8. Let $\mathcal{S} \subset \mathbb{R}^2$ denote a triangle with the notation defined in Fact 5.2.7. Then, the following statements hold:

i) $2r \le R$, $\frac{5}{2} \le \frac{R}{r} + \frac{r}{R}$, $2 \le \frac{3R}{4r} + \frac{r}{R}$, $\sqrt{R(R-2r)} \le R - r$, $(4R+r)^2(2R-r) < 32R^3$.

ii)

$$2 \le \frac{2\sqrt[4]{3}}{3}\sqrt{\frac{s}{r}} \le \left\{\begin{matrix} \frac{2\sqrt{3}}{9}\frac{s}{r} \\ \frac{2}{r^2 s}\left(\frac{4Rr + r^2}{3}\right)^{3/2} \\ 4\left(2 + \frac{R}{r}\right)\sqrt[3]{\frac{abc}{(a+b+2c)(b+c+2a)(c+a+2b)}} - 2 \end{matrix}\right\} \le \frac{R}{r} \le \frac{2}{27}\left(\frac{s}{r}\right)^2.$$

iii) $1 < \frac{2R}{s} + \frac{r}{2R}$, $\quad s \le \frac{\sqrt{16R^4 - (13/2)r^3R}}{2R - r} < \frac{4R^2}{2R - r}$,

$$s \le 2R + (3\sqrt{3} - 4)r \le \frac{\sqrt{3}}{4}\left(\frac{11}{2}R + r\right) \le \sqrt{3}(2R - r),$$

$$\frac{2rs}{R} \le 3\sqrt{3}r \le \frac{3\sqrt{6Rr}}{2} \le s = \frac{\text{area}(\mathcal{S})}{r} = \frac{1}{2}(a+b+c) = \frac{abc}{4Rr} \le \sqrt{\frac{(4R+r)^2R}{2(2R-r)}}$$
$$\le \sqrt{4R^2 + 4Rr + 3r^2} \le 2R + (3\sqrt{3} - 4)r \le \frac{\sqrt{3}}{3}(4R + r) \le \frac{3\sqrt{3}}{2}R \le \sqrt{7R^2 - r^2} \le \frac{3\sqrt{3}R^2}{4r}.$$

iv)
$$\left(\frac{2rs}{R}\right)^2 \le 27r^2 \le \frac{(4R+r)^2r}{2R-r} \le 3r(4R+r) \le \frac{27}{2}Rr \le 14Rr - r^2 \le \frac{(4R+r)^2r}{R+r}$$
$$\le 16Rr - 5r^2 \le \frac{4r(12R^2 - 11Rr + r^2)}{3R - 2r} \le \tfrac{1}{2}r[20R - r + \sqrt{3(12R+r)(4R-5r)}]$$
$$\le 2R^2 + 10Rr - r^2 - 2(R-2r)\sqrt{R(R-2r)} \le s^2 \le 2R^2 + 10Rr - r^2 + 2(R-2r)\sqrt{R(R-2r)}$$
$$\le \frac{(4R+r)^2R}{2(2R-r)} \le 4R^2 + 4Rr + 3r^2 \le [2R + (3\sqrt{3}-4)r]^2 \le 4R^2 + 5Rr + r^2$$
$$\le 4R^2 + 6Rr - r^2 \le \frac{9}{2}R^2 + 4Rr + r^2 \le \frac{9}{2}(R^2 + Rr) \le \frac{14}{3}(R^2 + Rr) - r^2$$
$$\le \frac{1}{3}(4R+r)^2 \le \frac{27}{4}R^2 \le 7R^2 - r^2 \le (4R+r)(2R-r) \le \frac{27R^4}{16r^2}.$$

v)
$$8Rr + \tfrac{11\sqrt{3}}{9}\text{area}(\mathcal{S}) \le s^2 \le 4R^2 + \tfrac{11\sqrt{3}}{9}\text{area}(\mathcal{S}),$$
$$\tfrac{1}{2}[(a-b)^2+(b-c)^2+(c-a)^2] + 3\sqrt{3}\text{area}(\mathcal{S}) \le s^2 \le 2[(a-b)^2+(b-c)^2+(c-a)^2] + 3\sqrt{3}\text{area}(\mathcal{S}).$$

vi)
$$\frac{3r(4R+r)}{(7R-5r)^2} \le \frac{r}{2R-r} \le \frac{3r}{4R+r} \le \frac{r(4R+r)}{(2R-r)(2R+5r)} \le \frac{r(16R+3r)}{(4R-r)(4R+7r)}$$
$$\le \frac{r}{R+r} \le \frac{r(16R-5r)}{(4R+r)^2} \le \frac{4r(12R^2-11Rr+r^2)}{(3R-2r)(4R+r)^2} \le \left(\frac{s}{4R+r}\right)^2$$
$$\le \frac{R}{4R-2r} \le \frac{4R^2+4Rr+3r^2}{(4R+r)^2} \le \frac{1}{3} \le \frac{4R+r}{27r} \le \frac{R^2}{4r(R+r)}.$$

vii)
$$\frac{27}{2}Rr + r^2 \le 14Rr \le 16Rr - 4r^2 \le r^2 + s^2 \le 4(R^2+Rr+r^2) \le 4R^2 + 6Rr$$
$$\le \frac{14}{3}(R^2+Rr) \le \frac{27}{4}R^2 + r^2 \le 7R^2 \le \frac{27R^4}{16r^2} + r^2.$$

viii) $(1+3\sqrt{3})r \le r+s \le \left(\frac{1+3\sqrt{3}}{2}\right)R \approx 3.098R.$

ix) Define $\delta \triangleq 1 - \sqrt{1 - \frac{2r}{R}}$. Then, $\delta(4-\delta)^3 \le \frac{4s^2}{R^2} \le (2-\delta)(2+\delta)^3.$

x) Let $p \ge 0$. Then, $\left(\frac{r}{R}\right)^p + \left(\frac{s}{R}\right)^p \le \frac{1+3^{3p/2}}{2^p}.$

xi) $2 \le \frac{a}{b} + \frac{b}{a} \le \frac{R}{r}.$

xii) $2 \le \frac{2}{3}\left(\frac{a}{b}+\frac{b}{c}+\frac{c}{a}\right) \le \frac{a}{b}+\frac{b}{c}+\frac{c}{a} - 1 \le \frac{1}{2}\left(1 + \frac{a^2}{bc} + \frac{b^2}{ca} + \frac{c^2}{ab}\right) \le \frac{R}{r}.$

xiii) $2 \le \frac{2abc}{(a+b-c)(b+c-a)(c+a-b)} = \frac{R}{r}.$

xiv) $\frac{a(4r-R)}{R} \le 2\sqrt{(s-b)(s-c)} \le a \le \frac{R}{r}\sqrt{(s-b)(s-c)}.$

xv) $r \le \frac{1}{2}\left(\frac{2}{1+\sqrt{5}}\right)^{5/2}(a+b) \approx 0.1501(a+b).$

xvi) If $a \le b \le c$, then $\frac{1}{a} - \frac{1}{c} < \frac{1}{2r}.$

xvii) $3\sqrt{3}r^2 \le \operatorname{area}(\mathcal{S}) = rs \le \frac{3\sqrt{3}}{2}Rr = \frac{3\sqrt{3}}{4}\frac{abc}{a+b+c} \le \frac{1}{2}Rs = \frac{abc}{8r} \le \frac{3\sqrt{3}}{4}R^2.$

xviii) $\operatorname{area}(\mathcal{S}) \le \frac{1}{2}\left(\frac{2}{1+\sqrt{5}}\right)^{5/2}(a+b)s.$

xix) $\operatorname{area}(\mathcal{S}) \le \frac{bc}{4}\sqrt{\frac{s}{s-a}} \le \frac{bcs}{2(b+c-a)} = \frac{bcs}{4(s-a)}.$

xx) $\operatorname{area}(\mathcal{S}) \le \frac{1}{2}(b+c)\sqrt{s(s-a)}.$

xxi) $\operatorname{area}(\mathcal{S}) \le \frac{\sqrt{3}}{6}(ab+bc+ca) - \frac{\sqrt{3}}{12}(a^2+b^2+c^2) \le \frac{\sqrt{3}}{12}(ab+bc+ca).$

xxii) $\operatorname{area}(\mathcal{S}) \le \begin{cases} \frac{1}{4}\min,\{a^2+b^2, b^2+c^2, c^2+a^2\} \le \frac{\sqrt{3}}{24}(a+b+c)^2 \\ \frac{1}{12}\min,\{a^2+b^2+4c^2, b^2+c^2+4a^2, c^2+a^2+4b^2\}. \end{cases}$

xxiii)
$$0 < \frac{3\sqrt{3}}{4}\left(\frac{abc}{\frac{a^2}{b+c-a} + \frac{b^2}{c+a-b} + \frac{c^2}{a+b-c}}\right) \le \frac{3(a+b-c)(b+c-a)(c+a-b)}{4\sqrt{a^2+b^2+c^2}}$$
$$\le \frac{\sqrt[4]{3}[(a+b-c)(b+c-a)(c+a-b)(a+b+c)]^{3/4}}{4\sqrt{a^2+b^2+c^2}}$$

$$\le \operatorname{area}(\mathcal{S}) = rs = \frac{abc}{4R} = \sqrt{s(s-a)(s-b)(s-c)} = \sqrt{a^2b^2+b^2c^2+c^2a^2-\tfrac{1}{2}(a^4+b^4+c^4)}$$

$$\le \left\{ \begin{array}{c} \dfrac{3abc}{4\sqrt{a^2+b^2+c^2}} \\ \frac{\sqrt{3}}{12}[2(ab+bc+ca)-(a^2+b^2+c^2)] = \frac{\sqrt{3}}{9}(s^2-\frac{1}{2}[(a-b)^2+(b-c)^2+(c-a)^2]) \end{array} \right\}$$

$$\le \frac{3\sqrt{3}abc}{8s} \le \left\{ \begin{array}{c} \frac{\sqrt{3}}{4}\sqrt[3]{a^2b^2c^2} \\ \dfrac{\sqrt{3}abcs}{2(ab+bc+ca)} \end{array} \right\} \le \tfrac{\sqrt{3}}{12}(a\sqrt{bc}+b\sqrt{ca}+c\sqrt{ab}) \le \tfrac{1}{4}\sqrt{2abcs}$$

$$\le \tfrac{\sqrt{3}}{12}(ab+bc+ca) \le \left\{ \begin{array}{c} \frac{\sqrt{3}}{9}s^2 \\ \frac{1}{4}\sqrt{a^2b^2+b^2c^2+c^2a^2} \end{array} \right\} \le \tfrac{\sqrt{3}}{12}(a^2+b^2+c^2) \le \left\{ \begin{array}{c} \frac{\sqrt{3}}{6}s^2 \\ \frac{1}{4}\sqrt{a^4+b^4+c^4} \end{array} \right\}.$$

xxiv)
$$0 < 3\sqrt{3}r^2 \le \sqrt{3}r\sqrt{4Rr+r^2} \le \frac{3r\sqrt{6Rr}}{2} \le r\sqrt{16Rr-5r^2} \le 2r\sqrt{\frac{r(12R^2-11Rr+r^2)}{3R-2r}}$$

$$\le r\sqrt{2R^2+10Rr-r^2-2(R-2r)\sqrt{R(R-2r)}} \le \operatorname{area}(\mathcal{S}) = rs = \frac{1}{2}(a+b+c)r = \frac{abc}{4R}$$

$$\le r\sqrt{2R^2+10Rr-r^2+2(R-2r)\sqrt{R(R-2r)}} \le r\sqrt{4R^2+4Rr+3r^2} \le 2Rr+(3\sqrt{3}-4)r^2$$

$$\le r\sqrt{4R^2+5Rr+r^2} \le r\sqrt{4R^2+\tfrac{11}{2}Rr}$$

$$\le \left\{ \begin{array}{c} r\sqrt{4R^2+6Rr-r^2} \le r\sqrt{\dfrac{14R^2+14Rr-3r^2}{3}} \le r\sqrt{5R^2+4Rr-r^2} \\ \le \frac{\sqrt{3}}{3}(4Rr+r^2) \le r\sqrt{6R^2+3r^2} \le \frac{3\sqrt{3}}{2}Rr \\ \frac{3\sqrt{3}}{25}(R+3r)^2 \end{array} \right\}$$

$$\le \tfrac{\sqrt{3}}{3}(R+r)^2 \le \tfrac{12\sqrt{3}}{25}(R+\tfrac{1}{2}r)^2 \le \left\{ \begin{array}{c} (R+\frac{1}{2}r)^2 \\ \frac{3\sqrt{3}}{4}R^2 \end{array} \right\} \le \tfrac{25}{16}R^2.$$

xxv) $(s-a)(s-b) \le ab$, $8(s-a)(s-b)(s-c) \le abc$, $(a+b-c)(b+c-a)(c+a-b) \le abc$.

xxvi) $(s-a)(s-b)+(s-b)(s-c)+(s-c)(s-a) \le \frac{1}{4}(ab+bc+ca)$.

xxvii) $\dfrac{9}{s} \le \dfrac{1}{s-a} + \dfrac{1}{s-b} + \dfrac{1}{s-c} = \dfrac{4R+r}{rs}$.

xxviii) $\sqrt{s} \le \sqrt{s-a} + \sqrt{s-b} + \sqrt{s-c} \le \sqrt{3s}$.

xxix) $\sqrt{a(s-a)} + \sqrt{b(s-b)} + \sqrt{c(s-c)} \le \sqrt{2}s$.

xxx) $a(s-a)+b(s-b)+c(s-c) \le 9Rr$.

xxxi) $abc < a^2(s-a)+b^2(s-b)+c^2(s-c) \le \frac{3}{2}abc$.

xxxii) $a^3(s-a)+b^3(s-b)+c^3(s-c) \le abcs$, $3\sqrt{3}r^3s^3 \le s^3(s-a)(s-b)(s-c) \le \frac{27}{64}(abc)^2 \le \frac{3\sqrt{3}R^3s^3}{8}$.

xxxiii) Let $n \ge 0$. Then, $a^n+b^n+c^n \le 2^{n+1}R^n + 2^n(3^{1+n/2}-2^{1+n})r^n$.

xxxiv)
$$\frac{abc}{R^2} \le 6\sqrt{3}r \le 3\sqrt{3}\sqrt{\frac{abc}{a+b+c}} \le 3\sqrt[3]{abc} \le a+b+c$$

$$< \left\{ \begin{array}{c} \sqrt{3(a^2+b^2+c^2)} \\ 4R+(6\sqrt{3}-8)r \le \frac{2\sqrt{3}}{3}(4R+r) \end{array} \right\} \le 3\sqrt{3}R \le \frac{abc}{4r^2}.$$

xxxv)

$$36r^2 \le 2\sqrt{3}r(a+b+c) \le 18Rr \le \left\{\begin{matrix} \dfrac{9(a^2\sqrt{bc}+b^2\sqrt{ca}+c^2\sqrt{ab})}{(\sqrt{a}+\sqrt{b}+\sqrt{c})^2} \le \dfrac{2s}{3}(a+b+c) \\ 24Rr-12r^2 \le 4R^2+12Rr-4r^2-4(R-2r)\sqrt{R^2-2Rr} \end{matrix}\right\}$$

$$\le a^2+b^2+c^2 \le \left\{\begin{matrix} 4R^2+12Rr-4r^2+4(R-2r)\sqrt{R^2-2Rr} \le \dfrac{72R^4}{9R^2-4r^2} \le 8R^2+4r^2 \\ (s-\sqrt{3}r)(a+b+c) \end{matrix}\right\}$$

$$\le 8R^2+\frac{4\sqrt{3}}{9}rs \le 9R^2, \quad a^2+b^2+c^2 \le \sqrt{64R^4+48r^2s^2}.$$

xxxvi)

$$4\sqrt{3}rs \le \left\{\begin{matrix} 3\sqrt[3]{(abc)^2} \le a\sqrt{bc}+b\sqrt{ca}+c\sqrt{ab} \le ab+bc+ca \\ 4\sqrt{3}rs+(a-b)^2+(b-c)^2+(c-a)^2 \end{matrix}\right\}$$

$$\le 4r\sqrt{4R^2+4Rr+3r^2}\sqrt{3+\frac{4(R-2r)}{4R+r}}+(a-b)^2+(b-c)^2+(c-a)^2$$

$$\le a^2+b^2+c^2 \le 4\sqrt{3}rs+3[(a-b)^2+(b-c)^2+(c-a)^2].$$

If, in addition, $\mathcal{S}$ is acute, then

$$4(R+r)^2 \le a^2+b^2+c^2 \le 4\sqrt{3}rs+\tfrac{2-\sqrt{3}}{3-2\sqrt{2}}[(a-b)^2+(b-c)^2+(c-a)^2]$$

$$\le 4\sqrt{3}rs+2[(a-b)^2+(b-c)^2+(c-a)^2].$$

xxxvii) $$\frac{16r^2s^2}{3R^2} \le 36r^2 \le \frac{16}{3}r^2s^2\left(\frac{1}{a^2}+\frac{1}{b^2}+\frac{1}{c^2}\right) \le a^2+b^2+c^2 \le 9R^2 \le 9R^4\left(\frac{1}{a^2}+\frac{1}{b^2}+\frac{1}{c^2}\right).$$

xxxviii) $$54Rr = \frac{27abc}{2s} \le (a+b+c)^2 \le \frac{9\sqrt{3}R}{2s}(a^2+b^2+c^2).$$

xxxix)

$$36r^2 \le 4\sqrt{3}rs \le 18Rr \le 20Rr-4r^2 \le 2R^2+14Rr-2(R-2r)\sqrt{R^2-2Rr}$$

$$\le ab+bc+ca \le 2R^2+14Rr+2(R-2r)\sqrt{R^2-2Rr} \le 4(R+r)^2 \le 9R^2.$$

xl)

$$8r(R-2r) \le 4(R-2r)(R+r-\sqrt{R(R-2r)}) \le (a-b)^2+(b-c)^2+(c-a)^2$$

$$\le 4(R-2r)(R+r+\sqrt{R(R-2r)}) \le 8R(R-2r).$$

xli)

$$192\sqrt{3}r^3 \le 64r^2s \le 96\sqrt{3}r^2R \le 12\sqrt{3}r^2(9R-2r) \le 4rs(9R-2r)$$

$$\le (a+b)(b+c)(c+a) \le 4s(2R^2+3Rr+2r^2)$$

$$\le \left\{\begin{matrix} 4(2R+(3\sqrt{3}-4)r)(2R^2+3Rr+2r^2) \\ 8R(R+2r)s \end{matrix}\right\}$$

$$\le 6\sqrt{3}R(2R^2+3Rr+2r^2) \le 24\sqrt{3}R^3.$$

xlii) $$24\sqrt{3}r^3 \le 12\sqrt{3}Rr^2 \le abc \le \left\{\begin{matrix} 4Rr[2R+(3\sqrt{3}-4)r] \le 6\sqrt{3}R^2r \\ \tfrac{8}{27}s^3 \end{matrix}\right\} \le 3\sqrt{3}R^3.$$

xliii)

$$72\sqrt{3}r^3 \le 24r^2s \le 36\sqrt{3}Rr^2 \le \left\{\begin{matrix} 4\sqrt{3}rs(s-\sqrt{3}r) \\ 12Rrs \\ 12\sqrt{3}r^2(5R-4r) \end{matrix}\right\} \le 8rs(2R-r) \le 4rs(5R-4r)$$

$$\le a^3+b^3+c^3 \le 4R(2R-r)s \le \left\{\begin{matrix} 4R[2R+(3\sqrt{3}-4)r](2R-r) \le 6\sqrt{3}R^2(2R-r) \\ 18R^2s-8r^2s-20Rrs. \end{matrix}\right.$$

xliv)

$$\frac{\sqrt{3}}{R} \le \frac{2\sqrt{3}(5R-r)}{9R^2} \le \begin{Bmatrix} \dfrac{5R-r}{R[2R+(3\sqrt{3}-4)r]} \le \dfrac{5R-r}{Rs} \\ \dfrac{3\sqrt{3}}{2(R+r)} \end{Bmatrix}$$

$$\le \frac{1}{a}+\frac{1}{b}+\frac{1}{c} \le \begin{Bmatrix} \dfrac{(R+r)^2}{Rrs} \\ \dfrac{\sqrt{3}}{2r} \end{Bmatrix} \le \frac{\sqrt{3}(R+r)^2}{9Rr^2} \le \frac{\sqrt{3}R}{4r^2},$$

$$\frac{R^2+7Rr-(R-2r)\sqrt{R(R-2r)}}{2Rrs} \le \frac{1}{a}+\frac{1}{b}+\frac{1}{c} \le \frac{R^2+7Rr+(R-2r)\sqrt{R(R-2r)}}{2Rrs}.$$

xlv) $\dfrac{2(5R-r)}{R} \le (a+b+c)\left(\dfrac{1}{a}+\dfrac{1}{b}+\dfrac{1}{c}\right) \le \dfrac{2(R+r)^2}{Rr} \le \dfrac{9R}{2r}.$

xlvi) $a^4+b^4+c^4 \le \frac{8}{3}R(R-r)(4R+r)^2 \le 54R^3(R-r).$

xlvii) $\dfrac{16r^2s^2}{9R^2}(a^2+b^2+c^2) \le 16r^2s^2 \le a^2b^2+b^2c^2+c^2a^2 \le 4R^2s^2 \le 3R^2(a^2+b^2+c^2).$

xlviii) $\dfrac{1}{R^2} \le \dfrac{1}{2Rr} \le \dfrac{1}{3}\left(\dfrac{1}{a}+\dfrac{1}{b}+\dfrac{1}{c}\right)^2 \le \dfrac{1}{a^2}+\dfrac{1}{b^2}+\dfrac{1}{c^2} \le \dfrac{(R^2+r^2)^2+Rr^3}{R^2r^3(16R-5r)} \le \dfrac{1}{4r^2}.$

xlix) $2 \le \dfrac{7R-2r}{3R} \le \dfrac{a+b}{3c}+\dfrac{b+c}{3a}+\dfrac{c+a}{3b} \le \dfrac{2R^2+Rr+2r^2}{3Rr} \le \dfrac{R}{r}.$

l) $\dfrac{(5R-2r)s}{R} \le \dfrac{ab}{s-c}+\dfrac{bc}{s-a}+\dfrac{ca}{s-b}.$

li) $\dfrac{R^2+3Rr+2r^2}{2R^2+3Rr+2r^2} \le \dfrac{s-a}{b+c}+\dfrac{s-b}{c+a}+\dfrac{s-c}{a+b} \le \dfrac{6R}{9R-2r}.$

lii) $\min\{(a-b)^2,(b-c)^2,(c-a)^2\} \le \begin{cases} 2r(R-2r) \le \dfrac{R^2}{4} \\ \dfrac{s^2-8Rr-2r^2}{5}. \end{cases}$

liii) $2 \le \dfrac{16a^2b^2c^2}{[2a^2-(b-c)^2][2b^2-(c-a)^2][2c^2-(a-b)^2]} \le \dfrac{R}{r}.$

liv) $18r^2 \le \dfrac{a^2bc}{(b+c)(b+c-a)}+\dfrac{b^2ca}{(c+a)(c+a-b)}+\dfrac{c^2ab}{(a+b)(a+b-c)}.$

lv) Let $\alpha \triangleq (\log 9-\log 4)/(\log 4-\log 3) \approx 2.8$, and let $p \in [-1,\alpha]$, where $p \ne 0$. Then,

$$2\sqrt{3}r \le [\tfrac{1}{3}(a^p+b^p+c^p)]^{1/p} \le \sqrt{3}R.$$

lvi) Let $p \in [0,1]$. Then, $\dfrac{1}{a^p}+\dfrac{1}{b^p}+\dfrac{1}{c^p} \le 3^{1-2p}\left(\dfrac{s}{Rr}\right)^p.$

lvii) Let $\alpha,\beta,\gamma \ge 0$, and assume that $\alpha+\beta+\gamma = 1$. Then, $2r \le \min\{a,b,c\} \le a^\alpha b^\beta c^\gamma \le \max\{a,b,c\} \le 2R$. Furthermore, $2r \le \sqrt[3]{4Rr^2} \le \sqrt{2Rr} \le \frac{\sqrt{3}}{3}\sqrt[3]{abc} \le \sqrt[3]{2R^2r} \le R.$

lviii) $a < b+c \le a+\sqrt{2b^2+2c^2-a^2}.$

lix) $a < b+c \le a+\dfrac{2bc}{a}.$

lx) Let $a \ge b \ge c$. Then, $\dfrac{b^2+2ac}{a+b+c} \le \frac{2}{3}[2R+(3\sqrt{3}-4)r] \le \sqrt{3}R.$

lxi) Let $a \ge b \ge c$. Then, $b^2+2ac \le \frac{4}{3}(4R^2+4Rr+3r^2).$

lxii) $5184r^6 \le a^6 + b^6 + c^6, \quad 576r^6 \le \dfrac{(abc)^3}{a^3+b^3+c^3}.$

lxiii) $a^3 + b^3 + c^3 + 3abc \le \frac{2}{3}(a+b+c)(a^2+b^2+c^2).$

lxiv) $abc(a^2+b^2+c^2-6R^2) \le (a^3+b^3+c^3)R^2.$

lxv) $(a^2+b^2-c^2)(b^2+c^2-a^2)(c^2+a^2-b^2) \le \left(\dfrac{4\sqrt{3}rs}{3}\right)^3.$

lxvi) $a^9+b^9+c^9 \le 3\left(\dfrac{sR}{3r}\right)^9.$

lxvii) $8\left(\dfrac{r}{s}\right)^2 < \log\left(\dfrac{b+c}{a}\right)\log\left(\dfrac{c+a}{b}\right)\log\left(\dfrac{a+b}{c}\right) < \dfrac{2r}{R} \le 1.$

lxviii) $4 - 4\left[\left(\dfrac{a}{b+c}\right)^2 + \left(\dfrac{a}{b+c}\right)^2 + \left(\dfrac{a}{b+c}\right)^2\right] \le \dfrac{2r}{R} \le 1.$

lxix) Let $p > 0$. Then, $3\left(\frac{4\sqrt{3}rs}{3}\right)^{p/2} \le a^p + b^p + c^p.$

lxx) Let $p \in \mathbb{R}$. Then, $a^p(s-a) + b^p(s-b) + c^p(s-c) \le \frac{1}{2}abc(a^{p-2}+b^{p-2}+c^{p-2}).$

lxxi) $\dfrac{1}{2r}\sqrt{4 - \dfrac{9r}{4R+r}} \le \dfrac{1}{a+b-c} + \dfrac{1}{b+c-a} + \dfrac{1}{c+a-b}.$

lxxii) $3\sqrt{3}R \le 3R\sqrt{4 - \dfrac{9r}{4R+r}} \le \dfrac{c^2}{a+b-c} + \dfrac{a^2}{b+c-a} + \dfrac{b^2}{c+a-b}.$

lxxiii) $\dfrac{1}{8Rr}\left(5 - \dfrac{9r}{4R+r}\right) \le \dfrac{1}{c(a+b-c)} + \dfrac{1}{a(b+c-a)} + \dfrac{1}{b(c+a-b)}.$

lxxiv) $\dfrac{1}{r^2}\left(\dfrac{1}{2} - \dfrac{9r}{4(4R+r)}\right) \le \dfrac{1}{(a+b-c)^2} + \dfrac{1}{(b+c-a)^2} + \dfrac{1}{(c+a-b)^2}.$

lxxv) Let $x, y, z \in \mathbb{R}$, and assume that $x+y$, $y+z$, and $z+x$ are positive. Then,

$$2\sqrt{3}rs \le \frac{x}{y+z}a^2 + \frac{y}{z+x}b^2 + \frac{z}{x+y}c^2.$$

lxxvi) $4\sqrt{3}rs + \frac{1}{2}(|a-b|+|b-c|+|c-a|)^2 \le a^2+b^2+c^2 \le 4\sqrt{3}rs + \frac{3}{2}(|a-b|+|b-c|+|c-a|)^2.$

lxxvii) $\dfrac{3}{2} \le \dfrac{5s^2+r^2+4Rr}{3(s^2+r^2+2Rr)} \le \dfrac{2(s^2-r^2-Rr)}{s^2+r^2+2Rr} < 2, \quad \dfrac{3}{2} \le \dfrac{s^2+r^2+4Rr}{12Rr} \le \dfrac{s^2+r^2-8Rr}{4Rr}.$

lxxviii) $18 \le \dfrac{a+2b+3c}{-a+b+c} + \dfrac{b+2c+3a}{-b+c+a} + \dfrac{c+2a+3b}{-c+a+b}.$

lxxix) $\dfrac{1}{3} \le \dfrac{a^2}{4a^2+5bc} + \dfrac{b^2}{4b^2+5ca} + \dfrac{c^2}{4c^2+5bc}.$

lxxx) $\frac{a^4+b^4}{a^2+ab+b^2} + \frac{b^4+c^4}{b^2+bc+c^2} + \frac{c^4+a^4}{c^2+ca+a^2} < 2(a^2+b^2+c^2) < 4(ab+bc+ca) < \frac{14}{3}(ab+bc+ca).$

lxxxi) Let $c = \max\{a, b, c\}$. Then, $\frac{3ab(a+b+2c)}{(a+c)(b+c)} \le a+b+c.$

lxxxii) $\dfrac{sr}{R}\left(\dfrac{5R-r}{4R+r}\right) \le \dfrac{(s-a)(s-b)}{c} + \dfrac{(s-b)(s-c)}{a} + \dfrac{(s-c)(s-a)}{b}.$

lxxxiii) $\dfrac{r}{R} \le \dfrac{8Rr}{s^2+r^2+2Rr} \le \dfrac{1}{2}.$

lxxxiv)

$$(R+r)(s^2+r^2+4Rr) \le 3R(s^2-r^2-4Rr), \quad 2r(R+r)(4R+r) \le 3R(s^2-2r^2-8Rr),$$

$$16s^2r(R+r) \le 3[(s^2+r^2+4Rr)^2 - 16s^2Rr], \quad 2(R+r)s^2 \le 3R[(4R+r)^2 - 2s^2],$$

$$(R+r)(s^2+r^2-8Rr) \le 3R(8R^2+r^2-s^2), \quad (R+r)[s^2+(4R+r)^2] \le 3R[(4R+r)^2 - s^2].$$

lxxxv) $\frac{b}{c}+\frac{c}{b}+1\le\frac{3R}{2r},\quad \frac{a+b}{c}+\frac{b+c}{a}+\frac{c+a}{b}\le\frac{3R}{r},\quad 432r^4\le a^2b^2+b^2c^2+c^2a^2\le 27R^4.$

lxxxvi) $\left(\frac{a}{b+c}+\frac{b}{c+a}+\frac{c}{a+b}\right)^2\le\frac{9R}{8r},\quad \sqrt{\frac{a}{s-a}}+\sqrt{\frac{b}{s-b}}+\sqrt{\frac{c}{s-c}}\le\frac{3\sqrt{2}R}{2r}.$

lxxxvii) $\frac{1}{R^2}\le\frac{1}{ab}+\frac{1}{bc}+\frac{1}{ca}\le\frac{1}{4r^2},\quad \frac{1}{3R^4}\le\frac{1}{a^2b^2}+\frac{1}{b^2c^2}+\frac{1}{c^2a^2}\le\frac{1}{48r^4}.$

lxxxviii) $8\le\frac{(a+b)(b+c)(c+a)}{abc}\le\frac{4R}{r}.$

Source: *i)* is given in [47, p. 112], [1730], and [2062, p. 165]; *ii)* is given in [47, p. 60], [367, 889], [1964], and [1938, p. 68]; *iii)* is given in [47, p. 60] and [1730]; *iv)–viii)* are given in [917] and [2062, pp. 166, 177, 189, 190]; *ix)* is given in [2927]; *x)* is given in [2062, p. 165]; *xi)* and *xii)* are given in [2791]; *xiii)* follows from Fact 5.2.7 and *ix)*; *xiv)* is given in [2244] and [2341]; *xv)* and *xvii)* are given in [215]; *xvi)* is given in [316]; *xx)* is given in [1533]; *xxii)* is given in [535, p. 42] and [2062, p. 110]; *xxiii)* is given in [47, p. 91], [110, p. 145], [535, p. 42], and [1938, pp. 73, 74]; *xxiv)* is given in [535, pp. 49, 50], [1860, p. 122], [1938, p. 168], [2062, pp. 50, 150, 168, 189], and [2377, 2922, 2926]; *xxv)* and *xxvi)* are given in [917] and [1938, p. 58]; *xxvii)* and *xxviii)* are given in [47, p. 18]; *xxix)* is given in [2062, p. 679]; *xxx)* is given in [806, p. 207]; *xxxi)* is given in [2062, p. 683]; *xxxii)* is given in [47, p. 38] and [917]; *xxxiii)* is given in [2377]; *xxxiv)* and *xxxv)* are given in [497], [2062, pp. 170, 171], and [2292]; *xxxvi)* is given in [933, 1906, 1905, 1907, 2292]; *xxxvii)–lii)* are given in [395, 917] and [2062, pp. 172–174, 177, 178, 192]; *liii)* is given in [2262]; *liv)* is given in [1822]; *lv)–lvii)* are given in [2062, p. 178]; *lviii)* follows from an inequality involving the medians of a triangle in Fact 5.2.12; *lix)* is given in [1938, p. 54]; *lx)* and *lxi)* are given in [2062, p. 177]; *lxii)* and *lxvi)* are given in [2467, 2925]; *lxiii)–lxv)* are given in [917]; *lxvii)* is given in [2002]; *lxviii)* is given in [109, pp. 314, 315]; *lxix)* and *lxx)* are given in [917]; *lxxi)–lxxiv)* are given in [1907]. *lxxv)* is given in [1956, 2739]; *lxxvi)* is given in [1956]; *lxxvii)* and *lxxviii)* are given in [397]; *lxxix)* is given in [751]; *lxxx)* is given in [1732]; *lxxxi)* is given in [109, pp. 110, 111]; *lxxxii)* is given in [109, pp. 198, 199]; *lxxxiii)* is given in [373]; *lxxxiv)* is given in [392]; *lxxxv)* is given in [395, 2995]; *lxxxvi)–lxxxviii)* are given in [395]. **Remark:** In *xiii)*, the inequality $(a+b-c)(b+c-a)(c+a-b)\le abc$ is *Padoa's inequality.* See Fact 5.2.25. **Remark:** $\operatorname{area}(\mathcal{S})\le\frac{\sqrt{3}}{12}(a^2+b^2+c^2)$ in *xxiii)* is *Weitzenbock's inequality.* See [47, pp. 84, 85]. **Remark:** $\operatorname{area}(\mathcal{S})\le\frac{\sqrt{3}}{12}[2(ab+bc+ca)-(a^2+b^2+c^2)]$ in *xxiii)* can be rewritten as $4\sqrt{3}\operatorname{area}(\mathcal{S})+(a-b)^2+(b-c)^2+(c-a)^2\le a^2+b^2+c^2$. This is the *Hadwiger-Finsler inequality.* This is interpolated in *xxxvi)*. Extensions are given in [1690]. See [993, p. 174] and [1907]. **Remark:** The adjacent upper and lower bounds for s^2 in *iv)* are *Blundon's inequalities.* See [119]. According to *lvi)*, these inequalities are necessary and sufficient for the existence of a triangle with semiperimeter s, inner radius r, and outer radius R. **Remark:** In *xxiv)*, $\operatorname{area}(\mathcal{S})\le(R+\frac{1}{2}r)^2$ in *iii)* is *Mircea's inequality.* The adjacent upper and lower bounds for area($\mathcal{S}$) are due to É. Rouché. See [2922]. The lower bounds $3\sqrt{3}r^2$ and $r\sqrt{16Rr-5r^2}$ and the upper bounds $r\sqrt{4R^2+4Rr+3r^2}$ and $2Rr+(3\sqrt{3}-4)r^2$ are due to W. J. Blundon. See [927, 2377]. **Remark:** $2r\le R$ is *Euler's inequality.* See [47, pp. 56–60]. If $\mathcal{S}$ is a right triangle, then $(1+\sqrt{2})r\le R$. See [2366, p. 19]. The interpolation in *xi)* is *Bandila's inequality.* **Remark:** $a^2+b^2+c^2\le 9R^2$ in *xxxv)* is *Leibniz's inequality.* See [1938, pp. 68–70]. **Remark:** The upper bounds for $a+b+c$ and $a^2+b^2+c^2$ in *xxxiv)* and *xxxv)* are due to W. J. Blundon. See [497, 535] and [2062, pp. 170, 171]. Versions of the adjacent upper and lower bounds in *xxxv)* given in [497] and quoted in [535, p. 53], [917], and [2062, pp. 170, 171] are incorrect. **Credit:** *i)* is due to L. Bankoff. See [47, p. 112] and [2062, p. 165].

Fact 5.2.9. Let $\mathcal{S}\subset\mathbb{R}^2$ denote a triangle with the notation defined in Fact 5.2.7. Then, the following statements hold:

i) $\sin A + \sin B + \sin C = 4(\cos \frac{A}{2})(\cos \frac{B}{2}) \cos \frac{C}{2} = \frac{s}{R}$.

ii) $\sin A + \sin B - \sin C = 4(\sin \frac{A}{2})(\sin \frac{B}{2}) \cos \frac{C}{2}$.

iii) $\sin^2 A + \sin^2 B - \sin^2 C = 2(\sin A)(\sin B) \cos C$.

iv) $(\sin^2 A + \sin^2 B - \sin^2 C)(\sin^2 B + \sin^2 C - \sin^2 A)(\sin^2 C + \sin^2 A - \sin^2 B)$
$= 8(\sin^2 A)(\sin^2 B)(\sin^2 C)(\cos A)(\cos B) \cos C = \frac{r^2 s^2(s^2 - 4R^2 - 4Rr - r^2)}{2R^6}$.

v) $a \sin A + b \sin B + c \sin C = \frac{a^2 + b^2 + c^2}{2R} = \frac{s^2 - 4Rr - r^2}{R}$.

vi) $(\sin A)(\sin B) \sin C = \frac{rs}{2R^2}$.

vii) $4(\sin A)(\sin B) \sin C = \sin(A + B - C) + \sin(B + C - A) + \sin(C + A - B) - \sin(A + B + C)$.

viii) $4(\sin A)(\cos B) \cos C = \sin(A + B - C) + \sin(A - B + C) + \sin(A - B - C) + \sin(A + B + C)$.

ix) $(\sin A)(\cos B) \cos C + (\sin B)(\cos C) \cos A + (\sin C)(\cos A) \cos B = \sin(A + B - C)$
$+ (\sin A)(\sin B) \sin C$.

x) $(\sin A) \sin B + (\sin B) \sin C + (\sin C) \sin A = \frac{s^2 + 4Rr + r^2}{4R^2}$.

xi) $\sin^2 A + \sin^2 B + \sin^2 C = \frac{s^2 - 4Rr - r^2}{2R^2}$.

xii) $\sin^3 A + \sin^3 B + \sin^3 C = \frac{s(s^2 - 6Rr - 3r^2)}{4R^3}$.

xiii) $\sin^4 A + \sin^4 B + \sin^4 C = \frac{s^4 - (8Rr + 6r^2)s^2 + (4R + r)^2 r^2}{8R^4}$.

xiv) $(\sin A + \sin B)(\sin B + \sin C)(\sin C + \sin A) = \frac{s(s^2 + 2Rr + r^2)}{4R^3}$.

xv) $\sin^2 \frac{A}{2} + \sin^2 \frac{B}{2} + \sin^2 \frac{C}{2} = 1 - \frac{r}{2R}$.

xvi) $(a + b)^2 \sin^2 \frac{C}{2} + (b + c)^2 \sin^2 \frac{A}{2} + (c + a)^2 \sin^2 \frac{B}{2} = \frac{2Rs^2 + 4Rr^2 + r^3 + rs^2}{2R}$.

xvii) $\sin^4 \frac{A}{2} + \sin^4 \frac{B}{2} + \sin^4 \frac{C}{2} = \frac{8R^2 + r^2 - s^2}{8R^2}$.

xviii) $(\sin^2 \frac{A}{2}) \sin^2 \frac{B}{2} + (\sin^2 \frac{B}{2}) \sin^2 \frac{C}{2} + (\sin^2 \frac{C}{2}) \sin^2 \frac{A}{2} = \frac{s^2 + r^2 - 8Rr}{16R^2}$.

xix) $\frac{\sin A + \sin B}{\sin C} + \frac{\sin B + \sin C}{\sin A} + \frac{\sin C + \sin A}{\sin B} = \frac{s^2 + r^2 - 2Rr}{2Rr}$.

xx) $\left(\frac{\sin A + \sin B}{\cos A + \cos B}\right)\left(\frac{\sin B + \sin C}{\cos B + \cos C}\right) \frac{\sin C + \sin A}{\cos C + \cos A} = \frac{s}{r}$.

xxi) $(\sin \frac{A}{2})(\sin \frac{B}{2}) \sin \frac{C}{2} = \frac{1}{4}(\cos A + \cos B + \cos C - 1) = \frac{r}{4R}$.

xxii) $(1 - \cos A)(1 - \cos B)(1 - \cos B) = 8(\sin^2 \frac{A}{2})(\sin^2 \frac{B}{2}) \sin^2 \frac{C}{2} = \frac{r^2}{2R^2}$.

xxiii) $\sin 2A + \sin 2B + \sin 2C = 4(\sin A)(\sin B) \sin C = \frac{2rs}{R^2}$.

xxiv) $(\sin 2A)(\sin 2B) \sin 2C = \frac{rs[s^2 - (2R + r)^2]}{R^4}$.

xxv) $\sin A, \sin B, \sin C$ are the roots of $4R^2x^3 - 4rsx^2 + (s^2 + 4Rr + r^2)x - 2rs = 0$.

xxvi) $\sin^2 \frac{A}{2}, \sin^2 \frac{B}{2}, \sin^2 \frac{C}{2}$ are the roots of $16R^2x^3 - 8R(2R - r)x^2 + (s^2 - 8Rr + r^2)x - r^2 = 0$.

xxvii) $\cos A + \cos B + \cos C = 4(\sin \frac{A}{2})(\sin \frac{B}{2}) \sin \frac{C}{2} + 1 = 1 + \frac{r}{R}$.

xxviii) $\cos A + \cos B - \cos C = 4(\cos \frac{A}{2})(\cos \frac{B}{2}) \sin \frac{C}{2} - 1$.

xxix) $(\cos A)(\cos B)\cos C = \dfrac{s^2-(2R+r)^2}{4R^2}.$

xxx) $(\cos A)\cos B + (\cos B)\cos C + (\cos C)\cos A = \dfrac{s^2-4R^2+r^2}{4R^2}.$

xxxi) $\cos^2 A + \cos^2 B + \cos^2 C = 1 - 2(\cos A)(\cos B)\cos C = \dfrac{6R^2+4Rr+r^2-s^2}{2R^2}.$

xxxii) $\cos^3 A + \cos^3 B + \cos^3 C = \dfrac{(2R+r)^3-3s^2r}{4R^3} - 1.$

xxxiii) $(\cos A + \cos B)(\cos B + \cos C)(\cos C + \cos A) = \dfrac{rs^2+2Rr^2+r^3}{4R^3}.$

xxxiv) $\cos^2 \frac{A}{2} + \cos^2 \frac{B}{2} + \cos^2 \frac{C}{2} = 2 + \dfrac{r}{2R}.$

xxxv) $(\cos \frac{A}{2})(\cos \frac{B}{2})\cos \frac{C}{2} = \frac{1}{4}(\sin A + \sin B + \sin C) = \dfrac{s}{4R}.$

xxxvi) $(1+\cos A)(1+\cos B)(1+\cos B) = 8(\cos^2 \frac{A}{2})(\cos^2 \frac{B}{2})\cos^2 \frac{C}{2} = \dfrac{s^2}{2R^2}.$

xxxvii) $\cos 2A + \cos 2B + \cos 2C = -4(\cos A)(\cos B)\cos C - 1 = \dfrac{3R^2+4Rr+r^2-s^2}{R^2}.$

xxxviii) $(\cos^2 \frac{A}{2})\cos^2 \frac{B}{2} + (\cos^2 \frac{B}{2})\cos^2 \frac{C}{2} + (\cos^2 \frac{C}{2})\cos^2 \frac{A}{2} = \dfrac{s^2+(4R+r)^2}{16R^2}.$

xxxix) $\dfrac{\cos A + \cos B}{\cos C} + \dfrac{\cos B + \cos C}{\cos A} + \dfrac{\cos C + \cos A}{\cos B} = \dfrac{(R+r)(s^2+r^2-4R^2)}{R[s^2-(2R+r)^2]} - 3.$

xl) $2(\cos \frac{A}{2})(\cos \frac{B}{2})\cos \frac{C}{2} - 2(\sin \frac{A}{2})(\sin \frac{B}{2})\sin \frac{C}{2} = 1 + (\cos \frac{A}{2} - \sin \frac{A}{2})(\cos \frac{B}{2} - \sin \frac{B}{2})(\cos \frac{C}{2} - \sin \frac{C}{2}).$

xli) $(\cos^3 \frac{A}{2})\sin \frac{B-C}{2} + (\cos^3 \frac{B}{2})\sin \frac{C-A}{2} + (\cos^3 \frac{C}{2})\sin \frac{A-B}{2} = 0.$

xlii) $(\cos \frac{A}{4})(\cos \frac{B}{4})\cos \frac{C}{4} = \frac{\sqrt{2}}{8} + \frac{1}{4}[\cos(\frac{\pi}{4} - \frac{A}{2}) + \cos(\frac{\pi}{4} - \frac{B}{2}) + \cos(\frac{\pi}{4} - \frac{C}{2})] = \frac{\sqrt{2}}{8}[1 + \cos \frac{A}{2} + \cos \frac{B}{2} + \cos \frac{C}{2} + \sin \frac{A}{2} + \sin \frac{B}{2} + \sin \frac{C}{2}].$

xliii) $(\cos A)(\sin B)\sin C = \frac{1}{4}(\cos 2A - \cos 2B - \cos 2C + 1).$

xliv) $\cos A, \cos B, \cos C$ are the roots of $4R^2x^3 - 4R(R+r)x^2 + (s^2+4R^2+r^2)x + (2R+r)^2 - s^2 = 0.$

xlv) $\cos^2 \frac{A}{2}, \cos^2 \frac{B}{2}, \cos^2 \frac{C}{2}$ are the roots of $16R^2x^3 - 8R(4R+r)x^2 + [s^2+(4R+r)^2]x - s^2 = 0.$

xlvi) $a\cos A + b\cos B + c\cos C = 4R(\sin A)(\sin B)\sin C.$

xlvii) $\tan A + \tan B + \tan C = (\tan A)(\tan B)\tan C = \dfrac{2rs}{s^2-(2R+r)^2}.$

xlviii) $(\tan A)\tan B + (\tan B)\tan C + (\tan C)\tan A = \dfrac{s^2-r^2-4Rr}{s^2-(2R+r)^2}.$

xlix) $(\tan A)(\tan B)\tan C = \dfrac{2rs}{s^2-(2R+r)^2}.$

l) $\tan^2 A + \tan^2 B + \tan^2 C = \dfrac{4s^2r^2 - 2(s^2-r^2-4Rr)[s^2-(2R+r)^2]}{[s^2-(2R+r)^2]^2}.$

li) $\tan^3 A + \tan^3 B + \tan^3 C = \dfrac{8rs(s^2r^2 - 3R^2[s^2-(2R+r)^2])}{[s^2-(2R+r)^2]^3}.$

lii) $(\tan A + \tan B)(\tan B + \tan C)(\tan C + \tan A) = \dfrac{8R^2rs}{[s^2-(2R+r)^2]^2}.$

liii) $\tan \frac{A}{2} + \tan \frac{B}{2} + \tan \frac{C}{2} = \dfrac{4R+r}{s}.$

liv) $a\tan \frac{A}{2} + b\tan \frac{B}{2} + c\tan \frac{C}{2} = 4R - 2r.$

lv) $(a+b)\tan \frac{C}{2} + (b+c)\tan \frac{A}{2} + (c+a)\tan \frac{B}{2} = 4(R+r).$

lvi) $\tan^2 \frac{A}{2} + \tan^2 \frac{B}{2} + \tan^2 \frac{C}{2} = \dfrac{(4R+r)^2 - 2s^2}{s^2}.$

lvii) $(\tan \frac{A}{2})(\tan \frac{B}{2}) \tan \frac{C}{2} = \dfrac{r}{s}.$

lviii) $(\tan \frac{A}{2} + \tan \frac{B}{2})(\tan \frac{B}{2} + \tan \frac{C}{2})(\tan \frac{C}{2} + \tan \frac{A}{2}) = \dfrac{4R}{s}.$

lix) $(\tan \frac{A}{2}) \tan \frac{B}{2} + (\tan \frac{B}{2}) \tan \frac{C}{2} + (\tan \frac{C}{2}) \tan \frac{A}{2} = 1.$

lx) $\tan^3 \frac{A}{2} + \tan^3 \frac{B}{2} + \tan^3 \frac{C}{2} = \dfrac{(4R+r)^3 - 12s^2R}{s^3}.$

lxi)

$$\frac{\tan \frac{A}{2} + \tan \frac{B}{2}}{\tan \frac{C}{2}} + \frac{\tan \frac{B}{2} + \tan \frac{C}{2}}{\tan \frac{A}{2}} + \frac{\tan \frac{C}{2} + \tan \frac{A}{2}}{\tan \frac{B}{2}} = \frac{\cot \frac{A}{2} + \cot \frac{B}{2}}{\cot \frac{C}{2}} + \frac{\cot \frac{B}{2} + \cot \frac{C}{2}}{\cot \frac{A}{2}} + \frac{\cot \frac{C}{2} + \cot \frac{A}{2}}{\cot \frac{B}{2}}$$
$$= \frac{4R - 2r}{r} = \frac{a}{s-a} + \frac{b}{s-b} + \frac{c}{s-c}.$$

lxii) $\tan A, \tan B, \tan C$ are the roots of $[s^2 - (2R+r)^2]x^3 - 2rsx^2 + (s^2 - 4Rr - r^2)x - 2rs = 0.$

lxiii) $\tan \frac{A}{2}, \tan \frac{B}{2}, \tan \frac{C}{2}$ are the roots of $sx^3 - (4R+r)x^2 + sx - r = 0.$

lxiv) $\csc A + \csc B + \csc C = \dfrac{s^2 + r^2 + 4Rr}{2rs}.$

lxv) $\csc^2 A + \csc^2 B + \csc^2 C = \dfrac{(s^2 + r^2 + 4Rr)^2 - 16s^2Rr}{4s^2R^2}.$

lxvi) $(\csc A) \csc B + (\csc B) \csc C + (\csc C) \csc A = \dfrac{2R}{r}.$

lxvii) $(\csc A)(\csc B) \csc C = \dfrac{2R^2}{rs}.$

lxviii) $\csc^2 \frac{A}{2} + \csc^2 \frac{B}{2} + \csc^2 \frac{C}{2} = \dfrac{s^2 + r^2 - 8Rr}{r^2}.$

lxix) $(\csc^2 \frac{A}{2}) \csc^2 \frac{B}{2} + (\csc^2 \frac{B}{2}) \csc^2 \frac{C}{2} + (\csc^2 \frac{C}{2}) \csc^2 \frac{A}{2} = \dfrac{8R(2R-r)}{r^2}.$

lxx) $\csc A, \csc B, \csc C$ are the roots of $2rsx^3 - (s^2 + r^2 + 4Rr)x^2 + 4Rsx - 4R^2 = 0.$

lxxi) $\csc^2 \frac{A}{2}, \csc^2 \frac{B}{2}, \csc^2 \frac{C}{2}$ are the roots of $r^2x^3 - (s^2 + r^2 - 8Rr)x^2 + 8R(2R-r)x - 16R^2 = 0.$

lxxii) $\sec A + \sec B + \sec C = \dfrac{s^2 + r^2 - 4R^2}{s^2 - (2R-r)^2}.$

lxxiii) $\sec^2 A + \sec^2 B + \sec^2 C = \dfrac{(s^2 + r^2 - 4R^2)^2 - 8R(R+r)[s^2 - (2R+r)^2]}{[s^2 - (2R+r)^2]^2}.$

lxxiv) $(\sec A) \sec B + (\sec B) \sec C + (\sec C) \sec A = \dfrac{4R(R+r)}{s^2 - (2R+r)^2}.$

lxxv) $\sec^2 \frac{A}{2} + \sec^2 \frac{B}{2} + \sec^2 \frac{C}{2} = \dfrac{s^2 + (4R+r)^2}{s^2}.$

lxxvi) $(\sec^2 \frac{A}{2}) \sec^2 \frac{B}{2} + (\sec^2 \frac{B}{2}) \sec^2 \frac{C}{2} + (\sec^2 \frac{C}{2}) \sec^2 \frac{A}{2} = \dfrac{8R(4R+r)}{s^2}.$

lxxvii) $\sec A, \sec B, \sec C$ are the roots of $[s^2 - (2R+r)^2]x^3 - (s^2 + r^2 - 4R^2)x^2 + 4R(R+r)x - 4R^2 = 0.$

lxxviii) $\sec^2 \frac{A}{2}, \sec^2 \frac{B}{2}, \sec^2 \frac{C}{2}$ are the roots of $s^3x^3 - [s^2 + (4R+r)^2]x^2 + 8R(4R+r)x - 16R^2 = 0.$

lxxix) $\cot A + \cot B + \cot C = \dfrac{a^2 + b^2 + c^2}{4rs} = \dfrac{s^2 - r^2 - 4Rr}{2rs}.$

lxxx) $(\cot A) \cot B + (\cot B) \cot C + (\cot C) \cot A = 1.$

lxxxi) $(\cot A)(\cot B)\cot C = \dfrac{s^2 - (2R+r)^2}{2rs}$.

lxxxii) $\cot^2 A + \cot^2 B + \cot^2 C = \dfrac{(s^2 - r^2 - 4Rr)^2}{4s^2r^2} - 2$.

lxxxiii) $(\cot A + \cot B)(\cot B + \cot C)(\cot C + \cot A) = \dfrac{2R^2}{rs}$.

lxxxiv) $\cot^3 A + \cot^3 B + \cot^3 C = \dfrac{(s^2 - r^2 - 4Rr)^3 - 48s^2R^2r^2}{8s^3r^3}$.

lxxxv) $\dfrac{1 - (\cot^2 A)(\cot B)\cot C}{1 + \cot^2 A} + \dfrac{1 - (\cot^2 B)(\cot C)\cot A}{1 + \cot^2 B} + \dfrac{1 - (\cot^2 C)(\cot A)\cot B}{1 + \cot^2 C} = 2$.

lxxxvi) $\cot \frac{A}{2} + \cot \frac{B}{2} + \cot \frac{C}{2} = (\cot \frac{A}{2})(\cot \frac{B}{2})\cot \frac{C}{2} = \dfrac{s}{r}$.

lxxxvii) $(\cot \frac{A}{2})\cot \frac{B}{2} + (\cot \frac{B}{2})\cot \frac{C}{2} + (\cot \frac{C}{2})\cot \frac{A}{2} = \dfrac{4R + r}{r}$.

lxxxviii) $\cot^2 \frac{A}{2} + \cot^2 \frac{B}{2} + \cot^2 \frac{C}{2} = \dfrac{s^2 - 2r(4R + r)}{r^2}$.

lxxxix) $\cot^3 \frac{A}{2} + \cot^3 \frac{B}{2} + \cot^3 \frac{C}{2} = \dfrac{s(s^2 - 12Rr)}{r^3}$.

xc) $(\cot \frac{A}{2} + \cot \frac{B}{2})(\cot \frac{B}{2} + \cot \frac{C}{2})(\cot \frac{C}{2} + \cot \frac{A}{2}) = \dfrac{4Rs}{r^2}$.

xci) $\cot 2A + \cot 2B + \cot 2C = \dfrac{s^2 - r^2 - 4Rr}{4rs} + \dfrac{rs}{(2R+r)^2 - s^2}$.

xcii) $(\cot 2A)(\cot 2B)\cot 2C = \dfrac{[2s^2 - (2R+r)^2 - r^2 - 4Rr]^2 - 16s^2r^2}{16rs[s^2 - (2R+r)^2]}$.

xciii) $\cot A, \cot B, \cot C$ are the roots of $2rsx^3 - (s^2 - r^2 - 4Rr)x^2 + 2rsx + (2R+r)^2 - s^2 = 0$.

xciv) $\cot \frac{A}{2}, \cot \frac{B}{2}, \cot \frac{C}{2}$ are the roots of $rx^3 - sx^2 + 2rsx + (4R + r)^2 - s = 0$.

Source: [47, p. 124], [107, p. 117], [108, pp. 11, 99], [161, 292], [644, pp. 166, 182], [806, p. 79], [1026], [1860, p. 135], [2062, pp. 54–60, 89, 90], [2162], and [2413]. **Remark:** The first equality in *xxviii)* is the *Cayley cosine cubic*. See [1307].

Fact 5.2.10. Let $\mathcal{S} \subset \mathbb{R}^2$ denote a triangle with the notation defined in Fact 5.2.7. Then, the following statements hold:

i) $a + (2 - \sqrt{2 - 2\cos A})\min\{b, c\} \le b + c \le a + (2 - \sqrt{2 - 2\cos A})\max\{b, c\}$.

ii) If $a < \frac{1}{2}(b + c)$, then $A < \frac{1}{2}(B + C)$.

iii) $\frac{\pi}{3}(a + b + c) \le aA + bB + cC \le \frac{1}{2}(\pi - \min\{A, B, C\})(a + b + c) < \frac{\pi}{2}(a + b + c)$.

iv) If $A \le B \le C$, then

$$\frac{\pi}{3} \le \frac{aA + bB + cC}{a + b + c} \le \left\{ \begin{array}{c} \dfrac{\pi - A}{2} \\ \frac{\pi}{2}\left[1 - (\tan \frac{A}{2})\tan \frac{B}{2}\right] \end{array} \right\} < \frac{\pi}{2}.$$

v) $2 \le \dfrac{R}{r} \le 2\left(\dfrac{\pi}{9}\right)^3\left(\dfrac{1}{A} + \dfrac{1}{B} + \dfrac{1}{C}\right)^3$, $64ABC \le (A + \pi)(B + \pi)(C + \pi)$.

vi) $\dfrac{9}{\pi} \le \dfrac{8}{\pi} + \dfrac{3}{\pi^3}(A^2 + B^2 + C^2) \le \dfrac{1}{A} + \dfrac{1}{B} + \dfrac{1}{C}$.

vii) $\pi < \dfrac{3\sqrt{3}}{4} + \dfrac{A^3}{A^2 + \pi^2} + \dfrac{B^3}{B^2 + \pi^2} + \dfrac{C^3}{C^2 + \pi^2}$.

viii) $2R\left(\frac{\pi^2 - A^2}{\pi^2 + A^2} + \frac{\pi^2 - B^2}{\pi^2 + B^2} + \frac{\pi^2 - C^2}{\pi^2 + C^2}\right) \le \frac{a}{A} + \frac{b}{B} + \frac{c}{C}.$

ix) $\frac{3}{\pi}(a+b+c) \le \frac{a}{A} + \frac{b}{B} + \frac{c}{C} \le 2a\left(\frac{1}{B} + \frac{1}{C}\right) + 2b\left(\frac{1}{C} + \frac{1}{A}\right) + 2c\left(\frac{1}{A} + \frac{1}{B}\right).$

x) $\frac{6}{\pi}(a+b+c) \le \frac{a+b}{C} + \frac{b+c}{A} + \frac{c+a}{B}.$

xi) $\frac{3}{\pi}(a+b+c) \le \frac{a+b-c}{C} + \frac{b+c-a}{A} + \frac{c+a-b}{B}.$

xii) $\frac{9}{\pi} \le \frac{a+b-c}{cC} + \frac{b+c-a}{aA} + \frac{c+a-b}{bB}.$

xiii) $$\left(\frac{\pi}{3}\right)^3 \frac{2r}{R} \le ABC \le \left\{\begin{matrix} \left\{\begin{matrix} \left(\frac{\pi}{3}\right)^3 \left(\frac{2r}{R}\right)^{2/3} \\ \left(\frac{\pi}{2s}\right)^3 abc \end{matrix}\right\} \le \left(\frac{\pi}{3}\right)^3 \approx 1.14838 < \frac{29}{25} < \frac{10\pi}{27} < \frac{7}{6} \\ \frac{\pi r}{R} \end{matrix}\right\} \le \frac{\pi}{2}.$$

xiv) $\frac{6s}{\pi} \le 3\sqrt[3]{\frac{abc}{ABC}} \le \frac{a}{\sqrt{BC}} + \frac{b}{\sqrt{CA}} + \frac{c}{\sqrt{AB}}.$

xv)

$$\left.\begin{matrix} 2 \le \frac{27ABCR}{\pi^3 r} \\ \left(1 - \frac{\pi^2}{12}\right)\frac{R}{r} \end{matrix}\right\} \le (1 - \tfrac{1}{24}[(A-B)^2 + (B-C)^2 + (C-A)^2])\frac{R}{r}$$

$$\le e^{-\frac{1}{24}[(A-B)^2+(B-C)^2+(C-A)^2]}\frac{R}{r} \le \left(\frac{2\pi^3}{27ABC}\right)^2 \frac{r}{R} \le \frac{2\pi^3}{27ABC} \le \frac{R}{r}.$$

xvi) $\frac{16\sqrt{3}}{9}\left(\frac{\pi}{3}\right)^3 \frac{s}{R} \le (\pi - A)(\pi - B)(\pi - C),\ \frac{s}{R} < 4e^{-(A^2+B^2+C^2)/8}.$

xvii) $(A-B)^2 + (B-C)^2 + (C-A)^2 < 2\pi^2 \approx 19.739.$

xviii) $(aB - bA)^2 + (bC - cB)^2 + (cA - aC)^2 < \frac{\pi^2}{4}(a+b+c)^2.$

xix) $\frac{2bc\cos A}{b+c} < b + c - a < \frac{2bc}{a}.$

xx) Let $c \le b \le a$. Then, $\frac{C}{c} \le \frac{B}{b} \le \frac{A}{a}.$

xxi) Let $c < b < a$. Then, $Ac + Ba + Cb < Ab + Bc + Ca.$

xxii) $\frac{\sqrt{3}\pi}{3R} \le \frac{A}{a} + \frac{B}{b} + \frac{C}{c} \le \frac{3\pi}{2s} \le \frac{\sqrt{3}\pi}{6r}.$

xxiii) $\frac{A}{A+B} < \frac{a}{c}.$

xxiv) $\frac{12s}{\pi} \le \frac{ab}{C(s-C)} + \frac{bc}{A(s-A)} + \frac{ca}{B(s-B)}.$

xxv) $(a+b)\frac{C}{c} + (b+c)\frac{A}{a} + (c+a)\frac{B}{b} \le 2(A+B+C).$

xxvi) $(b+c)\sin\tfrac{A}{2} \le a,\ \sqrt{(b+c)^2 - a^2} \le (b+c)\cos\tfrac{A}{2}.$

xxvii) $0 \le (a-b)\left(\frac{A}{a} - \frac{B}{b}\right) + (b-c)\left(\frac{B}{b} - \frac{C}{c}\right) + (c-a)\left(\frac{C}{c} - \frac{A}{a}\right).$

xxviii)

$$\frac{3\sqrt{3}r^2}{2R^2} \le (\sin A)(\sin B)\sin C = \tfrac{1}{4}(\sin 2A + \sin 2B + \sin 2C)$$

$$\le \left\{ \begin{array}{c} \dfrac{2Rr + (3\sqrt{3}-4)r^2}{2R^2} \le \dfrac{3\sqrt{3}r}{4R} \le \left(\dfrac{3\sqrt{3}}{2\pi}\right)^3 ABC \\ \tfrac{1}{8}(\sin A + \sin B)(\sin B + \sin C)(\sin C + \sin A) \end{array} \right\}$$

$$\le (\cos \tfrac{A}{2})(\cos \tfrac{B}{2})\cos \tfrac{C}{2} = \tfrac{1}{4}(\sin A + \sin B + \sin C)$$

$$\le [\sin \tfrac{1}{2}\sqrt[3]{(\pi - A)(\pi - B)(\pi - C)}]^3 \le \tfrac{3\sqrt{3}}{8}.$$

xxix) $(\sin A)(\sin B)\sin C \le \left(\dfrac{s}{3R}\right)^3 \le \left[\dfrac{2}{3} + \left(\sqrt{3} - \dfrac{4}{3}\right)\dfrac{r}{R}\right]^3 \le \dfrac{3\sqrt{3}}{8}.$

xxx)

$$\left. \begin{array}{r} \tfrac{3\sqrt{3}}{2} - \tfrac{1}{2}[(A-B)^2 + (B-C)^2 + (C-A)^2] \\ \dfrac{3\sqrt{3}r}{R} \end{array} \right\} \le \left\{ \begin{array}{c} \sin A + \sin B + \sin C = \dfrac{s}{R} \\ \dfrac{\sqrt{3}}{2}\left(1 + \dfrac{4r}{R}\right) \end{array} \right\}$$

$$\le \left\{ \begin{array}{c} 2 + (3\sqrt{3} - 4)\dfrac{r}{R} \\ \tfrac{\sqrt{3}}{2}(\cos \tfrac{A-B}{2} + \cos \tfrac{B-C}{2} + \cos \tfrac{C-A}{2}) \\ \le \tfrac{\sqrt{3}}{2}(1 + \tfrac{r}{R} + \sin \tfrac{A}{2} + \sin \tfrac{B}{2} + \sin \tfrac{C}{2}) \end{array} \right\}$$

$$\le \cos \tfrac{A}{2} + \cos \tfrac{B}{2} + \cos \tfrac{C}{2}$$

$$\le \tfrac{\sqrt{3}}{2}(\cos \tfrac{A-B}{4} + \cos \tfrac{B-C}{4} + \cos \tfrac{C-A}{4}) \le \tfrac{3\sqrt{3}}{2}.$$

xxxi)

$$\frac{2\sqrt{3}rs}{R} \le 9r \le 12r - \frac{6r^2}{R} \le a\sin A + b\sin B + c\sin C$$

$$= \frac{a^2 + b^2 + c^2}{2R} \le 4R + \frac{2r^2}{R} \le 4R + r \le \frac{9R}{2}.$$

xxxii) Let x, y, z be positive numbers. Then,

$$x\sin A + y\sin B + z\sin C \le \tfrac{1}{2}(xy + yz + zx)\sqrt{\frac{1}{xy} + \frac{1}{yz} + \frac{1}{zx}} \le \frac{\sqrt{3}}{2}\left(\frac{yz}{x} + \frac{zx}{y} + \frac{xy}{z}\right),$$

$$4[xy(\sin A)\sin B + yz(\sin B)\sin C + zx(\sin C)\sin A] \le (x + y + z)^2,$$

$$2\sqrt{3}(xy\sin C + yz\sin A + zx\sin B) \le (x + y + z)^2,$$

$$x\cos A + y\cos B + z\cos C \le \frac{1}{2}\left(\frac{yz}{x} + \frac{zx}{y} + \frac{xy}{z}\right).$$

xxxiii) $\dfrac{9r}{2R} \le \dfrac{3r(2R-r)}{R^2} \le \sin^2 A + \sin^2 B + \sin^2 C \le 2 + \left(\dfrac{r}{R}\right)^2 \le \cos^2 \tfrac{A}{2} + \cos^2 \tfrac{B}{2} + \cos^2 \tfrac{C}{2} \le \dfrac{9}{4}.$

xxxiv) $\dfrac{9\sqrt{3}r^2}{2R^2} \le \dfrac{3\sqrt{3}r^2(5R-4r)}{2R^3} \le \dfrac{rs(5R-4r)}{2R^3} \le \sin^3 A + \sin^3 B + \sin^3 C = \dfrac{s(s^2 - 6Rr - 3r^2)}{4R^3}$

$$\le \frac{s(2R-r)}{2R^2} \le \frac{(2R-r)[2R + (3\sqrt{3}-4)r]}{2R^2} \le \frac{3\sqrt{3}(2R-r)}{4R} < \frac{3\sqrt{3}}{2}.$$

xxxv) $\sin^3 A + \sin^3 B + \sin^3 C \le (\sin A + \sin B + \sin C)(\sin \tfrac{A}{2} + \sin \tfrac{B}{2} + \sin \tfrac{C}{2}).$

xxxvi) $\sin A + \sin B + \sin C \le \sqrt{3}(\sin \tfrac{A}{2} + \sin \tfrac{B}{2} + \sin \tfrac{C}{2}).$

xxxvii) $9\sqrt{3}(\cos A)(\cos B)\cos C \le \sin^3 A + \sin^3 B + \sin^3 C \le \tfrac{3\sqrt{3}}{2}(\sin^2 \tfrac{A}{2} + \sin^2 \tfrac{B}{2} + \sin^2 \tfrac{C}{2}).$

xxxviii) $\sin^4 A + \sin^4 B + \sin^4 C \le 2 - \frac{1}{2}\left(\frac{r}{R}\right)^2 - 3\left(\frac{r}{R}\right)^4 \le 2 - 5\left(\frac{r}{R}\right)^4.$

xxxix) Let $p > 0$. Then,

$$3^{1+p/2}\left(\frac{r}{R}\right)^p \le 3^{1+p/2}\left(\frac{r^2}{2R^2}\right)^{p/3} \le 3[(\sin A)(\sin B)\sin C]^{p/3} \le \sin^p A + \sin^p B + \sin^p C.$$

In particular,

$$3\sqrt{3}\left(\frac{r}{R}\right)^3 \le \frac{3\sqrt{3}}{2}\left(\frac{r}{R}\right)^2 \le (\sin A)(\sin B)\sin C \le \frac{1}{3}(\sin^3 A + \sin^3 B + \sin^3 C).$$

xl) $0 \le \sin 2A + \sin 2B + \sin 2C \le \sin A + \sin B + \sin C < \sin(\frac{A}{2} + B) + \sin(\frac{B}{2} + C) + \sin(\frac{C}{2} + A).$

xli) $\frac{\sqrt{3}s}{3R} \le \frac{1}{2} - \frac{r}{4R} + \frac{\sqrt{3}s}{12R} \le \sin\frac{A}{2} + \sin\frac{B}{2} + \sin\frac{C}{2} \le \sqrt{2 + \frac{r}{2R}} \le \frac{3}{2}.$

xlii) Let $p \ge 1$. Then, $\sin\frac{\pi}{p} \le \sin\frac{A}{p} + \sin\frac{B}{p} + \sin\frac{C}{p} \le \frac{\pi}{p}.$

xliii)

$$\begin{aligned}\frac{3}{4} &\le \frac{2Rr - r^2}{R^2} \le \sin^2\tfrac{A}{2} + \sin^2\tfrac{B}{2} + \sin^2\tfrac{C}{2} = 1 - \frac{r}{2R} \le 1 - \left(\frac{r}{R}\right)^2\\ &\le \frac{7}{4} - \frac{2r}{R} \le \frac{15}{8} - \frac{9r}{4R} \le 2 - \frac{2r}{R} - \frac{1}{4}(\cos\tfrac{A-B}{2})(\cos\tfrac{B-C}{2})\cos\tfrac{C-A}{2}\\ &\le \sin^2\tfrac{A}{2} + \sin^2\tfrac{B}{2} + \sin^2\tfrac{C}{2} + 1 - \frac{2r}{R} \le \frac{17}{8} - \frac{11r}{4R} < 2 - \frac{2r}{R}.\end{aligned}$$

xliv) $\frac{3r}{R} - \frac{1}{2} \le (\cos\frac{A-B}{2})(\cos\frac{B-C}{2})\cos\frac{C-A}{2} \le \frac{1}{2} + \frac{r}{R} < 1.$

xlv)

$$\left.\begin{aligned}&\frac{1}{4\pi}ABC\\ \frac{r}{2R} - \frac{1}{8} \le\ &\frac{3r}{8R} - \frac{1}{16}\end{aligned}\right\} \le (\sin\tfrac{A}{2})(\sin\tfrac{B}{2})\sin\tfrac{C}{2} = \frac{r}{4R}$$

$$\le \left\{\begin{aligned}&\left(\frac{3}{2\pi}\right)^3 ABC \le (\sin\tfrac{1}{2}\sqrt[3]{ABC})^3 \le \frac{1}{24} + \frac{r}{6R}\\ &\frac{r^2 + s^2 + 4Rr}{72R^2} \le \tfrac{1}{16}(\cos^2\tfrac{A-B}{2} + \cos^2\tfrac{B-C}{2} + \cos^2\tfrac{C-A}{2} - 1)\\ &\qquad = \tfrac{1}{8}(\cos\tfrac{A-B}{2})(\cos\tfrac{B-C}{2})\cos\tfrac{C-A}{2}\end{aligned}\right\}$$

$$\begin{aligned}&\le \frac{1}{16} + \frac{r}{8R} \le \frac{1}{12} + \frac{r}{12R} = \frac{1}{12}(\cos A + \cos B + \cos C)\\ &\le \frac{1}{8} \le \frac{1}{2} - \frac{3r}{4R}.\end{aligned}$$

xlvi)

$$\begin{aligned}\frac{r}{4R} &\le \tfrac{1}{16}(\cos^2\tfrac{A-B}{2} + \cos^2\tfrac{B-C}{2} + \cos^2\tfrac{C-A}{2} - 1)\\ &= \tfrac{1}{8}(\cos\tfrac{A-B}{2})(\cos\tfrac{B-C}{2})\cos\tfrac{C-A}{2} \le \frac{r^2 + s^2}{56R^2}\\ &\le \left\{\begin{aligned}&\frac{1}{12} + \frac{r}{12R} = \tfrac{1}{12}(\cos A + \cos B + \cos C)\\ &\tfrac{1}{8}e^{-\frac{1}{24}[(A-B)^2 + (B-C)^2 + (C-A)^2]}\end{aligned}\right\} \le \frac{1}{8}.\end{aligned}$$

xlvii)

$$\frac{9r}{2R} \le \frac{r(5R - r)}{R^2} \le (\sin A)\sin B + (\sin B)\sin C + (\sin C)\sin A$$

$$= \frac{s^2 + r^2 + 4Rr}{4R^2} \le \left\{ \begin{array}{c} \frac{(R+r)^2}{R^2} \le 1 + \frac{5r}{2R} \le \frac{5}{4} + \frac{2r}{R} \\ \frac{s^2}{3R^2} \end{array} \right\} \le \frac{9}{4}.$$

xlviii)
$$\frac{2rs(6Rr - 2R^2 - 3r^2)}{R^4} \le (\sin 2A)(\sin 2B) \sin 2C \le \frac{2r^3 s}{R^4}$$
$$\le \frac{2r^3[2R + (3\sqrt{3} - 4)r]}{R^4} \le \frac{3\sqrt{3}r^3}{R^3} \le \frac{3\sqrt{3}}{8}.$$

xlix) $\frac{9\sqrt{3}r}{4s} \le (\sin \frac{A}{2}) \sin \frac{B}{2} + (\sin \frac{B}{2}) \sin \frac{C}{2} + (\sin \frac{C}{2}) \sin \frac{A}{2} \le \frac{3}{4}.$

l) $36\left(\frac{r}{R}\right)^4 \le 6\left(\frac{r}{R}\right)^2 + 12\left(\frac{r}{R}\right)^4 \le \sin^2 2A + \sin^2 2B + \sin^2 2C.$

li) $(\sin 2A) \sin 2B + (\sin 2B) \sin 2C + (\sin 2C) \sin 2A \le 5\left(\frac{r}{R}\right)^2 + 8\left(\frac{r}{R}\right)^3 \le 9\left(\frac{r}{R}\right)^2.$

lii) $\sin 2A + \sin 2B + \sin 2C \le \sin A + \sin B + \sin C + \sin 3A + \sin 3B + \sin 3C.$

liii) Let $n \ge 1$. Then, $a^n \sin 2A + b^n \sin 2B + c^n \sin 2C \le \frac{a^{n+1} + b^{n+1} + c^{n+1}}{2R}.$

liv) $(\sin \frac{A}{2}) \sin^2 \frac{B}{2} + (\sin \frac{B}{2}) \sin^2 \frac{C}{2} + (\sin \frac{C}{2}) \sin^2 \frac{A}{2} + (\sin \frac{A}{2})(\sin \frac{B}{2}) \sin \frac{C}{2} \le \frac{1}{2}.$

lv) $|\sin(A - B) \cos^3 C + \sin(B - C) \cos^3 A + \sin(C - A) \cos^3 B| < \frac{3\sqrt{3}}{8}.$

lvi) $2 \le 8(\sin \frac{A}{2})(\sin \frac{B}{2}) \sin \frac{C}{2} + \tan^2 \frac{A}{2} + \tan^2 \frac{B}{2} + \tan^2 \frac{C}{2}.$

lvii) $\frac{9\sqrt{3}}{2} \le (\sin \frac{A}{2} + \sin \frac{B}{2} + \sin \frac{C}{2})(\cot \frac{A}{2} + \cot \frac{B}{2} + \cot \frac{C}{2}).$

lviii) $\sqrt{\sin A} + \sqrt{\sin B} + \sqrt{\sin C} \le \sqrt{\frac{3s}{R}} \le \sqrt{3\left[2 + (3\sqrt{3} - 4)\frac{r}{R}\right]} \le 3\sqrt[4]{\frac{3}{4}}.$

lix) $3\sqrt{\frac{3r}{2R}} \le \sqrt{(\sin A) \sin B} + \sqrt{(\sin B) \sin C} + \sqrt{(\sin C) \sin A} \le \frac{s}{R}.$

lx) $\frac{4}{27}(a + b + c)^3 \le \frac{a^3}{\sin^2 A} + \frac{b^3}{\sin^2 B} + \frac{c^3}{\sin^2 C} \le 12\sqrt{3}R^3.$

lxi) $(\sin \frac{A}{2}) \sqrt{(\sin B) \sin C} + (\sin \frac{B}{2}) \sqrt{(\sin C) \sin A} + (\sin \frac{C}{2}) \sqrt{(\sin A) \sin B} \le \frac{s}{2R}.$

lxii) $(\cos \frac{A}{2}) \sqrt{(\sin B) \sin C} + (\cos \frac{B}{2}) \sqrt{(\sin C) \sin A} + (\cos \frac{C}{2}) \sqrt{(\sin A) \sin B} \le \frac{s}{R}.$

lxiii)
$$\frac{9r^2}{2R^2} - 1 \le \sqrt{3}(\sin A)(\sin B) \sin C - 1 \le \frac{9r}{4R} - 1 \le \frac{3r}{R} - \frac{3r^2}{2R^2} - 1 \le (\cos A)(\cos B) \cos C$$
$$= \frac{s^2 - (2R + r)^2}{4R^2} \le \frac{2r^2 s^2}{27R^4} \le (1 - \cos A)(1 - \cos B)(1 - \cos C)$$
$$= 8(\sin^2 \tfrac{A}{2})(\sin^2 \tfrac{B}{2}) \sin^2 \tfrac{C}{2} = \frac{r^2}{2R^2} \le \frac{\sqrt{3}rs}{18R^2} \le \frac{r}{4R} \le \frac{1}{8}.$$

lxiv)
$$(\cos A)(\cos B) \cos C \le (2 - 3\sqrt{3}r/s)(\cos A)(\cos B) \cos C$$
$$\le (2 - 2r/R)(\cos A)(\cos B) \cos C$$
$$\le (1 - \cos A)(1 - \cos B)(1 - \cos C).$$

lxv)
$$\left. \begin{array}{r} 1 + 4(\cos A)(\cos B) \cos C \\ \frac{3r}{R} \end{array} \right\} \le 2[(\sin \tfrac{A}{2}) \sin \tfrac{B}{2} + (\sin \tfrac{B}{2}) \sin \tfrac{C}{2} + (\sin \tfrac{C}{2}) \sin \tfrac{A}{2}]$$
$$\le \sqrt{(\sin A) \sin B + (\sin B) \sin C + (\sin C) \sin A}$$

$$\leq 1 + \frac{r}{R} = \cos A + \cos B + \cos C$$
$$= 1 + 4(\sin \tfrac{A}{2})(\sin \tfrac{B}{2}) \sin \tfrac{A}{2} \leq \sin \tfrac{A}{2} + \sin \tfrac{B}{2} + \sin \tfrac{C}{2}$$
$$\leq \sqrt{\cos^2 \tfrac{A}{2} + \cos^2 \tfrac{B}{2} + \cos^2 \tfrac{C}{2}} \leq \frac{3}{2} \leq 2 - \frac{r}{R}.$$

lxvi) $\frac{9r}{4R} - 1 \leq (\cos A)(\cos B) \cos C \leq \frac{r}{2R}\left(1 - \frac{r}{R}\right)$.

lxvii) $\max\{1, \frac{3r}{R}\} \leq 1 + \frac{r}{R} = \cos A + \cos B + \cos C \leq \frac{3}{2}$.

lxviii) $a \cos A + b \cos B + c \cos C \leq \dfrac{9abc}{2(ab + bc + ca)} \leq \frac{3}{2}\sqrt[3]{abc}$.

lxix) Let $n \geq 1$. Then,

$$a^n \cos A + b^n \cos B + c^n \cos C \leq \tfrac{1}{3}(a^n + b^n + c^n)(\cos A + \cos B + \cos C) \leq \tfrac{1}{2}(a^n + b^n + c^n).$$

lxx) Let $n \geq 1$. Then,

$$\frac{abc(a^{n-1} + b^{n-1} + c^{n-1})}{2R^2} \leq a^n \cos(B - C) + b^n \cos(C - A) + c^n \cos(A - B) + \tfrac{1}{2}(a^n + b^n + c^n).$$

lxxi)
$$\frac{3}{4} \leq \frac{2Rr - r^2}{R^2} \leq 1 - \frac{r}{2R} = \sin^2 \tfrac{A}{2} + \sin^2 \tfrac{B}{2} + \sin^2 \tfrac{C}{2} \leq 1 - \left(\frac{r}{R}\right)^2$$
$$\leq \cos^2 A + \cos^2 B + \cos^2 C \leq \frac{3(R - r)^2}{R^2} < 3.$$

lxxii) $\dfrac{3}{8} \leq \dfrac{2R^3 - 3Rr^2 - 4r^3}{2R^3} \leq \cos^3 A + \cos^3 B + \cos^3 C \leq \dfrac{4R^2 + 12Rr - 34r^2}{4R^2}$.

lxxiii) $\operatorname{area}(\mathcal{S}) \leq R(a \cos^2 A + b \cos^2 B + c \cos^2 C)$.

lxxiv) $4 \leq 2(\cos A + \cos B + \cos C) + \tan^2 \frac{A}{2} + \tan^2 \frac{B}{2} + \tan^2 \frac{C}{2}$.

lxxv) $3(\sin^2 A + \sin^2 B + \sin^2 C) \leq 6 + 2(\cos^3 A + \cos^3 B + \cos^3 C)$.

lxxvi) $\cos^3 A + \cos^3 B + \cos^3 C + 3(\cos A)(\cos B) \cos C \leq \cos^2 A + \cos^2 B + \cos^2 C$.

lxxvii) $(\cos A) \cos^2 B + (\cos B) \cos^2 C + (\cos C) \cos^2 A + (\cos A)(\cos B) \cos C \leq \frac{1}{2}$.

lxxviii) $\cos^2 2A + \cos^2 2B + \cos^2 2C \leq 3 - 6\left(\frac{r}{R}\right)^2 - 12\left(\frac{r}{R}\right)^4 \leq 3 - 36\left(\frac{r}{R}\right)^4$.

lxxix) Let $n \geq 1$. Then, $\frac{3}{2^n} \leq \cos^n A + \cos^n B + \cos^n C$ and

$$0 \leq (\cos 2^n A)(\cos 2^n B) \cos 2^n C + (1 + \cos 2^n A)(1 + \cos 2^n B)(1 + \cos 2^n C).$$

lxxx) $\sin(A - B) + \sin(B - C) + \sin(C - A) \leq \dfrac{7R - 2r - 2s}{2R}$.

lxxxi) $\cos(A - B) + \cos(B - C) + \cos(C - A) \leq \dfrac{(R + r)(R + 2r)}{R^2}$.

lxxxii) $\dfrac{7r - 2R}{2R} \leq \dfrac{4r}{R} - \dfrac{r^2}{R^2} - 1 \leq (\cos A) \cos B + (\cos B) \cos C + (\cos C) \cos A \leq \dfrac{r}{R} + \dfrac{r^2}{R^2} \leq \dfrac{3r}{2R} \leq \dfrac{3}{4}$.

lxxxiii) Let $C < B < A$. Then,

$$2R \cos A < R - \sqrt{R(R - 2r)} < 2R \cos B < R + \sqrt{R(R - 2r)} < 2R \cos C.$$

lxxxiv) $2 \leq \dfrac{1}{1 + \cos A} + \dfrac{1}{1 + \cos A} + \dfrac{1}{1 + \cos A} \leq \dfrac{R}{r}$.

lxxxv) $\frac{27r^3}{8R^3} \leq (\cos^2 \frac{A}{2} + \cos^2 \frac{B}{2} - \cos^2 \frac{C}{2})(\cos^2 \frac{B}{2} + \cos^2 \frac{C}{2} - \cos^2 \frac{A}{2})(\cos^2 \frac{C}{2} + \cos^2 \frac{A}{2} - \cos^2 \frac{B}{2}) \leq \frac{27r}{32R}$.

lxxxvi) $9r \leq a \cos \frac{A}{2} + b \cos \frac{B}{2} + c \cos \frac{C}{2} \leq \sqrt{3}s \leq 2\sqrt{3}R + (9 - 4\sqrt{3})r \leq \dfrac{9R}{2}$.

lxxxvii) $3\sqrt{\frac{rs}{R}} \le \sqrt{a}\cos\frac{A}{2} + \sqrt{b}\cos\frac{B}{2} + \sqrt{c}\cos\frac{C}{2}.$

lxxxviii) $\frac{8\sqrt{3}s}{3} \le \frac{a+b}{\cos\frac{C}{2}} + \frac{b+c}{\cos\frac{A}{2}} + \frac{c+a}{\cos\frac{B}{2}}.$

lxxxix) $\frac{3}{s} \le \frac{9\sqrt{3}R}{2s^2} \le \frac{1}{(a+b)\cos^2\frac{C}{2}} + \frac{1}{(b+c)\cos^2\frac{A}{2}} + \frac{1}{(c+a)\cos^2\frac{B}{2}} \le \frac{\sqrt{3}}{3r}.$

xc) $\left.\begin{matrix} 2 \\ \frac{9r}{2R} \end{matrix}\right\} \le \cos^2\frac{A}{2} + \cos^2\frac{B}{2} + \cos^2\frac{C}{2} < \frac{9}{4} \le \frac{s^2}{6Rr}.$

xci) $\cos^4\frac{A}{2} + \cos^4\frac{B}{2} + \cos^4\frac{C}{2} < \frac{s^2}{8Rr}.$

xcii) $\frac{27}{8s} \le \frac{\cos^2\frac{A}{2}}{a} + \frac{\cos^2\frac{B}{2}}{b} + \frac{\cos^2\frac{C}{2}}{c}.$

xciii) $1 \le 2(\cos\frac{A}{2})(\cos\frac{B}{2})\cos\frac{C}{2} + (\cos\frac{A}{2})\cos\frac{B}{2} + (\cos\frac{B}{2})\cos\frac{C}{2} + (\cos\frac{C}{2})\cos\frac{A}{2}.$

xciv) $(\cos\frac{A}{2})(\cos\frac{B}{2})\cos\frac{C}{2} \le \frac{3\sqrt{3}}{8}.$

xcv) $\frac{27r^2}{2R^2} \le (1+\cos A)(1+\cos B)(1+\cos C) = 8(\cos^2\frac{A}{2})(\cos^2\frac{B}{2})\cos^2\frac{C}{2} = \frac{s^2}{2R^2} \le \frac{27}{8}.$

xcvi) Let $p \in \mathbb{R}$. Then, $\cos A + p(\cos B + \cos C) \le 1 + \frac{p^2}{2}.$

xcvii) $2 < 2 + (3\sqrt{3}-4)\frac{r}{R} \le \cos\frac{A}{2} + \cos\frac{B}{2} + \cos\frac{C}{2} \le \frac{\sqrt{6}}{2}\sqrt{3+\cos A+\cos B+\cos C} \le \frac{3\sqrt{3}}{2}.$

xcviii) $0 < \tan^2 A + \tan^2 B + \tan^2 C.$

xcix) $(\sec A)(\sec B)\sec C \le \frac{2}{7}[1 + (\tan^2 A)(\tan^2 B)\tan^2 C].$

c) $(1+\sec A)^2 + (1+\sec B)^2 + (1+\sec C)^2 \le (\tan A + \tan B + \tan C)^2.$

ci)
$$\frac{2\sqrt{3}r}{R} \le \frac{2\sqrt{3}(4R+r)}{9R} \le \sqrt{3} \le \frac{4R+r}{2R+(3\sqrt{3}-4)r} \le \sqrt{3+\frac{4(R-2r)}{4R+r}} = \sqrt{\frac{16R-5r}{4R+r}}$$
$$\le \tan\frac{A}{2} + \tan\frac{B}{2} + \tan\frac{C}{2} \le \frac{9R}{2s} \le \frac{s}{3r} \le \left\{\begin{matrix} \frac{\sqrt{3}(4R+r)}{9r} \le \frac{\sqrt{3}R}{2r} \\ \cot A + \cot B + \cot C \end{matrix}\right\} \le \frac{9R^2}{4rs}.$$

cii) $4\sqrt{3} \le 4(\tan\frac{A}{2} + \tan\frac{B}{2} + \tan\frac{C}{2}) \le \sqrt{3} + \cot\frac{A}{2} + \cot\frac{B}{2} + \cot\frac{C}{2} \le 2(\csc\frac{A}{2} + \csc\frac{B}{2} + \csc\frac{C}{2}).$

ciii) $3R \le a\tan\frac{A}{2} + b\tan\frac{B}{2} + c\tan\frac{C}{2} = 4R - 2r \le 5R - 4r.$

civ) $\frac{4rs^2}{4R+r} \le a^2(\tan\frac{B}{2})\tan\frac{C}{2} + b^2(\tan\frac{C}{2})\tan\frac{A}{2} + c^2(\tan\frac{A}{2})\tan\frac{B}{2}.$

cv) $\frac{2\sqrt{3}r}{9R} \le \frac{r}{2R+(3\sqrt{3}-4)r} \le (\tan\frac{A}{2})(\tan\frac{B}{2})\tan\frac{C}{2} \le \frac{\sqrt{3}}{9}.$

cvi) $1 \le 2 - \frac{2r}{R} \le \tan^2\frac{A}{2} + \tan^2\frac{B}{2} + \tan^2\frac{C}{2} = \frac{(4R+r)^2 - 2s^2}{s^2} \le \frac{16R^2 - 24Rr + 11r^2}{16Rr - 5r^2}.$

cvii) $\frac{3r}{s} \le \frac{\sqrt{3}}{3} \le \tan^3\frac{A}{2} + \tan^3\frac{B}{2} + \tan^3\frac{C}{2} \le \frac{3\sqrt{3}R^3}{8r^3} - \frac{8\sqrt{3}}{3}.$

cviii)
$$\frac{1}{3} \le \frac{a}{b+c}\left(\tan^4\frac{B}{2} + \tan^4\frac{C}{2}\right) + \frac{b}{c+a}\left(\tan^4\frac{C}{2} + \tan^4\frac{A}{2}\right) + \frac{c}{a+b}\left(\tan^4\frac{A}{2} + \tan^4\frac{B}{2}\right)$$
$$\le \frac{1}{3} + \frac{8}{3}\left[\left(\frac{R}{2r}\right)^2 - 1\right].$$

cix) $\frac{a}{b+c}\tan^2\frac{B}{2}\tan^2\frac{C}{2}+\frac{b}{c+a}\tan^2\frac{C}{2}\tan^2\frac{A}{2}+\frac{c}{a+b}\tan^2\frac{A}{2}\tan^2\frac{B}{2}\le\frac{1}{6}.$

cx) $\frac{1}{9}\le\tan^6\frac{A}{2}+\tan^6\frac{B}{2}+\tan^6\frac{C}{2}.$

cxi) $\sqrt{\tan\frac{A}{2}}+\sqrt{\tan\frac{B}{2}}+\sqrt{\tan\frac{C}{2}}\le 3\sqrt{\frac{4R+r}{3s}}\le\sqrt{\frac{\sqrt{3}(4R+r)}{3r}}\le 3\sqrt{\frac{\sqrt{3}R}{6r}}.$

cxii) $21-12\sqrt{3}=3\tan^2\frac{\pi}{12}\le\tan^2\frac{A}{4}+\tan^2\frac{B}{4}+\tan^2\frac{C}{4}\le 1.$

cxiii) $0<(\tan\frac{1}{2}\sqrt[3]{ABC})^3\le(\tan\frac{A}{2})(\tan\frac{B}{2})\tan\frac{C}{2}\le\frac{\sqrt{3}}{9}.$

cxiv) $0<A\cos B+(\sin A)\cos C.$

cxv) $\sin^2 A+(\sin B)\sin C\le\frac{25}{16}.$

cxvi) $(\sin B)\sin C\le\cos^2\frac{A}{2}.$

cxvii) Let $p\in\mathbb{R}$. Then, $p(2-p)\sin\frac{A}{2}\le\cos\frac{B-C}{2}.$

cxviii) Let $p>0$. Then, $\sin\frac{A-B}{2}+\sin\frac{A-C}{2}+\frac{1}{p}\sin\frac{3A}{2}\le\frac{p^2+2}{2p}.$

cxix) $\sin^2 B+\sin^2 C\le 1+2(\sin B)(\sin C)\cos A.$

cxx) $2<\frac{\sin A}{A}+\frac{\sin B}{B}+\frac{\sin C}{C}\le\frac{9\sqrt{3}}{2\pi}.$

cxxi) $0<\frac{\sin A}{(A-B)(C-A)}+\frac{\sin B}{(B-C)(A-B)}+\frac{\sin C}{(C-A)(B-C)}<\frac{1}{2}.$

cxxii) $6\le\frac{7R-2r}{R}\le\frac{\sin A+\sin B}{\sin C}+\frac{\sin B+\sin C}{\sin A}+\frac{\sin C+\sin A}{\sin B}\le\frac{2R^2+Rr+2r^2}{Rr}\le\frac{3R}{r}.$

cxxiii) $\frac{\sin A}{(\sin B+\sin C)^2}+\frac{\sin B}{(\sin C+\sin A)^2}+\frac{\sin C}{(\sin A+\sin B)^2}\le\frac{3}{4}(\csc A)(\csc B)\csc C.$

cxxiv) $9\le\frac{(\sin A)\sin B}{\sin^2\frac{C}{2}}+\frac{(\sin B)\sin C}{\sin^2\frac{A}{2}}+\frac{(\sin C)\sin A}{\sin^2\frac{B}{2}}.$

cxxv) $\frac{\sin\frac{A}{2}}{1+\sin\frac{A}{2}}+\frac{\sin\frac{B}{2}}{1+\sin\frac{B}{2}}+\frac{\sin\frac{C}{2}}{1+\sin\frac{C}{2}}\le 1\le\frac{1}{2+2\sin\frac{A}{2}}+\frac{1}{2+2\sin\frac{B}{2}}+\frac{1}{2+2\sin\frac{C}{2}}.$

cxxvi) $\frac{9}{2}\le 5-\frac{r}{R}\le\frac{(\cos\frac{A}{2})\cos\frac{B}{2}}{\sin\frac{C}{2}}+\frac{(\cos\frac{B}{2})\cos\frac{C}{2}}{\sin\frac{A}{2}}+\frac{(\cos\frac{C}{2})\cos\frac{A}{2}}{\sin\frac{B}{2}}\le\frac{(R+r)^2}{Rr}.$

cxxvii) $2\sqrt{3}\le\frac{\sin\frac{A}{2}}{(\sin\frac{B}{2})\cos\frac{C}{2}}+\frac{\sin\frac{B}{2}}{(\sin\frac{C}{2})\cos\frac{A}{2}}+\frac{\sin\frac{C}{2}}{(\sin\frac{A}{2})\cos\frac{B}{2}}.$

cxxviii) $\frac{3}{2}\le\frac{\cos\frac{A+B}{2}}{\cos\frac{A-B}{2}}+\frac{\cos\frac{B+C}{2}}{\cos\frac{B-C}{2}}+\frac{\cos\frac{C+A}{2}}{\cos\frac{C-A}{2}}<2.$

cxxix) $\frac{3}{2}\le\frac{\cos A}{\cos(B-C)}+\frac{\cos B}{\cos(C-A)}+\frac{\cos C}{\cos(A-B)}.$

cxxx) $-\frac{3\sqrt{3}}{2}\le\frac{\cos A-\cos B}{A-B}+\frac{\cos B-\cos C}{B-C}+\frac{\cos C-\cos A}{C-A}<-\frac{4}{\pi}.$

cxxxi) $\frac{A-B}{\cos A-\cos B}+\frac{B-C}{\cos B-\cos C}+\frac{C-A}{\cos C-\cos A}\le-2\sqrt{3}.$

cxxxii) $\frac{3\sqrt{3}}{2}\le\frac{\cos\frac{A}{2}}{\cos\frac{B-C}{2}}+\frac{\cos\frac{B}{2}}{\cos\frac{C-A}{2}}+\frac{\cos\frac{C}{2}}{\cos\frac{A-B}{2}}\le\frac{1}{2}(\cot\frac{A}{2})(\cot\frac{B}{2})\cot\frac{C}{2}.$

cxxxiii) $\sqrt{3}\le\frac{1+(\cos A)(\cos B)\cos C}{(\sin A)(\sin B)\sin C}.$

cxxxiv) $\dfrac{\sqrt{(\cos\frac{A}{2})\cos\frac{B}{2}}}{\sin\frac{C}{2}} + \dfrac{\sqrt{(\cos\frac{B}{2})\cos\frac{C}{2}}}{\sin\frac{A}{2}} + \dfrac{\sqrt{(\cos\frac{C}{2})\cos\frac{A}{2}}}{\sin\frac{B}{2}} \le \dfrac{2R}{r}[(\sin A)\sin\frac{A}{2} + (\sin B)\sin\frac{B}{2} + (\sin C)\sin\frac{C}{2}].$

cxxxv) $\dfrac{\sqrt{(\tan\frac{A}{2})\tan\frac{B}{2}}}{\cos\frac{C}{2}} + \dfrac{\sqrt{(\tan\frac{B}{2})\tan\frac{C}{2}}}{\cos\frac{A}{2}} + \dfrac{\sqrt{(\tan\frac{C}{2})\tan\frac{A}{2}}}{\cos\frac{B}{2}} \le 2.$

cxxxvi) $\dfrac{3r}{2R} \le \dfrac{\tan\frac{A}{2}}{\csc B + \csc C} + \dfrac{\tan\frac{B}{2}}{\csc C + \csc A} + \dfrac{\tan\frac{C}{2}}{\csc A + \csc B}.$

cxxxvii) Let $a \le b \le c$. Then,

$$2\cos^2\frac{C}{2} \le \frac{2}{1 + (\tan\frac{B}{2})\tan\frac{C}{2}} \le \frac{3}{2} \le \frac{a}{b+c} + \frac{b}{c+a} + \frac{c}{a+b} \le \frac{2}{1 + (\tan\frac{A}{2})\tan\frac{B}{2}} \le 2\cos^2\frac{A}{2}.$$

cxxxviii) $2 \le \dfrac{\tan\frac{A}{2} + \tan\frac{B}{2}}{\sin A + \sin B} + \dfrac{\tan\frac{B}{2} + \tan\frac{C}{2}}{\sin B + \sin C} + \dfrac{\tan\frac{C}{2} + \tan\frac{A}{2}}{\sin C + \sin A}.$

cxxxix)

$$\begin{aligned} 6 &\le \frac{\tan\frac{A}{2} + \tan\frac{B}{2}}{\tan\frac{C}{2}} + \frac{\tan\frac{B}{2} + \tan\frac{C}{2}}{\tan\frac{A}{2}} + \frac{\tan\frac{C}{2} + \tan\frac{A}{2}}{\tan\frac{B}{2}} \\ &= \frac{\cot\frac{A}{2} + \cot\frac{B}{2}}{\cot\frac{C}{2}} + \frac{\cot\frac{B}{2} + \cot\frac{C}{2}}{\cot\frac{A}{2}} + \frac{\cot\frac{C}{2} + \cot\frac{A}{2}}{\cot\frac{B}{2}} = \frac{4R - 2r}{8R}. \end{aligned}$$

cxl) Let α and β be positive numbers. Then,

$$\frac{s^2}{\alpha(4R+r)^2 + (\beta - 2\alpha)s^2} \le \frac{\tan\frac{A}{2}\tan\frac{B}{2}}{\alpha\tan\frac{A}{2} + \beta\tan\frac{B}{2}} + \frac{\tan\frac{B}{2}\tan\frac{C}{2}}{\alpha\tan\frac{B}{2} + \beta\tan\frac{C}{2}} + \frac{\tan\frac{C}{2}\tan\frac{A}{2}}{\alpha\tan\frac{C}{2} + \beta\tan\frac{A}{2}},$$

$$\frac{s^2}{\alpha s^2 + \beta(4R+r)r} \le \frac{\tan\frac{A}{2}\tan\frac{B}{2}}{\alpha + \beta\tan\frac{C}{2}} + \frac{\tan\frac{B}{2}\tan\frac{C}{2}}{\alpha + \beta\tan\frac{A}{2}} + \frac{\tan\frac{C}{2}\tan\frac{A}{2}}{\alpha + \beta\tan\frac{B}{2}}.$$

cxli) $\dfrac{8}{3 + 2\cos C} \le \csc A + \csc B.$

cxlii) $\dfrac{2}{\sin A + \sin B} + \dfrac{2}{\sin B + \sin C} + \dfrac{2}{\sin C + \sin A} \le \csc A + \csc B + \csc C.$

cxliii)

$$2\sqrt{3} \le \frac{4\sqrt{3}(5R - r)}{9R} \le \frac{4\sqrt{3}(2R - r)}{3R} \le \frac{2(5R - r)}{2R + (3\sqrt{3} - 4)r} \le \frac{2(5R - r)}{s}$$

$$\le \csc A + \csc B + \csc C$$

$$\le \left\{ \begin{array}{c} \dfrac{2(R+r)^2}{rs} \\ \left\{ \begin{array}{c} \dfrac{2s}{3r} \\ \frac{\sqrt{3}}{3} + \frac{5}{3}(\cot A + \cot B + \cot C) \end{array} \right\} \le \left\{ \begin{array}{c} \dfrac{\sqrt{3}R}{r} \\ 2(\cot A + \cot B + \cot C) \end{array} \right\} \end{array} \right\}$$

$$\le \frac{2(2R^2 + r^2)}{rs} \le \frac{2\sqrt{3}(R+r)^2}{9r^2} \le \frac{2\sqrt{3}(2R^2 + r^2)}{9r^2} \le \frac{\sqrt{3}R^2}{2r^2} \le \frac{9R^3}{4r^2 s}.$$

cxliv) $\csc A + \csc B + \csc C \le \dfrac{R}{r}(\tan\frac{A}{2} + \tan\frac{B}{2} + \tan\frac{C}{2}) \le \dfrac{9R^2}{2rs} \le \dfrac{Rs}{3r^2} \le \dfrac{\sqrt{3}R(4R + r)}{9r^2} \le \dfrac{\sqrt{3}R^2}{2r^2}.$

cxlv) $\dfrac{2\sqrt{3}(2R-r)}{3R} \le \dfrac{3(2R-r)}{2R+(3\sqrt{3}-4)r} \le \dfrac{3(2R-r)}{s} \le \cot A + \cot B + \cot C.$

cxlvi) $\left.\begin{matrix} 4 \\ \dfrac{4(5R-r)^2}{4R^2+4Rr+3r^2} - \dfrac{4R}{r} \end{matrix}\right\} \le \dfrac{27R^2}{s^2} \le \csc^2 A + \csc^2 B + \csc^2 C \le \dfrac{4(R+r)^4}{r^3(16R-5r)} - \dfrac{4R}{r}.$

cxlvii) $6 - \dfrac{3\sqrt{3}}{2} + \dfrac{s}{R} \le 6 \le \left\{\begin{matrix} \dfrac{9\sqrt{3}R}{s} \\ 6\sqrt[3]{\dfrac{R}{2r}} \end{matrix}\right\} \le \csc \frac{A}{2} + \csc \frac{B}{2} + \csc \frac{C}{2}.$

cxlviii) $12 \le \csc^2 \frac{A}{2} + \csc^2 \frac{B}{2} + \csc^2 \frac{C}{2}.$

cxlix) $6 \le (\csc \frac{A}{2}) \cos \frac{B-C}{2} + (\csc \frac{B}{2}) \cos \frac{C-A}{2} + (\csc \frac{C}{2}) \cos \frac{A-B}{2}.$

cl) $(\csc \frac{A}{2})(\csc \frac{B}{2}) \csc \frac{C}{2} \le \frac{2}{7}[1 + (\cot^2 \frac{A}{2})(\cot^2 \frac{B}{2}) \cot^2 \frac{C}{2}].$

cli) $12 \le (\csc \frac{A}{2}) \csc \frac{B}{2} + (\csc \frac{B}{2}) \csc \frac{C}{2} + (\csc \frac{C}{2}) \csc \frac{A}{2}.$

clii) $\sqrt{\csc \frac{A}{2}} + \sqrt{\csc \frac{B}{2}} + \sqrt{\csc \frac{C}{2}} \le 2\sqrt{\frac{R}{r}}(\sin \frac{A}{2} + \sin \frac{B}{2} + \sin \frac{C}{2}).$

cliii) $2\sqrt{3} \le \sec \frac{A}{2} + \sec \frac{B}{2} + \sec \frac{C}{2} \le \frac{9}{4}(\sec \frac{A}{2})(\sec \frac{B}{2}) \sec \frac{C}{2} \le \csc A + \csc B + \csc C \le \dfrac{a^2+b^2+c^2}{2\,\mathrm{area}(\mathcal{S})}.$

cliv) $4 \le \sec^2 \frac{A}{2} + \sec^2 \frac{B}{2} + \sec^2 \frac{C}{2}.$

clv) Let $p \ge 1$. Then, $4\left(\dfrac{2\sqrt{3}}{3}\right)^p s \le (a+b)\sec^p \frac{C}{2} + (b+c)\sec^p \frac{A}{2} + (c+a)\sec^p \frac{B}{2}.$

clvi) $12 \le (\sec A) \sec B + (\sec B) \sec C + (\sec C) \sec A.$

clvii)
$$\sqrt{3} \le \frac{\sqrt{3}(a+b+c)(a^2+b^2+c^2)}{9abc} \le \sqrt{3} + \frac{(a-b)^2+(b-c)^2+(c-a)^2}{4rs}$$
$$\le \cot A + \cot B + \cot C = \frac{a^2+b^2+c^2}{4rs} \le \sqrt{3} + \frac{3[(a-b)^2+(b-c)^2+(c-a)^2]}{4rs}.$$

clviii) $1 \le \dfrac{9(2R-r)^2}{4R^2+4Rr+3r^2} - 2 \le \dfrac{s^2}{9r^2} - 2 \le \cot^2 A + \cot^2 B + \cot^2 C \le \dfrac{(2R^2+r^2)^2}{r^3(16R-5r)} - 2 \le \dfrac{3R^3}{8r^3} - 2.$

clix) Let $n \ge 1$. Then, $\frac{3}{\sqrt{3^n}} \le \cot^n A + \cot^n B + \cot^n C.$

clx) $\dfrac{9}{4} \le \dfrac{1}{(\cot A + \cot B)^2} + \dfrac{1}{(\cot B + \cot C)^2} + \dfrac{1}{(\cot C + \cot A)^2}.$

clxi) $\dfrac{12Rr - 4R^2 - 6r^2}{3\sqrt{3}Rr} \le \dfrac{6Rr - 2R^2 - 3r^2}{2Rr + (3\sqrt{3}-4)r^2} \le \dfrac{6Rr - 2R^2 - 3r^2}{rs} \le (\cot A)(\cot B)\cot C \le \dfrac{r}{s} \le \dfrac{\sqrt{3}}{9}.$

clxii) $3\sqrt{3} \le \cot \frac{A}{2} + \cot \frac{B}{2} + \cot \frac{C}{2} \le \dfrac{2R}{r} + 3\sqrt{3} - 4 \le \dfrac{3\sqrt{3}R}{2r}.$

clxiii) $9 \le \dfrac{9R}{2r} \le \left\{\begin{matrix} \dfrac{s^2}{3r^2} \\ \dfrac{8R-7r}{r} \end{matrix}\right\} \le \cot^2 \frac{A}{2} + \cot^2 \frac{B}{2} + \cot^2 \frac{C}{2} \le \frac{(2R-r)^2}{r^2}.$

clxiv) Let $n \ge 1$. Then, $3\sqrt{3^n} \le \cot^n \frac{A}{2} + \cot^n \frac{B}{2} + \cot^n \frac{C}{2}.$

clxv) $9 \le (\cot \frac{A}{2}) \cot \frac{B}{2} + (\cot \frac{B}{2}) \cot \frac{C}{2} + (\cot \frac{C}{2}) \cot \frac{A}{2} = 1 + \dfrac{4R}{r} \le \dfrac{9R}{2r} \le \dfrac{s^2}{3r^2}.$

clxvi) $\frac{\sqrt{3}}{2}(\cot \frac{A}{2} + \cot \frac{B}{2} + \cot \frac{C}{2}) \le (\cos \frac{A}{2}) \cot \frac{A}{2} + (\cos \frac{B}{2}) \cot \frac{B}{2} + (\cos \frac{C}{2}) \cot \frac{C}{2}.$

clxvii) $3\sqrt{3} \le (\cot\frac{A}{2})(\cot\frac{B}{2})(\cot\frac{C}{2})$.

clxviii) $\sqrt{\cot\frac{A}{2}} + \sqrt{\cot\frac{B}{2}} + \sqrt{\cot\frac{C}{2}} \le \sqrt{\frac{3s}{r}} \le \sqrt{3\left(\frac{2R}{r} + 3\sqrt{3} - 4\right)} \le 3\sqrt{\frac{\sqrt{3}R}{2r}}$.

clxix) $3(2+\sqrt{3}) \le 6 + \frac{s}{r} \le 3(2-\sqrt{3}) + \frac{2s}{r} \le \cot\frac{A}{4} + \cot\frac{B}{4} + \cot\frac{C}{4}$.

clxx) Let A denote the largest angle of $\mathcal{S}$. Then,

$$s \le \begin{Bmatrix} \sqrt{\dfrac{(4R+r)^2R}{2(2R-r)}} \\ 2R + \left(3\sqrt{3} - 4 - \dfrac{9(2-\sqrt{3})(\sqrt{3}-\cot\frac{A}{2})^2}{8\cot^2\frac{A}{2}}\right)r \le 2R + \left(3\sqrt{3} - 4 - \dfrac{(\sqrt{3}-\cot\frac{A}{2})^3}{4\cot\frac{A}{2}}\right)r \end{Bmatrix}$$

$$\le 2R + (3\sqrt{3} - 4)r.$$

clxxi) $\frac{3r}{R} \le \frac{a^2\cot\frac{A}{2} + b^2\cot\frac{B}{2} + c^2\cot\frac{C}{2}}{a^2\tan\frac{A}{2} + b^2\tan\frac{B}{2} + c^2\tan\frac{C}{2}} \le 3 \le \frac{3R}{2r}$.

clxxii) $3\sqrt{3} \le \cot\frac{1}{4}(A+B) + \cot\frac{1}{4}(B+C) + \cot\frac{1}{4}(C+A) \le \frac{3\sqrt{3}}{2} + \frac{s}{2r}$.

clxxiii) $AB\log C + BC\log A + CA\log B \le \frac{\pi^2}{3}\log\frac{\pi}{3}$.

Source: [47, pp. 18, 27, 98, 127, 141, 144], [107, pp. 31, 32, 55, 176, 180, 295, 296, 585, 607, 608], [108, p. 446], [109, pp. 136, 137, 270, 271, 455–457], [120, 142, 158, 159], [288, p. 231], [321, 367, 373, 364], [535, pp. 15, 18, 21, 23–27, 29, 31–33, 35–41], [626], [806, pp. 45, 210, 211, 253], [1234, 1532, 1681, 1728], [1860, pp. 22, 122, 123, 150], [1901], [1938, pp. 31, 32, 68, 101], [1969, pp. 202, 203], [1971, pp. 281, 282], [2062, pp. 96, 97, 127, 135, 154–159, 164, 165, 168, 170, 180, 182, 184–192, 683, 684], [2129, 2162, 2278], [2366, pp. 14, 19], [2367], [2604, 2605, 2791, 2928, 2995, 3001]. **Remark:** The left-hand inequality in *xxii)* is *Klamkin's inequality*. See Fact 2.3.69 and [2791]. The first and third terms are *Vasic's inequality*. See [2791]. The right-most inequality in *iv)* follows from *v)* with $x = y = z = 1$.

Fact 5.2.11. Let $\mathcal{S} \subset \mathbb{R}^2$ denote a triangle with the notation defined in Fact 5.2.7. Furthermore, let l_a denote the length of the *angle bisector* from vertex A to side a, let h_a denote the length of the *altitude* from vertex A to side a, which intersects the line containing side a in a right angle, and let m_a denote the length of the *median* from vertex A to the center of side a. Then, the following statements hold:

i) $m_a = \frac{1}{2}\sqrt{2b^2 + 2c^2 - a^2} = \sqrt{s(s-a) + \frac{1}{4}(b-c)^2}$.

ii) $h_a = \frac{bc}{2R} = 2R(\sin B)\sin C = \frac{a}{\cot B + \cot C} = l_a\cos\frac{B-C}{2}$.

iii) $l_a = \frac{2bc}{b+c}\cos\frac{A}{2} = \frac{2\,\mathrm{area}(\mathcal{S})}{(b+c)\sin\frac{A}{2}} = \frac{2\sqrt{bc}}{b+c}\sqrt{s(s-a)}$.

iv) $l_a = \frac{\sqrt{bc}}{b+c}\sqrt{(b+c)^2 - a^2} = \frac{2abc\sin\frac{B-C}{2}}{b^2 - c^2} = h_a\sec\frac{B-C}{2} = \frac{bc}{2R\cos\frac{B-C}{2}}$.

v) $\mathrm{area}(\mathcal{S}) = \frac{1}{2}h_a a = \frac{a^2}{2(\cot B + \cot C)} = \frac{h_a h_b}{2\sin C} = \frac{R h_b h_c}{a} = \frac{(a+b)(b+c)(c+a)}{4abc(a+b+c)} l_a l_b l_c$.

vi) $\mathrm{area}(\mathcal{S}) = \sqrt{\frac{1}{2}R h_a h_b h_c} = \frac{1}{2}\sqrt{ab h_a h_b} = \frac{1}{2}\sqrt[3]{abc h_a h_b h_c} = \frac{a+b}{2(h_a^{-1} + h_b^{-1})}$.

vii) Let $m \triangleq \frac{1}{2}(m_{\rm a} + m_{\rm b} + m_{\rm c})$. Then, $m_{\rm a} < m_{\rm b} + m_{\rm c}$, and

$$\text{area}(\mathcal{S}) = \tfrac{4}{3}\sqrt{m(m-m_{\rm a})(m-m_{\rm b})(m-m_{\rm c})}.$$

Consequently, the medians $m_{\rm a}, m_{\rm b}, m_{\rm c}$ are the sides of a triangle whose area is $\frac{3}{4}$ area($\mathcal{S}$).

viii) Let $h \triangleq \frac{1}{2}(h_{\rm a}^{-1} + h_{\rm b}^{-1} + h_{\rm c}^{-1})$. Then, $h_{\rm a}^{-1} < h_{\rm b}^{-1} + h_{\rm c}^{-1}$, and

$$\text{area}(\mathcal{S}) = \frac{1}{4\sqrt{h(h-h_{\rm a}^{-1})(h-h_{\rm b}^{-1})(h-h_{\rm c}^{-1})}}.$$

ix) $m_{\rm a}^2 + m_{\rm b}^2 + m_{\rm c}^2 = \frac{3}{4}(a^2+b^2+c^2)$.

x) $h_{\rm a} + h_{\rm b} + h_{\rm c} = \dfrac{s^2+r^2+4Rr}{2R} = \dfrac{ab+bc+ca}{2R} = 2rs\left(\dfrac{1}{a}+\dfrac{1}{b}+\dfrac{1}{c}\right).$

xi) $h_{\rm a}h_{\rm b} + h_{\rm b}h_{\rm c} + h_{\rm c}h_{\rm a} = \dfrac{2rs^2}{R}, \quad ah_{\rm a} + bh_{\rm b} + ch_{\rm c} = 6\text{area}(\mathcal{S}).$

xii) $h_{\rm a}^2 + h_{\rm b}^2 + h_{\rm c}^2 = \dfrac{(s^2+r^2+4Rr)^2 - 16s^2Rr}{4R^2}, \quad \dfrac{1}{h_{\rm a}} + \dfrac{1}{h_{\rm b}} + \dfrac{1}{h_{\rm c}} = \dfrac{1}{r}.$

xiii) $h_{\rm a}^3 + h_{\rm b}^3 + h_{\rm c}^3 = \dfrac{(s^2+r^2+4Rr)^3 - 24s^2Rr(s^2+r^2+4Rr) + 48s^2R^2r^2)}{8R^3}.$

xiv) $h_{\rm a}h_{\rm b}h_{\rm c} = \dfrac{2r^2s^2}{R}, \quad (h_{\rm a}+h_{\rm b})(h_{\rm b}+h_{\rm c})(h_{\rm c}+h_{\rm a}) = \dfrac{rs^2(s^2+r^2+2Rr)}{R^2}.$

xv) $\dfrac{h_{\rm a}+h_{\rm b}}{h_{\rm c}} + \dfrac{h_{\rm b}+h_{\rm c}}{h_{\rm a}} + \dfrac{h_{\rm c}+h_{\rm a}}{h_{\rm b}} = \dfrac{a+b}{c} + \dfrac{b+c}{a} + \dfrac{c+a}{b} = \dfrac{s^2+r^2-2Rr}{2Rr}.$

xvi) $\dfrac{(h_{\rm a}+h_{\rm b})(h_{\rm b}+h_{\rm c})(h_{\rm c}+h_{\rm a})}{(a+b)(b+c)(c+a)} = \dfrac{rs}{2R^2}, \quad r = \dfrac{h_{\rm a}h_{\rm b}h_{\rm c}}{h_{\rm a}h_{\rm b} + h_{\rm b}h_{\rm c} + h_{\rm c}h_{\rm a}}.$

xvii) $\dfrac{1}{h_{\rm a}h_{\rm b}} + \dfrac{1}{h_{\rm b}h_{\rm c}} + \dfrac{1}{h_{\rm c}h_{\rm a}} = \dfrac{s^2+r^2+4Rr}{4r^2s^2}, \quad \dfrac{1}{h_{\rm a}^2} + \dfrac{1}{h_{\rm b}^2} + \dfrac{1}{h_{\rm c}^2} = \dfrac{s^2-r^2-4Rr}{2r^2s^2}.$

xviii) $h_{\rm a}, h_{\rm b}, h_{\rm c}$ are the roots of the cubic equation

$$2Rx^3 - (s^2+r^2+4Rr)x^2 + 4s^2rx - 4s^2r^2 = 0.$$

xix) $1/h_{\rm a}, 1/h_{\rm b}, 1/h_{\rm c}$ are the roots of the cubic equation

$$4s^2r^2x^3 - 4s^2rx^2 + (s^2+r^2+4Rr)x - 2R = 0.$$

xx) $\dfrac{(b+c)^2l_{\rm a}^2}{bc} = (b+c)^2 - a^2, \quad \dfrac{(b+c)^2l_{\rm a}^2}{bc} + \dfrac{(c+a)^2l_{\rm b}^2}{ca} + \dfrac{(a+b)^2l_{\rm c}^2}{ab} = (a+b+c)^2.$

xxi) $al_{\rm a}^2 + bl_{\rm b}^2 + cl_{\rm c}^2 \le 2rs(4R+r), \quad \dfrac{l_{\rm a}l_{\rm b}}{l_{\rm c}} + \dfrac{l_{\rm b}l_{\rm c}}{l_{\rm a}} + \dfrac{l_{\rm c}l_{\rm a}}{l_{\rm b}} = \dfrac{2(2Rs^2+4Rr^2+r^3+rs^2)}{s^2+2Rr+r^2}.$

xxii) $\dfrac{\sin\frac{A-B}{2}}{l_{\rm c}} + \dfrac{\sin\frac{1}{2}(B-C)}{l_{\rm a}} + \dfrac{\sin\frac{1}{2}(C-A)}{l_{\rm c}} = 0, \quad m_{\rm a}^2 - l_{\rm a}^2 = \left(\dfrac{b-c}{2}\right)^2\left[2 - \left(\dfrac{a}{b+c}\right)^2\right].$

xxiii) If $\mathcal{S}$ is not equilateral, then $R = \dfrac{l_{\rm a}^2\sqrt{m_{\rm a}^2-h_{\rm a}^2}}{2h_{\rm a}\sqrt{l_{\rm a}^2-h_{\rm a}^2}}.$

xxiv) $m_{\rm a}^2l_{\rm a}^2 = s^2(s-a)^2 + \dfrac{(b-c)^2[\text{area}(\mathcal{S})]^2}{(b+c)^2}.$

xxv) If $\mathcal{S}$ is equilateral, then $l_{\rm a} = h_{\rm a} = m_{\rm a} = \frac{\sqrt{3}}{2}a.$

xxvi) Assume that $C = \frac{\pi}{2}$, and let c_0 denote the length of the portion of c between the intersection of h_c and B. Then,

$$a = c\sin A = c\cos A, \quad a^2 + b^2 = c^2, \quad \operatorname{area}(\mathcal{S}) = \tfrac{1}{2}ab,$$

$$r = \frac{ab}{2s}, \quad R = \tfrac{1}{2}c, \quad c = \frac{a^2}{c_0} = \frac{ab}{h_c}.$$

Furthermore, if $n \ge 3$, then $a^n + b^n < c^n$. Finally, let $\mathcal{S}_1$ and $\mathcal{S}_2$ denote the triangles formed by partitioning $\mathcal{S}$ by h_c, and let r_1 and r_2 denote the inradii of $\mathcal{S}_1$ and $\mathcal{S}_2$, respectively. Then, $r_1 + r_2 = h_c$.

xxvii) The three bisectors of the angles of $\mathcal{S}$ intersect at a single point inside $\mathcal{S}$. This point is the *incenter* and is denoted by I. The incenter is the center of the incircle.

xxviii) The three lines that are perpendicular bisectors of the sides of $\mathcal{S}$ intersect at a single point. This point is the *circumcenter* and is denoted by O. The circumcenter is the center of the circumcircle.

xxix) The three altitudes of $\mathcal{S}$ intersect at a single point. This point is the *orthocenter* and is denoted by H.

xxx) The three medians of $\mathcal{S}$ intersect at a single point inside $\mathcal{S}$. This point is the *centroid* and is denoted by G. The distance from each vertex to the centroid is two-thirds of the length of the corresponding median.

xxxi) The circumcenter, centroid, and orthocenter are colinear. The centroid lies between the circumcenter and the orthocenter.

xxxii) The distance from the circumcenter O to the centroid G is $\mathrm{OG} = \sqrt{R^2 - (a^2 + b^2 + c^2)/9}$. Furthermore, $3\mathrm{OG}^2 + \mathrm{GA}^2 + \mathrm{GB}^2 + \mathrm{GC}^2 = 3R^2$.

xxxiii) The distance from the centroid G to the orthocenter H is $\mathrm{GH} = 2\sqrt{R^2 - (a^2 + b^2 + c^2)/9}$.

xxxiv) The distance from the circumcenter O to the orthocenter H is $\mathrm{OH} = \sqrt{9R^2 - (a^2 + b^2 + c^2)}$ $= R\sqrt{1 - 8(\cos A)(\cos B)\cos C}$.

xxxv) The distance from the incenter I to the circumcenter O is $\mathrm{IO} = \sqrt{R(R - 2r)}$.

xxxvi) The distance from the incenter I to the centroid G is $\mathrm{IG} = \frac{1}{3}\sqrt{s^2 + 5r^2 - 16Rr}$.

xxxvii) The distance from the incenter I to the orthocenter H is $\mathrm{IH} = \sqrt{4R^2 + 4Rr + 3r^2 - s^2}$ $= \sqrt{2r^2 - 4R^2(\cos A)(\cos B)\cos C}$.

xxxviii) If $\mathcal{S}$ is equilateral, then the incenter, circumcenter, orthocenter, and centroid coincide.

xxxix) Let GA, GB, GC denote the distances from the centroid G to the vertices A, B, C, respectively. Then, $\mathrm{GA}^2 + \mathrm{GB}^2 + \mathrm{GC}^2 = \frac{1}{3}(a^2 + b^2 + c^2)$.

xl) Let M be a point in the plane containing $\mathcal{S}$ with centroid G. Then,

$$\mathrm{MA}^2 + \mathrm{MB}^2 + \mathrm{MC}^2 = \mathrm{GA}^2 + \mathrm{GB}^2 + \mathrm{GC}^2 + 3\mathrm{MG}^2.$$

xli) The distance from the incenter I to the vertex A is $\mathrm{IA} = r\csc\frac{A}{2}$.

xlii) Let IA, IB, IC denote the distances from the incenter I to the vertices A, B, C, respectively. Then,

$$2r \le \max\{\mathrm{IA}, \mathrm{IB}, \mathrm{IC}\}, \quad 8r^3 \le (\mathrm{IA})(\mathrm{IB})\mathrm{IC},$$

$$6r \le 3\sqrt[3]{4Rr^2} \le \mathrm{IA} + \mathrm{IB} + \mathrm{IC} \le 2(R + r) \le \sqrt{12(R^2 - Rr + r^2)} \le 4R - 2r.$$

xliii) Let Ta, Tb, Tc denote the points of tangency of the incircle on sides a, b, c, respectively, let Ma, Mb, Mc denote the midpoints of sides a, b, c, respectively, let IMa, IMb, IMc be the distances from the incenter I to Ma, Mb, Mc, respectively, and let MaTa denote the distance between Ma and Ta. Then, $\mathrm{MaTa} = \frac{1}{2}(b - c)$ and $r(2R - r) \le \mathrm{IMa}^2 + \mathrm{IMb}^2 + \mathrm{IMc}^2$.

xliv) Let Oa, Ob, Oc be the distances from the circumcenter O to the sides a, b, c, respectively. Then, $R < \mathrm{Oa} + \mathrm{Ob} + \mathrm{Oc}$. Furthermore, $R + r = \pm\mathrm{Oa} \pm \mathrm{Ob} \pm \mathrm{Oc}$, where "−" is used in the case where the corresponding line segment lies entirely outside $\mathcal{S}$. Finally, if $\mathcal{S}$ is acute, then $R + r = \mathrm{Oa} + \mathrm{Ob} + \mathrm{Oc}$.

xlv) Let a_1 and a_2 denote the lengths of the portions of side a determined by the bisector of A. Then, $\min\{b/c, c/b\} = \min\{a_1/a_2, a_2/a_1\}$.

xlvi) Let P be a point inside $\mathcal{S}$, and let Pa, Pb, Pc be the distances from P to the sides a, b, c, respectively. Then,

$$\sqrt{\mathrm{Pa}} + \sqrt{\mathrm{Pb}} + \sqrt{\mathrm{Pc}} \le \sqrt{\frac{a^2+b^2+c^2}{2R}} \le 3\sqrt{\frac{R}{2}}.$$

In particular, if P is the incenter I, then $\mathrm{Pa} = \mathrm{Pb} = \mathrm{Pc} = r$, and thus

$$18rR \le a^2 + b^2 + c^2.$$

Furthermore, equality holds if and only if P is the incenter and $\mathcal{S}$ is equilateral; that is, $R = 2r$ and $a = \sqrt{3}R$.

xlvii) Let P be a point inside $\mathcal{S}$, and let Pa, Pb, Pc be the distances from P to the sides a, b, c, respectively. Then,

$$\frac{a+b+c}{r} \le \frac{a}{\mathrm{Pa}} + \frac{b}{\mathrm{Pb}} + \frac{c}{\mathrm{Pc}}.$$

Equality holds if and only if P is the incenter. Furthermore,

$$2sr^2 \le a\mathrm{Pa}^2 + b\mathrm{Pb}^2 + c\mathrm{Pc}^2.$$

Equality holds if and only if P is the incenter. In addition,

$$2rs \le a\mathrm{Pa} + b\mathrm{Pb} + c\mathrm{Pc}.$$

If $\mathcal{S}$ is equilateral, then

$$\frac{3}{r} \le \frac{1}{\mathrm{Pa}} + \frac{1}{\mathrm{Pb}} + \frac{1}{\mathrm{Pc}}.$$

If P is the centroid, then

$$\mathrm{Pa} + \mathrm{Pb} + \mathrm{Pc} \le 3r, \quad \frac{1}{3}(\sin A + \sin B + \sin C) \le \frac{\mathrm{Pa}}{a} + \frac{\mathrm{Pb}}{b} + \frac{\mathrm{Pc}}{c}, \quad \frac{6}{R} \le \frac{1}{\mathrm{Pa}} + \frac{1}{\mathrm{Pb}} + \frac{1}{\mathrm{Pc}}.$$

xlviii) Let P be a point inside $\mathcal{S}$, and let Pa, Pb, Pc be the distances from P to the sides a, b, c, respectively. Then,

$$\operatorname{area}(\mathcal{S}) \le \frac{3}{2}R\sqrt{\mathrm{Pa}^2 + \mathrm{Pb}^2 + \mathrm{Pc}^2}.$$

xlix) Let P be a point inside $\mathcal{S}$, let PA, PB, PC denote the distances from P to the vertices A, B, C, respectively, and let Pa, Pb, Pc denote the distances from P to the sides a, b, c, respectively. Furthermore, define $u = \mathrm{Pa}$, $v = \mathrm{Pb}$, $w = \mathrm{Pc}$, $p = \mathrm{PA}$, $q = \mathrm{PB}$, and $r = \mathrm{PC}$. Then,

$$2u + \frac{(v+w)^2}{p} \le q + r, \quad \sqrt{a^2 + 4u^2} \le q + r,$$

$$4(uv + vw + wu) \le 2(up + vq + wr) \le pq + qr + rp,$$

$$u^2 + v^2 + w^2 + 3(uv + vw + wu) \le pq + qr + rp,$$

$$2(u + v + w) \le p + q + r \le \frac{1}{2}\left(\frac{pq}{w} + \frac{qr}{u} + \frac{rp}{v}\right), \quad (u + v + w)^3 \le uh_{\mathrm{a}}^2 + vh_{\mathrm{b}}^2 + wh_{\mathrm{c}}^2,$$

$$2\left(\frac{1}{p}+\frac{1}{q}+\frac{1}{r}\right) \le \frac{1}{u}+\frac{1}{v}+\frac{1}{w}, \qquad \frac{36}{p+q+r} \le \frac{p}{vw}+\frac{q}{wu}+\frac{r}{uv},$$

$$4(l_{\rm a}l_{\rm b}+l_{\rm b}l_{\rm c}+l_{\rm c}l_{\rm a}) \le uv+vw+wu, \qquad uvw \le (\sin\tfrac{A}{2})(\sin\tfrac{B}{2})(\sin\tfrac{C}{2})pqr.$$

If $\alpha \in [-2,2]$, then

$$2+\frac{\alpha}{4} \le \frac{p^2+q^2+\alpha w^2}{a^2}+\frac{q^2+r^2+\alpha u^2}{b^2}+\frac{r^2+p^2+\alpha v^2}{c^2}.$$

If $x,y,z \ge 0$, then $2(yzu+zxv+xyw) \le x^2p+y^2q+z^2r$. If $\mathcal{S}$ is equilateral, then

$$a^2 = \tfrac{1}{2}(p^2+q^2+r^2)+2\sqrt{3\mu(\mu-p)(\mu-q)(\mu-r)},$$

where $\mu \triangleq \frac{1}{2}(p+q+r)$.

l) Let P be a point inside $\mathcal{S}$, and let PA, PB, PC denote the distances from P to the vertices A, B, C, respectively. Then,

$$abc \le a\mathrm{PBPC}+b\mathrm{PCPA}+c\mathrm{PAPB}.$$

Equality holds if and only if P = O. Furthermore,

$$abc \le a\mathrm{PA}^2+b\mathrm{PB}^2+c\mathrm{PC}^2 \le a^2\mathrm{PA}+b^2\mathrm{PB}+c^2\mathrm{PC}.$$

Equality holds in the first inequality if and only if P = I. In addition,

$$\frac{16}{9}[\mathrm{area}(\mathcal{S})]^2 \le 8(R+r)Rr^2 \le \frac{(abc)^2}{a^2+b^2+c^2} \le (\mathrm{PAPB})^2+(\mathrm{PBPC})^2+(\mathrm{PCPA})^2,$$

$$48r^3 \le \mathrm{PA}^3+\mathrm{PB}^3+\mathrm{PC}^3+3\mathrm{PAPBPC},$$

$$12Rr^2 \le (\mathrm{PAPB})^{3/2}+(\mathrm{PBPC})^{3/2}+(\mathrm{PCPA})^{3/2}.$$

Now, let $\alpha \ge \frac{2(\log 3-\log 2)}{3\log 3-4\log 2} \approx 1.54980$. Then,

$$3\left(\frac{4\sqrt{3}}{9}\mathrm{area}(\mathcal{S})\right)^{\alpha} \le (\mathrm{PAPB})^{\alpha}+(\mathrm{PBPC})^{\alpha}+(\mathrm{PCPA})^{\alpha}.$$

Finally, let $\alpha \ge \frac{9}{8}$. Then,

$$3(4r^2)^{\alpha} \le (\mathrm{PAPB})^{\alpha}+(\mathrm{PBPC})^{\alpha}+(\mathrm{PCPA})^{\alpha}.$$

li) Let P be a point in the plane of $\mathcal{S}$, and let PA, PB, PC denote the distances from P to the vertices A, B, C, respectively. Then,

$$a^2+b^2+c^2 \le 3(\mathrm{PA}^2+\mathrm{PB}^2+\mathrm{PC}^2).$$

Now, assume that $c \le b \le a$. Then,

$$bc \le \mathrm{PAPB}+\mathrm{PBPC}+\mathrm{PCPA}.$$

lii) A triangle $\mathcal{S}$ with semiperimeter s, inner radius r, and outer radius R exists if and only if

$$|s^2-2R^2-10Rr+r^2| \le 2(R-2r)\sqrt{R^2-2Rr}.$$

liii) Let $\theta \triangleq \min\{|A-B|,|A-C|,|B-C|\}$. Then,

$$2R^2+10Rr-r^2-2(R-2r)\sqrt{R(R-2r)}\cos\theta$$
$$\le s^2 \le 2R^2+10Rr-r^2+2(R-2r)\sqrt{R(R-2r)}\cos\theta.$$

liv) A triangle $\mathcal{S}$ with values $u = \cos A$, $v = \cos B$, and $v = \cos C$ exists if and only if $u+v+w \geq 1$, $uvw \geq -1$, and $u^2 + v^2 + w^2 + 2uvw = 1$.

lv) The triangle $\mathcal{S}$ is similar to the triangle $\mathcal{S}'$ with sides of length a', b', c' if and only if

$$\sqrt{aa'} + \sqrt{bb'} + \sqrt{cc'} = \sqrt{(a+b+c)(a'+b'+c')}.$$

lvi) Let P be a point in the plane containing $\mathcal{S}$, assume that P does not coincide with a vertex of $\mathcal{S}$, and let $\mathcal{S}'$ denote the triangle whose vertices A′, B′, C′ are the centroids of the triangles PAB, PBC, PCA. Then, $\mathcal{S}$ and $\mathcal{S}'$ are similar.

lvii) Let P be a point in the plane containing $\mathcal{S}$, let OP denote the distance from the circumcenter O to P, let A′, B′, C′ denote the points closest to P that are on the lines containing the sides of $\mathcal{S}$, and let $\mathcal{S}'$ denote the triangle with vertices A′, B′, C′. Then,

$$\text{area}(\mathcal{S}') = \frac{1}{4}\left|1 - \frac{\text{OP}^2}{R^2}\right| \text{area}(\mathcal{S}).$$

Furthermore, A′, B′, C′ lie on a single line if and only if P lies on the circle that circumscribes $\mathcal{S}$.

lviii) Assume that $b = c$ so that $\mathcal{S}$ is isosceles. Then,

$$h_\text{a} = \sqrt{b^2 - \tfrac{1}{4}a^2}, \quad \text{area}(\mathcal{S}) = \tfrac{1}{2}ah_\text{a} = \tfrac{1}{2}b^2\sin\theta, \quad r = \frac{a\sqrt{a^2 + 4h_\text{a}^2} - a}{4h_\text{a}}, \quad R = \frac{b^2}{\sqrt{4b^2 - a^2}}.$$

Furthermore, the distance from side a to the centroid of $\mathcal{S}$ is $h_\text{a}/3$.

lix) Assume that $\mathcal{S}$ lies in the x-y plane, let P be a point inside $\mathcal{S}$, let (x_A, y_A), (x_B, y_B), (x_C, y_C), (x_P, y_P) denote, respectively, the coordinates of A, B, C, P, and let PBC, PCA, PAB denote the three subtriangles of $\mathcal{S}$ with the listed vertices. Furthermore, define

$$\lambda_{\text{A,P}} \triangleq \frac{\text{area(PBC)}}{\text{area}(\mathcal{S})}, \quad \lambda_{\text{B,P}} \triangleq \frac{\text{area(PCA)}}{\text{area}(\mathcal{S})}, \quad \lambda_{\text{C,P}} \triangleq \frac{\text{area(PAB)}}{\text{area}(\mathcal{S})}.$$

Then,

$$\begin{bmatrix} x_\text{P} \\ y_\text{P} \end{bmatrix} = \lambda_{\text{A,P}}\begin{bmatrix} x_\text{A} \\ y_\text{A} \end{bmatrix} + \lambda_{\text{B,P}}\begin{bmatrix} x_\text{B} \\ y_\text{B} \end{bmatrix} + \lambda_{\text{C,P}}\begin{bmatrix} x_\text{C} \\ y_\text{C} \end{bmatrix}.$$

Finally, the coordinates of the incenter I, circumcenter O, orthocenter H, and centroid G are given by

$$\lambda_{\text{A,I}} = \frac{a}{a+b+c}, \quad \lambda_{\text{B,I}} = \frac{b}{a+b+c}, \quad \lambda_{\text{C,I}} = \frac{c}{a+b+c},$$

$$\lambda_{\text{A,O}} = \frac{a^2(b^2+c^2-a^2)}{\alpha} = \frac{2R^2\cos A}{bc}, \quad \lambda_{\text{B,O}} = \frac{b^2(c^2+a^2-b^2)}{\alpha} = \frac{2R^2\cos B}{ac},$$

$$\lambda_{\text{C,O}} = \frac{c^2(a^2+b^2-c^2)}{\alpha} = \frac{2R^2\cos C}{ab},$$

$$\lambda_{\text{A,H}} = \frac{(a^2+b^2-c^2)(c^2+a^2-b^2)}{\beta} = \frac{4R^2(\cos B)\cos C}{bc},$$

$$\lambda_{\text{B,H}} = \frac{(b^2+c^2-a^2)(a^2+b^2-c^2)}{\beta} = \frac{4R^2(\cos A)\cos C}{ac},$$

$$\lambda_{\text{C,H}} = \frac{(c^2+a^2-b^2)(b^2+c^2-a^2)}{\beta} = \frac{4R^2(\cos A)\cos B}{ab},$$

$$\lambda_{\text{A,G}} = \lambda_{\text{B,G}} = \lambda_{\text{C,G}} = 1/3,$$

where

$$\alpha \triangleq a^2(b^2+c^2-a^2)+b^2(c^2+a^2-b^2)+c^2(a^2+b^2-c^2),$$

$$\beta \triangleq (a^2+b^2-c^2)(c^2+a^2-b^2)+(b^2+c^2-a^2)(a^2+b^2-c^2)+(c^2+a^2-b^2)(b^2+c^2-a^2).$$

lx) The distances OI, OG, IG satisfy

$$\mathrm{OI}^2 = \mathrm{OG}^2 + 2\mathrm{IG}^2 + \tfrac{2}{3}r(R-2r),$$

$$\mathrm{IG} \le 2\mathrm{OG}, \quad \mathrm{OG} \le \mathrm{OI}, \quad 3\mathrm{IG} \le 2\mathrm{OI}$$

$$4\mathrm{OG}^2 + 8\mathrm{IG}^2 \le 4\mathrm{OI}^2 \le 16\mathrm{OG}^2 + 5\mathrm{IG}^2.$$

Furthermore, the triangle with vertices O, G, and I is obtuse.

lxi) Let D, E, F be arbitrary points on the sides a, b, c of $\mathcal{S}$, respectively. Then,

$$\min\{\mathrm{area(CDE)}, \mathrm{area(DBF)}, \mathrm{area(AEF)}\} \le \frac{3}{\frac{1}{\mathrm{area(CDE)}}+\frac{1}{\mathrm{area(DBF)}}+\frac{1}{\mathrm{area(AEF)}}} \le \mathrm{area(DEF)}.$$

lxii) Let D be the point on side a whose distance from A along the perimeter is s, let E be the point on side b whose distance from B along the perimeter is s, and let F be the point on side c whose distance from C along the perimeter is s. Then, the line segments AD, BE, and CF intersect at a point inside $\mathcal{S}$. Denote this point by N, and let $\mathcal{S}_\mathrm{N}$ denote the triangle whose vertices are D, E, F. Then, O, I, and N are collinear. Furthermore,

$$\mathrm{area}(\mathcal{S}_\mathrm{N}) = \frac{2r^2 s}{abc}\mathrm{area}(\mathcal{S}).$$

lxiii) Let D be the point on side a at which the inscribed circle intersects side a, let E be the point on side b at which the inscribed circle intersects side b, and let F be the point on side c at which the inscribed circle intersects side c. Then, the line segments AD, BE, and CF intersect at a point inside $\mathcal{S}$. Denote this point by $\mathrm{G_e}$, and let $\mathcal{S}_{\mathrm{G_e}}$ denote the triangle whose vertices are D, E, F. Then, $\mathrm{area}(\mathcal{S}_{\mathrm{G_e}}) = \mathrm{area}(\mathcal{S}_\mathrm{N})$.

Source: [47, pp. 90, 96, 99], [59, pp. 64, 65, 70, 83, 85, 101, 102], [107, pp. 13, 25, 102, 103, 115, 462, 463, 608–610], [108, pp. 13, 108], [109, pp. 151, 315], [110, pp. 143, 144], [157, 286, 396], [806, p. 208], [993, pp. 165, 181, 183, 322], [1026], [1158, p. 209], [1295, 1521, 1559, 1681], [1757, pp. 255, 256, 259], [1864, 1865, 1866, 1867], [1938, pp. 58, 68, 69], [1994], [2062, pp. 2, 62, 63, 163, 220, 223], [2294, p. 272], [2375, 2457, 2922, 2925]. **Remark:** The centroid of a triangle with uniform mass distribution coincides with the *center of mass* of the triangle, which is also called the *barycenter*. **Remark:** The equality in *xliii*) for the acute case is *Carnot's theorem*. See [47, p. 99]. The assumption that $\mathcal{S}$ is acute is omitted in [59, p. 83]. **Remark:** $2(u+v+w) \le p+q+r$ in *xlvii*) is the *Erdös-Mordell theorem*. See [47, pp. 93–97] and [1938, p. 81]. $u^2+v^2+w^2+3(uv+vw+wu) \le pq+qr+rp$ is *Oppenheim's geometric inequality*. See [1864]. The last statement is given in [396]. **Remark:** *li*), which is due to É. Rouché, is the *fundamental triangle inequality*. See [2062, p. 2] and [2922]. **Remark:** *xxxiii*) corrects a misprint in [59, p. 102]. **Remark:** In *lv*), $\lambda_{\mathrm{A,P}}, \lambda_{\mathrm{B,P}}, \lambda_{\mathrm{C,P}}$ are called *barycentric coordinates* (also called *areal coordinates*). See [1698, 2409, 2858, 2860]. **Remark:** The first and third terms in *l*) comprise the *Erdös-Debrunner inequality*. Refined bounds are discussed in [1521]. **Remark:** In *lxi*), N is the *Nagel point*. In *lxii*), $\mathrm{G_e}$ is the *Gergonne point*. **Problem:** Apply l'Hôpital's rule given by Fact 12.17.13 to *xxiii*) to obtain the expression for R for equilaterals.

Fact 5.2.12. Let $\mathcal{S} \subset \mathbb{R}^2$ denote a triangle with the notation defined in Fact 5.2.7. Then, the following statements hold:

i) The medians m_a, m_b, m_c satisfy

$$\left.\begin{array}{r}\frac{1}{2}(b+c-a) \le \sqrt{s(s-a)} \\ b-\frac{1}{2}a \\ c-\frac{1}{2}a \\ \frac{1}{2}|b-c|\end{array}\right\} \le m_a \le \begin{cases}\dfrac{Rs}{a} \\ \frac{1}{2}(b+c),\end{cases}$$

$$\sqrt{s(s-a)} + \sqrt{s(s-b)} \le \sqrt{2(s^2 - m_c^2)} \le \sqrt{3}s - m_c,$$

$$\left.\begin{array}{r} m_a + m_b \le \frac{3}{4}(a+b+c) \\ 9r \le \dfrac{ab+bc+ca}{2R} \le \dfrac{a^2+b^2+c^2}{2R}\end{array}\right\} \le m_a + m_b + m_c \le \begin{cases} a+b+c \\ 4R + r \le \frac{9}{2}R,\end{cases}$$

$$m_a + m_b + m_c \le \sqrt{4s^2 - 16Rr + 5r^2} \le 2s - 3(2\sqrt{3} - 3)r,$$

$$m_a + m_b + m_c \le \tfrac{1}{2}\sqrt{7(a^2+b^2+c^2) + 2(ab+bc+ca)},$$

$$a + b \le m_a + m_b + m_c \le a + b + c - \tfrac{6-3\sqrt{3}}{2}\min\{a,b,c\},$$

$$m_a + m_b \le \tfrac{3}{4}(a+b+c) \le m_a + m_b + m_c - \tfrac{6-3\sqrt{3}}{2}\min\{m_a, m_b, m_c\},$$

$$m_a + m_b + m_c \le \frac{(a^2+b^2+c^2)(am_a + bm_b + cm_c)}{3abc}, \quad \frac{3\sqrt{3}r}{2R} \le \frac{m_a}{b+c} + \frac{m_b}{c+a} + \frac{m_c}{a+b},$$

$$am_a + bm_b + cm_c \le \tfrac{\sqrt{3}}{2}(a^2+b^2+c^2), \quad m_a + m_b + m_c \le \frac{3}{4}\left(\frac{a^2}{m_a} + \frac{b^2}{m_b} + \frac{c^2}{m_c}\right),$$

$$6rs \le rs\max\left\{\frac{a+b}{c} + \frac{b+c}{a} + \frac{c+a}{b}, \frac{m_a+m_b}{m_c} + \frac{m_b+m_c}{m_a} + \frac{m_c+m_a}{m_b}\right\} \le am_a + bm_b + cm_c,$$

$$(\max\{a,b,c\})^2 - (\min\{a,b,c\})^2 \le am_a + bm_b + cm_c \le s\sqrt{4(\max\{a,b,c\})^2 - (\min\{a,b,c\})^2},$$

$$\frac{3\sqrt{3}}{2} \le \frac{3\sqrt{3}}{2} + \frac{3(2-\sqrt{3})}{2}\left[\frac{(a-b)^2}{ab} + \frac{(b-c)^2}{bc} + \frac{(c-a)^2}{ca}\right] \le \frac{m_a}{a} + \frac{m_b}{b} + \frac{m_c}{c},$$

$$a\cos A + b\cos B + c\cos C \le \tfrac{2}{3}(m_a\sin A + m_b\sin B + m_c\sin C) \le s,$$

$$\tfrac{9}{20}(ab+bc+ca) < m_am_b + m_bm_c + m_cm_a < \tfrac{5}{4}(ab+bc+ca),$$

$$\tfrac{1}{4}\max\{(a-b)^2, (b-c)^2, (c-a)^2\} < m_am_b + m_bm_c + m_cm_a < \tfrac{3}{4}\max\{(a+b)^2, (b+c)^2, (c+a)^2\},$$

$$9r(R+r) \le m_am_b + m_bm_c + m_cm_a \le 5R^2 + 2Rr + 3r^2 \le 5R^2 + \tfrac{9}{4}Rr + \tfrac{5}{2}r^2,$$

$$m_a\sec\tfrac{A}{2} + m_b\sec\tfrac{B}{2} + m_c\sec\tfrac{C}{2} \le \frac{2R}{rs}(m_am_b\sin\tfrac{C}{2} + m_bm_c\sin\tfrac{A}{2} + m_cm_a\sin\tfrac{B}{2}),$$

$$27r^2 \le \tfrac{27}{2}Rr \le \left\{\begin{array}{c}9r(2R-r) \\ s^2\end{array}\right\} \le m_a^2 + m_b^2 + m_c^2 \le 6R^2 + 3r^2 \le \tfrac{27}{4}R^2,$$

$$\frac{3}{4}(a+b+c) \le \frac{m_a^2}{a} + \frac{m_b^2}{b} + \frac{m_c^2}{c}, \quad \frac{9}{4} \le \frac{m_a^2}{a^2} + \frac{m_b^2}{b^2} + \frac{m_c^2}{c^2},$$

$$\frac{m_a^2}{b^2+c^2} + \frac{m_b^2}{c^2+a^2} + \frac{m_c^2}{a^2+b^2} \le \frac{9}{8}, \quad \frac{81r^2}{4} \le \frac{m_a^2m_b^2}{ab} + \frac{m_b^2m_c^2}{bc} + \frac{m_c^2m_a^2}{ca},$$

$$\frac{9}{4} \le \frac{m_a^2}{bc} + \frac{m_b^2}{ca} + \frac{m_c^2}{ab} = \frac{s^2 + 2Rr + 5r^2}{8Rr} \le \frac{4R^2 + 6Rr + 8r^2}{8Rr} \le \frac{2R+5r}{4r} \le \frac{9R}{8r},$$

$$r(m_a^2+m_b^2+m_c^2) \le m_a m_b m_c \le \frac{(a^2+b^2+c^2)(am_a+bm_b+cm_c)}{4(a+b+c)},$$

$$a(b+c)m_a^2+b(c+a)m_b^2+c(a+b)m_c^2 \le 12Rm_am_bm_c, \quad rs^2 \le m_am_bm_c \le \tfrac{1}{2}Rs^2,$$

$$9Rrs \le am_a^2+bm_b^2+cm_c^2 = \tfrac{1}{2}s(s^2+2Rr+5r^2), \quad r \le \frac{m_am_bm_c}{m_a^2+m_b^2+m_c^2} \le \tfrac{1}{2}R,$$

$$\operatorname{area}(\mathcal{S}) \le \frac{s\sqrt{4Rr+r^2}}{3} \le \frac{\sqrt{m_a^2m_b^2+m_b^2m_c^2+m_c^2m_a^2}}{3},$$

$$\operatorname{area}(\mathcal{S}) \le \tfrac{\sqrt{3}}{3}(m_am_bm_c)^{2/3} \le \tfrac{\sqrt{3}}{27}(m_a+m_b+m_c)^2,$$

$$m_a(cb-a^2)+m_b(ac-b^2)+m_c(ba-c^2) \le 0,$$

$$\frac{c^2-(a-b)^2}{2(a+b)} \le a+b-2m_c < \frac{c^2+(a-b)^2}{4m_c},$$

$$3\left(\frac{s-a}{m_a}+\frac{s-b}{m_b}+\frac{s-c}{m_c}\right) \le \frac{m_a}{s-a}+\frac{m_b}{s-b}+\frac{m_c}{s-c},$$

$$\frac{5}{s} < \frac{1}{m_a}+\frac{1}{m_b}+\frac{1}{m_c} \le \frac{2}{3}\left(\frac{1}{R}+\frac{1}{r}\right) \le \frac{1}{r}, \quad \frac{ab}{m_a}+\frac{bc}{m_b}+\frac{ca}{m_c} \le 6R,$$

$$\frac{a^2+b^2}{m_c}+\frac{b^2+c^2}{m_a}+\frac{c^2+a^2}{m_b} \le 12R, \quad \frac{1}{m_am_b}+\frac{1}{m_bm_c}+\frac{1}{m_cm_a} \le \frac{\sqrt{3}}{rs},$$

$$4 \le \frac{a^2}{m_bm_c}+\frac{b^2}{m_cm_a}+\frac{c^2}{m_am_b}, \quad \frac{1}{R^2} \le \frac{\sin^2 A}{m_bm_c}+\frac{\sin^2 B}{m_cm_a}+\frac{\sin^2 C}{m_am_b},$$

$$\frac{m_a}{m_bm_c}+\frac{m_b}{m_cm_a}+\frac{m_c}{m_am_b} \le 2R\left(\frac{1}{ab}+\frac{1}{bc}+\frac{1}{ca}\right), \quad 4 \le \frac{a^2}{m_a^2}+\frac{b^2}{m_b^2}+\frac{c^2}{m_c^2},$$

$$\frac{a^2}{m_b^2+m_c^2}+\frac{b^2}{m_c^2+m_a^2}+\frac{c^2}{m_a^2+m_b^2} \le 2, \quad 27r^2 \le \frac{m_a^4}{m_bm_c}+\frac{m_b^4}{m_cm_a}+\frac{m_c^4}{m_am_b}.$$

If $b < c$, then

$$\tfrac{1}{2}(c-b) < m_b - m_c < \tfrac{3}{2}(c-b).$$

The distance from the vertex A to the centroid G is $\mathrm{AG} = \frac{2}{3}m_a$.

ii) The altitudes h_a, h_b, h_c satisfy

$$\min\{h_a,h_b,h_c\} \le 3r, \quad h_a \le \begin{cases} \sqrt{s(s-a)} \le m_a \\ \sqrt{bc}\cos\frac{A}{2} \\ \dfrac{a}{2\tan\frac{A}{2}}, \end{cases}$$

$$9r \le \left\{\begin{matrix} 10r - \dfrac{2r^2}{R} \\ 2s\sqrt{\dfrac{2r}{R}}+(9-6\sqrt{3})r \le \dfrac{2\sqrt{3}s^2}{9R}+(9-3\sqrt{3})r \end{matrix}\right\}$$

$$\le \frac{2s^2}{5R}+\frac{18}{5}r \le h_a+h_b+h_c \le \frac{2s^2}{3R} \le \left\{\begin{matrix} 3(R+r) \\ \dfrac{a^2+b^2+c^2}{2R} \\ \sqrt{3}s \end{matrix}\right\} \le \frac{9R}{2},$$

$$0 \le \sqrt{3}s - (h_a + h_b + h_c) \le \tfrac{9}{2}(R - 2r),$$

$$h_a + h_b + h_c \le \frac{2(R+r)^2}{R} \le 2R + 5r \le 3(R+r) \le \frac{9R}{2},$$

$$\frac{2s}{R} \le \frac{2(4R^2 + 6Rr - r^2)}{Rs} \le \frac{h_a}{s-a} + \frac{h_b}{s-b} + \frac{h_c}{s-c} \le \frac{2(5R^2 + 3Rr + r^2)}{Rs},$$

$$3\sqrt{3r} \le \sqrt{h_a} + \sqrt{h_b} + \sqrt{h_c} \le \begin{cases} \sqrt{3\sqrt{3}s} \\ \left(1 + \frac{r}{R}\right)\sqrt{6R}, \end{cases}$$

$$\frac{54r^3}{R} \le \left\{ \begin{matrix} 27r^2 \\ \dfrac{6r^2(4R+r)}{R} \end{matrix} \right\} \le \frac{2r^2(16R - 5r)}{R} \le h_a h_b + h_b h_c + h_c h_a \le \left\{ \begin{matrix} \dfrac{2r(4R^2 + 4Rr + 3r^2)}{R} \\ 3\sqrt{3}rs \end{matrix} \right\}$$

$$\le \frac{2r(4R+r)^2}{3R} \le 3[(s-a)(s-b) + (s-b)(s-c) + (s-c)(s-a)] \le \frac{27Rr}{2} \le s^2 \le \frac{27R^2}{4},$$

$$\frac{54r^4}{R} \le 27r^3 \le \frac{2r^3(16R - 5r)}{R} \le h_a h_b h_c \le \left\{ \begin{matrix} \dfrac{2r^2(4R^2 + 4Rr + 3r^2)}{R} \\ 3\sqrt{3}(s-a)(s-b)(s-c) \end{matrix} \right\}$$

$$\le \left\{ \begin{matrix} \sqrt{3\sqrt{3}r^3 s^3} \\ \le \dfrac{2r^2(4R+r)^2}{3R} \le \dfrac{27Rr^2}{2} \end{matrix} \right\} \le rs^2 \le \frac{3\sqrt{3}}{8}abc \le \frac{27R^3}{8},$$

$$0 \le \sqrt{3}\sqrt[3]{abc} - 2\sqrt[3]{h_a h_b h_c} \le 3(R - 2r),$$

$$27r^2 \le h_a^2 + h_b^2 + h_c^2 \le \frac{(R+3r)s^2 - (4R+r)^2 r}{R} \le \left\{ \begin{matrix} s^2 \le \dfrac{3}{4}(a^2 + b^2 + c^2) \\ \dfrac{4R^2 + 7r^2 + 8r^3}{R} \le 4R^2 + 11r^2 \end{matrix} \right\} \le \frac{27R^2}{4},$$

$$\sqrt{a^2 + b^2 - h_c^2} + \sqrt{b^2 + c^2 - h_a^2} + \sqrt{c^2 + a^2 - h_b^2} \le 6R,$$

$$\frac{432r^4}{R} \le (h_a + h_b)(h_b + h_c)(h_c + h_a) \le 27R^3, \quad \frac{h_a}{a} + \frac{h_b}{b} + \frac{h_c}{c} \le \frac{s}{2r} \le \frac{3}{4}\left(\frac{a}{h_a} + \frac{b}{h_b} + \frac{b}{h_b}\right),$$

$$\frac{9r}{s} \le \frac{1}{3}\left(\frac{h_a}{s-a} + \frac{h_b}{s-b} + \frac{h_c}{s-c}\right) \le \frac{s-a}{h_a} + \frac{s-b}{h_b} + \frac{s-c}{h_c} \le \frac{9R}{2s},$$

$$\frac{4\sqrt{3}(7R - 2r)}{3(9R - 2r)} \le \frac{a}{h_b + h_c} + \frac{b}{h_c + h_a} + \frac{c}{h_a + h_b} < \sqrt{\frac{3R}{2r}},$$

$$3R \le \frac{a^2}{h_b + h_c} + \frac{b^2}{h_c + h_a} + \frac{c^2}{h_a + h_b} < 4R, \quad 2 \le \frac{a^2}{h_b^2 + h_c^2} + \frac{b^2}{h_c^2 + h_a^2} + \frac{c^2}{h_a^2 + h_b^2},$$

$$\frac{4}{3R^2} \le \frac{2}{3Rr} \le \frac{5R - r}{r(4R^2 + 4Rr + 3r^2)} \le \frac{1}{h_a h_b} + \frac{1}{h_b h_c} + \frac{1}{h_c h_a} \le \frac{1}{3r^2} \le \frac{(R+r)^2}{r^3(16R - 5r)} \le \frac{R}{6r^3},$$

$$\frac{2}{R} \le \frac{1}{h_a} + \frac{1}{h_b} + \frac{1}{h_c} = \frac{1}{r} \le \sqrt{\frac{h_a}{h_b^3}} + \sqrt{\frac{h_b}{h_c^3}} + \sqrt{\frac{h_c}{h_a^3}}, \quad 4 \le \frac{a^2}{h_b h_c} + \frac{b^2}{h_c h_a} + \frac{c^2}{h_a h_b},$$

$$\frac{4}{3R^2} \le \frac{2}{3Rr} \le \left\{ \begin{array}{c} \dfrac{\sqrt{3}}{\text{area}(\mathcal{S})} \\ \dfrac{3(2R-r)}{r(4R^2+4Rr+3r^2)} \end{array} \right\} \le \frac{1}{h_a^2} + \frac{1}{h_b^2} + \frac{1}{h_c^2} \le \frac{2R^2+r^2}{r^3(16R-5r)} \le \frac{R}{6r^3},$$

$$6 \le \frac{7R-2r}{R} \le \frac{h_a+h_b}{h_c} + \frac{h_b+h_c}{h_a} + \frac{h_c+h_a}{h_b} \le \frac{2R^2+Rr+2r^2}{Rr} + \frac{7R-2r}{2r},$$

$$\left(\frac{h_a+h_b}{a+b}\right)\left(\frac{h_b+h_c}{b+c}\right)\left(\frac{h_c+h_a}{c+a}\right) \le \frac{3\sqrt{3}}{8}, \quad \frac{3}{r} \le \frac{1}{h_a-2r} + \frac{1}{h_b-2r} + \frac{1}{h_c-2r},$$

$$\frac{3}{2} \le \frac{115R-38r}{75R-22r} \le \frac{h_a-r}{h_a+r} + \frac{h_b-r}{h_b+r} + \frac{h_c-r}{h_c+r} < \frac{5}{3},$$

$$6 \le \frac{3(19R-6r)}{9R-2r} \le \frac{h_a+r}{h_a-r} + \frac{h_b+r}{h_b-r} + \frac{h_c+r}{h_c-r} < 7,$$

$$\frac{3}{5} \le \frac{h_a-2r}{h_a+2r} + \frac{h_b-2r}{h_b+2r} + \frac{h_c-2r}{h_c+2r} < 1.$$

If $c \le b \le a$, then $c + h_c \le b + h_b \le a + h_a$. If $r \in [-1, \alpha]$, where $\alpha \triangleq (\log 9 - \log 4)/(\log 4 - \log 3) \approx 2.81$ and $\alpha \ne 0$, then

$$0 \le \sqrt{3}\left(\frac{a^r+b^r+c^r}{3}\right)^{1/r} - 2\left(\frac{h_a^r+h_b^r+h_c^r}{3}\right)^{1/r} \le 3(R-2r).$$

If $n \ge 3$ is an integer, then

$$h_a^n + h_b^n + h_c^n < \frac{2^n-1}{2^n}(a^n+b^n+c^n).$$

If $\mathcal{S}$ is acute, then

$$s \le \sqrt{3}\max\{h_a, h_b, h_c\}, \quad R + r \le \max\{h_a, h_b, h_c\}.$$

If $\mathcal{S}$ is not obtuse, then

$$\frac{h_a^2}{a^2} + \frac{h_b^2}{b^2} + \frac{h_c^2}{c^2} \le \frac{9}{4}.$$

If $C \le B \le A$, then

$$\frac{h_c}{h_b} + \frac{h_b}{h_a} + \frac{h_a}{h_c} \le \frac{h_a}{h_b} + \frac{h_b}{h_c} + \frac{h_c}{h_a}.$$

iii) The angle bisectors l_a, l_b, l_c satisfy

$$l_a \le \sqrt{bc}\cos\tfrac{A}{2} < \sqrt{bc}, \quad l_a \le \sqrt{s(s-a)},$$

$$l_a + l_b + l_c \le \left\{ \begin{array}{l} \sqrt{3}s \\ 3(R+r), \end{array} \right. \qquad 18\sqrt{3}r^2 \le al_a + bl_b + cl_c < 9R^2,$$

$$l_a l_b + l_b l_c + l_c l_a \le (ab+bc+ca)\left(\frac{1}{2} + \frac{2abc}{(a+b)(b+c)(c+a)}\right) \le \frac{3(ab+bc+ca)}{4} \le s^2,$$

$$al_b l_c + bl_c l_a + cl_a l_b \le \frac{Rs^3}{3r}, \qquad \left. \begin{array}{r} 3\sqrt{3}rs \\ \frac{8}{9}s^2 \end{array} \right\} \le l_a^2 + l_b^2 + l_c^2 \le s^2 + \tfrac{1}{2}(2r-R) \le s^2,$$

$$\frac{16}{9}(2R-r) < \frac{l_a l_b}{l_c} + \frac{l_b l_c}{l_a} + \frac{l_c l_a}{l_b} \le 3(2R-r), \quad \frac{3}{2} \le \sqrt{1-\frac{l_a^2}{bc}} + \sqrt{1-\frac{l_b^2}{ca}} + \sqrt{1-\frac{l_c^2}{ab}} < 3,$$

$$\left.\begin{matrix} \dfrac{1}{a}+\dfrac{1}{b}+\dfrac{1}{c} \\ \dfrac{6\sqrt{3}}{a+b+c} \end{matrix}\right\} \le \frac{1}{l_a}+\frac{1}{l_b}+\frac{1}{l_c} < \frac{3R}{rs}, \quad \frac{4\sqrt{3}s}{9abc} \le \frac{1}{al_a}+\frac{1}{bl_b}+\frac{1}{cl_c},$$

$$\frac{4\sqrt{3}}{9}\sqrt{\frac{2s}{(abc)^3}}(al_a+bl_b+cl_c) \le \frac{1}{al_a}+\frac{1}{bl_b}+\frac{1}{cl_c},$$

$$\frac{l_a}{a}+\frac{l_b}{b}+\frac{l_c}{c} \le \frac{3}{4}\left(\frac{a}{l_a}+\frac{b}{l_b}+\frac{c}{l_c}\right), \quad \frac{l_a}{a}+\frac{l_b}{b}+\frac{l_c}{c} \le \frac{s}{2r},$$

$$\begin{aligned} \frac{3\sqrt{3}}{2}+\left(\frac{8}{3}-\frac{3\sqrt{3}}{2}\right)\left(1-\frac{2r}{R}\right) &\le \frac{l_a}{a}+\frac{l_b}{b}+\frac{l_c}{c} \\ &\le \frac{3\sqrt{3}}{2}+\frac{3\sqrt{2}}{2}\left(\frac{R}{2r}-1\right) \le \frac{3\sqrt{3}}{2}+2\sqrt{3}\left(\frac{R}{2r}-1\right), \end{aligned}$$

$$\begin{aligned} 2\sqrt{3}+\frac{3}{2}\left(1-\frac{2r}{R}\right) \le \frac{1}{2}\left(\frac{s}{r}+\sqrt{3}\right) &\le \frac{a}{l_a}+\frac{b}{l_b}+\frac{c}{l_c} \\ &\le \frac{\sqrt{2}}{2}\left(\frac{s}{r}+2\sqrt{6}-3\sqrt{3}\right) \le 2\sqrt{3}+2\sqrt{2}\left(\frac{R}{2r}-1\right), \end{aligned}$$

$$3\left(\frac{s-a}{l_a}+\frac{s-b}{l_b}+\frac{s-c}{l_c}\right) \le \frac{l_a}{s-a}+\frac{l_b}{s-b}+\frac{l_c}{s-c},$$

$$\frac{9}{s^2} \le \frac{1}{l_a^2}+\frac{1}{l_b^2}+\frac{1}{l_c^2}, \quad \frac{27}{s^4} \le \frac{1}{l_a^4}+\frac{1}{l_b^4}+\frac{1}{l_c^4}, \quad 8 \le \frac{b^2+c^2}{l_a^2}+\frac{c^2+a^2}{l_b^2}+\frac{a^2+b^2}{l_c^2},$$

$$l_a l_b l_c \le \sqrt{s^3(s-a)(s-b)(s-c)} \le \tfrac{3\sqrt{3}}{8}abc, \quad \frac{l_a^2}{bc}+\frac{l_b^2}{ca}+\frac{l_c^2}{ab} \le \cos^2\tfrac{A}{2}+\cos^2\tfrac{B}{2}+\cos^2\tfrac{C}{2},$$

$$\frac{8Rr^2s^2}{2R^2+3Rr+2r^2} \le l_a l_b l_c \le \begin{cases} \dfrac{8Rrs^2}{9R-2r} \le rs^2 \\ \dfrac{8Rr^2(4R^2+4Rr+3r^2)}{2R^2+3Rr+2r^2}, \end{cases}$$

$$(l_a l_b)^2+(l_b l_c)^2+(l_c l_a)^2 \le rs^2(4R+r).$$

If x, y, z are positive numbers, then

$$\frac{54}{s^4} \le \frac{y+z}{xl_a^4}+\frac{z+x}{yl_b^4}+\frac{x+y}{zl_c^4}.$$

If $a \le b \le c$, then

$$a+l_a \le b+l_b.$$

If $p > 0$, then

$$3^{p+1}r^p \le 3^{p+1}\left(\frac{Rr^2}{2}\right)^{p/3} \le l_a^p+l_b^p+l_c^p.$$

If $p \in (0,1]$, then

$$l_a^p + l_b^p + l_c^p \le 3^{1-p/2}s^p \le 3\left(\frac{3R}{2}\right)^p.$$

If $p \in \mathbb{R}$, then

$$a^p l_a^2 + b^p l_b^2 + c^p l_c^2 \le \tfrac{1}{2}abcs(a^{p-2} + b^{p-2} + c^{p-2}).$$

iv) The medians m_a, m_b, m_c, altitudes h_a, h_b, h_c, and angle bisectors l_a, l_b, l_c satisfy

$$h_a \le l_a \le m_a, \quad 9r \le h_a + h_b + h_c \le l_a + l_b + l_c \le m_a + m_b + m_c \le \frac{9R}{2},$$

$$6rs \le al_a + bl_b + cl_c \le am_a + bm_b + cm_c \le \frac{2\sqrt{3}s^2}{3} \le \begin{cases} \frac{\sqrt{3}}{2}(a^2 + b^2 + c^2) \\ 3Rs, \end{cases}$$

$$m_a + m_b + m_c + \min\{a,b,c\} \le l_a + l_b + l_c + \max\{a,b,c\}, \quad l_a - h_a \le R - 2r,$$

$$m_a + m_b + m_c \le 2\left(\frac{m_a m_b}{h_c}\sin\tfrac{C}{2} + \frac{m_b m_c}{h_a}\sin\tfrac{A}{2} + \frac{m_c m_a}{h_b}\sin\tfrac{B}{2}\right),$$

$$m_a h_a + m_b h_b + m_c h_c \le m_a m_b \sin\tfrac{C}{2} + m_b m_c \sin\tfrac{A}{2} + m_c m_a \sin\tfrac{B}{2},$$

$$m_a h_a \sec\tfrac{A}{2} + m_b h_b \sec\tfrac{B}{2} + m_c h_c \sec\tfrac{C}{2} \le m_a m_b \sec\tfrac{C}{2} + m_b m_c \sec\tfrac{A}{2} + m_c m_a \sec\tfrac{B}{2},$$

$$\sqrt{\frac{h_a h_b}{m_a m_b}} + \sqrt{\frac{h_b h_c}{m_b m_c}} + \sqrt{\frac{h_c h_a}{m_c m_a}} \le \frac{h_a}{\sqrt{m_b m_c}} + \frac{h_b}{\sqrt{m_c m_a}} + \frac{h_c}{\sqrt{m_a m_b}} \le 2(\sin\tfrac{A}{2} + \sin\tfrac{B}{2} + \sin\tfrac{C}{2}),$$

$$m_a\sqrt{(\cos\tfrac{B}{2})\cos\tfrac{C}{2}} + m_b\sqrt{(\cos\tfrac{C}{2})\cos\tfrac{A}{2}} + m_c\sqrt{(\cos\tfrac{A}{2})\cos\tfrac{B}{2}} \le 2\left(\frac{m_a m_b}{h_c}\sin C + \frac{m_b m_c}{h_a}\sin A + \frac{m_c m_a}{h_b}\sin B\right),$$

$$\sqrt{m_a h_a(\cos\tfrac{B}{2})\cos\tfrac{C}{2}} + \sqrt{m_b h_b(\cos\tfrac{C}{2})\cos\tfrac{A}{2}} + \sqrt{m_c h_c(\cos\tfrac{A}{2})\cos\tfrac{B}{2}} \le m_a m_b \sin C + m_b m_c \sin A + m_c m_a \sin B,$$

$$m_a\sqrt{\sin\tfrac{A}{2}} + m_b\sqrt{\sin\tfrac{B}{2}} + m_c\sqrt{\sin\tfrac{C}{2}} \le \sqrt{\frac{r}{R}}\left(\frac{m_a m_b}{h_c} + \frac{m_b m_c}{h_a} + \frac{m_c m_a}{h_b}\right),$$

$$m_a h_a\sqrt{\sin\tfrac{A}{2}} + m_b h_b\sqrt{\sin\tfrac{B}{2}} + m_c h_c\sqrt{\sin\tfrac{C}{2}} \le \sqrt{\frac{r}{R}}(m_a m_b + m_b m_c + m_c m_a),$$

$$l_a + l_b + m_c \le \sqrt{s(s-a)} + \sqrt{s(s-b)} + m_c \le \sqrt{2(s^2 - m_c^2)} + m_c \le \sqrt{3}s,$$

$$\frac{1}{2} \le \frac{l_a + l_b + m_c}{a+b+c} \le \frac{\sqrt{3}}{2}, \quad \frac{1}{4} \le \frac{h_a + m_b + l_c}{a+b+c} \le \frac{\sqrt{3}}{2}, \quad \frac{3}{8} \le \frac{h_a + m_b + m_c}{a+b+c} \le 1,$$

$$h_a + h_b + h_c \le \frac{2(m_a^2 + m_b^2 + m_c^2)}{3R}, \quad m_a h_a + m_b h_b + m_c h_c \le s^2,$$

$$s^2 \le s(s-a) \le m_a l_a \le s(s-a) + \frac{[\mathrm{area}(\mathcal{S})]^2(b-c)^2}{2s(s-a)(b+c)^2} \le s(s-a) + \frac{1}{8}(b-c)^2,$$

$$s^2 \le m_a l_a + m_b l_b + m_c l_c \le \left\{\begin{matrix} s^2 + \frac{1}{8}[(a-b)^2 + (b-c)^2 + (c-a)^2] = \frac{5}{4}s^2 - 3Rr - \frac{3}{4}r^2 \\ s^2 + 2Rr - 4r^2 \end{matrix}\right\} \le 5R^2 + 2Rr + 3r^2 \le 6R^2 + 3r^2 \le \frac{27R^2}{4},$$

$$am_a l_a + bm_b l_b + cm_c l_c \le \tfrac{1}{4}s(s^2 + 18Rr + 9r^2), \quad l_a l_b l_c \le m_a m_b m_c, \quad \frac{R}{2r} h_a h_b h_c \le m_a m_b m_c,$$

$$(b+c)m_a l_a + (c+a)m_b l_b + (a+b)m_c l_c \le \tfrac{1}{2}s(5s^2 - 22Rr - 7r^2),$$

$$h_a^2 + h_b^2 + h_c^2 \le l_a^2 + l_b^2 + l_c^2 \le s^2 \le m_a^2 + m_b^2 + m_c^2, \quad l_a^6 + l_b^6 + l_c^6 \le s^4(s^2 - 12Rr) \le m_a^6 + m_b^6 + m_c^6,$$

$$\frac{(b+c)^2}{4bc} \le \frac{m_a}{l_a}, \quad \frac{b^2+c^2}{2bc} \le \frac{m_a}{h_a} \le \frac{R}{2r}, \quad 3 \le \sqrt{\frac{m_a}{h_a}} + \sqrt{\frac{m_b}{h_b}} + \sqrt{\frac{m_c}{h_c}} \le 3\sqrt{\frac{R}{2r}},$$

$$\frac{m_a}{h_b} + \frac{m_b}{h_c} + \frac{m_c}{h_a} \le \frac{9R^2 - 14Rr + 4r^2}{2Rr} \le \frac{9R - 12r}{2r}, \quad \frac{6r}{3R - 4r} \le \frac{h_a}{m_b} + \frac{h_b}{m_c} + \frac{h_c}{m_a} \le 3,$$

$$\left.\begin{matrix} 1 + \dfrac{4r}{R} \\ \dfrac{2\sqrt{3}s}{3R} \\ \dfrac{9}{4}\dfrac{(9R-2r)r}{\sqrt{(13R^2 + 14Rr - 8r^2)R}} \end{matrix}\right\} \le \frac{h_a}{l_a} + \frac{h_b}{l_b} + \frac{h_c}{l_c} \le \frac{\sqrt{6R^2 + 5Rr + 2r^2}}{R}$$

$$\le \frac{1}{2R}\sqrt{\frac{48R^3 + 16R^2r - 7Rr^2 - 2r^3}{2R - r}}$$

$$\le \frac{\sqrt{6}[R + (\sqrt{6} - 2)r]}{R} \le \frac{5R + 2r}{2R} \le 3,$$

$$\frac{6r}{R} \le \frac{l_a}{m_a} + \frac{l_b}{m_b} + \frac{l_c}{m_c} \le 3, \quad 3 \le \left\{\begin{matrix} \dfrac{13}{4} - \dfrac{r}{2R} \\ \dfrac{5}{3} + \dfrac{8s^2}{81Rr} \end{matrix}\right\} \le \frac{s^2 + 10Rr + r^2}{8Rr} \le \frac{m_a}{l_a} + \frac{m_b}{l_b} + \frac{m_c}{l_c} \le \frac{3R}{2r},$$

$$3 \le \frac{9R}{\sqrt{6R^2 + 5Rr + 2r^2}} \le \frac{l_a}{h_a} + \frac{l_b}{h_b} + \frac{l_c}{h_c} \le 4\sqrt{\frac{R(13R^2 + 14Rr - 8r^2)}{(9R - 2r)r}} \le \sqrt{3\left(1 + \frac{R}{r}\right)} \le \frac{3R}{2r},$$

$$3^{5/4}\sqrt{\text{area}(\mathcal{S})} \le \frac{m_a l_b}{h_c} + \frac{m_b l_c}{h_a} + \frac{m_c l_a}{h_b}, \quad 0 \le m_a^2 - l_a^2 \le \tfrac{1}{2}(b - c)^2,$$

$$\sqrt{l_a} + \sqrt{l_b} + \sqrt{l_c} \le \sqrt{m_a} + \sqrt{m_b} + \sqrt{m_c},$$

$$\frac{l_\mathrm{a}l_\mathrm{b}l_\mathrm{c}}{s} \le \frac{2l_\mathrm{a}l_\mathrm{b}l_\mathrm{c}}{3\sqrt[3]{abc}} \le \frac{l_\mathrm{a}l_\mathrm{b}l_\mathrm{c}}{9}\left(\frac{2}{a}+\frac{2}{b}+\frac{2}{c}\right) \\ \frac{\sqrt{3}}{3}\sqrt[3]{h_\mathrm{a}h_\mathrm{b}h_\mathrm{c}l_\mathrm{a}l_\mathrm{b}l_\mathrm{c}} \Bigg\} \le \operatorname{area}(\mathcal{S}) \le \begin{cases} \frac{\sqrt{3}}{3}\sqrt[6]{m_\mathrm{a}m_\mathrm{b}m_\mathrm{c}}\sqrt{l_\mathrm{a}l_\mathrm{b}l_\mathrm{c}} \le \frac{\sqrt{3}}{3}\sqrt[3]{m_\mathrm{a}m_\mathrm{b}m_\mathrm{c}l_\mathrm{a}l_\mathrm{b}l_\mathrm{c}} \\ \sqrt[3]{\frac{\sqrt{3}}{12}}(abcm_\mathrm{a}m_\mathrm{b}m_\mathrm{c}l_\mathrm{a}l_\mathrm{b}l_\mathrm{c})^{2/9} \\ \dfrac{8s^2l_\mathrm{a}l_\mathrm{b}l_\mathrm{c}}{27abc} \\ \frac{1}{2}\sqrt[3]{abcl_\mathrm{a}l_\mathrm{b}l_\mathrm{c}} \le \frac{1}{2}\sqrt[3]{abcm_\mathrm{a}m_\mathrm{b}m_\mathrm{c}} \\ \dfrac{\sqrt{m_\mathrm{a}m_\mathrm{b}m_\mathrm{c}l_\mathrm{a}l_\mathrm{b}l_\mathrm{c}}}{s} \le \dfrac{m_\mathrm{a}m_\mathrm{b}m_\mathrm{c}}{s} \\ \sqrt[3]{\dfrac{(abc)^2m_\mathrm{a}m_\mathrm{b}m_\mathrm{c}}{(a+b)(b+c)(c+a)}} \\ \dfrac{\sqrt{3}}{12}\left(\dfrac{(a+b)(b+c)(c+a)l_\mathrm{a}l_\mathrm{b}l_\mathrm{c}}{abc}\right)^{2/3} \\ \dfrac{3abcm_\mathrm{a}m_\mathrm{b}m_\mathrm{c}}{a(b+c)m_\mathrm{b}m_\mathrm{c}+b(c+a)m_\mathrm{c}m_\mathrm{a}+c(a+b)m_\mathrm{a}m_\mathrm{b}}. \end{cases}$$

If $a \le b \le c$, then $m_\mathrm{c} \le m_\mathrm{b} \le m_\mathrm{a}$, $h_\mathrm{c} \le h_\mathrm{b} \le h_\mathrm{a}$, and $l_\mathrm{c} \le l_\mathrm{b} \le l_\mathrm{a}$.

Source: [47, pp. 45–47, 51, 72, 109], [54, 56], [59, p. 70], [107, pp. 47, 99], [108, pp. 3, 26, 66, 67, 166, 167], [109, p. 343], [160], [289, pp. 7, 13, 69, 90], [317, 322, 395, 396, 401], [806, p. 207], [813], [993, p. 185], [1026, 1533, 1681], [1938, pp. 53, 54, 66, 70, 98, 108, 109, 141], [1994, 2011], [2062, pp. 110, 163, 200–206, 210–223, 229, 231, 680, 683, 684], [2376, 2444, 2602, 2603, 2924, 3013].

Fact 5.2.13. Let $\mathcal{S} \subset \mathbb{R}^2$ denote a triangle with the notation defined in Fact 5.2.7. Then, the following statement holds:

i) $[A\ B\ C] \overset{\mathrm{s}}{\prec} [\pi\ 0\ 0]$, and $[\frac{\pi}{3}\ \frac{\pi}{3}\ \frac{\pi}{3}] \overset{\mathrm{s}}{\prec} [A\ B\ C]$.

Furthermore, the following statements are equivalent:

ii) $\mathcal{S}$ is acute.

iii) $1 < (\tan A)\tan B$.

iv) $[A\ B\ C] \overset{\mathrm{s}}{\prec} [\frac{\pi}{2}\ \frac{\pi}{2}\ 0]$, and $[\frac{\pi}{3}\ \frac{\pi}{3}\ \frac{\pi}{3}] \overset{\mathrm{s}}{\prec} [A\ B\ C]$.

v) $2R + r < s$.

vi) $8R^2 < a^2 + b^2 + c^2$.

vii) $\dfrac{a^2}{A} + \dfrac{b^2}{B} + \dfrac{c^2}{C} \le \dfrac{3}{\pi}(a^2 + b^2 + c^2)$.

viii) $0 < (a^2 + b^2 - c^2)(b^2 + c^2 - a^2)(c^2 + a^2 - b^2)$.

ix) $0 < (5m_\mathrm{a}^2 - m_\mathrm{b}^2 - m_\mathrm{c}^2)(5m_\mathrm{b}^2 - m_\mathrm{c}^2 - m_\mathrm{a}^2)(5m_\mathrm{c}^2 - m_\mathrm{a}^2 - m_\mathrm{b}^2)$.

Furthermore, the following statements are equivalent:

x) $\mathcal{S}$ is right.

xi) $2R + r = s$.

xii) $8R^2 = a^2 + b^2 + c^2$.

xiii) $\cos\frac{A}{2}\cos\frac{B}{2}\cos\frac{C}{2} = \frac{1}{2} + \sin\frac{A}{2}\sin\frac{B}{2}\sin\frac{C}{2}$.

xiv) $(5m_\mathrm{a}^2 - m_\mathrm{b}^2 - m_\mathrm{c}^2)(5m_\mathrm{b}^2 - m_\mathrm{c}^2 - m_\mathrm{a}^2)(5m_\mathrm{c}^2 - m_\mathrm{a}^2 - m_\mathrm{b}^2) = 0$.

Furthermore, the following statements are equivalent:

xv) $\mathcal{S}$ is obtuse.

xvi) $[A\ B\ C] \overset{\rm s}{\prec} [\pi\ 0\ 0]$, and $[\frac{\pi}{2}\ \frac{\pi}{4}\ \frac{\pi}{4}] \overset{\rm s}{\prec} [A\ B\ C]$.

xvii) $s < 2R + r$.

xviii) $a^2 + b^2 + c^2 < 8R^2$.

xix) $(5m_{\rm a}^2 - m_{\rm b}^2 - m_{\rm c}^2)(5m_{\rm b}^2 - m_{\rm c}^2 - m_{\rm a}^2)(5m_{\rm c}^2 - m_{\rm a}^2 - m_{\rm b}^2) < 0$.

The following statements hold:

xx) If $\mathcal{S}$ is acute, $n \geq 1$, and p is a nonnegative number, then

$$\tan A > 0, \quad \min\{|A-B|, |B-C|, |C-A|\} \leq \tfrac{\pi}{6}, \quad ABC \leq \tfrac{\pi^3}{27}, \quad \frac{12R}{\pi} < \frac{a}{A} + \frac{b}{B} + \frac{c}{C},$$

$$\tfrac{4}{\pi^2}ABC \leq (\sin A)(\sin B)\sin C \leq \left(\tfrac{3\sqrt{3}}{2\pi}\right)^3 ABC \leq \tfrac{3\sqrt{3}}{8},$$

$$\frac{5}{4} + \frac{r}{2R} \leq \frac{4}{3} + \frac{r}{3R} \leq \sin\tfrac{A}{2} + \sin\tfrac{B}{2} + \sin\tfrac{C}{2}, \quad \tfrac{1}{\pi^2}ABC \leq (\sin\tfrac{A}{2})(\sin\tfrac{B}{2})\sin\tfrac{C}{2} \leq \left(\tfrac{3}{2\pi}\right)^3 ABC \leq \tfrac{1}{8},$$

$$\tfrac{1}{2} < (\sin\tfrac{A}{2})\sin\tfrac{B}{2} + (\sin\tfrac{B}{2})\sin\tfrac{C}{2} + (\sin\tfrac{C}{2})\sin\tfrac{A}{2} \leq \tfrac{3}{4}, \quad 2 < \sin^2 A + \sin^2 B + \sin^2 C \leq \tfrac{9}{4},$$

$$\cos A + \cos B + \cos C \leq \tfrac{3}{2}, \quad \tfrac{3}{4} \leq \cos^2 A + \cos^2 B + \cos^2 C < 1,$$

$$0 \leq (\cos A)(\cos B)\cos C \leq \tfrac{1}{8}, \quad \tfrac{1}{2} < (\cos\tfrac{A}{2})(\cos\tfrac{B}{2})\cos\tfrac{C}{2} \leq \tfrac{3\sqrt{3}}{8},$$

$$R(a\cos^3 A + b\cos^3 B + c\cos^3 C) \leq \operatorname{area}(\mathcal{S}),$$

$$\tfrac{1}{2} \leq \cos^3 A + \cos^3 B + \cos^3 C + (\cos A)(\cos B)\cos C, \quad 9 \leq \tan^2 A + \tan^2 B + \tan^2 C,$$

$$3 + \tfrac{3n}{2} < \tan^n A + \tan^n B + \tan^n C, \quad 3^{1+p/2} < \tan^p A + \tan^p B + \tan^p C,$$

$$(\tan A + \tan B)(\tan B + \tan C)(\tan C + \tan A) > 0,$$

$$0 < \left\{\begin{matrix} 2(\sin 2A + \sin 2B + \sin 2C) \\ 3(\cos A + \cos B + \cos C) \\ 4 \end{matrix}\right\} \leq 2(\sin A + \sin B + \sin C) \leq 3\sqrt{3}$$

$$\leq \left\{\begin{matrix} \tfrac{27}{8}(\csc A)(\csc B)\csc C = \dfrac{27R^2}{4rs} \\ 2(\cot\tfrac{A}{2} + \cot\tfrac{B}{2} + \cot\tfrac{C}{2}) - 3\sqrt{3} = 2(\cot\tfrac{A}{2})(\cot\tfrac{B}{2})\cot\tfrac{C}{2} - 3\sqrt{3} \end{matrix}\right\}$$

$$\leq (\tan A)(\tan B)\tan C,$$

$$\frac{s}{r} = \frac{\sin A}{1-\cos A} + \frac{\sin B}{1-\cos B} + \frac{\sin C}{1-\cos C} = \cot\tfrac{A}{2} + \cot\tfrac{B}{2} + \cot\tfrac{C}{2}$$

$$\leq \frac{(\sin A)(\sin B)\sin C}{(\cos A)(\cos B)\cos C} = \tan A + \tan B + \tan C = (\tan A)(\tan B)\tan C,$$

$$\tfrac{9}{2} \leq (\sin A)\tan A + (\sin B)\tan B + (\sin C)\tan C,$$

$$\tfrac{3\sqrt{3}}{8} \leq \tfrac{1}{6}[(\sin^2 A)\tan A + (\sin^2 B)\tan B + (\sin^2 C)\tan C],$$

$$\sqrt{3} \leq \tan\tfrac{\pi-A}{4} + \tan\tfrac{\pi-B}{4} + \tan\tfrac{\pi-C}{4} \leq \tan\tfrac{A}{2} + \tan\tfrac{B}{2} + \tan\tfrac{C}{2} \leq \cot A + \cot B + \cot C,$$

$$9 \leq 3 + \sec A + \sec B + \sec C \leq (\tan A)\tan B + (\tan B)\tan C + (\tan C)\tan A,$$

$$12 \leq \sec^2 A + \sec^2 B + \sec^2 C, \quad (\cot A)(\cot B)\cot C \leq \tfrac{\sqrt{3}}{9},$$

$$\frac{3}{\pi} < \frac{\sin A}{\pi - A} + \frac{\sin B}{\pi - B} + \frac{\sin C}{\pi - C} \leq \frac{3\sqrt{3}}{\pi},$$

$$\frac{4}{\pi} < \frac{\sin A - \sin B}{A - B} + \frac{\sin B - \sin C}{B - C} + \frac{\sin C - \sin A}{C - A} \leq \frac{3}{2},$$

$$-\frac{3\sqrt{3}}{2} \le -(\cos A + \cos B + \cos C)$$
$$\le \frac{\cos A - \cos B}{A - B} + \frac{\cos B - \cos C}{B - C} + \frac{\cos C - \cos A}{C - A}$$
$$< \left\{ \begin{matrix} -(\sin A + \sin B + \sin C) \\ -1 - \frac{4}{\pi} \end{matrix} \right\} < 0,$$

$$\csc^2 \tfrac{A}{2} + \csc^2 \tfrac{B}{2} + \csc^2 \tfrac{C}{2} \le \frac{\tan A - \tan B}{A - B} + \frac{\tan B - \tan C}{B - C} + \frac{\tan C - \tan A}{C - A}$$
$$\le \sec^2 A + \sec^2 B + \sec^2 C,$$

$$-(\csc^2 A + \csc^2 B + \csc^2 C) \le \frac{\cot A - \cot B}{A - B} + \frac{\cot B - \cot C}{B - C} + \frac{\cot C - \cot A}{C - A}$$
$$\le -(\sec^2 \tfrac{A}{2} + \sec^2 \tfrac{B}{2} + \sec^2 \tfrac{C}{2}),$$

$$6 \le \csc \tfrac{A}{2} + \csc \tfrac{B}{2} + \csc \tfrac{C}{2}$$
$$\le \frac{A - B}{\sin A - \sin B} + \frac{B - C}{\sin B - \sin C} + \frac{C - A}{\sin C - \sin A}$$
$$\le \frac{2}{\cos A + \cos B} + \frac{2}{\cos B + \cos C} + \frac{2}{\cos C + \cos A},$$

$$-\left(\frac{2}{\sin A + \sin B} + \frac{2}{\sin B + \sin C} + \frac{2}{\sin C + \sin A}\right)$$
$$\le \frac{A - B}{\cos A - \cos B} + \frac{B - C}{\cos B - \cos C} + \frac{C - A}{\cos C - \cos A}$$
$$\le -(\sec \tfrac{A}{2} + \sec \tfrac{B}{2} + \sec \tfrac{C}{2}) \le -2\sqrt{3},$$

$$0 < \frac{2}{\sec^2 A + \sec^2 B} + \frac{2}{\sec^2 B + \sec^2 C} + \frac{2}{\sec^2 C + \sec^2 A}$$
$$\le \frac{A - B}{\tan A - \tan B} + \frac{B - C}{\tan B - \tan C} + \frac{C - A}{\tan C - \tan A}$$
$$\le \left\{ \begin{matrix} \frac{3}{4} \\ \sin^2 \frac{A}{2} + \sin^2 \frac{B}{2} + \sin^2 \frac{C}{2} \end{matrix} \right\} < 1,$$

$$-\frac{9}{4} \le -(\cos^2 \tfrac{A}{2} + \cos^2 \tfrac{B}{2} + \cos^2 \tfrac{C}{2})$$
$$\le \frac{A - B}{\cot A - \cot B} + \frac{B - C}{\cot B - \cot C} + \frac{C - A}{\cot C - \cot A}$$
$$\le \begin{cases} -\left(\frac{2}{\csc^2 A + \csc^2 B} + \frac{2}{\csc^2 B + \csc^2 C} + \frac{2}{\csc^2 C + \csc^2 A}\right) \\ -1, \end{cases}$$

$$\left. \begin{matrix} 0 \\ \frac{\cos A + \cos B}{\sec^2 A + \sec^2 B} + \frac{\cos B + \cos C}{\sec^2 B + \sec^2 C} + \frac{\cos C + \cos A}{\sec^2 A + \sec^2 B} \end{matrix} \right\}$$

$$\leq \frac{\sin A - \sin B}{\tan A - \tan B} + \frac{\sin B - \sin C}{\tan B - \tan C} + \frac{\sin C - \sin A}{\tan C - \tan A}$$
$$\leq \begin{cases} \frac{3}{8} \\ \sin^3 A + \sin^3 B + \sin^3 C \leq \frac{\sqrt{2}}{2}, \end{cases}$$

$$24 \leq \frac{\sec^2 A + \sec^2 B}{\cos A + \cos B} + \frac{\sec^2 B + \sec^2 C}{\cos B + \cos C} + \frac{\sec^2 A + \sec^2 B}{\cos C + \cos A}$$
$$\leq \frac{\tan A - \tan B}{\sin A - \sin B} + \frac{\tan B - \tan C}{\sin B - \sin C} + \frac{\tan C - \tan A}{\sin C - \sin A}$$
$$\leq \csc^3 \tfrac{A}{2} + \csc^3 \tfrac{B}{2} + \csc^3 \tfrac{C}{2},$$

$$\left.\begin{matrix} 1 \\ \cos^3 \tfrac{A}{2} + \cos^3 \tfrac{B}{2} + \cos^3 \tfrac{C}{2} \end{matrix}\right\} \leq \frac{\cos A - \cos B}{\cot A - \cot B} + \frac{\cos B - \cos C}{\cot B - \cot C} + \frac{\cos C - \cos A}{\cot C - \cot A}$$
$$\leq \begin{cases} \frac{9\sqrt{3}}{8} \\ \dfrac{\sin A + \sin B}{\csc^2 A + \csc^2 B} + \dfrac{\sin B + \sin C}{\csc^2 B + \csc^2 C} + \dfrac{\sin C + \sin A}{\csc^2 A + \csc^2 B}, \end{cases}$$

$$\tfrac{8\sqrt{3}}{3} \leq \sec^3 \tfrac{A}{2} + \sec^3 \tfrac{B}{2} + \sec^3 \tfrac{C}{2}$$
$$\leq \frac{\cot A - \cot B}{\cos A - \cos B} + \frac{\cot B - \cot C}{\cos B - \cos C} + \frac{\cot C - \cot A}{\cos C - \cos A}$$
$$\leq \frac{\csc^2 A + \csc^2 B}{\sin A + \sin B} + \frac{\csc^2 B + \csc^2 C}{\sin B + \sin C} + \frac{\csc^2 A + \csc^2 B}{\sin C + \sin A}.$$

$$\frac{9}{4} \leq \left(\frac{(\sin A)\sin B}{\sin C}\right)^2 + \left(\frac{(\sin B)\sin C}{\sin A}\right)^2 + \left(\frac{(\sin C)\sin A}{\sin B}\right)^2,$$

$$\frac{3}{4} \leq \left(\frac{(\cos A)\cos B}{\cos C}\right)^2 + \left(\frac{(\cos B)\cos C}{\cos A}\right)^2 + \left(\frac{(\cos C)\cos A}{\cos B}\right)^2,$$

$$\frac{\sqrt{3}}{4}\sqrt[3]{(a^2 + b^2 - c^2)(b^2 + c^2 - a^2)(c^2 + a^2 - b^2)} \leq \operatorname{area}(\mathcal{S}),$$

$$3 \leq \frac{3R}{2r} \leq \frac{a^2}{b^2 + c^2 - a^2} + \frac{b^2}{c^2 + a^2 - b^2} + \frac{c^2}{a^2 + b^2 - c^2},$$

$$\sqrt{a^2 + b^2 - c^2} + \sqrt{b^2 + c^2 - a^2} + \sqrt{c^2 + a^2 - b^2} \leq a + b + c,$$

$$\frac{27}{(a + b + c)^2} \leq \frac{1}{2Rr} = \frac{a + b + c}{abc} \leq \frac{1}{a^2 + b^2 - c^2} + \frac{1}{b^2 + c^2 - a^2} + \frac{1}{c^2 + a^2 - b^2},$$

$$3(a + b + c) \leq \pi\left(\frac{a}{A} + \frac{b}{B} + \frac{c}{C}\right), \quad 3(a^2 + b^2 + c^2) \leq \pi\left(\frac{a^2}{A^2} + \frac{b^2}{B^2} + \frac{c^2}{C^2}\right).$$

xxi) If $\mathcal{S}$ is a right triangle with $C = \frac{\pi}{2}$, then

$$\cos 2B = \frac{a^2 - b^2}{a^2 + b^2}, \quad \tan 2B = \frac{2ab}{a^2 + b^2}, \quad \tan \frac{A}{2} = \sqrt{\frac{c - b}{c + b}}, \quad R + r = \tfrac{1}{2}(a + b).$$

xxii) If $\mathcal{S}$ is obtuse, then

$$r\sqrt{2R^2 + 8Rr + 3r^2} \leq \operatorname{area}(\mathcal{S}), \quad ABC < \tfrac{\pi^3}{32}, \quad 1 < \cos^2 A + \cos^2 B + \cos^2 C < 3,$$
$$0 < (\sin A)(\sin B)\sin C < \tfrac{16}{\pi^3}ABC \leq \tfrac{1}{2}, \quad -1 < (\cos A)(\cos B)\cos C < 0,$$
$$\tan A + \tan B + \tan C < 0 \leq \sin 2A + \sin 2B + \sin 2C \leq \sin A + \sin B + \sin C \leq 1 + \sqrt{2},$$

$$0 < \sin^2 A + \sin^2 B + \sin^2 C \le 2, \quad \tfrac{1}{4\pi}ABC \le (\sin\tfrac{A}{2})(\sin\tfrac{B}{2})\sin\tfrac{C}{2} \le \tfrac{8(\sqrt{2}-1)}{\pi^3}ABC \le \tfrac{\sqrt{2}-1}{4},$$

$$\tfrac{1}{6}[(\sin^2 A)\tan A + (\sin^2 B)\tan B + (\sin^2 C)\tan C] < 0 < (\sin A)(\sin B)\sin C < \tfrac{\sqrt{3}}{3},$$

$$0 < (\sin\tfrac{A}{2})\sin\tfrac{B}{2} + (\sin\tfrac{B}{2})\sin\tfrac{C}{2} + (\sin\tfrac{C}{2})\sin\tfrac{A}{2} \le \tfrac{2-\sqrt{2}}{4} + \sqrt{\tfrac{2-\sqrt{2}}{2}},$$

$$0 < (\cos\tfrac{A}{2})(\cos\tfrac{B}{2})\cos\tfrac{C}{2} < \tfrac{1+\sqrt{2}}{4}, \quad (\tan A)(\tan B)(\tan C) < 0,$$

$$3 < \sec^2 A + \sec^2 B + \sec^2 C, \quad (\cot A)(\cot B)\cot C < 0,$$

$$(\cot\tfrac{A}{2})(\cot\tfrac{B}{2})\cot\tfrac{C}{2} = \cot\tfrac{A}{2} + \cot\tfrac{B}{2} + \cot\tfrac{C}{2} = (1+\sqrt{2})\frac{R}{r}.$$

Source: [47, pp. 98, 109, 141, 144], [107, pp. 18, 23, 117, 118, 142], [108, pp. 6, 50, 77, 78, 289, 290, 427], [109, pp. 162, 164], [52, 319], [535, pp. 18, 20, 24–27, 28, 29, 31, 35, 95–97, 102], [806, p. 253], [993, p. 337], [1860, pp. 20, 117, 118], [1900, 1902, 2168, 2278], [2062, pp. 50, 133–136, 170, 241–243, 232, 241, 252], [2347, 3001]. **Remark:** The lower bound for area($\mathcal{S}$) in *xix*) is *Ono's inequality*. **Remark:** For acute triangles, the upper bounds for R/r given by

$$\frac{R}{r} \le \frac{2}{3}\left(\frac{a^2}{b^2+c^2-a^2} + \frac{b^2}{c^2+a^2-b^2} + \frac{c^2}{a^2+b^2-c^2}\right), \quad \frac{R}{r} \le 2\left(\frac{\pi}{9}\right)^3\left(\frac{1}{A} + \frac{1}{B} + \frac{1}{C}\right)^3,$$

where the second bound is given by *v*) of Fact 5.2.10, cannot be ordered.

Fact 5.2.14. Let x, y, z be positive numbers, and define $f\colon \mathbb{R} \mapsto \mathbb{R}$ by $f(p) = px^2 + (1-p)y^2 - p(1-p)z^2$. Then, the following statements are equivalent:

i) x, y, z represent the lengths of the sides of a triangle.

ii) $z < x + y$, $x < y + z$, and $y < z + x$.

iii) $(x + y - z)(y + z - x)(z + x - y) > 0$.

iv) $x > |y - z|$, $y > |z - x|$, and $z > |x - y|$.

v) $|y - z| < x < y + z$.

vi) $2(x^4 + y^4 + z^4) < (x^2 + y^2 + z^2)^2$.

vii) $(x^2 + y^2 + z^2)^2 < 4(x^2y^2 + y^2z^2 + z^2x^2)$.

viii) There exist positive numbers a, b, c such that $x = a + b$, $y = b + c$, and $z = c + a$.

ix) For all $p \in \mathbb{R}$, $f(p) > 0$.

x) $f[(y^2 + z^2 - x^2)/(2z^2)] > 0$.

If these statements hold, then a, b, c in *viii*) are given by

$$a = \tfrac{1}{2}(z + x - y), \quad b = \tfrac{1}{2}(x + y - z), \quad c = \tfrac{1}{2}(y + z - x).$$

Source: [993, p. 164]. To prove the equivalence of *iii*) and *vi*), note that $(x^2 + y^2 + z^2)^2 - 2(x^4 + y^4 + z^4) = (x + y + z)(x + y - z)(x - y + z)(-x + y + z)$. *v*) and *vi*) are given in [1371, p. 125]; *vii*) is given in [47, p. 71] and [1158, p. 38]. To prove *iii*) $\Longrightarrow$ *ix*), note that

$$\begin{aligned} f(p) &= z^2p^2 + (x^2 - y^2 - z^2)p + y^2 \\ &= z^2\left(p + \frac{x^2 - y^2 - z^2}{2z^2}\right)^2 + \frac{(x+y+z)(x+y-z)(y+z-x)(z+x-y)}{4z^2}. \end{aligned}$$

To prove *x*) $\Longrightarrow$ *iii*), note that

$$0 < f[(y^2 + z^2 - x^2)/(2c^2)] = (x + y + z)(x + y - z)(y + z - x)(z + x - y).$$

Remark: The expressions $x = a + b$, $y = b + c$, and $z = c + a$ can be used to recast inequalities involving the lengths x, y, z of the sides of a triangle into inequalities involving arbitrary positive numbers a, b, c. Conversely, each inequality that holds for arbitrary positive numbers a, b, c necessarily holds in the case where a, b, c are the lengths of the sides of a triangle, while the expressions $a = \frac{1}{2}(z + x - y)$, $b = \frac{1}{2}(x + y - z)$, and $c = \frac{1}{2}(y + z - x)$ can be used to obtain possibly different inequalities that are valid only for x, y, z that represent the lengths of the sides of a triangle. **Related:** Fact 10.10.7.

Fact 5.2.15. Let $\mathcal{S} \subset \mathbb{R}^2$ denote a triangle with the notation defined in Fact 5.2.7, and define $x = \frac{1}{2}(c + a - b)$, $y = \frac{1}{2}(a + b - c)$, and $z = \frac{1}{2}(b + c - a)$. Then,

$$a = x + y, \quad b = y + z, \quad c = z + x,$$

$$(a - b)^2 + (b - c)^2 + (c - a)^2 = (x - y)^2 + (y - z)^2 + (z - x)^2,$$

$$s = x + y + z, \quad \text{area}(\mathcal{S}) = \sqrt{(x + y + z)xyz},$$

$$r = \sqrt{\frac{xyz}{x + y + z}}, \quad R = \frac{(x + y)(y + z)(z + x)}{4\sqrt{(x + y + z)xyz}}.$$

Source: [1938, pp. 56, 57, 74]. **Remark:** The numbers x, y, z determine the points along the sides of the triangle at which the incircle is tangent to the sides of the triangle. See [47, p. 57].

Fact 5.2.16. Let $n \ge 2$ be an integer, let x, y, z be positive numbers, and assume that $x^n + y^n = z^n$. Then, x, y, z represent the lengths of the sides of a triangle. **Source:** [1371, p. 112]. **Remark:** If x and y are positive numbers and $n \ge 1$, then x, y, and $\sqrt[n]{x^n + y^n}$ represent the lengths of the sides of a triangle.

Fact 5.2.17. Let a, b, c be positive numbers that represent the lengths of the sides of a triangle. Then, $1/(a + b)$, $1/(b + c)$, $1/(c + a)$ represent the lengths of the sides of a triangle. **Source:** [1757, p. 44]. **Related:** Fact 5.2.14 and Fact 5.2.18.

Fact 5.2.18. Let a, b, c be positive numbers that represent the lengths of the sides of a triangle whose circumradius is R. Then, $\sqrt{a}$, $\sqrt{b}$, $\sqrt{c}$ represent the lengths of the sides of a triangle whose circumradius R_2 satisfies $\sqrt{3}R_2^2 \le R$. **Source:** [1371, p. 99] and Fact 5.2.19. **Related:** Fact 5.2.14 and Fact 5.2.17.

Fact 5.2.19. Let a, b, c be positive numbers that represent the lengths of the sides of a triangle whose circumradius is R, and let $p > 1$. Then, $a^{1/p}$, $b^{1/p}$, and $c^{1/p}$, represent the lengths of the sides of a triangle whose circumradius R_p satisfies $(\sqrt{3}R_p)^p \le \sqrt{3}R$. **Source:** [1860, p. 119]. **Credit:** A. Oppenheim. **Related:** Fact 5.2.14 and Fact 5.2.17.

Fact 5.2.20. Let a, b, c be positive numbers that represent the lengths of the sides of triangle $\mathcal{S}_1$. Then, $a + b$, $b + c$, and $c + a$ represent the lengths of the sides of triangle $\mathcal{S}_2$. Furthermore, $4\,\text{area}(\mathcal{S}_1) \le \text{area}(\mathcal{S}_2)$. **Source:** [806, p. 207].

Fact 5.2.21. Let x, y, z be positive numbers. Then, the following statements hold:

i) x, y, z represent the lengths of the sides of a triangle if and only if $x^4 + y^4 + z^4 < 2(x^2y^2 + y^2z^2 + z^2x^2$.

ii) If $z^2 \ge x^2 + y^2$, then $z = \max\{x, y, z\}$.

iii) Let $z = \max\{x, y, z\}$. Then, x, y, z represent the lengths of the sides of a triangle if and only if $z < x + y$.

iv) Let $z = \max\{x, y, z\}$. Then, x, y, z represent the lengths of the sides of an obtuse triangle if and only if $\sqrt{x^2 + y^2} < z < x + y$.

v) Let $z = \max\{x, y, z\}$. Then, x, y, z represent the lengths of the sides of a right triangle if and only if $z^2 = x^2 + y^2$.

vi) Let $z = \max\{x, y, z\}$. Then, x, y, z represent the lengths of the sides of an acute triangle if and only if $z < \sqrt{x^2 + y^2}$.

vii) x, y, z represent the lengths of the sides of an acute triangle if and only if $x^2 < y^2 + z^2$, $y^2 < z^2 + x^2$, and $z^2 < x^2 + y^2$.

viii) If x, y, z represent the lengths of the sides of an acute triangle, then x^2, y^2, z^2 represent the lengths of the sides of a triangle.

ix) If, for every positive integer n, x^n, y^n, z^n represent the lengths of the sides of a triangle $\mathcal{S}$, then $\mathcal{S}$ is isosceles.

x) $\sqrt{x+y}$, $\sqrt{y+z}$, $\sqrt{z+x}$ represent the sides of a triangle whose area is $\frac{1}{2}\sqrt{xy+yz+zx}$.

Source: [47, pp. 88, 89] and [1007]. To prove sufficiency in *vii)*, note that x^2, y^2, z^2 represent the lengths of the sides of a triangle. It thus follows from Fact 5.2.18 that x, y, z represent the lengths of the sides of a triangle. However, z is shorter than the hypotenuse of a right triangle whose remaining sides are x and y, and likewise for x and y. Therefore, the triangle is acute. *ix)* is given in [1158, p. 38]. *x)* is given in [1956]. See Fact 2.3.40.

Fact 5.2.22. Let $\mathcal{S}_1$ and $\mathcal{S}_2$ be triangles with sides a_1, b_1, c_1 and a_2, b_2, c_2, respectively. Then,

$$16\,\text{area}(\mathcal{S}_1)\,\text{area}(\mathcal{S}_2) \le a_1^2(-a_2^2 + b_2^2 + c_2^2) + b_1^2(a_2^2 - b_2^2 + c_2^2) + c_1^2(a_2^2 + b_2^2 - c_2^2).$$

Furthermore, equality holds if and only if the triangles are similar. **Source:** [47, p. 108]. **Remark:** This is the *Neuberg-Pedoe inequality*.

Fact 5.2.23. Let a, b, c be positive numbers that represent the lengths of the sides of a triangle, and let $f\colon [0,\infty) \mapsto [0,\infty)$ be nonincreasing and subadditive. Then, $f(a), f(b), f(c)$ denote the lengths of the sides of a triangle. **Source:** [47, p. 120]. **Remark:** "Subadditive" means that, for all $x, y \ge 0$, $f(x+y) \le f(x) + f(y)$.

Fact 5.2.24. Let $x, y, z > 0$ represent the lengths of the sides of a triangle. Then,

$$\left|\frac{x-y}{x+y} + \frac{y-z}{y+z} + \frac{z-x}{z+x}\right| < \frac{1}{8}, \quad \left|\frac{x}{y} + \frac{y}{z} + \frac{z}{x} - \left(\frac{y}{x} + \frac{z}{y} + \frac{x}{z}\right)\right| < 1.$$

Source: [993, pp. 181, 183].

Fact 5.2.25. Let x, y, z be positive numbers that represent the lengths of the sides of a triangle $\mathcal{S}$. Then,

$$1 \le \frac{x^2+y^2+z^2}{xy+yz+zx} < 2, \quad 2 < \frac{(x+y+z)^2}{x^2+y^2+z^2} \le 3, \quad 3 \le \frac{(x+y+z)^2}{xy+yz+zx} < 4,$$

$$\frac{1}{4} \le \frac{(x+y)(y+z)(z+x)}{(x+y+z)^3} \le \frac{8}{27}, \quad \frac{13}{27} \le \frac{(x+y+z)(x^2+y^2+z^2)+4xyz}{(x+y+z)^3} < \frac{1}{2},$$

$$\begin{aligned}
0 &\le 6(x+y-z)(y+z-x)(z+x-y) \le 2\sqrt[4]{3}[(x+y-z)(y+z-x)(z+x-y)(x+y+z)]^{3/4}\\
&\le 2\sqrt{(x^2+y^2+z^2)(x+y-z)(y+z-x)(z+x-y)(x+y+z)} = 8\sqrt{x^2+y^2+z^2}\,\text{area}(\mathcal{S}) \le 6xyz\\
&\le \left\{\begin{array}{c} \frac{2\sqrt{3}}{3}(xy+yz+zx)^{3/2} \le (x+y)(y+z)(z+x)\\ \dfrac{6\sqrt{3}xyz\sqrt{x^2+y^2+z^2}}{x+y+z} \le 2\sqrt{3}\sqrt[3]{x^2y^2z^2}\sqrt{x^2+y^2+z^2} \le 2\sqrt{xyz(x+y+z)(x^2+y^2+z^2)} \end{array}\right\}\\
&\le \left\{\begin{array}{c} x(y^2+z^2)+y(z^2+x^2)+z(x^2+y^2)\\ \frac{2}{9}(x+y+z)^3 \end{array}\right\} \le \left\{\begin{array}{c} 3xyz+x^3+y^3+z^3\\ \frac{2\sqrt{3}}{9}(x+y+z)^2\sqrt{x^2+y^2+z^2} \end{array}\right\}\\
&\le \tfrac{2}{3}(x+y+z)(x^2+y^2+z^2) \le \tfrac{2\sqrt{3}}{3}(x^2+y^2+z^2)^{3/2} \le \left\{\begin{array}{c} 2(x^3+y^3+z^3)\\ \frac{\sqrt{3}}{3}(x+y+z)^2\sqrt{x^2+y^2+z^2} \end{array}\right\}\\
&< 2xyz + 2[x(y^2+z^2)+y(z^2+x^2)+z(x^2+y^2)] \le 2xyz + 4(x^3+y^3+z^3)
\end{aligned}$$

$$\leq 6xyz + 4[x(y^2+z^2) + y(z^2+x^2) + z(x^2+y^2)] \leq 5[x(y^2+z^2) + y(z^2+x^2) + z(x^2+y^2)],$$

$$\left.\begin{array}{r} x^3(y+z-x) + y^3(z+x-y) + z^3(x+y-z) \leq xyz(x+y+z) \\ \frac{1}{2}(x^4+y^4+z^4) \end{array}\right\} \leq x^2y^2 + y^2z^2 + z^2x^2$$

$$\leq x^3y + y^3z + z^3x \leq x^4 + y^4 + z^4,$$

$$(x+y-z)(x-y+z)(y+z)^2 \leq 4x^2yz, \quad \max\{x,y,z\} \leq \sqrt[3]{\tfrac{1}{2}(x^3+y^3+z^3+3xyz)},$$

$$(x+y-z)^x(y+z-x)^y(z+x-y)^z \leq x^xy^yz^z, \quad (x+y+z)^4 \leq 9(xy+yz+zx)(x^2+y^2+z^2),$$

$$9(x^4+y^4+z^4) \leq (x+y)^4 + (y+z)^4 + (z+x)^4, \quad (xy+yz+zx)(x^2+y^2+z^2) \leq 3(x^3y+y^3z+z^3x),$$

$$xyz(x+y+z)(x^2+y^2+z^2) \leq (x^2y+y^2z+z^2x)^2,$$

$$\sqrt{x+y-z} + \sqrt{y+z-x} + \sqrt{z+x-y} \leq \sqrt{x} + \sqrt{y} + \sqrt{z},$$

$$0 \leq x^2(x+y)(y-z) + y^2(y+z)(z-x) + z^2(z+x)(x-y),$$

$$(x^2+y^2+z^2)(x^2y^2+y^2z^2+z^2x^2) \leq 3(x^2y^4+y^2z^4+z^2x^4),$$

$$3(x+y-z)(y+z-x)(z+x-y)(x+y+z) \leq (x^2+y^2+z^2)^2,$$

$$0 \leq (2x^2-yz)(y-z)^2 + (2y^2-zx)(z-x)^2 + (2z^2-xy)(x-y)^2,$$

$$(x^2+y^2+z^2)[2xyz + (x+y-z)(y+z-x)(z+x-y)] \leq xyz(x+y+z)^2.$$

$$\frac{9}{x+y+z} \leq \frac{1}{x} + \frac{1}{y} + \frac{1}{z} \leq \frac{3}{x+y+z}\min\left\{\frac{x}{y}+\frac{y}{z}+\frac{z}{x}, \frac{y}{x}+\frac{z}{y}+\frac{x}{z}\right\}$$

$$\leq \frac{3}{2(x+y+z)}\left(\frac{x}{y}+\frac{y}{z}+\frac{z}{x}+\frac{y}{x}+\frac{z}{y}+\frac{x}{z}\right) \leq \frac{3}{x+y+z}\max\left\{\frac{x}{y}+\frac{y}{z}+\frac{z}{x}, \frac{y}{x}+\frac{z}{y}+\frac{x}{z}\right\}$$

$$\leq \frac{1}{x+y-z} + \frac{1}{y+z-x} + \frac{1}{z+x-y},$$

$$\frac{x(y+z)}{x^2+yz} + \frac{y(z+x)}{y^2+zx} + \frac{z(x+y)}{z^2+xy} \leq 3, \quad 6 \leq \frac{(x+y)^2}{xy+z^2} + \frac{(y+z)^2}{yz+x^2} + \frac{(z+x)^2}{zx+y^2},$$

$$\frac{9}{8} \leq \frac{(x^3+y^3+z^3)^2}{(x^2+y^2)(y^2+z^2)(z^2+x^2)} < 2, \quad 3 \leq \frac{z}{x+y-z} + \frac{x}{y+z-x} + \frac{y}{z+x-y},$$

$$3 \leq \frac{x+y-z}{z} + \frac{y+z-x}{x} + \frac{z+x-y}{y}, \quad 1 \leq \frac{x}{3x-y+z} + \frac{y}{3y-z+x} + \frac{z}{3z-x+y},$$

$$\left.\begin{array}{r} \frac{1}{3}(x+y+z)\left(\frac{1}{x}+\frac{1}{y}+\frac{1}{z}\right) \\ \frac{1}{2}(x+y+z)\sqrt{\frac{1}{x^2}+\frac{1}{y^2}+\frac{1}{z^2}} \end{array}\right\} \leq \frac{x}{y}+\frac{y}{z}+\frac{z}{x}, \quad 0 \leq x^2\left(\frac{y}{z}-1\right) + y^2\left(\frac{z}{x}-1\right) + z^2\left(\frac{x}{y}-1\right),$$

$$0 \leq \frac{x-y}{y+z} + \frac{y-z}{z+x} + \frac{z-x}{x+y} < 1, \quad \frac{4}{3} \leq \sqrt{\frac{x}{y+3z}} + \sqrt{\frac{y}{z+3x}} + \sqrt{\frac{z}{x+3y}},$$

$$4 \leq \frac{3x+y}{2x+z} + \frac{3y+z}{2y+x} + \frac{3z+x}{2z+y}, \quad \frac{1}{x}+\frac{1}{y}+\frac{1}{z} \leq \frac{1}{x+y-z} + \frac{1}{y+z-x} + \frac{1}{z+x-y},$$

$$0 \leq \frac{xy-z^2}{x^2+3y^2+z^2} + \frac{yz-x^2}{y^2+3z^2+x^2} + \frac{zx-y^2}{z^2+3x^2+y^2},$$

$$6\left(\frac{x}{y+z}+\frac{y}{z+x}+\frac{z}{x+y}\right)\le (x+y+z)\left(\frac{1}{x}+\frac{1}{y}+\frac{1}{z}\right),$$

$$\frac{9}{2}+\frac{x}{y+z}+\frac{y}{z+x}+\frac{z}{x+y}\le \frac{x+y}{z}+\frac{y+z}{x}+\frac{z+x}{y},$$

$$x+y+z\le \sqrt{\frac{xyz}{-x+y+z}}+\sqrt{\frac{xyz}{x-y+z}}+\sqrt{\frac{xyz}{x+y-z}},$$

$$xy+yz+zx\le \frac{(x+y-z)^4}{y(y+z-x)}+\frac{(y+z-x)^4}{z(z+x-y)}+\frac{(z+x-y)^4}{x(x+y-z)},$$

$$7\le \frac{x+y}{z}+\frac{y+z}{x}+\frac{z+x}{y}+\frac{(x+y-z)(y+z-x)(z+x-y)}{xyz},$$

$$\frac{1}{8xyz+(x+y-z)^3}+\frac{1}{8xyz+(y+z-x)^3}+\frac{1}{8xyz+(z+x-y)^3}\le \frac{1}{3xyz},$$

$$\frac{x^2+y^2-z^2}{xy}+\frac{y^2+z^2-x^2}{yz}+\frac{z^2+x^2-y^2}{zx}\le 3\le \frac{1}{3}(x^2+y^2+z^2)\left(\frac{1}{x^2}+\frac{1}{y^2}+\frac{1}{z^2}\right)\le \frac{x^2}{y^2}+\frac{y^2}{z^2}+\frac{z^2}{x^2}.$$

If $x\le y\le z$, then

$$1\le \frac{z}{y}<2,\quad 1\le \min\left\{\frac{y}{x},\frac{z}{y}\right\}<\frac{1+\sqrt{5}}{2}\approx 1.618.$$

If $r\in[0,1]$, then

$$0\le x^2y(x^r-y^r)+y^2z(y^r-z^r)+z^2x(z^r-x^r).$$

If $r\in[0,2]$, then

$$0\le x^ry(x-y)+y^rz(y-z)+z^rx(z-x),$$

$$(x+y+z)(x^{r-1}y+y^{r-1}z+z^{r-1}x)\le 3(x^ry+y^rz+z^rx).$$

If $\mathcal{S}$ is isosceles, then

$$1\le \frac{x^2+y^2+z^2}{xy+yz+zx}<\frac{6}{5},\quad \frac{8}{3}<\frac{(x+y+z)^2}{x^2+y^2+z^2}\le 3,$$

$$3\le \frac{(x+y+z)^2}{xy+yz+zx}<\frac{16}{5},\quad \frac{9}{32}\le \frac{(x+y)(y+z)(z+x)}{(x+y+z)^3}\le \frac{8}{27}.$$

If $\mathcal{S}$ is obtuse, then

$$\frac{4}{1+2\sqrt{2}}\le \frac{x^2+y^2+z^2}{xy+yz+zx}<2,\quad 2<\frac{(x+y+z)^2}{x^2+y^2+z^2}\le \frac{(1+\sqrt{2})^2}{2},$$

$$\frac{2(1+\sqrt{2})^2}{1+2\sqrt{2}}\le \frac{(x+y+z)^2}{xy+yz+zx}<4,\quad \frac{1}{4}\le \frac{(x+y)(y+z)(z+x)}{(x+y+z)^3}\le \frac{\sqrt{2}(2+\sqrt{2})^2}{4(1+\sqrt{2})^3}.$$

If $\mathcal{S}$ is isosceles and obtuse, then

$$\frac{4}{1+2\sqrt{2}}\le \frac{x^2+y^2+z^2}{xy+yz+zx}<\frac{6}{5},\quad \frac{8}{3}<\frac{(x+y+z)^2}{x^2+y^2+z^2}\le \frac{(1+\sqrt{2})^2}{2},$$

$$\frac{2(1+\sqrt{2})^2}{1+2\sqrt{2}}\le \frac{(x+y+z)^2}{xy+yz+zx}<\frac{16}{5},\quad \frac{9}{32}\le \frac{(x+y)(y+z)(z+x)}{(x+y+z)^3}\le \frac{\sqrt{2}(2+\sqrt{2})^2}{4(1+\sqrt{2})^3}.$$

If $\mathcal{S}$ is either acute or right, then

$$\sqrt{x^2+y^2-z^2}+\sqrt{y^2+z^2-x^2}+\sqrt{z^2+x^2-y^2}\le x+y+z.$$

If $\mathcal{S}$ is right and the hypotenuse is z, then

$$\tfrac{2+\sqrt{2}}{2}(x+y)\le x+y+z,\quad 2\sqrt{2}xy\le(x+y)z.$$

If $\mathcal{S}$ is not isosceles, then

$$5<\left|\frac{x+y}{x-y}+\frac{y+z}{y-z}+\frac{z+x}{z-x}\right|,\quad 3<\left|\frac{x^2+y^2}{x^2-y^2}+\frac{y^2+z^2}{y^2-z^2}+\frac{z^2+x^2}{z^2-x^2}\right|.$$

Source: [47, p. 143], [108, pp. 15, 27, 117, 118, 167, 168], [109, pp. 330, 331], [287, 313], [748, pp. 70, 113, 345, 378, 379], [752, pp. 30, 40, 41], [806, pp. 115, 125, 208, 209, 215], [1470], [1860, pp. 5, 7, 19], [1823], [1938, pp. 34, 56, 65, 113], [1969, pp. 72, 199], [1971, pp. 103, 278], [2062, pp. 140, 144, 147], and [2845]. The inequality $x(y^2+z^2)+y(z^2+x^2)+z(x^2+y^2)\le 3xyz+x^3+y^3+z^3$ is given in [993, p. 183] and follows from

$$3\le\frac{x}{y+z-x}+\frac{y}{z+x-y}+\frac{z}{x+y-z},$$

which is given in [993, p. 186] and [1938, p. 19]. The inequalities given by

$$x(y^2+z^2)+y(z^2+x^2)+z(x^2+y^2)\le 2(x^3+y^3+z^3)\le 2xyz+2[x(y^2+z^2)+y(z^2+x^2)+z(x^2+y^2)]$$

are equivalent to

$$\frac{3}{2}\le\frac{x}{y+z}+\frac{y}{z+x}+\frac{z}{x+y}<2,$$

which is given in [1757, p. 267]. The left-hand inequality is Nesbitt's inequality given by Fact 2.3.60 and Fact 2.3.61. To prove the right-hand inequality, assume without loss of generality that $x\le y\le z$. Therefore,

$$\frac{x}{y+z}+\frac{y}{z+x}\le\frac{x}{y+x}+\frac{y}{y+x}=1.$$

Finally, add this inequality to the inequality $\frac{z}{x+y}<1$. The inequality $\frac{1}{2}(x^4+y^4+z^4)\le x^2y^2+y^2z^2+z^2x^2$ is equivalent to the median inequality $m_{\rm a}<m_{\rm b}+m_{\rm c}$ given by Fact 5.2.7. The inequality $x^2y^2+y^2z^2+z^2x^2\le x^3y+y^3z+z^3x$ is given in [993, p. 183] and [1938, p. 15]. This inequality orders the upper bounds for $xyz(x+y+z)$ given by Fact 2.3.66. The inequalities for isosceles triangles are given in [507], [1969, p. 199], and [1971, p. 278]. To prove $\min\left\{\frac{y}{x},\frac{z}{y}\right\}<\frac{1}{2}(1+\sqrt{5})$, suppose that $\frac{1}{2}(1+\sqrt{5})\le y/x$ and $\frac{1}{2}(1+\sqrt{5})\le z/y$. Then, $x\le\frac{1}{2}(\sqrt{5}-1)y$. Therefore, $x+y\le\frac{1}{2}(\sqrt{5}-1)y+y=\frac{1}{2}(1+\sqrt{5})y\le z$, which is a contradiction. **Remark:** The 12th string is *Walker's inequality*. See [603, p. 259]. **Remark:** The inequality $6xyz\le\frac{2\sqrt{3}}{3}(x^2+y^2+z^2)^{3/2}$ implies that $27x^2y^2z^2\le(x^2+y^2+z^2)^3$, which is Leibniz's inequality given by Fact 5.2.8. Using Fact 5.2.18, Leibniz's inequality yields $6xyz\le\frac{2}{9}(x+y+z)^2$. **Remark:** The inequality $(x+y-z)(y+z-x)(z+x-y)\le 6xyz$ is *Padoa's inequality*. See [47, p. 14]. This inequality is also given by Fact 5.2.8. **Remark:** Fact 5.2.21 implies that, for obtuse triangles, the first three strings hold with x,y,z replaced by x^2,y^2,z^2.

5.3 Facts on Polygons and Polyhedra

Fact 5.3.1. Let $\mathcal{S}\subset\mathbb{R}^2$ be a convex quadrilateral whose sides have lengths and names a,b,c,d labeled counterclockwise, define the semiperimeter $s\triangleq\frac{1}{2}(a+b+c+d)$, let A,B,C,D denote the angles and vertices of $\mathcal{S}$ labeled counterclockwise with A and B connected by a, let p,q denote the names and lengths of the diagonals of $\mathcal{S}$, where p connects A and C and q connects B and D, let m_{ac} denote the length of the *median* from the midpoint of a to the midpoint of c, let E denote the point at which the diagonals intersect, and let θ denote the angle formed by AEB. Then, the following statements hold:

i) $A + B + C + D = 2\pi$.

ii) $\sin A + \sin B + \sin C + \sin D = 4(\sin \frac{A+B}{2})(\sin \frac{A+C}{2}) \sin \frac{A+D}{2}$.

iii) If $\mathcal{S}$ has no right angle, then $\dfrac{(\tan A)\tan B - (\tan C)\tan D}{(\tan A)\tan C - (\tan B)\tan D} = \dfrac{\tan(A+C)}{\tan(A+B)}$.

iv) If $\mathcal{S}$ has no right angle, then $\dfrac{\tan A + \tan B + \tan C + \tan D}{\cot A + \cot B + \cot C + \cot D} = (\tan A)(\tan B)(\tan C)\tan D$.

v) $(\cos A)(\cos B)(\cos C)\cos D = \frac{1}{8}[1 + \cos 2A + \cos 2B + \cos 2C + \cos 2D + \cos 2(A+D) + \cos 2(B+D) + \cos 2(C+D)] = \frac{1}{8}(1 + \cos 2A + \cos 2B + \cos 2C + \cos 2D) + \frac{1}{16}[\cos 2(A+B) + \cos 2(A+C) + \cos 2(A+D) + \cos 2(B+C) + \cos 2(B+D) + \cos 2(C+D)]$.

vi) $(\cos A)(\sin B)(\sin C)\sin D + (\cos B)(\sin C)(\sin D)\sin A + (\cos C)(\sin D)(\sin A)\sin B + (\cos D)(\sin A)(\sin B)\sin C + \frac{1}{4}(\sin 2A + \sin 2B + \sin 2C + \sin 2D) = 0$.

vii) $(\cos A)\cos C + (\cos B)\cos D = (\sin A)\sin C + (\sin B)\sin D \le 1$.

viii) $2(\cos A)\cos C + 2(\cos B)\cos D \le \cos^2 A + \cos^2 B + \cos^2 C + \cos^2 D$.

ix) $pq \le ac + bd, \quad b^2 + d^2 - a^2 - c^2 = 2pq\cos\theta$.

x)

$$\begin{aligned}\operatorname{area}(\mathcal{S}) &= \tfrac{1}{2}pq\sin\theta = \tfrac{1}{4}(b^2+d^2-a^2-c^2)\tan\theta = \tfrac{1}{4}\sqrt{(2pq)^2-(b^2+d^2-a^2-c^2)^2}\\ &= \sqrt{(s-a)(s-b)(s-c)(s-d) - abcd\cos^2\tfrac{1}{2}(A+C)}\\ &= \sqrt{(s-a)(s-b)(s-c)(s-d) - \tfrac{1}{2}abcd[1+\cos(A+C)]}\\ &= \sqrt{(s-a)(s-b)(s-c)(s-d) - \tfrac{1}{4}(ac+bd+pq)(ac+bd-pq)}.\end{aligned}$$

xi)

$$\operatorname{area}(\mathcal{S}) \le \begin{cases}\tfrac{1}{2}(ab+cd)\\ \tfrac{1}{2}(ac+bd)\\ \tfrac{1}{2}(ad+bc)\\ \tfrac{1}{4}(a+c)(b+d)\\ \tfrac{1}{4}s^2\\ \tfrac{1}{2}pq \le \tfrac{1}{4}(p^2+q^2) \le \tfrac{1}{4}(a^2+b^2+c^2+d^2)\\ \tfrac{1}{2}\sqrt{(a^2+c^2)(b^2+d^2)}\\ \tfrac{1}{2}\sqrt[3]{(ad+bc)(ac+bd)(ab+cd)}.\end{cases}$$

xii) The following statements are equivalent:

a) The diagonals of $\mathcal{S}$ are perpendicular.

b) $a^2 + c^2 = b^2 + d^2$.

c) $m_{ac} = m_{bd}$.

xiii) The diagonals of $\mathcal{S}$ are perpendicular and $p = q$ if and only if $\operatorname{area}(\mathcal{S}) = \frac{1}{4}(p^2 + q^2)$.

xiv) If $\mathcal{S}$ has an inscribed circle of radius r that contacts a, b, c, d, then

$$\operatorname{area}(\mathcal{S}) = rs = \sqrt{abcd}\sin\tfrac{1}{2}(A+C).$$

xv) The following statements are equivalent:

a) $\mathcal{S}$ is cyclic.

b) $A + C = \pi$.

c) $pq = ac + bd$.

xvi) Assume that $\mathcal{S}$ is cyclic, and let R denote the radius of the circumscribed circle. Then,

$$\begin{aligned}\operatorname{area}(\mathcal{S}) &= \tfrac{1}{2}(ac+bd)\sin\theta = \tfrac{1}{2}(ad+bc)\sin A = \sqrt{(s-a)(s-b)(s-c)(s-d)}\\ &= \frac{1}{4}\sqrt{(a^2+b^2+c^2+d^2)^2+8abcd-2(a^4+b^4+c^4+d^4)}\\ &= \frac{1}{4R}\sqrt{(ad+bc)(ac+bd)(ab+cd)} \le 2R^2,\end{aligned}$$

$$2|p-q| \le |a-c| + |b-d|,$$

$$p = \sqrt{\frac{(ac+bd)(ad+bc)}{ab+cd}}, \quad q = \sqrt{\frac{(ac+bd)(ab+cd)}{ad+bc}}, \quad \frac{p}{q} = \frac{ad+bc}{ab+cd},$$

$$R = \tfrac{1}{4}\sqrt{\frac{(ad+bc)(ac+bd)(ab+cd)}{(s-a)(s-b)(s-c)(s-d)}}, \quad \tan\tfrac{1}{2}A = \sqrt{\frac{(s-a)(s-d)}{(s-b)(s-c)}},$$

$$\sin A = \frac{2\sqrt{(s-a)(s-b)(s-c)(s-d)}}{ad+bc}, \quad \cos A = \frac{a^2+d^2-b^2-c^2}{2(ad+bc)}.$$

xvii) Assume that $\mathcal{S}$ is both inscribable and circumscribable by circles of radii r and R, respectively, and let t denote the distance between the centers of the inscribed and circumscribed circles. Then,

$$2pq \le s^2, \quad a+c = b+d,$$

$$4r^2 \le \operatorname{area}(\mathcal{S}) = rs = \sqrt{abcd} = \sqrt{p^2q^2-(ac-bd)^2}$$
$$\le \begin{cases} 2R^2\\ \tfrac{1}{6}(ab+ac+ad+bc+bd+cd)\\ \frac{1}{\sqrt[3]{16}}(abc+abd+acd+bcd)^{2/3},\end{cases}$$

$$R = \frac{1}{4}\sqrt{\frac{(ad+bc)(ac+bd)(ab+cd)}{abcd}}, \quad \frac{1}{r^2} = \frac{1}{(R-t)^2} + \frac{1}{(R+t)^2}.$$

xviii) $p^2 + q^2 = 2(m_{ac}^2 + m_{bd}^2)$.

xix) $\mathcal{S}$ is a parallelogram if and only if $p^2+q^2 = a^2+b^2+c^2+d^2$.

xx) The midpoints of the sides of $\mathcal{S}$ are the vertices of a parallelogram. The perimeter of the parallelogram is $p+q$, and the area of the parallelogram is $\tfrac{1}{2}\operatorname{area}(\mathcal{S})$.

xxi) area(AEB)area(CED) = area(BEC)area(DEA).

Source: [47, p. 71], [111, pp. 37, 38], [993, p. 293], and [3024, pp. 322, 323]. *v)* and *vi)* are given in [292]; *vii)* and *viii)* are given in [1860, pp. 22, 150]; the inequality in *ix)* is given in [110, p. 130]; the first expression in *x)* is proved in [1322], see also [2057]; the bounds for area($\mathcal{S}$) in *xi)* are given in [47, pp. 39, 64]; *xii)* is given in [993, p. 293]; *xiii)* is given in [47, p. 71]; *xv)* is given in [110, p. 130]; in *xvi)*, the last expression for area($\mathcal{S}$) and the upper bound $2R^2$ are given in [47, pp. 64, 66]; the inequality in *xvi)* is given in [1158, p. 38]; in *xvii)*, the first inequality is given in [47, p. 66]; *xviii)* is given in [993, p. 298]; *xix)* is given in [993, p. 294]; *xx)* is given in [59, p. 124] and [993, p. 291]. **Remark:** $pq \le ac + bc$ in *ix)* is *Ptolemy's inequality*, which holds for nonconvex quadrilaterals. The equality case in *xv)* is *Ptolemy's theorem*. See [110, p. 130]. An extension to $2n$ points on the circumference of a circle is given in [668]. **Remark:** The fourth expression for area($\mathcal{S}$) in *x)* in the circumscribable case ($A + C = \pi$) is *Brahmagupta's formula*. The limiting case $d = 0$ yields Heron's formula. See Fact 5.2.7. **Remark:** The fifth expression for area($\mathcal{S}$) in x) is *Bretschneider's formula*, see [2057]. **Remark:** *xviii*) is *Euler's theorem.* **Remark:** For each quadrilateral, there exists a quadrilateral with the same side lengths and whose vertices lie on a circle. The area of the latter quadrilateral is maximum over all quadrilaterals with the same side lengths. See [2223]. **Remark:** A quadrilateral is convex if and only if its diagonals intersect. See [1913]. **Remark:** A convex quadrilateral $\mathcal{S}$ is *cyclic* if it has a circumscribed circle that passes through A, B, C, D. See [562]. **Remark:** The first statement in *xxi*) is true if $\mathcal{S}$ is either concave or crossed; that is, $\mathcal{S}$ is the union of triangles with one common vertex. See [2266]. If $\mathcal{S}$ is crossed, then the area of the parallelogram is half of the difference of the areas of the triangles comprising $\mathcal{S}$. **Related:** Fact 11.8.13.

Fact 5.3.2. Let a, b, c, d be positive numbers that represent the lengths of the sides of a quadrilateral. Then, $d < a + b + c$, $d^2 < 3(a^2 + b^2 + c^2)$, and $d^4 \le 27(a^4 + b^4 + c^4)$. **Source:** [289, pp. 24, 44, 139, 140, 201]. **Remark:** The quadrilateral need not be planar.

Fact 5.3.3. Let a, b, c, d be positive numbers that represent the lengths of the sides of a quadrilateral. Then, $\sqrt{a}/(4 + \sqrt{a})$, $\sqrt{b}/(4 + \sqrt{b})$, $\sqrt{c}/(4 + \sqrt{c})$, and $\sqrt{d}/(4 + \sqrt{d})$, represent the lengths of the sides of a quadrilateral. **Source:** [1860, p. 5].

Fact 5.3.4. Let a_1, a_2, a_3, a_4 be positive numbers that represent the lengths of the sides of a quadrilateral, and define $s \triangleq \frac{1}{2}(a_1 + a_2 + a_3 + a_4)$. Then,

$$\sum_{i=1}^{4} \frac{1}{s + a_i} \le \frac{2}{9} \sum_{1 \le i < j \le 4} \frac{1}{\sqrt{(s - a_i)(s - a_j)}}.$$

Source: [806, p. 210]. **Remark:** Equality holds for squares.

Fact 5.3.5. Consider a convex quadrilateral whose sides have lengths a, b, c, d, whose diagonals have lengths f and g, and where the distance between the midpoints of the diagonals is h. Then,

$$a^2 + b^2 + c^2 + d^2 = f^2 + g^2 + 4h^2.$$

Furthermore,

$$f^2 + g^2 \le a^2 + b^2 + c^2 + d^2 \le 2(f^2 + fg + g^2).$$

Source: [1860, pp. 24, 154, 155]. **Remark:** The upper bound is approached as one of the sides approaches zero and the diagonals become aligned.

Fact 5.3.6. Let $\mathcal{S} \subset \mathbb{R}^2$ be a hexagon inscribed around an ellipse, and let A, B, C, D, E, F denote the vertices of $\mathcal{S}$ labeled counterclockwise. Then, the three lines that pass through the pairs of points (A, D), (B, E), and (C, F) intersect at one point inside the ellipse. **Remark:** This is *Brianchon's theorem.*

Fact 5.3.7. Let $\mathcal{S}$ be a convex pentagon with sides of length $a_1, \dots, a_5$ and diagonals of length $d_1, \dots, d_5$. Then,

$$\frac{1}{2} \le \frac{a_1 + a_2 + a_3 + a_4 + a_5}{d_1 + d_2 + d_3 + d_4 + d_5} < 1.$$

Source: [1938, p. 53].

Fact 5.3.8. Let $\mathcal{S}$ be a convex hexagon with successively labeled vertices A, B, C, D, E, F. Then,

$$2\sqrt{3}\,\mathrm{area}(\mathcal{S}) \le \mathrm{AC}(\mathrm{BD} + \mathrm{BF} - \mathrm{DF}) + \mathrm{CE}(\mathrm{BD} + \mathrm{DF} - \mathrm{BF}) + \mathrm{AE}(\mathrm{BF} + \mathrm{DF} - \mathrm{BD}).$$

Source: [108, pp. 56, 324–326].

Fact 5.3.9. Let $n \ge 3$, and let $\mathcal{S}$ be an n-sided convex polygon. Furthermore, let $\mathcal{L} = \{L_1, \dots, L_m\}$ denote the set of lines that pass through all pairs of nonadjacent vertices. Then, $m = \frac{1}{2}n(n-3)$. Now, assume that no pair of lines in $\mathcal{L}$ is parallel and no triple of lines in $\mathcal{L}$ intersect at a single point in the interior of $\mathcal{S}$. Then, the number of points in the interior of $\mathcal{S}$ where the lines in $\mathcal{L}$ intersect is $\binom{n}{4}$, and the number of points outside of $\mathcal{S}$ where the lines in $\mathcal{L}$ intersect is $\frac{1}{12}n(n-3)(n-4)(n-5)$.

Finally, let $\mathcal{P}$ denote the set of polygons each of whose sides is either a side of $\mathcal{S}$ or a subset of $\cup_{i=1}^n L_i$ and whose interiors are disjoint from the sides of $\mathcal{S}$ and $\cup_{i=1}^n L_i$. Then, the number of polygons in $\mathcal{P}$ that are contained in $\mathcal{S}$ is $\frac{1}{24}(n-1)(n-2)(n^2-3n+12)$, and the number of polygons in $\mathcal{P}$ that are outside of $\mathcal{S}$ is $\frac{1}{8}(n^4-6n^3+23n^2-26n+8)$. **Source:** [771, p. 74].

Fact 5.3.10. Let $\mathcal{S} \subset \mathbb{R}^2$ denote the polygon with vertices $\left[\begin{smallmatrix} x_1 \\ y_1 \end{smallmatrix}\right], \ldots, \left[\begin{smallmatrix} x_n \\ y_n \end{smallmatrix}\right] \in \mathbb{R}^2$ arranged in counterclockwise order, and assume that the interior of the polygon is either empty or simply connected. Then,

$$\operatorname{area}(\mathcal{S}) = \tfrac{1}{2}\det\begin{bmatrix} x_1 & x_2 \\ y_1 & y_2 \end{bmatrix} + \tfrac{1}{2}\det\begin{bmatrix} x_2 & x_3 \\ y_2 & y_3 \end{bmatrix} + \cdots + \tfrac{1}{2}\det\begin{bmatrix} x_{n-1} & x_n \\ y_{n-1} & y_n \end{bmatrix} + \tfrac{1}{2}\det\begin{bmatrix} x_n & x_1 \\ y_n & y_1 \end{bmatrix}.$$

Source: [110, p. 100]. **Remark:** The polygon need not be convex, while "counterclockwise" is determined with respect to a point in the interior of the polygon. *Simply connected* means that the polygon has no holes. See [2552]. **Related:** Fact 11.8.13.

Fact 5.3.11. Let $\mathcal{S}$ be a regular polygon with n sides each of which has length a, let r denote the distance from the center of $\mathcal{S}$ to each side, and let R denote the distance from the center of $\mathcal{S}$ to each vertex. Then,

$$r = \frac{a}{2}\cot\frac{\pi}{n} = \frac{a}{2}\tan\frac{(n-2)\pi}{2n}, \quad R = \frac{a}{2}\csc\frac{\pi}{n},$$
$$\operatorname{area}(\mathcal{S}) = \frac{a^2}{4}n\cot\frac{\pi}{n} = nr^2\tan\frac{\pi}{n} = \frac{1}{2}nR^2\sin\frac{2\pi}{n}.$$

Source: [1757, p. 71].

Fact 5.3.12. Let $\mathcal{S}$ be a regular pentagon each of whose sides has length a, let r denote the distance from the center of $\mathcal{S}$ to each side, let R denote the distance from the center of $\mathcal{S}$ to each vertex, and let d denote the distance between each pair of nonadjacent vertices. Then,

$$r = \frac{\tau a}{2\sqrt{3-\tau}} = \frac{a}{10}\sqrt{25+10\sqrt{5}}, \quad R = \frac{a}{\sqrt{3-\tau}} = \frac{a}{10}\sqrt{50+10\sqrt{5}},$$
$$d = \frac{a}{2}(1+\sqrt{5}), \quad \operatorname{area}(\mathcal{S}) = \frac{5\tau a^2}{4\sqrt{3-\tau}} = \frac{5a^2}{4\sqrt{5-2\sqrt{5}}} = \frac{a^2}{4}\sqrt{5(5+2\sqrt{5})}.$$

Source: [2764, p. 146].

Fact 5.3.13. Let $\mathcal{S}$ be a regular polygon with n sides each of which has length a, let P denote a point inside $\mathcal{S}$, and let $d_1, \ldots, d_n$ denote the distances from P to each of the sides of $\mathcal{S}$. Then,

$$\frac{2\pi}{a} < \sum_{i=1}^n \frac{1}{d_i}.$$

Source: [112, p. 19].

Fact 5.3.14. Let $\mathcal{S}$ be a polygon with sides $a_1, \ldots, a_n$, and define $p \triangleq \sum_{i=1}^n a_i$. Then,

$$\frac{n}{n-1} \le \sum_{i=1}^n \frac{a_i}{p-a_i} \le 1 + \max_{i\in\{1,\ldots,n\}} \frac{a_i}{p-a_i} < 2.$$

Credit: O. Bagdasar.

5.4 Facts on Polytopes

Fact 5.4.1. Let $\mathcal{S} \subset \mathbb{R}^3$ denote the tetrahedron with vertices $x, y, z, w \in \mathbb{R}^3$. Then,

$$\operatorname{volume}(\mathcal{S}) = \tfrac{1}{6}\left|(x-w)^{\mathsf{T}}[(y-w)\times(z-w)]\right|.$$

Source: The volume of the unit simplex $\mathcal{S} \subset \mathbb{R}^3$ with vertices $(0,0,0), (1,0,0), (0,1,0), (0,0,1)$ is $1/6$. Now, Fact 5.4.5 implies that the volume of $A\mathcal{S}$ is $(1/6)|\det A|$. **Remark:** The connection between the *signed volume* of a simplex and the determinant is discussed in [1768, pp. 32, 33].

Fact 5.4.2. Let $\mathcal{S} \subset \mathbb{R}^3$ denote a tetrahedron with vertices $\mathrm{V}_1, \mathrm{V}_2, \mathrm{V}_3, \mathrm{V}_4$, edge lengths $a_1 \triangleq \mathrm{V}_1\mathrm{V}_2$, $a_2 \triangleq \mathrm{V}_1\mathrm{V}_3$, $a_3 \triangleq \mathrm{V}_1\mathrm{V}_4$, $a_4 \triangleq \mathrm{V}_2\mathrm{V}_3$, $a_5 \triangleq \mathrm{V}_2\mathrm{V}_4$, $a_6 \triangleq \mathrm{V}_3\mathrm{V}_4$, median lengths m_1, m_2, m_3, m_4, and bimedian lengths b_1, b_2, b_3. Then,

$$\sum_{i=1}^{4} m_i^2 = \frac{4}{9}\sum_{i=1}^{6} a_i^2, \quad \sum_{i=1}^{3} b_i^2 = \frac{1}{4}\sum_{i=1}^{6} a_i^2.$$

Next, let f_1, f_2, f_3, f_4 denote the faces of $\mathcal{S}$, and, for all $i, j \in \{1,2,3,4\}$, let $\theta_{i,j}$ denote the angle between f_i and f_j. Then,

$$\sum_{1\le i<j\le 4} \cos\theta_{i,j} \le 2, \quad \sum_{1\le i<j\le 4} \cos^2\frac{\theta_{i,j}}{2} \le 4, \quad \prod_{1\le i<j\le 4} \cos\theta_{i,j} \le \frac{1}{36}, \quad \prod_{1\le i<j\le 4} \cos\frac{\theta_{i,j}}{2} \le \frac{8}{27}.$$

Next, let r and R denote the radii of the inscribed and circumscribed spheres of $\mathcal{S}$, respectively. Then,

$$2\sqrt{6}r \le \max\{a_1, \ldots, a_6\}, \quad 3r \le R, \quad \sum_{i=1}^{6} a_i^2 \le 16R^2.$$

Finally, for all $i \in \{1,2,3,4\}$, let h_i denote the altitude from V_i to the opposite face. Then,

$$9 \le \sum_{1\le i<j\le 4} \frac{(\mathrm{V}_i\mathrm{V}_j)^2}{h_ih_j} \le \left(\frac{R}{r}\right)^2, \quad \frac{1}{h_1} + \frac{1}{h_2} + \frac{1}{h_3} + \frac{1}{h_4} = \frac{1}{r}, \quad \sum_{1\le i<j\le 4} \frac{1}{h_ih_j} \le \frac{3}{8r^2}.$$

Source: [108, pp. 427–433, 447, 450] and [1063]. **Remark:** Each median connects a vertex to the centroid of the opposite face. Each bimedian connects the midpoint of an edge to the midpoint of the nonadjoining edge.

Fact 5.4.3. Let $x, y, z \in \mathbb{R}^3$ be linearly independent vectors, and define the parallelepiped $\mathcal{S} \triangleq \{\alpha x + \beta y + \gamma z\colon \alpha, \beta, \gamma \in [0,1]\}$. Then,

$$\mathrm{volume}(\mathcal{S}) = |\det[x\ \ y\ \ z]|.$$

Remark: The parallelotope $\mathcal{S}$ has vertices $0, x, y, z, x+y, x+z, y+z, x+y+z$. **Related:** Fact 3.11.19.

Fact 5.4.4. Let $A \in \mathbb{R}^{n\times m}$, assume that $\operatorname{rank} A = m$, and define the parallelotope

$$\mathcal{S} = \left\{\sum_{i=1}^{m} \alpha_i \mathrm{col}_i(A)\colon 0 \le \alpha_i \le 1 \text{ for all } i \in \{1, \ldots, m\}\right\}.$$

Then,

$$\mathrm{volume}(\mathcal{S}) = (\det A^{\mathsf{T}}A)^{1/2}.$$

If, in addition, $m = n$, then

$$\mathrm{volume}(\mathcal{S}) = |\det A|.$$

Remark: volume($\mathcal{S}$) denotes the m-dimensional volume of $\mathcal{S}$. If $m = 2$, then volume($\mathcal{S}$) is the area of a parallelogram. See [970, p. 202]. **Related:** Fact 3.11.19.

Fact 5.4.5. Let $\mathcal{S} \subset \mathbb{R}^n$ and $A \in \mathbb{R}^{n\times n}$. Then,

$$\mathrm{volume}(A\mathcal{S}) = |\det A|\mathrm{volume}(\mathcal{S}).$$

Remark: See [2036, p. 468].

Fact 5.4.6. Let $\mathcal{S} \subset \mathbb{R}^n$ be a simplex, and assume that $\mathcal{S}$ is inscribed in a sphere of radius R. Then,

$$\operatorname{volume}(\mathcal{S}) \le \sqrt{\frac{(n+1)^{n+1}}{n^n}}\frac{R^n}{n!}.$$

Furthermore, equality holds if and only if $\mathcal{S}$ is a regular polytope. **Source:** [1041, p. 66-13] and [2790]. **Remark:** The definition, enumeration, and classification of regular polytopes is discussed in [1261, 1262, 2307].

Fact 5.4.7. Let $x_1, \dots, x_{n+1} \in \mathbb{R}^n$, define the polytope $\mathcal{S} \triangleq \operatorname{conv}\{x_1, \dots, x_{n+1}\}$, and let $A \in \mathbb{R}^{(n+1)\times(n+1)}$, where, for all $i, j \in \{1, \dots, n+1\}$, $A_{(i,j)} \triangleq \|x_i - x_j\|_2$. Then,

$$\operatorname{volume}(\mathcal{S}) = \frac{1}{2^{n/2}n!}\left|\det\begin{bmatrix} 0 & 1_{1\times(n+1)} \\ 1_{(n+1)\times 1} & A \end{bmatrix}\right|^{1/2}.$$

Now, assume that $n = 2$, and define $a \triangleq \|x_1 - x_2\|_2$, $b \triangleq \|x_2 - x_3\|_2$, $c \triangleq \|x_3 - x_1\|_2$, and $s \triangleq \frac{1}{2}(a+b+c)$. Then,

$$\operatorname{area}(\mathcal{S}) = \frac{1}{4}\left|\det\begin{bmatrix} 0 & 1 & 1 & 1 \\ 1 & 0 & a^2 & c^2 \\ 1 & a^2 & 0 & b^2 \\ 1 & c^2 & b^2 & 0 \end{bmatrix}\right|^{1/2} = \sqrt{s(s-a)(s-b)(s-c)}.$$

Source: [496, pp. 97–99], [517, pp. 234, 235], [1307], and [2979, pp. 220, 221]. **Remark:** The volume of $\mathcal{S}$ is given in terms of the *Cayley-Menger determinant.* **Remark:** The area of the triangle with side lengths a, b, c is given by Heron's formula. See Fact 5.2.7. **Related:** Fact 3.16.23.

Fact 5.4.8. Let $\mathcal{V}$ be a convex polyhedron in $\mathbb{R}^3$, and let F, V, and E denote the number of faces, vertices, and edges of $\mathcal{V}$, respectively. Then, $F + V - E = 2$. Furthermore,

$$F \le 2V - 4, \quad E \le 3V - 6, \quad E \le 3F - 6, \quad V \le 2F + 4.$$

Finally, let S denote the sum of the angles on each facet. Then, $2\pi V = S + 4\pi$. **Source:** The last statement is given in [1158, p. 291]. **Remark:** The first equality gives the *Euler characteristic* for a convex polyhedron in three dimensions, such as either a tetrahedron, cube (hexahedron), octahedron, dodecahedron, or icosahedron. For the cube, $F = 6$, $V = 8$, and $E = 12$. See [47, p. 115] and [2307].

5.5 Facts on Circles, Ellipses, Spheres, and Ellipsoids

Fact 5.5.1. Let $n \ge 1$. Then, the following statements hold:

i) Let $C_1, \dots, C_n$ be circles in a plane. Then, the maximum number of connected, open sets whose boundaries are subsets of $\cup_{i=1}^n C_i$ but whose interiors do not intersect $\cup_{i=1}^n C_i$ is $n^2 - n + 2$.

ii) Let $S_1, \dots, S_n$ be spheres in $\mathbb{R}^3$. Then, the maximum number of connected, open sets whose boundaries are subsets of $\cup_{i=1}^n S_i$ but whose interiors do not intersect $\cup_{i=1}^n S_i$ is $n(n^2-3n+8)$.

iii) Let $m \ge 2$, and let $H_1, \dots, H_n$ be hyperspheres in $\mathbb{R}^m$. Then, the maximum number of connected open sets whose whose boundaries are subsets of $\cup_{i=1}^n H_i$ but whose interiors do not intersect $\cup_{i=1}^n H_i$ is $\binom{n-1}{m} + \sum_{i=0}^m \binom{n}{i}$.

Source: [771, p. 73].

Fact 5.5.2. Let a be a complex number, let $b \in (0, |a|^2)$, and define

$$\mathcal{S} \triangleq \{z \in \mathbb{C}\colon |z|^2 - \overline{a}z - a\overline{z} + b = 0\}.$$

Then, $\mathcal{S}$ is the circle with center at a and radius $\sqrt{|a|^2 - b}$. That is,

$$\mathcal{S} = \{z \in \mathbb{C}\colon |z - a| = \sqrt{|a|^2 - b}\}.$$

Source: [110, pp. 84, 85].

Fact 5.5.3. Let $\mathcal{S}$ be a triangle, and let $\mathcal{E}$ be an ellipse contained in $\mathcal{S}$. Then,

$$\frac{\operatorname{area}(\mathcal{E})}{\operatorname{area}(\mathcal{S})} \le \frac{\sqrt{3}\pi}{9}.$$

Equality holds if and only if $\mathcal{E}$ is tangent to the sides of $\mathcal{S}$ at their midpoints. In addition, there exists a unique ellipse $\mathcal{E}_0$ satisfying these conditions. Now, let $z_1, z_2, z_3 \in \mathbb{C}$ be the vertices of $\mathcal{S}$, and, for $z \in \mathbb{C}$, define the cubic polynomial $p(z) \triangleq (z - z_1)(z - z_2)(z - z_3)$. Then, the roots of p' are the foci of $\mathcal{E}_0$, and the root $\frac{1}{3}(z_1 + z_2 + z_3)$ of p'' is the centroid of the triangle and the center of $\mathcal{E}_0$. Finally, the following statements are equivalent:

i) $\mathcal{S}$ is equilateral.

ii) $\mathcal{E}_0$ is a circle.

iii) p' has a repeated root.

Source: [662, 1204, 1576, 2049]. **Remark:** The center of $\mathcal{E}_0$ is the midpoint of the line segment connecting to the foci of $\mathcal{E}$. **Remark:** $\mathcal{E}_0$ is the *Steiner inellipse*. **Remark:** Extensions to polygons are considered in [2199]. **Credit:** J. Siebeck. **Related:** Fact 12.16.5.

Fact 5.5.4. Let $\mathcal{E}$ be an ellipse, and let $\mathcal{S}$ be a triangle contained in $\mathcal{E}$. Then,

$$\frac{\operatorname{area}(\mathcal{S})}{\operatorname{area}(\mathcal{E})} \le \frac{3\sqrt{3}}{4\pi}.$$

Furthermore, equality holds if and only if all of the vertices of $\mathcal{S}$ lie on the boundary of $\mathcal{E}$ and the centroid of $\mathcal{S}$ coincides with the center of $\mathcal{E}$. If these conditions hold, then $\mathcal{S}$ is isosceles. If, in addition, $\mathcal{E}$ is a circle, then $\mathcal{S}$ is equilateral. **Source:** [662].

Fact 5.5.5. Let $\mathcal{S}$ be a convex quadrilateral. Then, there exists an ellipse that is inscribed in $\mathcal{S}$. Furthermore, the set of centers of all ellipses inscribed in $\mathcal{S}$ is the line segment connecting the midpoints of the diagonals of $\mathcal{S}$. **Source:** [1462, 1464]. **Remark:** The ellipse $\mathcal{E}$ is *inscribed* in $\mathcal{S}$ if it is tangent to each side of $\mathcal{S}$.

Fact 5.5.6. Let $\mathcal{S}$ be a convex quadrilateral. Then, there exist infinitely many ellipses that circumscribe $\mathcal{S}$. **Source:** [2102]. **Remark:** The ellipse $\mathcal{E}$ *circumscribes* $\mathcal{S}$ if all of the vertices of $\mathcal{S}$ lie on $\mathcal{E}$.

Fact 5.5.7. Let $\mathcal{S}$ be a parallelogram, and let $\mathcal{E}$ be an ellipse contained in $\mathcal{S}$. Then,

$$\frac{\operatorname{area}(\mathcal{E})}{\operatorname{area}(\mathcal{S})} \le \frac{\pi}{4}.$$

Equality holds if and only if $\mathcal{E}$ is tangent to the sides of $\mathcal{S}$ at their midpoints. In addition, there exists a unique ellipse $\mathcal{E}_0$ satisfying these conditions. Finally, $\mathcal{S}$ is square if and only if $\mathcal{E}_0$ is a circle. **Source:** [1463].

Fact 5.5.8. Let a, b be positive numbers, and define the ellipse

$$\mathcal{S} \triangleq \left\{ x \in \mathbb{R}^2 \colon \frac{x_{(1)}^2}{a^2} + \frac{x_{(2)}^2}{b^2} \le 1 \right\}.$$

Then, $\operatorname{area}(\mathcal{S}) = \pi ab$. **Related:** This is a special case of Fact 5.5.9. See Fact 13.9.5 for the arc length of an ellipse.

Fact 5.5.9. Let a, c be positive numbers, let b be a real number, assume that $b^2 < ac$, and define the ellipse

$$\mathcal{S} \triangleq \{x \in \mathbb{R}^2 \colon a x_{(1)}^2 + b x_{(1)} x_{(2)} + c x_{(2)}^2 \le 1\}.$$

Then, $\text{area}(\mathcal{S}) = \frac{2\pi}{\sqrt{4ac-b^2}}$.

Fact 5.5.10. Let $\mathcal{S}$ denote the spherical triangle on the surface of the unit sphere whose vertices are $x, y, z \in \mathbb{R}^3$, and let A, B, C denote the angles of $\mathcal{S}$ located at the points x, y, z, respectively. Furthermore, let a, b, c denote the planar angles subtended by the pairs (y, z), (x, z), (x, y) from the center of the sphere, respectively; that is, the lengths of the sides of the spherical triangle opposite A, B, C, respectively. Finally, define the solid angle Ω to be the area of $\mathcal{S}$. Then,

$$\Omega = A + B + C - \pi.$$

Furthermore,

$$\tan\frac{\Omega}{4} = \sqrt{(\tan\tfrac{s}{2})(\tan\tfrac{s-a}{2})(\tan\tfrac{s-b}{2})\tan\tfrac{s-c}{2}},$$

$$\tan\frac{\Omega}{2} = \frac{|\det\,[x\;\,y\;\,z]|}{1 + x^{\mathsf{T}}y + y^{\mathsf{T}}z + z^{\mathsf{T}}x} = \frac{\sqrt{1 - \cos^2 a - \cos^2 b - \cos^2 c + 2(\cos a)(\cos b)\cos c}}{1 + \cos a + \cos b + \cos c}.$$

Source: [1000] and [3024, pp. 218–220]. **Remark:** Spherical triangles are discussed in [1035, pp. 253–260], [1527, Chapter 2], [2769], [2871, pp. 904–907], and [2889, pp. 26–29]. A linear algebraic approach is given in [274].

Fact 5.5.11. Let $\mathcal{S}$ denote a circular cap on the surface of the unit sphere, where the angle subtended by cross sections of the cone with apex at the center of the sphere is 2θ. Furthermore, define the solid angle Ω to be the area of $\mathcal{S}$. Then,

$$\Omega = 2\pi(1 - \cos\theta).$$

Fact 5.5.12. Let $\mathcal{S}$ denote a region on the surface of the unit sphere subtended by the sides of a right rectangular pyramid with apex at the center of the sphere, where the subtended planar angles of the edges of the pyramid are θ and ϕ. Furthermore, define the solid angle Ω to be the area of $\mathcal{S}$. Then,

$$\Omega = 4\,\text{asin}[(\sin\tfrac{\theta}{2})\sin\tfrac{\phi}{2}].$$

Remark: The solid angle of a polygonal cone is considered in [1296].

Fact 5.5.13. Let $n \geq 1$, let $r > 0$, and define the hypersphere of radius r by $\mathcal{S}_n \triangleq \{x \in \mathbb{R}^n\colon\ \|x\|_2 = r\}$ and the hyperball of radius r by $\mathcal{B}_n \triangleq \{x \in \mathbb{R}^n\colon\ \|x\|_2 \leq r\}$. Then,

$$\text{area}(\mathcal{S}_n) = \frac{2\pi^{n/2}r^{n-1}}{\Gamma\left(\frac{n}{2}\right)}, \quad \text{volume}(\mathcal{B}_n) = \frac{\pi^{n/2}r^n}{\Gamma\left(\frac{n}{2}+1\right)}.$$

In particular,

$$\begin{aligned} \text{area}(\mathcal{S}_1) &= 2, & \text{volume}(\mathcal{B}_1) &= 2r,\\ \text{area}(\mathcal{S}_2) &= 2\pi r, & \text{volume}(\mathcal{B}_2) &= \pi r^2,\\ \text{area}(\mathcal{S}_3) &= 4\pi r^2, & \text{volume}(\mathcal{B}_3) &= \tfrac{4}{3}\pi r^3,\\ \text{area}(\mathcal{S}_4) &= 2\pi^2 r^3, & \text{volume}(\mathcal{B}_4) &= \tfrac{1}{2}\pi^2 r^4,\\ \text{area}(\mathcal{S}_5) &= \tfrac{8}{3}\pi^2 r^4, & \text{volume}(\mathcal{B}_5) &= \tfrac{8}{15}\pi^2 r^5. \end{aligned}$$

Furthermore,

$$\frac{\text{area}(\mathcal{S}_{n+2})}{\text{area}(\mathcal{S}_n)} = \frac{2\pi r^2}{n}, \quad \frac{\text{volume}(\mathcal{B}_{n+2})}{\text{volume}(\mathcal{B}_n)} = \frac{2\pi r^2}{n+2}, \quad \frac{\text{volume}(\mathcal{B}_n)}{\text{area}(\mathcal{S}_n)} = \frac{r}{n}, \quad \frac{\text{area}(\mathcal{S}_{n+2})}{\text{volume}(\mathcal{B}_n)} = 2\pi r.$$

Finally,

$$\text{volume}(\mathcal{B}_n) = \begin{cases} \dfrac{\pi^{n/2}r^n}{(n/2)!}, & n \text{ even}, \\[2ex] \dfrac{2^n\pi^{(n-1)/2}[(n-1)/2]!r^n}{n!}, & n \text{ odd}. \end{cases}$$

Fact 5.5.14. Let $A \in \mathbb{R}^{n\times n}$, assume that A is positive definite, and define the hyperellipsoidal solid

$$\mathcal{E} \triangleq \{x \in \mathbb{R}^n\colon\ x^{\mathsf{T}}Ax \le 1\}.$$

Then,

$$\text{volume}(\mathcal{E}) = \frac{\text{volume}(\mathcal{B}_n)}{\sqrt{\det A}},$$

where volume$(\mathcal{B}_n)$ is the volume of the hyperball in $\mathbb{R}^n$. In particular, the area of the ellipse $\{x \in \mathbb{R}^2\colon\ x^{\mathsf{T}}Ax \le 1\}$ is $\pi/\det A$. **Source:** [1600, p. 36]. **Related:** Fact 5.5.15 and Fact 14.13.13.

Fact 5.5.15. Let $\alpha_1,\ldots,\alpha_n > 0$, define $\beta \triangleq \sum_{i=1}^n \frac{1}{\alpha_i}$, let $r > 0$, and define

$$\mathcal{S} \triangleq \left\{x \in \mathbb{R}^n\colon\ \sum_{i=1}^n |x_{(i)}|^{\alpha_i} \le r\right\}.$$

Then,

$$\text{volume}(\mathcal{S}) = \frac{2^n \prod_{i=1}^n \Gamma(1+\frac{1}{\alpha_i})}{\Gamma(1+\beta)} r^{\beta}.$$

Source: [1141]. **Related:** Fact 5.5.14 and Fact 14.13.13.

Chapter Six
Polynomial Matrices and Rational Transfer Functions

In this chapter we consider matrices whose entries are either polynomials or rational functions. The decomposition of a polynomial matrix in terms of a Smith matrix provides the foundation for developing canonical matrices in Chapter 7. In this chapter we also present some basic properties of eigenvalues and eigenvectors as well as the minimal and characteristic polynomials of a square matrix. Finally, we consider the extension of the Smith matrix to the Smith-McMillan matrix for rational transfer functions.

6.1 Polynomials

A function $p\colon\ \mathbb{C} \mapsto \mathbb{C}$ of the form

$$p(s) = \beta_k s^k + \beta_{k-1}s^{k-1} + \cdots + \beta_1 s + \beta_0, \tag{6.1.1}$$

where $k \in \mathbb{N}$ and $\beta_0, \ldots, \beta_k \in \mathbb{F}$, is a *polynomial.* The set of polynomials is denoted by $\mathbb{F}[s]$. If the coefficient $\beta_k \in \mathbb{F}$ is nonzero, then the *degree* of p, denoted by $\deg p$, is k. If, in addition, $\beta_k = 1$, then p is *monic*. If $k = 0$, then p is *constant*. The degree of a nonzero constant polynomial is zero, while the degree of the zero polynomial is defined to be $-\infty$.

Let p_1 and p_2 be polynomials. Then,

$$\deg p_1 p_2 = \deg p_1 + \deg p_2. \tag{6.1.2}$$

If either $p_1 = 0$ or $p_2 = 0$, then $\deg p_1 p_2 = \deg p_1 + \deg p_2 = -\infty$. If p_2 is a nonzero constant, then $\deg p_2 = 0$, and thus $\deg p_1 p_2 = \deg p_1$. Furthermore,

$$\deg(p_1 + p_2) \le \max\{\deg p_1, \deg p_2\}. \tag{6.1.3}$$

Therefore, $\deg(p_1 + p_2) = \max\{\deg p_1, \deg p_2\}$ if and only if either *i*) $\deg p_1 \ne \deg p_2$ or *ii*) $p_1 = p_2 = 0$ or *iii*) $r \triangleq \deg p_1 = \deg p_2 \ne -\infty$ and the sum of the coefficients of s^r in p_1 and p_2 is not zero. Equivalently, $\deg(p_1 + p_2) < \max\{\deg p_1, \deg p_2\}$ if and only if $r \triangleq \deg p_1 = \deg p_2 \ne -\infty$ and the sum of the coefficients of s^r in p_1 and p_2 is zero.

Let $p \in \mathbb{F}[s]$ be a polynomial of degree $k \ge 1$. Then, it follows from the *fundamental theorem of algebra* that p has k possibly repeated complex roots $\lambda_1, \ldots, \lambda_k$ and thus can be factored as

$$p(s) = \beta \prod_{i=1}^{k} (s - \lambda_i), \tag{6.1.4}$$

where $\beta \in \mathbb{F}$. The multiplicity of a root $\lambda \in \mathbb{C}$ of p is denoted by $\mathrm{mult}_p(\lambda)$. If λ is not a root of p, then $\mathrm{mult}_p(\lambda) = 0$. The multiset consisting of the roots of p including multiplicity is $\mathrm{mroots}(p) = \{\lambda_1, \ldots, \lambda_k\}_{\mathrm{ms}}$, while the set of roots of p ignoring multiplicity is $\mathrm{roots}(p) = \{\hat{\lambda}_1, \ldots, \hat{\lambda}_l\}$, where $\sum_{i=1}^{l} \mathrm{mult}_p(\hat{\lambda}_i) = k$. If $\mathbb{F} = \mathbb{R}$, then the multiplicity of a root λ_i whose imaginary part is nonzero is equal to the multiplicity of its complex conjugate $\overline{\lambda_i}$. Hence, $\mathrm{mroots}(p)$ is *self-conjugate*; that is, $\mathrm{mroots}(p) = \overline{\mathrm{mroots}(p)}$.

Let $p \in \mathbb{F}[s]$, and let $\operatorname{mroots}(A) = \{\lambda_1, \ldots, \lambda_n\}_{\rm ms}$. The *spread* of p is defined by

$$\delta(p) \triangleq \max_{i,j \in \{1,\ldots,n\}} |\lambda_i - \lambda_j|. \tag{6.1.5}$$

Then, the *root moduli* of p are the nonnegative numbers $\rho_n(p) \le \cdots \le \rho_1(p)$ such that $\{\rho_1(p), \ldots, \rho_n(p)\}_{\rm ms} = \{|\lambda_1|, \ldots, |\lambda_n|\}_{\rm ms}$. In particular, define the *minimum root modulus* of p by

$$\rho_{\min}(p) \triangleq \rho_n(p) \tag{6.1.6}$$

and the *root radius* of p by

$$\rho_{\max}(p) \triangleq \rho_1(p). \tag{6.1.7}$$

The *root real parts* of p are the real numbers $\alpha_n(p) \le \cdots \le \alpha_1(p)$ such that $\{\alpha_1(p), \ldots, \alpha_n(p)\}_{\rm ms} = \{\operatorname{Re} \lambda_1, \ldots, \operatorname{Re} \lambda_n\}_{\rm ms}$. In particular, define the *minimum root real part* of A by

$$\alpha_{\min}(p) \triangleq \alpha_n(p) \tag{6.1.8}$$

and the *root real abscissa* of p by

$$\alpha_{\max}(p) \triangleq \alpha_1(p). \tag{6.1.9}$$

The *root imaginary parts* of p are the real numbers $\beta_n(p) \le \cdots \le \beta_1(p)$ such that $\{\beta_1(p), \ldots, \beta_n(p)\}_{\rm ms} = \{\operatorname{Im} \lambda_1, \ldots, \operatorname{Im} \lambda_n\}_{\rm ms}$. In particular, define the *minimum root imaginary part* of A by

$$\beta_{\min}(p) \triangleq \beta_n(p) \tag{6.1.10}$$

and the *root imaginary abscissa* of p by

$$\beta_{\max}(p) \triangleq \beta_1(p). \tag{6.1.11}$$

Let $p \in \mathbb{F}[s]$. If $p(-s) = p(s)$ for all $s \in \mathbb{C}$, then p is *even*, while, if $p(-s) = -p(s)$ for all $s \in \mathbb{C}$, then p is *odd*. If p is either odd or even, then $\operatorname{mroots}(p) = -\operatorname{mroots}(p)$. If $p \in \mathbb{R}[s]$ and there exists a polynomial $q \in \mathbb{R}[s]$ such that $p(s) = q(s)q(-s)$ for all $s \in \mathbb{C}$, then p has a *spectral factorization*. If p has a spectral factorization, then p is even and $\deg p$ is an even integer.

Proposition 6.1.1. Let $p \in \mathbb{R}[s]$. Then, the following statements are equivalent:

i) p has a spectral factorization.

ii) p is even, and every imaginary root of p has even multiplicity.

iii) For all $\omega \in \mathbb{R}$, $p(\omega \jmath) \in [0, \infty)$.

Proof. *i)* $\Longleftrightarrow$ *ii)* is immediate. To prove *i)* $\Longrightarrow$ *iii)*, note that, for all $\omega \in \mathbb{R}$,

$$p(\omega \jmath) = q(\omega \jmath)q(-\omega \jmath) = |q(\omega \jmath)|^2 \ge 0.$$

Conversely, to prove *iii)* $\Longrightarrow$ *i)* write $p = p_1 p_2$, where every root of p_1 is imaginary and none of the roots of p_2 are imaginary. Now, let z be a root of p_2. Then, $-z$, $\overline{z}$, and $-\overline{z}$ are also roots of p_2 with the same multiplicity as z. Hence, there exists a polynomial $p_{20} \in \mathbb{R}[s]$ such that $p_2(s) = p_{20}(s)p_{20}(-s)$ for all $s \in \mathbb{C}$.

Next, assuming that p has at least one imaginary root, write $p_1(s) = \prod_{i=1}^k (s^2 + \omega_i^2)^{m_i}$, where $0 \le \omega_1 < \cdots < \omega_k$ and $m_i \triangleq \operatorname{mult}_p(\omega_i \jmath)$. Let ω_{i_0} denote the smallest element of the set $\{\omega_1, \ldots, \omega_k\}$ such that m_i is odd. Then, it follows that $p_1(\omega \jmath) = \prod_{i=1}^k (\omega_i^2 - \omega^2)^{m_i} < 0$ for all $\omega \in (\omega_{i_0}, \omega_{i_0+1})$, where $\omega_{k+1} \triangleq \infty$. However, note that $p_1(\omega \jmath) = p(\omega \jmath)/p_2(\omega \jmath) = p(\omega \jmath)/|p_{20}(\omega \jmath)|^2 \ge 0$ for all $\omega \in \mathbb{R}$, which is a contradiction. Therefore, m_i is even for all $i \in \{1, \ldots, k\}$, and thus $p_1(s) = p_{10}(s)p_{10}(-s)$ for all $s \in \mathbb{C}$, where $p_{10}(s) \triangleq \prod_{i=1}^k (s^2 + \omega_i^2)^{m_i/2}$. Consequently, $p(s) = p_{10}(s)p_{20}(s)p_{10}(-s)p_{20}(-s)$ for all $s \in \mathbb{C}$. Finally, if p has no imaginary roots, then $p_1 = 1$, and $p(s) = p_{20}(s)p_{20}(-s)$ for all $s \in \mathbb{C}$. $\square$

The following division algorithm is essential to the study of polynomials.

Lemma 6.1.2. Let $p_1, p_2 \in \mathbb{F}[s]$, and assume that p_2 is not the zero polynomial. Then, there exist unique polynomials $q, r \in \mathbb{F}[s]$ such that $\deg r < \deg p_2$ and

$$p_1 = qp_2 + r. \tag{6.1.12}$$

Proof. Define $n \triangleq \deg p_1$ and $m \triangleq \deg p_2$. If $n < m$, then $q = 0$ and $r = p_1$. Hence, $\deg r = \deg p_1 = n < m = \deg p_2$.

Now, assume that $n \geq m \geq 0$, and write $p_1(s) = \beta_n s^n + \cdots + \beta_0$ and $p_2(s) = \gamma_m s^m + \cdots + \gamma_0$. If $n = 0$, then $m = 0$, $\gamma_0 \neq 0$, $q = \beta_0/\gamma_0$, and $r = 0$. Hence, $-\infty = \deg r < 0 = \deg p_2$.

If $n = 1$, then either $m = 0$ or $m = 1$. If $m = 0$, then $p_2(s) = \gamma_0 \neq 0$, and (6.1.12) holds with $q(s) = p_1(s)/\gamma_0$ and $r = 0$, in which case $-\infty = \deg r < 0 = \deg p_2$. If $m = 1$, then (6.1.12) holds with $q(s) = \beta_1/\gamma_1$ and $r(s) = \beta_0 - \beta_1\gamma_0/\gamma_1$. Hence, $\deg r \leq 0 < 1 = \deg p_2$.

Now, suppose that $n = 2$. Then, $\hat{p}_1(s) = p_1(s) - (\beta_2/\gamma_m)s^{2-m}p_2(s)$ has degree 1. Applying (6.1.12) with p_1 replaced by $\hat{p}_1$, it follows that there exist polynomials $q_1, r_1 \in \mathbb{F}[s]$ such that $\hat{p}_1 = q_1 p_2 + r_1$ and such that $\deg r_1 < \deg p_2$. It thus follows that $p_1(s) = q_1(s)p_2(s) + r_1(s) + (\beta_2/\gamma_m)s^{2-m}p_2(s) = q(s)p_2(s) + r(s)$, where $q(s) = q_1(s) + (\beta_2/\gamma_m)s^{n-m}$ and $r = r_1$, which verifies (6.1.12). Similar arguments apply to successively larger values of n.

To prove uniqueness, suppose there exist polynomials $\hat{q}$ and $\hat{r}$ such that $\deg \hat{r} < \deg p_2$ and $p_1 = \hat{q}p_2 + \hat{r}$. Then, it follows that $(\hat{q} - q)p_2 = r - \hat{r}$. Next, note that $\deg(r - \hat{r}) < \deg p_2$. If $\hat{q} \neq q$, then $\deg p_2 \leq \deg[(\hat{q} - q)p_2]$ so that $\deg(r - \hat{r}) < \deg[(\hat{q} - q)p_2]$, which is a contradiction. Thus, $\hat{q} = q$, and, hence, $r = \hat{r}$. □

In Lemma 6.1.2, q is the *quotient* of p_1 and p_2, while r is the *remainder*. If $r = 0$, then p_2 *divides* p_1,; equivalently, p_1 is a *multiple* of p_2. Note that, if $p_2(s) = s - \alpha$, where $\alpha \in \mathbb{F}$, then r is constant and is given by $r(s) = p_1(\alpha)$.

If a polynomial $p_3 \in \mathbb{F}[s]$ divides two polynomials $p_1, p_2 \in \mathbb{F}[s]$, then p_3 is a *common divisor* of p_1 and p_2. Given polynomials $p_1, p_2 \in \mathbb{F}[s]$, there exists a unique monic polynomial $p_3 \in \mathbb{F}[s]$, the *greatest common divisor* of p_1 and p_2, such that p_3 is a common divisor of p_1 and p_2 and such that every common divisor of p_1 and p_2 divides p_3. In addition, there exist polynomials $q_1, q_2 \in \mathbb{F}[s]$ such that the greatest common divisor p_3 of p_1 and p_2 is given by $p_3 = q_1 p_1 + q_2 p_2$. See [2221, p. 113] for proofs of these results. Finally, p_1 and p_2 are *coprime* if their greatest common divisor is $p_3 = 1$, while a polynomial $p \in \mathbb{F}[s]$ is *irreducible* if there do not exist nonconstant polynomials $p_1, p_2 \in \mathbb{F}[s]$ such that $p = p_1 p_2$. For example, if $\mathbb{F} = \mathbb{R}$, then $p(s) = s^2 + s + 1$ is irreducible.

If a polynomial $p_3 \in \mathbb{F}[s]$ is a multiple of two polynomials $p_1, p_2 \in \mathbb{F}[s]$, then p_3 is a *common multiple* of p_1 and p_2. Given nonzero polynomials p_1 and p_2, there exists (see [2221, p. 113]) a unique monic polynomial $p_3 \in \mathbb{F}[s]$ that is a common multiple of p_1 and p_2 and that divides every common multiple of p_1 and p_2. The polynomial p_3 is the *least common multiple* of p_1 and p_2.

The polynomial $p \in \mathbb{F}[s]$ given by (6.1.1) can be evaluated with a square matrix argument $A \in \mathbb{F}^{n\times n}$ by defining

$$p(A) \triangleq \beta_k A^k + \beta_{k-1}A^{k-1} + \cdots + \beta_1 A + \beta_0 I. \tag{6.1.13}$$

6.2 Polynomial Matrices

The set $\mathbb{F}[s]^{n\times m}$ of *polynomial matrices* consists of matrix functions $P\colon\ \mathbb{C} \mapsto \mathbb{C}^{n\times m}$ whose entries are elements of $\mathbb{F}[s]$. A polynomial matrix $P \in \mathbb{F}[s]^{n\times m}$ can thus be written as

$$P(s) = s^k B_k + s^{k-1}B_{k-1} + \cdots + sB_1 + B_0, \tag{6.2.1}$$

where $B_0, \ldots, B_k \in \mathbb{F}^{n\times m}$. If B_k is nonzero, then the *degree* of P, denoted by $\deg P$, is k, whereas, if $P = 0$, then $\deg P = -\infty$. If $n = m$ and B_k is nonsingular, then P is *regular*, while, if $B_k = I$, then P

is *monic*.

The following result, which generalizes Lemma 6.1.2, provides a division algorithm for polynomial matrices.

Lemma 6.2.1. Let $P_1, P_2 \in \mathbb{F}[s]^{n\times n}$, where P_2 is regular. Then, there exist unique polynomial matrices $Q, R, \hat{Q}, \hat{R} \in \mathbb{F}[s]^{n\times n}$ such that $\deg R < \deg P_2$, $\deg \hat{R} < \deg P_2$,

$$P_1 = QP_2 + R, \tag{6.2.2}$$

$$P_1 = P_2\hat{Q} + \hat{R}. \tag{6.2.3}$$

Proof. See [1186, p. 90] and [2221, pp. 134–135]. □

If $R = 0$, then P_2 *right divides* P_1, while, if $\hat{R} = 0$, then P_2 *left divides* P_1.

Let the polynomial matrix $P \in \mathbb{F}[s]^{n\times m}$ be given by (6.2.1). Then, P can be evaluated with a square matrix argument in two different ways, either from the right or from the left. For all $A \in \mathbb{C}^{m\times m}$, define

$$P_{\rm R}(A) \triangleq B_kA^k + B_{k-1}A^{k-1} + \cdots + B_1A + B_0, \tag{6.2.4}$$

while, for all $A \in \mathbb{C}^{n\times n}$, define

$$P_{\rm L}(A) \triangleq A^kB_k + A^{k-1}B_{k-1} + \cdots + AB_1 + B_0. \tag{6.2.5}$$

$P_{\rm R}(A)$ and $P_{\rm L}(A)$ are *matrix polynomials*.

If $n = m$, then $P_{\rm R}(A)$ and $P_{\rm L}(A)$ can be evaluated for all $A \in \mathbb{F}^{n\times n}$, although these matrices may be different.

The following result is useful.

Lemma 6.2.2. Let $Q, \hat{Q} \in \mathbb{F}[s]^{n\times n}$ and $A \in \mathbb{F}^{n\times n}$. Furthermore, define $P, \hat{P} \in \mathbb{F}[s]^{n\times n}$ by $P(s) \triangleq Q(s)(sI - A)$ and $\hat{P}(s) \triangleq (sI - A)\hat{Q}(s)$. Then, $P_{\rm R}(A) = 0$ and $\hat{P}_{\rm L}(A) = 0$.

Let $p \in \mathbb{F}[s]$ be given by (6.1.1), and define

$$P(s) \triangleq p(s)I_n = s^k\beta_kI_n + s^{k-1}\beta_{k-1}I_n + \cdots + s\beta_1I_n + \beta_0I_n \in \mathbb{F}[s]^{n\times n}.$$

For $A \in \mathbb{C}^{n\times n}$ it follows that $p(A) = P(A) = P_{\rm R}(A) = P_{\rm L}(A)$.

The following result specializes Lemma 6.2.1 to polynomial matrix divisors of degree 1.

Corollary 6.2.3. Let $P \in \mathbb{F}[s]^{n\times n}$ and $A \in \mathbb{F}^{n\times n}$. Then, there exist unique polynomial matrices $Q, \hat{Q} \in \mathbb{F}[s]^{n\times n}$ and unique matrices $R, \hat{R} \in \mathbb{F}^{n\times n}$ such that

$$P(s) = Q(s)(sI - A) + R, \tag{6.2.6}$$

$$P(s) = (sI - A)\hat{Q}(s) + \hat{R}. \tag{6.2.7}$$

Furthermore, $R = P_{\rm R}(A)$ and $\hat{R} = P_{\rm L}(A)$.

Proof. In Lemma 6.2.1 set $P_1 = P$ and $P_2(s) = sI - A$. Since $\deg P_2 = 1$, it follows that $\deg R = \deg \hat{R} = 0$, and thus R and $\hat{R}$ are constant. The last statement follows from Lemma 6.2.2. □

Definition 6.2.4. Let $P \in \mathbb{F}[s]^{n\times m}$. Then, rank P is defined by

$$\operatorname{rank} P \triangleq \max_{s\in\mathbb{C}} \operatorname{rank} P(s). \tag{6.2.8}$$

Let $P \in \mathbb{F}[s]^{n\times n}$. Then, $P(s) \in \mathbb{C}^{n\times n}$ for all $s \in \mathbb{C}$. Furthermore, $\det P$ is a polynomial in s; that is, $\det P \in \mathbb{F}[s]$.

Definition 6.2.5. Let $P \in \mathbb{F}[s]^{n\times n}$. Then, P is *nonsingular* if $\det P$ is not the zero polynomial; otherwise, P is *singular*.

Proposition 6.2.6. Let $P \in \mathbb{F}[s]^{n\times n}$, and assume that P is regular. Then, P is nonsingular.

Let $P \in \mathbb{F}[s]^{n\times n}$. If P is nonsingular, then the *inverse* P^{-1} of P can be constructed according to (3.8.22). In general, the entries of P^{-1} are rational functions of s (see Definition 6.7.1). For example, if $P(s) = \begin{bmatrix} s+2 & s+1 \\ s-2 & s-1 \end{bmatrix}$, then $P^{-1}(s) = \frac{1}{2s}\begin{bmatrix} s-1 & -s-1 \\ -s+2 & s+2 \end{bmatrix}$. In certain cases, P^{-1} is also a polynomial matrix. For example, if $P(s) = \begin{bmatrix} s & 1 \\ s^2+s-1 & s+1 \end{bmatrix}$, then $P^{-1}(s) = \begin{bmatrix} s+1 & -1 \\ -s^2-s+1 & s \end{bmatrix}$.

The following result extends Proposition 3.8.7 from constant matrices to polynomial matrices.

Proposition 6.2.7. Let $P \in \mathbb{F}[s]^{n\times m}$. Then, $\operatorname{rank} P$ is the size of the largest nonsingular polynomial matrix that is a submatrix of P.

Proof. For all $s \in \mathbb{C}$ it follows from Proposition 3.8.7 that $\operatorname{rank} P(s)$ is the size of the largest nonsingular submatrix of $P(s)$. Now, let $s_0 \in \mathbb{C}$ be such that $\operatorname{rank} P(s_0) = \operatorname{rank} P$. Then, $P(s_0)$ has a nonsingular submatrix of maximal size $\operatorname{rank} P$. Therefore, P has a nonsingular polynomial submatrix of maximal size $\operatorname{rank} P$. □

A polynomial matrix can be transformed by performing elementary row and column operations of the following types:

i) Multiply either a row or a column by a nonzero constant.

ii) Interchange either two rows or two columns.

iii) Add a polynomial multiple of one (row, column) to another (row, column).

These operations correspond, respectively, to left multiplication and right multiplication by the elementary matrices

$$I_n + (\alpha-1)E_{i,i} = \begin{bmatrix} I_{i-1} & 0 & 0 \\ 0 & \alpha & 0 \\ 0 & 0 & I_{n-i} \end{bmatrix}, \tag{6.2.9}$$

where $\alpha \in \mathbb{F}$ is nonzero,

$$I_n + E_{i,j} + E_{j,i} - E_{i,i} - E_{j,j} = \begin{bmatrix} I_{i-1} & 0 & 0 & 0 & 0 \\ 0 & 0 & 0 & 1 & 0 \\ 0 & 0 & I_{j-i-1} & 0 & 0 \\ 0 & 1 & 0 & 0 & 0 \\ 0 & 0 & 0 & 0 & I_{n-j} \end{bmatrix}, \tag{6.2.10}$$

where $i \neq j$, and the *elementary polynomial matrix*

$$I_n + pE_{i,j} = \begin{bmatrix} I_{i-1} & 0 & 0 & 0 & 0 \\ 0 & 1 & 0 & p & 0 \\ 0 & 0 & I_{j-i-1} & 0 & 0 \\ 0 & 0 & 0 & 1 & 0 \\ 0 & 0 & 0 & 0 & I_{n-j} \end{bmatrix}, \tag{6.2.11}$$

where $i \neq j$ and $p \in \mathbb{F}[s]$. The matrices shown in (6.2.10) and (6.2.11) illustrate the case $i < j$. Applying these operations sequentially corresponds to forming products of elementary matrices and elementary polynomial matrices. Note that the elementary polynomial matrix $I + pE_{i,j}$ is nonsingular, and that $(I + pE_{i,j})^{-1} = I - pE_{i,j}$. Therefore, the inverse of an elementary polynomial matrix is an elementary polynomial matrix.

6.3 The Smith Form and Similarity Invariants

Definition 6.3.1. Let $P \in \mathbb{F}[s]^{n\times n}$. Then, P is *unimodular* if P is the product of elementary matrices and elementary polynomial matrices.

The following result provides a canonical matrix, known as the *Smith matrix*, for polynomial matrices under unimodular transformation.

Theorem 6.3.2. Let $P \in \mathbb{F}[s]^{n\times m}$, and let $r \triangleq \operatorname{rank} P$. Then, there exist unimodular matrices $S_1 \in \mathbb{F}[s]^{n\times n}$ and $S_2 \in \mathbb{F}[s]^{m\times m}$ and unique monic polynomials $p_1, \ldots, p_r \in \mathbb{F}[s]$ such that p_i divides p_{i+1} for all $i \in \{1, \ldots, r-1\}$ and such that

$$P = S_1 \begin{bmatrix} p_1 & & & 0 \\ & \ddots & & \\ & & p_r & \\ 0 & & & 0_{(n-r)\times(m-r)} \end{bmatrix} S_2. \tag{6.3.1}$$

Furthermore, for all $i \in \{1, \ldots, r\}$, the polynomial

$$\Delta_i \triangleq \prod_{j=1}^{i} p_j \tag{6.3.2}$$

is the monic greatest common divisor of all $i \times i$ subdeterminants of P.

Proof. This result is obtained by applying elementary row and column operations to P. See [1573, pp. 390–392] and [2221, pp. 125–128]. □

The diagonal matrix in (6.3.1) is a *Smith matrix* and the *Smith form* of P.

Definition 6.3.3. Let $P \in \mathbb{F}[s]^{n\times m}$. Then, the monic polynomials $p_1, \ldots, p_r \in \mathbb{F}[s]$ in the Smith form of P are the *Smith polynomials* of P. The *Smith zeros* of P are the roots of $p_1, \ldots, p_r$; that is,

$$\operatorname{Szeros}(P) \triangleq \operatorname{roots}(p_r), \tag{6.3.3}$$

$$\operatorname{mSzeros}(P) \triangleq \bigcup_{i=1}^{r} \operatorname{mroots}(p_i). \tag{6.3.4}$$

Proposition 6.3.4. Let $P \in \mathbb{R}[s]^{n\times m}$, and assume that there exist unimodular matrices $S_1 \in \mathbb{F}[s]^{n\times n}$ and $S_2 \in \mathbb{F}[s]^{m\times m}$ and monic polynomials $p_1, \ldots, p_r \in \mathbb{F}[s]$ satisfying (6.3.1). Then, $\operatorname{rank} P = r$.

Proposition 6.3.5. Let $P \in \mathbb{F}[s]^{n\times m}$, and let $r \triangleq \operatorname{rank} P$. Then, r is the largest size of all nonsingular submatrices of P.

Proof. Let r_0 denote the largest size of all nonsingular submatrices of P, and let $P_0 \in \mathbb{F}[s]^{r_0\times r_0}$ be a nonsingular submatrix of P. First, assume that $r < r_0$. Then, there exists $s_0 \in \mathbb{C}$ such that $\operatorname{rank} P(s_0) = \operatorname{rank} P_0(s_0) = r_0$. Thus, $r = \operatorname{rank} P = \max_{s\in\mathbb{C}} \operatorname{rank} P(s) \ge \operatorname{rank} P(s_0) = r_0$, which is a contradiction. Next, assume that $r > r_0$. Then, it follows from (6.3.1) that there exists $s_0 \in \mathbb{C}$ such that $\operatorname{rank} P(s_0) = r$. Consequently, $P(s_0)$ has a nonsingular $r \times r$ submatrix. Let $\hat{P}_0 \in \mathbb{F}[s]^{r\times r}$ denote the corresponding submatrix of P. Thus, $\hat{P}_0$ is nonsingular, which implies that P has a nonsingular submatrix whose size is greater than r_0, which is a contradiction. Consequently, $r = r_0$. □

Proposition 6.3.6. Let $P \in \mathbb{F}[s]^{n\times m}$. Then, $\operatorname{rank} P(s) < \operatorname{rank} P$ if and only if $s \in \operatorname{Szeros}(P)$.

Proposition 6.3.7. Let $P \in \mathbb{F}[s]^{n\times m}$, and let $\mathcal{S} \subset \mathbb{C}$ be a countable set. Then,

$$\operatorname{rank} P = \max_{s\in\mathbb{C}\backslash\mathcal{S}} \operatorname{rank} P(s). \tag{6.3.5}$$

Proposition 6.3.8. Let $P \in \mathbb{F}[s]^{n\times n}$. Then, the following statements are equivalent:

i) P is unimodular.

ii) $\det P$ is a nonzero constant.

iii) The Smith form of P is the identity matrix.

iv) P is nonsingular, and P^{-1} is a polynomial matrix.

v) P is nonsingular, and P^{-1} is unimodular.

Proof. To prove *i)* $\Longrightarrow$ *ii)*, note that every elementary matrix and every elementary polynomial matrix has a constant nonzero determinant. Since P is a product of elementary matrices and elementary polynomial matrices, its determinant is a constant.

To prove *ii)* $\Longrightarrow$ *iii)*, note that it follows from (6.3.1) that $\operatorname{rank} P = n$ and $\det P = (\det S_1)(\det S_2)p_1 \cdots p_n$, where $S_1, S_2 \in \mathbb{F}^{n\times n}$ are unimodular and $p_1, \ldots, p_n$ are monic polynomials. From the result *i)* $\Longrightarrow$ *ii)*, it follows that $\det S_1$ and $\det S_2$ are nonzero constants. Since $\det P$ is a nonzero constant, it follows that $p_1 \cdots p_n = \det P/[(\det S_1)(\det S_2)]$ is a nonzero constant. Since $p_1, \ldots, p_n$ are monic polynomials, it follows that $p_1 = \cdots = p_n = 1$.

Next, to prove *iii)* $\Longrightarrow$ *iv)*, note that P is unimodular, and thus it follows that $\det P$ is a nonzero constant. Furthermore, since P^{A} is a polynomial matrix, it follows that $P^{-1} = (\det P)^{-1}P^{\mathsf{A}}$ is a polynomial matrix.

To prove *iv)* $\Longrightarrow$ *v)*, note that $\det P^{-1}$ is a polynomial. Since $\det P$ is a polynomial and $\det P^{-1} = 1/\det P$ it follows that $\det P$ is a nonzero constant. Hence, P is unimodular, and thus $P^{-1} = (\det P)^{-1}P^{\mathsf{A}}$ is unimodular.

Finally, to prove *v)* $\Longrightarrow$ *i)*, note that $\det P^{-1}$ is a nonzero constant, and thus $P = [\det P^{-1}]^{-1}[P^{-1}]^{\mathsf{A}}$ is a polynomial matrix. Furthermore, since $\det P = 1/\det P^{-1}$, it follows that $\det P$ is a nonzero constant. Hence, P is unimodular. □

Proposition 6.3.9. Let $A_1, B_1, A_2, B_2 \in \mathbb{F}^{n\times n}$, where A_2 is nonsingular, and define the polynomial matrices $P_1, P_2 \in \mathbb{F}[s]^{n\times n}$ by $P_1(s) \triangleq sA_1 + B_1$ and $P_2(s) \triangleq sA_2 + B_2$. Then, P_1 and P_2 have the same Smith polynomials if and only if there exist nonsingular matrices $S_1, S_2 \in \mathbb{F}^{n\times n}$ such that $P_2 = S_1P_1S_2$.

Proof. The sufficiency result is immediate. To prove necessity, note that it follows from Theorem 6.3.2 that there exist unimodular matrices $T_1, T_2 \in \mathbb{F}[s]^{n\times n}$ such that $P_2 = T_2P_1T_1$. Now, since P_2 is regular, it follows from Lemma 6.2.1 that there exist polynomial matrices $Q, \hat{Q} \in \mathbb{F}[s]^{n\times n}$ and constant matrices $R, \hat{R} \in \mathbb{F}^{n\times n}$ such that $T_1 = QP_2 + R$ and $T_2 = P_2\hat{Q} + \hat{R}$. Next, we have

$$
\begin{aligned}
P_2 = T_2P_1T_1 &= (P_2\hat{Q} + \hat{R})P_1T_1 = \hat{R}P_1T_1 + P_2\hat{Q}T_2^{-1}P_2 = \hat{R}P_1(QP_2 + R) + P_2\hat{Q}T_2^{-1}P_2 \\
&= \hat{R}P_1R + (T_2 - P_2\hat{Q})P_1QP_2 + P_2\hat{Q}T_2^{-1}P_2 = \hat{R}P_1R + T_2P_1QP_2 + P_2(-\hat{Q}P_1Q + \hat{Q}T_2^{-1})P_2 \\
&= \hat{R}P_1R + P_2(T_1^{-1}Q - \hat{Q}P_1Q + \hat{Q}T_2^{-1})P_2.
\end{aligned}
$$

Since P_2 is regular and has degree 1, it follows that, if $T_1^{-1}Q - \hat{Q}P_1Q + \hat{Q}T_2^{-1}$ is not zero, then $\deg P_2(T_1^{-1}Q - \hat{Q}P_1Q + \hat{Q}T_2^{-1})P_2 \ge 2$. However, since P_2 and $\hat{R}P_1R$ have degree less than 2, it follows that $T_1^{-1}Q - \hat{Q}P_1Q + \hat{Q}T_2^{-1} = 0$. Hence, $P_2 = \hat{R}P_1R$.

Next, to show that $\hat{R}$ and R are nonsingular, note that, for all $s \in \mathbb{C}$,

$$P_2(s) = \hat{R}P_1(s)R = s\hat{R}A_1R + \hat{R}B_1R,$$

which implies that $A_2 = S_1A_1S_2$, where $S_1 = \hat{R}$ and $S_2 = R$. Since A_2 is nonsingular, it follows that S_1 and S_2 are nonsingular. □

Definition 6.3.10. Let $A \in \mathbb{F}^{n\times n}$. Then, the *similarity invariants* of A are the Smith polynomials of $sI - A$.

The following result provides necessary and sufficient conditions for two matrices to be similar.

Theorem 6.3.11. Let $A, B \in \mathbb{F}^{n\times n}$. Then, A and B are similar if and only if they have the same similarity invariants.

Proof. To prove necessity, assume that A and B are similar. Then, the matrices $sI - A$ and $sI - B$ have the same Smith form and thus the same similarity invariants. To prove sufficiency, it follows from Proposition 6.3.9 that there exist nonsingular matrices $S_1, S_2 \in \mathbb{F}^{n\times n}$ such that $sI - A = S_1(sI - B)S_2$. Thus, $S_1 = S_2^{-1}$, and, hence, $A = S_1BS_1^{-1}$. □

Corollary 6.3.12. Let $A \in \mathbb{F}^{n\times n}$. Then, A and A^{T} are similar.

A stronger form of Corollary 6.3.12 is given by Corollary 7.4.9.

6.4 Eigenvalues

Let $A \in \mathbb{F}^{n\times n}$. Then, the polynomial matrix $sI - A \in \mathbb{F}[s]^{n\times n}$ is monic and has degree 1.

Definition 6.4.1. Let $A \in \mathbb{F}^{n\times n}$. Then, the *characteristic polynomial* of A is the polynomial $\chi_A \in \mathbb{F}[s]$ given by

$$\chi_A(s) \triangleq \det(sI - A). \tag{6.4.1}$$

Since $sI - A$ is a polynomial matrix, its determinant is the product of its Smith polynomials, which are the similarity invariants of A.

Proposition 6.4.2. Let $A \in \mathbb{F}^{n\times n}$, and let $p_1, \dots, p_n \in \mathbb{F}[s]$ denote the similarity invariants of A. Then,

$$\chi_A = \prod_{i=1}^{n} p_i. \tag{6.4.2}$$

Proposition 6.4.3. Let $A \in \mathbb{F}^{n\times n}$. Then, χ_A is monic and $\deg \chi_A = n$.

Let $A \in \mathbb{F}^{n\times n}$, and write the characteristic polynomial of A as

$$\chi_A(s) = s^n + \beta_{n-1}s^{n-1} + \cdots + \beta_1 s + \beta_0, \tag{6.4.3}$$

where $\beta_0, \dots, \beta_{n-1} \in \mathbb{F}$. The *eigenvalues* of A are the n possibly repeated roots $\lambda_1, \dots, \lambda_n \in \mathbb{C}$ of χ_A, which are the solutions of the *characteristic equation*

$$\chi_A(s) = 0. \tag{6.4.4}$$

It is often convenient to denote the eigenvalues of $A \in \mathbb{F}^{n\times n}$ by $\lambda_1, \dots, \lambda_n$. This notation, however, does not specify which eigenvalue of A is denoted by λ_i. If, however, every eigenvalue of A is real, then we denote the eigenvalues of A unambiguously by $\lambda_1(A), \dots, \lambda_n(A)$, where

$$\lambda_n(A) \le \cdots \le \lambda_1(A). \tag{6.4.5}$$

Furthermore, we define

$$\lambda_{\min}(A) \triangleq \lambda_n(A), \quad \lambda_{\max}(A) \triangleq \lambda_1(A) \tag{6.4.6}$$

and the *eigenvalue vector* of A by

$$\lambda(A) \triangleq \begin{bmatrix} \lambda_1(A) \\ \vdots \\ \lambda_n(A) \end{bmatrix}. \tag{6.4.7}$$

Definition 6.4.4. Let $A \in \mathbb{F}^{n\times n}$. The *algebraic multiplicity* of an eigenvalue λ of A, denoted by $\operatorname{amult}_A(\lambda)$, is the algebraic multiplicity of λ as a root of χ_A; that is,

$$\operatorname{amult}_A(\lambda) \triangleq \operatorname{mult}_{\chi_A}(\lambda). \tag{6.4.8}$$

The multiset consisting of the eigenvalues of A including their algebraic multiplicity, denoted by $\operatorname{mspec}(A)$, is the *multispectrum* of A; that is,

$$\operatorname{mspec}(A) \triangleq \operatorname{mroots}(\chi_A). \tag{6.4.9}$$

Ignoring algebraic multiplicity, $\operatorname{spec}(A)$ denotes the *spectrum* of A; that is,

$$\operatorname{spec}(A) \triangleq \operatorname{roots}(\chi_A). \tag{6.4.10}$$

Note that

$$\operatorname{Szeros}(sI - A) = \operatorname{spec}(A), \tag{6.4.11}$$

$$\operatorname{mSzeros}(sI - A) = \operatorname{mspec}(A). \tag{6.4.12}$$

We can thus write $\operatorname{mspec}(A) = \{\lambda_1, \ldots, \lambda_n\}_{\rm ms}$ to denote the multiset of repeated eigenvalues of A, and $\operatorname{spec}(A) = \{\lambda_1, \ldots, \lambda_r\}$ to denote the set of distinct eigenvalues of A. However, as noted above, this notation is generic.

If $\lambda \notin \operatorname{spec}(A)$, then $\lambda \notin \operatorname{roots}(\chi_A)$, and thus $\operatorname{amult}_A(\lambda) = \operatorname{mult}_{\chi_A}(\lambda) = 0$.

Let $A \in \mathbb{F}^{n\times n}$ and $\operatorname{mroots}(\chi_A) = \{\lambda_1, \ldots, \lambda_n\}_{\rm ms}$. Then,

$$\chi_A(s) = \prod_{i=1}^{n}(s - \lambda_i). \tag{6.4.13}$$

If $\mathbb{F} = \mathbb{R}$, then $\overline{\chi_A(s)}$ has real coefficients, and thus the eigenvalues of A occur in complex conjugate pairs; that is, $\overline{\operatorname{mroots}(\chi_A)} = \operatorname{mroots}(\chi_A)$. Now, let $\operatorname{spec}(A) = \{\lambda_1, \ldots, \lambda_r\}$, and, for all $i \in \{1, \ldots, r\}$, let n_i denote the algebraic multiplicity of λ_i. Then,

$$\chi_A(s) = \prod_{i=1}^{r}(s - \lambda_i)^{n_i}. \tag{6.4.14}$$

The following result gives some basic properties of the spectrum of a matrix.

Proposition 6.4.5. Let $A, B \in \mathbb{F}^{n\times n}$. Then, the following statements hold:

i) $\chi_{A^{\mathsf T}} = \chi_A$.

ii) For all $s \in \mathbb{C}$, $\chi_{-A}(s) = (-1)^n\chi_A(-s)$.

iii) $\operatorname{mspec}(A^{\mathsf T}) = \operatorname{mspec}(A)$.

iv) $\operatorname{mspec}(\overline{A}) = \overline{\operatorname{mspec}(A)}$.

v) $\operatorname{mspec}(A^*) = \overline{\operatorname{mspec}(A)}$.

vi) $0 \in \operatorname{spec}(A)$ if and only if $\det A = 0$.

vii) If either $k \in \mathbb{N}$ or both A is nonsingular and $k \in \mathbb{Z}$, then

$$\operatorname{mspec}(A^k) = \{\lambda^k\colon\ \lambda \in \operatorname{mspec}(A)\}_{\rm ms}. \tag{6.4.15}$$

viii) If $\alpha \in \mathbb{F}$, then $\chi_{\alpha A+I}(s) = \chi_A(s - \alpha)$.

ix) If $\alpha \in \mathbb{F}$, then $\operatorname{mspec}(\alpha I + A) = \alpha + \operatorname{mspec}(A)$.

x) If $\alpha \in \mathbb{F}$, then $\operatorname{mspec}(\alpha A) = \alpha \operatorname{mspec}(A)$.

xi) If A is Hermitian, then $\operatorname{spec}(A) \subset \mathbb{R}$.

xii) If A and B are similar, then $\chi_A = \chi_B$ and $\operatorname{mspec}(A) = \operatorname{mspec}(B)$.

Proof. To prove *i)*, note that $\det(sI - A^{\mathsf T}) = \det\,(sI - A)^{\mathsf T} = \det(sI - A)$. To prove *ii)*, note that $\chi_{-A}(s) = \det(sI + A) = (-1)^n\det(-sI - A) = (-1)^n\chi_A(-s)$. Next, *iii)* follows from *i)*, *iv)* follows from $\det(sI - \overline{A}) = \det(\overline{\overline{s}I - A}) = \overline{\det(\overline{s}I - A)}$, *v)* follows from *iii)* and *iv)*, and *vi)* follows from $\chi_A(0) = (-1)^n\det A$. To prove "$\supseteq$" in *vii)*, let $\lambda \in \operatorname{spec}(A)$ and let $x \in \mathbb{C}^n$ be an eigenvector of A associated with λ (see Section 6.5). Then, $A^2x = A(Ax) = A(\lambda x) = \lambda Ax = \lambda^2x$. Similarly, in the case where A is nonsingular, $Ax = \lambda x$ implies that $A^{-1}x = \lambda^{-1}x$, and thus $A^{-2}x = \lambda^{-2}x$. Similar arguments apply to arbitrary $k \in \mathbb{Z}$. The reverse inclusion follows from the Jordan decomposition given by Theorem 7.4.2. To prove *viii)*, note that $\chi_{\alpha I+A}(s) = \det[sI-(\alpha I+A)] = \det[(s-\alpha)I-A] = \chi_A(s-\alpha)$. *ix)* follows immediately. *x)* is true for $\alpha = 0$. For $\alpha \neq 0$, $\chi_{\alpha A}(s) = \det(sI-\alpha A) = \alpha^n\det[(s/\alpha)I-A] = \alpha^n\chi_A(s/\alpha)$. To prove *xi)*, assume that $A = A^*$, let $\lambda \in \operatorname{spec}(A)$, and let $x \in \mathbb{C}^n$ be an eigenvector of A associated with λ. Then, $\lambda = x^*Ax/x^*x$, which is real. Finally, *xii)* is immediate. □

The following result characterizes the coefficients of χ_A in terms of the eigenvalues of A.

Proposition 6.4.6. Let $A \in \mathbb{F}^{n\times n}$, and let $\operatorname{mspec}(A) = \{\lambda_1, \dots, \lambda_n\}_{\rm ms}$. Then, for all $i \in \{0, \dots, n-1\}$, the coefficient β_i of s^i in (6.4.3) is given by

$$\beta_i = (-1)^{n-i} \sum_{1\le j_1<\cdots<j_{n-i}\le n} \lambda_{j_1}\cdots\lambda_{j_{n-i}}. \tag{6.4.16}$$

Now, let γ_i denote the sum of all $i \times i$ principal subdeterminants of A. Then,

$$\gamma_{n-i} = (-1)^{n-i}\beta_i = \sum_{1\le j_1<\cdots<j_{n-i}\le n} \lambda_{j_1}\cdots\lambda_{j_{n-i}}. \tag{6.4.17}$$

In particular,

$$\beta_{n-1} = -\sum_{i=1}^{n} \lambda_i = -\gamma_1 = -\operatorname{tr} A = -\sum_{i=1}^{n} A_{(i,i)}, \tag{6.4.18}$$

$$\beta_{n-2} = \sum_{1\le j_1<j_2\le n} \lambda_{j_1}\lambda_{j_2} = \gamma_2 = \tfrac{1}{2}[(\operatorname{tr} A)^2 - \operatorname{tr} A^2] = \sum_{1\le i<j\le n} \det A_{(\{i,j\},\{i,j\})}, \tag{6.4.19}$$

$$\beta_1 = (-1)^{n-1}\sum_{1\le j_1<\cdots<j_{n-1}\le n} \lambda_{j_1}\cdots\lambda_{j_{n-1}} = (-1)^{n-1}\gamma_{n-1} = (-1)^{n-1}\operatorname{tr} A^{\mathsf{A}} = (-1)^{n-1}\sum_{i=1}^{n} \det A_{[i,i]}, \tag{6.4.20}$$

$$\beta_0 = (-1)^n \prod_{i=1}^{n} \lambda_i = (-1)^n = (-1)^n\gamma_n \det A. \tag{6.4.21}$$

Proof. The expression for β_i given by (6.4.16) follows from the factored form of $\chi_A(s)$ given by (6.4.13), while the expression for γ_i given by (6.4.17) follows from the cofactor expansion (3.8.16) of $\det(sI - A)$. For details, see [2036, p. 495]. (6.4.18) follows from (6.4.17) and the fact that the $(n-1)\times(n-1)$ principal subdeterminants of A are the diagonal entries $A_{(i,i)}$. Using

$$\sum_{i=1}^{n} \lambda_i^2 = \left(\sum_{i=1}^{n} \lambda_i\right)^2 - 2\sum_{1\le j_1<j_2\le n} \lambda_{j_1}\lambda_{j_2},$$

and (6.4.18) yields (6.4.19). Next, if A is nonsingular, then $\chi_{A^{-1}}(s) = (-s)^n(\det A^{-1})\chi_A(1/s)$. Using (6.4.3) with s replaced by $1/s$ and (6.4.18), it follows that $\operatorname{tr} A^{-1} = (-1)^{n-1}(\det A^{-1})\beta_1$, and, hence, (6.4.20) holds. Using continuity in the case where A is singular yields (6.4.20) for arbitrary A. Finally, $\beta_0 = \chi_A(0) = \det(0I - A) = (-1)^n \det A$, which verifies (6.4.21). □

From the definition of the adjugate of a matrix it follows that $(sI - A)^{\mathsf{A}} \in \mathbb{F}[s]^{n\times n}$ is a monic polynomial matrix of degree $n-1$ of the form

$$(sI - A)^{\mathsf{A}} = s^{n-1}I + s^{n-2}B_{n-2} + \cdots + sB_1 + B_0, \tag{6.4.22}$$

where $B_0, B_1, \dots, B_{n-2} \in \mathbb{F}^{n\times n}$. Since $(sI - A)^{\mathsf{A}}$ is regular, it follows from Proposition 6.2.6 that $(sI - A)^{\mathsf{A}}$ is a nonsingular polynomial matrix. The matrix $(sI - A)^{-1}$ is the *resolvent* of A, which is given by

$$(sI - A)^{-1} = \frac{1}{\chi_A(s)}(sI - A)^{\mathsf{A}}. \tag{6.4.23}$$

Therefore,

$$(sI - A)^{-1} = \frac{s^{n-1}}{\chi_A(s)}I + \frac{s^{n-2}}{\chi_A(s)}B_{n-2} + \cdots + \frac{s}{\chi_A(s)}B_1 + \frac{1}{\chi_A(s)}B_0. \tag{6.4.24}$$

The next result is the *Cayley-Hamilton theorem*, which shows that every matrix is a "root" of its characteristic polynomial. Note that, in (6.4.25), the characteristic polynomial χ_A is first defined by (6.4.1) and then evaluated according to (6.1.13). This procedure is not equivalent to replacing s by A in $\det(sI - A)$.

Theorem 6.4.7. Let $A \in \mathbb{F}^{n\times n}$. Then,

$$\chi_A(A) = 0. \tag{6.4.25}$$

Proof. Define $P, Q \in \mathbb{F}[s]^{n\times n}$ by $P(s) \triangleq \chi_A(s)I$ and $Q(s) \triangleq (sI - A)^{\mathsf{A}}$. Then, (6.4.23) implies that $P(s) = Q(s)(sI - A)$. It thus follows from Lemma 6.2.2 that $P_{\mathrm{R}}(A) = 0$. Furthermore, $\chi_A(A) = P(A) = P_{\mathrm{R}}(A)$. Hence, $\chi_A(A) = 0$. □

In the notation of (6.4.14), Theorem 6.4.7 implies that

$$\prod_{i=1}^{r}(\lambda_i I - A)^{n_i} = 0. \tag{6.4.26}$$

Lemma 6.4.8. Let $A \in \mathbb{F}^{n\times n}$. Then,

$$\frac{\mathrm{d}}{\mathrm{d}s}\chi_A(s) = \operatorname{tr}(sI - A)^{\mathsf{A}} = \sum_{i=1}^{n}\det(sI - A_{[i,i]}). \tag{6.4.27}$$

Proof. It follows from (6.4.20) that $\frac{\mathrm{d}}{\mathrm{d}s}\chi_A(s)\big|_{s=0} = \beta_1 = (-1)^{n-1}\operatorname{tr} A^{\mathsf{A}}$. Hence,

$$\frac{\mathrm{d}}{\mathrm{d}s}\chi_A(s) = \frac{\mathrm{d}}{\mathrm{d}z}\det[(s+z)I - A]\bigg|_{z=0} = \frac{\mathrm{d}}{\mathrm{d}z}\det[zI - (-sI + A)]\bigg|_{z=0}$$

$$= (-1)^{n-1}\operatorname{tr}(-sI + A)^{\mathsf{A}} = \operatorname{tr}(sI - A)^{\mathsf{A}}.$$ □

The following result, known as *Leverrier's algorithm*, provides a recursive formula for the coefficients $\beta_0, \dots, \beta_{n-1}$ of χ_A and $B_0, \dots, B_{n-2}$ of $(sI - A)^{\mathsf{A}}$.

Proposition 6.4.9. Let $A \in \mathbb{F}^{n\times n}$, let χ_A be given by (6.4.3), and let $(sI - A)^{\mathsf{A}}$ be given by (6.4.22). Then, $\beta_{n-1}, \dots, \beta_0$ and $B_{n-2}, \dots, B_0$ are given by

$$\beta_k = \tfrac{1}{k-n}\operatorname{tr} AB_k, \quad k = n-1, \dots, 0, \tag{6.4.28}$$

$$B_{k-1} = AB_k + \beta_k I, \quad k = n-1, \dots, 1, \tag{6.4.29}$$

where $B_{n-1} = I$.

Proof. Since $(sI - A)(sI - A)^{\mathsf{A}} = \chi_A(s)I$, it follows that

$$s^n I + s^{n-1}(B_{n-2} - A) + s^{n-2}(B_{n-3} - AB_{n-2}) + \cdots + s(B_0 - AB_1) - AB_0$$
$$= (s^n + \beta_{n-1}s^{n-1} + \cdots + \beta_1 s + \beta_0)I.$$

Equating coefficients of powers of s yields (6.4.29) along with $-AB_0 = \beta_0 I$. Taking the trace of this last equality yields $\beta_0 = -\frac{1}{n}\operatorname{tr} AB_0$, which confirms (6.4.28) for $k = 0$. Next, using (6.4.27) and (6.4.22), it follows that

$$\frac{\mathrm{d}}{\mathrm{d}s}\chi_A(s) = \sum_{k=1}^{n} k\beta_k s^{k-1} = \sum_{k=1}^{n}(\operatorname{tr} B_{k-1})s^{k-1},$$

where $B_{n-1} \triangleq I_n$ and $\beta_n \triangleq 1$. Equating powers of s, it follows that $k\beta_k = \operatorname{tr} B_{k-1}$ for all $k \in \{1, \dots, n\}$. Now, (6.4.29) implies that $k\beta_k = \operatorname{tr}(AB_k + \beta_k I)$ for all $k \in \{1, \dots, n-1\}$, which implies (6.4.28). □

Proposition 6.4.10. Let $A \in \mathbb{F}^{n\times m}$ and $B \in \mathbb{F}^{m\times n}$, and assume that $m \le n$. Then,

$$\chi_{AB}(s) = s^{n-m}\chi_{BA}(s). \tag{6.4.30}$$

Consequently,

$$\operatorname{mspec}(AB) = \operatorname{mspec}(BA) \cup \{0,\ldots,0\}_{\text{ms}}, \tag{6.4.31}$$

where the multiset $\{0,\ldots,0\}_{\text{ms}}$ contains $n-m$ 0's.

Proof. First note that

$$\begin{bmatrix} 0_{m\times m} & 0_{m\times n} \\ A & AB \end{bmatrix} = \begin{bmatrix} I_m & -B \\ 0_{n\times m} & I_n \end{bmatrix}\begin{bmatrix} BA & 0_{m\times n} \\ A & 0_{n\times n} \end{bmatrix}\begin{bmatrix} I_m & B \\ 0_{n\times m} & I_n \end{bmatrix},$$

which shows that $\left[\begin{smallmatrix} 0_{m\times m} & 0_{m\times n} \\ A & AB \end{smallmatrix}\right]$ and $\left[\begin{smallmatrix} BA & 0_{m\times n} \\ A & 0_{n\times n} \end{smallmatrix}\right]$ are similar. It thus follows from *xi*) of Proposition 6.4.5 that $s^m\chi_{AB}(s) = s^n\chi_{BA}(s)$, which implies (6.4.30). □

If $n = m$, then Proposition 6.4.10 specializes to the following result.

Corollary 6.4.11. Let $A, B \in \mathbb{F}^{n\times n}$. Then,

$$\chi_{AB} = \chi_{BA}. \tag{6.4.32}$$

Consequently,

$$\operatorname{mspec}(AB) = \operatorname{mspec}(BA). \tag{6.4.33}$$

Let $A \in \mathbb{F}^{n\times n}$, and let $\operatorname{mspec}(A) = \{\lambda_1,\ldots,\lambda_n\}_{\text{ms}}$. The *spread* of A is defined by

$$\delta(A) \triangleq \max_{i,j\in\{1,\ldots,n\}} |\lambda_i - \lambda_j|. \tag{6.4.34}$$

The *spectral moduli* of A are the nonnegative numbers $\rho_n(A) \le \cdots \le \rho_1(A)$ such that $\{\rho_1(A),\ldots,\rho_n(A)\}_{\text{ms}} = \{|\lambda_1|,\ldots,|\lambda_n|\}_{\text{ms}}$. In particular, define the *minimum spectral modulus* of A by

$$\rho_{\min}(A) \triangleq \rho_n(A) \tag{6.4.35}$$

and the *spectral radius* of A by

$$\rho_{\max}(A) \triangleq \rho_1(A). \tag{6.4.36}$$

Finally, define the vector $\rho(A)$ of spectral moduli of A by

$$\rho(A) \triangleq \begin{bmatrix} \rho_1(A) \\ \vdots \\ \rho_n(A) \end{bmatrix}. \tag{6.4.37}$$

Next, the *spectral real parts* of A are the real numbers $\alpha_n(A) \le \cdots \le \alpha_1(A)$ such that $\{\alpha_1(A),\ldots,\alpha_n(A)\}_{\text{ms}} = \{\operatorname{Re}\lambda_1,\ldots,\operatorname{Re}\lambda_n\}_{\text{ms}}$. In particular, define the *minimum spectral real part* of A by

$$\alpha_{\min}(A) \triangleq \alpha_n(A) \tag{6.4.38}$$

and the *spectral abscissa* of A by

$$\alpha_{\max}(A) \triangleq \alpha_1(A). \tag{6.4.39}$$

Finally, define the vector $\alpha(A)$ of spectral real parts of A by

$$\alpha(A) \triangleq \begin{bmatrix} \alpha_1(A) \\ \vdots \\ \alpha_n(A) \end{bmatrix}. \tag{6.4.40}$$

The *spectral imaginary parts* of A are the real numbers $\beta_n(A) \le \cdots \le \beta_1(A)$ such that $\{\beta_1(A),\ldots,\beta_n(A)\}_{\text{ms}} = \{\operatorname{Im}\lambda_1,\ldots,\operatorname{Im}\lambda_n\}_{\text{ms}}$. In particular, define the *minimum spectral imaginary part* of A by

$$\beta_{\min}(A) \triangleq \beta_n(A) \tag{6.4.41}$$

and the *spectral imaginary abscissa* of A by

$$\beta_{\max}(A) \triangleq \beta_1(A). \tag{6.4.42}$$

Finally, define the vector $\beta(A)$ of spectral imaginary parts of A by

$$\beta(A) \triangleq \begin{bmatrix} \beta_1(A) \\ \vdots \\ \beta_n(A) \end{bmatrix}. \tag{6.4.43}$$

Let $A \in \mathbb{F}^{n\times n}$. Then, $\nu_-(A)$, $\nu_0(A)$, and $\nu_+(A)$ denote the number of eigenvalues of A counting algebraic multiplicity having, respectively, negative, zero, and positive real part. Define the *inertia* of A by

$$\operatorname{In} A \triangleq \begin{bmatrix} \nu_-(A) \\ \nu_0(A) \\ \nu_+(A) \end{bmatrix} \tag{6.4.44}$$

and the *signature* of A by

$$\operatorname{sig} A \triangleq \nu_+(A) - \nu_-(A). \tag{6.4.45}$$

Note that $\alpha_{\max}(A) < 0$ if and only if $\nu_-(A) = n$, while $\alpha_{\max}(A) = 0$ if and only if $\nu_+(A) = 0$.

6.5 Eigenvectors

Let $A \in \mathbb{F}^{n\times n}$, and let $\lambda \in \mathbb{C}$ be an eigenvalue of A. Then, $\chi_A(\lambda) = \det(\lambda I - A) = 0$, and thus $\lambda I - A \in \mathbb{C}^{n\times n}$ is singular. Furthermore, $\mathcal{N}(\lambda I - A)$ is a nontrivial subspace of $\mathbb{C}^n$; that is, $\operatorname{def}(\lambda I - A) > 0$. If $x \in \mathcal{N}(\lambda I - A)$, that is, $Ax = \lambda x$, and $x \neq 0$, then x is an *eigenvector of* A *associated with* λ. By definition, all eigenvectors are nonzero. Note that, if A and λ are real, then there exists a real eigenvector associated with λ.

Definition 6.5.1. The *geometric multiplicity* of $\lambda \in \operatorname{spec}(A)$, denoted by $\operatorname{gmult}_A(\lambda)$, is the number of linearly independent eigenvectors associated with λ, that is,

$$\operatorname{gmult}_A(\lambda) \triangleq \operatorname{def}(\lambda I - A). \tag{6.5.1}$$

By convention, if $\lambda \notin \operatorname{spec}(A)$, then $\operatorname{gmult}_A(\lambda) \triangleq 0$.

Proposition 6.5.2. Let $A \in \mathbb{F}^{n\times n}$, and let $\lambda \in \operatorname{spec}(A)$. Then, the following statements hold:

i) $\operatorname{rank}(\lambda I - A) + \operatorname{gmult}_A(\lambda) = n$.

ii) $\operatorname{def} A = \operatorname{gmult}_A(0)$.

iii) $\operatorname{rank} A + \operatorname{gmult}_A(0) = n$.

The spectral properties of normal matrices deserve special attention.

Lemma 6.5.3. Let $A \in \mathbb{F}^{n\times n}$ be normal, let $\lambda \in \operatorname{spec}(A)$, and let $x \in \mathbb{C}^n$ be an eigenvector of A associated with λ. Then, x is an eigenvector of A^* associated with $\overline{\lambda} \in \operatorname{spec}(A^*)$.

Proof. Since $\lambda \in \operatorname{spec}(A)$, statement *v*) of Proposition 6.4.5 implies that $\overline{\lambda} \in \operatorname{spec}(A^*)$. Next, using $Ax = \lambda x$, $x^*A^* = \overline{\lambda}x^*$, and $AA^* = A^*A$, it follows that

$$\begin{aligned}(A^*x - \overline{\lambda}x)^*(A^*x - \overline{\lambda}x) &= x^*AA^*x - \overline{\lambda}x^*Ax - \lambda x^*A^*x + \lambda\overline{\lambda}x^*x \\ &= x^*A^*Ax - \lambda\overline{\lambda}x^*x - \lambda\overline{\lambda}x^*x + \lambda\overline{\lambda}x^*x \\ &= \lambda\overline{\lambda}x^*x - \lambda\overline{\lambda}x^*x = 0.\end{aligned}$$

Hence, $A^*x = \overline{\lambda}x$. □

Proposition 6.5.4. Let $A \in \mathbb{F}^{n\times n}$. Then, eigenvectors associated with distinct eigenvalues of A

are linearly independent. If, in addition, A is normal, then these eigenvectors are mutually orthogonal.

Proof. Let $\lambda_1, \lambda_2 \in \operatorname{spec}(A)$ be distinct with associated eigenvectors $x_1, x_2 \in \mathbb{C}^n$. Suppose that x_1 and x_2 are linearly dependent; that is, $x_1 = \alpha x_2$, where $\alpha \in \mathbb{C}$ and $\alpha \neq 0$. Then, $Ax_1 = \lambda_1 x_1 = \lambda_1 \alpha x_2$ and $Ax_1 = A\alpha x_2 = \alpha\lambda_2 x_2$. Hence, $\alpha(\lambda_1 - \lambda_2)x_2 = 0$, which contradicts $\alpha \neq 0$. Since pairwise linear independence does not imply the linear independence of larger sets, next, let $\lambda_1, \lambda_2, \lambda_3 \in \operatorname{spec}(A)$ be distinct with associated eigenvectors $x_1, x_2, x_3 \in \mathbb{C}^n$. Suppose that x_1, x_2, x_3 are linearly dependent. Then, there exist $a_1, a_2, a_3 \in \mathbb{C}$, not all zero, such that $a_1x_1 + a_2x_2 + a_3x_3 = 0$. If $a_1 = 0$, then $a_2x_2 + a_3x_3 = 0$. However, $\lambda_2 \neq \lambda_3$ implies that x_2 and x_3 are linearly independent, which in turn implies that $a_2 = 0$ and $a_3 = 0$. Since a_1, a_2, a_3 are not all zero, it follows that $a_1 \neq 0$. Therefore, $x_1 = \alpha x_2 + \beta x_3$, where $\alpha \triangleq -a_2/a_1$ and $\beta \triangleq -a_3/a_1$ are not both zero. Thus, $Ax_1 = A(\alpha x_2 + \beta x_3) = \alpha Ax_2 + \beta Ax_3 = \alpha\lambda_2 x_2 + \beta\lambda_3 x_3$. However, $Ax_1 = \lambda_1 x_1 = \lambda_1(\alpha x_2 + \beta x_3) = \alpha\lambda_1 x_2 + \beta\lambda_1 x_3$. Subtracting these equations yields $0 = \alpha(\lambda_1 - \lambda_2)x_2 + \beta(\lambda_1 - \lambda_3)x_3$. Since x_2 and x_3 are linearly independent, it follows that $\alpha(\lambda_1 - \lambda_2) = 0$ and $\beta(\lambda_1 - \lambda_3) = 0$. Since α and β are not both zero, it follows that either $\lambda_1 = \lambda_2$ or $\lambda_1 = \lambda_3$, which contradicts the assumption that $\lambda_1, \lambda_2, \lambda_3$ are distinct. The same arguments apply to sets of four or more eigenvectors.

Now, suppose that A is normal, and let $\lambda_1, \lambda_2 \in \operatorname{spec}(A)$ be distinct eigenvalues with associated eigenvectors $x_1, x_2 \in \mathbb{C}^n$. Then, by Lemma 6.5.3, $Ax_1 = \lambda_1 x_1$ implies that $A^*x_1 = \overline{\lambda}_1 x_1$. Consequently, $x_1^*A = \lambda_1 x_1^*$, which implies that $x_1^*Ax_2 = \lambda_1 x_1^*x_2$. Furthermore, $x_1^*Ax_2 = \lambda_2 x_1^*x_2$. It thus follows that $0 = (\lambda_1 - \lambda_2)x_1^*x_2$. Hence, $\lambda_1 \neq \lambda_2$ implies that $x_1^*x_2 = 0$. □

If $A \in \mathbb{R}^{n\times n}$ is symmetric, then Lemma 6.5.3 is not needed and the proof of Proposition 6.5.4 is simpler. In this case, it follows from x) of Proposition 6.4.5 that $\lambda_1, \lambda_2 \in \operatorname{spec}(A)$ are real, and thus associated eigenvectors $x_1 \in \mathcal{N}(\lambda_1 I - A)$ and $x_2 \in \mathcal{N}(\lambda_2 I - A)$ can be chosen to be real. Hence, $Ax_1 = \lambda_1 x_1$ and $Ax_2 = \lambda_2 x_2$ imply that $x_2^{\mathsf{T}}Ax_1 = \lambda_1 x_2^{\mathsf{T}}x_1$ and $x_1^{\mathsf{T}}Ax_2 = \lambda_2 x_1^{\mathsf{T}}x_2$. Since $x_1^{\mathsf{T}}Ax_2 = x_2^{\mathsf{T}}A^{\mathsf{T}}x_1 = x_2^{\mathsf{T}}Ax_1$ and $x_1^{\mathsf{T}}x_2 = x_2^{\mathsf{T}}x_1$, it follows that $(\lambda_1 - \lambda_2)x_1^{\mathsf{T}}x_2 = 0$. Since $\lambda_1 \neq \lambda_2$, it follows that $x_1^{\mathsf{T}}x_2 = 0$.

6.6 The Minimal Polynomial

Theorem 6.4.7 showed that every square matrix $A \in \mathbb{F}^{n\times n}$ is a root of its characteristic polynomial. However, there may be polynomials of degree less than n having A as a root. In fact, the following result shows that there exists a unique monic polynomial that has A as a root and that divides all polynomials that have A as a root.

Theorem 6.6.1. Let $A \in \mathbb{F}^{n\times n}$. Then, there exists a unique monic polynomial $\mu_A \in \mathbb{F}[s]$ of minimal degree such that $\mu_A(A) = 0$. Furthermore, $\deg \mu_A \le n$, and μ_A divides every polynomial $p \in \mathbb{F}[s]$ satisfying $p(A) = 0$.

Proof. Since $\chi_A(A) = 0$ and $\deg \chi_A = n$, it follows that there exists a minimal positive integer $n_0 \le n$ such that there exists a monic polynomial $p_0 \in \mathbb{F}[s]$ satisfying $p_0(A) = 0$ and $\deg p_0 = n_0$. Let $p \in \mathbb{F}[s]$ satisfy $p(A) = 0$. Then, by Lemma 6.1.2, there exist polynomials $q, r \in \mathbb{F}[s]$ such that $p = qp_0 + r$ and $\deg r < \deg p_0$. However, $p(A) = p_0(A) = 0$ implies that $r(A) = 0$. If $r \neq 0$, then r can be normalized to obtain a monic polynomial of degree less than n_0, which contradicts the definition n_0. Hence, $r = 0$, which implies that p_0 divides p. This proves existence.

Now, suppose there exist two monic polynomials $p_0, \hat{p}_0 \in \mathbb{F}[s]$ of degree n_0 and such that $p_0(A) = \hat{p}_0(A) = 0$. By the previous argument, p_0 divides $\hat{p}_0$, and vice versa. Therefore, p_0 is a constant multiple of $\hat{p}_0$. Since p_0 and $\hat{p}_0$ are both monic, it follows that $p_0 = \hat{p}_0$. This proves uniqueness. Denote this polynomial by μ_A. □

The *minimal polynomial* μ_A of A is the monic polynomial of smallest degree having A as a root.

The following result relates the characteristic and minimal polynomials of $A \in \mathbb{F}^{n\times n}$ to the similarity invariants of A. Note that $\operatorname{rank}(sI - A) = n$, so that A has n similarity invariants $p_1, \ldots, p_n \in$

$\mathbb{F}[s]$. In this case, (6.3.1) becomes

$$sI - A = S_1(s)\begin{bmatrix} p_1(s) & & 0 \\ & \ddots & \\ 0 & & p_n(s) \end{bmatrix} S_2(s), \tag{6.6.1}$$

where $S_1, S_2 \in \mathbb{F}[s]^{n\times n}$ are unimodular and p_i divides p_{i+1} for all $i \in \{1,\ldots,n-1\}$.

Proposition 6.6.2. Let $A \in \mathbb{F}^{n\times n}$, and let $p_1,\ldots,p_n \in \mathbb{F}[s]$ be the similarity invariants of A, where p_i divides p_{i+1} for all $i \in \{1,\ldots,n-1\}$. Then,

$$\chi_A = \prod_{i=1}^{n} p_i, \tag{6.6.2}$$

$$\mu_A = p_n. \tag{6.6.3}$$

Proof. Using Theorem 6.3.2 and (6.6.1), it follows that

$$\chi_A(s) = \det(sI - A) = [\det S_1(s)][\det S_2(s)]\prod_{i=1}^{n} p_i(s).$$

Since S_1 and S_2 are unimodular and χ_A and $p_1,\ldots,p_n$ are monic, it follows that $[\det S_1(s)][\det S_2(s)] = 1$, which proves (6.6.2).

To prove (6.6.3), first note that it follows from Theorem 6.3.2 that $\chi_A = \Delta_{n-1}p_n$, where $\Delta_{n-1} \in \mathbb{F}[s]$ is the greatest common divisor of all $(n-1)\times(n-1)$ subdeterminants of $sI - A$. Since the $(n-1)\times(n-1)$ subdeterminants of $sI - A$ are the entries of $\pm(sI - A)^{\mathsf{A}}$, it follows that Δ_{n-1} divides every entry of $(sI - A)^{\mathsf{A}}$. Hence, there exists a polynomial matrix $P \in \mathbb{F}[s]^{n\times n}$ such that $(sI - A)^{\mathsf{A}} = \Delta_{n-1}(s)P(s)$. Furthermore, since $(sI - A)^{\mathsf{A}}(sI - A) = \chi_A(s)I$, it follows that $\Delta_{n-1}(s)P(s)(sI - A) = \chi_A(s)I = \Delta_{n-1}(s)p_n(s)I$, and thus $P(s)(sI - A) = p_n(s)I$. Lemma 6.2.2 now implies that $p_n(A) = 0$.

Since $p_n(A) = 0$, it follows from Theorem 6.6.1 that μ_A divides p_n. Hence, let $q \in \mathbb{F}[s]$ be the monic polynomial satisfying $p_n = q\mu_A$. Furthermore, since $\mu_A(A) = 0$, it follows from Corollary 6.2.3 that there exists a polynomial matrix $Q \in \mathbb{F}[s]^{n\times n}$ such that $\mu_A(s)I = Q(s)(sI - A)$. Thus, $P(s)(sI - A) = p_n(s)I = q(s)\mu_A(s)I = q(s)Q(s)(sI - A)$, which implies that $P = qQ$. Thus, q divides every entry of P. However, since P is obtained by dividing $(sI - A)^{\mathsf{A}}$ by the greatest common divisor of all of its entries, it follows that the greatest common divisor of the entries of P is 1. Hence, $q = 1$, which implies that $p_n = \mu_A$, which proves (6.6.3). □

Proposition 6.6.2 shows that μ_A divides χ_A, which is also a consequence of Theorem 6.4.7 and Theorem 6.6.1. Proposition 6.6.2 also shows that $\mu_A = \chi_A$ if and only if $p_1 = \cdots = p_{n-1} = 1$; that is, if and only if $p_n = \chi_A$ is the unique nonconstant similarity invariant of A. Note that, more generally, (6.6.2) implies that $\sum_{i=1}^{n} \deg p_i = n$.

Finally, note that the similarity invariants of the $n\times n$ identity matrix I_n are given by $p_i(s) = s - 1$ for all $i \in \{1,\ldots,n\}$. Thus, $\chi_{I_n}(s) = (s-1)^n$ and $\mu_{I_n}(s) = s - 1$.

Proposition 6.6.3. Let $A \in \mathbb{F}^{n\times n}$, and assume that A and B are similar. Then,

$$\mu_A = \mu_B. \tag{6.6.4}$$

6.7 Rational Transfer Functions and the Smith-McMillan Form

We now turn our attention to rational functions.

Definition 6.7.1. The set $\mathbb{F}(s)$ of *rational functions* consists of functions $g\colon \mathbb{C}\backslash\mathcal{S} \mapsto \mathbb{C}$, where $g(s) = p(s)/q(s)$, $p, q \in \mathbb{F}[s]$, $q \neq 0$, and $\mathcal{S} \triangleq \operatorname{roots}(q)$. The rational function g is *strictly proper, proper, exactly proper, improper*, respectively, if $\deg p < \deg q$, $\deg p \le \deg q$, $\deg p = \deg q$, $\deg p > \deg q$. If p and q are coprime, then the *zeros* of g are the elements of $\operatorname{mroots}(p)$, while the *poles* of g are the elements of mroots(q). The set of proper rational functions is denoted by $\mathbb{F}(s)_{\text{prop}}$. The *relative degree* of $g \in \mathbb{F}(s)_{\text{prop}}$, denoted by reldeg g, is $\deg q - \deg p$.

Definition 6.7.2. The set $\mathbb{F}^{l \times m}(s)$ of *rational transfer functions* consists of matrices whose entries are elements of $\mathbb{F}(s)$. The rational transfer function $G \in \mathbb{F}^{l \times m}(s)$ is *strictly proper* if every entry of G is strictly proper, *proper* if every entry of G is proper, *exactly proper* if every entry of G is proper and at least one entry of G is exactly proper, and *improper* if at least one entry of G is improper. The set of proper rational transfer functions is denoted by $\mathbb{F}^{l \times m}_{\text{prop}}(s)$.

Definition 6.7.3. Let $G \in \mathbb{F}(s)^{l \times m}_{\text{prop}}$. Then, the *relative degree* of G, denoted by reldeg G, is defined by

$$\operatorname{reldeg} G \triangleq \min_{\substack{i=1,\ldots,l \\ j=1,\ldots,m}} \operatorname{reldeg} G_{(i,j)}. \tag{6.7.1}$$

Since

$$(sI - A)^{-1} = \frac{1}{\chi_A(s)} (sI - A)^{\mathsf{A}}, \tag{6.7.2}$$

it follows from (6.4.22) that $(sI - A)^{-1}$ is a strictly proper rational transfer function. In fact, for all $i \in \{1, \ldots, n\}$,

$$\operatorname{reldeg} [(sI - A)^{-1}]_{(i,i)} = 1, \tag{6.7.3}$$

and thus

$$\operatorname{reldeg} (sI - A)^{-1} = 1. \tag{6.7.4}$$

The following definition is an extension of Definition 6.2.4 to rational transfer functions.

Definition 6.7.4. Let $G \in \mathbb{F}^{l \times m}(s)$, and, for all $i \in \{1, \ldots, l\}$ and $j \in \{1, \ldots, m\}$, let $G_{(i,j)} = p_{ij}/q_{ij}$, where $q_{ij} \neq 0$, and $p_{ij}, q_{ij} \in \mathbb{F}[s]$ are coprime. Then, the *poles* of G are the elements of the set

$$\operatorname{poles}(G) \triangleq \bigcup_{i,j=1}^{l,m} \operatorname{roots}(q_{ij}), \tag{6.7.5}$$

and the *blocking zeros* of G are the elements of the set

$$\operatorname{bzeros}(G) \triangleq \bigcap_{i,j=1}^{l,m} \operatorname{roots}(p_{ij}). \tag{6.7.6}$$

Finally, the rank of G is the nonnegative integer

$$\operatorname{rank} G \triangleq \max_{s \in \mathbb{C} \setminus \operatorname{poles}(G)} \operatorname{rank} G(s). \tag{6.7.7}$$

The following result provides a canonical matrix, known as the *Smith-McMillan matrix*, for rational transfer functions under unimodular transformation.

Theorem 6.7.5. Let $G \in \mathbb{F}^{l \times m}(s)$, and let $r \triangleq \operatorname{rank} G$. Then, there exist unimodular matrices $S_1 \in \mathbb{F}[s]^{l \times l}$ and $S_2 \in \mathbb{F}[s]^{m \times m}$ and unique monic polynomials $p_1, \ldots, p_r, q_1, \ldots, q_r \in \mathbb{F}[s]$ such that p_i and q_i are coprime for all $i \in \{1, \ldots, r\}$, p_i divides p_{i+1} for all $i \in \{1, \ldots, r-1\}$, q_{i+1} divides q_i for all $i \in \{1, \ldots, r-1\}$, and

$$G = S_1 \begin{bmatrix} p_1/q_1 & & & \\ & \ddots & & 0_{r \times (m-r)} \\ & & p_r/q_r & \\ & 0_{(l-r) \times r} & & 0_{(l-r) \times (m-r)} \end{bmatrix} S_2. \tag{6.7.8}$$

Proof. Let n_{ij}/d_{ij} denote the (i, j) entry of G, where $n_{ij}, d_{ij} \in \mathbb{F}[s]$ are coprime, and let $d \in \mathbb{F}[s]$ denote the least common multiple of d_{ij} for all $i \in \{1, \ldots, l\}$ and $j \in \{1, \ldots, m\}$. From Theorem

6.3.2 it follows that the polynomial matrix dG has the Smith form $\operatorname{diag}(\hat{p}_1, \ldots, \hat{p}_r, 0, \ldots, 0)$, where $\hat{p}_1, \ldots, \hat{p}_r \in \mathbb{F}[s]$ and $\hat{p}_i$ divides $\hat{p}_{i+1}$ for all $i \in \{1, \ldots, r-1\}$. Now, divide this Smith form by d and express every rational function $\hat{p}_i/d$ in coprime form p_i/q_i so that p_i divides p_{i+1} for all $i \in \{1, \ldots, r-1\}$ and q_{i+1} divides q_i for all $i \in \{1, \ldots, r-1\}$. □

The diagonal matrix in (6.7.8) is a *Smith-McMillan matrix* and the *Smith-McMillan form of G*.

Proposition 6.7.6. Let $G \in \mathbb{F}^{l \times m}(s)$, and assume that there exist unimodular matrices $S_1 \in \mathbb{F}[s]^{l \times l}$ and $S_2 \in \mathbb{F}[s]^{m \times m}$ and monic polynomials $p_1, \ldots, p_r, q_1, \ldots, q_r \in \mathbb{F}[s]$ such that p_i and q_i are coprime for all $i \in \{1, \ldots, r\}$ and such that (6.7.8) holds. Then, $\operatorname{rank} G = r$.

Proposition 6.7.7. Let $G \in \mathbb{F}[s]^{n \times m}$, and let $r \triangleq \operatorname{rank} G$. Then, r is the largest size of all nonsingular submatrices of G.

Proposition 6.7.8. Let $G \in \mathbb{F}^{n \times m}(s)$, and let $\mathcal{S} \subset \mathbb{C}$ be a countable set such that $\operatorname{poles}(G) \subseteq \mathcal{S}$. Then,

$$\operatorname{rank} G = \max_{s \in \mathbb{C} \setminus \mathcal{S}} \operatorname{rank} G(s). \tag{6.7.9}$$

Let $g_1, \ldots, g_r \in \mathbb{F}^n(s)$. Then, $g_1, \ldots, g_r$ are *linearly independent* if $\alpha_1, \ldots, \alpha_r \in \mathbb{F}[s]$ and $\sum_{i=1}^r \alpha_i g_i = 0$ imply that $\alpha_1 = \cdots = \alpha_r = 0$. Equivalently, $g_1, \ldots, g_r$ are *linearly independent* if $\alpha_1, \ldots, \alpha_r \in \mathbb{F}(s)$ and $\sum_{i=1}^r \alpha_i g_i = 0$ imply that $\alpha_1 = \cdots = \alpha_r = 0$. In other words, the coefficients α_i can be either polynomials or rational functions.

Proposition 6.7.9. Let $G \in \mathbb{F}^{l \times m}(s)$. Then, $\operatorname{rank} G$ is equal to the number of linearly independent columns of G.

Since $G \in \mathbb{F}[s]^{l \times m} \subset \mathbb{F}^{l \times m}(s)$, Proposition 6.7.9 applies to polynomial matrices.

Definition 6.7.10. Let $G \in \mathbb{F}^{l \times m}(s)$, assume that $G \neq 0$, let $r \triangleq \operatorname{rank} G$, and let $p_1, \ldots, p_r, q_1, \ldots, q_r \in \mathbb{F}[s]$ be given by Theorem 6.7.5. Then, the *McMillan degree* Mcdeg G of G is defined by

$$\operatorname{Mcdeg} G \triangleq \sum_{i=1}^r \deg q_i. \tag{6.7.10}$$

Furthermore, the *transmission zeros* of G are the elements of the set

$$\operatorname{tzeros}(G) \triangleq \operatorname{roots}(p_r). \tag{6.7.11}$$

Proposition 6.7.11. Let $G \in \mathbb{F}^{l \times m}(s)$, assume that $G \neq 0$, and assume that G has the Smith-McMillan form (6.7.8). Then,

$$\operatorname{poles}(G) = \operatorname{roots}(q_1), \tag{6.7.12}$$

$$\operatorname{bzeros}(G) = \operatorname{roots}(p_1). \tag{6.7.13}$$

Note that

$$\operatorname{bzeros}(G) \subseteq \operatorname{tzeros}(G). \tag{6.7.14}$$

Furthermore, we define the multisets

$$\operatorname{mpoles}(G) \triangleq \bigcup_{i=1}^r \operatorname{mroots}(q_i), \tag{6.7.15}$$

$$\operatorname{mtzeros}(G) \triangleq \bigcup_{i=1}^r \operatorname{mroots}(p_i), \tag{6.7.16}$$

$$\operatorname{mbzeros}(G) \triangleq \operatorname{mroots}(p_1). \tag{6.7.17}$$

Hence,

$$\text{mbzeros}(G) \subseteq \text{mtzeros}(G). \tag{6.7.18}$$

If $G = 0$, then these multisets as well as the sets poles(G), tzeros(G), and bzeros(G) are empty.

Proposition 6.7.12. Let $G \in \mathbb{F}(s)^{l\times m}_{\text{prop}}$, assume that $G \neq 0$, let $z \in \mathbb{C}$, and assume that z is not a pole of G. Then, z is a transmission zero of G if and only if $\operatorname{rank} G(z) < \operatorname{rank} G$. Furthermore, z is a blocking zero of G if and only if $G(z) = 0$.

The following example shows that a pole of G can also be a transmission zero of G.

Example 6.7.13. Define $G \in \mathbb{R}(s)^{2\times 2}_{\text{prop}}$ by

$$G(s) = \begin{bmatrix} \dfrac{1}{(s+1)^2} & \dfrac{1}{(s+1)(s+2)} \\[2ex] \dfrac{1}{(s+1)(s+2)} & \dfrac{s+3}{(s+2)^2} \end{bmatrix}.$$

Then, $\operatorname{rank} G = 2$. Furthermore,

$$G(s) = S_1(s)\begin{bmatrix} \dfrac{1}{(s+1)^2(s+2)^2} & 0 \\[2ex] 0 & s+2 \end{bmatrix} S_2(s),$$

where $S_1, S_2 \in \mathbb{R}[s]^{2\times 2}$ are the unimodular matrices

$$S_1(s) = \begin{bmatrix} (s+2)(s^3+4s^2+5s+1) & 1 \\ (s+1)(s^3+5s^2+8s+3) & 1 \end{bmatrix}, \quad S_2(s) = \begin{bmatrix} -(s+2) & (s+1)(s^2+3s+1) \\ 1 & -s(s+2) \end{bmatrix}.$$

Hence, the McMillan degree of G is 4, the poles of G are -1 and -2, the transmission zero of G is -2, and G has no blocking zeros. Note that -2 is both a pole and a transmission zero of G. Note also that, although G is strictly proper, the Smith-McMillan form of G is improper. ◇

Let $G \in \mathbb{F}(s)^{l\times m}_{\text{prop}}$. A factorization of G of the form

$$G(s) = N(s)D^{-1}(s), \tag{6.7.19}$$

where $N \in \mathbb{F}[s]^{l\times m}$ and $D \in \mathbb{F}[s]^{m\times m}$, is a *right polynomial fraction description* of G. We say that N and D are *right coprime* if every $R \in \mathbb{F}[s]^{m\times m}$ that right divides both N and D is unimodular. In this case, (6.7.19) is a *coprime right polynomial fraction description* of G.

Theorem 6.7.14. Let $N \in \mathbb{F}[s]^{l\times m}$ and $D \in \mathbb{F}[s]^{m\times m}$. Then, the following statements are equivalent:

i) N and D are right coprime.

ii) There exist $X \in \mathbb{F}[s]^{m\times l}$ and $Y \in \mathbb{F}[s]^{m\times m}$ such that

$$XN + YD = I. \tag{6.7.20}$$

iii) For all $s \in \mathbb{C}$,

$$\operatorname{rank}\begin{bmatrix} N(s) \\ D(s) \end{bmatrix} = m. \tag{6.7.21}$$

Proof. See [2349, p. 297]. □

Equation (6.7.20) is the *Bezout identity.*

The following result shows that all coprime right polynomial fraction descriptions of a proper rational transfer function G are related by a unimodular transformation.

Proposition 6.7.15. Let $G \in \mathbb{F}(s)^{l\times m}_{\text{prop}}$, let $N, \hat{N} \in \mathbb{F}[s]^{l\times m}$, let $D, \hat{D} \in \mathbb{F}[s]^{m\times m}$, and assume that $G = ND^{-1} = \hat{N}\hat{D}^{-1}$. Then, there exists a unimodular matrix $R \in \mathbb{F}[s]^{m\times m}$ such that $N = \hat{N}R$ and $D = \hat{D}R$.

Proof. See [2349, p. 298]. □

The following result uses the Smith-McMillan form to show that every proper rational transfer function has a coprime right polynomial fraction description.

Proposition 6.7.16. Let $G \in \mathbb{F}(s)^{l\times m}_{\text{prop}}$. Then, G has a coprime right polynomial fraction description. If, in addition, $G(s) = N(s)D^{-1}(s)$, where $N \in \mathbb{F}[s]^{l\times m}$ and $D \in \mathbb{F}[s]^{m\times m}$, is a coprime right polynomial fraction description of G, then

$$\text{Szeros}(N) = \text{tzeros}(G), \tag{6.7.22}$$

$$\text{Szeros}(D) = \text{poles}(G). \tag{6.7.23}$$

Proof. Note that (6.7.8) can be written as

$$G = S_1 \begin{bmatrix} p_1/q_1 & & & 0 \\ & \ddots & & \\ & & p_r/q_r & \\ 0 & & & 0_{(l-r)\times(m-r)} \end{bmatrix} S_2$$

$$= S_1 \begin{bmatrix} p_1 & & & 0 \\ & \ddots & & \\ & & p_r & \\ 0 & & & 0_{(l-r)\times(m-r)} \end{bmatrix} \begin{bmatrix} q_1 & & & 0 \\ & \ddots & & \\ & & q_r & \\ 0 & & & I_{m-r} \end{bmatrix}^{-1} S_2$$

$$= S_1 \begin{bmatrix} p_1 & & & 0 \\ & \ddots & & \\ & & p_r & \\ 0 & & & 0_{(l-r)\times(m-r)} \end{bmatrix} \left(S_2^{-1} \begin{bmatrix} q_1 & & & 0 \\ & \ddots & & \\ & & q_r & \\ 0 & & & I_{m-r} \end{bmatrix} \right)^{-1},$$

which, by Theorem 6.7.14, is a right coprime polynomial fraction description of G. The last statement follows from Theorem 6.7.5 and Proposition 6.7.15. □

6.8 Facts on Polynomials and Rational Functions

Fact 6.8.1. Let $p \in \mathbb{R}[s]$, where $p(s) = \beta_n s^n + \beta_{n-1}s^{n-1} + \cdots + \beta_1 s + \beta_0$, assume that $\beta_0, \ldots, \beta_n$ are integers, let k and l be nonzero coprime integers, and assume that $p(k/l) = 0$. Then, $k|\beta_0$ and $l|\beta_n$. **Source:** [1352, p. 130]. **Related:** Fact 6.8.2

Fact 6.8.2. Let $p \in \mathbb{R}[s]$, where $p(s) = s^n + \beta_{n-1}s^{n-1} + \cdots + \beta_1 s + \beta_0$, assume that $\beta_0, \ldots, \beta_{n-1}$ are integers, and let λ be a real root of p. Then, λ is either an integer or an irrational number. **Remark:** This is *Gauss's lemma*. See [1174]. **Remark:** If k and n are positive integers, then $\sqrt[n]{k}$ is either an integer or an irrational number. **Related:** Fact 1.12.15 and Fact 6.8.1.

Fact 6.8.3. Let $p \in \mathbb{R}[s]$ be monic, and define $q(s) \triangleq s^n p(1/s)$, where $n \triangleq \deg p$. If $0 \notin \text{roots}(p)$, then $\deg(q) = n$ and

$$\text{mroots}(q) = \{1/\lambda\colon\ \lambda \in \text{mroots}(p)\}_{\text{ms}}.$$

If $0 \in \text{roots}(p)$ with multiplicity r, then $\deg(q) = n - r$ and

$$\text{mroots}(q) = \{1/\lambda\colon\ \lambda \neq 0 \text{ and } \lambda \in \text{mroots}(p)\}_{\text{ms}}.$$

Related: Fact 15.18.6 and Fact 15.18.7.

Fact 6.8.4. Let $p \in \mathbb{F}^n[s]$ be given by $p(s) = s^n + \beta_{n-1}s^{n-1} + \cdots + \beta_1 s + \beta_0$, let $\operatorname{mroots}(p) = \{\lambda_1, \ldots, \lambda_n\}_{\mathrm{ms}}$, and, for all $i \in \{1, \ldots, n\}$, define $\mu_i \triangleq \lambda_1^i + \cdots + \lambda_n^i$. Then, for all $k \in \{1, \ldots, n\}$,

$$k\beta_{n-k} + \mu_1\beta_{n-k+1} + \mu_2\beta_{n-k+2} + \cdots + \mu_k\beta_n = 0.$$

That is,

$$\begin{bmatrix} n & \mu_1 & \mu_2 & \mu_3 & \mu_4 & \cdots & \mu_n \\ 0 & n-1 & \mu_1 & \mu_2 & \mu_3 & \cdots & \mu_{n-1} \\ \vdots & \ddots & \ddots & \ddots & \ddots & \ddots & \vdots \\ \vdots & \ddots & \ddots & \ddots & \ddots & \ddots & \vdots \\ 0 & 0 & \cdots & 0 & 2 & \mu_1 & \mu_2 \\ 0 & 0 & \cdots & 0 & 0 & 1 & \mu_1 \end{bmatrix} \begin{bmatrix} \beta_0 \\ \beta_1 \\ \vdots \\ \beta_{n-1} \\ 1 \end{bmatrix} = 0.$$

Therefore,

$$\begin{bmatrix} 1 & 0 & 0 & \cdots & 0 & 0 \\ \beta_{n-1} & 1 & 0 & \cdots & 0 & 0 \\ \beta_{n-2} & \beta_{n-1} & 1 & \cdots & 0 & 0 \\ \vdots & \vdots & \ddots & \ddots & \vdots & \vdots \\ \beta_2 & \beta_3 & \beta_4 & \cdots & 1 & 0 \\ \beta_1 & \beta_2 & \beta_3 & \cdots & \beta_{n-1} & 1 \end{bmatrix} \begin{bmatrix} \mu_1 \\ \mu_2 \\ \mu_3 \\ \vdots \\ \mu_{n-1} \\ \mu_n \end{bmatrix} = - \begin{bmatrix} \beta_{n-1} \\ 2\beta_{n-2} \\ 3\beta_{n-3} \\ \vdots \\ (n-1)\beta_1 \\ n\beta_0 \end{bmatrix}.$$

Consequently,

$$\beta_{n-1} = -\mu_1, \quad \beta_{n-2} = \tfrac{1}{2}(\mu_1^2 - \mu_2), \quad \beta_{n-3} = \tfrac{1}{6}(-\mu_1^3 + 3\mu_1\mu_2 - 2\mu_3),$$
$$\mu_1 = -\beta_{n-1}, \quad \mu_2 = \beta_{n-1}^2 - 2\beta_{n-2}, \quad \mu_3 = \beta_{n-1}^3 + 3\beta_{n-2}\beta_{n-1} - 3\beta_{n-3}.$$

Source: [1448, p. 44], [1575], and [2043, p. 9]. **Remark:** These equations are a consequence of Newton's identities given by Fact 2.11.34. Note that, for all $i \in \{0, \ldots, n\}$, $\beta_i = (-1)^{n-i}E_{n-i}$, where E_i is the ith elementary symmetric polynomial of the roots of p.

Fact 6.8.5. Let $p, q \in \mathbb{F}[s]$ be monic. Then, p and q are coprime if and only if their least common multiple is pq.

Fact 6.8.6. Let $n + m \geq 1$, let $p, q \in \mathbb{F}[s]$, where $p(s) = a_n s^n + \cdots + a_1 s + a_0$, $q(s) = b_m s^m + \cdots + b_1 s + b_0$, and assume that $\deg p = n$ and $m \leq n$. Furthermore, define the Toeplitz matrices $[p]^{(m)} \in \mathbb{F}^{m\times(m+n)}$ and $[q]^{(n)} \in \mathbb{F}^{n\times(m+n)}$ by

$$[p]^{(m)} \triangleq \begin{bmatrix} a_n & a_{n-1} & \cdots & a_1 & a_0 & 0 & 0 & \cdots & 0 \\ 0 & a_n & a_{n-1} & \cdots & a_1 & a_0 & 0 & \cdots & 0 \\ \vdots & \ddots & \ddots & \ddots & \cdots & \ddots & \ddots & \ddots & \vdots \end{bmatrix},$$

$$[q]^{(n)} \triangleq \begin{bmatrix} b_m & b_{m-1} & \cdots & b_1 & b_0 & 0 & 0 & \cdots & 0 \\ 0 & b_m & b_{m-1} & \cdots & b_1 & b_0 & 0 & \cdots & 0 \\ \vdots & \ddots & \ddots & \ddots & \cdots & \ddots & \ddots & \ddots & \vdots \end{bmatrix},$$

and $R(p, q) \in \mathbb{F}^{(m+n)\times(m+n)}$ by

$$R(p, q) \triangleq \begin{bmatrix} [p]^{(m)} \\ [q]^{(n)} \end{bmatrix}.$$

Then, p and q are coprime if and only if $R(p,q)$ is nonsingular. **Source:** [1040, p. 162] and [2263, pp. 187–191]. **Remark:** $R(p,q)$ is the *Sylvester matrix*, and $\det R(p,q)$ is the *resultant* of p and q. **Remark:** The form of $R(p,q)$ appears in [2263, pp. 187–191]. This result is given in [1040, p. 162] in terms of $\begin{bmatrix} \hat{I}[p]^{(m)} \\ \hat{I}[q]^{(n)} \end{bmatrix}\hat{I}$ and in [3024, p. 85] in terms of $\begin{bmatrix} [p]^{(m)} \\ \hat{I}[q]^{(n)} \end{bmatrix}$. Interweaving the entries of the rows of $[p]^{(m)}$ and $[q]^{(n)}$ and taking the transpose yields a *step-down matrix* [855].

Fact 6.8.7. Let $p_1,\dots,p_n \in \mathbb{F}[s]$, and let $d \in \mathbb{F}[s]$ be the greatest common divisor of $p_1,\dots,p_n$. Then, there exist polynomials $q_1,\dots,q_n \in \mathbb{F}[s]$ such that

$$d = \sum_{i=1}^{n} q_i p_i.$$

In addition, $p_1,\dots,p_n$ are coprime if and only if there exist polynomials $q_1,\dots,q_n \in \mathbb{F}[s]$ such that

$$1 = \sum_{i=1}^{n} q_i p_i.$$

Source: [1084, p. 16]. **Remark:** The polynomial d is given by the *Bezout equation*.

Fact 6.8.8. Let $p,q \in \mathbb{F}[s]$, where $p(s) = a_n s^n + \cdots + a_1 s + a_0$ and $q(s) = b_n s^n + \cdots + b_1 s + b_0$, and define $[p]^{(n)}, [q]^{(n)} \in \mathbb{F}^{n\times 2n}$ and $R(p,q) \in \mathbb{F}^{2n\times 2n}$ as in Fact 6.8.6. Furthermore, define

$$\begin{bmatrix} A_1 & A_2 \\ B_1 & B_2 \end{bmatrix} \triangleq R(p,q),$$

where $A_1, A_2, B_1, B_2 \in \mathbb{F}^{n\times n}$, and define $\hat{p}(s) \triangleq s^n p(-s)$ and $\hat{q}(s) \triangleq s^n q(-s)$. Then,

$$\begin{bmatrix} A_1 & A_2 \\ B_1 & B_2 \end{bmatrix} = \begin{bmatrix} \hat{p}(N_n^{\mathsf{T}}) & p(N_n) \\ \hat{q}(N_n^{\mathsf{T}}) & q(N_n) \end{bmatrix},$$

$$A_1B_1 = B_1A_1, \quad A_2B_2 = B_2A_2, \quad A_1B_2 + A_2B_1 = B_1A_2 + B_2A_1.$$

Therefore,

$$\begin{bmatrix} I & 0 \\ -B_1 & A_1 \end{bmatrix}\begin{bmatrix} A_1 & A_2 \\ B_1 & B_2 \end{bmatrix} = \begin{bmatrix} A_1 & A_2 \\ 0 & A_1B_2 - B_1A_2 \end{bmatrix},$$

$$\begin{bmatrix} -B_2 & A_2 \\ 0 & I \end{bmatrix}\begin{bmatrix} A_1 & A_2 \\ B_1 & B_2 \end{bmatrix} = \begin{bmatrix} A_2B_1 - B_2A_1 & 0 \\ B_1 & B_2 \end{bmatrix},$$

$$\det R(p,q) = \det(A_1B_2 - B_1A_2) = \det(B_2A_1 - A_2B_1).$$

Now, define $B(p,q) \in \mathbb{F}^{n\times n}$ by $B(p,q) \triangleq (A_1B_2 - B_1A_2)\hat{I}$. Then, the following statements hold:

i) For all $s, \hat{s} \in \mathbb{C}$,

$$p(s)q(\hat{s}) - q(s)p(\hat{s}) = (s - \hat{s})\begin{bmatrix} 1 \\ s \\ \vdots \\ s^{n-1} \end{bmatrix}^{\mathsf{T}} B(p,q)\begin{bmatrix} 1 \\ \hat{s} \\ \vdots \\ \hat{s}^{n-1} \end{bmatrix}.$$

ii) $B(p,q) = (B_2A_1 - A_2B_1)\hat{I} = \hat{I}(A_1^{\mathsf{T}}B_2^{\mathsf{T}} - B_1^{\mathsf{T}}A_2^{\mathsf{T}}) = \hat{I}(B_1^{\mathsf{T}}A_2^{\mathsf{T}} - A_1^{\mathsf{T}}B_2^{\mathsf{T}})$.

iii) $\begin{bmatrix} 0 & B(p,q) \\ -B(p,q) & 0 \end{bmatrix} = QR^{\mathsf{T}}(p,q)QR(p,q)Q$, where $Q \triangleq \begin{bmatrix} 0 & \hat{I} \\ -\hat{I} & 0 \end{bmatrix}$.

iv) $|\det B(p,q)| = |\det R(p,q)| = |\det q[C(p)]|$.

v) $B(p,q)$ and $\hat{B}(p,q)$ are symmetric.

vi) $B(p,q)$ is a linear function of (p,q).

vii) $B(p,q) = -B(q,p)$.

Now, assume, in addition, that $\deg q \le \deg p = n$ and p is monic. Then, the following statements hold:

viii) $\operatorname{def} B(p,q)$ equals the degree of the greatest common divisor of p and q.

ix) p and q are coprime if and only if $B(p,q)$ is nonsingular.

x) If $B(p,q)$ is nonsingular, then $[B(p,q)]^{-1}$ is Hankel. In fact, $[B(p,q)]^{-1} = H_{n,n}(b/p)$, where $a,b \in \mathbb{F}[s]$ satisfy the Bezout equation $ap + bq = 1$ and $H_{n,n}$ is defined in Fact 6.8.10.

xi) If $q = q_1q_2$, where $q_1, q_2 \in \mathbb{F}[s]$, then $B(p,q) = B(p,q_1)q_2[C(p)] = q_1[C^{\mathsf{T}}(p)]B(p,q_2)$.

xii) $B(p,q) = B(p,q)C(p) = C^{\mathsf{T}}(p)B(p,q)$.

xiii) $B(p,q) = B(p,1)q[C(p)] = q[C^{\mathsf{T}}(p)]B(p,1)$, where $B(p,1)$ is the Hankel matrix

$$B(p,1) = \begin{bmatrix} a_1 & a_2 & \cdots & a_{n-1} & 1 \\ a_2 & a_3 & ⋰ & 1 & 0 \\ \vdots & ⋰ & ⋰ & ⋰ & \vdots \\ a_{n-1} & 1 & ⋰ & 0 & 0 \\ 1 & 0 & \cdots & 0 & 0 \end{bmatrix}.$$

In particular, for $n = 3$ and $q(s) = s$, it follows that

$$\begin{bmatrix} -a_0 & 0 & 0 \\ 0 & a_2 & 1 \\ 0 & 1 & 0 \end{bmatrix} = \begin{bmatrix} a_1 & a_2 & 1 \\ a_2 & 1 & 0 \\ 1 & 0 & 0 \end{bmatrix} \begin{bmatrix} 0 & 1 & 0 \\ 0 & 0 & 1 \\ -a_0 & -a_1 & -a_2 \end{bmatrix}.$$

xiv) If A_2 is nonsingular, then

$$\begin{bmatrix} A_1 & A_2 \\ B_1 & B_2 \end{bmatrix} = \begin{bmatrix} 0 & I \\ A_2^{-1}\hat{I} & B_2A_2^{-1} \end{bmatrix} \begin{bmatrix} B(p,q) & 0 \\ 0 & I \end{bmatrix} \begin{bmatrix} I & 0 \\ A_1 & A_2 \end{bmatrix}.$$

xv) If p has distinct roots $\lambda_1,\dots,\lambda_n$, then

$$V^{\mathsf{T}}(\lambda_1,\dots,\lambda_n)B(p,q)V(\lambda_1,\dots,\lambda_n) = \operatorname{diag}[q(\lambda_1)p'(\lambda_1),\dots,q(\lambda_n)p'(\lambda_n)].$$

Source: [1040, pp. 164–167], [1084, pp. 200–207], and [1362]. To prove *ii)*, note that A_1, A_2, B_1, B_2 are square and Toeplitz, and thus reverse symmetric; that is, $A_1 = A_1^{\hat{\mathsf{T}}}$. See Fact 4.23.5. **Remark:** $B(p,q)$ is the *Bezout matrix* of p and q. See [301, 1361, 1468], [2263, p. 189], [2768, 2898], and Fact 7.17.26. **Remark:** *xiii)* is the *Barnett factorization.* See [294, 2768]. The definitions of $B(p,q)$ and *ii)* are the *Gohberg-Semencul formulas.* See [1084, p. 206]. **Remark:** It follows from continuity that the expressions for $\det R(p,q)$ are valid whether or not either A_1 or B_2 is singular. See Fact 3.17.14. **Remark:** The inverse of a Hankel matrix is a Bezout matrix. See [1040, p. 174].

Fact 6.8.9. Let $p,q \in \mathbb{F}[s]$, where $p(s) = \alpha_1 s + \alpha_0$ and $q(s) = s^2 + \beta_1 s + \beta_0$. Then, p and q are coprime if and only if $\alpha_0^2 + \alpha_1^2\beta_0 \ne \alpha_0\alpha_1\beta_1$. **Source:** Fact 6.8.8.

Fact 6.8.10. Let $p,q \in \mathbb{F}[s]$, assume that q is monic, assume that $\deg p < \deg q = n$, and define $B(p,q)$ as in Fact 6.8.8. Furthermore, define $g \in \mathbb{F}(s)$ by

$$g(s) \triangleq \frac{p(s)}{q(s)} = \sum_{i=1}^{\infty} \frac{h_i}{s^i}.$$

Finally, define the Hankel matrix $H_{i,j}(g) \in \mathbb{R}^{i \times j}$ by

$$H_{i,j}(g) = \begin{bmatrix} h_1 & h_2 & h_{k+3} & \cdots & h_j \\ h_{k+2} & h_{k+3} & \iddots & \iddots & \vdots \\ h_{k+3} & \iddots & \iddots & \iddots & \vdots \\ \vdots & \iddots & \iddots & \iddots & \vdots \\ \vdots & \iddots & \iddots & \iddots & \vdots \\ h_i & \cdots & \cdots & \cdots & h_{j+i-1} \end{bmatrix}.$$

Then, the following statements are equivalent:

i) p and q are coprime.

ii) $H_{n,n}(g)$ is nonsingular.

iii) For all $i, j \ge n$, $\operatorname{rank} H_{i,j}(g) = n$.

iv) There exist $i, j \ge n$ such that $\operatorname{rank} H_{i,j}(g) = n$.

Furthermore, the following statements hold:

v) If p and q are coprime, then $[H_{n,n}(g)]^{-1} = B(q, a)$, where $a, b \in \mathbb{F}[s]$ satisfy the Bezout equation $ap + bq = 1$.

vi) $B(q, p) = B(q, 1)H_{n,n}(g)B(q, 1)$.

vii) $B(q, p)$ and $H_{n,n}(g)$ are congruent.

viii) $\operatorname{In} B(q, p) = \operatorname{In} H_{n,n}(g)$.

ix) $\det H_{n,n}(g) = \det B(q, p)$.

Source: [1084, pp. 215–221]. **Related:** Proposition 16.9.13.

Fact 6.8.11. Let $q \in \mathbb{R}[s]$, define $g \in \mathbb{F}(s)$ by $g \triangleq q'/q$, and define $B(q, q')$ as in Fact 6.8.8. Then, the following statements hold:

i) The number of distinct roots of q is $\operatorname{rank} B(q, q')$.

ii) q has n distinct roots if and only if $B(q, q')$ is nonsingular.

iii) The number of distinct real roots of q is $\operatorname{sig} B(q, q')$.

iv) q has n distinct, real roots if and only if $B(q, q')$ is positive definite.

v) The number of distinct complex roots of q is $2\nu_-[B(q, q')]$.

vi) q has n distinct, complex roots if and only if n is even and $\nu_-[B(q,q')] = n/2$.

vii) q has n real roots if and only if $B(q, q')$ is positive semidefinite.

Source: [1084, p. 252]. **Remark:** $q'(s) \triangleq (\mathrm{d}/\mathrm{d}s)q(s)$.

Fact 6.8.12. Let $q \in \mathbb{F}[s]$, where $q(s) = \sum_{i=0}^{n} b_i s^i$, and define $\operatorname{coeff}(q) \triangleq [b_n \ \cdots \ b_0]^{\mathsf{T}}$. Now, let $p \in \mathbb{F}[s]$, where $p(s) = \sum_{i=0}^{n} a_i s^i$. Then,

$$\operatorname{coeff}(pq) = A\operatorname{coeff}(q),$$

where $A \in \mathbb{F}^{2n\times(n+1)}$ is the Toeplitz matrix

$$A = \begin{bmatrix} a_n & 0 & 0 & \cdots & 0 \\ a_{n-1} & a_n & 0 & \cdots & 0 \\ \vdots & \ddots & \ddots & \ddots & \vdots \\ a_1 & \ddots & \ddots & \ddots & \vdots \\ a_0 & a_1 & \ddots & \ddots & a_n \\ 0 & a_0 & \ddots & \ddots & a_{n-1} \\ \vdots & \ddots & \ddots & \ddots & \vdots \\ 0 & 0 & \cdots & a_0 & a_1 \end{bmatrix}.$$

In particular, if $n = 3$, then

$$A = \begin{bmatrix} a_2 & 0 & 0 \\ a_1 & a_2 & 0 \\ a_0 & a_1 & a_2 \\ 0 & a_0 & a_1 \end{bmatrix}.$$

Fact 6.8.13. Let $\lambda_1, \ldots, \lambda_n \in \mathbb{C}$ be distinct and, for all $i \in \{1, \ldots, n\}$, define

$$p_i(s) \triangleq \prod_{\substack{j=1 \\ j\neq i}}^{n} \frac{s - \lambda_i}{\lambda_i - \lambda_j}.$$

Then, for all $i \in \{1, \ldots, n\}$,

$$p_i(\lambda_j) = \begin{cases} 1, & i = j, \\ 0, & i \neq j. \end{cases}$$

Remark: This is the *Lagrange interpolation formula.*

Fact 6.8.14. Let $A \in \mathbb{F}^{n\times n}$, and assume that $\det(I + A) \neq 0$. Then, there exists $p \in \mathbb{F}[s]$ such that $\deg p \leq n - 1$ and $(I + A)^{-1} = p(A)$. **Related:** Fact 6.8.15.

Fact 6.8.15. Let $A \in \mathbb{F}^{n\times n}$, let $q \in \mathbb{F}[s]$, and assume that $q(A)$ is nonsingular. Then, there exists $p \in \mathbb{F}[s]$ such that $\deg p \leq n - 1$ and $[q(A)]^{-1} = p(A)$. **Source:** Fact 7.16.9. **Related:** Fact 6.8.14.

Fact 6.8.16. Let $n \geq 2$ be an even integer, let $A \in \mathbb{R}^{n\times n}$, assume that A is skew symmetric, and let the components of $x_A \in \mathbb{R}^{n(n-1)/2}$ be the entries $A_{(i,j)}$ for all $i > j$. Then, there exists a polynomial $p\colon\ \mathbb{R}^{n(n-1)/2} \mapsto \mathbb{R}$ such that, for all $\alpha \in \mathbb{R}$ and $x \in \mathbb{R}^{n(n-1)/2}$,

$$p(\alpha x) = \alpha^{n/2} p(x), \quad \det A = p^2(x_A).$$

In particular,

$$\det\begin{bmatrix} 0 & a \\ -a & 0 \end{bmatrix} = a^2, \quad \det\begin{bmatrix} 0 & a & b & c \\ -a & 0 & d & e \\ -b & -d & 0 & f \\ -c & -e & -f & 0 \end{bmatrix} = (af - be + cd)^2.$$

Source: [1768, p. 224] and [2263, pp. 125–127]. **Remark:** p is the *Pfaffian*, and this is *Pfaff's theorem.* **Remark:** An extension to the product of a pair of skew-symmetric matrices is given in [948]. **Related:** Fact 4.10.35.

Fact 6.8.17. Let $G \in \mathbb{F}(s)^{n\times m}$, and let $G_{(i,j)} = n_{ij}/d_{ij}$, where $n_{ij} \in \mathbb{F}[s]$ and $d_{ij} \in \mathbb{F}[s]$ are coprime for all $i \in \{1, \ldots, n\}$ and $j \in \{1, \ldots, m\}$. Then, q_1 given by the Smith-McMillan form is the least common multiple of $d_{11}, d_{12}, \ldots, d_{nm}$.

Fact 6.8.18. Let $G \in \mathbb{F}(s)^{n\times m}$, assume that $\operatorname{rank} G = m$, and let $\lambda \in \mathbb{C}$, where λ is not a pole of G. Then, λ is a transmission zero of G if and only if there exists a nonzero vector $u \in \mathbb{C}^m$ such that $G(\lambda)u = 0$. Furthermore, if G is square, then λ is a transmission zero of G if and only if $\det G(\lambda) = 0$.

Fact 6.8.19. Let $G \in \mathbb{R}^{n\times m}(s)$, let $\omega \in \mathbb{R}$, and assume that $\omega\jmath$ is not a pole of G. Then,

$$\operatorname{Im} G(-\omega\jmath) = -\operatorname{Im} G(\omega\jmath).$$

Fact 6.8.20. Let $p \in \mathbb{R}[s]$, and assume that all of the coefficients of p are nonnegative. Then, p is increasing on $[0, \infty)$.

Fact 6.8.21. Let $n \geq 2$, let $p \in \mathbb{F}[s]$, where $p(s) = \sum_{i=0}^{n} \beta_i s^i$, let $\operatorname{mroots}(p) = \{\lambda_1, \ldots, \lambda_n\}_{\mathrm{m}}$, and define

$$\operatorname{dis(p)} \triangleq \beta_n^{2n-2} \prod (\lambda_i - \lambda_j)^2 = (-1)^{n(n-1)/2} \beta_n^{2n-2} \prod (\lambda_i - \lambda_j),$$

where the first product is taken over all $i, j \in \{1, \ldots, n\}$ such that $i < j$, and the second product is taken over all distinct $i, j \in \{1, \ldots, n\}$. Then, the following statements are equivalent:

i) p has distinct roots.

ii) $\operatorname{dis}(p) \neq 0$.

iii) $\operatorname{mroots}(p) \cap \operatorname{mroots}(p') = \varnothing$.

iv) p and p' are coprime.

Furthermore, the following statements hold:

v) If $p(s) = as^2 + bs + c$, then $\operatorname{dis}(p) = b^2 - 4ac$.

vi) If $p(s) = as^3 + bs^2 + cs + d$, then $\operatorname{dis}(p) = b^2c^2 - 4ac^2 - 4b^3d - 27a^2d^2 + 18abcd$.

vii) If $p(s) = s^3 + cs + d$, then $\operatorname{dis}(p) = -4c^3 - 27d^2$.

viii) If $p(s) = as^4 + bs^3 + cs^2 + ds + e$, then

$$\begin{aligned}\operatorname{dis}(p) = {} & 256a^3e^3 - 192a^2bde^2 - 128a^2c^2e^2 + 144a^2cd^2e - 27a^2d^4 + 144ab^2ce^2 \\ & - 6ab^2d^2e - 80abc^2de + 18abcd^3 + 16ac^4e - 4ac^3d^2 - 27b^4e^2 + 18b^3cde \\ & - 4b^3d^3 - 4b^2c^3e + b^2c^2d^2.\end{aligned}$$

Source: [128], [1083, pp. 70, 119]. **Remark:** $\operatorname{dis}(p)$ is the *discriminant* of p. **Related:** Fact 6.8.6

Fact 6.8.22. Let $p_1, p_2 \in \mathbb{C}[s]$, assume that p_1 and p_2 are coprime, and assume that p_1, p_2, and $p_1 + p_2$ are not constant. Then,

$$\max\{\deg p_1, \deg p_2\} + 1 \leq \operatorname{card}(\operatorname{roots}[p_1p_2(p_1 + p_2)]),$$
$$\deg[p_1p_2(p_1 + p_2)] + 3 \leq 3\operatorname{card}(\operatorname{roots}[p_1p_2(p_1 + p_2)]).$$

Source: [331]. **Remark:** This is *Stothers's theorem.* **Related:** Fact 6.8.23.

Fact 6.8.23. Let $p_1, p_2, p_3 \in \mathbb{C}[s]$, and assume that $p_1, p_2, p_3, p_1 + p_2 + p_3$ are coprime and not constant. Then,

$$\max\{\deg p_1, \deg p_2, \det p_3, \deg(p_1 + p_2 + p_3)\} + 3 \leq 2\operatorname{card}(\operatorname{roots}[p_1p_2p_3(p_1 + p_2 + p_3)]).$$

Furthermore,

$$\deg[p_1p_2p_3(p_1 + p_2 + p_3)] + 12 \leq 8\operatorname{card}(\operatorname{roots}[p_1p_2(p_1 + p_2)]).$$

If, in addition, $p_1, p_2, p_3, p_1 + p_2 + p_3$ are linearly dependent, then

$$\max\{\deg p_1, \deg p_2, \det p_3, \deg(p_1 + p_2 + p_3)\} + 5 \leq 2\operatorname{card}(\operatorname{roots}[p_1p_2p_3(p_1 + p_2 + p_3)]).$$

Source: [331]. **Related:** This result extends Fact 6.8.22.

6.9 Facts on the Characteristic and Minimal Polynomials

Fact 6.9.1. Let $A = \begin{bmatrix} a & b \\ c & d \end{bmatrix} \in \mathbb{R}^{2\times 2}$. Then, the following statements hold:

i) $\operatorname{mspec}(A) = \left\{\frac{1}{2}\left[a + d \pm \sqrt{(a-d)^2 + 4bc}\right]\right\}_{\mathrm{ms}} = \left\{\frac{1}{2}\left[\operatorname{tr} A \pm \sqrt{(\operatorname{tr} A)^2 - 4\det A}\right]\right\}_{\mathrm{ms}}$.

ii) $\chi_A(s) = s^2 - (\operatorname{tr} A)s + \det A$.

iii) $A^2 = (\operatorname{tr} A)A - (\det A)I$.

iv) $\operatorname{tr} A^2 = (\operatorname{tr} A)^2 - 2\det A$.

v) $\det A = \frac{1}{2}[(\operatorname{tr} A)^2 - \operatorname{tr} A^2]$.

vi) $(sI - A)^{\mathsf{A}} = sI + A - (\operatorname{tr} A)I$.

vii) $A^{\mathsf{A}} = (\operatorname{tr} A)I - A$.

viii) $\operatorname{mspec}(A^{\mathsf{A}}) = \operatorname{mspec}(A)$.

ix) $\chi_{A^{\mathsf{A}}}(s) = \chi_A(s)$.

x) $\operatorname{tr} A^{\mathsf{A}} = \operatorname{tr} A$.

xi) $\det A^{\mathsf{A}} = \det A$.

xii) If A is nonsingular, then $A^{-1} = (\det A)^{-1}[(\operatorname{tr} A)I - A]$ and $\operatorname{tr} A^{-1} = \frac{\operatorname{tr} A}{\det A}$.

xiii) If A is singular, then $A^2 = (\operatorname{tr} A)A$ and $\operatorname{tr} A^2 = (\operatorname{tr} A)^2$.

xiv) If $\operatorname{tr} A = 0$, then $\operatorname{tr} A^2 + 2\det A = 0$ and $A^2 = -(\det A)I = \frac{1}{2}(\operatorname{tr} A^2)I$.

xv) If A is singular and $\operatorname{tr} A = 0$, then $A = 0$.

Fact 6.9.2. Let $A \in \mathbb{R}^{3\times 3}$. Then, the following statements hold:

i) $\chi_A(s) = s^3 - (\operatorname{tr} A)s^2 + (\operatorname{tr} A^{\mathsf{A}})s - \det A$.

ii) $\operatorname{tr} A^{\mathsf{A}} = \frac{1}{2}[(\operatorname{tr} A)^2 - \operatorname{tr} A^2] = \sum_{i=1}^{3} \det A_{[i,i]}$.

iii) $A^3 = (\operatorname{tr} A)A^2 - (\operatorname{tr} A^{\mathsf{A}})A + (\det A)I = (\operatorname{tr} A)A^2 + \frac{1}{2}[\operatorname{tr} A^2 - (\operatorname{tr} A)^2]A + (\det A)I$.

iv) $\operatorname{tr} A^3 = (\operatorname{tr} A)\operatorname{tr} A^2 - (\operatorname{tr} A^{\mathsf{A}})\operatorname{tr} A + 3(\det A) = \frac{3}{2}(\operatorname{tr} A)\operatorname{tr} A^2 - \frac{1}{2}(\operatorname{tr} A)^3 + 3(\det A)$.

v) $\det A = \frac{1}{3}[\operatorname{tr} A^3 - (\operatorname{tr} A)\operatorname{tr} A^2 + (\operatorname{tr} A^{\mathsf{A}})\operatorname{tr} A] = \frac{1}{3}\operatorname{tr} A^3 - \frac{1}{2}(\operatorname{tr} A)\operatorname{tr} A^2 + \frac{1}{6}(\operatorname{tr} A)^3$.

vi) $(sI - A)^{\mathsf{A}} = s^2 I + s[A - (\operatorname{tr} A)I] + A^2 - (\operatorname{tr} A)A + \frac{1}{2}[(\operatorname{tr} A)^2 - \operatorname{tr} A^2]I$.

vii) $A^{\mathsf{A}} = A^2 - (\operatorname{tr} A)A + \frac{1}{2}[(\operatorname{tr} A)^2 - \operatorname{tr} A^2]I$.

viii) $\det A^{\mathsf{A}} = (\det A)^2$.

ix) If A is singular, then $A^3 = (\operatorname{tr} A)A^2 - (\operatorname{tr} A^{\mathsf{A}})A = (\operatorname{tr} A)A^2 + \frac{1}{2}[\operatorname{tr} A^2 - (\operatorname{tr} A)^2]A$.

x) If $\operatorname{tr} A = 0$, then $\operatorname{tr} A^{\mathsf{A}} = -\frac{1}{2}\operatorname{tr} A^2$ and $\operatorname{tr} A^3 = 3(\det A)$.

xi) If A is nonsingular, then

$$A^{-1} = (\det A)^{-1}[A^2 - (\operatorname{tr} A)A + (\operatorname{tr} A^{\mathsf{A}})I] = (\det A)^{-1}[A^2 - (\operatorname{tr} A)A + \tfrac{1}{2}[(\operatorname{tr} A)^2 - \operatorname{tr} A^2]I,$$

$$\operatorname{tr} A^{-1} = \frac{\operatorname{tr} A^2 - (\operatorname{tr} A)^2 + 3\operatorname{tr} A^{\mathsf{A}}}{\det A} = \frac{(\operatorname{tr} A)^2 - \operatorname{tr} A^2}{2\det A} = \frac{\operatorname{tr} A^{\mathsf{A}}}{\det A}.$$

Related: Fact 3.19.1 and Fact 9.5.18.

Fact 6.9.3. Let $A \in \mathbb{R}^{4\times 4}$. Then, the following statements hold:

i) $\chi_A(s) = s^4 - (\operatorname{tr} A)s^3 + \frac{1}{2}[(\operatorname{tr} A)^2 - \operatorname{tr} A^2]s^2 - (\operatorname{tr} A^{\mathsf{A}})s + \det A$.

ii) $\frac{1}{2}[(\operatorname{tr} A)^2 - \operatorname{tr} A^2] = \sum_{1\le i<j\le 4}^{4} \det A_{(\{i,j\},\{i,j\})}$.

iii) $\operatorname{tr} A^{\mathsf{A}} = \frac{1}{6}[(\operatorname{tr} A)^3 - 3(\operatorname{tr} A)\operatorname{tr} A^2 + 2\operatorname{tr} A^3] = \sum_{i=1}^{4} \det A_{[i,i]}$.

iv) $A^4 = (\operatorname{tr} A)A^3 - \frac{1}{2}[(\operatorname{tr} A)^2 - \operatorname{tr} A^2]A^2 + \frac{1}{6}[(\operatorname{tr} A)^3 - 3(\operatorname{tr} A)\operatorname{tr} A^2 + 2\operatorname{tr} A^3]A - (\det A)I$.

v) $\operatorname{tr} A^4 = (\operatorname{tr} A)\operatorname{tr} A^3 - \frac{1}{2}[(\operatorname{tr} A)^2 - \operatorname{tr} A^2]A^2 + \frac{1}{6}[(\operatorname{tr} A)^3 - 3(\operatorname{tr} A)\operatorname{tr} A^2 + 2\operatorname{tr} A^3]\operatorname{tr} A - 4(\det A)$.

vi) If $\operatorname{tr} A = 0$, then $\operatorname{tr} A^{\mathsf{A}} = \frac{1}{3}\operatorname{tr} A^3$ and $\operatorname{tr} A^4 = \frac{1}{2}(\operatorname{tr} A^2)^2 - 4(\det A)$.

Fact 6.9.4. Let $A \in \mathbb{F}^{n\times n}$, let $\chi_A(s) = s^n + \beta_{n-1}s^{n-1} + \cdots + \beta_0$, and let $\operatorname{mspec}(A) = \{\lambda_1, \ldots, \lambda_n\}_{\mathrm{ms}}$. Then,

$$A^{\mathsf{A}} = (-1)^{n-1}(A^{n-1} + \beta_{n-1}A^{n-2} + \cdots + \beta_1 I).$$

Furthermore,

$$\operatorname{tr} A^{\mathsf{A}} = (-1)^{n-1}\chi_A'(0) = (-1)^{n-1}\beta_1 = \sum_{1\le j_1<\cdots<j_{n-1}\le n} \lambda_{j_1}\cdots\lambda_{j_{n-1}} = \sum_{i=1}^{n} \det A_{[i,i]}.$$

Source: Use $A^{-1}\chi_A(A) = 0$. The second equality follows from either (6.4.20) or Lemma 6.4.8. **Related:** Fact 6.10.13.

Fact 6.9.5. Let $A \in \mathbb{F}^{n\times n}$, assume that A is nonsingular, and let $\chi_A(s) = s^n + \beta_{n-1}s^{n-1} + \cdots + \beta_0$. Then,

$$\chi_{A^{-1}}(s) = \frac{1}{\det A}(-s)^n\chi_A(1/s) = s^n + (\beta_1/\beta_0)s^{n-1} + \cdots + (\beta_{n-1}/\beta_0)s + 1/\beta_0.$$

Related: Fact 7.18.4.

Fact 6.9.6. Let $A \in \mathbb{F}^{n\times n}$, and assume that A and $-A$ are similar. Then, $\chi_A(s) = (-1)^n\chi_A(-s)$. Furthermore, if n is even, then χ_A is even, whereas, if n is odd, then χ_A is odd. **Remark:** A and A^{T} are similar. See Corollary 6.3.12 and Corollary 7.4.9.

Fact 6.9.7. Let $A \in \mathbb{F}^{n\times n}$. Then, for all $s \in \mathbb{C}$,

$$(sI - A)^{\mathsf{A}} = \chi_A(s)(sI - A)^{-1} = \sum_{i=0}^{n-1} \chi_A^{[i]}(s)A^i,$$

where $\chi_A(s) = s^n + \beta_{n-1}s^{n-1} + \cdots + \beta_1 s + \beta_0$ and, for all $i \in \{0, \ldots, n-1\}$, the polynomial $\chi_A^{[i]}$ is defined by

$$\chi_A^{[i]}(s) \triangleq s^{n-i} + \beta_{n-1}s^{n-1-i} + \cdots + \beta_{i+1}.$$

Furthermore,

$$\chi_A^{[n-1]}(s) = s + \beta_{n-1}, \quad \chi_A^{[n]}(s) = 1,$$

and, for all $i \in \{0, \ldots, n-1\}$ and with $\chi_A^{[0]} \triangleq \chi_A$, the polynomials $\chi_A^{[i]}$ satisfy the recursion

$$s\chi_A^{[i+1]}(s) = \chi_A^{[i]}(s) - \beta_i.$$

Source: [2912, p. 31].

Fact 6.9.8. For all $n \ge 1$, define $A \in \mathbb{F}^{n\times n}$ by

$$A_n \triangleq \begin{bmatrix} A_{(1,1)} & A_{(1,2)} & A_{(1,3)} & \cdots & A_{(1,n-1)} & A_{(1,n)} \\ -1 & A_{(2,2)} & A_{(2,3)} & \cdots & A_{(2,n-1)} & A_{(2,n)} \\ 0 & -1 & A_{(3,3)} & \ddots & A_{(3,n-1)} & A_{(3,n)} \\ \vdots & \ddots & \ddots & \ddots & \ddots & \vdots \\ 0 & 0 & 0 & \ddots & A_{(n-1,n-1)} & A_{(n-1,n)} \\ 0 & 0 & 0 & \cdots & -1 & A_{(n,n)} \end{bmatrix},$$

let $a_1 \in \mathbb{F}$, and, for all $n \ge 1$, define $a_{n+1} \triangleq \sum_{i=1}^n A_{(i,n)}a_i$. Then, for all $n \ge 1$, $a_{n+1} = (\det A_n)a_1$. **Source:** [1519]. **Remark:** $(\det A_n)a_1$ solves the recursion.

Fact 6.9.9. Let $n \geq 1$. Then, the nth derangement, Catalan, and Bell numbers are given by

$$d_n = \det \begin{bmatrix} 1 & 1 & 0 & 0 & \cdots & 0 & 0 \\ -1 & 1 & 2 & 0 & \cdots & 0 & 0 \\ 0 & 0 & -1 & 3 & \ddots & 0 & 0 \\ \vdots & \ddots & \ddots & \ddots & \ddots & \ddots & \vdots \\ 0 & 0 & 0 & 0 & \ddots & n-2 & n-1 \\ 0 & 0 & 0 & 0 & \cdots & -1 & n-1 \end{bmatrix},$$

$$C_n = \det \begin{bmatrix} C_0 & C_1 & C_2 & \cdots & C_{n-2} & C_{n-1} \\ -1 & C_0 & C_1 & \cdots & C_{n-3} & C_{n-2} \\ 0 & -1 & C_0 & \ddots & C_{n-4} & C_{n-3} \\ \vdots & \ddots & \ddots & \ddots & \ddots & \vdots \\ 0 & 0 & 0 & \ddots & C_0 & C_1 \\ 0 & 0 & 0 & \cdots & -1 & C_0 \end{bmatrix}, \quad \mathcal{B}_n = \det \begin{bmatrix} \binom{0}{0} & \binom{1}{0} & \binom{2}{0} & \cdots & \binom{n-2}{0} & \binom{n-1}{0} \\ -1 & \binom{1}{1} & \binom{2}{1} & \cdots & \binom{n-2}{1} & \binom{n-1}{1} \\ 0 & -1 & \binom{0}{0} & \ddots & \binom{n-2}{2} & \binom{n-1}{2} \\ \vdots & \ddots & \ddots & \ddots & \ddots & \vdots \\ 0 & 0 & 0 & \ddots & \binom{n-2}{n-2} & \binom{n-1}{n-2} \\ 0 & 0 & 0 & \cdots & -1 & \binom{n-1}{n-1} \end{bmatrix}.$$

Furthermore,

$$n! = \det \begin{bmatrix} 1 & 1 & 1 & \cdots & 1 & 1 \\ -1 & 0 & 0 & \cdots & 0 & 0 \\ 0 & -1 & 1 & \ddots & 1 & 1 \\ \vdots & \ddots & \ddots & \ddots & \ddots & \vdots \\ 0 & 0 & 0 & \ddots & n-3 & n-2 \\ 0 & 0 & 0 & \cdots & -1 & n-1 \end{bmatrix}.$$

Next, let $k \geq 1$. Then,

$$(k+n-1)^{\underline{n-1}} = \det \begin{bmatrix} 1 & 1 & 1 & \cdots & 1 & 1 \\ -1 & k & k & \cdots & k & k \\ 0 & -1 & k+1 & \ddots & k+1 & k+1 \\ \vdots & \ddots & \ddots & \ddots & \ddots & \vdots \\ 0 & 0 & 0 & \ddots & k+n-3 & k+n-3 \\ 0 & 0 & 0 & \cdots & -1 & k+n-2 \end{bmatrix}.$$

Finally, let T_n denote the nth Chebyshev polynomial of the first kind. Then,

$$T_n(s) = \det \begin{bmatrix} s & -1 & 0 & \cdots & 0 & 0 \\ -1 & 2s & -1 & \cdots & 0 & 0 \\ 0 & -1 & 2s & \ddots & 0 & 0 \\ \vdots & \ddots & \ddots & \ddots & \ddots & \vdots \\ 0 & 0 & 0 & \ddots & 2s & -1 \\ 0 & 0 & 0 & \cdots & -1 & 2s \end{bmatrix}.$$

Source: [1520] and Fact 6.9.8. **Related:** Fact 1.18.4, Fact 1.19.6, and Fact 13.2.6.

Fact 6.9.10. Define $A \in \mathbb{F}[s]^{(n+1)\times(n+1)}$ by

$$A(s) \triangleq \begin{bmatrix} a_n & a_{n-1} & a_{n-2} & \cdots & a_1 & a_0 \\ -1 & s & 0 & \cdots & 0 & 0 \\ 0 & -1 & s & \ddots & 0 & 0 \\ \vdots & \ddots & \ddots & \ddots & \ddots & \vdots \\ 0 & 0 & 0 & \ddots & s & 0 \\ 0 & 0 & 0 & \cdots & -1 & s \end{bmatrix}.$$

Then,

$$\det A(s) = \sum_{i=0}^{n} a_i s^i.$$

Source: Fact 6.9.8, [588, p. 95], and [1519]. **Remark:** $A(s)$ is an upper Hessenberg matrix of size $n+1$, and $\deg p = n$, where $p(s) \triangleq \det A(s)$. In contrast, a companion matrix whose characteristic polynomial is $\det A(s)$ is of size n.

Fact 6.9.11. Let $A \in \mathbb{R}^{n\times n}$, and assume that A is skew symmetric. If n is even, then χ_A is even, whereas, if n is odd, then χ_A is odd.

Fact 6.9.12. Let $A \in \mathbb{F}^{n\times n}$, and assume that A is skew Hermitian. Then, for all $s \in \mathbb{C}$,

$$\chi_A(-s) = (-1)^n \overline{p(\overline{s})}.$$

Fact 6.9.13. Let $A \in \mathbb{F}^{n\times n}$. Then, $\chi_{\mathcal{A}}$ is even for the matrices $\mathcal{A} \in \mathbb{F}^{2n\times 2n}$ given by $\begin{bmatrix} 0 & A \\ A^* & 0 \end{bmatrix}$, $\begin{bmatrix} A & 0 \\ 0 & -A \end{bmatrix}$, and $\begin{bmatrix} A & 0 \\ 0 & -A^* \end{bmatrix}$.

Fact 6.9.14. Let $A, B \in \mathbb{F}^{n\times n}$, and define $\mathcal{A} \triangleq \begin{bmatrix} 0 & A \\ B & 0 \end{bmatrix}$. Then, $\chi_{\mathcal{A}}(s) = \chi_{AB}(s^2) = \chi_{BA}(s^2)$. Consequently, $\chi_{\mathcal{A}}$ is even. **Source:** Fact 3.17.14 and Proposition 6.4.10.

Fact 6.9.15. Let $x, y, z, w \in \mathbb{F}^n$, and define $A \triangleq xy^{\mathsf{T}}$ and $B \triangleq xy^{\mathsf{T}} + zw^{\mathsf{T}}$. Then,

$$\chi_A(s) = s^{n-1}(s - x^{\mathsf{T}}y), \quad \chi_B(s) = s^{n-2}[s^2 - (x^{\mathsf{T}}y + z^{\mathsf{T}}w)s + x^{\mathsf{T}}yz^{\mathsf{T}}w - y^{\mathsf{T}}zx^{\mathsf{T}}w].$$

Related: Fact 7.12.16.

Fact 6.9.16. Let $x, y \in \mathbb{F}^{n-1}$, and define $A \in \mathbb{F}^{n\times n}$ by

$$A \triangleq \begin{bmatrix} 0 & x^{\mathsf{T}} \\ y & 0 \end{bmatrix}.$$

Then,

$$\chi_A(s) = s^{n-2}(s^2 - y^{\mathsf{T}}x).$$

Source: [2721].

Fact 6.9.17. Let $x, y, z, w \in \mathbb{F}^{n-1}$, and define $A \in \mathbb{F}^{n\times n}$ by

$$A \triangleq \begin{bmatrix} 1 & x^{\mathsf{T}} \\ y & zw^{\mathsf{T}} \end{bmatrix}.$$

Then,

$$\chi_A(s) = s^{n-3}[s^3 - (1 + w^{\mathsf{T}}z)s^2 + (w^{\mathsf{T}}z - x^{\mathsf{T}}y)s + w^{\mathsf{T}}zx^{\mathsf{T}}y - x^{\mathsf{T}}zw^{\mathsf{T}}y].$$

Source: [907]. **Remark:** Extensions are given in [2721].

Fact 6.9.18. Let $x \in \mathbb{R}^3$, and define $\theta \triangleq \sqrt{x^{\mathsf{T}}x}$. Then,

$$\chi_{K(x)}(s) = s^3 + \theta^2 s.$$

Hence,

$$\operatorname{mspec}[K(x)] = \{0, \jmath\theta, -\jmath\theta\}_{\rm ms}, \quad \operatorname{mspec}[K(x)K^{\mathsf{T}}(x)] = \{\theta^2, \theta^2, 0\}_{\rm ms}.$$

Now, assume, in addition, that $x \neq 0$. Then, x is an eigenvector corresponding to the eigenvalue 0; that is, $K(x)x = 0$. Furthermore, if either $x_{(1)} \neq 0$ or $x_{(2)} \neq 0$, then

$$\begin{bmatrix} x_{(1)}x_{(3)} + \jmath\theta x_{(2)} \\ x_{(2)}x_{(3)} - \jmath\theta x_{(1)} \\ -x_{(1)}^2 - x_{(2)}^2 \end{bmatrix}$$

is an eigenvector corresponding to the eigenvalue $\theta\jmath$. Finally, if $x_{(1)} = x_{(2)} = 0$, then $\left[\begin{smallmatrix} \jmath \\ 1 \\ 0 \end{smallmatrix}\right]$ is an eigenvector corresponding to the eigenvalue $\theta\jmath$. **Related:** Fact 4.12.1 and Fact 15.12.6.

Fact 6.9.19. Let $a, b \in \mathbb{R}^3$, where $a = [a_1 \ \ a_2 \ \ a_3]^{\mathsf{T}}$ and $b = [b_1 \ \ b_2 \ \ b_3]^{\mathsf{T}}$, and define the skew-symmetric matrix $A \in \mathbb{R}^{4\times4}$ by

$$A \triangleq \begin{bmatrix} K(a) & b \\ -b^{\mathsf{T}} & 0 \end{bmatrix}.$$

Then, the following statements hold:

i) $\det A = (a^{\mathsf{T}}b)^2$.

ii) $\chi_A(s) = s^4 + (a^{\mathsf{T}}a + b^{\mathsf{T}}b)s^2 + (a^{\mathsf{T}}b)^2$.

iii) $A^{\mathsf{A}} = -a^{\mathsf{T}}b\begin{bmatrix} K(b) & a \\ -a^{\mathsf{T}} & 0 \end{bmatrix}$.

iv) If $\det A \neq 0$, then $A^{-1} = -(a^{\mathsf{T}}b)^{-1}\begin{bmatrix} K(b) & a \\ -a^{\mathsf{T}} & 0 \end{bmatrix}$.

v) If $\det A = 0$, then $A^3 = -(a^{\mathsf{T}}a + b^{\mathsf{T}}b)^2 A$ and $A^+ = -(a^{\mathsf{T}}a + b^{\mathsf{T}}b)^{-2}A$.

Source: [2723]. **Related:** Fact 6.10.8 and Fact 15.12.18.

Fact 6.9.20. Let $A \in \mathbb{R}^{2n\times2n}$, and assume that A is Hamiltonian. Then, χ_A is even, and thus $\operatorname{mspec}(A) = -\operatorname{mspec}(A)$. **Related:** Fact 6.10.19 and Fact 7.10.25.

Fact 6.9.21. Let $A, B \in \mathbb{F}^{2\times2}$. Then,

$$AB + BA - (\operatorname{tr} A)B - (\operatorname{tr} B)A + [(\operatorname{tr} A)\operatorname{tr} B - \operatorname{tr} AB]I = 0,$$

$$\det(A + B) - \det A - \det B = (\operatorname{tr} A)\operatorname{tr} B - \operatorname{tr} AB.$$

Source: Apply the Cayley-Hamilton theorem to $A + xB$, differentiate with respect to x, and set $x = 0$. For the second equality, evaluate the Cayley-Hamilton theorem with $A + B$. See [273, 1065, 1066, 1791, 2313] and [2423, p. 37]. **Remark:** This is a *polarized Cayley-Hamilton theorem.* See [180, 181, 1791, 2313]. **Related:** Fact 6.9.22 and Fact 6.9.28.

Fact 6.9.22. Let $A, B, C \in \mathbb{F}^{2\times2}$. Then,

$$\begin{aligned} 2ABC = {} & (\operatorname{tr} A)BC + (\operatorname{tr} B)AC + (\operatorname{tr} C)AB - (\operatorname{tr} AC)B + [\operatorname{tr} AB - (\operatorname{tr} A)\operatorname{tr} B]C \\ & + [\operatorname{tr} BC - (\operatorname{tr} B)\operatorname{tr} C]A - [\operatorname{tr} ACB - (\operatorname{tr} AC)\operatorname{tr} B]I. \end{aligned}$$

Remark: An analogous formula exists for the product of six 3×3 matrices. See [180]. **Related:** Fact 6.9.21 and Fact 6.9.28.

Fact 6.9.23. Let $A, B \in \mathbb{F}^{2\times2}$. Then, $\det(A - B) = \det A + \det B - \operatorname{tr} AB^{\mathsf{A}}$. **Source:** [1611].

Fact 6.9.24. Let $A_1, \ldots, A_n \in \mathbb{F}^{2\times 2}$. Then,

$$\sum \det\left(\sum_{i=1}^{n} \alpha_i A_i\right) = 2^n \sum_{i=1}^{n} \det A_i,$$

where the first sum is taken over all n-tuples $(\alpha_1, \ldots, \alpha_n)$ whose components are ± 1. In particular, if $A, B \in \mathbb{F}^{2\times 2}$, then

$$\det(A + B) + \det(A - B) = 2(\det A + \det B).$$

Source: [365].

Fact 6.9.25. $A, B \in \mathbb{F}^{2\times 2}$, assume that $\det A = \det B$, and let $\alpha, \beta \in \mathbb{R}$. Then, $\det(\alpha A + \beta B) = \det(\beta A + \alpha B)$. **Source:** [372].

Fact 6.9.26. Let $A, B \in \mathbb{F}^{2\times 2}$, and let $z \in \mathbb{C}$. Then,

$$\det(A + zB) = \det A + z \operatorname{tr} B(AJ_2)^{\mathsf{T}} J_2 + z^2 \det B.$$

Source: [363].

Fact 6.9.27. Let $A, B \in \mathbb{F}^{3\times 3}$, and assume that $\operatorname{tr} A = \operatorname{tr} B = 0$. Then,

$$4 \operatorname{tr} A^2B^2 + 2 \operatorname{tr} (AB)^2 = (\operatorname{tr} A^2) \operatorname{tr} B^2 + 2(\operatorname{tr} AB)^2,$$

$$6 \operatorname{tr} A^2B^2AB + 6 \operatorname{tr} B^2A^2BA + 2(\operatorname{tr} AB) \operatorname{tr} (AB)^2 + 2(\operatorname{tr} A^3) \operatorname{tr} B^3$$
$$= 2(\operatorname{tr} AB) \operatorname{tr} A^2B^2 + (\operatorname{tr} A^2)(\operatorname{tr} AB) \operatorname{tr} B^2 + 2(\operatorname{tr} AB)^3 + 6(\operatorname{tr} A^2B) \operatorname{tr} AB^2.$$

Source: [183].

Fact 6.9.28. Let $A, B, C \in \mathbb{F}^{3\times 3}$. Then,

$$\sum [A'B'C' - (\operatorname{tr} A')B'C' + (\operatorname{tr} A')(\operatorname{tr} B')C' - (\operatorname{tr} A'B')C']$$
$$- [(\operatorname{tr} A)(\operatorname{tr} B)\operatorname{tr} C - (\operatorname{tr} A)\operatorname{tr} BC - (\operatorname{tr} B)\operatorname{tr} CA - (\operatorname{tr} C)\operatorname{tr} AB + \operatorname{tr} ABC + \operatorname{tr} CBA]I = 0,$$

where the sum is taken over all six permutations A', B', C' of A, B, C. **Related:** Fact 6.9.21 and Fact 6.9.22.

Fact 6.9.29. Let $A, B \in \mathbb{F}^{n\times n}$, assume that A and B commute, and define $f\colon\ \mathbb{C}^2 \mapsto \mathbb{C}$ by $f(r, s) \triangleq \det(rA - sB)$. Then, $f(B, A) = 0$. **Remark:** This is the *generalized Cayley-Hamilton theorem*. See [794, 1390] and [1391, p. 30]. The noncommuting case is considered in [2031].

Fact 6.9.30. Let $B_0, \ldots, B_k \in \mathbb{F}^{n\times n}$, define $P \in \mathbb{F}[s]^{n\times n}$ by $P(s) \triangleq s^kB_k + s^{k-1}B_{k-1} + \cdots + sB_1 + B_0$, define $p \in \mathbb{F}[s]$ by $p(s) \triangleq \det P(s)$, let $A \in \mathbb{F}^{n\times n}$, and assume that $P_{\mathrm{R}}(A) = 0$. Then, $p(A) = 0$. **Source:** [1579].

Fact 6.9.31. Let $A, B \in \mathbb{F}^{n\times n}$, let k be a positive integer, and assume that

$$\sum_{i=0}^{k} (-1)^i \binom{k}{i} A^{k-i}B^i = 0.$$

Then, $\chi_A = \chi_B$. Hence, $\operatorname{tr} A = \operatorname{tr} B$ and $\det A = \det B$. If, in addition, $n = k = 2$, then $AB = BA$. **Source:** [529]. **Remark:** Aside from $n = k = 2$, A and B do not necessarily commute.

Fact 6.9.32. Let $A, B, C \in \mathbb{R}^{n\times n}$, and define

$$\mathcal{A} \triangleq \begin{bmatrix} A & B \\ C & -A^{\mathsf{T}} \end{bmatrix}.$$

If B and C are symmetric, then $\mathcal{A}$ is Hamiltonian. If B and C are skew symmetric, then $\chi_{\mathcal{A}}$ is even, although $\mathcal{A}$ is not necessarily Hamiltonian. **Source:** For the second result, replace J_{2n} with $\left[\begin{smallmatrix} 0 & I_n \\ I_n & 0 \end{smallmatrix}\right]$.

Fact 6.9.33. Let $A \in \mathbb{R}^{n\times n}$, $R \in \mathbb{R}^{n\times n}$, and $B \in \mathbb{R}^{n\times m}$, and define $\mathcal{A} \in \mathbb{R}^{2n\times 2n}$ by

$$\mathcal{A} \triangleq \begin{bmatrix} A & BB^{\mathsf{T}} \\ R & -A^{\mathsf{T}} \end{bmatrix}.$$

Then, for all $s \notin \operatorname{spec}(A)$,

$$\chi_{\mathcal{A}}(s) = (-1)^n \chi_A(s)\chi_A(-s)\det[I + B^{\mathsf{T}}(-sI - A^{\mathsf{T}})^{-1}R(sI - A)^{-1}B].$$

Now, assume, in addition, that R is symmetric. Then, $\mathcal{A}$ is Hamiltonian, and $\chi_{\mathcal{A}}$ is even. If, in addition, R is positive semidefinite, then $(-1)^n\chi_{\mathcal{A}}$ has a spectral factorization. **Source:** Using (3.9.8) and (3.9.15), it follows that, for all $\pm s \notin \operatorname{spec}(A)$,

$$\begin{aligned}\chi_{\mathcal{A}}(s) &= \det(sI - A)\det[sI + A^{\mathsf{T}} - R(sI - A)^{-1}BB^{\mathsf{T}}]\\ &= (-1)^n\chi_A(s)\chi_A(-s)\det[I - B^{\mathsf{T}}(sI + A^{\mathsf{T}})^{-1}R(sI - A)^{-1}B].\end{aligned}$$

To prove the second statement, note that, for all $\omega \in \mathbb{R}$ such that $\omega \jmath \notin \operatorname{spec}(A)$, it follows that

$$\chi_{\mathcal{A}}(\omega \jmath) = (-1)^n \chi_A(\omega \jmath)\overline{\chi_A(\omega \jmath)}\det[I + B^{\mathsf{T}}(\omega \jmath I - A)^{-*}R(\omega \jmath I - A)^{-1}B].$$

Thus, $(-1)^n\chi_{\mathcal{A}}(\omega\jmath) \geq 0$. By continuity, $(-1)^n\chi_{\mathcal{A}}(\omega\jmath) \geq 0$ for all $\omega \in \mathbb{R}$. Now, Proposition 6.1.1 implies that $(-1)^n\chi_{\mathcal{A}}$ has a spectral factorization. **Remark:** Not all Hamiltonian matrices $\mathcal{A} \in \mathbb{R}^{2n\times 2n}$ have the property that $(-1)^n\chi_{\mathcal{A}}$ has a spectral factorization. Consider $\begin{bmatrix} 0 & 0 & 1 & 0 \\ 0 & 0 & 0 & 1 \\ -1 & 0 & 0 & 0 \\ 0 & -3 & 0 & 0 \end{bmatrix}$, whose spectrum is $\{\jmath, -\jmath, \sqrt{3}\jmath, -\sqrt{3}\jmath\}$. **Related:** Fact 4.28.6 and Proposition 16.17.8.

Fact 6.9.34. Let $A \in \mathbb{F}^{n\times n}$. Then, $\mu_A = \chi_A$ if and only if there exists a unique monic polynomial $p \in \mathbb{F}[s]$ of degree n such that $p(A) = 0$. **Source:** To prove necessity, note that, if $\hat{p} \neq p$ is monic, of degree n, and satisfies $\hat{p}(A) = 0$, then $p - \hat{p}$ is nonzero, has degree less than n, and satisfies $(p - \hat{p})(A) = 0$. Conversely, if $\mu_A \neq \chi_A$, then $\mu_A + \chi_A$ is monic, has degree n, and satisfies $(\mu_A + \chi_A)(A) = 0$.

Fact 6.9.35. Let $A \in \mathbb{F}^{n\times n}$, let $\mu_A(s) = s^m + \beta_{m-1}s^{m-1} + \cdots + \beta_1 s + \beta_0$, and define $k \triangleq \operatorname{ind} A$. Then, $\beta_0 = \beta_1 = \cdots = \beta_{k-1} = 0$ and $\beta_k \neq 0$. **Source:** [2238, p. 228].

Fact 6.9.36. Let $A \in \mathbb{F}^{n\times n}$, let $\operatorname{mspec}(A) = \{\lambda_1, \ldots, \lambda_n\}_{\mathrm{ms}}$, and let $p \in \mathbb{F}[s]$. Then, the following statements hold:

i) $\operatorname{mspec}[p(A)] = \{p(\lambda_1), \ldots, p(\lambda_n)\}_{\mathrm{ms}}$.

ii) $\operatorname{roots}(p) \cap \operatorname{spec}(A) = \varnothing$ if and only if $p(A)$ is nonsingular.

iii) μ_A divides p if and only if $p(A) = 0$.

Fact 6.9.37. Let $1 \leq n \leq m$, $A \in \mathbb{F}^{n\times m}$, $B \in \mathbb{F}^{m\times n}$, and $\chi_{AB}(s) = s^n + \beta_{n-1}s^{n-1} + \cdots + \beta_1 s + \beta_0$. Then, for all $i \in \{0, \ldots, n-1\}$,

$$\beta_i = (-1)^{n-i}\sum \det A_{(\mathcal{S}_1,\mathcal{S}_2)} \det B_{(\mathcal{S}_2,\mathcal{S}_1)},$$

where the sum is taken over all subsets $\mathcal{S}$ of $\{1, \ldots, m\}$ having $n-i$ elements. **Source:** [1428, 1646]. **Remark:** This is a *generalized Cauchy-Binet formula*. **Related:** Fact 3.16.8.

Fact 6.9.38. Let $A, B, M, X \in \mathbb{F}^{n\times n}$, and assume that $AM = MB$ and $\chi_A = \chi_B$. Then, $\det(A - MX) = \det(B - XM)$. **Source:** [1643].

6.10 Facts on the Spectrum

Fact 6.10.1. Let $A \in \mathbb{C}^{2\times 2}$, let $\operatorname{mspec}(A) = \{\alpha, \beta\}_{\mathrm{ms}}$, and let $n \geq 0$. Then,

$$A^n = \begin{cases} \frac{\alpha^n}{\alpha-\beta}(A - \beta I) + \frac{\beta^n}{\beta-\alpha}(A - \alpha I), & \alpha \neq \beta, \\ \alpha^{n-1}[nA - (n-1)\alpha I], & \alpha = \beta. \end{cases}$$

If, in addition, A is nonsingular, then, for all $n \in \mathbb{Z}$, A^n is given by the above expressions. **Source:** [1669, 2884].

Fact 6.10.2. Let $A = \begin{bmatrix} a & b \\ c & d \end{bmatrix} \in \mathbb{C}^{2\times 2}$, let $n \geq 1$, and, for all $k \in \{n-1, n\}$, define

$$\delta_k \triangleq \sum_{i=0}^{\lfloor k/2 \rfloor} (-1)^i \binom{k-i}{i} (\operatorname{tr} A)^{k-2i} (\det A)^i.$$

Then,

$$A^n = \begin{bmatrix} \delta_n - d\delta_{n-1} & b\delta_{n-1} \\ c\delta_{n-1} & \delta_n - a\delta_{n-1} \end{bmatrix}.$$

Source: [1999].

Fact 6.10.3. Let $\alpha \in \mathbb{C}$, and assume that $|\alpha| \leq 1$. Then,

$$\operatorname{spec}\left(\begin{bmatrix} 1 & \alpha \\ \overline{\alpha} & 1 \end{bmatrix}\right) \subseteq [1 - |\alpha|, 1 + |\alpha|].$$

Source: [2991, p. 250].

Fact 6.10.4. Let $A \in \mathbb{C}^{2\times 2}$, and assume that A is Hermitian. Then,

$$2|A_{(1,2)}| \leq \lambda_1(A) - \lambda_2(A).$$

Source: [2991, p. 250].

Fact 6.10.5. Let $A \in \mathbb{F}^{n\times n}$. Then, $\operatorname{rank} A = 1$ if and only if $\operatorname{gmult}_A(0) = n-1$. If these conditions hold, then $\operatorname{mspec}(A) = \{\operatorname{tr} A, 0, \ldots, 0\}_{\mathrm{ms}}$. **Source:** Proposition 6.5.2. **Related:** Fact 3.13.26.

Fact 6.10.6. Let $A \in \mathbb{R}^{n\times n}$, and assume that A is row stochastic. Then, $1 \in \operatorname{spec}(A)$.

Fact 6.10.7. Let $A \in \mathbb{F}^{3\times 3}$, assume that A is symmetric, let $\lambda_1, \lambda_2, \lambda_3 \in \mathbb{R}$ denote the eigenvalues of A, where $\lambda_1 \geq \lambda_2 \geq \lambda_3$, and define

$$p \triangleq \tfrac{1}{6} \operatorname{tr} [A - \tfrac{1}{3}(\operatorname{tr} A)I]^2, \quad q \triangleq \tfrac{1}{2} \det[A - \tfrac{1}{3}(\operatorname{tr} A)I].$$

Then, the following statements hold:

i) $0 \leq |q| \leq p^{3/2}$.

ii) $p = 0$ if and only if $\lambda_1 = \lambda_2 = \lambda_3 = \frac{1}{3} \operatorname{tr} A$.

iii) $p > 0$ if and only if

$$\begin{aligned} \lambda_1 &= \tfrac{1}{3} \operatorname{tr} A + 2\sqrt{p} \cos \phi, \\ \lambda_2 &= \tfrac{1}{3} \operatorname{tr} A + \sqrt{3p} \sin \phi - \sqrt{p} \cos \phi, \\ \lambda_3 &= \tfrac{1}{3} \operatorname{tr} A - \sqrt{3p} \sin \phi - \sqrt{p} \cos \phi, \end{aligned}$$

where $\phi \in [0, \pi/3]$ is given by

$$\phi = \tfrac{1}{3} \operatorname{acos} \frac{q}{p^{3/2}}.$$

iv) $\phi = 0$ if and only if $q = p^{3/2} > 0$. If these conditions hold, then

$$\lambda_1 = \tfrac{1}{3} \operatorname{tr} A + 2\sqrt{p}, \quad \lambda_2 = \lambda_3 = \tfrac{1}{3} \operatorname{tr} A - \sqrt{p}.$$

v) $\phi = \pi/6$ if and only if $p > 0$ and $q = 0$. If these conditions hold, then $\sin \phi = 1/2$, $\cos \phi = \sqrt{3}/2$, and

$$\lambda_1 = \tfrac{1}{3} \operatorname{tr} A + \sqrt{3p}, \quad \lambda_2 = \tfrac{1}{3} \operatorname{tr} A, \quad \lambda_3 = \tfrac{1}{3} \operatorname{tr} A - \sqrt{3p}.$$

vi) $\phi = \pi/3$ if and only if $q = -p^{3/2} < 0$. If these conditions hold, then $\sin\phi = \sqrt{3}/2$, $\cos\phi = 1/2$, and

$$\lambda_1 = \lambda_2 = \tfrac{1}{3}\operatorname{tr} A + \sqrt{p}, \quad \lambda_3 = \tfrac{1}{3}\operatorname{tr} A - 2\sqrt{p}.$$

Source: [2471]. **Remark:** This result is based on *Cardano's trigonometric solution* for the roots of a cubic polynomial. See [502], [1083, Lecture 4], and [2471]. **Remark:** $q^2 \le p^3$ follows from Fact 2.2.27. **Related:** Fact 2.21.2.

Fact 6.10.8. Let $a, b, c, d, \omega \in \mathbb{R}$, and define the skew-symmetric matrix $A \in \mathbb{R}^{4\times 4}$ given by

$$A \triangleq \begin{bmatrix} 0 & \omega & a & b \\ -\omega & 0 & c & d \\ -a & -c & 0 & \omega \\ -b & -d & -\omega & 0 \end{bmatrix}.$$

Then,

$$\chi_A(s) = s^4 + (2\omega^2 + a^2 + b^2 + c^2 + d^2)s^2 + [\omega^2 - (ad - bc)]^2, \quad \det A = [\omega^2 - (ad - bc)]^2.$$

Hence, A is singular if and only if $bc \le ad$ and $\omega = \sqrt{ad - bc}$. Furthermore, A has a repeated eigenvalue if and only if either *i)* A is singular or *ii)* $a = -d$ and $b = c$. In case *i)*, A has the repeated eigenvalue 0, whereas, in case *ii)*, A has the repeated eigenvalues $\sqrt{\omega^2 + a^2 + b^2}\,\jmath$ and $-\sqrt{\omega^2 + a^2 + b^2}\,\jmath$. Finally, cases *i)* and *ii)* cannot occur simultaneously. **Related:** Fact 4.10.35, Fact 6.9.19, Fact 15.12.16, and Fact 15.12.18.

Fact 6.10.9. Define $A, B \in \mathbb{R}^{n\times n}$ by

$$A \triangleq \begin{bmatrix} 1 & -2 & 0 & \cdots & 0 & 0 \\ 0 & 1 & -2 & \ddots & 0 & 0 \\ 0 & 0 & 1 & \ddots & 0 & 0 \\ \vdots & \vdots & \ddots & \ddots & \ddots & \vdots \\ 0 & 0 & 0 & \ddots & 1 & -2 \\ 0 & 0 & 0 & \cdots & 0 & 1 \end{bmatrix}, \quad B \triangleq \begin{bmatrix} 1 & -2 & 0 & \cdots & 0 & 0 \\ 0 & 1 & -2 & \ddots & 0 & 0 \\ 0 & 0 & 1 & \ddots & 0 & 0 \\ \vdots & \vdots & \ddots & \ddots & \ddots & \vdots \\ 0 & 0 & 0 & \ddots & 1 & -2 \\ \alpha & 0 & 0 & \cdots & 0 & 1 \end{bmatrix},$$

where $\alpha \triangleq -1/2^{n-1}$. Then, $\operatorname{spec}(A) = \{1\}$ and $\det B = 0$.

Fact 6.10.10. Let $A \in \mathbb{F}^{n\times n}$. Then, $|\alpha_{\max}(A)| \le \rho_{\max}(A)$.

Fact 6.10.11. Let $A \in \mathbb{F}^{n\times n}$, assume that A is nonsingular, and assume that $\rho_{\max}(I - A) < 1$. Then,

$$A^{-1} = \sum_{k=0}^{\infty} (I - A)^k.$$

Fact 6.10.12. Let $A \in \mathbb{F}^{n\times n}$ and $B \in \mathbb{F}^{m\times m}$. If $\operatorname{tr} A^k = \operatorname{tr} B^k$ for all $k \in \{1, \ldots, \max\{m, n\}\}$, then A and B have the same nonzero eigenvalues with the same algebraic multiplicity. Now, assume, in addition, that $n = m$. Then, $\operatorname{tr} A^k = \operatorname{tr} B^k$ for all $k \in \{1, \ldots, n\}$ if and only if $\operatorname{mspec}(A) = \operatorname{mspec}(B)$. **Source:** Fact 6.8.4. **Remark:** Since, for all $k \ge 1$, $\operatorname{tr}(AB)^k = \operatorname{tr}(BA)^k$, this result yields Proposition 6.4.10. **Remark:** Setting $B = 0_{n\times n}$ yields necessity in Fact 3.15.15.

Fact 6.10.13. Let $A \in \mathbb{F}^{n\times n}$, and let $\operatorname{mspec}(A) = \{\lambda_1, \ldots, \lambda_n\}_{\rm ms}$. Then,

$$\operatorname{mspec}(A^{\mathsf{A}}) = \left\{ \prod_{i\in\{2,\ldots,n\}} \lambda_i,\ \prod_{i\in\{1,3,\ldots,n\}} \lambda_i, \ldots, \prod_{i\in\{1,\ldots,n-1\}} \lambda_i \right\}_{\rm ms}.$$

Consequently,

$$\operatorname{mspec}(A^{\mathsf{A}}) = \begin{cases} \left\{\dfrac{\det A}{\lambda_1}, \ldots, \dfrac{\det A}{\lambda_n}\right\}_{\mathrm{ms}}, & \operatorname{rank} A = n, \\ \left\{\displaystyle\sum_{i=1}^{n} \det A_{[i,i]}, 0, \ldots, 0\right\}_{\mathrm{ms}}, & \operatorname{rank} A = n-1, \\ \{0\}, & \operatorname{rank} A \le n-2. \end{cases}$$

In particular, the following statements hold:

i) If $n = 2$, then $\operatorname{mspec}(A^{\mathsf{A}}) = \operatorname{mspec}(A)$.

ii) If $n = 3$, then $\operatorname{mspec}(A^{\mathsf{A}}) = \{\lambda_2\lambda_3, \lambda_1\lambda_3, \lambda_1\lambda_2\}_{\mathrm{ms}}$.

iii) If $n = 4$, then $\operatorname{mspec}(A^{\mathsf{A}}) = \{\lambda_2\lambda_3\lambda_4, \lambda_1\lambda_3\lambda_4, \lambda_1\lambda_2\lambda_3\}_{\mathrm{ms}}$.

Furthermore,

$$\operatorname{tr} A^{\mathsf{A}} = \sum_{i=1}^{n} \det A_{[i,i]} = (-1)^{n-1}\chi_A'(0) = \sum_{i=1}^{n} \prod_{j\in\{1,\ldots,n\}\setminus\{i\}} \lambda_j.$$

Finally, if A is singular and $\lambda_n = 0$, then

$$\operatorname{tr} A^{\mathsf{A}} = \sum_{i=1}^{n} \det A_{[i,i]} = (-1)^{n-1}\chi_A'(0) = \prod_{i=1}^{n-1} \lambda_i.$$

Source: [2263, p. 68]. The expression for $\operatorname{tr} A^{\mathsf{A}}$ is given by (6.4.20). **Remark:** If $\operatorname{rank} A = n-1$, then $\operatorname{mspec}(A) = \{0\}$ is possible. For example, $N_2^{\mathsf{A}} = -N_2$. See Fact 3.19.3. **Remark:** If $\operatorname{rank} A \le n-2$, then $2 \le n - \operatorname{rank} A = \operatorname{def} A \le \operatorname{amult}_A(0)$, and thus $\operatorname{tr} A^{\mathsf{A}} = \sum_{i=1}^{n} \det A_{[i,i]} = 0$. **Related:** Fact 3.19.1, Fact 3.19.3, Fact 4.10.9, Fact 6.9.4, and Fact 7.12.39.

Fact 6.10.14. Let $A \in \mathbb{F}^{n\times n}$, and assume that A is either upper triangular or lower triangular. Then,

$$\chi_A(s) = \prod_{i=1}^{n}(s - A_{(i,i)}), \quad \operatorname{mspec}(A) = \{A_{(1,1)}, \ldots, A_{(n,n)}\}_{\mathrm{ms}}.$$

Related: Fact 4.25.1.

Fact 6.10.15. Let $A \in \mathbb{F}^{n\times n}$, $B \in \mathbb{F}^{n\times m}$, and $C \in \mathbb{F}^{m\times m}$, and let $p \in \mathbb{F}[s]$. Then, there exists $\hat{B} \in \mathbb{F}^{n\times m}$ such that

$$p\left(\begin{bmatrix} A & B \\ 0 & C \end{bmatrix}\right) = \begin{bmatrix} p(A) & \hat{B} \\ 0 & p(C) \end{bmatrix}.$$

Fact 6.10.16. Let $A_1 \in \mathbb{F}^{n\times n}$, $A_{12} \in \mathbb{F}^{n\times m}$, and $A_2 \in \mathbb{F}^{m\times m}$, and define $A \in \mathbb{F}^{(n+m)\times(n+m)}$ by

$$A \triangleq \begin{bmatrix} A_1 & A_{12} \\ 0 & A_2 \end{bmatrix}.$$

Then, $\chi_A = \chi_{A_1}\chi_{A_2}$. Furthermore, there exist $B_1, B_2 \in \mathbb{F}^{n\times m}$ such that

$$\chi_{A_1}(A) = \begin{bmatrix} 0 & B_1 \\ 0 & \chi_{A_1}(A_2) \end{bmatrix}, \quad \chi_{A_2}(A) = \begin{bmatrix} \chi_{A_2}(A_1) & B_2 \\ 0 & 0 \end{bmatrix}.$$

Therefore,

$$\mathcal{R}[\chi_{A_2}(A)] \subseteq \mathcal{R}\left(\begin{bmatrix} I_n \\ 0 \end{bmatrix}\right) \subseteq \mathcal{N}[\chi_{A_1}(A)], \quad \chi_{A_2}(A_1)B_1 + B_2\chi_{A_1}(A_2) = 0.$$

Hence,

$$\chi_A(A) = \chi_{A_1}(A)\chi_{A_2}(A) = \chi_{A_2}(A)\chi_{A_1}(A) = 0.$$

Fact 6.10.17. Let $A_1 \in \mathbb{F}^{n\times n}$, $A_{12} \in \mathbb{F}^{n\times m}$, and $A_2 \in \mathbb{F}^{m\times m}$, assume that $\operatorname{spec}(A_1)$ and $\operatorname{spec}(A_2)$ are disjoint, and define $A \in \mathbb{F}^{(n+m)\times(n+m)}$ by

$$A \triangleq \begin{bmatrix} A_1 & A_{12} \\ 0 & A_2 \end{bmatrix}.$$

Furthermore, let $\mu_1, \mu_2 \in \mathbb{F}[s]$ satisfy

$$\mu_A = \mu_1\mu_2, \quad \operatorname{roots}(\mu_1) = \operatorname{spec}(A_1), \quad \operatorname{roots}(\mu_2) = \operatorname{spec}(A_2).$$

Then, there exist $B_1, B_2 \in \mathbb{F}^{n\times m}$ such that

$$\mu_1(A) = \begin{bmatrix} 0 & B_1 \\ 0 & \mu_1(A_2) \end{bmatrix}, \quad \mu_2(A) = \begin{bmatrix} \mu_2(A_1) & B_2 \\ 0 & 0 \end{bmatrix}.$$

Therefore,

$$\mathcal{R}[\mu_2(A)] \subseteq \mathcal{R}\left(\begin{bmatrix} I_n \\ 0 \end{bmatrix}\right) \subseteq \mathcal{N}[\mu_1(A)], \quad \mu_2(A_1)B_1 + B_2\mu_1(A_2) = 0.$$

Hence,

$$\mu_A(A) = \mu_1(A)\mu_2(A) = \mu_2(A)\mu_1(A) = 0.$$

Fact 6.10.18. Let $A_1, A_2, A_3, A_4, B_1, B_2 \in \mathbb{F}^{n\times n}$, and define $A \in \mathbb{F}^{4n\times 4n}$ by

$$A \triangleq \begin{bmatrix} A_1 & B_1 & 0 & 0 \\ 0 & A_2 & 0 & 0 \\ 0 & 0 & A_3 & 0 \\ 0 & 0 & B_2 & A_4 \end{bmatrix}.$$

Then,

$$\operatorname{mspec}(A) = \bigcup_{i=1}^{4} \operatorname{mspec}(A_i).$$

Fact 6.10.19. Let $A \in \mathbb{F}^{n\times n}$. Then,

$$\operatorname{mspec}\left(\begin{bmatrix} 0 & A \\ A & 0 \end{bmatrix}\right) = \operatorname{mspec}(A) \cup \operatorname{mspec}(-A), \quad \operatorname{mspec}\left(\begin{bmatrix} 0 & A \\ A^* & 0 \end{bmatrix}\right) = -\operatorname{mspec}\left(\begin{bmatrix} 0 & A \\ A^* & 0 \end{bmatrix}\right).$$

Related: Fact 6.9.20 and Fact 7.10.25.

Fact 6.10.20. Let $A \in \mathbb{F}^{n\times m}$ and $B \in \mathbb{F}^{m\times n}$, and assume that $m < n$. Then,

$$\operatorname{mspec}(I_n + AB) = \operatorname{mspec}(I_m + BA) \cup \{1, \dots, 1\}_{\mathrm{ms}}.$$

Fact 6.10.21. Let $a, b \in \mathbb{F}$, and define the symmetric, Toeplitz matrix $A \in \mathbb{F}^{n\times n}$ by

$$A \triangleq \begin{bmatrix} a & b & b & \cdots & b \\ b & a & b & \cdots & b \\ b & b & a & \cdots & b \\ \vdots & \vdots & \vdots & \ddots & \vdots \\ b & b & b & \cdots & a \end{bmatrix}.$$

Then,

$$\operatorname{mspec}(A) = \{a + (n-1)b, a - b, \dots, a - b\}_{\mathrm{ms}}, \quad \operatorname{mspec}(aI_n + b1_{n\times n}) = \{a + nb, a, \dots, a\}_{\mathrm{ms}},$$
$$A1_{n\times 1} = [a + (n-1)b]1_{n\times 1}, \quad A^2 + a_1A + a_0I = 0,$$

where $a_1 \triangleq -2a + (2-n)b$ and $a_0 \triangleq a^2 + (n-2)ab + (1-n)b^2$. **Remark:** For the remaining eigenvectors of A, see [2418, pp. 149, 317]. **Related:** Fact 3.16.19, Fact 10.9.1, and Fact 10.10.39.

Fact 6.10.22. Let $A \in \mathbb{F}^{n\times n}$. Then,

$$\operatorname{spec}(A) \subset \bigcup_{i=1}^{n} \left\{ s \in \mathbb{C}\colon\ |s - A_{(i,i)}| \le \sum_{\substack{j=1\\ j\ne i}}^{n} |A_{(i,j)}| \right\}.$$

Now, let $\lambda \in \operatorname{spec}(A)$, and let $r \triangleq \operatorname{gmult}_A(\lambda)$. Then, there exist $1 \le i_1 < \cdots < i_r \le n$ such that

$$\lambda \in \bigcap_{k=1}^{r} \left\{ s \in \mathbb{C}\colon\ |s - A_{(i_k,i_k)}| \le \sum_{\substack{j=1\\ j\ne i_k}}^{n} |A_{(i_k,j)}| \right\}.$$

Source: [590, 2786]. The last statement is given in [1972]. **Remark:** This is the *Gershgorin circle theorem.* **Remark:** This result yields Corollary 11.4.5 for $\|\cdot\|_{\mathrm{col}}$ and $\|\cdot\|_{\mathrm{row}}$.

Fact 6.10.23. Let $A \in \mathbb{F}^{n\times n}$, and assume that, for all $i \in \{1,\ldots,n\}$, $\sum_{j=1,j\ne i}^{n}|A_{(i,j)}| < |A_{(i,i)}|$. Then, A is nonsingular. **Source:** Fact 6.10.22 and [2979, p. 188]. **Remark:** This is the *diagonal dominance theorem*, and A is *diagonally dominant*. See [2397]. **Related:** Fact 6.10.25.

Fact 6.10.24. Let $A \in \mathbb{F}^{n\times n}$, assume that, for all $i \in \{1,\ldots,n\}$, $A_{(i,i)} \ne 0$, and assume that

$$\alpha_i \triangleq \frac{\sum_{j=1,j\ne i}^{n} |A_{(i,j)}|}{|A_{(i,i)}|} < 1.$$

Then,

$$|A_{(1,1)}| \prod_{i=2}^{n} (|A_{(i,i)}| - l_i + L_i) \le |\det A|,$$

where

$$l_i \triangleq \sum_{j=1}^{i-1} \alpha_j |A_{(i,j)}|, \quad L_i \triangleq \left| \frac{A_{(i,1)}}{A_{(1,1)}} \right| \sum_{j=i+1}^{n} |A_{(i,j)}|.$$

Source: [567]. **Remark:** Note that, for all $i \in \{1,\ldots,n\}$,

$$l_i = \sum_{j=1}^{i-1} \alpha_j |A_{(i,j)}| \le \sum_{j=1,j\ne i}^{n} \alpha_j |A_{(i,j)}| \le \sum_{j=1,j\ne i}^{n} |A_{(i,j)}| = \alpha_i |A_{(i,i)}| < |A_{(i,i)}|.$$

Hence, the lower bound for $|\det A|$ is positive.

Fact 6.10.25. Let $A \in \mathbb{F}^{n\times n}$, and, for all $i \in \{1,\ldots,n\}$, define $r_i \triangleq \sum_{j=1,j\ne i}^{n}|A_{(i,j)}|$ and $c_i \triangleq \sum_{j=1,j\ne i}^{n}|A_{(j,i)}|$. Furthermore, assume that at least one of the following statements holds:

i) For all distinct $i, j \in \{1,\ldots,n\}$, $r_i c_j < |A_{(i,i)}A_{(j,j)}|$.

ii) A is irreducible, for all $i \in \{1,\ldots,n\}$ it follows that $r_i \le |A_{(i,i)}|$, and there exists $i \in \{1,\ldots,n\}$ such that $r_i < |A_{(i,i)}|$.

iii) There exist positive integers $k_1,\ldots,k_n$ such that $\sum_{i=1}^{n}(1+k_i)^{-1} \le 1$ and such that, for all $i \in \{1,\ldots,n\}$, $k_i \max_{j\in\{1,\ldots,n\},j\ne i} |A_{(i,j)}| < |A_{(i,i)}|$.

iv) There exists $\alpha \in [0,1]$ such that, for all $i \in \{1,\ldots,n\}$, $r_i^{\alpha} c_i^{1-\alpha} < |A_{(i,i)}|$.

Then, A is nonsingular. **Source:** [214]. **Remark:** Each statement is stronger than Fact 6.10.23.

Fact 6.10.26. Let $A \in \mathbb{R}^{n\times n}$, assume that A is symmetric, and, for all $i \in \{1,\ldots,n\}$, define $\alpha_i \triangleq \sum_{j=1,j\ne i}^{n} |A_{(i,j)}|$. Then,

$$\operatorname{spec}(A) \subset \bigcup_{i=1}^{n} [A_{(i,i)} - \alpha_i, A_{(i,i)} + \alpha_i].$$

Now, for all $i \in \{1,\ldots,n\}$, let $\beta_i \triangleq \max\{0, \max_{j\in\{1,\ldots,n\},j\neq i} A_{(i,j)}\}$ and $\gamma_i \triangleq \min\{0, \min_{j\in\{1,\ldots,n\},j\neq i} A_{(i,j)}\}$. Then,

$$\operatorname{spec}(A) \subset \bigcup_{i=1}^{n}\left[\left(\sum_{j=1}^{n} A_{(i,j)}\right) - n\beta_i, \left(\sum_{j=1}^{n} A_{(i,j)}\right) - n\gamma_i\right].$$

Source: The first statement is the specialization of the Gershgorin circle theorem to real, symmetric matrices. See Fact 6.10.22. The second result is given in [291].

Fact 6.10.27. Let $A \in \mathbb{F}^{n\times n}$. Then,

$$\operatorname{spec}(A) \subset \bigcup_{\substack{i,j=1\\ i\neq j}}^{n}\left\{s \in \mathbb{C}\colon |s - A_{(i,i)}||s - A_{(j,j)}| \le \sum_{\substack{k=1\\ k\neq i}}^{n}|A_{(i,k)}|\sum_{\substack{k=1\\ k\neq j}}^{n}|A_{(j,k)}|\right\}.$$

Remark: The inclusion region is the *ovals of Cassini*. See [1448, p. 380]. **Credit:** A. Brauer.

Fact 6.10.28. Let $A \in \mathbb{F}^{n\times n}$. Then,

$$\rho_{\min}(A) \le \max_{i\in\{1,\ldots,n\}} |\operatorname{tr} A^i|^{1/i}, \quad \rho_{\max}(A) \le \max_{i\in\{1,\ldots,2n-1\}} |\operatorname{tr} A^i|^{1/i}, \quad \rho_{\max}(A) \le \tfrac{5}{n} \max_{i\in\{1,\ldots,n\}} |\operatorname{tr} A^i|^{1/i}.$$

Remark: These are *Turan's inequalities*. See [2061, p. 657].

Fact 6.10.29. Let $A \in \mathbb{F}^{n\times n}$, and, for all $j \in \{1,\ldots,n\}$, define $b_j \triangleq \sum_{i=1}^{n}|A_{(i,j)}|$. Then,

$$\sum_{j=1}^{n}\frac{|A_{(j,j)}|}{b_j} \le \operatorname{rank} A.$$

Source: [2263, p. 67]. **Remark:** Interpret $0/0$ as 0. **Related:** Fact 6.10.23.

Fact 6.10.30. Let $A_1,\ldots,A_r \in \mathbb{F}^{n\times n}$, assume that $A_1,\ldots,A_r$ are normal, and let $A \in \operatorname{conv}\{A_1,\ldots,A_r\}$. Then,

$$\operatorname{spec}(A) \subseteq \operatorname{conv} \bigcup_{i\in\{1,\ldots,r\}} \operatorname{spec}(A_i).$$

Source: [2829]. **Remark:** The spectrum of a polytope of matrices is considered in [992, 2075, 2282]. **Related:** Fact 10.17.8.

Fact 6.10.31. Let $A, B \in \mathbb{R}^{n\times n}$. Then,

$$\operatorname{mspec}\left(\begin{bmatrix} A & B\\ B & A\end{bmatrix}\right) = \operatorname{mspec}(A + B) \cup \operatorname{mspec}(A - B).$$

Source: [2418, p. 93]. **Related:** Fact 3.17.30.

Fact 6.10.32. Let $A, B \in \mathbb{R}^{n\times n}$. Then,

$$\operatorname{mspec}\left(\begin{bmatrix} A & B\\ -B & A\end{bmatrix}\right) = \operatorname{mspec}(A + \jmath B) \cup \operatorname{mspec}(A - \jmath B).$$

Now, assume, in addition, that A is symmetric and B is skew symmetric. Then, $\left[\begin{smallmatrix} A & B\\ B^{\mathsf{T}} & A\end{smallmatrix}\right]$ is symmetric, $A + \jmath B$ is Hermitian, and

$$\operatorname{mspec}\left(\begin{bmatrix} A & B\\ B^{\mathsf{T}} & A\end{bmatrix}\right) = \operatorname{mspec}(A + \jmath B) \cup \operatorname{mspec}(A + \jmath B).$$

Related: Fact 3.24.6 and Fact 10.19.2.

Fact 6.10.33. Let $A, B \in \mathbb{C}^{n\times n}$, and define

$$M \triangleq \begin{bmatrix} A & B\\ -\overline{B} & \overline{A}\end{bmatrix}.$$

Then, the following statements hold:

i) $\chi_A \in \mathbb{R}[s]$.

ii) mspec(A) is conjugate symmetric.

iii) For all $\lambda \in \operatorname{spec}(A)$, there exist $x, y \in \mathbb{C}^n$ such that $\left[\begin{smallmatrix} x \\ y \end{smallmatrix}\right]$ and $\left[\begin{smallmatrix} -\overline{y} \\ \overline{x} \end{smallmatrix}\right]$ are linearly independent eigenvectors of M associated with λ and $\overline{\lambda}$, respectively.

Source: [2991, p. 106]. **Related:** Fact 3.24.8

Fact 6.10.34. Let $A \in \mathbb{F}^{n\times n}$, $B \in \mathbb{F}^{m\times m}$, and $C \in \mathbb{F}^{n\times m}$, assume that A and B are Hermitian, and define $\mathcal{A}_0 \triangleq \left[\begin{smallmatrix} A & 0 \\ 0 & B \end{smallmatrix}\right]$ and $\mathcal{A} \triangleq \left[\begin{smallmatrix} A & C \\ C^* & B \end{smallmatrix}\right]$. Furthermore, define

$$\eta \triangleq \min_{\substack{i=1,\ldots,n \\ j=1,\ldots,m}} |\lambda_i(A) - \lambda_j(B)|.$$

Then, for all $i \in \{1, \ldots, n+m\}$,

$$|\lambda_i(\mathcal{A}) - \lambda_i(\mathcal{A}_0)| \le \frac{2\sigma_{\max}^2(C)}{\eta + \sqrt{\eta^2 + 4\sigma_{\max}(C)}}.$$

Source: [453, pp. 142–146] and [1797].

Fact 6.10.35. Let $A \in \mathbb{R}^{n\times n}$, let $b, c \in \mathbb{R}^n$, define $p \in \mathbb{R}[s]$ by $p(s) \triangleq c^{\mathsf{T}}(sI - A)^{\mathsf{A}}b$, and assume that p and $\det(sI - A)$ are coprime. Furthermore, for all $\alpha \in [0, \infty)$, define $A_\alpha \triangleq A - \alpha bc^{\mathsf{T}}$, and let $\lambda\colon\ [0, \infty) \to \mathbb{C}$ be a continuous function such that, for all $\alpha \in [0, \infty)$, $\lambda(\alpha) \in \operatorname{spec}(A_\alpha)$. Then, either $\lim_{\alpha\to\infty} |\lambda(\alpha)| = \infty$ or $\lim_{\alpha\to\infty} \lambda(\alpha) \in \operatorname{roots}(p)$. **Remark:** This result is a consequence of *root locus* analysis from classical control theory, which determines asymptotic pole locations under high-gain feedback. In particular, the loop transfer function is $L(s) = c(sI - A)^{-1}b$ with feedback gain $-\alpha$.

Fact 6.10.36. Let $A \in \mathbb{F}^{n\times n}$, where $n \ge 2$, and assume that there exist $\alpha \in [0, \infty)$ and $B \in \mathbb{F}^{n\times n}$ such that $A = \alpha I - B$ and $\rho_{\max}(B) \le \alpha$. Then, $\operatorname{spec}(A) \subset \{0\} \cup \text{ORHP}$. If, in addition, $\rho_{\max}(B) < \alpha$, then $\operatorname{spec}(A) \subset \text{ORHP}$, and thus A is nonsingular. **Source:** Let $\lambda \in \operatorname{spec}(A)$. Then, there exists $\mu \in \operatorname{spec}(B)$ such that $\lambda = \alpha - \mu$. Hence, $\operatorname{Re}\lambda = \alpha - \operatorname{Re}\mu$. Since $\operatorname{Re}\mu \le |\operatorname{Re}\mu| \le |\mu| \le \rho_{\max}(B)$, it follows that $\operatorname{Re}\lambda \ge \alpha - |\operatorname{Re}\mu| \ge \alpha - |\mu| \ge \alpha - \rho_{\max}(B) \ge 0$. Hence, $\operatorname{Re}\lambda \ge 0$. Now, suppose that $\operatorname{Re}\lambda = 0$. Then, since $\alpha - \lambda = \mu \in \operatorname{spec}(B)$, it follows that $\alpha^2 + |\lambda|^2 \le \rho_{\max}^2(B) \le \alpha^2$. Hence, $\lambda = 0$. By a similar argument, $\rho_{\max}(B) < \alpha$ implies that $\operatorname{Re}\lambda > 0$. **Remark:** Converses of these statements hold in the case where B is nonnegative. See Fact 6.11.13.

6.11 Facts on Graphs and Nonnegative Matrices

Fact 6.11.1. Let $\mathcal{G} = (\mathcal{X}, \mathcal{R})$ be a directed graph, where $\mathcal{X} = \{x_1, \ldots, x_n\}$, and let A be the adjacency matrix of $\mathcal{G}$. Then, the following statements hold:

i) The number of distinct walks from x_i to x_j of length $k \ge 1$ is $(A^k)_{(j,i)}$.

ii) Let k be an integer such that $1 \le k \le n-1$. Then, for distinct $x_i, x_j \in \mathcal{X}$, the number of distinct walks from x_i to x_j whose length is either less than or equal to k is $[(I+A)^k]_{(j,i)}$.

Fact 6.11.2. Let $\mathcal{G} = (\mathcal{X}, \mathcal{R})$ be a directed graph, where $\mathcal{X} = \{x_1, \ldots, x_n\}$, and let A be the adjacency matrix of $\mathcal{G}$. Then, every subdeterminant of A is either -1, 0, or 1. Now, assume that $\mathcal{G}$ is a symmetric graph. Then, the following statements hold:

i) $\mathcal{G}$ is bipartite if and only if, for all $\lambda \in \operatorname{spec}(A)$, it follows that $-\lambda \in \operatorname{spec}(A)$ and $\operatorname{am}_A(\lambda) = \operatorname{am}_A(-\lambda)$.

ii) mspec(A) is the union of the multispectra of the adjacency matrices of the connected components of $\mathcal{G}$.

iii) Assume that every node of $\mathcal{G}$ has degree k. Then, $\lambda_{\max}(A) = k$, and $\operatorname{am}_A(\lambda)$ is the number of connected components of $\mathcal{G}$. In particular, if k is a simple eigenvalue of A, then $\mathcal{G}$ is connected.

iv) Assume that $\mathcal{G}$ is connected, and let m denote the length of the longest path in $\mathcal{G}$. Then, $\text{card}[\text{spec}(A)] \geq m + 1$.

Source: [1182, pp. 272, 325–330].

Fact 6.11.3. Let $A \in \mathbb{F}^{n \times n}$, and consider the directed graph $\mathcal{G}(A) = (\mathcal{X}, \mathcal{R})$, where $\mathcal{X} = \{x_1, \ldots, x_n\}$. Then, the following statements are equivalent:

i) $\mathcal{G}(A)$ is directionally connected.

ii) There exists $k \geq 1$ such that $(I + |A|)^{k-1} >> 0$.

iii) $(I + |A|)^{n-1} >> 0$.

Source: [1448, pp. 358, 359] and [1451, p. 401]. **Remark:** $\mathcal{G}(A)$ is defined in Definition 4.2.4.

Fact 6.11.4. Let $\mathcal{G} = (\{x_1, \ldots, x_n\}, \mathcal{R})$ be a directed graph, and let $A \in \mathbb{R}^{n \times n}$ be the adjacency matrix of $\mathcal{G}$. Then, the following statements are equivalent:

i) $\mathcal{G}$ is directionally connected.

ii) $\mathcal{G}$ has no directed cuts.

iii) A is irreducible.

iv) $\sum_{i=0}^{n-1} A^i >> 0$.

v) $(I + A)^{n-1} >> 0$.

If, in addition, every node has a self-arc, then the following statement is equivalent to *i*)–*v*):

vi) A^{n-1} is positive.

Furthermore, the following statements are equivalent:

vii) $\mathcal{G}$ is not directionally connected.

viii) $\mathcal{G}$ has a directed cut.

ix) A is reducible.

x) $\sum_{i=0}^{n-1} A^i$ has at least one entry that is zero.

xi) $(I + A)^{n-1}$ has at least one entry that is zero.

If, in addition, every node has a self-arc, then the following statement is equivalent to *vii*)–*xi*):

xii) A^{n-1} has at least one zero entry.

Finally, suppose that A is reducible and there exist $k \geq 1$ and a permutation matrix $S \in \mathbb{R}^{n \times n}$ such that $SAS^\mathsf{T} = \left[\begin{smallmatrix} B & C \\ 0_{k \times (n-k)} & D \end{smallmatrix}\right]$, where $B \in \mathbb{F}^{(n-k) \times (n-k)}$, $C \in \mathbb{F}^{(n-k) \times k}$, and $D \in \mathbb{F}^{k \times k}$, and define $[i_1 \ \cdots \ i_n]^\mathsf{T} \triangleq S[1 \ \cdots \ n]^\mathsf{T}$. Then, $(\{x_{i_1}, \ldots, x_{i_{n-k}}\}, \{x_{i_{n-k+1}}, \ldots, x_{i_n}\})$ is a directed cut. **Source:** [1448, p. 362], and [2344, p. 9-3], and [2432, pp. 238, 239].

Fact 6.11.5. Let $A \in \mathbb{R}^{n \times n}$, where $n \geq 2$, and assume that A is nonnegative. Then, the following statements hold:

i) $\rho_{\max}(A) \in \text{spec}(A)$.

ii) There exists a nonnegative eigenvector $x \in \mathbb{R}^n$ associated with $\rho_{\max}(A)$.

iii) If $x \in \mathbb{R}^n$ is a positive eigenvector of A associated with $\lambda \in \text{spec}(A)$, then $\lambda = \rho_{\max}(A)$.

iv) If A has less than $n - 1$ zero entries, then A is irreducible.

Furthermore, the following statements are equivalent:

v) A is irreducible.

vi) $(I + A)^{n-1} >> 0$.

vii) For all $i, j \in \{1, \ldots, n\}$, there exists $k \geq 1$ such that $(A^k)_{(i,j)} > 0$.

viii) $\mathcal{G}(A)$ is directionally connected.

ix) A has a positive eigenvector and a unique unit-length nonnegative eigenvector.

If A is irreducible, then the following statements hold:

x) $\rho_{\max}(A) > 0$.

xi) $\rho_{\max}(A)$ is a simple eigenvalue of A.

xii) A has a unique positive eigenvector $x \in \mathbb{R}^n$ such that $\|x\|_2 = 1$.

xiii) If $x \in \mathbb{R}^n$ is a positive eigenvector of A, then $Ax = \rho_{\max}(A)x$.

xiv) Define $\{\lambda_1, \ldots, \lambda_k\}_{\rm ms} = \{\lambda \in \operatorname{mspec}(A)\colon\ |\lambda| = \rho_{\max}(A)\}_{\rm ms}$. Then, $\lambda_1, \ldots, \lambda_k$ are distinct, and $\{\lambda_1, \ldots, \lambda_k\} = \{e^{(2\pi i/k)J}\rho_{\max}(A)\colon\ i = 1, \ldots, k\}$. Furthermore, $\operatorname{mspec}(A) = e^{(2\pi/k)J}\operatorname{mspec}(A)$.

xv) If at least one diagonal entry of A is positive, then $\rho_{\max}(A)$ is the unique eigenvalue of A whose absolute value is $\rho_{\max}(A)$.

xvi) If A has at least m positive diagonal entries, then $A^{2n-m-1} >> 0$.

xvii) If $x, y \in \mathbb{R}^n$ are positive and satisfy $Ax = \rho_{\max}(A)x$ and $A^{\mathsf{T}}y = \rho_{\max}(A)y$, then

$$\lim_{k\to\infty} \frac{1}{k}\sum_{i=1}^{k}\left(\frac{1}{\rho_{\max}(A)}A\right)^{i} = \frac{1}{x^{\mathsf{T}}y}xy^{\mathsf{T}}.$$

If A is irreducible, then the *index of imprimitivity* of A is $\operatorname{card}\{\lambda \in \operatorname{spec}(A)\colon |\lambda| = \rho_{\max}(A)\}$. In addition, the following statements are equivalent:

xviii) There exists $k \geq 1$ such that $A^k >> 0$.

xix) $A^{n^2-2n+2} >> 0$.

xx) A is irreducible, and $\{\lambda \in \operatorname{spec}(A)\colon |\lambda| = \rho_{\max}(A)\} = \{\rho_{\max}(A)\}$.

xxi) A is irreducible, and $\mathcal{G}(A)$ is aperiodic.

xxii) A is irreducible, and the index of imprimitivity of A is 1.

A is *primitive* if *xvi)*–*xx)* hold. The following statement holds:

xxiii) If A is irreducible and $\operatorname{tr} A > 0$, then A is primitive.

If A is primitive, then the following statements hold:

xxiv) For all $k \geq 1$, A^k is primitive.

xxv) If $k \geq 1$ and $A^k >> 0$, then, for all $l \geq k$, $A^l >> 0$.

xxvi) There exists a positive integer $k \leq (n-1)n^n$ such that $A^k >> 0$.

xxvii) If $x, y \in \mathbb{R}^n$ are positive and satisfy $Ax = \rho_{\max}(A)x$ and $A^{\mathsf{T}}y = \rho_{\max}(A)y$, then

$$\lim_{k\to\infty}\left(\frac{1}{\rho_{\max}(A)}A\right)^{k} = \frac{1}{x^{\mathsf{T}}y}xy^{\mathsf{T}}.$$

xxviii) If $x_0 \in \mathbb{R}^n$ is nonzero and nonnegative and $x, y \in \mathbb{R}^n$ are positive and satisfy $Ax = \rho_{\max}(A)x$ and $A^{\mathsf{T}}y = \rho_{\max}(A)y$, then

$$\lim_{k\to\infty}\frac{A^k x_0 - \rho_{\max}^k(A)y^{\mathsf{T}}x_0 x}{\|A^k x_0\|_2} = 0.$$

xxix) $\rho_{\max}(A) = \lim_{k\to\infty}(\operatorname{tr} A^k)^{1/k}$.

Source: [34, pp. 45–49], [283, p. 17], [422, pp. 26–28, 32, 55], [1040, Chapter 4], [1448, pp. 507–518, 524, 525], [2344, p. 9-3], and [2979, pp. 120–134]. For *xxx)*, see [2441] and [2785, p. 49]. **Remark:** This is the *Perron-Frobenius theorem.* **Remark:** *xix)* is due to H. Wielandt. See [2263, p. 157]. **Remark:** *xxi)* is given in [2344, p. 9-3]. **Example:** $\left[\begin{smallmatrix}1&0\\0&0\end{smallmatrix}\right]$, $\left[\begin{smallmatrix}1&1\\0&0\end{smallmatrix}\right]$, $\left[\begin{smallmatrix}1&1\\0&1\end{smallmatrix}\right]$, and $\left[\begin{smallmatrix}1&1\\0&2\end{smallmatrix}\right]$ are reducible. $\left[\begin{smallmatrix}1&0\\0&0\end{smallmatrix}\right]$ has two unit-length linearly independent nonnegative eigenvectors given by $\left[\begin{smallmatrix}1\\0\end{smallmatrix}\right]$ and $\left[\begin{smallmatrix}0\\1\end{smallmatrix}\right]$, but does not have a positive eigenvector. $\left[\begin{smallmatrix}1&1\\0&0\end{smallmatrix}\right]$ has two unit-length linearly independent eigenvectors given by $\left[\begin{smallmatrix}1\\0\end{smallmatrix}\right]$ and $\frac{1}{\sqrt{2}}\left[\begin{smallmatrix}1\\-1\end{smallmatrix}\right]$, and thus has a unique unit-length nonnegative eigenvector, but does not have a positive eigenvector. $\begin{bmatrix} 1 & 1 \\ 0 & 1 \end{bmatrix}$ has one unit-length linearly independent eigenvector given by $\begin{bmatrix} 1 \\ 0 \end{bmatrix}$, and thus has a unique unit-length nonnegative eigenvector given by $\begin{bmatrix} 1 \\ 0 \end{bmatrix}$, but does not have a positive eigenvector. $\begin{bmatrix} 1 & 1 \\ 0 & 2 \end{bmatrix}$ has two unit-length linearly independent nonnegative eigenvectors given by $\frac{1}{\sqrt{2}}\begin{bmatrix} 1 \\ 1 \end{bmatrix}$ and $\begin{bmatrix} 1 \\ 0 \end{bmatrix}$, and thus has a positive eigenvector but does not have a unique unit-length nonnegative eigenvector. $\begin{bmatrix} 0 & 1 \\ 1 & 0 \end{bmatrix}$ is irreducible but not primitive. For all $k \geq 1$, $\begin{bmatrix} 0 & 1 \\ 1 & 0 \end{bmatrix}^k$ is not positive. $\begin{bmatrix} 1 & 1 \\ 1 & 0 \end{bmatrix}$ is primitive. $\begin{bmatrix} 1 & 1 \\ 1 & 1 \end{bmatrix}$ is positive and thus primitive. **Remark:** For an arbitrary nonzero and nonnegative initial condition, *xxvii*) shows that the state $x_k = A^k x_0$ of the difference equation $x_{k+1} = Ax_k$ approaches a distribution given by the eigenvector associated with the positive eigenvalue of maximum absolute value. In demography, this eigenvector is interpreted as the *stable age distribution*. See [1605, pp. 47, 63]. **Example:** Let x and y be positive numbers such that $x + y < 1$, and define

$$A \triangleq \begin{bmatrix} x & y & 1-x-y \\ 1-x-y & x & y \\ y & 1-x-y & x \end{bmatrix}.$$

Then, $A1_{3\times1} = A^\mathsf{T}1_{3\times1} = 1_{3\times1}$, and thus $\lim_{k\to\infty} A^k = \frac{1}{3}1_{3\times3}$. See [517, p. 213]. **Related:** Fact 8.11.4, Fact 15.19.20, and Fact 15.22.24.

Fact 6.11.6. Let $A \in \mathbb{R}^{n\times n}$, where $n \geq 2$, and assume that A is nonnegative. Then, there exists a permutation matrix $S \in \mathbb{R}^{n\times n}$ such that SAS^T is upper block triangular and every diagonally located block is irreducible. If, in addition, A is a permutation matrix, then there exists a permutation matrix $S \in \mathbb{R}^{n\times n}$ such that SAS^T is block diagonal and every diagonally located block is an irreducible permutation matrix. **Source:** If A is either zero or irreducible, then the result holds with $S = I$. If A is either nonzero or reducible, then there exists an $n\times n$ permutation matrix S such that SAS^T is upper block triangular with square, nonnegative, diagonally located blocks B and C. If each matrix B and C is either zero or irreducible, then the result holds. If not, then either B or C can be transformed as needed. The last statement is given in [2980, p. 155] and Fact 7.18.14. **Example:** $\begin{bmatrix} 1 & 1 \\ 0 & 0 \end{bmatrix}$ and $\begin{bmatrix} 1 & 1 \\ 0 & 1 \end{bmatrix}$ are reducible, and every diagonally located block is irreducible. **Remark:** This result gives the *Frobenius normal form*. See [587, p. 27-6]. Note that all 1×1 matrices are defined to be irreducible. The only 1×1 irreducible permutation matrix is $[1]$.

Fact 6.11.7. Let $A \in \mathbb{R}^{n\times n}$. Then, the following statements hold:

i) If $n \geq 2$ and $A^2 \leq\leq 0$, then A is reducible.

ii) A^2 has at least one nonnegative entry.

Source: [1001] and [2979, pp. 130–132]. **Remark:** For all $n \geq 1$, there exists $B \in \mathbb{R}^{n\times n}$ such that B^2 has $n^2 - 1$ negative entries. See [1001] and [2979, pp. 131, 132].

Fact 6.11.8. Let $A \in \mathbb{R}^{n\times n}$, assume that A is positive, and assume that, for all $i, j \in \{1, \ldots, n\}$, $A^{\odot -1} = A^\mathsf{T}$. Then, $\rho_{\max}(A) \geq n$. **Source:** [2356].

Fact 6.11.9. Let $A \in \mathbb{R}^{n\times n}$, and assume that A is totally nonnegative. Then, $\operatorname{spec}(A) \subset [0, \infty)$. **Source:** [2979, p. 139].

Fact 6.11.10. Let $A \in \mathbb{R}^{n\times n}$, and assume that A is totally nonnegative. Then, the following statements are equivalent:

i) There exists $k \geq 1$ such that A^k is totally positive.

ii) A is nonsingular, and all of the entries on the subdiagonal and superdiagonal of A are positive.

A is *oscillatory* if *i*) and *ii*) hold. If A is oscillatory, then the following statements hold:

iii) A is primitive.

iv) The eigenvalues of A are distinct positive numbers.

v) A^{n-1} is totally positive.

Source: [2979, pp. 138, 139, 148].

Fact 6.11.11. Let $A \in \mathbb{R}^{n \times n}$, and assume that A is row stochastic. Then, $\rho_{\max}(A) = 1$. **Source:** Since $1_{n \times 1}$ is an eigenvector of A associated with the eigenvalue 1, the result follows from *ii*) of Fact 6.11.5. Alternatively, note that $\|A\|_{\infty,\infty} = \|A\|_{\text{row}} = 1$. Since $1 \in \text{spec}(A)$ and $\|\cdot\|_{\infty,\infty}$ is an induced norm, it follows from Corollary 11.4.5 that $1 \le \rho_{\max}(A) \le \|A\|_{\infty,\infty} = 1$. **Remark:** Fact 4.13.1 implies that, if $A \in \text{conv}\,\text{P}(n)$, then $\rho_{\max}(A) = 1$. **Related:** Fact 15.22.12.

Fact 6.11.12. Let $\mathcal{G} = (\mathcal{X}, \mathcal{R})$ be a directed graph, where $\mathcal{X} = \{x_1, \ldots, x_n\}$, and let $L_{\text{out}} \in \mathbb{R}^{n \times n}$ denote the outbound Laplacian of $\mathcal{G}$. Then, the following statements hold:

i) $0 \in \text{spec}(L_{\text{out}}) \subset \{0\} \cup \text{ORHP}$.

ii) $L_{\text{out}} 1_{n \times 1} = 0$.

iii) 0 is a semisimple eigenvalue of L_{out}.

iv) If $\mathcal{G}$ is directionally connected, then 0 is a simple eigenvalue of L_{out}.

v) The following statements are equivalent:

a) There exists a node $x \in \text{supp}(\mathcal{G})$ such that, for every node $y \in \text{supp}(\mathcal{G})$, there exists a walk from y to x.

b) 0 is a simple eigenvalue of L_{out}.

vi) $L_{\text{out}} + L_{\text{out}}^{\mathsf{T}}$ is positive semidefinite.

vii) L_{out} is symmetric if and only if $\mathcal{G}$ is symmetric.

Now, assume, in addition, that $\mathcal{G}$ is symmetric, and let $L \in \mathbb{R}^{n \times n}$ denote the Laplacian of $\mathcal{G}$. Then, the following statements hold:

viii) L is positive semidefinite.

ix) $0 \in \text{spec}(L) \subset \{0\} \cup [0, \infty)$.

x) $\mathcal{G}$ is connected if and only if $\text{rank}\, L = n - 1$.

xi) If $\mathcal{G}$ is connected, then 0 is a simple eigenvalue of L.

xii) 0 is a simple eigenvalue of L if and only if $\mathcal{G}$ has a spanning subgraph that is a tree.

Source: [606, pp. 40, 77] and [2032, p. 27]. For *xii*), see [2028, p. 147]. **Remark:** *v*) means that $\mathcal{G}$ has at least one globally reachable node. See [606, p. 40]. **Related:** Fact 15.20.7.

Fact 6.11.13. Let $A \in \mathbb{R}^{n \times n}$, where $n \ge 2$, and assume that A is a Z-matrix. Then, the following statements are equivalent:

i) There exist a nonnegative matrix $B \in \mathbb{R}^{n \times n}$ and $\alpha \ge \rho_{\max}(B)$ such that $A = \alpha I - B$.

ii) $\text{spec}(A) \subset \text{ORHP} \cup \{0\}$.

iii) $\text{spec}(A) \subset \text{CRHP}$.

iv) If $\lambda \in \text{spec}(A)$ is real, then $\lambda \ge 0$.

v) Every principal subdeterminant of A is nonnegative.

vi) If $D \in \mathbb{R}^{n \times n}$ is diagonal and positive definite, then $A + D$ is nonsingular.

A is an *M-matrix* if *i*)–*vi*) hold. The following statements are equivalent:

vii) A is a nonsingular M-matrix.

viii) There exist a nonnegative matrix $B \in \mathbb{R}^{n \times n}$ and $\alpha > \rho_{\max}(B)$ such that $A = \alpha I - B$.

ix) $\text{spec}(A) \subset \text{ORHP}$.

x) If $\lambda \in \text{spec}(A)$ is real, then $\lambda > 0$.

xi) A is nonsingular, and $A^{-1} \ge\ge 0$.

xii) There exists $x \in \mathbb{R}^n$ such that $x >> 0$ and $Ax >> 0$.

xiii) Every principal subdeterminant of A is positive.

xiv) Every leading principal subdeterminant of A is positive.

Source: *i*) $\Longrightarrow$ *ii*) follows from Fact 6.10.36. To prove *iii*) $\Longrightarrow$ *i*), let $\alpha \in (0, \infty)$ be sufficiently large that $B \triangleq \alpha I - A$ is nonnegative. Hence, for every $\mu \in \operatorname{spec}(B)$, it follows that $\lambda \triangleq \alpha - \mu \in \operatorname{spec}(A)$. Since $\operatorname{Re}\lambda \geq 0$, it follows that every $\mu \in \operatorname{spec}(B)$ satisfies $\operatorname{Re}\mu \leq \alpha$. Since B is nonnegative, it follows from *i*) of Fact 6.11.5 that $\rho_{\max}(B)$ is an eigenvalue of B. Hence, setting $\mu = \rho_{\max}(B)$ implies that $\rho_{\max}(B) \leq \alpha$. *iv*) and *v*) are proved in [423, pp. 149, 150]. The argument used to prove *i*) $\Longrightarrow$ *ii*) shows that *viii*) $\Longrightarrow$ *ix*). See [2979, pp. 140–142]. **Example:** $A = \left[\begin{smallmatrix} 0 & -1 \\ 0 & 0 \end{smallmatrix}\right] = I - \left[\begin{smallmatrix} 1 & 1 \\ 0 & 1 \end{smallmatrix}\right]$ is an M-matrix. **Related:** Fact 15.20.3 and Fact 15.20.5.

Fact 6.11.14. Let $A \in \mathbb{R}^{n \times n}$, where $n \geq 2$. Then, A is a Z-matrix if and only if every principal submatrix of A is a Z-matrix. Furthermore, A is an M-matrix if and only if every principal submatrix of A is an M-matrix. **Source:** [1450, p. 114] and [2979, p. 139].

Fact 6.11.15. Let $A \in \mathbb{R}^{n \times n}$, where $n \geq 2$, and assume that A is a Z-matrix. Then, the following statement holds:

i) If there exists $x \in \mathbb{R}^n$ such that $x >> 0$ and $Ax \geq\geq 0$, then A is an M-matrix.

Now, assume that A is an M-matrix. Then, the following statements hold:

ii) There exists a nonzero vector $x \in \mathbb{R}^n$ such that $x \geq\geq 0$ and $Ax \geq\geq 0$.

iii) If A is irreducible, then there exists a positive vector $x \in \mathbb{R}^n$ such that Ax is nonnegative.

Now, assume, in addition, that A is singular. Then, the following statements hold:

iv) $\operatorname{rank} A = n - 1$.

v) There exists a positive vector $x \in \mathbb{R}^n$ such that $Ax = 0$.

vi) A is group invertible.

vii) Every principal submatrix of A of size less than n and greater than 1 is a nonsingular M-matrix.

viii) If $x \in \mathbb{R}^n$ and Ax is nonnegative, then $Ax = 0$.

Source: To prove *ii*), Fact 6.11.13 implies that there exist $\alpha \in (0, \infty)$ and $B \in \mathbb{R}^{n \times n}$ such that $A = \alpha I - B$, B is nonnegative, and $\rho_{\max}(B) \leq \alpha$. Consequently, *ii*) of Fact 6.11.5 implies that there exists a nonzero nonnegative vector $x \in \mathbb{R}^n$ such that $Bx = \rho_{\max}(B)x$. Therefore, $Ax = [\alpha - \rho_{\max}(B)]x$ is nonnegative. *iv*)–*viii*) are given in [423, p. 156].

Fact 6.11.16. Let $A \in \mathbb{R}^{n \times n}$, where $n \geq 2$, and assume that A is a nonsingular M-matrix, B is a Z-matrix, and $A \leq\leq B$. Then, the following statements hold:

i) $\operatorname{tr} A^{-1}A^{\mathsf{T}} \leq n$.

ii) $\operatorname{tr} A^{-1}A^{\mathsf{T}} = n$ if and only if A is symmetric.

iii) B is a nonsingular M-matrix.

iv) $0 \leq B^{-1} \leq A^{-1}$.

v) $0 < \det A \leq \det B$.

Source: [1450, pp. 117, 370].

Fact 6.11.17. Let $A \in \mathbb{R}^{n \times n}$, where $n \geq 2$, assume that A is a Z-matrix. Then, the following statements hold:

i) $\alpha_{\min}(A) \in \operatorname{spec}(A)$.

ii) $\min_{i \in \{1,\ldots,n\}} \sum_{j=1}^n A_{(i,j)} \leq \alpha_{\min}(A)$.

Now, assume, in addition, that A is an M-matrix. Then, the following statements hold:

iii) If A is nonsingular, then $\alpha_{\min}(A) = \rho_{\min}(A)$.

iv) $\alpha^n_{\min}(A) \le \det A$.

v) If $B \in \mathbb{R}^{n\times n}$, B is an M-matrix, and $B \le\le A$, then $\alpha_{\min}(B) \le \alpha_{\min}(A)$.

Source: [1450, pp. 128–131]. **Related:** Fact 9.6.22.

Fact 6.11.18. Consider the nonnegative companion matrix $A \in \mathbb{R}^{n\times n}$ defined by

$$A \triangleq \begin{bmatrix} 0 & 1 & 0 & \cdots & 0 & 0 \\ 0 & 0 & 1 & \ddots & 0 & 0 \\ 0 & 0 & 0 & \ddots & 0 & 0 \\ \vdots & \vdots & \vdots & \ddots & \ddots & \vdots \\ 0 & 0 & 0 & \cdots & 0 & 1 \\ 1/n & 1/n & 1/n & \cdots & 1/n & 1/n \end{bmatrix}.$$

Then, A is irreducible, 1 is a simple eigenvalue of A with associated eigenvector $1_{n\times 1}$, and $|\lambda| < 1$ for all $\lambda \in \operatorname{spec}(A)\backslash\{1\}$. Furthermore, if $x \in \mathbb{R}^n$, then

$$\lim_{k\to\infty} A^k x = \left(\frac{2}{n(n+1)}\sum_{i=1}^n i x_{(i-1)}\right) 1_{n\times 1}.$$

Source: [1304, pp. 82, 83, 263–266] and Fact 6.11.5.

Fact 6.11.19. Let $A \in \mathbb{R}^{n\times m}$ and $b \in \mathbb{R}^m$. Then, the following statements are equivalent:

i) If $x \in \mathbb{R}^m$ and $Ax \ge\ge 0$, then $b^{\mathsf{T}}x \ge 0$.

ii) There exists $y \in \mathbb{R}^n$ such that $y \ge\ge 0$ and $A^{\mathsf{T}}y = b$.

Equivalently, exactly one of the following two statements holds:

iii) There exists $x \in \mathbb{R}^m$ such that $Ax \ge\ge 0$ and $b^{\mathsf{T}}x < 0$.

iv) There exists $y \in \mathbb{R}^n$ such that $y \ge\ge 0$ and $A^{\mathsf{T}}y = b$.

Source: [333, p. 47], [523, p. 24], and [2979, pp. 27, 28]. **Remark:** This is the *Farkas theorem*.

Fact 6.11.20. Let $A \in \mathbb{R}^{n\times m}$. Then, the following statements are equivalent:

i) There exists $x \in \mathbb{R}^m$ such that $Ax >> 0$.

ii) If $y \in \mathbb{R}^n$ is nonzero and $y \ge\ge 0$, then $A^{\mathsf{T}}y \ne 0$.

Equivalently, exactly one of the following two statements holds:

iii) There exists $x \in \mathbb{R}^m$ such that $Ax >> 0$.

iv) There exists a nonzero vector $y \in \mathbb{R}^n$ such that $y \ge\ge 0$ and $A^{\mathsf{T}}y = 0$.

Source: [333, p. 47] and [523, p. 23]. **Remark:** This is *Gordan's theorem*.

Fact 6.11.21. Let $A \in \mathbb{C}^{n\times n}$, let $B \in \mathbb{R}^{n\times n}$, assume that B is nonnegative, and assume that $|A| \le\le B$. Then, $\rho_{\max}(A) \le \rho_{\max}(B)$. Now, assume that A is irreducible, and let $\lambda \in \operatorname{spec}(A)$. Then, $|\lambda| = \rho_{\max}(B)$ if and only if there exists a diagonal unitary matrix $D \in \mathbb{C}^{n\times n}$ such that $\lambda DBD^{-1} = \rho_{\max}(B)A$. **Source:** [2979, p. 619].

Fact 6.11.22. Let $A \in \mathbb{C}^{n\times n}$. Then, $\rho_{\max}(A) \le \rho_{\max}(|A|)$. **Source:** [2036, p. 619].

Fact 6.11.23. Let $A, B \in \mathbb{R}^{n\times n}$, where $0 \le\le A \le\le B$. Then, the following statements hold:

i) $\rho_{\max}(A) \le \rho_{\max}(B)$.

ii) Assume that A is irreducible. Then, the following statements are equivalent:

a) $\rho_{\max}(A) = \rho_{\max}(B)$.

b) $A = B$.

iii) If $A + B$ is irreducible and $\rho_{\max}(A) = \rho_{\max}(B)$, then $A = B$.

iv) If $A \neq B$ and $A + B$ is irreducible, then $\rho_{\max}(A) < \rho_{\max}(B)$.

Source: [423, p. 27], [970, pp. 500, 501], and [2991, p. 165].

Fact 6.11.24. Let $A \in \mathbb{R}^{n\times n}$ and $B \in \mathbb{R}^{m\times m}$, and assume that A is nonnegative and B is a principal submatrix of A. Then, $\rho_{\max}(B) \le \rho_{\max}(A)$. If, in addition, A is irreducible and $m < n$, then $\rho_{\max}(B) < \rho_{\max}(A)$. **Source:** Fact 6.11.23, [2979, p. 128], and [2991, p. 170].

Fact 6.11.25. Let $A \in \mathbb{R}^{n\times n}$, assume that A is nonnegative, and define $f\colon [0,1] \mapsto [0,\infty)$ by $f(\alpha) \triangleq \rho_{\max}[\alpha A + (1-\alpha)A^{\mathsf{T}}]$. Then, f is nondecreasing on $[0, \frac{1}{2}]$ and nonincreasing on $[\frac{1}{2}, 1]$. Hence, for all $\alpha \in [0,1]$,

$$\rho_{\max}(A) \le \rho_{\max}[\alpha A + (1-\alpha)A^{\mathsf{T}}] \le \tfrac{1}{2}\rho_{\max}(A + A^{\mathsf{T}}).$$

Source: [279] and [2979, p. 147]. **Related:** Fact 9.6.18.

Fact 6.11.26. Let $A, B \in \mathbb{R}^{n\times n}$, assume that B is diagonal, assume that A and $A+B$ are nonnegative, and let $\alpha \in [0,1]$. Then,

$$\rho_{\max}[\alpha A + (1-\alpha)B] \le \alpha\rho_{\max}(A) + (1-\alpha)\rho_{\max}(A+B).$$

Source: [2344, p. 9-5].

Fact 6.11.27. Let $A \in \mathbb{R}^{n\times n}$, assume that $A >> 0$, let $\lambda \in \operatorname{spec}(A)$, and assume that $|\lambda| < \rho_{\max}(A)$. Then,

$$|\lambda| \le \frac{\alpha - \beta}{\alpha + \beta}\rho_{\max}(A),$$

where $\beta \triangleq \min\{A_{(i,j)}\colon\ i,j \in \{1,\ldots,n\}\}$ and $\alpha \triangleq \max\{A_{(i,j)}\colon\ i,j \in \{1,\ldots,n\}\}$. **Source:** [2979, pp. 143, 144]. **Remark:** This is *Hopf's theorem.* **Remark:** The equality case is considered in [1396].

Fact 6.11.28. Let $A \in \mathbb{R}^{n\times n}$, assume that A is nonnegative and irreducible, and let $x, y \in \mathbb{R}^n$, where $x >> 0$ and $y >> 0$ satisfy $Ax = \rho_{\max}(A)x$ and $A^{\mathsf{T}}y = \rho_{\max}(A)y$. Then,

$$\lim_{l\to\infty} \frac{1}{l}\sum_{k=1}^{l}\left(\frac{1}{\rho_{\max}(A)}A\right)^k = xy^{\mathsf{T}}.$$

If, in addition, A is primitive, then

$$\lim_{k\to\infty}\left(\frac{1}{\rho_{\max}(A)}A\right)^k = xy^{\mathsf{T}}.$$

Source: [970, p. 503] and [1448, p. 516].

Fact 6.11.29. Let $A \in \mathbb{R}^{n\times n}$, assume that A is nonnegative, and let $k \ge 1$ and $m \ge 1$. Then,

$$(\operatorname{tr} A^k)^m \le n^{m-1}\operatorname{tr} A^{km}.$$

Source: [1723]. **Remark:** This is the *JLL inequality*.

6.12 Notes

The proofs of Lemma 6.4.8 and Leverrier's algorithm Proposition 6.4.9 are based on [2314, pp. 432, 433], where it is called the *Souriau-Frame algorithm*. Alternative proofs of Leverrier's algorithm are given in [299, 1466]. The proof of Theorem 6.6.1 is based on [1448]. Polynomial-based approaches to linear algebra are given in [607, 1084], while polynomial matrices and rational transfer functions are studied in [1186, 2784]. The term *normal rank* is often used to refer to what we call the rank of a rational transfer function.

Chapter Seven
Matrix Decompositions

In this chapter we present several matrix decompositions, namely, the Smith, multicompanion, elementary multicompanion, Jordan, Schur, singular value, polar, and full-rank. The Smith, multicompanion, elementary multicompanion, Jordan, and singular value decompositions involve the transformation of a matrix into a unique canonical matrix.

7.1 Smith Decomposition

For rectangular matrices under a biequivalence transformation, the following result, which follows from Theorem 6.3.2, provides a canonical matrix given by a Smith matrix.

Theorem 7.1.1. Let $A \in \mathbb{F}^{n\times m}$ and $r \triangleq \operatorname{rank} A$. Then, there exist nonsingular matrices $S_1 \in \mathbb{F}^{n\times n}$ and $S_2 \in \mathbb{F}^{m\times m}$ such that

$$A = S_1 \begin{bmatrix} I_r & 0_{r\times(m-r)} \\ 0_{(n-r)\times r} & 0_{(n-r)\times(m-r)} \end{bmatrix} S_2. \tag{7.1.1}$$

The Smith matrix $\begin{bmatrix} I_r & 0_{r\times(m-r)} \\ 0_{(n-r)\times r} & 0_{(n-r)\times(m-r)} \end{bmatrix}$ in (7.1.1) is the *Smith form* of A. Note that the Smith polynomials $p_1, \ldots, p_r$ of a constant matrix whose rank is r are given by $p_1 = \cdots = p_r = 1$.

Proposition 7.1.2. Let $A, B \in \mathbb{F}^{n\times m}$. Then, the following statements hold:

i) A and B are left equivalent if and only if $\mathcal{N}(A) = \mathcal{N}(B)$.

ii) A and B are right equivalent if and only if $\mathcal{R}(A) = \mathcal{R}(B)$.

iii) The following statements are equivalent:

a) A and B are biequivalent.

b) $\operatorname{rank} A = \operatorname{rank} B$.

c) A and B have the same Smith form.

Proof. See [2314, pp. 179–181]. □

7.2 Reduced Row Echelon Decomposition

Definition 7.2.1. Let $A \in \mathbb{F}^{n\times m}$, and define $r \triangleq \operatorname{rank} A$. Then, A is a *reduced row echelon matrix* if the following statements hold:

i) For all $i \in \{1, \ldots, r\}$, the left-most nonzero entry of $\operatorname{row}_i(A)$ is 1, and all entries of A above this entry are zero.

ii) For all $1 \le i < j \le r$, the left-most nonzero entry of $\operatorname{row}_i(A)$ is to the left of the left-most nonzero entry of $\operatorname{row}_j(A)$.

iii) $\operatorname{row}_{r+1}(A) = \cdots = \operatorname{row}_n(A) = 0$.

As an example, the matrix

$$A = \begin{bmatrix} 0 & 1 & -2 & 0 & 4 \\ 0 & 0 & 0 & 1 & -3 \\ 0 & 0 & 0 & 0 & 0 \end{bmatrix}$$

is a reduced row echelon matrix.

For rectangular matrices under a left equivalence transformation, the following result provides a canonical matrix given by a reduced row echelon matrix.

Theorem 7.2.2. Let $A \in \mathbb{F}^{n\times m}$. Then, there exist a nonsingular matrix $B \in \mathbb{F}^{n\times n}$ and a unique reduced row echelon matrix $C \in \mathbb{F}^{n\times m}$ such that $\operatorname{rank} C = \operatorname{rank} A$ and $A = BC$.

Proof. See [1451, p. 11], where B is constructed as a product of elementary matrices. □

The reduced row echelon matrix C in Theorem 7.2.2 is the *reduced row echelon form* of A.

7.3 Multicompanion and Elementary Multicompanion Decompositions

For the monic polynomial $p(s) = s^n + \beta_{n-1}s^{n-1} + \cdots + \beta_1 s + \beta_0 \in \mathbb{F}[s]$ of degree $n \ge 1$, the *companion matrix* $C(p) \in \mathbb{F}^{n\times n}$ associated with p is defined to be

$$C(p) \triangleq \begin{bmatrix} 0 & 1 & 0 & \cdots & 0 & 0 \\ 0 & 0 & 1 & \ddots & 0 & 0 \\ 0 & 0 & 0 & \ddots & 0 & 0 \\ \vdots & \vdots & \vdots & \ddots & \ddots & \vdots \\ 0 & 0 & 0 & \cdots & 0 & 1 \\ -\beta_0 & -\beta_1 & -\beta_2 & \cdots & -\beta_{n-2} & -\beta_{n-1} \end{bmatrix}. \tag{7.3.1}$$

If $n = 1$, then $p(s) = s + \beta_0$ and $C(p) = -\beta_0$. Furthermore, if $n = 0$ and $p = 1$, then we define $C(p) \triangleq 0_{0\times 0}$. Note that, if $n \ge 1$, then $\operatorname{tr} C(p) = -\beta_{n-1}$ and $\det C(p) = (-1)^n\beta_0 = (-1)^n p(0)$.

It is easy to see that the characteristic polynomial of the companion matrix $C(p)$ associated with p is p. For example, let $n = 3$ so that

$$C(p) = \begin{bmatrix} 0 & 1 & 0 \\ 0 & 0 & 1 \\ -\beta_0 & -\beta_1 & -\beta_2 \end{bmatrix}, \tag{7.3.2}$$

and thus

$$sI - C(p) = \begin{bmatrix} s & -1 & 0 \\ 0 & s & -1 \\ \beta_0 & \beta_1 & s+\beta_2 \end{bmatrix}. \tag{7.3.3}$$

Adding s times the second column and s^2 times the third column to the first column leaves the determinant of $sI - C(p)$ unchanged and yields

$$\begin{bmatrix} 0 & -1 & 0 \\ 0 & s & -1 \\ p(s) & \beta_1 & s+\beta_2 \end{bmatrix}. \tag{7.3.4}$$

Hence, $\chi_{C(p)} = p$. If $n = 0$ and $p = 1$, then we define $\chi_{C(p)} \triangleq \chi_{0_{0\times 0}} = 1$. The following result shows that the characteristic polynomial of a companion matrix is also its minimal polynomial.

Proposition 7.3.1. Let $p \in \mathbb{F}[s]$ be a monic polynomial having degree n. Then, there exist unimodular matrices $S_1, S_2 \in \mathbb{F}[s]^{n\times n}$ such that

$$sI - C(p) = S_1(s)\begin{bmatrix} I_{n-1} & 0_{(n-1)\times 1} \\ 0_{1\times(n-1)} & p(s) \end{bmatrix} S_2(s). \tag{7.3.5}$$

Furthermore,

$$\chi_{C(p)} = \mu_{C(p)} = p. \tag{7.3.6}$$

Proof. Since $\chi_{C(p)} = p$, it follows that $\operatorname{rank}[sI - C(p)] = n$. Next, since $\det([sI - C(p)]_{[n,1]}) = (-1)^{n-1}$, it follows that $\Delta_{n-1} = 1$, where Δ_{n-1} is the greatest common divisor (which is monic by definition) of all $(n-1) \times (n-1)$ subdeterminants of $sI - C(p)$. Furthermore, since Δ_{i-1} divides Δ_i for all $i \in \{2, \ldots, n-1\}$, it follows that $\Delta_1 = \cdots = \Delta_{n-2} = 1$. Consequently, the similarity invariants $p_1, \ldots, p_n$ of $C(p)$ satisfy $p_1 = \cdots = p_{n-1} = 1$. Furthermore, it follows from Proposition 6.6.2 that $\chi_{C(p)} = \prod_{i=1}^n p_i = p_n$ and $\mu_{C(p)} = p_n$. Therefore, $\mu_{C(p)} = \chi_{C(p)} = p$. □

Next, we consider block-diagonal matrices all of whose diagonally located blocks are companion matrices. A matrix with this structure is a *multicompanion matrix.*

Lemma 7.3.2. Let $p_1, \ldots, p_n \in \mathbb{F}[s]$ be monic polynomials such that p_i divides p_{i+1} for all $i \in \{1, \ldots, n-1\}$ and $n = \sum_{i=1}^n \deg p_i$. Furthermore, define $C \triangleq \operatorname{diag}[C(p_1), \ldots, C(p_n)] \in \mathbb{F}^{n\times n}$. Then, there exist unimodular matrices $S_1, S_2 \in \mathbb{F}[s]^{n\times n}$ such that

$$sI - C = S_1(s)\begin{bmatrix} p_1(s) & & 0 \\ & \ddots & \\ 0 & & p_n(s) \end{bmatrix} S_2(s). \tag{7.3.7}$$

Proof. For all $i \in \{1, \ldots, n\}$, define $k_i \triangleq \deg p_i$, and note that

$$sI - C = \begin{bmatrix} sI_{k_1} - C(p_1) & & 0 \\ & \ddots & \\ 0 & & sI_{k_n} - C(p_n) \end{bmatrix}.$$

For all $i \in \{1, \ldots, n\}$, Proposition 7.3.1 implies that the Smith form of $sI_{k_i} - C(p_i)$ is $0_{0\times 0}$ for $k_i = 0$, p_i for $k_i = 1$, and $\operatorname{diag}(I_{k_i-1}, p_i)$ for $k_i \geq 2$. Note that $p_i = 1$ if and only if $i \leq n_0 \triangleq \sum_{i=1}^n \max\{0, k_i - 1\}$. By combining these Smith matrices into a block-diagonal matrix and rearranging the diagonal entries, it follows that there exist unimodular matrices $S_1, S_2 \in \mathbb{F}[s]^{n\times n}$ such that (7.3.7) holds. Since, for all $i \in \{1, \ldots, n-1\}$, p_i divides p_{i+1}, this diagonal matrix is the Smith form of $sI - C$. □

For square matrices under a similarity transformation, the following result provides a canonical matrix given by a multicompanion matrix.

Theorem 7.3.3. Let $A \in \mathbb{F}^{n\times n}$, and let $p_1, \ldots, p_n \in \mathbb{F}[s]$ denote the similarity invariants of A, where p_i divides p_{i+1} for all $i \in \{1, \ldots, n-1\}$. Then, there exists a nonsingular matrix $S \in \mathbb{F}^{n\times n}$ such that

$$A = S\begin{bmatrix} C(p_1) & & 0 \\ & \ddots & \\ 0 & & C(p_n) \end{bmatrix} S^{-1}. \tag{7.3.8}$$

Proof. Lemma 7.3.2 implies that the $n\times n$ matrix $sI - C$, where $C \triangleq \operatorname{diag}[C(p_1), \ldots, C(p_n)]$, has the Smith form $\operatorname{diag}(p_1, \ldots, p_n)$. Now, since $sI - A$ has the same similarity invariants as C, Theorem 6.3.11 implies that A and C are similar. □

The multicompanion matrix $\begin{bmatrix} C(p_1) & & 0 \\ & \ddots & \\ 0 & & C(p_n) \end{bmatrix}$ in (7.3.8) is the *multicompanion form* of A. Recall from Proposition 6.6.2 that $\chi_A = \prod_{i=1}^n p_i$ and $\mu_A = p_n$.

Corollary 7.3.4. Let $A \in \mathbb{F}^{n\times n}$. Then, $\mu_A = \chi_A$ if and only if A and $C(\chi_A)$ are similar.

Proof. Suppose that $\mu_A = \chi_A$. Then, it follows from Proposition 6.6.2 that $p_i = 1$ for all $i \in \{1, \ldots, n-1\}$ and $p_n = \chi_A$ is the unique nonconstant similarity invariant of A. Thus, $C(p_i) = 0_{0\times 0}$ for all $i \in \{1, \ldots, n-1\}$, and it follows from Theorem 7.3.3 that A is similar to $C(\chi_A)$. Conversely, it follows from (7.3.6) that $\mu_{C(\chi_A)} = \chi_A$. Next, since A and $C(\chi_A)$ are similar, it follows from Proposition 6.6.3 that $\mu_A = \mu_{C(\chi_A)}$. Hence, $\mu_A = \chi_A$. □

Corollary 7.3.5. Let $A \in \mathbb{F}^{n\times n}$ be a companion matrix. Then, $A = C(\chi_A)$ and $\mu_A = \chi_A$.

Note that, if $A = I_n$, then the similarity invariants of A are $p_i(s) = s - 1$ for all $i \in \{1,\dots,n\}$. Thus, $C(p_i) = 1$ for all $i \in \{1,\dots,n\}$, as expected.

Although the multicompanion matrix given by Theorem 7.3.3 provides a canonical matrix for all square A, in some cases it is possible to use similarity transformation to further decompose some of the companion blocks in the multicompanion matrix into multicompanion matrices. This procedure provides an alternative canonical matrix, which is also a multicompanion matrix. To show this, note that, if A_i is similar to B_i for all $i \in \{1,\dots,r\}$, then $\operatorname{diag}(A_1,\dots,A_r)$ is similar to $\operatorname{diag}(B_1,\dots,B_r)$. Therefore, it follows from Corollary 7.3.9 that, if $sI - A_i$ and $sI - B_i$ have the same Smith form for all $i \in \{1,\dots,r\}$, then $sI - \operatorname{diag}(A_1,\dots,A_r)$ and $sI - \operatorname{diag}(B_1,\dots,B_r)$ have the same Smith form. The following lemma is needed.

Lemma 7.3.6. Let $A = \operatorname{diag}(A_1, A_2)$, where $A_i \in \mathbb{F}^{n_i\times n_i}$ for $i = 1,2$. Then, μ_A is the least common multiple of μ_{A_1} and μ_{A_2}. In particular, if μ_{A_1} and μ_{A_2} are coprime, then $\mu_A = \mu_{A_1}\mu_{A_2}$.

Proof. Since $0 = \mu_A(A) = \operatorname{diag}[\mu_A(A_1), \mu_A(A_2)]$, it follows that $\mu_A(A_1) = 0$ and $\mu_A(A_2) = 0$. Therefore, Theorem 6.6.1 implies that μ_{A_1} and μ_{A_2} both divide μ_A. Consequently, the least common multiple q of μ_{A_1} and μ_{A_2} also divides μ_A. Since $q(A_1) = 0$ and $q(A_2) = 0$, it follows that $q(A) = 0$. Therefore, μ_A divides q. Hence, $q = \mu_A$. In the case where μ_{A_1} and μ_{A_2} are coprime, $\mu_A = \mu_{A_1}\mu_{A_2}$. □

Proposition 7.3.7. Let $p \in \mathbb{F}[s]$ be a monic polynomial of positive degree n, and let $p = p_1 \cdots p_r$, where $p_1,\dots,p_r \in \mathbb{F}[s]$ are monic and pairwise coprime polynomials. Then, the matrices $C(p)$ and $\operatorname{diag}[C(p_1),\dots,C(p_r)]$ are similar.

Proof. Let $\hat{p}_2 = p_2 \cdots p_r$ and $\hat{C} \triangleq \operatorname{diag}[C(p_1), C(\hat{p}_2)]$. Since p_1 and $\hat{p}_2$ are coprime, it follows from Lemma 7.3.6 that $\mu_{\hat{C}} = \mu_{C(p_1)}\mu_{C(\hat{p}_2)}$. Furthermore, $\chi_{\hat{C}} = \chi_{C(p_1)}\chi_{C(\hat{p}_2)} = \mu_{\hat{C}}$. Hence, Corollary 7.3.4 implies that $\hat{C}$ is similar to $C(\chi_{\hat{C}})$. However, $\chi_{\hat{C}} = p_1 \cdots p_r = p$, so that $\hat{C}$ is similar to $C(p)$. If $r > 2$, then the same argument can be used to decompose $C(\hat{p}_2)$ to show that $C(p)$ is similar to $\operatorname{diag}[C(p_1),\dots,C(p_r)]$. □

Proposition 7.3.7 can be used to decompose some of the companion blocks of a multicompanion matrix into multicompanion matrices. This procedure can be carried out for every companion block whose characteristic polynomial has either two or more nonconstant coprime factors. For example, suppose that $A \in \mathbb{R}^{10\times 10}$ has the similarity invariants $p_i(s) = 1$ for all $i \in \{1,\dots,7\}$, $p_8(s) = (s+1)^2$, $p_9(s) = (s+1)^2(s+2)$, and $p_{10}(s) = (s+1)^2(s+2)(s^2+3)$, so that, by Theorem 7.3.3, the multicompanion form of A is $\operatorname{diag}[C(p_8), C(p_9), C(p_{10})]$, where $C(p_8) \in \mathbb{R}^{2\times 2}$, $C(p_9) \in \mathbb{R}^{3\times 3}$, and $C(p_{10}) \in \mathbb{R}^{5\times 5}$. According to Proposition 7.3.7, the matrices $C(p_9)$ and $C(p_{10})$ can be further decomposed. For example, $C(p_9)$ is similar to $\operatorname{diag}[C(p_{9,1}), C(p_{9,2})]$, where $p_{9,1}(s) = (s+1)^2$ and $p_{9,2}(s) = s + 2$ are coprime. Furthermore, $C(p_{10})$ is similar to four different multicompanion matrices, three of which have two companion blocks and one of which has three companion blocks. Since $p_8(s) = (s+1)^2$ has no nonconstant coprime factors, however, it follows that $C(p_8)$ cannot be decomposed into smaller companion matrices.

By using a similarity transformation, the largest number of companion blocks in a multicompanion matrix is obtained by factoring each similarity invariant into *elementary divisors*, which are powers of nonconstant, monic, irreducible polynomials that are pairwise coprime. In the above example, this factorization is given by $p_9 = p_{9,1}p_{9,2}$, where $p_{9,1}(s) = (s+1)^2$ and $p_{9,2}(s) = s + 2$, and by $p_{10} = p_{10,1}p_{10,2}p_{10,3}$, where $p_{10,1}(s) = (s+1)^2$, $p_{10,2}(s) = s + 2$, and $p_{10,3}(s) = s^2 + 3$. The elementary divisors of A arising from p_8, p_9, and p_{10} are thus $(s+1)^2$, $(s+1)^2$, $s+2$, $(s+1)^2$, $s+2$, and s^2+3, which yields six companion blocks. Viewing $A \in \mathbb{C}^{n\times n}$ we can further factor $p_{10,3}(s) = (s + \sqrt{3}\jmath)(s - \sqrt{3}\jmath)$, which yields a total of seven companion blocks.

For square matrices under a similarity transformation, the following result provides a canonical matrix given by a multicompanion matrix. This result follows from Proposition 7.3.7 and Theorem 7.3.3.

Theorem 7.3.8. Let $A \in \mathbb{F}^{n\times n}$, and let $q_1^{l_1}, \ldots, q_h^{l_h} \in \mathbb{F}[s]$ be the elementary divisors of A, where $l_1, \ldots, l_h$ are positive integers. Then, there exists a nonsingular matrix $S \in \mathbb{F}^{n\times n}$ such that

$$A = S \begin{bmatrix} C(q_1^{l_1}) & & 0 \\ & \ddots & \\ 0 & & C(q_h^{l_h}) \end{bmatrix} S^{-1}. \tag{7.3.9}$$

The multicompanion matrix $\operatorname{diag}[C(q_1^{l_1}), \ldots, C(q_h^{l_h})]$ in (7.3.9) is the *elementary multicompanion form* of A. The multicompanion form of A is a canonical matrix given by a multicompanion matrix with the smallest possible number of companion blocks, whereas the elementary multicompanion form of A is a canonical matrix given by a multicompanion matrix with the largest possible number of companion blocks. Of course, A may be similar to a multicompanion matrix that is neither the multicompanion form of A nor the elementary multicompanion form of A, but this matrix is not uniquely specified and thus it is not a canonical matrix.

Corollary 7.3.9. Let $A, B \in \mathbb{F}^{n\times n}$. Then, the following statements are equivalent:

i) A and B are similar.

ii) A and B have the same similarity invariants.

iii) A and B have the same multicompanion form.

iv) A and B have the same elementary multicompanion form.

7.4 Jordan Decomposition

We now present an alternative form of the multicompanion matrix in (7.3.9). To do this we define the *Jordan matrix* $\mathfrak{J}_l(q)$, where l is a positive integer and $q(s) = s - \lambda \in \mathbb{C}[s]$, to be the $l \times l$ Toeplitz upper bidiagonal matrix

$$\mathfrak{J}_l(q) \triangleq \lambda I_l + N_l = \begin{bmatrix} \lambda & 1 & 0 & & & \\ 0 & \lambda & 1 & & 0 & \\ & & \ddots & \ddots & & \\ & & & \ddots & 1 & 0 \\ & 0 & & & \lambda & 1 \\ & & & & 0 & \lambda \end{bmatrix}. \tag{7.4.1}$$

The following result shows that $\mathfrak{J}_l(q)$ is similar to the companion matrix $C(q^l)$.

Lemma 7.4.1. Let $l \in \mathbb{P}$, let $\lambda \in \mathbb{C}$, and define $q(s) = s - \lambda \in \mathbb{C}[s]$. Then, q^l is the only elementary divisor of $\mathfrak{J}_l(q)$. Furthermore, $\mathfrak{J}_l(q)$ and $C(q^l)$ are similar.

Proof. Note that $\chi_{\mathfrak{J}_l(q)} = q^l$ and $\det([sI - \mathfrak{J}_l(q)]_{[l,1]}) = (-1)^{l-1}$. Hence, as in the proof of Proposition 7.3.1, $\chi_{\mathfrak{J}_l(q)} = \mu_{\mathfrak{J}_l(q)}$. Corollary 7.3.4 implies that $\mathfrak{J}_l(q)$ and $C(q^l)$ are similar. □

A block-diagonal matrix whose diagonally located blocks are Jordan matrices is a *multi-Jordan matrix*. For square matrices under a similarity transformation, the following result, which follows from Proposition 7.3.7 and Lemma 7.4.1, provides a canonical matrix given by a multi-Jordan matrix.

Theorem 7.4.2. Let $A \in \mathbb{C}^{n\times n}$, and let $q_1^{l_1}, \ldots, q_h^{l_h} \in \mathbb{C}[s]$ be the elementary divisors of A, where $l_1, \ldots, l_h$ are positive integers and each polynomial $q_1, \ldots, q_h \in \mathbb{C}[s]$ has degree 1. Then, there exists a nonsingular matrix $S \in \mathbb{C}^{n\times n}$ such that

$$A = S \begin{bmatrix} \mathfrak{J}_{l_1}(q_1) & & 0 \\ & \ddots & \\ 0 & & \mathfrak{J}_{l_h}(q_h) \end{bmatrix} S^{-1}. \tag{7.4.2}$$

The multi-Jordan matrix $\operatorname{diag}[\mathcal{J}_{l_1}(q_1), \ldots, \mathcal{J}_{l_h}(q_h)]$ in (7.4.2) is the *Jordan form* of A. The Jordan form of A is unique up to the ordering of the Jordan matrices. Uniqueness can be ensured by specifying a total ordering on $\mathbb{C}$. To illustrate the structure of the Jordan form, let $l_i = 3$ and $q_i(s) = s - \lambda_i$, where $\lambda_i \in \mathbb{C}$. Then, $\mathcal{J}_{l_i}(q_i)$ is the 3×3 matrix

$$\mathcal{J}_{l_i}(q_i) = \lambda_i I_3 + N_3 = \begin{bmatrix} \lambda_i & 1 & 0 \\ 0 & \lambda_i & 1 \\ 0 & 0 & \lambda_i \end{bmatrix} \tag{7.4.3}$$

so that $\operatorname{mspec}[\mathcal{J}_{l_i}(q_i)] = \{\lambda_i, \lambda_i, \lambda_i\}_{\rm ms}$. If $\mathcal{J}_{l_i}(q_i)$ is the unique diagonally located block of the Jordan form associated with the eigenvalue λ_i, then the algebraic multiplicity of λ_i is equal to 3, while its geometric multiplicity is equal to 1.

Corollary 7.4.3. Let $p \in \mathbb{F}[s]$, let $\lambda_1, \ldots, \lambda_r$ denote the distinct roots of p, and, for all $i \in \{1, \ldots, r\}$, let $l_i \triangleq \operatorname{mult}_p(\lambda_i)$ and $p_i(s) \triangleq s - \lambda_i$. Then, $C(p)$ is similar to $\operatorname{diag}[\mathcal{J}_{l_1}(p_1), \ldots, \mathcal{J}_{l_r}(p_r)]$.

For a real matrix $A \in \mathbb{R}^{n \times n}$, we now obtain a real decomposition that is analogous to (7.4.2). This can be done in two different ways, namely, in terms of either the coefficients of the quadratic irreducible elementary divisors of A or the real and imaginary parts of the nonreal eigenvalues of A. Note that every real elementary divisor $q_i^{l_i}$ is either of the form $(s - \lambda_i)^{l_i}$, where $\lambda_i \in \mathbb{R}$, or of the form $(s^2 - \beta_{1i}s - \beta_{0i})^{l_i}$, where $\beta_{0i}, \beta_{1i} \in \mathbb{R}$.

For $q(s) = s^2 - \beta_1 s - \beta_0 \in \mathbb{R}[s]$, define the $2l \times 2l$ real, tridiagonal matrix

$$\mathcal{H}_l(q) \triangleq \begin{bmatrix} 0 & 1 & & & & & \\ \beta_0 & \beta_1 & 1 & & & 0 & \\ & 0 & 0 & 1 & & & \\ & & \beta_0 & \beta_1 & 1 & & \\ & & & \ddots & \ddots & \ddots & \\ & 0 & & & \ddots & 0 & 1 \\ & & & & & \beta_0 & \beta_1 \end{bmatrix}. \tag{7.4.4}$$

For $q(s) = s - \lambda \in \mathbb{R}[s]$, define $\mathcal{H}_l(q) \triangleq \mathcal{J}_l(q)$. The matrix $\mathcal{H}_l(q)$ is a *hypercompanion matrix*. The following result shows that the hypercompanion matrix $\mathcal{H}_l(q)$ is similar to the companion matrix $C(q^l)$.

Lemma 7.4.4. Let $l \in \mathbb{P}$, and let $q(s) = s^2 - \beta_1 s - \beta_0 \in \mathbb{R}[s]$. Then, q^l is the only elementary divisor of $\mathcal{H}_l(q)$. Furthermore, $\mathcal{H}_l(q)$ and $C(q^l)$ are similar.

Proof. Note that $\chi_{\mathcal{H}_l(q)} = q^l$ and $\det([sI - \mathcal{H}_l(q)]_{[2l,1]}) = (-1)^{2l-1}$. Hence, as in the proof of Proposition 7.3.1, $\chi_{\mathcal{H}_l(q)} = \mu_{\mathcal{H}_l(q)}$. Corollary 7.3.4 now implies that $\mathcal{H}_l(q)$ is similar to $C(q^l)$. □

For real square matrices under a similarity transformation, the following result provides a canonical matrix given by a *multihypercompanion matrix*, which is a block-diagonal matrix whose diagonally located blocks are hypercompanion matrices.

Theorem 7.4.5. Let $A \in \mathbb{R}^{n \times n}$, and let $q_1^{l_1}, \ldots, q_h^{l_h} \in \mathbb{R}[s]$ be the real elementary divisors of A, where $l_1, \ldots, l_h$ are positive integers. Then, there exists a nonsingular matrix $S \in \mathbb{R}^{n \times n}$ such that

$$A = S \begin{bmatrix} \mathcal{H}_{l_1}(q_1) & & 0 \\ & \ddots & \\ 0 & & \mathcal{H}_{l_h}(q_h) \end{bmatrix} S^{-1}. \tag{7.4.5}$$

The multihypercompanion matrix $\operatorname{diag}[\mathcal{H}_{l_1}(q_1), \ldots, \mathcal{H}_{l_h}(q_h)]$ in (7.4.5) is the *multihypercompanion form* of A. Applying an additional real similarity transformation to each diagonally located block of a multihypercompanion matrix yields a *real Jordan matrix*. To do this, define the *real Jordan ma-*

trix $\mathcal{J}_l(q)$ for the positive integer l as follows. For $\lambda \in \mathbb{R}$ and $q(s) = s-\lambda \in \mathbb{R}[s]$, define $\mathcal{J}_l(q) \triangleq \mathcal{H}_l(q)$, whereas, for irreducible $q(s) = s^2 - \beta_1 s - \beta_0 \in \mathbb{R}[s]$ with a nonreal root $\lambda = \nu + \omega \jmath$, define the $2l \times 2l$ upper Hessenberg matrix

$$\mathcal{J}_l(q) \triangleq \begin{bmatrix} \nu & \omega & 1 & 0 & & & & \\ -\omega & \nu & 0 & 1 & \ddots & & 0 & \\ & 0 & \nu & \omega & 1 & \ddots & & \\ & & -\omega & \nu & 0 & \ddots & \ddots & \\ & & & & \ddots & \ddots & 1 & 0 \\ & & & & & \ddots & 0 & 1 \\ & 0 & & & & & \nu & \omega \\ & & & & & & -\omega & \nu \end{bmatrix}. \tag{7.4.6}$$

For real square matrices under a similarity transformation, the following result provides a canonical matrix given by a *multi-real-Jordan matrix*, which is a block-diagonal matrix whose diagonally located blocks are real Jordan matrices.

Theorem 7.4.6. Let $A \in \mathbb{R}^{n\times n}$, and let $q_1^{l_1}, \ldots, q_h^{l_h} \in \mathbb{R}[s]$ be the elementary divisors of A, where $l_1, \ldots, l_h$ are positive integers. Then, there exists a nonsingular matrix $S \in \mathbb{R}^{n\times n}$ such that

$$A = S \begin{bmatrix} \mathcal{J}_{l_1}(q_1) & & 0 \\ & \ddots & \\ 0 & & \mathcal{J}_{l_h}(q_h) \end{bmatrix} S^{-1}. \tag{7.4.7}$$

Proof. For the irreducible quadratic $q(s) = s^2 - \beta_1 s - \beta_0 \in \mathbb{R}[s]$, we show that $\mathcal{J}_l(q)$ and $\mathcal{H}_l(q)$ are similar. Writing $q(s) = (s-\lambda)(s-\overline{\lambda})$, Theorem 7.4.2 implies that $\mathcal{H}_l(q) \in \mathbb{R}^{2l\times 2l}$ is similar to $\operatorname{diag}(\lambda I_l + N_l, \overline{\lambda} I_l + N_l)$. Next, by using a permutation similarity transformation, $\mathcal{H}_l(q)$ is similar to

$$\begin{bmatrix} \lambda & 0 & 1 & 0 & & & & & \\ 0 & \overline{\lambda} & 0 & 1 & 0 & & & 0 & \\ & 0 & \lambda & 0 & 1 & 0 & & & \\ & & 0 & \overline{\lambda} & 0 & 1 & & & \\ & & & \ddots & \ddots & \ddots & & & \\ & & & & \ddots & \ddots & 1 & 0 \\ & & & & & \ddots & 0 & 1 \\ & 0 & & & & & \lambda & 0 \\ & & & & & & 0 & \overline{\lambda} \end{bmatrix}.$$

Finally, applying the similarity transformation $S \triangleq \operatorname{diag}(\hat{S}, \ldots, \hat{S})$ to the above matrix, where $\hat{S} \triangleq \left[\begin{smallmatrix} -\jmath & -\jmath \\ 1 & -1 \end{smallmatrix}\right]$ and $\hat{S}^{-1} = \frac{1}{2}\left[\begin{smallmatrix} \jmath & 1 \\ \jmath & -1 \end{smallmatrix}\right]$, yields $\mathcal{J}_l(q)$. □

The multi-real-Jordan matrix $[\mathcal{J}_{l_1}(q_1), \ldots, \mathcal{J}_{l_h}(q_h)]$ in (7.4.7) is the *multi-real-Jordan form* of A.

Example 7.4.7. Let $A, B \in \mathbb{R}^{4\times4}$ and $C \in \mathbb{C}^{4\times4}$ be given by

$$A = \begin{bmatrix} 0 & 1 & 0 & 0 \\ 0 & 0 & 1 & 0 \\ 0 & 0 & 0 & 1 \\ -16 & 0 & -8 & 0 \end{bmatrix}, \quad B = \begin{bmatrix} 0 & 1 & 0 & 0 \\ -4 & 0 & 1 & 0 \\ 0 & 0 & 0 & 1 \\ 0 & 0 & -4 & 0 \end{bmatrix}, \quad C = \begin{bmatrix} 2\jmath & 1 & 0 & 0 \\ 0 & 2\jmath & 0 & 0 \\ 0 & 0 & -2\jmath & 1 \\ 0 & 0 & 0 & -2\jmath \end{bmatrix}.$$

Then, A is a companion matrix, B is a hypercompanion matrix, and C is a multi-Jordan matrix. Furthermore, A, B, and C are similar. ◇

Example 7.4.8. Let $A, B, D \in \mathbb{R}^{6\times6}$ and $C \in \mathbb{C}^{6\times6}$ be given by

$$A = \begin{bmatrix} 0 & 1 & 0 & 0 & 0 & 0 \\ 0 & 0 & 1 & 0 & 0 & 0 \\ 0 & 0 & 0 & 1 & 0 & 0 \\ 0 & 0 & 0 & 0 & 1 & 0 \\ 0 & 0 & 0 & 0 & 0 & 1 \\ 27 & -18 & -3 & 20 & -15 & 6 \end{bmatrix}, \quad B = \begin{bmatrix} 0 & 1 & 0 & 0 & 0 & 0 \\ 3 & 2 & 0 & 0 & 0 & 0 \\ 0 & 0 & 0 & 1 & 0 & 0 \\ 0 & 0 & -3 & 2 & 1 & 0 \\ 0 & 0 & 0 & 0 & 0 & 1 \\ 0 & 0 & 0 & 0 & -3 & 2 \end{bmatrix},$$

$$C = \begin{bmatrix} -1 & 0 & 0 & 0 & 0 & 0 \\ 0 & 3 & 0 & 0 & 0 & 0 \\ 0 & 0 & 1-\sqrt{2}\jmath & 1 & 0 & 0 \\ 0 & 0 & 0 & 1-\sqrt{2}\jmath & 0 & 0 \\ 0 & 0 & 0 & 0 & 1+\sqrt{2}\jmath & 1 \\ 0 & 0 & 0 & 0 & 0 & 1+\sqrt{2}\jmath \end{bmatrix}, \quad D = \begin{bmatrix} -1 & 0 & 0 & 0 & 0 & 0 \\ 0 & 3 & 0 & 0 & 0 & 0 \\ 0 & 0 & 1 & \sqrt{2} & 0 & 0 \\ 0 & 0 & -\sqrt{2} & 1 & 1 & 0 \\ 0 & 0 & 0 & 0 & 1 & \sqrt{2} \\ 0 & 0 & 0 & 0 & -\sqrt{2} & 1 \end{bmatrix}.$$

Then, A is a companion matrix, B is a multihypercompanion matrix, C is a multi-Jordan matrix, and D is a multi-real-Jordan matrix. Furthermore, A, B, C, and D are similar. ◇

The next result shows that every matrix is similar to its transpose by means of a symmetric similarity transformation. This result, which improves Corollary 6.3.12, is due to F. G. Frobenius.

Corollary 7.4.9. Let $A \in \mathbb{F}^{n\times n}$. Then, there exists a symmetric, nonsingular matrix $S \in \mathbb{F}^{n\times n}$ such that $A = SA^{\mathsf{T}}S^{-1}$.

Proof. It follows from Theorem 7.4.2 that there exists a nonsingular matrix $\hat{S} \in \mathbb{C}^{n\times n}$ such that $A = \hat{S}B\hat{S}^{-1}$, where $B = \operatorname{diag}(B_1, \ldots, B_r)$ is the Jordan form of A, and $B_i \in \mathbb{C}^{n_i\times n_i}$ for all $i \in \{1, \ldots, r\}$. Now, define the symmetric nonsingular matrix $S \triangleq \hat{S}\tilde{I}\hat{S}^{\mathsf{T}}$, where $\tilde{I} \triangleq \operatorname{diag}(\hat{I}_{n_1}, \ldots, \hat{I}_{n_r})$ is symmetric and involutory. Furthermore, note that $\hat{I}_{n_i}B_i\hat{I}_{n_i} = B_i^{\mathsf{T}}$ for all $i \in \{1, \ldots, r\}$ so that $\tilde{I}B\tilde{I} = B^{\mathsf{T}}$, and thus $\tilde{I}B^{\mathsf{T}}\tilde{I} = B$. Hence, it follows that

$$SA^{\mathsf{T}}S^{-1} = S\hat{S}^{-\mathsf{T}}B^{\mathsf{T}}\hat{S}^{\mathsf{T}}S^{-1} = \hat{S}\tilde{I}\hat{S}^{\mathsf{T}}\hat{S}^{-\mathsf{T}}B^{\mathsf{T}}\hat{S}^{\mathsf{T}}\hat{S}^{-\mathsf{T}}\tilde{I}\hat{S}^{-1} = \hat{S}\tilde{I}B^{\mathsf{T}}\tilde{I}\hat{S}^{-1} = \hat{S}B\hat{S}^{-1} = A.$$

If A is real, then the real Jordan form can be used to show that S can be chosen to be real. □

An extension of Corollary 7.4.9 to the case where A is normal is given by Fact 7.10.10.

Corollary 7.4.10. Let $A \in \mathbb{F}^{n\times n}$. Then, there exist symmetric matrices $S_1, S_2 \in \mathbb{F}^{n\times n}$ such that S_2 is nonsingular and $A = S_1S_2$.

Proof. From Corollary 7.4.9 it follows that there exists a symmetric, nonsingular matrix $S \in \mathbb{F}^{n\times n}$ such that $A = SA^{\mathsf{T}}S^{-1}$. Now, let $S_1 \triangleq SA^{\mathsf{T}}$ and $S_2 \triangleq S^{-1}$. Note that S_2 is symmetric and nonsingular. Furthermore, $S_1^{\mathsf{T}} = AS = SA^{\mathsf{T}} = S_1$, which shows that S_1 is symmetric. □

Corollary 7.4.10 implies Corollary 7.4.9. To show this, note that, if $A = S_1S_2$, where S_1, S_2 are symmetric and S_2 is nonsingular, then $A = S_2^{-1}S_2S_1S_2 = S_2^{-1}A^{\mathsf{T}}S_2$.

7.5 Schur Decomposition

The *Schur decomposition* uses a unitary similarity transformation to transform an arbitrary square matrix into an upper triangular matrix.

Theorem 7.5.1. Let $A \in \mathbb{C}^{n\times n}$. Then, there exist a unitary matrix $S \in \mathbb{C}^{n\times n}$ and an upper triangular matrix $B \in \mathbb{C}^{n\times n}$ such that

$$A = SBS^*. \tag{7.5.1}$$

Proof. Let $\lambda_1 \in \mathbb{C}$ be an eigenvalue of A with associated eigenvector $x \in \mathbb{C}^n$ chosen such that $x^*x = 1$. Furthermore, let $S_1 \triangleq [x \;\; \hat{S}_1] \in \mathbb{C}^{n\times n}$ be unitary, where $\hat{S}_1 \in \mathbb{C}^{n\times(n-1)}$ satisfies $\hat{S}_1^*\hat{S}_1 = I_{n-1}$ and $x^*\hat{S}_1 = 0_{1\times(n-1)}$. Then, $S_1e_1 = x$, and $\mathrm{col}_1(S_1^{-1}AS_1) = S_1^{-1}Ax = \lambda_1 S_1^{-1}x = \lambda_1 e_1$. Consequently,

$$A = S_1\begin{bmatrix} \lambda_1 & C_1 \\ 0_{(n-1)\times 1} & A_1 \end{bmatrix}S_1^{-1},$$

where $C_1 \in \mathbb{C}^{1\times(n-1)}$ and $A_1 \in \mathbb{C}^{(n-1)\times(n-1)}$. Next, let $S_{20} \in \mathbb{C}^{(n-1)\times(n-1)}$ be a unitary matrix such that

$$A_1 = S_{20}\begin{bmatrix} \lambda_2 & C_2 \\ 0_{(n-2)\times 1} & A_2 \end{bmatrix}S_{20}^{-1},$$

where $C_2 \in \mathbb{C}^{1\times(n-2)}$ and $A_2 \in \mathbb{C}^{(n-2)\times(n-2)}$. Hence,

$$A = S_1S_2\begin{bmatrix} \lambda_1 & C_{11} & C_{12} \\ 0 & \lambda_2 & C_2 \\ 0 & 0 & A_2 \end{bmatrix}S_2^{-1}S_1,$$

where $C_1 = [C_{11} \;\; C_{12}]$, $C_{11} \in \mathbb{C}$, and $S_2 \triangleq \left[\begin{smallmatrix} 1 & 0 \\ 0 & S_{20} \end{smallmatrix}\right]$ is unitary. Proceeding in a similar manner yields (7.5.1) with $S \triangleq S_1S_2\cdots S_{n-1}$, where $S_1,\dots,S_{n-1} \in \mathbb{C}^{n\times n}$ are unitary. □

Since A and B in (7.5.1) are similar and B is upper triangular, Fact 6.10.14 implies that A and B have the same eigenvalues with the same algebraic multiplicities.

The *real Schur decomposition* uses a real orthogonal similarity transformation to transform a real matrix into an upper Hessenberg matrix with real 1×1 and 2×2 diagonally located blocks.

Corollary 7.5.2. Let $A \in \mathbb{R}^{n\times n}$, and let $\mathrm{mspec}(A) = \{\lambda_1,\dots,\lambda_r\}_{\mathrm{ms}} \cup \{\nu_1+\omega_1 J, \nu_1-\omega_1 J,\dots,\nu_l+\omega_l J, \nu_l-\omega_l J\}_{\mathrm{ms}}$, where $\lambda_1,\dots,\lambda_r \in \mathbb{R}$ and, for all $i \in \{1,\dots,l\}$, $\nu_i,\omega_i \in \mathbb{R}$ and $\omega_i \neq 0$. Then, there exists an orthogonal matrix $S \in \mathbb{R}^{n\times n}$ such that

$$A = SBS^{\mathsf{T}}, \tag{7.5.2}$$

where B is upper block triangular and the diagonally located blocks $B_1,\dots,B_r \in \mathbb{R}$ and $\hat{B}_1,\dots,\hat{B}_l \in \mathbb{R}^{2\times 2}$ of B satisfy $B_i \triangleq [\lambda_i]$ for all $i \in \{1,\dots,r\}$ and $\mathrm{spec}(\hat{B}_i) = \{\nu_i+\omega_i J, \nu_i-\omega_i J\}$ for all $i \in \{1,\dots,l\}$.

Proof. The proof is analogous to the proof of Theorem 7.4.6. See also [1448, p. 82]. □

Corollary 7.5.3. Let $A \in \mathbb{R}^{n\times n}$, and assume that the spectrum of A is real. Then, there exist an orthogonal matrix $S \in \mathbb{R}^{n\times n}$ and an upper triangular matrix $B \in \mathbb{R}^{n\times n}$ such that

$$A = SBS^{\mathsf{T}}. \tag{7.5.3}$$

The Schur decomposition reveals the structure of range-Hermitian matrices and thus, as a special case, normal matrices.

Corollary 7.5.4. Let $A \in \mathbb{F}^{n\times n}$, and define $r \triangleq \mathrm{rank}\, A$. Then, A is range Hermitian if and only if there exist a unitary matrix $S \in \mathbb{F}^{n\times n}$ and a nonsingular matrix $B \in \mathbb{F}^{r\times r}$ such that

$$A = S\begin{bmatrix} B & 0 \\ 0 & 0 \end{bmatrix}S^*. \tag{7.5.4}$$

In addition, A is normal if and only if there exist a unitary matrix $S \in \mathbb{C}^{n\times n}$ and a diagonal matrix $B \in \mathbb{C}^{r\times r}$ such that (7.5.4) holds.

Proof. Suppose that A is range Hermitian, and let $A = S\hat{B}S^*$, where $\hat{B}$ is upper triangular and $S \in \mathbb{F}^{n\times n}$ is unitary. Assume that A is singular, and choose S such that $\hat{B}_{(j,j)} = \hat{B}_{(j+1,j+1)} = \cdots = \hat{B}_{(n,n)} = 0$ and such that all other diagonal entries of $\hat{B}$ are nonzero. Thus, $\mathrm{row}_n(\hat{B}) = 0$, which implies that $e_n \in \mathcal{R}(\hat{B})^\perp$. Since A is range Hermitian, it follows that $\mathcal{R}(\hat{B}) = \mathcal{R}(\hat{B}^*)$, and thus $e_n \in \mathcal{R}(\hat{B}^*)^\perp$. Therefore, it follows from (3.5.13) that $e_n \in \mathcal{N}(\hat{B})$, which implies that $\mathrm{col}_n(\hat{B}) = 0$. In the case where $\hat{B}_{(n-1,n-1)} = 0$, it follows that $\mathrm{col}_{n-1}(\hat{B}) = 0$. Repeating this argument shows that $\hat{B}$ has the form $\left[\begin{smallmatrix} B & 0 \\ 0 & 0 \end{smallmatrix}\right]$, where $B \in \mathbb{F}^{r\times r}$ is nonsingular. The converse result is immediate.

Now, suppose that A is normal, and let $A = S\hat{B}S^*$, where $\hat{B} \in \mathbb{C}^{n\times n}$ is upper triangular and $S \in \mathbb{C}^{n\times n}$ is unitary. Since A is normal, it follows that $AA^* = A^*A$, which implies that $\hat{B}\hat{B}^* = \hat{B}^*\hat{B}$. Since $\hat{B}$ is upper triangular, it follows that $(\hat{B}^*\hat{B})_{(1,1)} = \overline{\hat{B}_{(1,1)}}\hat{B}_{(1,1)}$, whereas $(\hat{B}\hat{B}^*)_{(1,1)} = \mathrm{row}_1(\hat{B})[\mathrm{row}_1(\hat{B})]^* = \sum_{i=1}^n \hat{B}_{(1,i)}\overline{\hat{B}_{(1,i)}}$. Since $(\hat{B}^*\hat{B})_{(1,1)} = (\hat{B}\hat{B}^*)_{(1,1)}$, it follows that $\hat{B}_{(1,i)} = 0$ for all $i \in \{2,\ldots,n\}$. Continuing in a similar fashion row by row, it follows that $\hat{B}$ is diagonal. The converse result is immediate. □

Corollary 7.5.5. Let $A \in \mathbb{F}^{n\times n}$, assume that A is Hermitian, and define $r \triangleq \mathrm{rank}\, A$. Then, there exist a unitary matrix $S \in \mathbb{F}^{n\times n}$ and a diagonal matrix $B \in \mathbb{R}^{r\times r}$ such that (7.5.4) holds. In addition, A is positive semidefinite if and only if the diagonal entries of B are positive, and A is positive definite if and only if A is positive semidefinite and $r = n$.

Proof. Corollary 7.5.4 and *x*), *xi*) of Proposition 6.4.5 imply that there exist a unitary matrix $S \in \mathbb{F}^{n\times n}$ and a diagonal matrix $B \in \mathbb{R}^{r\times r}$ such that (7.5.4) holds. If A is positive semidefinite, then $x^*Ax \ge 0$ for all $x \in \mathbb{F}^n$. Choosing $x = Se_i$, it follows that $B_{(i,i)} = e_i^\mathsf{T} S^*ASe_i \ge 0$ for all $i \in \{1,\ldots,r\}$. If A is positive definite, then $r = n$ and $B_{(i,i)} > 0$ for all $i \in \{1,\ldots,n\}$. □

Proposition 7.5.6. Let $A \in \mathbb{F}^{n\times n}$ be Hermitian. Then, there exists a nonsingular matrix $S \in \mathbb{F}^{n\times n}$ such that

$$A = S\begin{bmatrix} -I_{\nu_-(A)} & 0 & 0 \\ 0 & 0_{\nu_0(A)\times\nu_0(A)} & 0 \\ 0 & 0 & I_{\nu_+(A)} \end{bmatrix}S^*. \tag{7.5.5}$$

Furthermore,

$$\mathrm{rank}\, A = \nu_+(A) + \nu_-(A), \tag{7.5.6}$$

$$\mathrm{def}\, A = \nu_0(A). \tag{7.5.7}$$

Proof. Since A is Hermitian, it follows from Corollary 7.5.5 that there exist a unitary matrix $\hat{S} \in \mathbb{F}^{n\times n}$ and a diagonal matrix $B \in \mathbb{R}^{n\times n}$ such that $A = \hat{S}B\hat{S}^*$. Choose $\hat{S}$ to arrange the diagonal entries of B such that $B = \mathrm{diag}(-B_1, 0_{\nu_0(A)\times\nu_0(A)}, B_2)$, where the diagonal matrices B_1, B_2 are both positive definite. Since the diagonal entries of B are the eigenvalues of A, it follows that $B_1 \in \mathbb{R}^{\nu_-(A)\times\nu_-(A)}$ and $B_2 \in \mathbb{R}^{\nu_+(A)\times\nu_+(A)}$. Now, define $\hat{B} \triangleq \mathrm{diag}(B_1, I_{\nu_0(A)}, B_2)$. Then, $B = \hat{B}^{1/2}D\hat{B}^{1/2}$, where $D \triangleq \mathrm{diag}(-I_{\nu_-(A)}, 0_{\nu_0(A)\times\nu_0(A)}, I_{\nu_+(A)})$. Hence, $A = \hat{S}\hat{B}^{1/2}D\hat{B}^{1/2}\hat{S}^* = SDS^*$, where $S \triangleq \hat{S}\hat{B}^{1/2}$. □

The following result is *Sylvester's law of inertia.*

Theorem 7.5.7. Let $A, B \in \mathbb{F}^{n\times n}$ be Hermitian. Then, A and B are congruent if and only if $\mathrm{In}\, A = \mathrm{In}\, B$.

Proof. To prove sufficiency, assume that $\mathrm{In}\, A = \mathrm{In}\, B = [n_1 \;\; n_2 \;\; n_3]^\mathsf{T}$. It thus follows from Proposition 7.5.6 that there exist nonsingular matrices $S_1 \in \mathbb{F}^{n\times n}$ and $S_2 \in \mathbb{F}^{n\times n}$

$$A = S_1\begin{bmatrix} -I_{n_1} & 0 & 0 \\ 0 & 0_{n_2\times n_2} & 0 \\ 0 & 0 & I_{n_3} \end{bmatrix}S_1^*, \quad B = S_2\begin{bmatrix} -I_{n_1} & 0 & 0 \\ 0 & 0_{n_2\times n_2} & 0 \\ 0 & 0 & I_{n_3} \end{bmatrix}S_2^*.$$

Therefore, $A = S_1S_2^{-1}B(S_1S_2^{-1})^*$, which shows that A and B are congruent.

Conversely, assume that A and B are congruent, and let $\hat{S} \in \mathbb{F}^{n\times n}$ be a nonsingular matrix such that $A = \hat{S}B\hat{S}^*$. Proposition 7.5.6 implies that there exists a nonsingular matrix $S \in \mathbb{F}^{n\times n}$ such that

$$B = S\begin{bmatrix} -I_{\nu_-(B)} & 0 & 0 \\ 0 & 0_{\nu_0(B)\times\nu_0(B)} & 0 \\ 0 & 0 & I_{\nu_+(B)} \end{bmatrix}S^*.$$

Therefore,

$$A = \hat{S}S\begin{bmatrix} -I_{\nu_-(B)} & 0 & 0 \\ 0 & 0_{\nu_0(B)\times\nu_0(B)} & 0 \\ 0 & 0 & I_{\nu_+(B)} \end{bmatrix}(\hat{S}S)^*.$$

Since A and $\begin{bmatrix} -I_{\nu_-(B)} & 0 & 0 \\ 0 & 0_{\nu_0(B)\times\nu_0(B)} & 0 \\ 0 & 0 & I_{\nu_+(B)} \end{bmatrix}$ are congruent, Fact 7.9.17 implies that $\operatorname{In} A = \operatorname{In} B$. □

Proposition 6.5.4 shows that eigenvectors associated with distinct eigenvalues of a normal matrix are mutually orthogonal. Thus, a normal matrix has at least as many mutually orthogonal eigenvectors as it has distinct eigenvalues. The next result, which follows from Corollary 7.5.4, shows that every $n \times n$ normal matrix has n mutually orthogonal eigenvectors. In fact, the converse is also true.

Corollary 7.5.8. Let $A \in \mathbb{C}^{n\times n}$. Then, A is normal if and only if A has n mutually orthogonal eigenvectors.

The following result concerns the *real normal form.*

Corollary 7.5.9. Let $A \in \mathbb{R}^{n\times n}$ be range symmetric. Then, there exist an orthogonal matrix $S \in \mathbb{R}^{n\times n}$ and a nonsingular matrix $B \in \mathbb{R}^{r\times r}$, where $r \triangleq \operatorname{rank} A$, such that

$$A = S\begin{bmatrix} B & 0 \\ 0 & 0 \end{bmatrix}S^{\mathsf{T}}. \tag{7.5.8}$$

In addition, assume that A is normal, and let $\operatorname{mspec}(A) = \{\lambda_1,\ldots,\lambda_r\}_{\rm ms} \cup \{\nu_1+\omega_1 J, \nu_1-\omega_1 J,\ldots,\nu_l+\omega_l J, \nu_l-\omega_l J\}_{\rm ms}$, where $\lambda_1,\ldots,\lambda_r \in \mathbb{R}$ and, for all $i \in \{1,\ldots,l,\}$ $\nu_i,\omega_i \in \mathbb{R}$ and $\omega_i \neq 0$. Then, there exists an orthogonal matrix $S \in \mathbb{R}^{n\times n}$ such that

$$A = SBS^{\mathsf{T}}, \tag{7.5.9}$$

where $B \triangleq \operatorname{diag}(B_1,\ldots,B_r,\hat{B}_1,\ldots,\hat{B}_l)$, $B_i \triangleq [\lambda_i]$ for all $i \in \{1,\ldots,r\}$, and $\hat{B}_i \triangleq \left[\begin{smallmatrix} \nu_i & \omega_i \\ -\omega_i & \nu_i \end{smallmatrix}\right]$ for all $i \in \{1,\ldots,l\}$.

7.6 Singular Value Decomposition, Polar Decomposition, and Full-Rank Factorization

We now consider the *singular value decomposition*, which, like the Smith decomposition but unlike the Jordan and Schur decompositions, applies to rectangular matrices. Let $A \in \mathbb{F}^{n\times m}$, where $A \neq 0$, and consider the positive-semidefinite matrices $AA^* \in \mathbb{F}^{n\times n}$ and $A^*A \in \mathbb{F}^{m\times m}$. It follows from Proposition 6.4.10 that AA^* and A^*A have the same nonzero eigenvalues with the same algebraic multiplicities. Since AA^* and A^*A are positive semidefinite, it follows that they have the same *positive* eigenvalues with the same algebraic multiplicities. Furthermore, since AA^* is Hermitian, it follows that the number of positive eigenvalues of AA^* (or A^*A), counting algebraic multiplicity, is equal to the rank of AA^* (or A^*A). Since $\operatorname{rank} A = \operatorname{rank} AA^* = \operatorname{rank} A^*A$, it thus follows that AA^* and A^*A both have r positive eigenvalues, where $r \triangleq \operatorname{rank} A$.

Definition 7.6.1. Let $A \in \mathbb{F}^{n\times m}$. Then, the *singular values* of A are the $\min\{n,m\}$ nonnegative numbers $\sigma_1(A),\ldots,\sigma_{\min\{n,m\}}(A)$, where, for all $i\in\{1,\ldots,\min\{n,m\}\}$,

$$\sigma_i(A) \triangleq \lambda_i^{1/2}(AA^*) = \lambda_i^{1/2}(A^*A). \tag{7.6.1}$$

For $A\in\mathbb{F}^{n\times m}$, define the *singular value vector* of A by

$$\sigma(A) \triangleq \begin{bmatrix}\sigma_1(A)\\ \vdots\\ \sigma_{\min\{n,m\}}(A)\end{bmatrix} \tag{7.6.2}$$

and the multiset of singular values of A by

$$\operatorname{msval}(A) \triangleq \{\sigma_1(A),\ldots,\sigma_{\min\{n,m\}}(A)\}_{\rm ms}. \tag{7.6.3}$$

In terms of the notation $\langle A\rangle$ defined by (10.5.6), it follows that, if $m\le n$, then $\sigma(A)=\lambda(\langle A\rangle)$, whereas, if $n\le m$, then $\sigma(A)=\lambda(\langle A^*\rangle)$.

Let $A\in\mathbb{F}^{n\times m}$. Since AA^* is positive semidefinite, it follows that

$$0\le\sigma_{\min\{n,m\}}(A)\le\cdots\le\sigma_1(A). \tag{7.6.4}$$

Define $r\triangleq\operatorname{rank}A$. If $1\le r<\min\{n,m\}$, then

$$0=\sigma_{\min\{n,m\}}(A)=\cdots=\sigma_{r+1}(A)<\sigma_r(A)\le\cdots\le\sigma_1(A), \tag{7.6.5}$$

whereas, if $r=\min\{m,n\}$, then

$$0<\sigma_{\min\{n,m\}}(A)=\sigma_r(A)\le\cdots\le\sigma_1(A). \tag{7.6.6}$$

Consequently, $\operatorname{rank}A$ is the number of positive singular values of A. For convenience, define

$$\sigma_{\max}(A)\triangleq\sigma_1(A) \tag{7.6.7}$$

and, if $n=m$,

$$\sigma_{\min}(A)\triangleq\sigma_n(A). \tag{7.6.8}$$

If $n\ne m$, then $\sigma_{\min}(A)$ is not defined. By convention, we define

$$\sigma_{\max}(0_{n\times m})=\sigma_{\min}(0_{n\times n})=0, \tag{7.6.9}$$

and, for all $i\in\{1,\ldots,\min\{n,m\}\}$,

$$\sigma_i(A)=\sigma_i(A^*)=\sigma_i(\overline{A})=\sigma_i(A^{\mathsf T}). \tag{7.6.10}$$

Now, suppose that $n=m$. If A is Hermitian, then, for all $i\in\{1,\ldots,n\}$,

$$\sigma_i(A)=\rho_i(A) \tag{7.6.11}$$

and

$$\{\sigma_1(A),\ldots,\sigma_n(A)\}_{\rm ms}=\{\rho_1(A),\ldots,\rho_n(A)\}_{\rm ms}=\{|\lambda_1(A)|,\ldots,|\lambda_n(A)|\}_{\rm ms}. \tag{7.6.12}$$

Finally, if A is positive semidefinite, then, for all $i\in\{1,\ldots,n\}$,

$$\sigma_i(A)=\rho_i(A)=\lambda_i(A). \tag{7.6.13}$$

Proposition 7.6.2. Let $A\in\mathbb{F}^{n\times m}$. If $n\le m$, then the following statements are equivalent:

i) $\operatorname{rank}A=n$.

ii) $\sigma_n(A)>0$.

If $m\le n$, then the following statements are equivalent:

iii) $\operatorname{rank}A=m$.

iv) $\sigma_m(A)>0$.

If $n = m$, then the following statements are equivalent:

v) A is nonsingular.

vi) $\sigma_{\min}(A) > 0$.

We now state the singular value decomposition.

Theorem 7.6.3. Let $A \in \mathbb{F}^{n\times m}$, assume that A is nonzero, let $r \triangleq \operatorname{rank} A$, and define $B \triangleq \operatorname{diag}[\sigma_1(A),\ldots,\sigma_r(A)]$. Then, there exist unitary matrices $S_1 \in \mathbb{F}^{n\times n}$ and $S_2 \in \mathbb{F}^{m\times m}$ such that

$$A = S_1\begin{bmatrix} B & 0_{r\times(m-r)} \\ 0_{(n-r)\times r} & 0_{(n-r)\times(m-r)} \end{bmatrix} S_2. \tag{7.6.14}$$

Furthermore, each column of S_1 is an eigenvector of AA^*, while each column of S_2^* is an eigenvector of A^*A.

Proof. For convenience, assume that $r < \min\{n,m\}$, since otherwise some of the zero matrices in (7.6.14) become empty matrices. By Corollary 7.5.5 there exists a unitary matrix $U \in \mathbb{F}^{n\times n}$ such that

$$AA^* = U\begin{bmatrix} B^2 & 0 \\ 0 & 0 \end{bmatrix}U^*.$$

Partition $U = [U_1\ \ U_2]$, where $U_1 \in \mathbb{F}^{n\times r}$ and $U_2 \in \mathbb{F}^{n\times(n-r)}$. Since $U^*U = I_n$, it follows that $U_1^*U_1 = I_r$ and $U_1^*U = [I_r\ \ 0_{r\times(n-r)}]$. Now, define $V_1 \triangleq A^*U_1B^{-1} \in \mathbb{F}^{m\times r}$, and note that

$$V_1^*V_1 = B^{-1}U_1^*AA^*U_1B^{-1} = B^{-1}U_1^*U\begin{bmatrix} B^2 & 0 \\ 0 & 0 \end{bmatrix}U^*U_1B^{-1} = I_r.$$

Next, since $U_2^*U = [0_{(n-r)\times r}\ \ I_{n-r}]$, it follows that

$$U_2^*AA^* = [0\ \ I]\begin{bmatrix} B^2 & 0 \\ 0 & 0 \end{bmatrix}U^* = 0.$$

However, since $\mathcal{R}(A) = \mathcal{R}(AA^*)$, it follows that $U_2^*A = 0$. Finally, let $V_2 \in \mathbb{F}^{m\times(m-r)}$ be such that $V \triangleq [V_1\ \ V_2] \in \mathbb{F}^{m\times m}$ is unitary. Hence,

$$\begin{aligned} U\begin{bmatrix} B & 0 \\ 0 & 0 \end{bmatrix}V^* &= [U_1\ \ U_2]\begin{bmatrix} B & 0 \\ 0 & 0 \end{bmatrix}\begin{bmatrix} V_1^* \\ V_2^* \end{bmatrix} = U_1BV_1^* = U_1BB^{-1}U_1^*A \\ &= U_1U_1^*A = (U_1U_1^* + U_2U_2^*)A = UU^*A = A, \end{aligned}$$

which yields (7.6.14) with $S_1 = U$ and $S_2 = V^*$. □

A corollary of the singular value decomposition is the *polar decomposition*.

Corollary 7.6.4. Let $A \in \mathbb{F}^{n\times n}$. Then, there exists a positive-semidefinite matrix $M \in \mathbb{F}^{n\times n}$ and a unitary matrix $S \in \mathbb{F}^{n\times n}$ such that

$$A = MS. \tag{7.6.15}$$

Proof. It follows from the singular value decomposition that there exist unitary matrices $S_1, S_2 \in \mathbb{F}^{n\times n}$ and a diagonal positive-definite matrix $B \in \mathbb{F}^{r\times r}$, where $r \triangleq \operatorname{rank} A$, such that $A = S_1\left[\begin{smallmatrix} B & 0 \\ 0 & 0 \end{smallmatrix}\right]S_2$. Hence,

$$A = S_1\begin{bmatrix} B & 0 \\ 0 & 0 \end{bmatrix}S_1^*S_1S_2 = MS,$$

where $M \triangleq S_1\left[\begin{smallmatrix} B & 0 \\ 0 & 0 \end{smallmatrix}\right]S_1^*$ is positive semidefinite and $S \triangleq S_1S_2$ is unitary. □

Proposition 7.6.5. Let $A \in \mathbb{F}^{n\times m}$, let $r \triangleq \operatorname{rank} A$, and define the Hermitian matrix $\mathcal{A} \triangleq \left[\begin{smallmatrix} 0 & A \\ A^* & 0 \end{smallmatrix}\right] \in \mathbb{F}^{(n+m)\times(n+m)}$. Then, $\operatorname{In}\mathcal{A} = [r\ \ 0\ \ r]^{\mathsf{T}}$, and the $2r$ nonzero eigenvalues of $\mathcal{A}$ are the r positive singular values of A and their negatives.

Proof. Since $\chi_{\mathcal{A}}(s) = \det(s^2 I - A^*A)$, it follows that

$$\operatorname{mspec}(\mathcal{A})\backslash\{0,\ldots,0\}_{\mathrm{ms}} = \{\sigma_1(A), -\sigma_1(A), \ldots, \sigma_r(A), -\sigma_r(A)\}_{\mathrm{ms}}.$$ □

The following result shows that every nonzero matrix has a *full-rank factorization* $A = BC$, which is unique up to a nonsingular factor. Note that B is left invertible and C is right invertible.

Proposition 7.6.6. Let $A \in \mathbb{F}^{n\times m}$, assume that A is nonzero, and let $r \triangleq \operatorname{rank} A$. Then, the following statements hold.

i) There exist $B \in \mathbb{F}^{n\times r}$ and $C \in \mathbb{F}^{r\times m}$ such that $\operatorname{rank} B = \operatorname{rank} C = r$ and $A = BC$.

ii) Let $B, \hat{B} \in \mathbb{F}^{n\times r}$ and $C, \hat{C} \in \mathbb{F}^{r\times m}$, and assume that $A = BC = \hat{B}\hat{C}$ and $\operatorname{rank} B = \operatorname{rank} \hat{B} = \operatorname{rank} C = \operatorname{rank} \hat{C} = r$. Then, there exists a nonsingular matrix $S \in \mathbb{F}^{r\times r}$ such that $B = \hat{B}S$ and $C = S^{-1}\hat{C}$.

Proof. *i)* follows from (7.6.14) with $B = S_1\begin{bmatrix} B \\ 0 \end{bmatrix}$ and $C = [I\ \ 0]S_2$. *ii)* holds with $S \triangleq \hat{C}C^{\mathsf{T}}(CC^{\mathsf{T}})^{-1}$ and $S^{-1} = (B^{\mathsf{T}}B)^{-1}B^{\mathsf{T}}\hat{B}$. □

Corollary 7.6.7. Let $A \in \mathbb{F}^{n\times m}$, and assume that $\operatorname{rank} A = 1$. Then, there exist nonzero $x \in \mathbb{F}^n$ and nonzero $y \in \mathbb{F}^m$ such that $A = xy^{\mathsf{T}}$. If, in addition, $\hat{x} \in \mathbb{F}^n$ and $\hat{y} \in \mathbb{F}^m$ satisfy $A = \hat{x}\hat{y}^{\mathsf{T}}$, then there exists nonzero $\alpha \in \mathbb{F}$ such that $\hat{x} = \alpha x$ and $\hat{y} = (1/\alpha)y$.

7.7 Eigenstructure Properties

Definition 7.7.1. Let $A \in \mathbb{F}^{n\times n}$ and $\lambda \in \mathbb{C}$. Then, the *index of λ with respect to A*, denoted by $\operatorname{ind}_A(\lambda)$, is the smallest nonnegative integer k such that

$$\mathcal{R}[(\lambda I - A)^k] = \mathcal{R}[(\lambda I - A)^{k+1}]. \tag{7.7.1}$$

That is,

$$\operatorname{ind}_A(\lambda) = \operatorname{ind}(\lambda I - A). \tag{7.7.2}$$

Note that $\lambda \notin \operatorname{spec}(A)$ if and only if $\operatorname{ind}_A(\lambda) = 0$. Hence, $0 \notin \operatorname{spec}(A)$ if and only if $\operatorname{ind} A = \operatorname{ind}_A(0) = 0$. Finally, note that $\operatorname{ind}_A(0) = \operatorname{ind} A$.

Proposition 7.7.2. Let $A \in \mathbb{F}^{n\times n}$ and $\lambda \in \mathbb{C}$. Then, $\operatorname{ind}_A(\lambda)$ is the smallest nonnegative integer k such that

$$\operatorname{rank}[(\lambda I - A)^k] = \operatorname{rank}[(\lambda I - A)^{k+1}]. \tag{7.7.3}$$

Furthermore, $\operatorname{ind} A$ is the smallest nonnegative integer k such that

$$\operatorname{rank}(A^k) = \operatorname{rank}(A^{k+1}). \tag{7.7.4}$$

Proof. Corollary 3.5.2 implies that $\mathcal{R}[(\lambda I - A)^k] \subseteq \mathcal{R}[(\lambda I - A)^{k+1}]$. Consequently, Lemma 3.1.5 implies that $\mathcal{R}[(\lambda I - A)^k] = \mathcal{R}[(\lambda I - A)^{k+1}]$ if and only if $\operatorname{rank}[(\lambda I - A)^k] = \operatorname{rank}[(\lambda I - A)^{k+1}]$. □

Proposition 7.7.3. Let $A \in \mathbb{F}^{n\times n}$ and $\lambda \in \operatorname{spec}(A)$. Then, the following statements hold:

i) The largest size of all of the Jordan blocks of A associated with λ is $\operatorname{ind}_A(\lambda)$.

ii) $\operatorname{mult}_{\mu_A}(\lambda) = \operatorname{ind}_A(\lambda)$.

iii) The number of Jordan blocks of A associated with λ is $\operatorname{gmult}_A(\lambda)$.

iv) The number of linearly independent eigenvectors of A associated with λ is $\operatorname{gmult}_A(\lambda)$.

v) $\operatorname{ind}_A(\lambda) \le \operatorname{amult}_A(\lambda)$.

vi) $\operatorname{ind}_A(\lambda) = \operatorname{amult}_A(\lambda)$ if and only if exactly one block is associated with λ.

vii) $\operatorname{gmult}_A(\lambda) \le \operatorname{amult}_A(\lambda)$.

viii) $\operatorname{gmult}_A(\lambda) = \operatorname{amult}_A(\lambda)$ if and only if every block associated with λ is of size equal to 1.

ix) $\mathrm{ind}_A(\lambda) + \mathrm{gmult}_A(\lambda) \le \mathrm{amult}_A(\lambda) + 1$.

x) $\mathrm{ind}_A(\lambda) + \mathrm{gmult}_A(\lambda) = \mathrm{amult}_A(\lambda) + 1$ if and only if at most one block associated with λ is of size greater than 1.

Definition 7.7.4. Let $A \in \mathbb{F}^{n\times n}$ and $\lambda \in \mathrm{spec}(A)$. Then, the following terminology is defined:

i) λ is *simple* if $\mathrm{amult}_A(\lambda) = 1$.

ii) A is *simple* if every eigenvalue of A is simple.

iii) λ is *cyclic* (or *nonderogatory*) if $\mathrm{gmult}_A(\lambda) = 1$.

iv) A is *cyclic* (or *nonderogatory*) if every eigenvalue of A is cyclic.

v) λ is *derogatory* if $\mathrm{gmult}_A(\lambda) > 1$.

vi) A is *derogatory* if A has at least one derogatory eigenvalue.

vii) λ is *semisimple* if $\mathrm{gmult}_A(\lambda) = \mathrm{amult}_A(\lambda)$.

viii) A is *semisimple* if every eigenvalue of A is semisimple.

ix) λ is *defective* if $\mathrm{gmult}_A(\lambda) < \mathrm{amult}_A(\lambda)$.

x) A is *defective* if A has at least one defective eigenvalue.

Proposition 7.7.5. Let $A \in \mathbb{F}^{n\times n}$ and $\lambda \in \mathrm{spec}(A)$. Then, λ is simple if and only if λ is cyclic and semisimple.

Proposition 7.7.6. Let $A \in \mathbb{F}^{n\times n}$ and $\lambda \in \mathrm{spec}(A)$. Then,

$$\mathrm{gmult}_A(\lambda) = \mathrm{def}(\lambda I - A) \le \mathrm{def}\,(\lambda I - A)^{\mathrm{ind}_A(\lambda)} = \mathrm{amult}_A(\lambda). \tag{7.7.5}$$

Theorem 7.4.2 yields the following result, which shows that the subspaces $\mathcal{N}[(\lambda I - A)^k]$, where $\lambda \in \mathrm{spec}(A)$ and $k = \mathrm{ind}_A(\lambda)$, provide a decomposition of $\mathbb{F}^n$.

Proposition 7.7.7. Let $A \in \mathbb{F}^{n\times n}$, let $\mathrm{spec}(A) = \{\lambda_1, \ldots, \lambda_r\}$, and, for all $i \in \{1, \ldots, r\}$, let $k_i \triangleq \mathrm{ind}_A(\lambda_i)$. Then, the following statements hold:

i) $\mathcal{N}[(\lambda_i I - A)^{k_i}] \cap \mathcal{N}[(\lambda_j I - A)^{k_j}] = \{0\}$ for all $i, j \in \{1, \ldots, r\}$ such that $i \ne j$.

ii) $\sum_{i=1}^r \mathcal{N}[(\lambda_i I - A)^{k_i}] = \mathbb{F}^n$.

Proposition 7.7.8. Let $A \in \mathbb{F}^{n\times n}$ and $\lambda \in \mathrm{spec}(A)$. Then, the following statements are equivalent:

i) λ is semisimple.

ii) $\mathrm{def}(\lambda I - A) = \mathrm{def}[(\lambda I - A)^2]$.

iii) $\mathcal{N}(\lambda I - A) = \mathcal{N}[(\lambda I - A)^2]$.

iv) $\mathrm{ind}_A(\lambda) = 1$.

Proof. To prove *i)* $\Longrightarrow$ *ii)*, suppose that λ is semisimple so that $\mathrm{gmult}_A(\lambda) = \mathrm{amult}_A(\lambda)$, and thus $\mathrm{def}(\lambda I - A) = \mathrm{amult}_A(\lambda)$. Then, it follows from Proposition 7.7.6 that $\mathrm{def}[(\lambda I - A)^k] = \mathrm{amult}_A(\lambda)$, where $k \triangleq \mathrm{ind}_A(\lambda)$. Therefore, Corollary 3.6.6 implies that $\mathrm{amult}_A(\lambda) = \mathrm{def}(\lambda I - A) \le \mathrm{def}[(\lambda I - A)^2] \le \mathrm{def}[(\lambda I - A)^k] = \mathrm{amult}_A(\lambda)$, which implies that $\mathrm{def}(\lambda I - A) = \mathrm{def}[(\lambda I - A)^2]$.

To prove *ii)* $\Longrightarrow$ *iii)*, note that Corollary 3.6.6 implies that $\mathcal{N}(\lambda I - A) \subseteq \mathcal{N}[(\lambda I - A)^2]$. Since, by *ii)*, these subspaces have equal dimension, Lemma 3.1.5 implies that these subspaces are equal.

To prove *iii)* $\Longrightarrow$ *iv)*, note that *iii)* implies *ii)*, and thus $\mathrm{rank}(\lambda I - A) = n - \mathrm{def}(\lambda I - A) = n - \mathrm{def}[(\lambda I - A)^2] = \mathrm{rank}[(\lambda I - A)^2]$. Therefore, since $\mathcal{R}(\lambda I - A) \subseteq \mathcal{R}[(\lambda I - A)^2]$, Corollary 3.6.6 implies that $\mathcal{R}(\lambda I - A) = \mathcal{R}[(\lambda I - A)^2]$. Finally, since $\lambda \in \mathrm{spec}(A)$, it follows from Definition 7.7.1 that $\mathrm{ind}_A(\lambda) = 1$.

Finally, to prove *iv)* $\Longrightarrow$ *i)*, note that *iv)* is equivalent to the fact that every Jordan block of A associated with λ has size 1, which is equivalent to the fact that the geometric multiplicity of λ is equal to the algebraic multiplicity of λ; that is, λ is semisimple. $\square$

Corollary 7.7.9. Let $A \in \mathbb{F}^{n\times n}$. Then, A is group invertible if and only if $\operatorname{ind} A \le 1$.

Proposition 7.7.10. Let $A, B \in \mathbb{F}^{n\times n}$. If A and B are similar, then the following statements hold:

i) $\operatorname{mspec}(A) = \operatorname{mspec}(B)$.

ii) For all $\lambda \in \operatorname{spec}(A)$, $\operatorname{gmult}_A(\lambda) = \operatorname{gmult}_B(\lambda)$.

Furthermore, if $n \le 3$ and *i)* and *ii)* hold, then A and B are similar.

The example

$$A = \begin{bmatrix} 0 & 1 & 0 & 0 \\ 0 & 0 & 0 & 0 \\ 0 & 0 & 0 & 1 \\ 0 & 0 & 0 & 0 \end{bmatrix}, \quad B = \begin{bmatrix} 0 & 1 & 0 & 0 \\ 0 & 0 & 1 & 0 \\ 0 & 0 & 0 & 0 \\ 0 & 0 & 0 & 0 \end{bmatrix} \tag{7.7.6}$$

shows that, for all $n \ge 4$, the converse of Proposition 7.7.10 does not hold.

Proposition 7.7.11. Let $A \in \mathbb{C}^{n\times n}$. Then, the following statements are equivalent:

i) A is semisimple.

ii) There exists a nonsingular matrix $S \in \mathbb{C}^{n\times n}$ such that SAS^{-1} is diagonal.

Now, assume that $A \in \mathbb{R}^{n\times n}$. Then, the following statements are equivalent:

iii) A is semisimple and $\operatorname{spec}(A) \subset \mathbb{R}$.

iv) There exists a nonsingular matrix $S \in \mathbb{R}^{n\times n}$ such that SAS^{-1} is diagonal.

In view of Proposition 7.7.11, a complex matrix A is *diagonalizable over* $\mathbb{C}$ if and only if A is semisimple. If, in addition, A is real, then A is *diagonalizable over* $\mathbb{R}$ if and only if A is semisimple and every eigenvalue of A is real. In the following result, "similar over $\mathbb{F}$" means that the entries of the similarity transformation are elements of $\mathbb{F}$.

Proposition 7.7.12. Let $A \in \mathbb{C}^{n\times n}$. Then, A is semisimple if and only if A is similar over $\mathbb{C}$ to a normal matrix. Now, assume that $A \in \mathbb{R}^{n\times n}$. Then, A is semisimple and $\operatorname{spec}(A) \subset \mathbb{R}$ if and only if A is similar over $\mathbb{R}$ to a real symmetric matrix.

The following result is an extension of Corollary 7.4.10.

Proposition 7.7.13. Let $A \in \mathbb{F}^{n\times n}$. Then, the following statements are equivalent:

i) A is semisimple, and $\operatorname{spec}(A) \subset \mathbb{R}$.

ii) There exists a positive-definite matrix $S \in \mathbb{F}^{n\times n}$ such that $A = SA^*S^{-1}$.

iii) There exist a Hermitian matrix $S_1 \in \mathbb{F}^{n\times n}$ and a positive-definite matrix $S_2 \in \mathbb{F}^{n\times n}$ such that $A = S_1S_2$.

Proof. To prove *i)* $\Longrightarrow$ *ii)*, let $\hat{S} \in \mathbb{F}^{n\times n}$ be a nonsingular matrix such that $A = \hat{S}B\hat{S}^{-1}$, where $B \in \mathbb{R}^{n\times n}$ is diagonal. Then, $B = \hat{S}^{-1}A\hat{S} = \hat{S}^*A^*\hat{S}^{-*}$. Hence, $A = \hat{S}B\hat{S}^{-1} = \hat{S}(\hat{S}^*A^*\hat{S}^{-*})\hat{S}^{-1} = (\hat{S}\hat{S}^*)A^*(\hat{S}\hat{S}^*)^{-1} = SA^*S^{-1}$, where $S \triangleq \hat{S}\hat{S}^*$ is positive definite. To show that *ii)* implies *iii)*, note that $A = SA^*S^{-1} = S_1S_2$, where $S_1 \triangleq SA^*$ and $S_2 = S^{-1}$. Since $S_1^* = (SA^*)^* = AS^* = AS = SA^* = S_1$, it follows that S_1 is Hermitian. Furthermore, since S is positive definite, it follows that S^{-1}, and hence S_2, is also positive definite. Finally, to prove *iii)* $\Longrightarrow$ *i)*, note that $A = S_1S_2 = S_2^{-1/2}(S_2^{1/2}S_1S_2^{1/2})S_2^{1/2}$. Since $S_2^{1/2}S_1S_2^{1/2}$ is Hermitian, Corollary 7.5.5 implies that $S_2^{1/2}S_1S_2^{1/2}$ is unitarily similar to a real diagonal matrix. Thus, A is semisimple and $\operatorname{spec}(A) \subset \mathbb{R}$. □

If a matrix is block triangular, then the following result shows that its eigenvalues and their algebraic multiplicity are determined by the diagonally located blocks. If, in addition, the matrix is block diagonal, then the geometric multiplicities of its eigenvalues are determined by the diagonally located blocks.

Proposition 7.7.14. Let $A \in \mathbb{F}^{n\times n}$, assume that A is partitioned as $A = \begin{bmatrix} A_{11} & \cdots & A_{1k} \\ \vdots & \ddots & \vdots \\ A_{k1} & \cdots & A_{kk} \end{bmatrix}$, where, for all $i, j \in \{1,\dots,k\}$, $A_{ij} \in \mathbb{F}^{n_i \times n_j}$, and let $\lambda \in \operatorname{spec}(A)$. Then, the following statements hold:

i) If A_{ii} is the unique nonzero block in the ith column of blocks, then

$$\operatorname{amult}_{A_{ii}}(\lambda) \le \operatorname{amult}_A(\lambda). \tag{7.7.7}$$

ii) If A is either upper block triangular or lower block triangular, then

$$\operatorname{amult}_A(\lambda) = \sum_{i=1}^{r} \operatorname{amult}_{A_{ii}}(\lambda), \quad \operatorname{mspec}(A) = \bigcup_{i=1}^{k} \operatorname{mspec}(A_{ii}). \tag{7.7.8}$$

iii) If A_{ii} is the unique nonzero block in the ith column of blocks, then

$$\operatorname{gmult}_{A_{ii}}(\lambda) \le \operatorname{gmult}_A(\lambda). \tag{7.7.9}$$

iv) If A is upper block triangular, then

$$\operatorname{gmult}_{A_{11}}(\lambda) \le \operatorname{gmult}_A(\lambda). \tag{7.7.10}$$

v) If A is lower block triangular, then

$$\operatorname{gmult}_{A_{kk}}(\lambda) \le \operatorname{gmult}_A(\lambda). \tag{7.7.11}$$

vi) If A is block diagonal, then

$$\operatorname{gmult}_A(\lambda) = \sum_{i=1}^{r} \operatorname{gmult}_{A_{ii}}(\lambda). \tag{7.7.12}$$

Proposition 7.7.15. Let $A \in \mathbb{F}^{n\times n}$, let $\operatorname{spec}(A) = \{\lambda_1,\dots,\lambda_r\}$, and, for all $i \in \{1,\dots,r\}$, let $k_i \triangleq \operatorname{ind}_A(\lambda_i)$. Then,

$$\mu_A(s) = \prod_{i=1}^{r} (s-\lambda_i)^{k_i}, \quad \deg \mu_A = \sum_{i=1}^{r} k_i. \tag{7.7.13}$$

Furthermore, the following statements are equivalent:

i) $\mu_A = \chi_A$.

ii) A is cyclic.

iii) For all $\lambda \in \operatorname{spec}(A)$, the Jordan form of A contains exactly one block associated with λ.

iv) A is similar to $C(\chi_A)$.

Proof. Let $A = SBS^{-1}$, where $B = \operatorname{diag}(B_1,\dots,B_{n_{\rm h}})$ denotes the Jordan form of A given by (7.4.2). Let $\lambda_i \in \operatorname{spec}(A)$, and let B_j be a Jordan block associated with λ_i. Then, the size of B_j is either less than or equal to k_i. Consequently, $(B_j - \lambda_i I)^{k_i} = 0$.

Next, let $p(s)$ denote the right-hand side of the first equality in (7.7.13). Then,

$$\begin{aligned} p(A) &= \prod_{i=1}^{r} (A-\lambda_i I)^{k_i} = S\left[\prod_{i=1}^{r} (B-\lambda_i I)^{k_i}\right] S^{-1} \\ &= S \operatorname{diag}\left(\prod_{i=1}^{r} (B_1-\lambda_i I)^{k_i},\dots,\prod_{i=1}^{r} (B_{n_{\rm h}}-\lambda_i I)^{k_i}\right) S^{-1} = 0. \end{aligned}$$

Therefore, it follows from Theorem 6.6.1 that μ_A divides p. Furthermore, note that, replacing k_i by $\hat{k}_i < k_i$ yields $p(A) \ne 0$. Hence, p is the minimal polynomial of A. The equivalence of *i)* and *ii)* is now immediate, while the equivalence of *ii)* and *iii)* follows from Theorem 7.4.6. The equivalence of *i)* and *iv)* is given by Corollary 7.3.4. □

Example 7.7.16. The standard nilpotent matrix N_n is in companion form, and thus is cyclic. In fact, N_n consists of a single Jordan block, and $\chi_{N_n}(s) = \mu_{N_n}(s) = s^n$. ◇

Example 7.7.17. The matrix $\begin{bmatrix} 1 & 1 \\ -1 & 1 \end{bmatrix}$ is normal but is neither symmetric nor skew symmetric, while the matrix $\begin{bmatrix} 0 & 1 \\ -1 & 0 \end{bmatrix}$ is normal but is neither symmetric nor semisimple with real eigenvalues. ◇

Example 7.7.18. The matrices $\begin{bmatrix} 1 & 0 \\ 2 & -1 \end{bmatrix}$ and $\begin{bmatrix} 1 & 1 \\ 0 & 2 \end{bmatrix}$ are diagonalizable over $\mathbb{R}$ but not normal, while the matrix $\begin{bmatrix} -1 & 1 \\ -2 & 1 \end{bmatrix}$ is diagonalizable but is neither normal nor diagonalizable over $\mathbb{R}$. ◇

Example 7.7.19. The product of the Hermitian matrices $\begin{bmatrix} 1 & 2 \\ 2 & 1 \end{bmatrix}$ and $\begin{bmatrix} 2 & 1 \\ 1 & -2 \end{bmatrix}$ has no real eigenvalues. ◇

Example 7.7.20. The matrices $\begin{bmatrix} 1 & 0 \\ 0 & 2 \end{bmatrix}$ and $\begin{bmatrix} 0 & 1 \\ -2 & 3 \end{bmatrix}$ are similar, whereas $\begin{bmatrix} 1 & 0 \\ 0 & 1 \end{bmatrix}$ and $\begin{bmatrix} 0 & 1 \\ -1 & 2 \end{bmatrix}$ have the same spectrum but are not similar. ◇

Proposition 7.7.21. Let $A \in \mathbb{F}^{n\times n}$. Then, the following statements hold:

i) A is singular if and only if $0 \in \operatorname{spec}(A)$.

ii) A is group invertible if and only if either A is nonsingular or $0 \in \operatorname{spec}(A)$ is semisimple.

iii) A is Hermitian if and only if A is normal and $\operatorname{spec}(A) \subset \mathbb{R}$.

iv) A is skew Hermitian if and only if A is normal and $\operatorname{spec}(A) \subset \mathrm{IA}$.

v) A is positive semidefinite if and only if A is normal and $\operatorname{spec}(A) \subset [0, \infty)$.

vi) A is positive definite if and only if A is normal and $\operatorname{spec}(A) \subset (0, \infty)$.

vii) A is unitary if and only if A is normal and $\operatorname{spec}(A) \subset \mathrm{UC}$.

viii) A is shifted unitary if and only if A is normal and

$$\operatorname{spec}(A) \subset \{\lambda \in \mathbb{C}\colon\ |\lambda - \tfrac{1}{2}| = \tfrac{1}{2}\}. \tag{7.7.14}$$

ix) A is involutory if and only if A is semisimple and $\operatorname{spec}(A) \subseteq \{-1, 1\}$.

x) A is skew involutory if and only if A is semisimple and $\operatorname{spec}(A) \subseteq \{-\jmath, \jmath\}$.

xi) A is idempotent if and only if A is semisimple and $\operatorname{spec}(A) \subseteq \{0, 1\}$.

xii) A is skew idempotent if and only if A is semisimple and $\operatorname{spec}(A) \subseteq \{0, -1\}$.

xiii) A is tripotent if and only if A is semisimple and $\operatorname{spec}(A) \subseteq \{-1, 0, 1\}$.

xiv) A is nilpotent if and only if $\operatorname{spec}(A) = \{0\}$.

xv) A is unipotent if and only if $\operatorname{spec}(A) = \{1\}$.

xvi) A is a projector if and only if A is normal and $\operatorname{spec}(A) \subseteq \{0, 1\}$.

xvii) A is a reflector if and only if A is normal and $\operatorname{spec}(A) \subseteq \{-1, 1\}$.

xviii) A is a skew reflector if and only if A is normal and $\operatorname{spec}(A) \subseteq \{-\jmath, \jmath\}$.

xix) A is an elementary projector if and only if A is normal and $\operatorname{mspec}(A) = \{0, 1, \ldots, 1\}_{\mathrm{ms}}$.

xx) A is an elementary reflector if and only if A is normal and $\operatorname{mspec}(A) = \{-1, 1, \ldots, 1\}_{\mathrm{ms}}$.

If, furthermore, $A \in \mathbb{F}^{2n\times 2n}$, then the following statements hold:

xxi) If A is Hamiltonian, then $\operatorname{mspec}(A) = \operatorname{mspec}(-A)$.

xxii) If A is symplectic, then $\operatorname{mspec}(A) = \operatorname{mspec}(A^{-1})$.

The following result is a consequence of Proposition 7.7.13 and Proposition 7.7.21.

Corollary 7.7.22. Let $A \in \mathbb{F}^{n\times n}$, and assume that A is either involutory, idempotent, skew idempotent, tripotent, a projector, or a reflector. Then, the following statements hold:

i) There exists a positive-definite matrix $S \in \mathbb{F}^{n\times n}$ such that $A = SA^*S^{-1}$.

ii) There exist a Hermitian matrix $S_1 \in \mathbb{F}^{n\times n}$ and a positive-definite matrix $S_2 \in \mathbb{F}^{n\times n}$ such that $A = S_1S_2$.

Proposition 7.7.23. Let $A, B \in \mathbb{F}^{n\times n}$. Then, the following statements hold:

i) Assume that A and B are normal. Then, A and B are unitarily similar if and only if mspec(A) = mspec(B).

ii) Assume that A and B are projectors. Then, A and B are unitarily similar if and only if $\operatorname{rank} A = \operatorname{rank} B$.

iii) Assume that A and B are (projectors, reflectors). Then, A and B are unitarily similar if and only if $\operatorname{tr} A = \operatorname{tr} B$.

iv) Assume that A and B are semisimple. Then, A and B are similar if and only if $\operatorname{mspec}(A) = \operatorname{mspec}(B)$.

v) Assume that A and B are (involutory, skew involutory, idempotent). Then, A and B are similar if and only if $\operatorname{tr} A = \operatorname{tr} B$.

vi) Assume that A and B are idempotent. Then, A and B are similar if and only if $\operatorname{rank} A = \operatorname{rank} B$.

vii) Assume that A and B are tripotent. Then, A and B are similar if and only if $\operatorname{rank} A = \operatorname{rank} B$ and $\operatorname{tr} A = \operatorname{tr} B$.

7.8 Pencils and the Kronecker Canonical Form

Let $A, B \in \mathbb{F}^{n\times m}$, and define the polynomial matrix $P_{A,B} \in \mathbb{F}[s]^{n\times m}$, called a *pencil*, by

$$P_{A,B}(s) \triangleq sB - A.$$

The pencil $P_{A,B}$ is *regular* if $\operatorname{rank} P_{A,B} = \min\{n, m\}$ (see Definition 6.2.4). Otherwise, $P_{A,B}$ is *singular*.

Let $A, B \in \mathbb{F}^{n\times m}$. Since $P_{A,B} \in \mathbb{F}^{n\times m}$ we define the *generalized spectrum* of $P_{A,B}$ by

$$\operatorname{spec}(A, B) \triangleq \operatorname{Szeros}(P_{A,B}) \tag{7.8.1}$$

and the *generalized multispectrum* of $P_{A,B}$ by

$$\operatorname{mspec}(A, B) \triangleq \operatorname{mSzeros}(P_{A,B}). \tag{7.8.2}$$

Furthermore, the elements of spec(A, B) are the *generalized eigenvalues* of $P_{A,B}$. The structure of a pencil is illuminated by the following result called the *Kronecker canonical form*.

Theorem 7.8.1. Let $A, B \in \mathbb{C}^{n\times m}$. Then, there exist nonsingular matrices $S_1 \in \mathbb{C}^{n\times n}$ and $S_2 \in \mathbb{C}^{m\times m}$ such that, for all $s \in \mathbb{C}$,

$$\begin{aligned} P_{A,B}(s) = S_1\operatorname{diag}(&sI_{r_1} - A_1, sB_2 - I_{r_2}, [sI_{k_1}{-}N_{k_1} \quad -e_{k_1}], \ldots, [sI_{k_p}{-}N_{k_p} \quad -e_{k_p}],\\ &[sI_{l_1}{-}N_{l_1} \quad -e_{l_1}]^{\mathsf T}, \ldots, [sI_{l_q}{-}N_{l_q} \quad -e_{l_q}]^{\mathsf T}, 0_{t\times u})S_2, \end{aligned} \tag{7.8.3}$$

where $A_1 \in \mathbb{C}^{r_1\times r_1}$ is in Jordan form, $B_2 \in \mathbb{R}^{r_2\times r_2}$ is nilpotent and in Jordan form, $k_1, \ldots, k_p, l_1, \ldots, l_q$ are positive integers, and $[sI_l - N_l \quad -e_l] \in \mathbb{C}^{l\times(l+1)}$. Furthermore,

$$\operatorname{rank} P_{A,B} = r_1 + r_2 + \sum_{i=1}^{p} k_i + \sum_{i=1}^{q} l_i. \tag{7.8.4}$$

Proof. See [151, Chapter 2], [1140, Chapter XII], [1573, pp. 395–398], [1737], [1762, pp. 128, 129], and [2539, Chapter VI]. □

In Theorem 7.8.1, note that

$$n = r_1 + r_2 + \sum_{i=1}^{p} k_i + \sum_{i=1}^{q} l_i + q + t, \quad m = r_1 + r_2 + \sum_{i=1}^{p} k_i + \sum_{i=1}^{q} l_i + p + u. \tag{7.8.5}$$

Proposition 7.8.2. Let $A, B \in \mathbb{C}^{n\times m}$, and consider the notation of Theorem 7.8.1. Then, $P_{A,B}$ is regular if and only if $t = u = 0$ and either $p = 0$ or $q = 0$.

Let $A, B \in \mathbb{F}^{n\times m}$, and let $\lambda \in \mathbb{C}$. Then,

$$\operatorname{rank} P_{A,B}(\lambda) = \operatorname{rank}(\lambda I - A_1) + r_2 + \sum_{i=1}^{p} k_i + \sum_{i=1}^{q} l_i. \tag{7.8.6}$$

Note that λ is a generalized eigenvalue of $P_{A,B}$ if and only if $\operatorname{rank} P_{A,B}(\lambda) < \operatorname{rank} P_{A,B}$. Consequently, λ is a generalized eigenvalue of $P_{A,B}$ if and only if λ is an eigenvalue of A_1; that is,

$$\operatorname{spec}(A, B) = \operatorname{spec}(A_1). \tag{7.8.7}$$

Furthermore,

$$\operatorname{mspec}(A, B) = \operatorname{mspec}(A_1). \tag{7.8.8}$$

The *generalized algebraic multiplicity* $\operatorname{amult}_{A,B}(\lambda)$ of $\lambda \in \operatorname{spec}(A, B)$ is defined by

$$\operatorname{amult}_{A,B}(\lambda) \triangleq \operatorname{amult}_{A_1}(\lambda). \tag{7.8.9}$$

The *generalized geometric multiplicity* $\operatorname{gmult}_{A,B}(\lambda)$ of $\lambda \in \operatorname{spec}(A, B)$ is defined by

$$\operatorname{gmult}_{A,B}(\lambda) \triangleq \operatorname{gmult}_{A_1}(\lambda). \tag{7.8.10}$$

For all $\lambda \in \operatorname{spec}(A, B)$,

$$\operatorname{gmult}_{A_1}(\lambda) = \operatorname{rank} P_{A,B} - \operatorname{rank} P_{A,B}(\lambda).$$

Now, assume that $A, B \in \mathbb{F}^{n\times n}$, and thus A and B are square, which, from (7.8.5), is equivalent to $q + t = p + u$. Then, the *characteristic polynomial* $\chi_{A,B} \in \mathbb{F}[s]$ of (A, B) is defined by

$$\chi_{A,B}(s) \triangleq \det P_{A,B}(s) = \det(sB - A).$$

Proposition 7.8.3. Let $A, B \in \mathbb{F}^{n\times n}$. Then, the following statements hold:

i) $P_{A,B}$ is singular if and only if $\chi_{A,B} = 0$.

ii) $P_{A,B}$ is singular if and only if $\deg \chi_{A,B} = -\infty$.

iii) $P_{A,B}$ is regular if and only if $\chi_{A,B}$ is not the zero polynomial.

iv) $P_{A,B}$ is regular if and only if $0 \le \deg \chi_{A,B} \le n$.

v) If $P_{A,B}$ is regular, then $\operatorname{mult}_{\chi_{A,B}}(0) = n - \deg \chi_{B,A}$.

vi) $\deg \chi_{A,B} = n$ if and only if B is nonsingular.

vii) If B is nonsingular, then $\chi_{A,B} = (\det B)\chi_{B^{-1}A}$, $\operatorname{spec}(A, B) = \operatorname{spec}(B^{-1}A)$, and $\operatorname{mspec}(A, B) = \operatorname{mspec}(B^{-1}A)$.

viii) $\operatorname{roots}(\chi_{A,B}) = \operatorname{spec}(A, B)$.

ix) $\operatorname{mroots}(\chi_{A,B}) = \operatorname{mspec}(A, B)$.

x) If either A or B is nonsingular, then $P_{A,B}$ is regular.

xi) If all of the generalized eigenvalues of (A, B) are real, then $P_{A,B}$ is regular.

xii) If $P_{A,B}$ is regular, then $\mathcal{N}(A) \cap \mathcal{N}(B) = \{0\}$.

xiii) If $P_{A,B}$ is regular, then there exist nonsingular matrices $S_1, S_2 \in \mathbb{C}^{n\times n}$ such that, for all $s \in \mathbb{C}$,

$$P_{A,B}(s) = S_1\left(s\begin{bmatrix} I_r & 0 \\ 0 & B_2 \end{bmatrix} - \begin{bmatrix} A_1 & 0 \\ 0 & I_{n-r} \end{bmatrix}\right)S_2, \tag{7.8.11}$$

where $r \triangleq \deg \chi_{A,B}$, $A_1 \in \mathbb{C}^{r\times r}$ is in Jordan form, and $B_2 \in \mathbb{R}^{(n-r)\times(n-r)}$ is nilpotent and in Jordan form. Furthermore,

$$\chi_{A,B} = \chi_{A_1}, \quad \operatorname{roots}(\chi_{A,B}) = \operatorname{spec}(A_1), \quad \operatorname{mroots}(\chi_{A,B}) = \operatorname{mspec}(A_1). \tag{7.8.12}$$

Proof. See [1762, p. 128] and [2539, Chapter VI]. □

xiii) is the *Weierstrass canonical form* for a square, regular pencil.

Proposition 7.8.4. Let $A, B \in \mathbb{F}^{n\times n}$, and assume that A is positive semidefinite and B is Hermitian. Then, the following statements are equivalent:

i) $P_{A,B}$ is regular.

ii) There exists $\alpha \in \mathbb{F}$ such that $A + \alpha B$ is nonsingular.

iii) $\mathcal{N}(A) \cap \mathcal{N}(B) = \{0\}$.

iv) $\mathcal{N}(\left[\begin{smallmatrix} A \\ B \end{smallmatrix}\right]) = \{0\}$.

v) There exists nonzero $\alpha \in \mathbb{F}$ such that $\mathcal{N}(A) \cap \mathcal{N}(B + \alpha A) = \{0\}$.

vi) For all nonzero $\alpha \in \mathbb{F}$, $\mathcal{N}(A) \cap \mathcal{N}(B + \alpha A) = \{0\}$.

vii) All generalized eigenvalues of (A, B) are real.

If, in addition, B is positive semidefinite, then the following statement is equivalent to *i*)–*vii*):

viii) There exists $\beta \in \mathbb{R}$ such that $\beta B < A$.

Proof. *i*) $\Longrightarrow$ *ii*) and *ii*) $\Longrightarrow$ *iii*) are immediate. Next, Fact 3.13.17 and Fact 3.14.10 imply that *iii*), *iv*), *v*), and *vi*) are equivalent. Next, to prove *iii*) $\Longrightarrow$ *vii*), let $\lambda \in \mathbb{C}$ be a generalized eigenvalue of (A, B). Since $\lambda = 0$ is real, suppose $\lambda \neq 0$. Since $\det(\lambda B - A) = 0$, let nonzero $\theta \in \mathbb{C}^n$ satisfy $(\lambda B - A)\theta = 0$, and thus it follows that $\theta^*A\theta = \lambda\theta^*B\theta$. Furthermore, note that $\theta^*A\theta$ and $\theta^*B\theta$ are real. Now, suppose $\theta \in \mathcal{N}(A)$. Then, it follows from $(\lambda B - A)\theta = 0$ that $\theta \in \mathcal{N}(B)$, which contradicts $\mathcal{N}(A) \cap \mathcal{N}(B) = \{0\}$. Hence, $\theta \notin \mathcal{N}(A)$, and thus $\theta^*A\theta > 0$ and, consequently, $\theta^*B\theta \neq 0$. Hence, it follows that $\lambda = \theta^*A\theta/\theta^*B\theta$, and thus λ is real. Hence, all generalized eigenvalues of (A, B) are real.

To prove *vii*) $\Longrightarrow$ *i*), let $\lambda \in \mathbb{C}\backslash\mathbb{R}$ so that λ is not a generalized eigenvalue of (A, B). Consequently, $\chi_{A,B}(s)$ is not the zero polynomial, and thus (A, B) is regular.

To prove *i*)–*vii*) $\Longrightarrow$ *viii*), let $\theta \in \mathbb{R}^n$ be nonzero, and note that $\mathcal{N}(A) \cap \mathcal{N}(B) = \{0\}$ implies that either $A\theta \neq 0$ or $B\theta \neq 0$. Hence, either $\theta^{\mathsf{T}}A\theta > 0$ or $\theta^{\mathsf{T}}B\theta > 0$. Thus, $\theta^{\mathsf{T}}(A + B)\theta > 0$, which implies that $A + B > 0$ and hence $-B < A$.

Finally, to prove *viii*) $\Longrightarrow$ *i*)–*vii*), let $\beta \in \mathbb{R}$ satisfy $\beta B < A$, so that, for all nonzero $\theta \in \mathbb{R}^n$, $\beta\theta^{\mathsf{T}}B\theta < \theta^{\mathsf{T}}A\theta$. Next, suppose $\hat{\theta} \in \mathcal{N}(A) \cap \mathcal{N}(B)$ is nonzero. Then, $A\hat{\theta} = 0$ and $B\hat{\theta} = 0$, and thus $\hat{\theta}^{\mathsf{T}}B\hat{\theta} = 0$ and $\hat{\theta}^{\mathsf{T}}A\hat{\theta} = 0$, which contradicts $\beta\hat{\theta}^{\mathsf{T}}B\hat{\theta} < \hat{\theta}^{\mathsf{T}}A\hat{\theta}$. Thus, $\mathcal{N}(A) \cap \mathcal{N}(B) = \{0\}$. □

7.9 Facts on the Inertia

Fact 7.9.1. Let $A \in \mathbb{F}^{n\times n}$, and assume that A is idempotent. Then,

$$\operatorname{rank} A = \operatorname{sig} A = \operatorname{tr} A, \quad \operatorname{In} A = \begin{bmatrix} 0 \\ n - \operatorname{tr} A \\ \operatorname{tr} A \end{bmatrix}.$$

Fact 7.9.2. Let $A \in \mathbb{F}^{n\times n}$, and assume that A is involutory. Then,

$$\operatorname{rank} A = n, \quad \operatorname{sig} A = \operatorname{tr} A, \quad \operatorname{In} A = \begin{bmatrix} \frac{1}{2}(n - \operatorname{tr} A) \\ 0 \\ \frac{1}{2}(n + \operatorname{tr} A) \end{bmatrix}.$$

Fact 7.9.3. Let $A \in \mathbb{F}^{n\times n}$, and assume that A is tripotent. Then,

$$\operatorname{rank} A = \operatorname{tr} A^2, \quad \operatorname{sig} A = \operatorname{tr} A, \quad \operatorname{In} A = \begin{bmatrix} \frac{1}{2}(\operatorname{tr} A^2 - \operatorname{tr} A) \\ n - \operatorname{tr} A^2 \\ \frac{1}{2}(\operatorname{tr} A^2 + \operatorname{tr} A) \end{bmatrix}.$$

Fact 7.9.4. Let $A \in \mathbb{F}^{n\times n}$, and assume that A is either skew Hermitian, skew involutory, or nilpotent. Then,

$$\operatorname{sig} A = \nu_-(A) = \nu_+(A) = 0, \quad \operatorname{In} A = \begin{bmatrix} 0 \\ n \\ 0 \end{bmatrix}.$$

Fact 7.9.5. Let $A \in \mathbb{F}^{n\times n}$, assume that A is group invertible, and assume that $\operatorname{spec}(A) \cap \mathrm{IA} \subseteq \{0\}$. Then,

$$\operatorname{rank} A = \nu_-(A) + \nu_+(A), \quad \operatorname{def} A = \nu_0(A) = \operatorname{amult}_A(0).$$

Fact 7.9.6. Let $A \in \mathbb{F}^{n\times n}$, and assume that A is Hermitian. Then,

$$\operatorname{rank} A = \nu_-(A) + \nu_+(A), \quad \operatorname{In} A = \begin{bmatrix} \nu_-(A) \\ \nu_0(A) \\ \nu_+(A) \end{bmatrix} = \begin{bmatrix} \frac{1}{2}(\operatorname{rank} A - \operatorname{sig} A) \\ \operatorname{def} A \\ \frac{1}{2}(\operatorname{rank} A + \operatorname{sig} A) \end{bmatrix}.$$

Fact 7.9.7. Let $A, B \in \mathbb{F}^{n\times n}$, and assume that A and B are Hermitian. Then, $\operatorname{In} A = \operatorname{In} B$ if and only if $\operatorname{rank} A = \operatorname{rank} B$ and $\operatorname{sig} A = \operatorname{sig} B$.

Fact 7.9.8. Let $A \in \mathbb{F}^{n\times n}$, assume that A is Hermitian, and let A_0 be a principal submatrix of A. Then, $\nu_-(A_0) \le \nu_-(A)$ and $\nu_+(A_0) \le \nu_+(A)$. **Source:** [1546] and [2991, p. 259].

Fact 7.9.9. Let $A \in \mathbb{F}^{n\times n}$, and assume that A is positive semidefinite. Then,

$$\operatorname{rank} A = \operatorname{sig} A = \nu_+(A), \quad \operatorname{In} A = \begin{bmatrix} 0 \\ \operatorname{def} A \\ \operatorname{rank} A \end{bmatrix}.$$

If, in addition, A is positive definite, then $\operatorname{In} A = [0 \;\; 0 \;\; n]^\mathsf{T}$.

Fact 7.9.10. Let $A \in \mathbb{F}^{n\times n}$. Then, the following statements are equivalent:

i) A is an elementary projector.

ii) A is a projector, and $\operatorname{tr} A = n - 1$.

iii) A is a projector, and $\operatorname{In} A = \left[\begin{smallmatrix} 0 \\ 1 \\ n-1 \end{smallmatrix}\right]$.

Furthermore, the following statements are equivalent:

iv) A is an elementary reflector.

v) A is a reflector, and $\operatorname{tr} A = n - 2$.

vi) A is a reflector, and $\operatorname{In} A = \left[\begin{smallmatrix} 1 \\ 0 \\ n-1 \end{smallmatrix}\right]$.

Source: Proposition 7.7.21.

Fact 7.9.11. Let $A, B \in \mathbb{F}^{n\times n}$, and assume that A and B are projectors. Then,

$$\operatorname{In}(A - B) = \begin{bmatrix} \operatorname{rank}\,[A \;\; B] - \operatorname{rank} A \\ \operatorname{rank} A + \operatorname{rank} B + n - 2\operatorname{rank}\,[A \;\; B] \\ \operatorname{rank}\,[A \;\; B] - \operatorname{rank} B \end{bmatrix} = \begin{bmatrix} \operatorname{rank}(B - AB) \\ n - \operatorname{rank}(B - AB) - \operatorname{rank}(A - AB) \\ \operatorname{rank}(A - AB) \end{bmatrix}.$$

Furthermore, the following statements are equivalent:

i) $\operatorname{rank}(A - B) = \operatorname{rank} A + \operatorname{rank} B$.

ii) $\nu_+(A - B) = \operatorname{rank} A$.

iii) $\nu_-(A - B) = \operatorname{rank} B$.

iv) $\mathcal{R}(A) \cap \mathcal{R}(B) = \{0\}$.

In addition, the following statements are equivalent:

v) $\operatorname{rank}(A - B) = \operatorname{rank} A - \operatorname{rank} B$.

vi) $\nu_+(A-B) = \operatorname{rank} A - \operatorname{rank} B$.

vii) $\nu_-(A-B) = 0$.

viii) $\mathcal{R}(B) \subseteq \mathcal{R}(A)$.

Source: [2659].

Fact 7.9.12. Let $A, B \in \mathbb{F}^{n\times n}$, and assume that A and B are projectors. Then,

$$\operatorname{In}(I-A-B) = \begin{bmatrix} \operatorname{rank} AB \\ \operatorname{rank} A + \operatorname{rank} B - 2\operatorname{rank} AB \\ n - \operatorname{rank} A - \operatorname{rank} B + \operatorname{rank} AB \end{bmatrix}.$$

Furthermore, the following statements hold:

i) $I - A - B$ is nonsingular if and only if $\operatorname{rank} A = \operatorname{rank} B = \operatorname{rank} AB$.

ii) $A + B = I$ if and only if $\operatorname{rank} A + \operatorname{rank} B = n$ and $AB = 0$.

iii) $A + B < I$ if and only if $A = B = 0$.

iv) $I < A + B$ if and only if $A = B = I$.

v) $A + B \le I$ if and only if $AB = 0$.

vi) $I \le A + B$ if and only if $\operatorname{rank} A + \operatorname{rank} B = n + \operatorname{rank} AB$.

Source: [2659].

Fact 7.9.13. Let $A \in \mathbb{F}^{n\times n}$. Then, the following statements are equivalent:

i) $A + A^*$ is positive definite.

ii) For all Hermitian matrices $B \in \mathbb{F}^{n\times n}$, $\operatorname{In} B = \operatorname{In} AB$.

Source: [613].

Fact 7.9.14. Let $A, B \in \mathbb{F}^{n\times n}$, assume that AB and B are Hermitian, and assume that $\operatorname{spec}(A) \cap [0,\infty) = \varnothing$. Then, $\operatorname{In}(-AB) = \operatorname{In} B$. **Source:** [613].

Fact 7.9.15. Let $A, B \in \mathbb{F}^{n\times n}$, assume that A and B are Hermitian and nonsingular, and assume that $\operatorname{spec}(AB) \cap [0,\infty) = \varnothing$. Then, $\nu_+(A) + \nu_+(B) = n$. **Source:** Fact 7.9.14 and [613]. **Remark:** Weaker versions are given in [1538, 2125].

Fact 7.9.16. Let $A \in \mathbb{F}^{n\times n}$, assume that A is Hermitian, and let $S \in \mathbb{F}^{m\times n}$. Then,

$$\nu_-(SAS^*) + \nu_+(SAS^*) = \operatorname{rank} SAS^* \le \min\{\operatorname{rank} A, \operatorname{rank} S\},$$

$$\nu_-(A) + \operatorname{rank} S - n \le \nu_-(SAS^*) \le \nu_-(A),$$

$$\nu_+(A) + \operatorname{rank} S - n \le \nu_+(SAS^*) \le \nu_+(A),$$

$$m + \nu_0(A) \le n + \nu_0(SAS^*),$$

$$\operatorname{rank} SAS^* \le \operatorname{rank} A \le 2(n - \operatorname{rank} S) + \operatorname{rank} SAS^* \le 2n - \operatorname{rank} S.$$

Furthermore, consider the following statements:

i) $\operatorname{rank} S = n$.

ii) $\operatorname{rank} SAS^* = \operatorname{rank} A$.

iii) $\nu_-(SAS^*) = \nu_-(A)$ and $\nu_+(SAS^*) = \nu_+(A)$.

iv) $m + \nu_0(A) = n + \nu_0(SAS^*)$.

Then, *i)* $\Longrightarrow$ *ii)* $\Longleftrightarrow$ *iii)* $\Longrightarrow$ *iv)*. Finally, the following statements hold:

v) If $\operatorname{rank} S = m$, then

$$\operatorname{rank} SAS^* \le \operatorname{rank} A \le 2(n-m) + \operatorname{rank} SAS^* \le 2n - m,$$

$$\nu_0(SAS^*) \le n - m + \nu_0(A) \le 2(n-m) + \nu_0(SAS^*).$$

vi) If SAS^* is positive semidefinite, then $\operatorname{rank} S \le \nu_0(A) + \nu_+(A)$.

vii) If SAS^* is positive definite, then $\operatorname{rank} S = m = \operatorname{rank} SAS^* = \nu_+(SAS^*) \le \nu_+(A)$.

Source: [970, pp. 430, 431] and [1084, p. 194]. The first three strings are given in [2184]. *vi*) follows from the second string. Adding the second and third strings yields the fifth string. **Remark:** *vi*) and *vii*) are given in [1448, p. 192].

Fact 7.9.17. Let $S \in \mathbb{F}^{n\times n}$, assume that S is nonsingular, let n_1, n_2, and n_3 be nonnegative integers such that $n_1 + n_2 + n_3 = n$, define $D \triangleq \operatorname{diag}(-I_{n_1}, 0_{n_2\times n_2}, I_{n_3})$, and define $A \in \mathbb{F}^{n\times n}$ by $A \triangleq SDS^*$. Then, $\operatorname{In}(A) = [n_1 \;\; n_2 \;\; n_3]^\mathsf{T}$. **Source:** Note that $\nu_0(A) = \operatorname{def} A = \operatorname{def} D = n_2$. Next, define $\hat{S} \triangleq [0_{n_3\times(n-n_3)} \;\; I_{n_3}]S^{-1}$ so that $\hat{S}A\hat{S}^* = I_{n_3}$. Since $\hat{S}A\hat{S}^*$ is positive definite, it follows from *vii*) and the second inequality in the third string of Fact 7.9.16 that $\operatorname{rank} \hat{S} = n_3 = \operatorname{rank} \hat{S}A\hat{S}^* = \nu_+(\hat{S}A\hat{S}^*) \le \nu_+(A) = \nu_+(SDS^*) \le \nu_+(D) = n_3$. Hence, $\nu_+(A) = n_3$. Likewise, $\nu_-(A) = n_1$. **Remark:** This result completes the proof of Theorem 7.5.7.

Fact 7.9.18. Let $A, S \in \mathbb{F}^{n\times n}$, and assume that A is Hermitian and S is nonsingular. Then, there exist $\alpha_1, \ldots, \alpha_n \in [\lambda_{\min}(SS^*), \lambda_{\max}(SS^*)]$ such that, for all $i \in \{1, \ldots, n\}$, $\lambda_i(SAS^*) = \alpha_i\lambda_i(A)$. **Source:** [2893]. **Remark:** This is a quantitative version of Sylvester's law of inertia given by Theorem 7.5.7. **Credit:** A. Ostrowski.

Fact 7.9.19. Let $A, S \in \mathbb{F}^{n\times n}$, assume that A is Hermitian. Then, the following statements are equivalent:

i) $\operatorname{In}(SAS^*) = \operatorname{In} A$.

ii) $\operatorname{rank}(SAS^*) = \operatorname{rank} A$.

iii) $\mathcal{R}(A) \cap \mathcal{N}(A) = \{0\}$.

Source: [228].

Fact 7.9.20. Let $A, B \in \mathbb{R}^{n\times n}$, and assume that A and B are projectors. Then,

$$\nu_+(A-B) = \operatorname{rank}(A+B) - \operatorname{rank} B, \quad \nu_-(A-B) = \operatorname{rank}(A+B) - \operatorname{rank} A, \quad \operatorname{sig}(A-B) = \operatorname{rank} A - \operatorname{rank} B.$$

Source: [2661].

Fact 7.9.21. Let $A \in \mathbb{R}^{n\times m}$. Then,

$$\operatorname{In}\begin{bmatrix} 0 & A \\ A^* & 0 \end{bmatrix} = \operatorname{In}\begin{bmatrix} AA^* & 0 \\ 0 & -A^*A \end{bmatrix} = \operatorname{In}\begin{bmatrix} AA^+ & 0 \\ 0 & -A^+A \end{bmatrix} = \begin{bmatrix} \operatorname{rank} A \\ n + m - 2\operatorname{rank} A \\ \operatorname{rank} A \end{bmatrix}.$$

Source: [970, pp. 432, 434].

Fact 7.9.22. Let $A \in \mathbb{C}^{n\times n}$, assume that A is Hermitian, and let $B \in \mathbb{C}^{n\times m}$. Then,

$$\operatorname{In}\begin{bmatrix} A & B \\ B^* & 0 \end{bmatrix} \ge\ge \begin{bmatrix} \operatorname{rank} B \\ m - \operatorname{rank} B \\ \operatorname{rank} B \end{bmatrix}.$$

Furthermore, if $\mathcal{R}(A) \subseteq \mathcal{R}(B)$, then

$$\operatorname{In}\begin{bmatrix} A & B \\ B^* & 0 \end{bmatrix} = \begin{bmatrix} \operatorname{rank} B \\ n + m - 2\operatorname{rank} B \\ \operatorname{rank} B \end{bmatrix}.$$

Finally, if $\operatorname{rank} B = n$, then

$$\operatorname{In}\begin{bmatrix} A & B \\ B^* & 0 \end{bmatrix} = \begin{bmatrix} n \\ m - n \\ n \end{bmatrix}.$$

Source: [970, pp. 433, 434] and [1921]. **Related:** Fact 10.19.21.

Fact 7.9.23. Let $A \in \mathbb{C}^{n\times n}$, let $C \in \mathbb{C}^{m\times m}$, assume that A and C are Hermitian, and let $B \in \mathbb{C}^{n\times m}$. Then,

$$\operatorname{In}\begin{bmatrix} A & 0 \\ 0 & C \end{bmatrix} = \operatorname{In} A + \operatorname{In} C, \quad \operatorname{In}\begin{bmatrix} 0 & B \\ B^* & 0 \end{bmatrix} = \begin{bmatrix} \operatorname{rank} B \\ n+m-2\operatorname{rank} B \\ \operatorname{rank} B \end{bmatrix},$$

$$\max\{\nu_-(A), \nu_-(C)\} \le \nu_-\left(\begin{bmatrix} A & B \\ B^* & C \end{bmatrix}\right), \quad \max\{\nu_+(A), \nu_+(C)\} \le \nu_+\left(\begin{bmatrix} A & B \\ B^* & C \end{bmatrix}\right).$$

$$\max\{\nu_-(A) + \nu_0(A), \nu_-(C) + \nu_0(C)\} \le \nu_-\left(\begin{bmatrix} A & B \\ B^* & C \end{bmatrix}\right) + \nu_0\left(\begin{bmatrix} A & B \\ B^* & C \end{bmatrix}\right),$$

$$\max\{\nu_+(A) + \nu_0(A), \nu_+(C) + \nu_0(C)\} \le \nu_+\left(\begin{bmatrix} A & B \\ B^* & C \end{bmatrix}\right) + \nu_0\left(\begin{bmatrix} A & B \\ B^* & C \end{bmatrix}\right).$$

Source: [2656]. **Remark:** The last four inequalities are *Poincaré's inequalities*. See [824].

Fact 7.9.24. Let $A \in \mathbb{C}^{n\times n}$, let $C \in \mathbb{C}^{m\times m}$, assume that A and C are positive semidefinite, let $B \in \mathbb{C}^{n\times m}$, and define $\mathcal{A} \triangleq \begin{bmatrix} A & B \\ B^* & -C \end{bmatrix}$. Then,

$$\operatorname{In} \mathcal{A} = \begin{bmatrix} \operatorname{rank}[B^* \ \ C] \\ n+m-\operatorname{rank}[A \ \ B] - \operatorname{rank}[B^* \ \ C] \\ \operatorname{rank}[A \ \ B] \end{bmatrix}.$$

Source: [2662].

Fact 7.9.25. Let $A, B \in \mathbb{F}^{n\times n}$, and assume that A and B are projectors. Then,

$$\operatorname{In}\begin{bmatrix} A & B \\ B & A \end{bmatrix} = \begin{bmatrix} \operatorname{rank}[A \ \ B] - \operatorname{rank} A \\ 2n - 3\operatorname{rank}[A \ \ B] + \operatorname{rank} A + \operatorname{rank} B \\ 2\operatorname{rank}[A \ \ B] - \operatorname{rank} B \end{bmatrix}.$$

Source: [2667]. **Related:** Fact 4.18.22.

Fact 7.9.26. Let $A \in \mathbb{F}^{n\times n}$. Then, there exist a nonsingular matrix $S \in \mathbb{F}^{n\times n}$ and a skew-Hermitian matrix $B \in \mathbb{F}^{n\times n}$ such that

$$A = S\left(\begin{bmatrix} I_{\nu_-(A+A^*)} & 0 & 0 \\ 0 & 0_{\nu_0(A+A^*)\times\nu_0(A+A^*)} & 0 \\ 0 & 0 & -I_{\nu_+(A+A^*)} \end{bmatrix} + B\right)S^*.$$

Source: Write $A = \frac{1}{2}(A + A^*) + \frac{1}{2}(A - A^*)$, and apply Proposition 7.5.6 to $\frac{1}{2}(A + A^*)$.

Fact 7.9.27. Let $A \in \mathbb{F}^{n\times n}$, $B \in \mathbb{F}^{n\times m}$, and $C \in \mathbb{F}^{m\times m}$, and assume that A is positive definite and C is negative definite. Then,

$$\operatorname{In}\begin{bmatrix} A & B & 0 \\ B^* & C & 0 \\ 0 & 0 & 0_{l\times l} \end{bmatrix} = \begin{bmatrix} m \\ l \\ n \end{bmatrix}.$$

Source: Fact 7.9.6 and [1546].

7.10 Facts on Matrix Transformations for One Matrix

Fact 7.10.1. Define $S \in \mathbb{C}^{3\times 3}$ by

$$S \triangleq \frac{1}{\sqrt{2}}\begin{bmatrix} 0 & 1 & 1 \\ 0 & -\jmath & \jmath \\ \sqrt{2} & 0 & 0 \end{bmatrix}.$$

Then, S is unitary, and $K(e_3) = S\operatorname{diag}(0, \jmath, -\jmath)S^{-1}$. **Related:** Fact 7.10.2.

Fact 7.10.2. Let $x \in \mathbb{R}^3$, assume that either $x_{(1)} \neq 0$ or $x_{(2)} \neq 0$, and define

$$\begin{bmatrix} a \\ b \\ c \end{bmatrix} \triangleq \frac{1}{\|x\|_2}x, \quad \alpha \triangleq \sqrt{a^2+b^2+a^2c^2+b^2c^2+(1-c^2)^2}, \quad S \triangleq \begin{bmatrix} a & \frac{-b+ac\jmath}{\alpha} & \frac{-b-ac\jmath}{\alpha} \\ b & \frac{a+bc\jmath}{\alpha} & \frac{a-bc\jmath}{\alpha} \\ c & \frac{-(1-c^2)\jmath}{\alpha} & \frac{(1-c^2)\jmath}{\alpha} \end{bmatrix}.$$

Then, $\alpha \geq \sqrt{a^2+b^2} > 0$, S is unitary, and $K(x) = S\operatorname{diag}(0, \|x\|_2\jmath, -\|x\|_2\jmath)S^{-1}$. **Source:** [1656, p. 154]. **Related:** If $x_{(1)} = x_{(2)} = 0$, then $a = b = 0$, $c = 1$, and $\alpha = 0$. See Fact 7.10.1. **Problem:** Find a decomposition of $K(x)$ that applies to all $x \in \mathbb{R}^3$.

Fact 7.10.3. Let $A \in \mathbb{F}^{n\times n}$, and assume that there exists a nonsingular matrix $S \in \mathbb{F}^{n\times n}$ such that $S^{-1}AS$ is upper triangular. Then, for all $r \in \{1,\ldots,n\}$, $\mathcal{R}\left(S\left[\begin{smallmatrix} I_r \\ 0 \end{smallmatrix}\right]\right)$ is an invariant subspace of A. **Remark:** Analogous results hold for lower triangular matrices and block-triangular matrices.

Fact 7.10.4. Let $A \in \mathbb{F}^{n\times n}$. Then, there exist unique matrices $B, C \in \mathbb{F}^{n\times n}$ such that the following statements hold:

i) B is semisimple.

ii) C is nilpotent.

iii) $A = B + C$.

iv) $BC = CB$.

Furthermore, the following statements hold:

v) $B = (A^{\mathsf{D}})^{\mathsf{D}} = (A^{\mathsf{D}})^{\#} = A^2A^{\mathsf{D}}$.

vi) $C = A - A^2A^{\mathsf{D}}$.

vii) $\operatorname{mspec}(A) = \operatorname{mspec}(B)$.

Source: [1401, p. 112], [1451, pp. 181, 189], and [1474, p. 74]. *v)* is given by *xiv)* of Proposition 8.2.2; *vi)* follows from Fact 7.19.5. **Remark:** This is the *S-N decomposition* (also called the *Jordan-Chevalley decomposition*).

Fact 7.10.5. Let $A \in \mathbb{F}^{n\times n}$. Then, the following statements are equivalent:

i) A is similar to a skew-Hermitian matrix.

ii) A is semisimple, and $\operatorname{spec}(A) \subset \mathrm{I}\mathbb{A}$.

Related: Fact 15.19.12.

Fact 7.10.6. Let $A \in \mathbb{F}^{n\times n}$, and let $r \triangleq \operatorname{rank} A$. Then, A is group invertible if and only if there exist a nonsingular matrix $B \in \mathbb{F}^{r\times r}$ and a nonsingular matrix $S \in \mathbb{F}^{n\times n}$ such that

$$A = S\begin{bmatrix} B & 0 \\ 0 & 0 \end{bmatrix}S^{-1}.$$

Fact 7.10.7. Let $A \in \mathbb{F}^{n\times n}$, and let $r \triangleq \operatorname{rank} A$. Then, A is range Hermitian if and only if there exist a nonsingular matrix $S \in \mathbb{F}^{n\times n}$ and a nonsingular matrix $B \in \mathbb{F}^{r\times r}$ such that

$$A = S\begin{bmatrix} B & 0 \\ 0 & 0 \end{bmatrix}S^{*}.$$

Remark: S need not be unitary for sufficiency. See Corollary 7.5.4. **Source:** Use the QR decomposition Fact 7.17.11 to let $S \triangleq \hat{S}R$, where $\hat{S}$ is unitary and R is upper triangular. See [2658].

Fact 7.10.8. Let $A \in \mathbb{F}^{n\times n}$. Then, there exists involutory $S \in \mathbb{F}^{n\times n}$ such that $A^{\mathsf{T}} = SAS^{\mathsf{T}}$. **Source:** [920] and [1215]. **Remark:** Note A^{T} rather than A^*. **Remark:** Unitary similarity between a matrix and its transpose is discussed in [1144].

Fact 7.10.9. Let $A \in \mathbb{F}^{n\times n}$. Then, there exists a nonsingular matrix $S \in \mathbb{F}^{n\times n}$ such that $A =$

SA^*S^{-1} if and only if there exist Hermitian matrices $S_1, S_2 \in \mathbb{F}^{n\times n}$ such that $A = S_1S_2$. **Source:** [2983, pp. 215, 216] and [2991, pp. 262, 263]. **Remark:** For normal A, an analogous result in Hilbert space is given in [1531]. **Related:** Corollary 6.3.12, Corollary 7.4.9, Proposition 7.7.13, and Fact 7.17.26.

Fact 7.10.10. Let $A \in \mathbb{F}^{n\times n}$, and assume that A is normal. Then, there exists a symmetric, nonsingular matrix $S \in \mathbb{F}^{n\times n}$ such that $A^\mathsf{T} = SAS^{-1}$ and such that $S^{-1} = \overline{S}$. **Source:** For $\mathbb{F} = \mathbb{C}$, let $A = UBU^*$, where U is unitary and B is diagonal. Then, $A^\mathsf{T} = SA\overline{S} = SAS^{-1}$, where $S \triangleq \overline{U}U^{-1}$. For $\mathbb{F} = \mathbb{R}$, use the real normal form and let $S \triangleq U\tilde{I}U^\mathsf{T}$, where U is orthogonal and $\tilde{I} \triangleq \operatorname{diag}(\hat{I}, \ldots, \hat{I})$. **Related:** Corollary 7.4.9.

Fact 7.10.11. Let $A \in \mathbb{R}^{n\times n}$, and assume that A is normal. Then, there exists a reflector $S \in \mathbb{R}^{n\times n}$ such that $A^\mathsf{T} = SAS^{-1}$. Consequently, A and A^T are orthogonally similar. Finally, if A is skew symmetric, then A and $-A$ are orthogonally similar. **Source:** Specialize Fact 7.10.10 to the case $\mathbb{F} = \mathbb{R}$.

Fact 7.10.12. Let $A \in \mathbb{F}^{n\times n}$. Then, there exists a reverse-symmetric, nonsingular matrix $S \in \mathbb{F}^{n\times n}$ such that $A^{\hat{\mathsf{T}}} = SAS^{-1}$. **Source:** Corollary 7.4.9 and [1775].

Fact 7.10.13. Let $A \in \mathbb{F}^{n\times n}$. Then, there exist reverse-symmetric matrices $S_1, S_2 \in \mathbb{F}^{n\times n}$ such that S_2 is nonsingular and $A = S_1S_2$. **Source:** Corollary 7.4.10 and [1775].

Fact 7.10.14. Let $A \in \mathbb{R}^{n\times n}$, and assume that A is not of the form aI, where $a \in \mathbb{R}$. Then, A is similar to a matrix with diagonal entries $0, \ldots, 0, \operatorname{tr} A$. **Source:** [2263, p. 77]. **Credit:** P. M. Gibson.

Fact 7.10.15. Let $A \in \mathbb{R}^{n\times n}$, and assume that A is not zero. Then, A is similar to a matrix whose diagonal entries are all nonzero. **Source:** [2263, p. 79]. **Credit:** M. Marcus and R. Purves.

Fact 7.10.16. Let $A \in \mathbb{R}^{n\times n}$, and assume that A is symmetric. Then, there exists an orthogonal matrix $S \in \mathbb{R}^{n\times n}$ such that $-1 \notin \operatorname{spec}(S)$ and SAS^T is diagonal. **Source:** [2263, p. 101]. **Credit:** P. L. Hsu.

Fact 7.10.17. Let $A \in \mathbb{R}^{n\times n}$, and assume that A is symmetric. Then, there exist a diagonal matrix $B \in \mathbb{R}^{n\times n}$ and a skew-symmetric matrix $C \in \mathbb{R}^{n\times n}$ such that

$$A = [2(I + C)^{-1} - I]B[2(I + C)^{-1} - I]^\mathsf{T}.$$

Source: Fact 7.10.16 and [2263, p. 101].

Fact 7.10.18. Let $A \in \mathbb{F}^{n\times n}$. Then, there exists a unitary matrix $S \in \mathbb{F}^{n\times n}$ such that S^*AS has equal diagonal entries. In particular, $I \odot (S^*AS) = \frac{1}{n}(\operatorname{tr} A)I$. **Source:** Fact 7.10.19, [1048], and [2263, p. 78]. **Credit:** W. V. Parker. See [1130].

Fact 7.10.19. Let $A \in \mathbb{F}^{n\times n}$. Then, the following statements are equivalent:

i) $\operatorname{tr} A = 0$.

ii) There exist $B, C \in \mathbb{F}^{n\times n}$ such that $A = [B, C]$.

iii) A is unitarily similar to a matrix whose diagonal entries are zero.

Source: [29, 1130], [1301, p. 146], and [1596, 1625]. **Remark:** This is *Shoda's theorem*. **Related:** Fact 7.10.20.

Fact 7.10.20. Let $R \in \mathbb{F}^{n\times n}$, and assume that R is Hermitian. Then, the following statements are equivalent:

i) $\operatorname{tr} R < 0$.

ii) R is unitarily similar to a matrix all of whose diagonal entries are negative.

iii) There exists an asymptotically stable matrix $A \in \mathbb{F}^{n\times n}$ such that $R = A + A^*$.

Source: [266]. **Related:** Fact 7.10.19.

Fact 7.10.21. Let $A \in \mathbb{F}^{n\times m}$. Then, there exists $S \in \mathbb{F}^{n\times m}$ such that $S^*S = I_m$ and such that $AA^* =$

SA^*AS^*. Now, assume that $n = m$. Then, AA^* and A^*A are unitarily similar. **Source:** Fact 10.8.4.

Fact 7.10.22. Let $A \in \mathbb{F}^{n\times n}$. Then, A is symmetric if and only if there exists a unitary matrix $S \in \mathbb{F}^{n\times n}$ such that $A = SBS^\mathrm{T}$, where $B \triangleq \operatorname{diag}[\sigma_1(A),\dots,\sigma_n(A)]$. **Source:** [1451, pp. 153, 263]. **Remark:** A is symmetric, complex, and T-congruent to B. **Remark:** This is the *Autonne-Takagi factorization.*

Fact 7.10.23. Let $A \in \mathbb{F}^{n\times n}$. Then,

$$\begin{bmatrix} A & 0 \\ 0 & -A \end{bmatrix} = \frac{1}{\sqrt{2}}\begin{bmatrix} I & I \\ -I & I \end{bmatrix}\begin{bmatrix} 0 & A \\ A & 0 \end{bmatrix}\frac{1}{\sqrt{2}}\begin{bmatrix} I & -I \\ I & I \end{bmatrix}.$$

Hence, $\left[\begin{smallmatrix} A & 0 \\ 0 & -A \end{smallmatrix}\right]$ and $\left[\begin{smallmatrix} 0 & A \\ A & 0 \end{smallmatrix}\right]$ are unitarily similar. **Related:** Fact 4.13.31.

Fact 7.10.24. Let n be a positive integer. Then,

$$J_{2n} = S\begin{bmatrix} \jmath I_n & 0 \\ 0 & -\jmath I_n \end{bmatrix}S^*, \quad S \triangleq \frac{1}{\sqrt{2}}\begin{bmatrix} I & I \\ \jmath I & -\jmath I \end{bmatrix}.$$

Hence, $\operatorname{mspec}(J_{2n}) = \{\jmath,-\jmath,\dots,\jmath,-\jmath\}_{\mathrm{ms}}$, and $\det J_{2n} = 1$. **Source:** Fact 3.24.6. **Remark:** By Fact 4.28.3, J_{2n} is Hamiltonian, and thus, by Fact 6.9.20, $\operatorname{mspec}(J_{2n}) = -\operatorname{mspec}(J_{2n})$. **Remark:** S is unitary. See Fact 4.13.31.

Fact 7.10.25. Let n be a positive integer. Then,

$$\hat{I}_n = \begin{cases} S\begin{bmatrix} I_{n/2} & 0 \\ 0 & -I_{n/2} \end{bmatrix}S^\mathrm{T}, & n \text{ even}, \\ S\begin{bmatrix} I_{(n-1)/2} & 0 & 0 \\ 0 & 1 & 0 \\ 0 & 0 & -I_{(n-1)/2} \end{bmatrix}S^\mathrm{T}, & n \text{ odd}, \end{cases} \qquad S \triangleq \begin{cases} \dfrac{1}{\sqrt{2}}\begin{bmatrix} I_{n/2} & -\hat{I}_{n/2} \\ \hat{I}_{n/2} & I_{n/2} \end{bmatrix}, & n \text{ even}, \\ \dfrac{1}{\sqrt{2}}\begin{bmatrix} I_{(n-1)/2} & 0 & -\hat{I}_{(n-1)/2} \\ 0 & \sqrt{2} & 0 \\ \hat{I}_{(n-1)/2} & 0 & I_{(n-1)/2} \end{bmatrix}, & n \text{ odd}. \end{cases}$$

Therefore,

$$\operatorname{mspec}(\hat{I}_n) = \begin{cases} \{-1,1,\dots,-1,1\}_{\mathrm{ms}}, & n \text{ even}, \\ \{1,-1,1,\dots,-1,1\}_{\mathrm{ms}}, & n \text{ odd}. \end{cases}$$

Remark: For even n, Fact 4.28.3 implies that $\hat{I}_n$ is Hamiltonian, and thus, by Fact 6.9.20, $\operatorname{mspec}(\hat{I}_n) = -\operatorname{mspec}(\hat{I}_n)$. See [2844]. **Remark:** S is orthogonal. See Fact 4.13.31.

Fact 7.10.26. Let $A \in \mathbb{F}^{n\times n}$, assume that A is idempotent, and let $r \triangleq \operatorname{rank} A$. Then, there exist a unitary matrix $S \in \mathbb{F}^{n\times n}$ and positive numbers $a_1,\dots,a_k$ such that

$$A = S\operatorname{diag}\left(\begin{bmatrix} 1 & a_1 \\ 0 & 0 \end{bmatrix},\dots,\begin{bmatrix} 1 & a_k \\ 0 & 0 \end{bmatrix}, I_{r-k}, 0_{(n-r-k)\times(n-r-k)}\right)S^*.$$

Source: [919]. **Remark:** This result provides a canonical matrix for idempotent matrices under unitary similarity. **Remark:** See [1134].

Fact 7.10.27. Let $A \in \mathbb{F}^{n\times n}$ be nonzero, define $r \triangleq \operatorname{rank} A$, define $B \triangleq \operatorname{diag}[\sigma_1(A),\dots,\sigma_r(A)]$, let $S, T \in \mathbb{F}^{n\times n}$ be unitary matrices such that

$$A = S\begin{bmatrix} B & 0_{r\times(n-r)} \\ 0_{(n-r)\times r} & 0_{(n-r)\times(n-r)} \end{bmatrix}T,$$

and define $K \in \mathbb{F}^{r\times r}$ and $L \in \mathbb{F}^{r\times(n-r)}$ by $[K \;\; L] \triangleq [I_r \;\; 0_{r\times(n-r)}]TS$. Then,

$$KK^* + LL^* = I_r, \quad A = S\begin{bmatrix} BK & BL \\ 0_{(n-r)\times r} & 0_{(n-r)\times(n-r)} \end{bmatrix}S^*.$$

Furthermore, the following statements hold:

i) A is group invertible if and only if K is nonsingular.

ii) A is idempotent if and only if $BK = I_r$.

iii) A is tripotent if and only if BK is involutory.

iv) $A^2 = 0$ if and only if $K = 0$.

v) A is range Hermitian if and only if $L = 0$.

vi) A is range disjoint if and only if $\operatorname{rank} L = r$.

vii) A is range spanning if and only if $\operatorname{rank} L = n - r$.

viii) A is Hermitian if and only if $L = 0$ and $BK = K^*B$.

ix) A is normal if and only if $L = 0$ and $BK = KB$.

x) A is a projector if and only if $B = K = I_r$.

xi) A is a partial isometry if and only if $B = I_r$.

xii) Let $k \geq 1$. Then, $A^k = A$ if and only if $(BK)^{k-1} = I_r$.

xiii) $A^2 = 0$ if and only if $K = 0$.

xiv) $A^2 = A^*$ if and only if $L = 0$ and $B = K^3 = I_r$.

xv) $A^4 = A$ if and only if $(BK)^3 = I_r$.

xvi) $A^2 = A^+$ if and only if $L = 0$ and $(BK)^3 = I_r$.

xvii) A is semicontractive if and only if B is semicontractive.

xviii) A is contractive if and only if B is contractive.

xix) A is Hermitian and tripotent if and only if $L = 0$, $B = I_r$, and K is Hermitian.

xx) $\mathcal{N}(A) \subseteq \mathcal{R}(A)$ if and only if $L^*L = I_{n-r}$.

xxi) $\mathcal{R}(A) \subseteq \mathcal{N}(A)$ if and only if $LL^* = I_r$.

xxii) The following statements are equivalent:

a) $\mathcal{R}(A) \cap \mathcal{N}(A) = \{0\}$; *b)* $\mathcal{R}(L) \subseteq \mathcal{R}(K)$; *c)* $\operatorname{rank} K = r$.

Source: [234, 236, 240, 247, 257, 1338]. **Remark:** $L = 0$ if and only if K is unitary. **Related:** Fact 8.3.23, Fact 8.5.13, and Fact 8.5.5.

Fact 7.10.28. Let $A \in \mathbb{F}^{n\times n}$, assume that A is idempotent, and let $r \triangleq \operatorname{rank} A$. Then, there exists $B \in \mathbb{F}^{r\times(n-r)}$ and a unitary matrix $S \in \mathbb{F}^{n\times n}$ such that

$$A = S\begin{bmatrix} I_r & B \\ 0 & 0_{(n-r)\times(n-r)} \end{bmatrix}S^*.$$

Source: Use Fact 7.10.27. See [1133, p. 46].

Fact 7.10.29. Let $A \in \mathbb{F}^{n\times n}$, assume that A is unitary, and partition A as

$$A = \begin{bmatrix} A_{11} & A_{12} \\ A_{21} & A_{22} \end{bmatrix},$$

where $A_{11} \in \mathbb{F}^{m\times k}$, $A_{12} \in \mathbb{F}^{m\times q}$, $A_{21} \in \mathbb{F}^{p\times k}$, $A_{22} \in \mathbb{F}^{p\times q}$, and $m + p = k + q = n$. Then, there exist unitary matrices $U, V \in \mathbb{F}^{n\times n}$ and nonnegative integers l, r such that

$$A = U\begin{bmatrix} I_r & 0 & 0 & 0 & 0 & 0 \\ 0 & \Gamma & 0 & 0 & \Sigma & 0 \\ 0 & 0 & 0 & 0 & 0 & I_{m-r-l} \\ 0 & 0 & 0 & I_{q-m+r} & 0 & 0 \\ 0 & \Sigma & 0 & 0 & -\Gamma & 0 \\ 0 & 0 & I_{k-r-l} & 0 & 0 & 0 \end{bmatrix}V,$$

where $\Gamma, \Sigma \in \mathbb{R}^{l\times l}$ are diagonal and satisfy

$$0 < \Gamma_{(l,l)} \le \cdots \le \Gamma_{(1,1)} < 1, \quad 0 < \Sigma_{(1,1)} \le \cdots \le \Sigma_{(l,l)} < 1, \quad \Gamma^2 + \Sigma^2 = I_m.$$

Source: [1133, p. 12] and [2539, p. 37]. **Remark:** This is the *CS decomposition*. See [2183, 2185]. The diagonal entries of Σ and Γ can be interpreted as the sines and cosines, respectively, of the principal angles between a pair of subspaces $\mathcal{S}_1 = \mathcal{R}(X_1)$ and $\mathcal{S}_2 = \mathcal{R}(Y_1)$ such that $[X_1\ X_2]$ and $[Y_1\ Y_2]$ are unitary and $A = [X_1\ X_2]^*[Y_1\ Y_2]$; see [1133, pp. 25–29], [1134, 1595], [2539, pp. 40–43], [2847], and Fact 3.12.21. **Related:** Fact 7.10.30 and Fact 7.13.27.

Fact 7.10.30. Let $A \in \mathbb{F}^{n\times n}$, define $r \triangleq \operatorname{rank} A$, let $P \in \mathbb{F}^{n\times n}$ and $Q \in \mathbb{F}^{n\times n}$ be the projectors onto $\mathcal{R}(A)$ and $\mathcal{R}(A^*)$, respectively, let $k \ge 0$ be the multiplicity of the singular value 1 of PQ, define $l \triangleq r - \operatorname{rank} PQ \ge 0$, and, for all $i \in \{1, \dots, r\}$, define $\theta_i \triangleq \operatorname{acos} \sigma_i(PQ)$. Then, the following statements hold:

i) $\dim[\mathcal{R}(A) \cap \mathcal{R}(A^*)] = k$.

ii) $\dim[\mathcal{R}(A) \cap \mathcal{N}(A)] = l$.

iii) $\operatorname{def} PQ = n + l - r$.

iv) $k \le \operatorname{rank} PQ = r - l \le r$.

v) If $k = 0$, then $1 > \sigma_1(PQ) \ge \cdots \ge \sigma_{r-l}(PQ) > \sigma_{r-l+1}(PQ) = \cdots = \sigma_n(PQ) = 0$.

vi) If $1 \le k < r$, then $1 = \sigma_1(PQ) = \cdots = \sigma_k(PQ) > \sigma_{k+1}(PQ) \ge \cdots \ge \sigma_{r-l}(PQ) > \sigma_{r-l+1}(PQ) = \cdots = \sigma_n(PQ) = 0$.

vii) If $k = r$, then $l = 0$ and $1 = \sigma_1(PQ) = \cdots = \sigma_r(PQ) > \sigma_{r+1}(PQ) = \cdots = \sigma_n(PQ) = 0$.

viii) $\theta_1 = \cdots = \theta_k = 0$, $\theta_{k+1}, \dots, \theta_{r-l} \in (0, \frac{\pi}{2})$, and $\theta_{r-l+1} = \cdots = \theta_r = \frac{\pi}{2}$.

ix) $\sigma_{\max}(P - Q) = \max\{\sin\theta_i \colon i \in \{1, \dots, r\}\}$.

x) There exist a unitary matrix $U \in \mathbb{F}^{n\times n}$ and a nonsingular matrix $M \in \mathbb{F}^{r\times r}$ such that

$$A = U\begin{bmatrix} MC & MS \\ 0_{(n-r)\times r} & 0_{(n-r)\times(n-r)} \end{bmatrix}U^*,$$

where $C \triangleq \operatorname{diag}(0_{l\times l}, \cos\theta_{k+1}, \dots, \cos\theta_{r-l}, I_k) \in \mathbb{R}^{r\times r}$ and $S \triangleq \operatorname{diag}(\sin\theta_{k+1}, \dots, \sin\theta_{r-l}, I_l, 0_{k\times(n+k-2r)}) \in \mathbb{R}^{r\times(n-r)}$. Furthermore, $C^2 + SS^* = I_r$.

xi) A is a partial isometry if and only if M is unitary.

xii) A is idempotent if and only if $CM = I_r$.

xiii) A is a projector if and only if $\theta_1 = \cdots = \theta_r = 0$ and $M = I_r$.

xiv) A is range Hermitian if and only if $\theta_1 = \cdots = \theta_r = 0$.

xv) A is normal if and only if $\theta_1 = \cdots = \theta_r = 0$ and M is normal.

xvi) The following conditions are equivalent:

a) A is group invertible.

b) $l = 0$.

c) C is nonsingular.

If these statements hold, then

$$A^{\#} = U\begin{bmatrix} (MC)^{-1} & (CMC)^{-1}S \\ 0_{(n-r)\times r} & 0_{(n-r)\times(n-r)} \end{bmatrix}U^*.$$

xvii) A is (contractive, semicontractive) if and only if M is.

Source: [402, 404]. **Remark:** $\theta_1, \ldots, \theta_r$ are the principal angles between the subspaces $\mathcal{R}(A)$ and $\mathcal{R}(A^*)$. **Related:** Fact 7.12.42.

Fact 7.10.31. Let $A \in \mathbb{F}^{n\times n}$, and define $r \triangleq \operatorname{rank} A$ and $\alpha \triangleq [\prod_{i=1}^r \sigma_i(A)]^{1/r}$. Then, there exist $S_1 \in \mathbb{F}^{n\times r}$, $B \in \mathbb{R}^{r\times r}$, and $S_2 \in \mathbb{F}^{r\times n}$ such that S_1 is left inner, S_2 is right inner, B is upper triangular, $I \odot B = \alpha I$, and $A = S_1BS_2$. **Source:** [1534]. **Remark:** B is real. **Remark:** This is the *geometric mean decomposition.*

Fact 7.10.32. Let $A \in \mathbb{F}^{n\times n}$. Then, there exist upper triangular matrices $S_1 \in \mathbb{F}^{n\times n}$ and $S_2 \in \mathbb{F}^{n\times n}$ and a permutation matrix $B \in \mathbb{R}^{n\times n}$ such that $A = S_1BS_2$. **Source:** [2336, pp. 54, 55]. **Remark:** This is the *Bruhat decomposition.*

Fact 7.10.33. Let $A \in \mathbb{C}^{n\times n}$. Then, there exists $B \in \mathbb{R}^{n\times n}$ such that $A\overline{A}$ and B^2 are similar. **Source:** [913].

7.11 Facts on Matrix Transformations for Two or More Matrices

Fact 7.11.1. Let $A \in \mathbb{R}^{n\times n}$, let $\lambda \in \operatorname{spec}(A)$, and assume that $\lambda \notin \mathbb{R}$. Then, the number and size of the Jordan blocks associated with λ in the Jordan form of A are equal to the number and size of the Jordan blocks associated with $\overline{\lambda}$ in the Jordan form of A.

Fact 7.11.2. Let $q(s) \triangleq s^2 - \beta_1 s - \beta_0 \in \mathbb{R}[s]$ be irreducible, and let $\lambda = \nu + \omega \jmath$ denote a root of q so that $\beta_1 = 2\nu$ and $\beta_0 = -(\nu^2 + \omega^2)$. Then,

$$\mathcal{H}_1(q) = \begin{bmatrix} 0 & 1 \\ \beta_0 & \beta_1 \end{bmatrix} = \begin{bmatrix} 1 & 0 \\ \nu & \omega \end{bmatrix}\begin{bmatrix} \nu & \omega \\ -\omega & \nu \end{bmatrix}\begin{bmatrix} 1 & 0 \\ -\nu/\omega & 1/\omega \end{bmatrix} = S\mathcal{J}_1(q)S^{-1}.$$

The transformation matrix $S = \left[\begin{smallmatrix} 1 & 0 \\ \nu & \omega \end{smallmatrix}\right]$ is not unique; an alternative choice is $S = \left[\begin{smallmatrix} \omega & \nu \\ 0 & \nu^2+\omega^2 \end{smallmatrix}\right]$. Furthermore,

$$\mathcal{H}_2(q) = \begin{bmatrix} 0 & 1 & 0 & 0 \\ \beta_0 & \beta_1 & 1 & 0 \\ 0 & 0 & 0 & 1 \\ 0 & 0 & \beta_0 & \beta_1 \end{bmatrix} = S\begin{bmatrix} \nu & \omega & 1 & 0 \\ -\omega & \nu & 0 & 1 \\ 0 & 0 & \nu & \omega \\ 0 & 0 & -\omega & \nu \end{bmatrix}S^{-1} = S\mathcal{J}_2(q)S^{-1},$$

where

$$S \triangleq \begin{bmatrix} \omega & \nu & \omega & \nu \\ 0 & \nu^2+\omega^2 & \omega & \nu^2+\omega^2+\nu \\ 0 & 0 & -2\omega\nu & 2\omega^2 \\ 0 & 0 & -2\omega(\nu^2+\omega^2) & 0 \end{bmatrix}.$$

Fact 7.11.3. Let $q(s) \triangleq s^2 - 2\nu s + \nu^2 + \omega^2 \in \mathbb{R}[s]$ with roots $\lambda = \nu + \omega\jmath$ and $\overline{\lambda} = \nu - \omega\jmath$. Then,

$$\mathcal{H}_1(q) = \begin{bmatrix} \nu & \omega \\ -\omega & \nu \end{bmatrix} = \frac{1}{\sqrt{2}}\begin{bmatrix} 1 & 1 \\ \jmath & -\jmath \end{bmatrix}\begin{bmatrix} \lambda & 0 \\ 0 & \overline{\lambda} \end{bmatrix}\frac{1}{\sqrt{2}}\begin{bmatrix} 1 & -\jmath \\ 1 & \jmath \end{bmatrix},$$

$$\mathcal{H}_2(q) = \begin{bmatrix} \nu & \omega & 1 & 0 \\ -\omega & \nu & 0 & 1 \\ 0 & 0 & \nu & \omega \\ 0 & 0 & -\omega & \nu \end{bmatrix} = S\begin{bmatrix} \lambda & 1 & 0 & 0 \\ 0 & \lambda & 0 & 0 \\ 0 & 0 & \overline{\lambda} & 1 \\ 0 & 0 & 0 & \overline{\lambda} \end{bmatrix}S^{-1},$$

where

$$S \triangleq \frac{1}{\sqrt{2}}\begin{bmatrix} 1 & 0 & 1 & 0 \\ \jmath & 0 & -\jmath & 0 \\ 0 & 1 & 0 & 1 \\ 0 & \jmath & 0 & -\jmath \end{bmatrix}, \quad S^{-1} = \frac{1}{\sqrt{2}}\begin{bmatrix} 1 & -\jmath & 0 & 0 \\ 0 & 0 & 1 & -\jmath \\ 1 & \jmath & 0 & 0 \\ 0 & 0 & 1 & \jmath \end{bmatrix}.$$

Fact 7.11.4. Left equivalence, right equivalence, biequivalence, unitary left equivalence, unitary right equivalence, and unitary biequivalence are equivalence relations on $\mathbb{F}^{n\times m}$. Similarity, congruence, and unitary similarity are equivalence relations on $\mathbb{F}^{n\times n}$.

Fact 7.11.5. Let $A, B \in \mathbb{F}^{n\times m}$. Then, A and B are in the same equivalence class of $\mathbb{F}^{n\times m}$ induced by biequivalent transformations if and only if A and B are biequivalent to $\left[\begin{smallmatrix} I & 0 \\ 0 & 0 \end{smallmatrix}\right]$. Now, let $n = m$. Then, A and B are in the same equivalence class of $\mathbb{F}^{n\times n}$ induced by similarity transformations if and only if A and B have the same Jordan form.

Fact 7.11.6. Let $A, B \in \mathbb{C}^{n\times n}$. Then, the following statements are equivalent:

i) A and B are similar.

ii) $\operatorname{spec}(A) = \operatorname{spec}(B)$, and, for all $\lambda \in \operatorname{spec}(A)$ and $i \in \{1, \ldots, n\}$, $\operatorname{def}[(\lambda I - A)^i] = \operatorname{def}[(\lambda I - B)^i]$.

Source: [2166, pp. 63–65].

Fact 7.11.7. Let $A, B \in \mathbb{F}^{n\times n}$, and assume that A and B are similar. Then, A is semisimple if and only if B is.

Fact 7.11.8. Let $A \in \mathbb{F}^{n\times n}$, and assume that A is normal. Then, A is unitarily similar to its Jordan form. **Related:** Fact 7.11.10.

Fact 7.11.9. Let $A, B \in \mathbb{R}^{n\times n}$. Then, the following statements are equivalent:

i) There exists a nonsingular matrix $S \in \mathbb{C}^{n\times n}$ such that $A = SBS^{-1}$.

ii) There exists a nonsingular matrix $S \in \mathbb{R}^{n\times n}$ such that $A = SBS^{-1}$.

Furthermore, the following statements are equivalent:

iii) There exists an orthogonal matrix $S \in \mathbb{C}^{n\times n}$ such that $A = SBS^{-1}$.

iv) There exists a unitary matrix $S \in \mathbb{R}^{n\times n}$ such that $A = SBS^{-1}$.

Source: [2979, pp. 181, 182].

Fact 7.11.10. Let $A, B \in \mathbb{F}^{n\times n}$, and assume that A and B are normal. Then, the following statements are equivalent:

i) $\operatorname{mspec}(A) = \operatorname{mspec}(B)$.

ii) A and B are similar.

iii) A and B are unitarily similar.

Source: Since A and B are similar, it follows that $\operatorname{mspec}(A) = \operatorname{mspec}(B)$. Since A and B are normal, it follows that they are unitarily similar to the same diagonal matrix. See [1140, p. 8], [1302, p. 104], and [2991, p. 303]. **Related:** Fact 7.11.8.

Fact 7.11.11. Let $A, B \in \mathbb{F}^{n\times n}$. Then, the following statements are equivalent:

i) A and B are unitarily similar.

ii) For all $k_1, \ldots, k_r, l_1, \ldots, l_r \in \mathbb{N}$ such that $\sum_{i,j=1}^r (k_i + l_j) \le r$, where $r \triangleq \sqrt[n]{2n^2/(n-1) + 1/4} + n/2 - 2$, it follows that

$$\operatorname{tr} A^{k_1}A^{l_1*} \cdots A^{k_r}A^{l_r*} = \operatorname{tr} B^{k_1}B^{l_1*} \cdots B^{k_r}B^{l_r*}.$$

Source: [481], [1451, pp. 97, 98], [1580, pp. 71, 72], and [2209, 2429]. **Remark:** This is *Specht's theorem.* **Remark:** In the case $n = 2$, it suffices to check three equalities, specifically, $\operatorname{tr} A = \operatorname{tr} B$, $\operatorname{tr} A^2 = \operatorname{tr} B^2$, and $\operatorname{tr} A^*A = \operatorname{tr} B^*B$. In the case $n = 3$, it suffices to check 7 equalities. See [481], [1451, p. 98], and [2429]. **Related:** Fact 7.11.15.

Fact 7.11.12. Let $A, B \in \mathbb{F}^{n\times n}$, assume that A and B are idempotent, assume that $\rho_{\max}(A-B) < 1$, and define

$$S \triangleq (AB + A_\perp B_\perp)[I - (A - B)^2]^{-1/2}.$$

Then, the following statements hold:

i) S is nonsingular.

ii) If $A = B$, then $S = I$.

iii) $S^{-1} = (BA + B_\perp A_\perp)[I - (B - A)^2]^{-1/2}$.

iv) A and B are similar. In fact, $A = SBS^{-1}$.

v) If A and B are projectors, then S is unitary and A and B are unitarily similar.

Source: [1399, p. 412]. **Remark:** $[I - (A - B)^2]^{-1/2}$ is defined by *ix)* of Fact 13.4.15.

Fact 7.11.13. Let $A, B \in \mathbb{F}^{n\times n}$, and assume that A and B are idempotent. Then, the following statements are equivalent:

i) A and B are unitarily similar.

ii) $\operatorname{tr} A = \operatorname{tr} B$ and, for all $i \in \{1, \dots, \lfloor n/2 \rfloor\}$, $\operatorname{tr}(AA^*)^i = \operatorname{tr}(BB^*)^i$.

iii) $\chi_{AA^*} = \chi_{BB^*}$.

Source: Fact 7.10.26 and [919].

Fact 7.11.14. Let $A, B \in \mathbb{F}^{n\times n}$, and assume that either A or B is nonsingular. Then, AB and BA are similar. **Source:** If A is nonsingular, then $AB = A(BA)A^{-1}$, whereas, if B is nonsingular, then $BA = B(AB)B^{-1}$.

Fact 7.11.15. Let $A, B \in \mathbb{F}^{n\times n}$, and assume that A and B are projectors. Then, AB and BA are unitarily similar. **Source:** [2291]. The result follows from Fact 7.11.11.

Fact 7.11.16. Let $A \in \mathbb{F}^{n\times n}$, and assume that A is idempotent. Then, A and A^* are unitarily similar. **Source:** Use Fact 7.10.26 and the fact that $\begin{bmatrix} 1 & a \\ 0 & 0 \end{bmatrix}$ and $\begin{bmatrix} 1 & 0 \\ a & 0 \end{bmatrix}$ are unitarily similar. See [919]. Alternatively, the result follows from Fact 7.11.15 and Fact 8.7.2.

Fact 7.11.17. Let $A, B \in \mathbb{F}^{n\times n}$, where A and B are Hermitian. Then, AB and BA are similar. **Source:** [2991, p. 264].

Fact 7.11.18. Let $A \in \mathbb{F}^{n\times n}$. Then, A is idempotent if and only if there exists a projector $B \in \mathbb{F}^{n\times n}$ such that A and B are similar.

Fact 7.11.19. Let $A, B \in \mathbb{F}^{n\times n}$, assume that A and B are idempotent, and assume that $A + B - I$ is nonsingular. Then, A and B are similar. In particular, $A = (A + B - I)^{-1}B(A + B - I)$.

Fact 7.11.20. Let $A_1, \dots, A_r \in \mathbb{F}^{n\times n}$, and assume that, for all $i, j \in \{1, \dots, r\}$, $A_iA_j = A_jA_i$. Then,

$$\dim \operatorname{span} \left\{ \prod_{i=1}^{r} A_i^{n_i} \colon\ 0 \le n_i \le n - 1 \text{ for all } i \in \{1, \dots, r\} \right\} \le \lfloor n^2/4 \rfloor + 1.$$

Source: [1582, 1722] and [2166, p. 114]. **Remark:** This result bounds the dimension of a commutative subalgebra. **Credit:** I. Schur.

Fact 7.11.21. Let $A, B \in \mathbb{F}^{n\times n}$, and assume that $AB = BA$. Then,

$$\dim \operatorname{span} \{A^iB^j \colon\ 0 \le i \le n - 1,\ 0 \le j \le n - 1\} \le n.$$

Source: [306, 1432, 1722] and [2166, p. 219]. **Remark:** This result bounds the dimension of a commutative subalgebra. **Credit:** M. Gerstenhaber.

Fact 7.11.22. Let $n \ge 2$, let $A, B \in \mathbb{C}^{n\times n}$, define $\mathcal{S}_0(A, B) \triangleq \{\alpha I \colon \alpha \in \mathbb{C}\}$, for all $k \ge 1$, define $\mathcal{S}_k(A, B) \triangleq \operatorname{span}\{X_1 \cdots X_k \colon X_1, \dots, X_k \in \{A, B, I\}\}$, and define $\ell(A, B) \triangleq \max_{k\ge 0} \dim \mathcal{S}_k(A, B)$.

Then, the following statements hold:

i) There exist $C, D \in \mathbb{C}^{n\times n}$ such that $\ell(C, D) = 2n - 2$.

ii) $\ell(A,B) \le \lceil (n^2+2)/3 \rceil$.

iii) If $AB = BA$, then $\ell(A,B) \le n-1$.

iv) If either A or B is cyclic, then $\ell(A,B) \le 3n-3$.

v) If either A or B is simple, then $\ell(A,B) \le 2n-2$.

vi) If $n \le 6$, then $\ell(A,B) \le 2n-2$.

Source: [1727, 2197, 2208]. **Remark:** $\cup_{i=0}^{\infty}\mathcal{R}[\mathcal{W}_i(A,B)]$ is the *unital algebra generated by* A, B. **Remark:** The bound $2n-2$ is conjectured to hold for all $n \ge 2$ and all $A, B \in \mathbb{C}^{n\times n}$. See [1727].

Fact 7.11.23. Let $A, B \in \mathbb{F}^{n\times m}$. Then, the following statements hold:

i) The matrices A and B are unitarily left equivalent if and only if $A^*A = B^*B$.

ii) The matrices A and B are unitarily right equivalent if and only if $AA^* = BB^*$.

iii) The matrices A and B are unitarily biequivalent if and only if A and B have the same singular values with the same multiplicity.

Source: [1455] and [2314, pp. 372, 373]. **Remark:** In [1455] A and B need not be the same size. **Remark:** The singular value decomposition provides a canonical matrix under unitary biequivalence in analogy with the Smith form under biequivalence. **Remark:** Note that $AA^* = BB^*$ implies that $\mathcal{R}(A) = \mathcal{R}(B)$, which implies right equivalence. This is an alternative proof of the fact that unitary right equivalence implies right equivalence.

Fact 7.11.24. Let $A, B \in \mathbb{F}^{n\times n}$. Then, the following statements hold:

i) $A^*A = B^*B$ if and only if there exists a unitary matrix $S \in \mathbb{F}^{n\times n}$ such that $A = SB$.

ii) $A^*A \le B^*B$ if and only if there exists $S \in \mathbb{F}^{n\times n}$ such that $A = SB$ and $S^*S \le I$.

iii) $A^*B + B^*A = 0$ if and only if there exists a unitary matrix $S \in \mathbb{F}^{n\times n}$ such that $(I-S)A = (I+S)B$.

iv) $A^*B + B^*A \ge 0$ if and only if there exists $S \in \mathbb{F}^{n\times n}$ such that $(I-S)A = (I+S)B$ and $S^*S \le I$.

Source: [1448, p. 406] and [2298]. **Remark:** *iii)* and *iv)* follow from *i)* and *ii)* by replacing A and B with $A-B$ and $A+B$, respectively.

Fact 7.11.25. Let $A \in \mathbb{F}^{n\times n}$, $B \in \mathbb{F}^{m\times m}$, and $C \in \mathbb{F}^{n\times m}$. Then, there exist $X, Y \in \mathbb{F}^{n\times m}$ satisfying

$$AX + YB + C = 0$$

if and only if

$$\operatorname{rank}\begin{bmatrix} A & 0 \\ 0 & B \end{bmatrix} = \operatorname{rank}\begin{bmatrix} A & C \\ 0 & B \end{bmatrix}.$$

Source: [2263, pp. 194, 195] and [2835]. **Remark:** $AX + YB + C = 0$ is a generalization of Sylvester's equation. See Fact 7.11.26. **Remark:** An explicit expression for all solutions is given by Fact 8.9.10, which applies to the case where A and B are not necessarily square and thus X and Y are not necessarily the same size. **Credit:** W. E. Roth.

Fact 7.11.26. Let $A \in \mathbb{F}^{n\times n}$, $B \in \mathbb{F}^{m\times m}$, and $C \in \mathbb{F}^{n\times m}$. Then, there exists $X \in \mathbb{F}^{n\times m}$ satisfying

$$AX + XB + C = 0$$

if and only if $\left[\begin{smallmatrix} A & 0 \\ 0 & -B \end{smallmatrix}\right]$ and $\left[\begin{smallmatrix} A & C \\ 0 & -B \end{smallmatrix}\right]$ are similar. If these conditions hold, then

$$\begin{bmatrix} A & C \\ 0 & -B \end{bmatrix} = \begin{bmatrix} I & X \\ 0 & I \end{bmatrix}\begin{bmatrix} A & 0 \\ 0 & -B \end{bmatrix}\begin{bmatrix} I & -X \\ 0 & I \end{bmatrix}.$$

Source: [2835] and [2979, pp. 41, 42]. For sufficiency, see [1738, pp. 422–424] and [2263, pp. 194, 195]. **Remark:** $AX + XB + C = 0$ is *Sylvester's equation.* See Proposition 9.2.5, Corollary 9.2.6, and Proposition 15.10.3 for the case where X is unique. **Credit:** W. E. Roth. See [478].

Fact 7.11.27. Let $A, B \in \mathbb{F}^{n\times n}$, and assume that A and B are idempotent. Then, the matrices $\begin{bmatrix} A+B & A \\ 0 & -A-B \end{bmatrix}$ and $\begin{bmatrix} A+B & 0 \\ 0 & -A-B \end{bmatrix}$ are similar and satisfy

$$\begin{bmatrix} A+B & -A \\ 0 & -A-B \end{bmatrix} = \begin{bmatrix} I & X \\ 0 & I \end{bmatrix}\begin{bmatrix} A+B & 0 \\ 0 & -A-B \end{bmatrix}\begin{bmatrix} I & -X \\ 0 & I \end{bmatrix},$$

where $X \triangleq \frac{1}{4}(I + A - B)$. **Credit:** Y. Tian. **Related:** Fact 7.11.26.

Fact 7.11.28. Let $A \in \mathbb{F}^{n\times n}$, $B \in \mathbb{F}^{m\times m}$, and $C \in \mathbb{F}^{n\times m}$, and assume that A and B are nilpotent. Then, the matrices $\begin{bmatrix} A & C \\ 0 & B \end{bmatrix}$ and $\begin{bmatrix} A & 0 \\ 0 & B \end{bmatrix}$ are similar if and only if

$$\operatorname{rank}\begin{bmatrix} A & C \\ 0 & B \end{bmatrix} = \operatorname{rank} A + \operatorname{rank} B, \quad AC + CB = 0.$$

Source: [2643].

Fact 7.11.29. Let $A \in \mathbb{F}^{n\times m}$ and $B \in \mathbb{F}^{m\times n}$. Then, $\begin{bmatrix} AB & 0 \\ B & 0 \end{bmatrix}$ and $\begin{bmatrix} 0 & 0 \\ B & BA \end{bmatrix}$ are similar. **Source:** [2979, p. 32].

Fact 7.11.30. Let $A, B, C \in \mathbb{F}^{n\times n}$, assume that $\operatorname{spec}(A) \cap \operatorname{spec}(B) = \varnothing$, and assume that $[A + B, C] = 0$ and $[AB, C] = 0$. Then, $[A, C] = [B, C] = 0$. **Source:** This result follows from Corollary 9.2.6. **Credit:** M. R. Embry. See [478].

7.12 Facts on Eigenvalues and Singular Values for One Matrix

Fact 7.12.1. Let $A \in \mathbb{F}^{n\times n}$, and assume that A is singular. If A is either simple or cyclic, then $\operatorname{rank} A = n - 1$.

Fact 7.12.2. Let $A \in \mathbb{R}^{n\times n}$, and assume that $A \in \mathrm{SO}(n)$. Then, $\operatorname{amult}_A(-1)$ is even. Now, assume that $n = 3$. Then, the following statements hold:

i) Either $\operatorname{amult}_A(1) = 1$ or $\operatorname{amult}_A(1) = 3$.

ii) $\operatorname{tr} A \geq -1$.

iii) $\operatorname{tr} A = -1$ if and only if $\operatorname{mspec}(A) = \{1, -1, -1\}_{\mathrm{ms}}$.

Fact 7.12.3. Let $A \in \mathbb{F}^{n\times n}$, let $\alpha \in \mathbb{F}$, and assume that $A^2 = \alpha A$. Then, $\operatorname{spec}(A) \subseteq \{0, \alpha\}$.

Fact 7.12.4. Let $A \in \mathbb{F}^{n\times n}$, assume that A is Hermitian, and let $\alpha \in \mathbb{R}$. Then, $A^2 = \alpha A$ if and only if $\operatorname{spec}(A) \subseteq \{0, \alpha\}$. **Related:** Fact 4.10.22.

Fact 7.12.5. Let $A \in \mathbb{F}^{n\times n}$. Then, the following statements are equivalent:

i) $A^2 = -A$.

ii) $\operatorname{rank} A = -\operatorname{tr} A$ and $\operatorname{rank}(A + I) = n + \operatorname{tr} A$.

Source: [2238, p. 530] and [2714].

Fact 7.12.6. Let $A \in \mathbb{F}^{n\times n}$, assume that A is Hermitian, let $x_1, \ldots, x_n \in \mathbb{F}^n$ be eigenvectors associated with $\lambda_1(A), \ldots, \lambda_n(A)$, respectively, let $k, l \in \{1, \ldots, n\}$, where $k \leq l$, and let $x \in \operatorname{span}\{x_k, \ldots, x_l\}$, where $x \neq 0$. Then,

$$\lambda_l(A) \leq \frac{x^*Ax}{x^*x} \leq \lambda_k(x).$$

Source: [2991, p. 266]. **Related:** This result implies Lemma 10.4.3.

Fact 7.12.7. Let $A \in \mathbb{F}^{n\times n}$, assume that A is Hermitian, let $k \in \{1, \ldots, n\}$, and let $\mathcal{W}$ denote the set of subspaces of $\mathbb{F}^n$ whose dimension is k. Then,

$$\lambda_k(A) = \max_{\mathcal{S}\in\mathcal{W}} \min_{\{x\in\mathcal{S}:\ x\neq 0\}} \frac{x^*Ax}{x^*x} = \max_{\mathcal{S}^\perp\in\mathcal{W}} \min_{\{x\in\mathcal{S}:\ x\neq 0\}} \frac{x^*Ax}{x^*x}.$$

Source: [2991, pp. 268, 269]. **Remark:** This is the *min-max theorem*. **Credit:** R. Courant and E. S. Fischer.

Fact 7.12.8. Let $A \in \mathbb{F}^{n\times n}$, and assume that A is Hermitian. Then,

$$\max_{x\in\mathbb{F}^n\backslash\{0\}} \frac{|x^*Ax|}{x^*x} = \sigma_{\max}(A).$$

Furthermore, if A is either positive definite or negative definite, then

$$\min_{x\in\mathbb{F}^n\backslash\{0\}} \frac{|x^*Ax|}{x^*x} = \sigma_{\min}(A).$$

Finally, A is neither positive definite nor negative definite, then

$$\min_{x\in\mathbb{F}^n\backslash\{0\}} \frac{|x^*Ax|}{x^*x} = 0.$$

Related: Fact 11.15.1.

Fact 7.12.9. Let $A \in \mathbb{F}^{n\times n}$, and assume that A is Hermitian. Then,

$$\alpha_{\max}(A) = \lambda_{\max}(A), \quad \rho_{\max}(A) = \sigma_{\max}(A) = \max\{|\lambda_{\min}(A)|, \lambda_{\max}(A)\}.$$

If, in addition, A is positive semidefinite, then

$$\rho_{\max}(A) = \sigma_{\max}(A) = \alpha_{\max}(A) = \lambda_{\max}(A).$$

Related: Fact 7.13.8.

Fact 7.12.10. Let $A \in \mathbb{F}^{n\times n}$, and assume that A is skew Hermitian. Then, the eigenvalues of A are imaginary. **Source:** Let $\lambda \in \operatorname{spec}(A)$. Since $0 \le AA^* = -A^2$, it follows that $-\lambda^2 \ge 0$, and thus $\lambda^2 \le 0$.

Fact 7.12.11. Let $A \in \mathbb{F}^{n\times n}$. Then, the following statements are equivalent:

i) A is idempotent.

ii) $\operatorname{rank}(I - A) \le \operatorname{tr}(I - A)$, A is group invertible, and every eigenvalue of A is nonnegative.

iii) A and $I - A$ are group invertible, and every eigenvalue of A is nonnegative.

iv) A is group invertible, and there exists a positive integer k such that $A^k = A^{k+1}$.

Source: [1336].

Fact 7.12.12. Let $A \in \mathbb{F}^{n\times n}$, and define $k \triangleq n - \operatorname{amult}_A(0)$. Then,

$$|\operatorname{tr} A|^2 \le \left(\sum_{i=1}^k \rho_i(A)\right)^2 \le k\sum_{i=1}^k \rho_i^2(A).$$

Source: Fact 2.11.17.

Fact 7.12.13. Let $A \in \mathbb{F}^{n\times n}$, and assume that A has exactly k nonzero eigenvalues. Then,

$$\left.\begin{matrix} |\operatorname{tr} A|^2 \\ k|\operatorname{tr} A^2| \le k\operatorname{tr}(A^{2*}A^2)^{1/2} \end{matrix}\right\} \le k\operatorname{tr} A^*A \le (\operatorname{rank} A)\operatorname{tr} A^*A.$$

Furthermore, the upper left-most inequality is an equality if and only if A is normal and all of the nonzero eigenvalues of A have the same absolute value. Moreover, the right-most inequality is an equality if and only if A is group invertible. If, in addition, $\operatorname{spec}(A) \subset \mathbb{R}$, then

$$(\operatorname{tr} A)^2 \le k\operatorname{tr} A^2 \le \left\{\begin{matrix} (\operatorname{rank} A)\operatorname{tr} A^2 \\ k\operatorname{tr} A^*A \end{matrix}\right\} \le (\operatorname{rank} A)\operatorname{tr} A^*A.$$

Source: The upper left-hand inequality in the first string is given in [2905]. The lower left-hand inequality in the first string is given by Fact 11.13.2. In the case where all of the eigenvalues of A are real, the inequality $(\operatorname{tr} A)^2 \le k\operatorname{tr} A^2$ follows from Fact 7.12.12. See [2991, p. 264]. **Remark:**

$|\operatorname{tr} A|^2 \le k|\operatorname{tr} A^2|$ does not necessarily hold. Consider $\operatorname{mspec}(A) = \{1, 1, \jmath, -\jmath\}_{\mathrm{ms}}$. **Related:** Fact 4.10.22, Fact 10.13.3, Fact 10.13.5, and Fact 10.13.6.

Fact 7.12.14. Let $A \in \mathbb{R}^{n\times n}$ and $\operatorname{mspec}(A) = \{\lambda_1, \dots, \lambda_n\}_{\mathrm{ms}}$. Then,

$$\sum_{i=1}^n (\operatorname{Re} \lambda_i)(\operatorname{Im} \lambda_i) = 0, \quad \operatorname{tr} A^2 = \sum_{i=1}^n (\operatorname{Re} \lambda_i)^2 - \sum_{i=1}^n (\operatorname{Im} \lambda_i)^2.$$

Fact 7.12.15. Let $n \ge 2$, let $a_1, \dots, a_n > 0$, and define the symmetric matrix $A \in \mathbb{R}^{n\times n}$ by $A_{(i,j)} \triangleq a_i + a_j$ for all $i, j \in \{1, \dots, n\}$. Then, $\operatorname{rank} A \le 2$ and $\operatorname{mspec}(A) = \{\lambda, \mu, 0, \dots, 0\}_{\mathrm{ms}}$, where

$$\lambda \triangleq \sum_{i=1}^n a_i + \sqrt{n\sum_{i=1}^n a_i^2}, \quad \mu \triangleq \sum_{i=1}^n a_i - \sqrt{n\sum_{i=1}^n a_i^2}.$$

Furthermore, the following statements hold:

i) $\lambda > 0$.

ii) $\mu \le 0$.

Moreover, the following statements are equivalent:

iii) $\mu < 0$.

iv) At least two of the numbers $a_1, \dots, a_n > 0$ are distinct.

v) $\operatorname{rank} A = 2$.

If *iii)*–*v)* hold, then $\lambda_{\min}(A) = \mu < 0 < \operatorname{tr} A = 2\sum_{i=1}^n a_i < \lambda_{\max}(A) = \lambda$. **Source:** $A = a1_{1\times n} + 1_{n\times 1}a^{\mathsf{T}}$, where $a \triangleq [a_1 \ \cdots \ a_n]^{\mathsf{T}}$. Then, Fact 3.14.18 implies that $\operatorname{rank} A \le \operatorname{rank}(a1_{1\times n}) + \operatorname{rank}(1_{n\times 1}a^{\mathsf{T}}) = 2$. Furthermore, $\operatorname{mspec}(A)$ follows from Fact 7.12.16, while Fact 2.11.81 implies that $\mu \le 0$. **Related:** Fact 10.9.11.

Fact 7.12.16. Let $x, y \in \mathbb{R}^n$. Then,

$$\operatorname{mspec}(xy^{\mathsf{T}} + yx^{\mathsf{T}}) = \left\{x^{\mathsf{T}}y + \sqrt{x^{\mathsf{T}}xy^{\mathsf{T}}y}, x^{\mathsf{T}}y - \sqrt{x^{\mathsf{T}}xy^{\mathsf{T}}y}, 0, \dots, 0\right\}_{\mathrm{ms}},$$

$$\alpha_{\max}(xy^{\mathsf{T}} + yx^{\mathsf{T}}) = x^{\mathsf{T}}y + \sqrt{x^{\mathsf{T}}xy^{\mathsf{T}}y}, \quad \rho_{\max}(xy^{\mathsf{T}} + yx^{\mathsf{T}}) = \begin{cases} x^{\mathsf{T}}y + \sqrt{x^{\mathsf{T}}xy^{\mathsf{T}}y}, & x^{\mathsf{T}}y \ge 0, \\ \left|x^{\mathsf{T}}y - \sqrt{x^{\mathsf{T}}xy^{\mathsf{T}}y}\right|, & x^{\mathsf{T}}y \le 0. \end{cases}$$

Now, assume that x and y are nonzero, and define $v_1, v_2 \in \mathbb{R}^n$ by

$$v_1 \triangleq \frac{1}{\|x\|}x + \frac{1}{\|y\|}y, \quad v_2 \triangleq \frac{1}{\|x\|}x - \frac{1}{\|y\|}y.$$

Then, v_1 and v_2 are eigenvectors of $xy^{\mathsf{T}} + yx^{\mathsf{T}}$ corresponding to $x^{\mathsf{T}}y + \sqrt{x^{\mathsf{T}}xy^{\mathsf{T}}y}$ and $x^{\mathsf{T}}y - \sqrt{x^{\mathsf{T}}xy^{\mathsf{T}}y}$, respectively. **Source:** [835, p. 539]. **Example:** $\operatorname{mspec}\left(\begin{bmatrix} 0_{n\times n} & 1_{n\times 1} \\ 1_{1\times n} & 0 \end{bmatrix}\right) = \{-\sqrt{n}, 0, \dots, 0, \sqrt{n}\}_{\mathrm{ms}}$. **Problem:** Extend this result to $\mathbb{C}$ and $xy^{\mathsf{T}} + zw^{\mathsf{T}}$. See Fact 6.9.15.

Fact 7.12.17. Let $A \in \mathbb{F}^{n\times n}$, and let $\operatorname{mspec}(A) = \{\lambda_1, \dots, \lambda_n\}_{\mathrm{ms}}$. Then,

$$\operatorname{mspec}[(I + A)^2] = \{(1 + \lambda_1)^2, \dots, (1 + \lambda_n)^2\}_{\mathrm{ms}}.$$

If A is nonsingular, then

$$\operatorname{mspec}(A^{-1}) = \{\lambda_1^{-1}, \dots, \lambda_n^{-1}\}_{\mathrm{ms}}.$$

Finally, if $I + A$ is nonsingular, then

$$\operatorname{mspec}[(I + A)^{-1}] = \{(1 + \lambda_1)^{-1}, \dots, (1 + \lambda_n)^{-1}\}_{\mathrm{ms}},$$

$$\operatorname{mspec}[A(I + A)^{-1}] = \{\lambda_1(1 + \lambda_1)^{-1}, \dots, \lambda_n(1 + \lambda_n)^{-1}\}_{\mathrm{ms}}.$$

Source: Fact 7.12.18.

Fact 7.12.18. Let $p, q \in \mathbb{F}[s]$, assume that p and q are coprime, and define $g \triangleq p/q \in \mathbb{F}(s)$. Furthermore, let $A \in \mathbb{F}^{n\times n}$, let $\mathrm{mspec}(A) = \{\lambda_1, \ldots, \lambda_n\}_{\mathrm{ms}}$, assume that $\mathrm{roots}(q) \cap \mathrm{spec}(A) = \emptyset$, and define $g(A) \triangleq p(A)[q(A)]^{-1}$. Then, $\mathrm{mspec}[g(A)] = \{g(\lambda_1), \ldots, g(\lambda_n)\}_{\mathrm{ms}}$. **Source:** *ii*) of Fact 6.9.36 implies that $q(A)$ is nonsingular.

Fact 7.12.19. Let $x \in \mathbb{F}^n$ and $y \in \mathbb{F}^m$. Then,

$$\sigma_{\max}(xy^*) = \sqrt{x^*xy^*y}.$$

If, in addition, $m = n$, then

$$\mathrm{mspec}(xy^*) = \{x^*y, 0, \ldots, 0\}_{\mathrm{ms}}, \quad \mathrm{mspec}(I + xy^*) = \{1 + x^*y, 1, \ldots, 1\}_{\mathrm{ms}},$$

$$\rho_{\max}(xy^*) = |x^*y|, \quad \alpha_{\max}(xy^*) = \max\{0, \mathrm{Re}\, x^*y\}.$$

Related: Fact 11.8.25.

Fact 7.12.20. Let $A \in \mathbb{F}^{n\times n}$, and assume that $\mathrm{rank}\, A = 1$. Then, $\sigma_{\max}(A) = (\mathrm{tr}\, AA^*)^{1/2}$.

Fact 7.12.21. Let $x, y \in \mathbb{F}^n$, and assume that $x^*y \neq 0$. Then, $\sigma_{\max}[(x^*y)^{-1}xy^*] \geq 1$.

Fact 7.12.22. Let $A \in \mathbb{F}^{n\times m}$ and $\alpha \in \mathbb{F}$. Then, for all $i \in \{1, \ldots, \min\{n, m\}\}$, $\sigma_i(\alpha A) = |\alpha|\sigma_i(A)$.

Fact 7.12.23. Let $A \in \mathbb{F}^{n\times m}$. Then, for all $i \in \{1, \ldots, \mathrm{rank}\, A\}$, $\sigma_i(A) = \sigma_i(A^*)$.

Fact 7.12.24. Let $A \in \mathbb{F}^{n\times n}$, and let $\lambda \in \mathrm{spec}(A)$. Then, the following statements hold:

i) $\sigma_{\min}(A) \leq |\lambda| \leq \sigma_{\max}(A)$.

ii) $\lambda_{\min}[\frac{1}{2}(A + A^*)] \leq \mathrm{Re}\, \lambda \leq \lambda_{\max}[\frac{1}{2}(A + A^*)]$.

iii) $\lambda_{\min}[\frac{1}{2j}(A - A^*)] \leq \mathrm{Im}\, \lambda \leq \lambda_{\max}[\frac{1}{2j}(A - A^*)]$.

iv) $\sigma_n[\frac{1}{2j}(A - A^*)] \leq |\mathrm{Im}\, \lambda_i| \leq \sigma_1[\frac{1}{2j}(A - A^*)]$.

Remark: *i*) is *Browne's theorem*, *ii*) is *Bendixson's theorem*, and *iii*) is *Hirsch's theorem*. See [688, p. 17] and [1952, pp. 140–144]. *iv*) is given in [1173]. **Related:** Fact 7.12.29, Fact 7.13.9, Fact 11.13.6, and Fact 11.13.7.

Fact 7.12.25. Let $A \in \mathbb{F}^{n\times n}$. Then,

$$\lambda[\tfrac{1}{2}(A + A^*)] \leq\leq \sigma(A).$$

Hence, for all $i \in \{1, \ldots, n\}$,

$$\lambda_i[\tfrac{1}{2}(A + A^*)] \leq \sigma_i(A), \quad -\sigma_{\max}(A) \leq \lambda_i[\tfrac{1}{2}(A + A^*)] \leq \sigma_{\max}(A).$$

In particular,

$$\lambda_{\min}[\tfrac{1}{2}(A + A^*)] \leq \sigma_{\min}(A), \quad -\sigma_{\max}(A) \leq \lambda_{\min}[\tfrac{1}{2}(A + A^*)] \leq \lambda_{\max}[\tfrac{1}{2}(A + A^*)] \leq \sigma_{\max}(A).$$

Furthermore,

$$\lambda[\tfrac{1}{2}(A + A^*)] \overset{\mathrm{w}}{\prec} \sigma(A), \quad \mathrm{Re}\,\mathrm{tr}\, A = \mathrm{tr}[\tfrac{1}{2}(A + A^*)] \leq \mathrm{tr}\,\langle A\rangle.$$

Source: [1399, p. 447], [1450, p. 151], [2479], and [2991, pp. 288, 289, 361]. **Remark:** This result generalizes $\mathrm{Re}\, z \leq |z|$, where $z \in \mathbb{C}$. **Remark:** For all $i \in \{1, \ldots, n\}$, $|\lambda_i[\frac{1}{2}(A + A^*)]| \leq \sigma_{\max}(A)$. However, $|\lambda_i[\frac{1}{2}(A + A^*)]| \leq \sigma_i(A)$ given in [1969, p. 240] and [1971, p. 327] is erroneous. Consider $A = \begin{bmatrix} -1 & 0 \\ 0 & 0 \end{bmatrix}$. **Related:** Fact 7.12.26, Fact 10.13.18, and Fact 10.21.9.

Fact 7.12.26. Let $A \in \mathbb{F}^{n\times n}$. Then,

$$\mathrm{d}(|A|) \overset{\mathrm{w}}{\prec} \sigma(A), \quad \rho(A) \overset{\mathrm{w}}{\prec} \sigma(A), \quad \rho[\tfrac{1}{2}(A + A^*)] \overset{\mathrm{w}}{\prec} \sigma(A).$$

If, in addition, A is Hermitian, then

$$\mathrm{d}(A) \overset{\mathrm{s}}{\prec} \lambda(A).$$

Source: Fact 7.12.25, [1969, pp. 228, 240], [1971, pp. 314, 327], [2979, p. 82], and [2991, pp. 349, 351, 361]. **Related:** Fact 7.12.25, Fact 10.21.10, and Fact 10.21.11.

Fact 7.12.27. Let $A \in \mathbb{F}^{n\times n}$. Then,

$$\alpha(A) \overset{\mathrm{s}}{\prec} \lambda[\tfrac{1}{2}(A + A^*)].$$

In particular, for all $i \in \{1, \ldots, n\}$,

$$\lambda_{\min}[\tfrac{1}{2}(A + A^*)] \le \alpha_n(A) \le \alpha_i(A) \le \alpha_{\max}(A) = \alpha_1(A) \le \lambda_{\max}[\tfrac{1}{2}(A + A^*)].$$

Furthermore,

$$\sum_{i=1}^{n} \alpha_i(A) = \operatorname{Re}\operatorname{tr} A = \operatorname{tr} \tfrac{1}{2}(A + A^*) = \sum_{i=1}^{n} \lambda_i[\tfrac{1}{2}(A + A^*)] \le |\operatorname{tr} A| \le \sum_{i=1}^{n} \rho_i(A) \le \sum_{i=1}^{n} \sigma_i(A).$$

Source: [449, p. 74] and [2991, p. 360]. **Credit:** K. Fan. **Related:** Fact 7.12.26, Fact 10.13.2, and *vi)* and *xvii)* of Fact 15.16.7.

Fact 7.12.28. Let $A \in \mathbb{F}^{n\times n}$. Then,

$$-\sigma_{\max}(A) \le \lambda_{\min}[\tfrac{1}{2}(A + A^*)] \le \alpha_{\max}(A) \le \left\{ \begin{matrix} |\alpha_{\max}(A)| \le \rho_{\max}(A) \\ \lambda_{\max}[\tfrac{1}{2}(A + A^*)] \end{matrix} \right\} \le \sigma_{\max}(A).$$

Source: Fact 7.12.25 and Fact 7.12.27.

Fact 7.12.29. Let $A \in \mathbb{F}^{n\times n}$, and let $\operatorname{mspec}(A) = \{\lambda_1, \ldots, \lambda_n\}_{\mathrm{ms}}$. Then, for all $k \in \{1, \ldots, n\}$,

$$\sum_{i=1}^{k} [\sigma_{n-i+1}^2(A) - |\lambda_i|^2] \le 2\sum_{i=1}^{k} (\sigma_i^2[\tfrac{1}{2j}(A - A^*)] - |\operatorname{Im} \lambda_i|^2),$$

$$2\sum_{i=1}^{k} (\sigma_{n-i+1}^2[\tfrac{1}{2j}(A - A^*)] - |\operatorname{Im} \lambda_i|^2) \le \sum_{i=1}^{k} [\sigma_i^2(A) - |\lambda_i|^2],$$

$$\sum_{i=1}^{n} [\sigma_i^2(A) - |\lambda_i|^2] = 2\sum_{i=1}^{n} (\sigma_i^2[\tfrac{1}{2j}(A - A^*)] - |\operatorname{Im} \lambda_i|^2).$$

Source: [1173]. **Related:** Fact 11.13.6.

Fact 7.12.30. Let $A \in \mathbb{F}^{n\times n}$. Then, the following statements are equivalent:

i) A is normal.

ii) $\{\sigma_1(A), \ldots, \sigma_n(A)\}_{\mathrm{ms}} = \{\rho_1(A), \ldots, \rho_n(A)\}_{\mathrm{ms}}$.

iii) For all $k \in \{1, \ldots, n\}$, $\prod_{i=1}^{k} \sigma_i(A) = \prod_{i=1}^{k} \rho_i(A)$.

iv) $\sum_{i=1}^{n} \rho_i^2(A) = \sum_{i=1}^{n} \sigma_i^2(A)$.

v) $\sum_{i=1}^{n} \alpha_i^2(A) = \frac{1}{4}\operatorname{tr}(A + A^*)^2$.

vi) $\sum_{i=1}^{n} \beta_i^2(A) = -\frac{1}{4}\operatorname{tr}(A - A^*)^2$.

vii) $\{\rho_1^2(A), \ldots, \rho_n^2(A)\}_{\mathrm{ms}} = \operatorname{mspec}(A^*A)$.

viii) $\{\alpha_1(A), \ldots, \alpha_n(A)\}_{\mathrm{ms}} = \operatorname{mspec}[\frac{1}{2}(A + A^*)]$.

ix) $\{\beta_1(A), \ldots, \beta_n(A)\}_{\mathrm{ms}} = \operatorname{mspec}[\frac{1}{2j}(A + A^*)]$.

x) There exists a permutation σ of $(1, \ldots, n)$ such that $\operatorname{mspec}(AA^*) = \{\lambda_1\overline{\lambda_{\sigma(1)}}, \ldots, \lambda_n\overline{\lambda_{\sigma(n)}}\}_{\mathrm{ms}}$.

xi) $\rho(A) \overset{\mathrm{s}}{\prec} \sigma(A)$.

xii) $[\sigma(A)]^{\odot 2} \overset{\mathrm{s}}{\prec} \sigma(A^2)$.

xiii) For all $x \in \mathbb{F}^n$, $\|Ax\|_2 = \|A^*x\|_2$.

If these statements hold, then the following statements hold:

xiv) $[\operatorname{Re} A_{(1,1)} \ \cdots \ \operatorname{Re} A_{(n,n)}] \overset{\mathrm{s}}{\prec} [\alpha_1(A) \ \cdots \ \alpha_n(A)]$.

xv) For all $x \in \mathbb{F}^n$ such that $\|x\|_2 = 1$, $|x^*Ax|^2 \le \frac{1}{2}(\|Ax\|_2^2 + |x^*A^2x|) \le \|Ax\|_2^2$.

Source: [940, 987, 1242] and [2991, pp. 294, 355, 361]. **Related:** Fact 4.10.12, Fact 4.13.6, Fact 7.15.16, Fact 10.21.10, and Fact 11.13.6.

Fact 7.12.31. Let $A \in \mathbb{F}^{n\times n}$, assume that A is normal, let $\alpha,\beta \in \mathbb{F}$, and let $p \in [1,\infty)$. If $p \in [1,2]$, then

$$[(|\alpha| + |\beta|)^p + ||\alpha| - |\beta||^p]\sigma_{\max}^p(A) \le \sigma_{\max}^p(\alpha A + \beta A^*) + \sigma_{\max}^p(\alpha A - \beta A^*).$$

If $p \in [2,\infty)$, then

$$2(|\alpha|^p + |\beta|^p)\sigma_{\max}^p(A) \le \sigma_{\max}^p(\alpha A + \beta A^*) + \sigma_{\max}^p(\alpha A - \beta A^*).$$

Source: [940].

Fact 7.12.32. Let $A \in \mathbb{F}^{n\times n}$. Then, $\rho(A) \overset{\mathrm{slog}}{\prec} \sigma(A)$. Therefore,

$$|\det A| = \prod_{i=1}^n \rho_i(A) = \prod_{i=1}^n \sigma_i(A),$$

and, for all $k \in \{1,\dots,n\}$,

$$\prod_{i=k}^n \sigma_i(A) \le \prod_{i=k}^n \rho_i(A).$$

In particular, $\sigma_{\min}(A) \le \rho_{\min}(A)$. **Source:** Fact 3.25.15, [449, p. 43], [1399, p. 445], [1450, p. 171], [2977, p. 19], and [2991, p. 353]. **Credit:** H. Weyl. **Related:** Fact 10.22.28 and Fact 11.15.20.

Fact 7.12.33. Let $A \in \mathbb{F}^{n\times n}$. Then,

$$\sigma_{\min}(A) \le \sigma_{\max}^{1/n}(A)\sigma_{\min}^{(n-1)/n}(A) \le \rho_{\min}(A) \le \rho_{\max}(A) \le \sigma_{\min}^{1/n}(A)\sigma_{\max}^{(n-1)/n}(A) \le \sigma_{\max}(A),$$

$$\sigma_{\min}^n(A) \le \sigma_{\max}(A)\sigma_{\min}^{n-1}(A) \le |\det A| \le \sigma_{\min}(A)\sigma_{\max}^{n-1}(A) \le \sigma_{\max}^n(A).$$

Source: Fact 7.12.32 and [1399, p. 445]. **Related:** Fact 10.15.1 and Fact 15.21.19.

Fact 7.12.34. Let $\beta_0,\dots,\beta_{n-1} \in \mathbb{F}$, define $A \in \mathbb{F}^{n\times n}$ by

$$A \triangleq \begin{bmatrix} 0 & 1 & 0 & \cdots & 0 & 0 \\ 0 & 0 & 1 & \ddots & 0 & 0 \\ 0 & 0 & 0 & \ddots & 0 & 0 \\ \vdots & \vdots & \vdots & \ddots & \ddots & \vdots \\ 0 & 0 & 0 & \cdots & 0 & 1 \\ -\beta_0 & -\beta_1 & -\beta_2 & \cdots & -\beta_{n-2} & -\beta_{n-1} \end{bmatrix},$$

and define $\alpha \triangleq 1 + \sum_{i=0}^{n-1} |\beta_i|^2$. Then, $\sigma_2(A) = \cdots = \sigma_{n-1}(A) = 1$ and

$$\sigma_1(A) = \sqrt{\frac{1}{2}\left(\alpha + \sqrt{\alpha^2 - 4|\beta_0|^2}\right)}, \quad \sigma_n(A) = \sqrt{\frac{1}{2}\left(\alpha - \sqrt{\alpha^2 - 4|\beta_0|^2}\right)}.$$

In particular, $\sigma_1(N_n) = \cdots = \sigma_{n-1}(N_n) = 1$ and $\sigma_{\min}(N_n) = 0$. **Source:** [1389, p. 523] and [1601, 1628]. **Related:** Fact 8.3.31 and Fact 15.21.19.

Fact 7.12.35. Let $\beta \in \mathbb{C}$. Then,

$$\sigma_{\max}\left(\begin{bmatrix} 1 & 2\beta \\ 0 & 1 \end{bmatrix}\right) = |\beta| + \sqrt{1+|\beta|^2}, \quad \sigma_{\min}\left(\begin{bmatrix} 1 & 2\beta \\ 0 & 1 \end{bmatrix}\right) = \sqrt{1+|\beta|^2} - |\beta|.$$

Source: [1801]. **Remark:** Singular-value inequalities for block-triangular matrices are given in [1801].

Fact 7.12.36. Let $A \in \mathbb{F}^{n\times m}$. Then,

$$\sigma_{\max}\left(\begin{bmatrix} I & 2A \\ 0 & I \end{bmatrix}\right) = \sigma_{\max}(A) + \sqrt{1+\sigma_{\max}^2(A)}.$$

Source: [1389, p. 116].

Fact 7.12.37. For all $i \in \{1,\dots,l\}$, let $A_i \in \mathbb{F}^{n_i\times m_i}$. Then,

$$\sigma_{\max}[\operatorname{diag}(A_1,\dots,A_l)] = \max\{\sigma_{\max}(A_1),\dots,\sigma_{\max}(A_l)\}.$$

Fact 7.12.38. Let $A \in \mathbb{F}^{n\times m}$, and define $r \triangleq \operatorname{rank} A$. Then, for all $i \in \{1,\dots,r\}$,

$$\lambda_i(AA^*) = \lambda_i(A^*A) = \sigma_i(AA^*) = \sigma_i(A^*A) = \sigma_i^2(A).$$

In particular,

$$\sigma_{\max}(AA^*) = \sigma_{\max}^2(A),$$

and, if $n = m$, then

$$\sigma_{\min}(AA^*) = \sigma_{\min}^2(A).$$

Furthermore, for all $i \in \{1,\dots,r\}$,

$$\sigma_i(AA^*A) = \sigma_i^3(A).$$

Related: Fact 10.10.24.

Fact 7.12.39. Let $A \in \mathbb{F}^{n\times n}$. Then, for all $i \in \{1,\dots,n\}$,

$$\sigma_i(A^{\mathsf{A}}) = \prod_{\substack{j=1\\ j\neq n+1-i}}^{n} \sigma_j(A).$$

Source: Fact 6.10.13 and [2263, p. 149].

Fact 7.12.40. Let $A \in \mathbb{F}^{n\times n}$. Then, $\sigma_1(A) = \sigma_n(A)$ if and only if there exist $\lambda \in \mathbb{F}$ and a unitary matrix $B \in \mathbb{F}^{n\times n}$ such that $A = \lambda B$. **Source:** [2263, pp. 149, 165].

Fact 7.12.41. Let $A \in \mathbb{F}^{n\times n}$, and assume that A is idempotent. Then, the following statements hold:

i) If σ is a singular value of A, then either $\sigma = 0$ or $\sigma \geq 1$.

ii) If $A \neq 0$, then $\sigma_{\max}(A) \geq 1$.

iii) $\sigma_{\max}(A) = 1$ if and only if A is a projector.

iv) If $1 \leq \operatorname{rank} A \leq n-1$, then $\sigma_{\max}(A) = \sigma_{\max}(A_\perp)$.

v) If $A \neq 0$, then

$$\sigma_{\max}(A) = \sigma_{\max}(A + A^* - I) = \sigma_{\max}(A + A^*) - 1,$$
$$\sigma_{\max}(I - 2A) = \sigma_{\max}(A) + [\sigma_{\max}^2(A) - 1]^{1/2}.$$

Source: [1134, 1469, 1504]. *iv)* is given in [1133, p. 61] and follows from Fact 7.12.42. **Problem:** Use Fact 7.10.28 to prove *iv)*.

Fact 7.12.42. Let $A \in \mathbb{F}^{n\times n}$, assume that A is idempotent, and assume that $1 \le \operatorname{rank} A \le n-1$. Then,

$$\sigma_{\max}(A) = \sigma_{\max}(A + A^* - I) = \frac{1}{\sin\theta},$$

where $\theta \in (0, \frac{\pi}{2}]$ is defined by

$$\cos\theta = \max\{|x^*y|\colon (x,y) \in \mathcal{R}(A) \times \mathcal{N}(A) \text{ and } x^*x = y^*y = 1\}.$$

Source: [1134, 1504]. **Remark:** θ is the minimal principal angle. See Fact 3.12.21 and Fact 7.13.27. **Remark:** Note that $\mathcal{N}(A) = \mathcal{R}(A_\perp)$. See Fact 4.15.5. **Remark:** This result yields *iii)* of Fact 7.12.41. **Credit:** V. E. Ljance. **Related:** Fact 7.10.30 and Fact 12.12.28.

Fact 7.12.43. Let $A \in \mathbb{R}^{n\times n}$, where $n \ge 2$, be the tridiagonal matrix

$$A \triangleq \begin{bmatrix} b_1 & c_1 & 0 & \cdots & 0 & 0 \\ a_1 & b_2 & c_2 & \cdots & 0 & 0 \\ 0 & a_2 & b_3 & \ddots & 0 & 0 \\ \vdots & \ddots & \ddots & \ddots & \ddots & \vdots \\ 0 & 0 & 0 & \ddots & b_{n-1} & c_{n-1} \\ 0 & 0 & 0 & \cdots & a_{n-1} & b_n \end{bmatrix},$$

and assume that, for all $i \in \{1,\ldots,n-1\}$, $a_ic_i > 0$. Then, A is simple, and every eigenvalue of A is real. Hence, $\operatorname{rank} A \ge n-1$. **Source:** SAS^{-1} is symmetric, where $S \triangleq \operatorname{diag}(d_1,\ldots,d_n)$, $d_1 \triangleq 1$, and, for all $i \in \{1,\ldots,n-1\}$, $d_{i+1} \triangleq (c_i/a_i)^{1/2}d_i$. For a proof that A is simple, see [1040, p. 198]. **Related:** Fact 7.12.44.

Fact 7.12.44. Let $A \in \mathbb{R}^{n\times n}$, where $n \ge 2$, be the tridiagonal matrix

$$A \triangleq \begin{bmatrix} b_1 & c_1 & 0 & \cdots & 0 & 0 \\ a_1 & b_2 & c_2 & \cdots & 0 & 0 \\ 0 & a_2 & b_3 & \ddots & 0 & 0 \\ \vdots & \ddots & \ddots & \ddots & \ddots & \vdots \\ 0 & 0 & 0 & \ddots & b_{n-1} & c_{n-1} \\ 0 & 0 & 0 & \cdots & a_{n-1} & b_n \end{bmatrix},$$

and assume that, for all $i \in \{1,\ldots,n-1\}$, $a_ic_i \ne 0$. Then, A is irreducible. Furthermore, let k_+ and k_- denote, respectively, the number of positive and negative numbers in the n-tuple

$$(1,\ a_1c_1,\ a_1a_2c_1c_2,\ \ldots,\ a_1a_2\cdots a_{n-1}c_1c_2\cdots c_{n-1}).$$

Then, A has at least $|k_+ - k_-|$ distinct real eigenvalues, at least $\max\{0, n-3\min\{k_+,k_-\}\}$ of which are simple. **Source:** [2794]. **Remark:** Note that $k_+ + k_- = n$ and $|k_+ - k_-| = n - 2\min\{k_+,k_-\}$. **Remark:** This result implies Fact 7.12.43.

Fact 7.12.45. Let $A \in \mathbb{R}^{n\times n}$ be the tridiagonal matrix

$$A \triangleq \begin{bmatrix} 0 & 1 & 0 & & & & \\ n-1 & 0 & 2 & & & 0 & \\ 0 & n-2 & 0 & \ddots & & & \\ & \ddots & \ddots & \ddots & \ddots & & \\ & & \ddots & \ddots & 0 & n-2 & 0 \\ & 0 & & \ddots & 2 & 0 & n-1 \\ & & & & 0 & 1 & 0 \end{bmatrix}.$$

Then,

$$\chi_A(s) = \prod_{i=1}^{n}[s-(n+1-2i)].$$

Hence,

$$\operatorname{spec}(A) = \begin{cases} \{n-1, -(n-1), \ldots, 1, -1\}, & n \text{ even}, \\ \{n-1, -(n-1), \ldots, 2, -2, 0\}, & n \text{ odd}. \end{cases}$$

Source: [2594].

Fact 7.12.46. Let $A \in \mathbb{R}^{n\times n}$, where $n \geq 1$, be the tridiagonal, Toeplitz matrix

$$A \triangleq \begin{bmatrix} b & c & 0 & \cdots & 0 & 0 \\ a & b & c & \cdots & 0 & 0 \\ 0 & a & b & \ddots & 0 & 0 \\ \vdots & \vdots & \ddots & \ddots & \ddots & \vdots \\ 0 & 0 & 0 & \ddots & b & c \\ 0 & 0 & 0 & \cdots & a & b \end{bmatrix},$$

and assume that $ac \geq 0$. Then,

$$\operatorname{spec}(A) = \left\{ b + 2\sqrt{ac}\cos\frac{i\pi}{n+1} \colon\ i \in \{1, \ldots, n\} \right\}.$$

Remark: See [1389, p. 522]. **Related:** Fact 4.23.9.

Fact 7.12.47. Let $A \in \mathbb{R}^{n\times n}$, where $n \geq 1$, be the tridiagonal, Toeplitz matrix

$$A \triangleq \begin{bmatrix} 0 & 1/2 & 0 & \cdots & 0 & 0 \\ 1/2 & 0 & 1/2 & \cdots & 0 & 0 \\ 0 & 1/2 & 0 & \ddots & 0 & 0 \\ \vdots & \vdots & \ddots & \ddots & \ddots & \vdots \\ 0 & 0 & 0 & \ddots & 0 & 1/2 \\ 0 & 0 & 0 & \cdots & 1/2 & 0 \end{bmatrix}.$$

Then,

$$\operatorname{spec}(A) = \left\{ \cos\frac{i\pi}{n+1} \colon\ i \in \{1, \ldots, n\} \right\}.$$

Furthermore, the associated eigenvectors $v_1, \ldots, v_n$ are given by

$$v_i = \sqrt{\frac{2}{n+1}} \begin{bmatrix} \sin \frac{i\pi}{n+1} \\ \sin \frac{2i\pi}{n+1} \\ \vdots \\ \sin \frac{ni\pi}{n+1} \end{bmatrix},$$

which are mutually orthogonal and satisfy $\|v_i\|_2 = 1$ for all $i \in \{1, \ldots, n\}$. **Source:** [1633].

Fact 7.12.48. Let $A \in \mathbb{F}^{n\times n}$, and assume that A has real eigenvalues. Then,

$$\begin{aligned} \tfrac{1}{n}\operatorname{tr} A - \sqrt{\tfrac{n-1}{n}\left[\operatorname{tr} A^2 - \tfrac{1}{n}(\operatorname{tr} A)^2\right]} \le \lambda_{\min}(A) &\le \tfrac{1}{n}\operatorname{tr} A - \sqrt{\tfrac{1}{n^2-n}\left[\operatorname{tr} A^2 - \tfrac{1}{n}(\operatorname{tr} A)^2\right]} \\ &\le \tfrac{1}{n}\operatorname{tr} A + \sqrt{\tfrac{1}{n^2-n}\left[\operatorname{tr} A^2 - \tfrac{1}{n}(\operatorname{tr} A)^2\right]} \\ &\le \lambda_{\max}(A) \le \tfrac{1}{n}\operatorname{tr} A + \sqrt{\tfrac{n-1}{n}\left[\operatorname{tr} A^2 - \tfrac{1}{n}(\operatorname{tr} A)^2\right]}. \end{aligned}$$

Furthermore, for all $i \in \{1, \ldots, n\}$,

$$\left|\lambda_i(A) - \tfrac{1}{n}\operatorname{tr} A\right| \le \sqrt{\tfrac{n-1}{n}\left[\operatorname{tr} A^2 - \tfrac{1}{n}(\operatorname{tr} A)^2\right]}.$$

Finally, if $n = 2$, then

$$\tfrac{1}{n}\operatorname{tr} A - \sqrt{\tfrac{1}{n}\operatorname{tr} A^2 - \tfrac{1}{n^2}(\operatorname{tr} A)^2} = \lambda_{\min}(A) \le \lambda_{\max}(A) = \tfrac{1}{n}\operatorname{tr} A + \sqrt{\tfrac{1}{n}\operatorname{tr} A^2 - \tfrac{1}{n^2}(\operatorname{tr} A)^2}.$$

Source: [2905, 2906]. **Remark:** See [2825]. **Related:** Fact 2.11.37 and Fact 7.12.49.

Fact 7.12.49. Let $A \in \mathbb{F}^{n\times n}$, assume that A has real eigenvalues, and define

$$a \triangleq \frac{1}{n}\operatorname{tr} A^2 - \left(\frac{1}{n}\operatorname{tr} A\right)^2, \quad b \triangleq \frac{1}{n}\operatorname{tr} A^3 - \frac{3}{n^2}(\operatorname{tr} A)\operatorname{tr} A^2 + \frac{2}{n^3}(\operatorname{tr} A)^3.$$

Then,

$$\lambda_{\min}(A) \le \frac{1}{n}\operatorname{tr} A + \frac{b - \sqrt{b^2 + 4a^3}}{2a} \le \frac{1}{n}\operatorname{tr} A + \frac{b + \sqrt{b^2 + 4a^3}}{2a} \le \lambda_{\max}(A),$$

$$\frac{\sqrt{b^2 + 4a^3}}{a} \le \lambda_{\max}(A) - \lambda_{\min}(A), \quad 1 + \frac{2\sqrt{b^2 + 4a^3}}{b + 2a(\operatorname{tr} A)/n - \sqrt{b^2 + 4a^3}} \le \frac{\lambda_{\max}(A)}{\lambda_{\min}(A)}.$$

Source: [2433]. **Related:** Fact 7.12.48.

Fact 7.12.50. Let $A \in \mathbb{F}^{n\times n}$. Then,

$$\tfrac{1}{n}|\operatorname{tr} A| - \sqrt{\tfrac{n-1}{n}\left(\operatorname{tr} AA^* - \tfrac{1}{n}|\operatorname{tr} A|^2\right)} \le \rho_{\min}(A) \le \sqrt{\tfrac{1}{n}\operatorname{tr} AA^*},$$

$$\tfrac{1}{n}|\operatorname{tr} A| \le \rho_{\max}(A) \le \tfrac{1}{n}|\operatorname{tr} A| + \sqrt{\tfrac{n-1}{n}\left(\operatorname{tr} AA^* - \tfrac{1}{n}|\operatorname{tr} A|^2\right)}.$$

Source: Theorem 3.1 of [2905].

Fact 7.12.51. Let $A \in \mathbb{F}^{n\times n}$, where $n \ge 2$, be the bidiagonal matrix

$$A \triangleq \begin{bmatrix} a_1 & b_1 & 0 & \cdots & 0 & 0 \\ 0 & a_2 & b_2 & \cdots & 0 & 0 \\ 0 & 0 & a_3 & \ddots & 0 & 0 \\ \vdots & \vdots & \ddots & \ddots & \ddots & \vdots \\ 0 & 0 & 0 & \ddots & a_{n-1} & b_{n-1} \\ 0 & 0 & 0 & \cdots & 0 & a_n \end{bmatrix},$$

and assume that $a_1, \ldots, a_n, b_1, \ldots, b_{n-1}$ are nonzero. Then, the following statements hold:

i) The singular values of A are distinct.

ii) If $B \in \mathbb{F}^{n\times n}$ is bidiagonal and $|B| = |A|$, then A and B have the same singular values.

iii) If $B \in \mathbb{F}^{n\times n}$ is bidiagonal, $|A| \le |B|$, and $|A| \ne |B|$, then $\sigma_{\max}(A) < \sigma_{\max}(B)$.

iv) If $B \in \mathbb{F}^{n\times n}$ is bidiagonal, $|I \odot A| \le |I \odot B|$, and $|I \odot A| \ne |I \odot B|$, then $\sigma_{\min}(A) < \sigma_{\min}(B)$.

v) If $B \in \mathbb{F}^{n\times n}$ is bidiagonal, $|N_n \odot A| \le |N_n \odot B|$, and $|N_n \odot A| \ne |N_n \odot B|$, then $\sigma_{\min}(B) < \sigma_{\min}(A)$.

Source: [1990, p. 17-5].

7.13 Facts on Eigenvalues and Singular Values for Two or More Matrices

Fact 7.13.1. Let $A, B \in \mathbb{F}^{n\times n}$, and assume that A and B are idempotent. Then, the following statements are equivalent:

i) $\operatorname{mspec}(A) = \operatorname{mspec}(B)$.

ii) $\operatorname{rank} A = \operatorname{rank} B$.

iii) $\operatorname{tr} A = \operatorname{tr} B$.

Fact 7.13.2. Let $A, B \in \mathbb{F}^{n\times n}$, and assume that A and B are projectors. Then, the following statements hold:

i) $\operatorname{amult}_{AB}(1) = \operatorname{def}(I - AB) \le \operatorname{tr} AB$.

ii) $\operatorname{rank}(I - AB) = \operatorname{def} AB + \operatorname{card}[\operatorname{mspec}(AB) \cap (0, 1)] = n - \operatorname{amult}_{AB}(1)$.

iii) $\operatorname{rank}(AB + BA) = \operatorname{rank} AB + \operatorname{card}[\operatorname{mspec}(AB) \cap (0, 1)]$.

iv) $\operatorname{rank}[AB - (AB)^2] = \operatorname{card}[\operatorname{mspec}(AB) \cap (0, 1)]$.

v) $\operatorname{rank}(I - ABA) = \operatorname{def} AB + \operatorname{card}[\operatorname{mspec}(AB) \cap (0, 1)]$.

vi) $\operatorname{rank}(AB)^2 = \operatorname{rank} ABA = \operatorname{rank} AB$.

vii) $\mathcal{R}(AB) \subseteq \mathcal{R}(AB + BA) = \mathcal{R}(AB) + \mathcal{R}(BA)$.

viii) If $AB + BA$ is idempotent, then $AB = BA = 0$.

ix) $\operatorname{rank}(AB - BA) = 2\operatorname{card}[\operatorname{mspec}(AB) \cap (0, 1)]$.

x) If $AB + BA$ and $AB - BA$ are nonsingular, then $\operatorname{rank} AB = n/2$.

Source: [245]. **Related:** Fact 8.8.3.

Fact 7.13.3. Let $A, B \in \mathbb{F}^{n\times n}$, assume that A and B are projectors, and define $r \triangleq \operatorname{rank} A$. Then, there exists a unitary matrix $S \in \mathbb{F}^{n\times n}$ such that

$$A = S\begin{bmatrix} I_r & 0 \\ 0 & 0 \end{bmatrix}S^*, \quad B = S\begin{bmatrix} B_{11} & B_{12} \\ B_{12}^* & B_{22} \end{bmatrix}S^*,$$

where $B_{11} \in \mathbb{F}^{r\times r}$, $B_{12} \in \mathbb{F}^{r\times(n-r)}$, and $B_{22} \in \mathbb{F}^{(n-r)\times(n-r)}$. Furthermore, the following statements hold:

i) $\operatorname{rank} B = \operatorname{rank} B_{11} - \operatorname{rank} B_{12} + \operatorname{rank} B_{22}$.

ii) $\operatorname{rank} AB = \operatorname{rank} BA = \operatorname{rank} B_{11}$.

iii) $\operatorname{rank}(I - AB) = n - \operatorname{rank} B_{11} + \operatorname{rank} B_{12}$.

iv) $\operatorname{rank}(A + B) = r + \operatorname{rank} B_{22}$.

v) $\operatorname{rank}(A - B) = r - \operatorname{rank} B_{11} + \operatorname{rank} B_{12} + \operatorname{rank} B_{22}$.

vi) $\operatorname{rank}(AB + BA) = \operatorname{rank} B_{11} + \operatorname{rank} B_{12}$.

vii) $\operatorname{rank}(AB - BA) = 2 \operatorname{rank} B_{12}$.

viii) $A + A_\perp(A_\perp B)^+$ is the projector onto $\mathcal{R}(A) + \mathcal{R}(B)$.

ix) $A - A(AB_\perp)^+$ is the projector onto $\mathcal{R}(A) \cap \mathcal{R}(B)$.

x) $\mathcal{R}(A) \cap \mathcal{R}(B) = \{0\}$ if and only if $\operatorname{rank} B_{11} = \operatorname{rank} B_{12}$.

xi) $\mathcal{R}(A) + \mathcal{R}(B) = \mathbb{F}^n$ if and only if $\operatorname{rank} B_{22} = n - r$.

xii) $\mathcal{R}(A) \perp \mathcal{R}(B)$ if and only if $B_{11} = 0$.

xiii) $\mathcal{R}(A)$ and $\mathcal{R}(B)$ are complementary subspaces if and only if $\operatorname{rank} B_{11} = \operatorname{rank} B_{12}$ and $\operatorname{rank} B_{22} = n - r$.

xiv) $\mathcal{R}(A)^\perp = \mathcal{R}(B)$ if and only if $B_{11} = 0$ and $\operatorname{rank} B_{22} = n - r$.

xv) $\operatorname{amult}_{B_{11}}(1) = \operatorname{rank} B_{11} - \operatorname{rank} B_{12}$ and $\operatorname{amult}_{B_{22}}(1) = \operatorname{rank} B_{22} - \operatorname{rank} B_{12}$

xvi) $\operatorname{spec}(AB) \subset [0, 1]$.

xvii) $\operatorname{def}(AB) = n - \operatorname{rank} B_{11} = \dim(\mathcal{N}(A) + [\mathcal{R}(A) \cap \mathcal{N}(B)])$.

xviii) $\operatorname{amult}_{AB}(1) = \operatorname{rank} B_{11} - \operatorname{rank} B_{12} = \dim[\mathcal{R}(A) \cap \mathcal{R}(B)]$.

xix) $\operatorname{card}[\operatorname{mspec}(AB) \cap (0, 1)] = \operatorname{rank} B_{12}$.

xx) $\operatorname{rank}(AB - BA) = \operatorname{card}[\operatorname{mspec}[(A - B)^2] \cap (0, 1)]$.

xxi) $\operatorname{spec}(A - B) \subset [-1, 1]$.

xxii) $\operatorname{amult}_{A-B}(1) = \operatorname{def}(I - A + B) = r - \operatorname{rank} B_{11} = \dim[\mathcal{R}(A) \cap \mathcal{N}(B)]$.

xxiii) $\operatorname{amult}_{A-B}(-1) = \operatorname{def}(I + A - B) = \operatorname{rank} B_{22} - \operatorname{rank} B_{12} = \dim[\mathcal{N}(A) \cap \mathcal{R}(B)]$.

xxiv) $\operatorname{def}(A - B) = n - r + \operatorname{rank} B_{11} - \operatorname{rank} B_{12} - \operatorname{rank} B_{22}$.

xxv) $\operatorname{card}(\operatorname{mspec}(A - B) \cap [(-1, 0) \cup (0, 1)]) = 2 \operatorname{rank} B_{12}$.

xxvi) $\operatorname{spec}(A + B) \subset [0, 2]$.

xxvii) $\operatorname{amult}_{A+B}(2) = \operatorname{def}(2I - A - B) = \operatorname{rank} B_{11} - \operatorname{rank} B_{22}$.

xxviii) $\operatorname{amult}_{A+B}(1) = \operatorname{def}(I - A - B) = r - \operatorname{rank} B_{11} - \operatorname{rank} B_{12} + \operatorname{rank} B_{22}$.

xxix) $\operatorname{def}(A + B) = n - r - \operatorname{rank} B_{22} = \dim[\mathcal{N}(A) \cap \mathcal{N}(B)]$.

xxx) $\operatorname{card}(\operatorname{mspec}(A + B) \cap [(0, 1) \cup (1, 2)]) = 2 \operatorname{rank} B_{12}$.

xxxi) $\operatorname{spec}(AB - BA) \subset \jmath[-1, 1]$.

xxxii) $\operatorname{def}(AB - BA) = n - 2 \operatorname{rank} B_{12}$.

xxxiii) $\operatorname{card}(\operatorname{mspec}(AB - BA) \cap \jmath[[-1, 0) \cup (0, 1)]) = 2 \operatorname{rank} B_{12}$.

xxxiv) $(A - B)^2 + (I - A - B)^2 = I$.

xxxv) $AB - BA = (I - A - B)(A - B) = (B - A)(I - A - B)$.

Source: [243]. **Related:** Fact 7.13.4.

Fact 7.13.4. Let $A, B \in \mathbb{F}^{n \times n}$, and assume that A and B are projectors. Then,

$$\operatorname{spec}(A + B) \subset [0, 2], \quad \operatorname{spec}(A - B) \subset [-1, 1],$$

$$\operatorname{spec}(AB) \subset [0, 1], \quad \operatorname{spec}(AB + BA) \subset [-\tfrac{1}{4}, 2], \quad \operatorname{spec}([A, B]) \subset \jmath[-1, 1].$$

Source: [83], [246, 251], [1133, p. 53], [2081], and [2263, p. 147]. **Remark:** Let $A = \begin{bmatrix} 1 & 0 \\ 0 & 0 \end{bmatrix}$

and $B = \frac{1}{4}\begin{bmatrix} 1 & \sqrt{3} \\ \sqrt{3} & 3 \end{bmatrix}$. Then, $-\frac{1}{4} \in \operatorname{spec}(AB + BA)$. **Credit:** The first inclusion is due to S. N. Afriat. **Related:** Fact 7.13.3, Fact 8.5.14, and Fact 10.11.10 .

Fact 7.13.5. Let $A, B \in \mathbb{F}^{n\times n}$, and assume that A and B are projectors. Then, the following statements are equivalent:

i) AB is a projector.

ii) $\operatorname{spec}(AB) \subseteq \{0, 1\}$.

iii) $\operatorname{spec}(A + B) \subset \{0\} \cup [1, 2]$.

iv) $\operatorname{spec}(A - B) \subseteq \{-1, 0, 1\}$.

Source: [245, 1134, 1252] and [2238, p. 336]. **Related:** Fact 4.18.7, Fact 7.13.6, Fact 8.8.5, and Fact 8.11.13.

Fact 7.13.6. Let $A, B \in \mathbb{F}^{n\times n}$, and assume that A and B are projectors. Then, the following statements are equivalent:

i) AB is a projector.

ii) AB is Hermitian.

iii) AB is normal.

iv) AB is range Hermitian.

v) $2AB - I$ is involutory.

vi) $2AB - I$ is unitary.

vii) $[A, B]$ is idempotent.

viii) $[A, B]$ is nilpotent.

ix) $[A, B]^2 = 0$.

x) $[ABA, BAB] = 0$.

xi) $\mathcal{R}(BABAB) \subseteq \mathcal{R}(AB)$.

xii) $\operatorname{amult}_{AB}(1) = \operatorname{tr} AB$.

xiii) $\operatorname{def}(I - AB) = \operatorname{tr} AB$.

xiv) $\operatorname{rank}(AB + BA) = \operatorname{rank} AB$.

xv) $\operatorname{spec}(AB) \cap (0, 1) = \emptyset$.

xvi) $\operatorname{rank}(A + B) = \operatorname{rank} AB + \operatorname{rank}(A - B)$.

xvii) $\operatorname{rank}(A + B) + \operatorname{rank} AB = \operatorname{rank} A + \operatorname{rank} B$.

xviii) $\operatorname{rank}(2I - A - B) = \operatorname{def} AB$.

xix) $\operatorname{rank}(I - BAB) = \operatorname{def} AB$.

xx) $\operatorname{rank}\, [AB \;\; BA] = \operatorname{rank} AB$.

xxi) $\operatorname{tr}\, (AB)^2 = \operatorname{tr} AB$.

xxii) $\operatorname{tr}\, (I - AB)^2 = \operatorname{tr}(I - AB)$.

xxiii) $\operatorname{tr}(I - AB) = \operatorname{rank}(I - AB)$.

Source: [245, 2705, 2706]. **Related:** Fact 7.13.5 and Fact 8.8.5.

Fact 7.13.7. Let $A \in \mathbb{F}^{n\times n}$ and $B \in \mathbb{F}^{n\times m}$, and define $r \triangleq \operatorname{rank} B$ and $\mathcal{A} \triangleq \begin{bmatrix} A & B \\ B^* & 0 \end{bmatrix}$. Then, $\nu_-(\mathcal{A}) \ge r$, $\nu_0(\mathcal{A}) \ge 0$, and $\nu_+(\mathcal{A}) \ge r$. If, in addition, $n = m$ and B is nonsingular, then $\operatorname{In} \mathcal{A} = \begin{bmatrix} n \\ 0 \\ n \end{bmatrix}$. **Source:** [1458]. **Related:** Proposition 7.6.5.

Fact 7.13.8. Let $A, B \in \mathbb{F}^{n\times n}$. Then,

$$\rho_{\max}(A + B) \le \sigma_{\max}(A + B) \le \sigma_{\max}(A) + \sigma_{\max}(B).$$

If, in addition, A and B are Hermitian, then

$$\rho_{\max}(A+B) = \sigma_{\max}(A+B) \le \sigma_{\max}(A) + \sigma_{\max}(B) = \rho_{\max}(A) + \rho_{\max}(B),$$
$$\lambda_{\min}(A) + \lambda_{\min}(B) \le \lambda_{\min}(A+B) \le \lambda_{\max}(A+B) \le \lambda_{\max}(A) + \lambda_{\max}(B).$$

Source: Use Lemma 10.4.3 for the last string of inequalities. **Related:** Fact 7.12.9.

Fact 7.13.9. Let $A, B \in \mathbb{F}^{n\times n}$ and $\lambda \in \operatorname{spec}(A+B)$. Then,

$$\tfrac{1}{2}\lambda_{\min}(A^* + A) + \tfrac{1}{2}\lambda_{\min}(B^* + B) \le \operatorname{Re}\lambda \le \tfrac{1}{2}\lambda_{\max}(A^* + A) + \tfrac{1}{2}\lambda_{\max}(B^* + B).$$

Source: [688, p. 18]. **Related:** Fact 7.12.24.

Fact 7.13.10. Let $A \in \mathbb{F}^{n\times n}$. Then, the following statements hold:

i) $\operatorname{tr} AX = 0$ for all Hermitian $X \in \mathbb{F}^{n\times n}$ if and only if $A = 0$.

ii) $\operatorname{Im}\operatorname{tr} AX = 0$ for all Hermitian $X \in \mathbb{F}^{n\times n}$ if and only if A is Hermitian.

iii) If A is Hermitian and $\operatorname{Re}\operatorname{tr} AX \le \operatorname{tr} A$ for all unitary $X \in \mathbb{F}^{n\times n}$, then A is positive semidefinite.

Source: [2991, p. 258].

Fact 7.13.11. Let $A, B \in \mathbb{F}^{n\times n}$, assume that either A and B are Hermitian or A and B are skew Hermitian, and let $k \ge 1$. Then, $\operatorname{tr} A^kB^k$ and $\operatorname{tr}(AB)^k$ are real. **Source:** $\operatorname{tr} AB = \operatorname{tr} A^*B^* = \operatorname{tr}(BA)^* = \overline{\operatorname{tr} BA} = \overline{\operatorname{tr} AB}$. See [103, 2954] and [2991, pp. 260, 264].

Fact 7.13.12. Let $A, B \in \mathbb{F}^{n\times n}$ be normal, and let $\operatorname{mspec}(A) = \{\lambda_1, \ldots, \lambda_n\}_{\mathrm{ms}}$ and $\operatorname{mspec}(B) = \{\mu_1, \ldots, \mu_n\}_{\mathrm{ms}}$. Then,

$$\min \operatorname{Re}\sum_{i=1}^n \lambda_i\mu_{\sigma(i)} \le \operatorname{Re}\operatorname{tr} AB \le \max \operatorname{Re}\sum_{i=1}^n \lambda_i\mu_{\sigma(i)},$$

where "max" and "min" are taken over all permutations σ of the eigenvalues of B. Now, assume that A and B are Hermitian. Then, $\operatorname{tr} AB$ is real, and

$$\sum_{i=1}^n \lambda_i(A)\lambda_{n-i+1}(B) \le \operatorname{tr} AB \le \sum_{i=1}^n \lambda_i(A)\lambda_i(B).$$

Furthermore, the last inequality is an equality if and only if there exists a unitary matrix $S \in \mathbb{F}^{n\times n}$ such that $A = S\operatorname{diag}[\lambda_1(A), \ldots, \lambda_n(A)]S^*$ and $B = S\operatorname{diag}[\lambda_1(B), \ldots, \lambda_n(B)]S^*$. **Source:** [1946]. For the second string of inequalities, use Fact 2.12.8. For the last statement, see [523, p. 10] and [1792]. **Credit:** The upper bound for $\operatorname{tr} AB$ is due to K. Fan. **Related:** Fact 7.13.13, Fact 7.13.16, Proposition 10.4.13, Fact 10.14.35, and Fact 10.22.25.

Fact 7.13.13. Let $A, B \in \mathbb{F}^{n\times n}$, and assume that B is Hermitian. Then,

$$\sum_{i=1}^n \lambda_i[\tfrac{1}{2}(A + A^*)]\lambda_{n-i+1}(B) \le \operatorname{Re}\operatorname{tr} AB \le \sum_{i=1}^n \lambda_i[\tfrac{1}{2}(A + A^*)]\lambda_i(B).$$

Source: Apply the second string of inequalities in Fact 7.13.12. **Remark:** For A, B real, these inequalities are given in [1663]. For the complex case, see [1760]. **Related:** Proposition 10.4.13 in the case where B is positive semidefinite.

Fact 7.13.14. Let $A \in \mathbb{F}^{n\times m}$ and $B \in \mathbb{F}^{m\times n}$, and define $r \triangleq \min\{m, n\}$. Then,

$$|\operatorname{tr} AB| \le \sum_{i=1}^r \sigma_i(A)\sigma_i(B).$$

Now, assume that $n = m$. Then,

$$\max \operatorname{Re}\operatorname{tr} UA^*VB = \max |\operatorname{tr} UA^*VB| = \sum_{i=1}^{r} \sigma_i(A)\sigma_i(B),$$

where "max" is taken over all unitary matrices $U, V \in \mathbb{F}^{n\times n}$. **Source:** [541], [1969, pp. 514, 515], [1971, pp. 789, 790], and [2263, p. 148]. **Remark:** Applying Fact 7.13.12 to $\begin{bmatrix} 0 & A \\ A^* & 0 \end{bmatrix}$ and $\begin{bmatrix} 0 & B^* \\ B & 0 \end{bmatrix}$ and using Proposition 7.6.5 yields the weaker result $|\operatorname{Re}\operatorname{tr} AB| \le \sum_{i=1}^{r} \sigma_i(A)\sigma_i(B)$. **Remark:** See [523, p. 14] and [1665]. **Credit:** The equalities are due to J. von Neumann and K. Fan. **Related:** Fact 7.13.15, Fact 11.14.1, and Fact 11.16.2.

Fact 7.13.15. Let $A, B \in \mathbb{F}^{n\times n}$, and assume that B is positive semidefinite. Then,

$$|\operatorname{tr} AB| \le \sum_{i=1}^{n} \sigma_i(A)\lambda_i(B) \le \sigma_{\max}(A)\operatorname{tr} B.$$

Source: Fact 7.13.14. **Related:** An extension of this result is given by Fact 11.16.2.

Fact 7.13.16. Let $A, B \in \mathbb{R}^{n\times n}$, assume that B is symmetric, and define $C \triangleq \frac{1}{2}(A + A^{\mathsf T})$. Then,

$$\lambda_{\min}(C)\operatorname{tr} B - \lambda_{\min}(B)[n\lambda_{\min}(C) - \operatorname{tr} A] \le \operatorname{tr} AB \le \lambda_{\max}(C)\operatorname{tr} B - \lambda_{\min}(B)[n\lambda_{\max}(C) - \operatorname{tr} A].$$

Source: [1014]. **Remark:** Extensions are given in [2201]. **Related:** Fact 7.13.12, Proposition 10.4.13, and Fact 10.14.35.

Fact 7.13.17. Let $A, B \in \mathbb{F}^{n\times n}$, assume that A and B are Hermitian, and assume that $\operatorname{tr} A$ and $\operatorname{tr} B$ are positive. Then,

$$\frac{\operatorname{tr} (A + B)^2}{\operatorname{tr}(A + B)} \le \frac{\operatorname{tr} A^2}{\operatorname{tr} A} + \frac{\operatorname{tr} B^2}{\operatorname{tr} B}.$$

Furthermore, equality holds if and only if there exists $\alpha \in (0, \infty)$ such that $A = \alpha B$. **Source:** Note that

$$\operatorname{tr} (A + B)^2 + \operatorname{tr} \left(\sqrt{\frac{\operatorname{tr} B}{\operatorname{tr} A}}A - \sqrt{\frac{\operatorname{tr} A}{\operatorname{tr} B}}B \right)^2 = (\operatorname{tr} A + \operatorname{tr} B)\left(\frac{\operatorname{tr} A^2}{\operatorname{tr} A} + \frac{\operatorname{tr} B^2}{\operatorname{tr} B}\right).$$

See [2991, p. 264].

Fact 7.13.18. Let $A, B, Q, S_1, S_2 \in \mathbb{R}^{n\times n}$, assume that A and B are symmetric, and assume that Q, S_1, and S_2 are orthogonal. Furthermore, assume that $S_1^{\mathsf T}AS_1$ and $S_2^{\mathsf T}BS_2$ are diagonal with the diagonal entries arranged downward in nonincreasing order, and define the orthogonal matrices $Q_1, Q_2 \in \mathbb{R}^{n\times n}$ by $Q_1 \triangleq S_1\hat{I}_nS_1^{\mathsf T}$ and $Q_2 \triangleq S_1S_2^{\mathsf T}$. Then,

$$\operatorname{tr} AQ_1BQ_1^{\mathsf T} \le \operatorname{tr} AQBQ^{\mathsf T} \le \operatorname{tr} AQ_2BQ_2^{\mathsf T}.$$

Source: [330, 1792]. **Related:** Fact 7.13.16 and Fact 7.13.19.

Fact 7.13.19. Let $A \in \mathbb{F}^{n\times n}$, assume that A is Hermitian, let $k \le \operatorname{rank} A$, let $S, S_1, S_2 \in \mathbb{F}^{n\times k}$, assume that S, S_1, and S_2 are left inner, and assume that $AS_1 = S_1 \operatorname{diag}[\lambda_{n-k+1}(A), \ldots, \lambda_n(A)]$ and $AS_2 = S_2 \operatorname{diag}[\lambda_1(A), \ldots, \lambda_k(A)]$. Then,

$$\operatorname{tr} S_1^*AS_1 \le \operatorname{tr} S^*AS \le \operatorname{tr} S_2^*AS_2.$$

Source: [775]. **Remark:** This is the *Fan trace minimization principle*. **Related:** Fact 7.13.18 and Fact 11.17.16.

Fact 7.13.20. Let $A, B \in \mathbb{F}^{n\times n}$, and assume that A and B are Hermitian. Then,

$$\max_{S\in \mathrm{U}(n)} \operatorname{tr} AS^*BS = \sum_{i=1}^{n} \lambda_i(A)\lambda_i(B).$$

Source: [775]. **Related:** Fact 7.13.14, Fact 7.13.19, and Fact 11.17.15.

Fact 7.13.21. Let $A_1, \ldots, A_m, B_1, \ldots, B_m \in \mathbb{F}^{n\times n}$, and assume that $A_1, \ldots, A_m$ are unitary. Then,

$$|\operatorname{tr} A_1B_1 \cdots A_mB_m| \le \sum_{i=1}^{n} \sigma_i(B_1)\cdots\sigma_i(B_m).$$

Source: [1969, p. 516] and [1971, p. 791]. **Credit:** K. Fan. **Related:** Fact 7.13.18.

Fact 7.13.22. Let $A_1, \ldots, A_m \in \mathbb{F}^{n\times n}$. Then, for all $k \in \{1, \ldots, n\}$,

$$\sum_{i=1}^{k} \sigma_i(A_1\cdots A_m) \le \sum_{i=1}^{k} \sigma_i(A_1)\cdots\sigma_i(A_m).$$

Source: [1969, p. 250] and [1971, p. 342].

Fact 7.13.23. Let $A, B \in \mathbb{R}^{n\times n}$, and assume that $AB = BA$. Then,

$$\rho_{\max}(AB) \le \rho_{\max}(A)\rho_{\max}(B), \quad \rho_{\max}(A + B) \le \rho_{\max}(A) + \rho_{\max}(B).$$

Source: Fact 7.19.5. **Remark:** If $AB \ne BA$, then both of these inequalities may be violated. Consider $A = \left[\begin{smallmatrix} 0 & 1 \\ 0 & 0 \end{smallmatrix}\right]$ and $B = \left[\begin{smallmatrix} 0 & 0 \\ 1 & 0 \end{smallmatrix}\right]$.

Fact 7.13.24. Let $A, B \in \mathbb{C}^{n\times n}$, assume that A and B are normal, and let $\operatorname{mspec}(A) = \{\lambda_1, \ldots, \lambda_n\}_{\rm ms}$ and $\operatorname{mspec}(B) = \{\mu_1, \ldots, \mu_n\}_{\rm ms}$. Then,

$$|\det(A + B)| \le \min\left\{\prod_{i=1}^{n} \max_{j\in\{1,\ldots,n\}} |\lambda_i + \mu_j|, \prod_{j=1}^{n} \max_{i\in\{1,\ldots,n\}} |\lambda_i + \mu_j|\right\}.$$

Source: [2286]. **Remark:** Equality is discussed in [339]. **Related:** Fact 11.16.22.

Fact 7.13.25. Let $A, B \in \mathbb{F}^{n\times n}$, assume that B is nonsingular, assume that $\operatorname{spec}(B^{-1}A) \subset \mathrm{ORHP}$, and let $p \ge 2/n$. Then,

$$|\det A|^p + |\det B|^p \le |\det(A + B)|^p.$$

Source: [2972]. **Related:** Corollary 10.4.15 and Fact 10.16.8.

Fact 7.13.26. Let $A \in \mathbb{F}^{n\times m}$ and $B \in \mathbb{F}^{m\times m}$, and assume that $n \le m$. Then,

$$\det(ABB^*A^*) \le \det(AA^*)\prod_{i=1}^{n} \sigma_i^2(B) = \prod_{i=1}^{n} \sigma_i^2(A)\sigma_i^2(B).$$

Source: [970, p. 218].

Fact 7.13.27. Let $A, B \in \mathbb{F}^{n\times n}$, assume that A and B are nonzero projectors, and define the minimal principal angle $\theta \in [0, \frac{\pi}{2}]$ by

$$\cos\theta = \max\{|x^*y| \colon (x, y) \in \mathcal{R}(A) \times \mathcal{R}(B) \text{ and } x^*x = y^*y = 1\}.$$

Then, the following statements hold:

i) $\sigma_{\max}(AB) = \sigma_{\max}(BA) = \cos\theta$.

ii) $\sigma_{\max}(A + B) = 1 + \sigma_{\max}(AB) = 1 + \cos\theta$.

iii) $1 \le \sigma_{\max}(AB) + \sigma_{\max}(A - B)$.

iv) $\sigma_{\max}(A - B) = \max\{\sigma_{\max}(AB_\perp), \sigma_{\max}(BA_\perp)\} \le 1$.

v) $\theta > 0$ if and only if $\mathcal{R}(A) \cap \mathcal{R}(B) = \{0\}$.

Furthermore, the following statements are equivalent:

vi) $\sigma_{\max}(A - B) < 1$.

vii) $\operatorname{rank} A = \operatorname{rank} B = \operatorname{rank} AB$.

viii) $\mathcal{R}(A) \cap \mathcal{N}(B) = \{0\}$ and $\mathcal{N}(A) \cap \mathcal{R}(B) = \{0\}$.

ix) $\mathcal{R}(A) = \mathcal{R}(AB)$ and $\mathcal{R}(B) = \mathcal{R}(BA)$.

x) $\operatorname{rank}(A + B) = \operatorname{rank}(AB + BA)$.

xi) $1 \notin \operatorname{spec}(A + B)$.

If these statements hold, then A and B are unitarily similar. Furthermore, the following statements are equivalent:

xii) $A - B$ is nonsingular.

xiii) $\mathcal{R}(A)$ and $\mathcal{R}(B)$ are complementary subspaces.

xiv) $\sigma_{\max}(A + B - I) < 1$.

Now, assume that $A - B$ is nonsingular. Then, the following statements hold:

xv) $\theta > 0$.

xvi) $\sigma_{\max}(AB) < 1$.

xvii) $\sigma_{\max}[(A - B)^{-1}] = \frac{1}{\sqrt{1-\sigma_{\max}^2(AB)}} = 1/\sin\theta$.

xviii) $\sigma_{\min}(A - B) = \sin\theta$.

xix) $\sigma_{\min}^2(A - B) + \sigma_{\max}^2(AB) = 1$.

xx) $I - AB$ is nonsingular.

xxi) If $\operatorname{rank} A = \operatorname{rank} B$, then $\sigma_{\max}(A - B) = \sin\theta_{\max}$, where $\theta_{\max}$ is the maximum principal angle defined in Fact 7.10.29.

xxii) $\sqrt{\operatorname{tr}(A - B)^2} = \sqrt{\operatorname{rank} A + \operatorname{rank} B - 2\operatorname{tr} AB} \le \operatorname{rank}(A - B)$.

Source: *i)* is given in [1504]; *ii)* is given in [1134]; *iii)* follows from the first inequality in Fact 10.22.17; *iv)* is given in [779] and [1590, p. 56]; *v)* is given in [1187, p. 393]; *vi)–xi)* are given in [249], see also [970, p. 195] and [1187, p. 389]; Fact 4.18.19 implies that *xi)* and *xii)* are equivalent; *xiv)* is given in [602], see also [1133, p. 236]; *xxi)* follows from [2539, pp. 92, 93]; *xxii)* is given in [249]. **Remark:** The equality in *iv)* is the *Krein-Krasnoselskii-Milman equality*. See [779]. **Remark:** The projectors $A = \begin{bmatrix} 1 & 0 \\ 0 & 0 \end{bmatrix}$ and $B = \begin{bmatrix} 0 & 0 \\ 0 & 1 \end{bmatrix}$ satisfy $A = \hat{I}_2 B \hat{I}_2$ and thus are unitarily similar, whereas $\sigma_{\max}(A - B) = 1$. Hence, the converse of the statement after *xi)* does not hold. **Remark:** Additional results for $A - B$ nonsingular are given in Fact 4.18.19. **Related:** Fact 3.12.21, Fact 7.12.42, Fact 7.13.28, and Fact 8.8.17.

Fact 7.13.28. Let $A \in \mathbb{F}^{n\times n}$, assume that A is idempotent, let $P, Q \in \mathbb{F}^{n\times n}$, where P is the projector onto $\mathcal{R}(A)$ and Q is the projector onto $\mathcal{N}(A)$, and define the minimal principal angle $\theta \in [0, \frac{\pi}{2}]$ by

$$\cos\theta = \max\{|x^*y|\colon (x, y) \in \mathcal{R}(P) \times \mathcal{R}(Q) \text{ and } x^*x = y^*y = 1\}.$$

Then, the following statements hold:

i) $P - Q$ is nonsingular.

ii) $(P - Q)^{-1} = A + A^* - I = A - A_{\perp}^*$.

iii) $\sigma_{\max}(A) = \frac{1}{\sqrt{1-\sigma_{\max}^2(PQ)}} = \sigma_{\max}[(P - Q)^{-1}] = \sigma_{\max}(A + A^* - I) = 1/\sin\theta$.

iv) $\sigma_{\min}^2(P - Q) = 1 - \sigma_{\max}^2(PQ)$.

v) $\sigma_{\max}(PQ) = \sigma_{\max}(QP) = \sigma_{\max}(P + Q - I) < 1$.

vi) $A = (I - PQ)^{-1}P(I - PQ)$.

Source: [2296] and Fact 7.13.27. The nonsingularity of $P - Q$ follows from Fact 4.18.19; *ii)* is given by Fact 4.18.19 and Fact 8.7.3; the first equality in *iii)* is given in [602], see also [1134]; *vi)* is given in [15, 531]. **Remark:** $A_{\perp}^*$ is the idempotent matrix onto $\mathcal{R}(A)^{\perp}$ along $\mathcal{N}(A)^{\perp}$. **Remark:** $P = AA^+$ and $Q = (A^+A)_{\perp}$. **Related:** Fact 4.15.5, Fact 4.18.19, Fact 7.13.27, and Fact 8.8.17.

Fact 7.13.29. Let $A, B \in \mathbb{F}^{n\times n}$, and assume that A and B are idempotent. Then, $A - B$ is idempotent if and only if $A - B$ is group invertible and every eigenvalue of $A - B$ is nonnegative. **Source:** [1336]. **Remark:** Conditions under which a matrix is a difference of idempotents are given in [1336]. **Credit:** T. Makelainen and G. P. H. Styan. **Related:** Fact 4.16.13.

Fact 7.13.30. Let $A \in \mathbb{F}^{n\times n}$, $B \in \mathbb{F}^{n\times m}$, and $C \in \mathbb{F}^{m\times m}$, define $\mathcal{A} \triangleq \begin{bmatrix} A & B \\ B^* & C \end{bmatrix} \in \mathbb{F}^{(n+m)\times(n+m)}$, and assume that $\mathcal{A}$ is Hermitian. Then, for all $i \in \{1,\dots,n\}$ and $j \in \{1,\dots,n\}$ such that $i + j \le n+1$,

$$\lambda_{i+j-1}(\mathcal{A}) + \lambda_{\min}(\mathcal{A}) \le \lambda_i(A) + \lambda_j(C).$$

In particular,

$$\lambda_{\max}(\mathcal{A}) + \lambda_{\min}(\mathcal{A}) \le \lambda_{\max}(A) + \lambda_{\max}(C).$$

Source: [486, p. 56] and [2979, p. 74].

Fact 7.13.31. Let $M \in \mathbb{R}^{r\times r}$, assume that M is positive definite, let $C, K \in \mathbb{R}^{r\times r}$, assume that C and K are positive semidefinite, and consider the equation

$$M\ddot{q} + C\dot{q} + Kq = 0.$$

Then, $x(t) \triangleq \begin{bmatrix} q(t) \\ \dot{q}(t) \end{bmatrix}$ satisfies $\dot{x}(t) = Ax(t)$, where A is the $2r \times 2r$ matrix

$$A \triangleq \begin{bmatrix} 0 & I \\ -M^{-1}K & -M^{-1}C \end{bmatrix}.$$

Furthermore, the following statements hold:

i) A, K, and M satisfy

$$\det A = \frac{\det K}{\det M}.$$

ii) A and K satisfy $\operatorname{rank} A = r + \operatorname{rank} K$.

iii) A is nonsingular if and only if K is positive definite. If these conditions hold, then

$$A^{-1} = \begin{bmatrix} -K^{-1}C & -K^{-1}M \\ I & 0 \end{bmatrix}.$$

iv) If K is singular, then A is not group invertible. In particular, every Jordan block of A associated with the zero eigenvalue is of size 2.

v) $\chi_A(s) = \det(s^2 I + sM^{-1}C + M^{-1}K)$.

vi) Let $\lambda \in \mathbb{C}$. Then, $\lambda \in \operatorname{spec}(A)$ if and only if $\det(\lambda^2 M + \lambda C + K) = 0$.

vii) If $\lambda \in \operatorname{spec}(A)$, $\operatorname{Re}\lambda = 0$, and $\operatorname{Im}\lambda \ne 0$, then λ is semisimple.

viii) $\operatorname{mspec}(A) \subset \mathrm{CLHP}$.

ix) If $C = 0$, then $\operatorname{spec}(A) \subset \mathrm{IA}$.

x) If C and K are positive definite, then $\operatorname{spec}(A) \subset \mathrm{OLHP}$.

xi) $\hat{x}(t) \triangleq \begin{bmatrix} K^{1/2}q(t) \\ M^{1/2}\dot{q}(t) \end{bmatrix}$ satisfies $\dot{x}(t) = \hat{A}x(t)$, where

$$\hat{A} \triangleq \begin{bmatrix} 0 & K^{1/2}M^{-1/2} \\ -M^{-1/2}K^{1/2} & -M^{-1/2}CM^{-1/2} \end{bmatrix}.$$

If, in addition, $C = 0$, then $\hat{A}$ is skew symmetric.

xii) $\hat{x}(t) \triangleq \begin{bmatrix} M^{1/2}q(t) \\ M^{1/2}\dot{q}(t) \end{bmatrix}$ satisfies $\dot{x}(t) = \hat{A}x(t)$, where

$$\hat{A} \triangleq \begin{bmatrix} 0 & I \\ -M^{-1/2}KM^{-1/2} & -M^{-1/2}CM^{-1/2} \end{bmatrix}.$$

If, in addition, $C = 0$, then $\hat{A}$ is Hamiltonian.

Remark: M, C, and K are mass, damping, and stiffness matrices, respectively. See [431]. *iv)* shows that the absence of a stiffness implies that the structure has a rigid-body mode. *vii)* shows that the only type of instability is a rigid-body mode. **Related:** Fact 7.13.32, Fact 7.15.32, and Fact 15.19.38. **Problem:** Prove *vii)*.

Fact 7.13.32. Let $A, B \in \mathbb{R}^{n\times n}$, and assume that A and B are positive semidefinite. Then, every eigenvalue λ of $\begin{bmatrix} 0 & A \\ -B & 0 \end{bmatrix}$ satisfies $\operatorname{Re}\lambda = 0$. If, in addition, $\operatorname{rank} A + \operatorname{rank} B$ is odd, then $\begin{bmatrix} 0 & A \\ -B & 0 \end{bmatrix}$ is not group invertible. Now, let $C \in \mathbb{R}^{n\times n}$, and assume that C is (positive semidefinite, positive definite). Then, every eigenvalue of $\begin{bmatrix} 0 & A \\ -B & -C \end{bmatrix}$ satisfies $(\operatorname{Re}\lambda \le 0, \operatorname{Re}\lambda < 0)$. **Source:** To prove the second statement, note that $\operatorname{rank}\begin{bmatrix} 0 & A \\ -B & 0 \end{bmatrix}^2$ is even. **Example:** If $A \triangleq \begin{bmatrix} 1 & 0 \\ 0 & 0 \end{bmatrix}$ and $B \triangleq I_2$, then $3 = \operatorname{rank}\begin{bmatrix} 0 & A \\ -B & 0 \end{bmatrix} > \operatorname{rank}\begin{bmatrix} 0 & A \\ -B & 0 \end{bmatrix}^2 = 2$. **Problem:** Determine the structure of the Jordan blocks associated with the zero eigenvalue. **Problem:** Consider the case where A and B are of different size as well as the matrix $\begin{bmatrix} -C & A \\ -B & -C \end{bmatrix}$.

Fact 7.13.33. Let $A_0, \ldots, A_{r-1} \in \mathbb{F}^{n\times n}$, and define $\mathcal{A} \in \mathbb{F}^{rn\times rn}$ by

$$\mathcal{A} \triangleq \begin{bmatrix} 0 & I_n & 0 & \cdots & 0 & 0 \\ 0 & 0 & I_n & \ddots & 0 & 0 \\ 0 & 0 & 0 & \ddots & 0 & 0 \\ \vdots & \vdots & \vdots & \ddots & \ddots & \vdots \\ 0 & 0 & 0 & \cdots & 0 & I_n \\ -A_0 & -A_1 & -A_2 & \cdots & -A_{r-2} & -A_{r-1} \end{bmatrix}.$$

Then,

$$\chi_{\mathcal{A}}(s) = \det(s^r I_n + s^{r-1}A_{r-1} + \cdots + sA_1 + A_0).$$

7.14 Facts on Matrix Pencils

Fact 7.14.1. Let $A, B \in \mathbb{F}^{n\times n}$, assume that $P_{A,B}$ is a regular pencil, let $\mathcal{S}, \mathcal{T} \subseteq \mathbb{F}^n$, assume that $\mathcal{S}, \mathcal{T}$ are subspaces of the same dimension, define $k \triangleq \dim \mathcal{S} = \dim \mathcal{T}$, let $S, T \in \mathbb{F}^{n\times k}$, and assume that $\mathcal{R}(S) = \mathcal{S}$ and $\mathcal{R}(T) = \mathcal{T}$. Then, there exist $M_1, M_2 \in \mathbb{F}^{k\times k}$ such that $AS = TM_1$ and $BS = TM_2$. If, in addition, B is nonsingular, then $\mathcal{S}$ is an invariant subspace of $B^{-1}A$. **Source:** [1705, p. 68]. **Remark:** $\mathcal{S}$ is a *right deflating subspace* of $P_{A,B}$, and $\mathcal{T}$ is a *left deflating subspace* of $P_{A,B}$. See also [2539, pp. 303, 304]. **Related:** Fact 3.11.21.

7.15 Facts on Eigenstructure for One Matrix

Fact 7.15.1. Let $A \in \mathbb{F}^{n\times n}$. Then, $\operatorname{def} A = \operatorname{gmult}_A(0) \le \operatorname{amult}_A(0)$. If, in addition, A is group invertible, then $\operatorname{def} A = \operatorname{gmult}_A(0) = \operatorname{amult}_A(0)$.

Fact 7.15.2. Let $A \in \mathbb{F}^{n\times n}$. Then, $\operatorname{amult}_{A^*A}(0) = \operatorname{gmult}_{A^*A}(0) = \operatorname{gmult}_A(0)$. **Source:** [2991, p. 85].

Fact 7.15.3. Let $A \in \mathbb{F}^{n\times n}$, and let $p \in \mathbb{F}[s]$. Then, the following statements are equivalent:

i) μ_A divides p.

ii) $\operatorname{spec}(A) \subseteq \operatorname{roots}(p)$, and, for all $\lambda \in \operatorname{spec}(A)$, $\operatorname{ind}_A(\lambda) \le \operatorname{mult}_p(\lambda)$.

Source: Proposition 7.7.3.

Fact 7.15.4. Let $A \in \mathbb{F}^{n\times n}$, let $\lambda \in \operatorname{spec}(A)$, assume that λ is cyclic, let $i \in \{1,\dots,n\}$ satisfy $\operatorname{rank}(A-\lambda I)_{[i,\cdot]} = n-1$, and define $x \in \mathbb{C}^n$ by

$$x \triangleq \begin{bmatrix} \det(A-\lambda I)_{[i,1]} \\ -\det(A-\lambda I)_{[i,2]} \\ \vdots \\ (-1)^{n+1}\det(A-\lambda I)_{[i,n]} \end{bmatrix}.$$

Then, x is an eigenvector of A associated with λ. **Source:** [2737].

Fact 7.15.5. Let $A \in \mathbb{F}^{n\times n}$, where $n \ge 2$. Then, the following statements are equivalent:

i) $\operatorname{rank} A = 1$.

ii) $\operatorname{gmult}_A(0) = n-1$.

Now, assume that these statements hold, and let $x, y \in \mathbb{F}^n$ satisfy $A = xy^{\mathsf{T}}$. Then, the following statements are equivalent:

iii) $\operatorname{tr} A = 0$.

iv) $\operatorname{amult}_A(0) = n$.

v) $\operatorname{spec}(A) = \{0\}$.

vi) $A^2 = 0$.

vii) A is nilpotent.

viii) A is not group invertible.

ix) A is defective.

x) $\operatorname{ind} A = 2$.

xi) $y^{\mathsf{T}}x = 0$.

xii) A is similar to $\operatorname{diag}(N_2, 0_{(n-2)\times(n-2)})$.

In addition, the following statements are equivalent:

xiii) $\operatorname{tr} A \ne 0$.

xiv) $\operatorname{amult}_A(0) = n-1$.

xv) $\{0\} \subset \operatorname{spec}(A)$.

xvi) $A^2 \ne 0$.

xvii) A is group invertible.

xviii) A is semisimple.

xix) $\operatorname{ind} A = 1$.

xx) $y^{\mathsf{T}}x \ne 0$.

xxi) A is similar to $\operatorname{diag}(\operatorname{tr} A, 0_{(n-1)\times(n-1)})$.

Fact 7.15.6. Let $A \in \mathbb{F}^{n\times n}$. Then, the following statements are equivalent:

i) A is group invertible.

ii) $\mathcal{R}(A) = \mathcal{R}(A^2)$.

iii) $\operatorname{rank} A = \operatorname{rank} A^2$.

iv) $\operatorname{ind} A \le 1$.

v) $\operatorname{rank} A = \sum_{i=1}^r \operatorname{amult}_A(\lambda_i)$, where $\{\lambda_1,\dots,\lambda_r\} = \operatorname{spec}(A)\backslash\{0\}$.

vi) $\mathcal{R}(A) + \mathcal{N}(A) = \mathbb{F}^n$.

vii) $\mathcal{R}(A) \cap \mathcal{N}(A) = \{0\}$.

Related: Corollary 4.8.10.

Fact 7.15.7. Let $A \in \mathbb{F}^{n\times n}$, assume that A is diagonalizable over $\mathbb{F}$ with eigenvalues $\lambda_1, \ldots, \lambda_n$, and let $B \triangleq \operatorname{diag}(\lambda_1, \ldots, \lambda_n)$. If $x_1, \ldots, x_n \in \mathbb{F}^n$ are linearly independent eigenvectors of A associated with $\lambda_1, \ldots, \lambda_n$, respectively, then $A = SBS^{-1}$, where $S \triangleq [x_1 \ \cdots \ x_n]$. Conversely, if $S \in \mathbb{F}^{n\times n}$ is nonsingular and $A = SBS^{-1}$, then, for all $i \in \{1, \ldots, n\}$, $\operatorname{col}_i(S)$ is an associated eigenvector.

Fact 7.15.8. Let $A \in \mathbb{C}^{n\times n}$, and, for all $k \in \{1, \ldots, n\}$, define $\mathcal{A}_k \in \mathbb{C}^{k\times k}$, where, for all $i, j \in \{1, \ldots, k\}$, $(\mathcal{A}_k)_{(i,j)} = \operatorname{tr} A^{i+j-2}$. Then, $\operatorname{card}[\operatorname{spec}(A)] = \max\{k \in \{1, \ldots, n\}\colon \det \mathcal{A}_k \neq 0\}$. In particular, A has n distinct eigenvalues if and only if $\mathcal{A}_n$ is nonsingular. Furthermore, the following statements are equivalent:

i) A is diagonalizable over $\mathbb{C}$.

ii) $\operatorname{card}[\operatorname{spec}(A)] = \deg \mu_A$.

iii) $\det \mathcal{A}_{\deg \mu_A} \neq 0$.

Source: [995, 996].

Fact 7.15.9. Let $A, S \in \mathbb{F}^{n\times n}$, assume that S is nonsingular, let $\lambda \in \mathbb{C}$, and assume that $\operatorname{col}_1(S^{-1}AS) = \lambda e_1$. Then, $\lambda \in \operatorname{spec}(A)$, and $\operatorname{col}_1(S)$ is an associated eigenvector.

Fact 7.15.10. Let $A \in \mathbb{F}^{n\times n}$. Then, A is cyclic if and only if there exists $b \in \mathbb{F}^n$ such that $[b \ Ab \ \cdots \ A^{n-1}b]$ is nonsingular. **Source:** Fact 16.21.14. **Remark:** (A, b) is controllable. See Corollary 16.6.3.

Fact 7.15.11. Let $A \in \mathbb{F}^{n\times n}$, and define the positive integer $m \triangleq \max_{\lambda\in\operatorname{spec}(A)} \operatorname{gmult}_A(\lambda)$. Then, m is the smallest integer such that there exists $B \in \mathbb{F}^{n\times m}$ such that $\operatorname{rank}[B \ AB \ \cdots \ A^{n-1}B] = n$. **Source:** Fact 16.21.14. **Remark:** (A, B) is controllable. See Corollary 16.6.3.

Fact 7.15.12. Let $A \in \mathbb{C}^{n\times n}$. Then, there exist $v_1, \ldots, v_n \in \mathbb{C}^n$ such that the following statements hold:

i) $v_1, \ldots, v_n \in \mathbb{C}^n$ are linearly independent.

ii) If $\lambda \in \operatorname{spec}(A)$ and A has a $k \times k$ Jordan block associated with λ, then there exist distinct integers $i_1, \ldots, i_k$ such that $Av_{i_1} = \lambda v_{i_1}, Av_{i_2} = \lambda v_{i_2} + v_{i_1}, \ldots, Av_{i_k} = \lambda v_{i_k} + v_{i_{k-1}}$.

iii) Let λ and $v_{i_1}, \ldots, v_{i_k}$ be given by *ii)*. Then, $\operatorname{span}\{v_{i_1}, \ldots, v_{i_k}\} = \mathcal{N}[(\lambda I - A)^k]$.

Remark: $v_1, \ldots, v_n$ are *generalized eigenvectors* of A. **Remark:** $(v_{i_1}, \ldots, v_{i_k})$ is a *Jordan chain* of A associated with λ. See [1738, pp. 229–231]. **Related:** Fact 16.20.5.

Fact 7.15.13. Let $A \in \mathbb{R}^{n\times n}$. Then, A is cyclic and semisimple if and only if A is simple.

Fact 7.15.14. Let $A = \operatorname{revdiag}(a_1, \ldots, a_n) \in \mathbb{R}^{n\times n}$. Then, A is semisimple if and only if, for all $i \in \{1, \ldots, n\}$, a_i and a_{n+1-i} are either both zero or both nonzero. **Source:** [1301, p. 116], [1604], and [2263, pp. 68, 86].

Fact 7.15.15. Let $A \in \mathbb{F}^{n\times n}$. Then, A has at least m real eigenvalues and m associated linearly independent eigenvectors if and only if there exists a positive-semidefinite matrix $S \in \mathbb{F}^{n\times n}$ such that $\operatorname{rank} S = m$ and $AS = SA^*$. **Source:** [2263, pp. 68, 86]. **Credit:** M. A. Drazin and E. V. Haynsworth. **Related:** The case $m = n$ is given by Proposition 7.7.13.

Fact 7.15.16. Let $A \in \mathbb{F}^{n\times n}$, and let $\operatorname{mspec}(A) = \{\lambda_1, \ldots, \lambda_n\}_{\rm ms}$, where $|\lambda_1| \geq \cdots \geq |\lambda_n|$. Then, the following statements are equivalent:

i) A is normal.

ii) There exists $p \in \mathbb{F}[s]$ such that $A = p(A^*)$.

iii) Every eigenvector of A is also an eigenvector of A^*.

iv) There exist nonzero mutually orthogonal vectors $x_1, \ldots, x_n \in \mathbb{C}^n$ such that $A = \sum_{i=1}^n \lambda_i x_i x_i^*$.

v) $AA^* - A^*A$ is either positive semidefinite or negative semidefinite.

vi) For all $x \in \mathbb{F}^n$, $x^*A^*Ax = x^*AA^*x$.

vii) For all $x, y \in \mathbb{F}^n$, $x^*A^*Ay = x^*AA^*y$.

If these statements hold, then $\rho_{\max}(A) = \sigma_{\max}(A)$. **Source:** [1242], [2263, p. 146], and [2991, p. 86]. **Related:** Fact 4.10.12, Corollary 7.5.8, Fact 7.12.30, Fact 10.13.2, and Fact 11.9.16

Fact 7.15.17. Let $A \in \mathbb{F}^{n\times n}$. Then, the following statements are equivalent:

i) A is (simple, cyclic, derogatory, semisimple, defective, diagonalizable over $\mathbb{F}$).

ii) There exists $\alpha \in \mathbb{F}$ such that $A + \alpha I$ is (simple, cyclic, derogatory, semisimple, defective, diagonalizable over $\mathbb{F}$).

iii) For all $\alpha \in \mathbb{F}$, $A + \alpha I$ is (simple, cyclic, derogatory, semisimple, defective, diagonalizable over $\mathbb{F}$).

Fact 7.15.18. Let $x, y \in \mathbb{F}^n$, assume that $x^{\mathsf{T}}y \neq 1$, and define the elementary matrix $A \triangleq I - xy^{\mathsf{T}}$. Then, A is semisimple if and only if either $xy^{\mathsf{T}} = 0$ or $x^{\mathsf{T}}y \neq 0$. **Source:** Fact 7.15.5 and Fact 7.15.17.

Fact 7.15.19. Let $A \in \mathbb{F}^{n\times n}$, and assume that A is nilpotent. Then, A is nonzero if and only if A is defective.

Fact 7.15.20. Let $A \in \mathbb{F}^{n\times n}$, and assume that A is either involutory or skew involutory. Then, A is semisimple.

Fact 7.15.21. Let $A \in \mathbb{R}^{n\times n}$, and assume that A is either involutory, idempotent, or tripotent. Then, A is diagonalizable over $\mathbb{R}$.

Fact 7.15.22. Let $A \in \mathbb{F}^{n\times n}$, let $\lambda \in \operatorname{spec}(A)$, assume that $\lambda \neq 0$, assume that λ is the unique eigenvalue of A on the circle $\{z \in \mathbb{C}\colon |z| = |\lambda|\}$, and let k be a positive integer. Then, $\operatorname{amult}_A(\lambda) = \operatorname{amult}_{A^k}(\lambda^k)$ and $\operatorname{gmult}_A(\lambda) = \operatorname{gmult}_{A^k}(\lambda^k)$. Now, assume that A is nonsingular, and let k be a nonzero integer. Then, $\operatorname{amult}_A(\lambda) = \operatorname{amult}_{A^k}(\lambda^k)$ and $\operatorname{gmult}_A(\lambda) = \operatorname{gmult}_{A^k}(\lambda^k)$. In particular, $\operatorname{amult}_A(\lambda) = \operatorname{amult}_{A^{-1}}(1/\lambda)$ and $\operatorname{gmult}_A(\lambda) = \operatorname{gmult}_{A^{-1}}(1/\lambda)$. **Related:** Fact 8.10.7.

Fact 7.15.23. Let $A \in \mathbb{F}^{n\times n}$. Then, the following statements are equivalent:

i) A is semisimple.

ii) A is group invertible, and, for all positive integers k, A^k is semisimple.

iii) A is group invertible, and there exists a positive integer k such that A^k is semisimple.

Now, assume that A is nonsingular. Then, the following statements are equivalent:

iv) A is semisimple.

v) For every integer k, A^k is semisimple.

vi) There exists an integer k such that A^k is semisimple.

Fact 7.15.24. Let $A \in \mathbb{F}^{n\times n}$. Then, the following statements hold:

i) If $\operatorname{spec}(A) \subseteq \{0, 1\}$, A is group invertible, and k is a positive integer, then A^k and A are similar.

ii) If $\operatorname{spec}(A) = \{1\}$ and k is an integer, then A^k and A are similar.

Fact 7.15.25. Let $A \in \mathbb{F}^{n\times n}$. Then, the following statements hold:

i) If $\operatorname{spec}(A) \subseteq \{-1, 0, 1\}$, A is group invertible, and k is an odd positive integer, then A^k and A are similar.

ii) If $\operatorname{spec}(A) \subseteq \{-1, 1\}$ and k is an odd integer, then A^k and A are similar.

Fact 7.15.26. Let $A \in \mathbb{F}^{n\times n}$. Then, the following statements hold:

i) If $\operatorname{spec}(A) \subseteq \{0, -\frac{1}{2} + \frac{1}{2}\sqrt{3}\jmath, -\frac{1}{2} - \frac{1}{2}\sqrt{3}\jmath, 1\}$, A is group invertible, and k is a positive integer, then A^{3k+1} and A are similar.

ii) If $\operatorname{spec}(A) \subseteq \{-\frac{1}{2} + \frac{1}{2}\sqrt{3}\jmath, -\frac{1}{2} - \frac{1}{2}\sqrt{3}\jmath, 1\}$ and k is an integer, then A^{3k+1} and A are similar.

Fact 7.15.27. Let $A \in \mathbb{F}^{n\times n}$ and let m be a positive integer. Then, the following statements hold:

i) If $\operatorname{spec}(A) \subseteq \{e^{(2\pi i/m)\jmath}: i = 0, \ldots, m-1\} \cup \{0\}$, A is group invertible, and k is a nonnegative integer, then A^{km+1} and A are similar.

ii) If $\operatorname{spec}(A) \subseteq \{e^{(2\pi i/m)\jmath}: i = 0, \ldots, m-1\}$ and k is an integer, then A^{km+1} and A are similar.

Fact 7.15.28. Let $A \in \mathbb{F}^{n\times n}$. Then, the following statements hold:

i) If there exist distinct positive integers k and l such that A^k and A^l are similar, then $\operatorname{ind} A \le \min\{k, l\}$.

ii) If there exist distinct positive integers k and l such that $A^k = A^l$, then $\operatorname{ind} A \le \min\{k, l\}$ and $(A^{\mathsf{D}})^{\mathsf{D}}$ is semisimple.

iii) If A is group invertible and there exist distinct positive integers k and l such that $A^k = A^l$, then A is semisimple and $A = A^{|l-k|+1}$.

iv) If A is nonsingular and there exist distinct nonzero integers k and l such that $A^k = A^l$, then A is semisimple and $A^{l-k} = I$.

Source: Fact 7.15.23.

Fact 7.15.29. Let $A \in \mathbb{F}^{n\times n}$, assume that A is group invertible, and assume that $A^3 = A^2$. Then, A is idempotent.

Fact 7.15.30. Let $A \in \mathbb{C}^{n\times n}$, and let $x \in \mathbb{C}^n$ be an eigenvector of A associated with the eigenvalue λ. If A is nonsingular, then x is an eigenvector of A^{A} associated with the eigenvalue $(\det A)/\lambda$. If $\operatorname{rank} A = n-1$, then x is an eigenvector of A^{A} associated with the eigenvalue $\operatorname{tr} A^{\mathsf{A}}$ or 0. Finally, if $\operatorname{rank} A \le n-2$, then x is an eigenvector of A^{A} associated with the eigenvalue 0. **Source:** Use Fact 7.16.1 and $A^{\mathsf{A}}A = AA^{\mathsf{A}}$. See [791]. **Related:** Fact 3.19.3 and Fact 8.3.21.

Fact 7.15.31. Let $q(t)$ denote the displacement of a mass $m > 0$ connected to a wall by means of a dashpot $c \ge 0$ and subject to a force $f(t)$. Then, $q(t)$ satisfies

$$m\ddot{q}(t) + c\dot{q}(t) = f(t),$$

and thus

$$\ddot{q}(t) + \frac{c}{m}\dot{q}(t) = \frac{1}{m}f(t).$$

Next, $x_1(t) \triangleq q(t)$ and $x_2(t) \triangleq \dot{q}(t)$ satisfy

$$\begin{bmatrix} \dot{x}_1(t) \\ \dot{x}_2(t) \end{bmatrix} = \begin{bmatrix} 0 & 1 \\ 0 & -\frac{c}{m} \end{bmatrix}\begin{bmatrix} x_1(t) \\ x_2(t) \end{bmatrix} + \begin{bmatrix} 0 \\ \frac{1}{m} \end{bmatrix} f(t).$$

The eigenvalues of the companion matrix $A_{\rm c} \triangleq \left[\begin{smallmatrix} 0 & 1 \\ 0 & -\frac{c}{m} \end{smallmatrix}\right]$ are given by $\operatorname{mspec}(A_{\rm c}) = \{0, -\frac{c}{m}\}_{\rm ms}$. The matrix $A_{\rm c}$ has a repeated eigenvalue if and only if $c = 0$, in which case $\operatorname{mspec}(A_{\rm c}) = \{0, 0\}_{\rm ms}$, $A_{\rm c} = \left[\begin{smallmatrix} 0 & 1 \\ 0 & 0 \end{smallmatrix}\right]$ is defective and nilpotent, and $A_{\rm c}$ is in Jordan form. Finally, if $c > 0$, then $A_{\rm c} = SA_{\rm J}S^{-1}$, where

$$A_{\rm J} \triangleq \begin{bmatrix} 0 & 0 \\ 0 & -\frac{c}{m} \end{bmatrix}, \quad S \triangleq \begin{bmatrix} 1 & -\frac{m}{c} \\ 0 & 1 \end{bmatrix}, \quad S^{-1} \triangleq \begin{bmatrix} 1 & \frac{m}{c} \\ 0 & 1 \end{bmatrix}.$$

Remark: If $c > 0$, then this structure is a *damped rigid body*, whereas, if $c = 0$, then this structure is an *undamped rigid body*.

Fact 7.15.32. Let $q(t)$ denote the displacement of a mass $m > 0$ connected to a wall by means of a spring $k > 0$ and a dashpot $c \ge 0$ and subject to a force $f(t)$. Then, $q(t)$ satisfies

$$m\ddot{q}(t) + c\dot{q}(t) + kq(t) = f(t),$$

and thus

$$\ddot{q}(t) + \frac{c}{m}\dot{q}(t) + \frac{k}{m}q(t) = \frac{1}{m}f(t).$$

Now, define the *natural frequency* $\omega_{\rm n} \triangleq \sqrt{k/m}$ and *damping ratio* $\zeta \triangleq c/(2\sqrt{km})$ to obtain

$$\ddot{q}(t) + 2\zeta\omega_{\rm n}\dot{q}(t) + \omega_{\rm n}^2 q(t) = \frac{1}{m}f(t).$$

Next, $x_1(t) \triangleq q(t)$ and $x_2(t) \triangleq \dot{q}(t)$ satisfy

$$\begin{bmatrix} \dot{x}_1(t) \\ \dot{x}_2(t) \end{bmatrix} = \begin{bmatrix} 0 & 1 \\ -\omega_{\rm n}^2 & -2\zeta\omega_{\rm n} \end{bmatrix}\begin{bmatrix} x_1(t) \\ x_2(t) \end{bmatrix} + \begin{bmatrix} 0 \\ \frac{1}{m} \end{bmatrix} f(t).$$

The eigenvalues of the companion matrix $A_{\rm c} \triangleq \left[\begin{smallmatrix} 0 & 1 \\ -\omega_{\rm n}^2 & -2\zeta\omega_{\rm n} \end{smallmatrix}\right]$ are given by

$$\operatorname{mspec}(A_{\rm c}) = \begin{cases} \{-\zeta\omega_{\rm n} - \omega_{\rm d}\jmath, -\zeta\omega_n + \omega_{\rm d}\jmath\}_{\rm ms}, & 0 \le \zeta \le 1, \\ \left\{(-\zeta - \sqrt{\zeta^2 - 1})\omega_{\rm n}, (-\zeta + \sqrt{\zeta^2 - 1})\omega_{\rm n}\right\}, & \zeta > 1, \end{cases}$$

where $\omega_{\rm d} \triangleq \omega_{\rm n}\sqrt{1 - \zeta^2}$ is the *damped natural frequency*. The matrix $A_{\rm c}$ has a repeated eigenvalue if and only if $\zeta = 1$, in which case $\operatorname{mspec}(A_{\rm c}) = \{-\omega_{\rm n}, -\omega_{\rm n}\}_{\rm ms}$ and $A_{\rm c}$ is defective. In particular, in the case where $\zeta = 1$, it follows that $A_{\rm c} = SA_{\rm J}S^{-1}$, where $S \triangleq \left[\begin{smallmatrix} -1 & 0 \\ \omega_{\rm n} & -1 \end{smallmatrix}\right]$ and $A_{\rm J}$ is the Jordan form matrix $A_{\rm J} \triangleq \left[\begin{smallmatrix} -\omega_{\rm n} & 1 \\ 0 & -\omega_{\rm n} \end{smallmatrix}\right]$. If $A_{\rm c}$ is not defective, that is, if $\zeta \ne 1$, then the Jordan form $A_{\rm J}$ of $A_{\rm c}$ is given by

$$A_{\rm J} \triangleq \begin{cases} \begin{bmatrix} -\zeta\omega_{\rm n} + \omega_{\rm d}\jmath & 0 \\ 0 & -\zeta\omega_{\rm n} - \omega_{\rm d}\jmath \end{bmatrix}, & 0 \le \zeta < 1, \\ \begin{bmatrix} \left(-\zeta - \sqrt{\zeta^2 - 1}\right)\omega_{\rm n} & 0 \\ 0 & \left(-\zeta + \sqrt{\zeta^2 - 1}\right)\omega_{\rm n} \end{bmatrix}, & \zeta > 1. \end{cases}$$

In the case where $0 \le \zeta < 1$, define the real normal form

$$A_{\rm n} \triangleq \begin{bmatrix} -\zeta\omega_{\rm n} & \omega_{\rm d} \\ -\omega_{\rm d} & -\zeta\omega_{\rm n} \end{bmatrix}.$$

The matrices $A_{\rm c}$, $A_{\rm J}$, and $A_{\rm n}$ are related by the similarity transformations $A_{\rm c} = S_1 A_{\rm J} S_1^{-1} = S_2 A_{\rm n} S_2^{-1}$ and $A_{\rm J} = S_3 A_{\rm n} S_3^{-1}$, where

$$S_1 \triangleq \begin{bmatrix} 1 & 1 \\ -\zeta\omega_{\rm n} + \omega_{\rm d}\jmath & -\zeta\omega_{\rm n} - \omega_{\rm d}\jmath \end{bmatrix}, \quad S_1^{-1} = \frac{\jmath}{2\omega_{\rm d}}\begin{bmatrix} -\zeta\omega_{\rm n} - \omega_{\rm d}\jmath & -1 \\ \zeta\omega_{\rm n} - \omega_{\rm d}\jmath & 1 \end{bmatrix},$$

$$S_2 \triangleq \frac{1}{\omega_{\rm d}}\begin{bmatrix} 1 & 0 \\ -\zeta\omega_{\rm n} & \omega_{\rm d} \end{bmatrix}, \quad S_2^{-1} = \begin{bmatrix} \omega_{\rm d} & 0 \\ \zeta\omega_{\rm n} & 1 \end{bmatrix}, \quad S_3 \triangleq \frac{1}{2\omega_{\rm d}}\begin{bmatrix} 1 & -\jmath \\ 1 & \jmath \end{bmatrix}, \quad S_3^{-1} = \omega_{\rm d}\begin{bmatrix} 1 & 1 \\ \jmath & -\jmath \end{bmatrix}.$$

In the case where $\zeta = 1$, $A_{\rm c}$ and $A_{\rm J}$ are related by $A_{\rm c} = S_4 A_{\rm J} S_4^{-1}$, where

$$S_4 \triangleq \begin{bmatrix} -1 & 0 \\ \omega_{\rm n} & -1 \end{bmatrix}, \quad S_4^{-1} = \begin{bmatrix} -1 & 0 \\ -\omega_{\rm n} & -1 \end{bmatrix}.$$

In the case where $\zeta > 1$, $A_{\rm c}$ and $A_{\rm J}$ are related by $A_{\rm c} = S_5 A_{\rm J} S_5^{-1}$, where

$$S_5 \triangleq \begin{bmatrix} 1 & 1 \\ -\zeta\omega_{\rm n} + \omega_{\rm n}\sqrt{\zeta^2 - 1} & -\zeta\omega_{\rm n} - \omega_{\rm n}\sqrt{\zeta^2 - 1} \end{bmatrix}, \quad S_5^{-1} = \frac{1}{-2\omega_{\rm n}\sqrt{\zeta^2 - 1}}\begin{bmatrix} -\zeta\omega_{\rm n} - \omega_{\rm n}\sqrt{\zeta^2 - 1} & -1 \\ \zeta\omega_{\rm n} - \omega_{\rm n}\sqrt{\zeta^2 - 1} & 1 \end{bmatrix}.$$

Finally, define the energy-coordinates matrix

$$A_{\mathrm{e}} \triangleq \begin{bmatrix} 0 & \omega_{\mathrm{n}} \\ -\omega_{\mathrm{n}} & -2\zeta\omega_{\mathrm{n}} \end{bmatrix}.$$

Then, $A_{\mathrm{e}} = S_6 A_{\mathrm{c}} S_6^{-1}$, where

$$S_6 \triangleq \sqrt{\frac{m}{2}} \begin{bmatrix} 1 & 0 \\ 0 & \frac{1}{\omega_{\mathrm{n}}} \end{bmatrix}.$$

Remark: If $c > 0$, then this structure is a *damped oscillator*, whereas, if $c = 0$, then this structure is an *undamped oscillator*. **Related:** Fact 7.13.31.

7.16 Facts on Eigenstructure for Two or More Matrices

Fact 7.16.1. Let $A, B \in \mathbb{C}^{n \times n}$, and assume that $AB = BA$. Furthermore, let $x \in \mathbb{C}^n$ be an eigenvector of A associated with the eigenvalue $\lambda \in \mathbb{C}$, and assume that $Bx \neq 0$. Then, Bx is an eigenvector of A associated with the eigenvalue $\lambda \in \mathbb{C}$. **Source:** $A(Bx) = BAx = B(\lambda x) = \lambda(Bx)$.

Fact 7.16.2. Let $A, B \in \mathbb{C}^{n \times n}$, and assume that $\operatorname{rank}[A, B] \le 1$. Then, there exists a nonzero vector $x \in \mathbb{C}^n$ that is an eigenvector of both A and B. **Source:** [1448, p. 51] and [2991, p. 77]. **Example:** Let $A = \begin{bmatrix} 1 & -1 \\ 1 & -1 \end{bmatrix}$, $B = \begin{bmatrix} 1 & 0 \\ 1 & 0 \end{bmatrix}$, and $x = \begin{bmatrix} 1 \\ 1 \end{bmatrix}$. Then, $\operatorname{rank}[A, B] = 1$, $Ax = 0$, and $Bx = x$. **Related:** Fact 7.16.15, Fact 7.19.2, Fact 7.19.6, and Fact 11.10.52.

Fact 7.16.3. Let $A, B \in \mathbb{C}^{n \times n}$, and assume that A is simple. Then, $AB = BA$ if and only if A and B have a common set of n linearly independent eigenvectors. **Source:** [2991, p. 76].

Fact 7.16.4. Let $A, B \in \mathbb{F}^{n \times n}$, and assume that $[A, [A, B]] = 0$. Then, $[A, B]$ is nilpotent. **Credit:** N. Jacobson. See [1055], [1201], and [1448, p. 98]. **Related:** Fact 7.16.5.

Fact 7.16.5. Let $A, B \in \mathbb{F}^{n \times n}$, assume that A is semisimple, and assume that $[A, [A, B]] = 0$. Then, $[A, B] = 0$. **Source:** [1201]. **Credit:** H. Shapiro. **Related:** Fact 7.16.4.

Fact 7.16.6. Let $A, B \in \mathbb{F}^{n \times n}$, assume that A is cyclic, and assume that $[A, [A, B]] = 0$. Then, there exists nonsingular $S \in \mathbb{F}^{n \times n}$ such that SAS^{-1} and SBS^{-1} are upper triangular. **Source:** [1201]. **Credit:** H. Shapiro.

Fact 7.16.7. Let $A, B \in \mathbb{F}^{n \times n}$, assume that A^2 is cyclic, and assume that $A[A, B] = -[A, B]A$. Then, $[A, B] = 0$. **Source:** [1201]. **Remark:** If A^2 is cyclic, then A is cyclic, but the converse is false.

Fact 7.16.8. Let $A \in \mathbb{F}^{n \times n}$. Then, the following statements are equivalent:

i) A is cyclic.

ii) Every matrix $B \in \mathbb{F}^{n \times n}$ satisfying $AB = BA$ is a polynomial in A.

iii) For all $C, D \in \{B \in \mathbb{F}^{n \times n} \colon AB = BA\}$, $CD = DC$.

Source: [1201] and [1450, p. 275]. The equivalence of *ii)* and *iii)* follows from Fact 7.16.9. **Related:** Fact 3.23.9, Fact 7.16.9, Fact 7.16.10, and Fact 9.5.3.

Fact 7.16.9. Let $A, B \in \mathbb{F}^{n \times n}$. Then, B is a polynomial in A if and only if B commutes with every matrix that commutes with A. **Source:** [1450, pp. 276–278]. **Related:** Fact 3.23.9, Fact 6.8.15, Fact 7.16.8, Fact 7.16.10, Fact 9.5.3.

Fact 7.16.10. Let $A \in \mathbb{F}^{n \times n}$, assume that A is simple, let $B \in \mathbb{F}^{n \times n}$, and assume that $AB = BA$. Then, B is a polynomial in A whose degree is less than or equal to $n - 1$. **Source:** [2983, p. 59] and [2991, pp. 73, 74]. **Related:** Fact 7.16.8.

Fact 7.16.11. Let $A, B \in \mathbb{F}^{n \times n}$, and define $D \triangleq AB - BA^{\mathsf{T}}$. Then, the following statements hold:

i) If $AD = DA^{\mathsf{T}}$, then D is singular.

ii) If $AD = DA^{\mathsf{T}}$ and A is semisimple, then $D = 0$.

iii) If $AD = DA^{\mathsf{T}}$ and $DA = A^{\mathsf{T}}D$, then D is nilpotent.

iv) If $AD = DA^{\mathsf{T}}$ and A is cyclic, then D is symmetric and $\mathrm{card}[\mathrm{spec}(A)] \le \mathrm{gmult}_D(0)$.

Source: [1201]. **Credit:** *iv*) is due to O. Taussky and H. Zassenhaus.

Fact 7.16.12. Let $A \in \mathbb{C}^{n\times n}$, $B \in \mathbb{C}^{m\times m}$, and $C \in \mathbb{C}^{n\times m}$, assume that $\mathrm{rank}\, C = m$, and assume that $AC = CB$. Then, $\mathrm{mspec}(B) \subseteq \mathrm{mspec}(A)$. **Source:** [2991, p. 77].

Fact 7.16.13. Let $A, B \in \mathbb{C}^{n\times n}$, and assume that A and B are Hermitian. Then, $\mathrm{spec}([A, B]) \subset \mathrm{I}\mathbb{A}$. **Source:** [2991, p. 305].

Fact 7.16.14. Let $A, B \in \mathbb{C}^{n\times n}$, assume that A and B are nonsingular, and assume that $[A, B]$ is singular. Then, $1 \in \mathrm{spec}(A^{-1}B^{-1}AB)$. **Source:** [2991, p. 77].

Fact 7.16.15. Let $A, B \in \mathbb{C}^{n\times n}$. Then, the following statements are equivalent:

i) $\cap_{k,l=1}^{n-1} \mathcal{N}([A^k, B^l]) \ne \{0\}$.

ii) $\sum_{k,l=1}^{n-1} [A^k, B^l]^*[A^k, B^l]$ is singular.

iii) A and B have a common eigenvector.

Source: [1161]. **Credit:** D. Shemesh. **Related:** Fact 7.16.2, Fact 7.19.2, and Fact 11.10.52.

Fact 7.16.16. Let $A, B \in \mathbb{F}^{n\times n}$. Then, the following statements hold:

i) Assume that A and B are Hermitian. Then, AB is Hermitian if and only if $AB = BA$.

ii) A is normal if and only if, for all $C \in \mathbb{F}^{n\times n}$ such that $AC = CA$, it follows that $A^*C = CA^*$.

iii) Assume that B is Hermitian and $AB = BA$. Then, $A^*B = BA^*$.

iv) Assume that A and B are normal and $AB = BA$. Then, AB is normal.

v) Assume that A and B are normal. Then, AB is normal if and only if BA is normal.

vi) Assume that A is positive semidefinite and B is normal. Then, AB is normal if and only if $AB = BA$.

vii) Assume that A and B are normal and either A or B has the property that distinct eigenvalues have unequal absolute values. Then, AB is normal if and only if $AB = BA$.

Source: [798, 1168], [1305, p. 157], [2263, p. 102], [2875], and [2991, pp. 295, 305, 318]. **Credit:** *v*) is due to F. R. Gantmacher and M. G. Krein. See [1168]. **Related:** Fact 7.19.8.

Fact 7.16.17. Let $A, B, C \in \mathbb{F}^{n\times n}$, assume that A and B are normal, and assume that $AC = CB$. Then, $A^*C = CB^*$. **Source:** Consider $\left[\begin{smallmatrix} A & 0 \\ 0 & B \end{smallmatrix}\right]$ and $\left[\begin{smallmatrix} 0 & C \\ 0 & 0 \end{smallmatrix}\right]$ in *ii*) of Fact 7.16.16. See [1302, p. 104] and [1305, p. 321]. **Remark:** This is the *Putnam-Fuglede theorem.*

Fact 7.16.18. Let $A, B \in \mathbb{F}^{n\times n}$, and assume that A is dissipative and B is range Hermitian. Then, $\mathrm{ind}\, B = \mathrm{ind}\, AB$. **Source:** [433].

Fact 7.16.19. Let $A \in \mathbb{F}^{n\times n}$, $B \in \mathbb{F}^{n\times m}$, and $C \in \mathbb{F}^{m\times m}$. Then,

$$\max\{\mathrm{ind}\, A, \mathrm{ind}\, C\} \le \mathrm{ind}\begin{bmatrix} A & B \\ 0 & C \end{bmatrix} \le \mathrm{ind}\, A + \mathrm{ind}\, C.$$

If C is nonsingular, then

$$\mathrm{ind}\begin{bmatrix} A & B \\ 0 & C \end{bmatrix} = \mathrm{ind}\, A,$$

whereas, if A is nonsingular, then

$$\mathrm{ind}\begin{bmatrix} A & B \\ 0 & C \end{bmatrix} = \mathrm{ind}\, C.$$

Source: [584, 2037]. **Remark:** The eigenstructure of a partitioned Hamiltonian matrix is considered in Fact 16.25.1. **Related:** Fact 8.12.3.

Fact 7.16.20. Let $A, B \in \mathbb{R}^{n\times n}$, and assume that A and B are skew symmetric. Then, there exists an orthogonal matrix $S \in \mathbb{R}^{n\times n}$ such that

$$A = S\begin{bmatrix} 0_{(n-l)\times(n-l)} & A_{12} \\ -A_{12}^{\mathsf{T}} & A_{22}\end{bmatrix}S^{\mathsf{T}}, \quad B = S\begin{bmatrix} B_{11} & B_{12} \\ -B_{12}^{\mathsf{T}} & 0_{l\times l}\end{bmatrix}S^{\mathsf{T}},$$

where $l \triangleq \lfloor n/2 \rfloor$. Consequently,

$$\operatorname{mspec}(AB) = \operatorname{mspec}(-A_{12}B_{12}^{\mathsf{T}}) \cup \operatorname{mspec}(-A_{12}^{\mathsf{T}}B_{12}),$$

and thus every nonzero eigenvalue of AB has even algebraic multiplicity. **Source:** [72].

Fact 7.16.21. Let $A, B \in \mathbb{R}^{n\times n}$, and assume that A and B are skew symmetric. If n is even, then there exists a monic polynomial p of degree $n/2$ such that $\chi_{AB}(s) = p^2(s)$ and $p(AB) = 0$. If n is odd, then there exists a monic polynomial $p(s)$ of degree $(n-1)/2$ such that $\chi_{AB}(s) = sp^2(s)$ and $ABp(AB) = 0$. Consequently, if n is (even, odd), then χ_{AB} is (even, odd) and (every, every nonzero) eigenvalue of AB has even algebraic multiplicity and geometric multiplicity of at least 2. **Source:** [918, 1216].

7.17 Facts on Matrix Factorizations

Fact 7.17.1. Let $A \in \mathbb{F}^{n\times m}$ and $r \ge \operatorname{rank} A$. Then, there exist $x_1, \ldots, x_r \in \mathbb{F}^n$ and $y_1, \ldots, y_r \in \mathbb{F}^{1\times m}$ such that $A = \sum_{i=1}^r x_i y_i = XY$, where $X \triangleq [x_1 \ \cdots \ x_r] \in \mathbb{F}^{n\times r}$ and $Y \triangleq \begin{bmatrix} y_1 \\ \vdots \\ y_r \end{bmatrix} \in \mathbb{F}^{r\times m}$. Furthermore, $\operatorname{rank} X = \operatorname{rank} Y = \operatorname{rank} A$ if and only if $r = \operatorname{rank} A$. **Remark:** The last statement gives a full-rank factorization of A. See Proposition 7.6.6.

Fact 7.17.2. Let $A \in \mathbb{F}^{n\times n}$. Then, A is normal if and only if there exists a unitary matrix $S \in \mathbb{F}^{n\times n}$ such that $A^* = AS$. Now, assume that A is nonsingular. Then, A is normal if and only if $A^{-1}A^*$ is unitary. **Source:** [2263, pp. 102, 113] and [2991, p. 304].

Fact 7.17.3. Let $A \in \mathbb{C}^{n\times n}$. Then, there exists a nonsingular matrix $S \in \mathbb{C}^{n\times n}$ such that SAS^{-1} is symmetric. **Source:** [1448, p. 209]. **Related:** Corollary 7.4.9. **Remark:** The coefficient of the last matrix in [1448, p. 209] should be $\jmath/2$.

Fact 7.17.4. Let $A \in \mathbb{C}^{n\times n}$, and assume that A^2 is normal. Then, the following statements hold:

i) There exists a unitary matrix $S \in \mathbb{C}^{n\times n}$ such that SAS^{-1} is symmetric.

ii) There exists a symmetric unitary matrix $S \in \mathbb{C}^{n\times n}$ such that $A^{\mathsf{T}} = SAS^{-1}$.

Source: [2793].

Fact 7.17.5. Let $A \in \mathbb{F}^{n\times n}$, and assume that A is nonsingular. Then, A^{-1} and A^* are similar if and only if there exists a nonsingular matrix $B \in \mathbb{F}^{n\times n}$ such that $A = B^{-1}B^*$. Furthermore, A is unitary if and only if there exists a normal, nonsingular matrix $B \in \mathbb{F}^{n\times n}$ such that $A = B^{-1}B^*$. **Source:** [876]. Sufficiency in the second statement follows from Fact 4.13.6.

Fact 7.17.6. Let $A, B \in \mathbb{F}^{n\times n}$. Then, the following statements hold:

i) AB and BA have the same nonsingular Jordan blocks.

ii) If either A or B is nonsingular, then AB and BA are similar.

iii) If $\operatorname{rank} AB = \operatorname{rank} BA = \operatorname{rank} A$, then AB and BA are similar.

iv) If A and B are normal, then $\operatorname{rank} AB = \operatorname{rank} BA$.

v) If A and B are normal and $\operatorname{rank} A \in \{0, 1, 2, n\}$, then AB and BA are similar.

vi) A and B are Hermitian, then AB and BA are similar.

vii) If A is normal and B is positive semidefinite, then AB and BA are similar.

viii) A and B are similar if and only if $(\operatorname{rank}\,(AB)^i)_{i=0}^\infty = (\operatorname{rank}\,(BA)^i)_{i=0}^\infty$.

ix) If $A + A^*$ is positive semidefinite, $\operatorname{rank}(A + A^*) = \operatorname{rank} A$, and B is range Hermitian, then AB

and BA are similar.

x) If A and B are normal and either $n \le 2$ or $\operatorname{rank} A \le 1$, then AB and BA are unitarily similar.

Source: [1143].

Fact 7.17.7. Let $A \in \mathbb{F}^{m\times m}$ and $B \in \mathbb{F}^{n\times n}$. Then, there exist $C \in \mathbb{F}^{m\times n}$ and $D \in \mathbb{F}^{n\times m}$ such that $A = CD$ and $B = DC$ if and only if both of the following statements hold:

i) The Jordan blocks associated with nonzero eigenvalues are identical in A and B.

ii) Let $n_1 \ge n_2 \ge \cdots \ge n_r$ denote the sizes of the Jordan blocks of A associated with $0 \in \operatorname{spec}(A)$, and let $m_1 \ge m_2 \ge \cdots \ge m_r$ denote the sizes of the Jordan blocks of B associated with $0 \in \operatorname{spec}(B)$, where $n_i = 0$ or $m_i = 0$ as needed. Then, $|n_i - m_i| \le 1$ for all $i \in \{1, \dots, r\}$.

Source: [1547, 1858]. **Example:** Let $C \triangleq \begin{bmatrix} 0&0&1\\0&0&0\\0&1&0 \end{bmatrix}$ and $D \triangleq \begin{bmatrix} 0&0&1\\1&0&0\\0&0&0 \end{bmatrix}$. Then, $A = CD = \begin{bmatrix} 0&0&0\\0&0&0\\1&0&0 \end{bmatrix}$ and $B = DC = \begin{bmatrix} 0&1&0\\0&0&1\\0&0&0 \end{bmatrix}$. Then, $A^2 = B^3 = 0$, whereas $B^2 \ne 0$. **Related:** Fact 7.17.8.

Fact 7.17.8. Let $A, B \in \mathbb{F}^{n\times n}$, and assume that A and B are nonsingular. Then, A and B are similar if and only if there exist nonsingular matrices $C, D \in \mathbb{F}^{n\times n}$ such that $A = CD$ and $B = DC$. **Source:** Sufficiency follows from Fact 7.11.14. Necessity is a special case of Fact 7.17.7.

Fact 7.17.9. Let $A, B \in \mathbb{F}^{n\times n}$, and assume that A and B are nonsingular. Then, $\det A = \det B$ if and only if there exist nonsingular matrices $C, D, E \in \mathbb{R}^{n\times n}$ such that $A = CDE$ and $B = EDC$. **Credit:** K. Shoda and O. Taussky-Todd. See [569].

Fact 7.17.10. Let $A \in \mathbb{F}^{n\times m}$, and assume that $\operatorname{rank} A = m$. Then, there exist a unique matrix $B \in \mathbb{F}^{n\times m}$ and a matrix $C \in \mathbb{F}^{m\times m}$ such that $B^*B = I_m$, C is upper triangular with positive diagonal entries, and $A = BC$. **Source:** [1448, p. 15] and [2314, p. 206]. **Remark:** $C \in \mathrm{UT}_+(n)$. See Fact 4.31.11. **Remark:** This factorization is a consequence of *Gram-Schmidt orthonormalization.*

Fact 7.17.11. Let $A \in \mathbb{F}^{n\times n}$. Then, there exist $B, C \in \mathbb{F}^{n\times n}$ such that B is unitary, C is upper triangular, and $A = BC$. If, in addition, A is nonsingular, then there exist unique matrices $B, C \in \mathbb{F}^{n\times n}$ such that B is unitary, C is upper triangular with positive diagonal entries, and $A = BC$. **Source:** [1448, p. 112] and [2314, p. 362]. **Remark:** This is the *QR decomposition.* The orthogonal matrix B is constructed as a product of elementary reflectors.

Fact 7.17.12. Let $A \in \mathbb{F}^{n\times n}$, let $r \triangleq \operatorname{rank} A$, and assume that the first r leading principal subdeterminants of A are nonzero. Then, there exist $B, C \in \mathbb{F}^{n\times n}$ such that B is lower triangular, C is upper triangular, and $A = BC$. Either B or C can be chosen to be nonsingular. Furthermore, both B and C are nonsingular if and only if A is nonsingular. **Source:** [1448, p. 160]. **Remark:** This is the *LU decomposition.* **Remark:** All LU factorizations of a singular matrix are characterized in [924]. **Related:** Fact 10.10.42.

Fact 7.17.13. Let $\theta \in (-\pi, \pi)$. Then,

$$\begin{bmatrix} \cos\theta & -\sin\theta \\ \sin\theta & \cos\theta \end{bmatrix} = \begin{bmatrix} 1 & -\tan\frac{\theta}{2} \\ 0 & 1 \end{bmatrix} \begin{bmatrix} 1 & 0 \\ \sin\theta & 1 \end{bmatrix} \begin{bmatrix} 1 & -\tan\frac{\theta}{2} \\ 0 & 1 \end{bmatrix}.$$

Remark: This is a *ULU factorization* involving three *shear factors.* The matrix $-I_2$ requires four shear factors. All shear factors may be different. See [2555, 2684].

Fact 7.17.14. Let $A \in \mathbb{F}^{n\times n}$. Then, A is nonsingular if and only if A is a product of elementary matrices.

Fact 7.17.15. Let $A \in \mathbb{F}^{n\times n}$, assume that A is a projector, and let $r \triangleq \operatorname{rank} A$. Then, there exist nonzero vectors $x_1, \dots, x_{n-r} \in \mathbb{F}^n$ such that $x_i^*x_j = 0$ for all $i \ne j$ and such that

$$A = \prod_{i=1}^{n-r} \left[I - (x_i^*x_i)^{-1}x_ix_i^*\right].$$

Source: A is unitarily similar to $\operatorname{diag}(1,\dots,1,0,\dots,0)$, which is a product of elementary projectors. **Remark:** Every projector is the product of mutually orthogonal elementary projectors.

Fact 7.17.16. Let $A \in \mathbb{F}^{n\times n}$, and assume that $A \neq I$. Then, A is a reflector if and only if there exist a positive integer $m \le n$ and nonzero vectors $x_1,\dots,x_m \in \mathbb{F}^n$ such that $x_i^*x_j = 0$ for all $i \neq j$ and such that

$$A = \prod_{i=1}^{m}\left[I - 2(x_i^*x_i)^{-1}x_ix_i^*\right].$$

If these conditions hold, then m is the algebraic multiplicity of $-1 \in \operatorname{spec}(A)$. **Source:** A is unitarily similar to $\operatorname{diag}(\pm 1,\dots,\pm 1)$, which can be written as the product of elementary reflectors. **Remark:** Every reflector is the product of mutually orthogonal elementary reflectors.

Fact 7.17.17. Let $A \in \mathbb{R}^{n\times n}$. Then, A is orthogonal if and only if there exist a positive integer m and nonzero vectors $x_1,\dots,x_m \in \mathbb{R}^n$ such that $\det A = (-1)^m$ and

$$A = \prod_{i=1}^{m}\left[I - 2(x_i^{\mathsf T}x_i)^{-1}x_ix_i^{\mathsf T}\right].$$

If these conditions hold, then the minimal number of factors is $\operatorname{rank}(A-I)$. **Source:** [218, p. 24] and [2388, 2761]. The minimal number of factors is given in [2222, p. 143]. **Remark:** Every orthogonal matrix is the product of elementary reflectors. **Credit:** E. Cartan and J. Dieudonné. **Related:** Fact 4.11.5 and Fact 4.19.4. **Problem:** Extend this result to complex matrices.

Fact 7.17.18. Let $A \in \mathbb{R}^{n\times n}$, where $n \ge 2$. Then, A is orthogonal and $\det A = 1$ if and only if there exist a positive integer m such that $1 \le m \le n(n-1)/2$, $\theta_1,\dots,\theta_m \in \mathbb{R}$, and $j_1,\dots,j_m$, $k_1,\dots,k_m \in \{1,\dots,n\}$ such that

$$A = \prod_{i=1}^{m} P(\theta_i, j_i, k_i),$$

where

$$P(\theta, j, k) \triangleq I_n + [(\cos\theta) - 1](E_{j,j} + E_{k,k}) + (\sin\theta)(E_{j,k} - E_{k,j}).$$

Source: [1018]. **Remark:** $P(\theta, j, k)$ is a *plane* or *Givens rotation*. See Fact 4.11.5. **Remark:** Assume that $\det A = -1$, and let $B \in \mathbb{R}^{n\times n}$ be an elementary reflector. Then, $AB \in \mathrm{SO}(n)$. Therefore, the factorization given above holds with an additional elementary reflector. **Remark:** Related results are given in [1784]. **Problem:** Extend this result to complex matrices.

Fact 7.17.19. Let $A \in \mathbb{F}^{n\times n}$. Then, $A^{2*}A = A^*A^2$ if and only if there exist a projector $B \in \mathbb{F}^{n\times n}$ and a Hermitian matrix $C \in \mathbb{F}^{n\times n}$ such that $A = BC$. **Source:** [2291].

Fact 7.17.20. Let $A \in \mathbb{R}^{n\times n}$. Then, $|\det A| = 1$ if and only if A is the product of $n+2$ or fewer involutory matrices that have exactly one negative eigenvalue. In addition, the following statements hold:

i) If $n = 2$, then 3 or fewer factors are needed.

ii) If $A \neq \alpha I$ for all $\alpha \in \mathbb{R}$ and $\det A = (-1)^n$, then n or fewer factors are needed.

iii) If $\det A = (-1)^{n+1}$, then $n+1$ or fewer factors are needed.

Source: [656, 2289]. **Remark:** See [915] for the minimal number of factors for a unitary matrix.

Fact 7.17.21. Let $A \in \mathbb{C}^{n\times n}$, and, for all $k \in \mathbb{P}$, define $r_0 \triangleq n$ and $r_k \triangleq \operatorname{rank} A^k$. Then, the following statements are equivalent:

i) There exists $B \in \mathbb{C}^{n\times n}$ such that $A = B^2$.

ii) The sequence $(r_k - r_{k+1})_{k=0}^{\infty}$ does not contain two components that are the same odd integer and, if $r_0 - r_1$ is odd, then $1 + 2r_1 \le r_0 + r_2$.

Now, assume that $A \in \mathbb{R}^{n\times n}$. Then, there exists $B \in \mathbb{R}^{n\times n}$ such that $A = B^2$ if and only if *ii)* holds and, for every negative eigenvalue λ of A and for every positive integer k, the Jordan form of A has an even number of $k \times k$ blocks associated with λ. **Source:** [1450, p. 472]. **Remark:** For all $l \geq 2$, $A \triangleq N_l$ does not have a square root. **Remark:** Uniqueness is discussed in [1545]. Square roots of A that are functions of A are defined in [1386]. **Remark:** The principal square root is considered in Theorem 12.9.1. **Remark:** mth roots are considered in [730, 1391, 2270, 2597]. **Related:** Fact 15.19.36.

Fact 7.17.22. Let $A \in \mathbb{C}^{n\times n}$, and assume that A is group invertible. Then, there exists $B \in \mathbb{C}^{n\times n}$ such that $A = B^2$. **Source:** Fact 7.17.21 and [795, 2270].

Fact 7.17.23. Let $A \in \mathbb{F}^{n\times n}$, and assume that A is nonsingular and has no negative eigenvalues. Furthermore, define $(P_k)_{k=0}^{\infty} \subset \mathbb{F}^{n\times n}$ and $(Q_k)_{k=0}^{\infty} \subset \mathbb{F}^{n\times n}$ by $P_0 \triangleq A$, and $Q_0 \triangleq I$, and, for all $k \geq 1$,

$$P_{k+1} \triangleq \tfrac{1}{2}(P_k + Q_k^{-1}), \quad Q_{k+1} \triangleq \tfrac{1}{2}(Q_k + P_k^{-1}).$$

Then, $B \triangleq \lim_{k\to\infty} P_k$ exists, satisfies $B^2 = A$, and is the unique square root of A satisfying $\operatorname{spec}(B) \subset$ ORHP. Furthermore,

$$\lim_{k\to\infty} Q_k = B^{-1}.$$

Source: [875, 1385]. **Remark:** All indicated inverses exist. **Remark:** This sequence is related to Newton's iteration for the matrix sign function. See Fact 12.15.2. **Remark:** An alternative algorithm for 3×3 matrices is given in [175]. **Related:** Fact 10.10.37.

Fact 7.17.24. Let $A \in \mathbb{F}^{n\times n}$, assume that A is positive semidefinite, and let $r \triangleq \operatorname{rank} A$. Then, there exists $B \in \mathbb{F}^{n\times r}$ such that $A = BB^*$.

Fact 7.17.25. Let $A \in \mathbb{F}^{n\times n}$, and let $k \geq 1$. Then, there exists a unique matrix $B \in \mathbb{F}^{n\times n}$ such that $A = B(B^*B)^k$. **Source:** [2250].

Fact 7.17.26. Let $A \in \mathbb{F}^{n\times n}$. Then, there exist symmetric matrices $B, C \in \mathbb{F}^{n\times n}$, at least one of which is nonsingular, such that $A = BC$. **Source:** [2263, p. 82]. **Remark:** Note that

$$\begin{bmatrix} 0 & 1 & 0 \\ 0 & 0 & 1 \\ -\beta_0 & -\beta_1 & -\beta_2 \end{bmatrix} = \begin{bmatrix} \beta_1 & \beta_2 & 1 \\ \beta_2 & 1 & 0 \\ 1 & 0 & 0 \end{bmatrix}^{-1} \begin{bmatrix} -\beta_0 & 0 & 0 \\ 0 & \beta_2 & 1 \\ 0 & 1 & 0 \end{bmatrix}$$

and use Theorem 7.3.3. **Remark:** B and C are symmetric for $\mathbb{F} = \mathbb{C}$. **Remark:** The equality is a *Bezout matrix factorization*; see Fact 6.8.8. See [527, 528, 1303]. **Credit:** F. G. Frobenius. **Related:** Corollary 6.3.12, Corollary 7.4.9, Proposition 7.7.13, and Fact 7.10.9.

Fact 7.17.27. Let $A \in \mathbb{C}^{n\times n}$. Then, $\det A$ is real if and only if A is the product of four Hermitian matrices. Furthermore, four is the smallest number for which the previous statement is true. **Source:** [2919].

Fact 7.17.28. Let $A \in \mathbb{R}^{n\times n}$. Then, the following statements hold:

i) A is the product of two positive-semidefinite matrices if and only if A is similar to a positive-semidefinite matrix.

ii) If A is nilpotent, then A is the product of three positive-semidefinite matrices.

iii) If A is singular, then A is the product of four positive-semidefinite matrices.

iv) $\det A \geq 0$ if and only if A is the product of five positive-semidefinite matrices.

v) $\det A > 0$ and $A \neq \alpha I$ for all $\alpha \leq 0$ if and only if A is the product of four positive-definite matrices.

vi) $\det A > 0$ if and only if A is the product of five positive-definite matrices.

Source: [263, 797, 1303, 2918, 2919]. **Remark:** See [797, 2919] for extensions to complex matrices. **Example:**

$$\begin{bmatrix} -1 & 0 \\ 0 & -1 \end{bmatrix} = \begin{bmatrix} 2 & 0 \\ 0 & 1/2 \end{bmatrix}\begin{bmatrix} 5 & 7 \\ 7 & 10 \end{bmatrix}\begin{bmatrix} 13/2 & -5 \\ -5 & 4 \end{bmatrix}\begin{bmatrix} 8 & 5 \\ 5 & 13/4 \end{bmatrix}\begin{bmatrix} 25/8 & -11/2 \\ -11/2 & 10 \end{bmatrix}.$$

Fact 7.17.29. Let $A \in \mathbb{R}^{n\times n}$. Then, the following statements hold:

i) $A = BC$, where $B \in \mathbb{R}^{n\times n}$ is symmetric and $C \in \mathbb{R}^{n\times n}$ is positive semidefinite, if and only if A^2 is diagonalizable over $\mathbb{R}$ and $\operatorname{spec}(A) \subset [0,\infty)$.

ii) $A = BC$, where $B \in \mathbb{R}^{n\times n}$ is symmetric and $C \in \mathbb{R}^{n\times n}$ is positive definite, if and only if A is diagonalizable over $\mathbb{R}$.

iii) $A = BC$, where $B, C \in \mathbb{R}^{n\times n}$ are positive semidefinite, if and only if $A = DE$, where $D \in \mathbb{R}^{n\times n}$ is positive semidefinite and $E \in \mathbb{R}^{n\times n}$ is positive definite.

iv) $A = BC$, where $B \in \mathbb{R}^{n\times n}$ is positive semidefinite and $C \in \mathbb{R}^{n\times n}$ is positive definite, if and only if A is diagonalizable over $\mathbb{R}$ and $\operatorname{spec}(A) \subset [0,\infty)$.

v) $A = BC$, where $B, C \in \mathbb{R}^{n\times n}$ are positive definite, if and only if A is diagonalizable over $\mathbb{R}$ and $\operatorname{spec}(A) \subset (0,\infty)$.

Source: [797, 1441, 2910, 2918].

Fact 7.17.30. Let $A \in \mathbb{F}^{n\times n}$. Then, A is either singular or I_n if and only if A is the product of n or fewer idempotent matrices in $\mathbb{F}^{n\times n}$, each of whose rank is equal to $\operatorname{rank} A$. Furthermore, $\operatorname{rank}(A-I) \le k \operatorname{def} A$, where $k \ge 1$, if and only if A is the product of k idempotent matrices. **Source:** [165, 271, 841, 999]. **Example:**

$$\begin{bmatrix} 0 & 1 \\ 0 & 0 \end{bmatrix} = \begin{bmatrix} 1 & 1/2 \\ 0 & 0 \end{bmatrix}\begin{bmatrix} 0 & 1/2 \\ 0 & 1 \end{bmatrix}, \quad \begin{bmatrix} 2 & 0 \\ 0 & 0 \end{bmatrix} = \begin{bmatrix} 1 & 1 \\ 0 & 0 \end{bmatrix}\begin{bmatrix} 1 & 0 \\ 1 & 0 \end{bmatrix}.$$

Fact 7.17.31. Let $A \in \mathbb{R}^{n\times n}$, and assume that A is singular and is not a 2×2 nilpotent matrix. Then, there exist nilpotent matrices $B, C \in \mathbb{R}^{n\times n}$ such that $A = BC$ and $\operatorname{rank} A = \operatorname{rank} B = \operatorname{rank} C$. **Source:** [2499, 2917]. See also [2567].

Fact 7.17.32. Let $A \in \mathbb{F}^{n\times n}$, define $r \triangleq \operatorname{rank} A$, let $B \in \mathbb{F}^{n\times r}$, let $C \in \mathbb{F}^{r\times n}$, and assume that $A = BC$. Then, A is idempotent if and only if $CB = I$. **Source:** [1275, p. 16].

Fact 7.17.33. Let $A \in \mathbb{F}^{n\times n}$, and assume that A is idempotent. Then, there exist $B, C \in \mathbb{F}^{n\times n}$ such that B is positive definite, C is positive semidefinite, and $A = BC$. **Source:** *iv)* of Fact 7.17.29 and [2708].

Fact 7.17.34. Let $A \in \mathbb{R}^{n\times n}$, and assume that A is nonsingular. Then, A is similar to A^{-1} if and only if A is the product of two involutory matrices. If, in addition, A is orthogonal, then A is the product of two reflectors. **Source:** [269, 912, 2908, 2909] and [2263, p. 108]. **Example:** $\begin{bmatrix} \cos\theta & \sin\theta \\ -\sin\theta & \cos\theta \end{bmatrix} = \begin{bmatrix} 1 & 0 \\ 0 & -1 \end{bmatrix}\begin{bmatrix} \cos\theta & \sin\theta \\ \sin\theta & -\cos\theta \end{bmatrix}$.

Fact 7.17.35. Let $A \in \mathbb{R}^{n\times n}$. Then, $|\det A| = 1$ if and only if A is the product of four or fewer involutory matrices. **Source:** [270, 1281, 2498].

Fact 7.17.36. Let $A \in \mathbb{R}^{n\times n}$, where $n \ge 2$. Then, A is the product of two commutators. **Source:** [2919].

Fact 7.17.37. Let $A \in \mathbb{R}^{n\times n}$, and assume that $\det A = 1$. Then, there exist nonsingular matrices $B, C \in \mathbb{R}^{n\times n}$ such that $A = BCB^{-1}C^{-1}$. **Source:** [2430]. **Remark:** This is a *multiplicative commutator*. **Remark:** For nonsingular A, B, note that $[A, B] = 0$ if and only if $ABA^{-1}B^{-1} = I$. **Credit:** K. Shoda. **Related:** Fact 7.17.38.

Fact 7.17.38. Let $A \in \mathbb{R}^{n\times n}$, assume that A is orthogonal, and assume that $\det A = 1$. Then, there exist reflectors $B, C \in \mathbb{R}^{n\times n}$ such that $A = BCB^{-1}C^{-1}$. **Source:** [2610]. **Related:** Fact 7.17.37.

Fact 7.17.39. Let $A \in \mathbb{F}^{n\times n}$, and assume that A is nonsingular. Then, there exist an involutory matrix $B \in \mathbb{F}^{n\times n}$ and a symmetric matrix $C \in \mathbb{F}^{n\times n}$ such that $A = BC$. **Source:** [1215].

Fact 7.17.40. Let $A \in \mathbb{F}^{n\times n}$, and assume that n is even. Then, the following statements are equivalent:

i) A is the product of two skew-symmetric matrices.

ii) Every elementary divisor of A has even algebraic multiplicity.

iii) There exists $B \in \mathbb{F}^{n/2\times n/2}$ such that A is similar to $\begin{bmatrix} B & 0 \\ 0 & B \end{bmatrix}$.

Remark: In *i)* the factors are skew symmetric in the case where A is complex. **Source:** [1216, 2919].

Fact 7.17.41. Let $A \in \mathbb{C}^{n\times n}$, and assume that $n \ge 4$ and n is even. Then, A is the product of five skew-symmetric matrices in $\mathbb{C}^{n\times n}$. **Source:** [1720, 1721].

Fact 7.17.42. Let $A \in \mathbb{F}^{n\times n}$. Then, there exist a symmetric matrix $B \in \mathbb{F}^{n\times n}$ and a skew-symmetric matrix $C \in \mathbb{F}^{n\times n}$ such that $A = BC$ if and only if A is similar to $-A$. **Source:** [2322].

Fact 7.17.43. Let $A \in \mathbb{F}^{n\times n}$. Then, A is diagonalizable over $\mathbb{F}$ with (nonnegative, positive) eigenvalues if and only if there exist (positive-semidefinite, positive-definite) matrices $B, C \in \mathbb{F}^{n\times n}$ such that $A = BC$. **Source:** To prove sufficiency, use Theorem 10.3.6 and $A = S^{-1}(SBS^*)(S^{-*}CS^{-1})S$.

Fact 7.17.44. Let $A \in \mathbb{R}^{n\times n}$, and assume that A is a permutation matrix. Then, there exist symmetric permutation matrices $A_1, \ldots, A_{n-1} \in \mathbb{R}^{n\times n}$ such that $A = A_1 \cdots A_{n-1}$. **Source:** [2991, p. 163].

Fact 7.17.45. Let $A \in \mathbb{F}^{n\times n}$. Then, there exist Toeplitz matrices $A_1, \ldots, A_{2n+5} \in \mathbb{F}^{n\times n}$ and Hankel matrices $B_1, \ldots, B_{2n+5} \in \mathbb{F}^{n\times n}$ such that $A = A_1 \cdots A_{2n+5} = B_1 \cdots B_{2n+5}$. **Source:** [2956].

7.18 Facts on Companion, Vandermonde, Circulant, Permutation, and Hadamard Matrices

Fact 7.18.1. Define $A \in \mathbb{F}^{n\times n}$ by

$$A \triangleq \begin{bmatrix} 0 & \alpha_1 & 0 & \cdots & 0 & 0 \\ 0 & 0 & \alpha_2 & \ddots & 0 & 0 \\ 0 & 0 & 0 & \ddots & 0 & 0 \\ \vdots & \vdots & \vdots & \ddots & \ddots & \vdots \\ 0 & 0 & 0 & \cdots & 0 & \alpha_{n-1} \\ -\beta_0 & -\beta_1 & -\beta_2 & \cdots & -\beta_{n-2} & -\beta_{n-1} \end{bmatrix},$$

and let $\chi_A(s) = s^n + a_{n-1}s^{n-1} + \cdots + a_1 s + a_0$. Then,

$$a_{n-1} = \beta_{n-1}, \quad a_{n-2} = \alpha_{n-1}\beta_{n-2}, \quad a_{n-3} = \alpha_{n-1}\alpha_{n-2}\beta_{n-3}, \quad \ldots, \quad a_0 = \alpha_{n-1}\alpha_{n-2}\cdots\alpha_1\beta_0.$$

Now, assume that $\alpha_1, \ldots, \alpha_{n-1}$ are nonzero, and define $S \triangleq \operatorname{diag}(1, \alpha_1, \alpha_1\alpha_2 \ldots, \alpha_1\alpha_2\cdots\alpha_{n-1})$. Then, $C(\chi_A) = SAS^{-1}$. **Source:** [300, pp. 370, 371] and [2966]. **Remark:** A is a *Leslie matrix*, which has applications to population biology in the case where A is nonnegative. See [655, pp. 8–10]. In [2966], A is called a *generalized Frobenius companion matrix.*

Fact 7.18.2. Let $\alpha_1, \ldots, \alpha_n, \beta_1, \ldots, \beta_n, \gamma_1, \ldots, \gamma_n, a_0, \ldots, a_{n-1} \in \mathbb{F}$, assume that $\alpha_1, \ldots, \alpha_n$ are nonzero, and define $A \in \mathbb{F}^{n\times n}$ by

$$A \triangleq \begin{bmatrix} -\frac{\beta_1}{\alpha_1} & \frac{1}{\alpha_1} & 0 & 0 & \cdots & 0 & 0 \\ \frac{\gamma_2}{\alpha_2} & -\frac{\beta_2}{\alpha_2} & \frac{1}{\alpha_2} & 0 & \cdots & 0 & 0 \\ 0 & \frac{\gamma_3}{\alpha_3} & -\frac{\beta_3}{\alpha_3} & \frac{1}{\alpha_3} & \ddots & 0 & 0 \\ \vdots & \vdots & \ddots & \ddots & \ddots & \ddots & \vdots \\ \vdots & \vdots & \ddots & \ddots & \ddots & \ddots & \vdots \\ 0 & 0 & 0 & \cdots & \frac{\gamma_{n-1}}{\alpha_{n-1}} & -\frac{\beta_{n-1}}{\alpha_{n-1}} & \frac{1}{\alpha_{n-1}} \\ -\frac{a_0}{\alpha_n} & -\frac{a_1}{\alpha_n} & -\frac{a_2}{\alpha_n} & \cdots & -\frac{a_{n-3}}{\alpha_n} & \frac{\gamma_n - a_{n-2}}{\alpha_n} & -\frac{\beta_n + a_{n-1}}{\alpha_n} \end{bmatrix}.$$

Furthermore, define $p_0, \ldots, p_n \in \mathbb{F}[s]$ by $p(s) \triangleq 1$, $p_1(s) \triangleq \alpha_1 s + \beta_2$, and, for all $i \in \{2, \ldots, n\}$, $p_i(s) \triangleq (\alpha_i s + \beta_i)p_{i-1}(s) - \gamma_i p_{i-2}(s)$. Then,

$$\chi_A(s) = \left(\frac{1}{\prod_{i=1}^{n} \alpha_i}\right)[p_n(s) + a_{n-1}p_{n-1}(s) + \cdots + a_1 p_1(s) + a_0 p_0(s)].$$

Source: [300, pp. 369–371]. **Remark:** A is a *comrade matrix*. Setting $\alpha_i = 1$, $\beta_i = 0$, and $\gamma_i = 0$ for all $i \in \{1, \ldots, n\}$, yields a companion matrix. Special cases include Leslie, Schwarz, and Routh matrices given, respectively, by Fact 7.18.1, Fact 15.19.25, and Fact 15.19.27. **Remark:** Choosing $\alpha_i = 1$, $\beta_i = 0$, and $\gamma_i = 1$ for all $i \in \{1, \ldots, n\}$ yields a *colleague matrix*. **Remark:** Chebyshev, Legendre, Laguerre, and Hermite polynomials can be obtained with suitable choices of α_i, β_i, and γ_i. **Remark:** An alternative approach to forming the characteristic polynomial is given by the *confederate matrix*, which is a lower Hessenberg matrix. Companion, confederate, comrade, and colleague matrices are *congenial* matrices. See [295, 297, 300].

Fact 7.18.3. Let $p \in \mathbb{F}[s]$, where $p(s) = s^n + \beta_{n-1}s^{n-1} + \cdots + \beta_0$, and define $C_{\mathrm{b}}(p), C_{\mathrm{r}}(p), C_{\mathrm{t}}(p), C_{\mathrm{l}}(p) \in \mathbb{F}^{n\times n}$ by

$$C_{\mathrm{b}}(p) \triangleq \begin{bmatrix} 0 & 1 & 0 & \cdots & 0 & 0 \\ 0 & 0 & 1 & \ddots & 0 & 0 \\ 0 & 0 & 0 & \ddots & 0 & 0 \\ \vdots & \vdots & \vdots & \ddots & \ddots & \vdots \\ 0 & 0 & 0 & \cdots & 0 & 1 \\ -\beta_0 & -\beta_1 & -\beta_2 & \cdots & -\beta_{n-2} & -\beta_{n-1} \end{bmatrix}, \quad C_{\mathrm{r}}(p) \triangleq \begin{bmatrix} 0 & 0 & 0 & \cdots & 0 & -\beta_0 \\ 1 & 0 & 0 & \cdots & 0 & -\beta_1 \\ 0 & 1 & 0 & \cdots & 0 & -\beta_2 \\ \vdots & \ddots & \ddots & \ddots & \vdots & \vdots \\ 0 & 0 & 0 & \ddots & 0 & -\beta_{n-2} \\ 0 & 0 & 0 & \cdots & 1 & -\beta_{n-1} \end{bmatrix},$$

$$C_{\rm t}(p) \triangleq \begin{bmatrix} -\beta_{n-1} & -\beta_{n-2} & \cdots & -\beta_2 & -\beta_1 & -\beta_0 \\ 1 & 0 & \cdots & 0 & 0 & 0 \\ \vdots & \ddots & \ddots & \vdots & \vdots & \vdots \\ 0 & 0 & \ddots & 0 & 0 & 0 \\ 0 & 0 & \ddots & 1 & 0 & 0 \\ 0 & 0 & \cdots & 0 & 1 & 0 \end{bmatrix}, \quad C_{\rm l}(p) \triangleq \begin{bmatrix} -\beta_{n-1} & 1 & \cdots & 0 & 0 & 0 \\ -\beta_{n-2} & 0 & \ddots & 0 & 0 & 0 \\ \vdots & \vdots & \ddots & \ddots & \ddots & \vdots \\ -\beta_2 & 0 & \cdots & 0 & 1 & 0 \\ -\beta_1 & 0 & \cdots & 0 & 0 & 1 \\ -\beta_0 & 0 & \cdots & 0 & 0 & 0 \end{bmatrix}.$$

Then,

$$C_{\rm r}(p) = C_{\rm b}^{\sf T}(p), \quad C_{\rm l}(p) = C_{\rm t}^{\sf T}(p),$$
$$C_{\rm t}(p) = \hat{I}C_{\rm b}(p)\hat{I}, \quad C_{\rm l}(p) = \hat{I}C_{\rm r}(p)\hat{I},$$
$$C_{\rm l}(p) = C_{\rm b}^{\hat{\sf T}}(p), \quad C_{\rm t}(p) = C_{\rm r}^{\hat{\sf T}}(p),$$
$$\chi_{C_{\rm b}(p)} = \chi_{C_{\rm r}(p)} = \chi_{C_{\rm t}(p)} = \chi_{C_{\rm l}(p)} = p.$$

Furthermore,

$$C_{\rm r}(p) = SC_{\rm b}(p)S^{-1}, \quad C_{\rm l}(p) = \hat{S}C_{\rm t}(p)\hat{S}^{-1},$$

where $S, \hat{S} \in \mathbb{F}^{n\times n}$ are the Hankel matrices

$$S \triangleq \begin{bmatrix} \beta_1 & \beta_2 & \cdots & \beta_{n-1} & 1 \\ \beta_2 & \beta_3 & .\cdot^{\cdot} & 1 & 0 \\ \vdots & .\cdot^{\cdot} & .\cdot^{\cdot} & .\cdot^{\cdot} & \vdots \\ \beta_{n-1} & 1 & .\cdot^{\cdot} & 0 & 0 \\ 1 & 0 & \cdots & 0 & 0 \end{bmatrix}, \quad \hat{S} \triangleq \hat{I}S\hat{I} = \begin{bmatrix} 0 & 0 & \cdots & 0 & 1 \\ 0 & 0 & .\cdot^{\cdot} & 1 & \beta_{n-1} \\ \vdots & .\cdot^{\cdot} & .\cdot^{\cdot} & .\cdot^{\cdot} & \vdots \\ 0 & 1 & .\cdot^{\cdot} & \beta_3 & \beta_2 \\ 1 & \beta_{n-1} & \cdots & \beta_2 & \beta_1 \end{bmatrix}.$$

Finally, defining $\beta_n \triangleq 1$,

$$\sum_{i=0}^{n-1} \beta_{i+1} \sum_{j=0}^{i} C_{\rm r}(p)^j E_{1,n} C_{\rm r}(p)^{i-j} = I.$$

Source: The last statement is given in [1815]. **Remark:** $(C_{\rm b}(p), C_{\rm r}(p), C_{\rm t}(p), C_{\rm l}(p))$ are the (*bottom, right, top, left*) companion matrices. Note that $C_{\rm b}(p) = C(p)$. See [300, p. 282] and [1573, p. 659]. **Remark:** $S = B(p, 1)$, where $B(p, 1)$ is a Bezout matrix. See Fact 6.8.8.

Fact 7.18.4. Let $p \in \mathbb{F}[s]$, where $p(s) = s^n + \beta_{n-1}s^{n-1} + \cdots + \beta_0$, assume that $\beta_0 \neq 0$, and define $C_{\rm b}(p)$ as in Fact 7.18.3. Then,

$$C_{\rm b}^{-1}(p) = C_{\rm t}(\hat{p}) = \begin{bmatrix} -\beta_1/\beta_0 & \cdots & -\beta_{n-2}/\beta_0 & -\beta_{n-1}/\beta_0 & -1/\beta_0 \\ 1 & \cdots & 0 & 0 & 0 \\ \vdots & \ddots & \vdots & \vdots & \vdots \\ 0 & \cdots & 1 & 0 & 0 \\ 0 & \cdots & 0 & 1 & 0 \end{bmatrix},$$

where $\hat{p}(s) \triangleq \beta_0^{-1}s^n p(1/s)$. **Related:** Fact 6.9.5.

Fact 7.18.5. Let $n \ge 2$, let $\lambda_1, \ldots, \lambda_n \in \mathbb{F}$, and define the *Vandermonde matrix* $V(\lambda_1, \ldots, \lambda_n) \in \mathbb{F}^{n \times n}$ by

$$V(\lambda_1, \ldots, \lambda_n) \triangleq \begin{bmatrix} 1 & 1 & \cdots & 1 \\ \lambda_1 & \lambda_2 & \cdots & \lambda_n \\ \lambda_1^2 & \lambda_2^2 & \cdots & \lambda_n^2 \\ \lambda_1^3 & \lambda_2^3 & \cdots & \lambda_n^3 \\ \vdots & \vdots & \iddots & \vdots \\ \lambda_1^{n-1} & \lambda_2^{n-1} & \cdots & \lambda_n^{n-1} \end{bmatrix}.$$

Then,

$$\det V(\lambda_1, \ldots, \lambda_n) = \prod_{1 \le i < j \le n} (\lambda_j - \lambda_i) = \prod_{i=1}^{n} \prod_{j=i+1}^{n} (\lambda_j - \lambda_i).$$

Thus, $V(\lambda_1, \ldots, \lambda_n)$ is nonsingular if and only if $\lambda_1, \ldots, \lambda_n$ are distinct. **Remark:** This result yields Proposition 6.5.4. Let $x_1, \ldots, x_k$ be eigenvectors of $V(\lambda_1, \ldots, \lambda_n)$ associated with distinct eigenvalues $\lambda_1, \ldots, \lambda_k$ of $V(\lambda_1, \ldots, \lambda_n)$. Suppose that $\alpha_1 x_1 + \cdots + \alpha_k x_k = 0$ so that $V^i(\lambda_1, \ldots, \lambda_n)(\alpha_1 x_1 + \cdots + \alpha_k x_k) = \alpha_1 \lambda_1^i x_i + \cdots + \alpha_k \lambda_k^i x_k = 0$ for all $i \in \{0, 1, \ldots, k-1\}$. Let $X \triangleq [x_1 \ \cdots \ x_k] \in \mathbb{F}^{n \times k}$ and $D \triangleq \operatorname{diag}(\alpha_1, \ldots, \alpha_k)$. Then, $XDV^{\mathsf{T}}(\lambda_1, \ldots, \lambda_k) = 0$, which implies that $XD = 0$. Hence, $\alpha_i x_i = 0$ for all $i \in \{1, \ldots, k\}$, and thus $\alpha_1 = \cdots = \alpha_k = 0$. **Remark:** Connections between the Vandermonde matrix and the Pascal matrix, *Stirling matrix*, *Bernoulli matrix*, *Bernstein matrix*, and companion matrices are discussed in [9]. See also Fact 15.12.4.

Fact 7.18.6. Let $n \ge 2$, define $V(1, \ldots, n)$ by Fact 7.18.5, and define $v_n \triangleq \det V(1, \ldots, n)$. Then,

$$v_n = \prod_{i=1}^{n-1} i!.$$

In particular, $(v_i)_{i=2}^{8} = (1, 2, 12, 288, 34560, 24883200, 125411328000)$. Furthermore, let $k_1 < \cdots < k_n$ be integers. Then, v_n divides $\det V(k_1, \ldots, k_n)$ **Source:** [2991, p. 145]. **Remark:** v_n is a *superfactorial*. See [14]. **Example:** $[\det V(-1, 4, 7)]/v_3 = 60$ and $[\det V(-4, -1, 5, 8)]/v_4 = 4374$.

Fact 7.18.7. Let $n \ge 2$ and $1 \le k \le n$, and define $V(1, \ldots, n)$ by Fact 7.18.5. Then,

$$\begin{bmatrix} n \\ k \end{bmatrix} = \frac{\det V(1, \ldots, n)_{[k,n]}}{\det V(1, \ldots, n-1)}, \quad \begin{Bmatrix} n \\ k \end{Bmatrix} = \frac{\det V(1, \ldots, n)_{[\{k, \ldots, n-1\}, \{k+1, \ldots, n\}]}}{\det V(1, \ldots, k)}.$$

Source: [771, p. 227].

Fact 7.18.8. Let n and k be positive integers, let $a, b \in \mathbb{R}^n$, and define $A, B \in \mathbb{R}^{n \times n}$ by $A \triangleq a 1_{1 \times n}$ and $B \triangleq 1_{n \times 1} b^{\mathsf{T}}$. Then,

$$\det (A + B)^{\odot k} = \begin{cases} (-1)^{\lfloor n/2 \rfloor} \left[\prod_{i=1}^{k} \binom{k}{i} \right] \prod_{1 \le i < j \le n} (a_{(j)} - a_{(i)})(b_{(j)} - b_{(i)}), & k = n-1, \\ 0, & k \le n-2. \end{cases}$$

Source: [2943]. **Example:** For $A = \begin{bmatrix} 1 & 1 & 1 \\ 2 & 2 & 2 \\ 3 & 3 & 3 \end{bmatrix}$ and $B = \begin{bmatrix} 4 & 5 & 6 \\ 4 & 5 & 6 \\ 4 & 5 & 6 \end{bmatrix}$, $\det(A + B) = 0$ and $\det (A + B)^{\odot 2} = -8$.

Fact 7.18.9. Let $p \in \mathbb{F}[s]$, where $p(s) = s^n + \beta_{n-1} s^{n-1} + \cdots + \beta_1 s + \beta_0$, assume that p has distinct roots $\lambda_1, \ldots, \lambda_n \in \mathbb{C}$, and define $V \triangleq V(\lambda_1, \ldots, \lambda_n)$. Then,

$$C(p) = V \operatorname{diag}(\lambda_1, \ldots, \lambda_n) V^{-1}.$$

Consequently, for all $i \in \{1, \ldots, n\}$, λ_i is an eigenvalue of $C(p)$ with associated eigenvector $\operatorname{col}_i(V)$. Finally,

$$(VV^{\mathsf{T}})^{-1} C V V^{\mathsf{T}} = C^{\mathsf{T}}.$$

Source: [295]. **Remark:** The case where $C(p)$ has repeated eigenvalues is considered in [295].

Fact 7.18.10. Let $a_1,\dots,a_n$ be distinct complex numbers, let x be a nonzero complex number, and define $A \triangleq V(xa_1,\dots,xa_n)[V(a_1,\dots,a_n)]^{-1}$. Then, $\operatorname{mspec}(A) = \{x^{n-1}, x^{n-2},\dots,x,1\}_{\rm ms}$. **Source:** [801].

Fact 7.18.11. Let $A \in \mathbb{F}^{n\times n}$. Then, A is cyclic if and only if A is similar to a companion matrix. **Source:** This result follows from Corollary 7.4.3. Alternatively, let $\operatorname{spec}(A) = \{\lambda_1,\dots,\lambda_r\}$ and $A = SBS^{-1}$, where $S \in \mathbb{C}^{n\times n}$ is nonsingular and $B = \operatorname{diag}(B_1,\dots,B_r)$ is the Jordan form of A, where, for all $i \in \{1,\dots,r\}$, $B_i \in \mathbb{C}^{n_i\times n_i}$ and $\lambda_i,\dots,\lambda_i$ are the diagonal entries of B_i. Now, define $R \in \mathbb{C}^{n\times n}$ by $R \triangleq [R_1 \ \cdots \ R_r] \in \mathbb{C}^{n\times n}$, where, for all $i \in \{1,\dots,r\}$, $R_i \in \mathbb{C}^{n\times n_i}$ is the matrix

$$R_i \triangleq \begin{bmatrix} 1 & 0 & \cdots & 0 \\ \lambda_i & 1 & \cdots & 0 \\ \vdots & \vdots & .\cdot\cdot^{\cdot}\!\!. & \vdots \\ \lambda_i^{n-2} & \binom{n-2}{1}\lambda_i^{n-3} & \cdots & \binom{n-2}{n_i-1}\lambda_i^{n-n_i-1} \\ \lambda_i^{n-1} & \binom{n-1}{1}\lambda_i^{n-2} & \cdots & \binom{n-1}{n_i-1}\lambda_i^{n-n_i} \end{bmatrix}.$$

Then, since $\lambda_1,\dots,\lambda_r$ are distinct, it follows that R is nonsingular. Furthermore, $C = RBR^{-1}$ is in companion form, and thus $A = SR^{-1}CRS$. If $n_i = 1$ for all $i \in \{1,\dots,r\}$, then R is a Vandermonde matrix. See Fact 7.18.5 and Fact 7.18.9.

Fact 7.18.12. Let $\lambda_1,\dots,\lambda_n \in \mathbb{F}$ and, for all $i \in \{1,\dots,n\}$, define

$$p_i(s) \triangleq \prod_{\substack{j=1\\ j\neq i}}^{n}(s-\lambda_j).$$

Furthermore, define $A \in \mathbb{F}^{n\times n}$ by

$$A \triangleq \begin{bmatrix} p_1(0) & \frac{1}{1!}p_1'(0) & \cdots & \frac{1}{(n-1)!}p_1^{(n-1)}(0) \\ \vdots & .\cdot\cdot^{\cdot}\!\!. & .\cdot\cdot^{\cdot}\!\!. & \vdots \\ p_n(0) & \frac{1}{1!}p_n'(0) & \cdots & \frac{1}{(n-1)!}p_n^{(n-1)}(0) \end{bmatrix}.$$

Then,

$$\operatorname{diag}[p_1(\lambda_1),\dots,p_n(\lambda_n)] = AV(\lambda_1,\dots,\lambda_n).$$

Source: [1040, p. 159]. **Remark:** p' is the derivative of p.

Fact 7.18.13. Let $n \ge 1$, let $a_0,\dots,a_{n-1} \in \mathbb{F}$, and define $\operatorname{circ}(a_0,\dots,a_{n-1}) \in \mathbb{F}^{n\times n}$ by

$$\operatorname{circ}(a_0,\dots,a_{n-1}) \triangleq \begin{bmatrix} a_0 & a_1 & a_2 & \cdots & a_{n-2} & a_{n-1} \\ a_{n-1} & a_0 & a_1 & \cdots & a_{n-3} & a_{n-2} \\ a_{n-2} & a_{n-1} & a_0 & \ddots & a_{n-4} & a_{n-3} \\ \vdots & \vdots & \ddots & \ddots & \ddots & \vdots \\ a_2 & a_3 & a_4 & \ddots & a_0 & a_1 \\ a_1 & a_2 & a_3 & \cdots & a_{n-1} & a_0 \end{bmatrix}.$$

A matrix of this form is *circulant*. Furthermore, for all $n \ge 2$, the $n\times n$ *primary circulant matrix* $\operatorname{circ}(0,1,0,\dots,0)$ is the cyclic permutation matrix P_n defined by (4.1.1). Note that $\operatorname{circ}(1) = P_1 = 1$.

Finally, define $p(s) \triangleq a_{n-1}s^{n-1} + \cdots + a_1 s + a_0 \in \mathbb{F}[s]$, and let $\theta \triangleq e^{(2\pi/n)\jmath}$. Then, the following statements hold:

i) $p(P_n) = \operatorname{circ}(a_0, \ldots, a_{n-1})$.

ii) $P_n = C(q)$, where $q \in \mathbb{F}[s]$ is defined by $q(s) \triangleq s^n - 1$ and $C(q)$ is the companion matrix associated with q.

iii) $q(s) = (s-1)q_0(s)$, where $q_0(s) = \sum_{i=0}^{n-1} s^{n-1-i}$.

iv) $\operatorname{roots}(q_0) = \{\theta, \theta^2, \ldots, \theta^{n-1}\}$.

v) $\operatorname{spec}(P_n) = \operatorname{roots}(q) = \{1, \theta, \theta^2, \ldots, \theta^{n-1}\}$.

vi) $\det P_n = (-1)^{n-1}$.

vii) $\operatorname{mspec}[\operatorname{circ}(a_0, \ldots, a_{n-1})] = \{p(1), p(\theta), p(\theta^2), \ldots, p(\theta^{n-1})\}_{\mathrm{ms}}$.

viii) If $A, B \in \mathbb{F}^{n\times n}$ are circulant and $\alpha, \beta \in \mathbb{F}$, then $\alpha A + \beta B$ is circulant.

ix) If $A, B \in \mathbb{F}^{n\times n}$ are circulant, then AB is circulant and $AB = BA$.

x) If A is circulant, then $\overline{A}$, A^{T}, and A^* are circulant.

xi) If A is circulant and k is a nonnegative integer, then A^k is circulant.

xii) If A is nonsingular and circulant, then A^{-1} is circulant.

xiii) $A \in \mathbb{F}^{n\times n}$ is circulant if and only if $A = P_n A P_n^{\mathsf{T}}$.

xiv) P_n is an orthogonal matrix, and $P_n^n = I_n$.

xv) If $A \in \mathbb{F}^{n\times n}$ is circulant, then A is reverse symmetric, Toeplitz, and normal.

xvi) If $A \in \mathbb{F}^{n\times n}$ is circulant and $A \notin \{\alpha I_n\colon \alpha \in \mathbb{F}\}$, then A is irreducible.

xvii) $A \in \mathbb{F}^{n\times n}$ is normal if and only if A is unitarily similar to a circulant matrix.

Next, define the *Fourier matrix* $S \in \mathbb{C}^{n\times n}$ by

$$S \triangleq n^{-1/2}V(1, \theta, \ldots, \theta^{n-1}) = \frac{1}{\sqrt{n}}\begin{bmatrix} 1 & 1 & 1 & \cdots & 1 \\ 1 & \theta & \theta^2 & \cdots & \theta^{n-1} \\ 1 & \theta^2 & \theta^4 & \cdots & \theta^{n-2} \\ \vdots & \vdots & \vdots & \iddots & \vdots \\ 1 & \theta^{n-1} & \theta^{n-2} & \cdots & \theta \end{bmatrix}.$$

Then, the following statements hold:

xviii) S is symmetric and unitary.

xix) If $n \ge 3$, then S is not Hermitian.

xx) $S^4 = I_n$ and $\operatorname{spec}(S) \subseteq \{1, -1, \jmath, -\jmath\}$.

xxi) $\operatorname{Re} S$ and $\operatorname{Im} S$ are symmetric, commute, and satisfy $(\operatorname{Re} S)^2 + (\operatorname{Im} S)^2 = I_n$.

xxii) $S^{-1}P_n S = \operatorname{diag}(1, \theta, \ldots, \theta^{n-1})$.

xxiii) $S^{-1}\operatorname{circ}(a_0, \ldots, a_{n-1})S = \operatorname{diag}[p(1), p(\theta), \ldots, p(\theta^{n-1})]$.

Source: Fact 3.16.2, [34, pp. 81–98], [839, p. 81], [2983, pp. 106–110], and [2991, pp. 138–142]. **Remark:** Circulant matrices play a role in digital signal processing, specifically, in the efficient implementation of the *fast Fourier transform.* See [2036, pp. 356–380], [2335], and [2776, pp. 206, 207]. **Remark:** S is a *Fourier matrix* and a Vandermonde matrix. **Remark:** If a real Toeplitz matrix is normal, then it must be either symmetric, skew symmetric, circulant, or skew circulant. See [167, 1019]. **Remark:** A unified treatment of the solutions of quadratic, cubic, and quartic equations using circulant matrices is given in [1577]. **Remark:** $\{I, P_k, P_k^2, \ldots, P_k^{k-1}\}$ is the multiplication group $\mathrm{C}(k)$. See Proposition 4.6.6 and Fact 4.31.17. **Remark:** Circulant matrices are generalized by *cycle matrices*, which correspond to visual geometric symmetries. See [1162].

Fact 7.18.14. Let $A \in \mathbb{R}^{n\times n}$, and assume that A is a permutation matrix. Then, there exists a permutation matrix $S \in \mathbb{R}^{n\times n}$ such that

$$A = S\,\mathrm{diag}(P_{n_1},\dots,P_{n_r})S^{-1},$$

where, for all $i \in \{1,\dots,r\}$, P_{n_i} is the $n_i \times n_i$ cyclic permutation matrix. Furthermore, the cyclic permutation matrices $P_{n_1},\dots,P_{n_r}$ are unique up to a relabeling. Consequently,

$$\mathrm{mspec}(A) = \bigcup_{i=1}^{r}\{1,\theta_i,\dots,\theta_i^{n_i-1}\}_{\mathrm{ms}},$$

where $\theta_i \triangleq e^{(2\pi/n_i)J}$. Hence, $\det A = (-1)^{n-r}$. Finally, $A^{n!} = I$, and $m \triangleq \mathrm{lcm}\,\{n_1,\dots,n_r\}$ is the smallest positive integer such that $A^m = I$. **Source:** [839, p. 29]. The last statement follows from [966, pp. 32, 33] and [2991, p. 161]. **Remark:** This result provides a canonical matrix for permutation matrices under unitary similarity with a permutation matrix. A related result for nonnegative matrices is given by Fact 6.11.6. **Remark:** It follows that A can be written as the product

$$A = S\begin{bmatrix} P_{n_1} & 0\\ 0 & I\end{bmatrix}\cdots\begin{bmatrix} I & 0 & 0\\ 0 & P_{n_i} & 0\\ 0 & 0 & I\end{bmatrix}\cdots\begin{bmatrix} I & 0\\ 0 & P_{n_r}\end{bmatrix}S^{-1},$$

where the factors represent disjoint cycles. This factorization reveals the *cycle decomposition* for an element of the symmetric group S_n (see Fact 4.31.16) on a set having n elements, where S_n is represented by the multiplication group $\mathrm{P}(n)$ of $n \times n$ permutation matrices. See [966, pp. 29–32], [2345, p. 18], and Fact 4.31.14. **Remark:** The number of possible canonical matrices for $n \times n$ permutation matrices is p_n, where p_n is the number of integral partitions of n. For example, $p_1 = 1$, $p_2 = 2$, $p_3 = 3$, $p_4 = 5$, and $p_5 = 7$. For all n, p_n is given by the generating function

$$1 + \sum_{n=1}^{\infty} p_n x^n = \frac{1}{(1-x)(1-x^2)(1-x^3)\cdots}.$$

See [118, p. 50] and [3024, p. 138]. **Remark:** The number of $n \times n$ permutation matrices with exactly k blocks is the number of permutations of an n-tuple that have exactly k cycles, which is given by the cycle number $\left[{n \atop k}\right]$. See Fact 1.19.1.

Fact 7.18.15. Let $A \in \mathbb{R}^{n\times n}$, and assume that A is a permutation matrix. Then, the following statements are equivalent:

i) A is irreducible.

ii) There exists a permutation matrix $S \in \mathbb{R}^{n\times n}$ such that SAS^{-1} is the $n\times n$ cyclic permutation matrix P_n.

iii) $\mathrm{spec}(A) = \{1, e^{(2\pi/n)J}, e^{(4\pi/n)J},\dots, e^{[(n-1)2\pi/n]J}\}$.

iv) $\{k \ge 1\colon A^k = I\} = \{n, 2n, 3n, \dots\}$.

Source: [2418, p. 177] and [2991, pp. 157, 161].

Fact 7.18.16. Let $A \in \mathbb{R}^{n\times n}$, assume that every entry of A is either 1 or -1, and assume that $AA^{\mathsf{T}} = nI$. (A is a *Hadamard matrix*.) Then, the following statements hold:

i) Either $n = 1$, $n = 2$, or n is a multiple of 4.

ii) $n^{-1/2}A$ is orthogonal.

iii) $|\det A| = n^{n/2}$.

iv) $\left[\begin{smallmatrix} A & A\\ A & -A\end{smallmatrix}\right]$ is a Hadamard matrix.

v) If $B \in \mathbb{R}^{m\times m}$ is a Hadamard matrix, then $A \otimes B$ is a Hadamard matrix.

vi) For every positive integer k, there exists a Hadamard matrix of size 2^k.

Source: [1445, p. 10], [2403, pp. 333–335], and [2991, pp. 150–154]. **Remark:** $\begin{bmatrix} 1 & 1 \\ 1 & -1 \end{bmatrix}$ is a Hadamard matrix. **Remark:** It is not known whether there exists a Hadamard matrix for every integer n that is divisible by 4. See [1445, p. 9].

Fact 7.18.17. Let $\mathcal{S}_n \subset \mathbb{R}^{n\times n}$ denote the set of matrices A such that every entry of A is either -1, 0, or 1, $1_{1\times n}A = 1_{1\times n}A^{\mathsf{T}} = 1_{1\times n}$, and, in every row and every column of A, every pair of nonzero entries that are either adjacent or separated by zeros have opposite signs. Then,

$$\operatorname{card}(\mathcal{S}_n) = \prod_{i=0}^{n-1} \frac{(3i+1)!}{(i+n)!}.$$

Source: [570, 2971]. **Remark:** This is the *alternating sign matrix conjecture*. **Remark:** The set of $n\times n$ permutation matrices, whose cardinality is $n!$, is a proper subset of $\mathcal{S}_n$. **Remark:** $(\operatorname{card}(\mathcal{S}_i))_{i=1}^{10} = (1, 2, 7, 42, 429, 7436, 218348, 10850216, 911835460, 129534272700)$.

7.19 Facts on Simultaneous Transformations

Fact 7.19.1. Let $\mathcal{S} \subset \mathbb{C}^{n\times n}$, and assume that there exists $S \in \mathbb{C}^{n\times n}$ such that, for all $A \in \mathcal{S}$, SAS^{-1} is upper triangular. Then, there exists a unitary matrix $U \in \mathbb{C}^{n\times n}$ such that, for all $A \in \mathcal{S}$, UAU^{-1} is upper triangular. If, in addition, every matrix in $\mathcal{S}$ is normal, then there exists a unitary matrix $U \in \mathbb{C}^{n\times n}$ such that, for all $A \in \mathcal{S}$, UAU^{-1} is diagonal. **Source:** [2432, p. 153].

Fact 7.19.2. Let $A, B \in \mathbb{F}^{n\times n}$, and assume that there exists a nonsingular matrix $S \in \mathbb{F}^{n\times n}$ such that SAS^{-1} and SBS^{-1} are upper triangular. Then, A and B have a common eigenvector with corresponding eigenvalues $(SAS^{-1})_{(1,1)}$ and $(SBS^{-1})_{(1,1)}$. **Source:** [1161]. **Related:** Fact 7.16.2 and Fact 7.16.15.

Fact 7.19.3. Let $A, B \in \mathbb{C}^{n\times n}$, and assume that $P_{A,B}$ is regular. Then, there exist unitary matrices $S_1, S_2 \in \mathbb{C}^{n\times n}$ such that S_1AS_2 and S_1BS_2 are upper triangular. **Source:** [2539, p. 276].

Fact 7.19.4. Let $A, B \in \mathbb{R}^{n\times n}$, and assume that $P_{A,B}$ is regular. Then, there exist orthogonal matrices $S_1, S_2 \in \mathbb{R}^{n\times n}$ such that S_1AS_2 is upper triangular and S_1BS_2 is upper Hessenberg with 2×2 diagonally located blocks. **Source:** [2539, p. 290]. **Credit:** C. Moler and G. W. Stewart.

Fact 7.19.5. Let $\mathcal{S} \subset \mathbb{F}^{n\times n}$, and assume that $AB = BA$ for all $A, B \in \mathcal{S}$. Then, there exists a unitary matrix $S \in \mathbb{F}^{n\times n}$ such that, for all $A \in \mathcal{S}$, SAS^* is upper triangular. **Source:** [1448, p. 81], [2290], [2432, pp. 153, 154], and [2991, pp. 75–77]. **Related:** Fact 7.19.10.

Fact 7.19.6. Let $A, B \in \mathbb{C}^{n\times n}$, and assume that at least one of the following conditions holds:

i) $[A, [A, B]] = [B, [A, B]] = 0$

ii) $A[A, B] = [A, B]B = 0$.

iii) $A[A, B] = B[A, B] = 0$.

iv) $\operatorname{rank}\,[A, B] \le 1$.

Then, there exists a nonsingular matrix $S \in \mathbb{C}^{n\times n}$ such that SAS^{-1} and SBS^{-1} are upper triangular. **Source:** [1161, 1719, 2290, 2440]. **Related:** Fact 7.16.2.

Fact 7.19.7. Let $A, B \in \mathbb{C}^{n\times n}$, and assume that A and B are idempotent. Then, $[A, B]$ is nilpotent if and only if there exists a unitary matrix $S \in \mathbb{C}^{n\times n}$ such that SAS^* and SBS^* are upper triangular. **Source:** [2581]. **Remark:** Necessity follows from Fact 4.22.13. **Related:** Fact 7.19.5.

Fact 7.19.8. Let $\mathcal{S} \subset \mathbb{F}^{n\times n}$, and assume that every matrix $A \in \mathcal{S}$ is normal. Then, $AB = BA$ for all $A, B \in \mathcal{S}$ if and only if there exists a unitary matrix $S \in \mathbb{F}^{n\times n}$ such that, for all $A \in \mathcal{S}$, SAS^* is diagonal. **Source:** [1448, pp. 103, 172]. **Related:** Fact 7.16.16, Fact 7.19.9, and Fact 10.20.2.

Fact 7.19.9. Let $\mathcal{S} \subset \mathbb{F}^{n\times n}$, and assume that every matrix $A \in \mathcal{S}$ is diagonalizable over $\mathbb{F}$. Then, $AB = BA$ for all $A, B \in \mathcal{S}$ if and only if there exists a nonsingular matrix $S \in \mathbb{F}^{n\times n}$ such that, for all $A \in \mathcal{S}$, SAS^{-1} is diagonal. **Source:** Fact 7.19.8 and [1448, p. 52].

Fact 7.19.10. Let $A, B \in \mathbb{F}^{n\times n}$, and assume that $\{x \in \mathbb{F}^n\colon\ x^*Ax = x^*Bx = 0\} = \{0\}$. Then, there exists a nonsingular matrix $S \in \mathbb{F}^{n\times n}$ such that SAS^* and SBS^* are upper triangular. **Source:** [2263, p. 96]. **Remark:** A and B need not be Hermitian. **Related:** Fact 7.19.5 and Fact 10.20.8. **Remark:** In the case where $P_{A,B}$ is a regular pencil, simultaneous triangularization of A and B by means of a biequivalence transformation is given in Proposition 7.8.3.

7.20 Facts on Additive Decompositions

Fact 7.20.1. Let $A \in \mathbb{F}^{n\times n}$, assume that A is nonsingular, let $B, C, D, E \in \mathbf{H}^n$, and assume that $A = B + \jmath C$ and $A^{-1} = D + \jmath E$. Then, $\operatorname{In} D = \operatorname{In} B$. In particular, A is dissipative if and only if A^{-1} is dissipative. **Source:** [1501]. **Related:** Fact 10.10.36, Fact 10.15.2, and Fact 10.21.18.

Fact 7.20.2. Let $A \in \mathbb{F}^{n\times n}$. Then, there exist a unique Hermitian matrix $B \in \mathbb{F}^{n\times n}$ and a unique skew-Hermitian matrix $C \in \mathbb{F}^{n\times n}$ such that $A = B + C$. Now, define $\hat{B} \triangleq \operatorname{Re} A$ and $\hat{C} \triangleq \operatorname{Im} A$ so that $A = \hat{B} + \jmath\hat{C}$. Then,

$$B = \tfrac{1}{2}(A + A^*) = \tfrac{1}{2}(\hat{B} + \hat{B}^{\mathsf{T}}) + \jmath\tfrac{1}{2}(\hat{C} - \hat{C}^{\mathsf{T}}),$$
$$C = \tfrac{1}{2}(A - A^*) = \tfrac{1}{2}(\hat{B} - \hat{B}^{\mathsf{T}}) + \jmath\tfrac{1}{2}(\hat{C} + \hat{C}^{\mathsf{T}}).$$

Furthermore, A is normal if and only if $BC = CB$. **Related:** Fact 4.10.28, Fact 7.20.3, and Fact 15.14.4.

Fact 7.20.3. Let $A \in \mathbb{F}^{n\times n}$. Then, there exist unique Hermitian matrices $B, C \in \mathbb{C}^{n\times n}$ such that $A = B + \jmath C$. Now, define $\hat{B} \triangleq \operatorname{Re} A$ and $\hat{C} \triangleq \operatorname{Im} A$ so that $A = \hat{B} + \jmath\hat{C}$. Then,

$$B = \tfrac{1}{2}(A + A^*) = \tfrac{1}{2}(\hat{B} + \hat{B}^{\mathsf{T}}) + \jmath\tfrac{1}{2}(\hat{C} - \hat{C}^{\mathsf{T}}),$$
$$C = \tfrac{1}{2\jmath}(A - A^*) = \tfrac{1}{2}(\hat{C} + \hat{C}^{\mathsf{T}}) - \jmath\tfrac{1}{2}(\hat{B} - \hat{B}^{\mathsf{T}}).$$

Furthermore, A is normal if and only if $BC = CB$. **Remark:** $B + \jmath C$ is the *Cartesian decomposition* of A. **Related:** Fact 4.10.28, Fact 7.20.2, and Fact 15.14.4.

Fact 7.20.4. Let $A \in \mathbb{R}^{(2n+1)\times(2n+1)}$. Then, there exist $a \in \mathbb{R}$, $N \in \mathbb{R}^{(2n+1)\times(2n+1)}$, and $S \in \mathbb{R}^{(2n+1)\times(2n+1)}$ such that $\operatorname{rank} N \le n$, S is skew symmetric, and $A = aI + N + S$. **Source:** [2838].

Fact 7.20.5. Let $A \in \mathbb{C}^{n\times n}$. Then, there exist unitary matrices $B, C \in \mathbb{C}^{n\times n}$ such that

$$A = \tfrac{1}{2}\sigma_{\max}(A)(B + C).$$

Source: [1803, 2976].

Fact 7.20.6. Let $A \in \mathbb{R}^{n\times n}$. Then, there exist orthogonal matrices $B, C, D, E \in \mathbb{R}^{n\times n}$ such that

$$A = \tfrac{1}{2}\sigma_{\max}(A)(B + C + D - E).$$

Source: [1803]. See also [2976]. **Remark:** $[1/\sigma_{\max}(A)]A$ is expressed as an affine combination of B, C, D, E where the sum of the coefficients $\tfrac{1}{2}, \tfrac{1}{2}, \tfrac{1}{2}, -\tfrac{1}{2}$ is 1.

Fact 7.20.7. Let $A \in \mathbb{R}^{n\times n}$, assume that $\sigma_{\max}(A) \le 1$, and define $r \triangleq \operatorname{rank}(I - A^*A)$. Then, A is a convex combination of $h(r)$ or fewer orthogonal matrices, where

$$h(r) \triangleq \begin{cases} 1 + r, & r \le 4,\\ 3 + \lfloor \log_2 r \rfloor, & r > 4. \end{cases}$$

Source: [1803].

Fact 7.20.8. Let $n \ge 2$, and let $A \in \mathbb{R}^{n\times n}$. Then, A is the sum of a finite number of real, orthogonal matrices. **Source:** [2026].

Fact 7.20.9. Let $A \in \mathbb{F}^{n\times n}$. Then, the following statements are equivalent:

i) A is positive semidefinite, $\operatorname{tr} A$ is a positive integer, and $\operatorname{rank} A \le \operatorname{tr} A$.

ii) There exist nonzero projectors $B_1, \ldots, B_l \in \mathbb{F}^{n\times n}$, where $l \le \operatorname{tr} A$, such that $A = \sum_{i=1}^{l} B_i$.

Source: [1049, 2920]. **Remark:** For some A, the smallest number of nonzero projectors is $\operatorname{tr} A$. **Related:** Fact 7.20.12.

Fact 7.20.10. Let $A \in \mathbb{F}^{n\times n}$, assume that A is Hermitian, $0 \le A \le I$, and $\operatorname{tr} A$ is a rational number. Then, A is the average of a finite set of projectors in $\mathbb{F}^{n\times n}$. **Source:** [726]. **Remark:** The required number of projectors can be arbitrarily large.

Fact 7.20.11. Let $A \in \mathbb{F}^{n\times n}$, assume that A is Hermitian, and assume that $0 \le A \le I$. Then, A is a convex combination of $\lfloor \log_2 n \rfloor + 2$ projectors in $\mathbb{F}^{n\times n}$. **Source:** [726].

Fact 7.20.12. Let $A \in \mathbb{F}^{n\times n}$. Then, the following statements are equivalent:

i) $\operatorname{tr} A$ is an integer, and $\operatorname{rank} A \le \operatorname{tr} A$.

ii) There exist idempotent matrices $B_1, \ldots, B_m \in \mathbb{F}^{n\times n}$ such that $A = \sum_{i=1}^m B_i$.

iii) There exist a positive integer m and idempotent matrices $B_1, \ldots, B_m \in \mathbb{F}^{n\times n}$ such that, for all $i \in \{1, \ldots, m\}$, $\operatorname{rank} B_i = 1$ and $\mathcal{R}(B_i) \subseteq \mathcal{R}(A)$, and such that $A = \sum_{i=1}^m B_i$.

iv) There exist idempotent matrices $B_1, \ldots, B_l \in \mathbb{F}^{n\times n}$, where $l \triangleq \operatorname{tr} A$, such that $A = \sum_{i=1}^l B_i$.

Source: [1337, 2503, 2920]. **Remark:** The minimal number of idempotent matrices is discussed in [2824]. **Related:** Fact 7.20.13.

Fact 7.20.13. Let $A \in \mathbb{F}^{n\times n}$, and assume that $2(\operatorname{rank} A - 1) \le \operatorname{tr} A \le 2n$. Then, there exist idempotent matrices $B, C, D, E \in \mathbb{F}^{n\times n}$ such that $A = B + C + D + E$. **Source:** [1764]. **Related:** Fact 7.20.15.

Fact 7.20.14. Let $A \in \mathbb{F}^{n\times n}$. If $n = 2$ or $n = 3$, then there exist $b, c \in \mathbb{F}$ and idempotent matrices $B, C \in \mathbb{F}^{n\times n}$ such that $A = bB + cC$. Furthermore, if $n \ge 4$, then there exist $b, c, d \in \mathbb{F}$ and idempotent matrices $B, C, D \in \mathbb{F}^{n\times n}$ such that $A = bB + cC + dD$. **Source:** [2287].

Fact 7.20.15. Let $A \in \mathbb{C}^{n\times n}$, and assume that A is Hermitian. If $n = 2$ or $n = 3$, then there exist $b, c \in \mathbb{C}$ and projectors $B, C \in \mathbb{C}^{n\times n}$ such that $A = bB + cC$. Furthermore, if $4 \le n \le 7$, then there exist $b, c, d \in \mathbb{F}$ and projectors $B, C, D \in \mathbb{F}^{n\times n}$ such that $A = bB + cC + dD$. If $n \ge 8$, then there exist $b, c, d, e \in \mathbb{C}$ and projectors $B, C, D, E \in \mathbb{C}^{n\times n}$ such that $A = bB + cC + dD + eE$. **Source:** [2110]. **Related:** Fact 7.20.13.

7.21 Notes

Much of the development in this chapter is based on [2221]. Additional discussions of the Smith and Smith-McMillan forms are given in [1573] and [2999]. The Smith form of a matrix whose entries are integers is considered in [1182, pp. 494–505]. The multicompanion form and the elementary multicompanion form are known as *rational canonical forms* [966, pp. 472–488], while the multicompanion form is traditionally called the *Frobenius canonical form* [302]. The derivation of the Jordan form by means of the elementary multicompanion form and the hypercompanion form follows [2221]. Corollary 7.4.9, Corollary 7.4.10, and Proposition 7.7.13 are given in [527, 528, 2591, 2592, 2595]. Corollary 7.4.10 is due to F. G. Frobenius. Canonical forms for congruence transformations are given in [1780, 2617].

The companion matrix is sometimes called a *Frobenius matrix*, see [9]. It is sometimes useful to define block-companion form matrices in which the scalars are replaced by matrix blocks [1186, 1187, 1189].

A variation of the Jordan form called the *Weyr form* is discussed in [144, 2166, 2431, 2432]. For Weyr form differs from the Jordan form in terms of the nilpotent component of the canonical form.

Matrix pencils are discussed in [187, 341, 487, 1692, 2738, 2759]. Computational algorithms for the Kronecker canonical form are given in [1856, 2772]. Applications to linear system theory are discussed in [688, pp. 52–55] and [1583]. Since a pencil is a polynomial matrix, the terminology "regular pencil" conflicts with "regular polynomial matrix."

The polar decomposition is applied to the elastic deformation of solids in [2203, pp. 140–142].

Chapter Eight
Generalized Inverses

Generalized inverses provide a useful extension of the matrix inverse to singular matrices and to rectangular matrices that are neither left nor right invertible.

8.1 Moore-Penrose Generalized Inverse

Let $A \in \mathbb{F}^{n\times m}$. If A is nonzero, then it follows from the singular value decomposition Theorem 7.6.3 that there exist orthogonal matrices $S_1 \in \mathbb{F}^{n\times n}$ and $S_2 \in \mathbb{F}^{m\times m}$ such that

$$A = S_1 \begin{bmatrix} B & 0_{r\times(m-r)} \\ 0_{(n-r)\times r} & 0_{(n-r)\times(m-r)} \end{bmatrix} S_2, \tag{8.1.1}$$

where $B \triangleq \operatorname{diag}[\sigma_1(A),\ldots,\sigma_r(A)]$, $r \triangleq \operatorname{rank} A$, and $\sigma_1(A) \ge \sigma_2(A) \ge \cdots \ge \sigma_r(A) > 0$ are the positive singular values of A. In (8.1.1), some of the border zero matrices may be empty. Then, the *(Moore-Penrose) generalized inverse* A^+ of A is the $m \times n$ matrix

$$A^+ \triangleq S_2^* \begin{bmatrix} B^{-1} & 0_{r\times(n-r)} \\ 0_{(m-r)\times r} & 0_{(m-r)\times(n-r)} \end{bmatrix} S_1^*. \tag{8.1.2}$$

If $A = 0_{n\times m}$, then $A^+ \triangleq 0_{m\times n}$, while, if $m = n$ and $\det A \neq 0$, then $A^+ = A^{-1}$. It is helpful to remember that A^+ and A^* are the same size. Note that A^+ satisfies

$$AA^+A = A, \tag{8.1.3}$$

$$A^+AA^+ = A^+, \tag{8.1.4}$$

$$(AA^+)^* = AA^+, \tag{8.1.5}$$

$$(A^+A)^* = A^+A. \tag{8.1.6}$$

Hence, for each $A \in \mathbb{F}^{n\times m}$ there exists $X \in \mathbb{F}^{m\times n}$ satisfying the four statements

$$AXA = A, \tag{8.1.7}$$

$$XAX = X, \tag{8.1.8}$$

$$(AX)^* = AX, \tag{8.1.9}$$

$$(XA)^* = XA. \tag{8.1.10}$$

We now show that X is uniquely defined by (8.1.7)–(8.1.10).

Theorem 8.1.1. Let $A \in \mathbb{F}^{n\times m}$. Then, $X = A^+$ is the unique matrix $X \in \mathbb{F}^{m\times n}$ satisfying (8.1.7)–(8.1.10).

Proof. Suppose there exists $X \in \mathbb{F}^{m\times n}$ satisfying (8.1.7)–(8.1.10). Then,

$$\begin{aligned} X &= XAX = X(AX)^* = XX^*A^* = XX^*(AA^+A)^* = XX^*A^*A^{+*}A^* = X(AX)^*(AA^+)^* \\ &= XAXAA^+ = XAA^+ = (XA)^*A^+ = A^*X^*A^+ = (AA^+A)^*X^*A^+ = A^*A^{+*}A^*X^*A^+ \\ &= (A^+A)^*(XA)^*A^+ = A^+AXAA^+ = A^+AA^+ = A^+. \end{aligned}$$

□

Given $A \in \mathbb{F}^{n\times m}$, $X \in \mathbb{F}^{m\times n}$ is a *(1)-inverse* of A if (8.1.7) holds, a *(1,2)-inverse* of A if (8.1.7) and (8.1.8) hold, and so forth.

Proposition 8.1.2. Let $A \in \mathbb{F}^{n\times m}$, assume that A is left invertible, and let $X \in \mathbb{F}^{m\times n}$. Then, the following statements are equivalent:

i) X is a left inverse of A.

ii) X is a (1)-inverse of A.

iii) X is a (1,2,4)-inverse of A.

Proposition 8.1.3. Let $A \in \mathbb{F}^{n\times m}$, assume that A is right invertible, and let $X \in \mathbb{F}^{m\times n}$. Then, the following statements are equivalent:

i) X is a right inverse of A.

ii) X is a (1)-inverse of A.

iii) X is a (1,2,3)-inverse of A.

Proof. To prove *i)* $\Longrightarrow$ *ii)*, note that $AX = I_n$, and thus $AXA = A$, which implies that X is a (1)-inverse of A. Conversely, since X is a (1)-inverse of A, it follows that $AXA = A$. Then, letting $A^{\mathsf{R}} \in \mathbb{F}^{m\times n}$ denote a right inverse of A, it follows that $I_n = AA^{\mathsf{R}} = AXAA^{\mathsf{R}} = AX$. Hence, X is a right inverse of A. Finally, to prove *ii)* $\Longrightarrow$ *iii)*, note that, since X is a right inverse of A, it is also a (2,3)-inverse of A. □

It can now be seen that A^+ is a particular (left, right) inverse if A is (left, right) invertible.

Corollary 8.1.4. Let $A \in \mathbb{F}^{n\times m}$. If A is left invertible, then A^+ is a left inverse of A. Furthermore, if A is right invertible, then A^+ is a right inverse of A.

The following result provides an explicit expression for A^+ in the case where A is either left invertible or right invertible. It is helpful to note that A is (left, right) invertible if and only if (A^*A, AA^*) is positive definite.

Proposition 8.1.5. Let $A \in \mathbb{F}^{n\times m}$. If A is left invertible, then

$$A^+ = (A^*A)^{-1}A^* \tag{8.1.11}$$

and A^+ is a left inverse of A. If A is right invertible, then

$$A^+ = A^*(AA^*)^{-1} \tag{8.1.12}$$

and A^+ is a right inverse of A.

Proof. It suffices to verify (8.1.7)–(8.1.10) with $X = A^+$. □

Corollary 8.1.6. Let $x \in \mathbb{F}^n$, and assume that x is nonzero. Then,

$$x^+ = \|x\|_2^{-2}x^*. \tag{8.1.13}$$

Proposition 8.1.7. Let $A \in \mathbb{F}^{n\times m}$. Then, the following statements hold:

i) $A = 0$ if and only if $A^+ = 0$.

ii) $(A^+)^+ = A$.

iii) $\overline{A}^+ = \overline{A^+}$.

iv) $A^{+\mathsf{T}} \triangleq (A^{\mathsf{T}})^+ = (A^+)^{\mathsf{T}}$.

v) $A^{+*} \triangleq (A^*)^+ = (A^+)^*$.

vi) $\mathcal{R}(A) = \mathcal{R}(AA^*) = \mathcal{R}(AA^+) = \mathcal{R}(A^{+*}) = \mathcal{N}[(AA^+)_\perp] = \mathcal{N}(A^*)^\perp$.

vii) $\mathcal{R}(A^*) = \mathcal{R}(A^*A) = \mathcal{R}(A^+A) = \mathcal{R}(A^+) = \mathcal{N}[(A^+A)_\perp] = \mathcal{N}(A)^\perp$.

viii) $\mathcal{N}(A) = \mathcal{N}(A^+A) = \mathcal{N}(A^*A) = \mathcal{N}(A^{+*}) = \mathcal{R}[(A^+A)_\perp] = \mathcal{R}(A^*)^\perp$.

ix) $\mathcal{N}(A^*) = \mathcal{N}(AA^+) = \mathcal{N}(AA^*) = \mathcal{N}(A^+) = \mathcal{R}[(AA^+)_\perp] = \mathcal{R}(A)^\perp$.

x) AA^+ and A^+A are positive semidefinite.

xi) $\operatorname{spec}(AA^+) \subseteq \{0, 1\}$ and $\operatorname{spec}(A^+A) \subseteq \{0, 1\}$.

xii) AA^+ is the projector onto $\mathcal{R}(A)$.

xiii) A^+A is the projector onto $\mathcal{R}(A^*)$.

xiv) $(A^+A)_\perp$ is the projector onto $\mathcal{N}(A)$.

xv) $(AA^+)_\perp$ is the projector onto $\mathcal{N}(A^*)$.

xvi) $x \in \mathcal{R}(A)$ if and only if $x = AA^+x$.

xvii) $\operatorname{rank} A = \operatorname{rank} A^+ = \operatorname{rank} AA^+ = \operatorname{rank} A^+A = \operatorname{tr} AA^+ = \operatorname{tr} A^+A$.

xviii) If A is idempotent, then $\operatorname{rank} A = \operatorname{tr} AA^+ = \operatorname{tr} A^2A^+ = \operatorname{tr} AA^+A = \operatorname{tr} A$.

xix) $\operatorname{rank} (AA^+)_\perp = n - \operatorname{rank} A$.

xx) $\operatorname{rank} (A^+A)_\perp = m - \operatorname{rank} A$.

xxi) $A = AA^*A^{+*} = A^{+*}A^*A = A(A^*A)^+A^*A = AA^*A(A^*A)^+ = AA^*(AA^*)^+A = (AA^*)^+AA^*A$.

xxii) $A^* = A^*AA^+ = A^+AA^*$.

xxiii) $A^+ = A^*(AA^*)^+ = (A^*A)^+A^* = A^+A^{+*}A^* = A^*A^{+*}A^+ = A^*(A^*AA^*)^+A^*$.

xxiv) $A^{+*} = (AA^*)^+A = A(A^*A)^+ = AA^+A^{+*} = A^{+*}A^+A = A(AA^*A)^+A$.

xxv) $(AA^*)^+ = A^{+*}A^+$.

xxvi) $(A^*A)^+ = A^+A^{+*}$.

xxvii) $AA^+ = (AA^+)^* = A^{+*}A^* = A(A^*A)^+A^* = AA^*(AA^*)^+ = (AA^*)^+AA^* = ([(AA^+)_\perp]^+(AA^+)_\perp)_\perp$.

xxviii) $A^+A = (A^+A)^* = A^*A^{+*} = A^*(AA^*)^+A = A^*A(A^*A)^+ = (A^*A)^+A^*A = ([(A^+A)_\perp]^+(A^+A)_\perp)_\perp$.

xxix) If $S_1 \in \mathbb{F}^{n\times n}$ and $S_2 \in \mathbb{F}^{m\times m}$ are unitary, then $(S_1AS_2)^+ = S_2^*A^+S_1^*$.

xxx) A is (range Hermitian, normal, Hermitian, skew Hermitian, positive semidefinite, positive definite) if and only if A^+ is.

xxxi) The following statements are equivalent:

a) A is range Hermitian.

b) $\mathcal{R}(A) = \mathcal{R}(A^+)$.

c) $AA^+ = A^+A$.

xxxii) If A is a projector, then $A^+ = A$.

xxxiii) If A is normal and $k \geq 1$, then $(A^k)^+ = (A^+)^k$.

xxxiv) If $B \in \mathbb{F}^{n\times m}$, then $A = B$ if and only if $A^+ = B^+$.

xxxv) If $B \in \mathbb{F}^{m\times l}$ and $AB = 0$, then $B^+A^+ = 0$.

xxxvi) If $B \in \mathbb{F}^{n\times l}$, then $A^*B = 0$ if and only if $A^+B = 0$.

xxxvii) If $B \in \mathbb{F}^{n\times l}$, then $\mathcal{R}(AA^+B)$ is the projection of $\mathcal{R}(B)$ into $\mathcal{R}(A)$.

xxxviii) If $B \in \mathbb{F}^{m\times l}$, $C \in \mathbb{F}^{n\times l}$, and $A^*AB = A^*C$, then $AB = AA^+C$.

xxxix) If $B \in \mathbb{F}^{l\times m}$, then $AB^* = 0$ if and only if $A^+AB^+B = 0$.

xl) If $S \in \mathbb{F}^{m\times m}$ is nonsingular, then $AS(AS)^+ = AA^+$.

Proof. The last three statements are given in [2238, pp. 186, 188, 312]. □

Theorem 3.7.5 shows that the equation $Ax = b$, where $A \in \mathbb{F}^{n\times m}$ and $b \in \mathbb{F}^n$, has a solution $x \in \mathbb{F}^m$ if and only if $\operatorname{rank} A = \operatorname{rank} [A \;\; b]$. In particular, $Ax = b$ has a unique solution $x \in \mathbb{F}^m$ if and only if $\operatorname{rank} A = \operatorname{rank} [A \;\; b] = m$, while $Ax = b$ has infinitely many solutions if and only if $\operatorname{rank} A = \operatorname{rank} [A \;\; b] < m$, and at least one solution if and only if $b \in \mathcal{R}(A)$. The following result expresses the last condition in terms of the (1)-inverse and the generalized inverse.

Proposition 8.1.8. Let $A \in \mathbb{F}^{n\times m}$ and $b \in \mathbb{F}^n$. Then, the following statements are equivalent:

i) $b \in \mathcal{R}(A)$.

ii) If $B \in \mathbb{F}^{m\times n}$ is a (1)-inverse of A, then $ABb = b$.

iii) There exists a (1)-inverse $B \in \mathbb{F}^{m\times n}$ of A such that $ABb = b$.

iv) $AA^+b = b$.

Proof. To show that *i*) $\Longrightarrow$ *ii*), let $B \in \mathbb{F}^{m\times n}$ be a (1)-inverse of A, and let $x \in \mathbb{F}^m$ satisfy $Ax = b$. Then, $ABb = ABAx = Ax = b$. *ii*) $\Longrightarrow$ *iii*) is immediate. Next, *iii*) implies that $x \triangleq Bb$ satisfies $Ax = b$, and thus *i*) holds. Next, to show that *i*) $\Longrightarrow$ *iv*), let $x \in \mathbb{F}^m$ satisfy $Ax = b$. Then, $AA^+b = AA^+Ax = Ax = b$. Finally, *iv*) implies that $x \triangleq A^+b$ satisfies $Ax = b$, and thus *i*) holds. □

Proposition 8.1.8 shows that, if B is a (1)-inverse of A, then Bb is a solution of $Ax = b$. However, it is not true that every solution x of $Ax = b$ is of the form $x = Bb$ for some (1)-inverse B of A. For example, let $A = [1\ \ 0]$ and $b = 0$ so that every (1)-inverse B of A is of the form $B = \begin{bmatrix} 1 \\ \beta \end{bmatrix}$ for some $\beta \in \mathbb{F}$. Then, $x = \begin{bmatrix} 0 \\ 1 \end{bmatrix}$ satisfies $Ax = b$, but $Bb = 0 \neq x$ for every (1)-inverse B of A.

The following result assumes that $Ax = b$ has at least one solution and characterizes all solutions in terms of the (1)-inverse and the generalized inverse. The case where A is right invertible is considered in Theorem 3.7.5.

Proposition 8.1.9. Let $A \in \mathbb{F}^{n\times m}$ and $b \in \mathbb{F}^n$, assume that $b \in \mathcal{R}(A)$, let $B \in \mathbb{F}^{m\times n}$ be a (1)-inverse of A, and let $x \in \mathbb{F}^m$. Then, the following statements are equivalent:

i) $Ax = b$.

ii) $BAx = Bb$.

iii) $x = Bb + (I - BA)x$.

iv) There exists $y \in \mathbb{F}^m$ such that $x = Bb + (I - BA)y$.

v) $A^+Ax = A^+b$.

vi) $x = A^+b + (I - A^+A)x$.

vii) There exists $y \in \mathbb{F}^m$ such that $x = A^+b + (I - A^+A)y$.

Now, assume that these statements hold and $\operatorname{rank} A = m$. Then, the following statements hold:

viii) If $A^{\mathsf{L}} \in \mathbb{F}^{m\times n}$ is a left inverse of A, then $x = A^{\mathsf{L}}b$.

ix) B is a (1)-inverse of A, and $x = Bb$.

x) $x = A^+b$.

Proof. Immediately, *i*) $\Longrightarrow$ *ii*) $\Longrightarrow$ *iii*) $\Longrightarrow$ *iv*). To show that *iv*) $\Longrightarrow$ *i*), note that $Ax = ABb + A(I - BA)y = ABb = b$. Next, it follows from Proposition 8.1.8 that $AA^+b = b$. Therefore, setting $B = A^+$ yields *i*) $\Longrightarrow$ *v*) $\Longrightarrow$ *vi*) $\Longrightarrow$ *vii*) $\Longrightarrow$ *i*).

viii) is immediate. To prove *ix*), note that Proposition 8.1.2 implies that B is a left inverse of A. Finally, *x*) follows from the fact that A^+ is a left inverse of A. □

Let $y \in \mathbb{F}^m$ and $x = A^+b + (I - A^+A)y$. Then, $x^*x = b^*A^{+*}A^+b + y^*(I - A^+A)y$. Therefore, x^*x is minimized by $y = 0$. Connections to least squares solutions are discussed in Fact 11.17.10.

The following result extends Proposition 3.7.10.

Proposition 8.1.10. Let $A \in \mathbb{F}^{n\times m}$ and $\mathcal{S} \subseteq \mathbb{F}^n$. Then,

$$A^{\mathsf{inv}}(\mathcal{S}) = A^+[\mathcal{S} \cap \mathcal{R}(A)] + \mathcal{R}(I - A^+A). \tag{8.1.14}$$

Hence,

$$AA^{\mathsf{inv}}(\mathcal{S}) = \mathcal{S} \cap \mathcal{R}(A). \tag{8.1.15}$$

If A is right invertible, then

$$AA^{\mathsf{inv}}(\mathcal{S}) = \mathcal{S}, \tag{8.1.16}$$

$$A^+\mathcal{S} \subseteq A^+\mathcal{S} + \mathcal{R}(I - A^+A) = A^{\mathsf{inv}}(\mathcal{S}). \tag{8.1.17}$$

Proof. Let $x \in A^{\mathrm{inv}}(\mathcal{S})$. Then, $y \triangleq Ax \in \mathcal{S}$. It thus follows from *vii*) of Proposition 8.1.9 that there exists $z \in \mathbb{F}^m$ such that $x = A^+y + (I - A^+A)z \in A^+[\mathcal{S} \cap \mathcal{R}(A)] + \mathcal{R}(I - A^+A)$. Conversely, let $x \in A^+[\mathcal{S} \cap \mathcal{R}(A)] + \mathcal{R}(I - A^+A)$. Then, there exist $y \in \mathcal{S} \cap \mathcal{R}(A)$ and $z \in \mathcal{R}(I - A^+A)$ such that $x = A^+y + (I - A^+A)z$. Hence, $Ax = AA^+y + A(I - A^+A)z = y$. □

The last statement of Proposition 8.1.10 is given by (3.7.9) of Proposition 3.7.10.

The following corollary of Proposition 8.1.10 characterizes the solution set of $Ax = b$.

Corollary 8.1.11. Let $A \in \mathbb{F}^{n\times m}$ and $b \in \mathbb{F}^n$. Then, the set of solutions x to the equation $Ax = b$ is given by

$$A^{\mathrm{inv}}(\{b\}) = \begin{cases} A^+b + \mathcal{R}(I - A^+A), & b \in \mathcal{R}(A), \\ \varnothing, & b \notin \mathcal{R}(A). \end{cases} \tag{8.1.18}$$

The following result provides an expression for the generalized inverse in terms of the full-rank factorization given by Proposition 7.6.6.

Proposition 8.1.12. Let $A \in \mathbb{F}^{n\times m}$, $B \in \mathbb{F}^{n\times r}$, and $C \in \mathbb{F}^{r\times m}$, and assume that $\operatorname{rank} A = \operatorname{rank} B = r$ and $A = BC$. Then,

$$A^+ = C^+B^+ = C^*(CC^*)^{-1}(B^*B)^{-1}B^*, \tag{8.1.19}$$

$$AA^+ = BB^+, \quad A^+A = C^+C. \tag{8.1.20}$$

If $m = r$, then

$$A^+ = C^{-1}(B^*B)^{-1}B^*. \tag{8.1.21}$$

If $n = r$, then

$$A^+ = C^*(CC^*)^{-1}B^{-1}. \tag{8.1.22}$$

If $n = m = r$, then

$$A^{-1} = C^{-1}B^{-1}. \tag{8.1.23}$$

Definition 8.1.13. Let $A \in \mathbb{F}^{n\times m}$, $B \in \mathbb{F}^{n\times l}$, $C \in \mathbb{F}^{k\times m}$, and $D \in \mathbb{F}^{k\times l}$, and define $\mathcal{A} \triangleq \left[\begin{smallmatrix} A & B \\ C & D \end{smallmatrix}\right] \in \mathbb{F}^{(n+k)\times(m+l)}$. Then, the *Schur complement* $A|\mathcal{A}$ of A with respect to $\mathcal{A}$ is defined by

$$A|\mathcal{A} \triangleq D - CA^+B. \tag{8.1.24}$$

Likewise, the *Schur complement* $D|\mathcal{A}$ of D with respect to $\mathcal{A}$ is defined by

$$D|\mathcal{A} \triangleq A - BD^+C. \tag{8.1.25}$$

8.2 Drazin Generalized Inverse

We now introduce a different type of generalized inverse, which applies only to square matrices yet is more useful in certain applications. Let $A \in \mathbb{F}^{n\times n}$. Then, A has a decomposition

$$A = S\begin{bmatrix} J_1 & 0 \\ 0 & J_2 \end{bmatrix}S^{-1}, \tag{8.2.1}$$

where $S \in \mathbb{F}^{n\times n}$ is nonsingular, $J_1 \in \mathbb{F}^{m\times m}$ is nonsingular, and $J_2 \in \mathbb{F}^{(n-m)\times(n-m)}$ is nilpotent. Then, the *Drazin generalized inverse* A^{D} of A is the matrix

$$A^{\mathrm{D}} \triangleq S\begin{bmatrix} J_1^{-1} & 0 \\ 0 & 0 \end{bmatrix}S^{-1}. \tag{8.2.2}$$

Let $A \in \mathbb{F}^{n\times n}$. Then, it follows from Definition 7.7.1 that $\operatorname{ind} A = \operatorname{ind}_A(0)$. Furthermore, A is nonsingular if and only if $\operatorname{ind} A = 0$, whereas $\operatorname{ind} A = 1$ if and only if A is singular and the zero eigenvalue of A is semisimple. In particular, $\operatorname{ind} 0_{n\times n} = 1$. Note that $\operatorname{ind} A$ is the largest size of all of the Jordan blocks of A associated with the zero eigenvalue of A.

It can be seen that A^{D} satisfies

$$A^{\mathsf{D}}AA^{\mathsf{D}} = A^{\mathsf{D}}, \tag{8.2.3}$$

$$AA^{\mathsf{D}} = A^{\mathsf{D}}A, \tag{8.2.4}$$

$$A^{k+1}A^{\mathsf{D}} = A^k, \tag{8.2.5}$$

where $k = \operatorname{ind} A$. Hence, for all $A \in \mathbb{F}^{n\times n}$ such that $\operatorname{ind} A = k$ there exists $X \in \mathbb{F}^{n\times n}$ satisfying the three statements

$$XAX = X, \tag{8.2.6}$$

$$AX = XA, \tag{8.2.7}$$

$$A^{k+1}X = A^k. \tag{8.2.8}$$

We now show that X is uniquely defined by (8.2.6)–(8.2.8).

Theorem 8.2.1. Let $A \in \mathbb{F}^{n\times n}$, and define $k \triangleq \operatorname{ind} A$. Then, $X = A^{\mathsf{D}}$ is the unique matrix $X \in \mathbb{F}^{n\times n}$ satisfying (8.2.6)–(8.2.8).

Proof. Let $X \in \mathbb{F}^{n\times n}$ satisfy (8.2.6)–(8.2.8). If $k = 0$, then it follows from (8.2.8) that $X = A^{-1}$. Hence, let $A = S\begin{bmatrix} J_1 & 0 \\ 0 & J_2 \end{bmatrix}S^{-1}$, where $k = \operatorname{ind} A \geq 1$, $S \in \mathbb{F}^{n\times n}$ is nonsingular, $J_1 \in \mathbb{F}^{m\times m}$ is nonsingular, and $J_2 \in \mathbb{F}^{(n-m)\times(n-m)}$ is nilpotent. Now, let $\hat{X} \triangleq S^{-1}XS = \begin{bmatrix} \hat{X}_1 & \hat{X}_{12} \\ \hat{X}_{21} & \hat{X}_2 \end{bmatrix}$ be partitioned conformably with $\hat{A} \triangleq S^{-1}AS = \begin{bmatrix} J_1 & 0 \\ 0 & J_2 \end{bmatrix}$. Since, by (8.2.7), $\hat{A}\hat{X} = \hat{X}\hat{A}$, it follows that $J_1\hat{X}_1 = \hat{X}_1J_1$, $J_1\hat{X}_{12} = \hat{X}_{12}J_2$, $J_2\hat{X}_{21} = \hat{X}_{21}J_1$, and $J_2\hat{X}_2 = \hat{X}_2J_2$. Since $J_2^k = 0$, it follows that $J_1\hat{X}_{12}J_2^{k-1} = 0$, and thus $\hat{X}_{12}J_2^{k-1} = 0$. By repeating this argument, it follows that $J_1\hat{X}_{12}J_2 = 0$, and thus $\hat{X}_{12}J_2 = 0$, which implies that $J_1\hat{X}_{12} = 0$, and thus $\hat{X}_{12} = 0$. Similarly, $\hat{X}_{21} = 0$, so that $\hat{X} = \begin{bmatrix} \hat{X}_1 & 0 \\ 0 & \hat{X}_2 \end{bmatrix}$. Now, (8.2.8) implies that $J_1^{k+1}\hat{X}_1 = J_1^k$, and hence $\hat{X}_1 = J_1^{-1}$. Next, (8.2.6) implies that $\hat{X}_2J_2\hat{X}_2 = \hat{X}_2$, which, together with $J_2\hat{X}_2 = \hat{X}_2J_2$, yields $\hat{X}_2^2J_2 = \hat{X}_2$. Consequently, $0 = \hat{X}_2^2J_2^k = \hat{X}_2J_2^{k-1}$, and thus, by repeating this argument, $\hat{X}_2 = 0$. Hence, $A^{\mathsf{D}} = S\begin{bmatrix} J_1^{-1} & 0 \\ 0 & 0 \end{bmatrix}S^{-1} = S\begin{bmatrix} \hat{X}_1 & 0 \\ 0 & 0 \end{bmatrix}S^{-1} = S\hat{X}S^{-1} = X$. □

Proposition 8.2.2. Let $A \in \mathbb{F}^{n\times n}$, and define $k \triangleq \operatorname{ind} A$. Then, the following statements hold:

i) $\overline{A}^{\mathsf{D}} = \overline{A^{\mathsf{D}}}$.

ii) $A^{\mathsf{DT}} \triangleq A^{\mathsf{TD}} \triangleq (A^{\mathsf{T}})^{\mathsf{D}} = (A^{\mathsf{D}})^{\mathsf{T}}$.

iii) $A^{\mathsf{D}*} \triangleq A^{*\mathsf{D}} \triangleq (A^*)^{\mathsf{D}} = (A^{\mathsf{D}})^*$.

iv) If $r \in \mathbb{P}$, then $A^{\mathsf{D}r} \triangleq A^{r\mathsf{D}} \triangleq (A^{\mathsf{D}})^r = (A^r)^{\mathsf{D}}$.

v) $\mathcal{R}(A^k) = \mathcal{R}(A^{\mathsf{D}}) = \mathcal{R}(AA^{\mathsf{D}}) = \mathcal{N}(I - AA^{\mathsf{D}})$.

vi) $\mathcal{N}(A^k) = \mathcal{N}(A^{\mathsf{D}}) = \mathcal{N}(AA^{\mathsf{D}}) = \mathcal{R}(I - AA^{\mathsf{D}})$.

vii) $\operatorname{rank} A^k = \operatorname{rank} A^{\mathsf{D}} = \operatorname{rank} AA^{\mathsf{D}} = \operatorname{def}(I - AA^{\mathsf{D}})$.

viii) $\operatorname{def} A^k = \operatorname{def} A^{\mathsf{D}} = \operatorname{def} AA^{\mathsf{D}} = \operatorname{rank}(I - AA^{\mathsf{D}})$.

ix) AA^{D} is the idempotent matrix onto $\mathcal{R}(A^{\mathsf{D}})$ along $\mathcal{N}(A^{\mathsf{D}})$.

x) $A^{\mathsf{D}} = 0$ if and only if A is nilpotent.

xi) A^{D} is group invertible.

xii) $\operatorname{ind} A^{\mathsf{D}} = 0$ if and only if A is nonsingular.

xiii) $\operatorname{ind} A^{\mathsf{D}} = 1$ if and only if A is singular.

xiv) $(A^{\mathsf{D}})^{\mathsf{D}} = (A^{\mathsf{D}})^{\#} = A^2A^{\mathsf{D}}$.

xv) The following statements are equivalent:

a) A is group invertible.

b) $(A^{\mathsf{D}})^{\mathsf{D}} = A$.

c) $(A^{\mathsf{D}})^{\#} = A$.

d) $A^2A^{\mathsf{D}} = A$.

xvi) $[(A^{\mathsf{D}})^{\mathsf{D}}]^{\mathsf{D}} = [(A^{\mathsf{D}})^{\#}]^{\mathsf{D}} = (A^2A^{\mathsf{D}})^{\mathsf{D}} = A^{\mathsf{D}}$.

xvii) If A is idempotent, then $k = 1$ and $A^{\mathsf{D}} = A$.

xviii) $A = A^{\mathsf{D}}$ if and only if A is tripotent.

xix) $A - A^2A^{\mathsf{D}}$ is nilpotent, and $\operatorname{ind}(A - A^2A^{\mathsf{D}}) = k$.

xx) $(A - A^2A^{\mathsf{D}})A^2A^{\mathsf{D}} = A^2A^{\mathsf{D}}(A - A^2A^{\mathsf{D}}) = 0$.

xxi) $(A - A^2A^{\mathsf{D}})A^{\mathsf{D}} = A^{\mathsf{D}}(A - A^2A^{\mathsf{D}}) = 0$.

xxii) $\operatorname{rank} A^2A^{\mathsf{D}} = n - \operatorname{amult}_A(0)$, and $\operatorname{def} A^2A^{\mathsf{D}} = \operatorname{amult}_A(0)$.

xxiii) $\operatorname{rank}(A - A^2A^{\mathsf{D}}) = \operatorname{amult}_A(0) - \operatorname{gmult}_A(0)$, and $\operatorname{def}(A - A^2A^{\mathsf{D}}) = n - \operatorname{amult}_A(0) + \operatorname{gmult}_A(0)$.

xxiv) Let $X, Y \in \mathbb{F}^{n\times n}$ be such that $XY = YX = 0$, $\operatorname{ind} X \le 1$, Y is nilpotent, and $\operatorname{ind} Y = k$. Then, $X = A^2A^{\mathsf{D}}$ and $Y = A - A^2A^{\mathsf{D}}$.

Let $A \in \mathbb{F}^{n\times n}$, and assume that $\operatorname{ind} A \le 1$ so that, by Corollary 7.7.9, A is group invertible. In this case, the Drazin generalized inverse A^{D} is denoted by $A^{\#}$, which is the *group generalized inverse* of A. Therefore, $A^{\#}$ satisfies

$$A^{\#}AA^{\#} = A^{\#}, \tag{8.2.9}$$

$$AA^{\#} = A^{\#}A, \tag{8.2.10}$$

$$AA^{\#}A = A, \tag{8.2.11}$$

while $A^{\#}$ is the unique matrix $X \in \mathbb{F}^{n\times n}$ satisfying

$$XAX = X, \tag{8.2.12}$$

$$AX = XA, \tag{8.2.13}$$

$$AXA = A. \tag{8.2.14}$$

Proposition 8.2.3. Let $A \in \mathbb{F}^{n\times n}$, and assume that A is group invertible. Then, the following statements hold:

i) $\overline{A}^{\#} = \overline{A^{\#}}$.

ii) $A^{\#\mathsf{T}} \triangleq A^{\mathsf{T}\#} \triangleq (A^{\mathsf{T}})^{\#} = (A^{\#})^{\mathsf{T}}$.

iii) $A^{\#*} \triangleq A^{*\#} \triangleq (A^*)^{\#} = (A^{\#})^*$.

iv) If $r \in \mathbb{P}$, then $A^{\#r} \triangleq A^{r\#} \triangleq (A^{\#})^r = (A^r)^{\#}$.

v) $\mathcal{R}(A) = \mathcal{R}(AA^{\#}) = \mathcal{N}(I - AA^{\#}) = \mathcal{R}(AA^+) = \mathcal{N}(I - AA^+)$.

vi) $\mathcal{N}(A) = \mathcal{N}(AA^{\#}) = \mathcal{R}(I - AA^{\#}) = \mathcal{N}(A^+A) = \mathcal{R}(I - A^+A)$.

vii) $\operatorname{rank} A = \operatorname{rank} A^{\#} = \operatorname{rank} AA^{\#} = \operatorname{rank} A^{\#}A$.

viii) $\operatorname{def} A = \operatorname{def} A^{\#} = \operatorname{def} AA^{\#} = \operatorname{def} A^{\#}A$.

ix) $AA^{\#}$ is the idempotent matrix onto $\mathcal{R}(A)$ along $\mathcal{N}(A)$.

x) $A^{\#} = 0$ if and only if $A = 0$.

xi) $A^{\#}$ is group invertible.

xii) $(A^{\#})^{\#} = A$.

xiii) If $X \in \mathbb{F}^{n\times n}$ is a (1)-inverse of A^3, then $A^{\#} = AXA$.

xiv) If $S \in \mathbb{F}^{n\times n}$ is nonsingular, then $(SAS^{-1})^{\#} = SA^{\#}S^{-1}$.

An alternative expression for the idempotent matrix onto $\mathcal{R}(A)$ along $\mathcal{N}(A)$ is given by Proposition 4.8.11.

8.3 Facts on the Moore-Penrose Generalized Inverse for One Matrix

Fact 8.3.1. Let $A \in \mathbb{F}^{n\times m}$. Then, A has a unique (1)-inverse if and only if $n = m$ and A is nonsingular. **Source:** [281, p. 8].

Fact 8.3.2. Let $A \in \mathbb{F}^{n\times m}$, and let $B \in \mathbb{F}^{m\times n}$ be a (1)-inverse of A. Then, AB and BA are idempotent.

Fact 8.3.3. Let $A \in \mathbb{R}^{n\times m}$, and let $B \in \mathbb{R}^{m\times n}$ be a (1)-inverse of A. Then, the following statements are equivalent:

i) $AA^{\mathsf{T}}B^{\mathsf{T}}B = AB$.

ii) $\operatorname{tr} AA^{\mathsf{T}}B^{\mathsf{T}}B = \operatorname{tr} AB$.

iii) $A^{\mathsf{T}}B^{\mathsf{T}} = BA$.

Source: [2703]. **Related:** Fact 3.15.37.

Fact 8.3.4. Let $A \in \mathbb{F}^{n\times m}$, let $B \in \mathbb{F}^{m\times n}$ be a (1)-inverse of A, and define $G \in \mathbb{F}^{m\times n}$ by $G \triangleq BAB$. Then, G is a (1,2)-inverse of A. **Source:** [281, p. 8].

Fact 8.3.5. Let $A \in \mathbb{F}^{n\times m}$, define $r \triangleq \operatorname{rank} A$, let $F \in \mathbb{F}^{n\times r}$, let $G \in \mathbb{F}^{r\times m}$, assume that $A = FG$, let $B \in \mathbb{F}^{r\times n}$, and let $C \in \mathbb{F}^{m\times r}$. Then, the following statements are equivalent:

i) BAC is nonsingular.

ii) BF and GC are nonsingular.

Now, assume that these statements hold, and define $X \triangleq C(BAC)^{-1}B$. Then, the following statements hold:

iii) X is a (1,2)-inverse of A.

iv) If $B = F^*$, then X is a (1,2,3)-inverse of A.

v) If $C = G^*$, then X is a (1,2,4)-inverse of A.

vi) If $B = F^*$ and $C = G^*$, then $X = A^+$.

Source: [2238, p. 220].

Fact 8.3.6. Let $A \in \mathbb{F}^{n\times m}$ and $B \in \mathbb{F}^{m\times n}$. Then, B is a (1,3)-inverse of A if and only if $AB = AA^+$. If these conditions hold, then the following statements hold:

i) $I - AB$ is a projector.

ii) $(I - AB)^*A = 0$.

iii) Let $b \in \mathbb{F}^n$. Then, for all $x \in \mathbb{F}^m$, $\|ABb - b\|_2 \le \|Ax - b\|_2$.

Source: [2238, p. 285]. **Remark:** Bb is a least-squares solution of $Ax = b$. See Fact 11.17.6.

Fact 8.3.7. Let $A \in \mathbb{F}^{n\times m}$ and $B \in \mathbb{F}^{m\times n}$. Then, B is a (1,4)-inverse of A if and only if $BA = A^+A$. If these conditions hold, then the following statements hold:

i) $I - BA$ is a projector.

ii) $(I - BA)A^* = 0$.

iii) Let $b \in \mathcal{R}(A)$. Then, $Bb = A^+b$, and, for all $x \in \mathbb{F}^m$ such that $Ax = b$ and $x \ne Bb$, $\|Bb\|_2 < \|x\|_2$.

Source: [2238, p. 284]. **Remark:** Bb is a minimum-norm solution of $Ax = b$. See Fact 11.17.7.

Fact 8.3.8. Let $A \in \mathbb{F}^{n\times m}$ and $B, C \in \mathbb{F}^{m\times n}$, assume that B is a (1,3)-inverse of A, and assume that C is a (1,4)-inverse of A. Then, $A^+ = CAB$. **Source:** [360, p. 48] and [2238, p. 285]. **Credit:** N. S. Urquhart.

Fact 8.3.9. Let $A \in \mathbb{F}^{n\times m}$ and $D \in \mathbb{F}^{m\times n}$. Then, the following statements are equivalent:

i) D is a (2)-inverse of A.

ii) There exist projectors $C \in \mathbb{F}^{n\times n}$ and $B \in \mathbb{F}^{m\times m}$ such that $D = (CAB)^+$.

Source: [2238, p. 297].

Fact 8.3.10. Let $A \in \mathbb{F}^{n\times m}$. Then, the following statements hold:

i) If $B \in \mathbb{F}^{n\times n}$ is a (1,2)-inverse of AA^*, then A^*B is a (1,2)-inverse of A.

ii) If $B \in \mathbb{F}^{m\times m}$ is a (1,2)-inverse of A^*A, then BA^* is a (1,2)-inverse of A.

Source: [2238, p. 289].

Fact 8.3.11. Let $A \in \mathbb{F}^{n\times m}$, $B \in \mathbb{F}^{k\times n}$, and $C \in \mathbb{F}^{m\times l}$, and assume that B is left inner and C is right inner. Then, $(BAC)^+ = C^*A^+B^*$. **Source:** [1343, p. 506].

Fact 8.3.12. Let $A \in \mathbb{F}^{n\times m}$, assume that A is right invertible, let $A^{\mathsf{R}} \in \mathbb{F}^{m\times n}$, and assume that A^{R} is a right inverse of A. Then, $A^{\mathsf{R}} = A^+$ if and only if $A^{\mathsf{R}}A$ is Hermitian.

Fact 8.3.13. Let $A \in \mathbb{F}^{n\times m}$, assume that A is left invertible, let $A^{\mathsf{L}} \in \mathbb{F}^{m\times n}$, and assume that A^{L} is a left inverse of A. Then, $A^{\mathsf{L}} = A^+$ if and only if AA^{L} is Hermitian.

Fact 8.3.14. Let $A \in \mathbb{F}^{n\times m}$, assume that A is left invertible, and let $A^{\mathsf{L}} \in \mathbb{F}^{m\times n}$ be a left inverse of A. Then, $B \in \mathbb{F}^{m\times n}$ is a left inverse of A if and only if there exists $S \in \mathbb{F}^{m\times n}$ such that $B = A^{\mathsf{L}} + S(I_n - AA^+)$. **Related:** Fact 3.18.3, Fact 11.17.4, and [2238, p. 150].

Fact 8.3.15. Let $A \in \mathbb{F}^{n\times m}$, assume that A is right invertible, and let $A^{\mathsf{R}} \in \mathbb{F}^{m\times n}$ be a right inverse of A. Then, $B \in \mathbb{F}^{m\times n}$ is a right inverse of A if and only if there exists $S \in \mathbb{F}^{m\times n}$ such that $B = A^{\mathsf{R}} + (I_m - A^+A)S$. **Related:** Fact 3.18.4 and Fact 11.17.5.

Fact 8.3.16. Let $A \in \mathbb{F}^{n\times m}$, $b \in \mathbb{F}^n$, and $y \in \mathbb{F}^m$, assume that A is right invertible, and define $x = A^+b + (I - A^+A)y$, which satisfies $Ax = b$. Furthermore, let $z \triangleq (I - A^+A)y$, and assume that there exists $S \in \mathbb{F}^{m\times n}$ such that $z^{\mathsf{T}}Sb \neq 0$, and define

$$A^{\mathsf{R}} = A^+ + \frac{1}{z^{\mathsf{T}}Sb}zz^{\mathsf{T}}S.$$

Then, $A^{\mathsf{R}} \in \mathbb{F}^{m\times n}$ is a right inverse of A, and $x = A^{\mathsf{R}}b$. **Related:** See the comments following Theorem 3.7.5. **Problem:** Assuming that A is right invertible, find necessary and sufficient conditions under which every solution x of $Ax = b$ is given by $x = A^{\mathsf{R}}b$ for some right inverse of A.

Fact 8.3.17. Let $A \in \mathbb{F}^{n\times m}$, and assume that $\operatorname{rank} A = 1$. Then,

$$A^+ = (\operatorname{tr} AA^*)^{-1}A^*.$$

Consequently, if $x \in \mathbb{F}^n$ and $y \in \mathbb{F}^n$ are nonzero, then

$$(xy^*)^+ = (x^*xy^*y)^{-1}yx^* = \frac{1}{\|x\|_2^2\|y\|_2^2}yx^*.$$

In particular,

$$1_{n\times m}^+ = \frac{1}{nm}1_{m\times n}.$$

Fact 8.3.18. Let $x \in \mathbb{F}^n$, and assume that x is nonzero. Then, the projector $A \in \mathbb{F}^{n\times n}$ onto $\operatorname{span}\{x\}$ is given by

$$A = (x^*x)^{-1}xx^*.$$

Fact 8.3.19. Let $x, y \in \mathbb{F}^n$, assume that x, y are nonzero, and assume that $x^*y = 0$. Then, the projector $A \in \mathbb{F}^{n\times n}$ onto $\operatorname{span}\{x, y\}$ is given by

$$A = (x^*x)^{-1}xx^* + (y^*y)^{-1}yy^*.$$

Fact 8.3.20. Let $x, y \in \mathbb{F}^n$, and assume that x, y are linearly independent. Then, the projector $A \in \mathbb{F}^{n\times n}$ onto $\operatorname{span}\{x, y\}$ is given by

$$A = (x^*xy^*y - |x^*y|^2)^{-1}(y^*yxx^* - y^*xyx^* - x^*yxy^* + x^*xyy^*).$$

Furthermore, define $z \triangleq [I - (x^*x)^{-1}xx^*]y$. Then,

$$A = (x^*x)^{-1}xx^* + (z^*z)^{-1}zz^*.$$

Remark: For $\mathbb{F} = \mathbb{R}$, this result is given in [2474, p. 178].

Fact 8.3.21. Let $n \geq 2$, let $A \in \mathbb{F}^{n\times n}$, assume that $\operatorname{rank} A = n - 1$, let $x \in \mathcal{N}(A)$ be nonzero, let $y \in \mathcal{N}(A^*)$ be nonzero, let $\alpha = 1$ if $\operatorname{spec}(A) = \{0\}$ and the product of the nonzero eigenvalues of A otherwise, and define $k \triangleq \operatorname{amult}_A(0)$. Then,

$$A^{\mathsf{A}} = \frac{(-1)^{k+1}\alpha}{y^*(A^{k-1})^+x}xy^*.$$

In particular,

$$N_n^{\mathsf{A}} = (-1)^{n+1}E_{1,n}.$$

If, in addition, $k = 1$, then

$$A^{\mathsf{A}} = \frac{\alpha}{y^*x}xy^*.$$

Source: [1924, pp. 40, 41] and Fact 4.22.4. **Remark:** This result provides an expression for *ii*) of Fact 3.19.3. **Remark:** If A is range Hermitian, then $\mathcal{N}(A) = \mathcal{N}(A^*)$ and $y^*x \neq 0$, and thus Fact 7.15.5 implies that A^{A} is semisimple. **Related:** Fact 7.15.30.

Fact 8.3.22. Let $A \in \mathbb{F}^{n\times m}$, and assume that $\operatorname{rank} A = n - 1$. Then,

$$A^+ = \frac{1}{\det[AA^* + (AA^*)^{\mathsf{A}}]}A^*[AA^* + (AA^*)^{\mathsf{A}}]^{\mathsf{A}}.$$

Source: [772]. **Remark:** Extensions to matrices of arbitrary rank are given in [772]. Expressions for the generalized inverse in terms of submatrices are given in [2822].

Fact 8.3.23. Let $A \in \mathbb{F}^{n\times n}$, assume that A is nonzero, and define $r \triangleq \operatorname{rank} A$, and $B \triangleq \operatorname{diag}[\sigma_1(A), \ldots, \sigma_r(A)]$. Then, there exist $S \in \mathbb{F}^{n\times n}$, $K \in \mathbb{F}^{r\times r}$, and $L \in \mathbb{F}^{r\times(n-r)}$ such that S is unitary, $KK^* + LL^* = I_r$, and

$$A = S\begin{bmatrix} BK & BL \\ 0_{(n-r)\times r} & 0_{(n-r)\times(n-r)} \end{bmatrix}S^*.$$

Furthermore,

$$A^+ = S\begin{bmatrix} K^*B^{-1} & 0_{r\times(n-r)} \\ L^*B^{-1} & 0_{(n-r)\times(n-r)} \end{bmatrix}S^*.$$

Source: Fact 7.10.27 and [240, 1338]. **Related:** Fact 8.5.13.

Fact 8.3.24. Let $A \in \mathbb{F}^{n\times n}$. Then,

$$\operatorname{rank}[A, A^+] = 2\operatorname{rank}[A\ \ A^*] - 2\operatorname{rank} A,$$

$$\operatorname{rank}[AA^+, A^+A] = 2\operatorname{rank}[A\ \ A^*] + 2\operatorname{rank} A^2 - 4\operatorname{rank} A.$$

If, in addition, $k \geq 1$, then

$$\operatorname{rank}[A^k, A^+] = \operatorname{rank}\begin{bmatrix} A^k \\ A^* \end{bmatrix} + \operatorname{rank}[A^k\ \ A^*] - 2\operatorname{rank} A.$$

Source: [2657, 2672].

Fact 8.3.25. Let $A \in \mathbb{F}^{n\times m}$ and $B \in \mathbb{F}^{n\times n}$. Then, the following statements are equivalent:

i) $B = AA^+$.

ii) $\mathcal{R}(B) = \mathcal{R}(A)$ and $B = BB^*$.

iii) $\mathcal{R}(B) \subseteq \mathcal{R}(A)$ and $A^*B = A^*$.

iv) $\mathcal{R}(B^*) \subseteq \mathcal{R}(A)$ and $BA = A$.

v) For all $X \in \mathbb{F}^{n\times n}$ such that $XA = A$, $\operatorname{tr} BB^* \le \operatorname{tr} XX^*$.

vi) $A^*B = A^*$, $BA = A$, and $\operatorname{rank} B = \operatorname{rank} A$.

Source: [2657].

Fact 8.3.26. Let $A \in \mathbb{F}^{n\times m}$, and assume that $\operatorname{rank} A = m$. Then, $(AA^*)^+ = A(A^*A)^{-2}A^*$. **Related:** Fact 8.4.18.

Fact 8.3.27. Let $A \in \mathbb{F}^{n\times m}$. Then,

$$A^+ = \lim_{\alpha\downarrow 0} A^*(AA^* + \alpha I)^{-1} = \lim_{\alpha\downarrow 0} (A^*A + \alpha I)^{-1}A^*,$$

$$A^+A = \lim_{\alpha\downarrow 0} A^*(AA^* + \alpha I)^{-1}A, \quad AA^+ = \lim_{\alpha\downarrow 0} A(A^*A + \alpha I)^{-1}A^*.$$

Fact 8.3.28. Let $A \in \mathbb{F}^{n\times m}$, let $\chi_{AA^*}(s) = s^n + \beta_{n-1}s^{n-1} + \cdots + \beta_1 s + \beta_0$, and let k denote the largest integer in $\{0, \ldots, n-1\}$ such that $\beta_{n-k} \ne 0$. Then,

$$A^+ = -\beta_{n-k}^{-1}A^*[(AA^*)^{k-1} + \beta_{n-1}(AA^*)^{k-2} + \cdots + \beta_{n-k+1}I].$$

Source: [863].

Fact 8.3.29. Let $A \in \mathbb{F}^{n\times n}$, and assume that A is Hermitian. Then, $\operatorname{In} A = \operatorname{In} A^+ = \operatorname{In} A^{\mathsf{D}}$. If, in addition, A is nonsingular, then $\operatorname{In} A = \operatorname{In} A^{-1}$.

Fact 8.3.30. Let $A \in \mathbb{F}^{n\times n}$. Then, the following statements are equivalent:

i) A is Hermitian.

ii) $A^* = A^2A^+$.

Source: [257].

Fact 8.3.31. Let $A \in \mathbb{F}^{n\times m}$, and define $r \triangleq \operatorname{rank} A$. Then, the following statements are equivalent:

i) A is a partial isometry.

ii) A^*A is a projector.

iii) $AA^*A = A$.

iv) $A^*AA^* = A^*$.

v) $A^+ = A^*$.

vi) $\sigma_1(A) = \sigma_r(A) = 1$.

In particular, $N_n^+ = N_n^{\mathsf{T}}$. **Source:** [360, pp. 219–220]. **Remark:** The partial isometry A preserves lengths and distances with respect to the Euclidean norm on $\mathcal{R}(A^*)$. See [360, p. 219]. **Related:** Fact 7.12.34 and Fact 8.6.1.

Fact 8.3.32. Let $A \in \mathbb{F}^{n\times n}$. Then, the following statements are equivalent:

i) $A^+A^* = A^*A^+$.

ii) $AA^+A^*A = AA^*A^+A$.

iii) $AA^*A^2 = A^2A^*A$.

If these statements hold, then A is *star-dagger*. If A is star-dagger, then $A^2(A^+)^2$ and $(A^+)^2A^2$ are positive semidefinite. **Source:** [1338, 2628]. **Related:** Fact 8.6.1.

Fact 8.3.33. Let $A \in \mathbb{F}^{n\times m}$, assume that A is nonzero, and define $r \triangleq \operatorname{rank} A$. Then, for all $i \in \{1, \ldots, r\}$, the singular values of A^+ are given by

$$\sigma_i(A^+) = \sigma_{r+1-i}^{-1}(A).$$

In particular,

$$\sigma_r(A) = 1/\sigma_{\max}(A^+).$$

If, in addition, $A \in \mathbb{F}^{n\times n}$ and A is nonsingular, then

$$\sigma_{\min}(A) = 1/\sigma_{\max}(A^{-1}).$$

Fact 8.3.34. Let $A \in \mathbb{F}^{n\times m}$. Then, $X = A^+$ is the unique matrix satisfying

$$\operatorname{rank}\begin{bmatrix} A & AA^+ \\ A^+A & X \end{bmatrix} = \operatorname{rank} A.$$

Source: [1043]. **Related:** Fact 3.22.11 and Fact 8.10.11.

Fact 8.3.35. Let $A \in \mathbb{F}^{n\times n}$, and assume that A is centrohermitian. Then, A^+ is centrohermitian. **Source:** [1776].

Fact 8.3.36. Let $A \in \mathbb{F}^{n\times m}$. Then, $I + A^+A$ and $I + AA^+$ are positive definite, and $A^+ = 4(I + A^+A)^{-1}A^+(I + AA^+)^{-1}$.

Fact 8.3.37. Let $A \in \mathbb{F}^{n\times n}$, and assume that A is unitary. Then,

$$\lim_{k\to\infty} \frac{1}{k}\sum_{i=0}^{k-1} A^i = I - (A - I)(A - I)^{\#} = I - (A - I)(A - I)^+.$$

Source: Fact 15.22.14 and Fact 15.22.17. Since $A - I$ is normal, it is range Hermitian. Fact 8.10.16 implies that $(A - I)^{\#} = (A - I)^+$. See [1301, p. 185]. **Remark:** $I - (A - I)(A - I)^+$ is the projector onto $\mathcal{N}[(A - I)^*] = \mathcal{N}(A - I) = \{x\colon\ Ax = x\}$. See Fact 4.9.6. **Remark:** This is the *ergodic theorem*.

Fact 8.3.38. Let $A \in \mathbb{F}^{n\times m}$, and define the sequence $(B_i)_{i=1}^{\infty}$ by $B_{i+1} \triangleq 2B_i - B_iAB_i$, where $B_0 \triangleq \alpha A^*$ and $\alpha \in (0, 2/\sigma_{\max}^2(A))$. Then, $\lim_{i\to\infty} B_i = A^+$. **Source:** [300, p. 259] and [624, p. 250]. **Remark:** This is a Newton-Raphson algorithm. **Remark:** B_0 satisfies $\rho_{\max}(I - B_0A) < 1$. **Remark:** For the case where A is square and nonsingular, see Fact 3.20.22. **Credit:** A. Ben-Israel. **Problem:** Determine whether convergence holds for all $B_0 \in \mathbb{F}^{n\times n}$ satisfying $\rho_{\max}(I - B_0A) < 1$.

Fact 8.3.39. Let $A \in \mathbb{F}^{n\times m}$, let $(A_i)_{i=1}^{\infty} \subset \mathbb{F}^{n\times m}$, and assume that $\lim_{i\to\infty} A_i = A$. Then, the following statements are equivalent:

i) $\lim_{i\to\infty} A_i^+ = A^+$.

ii) There exists a positive integer k such that, for all $i > k$, $\operatorname{rank} A_i = \operatorname{rank} A$.

iii) $(\sigma_{\max}(A_i^+))_{i=1}^{\infty}$ is bounded.

iv) $\lim_{i\to\infty} A_iA_i^+ = AA^+$.

Source: [403], [624, pp. 218, 219], [1275, p. 211], and [2403, pp. 199, 200].

8.4 Facts on the Moore-Penrose Generalized Inverse for Two or More Matrices

Fact 8.4.1. Let $A, B \in \mathbb{F}^{n\times m}$. Then, $\mathcal{N}(I - B^+A) = \mathcal{N}(A - B) \cap \mathcal{R}(B^+)$. **Source:** Let $x \in \mathcal{N}(I - B^+A)$. Then, $x = B^+Ax \in \mathcal{R}(B^+) \subseteq \mathcal{N}(A - B) \cap \mathcal{R}(B^+)$. Now, let $x \in \mathcal{N}(A - B) \cap \mathcal{R}(B^+)$. Then, $Ax = Bx$, and there exists $y \in \mathbb{F}^n$ such that $x = B^+y$. Therefore, $x - B^+Ax = x - B^+Bx = x - B^+BB^+y = x - B^+y = 0$. **Remark:** This result generalizes (34) of [620].

Fact 8.4.2. Let $A \in \mathbb{F}^{n\times m}$ and $B \in \mathbb{F}^{n\times l}$, and define $P \triangleq AA^+$ and $Q \triangleq BB^+$. Then,

$$\begin{aligned}
\operatorname{rank} A^*B &= \operatorname{rank} A^*Q = \operatorname{rank} PB = \operatorname{rank} PQ \\
&= \operatorname{rank} B - \dim[\mathcal{N}(P) \cap \mathcal{R}(Q)] \\
&= \tfrac{1}{2}\operatorname{rank}\,[P, Q] + \dim[\mathcal{R}(A) \cap \mathcal{R}(B)] \\
&= \operatorname{rank} A + \operatorname{rank} B - \dim(\mathcal{R}(B) + [\mathcal{R}(A) \cap \mathcal{N}(B^*)]) \\
&= \operatorname{rank}(I - P - Q) - \operatorname{rank}\ P_\perp Q_\perp \\
&= \operatorname{rank}(PQ + QP) - \tfrac{1}{2}\operatorname{rank}\,[P, Q]
\end{aligned}$$

$$\begin{aligned}
&= n + \operatorname{rank}[P, Q] - \operatorname{rank} PQ_\perp - \operatorname{rank} P_\perp Q - \operatorname{rank} P_\perp Q_\perp \\
&= \operatorname{rank} A + \operatorname{rank} B + \operatorname{rank} P_\perp Q_\perp - n \\
&= \operatorname{rank} A + \operatorname{rank} B - \operatorname{rank}[A \;\; B] + \tfrac{1}{2}\operatorname{rank}[P, Q] \\
&= \tfrac{1}{2}[\operatorname{rank} A + \operatorname{rank} B + \operatorname{rank}(I - P - Q) - n] \\
&= n + \tfrac{1}{2}\operatorname{rank}[P, Q] - \operatorname{rank}(I - PQ) \\
&= \tfrac{1}{2}[\operatorname{rank} A + \operatorname{rank} B + \operatorname{rank}[P, Q] - \operatorname{rank}(P - Q)],
\end{aligned}$$

$$\begin{aligned}
\operatorname{rank}[A \;\; B] &= \operatorname{rank} A + \operatorname{rank} P_\perp B = \operatorname{rank} B + \operatorname{rank} Q_\perp A \\
&= \operatorname{rank}(P + Q - PQ) \\
&= \operatorname{rank}(P - Q) + \dim[\mathcal{R}(A) \cap \mathcal{R}(B)] \\
&= \tfrac{1}{2}[\operatorname{rank} A + \operatorname{rank} B + \operatorname{rank}(P - Q)] \\
&= n - \operatorname{rank} P_\perp Q_\perp + \tfrac{1}{2}\operatorname{rank}[P, Q] \\
&= \operatorname{rank}(PQ + QP) + \dim([\mathcal{N}(A^*) \cap \mathcal{R}(B)] + [\mathcal{R}(A) \cap \mathcal{N}(B^*)]) \\
&= n + \operatorname{rank}[P, Q] - \dim([\mathcal{R}(A) + \mathcal{N}(B^*)] \cap [\mathcal{N}(A^*) + \mathcal{R}(B)] \cap [\mathcal{N}(A^*) + \mathcal{N}(B^*)]),
\end{aligned}$$

$$\operatorname{rank} A + \tfrac{1}{2}\operatorname{rank}[P, Q] \le \operatorname{rank}[A \;\; B],$$

$$\tfrac{1}{2}\operatorname{rank}[P, Q] = \operatorname{rank} A^* Q_\perp P_\perp B \le \operatorname{rank} A^* B \le \operatorname{rank}(PQ + QP).$$

Furthermore, the following statements are equivalent:

i) $\operatorname{rank} A^*B = \operatorname{rank} A$.

ii) $\mathcal{R}(A) \cap \mathcal{N}(B^*) = \{0\}$.

iii) $\mathcal{R}(A) \subseteq \mathcal{N}(A^*) + \mathcal{R}(B)$.

iv) $P = PQ(PQ)^+$.

v) P is the projector onto $\mathcal{R}(P) \cap [\mathcal{N}(P) + \mathcal{R}(Q)]$.

Finally, the following statements are equivalent:

vi) $\operatorname{rank} A^*B = \operatorname{rank} B$.

vii) $\mathcal{N}(A^*) \cap \mathcal{R}(B) = \{0\}$.

viii) $\mathcal{N}(A^*) \subseteq \mathcal{R}(A) + \mathcal{N}(B^*)$.

ix) $P_\perp = P_\perp Q_\perp (P_\perp Q_\perp)^+$.

x) $P_\perp$ is the projector onto $\mathcal{N}(P) \cap [\mathcal{R}(P) + \mathcal{N}(Q)]$.

Proof. See [252], [2238, p. 312], and [3014].

Fact 8.4.3. Let $A \in \mathbb{F}^{n\times m}$ and $B \in \mathbb{F}^{n\times l}$. Then, $\mathcal{R}(A) \subseteq \mathcal{R}(B)$ if and only if $BB^+A = A$. **Source:** [31, p. 35].

Fact 8.4.4. Let $A \in \mathbb{F}^{n\times m}$ and $B \in \mathbb{F}^{m\times n}$. Then, the following statements are equivalent:

i) $AB = 0$.

ii) $B = (A^+A)_\perp B$.

Source: [2238, p. 311].

Fact 8.4.5. Let $A, B \in \mathbb{F}^{n\times n}$, assume that B is Hermitian, and assume that $AB = BA$. Then, the following statements hold:

i) $A(B^+B)_\perp = (B^+B)_\perp A$.

ii) $(A^+A)_\perp B = B(A^+A)_\perp$.

Source: [2238, p. 311].

Fact 8.4.6. Let $A \in \mathbb{F}^{n\times m}$ and $B \in \mathbb{F}^{m\times n}$. Then, the following statements are equivalent:

i) $B = A^+$.

ii) $A^*AB = A^*$ and $B^*BA = B^*$.

iii) $BAA^* = A^*$ and $ABB^* = B^*$.

Remark: See [1343, pp. 503, 513].

Fact 8.4.7. Let $A, B \in \mathbb{F}^{n\times m}$. Then, the following statements are equivalent:

i) $\operatorname{rank}\,[A\ \ B] = \operatorname{rank}\begin{bmatrix} A \\ B \end{bmatrix}$.

ii) $\operatorname{rank}(AA^* + BB^*) = \operatorname{rank}(A^*A + B^*B)$.

iii) $\operatorname{rank}(AA^+ + BB^+) = \operatorname{rank}(A^+A + B^+B)$.

iv) $\operatorname{rank}(AA^+ - BB^+) = \operatorname{rank}(A^+A - B^+B)$.

v) $\begin{bmatrix} AA^* & B \\ B^* & 0 \end{bmatrix} = \operatorname{rank}\begin{bmatrix} A^*A & B^* \\ B & 0 \end{bmatrix}$.

vi) $\begin{bmatrix} BB^* & A \\ A^* & 0 \end{bmatrix} = \operatorname{rank}\begin{bmatrix} B^*B & A^* \\ A & 0 \end{bmatrix}$.

vii) $\operatorname{rank}\begin{bmatrix} AA^+ & BB^+ \\ BB^+ & AA^+ \end{bmatrix} = \operatorname{rank}\begin{bmatrix} A^+A & B^+B \\ B^+B & A^+A \end{bmatrix}$.

viii) $\operatorname{rank}\,[I - AA^+\ \ I - BB^+] + m = \operatorname{rank}\,[I - A^+A\ \ I - B^+B] + n$.

ix) $\dim[\mathcal{R}(A) \cap \mathcal{R}(B)] = \dim[\mathcal{R}(A^*) \cap \mathcal{R}(B^*)]$.

x) $\dim[\mathcal{N}(A) \cap \mathcal{N}(B)] + m = \dim[\mathcal{N}(A^*) \cap \mathcal{N}(B^*)] + n$.

Related: Fact 3.14.11. **Credit:** Y. Tian.

Fact 8.4.8. Let $A \in \mathbb{F}^{n\times m}$ and $B \in \mathbb{F}^{n\times l}$. Then,

$$\mathcal{R}(A) \cap \mathcal{R}(B) = \mathcal{R}[AA^+(AA^+ + BB^+)^+BB^+],$$

$$\dim[\mathcal{R}(A) \cap \mathcal{R}(B)] = \operatorname{rank} AA^+(AA^+ + BB^+)^+BB^+ = \operatorname{rank} A + \operatorname{rank} B - \operatorname{rank}\,[A\ \ B].$$

Now, assume that $A^*B = 0$. Then, $\mathcal{R}(A) \cap \mathcal{R}(B) = \{0\}$, $AA^+(AA^+ + BB^+)^+BB^+ = 0$, and $\operatorname{rank}\,[A\ \ B] = \operatorname{rank} A + \operatorname{rank} B$. **Source:** Fact 8.8.19. For the second statement, Proposition 8.1.7 implies that $A^+B = 0$, and thus Fact 8.8.19 implies that $AA^+(AA^+ + BB^+)^+BB^+ = AA^+BB^+(AA^+ + BB^+)^+AA^+BB^+ = 0$. Alternatively, Fact 3.13.37 implies that $\mathcal{R}(A)$ and $\mathcal{R}(B)$ are mutually orthogonal subspaces, and thus $\mathcal{R}(A) \cap \mathcal{R}(B) = \{0\}$. For the last statement, use Fact 3.14.8, Fact 3.14.18, and Fact 8.4.8. See [2238, p. 306]. **Remark:** See Theorem 3.1.3, Fact 3.13.37, Fact 3.14.15, Fact 8.8.19, Fact 8.9.5, Fact 8.9.26, and Fact 10.24.20.

Fact 8.4.9. Let $A \in \mathbb{F}^{n\times m}$ and $B \in \mathbb{F}^{n\times l}$. Then,

$$\mathcal{R}(A) + \mathcal{R}(B) = \mathcal{R}([A\ \ B][A\ \ B]^+).$$

Source: [2238, p. 306]. **Related:** Fact 8.9.20 and Fact 8.9.23.

Fact 8.4.10. Let $A \in \mathbb{F}^{n\times m}$, and let $x \in \mathbb{F}^n$ and $y \in \mathbb{F}^m$ be nonzero. Furthermore, define

$$d \triangleq A^+x, \quad e \triangleq A^{+*}y, \quad f \triangleq (I - AA^+)x, \quad g \triangleq (I - A^+A)y,$$

$$\delta \triangleq d^*d, \quad \eta \triangleq e^*e, \quad \phi \triangleq f^*f, \quad \psi \triangleq g^*g,$$

$$\lambda \triangleq 1 + y^*A^+x, \quad \mu \triangleq |\lambda|^2 + \delta\psi, \quad \nu \triangleq |\lambda|^2 + \eta\phi.$$

Then,

$$\operatorname{rank}(A + xy^*) = \operatorname{rank} A - 1$$

if and only if

$$x \in \mathcal{R}(A), \quad y \in \mathcal{R}(A^*), \quad \lambda = 0.$$

If these conditions hold, then

$$(A + xy^*)^+ = A^+ - \delta^{-1}dd^*A^+ - \eta^{-1}A^+ee^* + (\delta\eta)^{-1}d^*A^+ede^*.$$

Furthermore,

$$\operatorname{rank}(A + xy^*) = \operatorname{rank} A$$

if and only if exactly one of the following statements holds:

$$\begin{cases} x \in \mathcal{R}(A), & y \in \mathcal{R}(A^*), \quad \lambda \neq 0,\\ x \in \mathcal{R}(A), & y \notin \mathcal{R}(A^*),\\ x \notin \mathcal{R}(A), & y \in \mathcal{R}(A^*).\end{cases}$$

If these conditions hold, then, respectively,

$$\begin{cases} (A + xy^*)^+ = A^+ - \lambda^{-1}de^*,\\ (A + xy^*)^+ = A^+ - \mu^{-1}(\psi dd^*A^+ + \delta ge^*) + \mu^{-1}(\lambda gd^*A^+ - \overline{\lambda}de^*),\\ (A + xy^*)^+ = A^+ - \nu^{-1}(\phi A^+ee^* + \eta df^*) + \nu^{-1}(\lambda A^+ef^* - \overline{\lambda}de^*).\end{cases}$$

Finally, assume that A does not have full rank. Then,

$$\operatorname{rank}(A + xy^*) = \operatorname{rank} A + 1$$

if and only if

$$x \notin \mathcal{R}(A), \quad y \notin \mathcal{R}(A^*).$$

If these conditions hold, then

$$(A + xy^*)^+ = A^+ - \phi^{-1}df^* - \psi^{-1}ge^* + \lambda(\phi\psi)^{-1}gf^*.$$

Source: [227]. To prove sufficiency in the first alternative of the third statement, let $\hat{x} \in \mathbb{F}^m$ and $\hat{y} \in \mathbb{F}^n$ satisfy $x = A\hat{x}$ and $y = A^*\hat{y}$. Then, $A + xy^* = A(I + \hat{x}y^*)$. Since $\alpha \neq 0$ it follows that $-1 \neq y^*A^+x = \hat{y}^*AA^+A\hat{x} = \hat{y}^*A\hat{x} = y^*\hat{x}$. It now follows that $I + \hat{x}y^*$ is an elementary matrix and thus, by Fact 4.10.19, is nonsingular. **Remark:** An equivalent version of the first statement is given in [734] and [1467, p. 33]. A detailed treatment of the generalized inverse of an outer-product perturbation is given in [2821, pp. 152–157]. **Remark:** The equality $\operatorname{rank}(A + xy^*) = \operatorname{rank} A - 1$ is the Wedderburn rank-one reduction formula. See [734] and Fact 8.9.13. **Related:** Fact 3.13.32.

Fact 8.4.11. Let $A \in \mathbb{F}^{n\times n}$, assume that A is Hermitian and nonsingular, and let $x \in \mathbb{F}^n$ and $y \in \mathbb{F}^n$ be nonzero. Then, $A + xy^*$ is singular if and only if $y^*A^{-1}x + 1 = 0$. If these conditions hold, then

$$(A + xy^*)^+ = (I - aa^+)A^{-1}(I - bb^+),$$

where $a \triangleq A^{-1}x$ and $b \triangleq A^{-1}y$. **Source:** [2403, pp. 197, 198].

Fact 8.4.12. Let $A \in \mathbb{F}^{n\times n}$, assume that A is Hermitian, let $b \in \mathbb{F}^n$, and define $S \triangleq I - A^+A$. Then,

$$(A + bb^*)^+ = \begin{cases} [I - (b^*(A^+)^2b)^{-1}A^+bb^*A^+]A^+[I - (b^*A^{+2}b)^{-1}A^+bb^*A^+], & 1+b^*A^+b = 0,\\ A^+ - (1 + b^*A^+b)^{-1}A^+bb^*A^+, & 1+b^*A^+b \neq 0,\\ [I - (b^*Sb)^{-1}Sbb^*]A^+[I - (b^*Sb)^{-1}bb^*S] + (b^*Sb)^{-2}Sbb^*S, & b^*Sb \neq 0.\end{cases}$$

Source: [2048].

Fact 8.4.13. Let $A \in \mathbb{F}^{n\times n}$, assume that A is positive semidefinite, let $C \in \mathbb{F}^{m\times m}$, assume that C is positive definite, and let $B \in \mathbb{F}^{n\times m}$. Then,

$$(A + BCB^*)^+ = A^+ - A^+B(C^{-1} + B^*A^+B)^{-1}B^*A^+$$

if and only if $AA^+B = B$. **Source:** [2146]. **Remark:** $AA^+B = B$ is equivalent to $\mathcal{R}(B) \subseteq \mathcal{R}(A)$. See Fact 8.4.3. **Remark:** Extensions of the matrix inversion lemma are considered in [849, 1047, 2048, 2310] and [1343, pp. 426–428, 447, 448].

Fact 8.4.14. Let $A \in \mathbb{F}^{n\times m}$ and $B \in \mathbb{F}^{m\times l}$. Then, $AB = 0$ if and only if $B^+A^+ = 0$. **Source:** *ix*) $\Longrightarrow$ *i*) of Fact 8.4.26.

Fact 8.4.15. Let $A \in \mathbb{F}^{n\times m}$ and $B \in \mathbb{F}^{n\times l}$. Then, $A^+B = 0$ if and only if $A^*B = 0$. **Source:** Proposition 8.1.7.

Fact 8.4.16. Let $A \in \mathbb{F}^{n\times m}$, let $B \in \mathbb{F}^{m\times p}$, and assume that $\operatorname{rank} B = m$. Then,

$$AB(AB)^+ = AA^+.$$

Source: [2403, p. 215].

Fact 8.4.17. Let $A \in \mathbb{F}^{n\times m}$, let $B \in \mathbb{F}^{m\times m}$, and assume that B is positive definite. Then,

$$ABA^*(ABA^*)^+A = A.$$

Source: [2403, p. 215].

Fact 8.4.18. Let $A \in \mathbb{F}^{n\times m}$, assume that $\operatorname{rank} A = m$, let $B \in \mathbb{F}^{m\times m}$, and assume that B is positive definite. Then,

$$(ABA^*)^+ = A(A^*A)^{-1}B^{-1}(A^*A)^{-1}A^*.$$

Source: Fact 8.3.26.

Fact 8.4.19. Let $A \in \mathbb{F}^{n\times m}$, let $S \in \mathbb{F}^{m\times m}$, assume that S is nonsingular, and define $B \triangleq AS$. Then, $BB^+ = AA^+$. **Source:** [2418, p. 144].

Fact 8.4.20. Let $A \in \mathbb{F}^{n\times r}$ and $B \in \mathbb{F}^{r\times m}$, and assume that $\operatorname{rank} A = r$ and $\operatorname{rank} B = m$. Then, B^+A^+ is a left inverse of AB, and

$$B^+A^+ = (B^*B)^{-1}B^*(A^*A)^{-1}A^*.$$

If, in addition, $m = r$, then

$$(AB)^+ = B^+A^+ = B^{-1}(A^*A)^{-1}A^*.$$

Remark: A and B are left invertible. **Related:** Fact 3.18.9.

Fact 8.4.21. Let $A \in \mathbb{F}^{n\times r}$ and $B \in \mathbb{F}^{r\times m}$, and assume that $\operatorname{rank} A = n$ and $\operatorname{rank} B = r$. Then, B^+A^+ is a right inverse of AB, and

$$B^+A^+ = B^*(BB^*)^{-1}A^*(AA^*)^{-1}.$$

If, in addition, $n = r$, then

$$(AB)^+ = B^+A^+ = B^*(BB^*)^{-1}A^{-1}.$$

Remark: A and B are right invertible. **Related:** Fact 3.18.10.

Fact 8.4.22. Let $A \in \mathbb{F}^{n\times m}$ and $B \in \mathbb{F}^{m\times l}$, and define $B_1 \triangleq A^+AB$ and $A_1 \triangleq AB_1B_1^+$. Then,

$$(AB)^+ = (A_1B_1)^+ = B_1^+A_1^+.$$

Source: [2238, p. 188] and [2403, pp. 191, 192]. **Remark:** Products of generalized inverses are considered in [2663].

Fact 8.4.23. Let $A \in \mathbb{F}^{n\times m}$ and $B \in \mathbb{F}^{m\times l}$. Then,

$$(AB)^+ = (A^+AB)^+(ABB^+)^+ = (A^{+*}B)^+(B^+A^+)^*(AB^{+*})^+, \quad (AB)^+AB = (A^+AB)^+A^+AB.$$

Furthermore, if $\mathcal{R}(B) = \mathcal{R}(A^*)$, then

$$A^+AB = B, \quad ABB^+ = A, \quad (AB)^+ = B^+A^+.$$

Source: [2238, p. 312], [2403, p. 192], [2652, 2665]. **Credit:** R. E. Cline and T. N. E. Greville.

Fact 8.4.24. Let $A \in \mathbb{F}^{n\times m}$ and $B \in \mathbb{F}^{m\times l}$, and assume that at least one of the following statements holds:

i) $(AB)^+ = B^+(A^+ABB^+)^+A^+$.

ii) $(AB)^+ = B^*(A^*ABB^*)^+A^*$.

iii) $(AB)^+ = B^+A^+ - B^+[(I - BB^+)(I - A^+A)]^+A^+$.

iv) $\mathcal{R}(AA^*AB) = \mathcal{R}(AB)$ and $\mathcal{R}[(ABB^*B)^*] = \mathcal{R}[(AB)^*]$.

Then, all four statements hold. **Source:** [2637].

Fact 8.4.25. Let $A \in \mathbb{F}^{n\times m}$ and $B \in \mathbb{F}^{m\times l}$.

$$\operatorname{rank}[(ABB^+)^+ - BB^+A^+] = \operatorname{rank}\,[A^*AB \;\; B] - \operatorname{rank} B,$$

$$\operatorname{rank}[(A^+AB)^+ - B^+A^+A] = \operatorname{rank}\begin{bmatrix} A \\ ABB^* \end{bmatrix} - \operatorname{rank} A,$$

$$\operatorname{rank}[(A^+ABB^+)^+ - A^+ABB^+] = \operatorname{rank}\,[A^* \;\; B] + \operatorname{rank} AB - \operatorname{rank} A - \operatorname{rank} B.$$

Source: [2657].

Fact 8.4.26. Let $A \in \mathbb{F}^{n\times m}$ and $B \in \mathbb{F}^{m\times l}$. Then, the following statements are equivalent:

i) $(AB)^+ = B^+A^+$.

ii) $\mathcal{R}(A^*AB) \subseteq \mathcal{R}(B)$ and $\mathcal{R}(BB^*A^*) \subseteq \mathcal{R}(A^*)$.

iii) $\mathcal{R}(AA^*AB) = \mathcal{R}(AB)$ and $\mathcal{R}(BB^*A^*) \subseteq \mathcal{R}(A^*)$.

iv) $\mathcal{R}(A^*AB) \subseteq \mathcal{R}(B)$ and $\mathcal{R}[(ABB^*B)^*] = \mathcal{R}[(AB)^*)]$.

v) $(AB)^+ = B^+A^+ABB^+A^*$.

vi) $(AB)^+ = B^+(ABB^+)^+$ and $(ABB^+)^+ = BB^+A^+$.

vii) $(AB)^+ = (A^+AB)^+A^+$ and $(A^+AB)^+ = B^+A^+A$.

viii) $(AB)^+ = B^+(A^+ABB^+)^+A^+$ and $(A^+ABB^+)^+ = BB^+A^+A$.

ix) $(AB)^+ = (A^*AB)^+A^*$ and $(A^*AB)^+ = B^+(A^*A)^+$.

x) $(AB)^+ = B^*(ABB^*)^+$ and $(ABB^*)^+ = (BB^*)^+A^+$.

xi) $(AB)^+ = B^*(A^*ABB^*)^+A^*$ and $(A^*ABB^*)^+ = (BB^*)^+(A^*A)^+$.

xii) $AB(AB)^+ = ABB^+A^+$ and $(AB)^+AB = B^+A^+AB$.

xiii) $AB(AB)^+A = ABB^+$ and $A^+AB = B(AB)^+AB$.

xiv) $(ABB^+)^+ = BB^+A^+$ and $(A^+AB)^+ = B^+A^+A$.

xv) $B^+(ABB^+)^+ = B^+A^+$ and $(A^+AB)^+A = B^+A^+$.

xvi) $A^+ABB^+ = BB^+A^+A$.

xvii) A^*ABB^+ and A^+ABB^* are Hermitian.

xviii) $A^*AB = BB^+A^*AB$ and $ABB^* = ABB^*A^+A$.

xix) $A^+AB = B(AB)^+AB$ and $BB^+A^* = A^*AB(AB)^+$.

xx) $A^*ABB^* = BB^+A^*ABB^*A^+A$.

xxi) $(A^{+*}B)^+ = B^+A^*$.

xxii) $(AB^{+*})^+ = B^*A^+$.

Source: [31, p. 53], [1231], [2238, pp. 187, 188], [2403, pp. 190, 191], [2639, 2652], and [2853]. **Remark:** Conditions under which B^+A^+ is a (1)-inverse of AB are given in [2639]. **Credit:** The equivalence of *i)* and *ii)* is due to T. N. E. Greville.

Fact 8.4.27. Let $A \in \mathbb{F}^{n\times n}$ and $B \in \mathbb{F}^{m\times n}$, and assume that A is idempotent. Then,

$$A^*(BA)^+ = (BA)^+.$$

Source: [1343, p. 514].

Fact 8.4.28. Let $A, B \in \mathbb{F}^{n\times n}$, assume that A and B are idempotent, and assume that $AB = BA$. Then, $A + B - \frac{3}{2}AB$ is a 1-inverse of $A + B$, and $A - B - AB$ is a 1-inverse of $A - B$. **Source:** [235]. **Related:** Fact 8.8.5.

Fact 8.4.29. Let $A \in \mathbb{F}^{n\times m}$ and $B \in \mathbb{F}^{m\times l}$, and assume that $\operatorname{rank} B = m$. Then, $AB(AB)^+ = AA^+$. **Source:** [2238, p. 188].

Fact 8.4.30. Let $A \in \mathbb{F}^{n\times m}$, $B \in \mathbb{F}^{m\times n}$, and $C \in \mathbb{F}^{m\times n}$, and assume that $BAA^* = A^*$ and $A^*AC = A^*$. Then, $A^+ = BAC$. **Source:** [31, p. 36]. **Credit:** H. P. Decell.

Fact 8.4.31. Let $A, B \in \mathbb{F}^{n\times n}$, and assume that $A + B$ is nonsingular. Then, the following statements are equivalent:

i) $\operatorname{rank} A + \operatorname{rank} B = n$.

ii) $A(A + B)^{-1}B = 0$.

iii) $B(A + B)^{-1}A = 0$.

iv) $A(A + B)^{-1}A = A$.

v) $B(A + B)^{-1}B = B$.

vi) $A(A + B)^{-1}B + B(A + B)^{-1}A = 0$.

vii) $A(A + B)^{-1}A + B(A + B)^{-1}B = A + B$.

viii) $(A + B)^{-1} = [(I - BB^+)A(I - B^+B)]^+ + [(I - AA^+)B(I - A^+A)]^+$.

Source: [2666]. **Related:** Fact 3.14.11 and Fact 10.24.24.

Fact 8.4.32. Let $A \in \mathbb{F}^{n\times m}$ and $B \in \mathbb{F}^{n\times p}$. Then, the following statements are equivalent:

i) $[AA^+, BB^+] = 0$.

ii) $\operatorname{rank}\,[A\ \ B] = \operatorname{rank} A + \operatorname{rank} B - \operatorname{rank} A^*B$.

Source: [2657]. **Related:** Fact 4.18.7.

Fact 8.4.33. Let $A \in \mathbb{F}^{n\times m}$ and $B \in \mathbb{F}^{n\times p}$. Then,

$$(AA^* + BB^*)^+ = (I - C^{+*}B^*)A^{+*}EA^+(I - BC^+) + (CC^*)^+,$$

where

$$C \triangleq (I - AA^+)B, \quad E \triangleq I - A^*B(I - C^+C)[I + (I - C^+C)B^*(AA^*)^+B(I - C^+C)]^{-1}(A^+B)^*.$$

Source: [536, p. 16] and [2403, p. 196]. **Related:** Fact 8.9.20.

Fact 8.4.34. Let $A, B \in \mathbb{F}^{n\times m}$, and assume that $BA^* = 0$. Then,

$$(A + B)^+ = A^+ + (I - A^+B)(C^+ + D),$$

where

$$C \triangleq (I - AA^+)B, \quad D \triangleq (I - C^+C)[I + (I - C^+C)B^*(AA^*)^+B(I - C^+C)]^{-1}B^*(AA^*)^+(I - BC^+).$$

Source: [536, p. 17] and [2403, p. 196]. **Remark:** As noted in Fact 8.9.25, if $A^*B = 0$, then $C = B$ and $D = 0$. This observation yields Fact 8.4.35.

Fact 8.4.35. Let $A, B \in \mathbb{F}^{n\times m}$, and assume that $A^*B = 0$ and $BA^* = 0$. Then,

$$(A + B)^+ = A^+ + B^+.$$

Source: Fact 3.13.38, Fact 8.4.34, Fact 8.4.37, [762], [1343, p. 513], and [2403, p. 197]. **Credit:** R. Penrose.

Fact 8.4.36. Let $A, B \in \mathbb{F}^{n\times m}$, assume that $\operatorname{rank}(A + B) = \operatorname{rank} A$, and assume that $I + A^+B$ is nonsingular. Then,

$$(A+B)^+ = (A^+A+X^*)[A^+A - X(I+X^*X)^{-1}X^*](I+A^+B)^{-1}[AA^+ - Y^*(I+YY^*)^{-1}Y](AA^+ + Y^*),$$

where

$$X \triangleq (I+A^+B)^{-1}A^+B(I-A^+A), \quad Y \triangleq (I-AA^+)BA^+(I+BA^+)^{-1}.$$

Source: [1275, p. 220].

Fact 8.4.37. Let $A, B \in \mathbb{F}^{n\times m}$, and assume that $\operatorname{rank}(A+B) = \operatorname{rank} A + \operatorname{rank} B$. Then,

$$(A+B)^+ = (I - C^+B)A^+(I - BC^+) + C^+,$$

where

$$C \triangleq (I - AA^+)B(I - A^+A).$$

Furthermore,

$$(A+B)^+ = (I - S)A^+(I - T) + SB^+T,$$

where

$$S \triangleq [B^+B(I - A^+A)]^+, \quad T \triangleq [(I - AA^+)BB^+]^+.$$

Source: [762] and [2238, p. 197].

Fact 8.4.38. Let $A, B \in \mathbb{F}^{n\times m}$. Then,

$$(A+B)^+ = (I + A^+B)^+(A^+ + A^+BA^+)(I + BA^+)^+$$

if and only if $AA^+B = B = BA^+A$. Furthermore, if $n = m$ and A is nonsingular, then

$$(A+B)^+ = (I + A^{-1}B)^+(A^{-1} + A^{-1}BA^{-1})(I + BA^{-1})^+.$$

Source: [762].

Fact 8.4.39. Let $A, B \in \mathbb{F}^{n\times m}$. Then,

$$\begin{aligned} A^+ - B^+ &= B^+(B - A)A^+ + (I - B^+B)(A^* - B^*)A^{+*}A^+ + B^+B^{+*}(A^* - B^*)(I - AA^+) \\ &= A^+(B - A)B^+ + (I - A^+A)(A^* - B^*)B^{+*}B^+ + A^+A^{+*}(A^* - B^*)(I - BB^+). \end{aligned}$$

Furthermore, if B is left invertible, then

$$A^+ - B^+ = B^+(B - A)A^+ + B^+B^{+*}(A^* - B^*)(I - AA^+),$$

while, if B is right invertible, then

$$A^+ - B^+ = A^+(B - A)B^+ + (I - A^+A)(A^* - B^*)B^{+*}B^+.$$

Source: [624, p. 224].

Fact 8.4.40. Let $A \in \mathbb{F}^{n\times m}$, $B \in \mathbb{F}^{l\times k}$, and $C \in \mathbb{F}^{n\times k}$. Then, there exists $X \in \mathbb{F}^{m\times l}$ satisfying $AXB = C$ if and only if $AA^+CB^+B = C$. Furthermore, X satisfies $AXB = C$ if and only if there exists $Y \in \mathbb{F}^{m\times l}$ such that

$$X = A^+CB^+ + Y - A^+AYBB^+.$$

Finally, if $Y = 0$, then $\operatorname{tr} X^*X$ is minimized. **Source:** Proposition 8.1.8. See [1924, p. 37] and, for Hermitian X, see [1608].

Fact 8.4.41. Let $A \in \mathbb{F}^{n\times m}$, and assume that $\operatorname{rank} A = m$. Then, $A^{\mathsf{L}} \in \mathbb{F}^{m\times n}$ is a left inverse of A if and only if there exists $B \in \mathbb{F}^{m\times n}$ such that

$$A^{\mathsf{L}} = A^+ + B(I - AA^+).$$

Source: Use Fact 8.4.40 with $A = C = I_n$.

Fact 8.4.42. Let $A \in \mathbb{F}^{n\times m}$, and assume that $\operatorname{rank} A = n$. Then, $A^{\mathrm{R}} \in \mathbb{F}^{m\times n}$ is a right inverse of A if and only if there exists $B \in \mathbb{F}^{m\times n}$ such that

$$A^{\mathrm{R}} = A^+ + (I - A^+A)B.$$

Source: Use Fact 8.4.40 with $B = C = I_n$.

Fact 8.4.43. Let $A, B \in \mathbb{F}^{n\times m}$. Then, the following statements are equivalent:

i) $A \overset{*}{\leq} B$.

ii) $A^+A = A^+B$ and $AA^+ = BA^+$.

iii) $B - A \overset{*}{\leq} B$.

If these statements hold, then the following statements hold:

iv) AB^+, B^+A, A^+B, and BA^+ are Hermitian.

v) $BA^+B = A$ and $B^+AB^+ = A^+$.

Source: [1339, 2058] and [2059, pp. 127–131]. **Related:** Fact 4.30.8 and Fact 8.5.17.

Fact 8.4.44. Let $A, B \in \mathbb{F}^{n\times m}$. Then, the following statements are equivalent:

i) $A \overset{\mathrm{rs}}{\leq} B$.

ii) $B^+A \overset{\mathrm{rs}}{\leq} B^+B$ and $\mathcal{R}(A) \subseteq \mathcal{R}(B)$.

iii) $AB^+ \overset{\mathrm{rs}}{\leq} BB^+$ and $\mathcal{R}(A^*) \subseteq \mathcal{R}(B^*)$.

iv) $B^+AB^+ \overset{\mathrm{rs}}{\leq} B^+$, $\mathcal{R}(A) \subseteq \mathcal{R}(B)$, and $\mathcal{R}(A^*) \subseteq \mathcal{R}(B^*)$.

Furthermore, the following statements are equivalent:

v) $B^+A \overset{\mathrm{rs}}{\leq} B^+B$ and $\mathcal{R}(A^*) \subseteq \mathcal{R}(B^*)$.

vi) $B^+AB^+ \overset{\mathrm{rs}}{\leq} B^+$ and $\mathcal{R}(A^*) \subseteq \mathcal{R}(B^*)$.

Furthermore, the following statements are equivalent:

vii) $AB^+ \overset{\mathrm{rs}}{\leq} BB^+$ and $\mathcal{R}(A) \subseteq \mathcal{R}(B)$.

viii) $B^+AB^+ \overset{\mathrm{rs}}{\leq} B^+$ and $\mathcal{R}(A) \subseteq \mathcal{R}(B)$.

Source: [1880]. **Related:** Fact 4.30.4.

Fact 8.4.45. Let $A, C \in \mathbb{F}^{n\times m}$, and let $B, D \in \mathbb{F}^{n\times l}$. Then, the following statements hold:

i) If $A \overset{\mathrm{rs}}{\leq} C$, $B \overset{\mathrm{rs}}{\leq} D$, and $\mathcal{R}(C) \cap \mathcal{R}(D) = \{0\}$, then $[A \;\; B] \overset{\mathrm{rs}}{\leq} [C \;\; D]$.

ii) Assume that $A \overset{\mathrm{rs}}{\leq} C$ and $\mathcal{R}(A) \subseteq \mathcal{R}(C)$. Then, $[A \;\; B] \overset{\mathrm{rs}}{\leq} [C \;\; D]$ if and only if $B = AC^+D$.

Source: [1880].

Fact 8.4.46. Let $A, C \in \mathbb{F}^{n\times m}$, and let $B, D \in \mathbb{F}^{l\times m}$. Then, the following statements hold:

i) Assume that $A \overset{\mathrm{rs}}{\leq} C$, $B \overset{\mathrm{rs}}{\leq} D$, and $\mathcal{R}(C^*) \cap \mathcal{R}(D^*) = \{0\}$. Then, $\left[\begin{smallmatrix} A \\ B \end{smallmatrix}\right] \overset{\mathrm{rs}}{\leq} \left[\begin{smallmatrix} C \\ D \end{smallmatrix}\right]$.

ii) Assume that $A \overset{\mathrm{rs}}{\leq} C$ and $\mathcal{R}(D^*) \subseteq \mathcal{R}(C^*)$. Then, $\left[\begin{smallmatrix} A \\ B \end{smallmatrix}\right] \overset{\mathrm{rs}}{\leq} \left[\begin{smallmatrix} C \\ D \end{smallmatrix}\right]$ if and only if $B = DC^+A$.

Source: [1880].

Fact 8.4.47. Let $A_1, \ldots, A_k \in \mathbb{F}^{n\times m}$, and assume that, for all distinct $i, j \in \{1, \ldots, k\}$, $A_i^*A_j = 0$. Then,

$$\left(\sum_{i=1}^{k} A_i\right)^+ = \sum_{i=1}^{k} A_i^+.$$

Source: [2238, p. 186].

8.5 Facts on the Moore-Penrose Generalized Inverse for Range-Hermitian, Range-Disjoint, and Range-Spanning Matrices

Fact 8.5.1. Let $A \in \mathbb{F}^{n\times n}$ and $k \geq 1$. Then, the following statements are equivalent:

i) $A^k = A^+$.

ii) $A^{k+2} = A$, and A is range Hermitian.

Furthermore, if $k \geq 2$ and $A^k = A^*$, then *i)* and *ii)* hold. **Source:** [233].

Fact 8.5.2. Let $A \in \mathbb{F}^{n\times n}$. Then,

$$\operatorname{rank}\,[A, A^+] = \operatorname{rank}\,[AA^+ \;\; A^+A] = 2\operatorname{rank}\,[A \;\; A^*] - 2\operatorname{rank} A = \operatorname{rank}(A - A^2A^+) = \operatorname{rank}(A - A^+A^2).$$

Furthermore, the following statements are equivalent:

i) A is range Hermitian.

ii) $AA^+ = A^+A$.

iii) $\operatorname{rank}\,[A \;\; A^*] = \operatorname{rank} A$.

iv) $A = A^2A^+$.

v) $A = A^+A^2$.

Source: [2673]. **Related:** Fact 4.9.6, Fact 8.5.3, and Fact 8.5.8.

Fact 8.5.3. Let $A \in \mathbb{F}^{n\times n}$. Then, the following statements are equivalent:

i) A is range Hermitian.

ii) $\mathcal{R}(A) = \mathcal{R}(A^+)$.

iii) $A^+A = AA^+$.

iv) $(I - A^+A)_\perp = AA^+$.

v) $A = A^2A^+$.

vi) $A = A^+A^2$.

vii) $AA^+ = A^2(A^+)^2$.

viii) $(AA^+)^2 = A^2(A^+)^2$.

ix) $(A^+A)^2 = (A^+)^2A^2$.

x) $\operatorname{ind} A \leq 1$, and $(A^+)^2 = (A^2)^+$.

xi) $\operatorname{ind} A \leq 1$, and $AA^+A^*A = A^*A^2A^+$.

xii) $A^2A^+ + A^*A^{+*}A = 2A$.

xiii) $A^2A^+ + (A^2A^+)^* = A + A^*$.

xiv) $\mathcal{R}(A - A^+) = \mathcal{R}(A - A^3)$.

xv) $\mathcal{R}(A + A^+) = \mathcal{R}(A + A^3)$.

Source: [719, 2628, 2647, 2718] and Fact 8.10.16. **Related:** Fact 4.9.6, Fact 8.5.2, and Fact 8.5.8.

Fact 8.5.4. Let $A \in \mathbb{F}^{n\times n}$. Then, the following statements are equivalent:

i) $A = A^+$.

ii) $A^2 = AA^+$.

iii) $A^+ = A^2A^+$.

iv) $AA^{2*} = A$.

v) $A^{2*}A = A$.

vi) $A^2 = AA^+ = A^+A$.

vii) $(A^+)^2 = AA^+ = A^+A$.

viii) A is range Hermitian, and A^2 is idempotent.

ix) A is range Hermitian, and $(A^+)^2$ is idempotent.

x) A is range Hermitian, and $A^*A = A^*A^+$.

xi) A is range Hermitian, and $AA^* = A^+A^*$.

xii) A is range Hermitian, and $A^*AA^* = A^*A^+A^*$.

xiii) A is tripotent and range Hermitian.

xiv) A^* is tripotent and range Hermitian.

xv) A^+ is tripotent and range Hermitian.

xvi) A is tripotent, and A^2 is Hermitian.

xvii) A is group invertible, and A^2 is a projector.

xviii) A is group invertible, and $(A^+)^2$ is a projector.

xix) A is group invertible, and $A^{\mathrm{c}} = A$.

xx) A is group invertible, and $A^\# = A^2A^+$.

xxi) $\operatorname{rank}\begin{bmatrix} AA^*A & A \\ A & A^* \end{bmatrix} = \operatorname{rank} A$.

xxii) $\mathcal{R}\left(\begin{bmatrix} A \\ A^* \end{bmatrix}\right) = \mathcal{R}\left(\begin{bmatrix} AA^*A \\ A \end{bmatrix}\right)$.

xxiii) There exists a unitary matrix $S \in \mathbb{F}^{n\times n}$ and an involutory matrix $B \in \mathbb{F}^{r\times r}$, where $r \triangleq \operatorname{rank} A$, such that $A = S\left[\begin{smallmatrix} B & 0 \\ 0 & 0 \end{smallmatrix}\right]S^*$.

Source: [360, p. 49] and [233, 257].

Fact 8.5.5. Let $A \in \mathbb{F}^{n\times n}$. Then, the following statements are equivalent:

i) A is Hermitian and tripotent.

ii) A is a partial isometry and tripotent.

iii) A is Hermitian, and $A = A^+$.

iv) A is a partial isometry, tripotent, and either $\frac{1}{2}(AA^+ - A)$ or $\frac{1}{2}(AA^+ + A)$ is idempotent.

v) A is tripotent, range Hermitian, and semicontractive.

vi) A is group invertible, and $A = A^+ = A^* = A^\# = A^{\mathrm{c}}$.

Source: [257]. **Related:** Fact 7.10.27.

Fact 8.5.6. Let $A \in \mathbb{F}^{n\times n}$. Then, the following statements are equivalent:

i) $A + A^+ = 2AA^+$.

ii) $A + A^+ = 2A^+A$.

iii) $A + A^+ = AA^+ + A^+A$.

iv) A is range Hermitian, and $A^2 + AA^+ = 2A$.

v) A is range Hermitian, and $(I - A)^2A = 0$.

Source: [2707, 2717].

Fact 8.5.7. Let $A \in \mathbb{F}^{n\times n}$ and $k \geq 3$. Then, the following statements are equivalent:

i) A is normal, and $A^{k+1} = A$.

ii) A is a partial isometry, and $A^{k+1} = A$.

iii) A is a partial isometry, and $A^k = AA^*$.

iv) A is group invertible and a partial isometry, and $A^{k-1} = A^\#$.

v) A is group invertible, and $A^{k-1} = A^* = A^\#$.

vi) $A^{k-1} = A^*$.

The following statements are equivalent:

vii) A is range Hermitian, and $A^{k+1} = A$.

viii) $A^k = AA^+$.

The following statements are equivalent:

ix) A is range Hermitian and a partial isometry.

x) $A = A^*A^2$.

xi) $A = A^2A^*$.

Now, let $k \geq 1$. Then, following statements are equivalent:

xii) $A^k = A$.

xiii) A is group invertible, and $A^{k+1} = A^2$.

Source: [257].

Fact 8.5.8. Let $A \in \mathbb{F}^{n\times n}$, let $r \triangleq \operatorname{rank} A$, let $B \in \mathbb{F}^{n\times r}$ and $C \in \mathbb{F}^{r\times n}$, and assume that $A = BC$ and $\operatorname{rank} B = \operatorname{rank} C = r$. Then, the following statements are equivalent:

i) A is range Hermitian.

ii) $BB^+ = C^+C$.

iii) $\mathcal{N}(B^*) = \mathcal{N}(C)$.

iv) $B = C^+CB$ and $C = CBB^+$.

v) $B^+ = B^+C^+C$ and $C = CBB^+$.

vi) $B = C^+CB$ and $C^+ = BB^+C^+$.

vii) $B^+ = B^+C^+C$ and $C^+ = BB^+C^+$.

Source: [950]. **Remark:** $A = BC$ is a full-rank factorization. **Related:** Fact 4.9.6, Fact 8.5.2, and Fact 8.5.3.

Fact 8.5.9. Let $A, B \in \mathbb{F}^{n\times n}$, and assume that A is range Hermitian. Then, $AB = BA$ if and only if $A^+B = BA^+$. **Source:** [2627].

Fact 8.5.10. Let $A, B \in \mathbb{F}^{n\times n}$, and assume that A and B are range Hermitian. Then, the following statements are equivalent:

i) $AB = BA$.

ii) $A^+B = BA^+$.

iii) $AB^+ = B^+A$.

iv) $A^+B^+ = B^+A^+$.

Source: [2627].

Fact 8.5.11. Let $A, B \in \mathbb{F}^{n\times n}$, assume that A and B are range Hermitian, and assume that $(AB)^+ = A^+B^+$. Then, AB is range Hermitian. **Source:** [1335]. **Related:** Fact 10.24.23.

Fact 8.5.12. Let $A, B \in \mathbb{F}^{n\times n}$, and assume that A and B are range Hermitian. Then, the following statements are equivalent:

i) AB is range Hermitian.

ii) $AB(I - A^+A) = 0$ and $(I - B^+B)AB = 0$.

iii) $\mathcal{N}(A) \subseteq \mathcal{N}(AB)$ and $\mathcal{R}(AB) \subseteq \mathcal{R}(B)$.

iv) $\mathcal{N}(AB) = \mathcal{N}(A) + \mathcal{N}(B)$ and $\mathcal{R}(AB) = \mathcal{R}(A) \cap \mathcal{R}(B)$.

Source: [1335, 1658].

Fact 8.5.13. Let $A \in \mathbb{F}^{n\times n}$ be nonzero, and define $r \triangleq \operatorname{rank} A$ and $B \triangleq \operatorname{diag}[\sigma_1(A), \ldots, \sigma_r(A)]$.

Then, there exist $S \in \mathbb{F}^{n\times n}$, $K \in \mathbb{F}^{r\times r}$, and $L \in \mathbb{F}^{r\times(n-r)}$ such that S is unitary, $KK^* + LL^* = I_r$, and

$$A = S\begin{bmatrix} BK & BL \\ 0_{(n-r)\times r} & 0_{(n-r)\times(n-r)} \end{bmatrix}S^*.$$

Now, define the projectors $P \triangleq AA^+$ and $Q \triangleq A^+A$. Then, the following statements hold:

i) $PQ = QP$ if and only if K is a partial isometry.

ii) $\operatorname{rank} PQ = \operatorname{rank} K$.

iii) $\operatorname{rank}(P + Q) = r + \operatorname{rank} L$.

iv) $\operatorname{rank}(P - Q) = 2 \operatorname{rank} L$.

v) $\operatorname{rank}(PQ + QP) = 2 \operatorname{rank} K + \operatorname{rank} L - r$.

vi) $\operatorname{rank}(PQ - QP) = 2(\operatorname{rank} K + \operatorname{rank} L - r)$.

vii) $\operatorname{rank}(I - PQ) = n - r + \operatorname{rank} L$.

viii) $\operatorname{rank}(P + Q - PQ) = r + \operatorname{rank} L$.

ix) $\operatorname{rank}(I - P - Q) = n - 2(r - \operatorname{rank} K)$.

Furthermore, the following statements are equivalent:

x) A is group invertible.

xi) K is nonsingular.

xii) $\operatorname{rank} P = \operatorname{rank} PQ$.

xiii) $\operatorname{rank}(P + Q) = \operatorname{rank}(PQ + QP)$.

xiv) $\operatorname{rank}(P - Q) = \operatorname{rank}(PQ - QP)$.

xv) $\sigma_{\max}(P - Q) < 1$.

xvi) $\mathcal{R}(A) \cap \mathcal{N}(A) = \{0\}$.

xvii) $\mathcal{R}(A^*) \cap \mathcal{N}(A^*) = \{0\}$.

xviii) $\mathcal{R}(A) = \mathcal{R}(PQ)$.

xix) $\mathcal{R}(A^*) = \mathcal{R}(QP)$.

xx) $1 \notin \operatorname{spec}(P + Q)$.

xxi) $1 \notin \operatorname{spec}(P - Q)$.

xxii) $-1 \notin \operatorname{spec}(P - Q)$.

If *x)*–*xxii)* hold, then

$$A^\# = S\begin{bmatrix} (BK)^{-1} & (KBK)^{-1}L \\ 0_{(n-r)\times r} & 0_{(n-r)\times(n-r)} \end{bmatrix}S^*.$$

Finally, let $\alpha, \beta \in \mathbb{R}$ be nonzero. Then, $\alpha P + \beta Q$ is a projector if and only if at least one of the following statements holds:

xxiii) $\alpha = \beta = 1$ and $A^2 = 0$.

xxiv) $\alpha = -\beta = 1$ and A is range Hermitian.

xxv) $-\alpha = \beta = 1$ and A is range Hermitian.

Source: [240, 247, 250, 251, 1338]. **Related:** Fact 4.18.20, Fact 7.10.27, and Fact 8.3.23.

Fact 8.5.14. Let $A \in \mathbb{F}^{n\times n}$, assume that A is nonzero, and define the projectors $P \triangleq AA^+$ and $Q \triangleq A^+A$. Then, the following statements are equivalent:

i) A is range disjoint.

ii) A^+ is range disjoint.

iii) $\mathcal{R}(P) \cap \mathcal{R}(Q) = \{0\}$.

iv) PQ is range disjoint.

v) $\operatorname{rank}(P - Q) = \operatorname{rank} P + \operatorname{rank} Q$.

vi) $\mathcal{R}(P - Q) = \mathcal{R}(P) + \mathcal{R}(Q)$.

vii) $\operatorname{rank} PQ_\perp + \operatorname{rank} P_\perp Q = \operatorname{rank} P + \operatorname{rank} Q$.

viii) $\mathcal{R}(PQ_\perp) + \mathcal{R}(P_\perp Q) = \mathcal{R}(P) + \mathcal{R}(Q)$, and $\mathcal{R}(PQ_\perp)$ and $\mathcal{R}(P_\perp Q)$ are mutually orthogonal.

ix) $\operatorname{rank}(P - Q) = \operatorname{rank}(P + Q)$.

x) $\mathcal{R}(P - Q) = \mathcal{R}(P + Q)$.

xi) $\operatorname{rank}(PQ + QP) = \operatorname{rank} PQ + \operatorname{rank} QP$.

xii) $I - PQ$ is nonsingular.

xiii) $\operatorname{rank} P_\perp Q = \operatorname{rank} A$.

xiv) $\operatorname{rank} Q_\perp P = \operatorname{rank} A$.

xv) $\operatorname{rank}(P + Q) = 2 \operatorname{rank} A$.

xvi) $\operatorname{rank}(P + Q - PQ) = 2 \operatorname{rank} A$.

xvii) $(Q_\perp P)^+ = P(P + Q - QP)^+$.

xviii) $(Q_\perp P)^+ A = A$.

xix) $A(QP_\perp)^+ = A$.

xx) $(QP_\perp)^+ = [I - (Q_\perp P)^{+*}]Q$.

xxi) $(Q_\perp P)^+ = P[I - (QP_\perp)^{+*}]$.

xxii) $(QP_\perp)^+ A^+ (Q_\perp P)^+$ is a 1-inverse of A.

xxiii) $(QP_\perp)^+ + (Q_\perp P)^{+*}$ is the projector onto $\mathcal{R}(A) + \mathcal{R}(A^*)$.

xxiv) $(QP_\perp)^{+*} + (Q_\perp P)^+$ is the projector onto $\mathcal{R}(A) + \mathcal{R}(A^*)$.

xxv) $[\mathcal{R}(P) + \mathcal{N}(Q)] \cap [\mathcal{N}(P) + \mathcal{N}(Q)] = \mathcal{R}(P) + \mathcal{N}(Q)$.

xxvi) $[\mathcal{R}(P) + \mathcal{R}(Q)] \cap [\mathcal{N}(P) + \mathcal{N}(Q)] = \mathcal{R}(P + Q)$.

xxvii) $\operatorname{rank} [A \;\; A^*] = 2 \operatorname{rank} A$.

xxviii) $\operatorname{rank}(AA^* + A^*A) = 2 \operatorname{rank} A$.

xxix) $A^*(AA^* + A^*A)^+ A$ is idempotent.

xxx) $A^*(AA^* + A^*A)^+ A = A^+ A$.

xxxi) $A^*(AA^* + A^*A)^+ A^* = 0$.

xxxii) $2 \notin \operatorname{spec}(P + Q)$.

Furthermore, the following statements hold:

xxxiii) If $A \neq 0$, then $P + Q \neq 0$.

xxxiv) $P - Q \neq I$, $PQ + QP \neq I$, and $P + Q \neq PQ$.

xxxv) $P + Q = I$ if and only if $A^2 = 0$ and A is range spanning.

xxxvi) The following statements are equivalent:

a) $A^2 = 0$.

b) $PQ = 0$.

c) $P_\perp - Q$ is a projector.

d) $PQP = 0$.

e) $PQ + QP$ is idempotent.

f) $PQ + QP = 0$.

g) $\operatorname{spec}(PQ + QP) \cap (0, 2] = \varnothing$.

xxxvii) The following statements are equivalent:

a) A is range Hermitian.

b) $P - Q = 0$.

c) $P_\perp + Q$ is a projector.

d) $P + Q_\perp$ is a projector.

e) $P_\perp - Q_\perp$ is a projector.

f) $\operatorname{rank} A = \operatorname{tr} PQ$.

xxxviii) The following statements are equivalent:

a) A is nonsingular.

b) $PQ = I$.

c) PQ is nonsingular.

d $PQP = I$.

e) PQP is nonsingular.

xxxix) The following statements are equivalent:

a) A is range spanning.

b) $P + Q$ is nonsingular.

c) $P + Q - PQ$ is nonsingular.

d) $(Q_\perp P)^+ = (I - QP)^+ Q_\perp$.

xl) The following statements are equivalent:

a) A is group invertible and range spanning.

b) PQ is range spanning.

c) $PQ + QP$ is nonsingular.

xli) The following statements are equivalent:

a) A is range disjoint and range spanning.

b) $P - Q$ is nonsingular.

c) $(I - QP)^+ Q_\perp = P(P + Q - QP)^+$.

xlii) $P - Q$ is nonsingular if and only if A is range spanning and at least one of the following statements holds:

a) $\operatorname{spec}(P + Q) \subset [0, 2)$.

b) $\operatorname{rank}(P - Q) = \operatorname{rank} P + \operatorname{rank} Q$.

c) $\mathcal{R}(P - Q) = \mathcal{R}(P) + \mathcal{R}(Q)$.

d) $\operatorname{rank} PQ_\perp + \operatorname{rank} P_\perp Q = \operatorname{rank} P + \operatorname{rank} Q$.

e) $\mathcal{R}(PQ_\perp) + \mathcal{R}(P_\perp Q) = \mathcal{R}(P) + \mathcal{R}(Q)$ and $\mathcal{R}(PQ_\perp) \cap \mathcal{R}(P_\perp Q) = \{0\}$.

f) $(Q_\perp P)^+ = P(P + Q - QP)^+$.

xliii) The following statements are equivalent:

a) $PQ = QP$.

b) $P_\perp Q$ is a projector.

c) $PQ_\perp$ is a projector.

d) $P_\perp Q_\perp$ is a projector.

e) PQP is idempotent.

f) $I - PQ$ is idempotent.

g) $PQ - QP$ is idempotent.

h) $P + Q - PQ$ is idempotent.

xliv) The following statements are equivalent:

a) A is range spanning, and $PQ = QP$.

b) $P_\perp + Q_\perp$ is a projector.

c) $P - Q_\perp$ is a projector.

xlv) A is nilpotent if and only if $\operatorname{tr} PQ = 0$.

xlvi) Assume that A is range disjoint. Then, $(A + A^*)^+ = A^+ + A^{+*}$ if and only if $A^2 = 0$.

xlvii) A is range spanning and $PQ = QP$ if and only if $P + Q = I + PQ$.

xlviii) A is group invertible, range disjoint, and range spanning if and only if $[P, Q]$ is nonsingular.

xlix) Let $B \in \mathbb{F}^{n\times n}$, and assume that $AB = BA$ and either A or B is range disjoint. Then, AB is range disjoint.

l) $\operatorname{spec}(PQ) \subset [0,1]$, $\operatorname{spec}(P - Q) \subset [-1,1]$, $\operatorname{spec}(P + Q) \subset [0,2]$, $\operatorname{spec}(I - PQ) \subset [0,1]$, $\operatorname{spec}(PQ + QP) \subset [-\frac{1}{4}, 2]$, $\operatorname{spec}([P,Q]) \subset \jmath[-1,1]$, $\operatorname{spec}(P + Q - PQ) \subset [0,1]$.

Source: [234, 250, 251]. **Related:** Fact 7.13.4.

Fact 8.5.15. Let $A \in \mathbb{C}^{n\times n}$. Then, the following statements are equivalent:

i) A is group invertible.

ii) There exists $A^{\mathrm{c}} \in \mathbb{F}^{n\times n}$ such that $AA^{\mathrm{c}} = AA^+$ and $\mathcal{R}(A^{\mathrm{c}}) \subseteq \mathcal{R}(A)$.

iii) There exists a unique matrix $A^{\mathrm{c}} \in \mathbb{F}^{n\times n}$ such that $AA^{\mathrm{c}} = AA^+$ and $\mathcal{R}(A^{\mathrm{c}}) \subseteq \mathcal{R}(A)$.

Now, assume that A is group invertible, define $r \triangleq \operatorname{rank} A$, and define $B \in \mathbb{R}^{r\times r}$, $K \in \mathbb{F}^{r\times r}$, $L \in \mathbb{F}^{r\times(n-r)}$, and $S \in \mathbb{F}^{n\times n}$ as in Fact 8.5.13. Then, the following statements hold:

iv) If A is nonzero, then K is nonsingular, and

$$A^{\mathrm{c}} = S\begin{bmatrix} (BK)^{-1} & 0_{r\times(n-r)} \\ 0_{(n-r)\times r} & 0_{(n-r)\times(n-r)} \end{bmatrix} S^*.$$

v) $A^{\mathrm{c}} = 0$ if and only if $A = 0$.

vi) The following statements are equivalent:

a) A is range Hermitian.

b) $A^\# = A^{\mathrm{c}}$.

c) $A^+ = A^{\mathrm{c}}$.

d) $(A^{\mathrm{c}})^{\mathrm{c}} = A$.

e) $A^{\mathrm{c}}A = AA^{\mathrm{c}}$.

f) $(A^+)^{\mathrm{c}} = A$.

g) $(A^+)^{\mathrm{c}} = (A^{\mathrm{c}})^+$.

vii) $A^{\mathrm{c}} = AA^+$ if and only if A is idempotent.

viii) $A^{\mathrm{c}} = A$ if and only if A is tripotent and range Hermitian.

ix) $A^{\mathrm{c}} = A^*$ if and only if A is a partial isometry and is range Hermitian.

x) $A^{\mathrm{c}} = A^\# AA^+$.

xi) A^{c} is range Hermitian.

xii) A^{c} is a (1,2)-inverse of A.

xiii) $(A^{\mathrm{c}})^+ = (A^{\mathrm{c}})^{\mathrm{c}} = A^2A^+$.

xiv) $(A^{\mathrm{c}})^2A = A^\#$.

xv) If $k \geq 1$, then $(A^{\rm c})^k = (A^k)^{\rm c}$.

xvi) $A^{\rm c}A = A^{\#}A$.

xvii) If A is a projector and $B \in \mathbb{F}^{n\times n}$ is a projector, then $(AB)^{\rm c} = (ABA)^+$.

xviii) If $B \in \mathbb{F}^{n\times n}$ is a (1)-inverse of A, then $ABA^{\rm c} = A^{\rm c}$.

xix) $A^{\rm c}$ is a (2)-inverse of B if and only if $A^+BA^{\#} = A^{\rm c}$.

Source: [247, 1934]. **Remark:** $A^{\rm c}$ is the *core inverse* of A.

Fact 8.5.16. Let $A, B \in \mathbb{F}^{n\times n}$, assume that A and B are group invertible, and let $A \overset{\#}{\leq} B$ denote $A^{\#}A = A^{\#}B = BA^{\#}$. Then, "$\overset{\#}{\leq}$" is a partial ordering on $\{A \in \mathbb{F}^{n\times n}\colon A$ is group invertible$\}$. Furthermore, the following statements are equivalent:

i) $A \overset{\#}{\leq} B$.

ii) $A^2 = AB = BA$.

iii) $B - A \overset{\#}{\leq} B$.

iv) $A \overset{\rm rs}{\leq} B$ and $AB = BA$.

If these statements hold, then the following statements hold:

v) $BA^{\#}B = A$ and $B^{\#}AB^{\#} = A^{\#}$.

vi) $AB = BA$.

vii) $AB^{\#} = B^{\#}A$.

viii) For all $k \geq 1$, $A^k \overset{\#}{\leq} B^k$.

Source: [2024, 2058] and [2059, Chapter 4]. **Remark:** "$\overset{\#}{\leq}$" is the *sharp partial ordering*.

Fact 8.5.17. Let $A, B \in \mathbb{F}^{n\times n}$, assume that A and B are group invertible, and let $A \overset{\rm c}{\leq} B$ denote $A^{\rm c}A = A^{\rm c}B$ and $AA^{\rm c} = BA^{\rm c}$. Then, "$\overset{\rm c}{\leq}$" is a partial ordering on $\{A \in \mathbb{F}^{n\times n}\colon A$ is group invertible$\}$. Furthermore, the following statements are equivalent:

i) $A \overset{\rm c}{\leq} B$.

ii) $A^2 = BA$ and $A^+A = A^+B$.

iii) There exist an idempotent matrix $P \in \mathbb{F}^{n\times n}$ and a projector $Q \in \mathbb{F}^{n\times n}$ such that $A = BP = QB$ and $PA^{\#} = A^{\#}$.

If *i)*–*iii)* hold, then the following statements hold:

iv) $BA^{\rm c}B = A$, $B^{\rm c}AB^{\rm c} = A^{\rm c}$, and $B^{\rm c}BA^{\rm c} = A^{\rm c}BB^{\rm c} = A^{\rm c}$.

v) The following statements are equivalent:

a) A is range Hermitian.

b) $A^{\rm c}B = BA^{\rm c}$.

c) $AB^{\rm c} = B^{\rm c}A$.

If A is range Hermitian, then the following statements are equivalent:

vi) $A \overset{*}{\leq} B$.

vii) $AB = BA = A^2$.

viii) $(AB)^+ = B^+A^+ = A^+B^+$ and $A = AA^+B$.

ix) $(AB)^+ = B^+A^+ = A^+B^+$ and $A = BAA^+$.

If either A is range Hermitian, $A^{\rm c}B = BA^{\rm c}$, or $AB^{\rm c} = B^{\rm c}A$, then the following statements are equivalent:

x) $A \overset{\rm c}{\leq} B$.

xi) $A \overset{*}{\leq} B$.

xii) $A \overset{\#}{\leq} B$.

If A is range Hermitian, then the following statements are equivalent:

xiii) $A \overset{\rm c}{\leq} B$.

xiv) $A^2 \overset{*}{\leq} B^2$ and $AB = BA = A^2$.

xv) $A^2 \overset{\rm c}{\leq} B^2$ and $AB = BA = A^2$.

If B is range Hermitian, then the following statements hold:

xvi) If $A \overset{*}{\leq} B$, then $ABB^+ = A = BB^+A$.

xvii) If $ABB^+ = BB^+A$, then $BB^+AA^+ = AA^+BB^+$, $BB^+A^+A = A^+ABB^+$, B^+A^+ is a (1,2,3)-inverse of AB, and A^+B^+ is a (1,2,4)-inverse of BA.

If B is range Hermitian and $A \overset{*}{\leq} B$, then the following statements hold:

xviii) $(AB)^+ = B^+A^+$ if and only if $BB^*A^+A = A^+ABB^*$.

xix) $(BA)^+ = A^+B^+$ if and only if $B^*BAA^+ = AA^+B^*B$.

xx) If B is a partial isometry, then $(AB)^+ = B^+A^+$ and $(BA)^+ = A^+B^+$.

If $A \overset{\rm c}{\leq} B$ and B is range Hermitian, then the following statements hold:

xxi) $B^{\rm c}A^{\rm c}$ is a (1,2,3)-inverse of AB.

xxii) $(AB)^{\rm c} = (AB)^+ = B^+A^+ = B^{\rm c}A^{\rm c}$.

xxiii) If A is range Hermitian, then $(AB)^+ = B^{\rm c}A^{\rm c}$.

If A is group invertible, B is range Hermitian, and $A \overset{\#}{\leq} B$, then the following statements hold:

xxiv) B^+A^+ is a (1,2,3)-inverse of AB, and A^+B^+ is a (1,2,4)-inverse of BA.

xxv) $(AB)^+ = B^+A^+$ if and only if $BB^*A^+A = A^+ABB^*$.

xxvi) $(BA)^+ = A^+B^+$ if and only if $B^*BAA^+ = AA^+B^*B$.

The following statements hold:

xxvii) If A is range Hermitian and $A \overset{\rm rs}{\leq} B$, then B^+A^+ is a (1,2,3)-inverse of AB and A^+B^+ is a (1,2,4)-inverse of BA.

xxviii) If A is range Hermitian and B is group invertible, then $A \overset{*}{\leq} B$ if and only if $A \overset{\#}{\leq} B$.

xxix) If A and B are partial isometries, then $A \overset{*}{\leq} B$ if and only if $A \overset{\rm rs}{\leq} B$.

xxx) If A and B are group invertible and $A \overset{\#}{\leq} B$, then AB is group invertible and $(AB)^\# = B^\#A^\# = A^\#B^\#$.

Source: [405, 1934]. **Remark:** "$\overset{\rm c}{\leq}$" is the *core partial ordering*. **Remark:** "$\overset{*}{\leq}$" is the star partial ordering. See Fact 4.30.8 and Fact 8.4.43.

8.6 Facts on the Moore-Penrose Generalized Inverse for Normal Matrices, Hermitian Matrices, and Partial Isometries

Fact 8.6.1. Let $A \in \mathbb{F}^{n\times n}$. Then, the following statements are equivalent:

i) A is normal.

ii) A^+ is normal.

iii) $AA^*A^+ = A^+AA^*$.

iv) A is range Hermitian, and $A^+A^* = A^*A^+$.

v) $A(AA^*A)^+ = (AA^*A)^+A$.

vi) $AA^+A^*A^2A^+ = AA^*$.

vii) $A(A^* + A^+) = (A^* + A^+)A$.

viii) $A^*A(AA^*)^+A^*A = AA^*$.

ix) $2AA^*(AA^* + A^*A)^+AA^* = AA^*$.

x) There exists $X \in \mathbb{F}^{n\times n}$ such that $AA^*X = A^*A$ and $A^*AX = AA^*$.

xi) There exists $X \in \mathbb{F}^{n\times n}$ such that $AX = A^*$ and $A^{+*}X = A^+$.

xii) A^+A^* is a partial isometry.

Source: [719] and [1242]. **Related:** Fact 4.10.12, Fact 4.13.6, Fact 7.17.5, Fact 8.3.32, and Fact 8.10.17.

Fact 8.6.2. Let $A \in \mathbb{F}^{n\times n}$. Then, the following statements are equivalent:

i) A is Hermitian.

ii) $AA^+ = A^*A^+$.

iii) $A^2A^+ = A^*$.

iv) $AA^*A^+ = A$.

Source: [240].

Fact 8.6.3. Let $A \in \mathbb{F}^{n\times n}$. Then, $2\,\mathrm{rank}\,A \le \mathrm{tr}\,A^*A + \mathrm{tr}\,(A^*A)^+$. Furthermore, equality holds if and only if A is a partial isometry. **Source:** [234].

8.7 Facts on the Moore-Penrose Generalized Inverse for Idempotent Matrices

Fact 8.7.1. Let $A \in \mathbb{F}^{n\times n}$, and assume that A is idempotent. Then, $A^+ = A^+A^2A^+ = A^*A^{*+}AA^+$. **Source:** [778]. **Remark:** A^+ is the product of the projector A^+A onto $\mathcal{N}(A)^\perp = \mathcal{R}(A^*)$ and the projector AA^+ onto $\mathcal{R}(A)$. **Credit:** T. N. E. Greville. **Related:** Fact 8.7.2.

Fact 8.7.2. Let $A \in \mathbb{F}^{n\times n}$, and consider the following statements:

i) A is idempotent.

ii) $\mathrm{rank}\,A = \mathrm{tr}\,A$.

iii) $\mathrm{rank}\,A \le \mathrm{tr}\,A^2A^+A^*$.

Then, *i*) $\Longrightarrow$ *ii*) $\Longrightarrow$ *iii*). Furthermore, the following statements are equivalent:

iv) A is idempotent.

v) $\mathrm{rank}\,A + \mathrm{rank}(I - A) = n$.

vi) $\mathrm{rank}\,A = \mathrm{tr}\,A$ and $\mathrm{rank}(I - A) = \mathrm{tr}(I - A)$.

vii) $\mathrm{rank}\,A \le \mathrm{tr}\,A$ and $\mathrm{rank}(I - A) \le \mathrm{tr}(I - A)$.

viii) $\mathrm{rank}\,A = \mathrm{tr}\,A$ and there exist distinct $k, l \ge 1$ such that $A^k = A^l$.

ix) $\mathrm{rank}\,A = \mathrm{tr}\,A = \mathrm{tr}\,A^2A^+A^*$.

x) $\mathrm{rank}\,A + \mathrm{tr}\,A^2A^+A^* = \mathrm{tr}(A + A^*)$.

xi) There exist projectors $B, C \in \mathbb{F}^{n\times n}$ such that $A = (BC)^+$.

xii) $A^*A^+ = A^+$.

xiii) $A^+A^* = A^+$.

xiv) $A^*A^+A = A^+A$.

xv) $AA^+A^* = AA^+$.

xvi) $A^+A + (I - A)(I - A)^+ = I$.

xvii) $AA^+ + (I - A)^+(I - A) = I$.

Finally, let B and C satisfy *x)*. Then, $\mathcal{R}(B) = \mathcal{R}(A^*)$ and $\mathcal{R}(C) = \mathcal{R}(A)$. **Source:** [234, 718, 1134, 1247, 1607] and [2418, p. 166]. The last statement is given in [2238, p. 298]. **Remark:** Note that A^*A^+A is a projector, and $\mathcal{R}(A^*A^+A) = \mathcal{R}(A^*) = \mathcal{R}(A^+A)$. **Remark:** $\mathcal{N}(A) = \mathcal{R}(I - A^+A) = \mathcal{R}(I - A) = \mathcal{R}[(I - A)(I - A^+)]$. **Remark:** *xvi)* states that the projector onto the null space of A is the projector onto the range of $I - A$, while *xvii)* states that the projector onto the range of A is the projector onto the null space of $I - A$. **Remark:** *xi)*, which is due to R. Penrose, follows from Fact 8.8.9 by setting $A = I$. **Credit:** Sufficiency of *xiv)*–*xvii)* is due to G. Trenkler. **Related:** Fact 4.18.19 and Fact 7.13.28.

Fact 8.7.3. Let $A \in \mathbb{F}^{n\times n}$, and assume that A is idempotent. Then, $A + A^* - I$ is nonsingular, and $(A + A^* - I)^{-1} = AA^+ + A^+A - I$. **Source:** Fact 8.7.2. **Related:** Fact 4.18.19, Fact 7.13.28, and [2036, p. 457] for a geometric interpretation of this equality.

Fact 8.7.4. Let $A \in \mathbb{F}^{n\times n}$, and assume that A is idempotent. Then, $2A(A + A^*)^+A^*$ is the projector onto $\mathcal{R}(A) \cap \mathcal{R}(A^*)$. **Source:** [2702].

Fact 8.7.5. Let $A \in \mathbb{F}^{n\times n}$, and assume that A is idempotent. Then, $I - A^+$ is range Hermitian. **Source:** [258].

Fact 8.7.6. Let $A \in \mathbb{F}^{n\times n}$. Then, the following statements are equivalent:

i) A^+ is idempotent.

ii) $AA^*A = A^2$.

If A is range Hermitian, then the following statements are equivalent:

iii) A^+ is idempotent.

iv) $AA^* = A^*A = A$.

The following statements are equivalent:

v) A^+ is a projector.

vi) A is a projector.

vii) A is idempotent, and A and A^+ are similar.

viii) A is idempotent, and $A = A^+$.

ix) A is idempotent, and $AA^+ = AA^*$.

x) $A^+ = A$, and $A^2 = A^*$.

xi) A and A^+ are idempotent.

xii) $A = AA^+$.

Source: [2418, pp. 167, 168] and [2628, 2710, 2869]. **Related:** Fact 4.17.4.

Fact 8.7.7. Let $A \in \mathbb{F}^{n\times n}$, and assume that A and B are idempotent. Then,

$$\sigma_{\max}(AA^+ - BB^+) \le \sigma_{\max}(A - B), \quad \sigma_{\max}(A^+ - B^+) \le 2\sigma_{\max}(A - B).$$

Source: [778] and [1590, p. 58].

Fact 8.7.8. Let $A \in \mathbb{F}^{n\times n}$. Then, A is idempotent if and only if there exist $B \in \mathbb{F}^{n\times n}$ and $C \in \mathbb{F}^{l\times n}$ such that $A = B(CB)^+C$. If these conditions hold, then

$$\mathcal{R}(A) = \mathcal{R}(AA^*B^*) = \mathcal{R}(AA^*B^*B) = \mathcal{R}(A) \cap [(AA^*)^+(\mathcal{R}(A) \cap \mathcal{N}(B))]^\perp,$$
$$\mathcal{N}(A) = \mathcal{N}(A^*B^*B) = \mathcal{N}(AA^*B^*B) = \mathcal{N}(B) + (B^*B)^+[\mathcal{R}(A) + \mathcal{N}(B)]^\perp.$$

Source: [659]. **Related:** Fact 8.7.9.

Fact 8.7.9. Let $A \in \mathbb{F}^{n\times m}$ and $B \in \mathbb{F}^{l\times n}$, assume that $\mathcal{R}(A)$ and $\mathcal{N}(B)$ are complementary subspaces, and let $C \in \mathbb{F}^{n\times n}$ be the idempotent matrix onto $\mathcal{R}(A)$ along $\mathcal{N}(B)$. Then, $C = A(BA)^+B$. If, in addition, A and B are projectors, then $C = (BA)^+$. **Source:** [659]. **Related:** Fact 4.15.3, Fact 8.7.8, and Fact 8.8.14.

8.8 Facts on the Moore-Penrose Generalized Inverse for Projectors

Fact 8.8.1. Let $A \in \mathbb{F}^{n\times n}$. Then, the following statements are equivalent:

i) A is a projector.

ii) A is idempotent, and $\operatorname{rank} A = \operatorname{tr} A^+$.

iii) A is idempotent, and $\operatorname{tr} A = \operatorname{tr} AA^*$.

iv) A is idempotent, and $A^+ = A$.

Source: [234].

Fact 8.8.2. Let $A \in \mathbb{F}^{n\times n}$. Then, the following statements are equivalent:

i) A is a projector.

ii) $A = A^+A$.

iii) $A = AA^+$.

iv) $A = A^2 = A^+$.

Source: [247].

Fact 8.8.3. Let $A, B \in \mathbb{F}^{n\times n}$, and assume that A and B are projectors. Then, the following statements hold:

i) $(AB)^+ = B(AB)^+$.

ii) $(AB)^+ = (AB)^+A$.

iii) $(AB)^+ = B(AB)^+A$.

iv) $(AB)^+ = BA - B(B_\perp A_\perp)^+A$.

v) $(AB)^+$, $B(AB)^+$, $(AB)^+A$, $B(AB)^+A$, and $BA - B(B_\perp A_\perp)^+A$ are idempotent.

vi) $AB = A(AB)^+B$.

vii) $(AB)^2 = AB + AB(B_\perp A_\perp)^+AB$.

viii) $AB = AB(AB)^+BA(BA)^+ = (BA)^+AB$.

ix) $AB(AB + BA)^+BA = A(A + B)^+B$.

x) $(ABA)^+ = (BA)^+(AB)^+ = A[I - (A_\perp B_\perp)^+]B[I - (B_\perp A_\perp)^+]A$.

xi) $(AB)^+ = (AB)^+A = B(AB)^+$.

xii) $(AB)^+AB = BA(BA)^+ = (AB)^+B = B(BA)^+$.

xiii) $[(AB)^2]^+ = B[I - (B_\perp A_\perp)^+]A[I - (A_\perp B_\perp)^+]B[I - (B_\perp A_\perp)^+]A$.

xiv) $(AB)^\# = (BA)^+(AB)^+(BA)^+ = A[I - (A_\perp B_\perp)^+]B[I - (B_\perp A_\perp)^+]A[I - (A_\perp B_\perp)^+]B$.

Source: To prove *i)* note that $\mathcal{R}[(AB)^+] = \mathcal{R}[(AB)^*] = \mathcal{R}(BA)$, and thus $\mathcal{R}[B(AB)^+] = \mathcal{R}[B(AB)^*] = \mathcal{R}(BA)$. Hence, $\mathcal{R}[(AB)^+] = \mathcal{R}[B(AB)^+]$. It now follows from Fact 4.18.3 that $(AB)^+ = B(AB)^+$. *iv)* follows from Fact 8.4.24; *v)* and *vi)* follow from *iii)*; *vii)* follows from *iv)* and *vi)*; *viii)*–*xiv)* are given in [245, 2657]. **Remark:** The fact that the first expression in *v)* is idempotent is given by Fact 8.7.2. **Remark:** See [2637, 2869]. **Related:** Fact 7.13.2, Fact 8.4.23, Fact 8.7.6, and Fact 8.8.23.

Fact 8.8.4. Let $A, B \in \mathbb{F}^{n\times n}$, and assume that A and B are projectors. Then, the following statements hold:

i) $A(A - B)^+B = B(A - B)^+A = 0$.

ii) $A - B = A(A - B)^+A - B(B - A)^+B$.

iii) $(A - B)^+ = (A - AB)^+ + (AB - B)^+$.

iv) $(A - B)^+ = (A - BA)^+ + (BA - B)^+$.

v) $(A - B)^+ = A - B + B(A - BA)^+ - (B - BA)^+A$.

vi) $(A-B)^+ = A-B+(A-AB)^+B - A(B-AB)^+.$

vii) $(I-A-B)^+ = (A_\perp B_\perp)^+ - (AB)^+.$

viii) $(I-A-B)^+ = (B_\perp A_\perp)^+ - (BA)^+.$

ix) $\mathcal{R}([A,B]) = \mathcal{R}[(A-B)^+ - (A-B)].$

x) $[A\ \ B][A\ \ B]^+ = (A+B)(A+B)^+ = \begin{bmatrix} A \\ B \end{bmatrix}^+ \begin{bmatrix} A \\ B \end{bmatrix}.$

xi) $\operatorname{rank}[AB-(AB)^2] = \operatorname{rank} B(B_\perp A_\perp)^+A = \operatorname{rank} B_\perp(BA)^+A_\perp.$

xii) $\operatorname{rank}[AB,(AB)^+] = 2\operatorname{rank}[AB\ \ BA] - 2\operatorname{rank} AB.$

xiii) $\operatorname{rank}(A-B) = \operatorname{tr}(A-AB)^+ + \operatorname{tr}(B-BA)^+.$

Now, let $C, D \in \mathbb{F}^{n\times n}$ denote the projectors onto $\mathcal{R}(A)\cap[\mathcal{R}(A)\cap\mathcal{R}(B)]^\perp$ and $\mathcal{R}(B)\cap[\mathcal{R}(B)\cap\mathcal{R}(A)]^\perp$, respectively. Then, the following statements hold:

xiv) $\sigma_{\max}(A-B) = \sigma_{\max}(C-D).$

xv) $\operatorname{tr}(A-B)^2 = \operatorname{tr}(C-D)^2.$

xvi) $\operatorname{rank}(A-B) = \dim[\mathcal{R}(A)+\mathcal{R}(B)] - \dim[\mathcal{R}(A)\cap\mathcal{R}(B)].$

xvii) $\operatorname{rank}(A-B) = \operatorname{rank}[A + A_\perp(A_\perp B)^+] - \operatorname{rank}[A - A(AB_\perp)^+].$

Finally, let $C, D \in \mathbb{F}^{n\times n}$ denote the projectors onto $\mathcal{R}(A)\cap\mathcal{R}(I-B)$ and $\mathcal{R}(I-A)\cap\mathcal{R}(B)$, respectively. Then, the following statement holds:

xviii) $C = A - A(AB)^\perp$ and $D = A_\perp - A_\perp(A_\perp B_\perp)^+.$

xix) $\operatorname{rank}(C-D) = \dim[\mathcal{R}(C)+\mathcal{R}(D)] = \operatorname{rank}[C + D_\perp(C_\perp D)^+].$

Source: [249, 718, 2636]. **Related:** Fact 4.16.5.

Fact 8.8.5. Let $A, B \in \mathbb{F}^{n\times n}$, and assume that A and B are projectors. Then, the following statements are equivalent:

i) AB is a projector.

ii) $(AB)^+ = AB.$

iii) $(AB)^+ = BA.$

iv) $(A-B)^+ = A-B.$

v) $B(A-BA)^+ = (B-BA)^+A.$

vi) $A-B$ is tripotent.

vii) $[AB,(AB)^+] = 0.$

viii) $[AB(AB)^+,(AB)^+AB] = 0.$

ix) $[AB(AB)^*,(AB)^+AB] = 0.$

x) $[AB(AB)^+,(AB)^*AB] = 0.$

xi) $[(AB)^+]^2 = [(AB)^2]^+.$

xii) $I-AB$ is normal.

xiii) $I-AB$ is a partial isometry.

xiv) $I-AB$ is star-dagger.

xv) $(I-AB)^+$ is idempotent.

xvi) $[(AB)^2]^+ = [(AB)^+]^2.$

xvii) $(I-AB)^+ = I-AB.$

xviii) $(ABA)^+ = ABA.$

xix) $AB(AB)^+ = ABA.$

xx) $(AB)^+ = A(AB)^+.$

xxi) $(AB)^+ = B(BA)^+$.

xxii) $(AB - BA)^+ = (AB)^+ - (BA)^+$.

xxiii) $(ABA)^+$ is an orthogonal projector.

xxiv) $\operatorname{tr}(ABA)^+ = \operatorname{tr}(AB)^+$.

xxv) $\operatorname{rank}[I - (AB)^+(BA)^+] = \operatorname{def} AB$.

xxvi) $(AB)^+ = (AB)^\#$.

xxvii) $A + B - \frac{3}{2}AB$ is a 1-inverse of $A + B$.

xxviii) $A - B - AB$ is a 1-inverse of $A - B$.

xxix) $B = AB(AB)^+ + A_\perp - A_\perp(A_\perp B_\perp)^+$.

xxx) $AB(AB)^+ = A - A(AB_\perp)^+$.

xxxi) $I - A - B = (I - A - B)^+$.

xxxii) $A + B - AB = (A + B - AB)^+$.

xxxiii) $\operatorname{rank} ABA = \operatorname{tr}(ABA)^+$.

xxxiv) $\operatorname{tr} ABA = \operatorname{tr}(ABA)^+$.

xxxv) $\operatorname{rank}(AB - BA) = \operatorname{tr}(AB - BA)^+$.

xxxvi) $\operatorname{rank}(A + B - AB) = \operatorname{tr}(A + B - AB)^+$.

Source: [235, 245, 252, 261, 718, 2636, 2657, 2705, 2706]. **Remark:** Star-dagger is defined in Fact 8.3.32. **Related:** Fact 4.18.7, Fact 7.13.5, Fact 7.13.6, Fact 8.4.28, and Fact 8.11.13.

Fact 8.8.6. Let $A, B \in \mathbb{F}^{n\times n}$, assume that A and B are projectors. Then, the following statements are equivalent:

i) $A + B$ is a projector.

ii) $(A + B)^+ = A + B$.

iii) $AB + BA = (AB + BA)^+$.

iv) $\operatorname{rank}(I - A - B) = \operatorname{tr}(I - A - B)^+$.

v) AB is a projector, and $\operatorname{rank}(A + B) = \operatorname{tr}(A + B)^+$.

Source: [261, 2657]. **Related:** Fact 4.18.9.

Fact 8.8.7. Let $A, B \in \mathbb{F}^{n\times n}$, assume that A and B are projectors. Then, the following statements are equivalent:

i) $A - B$ is a projector.

ii) $\operatorname{rank}(A - B) = \operatorname{tr}(A - B)^+$.

Source: [261]. **Related:** Fact 4.18.10.

Fact 8.8.8. Let $A, B \in \mathbb{F}^{n\times n}$, assume that A and B are projectors, let $P \in \mathbb{F}^{n\times n}$ be the projector onto $\mathcal{N}([A, B])$, and let $\alpha, \beta \in \mathbb{F}$, where α and β are nonzero. Then, the following statements hold:

i) $\operatorname{rank}[A, B] \le \operatorname{rank}(\alpha A + \beta B)$.

ii) If $[A, B]$ is nonsingular, then $\alpha A + \beta B$.

iii) $\operatorname{rank}[A, B] = \operatorname{rank}(\alpha A + \beta B)P_\perp$.

iv) $\alpha A + \beta B$ is nonsingular if and only if $\operatorname{rank}(\alpha A + \beta B)P + \operatorname{rank}(\alpha A + \beta B) = n$.

v) $A - B$ is nonsingular if and only if $(A + B)P = P$.

Now, assume that $AB = BA$. Then, the following statements hold:

vi) $\operatorname{spec}(\alpha A + \beta B) \subset \{0, \alpha, \beta, \alpha + \beta\}$.

vii) If $\alpha + \beta \ne 0$, then $\operatorname{rank}(\alpha A + \beta B) = \operatorname{rank}(A + B)$.

viii) $(\alpha A + \beta B)^+ = [(\alpha + \beta)^+ - \frac{1}{\alpha} - \frac{1}{\beta}]AB + \frac{1}{\alpha}A + \frac{1}{\beta}B$.

Source: [406]. **Related:** Fact 4.18.20.

Fact 8.8.9. Let $A \in \mathbb{F}^{n\times m}$ and $B \in \mathbb{F}^{m\times n}$. Then, $BAB = B$ if and only if there exist projectors $C \in \mathbb{F}^{n\times n}$ and $D \in \mathbb{F}^{m\times m}$ such that $B = (CAD)^+$. **Source:** [1232].

Fact 8.8.10. Let $A \in \mathbb{F}^{n\times n}$. Then, the following statements are equivalent:

i) $A^2 = AA^*A$.

ii) A is the product of two projectors.

iii) $A = A(A^+)^2A$.

Credit: T. Crimmins. See [2291].

Fact 8.8.11. Let $A \in \mathbb{F}^{n\times n}$, and assume that A is idempotent. Then, A is the idempotent matrix onto $\mathcal{R}(AA^+)$ along $\mathcal{R}(I - A^+A)$. **Source:** Use Fact 8.10.2 with $A^\# = A$, or note that $\mathcal{R}(A) = \mathcal{R}(AA^+)$ and $\mathcal{N}(A) = \mathcal{R}(I - A^+A)$.

Fact 8.8.12. Let $A, B \in \mathbb{F}^{n\times n}$, and assume that A and B are projectors. Then, the following statements hold:

i) $(A_\perp B)^+ = (I - BA)^+B(I - BA) = (I - BA)^+BA_\perp$.

ii) $(A_\perp B)^+$ is the idempotent matrix onto $\mathcal{R}(B) \cap [\mathcal{N}(A) + \mathcal{N}(B)]$ along $\mathcal{R}(A) + [\mathcal{N}(A) \cap \mathcal{N}(B)]$.

iii) If $\mathcal{R}(A) + \mathcal{R}(B) = \mathbb{F}^n$, then $(A_\perp B)^+$ is the idempotent matrix onto $\mathcal{R}(B) \cap [\mathcal{N}(A) + \mathcal{N}(B)]$ along $\mathcal{R}(A)$.

iv) If $\mathcal{R}(A) \cap \mathcal{R}(B) = \{0\}$, then $I - BA$ is nonsingular and $(A_\perp B)^+ = (I - BA)^{-1}B(I - BA) = (I - BA)^{-1}BA_\perp = (I - AB)^{-1}A_\perp$ is the idempotent matrix onto $\mathcal{R}(B)$ along $\mathcal{R}(A) + [\mathcal{N}(A) \cap \mathcal{N}(B)]$.

v) $(I - BA)^+B_\perp$ is the idempotent matrix onto $(\mathcal{R}(B) \cap [\mathcal{N}(A) + \mathcal{N}(B)]) + [\mathcal{N}(A) \cap \mathcal{N}(B)]$ along $\mathcal{R}(B)$.

vi) $A(A + B - BA)^+$ is the idempotent matrix onto $\mathcal{R}(A)$ along $(\mathcal{R}(A) \cap [\mathcal{N}(A) + \mathcal{N}(B)]) + [\mathcal{N}(A) \cap \mathcal{N}(B)]$.

vii) $\mathcal{R}[(I - BA)^+B_\perp] = \mathcal{N}[B(A + B - AB)]^+$ and $\mathcal{N}[(I - BA)^+B_\perp] = \mathcal{R}[B(A + B - AB)]^+$.

viii) $\mathcal{R}(A) + \mathcal{R}(B) = \mathbb{F}^n$ if and only if $(A_\perp B)^+ = (I - AB)^+A_\perp$.

ix) $\mathcal{R}(A) \cap \mathcal{R}(B) = \{0\}$ if and only if $(A_\perp B)^+ = B(A + B - BA)^+$.

x) $\mathcal{R}(A)$ and $\mathcal{R}(B)$ are complementary subspaces if and only if $(I - AB)^+A_\perp = B(A + B - BA)^+$.

xi) $(A_\perp B)^+ = (I - BAB)^+BA_\perp$.

xii) $(A_\perp B)^+ = B(B - A)^+$.

xiii) $(B - A)^+ = (A_\perp B)^+ - (BA_\perp)^+$.

xiv) Let C denote the projector onto $[\mathcal{R}(A) \cap \mathcal{R}(B)] + [\mathcal{N}(A) \cap \mathcal{N}(B)]$. Then, $(B - A)^+ = (A_\perp B)^+ + (BA_\perp)^+ - I + C$.

xv) Let C denote the projector onto $[\mathcal{R}(A) \cap \mathcal{R}(B)] + [\mathcal{N}(A) \cap \mathcal{N}(B)]$, and let D denote the projector onto $[\mathcal{R}(A) + \mathcal{R}(B)] \cap [\mathcal{N}(A) + \mathcal{N}(B)]$. Then, $C + D = I$.

xvi) $(B - A)^+ = (A + B)^+(B - A)(A + B)^+$.

xvii) $(A + B)^+ = (B - A)^+(A + B)(B - A)^+ + A(A + B)^+B$.

xviii) $(AB)^+ = (A + B - I)^+A = B(A + B - I)^+$.

xix) $[(A + B - I)^+]^2A = (ABA)^+ = A[(A + B - I)^+]^2$.

xx) $(BA)^+ = (BA)^+B = A(A + B - I)^+B = A[(A + B - I)^+]^2B$.

xxi) For all $k \geq 0$, $A[(A + B - I)^+]^{2k+1}B = A[(A + B - I)^+]^{2k+2}B$.

xxii) $(BA)^+ = AB(AB)^\#$.

xxiii) $(AB)^\# = A[(A + B - I)^+]^3B = A[(A + B - I)^+]^4B$.

xxiv) $A[(A + B - I)^+]^3B = (BA)^+(A + B - I)^+(BA)^+$.

xxv) $A[(A+B-I)^+]^2B + B[(A+B-I)^+]^2A = (A+B-I)^+ + (A+B-I)(A+B-I)^+$.

xxvi) $(A_\perp B)^+A = A(BA_\perp)^+ = 0$.

xxvii) $(A_\perp B)^+ = B(A_\perp B)^+ = (A_\perp B)^+B$.

xxviii) $(A+B)(A-B)^+ + (A-B)^+(A+B) = 2(A-B)^+$.

xxix) $(A-B)(A-B)^+ = [(A+B)(A-B)^+]^2 = (AB_\perp)^+ + (BA_\perp)^+ = (B_\perp A)^+ + (A_\perp B)^+$
$= AB_\perp(AB_\perp)^+ + A_\perp B(A_\perp B)^+$.

Source: Fact 4.18.19, Fact 8.8.14, and [248].

Fact 8.8.13. Let $A, B \in \mathbb{F}^{n\times n}$, assume that A and B are projectors. Then, $(AB)^+$ is the idempotent matrix onto $\mathcal{R}(BA)$ along $\mathcal{N}(BA) = \mathcal{R}(AB)^\perp$. **Source:** [778]. **Remark:** Corollary 10.3.7 implies that BA is diagonalizable over $\mathbb{F}$ and thus group invertible. Corollary 4.8.10 thus implies that $\mathcal{R}(BA)$ and $\mathcal{N}(BA)$ are complementary subspaces. **Credit:** T. N. E. Greville. **Related:** Fact 8.8.14.

Fact 8.8.14. Let $A, B \in \mathbb{F}^{n\times n}$, assume that A and B are projectors, and assume that $\mathcal{R}(A)$ and $\mathcal{R}(B)$ are complementary subspaces. Then, $(B_\perp A)^+$ is the idempotent matrix onto $\mathcal{R}(A)$ along $\mathcal{R}(B)$. **Source:** Fact 8.8.15, [1246], and [1504]. **Remark:** Fact 8.8.3 implies that $(B_\perp A)^+$ is idempotent. Fact 4.18.19 implies that $I-BA$ is nonsingular and $(B_\perp A)^+ = (I-AB)^{-1}A(I-AB) = A(I-BA)^{-1}(I-AB) = (I-BA)^{-1}(I-B) = B(A+B-BA)^{-1}$. **Related:** Proposition 4.8.6, Fact 4.16.4, Fact 8.8.13, and Fact 8.8.15.

Fact 8.8.15. Let $A, B \in \mathbb{F}^{n\times n}$, and assume that $\mathcal{R}(A)$ and $\mathcal{R}(B)$ are complementary subspaces. Furthermore, define $P \triangleq AA^+$ and $Q \triangleq BB^+$. Then, $(P_\perp Q)^+$ is the idempotent matrix onto $\mathcal{R}(A)$ along $\mathcal{R}(B)$. **Source:** [1232]. **Related:** Fact 4.16.4, Fact 4.18.19, Fact 8.8.14, and Fact 8.10.2.

Fact 8.8.16. Let $A, B \in \mathbb{F}^{n\times n}$, and assume that A and B are projectors. Then, the following statements hold:

i) $A + A_\perp(A_\perp B)^+$ is the projector onto $\mathcal{R}(A) + \mathcal{R}(B)$.

ii) $A + A_\perp(A_\perp B)^+ = (A+B)(A+B)^+ = A + (AB_\perp)^+AB_\perp$.

iii) $A - A(AB_\perp)^+$ is the projector onto $\mathcal{R}(A) \cap \mathcal{R}(B)$.

iv) $A - A(AB_\perp)^+ = A - AB_\perp(AB_\perp)^+ = A - (B_\perp A)^+B_\perp A = 2A(A+B)^+B = 2[A - A(A+B)^+B]$.

Source: [2239]. **Related:** Fact 4.18.11 and Fact 8.8.17.

Fact 8.8.17. Let $A, B \in \mathbb{F}^{n\times n}$, and assume that A and B are projectors. Then, the following statements hold:

i) $\mathcal{R}[A - A(AB_\perp)^+] = \mathcal{N}\left(\begin{bmatrix} A_\perp \\ B_\perp \end{bmatrix}\right)$.

ii) $\sigma_{\max}(AB) = \sigma_{\max}[B - A_\perp(A_\perp B)^+]$.

iii) If $\mathcal{R}(A) \cap \mathcal{R}(B) = \{0\}$, then

$$\sigma_{\max}^2[(A_\perp B)^+] = \frac{1}{1 - \sigma_{\max}^2(AB)}.$$

iv) If $A \neq 0$ and $\mathcal{R}(A) \cap \mathcal{R}(B) = \{0\}$, then

$$\sigma_{\max}^2(AB) < \frac{\sigma_{\max}^2[(A_\perp B)^+]}{1 + \sigma_{\max}^2[(A_\perp B)^+]} < 1.$$

v) $(A-B)(A-B)^+ + [(A-B)(A-B)^+]_\perp[[(A-B)(A-B)^+]_\perp(I-A-B)(I-A-B)^+]^+ = I$.

vi) $(A-B)(A-B)^+(I-A-B)(I-A-B)^+ = [A,B][A,B]^+$.

vii) $(A-B)(A-B)^+ + (I-A-B)(I-A-B)^+ - [A,B][A,B]^+ = I$.

viii) $(A+B)(A+B)^+ + [(A-B)(A-B)^+]_\perp[[(A+B)(A+B)^+]_\perp(I-A-B)(I-A-B)^+]^+ = I$.

ix) $(A+B)(A+B)^+(I-A-B)(I-A-B)^+ = (AB+BA)(AB+BA)^+$.

x) $(A-B)(A-B)^+ + (I-A-B)(I-A-B)^+ - (AB+BA)(AB+BA)^+ = I.$

In addition, the following statements are equivalent:

xi) $\mathcal{R}([A,B]) = \mathcal{R}(A-B).$

xii) $\mathcal{R}(AB+BA) = \mathcal{R}(A+B).$

xiii) $I-A-B$ is nonsingular.

xiv) $A = AB(AB)^+$ and $B = BA(BA)^+.$

Finally, the following statements are equivalent:

xv) $\mathcal{R}([A,B])$ is nonsingular.

xvi) $A-B$ and $I-A-B$ are nonsingular.

xvii) $[\mathcal{R}(A)+\mathcal{R}(B)]\cap[\mathcal{N}(A)+\mathcal{N}(B)] = \mathbb{F}^n$ and $[\mathcal{R}(A)+\mathcal{N}(B)]\cap[\mathcal{N}(A)+\mathcal{R}(B)] = \mathbb{F}^n.$

xviii) $\mathcal{R}(A)+\mathcal{R}(B) = \mathbb{F}^n$, $\mathcal{R}(A)+\mathcal{N}(B) = \mathbb{F}^n$, $\mathcal{N}(A)+\mathcal{R}(B) = \mathbb{F}^n$, and $\mathcal{N}(A)+\mathcal{N}(B) = \mathbb{F}^n.$

Source: [242, 244]. **Related:** Fact 4.18.11, Fact 4.18.12, Fact 7.13.27, and Fact 7.13.28.

Fact 8.8.18. Let $A, B \in \mathbb{F}^{n\times n}$, and assume that A and B are complementary projectors. Then, the following statements hold:

i) $(B_\perp A)^+ = AB_\perp - A(A_\perp B)^+ B_\perp.$

ii) $\operatorname{rank}[(B_\perp A)^+ - AB_\perp] = \operatorname{rank} AB.$

iii) The following statements are equivalent:

a) $(B_\perp A)^+ = A.$

b) $AB = 0.$

c) $\mathcal{R}(A)^\perp = \mathcal{R}(B).$

Source: [2657].

Fact 8.8.19. Let $A, B \in \mathbb{F}^{n\times n}$, and assume that A and B are projectors. Then,

$$\operatorname{glb}(\{A,B\}) = 2A(A+B)^+B = 2B(A+B)^+A.$$

Furthermore, $\operatorname{glb}(\{A,B\})$ is the projector onto $\mathcal{R}(A)\cap\mathcal{R}(B)$; that is, $\mathcal{R}(A)\cap\mathcal{R}(B) = \mathcal{R}[2A(A+B)^+B]$. In addition,

$$\begin{aligned}\operatorname{glb}(\{A,B\}) &= \lim_{k\to\infty} A(BA)^k = A\#B\\ &= 2AB(A+B)^+A = 2BA(A+B)^+B\\ &= 2A(A+B)^+BA = 2B(A+B)^+AB\\ &= 2AB(A+B)^+AB = 2BA(A+B)^+BA\\ &= 2[A - A(A+B)^+A] = 2[B - B(A+B)^+B]\\ &= A - A(AB_\perp)^+ = B - B(BA_\perp)^+\\ &= A - (B_\perp A)^+B_\perp A = B - (A_\perp B)^+A_\perp B\\ &= 2AA^+(AA^+ + BB^+)^+BB^+ = 2BB^+(AA^+ + BB^+)^+AA^+\\ &= \left(\begin{bmatrix}A_\perp\\ B_\perp\end{bmatrix}^+\begin{bmatrix}A_\perp\\ B_\perp\end{bmatrix}\right)_\perp.\end{aligned}$$

Source: [84], [1302, pp. 64, 65, 121, 122], [1381, p. 191], [2238, pp. 304–306], and Fact 4.18.12. **Related:** Fact 8.4.8, Fact 10.11.68, and Fact 10.24.20.

Fact 8.8.20. Let $A, B \in \mathbb{F}^{n\times n}$, and assume that A and B are projectors. Then,

$$\operatorname{lub}(\{A,B\}) = (A+B)(A+B)^+ = (A+B)^+(A+B).$$

Furthermore, $\operatorname{lub}(\{A,B\})$ is the projector onto $\mathcal{R}(A+B)$; that is, $\mathcal{R}(A+B) = \mathcal{R}[(A+B)(A+B)^+]$. In addition,

$$\begin{aligned}\operatorname{lub}(\{A,B\}) &= I - \lim_{k\to\infty} A_\perp(B_\perp A_\perp)^k\\ &= I - 2A_\perp(A_\perp + B_\perp)^+B_\perp = I - 2B_\perp(B_\perp + A_\perp)^+A_\perp\\ &= A + (AB_\perp)^+AB_\perp = B + (BA_\perp)^+BA_\perp\\ &= A + (BA_\perp)^+BA_\perp = B + (AB_\perp)^+AB_\perp\\ &= A + A_\perp(A_\perp B)^+ = B + B_\perp(B_\perp A)^+\\ &= [A\ \ B][A\ \ B]^+.\end{aligned}$$

Source: For the first equality, use Fact 3.12.16 and Fact 10.7.3. For the second equality, see [84], [1302, pp. 64, 65, 121, 122], and [2238, pp. 303–306]. **Remark:** Fact 4.10.31 implies that $\mathcal{R}(A) + \mathcal{R}(B) = \mathcal{R}([A\ \ B]) = \mathcal{R}(A + B) = \operatorname{span}[\mathcal{R}(A) \cup \mathcal{R}(B)]$. **Related:** Fact 4.18.11, Fact 8.4.9, and Fact 10.24.20.

Fact 8.8.21. Let $A, B \in \mathbb{F}^{n\times n}$, and assume that A and B are projectors. Then, the following statements hold:

i) $\operatorname{glb}(\{A,B\}) \le \operatorname{lub}(\{A,B\})$.

ii) $\operatorname{glb}(\{A,B\}) + \operatorname{lub}(\{A_\perp,B_\perp\}) = I$.

iii) $\operatorname{lub}(\{A_\perp,B_\perp\}) = \operatorname{glb}(\{A,B\})_\perp$ and $\operatorname{glb}(\{A_\perp,B_\perp\}) = \operatorname{lub}(\{A,B\})_\perp$.

iv) $\operatorname{rank}\operatorname{glb}(\{A,B\}) = \operatorname{rank} AA^+(AA^+ + BB^+)^+BB^+ = \operatorname{rank} A + \operatorname{rank} B - \operatorname{rank}\,[A\ \ B]$.

v) $\operatorname{rank}\operatorname{lub}(\{A,B\}) = \operatorname{rank}\,[A\ \ B]$.

vi) $\operatorname{rank}\operatorname{glb}(\{A,B\}) + \operatorname{rank}\operatorname{lub}(\{A,B\}) = \operatorname{rank} A + \operatorname{rank} B$.

vii) $\operatorname{tr}\operatorname{glb}(\{A,B\}) + \operatorname{tr}\operatorname{lub}(\{A,B\}) = \operatorname{tr} A + \operatorname{tr} B$.

If $A \le B$, then the following statement holds:

viii) $B = \operatorname{lub}[\{A, \operatorname{glb}(\{A_\perp, B\})\}]$.

If $AB = BA$. Then, the following statements hold:

ix) $2AB \le A + B$.

x) $\operatorname{glb}(\{A,B\}) = AB = BA$.

xi) $\operatorname{lub}(\{A,B\}) = A + B - AB = B + A - BA$.

If $AB = BA = 0$, then the following statements hold:

xii) $\operatorname{glb}(\{A,B\}) = 0$.

xiii) $\operatorname{lub}(\{A,B\}) = A + B$.

Source: [2238, pp. 306, 311], Fact 8.4.8, and Fact 8.4.9. *vi)* is equivalent to Theorem 3.1.3.

Fact 8.8.22. Let $A \in \mathbb{F}^{n\times n}$, $B \in \mathbb{F}^{n\times m}$, and $C \in \mathbb{F}^{m\times m}$, define $\mathcal{A} \triangleq \begin{bmatrix} A & B\\ B^* & C\end{bmatrix}$, and assume that $\mathcal{A}$ is a projector. Then,

$$\operatorname{rank}\mathcal{A} = \operatorname{rank} A + \operatorname{rank} C - \operatorname{rank} B,$$

$$\operatorname{rank}(A - B^*C^+B) = \operatorname{rank} C - \operatorname{rank} B^*C^+B.$$

Source: [2657].

Fact 8.8.23. Let $k \ge 1$, let $A_1,\dots,A_k \in \mathbb{F}^{n\times n}$, assume that $A_1,\dots,A_k$ are projectors, and define $B_1,\dots,B_{k-1} \in \mathbb{F}^{n\times n}$ by

$$B_i \triangleq (A_1\cdots A_{k-i+1})^+A_1\cdots A_{k-i},\quad i = 1,\dots,k-2;\quad B_{k-1} \triangleq A_2\cdots A_k(A_1\cdots A_k)^+.$$

Then, $B_1,\dots,B_{k-1}$ are idempotent, and

$$(A_1\cdots A_k)^+ = B_1\cdots B_{k-1}.$$

Source: [2651]. **Remark:** For $k = 2$, the result B_1 is idempotent is given by *vi)* of Fact 8.8.3.

8.9 Facts on the Moore-Penrose Generalized Inverse for Partitioned Matrices

Fact 8.9.1. Let $A \in \mathbb{F}^{n\times m}$ and $B \in \mathbb{F}^{n\times l}$. Then,

$$\begin{aligned}\mathcal{R}(A) \cap \mathcal{R}(B) &= [A\ \ 0_{n\times l}]\mathcal{N}([A\ \ B]) \cap [0_{n\times m}\ \ B]\mathcal{N}([A\ \ B])\\ &= [A\ \ 0_{n\times l}]\mathcal{R}(I_{m+l} - [A\ \ B]^+[A\ \ B]) \cap [0_{n\times m}\ \ B]\mathcal{R}(I_{m+l} - [A\ \ B]^+[A\ \ B]).\end{aligned}$$

Source: Fact 3.14.14.

Fact 8.9.2. Let $A \in \mathbb{F}^{n\times m}$ and $B \in \mathbb{F}^{n\times l}$. Then,

$$\mathcal{R}(A) \cap \mathcal{R}(B) = \mathcal{N}\left(\begin{bmatrix} I - AA^+ \\ I - BB^+ \end{bmatrix}\right), \quad \dim[\mathcal{R}(A) \cap \mathcal{R}(B)] = \operatorname{def}\begin{bmatrix} I - AA^+ \\ I - BB^+ \end{bmatrix}.$$

Related: Fact 4.18.12.

Fact 8.9.3. Let $A \in \mathbb{F}^{n\times m}$ and $B \in \mathbb{F}^{l\times m}$. Then,

$$\mathcal{N}(A) + \mathcal{N}(B) = \mathcal{R}(I - A^+A) + \mathcal{R}(I - B^+B) = \mathcal{R}([I - A^+A\ \ \ I - B^+B]),$$

$$\dim[\mathcal{N}(A) + \mathcal{N}(B)] = \operatorname{rank}\,[I - A^+A\ \ \ I - B^+B],$$

$$\mathcal{N}\left(\begin{bmatrix} A \\ B \end{bmatrix}\right) = \mathcal{N}(A) \cap \mathcal{N}(B) = \mathcal{R}(I - A^+A) \cap \mathcal{R}(I - B^+B),$$

$$\operatorname{def}\begin{bmatrix} A \\ B \end{bmatrix} = \operatorname{def}\,[I - A^+A\ \ \ I - B^+B] - \operatorname{def} A - \operatorname{def} B.$$

Source: The last equality follows from Fact 3.14.14.

Fact 8.9.4. Let $A \in \mathbb{F}^{n\times m}$ and $B \in \mathbb{F}^{m\times l}$. Then,

$$\mathcal{N}(A) + \mathcal{R}(B) = \mathcal{R}(I - A^+A) + \mathcal{R}(B) = \mathcal{R}([I - A^+A\ \ \ B]),$$

$$\dim[\mathcal{N}(A) + \mathcal{R}(B)] = \operatorname{rank}\,[I - A^+A\ \ \ B] = \operatorname{rank}\begin{bmatrix} A \\ I - BB^+ \end{bmatrix} + \operatorname{rank} B - \operatorname{rank} A,$$

$$\mathcal{N}(A) \cap \mathcal{R}(B) = \mathcal{N}(A) \cap \mathcal{N}(I - BB^+) = \mathcal{N}\left(\begin{bmatrix} A \\ I - BB^+ \end{bmatrix}\right),$$

$$\dim[\mathcal{N}(A) \cap \mathcal{R}(B)] = \operatorname{def}\begin{bmatrix} A \\ I - BB^+ \end{bmatrix} = m - \operatorname{rank}\begin{bmatrix} A \\ I - BB^+ \end{bmatrix},$$

$$\operatorname{rank} A + \operatorname{rank} B \le \operatorname{rank}\begin{bmatrix} A \\ I - BB^+ \end{bmatrix} + \operatorname{rank} B = m - \dim[\mathcal{N}(A) \cap \mathcal{R}(B)] + \operatorname{rank} B = m + \operatorname{rank} AB.$$

Source: The fourth equality follows from Fact 8.9.15. The last equality follows from (3.6.19). **Remark:** This result provides an alternative proof of Sylvester's inequality given by Proposition 3.6.11. **Related:** Fact 3.13.21.

Fact 8.9.5. Let $A \in \mathbb{F}^{n\times m}$ and $B \in \mathbb{F}^{n\times l}$. Then, the following statements are equivalent:

i) $\operatorname{rank}\,[A\ \ B] = \operatorname{rank} A + \operatorname{rank} B$.

ii) $\mathcal{R}(A) \cap \mathcal{R}(B) = \{0\}$.

iii) $\operatorname{rank}(AA^* + BB^*) = \operatorname{rank} A + \operatorname{rank} B$.

iv) $A^*(AA^* + BB^*)^+A$ is idempotent.

v) $A^*(AA^* + BB^*)^+A = A^+A$.

vi) $A^*(AA^* + BB^*)^+B = 0.$

Source: [1924, pp. 56, 57]. **Remark:** Additional equivalent statements are given in Fact 8.9.26. **Related:** Fact 3.14.15, Fact 8.4.8, and Fact 8.9.21.

Fact 8.9.6. Let $A \in \mathbb{F}^{n\times m}$ and $B \in \mathbb{F}^{n\times l}$, and define the projectors $P \triangleq AA^+$ and $Q \triangleq BB^+$. Then, the following statements are equivalent:

i) $\operatorname{rank}\,[A\ \ B] = \operatorname{rank} A + \operatorname{rank} B = n.$

ii) $P - Q$ is nonsingular.

If these statements hold, then

$$\begin{aligned}(P-Q)^{-1} &= (P-PQ)^+ + (PQ-Q)^+ = (P-QP)^+ + (QP-Q)^+\\ &= P - Q + Q(P-QP)^+ - (Q-QP)^+P.\end{aligned}$$

Source: [718].

Fact 8.9.7. Let $A \in \mathbb{F}^{n\times m}$, $B \in \mathbb{F}^{n\times l}$, $C \in \mathbb{F}^{k\times m}$, and $D \in \mathbb{F}^{k\times l}$, and define $X \triangleq B - AA^+B$, $Y \triangleq C - CA^+A$, and $Z \triangleq (I_k - YY^+)(D - CA^+B)(I_l - X^+X)$. Then,

$$\begin{aligned}\operatorname{rank}\,[A\ \ B] &= \operatorname{rank} A + \operatorname{rank}(B - AA^+B) = \operatorname{rank} B + \operatorname{rank}(A - BB^+A)\\ &= \operatorname{rank} A + \operatorname{rank} B - \dim[\mathcal{R}(A)\cap\mathcal{R}(B)],\end{aligned}$$

$$\begin{aligned}\operatorname{rank}\begin{bmatrix}A\\C\end{bmatrix} &= \operatorname{rank} A + \operatorname{rank}(C - CA^+A) = \operatorname{rank} C + \operatorname{rank}(A - AC^+C)\\ &= \operatorname{rank} A + \operatorname{rank} C - \dim[\mathcal{R}(A^*)\cap\mathcal{R}(C^*)],\end{aligned}$$

$$\operatorname{rank}\begin{bmatrix}0 & B\\ C & D\end{bmatrix} = \operatorname{rank} B + \operatorname{rank} C + \operatorname{rank}\,(I_k - CC^+)D(I_l - B^+B),$$

$$\operatorname{rank}\begin{bmatrix}A & 0\\ C & D\end{bmatrix} = \operatorname{rank} A + \operatorname{rank} D + \operatorname{rank}\,(I_k - DD^+)C(I_m - A^+A),$$

$$\operatorname{rank}\begin{bmatrix}A & B\\ 0 & D\end{bmatrix} = \operatorname{rank} A + \operatorname{rank} D + \operatorname{rank}\,(I_n - AA^+)B(I_l - D^+D),$$

$$\operatorname{rank}\begin{bmatrix}A & B\\ C & 0\end{bmatrix} = \operatorname{rank} B + \operatorname{rank} C + \operatorname{rank}\,(I_n - BB^+)A(I_m - C^+C),$$

$$\begin{bmatrix}A & B\\ C & D\end{bmatrix} = \begin{bmatrix}I & 0\\ CA^+ & I\end{bmatrix}\begin{bmatrix}A & B - AA^+B\\ C - CA^+A & D - CA^+B\end{bmatrix}\begin{bmatrix}I & A^+B\\ 0 & I\end{bmatrix},$$

$$\begin{aligned}\operatorname{rank}\begin{bmatrix}A & B\\ C & D\end{bmatrix} &= \operatorname{rank} A + \operatorname{rank}\begin{bmatrix}0 & B - AA^+B\\ C - CA^+A & D - CA^+B\end{bmatrix}\\ &= \operatorname{rank}\begin{bmatrix}A\\C\end{bmatrix} + \operatorname{rank}\,[A\ \ B] - \operatorname{rank} A + \operatorname{rank} Z\\ &= \operatorname{rank} A + \operatorname{rank} X + \operatorname{rank} Y + \operatorname{rank} Z\\ &\le \operatorname{rank} A + \operatorname{rank} B + \operatorname{rank} C + \operatorname{rank}(D - CA^+B),\end{aligned}$$

$$\operatorname{rank}\begin{bmatrix}AA^* & B\\ B^* & 0\end{bmatrix} = \operatorname{rank}\begin{bmatrix}-AA^* & B\\ B^* & 0\end{bmatrix} = \operatorname{rank}\,[A\ \ B] + \operatorname{rank} B,$$

$$\operatorname{rank}\begin{bmatrix} A^*AA^* & A^*B \\ CA^* & D \end{bmatrix} = \operatorname{rank} A + \operatorname{rank}(D - CA^+B).$$

Furthermore, the following statements hold:

i)

$$\operatorname{rank} A + \operatorname{rank}(D - CA^+B) \le \operatorname{rank}\begin{bmatrix} A & B \\ C & D \end{bmatrix}.$$

ii) If $AA^+B = B$ and $CA^+A = C$, then

$$\operatorname{rank} A + \operatorname{rank}(D - CA^+B) = \operatorname{rank}\begin{bmatrix} A & B \\ C & D \end{bmatrix}.$$

iii) $\operatorname{rank} A + \operatorname{rank}(D - CA^+B) = \operatorname{rank}\left[\begin{smallmatrix} A & B \\ C & D \end{smallmatrix}\right]$ if and only if the following statements hold:

a) $\mathcal{N}(D - CA^+B) \subseteq \mathcal{N}[(I - AA^+)B]$.

b) $\mathcal{N}[(D - CA^+B)^*] \subseteq \mathcal{N}[(I - A^+A)C^*]$.

c) $(I - AA^+)B(D - CA^+B)^+C(I - A^+A) = 0$.

iv) If $n = m$ and A is nonsingular, then

$$n + \operatorname{rank}(D - CA^{-1}B) = \operatorname{rank}\begin{bmatrix} A & B \\ C & D \end{bmatrix}.$$

v) If $k = l$ and D is nonsingular, then

$$k + \operatorname{rank}(A - BD^{-1}C) = \operatorname{rank}\begin{bmatrix} A & B \\ C & D \end{bmatrix}.$$

vi) Let $A = A_{\mathrm{L}}A_{\mathrm{R}}$, $X = X_{\mathrm{L}}X_{\mathrm{R}}$, $Y = Y_{\mathrm{L}}Y_{\mathrm{R}}$, and $Z = Z_{\mathrm{L}}Z_{\mathrm{R}}$ be full rank factorizations. Then,

$$\begin{bmatrix} A & B \\ C & D \end{bmatrix} = \begin{bmatrix} A_{\mathrm{L}} & 0 & 0 & X_{\mathrm{L}} \\ CA_{\mathrm{R}}^+ & Z_{\mathrm{L}} & Y_{\mathrm{L}} & (I_k - YY^+)(D - CA^+B)X_{\mathrm{R}}^+ \end{bmatrix}\begin{bmatrix} A_{\mathrm{R}} & A_{\mathrm{L}}^+B \\ 0 & Z_{\mathrm{R}} \\ Y_{\mathrm{R}} & Y_{\mathrm{L}}^+(D - CA^+B) \\ 0 & X_{\mathrm{R}} \end{bmatrix}$$

is a full-rank factorization.

Source: The first expression for $\left[\begin{smallmatrix} A & B \\ C & D \end{smallmatrix}\right]$ follows from Fact 3.14.11. The inequality follows from the previous equality and Fact 3.14.20. See [212, 639, 1966, 2035], [2238, p. 194], [2634, 2638, 2656, 2941], Fact 3.14.15, and Fact 3.14.17. **Remark:** Since $\operatorname{rank} C \le \operatorname{rank}\left[\begin{smallmatrix} A & 0 \\ C & D \end{smallmatrix}\right]$, it follows that

$$\operatorname{rank} C \le \operatorname{rank} A + \operatorname{rank} D + \operatorname{rank}(I_n - DD^+)C(I_m - A^+A).$$

Using Fact 3.14.20, this inequality can be interpolated by noting that

$$\begin{aligned} \operatorname{rank} C &\le \operatorname{rank} CA^+A + \operatorname{rank} C(I_m - A^+A) + \operatorname{rank} DD^+C + \operatorname{rank}(I_k - DD^+)C - \operatorname{rank} C \\ &\le \operatorname{rank} A^+A + \operatorname{rank} C(I_m - A^+A) + \operatorname{rank} DD^+ + \operatorname{rank}(I_k - DD^+)C - \operatorname{rank} C \\ &= \operatorname{rank} A + \operatorname{rank} C(I_m - A^+A) + \operatorname{rank} D + \operatorname{rank}(I_k - DD^+)C - \operatorname{rank} C \\ &\le \operatorname{rank} A + \operatorname{rank} D + \operatorname{rank}(I_k - DD^+)C(I_m - A^+A). \end{aligned}$$

Related: Proposition 3.9.3, Fact 8.9.34, and Proposition 10.2.4.

Fact 8.9.8. Let $A \in \mathbb{F}^{n\times n}$, $B \in \mathbb{F}^{n\times m}$, and $C \in \mathbb{F}^{m\times m}$, and define $X \triangleq B - AA^+B$. Then,

$$\begin{aligned} \nu_-\left(\begin{bmatrix} A & B \\ B^* & 0 \end{bmatrix}\right) &= \operatorname{rank} B + \nu_-[(I - BB^+)A(I - BB^+)] \\ &= \operatorname{rank}\begin{bmatrix} A & B \\ B^* & 0 \end{bmatrix} - \operatorname{rank} B - \nu_+[(I - BB^+)A(I - BB^+)], \end{aligned}$$

$$\nu_+\left(\begin{bmatrix} A & B \\ B^* & 0 \end{bmatrix}\right) = \operatorname{rank} B + \nu_+[(I - BB^+)A(I - BB^+)]$$
$$= \operatorname{rank}\begin{bmatrix} A & B \\ B^* & 0 \end{bmatrix} - \operatorname{rank} B - \nu_-[(I - BB^+)A(I - BB^+)],$$

$$\nu_-\left(\begin{bmatrix} A & B \\ B^* & C \end{bmatrix}\right) = \operatorname{rank}\,[A\ \ B] - \nu_+(A) + \nu_+[(I - X^+X)(C - B^*A^+B)(I - X^+X)]$$
$$= \operatorname{rank}\begin{bmatrix} A & B \\ B^* & C \end{bmatrix} - \operatorname{rank}\,[A\ \ B] + \nu_-(A) - \nu_-[(I - X^+X)(C - B^*A^+B)(I - X^+X)],$$

$$\nu_+\left(\begin{bmatrix} A & B \\ B^* & C \end{bmatrix}\right) = \operatorname{rank}\,[A\ \ B] - \nu_-(A) + \nu_-[(I - X^+X)(C - B^*A^+B)(I - X^+X)]$$
$$= \operatorname{rank}\begin{bmatrix} A & B \\ B^* & C \end{bmatrix} - \operatorname{rank}\,[A\ \ B] + \nu_+(A) - \nu_+[(I - X^+X)(C - B^*A^+B)(I - X^+X)].$$

Source: [2656].

Fact 8.9.9. Let $A \in \mathbb{F}^{n\times n}$, $B \in \mathbb{F}^{n\times m}$, and $C \in \mathbb{F}^{m\times m}$, define $\mathcal{A} \triangleq \left[\begin{smallmatrix} A & B \\ B^* & C \end{smallmatrix}\right]$, and assume that $\mathcal{A}$ is Hermitian and $B = AA^+B$. Then, $\operatorname{In}\mathcal{A} = \operatorname{In}A + \operatorname{In}(A|\mathcal{A})$. **Remark:** This is the *Haynsworth inertia additivity formula.* See [2273]. **Remark:** If $\mathcal{A}$ is positive semidefinite, then $B = AA^+B$. See Proposition 10.2.5 and [2991, p. 257].

Fact 8.9.10. Let $A \in \mathbb{F}^{n\times m}$, $B \in \mathbb{F}^{k\times l}$, and $C \in \mathbb{F}^{n\times l}$. Then,

$$\min_{X\in\mathbb{F}^{m\times l},Y\in\mathbb{F}^{n\times k}} \operatorname{rank}(AX + YB + C) = \operatorname{rank}\begin{bmatrix} A & C \\ 0 & B \end{bmatrix} - \operatorname{rank} A - \operatorname{rank} B.$$

Furthermore, X, Y is a minimizing solution if and only if there exist $U \in \mathbb{F}^{m\times k}$, $U_1 \in \mathbb{F}^{m\times l}$, and $U_2 \in \mathbb{F}^{n\times k}$ such that

$$X = -A^+C + UB + (I_m - A^+A)U_1, \quad Y = (AA^+ - I)CB^+ - AU + U_2(I_k - BB^+).$$

Finally, all such matrices $X \in \mathbb{F}^{m\times l}$ and $Y \in \mathbb{F}^{n\times k}$ satisfy

$$AX + YB + C = 0$$

if and only if

$$\operatorname{rank}\begin{bmatrix} A & C \\ 0 & B \end{bmatrix} = \operatorname{rank} A + \operatorname{rank} B.$$

Source: [2633, 2668]. **Related:** Fact 7.11.25. Note that A and B are square in Fact 7.11.25.

Fact 8.9.11. Let $A \in \mathbb{F}^{n\times n}$, $B \in \mathbb{F}^{n\times m}$, and $C \in \mathbb{F}^{m\times m}$, and assume that $\left[\begin{smallmatrix} A & B \\ B^* & C \end{smallmatrix}\right]$ is a projector. Then,

$$\operatorname{rank}(C - B^*A^+B) = \operatorname{rank} C - \operatorname{rank} B^*A^+B.$$

Source: [226, 2646].

Fact 8.9.12. Let $A \in \mathbb{F}^{n\times m}$, $B \in \mathbb{F}^{n\times l}$, $C \in \mathbb{F}^{l\times m}$, and $D \in \mathbb{F}^{l\times l}$, and assume that D is nonsingular. Then,

$$\operatorname{rank} A = \operatorname{rank}(A - BD^{-1}C) + \operatorname{rank} BD^{-1}C$$

if and only if there exist $X \in \mathbb{F}^{m\times l}$ and $Y \in \mathbb{F}^{l\times n}$ such that $B = AX$, $C = YA$, and $D = YAX$. **Source:** [734]. **Related:** Fact 8.4.10 for the case $l = 1$.

Fact 8.9.13. Let $A \in \mathbb{F}^{n\times m}$, $B \in \mathbb{F}^{n\times l}$, and $C \in \mathbb{F}^{l\times m}$, and assume that $YAX \in \mathbb{F}^{l\times l}$ is nonsingular. Then,

$$\operatorname{rank}[A - B(YAX)^{-1}C] = \operatorname{rank} A - \operatorname{rank} B(YAX)^{-1}C.$$

If, in addition, $l = 1$, then

$$\operatorname{rank}\left(A - \frac{1}{YAX}BC\right) = \operatorname{rank} A - \operatorname{rank} \frac{1}{YAX}BC.$$

Source: Fact 8.4.10 and Fact 8.9.12. **Remark:** The second equality is the *Wedderburn rank-one reduction formula*. See [734] and [1451, p. 14].

Fact 8.9.14. Let $A_{11} \in \mathbb{F}^{n\times m}$, $A_{12} \in \mathbb{F}^{n\times l}$, $A_{21} \in \mathbb{F}^{k\times m}$, and $A_{22} \in \mathbb{F}^{k\times l}$, and define $A \triangleq \begin{bmatrix} A_{11} & A_{12} \\ A_{21} & A_{22} \end{bmatrix} \in \mathbb{F}^{(n+k)\times(m+l)}$ and $B \triangleq AA^+ = \begin{bmatrix} B_{11} & B_{12} \\ B_{12}^{\mathsf{T}} & B_{22} \end{bmatrix}$, where $B_{11} \in \mathbb{F}^{n\times n}$, $B_{12} \in \mathbb{F}^{n\times k}$, and $B_{22} \in \mathbb{F}^{k\times k}$. Then,

$$\operatorname{rank} B_{12} = \operatorname{rank}\,[A_{11}\ \ A_{12}] + \operatorname{rank}\,[A_{21}\ \ A_{22}] - \operatorname{rank} A.$$

Source: [2675]. **Related:** Fact 4.15.22 and Fact 4.17.13.

Fact 8.9.15. Let $A \in \mathbb{F}^{n\times m}$ and $B \in \mathbb{F}^{m\times l}$. Then,

$$\begin{aligned}\operatorname{rank}\begin{bmatrix} 0_{n\times l} & A \\ B & I_m \end{bmatrix} &= \operatorname{rank} A + \operatorname{rank}\,[I - A^+A\ \ B] = \operatorname{rank}\begin{bmatrix} A \\ I - BB^+ \end{bmatrix} + \operatorname{rank} B \\ &= m + \operatorname{rank} AB = \operatorname{rank} A + \operatorname{rank} B + \operatorname{rank}\,(I - BB^+)(I - A^+A),\end{aligned}$$

$$\begin{aligned}\max\{0, \operatorname{rank} A + \operatorname{rank} B - m\} &\le \operatorname{rank} A + \operatorname{rank} B - \operatorname{rank}\,[A^*\ \ B] \\ &\le \operatorname{rank} AB \le \min\{\operatorname{rank} A, \operatorname{rank} B\} \le \min\{n, m, l\}.\end{aligned}$$

The following statements are equivalent:

i) $\operatorname{rank} AB = \operatorname{rank} A$.

ii) $\mathcal{R}(AB) = \mathcal{R}(A)$.

iii) $[B\ \ I - A^+A]$ is right invertible.

The following statements are equivalent:

iv) $\operatorname{rank} AB = \operatorname{rank} B$.

v) $\mathcal{N}(AB) = \mathcal{N}(B)$.

vi) $\begin{bmatrix} A \\ I - BB^+ \end{bmatrix}$ is left invertible.

The following statements are equivalent:

vii) $\operatorname{rank}\begin{bmatrix} 0_{n\times l} & A \\ B & I_m \end{bmatrix} = \operatorname{rank} A + \operatorname{rank} B$.

viii) $\operatorname{rank} AB = \operatorname{rank} A + \operatorname{rank} B - m$.

ix) $\operatorname{rank} AB = \operatorname{rank} A + \operatorname{rank} B - \operatorname{rank}\,[A^*\ \ B]$, and $\operatorname{rank}\,[A^*\ \ B]$.

x) There exist $X \in \mathbb{F}^{l\times m}$ and $Y \in \mathbb{F}^{m\times n}$ such that $BX + YA = I$.

xi) $(I - BB^+)(I - A^+A) = 0$.

xii) $\mathcal{N}(A) \subseteq \mathcal{R}(B)$.

xiii) $\mathcal{N}(B^*) \subseteq \mathcal{R}(A^*)$.

Source: [1966]. Note that

$$\begin{bmatrix} 0_{n\times l} & A \\ B & I_m \end{bmatrix} = \begin{bmatrix} I & 0 \\ A^+ & I \end{bmatrix}\begin{bmatrix} 0 & A \\ B & I - A^+A \end{bmatrix} = \begin{bmatrix} 0 & A \\ B & I - BB^+ \end{bmatrix}\begin{bmatrix} I & B^+ \\ 0 & I \end{bmatrix} = \begin{bmatrix} -I & A \\ 0 & I \end{bmatrix}\begin{bmatrix} AB & 0 \\ B & I \end{bmatrix}$$

and that the rows of $[0\ \ A]$ are orthogonal to the rows of $[B\ \ I - A^+A]$. **Remark:** The generalized inverses can be replaced by arbitrary (1)-inverses. **Credit:** Y. Tian.

Fact 8.9.16. Let $A \in \mathbb{F}^{n\times m}$, $B \in \mathbb{F}^{m\times l}$, and $C \in \mathbb{F}^{l\times k}$. Then,

$$\begin{aligned}\operatorname{rank}\begin{bmatrix} 0 & AB \\ BC & B \end{bmatrix} &= \operatorname{rank} B + \operatorname{rank} ABC \\ &= \operatorname{rank} AB + \operatorname{rank} BC + \operatorname{rank}\,[(I - BC)(BC)^+]B[(I - (AB)^+(AB)],\end{aligned}$$

$$\begin{aligned}&\max\{0, \operatorname{rank} A + \operatorname{rank} B + \operatorname{rank} C - m - l\}\\ &\quad\le \max\{0, \operatorname{rank} A + \operatorname{rank} B + \operatorname{rank} C - \operatorname{rank}[A^* \;\; B] - \operatorname{rank}[B^* \;\; C]\}\\ &\quad\le \max\{0, \operatorname{rank} AB + \operatorname{rank} BC - \operatorname{rank} B\} \le \operatorname{rank} ABC \le \min\{\operatorname{rank} AB, \operatorname{rank} BC\}\\ &\quad\le \min\{\operatorname{rank} A, \operatorname{rank} B, \operatorname{rank} C\} \le \min\{n, m, l, k\}.\end{aligned}$$

Furthermore, the following statements are equivalent:

i) $\operatorname{rank} ABC = \operatorname{rank} B$.

ii) $\operatorname{rank} AB = \operatorname{rank} BC = \operatorname{rank} B$.

iii) $\mathcal{R}[(AB)^*] = \mathcal{R}(BC) = \mathcal{R}(B)$.

The following statements are equivalent:

iv) $\operatorname{rank}\begin{bmatrix} 0 & AB \\ BC & B\end{bmatrix} = \operatorname{rank} AB + \operatorname{rank} BC$.

v) $\operatorname{rank} ABC = \operatorname{rank} AB + \operatorname{rank} BC - \operatorname{rank} B$.

vi) There exist $X \in \mathbb{F}^{k\times l}$ and $Y \in \mathbb{F}^{m\times n}$ such that $BCX + YAB = B$.

vii) $[(I - BC)(BC)^+]B[(I - (AB)^+(AB)] = 0$.

The following statements are equivalent:

viii) $\operatorname{rank} ABC = \operatorname{rank} A + \operatorname{rank} B + \operatorname{rank} C - \operatorname{rank}[A^* \;\; B] - \operatorname{rank}[B^* \;\; C]$.

ix) $\operatorname{rank} ABC = \operatorname{rank} AB + \operatorname{rank} BC - \operatorname{rank} B$, $\operatorname{rank}[A^* \;\; B] = \operatorname{rank} A + \operatorname{rank} B + \operatorname{rank} AB$, and $\operatorname{rank}[B^* \;\; C] = \operatorname{rank} B + \operatorname{rank} C - \operatorname{rank} BC$.

The following statements are equivalent:

x) $\operatorname{rank} ABC = \operatorname{rank} A + \operatorname{rank} B + \operatorname{rank} C - m - l$.

xi) $\operatorname{rank} ABC = \operatorname{rank} AB + \operatorname{rank} BC - \operatorname{rank} B$, $\operatorname{rank} AB = \operatorname{rank} A + \operatorname{rank} B - m$, $\operatorname{rank} BC = \operatorname{rank} B + \operatorname{rank} C - l$.

xii) $\operatorname{rank} ABC = \operatorname{rank} A + \operatorname{rank} B + \operatorname{rank} C - \operatorname{rank}[A^* \;\; B] - \operatorname{rank}[B^* \;\; C]$, $\operatorname{rank}[A^* \;\; B] = m$, and $\operatorname{rank}[B^* \;\; C] = l$.

Source: [1966, 2675] and Fact 7.11.25. **Credit:** Y. Tian. **Related:** Fact 3.14.20.

Fact 8.9.17. Let $x, y \in \mathbb{R}^3$, and assume that x and y are linearly independent. Then,

$$[x \;\; y]^+ = \begin{bmatrix} x^+(I_3 - y\phi^\mathsf{T}) \\ \phi^\mathsf{T}\end{bmatrix},$$

where $x^+ = (x^\mathsf{T}x)^{-1}x^\mathsf{T}$, $\alpha \triangleq y^\mathsf{T}(I - xx^+)y$, and $\phi \triangleq \alpha^{-1}(I - xx^+)y$. Now, let $x, y, z \in \mathbb{R}^3$, and assume that x and y are linearly independent. Then,

$$[x \;\; y \;\; z]^+ = \begin{bmatrix} (I_2 - \beta ww^\mathsf{T})[x \;\; y]^+ \\ \beta w^\mathsf{T}[x \;\; y]^+\end{bmatrix},$$

where $w \triangleq [x \;\; y]^+z$ and $\beta \triangleq 1/(1 + w^\mathsf{T}w)$. **Source:** [2701].

Fact 8.9.18. Let $x, y \in \mathbb{R}^3$. Then, the following statements hold:

i) If $x \ne 0$, then $K^+(x) = -(x^\mathsf{T}x)^{-1}K(x)$.

ii) $K(x)K(x)^+ = I - xx^+$.

iii) $K(x)K(x)^+y = y$ if and only if $x^\mathsf{T}y = 0$.

iv) $(K(x) + \jmath I)^+ = -\frac{1}{4}[K(x) + 3xx^+\jmath + \jmath I]$.

v) If $x^\mathsf{T}y \ne 0$, then $[K(x)K(y)]^+ = \frac{1}{x^\mathsf{T}y}(xx^+ + yy^+ - I) - \frac{1}{x^\mathsf{T}xy^\mathsf{T}y}yx^\mathsf{T}$.

vi) If $x^\mathsf{T}y = 0$ and $x^\mathsf{T}x + y^\mathsf{T}y \neq 0$, then

$$\begin{bmatrix} K(x) & y \\ -y^\mathsf{T} & 0 \end{bmatrix}^+ = \frac{-1}{x^\mathsf{T}x + y^\mathsf{T}y}\begin{bmatrix} K(x) & y \\ -y^\mathsf{T} & 0 \end{bmatrix}.$$

Source: [2699, 2723]. **Related:** Fact 4.12.1.

Fact 8.9.19. Let $A \in \mathbb{F}^{n\times m}$ and $b \in \mathbb{F}^n$. Then,

$$[A\ \ b]^+ = \begin{bmatrix} A^+(I_n - b\phi^*) \\ \phi^* \end{bmatrix}, \quad [b\ \ A]^+ = \begin{bmatrix} \phi^* \\ A^+(I_n - b\phi^*) \end{bmatrix},$$

where

$$\phi \triangleq \begin{cases} (b - AA^+b)^{+*}, & b \neq AA^+b, \\ \gamma^{-1}(AA^*)^+b, & b = AA^+b. \end{cases}$$

and $\gamma \triangleq 1 + b^*(AA^*)^+b$. **Source:** [31, p. 44], [1040, p. 270], and [2423, p. 148]. **Credit:** T. N. E. Greville.

Fact 8.9.20. Let $A \in \mathbb{F}^{n\times m}$ and $B \in \mathbb{F}^{n\times l}$. Then,

$$[A\ \ B]^+ = \begin{bmatrix} A^*(AA^* + BB^*)^+ \\ B^*(AA^* + BB^*)^+ \end{bmatrix}, \quad [A\ \ B][A\ \ B]^+ = (AA^* + BB^*)(AA^* + BB^*)^+.$$

Source: [2645]. **Related:** Fact 8.4.33.

Fact 8.9.21. Let $A \in \mathbb{F}^{n\times m}$ and $B \in \mathbb{F}^{n\times l}$. Then,

$$[A\ \ B]^+ = \begin{bmatrix} A^+ - A^+B(C^+ + D) \\ C^+ + D \end{bmatrix},$$

where

$$C \triangleq (I - AA^+)B, \quad D \triangleq (I - C^+C)[I + (I - C^+C)B^*(AA^*)^+B(I - C^+C)]^{-1}B^*(AA^*)^+(I - BC^+).$$

Furthermore,

$$[A\ \ B]^+ = \begin{cases} \begin{bmatrix} A^*(AA^* + BB^*)^{-1} \\ B^*(AA^* + BB^*)^{-1} \end{bmatrix}, & \operatorname{rank}\,[A\ \ B] = n, \\ \begin{bmatrix} A^*A & A^*B \\ B^*A & B^*B \end{bmatrix}^{-1}\begin{bmatrix} A^* \\ B^* \end{bmatrix}, & \operatorname{rank}\,[A\ \ B] = m + l, \\ \begin{bmatrix} A^*(AA^*)^{-1}(I - BE) \\ E \end{bmatrix}, & \operatorname{rank} A = n, \end{cases}$$

where $E \triangleq [I + B^*(AA^*)^{-1}B]^{-1}B^*(AA^*)^{-1}$. Finally, define $M \triangleq [I + B^*(AA^*)^+B]^{-1}$. Then, the following statements hold:

i) $[A\ \ B]^+ = \begin{bmatrix} A^+ - A^+BMB^*(AA^*)^+ \\ C^+ + MB^*(AA^*)^+ \end{bmatrix}$ if and only if $C^+CB^*(AA^*)^+B = 0$.

ii) $[A\ \ B]^+ = \begin{bmatrix} A^+ - A^+BMB^*(AA^*)^+ \\ MB^*(AA^*)^+ \end{bmatrix}$ if and only if $C = 0$.

iii) $[A\ \ B]^+ = \begin{bmatrix} A^+ - A^+BC^+ \\ C^+ \end{bmatrix}$ if and only if $C^+CB^*(AA^*)^+B = B^*(AA^*)^+B$.

iv) $[A\ \ B]^+ = \begin{bmatrix} A^+ \\ B^+ \end{bmatrix}$ if and only if $C = B$.

Source: [761], [536, pp. 14–18] and [2403, pp. 193–195]. **Remark:** If $[A\ \ B]$ is square and nonsingular and $A^*B = 0$, then the second expression yields Fact 3.22.9. See Fact 8.4.34. **Related:** Fact 8.9.5 and Fact 8.9.26.

Fact 8.9.22. Let $A \in \mathbb{F}^{n\times m}$ and $B \in \mathbb{F}^{n\times l}$, and assume that $\mathcal{R}(B) \subseteq \mathcal{R}(A)$. Then,

$$[A\ \ B]^+ = \begin{bmatrix} A^+ - A^+BM^{-1}B^*(AA^*)^+ \\ M^{-1}B^*(AA^*)^+ \end{bmatrix},$$

where $M \triangleq I + B^*(AA^*)^+B$. **Source:** [1880].

Fact 8.9.23. Let $A \in \mathbb{F}^{n\times m}$ and $B \in \mathbb{F}^{n\times l}$. Then,

$$[A\ \ B][A\ \ B]^+ = AA^+ + B_1B_1^* - B_1A_1^+(I - AA^+),$$

where $A_1 \triangleq (I - BB^+)AA^+$ and $B_1 \triangleq (I - AA^+)BB^+$. **Source:** [2679].

Fact 8.9.24. Let $A \in \mathbb{F}^{n\times m}$ and $B \in \mathbb{F}^{n\times l}$. Then,

$$\begin{bmatrix} A \\ B \end{bmatrix}^+ = [A^+ - (E^+ + T)BA^+ \quad E^+ + T],$$

where

$$E \triangleq B(I - A^+A), \quad T \triangleq (I - E^+B)(A^*A)^+B^*[I + (I - EE^+)B(A^*A)^+B^*(I - EE^+)]^{-1}(I - EE^+).$$

Source: [2238, p. 188].

Fact 8.9.25. Let $A \in \mathbb{F}^{n\times m}$ and $B \in \mathbb{F}^{n\times l}$. Then,

$$\operatorname{rank}\left([A\ \ B]^+ - \begin{bmatrix} A^+ \\ B^+ \end{bmatrix}\right) = \operatorname{rank}\,[AA^*B \quad BB^*A].$$

Hence, $A^*B = 0$ if and only if

$$[A\ \ B]^+ = \begin{bmatrix} A^+ \\ B^+ \end{bmatrix}.$$

Source: [2637]. **Remark:** If $A^*B = 0$, then $C = B$ and $D = 0$, where C and D are defined in Fact 8.9.21. **Related:** Fact 8.9.26.

Fact 8.9.26. Let $A \in \mathbb{F}^{n\times m}$ and $B \in \mathbb{F}^{n\times l}$. Then, the following statements are equivalent:

i) $[A\ \ B][A\ \ B]^+ = \frac{1}{2}(AA^+ + BB^+)$.

ii) $\mathcal{R}(A) = \mathcal{R}(B)$.

Furthermore, the following statements are equivalent:

iii) $[A\ \ B]^+ = \frac{1}{2}\begin{bmatrix} A^+ \\ B^+ \end{bmatrix}$.

iv) $AA^* = BB^*$.

Furthermore, the following statements are equivalent:

v) $[A\ \ B]^+ = \begin{bmatrix} A^+ \\ B^+ \end{bmatrix}$.

vi) $\mathcal{R}(A) \subseteq \mathcal{N}(B^*)$.

vii) $B^*A = 0$.

Finally, the following statements are equivalent:

viii) $\mathcal{R}\left(\begin{bmatrix} A^+ \\ B^+ \end{bmatrix}\right) = \mathcal{R}\left(\begin{bmatrix} A^* \\ B^* \end{bmatrix}\right)$.

ix) $\mathcal{R}(A) \cap \mathcal{R}(B) = \{0\}$.

x) $AA^+(AA^+ + BB^+)^+BB^+ = 0$.

xi) $[A\ \ B]^+ = \begin{bmatrix} [(I - BB^+)A]^+ \\ [(I - AA^+)B]^+ \end{bmatrix}$.

xii) $[A\ \ B]^+ = \begin{bmatrix} A^+ - A^+B[(I - AA^+)B]^+ \\ B^+ - B^+A[(I - BB^+)A]^+ \end{bmatrix}$.

xiii) $[(I - BB^+)A]^+ = A^+ - A^+B[(I - AA^+)B]^+$.

xiv) $[(I - AA^+)B]^+ = B^+ - B^+A[(I - BB^+)A]^+$.

If these conditions hold, then the following equalities hold:

xv) $[(I - BB^+)A]^+B = 0$ and $[(I - AA^+)B]^+A = 0$.

xvi) $[(I - BB^+)A]^+A = A^+A$ and $[(I - AA^+)B]^+B = B^+B$.

xvii) $[A\ \ B]^+[A\ \ B] = \begin{bmatrix} A^+A & 0 \\ 0 & B^+B \end{bmatrix}$.

Source: [222, 2631, 2654] and Fact 8.4.8. **Related:** Fact 8.9.25. Additional conditions that are equivalent to $\mathcal{R}(A) \cap \mathcal{R}(B) = \{0\}$ are given by Fact 8.9.5.

Fact 8.9.27. Let $A, B \in \mathbb{F}^{n\times m}$. Then, the following statements are equivalent:

i) $\mathcal{R}\left(\begin{bmatrix} A \\ A^*A \end{bmatrix}\right) = \mathcal{R}\left(\begin{bmatrix} B \\ B^*B \end{bmatrix}\right)$.

ii) $\mathcal{R}\left(\begin{bmatrix} A \\ A^+A \end{bmatrix}\right) = \mathcal{R}\left(\begin{bmatrix} B \\ B^+B \end{bmatrix}\right)$.

iii) $A = B$.

Source: [2669].

Fact 8.9.28. Let $A \in \mathbb{F}^{n\times m}$ and $B \in \mathbb{F}^{k\times l}$. Then,

$$\begin{bmatrix} A & 0 \\ 0 & B \end{bmatrix}^+ = \begin{bmatrix} A^+ & 0 \\ 0 & B^+ \end{bmatrix}.$$

Fact 8.9.29. Let $A \in \mathbb{F}^{n\times m}$. Then,

$$\begin{bmatrix} I_n & A \\ 0_{m\times n} & 0_{m\times m} \end{bmatrix}^+ = \begin{bmatrix} (I_n + AA^*)^{-1} & 0_{n\times m} \\ A^*(I_n + AA^*)^{-1} & 0_{m\times m} \end{bmatrix}.$$

Source: [35, 2710].

Fact 8.9.30. Let $A \in \mathbb{F}^{n\times n}$, let $B \in \mathbb{F}^{n\times m}$, and assume that $BB^* = I$. Then,

$$\begin{bmatrix} A & B \\ B^* & 0 \end{bmatrix}^+ = \begin{bmatrix} 0 & B \\ B^* & -B^*AB \end{bmatrix}.$$

Source: [970, p. 237].

Fact 8.9.31. Let $A \in \mathbb{F}^{n\times n}$, assume that A is positive semidefinite, and let $B \in \mathbb{F}^{n\times m}$. Then,

$$\begin{bmatrix} A & B \\ B^* & 0 \end{bmatrix}^+ = \begin{bmatrix} C^+ - C^+BD^+B^*C^+ & C^+BD^+ \\ (C^+BD^+)^* & DD^+ - D^+ \end{bmatrix},$$

where $C \triangleq A + BB^*$ and $D \triangleq B^*C^+B$. **Source:** [1924, p. 58]. **Remark:** Representations of the generalized inverse of a partitioned matrix are given in [232, 284, 358], [360, Chapter 5], [608, 624, 652, 653, 654, 1248], [1275, pp. 161–165], [1329, 1331, 1489, 1814, 2034, 2035, 2037, 2038, 2041, 2140, 2301, 2327, 2624, 2655, 2677, 2678, 2855]. **Problem:** Show that the generalized inverses in this result and in Fact 8.9.30 are identical in the case where A is positive semidefinite and $BB^* = I$.

Fact 8.9.32. Let $A, B \in \mathbb{F}^{n\times m}$. Then,

$$(A+B)^+ = \tfrac{1}{2}[I_m \;\; I_m]\begin{bmatrix} A & B \\ B & A \end{bmatrix}^+ \begin{bmatrix} I_n \\ I_n \end{bmatrix}.$$

Source: [2624, 2629, 2666]. **Related:** Fact 3.22.5 and Fact 3.24.7.

Fact 8.9.33. Let $A_1, \ldots, A_k \in \mathbb{F}^{n\times m}$. Then,

$$(A_1 + \cdots + A_k)^+ = \tfrac{1}{k}[I_m \;\; \cdots \;\; I_m]\begin{bmatrix} A_1 & A_2 & \cdots & A_k \\ A_k & A_1 & \cdots & A_{k-1} \\ \vdots & \vdots & \ddots & \vdots \\ A_2 & A_3 & \cdots & A_1 \end{bmatrix}^+ \begin{bmatrix} I_n \\ \vdots \\ I_n \end{bmatrix}.$$

Source: [2629]. **Remark:** The partitioned matrix is block circulant. See Fact 3.22.7 and Fact 8.12.5.

Fact 8.9.34. Let $A \in \mathbb{F}^{n\times n}$, $x, y \in \mathbb{F}^n$, and $a \in \mathbb{F}$, and assume that $x \in \mathcal{R}(A)$. Then,

$$\begin{bmatrix} A & x \\ y^\mathsf{T} & a \end{bmatrix} = \begin{bmatrix} I & 0 \\ y^\mathsf{T} & 1 \end{bmatrix}\begin{bmatrix} A & 0 \\ y^\mathsf{T} - y^\mathsf{T}A & a - y^\mathsf{T}A^+x \end{bmatrix}\begin{bmatrix} I & A^+x \\ 0 & 1 \end{bmatrix}.$$

Remark: This factorization holds in the case where A is singular and $a = 0$. See Fact 3.17.11, Fact 3.21.4, and Fact 8.9.7, and note that $x = AA^+x$. **Problem:** Obtain a factorization in the case where $x \notin \mathcal{R}(A)$ (and thus x is nonzero and A is singular) and $a = 0$.

Fact 8.9.35. Let $A \in \mathbb{F}^{n\times m}$, assume that $A = \begin{bmatrix} A_1 \\ \vdots \\ A_k \end{bmatrix}$, and define $B \triangleq [A_1^+ \;\; \cdots \;\; A_k^+]$. Then, the following statements hold:

i) $\det AB = 0$ if and only if $\operatorname{rank} A < n$.

ii) $0 < \det AB \le 1$ if and only if $\operatorname{rank} A = n$.

iii) If $\operatorname{rank} A = n$, then

$$\det AB = \frac{\det AA^*}{\prod_{i=1}^k \det A_iA_i^*}, \qquad \det AA^* \le \prod_{i=1}^k \det A_iA_i^*.$$

iv) $\det AB = 1$ if and only if $AB = I$.

v) AB is group invertible.

vi) Every eigenvalue of AB is nonnegative.

vii) $\operatorname{rank} A = \operatorname{rank} B = \operatorname{rank} AB = \operatorname{rank} BA$.

Now, assume that $\operatorname{rank} A = \sum_{i=1}^k \operatorname{rank} A_i$, and let β denote the product of the positive eigenvalues of AB. Then, the following statements hold:

viii) $0 < \beta \le 1$.

ix) $\beta = 1$ if and only if $B = A^+$.

Source: [1765, 2563]. **Remark:** *iii)* yields Hadamard's inequality given by Fact 10.15.10 in the case where A is square and each A_i has a single row.

Fact 8.9.36. Let $A \in \mathbb{F}^{n\times m}$ and $B \in \mathbb{F}^{n\times l}$. Then,

$$\det\begin{bmatrix} A^*A & A^*B \\ B^*A & B^*B \end{bmatrix} = (\det A^*A)\det B^*(I - AA^+)B = (\det B^*B)\det A^*(I - BB^+)A.$$

Related: Fact 3.17.28.

Fact 8.9.37. Let $A \in \mathbb{F}^{n\times n}$, $B \in \mathbb{F}^{n\times m}$, $C \in \mathbb{F}^{m\times n}$, and $D \in \mathbb{F}^{m\times m}$, assume that either $\operatorname{rank}\,[A\ \ B] = \operatorname{rank} A$ or $\operatorname{rank}\left[\begin{smallmatrix} A \\ C \end{smallmatrix}\right] = \operatorname{rank} A$, and let $A^- \in \mathbb{F}^{n\times n}$ be a (1)-inverse of A. Then,

$$\det\begin{bmatrix} A & B \\ C & D \end{bmatrix} = (\det A)\det(D - CA^-B).$$

Source: [300, p. 266].

Fact 8.9.38. Let $A \triangleq \left[\begin{smallmatrix} A_{11} & A_{12} \\ A_{21} & A_{22} \end{smallmatrix}\right] \in \mathbb{F}^{(n+m)\times(n+m)}$, $B \in \mathbb{F}^{(n+m)\times l}$, $C \in \mathbb{F}^{l\times(n+m)}$, $D \in \mathbb{F}^{l\times l}$, and $\mathcal{A} \triangleq \left[\begin{smallmatrix} A & B \\ C & D \end{smallmatrix}\right]$, and assume that A, A_{11}, and $A_{11}|A$ are nonsingular. Then,

$$A|\mathcal{A} = (A_{11}|A)|(A_{11}|\mathcal{A}).$$

Source: [2238, pp. 195, 196] and [2263, pp. 18, 19]. **Remark:** Partitioning $B = \left[\begin{smallmatrix} B_1 \\ B_2 \end{smallmatrix}\right]$ and $C = [C_1\ \ C_2]$ conformably with A, the equality can be written as

$$D - CA^{-1}B = D - C_1A_{11}^{-1}B_1 - (C_2 - C_1A_{11}^{-1}A_{12})(A_{22} - A_{21}A_{11}^{-1}A_{12})^{-1}(B_2 - A_{21}A_{11}^{-1}B_1).$$

Remark: This is the *Crabtree-Haynsworth quotient formula*. See [1458]. **Remark:** Extensions are given in [2238, pp. 195, 196] and [2988].

Fact 8.9.39. Let $A, B \in \mathbb{F}^{n\times m}$. Then, the following statements are equivalent:

i) $A \overset{\text{rs}}{\leq} B$.

ii) $BB^+A = AB^+B = AB^+A = A$.

iii) $\operatorname{rank} A = \operatorname{rank}\,[A\ \ B] = \operatorname{rank}\left[\begin{smallmatrix} A \\ B \end{smallmatrix}\right]$ and $AB^+A = A$.

iv) $\mathcal{R}(A) \subseteq \mathcal{R}(B)$, $\mathcal{N}(B) \subseteq \mathcal{N}(A)$, and $AB^+A = A$.

Source: [402] and [2418, p. 45]. **Related:** Fact 10.24.8.

8.10 Facts on the Drazin and Group Generalized Inverses for One Matrix

Fact 8.10.1. Let $A \in \mathbb{F}^{n\times n}$. Then, the following statements hold:

i) $\operatorname{ind} A = \operatorname{ind}(A - A^2A^{\mathrm{D}})$.

ii) For all $k \geq 1$, $A^k - A^{k+1}A^{\mathrm{D}} = (A - A^2A^{\mathrm{D}})^k$.

iii) Let $X \in \mathbb{F}^{n\times n}$. Then, $X = A^{\mathrm{D}}$ if and only if the following statements hold:

a) $XAX = X$.

b) $AX = XA$.

c) $A - A^2X$ is nilpotent.

Source: [1657].

Fact 8.10.2. Let $A \in \mathbb{F}^{n\times n}$, and define the projectors $P \triangleq AA^+$ and $Q \triangleq I - A^+A$. Then, P and Q are complementary projectors if and only if A is group invertible. If these conditions hold, then $AA^{\#}$ is the idempotent matrix onto $\mathcal{R}(P)$ along $\mathcal{R}(Q)$. **Source:** Corollary 4.8.10. **Remark:** P and Q are the projectors onto $\mathcal{R}(A)$ and $\mathcal{N}(A)$, respectively. **Related:** Fact 8.10.3.

Fact 8.10.3. Let $A \in \mathbb{F}^{n\times n}$. Then, the following statements are equivalent:

i) A is group invertible.

ii) $\operatorname{rank}\,[AA^+\ \ I - A^+A] = \operatorname{rank} AA^+ + \operatorname{rank}(I - A^+A) = n$.

In this case, $AA^{\#} = (A^+A^2A^+)^+$. **Source:** Proposition 4.8.6 and Corollary 4.8.10. The expression for the idempotent matrix $AA^{\#}$ follows from Fact 8.8.14. **Remark:** This result shows that the range and null space of a group-invertible matrix are complementary subspaces.

Fact 8.10.4. Let $A \in \mathbb{F}^{n\times m}$, let $(A_i)_{i=1}^{\infty} \subset \mathbb{F}^{n\times m}$, for all $i \geq 1$, assume that A_i is group invertible, and assume that $\lim_{i\to\infty} A_i = A$. Then, the following statements are equivalent:

i) A is group invertible, and $\lim_{i\to\infty} A_i^{\#} = A^{\#}$.

ii) $(\sigma_{\max}(A_i^{\#}))_{i=1}^{\infty}$ is bounded.

iii) $\lim_{i\to\infty} A_i A_i^{\#} = AA^{\#}$.

Source: [403]. **Related:** Fact 8.10.6.

Fact 8.10.5. Let $A \in \mathbb{F}^{n\times n}$, and assume that A is group invertible and $\operatorname{rank} A = 1$. Then,

$$A^{\#} = \frac{1}{\operatorname{tr} A^2} A.$$

Consequently, if $x, y \in \mathbb{F}^n$ satisfy $x^*y \neq 0$, then

$$(xy^*)^{\#} = (x^*y)^{-2}xy^*.$$

In particular, $1_{n\times n}^{\#} = n^{-2}1_{n\times n}$.

Fact 8.10.6. Let $A \in \mathbb{F}^{n\times m}$, let $(A_i)_{i=1}^{\infty} \subset \mathbb{F}^{n\times m}$, and assume that $\lim_{i\to\infty} A_i = A$. Then, the following statements are equivalent:

i) $\lim_{i\to\infty} A_i^{\mathsf{D}} = A^{\mathsf{D}}$.

ii) $(\sigma_{\max}(A_i^{\mathsf{D}}))_{i=1}^{\infty}$ is bounded.

Source: [403]. **Related:** Fact 8.10.4.

Fact 8.10.7. Let $A \in \mathbb{F}^{n\times n}$, let $\lambda \in \operatorname{spec}(A)$, assume that $\lambda \neq 0$, and let $k \geq 1$. Then, $\operatorname{amult}_A(\lambda) = \operatorname{amult}_{A^{k\mathsf{D}}}(1/\lambda^k)$ and $\operatorname{gmult}_A(\lambda) = \operatorname{gmult}_{A^{k\mathsf{D}}}(1/\lambda^k)$. **Related:** Fact 7.15.23.

Fact 8.10.8. Let $A \in \mathbb{F}^{n\times n}$. Then, there exists $p \in \mathbb{F}[s]$ such that $p(A) = A^{\mathsf{D}}$. **Source:** [2238, p. 227].

Fact 8.10.9. Let $A \in \mathbb{F}^{n\times n}$. Then, A^{D} is group invertible. Furthermore, $(A^{\mathsf{D}})^{\#} = A^2A^{\mathsf{D}}$. **Source:** [1275, p. 63].

Fact 8.10.10. Let $A \in \mathbb{F}^{n\times n}$, define $k \triangleq \operatorname{ind} A$, and let $b \in \mathbb{F}^n$. Then, the following statements are equivalent:

i) $x = A^{\mathsf{D}}b$ satisfies $Ax = b$.

ii) $b \in \mathcal{R}(A^k)$.

iii) There exists $x \in \mathcal{R}([b \;\; Ab \;\; \cdots \;\; A^{n-1}b])$ such that $Ax = b$.

Now, assume that these statements hold, and define $m \triangleq \deg \mu_A$. Then, there exists a unique vector $x \in \mathcal{R}([b \;\; Ab \;\; \cdots \;\; A^{m-k-1}b])$ such that $Ax = b$. **Source:** [2238, pp. 228, 229].

Fact 8.10.11. Let $A \in \mathbb{F}^{n\times n}$. Then, $X = A^{\mathsf{D}}$ is the unique matrix satisfying

$$\operatorname{rank}\begin{bmatrix} A & AA^{\mathsf{D}} \\ A^{\mathsf{D}}A & X \end{bmatrix} = \operatorname{rank} A.$$

Source: [2854, 2993]. **Related:** Fact 3.22.11 and Fact 8.3.34.

Fact 8.10.12. Let $A \in \mathbb{F}^{n\times n}$, and let $k \triangleq \operatorname{ind} A$. Then,

$$A^{\mathsf{D}} = A^k(A^{2k+1})^+A^k.$$

If, in particular, $\operatorname{ind} A \leq 1$, then

$$A^{\#} = A(A^3)^+A, \quad (A^{\#})^+ = A^+A^3A^+.$$

Source: [360, pp. 165, 174] and [403].

Fact 8.10.13. Let $A \in \mathbb{F}^{n\times n}$, assume that A is group invertible, and let $S, B \in \mathbb{F}^{n\times n}$, where S is nonsingular, B is a Jordan form of A, and $A = SBS^{-1}$. Then, $A^{\#} = SB^{\#}S^{-1} = SB^+S^{-1}$. **Source:** Since B is range Hermitian, it follows from Fact 8.10.16 that $B^{\#} = B^+$. See [360, p. 158].

Fact 8.10.14. Let $A \in \mathbb{F}^{n\times n}$ and $k \geq 1$. Then, the following statements are equivalent:

i) $k \geq \operatorname{ind} A$.

ii) $\lim_{\alpha\to 0}\alpha^k(A+\alpha I)^{-1}$ exists.

iii) $\lim_{\alpha\to 0}(A^{k+1}+\alpha I)^{-1}A^k$ exists.

If these statements hold, then

$$A^{\mathrm{D}} = \lim_{\alpha\to 0}(A^{k+1}+\alpha I)^{-1}A^k,$$

$$\lim_{\alpha\to 0}\alpha^k(A+\alpha I)^{-1} = \begin{cases}(-1)^{k-1}(I-AA^{\mathrm{D}})A^{k-1}, & k=\operatorname{ind}A>0,\\ A^{-1}, & k=\operatorname{ind}A=0,\\ 0, & k>\operatorname{ind}A.\end{cases}$$

Source: [2037].

Fact 8.10.15. Let $A\in\mathbb{F}^{n\times n}$. Then, A is group invertible if and only if $\lim_{\alpha\to 0}(A+\alpha I)^{-1}A$ exists. If these conditions hold, then

$$\lim_{\alpha\to 0}(A+\alpha I)^{-1}A = AA^{\#}.$$

Source: [624, p. 138].

Fact 8.10.16. Let $A\in\mathbb{F}^{n\times n}$. Then, the following statements are equivalent:

i) A is range Hermitian.

ii) $A^+ = A^{\mathrm{D}}$.

iii) A is group invertible, and $A^+ = A^{\#}$.

iv) A is group invertible, and $A(A^+)^2 = A^{\#}$.

v) A is group invertible, and $AA^{\#}A^+ = A^{\#}$.

vi) A is group invertible, and $A^*AA^{\#} = A^*$.

vii) A is group invertible, and $A^+AA^{\#} = A^+$.

viii) A is group invertible, and $A^{\#}A^+A = A^+$.

ix) A is group invertible, and $AA^{\#} = A^+A$.

x) A is group invertible, and $A^*A^+ = A^*A^{\#}$.

xi) A is group invertible, and $A^+A^* = A^{\#}A^*$.

xii) A is group invertible, and $(A^+)^2 = A^+A^{\#}$.

xiii) A is group invertible, and $(A^+)^2 = A^{\#}A^+$.

xiv) A is group invertible, and $(A^+)^2 = A^{2\#}$.

xv) A is group invertible, and $A^+A^{\#} = A^{2\#}$.

xvi) A is group invertible, and $A^{\#}A^+ = A^{2\#}$.

xvii) A is group invertible, and $A^+A^{\#} = A^{\#}A^+$.

xviii) A is group invertible, and $AA^+A^* = A^*AA^+$.

xix) A is group invertible, and $AA^+A^{\#} = A^+A^{\#}A$.

xx) A is group invertible, and $AA^+A^{\#} = A^{\#}AA^+$.

xxi) A is group invertible, and $AA^{\#}A^* = A^*AA^{\#}$.

xxii) A is group invertible, and $AA^{\#}A^+ = A^+AA^{\#}$.

xxiii) A is group invertible, and $AA^{\#}A^+ = A^{\#}A^+A$.

xxiv) A is group invertible, and $A^*A^+A = A^+AA^*$.

xxv) A is group invertible, and $A^+AA^{\#} = A^{\#}A^+A$.

xxvi) A is group invertible, and $(A^+)^2A^{\#} = A^+A^{\#}A^+$.

xxvii) A is group invertible, and $(A^+)^2A^{\#} = A^{\#}(A^+)^2$.

xxviii) A is group invertible, and $A^+A^\#A^+ = A^\#(A^+)^2$.

xxix) A is group invertible, and $A^+A^{2\#} = A^\#A^+A^\#$.

xxx) A is group invertible, and $A^+A^{2\#} = A^{2\#}A^+$.

xxxi) A is group invertible, and $A^{2\#}A^+ = A^\#A^+A^\#$.

xxxii) A is group invertible, and $A^*A^\#A + AA^\#A^* = 2A^*$.

xxxiii) A is group invertible, and $A^+A^\#A + AA^\#A^+ = 2A^+$.

Source: [240, 719]. **Related:** Fact 8.5.3.

Fact 8.10.17. Let $A \in \mathbb{F}^{n\times n}$. Then, the following statements are equivalent:

i) A is normal.

ii) A is group invertible, and $A^*A^+ = A^\#A^*$.

iii) A is group invertible, and $A^*A^\# = A^+A^*$.

iv) A is group invertible, and $A^*A^\# = A^\#A^*$.

v) A is group invertible, and $AA^*A^\# = A^*A^\#A$.

vi) A is group invertible, and $AA^*A^\# = A^\#AA^*$.

vii) A is group invertible, and $AA^\#A^* = A^\#A^*A$.

viii) A is group invertible, and $A^*AA^\# = A^\#A^*A$.

ix) A is group invertible, and $A^{2*}A^\# = A^*A^\#A^*$.

x) A is group invertible, and $A^*A^+A^\# = A^\#A^*A^+$.

xi) A is group invertible, and $A^*A^\#A^* = A^\#A^{2*}$.

xii) A is group invertible, and $A^*A^\#A^+ = A^+A^*A^\#$.

xiii) A is group invertible, and $A^*A^{2\#} = A^\#A^*A^\#$.

xiv) A is group invertible, and $A^+A^*A^\# = A^\#A^+A^*$.

xv) A is group invertible, and $A^+A^\#A^* = A^\#A^*A^+$.

xvi) A is group invertible, and $A^\#A^*A^\# = A^{2\#}A^*$.

Source: [240, 719]. **Related:** Fact 4.10.12, Fact 4.13.6, Fact 7.17.5, and Fact 8.6.1.

Fact 8.10.18. Let $A \in \mathbb{F}^{n\times n}$. Then, the following statements are equivalent:

i) A is Hermitian.

ii) A is group invertible, and $AA^\# = A^*A^+$.

iii) A is group invertible, and $AA^\# = A^*A^\#$.

iv) A is group invertible, and $AA^\# = A^+A^*$.

v) A is group invertible, and $A^+A = A^\#A^*$.

vi) A is group invertible, and $A^*AA^\# = A$.

vii) A is group invertible, and $A^{2*}A^\# = A^*$.

viii) A is group invertible, and $A^*(A^+)^2 = A^\#$.

ix) A is group invertible, and $A^*A^+A^\# = A^+$.

x) A is group invertible, and $A^*A^+A^\# = A^\#$.

xi) A is group invertible, and $A^*A^{2\#} = A^\#$.

xii) A is group invertible, and $A^\#A^*A^\# = A^+$.

Source: [240].

Fact 8.10.19. Let $A \in \mathbb{F}^{n\times n}$. Then, the following statements are equivalent:

i) A is a projector.

ii) A is idempotent, and $A^* = A^\#$.

Source: [1249].

Fact 8.10.20. Let $A \in \mathbb{R}^{n\times n}$. Then, the following statements are equivalent:

i) A is idempotent.

ii) A is group invertible, and $A^{2\#} = A$.

Remark: The result is false if A is complex.

Fact 8.10.21. Let $A \in \mathbb{F}^{n\times n}$. Then, the following statements are equivalent:

i) A is tripotent.

ii) A is group invertible, and $A^\# = A$.

iii) A is group invertible, and $A^{3\#} = A$.

iv) A is group invertible, and $AA^\# = A^2$.

v) A is group invertible, and $\frac{1}{2}(AA^\# - A)$ is idempotent.

Source: [257].

Fact 8.10.22. Let $A \in \mathbb{F}^{n\times n}$. Then, the following statements are equivalent:

i) A is a generalized projector.

ii) $A^* = A^2$.

iii) $A = A^{2*}$.

iv) $A = A^*A^+$.

v) $A = A^+A^*$.

vi) $A^* = A^2 = A^+$.

vii) A is group invertible, and $A = A^*A^\#$.

viii) A is group invertible, and $A = A^\#A^*$.

Now, assume that these statements hold. Then, the following statements hold:

ix) A^*A is a projector.

x) For all $k \ge 0$, $A = A^{3k+1}$ and $A^* = A^2 = A^\# = A^{\mathrm{D}} = A^+ = A^2 = A^{3k+2}$.

Source: [233, 237, 1249]. **Related:** Fact 8.10.23.

Fact 8.10.23. Let $A \in \mathbb{F}^{n\times n}$. Then, the following statements are equivalent:

i) $A^+ = A^2$.

ii) $A = (A^2)^+$.

iii) $A = (A^+)^2$.

iv) $A^{2*} = A^{*+}$.

v) A is group invertible, and $A = A^+A^\#$.

vi) A is group invertible, and $A = A^\#A^+$.

vii) A is group invertible, and $A^{2\#} = (A^\#)^+$.

viii) A^3 is a projector, and $\mathcal{R}(A^3) = \mathcal{R}(A)$.

Source: [225, 233, 237, 1249]. **Remark:** If *i*)–*viii*) hold, then A is a *hypergeneralized projector*. **Related:** Fact 8.10.22.

Fact 8.10.24. Let $A \in \mathbb{F}^{n\times n}$. Then, the following statements are equivalent:

i) A is a projector.

ii) A is an idempotent generalized projector.

iii) A is an idempotent hypergeneralized projector.

Furthermore, the following statements hold:

iv) A is a projector if and only if there exists a generalized projector $B \in \mathbb{F}^{n\times n}$ such that $A = B^*B$.

v) If A is a generalized projector, then A and $I - A$ are normal.

vi) Assume that A is a generalized projector. Then, $I - A$ is a generalized projector if and only if A is a projector.

Source: [224, 233, 225, 1249]. *v*) follows from Fact 4.9.6.

Fact 8.10.25. Let $A \in \mathbb{F}^{n\times n}$, assume that A is a hypergeneralized projector, and define $B \in \mathbb{F}^{n\times n}$ by $B \triangleq \frac{1}{3}(A + A^2 + A^3)$. Then, the following statements hold:

i) B is idempotent.

ii) B is a projector if and only if $I - A$ is range Hermitian. If these conditions hold, then $B = I - (I - A)(I - A)^+$.

iii) $AB = BA = B$.

iv) $\mathcal{R}(B) = \mathcal{N}(I - A)$ and $\mathcal{N}(B) = \mathcal{R}(I - A)$.

Source: [1249].

Fact 8.10.26. Let $A, B \in \mathbb{F}^{n\times n}$, and assume that A and B are (idempotent, projectors, generalized projectors). Then, so is $A + B$ if and only if $AB = BA = 0$. **Source:** [1249].

Fact 8.10.27. Let $A, B \in \mathbb{F}^{n\times n}$, assume that A and B are hypergeneralized projectors, and assume that $AB = BA = 0$. Then, $A + B$ is a hypergeneralized projector. **Source:** [1249].

8.11 Facts on the Drazin and Group Generalized Inverses for Two or More Matrices

Fact 8.11.1. Let $A \in \mathbb{F}^{n\times r}$ and $B \in \mathbb{F}^{r\times n}$, and assume that $\operatorname{rank} AB = r$. Then, AB is group invertible if and only if BA is nonsingular. If these conditions hold, then $(AB)^\# = A(BA)^{-2}B$. **Source:** [360, p. 157] and [2238, p. 231]. **Credit:** R. E. Cline. **Related:** Fact 8.11.6.

Fact 8.11.2. Let $A, B \in \mathbb{F}^{n\times n}$, and assume that AB and BA are group invertible. Then,

$$(AB)^\# = A(BA)^{2\#}B.$$

Source: [873].

Fact 8.11.3. Let $A, B \in \mathbb{F}^{n\times n}$, assume that A and B are group invertible, and consider the following statements:

i) $ABA = B$.

ii) $BAB = A$.

iii) $A^2 = B^2$.

Then, if two of the above statements hold, then the third statement holds. Furthermore, if *i*)–*iii*) hold, then the following statements hold:

iv) A and B are group invertible.

v) $A^\# = A^3$ and $B^\# = B^3$.

vi) $A^5 = A$ and $B^5 = B$.

vii) $A^4 = B^4 = (AB)^4$.

viii) If A and B are nonsingular, then $A^4 = B^4 = (AB)^4 = I$.

Source: [1015].

Fact 8.11.4. Let $A \in \mathbb{R}^{n\times n}$, where $n \ge 2$, assume that A is positive, define $B \triangleq \rho_{\max}(A)I - A$, let $x, y \in \mathbb{R}^n$ be positive, and assume that $Ax = \rho_{\max}(A)x$ and $A^\mathsf{T}y = \rho_{\max}(A)y$. Then, the following statements hold:

i) $B + \frac{1}{x^\mathsf{T}y}xy^\mathsf{T}$ is nonsingular.

ii) $B^\# = (B + \frac{1}{x^\mathsf{T}y}xy^\mathsf{T})^{-1}(I - \frac{1}{x^\mathsf{T}y}xy^\mathsf{T})$.

iii) $I - BB^\# = \frac{1}{x^\mathsf{T}y}xy^\mathsf{T}$.

iv) $B^\# = \lim_{k\to\infty}\left(\sum_{i=0}^{k-1}\frac{1}{\rho_{\max}^i(A)}A^i - \frac{k}{x^\mathsf{T}y}xy^\mathsf{T}\right)$.

Source: [2344, p. 9-4]. **Related:** Fact 6.11.5.

Fact 8.11.5. Let $A, B \in \mathbb{F}^{n\times n}$, and assume that A and B are similar. Then, A^D and B^D are similar.

Fact 8.11.6. Let $A \in \mathbb{F}^{n\times m}$ and $B \in \mathbb{F}^{m\times n}$. Then, $(AB)^\mathsf{D} = A(BA)^{2\mathsf{D}}B$. **Source:** [624, pp. 149, 150] and [657]. **Remark:** This is *Cline's formula*. **Related:** Fact 8.11.1.

Fact 8.11.7. Let $A, B \in \mathbb{F}^{n\times n}$, and assume that $AB = BA$. Then,

$$(AB)^\mathsf{D} = B^\mathsf{D}A^\mathsf{D} = A^\mathsf{D}B^\mathsf{D}, \quad A^\mathsf{D}B = BA^\mathsf{D}, \quad AB^\mathsf{D} = B^\mathsf{D}A.$$

Furthermore, $\operatorname{ind} AB \le \max\{\operatorname{ind} A, \operatorname{ind} B\}$. **Source:** [624, pp. 149, 150] and [2238, pp. 227, 228].

Fact 8.11.8. Let $A, B \in \mathbb{F}^{n\times n}$, and assume that $AB = BA = 0$. Then,

$$(A + B)^\mathsf{D} = A^\mathsf{D} + B^\mathsf{D}.$$

Source: [1340]. **Credit:** M. P. Drazin.

Fact 8.11.9. Let $A, B \in \mathbb{F}^{n\times n}$, and assume that A and B are idempotent. Then, the following statements hold:

i) If $AB = 0$, then $(A + B)^\mathsf{D} = A + B - 2BA$ and $(A - B)^\mathsf{D} = A - B$.

ii) If $AB = 0$, B is nilpotent, and $k \triangleq \operatorname{ind} B$, then $(A + B)^\mathsf{D} = \sum_{i=0}^{k-1} B^iA^{(i+1)\mathsf{D}}$.

iii) If $BA = 0$, then $(A + B)^\mathsf{D} = A + B - 2AB$ and $(A - B)^\mathsf{D} = A - B$.

iv) If $BA = 0$, B is nilpotent, and $k \triangleq \operatorname{ind} B$, then $(A + B)^\mathsf{D} = \sum_{i=0}^{k-1} A^{(i+1)\mathsf{D}}B^i$.

v) If $AB = A$, then $(A + B)^\mathsf{D} = \frac{1}{4}A + B - \frac{3}{4}BA$ and $(A - B)^\mathsf{D} = BA - B$.

vi) If $AB = B$, then $(A + B)^\mathsf{D} = A + \frac{1}{4}B - \frac{3}{4}BA$ and $(A - B)^\mathsf{D} = A - BA$.

vii) If $BA = A$, then $(A + B)^\mathsf{D} = \frac{1}{4}A + B - \frac{3}{4}AB$ and $(A - B)^\mathsf{D} = AB - B$.

viii) If $BA = B$, then $(A + B)^\mathsf{D} = A + \frac{1}{4}B - \frac{3}{4}AB$ and $(A - B)^\mathsf{D} = A - AB$.

ix) If $AB = BA$, then $(A + B)^\mathsf{D} = A + B - \frac{3}{2}AB$ and $(A - B)^\mathsf{D} = A - B$.

x) If $ABA = 0$, then $(A + B)^\mathsf{D} = A + B - 2AB - 2BA + 3BAB$ and $(A - B)^\mathsf{D} = A - B - BAB$.

xi) If $BAB = 0$, then $(A + B)^\mathsf{D} = A + B - 2AB - 2BA + 3ABA$ and $(A - B)^\mathsf{D} = A - B + ABA$.

xii) If $ABA = A$, then $(A + B)^\mathsf{D} = \frac{1}{8}(A + B)^2 + \frac{7}{8}BA_\perp B$ and $(A - B)^\mathsf{D} = -BA_\perp B$.

xiii) If $BAB = B$, then $(A + B)^\mathsf{D} = \frac{1}{8}(A + B)^2 + \frac{7}{8}AB_\perp A$ and $(A - B)^\mathsf{D} = AB_\perp A$.

xiv) If $ABA = B$, then $(A + B)^\mathsf{D} = A - \frac{1}{2}B$ and $(A - B)^\mathsf{D} = A - B$.

xv) If $BAB = A$, then $(A + B)^\mathsf{D} = -\frac{1}{2}A + B$ and $(A - B)^\mathsf{D} = A - B$.

xvi) If $ABA = AB$, then $(A+B)^\mathsf{D} = A+B-2BA-\frac{3}{4}AB+\frac{5}{4}BAB$ and $(A-B)^\mathsf{D} = A-B-AB+BAB$.

xvii) If $ABA = BA$, then $(A+B)^\mathsf{D} = A+B-2AB-\frac{3}{4}BA+\frac{5}{4}BAB$ and $(A-B)^\mathsf{D} = A-B-BA+BAB$.

Source: [653, 872, 870]. **Related:** Fact 8.11.12.

Fact 8.11.10. Let $A, B \in \mathbb{F}^{n\times n}$, and assume that A and B are idempotent. Then, the following statements hold:

i) $(A - B)^\mathsf{D} = (I - AB)^\mathsf{D}(A - AB) + (A + B - AB)^\mathsf{D}(AB - B)$.

ii) $[A, B]^\mathsf{D} = (ABA)^\mathsf{D}(A - B)^\mathsf{D} - (A - B)^\mathsf{D}(ABA)^\mathsf{D}$.

iii) $(AB + BA)^\mathsf{D} = (A + B)^\mathsf{D}(A + B - I)^\mathsf{D}$.

iv) $(ABA)^\mathsf{D} = (I - A - B)^{2\mathsf{D}}A$.

v) $(AB)^{\mathrm{D}} = (ABA)^{2\mathrm{D}}B = (I - A - B)^{4\mathrm{D}}AB.$

vi) $(A - AB)^{\mathrm{D}} = (I - ABA)^{2\mathrm{D}}(A - AB) = (A - ABA)^{2\mathrm{D}}(I - AB).$

vii) $(A - BA)^{\mathrm{D}} = (A - BA)(I - ABA)^{2\mathrm{D}} = (I - BA)(A - ABA)^{2\mathrm{D}}.$

viii) $(I - AB)^{\mathrm{D}}A = A(A - BA)^{\mathrm{D}} = (A - AB)^{\mathrm{D}}A = A(I - ABA)^{\mathrm{D}} = (I - ABA)^{\mathrm{D}}A = A(I - BA)^{\mathrm{D}} = (A - ABA)^{\mathrm{D}} = A(A - B)^{2\mathrm{D}} = (A - B)^{2\mathrm{D}}A = A(I - ABA)^{\mathrm{D}} = (I - ABA)^{\mathrm{D}}A.$

ix) $(I - ABA)^{\mathrm{D}} = I - A + A(A - B)^{2\mathrm{D}}.$

Source: [872].

Fact 8.11.11. Let $A, B \in \mathbb{F}^{n\times n}$, and assume that A and B are projectors. Then, the following statements hold:

i) If $AB = A$, then $(A + B)^{\mathrm{D}} = -\frac{1}{2}A + B$ and $(A - B)^{\mathrm{D}} = A - B$.

ii) If $AB = B$, then $(A + B)^{\mathrm{D}} = A - \frac{1}{2}B$ and $(A - B)^{\mathrm{D}} = A - B$.

iii) If $AB = 0$, then $(A - B)^{\mathrm{D}} = A - B$.

iv) The following statements are equivalent:

a) $AB = BA$.

b) $(AB)^{\mathrm{D}} = AB$.

c) $(A - B)^{\mathrm{D}} = A - B$.

v) $AB = BA = 0$ if and only if $(A + B)^{\mathrm{D}} = A + B$.

vi) $(A - B)^{\mathrm{D}} = (A - B)^2[(A - BA)^{\mathrm{D}} - (B - BA)^{\mathrm{D}}].$

vii) $A[(A + B)^{\mathrm{D}} - (A - B)^{\mathrm{D}}](A - B)^2 = 0.$

viii) $(AB)^{\mathrm{D}} = (ABA)^{\mathrm{D}} - A(B_\perp B_\perp)^{\mathrm{D}}.$

ix) $(AB)^{\mathrm{D}}AB = (ABA)^{\mathrm{D}}AB.$

Source: [653, 871, 872]. **Related:** Fact 8.11.12.

Fact 8.11.12. Let $A, B \in \mathbb{F}^{n\times n}$, assume that A and B are group invertible, let $\alpha,\beta \in \mathbb{F}$, and assume that α and β are nonzero. Then, the following statements hold:

i) If $ABB^\# = BAA^\#$, then $\alpha A + \beta B$ is group invertible. If, in addition, $\alpha + \beta \neq 0$, then

$$(\alpha A+\beta B)^\# = \frac{1}{\alpha+\beta}(A^\#+B^\#-A^\#BB^\#)+\left(\frac{1}{\alpha}-\frac{1}{\alpha+\beta}\right)(I-BB^\#)A^\#+\left(\frac{1}{\beta}-\frac{1}{\alpha+\beta}\right)(I-AA^\#)B^\#.$$

ii) If $ABB^\# = BAA^\#$, then $A - B$ is group invertible and $(A - B)^\# = (A - B)(A^\# - B^\#)^2$.

iii) If $BB^\#A = AA^\#B$, then $\alpha A + \beta B$ is group invertible. If, in addition, $\alpha + \beta \neq 0$, then

$$(\alpha A+\beta B)^\# = \frac{1}{\alpha+\beta}(A^\#+B^\#-B^\#BA^\#)+\left(\frac{1}{\alpha}-\frac{1}{\alpha+\beta}\right)A^\#(I-BB^\#)+\left(\frac{1}{\beta}-\frac{1}{\alpha+\beta}\right)B^\#(I-AA^\#).$$

iv) If $BB^\#A = AA^\#B$, then $A - B$ is group invertible and $(A - B)^\# = (A^\# - B^\#)^2(A - B)$.

v) If $ABB^\# = BAA^\#$ and $BB^\#A = AA^\#B$, then $\alpha A + \beta B$ is group invertible. If, in addition, $\alpha + \beta \neq 0$, then

$$(\alpha A + \beta B)^\# = \frac{1}{\alpha+\beta}A^\#BB^\# + \frac{1}{\alpha}(I - BB^\#)A + \frac{1}{\beta}(I - AA^\#)B^\#.$$

vi) If $ABB^\# = BAA^\#$ and $BB^\#A = AA^\#B$, then $A - B$ and AB are group invertible, and $(A - B)^\# = A^\# - B^\#$ and $(AB)^\# = (BA)^\# = (A^\#BB^\#)^2$.

vii) If $BA^\#A = A$ and $\alpha + \beta \neq 0$, then $\alpha A + \beta B$ is group invertible and

$$(\alpha A+\beta B)^\#=\frac{\alpha}{(\alpha+\beta)^2}A^\#+\frac{\beta}{(\alpha+\beta)^2}B^\#+\frac{\alpha^2+2\alpha\beta}{\beta(\alpha+\beta)^2}(I-AA^\#)B^\#(I-AA^\#)-\frac{\alpha}{(\alpha+\beta)^2}A^\#(B-A)B^\#.$$

viii) Assume that $AB = BA$. Then, $\alpha A + \beta B$ is group invertible if and only if $\alpha ABB^\# + \beta BAA^\#$ is group invertible. If these conditions hold, then

$$(\alpha A + \beta B)^\# = (\alpha ABB^\# + \beta BAA^\#)^\# + \frac{1}{\alpha}A^\#(I - BB^\#) + \frac{1}{\beta}B^\#(I - AA^\#).$$

ix) If $AB = BA$, then AB is group invertible, $(AB)^\# = B^\#A^\# = A^\#B^\#$, $A^\#B = BA^\#$, and $B^\#A = AB^\#$.

x) If A and B are idempotent and $AB = BA$, then $A + B$ is group invertible, $(A + B)^\# = A + B - \frac{3}{2}AB$, and $(A - B)^\# = A - B$.

xi) If A and B are tripotent and $AB^2 = BA^2$, then $A + B$ is group invertible, $(A + B)^\# = A + B - \frac{1}{2}AB^2 - \frac{1}{2}A^2B - \frac{1}{2}B^2A$, and $(A - B)^\# = (A - B)^3 = A - B + B^2A - A^2B + BAB - ABA$.

xii) If A and B are tripotent and $A^2B = B^2A$, then $A + B$ is group invertible, $(A + B)^\# = A + B - \frac{1}{2}AB^2 - \frac{1}{2}B^2A - \frac{1}{2}BA^2$, and $(A - B)^\# = (A - B)^3 = A - B + BA^2 - AB^2 + BAB - ABA$.

Source: [1879]. **Related:** Fact 8.11.9.

Fact 8.11.13. Let $A, B \in \mathbb{F}^{n\times n}$, and assume that A and B are projectors. Then,

$$(AB)^\# = (BA)^+(AB)^+(BA)^+ = [(BA)^2]^+.$$

Furthermore, the following statements are equivalent:

i) AB is a projector.

ii) $(AB)^\#$ is Hermitian.

iii) $(AB)^\#$ is idempotent.

iv) $\operatorname{rank} AB = \operatorname{tr}(AB)^\#$.

v) $AB(AB)^\# = BA(AB)^\#$.

vi) $(I - AB)^\# = I - (AB)^\#$.

vii) $\operatorname{tr}[I - (AB)^\#] = \operatorname{rank}[I - (AB)^\#]$.

Source: [245]. **Related:** Fact 8.8.5.

Fact 8.11.14. Let $A, B \in \mathbb{F}^{n\times n}$, assume that A and B are idempotent, assume that $A - B$ is group invertible, and define $F \triangleq A(A - B)^\#$, $G \triangleq (A - B)^\#A$, and $H \triangleq (A - B)^\#(A - B)$. Then, the following statements hold:

i) F, G, and H are idempotent.

ii) $F = (A - B)^\#B_\perp$ and $\mathcal{R}(F) = \mathcal{R}(AH)$.

iii) $G = B_\perp(A - B)^\#$ and $\mathcal{N}(G) = \mathcal{N}(AH)$.

iv) $B(A - B)^\# = (A - B)^\#A_\perp$ and $A_\perp(A - B)^\# = (A - B)^\#B$.

v) $(A - B)^\# = F + G - H$.

vi) $AH = H$ if and only if $A + B$ is group invertible and $(A + B)^\# = (A - B)^\#(A + B)(A - B)^\# = (2G - H)(F + G - H)$.

vii) $FA = AG = AH = HA$.

viii) $BHB = BH = HB = HBH$.

ix) $B_\perp F_\perp = G_\perp B_\perp = B_\perp F_\perp B_\perp = A_\perp H_\perp$.

x) $(I - AB)^\# = FG + F_\perp A_\perp$.

xi) $(I - ABA)^\# = FG + A_\perp$.

xii) $(A - ABA)^\# = FG$.

xiii) $(A - AB)^\# = (FG)^2B_\perp$.

xiv) $(A - BA)^\# = B_\perp(FG)^2$.

Source: [873]. **Related:** Fact 4.16.10.

8.12 Facts on the Drazin and Group Generalized Inverses for Partitioned Matrices

Fact 8.12.1. Let $A \in \mathbb{F}^{n\times m}$, $B \in \mathbb{F}^{m\times n}$, and define $\mathcal{A} \triangleq \begin{bmatrix} 0 & A \\ B & 0 \end{bmatrix}$. Then, $\operatorname{ind}\mathcal{A} \le 2\operatorname{ind}AB + 1$, and

$$\mathcal{A}^{\mathsf{D}} = \begin{bmatrix} 0 & (AB)^{\mathsf{D}}A \\ B(AB)^{\mathsf{D}} & 0 \end{bmatrix} = \begin{bmatrix} 0 & A(BA)^{\mathsf{D}} \\ (BA)^{\mathsf{D}}B & 0 \end{bmatrix}.$$

Source: [657, 873].

Fact 8.12.2. Let $A \in \mathbb{F}^{n\times m}$, $B \in \mathbb{F}^{m\times n}$, and define $\mathcal{A} \triangleq \begin{bmatrix} 0 & A \\ B & 0 \end{bmatrix}$. Then, the following statements are equivalent:

i) $\mathcal{A}$ is group invertible.

ii) AB and BA are group invertible, and $[I - AB(AB)^{\#}]A = B[I - AB(AB)^{\#}] = 0$.

If these statements hold, then

$$\mathcal{A}^{\#} = \begin{bmatrix} 0 & (AB)^{\#}A \\ B(AB)^{\#} & 0 \end{bmatrix} = \begin{bmatrix} 0 & A(BA)^{\#} \\ (BA)^{\#}B & 0 \end{bmatrix}.$$

Source: [873].

Fact 8.12.3. Let $A \in \mathbb{F}^{n\times n}$, $B \in \mathbb{F}^{n\times m}$, and $C \in \mathbb{F}^{m\times m}$, and define $\mathcal{A} \triangleq \begin{bmatrix} A & B \\ 0 & C \end{bmatrix}$. Then, $\mathcal{A}$ is group invertible if and only if A and C are group invertible and $(I - AA^{\#})B(I - CC^{\#}) = 0$. If these conditions hold, then

$$\mathcal{A}^{\#} = \begin{bmatrix} A^{\#} & A^{2\#}B(I - CC^{\#}) + (I - AA^{\#})BC^{2\#} - A^{\#}BC^{\#} \\ 0 & C^{\#} \end{bmatrix}.$$

Source: [402, 2037]. **Credit:** C. G. Cao. **Related:** Fact 7.16.19.

Fact 8.12.4. Let $A \in \mathbb{F}^{n\times n}$, $B \in \mathbb{F}^{n\times m}$, $C \in \mathbb{F}^{m\times n}$, and $D \in \mathbb{F}^{m\times m}$, and define $\mathcal{A} \triangleq \begin{bmatrix} A & B \\ C & D \end{bmatrix}$ and $S \triangleq D - CA^{\mathsf{D}}B$. Then, the following statements hold:

i) If S is nonsingular, $(I - AA^{\mathsf{D}})B = 0$, and $C(I - AA^{\mathsf{D}}) = 0$, then

$$\mathcal{A}^{\mathsf{D}} = \begin{bmatrix} A^{\mathsf{D}} + A^{\mathsf{D}}BS^{-1}CA^{\mathsf{D}} & -A^{\mathsf{D}}BS^{-1} \\ -S^{-1}CA^{\mathsf{D}} & S^{-1} \end{bmatrix}.$$

ii) If $S = 0$, $(I - AA^{\mathsf{D}})B = 0$, and $C(I - AA^{\mathsf{D}}) = 0$, then

$$\mathcal{A}^{\mathsf{D}} = \begin{bmatrix} I \\ CA^{\mathsf{D}} \end{bmatrix} A(WA)^{2\mathsf{D}}[I \;\; A^{\mathsf{D}}B],$$

where $W \triangleq AA^{\mathsf{D}} + A^{\mathsf{D}}BCA^{\mathsf{D}}$.

iii) Assume that A is nonsingular. Then, $\operatorname{rank}\mathcal{A} = n$ if and only if $S = 0$.

iv) Assume that A is nonsingular and $S = 0$. Then, $\mathcal{A}$ is group invertible if and only if $W \triangleq I + A^{-1}BCA^{-1}$ is nonsingular. If these conditions hold, then

$$\mathcal{A}^{\#} = \begin{bmatrix} I \\ CA^{-1} \end{bmatrix} (WAW)^{-1}[I \;\; A^{-1}B].$$

v) If $\operatorname{ind}A = 1$, $\operatorname{ind}S \le 1$, and $(I - AA^{\#})B = 0$, then

$$\mathcal{A} = \begin{bmatrix} A^{\#} + A^{\#}BS^{\#}CA^{\#} & -A^{\#}BS^{\#} \\ -S^{\#}CA^{\#} & S^{\#} \end{bmatrix} \begin{bmatrix} I - A^{\#}BS^{\#}C(I - AA^{\#}) & A^{\#}B(I - SS^{\#}) \\ S^{\#}C(I - AA^{\#}) & I \end{bmatrix}.$$

Source: [653].

Fact 8.12.5. Let $A_1, \ldots, A_k \in \mathbb{F}^{n\times n}$. Then,

$$(A_1 + \cdots + A_k)^{\mathsf{D}} = \frac{1}{k}[I_n \ \ \cdots \ \ I_n]\begin{bmatrix} A_1 & A_2 & \cdots & A_k \\ A_k & A_1 & \cdots & A_{k-1} \\ \vdots & \vdots & \ddots & \vdots \\ A_2 & A_3 & \cdots & A_1 \end{bmatrix}^{\mathsf{D}} \begin{bmatrix} I_n \\ \vdots \\ I_n \end{bmatrix}.$$

Source: [2629]. **Related:** Fact 8.9.33.

8.13 Notes

A brief history of the generalized inverse is given in [359] and [360, p. 4]. The proof of the uniqueness of A^+ is given in [1924, p. 32]. Additional books on generalized inverses include [360, 536, 2299, 2821]. The terminology "range Hermitian" is used in [360]; the terminology "EP" is more common. Generalized inverses are widely used in least squares methods; see [489, 624, 1766]. Applications to singular differential equations are considered in [623]. Applications to Markov chains are discussed in [1490].

Chapter Nine
Kronecker and Schur Algebra

In this chapter we introduce Kronecker matrix algebra, which is useful for solving linear matrix equations.

9.1 Kronecker Product

For $A \in \mathbb{F}^{n\times m}$ define the *vec* operator as

$$\operatorname{vec} A \triangleq \begin{bmatrix} \operatorname{col}_1(A) \\ \vdots \\ \operatorname{col}_m(A) \end{bmatrix} \in \mathbb{F}^{nm}, \tag{9.1.1}$$

which is the column vector of size $nm \times 1$ obtained by stacking the columns of A. We recover A from $\operatorname{vec} A$ by writing

$$A = \operatorname{vec}^{-1}(\operatorname{vec} A). \tag{9.1.2}$$

Note that, if $x \in \mathbb{F}^n$, then $\operatorname{vec} x = \operatorname{vec} x^{\mathsf{T}} = x$.

Proposition 9.1.1. Let $A \in \mathbb{F}^{n\times m}$ and $B \in \mathbb{F}^{m\times n}$. Then,

$$\operatorname{tr} AB = (\operatorname{vec} A^{\mathsf{T}})^{\mathsf{T}} \operatorname{vec} B = (\operatorname{vec} B^{\mathsf{T}})^{\mathsf{T}} \operatorname{vec} A. \tag{9.1.3}$$

Proof. Note that

$$\operatorname{tr} AB = \sum_{i=1}^{n} \operatorname{row}_i(A)\operatorname{col}_i(B) = \sum_{i=1}^{n} [\operatorname{col}_i(A^{\mathsf{T}})]^{\mathsf{T}}\operatorname{col}_i(B)$$

$$= \begin{bmatrix} \operatorname{col}_1^{\mathsf{T}}(A^{\mathsf{T}}) & \cdots & \operatorname{col}_n^{\mathsf{T}}(A^{\mathsf{T}}) \end{bmatrix} \begin{bmatrix} \operatorname{col}_1(B) \\ \vdots \\ \operatorname{col}_n(B) \end{bmatrix} = (\operatorname{vec} A^{\mathsf{T}})^{\mathsf{T}} \operatorname{vec} B. \qquad \square$$

Next, we introduce the Kronecker product.

Definition 9.1.2. Let $A \in \mathbb{F}^{n\times m}$ and $B \in \mathbb{F}^{l\times k}$. Then, the *Kronecker product* $A \otimes B \in \mathbb{F}^{nl\times mk}$ of A and B is the partitioned matrix

$$A \otimes B \triangleq \begin{bmatrix} A_{(1,1)}B & A_{(1,2)}B & \cdots & A_{(1,m)}B \\ \vdots & \vdots & \ddots & \vdots \\ A_{(n,1)}B & A_{(n,2)}B & \cdots & A_{(n,m)}B \end{bmatrix}. \tag{9.1.4}$$

Unlike matrix multiplication, the Kronecker product $A \otimes B$ does not entail a restriction on either the size of A or the size of B.

The following results are immediate consequences of the definition of the Kronecker product.

Proposition 9.1.3. Let $\alpha \in \mathbb{F}$, $A \in \mathbb{F}^{n\times m}$, and $B \in \mathbb{F}^{l\times k}$. Then,

$$\alpha \otimes A = A \otimes \alpha = \alpha A, \quad A \otimes (\alpha B) = (\alpha A) \otimes B = \alpha (A \otimes B), \tag{9.1.5}$$

$$\overline{A \otimes B} = \overline{A} \otimes \overline{B}, \quad (A \otimes B)^{\mathsf{T}} = A^{\mathsf{T}} \otimes B^{\mathsf{T}}, \quad (A \otimes B)^* = A^* \otimes B^*. \tag{9.1.6}$$

Proposition 9.1.4. Let $A, B \in \mathbb{F}^{n\times m}$ and $C \in \mathbb{F}^{l\times k}$. Then,

$$(A + B) \otimes C = A \otimes C + B \otimes C, \tag{9.1.7}$$

$$C \otimes (A + B) = C \otimes A + C \otimes B. \tag{9.1.8}$$

The next result shows that the Kronecker product is associative.

Proposition 9.1.5. Let $A \in \mathbb{F}^{n\times m}$, $B \in \mathbb{F}^{l\times k}$, and $C \in \mathbb{F}^{p\times q}$. Then,

$$A \otimes (B \otimes C) = (A \otimes B) \otimes C. \tag{9.1.9}$$

We thus write $A \otimes B \otimes C$ for $A \otimes (B \otimes C)$ and $(A \otimes B) \otimes C$.

The next result shows how matrix multiplication interacts with the Kronecker product.

Proposition 9.1.6. Let $A \in \mathbb{F}^{n\times m}$, $B \in \mathbb{F}^{l\times k}$, $C \in \mathbb{F}^{m\times q}$, and $D \in \mathbb{F}^{k\times p}$. Then,

$$(A \otimes B)(C \otimes D) = AC \otimes BD. \tag{9.1.10}$$

Proof. Note that the ij block of $(A \otimes B)(C \otimes D)$ is given by

$$[(A \otimes B)(C \otimes D)]_{ij} = \begin{bmatrix} A_{(i,1)}B & \cdots & A_{(i,m)}B \end{bmatrix} \begin{bmatrix} C_{(1,j)}D \\ \vdots \\ C_{(m,j)}D \end{bmatrix}$$

$$= \sum_{k=1}^{m} A_{(i,k)}C_{(k,j)}BD = (AC)_{(i,j)}BD = (AC \otimes BD)_{ij}. \qquad \square$$

Next, we consider the inverse of a Kronecker product.

Proposition 9.1.7. Assume that $A \in \mathbb{F}^{n\times n}$ and $B \in \mathbb{F}^{m\times m}$ are nonsingular. Then, $A \otimes B$ is nonsingular, and

$$(A \otimes B)^{-1} = A^{-1} \otimes B^{-1}. \tag{9.1.11}$$

Proof. Note that $(A \otimes B)(A^{-1} \otimes B^{-1}) = AA^{-1} \otimes BB^{-1} = I_n \otimes I_m = I_{nm}$. $\square$

Proposition 9.1.8. Let $x \in \mathbb{F}^n$ and $y \in \mathbb{F}^m$. Then,

$$xy^{\mathrm{T}} = x \otimes y^{\mathrm{T}} = y^{\mathrm{T}} \otimes x, \quad \operatorname{vec} xy^{\mathrm{T}} = y \otimes x. \tag{9.1.12}$$

The following result concerns the vec of the product of three matrices.

Proposition 9.1.9. Let $A \in \mathbb{F}^{n\times m}$, $B \in \mathbb{F}^{m\times l}$, and $C \in \mathbb{F}^{l\times k}$. Then,

$$\operatorname{vec}(ABC) = (C^{\mathrm{T}} \otimes A)\operatorname{vec} B. \tag{9.1.13}$$

Proof. Using (9.1.10) and (9.1.12), it follows that

$$\operatorname{vec} ABC = \operatorname{vec} \sum_{i=1}^{l} A\operatorname{col}_i(B)e_i^{\mathrm{T}}C = \sum_{i=1}^{l} \operatorname{vec}[A\operatorname{col}_i(B)(C^{\mathrm{T}}e_i)^{\mathrm{T}}]$$

$$= \sum_{i=1}^{l} (C^{\mathrm{T}}e_i) \otimes [A\operatorname{col}_i(B)] = (C^{\mathrm{T}} \otimes A) \sum_{i=1}^{l} e_i \otimes \operatorname{col}_i(B)$$

$$= (C^{\mathrm{T}} \otimes A) \sum_{i=1}^{l} \operatorname{vec}[\operatorname{col}_i(B)e_i^{\mathrm{T}}] = (C^{\mathrm{T}} \otimes A)\operatorname{vec} B. \qquad \square$$

The following result concerns the eigenvalues and eigenvectors of the Kronecker product of two matrices.

Proposition 9.1.10. Let $A \in \mathbb{F}^{n\times n}$ and $B \in \mathbb{F}^{m\times m}$. Then,

$$\operatorname{mspec}(A \otimes B) = \{\lambda\mu\colon\ \lambda \in \operatorname{mspec}(A), \mu \in \operatorname{mspec}(B)\}_{\rm ms}. \tag{9.1.14}$$

If, in addition, $x \in \mathbb{C}^n$ is an eigenvector of A associated with $\lambda \in \operatorname{spec}(A)$ and $y \in \mathbb{C}^m$ is an eigenvector of B associated with $\mu \in \operatorname{spec}(B)$, then $x \otimes y$ is an eigenvector of $A \otimes B$ associated with $\lambda\mu$.

Proof. (9.1.10) implies $(A \otimes B)(x \otimes y) = (Ax) \otimes (By) = (\lambda x) \otimes (\mu y) = \lambda\mu(x \otimes y)$. □

Using the Minkowski product, (9.2.5) can be written as

$$\operatorname{mspec}(A \otimes B) = \operatorname{mspec}(A)\operatorname{mspec}(B). \tag{9.1.15}$$

Proposition 9.1.10 shows that $\operatorname{mspec}(A \otimes B) = \operatorname{mspec}(B \otimes A)$. Hence, $\det(A \otimes B) = \det(B \otimes A)$ and $\operatorname{tr}(A \otimes B) = \operatorname{tr}(B \otimes A)$.

Proposition 9.1.11. Let $A \in \mathbb{F}^{n\times n}$ and $B \in \mathbb{F}^{m\times m}$. Then,

$$\det(A \otimes B) = \det(B \otimes A) = (\det A)^m(\det B)^n. \tag{9.1.16}$$

Proof. Let $\operatorname{mspec}(A) = \{\lambda_1, \dots, \lambda_n\}_{\rm ms}$ and $\operatorname{mspec}(B) = \{\mu_1, \dots, \mu_m\}_{\rm ms}$. Then, Proposition 9.1.10 implies that

$$\begin{aligned}\det(A \otimes B) &= \prod_{i,j=1}^{n,m} \lambda_i\mu_j = \left(\lambda_1^m \prod_{j=1}^{m} \mu_j\right) \cdots \left(\lambda_n^m \prod_{j=1}^{m} \mu_j\right)\\ &= (\lambda_1 \cdots \lambda_n)^m(\mu_1 \cdots \mu_m)^n = (\det A)^m(\det B)^n.\end{aligned}$$ □

Proposition 9.1.12. Let $A \in \mathbb{F}^{n\times n}$ and $B \in \mathbb{F}^{m\times m}$. Then,

$$\operatorname{tr}(A \otimes B) = \operatorname{tr}(B \otimes A) = (\operatorname{tr} A)(\operatorname{tr} B). \tag{9.1.17}$$

Proof. Note that

$$\operatorname{tr}(A \otimes B) = \operatorname{tr}(A_{(1,1)}B) + \cdots + \operatorname{tr}(A_{(n,n)}B) = [A_{(1,1)} + \cdots + A_{(n,n)}] \operatorname{tr} B = (\operatorname{tr} A)(\operatorname{tr} B).$$ □

Next, define the *Kronecker permutation matrix* $P_{n,m} \in \mathbb{F}^{nm\times nm}$ by

$$P_{n,m} \triangleq \sum_{i,j=1}^{n,m} E_{i,j,n\times m} \otimes E_{j,i,m\times n}. \tag{9.1.18}$$

Proposition 9.1.13. Let $A \in \mathbb{F}^{n\times m}$. Then,

$$\operatorname{vec} A^{\mathsf T} = P_{n,m}\operatorname{vec} A. \tag{9.1.19}$$

9.2 Kronecker Sum and Linear Matrix Equations

Next, we define the Kronecker sum of two square matrices.

Definition 9.2.1. Let $A \in \mathbb{F}^{n\times n}$ and $B \in \mathbb{F}^{m\times m}$. Then, the *Kronecker sum* $A \oplus B \in \mathbb{F}^{nm\times nm}$ of A and B is

$$A \oplus B \triangleq A \otimes I_m + I_n \otimes B. \tag{9.2.1}$$

Proposition 9.2.2. Let $\alpha \in \mathbb{F}$, $A \in \mathbb{F}^{n\times n}$, and $B \in \mathbb{F}^{m\times m}$. Then,

$$(\alpha A) \oplus (\alpha B) = \alpha(A \oplus B), \tag{9.2.2}$$

$$\overline{A \oplus B} = \overline{A} \oplus \overline{B}, \quad (A \oplus B)^{\mathsf T} = A^{\mathsf T} \oplus B^{\mathsf T}, \quad (A \oplus B)^* = A^* \oplus B^*. \tag{9.2.3}$$

Proposition 9.2.3. Let $A \in \mathbb{F}^{n\times n}$, $B \in \mathbb{F}^{m\times m}$, and $C \in \mathbb{F}^{l\times l}$. Then,

$$A \oplus (B \oplus C) = (A \oplus B) \oplus C. \tag{9.2.4}$$

Hence, we write $A \oplus B \oplus C$ for $A \oplus (B \oplus C)$ and $(A \oplus B) \oplus C$.

Proposition 9.1.10 shows that, if $\lambda \in \operatorname{spec}(A)$ and $\mu \in \operatorname{spec}(B)$, then $\lambda\mu \in \operatorname{spec}(A \otimes B)$. The following result involving the Kronecker sum is analogous.

Proposition 9.2.4. Let $A \in \mathbb{F}^{n\times n}$ and $B \in \mathbb{F}^{m\times m}$. Then,

$$\operatorname{mspec}(A \oplus B) = \{\lambda + \mu\colon\ \lambda \in \operatorname{mspec}(A),\ \mu \in \operatorname{mspec}(B)\}_{\mathrm{ms}}. \tag{9.2.5}$$

Now, let $x \in \mathbb{C}^n$ be an eigenvector of A associated with $\lambda \in \operatorname{spec}(A)$, and let $y \in \mathbb{C}^m$ be an eigenvector of B associated with $\mu \in \operatorname{spec}(B)$. Then, $x \otimes y$ is an eigenvector of $A \oplus B$ associated with $\lambda + \mu$.

Proof. Using (9.1.10), we have

$$\begin{aligned}(A \oplus B)(x \otimes y) &= (A \otimes I_m)(x \otimes y) + (I_n \otimes B)(x \otimes y)\\ &= (Ax \otimes y) + (x \otimes By) = (\lambda x \otimes y) + (x \otimes \mu y)\\ &= \lambda(x \otimes y) + \mu(x \otimes y) = (\lambda + \mu)(x \otimes y).\end{aligned}$$

□

Using the Minkowski sum, (9.2.5) can be written as

$$\operatorname{mspec}(A \oplus B) = \operatorname{mspec}(A) + \operatorname{mspec}(B). \tag{9.2.6}$$

The next result concerns the existence and uniqueness of solutions to *Sylvester's equation*. See Fact 7.11.26 and Proposition 15.10.3.

Proposition 9.2.5. Let $A \in \mathbb{F}^{n\times n}$, $B \in \mathbb{F}^{m\times m}$, and $C \in \mathbb{F}^{n\times m}$. Then, $X \in \mathbb{F}^{n\times m}$ satisfies

$$AX + XB + C = 0 \tag{9.2.7}$$

if and only if X satisfies

$$(B^{\mathsf{T}} \oplus A)\operatorname{vec} X + \operatorname{vec} C = 0. \tag{9.2.8}$$

Consequently, $B^{\mathsf{T}} \oplus A$ is nonsingular if and only if there exists a unique matrix $X \in \mathbb{F}^{n\times m}$ satisfying (9.2.7). If these conditions hold, then X is given by

$$X = -\operatorname{vec}^{-1}[(B^{\mathsf{T}} \oplus A)^{-1}\operatorname{vec} C]. \tag{9.2.9}$$

Furthermore, $B^{\mathsf{T}} \oplus A$ is singular and $\operatorname{rank} B^{\mathsf{T}} \oplus A = \operatorname{rank}[B^{\mathsf{T}} \oplus A \quad \operatorname{vec} C]$ if and only if there exist infinitely many matrices $X \in \mathbb{F}^{n\times m}$ satisfying (9.2.7). In this case, for each $X \in \mathbb{F}^{n\times m}$ satisfying (9.2.7), the set of solutions of (9.2.7) is given by $\operatorname{vec}^{-1}[\operatorname{vec} X + \mathcal{N}(B^{\mathsf{T}} \oplus A)]$.

Proof. Note that (9.2.7) is equivalent to

$$\begin{aligned}0 &= \operatorname{vec}(AXI + IXB) + \operatorname{vec} C = (I \otimes A)\operatorname{vec} X + (B^{\mathsf{T}} \otimes I)\operatorname{vec} X + \operatorname{vec} C\\ &= (B^{\mathsf{T}} \otimes I + I \otimes A)\operatorname{vec} X + \operatorname{vec} C = (B^{\mathsf{T}} \oplus A)\operatorname{vec} X + \operatorname{vec} C,\end{aligned}$$

which yields (9.2.8). The remaining results follow from Corollary 3.7.8. □

For the following corollary, note Fact 7.11.26.

Corollary 9.2.6. Let $A \in \mathbb{F}^{n\times n}$, $B \in \mathbb{F}^{m\times m}$, and $C \in \mathbb{F}^{n\times m}$, and assume that $\operatorname{spec}(A) \cap \operatorname{spec}(-B) = \varnothing$. Then, there exists a unique matrix $X \in \mathbb{F}^{n\times m}$ satisfying (9.2.7). Furthermore, the matrices $\begin{bmatrix} A & 0\\ 0 & -B\end{bmatrix}$ and $\begin{bmatrix} A & C\\ 0 & -B\end{bmatrix}$ are similar and satisfy

$$\begin{bmatrix} A & C\\ 0 & -B\end{bmatrix} = \begin{bmatrix} I & X\\ 0 & I\end{bmatrix}\begin{bmatrix} A & 0\\ 0 & -B\end{bmatrix}\begin{bmatrix} I & -X\\ 0 & I\end{bmatrix}. \tag{9.2.10}$$

9.3 Schur Product

An alternative form of vector and matrix multiplication is given by the *Schur product.* If $A \in \mathbb{F}^{n\times m}$ and $B \in \mathbb{F}^{n\times m}$, then $A \odot B \in \mathbb{F}^{n\times m}$ is defined by

$$(A \odot B)_{(i,j)} \triangleq A_{(i,j)}B_{(i,j)}. \tag{9.3.1}$$

Hence, $A \odot B$ is formed by means of entry-by-entry multiplication. For matrices $A, B, C \in \mathbb{F}^{n\times m}$, the commutative, associative, and distributive equalities

$$A \odot B = B \odot A, \tag{9.3.2}$$

$$A \odot (B \odot C) = (A \odot B) \odot C, \tag{9.3.3}$$

$$A \odot (B + C) = A \odot B + A \odot C \tag{9.3.4}$$

hold. For all $A \in \mathbb{F}^{n\times m}$,

$$A \odot 1_{n\times m} = 1_{n\times m} \odot A = A. \tag{9.3.5}$$

Furthermore, if A is square, then $I \odot A$ is the diagonal part of A.

Next, let $A \in \mathbb{F}^{n\times m}$ and $\alpha \in [0, \infty)$. Then, the *Schur power* $A^{\odot\alpha} \in \mathbb{F}^{n\times m}$ is defined by

$$\left(A^{\odot\alpha}\right)_{(i,j)} \triangleq (A_{(i,j)})^{\alpha}. \tag{9.3.6}$$

Thus, $A^{\odot 2} = A \odot A$ and $A^{\odot 0} = 1_{n\times m}$. Furthermore, $\alpha < 0$ is allowed in the case where A has no zero entries. In particular, $A^{\odot -1}$ is the matrix whose entries are the reciprocals of the entries of A.

Proposition 9.3.1. Let $A, B \in \mathbb{F}^{n\times m}$. Then, the following statements hold:

i) $(A \odot B)^{\mathsf{T}} = A^{\mathsf{T}} \odot B^{\mathsf{T}}$, $\overline{A \odot B} = \overline{A} \odot \overline{B}$, $(A \odot B)^* = A^* \odot B^*$.

ii) If either A or B is (diagonal, lower bidiagonal, upper bidiagonal, tridiagonal, lower Hessenberg, upper Hessenberg), then so is $A \odot B$.

iii) If A and B are (Toeplitz, Hankel), then so is $A \odot B$.

iv) Let $\alpha \in [0, \infty)$. Then, $(A \odot B)^{\odot\alpha} = A^{\odot\alpha} \odot B^{\odot\alpha}$.

v) Assume that A and B have no zero entries, and let $\alpha \in \mathbb{R}$. Then, $(A \odot B)^{\odot\alpha} = A^{\odot\alpha} \odot B^{\odot\alpha}$.

Now, assume that $n = m$. Then, the following statements hold:

vi) If A and B are (symmetric, Hermitian), then so is $A \odot B$.

vii) If A and B are (skew symmetric, skew Hermitian), then $A \odot B$ is (symmetric, Hermitian).

viii) If A is symmetric and B is skew symmetric, then $A \odot B$ is skew symmetric.

ix) If A is Hermitian and B is skew Hermitian, then $A \odot B$ is skew Hermitian.

The following result shows that $A \odot B$ is a submatrix of $A \otimes B$.

Proposition 9.3.2. Let $A, B \in \mathbb{F}^{n\times m}$. Then,

$$A \odot B = (A \otimes B)_{(\{1,n+2,2n+3,\ldots,n^2\},\{1,m+2,2m+3,\ldots,m^2\})}. \tag{9.3.7}$$

If, in addition, $n = m$, then

$$A \odot B = (A \otimes B)_{(\{1,n+2,2n+3,\ldots,n^2\})}, \tag{9.3.8}$$

and thus $A \odot B$ is a principal submatrix of $A \otimes B$.

Proof. See [1450, p. 304] and [1951]. □

9.4 Facts on the Kronecker Product

Fact 9.4.1. Let $x, y \in \mathbb{F}^n$. Then, $x \otimes y = (x \otimes I_n)y = (I_n \otimes y)x$.

Fact 9.4.2. Let $x, y, w, z \in \mathbb{F}^n$. Then, $x^{\mathsf{T}}wy^{\mathsf{T}}z = (x^{\mathsf{T}} \otimes y^{\mathsf{T}})(w \otimes z) = (x \otimes y)^{\mathsf{T}}(w \otimes z)$.

Fact 9.4.3. Let $A \in \mathbb{F}^{n\times m}$ and $B \in \mathbb{F}^{1\times m}$. Then, $A(B\otimes I_m) = B\otimes A$.

Fact 9.4.4. Let $A \in \mathbb{F}^{n\times n}$ and $B \in \mathbb{F}^{m\times m}$, and assume that A and B are (diagonal, upper triangular, lower triangular). Then, so is $A\otimes B$.

Fact 9.4.5. Let $A \in \mathbb{F}^{n\times n}$, $B \in \mathbb{F}^{m\times m}$, and $l \ge 0$. Then, $(A\otimes B)^l = A^l\otimes B^l$.

Fact 9.4.6. Let $A \in \mathbb{F}^{n\times m}$. Then, $\operatorname{vec} A = (I_m\otimes A)\operatorname{vec} I_m = (A^{\mathsf{T}}\otimes I_n)\operatorname{vec} I_n$.

Fact 9.4.7. Let $A \in \mathbb{F}^{n\times m}$ and $B \in \mathbb{F}^{m\times l}$. Then,

$$\operatorname{vec} AB = (I_l\otimes A)\operatorname{vec} B = (B^{\mathsf{T}}\otimes A)\operatorname{vec} I_m = \sum_{i=1}^{m} \operatorname{col}_i(B^{\mathsf{T}})\otimes \operatorname{col}_i(A).$$

Fact 9.4.8. Let $A \in \mathbb{F}^{n\times m}$, $B \in \mathbb{F}^{m\times l}$, and $C \in \mathbb{F}^{l\times n}$. Then,

$$\operatorname{tr} ABC = (\operatorname{vec} A)^{\mathsf{T}}(B\otimes I_n)\operatorname{vec} C^{\mathsf{T}}.$$

Fact 9.4.9. Let $A, B, C \in \mathbb{F}^{n\times n}$, and assume that C is symmetric. Then,

$$(\operatorname{vec} C)^{\mathsf{T}}(A\otimes B)\operatorname{vec} C = (\operatorname{vec} C)^{\mathsf{T}}(B\otimes A)\operatorname{vec} C.$$

Fact 9.4.10. Let $A \in \mathbb{F}^{n\times m}$, $B \in \mathbb{F}^{m\times l}$, $C \in \mathbb{F}^{l\times k}$, and $D \in \mathbb{F}^{k\times n}$. Then,

$$\operatorname{tr} ABCD = (\operatorname{vec} A)^{\mathsf{T}}(B\otimes D^{\mathsf{T}})\operatorname{vec} C^{\mathsf{T}}.$$

Fact 9.4.11. Let $A \in \mathbb{F}^{n\times m}$, $B \in \mathbb{F}^{m\times l}$, and $k \ge 1$. Then,

$$(AB)^{\otimes k} = A^{\otimes k}B^{\otimes k},$$

where $A^{\otimes k} \triangleq A\otimes A\otimes\cdots\otimes A$, with A appearing k times.

Fact 9.4.12. Let $A, C \in \mathbb{F}^{n\times m}$ and $B, D \in \mathbb{F}^{l\times k}$, assume that A is (left equivalent, right equivalent, biequivalent) to C, and assume that B is (left equivalent, right equivalent, biequivalent) to D. Then, $A\otimes B$ is (left equivalent, right equivalent, biequivalent) to $C\otimes D$.

Fact 9.4.13. Let $A, B, C, D \in \mathbb{F}^{n\times n}$, assume that A is (similar, congruent, unitarily similar) to C, and assume that B is (similar, congruent, unitarily similar) to D. Then, $A\otimes B$ is (similar, congruent, unitarily similar) to $C\otimes D$.

Fact 9.4.14. Let $A, C \in \mathbb{F}^{n\times m}$ and $B, D \in \mathbb{F}^{k\times l}$. Then, the following statements are equivalent:

i) $A\otimes B = C\otimes D \ne 0$.

ii) A and B are nonzero, and there exist $a, b \in \mathbb{F}$ such that $A = aC$, $B = bD$, and $ab = 1$.

Source: [2991, p. 120].

Fact 9.4.15. Let $A \in \mathbb{F}^{n\times m}$ and $B \in \mathbb{F}^{l\times k}$. Then, $\mathcal{R}(A\otimes B) = \mathcal{R}(A\otimes I_l)\cap\mathcal{R}(I_n\otimes B)$. **Source:** [2641].

Fact 9.4.16. For all $i \in \{1,\ldots,k\}$, let $A_i \in \mathbb{F}^{n_i\times m_i}$ and $B_i \in \mathbb{F}^{n_i\times p_i}$ be nonzero matrices. Then, the following statements are equivalent:

i) For all $i \in \{1,\ldots,k\}$, $\mathcal{R}(A_i) \subseteq \mathcal{R}(B_i)$.

ii) $\mathcal{R}(A_1\otimes\cdots\otimes A_k) \subseteq \mathcal{R}(B_1\otimes\cdots\otimes B_k)$.

Source: [2649].

Fact 9.4.17. Let $A \in \mathbb{F}^{n\times n}$ and $B \in \mathbb{F}^{m\times m}$, and let $\gamma \in \operatorname{spec}(A\otimes B)$. Then,

$$\sum \operatorname{gmult}_A(\lambda)\operatorname{gmult}_B(\mu) \le \operatorname{gmult}_{A\otimes B}(\gamma) \le \operatorname{amult}_{A\otimes B}(\gamma) = \sum \operatorname{amult}_A(\lambda)\operatorname{amult}_B(\mu),$$

where both sums are taken over all $\lambda \in \operatorname{spec}(A)$ and $\mu \in \operatorname{spec}(B)$ such that $\lambda\mu = \gamma$.

Fact 9.4.18. Let $A \in \mathbb{F}^{n\times n}$ and $B \in \mathbb{F}^{m\times m}$. Then, $\rho_{\max}(A\otimes B) = \rho_{\max}(A)\rho_{\max}(B)$. Furthermore, $\rho_{\max}(A\otimes A) = \rho_{\max}^2(A)$.

Fact 9.4.19. Let $A \in \mathbb{F}^{n\times m}$ and $B \in \mathbb{F}^{l\times k}$. Then, $\operatorname{msval}(A\otimes B) = \operatorname{msval}(A)\operatorname{msval}(B)$. **Source:** [2979, pp. 36, 37].

Fact 9.4.20. Let $A \in \mathbb{F}^{n\times m}$ and $B \in \mathbb{F}^{l\times k}$. Then, $\operatorname{rank}(A\otimes B) = (\operatorname{rank} A)(\operatorname{rank} B) = \operatorname{rank}(B\otimes A)$. Consequently, $A\otimes B = 0$ if and only if either $A = 0$ or $B = 0$. **Source:** Fact 9.4.19. **Related:** Fact 10.25.15.

Fact 9.4.21. Let $A \in \mathbb{F}^{n\times n}$ and $B \in \mathbb{F}^{m\times m}$, and let $\gamma \in \operatorname{spec}(A\otimes B)$. Then, $\operatorname{ind}_{A\otimes B}(\gamma) = 1$ if and only if $\operatorname{ind}_A(\lambda) = 1$ and $\operatorname{ind}_B(\mu) = 1$ for all $\lambda \in \operatorname{spec}(A)$ and $\mu \in \operatorname{spec}(B)$ such that $\lambda\mu = \gamma$.

Fact 9.4.22. Let $A \in \mathbb{F}^{n\times n}$ and $B \in \mathbb{F}^{n\times n}$, and assume that A and B are (group invertible, range Hermitian, range symmetric, Hermitian, symmetric, normal, positive semidefinite, positive definite, unitary, orthogonal, projectors, reflectors, involutory, idempotent, tripotent, nilpotent, semisimple). Then, so is $A\otimes B$. **Related:** Fact 9.4.37.

Fact 9.4.23. Let $A_1,\dots,A_l \in \mathbb{F}^{n\times n}$, and assume that $A_1,\dots,A_l$ are skew Hermitian. If l is (even, odd), then $A_1\otimes\cdots\otimes A_l$ is (Hermitian, skew Hermitian).

Fact 9.4.24. Let $A_{i,j} \in \mathbb{F}^{n_i\times n_j}$ for all $i \in \{1,\dots,k\}$ and $j \in \{1,\dots,l\}$. Then,

$$\begin{bmatrix} A_{11} & A_{22} & \cdots & A_{1l} \\ A_{21} & A_{22} & \iddots & A_{2l} \\ \vdots & \iddots & \iddots & \vdots \\ A_{k1} & A_{k2} & \cdots & A_{kl} \end{bmatrix} \otimes B = \begin{bmatrix} A_{11}\otimes B & A_{22}\otimes B & \cdots & A_{1l}\otimes B \\ A_{21}\otimes B & A_{22}\otimes B & \iddots & A_{2l}\otimes B \\ \vdots & \iddots & \iddots & \vdots \\ A_{k1}\otimes B & A_{k2}\otimes B & \cdots & A_{kl}\otimes B \end{bmatrix}.$$

Fact 9.4.25. Let $x \in \mathbb{F}^k$, and, for all $i \in \{1,\dots,l\}$, let $A_i \in \mathbb{F}^{n\times n_i}$. Then,

$$x\otimes [A_1 \ \cdots \ A_l] = [x\otimes A_1 \ \cdots \ x\otimes A_l].$$

Fact 9.4.26. Let $x \in \mathbb{F}^m$, $A \in \mathbb{F}^{n\times m}$, and $B \in \mathbb{F}^{m\times l}$. Then,

$$(A\otimes x)B = (A\otimes x)(B\otimes 1) = (AB)\otimes x.$$

Fact 9.4.27. Let $A \in \mathbb{F}^{n\times n}$ and $B \in \mathbb{F}^{m\times m}$. Then, the eigenvalues of $\sum_{i,j=1,1}^{k,l} \gamma_{ij}A^i\otimes B^j$ are of the form $\sum_{i,j=1,1}^{k,l} \gamma_{ij}\lambda^i\mu^j$, where $\lambda \in \operatorname{spec}(A)$ and $\mu \in \operatorname{spec}(B)$, and an associated eigenvector is given by $x\otimes y$, where $x \in \mathbb{F}^n$ is an eigenvector of A associated with $\lambda \in \operatorname{spec}(A)$ and $y \in \mathbb{F}^n$ is an eigenvector of B associated with $\mu \in \operatorname{spec}(B)$. **Source:** Let $Ax = \lambda x$ and $By = \mu y$. Then, $\gamma_{ij}(A^i\otimes B^j)(x\otimes y) = \gamma_{ij}\lambda^i\mu^j(x\otimes y)$. See [1097], [1738, p. 411], and [1914, p. 83].

Fact 9.4.28. Let $A, B \in \mathbb{F}^{n\times m}$. Then,

$$\left.\begin{matrix} (\operatorname{rank} A)\,(\operatorname{rank}\,[A\ \ B] + \operatorname{rank}\left[\begin{smallmatrix} A \\ B \end{smallmatrix}\right]) - 2(\operatorname{rank} A)^2 \\ (\operatorname{rank} A)\operatorname{rank}\,[A\ \ B] + (\operatorname{rank} B)\operatorname{rank}\left[\begin{smallmatrix} A \\ B \end{smallmatrix}\right] - (\operatorname{rank} A)^2 - (\operatorname{rank} B)^2 \end{matrix}\right\}$$

$$\le \operatorname{rank}(A\otimes B + B\otimes A) \le \begin{cases} 2(\operatorname{rank} A)\operatorname{rank}\,[A\ \ B] - (\operatorname{rank} A)^2 \\ 2(\operatorname{rank} A)\operatorname{rank}\left[\begin{smallmatrix} A \\ B \end{smallmatrix}\right] - (\operatorname{rank} A)^2. \end{cases}$$

Now, assume that $n = m$ and B is nonsingular. Then,

$$2n\operatorname{rank} A - 2(\operatorname{rank} A)^2 \le \operatorname{rank}(A\otimes B + B\otimes A) \le 2n\operatorname{rank} A - (\operatorname{rank} A)^2.$$

Credit: Y. Tian.

Fact 9.4.29. Let $A \in \mathbb{F}^{n\times m}$, $B \in \mathbb{F}^{l\times k}$, $C \in \mathbb{F}^{n\times p}$, $D \in \mathbb{F}^{l\times q}$. Then,

$$\begin{aligned} &(\operatorname{rank} A)\operatorname{rank}\,[B\ \ D] + (\operatorname{rank} D)(\operatorname{rank} C - \operatorname{rank} A) \\ &\qquad \le \operatorname{rank}\,[A\otimes B\ \ C\otimes D] \end{aligned}$$

$$\leq \begin{cases} (\operatorname{rank} A)\operatorname{rank}\,[B\;\; D] + (\operatorname{rank} D)\operatorname{rank}\,[A\;\; C] - (\operatorname{rank} A)\operatorname{rank} D \\ (\operatorname{rank} B)\operatorname{rank}\,[A\;\; C] + (\operatorname{rank} C)\operatorname{rank}\,[B\;\; D] - (\operatorname{rank} B)\operatorname{rank} C. \end{cases}$$

If, in addition, either $\operatorname{rank}\,[A\;\; C] = \operatorname{rank} A + \operatorname{rank} C$ or $\operatorname{rank}\,[B\;\; D] = \operatorname{rank} B + \operatorname{rank} D$, then

$$\operatorname{rank}\,[A \otimes B\;\; C \otimes D] = (\operatorname{rank} A)\operatorname{rank} B + (\operatorname{rank} C)\operatorname{rank} D.$$

Source: [2644, 2650].

Fact 9.4.30. Let $A \in \mathbb{F}^{n\times n}$ and $B \in \mathbb{F}^{m\times m}$. Then,

$$\operatorname{rank}(I - A \otimes B) \leq nm - [n - \operatorname{rank}(I - A)][m - \operatorname{rank}(I - B)].$$

Source: [743].

Fact 9.4.31. Let $A \in \mathbb{F}^{n\times n}$ and $B \in \mathbb{F}^{m\times m}$. Then, $\operatorname{ind} A \otimes B = \max\{\operatorname{ind} A, \operatorname{ind} B\}$.

Fact 9.4.32. Let $A \in \mathbb{F}^{n\times m}$ and $B \in \mathbb{F}^{m\times n}$. Then, $|n - m|\min\{n, m\} \leq \operatorname{amult}_{A\otimes B}(0)$. **Source:** [1450, p. 249].

Fact 9.4.33. Let $A \in \mathbb{F}^{n\times m}$ and $B \in \mathbb{F}^{l\times k}$, and assume that $nl = mk$ and $n \neq m$. Then, $A \otimes B$ and $B \otimes A$ are singular. **Source:** [1450, p. 250].

Fact 9.4.34. Let $A \in \mathbb{F}^{n\times m}$ and $B \in \mathbb{F}^{l\times k}$. Then, A and B are left invertible if and only if $A \otimes B$ is left invertible. If these conditions hold, A^{L} is a left inverse of A, and B^{L} is a left inverse of B, then $B^{\mathsf{L}} \otimes A^{\mathsf{L}}$ is a left inverse of $A \otimes B$. **Remark:** If A and B have full column rank, then so does $A \otimes B$. **Related:** Fact 3.18.9.

Fact 9.4.35. Let $A \in \mathbb{F}^{n\times m}$ and $B \in \mathbb{F}^{l\times k}$. Then, A and B are right invertible if and only if $A \otimes B$ is right invertible. If these conditions hold, A^{R} is a right inverse of A, and B^{R} is a right inverse of B, then $B^{\mathsf{R}} \otimes A^{\mathsf{R}}$ is a right inverse of $A \otimes B$. **Remark:** If A and B have full row rank, then so does $A \otimes B$. **Related:** Fact 3.18.10.

Fact 9.4.36. Let $A \in \mathbb{F}^{n\times m}$ and $B \in \mathbb{F}^{l\times k}$. Then, $(A \otimes B)^+ = A^+ \otimes B^+$.

Fact 9.4.37. Let $A \in \mathbb{F}^{n\times n}$ and $B \in \mathbb{F}^{m\times m}$. Then, $(A \otimes B)^{\mathsf{D}} = A^{\mathsf{D}} \otimes B^{\mathsf{D}}$. Now, assume that A and B are group invertible. Then, $A \otimes B$ is group invertible, and $(A \otimes B)^{\#} = A^{\#} \otimes B^{\#}$. **Related:** Fact 9.4.22.

Fact 9.4.38. The Kronecker permutation matrix $P_{n,m} \in \mathbb{R}^{nm\times nm}$ has the following properties:

i) $P_{n,m}$ is a permutation matrix.

ii) $P_{n,m}^{\mathsf{T}} = P_{n,m}^{-1} = P_{m,n}$.

iii) $P_{n,m}$ is orthogonal.

iv) $P_{n,m}P_{m,n} = I_{nm}$.

v) $P_{n,n}$ is orthogonal, symmetric, and involutory.

vi) $P_{n,n}$ is a reflector.

vii) $P_{n,m} = \sum_{i=1}^{n} e_{i,n} \otimes I_m \otimes e_{i,n}^{\mathsf{T}}$.

viii) $P_{np,m} = (I_n \otimes P_{p,m})(P_{n,m} \otimes I_p) = (I_p \otimes P_{n,m})(P_{p,m} \otimes I_n)$.

ix) $\operatorname{sig} P_{n,n} = \operatorname{tr} P_{n,n} = n$.

x) The inertia of $P_{n,n}$ is given by $\operatorname{In} P_{n,n} = [\frac{1}{2}(n^2 - n)\;\; 0\;\; \frac{1}{2}(n^2 + n)]^{\mathsf{T}}$.

xi) $\det P_{n,n} = (-1)^{(n^2-n)/2}$.

xii) $P_{1,m} = I_m$ and $P_{n,1} = I_n$.

xiii) If $x \in \mathbb{F}^n$ and $y \in \mathbb{F}^m$, then $P_{n,m}(y \otimes x) = x \otimes y$.

xiv) If $A \in \mathbb{F}^{n\times m}$ and $b \in \mathbb{F}^k$, then $P_{k,n}(A \otimes b) = b \otimes A$ and $P_{n,k}(b \otimes A) = A \otimes b$.

xv) If $A \in \mathbb{F}^{n\times m}$ and $B \in \mathbb{F}^{l\times k}$, then

$$P_{l,n}(A \otimes B)P_{m,k} = B \otimes A,$$

$$\operatorname{vec}(A \otimes B) = (I_m \otimes P_{k,n} \otimes I_l)[(\operatorname{vec} A) \otimes (\operatorname{vec} B)],$$

$$\operatorname{vec}(A^{\mathsf{T}} \otimes B) = (P_{nk,m} \otimes I_l)[(\operatorname{vec} A) \otimes (\operatorname{vec} B)],$$

$$\operatorname{vec}(A \otimes B^{\mathsf{T}}) = (I_m \otimes P_{l,nk})[(\operatorname{vec} A) \otimes (\operatorname{vec} B)].$$

xvi) If $A \in \mathbb{F}^{n\times m}$, $B \in \mathbb{F}^{l\times k}$, and $nl = mk$, then

$$\operatorname{tr}(A \otimes B) = [\operatorname{vec}(I_m) \otimes (I_k)]^{\mathsf{T}}[(\operatorname{vec} A) \otimes (\operatorname{vec} B^{\mathsf{T}})].$$

xvii) If $A \in \mathbb{F}^{n\times n}$ and $B \in \mathbb{F}^{l\times l}$, then

$$P_{l,n}(A \otimes B)P_{n,l} = P_{l,n}(A \otimes B)P_{l,n}^{-1} = B \otimes A.$$

Hence, $A \otimes B$ and $B \otimes A$ are similar.

xviii) If $A \in \mathbb{F}^{n\times m}$ and $B \in \mathbb{F}^{m\times n}$, then $\operatorname{tr} AB = \operatorname{tr}[P_{m,n}(A \otimes B)]$.

xix) $P_{np,m} = P_{n,pm}P_{p,nm} = P_{p,nm}P_{n,pm}$.

xx) $P_{np,m}P_{pm,n}P_{mn,p} = I$.

Now, let $A \in \mathbb{F}^{n\times m}$, define $r \triangleq \operatorname{rank} A$, and define $K \triangleq P_{n,m}(A^* \otimes A)$. Then, the following statements hold:

xxi) K is Hermitian, $\operatorname{rank} K = r^2$, $\operatorname{tr} K = \operatorname{tr} A^*A$, $K^2 = (AA^*) \otimes (A^*A)$.

xxii) $\operatorname{mspec}(K) = \{\sigma_1^2(A), \ldots, \sigma_r^2(A)\} \cup \{\pm\sigma_i(A)\sigma_j(A)\colon\ i < j, i, j \in \{1, \ldots, r\}\}$.

Source: [2403, pp. 308–311, 342, 343].

Fact 9.4.39. Define $\Psi_n \in \mathbb{R}^{n^2\times n^2}$ by $\Psi_n \triangleq \frac{1}{2}(I_{n^2} + P_{n,n})$, let $x, y \in \mathbb{F}^n$, and let $A, B \in \mathbb{F}^{n\times n}$. Then, the following statements hold:

i) Ψ_n is a projector.

ii) $\Psi_n = \Psi_n P_{n,n} = P_{n,n}\Psi_n$.

iii) $\Psi_n(x \otimes y) = \frac{1}{2}(x \otimes y + y \otimes x)$.

iv) $\Psi_n \operatorname{vec}(A) = \frac{1}{2}\operatorname{vec}(A + A^{\mathsf{T}})$.

v) $\Psi_n(A \otimes B)\Psi_n = \Psi_n(B \otimes A)\Psi_n$.

vi) $\Psi_n(A \otimes A)\Psi_n = \Psi_n(A \otimes A) = (A \otimes A)\Psi_n$.

vii) $\Psi_n(A \otimes B + B \otimes A)\Psi_n = \Psi_n(A \otimes B + B \otimes A) = (A \otimes B + B \otimes A)\Psi_n = 2\Psi_n(B \otimes A)\Psi_n$.

viii) $(A \otimes A)\Psi_n(A^{\mathsf{T}} \otimes A^{\mathsf{T}}) = \Psi_n(AA^{\mathsf{T}} \otimes AA^{\mathsf{T}})$.

Source: [2403, p. 312].

Fact 9.4.40. For all $i \in \{1, \ldots, p\}$, let $A_i \in \mathbb{F}^{n_i\times n_i}$. Then,

$$\operatorname{mspec}(A_1 \otimes \cdots \otimes A_p) = \{\lambda_1 \cdots \lambda_p\colon\ \lambda_i \in \operatorname{mspec}(A_i) \text{ for all } i \in \{1, \ldots, p\}\}_{\mathrm{ms}}.$$

Finally, for all $i \in \{1, \ldots, p\}$, let $x_i \in \mathbb{C}^{n_i}$ be an eigenvector of A_i associated with $\lambda_i \in \operatorname{spec}(A_i)$. Then, $x_1 \otimes \cdots \otimes x_p$ is an eigenvector of $A_1 \otimes \cdots \otimes A_p$ associated with the eigenvalue $\lambda_1 \cdots \lambda_p$.

Fact 9.4.41. Let $A_1, \ldots, A_k \in \mathbb{F}^{n\times m}$, and define the *k-tensor* $f\colon \bigtimes_{i=1}^k \mathbb{F}^{m\times n} \mapsto \mathbb{F}$ on $\mathbb{F}^{m\times n}$ by

$$f(X_1, \ldots, X_k) \triangleq \operatorname{tr}[(A_1 \otimes \cdots \otimes A_k)(X_1 \otimes \cdots \otimes X_k)].$$

Then, for all $X_1, \ldots, X_k \in \mathbb{F}^{m\times n}$,

$$f(X_1, \ldots, X_k) = \operatorname{tr}[(A_1X_1) \otimes \cdots \otimes (A_kX_k)] = \prod_{i=1}^k \operatorname{tr} A_iX_i.$$

Now, let σ be a permutation of $(1,\dots,k)$, and define $f_\sigma\colon \bigtimes_{i=1}^k \mathbb{F}^{m\times n} \mapsto \mathbb{F}$ by

$$f_\sigma(X_1,\dots,X_k) \triangleq f(X_{\sigma(1)},\dots,X_{\sigma(k)}) = \prod_{i=1}^k \operatorname{tr} A_iX_{\sigma(i)}.$$

Furthermore, let g be the l-tensor on $\mathbb{F}^{m\times n}$ defined by

$$g(X_1,\dots,X_l) \triangleq \operatorname{tr}[(B_1\otimes\cdots\otimes B_l)(X_1\otimes\cdots\otimes X_l)],$$

where $B_1,\dots,B_l \in \mathbb{F}^{n\times m}$. Then, $f\otimes g$ defined by

$$\begin{aligned}(f\otimes g)(X_1,\dots,X_{k+l}) &\triangleq \operatorname{tr}[(A_1\otimes\cdots\otimes A_k\otimes B_1\otimes\cdots\otimes B_l)(X_1\otimes\cdots\otimes X_{k+l})]\\ &= \operatorname{tr}[(A_1X_1)\otimes\cdots\otimes(A_kX_k)\otimes(B_1X_{k+1})\otimes\cdots\otimes(B_lX_{k+l})]\\ &= \left(\prod_{i=1}^k \operatorname{tr} A_iX_i\right)\prod_{i=1}^l \operatorname{tr} B_iX_{k+i}\end{aligned}$$

is a $(k+l)$-tensor on $\mathbb{F}^{m\times n}$. Next, f is an *alternating k-tensor* on $\mathbb{F}^{m\times n}$ if, for every permutation σ of $(1,\dots,k)$, $f_\sigma = \operatorname{sign}(\sigma)f$. If $k=1$, then f is alternating. Now, define the k-tensor $\mathcal{A}(f)$ on $\mathbb{F}^{m\times n}$ by

$$\mathcal{A}(f) \triangleq \sum \operatorname{sign}(\sigma)f_\sigma,$$

where the sum is taken over all $k!$ permutations σ of $(1,\dots,k)$. Then, $\mathcal{A}(f)$ is alternating. Finally, define the alternating $(k+l)$-tensor $f\wedge g$ on $\mathbb{F}^{m\times n}$ by

$$f\wedge g \triangleq \frac{1}{k!l!}\mathcal{A}(f\otimes g).$$

Now, let f, g, and h be k-, l-, and j-tensors, respectively, on $\mathbb{F}^{m\times n}$. Then, the following statements hold:

i) Let $n=1$ and $k=m$, and define $A \triangleq \begin{bmatrix} A_1\\ \vdots\\ A_m\end{bmatrix}$. Then, for all $X_1,\dots,X_m \in \mathbb{F}^m$, $\mathcal{A}(f)(X_1,\dots,X_m) = \det AX$, where $X \triangleq [X_1\ \cdots\ X_m]$.

ii) If f is alternating, then $\mathcal{A}(f) = k!f$.

iii) The following statements are equivalent:

a) f is alternating.

b) For all $X_1,\dots,X_k \in \mathbb{F}^{m\times n}$ that are not distinct, $f(X_1,\dots,X_k)=0$.

iv) $\mathcal{A}(f\otimes g) = \frac{1}{k!}\mathcal{A}[\mathcal{A}(f)\otimes g] = \frac{1}{l!}\mathcal{A}[f\otimes\mathcal{A}(g)]$.

v) $\mathcal{A}(g\otimes f) = (-1)^{kl}\mathcal{A}(f\otimes g)$ and $g\wedge f = (-1)^{kl}f\wedge g$.

vi) If $\alpha,\beta\in\mathbb{F}$, then $(\alpha f)\wedge\beta g = \alpha\beta(f\wedge g)$.

vii) If $k=l=1$, then $f\wedge g = f\otimes g - g\otimes f$.

viii) $f\otimes g\otimes h \triangleq f\otimes(g\otimes h) = (f\otimes g)\otimes h$ is a $(k+l+j)$-tensor on $\mathbb{F}^{m\times n}$.

ix) $f\wedge g\wedge h \triangleq f\wedge(g\wedge h) = (f\wedge g)\wedge h$ is an alternating $(k+l+j)$-tensor on $\mathbb{F}^{m\times n}$. Furthermore,

$$\begin{aligned}f\wedge g\wedge h &= \frac{1}{l!j!}f\wedge\mathcal{A}(g\otimes h) = \frac{1}{k!(l+j)!}\mathcal{A}[f\otimes(g\wedge h)]\\ &= \frac{1}{k!l!}\mathcal{A}(f\otimes g)\wedge h = \frac{1}{(k+l)!j!}\mathcal{A}[(f\wedge g)\otimes h]\\ &= \frac{1}{k!l!j!}\mathcal{A}(f\otimes g\otimes h).\end{aligned}$$

x) If $k = l = j = 1$, then

$$f \wedge g \wedge h = f \otimes g \otimes h + g \otimes h \otimes f + h \otimes f \otimes g - h \otimes g \otimes f - f \otimes h \otimes g - g \otimes f \otimes h.$$

Source: [2742, pp. 18–33]. **Remark:** $f \wedge g$ is the *wedge product* of f and g. Alternating tensors play a central role in geometric algebra, see [784, 908, 925, 1178, 1268, 1375, 1376, 1759, 1892, 1977, 2265], as well as in integration on manifolds, including Green's theorem and Stokes's theorem, see [553, 1265, 2101, 2505, 2742, 2837, 2856]. **Remark:** The determinant function is an alternating m-tensor on the columns of $m \times m$ matrices. **Remark:** f is a *covariant tensor* on $\mathbb{F}^{m\times n}$ and a *contravariant tensor* on $\mathbb{F}^{n\times m}$. See [553, pp. 218–262].

Fact 9.4.42. Let $A, B, C \in \mathbb{F}^{n\times m}$, and define $A \wedge B \triangleq A \otimes B - B \otimes A$ and

$$A \wedge B \wedge C \triangleq A \otimes B \otimes C + B \otimes C \otimes A + C \otimes A \otimes B - C \otimes B \otimes A - A \otimes C \otimes B - B \otimes A \otimes C.$$

Then, the following statements hold:

i) $A \wedge B = 0$ if and only if A and B are linearly dependent.

ii) $A \wedge B \wedge C = 0$ if and only if A, B, C are linearly dependent.

iii) For all $\alpha \in \mathbb{F}$, $(\alpha A) \wedge B = A \wedge (\alpha B) = \alpha(A \wedge B)$.

iv) For all $\alpha \in \mathbb{F}$, $(\alpha A) \wedge B \wedge C = A \wedge (\alpha B) \wedge C = A \wedge B \wedge (\alpha C) = \alpha(A \wedge B \wedge C)$.

v) $A \wedge B = -(B \wedge A)$.

vi) $A \wedge B \wedge C = -B \wedge A \wedge C = -A \wedge C \wedge B = -C \wedge B \wedge A = B \wedge C \wedge A = C \wedge A \wedge B$.

vii) $A \wedge B \wedge C = \frac{1}{2}\sum \operatorname{sign}(\sigma)[A \otimes (B \wedge C)]_\sigma$, where the sum is taken over all permutations σ of A, B, C, and, writing A_1, A_2, A_3 for A, B, C, respectively, $[A_1 \otimes (A_2 \wedge A_3)]_\sigma$ denotes $[A_{\sigma(1)} \otimes (A_{\sigma(2)} \wedge A_{\sigma(3)})]_\sigma$.

Remark: A, B, C represent 1-tensors on $\mathbb{F}^{m\times n}$. See Fact 9.4.41.

9.5 Facts on the Kronecker Sum

Fact 9.5.1. Let $A \in \mathbb{F}^{n\times n}$. Then, $(A \oplus A)^2 = A^2 \oplus A^2 + 2A \otimes A$.

Fact 9.5.2. Let $A \in \mathbb{F}^{n\times n}$. Then, $\det(A \oplus A) = (-1)^n \det \chi_A(-A)$. **Source:** [2991, p. 121].

Fact 9.5.3. Let $A \in \mathbb{F}^{n\times n}$. Then,

$$n \le \operatorname{def}(A^\mathsf{T} \oplus -A) = \dim \{X \in \mathbb{F}^{n\times n}\colon AX = XA\},$$

$$\operatorname{rank}(A^\mathsf{T} \oplus -A) = \dim \{[A, X]\colon X \in \mathbb{F}^{n\times n}\} \le n^2 - n.$$

Source: Fact 3.23.9. **Remark:** $\operatorname{def}(A^\mathsf{T} \oplus -A)$ is the dimension of the *centralizer* (also called the *commutant*) of A. See Fact 3.23.9. **Related:** Fact 7.16.8 and Fact 7.16.9. **Problem:** Characterize $\operatorname{rank}(A^\mathsf{T} \oplus -A)$ in terms of the eigenstructure of A.

Fact 9.5.4. Let $A \in \mathbb{F}^{n\times n}$, assume that A is nilpotent, and assume that $A^\mathsf{T} \oplus -A = 0$. Then, $A = 0$. **Source:** Note that $A^\mathsf{T} \otimes A^k = I \otimes A^{k+1}$, and use Fact 9.4.20.

Fact 9.5.5. Let $A \in \mathbb{F}^{n\times n}$, and assume that, for all $X \in \mathbb{F}^{n\times n}$, $AX = XA$. Then, there exists $\alpha \in \mathbb{F}$ such that $A = \alpha I$. **Source:** It follows from Proposition 9.2.4 that all of the eigenvalues of A are equal. Hence, there exists $\alpha \in \mathbb{F}$ such that $A = \alpha I + B$, where B is nilpotent. Now, Fact 9.5.4 implies that $B = 0$.

Fact 9.5.6. Let $A \in \mathbb{F}^{n\times n}$ and $B \in \mathbb{F}^{m\times m}$, and let $\gamma \in \operatorname{spec}(A \oplus B)$. Then,

$$\sum \operatorname{gmult}_A(\lambda)\operatorname{gmult}_B(\mu) \le \operatorname{gmult}_{A\oplus B}(\gamma) \le \operatorname{amult}_{A\oplus B}(\gamma) = \sum \operatorname{amult}_A(\lambda)\operatorname{amult}_B(\mu),$$

where both sums are taken over all $\lambda \in \operatorname{spec}(A)$ and $\mu \in \operatorname{spec}(B)$ such that $\lambda + \mu = \gamma$.

Fact 9.5.7. Let $A \in \mathbb{F}^{n\times n}$ and $B \in \mathbb{F}^{m\times m}$. Then, $\alpha_{\max}(A \oplus B) = \alpha_{\max}(A) + \alpha_{\max}(B)$. Furthermore, $\alpha_{\max}(A \oplus A) = 2\alpha_{\max}(A)$.

Fact 9.5.8. Let $A \in \mathbb{F}^{n\times n}$ and $B \in \mathbb{F}^{m\times m}$, and let $\gamma \in \operatorname{spec}(A \oplus B)$. Then, $\operatorname{ind}_{A\oplus B}(\gamma) = 1$ if and only if $\operatorname{ind}_A(\lambda) = 1$ and $\operatorname{ind}_B(\mu) = 1$ for all $\lambda \in \operatorname{spec}(A)$ and $\mu \in \operatorname{spec}(B)$ such that $\lambda + \mu = \gamma$.

Fact 9.5.9. Let $A \in \mathbb{F}^{n\times n}$ and $B \in \mathbb{F}^{m\times m}$, and assume that A and B are (group invertible, range Hermitian, Hermitian, symmetric, skew Hermitian, skew symmetric, normal, positive semidefinite, positive definite, semidissipative, dissipative, nilpotent, semisimple). Then, so is $A \oplus B$.

Fact 9.5.10. Let $A \in \mathbb{F}^{n\times n}$ and $B \in \mathbb{F}^{m\times m}$. Then,

$$P_{m,n}(A \oplus B)P_{n,m} = P_{m,n}(A \oplus B)P_{m,n}^{-1} = B \oplus A.$$

Hence, $A \oplus B$ and $B \oplus A$ are similar, and thus

$$\operatorname{rank}(A \oplus B) = \operatorname{rank}(B \oplus A).$$

Source: *xiii*) of Fact 9.4.38.

Fact 9.5.11. Let $A \in \mathbb{F}^{n\times n}$ and $B \in \mathbb{F}^{m\times m}$. Then,

$$n \operatorname{rank} B + m \operatorname{rank} A - 2(\operatorname{rank} A)(\operatorname{rank} B) \le \operatorname{rank}(A \oplus B) \le \begin{cases} nm - [n - \operatorname{rank}(I + A)][m - \operatorname{rank}(I - B)] \\ nm - [n - \operatorname{rank}(I - A)][m - \operatorname{rank}(I + B)]. \end{cases}$$

If, in addition, $-A$ and B are idempotent, then

$$\operatorname{rank}(A \oplus B) = n \operatorname{rank} B + m \operatorname{rank} A - 2(\operatorname{rank} A)(\operatorname{rank} B).$$

Equivalently,

$$\operatorname{rank}(A \oplus B) = (\operatorname{rank}\,(-A)_\perp) \operatorname{rank} B + (\operatorname{rank} B_\perp) \operatorname{rank} A.$$

Source: [743]. **Remark:** The second inequality may be strict if $-A$ and B are idempotent.

Fact 9.5.12. Let $A \in \mathbb{F}^{n\times n}$ and $B \in \mathbb{F}^{m\times m}$, assume that A is positive definite, define $p(s) \triangleq \det(I - sA)$, and let $\operatorname{mroots}(p) = \{\lambda_1, \ldots, \lambda_n\}_{\rm ms}$. Then,

$$\det(A \oplus B) = (\det A)^m \prod_{i=1}^{n} \det(\lambda_i B + I).$$

Source: Specialize Fact 9.5.13. **Remark:** In the case where $\operatorname{rank} C = 1$, an expression for $\det(\operatorname{vec}^{-1}[(A \oplus B)^{-1} \operatorname{vec} C])$ is given in [2390].

Fact 9.5.13. Let $A, C \in \mathbb{F}^{n\times n}$ and $B, D \in \mathbb{F}^{m\times m}$, assume that A is positive definite, assume that C is positive semidefinite, define $p(s) \triangleq \det(C - sA)$, and let $\operatorname{mroots}(p) = \{\lambda_1, \ldots, \lambda_n\}_{\rm ms}$. Then,

$$\det(A \otimes B + C \otimes D) = (\det A)^m \prod_{i=1}^{n} \det(\lambda_i D + B).$$

Source: [2043, pp. 40, 41]. **Remark:** The Kronecker product definition in [2043] follows the convention of [1914], where "$A \otimes B$" denotes $B \otimes A$.

Fact 9.5.14. Let $A, C \in \mathbb{F}^{n\times n}$ and $B, D \in \mathbb{F}^{m\times m}$, assume that $\operatorname{rank} C = 1$, and assume that A is nonsingular. Then,

$$\det(A \otimes B + C \otimes D) = (\det A)^m (\det B)^{n-1} \det[B + (\operatorname{tr} CA^{-1})D].$$

Source: [2043, p. 41].

Fact 9.5.15. Let $A \in \mathbb{F}^{n\times n}$ and $B \in \mathbb{F}^{m\times m}$. Then, the following statements are equivalent:

i) $\operatorname{spec}(A) \cap \operatorname{spec}(-B) = \varnothing$.

ii) $A^{\mathsf T} \oplus B$ is nonsingular.

iii) $A \oplus B$ is nonsingular.

iv) For all $C \in \mathbb{F}^{n\times m}$, there exists a unique solution $X \in \mathbb{F}^{n\times m}$ of $AX + XB + C = 0$.

v) For all $C \in \mathbb{F}^{n\times m}$, there exists a solution $X \in \mathbb{F}^{n\times m}$ of $AX + XB + C = 0$.

vi) There exists $C \in \mathbb{F}^{n\times m}$ such that $AX + XB + C = 0$ has a unique solution $X \in \mathbb{F}^{n\times m}$.

vii) For all $C \in \mathbb{F}^{n\times m}$, $\left[\begin{smallmatrix} A & 0 \\ 0 & -B \end{smallmatrix}\right]$ and $\left[\begin{smallmatrix} A & C \\ 0 & -B \end{smallmatrix}\right]$ are similar.

Source: The equivalence of *i)*–*vi)* follows from Corollary 3.7.9 and Proposition 9.2.5. The equivalence of *v)* and *vii)* follows from Fact 7.11.26.

Fact 9.5.16. Let $A \in \mathbb{F}^{n\times n}$, $B \in \mathbb{F}^{m\times m}$, and $C \in \mathbb{F}^{n\times m}$, and assume that $\det(B^{\mathsf{T}} \oplus A) \neq 0$. Then, $X \in \mathbb{F}^{n\times m}$ satisfies

$$A^2X + 2AXB + XB^2 + C = 0$$

if and only if

$$X = -\operatorname{vec}^{-1}[(B^{\mathsf{T}} \oplus A)^{-2} \operatorname{vec} C].$$

Fact 9.5.17. For all $i \in \{1,\dots,p\}$, let $A_i \in \mathbb{F}^{n_i\times n_i}$. Then,

$$\operatorname{mspec}(A_1 \oplus \cdots \oplus A_p) = \{\lambda_1 + \cdots + \lambda_p\colon\ \lambda_i \in \operatorname{mspec}(A_i) \text{ for all } i \in \{1,\dots,p\}\}_{\rm ms}.$$

If, in addition, for all $i \in \{1,\dots,p\}$, $x_i \in \mathbb{C}^{n_i}$ is an eigenvector of A_i associated with $\lambda_i \in \operatorname{spec}(A_i)$, then $x_1 \oplus \cdots \oplus x_p$ is an eigenvector of $A_1 \oplus \cdots \oplus A_p$ associated with $\lambda_1 + \cdots + \lambda_p$.

Fact 9.5.18. Let $A \in \mathbb{F}^{n\times m}$, and, for $1 \le k \le \min\{n,m\}$, define the kth *compound* $A^{(k)}$ to be the $\binom{n}{k} \times \binom{m}{k}$ matrix whose entries are $k \times k$ subdeterminants of A, where the k-element subsets of rows and columns of A are ordered lexicographically. (Example: For $k = 3$, $n = 4$, and $m = 5$, the four 3-element subsets of the rows of A are ordered as $\{1,2,3\}, \{1,2,4\}, \{1,3,4\}, \{2,3,4\}$, and the ten 3-element subsets of the columns of A are ordered as $\{1,2,3\}, \{1,2,4\}, \{1,2,5\}, \{1,3,4\}, \{1,3,5\}, \{1,4,5\}, \{2,3,4\}, \{2,3,5\}, \{2,4,5\}, \{3,4,5\}$.) Specifically, $(A^{(k)})_{(i,j)}$ is the determinant of the $k\times k$ submatrix of A obtained from the ith subset of k rows of A and the jth subset of k columns of A. Furthermore, for all $x, y \in \mathbb{F}^n$, define $x \wedge y \in \mathbb{F}^{\binom{n}{2}}$ by

$$(x \wedge y)_{(p)} \triangleq \det \begin{bmatrix} x_{(i_p)} & y_{(i_p)} \\ x_{(j_p)} & x_{(j_p)} \end{bmatrix},$$

where, for all $p \in \{1,\dots,\binom{n}{2}\}$, the values of $1 \le i_p < j_p \le n$ are given by the pth component of the $\binom{n}{2}$-tuple whose components (i,j) are ordered lexicographically. (Example: For $n = 4$, this 6-tuple is $((1,2),(1,3),(1,4),(2,3),(2,4),(3,4))$.) Then, the following statements hold:

i) $A^{(0)} = 1$, $A^{(1)} = A$.

ii) $A^{(2)}(x \wedge y) = (Ax) \wedge (Ay)$.

iii) If $\alpha \in \mathbb{F}$, then $(\alpha A)^{(k)} = \alpha^k A^{(k)}$.

iv) $(A^{\mathsf{T}})^{(k)} = (A^{(k)})^{\mathsf{T}}$, $\overline{A}^{(k)} = \overline{A^{(k)}}$, $(A^*)^{(k)} = (A^{(k)})^*$, $(A^+)^{(k)} = (A^{(k)})^+$.

v) If $k \le \operatorname{rank} A$, then $\operatorname{rank} A^{(k)} = \binom{\operatorname{rank} A}{k}$. In particular, if $k = \operatorname{rank} A$, then $\operatorname{rank} A^{(k)} = 1$.

vi) If $k > \operatorname{rank} A$, then $A^{(k)} = 0$.

vii) If $B \in \mathbb{F}^{m\times l}$ and $1 \le k \le \min\{n,m,l\}$, then $(AB)^{(k)} = A^{(k)}B^{(k)}$.

viii) If $B \in \mathbb{F}^{m\times n}$, then $\det AB = A^{(n)}B^{(n)}$.

ix) $\operatorname{msval}(A^{(k)}) = \{\sigma_{i_1}(A)\cdots\sigma_{i_k}(A)\colon\ 1 \le i_1 < \cdots < i_k \le \min\{n,m\}\}_{\rm ms}$.

Now, assume that $m = n$, let $1 \le k \le n$, and let $\operatorname{mspec}(A) = \{\lambda_1,\dots,\lambda_n\}_{\rm ms}$. Then, the following statements hold:

x) If A is (diagonal, lower triangular, upper triangular, symmetric, Hermitian, positive semidefinite, positive definite, normal, unitary), then so is $A^{(k)}$.

xi) Assume that A is skew Hermitian. If k is odd, then $A^{(k)}$ is skew Hermitian. If k is even, then $A^{(k)}$ is Hermitian.

xii) Assume that A is diagonal, upper triangular, or lower triangular, and let $1 \le i_1 < \cdots < i_k \le n$. Then, the $(i_1+\cdots+i_k, i_1+\cdots+i_k)$ entry of $A^{(k)}$ is $A_{(i_1,i_1)}\cdots A_{(i_k,i_k)}$. In particular, $I_n^{(k)} = I_{\binom{n}{k}}$.

xiii) $\det A^{(k)} = (\det A)^{\binom{n-1}{k-1}}$, $A^{(n)} = \det A$.

xiv) $SA^{(n-1)\mathsf{T}}S = A^{\mathsf{A}}$, where $S \triangleq \operatorname{diag}(1,-1,1,\ldots)$.

xv) $\det A^{(n-1)} = \det A^{\mathsf{A}} = (\det A)^{n-1}$, $\operatorname{tr} A^{(n-1)} = \operatorname{tr} A^{\mathsf{A}}$.

xvi) If A is nonsingular, then $(A^{(k)})^{-1} = (A^{-1})^{(k)}$.

xvii) $\operatorname{mspec}(A^{(k)}) = \{\lambda_{i_1}\cdots\lambda_{i_k}\colon 1 \le i_1 < \cdots < i_k \le n\}_{\mathrm{ms}}$. In particular,

$$\operatorname{mspec}(A^{(2)}) = \{\lambda_i\lambda_j\colon\ i,j = 1,\ldots,n,\ i<j\}_{\mathrm{ms}}.$$

xviii) $\operatorname{tr} A^{(k)} = \sum_{1\le i_1<\cdots<i_k\le n}\lambda_{i_1}\cdots\lambda_{i_k}$.

xix) If A has exactly k nonzero eigenvalues, then $A^{(k)}$ has exactly one nonzero eigenvalue.

xx) If $k < n$ and A has exactly k nonzero eigenvalues, then $\operatorname{spec}(A^{(k+1)}) = \{0\}$, and thus $A^{(k+1)}$ is nilpotent.

xxi) If $B \in \mathbb{F}^{n\times n}$, then $\det(A+B) = [A\ \ I]^{(n)}\begin{bmatrix} I \\ B\end{bmatrix}^{(n)}$.

xxii) The characteristic polynomial of A is given by

$$\chi_A(s) = s^n + \sum_{i=1}^{n-1}(-1)^{n+i}[\operatorname{tr} A^{(n-i)}]s^i + (-1)^n\det A.$$

xxiii) $\det(A+I) = \sum_{i=0}^{n}\operatorname{tr} A^{(i)} = 1 + \sum_{i=1}^{n-1}\operatorname{tr} A^{(i)} + \det A$.

Now, for all $i \in \{0,1,\ldots,k\}$, define $A^{(k,i)}$ by

$$(A+sI)^{(k)} = s^kA^{(k,0)} + s^{k-1}A^{(k,1)} + \cdots + sA^{(k,k-1)} + A^{(k,k)}.$$

Then, the following statements hold:

xxiv) $A^{(k,0)} = I$, $A^{(k,k)} = A^{(k)}$.

xxv) If $B \in \mathbb{F}^{n\times n}$ and $\alpha,\beta \in \mathbb{F}$, then $(\alpha A + \beta B)^{(k,1)} = \alpha A^{(k,1)} + \beta B^{(k,1)}$.

xxvi) $\operatorname{mspec}(A^{(k,1)}) = \{\lambda_{i_1}+\cdots+\lambda_{i_k}\colon 1 \le i_1 < \cdots < i_k \le n\}_{\mathrm{ms}}$.

xxvii) $\operatorname{tr} A^{(k,1)} = \binom{n-1}{k-1}\operatorname{tr} A$.

xxviii) $\operatorname{mspec}(A^{(2,1)}) = \{\lambda_i + \lambda_j\colon\ i,j = 1,\ldots,n,\ i<j\}_{\mathrm{ms}}$.

xxix) $\operatorname{mspec}[(A^{(2,1)})^2 - 4A^{(2)}] = \{(\lambda_i - \lambda_j)^2\colon\ i,j = 1,\ldots,n,\ i<j\}_{\mathrm{ms}}$.

Source: [989], [1040, pp. 142–155], [1448, p. 11], [1868], [1947, pp. 116–130], [1969, pp. 502–506], [1971, pp. 775–780], [2263, p. 124], [2265], [2979, pp. 46–48], and [2991, pp. 122–124]. **Remark:** *vii*) follows from the Binet-Cauchy theorem given by Fact 3.16.8. *viii*) is a special case of *vii*), and is given by Fact 3.16.9, which is a special case of Fact 3.16.8. See [1969, p. 503] and [1971, p. 776]. *xiii*) is the *Sylvester-Franke theorem.* See [1947, p. 130]. **Remark:** $A^{(k,1)}$ is the kth *additive compound* of A. **Remark:** $(A^{(2,1)})^2 - 4A^{(2)}$ is the *discriminant* of A, which is singular if and only if A has a repeated eigenvalue. **Remark:** Induced norms of compound matrices are considered in [986]. **Related:** Fact 6.9.2, Fact 10.16.50, Fact 10.25.66, and Fact 15.18.16.

Fact 9.5.19. Let $A \in \mathbb{F}^{n\times n}$, let $k \le n$, and define the kth *adjugate* A^{A_k} to be the $\binom{n}{k}\times\binom{n}{k}$ matrix whose entries are signed $k\times k$ subdeterminants of A, ordered lexicographically. Specifically, let $\mathcal{S}_1, \mathcal{S}_2 \subseteq \{1,\ldots,n\}$, and assume that $\operatorname{card}(\mathcal{S}_1) = \operatorname{card}(\mathcal{S}_2) = k$. Then, the corresponding entry of A^{A_k} is given by $(-1)^\eta A_{[\mathcal{S}_2,\mathcal{S}_1]}$, where η denotes the sum of the elements of $\mathcal{S}_1\cup\mathcal{S}_2$. Then, the following statements hold:

i) $A^{\mathsf{A}_0} = \det A$, $A^{\mathsf{A}_1} = A^{\mathsf{A}}$, $A^{\mathsf{A}_n} = 1$.

ii) $A^{\mathsf{A}_k}A^{(k)} = A^{(k)}A^{\mathsf{A}_k} = (\det A)I_{\binom{n}{k}}$.

iii) $\det A^{(n-k)} = \det A^{\mathsf{A}_k}$.

iv) If A is nonsingular, then $A^{\mathsf{A}_k} = (\det A)[A^{(k)}]^{-1}$.

v) If $B \in \mathbb{F}^{n\times n}$, then $(AB)^{\mathsf{A}_k} = B^{\mathsf{A}_k}A^{\mathsf{A}_k}$.

vi) If $\alpha,\beta \in \mathbb{F}$, then $\det(\alpha A + \beta B) = \sum_{i=0}^{n} \alpha^{n-i}\beta^i \operatorname{tr}[A^{\mathsf{A}_i}B^{(i)}]$.

vii) $\det(A + I) = \sum_{i=0}^{n} \operatorname{tr} A^{\mathsf{A}_i} = \sum_{i=0}^{n} \operatorname{tr} A^{(i)}$.

viii) If $b, c \in \mathbb{F}^n$, then $\det(A + bc^{\mathsf{T}}) = \det A + c^{\mathsf{T}}A^{\mathsf{A}}b$.

ix) If $B, C \in \mathbb{F}^{n\times 2}$ and $\operatorname{rank} B = \operatorname{rank} C = 2$, then $\det(A+BC^{\mathsf{T}}) = \det A+\operatorname{tr} C^{\mathsf{T}}A^{\mathsf{A}}B+C^{(2)\mathsf{T}}A^{\mathsf{A}_2}B^{(2)}$.

Source: [861], [1451, p. 29], [2265]. **Remark:** $A^{(k)}$ is the kth compound of A. See Fact 9.5.18.

Fact 9.5.20. Let $A, B \in \mathbb{F}^{n\times n}$, and define the *bialternate product* $A \cdot B$ of A and B to be the $\binom{n}{2} \times \binom{n}{2}$ matrix whose (p, q) entry is

$$(A \cdot B)_{(p,q)} = \frac{1}{2}\left(\det\begin{bmatrix} A_{(i,k)} & A_{(i,l)} \\ B_{(j,k)} & B_{(j,l)} \end{bmatrix} + \det\begin{bmatrix} B_{(i,k)} & B_{(i,l)} \\ A_{(j,k)} & A_{(j,l)} \end{bmatrix}\right),$$

where the values of $1 \le i < j \le n$ corresponding to p are the pth component of the $\binom{n}{2}$-tuple with components (i, j) ordered lexicographically, and likewise for q and $1 \le k < l \le n$. (Example: For $n = 3$, the ordering of the rows of $A \cdot B$ corresponds to the ordering $(1,2), (1,3), (2,3)$ of (i, j).) Then, the following statements hold:

i) If $\alpha,\beta \in \mathbb{F}$, then $(\alpha A) \cdot (\beta B) = \alpha\beta(A \cdot B)$.

ii) $2A \cdot I = A \cdot I + I \cdot A = A^{(2,1)}$, $A \cdot A = A^{(2)}$.

iii) $2\operatorname{tr}(A \cdot I) = (n-1)\operatorname{tr} A$, $2\operatorname{tr}(A \cdot A) = 2\operatorname{tr} A^{(2)} = (\operatorname{tr} A)^2 - \operatorname{tr} A^2$.

iv) If $k \ge 0$, then $(A \cdot A)^k = A^k \cdot A^k$.

v) If A is nonsingular, then $A \cdot A$ is nonsingular and $(A \cdot A)^{-1} = A^{-1} \cdot A^{-1}$.

vi) $(A \cdot B)^{\mathsf{T}} = A^{\mathsf{T}} \cdot B^{\mathsf{T}}$, $\overline{A \cdot B} = \overline{A} \cdot \overline{B}$, $(A \cdot B)^* = A^* \cdot B^*$, $A \cdot B = B \cdot A$.

vii) $2A \cdot B = (A + B)^{(2)} - A^{(2)} - B^{(2)}$, $AB \cdot AB = (AB)^{(2)} = A^{(2)}B^{(2)} = (A \cdot A)(B \cdot B)$.

viii) $2\operatorname{tr}(A \cdot B) = (\operatorname{tr} A)\operatorname{tr} B - \operatorname{tr}(AB)$, $4\operatorname{tr}(A \cdot I)(B \cdot I) = (\operatorname{tr} A)\operatorname{tr} B + (n-2)\operatorname{tr} AB$.

ix) If $C \in \mathbb{F}^{n\times n}$, then $A \cdot (B + C) = A \cdot B + A \cdot C$.

x) If $C \in \mathbb{F}^{n\times n}$, then $(A \cdot B)(C \cdot C) = AC \cdot BC$.

xi) If $C, D \in \mathbb{F}^{n\times n}$, then $(A \cdot A)(C \cdot D) = AC \cdot AD$.

xii) If $C, D \in \mathbb{F}^{n\times n}$, then $(A \cdot B)(C \cdot D) = \frac{1}{2}(AC \cdot BD + AD \cdot BC)$.

xiii) Let $\operatorname{mspec}(A) = \{\lambda_1, \ldots, \lambda_n\}_{\mathrm{ms}}$, let $k \ge 0$, and, for all $p, q \in \{0, 1, \ldots, k\}$, let $\alpha_{p,q} \in \mathbb{F}$. Then,

$$\operatorname{mspec}\left(\sum_{p,q=0}^{k} \alpha_{p,q}A^p \cdot A^q\right) = \left\{\frac{1}{2}\sum_{p,q=0}^{k} \alpha_{p,q}(\lambda_i^p\lambda_j^q + \lambda_i^q\lambda_j^p) \colon 1 \le i < j \le n\right\}_{\mathrm{ms}}.$$

xiv) $\operatorname{mspec}(2A \cdot I) = \operatorname{mspec}(A \cdot I + I \cdot A) = \{\lambda_i + \lambda_j \colon 1 \le i < j \le n\}_{\mathrm{ms}}$.

Source: [989, 1097, 1213], [1562, pp. 313–320], and [1914, pp. 84, 85]. **Remark:** $A^{(k)}$ is the kth compound of A. See Fact 9.5.18.

Fact 9.5.21. Let $A, B \in \mathbb{F}^{n\times n}$, and define the *permanental bialternate product* $A \times B$ of A and B

to be the $\binom{n+1}{2}\times\binom{n+1}{2}$ matrix whose (p,q) entry is

$$(A\times B)_{(p,q)} = \frac{1}{\sqrt{(1+\delta_{i,j})(1+\delta_{k,l})}}\left(\det\begin{bmatrix} A_{(i,k)} & A_{(i,l)} \\ -B_{(j,k)} & B_{(j,l)}\end{bmatrix} + \det\begin{bmatrix} B_{(i,k)} & B_{(i,l)} \\ -A_{(j,k)} & A_{(j,l)}\end{bmatrix}\right),$$

where the values of $1\le i\le j\le n$ corresponding to p are the pth component of the $\binom{n+1}{2}$-tuple with components (i,j) ordered lexicographically, and likewise for q and $1\le k\le l\le n$. (Example: For $n=3$, the ordering of the rows of $A\times B$ corresponds to the ordering $(1,1),(1,2),(1,3),(2,2),(2,3),(3,3)$ of (i,j).) With the same indices, define $A^{[2]}$ to be the $\binom{n+1}{2}\times\binom{n+1}{2}$ matrix whose (p,q) entry is

$$A^{[2]}_{(p,q)} \triangleq \frac{1}{\sqrt{(1+\delta_{i,j})(1+\delta_{k,l})}}\det\begin{bmatrix} A_{(i,k)} & A_{(i,l)} \\ -A_{(j,k)} & A_{(j,l)}\end{bmatrix}.$$

Furthermore, let $x,y\in\mathbb{F}^n$, and define $x\vee y\in\mathbb{F}^{\binom{n+1}{2}}$ by

$$(x\vee y)_{(p)} \triangleq \det\begin{bmatrix} x_{(i)} & y_{(i)} \\ x_{(j)} & x_{(j)}\end{bmatrix},$$

where the values of $1\le i\le j\le n$ corresponding to p are given by the pth component of the $\binom{n}{2}$-tuple with components (i,j) ordered lexicographically. Then, the following statements hold:

i) If $\alpha,\beta\in\mathbb{F}$, then $(\alpha A)\times(\beta B) = \alpha\beta(A\times B)$.

ii) $A^{[2]}(x\vee y) = (Ax)\vee(Ay)$.

iii) $2A\times I = A\times I + I\times A = A^{(2,1)}$, $A\times A = A^{[2]}$.

iv) $2\operatorname{tr}(A\times I) = (n+1)\operatorname{tr}A$, $2\operatorname{tr}(A\times A) = 2\operatorname{tr}A^{[2]} = (\operatorname{tr}A)^2 + \operatorname{tr}A^2$.

v) If $k\ge 0$, then $(A\times A)^k = A^k\times A^k$.

vi) If A is nonsingular, then $A\times A$ is nonsingular and $(A\times A)^{-1} = A^{-1}\times A^{-1}$.

vii) $(A\times B)^{\mathsf{T}} = A^{\mathsf{T}}\times B^{\mathsf{T}}$, $\overline{A\times B} = \overline{A}\times\overline{B}$, $(A\times B)^* = A^*\times B^*$.

viii) $A\times B = B\times A$, $2A\times B = (A+B)^{[2]} - A^{[2]} - B^{[2]}$.

ix) $AB\times AB = (A\times A)(B\times B)$.

x) $2\operatorname{tr}(A\times B) = (\operatorname{tr}A)\operatorname{tr}B + \operatorname{tr}(AB)$, $4\operatorname{tr}(A\times I)(B\times I) = (\operatorname{tr}A)\operatorname{tr}B + (n+2)\operatorname{tr}AB$.

xi) If $C\in\mathbb{F}^{n\times n}$, then $A\times(B+C) = A\times B + A\times C$.

xii) If $C\in\mathbb{F}^{n\times n}$, then $(A\times B)(C\times C) = AC\times BC$.

xiii) If $C,D\in\mathbb{F}^{n\times n}$, then $(A\times A)(C\times D) = AC\times AD$.

xiv) If $C,D\in\mathbb{F}^{n\times n}$, then $(A\times B)(C\times D) = \frac{1}{2}(AC\times BD + AD\times BC)$.

xv) Let $\operatorname{mspec}(A) = \{\lambda_1,\ldots,\lambda_n\}_{\mathrm{ms}}$, let $k\ge 0$, and, for all $p,q\in\{0,\ldots,k\}$, let $\alpha_{p,q}\in\mathbb{F}$. Then,

$$\operatorname{mspec}\left(\sum_{p,q=0}^{k}\alpha_{p,q}A^p\times A^q\right) = \left\{\frac{1}{2}\sum_{p,q=0}^{k}\alpha_{p,q}(\lambda_i^p\lambda_j^q + \lambda_i^q\lambda_j^p)\colon 1\le i\le j\le n\right\}_{\mathrm{ms}}.$$

xvi) $\operatorname{mspec}(2A\times I) = \operatorname{mspec}(A\times I + I\times A) = \{\lambda_i+\lambda_j\colon 1\le i\le j\le n\}_{\mathrm{ms}}$.

xvii) There exists an orthogonal matrix $S\in\mathbb{R}^{2n\times 2n}$ such that

$$A\otimes B + B\otimes A = 2S\begin{bmatrix} A\cdot B & 0 \\ 0 & A\times B\end{bmatrix}S^{\mathsf{T}}.$$

xviii) $\operatorname{tr}A\otimes B = (\operatorname{tr}A)\operatorname{tr}B = \operatorname{tr}A\cdot B + \operatorname{tr}A\times B$.

xix) $\det(A\otimes B + B\otimes A) = 2^{n^2}(\det A\cdot B)\det A\times B$.

Source: [989]. **Remark:** $A^{[2]}$ is the *2nd induced matrix*. See [989]. **Remark:** $A^{(k)}$ is the kth compound of A, and $A \cdot B$ is the bialternate product of A and B. See Fact 9.5.18 and Fact 9.5.20.

9.6 Facts on the Schur Product

Fact 9.6.1. Let $x, y, z \in \mathbb{F}^n$. Then, $x^{\mathsf{T}}(y \odot z) = z^{\mathsf{T}}(x \odot y) = y^{\mathsf{T}}(x \odot z)$.

Fact 9.6.2. Let $w, y \in \mathbb{F}^n$ and $x, z \in \mathbb{F}^m$. Then, $(wx^{\mathsf{T}}) \odot (yz^{\mathsf{T}}) = (w \odot y)(x \odot z)^{\mathsf{T}}$.

Fact 9.6.3. Let $A \in \mathbb{F}^{n\times n}$ and $d \in \mathbb{F}^n$. Then, $\operatorname{diag}(d)A = A \odot d1_{1\times n}$.

Fact 9.6.4. Let $x \in \mathbb{F}^n$ and $y \in \mathbb{F}^m$. Then, $xy^{\mathsf{T}} = 1_{n\times 1}y^{\mathsf{T}} \odot x1_{1\times m}$.

Fact 9.6.5. Let $A \in \mathbb{F}^{n\times m}$, $x \in \mathbb{F}^n$, and $y, z \in \mathbb{F}^m$. Then,

$$(A \odot xy^{\mathsf{T}})z = [A(y \odot z)] \odot x.$$

In particular,

$$(A \odot x1_{1\times m})z = Az \odot x, \quad (A \odot 1_{n\times 1}y^{\mathsf{T}})z = A(y \odot z).$$

Fact 9.6.6. Let $A \in \mathbb{F}^{n\times m}$, $B \in \mathbb{F}^{m\times l}$, $x \in \mathbb{F}^n$, $y \in \mathbb{F}^l$, and $a, b \in \mathbb{F}^m$. Then,

$$\begin{aligned}(A \odot 1_{n\times 1}a^{\mathsf{T}})(B \odot b1_{1\times l}) \odot (xy^{\mathsf{T}}) &= (A \odot 1_{n\times 1}b^{\mathsf{T}})(B \odot a1_{1\times l}) \odot (xy^{\mathsf{T}})\\ &= (A \odot xa^{\mathsf{T}})(B \odot by^{\mathsf{T}}) = (A \odot xb^{\mathsf{T}})(B \odot ay^{\mathsf{T}}).\end{aligned}$$

In particular,

$$(AB) \odot (xy^{\mathsf{T}}) = (A \odot x1_{1\times m})(B \odot 1_{m\times 1}y^{\mathsf{T}}).$$

Source: [1711].

Fact 9.6.7. Let $A, B \in \mathbb{F}^{n\times m}$, let $x \in \mathbb{F}^n$, and assume that x has no zero components. Then,

$$(A \odot 1_{n\times 1}x^{\mathsf{T}})(B \odot x^{\odot -1}1_{1\times n}) = AB.$$

Source: [1711].

Fact 9.6.8. Let $A \in \mathbb{F}^{n\times n}$, let $x, y \in \mathbb{F}^n$, and assume that x and y have no zero components. Then, A is nonsingular if and only if $A \odot xy^{\mathsf{T}}$ is nonsingular. If these conditions hold, then

$$(A \odot xy^{\mathsf{T}})^{-1} = A^{-1} \odot (yx^{\mathsf{T}})^{\odot -1}.$$

Source: [1711].

Fact 9.6.9. Let $x, a \in \mathbb{F}^n$, $y, b \in \mathbb{F}^m$, and $A \in \mathbb{F}^{n\times m}$. Then, $x^{\mathsf{T}}(A \odot ab^{\mathsf{T}})y = (a \odot x)^{\mathsf{T}}A(b \odot y)$.

Fact 9.6.10. Let $A \in \mathbb{F}^{n\times m}$, $B \in \mathbb{F}^{m\times n}$, $x \in \mathbb{F}^m$, and $y \in \mathbb{F}^n$. Then,

$$x^{\mathsf{T}}(A^{\mathsf{T}} \odot B)y = \operatorname{tr} \operatorname{diag}(x)A^{\mathsf{T}}\operatorname{diag}(y)B^{\mathsf{T}} = \operatorname{tr} A(B \odot xy^{\mathsf{T}}).$$

In particular,

$$\operatorname{tr} AB = 1_{1\times m}(A^{\mathsf{T}} \odot B)1_{n\times 1}.$$

Fact 9.6.11. Let $A, B \in \mathbb{F}^{n\times m}$, $D_1 \in \mathbb{F}^{n\times n}$, and $D_2 \in \mathbb{F}^{m\times m}$, and assume that D_1 and D_2 are diagonal. Then,

$$(D_1A) \odot (BD_2) = D_1(A \odot B)D_2.$$

If, in addition, $n = m$, then

$$(D_1A) \odot (BD_2) = D_1(A \odot B)D_2 = (D_1AD_2) \odot B = A \odot (D_1BD_2).$$

Fact 9.6.12. Let $A, B \in \mathbb{F}^{n\times m}$. Then, $\operatorname{tr}[(A \odot B)(A \odot B)^{\mathsf{T}}] = \operatorname{tr}[(A \odot A)(B \odot B)^{\mathsf{T}}]$.

Fact 9.6.13. Let $A, B \in \mathbb{F}^{n\times m}$ and $C \in \mathbb{F}^{m\times n}$. Then,

$$I_n \odot [A(B^{\mathsf{T}} \odot C)] = I_n \odot [(A \odot B)C] = I_n \odot [(A \odot C^{\mathsf{T}})B^{\mathsf{T}}].$$

Hence,

$$\operatorname{tr}[A(B^{\mathsf{T}} \odot C)] = \operatorname{tr}[(A \odot B)C] = \operatorname{tr}[(A \odot C^{\mathsf{T}})B^{\mathsf{T}}].$$

Fact 9.6.14. Let $A_1,\dots,A_k \in \mathbb{F}^{n\times n}$. Then,

$$\mathcal{R}[(A_1A_1^*)\odot\cdots\odot(A_kA_k^*)] = \operatorname{span}\{(A_1x_1)\odot\cdots\odot(A_kx_k)\colon\ x_1,\dots,x_k \in \mathbb{F}^n\}.$$

Furthermore, if $A_1,\dots,A_k$ are positive semidefinite, then

$$\begin{aligned}\mathcal{R}(A_1\odot\cdots\odot A_k) &= \operatorname{span}\{(A_1x_1)\odot\cdots\odot(A_kx_k)\colon\ x_1,\dots,x_k \in \mathbb{F}^n\}\\ &= \operatorname{span}\{(A_1x)\odot\cdots\odot(A_kx)\colon\ x \in \mathbb{F}^n\}.\end{aligned}$$

Source: [2284].

Fact 9.6.15. Let $A, B \in \mathbb{F}^{n\times m}$. Then, $\operatorname{rank}(A\odot B) \le \operatorname{rank}(A\otimes B) = (\operatorname{rank} A)(\operatorname{rank} B)$. **Source:** Proposition 9.3.2. **Related:** Fact 10.25.15.

Fact 9.6.16. Let $x \in \mathbb{R}^m$ and $A \in \mathbb{R}^{n\times m}$, and define $x^A \in \mathbb{R}^n$ by

$$x^A \triangleq \begin{bmatrix}\prod_{i=1}^m x_{(i)}^{A_{(1,i)}}\\ \vdots \\ \prod_{i=1}^m x_{(i)}^{A_{(n,i)}}\end{bmatrix},$$

where every component of x^A is assumed to be a real number. Then, the following statements hold:

i) If $a \in \mathbb{R}$, then $a^x = \begin{bmatrix} a^{x_{(1)}}\\ \vdots\\ a^{x_{(m)}}\end{bmatrix}$.

ii) $x^{-A} = (x^A)^{\odot -1}$.

iii) If $y \in \mathbb{R}^m$, then $(x\odot y)^A = x^A \odot y^A$.

iv) If $B \in \mathbb{R}^{n\times m}$, then $x^{A+B} = x^A\odot x^B$.

v) If $B \in \mathbb{R}^{l\times n}$, then $(x^A)^B = x^{BA}$.

vi) If $a \in \mathbb{R}$, then $(a^x)^A = a^{Ax}$.

vii) If $A^{\mathsf{L}} \in \mathbb{R}^{m\times n}$ is a left inverse of A and $y = x^A$, then $x = y^{A^{\mathsf{L}}}$.

viii) If $A \in \mathbb{R}^{n\times n}$ is nonsingular and $y = x^A$, then $x = y^{A^{-1}}$.

ix) Define $f(x) \triangleq x^A$. Then, $f'(x) = \operatorname{diag}(x^A)A\operatorname{diag}(x^{\odot -1})$.

x) Let $x_1,\dots,x_n \in \mathbb{R}$, let $a \in \mathbb{R}^n$, and assume that $0 < x_1 < \cdots < x_n$ and $a_{(1)} < \cdots < a_{(n)}$. Then, $\det\,[x_1^a\ \cdots\ x_n^a] > 0$.

Remark: These operations are used to model chemical reaction kinetics. See [1794]. **Source:** *x)* is given in [2315].

Fact 9.6.17. Let $A \in \mathbb{R}^{n\times n}$, and assume that A is nonsingular. Then,

$$(A\odot A^{-\mathsf{T}})1_{n\times 1} = 1_{n\times 1}, \quad 1_{1\times n}(A\odot A^{-\mathsf{T}}) = 1_{1\times n}.$$

Source: [1548].

Fact 9.6.18. Let $A \in \mathbb{R}^{n\times n}$, and assume that A is nonnegative. Then,

$$\rho_{\max}[(A\odot A^{\mathsf{T}})^{\odot 1/2}] \le \rho_{\max}(A) \le \rho_{\max}[\tfrac{1}{2}(A+A^{\mathsf{T}})].$$

Source: [2407]. **Related:** Fact 6.11.25.

Fact 9.6.19. Let $A, B, C, D \in \mathbb{R}^{n\times n}$, and assume that A, B, C, D are nonnegative. Then,

$$(A\odot B)(C\odot D) \le\le [(A\odot A)(C\odot C)]^{\odot 1/2}\odot[(B\odot B)(D\odot D)]^{\odot 1/2}.$$

Furthermore,

$$(A\odot B)^2 \le\le [(A\odot A)(B\odot B)]^{\odot 1/2}\odot[(B\odot B)(A\odot A)]^{\odot 1/2},$$

$$(A^{\odot 1/2} \odot B^{\odot 1/2})^2 \leq\leq (A^2)^{\odot 1/2} \odot (A^2)^{\odot 1/2}.$$

Source: [2384].

Fact 9.6.20. Let $A_1, \ldots, A_k \in \mathbb{R}^{n\times n}$, and assume that, for all $i \in \{1, \ldots, k\}$, A_i is nonnegative. Then,

$$\rho_{\max}(A_1 \odot A_2 \odot \cdots \odot A_k) \le \rho_{\max}^{1-2/k}(A_1A_2\cdots A_k)\rho_{\max}^{1/k}(A_1^{\odot 2}A_2^{\odot 2}\cdots A_n^{\odot 2}) \le \rho_{\max}(A_1A_2\cdots A_k).$$

Now, let $\alpha_1, \ldots, \alpha_k$ be positive numbers, and assume that $\sum_{i=1}^k \alpha_i \ge 1$. Then,

$$\rho_{\max}(A_1^{\odot\alpha_1} \odot \cdots \odot A_k^{\odot\alpha_k}) \le \prod_{i=1}^k \rho_{\max}^{\alpha_i}(A_i).$$

In particular, let $A \in \mathbb{R}^{n\times n}$, and assume that A is nonnegative. Then, for all $\alpha \ge 1$,

$$\rho_{\max}(A^{\odot\alpha}) \le \rho_{\max}^{\alpha}(A),$$

whereas, for all $\alpha \le 1$,

$$\rho_{\max}^{\alpha}(A) \le \rho_{\max}(A^{\odot\alpha}).$$

Furthermore,

$$\rho_{\max}(A^{\odot 1/2} \odot A^{\mathsf{T}\odot 1/2}) \le \rho_{\max}(A), \quad \rho_{\max}^{1/2}(A \odot A) \le \rho_{\max}(A) = \rho_{\max}^{1/2}(A \otimes A).$$

Now, let $B \in \mathbb{R}^{n\times n}$ be nonnegative. Then,

$$\rho_{\max}(A \odot B) \le \begin{cases} \left.\begin{cases} \rho_{\max}^{1/2}(A \odot A)\rho_{\max}^{1/2}(B \odot B) \le \rho_{\max}(A)\rho_{\max}(B) = \rho_{\max}(A \otimes B) \\ \sigma_{\max}(A \odot B) \le \rho_{\max}(A^{\mathsf{T}}B) \le \sigma_{\max}(A^{\mathsf{T}}B) \end{cases}\right\} \le \sigma_{\max}(A)\sigma_{\max}(B) \\ \rho_{\max}^{1/2}[(A \odot A)(B \odot B)] < \rho_{\max}^{1/2}(AB \odot BA) \le \rho_{\max}^{1/4}(AB \odot AB)\rho_{\max}^{1/4}(BA \odot BA) \le \rho_{\max}(AB), \end{cases}$$

$$\rho_{\max}(A \odot B) \le \rho_{\max}(A)\rho_{\max}(B) + \max_{i\in\{1,\ldots,n\}} [2A_{(i,i)}B_{(i,i)} - \rho_{\max}(A)B_{(i,i)} - \rho_{\max}(B)A_{(i,i)}] \le \rho_{\max}(A)\rho_{\max}(B),$$

$$\rho_{\max}(A^{\odot 1/2} \odot B^{\odot 1/2}) \le \begin{cases} \sqrt{\rho_{\max}(A)\rho_{\max}(B)} \\ \rho_{\max}^{1/2}(AB). \end{cases}$$

Finally, if A and B are positive, then

$$\rho_{\max}(A \odot B) < \begin{cases} \rho_{\max}(A)\rho_{\max}(B) \\ \rho_{\max}(AB). \end{cases}$$

Source: [193, 697, 988, 1013, 1459, 1484, 1584, 2218, 2384] and [2991, pp. 166, 170]. Fact 9.4.18 implies $\rho_{\max}(A) = \rho_{\max}^{1/2}(A \otimes A)$. **Remark:** Extensions are given in [1272, 1481, 1871]. **Related:** $\rho_{\max}(A \odot A) \le \rho_{\max}(A \otimes A)$ follows from Fact 6.11.23 and Proposition 9.3.2.

Fact 9.6.21. Let $A, B \in \mathbb{R}^{n\times n}$, assume that A and B are nonnegative, and let $k \ge 1$. Then,

$$\operatorname{tr}(A \odot B)^{2k} \le \operatorname{tr}[(A \odot A)(B \odot B)]^k \le \operatorname{tr}(AB)^{2k}.$$

Source: [193].

Fact 9.6.22. Let $A, B \in \mathbb{R}^{n\times n}$, and assume that A and B are nonsingular M-matrices. Then, the following statements hold:

i) $A \odot B^{-1}$ is a nonsingular M-matrix.

ii) If $n = 2$, then $\rho_{\min}(A \odot A^{-1}) = 1$.

iii) If $n \ge 3$, then $\frac{1}{n} < \rho_{\min}(A \odot A^{-1}) \le 1$.

iv) $\rho_{\min}(A)d_{\min}(B^{-1}) \le \rho_{\min}(A \odot B^{-1})$.

v) $[\rho_{\min}(A)\rho_{\min}(B)]^n \le |\det(A \odot B)|$.

vi) $|(A \odot B)^{-1}| \le\le A^{-1} \odot B^{-1}$.

Source: [1450, pp. 359, 370, 375, 380]. **Related:** Fact 6.11.17 and [1481].

Fact 9.6.23. Let $A, B \in \mathbb{F}^{n\times m}$. Then,

$$\rho_{\max}(A \odot B) \le \sqrt{\rho_{\max}(A \odot \overline{A})\rho_{\max}(B \odot \overline{B})}.$$

Consequently,

$$\left.\begin{array}{r}\rho_{\max}(A \odot A)\\ \rho_{\max}(A \odot A^{\mathsf{T}})\\ \rho_{\max}(A \odot A^*)\end{array}\right\} \le \rho_{\max}(A \odot \overline{A}).$$

Source: [2441]. **Related:** Fact 11.16.45.

Fact 9.6.24. Let $A, B \in \mathbb{R}^{n\times n}$, assume that A and B are nonnegative, and let $\alpha \in [0, 1]$. Then,

$$\rho_{\max}(A^{\odot\alpha} \odot B^{\odot(1-\alpha)}) \le \rho_{\max}^{\alpha}(A)\rho_{\max}^{1-\alpha}(B).$$

In particular,

$$\rho_{\max}(A^{\odot 1/2} \odot B^{\odot 1/2}) \le \sqrt{\rho_{\max}(A)\rho_{\max}(B)}.$$

Finally,

$$\rho_{\max}(A^{\odot 1/2} \odot A^{\odot 1/2\mathsf{T}}) \le \rho_{\max}(A^{\odot\alpha} \odot A^{\odot(1-\alpha)\mathsf{T}}) \le \rho_{\max}(A).$$

Source: [2441]. **Related:** Fact 11.16.49.

Fact 9.6.25. Let $A_{11}, A_{12}, A_{21}, A_{22}, B_{11}, B_{12}, B_{21}, B_{22} \in \mathbb{F}^{n\times n}$, define

$$\mathcal{A} \triangleq \begin{bmatrix} A_{11} & A_{12}\\ A_{21} & A_{22}\end{bmatrix}, \quad \mathcal{B} \triangleq \begin{bmatrix} B_{11} & B_{12}\\ B_{21} & B_{22}\end{bmatrix},$$

and define the *block Kronecker product* of $\mathcal{A}$ and $\mathcal{B}$ by

$$\mathcal{A} \circledast \mathcal{B} \triangleq \begin{bmatrix} A_{11}\otimes B_{11} & A_{12}\otimes B_{12}\\ A_{21}\otimes B_{21} & A_{22}\otimes B_{22}\end{bmatrix}.$$

Then, the following statements hold:

i) $\mathcal{A} \odot \mathcal{B}$ is a principal submatrix of $\mathcal{A} \circledast \mathcal{B}$.

ii) $\mathcal{A} \otimes \mathcal{B}$ is a principal submatrix of $\mathcal{A} \circledast \mathcal{B}$.

iii) If A and B are positive semidefinite, then so is $\mathcal{A} \circledast \mathcal{B}$.

Source: [2991, p. 239]. **Remark:** Generalizations of the Schur and Kronecker products to partitioned matrices include the block-Kronecker, strong-Kronecker, Khatri-Rao, Tracy-Singh, and block-Hadamard products. These are discussed in [851, 1269, 1454, 1494, 1668, 1870, 1873, 1875, 1877, 2161] and [2300, pp. 216, 217]. The Tracy-Singh sum and Khatri-Rao sum are considered in [26]. **Problem:** Determine $\operatorname{rank}(\mathcal{A} \circledast \mathcal{B})$.

Fact 9.6.26. Let $A \in \mathbb{F}^{n\times m}$ and $B \in \mathbb{F}^{l\times k}$, define $p \triangleq \operatorname{lcm}\{m, l\}$, and define the *semi-tensor product* of A and B by

$$A \ltimes B \triangleq (A \otimes I_{p/m})(B \otimes I_{p/l}) \in \mathbb{F}^{np/m\times kp/l}.$$

Then, the following statements hold:

i) If $l|m$, then $A \ltimes B = A(B \otimes I_{m/l})$.

ii) If $m|l$, then $A \ltimes B = (A \otimes I_{l/m})B$.

iii) If $m = l$, then $A \ltimes B = AB$.

iv) If $\alpha,\beta \in \mathbb{F}$ and $C \in \mathbb{F}^{t\times r}$, then $A \ltimes (\alpha B + \beta C) = \alpha A \ltimes B + \beta A \ltimes C$ and $(\alpha A + \beta B) \ltimes C = \alpha A \ltimes C + \beta B \ltimes C$.

v) If $C \in \mathbb{F}^{t\times r}$, then $A \ltimes (B \ltimes C) = (A \ltimes B) \ltimes C$.

Now, assume that A and B are square. Then, the following statements hold:

vi) $\operatorname{tr}(A \ltimes B) = \operatorname{tr}(B \ltimes A)$.

vii) If $l|m$, then $\det(A \ltimes B) = (\det A)(\det B)^{m/l}$.

viii) If $m|l$, then $\det(A \ltimes B) = (\det A)^{l/m} \det B$.

ix) If $m = l$, then $\det(A \ltimes B) = \det AB = (\det A)\det B$.

x) If either A or B is nonsingular, then A and B are similar.

xi) If A and B are nonsingular, then $A \ltimes B$ is nonsingular and $(A \ltimes B)^{-1} = B^{-1} \ltimes A^{-1}$.

Source: [715, 716].

9.7 Notes

A history of the Kronecker product is given in [1365]. Kronecker matrix algebra is discussed in [572, 1218, 1367, 1924, 2029, 2522, 2797]. Applications are discussed in [2302, 2303, 2777].

The fact that the Schur product is a principal submatrix of the Kronecker product is noted in [1951]. A variation of Kronecker matrix algebra for symmetric matrices can be developed in terms of the half-vectorization operator "vech" and the associated elimination and duplication matrices [1367, 1923, 2747, 2748].

The Schur product is also called the Hadamard product.

The Kronecker product is associated with tensor analysis and multilinear algebra; see [777, Chapter 10] and [921, 1159, 1227, 1947, 1948, 2029].

Chapter Ten
Positive-Semidefinite Matrices

This chapter focuses on positive-semidefinite and positive-definite matrices, which are defined in Chapter 4 in terms of quadratic forms. These matrices arise in numerous applications, such as covariance analysis in signal processing and controllability analysis in linear system theory, and they have many special properties.

10.1 Positive-Semidefinite and Positive-Definite Orderings

Let $A \in \mathbb{F}^{n\times n}$ be a Hermitian matrix. As shown in Corollary 7.5.5, A is unitarily similar to a real diagonal matrix whose diagonal entries are the eigenvalues of A. We denote these eigenvalues by $\lambda_1(A), \ldots, \lambda_n(A)$, where

$$\lambda_n(A) \le \cdots \le \lambda_1(A), \tag{10.1.1}$$

and, in addition,

$$\lambda_{\min}(A) \triangleq \lambda_n(A), \quad \lambda_{\max}(A) \triangleq \lambda_1(A). \tag{10.1.2}$$

Then, A is positive semidefinite if and only if $\lambda_{\min}(A) \ge 0$, whereas A is positive definite if and only if $\lambda_{\min}(A) > 0$.

Let $\mathbf{H}^n$, $\mathbf{N}^n$, and $\mathbf{P}^n$ denote, respectively, the Hermitian, positive-semidefinite, and positive-definite matrices in $\mathbb{F}^{n\times n}$. If $\mathbb{F}$ is not specified, then $\mathbb{F} = \mathbb{C}$. Hence, $\mathbf{P}^n \subset \mathbf{N}^n \subset \mathbf{H}^n$. If $A \in \mathbf{N}^n$, then we write $A \ge 0$, whereas, if $A \in \mathbf{P}^n$, then we write $A > 0$. If $A, B \in \mathbf{H}^n$, then $A - B \in \mathbf{N}^n$ is possible in the case where neither A nor B is positive semidefinite. In this case, we write $A \ge B$ or $B \le A$. Similarly, $A - B \in \mathbf{P}^n$ is denoted by $A > B$ or $B < A$. This notation is consistent with the case $n = 1$, where $\mathbf{H}^1 = \mathbb{R}$, $\mathbf{N}^1 = [0, \infty)$, and $\mathbf{P}^1 = (0, \infty)$.

Since $0 \in \mathbf{N}^n$, it follows that $\mathbf{N}^n$ is a pointed cone. Furthermore, if $A, -A \in \mathbf{N}^n$, then $x^*Ax = 0$ for all $x \in \mathbb{F}^n$, which implies that $A = 0$. Hence, $\mathbf{N}^n$ is a one-sided cone. Finally, $\mathbf{N}^n$ is a convex cone since, if $A, B \in \mathbf{N}^n$ and $\alpha, \beta > 0$, then $\alpha A + \beta B \in \mathbf{N}^n$. Likewise, $\mathbf{P}^n$ is a convex cone. The following result shows that the relation "$\le$" is a partial ordering on $\mathbf{H}^n$.

Proposition 10.1.1. The relation "$\le$" is reflexive, antisymmetric, and transitive on $\mathbf{H}^n$; that is, if $A, B, C \in \mathbf{H}^n$, then the following statements hold:

i) $A \le A$.

ii) If $A \le B$ and $B \le A$, then $A = B$.

iii) If $A \le B$ and $B \le C$, then $A \le C$.

Proof. Since $\mathbf{N}^n$ is a pointed, one-sided, convex cone, it follows from Proposition 3.1.7 that the relation "$\le$" is reflexive, antisymmetric, and transitive. □

Additional properties of "$\le$" and "$<$" are given by the following result.

Proposition 10.1.2. Let $A, B, C, D \in \mathbf{H}^n$. Then, the following statements hold:

i) If $A \ge 0$, then $\alpha A \ge 0$ for all $\alpha \ge 0$, and $\alpha A \le 0$ for all $\alpha \le 0$.

ii) If $A > 0$, then $\alpha A > 0$ for all $\alpha > 0$, and $\alpha A < 0$ for all $\alpha < 0$.

iii) $\alpha A + \beta B \in \mathbf{H}^n$ for all $\alpha, \beta \in \mathbb{R}$.

iv) If $A \geq 0$ and $B \geq 0$, then $\alpha A + \beta B \geq 0$ for all $\alpha, \beta \geq 0$.

v) If $A \geq 0$ and $B > 0$, then $A + B > 0$.

vi) $A^2 \geq 0$.

vii) $A^2 > 0$ if and only if $\det A \neq 0$.

viii) If $A \leq B$ and $B < C$, then $A < C$.

ix) If $A < B$ and $B \leq C$, then $A < C$.

x) If $A \leq B$ and $C \leq D$, then $A + C \leq B + D$.

xi) If $A \leq B$ and $C < D$, then $A + C < B + D$.

Furthermore, let $S \in \mathbb{F}^{m \times n}$. Then, the following statements hold:

xii) If $A \leq B$, then $SAS^* \leq SBS^*$.

xiii) If $A < B$ and $\operatorname{rank} S = m$, then $SAS^* < SBS^*$.

xiv) If $SAS^* \leq SBS^*$ and $\operatorname{rank} S = n$, then $A \leq B$.

xv) If $SAS^* < SBS^*$ and $\operatorname{rank} S = n$, then $m = n$ and $A < B$.

xvi) Assume that $A \leq B$. Then, $SAS^* < SBS^*$ if and only if $\operatorname{rank} S = m$ and $\mathcal{R}(S^*) \cap \mathcal{N}(B - A) = \{0\}$.

Proof. *i*)–*xi*) are immediate. To prove *xiii*), note that $A < B$ implies that $(B - A)^{1/2}$ is positive definite. Thus, $\operatorname{rank} S(A - B)^{1/2} = m$, which implies that $S(A - B)S^*$ is positive definite. To prove *xiv*), note that, since $\operatorname{rank} S = n$, it follows that S has a left inverse $S^{\mathrm{L}} \in \mathbb{F}^{n \times m}$. Thus, *xii*) implies that $A = S^{\mathrm{L}}SAS^*S^{\mathrm{L}*} \leq S^{\mathrm{L}}SBS^*S^{\mathrm{L}*} = B$. To prove *xv*), note that, since $S(B - A)S^*$ is positive definite, it follows that $\operatorname{rank} S = m$. Hence, $m = n$ and S is nonsingular. Thus, *xiii*) implies that $A = S^{-1}SAS^*S^{-*} < S^{-1}SBS^*S^{-*} = B$. *xvi*) is proved in [627]. □

The following result is an immediate consequence of Theorem 7.5.7.

Corollary 10.1.3. Let $A, B \in \mathbf{H}^n$, and assume that A and B are congruent. Then, A is positive semidefinite if and only if B is positive semidefinite. Furthermore, A is positive definite if and only if B is positive definite.

10.2 Submatrices and Schur Complements

We first consider some factorizations of a partitioned Hermitian matrix. Note that $A_{11}|A$ and $A_{22}|A$ are Schur complements defined by Definition 8.1.13.

Proposition 10.2.1. Let $A = \begin{bmatrix} A_{11} & A_{12} \\ A_{12}^* & A_{22} \end{bmatrix} \in \mathbf{H}^{n+m}$. If

$$A_{12} = A_{11}A_{11}^+A_{12}, \tag{10.2.1}$$

then

$$\begin{bmatrix} A_{11} & A_{12} \\ A_{12}^* & A_{22} \end{bmatrix} = \begin{bmatrix} I & 0 \\ A_{12}^*A_{11}^+ & I \end{bmatrix} \begin{bmatrix} A_{11} & 0 \\ 0 & A_{11}|A \end{bmatrix} \begin{bmatrix} I & A_{11}^+A_{12} \\ 0 & I \end{bmatrix}, \tag{10.2.2}$$

where

$$A_{11}|A = A_{22} - A_{12}^*A_{11}^+A_{12}. \tag{10.2.3}$$

Alternatively, if

$$A_{12} = A_{12}A_{22}^+A_{22}, \tag{10.2.4}$$

then

$$\begin{bmatrix} A_{11} & A_{12} \\ A_{12}^* & A_{22} \end{bmatrix} = \begin{bmatrix} I & A_{12}A_{22}^+ \\ 0 & I \end{bmatrix} \begin{bmatrix} A_{22}|A & 0 \\ 0 & A_{22} \end{bmatrix} \begin{bmatrix} I & 0 \\ A_{22}^+A_{12}^* & I \end{bmatrix}, \tag{10.2.5}$$

where

$$A_{22}|A = A_{11} - A_{12}A_{22}^{+}A_{12}^{*}. \tag{10.2.6}$$

The following result shows that, if A is positive semidefinite, then (10.2.1) and (10.2.4) hold.

Lemma 10.2.2. Let $A = \begin{bmatrix} A_{11} & A_{12} \\ A_{12}^{*} & A_{22} \end{bmatrix} \in \mathbf{N}^{n+m}$. Then, (10.2.1) and (10.2.4) are satisfied.

Proof. Since $A \geq 0$, it follows from Corollary 7.5.5 that $A = BB^{*}$, where $B = \begin{bmatrix} B_1 \\ B_2 \end{bmatrix} \in \mathbb{F}^{(n+m)\times r}$ and $r \triangleq \operatorname{rank} A$. Thus, $A_{11} = B_1B_1^{*}$, $A_{12} = B_1B_2^{*}$, and $A_{22} = B_2B_2^{*}$. Since A_{11} is Hermitian, it follows from *xxvii)* of Proposition 8.1.7 that A_{11}^{+} is also Hermitian. Next, defining $S \triangleq B_1 - B_1B_1^{*}(B_1B_1^{*})^{+}B_1$, it follows that $SS^{*} = 0$, and thus $\operatorname{tr} SS^{*} = 0$. Hence, Lemma 3.3.3 implies that $S = 0$, and thus $B_1 = B_1B_1^{*}(B_1B_1^{*})^{+}B_1$. Consequently, $B_1B_2^{*} = B_1B_1^{*}(B_1B_1^{*})^{+}B_1B_2^{*}$; that is, (10.2.1) holds. The proof of (10.2.4) is analogous. □

Corollary 10.2.3. Let $A = \begin{bmatrix} A_{11} & A_{12} \\ A_{12}^{*} & A_{22} \end{bmatrix} \in \mathbf{N}^{n+m}$. Then, the following statements hold:

i) $\mathcal{R}(A_{12}) \subseteq \mathcal{R}(A_{11})$.

ii) $\mathcal{R}(A_{12}^{*}) \subseteq \mathcal{R}(A_{22})$.

iii) $\operatorname{rank}\,[A_{11} \;\; A_{12}] = \operatorname{rank} A_{11}$.

iv) $\operatorname{rank}\,[A_{12}^{*} \;\; A_{22}] = \operatorname{rank} A_{22}$.

v) $A_{11}|A$ and $A_{22}|A$ are positive semidefinite.

Proof. *i)* and *ii)* follow from (10.2.1) and (10.2.4); *iii)* and *iv)* are consequences of *i)* and *ii)*; and *v)* follows from (10.2.2), (10.2.5), and *xiv)* of Proposition 10.1.2. □

Proposition 10.2.4. Let $A \triangleq \begin{bmatrix} A_{11} & A_{12} \\ A_{12}^{*} & A_{22} \end{bmatrix} \in \mathbf{N}^{n+m}$. Then,

$$\operatorname{rank} A = \operatorname{rank} A_{11} + \operatorname{rank} A_{11}|A \tag{10.2.7}$$

$$= \operatorname{rank} A_{22}|A + \operatorname{rank} A_{22} \tag{10.2.8}$$

$$\leq \operatorname{rank} A_{11} + \operatorname{rank} A_{22}. \tag{10.2.9}$$

Furthermore,

$$\det A = (\det A_{11})\det(A_{11}|A), \tag{10.2.10}$$

$$\det A = (\det A_{22})\det(A_{22}|A). \tag{10.2.11}$$

Proposition 10.2.5. Let $A \triangleq \begin{bmatrix} A_{11} & A_{12} \\ A_{12}^{*} & A_{22} \end{bmatrix} \in \mathbf{H}^{n+m}$. Then, the following statements are equivalent:

i) $A \geq 0$.

ii) $A_{11} \geq 0, A_{12} = A_{11}A_{11}^{+}A_{12}$, and $A_{12}^{*}A_{11}^{+}A_{12} \leq A_{22}$.

iii) $A_{22} \geq 0$, $A_{12} = A_{12}A_{22}^{+}A_{22}$, and $A_{12}A_{22}^{+}A_{12}^{*} \leq A_{11}$.

The following statements are also equivalent:

iv) $A > 0$.

v) $A_{11} > 0$ and $A_{12}^{*}A_{11}^{-1}A_{12} < A_{22}$.

vi) $A_{22} > 0$ and $A_{12}A_{22}^{-1}A_{12}^{*} < A_{11}$.

The following result follows from (3.9.17) and (3.9.18) or from (10.2.2) and (10.2.5).

Proposition 10.2.6. Let $A \triangleq \begin{bmatrix} A_{11} & A_{12} \\ A_{12}^{*} & A_{22} \end{bmatrix} \in \mathbf{H}^{n+m}$. Then, the following statements hold:

i) Assume that A_{11} is nonsingular. Then, A is nonsingular if and only if $A_{11}|A$ is nonsingular. If these conditions hold, then

$$A^{-1} = \begin{bmatrix} A_{11}^{-1} + A_{11}^{-1}A_{12}(A_{11}|A)^{-1}A_{12}^{*}A_{11}^{-1} & -A_{11}^{-1}A_{12}(A_{11}|A)^{-1} \\ -(A_{11}|A)^{-1}A_{12}^{*}A_{11}^{-1} & (A_{11}|A)^{-1} \end{bmatrix}, \tag{10.2.12}$$

where

$$A_{11}|A = A_{22} - A_{12}^*A_{11}^{-1}A_{12}. \tag{10.2.13}$$

ii) Assume that A_{22} is nonsingular. Then, A is nonsingular if and only if $A_{22}|A$ is nonsingular. If these conditions hold, then

$$A^{-1} = \begin{bmatrix} (A_{22}|A)^{-1} & -(A_{22}|A)^{-1}A_{12}A_{22}^{-1} \\ -A_{22}^{-1}A_{12}^*(A_{22}|A)^{-1} & A_{22}^{-1}A_{12}^*(A_{22}|A)^{-1}A_{12}A_{22}^{-1} + A_{22}^{-1} \end{bmatrix}, \tag{10.2.14}$$

where

$$A_{22}|A = A_{11} - A_{12}A_{22}^{-1}A_{12}^*. \tag{10.2.15}$$

iii) Assume that A is nonsingular, and let $A^{-1} = \begin{bmatrix} B_{11} & B_{12} \\ B_{12}^* & B_{22} \end{bmatrix} \in \mathbf{H}^{n+m}$. Then,

$$B_{11}|A^{-1} = A_{22}^{-1}, \quad B_{22}|A^{-1} = A_{11}^{-1}. \tag{10.2.16}$$

iv) If A is positive definite, then A is nonsingular and A^{-1}, A_{11}, A_{22}, $A_{11}|A$, and $A_{22}|A$ are positive definite.

Lemma 10.2.7. Let $A \in \mathbb{F}^{n\times n}$, $b \in \mathbb{F}^n$, and $a \in \mathbb{R}$, and define $\mathcal{A} \triangleq \begin{bmatrix} A & b \\ b^* & a \end{bmatrix}$. Then, the following statements are equivalent:

i) $\mathcal{A}$ is positive semidefinite.

ii) A is positive semidefinite, $b = AA^+b$, and $b^*A^+b \le a$.

iii) Either A is positive semidefinite, $a = 0$, and $b = 0$ or $a > 0$ and $bb^* \le aA$.

Furthermore, the following statements are equivalent:

iv) $\mathcal{A}$ is positive definite.

v) A is positive definite, and $b^*A^{-1}b < a$.

vi) $a > 0$ and $bb^* < aA$.

If *iv)*–*vi)* hold, then

$$\det \mathcal{A} = (\det A)(a - b^*A^{-1}b). \tag{10.2.17}$$

For the following result note that every matrix is a principal submatrix of itself, while the determinant of a matrix is also a principal subdeterminant of the matrix.

Proposition 10.2.8. Let $A \in \mathbf{H}^n$. Then, $\chi_A \in \mathbb{R}[s]$. Furthermore, the following statements are equivalent:

i) A is positive semidefinite.

ii) Every principal submatrix of A is positive semidefinite.

iii) Every principal subdeterminant of A is nonnegative.

iv) For all $i \in \{1,\dots,n\}$, the sum of all $i\times i$ principal subdeterminants of A is nonnegative.

v) For all $i \in \{1,\dots,n-1\}$, $(-1)^{n-i}\beta_i \ge 0$, where $\chi_A(s) = s^n + \beta_{n-1}s^{n-1} + \cdots + \beta_1 s + \beta_0$.

Proof. To prove *i)* $\Longrightarrow$ *ii)*, let $\hat{A} \in \mathbb{F}^{m\times m}$ be the principal submatrix of A obtained from A by retaining rows and columns $i_1,\dots,i_m$. Then, $\hat{A} = S^\mathsf{T}AS$, where $S \triangleq [e_{i_1} \ \cdots \ e_{i_m}] \in \mathbb{R}^{n\times m}$. Now, let $\hat{x} \in \mathbb{F}^m$. Since A is positive semidefinite, it follows that $\hat{x}^*\hat{A}\hat{x} = \hat{x}^*S^\mathsf{T}AS\hat{x} \ge 0$, and thus $\hat{A}$ is positive semidefinite. Next, *ii)* $\Longrightarrow$ *iii)* $\Longrightarrow$ *iv)* are immediate. To prove *iv)* $\Longrightarrow$ *i)*, Proposition 6.4.6 implies

$$\chi_A(s) = \sum_{i=0}^n \beta_i s^i = \sum_{i=0}^n (-1)^{n-i}\gamma_{n-i}s^i = (-1)^n \sum_{i=0}^n \gamma_{n-i}(-s)^i, \tag{10.2.18}$$

where, for all $i \in \{1,\dots,n\}$, γ_i is the sum of all $i\times i$ principal subdeterminants of A, and $\beta_n = \gamma_0 = 1$. By assumption, $\gamma_i \ge 0$ for all $i \in \{1,\dots,n\}$. Now, suppose there exists $\lambda \in \operatorname{spec}(A)$ such that $\lambda < 0$.

Then, $0 = (-1)^n\chi_A(\lambda) = \sum_{i=0}^{n}\gamma_{n-i}(-\lambda)^i > 0$, which is a contradiction. Hence, $\mathrm{spec}(A) \subset [0,\infty)$. The equivalence of *iv*) and *v*) follows from Proposition 6.4.6. □

Proposition 10.2.9. Let $A \in \mathbf{H}^n$. Then, the following statements are equivalent:

i) A is positive definite.

ii) Every principal submatrix of A is positive definite.

iii) Every principal subdeterminant of A is positive.

iv) Every leading principal submatrix of A is positive definite.

v) Every leading principal subdeterminant of A is positive.

vi) For all $i \in \{1,\ldots,n\}$, the sum of all $i \times i$ principal subdeterminants of A is positive.

vii) For all $i \in \{1,\ldots,n-1\}$, $(-1)^{n-i}\beta_i > 0$, where $\chi_A(s) = s^n + \beta_{n-1}s^{n-1} + \cdots + \beta_1 s + \beta_0$.

Proof. To prove *i*) $\Longrightarrow$ *ii*), let $\hat{A} \in \mathbb{F}^{m\times m}$ and S be as in the proof of Proposition 10.2.8, and let $\hat{x}$ be nonzero so that $S\hat{x}$ is nonzero. Since A is positive definite, it follows that $\hat{x}^*\hat{A}\hat{x} = \hat{x}^*S^{\mathsf{T}}AS\hat{x} > 0$, and hence $\hat{A}$ is positive definite.

Next, *ii*) $\Longrightarrow$ *iii*) $\Longrightarrow$ *v*) and *ii*) $\Longrightarrow$ *iv*) $\Longrightarrow$ *v*) are immediate. To prove *v*) $\Longrightarrow$ *i*), suppose that the leading principal submatrix $A_i \in \mathbb{F}^{i\times i}$ has positive determinant for all $i \in \{1,\ldots,n\}$. The result is true for $n = 1$. For $n \ge 2$, we show that, if A_i is positive definite, then so is A_{i+1}. Writing $A_{i+1} = \begin{bmatrix} A_i & b_i \\ b_i^* & a_i \end{bmatrix}$, it follows from Lemma 10.2.7 that $\det A_{i+1} = (\det A_i)(a_i - b_i^*A_i^{-1}b_i) > 0$, and hence $a_i - b_i^*A_i^{-1}b_i = \det A_{i+1}/\det A_i > 0$. Lemma 10.2.7 now implies that A_{i+1} is positive definite. Using this argument for all $i \in \{2,\ldots,n\}$ implies that A is positive definite. □

The example $A = \begin{bmatrix} 0 & 0 \\ 0 & -1 \end{bmatrix}$ shows that every principal subdeterminant of A, rather than just the leading principal subdeterminants of A, must be checked to determine whether A is positive semidefinite. A less obvious example is $A = \begin{bmatrix} 1 & 1 & 1 \\ 1 & 1 & 1 \\ 1 & 1 & 0 \end{bmatrix}$, whose eigenvalues are 0, $1+\sqrt{3}$, and $1-\sqrt{3}$. In this case, the principal subdeterminant $\det A_{[1,1]} = \det \begin{bmatrix} 1 & 1 \\ 1 & 0 \end{bmatrix} < 0$.

Note that *iii*) of Proposition 10.2.9 includes $\det A > 0$ since the determinant of A is also a subdeterminant of A. The matrix $A = \begin{bmatrix} 3/2 & -1 & 1 \\ -1 & 2 & 1 \\ 1 & 1 & 2 \end{bmatrix}$ has the property that every 1×1 and 2×2 subdeterminant is positive but is not positive definite. This example shows, if the requirement that the determinant of A be positive is omitted, then *iii*) $\Longrightarrow$ *ii*) of Proposition 10.2.9 is false.

10.3 Simultaneous Diagonalization

This section considers the simultaneous diagonalization of a pair of matrices $A, B \in \mathbf{H}^n$. There are two types of simultaneous diagonalization. *Cogredient diagonalization* involves a nonsingular matrix $S \in \mathbb{F}^{n\times n}$ such that SAS^* and SBS^* are both diagonal, whereas *contragredient diagonalization* involves a nonsingular matrix $S \in \mathbb{F}^{n\times n}$ such that SAS^* and $S^{-*}BS^{-1}$ are both diagonal. Both types of simultaneous transformation involve congruence transformations. We begin by assuming that one of the matrices is positive definite. The first result is cogredient diagonalization.

Theorem 10.3.1. Let $A, B \in \mathbf{H}^n$, and assume that A is positive definite. Then, there exists a nonsingular matrix $S \in \mathbb{F}^{n\times n}$ such that $SAS^* = I$ and SBS^* is diagonal.

Proof. Setting $S_1 = A^{-1/2}$, it follows that $S_1AS_1^* = I$. Now, since $S_1BS_1^*$ is Hermitian, Corollary 7.5.5 implies that there exists a unitary matrix $S_2 \in \mathbb{F}^{n\times n}$ such that $SBS^* = S_2S_1BS_1^*S_2^*$ is diagonal, where $S = S_2S_1$. Finally, $SAS^* = S_2S_1AS_1^*S_2^* = S_2IS_2^* = I$. □

An analogous result holds for contragredient diagonalization.

Theorem 10.3.2. Let $A, B \in \mathbf{H}^n$, and assume that A is positive definite. Then, there exists a nonsingular matrix $S \in \mathbb{F}^{n\times n}$ such that $SAS^* = I$ and $S^{-*}BS^{-1}$ is diagonal.

Proof. Setting $S_1 = A^{-1/2}$, it follows that $S_1AS_1^* = I$. Since $S_1^{-*}BS_1^{-1}$ is Hermitian, it follows that there exists a unitary matrix $S_2 \in \mathbb{F}^{n\times n}$ such that $S^{-*}BS^{-1} = S_2^{-*}S_1^{-*}BS_1^{-1}S_2^{-1} = S_2(S_1^{-*}BS_1^{-1})S_2^*$ is diagonal, where $S = S_2S_1$. Finally, $SAS^* = S_2S_1AS_1^*S_2^* = S_2IS_2^* = I$. □

Corollary 10.3.3. Let $A, B \in \mathbf{H}^n$, and assume that A is positive definite. Then, AB is diagonalizable over $\mathbb{F}$, every eigenvalue of AB is real, and $\operatorname{In}(AB) = \operatorname{In}(B)$.

Corollary 10.3.4. Let $A, B \in \mathbf{P}^n$. Then, there exists a nonsingular matrix $S \in \mathbb{F}^{n\times n}$ such that SAS^* and $S^{-*}BS^{-1}$ are equal and diagonal.

Proof. By Theorem 10.3.2 there exists a nonsingular matrix $S_1 \in \mathbb{F}^{n\times n}$ such that $S_1AS_1^* = I$ and $B_1 = S_1^{-*}BS_1^{-1}$ is diagonal. Defining $S \triangleq B_1^{1/4}S_1$ yields $SAS^* = S^{-*}BS^{-1} = B_1^{1/2}$. □

The transformation S of Corollary 10.3.4 is a *balancing transformation*. See Definition 16.9.28.

Next, we weaken the requirement in Theorem 10.3.1 and Theorem 10.3.2 that A be positive definite by assuming only that A is positive semidefinite. In this case, however, we assume that B is also positive semidefinite.

Theorem 10.3.5. Let $A, B \in \mathbf{N}^n$. Then, there exists a nonsingular matrix $S \in \mathbb{F}^{n\times n}$ such that $SAS^* = \begin{bmatrix} I & 0 \\ 0 & 0 \end{bmatrix}$ and SBS^* is diagonal.

Proof. Let the nonsingular matrix $S_1 \in \mathbb{F}^{n\times n}$ satisfy $S_1AS_1^* = \begin{bmatrix} I & 0 \\ 0 & 0 \end{bmatrix}$, and similarly partition $S_1BS_1^* = \begin{bmatrix} B_{11} & B_{12} \\ B_{12}^* & B_{22} \end{bmatrix}$, which is positive semidefinite. Letting $S_2 \triangleq \begin{bmatrix} I & -B_{12}B_{22}^+ \\ 0 & I \end{bmatrix}$, it follows from Lemma 10.2.2 that

$$S_2S_1BS_1^*S_2^* = \begin{bmatrix} B_{11} - B_{12}B_{22}^+B_{12}^* & 0 \\ 0 & B_{22} \end{bmatrix}.$$

Next, let U_1 and U_2 be unitary matrices such that $U_1(B_{11} - B_{12}B_{22}^+B_{12}^*)U_1^*$ and $U_2B_{22}U_2^*$ are diagonal. Then, defining $S_3 \triangleq \begin{bmatrix} U_1 & 0 \\ 0 & U_2 \end{bmatrix}$ and $S \triangleq S_3S_2S_1$, it follows that $SAS^* = \begin{bmatrix} I & 0 \\ 0 & 0 \end{bmatrix}$ and $SBS^* = S_3S_2S_1BS_1^*S_2^*S_3^*$ is diagonal. □

Theorem 10.3.6. Let $A, B \in \mathbf{N}^n$. Then, there exists a nonsingular matrix $S \in \mathbb{F}^{n\times n}$ such that $SAS^* = \begin{bmatrix} I & 0 \\ 0 & 0 \end{bmatrix}$ and $S^{-*}BS^{-1}$ is diagonal.

Proof. Let $S_1 \in \mathbb{F}^{n\times n}$ be a nonsingular matrix such that $S_1AS_1^* = \begin{bmatrix} I & 0 \\ 0 & 0 \end{bmatrix}$, and similarly partition $S_1^{-*}BS_1^{-1} = \begin{bmatrix} B_{11} & B_{12} \\ B_{12}^* & B_{22} \end{bmatrix}$, which is positive semidefinite. Letting $S_2 \triangleq \begin{bmatrix} I & B_{11}^+B_{12} \\ 0 & I \end{bmatrix}$, it follows that

$$S_2^{-*}S_1^{-*}BS_1^{-1}S_2^{-1} = \begin{bmatrix} B_{11} & 0 \\ 0 & B_{22} - B_{12}^*B_{11}^+B_{12} \end{bmatrix}.$$

Now, let U_1 and U_2 be unitary matrices such that the matrices $U_1B_{11}U_1^*$ and $U_2(B_{22} - B_{12}^*B_{11}^+B_{12})U_2^*$ are diagonal. Then, defining $S_3 \triangleq \begin{bmatrix} U_1 & 0 \\ 0 & U_2 \end{bmatrix}$ and $S \triangleq S_3S_2S_1$, it follows that $SAS^* = \begin{bmatrix} I & 0 \\ 0 & 0 \end{bmatrix}$ and $S^{-*}BS^{-1} = S_3^{-*}S_2^{-*}S_1^{-*}BS_1^{-1}S_2^{-1}S_3^{-1}$ is diagonal. □

Corollary 10.3.7. Let $A, B \in \mathbf{N}^n$. Then, AB is diagonalizable over $\mathbb{F}$, and every eigenvalue of AB is nonnegative. If, in addition, A and B are positive definite, then every eigenvalue of AB is positive.

Proof. It follows from Theorem 10.3.6 that there exists a nonsingular matrix $S \in \mathbb{F}^{n\times n}$ such that $A_1 = SAS^*$ and $B_1 = S^{-*}BS^{-1}$ are diagonal with nonnegative diagonal entries. Hence, $AB = S^{-1}A_1B_1S$ is semisimple and all of its eigenvalues are nonnegative. □

A more direct approach to showing that AB has nonnegative eigenvalues is to use Corollary 6.4.11 and note that $\lambda_i(AB) = \lambda_i(B^{1/2}AB^{1/2}) \geq 0$.

Corollary 10.3.8. Let $A, B \in \mathbf{N}^n$, and assume that $\operatorname{rank} A = \operatorname{rank} B = \operatorname{rank} AB$. Then, there exists a nonsingular matrix $S \in \mathbb{F}^{n\times n}$ such that $SAS^* = S^{-*}BS^{-1}$ and SAS^* is diagonal.

Proof. By Theorem 10.3.6 there exists a nonsingular matrix $S_1 \in \mathbb{F}^{n\times n}$ such that $S_1AS_1^* = \begin{bmatrix} I_r & 0 \\ 0 & 0 \end{bmatrix}$, where $r \triangleq \operatorname{rank} A$, and such that $B_1 = S_1^{-*}BS_1^{-1}$ is diagonal. Hence, $AB = S_1^{-1}\begin{bmatrix} I_r & 0 \\ 0 & 0 \end{bmatrix}B_1S_1$. Since $\operatorname{rank} A = \operatorname{rank} B = \operatorname{rank} AB = r$, it follows that $B_1 = \begin{bmatrix} \hat{B}_1 & 0 \\ 0 & 0 \end{bmatrix}$, where $\hat{B}_1 \in \mathbb{F}^{r\times r}$ is a diagonal matrix all of whose diagonally located entries are positive. Hence, $S_1^{-*}BS_1^{-1} = \begin{bmatrix} \hat{B}_1 & 0 \\ 0 & 0 \end{bmatrix}$. Now, define $S_2 \triangleq \begin{bmatrix} \hat{B}_1^{1/4} & 0 \\ 0 & I \end{bmatrix}$ and $S \triangleq S_2S_1$. Then, $SAS^* = S_2S_1AS_1^*S_2^* = \begin{bmatrix} \hat{B}_1^{1/2} & 0 \\ 0 & 0 \end{bmatrix} = S_2^{-*}S_1^{-*}BS_1^{-1}S_2^{-1} = S^{-*}BS^{-1}$. □

10.4 Eigenvalue Inequalities

Next, we turn our attention to inequalities for eigenvalues. We begin with several lemmas.

Lemma 10.4.1. Let $A \in \mathbf{H}^n$ and $\beta \in \mathbb{R}$. Then, the following statements hold:

i) $\beta I \le A$ if and only if $\beta \le \lambda_{\min}(A)$.

ii) $\beta I < A$ if and only if $\beta < \lambda_{\min}(A)$.

iii) $A \le \beta I$ if and only if $\lambda_{\max}(A) \le \beta$.

iv) $A < \beta I$ if and only if $\lambda_{\max}(A) < \beta$.

Proof. To prove *i)*, assume that $\beta I \le A$, and let $S \in \mathbb{F}^{n\times n}$ be a unitary matrix such that $B = SAS^*$ is diagonal. Then, $\beta I \le B$, which yields $\beta \le \lambda_{\min}(B) = \lambda_{\min}(A)$. Conversely, let $S \in \mathbb{F}^{n\times n}$ be a unitary matrix such that $B = SAS^*$ is diagonal. Since the diagonal entries of B are the eigenvalues of A, it follows that $\lambda_{\min}(A)I \le B$, which implies that $\beta I \le \lambda_{\min}(A)I \le S^*BS = A$. *ii)*, *iii)*, and *iv)* are proved in a similar manner. □

Corollary 10.4.2. Let $A \in \mathbf{H}^n$. Then,

$$\lambda_{\min}(A)I \le A \le \lambda_{\max}(A)I. \tag{10.4.1}$$

Proof. Use *i)* and *iii)* of Lemma 10.4.1 with $\beta = \lambda_{\min}(A)$ and $\beta = \lambda_{\max}(A)$, respectively. □

The following result concerns the maximum and minimum values of the *Rayleigh quotient*.

Lemma 10.4.3. Let $A \in \mathbf{H}^n$. Then,

$$\lambda_{\min}(A) = \min_{x\in\mathbb{F}^n\backslash\{0\}} \frac{x^*Ax}{x^*x}, \qquad \lambda_{\max}(A) = \max_{x\in\mathbb{F}^n\backslash\{0\}} \frac{x^*Ax}{x^*x}. \tag{10.4.2}$$

Proof. It follows from (10.4.1) that $\lambda_{\min}(A) \le x^*Ax/x^*x$ for all nonzero $x \in \mathbb{F}^n$. Letting $x \in \mathbb{F}^n$ be an eigenvector of A associated with $\lambda_{\min}(A)$, it follows that this lower bound is attained. An analogous argument yields the second equality. □

The following result is the *Cauchy interlacing theorem*.

Lemma 10.4.4. Let $A \in \mathbf{H}^n$, and let A_0 be an $(n-1)\times(n-1)$ principal submatrix of A. Then, for all $i \in \{1,\dots,n-1\}$,

$$\lambda_{i+1}(A) \le \lambda_i(A_0) \le \lambda_i(A). \tag{10.4.3}$$

Proof. Note that (10.4.3) is the chain of inequalities

$$\lambda_n(A) \le \lambda_{n-1}(A_0) \le \lambda_{n-1}(A) \le \cdots \le \lambda_2(A) \le \lambda_1(A_0) \le \lambda_1(A).$$

Suppose that this chain of inequalities does not hold. In particular, first suppose that the rightmost inequality that is not true is $\lambda_j(A_0) \le \lambda_j(A)$, so that $\lambda_j(A) < \lambda_j(A_0)$. Choose δ such that $\lambda_j(A) < \delta < \lambda_j(A_0)$ and such that δ is not an eigenvalue of A_0. If $j = 1$, then $A - \delta I$ is negative definite, while, if $j \ge 2$, then $\lambda_j(A) < \delta < \lambda_j(A_0) \le \lambda_{j-1}(A_0) \le \lambda_{j-1}(A)$, so that $A - \delta I$ has $j-1$ positive eigenvalues. Thus, $\nu_+(A - \delta I) = j - 1$. Furthermore, since $\delta < \lambda_j(A_0)$, it follows that $\nu_+(A_0 - \delta I) \ge j$.

Now, assume for convenience that the rows and columns of A are ordered so that A_0 is the $(n-1)\times(n-1)$ leading principal submatrix of A. Thus, $A = \begin{bmatrix} A_0 & \beta \\ \beta^* & \gamma \end{bmatrix}$, where $\beta \in \mathbb{F}^{n-1}$ and $\gamma \in \mathbb{F}$. Next, note the equality

$$A - \delta I = \begin{bmatrix} I & 0 \\ \beta^*(A_0-\delta I)^{-1} & 1 \end{bmatrix} \begin{bmatrix} A_0 - \delta I & 0 \\ 0 & \gamma - \delta - \beta^*(A_0-\delta I)^{-1}\beta \end{bmatrix} \begin{bmatrix} I & (A_0-\delta I)^{-1}\beta \\ 0 & 1 \end{bmatrix},$$

where $A_0 - \delta I$ is nonsingular since δ is chosen to not be an eigenvalue of A_0. Since the right-hand side of this equality involves a congruence transformation, and since $\nu_+(A_0 - \delta I) \ge j$, it follows from Theorem 7.5.7 that $\nu_+(A - \delta I) \ge j$. However, this inequality contradicts the fact that $\nu_+(A - \delta I) = j - 1$.

Finally, suppose that the right-most inequality in (10.4.3) that is not true is $\lambda_{j+1}(A) \le \lambda_j(A_0)$, so that $\lambda_j(A_0) < \lambda_{j+1}(A)$. Choose δ such that $\lambda_j(A_0) < \delta < \lambda_{j+1}(A)$ and such that δ is not an eigenvalue of A_0. Then, it follows that $\nu_+(A - \delta I) \ge j + 1$ and $\nu_+(A_0 - \delta I) = j - 1$. Using the congruence transformation as in the previous case, it follows that $\nu_+(A - \delta I) \le j$, which contradicts the fact that $\nu_+(A - \delta I) \ge j + 1$. □

The following result is the *eigenvalue interlacing theorem*.

Theorem 10.4.5. Let $A \in \mathbf{H}^n$, and let $A_0 \in \mathbf{H}^k$ be a $k \times k$ principal submatrix of A. Then, for all $i \in \{1, \ldots, k\}$,

$$\lambda_{i+n-k}(A) \le \lambda_i(A_0) \le \lambda_i(A). \tag{10.4.4}$$

Proof. For $k = n - 1$, the result is given by Lemma 10.4.4. Hence, let $k = n - 2$, and let A_1 denote an $(n-1) \times (n-1)$ principal submatrix of A such that the $(n-2) \times (n-2)$ principal submatrix A_0 of A is also a principal submatrix of A_1. Therefore, Lemma 10.4.4 implies that $\lambda_n(A) \le \lambda_{n-1}(A_1) \le \cdots \le \lambda_2(A_1) \le \lambda_2(A) \le \lambda_1(A_1) \le \lambda_1(A)$ and $\lambda_{n-1}(A_1) \le \lambda_{n-2}(A_0) \le \cdots \le \lambda_2(A_0) \le \lambda_2(A_1) \le \lambda_1(A_0) \le \lambda_1(A_1)$. Combining these inequalities yields $\lambda_{i+2}(A) \le \lambda_i(A_0) \le \lambda_i(A)$ for all $i = 1, \ldots, n - 2$, while proceeding in a similar manner with $k < n - 2$ yields (10.4.4). □

Corollary 10.4.6. Let $A \in \mathbf{H}^n$, and let $A_0 \in \mathbf{H}^k$ be a $k \times k$ principal submatrix of A. Then,

$$\lambda_{\min}(A) \le \lambda_{\min}(A_0) \le \lambda_{\max}(A_0) \le \lambda_{\max}(A), \tag{10.4.5}$$

$$\lambda_{\min}(A_0) \le \lambda_k(A). \tag{10.4.6}$$

The following corollary of both Lemma 10.4.3 and Theorem 10.4.5 compares the maximum and minimum eigenvalues with the maximum and minimum diagonal entries.

Corollary 10.4.7. Let $A \in \mathbf{H}^n$. Then,

$$\lambda_{\min}(A) \le \mathrm{d}_{\min}(A) \le \mathrm{d}_{\max}(A) \le \lambda_{\max}(A). \tag{10.4.7}$$

Lemma 10.4.8. Let $A, B \in \mathbf{H}^n$, and assume that $A \le B$ and $\mathrm{mspec}(A) = \mathrm{mspec}(B)$. Then, $A = B$.

Proof. Let $\alpha \ge 0$ satisfy $0 < \hat{A} \le \hat{B}$, where $\hat{A} \triangleq A + \alpha I$ and $\hat{B} \triangleq B + \alpha I$. Note that $\mathrm{mspec}(\hat{A}) = \mathrm{mspec}(\hat{B})$, and thus $\det \hat{A} = \det \hat{B}$. Next, it follows that $I \le \hat{A}^{-1/2}\hat{B}\hat{A}^{-1/2}$. Hence, *i)* of Lemma 10.4.1 implies that $\lambda_{\min}(\hat{A}^{-1/2}\hat{B}\hat{A}^{-1/2}) \ge 1$. Furthermore, $\det(\hat{A}^{-1/2}\hat{B}\hat{A}^{-1/2}) = \det \hat{B}/\det \hat{A} = 1$, which implies that, for all $i \in \{1, \ldots, n\}$, $\lambda_i(\hat{A}^{-1/2}\hat{B}\hat{A}^{-1/2}) = 1$. Hence, $\hat{A}^{-1/2}\hat{B}\hat{A}^{-1/2} = I$, and thus $\hat{A} = \hat{B}$. Hence, $A = B$. □

The following result is the *monotonicity theorem* or *Weyl's inequality*.

Theorem 10.4.9. Let $A, B \in \mathbf{H}^n$, and assume that $A \le B$. Then, for all $i \in \{1, \ldots, n\}$,

$$\lambda_i(A) \le \lambda_i(B). \tag{10.4.8}$$

If $A \ne B$, then there exists $i \in \{1, \ldots, n\}$ such that

$$\lambda_i(A) < \lambda_i(B). \tag{10.4.9}$$

If $A < B$, then (10.4.9) holds for all $i \in \{1, \ldots, n\}$.

Proof. Since $A \le B$, Corollary 10.4.2 implies that $\lambda_{\min}(A)I \le A \le B \le \lambda_{\max}(B)I$. Hence, it follows from *iii)* and *i)* of Lemma 10.4.1 that $\lambda_{\min}(A) \le \lambda_{\min}(B)$ and $\lambda_{\max}(A) \le \lambda_{\max}(B)$. Next, let $S \in \mathbb{F}^{n \times n}$ be a unitary matrix such that $SAS^* = \mathrm{diag}[\lambda_1(A), \ldots, \lambda_n(A)]$. Furthermore, for $2 \le i \le n-1$, let $A_0 = \mathrm{diag}[\lambda_1(A), \ldots, \lambda_i(A)]$, and let B_0 denote the $i \times i$ leading principal submatrices of SAS^* and

SBS^*, respectively. Since $A \le B$, it follows that $A_0 \le B_0$, which implies that $\lambda_{\min}(A_0) \le \lambda_{\min}(B_0)$. It now follows from (10.4.6) that

$$\lambda_i(A) = \lambda_{\min}(A_0) \le \lambda_{\min}(B_0) \le \lambda_i(SBS^*) = \lambda_i(B),$$

which proves (10.4.8). If $A \neq B$, then Lemma 10.4.8 implies that $\operatorname{mspec}(A) \neq \operatorname{mspec}(B)$, and thus there exists $i \in \{1, \ldots, n\}$ such that (10.4.9) holds. In the case where $A < B$, it follows that $\lambda_{\min}(A_0) < \lambda_{\min}(B_0)$, which implies (10.4.9) for all $i \in \{1, \ldots, n\}$. □

Corollary 10.4.10. Let $A, B \in \mathbf{H}^n$. Then, the following statements hold:

i) If $A \le B$, then $\operatorname{tr} A \le \operatorname{tr} B$.

ii) If $A \le B$ and $\operatorname{tr} A = \operatorname{tr} B$, then $A = B$.

iii) If $A < B$, then $\operatorname{tr} A < \operatorname{tr} B$.

iv) If $0 \le A \le B$, then $0 \le \det A \le \det B$.

v) If $0 \le A < B$, then $0 \le \det A < \det B$.

vi) If $0 < A \le B$ and $\det A = \det B$, then $A = B$.

Proof. *i*), *iii*), *iv*), and *v*) follow from Theorem 10.4.9. To prove *ii*), note that, since $A \le B$ and $\operatorname{tr} A = \operatorname{tr} B$, Theorem 10.4.9 implies that $\operatorname{mspec}(A) = \operatorname{mspec}(B)$. Now, Lemma 10.4.8 implies that $A = B$. A similar argument yields *vi*). □

The following result, which is a generalization of Theorem 10.4.9, is due to H. Weyl.

Theorem 10.4.11. Let $A, B \in \mathbf{H}^n$. If $i + j \ge n + 1$, then

$$\lambda_i(A) + \lambda_j(B) \le \lambda_{i+j-n}(A + B). \tag{10.4.10}$$

If $i + j \le n + 1$, then

$$\lambda_{i+j-1}(A + B) \le \lambda_i(A) + \lambda_j(B). \tag{10.4.11}$$

In particular, for all $i \in \{1, \ldots, n\}$,

$$\lambda_i(A) + \lambda_{\min}(B) \le \lambda_i(A + B) \le \lambda_i(A) + \lambda_{\max}(B), \tag{10.4.12}$$

$$\lambda_{\min}(A) + \lambda_{\min}(B) \le \lambda_{\min}(A + B) \le \lambda_{\min}(A) + \lambda_{\max}(B), \tag{10.4.13}$$

$$\lambda_{\max}(A) + \lambda_{\min}(B) \le \lambda_{\max}(A + B) \le \lambda_{\max}(A) + \lambda_{\max}(B). \tag{10.4.14}$$

Furthermore, if $\operatorname{rank} B \le r$, then, for all $i \in \{1, \ldots, n - r\}$,

$$\lambda_{i+r}(A) \le \lambda_i(A + B), \tag{10.4.15}$$

$$\lambda_{i+r}(A + B) \le \lambda_i(A). \tag{10.4.16}$$

If, in particular, $x \in \mathbb{F}^n$, then, for all $i \in \{1, \ldots, n - 1\}$,

$$\lambda_{i+1}(A) \le \lambda_{i+1}(A + xx^*) \le \lambda_i(A) \le \lambda_i(A + xx^*). \tag{10.4.17}$$

Finally,

$$\nu_-(A + B) \le \nu_-(A) + \nu_-(B), \tag{10.4.18}$$

$$\nu_+(A + B) \le \nu_+(A) + \nu_+(B). \tag{10.4.19}$$

Proof. See [862], [1451, pp. 239–242], [2403, pp. 114, 115], and [2979, pp. 53, 54]. □

Lemma 10.4.12. Let $A, B, C \in \mathbf{H}^n$. If $A \le B$ and C is positive semidefinite, then

$$\operatorname{tr} AC \le \operatorname{tr} BC. \tag{10.4.20}$$

If $A < B$ and C is positive definite, then

$$\operatorname{tr} AC < \operatorname{tr} BC. \tag{10.4.21}$$

Proof. Since $C^{1/2}AC^{1/2} \le C^{1/2}BC^{1/2}$, *i*) of Corollary 10.4.10 implies that

$$\operatorname{tr} AC = \operatorname{tr} C^{1/2}AC^{1/2} \le \operatorname{tr} C^{1/2}BC^{1/2} = \operatorname{tr} BC.$$

(10.4.21) follows from *ii*) of Corollary 10.4.10 in a similar fashion. □

Proposition 10.4.13. Let $A, B \in \mathbb{F}^{n\times n}$, and assume that B is positive semidefinite. Then,

$$\tfrac{1}{2}\lambda_{\min}(A + A^*) \operatorname{tr} B \le \operatorname{Re} \operatorname{tr} AB \le \tfrac{1}{2}\lambda_{\max}(A + A^*) \operatorname{tr} B. \tag{10.4.22}$$

If, in addition, A is Hermitian, then

$$\lambda_{\min}(A) \operatorname{tr} B \le \operatorname{tr} AB \le \lambda_{\max}(A) \operatorname{tr} B. \tag{10.4.23}$$

Proof. It follows from Corollary 10.4.2 that $\frac{1}{2}\lambda_{\min}(A + A^*)I \le \frac{1}{2}(A + A^*)$, while Lemma 10.4.12 implies that $\frac{1}{2}\lambda_{\min}(A + A^*) \operatorname{tr} B = \operatorname{tr} \frac{1}{2}\lambda_{\min}(A + A^*)IB \le \operatorname{tr} \frac{1}{2}(A + A^*)B = \operatorname{Re} \operatorname{tr} AB$, which proves the left-hand inequality of (10.4.22). Similarly, the right-hand inequality holds. □

For results relating to Proposition 10.4.13, see Fact 7.13.12, Fact 7.13.13, Fact 7.13.16, and Fact 10.22.25.

Proposition 10.4.14. Let $A \in \mathbf{N}^n$, let $B \in \mathbf{P}^n$, and assume that $\det B = 1$. Then,

$$(\det A)^{1/n} \le \tfrac{1}{n}\operatorname{tr} AB. \tag{10.4.24}$$

Furthermore, equality holds if and only if $A = (\det A)^{1/n}B^{-1}$.

Proof. The arithmetic-mean–geometric-mean inequality given by Fact 2.11.81 implies that

$$(\det A)^{1/n} = (\det B^{1/2}AB^{1/2})^{1/n} = \left[\prod_{i=1}^{n} \lambda_i(B^{1/2}AB^{1/2})\right]^{1/n} \le \frac{1}{n}\sum_{i=1}^{n} \lambda_i(B^{1/2}AB^{1/2}) = \frac{1}{n} \operatorname{tr} AB.$$

Equality holds if and only if there exists $\beta > 0$ such that $B^{1/2}AB^{1/2} = \beta I$. If these conditions hold, then $\beta = (\det A)^{1/n}$ and $A = (\det A)^{1/n}B^{-1}$. □

For $A, B \in \mathbf{N}^n$, the following corollary of Proposition 10.4.14 includes *Minkowski's determinant theorem*, which is the second inequality in the string

$$(\det A + \det B)^{1/n} \le (\det A)^{1/n} + (\det B)^{1/n} \le [\det(A + B)]^{1/n}. \tag{10.4.25}$$

Corollary 10.4.15. Let $A, B \in \mathbf{N}^n$, and let p and q be real numbers such that $1 \le p \le n \le q$. Then,

$$\det A + \det B \le [(\det A)^{1/p} + (\det B)^{1/p}]^p \tag{10.4.26}$$

$$\le [(\det A)^{1/n} + (\det B)^{1/n}]^n \tag{10.4.27}$$

$$\le \begin{cases} [(\det A)^{1/q} + (\det B)^{1/q}]^q \\ \det(A + B). \end{cases} \tag{10.4.28}$$

Furthermore, the following statements hold:

i) If $p = 1$, then (10.4.26) is an equality.

ii) If $n = 1$, then (10.4.26), (10.4.27), and lower (10.4.28) are equalities.

iii) If either A or B is singular, then (10.4.26), (10.4.27), and upper (10.4.28) are equalities.

iv) If either $q = 1$, $A = 0$, $B = 0$, or $A + B$ is singular, then (10.4.26)–(10.4.28) are equalities.

Now, assume that $n \ge 2$. Then, the following statements hold:

v) If A is positive definite and $(\det A + \det B)^{1/n} = (\det A)^{1/n} + (\det B)^{1/n}$, then $\det B = 0$.

vi) The following statements are equivalent:

a) Either $A = 0$, $B = 0$, $A + B$ is singular, or there exists $\alpha \ge 0$ such that either $B = \alpha A$ or

$A = \alpha B$.

b) $(\det A)^{1/n} + (\det B)^{1/n} = [\det(A + B)]^{1/n}$.

vii) If $A + B$ is positive definite and $(\det A)^{1/n} + (\det B)^{1/n} = [\det(A + B)]^{1/n}$, then A and B are positive definite and there exists $\alpha > 0$ such that $B = \alpha A$.

viii) If A is positive definite and $\det A + \det B = \det(A + B)$, then $B = 0$.

ix) Either $A = 0$, $B = 0$, or $A + B$ is singular if and only if $\det A + \det B = \det(A + B)$.

x) Either $B = 0$ or $A + B$ is singular if and only if $\det A = \det(A + B)$.

Proof. Inequalities (10.4.26), (10.4.27), and upper (10.4.28) are consequences of the reverse power-sum inequality given by Fact 2.11.91 and Fact 11.8.21. To prove lower (10.4.28), note that, in the case where $A+B$ is singular, lower (10.4.28) is immediate. In the case where $A+B$ is positive definite, it follows from Proposition 10.4.14 that

$$\begin{aligned}(\det A)^{1/n} + (\det B)^{1/n} &\le \tfrac{1}{n}\operatorname{tr}[A[\det(A + B)]^{1/n}(A + B)^{-1}] + \tfrac{1}{n}\operatorname{tr}[B[\det(A + B)]^{1/n}(A + B)^{-1}]\\ &= [\det(A + B)]^{1/n}.\end{aligned}$$

Next, *v)* follows from Fact 2.11.91; *vi)* is given in [2991, p. 215]; *vii)* follows from *vi)*.

To prove *viii)*, note that (10.4.26) and (10.4.27) hold as equalities. Hence, *v)* implies that $\det B = 0$. Consequently, $\det A = \det(A + B)$. Since $0 < A \le A + B$, *vi)* of Corollary 10.4.10 implies that $B = 0$. *ix)* and *x)* are given in [2991, p. 215]. □

10.5 Exponential, Square Root, and Logarithm of Hermitian Matrices

Let $B \in \mathbb{R}^{n\times n}$ be diagonal, let $\mathcal{D} \subseteq \mathbb{R}$, let $f\colon\ \mathcal{D} \mapsto \mathbb{R}$, and assume that, for all $i \in \{1,\dots,n\}$, $B_{(i,i)} \in \mathcal{D}$. Then, we define

$$f(B) \triangleq \operatorname{diag}[f(B_{(1,1)}),\dots,f(B_{(n,n)})]. \tag{10.5.1}$$

Furthermore, let $A = SBS^* \in \mathbb{F}^{n\times n}$ be Hermitian, where $S \in \mathbb{F}^{n\times n}$ is unitary, $B \in \mathbb{R}^{n\times n}$ is diagonal, and assume that $\operatorname{spec}(A) \subset \mathcal{D}$. Then, we define $f(A) \in \mathbf{H}^n$ by

$$f(A) \triangleq Sf(B)S^*. \tag{10.5.2}$$

Hence, with an obvious extension of notation, $f\colon\ \{X \in \mathbf{H}^n\colon\ \operatorname{spec}(X) \subset \mathcal{D}\} \mapsto \mathbf{H}^n$. If $f\colon\ \mathcal{D} \mapsto \mathbb{R}$ is one-to-one, then its inverse $f^{-1}\colon\ \{X \in \mathbf{H}^n\colon\ \operatorname{spec}(X) \subset f(\mathcal{D})\} \mapsto \mathbf{H}^n$ exists. It remains to be shown, however, that the definition of $f(A)$ given by (10.5.2) is independent of the matrices S and B in the decomposition $A = SBS^*$. The following lemma is needed.

Lemma 10.5.1. Let $S \in \mathbb{F}^{n\times n}$ be unitary, let $D, \hat{D} \in \mathbb{R}^{n\times n}$ denote the diagonal matrices $D = \operatorname{diag}(\lambda_1 I_{n_1},\dots,\lambda_r I_{n_r})$ and $\hat{D} = \operatorname{diag}(\mu_1 I_{n_1},\dots,\mu_r I_{n_r})$, where $\lambda_1,\dots,\lambda_r,\mu_1,\dots,\mu_r \in \mathbb{R}$, and assume that $SD = DS$. Then, $S\hat{D} = \hat{D}S$.

Proof. Let $r = 2$, and partition $S = \begin{bmatrix} S_{11} & S_{12}\\ S_{21} & S_{22}\end{bmatrix}$. Then, it follows from $SD = DS$ that $\lambda_2 S_{12} = \lambda_1 S_{12}$ and $\lambda_1 S_{21} = \lambda_2 S_{21}$. Since $\lambda_1 \ne \lambda_2$, it follows that $S_{12} = 0$ and $S_{21} = 0$. Therefore, $S = \begin{bmatrix} S_{11} & 0\\ 0 & S_{22}\end{bmatrix}$, and thus $S\hat{D} = \hat{D}S$. A similar argument holds for $r \ge 3$. □

Proposition 10.5.2. Let $A = RBR^* = SCS^* \in \mathbf{H}^n$, where $R, S \in \mathbb{F}^{n\times n}$ are unitary and $B, C \in \mathbb{R}^{n\times n}$ are diagonal. Furthermore, let $\mathcal{D} \subseteq \mathbb{R}$, let $f\colon\ \mathcal{D} \mapsto \mathbb{R}$, and assume that all diagonal entries of B are contained in $\mathcal{D}$. Then, $Rf(B)R^* = Sf(C)S^*$.

Proof. Let $\operatorname{spec}(A) = \{\lambda_1,\dots,\lambda_r\}$. Then, the columns of R and S can be rearranged to obtain unitary matrices $\tilde{R}, \tilde{S} \in \mathbb{F}^{n\times n}$ such that $A = \tilde{R}D\tilde{R}^* = \tilde{S}D\tilde{S}^*$, where $D \triangleq \operatorname{diag}(\lambda_1 I_{n_1},\dots,\lambda_{n_r} I_{n_r})$. Hence, $UD = DU$, where $U \triangleq \tilde{S}^*\tilde{R}$. It thus follows from Lemma 10.5.1 that $U\hat{D} = \hat{D}U$, where $\hat{D} \triangleq f(D) = \operatorname{diag}[f(\lambda_1)I_{n_1},\dots,f(\lambda_{n_r})I_{n_r}]$. Hence, $\tilde{R}\hat{D}\tilde{R}^* = \tilde{S}\hat{D}\tilde{S}^*$, while rearranging the columns of $\tilde{R}$ and $\tilde{S}$ as well as the diagonal entries of $\hat{D}$ yields $Rf(B)R^* = Sf(C)S^*$. □

Let $A = SBS^* \in \mathbb{F}^{n\times n}$ be Hermitian, where $S \in \mathbb{F}^{n\times n}$ is unitary and $B \in \mathbb{R}^{n\times n}$ is diagonal. Then, the *matrix exponential* is defined by

$$e^A \triangleq Se^BS^* \in \mathbf{H}^n, \tag{10.5.3}$$

where, for all $i \in \{1,\dots,n\}$, $(e^B)_{(i,i)} \triangleq e^{B_{(i,i)}}$.

Let $A = SBS^* \in \mathbb{F}^{n\times n}$ be positive semidefinite, where $S \in \mathbb{F}^{n\times n}$ is unitary and $B \in \mathbb{R}^{n\times n}$ is diagonal with nonnegative entries. Then, for all $r \ge 0$ (not necessarily an integer), $A^r = SB^rS^*$ is positive semidefinite, where, for all $i \in \{1,\dots,n\}$, $(B^r)_{(i,i)} = [B_{(i,i)}]^r$. Note that $A^0 \triangleq I$. In particular, the positive-semidefinite matrix

$$A^{1/2} = SB^{1/2}S^* \tag{10.5.4}$$

is a square root of A since

$$A^{1/2}A^{1/2} = SB^{1/2}S^*SB^{1/2}S^* = SBS^* = A. \tag{10.5.5}$$

The uniqueness of the *positive-semidefinite square root* of A given by (10.5.4) follows from Theorem 12.9.1; see also [1450, p. 410] and [1767]. Uniqueness is shown in [970, pp. 265, 266], [1448, p. 405], [1655], and [2991, p. 81]. Hence, if $C \in \mathbb{F}^{n\times m}$, then C^*C is positive semidefinite, and we define

$$\langle C\rangle \triangleq (C^*C)^{1/2}. \tag{10.5.6}$$

If A is positive definite, then A^r is positive definite for all $r \in \mathbb{R}$, and, if $r \ne 0$, then $(A^r)^{1/r} = A$.

Now, assume that $A \in \mathbb{F}^{n\times n}$ is positive definite. Then, the *matrix logarithm* is defined by

$$\log A \triangleq S(\log B)S^* \in \mathbf{H}^n, \tag{10.5.7}$$

where, for all $i \in \{1,\dots,n\}$, $(\log B)_{(i,i)} \triangleq \log[B_{(i,i)}]$.

In chapters 10 and 11, the matrix exponential, square root, and logarithm are extended to matrices that are not necessarily Hermitian.

10.6 Matrix Inequalities

Lemma 10.6.1. Let $A, B \in \mathbf{H}^n$, and assume that $0 \le A \le B$. Then, $\mathcal{R}(A) \subseteq \mathcal{R}(B)$.

Proof. Let $x \in \mathcal{N}(B)$. Then, $x^*Bx = 0$, and thus $x^*Ax = 0$, which implies that $Ax = 0$. Hence, $\mathcal{N}(B) \subseteq \mathcal{N}(A)$, and thus $\mathcal{N}(A)^\perp \subseteq \mathcal{N}(B)^\perp$. Since A and B are Hermitian, it follows from Theorem 3.5.3 that $\mathcal{R}(A) = \mathcal{N}(A)^\perp$ and $\mathcal{R}(B) = \mathcal{N}(B)^\perp$. Hence, $\mathcal{R}(A) \subseteq \mathcal{R}(B)$. □

The following result is the *Douglas-Fillmore-Williams lemma* [930, 1050].

Theorem 10.6.2. Let $A \in \mathbb{F}^{n\times m}$ and $B \in \mathbb{F}^{n\times l}$. Then, the following statements are equivalent:

i) There exists $C \in \mathbb{F}^{l\times m}$ such that $A = BC$.

ii) There exists $\alpha > 0$ such that $AA^* \le \alpha BB^*$.

iii) $\mathcal{R}(A) \subseteq \mathcal{R}(B)$.

Proof. First we prove *i)* $\Longrightarrow$ *ii)*. Since $A = BC$, it follows that $AA^* = BCC^*B^*$. Since $CC^* \le \lambda_{\max}(CC^*)I$, it follows that $AA^* \le \alpha BB^*$, where $\alpha \triangleq \lambda_{\max}(CC^*)$. To prove *ii)* $\Longrightarrow$ *iii)*, let $x \in \mathcal{N}(B^*)$. Then, $0 \le x^*AA^*x = \alpha x^*BB^*x = 0$, and thus $x \in \mathcal{N}(A^*)$. Hence, $\mathcal{R}(B)^\perp = \mathcal{N}(B^*) \subseteq \mathcal{N}(A^*) = \mathcal{R}(A)^\perp$, which is equivalent to *iii)*. Finally, to prove *iii)* $\Longrightarrow$ *i)*, use Theorem 7.6.3 to write $B = S_1\left[\begin{smallmatrix} D & 0\\ 0 & 0\end{smallmatrix}\right]S_2$, where $S_1 \in \mathbb{F}^{n\times n}$ and $S_2 \in \mathbb{F}^{l\times l}$ are unitary and $D \in \mathbb{R}^{r\times r}$ is diagonal with positive diagonal entries, where $r \triangleq \operatorname{rank} B$. Since $\mathcal{R}(S_1^*A) \subseteq \mathcal{R}(S_1^*B)$ and $S_1^*B = \left[\begin{smallmatrix} D & 0\\ 0 & 0\end{smallmatrix}\right]S_2$, it follows that $S_1^*A = \left[\begin{smallmatrix} A_1\\ 0\end{smallmatrix}\right]$, where $A_1 \in \mathbb{F}^{r\times m}$. Consequently,

$$A = S_1\begin{bmatrix} A_1\\ 0\end{bmatrix} = S_1\begin{bmatrix} D & 0\\ 0 & 0\end{bmatrix}S_2S_2^*\begin{bmatrix} D^{-1} & 0\\ 0 & 0\end{bmatrix}\begin{bmatrix} A_1\\ 0\end{bmatrix} = BC,$$

where $C \triangleq S_2^*\begin{bmatrix} D^{-1} & 0 \\ 0 & 0 \end{bmatrix}\begin{bmatrix} A_1 \\ 0 \end{bmatrix} \in \mathbb{F}^{l\times m}$. □

Proposition 10.6.3. Let $(A_i)_{i=1}^{\infty} \subset \mathbf{N}^n$ satisfy $0 \le A_i \le A_j$ for all $i \le j$, and assume that there exists $B \in \mathbf{N}^n$ satisfying $A_i \le B$ for all $i \ge 1$. Then, $A \triangleq \lim_{i\to\infty} A_i$ exists and satisfies $0 \le A \le B$.

Proof. Let $k \in \{1,\dots,n\}$, and let $i < j$. Since $A_i \le A_j \le B$, it follows that the sequence $(A_{r(k,k)})_{r=1}^{\infty}$ is nondecreasing and bounded from above by $B_{(k,k)}$. Hence, $A_{(k,k)} \triangleq \lim_{r\to\infty} A_{r(k,k)}$ exists. Now, let $l \in \{1,\dots,n\}$, where $l \ne k$. Since $A_i \le A_j$, it follows that $(e_k+e_l)^{\mathsf{T}}A_i(e_k+e_l) \le (e_k+e_l)^{\mathsf{T}}A_j(e_k+e_l)$, which implies that $A_{i(k,l)} - A_{j(k,l)} \le \frac{1}{2}[A_{j(k,k)} - A_{i(k,k)} + A_{j(l,l)} - A_{i(l,l)}]$. Likewise, $(e_k-e_l)^{\mathsf{T}}A_i(e_k - e_l) \le (e_k-e_l)^{\mathsf{T}}A_j(e_k-e_l)$ implies that $A_{j(k,l)} - A_{i(k,l)} \le \frac{1}{2}[A_{j(k,k)} - A_{i(k,k)} + A_{j(l,l)} - A_{i(l,l)}]$. Hence, $|A_{j(k,l)} - A_{i(k,l)}| \le \frac{1}{2}[A_{j(k,k)} - A_{i(k,k)}] + \frac{1}{2}[A_{j(l,l)} - A_{i(l,l)}]$. Next, since $(A_{r(k,k)})_{r=1}^{\infty}$ and $(A_{r(l,l)})_{r=1}^{\infty}$ are convergent sequences and thus Cauchy sequences, it follows that $(A_{r(k,l)})_{r=1}^{\infty}$ is a Cauchy sequence. Consequently, $(A_{r(k,l)})_{r=1}^{\infty}$ is convergent, and thus $A_{(k,l)} \triangleq \lim_{i\to\infty} A_{i(k,l)}$ exists. Therefore, $(A_i)_{i=1}^{\infty}$ is convergent, and thus $A \triangleq \lim_{i\to\infty} A_i$ exists. Since $A_i \le B$ for all $i \ge 1$, it follows that $A \le B$. □

Proposition 10.6.4. Let $A \in \mathbf{P}^n$ and $p \in (0,\infty)$. Then,

$$A^{-1}(A-I) \le \log A \le p^{-1}(A^p - I), \tag{10.6.1}$$

$$\log A = \lim_{p\downarrow 0} p^{-1}(A^p - I). \tag{10.6.2}$$

Proof. Use Fact 2.15.3. □

Lemma 10.6.5. Let $A \in \mathbf{P}^n$. If $A \le I$, then $I \le A^{-1}$. Furthermore, if $A < I$, then $I < A^{-1}$.

Proof. Since $A \le I$, it follows from *xii*) of Proposition 10.1.2 that $I = A^{-1/2}AA^{-1/2} \le A^{-1/2}IA^{-1/2} = A^{-1}$. Similarly, $A < I$ implies that $I = A^{-1/2}AA^{-1/2} < A^{-1/2}IA^{-1/2} = A^{-1}$. □

Proposition 10.6.6. Let $A, B \in \mathbf{H}^n$, and assume that either A and B are positive definite or A and B are negative definite. If $A \le B$, then $B^{-1} \le A^{-1}$. If, in addition, $A < B$, then $B^{-1} < A^{-1}$.

Proof. Suppose that A and B are positive definite. Since $A \le B$, it follows that $B^{-1/2}AB^{-1/2} \le I$. Now, Lemma 10.6.5 implies that $I \le B^{1/2}A^{-1}B^{1/2}$, which implies that $B^{-1} \le A^{-1}$. If A and B are negative definite, then $A \le B$ is equivalent to $-B \le -A$. The case where $A < B$ is proved in a similar manner. □

The following result is the *Furuta inequality*.

Proposition 10.6.7. Let $A, B \in \mathbf{N}^n$, and assume that $0 \le A \le B$. Furthermore, let $p, q, r \in \mathbb{R}$ satisfy $p \ge 0$, $q \ge 1$, $r \ge 0$, and $p + 2r \le (1+2r)q$. Then,

$$A^{(p+2r)/q} \le (A^rB^pA^r)^{1/q}, \tag{10.6.3}$$

$$(B^rA^pB^r)^{1/q} \le B^{(p+2r)/q}. \tag{10.6.4}$$

Proof. See [1116] and [1124, pp. 129, 130]. □

Corollary 10.6.8. Let $A, B \in \mathbf{N}^n$, and assume that $0 \le A \le B$. Then,

$$A^2 \le (AB^2A)^{1/2}, \tag{10.6.5}$$

$$(BA^2B)^{1/2} \le B^2. \tag{10.6.6}$$

Proof. In Proposition 10.6.7 set $r = 1$, $p = 2$, and $q = 2$. □

Corollary 10.6.9. Let $A, B, C \in \mathbf{N}^n$, and assume that $0 \le A \le C \le B$. Then,

$$(CA^2C)^{1/2} \le C^2 \le (CB^2C)^{1/2}. \tag{10.6.7}$$

Proof. Use Corollary 10.6.8. See also [2820]. □

The following result provides representations for A^r, where $r \in (0,1)$.

Proposition 10.6.10. Let $A \in \mathbf{N}^n$ and $r \in (0,1)$. Then,

$$A^r = \left(\cos\frac{r\pi}{2}\right)I + \frac{\sin r\pi}{\pi}\int_0^\infty \left(\frac{x^{r+1}}{1+x^2}I - (A+xI)^{-1}x^r\right)\mathrm{d}x, \tag{10.6.8}$$

$$A^r = \frac{\sin r\pi}{\pi}\int_0^\infty (A+xI)^{-1}Ax^{r-1}\,\mathrm{d}x. \tag{10.6.9}$$

Proof. Let $t \ge 0$. As shown in [443], [449, p. 143],

$$\int_0^\infty \left(\frac{x^{r+1}}{1+x^2} - \frac{x^r}{t+x}\right)\mathrm{d}x = \frac{\pi}{\sin r\pi}\left(t^r - \cos\frac{r\pi}{2}\right).$$

Solving for t^r and replacing t by A yields (10.6.8). Likewise, replacing t by A in Fact 14.4.6 yields (10.6.9). □

The following result is the *Löwner-Heinz inequality.*

Corollary 10.6.11. Let $A, B \in \mathbf{N}^n$, assume that $0 \le A \le B$, and let $r \in [0,1]$. Then, $A^r \le B^r$. If, in addition, $A < B$ and $r \in (0,1]$, then $A^r < B^r$.

Proof. Let $A \le B$ and $r \in [0,1]$. In Proposition 10.6.7, replace p, q, r with $r, 1, 0$. The first result follows from (10.6.3).

Now, let $A < B$ and $r \in (0,1)$. Then, it follows from (10.6.8) of Proposition 10.6.10 as well as Proposition 10.6.6 that

$$B^r - A^r = \frac{\sin r\pi}{\pi}\int_0^\infty [(A+xI)^{-1} - (B+xI)^{-1}]x^r\,\mathrm{d}x > 0.$$

Hence, $A^r < B^r$.

Alternatively, let $A < B$ and $r \in (0,1)$. Then, Proposition 10.6.6 implies that, for all $x \ge 0$,

$$(A+xI)^{-1}A = I - x(A+xI)^{-1} < I - x(B+xI)^{-1} = (B+xI)^{-1}B.$$

It thus follows from (10.6.9) of Proposition 10.6.10 that

$$B^r - A^r = \frac{\sin r\pi}{\pi}\int_0^\infty [(B+xI)^{-1}B - (A+xI)^{-1}A]x^{r-1}\,\mathrm{d}x > 0.$$

Hence, $A^r < B^r$. Additional proofs are given in [1124, p. 127] and [2977, p. 2].

In the case where $A \le B$ and $r = 1/2$, let $\lambda \in \mathbb{R}$ be an eigenvalue of $B^{1/2} - A^{1/2}$, and let $x \in \mathbb{F}^n$ be an associated eigenvector. Then,

$$\begin{aligned}\lambda x^*(B^{1/2} + A^{1/2})x &= x^*(B^{1/2} + A^{1/2})(B^{1/2} - A^{1/2})x\\ &= x^*(B - B^{1/2}A^{1/2} + A^{1/2}B^{1/2} - A)x = x^*(B-A)x \ge 0.\end{aligned}$$

Since $B^{1/2} + A^{1/2}$ is positive semidefinite, it follows that either $\lambda \ge 0$ or $x^*(B^{1/2} + A^{1/2})x = 0$. In the latter case, $B^{1/2}x = A^{1/2}x = 0$, which implies that $\lambda = 0$. □

The Löwner-Heinz inequality does not extend to $r > 1$. In fact, $A \triangleq \begin{bmatrix} 2 & 1 \\ 1 & 1 \end{bmatrix}$ and $B \triangleq \begin{bmatrix} 1 & 0 \\ 0 & 0 \end{bmatrix}$ satisfy $A \ge B \ge 0$, whereas, for all $r > 1$, $A^r \not\ge B^r$. For details, see [1124, pp. 127, 128].

Many of the results given so far involve functions that are nondecreasing or increasing on suitable sets of matrices.

Definition 10.6.12. Let $\mathcal{D} \subseteq \mathbf{H}^n$ and $\phi\colon\ \mathcal{D} \mapsto \mathbf{H}^m$. Then, the following terminology is defined:

i) ϕ is *nondecreasing* if, for all $A, B \in \mathcal{D}$ such that $A \le B$, it follows that $\phi(A) \le \phi(B)$.

ii) ϕ is *increasing* if ϕ is nondecreasing and, for all $A, B \in \mathcal{D}$ such that $A < B$, it follows that $\phi(A) < \phi(B)$.

iii) ϕ is *strongly increasing* if ϕ is nondecreasing and, for all $A, B \in \mathcal{D}$ such that $A \le B$ and $A \ne B$, it follows that $\phi(A) < \phi(B)$.

iv) ϕ is (*nonincreasing, decreasing, strongly decreasing*) if $-\phi$ is (nondecreasing, increasing, strongly increasing).

Proposition 10.6.13. The following functions are nondecreasing:

i) $\phi\colon \mathbf{H}^n \mapsto \mathbf{H}^m$ defined by $\phi(A) \triangleq BAB^*$, where $B \in \mathbb{F}^{m\times n}$.

ii) $\phi\colon \mathbf{H}^n \mapsto \mathbb{R}$ defined by $\phi(A) \triangleq \operatorname{tr} AB$, where $B \in \mathbf{N}^n$.

iii) $\phi\colon \mathbf{N}^{n+m} \mapsto \mathbf{N}^n$ defined by $\phi(A) \triangleq A_{22}|A$, where $A \triangleq \begin{bmatrix} A_{11} & A_{12} \\ A_{12}^* & A_{22} \end{bmatrix}$.

iv) $\phi\colon \mathbf{N}^n \times \mathbf{N}^m \mapsto \mathbf{N}^{nm}$ defined by $\phi(A, B) \triangleq A^{r_1} \otimes B^{r_2}$, where $r_1, r_2 \in [0, 1]$ satisfy $r_1 + r_2 \le 1$.

v) $\phi\colon \mathbf{N}^n \times \mathbf{N}^n \mapsto \mathbf{N}^n$ defined by $\phi(A, B) \triangleq A^{r_1} \odot B^{r_2}$, where $r_1, r_2 \in [0, 1]$ satisfy $r_1 + r_2 \le 1$.

The following functions are increasing:

vi) $\phi\colon \mathbf{H}^n \mapsto \mathbb{R}$ defined by $\phi(A) \triangleq \lambda_i(A)$, where $i \in \{1, \ldots, n\}$.

vii) $\phi\colon \mathbf{N}^n \mapsto \mathbf{N}^n$ defined by $\phi(A) \triangleq A^r$, where $r \in [0, 1]$.

viii) $\phi\colon \mathbf{P}^n \mapsto -\mathbf{P}^n$ defined by $\phi(A) \triangleq -A^{-r}$, where $r \in [0, 1]$.

ix) $\phi\colon -\mathbf{P}^n \mapsto \mathbf{P}^n$ defined by $\phi(A) \triangleq (-A)^{-r}$, where $r \in [0, 1]$.

x) $\phi\colon \mathbf{H}^n \mapsto \mathbf{H}^m$ defined by $\phi(A) \triangleq BAB^*$, where $B \in \mathbb{F}^{m\times n}$ and $\operatorname{rank} B = m$.

xi) $\phi\colon \mathbf{P}^{n+m} \mapsto \mathbf{P}^n$ defined by $\phi(A) \triangleq A_{22}|A$, where $A \triangleq \begin{bmatrix} A_{11} & A_{12} \\ A_{12}^* & A_{22} \end{bmatrix}$.

xii) $\phi\colon \mathbf{P}^{n+m} \mapsto \mathbf{P}^n$ defined by $\phi(A) \triangleq -(A_{22}|A)^{-1}$, where $A \triangleq \begin{bmatrix} A_{11} & A_{12} \\ A_{12}^* & A_{22} \end{bmatrix}$.

xiii) $\phi\colon \mathbf{P}^n \mapsto \mathbf{H}^n$ defined by $\phi(A) \triangleq \log A$.

The following functions are strongly increasing:

xiv) $\phi\colon \mathbf{H}^n \mapsto [0, \infty)$ defined by $\phi(A) \triangleq \operatorname{tr} BAB^*$, where $B \in \mathbb{F}^{m\times n}$ and $\operatorname{rank} B = m$.

xv) $\phi\colon \mathbf{H}^n \mapsto \mathbb{R}$ defined by $\phi(A) \triangleq \operatorname{tr} AB$, where $B \in \mathbf{P}^n$.

xvi) $\phi\colon \mathbf{N}^n \mapsto [0, \infty)$ defined by $\phi(A) \triangleq \operatorname{tr} A^r$, where $r > 0$.

xvii) $\phi\colon \mathbf{N}^n \mapsto [0, \infty)$ defined by $\phi(A) \triangleq \det A$.

Proof. For the proof of *iii)*, see [1800]. To prove *xviii)*, let $A, B \in \mathbf{P}^n$, and assume that $A \le B$. Then, for all $r \in [0, 1]$, it follows from *vii)* that $r^{-1}(A^r - I) \le r^{-1}(B^r - I)$. Letting $r \downarrow 0$ and using Proposition 10.6.4 yields $\log A \le \log B$, which proves that log is nondecreasing. See [1124, p. 139] and [2586, p. 256]. To prove that log is increasing, assume that $A < B$, and let $\varepsilon > 0$ satisfy $A + \varepsilon I < B$. Then, $\log A < \log(A + \varepsilon I) \le \log B$. □

Finally, we consider convex functions defined with respect to matrix inequalities. The following definition generalizes Definition 1.6.6 in the case where $n = m = p = 1$.

Definition 10.6.14. Let $\mathcal{D} \subseteq \mathbb{F}^{n\times m}$ be a convex set, and let $\phi\colon \mathcal{D} \mapsto \mathbf{H}^p$. Then, the following terminology is defined:

i) ϕ is *convex* if, for all $\alpha \in [0, 1]$ and $A_1, A_2 \in \mathcal{D}$,

$$\phi[\alpha A_1 + (1 - \alpha)A_2] \le \alpha\phi(A_1) + (1 - \alpha)\phi(A_2). \tag{10.6.10}$$

ii) ϕ is *concave* if $-\phi$ is convex.

iii) ϕ is *strictly convex* if, for all $\alpha \in (0, 1)$ and distinct $A_1, A_2 \in \mathcal{D}$,

$$\phi[\alpha A_1 + (1 - \alpha)A_2] < \alpha\phi(A_1) + (1 - \alpha)\phi(A_2). \tag{10.6.11}$$

iv) ϕ is *strictly concave* if $-\phi$ is strictly convex.

Theorem 10.6.15. Let $\mathcal{S}_1, \mathcal{S}_2 \subseteq \mathbb{R}$, let $\phi : \mathcal{S}_1 \mapsto \mathcal{S}_2$, and assume that ϕ is continuous. Then, the following statements hold:

i) Let $\mathcal{S}_1 = \mathcal{S}_2 = (0,\infty)$, and assume that, for all $n \ge 1$, ϕ: $\mathbf{P}^n \mapsto \mathbf{P}^n$ is increasing. Then, ψ: $\mathbf{P}^n \mapsto \mathbf{P}^n$ defined by $\psi(x) = 1/\phi(x)$ is convex.

ii) Let $\mathcal{S}_1 = \mathcal{S}_2 = [0,\infty)$. Then, for all $n \ge 1$, ϕ: $\mathbf{N}^n \mapsto \mathbf{N}^n$ is increasing if and only if, for all $n \ge 1$, ϕ: $\mathbf{N}^n \mapsto \mathbf{N}^n$ is concave.

iii) Let $\mathcal{S}_1 = [0,\infty)$ and $\mathcal{S}_2 = \mathbb{R}$. Then, for all $n \ge 1$, ϕ: $\mathbf{N}^n \mapsto \mathbf{H}^n$ is convex and $\phi(0) \le 0$ if and only if, for all $n \ge 1$, ψ: $\mathbf{P}^n \mapsto \mathbf{H}^n$ defined by $\psi(x) = \phi(x)/x$ is increasing.

Proof. See [449, pp. 120–122]. □

Lemma 10.6.16. Let $\mathcal{D} \subseteq \mathbb{F}^{n\times m}$ and $\mathcal{S} \subseteq \mathbf{H}^p$ be convex sets, and let ϕ_1: $\mathcal{D} \mapsto \mathcal{S}$ and ϕ_2: $\mathcal{S} \mapsto \mathbf{H}^q$. Then, the following statements hold:

i) If ϕ_1 is convex and ϕ_2 is nondecreasing and convex, then $\phi_2 \circ \phi_1$: $\mathcal{D} \mapsto \mathbf{H}^q$ is convex.

ii) If ϕ_1 is concave and ϕ_2 is nonincreasing and convex, then $\phi_2 \circ \phi_1$: $\mathcal{D} \mapsto \mathbf{H}^q$ is convex.

iii) If $\mathcal{S}$ is symmetric, $\phi_2(-A) = -\phi_2(A)$ for all $A \in \mathcal{S}$, ϕ_1 is concave, and ϕ_2 is nonincreasing and concave, then $\phi_2 \circ \phi_1$: $\mathcal{D} \mapsto \mathbf{H}^q$ is convex.

iv) If $\mathcal{S}$ is symmetric, $\phi_2(-A) = -\phi_2(A)$ for all $A \in \mathcal{S}$, ϕ_1 is convex, and ϕ_2 is nondecreasing and concave, then $\phi_2 \circ \phi_1$: $\mathcal{D} \mapsto \mathbf{H}^q$ is convex.

Proof. To prove *i)* and *ii)*, let $\alpha \in [0,1]$ and $A_1, A_2 \in \mathcal{D}$. In both cases it follows that

$$\phi_2(\phi_1[\alpha A_1 + (1-\alpha)A_2]) \le \phi_2[\alpha\phi_1(A_1) + (1-\alpha)\phi_1(A_2)] \le \alpha\phi_2[\phi_1(A_1)] + (1-\alpha)\phi_2[\phi_1(A_2)].$$

iii) and *iv)* follow from *i)* and *ii)*, respectively. □

Proposition 10.6.17. The following functions are convex:

i) ϕ: $\mathbf{N}^n \mapsto \mathbf{N}^n$ defined by $\phi(A) \triangleq A^r$, where $r \in [1,2]$.

ii) ϕ: $\mathbf{N}^n \mapsto \mathbf{N}^n$ defined by $\phi(A) \triangleq A^2$.

iii) ϕ: $\mathbf{P}^n \mapsto \mathbf{P}^n$ defined by $\phi(A) \triangleq A^{-r}$, where $r \in [0,1]$.

iv) ϕ: $\mathbf{P}^n \mapsto \mathbf{P}^n$ defined by $\phi(A) \triangleq A^{-1}$.

v) ϕ: $\mathbf{P}^n \mapsto \mathbf{P}^n$ defined by $\phi(A) \triangleq A^{-1/2}$.

vi) ϕ: $\mathbf{N}^n \mapsto -\mathbf{N}^n$ defined by $\phi(A) \triangleq -A^r$, where $r \in [0,1]$.

vii) ϕ: $\mathbf{N}^n \mapsto -\mathbf{N}^n$ defined by $\phi(A) \triangleq -A^{1/2}$.

viii) ϕ: $\mathbf{N}^n \mapsto \mathbf{H}^m$ defined by $\phi(A) \triangleq \gamma BAB^*$, where $\gamma \in \mathbb{R}$ and $B \in \mathbb{F}^{m\times n}$.

ix) ϕ: $\mathbf{N}^n \mapsto \mathbf{N}^m$ defined by $\phi(A) \triangleq BA^rB^*$, where $B \in \mathbb{F}^{m\times n}$ and $r \in [1,2]$.

x) ϕ: $\mathbf{P}^n \mapsto \mathbf{N}^m$ defined by $\phi(A) \triangleq BA^{-r}B^*$, where $B \in \mathbb{F}^{m\times n}$ and $r \in [0,1]$.

xi) ϕ: $\mathbf{N}^n \mapsto -\mathbf{N}^m$ defined by $\phi(A) \triangleq -BA^rB^*$, where $B \in \mathbb{F}^{m\times n}$ and $r \in [0,1]$.

xii) ϕ: $\mathbf{P}^n \mapsto -\mathbf{P}^m$ defined by $\phi(A) \triangleq -(BA^{-r}B^*)^{-p}$, where $B \in \mathbb{F}^{m\times n}$, $\operatorname{rank} B = m$, and $r, p \in [0,1]$.

xiii) ϕ: $\mathbb{F}^{n\times m} \mapsto \mathbf{N}^n$ defined by $\phi(A) \triangleq ABA^*$, where $B \in \mathbf{N}^m$.

xiv) ϕ: $\mathbf{P}^n \times \mathbb{F}^{m\times n} \mapsto \mathbf{N}^m$ defined by $\phi(A,B) \triangleq BA^{-1}B^*$.

xv) ϕ: $\{A \in \mathbb{F}^{n\times n}\colon A + A^* > 0\} \mapsto \mathbf{P}^n$ defined by $\phi(A) \triangleq (A^{-1} + A^{-*})^{-1}$.

xvi) ϕ: $\mathbf{N}^n \times \mathbf{N}^n \mapsto \mathbf{N}^n$ defined by $\phi(A,B) \triangleq -A(A+B)^+B$.

xvii) ϕ: $\mathbf{N}^{n+m} \mapsto \mathbf{N}^n$ defined by $\phi(A) \triangleq -A_{22}|A$, where $A \triangleq \begin{bmatrix} A_{11} & A_{12} \\ A_{12}^* & A_{22} \end{bmatrix}$.

xviii) ϕ: $\mathbf{P}^{n+m} \mapsto \mathbf{P}^n$ defined by $\phi(A) \triangleq (A_{22}|A)^{-1}$, where $A \triangleq \begin{bmatrix} A_{11} & A_{12} \\ A_{12}^* & A_{22} \end{bmatrix}$.

xix) ϕ: $\mathbf{H}^n \mapsto [0,\infty)$ defined by $\phi(A) \triangleq \operatorname{tr} A^k$, where k is a nonnegative even integer.

xx) ϕ: $\mathbf{P}^n \mapsto (0,\infty)$ defined by $\phi(A) \triangleq \operatorname{tr} A^{-r}$, where $r > 0$.

xxi) ϕ: $\mathbf{P}^n \mapsto (-\infty,0)$ defined by $\phi(A) \triangleq -(\operatorname{tr} A^{-r})^{-p}$, where $r, p \in [0,1]$.

xxii) $\phi\colon\ \mathbf{N}^n\times\mathbf{N}^n\mapsto(-\infty,0]$ defined by $\phi(A,B)\triangleq-\operatorname{tr}(A^r+B^r)^{1/r}$, where $r\in[0,1]$.

xxiii) $\phi\colon\ \mathbf{N}^n\times\mathbf{N}^n\mapsto[0,\infty)$ defined by $\phi(A,B)\triangleq\operatorname{tr}(A^2+B^2)^{1/2}$.

xxiv) $\phi\colon\ \mathbf{N}^n\times\mathbf{N}^m\mapsto\mathbb{R}$ defined by $\phi(A,B)\triangleq-\operatorname{tr}A^rXB^pX^*$, where $X\in\mathbb{F}^{n\times m}$, $r,p\ge0$, and $r+p\le1$.

xxv) $\phi\colon\ \mathbf{N}^n\mapsto(-\infty,0)$ defined by $\phi(A)\triangleq-\operatorname{tr}A^rXA^pX^*$, where $X\in\mathbb{F}^{n\times n}$, $r,p\ge0$, and $r+p\le1$.

xxvi) $\phi\colon\ \mathbf{P}^n\times\mathbf{P}^m\times\mathbb{F}^{m\times n}\mapsto\mathbb{R}$ defined by $\phi(A,B,X)\triangleq(\operatorname{tr}A^{-p}XB^{-r}X^*)^q$, where $r,p\ge0$, $r+p\le1$, and $q\ge(2-r-p)^{-1}$.

xxvii) $\phi\colon\ \mathbf{P}^n\times\mathbb{F}^{n\times n}\mapsto[0,\infty)$ defined by $\phi(A,X)\triangleq\operatorname{tr}A^{-p}XA^{-r}X^*$, where $r,p\ge0$ and $r+p\le1$.

xxviii) $\phi\colon\ \mathbf{P}^n\times\mathbb{F}^{n\times n}\mapsto[0,\infty)$ defined by $\phi(A)\triangleq\operatorname{tr}A^{-p}XA^{-r}X^*$, where $r,p\in[0,1]$ and $X\in\mathbb{F}^{n\times n}$.

xxix) $\phi\colon\ \mathbf{P}^n\mapsto\mathbb{R}$ defined by $\phi(A)\triangleq-\operatorname{tr}([A^r,X][A^{1-r},X])$, where $r\in(0,1)$ and $X\in\mathbf{H}^n$.

xxx) $\phi\colon\ \mathbf{P}^n\mapsto\mathbf{H}^n$ defined by $\phi(A)\triangleq-\log A$.

xxxi) $\phi\colon\ \mathbf{P}^n\mapsto\mathbf{H}^m$ defined by $\phi(A)\triangleq A\log A$.

xxxii) $\phi\colon\ \mathbf{N}^n\backslash\{0\}\mapsto\mathbb{R}$ defined by $\phi(A)\triangleq-\log\operatorname{tr}A^r$, where $r\in[0,1]$.

xxxiii) $\phi\colon\ \mathbf{P}^n\mapsto\mathbb{R}$ defined by $\phi(A)\triangleq\log\operatorname{tr}A^{-1}$.

xxxiv) $\phi\colon\ \mathbf{P}^n\times\mathbf{P}^n\mapsto(0,\infty)$ defined by $\phi(A,B)\triangleq\operatorname{tr}[A(\log A-\log B)]$.

xxxv) $\phi\colon\ \mathbf{P}^n\times\mathbf{P}^n\to[0,\infty)$ defined by $\phi(A,B)\triangleq-e^{[1/(2n)]\operatorname{tr}(\log A+\log B)}$.

xxxvi) $\phi\colon\ \mathbf{H}^n\mapsto\mathbb{R}$ defined by $\phi(A)\triangleq\log\operatorname{tr}e^A$.

xxxvii) $\phi\colon\ \mathbf{N}^n\mapsto(-\infty,0]$ defined by $\phi(A)\triangleq-(\det A)^{1/n}$.

xxxviii) $\phi\colon\ \mathbf{P}^n\mapsto(0,\infty)$ defined by $\phi(A)\triangleq\log\det BA^{-1}B^*$, where $B\in\mathbb{F}^{m\times n}$ and $\operatorname{rank}B=m$.

xxxix) $\phi\colon\ \mathbf{P}^n\mapsto\mathbb{R}$ defined by $\phi(A)\triangleq-\log\det A$.

xl) $\phi\colon\ \mathbf{P}^n\mapsto(0,\infty)$ defined by $\phi(A)\triangleq\det A^{-1}$.

xli) $\phi\colon\ \mathbf{P}^n\mapsto\mathbb{R}$ defined by $\phi(A)\triangleq\log(\det A_k/\det A)$, where $k\in\{1,\ldots,n-1\}$ and A_k is the leading $k\times k$ principal submatrix of A.

xlii) $\phi\colon\ \mathbf{P}^n\mapsto\mathbb{R}$ defined by $\phi(A)\triangleq-\det A/\det A_{(\mathcal{S})}$, where $\mathcal{S}\subseteq\{1,\ldots,n\}$.

xliii) $\phi\colon\ \mathbf{N}^n\times\mathbf{N}^m\mapsto-\mathbf{N}^{nm}$ defined by $\phi(A,B)\triangleq-A^{r_1}\otimes B^{r_2}$, where $r_1,r_2\in[0,1]$ satisfy $r_1+r_2\le1$.

xliv) $\phi\colon\ \mathbf{P}^n\times\mathbf{N}^m\mapsto\mathbf{N}^{nm}$ defined by $\phi(A,B)\triangleq A^{-r}\otimes B^{1+r}$, where $r\in[0,1]$.

xlv) $\phi\colon\ \mathbf{N}^n\times\mathbf{N}^n\mapsto-\mathbf{N}^n$ defined by $\phi(A,B)\triangleq-A^{r_1}\odot B^{r_2}$, where $r_1,r_2\in[0,1]$ satisfy $r_1+r_2\le1$.

xlvi) $\phi\colon\ \mathbf{H}^n\mapsto\mathbb{R}$ defined by $\phi(A)\triangleq\sum_{i=1}^k\lambda_i(A)$, where $k\in\{1,\ldots,n\}$.

xlvii) $\phi\colon\ \mathbf{H}^n\mapsto\mathbb{R}$ defined by $\phi(A)\triangleq-\sum_{i=k}^n\lambda_i(A)$, where $k\in\{1,\ldots,n\}$.

Proof. *i)* and *iii)* are given in [88] and [449, p. 123]. Let $\alpha\in[0,1]$ for the remainder of the proof. To prove *ii)*, let $A_1,A_2\in\mathbf{H}^n$. Since

$$\alpha(1-\alpha)=(\alpha-\alpha^2)^{1/2}[(1-\alpha)-(1-\alpha)^2]^{1/2},$$

it follows that

$$\begin{aligned}0&\le[(\alpha-\alpha^2)^{1/2}A_1-[(1-\alpha)-(1-\alpha)^2]^{1/2}A_2]^2\\&=(\alpha-\alpha^2)A_1^2+[(1-\alpha)-(1-\alpha)^2]A_2^2-\alpha(1-\alpha)(A_1A_2+A_2A_1).\end{aligned}$$

Hence,

$$[\alpha A_1+(1-\alpha)A_2]^2\le\alpha A_1^2+(1-\alpha)A_2^2,$$

which shows that $\phi(A)=A^2$ is convex.

To prove *iv*), let $A_1, A_2 \in \mathbf{P}^n$. Then, $\begin{bmatrix} A_1^{-1} & I \\ I & A_1 \end{bmatrix}$ and $\begin{bmatrix} A_2^{-1} & I \\ I & A_2 \end{bmatrix}$ are positive semidefinite, and thus

$$\alpha\begin{bmatrix} A_1^{-1} & I \\ I & A_1 \end{bmatrix} + (1-\alpha)\begin{bmatrix} A_2^{-1} & I \\ I & A_2 \end{bmatrix} = \begin{bmatrix} \alpha A_1^{-1} + (1-\alpha)A_2^{-1} & I \\ I & \alpha A_1 + (1-\alpha)A_2 \end{bmatrix}$$

is positive semidefinite. It now follows from Proposition 10.2.5 that $[\alpha A_1 + (1-\alpha)A_2]^{-1} \le \alpha A_1^{-1} + (1-\alpha)A_2^{-1}$, which shows that $\phi(A) = A^{-1}$ is convex.

To prove *v*), note that $\phi(A) = A^{-1/2} = \phi_2[\phi_1(A)]$, where $\phi_1(A) \triangleq A^{1/2}$ and $\phi_2(B) \triangleq B^{-1}$. It follows from *vii*) that ϕ_1 is concave, while it follows from *iv*) that ϕ_2 is convex. Furthermore, *x*) of Proposition 10.6.13 implies that ϕ_2 is nonincreasing. It thus follows from *ii*) of Lemma 10.6.16 that $\phi(A) = A^{-1/2}$ is convex.

To prove *vi*), let $A \in \mathbf{P}^n$, and note that $\phi(A) = -A^r = \phi_2[\phi_1(A)]$, where $\phi_1(A) \triangleq A^{-r}$ and $\phi_2(B) \triangleq -B^{-1}$. It follows from *iii*) that ϕ_1 is convex, while it follows from *iv*) that ϕ_2 is concave. Furthermore, *x*) of Proposition 10.6.13 implies that ϕ_2 is nondecreasing. It thus follows from *iv*) of Lemma 10.6.16 that $\phi(A) = A^r$ is convex on $\mathbf{P}^n$. Continuity implies that $\phi(A) = A^r$ is convex on $\mathbf{N}^n$.

To prove *vii*), let $A_1, A_2 \in \mathbf{N}^n$. Then,

$$0 \le \alpha(1-\alpha)(A_1^{1/2} - A_2^{1/2})^2,$$

which is equivalent to

$$[\alpha A_1^{1/2} + (1-\alpha)A_2^{1/2}]^2 \le \alpha A_1 + (1-\alpha)A_2.$$

Using *viii*) of Proposition 10.6.13 yields

$$\alpha A_1^{1/2} + (1-\alpha)A_2^{1/2} \le [\alpha A_1 + (1-\alpha)A_2]^{1/2}.$$

Finally, multiplying by -1 shows that $\phi(A) = -A^{1/2}$ is convex.

The proof of *viii*) is immediate. *ix*), *x*), and *xi*) follow from *i*), *iii*), and *vi*), respectively.

To prove *xii*), note that $\phi(A) = -(BA^{-r}B^*)^{-p} = \phi_2[\phi_1(A)]$, where $\phi_1(A) = -BA^{-r}B^*$ and $\phi_2(C) = C^{-p}$. *x*) implies that ϕ_1 is concave, while *iii*) implies that ϕ_2 is convex. Furthermore, *ix*) of Proposition 10.6.13 implies that ϕ_2 is nonincreasing. It thus follows from *ii*) of Lemma 10.6.16 that $\phi(A) = -(BA^{-r}B^*)^{-p}$ is convex.

To prove *xiii*), let $A_1, A_2 \in \mathbb{F}^{n\times m}$, and let $B \in \mathbf{N}^m$. Then,

$$\begin{aligned} 0 &\le \alpha(1-\alpha)(A_1 - A_2)B(A_1 - A_2)^* \\ &= \alpha A_1 B A_1^* + (1-\alpha)A_2 B A_2^* - [\alpha A_1 + (1-\alpha)A_2]B[\alpha A_1 + (1-\alpha)A_2]^*. \end{aligned}$$

Thus,

$$[\alpha A_1 + (1-\alpha)A_2]B[\alpha A_1 + (1-\alpha)A_2]^* \le \alpha A_1 B A_1^* + (1-\alpha)A_2 B A_2^*,$$

which shows that $\phi(A) = ABA^*$ is convex.

To prove *xiv*), let $A_1, A_2 \in \mathbf{P}^n$ and $B_1, B_2 \in \mathbb{F}^{m\times n}$. Then, it follows from Proposition 10.2.5 that $\begin{bmatrix} B_1A_1^{-1}B_1^* & B_1 \\ B_1^* & A_1 \end{bmatrix}$ and $\begin{bmatrix} B_2A_2^{-1}B_2^* & B_2 \\ B_2^* & A_2 \end{bmatrix}$ are positive semidefinite, and thus

$$\alpha\begin{bmatrix} B_1A_1^{-1}B_1^* & B_1 \\ B_1^* & A_1 \end{bmatrix} + (1-\alpha)\begin{bmatrix} B_2A_2^{-1}B_2^* & B_2 \\ B_2^* & A_2 \end{bmatrix} = \begin{bmatrix} \alpha B_1A_1^{-1}B_1^* + (1-\alpha)B_2A_2^{-1}B_2^* & \alpha B_1 + (1-\alpha)B_2 \\ \alpha B_1^* + (1-\alpha)B_2^* & \alpha A_1 + (1-\alpha)A_2 \end{bmatrix}$$

is positive semidefinite. It thus follows from Proposition 10.2.5 that

$$[\alpha B_1 + (1-\alpha)B_2][\alpha A_1 + (1-\alpha)A_2]^{-1}[\alpha B_1 + (1-\alpha)B_2]^* \le \alpha B_1A_1^{-1}B_1^* + (1-\alpha)B_2A_2^{-1}B_2^*,$$

which shows that $\phi(A, B) = BA^{-1}B^*$ is convex.

xv) is given in [1987]. *xvi*) follows from Fact 10.24.20.

To prove *xvii*), let $A \triangleq \begin{bmatrix} A_{11} & A_{12} \\ A_{12}^* & A_{22} \end{bmatrix} \in \mathbf{P}^{n+m}$ and $B \triangleq \begin{bmatrix} B_{11} & B_{12} \\ B_{12} & B_{22} \end{bmatrix} \in \mathbf{P}^{n+m}$. Then, it follows from *xiv*) with A_1, B_1, A_2, B_2 replaced by $A_{22}, A_{12}, B_{22}, B_{12}$, respectively, that

$$[\alpha A_{12} + (1-\alpha)B_{12}][\alpha A_{22} + (1-\alpha)B_{22}]^{-1}[\alpha A_{12} + (1-\alpha)B_{12}]^* \le \alpha A_{12}A_{22}^{-1}A_{12}^* + (1-\alpha)B_{12}B_{22}^{-1}B_{12}^*.$$

Hence,

$$\begin{aligned}
-[\alpha A_{22} + (1-\alpha)B_{22}]|[\alpha A + (1-\alpha)B]& \\
&= [\alpha A_{12} + (1-\alpha)B_{12}][\alpha A_{22} + (1-\alpha)B_{22}]^{-1}[\alpha A_{12} + (1-\alpha)B_{12}]^* - [\alpha A_{11} + (1-\alpha)B_{11}] \\
&\le \alpha(A_{12}A_{22}^{-1}A_{12}^* - A_{11}) + (1-\alpha)(B_{12}B_{22}^{-1}B_{12}^* - B_{11}) \\
&= \alpha(-A_{22}|A) + (1-\alpha)(-B_{22}|B),
\end{aligned}$$

which shows that $\phi(A) \triangleq -A_{22}|A$ is convex. By continuity, the result holds for $A \in \mathbf{N}^{n+m}$.

To prove *xviii*), note that $\phi(A) = (A_{22}|A)^{-1} = \phi_2[\phi_1(A)]$, where $\phi_1(A) = A_{22}|A$ and $\phi_2(B) = B^{-1}$. It follows from *xv*) that ϕ_1 is concave, while it follows from *iv*) that ϕ_2 is convex. Furthermore, *x*) of Proposition 10.6.13 implies that ϕ_2 is nonincreasing. It thus follows from Lemma 10.6.16 that $\phi(A) \triangleq (A_{22}|A)^{-1}$ is convex.

xix) is given in [523, p. 106]. *xx*) is given in by Theorem 9 of [1817].

To prove *xxi*), note that $\phi(A) = -(\operatorname{tr} A^{-r})^{-p} = \phi_2[\phi_1(A)]$, where $\phi_1(A) = \operatorname{tr} A^{-r}$ and $\phi_2(B) = -B^{-p}$. *iii*) implies that ϕ_1 is convex and that ϕ_2 is concave. Furthermore, *ix*) of Proposition 10.6.13 implies that ϕ_2 is nondecreasing. It thus follows from *iv*) of Lemma 10.6.16 that $\phi(A) = -(\operatorname{tr} A^{-r})^{-p}$ is convex.

xxii) and *xxiii*) are proved in [633]. *xxiv*)–*xxviii*) are given by Corollary 1.1, Theorem 1, Corollary 2.1, Theorem 2, and Theorem 8, respectively, of [633]. A proof of *xxiv*) in the case where $p = 1 - r$ is given in [449, p. 273]. *xxix*) is proved in [449, p. 274] and [633]. *xxx*) is given in [454, p. 113]. *xxxi*) is given in [449, p. 123], [454, p. 113], and [1123].

To prove *xxxii*), note that $\phi(A) = -\log \operatorname{tr} A^r = \phi_2[\phi_1(A)]$, where $\phi_1(A) = \operatorname{tr} A^r$ and $\phi_2(x) = -\log x$. *vi*) implies that ϕ_1 is concave. Furthermore, ϕ_2 is convex and nonincreasing. It thus follows from *ii*) of Lemma 10.6.16 that $\phi(A) = -\log \operatorname{tr} A^r$ is convex.

xxxiii) is given in [2099]. *xxxiv*) is given in [449, p. 275]. *xxxv*) is given in [102]. *xxxvi*) is given in [1381, p. 144].

To prove *xxxvii*), let $A_1, A_2 \in \mathbf{N}^n$. From Corollary 10.4.15 it follows that $(\det A_1)^{1/n} + (\det A_2)^{1/n} \le [\det(A_1 + A_2)]^{1/n}$. Replacing A_1 and A_2 by αA_1 and $(1-\alpha)A_2$, respectively, and multiplying by -1 shows that $\phi(A) = -(\det A)^{1/n}$ is convex.

xxxviii) is proved in [2099]. *xxxix*) is a special case of *xxxvii*), which is due to K. Fan. See [788] and [789, p. 679]. To prove *xxxviii*), note that $\phi(A) = -n\log[(\det A)^{1/n}] = \phi_2[\phi_1(A)]$, where $\phi_1(A) = (\det A)^{1/n}$ and $\phi_2(x) = -n\log x$. It follows from *xix*) that ϕ_1 is concave. Since ϕ_2 is nonincreasing and convex, it follows from *ii*) of Lemma 10.6.16 that $\phi(A) = -\log\det A$ is convex.

To prove *xl*), note that $\phi(A) = \det A^{-1} = \phi_2[\phi_1(A)]$, where $\phi_1(A) = \log\det A^{-1}$ and $\phi_2(x) = e^x$. It follows from *xx*) that ϕ_1 is convex. Since ϕ_2 is nondecreasing and convex, it follows from *i*) of Lemma 10.6.16 that $\phi(A) = \det A^{-1}$ is convex.

xli) and *xlii*) are given in [788] and [789, pp. 684, 685]. Next, *xliii*) is given in [449, p. 273], [454, p. 114], and [2977, p. 9]. *xliv*) is given in [454, p. 114]. *xlv*) is given in [2977, p. 9]. Finally, *xlvi*) is given in [1969, p. 478] and [1971, p. 688]. *xlvii*) follows from *xlvi*). □

The following result follows from *xvii*) of Proposition 10.6.17 by setting $\alpha = 1/2$. Versions of this result appear in [639, 1355, 1800, 1869] and [2263, p. 152].

Corollary 10.6.18. Let $A \triangleq \begin{bmatrix} A_{11} & A_{12} \\ A_{12}^* & A_{22} \end{bmatrix} \in \mathbb{F}^{n+m}$ and $B \triangleq \begin{bmatrix} B_{11} & B_{12} \\ B_{12}^* & B_{22} \end{bmatrix} \in \mathbb{F}^{n+m}$, and assume that A and B

are positive semidefinite. Then,

$$A_{11}|A + B_{11}|B \le (A_{11} + B_{11})|(A + B). \tag{10.6.12}$$

The following result, which follows from *xlv*) and *xlvi*) of Proposition 10.6.17, yields inequalities for the eigenvalues of a pair of Hermitian matrices. These are the *Lidskii-Wielandt inequalities.*

Corollary 10.6.19. Let $A, B \in \mathbf{H}^n$. Then, for all $k \in \{1, \dots, n\}$ and $1 \le i_1 < \cdots < i_k \le n$,

$$\sum_{j=1}^{k} \lambda_{i_j}(A) + \sum_{j=n-k+1}^{n} \lambda_{i_j}(B) \le \sum_{i=1}^{k} \lambda_i(A + B) \le \sum_{i=1}^{k} [\lambda_i(A) + \lambda_i(B)] \tag{10.6.13}$$

with equality in both inequalities for $k = n$. Furthermore, for all $k \in \{1, \dots, n\}$,

$$\sum_{i=k}^{n} [\lambda_i(A) + \lambda_i(B)] \le \sum_{i=k}^{n} \lambda_i(A + B) \tag{10.6.14}$$

with equality for $k = 1$.

Proof. See [449, p. 71], [1450, p. 201], [1969, p. 688], [1971, p. 478], [2403, p. 116], and [2991, p. 356]. □

10.7 Facts on Range and Rank

Fact 10.7.1. Let $A, B \in \mathbb{F}^{n\times n}$, and assume that A and B are positive semidefinite. Then, there exists $\alpha > 0$ such that $A \le \alpha B$ if and only if $\mathcal{R}(A) \subseteq \mathcal{R}(B)$. If these conditions hold, then $\operatorname{rank} A \le \operatorname{rank} B$. **Source:** Theorem 10.6.2.

Fact 10.7.2. Let $A \in \mathbf{N}^n$. Then, $\mathcal{R}(A) = \mathcal{R}(A^{1/2}) = \mathcal{R}(A^2)$.

Fact 10.7.3. Let $A, B \in \mathbb{F}^{n\times n}$, assume that A is positive semidefinite, and assume that B is either positive semidefinite or skew Hermitian. Then, the following statements hold:

i) $\mathcal{N}(A + B) = \mathcal{N}(A) \cap \mathcal{N}(B)$.

ii) $\mathcal{R}(A + B) = \mathcal{R}(A) + \mathcal{R}(B) = \mathcal{R}([A\ \ B]) = \operatorname{span}[\mathcal{R}(A) \cup \mathcal{R}(B)]$.

iii) $\operatorname{rank}(A + B) = \dim[\mathcal{R}(A) + \mathcal{R}(B)] = \operatorname{rank}\,[A\ \ B] = \operatorname{rank}\left[\begin{smallmatrix} A \\ B \end{smallmatrix}\right]$.

Source: In *i*), "$\supseteq$" is given by Fact 3.13.15. To prove "$\subseteq$," let $x \in \mathcal{N}(A + B)$, which implies that $Ax + Bx = 0$, and thus $x^*Ax + x^*Bx = 0$. In the case where B is positive semidefinite, it follows that $x^*Ax = x^*Bx = 0$, and thus $Ax = Bx = 0$. In the case where B is skew Hermitian, it follows that $x^*Ax = -x^*Bx = 0$, and thus $Ax = 0$ and therefore $Bx = 0$. To prove *ii*) in the case where B is positive semidefinite, it follows from *i*) and Fact 3.12.17 that $\mathcal{R}(A + B) = \mathcal{N}(A + B)^\perp = [\mathcal{N}(A) \cap \mathcal{N}(B)]^\perp = \mathcal{N}(A)^\perp + \mathcal{N}(B)^\perp = \mathcal{R}(A) + \mathcal{R}(B)$. In the case where B is skew Hermitian, $\mathcal{R}(A + B) = \mathcal{N}(A - B)^\perp = [\mathcal{N}(A) \cap \mathcal{N}(B)]^\perp = \mathcal{N}(A)^\perp + \mathcal{N}(B)^\perp = \mathcal{R}(A) + \mathcal{R}(B)$. In both cases,

$$\mathcal{R}([A\ \ B]) = \mathcal{R}\left([A\ \ B]\left[\begin{smallmatrix} A \\ B^* \end{smallmatrix}\right]\right) = \mathcal{R}(A^2 + BB^*) = \mathcal{R}(A^2) + \mathcal{R}(BB^*) = \mathcal{R}(A) + \mathcal{R}(B) = \mathcal{R}(A + B).$$

Related: Fact 3.12.16, Fact 4.10.31, and Fact 4.18.11.

Fact 10.7.4. Let $A, B \in \mathbb{F}^{n\times n}$, and assume that A and B are positive semidefinite. Then, $(A + B)(A + B)^+$ is the projector onto the subspace $\mathcal{R}(A + B) = \mathcal{R}(A) + \mathcal{R}(B) = \mathcal{R}([A\ \ B]) = \operatorname{span}[\mathcal{R}(A) \cup \mathcal{R}(B)]$. **Source:** Fact 10.7.3. **Related:** Fact 8.8.20.

Fact 10.7.5. Let $A \in \mathbb{F}^{n\times n}$, and assume that $A + A^* \ge 0$. Then, the following statements hold:

i) $\mathcal{N}(A) = \mathcal{N}(A + A^*) \cap \mathcal{N}(A - A^*)$.

ii) $\mathcal{R}(A) = \mathcal{R}(A + A^*) + \mathcal{R}(A - A^*)$.

iii) $\operatorname{rank} A = \operatorname{rank}\,[A + A^*\ \ A - A^*]$.

Source: Fact 10.7.3.

Fact 10.7.6. Let $A_1,\ldots,A_r \in \mathbb{F}^{n\times n}$, and assume that $A_1,\ldots,A_r$ are positive semidefinite. Then,

$$\mathcal{R}\left(\sum_{i=1}^{r} A_i\right) = \sum_{i=1}^{r} \mathcal{R}(A_i) = \mathcal{R}([A_1 \ \cdots \ A_r]).$$

Source: Theorem 3.5.3, Fact 3.14.4, and Fact 10.7.2.

Fact 10.7.7. Let $A, B \in \mathbb{F}^{n\times n}$, and assume that A and B are positive semidefinite. Then,

$$\operatorname{rank}\begin{bmatrix} A & B \\ 0 & A \end{bmatrix} = \operatorname{rank}\begin{bmatrix} A & A+B \\ 0 & A \end{bmatrix} = \operatorname{rank} A + \operatorname{rank}(A+B).$$

Source: Use Theorem 10.3.5 to simultaneously diagonalize A and B.

Fact 10.7.8. Let $A, B \in \mathbb{F}^{n\times n}$, assume that A and B are positive semidefinite, and let $\mathcal{S} \subseteq \{1,\ldots,n\}$. Then,

$$\operatorname{rank}(AB)_{(\mathcal{S})} \le \operatorname{rank}(A^k)_{(\mathcal{S})} = \operatorname{rank}(A_{(\mathcal{S})})^k = \operatorname{rank} A_{(\mathcal{S})}.$$

Source: [2991, p. 226].

Fact 10.7.9. Let $A \in \mathbb{F}^{n\times n}$, and assume that A is either positive semidefinite or an irreducible, singular M-matrix. Then, the following statements hold:

i) If $\alpha \subset \{1,\ldots,n\}$, then $\operatorname{rank} A \le \operatorname{rank} A_{(\alpha)} + \operatorname{rank} A_{(\alpha^\sim)}$.

ii) If $\alpha,\beta \subseteq \{1,\ldots,n\}$, then $\operatorname{rank} A_{(\alpha\cup\beta)} \le \operatorname{rank} A_{(\alpha)} + \operatorname{rank} A_{(\beta)} - \operatorname{rank} A_{(\alpha\cap\beta)}$.

iii) If $1 \le k \le n-1$, then

$$k \sum_{\{\alpha:\ \operatorname{card}(\alpha)=k+1\}} \operatorname{rank} A_{(\alpha)} \le (n-k) \sum_{\{\alpha:\ \operatorname{card}(\alpha)=k\}} \operatorname{rank} A_{(\alpha)}.$$

If, in addition, A is either positive definite, a nonsingular M-matrix, or totally positive, then all three inequalities hold as equalities. **Source:** [1896]. **Related:** Fact 10.16.39.

Fact 10.7.10. Let $A \in \mathbb{F}^{n\times n}$ and $B \in \mathbb{F}^{n\times m}$, assume that A is Hermitian, $\operatorname{rank} A = \operatorname{rank} B = m$, and $\mathcal{R}(B) = \mathcal{R}(A)$. Then, B^*AB is nonsingular, and $B = AB(B^*AB)^{-1}B^*B$. **Source:** [278].

10.8 Facts on Unitary Matrices and the Polar Decomposition

Fact 10.8.1. Let $A \in \mathbb{F}^{n\times m}$. Then, $A\langle A\rangle = \langle A^*\rangle A$.

Fact 10.8.2. Let $A \in \mathbb{F}^{n\times m}$. Then, the following statements hold:

i) If $\operatorname{rank} A = m$, then $A\langle A\rangle^{-1}$ is left inner.

ii) If $\operatorname{rank} A = n$, then $\langle A^*\rangle^{-1}A$ is right inner.

iii) If $n = m$ and A is nonsingular, then $A\langle A\rangle^{-1}$ and $\langle A^*\rangle^{-1}A$ are unitary.

Fact 10.8.3. Let $A \in \mathbb{F}^{n\times m}$, where $m \le n$. Then, there exist $M \in \mathbb{F}^{m\times m}$ and $S \in \mathbb{F}^{n\times m}$ such that M is positive semidefinite, S satisfies $S^*S = I_m$, and $A = SM$. Furthermore, M is given uniquely by $M = \langle A\rangle$. If, in addition, $\operatorname{rank} A = m$, then M is positive definite and S is given uniquely by

$$S = A\langle A\rangle^{-1} = \frac{2}{\pi}A\int_0^\infty (t^2I + A^*A)^{-1}\,\mathrm{d}t.$$

Source: [1391, Chapter 8].

Fact 10.8.4. Let $A \in \mathbb{F}^{n\times m}$, where $n \le m$. Then, there exist $M \in \mathbb{F}^{n\times n}$ and $S \in \mathbb{F}^{n\times m}$ such that M is positive semidefinite, S satisfies $SS^* = I_n$, and $A = MS$. Furthermore, M is given uniquely by $M = \langle A^*\rangle$. If, in addition, $\operatorname{rank} A = n$, then S is given uniquely by

$$S = \langle A^*\rangle^{-1}A = \frac{2}{\pi}A^*\int_0^\infty (t^2I + AA^*)^{-1}\,\mathrm{d}t.$$

Source: [1391, Chapter 8].

Fact 10.8.5. Let $A \in \mathbb{F}^{n\times n}$, where A is nonsingular. Then, there exist unique matrices $M, S \in \mathbb{F}^{n\times n}$ such that $A = SM$, M is positive definite, and S is unitary. In particular, $M = \langle A\rangle$ and $S = A\langle A\rangle^{-1}$.

Fact 10.8.6. Let $A \in \mathbb{F}^{n\times n}$, where A is nonsingular. Then, there exist unique matrices $M, S \in \mathbb{F}^{n\times n}$ such that $A = MS$, M is positive definite, and S is unitary. In particular, $M = \langle A^*\rangle$ and $S = \langle A^*\rangle^{-1}A$. **Related:** Fact 10.8.1 and Fact 10.8.5.

Fact 10.8.7. Let $M_1, M_2 \in \mathbb{F}^{n\times n}$, and assume that M_1, M_2 are positive definite. Furthermore, let $S_1, S_2 \in \mathbb{F}^{n\times n}$, assume that S_1, S_2 are unitary, and assume that $M_1S_1 = S_2M_2$. Then, $S_1 = S_2$. **Source:** Let $A = M_1S_1 = S_2M_2$. Then, $S_1 = (S_2M_2^2S_2^*)^{-1/2}S_2M_2 = S_2$.

Fact 10.8.8. Let $A \in \mathbb{F}^{n\times n}$. Then, there exist a unitary matrix $S \in \mathbb{F}^{n\times n}$ and unique matrices $M_1, M_2 \in \mathbf{N}^n$ such that $A = M_1S = SM_2$. In particular, $M_1 = \langle A^*\rangle$ and $M_2 = \langle A\rangle$. **Remark:** If A is singular, then S is not uniquely determined.

Fact 10.8.9. Let $A, M, S \in \mathbb{F}^{n\times n}$, and assume that M is positive semidefinite, S is unitary, and $A = MS$. Then, the following statements are equivalent:

i) A is normal.

ii) $MS = SM$.

iii) $AS = SA$.

iv) $AM = MA$.

Source: [1448, p. 414] and [2991, p. 295].

Fact 10.8.10. Let $A, B \in \mathbb{F}^{n\times n}$, assume that A and B are unitary, and assume that $A + B$ is nonsingular. Then, the unitary factor in the polar decomposition of $A + B$ is $A(A^*B)^{1/2}$. **Source:** [1391, p. 216] and [2064]. **Remark:** $(A^*B)^{1/2}$ is the principal square root.

Fact 10.8.11. Let $A \in \mathbb{F}^{n\times n}$ be semicontractive, and define $B \in \mathbb{F}^{2n\times 2n}$ by

$$B \triangleq \begin{bmatrix} A & (I - AA^*)^{1/2} \\ (I - A^*A)^{1/2} & -A^* \end{bmatrix}.$$

Then, B is unitary. **Source:** [1084, p. 180] and [2991, p. 191].

Fact 10.8.12. Let $A \in \mathbb{F}^{n\times m}$, and define $B \in \mathbb{F}^{(n+m)\times(n+m)}$ by

$$B \triangleq \begin{bmatrix} (I + A^*A)^{-1/2} & -A^*(I + AA^*)^{-1/2} \\ (I + AA^*)^{-1/2}A & (I + AA^*)^{-1/2} \end{bmatrix}.$$

Then, B is unitary, $\det B = 1$, and $B^* = \tilde{I}B\tilde{I}$, where $\tilde{I} \triangleq \mathrm{diag}(I_m, -I_n)$. **Source:** [1324].

Fact 10.8.13. Let $A \in \mathbb{F}^{n\times m}$, assume that A is contractive, and define $B \in \mathbb{F}^{(n+m)\times(n+m)}$ by

$$B \triangleq \begin{bmatrix} (I - A^*A)^{-1/2} & A^*(I - AA^*)^{-1/2} \\ (I - AA^*)^{-1/2}A & (I - AA^*)^{-1/2} \end{bmatrix}.$$

Then, B is Hermitian, $\det B = 1$, and $B^*\tilde{I}B = \tilde{I}$, where $\tilde{I} \triangleq \mathrm{diag}(I_m, -I_n)$. **Source:** [1324].

10.9 Facts on Structured Positive-Semidefinite Matrices

Fact 10.9.1. Let $z \in \mathbb{C}$, assume that z is not real, and define the Hermitian, Toeplitz matrix $A \in \mathbb{C}^{n\times n}$ by

$$A \triangleq \begin{bmatrix} 1 & z & z & \cdots & z \\ \overline{z} & 1 & z & \cdots & z \\ \overline{z} & \overline{z} & 1 & \cdots & z \\ \vdots & \vdots & \vdots & \ddots & \vdots \\ \overline{z} & \overline{z} & \overline{z} & \cdots & 1 \end{bmatrix}.$$

Then, the following statements hold:

i) A is singular if and only if

$$\left(\frac{1-z}{1-\bar{z}}\right)^n = \frac{z}{\bar{z}}.$$

ii) Assume that $\operatorname{Im} z > 0$. Then, A is positive semidefinite if and only if

$$\frac{\arg z}{n} + \frac{(n-1)\pi}{n} \le \arg(z-1).$$

Source: [1789]. **Example:** Let $n = 2$ and $z = \alpha \jmath$, where $\alpha > 0$. Then, A is positive semidefinite if and only if $3\pi/4 < \pi/2 + \operatorname{atan} 1/\alpha$; that is, if and only if $\alpha < 1$. **Related:** Fact 6.10.21.

Fact 10.9.2. Let $\phi\colon \mathbb{R} \mapsto \mathbb{C}$, and assume that, for all $x_1, \ldots, x_n \in \mathbb{R}$, the matrix $A \in \mathbb{C}^{n\times n}$, where $A_{(i,j)} \triangleq \phi(x_i - x_j)$, is positive semidefinite. (The function ϕ is *positive semidefinite*.) Then, the following statements hold:

i) For all $x_1, x_2 \in \mathbb{R}$, it follows that $|\phi(x_1) - \phi(x_2)|^2 \le 2\phi(0)\operatorname{Re}[\phi(0) - \phi(x_1 - x_2)]$.

ii) The function $\psi\colon \mathbb{R} \mapsto \mathbb{C}$, where, for all $x \in \mathbb{R}$, $\psi(x) \triangleq \overline{\phi(x)}$, is positive semidefinite.

iii) For all $\alpha \in \mathbb{R}$, the function $\psi\colon \mathbb{R} \mapsto \mathbb{C}$, where, for all $x \in \mathbb{R}$, $\psi(x) \triangleq \phi(\alpha x)$, is positive semidefinite.

iv) The function $\psi\colon \mathbb{R} \mapsto \mathbb{C}$, where, for all $x \in \mathbb{R}$, $\psi(x) \triangleq |\phi(x)|^2$, is positive semidefinite.

v) The function $\psi\colon \mathbb{R} \mapsto \mathbb{C}$, where, for all $x \in \mathbb{R}$, $\psi(x) \triangleq \operatorname{Re}\phi(x)$, is positive semidefinite.

vi) If $\phi_1\colon \mathbb{R} \mapsto \mathbb{C}$ and $\phi_2\colon \mathbb{R} \mapsto \mathbb{C}$ are positive semidefinite, then $\phi_3\colon \mathbb{R} \mapsto \mathbb{C}$, where, for all $x \in \mathbb{R}$, $\phi_3(x) \triangleq \phi_1(x)\phi_2(x)$, is positive semidefinite.

vii) If $\phi_1\colon \mathbb{R} \mapsto \mathbb{C}$ and $\phi_2\colon \mathbb{R} \mapsto \mathbb{C}$ are positive semidefinite and α_1, α_2 are positive numbers, then $\phi_3\colon \mathbb{R} \mapsto \mathbb{C}$, where, for all $x \in \mathbb{R}$, $\phi_3(x) \triangleq \alpha_1\phi_1(x) + \alpha_2\phi_2(x)$, is positive semidefinite.

viii) Let $\phi\colon \mathbb{R} \mapsto \mathbb{C}$, and assume that ϕ is bounded and continuous. Furthermore, for all $x, y \in \mathbb{R}$, define $K\colon \mathbb{R}\times\mathbb{R} \mapsto \mathbb{C}$ by $K(x,y) \triangleq \phi(x-y)$. Then, ϕ is positive semidefinite if and only if, for every continuous integrable function $f\colon \mathbb{R} \mapsto \mathbb{C}$, it follows that

$$\int_{\mathbb{R}^2} K(x,y)f(x)\overline{f(y)}\,\mathrm{d}x\,\mathrm{d}y \ge 0.$$

Source: [454, pp. 141–144]. **Remark:** K is a *kernel function* associated with a reproducing kernel space. See [1160] for extensions to vector arguments, and [2400] and Fact 10.9.7 for applications.

Fact 10.9.3. Let $A \in \mathbb{C}^{n\times n}$, assume that A is positive definite, and let $B \in \mathbb{C}^{n\times n}$, where, for all $i, j \in \{1, \ldots, n\}$, $B_{(i,j)} \triangleq A_{(i,j)}/\sqrt{A_{(i,i)}A_{(j,j)}}$. Then, B is positive definite. **Source:** [2991, p. 204].

Fact 10.9.4. Let $a_1 < \cdots < a_n$ be positive numbers, and define $A \in \mathbb{R}^{n\times n}$ by $A_{(i,j)} \triangleq \min\{a_i, a_j\}$. Then, A is positive definite,

$$\det A = \prod_{i=1}^{n}(a_i - a_{i-1}),$$

and, for all $x \in \mathbb{R}^n$,

$$x^{\mathsf{T}}A^{-1}x = \sum_{i=1}^{n}\frac{(x_{(i)} - x_{(i-1)})^2}{a_i - a_{i-1}},$$

where $a_0 \triangleq 0$ and $x_{(0)} \triangleq 0$. **Remark:** A is a covariance matrix arising in Brownian motion. See [1379, p. 132] and [2911, p. 50]. **Related:** Fact 10.9.5.

Fact 10.9.5. Let $a_1, \ldots, a_n > 0$, and define $A \in \mathbb{R}^{n\times n}$ by the following expressions:

i) $A_{(i,j)} \triangleq \min\{a_i, a_j\}$.

ii) $A_{(i,j)} \triangleq \frac{1}{\max\{a_i,a_j\}}$.

iii) $A_{(i,j)} \triangleq \frac{a_i}{a_j}$, where $a_1 \le \cdots \le a_n$.

iv) $A_{(i,j)} \triangleq \frac{1}{a_i+a_j}$.

v) $A_{(i,j)} \triangleq \sqrt{a_i a_j}$.

vi) $A_{(i,j)} \triangleq 1/\sqrt{a_i a_j}$.

vii) $A_{(i,j)} \triangleq \frac{a_i a_j}{a_i+a_j}$.

viii) $A_{(i,j)} \triangleq \frac{a_i^p - a_j^p}{a_i-a_j}$, where $p \in [0,1]$.

ix) $A_{(i,j)} \triangleq \frac{a_i^p + a_j^p}{a_i+a_j}$, where $p \in [-1,1]$.

x) $A_{(i,j)} \triangleq \frac{\log a_i - \log a_j}{a_i-a_j}$.

Then, A is positive semidefinite. If, in addition, $\alpha > 0$, then $A^{\odot\alpha}$ is positive semidefinite. **Source:** [451, 452], [454, pp. 153, 178, 189], [455], and [922, p. 90]. **Remark:** A in *iii)* is the Schur product of the matrices defined in *i)* and *ii)*. **Related:** Fact 4.23.4, Fact 10.9.4, and Fact 10.9.6.

Fact 10.9.6. Let $A \in \mathbb{R}^{n\times n}$, and assume that, for all $i,j \in \{1,\dots,n\}$, $A_{(i,j)} = \min\{i,j\}$; that is,

$$A \triangleq \begin{bmatrix} 1 & 1 & 1 & \cdots & 1 & 1 \\ 1 & 2 & 2 & \cdots & 2 & 2 \\ 1 & 2 & 3 & \ddots & 3 & 3 \\ \vdots & \vdots & \ddots & \ddots & \ddots & \vdots \\ 1 & 2 & 3 & \ddots & n-1 & n-1 \\ 1 & 2 & 3 & \cdots & n-1 & n \end{bmatrix}.$$

Then, $\det A = 1$, A is positive definite,

$$A^{-1} = \begin{bmatrix} 2 & -1 & 0 & \cdots & 0 & 0 \\ -1 & 2 & -1 & \cdots & 0 & 0 \\ 0 & -1 & 2 & \ddots & 0 & 0 \\ \vdots & \vdots & \ddots & \ddots & \ddots & \vdots \\ 0 & 0 & 0 & \ddots & 2 & -1 \\ 0 & 0 & 0 & \cdots & -1 & 1 \end{bmatrix},$$

and

$$\operatorname{spec}(A) = \{\tfrac{1}{4}\csc^2\tfrac{\pi}{2(2n+1)}, \tfrac{1}{4}\csc^2\tfrac{3\pi}{2(2n+1)}, \dots, \tfrac{1}{4}\csc^2\tfrac{(2n-1)\pi}{2(2n+1)}\}.$$

Source: Six proofs that A is positive definite are given in [455]. **Remark:** $\operatorname{tr} A$ and $\operatorname{tr} A^{-1}$ yield equalities in Fact 2.16.10. Evaluating $\det A$ yields an equality in Fact 2.16.15. **Related:** Fact 10.9.5.

Fact 10.9.7. Let $a_1,\dots,a_n \in \mathbb{R}$, and define $A \in \mathbb{C}^{n\times n}$ by either of the following expressions:

i) $A_{(i,j)} \triangleq \frac{1}{1+\jmath(a_i-a_j)}$.

ii) $A_{(i,j)} \triangleq \frac{1}{1-\jmath(a_i-a_j)}$.

iii) $A_{(i,j)} \triangleq \frac{1}{1+(a_i-a_j)^2}$.

iv) $A_{(i,j)} \triangleq \frac{1}{1+|a_i-a_j|}$.

v) $A_{(i,j)} \triangleq e^{\jmath(a_i-a_j)}$.

vi) $A_{(i,j)} \triangleq \cos(a_i - a_j)$.

vii) $A_{(i,j)} \triangleq \frac{\sin[(a_i-a_j)]}{a_i-a_j}$.

viii) $A_{(i,j)} \triangleq \frac{a_i-a_j}{\sinh[(a_i-a_j)]}$.

ix) $A_{(i,j)} \triangleq \frac{\sinh p(a_i-a_j)}{\sinh(a_i-a_j)}$, where $p \in (0,1)$.

x) $A_{(i,j)} \triangleq \frac{\tanh[(a_i-a_j)]}{a_i-a_j}$.

xi) $A_{(i,j)} \triangleq \frac{\sinh[(a_i-a_j)]}{(a_i-a_j)[\cosh(a_i-a_j)+p]}$, where $p \in (-1,1]$.

xii) $A_{(i,j)} \triangleq \frac{1}{\cosh(a_i-a_j)+p}$, where $p \in (-1,1]$.

xiii) $A_{(i,j)} \triangleq \frac{\cosh p(a_i-a_j)}{\cosh(a_i-a_j)}$, where $p \in [-1,1]$.

xiv) $A_{(i,j)} \triangleq e^{-(a_i-a_j)^2}$.

xv) $A_{(i,j)} \triangleq e^{-|a_i-a_j|^p}$, where $p \in [0,2]$.

xvi) $A_{(i,j)} \triangleq \frac{1+p(a_i-a_j)^2}{1+q(a_i-a_j)^2}$, where $0 \le p \le q$.

xvii) $A_{(i,j)} \triangleq \operatorname{tr} e^{B+\jmath(a_i-a_j)C}$, where $B, C \in \mathbf{H}^n$ and $BC = CB$.

xviii) $A_{(i,j)} \triangleq 1/\det(A_i + A_j)$, where $A_1, \ldots, A_n \in \mathbf{P}^n$.

Then, A is positive semidefinite. Finally, if, α is a nonnegative number and A is defined by either *iv)*, *ix)*, *x)*, *xi)*, *xiii)*, or *xvi)*, then $A^{\odot\alpha}$ is positive semidefinite. **Source:** [454, pp. 141–144, 153, 177, 188], [477], [922, p. 90], and [1448, pp. 400, 401, 456, 457, 462, 463]. *viii)* is given in [2512]. **Remark:** *xv)* is related to the Bessis-Moussa-Villani conjecture in Fact 10.14.45 and Fact 10.14.46. **Related:** Fact 10.9.2. **Problem:** In each case, determine rank A.

Fact 10.9.8. Define $A \in \mathbb{R}^{n\times n}$ by either of the following expressions:

i) $A_{(i,j)} \triangleq \binom{i+j}{i}$.

ii) $A_{(i,j)} \triangleq (i+j)!$.

iii) $A_{(i,j)} \triangleq \min\{i,j\}$.

iv) $A_{(i,j)} \triangleq \gcd\{i,j\}$.

v) $A_{(i,j)} \triangleq \frac{i}{j}$.

vi) $A_{(i,j)} \triangleq \frac{1}{i+j-1}$.

Then, A is positive semidefinite. If, in addition, $\alpha \ge 0$, then $A^{\odot\alpha}$ is positive semidefinite. **Remark:** Fact 10.25.2 implies that $A^{\odot\alpha}$ is positive semidefinite for all $\alpha \in [0, n-2]$. **Remark:** *i)* is the *Pascal matrix*. See [9, 451, 973] and *xix)* of Fact 1.16.13. **Remark:** The determinant of *iv)* can be expressed in terms of the Euler totient function. See [155, 552] and Fact 1.20.4. **Remark:** *v)* is a special case of *iii)* of Fact 10.9.5. This is the *Lehmer matrix*. **Remark:** *vi)* is the Hilbert matrix, which is positive definite. See Fact 4.23.4. **Related:** Fact 3.16.28.

Fact 10.9.9. Let $a_1, \ldots, a_n \ge 0$, let $p \in \mathbb{R}$, assume that either $a_1, \ldots, a_n > 0$ or $p > 0$, and, for all $i, j \in \{1, \ldots, n\}$, define $A \in \mathbb{R}^{n\times n}$ by $A_{(i,j)} \triangleq (a_i a_j)^p$. Then, A is positive semidefinite. **Source:** Let $a \triangleq [a_1 \ \cdots \ a_n]^\mathsf{T}$ and $A \triangleq a^{\odot p} a^{\odot p\mathsf{T}}$.

Fact 10.9.10. Let $a \in \mathbb{C}^n$, assume that, for all $i \in \{1, \ldots, n\}$, $a_{(i)} + a_{(j)} \ne 0$, and, for all

$i, j \in \{1, \ldots, n\}$, define $A \in \mathbb{C}^{n\times n}$ by

$$A_{(i,j)} \triangleq \frac{1}{a_{(i)} + a_{(j)}}.$$

Then, A is positive definite if and only if $a >> 0$ and the components of a are distinct. Furthermore, A is totally positive if and only if A is positive definite and either $a = a^{\downarrow}$ or $a = a^{\uparrow}$. **Remark:** See [1042]. **Remark:** A is a Cauchy matrix. See Fact 10.9.20. **Related:** Fact 15.19.23.

Fact 10.9.11. Let $a_1, \ldots, a_n > 0$, let $\alpha > 0$, and, for all $i, j \in \{1, \ldots, n\}$, define $A \in \mathbb{R}^{n\times n}$ by

$$A_{(i,j)} \triangleq \frac{1}{(a_i + a_j)^{\alpha}}.$$

Then, A is positive semidefinite. **Source:** [451], [454, pp. 24, 25], and [2251]. **Remark:** For $\alpha = 1$, A is a Cauchy matrix. See Fact 4.27.5. **Related:** Fact 7.12.15 and Fact 10.9.10.

Fact 10.9.12. Let $a_1, \ldots, a_n > 0$, let $r \in [-1, 1]$, and, for all $i, j \in \{1, \ldots, n\}$, define $A \in \mathbb{R}^{n\times n}$ by

$$A_{(i,j)} \triangleq \frac{a_i^r + a_j^r}{a_i + a_j}.$$

Then, A is positive semidefinite. **Source:** [2977, p. 74].

Fact 10.9.13. Let $a_1, \ldots, a_n > 0$, let $q > 0$, let $p \in [-q, q]$, and, for all $i, j \in \{1, \ldots, n\}$, define $A \in \mathbb{R}^{n\times n}$ by

$$A_{(i,j)} \triangleq \frac{a_i^p + a_j^p}{a_i^q + a_j^q}.$$

Then, A is positive semidefinite. **Source:** Let $r = p/q$ and $b_i = a_i^q$. Then, $A_{(i,j)} = (b_i^r + b_j^r)/(b_i + b_j)$. Now, use Fact 10.9.12. See [1988] for the case where $q \geq p \geq 0$. **Remark:** The case where $q = 1$ and $p = 0$ yields a Cauchy matrix. In the case where $n = 2$, $A \geq 0$ yields Fact 2.2.31. **Problem:** Under what conditions is A positive definite?

Fact 10.9.14. Let $a_1, \ldots, a_n > 0$, let $p \in (-1, \infty)$, and define $A \in \mathbb{R}^{n\times n}$ by

$$A_{(i,j)} \triangleq \frac{1}{a_i^3 + p(a_i^2 a_j + a_i a_j^2) + a_j^3}.$$

Then, A is positive semidefinite. **Source:** [459].

Fact 10.9.15. Let $a_1, \ldots, a_n > 0$, $p \in [-1, 1]$, $q \in (-2, 2]$, and, for all $i, j \in \{1, \ldots, n\}$, define $A \in \mathbb{R}^{n\times n}$ by

$$A_{(i,j)} \triangleq \frac{a_i^p + a_j^p}{a_i^2 + q a_i a_j + a_j^2}.$$

Then, A is positive semidefinite. **Source:** [459, 2974] and [2977, p. 76].

Fact 10.9.16. Let $A \in \mathbb{F}^{n\times n}$, and assume that A is Hermitian, for all $i \in \{1, \ldots, n\}$, $A_{(i,i)} > 0$, and, for all $i, j \in \{1, \ldots, n\}$, $|A_{(i,j)}| < \frac{1}{n-1}\sqrt{A_{(i,i)}A_{(j,j)}}$. Then, A is positive definite. **Source:** Note that

$$x^*Ax = \sum_{i=1}^{n-1}\sum_{j=i+1}^{n} \begin{bmatrix} x_{(i)} \\ x_{(j)} \end{bmatrix}^* \begin{bmatrix} \frac{1}{n-1}A_{(i,i)} & A_{(i,j)} \\ \overline{A_{(i,j)}} & \frac{1}{n-1}A_{(j,j)} \end{bmatrix} \begin{bmatrix} x_{(i)} \\ x_{(j)} \end{bmatrix}.$$

Credit: A. Roup.

Fact 10.9.17. Let $A \in \mathbb{F}^{n\times n}$. Then, for all $i, j \in \{1, \ldots, n\}$, $|A_{(i,j)}|^2 < \langle A^*\rangle_{(i,i)}\langle A\rangle_{(j,j)}$. Now, let $B \in \mathbb{F}^{n\times n}$, and assume that B is semicontractive. Then,

$$|\det(A \odot B)|^2 \leq \prod_{i=1}^{n} \langle A^*\rangle_{(i,i)}\langle A\rangle_{(i,i)}.$$

Source: [1460].

Fact 10.9.18. Let $A \in \mathbb{R}^{n\times n}$, assume that A is positive semidefinite, assume that $A_{(i,i)} > 0$ for all $i \in \{1,\dots,n\}$, and define $B \in \mathbb{R}^{n\times n}$ by

$$B_{(i,j)} \triangleq \frac{A_{(i,j)}}{\mu_\alpha(A_{(i,i)}, A_{(j,j)})},$$

where, for positive scalars α, x, y,

$$\mu_\alpha(x,y) \triangleq \left[\tfrac{1}{2}(x^\alpha + y^\alpha)\right]^{1/\alpha}.$$

Then, B is positive semidefinite. If, in addition, A is positive definite, then B is positive definite. In particular, letting $\alpha \downarrow 0$, $\alpha = 1$, and $\alpha \to \infty$, the respective matrices $C, D, E \in \mathbb{R}^{n\times n}$ defined by

$$C_{(i,j)} \triangleq \frac{A_{(i,j)}}{\sqrt{A_{(i,i)}A_{(j,j)}}}, \quad D_{(i,j)} \triangleq \frac{2A_{(i,j)}}{A_{(i,i)} + A_{(j,j)}}, \quad E_{(i,j)} \triangleq \frac{A_{(i,j)}}{\max\{A_{(i,i)}, A_{(j,j)}\}}$$

are positive semidefinite. Finally, if A is positive definite, then C, D, and E are positive definite. **Source:** [2350]. **Remark:** The assumption that the diagonal entries of A are positive can be weakened. See [2350]. **Related:** Fact 2.2.58.

Fact 10.9.19. Let $\alpha, \beta, \gamma \in [0, \pi]$, and define $A \in \mathbb{R}^{3\times 3}$ by

$$A = \begin{bmatrix} 1 & \cos\alpha & \cos\gamma \\ \cos\alpha & 1 & \cos\beta \\ \cos\gamma & \cos\beta & 1 \end{bmatrix}.$$

Then,

$$\det A = 4[\sin\tfrac{1}{2}(\alpha+\beta+\gamma)][\sin\tfrac{1}{2}(-\alpha+\beta+\gamma)][\sin\tfrac{1}{2}(\alpha-\beta+\gamma)]\sin\tfrac{1}{2}(\alpha+\beta-\gamma).$$

Furthermore, A is positive semidefinite if and only if $\alpha \le \beta+\gamma, \beta \le \alpha+\gamma, \gamma \le \alpha+\beta$, and $\alpha+\beta+\gamma \le 2\pi$. Finally, A is positive definite if and only if all of these inequalities are strict. **Source:** [305, 955].

Fact 10.9.20. Let $\lambda_1, \dots, \lambda_n \in \mathbb{C}$, assume that, for all $i \in \{1,\dots,n\}$, $\operatorname{Re}\lambda_i < 0$, and, for all $i, j \in \{1,\dots,n\}$, define $A \in \mathbb{C}^{n\times n}$ by

$$A_{(i,j)} \triangleq \frac{-1}{\overline{\lambda_i} + \lambda_j}.$$

Then, A is positive definite. **Source:** Note that $A = 2B \odot (1_{n\times n} - C)^{\odot -1}$, where $B_{(i,j)} = \frac{1}{(\overline{\lambda_i}-1)(\lambda_j-1)}$ and $C_{(i,j)} = \frac{(\overline{\lambda_i}+1)(\lambda_j+1)}{(\overline{\lambda_i}-1)(\lambda_j-1)}$. Then, B is positive semidefinite and $(1_{n\times n} - C)^{\odot -1} = 1_{n\times n} + C + C^{\odot 2} + C^{\odot 3} + \cdots$.) **Remark:** A is the solution of a Lyapunov equation. See Fact 16.22.18 and Fact 16.22.19. **Remark:** A is a Cauchy matrix. See Fact 4.23.4, Fact 4.27.5, Fact 4.27.6, and Fact 10.9.20. **Remark:** A is the Gram matrix whose entries are inner products of the functions $f_i(t) = e^{-\lambda_i t}$. See [454, p. 3].

Fact 10.9.21. Let $\lambda_1, \dots, \lambda_n \in \text{OIUD}$ and $w_1, \dots, w_n \in \mathbb{C}$. Then, there exists an analytic function $\phi\colon \text{OIUD} \mapsto \text{OIUD}$ such that $\phi(\lambda_i) = w_i$ for all $i \in \{1,\dots,n\}$ if and only if $A \in \mathbb{C}^{n\times n}$ is positive semidefinite, where, for all $i, j \in \{1,\dots,n\}$,

$$A_{(i,j)} \triangleq \frac{1 - \overline{w_i}w_j}{1 - \overline{\lambda_i}\lambda_j}.$$

Source: [1996]. **Remark:** A is a *Pick matrix*.

Fact 10.9.22. Let $\alpha_0,\ldots,\alpha_n > 0$, and define the tridiagonal matrix $A \in \mathbb{R}^{n\times n}$ by

$$A \triangleq \begin{bmatrix} \alpha_0+\alpha_1 & -\alpha_1 & 0 & 0 & \cdots & 0 \\ -\alpha_1 & \alpha_1+\alpha_2 & -\alpha_2 & 0 & \cdots & 0 \\ 0 & -\alpha_2 & \alpha_2+\alpha_3 & -\alpha_3 & \cdots & 0 \\ \vdots & \vdots & \vdots & \vdots & \iddots & \vdots \\ 0 & 0 & 0 & 0 & \cdots & \alpha_{n-1}+\alpha_n \end{bmatrix}.$$

Then, for all $k = 2,\ldots,n$,

$$\det A_{(\{1,\ldots,k\})} = \left(\sum_{i=0}^{k} \alpha_i^{-1}\right)\alpha_0\alpha_1\cdots\alpha_k.$$

Furthermore, A is positive definite. **Source:** [302, p. 115]. **Remark:** A is a stiffness matrix arising in structural analysis. **Related:** Fact 4.24.4.

Fact 10.9.23. Let $A \in \mathbb{R}^{n\times n}$, where $n \ge 1$, be the tridiagonal, Toeplitz matrix

$$A \triangleq \begin{bmatrix} b & a & 0 & \cdots & 0 & 0 \\ a & b & a & \cdots & 0 & 0 \\ 0 & a & b & \ddots & 0 & 0 \\ \vdots & \vdots & \ddots & \ddots & \ddots & \vdots \\ 0 & 0 & 0 & \ddots & b & a \\ 0 & 0 & 0 & \cdots & a & b \end{bmatrix}.$$

Then,

$$\operatorname{spec}(A) = \left\{b + 2|a|\cos\frac{i\pi}{n+1}\colon\ i \in \{1,\ldots,n\}\right\}.$$

Furthermore, A is positive semidefinite if and only if $b + 2|a|\cos\frac{n\pi}{n+1} \ge 0$. Finally, A is positive definite if and only if $b + 2|a|\cos\frac{n\pi}{n+1} > 0$. **Related:** Fact 7.12.46.

10.10 Facts on Equalities and Inequalities for One Matrix

Fact 10.10.1. Let $A \in \mathbf{H}^n$. Then, the following statements are equivalent:

i) Neither A nor $-A$ is positive semidefinite.

ii) At least one of the following statements holds:

a) There exists an even integer $m \le n$ and a principal submatrix $B \in \mathbb{F}^{m\times m}$ such that $\det B < 0$.

b) There exist odd integers $k, l \le n$ and principal submatrices $C \in \mathbb{F}^{k\times k}$ and $D \in \mathbb{F}^{l\times l}$ such that $(\det C)\det D < 0$.

Source: [2991, p. 206].

Fact 10.10.2. Let $A \in \mathbb{F}^{n\times n}$, and assume that A is positive semidefinite. Then, $A^{\odot 2}$, $A^{\odot 3}$, and $|A|^{\odot 2}$ are positive semidefinite. If, in addition, $n \le 3$, then, $|A|$ is positive semidefinite. **Source:** [1953]. **Remark:** If $n \ge 4$, then this result does not hold. Let

$$A = \begin{bmatrix} 1 & \frac{1}{\sqrt{3}} & 0 & -\frac{1}{\sqrt{3}} \\ \frac{1}{\sqrt{3}} & 1 & \frac{1}{\sqrt{3}} & 0 \\ 0 & \frac{1}{\sqrt{3}} & 1 & \frac{1}{\sqrt{3}} \\ -\frac{1}{\sqrt{3}} & 0 & \frac{1}{\sqrt{3}} & 1 \end{bmatrix}.$$

Then, $\operatorname{mspec}(A) = \{1-\sqrt{6}/3, 1-\sqrt{6}/3, 1+\sqrt{6}/3, 1+\sqrt{6}/3\}_{\rm ms}$, whereas $\operatorname{mspec}(|A|) = \{1, 1, 1-$

$\sqrt{12}/3, 1+\sqrt{12}/3\}_{\rm ms}$. **Related:** Fact 11.12.7.

Fact 10.10.3. Let $A \in \mathbb{F}^{n\times n}$, assume that A is positive semidefinite, and let $\mathcal{S} \subseteq \{1,\dots,n\}$. Then,

$$(A_{(\mathcal{S})})^2 \le (A^2)_{(\mathcal{S})}, \quad (A^{1/2})_{(\mathcal{S})} \le (A_{(\mathcal{S})})^{1/2}.$$

Now, assume that A is positive definite. Then,

$$(A_{(\mathcal{S})})^{-1} \le (A^{-1})_{(\mathcal{S})}, \quad (A_{(\mathcal{S})})^{-1/2} \le (A^{-1/2})_{(\mathcal{S})}.$$

Source: [2991, p. 219].

Fact 10.10.4. Let $x \in \mathbb{F}^n$. Then, $xx^* \le x^*xI$.

Fact 10.10.5. Let $x \in \mathbb{F}^n$, assume that x is nonzero, and define $A \in \mathbb{F}^{n\times n}$ by $A \triangleq x^*xI - xx^*$. Then, A is positive semidefinite, $\operatorname{mspec}(A) = \{x^*x,\dots,x^*x,0\}_{\rm ms}$, and $\operatorname{rank} A = n-1$.

Fact 10.10.6. Let $x, y \in \mathbb{F}^n$, assume that x and y are linearly independent, and define $A \in \mathbb{F}^{n\times n}$ by

$$A \triangleq (x^*x + y^*y)I - xx^* - yy^*.$$

Then, A is positive definite. Now, let $\mathbb{F} = \mathbb{R}$. Then,

$$\operatorname{mspec}(A) = \Big\{x^{\mathsf T}x + y^{\mathsf T}y, \dots, x^{\mathsf T}x + y^{\mathsf T}y,\ \tfrac{1}{2}(x^{\mathsf T}x + y^{\mathsf T}y) + \sqrt{\tfrac{1}{4}(x^{\mathsf T}x - y^{\mathsf T}y)^2 + (x^{\mathsf T}y)^2},$$
$$\tfrac{1}{2}(x^{\mathsf T}x + y^{\mathsf T}y) - \sqrt{\tfrac{1}{4}(x^{\mathsf T}x - y^{\mathsf T}y)^2 + (x^{\mathsf T}y)^2}\Big\}_{\rm ms}.$$

Source: To show that A is positive definite, write $A = B+C$, where $B \triangleq x^*xI - xx^*$ and $C \triangleq y^*yI - yy^*$. Then, using Fact 10.10.5 it follows that $\mathcal{N}(B) = \operatorname{span}\{x\}$ and $\mathcal{N}(C) = \operatorname{span}\{y\}$. Now, it follows from Fact 10.7.3 that $\mathcal{N}(A) = \mathcal{N}(B) \cap \mathcal{N}(C) = \{0\}$. Therefore, A is nonsingular and thus positive definite. The expression for $\operatorname{mspec}(A)$ follows from Fact 6.9.15.

Fact 10.10.7. Let $x_1,\dots,x_n \in \mathbb{R}^3$, assume that $\operatorname{span}\{x_1,\dots,x_n\} = \mathbb{R}^3$, and define $A \in \mathbb{R}^{3\times 3}$ by

$$A \triangleq \sum_{i=1}^{n} (x_i^{\mathsf T}x_iI - x_ix_i^{\mathsf T}).$$

Then, A is positive definite. Furthermore,

$$\lambda_1(A) < \lambda_2(A) + \lambda_3(A), \quad \mathrm{d}_1(A) < \mathrm{d}_2(A) + \mathrm{d}_3(A).$$

Source: Suppose that $\mathrm{d}_1(A) = A_{(1,1)}$. Then, $\mathrm{d}_2(A) + \mathrm{d}_3(A) - \mathrm{d}_1(A) = 2\sum_{i=1}^n x_{i(3)}^2 > 0$. Now, let the matrix $S \in \mathbb{R}^{3\times 3}$ be such that $SAS^{\mathsf T} = \sum_{i=1}^n (\hat{x}_i^{\mathsf T}\hat{x}_iI - \hat{x}_i\hat{x}_i^{\mathsf T})$ is diagonal, where, for all $i \in \{1,\dots,n\}$, $\hat{x}_i \triangleq Sx_i$. Then, for all $i \in \{1,2,3\}$, $\mathrm{d}_i(A) = \lambda_i(A)$. **Remark:** A is the inertia matrix for a rigid body consisting of n discrete particles. For a homogeneous continuum body $\mathcal{B}$ whose density is ρ, the inertia matrix is given by

$$I = \rho \iiint_{\mathcal{B}} (r^{\mathsf T}rI - rr^{\mathsf T})\,\mathrm{d}x\mathrm{d}y\mathrm{d}z,$$

where $r \triangleq \left[\begin{smallmatrix} x\\ y\\ z\end{smallmatrix}\right]$. **Remark:** The eigenvalues and diagonal entries of A represent the lengths of the sides of triangles. See Fact 5.2.14 and [2196, p. 220].

Fact 10.10.8. Let $A \in \mathbb{F}^{2\times 2}$, assume that A is positive semidefinite and nonzero, and define $B \in \mathbb{F}^{2\times 2}$ by

$$B \triangleq (\operatorname{tr} A + 2\sqrt{\det A})^{-1/2}(A + \sqrt{\det A}I).$$

Then, $B = A^{1/2}$. **Source:** [1304, pp. 84, 266, 267]. **Related:** Fact 3.15.34.

Fact 10.10.9. Let $A \in \mathbb{F}^{n\times n}$, and assume that A is Hermitian. Then,

$$\operatorname{rank} A = \nu_-(A) + \nu_+(A), \quad \operatorname{def} A = \nu_0(A).$$

Fact 10.10.10. Let $A \in \mathbb{F}^{n \times n}$, and assume that A is positive semidefinite. Then, for all $i \in \{1, \ldots, n\}$, $A_{(i,i)} \geq 0$, and, for all $i, j \in \{1, \ldots, n\}$, $|A_{(i,j)}|^2 \leq A_{(i,i)}A_{(j,j)}$.

Fact 10.10.11. Let $A \in \mathbb{F}^{n \times n}$, assume that A is positive semidefinite, and assume that there exists $i \in \{1, \ldots, n\}$ such that $A_{(i,i)} = 0$. Then, $\operatorname{row}_i(A) = 0$ and $\operatorname{col}_i(A) = 0$.

Fact 10.10.12. Let $A \in \mathbb{F}^{n \times n}$. Then, $A \geq 0$ if and only if $A \geq -A$.

Fact 10.10.13. Let $A \in \mathbb{F}^{n \times n}$, and assume that A is Hermitian. Then, $A^2 \geq 0$.

Fact 10.10.14. Let $A \in \mathbb{F}^{n \times n}$, and assume that A is Hermitian. Then, $A \leq \frac{1}{2}[(A^2)^{1/2} + A]$.

Fact 10.10.15. Let $A \in \mathbb{F}^{n \times n}$, and assume that A is skew Hermitian. Then, $A^2 \leq 0$.

Fact 10.10.16. Let $A \in \mathbb{F}^{n \times n}$ and $\alpha > 0$. Then,

$$A^2 + A^{2*} \leq \alpha AA^* + \tfrac{1}{\alpha}A^*A.$$

Furthermore, equality holds if and only if $\alpha A = A^*$.

Fact 10.10.17. Let $A \in \mathbb{F}^{n \times n}$. Then,

$$(A - A^*)^2 \leq 0 \leq (A + A^*)^2 \leq 2(AA^* + A^*A).$$

Fact 10.10.18. Let $A \in \mathbb{F}^{n \times n}$ and $\alpha > 0$. Then,

$$A + A^* \leq \alpha I + \alpha^{-1}AA^*.$$

Equality holds if and only if $A = \alpha I$.

Fact 10.10.19. Let $A \in \mathbb{F}^{n \times n}$, and assume that A is positive definite. Then,

$$2I \leq A + A^{-1}.$$

Equality holds if and only if $A = I$. Furthermore,

$$2n \leq \operatorname{tr} A + \operatorname{tr} A^{-1}.$$

Fact 10.10.20. Let $A \in \mathbb{F}^{n \times n}$, and assume that A is positive definite. Then,

$$(1_{1 \times n}A^{-1}1_{n \times 1})^{-1}1_{n \times n} \leq A.$$

Source: Set $B = 1_{n \times n}$ in Fact 10.25.21. See [2985].

Fact 10.10.21. Let $A \in \mathbb{F}^{n \times n}$, and assume that A is positive definite. Then, $\begin{bmatrix} A & I \\ I & A^{-1} \end{bmatrix}$ is positive semidefinite.

Fact 10.10.22. Let $A \in \mathbb{F}^{n \times n}$, and assume that A is Hermitian. Then, $A^2 \leq A$ if and only if $0 \leq A \leq I$. Furthermore, $A^2 < A$ if and only if $0 < A < I$.

Fact 10.10.23. Let $A \in \mathbb{F}^{n \times n}$, and assume that A is Hermitian. Then, $A + \alpha I \geq 0$ if and only if $\alpha \geq -\lambda_{\min}(A)$. Furthermore,

$$A^2 + A + \tfrac{1}{4}I \geq 0.$$

Fact 10.10.24. Let $A \in \mathbb{F}^{n \times m}$. Then, the following statements are equivalent:

i) $\sigma_{\max}(A) \leq 1$.

ii) $\sigma_{\max}(A^*) \leq 1$.

iii) $\sigma_{\max}(AA^*) \leq 1$.

iv) $\sigma_{\max}(A^*A) \leq 1$.

v) $AA^* \leq I_n$.

vi) $A^*A \leq I_m$.

Source: Fact 7.12.38.

Fact 10.10.25. Let $A \in \mathbb{F}^{n \times n}$, and assume that either $AA^* \leq A^*A$ or $A^*A \leq AA^*$. Then, A is normal. **Source:** *ii)* of Corollary 10.4.10.

Fact 10.10.26. Let $A \in \mathbb{F}^{n\times n}$, and assume that A is a projector. Then, the following statements hold:

i) $0 \le A \le I$.

ii) For all $i \in \{1,\dots,n\}$, $0 \le A_{(i,i)} \le 1$.

iii) For all $i \in \{1,\dots,n\}$, $\sum_{j=1}^n |A_{(i,j)}| \le \sqrt{n(\operatorname{rank} A)A_{(i,i)}}$.

iv) $\sum_{i,j=1}^n |A_{(i,j)}| \le n\sqrt{\operatorname{rank} A}$.

v) For all $i \in \{1,\dots,n\}$, $\sum_{i,j=1, j\ne i}^n |A_{(i,j)}|^2 \le \frac{1}{4}$.

vi) For all $i,j \in \{1,\dots,n\}$, $|A_{(i,j)}|^2 \le \min\{A_{(i,i)}A_{(j,j)}, (1-A_{(i,i)})(1-A_{(j,j)})\}$.

vii) For all distinct $i,j \in \{1,\dots,n\}$, $|A_{(i,j)}| \le \frac{1}{2}$.

Source: [260].

Fact 10.10.27. Let $A \in \mathbb{F}^{n\times n}$, assume that A is (semisimple, Hermitian), and assume that there exists a nonnegative integer k such that $A^k = A^{k+1}$. Then, A is (idempotent, a projector).

Fact 10.10.28. Let $A \in \mathbb{F}^{n\times n}$, and assume that A is nonsingular. Then, $\langle A^{-1}\rangle = \langle A^*\rangle^{-1}$.

Fact 10.10.29. Let $A \in \mathbb{F}^{n\times m}$, and assume that A^*A is nonsingular. Then, $\langle A^*\rangle = A\langle A\rangle^{-1}A^*$.

Fact 10.10.30. Let $A \in \mathbb{F}^{n\times n}$. Then, A is unitary if and only if there exists a nonsingular matrix $B \in \mathbb{F}^{n\times n}$ such that $A = \langle B^*\rangle^{-1}B$. If, in addition, A is real, then $\det A = \operatorname{sign}(\det B)$. **Source:** For necessity, set $B = A$. **Related:** Fact 4.14.6.

Fact 10.10.31. Let $A \in \mathbb{F}^{n\times n}$. Then, the following statements hold:

i) A is positive semidefinite if and only if $A = \langle A\rangle$.

ii) A is normal if and only if $\langle A\rangle = \langle A^*\rangle$.

iii) A is normal if and only if $[A, \langle A\rangle] = 0$.

iv) If A is normal, then $\langle A + A^*\rangle = \langle A\rangle + \langle A^*\rangle$.

v) $\langle A\rangle$ and $\langle A^*\rangle$ are similar.

Source: [987, 1242] and [2991, pp. 290, 291]. **Related:** Fact 4.10.12.

Fact 10.10.32. Let $A \in \mathbb{F}^{n\times n}$. Then,

$$-\langle A\rangle - \langle A^*\rangle \le A + A^* \le \langle A\rangle + \langle A^*\rangle.$$

Source: Fact 10.12.27 and [1783].

Fact 10.10.33. Let $A \in \mathbb{F}^{n\times n}$, assume that A is normal, and let $\alpha,\beta \in (0,\infty)$. Then,

$$-\alpha\langle A\rangle - \beta\langle A^*\rangle \le \langle \alpha A + \beta A^*\rangle \le \alpha\langle A\rangle + \beta\langle A^*\rangle.$$

In particular,

$$-\langle A\rangle - \langle A^*\rangle \le \langle A + A^*\rangle \le \langle A\rangle + \langle A^*\rangle.$$

Source: [1783, 2987]. **Related:** Fact 10.12.27.

Fact 10.10.34. Let $A \in \mathbb{F}^{n\times n}$. Then, there exists a unitary matrix $S \in \mathbb{F}^{n\times n}$ such that

$$\tfrac{1}{2}(A + A^*) \le S^*\langle A\rangle S.$$

Source: Fact 10.11.13, Fact 10.21.9, and [2991, p. 289].

Fact 10.10.35. Let $A \in \mathbb{F}^{n\times n}$. The following statements hold:

i) If $A \in \mathbb{F}^{n\times n}$ is positive definite, then $I + A$ is nonsingular and the matrices $I - B$ and $I + B$ are positive definite, where $B \triangleq (I + A)^{-1}(I - A)$.

ii) If $I + A$ is nonsingular and the matrices $I - B$ and $I + B$ are positive definite, where $B \triangleq (I + A)^{-1}(I - A)$, then A is positive definite.

Source: [1009]. **Related:** Fact 4.13.24, Fact 4.13.25, Fact 4.13.26, Fact 4.28.12, Fact 15.22.10.

Fact 10.10.36. Let $A \in \mathbb{F}^{n\times n}$, assume that $\frac{1}{2\jmath}(A - A^*)$ is positive definite, and define

$$B \triangleq [\tfrac{1}{2\jmath}(A - A^*)]^{1/2}A^{-1}A^*[\tfrac{1}{2\jmath}(A - A^*)]^{-1/2}.$$

Then, B is unitary. **Source:** [1012]. **Remark:** A is *strictly dissipative* if $\frac{1}{2\jmath}(A - A^*)$ is negative definite. A is strictly dissipative if and only if $-\jmath A$ is dissipative. See [1010, 1011]. **Remark:** $A^{-1}A^*$ is similar to a unitary matrix. See Fact 4.13.6. **Related:** Fact 10.15.2 and Fact 10.21.18.

Fact 10.10.37. Let $A \in \mathbb{R}^{n\times n}$, assume that A is positive definite, assume that $A \le I$, and define the sequence $(B_k)_{k=0}^{\infty}$ by $B_0 \triangleq 0$ and $B_{k+1} \triangleq B_k + \frac{1}{2}(A - B_k^2)$. Then, $\lim_{k\to\infty} B_k = A^{1/2}$. **Source:** [353, p. 181]. **Related:** Fact 7.17.23.

Fact 10.10.38. Let $A \in \mathbb{R}^{n\times n}$, assume that A is nonsingular, and define the sequence $(B_k)_{k=0}^{\infty}$ by $B_0 \triangleq A$ and$B_{k+1} \triangleq \frac{1}{2}(B_k + B_k^{-\mathsf{T}})$. Then, $\lim_{k\to\infty} B_k = (AA^{\mathsf{T}})^{-1/2}A$. **Remark:** The limit is a unitary matrix. See Fact 10.10.30 and [300, p. 224].

Fact 10.10.39. Let $a, b \in \mathbb{R}$, and define the symmetric, Toeplitz matrix $A \in \mathbb{R}^{n\times n}$ by $A \triangleq aI_n + b1_{n\times n}$. Then, A is positive definite if and only if $a + nb > 0$ and $a > 0$. **Related:** Fact 3.16.18 and Fact 6.10.21.

Fact 10.10.40. Let $x_1, \ldots, x_n \in \mathbb{R}^m$, and define

$$\overline{x} \triangleq \frac{1}{n}\sum_{j=1}^{n} x_j, \quad S \triangleq \frac{1}{n}\sum_{j=1}^{n}(x_j - \overline{x})(x_j - \overline{x})^{\mathsf{T}}.$$

Then, for all $i \in \{1, \ldots, n\}$,

$$(x_i - \overline{x})(x_i - \overline{x})^{\mathsf{T}} \le (n-1)S.$$

Furthermore, equality holds if and only if all of the elements of $\{x_1, \ldots, x_n\}\backslash\{x_i\}$ are equal. **Source:** [1528, 2132, 2720]. **Related:** This result extends the Laguerre-Samuelson inequality given by Fact 2.11.37.

Fact 10.10.41. Let $x_1, \ldots, x_n \in \mathbb{F}^n$, and define $A \in \mathbb{F}^{n\times n}$ by $A_{(i,j)} \triangleq x_i^*x_j$ for all $i, j \in \{1, \ldots, n\}$, and $B \triangleq [x_1 \ \cdots \ x_n]$. Then, $A = B^*B$. Consequently, A is positive semidefinite and $\operatorname{rank} A = \operatorname{rank} B$. Conversely, let $A \in \mathbb{F}^{n\times n}$, and assume that A is positive semidefinite. Then, there exist $x_1, \ldots, x_n \in \mathbb{F}^n$ such that $A = B^*B$, where $B = [x_1 \ \cdots \ x_n]$. **Source:** The converse follows from Corollary 7.5.5. **Remark:** A is the *Gram matrix* of $x_1, \ldots, x_n$.

Fact 10.10.42. Let $A \in \mathbb{F}^{n\times n}$, and assume that A is positive semidefinite. Then, there exists $B \in \mathbb{F}^{n\times n}$ such that B is lower triangular, B has nonnegative diagonal entries, and $A = BB^*$. If, in addition, A is positive definite, then B is unique and has positive diagonal entries. **Remark:** This is the *Cholesky decomposition*. **Related:** Fact 7.17.12.

Fact 10.10.43. Let $A \in \mathbb{F}^{n\times m}$, and assume that $\operatorname{rank} A = m$. Then, $0 \le A(A^*A)^{-1}A^* \le I$.

Fact 10.10.44. Let $A \in \mathbb{F}^{n\times m}$. Then, $I - A^*A$ is positive definite if and only if $I - AA^*$ is positive definite. If these conditions hold, then

$$(I - A^*A)^{-1} = I + A^*(I - AA^*)^{-1}A.$$

Fact 10.10.45. Let $A \in \mathbb{F}^{n\times n}$, let $\alpha \in (0, \infty)$, and define $A_\alpha \in \mathbb{F}^{n\times n}$ by $A_\alpha \triangleq (\alpha I + A^*A)^{-1}A^*$. Then, the following statements are equivalent:

i) $AA_\alpha = A_\alpha A$.

ii) $AA^* = A^*A$.

Furthermore, the following statements are equivalent:

iii) $A_\alpha A^* = A^*A_\alpha$.

iv) $AA^*A^2 = A^2A^*A$.

Source: [2653]. **Remark:** A_α is a *regularized Tikhonov inverse*.

Fact 10.10.46. Let $A \in \mathbf{P}^{n\times n}$, and define $\alpha \triangleq \lambda_{\max}(A)$ and $\beta \triangleq \lambda_{\min}(A)$. Then,

$$A^{-1} \le \frac{\alpha+\beta}{\alpha\beta}I - \frac{1}{\alpha\beta}A \le \frac{(\alpha+\beta)^2}{4\alpha\beta}A^{-1}.$$

Source: [1970].

Fact 10.10.47. Let $A \in \mathbf{N}^n$. Then, the following statements hold:

i) If $\alpha \in [0,1]$, then $A^\alpha \le \alpha A + (1-\alpha)I$.

ii) If $\alpha \in [0,1]$ and A is positive definite, then

$$[\alpha A^{-1} + (1-\alpha)I]^{-1} \le A^\alpha \le \alpha A + (1-\alpha)I.$$

iii) If $\alpha \ge 1$, then $\alpha A + (1-\alpha)I \le A^\alpha$.

iv) If A is positive definite and either $\alpha \ge 1$ or $\alpha \le 0$, then

$$\alpha A + (1-\alpha)I \le A^\alpha \le [\alpha A^{-1} + (1-\alpha)I]^{-1}.$$

Source: [1124, pp. 122, 123] and [2991, p. 206]. **Remark:** This is a special case of Young's inequality. See Fact 2.2.53, Fact 2.2.50, Fact 10.11.73, Fact 10.14.8, Fact 10.14.33, Fact 10.14.34, and Fact 10.11.38.

Fact 10.10.48. Let $A \in \mathbf{N}^n$, let $\alpha_1, \ldots, \alpha_n > 0$, and assume that $\sum_{i=1}^n \frac{1}{\alpha_i} = 1$. Then,

$$A \le \operatorname{diag}(\alpha_1 A_{(1,1)}, \ldots, \alpha_n A_{(n,n)}).$$

Source: Assume that A is positive definite, and define $B \triangleq \operatorname{diag}(\alpha_1 A_{(1,1)}, \ldots, \alpha_n A_{(n,n)})$. Then, $C \triangleq B^{-1/2}AB^{-1/2}$ is positive definite. Furthermore, $\lambda_1(C) \le \operatorname{tr} C = 1$. Hence, $A \le B$.

Fact 10.10.49. Let $A \in \mathbf{P}^n$. Then,

$$\log A = \int_0^\infty \frac{1}{x+1}I - (xI+A)^{-1}\,dx = \int_0^1 (A-I)[x(A-I)+I]^{-1}\,dx = \int_1^\infty \frac{1}{x}(A-I)[(x-1)I+A]^{-1}\,dx.$$

Source: [1381, pp. 115, 116, 135]. **Related:** Fact 14.2.19.

Fact 10.10.50. Let $A \in \mathbf{P}^n$. Then, $I - A^{-1} \le \log A \le A - I$. Furthermore, if $A \ge I$, then $\log A$ is positive semidefinite, and, if $A > I$, then $\log A$ is positive definite. **Source:** Fact 2.15.2.

10.11 Facts on Equalities and Inequalities for Two or More Matrices

Fact 10.11.1. Let $A, B \in \mathbb{F}^{n\times n}$, and assume that A is positive semidefinite. Then, $AB = BA$ if and only if $A^{1/2}B = BA^{1/2}$. **Source:** [2991, p. 205].

Fact 10.11.2. Let $A \in \mathbb{F}^{n\times m}$, $B \in \mathbb{F}^{n\times l}$, and $\alpha > 0$. Then, the following statements are equivalent:

i) There exists $C \in \mathbb{F}^{l\times m}$ such that $A = BC$ and $\sigma_{\max}^2(C) \le \alpha$.

ii) $AA^* \le \alpha BB^*$.

If these statements hold, then $\mathcal{R}(A) \subseteq \mathcal{R}(B)$. **Related:** Theorem 10.6.2. See [2991, p. 187].

Fact 10.11.3. Let $(A_i)_{i=1}^\infty \subset \mathbf{H}^n$ and $(B_i)_{i=1}^\infty \subset \mathbf{H}^n$, assume that, for all $i \in \mathbb{P}$, $A_i \le B_i$, and assume that $A \triangleq \lim_{i\to\infty} A_i$ and $B \triangleq \lim_{i\to\infty} B_i$ exist. Then, $A \le B$.

Fact 10.11.4. Let $A, B \in \mathbf{N}^n$, and assume that $A \le B$. Then, $\mathcal{R}(A) \subseteq \mathcal{R}(B)$ and $\operatorname{rank} A \le \operatorname{rank} B$. Furthermore, $\mathcal{R}(A) = \mathcal{R}(B)$ if and only if $\operatorname{rank} A = \operatorname{rank} B$.

Fact 10.11.5. Let $A, B \in \mathbf{P}^n$ and $\mathcal{S} \subseteq \{1, \ldots, n\}$. Then,

$$[(A+B)_{(\mathcal{S})}]^{-1} \le \left\{ \begin{array}{c} [(A+B)^{-1}]_{(\mathcal{S})} \\ (A_{(\mathcal{S})})^{-1} + (B_{(\mathcal{S})})^{-1} \end{array} \right\} \le (A^{-1})_{(\mathcal{S})} + (B^{-1})_{(\mathcal{S})}.$$

Source: [2991, p. 223].

Fact 10.11.6. Let $A, B \in \mathbb{F}^{n\times n}$ and $\mathcal{S} \subseteq \{1,\dots,n\}$. Then,

$$(A^*)_{(\mathcal{S})}A_{(\mathcal{S})} \le (A^*A)_{(\mathcal{S})},$$

$$(AB)_{(\mathcal{S})}(B^*A^*)_{(\mathcal{S})} \le (ABB^*A^*)_{(\mathcal{S})},$$

$$A_{(\mathcal{S})}B_{(\mathcal{S})}(B^*)_{(\mathcal{S})}(A^*)_{(\mathcal{S})} \le A_{(\mathcal{S})}(BB^*)_{(\mathcal{S})}(A^*)_{(\mathcal{S})}.$$

Source: [2991, pp. 223, 318].

Fact 10.11.7. Let $A \in \mathbb{F}^{n\times m}$ and $B \in \mathbb{F}^{m\times l}$. Then, $(AB)^+AB \le B^+B$. **Source:** [2238, p. 312].

Fact 10.11.8. Let $A \in \mathbb{F}^{n\times m}$, let $B \in \mathbb{F}^{m\times m}$, and assume that B is a projector. Then, $AB = A$ if and only if $A^+A \le B$. **Source:** [2238, p. 312].

Fact 10.11.9. Let $A, B \in \mathbb{F}^{n\times n}$, and assume that A and B are projectors. Then, the following statements are equivalent:

i) $A \le B$.

ii) For all $x \in \mathbb{F}^n$, $\|Ax\|_2 \le \|Bx\|_2$.

iii) $\mathcal{R}(A) \subseteq \mathcal{R}(B)$.

iv) $AB = A$.

v) $BA = A$.

vi) $B - A$ is a projector.

Source: [1133, p. 43] and [2418, p. 24]. **Related:** Fact 4.18.3 and Fact 4.18.5.

Fact 10.11.10. Let $A, B \in \mathbf{N}^n$, and assume that A and B are semicontractive. Then,

$$-\tfrac{1}{4}I \le AB + BA.$$

Source: [2990, p. 81]. If A and B are projectors, then $AB + BA + \frac{1}{4}I = (A + B - \frac{1}{2}I)^2 \ge 0$. See [246]. **Related:** Fact 7.13.4, Fact 10.11.24, and Fact 10.22.32.

Fact 10.11.11. Let $A, B \in \mathbb{C}^{n\times n}$, assume that A and B are nonsingular, and assume that $\jmath(A^*B - B^*A)$ is positive semidefinite. Then, $A + \jmath B$ is nonsingular. **Source:** [1158, p. 71].

Fact 10.11.12. Let $A, B \in \mathbf{H}^n$. Then, the following statements hold:

i) $\lambda_{\min}(A) \le \lambda_{\min}(B)$ if and only if $\lambda_{\min}(A)I \le B$.

ii) $\lambda_{\max}(A) \le \lambda_{\max}(B)$ if and only if $A \le \lambda_{\max}(B)I$.

Fact 10.11.13. Let $A, B \in \mathbf{H}^n$, and consider the following statements:

i) $A \le B$.

ii) There exists a unitary matrix $S \in \mathbb{F}^{n\times n}$ such that $A \le SBS^*$.

iii) For all $i \in \{1,\dots,n\}$, $\lambda_i(A) \le \lambda_i(B)$.

Then, *i)* $\Longrightarrow$ *ii)* $\Longleftrightarrow$ *iii)*. **Source:** *ii)* $\Longrightarrow$ *iii)* follows from the monotonicity theorem given by Theorem 10.4.9 and the fact that, for all $i \in \{1,\dots,n\}$, $\lambda_i(B) = \lambda_i(SBS^*)$. *iii)* $\Longrightarrow$ *ii)* is given in [2991, pp. 287, 288].

Fact 10.11.14. Let $A, B \in \mathbb{F}^{n\times n}$, and assume that A is positive semidefinite and B is positive definite. Then, $0 \le A < B$ if and only if $\rho_{\max}(AB^{-1}) < 1$.

Fact 10.11.15. Let $A, B \in \mathbb{F}^{n\times n}$, and assume that A and B are positive semidefinite. Then, the following statements hold:

i) $I + AB$ and $I + BA$ are nonsingular.

ii) $(I + AB)^{-1}A = A(I + BA)^{-1}$.

iii) $(I + AB)^{-1}A$ is Hermitian.

Related: Fact 3.20.6.

Fact 10.11.16. Let $A, B \in \mathbb{F}^{n\times n}$, and assume that A is positive definite and B is Hermitian. Then,

$$2B \le BAB + A^{-1}.$$

Fact 10.11.17. Let $A, B \in \mathbf{P}^n$. Then,

$$(A^{-1} + B^{-1})^{-1} = A(A + B)^{-1}B = B(A + B)^{-1}A.$$

Remark: This equality holds for all nonsingular $A, B \in \mathbb{F}^{n\times n}$ such that $A + B$ is nonsingular.

Fact 10.11.18. Let $A, B \in \mathbf{P}^n$. Then,

$$(A + B)^{-1} \le \tfrac{1}{4}(A^{-1} + B^{-1}).$$

Equivalently,

$$A + B \le AB^{-1}A + BA^{-1}B.$$

In both inequalities, equality holds if and only if $A = B$. **Source:** [2983, p. 168]. **Related:** Fact 2.2.12.

Fact 10.11.19. Let $A, B \in \mathbb{F}^{n\times n}$, and assume that A is positive definite, B is Hermitian, and $A+B$ is nonsingular. Then,

$$(A + B)^{-1} + (A + B)^{-1}B(A + B)^{-1} \le A^{-1}.$$

If, in addition, B is nonsingular, then the inequality is strict. **Source:** This inequality is equivalent to $BA^{-1}B \ge 0$. See [2164].

Fact 10.11.20. Let $A, B \in \mathbf{P}^n$ and $\alpha \in [0, 1]$. Then,

$$\beta[\alpha A^{-1} + (1 - \alpha)B^{-1}] \le [\alpha A + (1 - \alpha)B]^{-1},$$

where

$$\beta \triangleq \min_{\mu\in \mathrm{mspec}(A^{-1}B)} \frac{4\mu}{(1 + \mu)^2}.$$

Source: [2072]. **Remark:** This is a reverse form of an inequality based on convexity.

Fact 10.11.21. Let $A \in \mathbb{F}^{n\times m}$ and $B \in \mathbb{F}^{m\times m}$, and assume that B is positive semidefinite. Then, $ABA^* = 0$ if and only if $AB = 0$.

Fact 10.11.22. Let $A, B \in \mathbf{N}^n$. Then, AB is positive semidefinite if and only if AB is normal. **Source:** [531, p. 1456].

Fact 10.11.23. Let $A, B \in \mathbf{H}^n$, and assume that either *i*) either A or B is positive definite, or *ii*) A and B are positive semidefinite. Then, AB is semisimple. **Source:** Theorem 10.3.2 and Theorem 10.3.6.

Fact 10.11.24. Let $A, B \in \mathbf{H}^n$, and assume that $A \odot I$ is positive definite and $AB + BA$ is (positive semidefinite, positive definite). Then, B is (positive semidefinite, positive definite). **Source:** [454, p. 8], [1768, p. 120], [2877], and [2991, p. 214]. Alternatively, use Corollary 15.10.4. **Related:** Fact 7.13.4, Fact 10.11.10, and Fact 10.22.32.

Fact 10.11.25. Let $A, B, C \in \mathbf{N}^n$, and assume that $A = B + C$. Then, the following statements are equivalent:

i) $\operatorname{rank} A = \operatorname{rank} B + \operatorname{rank} C$.

ii) There exists $S \in \mathbb{F}^{m\times n}$ such that $\operatorname{rank} S = m$, $\mathcal{R}(S) \cap \mathcal{N}(A) = \{0\}$, and either $B = AS^*(SAS^*)^{-1}SA$ or $C = AS^*(SAS^*)^{-1}SA$.

Source: [627, 735].

Fact 10.11.26. Let $A, B \in \mathbb{F}^{n\times n}$, and assume that A and B are Hermitian and nonsingular. Then, the following statements hold:

i) If every eigenvalue of AB is positive, then $\operatorname{In} A = \operatorname{In} B$.

ii) $\operatorname{In} A - \operatorname{In} B = \operatorname{In}(A - B) + \operatorname{In}(A^{-1} - B^{-1})$.

iii) If $\operatorname{In} A = \operatorname{In} B$ and $A \le B$, then $B^{-1} \le A^{-1}$.

Source: [97, 228, 2141]. **Remark:** An extension to singular A and B is given by Fact 10.24.15. **Credit:** The equality *ii)* is due to G. P. H. Styan. See [2141].

Fact 10.11.27. Let $A, B \in \mathbf{H}^n$, and assume that $A \le B$. Then, $A_{(i,i)} \le B_{(i,i)}$ for all $i \in \{1, \ldots, n\}$.

Fact 10.11.28. Let $A, B \in \mathbf{H}^n$, and assume that $A \le B$. Then, $\operatorname{sig} A \le \operatorname{sig} B$. **Source:** [860, p. 148].

Fact 10.11.29. Let $A, B \in \mathbf{H}^n$, and assume that $\langle A \rangle \le B$. Then, either $A \le B$ or $-A \le B$. **Source:** [2986].

Fact 10.11.30. Let $A, B \in \mathbb{F}^{n \times n}$, and assume that A is positive semidefinite and B is positive definite. Then, $A \le B$ if and only if $AB^{-1}A \le A$.

Fact 10.11.31. Let $A, B \in \mathbf{N}^n$, and assume that $A \le B$. Then, there exists $S \in \mathbb{F}^{n \times n}$ such that $A = S^*BS$ and $S^*S \le I$. **Source:** [970, p. 269].

Fact 10.11.32. Let $A, B \in \mathbb{F}^{n \times m}$, and assume that $AA^* \le BB^*$. Then, for all $i \in \{1, \ldots, \min\{n, m\}\}$, $\sigma_i(AA^*) \le \sigma_i(BB^*)$ and $\sigma_i(A) \le \sigma_i(B)$. In particular, $\sigma_{\max}(A) \le \sigma_{\max}(B)$. **Source:** Use Fact 7.12.38 and either Theorem 10.4.9 or Fact 11.10.30. **Related:** Fact 11.10.31.

Fact 10.11.33. Let $A, B, C, D \in \mathbf{N}^n$, and assume that $0 < D \le C$ and $BCB \le ADA$. Then, $B \le A$. **Source:** [186, 669].

Fact 10.11.34. Let $A \in \mathbb{F}^{n \times m}$ and $B \in \mathbb{F}^{k \times m}$. Then, there exist unitary matrices $S_1, S_2 \in \mathbb{F}^{m \times m}$ such that

$$\langle A + B \rangle \le S_1 \langle A \rangle S_1^* + S_2 \langle B \rangle S_2^*.$$

Source: [2991, pp. 289, 290]. **Remark:** This is a matrix version of the triangle inequality. See [93, 2613]. **Remark:** There exist $A \in \mathbb{F}^{n \times m}$ and $B \in \mathbb{F}^{k \times m}$ such that $\langle A + B \rangle \le \langle A \rangle + \langle B \rangle$ does not hold. See [2991, p. 291]. **Related:** Fact 10.14.50, Fact 10.25.52, and Fact 11.10.9.

Fact 10.11.35. Let $A, B \in \mathbb{F}^{n \times n}$, and let $p, q \in (0, \infty)$ satisfy $1/p + 1/q = 1$. Then,

$$\langle A - B \rangle^2 + \left\langle \sqrt{\tfrac{p}{q}} A + \sqrt{\tfrac{q}{p}} B \right\rangle^2 = p \langle A \rangle^2 + q \langle B \rangle^2.$$

Source: [2991, p. 291].

Fact 10.11.36. Let $A, B \in \mathbb{F}^{n \times n}$, $a, b \in (0, \infty)$, and $c \in (-\sqrt{ab}, \sqrt{ab}.)$ Then,

$$a \langle A \rangle^2 + b \langle B \rangle^2 + c(A^*B + B^*A) \ge 0.$$

Source: [2991, p. 291].

Fact 10.11.37. Let $A_1, \ldots, A_k \in \mathbb{F}^{n \times n}$, let $a_1, \ldots, a_k \in (0, \infty)$, and assume that $\sum_{i=1}^k a_i = 1$. Then,

$$\left\langle \sum_{i=1}^k a_i A_i \right\rangle^2 \le \sum_{i=1}^k a_i \langle A_i \rangle^2.$$

Source: [2991, p. 292].

Fact 10.11.38. Let $A \in \mathbb{F}^{n \times m}$ and $B \in \mathbb{F}^{l \times m}$, and let $p, q > 1$ satisfy $1/p + 1/q = 1$. Then, there exists a unitary matrix $S \in \mathbb{F}^{m \times m}$ such that

$$\langle AB^* \rangle \le S^* \left(\tfrac{1}{p} \langle A \rangle^p + \tfrac{1}{q} \langle B \rangle^q \right) S.$$

Furthermore,

$$\operatorname{tr} \langle AB^* \rangle \le \tfrac{1}{p} \operatorname{tr} \langle A \rangle^p + \tfrac{1}{q} \operatorname{tr} \langle B \rangle^q.$$

Source: [93, 95, 1418] and [2977, p. 28]. **Remark:** This is a matrix version of Young's inequality. **Related:** Fact 2.2.50, Fact 2.2.53, Fact 10.11.39 Fact 10.11.73, Fact 10.14.8, Fact 10.14.33, Fact

10.14.34, and Fact 11.16.30.

Fact 10.11.39. Let $A, B \in \mathbb{F}^{n\times n}$, and assume that A and B are positive definite. Then, there exists a unitary matrix $S \in \mathbb{F}^{n\times n}$ such that $\langle AB\rangle \le \frac{1}{2}S(A^2+B^2)S^*$. **Source:** [196, 468].

Fact 10.11.40. Let $A, B \in \mathbb{F}^{n\times n}$, and assume that A and B are projectors. Then, $ABA \le B$ if and only if $AB = BA$. **Source:** [2709].

Fact 10.11.41. Let $A, B \in \mathbb{F}^{n\times n}$, assume that A is positive semidefinite, $0 \le A \le I$, and B is positive definite, and define $\alpha \triangleq \lambda_{\min}(B)$ and $\beta \triangleq \lambda_{\max}(B)$. Then,

$$ABA \le \frac{(\alpha+\beta)^2}{4\alpha\beta}B.$$

Source: [542]. **Related:** Fact 2.12.10.

Fact 10.11.42. Let $A \in \mathbb{F}^{n\times n}$, define $r \triangleq \operatorname{rank} A$, assume that A is positive semidefinite, let $P \in \mathbb{F}^{n\times n}$ be the projector onto $\mathcal{R}(A)$, define $\alpha \triangleq \lambda_{\min}(A)$ and $\beta \triangleq \lambda_r(A)$, and let $B \in \mathbb{F}^{n\times m}$. Then,

$$B^*AB \le \frac{(\alpha+\beta)^2}{4\alpha\beta}B^*PB(B^*A^+B)^+B^*PB, \quad B^*A^2B \le \frac{(\alpha+\beta)^2}{4\alpha\beta}B^*AB(B^*PB)^+B^*AB,$$

$$B^*AB - B^*PB(B^*A^+B)^+B^*PB \le \frac{(\beta-\alpha)^2}{4}B^*A^+B, \quad B^*A^2B - B^*AB(B^*PB)^+B^*AB \le \frac{(\beta-\alpha)^2}{4}B^*PB,$$

$$B^*AB - B^*PB(B^*A^+B)^+B^*PB \le (\sqrt{\beta}-\sqrt{\alpha})^2B^*PB, \quad B^*A^2B - B^*AB(B^*PB)^+B^*AB \le (\sqrt{\beta}-\sqrt{\alpha})^2B^*AB.$$

Now, assume that A is positive definite. Then,

$$B^*AB \le \frac{(\alpha+\beta)^2}{4\alpha\beta}B^*B(B^*A^{-1}B)^{-1}B^*B, \quad B^*A^2B \le \frac{(\alpha+\beta)^2}{4\alpha\beta}B^*AB(B^*B)^{-1}B^*AB,$$

$$B^*AB - B^*B(B^*A^{-1}B)^{-1}B^*B \le \frac{(\beta-\alpha)^2}{4}B^*A^{-1}B, \quad B^*A^2B - B^*AB(B^*B)^{-1}B^*AB \le \frac{(\beta-\alpha)^2}{4}B^*B,$$

$$B^*AB - B^*B(B^*A^{-1}B)^{-1}B^*B \le (\sqrt{\beta}-\sqrt{\alpha})^2B^*B, \quad B^*A^2B - B^*AB(B^*PB)^{-1}B^*AB \le (\sqrt{\beta}-\sqrt{\alpha})^2B^*AB,$$

$$\det(B^*AB)\det(B^*A^{-1}B) \le \prod_{i=1}^{\min\{m,n-m\}} \frac{[\lambda_i(A)+\lambda_{n-i+1}(A)]^2}{4\lambda_i(A)\lambda_{n-i+1}(A)}.$$

Finally, if $2m \le n$, then

$$\frac{\operatorname{tr} B^*AB}{\operatorname{tr}(B^*A^{-1}B)^{-1}} \le \left(\frac{\sum_{i=1}^m[\lambda_i(A)+\lambda_{n-i+1}(A)]}{2\sum_{i=1}^m\sqrt{\lambda_i(A)\lambda_{n-i+1}(A)}}\right)^2.$$

Source: [1357]. The last two inequalities are given in [2826]. **Remark:** These are matrix extensions of the Kantorovich inequality. **Related:** Fact 2.11.134, Fact 10.18.8, and Fact 10.11.43.

Fact 10.11.43. Let $A \in \mathbb{F}^{n\times n}$, let $B \in \mathbb{F}^{n\times m}$, assume that A is positive definite, define $\alpha \triangleq \lambda_{\min}(A)$ and $\beta \triangleq \lambda_{\max}(A)$, and assume that B is left inner. Then,

$$B^*AB \le \frac{(\alpha+\beta)^2}{4\alpha\beta}(B^*A^{-1}B)^{-1}, \quad B^*A^2B \le \frac{(\alpha+\beta)^2}{4\alpha\beta}(B^*AB)^2,$$

$$B^*AB - (B^*A^{-1}B)^{-1} \le \frac{(\beta-\alpha)^2}{4}B^*A^{-1}B, \quad B^*A^2B - (B^*AB)^2 \le \frac{(\beta-\alpha)^2}{4}I,$$

$$B^*AB - (B^*A^{-1}B)^{-1} \le (\sqrt{\beta}-\sqrt{\alpha})^2I, \quad B^*A^2B - (B^*AB)^2 \le (\sqrt{\beta}-\sqrt{\alpha})^2B^*AB.$$

Source: [1357]. **Remark:** These are matrix extensions of the Kantorovich inequality. **Related:** Fact 2.11.134, Fact 10.18.8, and Fact 10.11.42.

Fact 10.11.44. Let $A, B \in \mathbb{F}^{n\times n}$, and assume that A and B are projectors. Then, $(A+B)^{1/2} \le A+B$ if and only if $AB = BA$. **Source:** [2696, p. 30].

Fact 10.11.45. Let $A, B \in \mathbf{N}^n$, and assume that $0 \le A \le B$. Then,

$$(A + \tfrac{1}{4}A^2)^{1/2} \le (B + \tfrac{1}{4}B^2)^{1/2}.$$

Source: [2063].

Fact 10.11.46. Let $A \in \mathbf{N}^n$, let $B \in \mathbb{F}^{l\times n}$, let $p, q \in \mathbb{R}[s]$, and assume that $p \ne 0$, $q \ne 0$, $p(0) = q(0) = 0$, and all of the coefficients of p and q are nonnegative. Then, $\operatorname{rank} Bp(A)B^* = \operatorname{rank} Bq(A)B^*$. **Source:** Note that $\operatorname{rank} Bp(A)^{1/2} = \operatorname{rank} Bq(A)^{1/2}$.

Fact 10.11.47. Let $A, B \in \mathbf{N}^n$, and assume that $A \le B$ and $AB = BA$. Then, $A^2 \le B^2$. **Source:** [229].

Fact 10.11.48. Let $A, B \in \mathbf{N}^n$, and assume that $BA^2B \le I$. Then, $B^{1/2}AB^{1/2} \le I$. **Source:** [2991, p. 214].

Fact 10.11.49. Let $A \in \mathbf{P}^n$ and $B, C \in \mathbf{N}^n$. Then, $2\operatorname{tr}\langle B^{1/2}C^{1/2}\rangle \le \operatorname{tr}(AB + A^{-1}C)$. Furthermore, there exists A such that equality holds if and only if $\operatorname{rank} B = \operatorname{rank} C = \operatorname{rank} B^{1/2}C^{1/2}$. **Source:** [80, 1057]. **Remark:** A matrix A for which equality holds is given in [80]. **Remark:** Applications to linear systems are given in [2896].

Fact 10.11.50. Let $A, B \in \mathbf{P}^n$, let $S \in \mathbb{F}^{n\times n}$, assume that S is nonsingular and $SAS^* = \operatorname{diag}(\alpha_1, \dots, \alpha_n)$ and $SBS^* = \operatorname{diag}(\beta_1, \dots, \beta_n)$, and define

$$C_l \triangleq S^{-1}\operatorname{diag}(\min\{\alpha_1,\beta_1\}, \dots, \min\{\alpha_n,\beta_n\})S^{-*},$$
$$C_u \triangleq S^{-1}\operatorname{diag}(\max\{\alpha_1,\beta_1\}, \dots, \max\{\alpha_n,\beta_n\})S^{-*}.$$

Then, $C_l \le A \le C_u$ and $C_l \le B \le C_u$. Furthermore, if $\hat{S} \in \mathbb{F}^{n\times n}$ is nonsingular, $SAS^* = \hat{S}A\hat{S}^*$, $SBS^* = \hat{S}B\hat{S}^*$, and $\hat{C}_l$ and $\hat{C}_u$ are defined as C_l and C_u with S replaced by $\hat{S}$, then $\hat{C}_l = C_l$ and $\hat{C}_u = C_u$. **Source:** [1805].

Fact 10.11.51. Let $A, B \in \mathbf{H}^n$, and define "glb" with respect to $\mathbf{H}^n$. Then, the following statements hold:

i) If $\operatorname{glb}(\{A, B\})$ exists, then either $A \le B$ or $B \le A$.

ii) If $A, B \in \mathbf{N}^n$ are projectors, then $\operatorname{glb}(\{A, B\}) = 2A(A + B)^+B$, which is the projector onto $\mathcal{R}(A) \cap \mathcal{R}(B)$.

iii) $\operatorname{glb}(\{A, B\})$ exists if and only if $\operatorname{glb}(\{A, \operatorname{glb}(\{AA^+, BB^+\})\})$ and $\operatorname{glb}(\{B, \operatorname{glb}(\{AA^+, BB^+\})\})$ are comparable. If these conditions hold, then

$$\operatorname{glb}(\{A, B\}) = \min\{\operatorname{glb}(\{A, \operatorname{glb}(\{AA^+, BB^+\})\}), \operatorname{glb}\{B, \operatorname{glb}(\{AA^+, BB^+\})\})\}.$$

iv) $\operatorname{glb}(\{A, B\})$ exists if and only if $\operatorname{sh}(A, B)$ and $\operatorname{sh}(B, A)$ are comparable, where $\operatorname{sh}(A, B) \triangleq \lim_{\alpha\to\infty} \alpha B(\alpha B + A)^+A$. If these conditions hold, then

$$\operatorname{glb}(\{A, B\}) = \min\{\operatorname{sh}(A, B), \operatorname{sh}(B, A)\}.$$

Source: [96, 957, 1169, 1570, 2078]. **Remark:** The shorted operator "sh" is defined in Fact 10.24.21. **Remark:** Let $A = \left[\begin{smallmatrix} 1 & 0 \\ 0 & 0 \end{smallmatrix}\right]$ and $B = \left[\begin{smallmatrix} 0 & 0 \\ 0 & 1 \end{smallmatrix}\right]$. Then, $C = 0$ is a lower bound for $\{A, B\}$. Furthermore, $D = \begin{bmatrix} -1 & \sqrt{2} \\ \sqrt{2} & -1 \end{bmatrix}$, which has eigenvalues $-1 - \sqrt{2}$ and $-1 + \sqrt{2}$, is also a lower bound for $\{A, B\}$ but is not comparable with C. Hence, neither C nor D is the greatest lower bound for A and B. Finally, since C is the unique positive-semidefinite lower bound for A and B, it follows that $\operatorname{glb}(\{A, B\})$ does not exist. Consequently, $\mathbf{H}^n$ with the ordering "$\le$" is not a lattice.

Fact 10.11.52. Let $A, B \in \mathbf{N}^n$, and define "glb" with respect to $\mathbf{N}^n$, and define $A_+ \triangleq \frac{1}{2}(A + \langle A \rangle)$. Then,

$$\text{glb}(\{A, B\}) = A - (A - B)_+ = B - (B - A)_+.$$

Source: [2623]. **Remark:** Let $A = \begin{bmatrix} 1 & 0 \\ 0 & 0 \end{bmatrix}$ and $B = \begin{bmatrix} 0 & 0 \\ 0 & 1 \end{bmatrix}$, and suppose that Z is the least upper bound for A and B. Hence, $A \le Z \le I$ and $B \le Z \le I$, and thus $Z = I$. Next, note that $X \triangleq \begin{bmatrix} 4/3 & 2/3 \\ 2/3 & 4/3 \end{bmatrix}$ satisfies $A \le X$ and $B \le X$. However, $Z \le X$ is false, and thus $\{A, B\}$ does not have a least upper bound. Therefore, $\mathbf{N}^n$ with the ordering "$\le$" is not a lattice. See [523, p. 11] and [1805]. **Remark:** The cone $\mathbf{N}^n$ is a partially ordered set under the spectral ordering. See Fact 10.11.57.

Fact 10.11.53. Let $A_1, \ldots, A_k \in \mathbf{P}^n$. Then,

$$n^2\left(\sum_{i=1}^k A_i\right)^{-1} \le \sum_{i=1}^k A_i^{-1}.$$

Remark: This is an extension of Fact 2.11.135.

Fact 10.11.54. Let $A, B \in \mathbf{H}^n$. Then, $[\frac{1}{2}(A + B)]^2 \le \frac{1}{2}(A^2 + B^2)$. **Source:** [2991, p. 292].

Fact 10.11.55. Let $A, B \in \mathbf{N}^n$, let p be a real number, and assume that either $p \in [1, 2]$ or A and B are positive definite and $p \in [-1, 0] \cup [1, 2]$. Then, $[\frac{1}{2}(A + B)]^p \le \frac{1}{2}(A^p + B^p)$. **Source:** [1716].

Fact 10.11.56. Let $A_1, \ldots, A_k \in \mathbf{N}^n$, and let $p, q \in \mathbb{R}$ satisfy $1 \le p \le q$. Then,

$$\left(\frac{1}{k}\sum_{i=1}^k A_i^p\right)^{1/p} \le \left(\frac{1}{k}\sum_{i=1}^k A_i^q\right)^{1/q}.$$

Source: [443].

Fact 10.11.57. Let $A, B \in \mathbf{N}^n$, and let $p, q \in \mathbb{R}$ satisfy $1 \le p \le q$. Then,

$$[\tfrac{1}{2}(A^p + B^p)]^{1/p} \le [\tfrac{1}{2}(A^q + B^q)]^{1/q}.$$

Furthermore,

$$\mu(A, B) \triangleq \lim_{r\to\infty} [\tfrac{1}{2}(A^r + B^r)]^{1/r}$$

exists and satisfies $A \le \mu(A, B)$ and $B \le \mu(A, B)$. Finally,

$$\lim_{r\to 0} [\tfrac{1}{2}(A^r + B^r)]^{1/r} = e^{\frac{1}{2}(\log A + \log B)}.$$

Source: [354, 443]. **Remark:** $\mu(A, B)$ is the least upper bound of A and B with respect to the spectral ordering. See [92, 102, 1589] and Fact 10.23.3. **Remark:** This result does not hold for $p = 1$ and $q = 1/3$. For example, let $A = \begin{bmatrix} 2 & 1 \\ 1 & 2 \end{bmatrix}^3 = \begin{bmatrix} 13 & 8 \\ 8 & 5 \end{bmatrix}$ and $B = \begin{bmatrix} 1 & 0 \\ 0 & 0 \end{bmatrix}$. **Related:** Fact 11.10.13 and Fact 12.17.11.

Fact 10.11.58. Let $A, B \in \mathbb{F}^{n\times n}$, assume that A and B are positive semidefinite, and assume that $A > B \ge 0$. If $r \in (0, 1]$, then

$$0 < (\sigma_{\max}^r(A) - [\sigma_{\max}(A) - \sigma_{\min}(A - B)]^r)I \le A^r - B^r,$$

$$0 < (\sigma_{\max}^r(A) - [\sigma_{\max}(A) - \sigma_{\min}(A - B)]^r)I \le ([\sigma_{\max}(B) + \sigma_{\min}(A - B)]^r - \sigma_{\max}^r(B))I \le A^r - B^r.$$

In particular, $0 < \sigma_{\min}(A - B)I \le A - B$. If $A > B > 0$, then

$$0 < \frac{\sigma_{\min}(A - B)}{[\sigma_{\max}(A) - \sigma_{\min}(A - B)]\sigma_{\max}(A)} I \le B^{-1} - A^{-1},$$

$$\begin{aligned} 0 &< (\log \sigma_{\max}(A) - \log[\sigma_{\max}(A) - \sigma_{\min}(A - B)])I \\ &\le (\log[\sigma_{\max}(B) + \sigma_{\min}(A - B)] - \log \sigma_{\max}(B))I \le \log A - \log B. \end{aligned}$$

Source: [1129, 2095].

Fact 10.11.59. Let $A_1, \ldots, A_k \in \mathbb{F}^{n\times n}$, assume that $A_1, \ldots, A_k$ are positive definite, let $\alpha_1, \ldots, \alpha_k$ be nonnegative numbers, and assume that $\sum_{i=1}^k \alpha_i = 1$. Then,

$$\lim_{r\to 0}\left(\sum_{i=1}^k \alpha_i A_i^r\right)^{1/r} = \exp\left(\sum_{i=1}^k \alpha_i \log A_i\right).$$

Source: [443, 462]. **Related:** Fact 10.11.57.

Fact 10.11.60. Let $A, B \in \mathbb{F}^{n\times n}$, assume that A and B are positive semidefinite, let $p \in (1, \infty)$, and let $\alpha \in [0, 1]$. Then,

$$\alpha^{1-1/p}A + (1-\alpha)^{1-1/p}B \le (A^p + B^p)^{1/p}.$$

Source: [102].

Fact 10.11.61. Let $A, B, C \in \mathbb{F}^{n\times n}$. Then,

$$A^*A + B^*B = (A - CB)^*(I + CC^*)^{-1}(A - CB) + (B + C^*A)^*(I + C^*C)^{-1}(B + C^*A).$$

Hence,

$$\left.\begin{array}{c}(A - CB)^*(I + CC^*)^{-1}(A - CB)\\ (B + C^*A)^*(I + C^*C)^{-1}(B + C^*A)\end{array}\right\} \le A^*A + B^*B.$$

Now, assume that $I - C^*C$ is nonsingular. Then,

$$A^*A - B^*B = (A - CB)^*(I - CC^*)^{-1}(A - CB) - (B - C^*A)^*(I - C^*C)^{-1}(B - C^*A).$$

If, in addition, $I - CC^*$ is positive definite, then

$$A^*A - B^*B \le (A - CB)^*(I - CC^*)^{-1}(A - CB).$$

Source: [1458, p. 36] and [3007]. The second equality corrects a misprint in [3007]. **Related:** Fact 10.16.32 and Fact 10.16.33.

Fact 10.11.62. Let $A \in \mathbb{F}^{n\times n}$, let $\alpha \in \mathbb{R}$, and assume that either A is nonsingular or $\alpha \ge 1$. Then,

$$(A^*A)^\alpha = A^*(AA^*)^{\alpha-1}A.$$

Source: Use the singular value decomposition and see [1088, 1120].

Fact 10.11.63. Let $A, B \in \mathbb{F}^{n\times n}$, assume that A is positive semidefinite, assume that B is semicontractive, and let $\alpha \in [0, 1]$. Then,

$$B^*A^\alpha B \le (B^*AB)^\alpha.$$

Source: [2991, p. 206].

Fact 10.11.64. Let $A, B \in \mathbb{F}^{n\times n}$, let $\alpha \in \mathbb{R}$, assume that A and B are positive semidefinite, and assume that either A and B are positive definite or $\alpha \ge 1$. Then, $(AB^2A)^\alpha = AB(BA^2B)^{\alpha-1}BA$. **Source:** Fact 10.11.62.

Fact 10.11.65. Let $A, B, C \in \mathbb{F}^{n\times n}$, assume that A is positive semidefinite, B is positive definite, and $B = C^*C$, and let $\alpha \in [0, 1]$. Then,

$$C^*(C^{-*}AC^{-1})^\alpha C \le \alpha A + (1-\alpha)B.$$

If, in addition, $\alpha \in (0, 1)$, then equality holds if and only if $A = B$. **Source:** [2030].

Fact 10.11.66. Let $A, B \in \mathbb{F}^{n\times n}$, assume that A is positive semidefinite, and let $p \in \mathbb{R}$. Furthermore, assume that either A and B are nonsingular or $p \ge 1$. Then,

$$(BAB^*)^p = BA^{1/2}(A^{1/2}B^*BA^{1/2})^{p-1}A^{1/2}B^*.$$

Source: [1120] and [1124, p. 129].

Fact 10.11.67. Let $A, B \in \mathbb{F}^{n\times n}$, assume that A and B are positive definite, and let $p \in \mathbb{R}$. Then,

$$(BAB)^p = BA^{1/2}(A^{1/2}B^2A^{1/2})^{p-1}A^{1/2}B.$$

Source: [1118, 1380].

Fact 10.11.68. Let $A, B \in \mathbb{F}^{n\times n}$, and assume that A and B are positive semidefinite. Furthermore, if A is positive definite, then define

$$A\#B \triangleq A^{1/2}(A^{-1/2}BA^{-1/2})^{1/2}A^{1/2},$$

whereas, if A is singular, then define

$$A\#B \triangleq \lim_{\varepsilon\downarrow 0}(A+\varepsilon I)\#B.$$

Then, the following statements hold:

i) $A\#B$ is positive semidefinite.

ii) $A\#A = A$.

iii) $A\#B = B\#A = \left[\frac{2}{\pi}\int_0^\infty \frac{1}{x}(xA + x^{-1}B)^{-1}\,\mathrm{d}x\right]^{-1}$.

iv) $\mathcal{R}(A\#B) = \mathcal{R}(A) \cap \mathcal{R}(B)$.

v) If $S \in \mathbb{F}^{m\times n}$ is right invertible, then $(SAS^*)\#(SBS^*) \le S(A\#B)S^*$.

vi) If $S \in \mathbb{F}^{n\times n}$ is nonsingular, then $(SAS^*)\#(SBS^*) = S(A\#B)S^*$.

vii) If $C, D \in \mathbf{P}^n$, $A \le C$, and $B \le D$, then $A\#B \le C\#D$.

viii) If $C, D \in \mathbf{P}^n$, then $(A\#C) + (B\#D) \le (A+B)\#(C+D)$.

ix) If $A \le B$, then $4A\#(B-A) = [A + A\#(4B-3A)]\#[-A + A\#(4B-3A)]$.

x) If $\alpha \in [0,1]$, then $\sqrt{\alpha}(A\#B) \pm \frac{1}{2}\sqrt{1-\alpha}(A-B) \le \frac{1}{2}(A+B)$. In particular, $A\#B \le \frac{1}{2}(A+B)$.

xi) $A\#B = \max\{X \in \mathbf{H}\colon \left[\begin{smallmatrix} A & X \\ X & B\end{smallmatrix}\right]$ is positive semidefinite$\}$.

xii) Let $X \in \mathbb{F}^{n\times n}$, and assume that X is Hermitian and $\left[\begin{smallmatrix} A & X \\ X & B\end{smallmatrix}\right]$ is positive semidefinite. Then, $-A\#B \le X \le A\#B$. Furthermore, $\left[\begin{smallmatrix} A & A\#B \\ A\#B & B\end{smallmatrix}\right]$ and $\left[\begin{smallmatrix} A & -A\#B \\ -A\#B & B\end{smallmatrix}\right]$ are positive semidefinite.

xiii) If $S \in \mathbb{F}^{n\times n}$ is unitary and $A^{1/2}SB^{1/2}$ is positive semidefinite, then $A\#B = A^{1/2}SB^{1/2}$.

If A is positive definite, then the following statements hold:

xiv) $(A\#B)A^{-1}(A\#B) = B$.

xv) For all $\alpha \in \mathbb{R}$, $A\#B = A^{1-\alpha}(A^{\alpha-1}BA^{-\alpha})^{1/2}A^{\alpha}$.

xvi) $A\#B = A(A^{-1}B)^{1/2} = (BA^{-1})^{1/2}A$.

xvii) $A\#B = (A+B)[(A+B)^{-1}A(A+B)^{-1}B]^{1/2}$.

xviii) If $C \in \mathbb{F}^{n\times n}$ is Hermitian, then $C \le A\#(CA^{-1}C) \le \frac{1}{2}(A + CA^{-1}C)$.

If A and B are positive definite, then the following statements hold:

xix) $A\#B$ is positive definite.

xx) $S \triangleq (A^{-1/2}BA^{-1/2})^{1/2}A^{1/2}B^{-1/2}$ is unitary, and $A\#B = A^{1/2}SB^{1/2}$.

xxi) $\det A\#B = \sqrt{(\det A)\det B}$, $\det(A\#B)^2 = \det AB$, $(A\#B)^{-1} = A^{-1}\#B^{-1}$.

xxii) Let $A_0 \triangleq A$ and $B_0 \triangleq B$, and, for all $k \in \mathbb{N}$, define $A_{k+1} \triangleq 2(A_k^{-1} + B_k^{-1})^{-1}$ and $B_{k+1} \triangleq \frac{1}{2}(A_k + B_k)$. Then, for all $k \in \mathbb{N}$,

$$A_k \le A_{k+1} \le A\#B \le B_{k+1} \le B_k, \quad \lim_{k\to\infty} A_k = \lim_{k\to\infty} B_k = A\#B.$$

xxiii) For all $\alpha \in (-1,1)$, $\left[\begin{smallmatrix} A & \alpha A\#B \\ \alpha A\#B & B\end{smallmatrix}\right]$ is positive definite.

xxiv) $\operatorname{rank}\begin{bmatrix} A & A\#B \\ A\#B & B \end{bmatrix} = \operatorname{rank}\begin{bmatrix} A & -A\#B \\ -A\#B & B \end{bmatrix} = n.$

xxv) Assume that $n = 2$, and let $\alpha \triangleq \sqrt{\det A}$ and $\beta \triangleq \sqrt{\det B}$. Then,

$$A\#B = \frac{\sqrt{\alpha\beta}}{\sqrt{\det(\alpha^{-1}A + \beta^{-1}B)}}(\alpha^{-1}A + \beta^{-1}B).$$

xxvi) If $0 < A \le B$, then $\phi\colon [0,\infty) \mapsto \mathbf{P}^n$ defined by $\phi(p) \triangleq A^{-p}\#B^p$ is nondecreasing.

xxvii)

$$A\#B = \frac{1}{\pi}\int_0^1 \frac{1}{\sqrt{x(1-x)}}[xA^{-1} + (1-x)B^{-1}]^{-1}\,\mathrm{d}x.$$

Furthermore, the following statements hold:

xxviii) If B is positive definite and $A \le B$, then $A^2\#B^{-2} \le A\#B^{-1} \le I$.

xxix) If $0 \le A \le B$, then $(BA^2B)^{1/2} \le B^{1/2}(B^{1/2}AB^{1/2})^{1/2}B^{1/2} \le B^2$.

Finally, let $X \in \mathbf{H}^n$. Then, the following statements are equivalent:

xxx) $\left[\begin{smallmatrix} A & X \\ X & B \end{smallmatrix}\right]$ is positive semidefinite.

xxxi) $XA^{-1}X \le B$.

xxxii) $XB^{-1}X \le A$.

xxxiii) $-A\#B \le X \le A\#B$.

Source: [90, 1046, 1224, 1767, 2692]. For *xiii)*, *xx)*, and *xxv)*, see [454, pp. 108, 109, 111]; *xviii)* is given in [1833]; *xxvi)* is given in [91]; *xxvii)* is given in [1381, p. 223]; *xxvi)* implies *xxviii)*, which, in turn, implies *xxix)*. **Remark:** The square roots in *xvi)* indicate a semisimple matrix with positive diagonal entries. **Remark:** $A\#B$ is the *geometric mean* of A and B. A related mean is defined in [1046]. Alternative means and their differences are considered in [38]. Geometric means for an arbitrary number of positive-definite matrices are discussed in [105, 1614, 2065, 2229]. **Remark:** Inverse problems are considered in [86]. **Remark:** *xxix)* interpolates (10.6.6). **Remark:** Compare *xiii)* and *xx)* with Fact 10.12.15. **Related:** Fact 8.8.19, Fact 12.15.4, Fact 14.3.13, and Fact 16.25.6. **Problem:** For singular A and B, express $A\#B$ in terms of generalized inverses.

Fact 10.11.69. Let $A, B \in \mathbb{F}^{n\times n}$, assume that A and B are positive definite, and define

$$L(A,B) \triangleq A^{1/2}(A^{-1/2}BA^{-1/2} - I)(\log A^{-1/2}BA^{-1/2})^{-1}A^{1/2}.$$

Then,

$$L(A,B) = \left(\int_0^\infty \frac{1}{x+1}(xA + B)^{-1}\mathrm{d}x\right)^{-1}, \quad A\#B \le L(A,B) \le \tfrac{1}{2}(A+B).$$

Source: [1381, p. 208]. **Remark:** $L(A,B)$ is the logarithmic mean of A and B. See Fact 2.2.63. **Remark:** The inequalities interpolate $A\#B \le \frac{1}{2}(A+B)$ given by *x)* of Fact 10.11.68. An alternative interpolation is given in [3019]. See Fact 2.2.65. **Related:** Fact 14.2.19.

Fact 10.11.70. Let $A, B \in \mathbb{F}^{n\times n}$, and assume that A and B are Hermitian. Then, there exist unitary matrices $U, V \in \mathbb{F}^{n\times n}$ such that

$$e^{2A}\#e^{2B} = e^{UAU^*+VBV^*}.$$

Source: [1613]. **Related:** Fact 15.15.37.

Fact 10.11.71. Let $A, B \in \mathbb{F}^{n\times n}$, and assume that A and B are Hermitian. Then, the following statements are equivalent:

i) $A \le B$.

ii) For all $t \ge 0$, $I \le e^{-tA}\#e^{tB}$.

iii) $\phi\colon [0,\infty) \mapsto \mathbf{P}^n$ defined by $\phi(t) \triangleq e^{-tA}\#e^{tB}$ is nondecreasing.

Source: [91].

Fact 10.11.72. Let $A, B \in \mathbf{P}^n$, let $\alpha \in [0, 1]$, and define

$$A\#_\alpha B \triangleq A^{1/2}(A^{-1/2}BA^{-1/2})^\alpha A^{1/2},$$

Then,

$$A\#_\alpha B = B\#_{1-\alpha}A, \quad (A\#_\alpha B)^{-1} = A^{-1}\#_\alpha B^{-1}.$$

Fact 10.11.73. Let $A, B \in \mathbf{P}^n$ and $\alpha \in [0, 1]$. Then,

$$[\alpha A^{-1} + (1-\alpha)B^{-1}]^{-1} \le A\#_{1-\alpha}B = B\#_\alpha A = A^{1/2}(A^{-1/2}BA^{-1/2})^{1-\alpha}A^{1/2} \le \alpha A + (1-\alpha)B,$$
$$[\alpha A + (1-\alpha)B]^{-1} \le A^{-1}\#_{1-\alpha}B^{-1} = B^{-1}\#_\alpha A^{-1} = A^{-1/2}(A^{-1/2}BA^{-1/2})^{\alpha-1}A^{-1/2} \le \alpha A^{-1} + (1-\alpha)B^{-1},$$
$$\operatorname{tr}[\alpha A + (1-\alpha)B]^{-1} \le \operatorname{tr} A^{-1}(A^{-1/2}BA^{-1/2})^{\alpha-1} \le \operatorname{tr}[\alpha A^{-1} + (1-\alpha)B^{-1}].$$

Remark: The left-hand inequality in the first string is *Young's inequality*. See [1124, p. 122] and Fact 2.2.53. Setting $B = I$ yields Fact 10.10.47. The second string interpolates the fact that $\phi(A) = A^{-1}$ is convex as shown by Proposition 10.6.17. **Remark:** Extensions are given in [37]. **Related:** Fact 2.2.50, Fact 10.11.72, Fact 10.14.8, Fact 10.14.33, Fact 10.14.34, and Fact 10.11.38.

Fact 10.11.74. Let $A, B \in \mathbb{F}^{n\times n}$, and assume that A and B are positive definite. Then,

$$\tfrac{2\alpha\beta}{(\alpha+\beta)^2}(A+B) \le 2(A^{-1}+B^{-1})^{-1} \le A\#B \le \tfrac{1}{2}(A+B) \le \tfrac{(\alpha+\beta)^2}{2\alpha\beta}(A^{-1}+B^{-1})^{-1},$$

where $\alpha \triangleq \min\{\lambda_{\min}(A), \lambda_{\min}(B)\}$ and $\beta \triangleq \max\{\lambda_{\max}(A), \lambda_{\max}(B)\}$. **Source:** Use the second string of inequalities of Fact 10.11.73 with $\alpha = 1/2$ along with results given in [2630] and [2983, p. 174]. See [2991, p. 216].

Fact 10.11.75. Let $A_1, \ldots, A_m \in \mathbb{F}^{n\times n}$, assume that A_1 is positive definite, assume that $A_2, \ldots, A_m$ are positive semidefinite, and, for all $i \in \{1, \ldots, n\}$, define $B_i \triangleq \sum_{j=1}^i A_j$. Then,

$$\sum_{i=1}^m B_i^{-1}A_iB_i^{-1} < 2A_1^{-1}, \quad \operatorname{tr}\sum_{i=1}^m A_iB_i^{-2} < 2\operatorname{tr} A_1^{-1}.$$

Now, assume that $A_2, \ldots, A_m$ are positive definite. Then,

$$\frac{1}{2}\sum_{i=1}^m\sum_{j=1}^i B_j \le \sum_{i=1}^m B_iA_i^{-1}B_i, \quad \frac{1}{2}\sum_{i=1}^m\sum_{j=1}^i \operatorname{tr} B_j \le \operatorname{tr}\sum_{i=1}^m A_i^{-1}B_i^2.$$

Source: [1833].

Fact 10.11.76. Let $(x_i)_{i=1}^\infty \subset \mathbb{R}^n$, assume that $\sum_{i=1}^\infty x_i$ exists, let $(A_i)_{i=1}^\infty \subset \mathbf{N}^n$, assume that, for all $i \ge 1$, $A_i \le A_{i+1}$, and assume that $\lim_{i\to\infty} \operatorname{tr} A_i = \infty$. Then,

$$\lim_{k\to\infty}(\operatorname{tr} A_k)^{-1}\sum_{i=1}^k A_ix_i = 0.$$

If, in addition A_i is positive definite for all $i \in \mathbb{P}$ and $\{\lambda_{\max}(A_i)/\lambda_{\min}(A_i)\}_{i=1}^\infty$ is bounded, then

$$\lim_{k\to\infty} A_k^{-1}\sum_{i=1}^k A_ix_i = 0.$$

Source: [75]. **Remark:** These equalities are matrix versions of Kronecker's lemma given by Fact 12.18.15. **Remark:** Extensions are given in [1298].

Fact 10.11.77. Let $A, B \in \mathbb{F}^{n\times n}$, assume that A and B are positive definite, assume that $A \le B$, and let $p \ge 1$. Then,

$$A^p \le K(\lambda_{\min}(A), \lambda_{\min}(A), p)B^p \le \left[\frac{\lambda_{\max}(A)}{\lambda_{\min}(A)}\right]^{p-1} B^p,$$

where

$$K(a,b,p) \triangleq \frac{a^pb - ab^p}{(p-1)(a-b)}\left[\frac{(p-1)(a^p-b^p)}{p(a^pb-ab^p)}\right]^p.$$

Source: [540, 1122] and [1124, pp. 193, 194]. **Remark:** $K(a,b,p)$ is the *Fan constant*.

Fact 10.11.78. Let $A, B \in \mathbb{F}^{n\times n}$, assume that A is positive definite and B is positive semidefinite, and let $p \ge 1$. Then, there exist unitary matrices $U, V \in \mathbb{F}^{n\times n}$ such that

$$\tfrac{1}{K(\lambda_{\min}(A),\lambda_{\min}(A),p)}U(BAB)^pU^* \le B^pA^pB^p \le K(\lambda_{\min}(A), \lambda_{\min}(A), p)V(BAB)^pV^*,$$

where $K(a,b,p)$ is the Fan constant defined in Fact 10.11.77. **Source:** [540]. **Related:** Fact 10.14.24, Fact 10.22.36, and Fact 11.10.50.

Fact 10.11.79. Let $A, B \in \mathbb{F}^{n\times n}$, assume that A is positive definite, B is positive semidefinite, and $B \le A$, and let $p \ge 1$ and $r \ge 1$. Then,

$$[A^{r/2}(A^{-1/2}B^pA^{-1/2})^rA^{r/2}]^{1/p} \le A^r.$$

In particular,

$$\langle A^{-1/2}B^pA^{1/2}\rangle^{2/p} \le A^2.$$

Source: [101].

Fact 10.11.80. Let $A, B \in \mathbb{F}^{n\times n}$, and assume that A and B are positive definite. Then, the following statements are equivalent:

i) $B \le A$.

ii) For all $r \in [0,\infty)$, $p \in [1,\infty)$, and $k \in \mathbb{N}$ such that $(k+1)(r+1) = p + r$,

$$B^{r+1} \le (B^{r/2}A^pB^{r/2})^{\frac{1}{k+1}}.$$

iii) For all $r \in [0,\infty)$, $p \in [1,\infty)$, and $k \in \mathbb{N}$ such that $(k+1)(r+1) = p + r$,

$$(A^{r/2}B^pA^{r/2})^{\frac{1}{k+1}} \le A^{r+1}.$$

Source: [1828]. **Related:** Fact 10.23.1.

Fact 10.11.81. Let $A, B \in \mathbb{F}^{n\times n}$, and assume that A is positive definite and B is positive semidefinite. Then, the following statements are equivalent:

i) $B \le A$.

ii) For all $p, q, r, t \in \mathbb{R}$ such that $p \ge 1$, $r \ge 0$, $t \ge 0$, and $q \in [1,2]$,

$$[A^{r/2}(A^{t/2}B^pA^{t/2})^qA^{r/2}]^{\frac{r+t+1}{r+qt+qp}} \le A^{r+t+1}.$$

iii) For all $p, q, r, \tau \in \mathbb{R}$ such that $p \ge 1$, $r \ge \tau$, $q \ge 1$, and $\tau \in [0,1]$,

$$[A^{r/2}(A^{-\tau/2}B^pA^{-\tau/2})^qA^{r/2}]^{\frac{r-\tau}{r-q\tau+qp}} \le A^{r-\tau}.$$

iv) For all $p, q, r, \tau \in \mathbb{R}$ such that $p \ge 1$, $r \ge \tau$, $\tau \in [0,1]$, and $q \ge 1$,

$$[A^{r/2}(A^{-\tau/2}B^pA^{-\tau/2})^qA^{r/2}]^{\frac{r-\tau+1}{r-q\tau+qp}} \le A^{r-\tau+1}.$$

In particular, if $B \le A$, $p \ge 1$, and $r \ge 1$, then

$$[A^{r/2}(A^{-1/2}B^pA^{-1/2})^rA^{r/2}]^{\frac{r-1}{pr}} \le A^{r-1}.$$

Source: *ii*) is given in [1088], *iii*) appears in [1125], and *iv*) appears in [1120]. See also [1088, 1089] and [1124, p. 133]. **Remark:** Setting $q = r$ and $\tau = 1$ in *iv*) yields Fact 10.11.79. **Remark:** *iv*) is the *generalized Furuta inequality*. This result interpolates Proposition 10.6.7 and Fact 10.11.79.

Fact 10.11.82. Let $A, B \in \mathbb{F}^{n\times n}$, and assume that A is positive definite, B is positive semidefinite, and $B \le A$. Furthermore, let $t \in [0, 1]$, $p \ge 1$, $r \ge t$, and $s \ge 1$, and define

$$F_{A,B}(r, s) \triangleq A^{-r/2}[A^{r/2}(A^{-t/2}B^pA^{-t/2})^sA^{r/2}]^{\frac{1-t+r}{(p-t)s+r}}A^{-r/2}.$$

Then,

$$F_{A,B}(r, s) \le F_{A,A}(r, s).$$

Equivalently,

$$[A^{r/2}(A^{-t/2}B^pA^{-t/2})^sA^{r/2}]^{\frac{1-t+r}{(p-t)s+r}} \le A^{1-t+r}.$$

Furthermore, if $r' \ge r$ and $s' \ge s$, then

$$F_{A,B}(r', s') \le F_{A,B}(r, s).$$

Source: [1120] and [1124, p. 143]. **Related:** This result extends *iv)* of Fact 10.11.81.

Fact 10.11.83. Let $A, B \in \mathbb{F}^{n\times n}$, let p and q be nonzero real numbers, and assume that $1/p+1/q = 1$. If $p \in (0, 1)$, then $q < 0$ and the following statements hold:

i) $p\langle A\rangle^2 + q\langle B\rangle^2 \le \langle A - B\rangle^2 + \langle(1 - p)A - B\rangle^2$.

ii) $p\langle A\rangle^2 + q\langle B\rangle^2 \le \langle A - B\rangle^2 + \langle A - (1 - q)B\rangle^2$.

iii) In *i*) and *ii*), equality holds if and only if $(1 - p)A = B$.

If $p \in (1, 2]$, then $q \in [2, \infty)$ and the following statements hold:

iv) $\langle A - B\rangle^2 + \langle(1 - p)A - B\rangle^2 \le p\langle A\rangle^2 + q\langle B\rangle^2$.

v) $p\langle A\rangle^2 + q\langle B\rangle^2 \le \langle A - B\rangle^2 + \langle A - (1 - q)B\rangle^2$.

vi) In *iv*) and *v*), equality holds if and only if either $p = q = 2$ or $(1 - p)A = B$.

vii) $\langle A - B\rangle^2 + \frac{2}{p}\langle(p - 1)A + B\rangle^2 \le p\langle A\rangle^2 + q\langle B\rangle^2 \le \langle A - B\rangle^2 + \frac{2}{q}\langle A + (q - 1)B\rangle^2$.

If $p \in [2, \infty)$, then $q \in (1, 2]$ and the following statements hold:

viii) $p\langle A\rangle^2 + q\langle B\rangle^2 \le \langle A - B\rangle^2 + \langle(1 - p)A - B\rangle^2$.

ix) $\langle A - B\rangle^2 + \langle A - (1 - q)B\rangle^2 \le p\langle A\rangle^2 + q\langle B\rangle^2$.

x) In *viii*) and *ix*), equality holds if and only if $(1 - p)A = B$.

xi) $\langle A - B\rangle^2 + \frac{2}{q}\langle A + (q - 1)B\rangle^2 \le p\langle A\rangle^2 + q\langle B\rangle^2 \le \langle A - B\rangle^2 + \frac{2}{p}\langle(p - 1)A + B\rangle^2$.

If $p > 1$, then $q > 1$ and the following statements hold:

xii) $\langle A - B\rangle^2 \le p\langle A\rangle^2 + q\langle B\rangle^2$.

xiii) In *xii*), equality holds if and only if $(1 - p)A = B$.

Finally, the following statements hold:

xiv) If $pq > 0$, then $\langle A - B\rangle \le p\langle A\rangle^2 + q\langle B\rangle^2$.

xv) If $pq < 0$, then $p\langle A\rangle^2 + q\langle B\rangle^2 \le \langle A - B\rangle$.

xvi) In *xiv*) and *xv*), equality holds if and only if $(1 - p)A = B$.

Source: [7, 1095, 1416, 3018]. **Remark:** This extends Bohr's inequality given by Fact 2.21.8. **Related:** Fact 10.11.84.

Fact 10.11.84. Let $A, B \in \mathbb{F}^{n\times n}$ and $\alpha \in \mathbb{R}$. If $\alpha \ne 0$, then

$$\langle A - B\rangle^2 + \frac{1}{\alpha}\langle \alpha A + B\rangle^2 = (1 + \alpha)\langle A\rangle^2 + (1 + \tfrac{1}{\alpha})\langle B\rangle^2.$$

In particular,

$$\langle A-B\rangle^2+\langle A+B\rangle^2=2\langle A\rangle^2+2\langle B\rangle^2.$$

If $\alpha\in[0,1]$, then

$$\langle A-B\rangle^2+\langle \alpha A+B\rangle^2\le(1+\alpha)\langle A\rangle^2+(1+\tfrac{1}{\alpha})\langle B\rangle^2.$$

If either $\alpha<0$ or $\alpha\ge 1$, then

$$(1+\alpha)\langle A\rangle^2+(1+\tfrac{1}{\alpha})\langle B\rangle^2\le\langle A-B\rangle^2+\langle \alpha A+B\rangle^2.$$

Finally,

$$\langle A+B\rangle^2-\langle A-B\rangle^2=4\,\mathrm{Re}(A^*B).$$

Source: [1095, 2093]. **Remark:** The first inequality is the generalized parallelogram law, and the second inequality is a matrix version of the parallelogram law. **Related:** Fact 2.21.8, Fact 10.11.83, and Fact 11.8.3.

Fact 10.11.85. Let $A_1,\dots,A_k\in\mathbb{F}^{n\times n}$, let $\alpha_1,\dots,\alpha_k$ be nonzero real numbers, and assume that $\sum_{i=1}^k\alpha_i=1$. Then,

$$\sum_{i=1}^k\frac{1}{\alpha_i}\langle A_i\rangle^2-\left\langle\sum_{i=1}^k A_i\right\rangle^2=\sum\frac{\alpha_i}{\alpha_j}\left\langle\frac{\alpha_j}{\alpha_i}A_i-A_j\right\rangle^2,$$

where the last summation is taken over all $i,j\in\{1,\dots,k\}$ such that $i\le j$. **Source:** [1095, 2989]. **Related:** Fact 2.21.14, Fact 10.11.86, and Fact 11.8.4.

Fact 10.11.86. Let $A_1,\dots,A_k\in\mathbb{F}^{n\times n}$, let $\alpha_1,\dots,\alpha_k\in[0,1]$, and assume that $\sum_{i=1}^k\alpha_i=1$. Then,

$$\left\langle\sum_{i=1}^k\alpha_iA_i\right\rangle^2\le\sum_{i=1}^k\alpha_i\langle A_i\rangle^2.$$

Source: [1095]. **Remark:** This is a convexity condition. **Related:** Fact 10.11.85.

Fact 10.11.87. Let $A,B\in\mathbb{F}^{n\times n}$, assume that A and B are nonsingular, let p and q be positive numbers, and assume that $1/p+1/q=1$. Then,

$$\langle\langle A\rangle^{-1}-B\langle B\rangle^{-1}\rangle^2\le\langle A\rangle^{-1}[p\langle A-B\rangle^2+q(\langle A\rangle-\langle B\rangle)^2]\langle A\rangle^{-1}.$$

Equality holds if and only if $(p-1)(A-B)\langle A\rangle^{-1}=B(\langle A\rangle^{-1}-\langle B\rangle^{-1})$. Furthermore,

$$\langle A\langle A\rangle^{-1}-B\langle B\rangle^{-1}\rangle\le(\langle A\rangle^{-1}[2\langle A-B\rangle^2+2(\langle A\rangle-\langle B\rangle)^2]\langle A\rangle^{-1})^{1/2}.$$

Source: [2212, 2360]. **Remark:** This is a matrix version of the Dunkl-Williams inequality given by Fact 11.7.11.

Fact 10.11.88. Let $A,B\in\mathbb{F}^{n\times n}$, assume that A and B are nonsingular, let $r\in\mathbb{R}$, let $p,q\in(0,\infty)$, and assume that $1/p+1/q=1$. Then,

$$\langle\langle A\rangle^{r-1}-B\langle B\rangle^{r-1}\rangle^2\le\langle A\rangle^{r-1}(p\langle A-B\rangle^2+q\langle\langle B\rangle^r\langle A\rangle^{1-r}-\langle B\rangle\rangle^2)\langle A\rangle^{r-1}.$$

Equality holds if and only if $(p-1)(A-B)\langle A\rangle^{r-1}=B(\langle A\rangle^{r-1}-\langle B\rangle^{r-1})$. **Source:** [810]. **Remark:** Setting $r=0$ yields Fact 10.11.87. **Related:** Fact 11.7.11.

Fact 10.11.89. Each of the following functions $\phi\colon(0,\infty)\mapsto(0,\infty)$ yields an increasing function $\phi\colon\mathbf{P}^n\mapsto\mathbf{P}^n$:

i) $\phi(x)=\frac{x^{p+1/2}}{x^{2p}+1}$, where $p\in[0,1/2]$.

ii) $\phi(x)=x(1+x)\log(1+1/x)$.

iii) $\phi(x)=\frac{1}{(1+x)\log(1+1/x)}$.

iv) $\phi(x)=\frac{x-1-\log x}{(\log x)^2}$.

v) $\phi(x) = \frac{x(\log x)^2}{x-1-\log x}$.

vi) $\phi(x) = \frac{x(x+2)\log(x+2)}{(x+1)^2}$.

vii) $\phi(x) = \frac{x(x+1)}{(x+2)\log(x+2)}$.

viii) $\phi(x) = \frac{(x^2-1)\log(1+x)}{x^2}$.

ix) $\phi(x) = \frac{x(x-1)}{(x+1)\log(x+1)}$.

x) $\phi(x) = \frac{(x-1)^2}{(x+1)\log x}$.

xi) $\phi(x) = \frac{p-1}{p}\left(\frac{x^p-1}{x^{p-1}-1}\right)$, where $p \in [-1, 2]$.

xii) $\phi(x) = \frac{x-1}{\log x}$.

xiii) $\phi(x) = \sqrt{x}$.

xiv) $\phi(x) = \frac{x}{x+1}$.

xv) $\phi(x) = \frac{x-1}{x^p-1}$, where $p \in (0, 1]$.

Source: [1128, 2229]. To obtain *xii)*, *xiii)*, and *xiv)*, set $p = 1, 1/2, -1$, respectively, in *xi)*.

Fact 10.11.90. Let $A, B \in \mathbb{F}^{n\times n}$, and assume that A and B are positive definite. Then,

$$\log AB^{-1} = A^{1/2}(\log A^{1/2}B^{-1}A^{1/2})A^{-1/2}.$$

Source: [1381, p. 134].

10.12 Facts on Equalities and Inequalities for Partitioned Matrices

Fact 10.12.1. Let $A \in \mathbb{F}^{n\times n}$, and assume that A is positive semidefinite. Then, the following statements hold:

i) $\begin{bmatrix} A & A \\ A & A \end{bmatrix}$ and $\begin{bmatrix} A & -A \\ -A & A \end{bmatrix}$ are positive semidefinite.

ii) If $\begin{bmatrix} \alpha & \beta \\ \overline{\beta} & \gamma \end{bmatrix} \in \mathbb{F}^{2\times 2}$ is positive semidefinite, then $\begin{bmatrix} \alpha A & \beta A \\ \overline{\beta} A & \gamma A \end{bmatrix}$ is positive semidefinite.

iii) If A and $\begin{bmatrix} \alpha & \beta \\ \overline{\beta} & \gamma \end{bmatrix}$ are positive definite, then $\begin{bmatrix} \alpha A & \beta A \\ \overline{\beta} A & \gamma A \end{bmatrix}$ is positive definite.

Source: Fact 9.4.22.

Fact 10.12.2. Let $A \in \mathbb{F}^{n\times n}$, $B \in \mathbb{F}^{n\times m}$, $C \in \mathbb{F}^{m\times m}$, assume that $\begin{bmatrix} A & B \\ B^* & C \end{bmatrix} \in \mathbb{F}^{(n+m)\times(n+m)}$ is positive semidefinite, and assume that $\begin{bmatrix} \alpha & \beta \\ \overline{\beta} & \gamma \end{bmatrix} \in \mathbb{F}^{2\times 2}$ is positive semidefinite. Then, the following statements hold:

i) $\begin{bmatrix} \alpha 1_{n\times n} & \beta 1_{n\times m} \\ \overline{\beta} 1_{m\times n} & \gamma 1_{m\times m} \end{bmatrix}$ is positive semidefinite.

ii) $\begin{bmatrix} \alpha A & \beta B \\ \overline{\beta} B^* & \gamma C \end{bmatrix}$ is positive semidefinite.

iii) If $\begin{bmatrix} A & B \\ B^* & C \end{bmatrix}$ is positive definite and α and γ are positive, then $\begin{bmatrix} \alpha A & \beta B \\ \overline{\beta} B^* & \gamma C \end{bmatrix}$ is positive definite.

Source: To prove *i)*, use Proposition 10.2.5. *ii)* and *iii)* follow from Fact 10.25.16.

Fact 10.12.3. Let $A \in \mathbb{F}^{n\times n}$, $B \in \mathbb{F}^{n\times m}$, $C \in \mathbb{F}^{m\times m}$, and assume that $\mathcal{A} \triangleq \begin{bmatrix} A & B \\ B^* & C \end{bmatrix} \in \mathbb{F}^{(n+m)\times(n+m)}$ is positive semidefinite. Then, the following statements are equivalent:

i) $\operatorname{rank} \mathcal{A} = \operatorname{rank} A + \operatorname{rank} C$.

ii) $\mathcal{A}^+ = \begin{bmatrix} A^+ + A^+B(A|\mathcal{A})^+B^*A^+ & -A^+B(A|\mathcal{A})^+ \\ -(A|\mathcal{A})^+B^*A^+ & (A|\mathcal{A})^+ \end{bmatrix}$.

Source: [2179, Theorem 4.6] and [1458, pp. 44–46].

Fact 10.12.4. Let $A = \begin{bmatrix} A_{11} & A_{12} \\ A_{12}^* & A_{22} \end{bmatrix} \in \mathbb{F}^{n\times n}$, and assume that A is positive semidefinite and A_{11} is positive definite. Then,

$$\lambda_{\min}(A) \le \lambda_{\min}(A_{11}|A) \le \lambda_{\min}(A_{22}).$$

Source: [2991, p. 273].

Fact 10.12.5. Let $A, B \in \mathbb{F}^{n\times n}$, assume that A and B are positive semidefinite, and assume that A and B are partitioned identically as $A = \begin{bmatrix} A_{11} & A_{12} \\ A_{12}^* & A_{22} \end{bmatrix}$ and $B = \begin{bmatrix} B_{11} & B_{12} \\ B_{12}^* & B_{22} \end{bmatrix}$. Then,

$$A_{22}|A + B_{22}|B \le (A_{22} + B_{22})|(A + B).$$

Now, assume that A_{22} and B_{22} are positive definite. Then, equality holds if and only if $A_{12}A_{22}^{-1} = B_{12}B_{22}^{-1}$. **Source:** *xvii*) of Proposition 10.6.17, Corollary 10.6.18, [1045, 2179], and [2991, p. 241]. **Remark:** The first inequality is an extension of Bergstrom's inequality, which corresponds to the case where A_{11} is a scalar. See Fact 10.19.4.

Fact 10.12.6. Let $A, B \in \mathbb{F}^{n\times n}$, assume that A and B are positive definite, and assume that A and B are partitioned identically as $A = \begin{bmatrix} A_{11} & A_{12} \\ A_{12}^* & A_{22} \end{bmatrix}$ and $B = \begin{bmatrix} B_{11} & B_{12} \\ B_{12}^* & B_{22} \end{bmatrix}$. Then,

$$4(A_{11} + B_{11})^{-1} \le (A_{22}|A)^{-1} + (B_{22}|B)^{-1}.$$

Source: [2991, p. 232].

Fact 10.12.7. Let $A, B \in \mathbb{F}^{n\times n}$, assume that A and B are positive semidefinite, assume that A and B are partitioned identically as $A = \begin{bmatrix} A_{11} & A_{12} \\ A_{12}^* & A_{22} \end{bmatrix}$ and $B = \begin{bmatrix} B_{11} & B_{12} \\ B_{12}^* & B_{22} \end{bmatrix}$, and assume that A_{11} and B_{11} are positive definite. Then,

$$\frac{\det A}{\det A_{11}} + \frac{\det B}{\det B_{11}} \le \frac{\det(A + B)}{\det(A_{11} + B_{11})} = \det[(A_{11} + B_{11})|(A + B)],$$

$$(A_{12} + B_{12})^*(A_{11} + B_{11})^{-1}(A_{12} + B_{12}) \le A_{12}^*A_{11}^{-1}A_{12} + B_{12}^*B_{11}^{-1}B_{12},$$

$$\operatorname{rank}[A_{12}^*A_{11}^{-1}A_{12} + B_{12}^*B_{11}^{-1}B_{12} - (A_{12} + B_{12})^*(A_{11} + B_{11})^{-1}(A_{12} + B_{12})] = \operatorname{rank}(A_{12} - A_{11}B_{11}^{-1}B_{12}).$$

Source: [2991, p. 243]. **Remark:** The first inequality follows from Fact 10.16.16.

Fact 10.12.8. Let $B, C \in \mathbb{F}^{n\times n}$, define $A \triangleq B + \jmath C$, assume that B and C are Hermitian, assume that A, B, and C are partitioned identically as $A = \begin{bmatrix} A_{11} & A_{12} \\ A_{21} & A_{22} \end{bmatrix}$, $B = \begin{bmatrix} B_{11} & B_{12} \\ B_{12}^* & B_{22} \end{bmatrix}$, and $C = \begin{bmatrix} C_{11} & C_{12} \\ C_{12}^* & C_{22} \end{bmatrix}$, and assume that A_{22}, B_{22}, and C_{22} are nonsingular. Then,

$$A_{22}|A = B_{22}|B + \jmath C_{22}|C + (B_{12}B_{22}^{-1} - C_{12}C_{22}^{-1})(B_{22}^{-1} - \jmath C_{22}^{-1})^{-1}(B_{12}B_{22}^{-1} - C_{12}C_{22}^{-1})^*.$$

Source: [1830].

Fact 10.12.9. Let $A \in \mathbb{F}^{n\times n}$, $B \in \mathbb{F}^{n\times m}$, and $C \in \mathbb{F}^{m\times m}$, define $\mathcal{A} \triangleq \begin{bmatrix} A & B \\ B^* & C \end{bmatrix}$, and assume that $\mathcal{A}$ is positive semidefinite. Then,

$$0 \le BC^+B^* \le A.$$

If, in addition, $\mathcal{A}$ is positive definite, then C is positive definite and

$$0 \le BC^{-1}B^* < A.$$

Source: Proposition 10.2.5.

Fact 10.12.10. Let $A, B, C \in \mathbb{F}^{n\times n}$, define $\mathcal{A} \triangleq \begin{bmatrix} A & B \\ B^* & C \end{bmatrix}$, and assume that $\mathcal{A}$ is positive semidefinite. Then,

$$-A - C \le B + B^* \le A + C, \quad |\operatorname{tr}(B + B^*)| \le \operatorname{tr}(A + C), \quad |\det(B + B^*)| \le \det(A + C).$$

Now, assume that $\mathcal{A}$ is positive definite. Then,

$$-A - C < B + B^* < A + C, \quad |\operatorname{tr}(B + B^*)| < \operatorname{tr}(A + C), \quad |\det(B + B^*)| < \det(A + C).$$

Source: Use Fact 10.12.9 with $S\mathcal{A}S^{\mathsf{T}}$, $S \triangleq [I \ I]$, and $S \triangleq [I \ -I]$. **Related:** Fact 10.12.11 and Fact 10.25.60.

Fact 10.12.11. Let $A, B, C \in \mathbb{F}^{n\times n}$, and assume that $\begin{bmatrix} A & B \\ B^* & C \end{bmatrix} \in \mathbb{F}^{2n\times 2n}$ is positive semidefinite. Then,

$$\sqrt{\operatorname{tr}(B+B^*)^2} \le \operatorname{tr}(A+C).$$

Source: [2991, p. 244]. **Related:** Fact 10.12.10.

Fact 10.12.12. Let $A, B, C \in \mathbb{F}^{n\times n}$, assume that $\begin{bmatrix} A & B \\ B^* & C \end{bmatrix} \in \mathbb{F}^{2n\times 2n}$ is positive semidefinite, and assume that $AB = BA$. Then, $B^*B \le A^{1/2}CA^{1/2}$. **Source:** [2985] and [2991, p. 224].

Fact 10.12.13. Let $A, B, C \in \mathbb{F}^{n\times n}$, and assume that $\begin{bmatrix} A & B \\ B^* & C \end{bmatrix} \in \mathbb{F}^{2n\times 2n}$ is positive semidefinite. Then,

$$|\operatorname{tr} B^2| \le \operatorname{tr} B^*B \le \left.\begin{matrix} |\operatorname{tr} B|^2 \\ \sqrt{(\operatorname{tr} A^2)\operatorname{tr} C^2} \\ \operatorname{tr} AC + |\operatorname{tr} B|^2 - \operatorname{tr} B^*B \end{matrix}\right\} \le (\operatorname{tr} A)\operatorname{tr} C.$$

If, in addition, $\begin{bmatrix} A & B^* \\ B & C \end{bmatrix} \in \mathbb{F}^{2n\times 2n}$ is positive semidefinite, then $\operatorname{tr} B^*B \le \operatorname{tr} AC$. **Source:** The uppermost inequality follows from Fact 10.12.64. In the middle string, the first inequality is given by Fact 11.13.2. $|\operatorname{tr} B^2| \le \sqrt{(\operatorname{tr} A^2)\operatorname{tr} C^2}$ is given in [1953]. $\operatorname{tr} B^*B \le (\operatorname{tr} A)\operatorname{tr} C$ is given in [2206]. Alternatively, use Fact 10.14.54 with $P = A$, $Q = B$, and $R = C$. The lowermost inequality and the last inequality are given in [439, 1834, 1840].

Fact 10.12.14. Let $A_{11} \in \mathbb{R}^{n\times n}$, $A_{12} \in \mathbb{R}^{n\times m}$, and $A_{22} \in \mathbb{R}^{m\times m}$, define $A \triangleq \begin{bmatrix} A_{11} & A_{12} \\ A_{12}^{\mathsf{T}} & A_{22} \end{bmatrix} \in \mathbb{R}^{(n+m)\times(n+m)}$, and assume that A is symmetric. Then, A is positive semidefinite if and only if, for all $B \in \mathbb{R}^{n\times m}$,

$$\operatorname{tr} BA_{12}^{\mathsf{T}} \le \operatorname{tr}(A_{11}^{1/2}BA_{22}B^{\mathsf{T}}A_{11}^{1/2})^{1/2}.$$

Source: [347].

Fact 10.12.15. Let $A \in \mathbb{F}^{n\times n}$, $B \in \mathbb{F}^{n\times m}$, and $C \in \mathbb{F}^{m\times m}$, and define $\mathcal{A} \triangleq \begin{bmatrix} A & B \\ B^* & C \end{bmatrix}$. Then, $\mathcal{A}$ is positive semidefinite if and only if A and C are positive semidefinite and there exists a semicontractive matrix $Y \in \mathbb{F}^{n\times m}$ such that $B = A^{1/2}YC^{1/2}$. **Source:** [1465] and [2991, p. 185]. **Related:** *xiii*) and *xx*) of Fact 10.11.68. Also, Fact 10.12.16 and Fact 10.24.25.

Fact 10.12.16. Let $A \in \mathbb{F}^{n\times n}$, $B \in \mathbb{F}^{n\times m}$, and $C \in \mathbb{F}^{m\times m}$, and assume that A and C are positive definite. Then, $\begin{bmatrix} A & B \\ B^* & C \end{bmatrix} \in \mathbb{F}^{(n+m)\times(n+m)}$ is positive semidefinite if and only if

$$\sigma_{\max}(A^{-1/2}BC^{-1/2}) \le 1.$$

Furthermore, $\begin{bmatrix} A & B \\ B^* & C \end{bmatrix} \in \mathbb{F}^{(n+m)\times(n+m)}$ is positive definite if and only if

$$\sigma_{\max}(A^{-1/2}BC^{-1/2}) < 1.$$

Source: [1953]. **Related:** Fact 10.12.15.

Fact 10.12.17. Let $A \in \mathbb{F}^{n\times n}$, $B \in \mathbb{F}^{n\times m}$, and $C \in \mathbb{F}^{m\times m}$, assume that A and C are positive definite, and assume that

$$\sigma_{\max}^2(B) \le \sigma_{\min}(A)\sigma_{\min}(C).$$

Then, $\begin{bmatrix} A & B \\ B^* & C \end{bmatrix} \in \mathbb{F}^{(n+m)\times(n+m)}$ is positive semidefinite. If, in addition,

$$\sigma_{\max}^2(B) < \sigma_{\min}(A)\sigma_{\min}(C),$$

then $\begin{bmatrix} A & B \\ B^* & C \end{bmatrix} \in \mathbb{F}^{(n+m)\times(n+m)}$ is positive definite. **Source:** Note that

$$\sigma_{\max}^2(A^{-1/2}BC^{-1/2}) \le \lambda_{\max}(A^{-1/2}BC^{-1}B^*A^{-1/2}) \le \sigma_{\max}(C^{-1})\lambda_{\max}(A^{-1/2}BB^*A^{-1/2})$$

$$\le \frac{1}{\sigma_{\min}(C)}\lambda_{\max}(B^*A^{-1}B) \le \frac{\sigma_{\max}(A^{-1})}{\sigma_{\min}(C)}\lambda_{\max}(B^*B) = \frac{1}{\sigma_{\min}(A)\sigma_{\min}(C)}\sigma_{\max}^2(B) \le 1.$$

The result now follows from Fact 10.12.16.

Fact 10.12.18. Let $A, B \in \mathbb{F}^{n\times n}$, and assume that A and B are Hermitian. Then, $-A \le B \le A$ if and only if $\left[\begin{smallmatrix} A & B \\ B & A \end{smallmatrix}\right]$ is positive semidefinite. Furthermore, $-A < B < A$ if and only if $\left[\begin{smallmatrix} A & B \\ B & A \end{smallmatrix}\right]$ is positive definite. **Source:** Note that

$$\frac{1}{\sqrt{2}}\begin{bmatrix} I & -I \\ I & I \end{bmatrix}\begin{bmatrix} A & B \\ B & A \end{bmatrix}\frac{1}{\sqrt{2}}\begin{bmatrix} I & I \\ -I & I \end{bmatrix} = \begin{bmatrix} A-B & 0 \\ 0 & A+B \end{bmatrix}.$$

Related: Fact 10.16.7.

Fact 10.12.19. Let $A \in \mathbb{F}^{n\times n}$, $B \in \mathbb{F}^{n\times m}$, and $C \in \mathbb{F}^{m\times m}$, assume that $\left[\begin{smallmatrix} A & B \\ B^* & C \end{smallmatrix}\right]$ is positive semidefinite, and let $r \triangleq \operatorname{rank} B$. Then, for all $k \in \{1, \dots, r\}$,

$$\prod_{i=1}^{k} \sigma_i(B) \le \prod_{i=1}^{k} \max\{\lambda_i(A), \lambda_i(C)\}, \quad \sum_{i=1}^{k} \sigma_i(B) \le \sum_{i=1}^{k} \max\{\lambda_i(A), \lambda_i(C)\}.$$

Source: [2985].

Fact 10.12.20. Let $A, B, C \in \mathbb{F}^{n\times n}$, and assume that $\left[\begin{smallmatrix} A & B \\ B^* & C \end{smallmatrix}\right]$ is positive semidefinite. Then, for all $k \in \{1, \dots, n\}$,

$$\prod_{i=1}^{k} \sigma_i(B) \le \prod_{i=1}^{k} \sqrt{\lambda_i(A)\lambda_i(C)}, \quad \sum_{i=1}^{k} \sigma_i(B) \le \sum_{i=1}^{k} \sqrt{\lambda_i(A)\lambda_i(C)},$$

$$\prod_{i=1}^{k} \rho_i(B) \le \prod_{i=1}^{k} \sqrt{\lambda_i(A)\lambda_i(C)}, \quad \sum_{i=1}^{k} \rho_i(B) \le \sum_{i=1}^{k} \sqrt{\lambda_i(A)\lambda_i(C)}.$$

Source: Fact 3.25.15 and [2991, p. 352].

Fact 10.12.21. Let $A, B, C \in \mathbb{F}^{n\times n}$, and assume that $\mathcal{A} \triangleq \left[\begin{smallmatrix} A & B \\ B^* & C \end{smallmatrix}\right] \in \mathbb{F}^{2n\times 2n}$ is Hermitian. Then,

$$\begin{bmatrix} \mathcal{A}_{(1,1)} \\ \vdots \\ \mathcal{A}_{(2n,2n)} \end{bmatrix} \overset{\text{s}}{\prec} \begin{bmatrix} \lambda(A) \\ \lambda(C) \end{bmatrix} \overset{\text{s}}{\prec} \lambda(\mathcal{A}).$$

Now, assume that B is either Hermitian or skew Hermitian. Then,

$$\begin{bmatrix} \lambda(A+C) \\ \lambda(A+C) \end{bmatrix}^{\downarrow} - [\lambda(\mathcal{A})]^{\downarrow} \overset{\text{s}}{\prec} \lambda(\mathcal{A}) \overset{\text{s}}{\prec} \begin{bmatrix} \lambda(A+C) \\ \lambda(A+C) \end{bmatrix}^{\downarrow} - [\lambda(\mathcal{A})]^{\uparrow},$$

$$[\lambda(\mathcal{A})]^{\downarrow} + [\lambda(\mathcal{A})]^{\uparrow} \overset{\text{s}}{\prec} \begin{bmatrix} \lambda(A+C) \\ \lambda(A+C) \end{bmatrix} \overset{\text{s}}{\prec} 2\lambda(\mathcal{A}) \overset{\text{s}}{\prec} \begin{bmatrix} \lambda(A+C) \\ \lambda(A+C) \end{bmatrix}^{\downarrow} + [\lambda(\mathcal{A})]^{\downarrow} - [\lambda(\mathcal{A})]^{\uparrow}.$$

In addition, assume that $\mathcal{A}$ is positive semidefinite. Then,

$$\begin{bmatrix} \mathcal{A}_{(1,1)} \\ \vdots \\ \mathcal{A}_{(2n,2n)} \end{bmatrix} \overset{\text{s}}{\prec} \begin{bmatrix} \lambda(A) \\ \lambda(C) \end{bmatrix} \overset{\text{s}}{\prec} \lambda(\mathcal{A}) \overset{\text{s}}{\prec} \begin{bmatrix} \lambda(A+C) \\ 0 \end{bmatrix}.$$

Source: [1850], [1969, p. 225], [1971, p. 308], and [2750].

Fact 10.12.22. Let $A \in \mathbb{F}^{n\times n}$, $B \in \mathbb{F}^{n\times m}$, and $C \in \mathbb{F}^{m\times m}$, define $\mathcal{A} \triangleq \left[\begin{smallmatrix} A & B \\ B^* & C \end{smallmatrix}\right]$, and assume that $\mathcal{A}$ is positive definite. Then,

$$\operatorname{tr} A^{-1} + \operatorname{tr} C^{-1} \le \operatorname{tr} \mathcal{A}^{-1}.$$

Furthermore, B is nonzero if and only if

$$\operatorname{tr} A^{-1} + \operatorname{tr} C^{-1} < \operatorname{tr} \mathcal{A}^{-1}.$$

Source: Proposition 10.2.6 and [2030].

Fact 10.12.23. Let $A \in \mathbb{F}^{n\times n}$, assume that A is positive semidefinite, let $\alpha > 0$, and define $\mathcal{A} \triangleq \begin{bmatrix} \alpha A & A \\ A & \alpha^{-1}A \end{bmatrix}$. Then, $\mathcal{A}$ is positive semidefinite, and $\operatorname{rank} \mathcal{A} = \operatorname{rank} A$. **Source:** [2991, p. 222].

Fact 10.12.24. Let $A \in \mathbb{F}^{n\times n}$, assume that A is positive definite, and define $\mathcal{A} \triangleq \begin{bmatrix} A & I \\ I & A^{-1} \end{bmatrix}$. Then, $\mathcal{A}$ is positive semidefinite, and $\operatorname{rank} \mathcal{A} = n$. **Source:** [2991, p. 222].

Fact 10.12.25. Let $A \in \mathbb{F}^{n\times m}$, and define

$$\mathcal{A} \triangleq \begin{bmatrix} \sigma_{\max}(A)I_m & A^* \\ A & \sigma_{\max}(A)I_n \end{bmatrix}.$$

Then, $\mathcal{A}$ is positive semidefinite, and $\operatorname{rank} \mathcal{A} = n + m - \operatorname{amult}_{\langle A\rangle}[\sigma_{\max}(A)]$. **Source:** [2991, p. 222].

Fact 10.12.26. Let $A \in \mathbb{F}^{n\times n}$, assume that A is positive semidefinite, and define

$$\mathcal{A} \triangleq \begin{bmatrix} \lambda_1(A)I_n & A \\ A & \lambda_1(A)I_n \end{bmatrix}.$$

Then, $\mathcal{A}$ is positive semidefinite, and $\operatorname{rank} \mathcal{A} = 2n - \operatorname{amult}_A[\lambda_1(A)]$.

Fact 10.12.27. Let $A \in \mathbb{F}^{n\times m}$, and define

$$\mathcal{A} \triangleq \begin{bmatrix} \langle A\rangle & A^* \\ A & \langle A^*\rangle \end{bmatrix}.$$

Then, $\mathcal{A}$ is positive semidefinite, and $\operatorname{rank} \mathcal{A} = \operatorname{rank} A$. If, in addition, $n = m$, then

$$-\langle A\rangle - \langle A^*\rangle \le A + A^* \le \langle A\rangle + \langle A^*\rangle.$$

Source: Fact 10.12.9. The rank equality follows from Proposition 10.2.4 and Fact 10.24.4. **Remark:** $\begin{bmatrix} \langle A^*\rangle & A^* \\ A & \langle A\rangle \end{bmatrix}$ is not necessarily positive semidefinite. See [2991, p. 317] and Fact 10.12.52. **Related:** Fact 10.10.32, Fact 10.24.4, and [2991, p. 222].

Fact 10.12.28. Let $A \in \mathbb{F}^{n\times m}$, let $\alpha \in [0, 1]$, and define

$$\mathcal{A} \triangleq \begin{bmatrix} \langle A\rangle^{2\alpha} & A^* \\ A & \langle A^*\rangle^{2(1-\alpha)} \end{bmatrix}.$$

Then, $\mathcal{A}$ is positive semidefinite. **Source:** [2991, p. 318].

Fact 10.12.29. Let $A \in \mathbb{F}^{n\times n}$, assume that A is normal, and define

$$\mathcal{A} \triangleq \begin{bmatrix} \langle A\rangle & A \\ A^* & \langle A\rangle \end{bmatrix}.$$

Then, $\mathcal{A}$ is positive semidefinite. **Source:** Use Fact 10.12.27 and $\langle A\rangle = \langle A^*\rangle$. See [1450, p. 213].

Fact 10.12.30. Let $A \in \mathbb{F}^{n\times n}$, and define

$$\mathcal{A} \triangleq \begin{bmatrix} I & A \\ A^* & I \end{bmatrix}.$$

Then, $\mathcal{A}$ is positive semidefinite if and only if $\sigma_{\max}(A) \le 1$. Furthermore, $\mathcal{A}$ is positive definite if and only if $\sigma_{\max}(A) < 1$. **Source:** Note that

$$\begin{bmatrix} I & A \\ A^* & I \end{bmatrix} = \begin{bmatrix} I & 0 \\ A^* & I \end{bmatrix} \begin{bmatrix} I & 0 \\ 0 & I - A^*A \end{bmatrix} \begin{bmatrix} I & A \\ 0 & I \end{bmatrix}.$$

Fact 10.12.31. Let $A \in \mathbb{F}^{n\times m}$, and define

$$\mathcal{A} \triangleq \begin{bmatrix} I_n & A \\ A^* & I_m \end{bmatrix}.$$

Then, $\mathcal{A}$ is (positive semidefinite, positive definite) if and only if A is (semicontractive, contractive). Furthermore,

$$\operatorname{In}\mathcal{A} = \begin{bmatrix} n \\ 0 \\ 0 \end{bmatrix} + \operatorname{In}(I_m - A^*A) = \begin{bmatrix} m \\ 0 \\ 0 \end{bmatrix} + \operatorname{In}(I_n - AA^*).$$

Source: Note that $\mathcal{A} = \left[\begin{smallmatrix} I_n & A \\ 0 & I_m \end{smallmatrix}\right]^* \left[\begin{smallmatrix} I_n & 0 \\ 0 & I_m - A^*A \end{smallmatrix}\right] \left[\begin{smallmatrix} I_n & A \\ 0 & I_m \end{smallmatrix}\right]$. See [2991, p. 259].

Fact 10.12.32. Let $A \in \mathbb{F}^{n\times n}$, assume that A is positive semidefinite, assume that $\operatorname{spec}(A) \subset [0,1]$, and define $\mathcal{A} \in \mathbb{F}^{2n\times 2n}$ by

$$\mathcal{A} \triangleq \begin{bmatrix} A & (A - A^2)^{1/2} \\ (A - A^2)^{1/2} & I - A \end{bmatrix}.$$

Then, $\mathcal{A}$ is a projector, and $\operatorname{rank}\mathcal{A} = n$. **Source:** [2648] and Fact 4.17.13.

Fact 10.12.33. Let $A \in \mathbb{F}^{n\times m}$ and $B \in \mathbb{F}^{n\times l}$, and define

$$\mathcal{A} \triangleq \begin{bmatrix} A^*A & A^*B \\ B^*A & B^*B \end{bmatrix}.$$

Then, $\mathcal{A}$ is positive semidefinite, and

$$0 \le A^*B(B^*B)^+B^*A \le A^*A.$$

If $m = l$, then

$$-A^*A - B^*B \le A^*B + B^*A \le A^*A + B^*B, \quad (A + B)^*(A + B) \le 2(A^*A + B^*B).$$

If $m = l = 1$, then

$$|A^*B|^2 \le A^*AB^*B.$$

Remark: The last inequality is the Cauchy-Schwarz inequality. See Fact 10.16.25. **Related:** Fact 10.25.57.

Fact 10.12.34. Let $A, B \in \mathbb{F}^{n\times m}$, and define

$$\mathcal{A} \triangleq \begin{bmatrix} I + A^*A & A^* + B^* \\ A + B & I + BB^* \end{bmatrix} = \begin{bmatrix} I & A^* \\ B & I \end{bmatrix} \begin{bmatrix} I & B^* \\ A & I \end{bmatrix}.$$

Then, $\mathcal{A}$ is positive semidefinite. Furthermore, there exists a semicontractive matrix $C \in \mathbb{F}^{m\times n}$ such that

$$A + B = (I + A^*A)^{1/2}C(I + BB^*)^{1/2}.$$

Source: Fact 10.12.15 and [2991, pp. 187, 222, 228]. **Related:** Fact 10.16.27.

Fact 10.12.35. Let $A, B \in \mathbb{F}^{n\times m}$, and define

$$\mathcal{A} \triangleq \begin{bmatrix} I + A^*A & I - A^*B \\ I - B^*A & I + B^*B \end{bmatrix}, \quad \mathcal{B} \triangleq \begin{bmatrix} I + A^*A & I + A^*B \\ I + B^*A & I + B^*B \end{bmatrix}.$$

Then, $\mathcal{A}$ and $\mathcal{B}$ are positive semidefinite, and

$$0 \le (I - A^*B)(I + B^*B)^{-1}(I - B^*A) \le I + A^*A,$$
$$0 \le (I + A^*B)(I + B^*B)^{-1}(I + B^*A) \le I + A^*A.$$

Related: Fact 10.16.28.

Fact 10.12.36. Let $A, B \in \mathbb{F}^{n\times m}$. Then,

$$\begin{aligned}(A + B)(I + B^*B)^{-1}(A + B)^* &\le (A + B)(I + B^*B)^{-1}(A + B)^* + (I - AB^*)(I + BB^*)^{-1}(I - BA^*) \\ &= I + AA^*.\end{aligned}$$

Source: Set $C = A$ in Fact 3.20.19. See also [2983, p. 185] and [2991, pp. 228–231].

Fact 10.12.37. Let $A, B \in \mathbb{F}^{n\times n}$, assume that A and B are positive semidefinite, and define

$$\mathcal{A} \triangleq \begin{bmatrix} A & A^{1/2}B^{1/2} \\ B^{1/2}A^{1/2} & B \end{bmatrix} = \begin{bmatrix} A^{1/2} \\ B^{1/2} \end{bmatrix} [A^{1/2} \;\; B^{1/2}].$$

Then, $\mathcal{A}$ is positive semidefinite

Fact 10.12.38. Let $A \in \mathbb{F}^{n\times n}$ and $B \in \mathbb{F}^{n\times m}$, assume that A is positive semidefinite, and define

$$\mathcal{A} \triangleq \begin{bmatrix} A & AB \\ B^*A & B^*AB \end{bmatrix} = \begin{bmatrix} A^{1/2} \\ B^*A^{1/2} \end{bmatrix} [A^{1/2} \;\; A^{1/2}B].$$

Then, $\mathcal{A}$ is positive semidefinite, and

$$0 \le AB(B^*AB)^+B^*A \le A.$$

If, in addition $n = m$, then

$$-A - B^*AB \le AB + B^*A \le A + B^*AB.$$

Now, let $\alpha \in (0, \infty)$. Then,

$$0 \le AB(\alpha I + B^*AB)^{-1}B^*A \le A.$$

Fact 10.12.39. Let $A \in \mathbb{F}^{n\times n}$ and $B \in \mathbb{F}^{n\times m}$, assume that A is positive definite, and define

$$\mathcal{A} \triangleq \begin{bmatrix} A & B \\ B^* & B^*A^{-1}B \end{bmatrix}.$$

Then,

$$\mathcal{A} = \begin{bmatrix} A^{1/2} \\ B^*A^{-1/2} \end{bmatrix} [A^{1/2} \;\; A^{-1/2}B] = \begin{bmatrix} I & 0 \\ 0 & B^* \end{bmatrix} \begin{bmatrix} A & I \\ I & A^{-1} \end{bmatrix} \begin{bmatrix} I & 0 \\ 0 & B \end{bmatrix},$$

and thus $\mathcal{A}$ is positive semidefinite. Furthermore,

$$0 \le B(B^*A^{-1}B)^+B^* \le A.$$

Furthermore, if $\operatorname{rank} B = m$, then

$$\operatorname{rank}[A - B(B^*A^{-1}B)^{-1}B^*] = n - m.$$

Now, assume that $n = m$. Then,

$$-A - B^*A^{-1}B \le B + B^* \le A + B^*A^{-1}B.$$

Source: Fact 10.12.9 and [2991, p. 223]. **Remark:** $I - A^{-1/2}B(B^*A^{-1}B)^+B^*A^{-1/2}$ is a projector. **Related:** Fact 10.25.58.

Fact 10.12.40. Let $A \in \mathbb{F}^{n\times n}$, $B \in \mathbb{F}^{n\times m}$, and $C \in \mathbb{F}^{m\times m}$, assume that A is positive definite, assume that C is positive semidefinite, and define $\mathcal{A} \triangleq \left[\begin{smallmatrix} A & B \\ B^* & C \end{smallmatrix}\right]$. Then, $\mathcal{A}$ is positive semidefinite if and only if $B^*A^{-1}B \le C$. **Source:** [2991, p. 220]. **Related:** Fact 10.12.65 and Fact 10.12.66.

Fact 10.12.41. Let $A, B \in \mathbb{F}^{n\times n}$, and assume that $\left[\begin{smallmatrix} A & B \\ B^* & A \end{smallmatrix}\right]$ is positive semidefinite. Then, $\left[\begin{smallmatrix} A & B^* \\ B & A \end{smallmatrix}\right]$ is positive semidefinite. **Source:** Consider a congruence transformation with $\left[\begin{smallmatrix} 0 & I \\ I & 0 \end{smallmatrix}\right]$. See [2991, p. 317]. **Remark:** If $\left[\begin{smallmatrix} A & B \\ B^* & \alpha I \end{smallmatrix}\right]$ is positive semidefinite, then $\left[\begin{smallmatrix} A & B^* \\ B & \alpha I \end{smallmatrix}\right]$ is positive semidefinite. Now, let $C \in \mathbb{F}^{n\times n}$, and assume that $\left[\begin{smallmatrix} A & B \\ B^* & C \end{smallmatrix}\right]$ is positive semidefinite. Then, $\left[\begin{smallmatrix} A & B^* \\ B & C \end{smallmatrix}\right]$ is not necessarily positive semidefinite. See [1840], [2991, p. 317], and Fact 10.12.52.

Fact 10.12.42. Let $x, y \in \mathbb{F}^n$. Then, $\left[\begin{smallmatrix} xx^*+x^*xI & yx^*+x^*yI \\ xy^*+y^*xI & yy^*+y^*yI \end{smallmatrix}\right]$ is positive semidefinite. **Source:** [1834, 1837]. **Remark:** $\left[\begin{smallmatrix} xx^* & xy^* \\ yx^* & yy^* \end{smallmatrix}\right]$ and $\left[\begin{smallmatrix} x^*xI & x^*yI \\ y^*xI & y^*yI \end{smallmatrix}\right]$ are positive semidefinite, but $\left[\begin{smallmatrix} xx^* & yx^* \\ xy^* & yy^* \end{smallmatrix}\right]$ is not necessarily positive semidefinite. See Fact 10.12.41.

Fact 10.12.43. Let $A \in \mathbb{R}^{n\times n}$ and $B \in \mathbb{R}^{n\times m}$, assume that A is symmetric, $m < n$, and rank $B = m$, and define $\mathcal{A} \triangleq \begin{bmatrix} 0 & B^\mathsf{T} \\ B & A \end{bmatrix}$. Then, the following statements are equivalent:

i) For all $x \in \mathcal{N}(B^\mathsf{T})$, $x^\mathsf{T}Ax > 0$.

ii) For all $i \in \{2m+1, \dots, n+m\}$, $\operatorname{sign}\det \mathcal{A}_{(\{1,\dots,i\})} = (-1)^m$.

Source: [1895, p. 48]. **Related:** Fact 10.12.44.

Fact 10.12.44. Let $A \in \mathbb{F}^{n\times n}$ and $B \in \mathbb{F}^{n\times m}$, assume that A is positive semidefinite and rank $B = m$, and define $\mathcal{A} \triangleq \begin{bmatrix} A & B \\ B^* & 0 \end{bmatrix}$. Then, the following statements are equivalent:

i) $\mathcal{A}$ is nonsingular.

ii) $\mathcal{N}(A) \cap \mathcal{N}(B^*) = \{0\}$.

iii) For all $x \in \mathcal{N}(B^*)$, $x^*Ax > 0$.

iv) $A + BB^*$ is positive definite.

Source: [558, p. 523]. **Remark:** $\mathcal{A}$ is the *KKT matrix*. **Related:** Fact 10.12.43.

Fact 10.12.45. Let $A \in \mathbb{F}^{n\times n}$ and $B \in \mathbb{F}^{n\times m}$, assume that A is positive definite and rank $B = m$, and define $\mathcal{A} \triangleq \begin{bmatrix} A & B \\ B^* & 0 \end{bmatrix}$. Then, $B^*A^{-1}B$ is positive definite, $\det\mathcal{A} = (-1)^m(\det A)\det B^*A^{-1}B$, $\mathcal{A}$ is nonsingular, and

$$\mathcal{A}^{-1} = \begin{bmatrix} A^{-1} - A^{-1}B(B^*A^{-1}B)^{-1}B^*A^{-1} & A^{-1}B(B^*A^{-1}B)^{-1} \\ (B^*A^{-1}B)^{-1}B^*A^{-1} & -(B^*A^{-1}B)^{-1} \end{bmatrix}.$$

Furthermore,

$$\mathcal{A}^2 = \begin{bmatrix} A^2 + BB^* & AB \\ B^*A & B^*B \end{bmatrix}$$

is positive definite, and $AB(B^*B)^{-1}B^*A < A^2 + BB^*$. **Source:** Proposition 3.9.3 and Proposition 3.9.7.

Fact 10.12.46. Let $A, B \in \mathbb{F}^{n\times n}$, and define

$$\mathcal{A} \triangleq \begin{bmatrix} \langle A\rangle + \langle B\rangle & A^* + B^* \\ A + B & \langle A^*\rangle + \langle B^*\rangle \end{bmatrix}.$$

Then, $\mathcal{A}$ is positive semidefinite. Furthermore,

$$(\det\langle A + B\rangle)^2 \le \det(\langle A\rangle + \langle B\rangle)\det(\langle A^*\rangle + \langle B^*\rangle).$$

In particular,

$$\det\langle A + A^*\rangle \le \det(\langle A\rangle + \langle A^*\rangle).$$

Source: [2991, p. 317].

Fact 10.12.47. Let $A, B \in \mathbf{H}^n$. Then, $\begin{bmatrix} A^2+B^2 & AB+BA \\ AB+BA & A^2+B^2 \end{bmatrix}$ is positive semidefinite. **Source:** [1840].

Fact 10.12.48. Let $A, B \in \mathbb{F}^{n\times n}$, assume that A is positive definite, and let $\mathcal{S} \subseteq \{1, \dots, n\}$. Then,

$$B^*_{(\mathcal{S})}(A^{-1})_{(\mathcal{S})}B_{(\mathcal{S})} \le (B^*A^{-1}B)_{(\mathcal{S})}.$$

Source: [2991, p. 221].

Fact 10.12.49. Let $A \in \mathbb{F}^{n\times n}$ and $B \in \mathbb{F}^{n\times m}$, assume that A is positive definite, and define

$$\mathcal{A} \triangleq \begin{bmatrix} B^*AB & B^*B \\ B^*B & B^*A^{-1}B \end{bmatrix}.$$

Then,

$$\mathcal{A} = \begin{bmatrix} B^*A^{1/2} \\ B^*A^{-1/2} \end{bmatrix} [A^{1/2}B \;\; A^{-1/2}B],$$

and thus $\mathcal{A}$ is positive semidefinite. Furthermore,

$$0 \le B^*B(B^*A^{-1}B)^+B^*B \le B^*AB.$$

Now, assume in addition that $n = m$. Then,

$$-B^*AB - B^*A^{-1}B \le 2B^*B \le B^*AB + B^*A^{-1}B.$$

Source: Fact 10.12.9. **Related:** Fact 10.16.26 and Fact 10.25.58.

Fact 10.12.50. Let $A, B, C \in \mathbb{F}^{n\times m}$, assume that A and B are positive definite, and assume that C is Hermitian. Then, $\left[\begin{smallmatrix} A+B & A \\ A & A+C \end{smallmatrix}\right]$ is positive semidefinite if and only if $-(A^{-1} + B^{-1})^{-1} \le C$. **Source:** [2991, p. 222].

Fact 10.12.51. Let $A, B \in \mathbb{F}^{n\times m}$, let $\alpha, \beta \in (0, \infty)$, and define

$$\mathcal{A} \triangleq \begin{bmatrix} \beta^{-1}I + \alpha A^*A & (A+B)^* \\ A+B & \alpha^{-1}I + \beta BB^* \end{bmatrix}.$$

Then,

$$\mathcal{A} = \begin{bmatrix} \beta^{-1/2}I & \alpha^{1/2}A^* \\ \beta^{1/2}B & \alpha^{-1/2}I \end{bmatrix}\begin{bmatrix} \beta^{-1/2}I & \beta^{1/2}B^* \\ \alpha^{1/2}A & \alpha^{-1/2}I \end{bmatrix} = \begin{bmatrix} \alpha A^*A & A^* \\ A & \alpha^{-1}I \end{bmatrix} + \begin{bmatrix} \beta^{-1}I & B^* \\ B & \beta BB^* \end{bmatrix},$$

and thus $\mathcal{A}$ is positive semidefinite. Furthermore,

$$(A+B)^*(\alpha^{-1}I + \beta BB^*)^{-1}(A+B) \le \beta^{-1}I + \alpha A^*A.$$

Now, assume in addition that $n = m$. Then,

$$-(\beta^{-1/2} + \alpha^{-1/2})I - \alpha A^*A - \beta BB^* \le A + B + (A+B)^* \le (\beta^{-1/2} + \alpha^{-1/2})I + \alpha A^*A + \beta BB^*.$$

Related: Fact 10.16.29 and Fact 10.25.61.

Fact 10.12.52. Let $A, B \in \mathbb{F}^{n\times m}$, and assume that $I - A^*A$ and thus $I - AA^*$ are nonsingular. Then,

$$I - B^*B = (I - B^*A)(I - A^*A)^{-1}(I - A^*B) - (A-B)^*(I - AA^*)^{-1}(A-B).$$

Now, assume that $I - A^*A$ is positive definite. Then,

$$I - B^*B \le (I - B^*A)(I - A^*A)^{-1}(I - A^*B).$$

Now, assume that $I - B^*B$ is positive definite. Then, $I - A^*B$ is nonsingular. Next, define

$$\mathcal{A} \triangleq \begin{bmatrix} (I - A^*A)^{-1} & (I - B^*A)^{-1} \\ (I - A^*B)^{-1} & (I - B^*B)^{-1} \end{bmatrix}, \quad \mathcal{A}_0 \triangleq \begin{bmatrix} (I - A^*A)^{-1} & (I - A^*B)^{-1} \\ (I - B^*A)^{-1} & (I - B^*B)^{-1} \end{bmatrix}.$$

Then, $\mathcal{A}$ and $\mathcal{A}_0$ are positive semidefinite. Furthermore,

$$-(I - A^*A)^{-1} - (I - B^*B)^{-1} \le (I - B^*A)^{-1} + (I - A^*B)^{-1} \le (I \quad A^*A)^{-1} + (I - B^*B)^{-1},$$

$$\operatorname{tr}(I - A^*B)^{-1}(I - B^*A)^{-1} \le \operatorname{tr}(I - A^*A)^{-1}(I - B^*B)^{-1}.$$

Source: For the first equality, set $D = -B^*$ and $C = -A^*$, and replace B with $-B$ in Fact 3.20.16. See [93, 2184] and [2991, p. 231]. The penultimate string follows from Fact 10.12.9. The last inequality is given in [1840]. **Remark:** The equality is *Hua's matrix equality*. This result does not assume that either $I - A^*A$ or $I - B^*B$ is positive semidefinite. The inequality and Fact 10.16.28 are *Hua's inequalities*. See [2184, 2933]. A generalization is given in Fact 3.20.17. **Remark:** The fact that $\mathcal{A}_0$ is positive semidefinite is noted in [1844]. See Fact 10.12.27. **Remark:** Extensions to the case where $I - A^*A$ is singular are considered in [2184]. **Related:** Fact 10.10.44 and Fact 10.16.28.

Fact 10.12.53. Let $X \in \mathbb{F}^{n\times m}$, define $U \in \mathbb{F}^{(n+m)\times(n+m)}$ by

$$U \triangleq \begin{bmatrix} (I+X^*X)^{-1/2} & -X^*(I+XX^*)^{-1/2} \\ (I+XX^*)^{-1/2}X & (I+XX^*)^{-1/2} \end{bmatrix},$$

and let $A \in \mathbb{F}^{n\times n}$, $B \in \mathbb{F}^{n\times m}$, $C \in \mathbb{F}^{m\times n}$, and $D \in \mathbb{F}^{m\times m}$. Then, the following statements hold:

i) Assume that D is nonsingular and $X = D^{-1}C$. Then,

$$\begin{bmatrix} A & B \\ C & D \end{bmatrix} = \begin{bmatrix} (A-BX)(I+X^*X)^{-1/2} & (B+AX^*)(I+XX^*)^{-1/2} \\ 0 & D(I+XX^*)^{1/2} \end{bmatrix} U.$$

ii) Assume that A is nonsingular and $X = CA^{-1}$. Then,

$$\begin{bmatrix} A & B \\ C & D \end{bmatrix} = U \begin{bmatrix} (I+X^*X)^{1/2}A & (I+X^*X)^{-1/2}(B+X^*D) \\ 0 & (I+XX^*)^{-1/2}(D-XB) \end{bmatrix}.$$

Source: [1324]. **Remark:** See Proposition 3.9.3 and Proposition 3.9.4.

Fact 10.12.54. Let $X \in \mathbb{F}^{n\times m}$, assume that $\sigma_{\max}(X) < 1$, define $U \in \mathbb{F}^{(n+m)\times(n+m)}$ by

$$U \triangleq \begin{bmatrix} (I-X^*X)^{-1/2} & X^*(I-XX^*)^{-1/2} \\ (I-XX^*)^{-1/2}X & (I-XX^*)^{-1/2} \end{bmatrix},$$

and let $A \in \mathbb{F}^{n\times n}$, $B \in \mathbb{F}^{n\times m}$, $C \in \mathbb{F}^{m\times n}$, and $D \in \mathbb{F}^{m\times m}$. Then, the following statements hold:

i) Assume that D is nonsingular and $X = D^{-1}C$. Then,

$$\begin{bmatrix} A & B \\ C & D \end{bmatrix} = \begin{bmatrix} (A-BX)(I-X^*X)^{-1/2} & (B+AX^*)(I-XX^*)^{-1/2} \\ 0 & D(I-XX^*)^{1/2} \end{bmatrix} U.$$

ii) Assume that A is nonsingular and $X = CA^{-1}$. Then,

$$\begin{bmatrix} A & B \\ C & D \end{bmatrix} = U \begin{bmatrix} (I-X^*X)^{1/2}A & (I-X^*X)^{-1/2}(B-X^*D) \\ 0 & (I-XX^*)^{-1/2}(D-XB) \end{bmatrix}.$$

Source: [1324]. **Related:** Proposition 3.9.3 and Proposition 3.9.4.

Fact 10.12.55. Let $A, B \in \mathbb{F}^{n\times m}$ and $C, D \in \mathbb{F}^{m\times m}$, assume that C and D are positive definite, and define

$$\mathcal{A} \triangleq \begin{bmatrix} AC^{-1}A^* + BD^{-1}B^* & A+B \\ (A+B)^* & C+D \end{bmatrix}.$$

Then, $\mathcal{A}$ is positive semidefinite, and

$$(A+B)(C+D)^{-1}(A+B)^* \le AC^{-1}A^* + BD^{-1}B^*.$$

Now, assume that $n = m$. Then,

$$-AC^{-1}A^* - BD^{-1}B^* - C - D \le A + B + (A+B)^* \le AC^{-1}A^* + BD^{-1}B^* + C + D.$$

Source: [1355, 1819] and [2263, p. 151]. **Remark:** Replacing A, B, C, D by αB_1, $(1-\alpha)B_2$, αA_1, $(1-\alpha)A_2$ yields *xiv)* of Proposition 10.6.17.

Fact 10.12.56. Let $A \in \mathbb{R}^{n\times n}$, assume that A is positive definite, and let $\mathcal{S} \subseteq \{1,\ldots,n\}$. Then,

$$(A_{(\mathcal{S})})^{-1} \le (A^{-1})_{(\mathcal{S})}.$$

Source: [1448, p. 474]. **Remark:** Generalizations of this result are given in [727].

Fact 10.12.57. Let $A, D \in \mathbb{F}^{n\times n}$, $B \in \mathbb{F}^{n\times m}$, and $C \in \mathbb{F}^{m\times m}$, and assume that $\left[\begin{smallmatrix} A & B \\ B^* & C \end{smallmatrix}\right] \in \mathbb{F}^{n\times n}$ is positive semidefinite, C is positive definite, and D is positive definite. Then, $\left[\begin{smallmatrix} A+D & B \\ B^* & C \end{smallmatrix}\right]$ is positive definite.

Fact 10.12.58. Let $A \in \mathbb{F}^{n\times n}$, $B \in \mathbb{F}^{n\times m}$, and $C \in \mathbb{F}^{m\times m}$, and assume that $\begin{bmatrix} A & B \\ B^* & C \end{bmatrix}$ is positive semidefinite. Then, there exist unitary matrices $U, V \in \mathbb{F}^{(n+m)\times(n+m)}$ such that

$$\begin{bmatrix} A & B \\ B^* & C \end{bmatrix} = U\begin{bmatrix} A & 0 \\ 0 & 0 \end{bmatrix}U^* + V\begin{bmatrix} 0 & 0 \\ 0 & C \end{bmatrix}V^*.$$

Now, let $f\colon [0,\infty) \mapsto \mathbb{R}$, and assume that f is concave. Then,

$$\operatorname{tr} f\left(\begin{bmatrix} A & B \\ B^* & C \end{bmatrix}\right) \le \operatorname{tr} f(A) + \operatorname{tr} f(C).$$

Source: [544, 546, 548]. **Related:** Fact 10.12.60 and Fact 11.12.6.

Fact 10.12.59. Let $A, B, C \in \mathbb{F}^{n\times n}$, and assume that $\begin{bmatrix} A & B \\ B^* & C \end{bmatrix}$ is positive semidefinite. Then, there exist unitary matrices $U, V \in \mathbb{F}^{(n+m)\times(n+m)}$ such that

$$\begin{bmatrix} A & B \\ B^* & C \end{bmatrix} = \tfrac{1}{2}U\begin{bmatrix} A + C + \jmath(B^* - B) & 0 \\ 0 & 0 \end{bmatrix}U^* + \tfrac{1}{2}V\begin{bmatrix} 0 & 0 \\ 0 & A + C + \jmath(B - B^*) \end{bmatrix}V^*.$$

Source: [548]. **Related:** Fact 10.12.58.

Fact 10.12.60. Let $A, B, C \in \mathbb{C}^{n\times n}$, and assume that B is Hermitian and $\begin{bmatrix} A & B \\ B & C \end{bmatrix}$ is positive semidefinite. Then, there exist left-inner matrices $U, V \in \mathbb{F}^{2n\times n}$ such that

$$\begin{bmatrix} A & B \\ B & C \end{bmatrix} = \tfrac{1}{2}[U(A + C)U^* + V(A + C)V^*].$$

Source: [546, 549]. **Related:** Fact 10.12.58.

Fact 10.12.61. Let $A \in \mathbb{F}^{(n+m+l)\times(n+m+l)}$, assume that A is Hermitian, assume that A is partitioned as

$$A = \begin{bmatrix} A_{11} & A_{12} & A_{13} \\ A_{12}^* & A_{22} & A_{23} \\ A_{13}^* & A_{23}^* & A_{33} \end{bmatrix},$$

and assume that $B \triangleq \begin{bmatrix} A_{11} & A_{12} \\ A_{12}^* & A_{22} \end{bmatrix}$ is positive definite. Then,

$$A_{23}^*A_{22}^{-1}A_{23} \le [A_{13}^* \;\; A_{23}^*]B^{-1}\begin{bmatrix} A_{13} \\ A_{23} \end{bmatrix}.$$

Equivalently,

$$B|A \le A_{22}\left|\begin{bmatrix} A_{22} & A_{23} \\ A_{23}^* & A_{33} \end{bmatrix}\right..$$

Source: [3007].

Fact 10.12.62. Let $A \in \mathbb{F}^{(n+m+l)\times(n+m+l)}$, assume that A is positive semidefinite, and assume that A has the form

$$A = \begin{bmatrix} A_{11} & A_{12} & 0 \\ A_{12}^* & A_{22} & A_{23} \\ 0 & A_{23}^* & A_{33} \end{bmatrix}.$$

Then, there exist positive-semidefinite matrices $B, C \in \mathbb{F}^{(n+m+l)\times(n+m+l)}$ such that $A = B + C$ and such that B and C have the form

$$B = \begin{bmatrix} B_{11} & B_{12} & 0 \\ B_{12}^* & B_{22} & 0 \\ 0 & 0 & 0 \end{bmatrix}, \quad C = \begin{bmatrix} 0 & 0 & 0 \\ 0 & C_{22} & C_{23} \\ 0 & C_{23}^* & C_{33} \end{bmatrix}.$$

Source: [1373].

Fact 10.12.63. For all $i, j \in \{1, \ldots, k\}$, let $A_{ij} \in \mathbb{F}^{n_i \times n_j}$, define

$$A \triangleq \begin{bmatrix} A_{11} & \cdots & A_{1k} \\ \vdots & \cdot^{\cdot^{\cdot}} & \vdots \\ A_{1k} & \cdots & A_{kk} \end{bmatrix},$$

and assume that A is square and (positive semidefinite, positive definite). Furthermore, define

$$\hat{A} \triangleq \begin{bmatrix} \hat{A}_{11} & \cdots & \hat{A}_{1k} \\ \vdots & \cdot^{\cdot^{\cdot}} & \vdots \\ \hat{A}_{1k} & \cdots & \hat{A}_{kk} \end{bmatrix},$$

where $\hat{A}_{ij} = 1_{1\times n_i} A_{ij} 1_{n_j \times 1}$ is the sum of the entries of A_{ij} for all $i, j \in \{1, \ldots, k\}$. Then, $\hat{A}$ is (positive semidefinite, positive definite). **Source:** $\hat{A} = BAB^{\mathsf{T}}$, where the entries of $B \in \mathbb{R}^{k \times \sum_{i=1}^k n_i}$ are 0's and 1's. See [87] and [2991, p. 224].

Fact 10.12.64. For all $i, j \in \{1, \ldots, k\}$, let $A_{ij} \in \mathbb{F}^{n \times n}$, define $A \in \mathbb{F}^{kn \times kn}$ by

$$A \triangleq \begin{bmatrix} A_{11} & \cdots & A_{1k} \\ \vdots & \cdot^{\cdot^{\cdot}} & \vdots \\ A_{1k}^* & \cdots & A_{kk} \end{bmatrix},$$

and assume that A is positive semidefinite. Then,

$$\begin{bmatrix} \operatorname{tr} A_{11} & \cdots & \operatorname{tr} A_{1k} \\ \vdots & \cdot^{\cdot^{\cdot}} & \vdots \\ \operatorname{tr} A_{1k}^* & \cdots & \operatorname{tr} A_{kk} \end{bmatrix} \ge 0, \quad \begin{bmatrix} \operatorname{tr} A_{11}^2 & \cdots & \operatorname{tr} A_{1k}^* A_{1k} \\ \vdots & \cdot^{\cdot^{\cdot}} & \vdots \\ \operatorname{tr} A_{1k}^* A_{1k} & \cdots & \operatorname{tr} A_{kk}^2 \end{bmatrix} \ge 0.$$

Furthermore,

$$\det A \le \frac{1}{k^{kn}} \det \begin{bmatrix} \operatorname{tr} A_{11} & \cdots & \operatorname{tr} A_{1k} \\ \vdots & \cdot^{\cdot^{\cdot}} & \vdots \\ \operatorname{tr} A_{1k}^* & \cdots & \operatorname{tr} A_{kk} \end{bmatrix}^k.$$

Source: [852, 1953, 2206] and [2991, p. 237]. The last inequality is given in [1853]. **Related:** Fact 10.16.50.

Fact 10.12.65. Let $A \in \mathbb{F}^{n \times n}$, $B \in \mathbb{F}^{n \times m}$, and $C \in \mathbb{F}^{m \times m}$, and assume that $\mathcal{A} \in \mathbb{F}^{(n+m)\times(n+m)}$ defined by $\mathcal{A} \triangleq \left[\begin{smallmatrix} A & B \\ B^* & C \end{smallmatrix}\right]$ is positive semidefinite. Then, for all $i \in \{1, \ldots, \min\{n, m\}\}$,

$$2\sigma_i(B) \le \sigma_i(\mathcal{A}).$$

Source: [474, 2587] and [2991, p. 355]. **Related:** Fact 10.12.40 and Fact 10.12.66.

Fact 10.12.66. Let $A \in \mathbb{F}^{n \times n}$, $B \in \mathbb{F}^{n \times m}$, and $C \in \mathbb{F}^{m \times m}$, and assume that $\mathcal{A} \triangleq \left[\begin{smallmatrix} A & B \\ B^* & C \end{smallmatrix}\right] \in \mathbb{F}^{(n+m)\times(n+m)}$ is positive semidefinite. Then,

$$\begin{aligned} \max\{\sigma_{\max}(A), \sigma_{\max}(B)\} &\le \sigma_{\max}(\mathcal{A}) \\ &\le \tfrac{1}{2}[\sigma_{\max}(A) + \sigma_{\max}(B) + \sqrt{[\sigma_{\max}(A) - \sigma_{\max}(B)]^2 + 4\sigma_{\max}^2(C)}] \\ &\le \sigma_{\max}(A) + \sigma_{\max}(B), \end{aligned}$$

$$\max\{\sigma_{\max}(A), \sigma_{\max}(B)\} \le \sigma_{\max}(\mathcal{A}) \le \max\{\sigma_{\max}(A), \sigma_{\max}(B)\} + \sigma_{\max}(C).$$

Furthermore, if $n = m$, then $\begin{bmatrix} \sigma_{\max}(A) & \sigma_{\max}(B) \\ \sigma_{\max}(B) & \sigma_{\max}(C) \end{bmatrix}$ and $\begin{bmatrix} \rho_{\max}(A) & \rho_{\max}(B) \\ \rho_{\max}(B) & \rho_{\max}(C) \end{bmatrix}$ are positive semidefinite. **Source:** [1465]. **Related:** Fact 10.12.40, Fact 10.12.65, and Fact 11.16.15.

10.13 Facts on the Trace for One Matrix

Fact 10.13.1. Let $A \in \mathbb{F}^{n\times n}$, and assume that A is positive semidefinite. Then, $A = 0$ if and only if $\operatorname{tr} A = 0$.

Fact 10.13.2. Let $A \in \mathbb{F}^{n\times n}$. Then,

$$|\operatorname{Re}\operatorname{tr} A| \le |\operatorname{tr} A| \le \sum_{i=1}^{n} \rho_i(A) \le \operatorname{tr}\langle A\rangle = \sum_{i=1}^{n} \sigma_i(A).$$

Furthermore, consider the following statements:

i) A is positive semidefinite

ii) $\operatorname{tr} A = \operatorname{tr}\langle A\rangle$.

iii) $|\operatorname{tr} A| = \operatorname{tr}\langle A\rangle$.

iv) There exists $\alpha \in \mathbb{C}$ such that $|\alpha| = 1$ and αA is positive semidefinite.

v) A is normal.

vi) $\sum_{i=1}^{n} \rho_i(A) = \operatorname{tr}\langle A\rangle$.

Then, *i)* $\Longleftrightarrow$ *ii)* $\Longrightarrow$ *iii)* $\Longleftrightarrow$ *iv)* $\Longrightarrow$ *v)* $\Longrightarrow$ *vi)*. **Source:** Fact 7.12.30, Fact 10.21.10, and [2991, pp. 312, 313]. **Remark:** $A = \begin{bmatrix} 1 & 0 \\ 0 & -1 \end{bmatrix}$ shows that *v)* does not imply *iv)*. **Related:** Fact 10.21.10 states that A is normal if and only if $\sum_{i=1}^{n} \rho_i^2(A) = \operatorname{tr}\langle A\rangle^2$. Additional necessary and sufficient conditions are given by Fact 7.12.30. See [1499].

Fact 10.13.3. Let $A \in \mathbb{F}^{n\times n}$. Then,

$$|\operatorname{tr} A^2| \le \begin{cases} \operatorname{tr}\langle A\rangle\langle A^*\rangle \\ \operatorname{tr}\langle A^2\rangle \le \operatorname{tr}\langle A\rangle^2 = \operatorname{tr} A^*A. \end{cases}$$

Source: For the upper inequality, see [1783, 2987]. For the lower inequalities, use Fact 10.21.9 and Fact 11.13.2. **Related:** Fact 7.12.13, Fact 10.13.5, and Fact 10.13.6.

Fact 10.13.4. Let $A \in \mathbb{F}^{n\times n}$. Then, $\operatorname{tr}\langle A\rangle = \operatorname{tr}\langle A^*\rangle$.

Fact 10.13.5. Let $A \in \mathbb{F}^{n\times n}$. Then, for all $k \in \{1,\dots,n\}$,

$$\sum_{i=1}^{k} \sigma_i(A^2) \le \sum_{i=1}^{k} \sigma_i^2(A).$$

Hence,

$$\operatorname{tr}(A^{2*}A^2)^{1/2} \le \operatorname{tr} A^*A.$$

Equivalently, $\operatorname{tr}\langle A^2\rangle \le \operatorname{tr}\langle A\rangle^2$. **Source:** Let $B = A$ and $r = 1$ in Proposition 11.6.2. See also Fact 11.13.2.

Fact 10.13.6. Let $A \in \mathbb{F}^{n\times n}$, and let k denote the number of nonzero eigenvalues of A. Then,

$$\left.\begin{matrix} |\operatorname{tr} A^2| \le \operatorname{tr}\langle A^2\rangle \\ \operatorname{tr}\langle A\rangle\langle A^*\rangle \\ \frac{1}{k}|\operatorname{tr} A|^2 \end{matrix}\right\} \le \operatorname{tr}\langle A\rangle^2.$$

Source: The upper bound for $|\operatorname{tr} A^2|$ is given by Fact 11.13.2. The upper bound for $\operatorname{tr}\langle A^2\rangle$ is given by Fact 10.13.5. To prove the next inequality, let $A = S_1DS_2$ denote the singular value decomposition of A. Then, $\operatorname{tr}\langle A\rangle\langle A^*\rangle = \operatorname{tr} S_3^*DS_3D$, where $S_3 \triangleq S_1S_2$, and $\operatorname{tr} A^*A = \operatorname{tr} D^2$. The result follows from Fact 7.13.12. The last inequality is given by Fact 7.12.13. **Related:** Fact 7.12.13 and Fact 11.13.2.

Fact 10.13.7. Let $A \in \mathbb{F}^{n\times n}$. Then, the following statements are equivalent:

i) A is Hermitian, and $\operatorname{tr} A > 0$.

ii) There exists matrices $B, S \in \mathbb{F}^{n\times n}$ such that S is nonsingular, SBS^{-1} is positive definite, and $A = B + B^*$.

Source: [2991, p. 265].

Fact 10.13.8. Let $A \in \mathbb{F}^{n\times n}$, assume that A is positive definite, let p and q be real numbers, and assume that $p \le q$. Then,

$$\left(\tfrac{1}{n}\operatorname{tr} A^p\right)^{1/p} \le \left(\tfrac{1}{n}\operatorname{tr} A^q\right)^{1/q}.$$

Furthermore,

$$\lim_{p\downarrow 0}\left(\tfrac{1}{n}\operatorname{tr} A^p\right)^{1/p} = \det A^{1/n}.$$

Source: Fact 2.11.86.

Fact 10.13.9. Let $A \in \mathbb{F}^{n\times n}$, and assume that A is positive definite. Then,

$$n^2 \le (\operatorname{tr} A)\operatorname{tr} A^{-1}.$$

Finally, equality holds if and only if $A = I_n$. **Related:** Fact 10.25.23. Bounds on $\operatorname{tr} A^{-1}$ are given in [211, 682, 2170, 2318].

Fact 10.13.10. Let $A \in \mathbb{F}^{n\times n}$, assume that A is positive definite, and let $\alpha > 0$. Then,

$$n\operatorname{tr} A \le (\operatorname{tr} A^{1+\alpha})\operatorname{tr} A^{-\alpha}.$$

In particular,

$$n\operatorname{tr} A \le (\operatorname{tr} A^{3/2})\operatorname{tr} A^{-1/2}.$$

Source: [2751].

Fact 10.13.11. Let $A \in \mathbb{F}^{n\times n}$, assume that A is Hermitian, and consider the following statements:

i) $\sqrt{(n-1)\operatorname{tr} A^2} \le \operatorname{tr} A$.

ii) A is positive semidefinite.

iii) $\operatorname{tr} A^2 \le (\operatorname{tr} A)^2 \le n\operatorname{tr} A^2$.

Then, *i)* $\Longrightarrow$ *ii)* $\Longrightarrow$ *iii)*. **Source:** [2991, pp. 205, 213]. The second inequality in *iii)* follows from Fact 2.11.16 with $k = 1$.

Fact 10.13.12. Define $\mathcal{S} \triangleq \{A \in \mathbf{N}^n\colon\ \operatorname{tr} A \le 1\}$, and define $f\colon \mathcal{S} \mapsto \mathbb{R}$ by $f(A) \triangleq (\operatorname{tr} A)\det(I-A)$. Then, f is concave. **Source:** [928]. **Related:** Fact 2.12.53.

Fact 10.13.13. Let $A \in \mathbb{F}^{n\times n}$, and assume that A is positive semidefinite. Then,

$$\operatorname{tr} e^A \le e^{\operatorname{tr} A} + n - 1.$$

Furthermore, equality holds if and only if $\operatorname{rank} A \le 1$. **Source:** [2991, p. 348].

Fact 10.13.14. Let $A \in \mathbb{F}^{n\times n}$, and assume that A is positive semidefinite. Then, the following statements hold:

i) Let $r \in [0, 1]$. Then, for all $k \in \{1, \dots, n\}$,

$$\sum_{i=k}^{n} \lambda_i^r(A) \le \sum_{i=k}^{n} \mathrm{d}_i^r(A).$$

In particular,

$$\operatorname{tr} A^r \le \sum_{i=1}^{n} [A_{(i,i)}]^r.$$

ii) Let $r \geq 1$. Then, for all $k \in \{1, \ldots, n\}$,

$$\sum_{i=1}^{k} \mathrm{d}_i^r(A) \leq \sum_{i=1}^{k} \lambda_i^r(A).$$

In particular,

$$\sum_{i=1}^{n} [A_{(i,i)}]^r \leq \operatorname{tr} A^r.$$

iii) If either $r = 0$ or $r = 1$, then

$$\operatorname{tr} A^r = \sum_{i=1}^{n} [A_{(i,i)}]^r.$$

iv) If $r \neq 0$ and $r \neq 1$, then

$$\operatorname{tr} A^r = \sum_{i=1}^{n} [A_{(i,i)}]^r$$

if and only if A is diagonal.

Source: Fact 3.25.8, Fact 10.21.11, [1922], and [1924, p. 217]. **Related:** Fact 10.21.11.

Fact 10.13.15. Let $A \in \mathbb{F}^{n \times n}$ and $k \geq 1$. Then,

$$\operatorname{tr} A^{*k}A^k \leq \operatorname{tr} (A^*A)^k \leq (\operatorname{tr} A^*A)^k.$$

In particular,

$$\operatorname{tr} A^{*2}A^2 \leq \operatorname{tr} (A^*A)^2 \leq (\operatorname{tr} A^*A)^2.$$

Source: Fact 10.13.16 and [2991, pp. 197, 371].

Fact 10.13.16. Let $A \in \mathbb{F}^{n \times n}$ and $p, q \in [0, \infty)$. Then,

$$\operatorname{tr} (A^{*p}A^p)^q \leq \operatorname{tr} (A^*A)^{pq}.$$

Furthermore, equality holds if and only if $\operatorname{tr} A^{*p}A^p = \operatorname{tr} (A^*A)^p$. **Source:** [2476] and [2991, p. 197].

Fact 10.13.17. Let $A \in \mathbb{F}^{n \times n}$, $p \in [2, \infty)$, and $q \in [1, \infty)$. Then, A is normal if and only if

$$\operatorname{tr} (A^{*p}A^p)^q = \operatorname{tr} (A^*A)^{pq}.$$

Source: [987, 2476]. **Related:** Fact 4.10.12.

Fact 10.13.18. Let $A \in \mathbb{F}^{n \times n}$, and define $B \triangleq \frac{1}{2}(A + A^*)$. Then, the following statements hold:

i) If B is positive semidefinite and $p \geq 0$, then $\operatorname{tr} B^p \leq \operatorname{tr} \langle A \rangle^p$.

ii) If B is positive definite and $p \leq 0$, then $\operatorname{tr} \langle A \rangle^p \leq \operatorname{tr} B^p$.

iii) If $k \geq 0$, then $\operatorname{tr} B^{2k+1} \leq \operatorname{tr} \langle A \rangle^{2k+1}$.

Source: [1142]. **Remark:** Setting $k = 0$ in *iii*) yields $\operatorname{tr} \frac{1}{2}(A + A^*) \leq \operatorname{tr} \langle A \rangle$, which is given by Fact 7.12.25.

10.14 Facts on the Trace for Two or More Matrices

Fact 10.14.1. Let $A, B \in \mathbb{F}^{n \times n}$, assume that A and B are Hermitian, and assume that, for all $i \in \{1, \ldots, 3n\}$, $\operatorname{tr} (A + B)^i = \operatorname{tr} A^i + \operatorname{tr} B^i$. Then, the following statements hold:

i) $AB = 0$.

ii) $\mathcal{R}(A + B) = \mathcal{R}(A) + \mathcal{R}(B)$.

iii) $\operatorname{rank}(A + B) = \operatorname{rank} A + \operatorname{rank} B$.

Source: [2236] and [2991, p. 286].

Fact 10.14.2. Let $A_1, \ldots, A_m \in \mathbb{F}^{n\times n}$, assume that $A_1, \ldots, A_m$ are normal, and assume that, for all $k \in \{1, \ldots, (m+1)n\}$, $\operatorname{tr}(\sum_{i=1}^m A_i)^k = \sum_{i=1}^m \operatorname{tr} A_i^k$. Then, for all distinct $i, j \in \{1, \ldots, m\}$, $A_iA_j = 0$. **Source:** [1842].

Fact 10.14.3. Let $A, B \in \mathbb{F}^{n\times n}$, and assume that A and B are Hermitian. Then,

$$|\operatorname{tr} AB| \le \sqrt{(\operatorname{tr} A^2)\operatorname{tr} B^2} \le \tfrac{1}{2}\operatorname{tr}(A^2 + B^2), \qquad \sqrt{\operatorname{tr}(A+B)^2} \le \sqrt{\operatorname{tr} A^2} + \sqrt{\operatorname{tr} B^2}.$$

The first inequality is an equality if and only if A and B are linearly dependent. The second inequality is an equality if and only if $\operatorname{tr} A^2 = \operatorname{tr} B^2$. All three terms are equal if and only if $A = B$. **Source:** Corollary 11.3.9 and [2991, p. 264]. **Related:** Fact 10.14.22.

Fact 10.14.4. Let $A, B \in \mathbb{F}^{n\times n}$, assume that A and B are Hermitian, and assume that $-A \le B \le A$. Then,

$$\operatorname{tr} B^2 \le \operatorname{tr} A^2.$$

Source: $0 \le \operatorname{tr}[(A-B)(A+B)] = \operatorname{tr} A^2 - \operatorname{tr} B^2$. See [2697]. **Remark:** For $0 \le B \le A$, this result is a special case of *xxi*) of Proposition 10.6.13.

Fact 10.14.5. Let $A, B \in \mathbb{F}^{n\times n}$, and assume that A and B are positive semidefinite. Then, $AB = 0$ if and only if $\operatorname{tr} AB = 0$.

Fact 10.14.6. Let $A, B \in \mathbb{F}^{n\times n}$, assume that A and B are positive semidefinite, let $p, q \in (1, \infty)$, and assume that $1/p + 1/q = 1$. Then,

$$\operatorname{tr} AB \le \operatorname{tr}\langle AB\rangle \le (\operatorname{tr} A^p)^{1/p}(\operatorname{tr} B^q)^{1/q} \le \operatorname{tr}\left(\tfrac{1}{p}A^p + \tfrac{1}{q}B^q\right).$$

Equality holds in at least one inequality if and only if A^{p-1} and B are linearly dependent. **Source:** [1922], [1924, pp. 219, 222], and [2991, p. 370]. **Remark:** This is a matrix version of Hölder's inequality. **Related:** Fact 10.14.8 and Fact 10.14.21.

Fact 10.14.7. Let $f\colon \mathbf{P}^n \mapsto (0, \infty)$ be given by one of the following definitions:

i) $f(A) = (\det A)^{1/n}$.

ii) $p \in (0, 1)$ and $f(A) = (\operatorname{tr} A^p)^{1/p}$.

iii) $p \in (0, 1)$ and $f(A) = (\operatorname{tr} A^{p-1})^{1/(p-1)}$.

iv) $p \in [0, 1]$ and $f(A) = \frac{\operatorname{tr} A^p}{\operatorname{tr} A^{p-1}}$.

Then, for all $A, B \in \mathbf{P}^n$, $f(A) + f(B) \le f(A+B)$. **Source:** [1779]. **Remark:** f is an anti-norm on $\mathbf{P}^n$. **Remark:** *i*) gives the Minkowski determinant theorem given by Corollary 10.4.15. **Related:** Fact 2.2.59, Fact 11.8.21, and Fact 11.10.4.

Fact 10.14.8. Let $A_1, \ldots, A_m \in \mathbb{F}^{n\times n}$, assume that $A_1, \ldots, A_m$ are positive semidefinite, let $p_1, \ldots, p_m \in [1, \infty)$, and assume that $\frac{1}{p_1} + \cdots + \frac{1}{p_m} = 1$. Then,

$$\operatorname{tr}\langle A_1 \cdots A_m\rangle \le \prod_{i=1}^m (\operatorname{tr} A_i^{p_i})^{1/p_i} \le \operatorname{tr}\sum_{i=1}^m \frac{1}{p_i}A_i^{p_i}.$$

Furthermore, the following statements are equivalent:

i) $\operatorname{tr}\langle A_1 \cdots A_m\rangle = \prod_{i=1}^m (\operatorname{tr} A_i^{p_i})^{1/p_i}$.

ii) $\operatorname{tr}\langle A_1 \cdots A_m\rangle = \operatorname{tr}\sum_{i=1}^m \frac{1}{p_i}A_i^{p_i}$.

iii) $A_1^{p_1} = \cdots = A_m^{p_m}$.

Source: [1940]. **Remark:** The first inequality is a matrix version of Hölder's inequality. The first and third terms are a matrix version of Young's inequality. See Fact 2.2.50, Fact 2.2.53, Fact 10.11.73, Fact 10.14.33, Fact 10.14.34, and Fact 10.11.38.

Fact 10.14.9. Let $A_1, \ldots, A_m \in \mathbb{F}^{n\times n}$, assume that $A_1, \ldots, A_m$ are positive semidefinite, let

$\alpha_1, \ldots, \alpha_m$ be nonnegative numbers, and assume that $\sum_{i=1}^m \alpha_i \ge 1$. Then,

$$\left|\operatorname{tr} \prod_{i=1}^m A_i^{\alpha_i}\right| \le \prod_{i=1}^m (\operatorname{tr} A_i)^{\alpha_i}.$$

Furthermore, if $\sum_{i=1}^m \alpha_i = 1$, then equality holds if and only if $A_2, \ldots, A_m$ are scalar multiples of A_1, whereas, if $\sum_{i=1}^m \alpha_i > 1$, then equality holds if and only if $A_2, \ldots, A_m$ are scalar multiples of A_1 and $\operatorname{rank} A_1 = 1$. **Source:** [709]. **Remark:** If $\sum_{i=1}^m \alpha_i = 1$, then $\prod_{i=1}^m (\operatorname{tr} A_i)^{\alpha_i} \le \sum_{i=1}^m \alpha_i \operatorname{tr} A_i$. See Fact 2.11.87. **Related:** Fact 10.14.6.

Fact 10.14.10. Let $A, B \in \mathbb{F}^{n\times n}$, let $p, q \in (1, \infty)$, assume that $1/p + 1/q = 1$, and let $r \in [1, \infty)$. Then,

$$\operatorname{tr} \langle AB\rangle^r \le (\operatorname{tr} \langle A\rangle^{rp})^{1/p} (\operatorname{tr} \langle B\rangle^{rq})^{1/q}.$$

Source: [2439]. **Related:** Fact 10.14.11.

Fact 10.14.11. Let $A_1, \ldots, A_m \in \mathbb{F}^{n\times n}$, let $p_1, \ldots, p_m \in (1, \infty)$, assume that $\sum_{i=1}^m 1/p_i = 1$, and let $r \in [1, \infty)$. Then,

$$\operatorname{tr} \langle A_1 A_2 \cdots A_m\rangle^r \le \prod_{i=1}^m (\operatorname{tr} \langle A_i\rangle^{rp_i})^{1/p_i} \le \operatorname{tr} \sum_{i=1}^m \tfrac{1}{p_i} \langle A_i\rangle^{rp_i}.$$

Now, assume that $A_1, \ldots, A_m$ are positive semidefinite. Then,

$$|\operatorname{tr} A_1 A_2 \cdots A_m|^m \le \prod_{i=1}^m \operatorname{tr} A_i^m.$$

Finally, if $\alpha_1, \ldots, \alpha_m \in [0, 1]$ and $\sum_{i=1}^m \alpha_i = 1$, then

$$\operatorname{tr} \langle A_1^{\alpha_1} A_2^{\alpha_2} \cdots A_m^{\alpha_m}\rangle^r \le \prod_{i=1}^m (\operatorname{tr} A_i^r)^{\alpha_i} \le \sum_{i=1}^m \alpha_i \operatorname{tr} A_i^r.$$

Source: [2439]. **Related:** Fact 10.14.10.

Fact 10.14.12. Let $A, B \in \mathbb{F}^{n\times n}$, let $p, q \in (1, \infty)$, assume that $1/p + 1/q = 1$, and let $r \in [1, \infty)$. Then,

$$\operatorname{tr} \langle AB^*\rangle^r \le \operatorname{tr}(\tfrac{1}{p}\langle A\rangle^{rp} + \tfrac{1}{q}\langle B\rangle^{rq}),$$

$$\operatorname{tr} \langle AB^*\rangle^r \le \operatorname{tr} (\tfrac{1}{p}\langle A\rangle^{p} + \tfrac{1}{q}\langle B\rangle^{q})^r.$$

Furthermore,

$$\operatorname{tr} \langle AB^*\rangle \le \tfrac{1}{2} \operatorname{tr}(A^*A + B^*B),$$

$$\operatorname{tr} (\tfrac{1}{2}\langle A\rangle^2 + \tfrac{1}{2}\langle B\rangle^2)^r \le \tfrac{1}{2} \operatorname{tr} \langle A\rangle^{2r} + \tfrac{1}{2} \operatorname{tr} \langle B\rangle^{2r}.$$

Source: [2439].

Fact 10.14.13. Let $A_1, \ldots, A_m, B_1, \ldots, B_m \in \mathbf{N}^{n\times n}$, let $p, q \in (1, \infty)$, and assume that $1/p + 1/q = 1$. Then,

$$\operatorname{tr} \sum_{i=1}^m A_i B_i \le \left(\operatorname{tr} \sum_{i=1}^m A_i^p\right)^{1/p} \left(\operatorname{tr} \sum_{i=1}^m B_i^p\right)^{1/p} \le \frac{1}{p} \operatorname{tr} \sum_{i=1}^m A_i^p + \frac{1}{q} \operatorname{tr} \sum_{i=1}^m B_i^q.$$

Source: [2439].

Fact 10.14.14. Let $A, B \in \mathbb{F}^{n\times n}$. Then, $|\operatorname{tr} AB|^2 \le (\operatorname{tr} A^*A) \operatorname{tr} BB^*$. **Source:** [2983, p. 25] and Corollary 11.3.9. **Related:** Fact 10.14.15.

Fact 10.14.15. Let $A \in \mathbb{F}^{n\times m}$ and $B \in \mathbb{F}^{m\times n}$, and let $k \ge 1$. Then,

$$\operatorname{Re} \operatorname{tr} (AB)^{2k} \le |\operatorname{tr} (AB)^{2k}| \le \operatorname{tr} (A^*ABB^*)^k \le \operatorname{tr} (A^*A)^k (BB^*)^k \le [\operatorname{tr} (A^*A)^k] \operatorname{tr} (BB^*)^k.$$

In particular,

$$|\operatorname{tr}(AB)^2| \le \operatorname{tr} A^*ABB^* \le (\operatorname{tr} A^*A)\operatorname{tr} BB^*.$$

Source: See [2954] for the case where $n = m$. If $n \ne m$, then A and B can be augmented with zeros. **Remark:** The term $|\operatorname{tr} AB|^2$ in Fact 10.14.14 cannot be ordered with $|\operatorname{tr}(AB)^2|$ and $\operatorname{tr} A^*ABB^*$.

Fact 10.14.16. Let $A, B \in \mathbb{F}^{n\times n}$, assume that A and B are Hermitian, and let $k \ge 1$. Then,

$$\operatorname{tr}(AB)^{2k} \le |\operatorname{tr}(AB)^{2k}| \le \operatorname{tr}\langle (AB)^{2k}\rangle \le \operatorname{tr}(A^2B^2)^k \le \begin{cases} (\operatorname{tr} A^2B^2)^k \\ \operatorname{tr} A^{2k}B^{2k} \le \sqrt{(\operatorname{tr} A^{4k})\operatorname{tr} B^{4k}}, \end{cases}$$

$$|\operatorname{tr}(A^kB^k)^2| \le \operatorname{tr} A^{2k}B^{2k}.$$

In particular, $\operatorname{tr}(AB)^2 \le \operatorname{tr} A^2B^2$. Furthermore, $\operatorname{tr}(AB)^2 = \operatorname{tr} A^2B^2$ if and only if A and B commute. Finally, if k is even, then

$$\operatorname{tr}(AB)^{2k} \le |\operatorname{tr}(AB)^{2k}| \le \operatorname{tr}\langle (AB)^{2k}\rangle \le \operatorname{tr}(A^2B^2)^k \le \begin{cases} (\operatorname{tr} A^2B^2)^k \\ |\operatorname{tr}(A^kB^k)^2| \le \operatorname{tr} A^{2k}B^{2k} \le \sqrt{(\operatorname{tr} A^{4k})\operatorname{tr} B^{4k}}. \end{cases}$$

Source: Fact 10.14.15, [103, 2954], and [2991, pp. 260, 261, 371]. **Remark:** Fact 7.13.11 implies that $\operatorname{tr}(AB)^{2k}$ and $\operatorname{tr}(A^2B^2)^k$ are real.

Fact 10.14.17. Let $A, B \in \mathbb{F}^{n\times n}$, assume that A and B are positive semidefinite, and let $k \ge 1$. Then,

$$\operatorname{tr}(AB)^k \le \sqrt{(\operatorname{tr} A^{2k})\operatorname{tr} B^{2k}} \le (\operatorname{tr} A^k)\operatorname{tr} B^k.$$

In particular,

$$\operatorname{tr} AB \le \sqrt{(\operatorname{tr} A^2)\operatorname{tr} B^2} \le (\operatorname{tr} A)\operatorname{tr} B.$$

Source: [2951].

Fact 10.14.18. Let $A, B \in \mathbb{F}^{n\times n}$, assume that A and B are positive semidefinite, and let $k, m \in \mathbb{P}$, where $m \le k$. Then,

$$\operatorname{tr}(A^mB^m)^k \le \operatorname{tr}(A^kB^k)^m.$$

In particular,

$$\operatorname{tr}(AB)^k \le \operatorname{tr} A^kB^k, \quad \operatorname{tr} AB \le (\operatorname{tr} A)\operatorname{tr} B.$$

If, in addition, k is even, then

$$\operatorname{tr}(AB)^k \le \operatorname{tr}(A^2B^2)^{k/2} \le \operatorname{tr} A^kB^k.$$

Finally, $\operatorname{tr}(AB)^k \le \operatorname{tr} A^kB^k$ if and only if either $n = 1$ or $[A, B] = 0$. **Source:** Fact 10.22.27, Fact 10.22.37, and [2991, pp. 207, 368, 369, 371]. **Remark:** Fact 7.13.11 implies that $\operatorname{tr}(AB)^k$ is real. **Remark:** $\operatorname{tr}(AB)^k \le \operatorname{tr} A^kB^k$ is the *Lieb-Thirring inequality*. See [449, p. 279]. $\operatorname{tr}(AB)^k \le \operatorname{tr}(A^2B^2)^{k/2}$ follows from Fact 10.14.24. See [2932, 2954].

Fact 10.14.19. Let $A, B \in \mathbb{F}^{n\times n}$, assume that A and B are positive semidefinite, and let $k \ge 1$. Then,

$$\operatorname{tr}(AB)^k \le \min\{\sigma_{\max}^k(A)\operatorname{tr} B^k, \sigma_{\max}^k(B)\operatorname{tr} A^k\}.$$

Source: [2439].

Fact 10.14.20. Let $A, B \in \mathbb{F}^{n\times n}$, assume that A is positive semidefinite, assume that B is Hermitian, and let $\alpha \in [0, 1]$. Then,

$$\operatorname{tr}(AB)^2 \le \operatorname{tr} A^{2\alpha}BA^{2-2\alpha}B \le \operatorname{tr} A^2B^2.$$

Source: [1111].

Fact 10.14.21. Let $A, B \in \mathbb{F}^{n\times n}$, and assume that A and B are positive semidefinite. Then,

$$\operatorname{tr} AB \le \operatorname{tr}(AB^2A)^{1/2} = \operatorname{tr}\langle AB\rangle \le \tfrac{1}{4}\operatorname{tr}(A+B)^2,$$
$$\operatorname{tr}(AB)^2 \le \operatorname{tr} A^2B^2 \le \tfrac{1}{16}\operatorname{tr}(A+B)^4.$$

Source: Fact 10.14.24 and Fact 11.10.45.

Fact 10.14.22. Let $A, B \in \mathbb{F}^{n\times n}$, and assume that A and B are positive semidefinite. Then,

$$\operatorname{tr} AB = \operatorname{tr} A^{1/2}BA^{1/2} = \operatorname{tr}[(A^{1/2}BA^{1/2})^{1/2}(A^{1/2}BA^{1/2})^{1/2}]$$
$$\le [\operatorname{tr}(A^{1/2}BA^{1/2})^{1/2}]^2 \le (\operatorname{tr} A)(\operatorname{tr} B) \le \tfrac{1}{4}(\operatorname{tr} A + \operatorname{tr} B)^2 \le \tfrac{1}{2}[(\operatorname{tr} A)^2 + (\operatorname{tr} B)^2],$$

$$\operatorname{tr} AB \le \operatorname{tr}(AB^2A)^{1/2} \le \sqrt{\operatorname{tr} A^2}\sqrt{\operatorname{tr} B^2} \le \tfrac{1}{4}(\sqrt{\operatorname{tr} A^2} + \sqrt{\operatorname{tr} B^2})^2 \le \tfrac{1}{2}(\operatorname{tr} A^2 + \operatorname{tr} B^2) \le \tfrac{1}{2}[(\operatorname{tr} A)^2 + (\operatorname{tr} B)^2].$$

Source: Fact 2.2.12. Note that

$$\operatorname{tr}(A^{1/2}BA^{1/2})^{1/2} = \sum_{i=1}^{n} \lambda_i^{1/2}(AB).$$

The second inequality follows from Proposition 11.3.6 with $p = q = 2$, $r = 1$, and A and B replaced by $A^{1/2}$ and $B^{1/2}$. **Related:** Fact 3.15.19.

Fact 10.14.23. Let $A, B \in \mathbb{F}^{n\times n}$, assume that A and B are positive semidefinite, and let $p \ge 1$. Then,

$$\operatorname{tr} AB \le \operatorname{tr}(A^{p/2}B^pA^{p/2})^{1/p}.$$

Source: [1108].

Fact 10.14.24. Let $A, B \in \mathbb{F}^{n\times n}$, assume that A and B are positive semidefinite, and let $p \ge 0$ and $r \ge 1$. Then,

$$\operatorname{tr}(A^{1/2}BA^{1/2})^{pr} \le \operatorname{tr}(A^{r/2}B^rA^{r/2})^p.$$

In particular,

$$\operatorname{tr}(A^{1/2}BA^{1/2})^{2p} \le \operatorname{tr}(AB^2A)^p, \quad \operatorname{tr} AB \le \operatorname{tr}(AB^2A)^{1/2} = \operatorname{tr}\langle AB\rangle.$$

Source: Fact 10.22.27 and Fact 10.22.37. **Remark:** This is the *Araki-Lieb-Thirring inequality*. See [163, 192] and [449, p. 258]. **Related:** Fact 10.11.78, Fact 10.22.36, and Fact 11.10.50.

Fact 10.14.25. Let $A, B \in \mathbb{F}^{n\times n}$, assume that A and B are positive semidefinite, and let $q \ge 0$ and $p \in [0, 1]$. Then,

$$\sigma_{\max}^{2pq}(A)\operatorname{tr} B^{pq} \le \operatorname{tr}(A^pB^pA^p)^q \le \operatorname{tr}(ABA)^{pq}.$$

Source: [192]. **Remark:** The right-hand inequality is equivalent to the Araki-Lieb-Thirring inequality given by Fact 10.14.24, where p and q correspond to $1/r$ and pr, respectively.

Fact 10.14.26. Let $A, B \in \mathbb{F}^{n\times n}$, and assume that A and B are positive semidefinite. Then,

$$\operatorname{tr} B^2(ABA)^2 \le \operatorname{tr} A^4B^4.$$

Source: [456].

Fact 10.14.27. Let $A, B \in \mathbb{F}^{n\times n}$, assume that A and B are positive semidefinite, and let $p, q \ge 0$. Then,

$$\operatorname{tr} AB^pAB^q \le \operatorname{tr} A^2B^{p+q}.$$

If, in addition, $p < q$, then

$$2\operatorname{tr} AB^pAB^q \le \operatorname{tr} AB^pAB^q + \operatorname{tr} AB^{2p}AB^{q-p} \le 2\operatorname{tr} A^2B^{p+q}.$$

Source: [456, 1353].

Fact 10.14.28. Let $A, B \in \mathbb{F}^{n\times n}$, assume that A and B are positive semidefinite, and let $p \ge r \ge 0$. Then,

$$[\operatorname{tr}(A^{1/2}BA^{1/2})^p]^{1/p} \le [\operatorname{tr}(A^{1/2}BA^{1/2})^r]^{1/r}.$$

Furthermore,

$$[\operatorname{tr}(AB)^2]^{1/2} \le (\operatorname{tr} A^2B^2)^{1/2} = (\operatorname{tr} AB^2A)^{1/2} \le \operatorname{tr}(AB^2A)^{1/2},$$

$$[\operatorname{tr}(A^{1/2}BA^{1/2})^2]^{1/2} \le \operatorname{tr} AB \le \begin{cases} \operatorname{tr}(AB^2A)^{1/2} \\ [\operatorname{tr}(A^{1/2}BA^{1/2})^{1/2}]^2. \end{cases}$$

Source: [829], Fact 2.11.90, Fact 10.14.20, and Fact 10.14.23.

Fact 10.14.29. Let $A, B \in \mathbb{F}^{n\times n}$, assume that A and B are positive semidefinite, assume that $B \le I$, and let $q \ge p > 0$. Then,

$$\operatorname{tr}(BA^pB)^{1/p} \le \operatorname{tr}(BA^qB)^{1/q}, \quad \operatorname{tr}(BA^pB)^{1/p} \le \operatorname{tr}(B^{p/q}A^pB^{p/q})^{1/p}.$$

Source: [2693].

Fact 10.14.30. Let $A, B \in \mathbb{F}^{n\times n}$, assume that A and B are positive semidefinite, assume that $A \le B$, and let $p, q \ge 0$. Then,

$$\operatorname{tr} A^pB^q \le \operatorname{tr} B^{p+q}.$$

If, in addition, A and B are positive definite, then this inequality holds for all $p, q \in \mathbb{R}$ satisfying $q \ge -1$ and $p + q \ge 0$. **Source:** [537].

Fact 10.14.31. Let $A, B \in \mathbb{F}^{n\times n}$, assume that A and B are positive semidefinite, assume that $A \le B$, let $f\colon [0,\infty) \mapsto [0,\infty)$, and assume that $f(0) = 0$, f is continuous, and f is increasing. Then,

$$\operatorname{tr} f(A) \le \operatorname{tr} f(B).$$

Now, let $p > 1$ and $q \ge \max\{-1, -p/2\}$, and, if $q < 0$, assume that A is positive definite. Then,

$$\operatorname{tr} f(A^{q/2}B^pA^{q/2}) \le \operatorname{tr} f(A^{p+q}).$$

Source: [1121].

Fact 10.14.32. Let $A, B \in \mathbb{F}^{n\times n}$, assume that A and B are positive semidefinite, and let $\alpha \in [0, 1]$. Then,

$$\operatorname{tr}(A + B) \le \operatorname{tr}\langle A - B\rangle + 2\operatorname{tr} A^\alpha B^{1-\alpha}.$$

Source: [1426]. **Remark:** This is the *Powers-Stormer inequality*. **Related:** Fact 2.2.45 and Fact 11.10.16.

Fact 10.14.33. Let $A, B \in \mathbb{F}^{n\times n}$, assume that A and B are positive semidefinite, let $\alpha \in [0, 1]$, and define $\delta \triangleq \min\{\alpha, 1-\alpha\}$. $\rho \triangleq \max\{\alpha, 1-\alpha\}$. Then,

$$\begin{aligned} \tfrac{1}{2}\operatorname{tr}(A + B - \langle A - B\rangle) &\le \operatorname{tr} A^\alpha B^{1-\alpha} \le \operatorname{tr}\langle A^\alpha B^{1-\alpha}\rangle \\ &\le \operatorname{tr}\langle A^\alpha B^{1-\alpha}\rangle + \delta(\sqrt{\operatorname{tr} A} - \sqrt{\operatorname{tr} B})^2 \\ &\le (\operatorname{tr} A)^\alpha(\operatorname{tr} B)^{1-\alpha} \le \operatorname{tr}[\alpha A + (1-\alpha)B] \\ &= [(\operatorname{tr} A^{2\alpha})\operatorname{tr} B^{2(1-\alpha)}]^{1/2} + \rho(\operatorname{tr} A + \operatorname{tr} B - \operatorname{tr}\langle A^{1/2}B^{1/2}\rangle). \end{aligned}$$

Furthermore, the first two inequalities are equalities if and only if A and B are linearly dependent, while the last inequality is an equality if and only if $A = B$. **Source:** Use Fact 10.14.6 or Fact 10.14.9 for the inequality involving the first and third terms, and use Fact 2.2.53 for the right-most inequality. The inequality involving the third and fifth terms is given in [1640]. The last inequality is given in [1356]. **Related:** Fact 2.2.50, Fact 2.2.53, Fact 10.11.38, Fact 10.11.73, Fact 10.14.8, Fact 10.14.33, and Fact 10.16.14.

Fact 10.14.34. Let $A, B \in \mathbb{F}^{n\times n}$, assume that A and B are positive definite, and let $\alpha \in [0,1]$. Then,

$$\left.\begin{array}{r}\operatorname{tr} A^{-\alpha}B^{\alpha-1}\\ \operatorname{tr}[\alpha A + (1-\alpha)B]^{-1}\end{array}\right\} \le (\operatorname{tr} A^{-1})^{\alpha}(\operatorname{tr} B^{-1})^{1-\alpha} \le \operatorname{tr}[\alpha A^{-1} + (1-\alpha)B^{-1}],$$

$$\operatorname{tr}[\alpha A + (1-\alpha)B]^{-1} \le \left\{\begin{array}{c}(\operatorname{tr} A^{-1})^{\alpha}(\operatorname{tr} B^{-1})^{1-\alpha}\\ \operatorname{tr} A^{-1}(A^{-1/2}BA^{-1/2})^{\alpha-1}\end{array}\right\} \le \operatorname{tr}[\alpha A^{-1} + (1-\alpha)B^{-1}].$$

Remark: In the first string of inequalities, the upper left inequality and right-hand inequality are equivalent to Fact 10.14.33. The lower left inequality is given by *xxxiii)* of Proposition 10.6.17. The second string of inequalities combines the lower left inequality in the first string of inequalities with the third string of inequalities in Fact 10.11.73. **Remark:** These inequalities interpolate the convexity of $\phi(A) = \operatorname{tr} A^{-1}$. See Fact 2.2.53. **Related:** Fact 2.2.50, Fact 2.2.53, Fact 10.11.38, Fact 10.11.73, Fact 10.14.8, and Fact 10.14.33.

Fact 10.14.35. Let $A, B \in \mathbb{F}^{n\times n}$, and assume that B is positive semidefinite. Then,

$$|\operatorname{tr} AB| \le \sigma_{\max}(A)\operatorname{tr} B.$$

Source: Proposition 10.4.13 and $\sigma_{\max}(A + A^*) \le 2\sigma_{\max}(A)$. **Related:** Fact 7.13.12.

Fact 10.14.36. Let $A, B \in \mathbb{F}^{n\times n}$, assume that A and B are positive semidefinite, and let $p \in [1,\infty)$. Then,

$$\operatorname{tr}(A^p + B^p) \le \operatorname{tr}(A+B)^p \le [(\operatorname{tr} A^p)^{1/p} + (\operatorname{tr} B^p)^{1/p}]^p \le 2^{p-1}(\operatorname{tr} A^p + \operatorname{tr} B^p).$$

Furthermore, the second inequality is an equality if and only if A and B are linearly dependent. **Source:** [537] and [1922]. **Remark:** The first inequality is the *McCarthy inequality*. The second inequality is a special case of the triangle inequality for the norm $\|\cdot\|_{\sigma p}$ and a matrix version of Minkowski's inequality. **Remark:** Fact 11.10.53 restates these inequalities in terms of the Schatten norm, and Fact 11.10.59 extends these inequalities to more than two matrices.

Fact 10.14.37. Let $A, B \in \mathbb{F}^{n\times n}$, assume that A and B are positive semidefinite, and let $p \in \mathbb{R}$. Then, the following statements hold:

i) If $p \in [0,1]$, then

$$2^{p-1}(\operatorname{tr} A^p + \operatorname{tr} B^p) \le [(\operatorname{tr} A^p)^{1/p} + (\operatorname{tr} B^p)^{1/p}]^p \le \operatorname{tr}(A+B)^p \le \operatorname{tr} A^p + \operatorname{tr} B^p.$$

ii) If $p \in [0,1] \cup [2,3]$, then

$$\operatorname{tr} A^p + \operatorname{tr} B^p + (2^p - 2)\operatorname{tr} A^{p/2}B^{p/2} \le \operatorname{tr}(A+B)^p,$$

$$\operatorname{tr} A^p + \operatorname{tr} B^p + (2^p - 2)\operatorname{tr}(A^{1/2}BA^{1/2})^{p/2} \le \operatorname{tr}(A+B)^p.$$

iii) If $p \ge 1$, then

$$2^{-1/p}\operatorname{tr}(A+B) + (1 - 2^{1-1/p})\operatorname{tr} A^{1/2}B^{1/2} \le \operatorname{tr}[\tfrac{1}{2}(A^p + B^p)]^{1/p}.$$

iv) If $p < 0$ and A and B are positive definite, then

$$\operatorname{tr}(A+B)^p \le 2^{p-1}(\operatorname{tr} A^p + \operatorname{tr} B^p).$$

v) If $p \in [0,1] \cup [2,3]$, then

$$\operatorname{tr}(A^p + B^p) + (2^p - 2)\operatorname{tr} A^{p/2}B^{p/2} \le \operatorname{tr}(A+B)^p.$$

vi) If either $p \in [1,2]$ or both A and B are positive definite and $p < 0$, then

$$\operatorname{tr}(A+B)^p \le \operatorname{tr}(A^p + B^p) + (2^p - 2)\operatorname{tr} A^{p/2}B^{p/2}.$$

vii) If $p \ge 1$, then

$$\frac{1}{2^{1/p}}\operatorname{tr}(A+B) + (1-2^{1-1/p})\operatorname{tr} A^{1/2}B^{1/2} \le \operatorname{tr}\,[\tfrac{1}{2}(A^p+B^p)]^{1/p}.$$

viii) If $p \in [0,1] \cup [2,\infty)$, then

$$\operatorname{tr}(A^p+B^p) + (2^p-2)\operatorname{tr}\,(A^{1/2}BA^{1/2})^{p/2} \le \operatorname{tr}\,(A+B)^p.$$

ix) If either $p \in [1,2]$ or both A and B are positive definite and $p<0$, then

$$\operatorname{tr}\,(A+B)^p \le \operatorname{tr}(A^p+B^p) + (2^p-2)\operatorname{tr}\,(A^{1/2}BA^{1/2})^{p/2}.$$

x) $12\operatorname{tr}\,(AB)^2 \le 4\operatorname{tr}(A^3B + A^2B^2 + AB^3) + 2\operatorname{tr}\,(AB)^2$.

Source: [188, 194, 2750]. The following cases are conjectures: $p \in (3,4) \cup (4,\infty)$ in *viii)*; $p \in (-2,0)$ in *ix)*.

Fact 10.14.38. Let $A, B \in \mathbb{F}^{n\times n}$, and assume that A and B are positive semidefinite. If $p \in [0,1]$, then

$$\operatorname{tr}\,[A^2 + (AB)^k(BA)^k]^p \le \operatorname{tr}\,[A^2 + (BA)^k(AB)^k]^p.$$

If $p \ge 1$, then

$$\operatorname{tr}\,[A^2 + (BA)^k(AB)^k]^p \le \operatorname{tr}\,[A^2 + (AB)^k(BA)^k]^p.$$

Source: [1850].

Fact 10.14.39. Let $A, B \in \mathbb{F}^{n\times n}$, and assume that A and B are positive semidefinite. If $p \in [0,1]$, then

$$\operatorname{tr}\,(I + A + B + AB)^p \le \operatorname{tr}\,(I + A + B + B^{1/2}AB^{1/2})^p.$$

If $p \ge 1$, then

$$\operatorname{tr}\,(I + A + B + B^{1/2}AB^{1/2})^p \le \operatorname{tr}\,(I + A + B + AB)^p.$$

Source: [1112] and [1381, pp. 266, 267].

Fact 10.14.40. Let $A, B \in \mathbb{F}^{n\times n}$, and assume that A and B are positive semidefinite. Then,

$$\lambda(A^2 + BA^2B) \overset{\mathrm{s}}{\prec} \lambda(A^2 + AB^2A).$$

If $p \in [0,1]$, then

$$\operatorname{tr}\,(A^2 + AB^2A)^p \le \operatorname{tr}\,(A^2 + BA^2B)^p.$$

If $p \ge 1$, then

$$\operatorname{tr}\,(A^2 + BA^2B)^p \le \operatorname{tr}\,(A^2 + AB^2A)^p.$$

Now, assume that A is positive definite. Then,

$$\operatorname{tr}\,(A^2 + BA^2B)^{-1} \le \operatorname{tr}\,(A^2 + AB^2A)^{-1}.$$

Source: [1112]. The last statement is given in [1845].

Fact 10.14.41. Let $A, B \in \mathbb{F}^{n\times n}$, and assume that A and B are positive definite. Then,

$$\operatorname{tr} A\#B \le \operatorname{tr} A^{1/2}B^{1/2}, \quad \operatorname{tr}\,(A\#B)^2 \le \operatorname{tr}\,(A^{1/2}B^{1/2})^2 \le \operatorname{tr} AB, \quad \operatorname{tr} A(A\#B) \le \operatorname{tr} A^{3/2}B^{1/2},$$

$$\operatorname{tr}[A + B + 2(A\#B)] \le \operatorname{tr}\,(A^{1/2} + B^{1/2})^2, \quad \operatorname{tr}\,[A + B + 2(A\#B)]^2 \le \operatorname{tr}\,(A^{1/2} + B^{1/2})^4,$$

$$\operatorname{tr}\log\,(A^{1/2} + B^{1/2})^2 \le \operatorname{tr}\log[A + B + 2(A\#B)],$$

$$\lambda(A\#B) \overset{\mathrm{slog}}{\prec} \lambda(A^{1/2}B^{1/2}), \quad \lambda(A^{1/2}(A\#B)A^{1/2}) \overset{\mathrm{slog}}{\prec} \lambda(A^{3/4}B^{1/2}A^{3/4}),$$

$$\sigma_{\max}[A + B + 2(A\#B)] \le \sigma_{\max}(A + B + A^{1/2}B^{1/2} + B^{1/2}A^{1/2}).$$

Furthermore, if $p \in [1,2]$, then

$$\operatorname{tr}[A+B+2(A\#B)]^p \le (2-p)\operatorname{tr}(A^{1/2}+B^{1/2})^2+(p-1)\operatorname{tr}(A^{1/2}+B^{1/2})^4.$$

Source: [476, 3020].

Fact 10.14.42. Let $A, B \in \mathbb{F}^{n\times n}$, and assume that A and B are positive definite. Then,

$$\sum_{i=1}^n \frac{1}{\lambda_i(A)+\lambda_{n+1-i}(B)} \le \operatorname{tr}(A+B)^{-1} \le \sum_{i=1}^n \frac{1}{\lambda_i(A)+\lambda_i(B)}.$$

Source: [2750] and [2991, p. 362].

Fact 10.14.43. Let $A, B \in \mathbb{F}^{n\times n}$, and assume that A is positive definite, B is positive semidefinite, and $B \le A$. Then,

$$\frac{\det B}{\det A} \le \frac{\operatorname{tr} B}{\operatorname{tr} A}.$$

Source: [2991, p. 215].

Fact 10.14.44. Let $A_1, \dots, A_m \in \mathbb{F}^{n\times n}$, and assume that $A_1, \dots, A_m$ are positive definite. Then,

$$\frac{\operatorname{tr}\left(\sum_{i=1}^m A_i\right)^2}{\operatorname{tr}\sum_{i=1}^m A_i} \le \sum_{i=1}^n \frac{\operatorname{tr} A_i^2}{\operatorname{tr} A_i}.$$

Source: [2998].

Fact 10.14.45. Let $A, B \in \mathbb{F}^{n\times n}$, assume that A and B are positive semidefinite, let $m \ge 1$, and define $p \in \mathbb{F}[s]$ by $p(s) = \operatorname{tr}(A+sB)^m$. Then, all of the coefficients of p are nonnegative. **Remark:** This is the *Bessis-Moussa-Villani trace conjecture*. See [1058, 1395, 1820] and Fact 10.14.46.

Fact 10.14.46. Let $A, B \in \mathbb{F}^{n\times n}$, assume that A is Hermitian and B is positive semidefinite, and define $f(t) = e^{A+tB}$. Then, for all $k \ge 0$ and $t \ge 0$, $(-1)^{k+1}f^{(k)}(t) \ge 0$. **Source:** Fact 10.17.24 and [1395, 1820]. **Related:** Fact 10.14.45.

Fact 10.14.47. Let $A, B \in \mathbb{F}^{n\times n}$, assume that A and B are Hermitian, and let $f\colon \mathbb{R} \mapsto \mathbb{R}$. Then, the following statements hold:

i) If f is convex, then there exist unitary matrices $S_1, S_2 \in \mathbb{F}^{n\times n}$ such that

$$f[\tfrac{1}{2}(A+B)] \le \tfrac{1}{2}[S_1(\tfrac{1}{2}[f(A)+f(B)])S_1^* + S_2(\tfrac{1}{2}[f(A)+f(B)])S_2^*].$$

ii) If f is convex and even, then there exist unitary matrices $S_1, S_2 \in \mathbb{F}^{n\times n}$ such that

$$f[\tfrac{1}{2}(A+B)] \le \tfrac{1}{2}[S_1 f(A)S_1^* + S_2 f(B)S_2^*].$$

iii) If f is convex and increasing, then there exists a unitary matrix $S \in \mathbb{F}^{n\times n}$ such that

$$f[\tfrac{1}{2}(A+B)] \le S(\tfrac{1}{2}[f(A)+f(B)])S^*.$$

iv) There exist unitary matrices $S_1, S_2 \in \mathbb{F}^{n\times n}$ such that

$$\langle A+B\rangle \le S_1\langle A\rangle S_1^* + S_2\langle B\rangle S_2^*.$$

v) If f is convex, then

$$\operatorname{tr} f[\tfrac{1}{2}(A+B)] \le \operatorname{tr}\tfrac{1}{2}[f(A)+f(B)].$$

Source: [538, 539, 546]. **Related:** Fact 10.14.48.

Fact 10.14.48. Let $f\colon \mathbb{R} \mapsto \mathbb{R}$, and assume that f is convex. Then, the following statements hold:

i) If $f(0) \le 0$, $A \in \mathbf{H}^n$, and $S \in \mathbb{F}^{n\times m}$ is contractive, then $\operatorname{tr} f(S^*AS) \le \operatorname{tr} S^*f(A)S$.

ii) If $A_1,\dots,A_k \in \mathbb{F}^{n\times n}$ are Hermitian and $S_1,\dots,S_k \in \mathbb{F}^{n\times m}$ satisfy $\sum_{i=1}^k S_i^*S_i = I$, then

$$\operatorname{tr} f\left(\sum_{i=1}^k S_i^*A_iS_i\right) \le \operatorname{tr}\sum_{i=1}^k S_i^*f(A_i)S_i.$$

iii) If $A \in \mathbf{H}^n$ and $S \in \mathbb{F}^{n\times n}$ is a projector, then $\operatorname{tr} Sf(SAS)S \le \operatorname{tr} Sf(A)S$.

Source: [539] and [2128, p. 36]. **Remark:** Special cases are considered in [1571]. **Remark:** The first result is due to L. G. Brown and H. Kosaki, the second result is due to F. Hansen and G. K. Pedersen, and the third result is due to F. A. Berezin. **Related:** *ii*) generalizes *v*) of Fact 10.14.47.

Fact 10.14.49. Let $A_1,\dots,A_m \in \mathbb{F}^{n\times n}$ and $r \in [1,\infty)$. Then,

$$\operatorname{tr}\left\langle\sum_{i=1}^m A_i\right\rangle^r \le m^{r-1}\operatorname{tr}\sum_{i=1}^m \langle A_i\rangle^r.$$

Source: [2439]. **Related:** Fact 10.14.50.

Fact 10.14.50. Let $A, B \in \mathbb{F}^{n\times m}$. Then, $\operatorname{tr}\langle A+B\rangle \le \operatorname{tr}\langle A\rangle + \operatorname{tr}\langle B\rangle$. **Source:** Fact 10.11.34, [2439], and [2991, p. 291]. **Related:** Fact 10.14.49 and Fact 10.14.51.

Fact 10.14.51. Let $A, B \in \mathbb{F}^{n\times n}$ and $r \in [1,\infty)$. Then,

$$\operatorname{tr}\langle A\rangle^r + \operatorname{tr}\langle B\rangle^r \le \operatorname{tr}\langle A+B\rangle^r + \operatorname{tr}\langle A-B\rangle^r.$$

Source: [2439]. **Related:** Fact 10.14.50.

Fact 10.14.52. Let $A, B \in \mathbb{F}^{n\times n}$, and assume that B is positive semidefinite and $A^*A \le B$. Then,

$$|\operatorname{tr} A| \le \operatorname{tr} B^{1/2}.$$

Source: Corollary 10.6.11 with $r = 2$ implies $(A^*A)^{1/2} \le \operatorname{tr} B^{1/2}$. It then follows from Fact 10.13.2 that $|\operatorname{tr} A| \le \sum_{i=1}^n \rho_i(A) \le \sum_{i=1}^n \sigma_i(A) = \operatorname{tr}(A^*A)^{1/2} \le \operatorname{tr} B^{1/2}$. See [347].

Fact 10.14.53. Let $A, B \in \mathbb{F}^{n\times n}$, assume that A is positive definite and B is positive semidefinite, let $\alpha \in [0,1]$, and let $\beta \ge 0$. Then,

$$\operatorname{tr}(-BA^{-1}B + \beta B^\alpha) \le \beta(1-\tfrac{\alpha}{2})\operatorname{tr}(\tfrac{\alpha\beta}{2}A)^{\alpha/(2-\alpha)}.$$

If, in addition, either A and B commute or B is a scalar multiple of a projector, then

$$-BA^{-1}B + \beta B^\alpha \le \beta(1-\tfrac{\alpha}{2})(\tfrac{\alpha\beta}{2}A)^{\alpha/(2-\alpha)}.$$

Source: [1312, 1313].

Fact 10.14.54. Let $A, P \in \mathbb{F}^{n\times n}$, $B, Q \in \mathbb{F}^{n\times m}$, and $C, R \in \mathbb{F}^{m\times m}$, and assume that $\begin{bmatrix} A & B\\ B^* & C\end{bmatrix}, \begin{bmatrix} P & Q\\ Q^* & R\end{bmatrix} \in \mathbb{F}^{(n+m)\times(n+m)}$ are positive semidefinite. Then,

$$|\operatorname{tr} BQ^*|^2 \le (\operatorname{tr} AP)\operatorname{tr} CR.$$

Source: [1783, 2987].

Fact 10.14.55. Let $A, B \in \mathbb{F}^{n\times n}$, and assume that A and B are projectors. Then,

$$(\operatorname{tr} AB)^2 \le (\operatorname{rank} AB)\operatorname{tr} ABAB.$$

Furthermore, equality holds if and only if there exists $\alpha > 0$ such that αAB is idempotent. **Source:** [239].

Fact 10.14.56. Let $A, B \in \mathbb{F}^{n\times m}$, let $X \in \mathbb{F}^{n\times n}$, and assume that X is positive definite. Then,

$$|\operatorname{tr} A^*B|^2 \le (\operatorname{tr} A^*XA)\operatorname{tr} B^*X^{-1}A.$$

Source: Use Fact 10.14.54 with $\begin{bmatrix} X & I\\ I & X^{-1}\end{bmatrix}$ and $\begin{bmatrix} AA^* & AB^*\\ BA^* & BB^*\end{bmatrix}$. See [1783, 2987].

Fact 10.14.57. Let $A, B, C \in \mathbb{F}^{n\times n}$, and assume that A and B are Hermitian and C is positive semidefinite. Then,

$$|\operatorname{tr} ABC^2 - \operatorname{tr} ACBC| \le \tfrac{1}{4}[\lambda_1(A) - \lambda_n(A)][\lambda_1(B) - \lambda_n(B)] \operatorname{tr} C^2.$$

Source: [541].

Fact 10.14.58. Let $A, B, C \in \mathbb{F}^{n\times n}$, assume that A, B, C are positive semidefinite, and assume that $B \le C$. Then,

$$\operatorname{tr} (A + B)^{-1}B \le \operatorname{tr} (A + C)^{-1}C.$$

Source: [2991, p. 213].

Fact 10.14.59. Let $A, B \in \mathbb{F}^{n\times n}$, and assume that A and B are positive definite. Then,

$$\operatorname{tr} (A - B)(B^{-1} - A^{-1}) \ge 0.$$

Source: [2991, p. 213].

Fact 10.14.60. Let $A, B, C, D, X \in \mathbb{F}^{n\times n}$, and assume that A and B are positive semidefinite, C and D are positive definite, and X is Hermitian. Then,

$$\operatorname{tr} X(A + C)^{-1}X(B + D)^{-1} \le \operatorname{tr} XA^{-1}XB^{-1},$$

$$4 \operatorname{tr} (C - D)[(A + C)^{-1} - (B + D)^{-1}] \le \operatorname{tr} (A - B)(B^{-1} - A^{-1}),$$

$$|\operatorname{tr} (C - D)(B + D)^{-1}(A - B)(A + C)^{-1}| \le \operatorname{tr}[(A - B)(B^{-1} - A^{-1}) + (C - D)((B + D)^{-1} - (A + C)^{-1})].$$

Source: [355, 1113].

Fact 10.14.61. Let $A_1, \dots, A_k \in \mathbf{P}^n$ and $\{j_1, \dots, j_k\} = \{1, \dots, k\}$. Then,

$$kn^2 \le \sum_{i=1}^{k} (\operatorname{tr} A_i) \operatorname{tr} A_{j_i}^{-1}.$$

Now, assume that $A_k \le \cdots \le A_1$. Then,

$$kn^2 \le \sum_{i=1}^{k} (\operatorname{tr} A_i) \operatorname{tr} A_i^{-1} \le \sum_{i=1}^{k} (\operatorname{tr} A_i) \operatorname{tr} A_{j_i}^{-1}, \quad kn \le \sum_{i=1}^{k} \operatorname{tr} A_i A_{j_i}^{-1}.$$

Source: [2681]. **Related:** Fact 10.14.62.

Fact 10.14.62. For all $i \in \{1, \dots, k\}$, let $A_i, B_i \in \mathbf{H}^n$, assume that $A_k \le \cdots \le A_1$ and $B_k \le \cdots \le B_1$, and let $\{j_1, \dots, j_k\} = \{1, \dots, k\}$. Then,

$$\sum_{i=1}^{k} \operatorname{tr} A_i B_{k-i+1} \le \sum_{i=1}^{k} \operatorname{tr} A_i B_{j_i} \le \sum_{i=1}^{k} \operatorname{tr} A_i B_i,$$

$$k \sum_{i=1}^{k} \operatorname{tr} A_i B_{k-i+1} \le \operatorname{tr} \left(\sum_{i=1}^{k} A_i \right) \sum_{i=1}^{k} B_i \le k \sum_{i=1}^{k} \operatorname{tr} A_i B_i,$$

$$\sum_{i=1}^{k} \operatorname{tr} (A_i + B_{k-i+1})^2 \le \sum_{i=1}^{k} \operatorname{tr} (A_i + B_{j_i})^2 \le \sum_{i=1}^{k} \operatorname{tr} (A_i + B_i)^2,$$

$$\sum_{i=1}^{k} \operatorname{tr} (A_i - B_i)^2 \le \sum_{i=1}^{k} \operatorname{tr} (A_i - B_{j_i})^2 \le \sum_{i=1}^{k} \operatorname{tr} (A_i - B_{k-i+1})^2.$$

Now, assume that A_k and B_k are positive semidefinite, and let $l \geq 0$. Then,

$$\sum_{i=1}^{k} (\mathrm{tr}\ A_i B_{k-i+1})^l \leq \sum_{i=1}^{k} (\mathrm{tr}\ A_i B_{j_i})^l \leq \sum_{i=1}^{k} (\mathrm{tr}\ A_i B_i)^l.$$

Source: [2681]. **Related:** Fact 10.14.61, Fact 10.16.36, and Fact 10.25.45.

10.15 Facts on the Determinant for One Matrix

Fact 10.15.1. Let $A \in \mathbb{F}^{n \times n}$, and assume that A is positive semidefinite. Then,

$$\lambda_{\min}(A) \leq \lambda_{\max}^{1/n}(A)\lambda_{\min}^{(n-1)/n}(A) \leq \lambda_n(A) \leq \lambda_1(A) \leq \lambda_{\min}^{1/n}(A)\lambda_{\max}^{(n-1)/n}(A) \leq \lambda_{\max}(A),$$

$$\lambda_{\min}^n(A) \leq \lambda_{\max}(A)\lambda_{\min}^{n-1}(A) \leq \det A \leq \lambda_{\min}(A)\lambda_{\max}^{n-1}(A) \leq \lambda_{\max}^n(A).$$

Source: Fact 7.12.33.

Fact 10.15.2. Let $A \in \mathbb{R}^{n \times n}$, and assume that $A + A^{\mathsf{T}}$ is positive semidefinite. Then,

$$[\tfrac{1}{2}(A + A^{\mathsf{T}})]^{\mathsf{A}} \leq \tfrac{1}{2}(A^{\mathsf{A}} + A^{\mathsf{AT}}).$$

Now, assume that $A + A^{\mathsf{T}}$ is positive definite. Then,

$$[\det \tfrac{1}{2}(A + A^{\mathsf{T}})][\tfrac{1}{2}(A + A^{\mathsf{T}})]^{-1} \leq (\det A)[\tfrac{1}{2}(A^{-1} + A^{-\mathsf{T}})].$$

Furthermore,

$$[\det \tfrac{1}{2}(A + A^{\mathsf{T}})][\tfrac{1}{2}(A + A^{\mathsf{T}})]^{-1} < (\det A)[\tfrac{1}{2}(A^{-1} + A^{-\mathsf{T}})]$$

if and only if $\mathrm{rank}(A - A^{\mathsf{T}}) \geq 4$. Finally, if $n \geq 4$ and $A - A^{\mathsf{T}}$ is nonsingular, then

$$(\det A)[\tfrac{1}{2}(A^{-1} + A^{-\mathsf{T}})] < [\det A - \det \tfrac{1}{2}(A - A^{\mathsf{T}})][\tfrac{1}{2}(A + A^{\mathsf{T}})]^{-1}.$$

Source: [1011, 1536]. **Remark:** This result does not hold for complex matrices. **Related:** Fact 7.20.1, Fact 10.10.36, Fact 10.15.2, and Fact 10.21.18.

Fact 10.15.3. Let $A \in \mathbb{F}^{n \times n}$, and assume that $A + A^*$ is positive semidefinite. Then,

$$\det \tfrac{1}{2}(A + A^*) \leq |\det A|.$$

Furthermore, if $A + A^*$ is positive definite, then equality holds if and only if A is Hermitian. **Source:** The inequality follows from Fact 7.12.25 and Fact 7.12.32. See [2991, p. 205]. **Remark:** This is the *Ostrowski-Taussky inequality*. **Related:** Fact 10.15.3.

Fact 10.15.4. Let $A \in \mathbb{F}^{n \times n}$, and assume that $A + A^*$ is positive semidefinite. Then,

$$[\det \tfrac{1}{2}(A + A^*)]^{2/n} + |\det \tfrac{1}{2}(A - A^*)|^{2/n} \leq |\det A|^{2/n}.$$

Furthermore, if $A + A^*$ is positive definite, then equality holds if and only if every eigenvalue of $(A + A^*)^{-1}(A - A^*)$ has the same absolute value. Finally, if $n \geq 2$, then

$$\det \tfrac{1}{2}(A + A^*) \leq \det \tfrac{1}{2}(A + A^*) + |\det \tfrac{1}{2}(A - A^*)| \leq |\det A|.$$

Source: [1012, 1537]. To prove the last result, use Fact 2.2.59. **Remark:** Setting $A = 1 + \jmath$ shows that the last result can fail for $n = 1$. **Remark:** $-A$ is semidissipative. **Remark:** The last result interpolates Fact 10.15.3. **Remark:** Extensions to the case where $A + A^*$ is not positive semidefinite are considered in [2611].

Fact 10.15.5. Let $A \in \mathbb{F}^{n \times m}$. Then, $|\det(I + A)| \leq \det(I + \langle A \rangle)$. **Source:** [2420]. **Related:** Fact 10.16.17.

Fact 10.15.6. Let $A \in \mathbb{F}^{n\times n}$, assume that A is positive semidefinite, and define

$$\alpha \triangleq \frac{1}{n}\operatorname{tr} A, \quad \beta \triangleq \frac{1}{n(n-1)}\sum_{\substack{i,j=1\\ i\neq j}}^{n} |A_{(i,j)}|.$$

Then,

$$|\det A| \le (\alpha-\beta)^{n-1}[\alpha+(n-1)\beta].$$

Furthermore, if $A = aI_n + b1_{n\times n}$, where $a + nb > 0$ and $a > 0$, then $\alpha = a + b, \beta = b$, and equality holds. **Source:** [2119]. **Related:** Fact 3.16.18 and Fact 10.10.39.

Fact 10.15.7. Let $A \in \mathbb{F}^{n\times n}$, assume that A is positive definite, and define

$$\beta \triangleq \frac{1}{n(n-1)}\sum_{\substack{i,j=1\\ i\neq j}}^{n} \frac{|A_{(i,j)}|}{\sqrt{A_{(i,i)}A_{(j,j)}}}.$$

Then,

$$|\det A| \le (1-\beta)^{n-1}[1+(n-1)\beta]\prod_{i=1}^{n} A_{(i,i)} \le \prod_{i=1}^{n} A_{(i,i)}.$$

Source: [2119]. **Remark:** This result interpolates Hadamard's inequality. See Fact 10.21.15 and [910].

Fact 10.15.8. Let $A \in \mathbb{R}^{n\times n}$, and assume that A is positive definite. Then,

$$\sum_{i=1}^{n} A_{(i,i)} \det A_{[i,i]} \le \det A + (n-1)\prod_{i=1}^{n} A_{(i,i)}.$$

Source: [1836].

Fact 10.15.9. Let $A \in \mathbb{F}^{n\times n}$, assume that A is positive definite, and let $\lambda_1 \triangleq \lambda_1(A)$ and $\lambda_n \triangleq \lambda_n(A)$. Then,

$$\left(\frac{n}{\operatorname{tr}(A+\lambda_n I)^{-1}} - \lambda_n\right)^n \le \det A \le \left(\frac{n}{\operatorname{tr}(A+\lambda_1 I)^{-1}} - \lambda_1\right)^n.$$

Source: [673].

Fact 10.15.10. Let $A \in \mathbb{F}^{n\times n}$. Then,

$$|\det A| \le \prod_{i=1}^{n}\left(\sum_{j=1}^{n} |A_{(i,j)}|^2\right)^{1/2} = \prod_{i=1}^{n} \|\operatorname{row}_i(A)\|_2.$$

Furthermore, equality holds if and only if AA^* is diagonal. Now, let $\alpha > 0$ be such that, for all $i, j \in \{1,\dots,n\}$, $|A_{(i,j)}| \le \alpha$. Then,

$$|\det A| \le \alpha^n n^{n/2}.$$

If, in addition, at least one entry of A has absolute value less than α, then

$$|\det A| < \alpha^n n^{n/2}.$$

Source: Replace A with AA^* in Fact 10.21.15. See [2432, p. 34]. **Related:** Fact 3.16.24 and Fact 8.9.35.

Fact 10.15.11. Let $A \in \mathbb{F}^{n\times n}$, and assume that A is positive definite. Then,

$$n + \operatorname{tr}\log A = n + \log\det A \le n(\det A)^{1/n} \le \operatorname{tr} A \le (n\operatorname{tr} A^2)^{1/2},$$

with equality if and only if $A = I$. **Remark:** $(\det A)^{1/n} \le (\operatorname{tr} A)/n$ follows from the arithmetic-mean–geometric-mean inequality.

Fact 10.15.12. Let $A \in \mathbb{F}^{n\times n}$, assume that A is Hermitian, and assume that, for all $i \in \{1,\dots,n-1\}$, $\det A_{(\{1,\dots,i\})} > 0$. Then, the following statements hold:

i) For all $i \in \{1,\dots,n-1\}$, $A_{(\{1,\dots,i\})}$ is positive definite.

ii) $\lambda_{n-1}(A)$ is positive.

iii) $\operatorname{sign}[\lambda_{\min}(A)] = \operatorname{sign}(\det A)$.

iv) A is (positive semidefinite, positive definite) if and only if $(\det A \ge 0, \det A > 0)$.

v) $\det A = 0$ if and only if $\operatorname{rank} A = n-1$.

Source: Note that $\det A = \lambda_{\min}(A)/[\lambda_1(A)\cdots\lambda_{n-1}(A)]$, and use Proposition 10.2.9 and Theorem 10.4.5. See [2403, p. 278].

Fact 10.15.13. Let $A \in \mathbb{R}^{n\times n}$, and assume that A is positive definite. Then,

$$\sum_{i=1}^{n}[\det A_{(\{1,\dots,i\})}]^{1/i} \le \left(1+\tfrac{1}{n}\right)^n \operatorname{tr} A < e\operatorname{tr} A.$$

Source: [70].

Fact 10.15.14. Let $A \in \mathbb{F}^{n\times n}$, assume that A is positive definite and Toeplitz, and, for all $i \in \{1,\dots,n\}$, define $A_i \triangleq A_{(\{1,\dots,i\})} \in \mathbb{F}^{i\times i}$. Then,

$$(\det A)^{1/n} \le (\det A_{n-1})^{1/(n-1)} \le \cdots \le (\det A_2)^{1/2} \le \det A_1.$$

Furthermore,

$$\frac{\det A}{\det A_{n-1}} \le \frac{\det A_{n-1}}{\det A_{n-2}} \le \cdots \le \frac{\det A_3}{\det A_2} \le \frac{\det A_2}{\det A_1}.$$

Source: [788] and [789, p. 682].

Fact 10.15.15. Let $A \in \mathbb{F}^{n\times n}$. Then, $0 \le \det(I + \overline{A}A) \le \det(I + A^*A)$. Furthermore, the second inequality is an equality if and only if A is symmetric. **Source:** [914, 1838].

10.16 Facts on the Determinant for Two or More Matrices

Fact 10.16.1. Let $A, B \in \mathbb{C}^{n\times n}$, and assume that $\operatorname{spec}(A) \subset [0,\infty)$ and $\operatorname{spec}(B) \subset [0,\infty)$. Then, the following statements hold:

i) If $\lambda(A) \overset{\mathrm{s}}{\prec} \lambda(B)$, then $\det B \le \det A$.

ii) If $\lambda(A) \overset{\mathrm{wlog}}{\prec} \lambda(B)$, then $\det(I + A) \le \det(I + B)$.

Source: Fact 3.25.14, [1846], and [2991, p. 347].

Fact 10.16.2. Let $A, B \in \mathbb{F}^{n\times n}$, and assume that A is positive semidefinite and B is skew Hermitian. Then,

$$(\det A)^{2/n} + |\det B|^{2/n} \le |\det(A + B)|^{2/n}.$$

If A is positive definite, then equality holds if and only if every eigenvalue of $A^{-1}B$ has the same absolute value. Finally, if $n \ge 2$, then

$$\det A \le \det A + |\det B| \le |\det(A + B)|.$$

Source: Replace A by $A + B$ in Fact 10.15.4. **Remark:** An extension of the first inequality to the case where A is Hermitian is given in [714, 2611]. Setting $A = 1$ and $B = \jmath$ shows that the second inequality in the last string does not hold in the case where $n = 1$.

Fact 10.16.3. Let $A, B \in \mathbb{F}^{n\times n}$, and assume that A is positive semidefinite and B is skew Hermitian. Then,

$$\det A \le |\det(A + B)| = (\det A)\prod_{i=1}^{n}\sqrt{1 + \sigma_i^2(A^{-1/2}BA^{-1/2})}.$$

If A is positive definite, then equality holds if and only if $B = 0$. Finally, if A and B are real, then

$$\det A \le \det(A + B).$$

Source: [100, 714], [1343, p. 447], and [2263, pp. 146, 163]. Suppose that A and B are real. If A is positive definite, then $A^{-1/2}BA^{-1/2}$ is skew symmetric, and thus $\det(A + B) = (\det A)\det(I + A^{-1/2}BA^{-1/2})$ is positive. If A is positive semidefinite, then continuity implies that $\det(A + B)$ is nonnegative. **Remark:** Extensions are given in [480].

Fact 10.16.4. Let $A, B \in \mathbb{F}^{n\times n}$, and assume that A and B are positive definite. Then,

$$\prod_{i=1}^{n}[\lambda_i(A) + \lambda_i(B)] \le \det(A + B) \le \prod_{i=1}^{n}[\lambda_i(A) + \lambda_{n+1-i}(B)].$$

Source: [2991, p. 362].

Fact 10.16.5. Let $A, B \in \mathbb{F}^{n\times n}$, and assume that A and B are Hermitian. Then,

$$|\det(A + \jmath B)| \le \prod_{i=1}^{n}[\sigma_i^2(A) + \sigma_{n-i+1}^2(B)]^{1/2}.$$

Now, assume that A and B are positive semidefinite. Then,

$$|\det(A + \jmath B)| \le \det(A + B) \le 2^{n/2}|\det(A + \jmath B)|,$$

$$\prod_{i=1}^{n}[\lambda_i^2(A) + \lambda_i^2(B)]^{1/2} \le |\det(A + \jmath B)| \le \prod_{i=1}^{n}[\lambda_i^2(A) + \lambda_{n-i+1}^2(B)]^{1/2}.$$

Source: [100, 336, 714, 1832, 2285] and [2979, p. 99]. **Related:** Fact 10.16.6.

Fact 10.16.6. Let $A, B \in \mathbb{C}^{n\times n}$, assume that A and B are Hermitian, and assume that $A^2 + B^2$ is positive definite. Then, $\jmath(AB - BA)$ is Hermitian,

$$\operatorname{spec}[\jmath(A^2 + B^2)^{-1/2}(AB - BA)(A^2 + B^2)^{-1/2}] \subset (-1, 1),$$

$$\begin{aligned}\det\begin{bmatrix} A & B \\ -B & A \end{bmatrix} &= [\det(A + \jmath B)]\det(A - \jmath B) = \det[A^2 + B^2 \pm \jmath(AB - BA)] \\ &= [\det(A^2 + B^2)]\det[I \pm \jmath(A^2 + B^2)^{-1/2}(AB - BA)(A^2 + B^2)^{-1/2}] > 0.\end{aligned}$$

Hence, $\left[\begin{smallmatrix} A & B \\ -B & A \end{smallmatrix}\right]$, $A + \jmath B$, $A - \jmath B$, and $A^2 + B^2 \pm \jmath(AB - BA)$ are nonsingular. **Related:** Fact 3.24.7 and Fact 10.16.5.

Fact 10.16.7. Let $A, B \in \mathbb{F}^{n\times n}$, assume that A and B are Hermitian, and assume that $-A \le B \le A$. Then, $|\det B| \le \det A$. **Source:** Let $S \in \mathbb{F}^{n\times n}$ be a nonsingular matrix such that $D_1 \triangleq SAS^*$ and $D_2 \triangleq SBS^*$ are diagonal. Then, $-D_1 \le D_2 \le D_1$, and thus $|D_2| \le D_1$ and $|\det D_2| = \det|D_2| \le \det D_1$. Therefore, $|\det B| = (\det S^{-1})(\det S^{-*})|\det D_2| \le (\det S^{-1})(\det S^{-*})\det D_1 = \det A$. **Related:** Fact 10.12.18.

Fact 10.16.8. Let $A, B \in \mathbb{F}^{n\times n}$, assume that A and B are positive semidefinite, and assume that $A \le B$. Then,

$$n\det A + \det B \le \det(A + B).$$

Source: [2263, pp. 154, 166]. **Remark:** Under weaker conditions, Corollary 10.4.15 implies that $\det A + \det B \le \det(A + B)$. **Related:** Fact 7.13.25.

Fact 10.16.9. Let $A, B \in \mathbb{F}^{n\times n}$, and assume that A and B are positive semidefinite. Then,

$$\det A + \det B \le \det A + \det B + (2^n - 2)\sqrt{\det AB} \le \det(A + B).$$

If A and B are positive definite, then

$$\begin{aligned}\det A + \det B &\le \det A + \det B + (2^n - 2)\sqrt{\det AB}\\ &\le \left(1 + \sum_{i=1}^{n-1} \frac{\det B_{(1,\ldots,i)}}{\det A_{(1,\ldots,i)}}\right)\det A + \left(1 + \sum_{i=1}^{n-1} \frac{\det A_{(1,\ldots,i)}}{\det B_{(1,\ldots,i)}}\right)\det B + (2^n - 2n)\sqrt{\det AB}\\ &\le \det(A+B).\end{aligned}$$

If $B \le A$, then

$$\det A + (2^n - 1)\det B \le \det A + \det B + (2^n - 2)\sqrt{\det AB} \le \det(A+B).$$

Source: [1835, 2179] and [2418, p. 231]. **Remark:** The last inequality in the second string is due to E. V. Haynsworth and D. J. Hartfiel. See [1839].

Fact 10.16.10. Let $A, B \in \mathbb{F}^{n\times n}$, and assume that A and B are positive semidefinite. Then,

$$|\det(A^{1/2}B^{1/2} + B^{1/2}A^{1/2})| \le \det(A+B).$$

Source: [1843].

Fact 10.16.11. Let $A, B \in \mathbb{F}^{n\times n}$, assume that A and B are positive semidefinite, and let $\alpha, \beta \in \mathbb{C}$. Then,

$$|\det(\alpha A + \beta B)| \le \det(|\alpha|A + |\beta|B).$$

Source: [2991, p. 215].

Fact 10.16.12. Let $A, B \in \mathbb{F}^{n\times n}$, assume that A and B are positive semidefinite, and let $p \in [0, 2]$. Then,

$$\det(A^2 + \langle BA\rangle^p) \le \det(A^2 + A^pB^p), \quad \det(A^2 + A^pB^p) \le \det(A^2 + \langle AB\rangle^p).$$

In particular,

$$\det(A^2 + \langle BA\rangle) \le \det(A^2 + AB), \quad \det(A^2 + AB) \le \det(A^2 + \langle AB\rangle).$$

Source: [1846]. **Remark:** The second inequality is a conjecture given in [1846].

Fact 10.16.13. Let $A, B \in \mathbb{F}^{n\times n}$, assume that B is Hermitian, and assume that $A^*BA < A + A^*$. Then, $\det A \ne 0$.

Fact 10.16.14. Let $A, B \in \mathbb{F}^{n\times n}$, assume that A and B are positive semidefinite, and let $\alpha \in [0, 1]$. Then,

$$(\det A)^\alpha(\det B)^{1-\alpha} = \det A^\alpha B^{1-\alpha} \le \det[\alpha A + (1-\alpha)B].$$

Now, assume that A and B are positive definite. Then, equality holds if and only if $A = B$. In addition,

$$\det A^\alpha B^{1-\alpha} \le \det A^\alpha B^{1-\alpha} + \delta^n \det(A + B - 2A\#B) \le \det[\alpha A + (1-\alpha)B],$$

where $\delta \triangleq \min\{\alpha, 1-\alpha\}$. **Source:** The first inequality is *xxxviii)* of Proposition 10.6.17. See [1448, p. 467] and [1451, p. 488]. The last inequality is given in [1640]. **Remark:** $A\#B$ is the geometric mean of A and B. See Fact 10.11.68. Note that $2A\#B \le A + B$. **Remark:** $\alpha = 1/2$ yields $\sqrt{(\det A)\det B} \le \det \frac{1}{2}(A+B)$. **Credit:** H. Bergstrom. **Related:** Fact 10.14.33.

Fact 10.16.15. Let $A, B \in \mathbb{F}^{n\times n}$, assume that A and B are positive semidefinite, assume that either $A \le B$ or $B \le A$, let $\alpha \in [0, 1]$, and let $p \ge 1$. Then,

$$(\det[\alpha A + (1-\alpha)B])^p \le \alpha(\det A)^p + (1-\alpha)(\det B)^p.$$

Source: [830, 2840].

Fact 10.16.16. Let $A, B \in \mathbb{F}^{n\times n}$, assume that A and B are positive definite, and let $\mathcal{S} \subseteq \{1,\dots,n\}$. Then,

$$\frac{\det A}{\det A_{(\mathcal{S})}} + \frac{\det B}{\det B_{(\mathcal{S})}} \le \frac{\det(A+B)}{\det(A_{(\mathcal{S})} + B_{(\mathcal{S})})}.$$

Source: *xli)* of Proposition 10.6.17, [2263, p. 145], and [2991, p. 243]. **Related:** Fact 10.12.7.

Fact 10.16.17. Let $A, B \in \mathbb{F}^{n\times m}$. Then, there exist unitary matrices $S_1, S_2 \in \mathbb{F}^{n\times n}$ such that

$$I + \langle A+B\rangle \le S_1(I+\langle A\rangle)^{1/2}S_2(I+\langle B\rangle)S_2^*(I+\langle A\rangle)^{1/2}S_1^*.$$

Therefore,

$$|\det(I+A+B)| \le \det(I+\langle A+B\rangle) \le \det(I+\langle A\rangle)\det(I+\langle B\rangle).$$

Source: [93, 2420, 2612]. **Related:** Fact 10.15.5 and Fact 10.16.18.

Fact 10.16.18. Let $A, B, C \in \mathbb{F}^{n\times n}$, assume that A and C are positive semidefinite, assume that B is Hermitian, and assume that $\left[\begin{smallmatrix} A & B \\ B & C \end{smallmatrix}\right]$ is positive semidefinite. Then,

$$\det(I+A+C) \le \det\left(I + \begin{bmatrix} A & B \\ B & C \end{bmatrix}\right) \le \det(I+A)\det(I+C).$$

Source: [546]. **Related:** Fact 10.16.17.

Fact 10.16.19. Let $A, B \in \mathbb{F}^{n\times m}$, assume that A and B are semicontractive, and let $\alpha \in [0,1]$. Then,

$$[\det(I+\langle A\rangle)]^\alpha[\det(I+\langle B\rangle)]^{1-\alpha} \le \det(I+\alpha\langle A\rangle)\det(I+(1-\alpha)\langle B\rangle),$$
$$[\det(I-\langle A\rangle)]^\alpha[\det(I-\langle B\rangle)]^{1-\alpha} \le \det(I-\alpha\langle A\rangle)\det(I-(1-\alpha)\langle B\rangle).$$

Source: [1831].

Fact 10.16.20. Let $A, B \in \mathbb{F}^{n\times n}$, assume that $A + A^* > 0$ and $B + B^* \ge 0$, and let $\alpha > 0$. Then, $\det(\alpha I + AB) \ne 0$, and thus AB has no negative eigenvalues. **Source:** [1284]. **Remark:** Equivalently, $-A$ is dissipative and $-B$ is semidissipative. **Problem:** Find a positive lower bound for $|\det(\alpha I + AB)|$ in terms of α, A, and B.

Fact 10.16.21. Let $A_1,\dots,A_k \in \mathbb{F}^{n\times n}$, assume that $A_1,\dots,A_k$ are positive semidefinite, and let $\alpha_1,\dots,\alpha_k \in \mathbb{C}$. Then,

$$\left|\det\sum_{i=1}^k \alpha_i A_i\right| \le \det\sum_{i=1}^k |\alpha_i|A_i.$$

Source: [2263, p. 144] and [2991, p. 362].

Fact 10.16.22. Let $A_1,\dots,A_k \in \mathbb{F}^{n\times n}$, assume that $A_1,\dots,A_k$ are contractive, let $\alpha_1,\dots,\alpha_k \in [0,1]$, and assume that $\sum_{i=1}^k \alpha_i = 1$. Then,

$$\left|\frac{\det(I+\sum_{i=1}^k \alpha_i A_i)}{\det(I-\sum_{i=1}^k \alpha_i A_i)}\right| \le \prod_{i=1}^k\left[\frac{\det(I+\langle A_i\rangle)}{\det(I-\langle A_i\rangle)}\right]^{\alpha_i}.$$

Source: [1831].

Fact 10.16.23. Let $A, B, C \in \mathbb{R}^{n\times n}$, let $D \triangleq A + \jmath B$, and assume that $CB + B^\mathsf{T}C^\mathsf{T} < D + D^*$. Then, $\det A \ne 0$.

Fact 10.16.24. Let $A, B \in \mathbb{F}^{n\times n}$, assume that A and B are positive semidefinite, and let $m \ge 1$. Then,

$$n(\det AB)^{m/n} \le \operatorname{tr} A^mB^m, \qquad \frac{n}{\sqrt[n]{\det A}} \le \operatorname{tr} A^{-1}.$$

Source: [829]. **Remark:** Assuming $\det B = 1$ and setting $m = 1$ yields Proposition 10.4.14.

Fact 10.16.25. Let $A, B \in \mathbb{F}^{n\times m}$. Then,

$$|\det A^*B|^2 \le (\det A^*A)(\det B^*B).$$

Source: Use Fact 10.12.33 or apply Fact 10.16.50 to $\left[\begin{smallmatrix} A^*A & B^*A \\ A^*B & B^*B \end{smallmatrix}\right]$. **Remark:** This is a determinantal version of the Cauchy-Schwarz inequality.

Fact 10.16.26. Let $A \in \mathbb{F}^{n\times n}$, assume that A is positive definite, and let $B \in \mathbb{F}^{m\times n}$. Then,

$$(\det BB^*)^2 \le (\det BAB^*)\det BA^{-1}B^*.$$

Source: Fact 10.12.49.

Fact 10.16.27. Let $A, B \in \mathbb{F}^{n\times n}$. Then,

$$|\det(A+B)|^2 + |\det(I - AB^*)|^2 \le \det(I + AA^*)\det(I + B^*B),$$
$$|\det(A-B)|^2 + |\det(I + AB^*)|^2 \le \det(I + AA^*)\det(I + B^*B).$$

Furthermore, the first inequality is an equality if and only if either $n = 1$, $A + B = 0$, or $AB^* = I$. **Source:** Fact 10.12.36, [2983, p. 184], and [2991, pp. 228–230]. **Related:** Fact 10.12.34.

Fact 10.16.28. Let $A, B \in \mathbb{F}^{n\times m}$, and assume that A and B are semicontractive. Then,

$$0 \le \det(I - A^*A)\det(I - B^*B) \le \left\{\begin{matrix} |\det(I - A^*B)|^2 \\ |\det(I + A^*B)|^2 \end{matrix}\right\} \le \det(I + A^*A)\det(I + B^*B).$$

Now, assume that $n = m$. Then,

$$0 \le \det(I - A^*A)\det(I - B^*B) \le |\det(I - A^*B)|^2 - |\det(A - B)|^2$$
$$\le |\det(I - A^*B)|^2 \le |\det(I - A^*B)|^2 + |\det(A + B)|^2 \le \det(I + A^*A)\det(I + B^*B),$$

$$0 \le \det(I - A^*A)\det(I - B^*B) \le |\det(I + A^*B)|^2 - |\det(A + B)|^2$$
$$\le |\det(I + A^*B)|^2 \le |\det(I + A^*B)|^2 + |\det(A - B)|^2 \le \det(I + A^*A)\det(I + B^*B).$$

Furthermore, for all $i \in \{1, \ldots, n\}$,

$$\sigma_i[(I - A^*A)(I - B^*B)] \le \sigma_i^2(I - A^*B), \quad \lambda_i[(I - A^*A)(I - B^*B)] \le \sigma_i^2(I - A^*B),$$
$$\sigma_i[(I - A^*A)(I - B^*B)] \le \sigma_i^2(I - AB^*), \quad \lambda_i[(I - A^*A)(I - B^*B)] \le \sigma_i^2(I - AB^*),$$
$$\lambda_i[(I - A^*A)^{1/2}(I - B^*B)^{1/2}] \le \sigma_i(I - A^*B), \quad \lambda_i[(I - A^*A)^{1/2}(I - B^*B)^{1/2}] \le \sigma_i(I - AB^*).$$

Now, assume that A and B are contractive. Then,

$$\begin{bmatrix} \det(I - A^*A)^{-1} & \det(I - A^*B)^{-1} \\ \det(I - B^*A)^{-1} & \det(I - B^*B)^{-1} \end{bmatrix} \ge 0.$$

Source: [93]. The third inequality follows from Fact 10.12.35. The second inequality for $n = m$ is given in [2184]. **Remark:** The second inequality and Fact 10.12.52 are *Hua's inequalities*. See [2564]. **Remark:** Extensions of the last inequality are given in [2933]. **Related:** Fact 10.12.52 and Fact 10.19.5.

Fact 10.16.29. Let $A, B \in \mathbb{F}^{n\times n}$ and $\alpha, \beta \in (0, \infty)$. Then,

$$|\det(A+B)|^2 \le \det(\beta^{-1}I + \alpha A^*A)\det(\alpha^{-1}I + \beta BB^*).$$

Source: Use Fact 10.12.51 or Fact 10.16.27 with A and B replaced by $\sqrt{\alpha\beta}A^*$ and $\sqrt{\alpha\beta}B^*$, respectively. See [2984].

Fact 10.16.30. Let $A \in \mathbb{F}^{n\times m}$, $B \in \mathbb{F}^{n\times l}$, $C \in \mathbb{F}^{n\times m}$, and $D \in \mathbb{F}^{n\times l}$. Then,

$$|\det(AC^* + BD^*)|^2 \le \det(AA^* + BB^*)\det(CC^* + DD^*).$$

Source: Use Fact 10.16.41 and $\mathcal{A}\mathcal{A}^* \ge 0$, where $\mathcal{A} \triangleq \left[\begin{smallmatrix} A & B \\ C & D \end{smallmatrix}\right]$. **Related:** Fact 3.17.24.

Fact 10.16.31. Let $A \in \mathbb{F}^{n\times m}$, $B \in \mathbb{F}^{n\times m}$, $C \in \mathbb{F}^{k\times m}$, and $D \in \mathbb{F}^{k\times m}$. Then,

$$|\det(A^*B + C^*D)|^2 \le \det(A^*A + C^*C)\det(B^*B + D^*D).$$

Source: Use Fact 10.16.41 and $\mathcal{A}^*\mathcal{A} \ge 0$, where $\mathcal{A} \triangleq \left[\begin{smallmatrix} A & B \\ C & D \end{smallmatrix}\right]$. **Related:** Fact 3.17.20.

Fact 10.16.32. Let $A, B, C \in \mathbb{F}^{n\times n}$. Then,

$$|\det(B + CA)|^2 \le \begin{cases} \det(A^*A + B^*B)\det(I + CC^*) \\ \det(BB^* + CC^*)\det(I + A^*A). \end{cases}$$

Source: [1458, 3007]. **Related:** Fact 10.11.61.

Fact 10.16.33. Let $A, B, C \in \mathbb{F}^{n\times n}$, and assume that C is semicontractive and $B^*B \le A^*A$. Then,

$$\det(A^*A - B^*B)\det(I - CC^*) + |\det(B - C^*A)|^2 \le |\det(A - CB)|^2.$$

Source: [3007] and Fact 10.11.61.

Fact 10.16.34. Let $A, B, C, D, X, Y \in \mathbb{F}^{n\times n}$, and assume that X and Y are positive definite. Then,

$$|\det(AXC^* + BYD^*)|^2 \le \det(AXA^* + BYB^*)\det(CXC^* + DYD^*).$$

In particular,

$$|\det(XC^* + BY)|^2 \le \det(X + BYB^*)\det(Y + CXC^*),$$
$$|\det(B + C^*)|^2 \le (I + BB^*)\det(I + CC^*),$$
$$|\det(AC^* + BD^*)|^2 \le \det(AA^* + BB^*)\det(CC^* + DD^*).$$

Source: [3007].

Fact 10.16.35. Let $A, B, C \in \mathbb{F}^{n\times n}$, and assume that A, B, C are positive semidefinite. Then,

$$\det(I + A + B) \le \det(I + A)(I + B),$$
$$(\det A)\det(A + B + C) \le \det(A + B)(A + C),$$
$$\det(A + B) + \det(B + C) \le \det(A + B + C) + \det B,$$
$$\det(A + B) + \det(B + C) + \det(C + A) \le \det(A + B + C) + \det A + \det B + \det C.$$

Now, assume that A, B, C are positive definite. Then,

$$\det(A + B)^{-1} + \det(A + C)^{-1} \le \det A^{-1} + \det(A + B + C)^{-1}.$$

Source: [1835, 2186, 2681]. **Remark:** Note the analogy between the fourth equality and Hlawka's identity given by Fact 2.21.11 and Fact 11.8.5. See [1835]. **Related:** Fact 2.13.3 and Fact 10.25.44.

Fact 10.16.36. For all $i \in \{1, \dots, k\}$, let $A_i, B_i \in \mathbf{H}^n$, assume that $A_k \le \cdots \le A_1$ and $B_k \le \cdots \le B_1$, assume that $A_k + B_k$ is positive semidefinite, and let $\{j_1, \dots, j_k\} = \{1, \dots, k\}$. Then,

$$\sum_{i=1}^k \det(A_i + B_{k-i+1}) \le \sum_{i=1}^k \det(A_i + B_{j_i}) \le \sum_{i=1}^k \det(A_i + B_i),$$

$$\prod_{i=1}^k \det(A_i + B_i) \le \prod_{i=1}^k \det(A_i + B_{j_i}) \le \prod_{i=1}^k \det(A_i + B_{k-i+1}).$$

Now, assume that $A_k + B_k$ is positive definite, and let $l \in \mathbb{Z}$. Then,

$$\sum_{i=1}^{k} \det (A_i + B_{k-i+1})^l \le \sum_{i=1}^{k} \det (A_i + B_{j_i})^l \le \sum_{i=1}^{k} \det (A_i + B_i)^l.$$

Source: [2681]. **Related:** Fact 10.14.62.

Fact 10.16.37. Let $A \in \mathbb{F}^{n\times n}$, $B \in \mathbb{F}^{n\times m}$, and $C \in \mathbb{F}^{m\times m}$, define $\mathcal{A} \triangleq \begin{bmatrix} A & B \\ B^* & C \end{bmatrix} \in \mathbb{F}^{(n+m)\times(n+m)}$, and assume that $\mathcal{A}$ is positive definite. Then,

$$\det \mathcal{A} = (\det A)\det(C - B^*A^{-1}B) \le (\det A)\det C \le \prod_{i=1}^{n+m} \mathcal{A}_{(i,i)}.$$

Now, suppose that $n = m$. Then, the first inequality is an equality if and only if $B = 0$. **Source:** The second inequality is obtained by successive application of the first inequality. See [2991, p. 216]. **Remark:** $\det \mathcal{A} \le (\det A)\det C$ is *Fischer's inequality*. The inequality comprised of the first and last terms is Hadamard's inequality. See Fact 10.21.15.

Fact 10.16.38. Let $A \triangleq \begin{bmatrix} A_{11} & A_{12} \\ A_{12}^* & A_{22} \end{bmatrix} \in \mathbb{F}^{(n+m)\times(n+m)}$, assume that A is nonsingular, and define $B \triangleq AA^* = \begin{bmatrix} B_{11} & B_{12} \\ B_{12}^* & B_{22} \end{bmatrix} \in \mathbf{P}^{(n+m)}$. Then,

$$|\det(A_{22} - B_{12}^*B_{11}^{-1}A_{12})|^2 \det B_{11} \le \det B.$$

Furthermore, equality holds if and only if $A_{12} = 0$. **Source:** [1841].

Fact 10.16.39. Let $A \in \mathbb{F}^{n\times n}$, $B \in \mathbb{F}^{n\times m}$, and $C \in \mathbb{F}^{m\times m}$, define $\mathcal{A} \triangleq \begin{bmatrix} A & B \\ B^* & C \end{bmatrix} \in \mathbb{F}^{(n+m)\times(n+m)}$, assume that $\mathcal{A}$ is positive definite, let $k \triangleq \min\{m, n\}$, and, for all $i \in \{1, \dots, n+m\}$, let $\lambda_i \triangleq \lambda_i(\mathcal{A})$. Then,

$$\prod_{i=1}^{n+m} \lambda_i \le (\det A)\det C \le \left(\prod_{i=k+1}^{n+m-k} \lambda_i\right) \prod_{i=1}^{k} [\tfrac{1}{2}(\lambda_i + \lambda_{n+m-i+1})]^2.$$

Source: The first inequality is given by Fact 10.16.37. The second inequality is given in [2103]. **Remark:** If $n = m$, then the second product is 1.

Fact 10.16.40. Let $A \in \mathbb{F}^{n\times n}$, and assume that A is positive semidefinite. Then, the following statements hold:

i) If $\alpha \subset \{1, \dots, n\}$, then $\det A \le [\det A_{(\alpha)}] \det A_{[\alpha]}$.

ii) If $\alpha, \beta \subseteq \{1, \dots, n\}$, then

$$\det A_{(\alpha\cup\beta)} \det A_{(\alpha\cap\beta)} \le [\det A_{(\alpha)}] \det A_{(\beta)}.$$

iii) If $1 \le k \le n-1$, then

$$\left(\prod_{\{\alpha:\ \mathrm{card}(\alpha)=k+1\}} \det A_{(\alpha)}\right)^{\binom{n-1}{k-1}} \le \left(\prod_{\{\alpha:\ \mathrm{card}(\alpha)=k\}} \det A_{(\alpha)}\right)^{\binom{n-1}{k}}.$$

Source: [1448, p. 485], [1451, p. 507], and [1896]. **Remark:** *i)* is Fischer's inequality, which is given in Fact 10.16.37; *ii)* is the *Hadamard-Fischer inequality*; *iii)* is *Szasz's inequality*. See [789, 868, p. 680], [1448, p. 479], and [1896]. **Related:** Fact 10.16.39.

Fact 10.16.41. Let $A, B, C \in \mathbb{F}^{n\times n}$, define $\mathcal{A} \triangleq \begin{bmatrix} A & B \\ B^* & C \end{bmatrix} \in \mathbb{F}^{2n\times 2n}$, and assume that $\mathcal{A}$ is positive semidefinite. Then,

$$0 \le \det \mathcal{A} \le (\det A)\det C - |\det B|^2 \le (\det A)\det C.$$

Furthermore, if $\mathcal{A}$ is positive definite, then $|\det B|^2 < (\det A)\det C$. **Source:** Assume that A is positive definite. Then, $\det \mathcal{A} = (\det A)\det(C - B^*A^{-1}B)$, and Minkowski's determinant theorem

Corollary 10.4.15 implies that

$$(\det \mathcal{A} + |\det B|^2)/\det A = \det \mathcal{A}/\det A + \det B^*A^{-1}B = \det(C - B^*A^{-1}B) + \det B^*A^{-1}B \le \det C.$$

Use continuity in the case where A is singular. If $\mathcal{A}$ is positive definite, then $0 \le B^*A^{-1}B < C$, and thus $|\det B|^2/\det A < \det C$. See [1005, 1849]. **Remark:** If $n = 1$, then the second inequality is an equality. **Remark:** $\mathcal{A} = \begin{bmatrix} 1&0&0&0\\0&2&2&0\\0&2&2&0\\0&0&0&2 \end{bmatrix}$ shows that the converse of the last statement is false. **Remark:** If B is nonsquare, then $\mathcal{A} = \begin{bmatrix} 1&1&1\\1&2&1\\1&1&1 \end{bmatrix}$ shows that it is not necessarily true that $\det B^*B \le (\det A)\det C$. See [2985]. **Remark:** $|\det B|^2 \le (\det A)\det C$ is proved in [2263, p. 142]. **Related:** Fact 10.16.50.

Fact 10.16.42. For all $i, j \in \{1, 2, 3\}$, let $A_{ij} \in \mathbb{F}^{n_i \times n_j}$, define $A \in \mathbb{F}^{(n_1+n_2+n_3)\times(n_1+n_2+n_3)}$ by

$$A \triangleq \begin{bmatrix} A_{11} & A_{12} & A_{13} \\ A_{12}^* & A_{22} & A_{23} \\ A_{13}^* & A_{23}^* & A_{33} \end{bmatrix},$$

and assume that A is positive semidefinite. Then,

$$(\det A)\det A_{22} \le \left(\det \begin{bmatrix} A_{11} & A_{12} \\ A_{12}^* & A_{22} \end{bmatrix}\right) \det \begin{bmatrix} A_{22} & A_{23} \\ A_{23}^* & A_{33} \end{bmatrix}.$$

If, in addition, A_{22} is positive definite, then equality holds if and only if $A_{13} = A_{12}A_{22}^{-1}A_{23}$. Now, assume that $n_1 = n_2 = n_3$. Then,

$$(\det A)\det A_{22} \le \left(\det \begin{bmatrix} A_{11} & A_{12} \\ A_{12}^* & A_{22} \end{bmatrix}\right) \det \begin{bmatrix} A_{22} & A_{23} \\ A_{23}^* & A_{33} \end{bmatrix} - \left|\det \begin{bmatrix} A_{12} & A_{13} \\ A_{22} & A_{23} \end{bmatrix}\right|^2.$$

If, in addition, A_{22} is positive definite, then equality holds if and only if $A_{13} = A_{12}A_{22}^{-1}A_{23}$. **Source:** [1849]. The first result follows from the Hadamard-Fischer inequality given by Fact 10.16.39. **Remark:** In the equality case, Fact 3.17.13 implies that $\det \begin{bmatrix} A_{12} & A_{13} \\ A_{22} & A_{23} \end{bmatrix} = 0$.

Fact 10.16.43. Let $B, C \in \mathbb{F}^{n\times n}$, assume that B and C are positive definite, and define

$$A = \begin{bmatrix} A_{11} & A_{12} \\ A_{12}^* & A_{22} \end{bmatrix} \triangleq B + \jmath C,$$

where $A_{11} \in \mathbb{F}^{n_1\times n_1}$ and $A_{22} \in \mathbb{F}^{n_2\times n_2}$. Then, the following statements hold:

i) Every principal submatrix of A is nonsingular.

ii) Define $S \triangleq A_{22} - A_{21}A_{22}^{-1}A_{22}$. Then, $\frac{1}{2}(S + S^*)$ and $\frac{1}{2\jmath}(S - S^*)$ are positive definite.

iii) A is nonsingular, and $A^{-1} = (B + CB^{-1}C)^{-1} - \jmath(C + BC^{-1}B)^{-1}$.

iv) Let $m \triangleq \min\{n_1, n_2\}$ and $\kappa \triangleq \max\left\{\frac{\lambda_{\max}(B)}{\lambda_{\min}(B)}, \frac{\lambda_{\max}(C)}{\lambda_{\min}(C)}\right\}$. Then,

$$\frac{(4\kappa)^m}{(1+\kappa)^{2m}}|\det A_{11}||\det A_{22}| \le 2^m|\det A_{11}||\det A_{22}|.$$

Source: [1500, 1807, 1830, 1832]. **Remark:** A is an *accretive-dissipative matrix.*

Fact 10.16.44. Let $A \in \mathbb{F}^{n\times n}$, $B \in \mathbb{F}^{n\times m}$, and $C \in \mathbb{F}^{m\times m}$, define $\mathcal{A} \triangleq \left[\begin{smallmatrix} A & B \\ B^* & C \end{smallmatrix}\right] \in \mathbb{F}^{(n+m)\times(n+m)}$, and assume that $\mathcal{A}$ is positive semidefinite and A is positive definite. Then,

$$B^*A^{-1}B \le \left[\frac{\lambda_{\max}(\mathcal{A}) - \lambda_{\min}(\mathcal{A})}{\lambda_{\max}(\mathcal{A}) + \lambda_{\min}(\mathcal{A})}\right]^2 C.$$

Source: [1783, 2987].

Fact 10.16.45. Let $A, B, C \in \mathbb{F}^{n\times n}$, define $\mathcal{A} \triangleq \begin{bmatrix} A & B \\ B^* & C \end{bmatrix} \in \mathbb{F}^{2n\times 2n}$, and assume that $\mathcal{A}$ is positive semidefinite. Then,

$$|\det B|^2 \le \left[\frac{\lambda_{\max}(\mathcal{A}) - \lambda_{\min}(\mathcal{A})}{\lambda_{\max}(\mathcal{A}) + \lambda_{\min}(\mathcal{A})}\right]^{2n} (\det A)\det C.$$

Hence,

$$|\det B|^2 \le \left[\frac{\lambda_{\max}(\mathcal{A}) - \lambda_{\min}(\mathcal{A})}{\lambda_{\max}(\mathcal{A}) + \lambda_{\min}(\mathcal{A})}\right]^{2} (\det A)\det C.$$

Now, define $\hat{\mathcal{A}} \triangleq \begin{bmatrix} \det A & \det B \\ \det B^* & \det C \end{bmatrix} \in \mathbb{F}^{2\times 2}$. Then,

$$|\det B|^2 \le \left[\frac{\lambda_{\max}(\hat{\mathcal{A}}) - \lambda_{\min}(\hat{\mathcal{A}})}{\lambda_{\max}(\hat{\mathcal{A}}) + \lambda_{\min}(\hat{\mathcal{A}})}\right]^{2} (\det A)\det C.$$

Source: [1783, 2987]. **Remark:** These bounds cannot be ordered.

Fact 10.16.46. Let $A \in \mathbb{F}^{n\times n}$, $B \in \mathbb{F}^{n\times m}$, and $C \in \mathbb{F}^{m\times m}$, define $\mathcal{A} \triangleq \begin{bmatrix} A & B \\ B^* & C \end{bmatrix} \in \mathbb{F}^{(n+m)\times(n+m)}$, assume that $\mathcal{A}$ is positive semidefinite, and assume that A and C are positive definite. Then,

$$\det(A|\mathcal{A})\det(C|\mathcal{A}) \le \det \mathcal{A}.$$

Source: [1458]. **Remark:** This is the *reverse Fischer inequality.*

Fact 10.16.47. Let $A \in \mathbb{F}^{n\times n}$, $B \in \mathbb{F}^{n\times m}$, $C \in \mathbb{F}^{m\times m}$, and define $\mathcal{A} \triangleq \begin{bmatrix} A & B \\ 0 & C \end{bmatrix} \in \mathbb{F}^{(n+m)\times(n+m)}$. Then,

$$|\det(I+\overline{A}A)\det(I+\overline{C}C)| \le \det(I+A^*A)\det(I+C^*C) \le \det(I+\mathcal{A}^*\mathcal{A}) \le \det(I+A^*A)\det(I+B^*B+C^*C).$$

Source: Fact 10.16.37 and [1838].

Fact 10.16.48. Let $A_1, \ldots, A_n \in \mathbf{N}^n$, let $\alpha_1, \ldots, \alpha_n \in [0, 1]$, and assume that $\sum_{i=1}^n \alpha_i = 1$. Then

$$\prod_{i=1}^n (\det A_i)^{\alpha_i} \le \det\left(\sum_{i=1}^n \alpha_i A_i\right).$$

In particular,

$$\sqrt[n]{\prod_{i=1}^n \det A_i} \le \det \frac{1}{n}\sum_{i=1}^n A_i.$$

Source: [1370]. **Related:** Fact 10.16.49.

Fact 10.16.49. Let $A_1, \ldots, A_k \in \mathbf{N}^n$. Then,

$$\sqrt[k]{\prod_{i=1}^k \det A_i} \le \frac{1}{k^n} \det \sum_{i=1}^k A_i.$$

Source: [1853]. **Related:** If $1 \le k \le n$, then this result follows from Fact 10.16.48 by setting $\alpha_1 = \cdots = \alpha_k = \frac{1}{k}$ and $\alpha_{k+1} = \cdots = \alpha_n = 0$. Setting $k = n$ yields the special case in Fact 10.16.48.

Fact 10.16.50. Let $A_{ij} \in \mathbb{F}^{n\times n}$ for all $i, j \in \{1, \ldots, m\}$, define $A \in \mathbb{F}^{mn\times mn}$ by

$$A \triangleq \begin{bmatrix} A_{11} & \cdots & A_{1m} \\ \vdots & \ddots & \vdots \\ A_{1m}^* & \cdots & A_{mm} \end{bmatrix},$$

assume that A is positive semidefinite, let $1 \le k \le n$, and define $\tilde{A}_k \in \mathbb{F}^{m\binom{n}{k}\times m\binom{n}{k}}$ by

$$\tilde{A}_k \triangleq \begin{bmatrix} A_{11}^{(k)} & \cdots & A_{1m}^{(k)} \\ \vdots & \therefore & \vdots \\ A_{1m}^{*(k)} & \cdots & A_{mm}^{(k)} \end{bmatrix}.$$

Then, $\tilde{A}_k$ is positive semidefinite. In particular,

$$\tilde{A}_n = \begin{bmatrix} \det A_{11} & \cdots & \det A_{1m} \\ \vdots & \therefore & \vdots \\ \det A_{1m}^* & \cdots & \det A_{mm} \end{bmatrix}$$

is positive semidefinite. Furthermore,

$$\det A \le \det \tilde{A}_n.$$

Now, assume that A is positive definite. Then, $\det A = \det \tilde{A}_n$ if and only if, for all distinct $i, j \in \{1, \ldots, m\}$, $A_{ij} = 0$. **Source:** The first statement is given in [852]. The inequality as well as the final statement are given in [2609]. See also [1848]. **Remark:** $B^{(k)}$ is the kth compound of B. See Fact 9.5.18. **Remark:** Every principal subdeterminant of $\tilde{A}_n$ is lower bounded by the determinant of a principal submatrix of A, which is positive semidefinite and thus has nonnegative determinant. Hence, this inequality implies that $\tilde{A}_n$ is positive semidefinite. **Remark:** A weaker result is given in [854] and quoted in [1950] in terms of elementary symmetric functions of the eigenvalues of each block. **Remark:** The example $A = \left[\begin{smallmatrix} 1&0&1&0\\0&1&0&0\\1&0&1&0\\0&0&0&1 \end{smallmatrix}\right]$ shows that $\tilde{A}$ can be positive definite while A is singular. **Remark:** The matrix whose (i, j) entry is $\det A_{ij}$ is a *determinantal compression* of A. See [853, 1953, 2609] and [2991, p. 221]. **Related:** Fact 10.12.64.

10.17 Facts on Convex Sets and Convex Functions

Fact 10.17.1. Let $f\colon \mathbb{R}^n \mapsto \mathbb{R}^n$, and assume that f is convex. Then, for all $\alpha \in \mathbb{R}$, the sets $\{x \in \mathbb{R}^n\colon f(x) \le \alpha\}$ and $\{x \in \mathbb{R}^n\colon f(x) < \alpha\}$ are convex. **Source:** [1059, p. 108]. **Remark:** The converse is false. Let $f(x) = x^3$.

Fact 10.17.2. Let $A \in \mathbb{F}^{n\times n}$, assume that A is Hermitian, let $\alpha \ge 0$, and define the set $\mathcal{S} \triangleq \{x \in \mathbb{F}^n\colon x^*Ax < \alpha\}$. Then, the following statements hold:

i) $\mathcal{S}$ is an open set.

ii) $\mathcal{S}$ is a blunt cone if and only if $\alpha = 0$.

iii) $\mathcal{S}$ is a nonempty set if and only if either $\alpha > 0$ or $\lambda_{\min}(A) < 0$.

iv) $\mathcal{S}$ is a convex set if and only if $A \ge 0$.

v) $\mathcal{S}$ is a nonempty, convex set if and only if $\alpha > 0$ and $A \ge 0$.

vi) The following statements are equivalent:

a) $\mathcal{S}$ is a bounded set.

b) $\mathcal{S}$ is a bounded, convex set.

c) $A > 0$.

vii) The following statements are equivalent:

a) $\mathcal{S}$ is a nonempty, bounded set.

b) $\mathcal{S}$ is a nonempty, bounded, convex set.

c) $\alpha > 0$ and $A > 0$.

Fact 10.17.3. Let $A \in \mathbb{F}^{n\times n}$, assume that A is Hermitian, let $\alpha \ge 0$, and define the set $\mathcal{S} \triangleq \{x \in$

$\mathbb{F}^n$: $x^*Ax \le \alpha\}$. Then, the following statements hold:

i) $\mathcal{S}$ is closed.

ii) $0 \in \mathcal{S}$, and thus $\mathcal{S}$ is nonempty.

iii) $\mathcal{S}$ is a pointed cone if and only if $\alpha = 0$ or $A \le 0$.

iv) $\mathcal{S}$ is a convex set if and only if $A \ge 0$.

v) The following statements are equivalent:

a) $\mathcal{S}$ is a bounded set.

b) $\mathcal{S}$ is bounded, convex set.

c) $A > 0$.

Fact 10.17.4. Let $A \in \mathbb{F}^{n\times n}$, assume that A is Hermitian, let $\alpha \ge 0$, and define the set $\mathcal{S} \triangleq \{x \in \mathbb{F}^n$: $x^*Ax = \alpha\}$. Then, the following statements hold:

i) $\mathcal{S}$ is closed.

ii) $\mathcal{S}$ is a nonempty set if and only if either $\alpha = 0$ or $\lambda_{\max}(A) > 0$.

iii) The following statements are equivalent:

a) $\mathcal{S}$ is a pointed cone.

b) $0 \in \mathcal{S}$.

c) $\alpha = 0$.

iv) $\mathcal{S} = \{0\}$ if and only if $\alpha = 0$ and either $A > 0$ or $A < 0$.

v) $\mathcal{S}$ is a bounded set if and only if either $A > 0$ or both $\alpha > 0$ and $A \le 0$.

vi) $\mathcal{S}$ is a nonempty, bounded set if and only if $A > 0$.

vii) The following statements are equivalent:

a) $\mathcal{S}$ is a convex set.

b) $\mathcal{S}$ is nonempty, convex set.

c) $\alpha = 0$ and either $A > 0$ or $A < 0$.

viii) If $\alpha > 0$, then the following statements are equivalent:

a) $\mathcal{S}$ is a nonempty set.

b) $\mathcal{S}$ is not a convex set.

c) $\lambda_{\max}(A) > 0$.

ix) The following statements are equivalent:

a) $\mathcal{S}$ is a bounded, convex set.

b) $\mathcal{S}$ is a nonempty, bounded, convex set.

c) $\alpha = 0$ and $A > 0$.

Fact 10.17.5. Let $A \in \mathbb{F}^{n\times n}$, assume that A is Hermitian, let $\alpha \ge 0$, and define the set $\mathcal{S} \triangleq \{x \in \mathbb{F}^n$: $x^*Ax \ge \alpha\}$. Then, the following statements hold:

i) $\mathcal{S}$ is a closed set.

ii) $\mathcal{S}$ is a pointed cone if and only if $\alpha = 0$.

iii) $\mathcal{S}$ is a nonempty set if and only if either $\alpha = 0$ or $\lambda_{\max}(A) > 0$.

iv) $\mathcal{S}$ is a bounded set if and only if $\mathcal{S} \subseteq \{0\}$.

v) The following statements are equivalent:

a) $\mathcal{S}$ is a nonempty, bounded set.

b) $\mathcal{S} = \{0\}$.

c) $\alpha = 0$ and $A < 0$.

vi) $\mathcal{S}$ is a convex set if and only if either $\mathcal{S} = \varnothing$ or $\mathcal{S} = \mathbb{F}^n$.

vii) $\mathcal{S}$ is convex and bounded if and only if $\mathcal{S} = \varnothing$.

viii) The following statements are equivalent:

a) $\mathcal{S}$ is a nonempty, convex set.

b) $\mathcal{S} = \mathbb{F}^n$.

c) $\alpha = 0$ and $A \geq 0$.

Fact 10.17.6. Let $A \in \mathbb{F}^{n\times n}$, assume that A is Hermitian, let $\alpha \geq 0$, and define the set $\mathcal{S} \triangleq \{x \in \mathbb{F}^n\colon x^*Ax > \alpha\}$. Then, the following statements hold:

i) $\mathcal{S}$ is an open set.

ii) $\mathcal{S}$ is a blunt cone if and only if $\alpha = 0$.

iii) $\mathcal{S}$ is a nonempty set if and only if $\lambda_{\max}(A) > 0$.

iv) The following statements are equivalent:

a) $\mathcal{S} = \varnothing$.

b) $\lambda_{\max}(A) \leq 0$.

c) $\mathcal{S}$ is a bounded set.

d) $\mathcal{S}$ is a convex set.

Fact 10.17.7. Let $f\colon \mathbb{R}^n \mapsto \mathbb{R}$, define $g\colon \mathbb{R}^n \times \mathbb{R}^n \mapsto \mathbb{R}$ by $g(x,y) \triangleq x^\mathsf{T}y - f(y)$, define $\mathcal{D} \triangleq \{x \in \mathbb{R}^n\colon g(x,\mathbb{R}^n) \text{ is bounded}\}$, and define $h\colon \mathcal{D} \mapsto \mathbb{R}$ by $h(x) \triangleq \sup_{y\in\mathbb{R}^n} g(x,y)$. Then, the following statements hold:

i) Define $f\colon \mathbb{R} \mapsto \mathbb{R}$ by $f(x) \triangleq |x|$. Then, $h\colon [-1,1] \mapsto \mathbb{R}$ is given by $h(x) = 0$.

ii) Let $p \in (1,\infty)$, and define $f\colon \mathbb{R} \mapsto \mathbb{R}$ by $f(x) \triangleq \frac{1}{p}|x|^p$. Then, $h\colon \mathbb{R} \mapsto \mathbb{R}$ is given by $h(x) = \frac{1}{q}|x|^q$, where $q \in (0,\infty)$ satisfies $1/p + 1/q = 1$.

iii) Define $f\colon \mathbb{R} \mapsto \mathbb{R}$ by $f(x) \triangleq e^x$. Then, $h\colon [0,\infty) \mapsto \mathbb{R}$ is given by $h(0) = 0$ and, for all $x > 0$, $h(x) = x\log x - x$.

iv) Let $a \in \mathbb{R}^n$ and $b \in \mathbb{R}$, and define $f(x) \triangleq a^\mathsf{T}x + b$. Then, $h\colon \{a\} \mapsto \mathbb{R}$ is given by $h(a) = -b$.

v) Let $A \in \mathbb{R}^{n\times n}$, assume that A is positive definite, and define $f\colon \mathbb{R}^n \mapsto \mathbb{R}$ by $f(x) \triangleq \frac{1}{2}x^\mathsf{T}Ax$. Then, $h\colon \mathbb{R}^n \mapsto \mathbb{R}$ is given by $h(x) = \frac{1}{2}x^\mathsf{T}A^{-1}x$.

Source: [523, pp. 49–63] and [1847]. **Remark:** h is the *Legendre-Fenchel transform* of f, and is also called the *convex conjugate* of f. **Remark:** h can be viewed as a mapping from $\mathbb{R}^n$ to $\mathbb{R} \cup \{\infty\}$, where $h(x) \triangleq \infty$ for all $x \in \mathbb{R}^n\backslash\mathcal{D}$.

Fact 10.17.8. Let $A \in \mathbb{C}^{n\times n}$, and define the *numerical range* of A by

$$\Theta_1(A) \triangleq \{x^*Ax\colon\ x \in \mathbb{C}^n \text{ and } x^*x = 1\}$$

and the set

$$\Theta(A) \triangleq \{x^*Ax\colon\ x \in \mathbb{C}^n\}.$$

Then, the following statements hold:

i) $\Theta_1(A)$ is a closed, bounded, convex subset of $\mathbb{C}$.

ii) $\Theta(A) = \{0\} \cup \operatorname{cone}\Theta_1(A)$.

iii) $\Theta(A)$ is a pointed, closed, convex cone contained in $\mathbb{C}$.

iv) $\Theta(A)$ is a closed, bounded interval contained in $\mathbb{R}$ if and only if there exist $\alpha,\beta \in \mathbb{C}$ and a Hermitian matrix $B \in \mathbb{C}^{n\times n}$ such that $A = \alpha B + \beta I$.

v) If A is Hermitian, then $\Theta_1(A)$ is a closed, bounded interval contained in $\mathbb{R}$.

vi) If A is Hermitian, then $\Theta(A)$ is either $(-\infty,0]$, $[0,\infty)$, or $\mathbb{R}$.

vii) $\Theta_1(A)$ satisfies

$$\operatorname{conv}\operatorname{spec}(A) \subseteq \Theta_1(A) \subseteq \operatorname{co}\{\nu_1 + \mu_1 J, \nu_1 + \mu_n J, \nu_n + \mu_1 J, \nu_n + \mu_n J\},$$

where

$$\nu_1 \triangleq \lambda_{\max}[\tfrac{1}{2}(A + A^*)], \quad \nu_n \triangleq \lambda_{\min}[\tfrac{1}{2}(A + A^*)],$$
$$\mu_1 \triangleq \lambda_{\max}[\tfrac{1}{2J}(A - A^*)], \quad \mu_n \triangleq \lambda_{\min}[\tfrac{1}{2J}(A - A^*)].$$

viii) If A is normal, then $\Theta_1(A) = \operatorname{conv}\operatorname{spec}(A)$.

ix) If $n \le 4$ and $\Theta_1(A) = \operatorname{conv}\operatorname{spec}(A)$, then A is normal.

x) $\Theta_1(A) = \operatorname{conv}\operatorname{spec}(A)$ if and only if either A is normal or there exist $A_1 \in \mathbb{F}^{n_1\times n_1}$ and $A_2 \in \mathbb{F}^{n_2\times n_2}$ such that $n_1 + n_2 = n$, $\Theta_1(A_1) \subseteq \Theta_1(A_2)$, A_2 is normal, and A is unitarily similar to $\begin{bmatrix} A_1 & 0 \\ 0 & A_2 \end{bmatrix}$.

xi) Let $\lambda \in \Theta_1(A)$ and assume that λ is the vertex of a triangle that contains $\Theta_1(A)$. Then, $\lambda \in \operatorname{spec}(A)$.

xii) If $\Theta_1(A)$ is a line segment, then A is normal.

xiii) If $\Theta_1(A) = [a,b] \subset \mathbb{R}$, then $a, b \in \operatorname{spec}(A)$.

xiv) If $\lambda \in \operatorname{spec}(A) \cap \operatorname{bd}\Theta_1(A)$, then $A\mathcal{N}(A - \lambda I) \subseteq \mathcal{N}(A - \lambda I)$ and $A\mathcal{R}(A^* - \overline{\lambda} I) \subseteq \mathcal{R}(A^* - \overline{\lambda} I)$.

Source: [1280], [1450, pp. 11, 52], [2432, Chapter 9], [2979, pp. 19, 193–196], and [2991, pp. 108, 109]. **Remark:** $\Theta_1(A)$ is called the *field of values* in [1450, p. 5]. **Remark:** *ix)* is an example of the *quartic barrier*. See [787], Fact 10.19.24, and Fact 15.18.5. **Related:** Fact 6.10.30 and Fact 10.17.8.

Fact 10.17.9. Let $A \in \mathbb{R}^{n\times n}$, and define the *real numerical range* of A by

$$\Psi_1(A) \triangleq \{x^{\mathsf{T}}Ax\colon\ x \in \mathbb{R}^n \text{ and } x^{\mathsf{T}}x = 1\}$$

and the set

$$\Psi(A) \triangleq \{x^{\mathsf{T}}Ax\colon\ x \in \mathbb{R}^n\}.$$

Then, the following statements hold:

i) $\Psi_1(A) = \Psi_1[\tfrac{1}{2}(A + A^{\mathsf{T}})]$.

ii) $\Psi_1(A) = [\lambda_{\min}[\tfrac{1}{2}(A + A^{\mathsf{T}})], \lambda_{\max}[\tfrac{1}{2}(A + A^{\mathsf{T}})]]$.

iii) If A is symmetric, then $\Psi_1(A) = [\lambda_{\min}(A), \lambda_{\max}(A)]$.

iv) $\Psi(A) = \{0\} \cup \operatorname{cone}\Psi_1(A)$.

v) $\Psi(A)$ is either $(-\infty, 0]$, $[0,\infty)$, or $\mathbb{R}$.

vi) $\Psi_1(A) = \Theta_1(A)$ if and only if A is symmetric.

Source: [1450, p. 83]. **Remark:** $\Theta_1(A)$ is defined in Fact 10.17.8.

Fact 10.17.10. Let $A, B \in \mathbb{C}^{n\times n}$, assume that A and B are Hermitian, and define

$$\Theta_1(A,B) \triangleq \left\{\begin{bmatrix} x^*Ax \\ x^*Bx \end{bmatrix}\colon\ x \in \mathbb{C}^n \text{ and } x^*x = 1\right\} \subseteq \mathbb{R}^2.$$

Then, $\Theta_1(A,B)$ is a convex set. **Source:** [2248]. **Related:** This result follows from Fact 10.17.8.

Fact 10.17.11. Let $A, B \in \mathbb{R}^{n\times n}$, assume that A and B are symmetric, and let α, β be real numbers. Then, the following statements are equivalent:

i) There exists $x \in \mathbb{R}^n$ such that $x^{\mathsf{T}}Ax = \alpha$ and $x^{\mathsf{T}}Bx = \beta$.

ii) There exists a positive-semidefinite matrix $X \in \mathbb{R}^{n\times n}$ such that $\operatorname{tr} AX = \alpha$ and $\operatorname{tr} BX = \beta$.

Source: [309, p. 84].

Fact 10.17.12. Let $A, B \in \mathbb{R}^{n\times n}$, assume that A and B are symmetric, and define

$$\Psi_1(A,B) \triangleq \left\{ \begin{bmatrix} x^\mathsf{T}Ax \\ x^\mathsf{T}Bx \end{bmatrix} \colon\ x \in \mathbb{R}^n \text{ and } x^\mathsf{T}x = 1 \right\} \subseteq \mathbb{R}^2, \quad \Psi(A,B) \triangleq \left\{ \begin{bmatrix} x^\mathsf{T}Ax \\ x^\mathsf{T}Bx \end{bmatrix} \colon\ x \in \mathbb{R}^n \right\} \subseteq \mathbb{R}^2.$$

Then, $\Psi(A,B)$ is a pointed, convex cone. If, in addition, $n \ge 3$, then $\Psi_1(A,B)$ is a convex set. **Source:** [309, pp. 84, 89] and [903, 2245, 2248]. **Remark:** $\Psi(A,B) = [\operatorname{cone}\Psi_1(A,B)] \cup \left\{\left[\begin{smallmatrix}0\\0\end{smallmatrix}\right]\right\}$. **Remark:** The set $\Psi(A,B)$ is not necessarily closed. See [903, 2188, 2189].

Fact 10.17.13. Let $A, B \in \mathbb{R}^{n\times n}$, assume that A and B are symmetric, let $\mathcal{A}, \mathcal{B}$ be closed subsets of $\mathbb{R}^n$, assume that $\mathcal{A}\cup\mathcal{B} = \mathbb{R}^n$, and assume that, for all $x \in \mathcal{A}$, $x^\mathsf{T}Ax \ge 0$ and, for all $x \in \mathcal{B}$, $x^\mathsf{T}Bx \ge 0$. Then, there exists $\alpha \in [0,1]$ such that $\alpha A + (1-\alpha)B$ is positive semidefinite. **Source:** [2245].

Fact 10.17.14. Let $A, B \in \mathbb{R}^{n\times n}$, and assume that A and B are symmetric. Then, the following statements are equivalent:

i) For all $x \in \mathbb{R}^n$, either $x^\mathsf{T}Ax > 0$ or $x^\mathsf{T}Bx > 0$.

ii) There exist $\alpha, \beta \in [0,\infty)$ such that $\alpha A + \beta B$ is positive definite.

Now, assume that there exists $y \in \mathbb{R}^n$ such that $y^\mathsf{T}By < 0$. Then, the following statements are equivalent:

iii) For all $x \in \mathbb{R}^n$, either $x^\mathsf{T}Ax \ge 0$ or $x^\mathsf{T}Bx > 0$.

iv) There exists $\alpha \in [0,\infty)$ such that $A + \alpha B$ is positive semidefinite.

Source: [2245]. **Remark:** The equivalence of *iii)* and *iv)* is the *S-lemma*.

Fact 10.17.15. Let $n \ge 2$, let $A, B \in \mathbb{R}^{n\times n}$, assume that A and B are symmetric, let $a, b \in \mathbb{R}^n$, let $a_0, b_0 \in \mathbb{R}$, assume that there exist real numbers α, β such that $\alpha A + \beta B > 0$, and define

$$\Psi \triangleq \left\{ \begin{bmatrix} x^\mathsf{T}Ax + a^\mathsf{T}x + a_0 \\ x^\mathsf{T}Bx + b^\mathsf{T}x + b_0 \end{bmatrix} \colon\ x \in \mathbb{R}^n \right\} \subseteq \mathbb{R}^2.$$

Then, Ψ is a closed, convex set. **Source:** [2248].

Fact 10.17.16. Let $A, B, C \in \mathbb{R}^{n\times n}$, where $n \ge 3$, assume that A, B, and C are symmetric, and define

$$\Phi_1(A,B,C) \triangleq \left\{ \begin{bmatrix} x^\mathsf{T}Ax \\ x^\mathsf{T}Bx \\ x^\mathsf{T}Cx \end{bmatrix} \colon\ x \in \mathbb{R}^n \text{ and } x^\mathsf{T}x = 1 \right\} \subseteq \mathbb{R}^3, \quad \Phi(A,B,C) \triangleq \left\{ \begin{bmatrix} x^\mathsf{T}Ax \\ x^\mathsf{T}Bx \\ x^\mathsf{T}Cx \end{bmatrix} \colon\ x \in \mathbb{R}^n \right\} \subseteq \mathbb{R}^3.$$

Then, $\Phi_1(A,B,C)$ is a convex set, and $\Phi(A,B,C)$ is a pointed, convex cone. **Source:** [573, 2245, 2248].

Fact 10.17.17. Let $A, B, C \in \mathbb{R}^{n\times n}$, where $n \ge 3$, assume that A, B, and C are symmetric, and define

$$\Phi(A,B,C) \triangleq \left\{ \begin{bmatrix} x^\mathsf{T}Ax \\ x^\mathsf{T}Bx \\ x^\mathsf{T}Cx \end{bmatrix} \colon\ x \in \mathbb{R}^n \right\} \subseteq \mathbb{R}^3.$$

Then, the following statements are equivalent:

i) There exist real numbers α, β, γ such that $\alpha A + \beta B + \gamma C$ is positive definite.

ii) $\Phi(A,B,C)$ is a pointed, one-sided, closed, convex cone, and $\{x \in \mathbb{R}^n\colon\ x^\mathsf{T}Ax = x^\mathsf{T}Bx = x^\mathsf{T}Cx = 0\} \subseteq \{0\}$.

Source: [2248].

Fact 10.17.18. Let $A \in \mathbb{F}^{n\times n}$, assume that A is Hermitian, let $b \in \mathbb{F}^n$ and $c \in \mathbb{R}$, and define $f\colon\ \mathbb{F}^n \mapsto \mathbb{R}$ by

$$f(x) \triangleq x^*Ax + \operatorname{Re}(b^*x) + c.$$

Then, the following statements hold:

i) f is convex if and only if A is positive semidefinite.

ii) f is strictly convex if and only if A is positive definite.

Now, assume that A is positive semidefinite. Then, f has a minimizer if and only if $b \in \mathcal{R}(A)$. If these conditions hold, then the following statements hold:

iii) The vector $x_0 \in \mathbb{F}^n$ is a minimizer of f if and only if x_0 satisfies $Ax_0 = -\frac{1}{2}b$.

iv) $x_0 \in \mathbb{F}^m$ minimizes f if and only if there exists a vector $y \in \mathbb{F}^m$ such that

$$x_0 = -\tfrac{1}{2}A^+b + (I - A^+A)y.$$

v) If $x_0 \in \mathbb{F}^m$ minimizes f, then

$$f(x_0) = c - x_0^*Ax_0 = c - \tfrac{1}{4}b^*A^+b.$$

vi) If A is positive definite, then $x_0 = -\frac{1}{2}A^{-1}b$ is the unique minimizer of f, and the minimum of f is given by

$$f(x_0) = c - x_0^*Ax_0 = c - \tfrac{1}{4}b^*A^{-1}b.$$

Source: Use Proposition 8.1.8 and note that, for each $x_0 \in \mathbb{F}^n$ satisfying $Ax_0 = -\frac{1}{2}b$, it follows that

$$\begin{aligned} f(x_0) &= (x - x_0)^*A(x - x_0) + c - x_0^*Ax_0 \\ &= (x - x_0)^*A(x - x_0) + c - \tfrac{1}{4}b^*A^+b. \end{aligned}$$

Remark: This is the *quadratic minimization lemma.* **Related:** Fact 11.17.10.

Fact 10.17.19. Let $A \in \mathbb{F}^{n\times n}$, assume that A is positive definite, and define $\phi\colon \mathbb{F}^{m\times n} \mapsto \mathbb{R}$ by $\phi(B) \triangleq \operatorname{tr} BAB^*$. Then, ϕ is strictly convex. **Source:** $\operatorname{tr} \alpha(1-\alpha)(B_1 - B_2)A(B_1 - B_2)^* > 0$.

Fact 10.17.20. Let $p, q \in \mathbb{R}$, and define $\phi\colon (0,\infty)\times(0,\infty) \mapsto (0,\infty)$ by $\phi(a,b) \triangleq a^pb^q$. Then, the following conditions are equivalent:

i) One of the following conditions holds:

a) $p \ge 1$, $q \le 0$, and $p + q \ge 1$.

b) $p \le 0$, $q \ge 1$, and $p + q \ge 1$.

c) $p \le 0$ and $q \le 0$.

ii) $\phi\colon (0,\infty)\times(0,\infty) \mapsto (0,\infty)$ defined by $\phi(a,b) \triangleq a^pb^q$ is convex.

Furthermore, the following conditions are equivalent:

iii) $p \in [0,1]$, $q \in [0,1]$, and $p + q \le 1$.

iv) ϕ is concave.

Source: [632].

Fact 10.17.21. Let $n \ge 1$, let $p, q \in \mathbb{R}$, and define $\phi\colon \mathbf{P}^n \times \mathbf{P}^n \to (0,\infty)$ by $\phi(A,B) \triangleq \operatorname{tr} A^pB^q$. Then, the following statements hold:

i) Assume that one of the following conditions holds:

a) $p \in [1,2]$, $q \in [-1,0]$, and $p + q \ge 1$.

b) $p \in [-1,0)$, $q \in [1,2]$, and $p + q \ge 1$.

c) $p \in [-1,0)$ and $q \in [-1,0)$.

Then, ϕ is convex.

ii) If $p \in (0,1)$, $q \in (0,1)$, and $p + q \le 1$, then ϕ is concave.

iii) If p and q do not satisfy the assumptions of either *i)* or *ii)*, then ϕ is neither convex nor concave.

Source: [346].

Fact 10.17.22. Let $n \geq 1$. Then, the following statements hold:

i) $\phi\colon \mathbf{P}^n \times \mathbf{P}^n \mapsto \mathbf{P}^n$ defined by $\phi(A, B) \triangleq A^{p/2}B^qA^{p/2}$ is convex if and only if $p = 2$ and $q \in [-1, 0)$.

ii) $\phi\colon \mathbf{P}^n \times \mathbf{P}^n \times \mathbf{P}^n \mapsto \mathbf{P}^n$ defined by $\phi(A, B, C) \triangleq A^{p/2}B^qA^{p/2}C^r$ is convex if and only if $p = 2$, $q < 0$, $r < 0$, and $q + r \in [-1, 0)$.

iii) If $p \in [-1, 0)$, $q \in [1, 2]$, $p+q \neq 0$, and $s \in [\min\{1/(q-1), 1/(p+1)\}, \infty)$, then $\phi\colon \mathbf{P}^n \times \mathbf{P}^n \mapsto (0, \infty)$ defined by $\phi(A, B) \triangleq \operatorname{tr}(A^{p/2}B^qA^{p/2})^s$ is convex.

iv) If $p \in [-1, 0)$ and $s \in [1/(p + 2), \infty)$, then $\phi\colon \mathbf{P}^n \times \mathbf{P}^n \mapsto (0, \infty)$ defined by $\phi(A, B) \triangleq \operatorname{tr}(A^{p/2}B^2A^{p/2})^s$ is convex.

v) If $p \in [0, 1]$, $q \in [0, 1]$, and $s \in [0, 1/(p + q]$, then $\phi\colon \mathbf{P}^n \times \mathbf{P}^n \mapsto (0, \infty)$ defined by $\phi(A, B) \triangleq \operatorname{tr}(A^{p/2}B^2A^{p/2})^s$ is convex.

Source: [632].

Fact 10.17.23. Let $A \in \mathbb{F}^{n\times n}$, assume that A is Hermitian, let $x \in \mathbb{F}^n$, and define $f\colon \mathbb{R} \mapsto \mathbb{R}$ by $f(t) \triangleq x^*e^{A+txx^*}x$. Then, f is increasing and strictly convex. **Source:** [2578].

Fact 10.17.24. Let $B \in \mathbb{F}^{n\times n}$, assume that B is Hermitian, let $\alpha_1, \ldots, \alpha_k \in (0, \infty)$, define $r \triangleq \sum_{i=1}^k \alpha_i$, assume that $r \leq 1$, let $q \in \mathbb{R}$, and define $\phi\colon \bigtimes_{i=1}^k \mathbf{P}^n \to [0, \infty)$ by

$$\phi(A_1, \ldots, A_k) \triangleq -\left(\operatorname{tr} e^{B+\sum_{i=1}^k \alpha_i \log A_i}\right)^q.$$

If $q \in (0, 1/r]$, then ϕ is convex. Furthermore, if $q < 0$, then $-\phi$ is convex. **Source:** [1817, 1891, 2003]. **Related:** Fact 10.14.46.

Fact 10.17.25. Let $A_1, \ldots, A_m \in \mathbb{R}^{n\times n}$, assume that $A_1, \ldots, A_m$ are positive definite, define $f\colon \mathbb{R}^n \mapsto \mathbb{R}$ by $f(x) \triangleq \prod_{i=1}^m x^\mathsf{T}A_ix$, and, for all $i, j \in \{1, \ldots, m\}$, define $\kappa_{i,j} \triangleq \lambda_{\max}(A_iA_j^{-1})/\lambda_{\min}(A_iA_j^{-1})$. Then, the following statements hold:

i) f is convex if and only if, for all $x, y \in \mathbb{R}^n$ such that $y \neq 0$,

$$\sum_{i=1}^m \frac{x^\mathsf{T}A_ix}{y^\mathsf{T}A_iy} + 2\sum_{i=1}^m \sum_{j=1, j\neq i}^m \frac{x^\mathsf{T}A_iyx^\mathsf{T}A_jy}{y^\mathsf{T}A_iyy^\mathsf{T}A_jy} \geq 0.$$

ii) If, for all $i \in \{1, \ldots, m\}$,

$$\sum_{j=1}^m \left(\frac{\sqrt{\kappa_{i,j}} - 1}{\sqrt{\kappa_{i,j}} + 1}\right)^2 \leq \frac{1}{2},$$

then f is convex.

iii) If, for all distinct $i, j \in \{1, \ldots, m\}$,

$$\kappa_{i,j} \leq \left(\frac{\sqrt{2m-2} + 1}{\sqrt{2m-2} - 1}\right)^2,$$

then f is convex.

iv) Let $A, B \in \mathbb{F}^{n\times n}$, assume that A and B are positive definite, define $g\colon \mathbb{R}^n \mapsto \mathbb{R}$ by $g(x) \triangleq x^\mathsf{T}Axx^\mathsf{T}Bx$, and define $\kappa \triangleq \lambda_{\max}(AB^{-1})/\lambda_{\min}(AB^{-1})$. Then, the following statements are equivalent:

a) g is convex.

b) $\dfrac{\sqrt{\kappa} - 1}{\sqrt{\kappa} + 1} \leq \dfrac{\sqrt{2}}{2}$.

c) $\kappa \leq 17 + 12\sqrt{2}$.

v) Let $A \in \mathbb{F}^{n\times n}$, assume that A is positive definite, and define $g\colon \mathbb{R}^n \mapsto \mathbb{R}$ by $g(x) \triangleq x^{\mathsf{T}}Axx^{\mathsf{T}}A^{-1}x$, and define $\kappa \triangleq \operatorname{cond}(A)$. Then, g is convex if and only if $\kappa \le 3 + 2\sqrt{2}$.

Source: [1847]. **Remark:** See [1400]. **Remark:** $(3 + 2\sqrt{2})^2 = 17 + 12\sqrt{2}$.

10.18 Facts on Quadratic Forms for One Matrix

Fact 10.18.1. Let $\mathcal{G} = (\mathcal{X}, \mathcal{R})$ be a symmetric graph, where $\mathcal{X} = \{x_1, \ldots, x_n\}$. Then, for all $z \in \mathbb{R}^n$, the Laplacian matrix L of $\mathcal{G}$ satisfies

$$z^{\mathsf{T}}Lz = \frac{1}{2}\sum (z_{(i)} - z_{(j)})^2,$$

where the sum is taken over the set $\{(i, j)\colon (x_i, x_j) \in \mathcal{R}\}$. **Source:** [591, pp. 29, 30] and [2028].

Fact 10.18.2. Let $A \in \mathbb{F}^{n\times n}$, and assume that A is Hermitian. Then, $\mathcal{N}(A) \subseteq \{x \in \mathbb{F}^n\colon x^*Ax = 0\}$. Furthermore, $\mathcal{N}(A) = \{x \in \mathbb{F}^n\colon x^*Ax = 0\}$ if and only if either $A \ge 0$ or $A \le 0$.

Fact 10.18.3. Let $x, y \in \mathbb{F}^n$. Then, $xx^* \le yy^*$ if and only if there exists $\alpha \in \mathbb{F}$ such that $|\alpha| \in [0, 1]$ and $x = \alpha y$.

Fact 10.18.4. Let $x, y \in \mathbb{F}^n$. Then, $xy^* + yx^* \ge 0$ if and only if x and y are linearly dependent. **Source:** Evaluate the product of the nonzero eigenvalues of $xy^* + yx^*$, and use the Cauchy-Schwarz inequality $|x^*y|^2 \le x^*xy^*y$.

Fact 10.18.5. Let $A \in \mathbb{F}^{n\times n}$, assume that A is positive definite, let $x \in \mathbb{F}^n$, and let $a \in [0, \infty)$. Then, the following statements are equivalent:

i) $xx^* \le aA$.

ii) $x^*A^{-1}x \le a$.

iii) $\begin{bmatrix} A & x \\ x^* & a \end{bmatrix} \ge 0$.

Source: Fact 3.17.3 and Proposition 10.2.5. Note that, if $a = 0$, then $x = 0$.

Fact 10.18.6. Let $A \in \mathbb{F}^{n\times n}$, assume that A is positive definite, and let $x, y \in \mathbb{F}^n$. Then,

$$2\operatorname{Re} x^*y \le x^*Ax + y^*A^{-1}y.$$

Furthermore, if $y = Ax$, then equality holds. Therefore,

$$x^*Ax = \max\{2\operatorname{Re} x^*z - z^*A^{-1}z\colon z \in \mathbb{F}^n\}.$$

Source: $(A^{1/2}x - A^{-1/2}y)^*(A^{1/2}x - A^{-1/2}y) \ge 0$. **Credit:** R. Bellman. See [1783, 2987].

Fact 10.18.7. Let $A \in \mathbb{F}^{n\times n}$, assume that A is positive definite, and let $x, y \in \mathbb{F}^n$. Then,

$$|x^*y|^2 \le (x^*Ax)(y^*A^{-1}y).$$

Source: Use Fact 10.12.33 with A replaced by $A^{1/2}x$ and B replaced by $A^{-1/2}y$. Alternatively, use Fact 10.12.24 and Fact 10.19.23.

Fact 10.18.8. Let $A \in \mathbb{F}^{n\times n}$, assume that A is positive definite, let $x \in \mathbb{F}^n$, and define $\alpha \triangleq \lambda_{\min}(A)$ and $\beta \triangleq \lambda_{\max}(A)$. Then,

$$(x^*x)^2 \le (x^*Ax)(x^*A^{-1}x) \le \frac{(\alpha+\beta)^2}{4\alpha\beta}(x^*x)^2 \le \frac{\beta}{\alpha}(x^*x)^2.$$

Source: [43], [1448, p. 443], [1451, p. 470], [1876], and [2991, pp. 249, 251]. **Remark:** The second inequality is the *Kantorovich inequality*. **Related:** Fact 2.11.134, Fact 10.11.42, and Fact 10.11.43.

Fact 10.18.9. Let $A \in \mathbb{F}^{n\times n}$, assume that A is nonsingular, let $x \in \mathbb{F}^n$, and define $\kappa \triangleq \rho_1(A)/\rho_n(A)$. Then,

$$|x^*Axx^*A^{-1}x| \le \tfrac{1}{4}(\kappa + \kappa^{-1} + 2)(x^*x)^2, \quad |x^*A^2xx^*A^{-2}x| \le \tfrac{1}{4}(\kappa + \kappa^{-1})^2(x^*x)^2.$$

Source: [1451, p. 474]. **Remark:** κ is the *spectral condition number* of A. **Related:** Fact 10.18.8.

Fact 10.18.10. Let $A \in \mathbb{F}^{n\times n}$, assume that A is positive definite, and define $\kappa \triangleq \lambda_1(A)/\lambda_n(A)$. If $x \in \mathbb{F}^n$, then

$$|x^*Axx^*A^{-1}x| \le \tfrac{1}{4}(\kappa + \kappa^{-1} + 2)(x^*x)^2.$$

If $x, y \in \mathbb{F}^n$ and $x^*y = 0$, then

$$|x^*Ay|^2 \le \left(\frac{\kappa - 1}{\kappa + 1}\right)^2 x^*Axy^*Ay.$$

Source: [1451, p. 474]. **Remark:** These are equivalent to the Kantorovich and Wielandt inequalities, respectively. **Related:** Fact 10.18.8 and Fact 10.18.21.

Fact 10.18.11. Let $A \in \mathbb{F}^{n\times n}$, assume that A is positive definite, let $x \in \mathbb{F}^n$, and define $\alpha \triangleq \lambda_{\min}(A)$ and $\beta \triangleq \lambda_{\max}(A)$. Then,

$$(x^*x)^{1/2}(x^*Ax)^{1/2} - x^*Ax \le \frac{(\alpha - \beta)^2}{4(\alpha + \beta)}x^*x, \quad (x^*x)(x^*A^2x) - (x^*Ax)^2 \le \tfrac{1}{4}(\alpha - \beta)^2(x^*x)^2.$$

Source: [2214]. **Remark:** Extensions are given in [1510, 2214].

Fact 10.18.12. Let $A \in \mathbb{F}^{n\times n}$, assume that A is positive semidefinite, let $r \triangleq \operatorname{rank} A$, let $x \in \mathbb{F}^n$, and assume that $x \notin \mathcal{N}(A)$. Then,

$$\frac{x^*Ax}{x^*x} - \frac{x^*x}{x^*A^+x} \le [\lambda_{\max}^{1/2}(A) - \lambda_r^{1/2}(A)]^2.$$

If, in addition, A is positive definite, then, for all nonzero $x \in \mathbb{F}^n$,

$$0 \le \frac{x^*Ax}{x^*x} - \frac{x^*x}{x^*A^{-1}x} \le [\lambda_{\max}^{1/2}(A) - \lambda_{\min}^{1/2}(A)]^2.$$

Source: [2071, 2214]. The left-hand inequality in the last string is given by Fact 10.18.8.

Fact 10.18.13. Let $A \in \mathbb{F}^{n\times n}$, assume that A is positive definite, let $y \in \mathbb{F}^n$, let $\alpha > 0$, and define $f\colon \mathbb{F}^n \mapsto \mathbb{R}$ by $f(x) \triangleq |x^*y|^2$. Then,

$$x_0 = \sqrt{\frac{\alpha}{y^*A^{-1}y}}A^{-1}y$$

minimizes $f(x)$ subject to $x^*Ax \le \alpha$. Furthermore, $f(x_0) = \alpha y^*A^{-1}y$. **Source:** [73].

Fact 10.18.14. Let $A \in \mathbb{F}^{n\times n}$ and $x \in \mathbb{F}^n$. Then,

$$|x^*Ax| \le x^*A^*Ax, \quad |x^*Ax|^2 \le (x^*x)x^*A^*Ax.$$

Source: [2991, pp. 291, 302].

Fact 10.18.15. Let $A \in \mathbb{F}^{n\times n}$ and $x \in \mathbb{F}^n$. Then,

$$x^*(A + A^*)x \le 2\sqrt{x^*xx^*A^*Ax}.$$

Source: [1142].

Fact 10.18.16. Let $A \in \mathbb{F}^{n\times n}$. Then, the following statements are equivalent:

i) A is normal.

ii) For all $x \in \mathbb{F}^n$, $|x^*Ax| \le x^*\langle A\rangle x$.

iii) For all $x \in \mathbb{F}^n$, $|Ax| = |A^*x|$.

iv) There exists $\alpha \in [0, \tfrac{1}{2}) \cup (\tfrac{1}{2}, 1]$ such that, for all $x \in \mathbb{F}^n$, $|x^*Ax| \le (x^*\langle A\rangle x)^\alpha (x^*\langle A^*\rangle x)^{1-\alpha}$.

v) For all $x, y \in \mathbb{F}^n$, $y^*A^*Ax = y^*AA^*x$.

Source: [2991, pp. 294, 295, 303, 315, 318].

Fact 10.18.17. Let $A \in \mathbb{F}^{n\times n}$, assume that A is positive semidefinite, and let $x \in \mathbb{F}^n$. Then,

$$(x^*Ax)^2 \le (x^*x)(x^*A^2x) \le (x^*x)\sqrt{(x^*Ax)(x^*A^3x)},$$
$$(\min \tfrac{1}{4}[\lambda_i(A) - \lambda_j(A)]^2)(x^*x)^2 \le (x^*x)(x^*A^2x) - (x^*Ax)^2 \le \tfrac{1}{4}[\lambda_1(A) - \lambda_n(A)]^2(x^*x)^2,$$

where the minimum is taken over all distinct $i, j \in \{1,\dots,n\}$. **Source:** Use the Cauchy-Schwarz inequality Corollary 11.1.7. The second string is given in [1344].

Fact 10.18.18. Let $A \in \mathbb{F}^{n\times n}$, assume that A is positive semidefinite, and let $x \in \mathbb{F}^n$. If $\alpha \in [0,1]$, then

$$x^*A^\alpha x \le (x^*x)^{1-\alpha}(x^*Ax)^\alpha.$$

Furthermore, if $\alpha > 1$, then

$$(x^*Ax)^\alpha \le (x^*x)^{\alpha-1}x^*A^\alpha x.$$

Remark: The first inequality is the *Hölder-McCarthy inequality*, which is equivalent to Young's inequality. See [1124, p. 125] and [1126]. Matrix versions of the second inequality are given in [1421]. **Related:** Fact 10.10.47 and Fact 10.11.68.

Fact 10.18.19. Let $A \in \mathbb{F}^{n\times n}$, assume that A is positive semidefinite, let $x \in \mathbb{F}^n$, and let $\alpha, \beta \in [1,\infty)$, where $\alpha \le \beta$. Then,

$$(x^*A^\alpha x)^{1/\alpha} \le (x^*A^\beta x)^{1/\beta}.$$

Now, assume that A is positive definite. Then,

$$x^*(\log A)x \le \log x^*Ax \le \tfrac{1}{\alpha}\log x^*A^\alpha x \le \tfrac{1}{\beta}\log x^*A^\beta x.$$

Source: [1085].

Fact 10.18.20. Let $A \in \mathbb{F}^{n\times n}$, $x, y \in \mathbb{F}^n$, and $\alpha \in (0,1)$. Then,

$$|x^*Ay| \le \|\langle A\rangle^\alpha x\|_2 \|\langle A^*\rangle^{1-\alpha} y\|_2.$$

Consequently,

$$|x^*Ay| \le (x^*\langle A\rangle x)^{1/2}(y^*\langle A^*\rangle y)^{1/2}.$$

Source: [1551] and [2991, p. 314].

Fact 10.18.21. Let $A \in \mathbb{F}^{n\times n}$, assume that A is positive semidefinite, let $x, y \in \mathbb{F}^n$, and assume that $x^*y = 0$. Then,

$$|x^*Ay|^2 \le \left[\frac{\lambda_{\max}(A) - \lambda_{\min}(A)}{\lambda_{\max}(A) + \lambda_{\min}(A)}\right]^2 (x^*Ax)(y^*Ay).$$

Furthermore, there exist $x, y \in \mathbb{F}^n$ satisfying $x^*y = 0$ and $x^*x = y^*y = 1$ for which equality holds. **Source:** [1448, p. 443], [1451, p. 471], and [1783, 2987]. **Remark:** This is the *Wielandt inequality*. **Related:** Fact 10.19.23.

Fact 10.18.22. Let $A \in \mathbb{F}^{n\times n}$, assume that A is Hermitian, let $x, y \in \mathbb{F}^n$, and assume that $x^*x = y^*y = 1$ and $x^*y = 0$. Then,

$$2|x^*Ay| \le \lambda_{\max}(A) - \lambda_{\min}(A).$$

Furthermore, there exist $x, y \in \mathbb{F}^n$ satisfying $x^*x = y^*y = 1$ and $x^*y = 0$ for which equality holds. **Source:** [1783, 2987] and [2991, p. 292]. **Remark:** $\lambda_{\max}(A) - \lambda_{\min}(A)$ is the spread of A. **Related:** Fact 6.10.4, Fact 10.21.5, Fact 11.11.7, and Fact 11.11.8. Fact 10.18.23 provides a lower bound in the case where A is positive semidefinite.

Fact 10.18.23. Let $A \in \mathbb{F}^{n\times n}$, assume that A is positive semidefinite, let $x, y \in \mathbb{F}^n$, and assume that $x^*y = 0$. Then,

$$(\min \tfrac{1}{2}[\lambda_i(A) - \lambda_j(A)])\sqrt{x^*xy^*y} \le |x^*Ay| \le \tfrac{1}{2}[\lambda_1(A) - \lambda_n(A)]\sqrt{x^*xy^*y},$$

where the minimum is taken over all $i, j \in \{1, \ldots, n\}$ such that $i < j$. **Source:** [1344]. **Related:** Fact 10.18.22 shows that the upper bound holds for Hermitian A.

Fact 10.18.24. Let $A \in \mathbb{F}^{n\times n}$, assume that A is positive semidefinite, let $x, y \in \mathbb{F}^n$, and assume that $x^*x = y^*y = 1$. Then,

$$|x^*Ax - y^*Ay| \le [\lambda_1(A) - \lambda_n(A)]\sqrt{1 - (x^*y)^2}.$$

Source: [1344].

Fact 10.18.25. Let $A \in \mathbb{F}^{n\times n}$, assume that A is positive semidefinite, let $x, y \in \mathbb{F}^n$, and assume that $x^*x = y^*y = 1$ and $x^*y = 0$. Then,

$$\tfrac{1}{4}[\lambda_{n-1}(A) + \lambda_n(A)]^2 \le x^*Axy^*Ay \le \tfrac{1}{4}[\lambda_1(A) + \lambda_2(A)]^2.$$

Source: [1344].

Fact 10.18.26. Let $A \in \mathbb{R}^{n\times n}$, assume that A is positive semidefinite, let $x, y \in \mathbb{R}^n$, assume that $x^\mathsf{T}x = y^\mathsf{T}y = 1$, and let $\theta \in [0, \pi]$ be the angle between x and y. Then,

$$x^\mathsf{T}Ay \le \tfrac{1}{2}(\cos\theta)(x^\mathsf{T}Ax + y^\mathsf{T}Ay) + \tfrac{1}{2}(\sin\theta)\delta(A).$$

Source: [1344]. **Problem:** Extend this result to complex matrices.

Fact 10.18.27. Let $A \in \mathbb{R}^{m\times m}$, assume that A is positive semidefinite, and assume that every entry of A is an integer. If every off-diagonal entry of A is even and, for all $k \in \{1,2,3,5,6,7,10,14,15\}$, there exists $x \in \mathbb{R}^m$ such that $k = x^\mathsf{T}Ax$, then, for all $n \ge 0$, there exists $x \in \mathbb{R}^m$ such that $n = x^\mathsf{T}Ax$. Furthermore, if, for all $k \in \{1,2,3,5,6,7,10,13,14,15,17,19,21,22,23,26,29,30,31,34,35,37,42,58,93,110,145,203,290\}$, there exists $x \in \mathbb{R}^m$ such that $k = x^\mathsf{T}Ax$, then, for all $n \ge 0$, there exists $x \in \mathbb{R}^m$ such that $n = x^\mathsf{T}Ax$. **Source:** [2887]. **Related:** Fact 1.11.27.

10.19 Facts on Quadratic Forms for Two or More Matrices

Fact 10.19.1. Let $A, B \in \mathbb{F}^{n\times n}$, assume that A and B are Hermitian, assume that $A + B$ is nonsingular, let $x, a, b \in \mathbb{F}^n$, and define $c \triangleq (A + B)^{-1}(Aa + Bb)$. Then,

$$(x - a)^*A(x - a) + (x - b)^*B(x - b) = (x - c)^*(A + B)(x - c) + (a - b)^*A(A + B)^{-1}B(a - b).$$

Source: [2418, p. 278].

Fact 10.19.2. Let $A, B \in \mathbb{R}^{n\times n}$, assume that A is symmetric and B is skew symmetric, and let $x, y \in \mathbb{R}^n$. Then,

$$\begin{bmatrix} x \\ y \end{bmatrix}^\mathsf{T} \begin{bmatrix} A & B \\ B^\mathsf{T} & A \end{bmatrix} \begin{bmatrix} x \\ y \end{bmatrix} = (x + \jmath y)^*(A + \jmath B)(x + \jmath y).$$

Related: Fact 6.10.32.

Fact 10.19.3. Let $A, B \in \mathbb{F}^{n\times n}$, assume that A is positive semidefinite, assume that AB is Hermitian, and let $x \in \mathbb{F}^n$. Then,

$$|x^*ABx| \le \rho_{\max}(B)x^*Ax.$$

Source: [1825]. **Remark:** This is an improvement of Reid's inequality. Related results are given in [1826]. **Credit:** P. R. Halmos.

Fact 10.19.4. Let $A, B \in \mathbb{F}^{n\times n}$, assume that A and B are positive definite, and let $x \in \mathbb{F}^n$. Then,

$$x^*(A + B)^{-1}x \le \frac{x^*A^{-1}xx^*B^{-1}x}{x^*A^{-1}x + x^*B^{-1}x} \le \tfrac{1}{4}(x^*A^{-1}x + x^*B^{-1}x).$$

In particular,

$$\frac{1}{(A^{-1})_{(i,i)}} + \frac{1}{(B^{-1})_{(i,i)}} \le \frac{1}{[(A + B)^{-1}]_{(i,i)}}.$$

Source: [1924, p. 201]. The right-hand inequality follows from Fact 2.2.12. **Remark:** This is *Bergstrom's inequality*. **Remark:** This is a special case of Fact 10.12.5, which is a special case of *xvii)* of Proposition 10.6.17.

Fact 10.19.5. Let $A, B \in \mathbb{F}^{n\times m}$, assume that $I - A^*A$ and $I - B^*B$ are positive semidefinite, and let $x \in \mathbb{C}^n$. Then,

$$x^*(I - A^*A)xx^*(I - B^*B)x \le |x^*(I - A^*B)x|^2.$$

Credit: M. Marcus. See [2184]. **Related:** Fact 10.16.28.

Fact 10.19.6. Let $A, B \in \mathbb{F}^n$, and assume that A is Hermitian and B is positive definite. Then,

$$\lambda_{\min}(AB^{-1}) = \min\{\lambda \in \mathbb{R}\colon\ \det(A - \lambda B) = 0\} = \min_{x\in\mathbb{F}^n\backslash\{0\}} \frac{x^*Ax}{x^*Bx},$$

$$\lambda_{\max}(AB^{-1}) = \max\{\lambda \in \mathbb{R}\colon\ \det(A - \lambda B) = 0\} = \max_{x\in\mathbb{F}^n\backslash\{0\}} \frac{x^*Ax}{x^*Bx}.$$

Source: Lemma 10.4.3 and [2991, pp. 273, 285].

Fact 10.19.7. Let $w, x, y, z \in \mathbb{F}^n$, let $A, B \in \mathbb{F}^{n\times n}$, and assume that A and B are positive definite. Then,

$$2w^*yx^*z \le w^*A^{-1}wx^*Bx + y^*Ayz^*B^{-1}z.$$

Source: [241].

Fact 10.19.8. Let $A \in \mathbb{F}^{n\times n}$, $B \in \mathbb{F}^{n\times m}$, $x \in \mathbb{F}^n$, and $y \in \mathbb{F}^m$, and assume that A is positive definite. Then,

$$|x^*By|^2 \le x^*Axy^*B^*A^{-1}By.$$

Source: [2991, p. 250].

Fact 10.19.9. Let $A, B \in \mathbb{F}^{n\times m}$, $x \in \mathbb{F}^n$, and $y \in \mathbb{F}^m$. Then,

$$|x^*(A + B)y|^2 \le x^*(I + AA^*)xy^*(I + B^*B)y.$$

Source: [2991, p. 250].

Fact 10.19.10. Let $A, B \in \mathbb{R}^{n\times n}$, assume that A is positive definite and B is symmetric, let $x, y \in \mathbb{R}^n$, and assume that $x^\mathsf{T}x = y^\mathsf{T}y = 1$. Then,

$$|x^\mathsf{T}ABy - x^\mathsf{T}Axx^\mathsf{T}By| \le \tfrac{1}{4}[\lambda_1(A) - \lambda_n(A)][\lambda_1(B) - \lambda_n(B)].$$

Source: [1344].

Fact 10.19.11. Let $A, B \in \mathbb{F}^{n\times n}$, and assume that A is positive definite and B is positive semidefinite. Then, for all nonzero $x \in \mathbb{F}^n$, $4(x^*x)(x^*Bx) < (x^*Ax)^2$ if and only if there exists $\alpha > 0$ such that $\alpha I + \alpha^{-1}B < A$. If these conditions hold, then $4B < A^2$, and hence $2B^{1/2} < A$. **Source:** Sufficiency follows from $\alpha x^*x + \alpha^{-1}x^*Bx < x^*Ax$. Necessity follows from Fact 10.19.12. The last statement follows from $(A - 2\alpha I)^2 \ge 0$ and $2B^{1/2} \le \alpha I + \alpha^{-1}B$.

Fact 10.19.12. Let $A, B, C \in \mathbb{F}^{n\times n}$, assume that A, B, C are positive semidefinite, and assume that, for all nonzero $x \in \mathbb{F}^n$, $4(x^*Cx)(x^*Bx) < (x^*Ax)^2$. Then, there exists $\alpha > 0$ such that $\alpha C + \alpha^{-1}B < A$. **Source:** [2224].

Fact 10.19.13. Let $A, B \in \mathbb{F}^{n\times n}$, where A is Hermitian and B is positive semidefinite. Then, $x^*Ax < 0$ for all nonzero $x \in \mathbb{F}^n$ such that $Bx = 0$ if and only if there exists $\alpha > 0$ such that $A < \alpha B$. **Source:** To prove necessity, suppose that, for every $\alpha > 0$, there exists a nonzero vector x such that $x^*Ax \ge \alpha x^*Bx$. Then, $Bx = 0$ implies that $x^*Ax \ge 0$. Sufficiency is immediate.

Fact 10.19.14. Let $A, B \in \mathbb{C}^{n\times n}$, and assume that A and B are Hermitian. Then, the following statements are equivalent:

i) There exist $\alpha, \beta \in \mathbb{R}$ such that $\alpha A + \beta B$ is positive definite.

ii) $\{x \in \mathbb{C}^n\colon\ x^*Ax = x^*Bx = 0\} = \{0\}$.

Remark: This is *Finsler's lemma*. See [185, 341, 1400, 1737, 2738, 2759]. **Related:** Fact 10.19.15, Fact 10.20.7, and Fact 10.20.8.

Fact 10.19.15. Let $A, B \in \mathbb{R}^{n\times n}$, and assume that A and B are symmetric. Then, the following statements are equivalent:

i) There exist $\alpha, \beta \in \mathbb{R}$ such that $\alpha A + \beta B$ is positive definite.

ii) Either $x^\mathsf{T}Ax > 0$ for all nonzero $x \in \{y \in \mathbb{F}^n\colon\ y^\mathsf{T}By = 0\}$ or $x^\mathsf{T}Ax < 0$ for all nonzero $x \in \{y \in \mathbb{F}^n\colon\ y^\mathsf{T}By = 0\}$.

Now, assume that $n \ge 3$. Then, the following statement is equivalent to *i*) and *ii*):

iii) $\{x \in \mathbb{R}^n\colon\ x^\mathsf{T}Ax = x^\mathsf{T}Bx = 0\} = \{0\}$.

Remark: This result is related to Finsler's lemma. See [185, 341, 2759]. **Related:** Fact 10.19.14, Fact 10.20.7, and Fact 10.20.8.

Fact 10.19.16. Let $A, B \in \mathbb{F}^{n\times n}$, assume that A and B are positive definite, and let $C, D \in \mathbb{F}^{n\times m}$. Then,

$$(C + D)^*(A + B)^{-1}(C + D) \le C^*A^{-1}C + D^*B^{-1}D.$$

In particular, if $x, y \in \mathbb{F}^n$, then

$$(x + y)^*(A + B)^{-1}(x + y) \le x^*A^{-1}x + y^*B^{-1}y.$$

Source: Use *xiv*) of Lemma 10.6.16. See [1451, p. 475], [2142], and [2991, p. 243]. **Related:** Fact 10.19.17 and Fact 10.25.59.

Fact 10.19.17. Let $A, B \in \mathbb{F}^{n\times n}$, assume that A and B are positive semidefinite, let $C, D \in \mathbb{F}^{n\times m}$, and assume that $\mathcal{R}(C) \subseteq \mathcal{R}(A)$ and $\mathcal{R}(D) \subseteq \mathcal{R}(B)$. Then,

$$(C + D)^*(A + B)^+(C + D) \le C^*A^+C + D^*B^+D.$$

In particular, if $x \in \mathcal{R}(A)$ and $y \in \mathcal{R}(B)$, then

$$(x + y)^*(A + B)^+(x + y) \le x^*A^+x + y^*B^+y.$$

Source: [2142]. **Related:** Fact 10.19.16.

Fact 10.19.18. Let $A, B \in \mathbb{F}^{n\times n}$, assume that A and B are positive semidefinite, let $C \in \mathbb{F}^{n\times m}$, assume that $\mathcal{R}(C) \subseteq \mathcal{R}(A) \cap \mathcal{R}(B)$, and let $\alpha \in [0, 1]$. Then,

$$C^*[\alpha A + (1 - \alpha)B]^+C \le \alpha C^*A^+C + (1 - \alpha)C^*B^+C.$$

Source: [2142]. **Related:** Fact 10.19.17.

Fact 10.19.19. Let $A_1, \ldots, A_k \in \mathbb{F}^{n\times n}$, assume that $A_1, \ldots, A_k$ are positive semidefinite, let $B_1, \ldots, B_k \in \mathbb{F}^{n\times m}$, assume that, for all $i \in \{1, \ldots, k\}$, $\mathcal{R}(B_i) \subseteq \mathcal{R}(A_i)$, let $\alpha_1, \ldots, \alpha_k \ge 0$, and assume that $\sum_{i=1}^k \alpha_i = 1$. Then,

$$\left(\sum_{i=1}^k \alpha_i B_i\right)^*\left(\sum_{i=1}^k \alpha_i A_i\right)^+\left(\sum_{i=1}^k \alpha_i B_i\right) \le \sum_{i=1}^k \alpha_i B_i^* A_i^+ B_i.$$

In particular, if, for all $i \in \{1, \ldots, k\}$, $x_i \in \mathcal{R}(A_i)$, then

$$\left(\sum_{i=1}^k \alpha_i x_i\right)^*\left(\sum_{i=1}^k \alpha_i A_i\right)^+\left(\sum_{i=1}^k \alpha_i x_i\right) \le \sum_{i=1}^k \alpha_i x_i^* A_i^+ x_i.$$

Source: [2142]. **Credit:** N. Gaffke and O. Krafft. **Related:** Fact 10.19.17.

Fact 10.19.20. Let $A, B \in \mathbb{C}^{n\times n}$, where A and B are Hermitian, and assume that $x^*(A + \jmath B)x \neq 0$ for all nonzero $x \in \mathbb{C}^n$. Then, there exists $t \in [0, \pi)$ such that $(\sin t)A + (\cos t)B$ is positive definite. **Source:** [793] and [2539, p. 282].

Fact 10.19.21. Let $A \in \mathbb{R}^{n\times n}$, assume that A is symmetric, and let $B \in \mathbb{R}^{n\times m}$. Then, the following statements are equivalent:

i) $x^\mathsf{T}Ax > 0$ for all nonzero $x \in \mathcal{N}(B^\mathsf{T})$.

ii) $\nu_+\left(\begin{bmatrix} A & B \\ B^\mathsf{T} & 0 \end{bmatrix}\right) = n.$

Furthermore, the following statements are equivalent:

iii) $x^\mathsf{T}Ax \geq 0$ for all $x \in \mathcal{N}(B^\mathsf{T})$.

iv) $\nu_-\left(\begin{bmatrix} A & B \\ B^\mathsf{T} & 0 \end{bmatrix}\right) = \operatorname{rank} B.$

Source: [661, 1921]. **Related:** Fact 7.9.22 and Fact 10.19.22.

Fact 10.19.22. Let $A \in \mathbb{R}^{n\times n}$, assume that A is symmetric, let $B \in \mathbb{R}^{n\times m}$, where $m \leq n$, and assume that $[I_m \ \ 0_{m\times(n-m)}]B$ is nonsingular. Then, the following statements are equivalent:

i) $x^\mathsf{T}Ax > 0$ for all nonzero $x \in \mathcal{N}(B^\mathsf{T})$.

ii) For all $i \in \{m+1, \ldots, n\}$, $\operatorname{sign}\det\begin{bmatrix} 0 & B^\mathsf{T} \\ B & A \end{bmatrix}_{(\{1,\ldots,i\})} = (-1)^m.$

Source: [200, p. 20], [1895, p. 312], and [1942]. **Related:** Fact 10.19.21.

Fact 10.19.23. Let $A \in \mathbb{F}^{n\times n}$, $B \in \mathbb{F}^{n\times m}$, and $C \in \mathbb{F}^{m\times m}$, define $\mathcal{A} \triangleq \left[\begin{smallmatrix} A & B \\ B^* & C \end{smallmatrix}\right]$, and assume that A and C are positive semidefinite. Then, the following statements are equivalent:

i) $\mathcal{A}$ is positive semidefinite.

ii) $|x^*By|^2 \leq (x^*Ax)(y^*Cy)$ for all $x \in \mathbb{F}^n$ and $y \in \mathbb{F}^m$.

iii) $2|x^*By| \leq x^*Ax + y^*Cy$ for all $x \in \mathbb{F}^n$ and $y \in \mathbb{F}^m$.

If, in addition, A and C are positive definite, then the following statement is equivalent to *i)–iii)*:

iv) $\rho_{\max}(B^*A^{-1}BC^{-1}) \leq 1.$

Finally, if $\mathcal{A}$ is positive semidefinite and nonzero, then, for all $x \in \mathbb{F}^n$ and $y \in \mathbb{F}^m$,

$$|x^*By|^2 \leq \left[\frac{\lambda_{\max}(\mathcal{A}) - \lambda_{\min}(\mathcal{A})}{\lambda_{\max}(\mathcal{A}) + \lambda_{\min}(\mathcal{A})}\right]^2 (x^*Ax)(y^*Cy).$$

Source: [1448, p. 473], [1783, 2987], and [2991, p. 246]. **Related:** Fact 10.18.7 and Fact 10.18.21.

Fact 10.19.24. Let $n \leq 4$, let $A \in \mathbb{R}^{n\times n}$, assume that A is symmetric, and assume that, for all nonnegative vectors $x \in \mathbb{R}^n$, $x^\mathsf{T}Ax \geq 0$. Then, there exist $B, C \in \mathbb{R}^{n\times n}$ such that B is positive semidefinite, C is symmetric and nonnegative, and $A = B + C$. **Remark:** If $n \geq 5$, then this result does not hold. Hence, this result is an example of the *quartic barrier*. See [787], Fact 10.17.8, and Fact 15.18.5. **Remark:** A is *copositive*.

Fact 10.19.25. Let $m > k \geq n \geq 1$, let $x_1, \ldots, x_m \in \mathbb{F}^n$, and assume that $\sum_{i=1}^k x_ix_i^*$ is positive definite. Then,

$$\sum_{j=k+1}^m x_j^*\left(\sum_{i=1}^j x_ix_i^*\right)^{-2} x_j \leq \operatorname{tr}\left(\sum_{i=1}^k x_ix_i^*\right)^{-1}.$$

Source: [1833]. **Credit:** T. W. Anderson and J. B. Taylor.

10.20 Facts on Simultaneous Diagonalization

Fact 10.20.1. Let $A, B \in \mathbb{F}^{n\times n}$, assume that A and B are Hermitian, and assume that $\mathcal{R}(A) \subseteq \mathcal{R}(B)$. Then, there exists a unitary matrix $S \in \mathbb{F}^{n\times n}$, such that

$$SAS^* = \begin{bmatrix} A_1 & 0 \\ 0 & 0 \end{bmatrix}, \quad SBS^* = \begin{bmatrix} B_1 & 0 \\ 0 & 0 \end{bmatrix},$$

where $A_1 \in \mathbf{H}^r$, $B_1 \triangleq \operatorname{diag}(\lambda_1(B), \ldots, \lambda_r(B))$, and $r \triangleq \operatorname{rank} B$.

Fact 10.20.2. Let $A, B \in \mathbb{F}^{n\times n}$, assume that A and B are Hermitian. Then, the following statements are equivalent:

i) There exists a unitary matrix $S \in \mathbb{F}^{n\times n}$ such that SAS^* and SBS^* are diagonal.

ii) $AB = BA$.

iii) AB is Hermitian.

If, in addition, A is nonsingular, then the following statement is equivalent to *i)*–*iii)*:

iv) $A^{-1}B$ is Hermitian.

Source: [360, p. 208], [970, pp. 188–190], and [1448, p. 229]. **Related:** The equivalence of *i)* and *ii)* is given by Fact 7.19.8.

Fact 10.20.3. Let $A, B \in \mathbb{F}^{n\times n}$, assume that A and B are Hermitian, and assume that A is nonsingular. Then, there exists a nonsingular matrix $S \in \mathbb{F}^{n\times n}$ such that SAS^* and SBS^* are diagonal if and only if $A^{-1}B$ is diagonalizable over $\mathbb{R}$. **Source:** [1448, p. 229] and [2263, p. 95].

Fact 10.20.4. Let $A \in \mathbb{F}^{n\times n}$. Then, the following statements are equivalent:

i) A is diagonalizable, and $\operatorname{spec}(A) \subset \mathbb{R}$.

ii) There exist positive-definite $B \in \mathbb{F}^{n\times n}$ and Hermitian $C \in \mathbb{F}^{n\times n}$ such that $A = BC$.

Now, assume that *i)* and *ii)* hold. Then, the following statements hold:

iii) $\operatorname{In}(A) = \operatorname{In}(C)$.

iv) A is positive definite if and only if A is Hermitian and C is positive definite.

Source: [2991, p. 265]. **Related:** This result provides the converse of Corollary 10.3.3.

Fact 10.20.5. Let $A, B \in \mathbb{F}^{n\times n}$, assume that A and B are symmetric, and assume that A is nonsingular. Then, there exists a nonsingular matrix $S \in \mathbb{F}^{n\times n}$ such that SAS^{T} and SBS^{T} are diagonal if and only if $A^{-1}B$ is diagonalizable. **Source:** [1448, p. 229] and [2759]. **Remark:** If $\mathbb{F} = \mathbb{C}$, then A and B may be complex symmetric.

Fact 10.20.6. Let $A, B \in \mathbb{F}^{n\times n}$, and assume that A and B are Hermitian. Then, there exists a nonsingular matrix $S \in \mathbb{F}^{n\times n}$ such that SAS^* and SBS^* are diagonal if and only if there exists a positive-definite matrix $M \in \mathbb{F}^{n\times n}$ such that $AMB = BMA$. **Source:** [185].

Fact 10.20.7. Let $A, B \in \mathbb{F}^{n\times n}$, assume that A and B are Hermitian, and assume that there exist $\alpha, \beta \in \mathbb{R}$ such that $\alpha A + \beta B$ is positive definite. Then, there exists a nonsingular matrix $S \in \mathbb{F}^{n\times n}$ such that SAS^* and SBS^* are diagonal. **Source:** [1448, p. 465]. **Remark:** This result extends a result of K. Weierstrass. See [2759]. **Remark:** Suppose that B is positive definite. Then, by necessity of Fact 10.20.3, it follows that $A^{-1}B$ is diagonalizable over $\mathbb{R}$, which proves *iii)* $\Longrightarrow$ *i)* of Proposition 7.7.13. **Related:** Fact 10.20.8.

Fact 10.20.8. Let $A, B \in \mathbb{F}^{n\times n}$, assume that A and B are Hermitian, assume that $\{x \in \mathbb{F}^n\colon\ x^*Ax = x^*Bx = 0\} = \{0\}$, and, if $\mathbb{F} = \mathbb{R}$, assume that $n \geq 3$. Then, there exists a nonsingular matrix $S \in \mathbb{F}^{n\times n}$ such that SAS^* and SBS^* are diagonal. **Source:** This result follows from Fact 7.19.10. See [1929] and [2263, p. 96]. **Remark:** For $\mathbb{F} = \mathbb{R}$, this result is due to E. Pesonen and J. Milnor. See [2759]. **Related:** Fact 7.19.10, Fact 10.19.14, Fact 10.19.15, and Fact 10.20.7.

10.21 Facts on Eigenvalues and Singular Values for One Matrix

Fact 10.21.1. Let $A = \begin{bmatrix} a & b \\ \bar{b} & c \end{bmatrix} \in \mathbb{F}^{2\times 2}$, assume that A is Hermitian, and let $\operatorname{mspec}(A) = \{\lambda_1, \lambda_2\}_{\rm ms}$, where $\lambda_2 \le \lambda_1$. Then,

$$2|b| \le \lambda_1 - \lambda_2.$$

Now, assume that A is positive semidefinite. Then,

$$\sqrt{2}|b| \le \left(\sqrt{\lambda_1} - \sqrt{\lambda_2}\right)\sqrt{\lambda_1 + \lambda_2}.$$

If $c > 0$, then

$$\frac{|b|}{\sqrt{c}} \le \sqrt{\lambda_1} - \sqrt{\lambda_2}.$$

If $a > 0$ and $c > 0$, then

$$\frac{|b|}{\sqrt{ac}} \le \frac{\lambda_1 - \lambda_2}{\lambda_1 + \lambda_2}.$$

Finally, if A is positive definite, then

$$\frac{|b|}{a} \le \frac{\lambda_1 - \lambda_2}{2\sqrt{\lambda_1\lambda_2}}, \quad 4|b| \le \frac{\lambda_1^2 - \lambda_2^2}{\sqrt{\lambda_1\lambda_2}}.$$

Source: [1783, 2987]. **Related:** Fact 10.21.4.

Fact 10.21.2. Let $A \in \mathbf{H}^n$, and let $\chi_A(s) = s^n + \beta_{n-1}s^{n-1} + \cdots + \beta_1 s + \beta_0$. Then, the following statements are equivalent:

i) A is negative semidefinite.

ii) For all $i \in \{1, \ldots, n-1\}$, $\beta_i \ge 0$.

Furthermore, the following statements are equivalent:

iii) A is negative definite.

iv) For all $i \in \{1, \ldots, n-1\}$, $\beta_i > 0$.

Related: Fact 15.18.3.

Fact 10.21.3. Let $A \in \mathbb{F}^{n\times m}$. If $m \le n$, then

$$\operatorname{msval}(A) = \operatorname{msval}(\langle A\rangle) = \operatorname{mspec}(\langle A\rangle).$$

If $n \le m$, then

$$\operatorname{msval}(A) = \operatorname{msval}(\langle A^*\rangle) = \operatorname{mspec}(\langle A^*\rangle).$$

Fact 10.21.4. Let $A \in \mathbb{F}^{n\times n}$, and assume that A is positive semidefinite. Then, the following statements hold:

i) If $i, j \in \{1, \ldots, n\}$, then $\frac{\lambda_1(A)\lambda_n(A)}{[\lambda_1(A)+\lambda_n(A)]^2}[A_{(i,i)} - A_{(j,j)}]^2 \le A_{(i,i)}A_{(j,j)} - |A_{(i,j)}|^2$.

ii) If $i \in \{1, \ldots, n\}$, then $|\lambda_i(A) - \frac{1}{n}\operatorname{tr} A| \le \delta(A)$.

iii) If $i, j \in \{1, \ldots, n\}$, then $|\lambda_i(A) - A_{(j,j)}| \le \delta(A)$.

iv) If $i, j \in \{1, \ldots, n\}$ and $i \ne j$, then $2|A_{(i,j)}| \le \delta(A)$.

v) If $i, j \in \{1, \ldots, n\}$, then $|A_{(i,i)} - A_{(j,j)}| \le \delta(A)$.

vi) If $i, j \in \{1, \ldots, n\}$, then $4|A_{(i,i)}A_{(j,j)}| \le [\lambda_1(A) + \lambda_2(A)]^2$.

vii) $\operatorname{tr} A^3 - \frac{1}{n}(\operatorname{tr} A^2)\operatorname{tr} A \le \frac{n}{4}[\lambda_1(A) + \lambda_n(A)][\lambda_1(A) - \lambda_n(A)]^2$.

Source: [1344]. **Remark:** $\delta(A) = \lambda_{\max}(A) - \lambda_{\min}(A)$ is the spread of A. **Related:** Fact 10.21.1.

Fact 10.21.5. Let $A \in \mathbb{F}^{n\times m}$, and assume that A is Hermitian. Then,

$$\left.\begin{array}{ll}\frac{2}{n}\sqrt{n\operatorname{tr}A^2-(\operatorname{tr}A)^2}, & n \text{ even}\\ \frac{2}{\sqrt{n^2-1}}\sqrt{n\operatorname{tr}A^2-(\operatorname{tr}A)^2}, & n \text{ odd}\end{array}\right\}\le \delta(A)\le\sqrt{\frac{2}{n}}\sqrt{n\operatorname{tr}A^2-(\operatorname{tr}A)^2}.$$

Source: [674, 1344]. **Remark:** $\delta(A)=\lambda_{\max}(A)-\lambda_{\min}(A)$ is the spread of A. **Related:** Fact 6.10.4, Fact 10.18.22, Fact 11.11.7, and Fact 11.11.8.

Fact 10.21.6. Let $A \in \mathbb{F}^{n\times n}$, assume that A is positive definite, and define $\kappa \triangleq \lambda_{\max}(A)/\lambda_{\min}(A)$. Then, for all distinct $i,j\in\{1,\ldots,n\}$,

$$\max\{|A_{(i,i)}-A_{(j,j)}|, |2\operatorname{Re}A_{(i,j)}|\}\le\frac{\kappa-1}{\kappa+1}(A_{(i,i)}+A_{(j,j)})\le A_{(i,i)}+A_{(j,j)},$$

$$\max\left\{\frac{A_{(i,i)}+2\operatorname{Re}A_{(i,j)}+A_{(j,j)}}{A_{(i,i)}-2\operatorname{Re}A_{(i,j)}+A_{(j,j)}},\frac{A_{(i,i)}-2\operatorname{Re}A_{(i,j)}+A_{(j,j)}}{A_{(i,i)}+2\operatorname{Re}A_{(i,j)}+A_{(j,j)}},\frac{A_{(i,i)}}{A_{(j,j)}}\right\}\le\kappa.$$

Source: [1344].

Fact 10.21.7. Let $A \in \mathbb{F}^{n\times m}$. Then, for all $i\in\{1,\ldots,\min\{n,m\}\}$, $\lambda_i(\langle A\rangle)=\sigma_i(A)$. Hence,

$$\operatorname{tr}\langle A\rangle=\sum_{i=1}^{\min\{n,m\}}\sigma_i(A).$$

Fact 10.21.8. Let $A \in \mathbb{F}^{n\times n}$, and define

$$\mathcal{A}\triangleq\begin{bmatrix}\sigma_{\max}(A)I & A^*\\ A & \sigma_{\max}(A)I\end{bmatrix}.$$

Then, $\mathcal{A}$ is positive semidefinite. Furthermore,

$$\langle A+A^*\rangle\le\left\{\begin{array}{c}\langle A\rangle+\langle A^*\rangle\le 2\sigma_{\max}(A)I\\ A^*A+I\end{array}\right\}\le[\sigma_{\max}^2(A)+1]I.$$

Source: [2985].

Fact 10.21.9. Let $A \in \mathbb{F}^{n\times n}$. Then, for all $i\in\{1,\ldots,n\}$,

$$\lambda_i[\tfrac{1}{2}(A+A^*)]\le\sigma_i(A).$$

Hence, $\operatorname{Re}\operatorname{tr}A\le\operatorname{tr}\langle A\rangle$. **Source:** [2479] and [2991, pp. 288, 289]. **Related:** Fact 7.12.25.

Fact 10.21.10. Let $A \in \mathbb{F}^{n\times n}$. If $p>0$, then, for all $k\in\{1,\ldots,n\}$,

$$\sum_{i=1}^k\rho_i^p(A)\le\sum_{i=1}^k\sigma_i^p(A).$$

In particular, for all $k\in\{1,\ldots,n\}$,

$$\sum_{i=1}^k\rho_i(A)\le\sum_{i=1}^k\sigma_i(A).$$

Hence,

$$|\operatorname{tr}A|\le\sum_{i=1}^n\rho_i(A)\le\sum_{i=1}^n\sigma_i(A)=\operatorname{tr}\langle A\rangle.$$

Furthermore, for all $k\in\{1,\ldots,n\}$,

$$\sum_{i=1}^k\rho_i^2(A)\le\sum_{i=1}^k\sigma_i^2(A).$$

Hence,

$$\operatorname{Re}\operatorname{tr} A^2 \le |\operatorname{tr} A^2| \le \sum_{i=1}^{n} \rho_i^2(A) \le \sum_{i=1}^{n} \sigma_i(A^2) = \operatorname{tr}\langle A^2\rangle \le \sum_{i=1}^{n} \sigma_i^2(A) = \operatorname{tr} A^*A.$$

Furthermore,

$$\sum_{i=1}^{n} \rho_i^2(A) = \operatorname{tr} A^*A$$

if and only if A is normal. Finally, $\operatorname{tr} A^2 = \operatorname{tr} A^*A$ if and only if A is Hermitian. **Source:** Fact 3.25.15, Fact 7.12.32, [449, p. 42], [1450, p. 176], [2977, p. 19], and [2991, pp. 86, 312, 355]. See Fact 10.13.5 for $\operatorname{tr}\langle A^2\rangle = \operatorname{tr}(A^{2*}A^2)^{1/2} \le \operatorname{tr} A^*A$. See Fact 4.10.13 and Fact 7.15.16. **Remark:** The first result is *Weyl's inequalities.* $\sum_{i=1}^n \rho_i^2(A) \le \operatorname{tr} A^*A$ is *Schur's inequality.* **Related:** Fact 7.12.26, Fact 7.12.30, Fact 10.13.2, Fact 10.13.14, Fact 10.21.11, and Fact 11.13.2.

Fact 10.21.11. Let $A \in \mathbb{F}^{n\times n}$, and assume that A is Hermitian. Then, for all $k \in \{1,\dots,n\}$,

$$\sum_{i=1}^{k} \mathrm{d}_i(A) \le \sum_{i=1}^{k} \lambda_i(A)$$

with equality for $k = n$; that is,

$$\operatorname{tr} A = \sum_{i=1}^{n} \mathrm{d}_i(A) = \sum_{i=1}^{n} \lambda_i(A).$$

Hence,

$$[\mathrm{d}_1(A) \ \cdots \ \mathrm{d}_n(A)] \overset{\mathrm{s}}{\prec} [\lambda_1(A) \ \cdots \ \lambda_n(A)],$$

and thus, for all $k \in \{1,\dots,n\}$,

$$\sum_{i=k}^{n} \lambda_i(A) \le \sum_{i=k}^{n} \mathrm{d}_i(A).$$

In particular,

$$\lambda_{\min}(A) \le \mathrm{d}_{\min}(A) \le \mathrm{d}_{\max}(A) \le \lambda_{\max}(A).$$

Furthermore, $[\mathrm{d}_1(A) \ \cdots \ \mathrm{d}_n(A)]^{\mathsf{T}}$ is an element of the convex hull of the $n!$ vectors obtained by permuting the components of $[\lambda_1(A) \ \cdots \ \lambda_n(A)]^{\mathsf{T}}$. **Source:** [449, p. 35], [1448, p. 193], [1969, p. 218], [1969, p. 300], and [2977, p. 18]. The last statement follows from Fact 4.11.6. **Remark:** This is *Schur's theorem.* **Related:** Fact 7.12.26, Fact 10.13.2, Fact 10.13.14, Fact 10.21.10, and Fact 11.13.2.

Fact 10.21.12. Let $A \in \mathbb{F}^{n\times n}$. Then,

$$[\rho_1^2(A) \ \cdots \ \rho_n^2(A)] \overset{\mathrm{w}}{\prec} [\sigma_1^2(A) \ \cdots \ \sigma_n^2(A)],$$

$$[\mathrm{d}_1^2(|A|) \ \cdots \ \mathrm{d}_n^2(|A|)] \overset{\mathrm{w}}{\prec} [\mathrm{d}_1(\langle A\rangle)\mathrm{d}_1(\langle A^*\rangle) \ \cdots \ \mathrm{d}_n(\langle A\rangle)\mathrm{d}_n(\langle A^*\rangle)] \overset{\mathrm{w}}{\prec} [\sigma_1^2(A) \ \cdots \ \sigma_n^2(A)],$$

$$[\mathrm{d}_1^2(|A|) \ \cdots \ \mathrm{d}_n^2(|A|)] \overset{\mathrm{wlog}}{\prec} [\mathrm{d}_1(\langle A\rangle)\mathrm{d}_1(\langle A^*\rangle) \ \cdots \ \mathrm{d}_n(\langle A\rangle)\mathrm{d}_n(\langle A^*\rangle)],$$

$$[\rho_1(A + A^*) \ \cdots \ \rho_n(A + A^*)] \overset{\mathrm{w}}{\prec} [\lambda_1(\langle A\rangle + \langle A^*\rangle) \ \cdots \ \lambda_n(\langle A\rangle + \langle A^*\rangle)],$$

$$[\rho_1(A \odot A^*) \ \cdots \ \rho_n(A \odot A^*)] \overset{\mathrm{w}}{\prec} [\lambda_1(\langle A\rangle\langle A^*\rangle) \ \cdots \ \lambda_n(\langle A\rangle\langle A^*\rangle)].$$

Source: [2991, p. 355].

Fact 10.21.13. Let $A \in \mathbb{F}^{n\times n}$, assume that A is Hermitian, let k denote the number of positive diagonal entries of A, and let l denote the number of positive eigenvalues of A. Then,

$$\sum_{i=1}^{k} \mathrm{d}_i^2(A) \le \sum_{i=1}^{l} \lambda_i^2(A).$$

Source: Write $A = B+C$, where B is positive semidefinite, C is negative semidefinite, and $\mathrm{mspec}(A) = \mathrm{mspec}(B) \cup \mathrm{mspec}(C)$. Furthermore, without loss of generality, assume that $A_{(1,1)}, \ldots, A_{(k,k)}$ are the positive diagonal entries of A. Then,

$$\sum_{i=1}^{k} \mathrm{d}_i^2(A) = \sum_{i=1}^{k} A_{(i,i)}^2 \le \sum_{i=1}^{k} (A_{(i,i)} - C_{(i,i)})^2 = \sum_{i=1}^{k} B_{(i,i)}^2 \le \sum_{i=1}^{n} B_{(i,i)}^2 \le \operatorname{tr} B^2 = \sum_{i=1}^{l} \lambda_i^2(A).$$

Remark: This can be written as $\operatorname{tr}(A + |A|)^{\odot 2} \le \operatorname{tr}(A + \langle A\rangle)^2$. **Credit:** Y. Li.

Fact 10.21.14. Let $x, y \in \mathbb{R}^n$, where $n \ge 2$. Then, the following statements are equivalent:

i) $x \overset{\mathrm{s}}{<} y$.

ii) x is an element of the convex hull of the vectors $y_1, \ldots, y_{n!} \in \mathbb{R}^n$, where each of these $n!$ vectors is formed by permuting the components of y.

iii) There exists a Hermitian matrix $A \in \mathbb{C}^{n\times n}$ such that $[A_{(1,1)} \ \cdots \ A_{(n,n)}]^{\mathsf{T}} = x$ and $\mathrm{mspec}(A) = \{y_{(1)}, \ldots, y_{(n)}\}_{\mathrm{ms}}$.

Remark: This is the *Schur-Horn theorem.* Schur's theorem given by Fact 10.21.11 is *iii)* $\Longrightarrow$ *i)*, while the result *i)* $\Longrightarrow$ *iii)* is given in [1446]. The equivalence of *ii)* is given by Fact 4.11.6. This result is discussed in [309, 450, 575]. An equivalent version is given by Fact 4.13.16.

Fact 10.21.15. Let $A \in \mathbb{F}^{n\times n}$, and assume that A is positive semidefinite. Then, for all $k \in \{1, \ldots, n\}$,

$$\prod_{i=k}^{n} \lambda_i(A) \le \prod_{i=k}^{n} \mathrm{d}_i(A).$$

In particular,

$$\det A \le \prod_{i=1}^{n} A_{(i,i)}.$$

Furthermore, equality holds if and only if either $\prod_{i=1}^{n} A_{(i,i)} = 0$ or A is diagonal. **Source:** [1124, pp. 21–24], [1448, pp. 200, 477], [2977, p. 18], and [2991, pp. 218, 355]. **Remark:** The case $k = 1$ is *Hadamard's inequality.* **Remark:** A geometric interpretation is given in [1138]. **Related:** Fact 10.15.10 and Fact 11.13.1. A refinement is given by Fact 10.15.7.

Fact 10.21.16. Let $A \in \mathbb{F}^{n\times n}$, and assume that A is positive definite. Then, for all $k \in \{1, \ldots, n\}$,

$$\sum_{i=k}^{n} \frac{1}{\mathrm{d}_i(A)} < \sum_{i=k}^{n} \frac{1}{\lambda_i(A)}.$$

Source: [2991, p. 355].

Fact 10.21.17. Let $A \in \mathbb{F}^{n\times n}$ and $\alpha > 0$. Then, for all $k \in \{1, \ldots, n\}$,

$$\prod_{i=1}^{k} [1 + \alpha\rho_i(A)] \le \prod_{i=1}^{k} [1 + \alpha\sigma_i(A)].$$

Source: [970, p. 222].

Fact 10.21.18. Let $A \in \mathbb{F}^{n\times n}$, define $H \triangleq \frac{1}{2}(A + A^*)$ and $S \triangleq \frac{1}{2}(A - A^*)$, and assume that H is positive definite. Then, the following statements hold:

i) A is nonsingular.

ii) $\frac{1}{2}(A^{-1} + A^{-*}) = (H + S^*H^{-1}S)^{-1}$.

iii) $\sigma_{\max}(A^{-1}) \le \sigma_{\max}(H^{-1})$.

iv) $\sigma_{\max}(A) \le \sigma_{\max}(H + S^*H^{-1}S)$.

Source: [1987]. **Related:** Fact 7.20.1, Fact 10.10.36, and Fact 10.15.2.

Fact 10.21.19. Let $A \in \mathbb{F}^{n\times n}$, and assume that A is Hermitian. Then, A is diagonal if and only if $\operatorname{mspec}(A) = \{A_{(1,1)}, \dots, A_{(n,n)}\}_{\rm ms}$. **Source:** Use Fact 10.21.15 with $A + \beta I > 0$.

Fact 10.21.20. Let $A \in \mathbb{F}^{n\times n}$, and assume that A is Hermitian. Then, for all $k \in \{1, \dots, n\}$,

$$\sum_{i=n+1-k}^{n} \lambda_i(A) = \min\{\operatorname{tr} S^*AS\colon\ S \in \mathbb{F}^{n\times k} \text{ and } S^*S = I_k\}$$
$$\le \max\{\operatorname{tr} S^*AS\colon\ S \in \mathbb{F}^{n\times k} \text{ and } S^*S = I_k\} = \sum_{i=1}^{k} \lambda_i(A).$$

Now, assume that A is positive definite. Then,

$$\sum_{i=1}^{k} [\lambda_i(A)]^{-1} = \min\{\operatorname{tr}(S^*AS)^{-1}\colon\ S \in \mathbb{F}^{n\times k} \text{ and } S^*S = I_k\}$$
$$\le \max\{\operatorname{tr}(S^*AS)^{-1}\colon\ S \in \mathbb{F}^{n\times k} \text{ and } S^*S = I_k\} = \sum_{i=n+1-k}^{n} \frac{1}{\lambda_i(A)}.$$

Source: [1448, p. 191] and [2991, p. 273]. **Remark:** This is the *minimum principle*.

Fact 10.21.21. Let $A \in \mathbb{F}^{n\times n}$, assume that A is Hermitian, and let $S \in \mathbb{F}^{n\times k}$ satisfy $S^*S = I_k$. Then, for all $i \in \{1, \dots, k\}$,

$$\lambda_{i+n-k}(A) \le \lambda_i(S^*AS) \le \lambda_i(A).$$

Consequently,

$$\sum_{i=1}^{k} \lambda_{i+n-k}(A) \le \operatorname{tr} S^*AS \le \sum_{i=1}^{k} \lambda_i(A), \quad \prod_{i=1}^{k} \lambda_{i+n-k}(A) \le \det S^*AS \le \prod_{i=1}^{k} \lambda_i(A).$$

Source: [1448, p. 190], [2403, p. 111], and [2991, p. 273]. **Remark:** This is the *Poincaré separation theorem*.

Fact 10.21.22. Let $A \in \mathbb{F}^{n\times n}$, and let $S \in \mathbb{F}^{n\times k}$ satisfy $S^*S = I_k$. Then, for all $i \in \{1, \dots, k\}$,

$$\sigma_i(S^*AS) \le \sigma_i(A).$$

Source: [2991, p. 273].

Fact 10.21.23. Let $n \ge 2$, let $A \in \mathbb{C}^{n\times n}$, and assume that A is positive definite. Then,

$$\sigma_{\max}(A^{\mathsf{A}}) \le \left(\frac{\operatorname{tr} A}{n-1}\right)^{n-1}.$$

Source: [1172, p. 29]. **Related:** Fact 11.9.15, Fact 11.15.16, and Fact 11.15.17.

10.22 Facts on Eigenvalues and Singular Values for Two or More Matrices

Fact 10.22.1. Let $A \in \mathbb{F}^n$, assume that A is Hermitian, and let $x \in \mathbb{F}^n$. Then, for all $i \in \{2, \dots, n-2\}$,

$$\lambda_{i+2}(A) \le \lambda_{i+1}(A + xx^*) \le \lambda_i(A) \le \lambda_i(A + xx^*).$$

Furthermore,

$$[\lambda_1(A + xx^*)\ \cdots\ \lambda_n(A + xx^*)] \overset{\rm s}{\prec} [\lambda_1(A) + x^*x\ \ \lambda_2(A)\ \cdots\ \lambda_n(A)].$$

Source: [2991, pp. 279, 361].

Fact 10.22.2. Let $A, B \in \mathbb{F}^n$, and assume that A and B are Hermitian. Then, for all $i \in \{1, \ldots, n\}$,

$$|\lambda_i(A) - \lambda_i(B)| \le \sigma_{\max}(A - B).$$

Source: [2979, p. 54].

Fact 10.22.3. Let $A, B \in \mathbb{F}^n$, and assume that A and B are Hermitian. Then,

$$\lambda(A) + [\lambda(B)]^{\uparrow} \overset{\mathrm{s}}{\prec} \lambda(A+B) \overset{\mathrm{s}}{\prec} \lambda(A) + \lambda(B),$$

$$2\lambda(A) \overset{\mathrm{s}}{\prec} \lambda(A+B) + \lambda(A-B), \quad \begin{bmatrix} \lambda[\frac{1}{2}(A+B)] \\ \lambda[\frac{1}{2}(A+B)] \end{bmatrix} \overset{\mathrm{s}}{\prec} \begin{bmatrix} \lambda(A) \\ \lambda(B) \end{bmatrix}.$$

Now, let $k \ge 1$. Then,

$$\lambda[A^2 + (BA)^k(AB)^k] \overset{\mathrm{s}}{\prec} \lambda[A^2 + (AB)^k(BA)^k].$$

Source: [449, p. 71] and [450, 843, 1850, 2750]. **Related:** Corollary 10.6.19.

Fact 10.22.4. Let $A, B \in \mathbb{F}^{n\times n}$, and assume that A and B are Hermitian. Then, for all $k \in \{1, \ldots, n\}$,

$$\sum_{i=1}^{k} \lambda_i(A) + \sum_{i=n-k+1}^{n} \lambda_i(B) \le \sum_{i=1}^{k} \lambda_i(A+B) \le \sum_{i=1}^{k} [\lambda_i(A) + \lambda_i(B)],$$

$$\sum_{i=1}^{k} [\lambda_i(A) - \lambda_i(B)] \le \sum_{i=1}^{k} \lambda_i(A-B) \le \sum_{i=1}^{k} \lambda_i(A) - \sum_{i=n-k+1}^{n} \lambda_i(B).$$

Source: Corollary 10.6.19. **Related:** Fact 15.17.6.

Fact 10.22.5. Let $A, B \in \mathbb{F}^{n\times n}$, assume that A and B are Hermitian, let $k \in \{1, \ldots, n\}$, and let $1 \le i_1 < \cdots < i_k \le n$. Then,

$$\sum_{j=1}^{k} \lambda_{i_j}(A) + \sum_{j=n-k+1}^{n} \lambda_j(B) \le \sum_{j=1}^{k} \lambda_{i_j}(A+B) \le \sum_{j=1}^{k} [\lambda_{i_j}(A) + \lambda_j(B)].$$

Source: [449, p. 69], [2403, pp. 115, 116], [2991, pp. 281–284], and Fact 10.22.3. **Remark:** The spectrum of a sum of Hermitian matrices is given by a subsequently confirmed conjecture from [1447]. See [450, 843, 1363, 1644, 1653, 1804, 1816].

Fact 10.22.6. Let $A, B \in \mathbb{F}^{n\times n}$, and assume that A and B are positive semidefinite. Then,

$$\begin{bmatrix} \lambda(A) \\ \lambda(B) \end{bmatrix} \overset{\mathrm{s}}{\prec} \begin{bmatrix} \lambda(A+B) \\ 0 \end{bmatrix}, \quad \begin{bmatrix} \lambda(A) & 0 \\ 0 & \lambda(B) \end{bmatrix} \overset{\mathrm{s}}{\prec} \begin{bmatrix} \lambda(A+B) \\ 0 \end{bmatrix}.$$

Source: [547], [1969, p. 242], [1971, p. 330], [2750], and [2991, p. 362].

Fact 10.22.7. Let $A, B \in \mathbb{F}^{n\times n}$, assume that A and B are positive semidefinite, and let $r \in \mathbb{R}$. If either $r \ge 1$ or both $r \le 0$ and A and B are positive definite, then

$$[(\lambda(A) + \lambda^{\uparrow}(B)]^{\odot r} \overset{\mathrm{w}}{\prec} [\lambda(A+B)]^{\odot r} \overset{\mathrm{w}}{\prec} [\lambda(A+B)]^{\odot r},$$

$$\begin{bmatrix} (\lambda[\frac{1}{2}(A+B)])^{\odot r} \\ (\lambda[\frac{1}{2}(A+B)])^{\odot r} \end{bmatrix} \overset{\mathrm{w}}{\prec} \begin{bmatrix} [\lambda(A)]^{\odot r} \\ [\lambda(B)]^{\odot r} \end{bmatrix} \overset{\mathrm{w}}{\prec} \begin{bmatrix} [\lambda(A+B)]^{\odot r} \\ 0 \end{bmatrix}.$$

If $r \in (0, 1)$, then

$$[\lambda(A+B)]^{\odot r} \overset{\mathrm{w}}{\prec} [\lambda(A+B)]^{\odot r} \overset{\mathrm{w}}{\prec} [(\lambda(A) + \lambda^{\uparrow}(B)]^{\odot r},$$

$$\begin{bmatrix} [\lambda(A+B)]^{\odot r} \\ 0 \end{bmatrix} \overset{\mathrm{w}}{\prec} \begin{bmatrix} [\lambda(A)]^{\odot r} \\ [\lambda(B)]^{\odot r} \end{bmatrix} \overset{\mathrm{w}}{\prec} \begin{bmatrix} (\lambda[\frac{1}{2}(A+B)])^{\odot r} \\ (\lambda[\frac{1}{2}(A+B)])^{\odot r} \end{bmatrix}.$$

Source: [2750].

Fact 10.22.8. Let $A, B \in \mathbb{F}^{n\times n}$. Then,

$$2\lambda(AA^* + BB^*) \overset{\text{s}}{\prec} \lambda(A^*A + B^*B - A^*B - B^*A) + \lambda(A^*A + B^*B + A^*B + B^*A).$$

Source: [2750].

Fact 10.22.9. Let $A, B \in \mathbb{F}^{n\times m}$, and assume that A^*B is either Hermitian or skew Hermitian. If $n < m$, then

$$\begin{bmatrix} \lambda(AA^* + BB^*) \\ 0 \end{bmatrix} \overset{\text{s}}{\prec} \lambda(A^*A + B^*B).$$

If $m = n$, then

$$\lambda(AA^* + BB^*) \overset{\text{s}}{\prec} \lambda(A^*A + B^*B).$$

If $m < n$, then

$$\lambda(AA^* + BB^*) \overset{\text{s}}{\prec} \begin{bmatrix} \lambda(A^*A + B^*B) \\ 0 \end{bmatrix}.$$

Source: [1850, 2750].

Fact 10.22.10. Let $A, B \in \mathbb{F}^n$, assume that A and B are positive definite.

$$\begin{aligned}&[\log\lambda_1(A) \ \cdots \ \log\lambda_n(A)] + [\log\lambda_n(B) \ \cdots \ \log\lambda_1(B)] \\ &\quad \overset{\text{s}}{\prec} [\log\lambda_1(AB) \ \cdots \ \log\lambda_n(AB)] \overset{\text{s}}{\prec} [\log\lambda_1(A) + \log\lambda_1(B) \ \cdots \ \log\lambda_n(A) + \log\lambda_n(B)].\end{aligned}$$

Source: [100].

Fact 10.22.11. Let $f\colon \mathbb{R} \mapsto \mathbb{R}$ be convex, define $f\colon \mathbf{H}^n \mapsto \mathbf{H}^n$ by (10.5.2), let $A, B \in \mathbb{F}^{n\times n}$, and assume that A and B are Hermitian, and let $\alpha \in [0, 1]$. Then,

$$\begin{aligned}&\begin{bmatrix}\lambda_1[f(\alpha A + (1-\alpha)B)] & \cdots & \lambda_n[f(\alpha A + (1-\alpha)B)]\end{bmatrix} \\ &\quad \overset{\text{w}}{\prec} \begin{bmatrix}\alpha\lambda_1[f(A)] + (1-\alpha)\lambda_1[f(B)] & \cdots & \alpha\lambda_n[f(A)] + (1-\alpha)\lambda_n[f(B)]\end{bmatrix}.\end{aligned}$$

If, in addition, f is either nonincreasing or nondecreasing, then, for all $i \in \{1, \dots, n\}$,

$$\lambda_i[f(\alpha A + (1-\alpha)B)] \le \lambda_i[\alpha f(A) + (1-\alpha)f(B)].$$

Source: [197]. **Remark:** Under the assumptions of the last statement, the inequality $\lambda_i[f(\alpha A + (1-\alpha)B)] \le \alpha\lambda_i[f(A)] + (1-\alpha)\lambda_i[f(B)]$ may not hold as shown by $A = \left[\begin{smallmatrix}2 & 0\\ 0 & 1\end{smallmatrix}\right]$ and $B = \left[\begin{smallmatrix}0 & 0\\ 0 & 1\end{smallmatrix}\right]$. **Remark:** Convexity of $f\colon \mathbb{R} \mapsto \mathbb{R}$ does not imply convexity of $f\colon \mathbf{H}^n \mapsto \mathbf{H}^n$.

Fact 10.22.12. Let $A, B \in \mathbb{F}^{n\times n}$, and assume that A and B are positive semidefinite. If $r \in [0, 1]$, then

$$\begin{bmatrix}\lambda_1[(A+B)^r] & \cdots & \lambda_n[(A+B)^r]\end{bmatrix} \overset{\text{w}}{\prec} \begin{bmatrix}\lambda_1(A^r + B^r) & \cdots & \lambda_n(A^r + B^r)\end{bmatrix},$$

and, for all $i \in \{1, \dots, n\}$,

$$2^{1-r}\lambda_i[(A+B)^r] \le \lambda_i(A^r + B^r).$$

If $r \ge 1$, then

$$\begin{bmatrix}\lambda_1(A^r + B^r) & \cdots & \lambda_n(A^r + B^r)\end{bmatrix} \overset{\text{w}}{\prec} \begin{bmatrix}\lambda_1[(A+B)^r] & \cdots & \lambda_n[(A+B)^r]\end{bmatrix},$$

and, for all $i \in \{1, \dots, n\}$,

$$\lambda_i(A^r + B^r) \le 2^{r-1}\lambda_i[(A+B)^r].$$

Source: This result follows from Fact 10.22.11. See [106, 195, 197].

Fact 10.22.13. Let $A, B \in \mathbb{F}^{n\times n}$, and assume that A and B are positive semidefinite. Then, for all $k \in \{1, \ldots, n\}$,

$$\prod_{i=1}^{k} |\lambda_i(A - B)| \le \prod_{i=1}^{k} \lambda_i(A + B).$$

Source: [2991, p. 362].

Fact 10.22.14. Let $A, B \in \mathbb{F}^{n\times n}$, and assume that A and B are Hermitian. Then,

$$[\sigma_1^2(A) + \sigma_n^2(B) \ \cdots \ \sigma_n^2(A) + \sigma_1^2(B)] \overset{\text{s}}{\prec} [\sigma_1^2(A + \jmath B) \ \cdots \ \sigma_n^2(A + \jmath B)],$$

$$\tfrac{1}{2}[\sigma_1^2(A + \jmath B) + \sigma_n^2(A + \jmath B) \ \cdots \ \sigma_n^2(A + \jmath B) + \sigma_1^2(A + \jmath B)] \overset{\text{s}}{\prec} [\sigma_1^2(A) + \sigma_1^2(B) \ \cdots \ \sigma_n^2(A) + \sigma_n^2(B)].$$

Now, assume that A and B are positive semidefinite. Then,

$$[\sigma_1^2(A + \jmath B) \ \cdots \ \sigma_n^2(A + \jmath B)] \overset{\text{s}}{\prec} [\sigma_1^2(A) + \sigma_1^2(B) \ \cdots \ \sigma_n^2(A) + \sigma_n^2(B)].$$

Furthermore, for all $k \in \{1, \ldots, n\}$,

$$\prod_{i=k}^{n} |\sigma_i(A) + \jmath\sigma_i(B)| \le \prod_{i=k}^{n} |\sigma_i(A + \jmath B)|.$$

Source: [100, 714] and [2979, pp. 97, 98]. **Related:** Fact 10.16.5 and Fact 11.10.69.

Fact 10.22.15. Let $A, B \in \mathbb{F}^{n\times n}$, and assume that A and B are positive semidefinite. Then, the following statements hold:

i) If $p \in [0, 1]$, then $\sigma_{\max}(A^p - B^p) \le \sigma_{\max}^p(A - B)$.

ii) If $p \ge \sqrt{2}$, then $\sigma_{\max}(A^p - B^p) \le p[\max\{\sigma_{\max}(A), \sigma_{\max}(B)\}]^{p-1}\sigma_{\max}(A - B)$.

iii) If a and b are positive numbers such that $aI \le A \le bI$ and $aI \le B \le bI$, then

$$\sigma_{\max}(A^p - B^p) \le b[b^{p-2} + (p - 1)a^{p-2}]\sigma_{\max}(A - B).$$

Source: [463, 1627].

Fact 10.22.16. Let $A, B \in \mathbb{F}^{n\times n}$, and assume that A and B are positive semidefinite. Then, for all $i \in \{1, \ldots, n\}$,

$$\sigma_i(A - B) \le \sigma_i\left(\begin{bmatrix} A & 0 \\ 0 & B \end{bmatrix}\right).$$

Source: [2587, 2975] and [2991, p. 362].

Fact 10.22.17. Let $A, B \in \mathbb{F}^{n\times n}$. Then,

$$\max\{\sigma_{\max}^2(A), \sigma_{\max}^2(B)\} - \sigma_{\max}(AB) \le \sigma_{\max}(A^*A - BB^*),$$

$$\sigma_{\max}(A^*A - BB^*) \le \max\{\sigma_{\max}^2(A), \sigma_{\max}^2(B)\} - \min\{\sigma_{\min}^2(A), \sigma_{\min}^2(B)\}.$$

Furthermore,

$$\max\{\sigma_{\max}^2(A), \sigma_{\max}^2(B)\} + \min\{\sigma_{\min}^2(A), \sigma_{\min}^2(B)\} \le \sigma_{\max}(A^*A + BB^*),$$

$$\sigma_{\max}(A^*A + BB^*) \le \max\{\sigma_{\max}^2(A), \sigma_{\max}^2(B)\} + \sigma_{\max}(AB).$$

Now, assume that A and B are positive semidefinite. Then,

$$\max\{\lambda_{\max}(A), \lambda_{\max}(B)\} - \sigma_{\max}(A^{1/2}B^{1/2}) \le \sigma_{\max}(A - B),$$

$$\sigma_{\max}(A - B) \le \max\{\lambda_{\max}(A), \lambda_{\max}(B)\} - \min\{\lambda_{\min}(A), \lambda_{\min}(B)\}.$$

Furthermore,

$$\max\{\lambda_{\max}(A), \lambda_{\max}(B)\} + \min\{\lambda_{\min}(A), \lambda_{\min}(B)\} \le \lambda_{\max}(A + B),$$

$$\lambda_{\max}(A+B) \le \max\{\lambda_{\max}(A), \lambda_{\max}(B)\} + \sigma_{\max}(A^{1/2}B^{1/2}).$$

Source: [1635, 2978]. **Related:** Fact 10.22.19 and Fact 11.15.10.

Fact 10.22.18. Let $A, B \in \mathbb{F}^{n\times n}$, and assume that A and B are positive semidefinite, and let $k \ge 1$. Then, for all $i \in \{1,\dots,n\}$,

$$2\sigma_i[A^{1/2}(A+B)^{k-1}B^{1/2}] \le \lambda_i[(A+B)^k].$$

Hence,

$$2\sigma_{\max}(A^{1/2}B^{1/2}) \le \lambda_{\max}(A+B), \quad \sigma_{\max}(A^{1/2}B^{1/2}) \le \max\{\lambda_{\max}(A), \lambda_{\max}(B)\}.$$

Source: Fact 10.22.17 and Fact 11.10.45.

Fact 10.22.19. Let $A, B \in \mathbb{F}^{n\times n}$, and assume that A and B are positive semidefinite. Then,

$$\begin{aligned}
&\max\{\lambda_{\max}(A), \lambda_{\max}(B)\} - \sigma_{\max}(A^{1/2}B^{1/2}) \\
&\quad \le \sigma_{\max}(A-B) \le \max\{\lambda_{\max}(A), \lambda_{\max}(B)\} \le \lambda_{\max}(A+B) \\
&\quad \le \frac{1}{2}\left[\lambda_{\max}(A) + \lambda_{\max}(B) + \sqrt{[\lambda_{\max}(A) - \lambda_{\max}(B)]^2 + 4\sigma_{\max}^2(A^{1/2}B^{1/2})}\right] \\
&\quad \le \left\{\begin{array}{c} \max\{\lambda_{\max}(A), \lambda_{\max}(B)\} + \sigma_{\max}(A^{1/2}B^{1/2}) \\ \lambda_{\max}(A) + \lambda_{\max}(B) \end{array}\right\} \le 2\max\{\lambda_{\max}(A), \lambda_{\max}(B)\}.
\end{aligned}$$

Furthermore,

$$\lambda_{\max}(A+B) = \lambda_{\max}(A) + \lambda_{\max}(B)$$

if and only if

$$\sigma_{\max}(A^{1/2}B^{1/2}) = \lambda_{\max}^{1/2}(A)\lambda_{\max}^{1/2}(B).$$

Source: [1629, 1632, 1635] and Fact 10.22.18. **Related:** Fact 10.22.17, Fact 10.22.19, Fact 11.10.30, Fact 11.10.78, and Fact 11.16.18.

Fact 10.22.20. Let $A, B \in \mathbb{F}^{n\times n}$, assume that A and B are positive semidefinite, and let $z \in \mathbb{C}$. Then, for all $k \in \{1,\dots,n\}$,

$$\prod_{i=1}^k \sigma_i(A-|z|B) \le \prod_{i=1}^k \sigma_i(A+zB) \le \prod_{i=1}^k \sigma_i(A+|z|B),$$

$$\sum_{i=1}^k \sigma_i(A-|z|B) \le \sum_{i=1}^k \sigma_i(A+zB) \le \sum_{i=1}^k \sigma_i(A+|z|B).$$

Source: [2975] and [2991, pp. 357, 358]. **Related:** Fact 11.10.24.

Fact 10.22.21. Let $A, B \in \mathbb{F}^{n\times n}$, and assume that A and B are positive semidefinite. Then,

$$\operatorname{tr} AB \le \operatorname{tr}(AB^2A)^{1/2} \le \tfrac{1}{4}\operatorname{tr}(A+B)^2, \quad \operatorname{tr}(AB)^2 \le \operatorname{tr} A^2B^2 \le \tfrac{1}{16}\operatorname{tr}(A+B)^4,$$

$$\begin{aligned}
&\sigma_{\max}(AB) \le \tfrac{1}{4}\sigma_{\max}[(A+B)^2] \\
&\quad \le \left\{\begin{array}{c} \tfrac{1}{2}\sigma_{\max}(A^2+B^2) \le \tfrac{1}{2}\sigma_{\max}(A^2) + \tfrac{1}{2}\sigma_{\max}(B^2) \\ \tfrac{1}{4}\sigma_{\max}^2(A+B) \le \tfrac{1}{4}[\sigma_{\max}(A) + \sigma_{\max}(B)]^2 \end{array}\right\} \le \tfrac{1}{2}\sigma_{\max}^2(A) + \tfrac{1}{2}\sigma_{\max}^2(B).
\end{aligned}$$

Source: Fact 11.10.45. The inequalities $\operatorname{tr} AB \le \operatorname{tr}(AB^2A)^{1/2}$ and $\operatorname{tr}(AB)^2 \le \operatorname{tr} A^2B^2$ follow from Fact 10.14.24. **Related:** Fact 10.22.22.

Fact 10.22.22. Let $A, B \in \mathbb{F}^{n\times n}$, assume that A and B are positive semidefinite, and let $k \geq 1$. Then, for all $i \in \{1,\dots,n\}$,

$$\lambda_i[(AB)^k] \leq \tfrac{1}{4}\lambda_i[(A+B)^{2k}], \quad \lambda_i[(A^2B^2)^k] \leq \tfrac{1}{16^k}\lambda_i[(A+B)^{4k}].$$

Therefore,

$$\operatorname{tr}(AB)^k \leq \tfrac{1}{4}\operatorname{tr}(A+B)^{2k}, \quad \operatorname{tr}(A^2B^2)^k \leq \tfrac{1}{16^k}\operatorname{tr}(A+B)^{4k}.$$

Source: [471, 2935]. **Related:** Fact 10.22.21.

Fact 10.22.23. Let $A, B \in \mathbb{F}^{n\times n}$, and assume that A and B are positive semidefinite. Then, for all $i, j \in \{0,1,\dots,n-1\}$ such that $i+j \leq n-1$,

$$\lambda_{n-i}(A)\lambda_{n-j}(B) \leq \lambda_{n-i-j}(AB),$$

and, for all $i, j \in \{1,\dots,n\}$ such that $i+j \leq n+1$,

$$\lambda_{i+j-1}(AB) \leq \lambda_i(A)\lambda_j(B).$$

Therefore, for all $i \in \{1,\dots,n\}$,

$$\lambda_i(A)\lambda_n(B) \leq \lambda_i(AB) \leq \lambda_i(A)\lambda_1(B).$$

In particular,

$$\lambda_{\min}(A)\lambda_{\min}(B) \leq \lambda_{\min}(AB) \leq \lambda_{\min}(A)\lambda_{\max}(B).$$

Source: [2403, pp. 126, 127], [2991, pp. 277, 278, 280]. **Related:** Fact 10.22.28 and Fact 10.22.36.

Fact 10.22.24. Let $A, B \in \mathbb{F}^{n\times n}$, and assume that A and B are positive definite. Then, for all $i \in \{1,\dots,n\}$,

$$\frac{\lambda_i^2(AB)}{\lambda_1(A)\lambda_1(B)} \leq \lambda_i(A)\lambda_i(B) \leq \frac{\lambda_i^2(AB)}{\lambda_n(A)\lambda_n(B)}.$$

Source: [2403, p. 137].

Fact 10.22.25. Let $A, B \in \mathbb{F}^{n\times n}$, assume that A is positive semidefinite, and assume that B is Hermitian. Then, for all $k \in \{1,\dots,n\}$,

$$\sum_{i=1}^k \lambda_i(A)\lambda_{n-i+1}(B) \leq \sum_{i=1}^k \lambda_i(AB), \quad \sum_{i=1}^k \lambda_{n-i+1}(AB) \leq \sum_{i=1}^k \lambda_i(A)\lambda_i(B).$$

In particular,

$$\sum_{i=1}^n \lambda_i(A)\lambda_{n-i+1}(B) \leq \operatorname{tr} AB \leq \sum_{i=1}^n \lambda_i(A)\lambda_i(B).$$

Source: [1664]. **Remark:** The bounds on $\operatorname{tr} AB$ are related to Fact 2.12.8. See [453, p. 140]. **Related:** Fact 7.13.12, Fact 7.13.13, Fact 7.13.16, and Proposition 10.4.13.

Fact 10.22.26. Let $A, B \in \mathbb{F}^{n\times n}$, assume that A and B are positive semidefinite, and let $1 \leq i_1 < \cdots < i_k \leq n$. Then,

$$\sum_{j=1}^k \lambda_{i_j}(A)\lambda_{n-j+1}(B) \leq \sum_{j=1}^k \lambda_{i_j}(AB) \leq \sum_{j=1}^k \lambda_{i_j}(A)\lambda_j(B).$$

Furthermore, for all $k \in \{1,\dots,n\}$,

$$\sum_{j=1}^k \lambda_{i_j}(A)\lambda_{n-i_j+1}(B) \leq \sum_{j=1}^k \lambda_j(AB).$$

In particular, for all $k \in \{1,\dots,n\}$,

$$\sum_{i=1}^{k} \lambda_i(A)\lambda_{n-i+1}(B) \le \sum_{i=1}^{k} \lambda_i(AB) \le \sum_{i=1}^{k} \lambda_i(A)\lambda_i(B).$$

Therefore,

$$\sum_{i=1}^{n} \lambda_i(A)\lambda_{n-i+1}(B) \le \operatorname{tr}(AB) \le \sum_{i=1}^{n} \lambda_i(A)\lambda_i(B).$$

Source: [2812]. The left-hand inequality in the penultimate string is given in [2403, p. 128]. **Related:** Fact 10.22.29 and Fact 11.16.36.

Fact 10.22.27. Let $A, B \in \mathbb{F}^{n\times n}$, and assume that A and B are positive semidefinite. If $0 \le p \le 1$, then

$$\sum_{i=1}^{n} \lambda_i^p(A)\lambda_{n-i+1}^p(B) \le \operatorname{tr} A^pB^p \le \operatorname{tr}(B^{1/2}AB^{1/2})^p \le \sum_{i=1}^{n} \lambda_i^p(A)\lambda_i^p(B).$$

If $p \ge 1$, then

$$\sum_{i=1}^{n} \lambda_i^p(A)\lambda_{n-i+1}^p(B) \le \operatorname{tr}(B^{1/2}AB^{1/2})^p \le \operatorname{tr} A^pB^p \le \sum_{i=1}^{n} \lambda_i^p(A)\lambda_i^p(B).$$

Now, suppose that A and B are positive definite. If $p \le -1$, then

$$\sum_{i=1}^{n} \lambda_i^p(A)\lambda_{n-i+1}^p(B) \le \operatorname{tr}(B^{1/2}AB^{1/2})^p \le \operatorname{tr} A^pB^p \le \sum_{i=1}^{n} \lambda_i^p(A)\lambda_i^p(B).$$

If $-1 \le p \le 0$, then

$$\sum_{i=1}^{n} \lambda_i^p(A)\lambda_{n-i+1}^p(B) \le \operatorname{tr} A^pB^p \le \operatorname{tr}(B^{1/2}AB^{1/2})^p \le \sum_{i=1}^{n} \lambda_i^p(A)\lambda_i^p(B).$$

Source: [2813]. See also [609, 1774, 1821, 2817]. **Related:** Fact 10.14.24. See Fact 10.14.15 for the indefinite case.

Fact 10.22.28. Let $A, B \in \mathbb{F}^{n\times n}$, and assume that A and B are positive semidefinite. Then, for all $k \in \{1,\dots,n\}$,

$$\prod_{i=1}^{k} \lambda_i(AB) \le \prod_{i=1}^{k} \sigma_i(AB) \le \prod_{i=1}^{k} \lambda_i(A)\lambda_i(B)$$

with equality for $k = n$. Furthermore, for all $k \in \{1,\dots,n\}$,

$$\prod_{i=k}^{n} \lambda_i(A)\lambda_i(B) \le \prod_{i=k}^{n} \sigma_i(AB) \le \prod_{i=k}^{n} \lambda_i(AB)$$

with equality for $k = 1$. In particular,

$$\lambda_{\max}(AB) \le \sigma_{\max}(AB) \le \lambda_{\max}(A)\lambda_{\max}(B),$$
$$\lambda_{\min}(A)\lambda_{\min}(B) \le \sigma_{\min}(AB) \le \lambda_{\min}(AB).$$

Source: Fact 7.12.32 and Fact 11.15.20. **Remark:** The last string and Fact 10.22.23 imply that

$$\lambda_{\min}(A)\lambda_{\min}(B) \le \sigma_{\min}(AB) \le \lambda_{\min}(AB) \le \lambda_{\min}(A)\lambda_{\max}(B).$$

Fact 10.22.29. Let $A, B \in \mathbb{F}^{n\times n}$, assume that A and B are positive semidefinite, and let $1 \le i_1 < \cdots < i_k \le n$. Then,

$$\prod_{j=1}^{k} \lambda_{i_j}(AB) \le \prod_{j=1}^{k} \lambda_{i_j}(A)\lambda_j(B)$$

with equality for $k = n$. Furthermore,

$$\prod_{j=1}^{k} \lambda_{i_j}(A)\lambda_{n-i_j+1}(B) \le \prod_{j=1}^{k} \lambda_j(AB)$$

with equality for $k = n$. In particular,

$$\prod_{i=1}^{k} \lambda_i(A)\lambda_{n-i+1}(B) \le \prod_{i=1}^{k} \lambda_i(AB) \le \prod_{i=1}^{k} \lambda_i(A)\lambda_i(B)$$

with equality for $k = n$. **Source:** [2812] and [2991, p. 363]. The first inequality is given in [2403, p. 127]. **Credit:** V. B. Lidskii. **Related:** Fact 10.22.26 and Fact 11.16.36.

Fact 10.22.30. Let $A, B \in \mathbb{F}^{n\times n}$, assume that A and B are positive semidefinite, and let $k \in \{1, \dots, n\}$. Then,

$$\prod_{i=k}^{n}[\lambda_i(A) + 1] + \prod_{i=k}^{n}[\lambda_i(B) + 1] \le 1 + \prod_{i=k}^{n}[\lambda_i(A + B) + 1].$$

Source: [2681].

Fact 10.22.31. Let $A, B \in \mathbb{F}^{n\times n}$, assume that A and B are positive definite, and let $\lambda \in \operatorname{spec}(AB)$. Then,

$$\frac{2}{n}\left[\frac{\lambda^2_{\min}(A)\lambda^2_{\min}(B)}{\lambda^2_{\min}(A) + \lambda^2_{\min}(B)}\right] < \lambda < \frac{n}{2}[\lambda^2_{\max}(A) + \lambda^2_{\max}(B)].$$

Source: [1477].

Fact 10.22.32. Let $A, B \in \mathbb{F}^{n\times n}$, assume that A and B are positive definite, and define

$$k_A \triangleq \frac{\lambda_{\max}(A)}{\lambda_{\min}(A)}, \quad k_B \triangleq \frac{\lambda_{\max}(B)}{\lambda_{\min}(B)}, \quad \gamma \triangleq \frac{(\sqrt{k_A} + 1)^2}{\sqrt{k_A}} - \frac{k_B(\sqrt{k_A} - 1)^2}{\sqrt{k_A}}.$$

If $\gamma < 0$, then

$$\left.\begin{matrix} \tfrac{1}{2}\lambda_{\max}(A)\lambda_{\max}(B)\gamma \\ -\tfrac{1}{4}\lambda_{\max}(A)\lambda_{\max}(B) \end{matrix}\right\} \le \lambda_{\min}(AB + BA) \le \lambda_{\max}(AB + BA) \le 2\lambda_{\max}(A)\lambda_{\max}(B),$$

whereas, if $\gamma > 0$, then

$$\tfrac{1}{2}\lambda_{\min}(A)\lambda_{\min}(B)\gamma \le \lambda_{\min}(AB + BA) \le \lambda_{\max}(AB + BA) \le 2\lambda_{\max}(A)\lambda_{\max}(B).$$

Furthermore, if $\sqrt{k_A k_B} < 1 + \sqrt{k_A} + \sqrt{k_B}$, then $AB + BA$ is positive definite. **Source:** [2127] and [2991, pp. 207, 208]. **Related:** Fact 7.13.4, Fact 10.11.10, and Fact 10.11.24.

Fact 10.22.33. Let $A, B \in \mathbb{F}^{n\times n}$, assume that A is positive definite, assume that B is positive semidefinite, and let $\alpha > 0$ and $\beta > 0$ satisfy $\alpha I \le A \le \beta I$. Then,

$$\sigma_{\max}(AB) \le \frac{\alpha + \beta}{2\sqrt{\alpha\beta}}\rho_{\max}(AB) \le \frac{\alpha + \beta}{2\sqrt{\alpha\beta}}\sigma_{\max}(AB).$$

In particular,

$$\sigma_{\max}(A) \le \frac{\alpha + \beta}{2\sqrt{\alpha\beta}}\rho_{\max}(A) \le \frac{\alpha + \beta}{2\sqrt{\alpha\beta}}\sigma_{\max}(A).$$

Source: [2686]. **Remark:** The left-hand inequality is tightest for $\alpha = \lambda_{\min}(A)$ and $\beta = \lambda_{\max}(A)$. **Credit:** J.-C. Bourin.

Fact 10.22.34. Let $A, B \in \mathbb{F}^{n\times n}$, and assume that A and B are positive semidefinite. Then, for all $i \in \{1,\dots,n\}$,

$$\sqrt{\sigma_i(AB)} \le \tfrac{1}{2}\lambda_i(A+B).$$

Source: [954].

Fact 10.22.35. Let $A, B \in \mathbb{F}^{n\times n}$, and assume that A and B are positive semidefinite. Then,

$$\sigma_{\max}(A^{1/2}B^{1/2}) \le \sigma_{\max}^{1/2}(AB).$$

Equivalently,

$$\lambda_{\max}(A^{1/2}BA^{1/2}) \le \lambda_{\max}^{1/2}(AB^2A).$$

Furthermore, $AB = 0$ if and only if $A^{1/2}B^{1/2} = 0$. **Source:** [1629, 1635]. **Related:** Fact 10.22.36.

Fact 10.22.36. Let $A, B \in \mathbb{F}^{n\times n}$, and assume that A and B are positive semidefinite. Then, the following statements hold:

i) If $q \in [0,1]$, then $\sigma_{\max}(A^qB^q) \le \sigma_{\max}^q(AB)$.

ii) If $q \in [0,1]$, then $\sigma_{\max}(B^qA^qB^q) \le \sigma_{\max}^q(BAB)$.

iii) If $q \in [0,1]$, then $\lambda_{\max}(A^qB^q) \le \lambda_{\max}^q(AB)$.

iv) If $q \ge 1$, then $\sigma_{\max}^q(AB) \le \sigma_{\max}(A^qB^q)$.

v) If $q \ge 1$, then $\lambda_{\max}^q(AB) \le \lambda_{\max}(A^qB^q)$.

vi) If $p \ge q > 0$, then $\sigma_{\max}^{1/q}(A^qB^q) \le \sigma_{\max}^{1/p}(A^pB^p)$.

vii) If $p \ge q > 0$, then $\lambda_{\max}^{1/q}(A^qB^q) \le \lambda_{\max}^{1/p}(A^pB^p)$

Source: [449, pp. 255–258] and [1117]. **Remark:** *iii)* is the *Cordes inequality.* **Related:** Fact 10.11.78, Fact 10.14.24, Fact 10.22.23, Fact 10.22.35, Fact 11.10.49, and Fact 11.10.50.

Fact 10.22.37. Let $A, B \in \mathbb{F}^{n\times n}$, assume that A and B are positive semidefinite, and let $p \ge r \ge 0$. Then,

$$[\lambda_1^{1/r}(A^rB^r) \ \ \cdots \ \ \lambda_n^{1/r}(A^rB^r)] \overset{\mathrm{s}}{\prec} [\lambda_1^{1/p}(A^pB^p) \ \ \cdots \ \ \lambda_n^{1/p}(A^pB^p)].$$

Furthermore, for all $q > 0$,

$$\det (A^qB^q)^{1/q} = \det AB.$$

Source: [449, p. 257], [2977, p. 20], and Fact 3.25.15.

Fact 10.22.38. Let $A, B \in \mathbb{F}^{n\times n}$, assume that A and B are positive semidefinite, let p and r be positive integers such that $p \le r$, let $k \in \{1,\dots,n\}$, and let $1 \le i_1 < \cdots < i_k \le n$. Then,

$$\prod_{j=1}^k \lambda_{i_j}(A)\lambda_{n-i_j+1}(B) \le \prod_{i=1}^k \lambda_i^r(A^{1/r}B^{1/r}) \le \prod_{i=1}^k \lambda_i^p(A^{1/p}B^{1/p}) \le \prod_{i=1}^k \lambda_i(AB),$$

$$\prod_{i=1}^k \lambda_i(AB) \le \prod_{i=1}^k \lambda_i^{1/p}(A^pB^p) \le \prod_{i=1}^k \lambda_i^{1/r}(A^rB^r) \le \prod_{i=1}^k \lambda_i(A)\lambda_i(B).$$

Furthermore, if $k = n$, then all of the above inequalities are equalities. Finally,

$$\lambda_n(A)\lambda_n(B) \le \lambda_n^r(A^{1/r}B^{1/r}) \le \lambda_n^p(A^{1/p}B^{1/p}) \le \lambda_n(AB),$$
$$\lambda_1(AB) \le \lambda_1^p(A^{1/p}B^{1/p}) \le \lambda_1^r(A^{1/r}B^{1/r}) \le \lambda_1(A)\lambda_1(B).$$

Source: [2991, pp. 366–368].

Fact 10.22.39. Let $A, B \in \mathbb{F}^{n \times n}$, and assume that A and B are positive semidefinite. Then,

$$\sigma_{\max}[(I + A)^{-1}AB(I + B)^{-1}] \le \frac{\sigma_{\max}(AB)}{[1 + \sigma_{\max}^{1/2}(AB)]^2}.$$

Source: [2756].

Fact 10.22.40. Let $A, B, C \in \mathbb{F}^{n \times n}$, and assume that A, B, and C are positive semidefinite. Then,

$$\operatorname{spec}(CABA + CA + BA) \subset [0, \infty).$$

If, in addition, A is positive definite and either B or C is positive definite, then

$$\operatorname{spec}(CABA + CA + BA) \subset (0, \infty).$$

Source: Assume that A is positive definite and either B or C is positive definite, and define $X \triangleq A^{1/2}CA^{1/2}$ and $Y \triangleq A^{1/2}BA^{1/2}$, at least one of which is positive definite. Then,

$$\begin{aligned}\operatorname{spec}(CABA + CA + BA) &= \operatorname{spec}[A^{1/2}(CAB + C + B)A^{1/2}] = \operatorname{spec}(XY + X + Y)\\ &= \operatorname{spec}(XY + X + Y + I) - \{1\} = \operatorname{spec}[(X + I)(Y + I)] - \{1\} \subset \mathbb{R}.\end{aligned}$$

Furthermore, it follows from Fact 10.22.23 that

$$\lambda_{\min}[(X + I)(Y + I)] - 1 \ge \lambda_{\min}(X + I)\lambda_{\min}(Y + I) - 1 > 0.$$

The first result follows from continuity. **Credit:** A. Ali and M. Lin.

Fact 10.22.41. Let $A, B \in \mathbb{F}^{n \times n}$, and assume that A and B are positive definite. Then,

$$[\lambda_1(\log A^{1/2}BA^{1/2}) \ \cdots \ \lambda_n(\log A^{1/2}BA^{1/2})] \overset{\text{slog}}{\prec} [\lambda_1(\log A + \log B) \ \cdots \ \lambda_n(\log A + \log B)].$$

Consequently,

$$\log\det AB = \operatorname{tr}(\log A + \log B) = \operatorname{tr}\log A^{1/2}BA^{1/2} = \log\det A^{1/2}BA^{1/2}.$$

Source: [196].

Fact 10.22.42. Let $A, B \in \mathbb{F}^{n \times n}$, and assume that A and B are positive semidefinite. Then, the following statements hold:

i) $\sigma_{\max}[\log(I + A)\log(I + B)] \le (\log[1 + \sigma_{\max}^{1/2}(AB)])^2$.

ii) $\sigma_{\max}[\log(I + B)\log(I + A)\log(I + B)] \le (\log[1 + \sigma_{\max}^{1/3}(BAB)])^3$.

iii) $\det[\log(I + A)\log(I + B)] \le \det[\log(I + \langle AB\rangle^{1/2})]^2$.

iv) $\det[\log(I + B)\log(I + A)\log(I + B)] \le \det(\log[I + (BAB)^{1/3}])^3$.

Source: [2756]. **Related:** Fact 15.17.8.

Fact 10.22.43. Let $A, B, C \in \mathbb{F}^{n \times n}$, and assume that A, B, and C are positive definite. Then,

$$[\log\lambda_1(AC) \ \cdots \ \log\lambda_n(AC)] \overset{\text{s}}{\prec} [\log\lambda_1(AB) + \log\lambda_1(B^{-1}C) \ \cdots \ \log\lambda_n(AB) + \log\lambda_n(B^{-1}C)].$$

Source: [2991, p. 371].

10.23 Facts on Alternative Partial Orderings

Fact 10.23.1. Let $A, B \in \mathbb{F}^{n \times n}$, and assume that A and B are positive definite. Then, the following statements are equivalent:

i) $\log B \le \log A$.

ii) For all $r \in (0, \infty)$, $B^r \le (B^{r/2}A^rB^{r/2})^{1/2}$.

iii) For all $r \in (0, \infty)$, $(A^{r/2}B^rA^{r/2})^{1/2} \le A^r$.

iv) For all $p, r \in (0, \infty)$ and $k \in \mathbb{N}$ such that $(k + 1)r = p + r$, $B^r \le (B^{r/2}A^pB^{r/2})^{\frac{1}{k+1}}$.

v) For all $p, r \in (0, \infty)$ and $k \in \mathbb{N}$ such that $(k+1)r = p + r$, $(A^{r/2}B^pA^{r/2})^{\frac{1}{k+1}} \le A^r$.

vi) For all $p, r \in [0, \infty)$, $B^r \le (B^{r/2}A^pB^{r/2})^{\frac{r}{r+p}}$.

vii) For all $p, r \in [0, \infty)$, $(A^{r/2}B^pA^{r/2})^{\frac{r}{r+p}} \le A^r$.

viii) For all $p, q, r, t \in \mathbb{R}$ such that $p \ge 0$, $r \ge 0$, $t \ge 0$, and $q \in [1, 2]$,

$$[A^{r/2}(A^{t/2}B^pA^{t/2})^qA^{r/2}]^{\frac{r+t}{r+qt+qp}} \le A^{r+t}.$$

Source: [1088, 1828, 2939] and [1124, pp. 139, 200]. **Remark:** $\log B \le \log A$ is the *chaotic order*. This order is weaker than the Löwner ordering since $B \le A$ implies that $\log B \le \log A$, but not vice versa. **Remark:** Additional statements are given in [1828].

Fact 10.23.2. Let $A, B \in \mathbb{F}^{n\times n}$, assume that A is positive definite and B is positive semidefinite, and let $\alpha > 0$. Then, the following statements are equivalent:

i) $B^\alpha \le A^\alpha$.

ii) For all $p, q, r, \tau \in \mathbb{R}$ such that $p \ge \alpha$, $r \ge \tau$, $q \ge 1$, and $\tau \in [0, \alpha]$,

$$[A^{r/2}(A^{-\tau/2}B^pA^{-\tau/2})^qA^{r/2}]^{\frac{r-\tau}{r-q\tau+qp}} \le A^{r-\tau}.$$

Source: [1088].

Fact 10.23.3. Let $A, B \in \mathbb{F}^{n\times n}$, and assume that A is positive definite and B is positive semidefinite. Then, the following statements are equivalent:

i) For all $k \ge 0$, $B^k \le A^k$.

ii) For all $\alpha > 0$, $B^\alpha \le A^\alpha$.

iii) For all $p, r \in \mathbb{R}$ such that $p > r \ge 0$, $(A^{-r/2}B^pA^{-r/2})^{\frac{2p-r}{p-r}} \le A^{2p-r}$.

iv) For all $p, q, r, \tau \in \mathbb{R}$ such that $p \ge \tau$, $r \ge \tau$, $q \ge 1$, and $\tau \ge 0$,

$$[A^{r/2}(A^{-\tau/2}B^pA^{-\tau/2})^qA^{r/2}]^{\frac{r-\tau}{r-q\tau+qp}} \le A^{r-\tau}.$$

Source: [1125]. **Remark:** A and B are related by the *spectral ordering*.

Fact 10.23.4. Let $A, B \in \mathbb{F}^{n\times n}$, and assume that A and B are positive semidefinite. Then, if two of the following statements hold, then the remaining statement also holds:

i) $A \overset{\rm rs}{\le} B$.

ii) $A^2 \overset{\rm rs}{\le} B^2$.

iii) $AB = BA$.

Source: [229, 1243, 1244]. **Remark:** The rank subtractivity partial ordering is defined in Fact 4.30.3.

Fact 10.23.5. Let $A, B, C \in \mathbb{F}^{n\times n}$, and assume that A, B, and C are positive semidefinite. Then, the following statements hold:

i) If $A^2 = AB$ and $B^2 = BA$, then $A = B$.

ii) If $A^2 = AB$ and $B^2 = BC$, then $A^2 = AC$.

Source: Fact 4.30.6 and Fact 4.30.7.

Fact 10.23.6. Let $A, B \in \mathbb{F}^{n\times n}$, assume that A and B are Hermitian, and let $A \overset{*}{\le} B$ denote $A^2 = AB$. Then, "$\overset{*}{\le}$" is a partial ordering on $\mathbf{H}^{n\times n}$. **Source:** Use Fact 4.30.8 or Fact 10.23.5 to show that "$\overset{*}{\le}$" is antisymmetric and transitive. **Remark:** "$\overset{*}{\le}$" is the star partial ordering. See Fact 4.30.8.

Fact 10.23.7. Let $A, B \in \mathbb{F}^{n\times n}$, and assume that A and B are positive semidefinite. Then, the following statements are equivalent:

i) $A \overset{*}{\le} B$.

ii) $B^+ \overset{*}{\le} A^+$.

iii) $A \overset{\rm rs}{\le} B$ and $A^2 \overset{\rm rs}{\le} B^2$.

Remark: See [1255, 1332]. **Remark:** The star partial ordering is defined in Fact 10.23.6.

Fact 10.23.8. Let $A, B \in \mathbb{F}^{n\times m}$, and let $A \overset{\rm GL}{\le} B$ denote the case where the following three statements hold:

i) $\langle A\rangle \le \langle B\rangle$.

ii) $\mathcal{R}(A^*) \subseteq \mathcal{R}(B^*)$.

iii) $AB^* = \langle A\rangle\langle B\rangle$.

Then, the following statements hold:

iv) "$\overset{\rm GL}{\le}$" is a partial ordering on $\mathbb{F}^{n\times m}$.

v) If $A \overset{\rm rs}{\le} B$, then $A \overset{\rm GL}{\le} B$.

vi) Assume that A and B are positive semidefinite. Then, $A \le B$ if and only if $A \overset{\rm GL}{\le} B$.

Furthermore, the following statements are equivalent:

vii) $A \overset{\rm GL}{\le} B$.

viii) $A^* \overset{\rm GL}{\le} B^*$.

ix) $\rho_{\max}(B^+A) \le 1$, $\mathcal{R}(A) \subseteq \mathcal{R}(B)$, $\mathcal{R}(A^*) \subseteq \mathcal{R}(B^*)$, and $AB^* = \langle A\rangle\langle B\rangle$.

Source: [1347]. **Remark:** "$\overset{\rm GL}{\le}$" is the *generalized Löwner partial ordering*. The polar decomposition links the Löwner, generalized Löwner, and star partial orderings. See [1347].

10.24 Facts on Generalized Inverses

Fact 10.24.1. Let $A \in \mathbb{F}^{n\times n}$. Then, the following statements are equivalent:

i) $A + A^* \ge 0$.

ii) $A^+ + A^{+*} \ge 0$.

If, in addition, A is group invertible, then the following statement is equivalent to *i)* and *ii)*:

iii) $A^\# + A^{\#*} \ge 0$.

Source: [2713].

Fact 10.24.2. Let $A \in \mathbb{F}^{n\times n}$, and assume that A is positive semidefinite. Then, the following statements hold:

i) $A^+ = A^{\rm D} = A^\# \ge 0$.

ii) $A^{+1/2} \triangleq (A^{1/2})^+ = (A^+)^{1/2}$.

iii) $A^{1/2} = A(A^+)^{1/2} = (A^+)^{1/2}A$.

iv) $AA^+ = A^{1/2}(A^{1/2})^+$.

v) $\begin{bmatrix} A & AA^+ \\ A^+A & A^+ \end{bmatrix}$ is positive semidefinite.

vi) $A^+A + AA^+ \le A + A^+$.

vii) $A^+A \odot AA^+ \le A \odot A^+$.

Source: [2985] and Fact 10.12.9. See Fact 10.25.60 for *v)–vii)*.

Fact 10.24.3. Let $A \in \mathbb{F}^{n\times n}$, and assume that A is positive semidefinite. Then,

$$\operatorname{rank} A \le (\operatorname{tr} A)\operatorname{tr} A^+.$$

Furthermore, equality holds if and only if $\operatorname{rank} A \le 1$. **Source:** [238].

Fact 10.24.4. Let $A \in \mathbb{F}^{n\times m}$. Then, $\langle A^*\rangle = A\langle A\rangle^+A^* = A\langle A\rangle A^+$. **Source:** Fact 10.8.4 and Fact

10.8.3. **Remark:** This result shows that the Schur complement of $\langle A\rangle$ in the partitioned matrix $\mathcal{A}$ defined in Fact 10.12.27 is zero.

Fact 10.24.5. Let $A \in \mathbb{F}^{n\times m}$, and define $S \in \mathbb{F}^{n\times n}$ by $S \triangleq \langle A^*\rangle + I_n - AA^+$. Then, S is positive definite, and

$$SAA^+S = \langle A^*\rangle AA^+\langle A^*\rangle = AA^*.$$

Source: [970, p. 432]. **Remark:** This result provides a congruence transformation between AA^+ and AA^*. **Related:** Fact 7.9.21.

Fact 10.24.6. Let $A, B \in \mathbb{F}^{n\times n}$, and assume that A and B are positive semidefinite. Then, the following statements are equivalent:

i) $A = B$.

ii) $A + AA^{|} = B + BB^{|}$.

iii) $\operatorname{rank} A = \operatorname{rank} B$ and $AB^+A = B$.

iv) $\operatorname{rank} A = \operatorname{rank} B$ and $2A(A + B)^+A = A$.

v) $\mathcal{R}\left(\begin{bmatrix} A \\ B \end{bmatrix}\right) = \mathcal{R}\left(\begin{bmatrix} B \\ A \end{bmatrix}\right)$.

Source: [2625].

Fact 10.24.7. Let $A, B \in \mathbb{F}^{n\times n}$, and assume that A and B are positive semidefinite. Then, $\mathcal{R}(A) \subseteq \mathcal{R}(A + B)$ and $A = (A + B)(A + B)^+A$.

Fact 10.24.8. Let $A, B \in \mathbb{F}^{n\times n}$, and assume that A and B are Hermitian. Then, the following statements are equivalent:

i) $A \overset{\rm rs}{\le} B$.

ii) $\mathcal{R}(A) \subseteq \mathcal{R}(B)$ and $AB^+A = A$.

Source: [1243, 1244]. **Related:** Fact 8.9.39.

Fact 10.24.9. Let $A, B \in \mathbb{F}^{n\times n}$, assume that A and B are Hermitian, assume that $\nu_-(A) = \nu_-(B)$, and consider the following statements:

i) $A \overset{*}{\le} B$.

ii) $A \overset{\rm rs}{\le} B$.

iii) $A \le B$.

iv) $\mathcal{R}(A) \subseteq \mathcal{R}(B)$ and $AB^+A \le A$.

Then, *i)* $\Longrightarrow$ *ii)* $\Longrightarrow$ *iii)* $\Longleftrightarrow$ *iv)*. If A and B are positive semidefinite, then the following statement is equivalent to *iii)* and *iv)*:

v) $\mathcal{R}(A) \subseteq \mathcal{R}(B)$ and $\rho_{\max}(B^+A) \le 1$.

Source: *i)* $\Longrightarrow$ *ii)* is given in [1339]. See [229, 1243, 1255], [2418, p. 229], and [2532]. **Related:** Fact 10.24.8.

Fact 10.24.10. Let $A, B \in \mathbb{F}^{n\times n}$, and assume that A and B are positive semidefinite. Then, the following statements are equivalent:

i) $A^2 \le B^2$.

ii) $\mathcal{R}(A) \subseteq \mathcal{R}(B)$ and $\sigma_{\max}(B^+A) \le 1$.

Source: [1255].

Fact 10.24.11. Let $A, B \in \mathbb{F}^{n\times n}$, assume that A and B are positive semidefinite, and assume that $A \le B$. Then, the following statements are equivalent:

i) $B^+ \le A^+$.

ii) $\operatorname{rank} A = \operatorname{rank} B$.

iii) $\mathcal{R}(A) = \mathcal{R}(B)$.

Furthermore, the following statements are equivalent:

iv) $A^+ \le B^+$.

v) $A^2 = AB$.

vi) $A^+ \overset{*}{\le} B^+$.

Source: [1332, 2044].

Fact 10.24.12. Let $A, B \in \mathbb{F}^{n\times n}$, and assume that A and B are positive semidefinite. Then, if two of the following statements hold, then the remaining statement also holds:

i) $A \le B$.

ii) $B^+ \le A^+$.

iii) $\operatorname{rank} A = \operatorname{rank} B$.

Source: [230, 2044, 2868, 2914].

Fact 10.24.13. Let $A, B \in \mathbb{F}^{n\times n}$, and assume that A and B are Hermitian. Then, if two of the following statements hold, then the remaining statement also holds:

i) $A \le B$.

ii) $B^+ \le A^+$.

iii) $\operatorname{In} A = \operatorname{In} B$.

Source: [228].

Fact 10.24.14. Let $A, B \in \mathbb{F}^{n\times n}$, assume that A and B are positive semidefinite, and assume that $A \le B$. Then,

$$0 \le AA^+ \le BB^+.$$

If, in addition, $\operatorname{rank} A = \operatorname{rank} B$, then

$$AA^+ = BB^+.$$

Fact 10.24.15. Let $A, B \in \mathbb{F}^{n\times n}$, assume that A and B are Hermitian, and assume that $\mathcal{R}(A) = \mathcal{R}(B)$. Then,

$$\operatorname{In} A - \operatorname{In} B = \operatorname{In}(A - B) + \operatorname{In}(A^+ - B^+).$$

Source: [2141]. **Related:** Fact 10.11.26.

Fact 10.24.16. Let $A, B \in \mathbb{F}^{n\times n}$, assume that A and B are positive semidefinite, and assume that $A \le B$. Then,

$$0 \le AB^+A \le A \le A + B[(I - AA^+)B(I - AA^+)]^+B \le B.$$

Source: [1332].

Fact 10.24.17. Let $A, B \in \mathbb{F}^{n\times n}$, and assume that A and B are positive semidefinite. Then,

$$\operatorname{spec}[(A + B)^+A] \subset [0, 1].$$

Source: Let C be positive definite and satisfy $B \le C$. Then, $(A + C)^{-1/2}C(A + C)^{-1/2} \le I$. The result now follows from Fact 10.24.19.

Fact 10.24.18. Let $A, B \in \mathbb{F}^{n\times n}$, and assume that A and B are positive semidefinite. Then, the following statements are equivalent:

i) For all $\alpha \in [0, 1]$, $[\alpha A + (1 - \alpha)B]^+ \le \alpha A^+ + (1 - \alpha)B^+$.

ii) $\mathcal{R}(A) = \mathcal{R}(B)$.

Now, let $x \in \mathbb{C}$. Then, the following statements are equivalent:

iii) For all $\alpha \in [0, 1]$, $x^*[\alpha A + (1 - \alpha)B]^+x \le \alpha x^*A^+x + (1 - \alpha)x^*B^+x$.

iv) $x \in [\mathcal{R}(A) \cap \mathcal{R}(B)] + [\mathcal{N}(A) \cap \mathcal{N}(B)]$.

Furthermore, the following statements are equivalent:

v) For all $\alpha \in [0,1]$, $x^*[\alpha A + (1-\alpha)B]^+x = \alpha x^*A^+x + (1-\alpha)x^*B^+x$.

vi) *iv)* holds and $(A^+ - B^+)x \in \mathcal{N}(A) + \mathcal{N}(B)$.

Source: [2142].

Fact 10.24.19. Let $A, B, C \in \mathbb{F}^{n\times n}$, assume that A, B, C are positive semidefinite, and assume that $B \le C$. Then, for all $i \in \{1,\dots,n\}$,

$$\lambda_i[(A+B)^+B] \le \lambda_i[(A+C)^+C].$$

Consequently,

$$\operatorname{tr}(A+B)^+B \le \operatorname{tr}(A+C)^+C.$$

Source: [2815]. **Related:** Fact 10.24.17.

Fact 10.24.20. Let $A, B \in \mathbb{F}^{n\times n}$, assume that A and B are positive semidefinite, and define

$$A\!:\!B \triangleq A(A+B)^+B.$$

Then, the following statements hold:

i) $A\!:\!B$ is positive semidefinite.

ii) $A\!:\!B = \lim_{\varepsilon\downarrow 0}(A+\varepsilon I)\!:\!(B+\varepsilon I)$.

iii) $A\!:\!A = \frac{1}{2}A$.

iv) $A\!:\!B = B\!:\!A = B - B(A+B)^+B = A - A(A+B)^+A$.

v) $A\!:\!B \le A$.

vi) $A\!:\!B \le B$.

vii) $A\!:\!B = -\begin{bmatrix} 0 & 0 & I \end{bmatrix}\begin{bmatrix} A & 0 & I \\ 0 & B & I \\ I & I & 0 \end{bmatrix}^+\begin{bmatrix} 0 \\ 0 \\ I \end{bmatrix}$.

viii) $A\!:\!B = (A^+ + B^+)^+$ if and only if $\mathcal{R}(A) = \mathcal{R}(B)$.

ix) $A(A+B)^+B = ACB$ for every (1)-inverse C of $A+B$.

x) $\operatorname{tr} A\!:\!B \le (\operatorname{tr} B)\!:\!(\operatorname{tr} A)$.

xi) $\operatorname{tr} A\!:\!B = (\operatorname{tr} B)\!:\!(\operatorname{tr} A)$ if and only if there exists $\alpha \in [0,\infty)$ such that either $A = \alpha B$ or $B = \alpha A$.

xii) $\det A\!:\!B \le (\det B)\!:\!(\det A)$.

xiii) $\mathcal{R}(A\!:\!B) = \mathcal{R}(A) \cap \mathcal{R}(B)$.

xiv) $\mathcal{N}(A\!:\!B) = \mathcal{N}(A) + \mathcal{N}(B)$.

xv) $\operatorname{rank} A\!:\!B = \operatorname{rank} A + \operatorname{rank} B - \operatorname{rank}(A+B)$.

xvi) Let $S \in \mathbb{F}^{p\times n}$, and assume that S is right invertible. Then, $S(A\!:\!B)S^* \le (SAS^*)\!:\!(SBS^*)$.

xvii) Let $S \in \mathbb{F}^{n\times n}$, and assume that S is nonsingular. Then, $S(A\!:\!B)S^* = (SAS^*)\!:\!(SBS^*)$.

xviii) For all positive numbers α, β, $(\alpha^{-1}A)\!:\!(\beta^{-1}B) \le \alpha A + \beta B$.

xix) Let $X \in \mathbb{F}^{n\times n}$, and assume that X is Hermitian and $\left[\begin{smallmatrix} A+B & A \\ A & A-X \end{smallmatrix}\right] \ge 0$. Then,

$$X \le A\!:\!B, \qquad \begin{bmatrix} A+B & A \\ A & A - A\!:\!B \end{bmatrix} \ge 0.$$

xx) $\phi\colon \mathbf{N}^n \times \mathbf{N}^n \mapsto -\mathbf{N}^n$ defined by $\phi(A,B) \triangleq -A\!:\!B$ is convex.

xxi) If A and B are projectors, then $2(A\!:\!B)$ is the projector onto $\mathcal{R}(A) \cap \mathcal{R}(B)$.

xxii) If $A+B$ is positive definite, then $A\!:\!B = A(A+B)^{-1}B$.

xxiii) $A\#B = [\frac{1}{2}(A+B)]\#[2(A:B)]$.

xxiv) If $C, D \in \mathbb{F}^{n\times n}$ are positive semidefinite, then

$$(A:B):C = A:(B:C), \quad A:C + B:D \le (A+B):(C+D).$$

xxv) If $C, D \in \mathbb{F}^{n\times n}$ are positive semidefinite, $A \le C$, and $B \le D$, then $A:B \le C:D$.

xxvi) If A and B are positive definite, then

$$A:B = (A^{-1} + B^{-1})^{-1} \le \tfrac{1}{2}(A\#B) \le \tfrac{1}{4}(A+B).$$

xxvii) If A and B are positive definite, then

$$\sum_{i=1}^{n} \frac{\lambda_i(A)\lambda_{n-i+1}(B)}{\lambda_i(A) + \lambda_{n-i+1}(B)} \le \operatorname{tr} A:B \le \sum_{i=1}^{n} \frac{\lambda_i(A)\lambda_i(B)}{\lambda_i(A) + \lambda_i(B)}.$$

xxviii) Let $x, y \in \mathbb{F}^n$. Then,

$$(x+y)^*(A:B)(x+y) \le x^*Ax + y^*By.$$

xxix) Let $x, y \in \mathbb{F}^n$. Then,

$$x^*(A:B)x \le y^*Ay + (x-y)^*B(x-y).$$

xxx) Let $x \in \mathbb{F}^n$. Then,

$$x^*(A:B)x = \inf_{y\in\mathbb{F}^n} [y^*Ay + (x-y)^*B(x-y)].$$

xxxi) Let $x \in \mathbb{F}^n$. Then,

$$x^*(A:B)x \le (x^*Ax):(x^*Bx).$$

Source: [81, 82, 85, 1224, 1696], [2299, p. 189], [2632, 2750], and [2977, p. 9]. **Remark:** $A:B$ is the *parallel sum* of A and B. **Remark:** A symmetric expression for the parallel sum of three or more positive-semidefinite matrices is given in [2632]. **Remark:** $A\#B$ is the geometric mean of A and B. See Fact 10.11.68. **Related:** Fact 8.4.8 and Fact 8.8.19.

Fact 10.24.21. Let $A, B \in \mathbb{F}^{n\times n}$, assume that A is positive semidefinite, and assume that B is a projector. Then,

$$\operatorname{sh}(A,B) \triangleq \max\{X \in \mathbf{N}^n\colon 0 \le X \le A \text{ and } \mathcal{R}(X) \subseteq \mathcal{R}(B)\}$$

exists. Furthermore,

$$\operatorname{sh}(A,B) = A - AB_\perp(B_\perp AB_\perp)^+B_\perp A.$$

That is,

$$\operatorname{sh}(A,B) = A \left\| \begin{bmatrix} A & AB_\perp \\ B_\perp A & B_\perp AB_\perp \end{bmatrix} \right. .$$

Finally,

$$\operatorname{sh}(A,B) = \lim_{\alpha\to\infty} (\alpha B):A.$$

Source: Existence of the maximum is proved in [85]. The expression for $\operatorname{sh}(A,B)$ is given in [1195]; a related expression involving the Schur complement is given in [81]. The last equality is shown in [85]. See also [96]. **Remark:** $\operatorname{sh}(A,B)$ is the *shorted operator*.

Fact 10.24.22. Let $B \in \mathbb{R}^{m\times n}$, define

$$\mathcal{S} \triangleq \{A \in \mathbb{R}^{n\times n}\colon A \ge 0 \text{ and } \mathcal{R}(B^\mathsf{T}BA) \subseteq \mathcal{R}(A)\},$$

and define $\phi\colon \mathcal{S} \mapsto -\mathbf{N}^m$ by $\phi(A) \triangleq -(BA^+B^\mathsf{T})^+$. Then, $\mathcal{S}$ is a convex cone, and ϕ is convex. **Source:** [1245]. **Related:** This result generalizes *xii)* of Proposition 10.6.17 in the case where $r = p = 1$.

Fact 10.24.23. Let $A, B \in \mathbb{F}^{n\times n}$, and assume that A and B are positive semidefinite. If $(AB)^+ = B^+A^+$, then AB is range Hermitian. Furthermore, the following statements are equivalent:

i) AB is range Hermitian.

ii) $(AB)^{\#} = B^+A^+$.

iii) $(AB)^+ = B^+A^+$.

Source: [2001]. **Related:** Fact 8.5.11.

Fact 10.24.24. Let $A, B \in \mathbb{F}^{n\times n}$, and assume that A and B are positive semidefinite. Then, the following statements are equivalent:

i) $A(A+B)^+B = 0$.

ii) $B(A+B)^+A = 0$.

iii) $A(A+B)^+A = A$.

iv) $B(A+B)^+B = B$.

v) $A(A+B)^+B + B(A+B)^+A = 0$.

vi) $A(A+B)^+A + B(A+B)^+B = A + B$.

vii) $\operatorname{rank}[A\ \ B] = \operatorname{rank} A + \operatorname{rank} B$.

viii) $\mathcal{R}(A) \cap \mathcal{R}(B) = \{0\}$.

ix) $(A+B)^+ = [(I - BB^+)A(I - B^+B)]^+ + [(I - AA^+)B(I - A^+A)]^+$.

Source: [2666]. **Related:** Fact 8.4.31.

Fact 10.24.25. Let $A \in \mathbb{F}^{n\times n}$ and $C \in \mathbb{F}^{m\times m}$, assume that A and C are positive semidefinite, let $B \in \mathbb{F}^{n\times m}$, and define $X \triangleq A^{+1/2}BC^{+1/2}$. Then, the following statements are equivalent:

i) $\begin{bmatrix} A & B \\ B^* & C \end{bmatrix}$ is positive semidefinite.

ii) $AA^+B = B$ and $XX^* \le I_n$.

iii) $BC^+C = B$ and $X^*X \le I_m$.

iv) $B = A^{1/2}XC^{1/2}$ and $X^*X \le I_m$.

v) There exists $Y \in \mathbb{F}^{n\times m}$ such that $B = A^{1/2}YC^{1/2}$ and $Y^*Y \le I_m$.

Source: [2977, p. 15]. **Related:** Fact 10.12.15.

10.25 Facts on the Kronecker and Schur Products

Fact 10.25.1. Let $A \in \mathbb{F}^{n\times n}$, and assume that every entry of A is nonzero. Then, A and $A^{\odot -1}$ are positive semidefinite if and only if there exists $x \in \mathbb{F}^n$ such that $A = xx^*$. **Source:** [1435, 1788].

Fact 10.25.2. Let $A \in \mathbb{F}^{n\times n}$, assume that A is positive semidefinite, assume that every entry of A is nonnegative, and let $\alpha \in [0, n-2]$. Then, $A^{\odot\alpha}$ is positive semidefinite. **Source:** [451, 1053]. **Remark:** $A^{\odot\alpha}$ may be positive semidefinite for all $\alpha \ge 0$. See Fact 10.9.8.

Fact 10.25.3. Let $A \in \mathbb{F}^{n\times n}$, assume that A is positive semidefinite, and let $k \ge 1$. Then, the following statements hold:

i) If $r \in [0,1]$, then $(A^r)^{\odot k} \le (A^{\odot k})^r$.

ii) If $r \in [1,2]$, then $(A^{\odot k})^r \le (A^r)^{\odot k}$.

iii) If A is positive definite and $r \in [0,1]$, then $(A^{\odot k})^{-r} \le (A^{-r})^{\odot k}$.

Source: [2977, p. 8].

Fact 10.25.4. Let $A \in \mathbb{F}^{n\times n}$, and assume that A is positive semidefinite. Then,

$$(I \odot A)^2 \le \tfrac{1}{2}(I \odot A^2 + A \odot A) \le I \odot A^2, \quad A \odot A \le I \odot A^2.$$

Hence,

$$\sum_{i=1}^n A_{(i,i)}^2 \le \sum_{i=1}^n \lambda_i^2(A).$$

Now, assume that A is positive definite. Then,

$$(A\odot A)^{-1} \le A^{-1}\odot A^{-1}, \quad (A\odot A^{-1})^{-1} \le I \le (A^{1/2}\odot A^{-1/2})^2 \le \tfrac{1}{2}(I + A\odot A^{-1}) \le A\odot A^{-1}.$$

If A is real, then

$$2\max_{i\in\{1,\ldots,n\}}\left(\sqrt{(A\odot A^{-1})_{(i,i)}} - 1\right) \le \sum_{i=1}^{n}[(A\odot A^{-1})_{(i,i)} - 1].$$

Furthermore,

$$(A\odot A^{-\mathsf{T}})1_{n\times 1} = 1_{n\times 1}, \quad 1 = \min\operatorname{spec}(A\odot A^{-\mathsf{T}}).$$

Next, let $\alpha \triangleq \lambda_{\min}(A)$ and $\beta \triangleq \lambda_{\max}(A)$. Then,

$$\frac{2\alpha\beta}{\alpha^2+\beta^2}I \le \frac{2\alpha\beta}{\alpha^2+\beta^2}(A^2\odot A^{-2})^{1/2} \le \frac{\alpha\beta}{\alpha^2+\beta^2}(I+A^2\odot A^{-2}) \le A\odot A^{-1} \le \frac{\alpha^2+\beta^2}{2\alpha\beta}I, \quad A\odot A^{-\mathsf{T}} \le \frac{\alpha^2+\beta^2}{2\alpha\beta}I.$$

Finally, define $\Phi(A) \triangleq A\odot A^{-1}$, and, for all $k\ge 1$, define $\Phi^{(k+1)}(A) \triangleq \Phi[\Phi^{(k)}(A)]$, where $\Phi^{(1)}(A) \triangleq \Phi(A)$. Then, for all $k \ge 1$,

$$\Phi^{(k)}(A) \ge I, \quad \lim_{k\to\infty}\Phi^{(k)}(A) = I.$$

Source: [1039, 1042, 1548, 2804, 2805], [1448, p. 475], [2403, pp. 304, 305], [2991, p. 251], and set $B = A^{-1}$ in Fact 10.25.49. The upper bound for $A\odot A^{-\mathsf{T}}$ is given in [1874]. **Remark:** $A\odot A^{-\mathsf{T}}$ is Hermitian in the case where $\mathbb{F} = \mathbb{C}$. **Remark:** The convergence result holds in the case where A is an *H-matrix* [1548]. $A\odot A^{-\mathsf{T}}$ is the *relative gain array*. **Related:** Fact 10.25.46 and Fact 10.25.55.

Fact 10.25.5. Let $A \in \mathbb{F}^{n\times n}$, and assume that A is positive definite. Then, for all $i \in \{1,\ldots,n\}$,

$$1 \le A_{(i,i)}(A^{-1})_{(i,i)},$$

$$2\max_{i=1,\ldots,n}\sqrt{A_{(i,i)}(A^{-1})_{(i,i)} - 1} \le \sum_{i=1}^{n}\sqrt{A_{(i,i)}(A^{-1})_{(i,i)} - 1},$$

$$2\max_{i=1,\ldots,n}\sqrt{A_{(i,i)}(A^{-1})_{(i,i)}} - 1 \le \sum_{i=1}^{n}\left[\sqrt{A_{(i,i)}(A^{-1})_{(i,i)}} - 1\right].$$

Source: [1041, p. 66-6].

Fact 10.25.6. Let $A \in \mathbb{F}^{n\times n}$, assume that A is positive definite, and define $\alpha \triangleq \lambda_n(A)$ and $\beta \triangleq \lambda_1(A)$. Then,

$$2I \le (A + A^{-1})\odot I \le \frac{\alpha^2+\beta^2}{\alpha\beta}I.$$

Source: [26]. **Remark:** The left-hand inequality is a special case of Fact 10.25.55.

Fact 10.25.7. Let $A, B, C \in \mathbb{F}^{n\times n}$, define $\mathcal{A} \triangleq \left[\begin{smallmatrix} A & B \\ B^* & C\end{smallmatrix}\right] \in \mathbb{F}^{2n\times 2n}$, and assume that $\mathcal{A}$ is positive semidefinite. Then,

$$-A\odot C \le B\odot B^* \le A\odot C.$$

Source: [2991, p. 236, 237].

Fact 10.25.8. Let $\mathcal{A} \triangleq \left[\begin{smallmatrix} A & B \\ B^* & C\end{smallmatrix}\right] \in \mathbb{F}^{(n+m)\times(n+m)}$, assume that $\mathcal{A}$ is positive definite, and partition $\mathcal{A}^{-1} = \left[\begin{smallmatrix} X & Y \\ Y^* & Z\end{smallmatrix}\right]$ conformably with $\mathcal{A}$. Then,

$$I \le \begin{bmatrix} A\odot A^{-1} & 0 \\ 0 & Z\odot Z^{-1}\end{bmatrix} \le \mathcal{A}\odot\mathcal{A}^{-1}, \quad I \le \begin{bmatrix} X\odot X^{-1} & 0 \\ 0 & C\odot C^{-1}\end{bmatrix} \le \mathcal{A}\odot\mathcal{A}^{-1}.$$

Source: [282].

Fact 10.25.9. Let $A \in \mathbb{F}^{n\times n}$, let $p,q \in \mathbb{R}$, assume that A is positive semidefinite, and assume that either p and q are nonnegative or A is positive definite. Then,

$$A^{(p+q)/2} \odot A^{(p+q)/2} \le A^p \odot A^q.$$

In particular, $I \le A \odot A^{-1}$. **Source:** [198].

Fact 10.25.10. Let $A \in \mathbb{F}^{n\times n}$, assume that A is positive semidefinite, and assume that $I_n \odot A = I_n$. Then, $\det A \le \lambda_{\min}(A \odot \overline{A})$. **Source:** [2842].

Fact 10.25.11. Let $A \in \mathbb{F}^{n\times n}$. Then, $-A^*A \odot I \le A^* \odot A \le A^*A \odot I$. **Source:** Use Fact 10.25.57 with $B = I$.

Fact 10.25.12. Let $A \in \mathbb{F}^{n\times n}$. Then, $\langle A \odot A^* \rangle \le \left\{ \begin{matrix} A^*A \odot I \\ \langle A \rangle \odot \langle A^* \rangle \end{matrix} \right\} \le \sigma_{\max}^2(A)I$. **Source:** [2985] and Fact 10.25.37.

Fact 10.25.13. Let $A \in \mathbb{F}^{n\times n}$ and $x \in \mathbb{C}^n$. Then, $|x^*(A \odot A^*)x| \le x^*(\langle A \rangle \odot \langle A^* \rangle)x$. **Source:** [2991, p. 314].

Fact 10.25.14. Let $A \triangleq \begin{bmatrix} A_{11} & A_{12} \\ A_{12}^* & A_{22} \end{bmatrix} \in \mathbb{F}^{(n+m)\times(n+m)}$ and $B \triangleq \begin{bmatrix} B_{11} & B_{12} \\ B_{12} & B_{22} \end{bmatrix} \in \mathbb{F}^{(n+m)\times(n+m)}$, and assume that A and B are positive semidefinite. Then,

$$(A_{11}|A) \odot (B_{11}|B) \le (A_{11}|A) \odot B_{22} \le (A_{11} \odot B_{11})|(A \odot B).$$

Source: [1800] and [2991, p. 241].

Fact 10.25.15. Let $A, B \in \mathbb{F}^{n\times n}$, and assume that A and B are positive semidefinite. Then,

$$\operatorname{rank}(A \odot B) \le \operatorname{rank}(A \otimes B) = (\operatorname{rank} A)(\operatorname{rank} B).$$

Now, assume that A is positive definite. Then,

$$\operatorname{rank} B \le \operatorname{rank}(A \odot B) = \operatorname{rank}(I \odot B).$$

Source: Fact 9.4.20, Fact 9.6.15, Fact 10.25.21, [2263, pp. 154, 166], and [2991, p. 238]. **Remark:** The first inequality is due to D. Z. Djokovic.

Fact 10.25.16. Let $A, B \in \mathbb{F}^{n\times n}$, and assume that A and B are positive semidefinite. Then, $A \odot B$ is positive semidefinite. If, in addition, A is positive definite and $I \odot B$ is positive definite, then $A \odot B$ is positive definite. **Source:** By Fact 9.4.22, $A \otimes B$ is positive semidefinite, and the Schur product $A \odot B$ is a principal submatrix of the Kronecker product. If A is positive definite, use Fact 10.25.33 to obtain $\det(A \odot B) > 0$. See [2403, p. 300] and [2991, pp. 234, 235]. **Remark:** The first result is *Schur's theorem.* The second result is *Schott's theorem.* See [1873], Fact 10.25.15, and Fact 10.25.33.

Fact 10.25.17. Let $A \in \mathbb{F}^{n\times n}$, assume that A is positive semidefinite, and define $e^{\odot A} \in \mathbb{F}^{n\times n}$ by $(e^{\odot A})_{(i,j)} \triangleq e^{A_{(i,j)}}$. Then, $e^{\odot A}$ is positive semidefinite. **Source:** Note that $e^{\odot A} = 1_{n\times n} + \frac{1}{2}A \odot A + \frac{1}{3!}A \odot A \odot A + \cdots$, and use Fact 10.25.16. See [922, p. 10].

Fact 10.25.18. Let $A \in \mathbb{F}^{n\times n}$, and assume that A is positive definite. Then, there exist positive-definite matrices $B, C \in \mathbb{F}^{n\times n}$ such that $A = B \odot C$. **Source:** [2263, pp. 154, 166]. **Credit:** D. Z. Djokovic.

Fact 10.25.19. Let $A, B \in \mathbb{F}^{n\times n}$. Then, A is positive semidefinite if and only if, for every positive-semidefinite matrix $B \in \mathbb{F}^{n\times n}$, $1_{1\times n}(A \odot B)1_{n\times 1} \ge 0$. **Source:** [1448, p. 459]. **Remark:** This is *Fejér's theorem.*

Fact 10.25.20. Let $A, B \in \mathbb{F}^{n\times n}$, and assume that A and B are positive definite. Then,

$$1_{1\times n}[(A - B) \odot (A^{-1} - B^{-1})]1_{n\times 1} \le 0.$$

Furthermore, equality holds if and only if $A = B$. **Source:** [304, p. 8-8].

Fact 10.25.21. Let $A, B \in \mathbb{F}^{n\times n}$, and assume that A is positive definite and B is positive semidefinite. Then,

$$(1_{1\times n}A^{-1}1_{n\times 1})^{-1}B \le A \odot B.$$

Source: [1044]. **Remark:** Setting $B = 1_{n\times n}$ yields Fact 10.10.20.

Fact 10.25.22. Let $A, B \in \mathbb{F}^{n\times n}$, and assume that A and B are positive definite. Then,

$$(1_{1\times n}A^{-1}1_{n\times 1}1_{1\times n}B^{-1}1_{n\times 1})^{-1}1_{n\times n} \le A \odot B.$$

Source: [2985].

Fact 10.25.23. Let $A \in \mathbb{F}^{n\times n}$, and assume that A is positive semidefinite. Then,

$$\operatorname{tr} A^{\odot 2} \le \operatorname{tr} A^2 \le (\operatorname{tr} A)^2, \quad (\operatorname{tr} A)^{1/2} \le \operatorname{tr} A^{1/2} \le \operatorname{tr}(A \odot I)^{1/2}.$$

Now, assume that A is positive definite. Then,

$$\frac{n^2}{\operatorname{tr} A} \le \operatorname{tr}(A \odot I)^{-1} \le \operatorname{tr} A^{-1}.$$

Source: [2991, p. 213]. **Related:** Fact 10.13.9.

Fact 10.25.24. Let $A, B \in \mathbb{F}^{n\times n}$. Then, A is positive semidefinite if and only if, for every positive-semidefinite matrix $B \in \mathbb{F}^{n\times n}$, $\operatorname{tr}(A \odot B) \ge 0$. **Source:** [2991, p. 237].

Fact 10.25.25. Let $A, B \in \mathbb{F}^{n\times n}$, and assume that A and B are positive semidefinite. Then, the following statements hold:

i) $\operatorname{tr} AB \le \operatorname{tr}(A \otimes B)$.

ii) $\operatorname{tr}(A \odot B) \le \frac{1}{2}\operatorname{tr}(A \odot A + B \odot B)$.

iii) $\operatorname{tr}(A \otimes B) \le \frac{1}{2}\operatorname{tr}(A \otimes A + B \otimes B)$.

iv) $\det(A \otimes B) \le \frac{1}{2}[\det(A \otimes A) + \det(B \otimes B)]$.

Source: [2991, p. 238].

Fact 10.25.26. Let $A \in \mathbb{F}^{n\times m}$ and $B \in \mathbb{F}^{k\times l}$. Then, $\langle A \otimes B\rangle = \langle A\rangle \otimes \langle B\rangle$. **Source:** [2991, p. 291].

Fact 10.25.27. Let $A, B \in \mathbb{F}^{n\times n}$, assume that A and B are Hermitian, and assume that either $A \le B$ or $B \le A$. Then,

$$2A \odot B \le A \odot A + B \odot B.$$

Source: [2991, p. 251].

Fact 10.25.28. Let $A, B \in \mathbb{F}^{n\times n}$, and assume that A and B are positive semidefinite. Then, the following statements hold:

i) If $p \ge 1$, then $\operatorname{tr}(A \odot B)^p \le \operatorname{tr} A^p \odot B^p$.

ii) If $0 \le p \le 1$, then $\operatorname{tr} A^p \odot B^p \le \operatorname{tr}(A \odot B)^p$.

iii) If A and B are positive definite and $p \le 0$, then $\operatorname{tr}(A \odot B)^p \le \operatorname{tr} A^p \odot B^p$.

Source: [2817].

Fact 10.25.29. Let $A, B \in \mathbb{F}^{n\times n}$, and assume that A and B are positive semidefinite. Then,

$$\lambda_{\min}(AB) \le \lambda_{\min}(A \odot B), \quad \lambda_{\min}(AB)I \le \lambda_{\min}(A \odot B)I \le A \odot B.$$

Source: [1541]. **Related:** This result interpolates an inequality in Fact 10.25.34.

Fact 10.25.30. Let $A, B \in \mathbb{F}^{n\times n}$, and assume that A is positive semidefinite and B is positive definite. Then,

$$\rho_{\max}(A \odot B) \le \rho_{\max}(B \odot B^{-1})\rho_{\max}(AB).$$

Source: [1459]. **Remark:** Fact 10.25.4 implies that $\rho_{\max}(B \odot B^{-1}) \ge \rho_{\min}(B \odot B^{-1}) \ge 1$.

Fact 10.25.31. Let $A, B \in \mathbb{F}^{n\times n}$, and assume that A and B are Hermitian. Then, for all $i \in \{1,\ldots,n\}$,

$$\min_{i,j\in\{1,\ldots,n\}} \lambda_i(A)\lambda_j(B) \le \lambda_{i+n^2-n}(A\otimes B) \le \lambda_i(A\odot B) \le \lambda_i(A\otimes B) \le \max_{i,j\in\{1,\ldots,n\}} \lambda_i(A)\lambda_j(B).$$

Now, assume that A and B are positive semidefinite. Then,

$$\lambda_n(A)\lambda_n(B) \le \lambda_{i+n^2-n}(A\otimes B) \le \lambda_i(A\odot B) \le \lambda_i(A\otimes B) \le \lambda_1(A)\lambda_1(B).$$

Source: Proposition 9.1.10, Proposition 9.3.2, and Theorem 10.4.5. For A, B positive semidefinite, see [1951] and [2991, p. 278].

Fact 10.25.32. Let $A, B \in \mathbb{F}^{n\times n}$, and assume that A and B are positive semidefinite. Then, for all $i \in \{1,\ldots,n\}$,

$$\mathrm{d}_{\min}(A)\lambda_{\min}(B) \le \mathrm{d}_i(A)\lambda_{\min}(B) \le \lambda_i(A\odot B) \le \mathrm{d}_i(A)\lambda_{\max}(B) \le \mathrm{d}_{\max}(A)\lambda_{\max}(B).$$

Source: [2403, pp. 303, 304] and [2991, pp. 274, 275].

Fact 10.25.33. Let $A, B \in \mathbb{F}^{n\times n}$, and assume that A and B are positive semidefinite. Then,

$$\det AB \le \left(\prod_{i=1}^n A_{(i,i)}\right)\det B \le \det(A\odot B) \le \prod_{i=1}^n A_{(i,i)}B_{(i,i)}.$$

Equivalently,

$$\det AB \le [\det(I\odot A)]\det B \le \det(A\odot B) \le \prod_{i=1}^n A_{(i,i)}B_{(i,i)}.$$

Furthermore,

$$2\det AB \le \left(\prod_{i=1}^n A_{(i,i)}\right)\det B + \left(\prod_{i=1}^n B_{(i,i)}\right)\det A \le \det(A\odot B) + (\det A)\det B.$$

Finally, the following statements hold:

i) If $I\odot A$ and B are positive definite, then $A\odot B$ is positive definite.

ii) If $I\odot A$ and B are positive definite and $\operatorname{rank} A = 1$, then equality holds in the right-hand equality.

iii) If A and B are positive definite, then equality holds in the right-hand equality if and only if B is diagonal.

Source: [1960], [2418, p. 253], [2955], and [2991, p. 242]. **Remark:** In the first string, the first and third inequalities follow from Hadamard's inequality Fact 10.21.15, while the second inequality is *Oppenheim's inequality*. See Fact 10.25.16. **Remark:** The right-hand inequality in the third string of inequalities is valid if A and B are M-matrices. See [89, 710].

Fact 10.25.34. Let $A, B \in \mathbb{F}^{n\times n}$, assume that A and B are positive semidefinite, let $k \in \{1,\ldots,n\}$, and let $r \in (0,1]$. Then,

$$\prod_{i=k}^n \lambda_i(A)\lambda_i(B) \le \prod_{i=k}^n \sigma_i(AB) \le \prod_{i=k}^n \lambda_i(AB) \le \prod_{i=k}^n \lambda_i^2(A\#B) \le \prod_{i=k}^n \lambda_i(A\odot B),$$

$$\prod_{i=k}^n \lambda_i(A)\lambda_i(B) \le \prod_{i=k}^n \sigma_i(AB) \le \prod_{i=k}^n \lambda_i(AB) \le \prod_{i=k}^n \lambda_i^{1/r}(A^rB^r)$$

$$\le \prod_{i=k}^n e^{\lambda_i(\log A+\log B)} \le \prod_{i=k}^n e^{\lambda_i[I\odot(\log A+\log B)]} \le \prod_{i=k}^n \lambda_i^{1/r}(A^r\odot B^r) \le \prod_{i=k}^n \lambda_i(A\odot B).$$

Consequently,

$$\lambda_{\min}(AB) \le \lambda_{\min}(A \odot B), \quad \det AB = \det (A\#B)^2 \le \det(A \odot B).$$

Source: [94, 1039, 2803], [2403, p. 305], [2977, p. 21], Fact 10.11.68, and Fact 10.22.28. **Remark:** Although $\det AB = \det (A\#B)^2$, the matrices AB and $(A\#B)^2$ do not necessarily have the same spectrum.

Fact 10.25.35. Let $A, B \in \mathbb{F}^{n\times n}$, assume that A and B are positive definite, let $k \in \{1, \dots, n\}$, and let $r > 0$. Then,

$$\prod_{i=k}^{n} \lambda_i^{-r}(A \odot B) \le \prod_{i=k}^{n} \lambda_i^{-r}(AB).$$

Source: [2802].

Fact 10.25.36. Let $A \in \mathbb{F}^{n\times n}$, assume that A is positive semidefinite, and define $\mathcal{S} \triangleq \{X \in \mathbf{N}^n : X \odot I = I\}$. Then,

$$\min_{X\in\mathcal{S}} \det(A \odot X) = \det A.$$

Source: [2991, p. 243]. **Remark:** X is a *correlation matrix*.

Fact 10.25.37. Let $A, B \in \mathbb{F}^{n\times n}$, let $C, D \in \mathbb{F}^{m\times m}$, assume that $A, B, C,$ and D are Hermitian, $A \le B$, $C \le D$, and that either A and C are positive semidefinite, A and D are positive semidefinite, or B and D are positive semidefinite. Then,

$$A \otimes C \le B \otimes D.$$

If, in addition, $n = m$, then

$$A \odot C \le B \odot D.$$

Source: [88, 230]. **Problem:** Determine conditions under which these inequalities are strict.

Fact 10.25.38. Let $A, B, C, D \in \mathbb{F}^{n\times n}$, assume that A, B, C, D are positive semidefinite, and assume that $A \le B$ and $C \le D$. Then,

$$0 \le A \otimes C \le B \otimes D, \quad 0 \le A \odot C \le B \odot D.$$

Source: Fact 10.25.37 and [2991, p. 238].

Fact 10.25.39. Let $A, B \in \mathbb{F}^{n\times n}$, and assume that A and B are positive semidefinite. Then, the following statements are equivalent:

i) $A \le B$

ii) $A \otimes I \le B \otimes I$.

iii) $A \otimes A \le B \otimes B$.

Source: [1873] and [2991, p. 237].

Fact 10.25.40. Let $A, B \in \mathbb{F}^{n\times n}$, assume that A and B are positive semidefinite, assume that $0 \le A \le B$, and let $k \ge 1$. Then, $A^{\odot k} \le B^{\odot k}$. **Source:** $0 \le (B-A)\odot(B+A)$ implies that $A\odot A \le B\odot B$; that is, $A^{\odot 2} \le B^{\odot 2}$.

Fact 10.25.41. Let $A_1, \dots, A_k, B_1, \dots, B_k \in \mathbb{F}^{n\times n}$, and assume that $A_1, \dots, A_k, B_1, \dots, B_k$ are positive semidefinite. Then,

$$(A_1 + B_1) \otimes \cdots \otimes (A_k + B_k) \le A_1 \otimes \cdots \otimes A_k + B_1 \otimes \cdots \otimes B_k.$$

Source: [2029, p. 143].

Fact 10.25.42. Let $A_1, A_2, B_1, B_2 \in \mathbb{F}^{n\times n}$, assume that A_1, A_2, B_1, B_2 are positive semidefinite, assume that $0 \le A_1 \le B_1$ and $0 \le A_2 \le B_2$, and let $\alpha \in [0, 1]$. Then,

$$[\alpha A_1 + (1-\alpha)B_1] \otimes [\alpha A_2 + (1-\alpha)B_2] \le \alpha(A_1 \otimes A_2) + (1-\alpha)(B_1 \otimes B_2).$$

Source: [2840].

Fact 10.25.43. Let $A \in \mathbb{F}^{n\times n}$ and $B \in \mathbb{F}^{m\times m}$, assume that A and B are positive semidefinite, let $r \in \mathbb{R}$, and assume that either A and B are positive definite or r is positive. Then, $(A\otimes B)^r = A^r \otimes B^r$.
Source: [2074].

Fact 10.25.44. Let $A, B, C \in \mathbb{F}^{n\times n}$, assume A, B, C are positive semidefinite, and let $k \ge 1$. Then,

$$(A+B)^{\otimes k} + (A+C)^{\otimes k} \le (A+B+C)^{\otimes k} + A^{\otimes k}.$$

Source: [2681]. **Related:** Fact 10.16.35.

Fact 10.25.45. For all $i \in \{1,\dots,k\}$, let $A_i \in \mathbf{H}^n$ and $B_i \in \mathbf{H}^m$, assume that $A_k \le \cdots \le A_1$ and $B_k \le \cdots \le B_1$, and let $\{j_1,\dots,j_k\} = \{1,\dots,k\}$. Then,

$$\sum_{i=1}^k A_i \otimes B_{k-i+1} \le \sum_{i=1}^k A_i \otimes B_{j_i} \le \sum_{i=1}^k A_i \otimes B_i,$$

$$k\sum_{i=1}^k A_i \otimes B_{k-i+1} \le \sum_{i=1}^k A_i \otimes \sum_{i=1}^k B_{j_i} \le k\sum_{i=1}^k A_i \otimes B_i.$$

Now, assume that $n = m$, assume that $A_k + B_k$ is positive semidefinite, and let $l \ge 1$. Then,

$$\sum_{i=1}^k (A_i + B_{k-i+1})^{\otimes l} \le \sum_{i=1}^k (A_i + B_{j_i})^{\otimes l} \le \sum_{i=1}^k (A_i + B_i)^{\otimes l}.$$

Source: [2681]. **Related:** Fact 10.14.62.

Fact 10.25.46. For all $i \in \{1,\dots,k\}$, let $A_i, B_i \in \mathbf{H}^n$, assume that $A_k \le \cdots \le A_1$ and $B_k \le \cdots \le B_1$, and let $\{j_1,\dots,j_k\} = \{1,\dots,k\}$. Then,

$$\sum_{i=1}^k A_i \odot B_{k-i+1} \le \sum_{i=1}^k A_i \odot B_{j_i} \le \sum_{i=1}^k A_i \odot B_i,$$

$$k\sum_{i=1}^k A_i \odot B_{k-i+1} \le \sum_{i=1}^k A_i \odot \sum_{i=1}^k B_{j_i} \le k\sum_{i=1}^k A_i \odot B_i.$$

Now, assume that $A_k + B_k$ is positive semidefinite, and let $l \ge 1$. Then,

$$\sum_{i=1}^k (A_i + B_{k-i+1})^{\odot l} \le \sum_{i=1}^k (A_i + B_{j_i})^{\odot l} \le \sum_{i=1}^k (A_i + B_i)^{\odot l}.$$

Now, assume that A_k and B_k are positive semidefinite. Then,

$$\sum_{i=1}^k \det(A_i \odot B_{k-i+1})^l \le \sum_{i=1}^k \det(A_i \odot B_{j_i})^l \le \sum_{i=1}^k \det(A_i \odot B_i)^l.$$

Finally, assume that A_k is positive definite. Then,

$$kI \le \sum_{i=1}^k A_i \odot A_i^{-1} \le \sum_{i=1}^k A_i \odot A_{j_i}^{-1}.$$

Source: [2681]. **Related:** Fact 10.14.62 and Fact 10.25.4.

Fact 10.25.47. Let $A \in \mathbb{F}^{n\times m}$ and $B \in \mathbb{F}^{k\times l}$. Then, $\langle A\otimes B\rangle = \langle A\rangle \otimes \langle B\rangle$.

Fact 10.25.48. Let $A, B \in \mathbb{F}^{n\times n}$, let $C, D \in \mathbb{F}^{m\times m}$, assume that A, B, C, D are positive semidefinite, let α and β be nonnegative numbers, and let $r \in [0, 1]$. Then,

$$\alpha(A^r \otimes C^{1-r}) + \beta(B^r \otimes D^{1-r}) \le (\alpha A + \beta B)^r \otimes (\alpha C + \beta D)^{1-r}.$$

Source: [1796].

Fact 10.25.49. Let $A, B \in \mathbb{F}^{n\times n}$, and assume that A and B are positive semidefinite. Then, the following statements hold:

i) If $r \in [0, 1]$, then $A^r \odot B^r \le (A \odot B)^r$.

ii) If $r \in [1, 2]$, then $(A \odot B)^r \le A^r \odot B^r$.

Now, define $\alpha \triangleq \lambda_{\min}(A \otimes B)$ and $\beta \triangleq \lambda_{\max}(A \otimes B)$. Then,

$$A^2 \odot B^2 - \tfrac{1}{4}(\beta-\alpha)^2 I \le (A \odot B)^2 \le \tfrac{1}{2}[A^2 \odot B^2 + (AB)^{\odot 2}] \le A^2 \odot B^2 \le \begin{cases} \frac{(\alpha+\beta)^2}{4\alpha\beta}(A \odot B)^2 \\ (A \odot B)^2 + \frac{1}{4}(\beta-\alpha)^2 I, \end{cases}$$

$$\begin{aligned} &A \odot B - \tfrac{1}{4}(\sqrt{\beta} - \sqrt{\alpha})^2 I \\ &\quad \le (A^{1/2} \odot B^{1/2})^2 \le \tfrac{1}{2}[A \odot B + (A^{1/2}B^{1/2})^{\odot 2}] \le A \odot B \le (A^2 \odot B^2)^{1/2} \le \begin{cases} \frac{\alpha+\beta}{2\sqrt{\alpha\beta}} A \odot B \\ A \odot B + \frac{(\beta-\alpha)^2}{4(\beta+\alpha)} I. \end{cases} \end{aligned}$$

Source: [88], [1448, p. 475], [2073, 2804], [2977, p. 8], and [2991, p. 252].

Fact 10.25.50. Let $A, B \in \mathbb{F}^{n\times n}$, and assume that A and B are positive definite. Then, the following statements hold:

i) If $r \in [0, 1]$, then $(A \odot B)^{-r} \le A^{-r} \odot B^{-r}$.

ii) If k, l are nonzero integers such that $k \le l$, then, $(A^k \odot B^k)^{1/k} \le (A^l \odot B^l)^{1/l}$.

In particular, for all $k \ge 1$,

$$A \odot B \le (A^k \odot B^k)^{1/k}, \quad A^{1/k} \odot B^{1/k} \le (A \odot B)^{1/k}.$$

Now, define $\alpha \triangleq \lambda_{\min}(A \otimes B)$ and $\beta \triangleq \lambda_{\max}(A \otimes B)$. Then,

$$(A^{-1} \odot B^{-1})^{-1} \le A \odot B \le \begin{cases} \frac{(\alpha+\beta)^2}{4\alpha\beta}(A^{-1} \odot B^{-1})^{-1} \\ (A^{-1} \odot B^{-1})^{-1} + (\sqrt{\beta} - \sqrt{\alpha})^2 I. \end{cases}$$

Source: [26, 2073].

Fact 10.25.51. Let $A, B \in \mathbb{F}^{n\times n}$, assume that A and B are positive semidefinite, and define $\alpha \triangleq \lambda_{\min}(A \oplus B)$ and $\beta \triangleq \lambda_{\max}(A \oplus B)$. Then,

$$(A^2 + B^2) \odot I \le \frac{(\alpha+\beta)^2}{4\alpha\beta}[(A + B) \odot I]^2, \quad (A^2 + B^2) \odot I \le [(A + B) \odot I]^2 + \frac{(\beta-\alpha)^2}{4} I.$$

Now, assume that A and B are positive definite. Then,

$$(A^{-1} + B^{-1}) \odot I \le \frac{(\alpha+\beta)^2}{4\alpha\beta}[(A + B) \odot I]^{-1}, \quad (A^{-1} + B^{-1}) \odot I \le [(A + B) \odot I]^{-1} + \frac{\sqrt{\beta} - \sqrt{\alpha}}{\alpha\beta} I.$$

Source: [26]. **Source:** $A \odot I + I \odot B = (A + B) \odot I$ is the *Hadamard sum* of A and B. See [26].

Fact 10.25.52. Let $A, B \in \mathbb{F}^{n\times n}$, and assume that A and B are Hermitian. Then, there exist unitary matrices $S_1, S_2 \in \mathbb{F}^{n\times n}$ such that

$$\langle A \odot B\rangle \le \tfrac{1}{2}[S_1(\langle A\rangle \odot \langle B\rangle)S_1^* + S_2(\langle A\rangle \odot \langle B\rangle)S_2^*].$$

Source: [196]. **Related:** Fact 10.11.34.

Fact 10.25.53. Let $A, B \in \mathbb{F}^{n\times n}$, and assume that A is positive definite, B is positive semidefinite, and $I \odot B$ is positive definite. Then, for all $i \in \{1,\dots,n\}$,

$$[(A\odot B)^{-1}]_{(i,i)} \le \frac{(A^{-1})_{(i,i)}}{B_{(i,i)}}.$$

Furthermore, if $\operatorname{rank} B = 1$, then equality holds. **Source:** [2955].

Fact 10.25.54. Let $A, B \in \mathbb{F}^{n\times n}$, assume that A and B are positive semidefinite, let $p, q \in \mathbb{R}$, and assume that at least one of the following statements holds:

i) $p \le q \le -1$, and A and B are positive definite.

ii) $p \le -1 < 1 \le q$, and A and B are positive definite.

iii) $1 \le p \le q$.

iv) $\frac{1}{2} \le p \le 1 \le q$.

v) $p \le -1 \le q \le -\frac{1}{2}$, and A and B are positive definite.

Then,

$$(A^p \odot B^p)^{1/p} \le (A^q \odot B^q)^{1/q}.$$

Source: [2074]. Consider case *iii*). Since $p/q \le 1$, it follows from Fact 10.25.49 that $A^p \odot B^p = (A^q)^{p/q} \odot (A^q)^{p/q} \le (A^q \odot B^q)^{p/q}$. Then, use Corollary 10.6.11 with p replaced by $1/p$. **Remark:** See [198] and [2977, p. 8].

Fact 10.25.55. Let $A, B \in \mathbb{F}^{n\times n}$, and assume that A and B are positive definite. Then,

$$2I \le A\odot B^{-1} + B\odot A^{-1}.$$

Source: [2804, 2985]. **Remark:** Setting $B = A$ yields an inequality given by Fact 10.25.4. Setting $B = I$ yields Fact 10.25.6.

Fact 10.25.56. Let $A, B \in \mathbb{F}^{n\times m}$, and define

$$\mathcal{A} \triangleq \begin{bmatrix} A^*A\odot B^*B & (A\odot B)^* \\ A\odot B & I \end{bmatrix}.$$

Then, $\mathcal{A}$ is positive semidefinite. Furthermore,

$$-A^*A\odot B^*B \le \left\{ \begin{matrix} (A\odot B)^*(A\odot B) \\ A^*B\odot B^*A \end{matrix} \right\} \le \tfrac{1}{2}(A^*A\odot B^*B + A^*B\odot B^*A) \le A^*A\odot B^*B.$$

Source: [1453, 2804, 2985]. **Remark:** $(A\odot B)^*(A\odot B) \le A^*A\odot B^*B$ is *Amemiya's inequality*. See [1873].

Fact 10.25.57. Let $A, B \in \mathbb{F}^{n\times m}$. Then,

$$-A^*A\odot B^*B \le A^*B\odot B^*A \le A^*A\odot B^*B, \quad |\det(A^*B\odot B^*A)| \le \det(A^*A\odot B^*B).$$

Source: Apply Fact 10.25.60 to $\left[\begin{smallmatrix} A^*A & A^*B \\ B^*A & B^*B \end{smallmatrix}\right]$. **Related:** Fact 10.12.33 and Fact 10.25.11.

Fact 10.25.58. Let $A, B \in \mathbb{F}^{n\times n}$, and assume that A is positive definite. Then,

$$-A\odot B^*A^{-1}B \le B\odot B^* \le A\odot B^*A^{-1}B, \quad |\det(B\odot B^*)| \le \det(A\odot B^*A^{-1}B).$$

Source: Fact 10.12.49 and Fact 10.25.60.

Fact 10.25.59. Let $A, B \in \mathbb{F}^{n\times n}$, assume that A and B are positive definite, and let $x, y \in \mathbb{F}^n$. Then,

$$(x\odot y)^*(A\odot B)^{-1}(x\odot y) \le (x^*A^{-1}x)\odot(y^*B^{-1}y).$$

Source: [2991, p. 243]. **Related:** Fact 10.19.16.

Fact 10.25.60. Let $A, B, C \in \mathbb{F}^{n\times n}$, define

$$\mathcal{A} \triangleq \begin{bmatrix} A & B \\ B^* & C \end{bmatrix},$$

and assume that $\mathcal{A}$ is positive semidefinite. Then,

$$-A \odot C \le B \odot B^* \le A \odot C, \quad |\det(B \odot B^*)| \le \det(A \odot C),$$
$$|\operatorname{tr}(B \odot B^*)| \le \operatorname{tr}(A \odot C), \quad \sqrt{\operatorname{tr}(B \odot B^*)^2} \le \operatorname{tr}(A \odot C).$$

If, in addition, $\mathcal{A}$ is positive definite, then

$$-A \odot C < B \odot B^* < A \odot C, \quad |\det(B \odot B^*)| < \det(A \odot C).$$

Source: [2985] and [2991, p. 244]. **Related:** Fact 10.12.9, Fact 10.12.10, and Fact 10.12.11.

Fact 10.25.61. Let $A, B \in \mathbb{F}^{n\times n}$ and $\alpha, \beta \in (0, \infty)$.

$$-(\beta^{-1/2}I + \alpha A^*A) \odot (\alpha^{-1/2}I + \beta BB^*) \le (A + B) \odot (A + B)^* \le (\beta^{-1/2}I + \alpha A^*A) \odot (\alpha^{-1/2}I + \beta BB^*).$$

Related: Fact 10.12.51.

Fact 10.25.62. Let $A, B \in \mathbb{F}^{n\times m}$, and define

$$\mathcal{A} \triangleq \begin{bmatrix} A^*A \odot I & (A \odot B)^* \\ A \odot B & BB^* \odot I \end{bmatrix}.$$

Then, $\mathcal{A}$ is positive semidefinite. Now, assume that $n = m$. Then,

$$-A^*A \odot I - BB^* \odot I \le A \odot B + (A \odot B)^* \le A^*A \odot I + BB^* \odot I,$$
$$-A^*A \odot BB^* \odot I \le A \odot A^* \odot B \odot B^* \le A^*A \odot BB^* \odot I.$$

Related: Fact 10.25.60.

Fact 10.25.63. Let $A, B \in \mathbb{F}^{n\times n}$, and assume that A and B are positive semidefinite. Then,

$$A \odot B \le \tfrac{1}{2}(A^2 + B^2) \odot I.$$

Source: Fact 10.25.62.

Fact 10.25.64. Let $A, B \in \mathbb{F}^{n\times m}$, assume that A is positive definite, and define

$$\mathcal{A} \triangleq \begin{bmatrix} A \odot (B^*A^{-1}B) & B^* \odot B \\ B^* \odot B & A \odot (B^*A^{-1}B) \end{bmatrix}.$$

Then, $\mathcal{A}$ is positive semidefinite. In particular, $\left[\begin{smallmatrix} A\odot A^{-1} & I \\ I & A\odot A^{-1} \end{smallmatrix}\right]$ is positive semidefinite. Furthermore,

$$\sqrt{\operatorname{tr}(B^* \odot B)^2} \le \operatorname{tr}[A \odot (B^*A^{-1}B)], \quad \det(B^* \odot B) \le \det[A \odot (B^*A^{-1}B)].$$

Finally, if $I \odot B^*B$ is positive definite, then

$$(B^* \odot B)(I \odot B^*B)^{-1}(B^* \odot B) \le I \odot B^*B.$$

Source: [2991, p. 244].

Fact 10.25.65. Let $A, B \in \mathbb{F}^{n\times n}$, assume that A and B are positive definite, and let $p, q \in (0, \infty)$ satisfy $p \le q$. Then,

$$I \odot (\log A + \log B) = \lim_{p\downarrow 0} \log (A^p \odot B^p)^{1/p} \le \log (A^p \odot B^p)^{1/p} \le \log (A^q \odot B^q)^{1/q}.$$

In particular,

$$I \odot (\log A + \log B) \le \log(A \odot B).$$

Source: [88, 2803] and [2977, p. 8]. **Remark:** $\log(A^p \odot B^p)^{1/p} = \frac{1}{p}\log(A^p \odot B^p)$. **Related:** Fact 15.15.22.

Fact 10.25.66. Let $A, B \in \mathbb{F}^{n\times n}$, assume that A and B are positive semidefinite, and let $k \geq 0$. Then,

$$A^{(k)} + B^{(k)} \leq (A + B)^{(k)}.$$

If either $\alpha \in [0, \infty)$ or both A is positive definite and $\alpha \in \mathbb{R}$, then

$$(A^\alpha)^{(k)} = (A^{(k)})^\alpha.$$

Source: [1868]. **Remark:** $A^{(k)}$ is the kth compound of A. See Fact 9.5.18.

Fact 10.25.67. Let $A, B \in \mathbb{F}^{n\times n}$, assume that A and B are positive definite, and let $C, D \in \mathbb{F}^{m\times n}$. Then,

$$(C \odot D)(A \odot B)^{-1}(C \odot D)^* \leq (CA^{-1}C^*) \odot (DB^{-1}D^*).$$

In particular,

$$(A \odot B)^{-1} \leq A^{-1} \odot B^{-1}, \quad (C \odot D)(C \odot D)^* \leq (CC^*) \odot (DD^*).$$

Source: Consider the Schur complement of the lower right block of the Schur product of the positive-semidefinite matrices $\left[\begin{smallmatrix} A & C^* \\ C & CA^{-1}C^* \end{smallmatrix}\right]$ and $\left[\begin{smallmatrix} B & D^* \\ D & DB^{-1}D^* \end{smallmatrix}\right]$. See [1959, 2818], [2977, p. 13], [2983, p. 198], and [2991, p. 240].

Fact 10.25.68. Let $A, B \in \mathbb{F}^{n\times n}$, assume that A and B are positive semidefinite, and let $p, q \in (1, \infty)$ satisfy $1/p + 1/q = 1$. Then,

$$(A \odot B) + (C \odot D) \leq (A^p + C^p)^{1/p} \odot (B^q + D^q)^{1/q}.$$

Source: Use *xxiv)* of Proposition 10.6.17 with $r = 1/p$. See [2977, p. 10].

Fact 10.25.69. Let $A_1, \ldots, A_m, B_1, \ldots, B_m \in \mathbb{F}^{n\times n}$, assume that $A_1 \geq \cdots \geq A_m \geq 0$ and $B_1 \geq \cdots \geq B_m \geq 0$, and, for all $i \in \{1, \ldots, m\}$, let $\alpha_i \geq 0$. Then,

$$\left(\sum_{i=1}^m \alpha_i A_i\right) \odot \left(\sum_{i=1}^m \alpha_i B_i\right) \leq \left(\sum_{i=1}^m \alpha_i\right) \sum_{i=1}^m \alpha_i (A_i \odot B_i).$$

Source: [1982]. **Remark:** This is an extension of the Chebyshev inequality given by Fact 2.12.7.

Fact 10.25.70. Let $A_1, \ldots, A_m, B_1, \ldots, B_m \in \mathbb{F}^{n\times n}$, assume that there exist $a, b \in (0, \infty)$ such that, for all $i \in \{1, \ldots, m\}$, $aI \leq A_i \leq bI$, and, for all $i \in \{1, \ldots, m\}$, assume that B_i is positive semidefinite. Then,

$$\left(\sum_{i=1}^m B^{1/2} A_i B^{1/2}\right) \odot \sum_{i=1}^m B^{1/2} A_i^{-1} B^{1/2} \leq \frac{a^2 + b^2}{2ab} \left(\sum_{i=1}^m B_i\right) \odot \sum_{i=1}^m B_i.$$

Source: [1982]. **Remark:** This is an extension of the Kantorovich inequality Fact 2.11.134.

Fact 10.25.71. Let $A, B, C, D \in \mathbb{F}^{n\times n}$, assume that A, B, C, and D are positive definite. Then,

$$(A\#C) \odot (B\#D) \leq (A \odot B)\#(C \odot D), \quad (A\#B) \odot (A\#B) \leq (A \odot B).$$

Source: [198].

Fact 10.25.72. Let $A \in \mathbb{R}^{n\times n}$, assume that A is nonnegative and positive semidefinite, and let $f\colon [0, \infty) \mapsto [0, \infty)$. Then, the following statements hold:

i) If f is concave and $f(0) = 0$, then $\lambda_{\max}[f(A)] \leq \lambda_{\max}[f \odot (A)]$ and $\operatorname{tr} f(A) \leq \operatorname{tr} f \odot (A)$.

ii) If f is convex and $f(0) = 0$, then $\lambda_{\max}[f \odot (A)] \leq \lambda_{\max}[f(A)]$ and $\operatorname{tr} f \odot (A) \leq \operatorname{tr} f(A)$.

iii) If $p \in [0, 1]$, then $\lambda_{\max}(A^p) \leq \lambda_{\max}(A^{\odot p})$ and $\operatorname{tr} A^p \leq \operatorname{tr} A^{\odot p}$.

iv) If $p \geq 1$, then $\lambda_{\max}(A^{\odot p}) \leq \lambda_{\max}(A^p)$ and $\operatorname{tr} A^{\odot p} \leq \operatorname{tr} A^p$.

Source: [1378]. **Remark:** $f \odot (A)$ denotes the $n \times n$ matrix whose i, j entry is $f(A_{(i,j)})$.

10.26 Notes

The ordering $A \leq B$ is called the *Löwner ordering*. Proposition 10.2.5 is given in [30] and [1703] with extensions in [347]. The proof of Proposition 10.2.8 is based on [583, p. 120], as suggested in [2579]. The proof given in [1139, p. 307] is incomplete.

Theorem 10.3.5 is due to R. W. Newcomb [2123]. Proposition 10.4.13 is given in [1425, 2082]. Special cases such as Fact 10.14.35 appear in numerous papers. The proofs of Lemma 10.4.4 and Theorem 10.4.5 are based on [2539]. Theorem 10.4.9 can also be obtained as a corollary of the *Fischer minimax theorem* given in [1448, 1969, 1971], which provides a geometric characterization of the eigenvalues of a symmetric matrix. Three proofs of Theorem 10.4.5 are given in [2991, pp. 269–271]. Theorem 10.3.6 appears in [2299, p. 121]. Theorem 10.6.2 is given in [85]. Additional inequalities appear in [2051].

Functions that are nondecreasing on $\mathbf{P}^n$ are characterized by the theory of *monotone matrix functions* [449, 922].

The literature on convex maps is extensive. *xiv*) of Proposition 10.6.17 is given in [1819]. *xxiv*) is the *Lieb concavity theorem*. See [449, p. 271] and [1817]. *xxxiv*) is due to T. Ando. *xlv*) and *xlvi*) are due to K. Fan. Some extensions to strict convexity are considered in [1969, 1971]. See also [88, 2099].

Products of positive-definite matrices are studied in [263, 264, 265, 267, 2918].

Essays on the legacy of Issai Schur appear in [1558]. Schur complements are discussed in [637, 639, 1355, 1800, 1869, 2179]. Majorization and eigenvalue inequalities for sums and products of matrices are discussed in [450].

Chapter Eleven
Norms

Norms are used to quantify vectors and matrices. This chapter introduces vector and matrix norms and their properties.

11.1 Vector Norms

Definition 11.1.1. A *norm* $\|\cdot\|$ on $\mathbb{F}^n$ is a function $\|\cdot\|\colon\ \mathbb{F}^n \mapsto [0,\infty)$ that satisfies the following statements:

i) $\|x\| = 0$ if and only if $x = 0$.

ii) For all $\alpha \in \mathbb{F}$ and $x \in \mathbb{F}^n$, $\|\alpha x\| = |\alpha|\|x\|$.

iii) For all $x, y \in \mathbb{F}^n$, $\|x + y\| \le \|x\| + \|y\|$.

iii) is the *triangle inequality*.

The norm $\|\cdot\|$ on $\mathbb{F}^n$ is *monotone* if, for all $x, y \in \mathbb{F}^n$, $|x| \le\le |y|$ implies that $\|x\| \le \|y\|$, while $\|\cdot\|$ is *absolute* if, for all $x \in \mathbb{F}^n$, $\||x|\| = \|x\|$.

Proposition 11.1.2. Let $\|\cdot\|$ be a norm on $\mathbb{F}^n$. Then, $\|\cdot\|$ is monotone if and only if $\|\cdot\|$ is absolute.

Proof. First, suppose that $\|\cdot\|$ is monotone. Let $x \in \mathbb{F}^n$, and define $y \triangleq |x|$. Then, $|y| = |x|$, and thus $|y| \le\le |x|$ and $|x| \le\le |y|$. Hence, $\|x\| \le \|y\|$ and $\|y\| \le \|x\|$, which implies that $\|x\| = \|y\|$. Thus, $\||x|\| = \|y\| = \|x\|$, which proves that $\|\cdot\|$ is absolute.

Conversely, suppose that $\|\cdot\|$ is absolute and, for convenience, let $n = 2$. Now, let $x, y \in \mathbb{F}^2$ satisfy $|x| \le\le |y|$. Then, there exist $\alpha_1, \alpha_2 \in [0,1]$ and $\theta_1, \theta_2 \in \mathbb{R}$ such that $x_{(i)} = \alpha_i e^{J\theta_i} y_{(i)}$ for $i = 1, 2$. Since $\|\cdot\|$ is absolute, it follows that

$$\begin{aligned}
\|x\| &= \left\|\begin{bmatrix}\alpha_1 e^{\theta_1 J} y_{(1)}\\ \alpha_2 e^{\theta_2 J} y_{(2)}\end{bmatrix}\right\| = \left\|\begin{bmatrix}\alpha_1 |y_{(1)}|\\ \alpha_2 |y_{(2)}|\end{bmatrix}\right\| \\
&= \left\|\tfrac{1}{2}(1-\alpha_1)\begin{bmatrix}-|y_{(1)}|\\ \alpha_2|y_{(2)}|\end{bmatrix} + \tfrac{1}{2}(1-\alpha_1)\begin{bmatrix}|y_{(1)}|\\ \alpha_2|y_{(2)}|\end{bmatrix} + \alpha_1\begin{bmatrix}|y_{(1)}|\\ \alpha_2|y_{(2)}|\end{bmatrix}\right\| \\
&\le \left[\tfrac{1}{2}(1-\alpha_1) + \tfrac{1}{2}(1-\alpha_1) + \alpha_1\right]\left\|\begin{bmatrix}|y_{(1)}|\\ \alpha_2|y_{(2)}|\end{bmatrix}\right\| = \left\|\begin{bmatrix}|y_{(1)}|\\ \alpha_2|y_{(2)}|\end{bmatrix}\right\| \\
&= \left\|\tfrac{1}{2}(1-\alpha_2)\begin{bmatrix}|y_{(1)}|\\ -|y_{(2)}|\end{bmatrix} + \tfrac{1}{2}(1-\alpha_2)\begin{bmatrix}|y_{(1)}|\\ |y_{(2)}|\end{bmatrix} + \alpha_2\begin{bmatrix}|y_{(1)}|\\ |y_{(2)}|\end{bmatrix}\right\| \le \left\|\begin{bmatrix}|y_{(1)}|\\ |y_{(2)}|\end{bmatrix}\right\| = \|y\|.
\end{aligned}$$

□

For $x \in \mathbb{F}^n$, a useful class of norms consists of the *Hölder norms* defined by

$$\|x\|_p \triangleq \begin{cases}\left(\displaystyle\sum_{i=1}^n |x_{(i)}|^p\right)^{1/p}, & 1 \le p < \infty,\\ \displaystyle\max_{i\in\{1,\ldots,n\}} |x_{(i)}|, & p = \infty.\end{cases} \tag{11.1.1}$$

Note that, for all $x \in \mathbb{C}^n$ and $p \in [1,\infty]$, $\|\overline{x}\|_p = \|x\|_p$. These norms depend on *Minkowski's inequality* given by the following result.

Lemma 11.1.3. Let $p \in [1, \infty]$ and $x, y \in \mathbb{F}^n$. Then,

$$\|x + y\|_p \le \|x\|_p + \|y\|_p. \tag{11.1.2}$$

If $p = 1$, then equality holds if and only if, for all $i \in \{1, \ldots, n\}$, there exists $\alpha_i \ge 0$ such that either $x_{(i)} = \alpha_i y_{(i)}$ or $y_{(i)} = \alpha_i x_{(i)}$. If $p \in (1, \infty)$, then equality holds if and only if there exists $\alpha \ge 0$ such that either $x = \alpha y$ or $y = \alpha x$.

Proof. See [340, 1952] and Fact 2.12.51. □

Proposition 11.1.4. Let $p \in [1, \infty]$. Then, $\|\cdot\|_p$ is a norm on $\mathbb{F}^n$.

For $p = 1$,

$$\|x\|_1 = \sum_{i=1}^{n} |x_{(i)}| \tag{11.1.3}$$

is the *absolute sum norm*; for $p = 2$,

$$\|x\|_2 = \left(\sum_{i=1}^{n} |x_{(i)}|^2 \right)^{1/2} = \sqrt{x^*x} \tag{11.1.4}$$

is the Euclidean norm defined in (3.3.17); and, for $p = \infty$,

$$\|x\|_\infty = \max_{i \in \{1, \ldots, n\}} |x_{(i)}| \tag{11.1.5}$$

is the *infinity norm*. Note that, for all $x \in \mathbb{F}^n$,

$$\lim_{p \to \infty} \|x\|_p = \|x\|_\infty. \tag{11.1.6}$$

The Hölder norms satisfy the following monotonicity property, which is related to the power-sum inequality given by Fact 2.11.90.

Proposition 11.1.5. Let $1 \le p \le q \le \infty$ and $x \in \mathbb{F}^n$. Then,

$$\|x\|_\infty \le \|x\|_q \le \|x\|_p \le \|x\|_1. \tag{11.1.7}$$

Assume, in addition, that $1 < p < q < \infty$. Then, x has at least two nonzero components if and only if

$$\|x\|_\infty < \|x\|_q < \|x\|_p < \|x\|_1. \tag{11.1.8}$$

Proof. If either $p = q$ or $x = 0$ or x has exactly one nonzero component, then $\|x\|_q = \|x\|_p$. Hence, to prove both (11.1.7) and (11.1.8), it suffices to prove (11.1.8) in the case where $1 < p < q < \infty$ and x has at least two nonzero components. Thus, let $n \ge 2$, let $x \in \mathbb{F}^n$ have at least two nonzero components, and define $f\colon [1, \infty) \to [0, \infty)$ by $f(\beta) \triangleq \|x\|_\beta$. Hence,

$$f'(\beta) = \tfrac{1}{\beta} \|x\|_\beta^{1-\beta} \sum_{i=1}^{n} \gamma_i,$$

where, for all $i \in \{1, \ldots, n\}$,

$$\gamma_i \triangleq \begin{cases} |x_i|^\beta (\log |x_{(i)}| - \log \|x\|_\beta), & x_{(i)} \ne 0, \\ 0, & x_{(i)} = 0. \end{cases}$$

If $x_{(i)} \ne 0$, then $\log |x_{(i)}| < \log \|x\|_\beta$. It thus follows that $f'(\beta) < 0$, which implies that f is decreasing on $[1, \infty)$. Hence, (11.1.8) holds. □

The following result is *Hölder's inequality*.

Proposition 11.1.6. Let $p, q \in [1, \infty]$ satisfy $1/p + 1/q = 1$, and let $x, y \in \mathbb{F}^n$. Then,

$$|x^*y| \le \|x\|_p \|y\|_q. \tag{11.1.9}$$

Furthermore, equality holds if and only if $|x^{\mathsf{T}}y| = |x|^{\mathsf{T}}|y|$ and

$$\begin{cases} |x| \odot |y| = \|y\|_\infty |x|, & p = 1, \\ \|y\|_q^{1/p} |x|^{\odot 1/q} = \|x\|_p^{1/q} |y|^{\odot 1/p}, & 1 < p < \infty, \\ |x| \odot |y| = \|x\|_\infty |y|, & p = \infty. \end{cases} \tag{11.1.10}$$

Proof. See [603, p. 127], [1448, p. 536], [1597, p. 71], Fact 2.12.23, and Fact 2.12.24. □

The case $p = q = 2$ is the *Cauchy-Schwarz inequality*.

Corollary 11.1.7. Let $x, y \in \mathbb{F}^n$. Then,

$$|x^*y| \le \|x\|_2 \|y\|_2. \tag{11.1.11}$$

Furthermore, equality holds if and only if x and y are linearly dependent.

Proof. Define $\theta \triangleq \arg x^*y \in (-\pi, \pi]$ satisfy $x^*y = e^{\theta J}|x^*y|$. Then,

$$0 \le \big\| \|x\|_2 y - \|y\|_2 e^{\theta J} x \big\|_2 = \|x\|_2 \|y\|_2 (\|x\|_2 \|y\|_2 - |x^*y|).$$ □

Let $x, y \in \mathbb{F}^n$, assume that x and y are both nonzero, let $p, q \in [1, \infty]$, and assume that $1/p + 1/q = 1$. Since $\|\overline{x}\|_p = \|x\|_p$, it follows that

$$|x^{\mathsf{T}}y| \le \|x\|_p \|y\|_q, \tag{11.1.12}$$

and, in particular,

$$|x^{\mathsf{T}}y| \le \|x\|_2 \|y\|_2. \tag{11.1.13}$$

The angle $\theta \in [0, \pi]$ between x and y, which is defined by (3.3.20), is thus given by

$$\theta = \operatorname{acos} \frac{\operatorname{Re} x^*y}{\|x\|_2 \|y\|_2}. \tag{11.1.14}$$

The norms $\|\cdot\|$ and $\|\cdot\|'$ on $\mathbb{F}^n$ are *equivalent* if, for all $x \in \mathbb{F}^n$, there exist $\alpha, \beta > 0$ such that

$$\alpha\|x\| \le \|x\|' \le \beta\|x\|. \tag{11.1.15}$$

Note that these inequalities can be written as

$$\tfrac{1}{\beta}\|x\|' \le \|x\| \le \tfrac{1}{\alpha}\|x\|'. \tag{11.1.16}$$

Hence, the word "equivalent" is justified. The following result shows that every pair of norms on $\mathbb{F}^n$ is equivalent.

Theorem 11.1.8. Let $\|\cdot\|$ and $\|\cdot\|'$ be norms on $\mathbb{F}^n$. Then, $\|\cdot\|$ and $\|\cdot\|'$ are equivalent.

Proof. See [1448, p. 272]. □

11.2 Matrix Norms

Definition 11.2.1. A *norm* $\|\cdot\|$ on $\mathbb{F}^{n\times m}$ is a function $\|\cdot\|\colon \mathbb{F}^{n\times m} \mapsto [0, \infty)$ that satisfies the following statements:

i) $\|A\| = 0$ if and only if $A = 0$.

ii) For all $\alpha \in \mathbb{F}$ and $A \in \mathbb{F}^{n\times m}$, $\|\alpha A\| = |\alpha|\|A\|$.

iii) For all $A, B \in \mathbb{F}^{n\times m}$, $\|A + B\| \le \|A\| + \|B\|$.

If $\|\cdot\|$ is a norm on $\mathbb{F}^{n\times m}$, then $\|\cdot\|'$ defined by $\|x\|' = \|\operatorname{vec}^{-1} x\|$, where $x \in \mathbb{F}^{nm}$ and $\operatorname{vec}^{-1} x \in \mathbb{F}^{n\times m}$, is a norm on $\mathbb{F}^{nm}$. Consequently, each matrix norm defines a vector norm. Conversely, if $\|\cdot\|$ is a norm on $\mathbb{F}^{nm}$, then $\|\cdot\|'$ defined by $\|A\|' \triangleq \|\operatorname{vec} A\|$, where $A \in \mathbb{F}^{n\times m}$ and $\operatorname{vec} A \in \mathbb{F}^{nm}$, is a norm on $\mathbb{F}^{n\times m}$.

Hölder norms are defined for matrices by choosing the vector norm $\|\cdot\| = \|\cdot\|_p$. Let $A \in \mathbb{F}^{n\times m}$. Then, the *Hölder matrix norm* of A is defined by

$$\|A\|_p \triangleq \begin{cases} \left(\displaystyle\sum_{j=1}^{m} \sum_{i=1}^{n} |A_{(i,j)}|^p \right)^{1/p}, & 1 \le p < \infty, \\ \displaystyle\max_{\substack{i\in\{1,\ldots,n\} \\ j\in\{1,\ldots,m\}}} |A_{(i,j)}|, & p = \infty. \end{cases} \tag{11.2.1}$$

Note that the same symbol $\|\cdot\|_p$ is used to denote the Hölder norm for both vectors and matrices. This notation is consistent since, if $m = 1$, then $\|A\|_p$ coincides with the vector Hölder norm. Furthermore, if $1 \le p \le \infty$, then

$$\|A\|_p = \|\operatorname{vec} A\|_p. \tag{11.2.2}$$

It follows from (11.1.7) that, if $1 \le p \le q \le \infty$, then

$$\|A\|_\infty \le \|A\|_q \le \|A\|_p \le \|A\|_1. \tag{11.2.3}$$

Finally, if $1 < p < q < \infty$ and A has at least two nonzero entries, then

$$\|A\|_\infty < \|A\|_q < \|A\|_p < \|A\|_1. \tag{11.2.4}$$

The Hölder norms with $p = 1, 2, \infty$ are the most commonly used. Let $A \in \mathbb{F}^{n\times m}$. For $p = 2$ we define the *Frobenius norm* $\|\cdot\|_{\mathrm{F}}$ by

$$\|A\|_{\mathrm{F}} \triangleq \|A\|_2. \tag{11.2.5}$$

Since $\|A\|_2 = \|\operatorname{vec} A\|_2$, it follows that

$$\|A\|_{\mathrm{F}} = \|A\|_2 = \|\operatorname{vec} A\|_2 = \|\operatorname{vec} A\|_{\mathrm{F}}. \tag{11.2.6}$$

It is easy to see that

$$\|A\|_{\mathrm{F}} = \sqrt{\operatorname{tr} A^*A}. \tag{11.2.7}$$

Let $A \in \mathbb{F}^{n\times m}$. The *mixed Hölder norm* $\|A\|_{p|q}$ of A is defined by applying $\|\cdot\|_p$ to each column of A and then applying $\|\cdot\|_q$ to the resulting vector. Hence,

$$\|A\|_{p|q} \triangleq \left\| \begin{bmatrix} \|\mathrm{col}_1(A)\|_p \\ \vdots \\ \|\mathrm{col}_m(A)\|_p \end{bmatrix} \right\|_q. \tag{11.2.8}$$

Therefore,

$$\|A\|_{p|q} = \begin{cases} \left[\displaystyle\sum_{j=1}^{m} \left(\sum_{i=1}^{n} |A_{(i,j)}|^p \right)^{q/p} \right]^{1/q}, & p, q \in (1, \infty), \\ \displaystyle\max_{j\in\{1,\ldots,m\}} \left(\sum_{i=1}^{n} |A_{(i,j)}|^p \right)^{1/p} & p \in (1, \infty),\ q = \infty, \\ \left[\displaystyle\sum_{j=1}^{m} \left(\max_{i\in\{1,\ldots,n\}} |A_{(i,j)}| \right)^{q} \right]^{1/q} & p = \infty,\ q \in (1, \infty), \\ \displaystyle\max_{\substack{i\in\{1,\ldots,n\} \\ j\in\{1,\ldots,m\}}} |A_{(i,j)}|, & p = q = \infty. \end{cases} \tag{11.2.9}$$

Furthermore, for all $p \in [1, \infty]$,

$$\|A\|_{p|p} = \|A\|_p. \tag{11.2.10}$$

Finally, note that $\|A^{\mathsf{T}}\|_{p|q}$ applies $\|\cdot\|_p$ to each row of A and then applies $\|\cdot\|_q$ to the resulting vector.

Let $\|\cdot\|$ be a norm on $\mathbb{F}^{n\times m}$. Then, $\|\cdot\|$ is *unitarily invariant* if, for all $A \in \mathbb{F}^{n\times m}$ and all unitary matrices $S_1 \in \mathbb{F}^{n\times n}$ and $S_2 \in \mathbb{F}^{m\times m}$, $\|S_1AS_2\| = \|A\|$. Now, let $m = n$. Then, $\|\cdot\|$ is *self-adjoint* if, for all $A \in \mathbb{F}^{n\times n}$, $\|A\| = \|A^*\|$. $\|\cdot\|$ is *normalized* if $\|I_n\| = 1$. Note that the Frobenius norm is not normalized since $\|I_n\|_{\mathrm{F}} = \sqrt{n}$. Finally, $\|\cdot\|$ is *weakly unitarily invariant* if, for all $A \in \mathbb{F}^{n\times n}$ and all unitary matrices $S \in \mathbb{F}^{n\times n}$, $\|SAS^*\| = \|A\|$.

Matrix norms can be defined in terms of singular values. Let $\sigma_1(A) \ge \sigma_2(A) \ge \cdots \ge \sigma_{\min\{n,m\}}$ denote the singular values of $A \in \mathbb{F}^{n\times m}$. The following result gives a weak majorization condition for the singular values of a sum of matrices.

Proposition 11.2.2. Let $A, B \in \mathbb{F}^{n\times m}$. Then, for all $k \in \{1, \ldots, \min\{n, m\}\}$,

$$\sum_{i=1}^{k}[\sigma_i(A) - \sigma_i(B)] \le \sum_{i=1}^{k}\sigma_i(A + B) \le \sum_{i=1}^{k}[\sigma_i(A) + \sigma_i(B)]. \tag{11.2.11}$$

In particular,

$$\sigma_{\max}(A) - \sigma_{\max}(B) \le \sigma_{\max}(A + B) \le \sigma_{\max}(A) + \sigma_{\max}(B), \tag{11.2.12}$$

$$\operatorname{tr}\langle A\rangle - \operatorname{tr}\langle B\rangle \le \operatorname{tr}\langle A + B\rangle \le \operatorname{tr}\langle A\rangle + \operatorname{tr}\langle B\rangle. \tag{11.2.13}$$

Proof. Define $\mathcal{A}, \mathcal{B} \in \mathbf{H}^{n+m}$ by $\mathcal{A} \triangleq \begin{bmatrix} 0 & A \\ A^* & 0 \end{bmatrix}$ and $\mathcal{B} \triangleq \begin{bmatrix} 0 & B \\ B^* & 0 \end{bmatrix}$. Then, Corollary 10.6.19 implies that, for all $k \in \{1, \ldots, n + m\}$,

$$\sum_{i=1}^{k}\lambda_i(\mathcal{A} + \mathcal{B}) \le \sum_{i=1}^{k}[\lambda_i(\mathcal{A}) + \lambda_i(\mathcal{B})].$$

Now, consider $k \le \min\{n, m\}$. Then, it follows from Proposition 7.6.5 that, for all $i \in \{1, \ldots, k\}$, $\lambda_i(\mathcal{A}) = \sigma_i(A)$. Setting $k = 1$ yields (11.2.12), while setting $k = \min\{n, m\}$ and using Fact 10.21.7 yields (11.2.13). □

Let $p \in [1, \infty]$ and $A \in \mathbb{F}^{n\times m}$, and define

$$\|A\|_{\sigma p} \triangleq \|\sigma(A)\|_p. \tag{11.2.14}$$

In other words,

$$\|A\|_{\sigma p} \triangleq \begin{cases} \left(\displaystyle\sum_{i=1}^{\min\{n,m\}}\sigma_i^p(A)\right)^{1/p}, & 1 \le p < \infty, \\ \sigma_{\max}(A), & p = \infty. \end{cases} \tag{11.2.15}$$

Note that, for all $p \in [1, \infty)$,

$$\|A\|_{\sigma p} = (\operatorname{tr}\langle A\rangle^p)^{1/p}. \tag{11.2.16}$$

Proposition 11.2.3. Let $p \in [1, \infty]$. Then, $\|\cdot\|_{\sigma p}$ is a norm on $\mathbb{F}^{n\times m}$.

Proof. Let $A, B \in \mathbb{F}^{n\times m}$. Then, Proposition 11.2.2 and Minkowski's inequality given by Fact 2.12.51 imply that

$$\begin{aligned}\|A + B\|_{\sigma p} &= \left(\sum_{i=1}^{\min\{n,m\}}\sigma_i^p(A + B)\right)^{1/p} \le \left(\sum_{i=1}^{\min\{n,m\}}[\sigma_i(A) + \sigma_i(B)]^p\right)^{1/p} \\ &\le \left(\sum_{i=1}^{\min\{n,m\}}\sigma_i^p(A)\right)^{1/p} + \left(\sum_{i=1}^{\min\{n,m\}}\sigma_i^p(B)\right)^{1/p} = \|A\|_{\sigma p} + \|B\|_{\sigma p}.\end{aligned}$$ □

For all $p \in [1,\infty)$, $\|\cdot\|_{\sigma p}$ is a *Schatten norm*. Let $A \in \mathbb{F}^{n\times m}$. Special cases are

$$\|A\|_{\sigma 1} = \|\sigma(A)\|_1 = \sigma_1(A) + \cdots + \sigma_{\min\{n,m\}}(A) = \operatorname{tr}\langle A\rangle, \tag{11.2.17}$$

$$\|A\|_{\sigma 2} = \|\sigma(A)\|_2 = [\sigma_1^2(A) + \cdots + \sigma_{\min\{n,m\}}^2(A)]^{1/2} = (\operatorname{tr} A^*A)^{1/2} = \|A\|_{\mathrm{F}}, \tag{11.2.18}$$

$$\|A\|_{\sigma\infty} = \|\sigma(A)\|_\infty = \sigma_1(A) = \sigma_{\max}(A), \tag{11.2.19}$$

which are the *trace norm*, Frobenius norm, and *spectral norm*, respectively.

The following result shows that the trace norm bounds the trace.

Proposition 11.2.4. Let $A \in \mathbb{F}^{n\times n}$. Then,

$$|\operatorname{Re}\operatorname{tr} A| \le |\operatorname{tr} A| \le \sum_{i=1}^n \rho_i(A) \le \operatorname{tr}\langle A\rangle = \sum_{i=1}^n \sigma_i(A) = \|A\|_{\sigma 1}. \tag{11.2.20}$$

Proof. See Fact 10.13.2. □

By applying Proposition 11.1.5 to the vector $\sigma(A)$, we obtain the following result.

Proposition 11.2.5. Let $p, q \in [1,\infty)$, where $p \le q$ and $A \in \mathbb{F}^{n\times m}$. Then,

$$\|A\|_{\sigma\infty} \le \|A\|_{\sigma q} \le \|A\|_{\sigma p} \le \|A\|_{\sigma 1}. \tag{11.2.21}$$

Assume, in addition, that $1 < p < q < \infty$ and $\operatorname{rank} A \ge 2$. Then,

$$\|A\|_{\sigma\infty} < \|A\|_{\sigma q} < \|A\|_{\sigma p} < \|A\|_{\sigma 1}. \tag{11.2.22}$$

Let $x \in \mathbb{F}^n = \mathbb{F}^{n\times 1}$. Then, $\sigma_{\max}(x) = (x^*x)^{1/2} = \|x\|_2$, and, since $\operatorname{rank} x \le 1$, it follows that, for all $p \in [1,\infty]$,

$$\|x\|_{\sigma p} = \|x\|_2. \tag{11.2.23}$$

Proposition 11.2.6. Let $\|\cdot\|$ be a norm on $\mathbb{F}^{n\times n}$, and let $A \in \mathbb{F}^{n\times n}$. Then,

$$\rho_{\max}(A) = \lim_{k\to\infty} \|A^k\|^{1/k}. \tag{11.2.24}$$

Proof. See [1448, pp. 299, 322] and [1451, p. 349]. □

11.3 Compatible Norms

The norms $(\|\cdot\|, \|\cdot\|', \|\cdot\|'')$ on $\mathbb{F}^{n\times l}$, $\mathbb{F}^{n\times m}$, and $\mathbb{F}^{m\times l}$, respectively, are *compatible* if, for all $A \in \mathbb{F}^{n\times m}$ and $B \in \mathbb{F}^{m\times l}$,

$$\|AB\| \le \|A\|'\|B\|''. \tag{11.3.1}$$

For $l = 1$, the norms $(\|\cdot\|, \|\cdot\|', \|\cdot\|'')$ on $\mathbb{F}^n$, $\mathbb{F}^{n\times m}$, and $\mathbb{F}^m$, respectively, are compatible if, for all $A \in \mathbb{F}^{n\times m}$ and $x \in \mathbb{F}^m$,

$$\|Ax\| \le \|A\|'\|x\|''. \tag{11.3.2}$$

Furthermore, the norm $\|\cdot\|$ on $\mathbb{F}^n$ is *compatible* with the norm $\|\cdot\|'$ on $\mathbb{F}^{n\times n}$ if, for all $A \in \mathbb{F}^{n\times n}$ and $x \in \mathbb{F}^n$,

$$\|Ax\| \le \|A\|'\|x\|. \tag{11.3.3}$$

Note that $\|I_n\|' \ge 1$. The norm $\|\cdot\|$ on $\mathbb{F}^{n\times n}$ is *submultiplicative* if, for all $A, B \in \mathbb{F}^{n\times n}$,

$$\|AB\| \le \|A\|\|B\|. \tag{11.3.4}$$

Hence, the norm $\|\cdot\|$ on $\mathbb{F}^{n\times n}$ is submultiplicative if and only if $(\|\cdot\|, \|\cdot\|, \|\cdot\|)$ are compatible. In this case, $\|I_n\| \ge 1$, and $\|\cdot\|$ is normalized if and only if $\|I_n\| = 1$.

Proposition 11.3.1. Let $\|\cdot\|'$ be a submultiplicative norm on $\mathbb{F}^{n\times n}$, and let $y \in \mathbb{F}^n$ be nonzero. Then, $\|x\| \triangleq \|xy^*\|'$ is a norm on $\mathbb{F}^n$, and $\|\cdot\|$ is compatible with $\|\cdot\|'$.

Proof. Note that $\|Ax\| = \|Axy^*\|' \le \|A\|'\|xy^*\|' = \|A\|'\|x\|$. □

Proposition 11.3.2. Let $\|\cdot\|$ be a submultiplicative norm on $\mathbb{F}^{n\times n}$, and let $A \in \mathbb{F}^{n\times n}$. Then,

$$\rho_{\max}(A) \le \|A\|. \tag{11.3.5}$$

Proof. Use Proposition 11.3.1 to construct a norm $\|\cdot\|'$ on $\mathbb{F}^n$ that is compatible with $\|\cdot\|$. Furthermore, let $A \in \mathbb{F}^{n\times n}$, let $\lambda \in \operatorname{spec}(A)$, and let $x \in \mathbb{C}^n$ be an eigenvector of A associated with λ. Then, $Ax = \lambda x$ implies that $|\lambda|\|x\|' = \|Ax\|' \le \|A\|\|x\|'$, and thus $|\lambda| \le \|A\|$, which implies (11.3.5). Alternatively, under the additional assumption that $\|\cdot\|$ is submultiplicative, it follows from Proposition 11.2.6 that

$$\rho_{\max}(A) = \lim_{k\to\infty} \|A^k\|^{1/k} \le \lim_{k\to\infty} \|A\|^{k/k} = \|A\|. \qquad \square$$

Proposition 11.3.3. Let $A \in \mathbb{F}^{n\times n}$ and $\varepsilon > 0$. Then, there exists a normalized submultiplicative norm $\|\cdot\|$ on $\mathbb{F}^{n\times n}$ such that

$$\rho_{\max}(A) \le \|A\| \le \rho_{\max}(A) + \varepsilon. \tag{11.3.6}$$

Proof. See [1448, p. 297], [1451, pp. 347, 348], [2403, p. 167], and [2979, pp. 13, 14]. □

It is shown in [2979, pp. 13, 14] that $\|\cdot\|$ can be chosen to be an induced norm.

Corollary 11.3.4. Let $A \in \mathbb{F}^{n\times n}$, and assume that $\rho_{\max}(A) < 1$. Then, there exists a submultiplicative norm $\|\cdot\|$ on $\mathbb{F}^{n\times n}$ such that $\|A\| < 1$.

We now identify several compatible norms. We begin with the Hölder norms.

Proposition 11.3.5. Let $A \in \mathbb{F}^{n\times m}$ and $B \in \mathbb{F}^{m\times l}$. If $p \in [1, 2]$, then

$$\|AB\|_p \le \|A\|_p\|B\|_p. \tag{11.3.7}$$

If $p \in [2, \infty]$ and q satisfies $1/p + 1/q = 1$, then

$$\|AB\|_p \le \|A\|_p\|B\|_q, \tag{11.3.8}$$

$$\|AB\|_p \le \|A\|_q\|B\|_p. \tag{11.3.9}$$

Proof. First, let $1 \le p \le 2$ so that $q \triangleq p/(p-1) \ge 2$. Using Hölder's inequality (11.1.9) along with (11.1.7) and the fact that $p \le q$ yields

$$\begin{aligned}
\|AB\|_p &= \left(\sum_{i,j=1}^{n,l} |\operatorname{row}_i(A)\operatorname{col}_j(B)|^p\right)^{1/p} = \left(\sum_{i,j=1}^{n,l} |\operatorname{col}_i(A^*)^*\operatorname{col}_j(B)|^p\right)^{1/p}\\
&\le \left(\sum_{i,j=1}^{n,l} \|\operatorname{col}_i(A^*)^*\|_p^p\|\operatorname{col}_j(B)\|_q^p\right)^{1/p} = \left(\sum_{i=1}^{n} \|\operatorname{col}_i(A^*)^*\|_p^p\right)^{1/p}\left(\sum_{j=1}^{l} \|\operatorname{col}_j(B)\|_q^p\right)^{1/p}\\
&\le \left(\sum_{i=1}^{n} \|\operatorname{col}_i(A^*)^*\|_p^p\right)^{1/p}\left(\sum_{j=1}^{l} \|\operatorname{col}_j(B)\|_p^p\right)^{1/p} = \|A^*\|_p\|B\|_p = \|A\|_p\|B\|_p.
\end{aligned}$$

Next, let $2 \le p \le \infty$ so that $q \triangleq p/(p-1) \le 2$. Using Hölder's inequality (11.1.9) along with (11.1.7) and the fact that $q \le p$ yields

$$\begin{aligned}
\|AB\|_p &\le \left(\sum_{i=1}^{n} \|\operatorname{col}_i(A^*)^*\|_p^p\right)^{1/p}\left(\sum_{j=1}^{l} \|\operatorname{col}_j(B)\|_q^p\right)^{1/p} \le \left(\sum_{i=1}^{n} \|\operatorname{col}_i(A^*)^*\|_p^p\right)^{1/p}\left(\sum_{j=1}^{l} \|\operatorname{col}_j(B)\|_q^q\right)^{1/q}\\
&= \|A^*\|_p\|B\|_q = \|A\|_p\|B\|_q.
\end{aligned}$$

Similarly, it can be shown that (11.3.9) holds. □

Proposition 11.3.6. Let $A \in \mathbb{F}^{n\times m}$, $B \in \mathbb{F}^{m\times l}$, and $p,q \in [1,\infty]$, define $r \triangleq pq/(p+q)$, and assume that $r \geq 1$. Then,

$$\|AB\|_{\sigma r} \leq \|A\|_{\sigma p}\|B\|_{\sigma q}. \tag{11.3.10}$$

Furthermore, if $r = 1$, then

$$|\operatorname{Re}\operatorname{tr} AB| \leq |\operatorname{tr} AB| \leq \sum_{i=1}^{n} \rho_i(AB) \leq \operatorname{tr}\langle AB\rangle = \sum_{i=1}^{n} \sigma_i(AB) = \|AB\|_{\sigma 1} \leq \|A\|_{\sigma p}\|B\|_{\sigma q}. \tag{11.3.11}$$

Proof. Proposition 11.6.2 and Hölder's inequality with $1/(p/r) + 1/(q/r) = 1$ imply

$$\begin{aligned}\|AB\|_{\sigma r} &= \left(\sum_{i=1}^{\min\{n,m,l\}} \sigma_i^r(AB)\right)^{1/r} \leq \left(\sum_{i=1}^{\min\{n,m,l\}} \sigma_i^r(A)\sigma_i^r(B)\right)^{1/r}\\ &\leq \left[\left(\sum_{i=1}^{\min\{n,m,l\}} \sigma_i^p(A)\right)^{r/p}\left(\sum_{i=1}^{\min\{n,m,l\}} \sigma_i^q(B)\right)^{r/q}\right]^{1/r} = \|A\|_{\sigma p}\|B\|_{\sigma q}.\end{aligned}$$

Finally, the inequalities involving the trace are given in Proposition 11.2.4. □

Corollary 11.3.7. Let $A \in \mathbb{F}^{n\times m}$ and $B \in \mathbb{F}^{m\times l}$. Then,

$$\|AB\|_{\sigma\infty} \leq \|AB\|_{\sigma 2} \leq \begin{Bmatrix} \|A\|_{\sigma\infty}\|B\|_{\sigma 2} \\ \|A\|_{\sigma 2}\|B\|_{\sigma\infty} \\ \|AB\|_{\sigma 1} \end{Bmatrix} \leq \|A\|_{\sigma 2}\|B\|_{\sigma 2}. \tag{11.3.12}$$

Equivalently,

$$\sigma_{\max}(AB) \leq \|AB\|_{\mathrm{F}} \leq \begin{Bmatrix} \sigma_{\max}(A)\|B\|_{\mathrm{F}} \\ \|A\|_{\mathrm{F}}\sigma_{\max}(B) \\ \operatorname{tr}\langle AB\rangle \end{Bmatrix} \leq \|A\|_{\mathrm{F}}\|B\|_{\mathrm{F}}. \tag{11.3.13}$$

Furthermore, for all $r \in [1,\infty]$,

$$\|AB\|_{\sigma 2r} \leq \|AB\|_{\sigma r} \leq \begin{Bmatrix} \|A\|_{\sigma r}\sigma_{\max}(B) \\ \sigma_{\max}(A)\|B\|_{\sigma r} \\ \|A\|_{\sigma 2r}\|B\|_{\sigma 2r} \end{Bmatrix} \leq \|A\|_{\sigma r}\|B\|_{\sigma r}. \tag{11.3.14}$$

In particular, setting $r = \infty$ yields

$$\sigma_{\max}(AB) \leq \sigma_{\max}(A)\sigma_{\max}(B). \tag{11.3.15}$$

Corollary 11.3.8. Let $A \in \mathbb{F}^{n\times m}$ and $B \in \mathbb{F}^{m\times l}$. Then,

$$\|AB\|_{\sigma 1} \leq \begin{cases} \sigma_{\max}(A)\|B\|_{\sigma 1} \\ \|A\|_{\sigma 1}\sigma_{\max}(B). \end{cases} \tag{11.3.16}$$

Note that the inequality $\|AB\|_{\mathrm{F}} \leq \|A\|_{\mathrm{F}}\|B\|_{\mathrm{F}}$ in (11.3.13) is equivalent to (11.3.7) with $p = 2$ as well as (11.3.8) and (11.3.9) with $p = q = 2$. The following result is a matrix version of the Cauchy-Schwarz inequality given by Corollary 11.1.7.

Corollary 11.3.9. Let $A \in \mathbb{F}^{n\times m}$ and $B \in \mathbb{F}^{n\times m}$. Then,

$$|\operatorname{tr} A^*B| \leq \|A\|_{\mathrm{F}}\|B\|_{\mathrm{F}}. \tag{11.3.17}$$

Equality holds if and only if A and B are linearly dependent.

The following result is an extension of the geometric series for scalars.

Proposition 11.3.10. Let $A \in \mathbb{F}^{n\times n}$, and assume that $\rho_{\max}(A) < 1$. Then, there exists a submultiplicative norm $\|\cdot\|$ on $\mathbb{F}^{n\times n}$ such that $\|A\| < 1$. Furthermore, $\sum_{i=0}^{\infty} A^i$ converges absolutely, and

$$(I - A)^{-1} = \sum_{i=0}^{\infty} A^i. \tag{11.3.18}$$

Furthermore,

$$\frac{1}{1 + \|A\|} \le \|(I - A)^{-1}\| \le \frac{1}{1 - \|A\|} + \|I\| - 1. \tag{11.3.19}$$

If, in addition, $\|\cdot\|$ is normalized, then

$$\frac{1}{1 + \|A\|} \le \|(I - A)^{-1}\| \le \frac{1}{1 - \|A\|}. \tag{11.3.20}$$

Proof. Corollary 11.3.4 implies that there exists a submultiplicative norm $\|\cdot\|$ on $\mathbb{F}^{n\times n}$ such that $\|A\| < 1$. It thus follows that

$$\sum_{i=0}^{\infty} \|A^i\| \le \|I\| - 1 + \sum_{i=0}^{\infty} \|A\|^i = \frac{1}{1 - \|A\|} + \|I\| - 1,$$

which proves that $\sum_{i=0}^{\infty} A^i$ converges absolutely. Next, to verify (11.3.18), note that

$$(I - A)\sum_{k=0}^{\infty} A^k = \sum_{k=0}^{\infty} A^k - \sum_{k=1}^{\infty} A^k = I + \sum_{k=1}^{\infty} A^k - \sum_{k=1}^{\infty} A^k = I,$$

which implies (11.3.18) and thus the right-hand inequality in (11.3.19). Furthermore,

$$1 \le \|I\| = \|(I - A)(I - A)^{-1}\| \le \|I - A\|\|(I - A)^{-1}\| \le (1 + \|A\|)\|(I - A)^{-1}\|,$$

which yields the left-hand inequality in (11.3.19). □

11.4 Induced Norms

In this section we consider the case where there exists a nonzero vector $x \in \mathbb{F}^m$ such that (11.3.3) holds as an equality. This statement characterizes a special class of norms on $\mathbb{F}^{n\times n}$, namely, the *induced norms*.

Definition 11.4.1. Let $\|\cdot\|''$ and $\|\cdot\|$ be norms on $\mathbb{F}^m$ and $\mathbb{F}^n$, respectively. Then, $\|\cdot\|'\colon \mathbb{F}^{n\times m} \mapsto \mathbb{F}$ defined by

$$\|A\|' = \max_{x\in\mathbb{F}^m\backslash\{0\}} \frac{\|Ax\|}{\|x\|''} \tag{11.4.1}$$

is an *induced norm* on $\mathbb{F}^{n\times m}$. In this case, $\|\cdot\|'$ is *induced by* $\|\cdot\|''$ and $\|\cdot\|$. If $m = n$ and $\|\cdot\|'' = \|\cdot\|$, then $\|\cdot\|'$ is *induced by* $\|\cdot\|$, and $\|\cdot\|'$ is an *equi-induced norm*.

The next result confirms that $\|\cdot\|'$ defined by (11.4.1) is a norm.

Theorem 11.4.2. Every induced norm is a norm. Furthermore, every equi-induced norm is normalized.

Proof. See [1448, p. 293]. □

Let $A \in \mathbb{F}^{n\times m}$. It can be seen that (11.4.1) is equivalent to

$$\|A\|' = \max_{x\in\{y\in\mathbb{F}^m\colon\ \|y\|''=1\}} \|Ax\|. \tag{11.4.2}$$

Theorem 12.4.11 implies that the maximum in (11.4.2) exists. Since, for all $x \neq 0$,

$$\|A\|' = \max_{x\in\mathbb{F}^m\setminus\{0\}} \frac{\|Ax\|}{\|x\|''} \geq \frac{\|Ax\|}{\|x\|''}, \tag{11.4.3}$$

it follows that, for all $x \in \mathbb{F}^m$,

$$\|Ax\| \leq \|A\|'\|x\|'' \tag{11.4.4}$$

so that $(\|\cdot\|, \|\cdot\|', \|\cdot\|'')$ are compatible. If $m = n$ and $\|\cdot\|'' = \|\cdot\|$, then the norm $\|\cdot\|$ is compatible with the induced norm $\|\cdot\|'$. The next result shows that compatible norms can be obtained from induced norms.

Proposition 11.4.3. Let $\|\cdot\|$, $\|\cdot\|'$, and $\|\cdot\|''$ be norms on $\mathbb{F}^l$, $\mathbb{F}^m$, and $\mathbb{F}^n$, respectively. Furthermore, let $\|\cdot\|'''$ be the norm on $\mathbb{F}^{m\times l}$ induced by $\|\cdot\|$ and $\|\cdot\|'$, let $\|\cdot\|''''$ be the norm on $\mathbb{F}^{n\times m}$ induced by $\|\cdot\|'$ and $\|\cdot\|''$, and let $\|\cdot\|'''''$ be the norm on $\mathbb{F}^{n\times l}$ induced by $\|\cdot\|$ and $\|\cdot\|''$. If $A \in \mathbb{F}^{n\times m}$ and $B \in \mathbb{F}^{m\times l}$, then

$$\|AB\|''''' \leq \|A\|''''\|B\|'''. \tag{11.4.5}$$

Proof. Note that, for all $x \in \mathbb{F}^l$, $\|Bx\|' \leq \|B\|'''\|x\|$, and, for all $y \in \mathbb{F}^m$, $\|Ay\|'' \leq \|A\|''''\|y\|'$. Hence, for all $x \in \mathbb{F}^l$, it follows that

$$\|ABx\|'' \leq \|A\|''''\|Bx\|' \leq \|A\|''''\|B\|'''\|x\|,$$

which implies that

$$\|AB\|''''' = \max_{x\in\mathbb{F}^l\setminus\{0\}} \frac{\|ABx\|''}{\|x\|} \leq \|A\|''''\|B\|'''. \qquad \square$$

Corollary 11.4.4. Every equi-induced norm is submultiplicative.

The following result is a consequence of Corollary 11.4.4 and Proposition 11.3.2.

Corollary 11.4.5. Let $\|\cdot\|$ be an equi-induced norm on $\mathbb{F}^{n\times n}$, and let $A \in \mathbb{F}^{n\times n}$. Then,

$$\rho_{\max}(A) \leq \|A\|. \tag{11.4.6}$$

By assigning $\|\cdot\|_p$ to $\mathbb{F}^m$ and $\|\cdot\|_q$ to $\mathbb{F}^n$, where $p \geq 1$ and $q \geq 1$, the *Hölder-induced norm* on $\mathbb{F}^{n\times m}$ is defined by

$$\|A\|_{q,p} \triangleq \max_{x\in\mathbb{F}^m\setminus\{0\}} \frac{\|Ax\|_q}{\|x\|_p}. \tag{11.4.7}$$

Proposition 11.4.6. Let $p, q, p', q' \in [1, \infty]$, where $p \leq p'$ and $q \leq q'$, and let $A \in \mathbb{F}^{n\times m}$. Then,

$$\|A\|_{q',p} \leq \|A\|_{q,p} \leq \|A\|_{q,p'}. \tag{11.4.8}$$

Proof. Use Proposition 11.1.5. $\square$

A subtlety of induced norms is that an induced norm may depend on the underlying field. In particular, the induced norm of a real matrix A computed over $\mathbb{R}$ may be smaller than the induced norm of A computed over $\mathbb{C}$. Although the chosen field is usually not made explicit, we do so in special cases for clarity. In particular, for all $A \in \mathbb{R}^{n\times m}$ and $p, q \in [1, \infty]$, $\|A\|_{p,q,\mathbb{F}}$ denotes the Hölder-induced norm of the real matrix A evaluated over $\mathbb{F}$.

Proposition 11.4.7. Let $A \in \mathbb{R}^{n\times m}$ and $p, q \in [1, \infty]$. Then,

$$\|A\|_{q,p,\mathbb{R}} \leq \|A\|_{q,p,\mathbb{C}} \leq c_p\|A\|_{q,p,\mathbb{R}}, \tag{11.4.9}$$

where

$$c_p \triangleq \max_{x,y\in\mathbb{R}^n\setminus\{0\}} \frac{\|x\|_p + \|y\|_p}{\|x + jy\|_p} = \begin{cases} \sqrt{2}, & 1 \leq p \leq 2, \\ 2^{1-1/p}, & 2 \leq p < \infty, \\ 2, & p = \infty. \end{cases} \tag{11.4.10}$$

If either A is nonnegative or $p \le q$, then

$$\|A\|_{q,p,\mathbb{R}} = \|A\|_{q,p,\mathbb{C}}. \tag{11.4.11}$$

Proof. See [864, pp. 347, 348], [1399, p. 716], and [1439, 1933, 2040, 2358]. □

Example 11.4.8. Let $A = \begin{bmatrix} 1 & -1 \\ 1 & 1 \end{bmatrix}$, $p = \infty$, and $q = 1$. Then, for all nonzero $x = [x_1 \;\; x_2]^\mathsf{T} \in \mathbb{C}^2$,

$$\frac{\|Ax\|_1}{\|x\|_\infty} = \frac{|x_1 - x_2| + |x_1 + x_2|}{\max\{|x_1|, |x_2|\}}.$$

Restricting $x \in \mathbb{R}^2$ implies that $\|A\|_{1,\infty,\mathbb{R}} = 2$. On the other hand, letting $x = [1 \;\; j]^\mathsf{T}$ yields $\|A\|_{1,\infty,\mathbb{C}} \ge \|Ax\|_1/\|x\|_\infty = 2\sqrt{2}$. Hence, $\|A\|_{1,\infty,\mathbb{R}} < \|A\|_{1,\infty,\mathbb{C}}$, and thus the first inequality in (11.4.9) is strict. Furthermore, the second inequality in (11.4.9) implies that $2\sqrt{2} \le \|A\|_{1,\infty,\mathbb{C}} \le 2\|A\|_{1,\infty,\mathbb{R}} = 4$. In fact, $\|A\|_{1,\infty,\mathbb{C}} = 2\sqrt{2}$. See [864, p. 176], [1399, p. 716], and [1439]. ◇

The following result gives explicit expressions for several Hölder-induced norms.

Proposition 11.4.9. Let $A \in \mathbb{F}^{n\times m}$. Then,

$$\|A\|_{2,2} = \sigma_{\max}(A). \tag{11.4.12}$$

If $p \in [1, \infty]$, then

$$\|A\|_{p,1} = \|A\|_{p|\infty} = \max_{j\in\{1,\dots,m\}} \|\mathrm{col}_j(A)\|_p. \tag{11.4.13}$$

Finally, if $p, q \in [1, \infty]$ satisfy $1/p + 1/q = 1$, then

$$\|A\|_{\infty,p} = \|A^\mathsf{T}\|_{q|\infty} = \max_{i\in\{1,\dots,n\}} \|\mathrm{row}_i(A)\|_q. \tag{11.4.14}$$

Proof. Since A^*A is Hermitian, Corollary 10.4.2 implies that, for all $x \in \mathbb{F}^m$,

$$x^*A^*Ax \le \lambda_{\max}(A^*A)x^*x,$$

which implies that, for all $x \in \mathbb{F}^m$, $\|Ax\|_2 \le \sigma_{\max}(A)\|x\|_2$, and thus $\|A\|_{2,2} \le \sigma_{\max}(A)$. Now, let $x \in \mathbb{F}^{n\times n}$ be an eigenvector associated with $\lambda_{\max}(A^*A)$ so that $\|Ax\|_2 = \sigma_{\max}(A)\|x\|_2$, which implies that $\sigma_{\max}(A) \le \|A\|_{2,2}$. Hence, (11.4.12) holds.

Next, note that, for all $x \in \mathbb{F}^m$,

$$\|Ax\|_p = \left\| \sum_{j=1}^m x_{(j)}\mathrm{col}_j(A) \right\|_p \le \sum_{j=1}^m |x_{(j)}|\|\mathrm{col}_j(A)\|_p \le \max_{j\in\{1,\dots,m\}} \|\mathrm{col}_j(A)\|_p\|x\|_1,$$

and hence $\|A\|_{p,1} \le \max_{j\in\{1,\dots,m\}} \|\mathrm{col}_j(A)\|_p$. Next, let $k \in \{1, \dots, m\}$ be such that $\|\mathrm{col}_k(A)\|_p = \max_{j\in\{1,\dots,m\}} \|\mathrm{col}_j(A)\|_p$. Now, since $\|e_k\|_1 = 1$, it follows that $\|Ae_k\|_p = \|\mathrm{col}_k(A)\|_p\|e_k\|_1$, which implies that

$$\max_{j\in\{1,\dots,m\}} \|\mathrm{col}_j(A)\|_p = \|\mathrm{col}_k(A)\|_p \le \|A\|_{p,1},$$

and hence (11.4.13) holds.

Next, for all $x \in \mathbb{F}^m$, it follows from Hölder's inequality (11.1.9) that

$$\|Ax\|_\infty = \max_{i\in\{1,\dots,n\}} |\mathrm{row}_i(A)x| \le \max_{i\in\{1,\dots,n\}} \|\mathrm{row}_i(A)\|_q\|x\|_p,$$

which implies that $\|A\|_{\infty,p} \le \max_{i\in\{1,\dots,n\}} \|\mathrm{row}_i(A)\|_q$. Next, let $k \in \{1, \dots, n\}$ be such that $\|\mathrm{row}_k(A)\|_q = \max_{i\in\{1,\dots,n\}} \|\mathrm{row}_i(A)\|_q$, and let nonzero $x \in \mathbb{F}^m$ satisfy $|\mathrm{row}_k(A)x| = \|\mathrm{row}_k(A)\|_q\|x\|_p$. Hence,

$$\|Ax\|_\infty = \max_{i\in\{1,\dots,n\}} |\mathrm{row}_i(A)x| \ge |\mathrm{row}_k(A)x| = \|\mathrm{row}_k(A)\|_q\|x\|_p,$$

which implies that

$$\max_{i\in\{1,\ldots,n\}} \|\text{row}_i(A)\|_q = \|\text{row}_k(A)\|_q \le \|A\|_{\infty,p},$$

and thus (11.4.14) holds. □

Let $A \in \mathbb{F}^{n\times m}$. Then,

$$\|A\|_{2|\infty} = \max_{i\in\{1,\ldots,m\}} \|\text{col}_i(A)\|_2 = \text{d}_{\max}^{1/2}(A^*A), \tag{11.4.15}$$

$$\|A^{\mathsf{T}}\|_{2|\infty} = \max_{i\in\{1,\ldots,n\}} \|\text{row}_i(A)\|_2 = \text{d}_{\max}^{1/2}(AA^*). \tag{11.4.16}$$

Therefore, Proposition 11.4.9 implies that

$$\|A\|_{1,1} = \|A\|_{1|\infty} = \max_{i\in\{1,\ldots,m\}} \|\text{col}_i(A)\|_1, \tag{11.4.17}$$

$$\|A\|_{2,1} = \|A\|_{2|\infty} = \max_{i\in\{1,\ldots,m\}} \|\text{col}_i(A)\|_2 = \text{d}_{\max}^{1/2}(A^*A), \tag{11.4.18}$$

$$\|A\|_{\infty,1} = \|A\|_{\infty|\infty} = \|A\|_\infty = \max_{\substack{i\in\{1,\ldots,n\}\\ j\in\{1,\ldots,m\}}} |A_{(i,j)}|, \tag{11.4.19}$$

$$\|A\|_{\infty,2} = \|A^{\mathsf{T}}\|_{2|\infty} = \max_{i\in\{1,\ldots,n\}} \|\text{row}_i(A)\|_2 = \text{d}_{\max}^{1/2}(AA^*), \tag{11.4.20}$$

$$\|A\|_{\infty,\infty} = \|A^{\mathsf{T}}\|_{1|\infty} = \max_{i\in\{1,\ldots,n\}} \|\text{row}_i(A)\|_1. \tag{11.4.21}$$

For convenience, we define the *column norm*

$$\|A\|_{\text{col}} \triangleq \|A\|_{1,1} = \|A\|_{1|\infty} \tag{11.4.22}$$

and the *row norm*

$$\|A\|_{\text{row}} \triangleq \|A\|_{\infty,\infty} = \|A^{\mathsf{T}}\|_{1|\infty}. \tag{11.4.23}$$

Note that

$$\|A^{\mathsf{T}}\|_{\text{col}} = \|A\|_{\text{row}}. \tag{11.4.24}$$

The following result follows from Corollary 11.4.5.

Corollary 11.4.10. Let $A \in \mathbb{F}^{n\times n}$. Then,

$$\rho_{\max}(A) \le \sigma_{\max}(A), \tag{11.4.25}$$

$$\rho_{\max}(A) \le \|A\|_{\text{col}}, \tag{11.4.26}$$

$$\rho_{\max}(A) \le \|A\|_{\text{row}}. \tag{11.4.27}$$

Proposition 11.4.11. Let $p, q \in [1,\infty]$ satisfy $1/p + 1/q = 1$, and let $A \in \mathbb{F}^{n\times m}$. Then,

$$\|A\|_{q,p} \le \|A\|_q. \tag{11.4.28}$$

Proof. For $p = 1$ and $q = \infty$, (11.4.28) follows from (11.4.19). For $q < \infty$ and $x \in \mathbb{F}^n$, it follows from Hölder's inequality (11.1.9) that

$$\|Ax\|_q = \left(\sum_{i=1}^n |\text{row}_i(A)x|^q\right)^{1/q} \le \left(\sum_{i=1}^n \|\text{row}_i(A)\|_q^q \|x\|_p^q\right)^{1/q} = \left(\sum_{i=1}^n\sum_{j=1}^m |A_{(i,j)}|^q\right)^{1/q} \|x\|_p = \|A\|_q\|x\|_p,$$

which implies (11.4.28). □

Next, we specialize Proposition 11.4.3 to the Hölder-induced norms.

Corollary 11.4.12. Let $p, q, r \in [1, \infty]$, and let $A \in \mathbb{F}^{n\times m}$ and $B \in \mathbb{F}^{m\times l}$. Then,

$$\|AB\|_{r,p} \le \|A\|_{r,q}\|B\|_{q,p}. \tag{11.4.29}$$

In particular,

$$\|AB\|_{\rm col} \le \|A\|_{\rm col}\|B\|_{\rm col}, \tag{11.4.30}$$

$$\sigma_{\max}(AB) \le \sigma_{\max}(A)\sigma_{\max}(B), \tag{11.4.31}$$

$$\|AB\|_{\rm row} \le \|A\|_{\rm row}\|B\|_{\rm row}, \tag{11.4.32}$$

$$\|AB\|_\infty \le \|A\|_\infty\|B\|_{\rm col}, \tag{11.4.33}$$

$$\|AB\|_\infty \le \|A\|_{\rm row}\|B\|_\infty, \tag{11.4.34}$$

$$\mathrm{d}_{\max}^{1/2}(B^*A^*AB) \le \mathrm{d}_{\max}^{1/2}(A^*A)\|B\|_{\rm col}, \tag{11.4.35}$$

$$\mathrm{d}_{\max}^{1/2}(B^*A^*AB) \le \sigma_{\max}(A)\mathrm{d}_{\max}^{1/2}(B^*B), \tag{11.4.36}$$

$$\mathrm{d}_{\max}^{1/2}(ABB^*A^*) \le \mathrm{d}_{\max}^{1/2}(AA^*)\sigma_{\max}(B), \tag{11.4.37}$$

$$\mathrm{d}_{\max}^{1/2}(ABB^*A^*) \le \|B\|_{\rm row}\mathrm{d}_{\max}^{1/2}(BB^*). \tag{11.4.38}$$

11.5 Induced Lower Bound

We now consider a variation of the induced norm.

Definition 11.5.1. Let $\|\cdot\|$ and $\|\cdot\|'$ denote norms on $\mathbb{F}^m$ and $\mathbb{F}^n$, respectively, and let $A \in \mathbb{F}^{n\times m}$. Then, $\ell\colon \mathbb{F}^{n\times m} \mapsto \mathbb{R}$ defined by

$$\ell(A) \triangleq \begin{cases} \displaystyle\min_{y\in\mathcal{R}(A)\backslash\{0\}}\ \max_{x\in\{z\in\mathbb{F}^m\colon Az=y\}} \frac{\|y\|'}{\|x\|}, & A \ne 0,\\ 0, & A = 0,\end{cases} \tag{11.5.1}$$

is the *lower bound induced by* $\|\cdot\|$ and $\|\cdot\|'$. Equivalently,

$$\ell(A) \triangleq \begin{cases} \displaystyle\min_{x\in\mathbb{F}^m\backslash\mathcal{N}(A)}\ \max_{z\in\mathcal{N}(A)} \frac{\|Ax\|'}{\|x+z\|}, & A \ne 0,\\ 0, & A = 0.\end{cases} \tag{11.5.2}$$

Proposition 11.5.2. Let $\|\cdot\|$ and $\|\cdot\|'$ be norms on $\mathbb{F}^m$ and $\mathbb{F}^n$, respectively, let $\|\cdot\|''$ be the norm induced by $\|\cdot\|$ and $\|\cdot\|'$, let $\|\cdot\|'''$ be the norm induced by $\|\cdot\|'$ and $\|\cdot\|$, and let ℓ be the lower bound induced by $\|\cdot\|$ and $\|\cdot\|'$. Then, the following statements hold:

i) For all $A \in \mathbb{F}^{n\times m}$, $\ell(A)$ exists; that is, for all $A \in \mathbb{F}^{n\times m}$, the minimum in (11.5.1) is attained.

ii) If $A \in \mathbb{F}^{n\times m}$, then $\ell(A) = 0$ if and only if $A = 0$.

iii) For all $A \in \mathbb{F}^{n\times m}$, there exists $x \in \mathbb{F}^m$ such that

$$\ell(A)\|x\| = \|Ax\|'. \tag{11.5.3}$$

iv) For all $A \in \mathbb{F}^{n\times m}$,

$$\ell(A) \le \|A\|''. \tag{11.5.4}$$

v) If $A \ne 0$ and B is a (1)-inverse of A, then

$$1/\|B\|''' \le \ell(A) \le \|B\|'''. \tag{11.5.5}$$

vi) If $A, B \in \mathbb{F}^{n\times m}$ and either $\mathcal{R}(A) \subseteq \mathcal{R}(A+B)$ or $\mathcal{N}(A) \subseteq \mathcal{N}(A+B)$, then

$$\ell(A) - \|B\|''' \le \ell(A+B). \tag{11.5.6}$$

vii) If $A, B \in \mathbb{F}^{n\times m}$ and either $\mathcal{R}(A+B) \subseteq \mathcal{R}(A)$ or $\mathcal{N}(A+B) \subseteq \mathcal{N}(A)$, then

$$\ell(A+B) \le \ell(A) + \|B\|'''. \tag{11.5.7}$$

viii) If $n = m$ and $A \in \mathbb{F}^{n\times n}$ is nonsingular, then

$$\ell(A) = 1/\|A^{-1}\|'''. \tag{11.5.8}$$

Proof. See [1223]. □

Proposition 11.5.3. Let $\|\cdot\|$, $\|\cdot\|'$, and $\|\cdot\|''$ be norms on $\mathbb{F}^l$, $\mathbb{F}^m$, and $\mathbb{F}^n$, respectively, let $\|\cdot\|'''$ denote the norm on $\mathbb{F}^{m\times l}$ induced by $\|\cdot\|$ and $\|\cdot\|'$, let $\|\cdot\|''''$ denote the norm on $\mathbb{F}^{n\times m}$ induced by $\|\cdot\|'$ and $\|\cdot\|''$, and let $\|\cdot\|'''''$ denote the norm on $\mathbb{F}^{n\times l}$ induced by $\|\cdot\|$ and $\|\cdot\|''$. If $A \in \mathbb{F}^{n\times m}$ and $B \in \mathbb{F}^{m\times l}$, then

$$\ell(A)\ell'(B) \le \ell''(AB). \tag{11.5.9}$$

In addition, the following statements hold:

i) If either $\operatorname{rank} B = \operatorname{rank} AB$ or $\operatorname{def} B = \operatorname{def} AB$, then

$$\ell''(AB) \le \|A\|''\ell(B). \tag{11.5.10}$$

ii) If $\operatorname{rank} A = \operatorname{rank} AB$, then

$$\ell''(AB) \le \ell(A)\|B\|''''. \tag{11.5.11}$$

iii) If $\operatorname{rank} B = m$, then

$$\|A\|''\ell(B) \le \|AB\|'''''. \tag{11.5.12}$$

iv) If $\operatorname{rank} A = m$, then

$$\ell(A)\|B\|'''' \le \|AB\|'''''. \tag{11.5.13}$$

Proof. See [1223]. □

By assigning $\|\cdot\|_p$ to $\mathbb{F}^m$ and $\|\cdot\|_q$ to $\mathbb{F}^n$, where $p \ge 1$ and $q \ge 1$, the *Hölder-induced lower bound* on $\mathbb{F}^{n\times m}$ is defined by

$$\ell_{q,p}(A) \triangleq \begin{cases} \min\limits_{y\in\mathcal{R}(A)\backslash\{0\}} \max\limits_{x\in\{z\in\mathbb{F}^m:\, Az=y\}} \frac{\|y\|_q}{\|x\|_p}, & A \ne 0,\\[2ex] 0, & A = 0. \end{cases} \tag{11.5.14}$$

The following result shows that $\ell_{2,2}(A)$ is the smallest positive singular value of A.

Proposition 11.5.4. Let $A \in \mathbb{F}^{n\times m}$, assume that A is nonzero, and let $r \triangleq \operatorname{rank} A$. Then,

$$\ell_{2,2}(A) = \sigma_r(A). \tag{11.5.15}$$

Proof. Use the singular value decomposition. □

Corollary 11.5.5. Let $A \in \mathbb{F}^{n\times m}$. If $n \le m$ and A is right invertible, then

$$\ell_{2,2}(A) = \sigma_{\min}(A) = \sigma_n(A). \tag{11.5.16}$$

If $m \le n$ and A is left invertible, then

$$\ell_{2,2}(A) = \sigma_{\min}(A) = \sigma_m(A). \tag{11.5.17}$$

Finally, if $n = m$ and A is nonsingular, then

$$\ell_{2,2}(A^{-1}) = \sigma_{\min}(A^{-1}) = \frac{1}{\sigma_{\max}(A)}. \tag{11.5.18}$$

Proof. Use Proposition 7.6.2 and Fact 8.3.33. □

In contrast to the submultiplicativity condition (11.4.4), which holds for the induced norm, the induced lower bound satisfies a supermultiplicativity condition. The following result is analogous to Proposition 11.4.3.

Proposition 11.5.6. Let $\|\cdot\|$, $\|\cdot\|'$, and $\|\cdot\|''$ be norms on $\mathbb{F}^l$, $\mathbb{F}^m$, and $\mathbb{F}^n$, respectively. Let $\ell(\cdot)$ be the lower bound induced by $\|\cdot\|$ and $\|\cdot\|'$, let $\ell'(\cdot)$ be the lower bound induced by $\|\cdot\|'$ and $\|\cdot\|''$, let $\ell''(\cdot)$ be the lower bound induced by $\|\cdot\|$ and $\|\cdot\|''$, let $A \in \mathbb{F}^{n\times m}$ and $B \in \mathbb{F}^{m\times l}$, and assume that either A or B is right invertible. Then,

$$\ell'(A)\ell(B) \le \ell''(AB). \tag{11.5.19}$$

Furthermore, if $1 \le p, q, r \le \infty$, then

$$\ell_{r,q}(A)\ell_{q,p}(B) \le \ell_{r,p}(AB). \tag{11.5.20}$$

In particular,

$$\sigma_m(A)\sigma_l(B) \le \sigma_l(AB). \tag{11.5.21}$$

Proof. See [1223] and [1738, pp. 369, 370]. □

11.6 Singular Value Inequalities

Proposition 11.6.1. Let $A \in \mathbb{F}^{n\times m}$ and $B \in \mathbb{F}^{m\times l}$. Then, for all $i \in \{1,\dots,\min\{n,m\}\}$ and $j \in \{1,\dots,\min\{m,l\}\}$ such that $i + j \le \min\{n,l\} + 1$,

$$\sigma_{i+j-1}(AB) \le \sigma_i(A)\sigma_j(B). \tag{11.6.1}$$

In particular, for all $i \in \{1,\dots,\min\{n,m,l\}\}$,

$$\sigma_i(AB) \le \sigma_{\max}(A)\sigma_i(B), \tag{11.6.2}$$

$$\sigma_i(AB) \le \sigma_i(A)\sigma_{\max}(B). \tag{11.6.3}$$

Proof. See [1450, p. 178] and [2979, pp. 78, 79]. □

Proposition 11.6.2. Let $A \in \mathbb{F}^{n\times m}$ and $B \in \mathbb{F}^{m\times l}$. If $r \ge 0$, then, for all $k \in \{1,\dots,\min\{n,m,l\}\}$,

$$\sum_{i=1}^{k} \sigma_i^r(AB) \le \sum_{i=1}^{k} \sigma_i^r(A)\sigma_i^r(B). \tag{11.6.4}$$

In particular, for all $k \in \{1,\dots,\min\{n,m,l\}\}$,

$$\sum_{i=1}^{k} \sigma_i(AB) \le \sum_{i=1}^{k} \sigma_i(A)\sigma_i(B). \tag{11.6.5}$$

If $r < 0$, $n = m = l$, and A and B are nonsingular, then

$$\sum_{i=1}^{n} \sigma_i^r(AB) \le \sum_{i=1}^{n} \sigma_i^r(A)\sigma_i^r(B). \tag{11.6.6}$$

Proof. The first statement follows from Proposition 11.6.3 and Fact 3.25.9. For the case where $r < 0$, use Fact 3.25.13. See [449, p. 94] and [1450, p. 177]. □

Proposition 11.6.3. Let $A \in \mathbb{F}^{n\times m}$ and $B \in \mathbb{F}^{m\times l}$. Then, for all $k \in \{1,\dots,\min\{n,m,l\}\}$,

$$\prod_{i=1}^{k} \sigma_i(AB) \le \prod_{i=1}^{k} \sigma_i(A)\sigma_i(B). \tag{11.6.7}$$

If, in addition, $n = m = l$, then

$$\prod_{i=1}^{n} \sigma_i(AB) = \prod_{i=1}^{n} \sigma_i(A)\sigma_i(B). \tag{11.6.8}$$

Proof. See [1450, p. 172]. □

Proposition 11.6.4. Let $A \in \mathbb{F}^{n\times m}$ and $B \in \mathbb{F}^{m\times l}$, and let $i \in \{\max\{m-n, 0\}, \ldots, m-1\}$ and $j \in \{\max\{m-l, 0\}, \ldots, m-1\}$ satisfy $i + j \le m - 1$. Then,

$$\sigma_{m-i}(A)\sigma_{m-j}(B) = \sigma_{m-i-j}(AB). \tag{11.6.9}$$

Consequently, if $m \le n$, then, for all $i \in \{1, \ldots, \min\{m, l\}\}$,

$$\sigma_{\min}(A)\sigma_i(B) = \sigma_m(A)\sigma_i(B) \le \sigma_i(AB). \tag{11.6.10}$$

Furthermore, if $m \le l$, then, for all $i \in \{1, \ldots, \min\{n, m\}\}$,

$$\sigma_i(A)\sigma_{\min}(B) = \sigma_i(A)\sigma_m(B) \le \sigma_i(AB). \tag{11.6.11}$$

Proof. Use Fact 10.22.23. To prove (11.6.10), note that Corollary 10.4.2 implies that $\sigma_m^2(A)I_m = \lambda_{\min}(A^*A)I_m \le A^*A$, which implies that $\sigma_m^2(A)B^*B \le B^*A^*AB$. Hence, it follows from the monotonicity theorem Theorem 10.4.9 that, for all $i \in \{1, \ldots, \min\{m, l\}\}$,

$$\sigma_m(A)\sigma_i(B) = \lambda_i[\sigma_m^2(A)B^*B]^{1/2} \le \lambda_i^{1/2}(B^*A^*AB) = \sigma_i(AB),$$

which proves (11.6.10). Similarly, for all $i \in \{1, \ldots, \min\{n, m\}\}$,

$$\sigma_i(A)\sigma_m(B) = \lambda_i[\sigma_m^2(B)AA^*]^{1/2} \le \lambda_i^{1/2}(ABB^*A^*) = \sigma_i(AB).$$ □

Specializing Proposition 11.6.1 and Proposition 11.6.4 yields the following four results.

Corollary 11.6.5. Let $A \in \mathbb{F}^{n\times m}$ and $B \in \mathbb{F}^{m\times l}$. Then,

$$\sigma_m(A)\sigma_{\min\{n,m,l\}}(B) \le \sigma_{\min\{n,m,l\}}(AB) \le \sigma_{\max}(A)\sigma_{\min\{n,m,l\}}(B), \tag{11.6.12}$$

$$\sigma_m(A)\sigma_{\max}(B) \le \sigma_{\max}(AB) \le \sigma_{\max}(A)\sigma_{\max}(B), \tag{11.6.13}$$

$$\sigma_{\min\{n,m,l\}}(A)\sigma_m(B) \le \sigma_{\min\{n,m,l\}}(AB) \le \sigma_{\min\{n,m,l\}}(A)\sigma_{\max}(B), \tag{11.6.14}$$

$$\sigma_{\max}(A)\sigma_m(B) \le \sigma_{\max}(AB) \le \sigma_{\max}(A)\sigma_{\max}(B). \tag{11.6.15}$$

Corollary 11.6.6. Let $A \in \mathbb{F}^{n\times n}$ and $B \in \mathbb{F}^{n\times l}$. Then, for all $i \in \{1, \ldots, \min\{n, l\}\}$,

$$\sigma_{\min}(A)\sigma_i(B) \le \sigma_i(AB) \le \sigma_{\max}(A)\sigma_i(B). \tag{11.6.16}$$

In particular,

$$\sigma_{\min}(A)\sigma_{\max}(B) \le \sigma_{\max}(AB) \le \sigma_{\max}(A)\sigma_{\max}(B). \tag{11.6.17}$$

Corollary 11.6.7. Let $A \in \mathbb{F}^{n\times m}$ and $B \in \mathbb{F}^{m\times m}$. Then, for all $i \in \{1, \ldots, \min\{n, m\}\}$,

$$\sigma_i(A)\sigma_{\min}(B) \le \sigma_i(AB) \le \sigma_i(A)\sigma_{\max}(B). \tag{11.6.18}$$

In particular,

$$\sigma_{\max}(A)\sigma_{\min}(B) \le \sigma_{\max}(AB) \le \sigma_{\max}(A)\sigma_{\max}(B). \tag{11.6.19}$$

Corollary 11.6.8. Let $A, B \in \mathbb{F}^{n\times n}$. Then, for all $i \in \{1, \ldots, n\}$,

$$\sigma_i(A)\sigma_{\min}(B) \le \sigma_i(AB) \le \sigma_i(A)\sigma_{\max}(B). \tag{11.6.20}$$

In particular,

$$\sigma_{\min}(A)\sigma_{\min}(B) \le \sigma_{\min}(AB) \le \sigma_{\min}(A)\sigma_{\max}(B), \tag{11.6.21}$$

$$\sigma_{\max}(A)\sigma_{\min}(B) \le \sigma_{\max}(AB) \le \sigma_{\max}(A)\sigma_{\max}(B). \tag{11.6.22}$$

Corollary 11.6.9. Let $A \in \mathbb{F}^{n\times m}$ and $B \in \mathbb{F}^{m\times l}$. If $m \le n$, then

$$\sigma_{\min}(A)\|B\|_{\mathrm{F}} = \sigma_m(A)\|B\|_{\mathrm{F}} \le \|AB\|_{\mathrm{F}}. \tag{11.6.23}$$

If $m \le l$, then

$$\|A\|_{\mathrm{F}}\sigma_{\min}(B) = \|A\|_{\mathrm{F}}\sigma_m(B) \le \|AB\|_{\mathrm{F}}. \tag{11.6.24}$$

Proposition 11.6.10. Let $A, B \in \mathbb{F}^{n\times m}$. Then, for all $i, j \in \{1, \ldots, \min\{n, m\}\}$ such that $i + j \le \min\{n, m\} + 1$,

$$\sigma_{i+j-1}(A + B) \le \sigma_i(A) + \sigma_j(B), \tag{11.6.25}$$

$$\sigma_{i+j-1}(A) - \sigma_j(B) \le \sigma_i(A + B). \tag{11.6.26}$$

Proof. See [1450, p. 178] and [2979, pp. 78, 79]. □

Corollary 11.6.11. Let $A, B \in \mathbb{F}^{n\times m}$. Then, for all $i \in \{1, \ldots, \min\{n, m\}\}$,

$$\sigma_i(A) - \sigma_{\max}(B) \le \sigma_i(A + B) \le \sigma_i(A) + \sigma_{\max}(B). \tag{11.6.27}$$

If, in addition, $n = m$, then

$$\sigma_{\min}(A) - \sigma_{\max}(B) \le \sigma_{\min}(A + B) \le \sigma_{\min}(A) + \sigma_{\max}(B). \tag{11.6.28}$$

Proof. Use Proposition 11.6.10 with $j = 1$. Alternatively, Lemma 10.4.3 and the Cauchy-Schwarz inequality given by Corollary 11.1.7 imply that, for all nonzero $x \in \mathbb{F}^n$,

$$\begin{aligned}\lambda_{\min}[(A + B)(A + B)^*] &\le \frac{x^*(AA^* + BB^* + AB^* + BA^*)x}{x^*x} = \frac{x^*AA^*x}{\|x\|_2^2} + \frac{x^*BB^*x}{\|x\|_2^2} + \mathrm{Re}\,\frac{2x^*AB^*x}{\|x\|_2^2}\\ &\le \frac{x^*AA^*x}{\|x\|_2^2} + \sigma_{\max}^2(B) + 2\frac{(x^*AA^*x)^{1/2}}{\|x\|_2}\sigma_{\max}(B).\end{aligned}$$

Minimizing with respect to x and using Lemma 10.4.3 yields

$$\begin{aligned}\sigma_n^2(A + B) = \lambda_{\min}[(A + B)(A + B)^*] &\le \lambda_{\min}(AA^*) + \sigma_{\max}^2(B) + 2\lambda_{\min}^{1/2}(AA^*)\sigma_{\max}(B)\\ &= [\sigma_n(A) + \sigma_{\max}(B)]^2,\end{aligned}$$

which proves the right-hand inequality of (11.6.27). Finally, the left-hand inequality follows from the right-hand inequality with A and B replaced by $A + B$ and $-B$, respectively. □

11.7 Facts on Vector Norms

Fact 11.7.1. Let $\|\cdot\|$ be a norm on $\mathbb{F}^n$, and define $f\colon \mathbb{F}^n \mapsto [0, \infty)$ by $f(x) = \|x\|$. Then, f is convex.

Fact 11.7.2. Let $\|\cdot\|$ and $\|\cdot\|'$ be norms on $\mathbb{F}^n$, and let $\alpha, \beta > 0$. Then, $\alpha\|\cdot\| + \beta\|\cdot\|'$ is also a norm on $\mathbb{F}^n$. Furthermore, $\max\{\|\cdot\|, \|\cdot\|'\}$ is a norm on $\mathbb{F}^n$. **Remark:** $\min\{\|\cdot\|, \|\cdot\|'\}$ is not necessarily a norm. See [109, p. 278].

Fact 11.7.3. Let $x, y \in \mathbb{F}^n$, and let $\|\cdot\|$ be a norm on $\mathbb{F}^n$. Then,

$$\left.\begin{matrix}\|x\|^2 + \|y\|^2\\ 2\|x\|^2 - 4\|x\|\|y\| + 2\|y\|^2\end{matrix}\right\} \le \|x + y\|^2 + \|x - y\|^2 \le 2\|x\|^2 + 4\|x\|\|y\| + 2\|y\|^2 \le 4(\|x\|^2 + \|y\|^2).$$

Source: [1124, pp. 9, 10] and [2112, p. 278].

Fact 11.7.4. Let $\|\cdot\|$ be a norm on $\mathbb{F}^n$. Then, there exists a unique $\alpha \in [1, 2]$ such that, for all $x, y \in \mathbb{F}^n$, at least one of which is nonzero,

$$\frac{2}{\alpha} \le \frac{\|x + y\|^2 + \|x - y\|^2}{\|x\|^2 + \|y\|^2} \le 2\alpha.$$

Furthermore, $\alpha = 1$ if and only if $\|\cdot\| = \|\cdot\|_2$. Finally, if $\|\cdot\| = \|\cdot\|_p$, then

$$\alpha = \begin{cases} 2^{(2-p)/p}, & 1 \le p \le 2, \\ 2^{(p-2)/p}, & p \ge 2. \end{cases}$$

Source: [605, p. 258] and [2061, p. 550]. **Remark:** This is the *von Neumann–Jordan* inequality. **Remark:** If $p = 2$, then $\alpha = 1$ and this result yields *v*) of Fact 11.8.3. **Related:** Fact 11.7.5.

Fact 11.7.5. Let $\|\cdot\|$ be a norm on $\mathbb{F}^n$. Then, the following statements are equivalent:

i) $\|\cdot\| = \|\cdot\|_2$.

ii) For all $x, y \in \mathbb{F}$, $\|x+y\|^2 + \|x-y\|^2 = 2\|x\|^2 + 2\|y\|^2$.

iii) For all $x, y \in \mathbb{F}$,

$$\left\| \frac{1}{\|x\|}x - \frac{1}{\|y\|}y \right\| \le \frac{2\|x-y\|}{\|x\| + \|y\|}.$$

iv) For all $x, y, z \in \mathbb{F}$, $\|x+y\|^2 + \|y+z\|^2 + \|z+x\|^2 = \|x\|^2 + \|y\|^2 + \|z\|^2 + \|x+y+z\|^2$.

Source: [2094, 1620]. **Credit:** The equivalence of *i*) and *iv*) is due to M. Fréchet. **Related:** Fact 11.7.4.

Fact 11.7.6. Let $x, y \in \mathbb{F}^n$, let $\alpha \in [0,1]$, and let $\|\cdot\|$ be a norm on $\mathbb{F}^n$. Then,

$$\|x+y\| \le \|\alpha x + (1-\alpha)y\| + \|(1-\alpha)x + \alpha y\| \le \|x\| + \|y\|.$$

Fact 11.7.7. Let $x, y \in \mathbb{F}^n$, and let $\|\cdot\|$ be a norm on $\mathbb{F}^n$. Then,

$$\big|\|x\| - \|y\|\big| \le \left\{ \begin{array}{c} \|x\| + \|y\| - \big|\|x+y\| - \|x-y\|\big| \\ \|x+y\| + \|x-y\| - \|x\| - \|y\| \le \min\{\|x+y\|, \|x-y\|\} \le \left\{ \begin{array}{c} \|x+y\| \\ \|x-y\| \end{array} \right\} \end{array} \right\} \le \|x\| + \|y\|.$$

Source: [1932]. **Related:** Fact 2.2.44.

Fact 11.7.8. Let $x, y \in \mathbb{F}^n$, let $\|\cdot\|$ be a norm on $\mathbb{F}^n$, assume that $\|x\| \ne \|y\|$, and let $p > 0$. Then,

$$\big|\|x\| - \|y\|\big| \le \frac{\big\|\|x\|^p x - \|y\|^p y\big\|}{\big|\|x\|^{p+1} - \|y\|^{p+1}\big|}\big|\|x\| - \|y\|\big| \le \|x-y\|.$$

Source: [2061, p. 516].

Fact 11.7.9. Let $x, y \in \mathbb{F}^n$, and let $\|\cdot\|$ be a norm on $\mathbb{F}^n$. Then,

$$\|x-y\| + \big|\|x\| - \|y\|\big| \le \sqrt{2\|x-y\|^2 + 2(\|x\| - \|y\|)^2} \le 2\|x-y\|,$$

If, in addition, x and y are nonzero, then

$$\left\| \frac{1}{\|x\|}x - \frac{1}{\|y\|}y \right\| \le \frac{\sqrt{2\|x-y\|^2 + 2(\|x\| - \|y\|)^2}}{\max\{\|x\|, \|y\|\}}.$$

Source: [2212].

Fact 11.7.10. Let $x, y \in \mathbb{F}^n$ be nonzero, and let $\|\cdot\|$ be a norm on $\mathbb{F}^n$. Then,

$$\|x+y\| \le \|x\| + \|y\| - \min\{\|x\|, \|y\|\}\left(2 - \left\| \frac{1}{\|x\|}x + \frac{1}{\|y\|}y \right\|\right) \le \|x\| + \|y\|,$$

$$\|x-y\| \le \|x\| + \|y\| - \min\{\|x\|, \|y\|\}\left(2 - \left\| \frac{1}{\|x\|}x + \frac{1}{\|y\|}y \right\|\right) \le \|x\| + \|y\|,$$

$$\|x\| + \|y\| - \max\{\|x\|, \|y\|\}\left(2 - \left\| \frac{1}{\|x\|}x + \frac{1}{\|y\|}y \right\|\right) \le \|x+y\| \le \|x\| + \|y\|,$$

$$\|x\| + \|y\| - \max\{\|x\|, \|y\|\}\left(2 - \left\|\frac{1}{\|x\|}x + \frac{1}{\|y\|}y\right\|\right) \le \|x - y\| \le \|x\| + \|y\|.$$

Source: [1931].

Fact 11.7.11. Let $x, y \in \mathbb{F}^n$ be nonzero, and let $\|\cdot\|$ be a norm on $\mathbb{F}^n$. Then,

$$\begin{aligned}\frac{(\|x\| + \|y\|)(\|x + y\| - \big|\|x\| - \|y\|\big|)}{4\min\{\|x\|, \|y\|\}} &\le \tfrac{1}{4}(\|x\| + \|y\|)\left\|\frac{1}{\|x\|}x + \frac{1}{\|y\|}y\right\| \\ &\le \tfrac{1}{2}\max\{\|x\|, \|y\|\}\left\|\frac{1}{\|x\|}x + \frac{1}{\|y\|}y\right\| \\ &\le \tfrac{1}{2}(\|x + y\| + \max\{\|x\|, \|y\|\} - \|x\| - \|y\|) \\ &\le \tfrac{1}{2}(\|x + y\| + \big|\|x\| - \|y\|\big|) \le \|x + y\|,\end{aligned}$$

$$\begin{aligned}\frac{\|x - y\| - \big|\|x\| - \|y\|\big|}{\min\{\|x\|, \|y\|\}} \le \left\|\frac{1}{\|x\|}x - \frac{1}{\|y\|}y\right\| &\le \left\{\begin{matrix}\dfrac{\|x - y\| + \big|\|x\| - \|y\|\big|}{\max\{\|x\|, \|y\|\}} \\[2ex] \dfrac{2\|x - y\|_2}{\|x\|_2 + \|y\|_2}\end{matrix}\right\} \\ &\le \left\{\begin{matrix}\dfrac{2\|x - y\|}{\max\{\|x\|, \|y\|\}} \\[2ex] \dfrac{2(\|x - y\| + \big|\|x\| - \|y\|\big|)}{\|x\| + \|y\|}\end{matrix}\right\} \le \frac{4\|x - y\|}{\|x\| + \|y\|}.\end{aligned}$$

Source: Fact 11.7.10, [1931, 1932, 2023, 2094], and [2061, p. 516]. **Remark:** In the second string, the first inequality is the *reverse Maligranda inequality*, the second inequality is the *Maligranda inequality*, the second and upper fourth terms are the *Massera-Schaffer inequality*, and the second and fifth terms are the Dunkl-Williams inequality. **Remark:** The term involving the Euclidean norm is present only for $\|\cdot\| = \|\cdot\|_2$. See Fact 2.21.8. **Remark:** Extensions to more than two vectors are given in [1588, 2211]. **Remark:** A matrix version is given by Fact 10.11.87. **Related:** Fact 10.11.87.

Fact 11.7.12. Let $x_1, \ldots, x_k \in \mathbb{F}^n$ be nonzero, and let $\|\cdot\|$ be a norm on $\mathbb{F}^n$. Then,

$$\left\|\sum_{i=1}^k x_i\right\| + \left(k - \left\|\sum_{i=1}^k \frac{1}{\|x_i\|}x_i\right\|\right)\min_{i\in\{1,\ldots,k\}}\|x_i\| \le \sum_{i=1}^k \|x_i\| \le \left\|\sum_{i=1}^k x_i\right\| + \left(k - \left\|\sum_{i=1}^k \frac{1}{\|x_i\|}x_i\right\|\right)\max_{i\in\{1,\ldots,k\}}\|x_i\|,$$

$$\begin{aligned}\max_{i\in\{1,\ldots,k\}}\left[\frac{1}{\|x_i\|}\left(\left\|\sum_{j=1}^k x_j\right\| - \sum_{j=1}^k \big|\|x_j\| - \|x_i\|\big|\right)\right] &\le \left\|\sum_{i=1}^k \frac{1}{\|x_i\|}x_i\right\| \\ &\le \min_{i\in\{1,\ldots,k\}}\left[\frac{1}{\|x_i\|}\left(\left\|\sum_{j=1}^k x_j\right\| + \sum_{j=1}^k \big|\|x_j\| - \|x_i\|\big|\right)\right].\end{aligned}$$

If $p \in (0, \infty)$, then

$$\begin{aligned}\left(\min_{i\in\{1,\ldots,k\}}\|x_i\|\right)\left(\sum_{i=1}^k \|x_i\|^{p-1} - \left\|\sum_{i=1}^k \frac{1}{\|x_i\|}x_i\right\|^p\right) &\le \sum_{i=1}^k \|x_i\|^p - k^{1-p}\left\|\sum_{i=1}^k x_i\right\|^p \\ &\le \left(\max_{i\in\{1,\ldots,k\}}\|x_i\|\right)\left(\sum_{i=1}^n \|x_i\|^{p-1} - \left\|\sum_{i=1}^k \frac{1}{\|x_i\|}x_i\right\|^p\right).\end{aligned}$$

If $\alpha_1,\ldots,\alpha_k \in \mathbb{F}$, then

$$\max_{i\in\{1,\ldots,k\}}\left(|\alpha_i|\left\|\sum_{j=1}^k x_j\right\| - \sum_{j=1}^k |\alpha_j-\alpha_i|\|x_j\|\right) \le \left\|\sum_{i=1}^k \alpha_i x_i\right\| \le \min_{i\in\{1,\ldots,k\}}\left(|\alpha_i|\left\|\sum_{j=1}^k x_j\right\| + \sum_{j=1}^k |\alpha_j-\alpha_i|\|x_j\|\right).$$

Source: [675, 939, 941, 1588, 2055, 2211].

Fact 11.7.13. Let $x,y\in\mathbb{F}^n$ be nonzero, let $p\in\mathbb{R}$, and let $\|\cdot\|$ be a norm on $\mathbb{F}^n$. Then, the following statements hold:

i) If $p\in(0,1]$, then

$$\left\|\frac{1}{\|x\|}x - \frac{1}{\|y\|}y\right\| \le \frac{2^{1+1/p}\|x-y\|}{(\|x\|^p+\|y\|^p)^{1/p}}.$$

ii) If $p\ge 1$, then

$$\left\|\frac{1}{\|x\|}x - \frac{1}{\|y\|}y\right\| \le \frac{4\|x-y\|}{(\|x\|^p+\|y\|^p)^{1/p}}.$$

Source: [25, 2094].

Fact 11.7.14. Let $x,y\in\mathbb{F}^n$, assume that x and y are nonzero, let $\|\cdot\|$ be a norm on $\mathbb{F}^n$, and let $p,q\in\mathbb{R}$. Then, the following statements hold:

i) If $p\le 0$, then

$$\big\|\|x\|^{p-1}x - \|y\|^{p-1}y\big\| \le \frac{(2-p)\max\{\|x\|^p,\|y\|^p\}}{\max\{\|x\|,\|y\|\}}\|x-y\|.$$

ii) If $p\in[0,1]$, then

$$\big\|\|x\|^{p-1}x - \|y\|^{p-1}y\big\| \le \frac{(2-p)\|x-y\|}{[\max\{\|x\|,\|y\|\}]^{1-p}}.$$

iii) If $p\in[0,1]$, then

$$\big\|\|x\|^{p-1}x - \|y\|^{p-1}y\big\| \le \frac{2\|x-y\|}{\|x\|^{1-p}+\|y\|^{1-p}}.$$

iv) If $p\ge 1$, then

$$\big\|\|x\|^{p-1}x - \|y\|^{p-1}y\big\| \le p[\max\{\|x\|,\|y\|\}]^{p-1}\|x-y\|.$$

v) If $p\in[0,1]$ and $q>0$, then

$$\big\|\|x\|^{p-1}x - \|y\|^{p-1}y\big\| \le \frac{2^{1+1/q}\|x-y\|}{(\|x\|^{(1-p)q}+\|y\|^{(1-p)q})^{1/q}}.$$

Source: [811, 1931, 2094].

Fact 11.7.15. Let $x,y\in\mathbb{F}^n$, let $\|\cdot\|$ be a norm on $\mathbb{F}^n$, let p and q be real numbers, and assume that $1<p\le q$. Then,

$$[\tfrac{1}{2}(\|x+\tfrac{1}{\sqrt{q-1}}y\|^q + \|x-\tfrac{1}{\sqrt{q-1}}y\|^q)]^{1/q} \le [\tfrac{1}{2}(\|x+\tfrac{1}{\sqrt{p-1}}y\|^p + \|x-\tfrac{1}{\sqrt{p-1}}y\|^p)]^{1/p}.$$

Source: [1146, p. 207]. **Remark:** This is *Bonami's inequality*. See Fact 2.2.32.

Fact 11.7.16. Let $A\in\mathbb{F}^{n\times n}$, assume that A is nonsingular, and let $\|\cdot\|$ be a norm on $\mathbb{F}^n$. Then, $\|x\|' \triangleq \|Ax\|$ is a norm on $\mathbb{F}^n$.

Fact 11.7.17. Let $A\in\mathbb{F}^{n\times n}$, where A is positive definite. Then, $\|x\|'\triangleq(x^*Ax)^{1/2}$ is a norm on $\mathbb{F}^n$.

Fact 11.7.18. Let $x\in\mathbb{R}^n$, and let $\|\cdot\|$ be a norm on $\mathbb{R}^n$. Then, $x^\mathsf{T}y>0$ for all $y\in\{z\in\mathbb{R}^n\colon\ \|z-x\|<\|x\|\}$.

Fact 11.7.19. Let $\|\cdot\|$ be a norm on $\mathbb{F}^n$, and assume that, for all $x\in\mathbb{F}^n$ and every permutation matrix $A\in\mathbb{R}^{n\times n}$, $\|Ax\|=\|x\|$. Then, $f\colon\mathbb{F}^n\mapsto\mathbb{R}$ defined by $f(x)=\|x\|$ is Schur-convex. **Source:** [2991, p. 376].

Fact 11.7.20. Let k and n be positive integers, assume that $k \le n$, let $p \ge 1$, and, for all $x \in \mathbb{F}^n$, define

$$\|x\| \triangleq \left[\sum_{i=1}^{k} (|x|^{\downarrow})_{(i)}^{p}\right]^{1/p}.$$

Then, $\|\cdot\|$ is a symmetric gauge function on $\mathbb{R}^n$ and an absolute and monotone norm on $\mathbb{F}^n$. **Source:** [449, p. 89] and [2991, pp. 373, 376]. **Remark:** Setting $p = 1$ yields the *Ky Fan k-norm.*

Fact 11.7.21. Let $x, y \in \mathbb{R}^n$, assume that x and y are nonzero, assume that $x^{\mathsf{T}}y = 0$, and let $\|\cdot\|$ be a norm on $\mathbb{R}^n$. Then, $\|x\| \le \|x + y\|$. **Source:** If $\|x + y\| < \|x\|$, then $x + y \in \mathbb{B}_{\|x\|}(0)$, and thus $y \in \mathbb{B}_{\|x\|}(-x)$. By Fact 11.7.18, $x^{\mathsf{T}}y < 0$. **Remark:** See [479, 1806] for related results on matrices.

Fact 11.7.22. Let $x, y \in \mathbb{F}^n$, and let $\|\cdot\|$ be a norm on $\mathbb{F}^n$. Then, the following statements hold:

i) If there exists $\beta \ge 0$ such that either $x = \beta y$ or $y = \beta x$, then $\|x + y\| = \|x\| + \|y\|$.

ii) If $\|x + y\| = \|x\| + \|y\|$ and x and y are linearly dependent, then there exists $\beta \ge 0$ such that either $x = \beta y$ or $y = \beta x$.

Remark: Let $x = \begin{bmatrix} 1 \\ 0 \end{bmatrix}$ and $y = \begin{bmatrix} 1 \\ 1 \end{bmatrix}$, which are linearly independent. Then, $\|x+y\|_\infty = \|x\|_\infty + \|y\|_\infty = 2$. **Problem:** If x and y are linearly independent and $p \in [1, \infty)$, then does $\|x + y\|_p < \|x\|_p + \|y\|_p$?

11.8 Facts on Vector p-Norms

Fact 11.8.1. For all $x \in \mathbb{R}^2$, define

$$\|x\| \triangleq \begin{cases} \|x\|_2, & x_{(1)} \le 0, \\ \|x\|_\infty, & x_{(1)} \ge 0. \end{cases}$$

Then, $\|\cdot\|$ is a norm on $\mathbb{R}^2$ that is neither absolute nor monotone. **Source:** [1233]. **Remark:** Let $x \triangleq \begin{bmatrix} -1/2 \\ -1/2 \end{bmatrix}$ and $y \triangleq \begin{bmatrix} 3/5 \\ 3/5 \end{bmatrix}$ so that $|x| \le\le |y|$ and $\|y\| = \frac{3}{5} < \frac{\sqrt{2}}{2} = \|x\|$.

Fact 11.8.2. Let $x, y \in \mathbb{F}^n$. Then, x and y are linearly dependent if and only if $|x|^{\odot 2}$ and $|y|^{\odot 2}$ are linearly dependent and $|x^*y| = |x|^{\mathsf{T}}|y|$. **Remark:** This clarifies the relationship between (11.1.10) with $p = 2$ and Corollary 11.1.7.

Fact 11.8.3. Let $x, y \in \mathbb{F}^n$. Then, the following statements hold:

i) $\|x\|_2^2 + \|y\|_2^2 = \|x + y\|_2^2 - 2\operatorname{Re} x^*y = \|x - y\|_2^2 + 2\operatorname{Re} x^*y$.

ii) $\|x + y\|_2 = \sqrt{\|x\|_2^2 + \|y\|_2^2 + 2\operatorname{Re} x^*y}$.

iii) $\|x - y\|_2 = \sqrt{\|x\|_2^2 + \|y\|_2^2 - 2\operatorname{Re} x^*y}$.

iv) The following statements are equivalent:

a) $\|x - y\|_2 = \|x + y\|_2$.

b) $\|x + y\|_2^2 = \|x\|_2^2 + \|y\|_2^2$.

c) $\|x - y\|_2^2 = \|x\|_2^2 + \|y\|_2^2$.

d) $\operatorname{Re} x^*y = 0$.

v) Let $\alpha \in \mathbb{R}$, and assume that $\alpha \ne 0$. Then,

$$\frac{1}{\alpha}\|\alpha x + y\|_2^2 + \|x - y\|_2^2 = (1 + \alpha)\|x\|_2^2 + (1 + \tfrac{1}{\alpha})\|y\|_2^2.$$

In particular,

$$\|x + y\|_2^2 + \|x - y\|_2^2 = 2\|x\|_2^2 + 2\|y\|_2^2.$$

vi) $\operatorname{Re} x^*y = \frac{1}{4}(\|x + y\|_2^2 - \|x - y\|_2^2) = \frac{1}{2}(\|x + y\|_2^2 - \|x\|_2^2 - \|y\|_2^2)$.

vii) If $\mathbb{F} = \mathbb{R}$, then $4x^{\mathsf{T}}y = \|x + y\|_2^2 - \|x - y\|_2^2 = 2(\|x + y\|_2^2 - \|x\|_2^2 - \|y\|_2^2)$.

viii) If $\mathbb{F} = \mathbb{C}$, then $\operatorname{Im} x^*y = \frac{1}{4}(\|x - \jmath y\|_2^2 - \|x + \jmath y\|_2^2)$.

ix) If $\mathbb{F} = \mathbb{C}$, then $4x^*y = \|x + y\|_2^2 - \|x - y\|_2^2 + (\|x - \jmath y\|_2^2 - \|x + \jmath y\|_2^2)\jmath$.

x) Let $\mathbb{F} = \mathbb{C}$ and $m \ge 3$. Then,

$$x^*y = \frac{1}{m}\sum_{i=0}^{m-1} e^{-(2i\pi/m)\jmath}\|x + e^{(2i\pi/m)\jmath}y\|_2^2.$$

xi) If $\mathbb{F} = \mathbb{C}$, then $2x^*y = \|x + y\|_2^2 - \|x\|_2^2 - \|y\|_2^2 + (\|x\|_2^2 + \|y\|_2^2 - \|x + \jmath y\|_2^2)\jmath$.

xii) $\|x + y\|_2 = \|x\|_2 + \|y\|_2$ if and only if there exists $\beta \ge 0$ such that either $x = \beta y$ or $y = \beta x$. If these conditions hold, then $\operatorname{Im} x^*y = 0$ and $\operatorname{Re} x^*y \ge 0$.

xiii) If x and y are linearly independent, then $\|x + y\|_2 < \|x\|_2 + \|y\|_2$.

xiv) $|x^*y|^2 + \frac{1}{2}\|x \otimes y - (x \otimes y)^*\|_2^2 = \|x\|_2^2\|y\|_2^2$.

xv) If $x, y \in \mathbb{R}^n$ and x and y are nonzero, then

$$x^\mathsf{T}y + \tfrac{1}{2}\|x\|_2\|y\|_2\left\|\frac{x}{\|x\|_2} - \frac{y}{\|y\|_2}\right\|_2^2 = \|x\|_2\|y\|_2,$$

$$1 - \frac{1}{2}\left[\frac{\big|\|x\|_2 - \|y\|_2\big| + \|x - y\|_2}{\max\{\|x\|_2, \|y\|_2\}}\right]^2 \le \frac{x^\mathsf{T}y}{\|x\|_2\|y\|_2} \le 1 - \frac{1}{2}\left[\frac{\|x - y\|_2 - \big|\|x\|_2 - \|y\|_2\big|}{\min\{\|x\|_2, \|y\|_2\}}\right]^2,$$

$$(\|x\|_2 + \|y\|_2 - \|x + y\|_2)^2 \le \|x - y\|_2^2 - (\|x\|_2 - \|y\|_2)^2 = \|x\|_2\|y\|_2\left\|\frac{x}{\|x\|_2} - \frac{y}{\|y\|_2}\right\|_2^2.$$

xvi) If x and y are nonzero, then

$$\operatorname{Re} y^*x = \|x\|_2\|y\|_2\left(1 - \frac{1}{2}\left\|\frac{x}{\|x\|_2} - \frac{y}{\|y\|_2}\right\|_2^2\right), \quad \operatorname{Im} y^*x = \|x\|_2\|y\|_2\left(1 - \frac{1}{2}\left\|\frac{x}{\|x\|_2} - \frac{\jmath y}{\|y\|_2}\right\|_2^2\right).$$

xvii) $|x^*y| \le |x|^\mathsf{T}|y| \le \|x\|_2\|y\|_2$.

xviii) $\|x + y\|_2\|x - y\|_2 \le \frac{1}{2}(\|x + y\|_2^2 + \|x - y\|_2^2) = \|x\|_2^2 + \|y\|_2^2$.

xix) If $\|x + y\|_2 \le 2$, then $(1 - \|x\|_2^2)(1 - \|y\|_2^2) \le |1 - x^*y|^2$.

xx) $(1 - \|x\|_2^2)(1 - \|y\|_2^2) + \|x + y\|_2^2 + \|x - y\|_2^2 = (1 + \|x\|_2^2)(1 + \|y\|_2^2)$.

xxi) $\|x - y\|_2^2 \le (1 + \|x\|_2^2)(1 + \|y\|_2^2)$.

xxii) $\dfrac{\|x + y\|_2}{1 + \|x + y\|_2} \le \dfrac{\|x\|_2}{1 + \|x\|_2} + \dfrac{\|y\|_2}{1 + \|y\|_2}$.

xxiii) If x and y are nonzero, then

$$\left\|\frac{1}{\|x\|_2}x - \|x\|_2 y\right\|_2 = \left\|\frac{1}{\|y\|_2}y - \|y\|_2 x\right\|_2.$$

xxiv) If x and y are nonzero, then

$$\tfrac{1}{2}(\|x\|_2 + \|y\|_2)\left\|\frac{1}{\|x\|_2}x - \frac{1}{\|y\|_2}y\right\|_2 \le \|x - y\|_2.$$

xxv) Let α be a nonzero real number. Then,

$$\|x\|_2^2\|y\|_2^2 - |x^*y|^2 \le \alpha^{-2}\|\alpha y - x\|_2^2\|x\|_2^2.$$

If, in addition, $\operatorname{Re} x^*y \ne 0$, then

$$\|x\|_2^2\|y\|_2^2 - |x^*y|^2 \le \alpha_0^{-2}\|\alpha_0 y - x\|_2^2\|x\|_2^2 \le \alpha^{-2}\|\alpha y - x\|_2^2\|x\|_2^2,$$

where $\alpha_0 \triangleq x^*x/(\operatorname{Re} x^*y)$.

xxvi) If $p \in [1,2]$, then

$$\left.\begin{array}{r}\|x+y\|_2^p \le 2^{p-1}(\|x\|_2^p + \|y\|_2^p)\\ (\|x\|_2 + \|y\|_2)^p + |\|x\|_2 - \|y\|_2|^p\end{array}\right\} \le \|x+y\|_2^p + \|x-y\|_2^p \le 2(\|x\|_2^p + \|y\|_2^p).$$

xxvii) If $p \ge 2$, then

$$2(\|x\|_2^p + \|y\|_2^p) \le \|x+y\|_2^p + \|x-y\|_2^p \le 2^{p-1}(\|x\|_2^p + \|y\|_2^p).$$

xxviii) If $p \in (1,2]$, $q \ge 2$, and $1/p + 1/q = 1$, then

$$\|x+y\|_2^q + \|x-y\|_2^q \le 2(\|x\|_2^p + \|y\|_2^p)^{q-1}.$$

xxix) If $p \ge 2$, $q \in (1,2]$, and $1/p + 1/q = 1$, then

$$2(\|x\|_2^p + \|y\|_2^p)^{q-1} \le \|x+y\|_2^q + \|x-y\|_2^q.$$

xxx) If $p, q > 1$ and $1/p + 1/q = 1$, then

$$\|x+y\|_2^2 \le p\|x\|_2^2 + q\|y\|_2^2.$$

Furthermore, equality holds if and only if $y = (p-1)x$.

xxxi) If $\alpha \in [0,1]$, then

$$\|x-y\|_2^2 + \|\alpha x + y\|_2^2 \le (1+\alpha)\|x\|_2^2 + (1+\tfrac{1}{\alpha})\|y\|_2^2.$$

xxxii) If either $\alpha < 0$ or $\alpha \ge 1$, then

$$(1+\alpha)\|x\|_2^2 + (1+\tfrac{1}{\alpha})\|y\|_2^2 \le \|x-y\|_2^2 + \|\alpha x + y\|_2^2.$$

xxxiii) Let $z \in \mathbb{F}$, let $\mathcal{S} \subseteq \mathbb{F}$ be a subspace, and let $x, y \in \mathcal{S}$. Then,

$$\|x-y\|_2 \le \sqrt{\|z-x\|_2^2 - \min_{w\in\mathcal{S}} \|z-w\|_2^2} + \sqrt{\|z-y\|_2^2 - \min_{w\in\mathcal{S}} \|z-w\|_2^2}.$$

Equality holds if and only if there exists $\alpha \in [0,1]$ such that $\min_{w\in\mathcal{S}} \|z-w\|_2 = \|z - [\alpha x + (1-\alpha)y]\|_2$.

xxxiv) Let $\mathcal{S} \subseteq \mathbb{F}^n$ be a subspace, let $A \in \mathbb{F}^{n\times n}$ be the projector onto $\mathcal{S}$, and assume that $y \in \mathcal{S}$. Then,

$$\|x-y\|_2^2 = \|x-Ax\|_2^2 + \|y-Ax\|_2^2.$$

Source: *ii*) and *iii*) is the *cosine law* (see Fact 11.10.1 for a matrix version); the equivalence of *c*) and *d*) in *iv*) is the *Pythagorean theorem*; the first equality in *v*) is the *generalized parallelogram law*, see [1095] and Fact 10.11.84; the second equality in *v*) is the *parallelogram law*, which relates the diagonals and the sides of a parallelogram; *vi*) follows from *ii*) and *iii*); *viii*) follows from *vi*) by replacing y with $\jmath y$; *ix*) is the *polarization identity* (see [2112, p. 276]); *x*) is a generalization of the polarization identity (see [828, p. 54]); *xi*) is given in [2238, p. 261]; *xiv*) is given in [2488]; *xv*) is given in [32, 2022]; *xvi*) is given in [33]; the first and third terms in *xvii*) are the Cauchy-Schwarz inequality; *xix*) is given by Lemma 1 in [2933] and implies Aczel's inequality given by Fact 2.12.37; *xxiv*) is the *Dunkl-Williams inequality*, which compares the distance between x and y with the distance between the projections of x and y into the unit sphere (see [967], [2061, p. 515], and [2983, p. 28]); *xxv*) is given in [1827]; *xxvi*)–*xxix*) are the *Clarkson inequalities* (see [1419], [2061, p. 536], and [2294, p. 253]); the lower left inequality in *xxvi*) is given in [940]; *xxx*) is *Bohr's inequality* (see [1984]); *xxxi*) and *xxxii*) are given in [1095]; *xxxiii*), which is the *Beppo Levi inequality*, is given in [2333]; *xxxiv*) is given in [2487, p. 67] **Remark:** In terms of the traditional inner product notation, $x^*y = \langle y, x\rangle$. **Remark:** Many of these results are extensions of results for complex scalars given by Fact 2.21.8. Note that, if $\mathbb{F} = \mathbb{R}$ and $n = 2$, then $\|\left[\begin{smallmatrix} x \\ y\end{smallmatrix}\right]\|_2 = |x + \jmath y|$. **Remark:** By replacing the Euclidean norm in *xvi*) with an arbitrary vector norm, it is possible to define an alternative notion of angle. For example, the nonzero vectors x and y are *orthogonal with respect to* $\|\cdot\|$ if $\left\|\frac{1}{\|x\|}x - \frac{1}{\|y\|}y\right\| = \sqrt{2}$. See [811, 902, 1075]. **Related:** Fact 14.8.5.

Fact 11.8.4. Let $x_1,\dots,x_k \in \mathbb{F}^n$, let $\alpha_1,\dots,\alpha_k$ be nonzero real numbers, and assume that $\sum_{i=1}^k \alpha_i = 1$. Then,

$$\sum_{i=1}^k \frac{1}{\alpha_i}\|x_i\|_2^2 - \left\|\sum_{i=1}^k x_i\right\|_2^2 = \sum \frac{\alpha_i}{\alpha_j}\left\|\frac{\alpha_j}{\alpha_i}x_i - x_j\right\|_2^2,$$

where the last summation is taken over all $i, j \in \{1,\dots,k\}$ such that $i \le j$. **Source:** [1095, 2989]. **Related:** Fact 2.21.14 and Fact 10.11.85.

Fact 11.8.5. Let $x, y, z \in \mathbb{F}^n$. Then, the following statements hold:

i) $\|x-z\|_2^2 + \|z-y\|_2^2 = \frac{1}{2}\|x-y\|_2^2 + 2\|z - \frac{1}{2}(x+y)\|_2^2$.

ii) $\|x+y\|_2^2 + \|y+z\|_2^2 + \|z+x\|_2^2 = \|x\|_2^2 + \|y\|_2^2 + \|z\|_2^2 + \|x+y+z\|_2^2$.

iii) $\|x-y\|_2^2 + \|y-z\|_2^2 + \|z-x\|_2^2 + \|x+y+z\|_2^2 = 3(\|x\|_2^2 + \|y\|_2^2 + \|z\|_2^2)$.

iv) $\|x+y-z\|_2^2 + \|y+z-x\|_2^2 + \|z+x-y\|_2^2 + \|x+y+z\|_2^2 = 4(\|x\|_2^2 + \|y\|_2^2 + \|z\|_2^2)$.

v) $\|x+y\|_2 + \|y+z\|_2 + \|z+x\|_2 \le \|x\|_2 + \|y\|_2 + \|z\|_2 + \|x+y+z\|_2$.

vi) $\|x\|_2 + \|y\|_2 + \|z\|_2 \le \|x+y-z\|_2 + \|y+z-x\|_2 + \|z+x-y\|_2$.

vii) $\operatorname{Re} x^*zz^*y \le \frac{1}{2}(\|x\|_2\|y\|_2 + \operatorname{Re} x^*y)\|z\|_2^2$, $|x^*zz^*y| \le \frac{1}{2}(\|x\|_2\|y\|_2 + |x^*y|)\|z\|_2^2$.

viii) $|\operatorname{Re}(x^*zz^*y - \frac{1}{2}x^*y)\|z\|_2^2)| \le |x^*zz^*y - \frac{1}{2}x^*y\|z\|_2^2| \le \frac{1}{2}\|x\|_2\|y\|_2\|z\|_2^2$.

ix) $|\operatorname{Re}(x^*zz^*y - \frac{1}{2}x^*y\|z\|_2^2)| \le \frac{1}{2}\|z\|_2^2\sqrt{\|x\|_2^2\|y\|_2^2 - (\operatorname{Im} x^*y)^2} \le \frac{1}{2}\|x\|_2\|y\|_2\|z\|_2^2$.

x) $|\operatorname{Im}(x^*zz^*y - \frac{1}{2}x^*y\|z\|_2^2)| \le \frac{1}{2}\|z\|_2^2\sqrt{\|x\|_2^2\|y\|_2^2 - (\operatorname{Re} x^*y)^2} \le \frac{1}{2}\|x\|_2\|y\|_2\|z\|_2^2$.

xi) $\|x\|_2^2|y^*z|^2 + \|y\|_2^2|z^*x|^2 + \|z\|_2^2|x^*y|^2 \le \|x\|_2^2\|y\|_2^2\|z\|_2^2 + 2|x^*yy^*zz^*x|^2 \le 3\|x\|_2^2\|y\|_2^2\|z\|_2^2$.

xii) $\|x\|_2^2|y^*z|^2 + \|y\|_2^2|z^*x|^2 \le \|x\|_2^2\|y\|_2^2\|z\|_2^2 + \|x\|_2\|y\|_2\|z\|_2^2|x^*y| \le 2\|x\|_2^2\|y\|_2^2\|z\|_2^2$.

xiii) $\sqrt{\|x+y\|_2} + \sqrt{\|y+z\|_2} + \sqrt{\|z+x\|_2} \le \sqrt{\|x\|_2} + \sqrt{\|y\|_2} + \sqrt{\|z\|_2} + \sqrt{\|x+y+z\|_2}$.

Remark: *i)* is the *Appolonius identity* (see [2238, p. 260]; *ii)* is *Hlawka's identity* (see [2128, p. 100]); *v)* is *Hlawka's inequality* (see [2061, p. 521], [2128, p. 100], and Fact 1.21.9); *vii)* is *Buzano's inequality* (see [937, 1090], [2527, p. 71], and Fact 2.13.2); *viii)* is an extension of Buzano's inequality (see [937]); *ix)* and *x)* are given in [2254]; *xi)* is given in [1897, 1908]; *xii)* is given in [1908]; *xiii)* is given in [2305]. **Remark:** As in Fact 11.8.3, some of these results are extensions of results for complex scalars given by Fact 2.21.8.

Fact 11.8.6. Let $w, x, y, z \in \mathbb{C}^n$. Then,

$$|w^{\mathrm{T}}x|^2 + |y^{\mathrm{T}}z|^2 + 2|\operatorname{Re}(w^*yx^{\mathrm{T}}\bar{z} - w^*\bar{z}x^{\mathrm{T}}y)| \le \|w\|_2^2\|x\|_2^2 + \|y\|_2^2\|z\|_2^2.$$

Source: [722, 1342]. **Remark:** This is a generalized Cauchy-Schwarz inequality. **Related:** Fact 11.10.2.

Fact 11.8.7. Let $x \in \mathbb{F}^n$, and let $p, q \in [1, \infty]$, where $p \le q$. Then,

$$\|x\|_q \le \|x\|_p \le n^{1/p-1/q}\|x\|_q.$$

In particular,

$$\|x\|_2 \le \|x\|_1 \le \sqrt{n}\|x\|_2, \quad \|x\|_\infty \le \|x\|_1 \le n\|x\|_\infty, \quad \|x\|_\infty \le \|x\|_2 \le \sqrt{n}\|x\|_\infty.$$

Source: [1388] and [1389, p. 107]. **Related:** Fact 2.11.90 and Fact 11.9.25.

Fact 11.8.8. Let $n \ge 3$, let $x_1,\dots,x_n \in \mathbb{F}^n$, and, for all $k \in \{1,\dots,n\}$, define $S_k \triangleq \sum \|x_{i_1} + \cdots + x_{i_k}\|_2$, where the sum is taken over all k-tuples $(i_1,\dots,i_k)$ such that $1 \le i_1 < \cdots < i_k \le n$. Then, for all $k \in \{2, \ldots, n-1\}$,

$$S_k \le \binom{n-2}{k-1} S_1 + \binom{n-2}{k-2} S_n.$$

In addition,

$$\sum_{i=2}^{n-1} S_i \le (2^{n-2} - 1)(S_1 + S_n).$$

Source: [2470]. **Example:** If $n = 3$, then $\|x_1 + x_2\|_2 + \|x_2 + x_3\|_2 + \|x_3 + x_1\|_2 \le \|x_1\|_2 + \|x_2\|_2 + \|x_3\|_2 + \|x_1 + x_2 + x_3\|_2$. If $n = 4$, then

$$\begin{aligned}\|x_1 + x_2\|_2 + \|x_1 + x_3\|_2 + \|x_1 + x_4\|_2 + \|x_2 + x_3\|_2 + \|x_2 + x_4\|_2 + \|x_3 + x_4\|_2 \\ \le 2(\|x_1\|_2 + \|x_2\|_2 + \|x_3\|_2 + \|x_4\|_2) + \|x_1 + x_2 + x_3 + x_4\|_2,\end{aligned}$$

$$\begin{aligned}\|x_1 + x_2 + x_3\|_2 + \|x_1 + x_2 + x_4\|_2 + \|x_1 + x_3 + x_4\|_2 + \|x_2 + x_3 + x_4\|_2 \\ \le \|x_1\|_2 + \|x_2\|_2 + \|x_3\|_2 + \|x_4\|_2 + 2\|x_1 + x_2 + x_3 + x_4\|_2.\end{aligned}$$

Remark: These inequalities concern the diagonals of a polygon. **Related:** Fact 2.21.25.

Fact 11.8.9. Let $x, y, z \in \mathbb{F}^n$, assume that x, y, z are nonzero, and define

$$\phi_{xy} \triangleq \operatorname{acos} \frac{\operatorname{Re} y^*x}{\|x\|_2\|y\|_2}, \quad \theta_{xy} \triangleq \operatorname{acos} \frac{|y^*x|}{\|x\|_2\|y\|_2}.$$

Then,

$$\phi_{xz} \le \phi_{xy} + \phi_{yz}, \quad \theta_{xz} \le \theta_{xy} + \theta_{yz},$$
$$\sin\phi_{xz} \le \sin\phi_{xy} + \sin\phi_{yz}, \quad \sin\theta_{xz} \le \sin\theta_{xy} + \sin\theta_{yz}.$$

Source: [1829]. **Remark:** The first inequality is *Krein's inequality*. **Related:** Fact 3.15.1 and Fact 5.1.5.

Fact 11.8.10. Let $y, x_1, \ldots, x_n \in \mathbb{F}^n$, assume that, for all $i, j \in \{1, \ldots, n\}$, $x_i^*x_j = \delta_{i,j}$, and let $k \in \{1, \ldots, n\}$. Then, the following statements hold:

i) $x_i^*(y - \sum_{i=1}^k x_i^*yx_i) = 0$.

ii) $\left\|y - \sum_{i=1}^k x_i^*yx_i\right\|_2^2 = \|y\|_2^2 - \sum_{i=1}^k |x_i^*y|^2$.

iii) $\sum_{i=1}^k |x_i^*y|^2 \le \|y\|_2^2$.

iv) $y = \sum_{i=1}^n x_i^*yx_i$.

v) $\|y\|_2 = \sum_{i=1}^n |x_i^*y|^2$.

vi) $\sum_{i=1}^n |x_i^*y|^2 = \|y\|_2^2$.

Source: [2238, pp. 264, 265]. **Remark:** *iii)* is *Bessel's inequality*; *vi)* is *Parseval's identity*.

Fact 11.8.11. Let $y, x_1, \ldots, x_n \in \mathbb{F}^n$, and assume that, for all $i \in \{1, \ldots, n\}$, $\sum_{j=1}^n |x_j^*x_i| \ne 0$. Then,

$$\sum_{i=1}^n \frac{|x_i^*y|^2}{\sum_{j=1}^n |x_j^*x_i|} \le \|y\|_2^2.$$

Source: [2527, p. 225]. **Remark:** This is *Selberg's inequality*.

Fact 11.8.12. Let $x_1, \ldots, x_m, y_1, \ldots, y_n \in \mathbb{F}^n$. Then,

$$\sum_{i,j=1}^{m,n} |y_j^*x_i|^2 \le \left(\sum_{i,j=1}^{m,m} |x_j^*x_i|^2\right)^{1/2} \left(\sum_{i,j=1}^{n,n} |y_j^*y_i|^2\right)^{1/2}.$$

Source: [2527, p. 225]. **Credit:** P. Enflo.

Fact 11.8.13. Let $x, y \in \mathbb{R}^3$, and let $\mathcal{S} \subset \mathbb{R}^3$ be the parallelogram with vertices 0, x, y, and $x + y$. Then, $\operatorname{area}(\mathcal{S}) = \|x \times y\|_2$. **Remark:** The parallelogram associated with the cross product can be interpreted as a bivector. See [926, pp. 86–88] and [1268, 1759]. **Related:** Fact 4.12.1, Fact 5.3.1, and Fact 5.3.10.

Fact 11.8.14. Let $x, y \in \mathbb{F}^n$, and assume that x and y are nonzero. Then,

$$\frac{\operatorname{Re} x^*y}{\|x\|_2\|y\|_2}(\|x\|_2 + \|y\|_2) \le \|x + y\|_2 \le \|x\|_2 + \|y\|_2.$$

Hence, if $\operatorname{Re} x^*y = \|x\|_2\|y\|_2$, then $\|x\|_2 + \|y\|_2 = \|x + y\|_2$. **Source:** [2061, p. 517]. **Remark:** This is a *reverse triangle inequality*. **Remark:** Setting $x = -y = 1$ shows that the first inequality can fail with $\operatorname{Re} x^*y$ replaced by $|\operatorname{Re} x^*y|$.

Fact 11.8.15. Let $x_1, \dots, x_n \in \mathbb{C}^n$, let $\alpha_1, \dots, \alpha_n \in \mathbb{R}$, and define $\alpha \triangleq \sum_{i=1}^n \alpha_i$. Then,

$$\sum_{i=1}^n \alpha_i \left\|\sum_{j=1}^n \alpha_j(x_j - x_i)\right\|_2^2 = \frac{\alpha}{2}\sum_{i,j=1}^n \alpha_i\alpha_j\|x_i - x_j\|_2^2.$$

Now, let $x \in \mathbb{C}^n$, and assume that α is nonzero. Then,

$$\sum_{i=1}^n \alpha_i\|x - x_i\|_2^2 = \alpha\left\|x - \frac{1}{\alpha}\sum_{i=1}^n \alpha_i x_i\right\|_2^2 + \frac{1}{2\alpha}\sum_{i,j=1}^n \alpha_i\alpha_j\|x_i - x_j\|_2^2,$$

$$\sum_{i=1}^n \alpha_i\|x - x_i\|_2^2 = \alpha\left\|x - \frac{1}{\alpha}\sum_{i=1}^n \alpha_i x_i\right\|_2^2 + \sum_{i=1}^n \alpha_i\left\|x_i - \frac{1}{\alpha}\sum_{j=1}^n \alpha_j x_j\right\|_2^2.$$

Source: [435, 1171]. **Remark:** The second equality is *Lagrange's second identity*. The third inequality is the *Huygens-Leibniz identity*.

Fact 11.8.16. Let $x_1, \dots, x_n \in \mathbb{F}^n$, and let $\alpha_1, \dots, \alpha_n$ be nonnegative numbers. Then,

$$\sum_{i=1}^n \alpha_i\left\|x_i - \sum_{j=1}^n \alpha_j x_j\right\|_2 \le \sum_{i=1}^n \alpha_i\|x_i\|_2 + \left[\left(\sum_{i=1}^n \alpha_i\right) - 2\right]\left\|\sum_{i=1}^n \alpha_i x_i\right\|_2.$$

In particular,

$$\sum_{i=1}^n \left\|\sum_{j=1, j\ne i}^n x_j\right\|_2 \le \sum_{i=1}^n \|x_i\|_2 + (n-2)\left\|\sum_{i=1}^n x_i\right\|_2.$$

Remark: The first inequality is the *generalized Hlawka inequality* (also called the *polygonal inequalities*). The second inequality is the *Djokovic inequality*. See [2584] and Fact 11.8.3.

Fact 11.8.17. Let $x_1, \dots, x_n, y_1, \dots, y_n \in \mathbb{F}^n$. Then,

$$\sum_{i,j=1}^n \|x_i - x_j\|_2^2 + \sum_{i,j=1}^n \|y_i - y_j\|_2^2 + 2\left\|\sum_{i=1}^n (x_i - y_i)\right\|_2^2 = 2\sum_{i,j=1}^n \|x_i - y_j\|_2^2.$$

Equivalently,

$$\sum_{i,j=1, i<j}^n \|x_i - x_j\|_2^2 + \sum_{i,j=1, i<j}^n \|y_i - y_j\|_2^2 + \left\|\sum_{i=1}^n (x_i - y_i)\right\|_2^2 = \sum_{i,j=1}^n \|x_i - y_j\|_2^2.$$

Source: [2093]. **Remark:** This is a generalized parallelogram law. Setting $x_1 = -x_2 = x$, $x_3 = -x_4 = y$, and $y_1 = y_2 = y_3 = y_4 = 0$ yields the parallelogram law $\|x + y\|_2^2 + \|x - y\|_2^2 = 2\|x\|_2^2 + 2\|y\|_2^2$ given by *v*) of Fact 11.8.3. **Related:** Fact 2.21.26 and Fact 11.10.61.

Fact 11.8.18. Let $x, y \in \mathbb{R}^n$, let α and δ be positive numbers, and let $p, q \in (0, \infty)$ satisfy $1/p + 1/q = 1$. Then,

$$\left(\frac{\alpha}{\alpha + \|y\|_2^q}\right)^{p-1} \delta^p \le |\delta - x^\mathsf{T} y|^p + \alpha^{p-1}\|x\|_2^p.$$

Equality holds if and only if $x = [\delta\|y\|_2^{q-2}/(\alpha + \|y\|_2^q)]y$. In particular,

$$\frac{\alpha\delta^2}{\alpha + \|y\|_2^2} \le (\delta - x^\mathsf{T} y)^2 + \alpha\|x\|_2^2.$$

Equality holds if and only if $x = [\delta/(\alpha + \|y\|_2^2)]y$. **Source:** [2583]. **Remark:** These are generalizations of Hua's inequality. See Fact 2.11.39 and Fact 11.8.19. **Credit:** The first inequality is due to J. Pecaric. The case $p = q = 2$ is due to S. S. Dragomir and G.-S. Yang.

Fact 11.8.19. Let $x_1, \dots, x_n, y \in \mathbb{R}^n$ and $\alpha \in (0, \infty)$. Then,

$$\frac{\alpha}{\alpha + n}\|y\|_2^2 \le \left\| y - \sum_{i=1}^n x_i \right\|_2^2 + \alpha \sum_{i=1}^n \|x_i\|_2^2.$$

Equality holds if and only if $x_1 = \cdots = x_n = [1/(\alpha + n)]y$. **Source:** [2583]. **Related:** This extends Hua's inequality. See Fact 2.11.39 and Fact 11.8.18.

Fact 11.8.20. Let $x \in \mathbb{F}^n$, and let $p, q \in [1, \infty]$ satisfy $1/p + 1/q = 1$. Then, $\|x\|_2 \le \sqrt{\|x\|_p\|x\|_q}$.

Fact 11.8.21. Let $x, y \in \mathbb{R}^n$, assume that x and y are nonnegative, let $p \in (0, 1]$, and define $\|x\|_p \triangleq (\sum_{i=1}^n x_{(i)}^p)^{1/p}$. Then,

$$\|x\|_p + \|y\|_p \le \|x + y\|_p.$$

Now, let $q \in (0, 1]$, and assume that $p \le q$. Then,

$$\|x\|_q \le \|x\|_p.$$

Remark: This notation is for convenience only since, for all $p \in (0, 1)$, $\|\cdot\|_p$ is not a norm but rather is an *anti-norm* on $\{x \in \mathbb{R}^n\colon x \geq\geq 0\}$. See [1779]. **Related:** Fact 2.2.59, Fact 2.11.91, Fact 10.14.7, Fact 11.10.4, Fact 11.10.54, and Fact 11.10.55.

Fact 11.8.22. Let $x, y \in \mathbb{F}^{n\times n}$. Then, $|\|x\|_2 - \|y\|_2| \le \|x - y\|_1$. **Source:** [1566, p. 12].

Fact 11.8.23. Let $x, y \in \mathbb{F}^{n\times n}$. If $p \in [1, 2]$, then

$$(\|x\|_p + \|y\|_p)^p + |\|x\|_p - \|y\|_p|^p \le \|x + y\|_p^p + \|x - y\|_p^p,$$
$$(\|x + y\|_p + \|x - y\|_p)^p + |\|x + y\|_p - \|x - y\|_p|^p \le 2^p(\|x\|_p^p + \|y\|_p^p).$$

If $p \in [2, \infty]$, then

$$\|x + y\|_p^p + \|x - y\|_p^p \le (\|x\|_p + \|y\|_{\sigma p})^p + |\|x\|_p - \|y\|_p|^p,$$
$$2^p(\|x\|_p^p + \|y\|_p^p) \le (\|x + y\|_p + \|x - y\|_p)^p + |\|x + y\|_p - \|x - y\|_p|^p.$$

Source: [262, 1818]. **Remark:** These are vector extensions of *Hanner's inequality*. These follow from integral inequalities on L_p by appropriate choice of measure. **Remark:** Equality holds for $p = 2$. The case where $p = 2$, $n = 1$, and $\mathbb{F} = \mathbb{C}$ is given by Fact 2.21.8. The case where $p = 2$ and $n \ge 1$ is given by Fact 11.8.3. **Remark:** Matrix versions are given in Fact 11.10.65.

Fact 11.8.24. Let $y \in \mathbb{F}^n$, let $\|\cdot\|$ be a norm on $\mathbb{F}^n$, let $\|\cdot\|'$ be the norm on $\mathbb{F}^{n\times n}$ induced by $\|\cdot\|$, and define

$$\|y\|_\mathrm{D} \triangleq \max_{x\in\{z\in\mathbb{F}^n\colon \|z\|=1\}} |y^*x|.$$

Then, $\|\cdot\|_{\mathrm{D}}$ is a norm on $\mathbb{F}^n$. Furthermore,

$$\|y\| = \max_{x\in\{z\in\mathbb{F}^n:\ \|z\|_{\mathrm{D}}=1\}} |y^*x|.$$

Hence, for all $x \in \mathbb{F}^n$,

$$|x^*y| \le \|x\|\|y\|_{\mathrm{D}}, \quad \|xy^*\|' = \|x\|\|y\|_{\mathrm{D}}.$$

Finally, let $p, q \in [1, \infty)$ satisfy $1/p + 1/q = 1$. Then,

$$\|\cdot\|_{p\mathrm{D}} = \|\cdot\|_q.$$

Hence, for all $x \in \mathbb{F}^n$,

$$|x^*y| \le \|x\|_p\|y\|_q, \quad \|xy^*\|_{p,p} = \|x\|_p\|y\|_q.$$

Source: [2539, p. 57]. **Remark:** $\|\cdot\|_{\mathrm{D}}$ is the *dual norm* of $\|\cdot\|$.

Fact 11.8.25. Let $x \in \mathbb{F}^n$ and $y \in \mathbb{F}^m$. Then,

$$\sigma_{\max}(xy^*) = \|xy^*\|_{\mathrm{F}} = \|x\|_2\|y\|_2, \quad \sigma_{\max}(xx^*) = \|xx^*\|_{\mathrm{F}} = \|x\|_2^2.$$

Related: Fact 7.12.19.

Fact 11.8.26. Let $x \in \mathbb{F}^n$, $y \in \mathbb{F}^m$, and $p \in (0, \infty)$. Then,

$$\|x \otimes y\|_p = \|\operatorname{vec}(x \otimes y^{\mathsf{T}})\|_p = \|\operatorname{vec}(xy^{\mathsf{T}})\|_p = \|xy^{\mathsf{T}}\|_p = \|x\|_p\|y\|_p.$$

Fact 11.8.27. Let $x \in \mathbb{C}^n$ and $p, q \in [1, \infty]$. Then, $\|x\|_{q,p,\mathbb{C}} = \|x\|_q$. Now, assume that $x \in \mathbb{R}^n$. Then, $\|x\|_{q,p,\mathbb{C}} = \|x\|_{q,p,\mathbb{R}} = \|x\|_q$. **Related:** Fact 11.9.43.

Fact 11.8.28. Let $x_1, \ldots, x_k \in \mathbb{F}^n$, let $\alpha_1, \ldots, \alpha_k \in (0, \infty)$, and assume that $\sum_{i=1}^k \alpha_i = 1$. Then,

$$|1_{1\times n}(x_1 \odot \cdots \odot x_k)| \le \prod_{i=1}^k \|x_i\|_{1/\alpha_i}.$$

Remark: This is the *generalized Hölder inequality*. See [603, p. 128].

Fact 11.8.29. Let $x_1, \ldots, x_m \in \mathbb{R}^n$, assume that $\sum_{i=1}^m x_i = 0$, and assume that, for all $i \in \{1, \ldots, m\}$, $\|x_i\|_2 \le 1$. Then, there exists a permutation σ of $\{1, \ldots, m\}$ such that, for all $k \in \{1, \ldots, m\}$, $\|\sum_{i=1}^k x_{\sigma(i)}\|_2 \le n$. **Source:** [1992, pp. 71–75].

11.9 Facts on Matrix Norms for One Matrix

Fact 11.9.1. Let $\mathcal{S} \subseteq \mathbb{F}^m$, assume that $\mathcal{S}$ is bounded, and let $A \in \mathbb{F}^{n\times m}$. Then, $A\mathcal{S}$ is bounded.

Fact 11.9.2. Let $A \in \mathbb{F}^{n\times n}$, assume that A is a idempotent, and assume that, for all $x \in \mathbb{F}^n$, $\|Ax\|_2 \le \|x\|_2$. Then, A is a projector. **Source:** [1133, p. 42].

Fact 11.9.3. Let $A \in \mathbb{F}^{n\times n}$, and assume that $\rho_{\max}(A) < 1$. Then, there exists a submultiplicative matrix norm $\|\cdot\|$ on $\mathbb{F}^{n\times n}$ such that $\|A\| < 1$. Furthermore, $\lim_{k\to\infty} A^k = 0$.

Fact 11.9.4. Let $A \in \mathbb{F}^{n\times n}$, assume that A is nonsingular, and let $\|\cdot\|$ be a submultiplicative norm on $\mathbb{F}^{n\times n}$. Then, $\|I_n\|/\|A\| \le \|A^{-1}\|$.

Fact 11.9.5. Let $A \in \mathbb{F}^{n\times n}$, assume that A is nonzero and idempotent, and let $\|\cdot\|$ be a submultiplicative norm on $\mathbb{F}^{n\times n}$. Then, $\|A\| \ge 1$.

Fact 11.9.6. Let $A \in \mathbb{F}^{n\times n}$, and let $\|\cdot\|$ be a unitarily invariant norm on $\mathbb{F}^{n\times n}$. Then, $\|A\| = \|A^*\| = \|\langle A\rangle\| = \|\langle A^*\rangle\|$. If, in addition, $r > 0$, then $\|\langle A\rangle^r\| = \|\langle A^*\rangle^r\|$. **Remark:** $\|\cdot\|$ is self-adjoint.

Fact 11.9.7. Let $A \in \mathbb{F}^{n\times m}$, let $\|\cdot\|$ be a norm on $\mathbb{F}^{n\times m}$, and define $\|A\|' \triangleq \|A^*\|$. Then, $\|\cdot\|'$ is a norm on $\mathbb{F}^{m\times n}$. If, in addition, $n = m$ and $\|\cdot\|$ is induced by $\|\cdot\|''$, then $\|\cdot\|'$ is induced by $\|\cdot\|''_{\mathrm{D}}$. **Source:** [1448, p. 309] and Fact 11.9.21. **Remark:** $\|\cdot\|'$ is the *adjoint norm* of $\|\cdot\|$. **Related:** Fact 11.8.24 defines the dual norm. **Problem:** Extend this result to nonsquare matrices and norms that are not equi-induced.

Fact 11.9.8. Let $A \in \mathbb{F}^{n\times n}$. Then,

$$\|A\|_{\mathrm{F}}^2 = \|\tfrac{1}{2}(A + A^*)\|_{\mathrm{F}}^2 + \|\tfrac{1}{2}(A - A^*)\|_{\mathrm{F}}^2, \quad \|A^*A - AA^*\|_{\mathrm{F}}^2 = 2[\operatorname{tr}(A^*A)^2 - \operatorname{tr} A^{2*}A^2].$$

Source: The second equality is given in [2476].

Fact 11.9.9. Let $A \in \mathbb{F}^{2\times 2}$. Then,

$$\sigma_{\max}(A) = \left[\frac{1}{2}\left(\|A\|_{\mathrm{F}}^2 + \sqrt{\|A\|_{\mathrm{F}}^4 - 4|\det A|^2}\right)\right]^{1/2}.$$

Source: [1485, p. 261].

Fact 11.9.10. Let $A \in \mathbb{F}^{n\times m}$. If $p \in (0, 2]$, then $\|A\|_{\sigma p} \le \|A\|_p$. If $p \ge 2$, then $\|A\|_p \le \|A\|_{\sigma p}$. **Source:** [2977, p. 50]. **Remark:** For $p \in (0, 1)$, $\|A\|_{\sigma p} \triangleq \|\sigma(A)\|_p$ is not a norm. See Fact 11.8.21.

Fact 11.9.11. Let $1 \le p \le \infty$. Then, $\|\cdot\|_{\sigma p}$ is unitarily invariant.

Fact 11.9.12. Let $A \in \mathbb{F}^{n\times m}$. Then,

$$\|A^*A\|_{\mathrm{F}} = \|AA^*\|_{\mathrm{F}} = \sqrt{\sum_{i=1}^{\min\{m,n\}} \sigma_i^4(A)} \le \sum_{i=1}^{\min\{m,n\}} \sigma_i^2(A) = \|A\|_{\mathrm{F}}^2, \quad \|A^*A\|_{\mathrm{F}} \begin{cases} = \|A\|_{\mathrm{F}}^2, & \operatorname{rank} A \le 1, \\ < \|A\|_{\mathrm{F}}^2, & \operatorname{rank} A \ge 2. \end{cases}$$

Related: Fact 11.10.31.

Fact 11.9.13. Let $A \in \mathbb{F}^{n\times m}$, and assume that A is positive definite. Then, $n^{3/2} \le \|A^{-1}\|_{\mathrm{F}} \operatorname{tr} A$. **Source:** [1476].

Fact 11.9.14. Let $A \in \mathbb{R}^{n\times n}$, and assume that A is positive definite. Then,

$$\frac{n^{3/2}\|A^{1/2}\|_{\mathrm{F}}}{\operatorname{tr} A^{1/2}} \le \|A\|_{\mathrm{F}}\|A^{-1}\|_{\mathrm{F}}, \quad \frac{2\operatorname{tr} A}{\sqrt[n]{\det A}} - n \le \|A\|_{\mathrm{F}}\|A^{-1}\|_{\mathrm{F}}, \quad 0 \le \frac{2\operatorname{tr} A^{3/2}}{(\operatorname{tr} A)\sqrt[2n]{\det A}} - n \le \|A\|_{\mathrm{F}}\|A^{-1}\|_{\mathrm{F}}.$$

Source: [1809, 2751]. **Related:** Fact 11.15.2.

Fact 11.9.15. Let $A \in \mathbb{F}^{n\times m}$. Then, $\|A^{\mathsf{A}}\|_{\mathrm{F}} \le n^{(2\ n)/2}\|A\|_{\mathrm{F}}^{n-1}$. Furthermore, equality holds if and only if either $n \ge 2$ or there exist $\alpha \in \mathbb{F}$ and a unitary matrix $B \in \mathbb{F}^{n\times n}$ such that $A = \alpha B$. **Source:** [2052] and [2263, pp. 151, 165]. **Related:** Fact 10.21.23.

Fact 11.9.16. Let $A \in \mathbb{F}^{n\times m}$, and assume that A is normal. Then,

$$\frac{1}{\sqrt{mn}}\sigma_{\max}(A) \le \|A\|_\infty \le \rho_{\max}(A) = \sigma_{\max}(A).$$

Source: Fact 7.15.16 and *xii*) of Fact 11.9.23.

Fact 11.9.17. Let $A \in \mathbb{R}^{n\times n}$, assume that A is symmetric, and assume that every diagonal entry of A is zero. Then, the following statements are equivalent:

i) For all $x \in \mathbb{R}^n$ such that $1_{1\times n}x = 0$, it follows that $x^{\mathsf{T}}Ax \le 0$.

ii) There exists $k \ge 1$ and $x_1, \ldots, x_n \in \mathbb{R}^k$ such that, for all $i, j \in \{1, \ldots, n\}$, $A_{(i,j)} = \|x_i - x_j\|_2^2$.

Source: [36]. **Remark:** A is a *Euclidean distance matrix*. **Credit:** I. J. Schoenberg.

Fact 11.9.18. Let $A \in \mathbb{F}^{n\times n}$, assume that A is normal, let $B \in \mathbb{F}^{n\times n}$ be the strictly lower triangular part of A, and let $C \in \mathbb{F}^{n\times n}$ be the strictly upper triangular part of A. Then,

$$\|B\|_{\mathrm{F}} \le \sqrt{n-1}\|C\|_{\mathrm{F}}, \quad \|C\|_{\mathrm{F}} \le \sqrt{n-1}\|B\|_{\mathrm{F}}.$$

Source: [2991, p. 321].

Fact 11.9.19. Let $A \in \mathbb{F}^{n\times n}$, and assume that A is semicontractive. Then,

$$\|A\|_{\mathrm{F}}^2 \le |\det A|^2 + n - 1.$$

Source: [2992].

Fact 11.9.20. Let $A \in \mathbb{F}^{n\times n}$. If $p \in [1,2]$, then

$$\|A\|_{\mathrm{F}} \le \|A\|_{\sigma p} \le n^{1/p-1/2}\|A\|_{\mathrm{F}}.$$

If $p \in [2,\infty]$, then

$$\|A\|_{\sigma p} \le \|A\|_{\mathrm{F}} \le n^{1/2-1/p}\|A\|_{\sigma p}.$$

Source: [453, p. 174].

Fact 11.9.21. Let $A \in \mathbb{F}^{n\times m}$, and let $p,q \in [1,\infty]$ satisfy $1/p + 1/q = 1$. Then, $\|A^*\|_{p,p} = \|A\|_{q,q}$. In particular, $\|A^*\|_{\mathrm{col}} = \|A\|_{\mathrm{row}}$. **Source:** Fact 11.9.7.

Fact 11.9.22. Let $A \in \mathbb{F}^{n\times m}$, and let $p,q \in [1,\infty]$ satisfy $1/p + 1/q = 1$. Then,

$$\left\|\begin{bmatrix} 0 & A \\ A^* & 0 \end{bmatrix}\right\|_{p,p} = \max\{\|A\|_{p,p}, \|A\|_{q,q}\}.$$

In particular,

$$\left\|\begin{bmatrix} 0 & A \\ A^* & 0 \end{bmatrix}\right\|_{\mathrm{col}} = \left\|\begin{bmatrix} 0 & A \\ A^* & 0 \end{bmatrix}\right\|_{\mathrm{row}} = \max\{\|A\|_{\mathrm{col}}, \|A\|_{\mathrm{row}}\}.$$

Fact 11.9.23. Let $A \in \mathbb{F}^{n\times m}$. Then, the following statements hold:

i) $\|A\|_{\mathrm{F}} \le \|A\|_1 \le \sqrt{mn}\|A\|_{\mathrm{F}}$.

ii) $\|A\|_\infty \le \|A\|_1 \le mn\|A\|_\infty$.

iii) $\|A\|_{\mathrm{col}} \le \|A\|_1 \le m\|A\|_{\mathrm{col}}$.

iv) $\|A\|_{\mathrm{row}} \le \|A\|_1 \le n\|A\|_{\mathrm{row}}$.

v) $\sigma_{\max}(A) \le \|A\|_1 \le \sqrt{mn\,\mathrm{rank}\,A}\,\sigma_{\max}(A)$.

vi) $\|A\|_\infty \le \|A\|_{\mathrm{F}} \le \sqrt{mn}\|A\|_\infty$.

vii) $\frac{1}{\sqrt{n}}\|A\|_{\mathrm{col}} \le \|A\|_{\mathrm{F}} \le \sqrt{m}\|A\|_{\mathrm{col}}$.

viii) $\frac{1}{\sqrt{m}}\|A\|_{\mathrm{row}} \le \|A\|_{\mathrm{F}} \le \sqrt{n}\|A\|_{\mathrm{row}}$.

ix) $\sigma_{\max}(A) \le \|A\|_{\mathrm{F}} \le \sqrt{\mathrm{rank}\,A}\,\sigma_{\max}(A)$.

x) $\frac{1}{n}\|A\|_{\mathrm{col}} \le \|A\|_\infty \le \|A\|_{\mathrm{col}}$.

xi) $\frac{1}{m}\|A\|_{\mathrm{row}} \le \|A\|_\infty \le \|A\|_{\mathrm{row}}$.

xii) $\frac{1}{\sqrt{mn}}\sigma_{\max}(A) \le \|A\|_\infty \le \sigma_{\max}(A)$.

xiii) $\frac{1}{m}\|A\|_{\mathrm{row}} \le \|A\|_{\mathrm{col}} \le n\|A\|_{\mathrm{row}}$.

xiv) $\frac{1}{\sqrt{m}}\sigma_{\max}(A) \le \|A\|_{\mathrm{col}} \le \sqrt{n}\sigma_{\max}(A)$.

xv) $\frac{1}{\sqrt{n}}\sigma_{\max}(A) \le \|A\|_{\mathrm{row}} \le \sqrt{m}\sigma_{\max}(A)$.

Source: [1448, p. 314] and [3008]. **Remark:** See [1389, p. 115] for equality cases.

Fact 11.9.24. Let $A \in \mathbb{F}^{n\times n}$, let $\|\cdot\|$ and $\|\cdot\|'$ be norms on $\mathbb{F}^n$, and define the induced norms

$$\|A\|'' \triangleq \max_{x\in\{y\in\mathbb{F}^m:\ \|y\|=1\}} \|Ax\|, \quad \|A\|''' \triangleq \max_{x\in\{y\in\mathbb{F}^m:\ \|y\|'=1\}} \|Ax\|'.$$

Then,

$$\max_{A\in\{X\in\mathbb{F}^{n\times n}:\ X\neq 0\}} \frac{\|A\|''}{\|A\|'''} = \max_{A\in\{X\in\mathbb{F}^{n\times n}:\ X\neq 0\}} \frac{\|A\|'''}{\|A\|''} = \max_{x\in\{y\in\mathbb{F}^n:\ y\neq 0\}} \frac{\|x\|}{\|x\|'} \max_{x\in\{y\in\mathbb{F}^n:\ y\neq 0\}} \frac{\|x\|'}{\|x\|}.$$

Source: [1448, p. 303]. **Remark:** This symmetry property is evident in Fact 11.9.23.

Fact 11.9.25. Let $A \in \mathbb{F}^{n\times n}$ and $p, q \in [1, \infty]$. Then,

$$\|A\|_{p,p} \le \begin{cases} n^{1/p-1/q}\|A\|_{q,q}, & p \le q, \\ n^{1/q-1/p}\|A\|_{q,q}, & q \le p. \end{cases}$$

Consequently,

$$n^{1/p-1}\|A\|_{\rm col} \le \|A\|_{p,p} \le n^{1-1/p}\|A\|_{\rm col},$$

$$n^{-|1/p-1/2|}\sigma_{\max}(A) \le \|A\|_{p,p} \le n^{|1/p-1/2|}\sigma_{\max}(A),$$

$$n^{-1/p}\|A\|_{\rm col} \le \|A\|_{p,p} \le n^{1/p}\|A\|_{\rm row}.$$

Source: [1388] and [1389, p. 112]. **Related:** Fact 11.8.7. **Problem:** Extend these inequalities to nonsquare matrices.

Fact 11.9.26. Let $A \in \mathbb{F}^{n\times m}$, $p, q \in [1, \infty]$, and $\alpha \in [0, 1]$, and define $r \triangleq pq/[(1-\alpha)p + \alpha q]$. Then,

$$\|A\|_{r,r} \le \|A\|_{p,p}^{\alpha}\|A\|_{q,q}^{1-\alpha}.$$

Source: [1388] and [1389, p. 113].

Fact 11.9.27. Let $A \in \mathbb{F}^{n\times m}$ and $p \in [1, \infty]$. Then,

$$\|A\|_{p,p} \le \|A\|_{\rm col}^{1/p}\|A\|_{\rm row}^{1-1/p}.$$

In particular,

$$\sigma_{\max}(A) \le \sqrt{\|A\|_{\rm col}\|A\|_{\rm row}} \le \tfrac{1}{2}(\|A\|_{\rm col} + \|A\|_{\rm row}) \le \max\{\|A\|_{\rm col}, \|A\|_{\rm row}\}.$$

Source: Set $\alpha = 1/p$, $p = 1$, and $q = \infty$ in Fact 11.9.26. See [1389, p. 113]. To prove the special case $p = 2$, note that $\sigma_{\max}^2(A) = \lambda_{\max}(A^*A) \le \|A^*A\|_{\rm col} \le \|A^*\|_{\rm col}\|A\|_{\rm col} = \|A\|_{\rm row}\|A\|_{\rm col}$.

Fact 11.9.28. Let $A \in \mathbb{F}^{n\times m}$. Then, $\|A\|_{2,1} \le \sigma_{\max}(A)$ and $\|A\|_{\infty,2} \le \sigma_{\max}(A)$. **Source:** Proposition 11.1.5.

Fact 11.9.29. Let $A \in \mathbb{F}^{n\times m}$ and $p \in [1, 2]$. Then,

$$\|A\|_{p,p} \le \|A\|_{\rm col}^{2/p-1}\sigma_{\max}^{2-2/p}(A).$$

Source: Let $\alpha = 2/p - 1$, $p = 1$, and $q = 2$ in Fact 11.9.26. See [1389, p. 113].

Fact 11.9.30. Let $A \in \mathbb{F}^{n\times n}$ and $p \in [1, \infty]$. Then,

$$\|A\|_{p,p} \le \||A|\|_{p,p} \le n^{\min\{1/p, 1-1/p\}}\|A\|_{p,p} \le \sqrt{n}\|A\|_{p,p}.$$

Remark: See [1389, p. 117].

Fact 11.9.31. Let $A \in \mathbb{F}^{n\times n}$ and $p, q, r, s \in [1, \infty]$. Then,

$$\|A\|_{q,p} \le \alpha(q, s, n)\alpha(r, p, m)\|A\|_{s,r},$$

where

$$\alpha(q, s, n) = \begin{cases} 1, & q \ge s, \\ n^{1/q-1/s}, & q < s. \end{cases}$$

Source: [1030, 1031].

Fact 11.9.32. Let $A \in \mathbb{F}^{n\times m}$ and $p, q \in [1, \infty]$. Then, $\|\overline{A}\|_{q,p} = \|A\|_{q,p}$.

Fact 11.9.33. Let $A \in \mathbb{F}^{n\times m}$ and $p, q \in [1, \infty]$. Then, $\|A^*\|_{q,p} = \|A\|_{p/(p-1),q/(q-1)}$.

Fact 11.9.34. Let $A \in \mathbb{F}^{n\times m}$ and $p, q \in [1, \infty]$. Then,

$$\|A\|_{q,p} \le \begin{cases} \|A\|_{p/(p-1)}, & 1/p + 1/q \le 1, \\ \|A\|_q, & 1/p + 1/q \ge 1. \end{cases}$$

Fact 11.9.35. Let $A \in \mathbb{F}^{n\times m}$, let $q, r \in [1, \infty]$, assume that $1 \le q < r$, define $p \triangleq qr/(r-q)$, and assume that $p \ge 2$. Then, $\|A\|_p \le \|A\|_{q,r}$. In particular, $\|A\|_\infty \le \|A\|_{\infty,\infty}$. **Source:** [1030, 1031]. **Credit:** G. H. Hardy and J. E. Littlewood.

Fact 11.9.36. Let $A \in \mathbb{R}^{n\times m}$. Then,

$$\|A^{\mathsf{T}}\|_{2|1} \le \sqrt{2}\|A\|_{1,\infty}, \quad \|A^{\mathsf{T}}\|_{1|2} \le \sqrt{2}\|A\|_{1,\infty}, \quad \|A\|_{4/3}^{3/4} \le \sqrt{2}\|A\|_{1,\infty}.$$

Source: [1146, p. 303]. **Credit:** The first and third results are due to J. E. Littlewood; the second result is due to W. Orlicz.

Fact 11.9.37. Let $A \in \mathbb{F}^{n\times n}$, and assume that A is positive semidefinite. Then,

$$\|A\|_{1,\infty} = \max_{x\in\{z\in\mathbb{F}^n:\ \|z\|_\infty=1\}} x^*Ax.$$

Credit: P. D. Tao. See [1389, p. 116] and [2328].

Fact 11.9.38. Let $A \in \mathbb{R}^{n\times m}$, let $p, q \in [1, \infty]$, and assume that $q \le p$. Then,

$$\|A\|_{q,p,\mathbb{R}} \le \|A\|_{q,p,\mathbb{C}} \le c_{q,p}\|A\|_{q,p,\mathbb{R}},$$

where

$$c_{q,p} \le \min\left\{\sqrt{2},\ \sqrt{\pi}^{1/q-1/p}\left(\frac{\Gamma[(1+p)/2]}{\Gamma[(2+p)/2]}\right)^{1/p}\left(\frac{\Gamma[(2+q)/2]}{\Gamma[(1+q)/2]}\right)^{1/q}\right\}.$$

Furthermore, if $p \in [1, 2]$, then

$$c_{p,2} = c_{2,p/(p-1)} = \frac{\sqrt{2}}{2}\left(\sqrt{\pi}\frac{\Gamma[(2+p)/2]}{\Gamma[(1+p)/2]}\right)^{1/p}.$$

In particular, $c_{1,2} = c_{2,\infty} = \sqrt{2}\pi/4$. **Source:** [864, p. 377] and [2040]. **Problem:** Compare these constants to the constants c_p given in Proposition 11.4.7.

Fact 11.9.39. Let $A \in \mathbb{F}^{n\times m}$, let $p, q \in [1, \infty]$, and assume that $1/p + 1/q = 1$. Then,

$$\|A\|_{1,p} \le \|A\|_{1|q}.$$

If, in addition, $A \in \mathbb{R}^{n\times m}$ and A is nonnegative, then

$$\|A\|_{1,p} = \|A\|_{1|q}.$$

Source: [2040].

Fact 11.9.40. Let $A \in \mathbb{F}^{n\times m}$ and $p, q, r, s \in [1, \infty]$. Then,

$$\|A\|_{p|q} \le \alpha(p, r, n)\alpha(q, s, m)\|A\|_{r|s},$$

where α is defined in Fact 11.9.31. **Source:** [1030, 1031].

Fact 11.9.41. Let $A \in \mathbb{F}^{n\times m}$ and $p, q \in [1, \infty]$. Then,

$$\|A\|_{q,p} = \max_{\substack{x\in\mathbb{F}^m, y\in\mathbb{F}^n \\ x,y\ne 0}} \frac{|y^*Ax|}{\|y\|_{q/(q-1)}\|x\|_p}.$$

Fact 11.9.42. Let $A \in \mathbb{F}^{n\times m}$, and let $p, q \in [1, \infty]$ satisfy $1/p + 1/q = 1$. Then,

$$\|A\|_{p,p} = \max_{\substack{x\in\mathbb{F}^m, y\in\mathbb{F}^n\\ x,y\neq 0}} \frac{|y^*Ax|}{\|y\|_q\|x\|_p} = \max_{\substack{x\in\mathbb{F}^m, y\in\mathbb{F}^n\\ x,y\neq 0}} \frac{|y^*Ax|}{\|y\|_{p/(p-1)}\|x\|_p}.$$

Related: See Fact 11.15.4 for the case $p = 2$.

Fact 11.9.43. Let $A \in \mathbb{F}^{n\times m}$ and $p, q \in [1, \infty]$. If $m < n$, then

$$\|[A \;\; 0_{n\times(n-m)}]\|_{q,p,\mathbb{F}} = \|A\|_{q,p,\mathbb{F}}.$$

If $n < m$, then

$$\left\| \begin{bmatrix} A \\ 0_{(m-n)\times m} \end{bmatrix} \right\|_{q,p,\mathbb{F}} = \|A\|_{q,p,\mathbb{F}}.$$

Related: Fact 11.8.27.

Fact 11.9.44. Let $A \in \mathbb{F}^{n\times n}$, and let $\|\cdot\|$ be a unitarily invariant norm on $\mathbb{F}^{n\times n}$. Then, $\|\langle A\rangle\| = \|A\|$. **Source:** [2991, p. 376].

Fact 11.9.45. Let $A, S \in \mathbb{F}^{n\times n}$, assume that S is nonsingular, and let $\|\cdot\|$ be a unitarily invariant norm on $\mathbb{F}^{n\times n}$. Then,

$$\|A\| \le \tfrac{1}{2}\|SAS^{-1} + S^{-*}AS^*\|.$$

Source: [121, 537].

Fact 11.9.46. For $A \in \mathbb{F}^{n\times n}$, define $\|A\| \triangleq \max\{|\operatorname{tr} AX| \colon X \in \mathbb{F}^{n\times n} \text{ and } \operatorname{tr} X^*X = 1\}$. Then, $\|\cdot\|$ is a unitarily invariant norm on $\mathbb{F}^{n\times n}$. **Source:** [2991, p. 377].

Fact 11.9.47. Let $A \in \mathbb{F}^{n\times m}$ and $p, r \in (0, \infty)$. and define $\|A\|_{\sigma p} \triangleq (\sum_{i=1}^{\min\{n,m\}} \sigma_i^p(A))^{1/p}$. Then, $\|\langle A\rangle^r\|_{\sigma p} = \|A\|_{\sigma rp}^r$. **Source:** [2436, 2438]. **Remark:** for $p \in (0, 1)$, $\|\cdot\|_{\sigma p}$ is not a norm. **Related:** Fact 11.8.21.

Fact 11.9.48. Let $A \in \mathbb{F}^{n\times n}$, let $\alpha_1 \ge \cdots \ge \alpha_n \ge 0$, assume that $\alpha_1, \ldots, \alpha_n$ are not all zero, and define

$$\|A\| \triangleq \sum_{i=1}^n \alpha_i\sigma_i(A).$$

Then, $\|\cdot\|$ is a unitarily invariant norm on $\mathbb{F}^{n\times n}$. **Source:** [2991, p. 377].

Fact 11.9.49. Let $A \in \mathbb{F}^{n\times m}$, and let $\|\cdot\|$ be a unitarily invariant norm on $\mathbb{F}^{n\times m}$. Then,

$$\lim_{p\to\infty} \|\langle A\rangle^p\|^{1/p} = \sigma_{\max}(A).$$

Now, assume that $\|\cdot\|$ is a normalized unitarily invariant norm on $\mathbb{F}^{n\times m}$, and define $f\colon (0, \infty) \to \mathbb{R}$ by $f(p) \triangleq \|\langle A\rangle^p\|^{1/p}$. Then, f is nonincreasing. **Source:** [1383].

Fact 11.9.50. Let $A \in \mathbb{F}^{n\times n}$, assume that A is positive semidefinite, and let $\|\cdot\|$ be a submultiplicative norm on $\mathbb{F}^{n\times n}$. Then,

$$\|A\|^{1/2} \le \|A^{1/2}\|.$$

Furthermore,

$$\sigma_{\max}^{1/2}(A) = \sigma_{\max}(A^{1/2}).$$

Fact 11.9.51. Let $A \in \mathbb{F}^{n\times n}$, assume that A is positive semidefinite, and let $\|\cdot\|$ be a unitarily invariant norm on $\mathbb{F}^{n\times n}$. If $r \in (0,1]$, then $\|A\|^p \le \|A^p\|$. Furthermore, if $r \in [1,\infty)$, then $\|A^r\| \le \|A\|^r$. **Source:** [1383, 2436, 2438].

Fact 11.9.52. Let $A_{11} \in \mathbb{F}^{n\times n}$, $A_{12} \in \mathbb{F}^{n\times m}$, and $A_{22} \in \mathbb{F}^{m\times m}$, assume that $\begin{bmatrix} A_{11} & A_{12} \\ A_{12}^* & A_{22} \end{bmatrix} \in \mathbb{F}^{(n+m)\times(n+m)}$ is positive semidefinite, let $\|\cdot\|$ and $\|\cdot\|'$ be unitarily invariant norms on $\mathbb{F}^{n\times n}$ and $\mathbb{F}^{m\times m}$, respectively, and let $p > 0$. Then,

$$\|\langle A_{12}\rangle^p\|'^2 \le \|A_{11}^p\|\|A_{22}^p\|'.$$

Source: [1453].

Fact 11.9.53. Let $A \in \mathbb{F}^{n\times n}$, let $\|\cdot\|$ be a norm on $\mathbb{F}^n$, let $\|\cdot\|_{\mathrm{D}}$ denote the dual norm on $\mathbb{F}^n$, and let $\|\cdot\|'$ denote the norm induced by $\|\cdot\|$ on $\mathbb{F}^{n\times n}$. Then,

$$\|A\|' = \max_{\substack{x,y\in\mathbb{F}^n\\ x,y\neq 0}} \frac{\mathrm{Re}\, y^*Ax}{\|y\|_{\mathrm{D}}\|x\|}.$$

Source: [1389, p. 115]. **Remark:** Fact 11.8.24 defines the dual norm. **Problem:** Extend this result to include Fact 11.9.41 as a special case.

Fact 11.9.54. Let $A \in \mathbb{F}^{n\times n}$, and assume that A is positive definite. Then,

$$\min_{x\in\mathbb{F}^n\setminus\{0\}} \frac{x^*Ax}{\|Ax\|_2\|x\|_2} = \frac{2\sqrt{\alpha\beta}}{\alpha+\beta}, \quad \min_{\alpha\geq 0} \sigma_{\max}(\alpha A - I) = \frac{\alpha-\beta}{\alpha+\beta},$$

where $\alpha \triangleq \lambda_{\max}(A)$ and $\beta \triangleq \lambda_{\min}(A)$. **Source:** [1279]. **Remark:** These are *antieigenvalues*.

Fact 11.9.55. Let $A \in \mathbb{F}^{n\times n}$, and define

$$\mathrm{nrad}(A) \triangleq \max\{|x^*Ax|\colon\ x \in \mathbb{C}^n \text{ and } x^*x \leq 1\}.$$

Then, the following statements hold:

i) $\mathrm{nrad}(A) = \max\{|z|\colon\ z \in \Theta(A)\} = \max_{\theta\in[-\pi,\pi]} \sigma_{\max}[\frac{1}{2}(e^{\theta J}A + e^{-\theta J}A^*)]$.

ii) $\rho_{\max}(A) \leq \mathrm{nrad}(A) \leq \mathrm{nrad}(|A|) = \frac{1}{2}\rho_{\max}(|A| + |A|^{\mathsf{T}})$.

iii) $\frac{1}{2}\sigma_{\max}(A) \leq \mathrm{nrad}(A) \leq \frac{1}{2}\sigma_{\max}(\langle A\rangle + \langle A^*\rangle) \leq \frac{1}{2}[\sigma_{\max}(A) + \sigma_{\max}^{1/2}(A^2)] \leq \sigma_{\max}(A) \leq 2\,\mathrm{nrad}(A)$.

iv) $\frac{1}{4}\sigma_{\max}(A^*A + AA^*) \leq [\mathrm{nrad}(A)]^2 \leq \frac{1}{2}\sigma_{\max}(A^*A + AA^*)$.

v) If $A^2 = 0$, then $\mathrm{nrad}(A) = \sigma_{\max}(A)$.

vi) If $\mathrm{nrad}(A) = \sigma_{\max}(A)$, then $\sigma_{\max}(A^2) = \sigma_{\max}^2(A)$.

vii) If A is normal, then $\mathrm{nrad}(A) = \rho_{\max}(A)$.

viii) $\mathrm{nrad}(A^k) \leq [\mathrm{nrad}(A)]^k$ for all $k \in \mathbb{N}$.

ix) $\mathrm{nrad}(\cdot)$ is a weakly unitarily invariant norm on $\mathbb{F}^{n\times n}$.

x) $\mathrm{nrad}(\cdot)$ is not a submultiplicative norm on $\mathbb{F}^{n\times n}$.

xi) $\|\cdot\| \triangleq \alpha\,\mathrm{nrad}(\cdot)$ is a submultiplicative norm on $\mathbb{F}^{n\times n}$ if and only if $\alpha \geq 4$.

xii) $\mathrm{nrad}(AB) \leq \mathrm{nrad}(A)\,\mathrm{nrad}(B)$ for all $A, B \in \mathbb{F}^{n\times n}$ such that A and B are normal.

xiii) $\mathrm{nrad}(A \odot B) \leq \alpha\,\mathrm{nrad}(A)\mathrm{nrad}(B)$ for all $A, B \in \mathbb{F}^{n\times n}$ if and only if $\alpha \geq 2$.

xiv) $\mathrm{nrad}(A \oplus B) = \max\{\mathrm{nrad}(A), \mathrm{nrad}(B)\}$ for all $A \in \mathbb{F}^{n\times n}$ and $B \in \mathbb{F}^{m\times m}$.

Source: [8, 983], [1280, pp. 109, 115], [1448, p. 331], [1450, pp. 43, 44], [1634, 1636, 2437], and [2991, pp. 109, 110]. **Remark:** nrad(A) is the *numerical radius* of A, while $\Theta(A)$ is the numerical range of A. See Fact 10.17.8. **Remark:** nrad($\cdot$) is not submultiplicative. The example $A = \left[\begin{smallmatrix} 0 & 1\\ 0 & 0\end{smallmatrix}\right]$, $B = \left[\begin{smallmatrix} 0 & 2\\ 2 & 0\end{smallmatrix}\right]$, where B is normal, $\mathrm{nrad}(A) = 1/2$, $\mathrm{nrad}(B) = 2$, and $\mathrm{nrad}(AB) = 2$, shows that *xii)* can be false if only one of the matrices A and B is normal, which corrects [1450, pp. 43, 73]. **Remark:** *viii)* is the *power inequality*.

Fact 11.9.56. Let $A \in \mathbb{F}^{n\times m}$, let $\gamma > \sigma_{\max}(A)$, and define $\beta \triangleq \sigma_{\max}(A)/\gamma$. Then,

$$\|A\|_{\mathrm{F}} \leq \sqrt{-[\gamma^2/(2\pi)]\log\det(I - \gamma^{-2}A^*A)} \leq \beta^{-1}\sqrt{-\log(1-\beta^2)}\|A\|_{\mathrm{F}}.$$

Source: [557].

Fact 11.9.57. Let $\|\cdot\|$ be a unitarily invariant norm on $\mathbb{F}^{n\times n}$. Then, $\|A\| = 1$ for all $A \in \mathbb{F}^{n\times n}$ such that $\mathrm{rank}\,A = 1$ if and only if $\|E_{1,1}\| = 1$. **Source:** $\|A\| = \|E_{1,1}\|\sigma_{\max}(A)$. **Remark:** These normalizations are used in [449] and [2539, p. 74].

Fact 11.9.58. Let $\|\cdot\|$ be a unitarily invariant norm on $\mathbb{F}^{n\times n}$. Then, the following statements are equivalent:

i) $\sigma_{\max}(A) \le \|A\|$ for all $A \in \mathbb{F}^{n\times n}$.

ii) $\|\cdot\|$ is submultiplicative.

iii) $\|A^2\| \le \|A\|^2$ for all $A \in \mathbb{F}^{n\times n}$.

iv) $\|A^k\| \le \|A\|^k$ for all $k \ge 1$ and $A \in \mathbb{F}^{n\times n}$.

v) $\|A \odot B\| \le \|A\|\|B\|$ for all $A, B \in \mathbb{F}^{n\times n}$.

vi) $\rho_{\max}(A) \le \|A\|$ for all $A \in \mathbb{F}^{n\times n}$.

vii) $\|Ax\|_2 \le \|A\|\|x\|_2$ for all $A \in \mathbb{F}^{n\times n}$ and $x \in \mathbb{F}^n$.

viii) $\|A\|_\infty \le \|A\|$ for all $A \in \mathbb{F}^{n\times n}$.

ix) $\|E_{1,1}\| \ge 1$.

x) $\sigma_{\max}(A) \le \|A\|$ for all $A \in \mathbb{F}^{n\times n}$ such that $\operatorname{rank} A = 1$.

xi) For all $A, B, C \in \mathbb{F}^{n\times n}$, $\|ABC\| \le \sigma_{\max}(A)\sigma_{\max}(C)\|B\|$.

Source: The equivalence of *i)*–*vii)* is given in [1450, pp. 211, 336]. Since $\|A\| = \|E_{1,1}\|\sigma_{\max}(A)$ for all $A \in \mathbb{F}^{n\times n}$ such that $\operatorname{rank} A = 1$, it follows that *vii)* and *viii)* are equivalent. To prove *ix)* $\Longrightarrow$ *x)*, let $A \in \mathbb{F}^{n\times n}$ satisfy $\operatorname{rank} A = 1$. Then, $\|A\| = \sigma_{\max}(A)\|E_{1,1}\| \ge \sigma_{\max}(A)$. To show *x)* $\Longrightarrow$ *ii)*, define $\|\cdot\|' \triangleq \|E_{1,1}\|^{-1}\|\cdot\|$. Since $\|E_{1,1}\|' = 1$, it follows from [449, p. 94] that $\|\cdot\|'$ is submultiplicative. Since $\|E_{1,1}\|^{-1} \le 1$, it follows that $\|\cdot\|$ is also submultiplicative. Alternatively, $\|A\|' = \sigma_{\max}(A)$ for all $A \in \mathbb{F}^{n\times n}$ having rank 1. Then, Corollary 3.10 of [2539, p. 80] implies that $\|\cdot\|'$, and thus $\|\cdot\|$, is submultiplicative. *xi)* is given in [2979, p. 101]. **Related:** Fact 11.10.34.

Fact 11.9.59. Let $\Phi\colon\ \mathbb{F}^n \mapsto [0,\infty)$, and assume that the following statements hold:

i) If $x \ne 0$, then $\Phi(x) > 0$.

ii) $\Phi(\alpha x) = |\alpha|\Phi(x)$ for all $\alpha \in \mathbb{F}$.

iii) $\Phi(x + y) \le \Phi(x) + \Phi(y)$ for all $x, y \in \mathbb{F}^n$.

iv) If $A \in \mathbb{R}^{n\times n}$ is a permutation matrix, then $\Phi(Ax) = \Phi(x)$ for all $x \in \mathbb{F}^n$.

v) $\Phi(|x|) = \Phi(x)$ for all $x \in \mathbb{F}^n$.

Then, the following statements hold:

vi) Φ is an absolute and monotone norm on $\mathbb{F}^n$.

vii) Φ is convex on $\mathbb{F}^n$.

viii) Φ is Schur-convex on $[0,\infty)^n$.

ix) For $A \in \mathbb{F}^{n\times m}$, where $n \le m$, define $\|A\| \triangleq \Phi[\sigma(A)]$. Then, $\|\cdot\|$ is a unitarily invariant norm on $\mathbb{F}^{n\times m}$.

x) If $\|\cdot\|$ is a unitarily invariant norm on $\mathbb{F}^{n\times m}$, where $n \le m$, then $\Phi\colon\ \mathbb{F}^n \mapsto [0,\infty)$ defined by

$$\Phi(x) \triangleq \left\|\begin{bmatrix} x_{(1)} & \cdots & 0 & 0_{n\times(m-n)} \\ \vdots & \ddots & \vdots & \vdots \\ 0 & \cdots & x_{(n)} & 0_{n\times(m-n)} \end{bmatrix}\right\|$$

satisfies *i)*–*v)*.

Finally, let $x, y \in \mathbb{F}^{1\times n}$. Then, the following statements are equivalent:

xi) $|x| \overset{\mathrm{w}}{\prec} |y|$.

xii) For all $\Phi\colon\ \mathbb{F}^n \mapsto [0,\infty)$ satisfying *i)*–*v)*, $\Phi(x) \le \Phi(y)$.

Source: [2539, pp. 75, 76] and [2991, pp. 373–376]. **Remark:** Φ is a *symmetric gauge function.* See Fact 3.25.16. **Credit:** J. von Neumann.

Fact 11.9.60. Let $\|\cdot\|$ and $\|\cdot\|'$ denote norms on $\mathbb{F}^m$ and $\mathbb{F}^n$, respectively, and define $\hat{\ell}\colon\ \mathbb{F}^{n\times m} \mapsto \mathbb{R}$ by

$$\hat{\ell}(A) \triangleq \min_{x\in\mathbb{F}^m\backslash\{0\}} \frac{\|Ax\|'}{\|x\|},$$

or, equivalently,

$$\hat{\ell}(A) \triangleq \min_{x\in\{y\in\mathbb{F}^m\colon\ \|y\|=1\}} \|Ax\|'.$$

Then, for all $A \in \mathbb{F}^{n\times m}$, the following statements hold:

i) $\hat{\ell}(A) \ge 0$.

ii) $\hat{\ell}(A) > 0$ if and only if $\operatorname{rank} A = m$.

iii) $\hat{\ell}(A) = \ell(A)$ if and only if either $A = 0$ or $\operatorname{rank} A = m$.

Source: [1738, pp. 369, 370]. **Remark:** $\hat{\ell}$ is a weaker version of ℓ.

Fact 11.9.61. Let $\|\cdot\|$ and $\|\cdot\|'$ denote norms on $\mathbb{F}^m$ and $\mathbb{F}^n$, respectively, let $\|\cdot\|'''$ denote the norm induced by $\|\cdot\|'$ and $\|\cdot\|$, and define $\hat{\ell}\colon\ \mathbb{F}^{n\times m} \mapsto \mathbb{R}$ by

$$\hat{\ell}(A) \triangleq \min_{x\in\mathcal{R}(A^*)\backslash\{0\}} \frac{\|Ax\|'}{\|x\|}.$$

Now, let $A \in \mathbb{F}^{n\times m}$. If A is nonzero, then

$$\frac{1}{\|A^+\|'''} \le \hat{\ell}(A).$$

If, in addition, $\operatorname{rank} A = m$, then

$$\frac{1}{\|A^+\|'''} = \hat{\ell}(A) = \ell(A).$$

Source: [2732].

Fact 11.9.62. Let $A \in \mathbb{F}^{n\times n}$, let $\|\cdot\|$ be a normalized, submultiplicative norm on $\mathbb{F}^{n\times n}$, and assume that $\|I - A\| < 1$. Then, A is nonsingular. **Related:** Fact 11.10.88.

Fact 11.9.63. Let $\|\cdot\|$ be a normalized, submultiplicative norm on $\mathbb{F}^{n\times n}$. Then, $\|\cdot\|$ is equi-induced if and only if $\|A\| \le \|A\|'$ for all $A \in \mathbb{F}^{n\times n}$ and all normalized submultiplicative norms $\|\cdot\|'$ on $\mathbb{F}^{n\times n}$. **Source:** [2543]. **Remark:** Not every normalized submultiplicative norm on $\mathbb{F}^{n\times n}$ is induced. See [684, 848]. For example, the norm $\|A\| \triangleq \max\{\|A\|_{\text{row}}, \|A\|_{\text{col}}\}$ on $\mathbb{F}^{n\times n}$ is normalized and submultiplicative but not induced. See [1451, p. 357].

11.10 Facts on Matrix Norms for Two or More Matrices

Fact 11.10.1. Let $A, B \in \mathbb{F}^{n\times m}$. Then,

$$\|A + B\|_{\text{F}} = \sqrt{\|A\|_{\text{F}}^2 + \|B\|_{\text{F}}^2 + 2\operatorname{tr} AB^*} \le \|A\|_{\text{F}} + \|B\|_{\text{F}}.$$

Therefore,

$$\|A - B\|_{\text{F}} = \sqrt{\|A\|_{\text{F}}^2 + \|B\|_{\text{F}}^2 - 2\operatorname{tr} AB^*} \le \|A\|_{\text{F}} + \|B\|_{\text{F}},$$

and thus

$$\|A + B\|_{\text{F}}^2 + \|A - B\|_{\text{F}}^2 = 2\|A\|_{\text{F}}^2 + 2\|B\|_{\text{F}}^2.$$

If, in addition, A is Hermitian and B is skew Hermitian, then $\operatorname{tr} AB^* = 0$, and thus

$$\|A + B\|_{\text{F}}^2 = \|A - B\|_{\text{F}}^2 = \|A\|_{\text{F}}^2 + \|B\|_{\text{F}}^2.$$

Finally, if A and B are nonzero, then

$$\frac{\operatorname{Re}\operatorname{tr} A^*B}{\|A\|_{\mathrm{F}}\|B\|_{\mathrm{F}}}(\|A\|_{\mathrm{F}} + \|B\|_{\mathrm{F}}) \le \|A + B\|_{\mathrm{F}} \le \|A\|_{\mathrm{F}} + \|B\|_{\mathrm{F}}.$$

Remark: The second equality is a matrix version of the cosine law given by Fact 11.8.3. **Related:** Fact 11.8.14.

Fact 11.10.2. Let $A, B, C, D \in \mathbb{C}^{n\times n}$. Then,

$$|\operatorname{tr} A^*B|^2 + |\operatorname{tr} C^*D|^2 + 2|\operatorname{Re}[(\operatorname{tr} A^*C)\operatorname{tr} B^*D - (\operatorname{tr} A^*D)\operatorname{tr} B^*C]| \le \|A\|_{\mathrm{F}}^2\|B\|_{\mathrm{F}}^2 + \|C\|_{\mathrm{F}}^2\|D\|_{\mathrm{F}}^2.$$

Source: [2957]. **Remark:** This is a generalized Cauchy-Schwarz inequality. **Related:** Fact 11.8.6.

Fact 11.10.3. Let $A, B \in \mathbb{F}^{n\times n}$, assume that A and B are positive semidefinite, let $f\colon [0,\infty) \mapsto [0,\infty)$, and assume that f is concave. Then, there exist unitary matrices $U, V \in \mathbb{F}^{n\times n}$ such that

$$f(A + B) \le Uf(A)U^* + Vf(B)V^*.$$

Now, let $\|\cdot\|$ be a unitarily invariant norm on $\mathbb{F}^{n\times n}$. Then,

$$\|f(A + B)\| \le \|f(A)\| + \|f(B)\|.$$

In particular,

$$\operatorname{tr} f(A + B) \le \operatorname{tr} f(A) + \operatorname{tr} f(B).$$

Source: [544, 546]. **Remark:** The last result is the *Rotfel'd trace inequality*. **Related:** Fact 10.12.58.

Fact 11.10.4. Let $A, B \in \mathbf{P}^n$, let $p > 0$, and let $\|\cdot\|$ be a unitarily invariant norm on $\mathbb{F}^{n\times n}$. Then,

$$\|A^{-p}\|^{-1/p} + \|A^{-p}\|^{-1/p} \le \|(A + B)^{-p}\|^{-1/p}.$$

Source: [1779]. **Remark:** $f(A) = \|A^{-p}\|^{-1/p}$ is an anti-norm on $\mathbf{P}^n$. **Related:** Fact 2.2.59, Fact 10.14.7, Fact 11.8.21, and Fact 11.9.47.

Fact 11.10.5. Let $A, B \in \mathbb{F}^{n\times n}$, and let $\|\cdot\|$ be a unitarily invariant norm on $\mathbb{F}^{n\times n}$. Then,

$$\|A + B\| \le \|\langle A\rangle + \langle B\rangle\|^{1/2}\|\langle A^*\rangle + \langle B^*\rangle\|^{1/2} \le \begin{cases} \sqrt{2}\|\langle A\rangle + \langle B\rangle\| \\ \frac{1}{2}(\|\langle A\rangle + \langle B\rangle\| + \|\langle A^*\rangle + \langle B^*\rangle\|). \end{cases}$$

Source: [1479, 1623]. **Remark:** $\|\langle A^*\rangle\| + \|\langle B^*\rangle\| \le 2\|\langle A\rangle + \langle B\rangle\|$. See [1479] and Fact 11.10.6.

Fact 11.10.6. Let $A, B \in \mathbb{F}^{n\times n}$, assume that A and B are positive semidefinite, and let $\|\cdot\|$ be a unitarily invariant norm on $\mathbb{F}^{n\times n}$. Then, the following statements hold:

i) If $A \le B$, then $\|A\| \le \|B\|$.

ii) $\|A\| + \|B\| \le \|A\| + \|A + B\| \le 2\|A + B\|$.

iii) $\|A\|_{\sigma 1} + \|B\|_{\sigma 1} \le \|A + B\|_{\sigma 1}$.

iv) $\|A\|_{\mathrm{F}} + \|B\|_{\mathrm{F}} \le \sqrt{2}\|A + B\|_{\mathrm{F}}$.

v) If $p \ge 1$, then $\|A\|_{\sigma p} + \|B\|_{\sigma p} \le 2^{(p-1)/p}\|A + B\|_{\sigma p}$.

Source: Fact 11.10.53 and Fact 11.8.7. *i)* follows from the fact that, for all $A \in \mathbb{F}^{n\times n}$, $\|A\|$ defines a symmetric gauge function, which is a monotone function of $\sigma(A)$. See Fact 11.9.59. Alternatively, *i)* follows from the Fan dominance theorem given by Fact 11.16.23. **Remark:** Letting $p \to \infty$ in *v)* yields *ii)* with $\|\cdot\| = \sigma_{\max}(\cdot)$.

Fact 11.10.7. Let $A, B \in \mathbb{F}^{n\times n}$. Then,

$$\|A + B\|_{\mathrm{F}} \le \sqrt[4]{2}\|\langle A\rangle + \langle B\rangle\|_{\mathrm{F}}.$$

If, in addition, A and B are nonzero, then

$$\|A+B\|_{\rm F} \le \left(2 - \frac{S\left(\frac{\|B\|_{\rm F}}{\|A\|_{\rm F}}\right) - 1}{S\left(\frac{\|B\|_{\rm F}^2}{\|A\|_{\rm F}^2}\right)}\right)^{1/4} \|\langle A\rangle + \langle B\rangle\|_{\rm F} \le \sqrt[4]{2}\|\langle A\rangle + \langle B\rangle\|_{\rm F}.$$

Source: [3016]. **Remark:** S is Specht's ratio. See Fact 12.17.5. **Related:** Fact 11.10.8.

Fact 11.10.8. Let $A, B \in \mathbb{F}^{n\times n}$, and let $\|\cdot\|$ be a unitarily invariant norm on $\mathbb{F}^{n\times n}$. Then,

$$\|A+B\| \le \sqrt{2}\|\langle A\rangle + \langle B\rangle\|.$$

Now, assume that A and B are positive semidefinite, let $S \in \mathbb{F}^{n\times n}$, and assume that S is unitary. Then,

$$\|A+SB\| \le \begin{cases} \sqrt{2}\|A+B\| \\ \|A+B+SBS^*\|, \end{cases} \qquad \|A+S\| \le \|A+I\| \le \begin{cases} \sqrt{2}\|A+I\| \\ \|A+2I\|. \end{cases}$$

Source: [1778, 1852]. **Remark:** The bound $\|A+S\| \le \|A+I\|$ is due to K. Fan and A. Hoffman.

Fact 11.10.9. Let $A, B \in \mathbb{F}^{n\times n}$, assume that A and B are normal, and let $\|\cdot\|$ be a unitarily invariant norm on $\mathbb{F}^{n\times n}$. Then,

$$\|A+B\| \le \|\langle A\rangle + \langle B\rangle\|, \quad \|A\odot B\| \le \|\langle A\rangle \odot \langle B\rangle\|.$$

Source: [196, 543, 551], [1450, p. 213], [1457, 1637, 2167], and [2991, p. 378]. **Remark:** Both inequalities can fail in the case where A and B are not both normal. Furthermore, there exist $A, B \in \mathbb{F}^{n\times n}$ such that $\langle A+B\rangle \le \langle A\rangle + \langle B\rangle$ does not hold. **Related:** Fact 10.11.34.

Fact 11.10.10. Let $A, B \in \mathbb{F}^{n\times n}$, assume that A and B are positive semidefinite, and let $\|\cdot\|$ be a unitarily invariant norm on $\mathbb{F}^{n\times n}$. If $r \in [0,1]$, then

$$\|(A+B)^r\| \le \|A^r + B^r\|.$$

Furthermore, if $r \in [1,\infty)$, then

$$\|A^r + B^r\| \le \|(A+B)^r\|.$$

Source: [106, 469, 1383, 1479].

Fact 11.10.11. Let $A, B \in \mathbb{F}^{n\times n}$, and let $\|\cdot\|$ be a unitarily invariant norm on $\mathbb{F}^{n\times n}$. If $p \in (0,1]$, then

$$\|\langle A+B\rangle^p\|^{1/p} \le 2^{1/p-1}(\|\langle A\rangle^p\|^{1/p} + \|\langle B\rangle^p\|^{1/p}).$$

If $p \in [1,\infty)$, then

$$\|\langle A+B\rangle^p\|^{1/p} \le \|\langle A\rangle^p\|^{1/p} + \|\langle B\rangle^p\|^{1/p}$$

and $\|\langle\cdot\rangle^p\|^{1/p}$ is a unitarily invariant norm on $\mathbb{F}^{n\times n}$. **Source:** [1383, 2436].

Fact 11.10.12. Let $A_1,\ldots,A_l \in \mathbf{N}^n$, let $p > 0$, let $\|\cdot\|$ be a unitarily invariant norm on $\mathbb{F}^{n\times n}$, and define $f\colon \ \mapsto [0,\infty)$ by $f(t) = \|(\sum_{i=1}^n A_i^t)^p\|$. Then, f is convex. **Source:** [1383].

Fact 11.10.13. Let $A, B \in \mathbb{F}^n$, assume that A and B are positive semidefinite, define $\mu(A,B) \triangleq \lim_{r\to\infty} [\frac{1}{2}(A^r + B^r)]^{1/r}$, and let $\|\cdot\|$ be a unitarily invariant norm on $\mathbb{F}^{n\times n}$. Then, the following statements hold:

i) Define $f\colon (0,1] \mapsto \mathbb{R}$ by $f(r) \triangleq \|(A^r + B^r)^{1/r}\|$. Then, f is nonincreasing.

ii) Assume that $\|\cdot\|$ is normalized, and define $f\colon (0,\infty) \mapsto \mathbb{R}$ by $f(r) \triangleq \|A^r + B^r\|^{1/r}$. Then, f is nonincreasing.

iii) $\lim_{r\to\infty} \|A^r + B^r\|^{1/r} = \sigma_{\max}[\mu(A,B)]$.

Source: [1383]. **Related:** Fact 10.11.57.

Fact 11.10.14. Let $A, B \in \mathbb{F}^{n\times n}$. Then,

$$\|\langle A\rangle - \langle B\rangle\|_{\mathrm{F}} \le \sqrt{2}\|A - B\|_{\mathrm{F}},$$
$$\|\langle A\rangle - \langle B\rangle\|_{\mathrm{F}}^2 + \|\langle A^*\rangle - \langle B^*\rangle\|_{\mathrm{F}}^2 \le 2\|A - B\|_{\mathrm{F}}^2.$$

If, in addition, A and B are normal, then

$$\|\langle A\rangle - \langle B\rangle\|_{\mathrm{F}} \le \|A - B\|_{\mathrm{F}}.$$

Source: [93, 164], [1391, pp. 217, 218], [1623, 1639], and [2991, pp. 319, 320].

Fact 11.10.15. Let $A, B \in \mathbb{F}^{n\times n}$, assume that A and B are positive semidefinite, and let $\|\cdot\|$ be a unitarily invariant norm on $\mathbb{F}^{n\times n}$. If $r \in [0, 1]$, then

$$\|A^r - B^r\| \le \|\langle A - B\rangle^r\|.$$

Furthermore, if $r \in [1, \infty)$, then

$$\|\langle A - B\rangle^r\| \le \|A^r - B^r\|.$$

In particular,

$$\|(A - B)^2\| \le \|A^2 - B^2\|.$$

Source: [449, pp. 293, 294] and [1631, 2438].

Fact 11.10.16. Let $A, B \in \mathbb{F}^{n\times n}$, assume that A and B are positive semidefinite, let $\alpha \ge 1$, and let $p \in [1/\alpha, \infty]$. Then,

$$\|A - B\|_{\sigma\alpha p}^{\alpha} \le \|A^{\alpha} - B^{\alpha}\|_{\sigma p}.$$

In particular, if $p \in [1/2, \infty]$, then

$$\|A - B\|_{\sigma 2p}^2 \le \|A^2 - B^2\|_{\sigma p}.$$

Source: [1381, p. 260] and [1624]. **Credit:** The case where $\alpha = 2$ and $p = 1$ is due to R. T. Powers and E. Stormer. **Related:** Fact 10.14.32.

Fact 11.10.17. Let $A, B \in \mathbb{F}^{n\times n}$ and $p \in [2, \infty]$. Then,

$$\|\langle A\rangle - \langle B\rangle\|_{\sigma p}^2 \le \|A + B\|_{\sigma p}\|A - B\|_{\sigma p}.$$

Source: [1639, 2438].

Fact 11.10.18. Let $A, B \in \mathbb{F}^{n\times n}$ and $p \ge 1$. Then,

$$\|\langle A\rangle - \langle B\rangle\|_{\sigma p} \le \max\{2^{1/p-1/2}, 1\}\sqrt{\|A + B\|_{\sigma p}\|A - B\|_{\sigma p}}.$$

Source: [93, 448]. **Credit:** F. Kittaneh, H. Kosaki, and R. Bhatia.

Fact 11.10.19. Let $A, B \in \mathbb{F}^{n\times n}$, and let $\|\cdot\|$ be a unitarily invariant norm on $\mathbb{F}^{n\times n}$. Then,

$$\|\langle A\rangle - \langle B\rangle\|^2 \le 2\|A + B\|\|A - B\|.$$

Source: [93, 448]. **Credit:** The case where $\|\cdot\| = \|\cdot\|_{\sigma 1}$ is due to H. J. Borchers and H. Kosaki. See [1639].

Fact 11.10.20. Let $A, B \in \mathbb{F}^{n\times n}$, and let $\|\cdot\|$ be a unitarily invariant norm on $\mathbb{F}^{n\times n}$. Then,

$$\|(\langle A\rangle - \langle B\rangle)^2\| \le \sigma_{\max}(A + B)\|A - B\|.$$

Source: [466, 2438].

Fact 11.10.21. Let $A, B \in \mathbb{F}^{n\times m}$, and assume that A and B are contractive, Then, for all $i \in \{1, \dots, n\}$,

$$2\sigma_i[(I - A^*B)^{-1}] \le \sigma_i[(I - A^*A)^{-1} + (I - B^*B)^{-1}].$$

Now, let $\|\cdot\|$ be a unitarily invariant norm on $\mathbb{F}^{m\times m}$. Then,

$$2\|(I-A^*B)^{-1}\| \le \|(I-A^*A)^{-1}+(I-B^*B)^{-1}\|.$$

Source: [1840, 1852].

Fact 11.10.22. Let $A,B\in\mathbb{F}^{n\times n}$. Then,

$$\sigma_{\max}(\langle A\rangle-\langle B\rangle) \le \frac{2}{\pi}\left[2+\log\frac{\sigma_{\max}(A)+\sigma_{\max}(B)}{\sigma_{\max}(A-B)}\right]\sigma_{\max}(A-B).$$

Credit: T. Kato. See [1639].

Fact 11.10.23. Let $A,B\in\mathbb{F}^{n\times n}$, assume that A and B are Hermitian, let $\|\cdot\|$ be a unitarily invariant norm on $\mathbb{F}^{n\times n}$, and let $k\in\mathbb{N}$. Then,

$$\|(A-B)^{2k+1}\| \le 2^{2k}\|A^{2k+1}-B^{2k+1}\|.$$

Source: [449, p. 294] and [1535].

Fact 11.10.24. Let $A,B\in\mathbb{F}^{n\times n}$, assume that A and B are positive semidefinite, let $\|\cdot\|$ be a unitarily invariant norm on $\mathbb{F}^{n\times n}$, and let $z\in\mathbb{F}$. Then,

$$\|A-|z|B\| \le \|A+zB\| \le \|A+|z|B\|.$$

In particular,

$$\|A-B\| \le \|A+B\|.$$

Source: [469, 474]. **Related:** Fact 10.22.20.

Fact 11.10.25. Let $\|\cdot\|$ be a normalized unitarily invariant norm on $\mathbb{F}^{n\times n}$. Then, $\|\cdot\|$ is submultiplicative. **Source:** [449, p. 94].

Fact 11.10.26. Let $A\in\mathbb{F}^{n\times m}$ and $B\in\mathbb{F}^{m\times l}$. Then, $\|AB\|_\infty \le m\|A\|_\infty\|B\|_\infty$. Furthermore, if $A=1_{n\times m}$ and $B=1_{m\times l}$, then equality holds.

Fact 11.10.27. $\|\cdot\|'_\infty \triangleq n\|\cdot\|_\infty$ is a submultiplicative norm on $\mathbb{F}^{n\times n}$. **Remark:** It is not necessarily true that $\|AB\|_\infty \le \|A\|_\infty\|B\|_\infty$. For example, let $A=B=\left[\begin{smallmatrix}1&1\\1&1\end{smallmatrix}\right]$.

Fact 11.10.28. Let $A,B\in\mathbb{F}^{n\times n}$, and let $\|\cdot\|$ be a submultiplicative norm on $\mathbb{F}^{n\times n}$. Then, $\|AB\| \le \|A\|\|B\|$. If $\|A\|\le 1$ and $\|B\|\le 1$, then $\|AB\|\le 1$, and, if $\|A\|<1$ and $\|B\|<1$, then $\|AB\|<1$. **Remark:** $\rho_{\max}(A)<1$ and $\rho_{\max}(B)<1$ do not imply that $\rho_{\max}(AB)<1$. Let $A=B^{\mathsf{T}}=\left[\begin{smallmatrix}0&2\\0&0\end{smallmatrix}\right]$.

Fact 11.10.29. Let $\|\cdot\|$ be a norm on $\mathbb{F}^{m\times m}$, and let

$$\delta > \sup\left\{\frac{\|AB\|}{\|A\|\|B\|}\colon\ A,B\in\mathbb{F}^{m\times m}, A,B\ne 0\right\}.$$

Then, $\|\cdot\|' \triangleq \delta\|\cdot\|$ is a submultiplicative norm on $\mathbb{F}^{m\times m}$. **Source:** [1448, p. 323].

Fact 11.10.30. Let $A,B\in\mathbb{F}^{n\times n}$, assume that A and B are Hermitian, assume that $-B\le A\le B$, and let $\|\cdot\|$ be a unitarily invariant norm on $\mathbb{F}^{n\times n}$. Then, $\|A\|\le\|B\|$. If, in addition, A and B are positive semidefinite, then $\|A-B\|\le\|A+B\|$. **Source:** [474]. For the second statement, note that $-2A\le 0\le 2B$, which implies that $-(A+B)\le A-B\le A+B$. **Related:** Fact 11.10.31.

Fact 11.10.31. Let $A,B\in\mathbb{F}^{n\times m}$, where $AA^*\le BB^*$. Then, $\|AA^*\|_{\rm F}\le\|BB^*\|_{\rm F}$ and $\|A\|_{\rm F}\le\|B\|_{\rm F}$. **Source:** Fact 11.10.30. **Related:** Fact 10.11.32 and Fact 11.9.12.

Fact 11.10.32. Let $A,B\in\mathbb{F}^{n\times n}$, assume that AB is normal, and let $\|\cdot\|$ be a unitarily invariant norm on $\mathbb{F}^{n\times n}$. Then, $\|AB\|\le\|BA\|$. **Source:** [449, p. 253] and [2991, p. 378].

Fact 11.10.33. Let $A,B\in\mathbb{F}^{n\times n}$, assume that A and B are positive semidefinite and nonzero, and let $\|\cdot\|$ be a submultiplicative unitarily invariant norm on $\mathbb{F}^{n\times n}$. Then,

$$\frac{\|AB\|}{\|A\|\|B\|} \le \frac{\|A+B\|}{\|A\|+\|B\|},\quad \frac{\|A\odot B\|}{\|A\|\|B\|} \le \frac{\|A+B\|}{\|A\|+\|B\|},$$

$$\frac{\|A \odot B\|}{\|A\|\|B\|} \le \frac{\|(A+B) \odot I\|}{\|A\| + \|B\|}, \quad \frac{\|A \otimes B\|}{\|A\|\|B\|} \le \frac{\|A \oplus B\|}{\|A\| + \|B\|}.$$

Source: [26, 1382]. **Related:** Fact 11.9.58.

Fact 11.10.34. Let $A, B \in \mathbb{F}^{n \times n}$, and let $\| \cdot \|$ be a unitarily invariant norm on $\mathbb{F}^{n \times n}$. Then,

$$\|AB\| \le \sigma_{\max}(A)\|B\|, \quad \|AB\| \le \|A\|\sigma_{\max}(B).$$

Consequently, if $C \in \mathbb{F}^{n \times n}$, then

$$\|ABC\| \le \sigma_{\max}(A)\|B\|\sigma_{\max}(C).$$

Source: [1631] and [2979, pp. 43, 101]. **Related:** Fact 11.9.58.

Fact 11.10.35. Let $A, B \in \mathbb{F}^{n \times m}$, and let $\| \cdot \|$ be a unitarily invariant norm on $\mathbb{F}^{m \times m}$. If $p > 0$, then

$$\|\langle A^*B \rangle^p\|^2 \le \|(A^*A)^p\|\|(B^*B)^p\|.$$

In particular,

$$\|(A^*BB^*A)^{1/4}\|^2 \le \|\langle A \rangle\|\|\langle B \rangle\|, \quad \|\langle A^*B \rangle\| = \|A^*B\|^2 \le \|A^*A\|\|B^*B\|.$$

Furthermore,

$$[\operatorname{tr}(A^*BB^*A)^{1/4}]^2 \le (\operatorname{tr}\langle A \rangle) \operatorname{tr}\langle B \rangle,$$

$$\left.\begin{matrix} |\operatorname{tr} A^*B| \\ \sqrt{|\operatorname{tr}(A^*B)^2|} \le \sqrt{\operatorname{tr} AA^*BB^*} \end{matrix}\right\} \le \operatorname{tr}\langle A^*B \rangle \le \|A\|_{\mathrm{F}}\|B\|_{\mathrm{F}}.$$

Source: [1453], Fact 11.9.44, Fact 11.14.1, and Fact 11.14.2.

Fact 11.10.36. Let $A, B \in \mathbb{F}^{n \times n}$, and assume that A and B are positive semidefinite. Then,

$$\|AB\|_{\mathrm{F}}^{1/2} \le \left\{ \begin{matrix} (2\|A\|_{\mathrm{F}}\|B\|_{\mathrm{F}})^{1/2} \le (\|A\|_{\mathrm{F}}^2 + \|B\|_{\mathrm{F}}^2)^{1/2} = \|(A^2 + B^2)^{1/2}\|_{\mathrm{F}} \\ 2\|AB\|_{\mathrm{F}}^{1/2} \le \|(A+B)^2\|_{\mathrm{F}}^{1/2} \end{matrix} \right\}$$

$$\le \|A + B\|_{\mathrm{F}} \le \|A\|_{\mathrm{F}} + \|B\|_{\mathrm{F}} \le \sqrt{2}(\|A\|_{\mathrm{F}}^2 + \|B\|_{\mathrm{F}}^2)^{1/2}.$$

Source: $2\|AB\|_{\mathrm{F}}^{1/2} \le \|(A+B)^2\|_{\mathrm{F}}^{1/2}$ follows from Fact 11.10.43.

Fact 11.10.37. Let $A, B \in \mathbb{F}^{n \times n}$, and assume that A and B are positive semidefinite. Then,

$$\|AB + BA\|_{\mathrm{F}} \le 2\|AB\|_{\mathrm{F}} \le \|A^2 + B^2\|_{\mathrm{F}},$$

$$\|A^2B + B^2A\|_{\mathrm{F}} \le \|A^2B + AB^2\|_{\mathrm{F}} \le \|A^3 + B^3\|_{\mathrm{F}},$$

$$\|A^3B + B^3A\|_{\mathrm{F}} \le \|A^3B + AB^3\|_{\mathrm{F}} \le \|A^4 + B^4\|_{\mathrm{F}}.$$

Now, let $p, q \in (0, \infty)$, and assume that $1/4 \le p/(p+q) \le 3/4$. Then,

$$\|A^pB^q + B^pA^q\|_{\mathrm{F}} \le \|A^pB^q + A^qB^p\|_{\mathrm{F}}.$$

In particular, if $p \in [1/3, 3]$, then

$$\|A^pB + B^pA\|_{\mathrm{F}} \le \|A^pB + AB^p\|_{\mathrm{F}}.$$

Source: [456, 1353] and use Fact 11.10.46.

Fact 11.10.38. Let $A, B \in \mathbb{F}^{n \times m}$, let $p, q \in [1, \infty]$, let $r \in [1, \infty)$, assume that $1/p + 1/q = 1/r$, and let $\| \cdot \|$ be a unitarily invariant norm on $\mathbb{F}^{m \times m}$. Then,

$$\|\langle A^*B \rangle^r\|^{1/r} \le \|\langle A \rangle^p\|\|^{1/p}\|\langle B \rangle^q\|^{1/q}.$$

In particular,

$$\|\langle A^*B\rangle^{1/2}\|^2 \le \|A\|\|B\|, \quad \|A^*B\| \le (\|A^*A\|\|B^*B\|)^{1/2}.$$

Source: [449, p. 95] and [1452]. **Remark:** This is Hölder's inequality for unitarily invariant norms. **Related:** Fact 2.12.23 and Fact 11.8.28.

Fact 11.10.39. Let $A, B \in \mathbb{F}^{n\times n}$, let $\|\cdot\|$ be a unitarily invariant norm on $\mathbb{F}^{n\times n}$, let $p, q \in (0, \infty)$, and assume that $\frac{1}{p} + \frac{1}{q} = 1$. Then,

$$\|AB\|^{pq} = \|AB\|^{p+q} \le \|\langle A\rangle^p \langle B\rangle^q\|.$$

Source: [2991, p. 378].

Fact 11.10.40. Let $A \in \mathbb{F}^{n\times n}$, assume that A is positive definite, and let $B \in \mathbb{F}^{n\times m}$. Then,

$$\|B^*ABB^*A^{-1}B\|_{\mathrm{F}} \le \frac{n}{2}\left(\frac{\lambda_1}{\lambda_n} + \frac{\lambda_n}{\lambda_1}\right)\|B^*B\|_{\mathrm{F}}^2.$$

Source: [2826].

Fact 11.10.41. Let $A, B \in \mathbb{R}^{n\times n}$, and assume that A and B are positive definite. Then,

$$(\operatorname{tr} A)\sqrt[n]{\det B} \le \|A\|_{\mathrm{F}}\|B\|_{\mathrm{F}}.$$

Source: [1809].

Fact 11.10.42. Let $A, B \in \mathbb{F}^{n\times n}$, assume that A and B are positive definite, and define

$$\operatorname{Cos}(A, B) \triangleq \frac{\operatorname{tr} AB}{\|A\|_{\mathrm{F}}\|B\|_{\mathrm{F}}}.$$

Then,

$$\operatorname{Cos}(A, I) \le 1, \quad \operatorname{Cos}(A, I)\operatorname{Cos}(B, I) \le \tfrac{1}{2}[\operatorname{Cos}(A, B) + 1],$$

$$\operatorname{Cos}(A, A^{-1}) \le \operatorname{Cos}(A, I)\operatorname{Cos}(A^{-1}, I) \le \tfrac{1}{2}[\operatorname{Cos}(A, A^{-1}) + 1] \le 1.$$

Source: [1809, 2751].

Fact 11.10.43. Let $A, B \in \mathbb{F}^{n\times n}$, and let $\|\cdot\|$ be a unitarily invariant norm on $\mathbb{F}^{n\times n}$. Then,

$$\|AB\| \le \tfrac{1}{4}\|(\langle A\rangle + \langle B^*\rangle)^2\|.$$

Source: [471].

Fact 11.10.44. Let $A, B \in \mathbb{F}^{n\times n}$, and assume that A and B are positive semidefinite. Then, for all $i \in \{1, \dots, n\}$,

$$\sigma_i(AB) \le \tfrac{1}{4}\sigma_i[(A + B)^2].$$

Now, let $\|\cdot\|$ be a unitarily invariant norm on $\mathbb{F}^{n\times n}$. Then,

$$\|\langle AB\rangle^{1/2}\| \le \tfrac{1}{2}\|A + B\|, \quad \|AB\| \le \tfrac{1}{4}\|(A + B)^2\|, \quad \|\langle A^{3/4}B^{3/4}\rangle^{2/3}\| \le \tfrac{1}{2}\|A + B\|.$$

Source: [471, 2935] and [2977, p. 77].

Fact 11.10.45. Let $A, B \in \mathbb{F}^{n\times n}$, assume that A and B are positive semidefinite, let $p \in [0, 1]$, and let $\|\cdot\|$ be a unitarily invariant norm on $\mathbb{F}^{n\times n}$. Then,

$$\|A^{1/2}B^{1/2}\| \le \tfrac{1}{2}\|A^pB^{1-p} + A^{1-p}B^p\| \le \tfrac{1}{2}\|A + B\|,$$

$$\|AB\| \le \tfrac{1}{4}\|(A + B)^2\| \le \tfrac{1}{2}\|A^2 + B^2\|,$$

$$\|AB + BA\| \le \|AB\| + \|BA\| = 2\|AB\| \le \|A^2 + B^2\|,$$

$$\|(A + B)^2\| \le \|A^2 + B^2\| + \|AB + BA\| \le 2\|A^2 + B^2\|.$$

Source: Let $p = 1/2$ and $X = I$ in Fact 11.10.82. The second inequality follows from Fact 11.10.44. **Remark:** Fact 11.9.6 implies that $\|\cdot\|$ is self adjoint, and thus $\|BA\| = \|(BA)^*\| = \|AB\|$. **Related:** Fact 10.22.18.

Fact 11.10.46. Let $A, B \in \mathbb{F}^{n\times n}$, assume that A and B are positive semidefinite, let $p, q \in (0, \infty)$, and let $\|\cdot\|$ be a unitarily invariant norm on $\mathbb{F}^{n\times n}$. Then,

$$\|A^pB^q + A^qB^p\| \le \|A^{p+q} + B^{p+q}\|.$$

Source: Use Fact 11.10.82 with $X = I$, A replaced by A^{p+q}, B replaced by B^{p+q}, and p replaced by $p/(p+q)$. **Remark:** See [1353].

Fact 11.10.47. Let $A, B \in \mathbb{F}^{n\times n}$, assume that A and B are positive semidefinite, let $\alpha \in [0, 1]$, and let $\|\cdot\|$ be a unitarily invariant norm on $\mathbb{F}^{n\times n}$. Then,

$$\|A^\alpha B^{1-\alpha} + B^{1-\alpha}A^\alpha\| \le 2\|\alpha A + (1-\alpha)B\|,$$
$$\|A^{1/2}B^{1/2}(A^\alpha B^{1-\alpha} + A^{1-\alpha}B^\alpha)\| \le \tfrac{1}{2}\|(A+B)^2\|.$$
$$\|B^{1/2}A^{1/2}(A^\alpha B^{1-\alpha} + A^{1-\alpha}B^\alpha)\| \le \tfrac{1}{2}\|(A+B)^2\|.$$

In particular,

$$\|(A^{1/2}B^{1/2})^2\| \le \tfrac{1}{4}\|(A+B)^2\|, \quad \|A^{1/2}B^{1/2}(A+B)\| \le \tfrac{1}{2}\|(A+B)^2\|, \quad \|B^{1/2}AB^{1/2}\| \le \tfrac{1}{4}\|(A+B)^2\|.$$

Source: [3017].

Fact 11.10.48. Let $A \in \mathbb{F}^{n\times m}$, $B \in \mathbb{F}^{m\times l}$, and $p, q, q', r \in [1, \infty]$, and assume that $1/q + 1/q' = 1$. Then,

$$\|AB\|_p \le \varepsilon_{pq}(n)\varepsilon_{pr}(l)\varepsilon_{q'r}(m)\|A\|_q\|B\|_r,$$

where

$$\varepsilon_{pq}(n) \triangleq \begin{cases} 1, & p \ge q, \\ n^{1/p-1/q}, & q \ge p. \end{cases}$$

Furthermore, there exist $A \in \mathbb{F}^{n\times m}$ and $B \in \mathbb{F}^{m\times l}$ such that equality holds. **Source:** [1191]. **Remark:** Related results are given in [1030, 1031, 1191, 1192, 1193, 1642, 2688].

Fact 11.10.49. Let $A, B \in \mathbb{F}^{n\times n}$, assume that A and B are positive semidefinite, let $\|\cdot\|$ be a unitarily invariant norm on $\mathbb{F}^{n\times n}$, and let $p \in [0, \infty)$. If $p \in [0, 1]$, then

$$\|A^pB^p\| \le \|AB\|^p.$$

If $p \in [1, \infty)$, then

$$\|AB\|^p \le \|A^pB^p\|.$$

Source: [458, 1117]. **Related:** Fact 10.22.36.

Fact 11.10.50. Let $A, B \in \mathbb{F}^{n\times n}$, assume that A and B are positive semidefinite, and let $\|\cdot\|$ be a unitarily invariant norm on $\mathbb{F}^{n\times n}$. If $p \in [0, 1]$, then

$$\|B^pA^pB^p\| \le \|(BAB)^p\|.$$

Furthermore, if $p \ge 1$, then

$$\|(BAB)^p\| \le \|B^pA^pB^p\|.$$

Source: [163] and [449, p. 258]. **Remark:** Extensions and a reverse inequality are given in Fact 10.11.78. **Related:** Fact 10.14.24 and Fact 10.22.36.

Fact 11.10.51. Let $A, B \in \mathbb{F}^{n\times n}$, assume that A and B are positive semidefinite, and let either $p = 1$ or $p \in [2, \infty]$. Then,

$$\|\langle AB\rangle^{1/2}\|_{\sigma p} \le \tfrac{1}{2}\|A + B\|_{\sigma p}.$$

Source: [196, 471]. **Remark:** This inequality holds for all Q-norms. See [449]. **Related:** Fact 10.22.18.

Fact 11.10.52. Let $A \in \mathbb{F}^{n\times m}$, $B \in \mathbb{F}^{m\times l}$, and $r \in \{1,2\}$. Then,

$$\|AB\|_{\sigma r} = \|A\|_{\sigma 2r}\|B\|_{\sigma 2r}$$

if and only if there exists $\alpha \geq 0$ such that $AA^* = \alpha B^*B$. Furthermore,

$$\|AB\|_{\infty} = \|A\|_{\infty}\|B\|_{\infty}$$

if and only if AA^* and B^*B have a common eigenvector associated with $\lambda_1(AA^*)$ and $\lambda_1(B^*B)$. **Source:** [2896]. **Related:** Fact 7.16.2.

Fact 11.10.53. Let $A, B \in \mathbb{F}^{n\times n}$, assume that A and B are positive semidefinite, let $p \in (0,\infty)$, and define $\|A\|_{\sigma p} \triangleq (\operatorname{tr} A^p)^{1/p}$. If $p \in (0,1]$, then

$$2^{p-1}(\|A\|_{\sigma p}^p + \|B\|_{\sigma p}^p) \leq (\|A\|_{\sigma p} + \|B\|_{\sigma p})^p \leq \|A+B\|_{\sigma p}^p \leq \|A\|_{\sigma p}^p + \|B\|_{\sigma p}^p.$$

If $p \in [1,\infty)$, then

$$\|A\|_{\sigma p}^p + \|B\|_{\sigma p}^p \leq \|A+B\|_{\sigma p}^p \leq (\|A\|_{\sigma p} + \|B\|_{\sigma p})^p \leq 2^{p-1}(\|A\|_{\sigma p}^p + \|B\|_{\sigma p}^p).$$

In particular,

$$\|A\|_{\sigma 1} + \|B\|_{\sigma 1} \leq \|A+B\|_{\sigma 1} \leq \|A\|_{\sigma 1} + \|B\|_{\sigma 1},$$

$$\tfrac{1}{2}(\|A\|_{\mathrm{F}} + \|B\|_{\mathrm{F}})^2 \leq \|A\|_{\mathrm{F}}^2 + \|B\|_{\mathrm{F}}^2 \leq \|A+B\|_{\mathrm{F}}^2 \leq (\|A\|_{\mathrm{F}} + \|B\|_{\mathrm{F}})^2 \leq 2(\|A\|_{\mathrm{F}}^2 + \|B\|_{\mathrm{F}}^2).$$

Source: Fact 10.14.36 and Fact 11.10.59. **Remark:** For all $p \in (0,1)$, $\|\cdot\|_{\sigma p}$ is an anti-norm on $\mathbf{N}^n$. See [545]. **Remark:** The first inequality in the second string is the McCarthy inequality given by Fact 10.14.36. **Related:** Fact 11.8.21, Fact 11.9.47, Fact 11.10.4, and Fact 11.10.6.

Fact 11.10.54. Let $A, B \in \mathbb{F}^{n\times n}$ and $p, q \in (0,\infty)$. Then, the following statements hold:

i) If $p \in (0,2]$, then

$$2^{p-1}(\|A\|_{\sigma p}^p + \|B\|_{\sigma p}^p) \leq \|A+B\|_{\sigma p}^p + \|A-B\|_{\sigma p}^p \leq 2(\|A\|_{\sigma p}^p + \|B\|_{\sigma p}^p).$$

ii) If $p \in [2,\infty)$, then

$$2(\|A\|_{\sigma p}^p + \|B\|_{\sigma p}^p) \leq \|A+B\|_{\sigma p}^p + \|A-B\|_{\sigma p}^p \leq 2^{p-1}(\|A\|_{\sigma p}^p + \|B\|_{\sigma p}^p).$$

iii) If $p \in (1,2]$ and $1/p + 1/q = 1$, then

$$\|A+B\|_{\sigma p}^q + \|A-B\|_{\sigma p}^q \leq 2(\|A\|_{\sigma p}^p + \|B\|_{\sigma p}^p)^{q/p}.$$

iv) If $p \in [2,\infty)$ and $1/p + 1/q = 1$, then

$$2(\|A\|_{\sigma p}^p + \|B\|_{\sigma p}^p)^{q/p} \leq \|A+B\|_{\sigma p}^q + \|A-B\|_{\sigma p}^q.$$

Source: [1420]. **Remark:** These are versions of the *Clarkson inequalities*. See Fact 2.21.8. **Remark:** See [1420] for extensions to unitarily invariant norms. See [472] for further extensions. **Remark:** For $p \in (0,1)$, $\|\cdot\|_{\sigma p}$ is defined in Fact 11.9.10.

Fact 11.10.55. Let $A, B \in \mathbb{F}^{n\times n}$ and $p \in (0,1]$. Then,

$$\|A+B\|_{\sigma p} \leq 2^{(1-p)/p}(\|A\|_{\sigma p} + \|B\|_{\sigma p}).$$

Source: [2438]. **Remark:** For $p \in (0,1)$, $\|\cdot\|_{\sigma p}$ is defined in Fact 11.9.10.

Fact 11.10.56. Let $A_1, \ldots, A_l \in \mathbf{H}^n$, let $p_1, \ldots, p_l \in (0,1]$, assume that $\sum_{i=1}^l p_i = 1$, let $r \in (1,2]$, and let $\|\cdot\|$ be a unitarily invariant norm on $\mathbb{F}^{n\times n}$. Then,

$$\left\|\left\langle \sum_{i=1}^l A_i \right\rangle^r\right\| \leq \left\|\sum_{i=1}^l \frac{1}{p_i}\langle A_i\rangle^r\right\|.$$

Source: [1984]. **Related:** Fact 2.21.23.

Fact 11.10.57. Let $A_1,\ldots,A_k \in \mathbb{F}^{n\times n}$, let $p \in [2,\infty)$, and let $\|\cdot\|$ be a unitarily invariant norm on $\mathbb{F}^{n\times n}$. Then,

$$\left\|\left(\sum_{i=1}^{k}\langle A_i\rangle^2\right)^{1/2}\right\| \le n^{1/2-1/p}\left\|\left(\sum_{i=1}^{k}\langle A_i\rangle^p\right)^{1/p}\right\|.$$

Source: [2436].

Fact 11.10.58. Let $A_1,\ldots,A_k \in \mathbb{F}^{n\times n}$ be positive semidefinite, and let $\|\cdot\|$ be a unitarily invariant norm on $\mathbb{F}^{n\times n}$. Then, the following statements hold:

i) If $p \in (0,1]$, then

$$\left\|\left(\sum_{i=1}^{k}A_i\right)^p\right\| \le \left\|\sum_{i=1}^{k}A_i^p\right\| \le k^{1-p}\left\|\left(\sum_{i=1}^{k}A_i\right)^p\right\|.$$

ii) If $p \in [1,\infty)$, then

$$\left\|\sum_{i=1}^{k}A_i^p\right\| \le \left\|\left(\sum_{i=1}^{k}A_i\right)^p\right\| \le k^{p-1}\left\|\sum_{i=1}^{k}A_i^p\right\|.$$

Source: [2436].

Fact 11.10.59. Let $A_1,\ldots,A_k \in \mathbb{F}^{n\times n}$ be positive semidefinite, and let $p \in [0,\infty)$. If $p \in [0,1]$, then

$$k^{p-1}\sum_{i=1}^{n}\|A_i\|_{\sigma p}^p \le \left(\sum_{i=1}^{k}\|A_i\|_{\sigma p}\right)^p \le \left\|\sum_{i=1}^{n}A_i\right\|_{\sigma p}^p \le \sum_{i=1}^{n}\|A_i\|_{\sigma p}^p.$$

If $p \in [1,\infty)$, then

$$\sum_{i=1}^{k}\|A_i\|_{\sigma p}^p \le \left\|\sum_{i=1}^{k}A_i\right\|_{\sigma p}^p \le \left(\sum_{i=1}^{k}\|A_i\|_{\sigma p}\right)^p \le k^{p-1}\sum_{i=1}^{k}\|A_i\|_{\sigma p}^p.$$

Source: [1423, 1629]. **Remark:** The first inequality in the second string extends the McCarthy inequality given by Fact 10.14.36 to more than two matrices. **Related:** Fact 2.2.59 and Fact 2.11.90.

Fact 11.10.60. Let $A, B \in \mathbb{F}^{n\times n}$, and let $\|\cdot\|$ be a unitarily invariant norm on $\mathbb{F}^{n\times n}$. Then, the following statements hold:

i) If $p \in (0,1]$, then

$$\|\langle A\rangle^p\|^{1/p} + \|\langle B\rangle^p\|^{1/p} \le 2^{1/p-1}(\|\langle A+B\rangle^p\|^{1/p} + \|\langle A-B\rangle^p\|^{1/p}).$$

ii) If $p \in [1,\infty)$, then

$$\|\langle A\rangle^p\|^{1/p} + \|\langle B\rangle^p\|^{1/p} \le \|\langle A+B\rangle^p\|^{1/p} + \|\langle A-B\rangle^p\|^{1/p}.$$

iii) If $p \in [1,\infty)$, then

$$\|\langle A+B\rangle^p + \langle A-B\rangle^p\|^{1/p} \le \|\langle A+B\rangle^p\|^{1/p} + \|\langle A-B\rangle^p\|^{1/p}.$$

iv) If $p \in [2,\infty)$, then

$$2\|\langle A\rangle^p + \langle B\rangle^p\| \le \|\langle A+B\rangle^p + \langle A-B\rangle^p\|.$$

Source: [2436].

Fact 11.10.61. Let $A_1,\ldots,A_k,B_1,\ldots,B_k \in \mathbb{F}^{n\times n}$. Then,

$$\sum_{i,j=1}^{k}\langle A_i-A_j\rangle^2 + \sum_{i,j=1}^{k}\langle B_i-B_j\rangle^2 + 2\left\langle\sum_{i=1}^{k}(A_i-B_i)\right\rangle^2 = 2\sum_{i,j=1}^{k}\langle A_i-B_j\rangle^2,$$

$$\sum_{i,j=1}^{k}\sigma_{\max}(A_i - A_j)^2 + \sum_{i,j=1}^{k}\sigma_{\max}(B_i - B_j)^2 + 2\sigma_{\max}\left(\sum_{i=1}^{k}(A_i - B_i)\right)^2 = 2\sum_{i,j=1}^{k}\sigma_{\max}(A_i - B_j)^2,$$

$$\sum_{i,j=1}^{k}\|A_i - A_j\|_{\mathrm{F}}^2 + \sum_{i,j=1}^{k}\|B_i - B_j\|_{\mathrm{F}}^2 + 2\left\|\sum_{i=1}^{k}(A_i - B_i)\right\|_{\mathrm{F}}^2 = 2\sum_{i,j=1}^{k}\|A_i - B_j\|_{\mathrm{F}}^2.$$

If $p \in (0, 2]$, then

$$\begin{aligned}2k^{p-2}\sum_{i,j=1}^{k}\|A_i - B_j\|_{\sigma p}^p &\le \sum_{i,j=1}^{k}\|A_i - A_j\|_{\sigma p}^p + \sum_{i,j=1}^{k}\|B_i - B_j\|_{\sigma p}^p + 2\left\|\sum_{i=1}^{k}(A_i - B_i)\right\|_{\sigma p}^p\\ &\le 2(k^2 - k + 1)^{1-p/2}\sum_{i,j=1}^{k}\|A_i - B_j\|_{\sigma p}^p.\end{aligned}$$

If $p \in [2, \infty)$, then

$$\begin{aligned}2(k^2 - k + 1)^{1-p/2}\sum_{i,j=1}^{k}\|A_i - B_j\|_{\sigma p}^p &\le \sum_{i,j=1}^{k}\|A_i - A_j\|_{\sigma p}^p + \sum_{i,j=1}^{k}\|B_i - B_j\|_{\sigma p}^p + 2\left\|\sum_{i=1}^{k}(A_i - B_i)\right\|_{\sigma p}^p\\ &\le 2k^{p-2}\sum_{i,j=1}^{k}\|A_i - B_j\|_{\sigma p}^p.\end{aligned}$$

Source: [2093, 2096]. **Related:** Fact 10.11.84.

Fact 11.10.62. Let $A_1, \ldots, A_k, B_1, \ldots, B_k \in \mathbb{F}^{n\times n}$, let $p \in [2, \infty)$, and let $\|\cdot\|$ be a unitarily invariant norm on $\mathbb{F}^{n\times n}$. Then,

$$\begin{aligned}k^{-1+1/p}\left\|\left(\sum_{i=1}^{k}\langle A_i + B_i\rangle^p\right)^{1/p}\right\| &\le \left\|\left(\sum_{i=1}^{k}\langle A_i\rangle^p\right)^{1/p}\right\| + \left\|\left(\sum_{i=1}^{k}\langle B_i\rangle^p\right)^{1/p}\right\|\\ &\le k^{1-1/p}\left(\left\|\left(\sum_{i=1}^{k}\langle A_i + B_i\rangle^p\right)^{1/p}\right\| + \left\|\left(\sum_{i=1}^{k}\langle A_i - B_i\rangle^p\right)^{1/p}\right\|\right),\end{aligned}$$

$$k^{-1+1/p}\left\|\sum_{i=1}^{k}\langle A_i + B_i\rangle^p\right\|^{1/p} \le \left\|\left(\sum_{i=1}^{k}\langle A_i\rangle^p\right)^{1/p}\right\| + \left\|\left(\sum_{i=1}^{k}\langle B_i\rangle^p\right)^{1/p}\right\|,$$

$$\left\|\sum_{i=1}^{k}\langle A_i\rangle^p\right\|^{1/p} + \left\|\sum_{i=1}^{k}\langle B_i\rangle^p\right\|^{1/p} \le k^{1-1/p}\left(\left\|\left(\sum_{i=1}^{k}\langle A_i + B_i\rangle^p\right)^{1/p}\right\| + \left\|\left(\sum_{i=1}^{k}\langle A_i - B_i\rangle^p\right)^{1/p}\right\|\right).$$

Source: [2436].

Fact 11.10.63. Let $A_1, \ldots, A_k, B_1, \ldots, B_k \in \mathbb{F}^{n\times n}$ be positive semidefinite, and let $\|\cdot\|$ be a unitarily invariant norm on $\mathbb{F}^{n\times n}$. Then, the following statements hold:

i) If $p \in [1, \infty)$, then

$$k^{-|1/p-1/2|}\left\|\sum_{i=1}^{k}\langle A_i + B_i\rangle^p\right\|^{1/p} \le \left\|\sum_{i=1}^{k}\langle A_i\rangle^p\right\|^{1/p} + \left\|\sum_{i=1}^{k}\langle B_i\rangle^p\right\|^{1/p}$$

$$\leq k^{|1/p-1/2|}\left(\left\|\sum_{i=1}^k \langle A_i + B_i\rangle^p\right\|^{1/p} + \left\|\sum_{i=1}^k \langle A_i - B_i\rangle^p\right\|^{1/p}\right).$$

ii) If $p \in (0, 1]$, then

$$2^{1-1/p}k^{-|1/p-1/2|}\left\|\sum_{i=1}^k \langle A_i + B_i\rangle^p\right\|^{1/p} \leq \left\|\sum_{i=1}^k \langle A_i\rangle^p\right\|^{1/p} + \left\|\sum_{i=1}^k \langle B_i\rangle^p\right\|^{1/p}$$
$$\leq 2^{1/p-1}k^{|1/p-1/2|}\left(\left\|\sum_{i=1}^k \langle A_i + B_i\rangle^p\right\|^{1/p} + \left\|\sum_{i=1}^k \langle A_i - B_i\rangle^p\right\|^{1/p}\right).$$

Source: [2436].

Fact 11.10.64. Let $A, B \in \mathbb{C}^{n\times m}$. If $p \in [1, 2]$, then

$$[\|A\|_{\sigma p}^2 + (p-1)\|B\|_{\sigma p}^2]^{1/2} \leq [\tfrac{1}{2}(\|A + B\|_{\sigma p}^p + \|A - B\|_{\sigma p}^p)]^{1/p}.$$

If $p \in [2, \infty]$, then

$$[\tfrac{1}{2}(\|A + B\|_{\sigma p}^p + \|A - B\|_{\sigma p}^p)]^{1/p} \leq [\|A\|_{\sigma p}^2 + (p-1)\|B\|_{\sigma p}^2]^{1/2}.$$

Source: [262, 342]. **Remark:** This is *Beckner's two-point inequality* (also called *optimal 2-uniform convexity*).

Fact 11.10.65. Let $A, B \in \mathbb{F}^{n\times n}$. If either $p \in [1, 4/3]$ or both $p \in (4/3, 2]$ and $A + B$ and $A - B$ are positive semidefinite, then

$$(\|A\|_{\sigma p} + \|B\|_{\sigma p})^p + |\|A\|_{\sigma p} - \|B\|_{\sigma p}|^p \leq \|A + B\|_{\sigma p}^p + \|A - B\|_{\sigma p}^p.$$

Furthermore, if either $p \in [4, \infty]$ or both $p \in [2, 4)$ and A and B are positive semidefinite, then

$$\|A + B\|_{\sigma p}^p + \|A - B\|_{\sigma p}^p \leq (\|A\|_{\sigma p} + \|B\|_{\sigma p})^p + |\|A\|_{\sigma p} - \|B\|_{\sigma p}|^p.$$

Source: [262, 1617]. **Remark:** These are matrix versions of *Hanner's inequality*. Vector versions are given in Fact 11.8.23.

Fact 11.10.66. Let $A, B \in \mathbb{C}^{n\times n}$, and assume that A and B are Hermitian. If $p \in [1, 2]$, then

$$2^{1/2-1/p}\|(A^2 + B^2)^{1/2}\|_{\sigma p} \leq \|A + \jmath B\|_{\sigma p} \leq \|(A^2 + B^2)^{1/2}\|_{\sigma p},$$
$$2^{1-2/p}(\|A\|_{\sigma p}^2 + \|B\|_{\sigma p}^2) \leq \|A + \jmath B\|_{\sigma p}^2 \leq 2^{2/p-1}(\|A\|_{\sigma p}^2 + \|B\|_{\sigma p}^2).$$

Furthermore, if $p \in [2, \infty)$, then

$$\|(A^2 + B^2)^{1/2}\|_{\sigma p} \leq \|A + \jmath B\|_{\sigma p} \leq 2^{1/2-1/p}\|(A^2 + B^2)^{1/2}\|_{\sigma p},$$
$$2^{2/p-1}(\|A\|_{\sigma p}^2 + \|B\|_{\sigma p}^2) \leq \|A + \jmath B\|_{\sigma p}^2 \leq 2^{1-2/p}(\|A\|_{\sigma p}^2 + \|B\|_{\sigma p}^2).$$

Source: [470].

Fact 11.10.67. Let $A, B \in \mathbb{C}^{n\times n}$, and assume that A and B are Hermitian. If $p \in [1, 2]$, then

$$2^{1-2/p}(\|A\|_{\sigma p}^p + \|B\|_{\sigma p}^p) \leq \|A + \jmath B\|_{\sigma p}^p.$$

If $p \in [2, \infty]$, then

$$\|A + \jmath B\|_{\sigma p}^p \leq 2^{1-2/p}(\|A\|_{\sigma p}^p + \|B\|_{\sigma p}^p).$$

In particular,

$$\|A + \jmath B\|_{\rm F}^2 = \|A\|_{\rm F}^2 + \|B\|_{\rm F}^2 = \|(A^2 + B^2)^{1/2}\|_{\rm F}^2.$$

Source: [470, 480].

Fact 11.10.68. Let $A, B \in \mathbb{C}^{n\times n}$, and assume that A is positive semidefinite and B is Hermitian. If $p \in [1,2]$, then

$$\|A\|_{\sigma p}^2 + 2^{1-2/p}\|B\|_{\sigma p}^2 \le \|A + \jmath B\|_{\sigma p}^2.$$

If $p \in [2,\infty]$, then

$$\|A + \jmath B\|_{\sigma p}^2 \le \|A\|_{\sigma p}^2 + 2^{1-2/p}\|B\|_{\sigma p}^2.$$

In particular,

$$(\operatorname{tr}\langle A\rangle)^2 + (\operatorname{tr}\langle B\rangle)^2 \le (\operatorname{tr}\langle A + \jmath B\rangle)^2,$$
$$\|A + \jmath B\|_{\mathrm{F}}^2 = \|A\|_{\mathrm{F}}^2 + \|B\|_{\mathrm{F}}^2,$$
$$\sigma_{\max}^2(A + \jmath B) \le \sigma_{\max}^2(A) + 2\sigma_{\max}^2(B).$$

In addition,

$$\|A\|_{\sigma 1}^2 + \|B\|_{\sigma 1}^2 \le \|A + \jmath B\|_{\sigma 1}^2.$$

Source: [480].

Fact 11.10.69. Let $A, B \in \mathbb{C}^{n\times n}$, and assume that A and B are positive semidefinite. If $p \in [1,2]$, then

$$\|A\|_{\sigma p}^2 + \|B\|_{\sigma p}^2 \le \|A + \jmath B\|_{\sigma p}^2.$$

If $p \in [2,\infty]$, then

$$\|A + \jmath B\|_{\sigma p}^2 \le \|A\|_{\sigma p}^2 + \|B\|_{\sigma p}^2.$$

Hence,

$$\|A\|_{\sigma 2}^2 + \|B\|_{\sigma 2}^2 = \|A + \jmath B\|_{\sigma 2}^2.$$

In particular,

$$(\operatorname{tr} A)^2 + (\operatorname{tr} B)^2 \le (\operatorname{tr}\langle A + \jmath B\rangle)^2,$$
$$\|A + \jmath B\|_{\mathrm{F}}^2 = \|A\|_{\mathrm{F}}^2 + \|B\|_{\mathrm{F}}^2,$$
$$\sigma_{\max}^2(A + \jmath B) \le \sigma_{\max}^2(A) + \sigma_{\max}^2(A).$$

Source: [480]. **Related:** Fact 10.22.14.

Fact 11.10.70. Let $A, B \in \mathbb{F}^{n\times n}$, assume that A and B are Hermitian, and let $\|\cdot\|$ be a unitarily invariant norm on $\mathbb{F}^{n\times n}$. Then,

$$\|A + B\| \le \sqrt{2}\|A + \jmath B\|.$$

Source: [475].

Fact 11.10.71. Let $A, B \in \mathbb{F}^{n\times n}$, and let $\|\cdot\|$ be a unitarily invariant norm on $\mathbb{F}^{n\times n}$. Then,

$$\|(A + B)(A + B)^*\| \le \|AA^* + BB^* + 2AB^*\| \le \begin{cases} \|(A - B)(A - B)^* + 4AB^*\| \\ (\|A\| + \|B\|)^2. \end{cases}$$

Source: [1983].

Fact 11.10.72. Let $A \in \mathbb{F}^{n\times n}$, $B \in \mathbb{F}^{m\times m}$, and $X \in \mathbb{F}^{n\times m}$, assume that A and B are nonsingular and either both Hermitian or both skew Hermitian, and let $\|\cdot\|$ be a unitarily invariant norm on $\mathbb{F}^{n\times m}$. Then,

$$\|X\| \le \tfrac{1}{2}\|AXB^{-1} + A^{-1}XB\|.$$

Source: Replace X by $A^{-1}XB^{-1}$ in Fact 11.10.79. See [1449].

Fact 11.10.73. Let $A, B \in \mathbb{F}^{n\times n}$, and let $\|\cdot\|$ be a unitarily invariant norm on $\mathbb{F}^{n\times n}$. Then, the following statements hold:

i) If B is Hermitian, then $\|A - \tfrac{1}{2}(A + A^*)\| \le \|A - B\|$.

ii) If B is skew Hermitian, then $\|A - \tfrac{1}{2}(A - A^*)\| \le \|A - B\|$.

iii) Let $A = MS$, where $M, S \in \mathbb{F}^{n\times n}$, M is positive semidefinite, and S is unitary. If B is unitary, then $\|A - S\| \le \|A - B\| \le \|A + S\|$.

iv) Let $A = S_1DS_2$, where $S_1, D, S_2 \in \mathbb{F}^{n\times n}$, S_1, S_2 are unitary, and D is nonnegative and diagonal. If B is unitary, then $\|A - S_1S_2\| \le \|A - B\| \le \|A + S_1S_2\|$.

v) If A is normal and B is nonsingular, then $\|A\| \le \|B^{-1}AB\|$.

vi) If A is positive semidefinite and B is unitary, then $\|A - I\| \le \|A - B\| \le \|A + I\|$.

Source: [449, p. 275], [2263, p. 150], and [2991, p. 378]. **Related:** Fact 11.16.26 and Fact 11.16.27

Fact 11.10.74. Let $A, B \in \mathbb{F}^{n\times m}$, let $X \in \mathbb{F}^{m\times m}$, and let $\|\cdot\|$ be a unitarily invariant norm on $\mathbb{F}^{n\times m}$. Then,

$$\|AA^+B - B\| \le \|AX - B\|.$$

Source: [2991, p. 377].

Fact 11.10.75. Let $A, B \in \mathbb{F}^{n\times m}$, and let $\|\cdot\|$ be a unitarily invariant norm on $\mathbb{F}^{n\times m}$. Then,

$$\begin{aligned}\|A^+ - B^+\| &\le [\sigma_{\max}(A^+)\sigma_{\max}(B^+) + \max\{\sigma_{\max}^2(A^+), \sigma_{\max}^2(B^+)\}]\|A - B\|\\ &\le 2\max\{\sigma_{\max}^2(A^+), \sigma_{\max}^2(B^+)\}\|A - B\|.\end{aligned}$$

If, in addition, $\operatorname{rank} A = \operatorname{rank} B$, then

$$\begin{aligned}\|A^+ - B^+\| &\le [\sigma_{\max}(A^+)\sigma_{\max}(B^+) + (\sigma_{\max}(A^+) + \sigma_{\max}(B^+))\max\{\sigma_{\max}(A^+), \sigma_{\max}(B^+)\}]\|A - B\|\\ &\le 3\sigma_{\max}(A^+)\sigma_{\max}(B^+)\|A - B\|.\end{aligned}$$

Source: [612].

Fact 11.10.76. Let $A, B \in \mathbb{F}^{n\times n}$, assume that A is nonsingular and B is Hermitian, and let $\|\cdot\|$ be a unitarily invariant norm on $\mathbb{F}^{n\times n}$. Then,

$$\|B\| \le \tfrac{1}{2}\|ABA^{-1} + A^{-1}BA\|.$$

Source: [780, 1093].

Fact 11.10.77. Let $A, M, S, B \in \mathbb{F}^{n\times n}$, assume that $A = MS$, M is positive semidefinite, and S and B are unitary, and let $\|\cdot\|$ be a unitarily invariant norm on $\mathbb{F}^{n\times n}$. Then, $\|A - S\| \le \|A - B\|$. **Source:** [449, p. 276] and [2263, p. 150]. **Remark:** $A = MS$ is the polar decomposition of A. See Corollary 7.6.4.

Fact 11.10.78. Let $A, B \in \mathbb{F}^{n\times n}$, assume that A and B are positive semidefinite, and let $\|\cdot\|$ be a unitarily invariant norm on $\mathbb{F}^{2n\times 2n}$. Then,

$$\left\|\begin{bmatrix} A+B & 0\\ 0 & 0\end{bmatrix}\right\| \le \left\|\begin{bmatrix} A & 0\\ 0 & B\end{bmatrix}\right\| + \left\|\begin{bmatrix} A^{1/2}B^{1/2} & 0\\ 0 & A^{1/2}B^{1/2}\end{bmatrix}\right\|.$$

In particular,

$$\sigma_{\max}(A + B) \le \max\{\sigma_{\max}(A), \sigma_{\max}(B)\} + \sigma_{\max}(A^{1/2}B^{1/2}),$$

and, for all $p \in [1, \infty)$,

$$\|A + B\|_{\sigma p} \le (\|A\|_{\sigma p}^p + \|B\|_{\sigma p}^p)^{1/p} + 2^{1/p}\|A^{1/2}B^{1/2}\|_{\sigma p}.$$

Source: [1629, 1632, 1637]. **Remark:** Fact 11.16.18 gives a refined upper bound for $\sigma_{\max}(A + B)$.

Fact 11.10.79. Let $A \in \mathbb{F}^{n\times n}$, $B \in \mathbb{F}^{m\times m}$, and $X \in \mathbb{F}^{n\times m}$, and let $\|\cdot\|$ be a unitarily invariant norm on $\mathbb{F}^{n\times m}$. Then,

$$\|A^*XB\| \le \tfrac{1}{2}\|AA^*X + XBB^*\|.$$

In particular, if $n = m$, then

$$\|A^*B\| \le \tfrac{1}{2}\|AA^* + BB^*\|.$$

Source: [121, 457, 468, 1119, 1449, 1626]. **Remark:** The first result is *McIntosh's inequality*. **Related:** Fact 11.16.31.

Fact 11.10.80. Let $A, X, B \in \mathbb{F}^{n\times n}$, assume that X is positive semidefinite, and let $\|\cdot\|$ be a unitarily invariant norm on $\mathbb{F}^{n\times n}$. Then,

$$\|A^*XB + B^*XA\| \le \|A^*XA + B^*XB\|.$$

In particular,

$$\|A^*B + B^*A\| \le \|A^*A + B^*B\|.$$

Source: [468, 1630]. **Remark:** See [1630] for the case of indefinite X.

Fact 11.10.81. Let $A_1, \dots, A_m \in \mathbb{F}^{n\times n}$, define $A_{m+1} \triangleq A_1$, and let $\|\cdot\|$ be a unitarily invariant norm on $\mathbb{F}^{n\times n}$. Then,

$$\left\|\sum_{i=1}^{m} A_i^*A_{i+1}\right\| \le \left\|\sum_{i=1}^{m} A_i^*A_i\right\|.$$

Source: [2167]. **Related:** Fact 11.10.80.

Fact 11.10.82. Let $A, X, B \in \mathbb{F}^{n\times n}$, assume that A and B are positive semidefinite, let $p \in [0,1]$, and let $\|\cdot\|$ be a unitarily invariant norm on $\mathbb{F}^{n\times n}$. Then,

$$\|A^pXB^{1-p} + A^{1-p}XB^p\| \le \|AX + XB\|,$$
$$\|A^pXB^{1-p} - A^{1-p}XB^p\| \le |2p-1|\|AX - XB\|.$$

Furthermore, for all $i \in \{1, \dots, n\}$,

$$\sigma_i(A^pB^{1-p} + A^{1-p}B^p) \le \sigma_i(A + B).$$

Source: [121, 458, 477, 1086] and [2979, pp. 87, 88]. See [190, 1640] for the last inequality. **Remark:** These are the *Heinz inequalities*. **Related:** Fact 11.10.46.

Fact 11.10.83. Let $A, B \in \mathbf{P}^n$, and define

$$V \triangleq (A^{-1/2}BA^{-1/2})^{1/2}A^{1/2}B^{-1/2}, \quad U \triangleq A^{1/2}B^{1/2}\langle A^{1/2}B^{1/2}\rangle^{-1}.$$

Then, U and V are unitary. If, in addition, $\|\cdot\|$ is a unitarily invariant norm, then

$$\|A + VBV^*\| \le \|A + B\| \le \|A + UBU^*\|.$$

Furthermore,

$$\det(A + UBU^*) \le \det(A + B) \le \det(A + VBV^*).$$

Source: [1779]. **Related:** Fact 10.11.68.

Fact 11.10.84. Let $A, B, C, D \in \mathbb{F}^{n\times n}$, and let $\|\cdot\|$ be a unitarily invariant norm on $\mathbb{F}^{n\times n}$. If $p \in [1, \infty)$, then

$$\|\langle A + B\rangle^p + \langle C + D\rangle^p\|^{1/p} \le 2^{|1/p-1/2|}\|\langle A\rangle^p + \langle C\rangle^p\|^{1/p} + \|\langle B\rangle^p + \langle D\rangle^p\|^{1/p}.$$

If $p, r \in [1, \infty]$, then

$$\|\langle A + B\rangle^p + \langle C + D\rangle^p\|_{\sigma r}^{1/p} \le 2^{(1/p)(1-1/r)}\|\langle A\rangle^p + \langle C\rangle^p\|_{\sigma r}^{1/p} + \|\langle B\rangle^p + \langle D\rangle^p\|_{\sigma r}^{1/p}.$$

If $p, r \in [1, \infty]$ and $p \le r$, then

$$\|(\langle A + B\rangle^p + \langle C + D\rangle^p)^{1/p}\|_{\sigma r} \le 2^{|1/p-1/2|}\|(\langle A\rangle^p + \langle C\rangle^p)^{1/p}\|_{\sigma r} + \|(\langle B\rangle^p + \langle D\rangle^p)^{1/p}\|_{\sigma r}.$$

If $p, r \in [1, \infty]$, then

$$\|(\langle A + B\rangle^p + \langle C + D\rangle^p)^{1/p}\|_{\sigma r} \le 2^{|1/p-1/r|}\|(\langle A\rangle^p + \langle C\rangle^p)^{1/p}\|_{\sigma r} + \|(\langle B\rangle^p + \langle D\rangle^p)^{1/p}\|_{\sigma r}.$$

Source: [1383].

Fact 11.10.85. Let $A, B, C, D \in \mathbb{F}^{n\times n}$, let $p, q \in [1,\infty]$, assume that $1/p + 1/q = 1$, and let $\|\cdot\|$ be a unitarily invariant norm on $\mathbb{F}^{n\times n}$. Then,

$$\|C^*A + D^*B\| \le 2^{|1/p-1/2|}\|\langle A\rangle^p + \langle B\rangle^p\|^{1/p}\|\langle C\rangle^q + \langle D\rangle^q\|^{1/q}.$$

If, in addition, $r \in [1,\infty]$, then

$$\|C^*A + D^*B\|_{\sigma r} \le 2^{1-1/r}\|\langle A\rangle^p + \langle B\rangle^p\|_{\sigma r}^{1/p}\|\langle C\rangle^q + \langle D\rangle^q\|_{\sigma r}^{1/q}.$$

Source: [1383].

Fact 11.10.86. Let $A, B \in \mathbb{F}^{n\times n}$, assume that A and B are positive semidefinite, and let $\|\cdot\|$ be a unitarily invariant norm on $\mathbb{F}^{n\times n}$. Then,

$$\|\log(I + A) - \log(I + B)\| \le \|\log(I + \langle A - B\rangle)\|,$$

$$\|\log(I + A + B)\| \le \|\log(I + A) + \log(I + B)\|.$$

Source: [106] and [449, p. 293]. **Related:** Fact 15.17.18.

Fact 11.10.87. Let $A, X, B \in \mathbb{F}^{n\times n}$, assume that A and B are positive definite, and let $\|\cdot\|$ be a unitarily invariant norm on $\mathbb{F}^{n\times n}$. Then,

$$\|(\log A)X - X(\log B)\| \le \|A^{1/2}XB^{-1/2} - A^{-1/2}XB^{1/2}\|.$$

Source: [477]. **Related:** Fact 15.17.19.

Fact 11.10.88. Let $A, B \in \mathbb{F}^{n\times n}$, assume that A is nonsingular, let $\|\cdot\|$ be a normalized submultiplicative norm on $\mathbb{F}^{n\times n}$, and assume that $\|A - B\| < 1/\|A^{-1}\|$. Then, B is nonsingular. **Related:** Fact 11.9.62.

Fact 11.10.89. Let $A, B \in \mathbb{F}^{n\times n}$, assume that A is nonsingular, let $\|\cdot\|$ be a normalized submultiplicative norm on $\mathbb{F}^{n\times n}$, let $\gamma > 0$, and assume that $\|A^{-1}\| < \gamma$ and $\|A - B\| < 1/\gamma$. Then, B is nonsingular and

$$\|B^{-1}\| \le \frac{\gamma}{1 - \gamma\|B - A\|}, \quad \|A^{-1} - B^{-1}\| \le \gamma^2\|A - B\|.$$

Source: [970, p. 148]. **Related:** Fact 11.9.62.

Fact 11.10.90. Let $A, B \in \mathbb{F}^{n\times n}$, let $\lambda \in \mathbb{C}$, assume that $\lambda I - A$ is nonsingular, let $\|\cdot\|$ be a normalized submultiplicative norm on $\mathbb{F}^{n\times n}$, let $\gamma > 0$, and assume that $\|(\lambda I - A)^{-1}\| < \gamma$ and $\|A - B\| < 1/\gamma$. Then, $\lambda I - B$ is nonsingular and

$$\|(\lambda I - B)^{-1}\| \le \frac{\gamma}{1 - \gamma\|B - A\|}, \quad \|(\lambda I - A)^{-1} - (\lambda I - B)^{-1}\| \le \frac{\gamma^2\|A - B\|}{1 - \gamma\|A - B\|}.$$

Source: [970, pp. 149, 150]. **Related:** Fact 11.10.89.

Fact 11.10.91. Let $A, B \in \mathbb{F}^{n\times n}$, assume that A and $A + B$ are nonsingular, and let $\|\cdot\|$ be a normalized submultiplicative norm on $\mathbb{F}^{n\times n}$. Then,

$$\|A^{-1} - (A + B)^{-1}\| \le \|A^{-1}\|\,\|(A + B)^{-1}\|\,\|B\|.$$

If, in addition, $\|A^{-1}B\| < 1$, then

$$\|A^{-1} + (A + B)^{-1}\| \le \frac{\|A^{-1}\|\|A^{-1}B\|}{1 - \|A^{-1}B\|}.$$

Furthermore, if $\|B\| < 1/\|A^{-1}\|$, then

$$\|A^{-1} - (A + B)^{-1}\| \le \frac{\|A^{-1}\|^2\|B\|}{1 - \|A^{-1}\|\|B\|}.$$

Fact 11.10.92. Let $A \in \mathbb{F}^{n\times n}$, assume that A is nonsingular, let $E \in \mathbb{F}^{n\times n}$, and let $\|\cdot\|$ be a normalized norm on $\mathbb{F}^{n\times n}$. Then, as $\|E\| \to 0$,

$$(A+E)^{-1} = A^{-1}(I+EA^{-1})^{-1} = A^{-1} - A^{-1}EA^{-1} + O(\|E\|^2).$$

Fact 11.10.93. Let $A \in \mathbb{F}^{n\times m}$ and $B \in \mathbb{F}^{l\times k}$. Then,

$$\|A\otimes B\|_{\rm col} = \|A\|_{\rm col}\|B\|_{\rm col}, \quad \|A\otimes B\|_{\infty} = \|A\|_{\infty}\|B\|_{\infty}, \quad \|A\otimes B\|_{\rm row} = \|A\|_{\rm row}\|B\|_{\rm row}.$$

Furthermore, if $p \in [1,\infty]$, then

$$\|A\otimes B\|_p = \|A\|_p\|B\|_p.$$

Fact 11.10.94. Let $A \in \mathbb{F}^{n\times m}$ and $B \in \mathbb{F}^{l\times k}$. If either $p,q \in [1,\infty]$ and $p \le q$ or A and B are nonnegative, then

$$\|A\otimes B\|_{q,p} \le \|A\|_{q,p}\|B\|_{q,p}.$$

If $p \in [1,2]$, then

$$\|A\otimes B\|_{p,2,\mathbb{C}} \le \Gamma\left(\tfrac{p+2}{2}\right)^{-1/p}\|A\|_{p,2,\mathbb{C}}\|B\|_{p,2,\mathbb{C}}.$$

Source: [2040].

Fact 11.10.95. Let $A \in \mathbb{F}^{n\times m}$, $B \in \mathbb{F}^{l\times k}$, and $p \in [1,\infty]$. Then,

$$\|A\otimes B\|_{\sigma p} = \|A\|_{\sigma p}\|B\|_{\sigma p}.$$

In particular,

$$\sigma_{\max}(A\otimes B) = \sigma_{\max}(A)\sigma_{\max}(B), \quad \|A\otimes B\|_{\rm F} = \|A\|_{\rm F}\|B\|_{\rm F}.$$

Source: [1399, p. 722].

Fact 11.10.96. Let $A, B \in \mathbb{F}^{n\times n}$, and let $\|\cdot\|$ be a unitarily invariant norm on $\mathbb{F}^{n\times n}$. Then,

$$\|A\odot B\|^2 \le \|A^*A\|\|B^*B\|.$$

If, in addition, $k \ge 1$, then

$$\|(A\odot B)^k\|_{\rm F}^2 \le \|(A^*A)^k\|_{\rm F}\|(B^*B)^k\|_{\rm F}.$$

Source: [1452, 2762].

11.11 Facts on Matrix Norms for Commutators

Fact 11.11.1. Let $A, B \in \mathbb{F}^{n\times n}$, and let $\|\cdot\|$ be a submultiplicative norm on $\mathbb{F}^{n\times n}$. Then, $\|\cdot\|' \triangleq 2\|\cdot\|$ is a submultiplicative norm on $\mathbb{F}^{n\times n}$ and satisfies $\|[A,B]\|' \le \|A\|'\|B\|'$.

Fact 11.11.2. Let $A \in \mathbb{F}^{n\times n}$. Then, the following statements are equivalent:

i) There exist projectors $Q, P \in \mathbb{R}^{n\times n}$ such that $A = [P,Q]$.

ii) $\sigma_{\max}(A) \le 1/2$, A and $-A$ are unitarily similar, and A is skew Hermitian.

Source: [1813]. **Remark:** Extensions are discussed in [1995]. **Remark:** For $\mathbb{F} = \mathbb{R}$, if A is skew symmetric, then A and $-A$ are orthogonally similar. See Fact 7.10.11. **Related:** Fact 4.22.10 considers idempotent matrices.

Fact 11.11.3. Let $A, B \in \mathbb{R}^{n\times n}$. Then, $\|AB - BA\|_{\rm F} \le \sqrt{2}\|A\|_{\rm F}\|B\|_{\rm F}$. **Source:** [532, 2807]. **Remark:** $\sqrt{2}$ is valid for all n. **Remark:** Extensions to complex matrices are given in [533].

Fact 11.11.4. Let $A, B \in \mathbb{F}^{n\times n}$, and assume that A and B are positive semidefinite. Then,

$$\|AB - BA\|_{\rm F}^2 + \|(A-B)^2\|_{\rm F}^2 \le \|A^2 - B^2\|_{\rm F}^2.$$

Source: [1631].

Fact 11.11.5. Let $A, B \in \mathbb{F}^{n\times n}$, let p be a positive number, and assume that either A is normal and $p \in [2,\infty]$, or A is Hermitian and $p \ge 1$. Then,

$$\|\langle A\rangle B - B\langle A\rangle\|_{\sigma p} \le \|AB - BA\|_{\sigma p}.$$

Source: [1].

Fact 11.11.6. Let $\|\cdot\|$ be a unitarily invariant norm on $\mathbb{F}^{n\times n}$, and let $A, X, B \in \mathbb{F}^{n\times n}$. Then,

$$\|AX - XB\| \le [\sigma_{\max}(A) + \sigma_{\max}(B)]\|X\|.$$

In particular,

$$\sigma_{\max}(AX - XA) \le 2\sigma_{\max}(A)\sigma_{\max}(X).$$

Now, assume that A and B are positive semidefinite. Then,

$$\|AX - XB\| \le \max\{\sigma_{\max}(A), \sigma_{\max}(B)\}\|X\|.$$

In particular,

$$\sigma_{\max}(AX - XA) \le \sigma_{\max}(A)\sigma_{\max}(X).$$

Finally, assume that A and X are positive semidefinite. Then,

$$\|AX - XA\| \le \tfrac{1}{2}\sigma_{\max}(A)\left\|\begin{bmatrix} X & 0 \\ 0 & X \end{bmatrix}\right\|.$$

In particular,

$$\sigma_{\max}(AX - XA) \le \tfrac{1}{2}\sigma_{\max}(A)\sigma_{\max}(X).$$

Source: [473]. **Remark:** Equality holds in the first inequality with $A = B = \left[\begin{smallmatrix} 1 & 0 \\ 0 & -1 \end{smallmatrix}\right]$ and $X = \left[\begin{smallmatrix} 0 & 1 \\ -1 & 0 \end{smallmatrix}\right]$. **Remark:** $\|\cdot\|$ can be extended to $\mathbb{F}^{2n\times 2n}$ by considering the n largest singular values of matrices in $\mathbb{F}^{2n\times 2n}$. See [449, pp. 90, 98].

Fact 11.11.7. Let $\|\cdot\|$ be a unitarily invariant norm on $\mathbb{F}^{n\times n}$, let $A, X \in \mathbb{F}^{n\times n}$, and assume that A is Hermitian. Then,

$$\|AX - XA\| \le \delta(A)\|X\|.$$

Source: [473]. **Remark:** $\delta(A) = \lambda_{\max}(A) - \lambda_{\min}(A)$ is the spread of A.

Fact 11.11.8. Let $\|\cdot\|$ be a unitarily invariant norm on $\mathbb{F}^{n\times n}$, let $A, X \in \mathbb{F}^{n\times n}$, and assume that A is normal. Then,

$$\|AX - XA\| \le \sqrt{2}\delta(A)\|X\|.$$

Furthermore, let $p \in [1, \infty]$. Then,

$$\|AX - XA\|_{\sigma p} \le 2^{|2-p|/(2p)}\delta(A)\|X\|_{\sigma p}.$$

In particular,

$$\|AX - XA\|_{\mathrm{F}} \le \delta(A)\|X\|_{\mathrm{F}}, \quad \sigma_{\max}(AX - XA) \le \sqrt{2}\delta(A)\sigma_{\max}(X).$$

Source: [473]. **Remark:** $\delta(A)$ is the spread of A.

11.12 Facts on Matrix Norms for Partitioned Matrices

Fact 11.12.1. Let $A, B, C \in \mathbb{F}^{n\times n}$, assume that $\left[\begin{smallmatrix} A & B \\ B^* & C \end{smallmatrix}\right]$ and $\left[\begin{smallmatrix} A & B^* \\ B & C \end{smallmatrix}\right]$ are positive semidefinite, and let $\|\cdot\|$ be a unitarily invariant norm on $\mathbb{F}^{n\times n}$. Then, $2\|B\| \le \|A + C\|$. **Source:** [1840].

Fact 11.12.2. Let $A, B \in \mathbb{F}^{n\times m}$, and let $\|\cdot\|$ be a unitarily invariant norm on $\mathbb{F}^{2n\times 2n}$. Then,

$$\left\|\begin{bmatrix} AA^* + BB^* & AB^* + BA^* \\ AB^* + BA^* & AA^* + BB^* \end{bmatrix}\right\| \le 2\left\|\begin{bmatrix} AA^* + BB^* & 0 \\ 0 & 0 \end{bmatrix}\right\|.$$

Now, assume that $n = m$. Then,

$$\frac{1}{2}\left\|\begin{bmatrix} A + B & 0 \\ 0 & A + B \end{bmatrix}\right\| \le \left\|\begin{bmatrix} A & 0 \\ 0 & B \end{bmatrix}\right\| \le \left\|\begin{bmatrix} \langle A\rangle & 0 \\ 0 & \langle B\rangle \end{bmatrix}\right\| \le \left\|\begin{bmatrix} \langle A\rangle + \langle B\rangle & 0 \\ 0 & 0 \end{bmatrix}\right\|,$$

$$\left\|\begin{bmatrix} AB & 0 \\ 0 & AB \end{bmatrix}\right\| \le \left\|\begin{bmatrix} A^*A & 0 \\ 0 & B^*B \end{bmatrix}\right\|, \quad \left\|\begin{bmatrix} A\odot B & 0 \\ 0 & A\odot B \end{bmatrix}\right\| \le \left\|\begin{bmatrix} A^*A & 0 \\ 0 & B^*B \end{bmatrix}\right\|.$$

If, in addition, A and B are Hermitian, then

$$\left\|\begin{bmatrix} A & 0 \\ 0 & A \end{bmatrix}\right\| \le \left\|\begin{bmatrix} A+B & 0 \\ 0 & A-B \end{bmatrix}\right\|.$$

Source: [449, p. 97], [2750], and [2991, p. 377].

Fact 11.12.3. Let $A, B, C \in \mathbb{F}^{n\times n}$, and assume that $\left[\begin{smallmatrix} A & B \\ B^* & C \end{smallmatrix}\right]$ is positive semidefinite. Then, the following statements hold:

i) Let $p > 0$, and let $\|\cdot\|$ be a unitarily invariant norm on $\mathbb{F}^{2n\times 2n}$. Then,

$$\left\|\begin{bmatrix} A & B \\ B^* & C \end{bmatrix}^p\right\| \le 2^{|p-1|}(\|(A+C)^p\| + \|\langle B - B^*\rangle^p\|).$$

ii) Assume that B is Hermitian, and let $\|\cdot\|$ be a unitarily invariant norm on $\mathbb{F}^{2n\times 2n}$. Then,

$$\left\|\begin{bmatrix} A & B \\ B^* & C \end{bmatrix}\right\| \le \left\|\begin{bmatrix} A+C & 0 \\ 0 & 0 \end{bmatrix}\right\|.$$

iii) Let $p \ge 1$, and let $\|\cdot\|$ be a unitarily invariant norm on $\mathbb{F}^{2n\times 2n}$. Then,

$$\left\|\begin{bmatrix} A & B \\ B^* & C \end{bmatrix}\right\|_{\sigma p} \le 2^{1-1/p}(\|A\|_{\sigma p}^p + \|C\|_{\sigma p}^p)^{1/p}.$$

iv) Let $\operatorname{nrad}(B)$ denote the numerical radius of B. Then,

$$\begin{aligned}\sigma_{\max}\left(\begin{bmatrix} A & B \\ B^* & C \end{bmatrix}\right) &\le \sigma_{\max}(A+C) + 2\operatorname{nrad}(B) \\ &\le \sigma_{\max}(A+C) + 2\sigma_{\max}(B) \\ &\le \sigma_{\max}(A) + \sigma_{\max}(C) + 2\sqrt{\sigma_{\max}(A)\sigma_{\max}(C)} \\ &\le 4\max\{\sigma_{\max}(A), \sigma_{\max}(C)\}.\end{aligned}$$

v)

$$\sigma_{\max}\left(\begin{bmatrix} A & B \\ B^* & C \end{bmatrix}\right) \le 2\max\{\sigma_{\max}(A), \sigma_{\max}(C)\}.$$

Source: [548]. **Related:** Fact 10.12.58 and Fact 10.12.59.

Fact 11.12.4. Let $A \in \mathbb{F}^{n\times m}$ be the partitioned matrix

$$A = \begin{bmatrix} A_{11} & A_{12} & \cdots & A_{1k} \\ A_{21} & A_{22} & \cdots & A_{2k} \\ \vdots & \vdots & \ddots & \vdots \\ A_{k1} & A_{k2} & \cdots & A_{kk} \end{bmatrix},$$

where $A_{ij} \in \mathbb{F}^{n_i\times n_j}$ for all $i, j \in \{1, \ldots, k\}$. Furthermore, define $\mu(A) \in \mathbb{R}^{k\times k}$ by

$$\mu(A) \triangleq \begin{bmatrix} \sigma_{\max}(A_{11}) & \sigma_{\max}(A_{12}) & \cdots & \sigma_{\max}(A_{1k}) \\ \sigma_{\max}(A_{21}) & \sigma_{\max}(A_{22}) & \cdots & \sigma_{\max}(A_{2k}) \\ \vdots & \vdots & \ddots & \vdots \\ \sigma_{\max}(A_{k1}) & \sigma_{\max}(A_{k2}) & \cdots & \sigma_{\max}(A_{kk}) \end{bmatrix}.$$

Finally, let $B \in \mathbb{F}^{n\times m}$ be partitioned conformally with A. Then, the following statements hold:

i) For all $\alpha \in \mathbb{F}$, $\mu(\alpha A) \le |\alpha|\mu(A)$.

ii) $\mu(A + B) \le \mu(A) + \mu(B)$.

iii) $\mu(AB) \le \mu(A)\mu(B)$.

iv) $\rho_{\max}(A) \le \rho_{\max}[\mu(A)]$.

v) $\sigma_{\max}(A) \le \sigma_{\max}[\mu(A)]$.

vi) If $k = 2$, then $\operatorname{nrad}(A) \le \operatorname{nrad}[\mu(A)]$.

Source: [881, 1633, 2177, 2473]. **Remark:** μ is a *matricial norm.* **Remark:** This is a norm-compression inequality.

Fact 11.12.5. Let $A \in \mathbb{F}^{n\times m}$ be the partitioned matrix

$$A = \begin{bmatrix} A_{11} & A_{12} & \cdots & A_{1k} \\ A_{21} & A_{22} & \cdots & A_{2k} \\ \vdots & \vdots & \ddots & \vdots \\ A_{k1} & A_{k2} & \cdots & A_{kk} \end{bmatrix},$$

where $A_{ij} \in \mathbb{F}^{n_i\times n_j}$ for all $i, j \in \{1, \ldots, k\}$. Then, the following statements hold:

i) If $p \in [1, 2]$, then

$$\sum_{i,j=1}^{k} \|A_{ij}\|_{\sigma p}^2 \le \|A\|_{\sigma p}^2 \le k^{4/p-2} \sum_{i,j=1}^{k} \|A_{ij}\|_{\sigma p}^2.$$

ii) If $p \in [2, \infty]$, then

$$k^{4/p-2} \sum_{i,j=1}^{k} \|A_{ij}\|_{\sigma p}^2 \le \|A\|_{\sigma p}^2 \le \sum_{i,j=1}^{k} \|A_{ij}\|_{\sigma p}^2.$$

iii) If $p \in [1, 2]$, then

$$\|A\|_{\sigma p}^p \le \sum_{i,j=1}^{k} \|A_{ij}\|_{\sigma p}^p \le k^{2-p} \|A\|_{\sigma p}^p.$$

iv) If $p \in [2, \infty)$, then

$$k^{2-p} \|A\|_{\sigma p}^p \le \sum_{i,j=1}^{k} \|A_{ij}\|_{\sigma p}^p \le \|A\|_{\sigma p}^p.$$

v) $\|A\|_{\sigma 2}^2 = \sum_{i,j=1}^{k} \|A_{ij}\|_{\sigma 2}^2$.

vi) For all $p \in [1, \infty]$,

$$\left(\sum_{i=1}^{k} \|A_{ii}\|_{\sigma p}^p \right)^{1/p} \le \|A\|_{\sigma p}.$$

vii) For all $i \in \{1, \ldots, k\}$, $\sigma_{\max}(A_{ii}) \le \sigma_{\max}(A)$.

Source: [277, 467].

Fact 11.12.6. For all $i, j \in \{1, \ldots, k\}$, let $A_{ij} \in \mathbb{F}^{n\times n}$ be Hermitian, define the partitioned matrix $A \in \mathbb{F}^{kn\times kn}$ by

$$A \triangleq \begin{bmatrix} A_{11} & A_{12} & \cdots & A_{1k} \\ A_{21} & A_{22} & \cdots & A_{2k} \\ \vdots & \vdots & \ddots & \vdots \\ A_{k1} & A_{k2} & \cdots & A_{kk} \end{bmatrix},$$

assume that A is positive semidefinite, and let $p \in [1, \infty]$. Then,

$$\|A\|_{\sigma p} \le \left\| \sum_{i=1}^{k} A_{ii} \right\|_{\sigma p}.$$

Source: [549]. **Remark:** This is *Hiroshima's theorem.* **Related:** Fact 10.12.58.

Fact 11.12.7. For all $i, j \in \{1, 2, 3\}$, let $A_{ij} \in \mathbb{F}^{n \times n}$, define $A \in \mathbb{F}^{3n \times 3n}$ by

$$A \triangleq \begin{bmatrix} A_{11} & A_{12} & A_{13} \\ A_{12}^* & A_{22} & A_{23} \\ A_{13}^* & A_{23}^* & A_{33} \end{bmatrix},$$

assume that A is positive semidefinite, and define $B \in \mathbb{F}^{3 \times 3}$ and $C \in \mathbb{R}^{3 \times 3}$ by

$$B \triangleq \begin{bmatrix} \operatorname{tr} A_{11} & \operatorname{tr} A_{12} & \operatorname{tr} A_{13} \\ \operatorname{tr} A_{12}^* & \operatorname{tr} A_{22} & \operatorname{tr} A_{23} \\ \operatorname{tr} A_{13}^* & \operatorname{tr} A_{23}^* & \operatorname{tr} A_{33} \end{bmatrix}, \quad C \triangleq \begin{bmatrix} \|A_{11}\|_{\sigma 1} & \|A_{12}\|_{\sigma 1} & \|A_{13}\|_{\sigma 1} \\ \|A_{12}^*\|_{\sigma 1} & \|A_{22}\|_{\sigma 1} & \|A_{23}\|_{\sigma 1} \\ \|A_{13}^*\|_{\sigma 1} & \|A_{23}^*\|_{\sigma 1} & \|A_{33}\|_{\sigma 1} \end{bmatrix}.$$

Then, the following statements hold:

i) B, $|B|$, and C are positive semidefinite.

ii) If $p \in [1, 2]$, then $\| \, |B| \, \|_{\sigma p} \le \|B\|_{\sigma p}$.

iii) If $p \in [2, \infty]$, then $\|B\|_{\sigma p} \le \| \, |B| \, \|_{\sigma p}$.

Source: [955, 1808, 1849]. **Remark:** Fact 10.10.2 implies that $|B|$ is positive semidefinite. **Related:** Fact 10.12.64.

Fact 11.12.8. Let $A, B \in \mathbb{F}^{n \times n}$, and define $\mathcal{A} \in \mathbb{F}^{kn \times kn}$ by

$$\mathcal{A} \triangleq \begin{bmatrix} A & B & B & \cdots & B \\ B & A & B & \cdots & B \\ B & B & A & \ddots & B \\ \vdots & \vdots & \ddots & \ddots & \vdots \\ B & B & B & \cdots & A \end{bmatrix}.$$

Then,

$$\sigma_{\max}(\mathcal{A}) = \max\{\sigma_{\max}(A + (k-1)B), \sigma_{\max}(A - B)\}.$$

Now, let $p \in [1, \infty)$. Then,

$$\|\mathcal{A}\|_{\sigma p} = (\|A + (k-1)B\|_{\sigma p}^p + (k-1)\|A - B\|_{\sigma p}^p)^{1/p}.$$

Source: [277].

Fact 11.12.9. Let $A \in \mathbb{F}^{n \times n}$, and define $\mathcal{A} \in \mathbb{F}^{kn \times kn}$ by

$$\mathcal{A} \triangleq \begin{bmatrix} A & A & A & \cdots & A \\ -A & A & A & \cdots & A \\ -A & -A & A & \ddots & A \\ \vdots & \vdots & \ddots & \ddots & \vdots \\ -A & -A & -A & \cdots & A \end{bmatrix}.$$

Then,

$$\sigma_{\max}(\mathcal{A}) = \sqrt{\frac{2}{1 - \cos(\pi/k)}}\,\sigma_{\max}(A).$$

Furthermore, define $\mathcal{A}_0 \in \mathbb{F}^{kn\times kn}$ by

$$\mathcal{A}_0 \triangleq \begin{bmatrix} 0 & A & A & \cdots & A \\ -A & 0 & A & \cdots & A \\ -A & -A & 0 & \ddots & A \\ \vdots & \vdots & \ddots & \ddots & \vdots \\ -A & -A & -A & \cdots & 0 \end{bmatrix}.$$

Then,

$$\sigma_{\max}(\mathcal{A}_0) = \sqrt{\frac{1+\cos(\pi/k)}{1-\cos(\pi/k)}}\sigma_{\max}(A).$$

Source: [277]. **Remark:** Extensions to Schatten norms are given in [277].

Fact 11.12.10. Let $A, B, C, D \in \mathbb{F}^{n\times n}$. Then,

$$\tfrac{1}{2}\max\{\sigma_{\max}(A+B+C+D), \sigma_{\max}(A-B-C+D)\} \le \sigma_{\max}\left(\begin{bmatrix} A & B \\ C & D \end{bmatrix}\right).$$

Now, let $p \in [1,\infty)$. Then,

$$\tfrac{1}{2}(\|A+B+C+D\|_{\sigma p}^p + \|A-B-C+D\|_{\sigma p}^p)^{1/p} \le \left\|\begin{bmatrix} A & B \\ C & D \end{bmatrix}\right\|_{\sigma p}.$$

Source: [277].

Fact 11.12.11. Let $\mathcal{A} \triangleq \left[\begin{smallmatrix} A & B \\ C & D \end{smallmatrix}\right] \in \mathbb{F}^{(n+m)\times(n+m)}$, and assume that $\mathcal{A}$ is normal. Then, $\|B\|_{\mathrm{F}} = \|C\|_{\mathrm{F}}$. **Source:** [2991, p. 323].

Fact 11.12.12. Let $A, B, C \in \mathbb{F}^{n\times n}$, define $\mathcal{A} \triangleq \left[\begin{smallmatrix} A & B \\ B^* & C \end{smallmatrix}\right]$, assume that $\mathcal{A}$ is positive semidefinite, let $p \in [1,\infty]$, and define

$$\mathcal{A}_0 \triangleq \begin{bmatrix} \|A\|_{\sigma p} & \|B\|_{\sigma p} \\ \|B\|_{\sigma p} & \|C\|_{\sigma p} \end{bmatrix}.$$

If $p \in [1,2]$, then

$$\|\mathcal{A}_0\|_{\sigma p} \le \|\mathcal{A}\|_{\sigma p}.$$

Furthermore, if $p \in [2,\infty]$, then

$$\|\mathcal{A}\|_{\sigma p} \le \|\mathcal{A}_0\|_{\sigma p}.$$

Hence, if $p = 2$, then

$$\|\mathcal{A}_0\|_{\sigma p} = \|\mathcal{A}\|_{\sigma p}.$$

Finally, if $A = C$, B is Hermitian, and p is an integer, then

$$\|\mathcal{A}\|_{\sigma p}^p = \|A+B\|_{\sigma p}^p + \|A-B\|_{\sigma p}^p,$$

$$\|\mathcal{A}_0\|_{\sigma p}^p = (\|A\|_{\sigma p} + \|B\|_{\sigma p})^p + |\|A\|_{\sigma p} - \|B\|_{\sigma p}|^p.$$

Source: [1616]. **Remark:** This is a norm-compression inequality. **Related:** Fact 11.16.15.

Fact 11.12.13. Let $A \in \mathbb{F}^{n\times n}$, $B \in \mathbb{F}^{n\times m}$, and $C \in \mathbb{F}^{m\times m}$, define $\mathcal{A} \triangleq \left[\begin{smallmatrix} A & B \\ B^* & C \end{smallmatrix}\right]$, assume that $\mathcal{A}$ is positive semidefinite, and let $p \ge 1$. Then, the following statements hold:

i) If $p \in [1,2]$, then

$$\|\mathcal{A}\|_{\sigma p}^p \le \|A\|_{\sigma p}^p + (2^p - 2)\|B\|_{\sigma p}^p + \|C\|_{\sigma p}^p.$$

ii) If $p \in [0,1] \cup [2,\infty)$, then

$$\|A\|_{\sigma p}^p + (2^p - 2)\|B\|_{\sigma p}^p + \|C\|_{\sigma p}^p \le \|\mathcal{A}\|_{\sigma p}^p.$$

iii) If $p = 2$, then

$$\|\mathcal{A}\|_{\sigma p}^{p} = \|A\|_{\sigma p}^{p} + (2^{p} - 2)\|B\|_{\sigma p}^{p} + \|C\|_{\sigma p}^{p}.$$

iv) Assume that C is positive definite. If $p \in (-\infty, -2] \cup [1, 2]$, then

$$\operatorname{tr}\begin{bmatrix} B^*C^{-1}B & B^* \\ B & C \end{bmatrix} \le \operatorname{tr}\begin{bmatrix} B^*C^{-1}B & 0 \\ 0 & C \end{bmatrix} + (2^p - 2)\|B\|_{\sigma p}^{p}.$$

v) Assume that C is positive definite. If $p \in [0, 1] \cup [2, 3]$, then

$$\operatorname{tr}\begin{bmatrix} B^*C^{-1}B & 0 \\ 0 & C \end{bmatrix} + (2^p - 2)\|B\|_{\sigma p}^{p} \le \operatorname{tr}\begin{bmatrix} B^*C^{-1}B & B^* \\ B & C \end{bmatrix}.$$

Source: [189, 194]. *ii*) for $p > 3$ is a conjecture.

Fact 11.12.14. Let $A \in \mathbb{F}^{n\times m}$ be the partitioned matrix

$$A = \begin{bmatrix} A_{11} & \cdots & A_{1k} \\ A_{21} & \cdots & A_{2k} \end{bmatrix},$$

where $A_{ij} \in \mathbb{F}^{n_i\times n_j}$ for all $i, j \in \{1, \ldots, k\}$. Then, the following statements are conjectured to hold:

i) If $p \in [1, 2]$, then

$$\left\| \begin{bmatrix} \|A_{11}\|_{\sigma p} & \cdots & \|A_{1k}\|_{\sigma p} \\ \|A_{21}\|_{\sigma p} & \cdots & \|A_{2k}\|_{\sigma p} \end{bmatrix} \right\|_{\sigma p} \le \|A\|_{\sigma p}.$$

ii) If $p \ge 2$, then

$$\|A\|_{\sigma p} \le \left\| \begin{bmatrix} \|A_{11}\|_{\sigma p} & \cdots & \|A_{1k}\|_{\sigma p} \\ \|A_{21}\|_{\sigma p} & \cdots & \|A_{2k}\|_{\sigma p} \end{bmatrix} \right\|_{\sigma p}.$$

Source: [191]. This result is true if either $p \ge 4$ or all blocks have rank 1. **Remark:** This is a norm-compression inequality.

11.13 Facts on Matrix Norms and Eigenvalues for One Matrix

Fact 11.13.1. Let $A \in \mathbb{F}^{n\times n}$. Then,

$$|\det A| \le \prod_{i=1}^{n} \|\mathrm{row}_i(A)\|_2, \quad |\det A| \le \prod_{i=1}^{n} \|\mathrm{col}_i(A)\|_2.$$

Source: Use Hadamard's inequality. See Fact 10.21.15.

Fact 11.13.2. Let $A \in \mathbb{F}^{n\times n}$. Then,

$$\begin{aligned} |\mathrm{Re}\,\operatorname{tr} A^2| \le |\operatorname{tr} A^2| \le \|\rho(A)\|_2^2 \le \|A^2\|_{\sigma 1} &= \operatorname{tr}\langle A^2\rangle = \|\sigma(A^2)\|_1 \\ &\le \|\sigma(A)\|_2^2 = \operatorname{tr} A^*A = \operatorname{tr}\langle A\rangle^2 = \|A\|_{\sigma 2}^2 = \|A\|_{\mathrm{F}}^2, \end{aligned}$$

$$\|A\|_{\mathrm{F}}^2 - \sqrt{\tfrac{n^3-n}{12}}\|[A, A^*]\|_{\mathrm{F}} \le \|\rho(A)\|_2^2 \le \sqrt{\|A\|_{\mathrm{F}}^4 - \tfrac{1}{2}\|[A, A^*]\|_{\mathrm{F}}^2} \le \|A\|_{\mathrm{F}}^2.$$

Consequently, A is normal if and only if $\|A\|_{\mathrm{F}}^2 = \|\rho(A)\|_2^2$. Furthermore,

$$\|\rho(A)\|_2^2 \le \sqrt{\|A\|_{\mathrm{F}}^4 - \tfrac{1}{4}(\operatorname{tr}|[A, A^*]|)^2} \le \|A\|_{\mathrm{F}}^2,$$

$$\|\rho(A)\|_2^2 \le \sqrt{\|A\|_{\mathrm{F}}^4 - \tfrac{n^2}{4}|\det[A, A^*]|^{2/n}} \le \|A\|_{\mathrm{F}}^2.$$

Finally, A is Hermitian if and only if $\operatorname{tr} A^2 = \|A\|_{\mathrm{F}}^2$. **Source:** Fact 10.13.2 and Fact 10.21.10. The lower bound involving the commutator is due to P. Henrici; the corresponding upper bound is given in [1704]. The bounds in the penultimate statement are given in [1704]. The last statement follows from Fact 4.10.13. **Remark:** $\operatorname{tr}(A + A^*)^2 \ge 0$ and $\operatorname{tr}(A - A^*)^2 \le 0$ yield $|\operatorname{tr} A^2| \le \|A\|_F^2$. **Remark:** $\sum_{i=1}^n \rho_i^2(A) \le \|A\|_F^2$ is *Schur's inequality*. See Fact 10.21.10. **Related:** Fact 7.12.13, Fact 10.13.5, Fact 11.13.4, and Fact 11.15.19.

Fact 11.13.3. Let $A \in \mathbb{F}^{n \times n}$. Then,

$$|\operatorname{tr} A^2| \le (\operatorname{rank} A)\sqrt{\|A\|_F^4 - \tfrac{1}{2}\|[A, A^*]\|_F^2}.$$

Source: [706].

Fact 11.13.4. Let $A \in \mathbb{F}^{n \times n}$, and define

$$\delta \triangleq \sqrt{(\|A\|_F^2 - \tfrac{1}{n}|\operatorname{tr} A|^2)^2 - \tfrac{1}{2}\|[A, A^*]\|_F^2} + \tfrac{1}{n}|\operatorname{tr} A|^2.$$

Then,

$$\|\rho(A)\|_2^2 \le \delta \le \sqrt{\|A\|_F^4 - \tfrac{1}{2}\|[A, A^*]\|_F^2} \le \|A\|_F^2,$$

$$\|\alpha(A)\|_2^2 \le \tfrac{1}{2}(\delta + \operatorname{Re}\operatorname{tr} A^2), \quad \|\beta(A)\|_2^2 \le \tfrac{1}{2}(\delta - \operatorname{Re}\operatorname{tr} A^2).$$

Source: [1482]. **Remark:** The first string interpolates the upper bound for $\sum_{i=1}^n \rho_i^2(A)$ in the second string in Fact 11.13.2.

Fact 11.13.5. Let $A \in \mathbb{F}^{n \times n}$ and $p \in (0, 2]$. Then,

$$\|\rho(A)\|_p^p \le \|\sigma(A)\|_p^p = \|A\|_{\sigma p}^p \le \|A\|_p^p.$$

Source: The left-hand inequality, which holds for all $p > 0$, follows from Weyl's inequality in Fact 10.21.10. The right-hand inequality is given by Fact 11.9.10. **Remark:** This is the *generalized Schur inequality*. **Remark:** Equality is discussed in [1499] for $p \in [1, 2)$.

Fact 11.13.6. Let $A \in \mathbb{F}^{n \times n}$. Then,

$$0 \le \|A\|_F^2 - \|\rho(A)\|_2^2 = \tfrac{1}{2}\|A + A^*\|_F^2 - 2\|\alpha(A)\|_2^2 = \tfrac{1}{2}\|A - A^*\|_F^2 - 2\|\beta(A)\|_2^2.$$

Furthermore, the following statements are equivalent:

i) $\|A\|_F^2 = \|\rho(A)\|_2^2$.

ii) $\|A + A^*\|_F^2 = 4\|\alpha(A)\|_2^2$.

iii) $\|A - A^*\|_F^2 = 4\|\beta(A)\|_2^2$.

iv) A is normal.

Source: Fact 7.12.29, [1172, pp. 11–17], and [2295, pp. 57, 58]. **Remark:** This is an extension of Browne's theorem. See Fact 7.12.24. **Related:** Fact 7.12.30.

Fact 11.13.7. Let $A \in \mathbb{F}^{n \times n}$. Then, the following statements hold:

i) $\rho_{\max}(A) \le n\|A\|_\infty$.

ii) $\max\{|\alpha_{\min}(A)|, |\alpha_{\max}(A)|\} \le \frac{n}{2}\|A + A^*\|_\infty$.

iii) $\max\{|\beta_{\min}(A)|, |\beta_{\max}(A)|\} \le \frac{\sqrt{n^2-n}}{2\sqrt{2}}\|A - A^*\|_\infty \le n\|A - A^*\|_\infty$.

iv) $|\det A| \le (\sqrt{n}\|A\|_\infty)^n$.

Now, assume that A is normal. Then, the following statements hold:

v) $\|A + \overline{A}\| \le 2\max\{|\alpha_{\min}(A)|, |\alpha_{\max}(A)|\}$.

vi) $\|A - \overline{A}\| \le 2\max\{|\beta_{\min}(A)|, |\beta_{\max}(A)|\}$.

Source: [1952, p. 140] and [2991, p. 323]. **Remark:** *i)* and *ii)* are *Hirsch's theorems*, while *iii)* is *Bendixson's theorem*. See Fact 7.12.24.

Fact 11.13.8. Let $A \in \mathbb{F}^{n \times n}$. Then, the following statements hold:

i) Let $p \in [1,2]$ and $q \in [2,\infty)$, assume that $p < q$, and define $r \triangleq pq/(q-p)$. Then, $\|\rho(A)\|_r \le 4\|A\|_{p,q}$.

ii) Let $p \in [1,2]$. Then, $\|\rho(A)\|_p \le 4\|A\|_{p,\infty}$.

iii) Let $p \in [1,2)$, and define $r \triangleq 2p/(2-p)$. Then, $\|\sigma(A)\|_r \le \sqrt{2}\|A\|_{p,2}$.

iv) Let $p \in [2,\infty)$ and $q \in (1,2]$, and assume that $1/p + 1/q = 1$. Then, $\|\lambda(A)\|_p \le \|A\|_{q|p}$.

Source: [1667, Chapter 2] and [2040].

Fact 11.13.9. Let $A \in \mathbb{R}^{n\times n}$, and assume that A is nonnegative. Then,

$$\min_{j\in\{1,\ldots,n\}} \|\mathrm{col}_j(A)\|_1 \le \rho_{\max}(A) \le \|A\|_{\mathrm{col}},$$

$$\min_{i\in\{1,\ldots,n\}} \|\mathrm{row}_i(A)\|_1 \le \rho_{\max}(A) \le \|A\|_{\mathrm{row}}.$$

Source: [2403, pp. 318, 319], [2979, p. 126], and [2991, pp. 166, 167]. **Remark:** The upper bounds are given by (11.4.26) and (11.4.27).

Fact 11.13.10. Let $A \in \mathbb{F}^{n\times n}$. Then, the following statements hold:

i) $\delta(A) \le \sqrt{2\|A\|_{\mathrm{F}}^2 - \frac{2}{n}|\mathrm{tr}\, A|^2} \le \sqrt{2}\|A\|_{\mathrm{F}}$.

ii) Let $\lambda \in \mathrm{spec}(A)$. Then, $\left|\lambda - \frac{\mathrm{tr}\, A}{n}\right| \le \sqrt{\frac{n-1}{n}\left(\|A\|_{\mathrm{F}}^2 - \frac{1}{n}|\mathrm{tr}\, A|^2\right)} \le \sqrt{\frac{n-1}{n}}\|A\|_{\mathrm{F}}$.

iii) $\delta(A) \le \max_{i,j\in\{1,\ldots,n\}}(|A_{(i,i)} - A_{(j,j)}| - A_{(i,i)} - A_{(j,j)} + \|\mathrm{row}_i(A)\|_1 + \|\mathrm{row}_j(A)\|_1)$.

iv) If A is normal, then $\sqrt{3}\|A - I \odot A\|_\infty \le \delta(A)$.

v) If A is Hermitian, then $2\|A - I \odot A\|_\infty \le \delta(A)$.

Source: [2916] and [2991, p. 324]. **Remark:** *i)* is due to L. Mirsky.

Fact 11.13.11. Let $A \in \mathbb{F}^{n\times n}$, let $\beta > \rho_{\max}(A)$, and let $\|\cdot\|$ be a submultiplicative norm on $\mathbb{F}^{n\times n}$. Then, there exists $\gamma > 1$ such that, for all $k \ge 0$, $\|A^k\| \le \gamma\beta^k$. **Remark:** If A is discrete-time asymptotically stable, then β can be chosen to satisfy $\rho_{\max}(A) < \beta < 1$ so that the bound converges to zero. **Related:** Fact 15.22.18.

Fact 11.13.12. Let $A \in \mathbb{F}^{n\times n}$, assume that A is Hermitian, let $\lambda \in \mathrm{spec}(A)$, let $x, z \in \mathbb{F}^n$, assume that x is an eigenvector of A associated with λ, assume that $\|x\|_2 = 1$, and assume that $Az \ne \lambda z$. Then,

$$|z^*x|^2 \le \frac{\|z\|_2^2\|Az\|_2^2 - (z^*Az)^2}{\|Az - \lambda z\|_2^2}.$$

Source: [438].

11.14 Facts on Matrix Norms and Eigenvalues for Two or More Matrices

Fact 11.14.1. Let $A, B \in \mathbb{F}^{n\times m}$, and let $p, q \in [1,\infty]$ satisfy $1/p + 1/q = 1$. Then,

$$|\mathrm{tr}\, A^*B| \le \|\rho(A^*B)\|_1 \le \|A^*B\|_{\sigma 1} = \|\sigma(A^*B\|_1 \le \sum_{i=1}^{\min\{m,n\}} \sigma_i(A)\sigma_i(B) \le \|A\|_{\sigma p}\|B\|_{\sigma q}.$$

In particular,

$$|\mathrm{tr}\, A^*B| \le \|A\|_{\mathrm{F}}\|B\|_{\mathrm{F}}.$$

Source: Fact 10.13.2 and Proposition 11.6.2. The last inequality in the first string is Hölder's inequality. **Related:** Fact 7.13.14, Fact 11.10.35, and Fact 11.16.2.

Fact 11.14.2. Let $A, B \in \mathbb{F}^{n\times m}$. Then,

$$|\mathrm{tr}\,(A^*B)^2| \le \|\rho(A^*B)\|_2^2 \le \|\sigma(A^*B)\|_2^2 = \mathrm{tr}\, AA^*BB^* = \|A^*B\|_{\mathrm{F}}^2 \le \|A\|_{\mathrm{F}}^2\|B\|_{\mathrm{F}}^2.$$

Source: Fact 10.21.10.

Fact 11.14.3. Let $A, B \in \mathbb{F}^{n\times n}$, and assume that A and B are Hermitian. Then,

$$\|\alpha(A + \jmath B)\|_2^2 \le \|A\|_{\rm F}^2, \quad \|\beta(A + \jmath B)\|_2^2 \le \|B\|_{\rm F}^2.$$

Source: [2263, p. 146].

Fact 11.14.4. Let $A, B \in \mathbb{F}^{n\times n}$, assume that A and B are Hermitian, and let $\|\cdot\|$ be a weakly unitarily invariant norm on $\mathbb{F}^{n\times n}$. Then,

$$\|\operatorname{diag}[\lambda(A) - \lambda(B)]\| \le \|A - B\| \le \|\operatorname{diag}[\lambda(A) - \lambda^{\uparrow}(B)]\|.$$

In particular,

$$\max_{i\in\{1,\ldots,n\}} |\lambda_i(A) - \lambda_i(B)| \le \sigma_{\max}(A - B) \le \max_{i\in\{1,\ldots,n\}} |\lambda_i(A) - \lambda_{n-i+1}(B)|,$$

$$\left(\sum_{i=1}^n [\lambda_i(A) - \lambda_i(B)]^2\right)^{1/2} \le \|A - B\|_{\rm F} \le \left(\sum_{i=1}^n [\lambda_i(A) - \lambda_{n-i+1}(B)]^2\right)^{1/2}.$$

Source: [93], [447, p. 38], [449, pp. 63, 69], [453, p. 38], [1590, p. 126], [1768, p. 134], [1799], and [2539, p. 202]. **Remark:** The first inequality is the *Lidskii-Mirsky-Wielandt theorem.* This result can be stated without norms using Fact 11.9.59. See [1799]. **Remark:** The case where A and B are normal is considered in Fact 11.14.6. **Related:** Fact 11.14.5 and Fact 11.16.40.

Fact 11.14.5. Let $A, B \in \mathbb{F}^{n\times n}$, assume that A and B are positive semidefinite, and let $\|\cdot\|$ be a weakly unitarily invariant norm on $\mathbb{F}^{n\times n}$. Then,

$$\|\operatorname{diag}[\lambda(A) + \lambda^{\uparrow}(B)]\| \le \|A + B\| \le \|\operatorname{diag}[\lambda(A) + \lambda(B)]\|.$$

In particular,

$$\max_{i\in\{1,\ldots,n\}} [\lambda_i(A) + \lambda_{n-i+1}(B)] \le \sigma_{\max}(A + B) \le \max_{i\in\{1,\ldots,n\}} [\lambda_i(A) + \lambda_i(B)],$$

$$\left(\sum_{i=1}^n [\lambda_i(A) + \lambda_{n-i+1}(B)]^2\right)^{1/2} \le \|A + B\|_{\rm F} \le \left(\sum_{i=1}^n [\lambda_i(A) + \lambda_i(B)]^2\right)^{1/2}.$$

Source: [1779]. **Remark:** This is the *Fan-Lidskii theorem.* **Related:** Fact 11.14.4.

Fact 11.14.6. Let $A, B \in \mathbb{F}^{n\times n}$, assume that A and B are normal, and let $\operatorname{spec}(A) = \{\lambda_1, \ldots, \lambda_q\}$ and $\operatorname{spec}(B) = \{\mu_1, \ldots, \mu_r\}$. Then,

$$\sigma_{\max}(A - B) \le \sqrt{2}\max\{|\lambda_i - \mu_j|\colon i \in \{1, \ldots, q\},\ j \in \{1, \ldots, r\}\}.$$

Source: [449, p. 164] and [453, p. 154]. **Related:** Fact 11.14.4.

Fact 11.14.7. Let $A, B \in \mathbb{F}^{n\times n}$, $\operatorname{mspec}(A) = \{\lambda_1, \ldots, \lambda_n\}_{\rm ms}$, and $\operatorname{mspec}(B) = \{\mu_1, \ldots, \mu_n\}_{\rm ms}$. Then, there exists a permutation σ of $(1, \ldots, n)$ such that

$$\max_{i\in\{1,\ldots,n\}} |\lambda_i - \mu_{\sigma(i)}| \le \frac{4}{\sqrt[n]{2}}[\sigma_{\max}(A) + \sigma_{\max}(B)]^{(n\ 1)/n}\sigma_{\max}^{1/n}(A - B).$$

Now, let $\|\cdot\|$ be an equi-induced norm on $\mathbb{F}^{n\times n}$. Then, there exists a permutation σ of $(1, \ldots, n)$ such that

$$\max_{i\in\{1,\ldots,n\}} |\lambda_i - \mu_{\sigma(i)}| \le 8\sqrt[n]{\frac{n}{4}}(\max\{\|A\|, \|B\|\})^{(n-1)/n}\|A - B\|^{1/n}.$$

Source: [461, 2230] and [2979, pp. 104, 105]. **Related:** Fact 12.13.3.

Fact 11.14.8. Let $A, B \in \mathbb{F}^{n\times n}$, let $\operatorname{mspec}(A) = \{\lambda_1, \ldots, \lambda_n\}_{\rm ms}$ and $\operatorname{mspec}(B) = \{\mu_1, \ldots, \mu_n\}_{\rm ms}$, and assume that at least one of the following statements holds:

i) A and B are Hermitian.

ii) A is Hermitian, and B is skew Hermitian.

iii) A is skew Hermitian, and B is Hermitian.

iv) A and B are unitary.

v) There exist nonzero $\alpha, \beta \in \mathbb{C}$ such that αA and βB are unitary.

vi) A, B, and $A - B$ are normal.

vii) $n = 2$ and A and B are normal.

Then, there exists a permutation σ of $(1, \ldots, n)$ such that

$$\max_{i \in \{1,\ldots,n\}} |\lambda_i - \mu_{\sigma(i)}| \le \sigma_{\max}(A - B).$$

Source: [453, pp. 52, 152].

Fact 11.14.9. Let $A, B \in \mathbb{F}^{n \times n}$, assume that A is normal, and let $\operatorname{mspec}(A) = \{\lambda_1, \ldots, \lambda_n\}_{\rm ms}$ and $\operatorname{mspec}(B) = \{\mu_1, \ldots, \mu_n\}_{\rm ms}$. Then, there exists a permutation σ of $(1, \ldots, n)$ such that

$$\left(\sum_{i=1}^{n} |\lambda_i - \mu_{\sigma(i)}|^2 \right)^{1/2} \le \sqrt{n} \|A - B\|_{\rm F}.$$

Source: [453, p. 173] and [2979, pp. 108, 109]. **Credit:** J. G. Sun.

Fact 11.14.10. Let $A, B \in \mathbb{F}^{n \times n}$, assume that A is Hermitian, and let $\operatorname{mspec}(B) = \{\mu_1, \ldots, \mu_n\}_{\rm ms}$. Then, there exists a permutation σ of $(1, \ldots, n)$ such that

$$\left(\sum_{i=1}^{n} |\lambda_i(A) - \mu_{\sigma(i)}|^2 \right)^{1/2} \le \sqrt{2} \|A - B\|_{\rm F}.$$

Source: [453, p. 174].

Fact 11.14.11. Let $A, B \in \mathbb{F}^{n \times n}$, assume that A is normal, and let $\operatorname{mspec}(A) = \{\lambda_1, \ldots, \lambda_n\}_{\rm ms}$ and $\operatorname{mspec}(B) = \{\mu_1, \ldots, \mu_n\}_{\rm ms}$. Then, there exists a permutation σ of $(1, \ldots, n)$ such that

$$\max_{i \in \{1,\ldots,n\}} |\lambda_i - \mu_{\sigma(i)}| \le n \|A - B\|_{\rm F}.$$

Source: [2979, p. 109].

Fact 11.14.12. Let $A, B \in \mathbb{F}^{n \times n}$, assume that A and B are normal, and let $\operatorname{mspec}(A) = \{\lambda_1, \ldots, \lambda_n\}_{\rm ms}$ and $\operatorname{mspec}(B) = \{\mu_1, \ldots, \mu_n\}_{\rm ms}$. Then, there exists a permutation σ of $(1, \ldots, n)$ such that

$$\max_{i \in \{1,\ldots,n\}} |\lambda_i - \mu_{\sigma(i)}| \le 3\sigma_{\max}(A - B).$$

Source: [453, pp. 153, 159]. **Remark:** Analogous inequalities for Schatten norms are given in [453, p. 159]. **Remark:** If $A - B$ is normal, then Fact 11.14.8 yields a stronger inequality.

Fact 11.14.13. Let $A, B \in \mathbb{F}^{n \times n}$, assume that A and B are normal, and let $\operatorname{mspec}(A) = \{\lambda_1, \ldots, \lambda_n\}_{\rm ms}$ and $\operatorname{mspec}(B) = \{\mu_1, \ldots, \mu_n\}_{\rm ms}$. Then, there exists a permutation σ of $(1, \ldots, n)$ such that

$$\left(\sum_{i=1}^{n} |\lambda_i - \mu_{\sigma(i)}|^2 \right)^{1/2} \le \|A - B\|_{\rm F}.$$

Source: [1448, p. 368], [1451, p. 407], [2263, pp. 160, 161], [2979, pp. 106, 107], and [2991, pp. 320, 321]. **Remark:** This is the *Hoffman-Wielandt theorem.* **Related:** Fact 11.14.4. The case where B is not necessarily normal is considered in Fact 11.14.9.

Fact 11.14.14. Let $A, B \in \mathbb{F}^{n \times n}$, assume that A and B are normal, and let $\operatorname{mspec}(A) = \{\lambda_1, \ldots, \lambda_n\}_{\rm ms}$ and $\operatorname{mspec}(B) = \{\mu_1, \ldots, \mu_n\}_{\rm ms}$. Then, there exists a permutation σ of $(1, \ldots, n)$ such that, for all $i \in \{1, \ldots, n\}$,

$$|\lambda_i - \mu_{\sigma(i)}| \le n\sigma_{\max}(A - B).$$

Source: [2991, p. 324].

Fact 11.14.15. Let $A, B \in \mathbb{F}^{n\times n}$, and assume that A is Hermitian and B is normal, and let $\operatorname{mspec}(B) = \{\mu_1, \ldots, \mu_n\}_{\rm ms}$, where $\operatorname{Re}\mu_n \le \cdots \le \operatorname{Re}\mu_1$. Then,

$$\left(\sum_{i=1}^{n} |\lambda_i(A) - \mu_i|^2\right)^{1/2} \le \|A - B\|_{\rm F}.$$

Source: [1448, p. 370] and [1451, pp. 407, 408]. **Related:** Fact 11.14.13.

Fact 11.14.16. Let $A, B \in \mathbb{F}^{n\times n}$, assume that A is Hermitian and B is skew Hermitian, let $\operatorname{mspec}(A) = \{\lambda_1, \ldots, \lambda_n\}_{\rm ms}$ and $\operatorname{mspec}(B) = \{\mu_1, \ldots, \mu_n\}_{\rm ms}$, assume that $|\lambda_1| \ge \cdots \ge |\lambda_n|$ and $|\mu_1| \ge \cdots \ge |\mu_n|$, and let $\|\cdot\|$ be a unitarily invariant norm. Then,

$$\|\operatorname{diag}(\lambda_1 - \mu_1, \ldots, \lambda_n - \mu_n)\| \le \|A - B\|.$$

Source: [2979, p. 109].

Fact 11.14.17. Let $A, B \in \mathbb{F}^{n\times n}$, assume that A and B are Hermitian, let $\operatorname{mspec}(A) = \{\lambda_1, \ldots, \lambda_n\}_{\rm ms}$ and $\operatorname{mspec}(B) = \{\mu_1, \ldots, \mu_n\}_{\rm ms}$, assume that $|\lambda_1| \ge \cdots \ge |\lambda_n|$ and $|\mu_1| \ge \cdots \ge |\mu_n|$, and let $\|\cdot\|$ be a unitarily invariant norm. Then,

$$\|\operatorname{diag}(\lambda_1 + \mu_1 \jmath, \ldots, \lambda_n + \mu_n \jmath)\| \le \sqrt{2}\|A + \jmath B\|.$$

Source: [2979, pp. 110, 111].

Fact 11.14.18. Let $A, B \in \mathbb{F}^{n\times n}$, and let $\|\cdot\|$ be an induced norm on $\mathbb{F}^{n\times n}$. Then,

$$|\det A - \det B| \le \begin{cases} \dfrac{\|A - B\|(\|A\|^n - \|B\|^n)}{\|A\| - \|B\|}, & \|A\| \ne \|B\|, \\ n\|A - B\|\|A\|^{n-1}, & \|A\| = \|B\|. \end{cases}$$

Source: [1078]. **Related:** Fact 2.21.8.

11.15 Facts on Matrix Norms and Singular Values for One Matrix

Fact 11.15.1. Let $A \in \mathbb{F}^{n\times m}$. Then,

$$\sigma_{\max}(A) = \max_{x\in\mathbb{F}^m\setminus\{0\}} \left(\frac{x^*A^*Ax}{x^*x}\right)^{1/2},$$

and thus

$$\|Ax\|_2 \le \sigma_{\max}(A)\|x\|_2.$$

Furthermore,

$$\lambda_{\min}^{1/2}(A^*A) = \min_{x\in\mathbb{F}^n\setminus\{0\}} \left(\frac{x^*A^*Ax}{x^*x}\right)^{1/2},$$

and thus

$$\lambda_{\min}^{1/2}(A^*A)\|x\|_2 \le \|Ax\|_2.$$

If, in addition, $m \le n$, then

$$\sigma_m(A) = \min_{x\in\mathbb{F}^n\setminus\{0\}} \left(\frac{x^*A^*Ax}{x^*x}\right)^{1/2},$$

and thus

$$\sigma_m(A)\|x\|_2 \le \|Ax\|_2.$$

Finally, if $m = n$, then

$$\sigma_{\min}(A) = \min_{x\in\mathbb{F}^n\setminus\{0\}} \left(\frac{x^*A^*Ax}{x^*x}\right)^{1/2},$$

and thus

$$\sigma_{\min}(A)\|x\|_2 \le \|Ax\|_2.$$

Source: Lemma 10.4.3. **Related:** Fact 7.12.8.

Fact 11.15.2. Let $A \in \mathbb{F}^{n\times m}$, and assume that A is nonsingular. Then,

$$\max\left\{1, \frac{\|A\|_F}{n|\det A|^{1/n}}\right\} \le \frac{\sigma_{\max}(A)}{\sigma_{\min}(A)} \le \frac{2}{|\det A|}\left(\frac{\|A\|_F}{\sqrt{n}}\right)^n \le \frac{\|A\|_F^n}{|\det A|},$$

$$\beta \triangleq |\det A|\left(\frac{\sqrt{n-1}}{\|A\|_F}\right)^{n-1} \le |\det A|\left(\frac{n-1}{\|A\|_F^2-\beta^2}\right)^{(n-1)/2} \le \sigma_{\min}(A).$$

Now, assume that $n \ge 3$. Then, for all $k \in \{2,\dots,n-1\}$,

$$\frac{\sigma_{\max}(A)}{\sigma_{\min}(A)} \le \frac{2^k}{|\det A|\prod_{i=2}^k \sigma_i(A)}\left(\frac{\|A\|_F}{\sqrt{n+k-1}}\right)^{n+k-1}.$$

Source: [1264, 1811, 2231, 3015]. **Related:** Fact 11.9.14.

Fact 11.15.3. Let $A \in \mathbb{F}^{n\times n}$. Then, the following statements are equivalent:

i) There exists $\alpha \ge 0$ such that, for all $x \in \mathbb{F}^n$, $\|Ax\|_2 = \alpha\|x\|_2$.

ii) There exists $\alpha \ge 0$ such that $A^*A = \alpha I$.

iii) For all $x \in \mathbb{F}^n$, $\|Ax\|_2 = \sigma_{\min}(A)\|x\|_2$.

iv) $A^*A = \sigma_{\min}(A)I$.

v) For all $x \in \mathbb{F}^n$, $\|Ax\|_2 = \sigma_{\max}(A)\|x\|_2$.

vi) $A^*A = \sigma_{\max}(A)I$.

vii) $\sigma_{\min}(A) = \sigma_{\max}(A)$.

Source: [2991, p. 111].

Fact 11.15.4. Let $A \in \mathbb{F}^{n\times m}$. Then,

$$\begin{aligned}\sigma_{\max}(A) &= \max\{|y^*Ax|:\ x \in \mathbb{F}^m,\ y \in \mathbb{F}^n,\ \|x\|_2 = \|y\|_2 = 1\}\\ &= \max\{|y^*Ax|:\ x \in \mathbb{F}^m,\ y \in \mathbb{F}^n,\ \|x\|_2 \le 1,\ \|y\|_2 \le 1\}.\end{aligned}$$

Related: Fact 11.9.42.

Fact 11.15.5. Let $x \in \mathbb{F}^n$ and $y \in \mathbb{F}^m$, and define $\mathcal{S} \triangleq \{A \in \mathbb{F}^{n\times m}:\ \sigma_{\max}(A) \le 1\}$. Then,

$$\max_{A\in\mathcal{S}} x^*Ay = \sqrt{x^*xy^*y}.$$

Fact 11.15.6. Let $\|\cdot\|$ be an equi-induced unitarily invariant norm on $\mathbb{F}^{n\times n}$. Then, $\|\cdot\| = \sigma_{\max}(\cdot)$.

Fact 11.15.7. Let $\|\cdot\|$ be an equi-induced self-adjoint norm on $\mathbb{F}^{n\times n}$. Then, $\|\cdot\| = \sigma_{\max}(\cdot)$.

Fact 11.15.8. Let $A \in \mathbb{F}^{n\times n}$. Then, $\sigma_{\min}(A)-1 \le \sigma_{\min}(A+I) \le \sigma_{\min}(A)+1$. **Source:** Proposition 11.6.10.

Fact 11.15.9. Let $A \in \mathbb{F}^{n\times n}$, assume that A is normal, and let $k \ge 0$. Then, $\sigma_{\max}(A^k) = \sigma_{\max}^k(A)$. **Remark:** Matrices that are not normal might also satisfy these statements. Consider $\begin{bmatrix}1&0&0\\0&0&0\\0&1&0\end{bmatrix}$.

Fact 11.15.10. Let $A \in \mathbb{F}^{n\times n}$. Then,

$$\sigma_{\max}^2(A) - \sigma_{\max}(A^2) \le \sigma_{\max}(A^*A - AA^*) \le \sigma_{\max}^2(A) - \sigma_{\min}^2(A),$$

$$\sigma_{\max}^2(A) + \sigma_{\min}^2(A) \le \sigma_{\max}(A^*A + AA^*) \le \sigma_{\max}^2(A) + \sigma_{\max}(A^2).$$

If $A^2 = 0$, then

$$\sigma_{\max}(A^*A - AA^*) = \sigma_{\max}^2(A).$$

Source: [1064, 1631, 1635]. **Remark:** If A is normal, then $\sigma^2_{\max}(A) \le \sigma_{\max}(A^2)$. Fact 11.15.9 implies that equality holds. **Related:** Fact 10.22.17.

Fact 11.15.11. Let $A \in \mathbb{F}^{n\times n}$. Then, the following statements are equivalent:

i) $\rho_{\max}(A) = \sigma_{\max}(A)$.

ii) $\sigma_{\max}(A^i) = \sigma^i_{\max}(A)$ for all $i \ge 1$.

iii) $\sigma_{\max}(A^n) = \sigma^n_{\max}(A)$.

Source: [1056] and [1450, p. 44]. **Remark:** Additional results are given in [1194]. **Credit:** *iii)* $\Longrightarrow$ *i)* is due to V. Ptak.

Fact 11.15.12. Let $A \in \mathbb{F}^{n\times n}$. Then, $\sigma_{\max}(A) \le \sigma_{\max}(|A|) \le \sqrt{\operatorname{rank} A}\,\sigma_{\max}(A)$. **Source:** [1389, p. 111].

Fact 11.15.13. Let $A \in \mathbb{F}^{n\times n}$, and let $p \in [2,\infty)$ be an even integer. Then,

$$\|A\|_{\sigma p} \le \||A|\|_{\sigma p}.$$

In particular,

$$\|A\|_{\mathrm{F}} \le \||A|\|_{\mathrm{F}}, \quad \sigma_{\max}(A) \le \sigma_{\max}(|A|).$$

Finally, let $\|\cdot\|$ be a unitarily invariant norm on $\mathbb{C}^{n\times m}$. Then, $\|A\| = \||A|\|$ for all $A \in \mathbb{C}^{n\times m}$ if and only if there exists $c > 0$ such that $\|\cdot\| = c\|\cdot\|_{\mathrm{F}}$. **Source:** [1452] and [1480].

Fact 11.15.14. Let $A \in \mathbb{R}^{n\times n}$, and assume that $r \triangleq \operatorname{rank} A \ge 2$. If $r \operatorname{tr} A^2 \le (\operatorname{tr} A)^2$, then

$$\sqrt{\frac{(\operatorname{tr} A)^2 - \operatorname{tr} A^2}{r(r-1)}} \le \rho_{\max}(A).$$

If $(\operatorname{tr} A)^2 \le r \operatorname{tr} A^2$, then

$$\frac{|\operatorname{tr} A|}{r} + \sqrt{\frac{r \operatorname{tr} A^2 - (\operatorname{tr} A)^2}{r^2(r-1)}} \le \rho_{\max}(A).$$

If $\operatorname{rank} A = 2$, then equality holds in both cases. Finally, if A is skew symmetric, then

$$\sqrt{\frac{3}{r(r-1)}}\|A\|_{\mathrm{F}} \le \rho_{\max}(A).$$

Source: [1461].

Fact 11.15.15. Let $A \in \mathbb{R}^{n\times n}$. Then,

$$\sqrt{\tfrac{1}{2(n^2-n)}(\|A\|^2_{\mathrm{F}} + \operatorname{tr} A^2)} \le \sigma_{\max}(A).$$

Furthermore, if $\|A\|_{\mathrm{F}} \le \operatorname{tr} A$, then

$$\sigma_{\max}(A) \le \tfrac{1}{n}\operatorname{tr} A + \sqrt{\tfrac{n-1}{n}[\|A\|^2_{\mathrm{F}} - \tfrac{1}{n}(\operatorname{tr} A)^2]}.$$

Source: [2025], which considers the complex case.

Fact 11.15.16. Let $n \ge 2$, $A \in \mathbb{C}^{n\times n}$, and $z \in \mathbb{C}$. Then,

$$\sigma_{\max}[(zI - A)^{\mathsf{A}}] \le \left(\tfrac{1}{n-1}[n|z| + \|A\|^2_{\mathrm{F}} - 2\operatorname{Re}(\bar{z}\operatorname{tr} A)]\right)^{\frac{n-1}{2}}.$$

Source: [1172, pp. 28, 29]. **Related:** Fact 10.21.23 and Fact 11.15.17.

Fact 11.15.17. Let $n \ge 2$, $A \in \mathbb{C}^{n\times n}$, and $\operatorname{mspec}(A) = \{\lambda_1, \ldots, \lambda_n\}_{\mathrm{ms}}$, and assume that $|\lambda_1| =$

$\min\{|\lambda_1|,\dots,|\lambda_n|\}$. Then,

$$\sigma_{\max}(A^{\mathsf{A}}) \le \sum_{i=0}^{n-1} \rho_{\max}^{n-i-1}(A)\sqrt{\frac{1}{(n-1)^i}\binom{n-1}{i}}\left(\|A\|_{\mathrm{F}}^2 - \sum_{j=1}^n |\lambda_j|^2\right)^{i/2},$$

$$\sigma_{\max}(A^{\mathsf{A}}) \le \left(\prod_{i=2}^n \max\{1,|\lambda_i|\}\right)\sum_{i=0}^{n-1} \sqrt{\frac{1}{(n-1)^i}\binom{n-1}{i}}\left(\|A\|_{\mathrm{F}}^2 - \sum_{j=1}^n |\lambda_j|^2\right)^{i/2}.$$

Source: [1172, pp. 30–32]. **Related:** Fact 10.21.23 and Fact 11.15.16.

Fact 11.15.18. Let $A \in \mathbb{F}^{n\times n}$. Then, the polynomial $p \in \mathbb{R}[s]$ defined by

$$p(s) \triangleq s^n - \|A\|_{\mathrm{F}}^2 s + (n-1)|\det A|^{2/(n-1)}$$

has either exactly one or exactly two positive roots $0 < \alpha \le \beta$. Furthermore, α and β satisfy

$$\alpha^{(n-1)/2} \le \sigma_{\min}(A) \le \sigma_{\max}(A) \le \beta^{(n-1)/2}.$$

Source: [2329].

Fact 11.15.19. Let $A \in \mathbb{F}^{n\times n}$ and $p \ge 0$. Then, $\rho(A)^{\odot p} \overset{\mathrm{w}}{\prec} \sigma(A)^{\odot p}$. In particular, $\rho(A) \overset{\mathrm{w}}{\prec} \sigma(A)$, and thus

$$|\operatorname{tr} A| \le \|\rho(A)\|_1 \le \operatorname{tr}\langle A\rangle = \|A\|_1.$$

Source: [449, p. 42]. **Remark:** This is *Weyl's majorant theorem.* **Related:** Fact 11.13.2.

Fact 11.15.20. Let $A \in \mathbb{F}^{n\times n}$. Then,

$$\rho(A)^{\odot 2} \overset{\mathrm{slog}}{\prec} \sigma(A^2) \overset{\mathrm{slog}}{\prec} \sigma(A)^{\odot 2}.$$

That is, for all $k \in \{1,\dots,n\}$,

$$\prod_{i=1}^k \rho_i^2(A) \le \prod_{i=1}^k \sigma_i(A^2) \le \prod_{i=1}^k \sigma_i^2(A),$$

$$\prod_{i=1}^n \rho_i^2(A) = \prod_{i=1}^n \sigma_i(A^2) = \prod_{i=1}^n \sigma_i^2(A) = |\det A|^2.$$

Source: [1450, p. 172], Fact 3.25.15, and Fact 7.12.32. **Credit:** H. Weyl. **Related:** Fact 10.13.2 and Fact 10.22.28.

Fact 11.15.21. Let $A \in \mathbb{F}^{n\times n}$ and $k \ge 1$. Then, $\sigma(A^k) \overset{\mathrm{slog}}{\prec} \sigma(A)^{\odot k}$. **Source:** [1810] and [2991, p. 371].

Fact 11.15.22. Let $A \in \mathbb{F}^{n\times n}$, and assume that A is normal. Then, $\|A\|_\infty \le \rho_{\max}(A)$. **Source:** [2991, p. 322].

Fact 11.15.23. Let $A \in \mathbb{F}^{n\times n}$, assume that A is normal, let $\mathcal{S} \subseteq \{1,\dots,n\}$, and define $k \triangleq \operatorname{card}(\mathcal{S})$. Then, $|1_{1\times n}A_{(\mathcal{S})}1_{n\times 1}| \le k\rho_{\max}(A)$. **Source:** [2991, p. 323].

Fact 11.15.24. Let $A \in \mathbb{F}^{n\times n}$, and define

$$r_i \triangleq \sum_{j=1}^n |A_{(i,j)}|, \quad c_i \triangleq \sum_{j=1}^n |A_{(j,i)}|, \quad r_{\min} \triangleq \min_{i\in\{1,\dots,n\}} r_i, \quad c_{\min} \triangleq \min_{i\in\{1,\dots,n\}} c_i,$$

$$\hat{r}_i \triangleq \sum_{\substack{j=1\\ j\ne i}}^n |A_{(i,j)}|, \quad \hat{c}_i \triangleq \sum_{\substack{j=1\\ j\ne i}}^n |A_{(j,i)}|, \quad \alpha \triangleq \min_{i\in\{1,\dots,n\}} (|A_{(i,i)}| - \hat{r}_i), \quad \beta \triangleq \min_{i\in\{1,\dots,n\}} (|A_{(i,i)}| - \hat{c}_i).$$

Then, the following statements hold:

i) If $\alpha > 0$, then A is nonsingular and $\|A^{-1}\|_{\rm row} < 1/\alpha$.

ii) If $\beta > 0$, then A is nonsingular and $\|A^{-1}\|_{\rm col} < 1/\beta$.

iii) If $\alpha > 0$ and $\beta > 0$, then A is nonsingular, and $\sqrt{\alpha\beta} \le \sigma_{\min}(A)$.

iv) $\min_{i=1,\ldots,n} \frac{1}{2}[2|A_{(i,i)}| - \hat{r}_i - \hat{c}_i] \le \sigma_{\min}(A)$.

v) $\min_{i=1,\ldots,n} \frac{1}{2}[(4|A_{(i,i)}|^2 + [\hat{r}_i - \hat{c}_i]^2)^{1/2} - \hat{r}_i - \hat{c}_i] \le \sigma_{\min}(A)$.

vi) $\left(\frac{n-1}{n}\right)^{(n-1)/2} |\det A| \max\left\{\frac{c_{\min}}{\prod_{i=1}^n c_i}, \frac{r_{\min}}{\prod_{i=1}^n r_i}\right\} \le \sigma_{\min}(A)$.

Source: Fact 11.9.27, [1442], [1450, pp. 227, 231], and [1540, 1550, 2782].

Fact 11.15.25. Let $A \in \mathbb{F}^{n\times n}$. Then, for all $i \in \{1,\ldots,n\}$,

$$\lim_{k\to\infty} \sigma_i^{1/k}(A^k) = \rho_i(A).$$

In particular,

$$\lim_{k\to\infty} \sigma_{\max}^{1/k}(A^k) = \rho_{\max}(A).$$

Source: [1450, p. 180] and [2979, p. 85]. **Credit:** T. Yamamoto. **Related:** The expression for $\rho_{\max}(A)$ is a special case of Proposition 11.2.6.

Fact 11.15.26. Let $A \in \mathbb{F}^{n\times n}$, and assume that A is nonzero. Then,

$$\frac{1}{\sigma_{\max}(A)} = \min_{B\in\{X\in\mathbb{F}^{n\times n}\colon\ \det(I-AX)=0\}} \sigma_{\max}(B).$$

Furthermore, there exists $B_0 \in \mathbb{F}^{n\times n}$ such that $\operatorname{rank} B_0 = 1$, $\det(I - AB_0) = 0$, and

$$\frac{1}{\sigma_{\max}(A)} = \sigma_{\max}(B_0).$$

Source: If $\sigma_{\max}(B) < 1/\sigma_{\max}(A)$, then $\rho_{\max}(AB) \le \sigma_{\max}(AB) < 1$, and thus $I - AB$ is nonsingular. Hence,

$$\begin{aligned}\frac{1}{\sigma_{\max}(A)} &= \min_{B\in\{X\in\mathbb{F}^{n\times n}\colon\ \sigma_{\max}(X)\ge 1/\sigma_{\max}(A)\}} \sigma_{\max}(B)\\ &= \min_{B\in\{X\in\mathbb{F}^{n\times n}\colon\ \sigma_{\max}(X)<1/\sigma_{\max}(A)\}^{\sim}} \sigma_{\max}(B)\\ &\le \min_{B\in\{X\in\mathbb{F}^{n\times n}\colon\ \det(I-AX)=0\}} \sigma_{\max}(B).\end{aligned}$$

Using the singular value decomposition, equality holds by constructing B_0 to have rank 1 and singular value $1/\sigma_{\max}(A)$. **Remark:** This result is related to the *small-gain theorem*. See [2999, pp. 276, 277].

11.16 Facts on Matrix Norms and Singular Values for Two or More Matrices

Fact 11.16.1. Let $A \in \mathbb{F}^{n\times m}$, $B \in \mathbb{F}^{k\times n}$, and $C \in \mathbb{F}^{m\times l}$, and assume that $n \le m$. Then, for all $i \in \{1,\ldots,n\}$,

$$\sigma_i(BA) \le \sigma_{\max}(B)\sigma_i(A), \quad \sigma_i(AC) \le \sigma_i(A)\sigma_{\max}(C).$$

Source: [2586, p. 54].

Fact 11.16.2. Let $A \in \mathbb{F}^{n\times m}$ and $B \in \mathbb{F}^{m\times n}$. Then,

$$|\operatorname{tr} AB| \le \|\rho(AB)\|_1 \le \|AB\|_{\sigma 1} = \|\sigma(AB)\|_1 \le \sum_{i=1}^{\min\{m,n\}} \sigma_i(A)\sigma_i(B) \le \sigma_{\max}(A)\|B\|_{\sigma 1}.$$

Source: Fact 11.14.1. **Remark:** Sufficient conditions for equality are given in [2418, p. 107]. **Related:** Fact 7.13.15.

Fact 11.16.3. Let $A, B \in \mathbb{F}^{n\times m}$, and assume that A and B are normal. Then,

$$\sigma_{\max}^2(AB) = \sigma_{\max}^2(\langle A\rangle\langle B\rangle) = \lambda_{\max}(\langle A\rangle^2\langle B\rangle^2).$$

Furthermore, if $k \ge 1$, then

$$\sigma_{\max}^k(AB) \le \sigma_{\max}(A^kB^k) = \sigma_{\max}(\langle A\rangle^k\langle B\rangle^k).$$

Finally, if $k \ge 1$, then

$$\sigma_{\max}^k(AB) \le \lambda_{\max}(\langle A\rangle^k\langle B\rangle^k).$$

Source: [1810]. **Remark:** If A is normal and $k \ge 1$, then $\langle A^k\rangle = \langle A\rangle^k$.

Fact 11.16.4. Let $A \in \mathbb{F}^{n\times m}$, $B \in \mathbb{F}^{m\times n}$, and $p \in [1,\infty)$, and assume that AB is normal. Then,

$$\|AB\|_{\sigma p} \le \|BA\|_{\sigma p}.$$

In particular,

$$\operatorname{tr}\langle AB\rangle \le \operatorname{tr}\langle BA\rangle, \quad \|AB\|_{\mathrm{F}} \le \|BA\|_{\mathrm{F}}, \quad \sigma_{\max}(AB) \le \sigma_{\max}(BA).$$

Source: [537]. **Credit:** B. Simon.

Fact 11.16.5. Let $A, B \in \mathbb{R}^{n\times n}$, assume that A is nonsingular, and assume that B is singular. Then, $\sigma_{\min}(A) \le \sigma_{\max}(A - B)$. Furthermore, if $\sigma_{\max}(A^{-1}) = \rho_{\max}(A^{-1})$, then there exists a singular matrix $C \in \mathbb{R}^{n\times n}$ such that $\sigma_{\max}(A - C) = \sigma_{\min}(A)$. **Source:** [2263, p. 151]. **Credit:** P. Franck.

Fact 11.16.6. Let $A \in \mathbb{C}^{n\times n}$, assume that A is nonsingular, let $\|\cdot\|$ and $\|\cdot\|'$ be norms on $\mathbb{C}^n$, let $\|\cdot\|''$ be the norm on $\mathbb{C}^{n\times n}$ induced by $\|\cdot\|$ and $\|\cdot\|'$, and let $\|\cdot\|'''$ be the norm on $\mathbb{C}^{n\times n}$ induced by $\|\cdot\|'$ and $\|\cdot\|$. Then,

$$\min\{\|B\|''\colon\ B \in \mathbb{C}^{n\times n} \text{ and } A + B \text{ is nonsingular}\} = 1/\|A^{-1}\|'''.$$

In particular,

$$\min\{\|B\|_{\mathrm{col}}\colon\ B \in \mathbb{C}^{n\times n} \text{ and } A + B \text{ is singular}\} = 1/\|A^{-1}\|_{\mathrm{col}},$$

$$\min\{\sigma_{\max}(B)\colon\ B \in \mathbb{C}^{n\times n} \text{ and } A + B \text{ is singular}\} = \sigma_{\min}(A),$$

$$\min\{\|B\|_{\mathrm{row}}\colon\ B \in \mathbb{C}^{n\times n} \text{ and } A + B \text{ is singular}\} = 1/\|A^{-1}\|_{\mathrm{row}}.$$

Source: [1387] and [1389, p. 111]. **Credit:** N. Gastinel. **Related:** The third equality is equivalent to the inequality in Fact 11.16.5.

Fact 11.16.7. Let $A, B \in \mathbb{F}^{n\times m}$, and assume that $\operatorname{rank} A = \operatorname{rank} B$ and $\alpha \triangleq \sigma_{\max}(A^+)\sigma_{\max}(A-B) < 1$. Then,

$$\sigma_{\max}(B^+) < \frac{1}{1-\alpha}\sigma_{\max}(A^+).$$

If, in addition, $n = m$, A and B are nonsingular, and $\sigma_{\max}(A - B) < \sigma_{\min}(A)$, then

$$\sigma_{\max}(B^{-1}) < \frac{\sigma_{\min}(A)}{\sigma_{\min}(A) - \sigma_{\max}(A - B)}\sigma_{\max}(A^{-1}).$$

Source: [1389, p. 400].

Fact 11.16.8. Let $A, B \in \mathbb{F}^{n\times n}$. Then,

$$\sigma_{\max}(I - [A, B]) \ge 1.$$

Source: Since $\operatorname{tr}[A, B] = 0$, it follows that there exists $\lambda \in \operatorname{spec}(I - [A, B])$ such that $\operatorname{Re}\lambda \ge 1$, and thus $|\lambda| \ge 1$. Hence, Corollary 11.4.5 implies that $\sigma_{\max}(I - [A, B]) \ge \rho_{\max}(I - [A, B]) \ge |\lambda| \ge 1$.

Fact 11.16.9. Let $A \in \mathbb{F}^{n\times m}$, and let $B \in \mathbb{F}^{k\times l}$ be a submatrix of A. Then, for all $i \in \{1,\ldots,\min\{k,l\}\}$,

$$\sigma_i(B) \le \sigma_i(A).$$

Furthermore, for all $i \in \{1,\ldots,\min\{k,l,k+l-n,k+l-m\}\}$,

$$\sigma_{n+m-(k+l)+i}(A) \le \sigma_i(B).$$

Source: Proposition 11.6.1, [1451, p. 451], [2979, pp. 79, 80], and [2991, p. 276].

Fact 11.16.10. Let $a_1,\ldots,a_n \in \mathbb{F}^n$ be linearly independent, and, for all $i \in \{1,\ldots,n\}$, define $A_i \triangleq I - (a_i^*a_i)^{-1}a_ia_i^*$. Then,

$$\sigma_{\max}(A_nA_{n-1}\cdots A_1) < 1.$$

Source: Assume that $n \ge 2$, and define $A \triangleq A_nA_{n-1}\cdots A_1$. Since $\sigma_{\max}(A_i) = 1$ for all $i \in \{1,\ldots,n\}$, it follows that $\sigma_{\max}(A) \le 1$. Suppose that $\sigma_{\max}(A) = 1$, and let $x \in \mathbb{F}^n$ satisfy $x^*x = 1$ and $\|Ax\|_2 = 1$. Then, for all $i \in \{1,\ldots,n\}$, $\|A_iA_{i-1}\cdots A_1x\|_2 = 1$. Consequently, $\|A_1x\|_2 = 1$, which implies that $a_1^*x = 0$, and thus $A_1x = x$. Hence, $\|A_iA_{i-1}\cdots A_2x\|_2 = 1$. Repeating this argument implies that, for all $i \in \{1,\ldots,n\}$, $a_i^*x = 0$. Since $a_1,\ldots,a_n$ are linearly independent, it follows that $x = 0$, which is a contradiction. **Credit:** J. C. Akers and D. Z. Djokovic.

Fact 11.16.11. Let $a_1,\ldots,a_n \in \mathbb{F}^n$ be linearly independent, and, for all $i \in \{1,\ldots,n\}$, define $A_i \triangleq (I + a_ia_i^*)^{-1}$. Then,

$$\sigma_{\max}(A_nA_{n-1}\cdots A_1) < 1.$$

Credit: A. A. Ali and M. Lin.

Fact 11.16.12. Let $A_1,\ldots,A_n \in \mathbb{F}^{n\times n}$, assume that, for all $i,j \in \{1,\ldots,n\}$, $[A_i,A_j] = 0$, and assume that, for all $i \in \{1,\ldots,n\}$, $\rho_{\max}(A_i) < \sigma_{\max}(A_i) = 1$. Then,

$$\sigma_{\max}(A_nA_{n-1}\cdots A_1) < 1.$$

Source: [2959].

Fact 11.16.13. Let

$$\mathcal{A} \triangleq \begin{bmatrix} A & B \\ C & D \end{bmatrix} \in \mathbb{F}^{(n+m)\times(n+m)},$$

and assume that $\mathcal{A}$ is unitary. Then, the following statements hold:

i) If $n < m$, then

$$\sigma_i(D) = \begin{cases} 1, & 1 \le i \le m-n, \\ \sigma_{i-m+n}(A), & m-n < i \le m. \end{cases}$$

ii) If $n = m$, then, for all $i \in \{1,\ldots,n\}$, $\sigma_i(D) = \sigma_i(A)$.

iii) If $n \le m$, then

$$|\det D| = \prod_{i=1}^{m} \sigma_i(D) = \prod_{i=1}^{n} \sigma_i(A) = |\det A|.$$

Source: [1212]. **Remark:** *iii)* follows from Fact 4.13.21 using Fact 7.12.32.

Fact 11.16.14. Let

$$\mathcal{A} \triangleq \begin{bmatrix} A & B \\ C & D \end{bmatrix} \in \mathbb{F}^{(n+m)\times(n+m)},$$

assume that $\mathcal{A}$ is nonsingular, and define $\left[\begin{smallmatrix} E & F \\ G & H \end{smallmatrix}\right] \in \mathbb{F}^{(n+m)\times(n+m)}$ by

$$\begin{bmatrix} E & F \\ G & H \end{bmatrix} \triangleq \mathcal{A}^{-1}.$$

Then, the following statements hold:

i) For all $i \in \{1,\ldots,\min\{n,m\}-1\}$,

$$\frac{\sigma_{n-i}(A)}{\sigma^2_{\max}(\mathcal{A})} \le \sigma_{m-i}(H) \le \frac{\sigma_{n-i}(A)}{\sigma^2_{\min}(\mathcal{A})}.$$

ii) Assume that $n < m$. Then, for all $i \in \{1,\ldots,m-n\}$,

$$\frac{1}{\sigma_{\max}(\mathcal{A})} \le \sigma_i(H) \le \frac{1}{\sigma_{\min}(\mathcal{A})}.$$

iii) Assume that $m < n$. Then, for all $i \in \{1,\ldots,n-m\}$,

$$\sigma_{\min}(\mathcal{A}) \le \sigma_i(A) \le \sigma_{\max}(\mathcal{A}).$$

iv) Assume that $n = m$. Then, for all $i \in \{1,\ldots,n\}$,

$$\frac{\sigma_i(A)}{\sigma^2_{\max}(\mathcal{A})} \le \sigma_i(H) \le \frac{\sigma_i(A)}{\sigma^2_{\min}(\mathcal{A})}.$$

v) Assume that $m < n$. Then,

$$\sigma_{\max}(H) \le \frac{\sigma_{n-m+1}(A)}{\sigma^2_{\min}(\mathcal{A})}.$$

vi) Assume that $m < n$. Then, $H = 0$ if and only if $\operatorname{def} A = m$.

Source: [1212]. **Remark:** *vi)* is a special case of the nullity theorem given by Fact 3.14.27.

Fact 11.16.15. Let $A \in \mathbb{F}^{n\times m}$, $B \in \mathbb{F}^{n\times l}$, $C \in \mathbb{F}^{k\times m}$, and $D \in \mathbb{F}^{k\times l}$. Then,

$$\sigma_{\max}\left(\begin{bmatrix} A & B \\ C & D \end{bmatrix}\right) \le \sigma_{\max}\left(\begin{bmatrix} \sigma_{\max}(A) & \sigma_{\max}(B) \\ \sigma_{\max}(C) & \sigma_{\max}(D) \end{bmatrix}\right).$$

Source: [1465, 1632]. **Credit:** J. Tomiyama. **Related:** Fact 10.12.66 and Fact 11.12.12.

Fact 11.16.16. Let $A \in \mathbb{F}^{n\times m}$, $B \in \mathbb{F}^{n\times l}$, and $C \in \mathbb{F}^{k\times m}$. Then, for all $X \in \mathbb{F}^{k\times l}$,

$$\max\left\{\sigma_{\max}([A \;\; B]), \sigma_{\max}\left(\begin{bmatrix} A \\ C \end{bmatrix}\right)\right\} \le \sigma_{\max}\left(\begin{bmatrix} A & B \\ C & X \end{bmatrix}\right).$$

Furthermore, there exists $X \in \mathbb{F}^{k\times l}$ such that equality holds. **Remark:** This is *Parrott's theorem*. See [825], [970, pp. 271, 272], and [2999, pp. 40–42].

Fact 11.16.17. Let $A \in \mathbb{F}^{n\times m}$ and $B \in \mathbb{F}^{n\times l}$. Then,

$$\max\{\sigma_{\max}(A), \sigma_{\max}(B)\} \le \sigma_{\max}([A \;\; B]) \le [\sigma^2_{\max}(A) + \sigma^2_{\max}(B)]^{1/2} \le \sqrt{2}\max\{\sigma_{\max}(A), \sigma_{\max}(B)\}.$$

Furthermore, if $n \le \min\{m, l\}$, then

$$[\sigma^2_n(A) + \sigma^2_n(B)]^{1/2} \le \sigma_n([A \;\; B]) \le \begin{cases} [\sigma^2_n(A) + \sigma^2_{\max}(B)]^{1/2} \\ [\sigma^2_{\max}(A) + \sigma^2_n(B)]^{1/2}. \end{cases}$$

Problem: Obtain analogous bounds for $\sigma_i([A \;\; B])$.

Fact 11.16.18. Let $A, B \in \mathbb{F}^{n\times n}$. Then,

$$\begin{aligned}\sigma_{\max}(A+B) \le{}& \frac{1}{2}\Big(\sigma_{\max}(A) + \sigma_{\max}(B) \\ &+ \sqrt{[\sigma_{\max}(A) - \sigma_{\max}(B)]^2 + 4\max\{\sigma^2_{\max}(\langle A\rangle^{1/2}\langle B\rangle^{1/2}), \sigma^2_{\max}(\langle A^*\rangle^{1/2}\langle B^*\rangle^{1/2})\}}\Big) \\ \le{}& \sigma_{\max}(A) + \sigma_{\max}(B).\end{aligned}$$

Source: [1632]. **Remark:** This result interpolates the triangle inequality for the maximum singular value. **Related:** Fact 10.22.19.

Fact 11.16.19. Let $A, B \in \mathbb{F}^{n\times n}$. Then,

$$\sigma_{\max}(A+B) = \sigma_{\max}(\langle A+B\rangle) \le \sigma_{\max}(A) + \sigma_{\max}(B) = \sigma_{\max}(\langle A\rangle) + \sigma_{\max}(\langle B\rangle).$$

Now, assume that A and B are normal. Then,

$$\sigma_{\max}(A+B) = \sigma_{\max}(\langle A+B\rangle) \le \sigma_{\max}(\langle A\rangle + \langle B\rangle) \le \sigma_{\max}(A) + \sigma_{\max}(B) \le \sigma_{\max}(\langle A\rangle) + \sigma_{\max}(\langle B\rangle).$$

If, in addition, $r \in [1,\infty)$, then

$$\sigma^r_{\max}(A+B) \le 2^{r-1}\sigma_{\max}(\langle A\rangle^r + \langle B\rangle^r).$$

Source: [454, p. 27] and [1636, 2437, 2439].

Fact 11.16.20. Let $A, B \in \mathbb{F}^{n\times n}$ and $\alpha > 0$. Then,

$$\sigma_{\min}(A+B) \le [(1+\alpha)\sigma^2_{\min}(A) + (1+\alpha^{-1})\sigma^2_{\max}(B)]^{1/2},$$
$$\sigma_{\max}(A+B) \le [(1+\alpha)\sigma^2_{\max}(A) + (1+\alpha^{-1})\sigma^2_{\max}(B)]^{1/2}.$$

Fact 11.16.21. Let $A, B \in \mathbb{F}^{n\times n}$. Then,

$$\sigma_{\min}(A) - \sigma_{\max}(B) \le |\det(A+B)|^{1/n} \le \prod_{i=1}^{n} |\sigma_i(A) + \sigma_{n-i+1}(B)|^{1/n} \le \sigma_{\max}(A) + \sigma_{\max}(B).$$

Source: [1467, p. 63] and [1798].

Fact 11.16.22. Let $A, B \in \mathbb{F}^{n\times n}$, and assume that $\sigma_{\max}(B) \le \sigma_{\min}(A)$. Then,

$$\begin{aligned}0 &\le [\sigma_{\min}(A) - \sigma_{\max}(B)]^n \le \prod_{i=1}^{n} |\sigma_i(A) - \sigma_{n-i+1}(B)| \\ &\le |\det(A+B)| \le \prod_{i=1}^{n} |\sigma_i(A) + \sigma_{n-i+1}(B)| \le [\sigma_{\max}(A) + \sigma_{\max}(B)]^n.\end{aligned}$$

Hence, if $\sigma_{\max}(B) < \sigma_{\min}(A)$, then A is nonsingular and, for all $\alpha \in [-1,1]$, $A + \alpha B$ is nonsingular. **Source:** [1798]. **Related:** Fact 7.13.24 and Fact 15.19.16.

Fact 11.16.23. Let $A, B \in \mathbb{F}^{n\times m}$, and define $r \triangleq \min\{n,m\}$. Then, the following statements are equivalent:

i) $\sigma(A) \overset{\mathrm{w}}{\prec} \sigma(B)$.

ii) For all unitarily invariant norms $\|\cdot\|$ on $\mathbb{F}^{n\times m}$, $\|A\| \le \|B\|$.

Source: [1450, pp. 205, 206] and [2991, p. 375]. **Remark:** This is the *Fan dominance theorem.*

Fact 11.16.24. Let $A, B \in \mathbb{F}^{n\times n}$. Then,

$$\sigma(A) + \sigma(B)^{\uparrow} \overset{\mathrm{w}}{\prec} \sigma(A+B) \overset{\mathrm{w}}{\prec} \begin{cases}\sigma(A) + \sigma(B)\\ \sigma(\langle A\rangle + \langle B\rangle).\end{cases}$$

If, in addition, A and B are Hermitian, then

$$\lambda(A) + \lambda(B)^{\uparrow} \overset{\mathrm{w}}{\prec} \lambda(A+B) \overset{\mathrm{w}}{\prec} \lambda(A) + \lambda(B).$$

Source: [1798, 2750] and [2991, pp. 357, 362]. **Related:** Fact 11.16.25.

Fact 11.16.25. Let $A, B \in \mathbb{F}^{n\times n}$. Then,

$$|\sigma(A) - \sigma(B)| \overset{\mathrm{w}}{\prec} \sigma(A-B).$$

Furthermore, if either $\sigma_{\max}(A) < \sigma_{\min}(B)$ or $\sigma_{\max}(B) < \sigma_{\min}(A)$, then

$$\sigma(A+B) \overset{\mathrm{w}}{\prec} |\sigma(A) - \sigma^{\uparrow}(B)|.$$

Source: Proposition 11.2.2, [1450, pp. 196, 197], [1798], [1969, p. 243], [1971, p. 330], and [2991, pp. 357, 361, 362]. **Related:** Fact 11.16.24.

Fact 11.16.26. Let $M, S, U \in \mathbb{F}^{n\times n}$, assume that M is positive semidefinite, and assume that S and U are unitary. Then,

$$\sigma(SM - S) \overset{\mathrm{w}}{\prec} \sigma(SM - U) \overset{\mathrm{w}}{\prec} \sigma(SM + S).$$

In particular,

$$\sigma(M - I) \overset{\mathrm{w}}{\prec} \sigma(M - U) \overset{\mathrm{w}}{\prec} \sigma(M + I).$$

Source: [2991, p. 359]. **Remark:** SM is a polar decomposition. **Related:** Fact 11.10.73.

Fact 11.16.27. Let $U, V, W, D \in \mathbb{F}^{n\times n}$, assume that D is nonnegative and diagonal, and assume that U, V, and W are unitary. Then,

$$\sigma(UDV - UV) \overset{\mathrm{w}}{\prec} \sigma(UDV - W) \overset{\mathrm{w}}{\prec} \sigma(UDV + UV).$$

Source: [2991, p. 361]. **Remark:** UDV is a singular value decomposition. **Related:** Fact 11.10.73.

Fact 11.16.28. Let $A, B, C \in \mathbb{F}^{n\times n}$, assume that B is Hermitian, and assume that C is skew Hermitian. Then,

$$\sigma[A - \tfrac{1}{2}(A + A^*)] \overset{\mathrm{w}}{\prec} \sigma(A - B), \quad \sigma[A - \tfrac{1}{2}(A - A^*)] \overset{\mathrm{w}}{\prec} \sigma(A - C).$$

Source: [2991, p. 359].

Fact 11.16.29. Let $A, B \in \mathbb{F}^{n\times m}$ and $\alpha \in [0, 1]$. Then, for all $i \in \{1, \dots, \min\{n, m\}\}$,

$$\sigma_i[\alpha A + (1-\alpha)B] \le \begin{cases} \sigma_i\left(\begin{bmatrix} A & 0 \\ 0 & B \end{bmatrix}\right) \\ \sigma_i\left(\begin{bmatrix} \sqrt{2\alpha}A & 0 \\ 0 & \sqrt{2(1-\alpha)}B \end{bmatrix}\right), \end{cases}$$

$$2\sigma_i(AB^*) \le \sigma_i(\langle A\rangle^2 + \langle B\rangle^2).$$

Furthermore,

$$\langle \alpha A + (1-\alpha)B\rangle^2 \le \alpha\langle A\rangle^2 + (1-\alpha)\langle B\rangle^2.$$

If, in addition, $n = m$, then, for all $i \in \{1, \dots, n\}$,

$$\tfrac{1}{2}\sigma_i(A + A^*) \le \sigma_i\left(\begin{bmatrix} A & 0 \\ 0 & A \end{bmatrix}\right).$$

Source: [1422]. **Related:** Fact 11.16.31.

Fact 11.16.30. Let $A \in \mathbb{F}^{n\times m}$ and $B \in \mathbb{F}^{l\times m}$, and let $p, q \in (0, \infty)$ satisfy $1/p + 1/q = 1$. Then, for all $i \in \{1, \dots, \min\{n, m, l\}\}$,

$$\sigma_i(AB^*) \le \sigma_i\left(\tfrac{1}{p}\langle A\rangle^p + \tfrac{1}{q}\langle B\rangle^q\right).$$

Source: [93, 95, 1418] and [2977, p. 28]. **Related:** Fact 2.2.50, Fact 2.2.53, Fact 10.11.73, Fact 10.14.8, Fact 10.14.33, and Fact 10.14.34.

Fact 11.16.31. Let $A \in \mathbb{F}^{n\times m}$ and $B \in \mathbb{F}^{l\times m}$. Then, for all $i \in \{1, \dots, \min\{n, m, l\}\}$,

$$\sigma_i(AB^*) \le \tfrac{1}{2}\sigma_i(A^*A + B^*B).$$

Source: Set $p = q = 2$ in Fact 10.11.38. See [468], [2979, p. 88], and [2991, p. 354]. **Related:** Fact 11.10.79 and Fact 11.16.29.

Fact 11.16.32. Let $A, B \in \mathbb{F}^{n\times n}$, and assume that A and B are semicontractive. Then, for all $i \in \{1,\dots,n\}$,

$$\sigma_i(2I - A^*A - B^*B) \le 2\sigma_i(I - A^*B).$$

Now, assume that A and B are contractive. Then, for all $i \in \{1,\dots,n\}$,

$$2\sigma_i[(I - A^*B)^{-1}] \le \sigma_i[(I - A^*A)^{-1} + (I - B^*B)^{-1}].$$

Now, let $\|\cdot\|$ be a unitarily invariant norm on $\mathbb{F}^{n\times n}$. Then,

$$2\|(I - A^*B)^{-1}\| \le \|(I - A^*A)^{-1} + (I - B^*B)^{-1}\|.$$

Source: [1840].

Fact 11.16.33. Let $A, B, C, D \in \mathbb{F}^{n\times m}$. Then, for all $i \in \{1,\dots,\min\{n,m\}\}$,

$$\sqrt{2}\sigma_i(\langle AB^* + CD^*\rangle) \le \sigma_i\left(\begin{bmatrix} A & B \\ C & D \end{bmatrix}\right).$$

Source: [1417].

Fact 11.16.34. Let $A, B, C, D, X \in \mathbb{F}^{n\times n}$, assume that A, B, C, D are positive semidefinite, and assume that $0 \le A \le C$ and $0 \le B \le D$. Then, for all $i \in \{1,\dots,n\}$,

$$\sigma_i(A^{1/2}XB^{1/2}) \le \sigma_i(C^{1/2}XD^{1/2}).$$

Source: [1422, 1627].

Fact 11.16.35. Let $A_1,\dots,A_k \in \mathbb{F}^{n\times n}$. Then, for all $l \in \{1,\dots,n\}$,

$$\sum_{i=1}^{l} \sigma_i\left(\prod_{j=1}^{k} A_j\right) \le \sum_{i=1}^{l}\prod_{j=1}^{k}\sigma_i(A_j).$$

Source: [709].

Fact 11.16.36. Let $A \in \mathbb{F}^{n\times m}$, $B \in \mathbb{F}^{m\times n}$, $1 \le k \le \min\{n,m\}$, and $1 \le i_1 < \cdots < i_k \le \min\{n,m\}$. Then,

$$\sum_{j=1}^{k} \sigma_{i_j}(A)\sigma_{n-j+1}(B) \le \sum_{j=1}^{k}\sigma_{i_j}(AB) \le \sum_{j=1}^{k}\sigma_{i_j}(A)\sigma_j(B),$$

$$\sum_{j=1}^{k} \sigma_{i_j}(A)\sigma_{n-i_j+1}(B) \le \sum_{j=1}^{k}\sigma_j(AB), \quad \sum_{j=1}^{k}\sigma_{i_j}^2(A)\sigma_{n-j+1}^2(B) \le \sum_{j=1}^{k}\sigma_{i_j}^2(AB),$$

$$\prod_{j=1}^{k} \sigma_{i_j}(A)\sigma_{n-i_j+1}(B) \le \prod_{j=1}^{k}\sigma_j(AB), \quad \prod_{j=1}^{k}\sigma_{i_j}(AB) \le \prod_{j=1}^{k}\sigma_{i_j}(A)\sigma_j(B).$$

If, in addition, $r > 0$, then

$$\sum_{j=1}^{k} \sigma_{i_j}^r(A)\sigma_{n-i_j+1}^r(B) \le \sum_{j=1}^{k}\sigma_j^r(AB), \quad \sum_{j=1}^{k}\sigma_{i_j}^r(A)\sigma_{n-j+1}^r(B) \le \sum_{j=1}^{k}\sigma_{i_j}^r(AB).$$

Source: [2812, 2814] and [2991, p. 364] **Remark:** Extensions to products of three or more matrices are given in [2814]. **Credit:** The second inequality in the last string is due to A. Horn. **Related:** Fact 10.22.26 and Fact 10.22.29.

Fact 11.16.37. Let $A, B \in \mathbb{F}^{n\times n}$. Then, the following statements hold:

i) $\sigma(A) \odot \sigma(B)^{\uparrow} \overset{\mathrm{w}}{\prec} \sigma(AB) \overset{\mathrm{w}}{\prec} \sigma(A) \odot \sigma(B)$.

ii) $[\sigma(A) \odot \sigma(B)^{\uparrow}]^{\odot 2} \overset{\mathrm{w}}{\prec} \sigma(AB)^{\odot 2}$.

iii) $\sigma(A) \odot \sigma(B)^{\uparrow} \overset{\mathrm{slog}}{\prec} \sigma(AB) \overset{\mathrm{slog}}{\prec} \sigma(A) \odot \sigma(B)$.

iv) $\sigma(A \odot B) \overset{\mathrm{w}}{\prec} \sigma(A) \odot \sigma(B)$.

v) If A and B are positive semidefinite, then $\lambda(AB) \overset{\mathrm{slog}}{\prec} \lambda(A) \odot \lambda(B)$.

vi) If A and B are Hermitian, then $\lambda(e^A) + \lambda(e^B)^{\uparrow} \overset{\mathrm{slog}}{\prec} \lambda(e^{A+B}) \overset{\mathrm{w}}{\prec} \lambda(e^A) \odot \lambda(e^B)$.

Source: [2750] and [2991, pp. 353, 364].

Fact 11.16.38. Let $A, B \in \mathbb{F}^{n\times n}$. Then,

$$\begin{bmatrix} \sigma[\frac{1}{2}(A+B)] \\ \sigma[\frac{1}{2}(A+B)] \end{bmatrix} \overset{\mathrm{w}}{\prec} \begin{bmatrix} \sigma(A) \\ \sigma(B) \end{bmatrix}, \quad \begin{bmatrix} \sigma[e^{\frac{1}{2}(A+B)}] \\ \sigma[e^{\frac{1}{2}(A+B)}] \end{bmatrix} \overset{\mathrm{slog}}{\prec} \begin{bmatrix} \sigma(e^A) \\ \sigma(e^B) \end{bmatrix},$$

$$\begin{bmatrix} \sigma(AB) \\ \sigma(AB) \end{bmatrix} \overset{\mathrm{slog}}{\prec} \begin{bmatrix} \sigma^{\odot 2}(A) \\ \sigma^{\odot 2}(B) \end{bmatrix}, \quad \begin{bmatrix} \sigma(A\odot B) \\ \sigma(A\odot B) \end{bmatrix} \overset{\mathrm{w}}{\prec} \begin{bmatrix} \sigma^{\odot 2}(A) \\ \sigma^{\odot 2}(B) \end{bmatrix}.$$

Now, let $p \ge 1$. Then,

$$\begin{bmatrix} \sigma^{\odot p}(AB) \\ \sigma^{\odot p}(AB) \end{bmatrix} \overset{\mathrm{w}}{\prec} \begin{bmatrix} \sigma^{\odot 2p}(A) \\ \sigma^{\odot 2p}(B) \end{bmatrix}, \quad \begin{bmatrix} \sigma^{\odot p}(A\odot B) \\ \sigma^{\odot p}(A\odot B) \end{bmatrix} \overset{\mathrm{w}}{\prec} \begin{bmatrix} \sigma^{\odot 2p}(A) \\ \sigma^{\odot 2p}(B) \end{bmatrix}.$$

If A and B are Hermitian, then

$$\begin{bmatrix} \lambda[\frac{1}{2}(A+B)] \\ \lambda[\frac{1}{2}(A+B)] \end{bmatrix} \overset{\mathrm{w}}{\prec} \begin{bmatrix} \lambda(A) \\ \lambda(B) \end{bmatrix}.$$

Finally, if A and B are positive semidefinite, then

$$\begin{bmatrix} \lambda[\frac{1}{2}(A+B)] \\ \lambda[\frac{1}{2}(A+B)] \end{bmatrix} \overset{\mathrm{w}}{\prec} \begin{bmatrix} \lambda(A) \\ \lambda(B) \end{bmatrix} \overset{\mathrm{w}}{\prec} \begin{bmatrix} \lambda(A+B) \\ 0 \end{bmatrix} \overset{\mathrm{w}}{\prec} \begin{bmatrix} \lambda(A)+\lambda(B) \\ 0 \end{bmatrix}.$$

Source: [2750].

Fact 11.16.39. Let $A \in \mathbb{F}^{n\times m}$, let $k \ge 1$ satisfy $k < \operatorname{rank} A$, and let $\|\cdot\|$ be a unitarily invariant norm on $\mathbb{F}^{n\times m}$. Then,

$$\min_{B\in\{X\in\mathbb{F}^{n\times m}:\ \operatorname{rank} X\le k\}} \|A-B\| = \|A-B_0\|,$$

where B_0 is formed by replacing the $(\operatorname{rank} A) - k$ smallest positive singular values in the singular value decomposition of A by 0's. Furthermore,

$$\sigma_{\max}(A-B_0) = \sigma_{k+1}(A), \quad \|A-B_0\|_{\mathrm{F}} = \sqrt{\sum_{i=k+1}^{r} \sigma_i^2(A)}.$$

Finally, B_0 is the unique solution if and only if $\sigma_{k+1}(A) < \sigma_k(A)$. **Source:** Use Fact 11.16.40 with $B_\sigma \triangleq \operatorname{diag}[\sigma_1(A),\ldots,\sigma_k(A),0_{(n-k)\times(m-k)}]$, $S_1 = I_n$, and $S_2 = I_m$. See [1196, p. 79] and [2539, p. 208]. **Remark:** This is the *Schmidt-Mirsky theorem.* For the Frobenius norm, this is the *Eckart-Young theorem.* See [1081] and [2539, p. 210]. **Related:** Fact 11.17.14.

Fact 11.16.40. Let $A, B \in \mathbb{F}^{n\times m}$, define $A_\sigma, B_\sigma \in \mathbb{F}^{n\times m}$ by

$$A_\sigma \triangleq \operatorname{diag}(\sigma_1(A),\ldots,\sigma_r(A),0_{(n-r)\times(m-r)}), \quad B_\sigma \triangleq \operatorname{diag}(\sigma_1(B),\ldots,\sigma_l(B),0_{(n-l)\times(m-l)}),$$

where $r \triangleq \operatorname{rank} A$ and $l \triangleq \operatorname{rank} B$, let $S_1 \in \mathbb{F}^{n\times n}$ and $S_2 \in \mathbb{F}^{m\times m}$ be unitary matrices, and let $\|\cdot\|$ be a unitarily invariant norm on $\mathbb{F}^{n\times m}$. Then,

$$\|A_\sigma - B_\sigma\| \le \|A - S_1BS_2\| \le \|A_\sigma + B_\sigma\|.$$

In particular,

$$\max_{i\in\{1,\dots,\max\{r,l\}\}} |\sigma_i(A) - \sigma_i(B)| \le \sigma_{\max}(A - B) \le \sigma_{\max}(A) + \sigma_{\max}(B).$$

Source: [2815] and [2991, p. 377]. **Remark:** In the case where $S_1 = I_n$ and $S_2 = I_m$, the left-hand inequality is *Mirsky's theorem*. See [2539, p. 204]. **Related:** Fact 11.14.4.

Fact 11.16.41. Let $A, B \in \mathbb{F}^{n\times m}$, and assume that $\operatorname{rank} A = \operatorname{rank} B$. Then,

$$\sigma_{\max}[AA^+(I - BB^+)] = \sigma_{\max}[BB^+(I - AA^+)] \le \min\{\sigma_{\max}(A^+), \sigma_{\max}(B^+)\}\sigma_{\max}(A - B).$$

Source: [1389, p. 400] and [2539, p. 141].

Fact 11.16.42. Let $A, B \in \mathbb{F}^{n\times m}$. Then, for all $k \in \{1,\dots,\min\{n,m\}\}$,

$$\sum_{i=1}^k \sigma_i(A\odot B) \le \sum_{i=1}^k \mathrm{d}_i^{1/2}(A^*A)\mathrm{d}_i^{1/2}(BB^*) \le \left\{\begin{matrix} \sum_{i=1}^k \mathrm{d}_i^{1/2}(A^*A)\sigma_i(B) \\ \sum_{i=1}^k \sigma_i(A)\mathrm{d}_i^{1/2}(BB^*) \end{matrix}\right\} \le \sum_{i=1}^k \sigma_i(A)\sigma_i(B),$$

$$\sum_{i=1}^k \sigma_i(A\odot B) \le \sum_{i=1}^k \mathrm{d}_i^{1/2}(AA^*)\mathrm{d}_i^{1/2}(B^*B) \le \left\{\begin{matrix} \sum_{i=1}^k \mathrm{d}_i^{1/2}(AA^*)\sigma_i(B) \\ \sum_{i=1}^k \sigma_i(A)\mathrm{d}_i^{1/2}(B^*B) \end{matrix}\right\} \le \sum_{i=1}^k \sigma_i(A)\sigma_i(B).$$

In particular,

$$\sigma_{\max}(A\odot B) \le \|A\|_{2,1}\|B\|_{\infty,2} \le \left\{\begin{matrix} \|A\|_{2,1}\sigma_{\max}(B) \\ \sigma_{\max}(A)\|B\|_{\infty,2} \end{matrix}\right\} \le \sigma_{\max}(A)\sigma_{\max}(B),$$

$$\sigma_{\max}(A\odot B) \le \|A\|_{\infty,2}\|B\|_{2,1} \le \left\{\begin{matrix} \|A\|_{\infty,2}\sigma_{\max}(B) \\ \sigma_{\max}(A)\|B\|_{2,1} \end{matrix}\right\} \le \sigma_{\max}(A)\sigma_{\max}(B).$$

Source: [104], [1450, pp. 332, 334], [1985, 2973], Fact 3.25.2, Fact 10.21.11, and Fact 11.9.28. **Remark:** $\mathrm{d}_i^{1/2}(A^*A)$ and $\mathrm{d}_i^{1/2}(AA^*)$ are the ith largest Euclidean norms of the columns and rows of A, respectively. **Remark:** Related results are given in [2749]. Equality is discussed in [713].

Fact 11.16.43. Let $A, B \in \mathbb{F}^{n\times m}$. Then,

$$\sigma_{\max}(A\odot B) \le \sqrt{n}\|A\|_\infty\sigma_{\max}(B).$$

Now, assume that $n = m$ and that either A is positive semidefinite and B is Hermitian or A and B are nonnegative and symmetric. Then,

$$\sigma_{\max}(A\odot B) \le \|A\|_\infty\sigma_{\max}(B).$$

Source: [1157].

Fact 11.16.44. Let $A, B \in \mathbb{R}^{n\times m}$, assume that B is positive semidefinite, and let β denote the smallest positive entry of $|B|$. Then,

$$\rho_{\max}(A\odot B) \le \frac{\|A\|_\infty\|B\|_\infty}{\beta}\sigma_{\max}(B), \quad \rho_{\max}(B) \le \rho_{\max}(|B|) \le \frac{\|B\|_\infty}{\beta}\rho_{\max}(B).$$

Source: [1157].

Fact 11.16.45. Let $A, B \in \mathbb{F}^{n\times m}$, and let $p \in [2,\infty)$ be an even integer. Then,

$$\|A\odot B\|_{\sigma p}^2 \le \|A\odot\overline{A}\|_{\sigma p}\|B\odot\overline{B}\|_{\sigma p}.$$

In particular,

$$\|A \odot B\|_{\mathrm{F}}^2 \le \|A \odot \overline{A}\|_{\mathrm{F}} \|B \odot \overline{B}\|_{\mathrm{F}},$$

$$\sigma_{\max}^2(A \odot B) \le \sigma_{\max}(A \odot \overline{A})\sigma_{\max}(B \odot \overline{B}) \le \sigma_{\max}^2(A)\sigma_{\max}^2(B).$$

If $B = \overline{A}$, then equality holds. Furthermore,

$$\|A \odot A\|_{\sigma p} \le \|A \odot \overline{A}\|_{\sigma p}.$$

In particular,

$$\|A \odot A\|_{\mathrm{F}} \le \|A \odot \overline{A}\|_{\mathrm{F}}, \quad \sigma_{\max}(A \odot A) \le \sigma_{\max}(A \odot \overline{A}).$$

Now, assume that $n = m$. Then,

$$\|A \odot A^{\mathsf{T}}\|_{\sigma p} \le \|A \odot \overline{A}\|_{\sigma p}.$$

In particular,

$$\|A \odot A^{\mathsf{T}}\|_{\mathrm{F}} \le \|A \odot \overline{A}\|_{\mathrm{F}}, \quad \sigma_{\max}(A \odot A^{\mathsf{T}}) \le \sigma_{\max}(A \odot \overline{A}).$$

Finally,

$$\|A \odot A^*\|_{\sigma p} \le \|A \odot \overline{A}\|_{\sigma p}.$$

In particular,

$$\|A \odot A^*\|_{\mathrm{F}} \le \|A \odot \overline{A}\|_{\mathrm{F}}, \quad \sigma_{\max}(A \odot A^*) \le \sigma_{\max}(A \odot \overline{A}).$$

Source: [1450, p. 340] and [1452, 2441]. **Related:** Fact 9.6.23 and Fact 11.16.42.

Fact 11.16.46. Let $A, B \in \mathbb{C}^{n\times m}$. Then,

$$\|A \odot B\|_{\mathrm{F}}^2 = \sum_{i=1}^n \sigma_i^2(A \odot B) = \operatorname{tr}(A \odot B)(\overline{A} \odot \overline{B})^{\mathsf{T}} = \operatorname{tr}(A \odot \overline{A})(B \odot \overline{B})^{\mathsf{T}}$$

$$\le \sum_{i=1}^n \sigma_i[(A \odot \overline{A})(B \odot \overline{B})^{\mathsf{T}}] \le \sum_{i=1}^n \sigma_i(A \odot \overline{A})\sigma_i(B \odot \overline{B}).$$

Source: [1480].

Fact 11.16.47. Let $A, B \in \mathbb{F}^{n\times n}$. Then,

$$\|(A \odot B)(\overline{A} \odot \overline{B})^{\mathsf{T}}\|_{\sigma 1} \le \|(A \odot \overline{A})(B \odot \overline{B})^{\mathsf{T}}\|_{\sigma 1},$$

$$\|(A \odot B)(\overline{A} \odot \overline{B})^{\mathsf{T}}\|_{\mathrm{F}} \le \|(A \odot \overline{A})(B \odot \overline{B})^{\mathsf{T}}\|_{\mathrm{F}},$$

$$\sigma_{\max}[(A \odot B)(\overline{A} \odot \overline{B})^{\mathsf{T}}] \le \sigma_{\max}[(A \odot \overline{A})(B \odot \overline{B})^{\mathsf{T}}].$$

Source: [958].

Fact 11.16.48. Let $A, B \in \mathbb{R}^{n\times n}$, and assume that A and B are nonnegative. Then,

$$\sigma_{\max}(A \odot B) \le \rho_{\max}(A^{\mathsf{T}}B).$$

Source: [2991, p. 170].

Fact 11.16.49. Let $A, B \in \mathbb{R}^{n\times n}$, assume that A and B are nonnegative, and let $\alpha \in [0, 1]$. Then,

$$\sigma_{\max}(A^{\odot\alpha} \odot B^{\odot(1-\alpha)}) \le \sigma_{\max}^{\alpha}(A)\sigma_{\max}^{1-\alpha}(B),$$

$$\sigma_{\max}(A^{\odot 1/2} \odot B^{\odot 1/2}) \le \sqrt{\sigma_{\max}(A)\sigma_{\max}(B)},$$

$$\sigma_{\max}(A^{\odot 1/2} \odot A^{\odot 1/2\mathsf{T}}) \le \sigma_{\max}(A^{\odot\alpha} \odot A^{\odot(1-\alpha)\mathsf{T}}) \le \sigma_{\max}(A).$$

Source: [2441]. **Related:** Fact 9.6.24.

Fact 11.16.50. Let $\|\cdot\|$ be a unitarily invariant norm on $\mathbb{C}^{n\times n}$, and let $A, X, B \in \mathbb{C}^{n\times n}$. Then,

$$\|A\odot X\odot B\| \le \tfrac{1}{2}\sqrt{n}\|A\odot X\odot \overline{A} + B\odot X\odot \overline{B}\|,$$

$$\|A\odot X\odot B\|^2 \le n\|A\odot X\odot \overline{A}\|\|B\odot X\odot \overline{B}\|,$$

$$\|A\odot X\odot B\|_{\mathrm{F}} \le \tfrac{1}{2}\|A\odot X\odot \overline{A} + B\odot X\odot \overline{B}\|_{\mathrm{F}}.$$

Source: [1480].

11.17 Facts on Linear Equations and Least Squares

Fact 11.17.1. Let $A \in \mathbb{R}^{n\times n}$, assume that A is nonsingular, let $b \in \mathbb{R}^n$, and let $\hat{x} \in \mathbb{R}^n$. Then,

$$\frac{1}{\kappa(A)}\frac{\|A\hat{x}-b\|}{\|b\|} \le \frac{\|\hat{x}-A^{-1}b\|}{\|A^{-1}b\|} \le \kappa(A)\frac{\|A\hat{x}-b\|}{\|b\|},$$

where $\kappa(A) \triangleq \|A\|\|A^{-1}\|$ and the vector and matrix norms are compatible. Equivalently, letting $\hat{b} \triangleq A\hat{x}$ and $x \triangleq A^{-1}b$, it follows that

$$\frac{1}{\kappa(A)}\frac{\|\hat{b}-b\|}{\|b\|} \le \frac{\|\hat{x}-x\|}{\|x\|} \le \kappa(A)\frac{\|\hat{b}-b\|}{\|b\|}.$$

Remark: This result estimates the accuracy of an approximate solution $\hat{x}$ to $Ax = b$. $\kappa(A)$ is the *condition number* of A. **Remark:** For $\|A\| = \sigma_{\max}(A)$, $\kappa(A) = \kappa(A^{-1})$. **Remark:** See [3008].

Fact 11.17.2. Let $A \in \mathbb{R}^{n\times n}$, assume that A is nonsingular, let $\tilde{A} \in \mathbb{R}^{n\times n}$, assume that $\|A^{-1}\tilde{A}\| < 1$, and let $b, \tilde{b} \in \mathbb{R}^n$. Furthermore, let $x \in \mathbb{R}^n$ satisfy $Ax = b$, and let $\hat{x} \in \mathbb{R}^n$ satisfy $(A+\tilde{A})\hat{x} = b + \tilde{b}$. Then,

$$\frac{\|\hat{x}-x\|}{\|x\|} \le \frac{\kappa(A)}{1-\|A^{-1}\tilde{A}\|}\left(\frac{\|\tilde{b}\|}{\|b\|} + \frac{\|\tilde{A}\|}{\|A\|}\right),$$

where $\kappa(A) \triangleq \|A\|\|A^{-1}\|$ and the vector and matrix norms are compatible. If, in addition, $\|A^{-1}\|\|\tilde{A}\| < 1$, then

$$\frac{1}{\kappa(A)+1}\frac{\|\tilde{b}-\tilde{A}x\|}{\|b\|} \le \frac{\|\hat{x}-x\|}{\|x\|} \le \frac{\kappa(A)}{1-\|A^{-1}\tilde{A}\|}\frac{\|\tilde{b}-\tilde{A}x\|}{\|b\|}.$$

Source: [905, 906].

Fact 11.17.3. Let $A, \hat{A} \in \mathbb{R}^{n\times n}$ satisfy $\|A^{+}\hat{A}\| < 1$, let $b \in \mathcal{R}(A)$, let $\hat{b} \in \mathbb{R}^n$, and assume that $b + \hat{b} \in \mathcal{R}(A+\hat{A})$. Furthermore, let $\hat{x} \in \mathbb{R}^n$ satisfy $(A+\hat{A})\hat{x} = b+\hat{b}$. Then, $x \triangleq A^{+}b + (I - A^{+}A)\hat{x}$ satisfies $Ax = b$ and

$$\frac{\|\hat{x}-x\|}{\|x\|} \le \frac{\kappa(A)}{1-\|A^{+}\hat{A}\|}\left(\frac{\|\hat{b}\|}{\|b\|} + \frac{\|\hat{A}\|}{\|A\|}\right),$$

where $\kappa(A) \triangleq \|A\|\|A^{-1}\|$ and the vector and matrix norms are compatible. **Source:** [905]. **Remark:** See [906] for a lower bound.

Fact 11.17.4. Let $A \in \mathbb{F}^{n\times m}$, assume that A is left invertible, and let $A^{\mathsf{L}} \in \mathbb{F}^{m\times n}$ be a left inverse of A. Then,

$$A^{+}A^{+*} \le A^{\mathsf{L}}A^{\mathsf{L}*}.$$

Therefore,

$$\sigma_{\max}(A^{+}) \le \sigma_{\max}(A^{\mathsf{L}}), \quad \|A^{+}\|_{\mathrm{F}} \le \|A^{\mathsf{L}}\|_{\mathrm{F}}.$$

Furthermore, $A^{+}A^{+*} = A^{\mathsf{L}}A^{\mathsf{L}*}$ if and only if $A^{\mathsf{L}} = A^{+}$. **Source:** Fact 8.3.14, Fact 10.11.32, and Fact 11.10.31.

Fact 11.17.5. Let $A \in \mathbb{F}^{n\times m}$, assume that A is right invertible, and let $A^{\mathsf{R}} \in \mathbb{F}^{m\times n}$ denote a right inverse of A. Then,

$$A^{+*}A^{+} \le A^{\mathsf{R}*}A^{\mathsf{R}}.$$

Therefore,

$$\sigma_{\max}(A^+) \le \sigma_{\max}(A^{\mathrm{R}}), \quad \|A^+\|_{\mathrm{F}} \le \|A^{\mathrm{R}}\|_{\mathrm{F}}.$$

Furthermore, $A^{+*}A^+ = A^{\mathrm{R}*}A^{\mathrm{R}}$ if and only if $A^{\mathrm{R}} = A^+$. **Source:** Fact 8.3.15, Fact 10.11.32, and Fact 11.10.31.

Fact 11.17.6. Let $A \in \mathbb{F}^{n\times m}$, let $b \in \mathbb{F}^n$, assume that $b \notin \mathcal{R}(A)$, and let $B \in \mathbb{F}^{m\times n}$. Then, the following statements are equivalent:

i) B is a (1,3)-inverse of A.

ii) For all $x \in \mathbb{F}^m$, $\|ABb - b\|_2 \le \|Ax - b\|_2$.

Source: [1275, p. 10] and [2238, p. 285]. **Remark:** Bb is a least-squares solution of $Ax = b$. See Fact 8.3.6.

Fact 11.17.7. Let $A \in \mathbb{F}^{n\times m}$, let $b \in \mathbb{F}^n$, assume that $b \in \mathcal{R}(A)$, and let $B \in \mathbb{F}^{m\times n}$. Then, the following statements are equivalent:

i) B is a (1,4)-inverse of A.

ii) For all $x \in \{x\colon Ax = b\}$ such that $x \ne Bb$, it follows that $\|Bb\|_2 < \|x\|_2$.

iii) If $C \in \mathbb{F}^{m\times n}$, $ACb = b$, and, for all $x \in \mathbb{F}^m$ such that $Ax = b$ and $x \ne Cb$, $\|CB\|_2 < \|x\|_2$, then C is a (1,4)-inverse of A.

Source: [281, pp. 8, 114], [1275, p. 9], and [2238, p. 284]. **Remark:** Bb is a minimum-norm solution of $Ax = b$. See Fact 8.3.7.

Fact 11.17.8. Let $x, x_1, \dots, x_n \in \mathbb{F}^n$, let $\alpha_1, \dots, \alpha_n$ be real numbers, define $\alpha \triangleq \sum_{i=1}^n \alpha_i$, and assume that $\alpha > 0$. Then,

$$\frac{1}{2\alpha}\sum_{i,j=1}^n \alpha_i\alpha_j\|x_i - x_j\|_2^2 \le \sum_{i=1}^n \alpha_i\|x - x_i\|_2^2.$$

Equality holds if and only if $x = \frac{1}{\alpha}\sum_{i=1}^n \alpha_i x_i$. **Source:** [1171]. **Remark:** This is a weighted least squares problem.

Fact 11.17.9. Let $A \in \mathbb{F}^{n\times m}$ and $b \in \mathbb{F}^n$, define $f\colon \mathbb{F}^m \mapsto \mathbb{R}$ by

$$f(x) \triangleq (Ax - b)^*(Ax - b) = \|Ax - b\|_2^2,$$

and let $B \in \mathbb{F}^{m\times n}$ be a (1,3)-inverse of A. Then, Bb minimizes f. Now, let $x_0 \in \mathbb{F}^m$. Then, the following statements are equivalent:

i) x_0 minimizes f.

ii) $f(x_0) = b^*(I - AB)b$.

iii) $Ax_0 = ABb$.

iv) There exists $y \in \mathbb{R}^m$ such that $x_0 = Bb + (I - BA)y$.

Source: [2403, pp. 233–236]. **Remark:** Existence of solutions of $Ax = b$ is not assumed.

Fact 11.17.10. Let $A \in \mathbb{F}^{n\times m}$ and $b \in \mathbb{F}^n$, and define $f\colon \mathbb{F}^m \mapsto \mathbb{R}$ by

$$f(x) \triangleq (Ax - b)^*(Ax - b) = \|Ax - b\|_2^2.$$

Then, f has a minimizer. Now, let $x_0 \in \mathbb{F}^m$. Then, the following statements are equivalent:

i) x_0 minimizes f.

ii) $A^*Ax_0 = A^*b$.

iii) There exists $y \in \mathbb{F}^m$ such that $x_0 = A^+b + (I - A^+A)y$.

If *i)*–*iii)* are satisfied, then, for all $x \in \mathbb{F}^n$,

$$f(x) = (x - x_0)^*A^*A(x - x_0) + b^*b - b^*AA^+b.$$

Therefore, if x_0 minimizes f, then

$$f(x_0) = b^*b - x_0^*A^*Ax_0 = b^*(I - AA^+)b.$$

Furthermore, if $y \in \mathbb{F}^m$ and $(I - A^+A)y$ is nonzero, then

$$\|A^+b\|_2 < \|A^+b + (I - A^+A)y\|_2 = \sqrt{\|A^+b\|_2^2 + \|(I - A^+A)y\|_2^2}.$$

Finally, A^+b is the unique minimizer of f if and only if A is left invertible. **Remark:** This is a *least squares problem.* See [31, 489, 2535]. The expression for x is identical to the expression given by *vii)* of Proposition 8.1.9 for solutions of $Ax = b$. Therefore, x satisfies $Ax = b$ if and only if x is optimal in the least-squares sense. However, unlike Proposition 8.1.9, consistency is not assumed; that is, there need not exist a solution to $Ax = b$. **Related:** Fact 10.17.18.

Fact 11.17.11. Let $A \in \mathbb{F}^{n\times m}$ and $B \in \mathbb{F}^{n\times l}$, and define $f\colon \mathbb{F}^{m\times l} \to \mathbb{R}$ by

$$f(X) \triangleq \operatorname{tr}[(AX - B)^*(AX - B)] = \|AX - B\|_{\mathrm{F}}^2.$$

Then, $X = A^+B$ minimizes f. **Remark:** This is the *orthogonal Procrustes problem.* See [1196, pp. 327, 328]. **Related:** Fact 11.17.12. **Problem:** Determine all minimizers.

Fact 11.17.12. Let $A \in \mathbb{F}^{n\times m}$ and $B \in \mathbb{F}^{l\times m}$, and define $f\colon \mathbb{F}^{l\times n} \to \mathbb{R}$ by

$$f(X) \triangleq \operatorname{tr}[(XA - B)^*(XA - B)] = \|XA - B\|_{\mathrm{F}}^2.$$

Then, $X = BA^+$ minimizes f. **Related:** Fact 11.17.11.

Fact 11.17.13. Let $A, B \in \mathbb{F}^{n\times m}$, and define $f\colon \mathrm{U}(m) \to \mathbb{R}$ by

$$f(X) \triangleq \operatorname{tr}[(AX - B)^*(AX - B)] = \|AX - B\|_{\mathrm{F}}^2.$$

Then, $X = S_1S_2$ minimizes f, where $S_1\left[\begin{smallmatrix} \hat{B} & 0 \\ 0 & 0 \end{smallmatrix}\right]S_2$ is the singular value decomposition of A^*B. **Source:** [300, p. 224], [1969, pp. 269, 270], and [1971, p. 375].

Fact 11.17.14. Let $A \in \mathbb{F}^{n\times m}$, $B \in \mathbb{F}^{n\times p}$, and $C \in \mathbb{F}^{q\times m}$, and let $k \ge 1$ satisfy $k < \operatorname{rank} A$. Then,

$$\min_{X\in\{Y\in\mathbb{F}^{p\times q}\colon\ \operatorname{rank} Y\le k\}} \|A - BXC\|_{\mathrm{F}} = \|A - BX_0C\|_{\mathrm{F}},$$

where $X_0 = B^+SC^+$ and S is formed by replacing all but the k largest singular values in the singular value decomposition of BB^+AC^+C by 0's. Furthermore, X_0 is a solution that minimizes $\|X\|_{\mathrm{F}}$. Finally, X_0 is the unique solution if and only if either $\operatorname{rank} BB^+AC^+C \le k$ or both $k \le BB^+AC^+C$ and $\sigma_{k+1}(BB^+AC^+C) < \sigma_k(BB^+AC^+C)$. **Source:** [1081]. **Related:** This result generalizes Fact 11.16.39.

Fact 11.17.15. Let $A, B \in \mathbb{F}^{n\times n}$, and assume that A and B are Hermitian. Then,

$$\min_{S\in\mathrm{U}(n)} \|A - S^*BS\|_{\mathrm{F}} = \sqrt{\sum_{i=1}^{n}[\lambda_i(A) - \lambda_i(B)]^2}.$$

Source: [775]. **Related:** Fact 7.13.20.

Fact 11.17.16. Let $A \in \mathbb{F}^{n\times m}$, assume that $m \le n$, let $k \le \operatorname{rank} A$, let $S, S_1, S_2 \in \mathbb{F}^{m\times k}$, assume that S, S_1, and S_2 are left inner, and assume that $A^*AS_1 = S_1\operatorname{diag}[\sigma_{m-k+1}^2(A), \ldots, \sigma_m^2(A)]$ and $A^*AS_2 = S_2\operatorname{diag}[\sigma_1^2(A), \ldots, \sigma_k^2(A)]$. Then, $\|AS_1\|_{\mathrm{F}} \le \|AS\|_{\mathrm{F}} \le \|AS_2\|_{\mathrm{F}}$. **Source:** [775]. **Related:** Fact 7.13.19.

Fact 11.17.17. Let $A, B \in \mathbb{R}^{n\times n}$, and define

$$f(X_1, X_2) \triangleq \operatorname{tr}[(X_1AX_2 - B)^{\mathsf{T}}(X_1AX_2 - B)] = \|X_1AX_2 - B\|_{\mathrm{F}}^2,$$

where $X_1, X_2 \in \mathbb{R}^{n\times n}$ are orthogonal. Then, $(X_1, X_2) = (V_2^{\mathsf{T}}U_1^{\mathsf{T}}, V_1^{\mathsf{T}}U_2^{\mathsf{T}})$ minimizes f, where $U_1\left[\begin{smallmatrix}\hat{A} & 0\\ 0 & 0\end{smallmatrix}\right]V_1$ is the singular value decomposition of A and $U_2\left[\begin{smallmatrix}\hat{B} & 0\\ 0 & 0\end{smallmatrix}\right]V_2$ is the singular value decomposition of B. **Source:** [1969, p. 270] and [1971, p. 375]. **Credit:** W. Kristof. **Related:** Fact 4.11.5. **Problem:** Extend this result to $\mathbb{C}$ and nonsquare matrices.

Fact 11.17.18. Let $A \in \mathbb{R}^{n\times m}$, let $b \in \mathbb{R}^n$, and assume that $\operatorname{rank}[A\ \ b] = m + 1$. Furthermore, consider the singular value decomposition of $[A\ \ b]$ given by

$$[A\ \ b] = U\begin{bmatrix}\Sigma\\ 0_{(n-m-1)\times(m+1)}\end{bmatrix}V,$$

where $U \in \mathbb{R}^{n\times n}$ and $V \in \mathbb{R}^{(m+1)\times(m+1)}$ are orthogonal and $\Sigma \triangleq \operatorname{diag}[\sigma_1(A), \ldots, \sigma_{m+1}(A)]$. Furthermore, define $\hat{A} \in \mathbb{R}^{n\times m}$ and $\hat{b} \in \mathbb{R}^n$ by

$$[\hat{A}\ \ \hat{b}] \triangleq U\begin{bmatrix}\Sigma_0\\ 0_{(n-m-1)\times(m+1)}\end{bmatrix}V,$$

where $\Sigma_0 \triangleq \operatorname{diag}[\sigma_1(A), \ldots, \sigma_m(A), 0]$. Finally, assume that $V_{(m+1,m+1)} \neq 0$, and define

$$\hat{x} \triangleq -\frac{1}{V_{(m+1,m+1)}}\begin{bmatrix}V_{(m+1,1)}\\ \vdots\\ V_{(m+1,m)}\end{bmatrix}.$$

Then, $\hat{A}\hat{x} = \hat{b}$. **Remark:** $\hat{x}$ is the *total least squares solution*. See [2773]. **Related:** The construction of $[\hat{A}\ \ \hat{b}]$ is based on Fact 11.16.39.

11.18 Notes

The equivalence of absolute and monotone norms given by Proposition 11.1.2 is given in [329]. More general monotonicity conditions are considered in [1544]. Induced lower bounds are treated in [1738, pp. 369, 370]. See also [2539, pp. 33, 80]. The induced norms (11.4.13) and (11.4.14) are given in [686] and [1389, p. 116]. Alternative norms for the convolution operator are given in [686, 2888]. Proposition 11.3.6 is given in [2311, p. 97]. Norm-related topics are discussed in [351]. Spectral perturbation theory in finite and infinite dimensions is treated in [1590], where the emphasis is on the regularity of the spectrum as a function of the perturbation rather than on bounds for finite perturbations. The trace norm is also called the *nuclear norm*.

Chapter Twelve
Functions, Limits, Sequences, Series, Infinite Products, and Derivatives

The norms discussed in Chapter 11 provide the foundation for the development in this chapter of some basic results in topology and analysis.

12.1 Open Sets and Closed Sets

Definition 12.1.1. Let $\|\cdot\|$ be a norm on $\mathbb{F}^n$, let $x \in \mathbb{F}^n$, and let $\varepsilon > 0$. Then, the *open ball of radius ε centered at x* is defined by

$$\mathbb{B}_\varepsilon(x) \triangleq \{y \in \mathbb{F}^n\colon\ \|x-y\| < \varepsilon\}, \tag{12.1.1}$$

and the *sphere of radius ε centered at x* is defined by

$$\mathbb{S}_\varepsilon(x) \triangleq \{y \in \mathbb{F}^n\colon\ \|x-y\| = \varepsilon\}. \tag{12.1.2}$$

It follows from Theorem 11.1.8 on the equivalence of norms on $\mathbb{F}^n$ that the following definitions are independent of the norm assigned to $\mathbb{F}^n$.

Definition 12.1.2. Let $x \in \mathcal{S} \subseteq \mathbb{F}^n$. Then, x is an *interior point* of $\mathcal{S}$ if there exists $\varepsilon > 0$ such that $\mathbb{B}_\varepsilon(x) \subseteq \mathcal{S}$. The *interior* of $\mathcal{S}$ is the set

$$\operatorname{int}\mathcal{S} \triangleq \{x \in \mathcal{S}\colon\ x \text{ is an interior point of } \mathcal{S}\}. \tag{12.1.3}$$

Finally, $\mathcal{S}$ is *open* if $\mathcal{S} = \operatorname{int}\mathcal{S}$.

Note that $\operatorname{int}\mathcal{S} \subseteq \mathcal{S}$. Hence, $\mathcal{S}$ is open if and only if every element of $\mathcal{S}$ is an interior point of $\mathcal{S}$.

Definition 12.1.3. Let $x \in \mathcal{S} \subseteq \mathcal{S}' \subseteq \mathbb{F}^n$. Then, x is an *interior point* of $\mathcal{S}$ *relative* to $\mathcal{S}'$ if there exists $\varepsilon > 0$ such that $\mathbb{B}_\varepsilon(x) \cap \mathcal{S}' \subseteq \mathcal{S}$; equivalently, $\mathbb{B}_\varepsilon(x) \cap \mathcal{S} = \mathbb{B}_\varepsilon(x) \cap \mathcal{S}'$. The *interior* of $\mathcal{S}$ *relative* to $\mathcal{S}'$ is the set

$$\operatorname{int}_{\mathcal{S}'}\mathcal{S} \triangleq \{x \in \mathcal{S}\colon\ x \text{ is an interior point of } \mathcal{S} \text{ relative to } \mathcal{S}'\}. \tag{12.1.4}$$

In particular, the *relative interior* of $\mathcal{S}$ is

$$\operatorname{relint}\mathcal{S} \triangleq \operatorname{int}_{\operatorname{affin}\mathcal{S}}\mathcal{S}. \tag{12.1.5}$$

Finally, $\mathcal{S}$ is *open relative* to $\mathcal{S}'$ if $\mathcal{S} = \operatorname{int}_{\mathcal{S}'}\mathcal{S}$, and $\mathcal{S}$ is *relatively open* if $\mathcal{S} = \operatorname{relint}\mathcal{S}$.

As an example, the interval $[0,1)$ is open relative to the interval $[0,2]$.

Proposition 12.1.4. Let $\mathcal{S} \subseteq \mathbb{F}^n$, let $\mathcal{S}' \subseteq \mathbb{F}^n$, and assume that $\mathcal{S} \subseteq \operatorname{int}\mathcal{S}' \subseteq \mathbb{F}^n$. Then, $\operatorname{int}_{\mathcal{S}'}\mathcal{S} = \operatorname{int}\mathcal{S}$. Furthermore, $\mathcal{S} \subseteq \mathbb{F}^n$ is open if and only if $\mathcal{S}$ is open relative to $\mathcal{S}'$. In particular, $\mathcal{S} \subseteq \mathbb{F}^n$ is open if and only if $\mathcal{S}$ is open relative to $\mathbb{F}^n$.

Proposition 12.1.5. Let $\mathcal{S} \subseteq \mathcal{S}' \subseteq \mathcal{S}'' \subseteq \mathbb{F}^n$. Then,

$$\operatorname{int}_{\mathcal{S}''}\mathcal{S} \subseteq \operatorname{int}_{\mathcal{S}'}\mathcal{S}. \tag{12.1.6}$$

In particular,

$$\operatorname{int}\mathcal{S} \subseteq \operatorname{relint}\mathcal{S}. \tag{12.1.7}$$

Definition 12.1.6. Let $\mathcal{S} \subseteq \mathbb{F}^n$ and $x \in \mathcal{S}$. Then, $x \in \mathbb{F}^n$ is a *closure point* of $\mathcal{S}$ if, for all $\varepsilon > 0$, $\mathbb{B}_\varepsilon(x) \cap \mathcal{S}$ is nonempty. The *closure* of $\mathcal{S}$ is the set

$$\operatorname{cl} \mathcal{S} \triangleq \{x \in \mathbb{F}^n\colon\ x \text{ is a closure point of } \mathcal{S}\}. \tag{12.1.8}$$

The vector $x \in \mathbb{F}^n$ is an *essential closure point* of $\mathcal{S}$ if, for all $\varepsilon > 0$, the set $[\mathbb{B}_\varepsilon(x)\backslash\{x\}] \cap \mathcal{S}$ is nonempty. The *essential closure* of $\mathcal{S}$ is the set

$$\operatorname{ecl} \mathcal{S} \triangleq \{x \in \mathbb{F}^n\colon\ x \text{ is an essential closure point of } \mathcal{S}\}. \tag{12.1.9}$$

Finally, $\mathcal{S}$ is *closed* if $\mathcal{S} = \operatorname{cl} \mathcal{S}$.

Note that $\mathcal{S} \subseteq \operatorname{cl} \mathcal{S}$. Hence, $\mathcal{S}$ is closed if and only if every closure point of $\mathcal{S}$ is an element of $\mathcal{S}$.

For example, let $\mathcal{S} \triangleq (0, 1)$. Then, $\operatorname{ecl} \mathcal{S} = \operatorname{cl} \mathcal{S} = [0, 1]$. As another example, let $\mathcal{S} \triangleq (0, 1) \cup \{2\}$. Then, $[0, 1] = \operatorname{ecl} \mathcal{S} \subset \operatorname{cl} \mathcal{S} = [0, 1] \cup \{2\}$. As another example, let $\mathcal{S} = \{1, 1/2, 1/3, \ldots\}$. Then, $\{0\} = \operatorname{ecl} \mathcal{S} \subset \operatorname{cl} \mathcal{S} = \mathcal{S} \cup \{0\}$. Finally, every affine subspace is closed.

Definition 12.1.7. Let $\mathcal{S} \subseteq \mathbb{F}^n$ and $x \in \mathcal{S}$. Then, $x \in \mathcal{S}$ is an *isolated point* of $\mathcal{S}$ if there exists $\varepsilon > 0$ such that $\mathbb{B}_\varepsilon(x) \cap \mathcal{S} = \{x\}$. The *isolated subset* of $\mathcal{S}$ is the set

$$\operatorname{iso} \mathcal{S} \triangleq \{x \in \mathcal{S}\colon\ x \text{ is an isolated point of } \mathcal{S}\}. \tag{12.1.10}$$

Proposition 12.1.8. Let $\mathcal{S} \subseteq \mathbb{F}^n$. Then,

$$\left.\begin{array}{r} \operatorname{iso} \mathcal{S} = \mathcal{S}\backslash(\operatorname{ecl} \mathcal{S}) \subseteq \mathcal{S} \\ \operatorname{cl}\operatorname{relint} \mathcal{S} \subseteq \operatorname{ecl} \mathcal{S} = \operatorname{cl}\operatorname{ecl} \mathcal{S} = (\operatorname{cl} \mathcal{S})\backslash(\operatorname{iso} \mathcal{S}) \end{array}\right\} \subseteq \operatorname{cl} \mathcal{S} = \mathcal{S} \cup \operatorname{ecl} \mathcal{S} = (\operatorname{iso} \mathcal{S}) \cup (\operatorname{ecl} \mathcal{S}). \tag{12.1.11}$$

Furthermore, the following statements hold:

i) $\operatorname{ecl} \mathcal{S}$ and $\operatorname{iso} \mathcal{S}$ are disjoint.

ii) $\operatorname{ecl} \mathcal{S}$ is closed.

iii) $\operatorname{ecl} \mathcal{S} \subseteq \mathcal{S}$ if and only if $\mathcal{S}$ is closed.

iv) $\operatorname{ecl} \mathcal{S} = \operatorname{cl} \mathcal{S}$ if and only if $\operatorname{iso} \mathcal{S}$ is empty.

v) $\operatorname{ecl} \mathcal{S} = \mathcal{S}$ if and only if $\mathcal{S}$ is closed and $\operatorname{iso} \mathcal{S}$ is empty.

vi) If $\mathcal{S}_0 \subseteq \mathcal{S}$, then $\operatorname{ecl} \mathcal{S}_0 \subseteq \operatorname{ecl} \mathcal{S}$.

vii) If $\mathcal{S}_0 \subseteq \mathcal{S}$ and $\mathcal{S}\backslash\mathcal{S}_0$ is finite, then $\operatorname{ecl} \mathcal{S}_0 = \operatorname{ecl} \mathcal{S}$.

viii) If $\mathcal{S}$ is finite, then $\operatorname{ecl} \mathcal{S}$ is empty.

Definition 12.1.9. Let $\mathcal{S} \subseteq \mathcal{S}' \subseteq \mathbb{F}^n$. Then, the *closure of* $\mathcal{S}$ *relative to* $\mathcal{S}'$ is the set

$$\operatorname{cl}_{\mathcal{S}'} \mathcal{S} \triangleq \{x \in \mathcal{S}'\colon\ x \text{ is a closure point of } \mathcal{S}\}. \tag{12.1.12}$$

The *essential closure of* $\mathcal{S}$ *relative to* $\mathcal{S}'$ is the set

$$\operatorname{ecl}_{\mathcal{S}'} \mathcal{S} \triangleq \{x \in \mathcal{S}'\colon\ x \text{ is an essential closure point of } \mathcal{S}\}. \tag{12.1.13}$$

Finally, $\mathcal{S}$ is *closed relative* to $\mathcal{S}'$ if $\mathcal{S} = \operatorname{cl}_{\mathcal{S}'} \mathcal{S}$, and $\mathcal{S}$ is *essentially closed relative* to $\mathcal{S}'$ if $\mathcal{S} = \operatorname{ecl}_{\mathcal{S}'} \mathcal{S}$.

As an example, the interval $(0, 1]$ is closed relative to the interval $(0, 2]$ and essentially closed relative to the interval $(0, 2]$. As another example, the set $(0, 1] \cup \{2\}$ is closed relative to $(0, \infty)$ but is not essentially closed relative to $(0, \infty)$.

Note that $\mathcal{S} \subseteq \mathbb{F}^n$ is closed if and only if $\mathcal{S}$ is closed relative to $\mathbb{F}^n$. Furthermore, $\mathcal{S} \subseteq \mathbb{F}^n$ is essentially closed if and only if $\mathcal{S}$ is essentially closed relative to $\mathbb{F}^n$.

The empty set is both open and closed, and $\operatorname{int} \varnothing = \operatorname{cl} \varnothing = \operatorname{int}_\varnothing \varnothing = \operatorname{cl}_\varnothing \varnothing = \operatorname{relint} \varnothing = \varnothing$.

Proposition 12.1.10. Let $\mathcal{S} \subseteq \mathcal{S}' \subseteq \mathcal{S}'' \subseteq \mathbb{F}^n$. Then,

$$\operatorname{cl}_{\mathcal{S}'} \mathcal{S} \subseteq \operatorname{cl}_{\mathcal{S}''} \mathcal{S}. \tag{12.1.14}$$

If, in addition, $\mathcal{S}'$ is closed, then

$$\operatorname{cl}_{\mathcal{S}'} \mathcal{S} = \operatorname{cl} \mathcal{S}. \tag{12.1.15}$$

In particular,

$$\operatorname{cl}_{\operatorname{affin} \mathcal{S}} \mathcal{S} = \operatorname{cl} \mathcal{S}. \tag{12.1.16}$$

Let $\mathcal{S} \subseteq \mathcal{S}' \subseteq \mathbb{F}^n$. Then,

$$\operatorname{cl}_{\mathcal{S}'} \mathcal{S} = (\operatorname{cl} \mathcal{S}) \cap \mathcal{S}', \tag{12.1.17}$$

$$\operatorname{int}_{\mathcal{S}'} \mathcal{S} = \mathcal{S}' \backslash \operatorname{cl}(\mathcal{S}' \backslash \mathcal{S}), \tag{12.1.18}$$

$$\operatorname{int} \mathcal{S} \subseteq \operatorname{int}_{\mathcal{S}'} \mathcal{S} \subseteq \mathcal{S} \subseteq \operatorname{cl}_{\mathcal{S}'} \mathcal{S} \subseteq \operatorname{cl} \mathcal{S}. \tag{12.1.19}$$

In particular,

$$\operatorname{cl} \mathcal{S} = (\operatorname{cl} \mathcal{S}) \cap \operatorname{affin} \mathcal{S}, \tag{12.1.20}$$

$$\operatorname{relint} \mathcal{S} = (\operatorname{affin} \mathcal{S}) \backslash \operatorname{cl}[(\operatorname{affin} \mathcal{S}) \backslash \mathcal{S}], \tag{12.1.21}$$

$$\operatorname{int} \mathcal{S} = [\operatorname{cl}(\mathcal{S}^{\sim})]^{\sim}, \tag{12.1.22}$$

$$\operatorname{int} \mathcal{S} \subseteq \operatorname{relint} \mathcal{S} \subseteq \mathcal{S} \subseteq \operatorname{cl} \mathcal{S}. \tag{12.1.23}$$

Definition 12.1.11. Let $\mathcal{S} \subseteq \mathcal{S}' \subseteq \mathbb{F}^n$. Then, the *boundary* of $\mathcal{S}$ *relative to* $\mathcal{S}'$ is the set

$$\operatorname{bd}_{\mathcal{S}'} \mathcal{S} \triangleq (\operatorname{cl}_{\mathcal{S}'} \mathcal{S}) \backslash \operatorname{int}_{\mathcal{S}'} \mathcal{S}. \tag{12.1.24}$$

In particular, the *boundary* of $\mathcal{S}$ is the set

$$\operatorname{bd} \mathcal{S} \triangleq (\operatorname{cl} \mathcal{S}) \backslash \operatorname{int} \mathcal{S}. \tag{12.1.25}$$

and the *relative boundary* of $\mathcal{S}$ is

$$\operatorname{relbd} \mathcal{S} \triangleq (\operatorname{cl} \mathcal{S}) \backslash \operatorname{relint} \mathcal{S}. \tag{12.1.26}$$

Note that, if $\mathcal{S} \subseteq \mathbb{F}^n$, then

$$\operatorname{cl} \mathcal{S} = (\operatorname{relint} \mathcal{S}) \cup \operatorname{relbd} \mathcal{S}. \tag{12.1.27}$$

Definition 12.1.12. Let $\mathcal{S} \subseteq \mathbb{F}^n$. Then, $\mathcal{S}$ is *solid* if $\operatorname{int} \mathcal{S}$ is nonempty. Furthermore, $\mathcal{S}$ is *completely solid* if $\mathcal{S}$ is solid and $\operatorname{cl} \operatorname{int} \mathcal{S} = \operatorname{cl} \mathcal{S}$.

The empty set is neither solid nor completely solid, and every nonempty, open set is both solid and completely solid.

Definition 12.1.13. Let $\mathcal{S} \subseteq \mathbb{F}^n$. Then, $\mathcal{S}$ is *bounded* if there exists $\delta > 0$ such that, for all $x, y \in \mathcal{S}$,

$$\|x - y\| < \delta. \tag{12.1.28}$$

The set $\mathcal{S}$ is *compact* if it is both closed and bounded.

Definition 12.1.14. Let $\mathcal{S} \subset \mathbb{F}^n$. Then, $\mathcal{S}$ is *disconnected* if there exist nonempty, disjoint subsets $\mathcal{S}_1, \mathcal{S}_2$ of $\mathcal{S}$ that are open relative to $\mathcal{S}$ and whose union is $\mathcal{S}$. Furthermore, $\mathcal{S}$ is *connected* if $\mathcal{S}$ is not disconnected.

The empty set is compact and connected.

12.2 Limits of Sequences

Proposition 12.2.1. Let $(x_i)_{i=1}^{\infty} \subset \mathbb{F}$. Then, there exists at most one $x \in \mathbb{F}$ such that, for all $\varepsilon > 0$, there exists $n \geq 1$ such that, for all $i > n$, $|x_i - x| < \varepsilon$.

Definition 12.2.2. Let $X \triangleq (x_i)_{i=1}^\infty \subset \mathbb{F}$. Then, X *converges* to the *limit* $x \in \mathbb{F}$ if, for all $\varepsilon > 0$, there exists $n \geq 1$ such that, for all $i > n$, $|x_i - x| < \varepsilon$. In this case, we write either $\lim X = \lim_{i\to\infty} x_i = x$ or $x_i \to x$ as $i \to \infty$. Finally, X *converges* and $\lim_{i\to\infty} x_i$ *exists* if there exists $x \in \mathbb{F}$ such that $(x_i)_{i=1}^\infty$ converges to x.

Definition 12.2.3. Let $X \triangleq (x_i)_{i=1}^\infty \subset \mathbb{R}$. Then, X has the *limit* ∞ if, for all $M > 0$, there exists $n \geq 1$ such that, for all $i > n$, $x_i > M$. In this case, we write either $\lim X = \lim_{i\to\infty} x_i = \infty$ or $x_i \to \infty$ as $i \to \infty$. Likewise, X has the limit $-\infty$ if, for all $M > 0$, there exists $n \geq 1$ such that, for all $i > n$, $x_i < -M$. In this case, we write either $\lim X = \lim_{i\to\infty} x_i = -\infty$ or $x_i \to -\infty$ as $i \to \infty$.

Note that $\lim X \in \mathbb{R} \cup \{-\infty, \infty\}$. However, the terminology "converges" and "$\lim_{i\to\infty} x_i$ exists" are used only in the case where $\lim X$ is finite.

Definition 12.2.4. Let $X \triangleq (x_i)_{i=1}^\infty \subset \mathbb{F}^n$. Then, X *converges* to the *limit* $x \in \mathbb{F}^n$ if $\lim_{i\to\infty} \|x - x_i\| = 0$, where $\|\cdot\|$ is a norm on $\mathbb{F}^n$. In this case, we write either $x = \lim_{i\to\infty} x_i$ or $x_i \to x$ as $i \to \infty$, where $i \in \mathbb{P}$. X *converges* if there exists $x \in \mathbb{F}^n$ such that $(x_i)_{i=1}^\infty$ converges to x. Now, let $\mathcal{A} \triangleq (A_i)_{i=1}^\infty \subset \mathbb{F}^{n\times m}$. Then, $\mathcal{A}$ *converges* to $A \in \mathbb{F}^{n\times m}$ if $\lim_{i\to\infty} \|A - A_i\| = 0$, where $\|\cdot\|$ is a norm on $\mathbb{F}^{n\times m}$. In this case, we write either $A = \lim_{i\to\infty} A_i$ or $A_i \to A$ as $i \to \infty$, where $i \in \mathbb{P}$. Finally, $\mathcal{A}$ *converges* and $\lim_{i\to\infty} A_i$ *exists* if there exists $A \in \mathbb{F}^{n\times m}$ such that $(A_i)_{i=1}^\infty$ converges to A.

Theorem 11.1.8 implies that convergence of a sequence is independent of the choice of norm.

Proposition 12.2.5. Let $\mathcal{S} \subseteq \mathbb{F}^n$, and let $x \in \mathbb{F}^n$. Then, $x \in \operatorname{cl} \mathcal{S}$ if and only if there exists a sequence $(x_i)_{i=1}^\infty \subseteq \mathcal{S}$ that converges to x. Furthermore, $x \in \operatorname{ecl} \mathcal{S}$ if and only if there exists a sequence $(x_i)_{i=1}^\infty \subseteq \mathcal{S}\backslash\{x\}$ that converges to x.

Proof. Suppose that $x \in \operatorname{ecl} \mathcal{S}$. Then, for all $i \in \mathbb{P}$, there exists $x_i \in \mathcal{S}\backslash\{x\}$ such that $\|x - x_i\| < 1/i$. Hence, $x - x_i \to 0$ as $i \to \infty$ and thus $(x_i)_{i=1}^\infty$ converges to x. Conversely, suppose that $(x_i)_{i=1}^\infty \subseteq \mathcal{S}\backslash\{x\}$ is such that $x_i \to x$ as $i \to \infty$, and let $\varepsilon > 0$. Then, there exists a positive integer p such that $\|x - x_i\| < \varepsilon$ for all $i > p$. Therefore, $x_{p+1} \in [\mathbb{B}_\varepsilon(x)\backslash\{x\}] \cap \mathcal{S}$, and thus $[\mathbb{B}_\varepsilon(x)\backslash\{x\}] \cap \mathcal{S}$ is nonempty. Hence, x is an essential closure point of $\mathcal{S}$. □

Let $X \triangleq (x_i)_{i=1}^\infty$, where, for all $i \in \mathbb{P}$, $x_i \in \mathbb{F}^n$. Recall from Chapter 1 that X can be viewed as the subset $\{x_1, x_2, \ldots\}$ of $\mathbb{F}^n$, where the multiplicity of the components of the sequence X is ignored. The following result gives necessary conditions for X to converge to $x \in \mathbb{F}^n$.

Proposition 12.2.6. Let $X \triangleq (x_i)_{i=1}^\infty \subset \mathbb{F}^n$, let $x \in \mathbb{F}^n$, and assume that X converges to x. Then, the following statements hold:

i) $\operatorname{ecl} X \subseteq \{x\} \subseteq \operatorname{cl} X = X \cup \{x\} = X \cup \operatorname{ecl} X$.

ii) $X = \operatorname{cl} X$ if and only if $x \in X$.

iii) The following statements are equivalent:

  a) $\operatorname{ecl} X = \varnothing$.

  b) X is finite.

  c) There exists a positive integer l such that, for all $i \geq l$, $x_i = x$.

iv) The following statements are equivalent:

  a) $\operatorname{ecl} X = \{x\}$.

  b) X is not finite.

  c) The subsequence $\hat{X}$ of X obtained by deleting all components of X that are equal to x converges to x.

To illustrate *iii*), let $X = (1, 1/2, 0, 0, 0, \ldots)$. Then, $\lim_{i\to\infty} x_i = 0$, $\operatorname{ecl} X = \varnothing$, $\operatorname{card}(X) = 3$, and $0 \in \operatorname{cl} X = X = \{0, 1/2, 1\}$. To illustrate *iv*), let $X = (0, 1, 1/2, 1/3, 1/4, \ldots)$. Then, $\lim_{i\to\infty} x_i = 0$, $\operatorname{ecl} X = \{0\}$, $\operatorname{card}(X)$ is not finite, and $\{i \in \mathbb{P}\colon x_i = 0\}$ is finite. Also to illustrate *iv*), let $X = (1, 0, 1/2, 0, 1/3, 0, 1/4, 0, \ldots)$. Then, $\lim_{i\to\infty} x_i = 0$, $\operatorname{ecl} X = \{0\}$, $\operatorname{card}(X)$ is not finite, $\{i \in \mathbb{P}\colon x_i = 0\}$

is not finite, and the subsequence $\hat{X}$ of X obtained by deleting all components of X that are equal to 0 converges to 0.

The following result gives a necessary and sufficient condition for $X \triangleq (x_i)_{i=1}^{\infty} \subset \mathbb{F}^n$ to converge to $x \in \mathbb{F}^n$.

Proposition 12.2.7. Let $X \triangleq (x_i)_{i=1}^{\infty} \subset \mathbb{F}^n$, and let $x \in \mathbb{F}^n$. Then, X converges to x if and only if, for all $\varepsilon > 0$, $\{i \in \mathbb{P}\colon \|x - x_i\| > \varepsilon\}$ is finite.

Let $X \triangleq (x_i)_{i=1}^{\infty} \subseteq \mathbb{R}^n$ and $Y \triangleq (y_i)_{i=1}^{\infty} \subseteq \mathbb{R}^n$. Then, Y is a *rearrangement of* X if there exists $\sigma\colon \mathbb{P} \mapsto \mathbb{P}$ such that σ is one-to-one and onto and such that, for all $i \in \mathbb{P}$, $y_i = x_{\sigma(i)}$. Note that Y is a rearrangement of X if and only if X is a rearrangement of Y.

Proposition 12.2.8. Let $X \triangleq (x_i)_{i=1}^{\infty} \subset \mathbb{F}^n$. Then, the following statements are equivalent:

i) X converges.

ii) Every subsequence of X converges to $\lim X$.

iii) Every rearrangement of X converges to $\lim X$.

iv) Every rearrangement of every subsequence of X converges to $\lim X$.

Proof. *i*) $\Longrightarrow$ *iii*) follows from Proposition 12.2.7. □

Theorem 12.2.9. Let $\mathcal{S} \subset \mathbb{F}^n$ be compact, and let $(x_i)_{i=1}^{\infty} \subseteq \mathcal{S}$. Then, there exists a subsequence $(x_{i_j})_{j=1}^{\infty}$ of $(x_i)_{i=1}^{\infty}$ such that $(x_{i_j})_{j=1}^{\infty}$ converges and $\lim_{j\to\infty} x_{i_j} \in \mathcal{S}$.

Proof. See [2112, p. 145]. □

The sequence $X = (x_i)_{i=1}^{\infty} \subset \mathbb{R}$ is (decreasing, nonincreasing, increasing, nondecreasing) if, for all $i \geq 1$, $(x_{i+1} < x_i,\ x_{i+1} \leq x_i,\ x_{i+1} > x_i,\ x_{i+1} \geq x_i)$. Furthermore, $(x_i)_{i=1}^{\infty} \subset \mathbb{R}$ is *bounded* if the set $\{x_1, x_2, \ldots\}$ is bounded. The same terminology is used for sequences of vectors and matrices with a specified partial ordering.

Proposition 12.2.10. Let $X = (x_i)_{i=1}^{\infty} \subset \mathbb{R}$. Then, the following statements hold:

i) X is bounded if and only if $\inf X \in \mathbb{R}$ and $\sup X \in \mathbb{R}$.

ii) Assume that X is nonincreasing. Then, $\lim X \in \mathbb{R} \cup \{-\infty\}$. Furthermore, $\lim X \in \mathbb{R}$ if and only if $\inf X \in \mathbb{R}$.

iii) Assume that $\sup X \in \mathbb{R}$. Then, $Y \triangleq (\sup\, (x_j)_{j=i}^{\infty})_{i=1}^{\infty} \subset \mathbb{R}$, and Y is nonincreasing. Furthermore, $\lim Y \in \mathbb{R}$ if and only if $\inf X \in \mathbb{R}$.

iv) Assume that X is nondecreasing. Then, $\lim X \in \mathbb{R} \cup \{\infty\}$. Furthermore, $\lim X \in \mathbb{R}$ if and only if $\sup X \in \mathbb{R}$.

v) Assume that $\inf X \in \mathbb{R}$. Then, $Y \triangleq (\inf\, (x_j)_{j=i}^{\infty})_{i=1}^{\infty} \subset \mathbb{R}$, and Y is nondecreasing. Furthermore, $\lim Y \in \mathbb{R}$ if and only if $\sup X \in \mathbb{R}$.

Definition 12.2.11. Let $X = (x_i)_{i=1}^{\infty} \subset \mathbb{R}$. Then, the *limit inferior* of $(x_i)_{i=1}^{\infty}$ is defined by

$$\liminf X \triangleq \liminf_{i\to\infty} x_i \triangleq \begin{cases} -\infty, & \inf X = -\infty, \\ \lim_{i\to\infty} \inf\, (x_j)_{j=i}^{\infty}, & \inf X \in \mathbb{R}. \end{cases} \tag{12.2.1}$$

Furthermore, the *limit superior* of $(x_i)_{i=1}^{\infty}$ is defined by

$$\limsup X \triangleq \limsup_{i\to\infty} x_i \triangleq \begin{cases} \lim_{i\to\infty} \sup\, (x_j)_{j=i}^{\infty}, & \sup X \in \mathbb{R}, \\ \infty, & \sup X = \infty. \end{cases} \tag{12.2.2}$$

Proposition 12.2.12. Let $X = (x_i)_{i=1}^{\infty} \subset \mathbb{R}$. Then, the following statements hold:

i) $\inf X \leq \liminf X \leq \limsup X \leq \sup X$.

ii) $\inf X \neq \infty$, and $\sup X \neq -\infty$.

iii) $\liminf X \in \mathbb{R} \cup \{-\infty, \infty\}$, and $\limsup X \in \mathbb{R} \cup \{-\infty, \infty\}$.

iv) $\inf X = -\infty$ if and only if $\liminf X = -\infty$.

v) If $\liminf X \in \mathbb{R}$, then $\inf X \in \mathbb{R}$.

vi) If $\inf X \in \mathbb{R}$, then $\liminf X \in \mathbb{R} \cup \{\infty\}$.

vii) $\sup X = \infty$ if and only if $\limsup X = \infty$.

viii) If $\limsup X \in \mathbb{R}$, then $\sup X \in \mathbb{R}$.

ix) If $\sup X \in \mathbb{R}$, then $\limsup X \in \mathbb{R} \cup \{-\infty\}$.

x) $\liminf X \in \mathbb{R}$ and $\limsup X \in \mathbb{R}$ if and only if X is bounded.

xi) X converges if and only if $\liminf X = \limsup X \in \mathbb{R}$.

xii) If X converges, then X is bounded and

$$-\infty < \inf X \le \liminf X = \lim X = \limsup X \le \sup X < \infty. \tag{12.2.3}$$

xiii) There exists a subsequence $(x_{i_j})_{j=1}^{\infty}$ of $(x_i)_{i=1}^{\infty}$ such that $\lim_{j\to\infty} x_{i_j} = \liminf_{i\to\infty} x_i$.

xiv) There exists a subsequence $(x_{i_j})_{j=1}^{\infty}$ of $(x_i)_{i=1}^{\infty}$ such that $\lim_{j\to\infty} x_{i_j} = \limsup_{i\to\infty} x_i$.

Definition 12.2.13. Let $X = (x_i)_{i=1}^{\infty} \subset \mathbb{R}$, and let $\alpha \ge 0$. Then, as $n \to \infty$,

$$x_n = O(n^{-\alpha}) \tag{12.2.4}$$

if $(i^{\alpha} x_i)_{i=1}^{\infty}$ is bounded. Furthermore, as $n \to \infty$,

$$x_n = o(n^{-\alpha}) \tag{12.2.5}$$

if $\lim_{n\to\infty} n^{\alpha} x_n = 0$.

Note that, as $n \to \infty$, $x_n = O(1)$ means that X is bounded, whereas $x_n = o(1)$ means that X converges to zero. Furthermore, if $\alpha > \beta \ge 0$ and, as $n \to \infty$, $x_n = O(n^{-\alpha})$, then, as $n \to \infty$, $x_n = o(n^{-\beta})$.

Definition 12.2.14. Let $X = (x_i)_{i=1}^{\infty} \subset \mathbb{R}$ and $Y = (y_i)_{i=1}^{\infty} \subset \mathbb{R}$, and assume that Y has a finite number of components that are zero. Then, as $n \to \infty$,

$$x_n \sim y_n \tag{12.2.6}$$

if $\lim_{n\to\infty} x_n/y_n = 1$.

Proposition 12.2.15. Let $X = (x_i)_{i=1}^{\infty} \subset \mathbb{R}$ and $Y = (y_i)_{i=1}^{\infty} \subset \mathbb{R}$, let α and β be nonnegative numbers, assume that $\beta \le \alpha$, assume that Y has a finite number of components that are zero, assume that $y_n = O(n^{-\beta})$, and assume that, as $n \to \infty$, $x_n = y_n + o(n^{-\alpha})$. Then, as $n \to \infty$, $x_n \sim y_n$.

Definition 12.2.16. Let $X = (x_i)_{i=1}^{\infty} \subset \mathbb{F}$ and $p \in [1, \infty)$. Then, $\|X\|_p \triangleq (\sum_{i=1}^{\infty} |x_i|^p)^{1/p}$, and

$$\ell_p \triangleq \{Y = (y_i)_{i=1}^{\infty} \subset \mathbb{F}\colon \|Y\|_p \in [0, \infty)\}. \tag{12.2.7}$$

Furthermore, $\|X\|_\infty \triangleq \sup_{i\ge 1} |x_i|$, and

$$\ell_\infty \triangleq \{Y = (y_i)_{i=1}^{\infty} \subset \mathbb{F}\colon \|Y\|_\infty \in [0, \infty)\}. \tag{12.2.8}$$

Finally,

$$c_0 \triangleq \{Y = (y_i)_{i=1}^{\infty} \subset \mathbb{F}\colon \lim_{i\to\infty} y_i = 0\}. \tag{12.2.9}$$

Proposition 12.2.17. Let $X = (x_i)_{i=1}^{\infty} \subset \mathbb{F}$, $Y = (y_i)_{i=1}^{\infty} \subset \mathbb{F}$, and $p, q \in [1, \infty]$. Then, the following statements hold:

i) $\|X\|_p = 0$ if and only if, for all $i \ge 1$, $x_i = 0$.

ii) If $\alpha \in \mathbb{F}$ and $X \in \ell_p$, then $\|\alpha X\|_p = |\alpha| \|X\|_p$.

iii) If $p \le q$, then $\|X\|_\infty \le \|X\|_q \le \|X\|_p \le \|X\|_1$.

iv) If $1 < p < q < \infty$, then $\ell_1 \subset \ell_p \subset \ell_q \subset c_0 \subset \ell_\infty$.

v) If $p \in (1, \infty)$ and $X \in \ell_p$, then $\lim_{r\to\infty} \|X\|_r = \|X\|_\infty$.

vi) If $X, Y \in \ell_p$, then $\|X + Y\|_p \le \|X\|_p + \|Y\|_p$ and thus $X + Y \in \ell_p$.

vii) If $X, Y \in c_0$, then $X + Y \in c_0$.

viii) For all $i \ge 1$, let $X_i \in \ell_p$ and assume that, for all $\varepsilon > 0$, there exists $n \ge 1$ such that, for all $k, l > n$, $\|X_k - X_l\|_p < \varepsilon$. Then, there exists $X \in \ell_p$ such that $\lim_{i\to\infty} \|X_i - X\|_p = 0$.

ix) For all $i \ge 1$, let $X_i \in c_0$ and assume that, for all $\varepsilon > 0$, there exists $n \ge 1$ such that, for all $k, l > n$, $\|X_k - X_l\|_\infty < \varepsilon$. Then, there exists $X \in c_0$ such that $\lim_{i\to\infty} \|X_i - X\|_\infty = 0$.

x) For all $i \ge 1$, let $X_i \in \ell_p$ and assume that $\sum_{i=1}^\infty \|X_i\|_p < \infty$. Then, there exists $X \in \ell_p$ such that $\lim_{i\to\infty} \| \sum_{j=1}^i X_j - X\|_p = 0$.

xi) For all $i \ge 1$, let $X_i \in c_0$ and assume that $\sum_{i=1}^\infty \|X_i\|_\infty < \infty$. Then, there exists $X \in c_0$ such that $\lim_{i\to\infty} \| \sum_{j=1}^i X_j - X\|_\infty = 0$.

xii) For all $i \ge 1$, let $X_i = (x_{i,j})_{j=1}^\infty \in \ell_p$, and assume that, for all $j \ge 1$, $x_{\infty,j} \triangleq \lim_{i\to\infty} x_{i,j}$ exists, define $X \triangleq (x_{\infty,j})_{j=1}^\infty$, and assume that $(\|X_i\|_p)_{i=1}^\infty$ is bounded. Then, $X \in \ell_p$.

For all $i \ge 1$, define $X_i = (x_{i,j})_{j=1}^\infty$, where $x_{i,j} \triangleq (\frac{i}{i+1})^{j-1}$, and let $p \in [1, \infty)$. Then, for all $i \ge 1$,

$$\|X_i\|_p = \frac{i+1}{\sqrt[p]{(i+1)^p - i^p}}.$$

Hence, $X_i \in \ell_p$. Note that, for all $j \ge 1$, $\lim_{i\to\infty} x_{i,j} = 1$. However, the limiting sequence $(1, 1, 1, \ldots)$ is not an element of ℓ_p. Furthermore, for all $i \ge 1$, $X_i \in c_0$, but $(1, 1, 1, \ldots)$ is not an element of c_0. Alternatively, define $X_i = (x_{i,j})_{j=1}^\infty$ by $x_{i,j} = \min\{i, j\}$. Then, for all $i \ge 1$, $\|X_i\|_\infty = i$, and, for all $j \ge 1$, $\lim_{i\to\infty} x_{i,j} = j$. However, the limiting sequence $(1, 2, 3, \ldots)$ is not an element of ℓ_∞. These examples show that, for all $p \in [1, \infty]$, the component-wise limit of a sequence of elements of ℓ_p is not necessarily an element of ℓ_p, and likewise for c_0. Finally, for all $i \ge 1$, define $X_i = (x_{i,j})_{j=1}^\infty \in c_0$, where, for all $j \in \{1, \ldots, i\}$, $x_{i,j} = 1$ and, for all $j \ge i + 1$, $x_{i,j} = 0$. Then, for all $i \ge 1$, $\|X_i\|_\infty = 1$ and, for all $j \ge 1$, $\lim_{i\to\infty} x_{i,j} = 1$. Since the limiting sequence $(1, 1, 1, \ldots)$ is not an element of c_0, it follows that *xii)* does not hold in the case where ℓ_p is replaced by c_0 and $\|\cdot\|_p$ is replaced by $\|\cdot\|_\infty$.

12.3 Series, Power Series, and Bi-power Series

Consider the sequence $(x_i)_{i=1}^\infty \subset \mathbb{F}^n$. Then, the sequence of *partial sums* $(\sum_{i=1}^k x_i)_{k=1}^\infty$ is a *series*. The limit of this series is denoted by $\sum_{i=1}^\infty x_i$. For convenience, we denote the series $(\sum_{i=1}^k x_i)_{k=1}^\infty$ by $\sum_{i=1}^\infty x_i$ whether or not $(\sum_{i=1}^k x_i)_{k=1}^\infty$ converges.

Definition 12.3.1. Let $(x_i)_{i=1}^\infty \subset \mathbb{F}^n$, and let $\|\cdot\|$ be a norm on $\mathbb{F}^n$. Then, the series $\sum_{i=1}^\infty x_i$ *converges* if $(\sum_{i=1}^k x_i)_{k=1}^\infty$ converges. Furthermore, $\sum_{i=1}^\infty x_i$ *converges absolutely* if the series $\sum_{i=1}^\infty \|x_i\|$ converges.

Proposition 12.3.2. Let $(x_i)_{i=1}^\infty \subset \mathbb{F}^n$, and assume that the series $\sum_{i=1}^\infty x_i$ converges absolutely. Then, the series $\sum_{i=1}^\infty x_i$ converges.

Definition 12.3.3. Let $(A_i)_{i=1}^\infty \subset \mathbb{F}^{n\times m}$, and let $\|\cdot\|$ be a norm on $\mathbb{F}^{n\times m}$. Then, the series $\sum_{i=1}^\infty A_i$ *converges* if $(\sum_{i=1}^k A_i)_{k=1}^\infty$ converges. Furthermore, $\sum_{i=1}^\infty A_i$ *converges absolutely* if the series $\sum_{i=1}^\infty \|A_i\|$ converges.

Proposition 12.3.4. Let $(A_i)_{i=1}^\infty \subset \mathbb{F}^{n\times m}$, and assume that the series $\sum_{i=1}^\infty A_i$ converges absolutely. Then, the series $\sum_{i=1}^\infty A_i$ converges.

Definition 12.3.5. Let $(\beta_i)_{i=0}^\infty \subset \mathbb{C}$, let $z_0 \in \mathbb{C}$, let $\mathcal{D} \subseteq \mathbb{C}$ denote the set of all $z \in \mathbb{C}$ such that

$\sum_{i=0}^{\infty}\beta_i(z-z_0)^i$ converges, and define $f\colon \mathcal{D}\mapsto\mathbb{C}$ by

$$f(z) \triangleq \sum_{i=0}^{\infty}\beta_i(z-z_0)^i. \tag{12.3.1}$$

Then, f is a *power series*, and $\mathcal{D}$ is the *domain of convergence of* f. For all $z\in\mathcal{D}$, the power series f *converges at* z. The power series f is *absolutely convergent at* z if $\sum_{i=0}^{n}|\beta_i||z-z_0|^i$ converges. Finally, f is a *generating function* for the sequence $(\beta_i)_{i=0}^{\infty}$.

Proposition 12.3.6. Let $(\beta_i)_{i=0}^{\infty}\subset\mathbb{C}$, let $z_0\in\mathbb{C}$, and define the power series f with domain of convergence $\mathcal{D}$ by (12.3.1). Then, exactly one of the following statements holds:

i) $\mathcal{D}=\{z_0\}$.

ii) There exist $\rho>0$ and $\mathcal{D}_\rho\subseteq\mathbb{S}_\rho(z_0)$ such that

$$\mathcal{D}=\{z\in\mathbb{C}\colon |z-z_0|<\rho\}\cup\mathcal{D}_\rho. \tag{12.3.2}$$

iii) $\mathcal{D}=\mathbb{C}$.

In cases *ii*) and *iii*), (12.3.1) converges absolutely for all $z\in\operatorname{int}\mathcal{D}$.

Proof. See [2113, pp. 67–69]. □

As an example, consider the power series $f(z)\triangleq\sum_{i=1}^{\infty}\frac{1}{i}z^i$. The domain of convergence $\mathcal{D}$ of f is CIUD$\backslash\{1\}$, which is case *ii*) of Proposition 12.3.6. Furthermore, f converges absolutely for all $z\in$ OIUD. However, for all $z\in$ UC, f does not converge absolutely.

Consider the power series f defined by (12.3.1). Then, the domain of convergence $\mathcal{D}$ of f is either the point z_0, the union of $\mathbb{B}_\rho(z_0)$ and a subset of $\mathbb{S}_\rho(z_0)$, or the entire complex plane. ρ is the *radius of convergence of* f. In case *iii*), we set $\rho=\infty$.

The following result gives the *Cauchy-Hadamard formula*, *d'Alembert's ratio test*, and *Cauchy's root test*, respectively, for the radius of convergence of a power series.

Proposition 12.3.7. Let $(\beta_i)_{i=1}^{\infty}\subset\mathbb{C}$, let $z_0\in\mathbb{C}$, and define the power series f by (12.3.1) with domain of convergence $\mathcal{D}$ and radius of convergence ρ. Then,

$$\rho=\frac{1}{\limsup_{i\to\infty}|\beta_i|^{1/i}}. \tag{12.3.3}$$

Furthermore, if either $\lim_{i\to\infty}|\beta_i|^{1/i}$ exists or is ∞, then

$$\rho=\frac{1}{\lim_{i\to\infty}|\beta_i|^{1/i}}. \tag{12.3.4}$$

Finally, if either $\lim_{i\to\infty}\left|\frac{\beta_i}{\beta_{i+1}}\right|$ exists or is ∞, then

$$\rho=\lim_{i\to\infty}\left|\frac{\beta_i}{\beta_{i+1}}\right|. \tag{12.3.5}$$

Definition 12.3.8. A *bi-sequence* $(x_i)_{i=-\infty}^{\infty}=(\ldots,x_{-2},x_{-1},x_0,x_1,x_2,\ldots)$ is a tuple with a countably infinite number of components. Now, let $\cdots<i_{-2}<i_{-1}<i_0<i_1<i_2<\cdots$. Then, $(x_{i_j})_{j=-\infty}^{\infty}$ is a *bisubsequence* of $(x_i)_{i=-\infty}^{\infty}$.

Definition 12.3.9. Let $(\beta_i)_{i=1-\infty}^{\infty}\subset\mathbb{C}$, let $z_0\in\mathbb{C}$, let $\mathcal{D}\subseteq\mathbb{C}$ denote the set of all $z\in\mathbb{C}$ such that $\sum_{i=1}^{\infty}\beta_{-i}(z-z_0)^{-i}$ and $\sum_{i=0}^{\infty}\beta_i(z-z_0)^i$ converge, and define $f\colon\mathcal{D}\mapsto\mathbb{C}$ by

$$f(z)\triangleq\sum_{i=-\infty}^{\infty}\beta_i(z-z_0)^i. \tag{12.3.6}$$

Then, f is a *bi-power series*; for all $z\in\mathcal{D}$, f *converges at* z; and $\mathcal{D}$ is the *domain of convergence of* f. The bi-power series f is *absolutely convergent at* z if $\sum_{i=-\infty}^{\infty}|\beta_i||(z-z_0)^i|$ converges. Finally, f is a *generating function* for the bi-sequence $(\beta_i)_{i=-\infty}^{\infty}$.

Proposition 12.3.10. Let $(\beta_i)_{i=-\infty}^{\infty} \subset \mathbb{C}$, and define the bi-power series f with domain of convergence $\mathcal{D}$ by (12.3.6). Then, exactly one of the following statements holds:

i) $\mathcal{D} = \varnothing$.

ii) $\mathcal{D} = \{z_0\}$.

iii) $\mathcal{D} = \mathbb{C}\backslash\{z_0\}$.

iv) $\mathcal{D} = \mathbb{C}$.

v) There exist $\rho > 0$ and $\mathcal{D}_\rho \subseteq \mathbb{S}_\rho(z_0)$ such that

$$\mathcal{D} = \mathbb{B}_\rho(z_0) \cup \mathcal{D}_\rho. \tag{12.3.7}$$

vi) There exist $\rho > 0$ and $\mathcal{D}_\rho \subseteq \mathbb{S}_\rho(z_0)$ such that

$$\mathcal{D} = \{z \in \mathbb{C}\colon 0 < |z - z_0| < \rho\} \cup \mathcal{D}_\rho. \tag{12.3.8}$$

vii) There exist $r > 0$, $\rho > r$, $\mathcal{D}_r \subseteq \mathbb{S}_r(z_0)$, and $\mathcal{D}_\rho \subseteq \mathbb{S}_\rho(z_0)$ such that

$$\mathcal{D} = \{z \in \mathbb{C}\colon r < |z - z_0| < \rho\} \cup \mathcal{D}_r \cup \mathcal{D}_\rho. \tag{12.3.9}$$

viii) There exist $r > 0$ and $\mathcal{D}_r \subseteq \mathbb{S}_r(z_0)$ such that

$$\mathcal{D} = \{z \in \mathbb{C}\colon r < |z - z_0|\} \cup \mathcal{D}_r. \tag{12.3.10}$$

ix) $\mathcal{D}$ is nonempty, and there exists $r > 0$ such that $\mathcal{D} \subseteq \mathbb{S}_r(z_0)$.

In cases *iii)*–*viii)*, (12.3.1) converges absolutely for all $z \in \operatorname{int} \mathcal{D}$.

Consider the bi-power series f defined by (12.3.6). Then, the domain of convergence $\mathcal{D}$ of f is either empty, the point z_0, the entire complex plane except for z_0, the entire complex plane, the union of $\mathbb{B}_\rho(z_0)$ and a subset of its boundary, the union of the punctured open disk $\mathbb{B}_\rho(z_0)\backslash\{z_0\}$ and a subset of its boundary, the union of the open annulus $\{z \in \mathbb{C}\colon r < |z - z_0| < \rho\}$ and a subset of its boundary, the union of the open outer disk $\{z \in \mathbb{C}\colon r < |z - z_0|\}$ and a subset of its boundary, or a nonempty subset of the circle $\mathbb{S}_r(z_0)$. r is the *inner radius of convergence of* f, and ρ is the *outer radius of convergence of* f. In cases i), *ii)*, *v)*, and *vi)*, we set $r = 0$; in cases *iii)*, *iv)*, and *viii)*, we set $\rho = \infty$; in case *ix)*, we set $\rho = r$.

Proposition 12.3.11. Let $(\beta_i)_{i=-\infty}^{\infty} \subset \mathbb{C}$, and define the bi-power series f by (12.3.6) with domain of convergence $\mathcal{D}$, assume that one of the statements *ii)*–*ix)* of Proposition 12.3.6 holds with inner radius of convergence r and outer radius of convergence ρ, where $r \le \rho$. Then,

$$r = \limsup_{i\to\infty} |\beta_{-i}|^{1/i}, \quad \rho = \frac{1}{\limsup_{i\to\infty} |\beta_i|^{1/i}}. \tag{12.3.11}$$

12.4 Continuity

Definition 12.4.1. Let $\mathcal{D} \subseteq \mathbb{F}^m$, $f\colon \mathcal{D} \mapsto \mathbb{F}^n$, and $x \in \mathcal{D}$. Then, f is *continuous* at x if, for every sequence $(x_i)_{i=1}^{\infty} \subseteq \mathcal{D}$ such that $\lim_{i\to\infty} x_i = x$, it follows that $\lim_{i\to\infty} f(x_i) = f(x)$. Furthermore, let $\mathcal{D}_0 \subseteq \mathcal{D}$. Then, f is *continuous* on $\mathcal{D}_0$ if, for all $x \in \mathcal{D}_0$, f is continuous at x. Finally, f is *continuous* if it is continuous on $\mathcal{D}$.

Definition 12.4.2. Let $\mathcal{D} \subseteq \mathbb{R}$, $f\colon \mathcal{D} \mapsto \mathbb{F}^n$, and $x_0 \in \operatorname{cl} \mathcal{D}$. Then, $\lim_{x\downarrow x_0} f(x)$ *exists* if there exists $\alpha \in \mathbb{F}^n$ such that, for every sequence $(x_i)_{i=1}^{\infty} \subseteq \mathcal{D} \cap [x_0, \infty)$ such that $\lim_{i\to\infty} x_i = x_0$, $\lim_{i\to\infty} f(x_i) = \alpha$. In this case, $\lim_{x\downarrow x_0} f(x) \triangleq \alpha$.

Definition 12.4.3. Let $\mathcal{D} \subseteq \mathbb{R}$, $f\colon \mathcal{D} \mapsto \mathbb{F}^n$, and $x_0 \in \operatorname{cl} \mathcal{D}$. Then, $\lim_{x\uparrow x_0} f(x)$ *exists* if there exists $\alpha \in \mathbb{F}^n$ such that, for every sequence $(x_i)_{i=1}^{\infty} \subseteq \mathcal{D} \cap (-\infty, x_0]$ such that $\lim_{i\to\infty} x_i = x_0$, $\lim_{i\to\infty} f(x_i) = \alpha$. In this case, $\lim_{x\uparrow x_0} f(x) \triangleq \alpha$.

Definition 12.4.4. Let $\mathcal{D} \subseteq \mathbb{R}$, $f\colon\ \mathcal{D} \mapsto \mathbb{F}^n$, and $x_0 \in \operatorname{cl} \mathcal{D}$. Then, $\lim_{x\to x_0} f(x)$ *exists* if there exists $\alpha \in \mathbb{F}^n$ such that, for every sequence $(x_i)_{i=1}^\infty \subseteq \mathcal{D}$ such that $\lim_{i\to\infty} x_i = x_0$, $\lim_{i\to\infty} f(x_i) = \alpha$. In this case, $\lim_{x\to x_0} f(x) \triangleq \alpha$.

Proposition 12.4.5. Let $\mathcal{D} \subseteq \mathbb{F}^m$, $f\colon\ \mathcal{D} \mapsto \mathbb{F}^n$, and $x_0 \in \mathcal{D}$. Then, f is continuous at x if and only if $\lim_{x\to x_0} f(x) = f(x_0)$.

Theorem 12.4.6. Let $\mathcal{D} \subseteq \mathbb{F}^n$ be convex, and let $f\colon\ \mathcal{D} \to \mathbb{F}$ be convex. Then, f is continuous on $\operatorname{relint} \mathcal{D}$.

Proof. See [333, p. 81] and [2319, p. 82]. □

Corollary 12.4.7. Let $A \in \mathbb{F}^{n\times m}$, and define $f\colon\ \mathbb{F}^m \to \mathbb{F}^n$ by $f(x) \triangleq Ax$. Then, f is continuous.

Proof. This is a consequence of Theorem 12.4.6. Alternatively, let $x \in \mathbb{F}^m$, and let $(x_i)_{i=1}^\infty \subset \mathbb{F}^m$ satisfy $x_i \to x$ as $i \to \infty$. Furthermore, let $\|\cdot\|$, $\|\cdot\|'$, and $\|\cdot\|''$ be compatible norms on $\mathbb{F}^n$, $\mathbb{F}^{n\times m}$, and $\mathbb{F}^m$, respectively. Since $\|Ax - Ax_i\| \le \|A\|'\|x - x_i\|''$, it follows that $Ax_i \to Ax$ as $i \to \infty$. □

The following result is a consequence of Corollary 12.4.7.

Proposition 12.4.8. Let $X \triangleq (x_i)_{i=1}^\infty \subset \mathbb{F}^n$ and $Y \triangleq (y_i)_{i=1}^\infty \subset \mathbb{F}^m$, assume that X and Y converge, define $x \triangleq \lim_{i\to\infty} x_i$ and $y \triangleq \lim_{i\to\infty} y_i$, let $A \in \mathbb{F}^{p\times n}$ and $B \in \mathbb{F}^{p\times m}$, and let $Z \triangleq (z_i)_{i=1}^\infty \subset \mathbb{F}^p$, where, for all $i \ge 0$, $z_i \triangleq Ax_i + By_i$. Then, Z converges, and

$$\lim_{i\to\infty} z_i = Ax + By. \tag{12.4.1}$$

The following result characterizes continuity of a function f in terms of properties of the inverse mapping f^{inv}.

Theorem 12.4.9. Let $\mathcal{D} \subseteq \mathbb{F}^m$ and $f\colon\ \mathcal{D} \mapsto \mathbb{F}^n$. Then, the following statements are equivalent:

i) f is continuous.

ii) For all open $\mathcal{S} \subseteq \mathbb{F}^n$, $f^{\mathsf{inv}}(\mathcal{S})$ is open relative to $\mathcal{D}$.

iii) For all closed $\mathcal{S} \subseteq \mathbb{F}^n$, $f^{\mathsf{inv}}(\mathcal{S})$ is closed relative to $\mathcal{D}$.

iv) For all $\mathcal{S} \subseteq \mathbb{F}^n$, $f^{\mathsf{inv}}(\operatorname{int} \mathcal{S}) \subseteq \operatorname{int}_{\mathcal{D}} f^{\mathsf{inv}}(\mathcal{S})$.

v) For all $\mathcal{S} \subseteq \mathbb{F}^n$, $\operatorname{cl} f^{\mathsf{inv}}(\mathcal{S}) \subseteq f^{\mathsf{inv}}(\operatorname{cl} \mathcal{S})$.

vi) For all $\mathcal{D}_0 \subseteq \mathcal{D}$, $f(\operatorname{cl} \mathcal{D}_0) \subseteq \operatorname{cl} f(\mathcal{D}_0)$.

Proof. See [874, pp. 9, 10] and [2112, pp. 87, 110]. □

Corollary 12.4.10. Let $A \in \mathbb{F}^{n\times m}$ and $\mathcal{S} \subseteq \mathbb{F}^n$. If $\mathcal{S}$ is open, then $A^{\mathsf{inv}}(\mathcal{S})$ is open. If $\mathcal{S}$ is closed, then $A^{\mathsf{inv}}(\mathcal{S})$ is closed.

Theorem 12.4.11. Let $\mathcal{D} \subset \mathbb{F}^m$ be compact, and let $f\colon\ \mathcal{D} \mapsto \mathbb{F}^n$ be continuous. Then, $f(\mathcal{D})$ is compact.

Proof. See [837, p. 125] and [2112, p. 146]. □

The following corollary of Theorem 12.4.11 shows that a continuous real-valued function defined on a compact set has a minimizer and a maximizer.

Corollary 12.4.12. Let $\mathcal{D} \subset \mathbb{F}^m$ be compact, and let $f\colon\ \mathcal{D} \mapsto \mathbb{R}$ be continuous. Then, there exist $x_0, x_1 \in \mathcal{D}$ such that, for all $x \in \mathcal{D}$, $f(x_0) \le f(x) \le f(x_1)$.

Corollary 12.4.13. Let $A \in \mathbb{F}^{n\times m}$, and let $\mathcal{S} \subseteq \mathbb{F}^m$ be compact. Then, $A\mathcal{S}$ is compact.

If $\mathcal{S} \subseteq \mathbb{F}^m$ is closed but not bounded, then $A\mathcal{S}$ is not necessarily closed. For example, consider $A = [1\ \ 0]$ and $\mathcal{S} = \{[\begin{smallmatrix} x \\ y \end{smallmatrix}] \in \mathbb{R}^2\colon x > 0, y = \frac{1}{x}\}$. Then, $A\mathcal{S} = (0, \infty)$.

Corollary 12.4.14. Let $\mathcal{D} \subseteq \mathbb{F}^m$ and $f\colon \mathcal{D} \mapsto \mathbb{F}^n$, assume that f is one-to-one and $f^{\mathsf{Inv}}\colon f(\mathcal{D}) \mapsto \mathcal{D}$ is continuous, and let $\mathcal{S} \subseteq f(\mathcal{D})$ be compact. Then, $f^{\mathsf{Inv}}(\mathcal{S})$ is compact.

Proof. The result follows from Theorem 12.4.11. □

Corollary 12.4.15. Let $A \in \mathbb{F}^{n\times m}$ be left invertible, and let $\mathcal{S} \subset \mathbb{F}^n$ be compact. Then, $A^{\mathsf{inv}}(\mathcal{S})$ is compact. Furthermore,

$$A^{\mathrm{inv}}(\mathcal{S}) = A^{+}[\mathcal{S} \cap \mathcal{R}(A)]. \tag{12.4.2}$$

Proof. Since A is left invertible, it follows that $f\colon \mathbb{F}^m \mapsto \mathcal{R}(A)$ defined by $f(x) \triangleq Ax$ is invertible. Let $A^{\mathrm{L}} \in \mathbb{F}^{m\times n}$ be a left inverse of A. Then, the inverse $f^{\mathrm{Inv}}\colon \mathcal{R}(A) \mapsto \mathbb{F}^n$ of f is given by $f^{\mathrm{Inv}}(y) = A^{\mathrm{L}}y$. Since f^{Inv} is linear, it follows from Corollary 12.4.7 that f^{Inv} is continuous. Corollary 12.4.14 thus implies that $A^{\mathrm{inv}}(\mathcal{S})$ is compact. The expression for $A^{\mathrm{inv}}(\mathcal{S})$ follows from Proposition 8.1.10. □

Definition 12.4.16. Let $\mathcal{D} \subseteq \mathbb{R}$ and $f\colon\ \mathcal{D} \mapsto \mathbb{F}^n$. Then, f is an *open mapping* if, for all $\mathcal{D}_0 \subseteq \mathcal{D}$ such that $\mathcal{D}_0$ is open relative to $\mathcal{D}$, $f(\mathcal{D}_0)$ is open.

Theorem 12.4.17. Let $\mathcal{D} \subseteq \mathbb{F}^m$ and $f\colon\ \mathcal{D} \mapsto \mathbb{F}^n$. Then, the following statements are equivalent:

i) f is an open mapping.

ii) For all $\mathcal{D}_0 \subseteq \mathcal{D}$, $f(\mathrm{int}_{\mathcal{D}}\, \mathcal{D}_0) \subseteq \mathrm{int}\, f(\mathcal{D}_0)$.

Proof. See [874, pp. 11, 12]. □

Theorem 12.4.18. Let $\mathcal{D} \subseteq \mathbb{F}^m$, $\mathcal{S} \subseteq \mathbb{F}^n$, and $f\colon\ \mathcal{D} \mapsto \mathcal{S}$, and assume that f is one-to-one and onto. Then, the following statements are equivalent:

i) f and f^{Inv} are continuous.

ii) f is continuous, and f is an open mapping.

iii) For all $\mathcal{D}_0 \subseteq \mathcal{D}$, $f(\mathrm{cl}_{\mathcal{D}}\, \mathcal{D}_0) = \mathrm{cl}\, f(\mathcal{D}_0)$.

If these conditions hold and $\mathcal{D}$ and $\mathcal{S}$ have nonempty interior, then $m = n$.

Proof. See [874, p. 12]. □

The following result, called *invariance of domain*, implies that a function that is continuous and one-to-one is an open mapping. This result assumes that the domain and range of f have the same dimension.

Theorem 12.4.19. Let $\mathcal{D} \subseteq \mathbb{F}^n$ be open, let $\mathcal{S} \subseteq \mathbb{F}^n$, and let $f\colon\ \mathcal{D} \mapsto \mathcal{S}$ be one-to-one, onto, and continuous. Then, $\mathcal{S}$ is open, and $f^{\mathrm{Inv}}\colon\ \mathcal{S} \mapsto \mathbb{F}^n$ is continuous.

Proof. See [2505, p. 3]. □

The following result is a variation of Theorem 12.4.19.

Theorem 12.4.20. Let $\mathcal{D} \subseteq \mathbb{F}^n$ be compact, and let $f\colon\ \mathcal{D} \mapsto \mathbb{F}^n$ be continuous and one-to-one. Then, $f(\mathcal{D})$ is compact, and $f^{\mathrm{Inv}}\colon\ f(\mathcal{D}) \mapsto \mathbb{R}^n$ is continuous.

Proof. See [837, p. 371]. □

The following result specializes Theorem 12.4.19 to the case $f(x) = Ax$, where $A \in \mathbb{F}^{n\times n}$. In this case, it follows from Corollary 3.7.7 that f is one-to-one if and only if A is nonsingular.

Corollary 12.4.21. Let $\mathcal{S} \subseteq \mathbb{F}^n$, assume that $\mathcal{S}$ is open, let $A \in \mathbb{F}^{n\times n}$, and assume that A is nonsingular. Then, $A\mathcal{S}$ is open.

The following result is the *open mapping theorem.* This is an extension of Corollary 12.4.21, where the domain and range of A need not have the same dimension. The nonsingularity condition of Corollary 12.4.21 is replaced by the assumption that A is right invertible; that is, $\mathrm{rank}\, A = n \le m$.

Theorem 12.4.22. Let $\mathcal{S} \subseteq \mathbb{F}^m$, let $A \in \mathbb{F}^{n\times m}$, assume that $\mathcal{S}$ is open, and assume that A is right invertible. Then, $A\mathcal{S}$ is open.

Definition 12.4.23. Let $\mathcal{S} \subseteq \mathbb{F}^n$. Then, $\mathcal{S}$ is *pathwise connected* if, for all $x_1, x_2 \in \mathcal{S}$, there exists a continuous function $f\colon\ [0,1] \mapsto \mathcal{S}$ such that $f(0) = x_1$ and $f(1) = x_2$.

Proposition 12.4.24. Let $\mathcal{S} \subseteq \mathbb{F}^n$ be pathwise connected. Then, $\mathcal{S}$ is connected.

Proof. See [837, p. 259]. □

The converse of Proposition 12.4.24 is false. Let $\mathcal{S} = \mathcal{S}_1 \cup \mathcal{S}_2$, where $\mathcal{S}_1 = \{0\} \times [0,1]$ and $\mathcal{S}_2 = \{\left[\begin{smallmatrix} x \\ y \end{smallmatrix}\right] \in \mathbb{R}^2\colon\ x \in (0,1], y = \sin \frac{1}{x}\}$. Then, $\mathcal{S}_1$ is not open relative to $\mathcal{S}$. In fact, $\mathrm{int}_{\mathcal{S}}\, \mathcal{S}_1 = \varnothing$. Thus, $\mathcal{S}$

is connected. However, $\mathcal{S}$ is not pathwise connected. See [837, p. 260].

Theorem 12.4.25. Let $\mathcal{D} \subset \mathbb{F}^m$ be connected, and let $f\colon\ \mathcal{D} \mapsto \mathbb{F}^n$ be continuous. Then, $f(\mathcal{D})$ is connected.

Proof. See [837, p. 259]. □

The following result is the *Schauder fixed-point theorem.*

Theorem 12.4.26. Let $\mathcal{D} \subseteq \mathbb{F}^m$ be nonempty, closed, and convex, let $f\colon\ \mathcal{D} \to \mathcal{D}$ be continuous, and assume that $f(\mathcal{D})$ is bounded. Then, there exists $x \in \mathcal{D}$ such that $f(x) = x$.

Proof. See [2836, p. 167]. □

The following corollary for the case of a bounded domain is the *Brouwer fixed-point theorem.*

Corollary 12.4.27. Let $\mathcal{D} \subseteq \mathbb{F}^m$ be nonempty, compact, and convex, and let $f\colon\ \mathcal{D} \to \mathcal{D}$ be continuous. Then, there exists $x \in \mathcal{D}$ such that $f(x) = x$.

Proof. See [2836, p. 163]. □

Definition 12.4.28. Let $\mathcal{D} \subseteq \mathbb{F}^m$, and let $f\colon \mathcal{D} \mapsto \mathbb{F}^n$. Then, f is *locally Lipschitz* if, for all $\xi \in \mathcal{D}$ and $\varepsilon > 0$, there exists $\mu > 0$ such that, for all $x, y \in \mathbb{B}_\varepsilon(\xi) \cap \mathcal{D}$, $\|f(x) - f(y)\|_2 \le \mu\|x - y\|_2$. Furthermore, f is *globally Lipschitz* if there exists $\mu > 0$ such that, for all $x, y \in \mathcal{D}$, $\|f(x) - f(y)\|_2 \le \mu\|x - y\|_2$.

The function $f\colon \mathbb{R} \mapsto \mathbb{R}$ defined by $f(x) = x^2$ is locally Lipschitz but not globally Lipschitz.

12.5 Derivatives

Let $\mathcal{D} \subseteq \mathbb{F}^m$, and let $x_0 \in \mathcal{D}$. Then, the *feasible cone of $\mathcal{D}$ with respect to x_0* is the set

$$\operatorname{fcone}(\mathcal{D}, x_0) \triangleq \{\xi \in \mathbb{F}^m\colon\ \text{there exists } \alpha_0 > 0 \text{ such that, for all } \alpha \in [0, \alpha_0), x_0 + \alpha\xi \in \mathcal{D}\}. \tag{12.5.1}$$

Note that $\operatorname{fcone}(\mathcal{D}, x_0)$ is a pointed cone, although it may consist of only $\xi = 0$ as can be seen from the example $x_0 = 0$ and

$$\mathcal{D} = \left\{x \in \mathbb{R}^2\colon\ 0 \le x_{(1)} \le 1,\ x_{(1)}^3 \le x_{(2)} \le x_{(1)}^2\right\}.$$

Now, let $\mathcal{D} \subseteq \mathbb{F}^m$ and $f\colon\ \mathcal{D} \to \mathbb{F}^n$. If $\xi \in \operatorname{fcone}(\mathcal{D}, x_0)$, then the *one-sided directional differential of f at x_0 in the direction ξ* is defined by

$$\mathrm{D}_+f(x_0;\xi) \triangleq \lim_{\alpha\downarrow 0} \tfrac{1}{\alpha}[f(x_0 + \alpha\xi) - f(x_0)] \tag{12.5.2}$$

in the case where the limit exists. Similarly, if $\xi \in \operatorname{fcone}(\mathcal{D}, x_0)$ and $-\xi \in \operatorname{fcone}(\mathcal{D}, x_0)$, then the *two-sided directional differential* $\mathrm{D}f(x_0;\xi)$ of f *at x_0 in the direction ξ* is defined by

$$\mathrm{D}f(x_0;\xi) \triangleq \lim_{\alpha\to 0} \tfrac{1}{\alpha}[f(x_0 + \alpha\xi) - f(x_0)] \tag{12.5.3}$$

in the case where the limit exists. If $\xi = e_i$ so that the direction ξ is one of the coordinate axes, then the *partial derivative* of f *with respect to $x_{(i)}$ at x_0*, denoted by $\frac{\partial f(x_0)}{\partial x_{(i)}}$, is given by

$$\frac{\partial f(x_0)}{\partial x_{(i)}} \triangleq \lim_{\alpha\to 0} \tfrac{1}{\alpha}[f(x_0 + \alpha e_i) - f(x_0)]. \tag{12.5.4}$$

Equivalently,

$$\frac{\partial f(x_0)}{\partial x_{(i)}} = \mathrm{D}f(x_0; e_i), \tag{12.5.5}$$

in the case where the two-sided directional differential $\mathrm{D}f(x_0; e_i)$ exists. Note that $\frac{\partial f(x_0)}{\partial x_{(i)}} \in \mathbb{F}^{n\times 1}$.

Proposition 12.5.1. Let $\mathcal{D} \subseteq \mathbb{F}^m$ be convex, let $f\colon\ \mathcal{D} \mapsto \mathbb{F}^n$ be convex, and let $x_0 \in \operatorname{int} \mathcal{D}$. Then, $\mathrm{D}_+f(x_0;\xi)$ exists for all $\xi \in \operatorname{fcone}(\mathcal{D}, x_0)$.

Proof. See [333, p. 83]. □

Note that $\mathrm{D}_+f(x_0;\xi) = \pm\infty$ is possible if x_0 is an element of the boundary of $\mathcal{D}$. For example, consider the convex, continuous function $f\colon\ [0,\infty) \mapsto \mathbb{R}$ given by $f(x) = 1 - \sqrt{x}$. In this case, $\mathrm{D}_+f(0;1) = -\infty$ and thus does not exist.

Next, we consider a stronger form of differentiation. Note that this result does not assume that x_0 is contained in the interior of the domain of f.

Proposition 12.5.2. Let $\mathcal{D} \subseteq \mathbb{F}^m$ and $f\colon\ \mathcal{D} \mapsto \mathbb{F}^n$, assume that $\mathcal{D}$ is a solid, convex set, and let $x_0 \in \mathcal{D}$. Then, there exists at most one matrix $F \in \mathbb{F}^{n\times m}$ satisfying

$$\lim_{\substack{x\to x_0\\ x\in\mathcal{D}\backslash\{x_0\}}} \|x - x_0\|^{-1}[f(x) - f(x_0) - F(x - x_0)] = 0. \tag{12.5.6}$$

Proof. See [2836, p. 170]. □

In (12.5.6) the limit is taken over all sequences that are contained in $\mathcal{D}$, do not include x_0, and converge to x_0. The restriction to sequences that do not include x_0 is necessitated by the fact that $\|x - x_0\|^{-1}[f(x) - f(x_0) - F(x - x_0)]$ is not defined for $x = x_0$. The restriction to sequences that are confined to $\mathcal{D}$ is not explicitly mentioned henceforth. Finally, note that $\mathcal{D}$ is not necessarily open, and x_0 may be an element of $\mathcal{D} \cap \operatorname{bd}\mathcal{D}$. However, x_0 must be an element of $\mathcal{D}$ since $f(x_0)$ plays a role in the limit.

Definition 12.5.3. Let $\mathcal{D} \subseteq \mathbb{F}^m$, let $f\colon\ \mathcal{D} \mapsto \mathbb{F}^n$, assume that $\mathcal{D}$ is a solid, convex set, let $x_0 \in \mathcal{D}$, and assume that there exists $F \in \mathbb{F}^{n\times m}$ satisfying (12.5.6). Then, f is *differentiable at* x_0, and the matrix F is the *derivative of* f *at* x_0. In this case, we write $f'(x_0) = F$ and

$$\lim_{x\to x_0} \|x - x_0\|^{-1}[f(x) - f(x_0) - f'(x_0)(x - x_0)] = 0. \tag{12.5.7}$$

f is *differentiable* if, for all $x \in \mathcal{D}$, f is differentiable at x.

We alternatively write $\frac{\mathrm{d}f(x_0)}{\mathrm{d}x}$ for $f'(x_0)$.

Proposition 12.5.4. Let $\mathcal{D} \subseteq \mathbb{F}^m$ be a solid, convex set, let $f\colon\ \mathcal{D} \mapsto \mathbb{F}^n$, let $x \in \mathcal{D}$, and assume that f is differentiable at x_0. Then, f is continuous at x_0.

Let $\mathcal{D} \subseteq \mathbb{F}^m$ be a solid, convex set, and let $f\colon\ \mathcal{D} \mapsto \mathbb{F}^n$. In terms of its scalar components, f can be written as $f = [f_1 \ \cdots \ f_n]^{\mathrm{T}}$, where $f_i\colon\ \mathcal{D} \mapsto \mathbb{F}$ for all $i \in \{1,\dots,n\}$ and $f(x) = [f_1(x) \ \cdots \ f_n(x)]^{\mathrm{T}}$ for all $x \in \mathcal{D}$. With this notation, if $f'(x_0)$ exists, then it can be written as

$$f'(x_0) = \begin{bmatrix} f_1'(x_0)\\ \vdots\\ f_n'(x_0)\end{bmatrix}, \tag{12.5.8}$$

where $f_i'(x_0) \in \mathbb{F}^{1\times m}$ is the *gradient of* f_i *at* x_0 and $f'(x_0)$ is the *Jacobian of* f *at* x_0. Furthermore, if $x \in \operatorname{int}\mathcal{D}$, then $f'(x_0)$ is related to the partial derivatives of f by

$$f'(x_0) = \begin{bmatrix} \dfrac{\partial f(x_0)}{\partial x_{(1)}} & \cdots & \dfrac{\partial f(x_0)}{\partial x_{(m)}}\end{bmatrix}, \tag{12.5.9}$$

where $\frac{\partial f(x_0)}{\partial x_{(i)}} \in \mathbb{F}^{n\times 1}$ for all $i \in \{1,\dots,m\}$. Finally, note that the (i,j) entry of the $n\times m$ matrix $f'(x_0)$ is $\frac{\partial f_i(x_0)}{\partial x_{(j)}}$. For example, if $x \in \mathbb{F}^n$ and $A \in \mathbb{F}^{n\times n}$, then

$$\frac{\mathrm{d}}{\mathrm{d}x}Ax = A. \tag{12.5.10}$$

Note that the existence of the partial derivatives of f at x_0 does not imply that $f'(x_0)$ exists. That is, f may not be differentiable at x_0 since $f'(x_0)$ given by (12.5.9) may not satisfy (12.5.7).

Let $\mathcal{D} \subseteq \mathbb{F}^m$ and $f\colon\ \mathcal{D} \mapsto \mathbb{F}^n$, and assume that $f'(x)$ exists for all $x \in \mathcal{D}$ and $f'\colon\ \mathcal{D} \mapsto \mathbb{F}^{n\times m}$ is continuous. Then, f is *continuously differentiable*, which is also called C^1. For all $x_0 \in \mathcal{D}$,

$f'(x_0) \in \mathbb{F}^{n\times m}$, and thus $f'(x_0)\colon\ \mathbb{F}^m \mapsto \mathbb{F}^n$. The *second derivative* of f at $x_0 \in \mathcal{D}$, denoted by $f''(x_0)$, is the derivative of $f'\colon\ \mathcal{D} \mapsto \mathbb{F}^{n\times m}$ at $x_0 \in \mathcal{D}$. By analogy with the first derivative, it follows that $f''(x_0)\colon\ \mathbb{F}^m \mapsto \mathbb{F}^{n\times m}$ is linear. Therefore, for all $\eta \in \mathbb{F}^m$, $f''(x_0)\eta \in \mathbb{F}^{n\times m}$, and, thus, for all $\eta, \hat{\eta} \in \mathbb{F}^m$, $[f''(x_0)\eta]\hat{\eta} \in \mathbb{F}^n$. Defining $f''(x_0)(\eta,\hat{\eta}) \triangleq [f''(x_0)\eta]\hat{\eta}$, it follows that $f''(x_0)\colon\ \mathbb{F}^m \times \mathbb{F}^m \mapsto \mathbb{F}^n$ is *bilinear*; that is, for all $\hat{\eta} \in \mathbb{F}^m$, the mapping $\eta \mapsto f''(x_0)(\eta,\hat{\eta})$ is linear, and, for all $\eta \in \mathbb{F}^m$, the mapping $\hat{\eta} \mapsto f''(x_0)(\eta,\hat{\eta})$ is linear. Letting $f = [f_1 \ \cdots \ f_n]^{\mathrm{T}}$, it follows that

$$f''(x_0)\eta = \begin{bmatrix} \eta^{\mathrm{T}} f_1''(x_0) \\ \vdots \\ \eta^{\mathrm{T}} f_n''(x_0) \end{bmatrix}, \qquad f''(x_0)(\eta,\hat{\eta}) = \begin{bmatrix} \eta^{\mathrm{T}} f_1''(x_0)\hat{\eta} \\ \vdots \\ \eta^{\mathrm{T}} f_n''(x_0)\hat{\eta} \end{bmatrix}, \tag{12.5.11}$$

where, for all $i \in \{1,\dots,n\}$, the matrix $f_i''(x_0)$ is the $m\times m$ *Hessian* of f_i at x_0. We write $f^{(2)}(x_0)$ for $f''(x_0)$ and $f^{(k)}(x_0)$ for the kth derivative of f at x_0. f is C^k if $f^{(k)}(x)$ exists for all $x \in \mathcal{D}$ and $f^{(k)}$ is continuous on $\mathcal{D}$. The following result is the *inverse function theorem* [974, p. 185].

Theorem 12.5.5. Let $\mathcal{D} \subseteq \mathbb{F}^n$ be open, let $f\colon\ \mathcal{D} \mapsto \mathbb{F}^n$, and assume that f is C^k. Furthermore, let $x_0 \in \mathcal{D}$ satisfy $\det f'(x_0) \neq 0$. Then, there exist open sets $\mathcal{N} \subset \mathbb{F}^n$ containing x_0 and $\mathcal{M} \subset \mathbb{F}^n$ containing $f(x_0)$ and a C^k function $g\colon\ \mathcal{M} \mapsto \mathcal{N}$ such that, for all $x \in \mathcal{N}$, $g[f(x)] = x$ and, for all $y \in \mathcal{M}$, $f[g(y)] = y$.

Let $S\colon\ [t_0,t_1] \mapsto \mathbb{F}^{n\times m}$, and assume that every entry of $S(t)$ is differentiable. Then, define $\dot{S}(t) \triangleq \frac{\mathrm{d}S(t)}{\mathrm{d}t} \in \mathbb{F}^{n\times m}$ for all $t \in [t_0,t_1]$ entrywise; that is, for all $i \in \{1,\dots,n\}$ and $j \in \{1,\dots,m\}$,

$$[\dot{S}(t)]_{(i,j)} \triangleq \frac{\mathrm{d}}{\mathrm{d}t} S_{(i,j)}(t). \tag{12.5.12}$$

If either $t = t_0$ or $t = t_1$, then either $\mathrm{d}^+/\mathrm{d}t$ or $\mathrm{d}^-/\mathrm{d}t$ (or just $\mathrm{d}/\mathrm{d}t$) denotes the right and left one-sided derivatives, respectively. Finally, define $\int_{t_0}^{t_1} S(t)\,\mathrm{d}t$ entrywise; that is, for all $i \in \{1,\dots,n\}$ and $j \in \{1,\dots,m\}$,

$$\left[\int_{t_0}^{t_1} S(t)\,\mathrm{d}t\right]_{(i,j)} \triangleq \int_{t_0}^{t_1} S_{(i,j)}(t)\,\mathrm{d}t. \tag{12.5.13}$$

12.6 Complex-Valued Functions

Complex-valued functions of a complex variable possess properties that warrant separate consideration. The following definition specializes Definition 12.5.3 to the case $\mathbb{F} = \mathbb{C}$ and $m = n = 1$.

Definition 12.6.1. Let $\mathcal{D} \subseteq \mathbb{C}$ be a solid, convex set, let $f\colon\ \mathcal{D} \mapsto \mathbb{C}$, let $z_0 \in \mathcal{D}$, and assume that

$$f'(z_0) \triangleq \lim_{z\to z_0} \frac{f(z) - f(z_0)}{z - z_0} \tag{12.6.1}$$

exists. Then, f is *differentiable at* z_0, and $f'(z_0)$ is the *derivative of f at z_0*.

Proposition 12.6.2. Let $\mathcal{D} \subseteq \mathbb{C}$ be a solid, convex set, let $f\colon\ \mathcal{D} \mapsto \mathbb{C}$, let $z_0 \in \mathcal{D}$, and assume that $f'(z_0)$ exists. Furthermore, let $\hat{\mathcal{D}} \triangleq \{(x,y) \in \mathbb{R}^2\colon\ x + y\jmath \in \mathcal{D}\}$, define $u(x,y) \triangleq \operatorname{Re} f(x + y\jmath)$, $v(x,y) \triangleq \operatorname{Im} f(x + y\jmath)$, and $\hat{f}\colon\ \hat{\mathcal{D}} \mapsto \mathbb{R}^2$ by

$$\hat{f}(x,y) \triangleq \begin{bmatrix} u(x,y) \\ v(x,y) \end{bmatrix}, \tag{12.6.2}$$

and let $z_0 = x_0 + y_0\jmath$, where $x_0, y_0 \in \mathbb{R}$. Then, $\hat{f}$ is differentiable at (x_0,y_0), and

$$\hat{f}'(x_0,y_0) = \begin{bmatrix} \dfrac{\partial u(x_0,y_0)}{\partial x} & \dfrac{\partial u(x_0,y_0)}{\partial y} \\[2ex] \dfrac{\partial v(x_0,y_0)}{\partial x} & \dfrac{\partial v(x_0,y_0)}{\partial y} \end{bmatrix}. \tag{12.6.3}$$

For convenience, we rewrite (12.6.3) as

$$\hat{f}'(x_0, y_0) = \begin{bmatrix} u_x(x_0, y_0) & u_y(x_0, y_0) \\ v_x(x_0, y_0) & v_y(x_0, y_0) \end{bmatrix}. \quad (12.6.4)$$

As an example, define $f\colon \mathbb{C} \mapsto \mathbb{C}$ by $f(z) = z^2$. Then, $f'(z) = 2z$ and

$$\hat{f}'(x, y) = \begin{bmatrix} 2x & -2y \\ 2y & 2x \end{bmatrix}.$$

Likewise, define $f\colon \mathbb{C} \mapsto \mathbb{C}$ by $f(z) = z^3$. Then, $f'(z) = 3z^2$ and

$$\hat{f}'(x, y) = \begin{bmatrix} 3x^2 - 3y^2 & -6xy \\ 6xy & 3x^2 - 3y^2 \end{bmatrix}.$$

The structure of $\hat{f}'(x, y)$ is explained by the following result.

Proposition 12.6.3. Let $\mathcal{D} \subseteq \mathbb{C}$, let $f\colon \mathcal{D} \mapsto \mathbb{C}$, and, for all $z = x + y\jmath \in \mathcal{D}$, where $x, y \in \mathbb{R}$, define $u(x, y) \triangleq \operatorname{Re} f(x + y\jmath)$ and $v(x, y) \triangleq \operatorname{Im} f(x + y\jmath)$. Then, the following statements hold:

i) Let $z_0 = x_0 + y_0\jmath \in \operatorname{int} \mathcal{D}$, and assume that $f'(z_0)$ exists. Then,

$$u_x(x_0, y_0) = v_y(x_0, y_0), \quad u_y(x_0, y_0) = -v_x(x_0, y_0). \quad (12.6.5)$$

ii) Let $z_0 = x_0 + y_0\jmath \in \operatorname{int} \mathcal{D}$, assume that there exists an open subset $\mathcal{D}_0 \subseteq \mathcal{D}$ such that $z_0 \in \mathcal{D}_0$ and such that, for all $z = x + y\jmath \in \mathcal{D}_0$, the partial derivatives $u_x(x, y)$, $u_y(x, y)$, $v_x(x, y)$, and $v_y(x, y)$ exist and are continuous at (x_0, y_0), and assume that (12.6.5) holds. Then, $f'(z_0)$ exists. In addition, $f'(z_0) = u_x(x_0, y_0) + v_x(x_0, y_0)\jmath$.

iii) Assume that there exists an open subset $\mathcal{D}_0 \subseteq \mathcal{D}$ such that, for all $z \in \mathcal{D}_0$, $f'(z)$ exists, assume that $u_{xx}(x_0, y_0)$, $u_{xy}(x_0, y_0)$, $u_{yx}(x_0, y_0)$, $u_{yy}(x_0, y_0)$, $v_{xx}(x_0, y_0)$, $v_{xy}(x_0, y_0)$, $v_{yx}(x_0, y_0)$, and $v_{yy}(x_0, y_0)$ exist, and assume that $u_{xy}(x_0, y_0) = u_{yx}(x_0, y_0)$ and $v_{xy}(x_0, y_0) = v_{yx}(x_0, y_0)$. Then,

$$u_{xx}(x_0, y_0) + u_{yy}(x_0, y_0) = 0, \quad v_{xx}(x_0, y_0) + v_{yy}(x_0, y_0) = 0. \quad (12.6.6)$$

Proof. See [582, pp. 63–68, 78, 79]. □

(12.6.5) is the *Cauchy-Riemann equations*. (12.6.6) shows that u and y are *harmonic functions*.

Define $f\colon \mathbb{C} \mapsto \mathbb{C}$ by $f(z) = \bar{z}$. Then, for all $(x, y) \in \mathbb{R}^2$, $u(x, y) = x$ and $v(x, y) = -y$. Consequently, for all $(x, y) \in \mathbb{R}^2$, $u_x(x, y) = 1 \neq -1 = v_y(x, y)$. It thus follows from Proposition 12.6.3 that, for all $z \in \mathbb{C}$, $f'(z)$ does not exist. However, in the notation of Proposition 12.6.2, $\hat{f}'(x, y)$ exists for all $(x, y) \in \mathbb{R}^2$ and is given by $\hat{f}'(x, y) = \begin{bmatrix} 1 & 0 \\ 0 & -1 \end{bmatrix}$.

As another example, define $f\colon \mathbb{C} \mapsto \mathbb{C}$ by $f(z) = |z|^2$. Then, for all $(x, y) \in \mathbb{R}^2$, $u(x, y) = x^2 + y^2$ and $v(x, y) = 0$. Consequently, $u_x(x, y) = 2x = 0 = v_y(x, y)$ and $u_y(x, y) = 2y = 0 = -v_x(x, y)$ if and only if $x = y = 0$. It thus follows from Proposition 12.6.3 that $f'(z)$ exists if and only if $z = 0$. However, in the notation of Proposition 12.6.2, $\hat{f}'(x, y)$ exists for all $(x, y) \in \mathbb{R}^2$ and is given by $\hat{f}'(x, y) = \begin{bmatrix} 2x & 2y \\ 0 & 0 \end{bmatrix}$.

Definition 12.6.4. Let $\mathcal{D} \subseteq \mathbb{C}$, let $f\colon \mathcal{D} \mapsto \mathbb{C}$, let $\mathcal{D}_0 \subseteq \mathcal{D}$, assume that $\mathcal{D}_0$ is open, assume that $f'(z)$ exists for all $z \in \mathcal{D}_0$, and let $z_0 \in \mathcal{D}_0$. Then, f is *analytic in* $\mathcal{D}_0$, and f is *analytic at* z_0.

The function $f\colon \mathbb{C} \mapsto \mathbb{C}$ defined by $f(z) = z^2$ is analytic in $\mathbb{C}$. The function $f\colon \mathbb{C}\backslash\{0\} \mapsto \mathbb{C}$ defined by $f(z) = 1/z$ is analytic in $\mathbb{C}\backslash\{0\}$. If $\mathcal{D} \subseteq \mathbb{C}$ is an open set, then the function $f\colon \mathcal{D} \mapsto \mathbb{C}$ defined by $f(z) = |z|^2$ is not analytic in $\mathcal{D}$ since $f'(z)$ exists only if $z = 0$. Finally, if $\mathcal{D} \subseteq \mathbb{C}$ is an open set, then the function $f\colon \mathcal{D} \mapsto \mathbb{C}$ defined by $f(z) = \overline{z}$ is not analytic since, for all $z \in \mathbb{C}$, $f'(z)$ does not exist.

Proposition 12.6.5. Let $\mathcal{D} \subseteq \mathbb{C}$, assume that $\mathcal{D}$ is open and connected, let $f\colon \mathcal{D} \mapsto \mathbb{C}$, assume that f is analytic in $\mathcal{D}$, let $\mathcal{D}_0 \subseteq \mathcal{D}$, assume that $\mathcal{D}_0$ is either a line segment of positive length or is open and nonempty, and assume that, for all $z \in \mathcal{D}_0$, $f(z) = 0$. Then, for all $z \in \mathcal{D}$, $f(z) = 0$.

Proof. See [582, pp. 83, 84]. □

Definition 12.6.6. Let $\mathcal{S}_1, \mathcal{S}_2 \subseteq \mathbb{C}$, assume that $\mathcal{S}_1$ and $\mathcal{S}_2$ are open, let $f\colon \mathcal{S}_1 \mapsto \mathcal{S}_2$, and assume that f is analytic in $\mathcal{S}_1$, one-to-one, and onto. Then, f is a *conformal mapping*.

Proposition 12.6.7. Let $\mathcal{D} \subseteq \mathbb{C}$, assume that $\mathcal{D}$ is open, let $f\colon \mathcal{D} \mapsto \mathbb{C}$, and assume that f is analytic in $\mathcal{D}$. Then, for all $n \ge 1$ and $z \in \mathcal{D}$, $f^{(n)}(z)$ exists. Furthermore, for all $n \ge 1$, $f^{(n)}$ is analytic in $\mathcal{D}$.

Proof. See [582, p. 168]. □

The following shows that an analytic function has a *power series expansion* in each open disk.

Proposition 12.6.8. Let $\mathcal{D} \subseteq \mathbb{C}$, assume that $\mathcal{D}$ is open, let $f\colon \mathcal{D} \mapsto \mathbb{C}$, assume that f is analytic in $\mathcal{D}$, let $z_0 \in \mathcal{D}$, let $\rho_0 > 0$ satisfy $\mathcal{D}_0 \triangleq \{z \in \mathbb{C}\colon |z - z_0| < \rho\} \subseteq \mathcal{D}$, and, for all $i \ge 0$, define

$$\beta_i \triangleq \frac{f^{(i)}(z_0)}{i!}. \tag{12.6.7}$$

Then, for all $z \in \mathcal{D}_0$,

$$f(z) = \sum_{i=0}^{\infty} \beta_i (z - z_0)^i. \tag{12.6.8}$$

Furthermore, (12.6.8) converges absolutely in $\mathcal{D}_0$.

Proof. See [582, p. 189]. □

The power series (12.6.8) is the *Taylor series of f at z_0*. The radius of convergence of the Taylor series (12.6.8) is characterized by Proposition 12.3.6.

Proposition 12.6.9. Let $\mathcal{D} \subseteq \mathbb{C}$, let $f\colon \mathcal{D} \mapsto \mathbb{C}$, let $z_0 \in \mathbb{C}$, let $r, R \in [0, \infty]$ satisfy $r < R$ and $\mathcal{D}_0 \triangleq \{z \in \mathbb{C}\colon r < |z - z_0| < R\} \subseteq \mathcal{D}$, and assume that f is analytic in $\mathcal{D}_0$. Then, there exists a unique bi-sequence $(\beta_i)_{i=-\infty}^{\infty} \subset \mathbb{C}$ such that, for all $z \in \mathcal{D}_0$,

$$f(z) = \sum_{i=-\infty}^{\infty} \beta_i (z - z_0)^i. \tag{12.6.9}$$

Furthermore, (12.6.9) converges absolutely in $\mathcal{D}_0$. Finally, for all $i \in \mathbb{Z}$,

$$\beta_i = \frac{1}{2\pi J} \oint_{\mathcal{C}} \frac{f(z)}{(z - z_0)^{i+1}}\, \mathrm{d}z, \tag{12.6.10}$$

where $\mathcal{C}$ is a counterclockwise circle centered at z_0 and lying in $\mathcal{D}_0$.

Proof. See [1136, pp. 165–168]. □

The series in (12.6.9) is the *Laurent series of f at z_0*. If $r = 0$ and R is finite, then $\mathcal{D}_0$ is a *punctured open disk*. If $r = 0$ and R is infinite, then $\mathcal{D}_0$ is a *punctured plane*. If $r > 0$ and R is finite, then $\mathcal{D}_0$ is an *open annulus*. If $r > 0$ and R is infinite, then $\mathcal{D}_0$ is an *open outer disk*. If $r = 0$ and f is analytic at z_0, then the Laurent series is a power series.

Define f, $\mathcal{D}$, z_0, r, R, $\mathcal{D}_0$, and $(\beta_i)_{i=-\infty}^{\infty}$ as in Proposition 12.6.9, and assume that $z_0 \notin \mathcal{D}$, and thus f is not defined at z_0. Then, (12.6.9) can be written as

$$f(z) = f_{\mathrm{outer}}(z) + f_{\mathrm{inner}}(z), \tag{12.6.11}$$

where $f_{\mathrm{outer}}(z) \triangleq \sum_{i=-\infty}^{-1} \beta_i (z - z_0)^i$ is analytic on $\mathcal{D}_{\mathrm{outer}} \triangleq \{z \in \mathbb{C}\colon r < |z - z_0|\}$ and $f_{\mathrm{inner}}(z) \triangleq$

$\sum_{i=0}^{\infty}\beta_i(z-z_0)^i$ is analytic on $\mathcal{D}_{\rm inner} \triangleq \{z \in \mathbb{C}\colon |z-z_0| < R\}$. If $r = 0$, then z_0 is an *isolated singularity of* f. If z_0 is an isolated singularity of f and, for all $i < 0$, $\beta_i = 0$, then $\tilde{f}\colon \mathcal{D}_0 \mapsto \mathbb{C}$ defined by $\tilde{f}(z) \triangleq f(z)$ for all $z \in \mathcal{D}_0$ and $\tilde{f}(z_0) \triangleq \beta_0$ is analytic on $\mathcal{D}_0$. In this case, z_0 is a *removable singularity of* f, and the radius of convergence ρ of (12.6.11) satisfies $R \le \rho$. If z_0 is an isolated singularity of f and $\{i < 0\colon \beta_i \ne 0\}$ is finite, then z_0 is a *pole of* f. If z_0 is a pole of f and $\max\{i < 0\colon \beta_i \ne 0\} = -1$, then z_0 is a *simple pole*. If z_0 is an isolated singularity of f and is neither a removable singularity nor a pole, then z_0 is an *essential singularity*.

As an example, define $f\colon \mathbb{C}\backslash\{1\} \mapsto \mathbb{C}$ by $f(z) = z/(z-1)$. Then, the Laurent series of f at 0 in OOUD $= \{z \in \mathbb{C}\colon |z| > 1\}$ is given by

$$f(z) = \sum_{i=1}^{\infty} \frac{1}{z^i}.$$

Note that 0 is not an isolated singularity of f since $r = 1$. On the other hand, the Laurent series of f centered at 1 in the punctured plane $\{z \in \mathbb{C}\colon |z-1| > 0\}$ is given by

$$f(z) = \frac{1}{z-1} + 1,$$

which shows that 1 is a pole of $f(z)$. As another example, the function $f\colon \mathbb{C}\backslash\{0\} \mapsto \mathbb{C}$ defined by $f(z) = e^{1/z}$ has the Laurent series

$$f(z) = \sum_{i=0}^{\infty} \frac{1}{i!}\frac{1}{z^i},$$

which shows that 0 is an essential singularity of f. Finally, the function $f\colon \mathbb{C}\backslash\{0\} \mapsto \mathbb{C}$ defined by $f(z) = (\sin z)/z$ has a removable singularity at 0 since $\tilde{f}\colon \mathbb{C} \mapsto \mathbb{C}$ defined by $\tilde{f}(z) \triangleq f(z)$ for all $z \in \mathbb{C}\backslash\{0\}$ and $\tilde{f}(0) \triangleq 1$ is analytic in $\mathbb{C}$.

For $n \ge 2$, consider the function $f\colon \mathbb{C} \mapsto \mathbb{C}$ defined by $f(z) = z^{1/n}$, which is the principal nth root. Note that, for all $z \in (-\infty, 0)$, f is not continuous at z. For example, it follows from (1.6.32) that $\lim_{\theta\uparrow\pi}(e^{\theta \jmath})^{1/2} = \jmath$ and $\lim_{\theta\downarrow-\pi}(e^{\theta \jmath})^{1/2} = -\jmath$. Therefore, there does not exist an open set $\mathcal{D} \subset \mathbb{C}$ such that $0 \in \mathcal{D}$ and f is analytic on $\mathcal{D}\backslash\{0\}$. Consequently, f does not have a Laurent series at $z = 0$.

As a final example, consider the function $f\colon \mathbb{C}\backslash\{1,2\} \mapsto \mathbb{C}$ defined by $f(z) = 1/[(z-1)(z-2)]$. Then, f has a power series in the region $\{z \in \mathbb{C}\colon |z| < 1\}$ and Laurent series in the regions $\{z \in \mathbb{C}\colon 1 < |z| < 2\}$ and $\{z \in \mathbb{C}\colon |z| > 2\}$ given by

$$f(z) = \begin{cases} \displaystyle\sum_{i=0}^{\infty}\left(1 - \frac{1}{2^{i+1}}\right)z^i, & |z| < 1, \\ \displaystyle -\sum_{i=0}^{\infty}\frac{1}{2^{i+1}}z^i - \sum_{i=1}^{\infty}\frac{1}{z^i}, & 1 < |z| < 2, \\ \displaystyle\sum_{i=1}^{\infty}(2^{i-1} - 1)\frac{1}{z^i}, & |z| > 2. \end{cases} \tag{12.6.12}$$

For details, see [582, pp. 203–205].

12.7 Infinite Products

Definition 12.7.1. Let $(x_i)_{i=1}^{\infty} \subset \mathbb{C}$. Then, the infinite product $\prod_{i=1}^{\infty} x_i$ *converges* if there exists $k \ge 1$ such that $\lim_{n\to\infty}\prod_{i=k}^{n} x_i$ exists and is nonzero. In this case,

$$\prod_{i=1}^{\infty} x_i \triangleq \begin{cases} 0, & 0 \in (x_i)_{i=1}^{\infty}, \\ \lim_{n\to\infty}\prod_{i=1}^{n} x_i, & 0 \notin (x_i)_{i=1}^{\infty}. \end{cases} \tag{12.7.1}$$

If $(x_i)_{i=1}^\infty \subset \mathbb{R}$ and $\lim_{n\to\infty} \prod_{i=1}^n x_i = \pm\infty$, then $\prod_{i=1}^\infty x_i \triangleq \pm\infty$, respectively.

For all $k \ge 1$, $\lim_{n\to\infty} \prod_{i=k}^n (-1)^i$ does not exist, and thus $\prod_{i=1}^\infty (-1)^i$ does not converge. For all $k \ge 1$, $\lim_{n\to\infty} \prod_{i=k}^n i = \infty$, and thus $\prod_{i=1}^\infty i$ does not converge. If $(x_i)_{i=1}^\infty = (0,1,2,3,\ldots)$, then $\lim_{n\to\infty} \prod_{i=1}^n x_i = 0$ and, for all $k \ge 2$, $\prod_{i=k}^\infty x_i = \infty$, and thus $\prod_{i=1}^\infty x_i$ does not converge. If $(x_i)_{i=1}^\infty = (0,1,0,1,\ldots)$, then, for all $k \ge 1$, $\lim_{n\to\infty} \prod_{i=k}^n x_i = 0$, and thus $\prod_{i=1}^\infty x_i$ does not converge. If $(x_i)_{i=1}^\infty = (1,1/2,1/3,1/4,\ldots)$, then, for all $k \ge 1$, $\lim_{n\to\infty} \prod_{i=k}^n x_i = 0$, and thus $\prod_{i=1}^\infty x_i$ does not converge. However, if $(x_i)_{i=1}^\infty = (0,1,1,1,\ldots)$, then $\lim_{n\to\infty} \prod_{i=2}^n x_i = 1$, and thus $\prod_{i=1}^\infty x_i$ converges; in fact, $\prod_{i=1}^\infty x_i = 0$.

Proposition 12.7.2. Let $(x_i)_{i=1}^\infty \subset \mathbb{C}$. Then, the following statements are equivalent:

i) $\prod_{i=1}^\infty x_i$ converges.

ii) $\lim_{i\to\infty} x_i = 1$ and $\sum_{\{i\in\mathbb{P}\colon\, x_i\neq 0\}} \log x_i$ converges.

iii) $\operatorname{card}(\{i \in \mathbb{P}\colon\, x_i = 0\})$ is finite and $\sum_{\{i\in\mathbb{P}\colon\, x_i\neq 0\}} \log x_i$ converges.

If $(x_i)_{i=1}^\infty \subset (0,\infty)$, then the following statements are equivalent:

iv) $\prod_{i=1}^\infty x_i$ converges.

v) $\sum_{i=1}^\infty \log x_i$ converges.

If $\prod_{i=1}^\infty x_i$ converges, then the following statements hold:

vi) $\operatorname{card}\{i \in \mathbb{P}\colon\, x_i = 0\}$ is finite.

vii) If $0 \notin (x_i)_{i=1}^\infty$, then $\prod_{i=1}^\infty x_i = e^{\sum_{i=1}^\infty \log x_i} \neq 0$.

The following statements are equivalent:

viii) $\prod_{i=1}^\infty (1 + |x_i|)$ converges.

ix) $\sum_{i=1}^\infty |x_i|$ converges.

Now, assume that $(x_i)_{i=1}^\infty \subset \mathbb{R}$. Then, the following statements are equivalent:

x) $\prod_{i=1}^\infty x_i = \infty$.

xi) $\operatorname{Re} \sum_{i=1}^\infty \log x_i = \infty$ and $\frac{1}{\pi} \operatorname{Im} \sum_{i=1}^\infty \log x_i$ is an even integer.

Furthermore, the following statements are equivalent:

xii) $\prod_{i=1}^\infty x_i = -\infty$.

xiii) $\operatorname{Re} \sum_{i=1}^\infty \log x_i = \infty$ and $\frac{1}{\pi} \operatorname{Im} \sum_{i=1}^\infty \log x_i$ is an odd integer.

Proof. See [116, pp. 595–597] and [1136, pp. 352, 353]. □

12.8 Functions of a Matrix

Let $(\beta_i)_{i=0}^\infty \subset \mathbb{C}$, let $f\colon \mathcal{D} \mapsto \mathbb{C}$ denote the power series

$$f(z) \triangleq \sum_{i=0}^{\infty} \beta_i z^i, \tag{12.8.1}$$

where $\mathcal{D}$ is the domain of convergence of f, and let ρ be the radius of convergence of f. Next, let $A \in \mathbb{C}^{n\times n}$ satisfy $\rho_{\max}(A) < \rho$, and define $f(A)$ by the infinite series

$$f(A) \triangleq \sum_{i=0}^{\infty} \beta_i A^i. \tag{12.8.2}$$

To show that the infinite series (12.8.2) converges, we express A as $A = SBS^{-1}$, where $S \in \mathbb{C}^{n \times n}$ is nonsingular and $B \in \mathbb{C}^{n \times n}$. It thus follows that

$$f(A) = Sf(B)S^{-1}. \tag{12.8.3}$$

Now, let $B = \operatorname{diag}(J_1, \ldots, J_r)$ be a Jordan form of A. Then,

$$f(A) = S\operatorname{diag}[f(J_1), \ldots, f(J_r)]S^{-1}. \tag{12.8.4}$$

Letting $\lambda \in \mathcal{D}$ be an eigenvalue of A and $J = \lambda I_k + N_k$ denote a $k \times k$ Jordan block, expanding the infinite series $\sum_{i=1}^{\infty} \beta_i J^i$ shows that $f(J)$ is the $k \times k$ upper triangular, Toeplitz matrix

$$f(J) = f(\lambda)I_k + f'(\lambda)N_k + \tfrac{1}{2}f''(\lambda)N_k^2 + \cdots + \frac{1}{(k-1)!}f^{(k-1)}(\lambda)N_k^{k-1}$$

$$= \begin{bmatrix} f(\lambda) & f'(\lambda) & \frac{1}{2}f''(\lambda) & \cdots & \frac{1}{(k-1)!}f^{(k-1)}(\lambda) \\ 0 & f(\lambda) & f'(\lambda) & \cdots & \frac{1}{(k-2)!}f^{(k-2)}(\lambda) \\ 0 & 0 & f(\lambda) & \cdots & \frac{1}{(k-3)!}f^{(k-3)}(\lambda) \\ \vdots & \vdots & \ddots & \ddots & \vdots \\ 0 & 0 & 0 & \cdots & f(\lambda) \end{bmatrix}. \tag{12.8.5}$$

Note that every entry of $f(J)$ exists since f converges in $\mathcal{D}$ and all of its derivatives exist in $\mathcal{D}$. Therefore, the infinite series (12.8.2) converges.

Alternatively, since $\rho_{\max}(A) < \rho$, Proposition 11.3.3 implies that there exists a normalized submultiplicative norm $\|\cdot\|$ on $\mathbb{F}^{n \times n}$ such that $\|A\| < \rho$. Therefore, since ρ is the radius of convergence of f, it follows that $\sum_{i=0}^{\infty} \|\beta_i A^i\| \le \sum_{i=0}^{\infty} |\beta_i| \|A\|^i$ exists. Hence, $\sum_{i=0}^{\infty} \beta_i A^i$ is absolutely convergent.

Next, we extend the definition $f(A)$ to functions $f\colon\ \mathcal{D} \subseteq \mathbb{C} \mapsto \mathbb{C}$ that are not necessarily of the form (12.8.1).

Definition 12.8.1. Let $f\colon\ \mathcal{D} \subseteq \mathbb{C} \mapsto \mathbb{C}$, let $A \in \mathbb{C}^{n \times n}$, where $\operatorname{spec}(A) \subset \mathcal{D}$, and assume that, for all $\lambda_i \in \operatorname{spec}(A)$, f is $k_i - 1$ times differentiable at λ_i, where $k_i \triangleq \operatorname{ind}_A(\lambda_i)$ is the largest size of all of the Jordan blocks associated with λ_i as given by Theorem 7.4.2. Then, f is *defined* at A, and $f(A)$ is given by (12.8.3) and (12.8.4), where $f(J_i)$ is defined by (12.8.5) with $k = k_i$ and $\lambda = \lambda_i$.

The following result shows that $f(A)$ in Definition 12.8.1 is well-defined in the sense that $f(A)$ is independent of the decomposition $A = SBS^{-1}$ used to define $f(A)$ in (12.8.3).

Theorem 12.8.2. Let $A \in \mathbb{F}^{n \times n}$, let $\operatorname{spec}(A) = \{\lambda_1, \ldots, \lambda_r\}$, and, for all $i \in \{1, \ldots, r\}$, let $k_i \triangleq \operatorname{ind}_A(\lambda_i)$. Furthermore, suppose that $f\colon\ \mathcal{D} \subseteq \mathbb{C} \mapsto \mathbb{C}$ is defined at A. Then, there exists a polynomial $p \in \mathbb{F}[s]$ such that $f(A) = p(A)$. Furthermore, there exists a unique polynomial p of degree less than $\sum_{i=1}^{r} k_i$ satisfying $f(A) = p(A)$ and such that, for all $i \in \{1, \ldots, r\}$ and $j \in \{0, 1, \ldots, k_i - 1\}$,

$$f^{(j)}(\lambda_i) = p^{(j)}(\lambda_i). \tag{12.8.6}$$

This polynomial is given by

$$p(s) = \sum_{i=1}^{r}\left(\left[\prod_{\substack{j=1\\ j\neq i}}^{r}(s-\lambda_j)^{n_j}\right]\left[\sum_{k=0}^{k_i-1}\frac{1}{k!}\frac{\mathrm{d}^k}{\mathrm{d}s^k}\frac{f(s)}{\prod_{\substack{l=1\\ l\neq i}}^{r}(s-\lambda_l)^{k_l}}\Bigg|_{s=\lambda_i}(s-\lambda_i)^k\right]\right). \tag{12.8.7}$$

If, in addition, A is simple, then p is given by

$$p(s) = \sum_{i=1}^{r} f(\lambda_i)\prod_{\substack{j=1\\ j\neq i}}^{r}\frac{s-\lambda_j}{\lambda_i-\lambda_j}. \tag{12.8.8}$$

Proof. See [799, pp. 263, 264]. □

The polynomial (12.8.7) is the *Lagrange-Hermite interpolation polynomial* for f.

The following result, called the *identity theorem*, is a special case of Theorem 12.8.2.

Theorem 12.8.3. Let $A \in \mathbb{F}^{n\times n}$, let $\operatorname{spec}(A) = \{\lambda_1, \ldots, \lambda_r\}$, and, for all $i \in \{1, \ldots, r\}$, let $k_i \triangleq \operatorname{ind}_A(\lambda_i)$. Furthermore, let $f\colon\ \mathcal{D} \subseteq \mathbb{C} \mapsto \mathbb{C}$ and $g\colon\ \mathcal{D} \subseteq \mathbb{C} \mapsto \mathbb{C}$ be analytic on a neighborhood of $\operatorname{spec}(A)$. Then, $f(A) = g(A)$ if and only if, for all $i \in \{1, \ldots, r\}$ and $j \in \{0, 1, \ldots, k_i - 1\}$,

$$f^{(j)}(\lambda_i) = g^{(j)}(\lambda_i). \tag{12.8.9}$$

Corollary 12.8.4. Let $A \in \mathbb{F}^{n\times n}$, and let $f\colon\ \mathcal{D} \subset \mathbb{C} \mapsto \mathbb{C}$ be analytic on a neighborhood of $\operatorname{mspec}(A)$. Then,

$$\operatorname{mspec}[f(A)] = f[\operatorname{mspec}(A)]. \tag{12.8.10}$$

12.9 Matrix Square Root and Matrix Sign Functions

Theorem 12.9.1. Let $A \in \mathbb{C}^{n\times n}$, and assume that A is group invertible and has no eigenvalues in $(-\infty, 0)$. Then, there exists a unique matrix $B \in \mathbb{C}^{n\times n}$ such that $\operatorname{spec}(B) \subset \mathrm{ORHP} \cup \{0\}$ and such that $B^2 = A$. If, in addition, A is real, then B is real.

Proof. See [1391, pp. 20, 31]. □

The matrix B given by Theorem 12.9.1 is the *principal square root* of A. This matrix is denoted by $A^{1/2}$. The existence of a square root that is not necessarily the principal square root is discussed in Fact 7.17.21.

The following result defines the *matrix sign function*.

Definition 12.9.2. Let $A \in \mathbb{C}^{n\times n}$, assume that A has no eigenvalues on the imaginary axis, and let

$$A = S\begin{bmatrix} J_1 & 0\\ 0 & J_2\end{bmatrix}S^{-1},$$

where $S \in \mathbb{C}^{n\times n}$ is nonsingular, $J_1 \in \mathbb{C}^{p\times p}$ and $J_2 \in \mathbb{C}^{q\times q}$ are Jordan matrices, and $\operatorname{spec}(J_1) \subset \mathrm{OLHP}$ and $\operatorname{spec}(J_2) \subset \mathrm{ORHP}$. Then, the *matrix sign* of A is defined by

$$\operatorname{Sign}(A) \triangleq S\begin{bmatrix} -I_p & 0\\ 0 & I_q\end{bmatrix}S^{-1}.$$

12.10 Vector and Matrix Derivatives

In this section we consider derivatives of scalar-valued functions with matrix arguments. Consider the linear function $f\colon\ \mathbb{F}^{m\times n} \mapsto \mathbb{F}$ given by $f(X) = \operatorname{tr} AX$, where $A \in \mathbb{F}^{n\times m}$ and $X \in \mathbb{F}^{m\times n}$. In terms of vectors $x = \operatorname{vec} X \in \mathbb{F}^{mn}$, we can define the linear function $\hat{f}(x) \triangleq f(X) = (\operatorname{vec} A^{\mathsf{T}})^{\mathsf{T}}x$. Consequently, for all $X_0 \in \mathbb{F}^{m\times n}$, the function $\frac{\mathrm{d}}{\mathrm{d}X}f(X_0)\colon\ \mathbb{F}^{m\times n} \mapsto \mathbb{F}$ can be represented for all $Y \in \mathbb{F}^{m\times n}$

by

$$\frac{\mathrm{d}}{\mathrm{d}X}f(X_0)Y = \hat{f}'(\operatorname{vec} X_0)\operatorname{vec} Y = (\operatorname{vec} A^{\mathsf{T}})^{\mathsf{T}}\operatorname{vec} Y = \operatorname{tr} AY. \tag{12.10.1}$$

Noting that $\hat{f}'(\operatorname{vec} X_0) = (\operatorname{vec} A^{\mathsf{T}})^{\mathsf{T}}$ and identifying $\frac{\mathrm{d}}{\mathrm{d}X}f(X_0)$ with the matrix A, we define the *matrix derivative* of $f\colon\ \mathcal{D} \subseteq \mathbb{F}^{m\times n} \mapsto \mathbb{F}$ by

$$\frac{\mathrm{d}}{\mathrm{d}X}f(X) \triangleq (\operatorname{vec}^{-1}[\hat{f}'(\operatorname{vec} X)]^{\mathsf{T}})^{\mathsf{T}}, \tag{12.10.2}$$

which is the $n \times m$ matrix A whose (i, j) entry is $\frac{\partial f(X)}{\partial X_{(j,i)}}$. Note the ordering of the indices. The matrix derivative is a representation of the derivative in the sense that

$$\lim_{\substack{X\to X_0\\ X\in\mathcal{D}\backslash\{X_0\}}} \frac{f(X) - f(X_0) - \operatorname{tr}[F(X - X_0)]}{\|X - X_0\|} = 0, \tag{12.10.3}$$

where F denotes $\frac{\mathrm{d}}{\mathrm{d}X}f(X_0)$ and $\|\cdot\|$ is a norm on $\mathbb{F}^{m\times n}$.

Proposition 12.10.1. Let $x \in \mathbb{F}^n$. Then, the following statements hold:

i) If $A \in \mathbb{F}^{n\times n}$, then

$$\frac{\mathrm{d}}{\mathrm{d}x}x^{\mathsf{T}}Ax = x^{\mathsf{T}}(A + A^{\mathsf{T}}). \tag{12.10.4}$$

ii) If $A \in \mathbb{F}^{n\times n}$ is symmetric, then

$$\frac{\mathrm{d}}{\mathrm{d}x}x^{\mathsf{T}}Ax = 2x^{\mathsf{T}}A. \tag{12.10.5}$$

iii) If $A \in \mathbb{F}^{n\times n}$ is Hermitian, then

$$\frac{\mathrm{d}}{\mathrm{d}x}x^*Ax = 2x^*A. \tag{12.10.6}$$

Proposition 12.10.2. Let $A \in \mathbb{F}^{n\times m}$ and $B \in \mathbb{F}^{l\times n}$. Then, the following statements hold:

i) For all $X \in \mathbb{F}^{m\times n}$,

$$\frac{\mathrm{d}}{\mathrm{d}X}\operatorname{tr} AX = A. \tag{12.10.7}$$

ii) For all $X \in \mathbb{F}^{m\times l}$,

$$\frac{\mathrm{d}}{\mathrm{d}X}\operatorname{tr} AXB = BA. \tag{12.10.8}$$

iii) For all $X \in \mathbb{F}^{l\times m}$,

$$\frac{\mathrm{d}}{\mathrm{d}X}\operatorname{tr} AX^{\mathsf{T}}B = A^{\mathsf{T}}B^{\mathsf{T}}. \tag{12.10.9}$$

iv) For all $X \in \mathbb{F}^{m\times l}$ and $k \geq 1$,

$$\frac{\mathrm{d}}{\mathrm{d}X}\operatorname{tr}(AXB)^k = kB(AXB)^{k-1}A. \tag{12.10.10}$$

v) For all $X \in \mathbb{F}^{m\times l}$,

$$\frac{\mathrm{d}}{\mathrm{d}X}\det AXB = B(AXB)^{\mathsf{A}}A. \tag{12.10.11}$$

vi) For all $X \in \mathbb{F}^{m\times l}$ such that AXB is nonsingular,

$$\frac{\mathrm{d}}{\mathrm{d}X}\log\det AXB = B(AXB)^{-1}A. \tag{12.10.12}$$

Proposition 12.10.3. Let $A \in \mathbb{F}^{n\times m}$ and $B \in \mathbb{F}^{m\times n}$. Then, the following statements hold:

i) For all $X \in \mathbb{F}^{m\times m}$ and $k \ge 1$,

$$\frac{\mathrm{d}}{\mathrm{d}X}\operatorname{tr} AX^kB = \sum_{i=0}^{k-1} X^{k-1-i}BAX^i. \tag{12.10.13}$$

ii) For all nonsingular $X \in \mathbb{F}^{m\times m}$,

$$\frac{\mathrm{d}}{\mathrm{d}X}\operatorname{tr} AX^{-1}B = -X^{-1}BAX^{-1}. \tag{12.10.14}$$

iii) For all nonsingular $X \in \mathbb{F}^{m\times m}$,

$$\frac{\mathrm{d}}{\mathrm{d}X}\det AX^{-1}B = -X^{-1}B(AX^{-1}B)^{\mathrm{A}}AX^{-1}. \tag{12.10.15}$$

iv) For all nonsingular $X \in \mathbb{F}^{m\times m}$,

$$\frac{\mathrm{d}}{\mathrm{d}X}\log\det AX^{-1}B = -X^{-1}B(AX^{-1}B)^{-1}AX^{-1}. \tag{12.10.16}$$

Proposition 12.10.4. The following statements hold:

i) Let $A, B \in \mathbb{F}^{n\times m}$. Then, for all $X \in \mathbb{F}^{m\times n}$,

$$\frac{\mathrm{d}}{\mathrm{d}X}\operatorname{tr} AXBX = AXB + BXA. \tag{12.10.17}$$

ii) Let $A \in \mathbb{F}^{n\times n}$ and $B \in \mathbb{F}^{m\times m}$. Then, for all $X \in \mathbb{F}^{n\times m}$,

$$\frac{\mathrm{d}}{\mathrm{d}X}\operatorname{tr} AXBX^{\mathsf{T}} = BX^{\mathsf{T}}A + B^{\mathsf{T}}X^{\mathsf{T}}A^{\mathsf{T}}. \tag{12.10.18}$$

iii) Let $A \in \mathbb{F}^{n\times n}$. Then, for all $X \in \mathbb{F}^{n\times m}$,

$$\frac{\mathrm{d}}{\mathrm{d}X}\operatorname{tr} X^{\mathsf{T}}AX = X^{\mathsf{T}}(A + A^{\mathsf{T}}). \tag{12.10.19}$$

iv) Let $A \in \mathbb{F}^{k\times l}$, $B \in \mathbb{F}^{l\times m}$, $C \in \mathbb{F}^{n\times l}$, $D \in \mathbb{F}^{l\times l}$, and $E \in \mathbb{F}^{l\times k}$, and let p be a positive integer. Then, for all $X \in \mathbb{F}^{m\times n}$,

$$\frac{\mathrm{d}}{\mathrm{d}X}\operatorname{tr} A(D + BXC)^pE = \sum_{i=1}^{p} C(D + BXC)^{p-i}EA(D + BXC)^{i-1}B. \tag{12.10.20}$$

v) Let $A \in \mathbb{F}^{k\times l}$, $B \in \mathbb{F}^{l\times m}$, $C \in \mathbb{F}^{n\times l}$, $D \in \mathbb{F}^{l\times l}$, and $E \in \mathbb{F}^{l\times k}$. Then, for all $X \in \mathbb{F}^{m\times n}$ such that $D + BXC$ is nonsingular,

$$\frac{\mathrm{d}}{\mathrm{d}X}\operatorname{tr} A(D + BXC)^{-1}E = -C(D + BXC)^{-1}EA(D + BXC)^{-1}B. \tag{12.10.21}$$

vi) Let $A \in \mathbb{F}^{k\times l}$, $B \in \mathbb{F}^{l\times m}$, $C \in \mathbb{F}^{n\times l}$, $D \in \mathbb{F}^{l\times l}$, and $E \in \mathbb{F}^{l\times k}$. Then, for all $X \in \mathbb{F}^{n\times m}$ such that $D + BX^{\mathsf{T}}C$ is nonsingular,

$$\frac{\mathrm{d}}{\mathrm{d}X}\operatorname{tr} A(D + BX^{\mathsf{T}}C)^{-1}E = -B^{\mathsf{T}}(D + BX^{\mathsf{T}}C)^{-\mathsf{T}}A^{\mathsf{T}}E^{\mathsf{T}}(D + BX^{\mathsf{T}}C)^{-\mathsf{T}}C^{\mathsf{T}}. \tag{12.10.22}$$

12.11 Facts on One Set

Fact 12.11.1. Let $x \in \mathbb{R}$, and assume that x is an irrational number. Then,

$$\operatorname{cl}\{nx + m\colon\ n, m \in \mathbb{Z}\} = \mathbb{R}.$$

Source: [2294, pp. 44, 45]. **Remark:** This is *Kronecker's theorem.*

Fact 12.11.2. Let $\mathcal{S} \subset \mathbb{F}^n$, let $\|\cdot\|$ be a norm on $\mathbb{F}^n$, assume that there exists $\delta > 0$ such that, for all $x, y \in \mathcal{S}$, $\|x - y\| < \delta$, and let $x_0 \in \mathcal{S}$. Then, $\mathcal{S} \subseteq \mathbb{B}_\delta(x_0)$.

Fact 12.11.3. Let $\mathcal{S} \subset \mathbb{R}^n$, assume that $\mathcal{S}$ is bounded, and let $\delta \triangleq \sup\{\|x - y\|_2 : x, y \in \mathcal{S}\}$. Then, there exists $x_0 \in \mathbb{R}^n$ such that $\mathcal{S} \subseteq \mathbb{B}_{\sqrt{\frac{n}{2n+2}}\delta}(x_0)$. **Source:** [1260, p. 49]. **Remark:** This is *Jung's theorem*. **Remark:** For $n = 2$, the bound is achieved in the case where $\mathcal{S}$ is an equilateral triangle.

Fact 12.11.4. Let $\mathcal{S} \subseteq \mathbb{F}^n$. Then, $\operatorname{cl}\mathcal{S}$ is the smallest closed set containing $\mathcal{S}$, and $\operatorname{int}\mathcal{S}$ is the largest open set contained in $\mathcal{S}$.

Fact 12.11.5. Let $\mathcal{S} \subseteq \mathbb{F}^n$. If $\mathcal{S}$ is (open, closed), then $\mathcal{S}^\sim$ is (closed, open).

Fact 12.11.6. Let $\mathcal{S} \subseteq \mathcal{S}' \subseteq \mathbb{F}^n$. If $\mathcal{S}$ is (open relative to $\mathcal{S}'$, closed relative to $\mathcal{S}'$), then $\mathcal{S}'\backslash\mathcal{S}$ is (closed relative to $\mathcal{S}'$, open relative to $\mathcal{S}'$).

Fact 12.11.7. Let $\mathcal{S} \subseteq \mathbb{F}^n$. Then,

$$(\operatorname{int}\mathcal{S})^\sim = \operatorname{cl}(\mathcal{S}^\sim), \quad \operatorname{bd}\mathcal{S} = \operatorname{cl}\operatorname{bd}\mathcal{S} = \operatorname{bd}(\mathcal{S}^\sim) = (\operatorname{cl}\mathcal{S}) \cap \operatorname{cl}(\mathcal{S}^\sim) = [(\operatorname{int}\mathcal{S}) \cup \operatorname{int}(\mathcal{S}^\sim)]^\sim.$$

Fact 12.11.8. Let $\mathcal{S} \subseteq \mathbb{F}^n$, and assume that $\operatorname{int}\mathcal{S} = \varnothing$. Then, $\operatorname{bd}\mathcal{S} = \operatorname{cl}\mathcal{S}$. Hence, if $\mathcal{S}$ is closed, then $\mathcal{S} = \operatorname{bd}\mathcal{S}$, whereas, if $\mathcal{S}$ is not closed, then $\mathcal{S} \subset \operatorname{bd}\mathcal{S}$.

Fact 12.11.9. Let $\mathcal{S} \subseteq \mathbb{F}^n$, and assume that $\mathcal{S}$ is either open or closed. Then, $\operatorname{int}\operatorname{bd}\mathcal{S}$ is empty. **Source:** [154, p. 68]. **Remark:** Let $\mathcal{S} = \{x \in \mathbb{R}\colon x$ is a rational number$\}$. Then, $\operatorname{bd}\mathcal{S} = \mathbb{R}$.

Fact 12.11.10. Let $\mathcal{S} \subseteq \mathbb{F}^n$, and assume that $\operatorname{int}\mathcal{S}$ is nonempty. Then, $\dim\mathcal{S} = n$.

Fact 12.11.11. Let $\mathcal{S} \subseteq \mathbb{F}^n$, and assume that $\mathcal{S}$ is an affine subspace. Then, $\mathcal{S}$ is closed. Furthermore, $\mathcal{S}$ is open if and only if either $\mathcal{S} = \varnothing$ or $\mathcal{S} = \mathbb{F}^n$.

Fact 12.11.12. Let $\mathcal{S} \subseteq \mathbb{F}^n$. Then, $\operatorname{cl}\mathcal{S} \subseteq \operatorname{affin}\operatorname{cl}\mathcal{S} = \operatorname{cl}\operatorname{affin}\mathcal{S} = \operatorname{affin}\mathcal{S}$. **Source:** [523, p. 7].

Fact 12.11.13. Let $n \ge 2$, let $1 \le k \le n-1$, and let $x_1, \ldots, x_k \in \mathbb{F}^n$. Then, $\operatorname{int}\operatorname{affin}\{x_1, \ldots, x_k\} = \varnothing$. **Related:** Fact 3.11.14.

Fact 12.11.14. Let $x \in \mathbb{F}^n$, and let $\varepsilon > 0$. Then, $\mathbb{B}_\varepsilon(x)$ is completely solid and convex.

Fact 12.11.15. Let $\mathcal{S} \subseteq \mathbb{F}^n$, and assume that $\mathcal{S}$ is a (cone, convex set, convex cone, affine subspace, subspace). Then, so are $\operatorname{cl}\mathcal{S}$, $\operatorname{int}\mathcal{S}$, and $\operatorname{relint}\mathcal{S}$. **Source:** [1260, p. 44], [2319, p. 45], and [2320, p. 64].

Fact 12.11.16. Let $\mathcal{S} \subseteq \mathbb{F}^n$, and assume that $\mathcal{S}$ is convex. Then, $\operatorname{relint}\operatorname{cl}\mathcal{S} = \operatorname{relint}\mathcal{S}$, $\operatorname{int}\operatorname{cl}\mathcal{S} = \operatorname{int}\mathcal{S}$, and $\mathcal{S} \subseteq \operatorname{cl}\operatorname{relint}\mathcal{S} = \operatorname{cl}\mathcal{S}$. **Source:** [1260, p. 44] and [2487, p. 95], and use Fact 12.12.4.

Fact 12.11.17. Let $\mathcal{S} \subseteq \mathbb{F}^n$, and assume that $\mathcal{S}$ is solid. Then, $\operatorname{conv}\mathcal{S}$ is completely solid.

Fact 12.11.18. Let $\mathcal{S} \subseteq \mathbb{F}^n$. Then, $\operatorname{conv}\operatorname{relint}\mathcal{S} \subseteq \operatorname{relint}\operatorname{conv}\mathcal{S}$, and $\operatorname{conv}\operatorname{int}\mathcal{S} \subseteq \operatorname{int}\operatorname{conv}\mathcal{S}$. If, in addition, $\mathcal{S}$ is completely solid, then $\operatorname{conv}\operatorname{int}\mathcal{S} = \operatorname{int}\operatorname{conv}\mathcal{S}$.

Fact 12.11.19. Let $\mathcal{S} \subseteq \mathbb{F}^n$, and assume that $\mathcal{S}$ is open. Then, $\operatorname{conv}\mathcal{S}$ is open. **Source:** Assume that $\mathcal{S}$ is nonempty, so that $\mathcal{S}$ is completely solid. Fact 12.11.18 implies that $\operatorname{conv}\mathcal{S} = \operatorname{conv}\operatorname{int}\mathcal{S} = \operatorname{int}\operatorname{conv}\mathcal{S}$.

Fact 12.11.20. Let $\mathcal{S} \subseteq \mathbb{F}^n$, and assume that $\mathcal{S}$ is convex. Then, the following statements are equivalent:

i) $\mathcal{S}$ is solid.

ii) $\mathcal{S}$ is completely solid.

iii) $\dim\mathcal{S} = n$.

iv) $\operatorname{affin}\mathcal{S} = \mathbb{F}^n$.

Fact 12.11.21. Let $\mathcal{S} \subseteq \mathbb{F}^n$. Then, the following statements hold:

i) $\operatorname{conv}\operatorname{cl}\mathcal{S} \subseteq \operatorname{cl}\operatorname{conv}\mathcal{S}$.

ii) If $\mathcal{S}$ is bounded, then $\operatorname{conv}\mathcal{S}$ is bounded.

iii) If $\mathcal{S}$ is either bounded or convex, then $\operatorname{conv}\operatorname{cl}\mathcal{S} = \operatorname{cl}\operatorname{conv}\mathcal{S}$.

iv) If $\mathcal{S}$ is convex, then $\operatorname{cl}\mathcal{S}$ is convex.

v) If $\mathcal{S}$ is compact, then $\operatorname{conv}\mathcal{S}$ is compact.

vi) $\operatorname{conv}\operatorname{relint}\mathcal{S} \subseteq \operatorname{relint}\operatorname{conv}\mathcal{S}$.

Source: [1260, p. 45] and [2487, pp. 128–130]. **Remark:** $\mathcal{S} = \{x \in \mathbb{R}^2\colon\ x_{(1)}^2 x_{(2)}^2 = 1$ for all $x_{(1)} > 0\}$ is closed, but $\operatorname{conv}\mathcal{S}$ is not closed. Hence, $\operatorname{conv}\operatorname{cl}\mathcal{S} \subset \operatorname{cl}\operatorname{conv}\mathcal{S}$. Likewise, $\mathcal{S} = \{[0\ \ 0]^{\mathsf{T}}\} \cup \{[x\ \ 1]^{\mathsf{T}} \in \mathbb{R}^2\colon\ x \in \mathbb{R}\}$ satisfies $\operatorname{conv}\operatorname{cl}\mathcal{S} \subset \operatorname{cl}\operatorname{conv}\mathcal{S}$.

Fact 12.11.22. Let $A \in \mathbb{F}^{n\times m}$, $\mathcal{S}_1 \subseteq \mathbb{F}^m$, and $\mathcal{S}_2 \subseteq \mathbb{F}^n$. Then,

$$\operatorname{relint}\operatorname{conv} A\mathcal{S}_1 = A\operatorname{relint}\operatorname{conv}\mathcal{S}_1,$$

$$\operatorname{relint}\operatorname{conv} A^{\mathsf{inv}}(\mathcal{S}_2) = A^{\mathsf{inv}}(\operatorname{relint}\operatorname{conv}[\mathcal{S}_2 \cap \mathcal{R}(A)]) \subseteq A^{\mathsf{inv}}(\operatorname{conv}\mathcal{S}_2).$$

Source: [2487, p. 135]. **Related:** Fact 3.11.17.

Fact 12.11.23. Let $\mathcal{S} \subseteq \mathbb{F}^n$. Then, $\operatorname{relbd}\mathcal{S}$ is closed. Now, assume that $\mathcal{S}$ is convex. Then, $\operatorname{relbd}\mathcal{S} = \operatorname{relbd}\operatorname{cl}\mathcal{S} = \operatorname{relbd}\operatorname{relint}\mathcal{S}$. Furthermore, $\operatorname{relbd}\mathcal{S}$ is empty if and only if $\mathcal{S}$ is an affine subspace. **Source:** [2487, p. 104].

Fact 12.11.24. Let $\mathcal{S}_1 \subseteq \mathbb{F}^m$, $\mathcal{S}_2 \subseteq \mathbb{F}^n$, and $A \in \mathbb{F}^{n\times m}$. Then,

$$\operatorname{relint} A\mathcal{S}_1 = A\operatorname{relint}\mathcal{S}_1,$$

$$\operatorname{relint} A^{\mathsf{inv}}(\mathcal{S}_2) = A^{\mathsf{inv}}(\operatorname{relint}[\mathcal{S}_2 \cap \mathcal{R}(A)]) \subseteq A^{\mathsf{inv}}(\operatorname{relint}\mathcal{S}_2).$$

If, in addition, $(\operatorname{relint}\mathcal{S}_2) \cap \mathcal{R}(A) \neq \varnothing$, then

$$\operatorname{relint} A^{\mathsf{inv}}(\mathcal{S}_2) = A^{\mathsf{inv}}(\operatorname{relint}\mathcal{S}_2).$$

Source: [2487, pp. 92, 93]. **Related:** Fact 12.11.25 and Fact 12.11.26.

Fact 12.11.25. Let $\mathcal{S}_1 \subseteq \mathbb{F}^m$, $\mathcal{S}_2 \subseteq \mathbb{F}^n$, and $A \in \mathbb{F}^{n\times m}$. Then,

$$\operatorname{relint}\operatorname{conv} A\mathcal{S}_1 = A\operatorname{relint}\operatorname{conv}\mathcal{S}_1,$$

$$\operatorname{relint}\operatorname{conv} A^{\mathsf{inv}}(\mathcal{S}_2) = A^{\mathsf{inv}}(\operatorname{relint}\operatorname{conv}[\mathcal{S}_2 \cap \mathcal{R}(A)]) \subseteq A^{\mathsf{inv}}(\operatorname{relint}\operatorname{conv}\mathcal{S}_2).$$

Source: [2487, p. 135]. **Related:** Fact 12.11.24.

Fact 12.11.26. Let $\mathcal{S}_1 \subseteq \mathbb{F}^m$, $\mathcal{S}_2 \subseteq \mathbb{F}^n$, and $A \in \mathbb{F}^{n\times m}$. Then,

$$A\operatorname{cl}\mathcal{S}_1 \subseteq \operatorname{cl} A\mathcal{S}_1,$$

$$\operatorname{cl} A^{\mathsf{inv}}(\mathcal{S}_2) = A^{\mathsf{inv}}(\operatorname{cl}[\mathcal{S}_2 \cap \mathcal{R}(A)]) \subseteq A^{\mathsf{inv}}(\operatorname{cl}\mathcal{S}_2).$$

Furthermore, the following statements hold:

i) If either A is left invertible or $\mathcal{S}_1$ is bounded, then $A\operatorname{cl}\mathcal{S}_1 = \operatorname{cl} A\mathcal{S}_1$.

ii) $A\operatorname{cl}\mathcal{S}_1 = \operatorname{cl} A\mathcal{S}_1$ if and only if $(\operatorname{cl}\mathcal{S}_1) + \mathcal{N}(A)$ is closed.

iii) If $\operatorname{relint}\mathcal{S}_2 \cap \mathcal{R}(A) \neq \varnothing$, then $\operatorname{cl} A^{\mathsf{inv}}(\mathcal{S}_2) = A^{\mathsf{inv}}(\operatorname{cl}\mathcal{S}_2)$.

Source: [2487, pp. 99, 100, 103]. **Related:** Fact 12.11.24.

Fact 12.11.27. Let $\mathcal{S} \subseteq \mathbb{F}^n$, and assume that $\mathcal{S}$ is solid. Then, $\dim\mathcal{S} = n$.

Fact 12.11.28. Let $\mathcal{S} \subseteq \mathbb{F}^m$, assume that $\mathcal{S}$ is solid, let $A \in \mathbb{F}^{n\times m}$, and assume that A is right invertible. Then, $A\mathcal{S}$ is solid. **Source:** Theorem 12.4.22. **Related:** Fact 3.13.4.

Fact 12.11.29. $\mathbf{N}^n$ is a closed and completely solid subset of $\mathbb{F}^{n(n+1)/2}$. Furthermore, $\operatorname{int}\mathbf{N}^n = \mathbf{P}^n$.

Fact 12.11.30. Let $\mathcal{D} \subseteq \mathbb{F}^n$, and let x_0 belong to a solid, convex subset of $\mathcal{D}$. Then,

$$\dim\operatorname{fcone}(\mathcal{D}, x_0) = n.$$

Fact 12.11.31. Let $\mathcal{S} \subset \mathbb{F}^n$, assume that $\mathcal{S}$ is equilibrated, solid, compact, and convex, and, for all $x \in \mathbb{F}^n$, define

$$\|x\| \triangleq \min\{\alpha \geq 0\colon\ x \in \alpha\mathcal{S}\} = \max\{\alpha \geq 0\colon\ \alpha x \in \mathcal{S}\}.$$

Then, $\|\cdot\|$ is a norm on $\mathbb{F}^n$, and $\mathrm{cl}\,\mathbb{B}_1(0) = \mathcal{S}$. Conversely, let $\|\cdot\|$ be a norm on $\mathbb{F}^n$. Then, $\mathrm{cl}\,\mathbb{B}_1(0)$ is equilibrated, solid, compact, and convex, and, in addition,

$$\|x\| = \min\{\alpha \geq 0\colon\ x \in \alpha\,\mathrm{cl}\,\mathbb{B}_1(0)\} = \max\{\alpha \geq 0\colon\ \alpha x \in \mathrm{cl}\,\mathbb{B}_1(0)\}.$$

Source: [1467, pp. 38, 39]. **Remark:** In all cases, $\mathbb{B}_1(0)$ is defined with respect to $\|\cdot\|$. **Remark:** $\mathcal{S}$ is *equilibrated* if, for all $\lambda \in \mathbb{F}$ such that $|\lambda| = 1$, $\lambda\mathcal{S} = \mathcal{S}$. If $\mathbb{F} = \mathbb{R}$, then $\mathcal{S}$ is equilibrated if and only if it is symmetric; that is, $\mathcal{S} = -\mathcal{S}$. In the case where $\mathbb{F} = \mathbb{C}$, let $z = x + y\jmath \in \mathbb{C}$, where $x, y \in \mathbb{R}$. Then, $f(z) \triangleq \sqrt{2x^2 + y^2}$ does not satisfy $f(\jmath z) = |\jmath|f(z)$, and thus f is not a norm on $\mathbb{C}$. In fact, if $\|\cdot\|$ is a norm on $\mathbb{C}$, then, for all $z \in \mathbb{C}$, $\|z\| = \|z \cdot 1\| = |z|\|1\|$. Hence, $\|\cdot\|$ is proportional to $|\cdot|$, whose unit ball is circular. **Credit:** H. Minkowski. **Related:** Fact 11.7.1.

Fact 12.11.32. Let $\mathcal{S} \subseteq \mathbb{R}^n$, assume that $\mathcal{S}$ is a nonempty, closed, convex set, and let $\mathcal{E} \subseteq \mathcal{S}$ denote the set of elements of $\mathcal{S}$ that cannot be represented as nontrivial convex combinations of two distinct elements of $\mathcal{S}$. Then, $\mathcal{E}$ is nonempty and $\mathcal{S} = \mathrm{conv}\,\mathcal{E}$. If, in addition, $n = 2$, then $\mathcal{E}$ is closed. **Source:** [970, pp. 482–484]. **Remark:** $\mathcal{E}$ is the set of *extreme points* of $\mathcal{S}$. **Remark:** The last result is the *Krein-Milman theorem*. $\mathcal{E}$ is not necessarily closed for $n \geq 3$. See [970, p. 483].

12.12 Facts on Two or More Sets

Fact 12.12.1. Let $\mathcal{S}_1 \subseteq \mathcal{S}_2 \subseteq \mathbb{F}^n$. Then, $\mathrm{cl}\,\mathcal{S}_1 \subseteq \mathrm{cl}\,\mathcal{S}_2$ and $\mathrm{int}\,\mathcal{S}_1 \subseteq \mathrm{int}\,\mathcal{S}_2$.

Fact 12.12.2. Let $\mathcal{S}_1, \mathcal{S}_2 \subseteq \mathbb{F}^n$. Then, the following statements hold:

i) $(\mathrm{int}\,\mathcal{S}_1) \cap (\mathrm{int}\,\mathcal{S}_2) = \mathrm{int}(\mathcal{S}_1 \cap \mathcal{S}_2)$.

ii) $(\mathrm{int}\,\mathcal{S}_1) \cup (\mathrm{int}\,\mathcal{S}_2) \subseteq \mathrm{int}(\mathcal{S}_1 \cup \mathcal{S}_2)$.

iii) $(\mathrm{cl}\,\mathcal{S}_1) \cup (\mathrm{cl}\,\mathcal{S}_2) = \mathrm{cl}(\mathcal{S}_1 \cup \mathcal{S}_2)$.

iv) $\mathrm{bd}(\mathcal{S}_1 \cup \mathcal{S}_2) \subseteq (\mathrm{bd}\,\mathcal{S}_1) \cup (\mathrm{bd}\,\mathcal{S}_2)$.

v) If $(\mathrm{cl}\,\mathcal{S}_1) \cap (\mathrm{cl}\,\mathcal{S}_2) = \varnothing$, then $\mathrm{bd}(\mathcal{S}_1 \cup \mathcal{S}_2) = (\mathrm{bd}\,\mathcal{S}_1) \cup (\mathrm{bd}\,\mathcal{S}_2)$.

vi) Assume that $\mathcal{S}_1$ and $\mathcal{S}_2$ are convex. Then, $\mathrm{cl}(\mathcal{S}_1 \cap \mathcal{S}_2) \subseteq (\mathrm{cl}\,\mathcal{S}_1) \cap \mathrm{cl}\,\mathcal{S}_2$. If, in addition, $(\mathrm{relint}\,\mathcal{S}_1) \cap \mathrm{relint}\,\mathcal{S}_2 \neq \varnothing$, then $\mathrm{cl}(\mathcal{S}_1 \cap \mathcal{S}_2) = (\mathrm{cl}\,\mathcal{S}_1) \cap \mathrm{cl}\,\mathcal{S}_2$.

Source: [154, p. 65]. The last statement is given in [2487, p. 97].

Fact 12.12.3. Let $\mathcal{S}_1, \mathcal{S}_2 \subseteq \mathbb{F}^n$, assume that either $\mathcal{S}_1$ is closed or $\mathcal{S}_2$ is closed, and assume that $\mathrm{int}\,\mathcal{S}_1 = \mathrm{int}\,\mathcal{S}_2 = \varnothing$. Then, $\mathrm{int}(\mathcal{S}_1 \cup \mathcal{S}_2)$ is empty. **Source:** [154, p. 69]. **Remark:** The set $\mathrm{int}(\mathcal{S}_1 \cup \mathcal{S}_2)$ is not necessarily empty if neither $\mathcal{S}_1$ nor $\mathcal{S}_2$ is closed. Consider the sets of rational and irrational numbers.

Fact 12.12.4. Let $\mathcal{S}_1, \mathcal{S}_2 \subseteq \mathbb{F}^n$, and assume that $\mathcal{S}_1$ is open, $\mathcal{S}_2$ is convex, and $\mathcal{S}_1 \subseteq \mathrm{cl}\,\mathcal{S}_2$. Then, $\mathcal{S}_1 \subseteq \mathcal{S}_2$. **Source:** [2403, pp. 72, 73]. **Remark:** The statement is false without the assumption that $\mathcal{S}_2$ is convex. Let $\mathcal{S}_1 = (0, 3/2)$ and $\mathcal{S}_2 = (0, 1) \cup (1, 2)$. **Related:** Fact 12.11.16.

Fact 12.12.5. Let $\mathcal{S}_1, \mathcal{S}_2 \subseteq \mathbb{F}^n$, and assume that $\mathcal{S}_1$ is open. Then, $\mathcal{S}_1 + \mathcal{S}_2$ is open. **Source:** Note that $\mathcal{S}_1 + \mathcal{S}_2 = \cup_{x \in \mathcal{S}_2}(\mathcal{S}_1 + \{x\})$ is the union of a collection of open sets. The result now follows from Fact 12.12.16. **Remark:** See [1267, p. 107].

Fact 12.12.6. Define $\mathcal{S}_1, \mathcal{S}_2 \subseteq \mathbb{R}$ by $\mathcal{S}_1 \triangleq \mathbb{Z}$ and $\mathcal{S}_2 \triangleq \{i + 1/i + \pi\colon i \in \mathbb{P}\}$. Then, $\mathcal{S}_1$ and $\mathcal{S}_2$ are closed, but $\mathcal{S}_1 + \mathcal{S}_2$ is not closed. **Remark:** The sum of two closed, convex sets in $\mathbb{R}^2$ is not necessarily closed. See [2487, p. 98]. The sum of two closed, convex cones in $\mathbb{R}^3$ is not necessarily closed. See [309, p. 65].

Fact 12.12.7. Let $\mathcal{S}_1, \mathcal{S}_2 \subseteq \mathbb{F}^n$. Then, $\mathrm{cl}\,\mathcal{S}_1 + \mathrm{cl}\,\mathcal{S}_2 \subseteq \mathrm{cl}(\mathcal{S}_1 + \mathcal{S}_2)$. Now, assume that at least one of the following statements holds:

i) Either $\mathcal{S}_1$ or $\mathcal{S}_2$ is bounded.

ii) There exist affine subspaces $\mathcal{S}_3 \subseteq \mathbb{F}^n$ and $\mathcal{S}_4 \subseteq \mathbb{F}^n$ such that $\mathcal{S}_1 \subseteq \mathcal{S}_3$, $\mathcal{S}_2 \subseteq \mathcal{S}_4$, and the subspaces $\mathcal{S}_3'$ and $\mathcal{S}_4'$ that are parallel to $\mathcal{S}_3$ and $\mathcal{S}_4$, respectively, satisfy $\mathcal{S}_3' \cap \mathcal{S}_4' = \{0\}$.

Then, $\operatorname{cl}\mathcal{S}_1 + \operatorname{cl}\mathcal{S}_2 = \operatorname{cl}(\mathcal{S}_1 + \mathcal{S}_2)$. **Source:** [2487, pp. 10, 98]. **Related:** Fact 12.12.8.

Fact 12.12.8. Let $\mathcal{S}_1, \mathcal{S}_2 \subseteq \mathbb{F}^n$, and assume that $\mathcal{S}_1$ is closed and $\mathcal{S}_2$ is compact. Then, $\mathcal{S}_1 + \mathcal{S}_2$ is closed. If, in addition, $\mathcal{S}_1$ is compact, then $\mathcal{S}_1 + \mathcal{S}_2$ is compact. **Source:** [309, p. 34], [962, p. 209], and [1260, p. 81]. **Remark:** This result follows from Fact 12.12.7.

Fact 12.12.9. Let $\mathcal{S}_1, \mathcal{S}_2 \subseteq \mathbb{F}^n$, and assume that $\mathcal{S}_1$ and $\mathcal{S}_2$ are convex. Then, $\operatorname{cl}\mathcal{S}_1 = \operatorname{cl}\mathcal{S}_2$ if and only if $\operatorname{relint}\mathcal{S}_1 = \operatorname{relint}\mathcal{S}_2$. **Source:** [2487, p. 95].

Fact 12.12.10. Let $\mathcal{S}_1, \mathcal{S}_2, \mathcal{S}_3 \subseteq \mathbb{F}^n$, assume that $\mathcal{S}_1$, $\mathcal{S}_2$, and $\mathcal{S}_3$ are closed, convex sets, assume that $\mathcal{S}_1 \cap \mathcal{S}_2 \neq \varnothing$, $\mathcal{S}_2 \cap \mathcal{S}_3 \neq \varnothing$, and $\mathcal{S}_3 \cap \mathcal{S}_1 \neq \varnothing$, and assume that $\mathcal{S}_1 \cup \mathcal{S}_2 \cup \mathcal{S}_3$ is convex. Then, $\mathcal{S}_1 \cap \mathcal{S}_2 \cap \mathcal{S}_3 \neq \varnothing$. **Source:** [309, p. 32].

Fact 12.12.11. Let $\mathcal{S}_1, \mathcal{S}_2, \mathcal{S}_3 \subseteq \mathbb{F}^n$, assume that $\mathcal{S}_1$ and $\mathcal{S}_2$ are convex sets, $\mathcal{S}_2$ is a closed set, and $\mathcal{S}_3$ is a bounded set, and assume that $\mathcal{S}_1 + \mathcal{S}_3 \subseteq \mathcal{S}_2 + \mathcal{S}_3$. Then, $\mathcal{S}_1 \subseteq \mathcal{S}_2$. **Source:** [523, p. 5]. **Credit:** H. Radstrom.

Fact 12.12.12. Let $\mathcal{S}_1, \ldots, \mathcal{S}_k \subseteq \mathbb{F}^n$, and let $\alpha_1, \ldots, \alpha_k \in \mathbb{F}$. Then,

$$\operatorname{relint}\operatorname{conv}\sum_{i=1}^k \alpha_i\mathcal{S}_i = \sum_{i=1}^k \alpha_i \operatorname{relint}\operatorname{conv}\mathcal{S}_i.$$

If, in addition, $\mathcal{S}_1, \ldots, \mathcal{S}_k \subseteq \mathbb{F}^n$ are convex, then

$$\operatorname{relint}\sum_{i=1}^k \alpha_i\mathcal{S}_i = \sum_{i=1}^k \alpha_i \operatorname{relint}\mathcal{S}_i.$$

Source: [2487, pp. 89, 90, 132]. **Related:** Fact 3.12.5.

Fact 12.12.13. Let $\mathcal{S}_1, \ldots, \mathcal{S}_k \subseteq \mathbb{F}^n$. Then,

$$\operatorname{relint}\operatorname{conv}\bigcup_{i=1}^k \mathcal{S}_i \subseteq \operatorname{conv}\bigcup_{i=1}^k \operatorname{relint}\mathcal{S}_i.$$

If, in addition, $\cap_{i=1}^k \operatorname{relint}\mathcal{S}_i \neq \varnothing$, then

$$\operatorname{relint}\operatorname{conv}\bigcup_{i=1}^k \mathcal{S}_i = \operatorname{conv}\bigcup_{i=1}^k \operatorname{relint}\mathcal{S}_i.$$

Source: [2487, pp. 134, 135].

Fact 12.12.14. Let $\mathcal{S}, \mathcal{S}_1, \mathcal{S}_2 \subseteq \mathbb{F}^n$. Then, the following statements hold:

i) $\operatorname{polar}\mathcal{S}$ is a closed, convex set containing the origin.

ii) $\operatorname{polar}\mathbb{F}^n = \{0\}$, and $\operatorname{polar}\{0\} = \mathbb{F}^n$.

iii) If $\alpha > 0$, then $\operatorname{polar}\alpha\mathcal{S} = \frac{1}{\alpha}\operatorname{polar}\mathcal{S}$.

iv) $\mathcal{S} \subseteq \operatorname{polar}\operatorname{polar}\mathcal{S}$.

v) If $\mathcal{S}$ is nonempty, then $\operatorname{polar}\operatorname{polar}\operatorname{polar}\mathcal{S} = \operatorname{polar}\mathcal{S}$.

vi) If $\mathcal{S}$ is nonempty, then $\operatorname{polar}\operatorname{polar}\mathcal{S} = \operatorname{cl}\operatorname{conv}(\mathcal{S} \cup \{0\})$.

vii) If $0 \in \mathcal{S}$ and $\mathcal{S}$ is a closed, convex set, then $\operatorname{polar}\operatorname{polar}\mathcal{S} = \mathcal{S}$.

viii) If $\mathcal{S}_1 \subseteq \mathcal{S}_2$, then $\operatorname{polar}\mathcal{S}_2 \subseteq \operatorname{polar}\mathcal{S}_1$.

ix) $\operatorname{polar}(\mathcal{S}_1 \cup \mathcal{S}_2) = (\operatorname{polar}\mathcal{S}_1) \cap (\operatorname{polar}\mathcal{S}_2)$.

x) If $\mathcal{S}$ is a convex cone, then $\operatorname{polar}\mathcal{S} = \operatorname{dcone}\mathcal{S}$.

xi) $\operatorname{dcone}\operatorname{dcone}\mathcal{S} = \operatorname{cl}\operatorname{coco}\mathcal{S}$.

Source: [309, pp. 143–147]. For *xi*), see [523, p. 54].

Fact 12.12.15. Let $\mathcal{S}_1, \mathcal{S}_2 \subseteq \mathbb{F}^n$, and assume that $\mathcal{S}_1$ and $\mathcal{S}_2$ are cones. Then,

$$\operatorname{dcone}(\mathcal{S}_1 + \mathcal{S}_2) = (\operatorname{dcone}\mathcal{S}_1) \cap (\operatorname{dcone}\mathcal{S}_1).$$

If, in addition, $\mathcal{S}_1$ and $\mathcal{S}_2$ are closed, convex sets, then

$$\operatorname{dcone}(\mathcal{S}_1 \cap \mathcal{S}_2) = \operatorname{cl}[(\operatorname{dcone}\mathcal{S}_1) + (\operatorname{dcone}\mathcal{S}_2)].$$

Source: [309, p. 147] and [523, pp. 58, 59].

Fact 12.12.16. Let $\mathcal{A}$ be a collection of open subsets of $\mathbb{F}^n$. Then, the union of all elements of $\mathcal{A}$ is open. If, in addition, $\mathcal{A}$ is finite, then the intersection of all elements of $\mathcal{A}$ is open. **Source:** [154, p. 50].

Fact 12.12.17. Let $\mathcal{A}$ be a collection of closed subsets of $\mathbb{F}^n$. Then, the intersection of all elements of $\mathcal{A}$ is closed. If, in addition, $\mathcal{A}$ is finite, then the union of all elements of $\mathcal{A}$ is closed. **Source:** [154, p. 50].

Fact 12.12.18. Let $\mathcal{S} \subseteq \mathbb{F}^n$. Then, $\operatorname{cl}\mathcal{S}$ is the intersection of all closed subsets of $\mathbb{F}^n$ that contain $\mathcal{S}$.

Fact 12.12.19. Let $\mathcal{S} \subseteq \mathbb{F}^n$, and define the following terminology:

i) $\mathcal{S}$ is G_δ if it is the intersection of a countable number of open subsets of $\mathbb{F}^n$.

ii) $\mathcal{S}$ is F_σ if it is the union of a countable number of closed subsets of $\mathbb{F}^n$.

Then, the following statements hold:

iii) The intersection of a countable number of G_δ sets is G_δ.

iv) The union of a countable number of F_σ sets is F_σ.

v) If $\mathcal{S}$ is either open or closed, then it is both G_δ and F_σ.

vi) $\mathcal{S}$ is G_δ if and only if $\mathcal{S}^\sim$ is F_σ.

vii) The intersection of all open sets that contain $\mathcal{S}$ is $\mathcal{S}$.

viii) The intersection of a countable collection of open sets that contain $\mathcal{S}$ is a G_δ set that contains $\mathcal{S}$.

ix) The union of all closed sets that are contained in $\mathcal{S}$ is $\mathcal{S}$.

x) The union of a countable collection of closed sets that are contained in $\mathcal{S}$ is an F_σ set that is contained in $\mathcal{S}$.

xi) The set $\mathbb{Q}$ of rational numbers is F_σ, and the set $\mathbb{R}\backslash\mathbb{Q}$ of irrational numbers is G_δ.

xii) Let $a, b \in \mathbb{R}$, assume that $a < b$, and let $f\colon [a,b] \mapsto \mathbb{R}$. Then, $\mathcal{S} \triangleq \{x \in [a,b]\colon f \text{ is continuous at } x\}$ is G_δ.

Fact 12.12.20. Let $\mathcal{A} = (A_i)_{i=1}^\infty$ be a nonincreasing sequence of nonempty, compact subsets of $\mathbb{R}^n$. Then, $\operatorname{glb}(\mathcal{A})$ is compact and nonempty. **Source:** [154, p. 56]. **Remark:** This is the *Cantor intersection theorem*, where $\operatorname{glb}(\mathcal{A}) = \cap_{i=1}^\infty A_i$.

Fact 12.12.21. Let $\|\cdot\|$ be a norm on $\mathbb{F}^n$, let $\mathcal{S} \subset \mathbb{F}^n$, assume that $\mathcal{S}$ is a subspace, let $y \in \mathbb{F}^n$, and define

$$\mu \triangleq \max_{x\in\{z\in\mathcal{S}\colon \|z\|=1\}} |y^*x|.$$

Then, there exists $w \in \mathcal{S}^\perp$ such that

$$\max_{x\in\{z\in\mathbb{F}^n\colon \|z\|=1\}} |(y+w)^*x| = \mu.$$

Source: [2539, p. 57]. **Remark:** This is a version of the *Hahn-Banach theorem*. **Problem:** Find a simple interpretation in $\mathbb{R}^2$.

Fact 12.12.22. Let $\mathcal{S} \subset \mathbb{R}^n$, assume that $\mathcal{S}$ is a convex cone, let $x \in \mathbb{R}^n$, and assume that $x \notin \operatorname{int} \mathcal{S}$. Then, there exists a nonzero vector $\lambda \in \mathbb{R}^n$ such that $\lambda^\mathsf{T} x \le 0$ and $\lambda^\mathsf{T} z \ge 0$ for all $z \in \mathcal{S}$. **Source:** [1769, p. 37], [2260, p. 443], [2319, pp. 95–101], and [2544, pp. 96–100]. **Remark:** This is a *separation theorem.* **Remark:** Every convex cone that is a proper subset of $\mathbb{R}^n$ is contained in a closed half space.

Fact 12.12.23. Let $\mathcal{S}_1, \mathcal{S}_2 \subset \mathbb{R}^n$, and assume that $\mathcal{S}_1$ and $\mathcal{S}_2$ are nonempty, convex sets. Then, the following statements are equivalent:

i) There exist a nonzero vector $\lambda \in \mathbb{R}^n$ and $\alpha \in \mathbb{R}$ such that $\lambda^\mathsf{T} x \le \alpha$ for all $x \in \mathcal{S}_1$, $\lambda^\mathsf{T} y \ge \alpha$ for all $y \in \mathcal{S}_2$, and either $\mathcal{S}_1$ or $\mathcal{S}_2$ is not contained in the affine hyperplane $\{x \in \mathbb{R}^n\colon\ \lambda^\mathsf{T} x = \alpha\}$.

ii) $\operatorname{int}_{\operatorname{affin} \mathcal{S}_1} \mathcal{S}_1$ and $\operatorname{int}_{\operatorname{affin} \mathcal{S}_2} \mathcal{S}_2$ are disjoint.

Source: [421, p. 82] and [1267, p. 148]. **Remark:** This is a *proper separation theorem.*

Fact 12.12.24. Let $\|\cdot\|$ be a norm on $\mathbb{F}^n$, let $y \in \mathbb{F}^n$, let $\mathcal{S} \subseteq \mathbb{F}^n$, and assume that $\mathcal{S}$ is nonempty and closed. Then, there exists $x_0 \in \mathcal{S}$ such that

$$\|y - x_0\| = \min_{x \in \mathcal{S}} \|y - x\|.$$

Now, assume that $\mathcal{S}$ is convex. Then, there exists a unique vector $x_0 \in \mathcal{S}$ such that

$$\|y - x_0\| = \min_{x \in \mathcal{S}} \|y - x\|.$$

Consequently, there exists $x_0 \in \mathcal{S}$ such that, for all $x \in \mathcal{S}\backslash\{x_0\}$, $\|y - x_0\| < \|y - x\|$. **Source:** [970, pp. 470, 471]. **Related:** Fact 12.12.26.

Fact 12.12.25. Let $\|\cdot\|$ be a norm on $\mathbb{F}^n$, let $y_1, y_2 \in \mathbb{F}^n$, let $\mathcal{S} \subseteq \mathbb{F}^n$, assume that $\mathcal{S}$ is a nonempty, closed, convex set, and let x_1 and x_2 denote the unique elements of $\mathcal{S}$ that are closest to y_1 and y_2, respectively. Then,

$$\|x_1 - x_2\| \le \|y_1 - y_2\|.$$

Source: [970, pp. 474, 475].

Fact 12.12.26. Let $\mathcal{S} \subseteq \mathbb{R}^n$, assume that $\mathcal{S}$ is a subspace, let $A \in \mathbb{F}^{n\times n}$ be the projector onto $\mathcal{S}$, and let $x \in \mathbb{F}^n$. Then,

$$\min_{y \in \mathcal{S}} \|x - y\|_2 = \|A_\perp x\|_2.$$

Source: [1133, p. 41] and [2539, p. 91]. **Related:** Fact 12.12.24.

Fact 12.12.27. Let $\mathcal{S}, \mathcal{S}_1, \mathcal{S}_2, \mathcal{S}_3 \subseteq \mathbb{F}^n$, let $\|\cdot\|$ be a norm on $\mathbb{F}^n$, and define

$$\operatorname{dist}(x, \mathcal{S}) \triangleq \inf_{y \in \mathcal{S}} \|x - y\|,$$

$$\mathcal{H}(\mathcal{S}_1, \mathcal{S}_2) \triangleq \max\left\{\sup_{x \in \mathcal{S}_1} \operatorname{dist}(x, \mathcal{S}_2), \sup_{y \in \mathcal{S}_2} \operatorname{dist}(y, \mathcal{S}_1)\right\} = \max\left\{\sup_{x \in \mathcal{S}_1} \inf_{y \in \mathcal{S}_2} \|x - y\|, \sup_{y \in \mathcal{S}_2} \inf_{x \in \mathcal{S}_1} \|x - y\|\right\}.$$

Then, the following statements hold:

i) $\mathcal{H}(\mathcal{S}_1, \mathcal{S}_2) = \mathcal{H}(\mathcal{S}_2, \mathcal{S}_1)$.

ii) $\mathcal{H}(\mathcal{S}_1, \mathcal{S}_2) \ge 0$.

iii) If $\mathcal{S}_1$ and $\mathcal{S}_2$ are bounded, then $\mathcal{H}(\mathcal{S}_1, \mathcal{S}_2)$ is finite.

iv) If $\mathcal{H}(\mathcal{S}_1, \mathcal{S}_2)$ is finite, then $\mathcal{S}_1$ and $\mathcal{S}_2$ are either both bounded or both unbounded.

v) $\mathcal{H}(\mathcal{S}_1, \mathcal{S}_2) = 0$ if and only if $\operatorname{cl} \mathcal{S}_1 = \operatorname{cl} \mathcal{S}_2$.

vi) $\mathcal{H}(\mathcal{S}_1, \mathcal{S}_3) \le \mathcal{H}(\mathcal{S}_1, \mathcal{S}_2) + \mathcal{H}(\mathcal{S}_2, \mathcal{S}_3)$.

Remark: The function "$\mathcal{H}$" is the *Hausdorff distance*, which is a metric on the set of nonempty, compact subsets of $\mathbb{F}^n$. See [344, pp. 85, 86].

Fact 12.12.28. Let $\mathcal{S}_1, \mathcal{S}_2, \mathcal{S}_3 \subseteq \mathbb{R}^n$, assume that $\mathcal{S}_1$, $\mathcal{S}_2$, and $\mathcal{S}_3$ are nonzero subspaces, let A_1 and A_2 be the projectors onto $\mathcal{S}_1$ and $\mathcal{S}_2$, respectively, define "dist" and "$\mathcal{H}$" as in Fact 12.12.27 with $\|\cdot\| = \|\cdot\|_2$, and define

$$\mathcal{G}(\mathcal{S}_1, \mathcal{S}_2) \triangleq \max\left\{\max_{x\in\mathcal{S}_1\cap\mathbb{S}_1(0)} \operatorname{dist}(x, \mathcal{S}_2), \max_{y\in\mathcal{S}_2\cap\mathbb{S}_1(0)} \operatorname{dist}(y, \mathcal{S}_1)\right\}$$
$$= \max\left\{\max_{\substack{x\in\mathcal{S}_1\\ \|x\|_2=1}} \min_{y\in\mathcal{S}_2} \|x-y\|_2, \max_{\substack{y\in\mathcal{S}_2\\ \|y\|_2=1}} \min_{x\in\mathcal{S}_1} \|x-y\|_2\right\}.$$

Then, the following statements hold:

i) $\mathcal{G}(\mathcal{S}_1, \mathcal{S}_2) = \mathcal{G}(\mathcal{S}_2, \mathcal{S}_1)$.

ii) $0 \le \mathcal{G}(\mathcal{S}_1, \mathcal{S}_2) \le 1$.

iii) $\mathcal{G}(\mathcal{S}_1, \mathcal{S}_2) = 0$ if and only if $\mathcal{S}_1 = \mathcal{S}_2$.

iv) $\mathcal{G}(\mathcal{S}_1, \mathcal{S}_2) \le \mathcal{H}(\mathcal{S}_1 \cap \mathbb{S}_1(0), \mathcal{S}_2 \cap \mathbb{S}_1(0)) \le 2\mathcal{G}(\mathcal{S}_1, \mathcal{S}_2)$.

v) If $\dim \mathcal{S}_1 = \dim \mathcal{S}_2$, then $\mathcal{G}(\mathcal{S}_1, \mathcal{S}_2) = \sin\theta$, where θ is the minimal principal angle defined in Fact 7.12.42.

vi) $\mathcal{G}(\mathcal{S}_1, \mathcal{S}_2) = \sigma_{\max}(A_1 - A_2)$.

vii) $\mathcal{G}(\mathcal{S}_1, \mathcal{S}_3) \le \mathcal{G}(\mathcal{S}_1, \mathcal{S}_2) + \mathcal{G}(\mathcal{S}_2, \mathcal{S}_3)$.

Source: [1187, Chapter 13], [1590, pp. 199, 200], and [2539, pp. 91–93]. **Remark:** The function "$\mathcal{G}$" is the *gap*, which is a metric on the set of subspaces in $\mathbb{F}^n$. See [1590, pp. 199, 200]. **Remark:** If $\|\cdot\|$ is a norm on $\mathbb{F}^{n\times n}$, then $\mathrm{d}(\mathcal{S}_1, \mathcal{S}_2) \triangleq \|A_1 - A_2\|$ is a metric on the set of subspaces of $\mathbb{F}^n$, yielding the *gap topology*. See [2539, p. 93]. **Related:** Fact 7.13.27.

12.13 Facts on Functions

Fact 12.13.1. Let $p \in \mathbb{R}[s]$, let $\alpha \in \mathbb{R}$, assume that $p(\alpha) \ne 0$, and define $m \triangleq \operatorname{card}[\operatorname{mroots}(p) \cap (\alpha, \infty)]$. Then, $\operatorname{sign}[p(\alpha)] = (-1)^m$.

Fact 12.13.2. Let $p \in \mathbb{C}[s]$, where $p(s) = s^n + a_{n-1}s^{n-1} + \cdots + a_0$, let $\operatorname{roots}(p) = \{\lambda_1, \ldots, \lambda_r\}$, and, for all $i \in \{1, \ldots, r\}$, let $\alpha_i \in \mathbb{R}$ satisfy $0 < \alpha_i < \min_{j\ne i} |\lambda_i - \lambda_j|$. Furthermore, for all $\varepsilon_0, \ldots, \varepsilon_{n-1} \in \mathbb{R}$, define

$$p_{\varepsilon_0,\ldots,\varepsilon_{n-1}}(s) \triangleq s^n + (a_{n-1} + \varepsilon_{n-1})s^{n-1} + \cdots + (a_1 + \varepsilon_1)s + a_0 + \varepsilon_0.$$

Then, there exists $\varepsilon > 0$ such that, for all $\varepsilon_0, \ldots, \varepsilon_{n-1}$ satisfying $|\varepsilon_i| < \varepsilon$ for all $i \in \{1, \ldots, n-1\}$, it follows that, for all $i \in \{1, \ldots, r\}$, the polynomial $p_{\varepsilon_0,\ldots,\varepsilon_{n-1}}$ has exactly $\operatorname{mult}_p(\lambda_i)$ roots in the open disk $\{s \in \mathbb{C}\colon |s - \lambda_i| < \alpha_i\}$. **Source:** [2047]. **Remark:** This result shows that the roots of a polynomial are continuous functions of the coefficients. **Remark:** $\lambda_1, \ldots, \lambda_r$ are the distinct roots of p.

Fact 12.13.3. Let $p, q \in \mathbb{C}[s]$, where $p(s) = s^n + a_{n-1}s^{n-1} + \cdots + a_1 s + a_0$ and $q(s) = s^n + b_{n-1}s^{n-1} + \cdots + b_1 s + b_0$, define $\gamma \triangleq 2\max_{i\in\{1,\ldots,n\}}\{|a_{n-i}|^{1/i}, |b_{n-i}|^{1/i}\}$, and let $\operatorname{mroots}(p) = \{\lambda_1, \ldots, \lambda_n\}_{\mathrm{ms}}$ and $\operatorname{mroots}(q) = \{\mu_1, \ldots, \mu_n\}_{\mathrm{ms}}$. Then, there exists a permutation σ of $(1, \ldots, n)$ such that

$$\max_{i\in\{1,\ldots,n\}} |\lambda_i - \mu_{\sigma(i)}| \le \frac{4}{\sqrt[n]{2}}\left(\sum_{i=1}^n |a_i - b_i|\gamma^i\right)^{1/n}.$$

Source: [461]. **Related:** Fact 11.14.7.

Fact 12.13.4. Let $f\colon [0, 1] \mapsto \mathbb{C}$, assume that f is continuous, and let $z_1, \ldots, z_n \in \mathbb{C}$. Then, there exists $\alpha \in [0, 1]$ such that

$$\frac{|f(1) - f(0)|^n}{2^{2n-1}} \le \prod_{i=1}^n |f(\alpha) - z_i|.$$

Source: [461]. **Related:** Fact 12.13.5.

Fact 12.13.5. Let $p \in \mathbb{C}[s]$, assume that p is monic, and define $n \triangleq \deg p$. Then,

$$\frac{1}{2^{2n-1}} \le \max_{x\in[0,1]} |p(x)|.$$

Source: [461]. **Credit:** P. L. Chebyshev. **Related:** Fact 12.13.4 and Fact 12.13.6.

Fact 12.13.6. Let $p \in \mathbb{C}[s]$, where $p(s) = \sum_{i=0}^{n} \beta_i s^i$. Then,

$$|\beta_n| + |\beta_0| \le \max_{s\in\{z\in\mathbb{C}\colon\ |z|=1\}} |p(s)|.$$

Furthermore, equality holds if and only if $p(s) = \beta_n s^n + \beta_0$. **Source:** [1512]. **Credit:** P. L. Visser. **Related:** Fact 12.13.5.

Fact 12.13.7. Let $p \in \mathbb{C}[s]$, define $n \triangleq \deg p$, and let $r > 1$. Then,

$$\max_{z\in\mathbb{S}_r(0)} |p(z)| \le r^n \max_{z\in\mathrm{UC}} |p(z)|.$$

If, in addition, $\rho_{\min}(p) \ge 1$, then

$$\max_{z\in\mathbb{S}_r(0)} |p(z)| \le \left\{ \begin{array}{c} \dfrac{r^n+1}{2}\max_{z\in\mathrm{UC}}|p(z)| - \dfrac{r^n-1}{2}\min_{z\in\mathrm{UC}}|p(z)| \\ \dfrac{r^n+\rho_{\min}(p)}{1+\rho_{\min}(p)}\max_{z\in\mathrm{UC}}|p(z)| \end{array} \right\} \le \frac{r^n+1}{2}\max_{z\in\mathrm{UC}}|p(z)|.$$

Source: [3011]. Use Fact 2.4.2 for the last inequality. **Related:** Fact 12.16.3.

Fact 12.13.8. Let $\mathbb{S}_1 \subseteq \mathbb{F}^n$, assume that $\mathbb{S}_1$ is compact, let $\mathbb{S}_2 \subset \mathbb{F}^m$, let $f\colon\ \mathbb{S}_1 \times \mathbb{S}_2 \to \mathbb{R}$, and assume that f is continuous. Furthermore, for each $x \in \mathbb{S}_1$, define $h_x\colon\ \mathbb{S}_2 \mapsto \mathbb{R}$ by $h_x(y) = f(x,y)$, and define $g\colon\ \mathbb{S}_2 \to \mathbb{R}$ by $g(y) \triangleq \max_{x\in\mathbb{S}_1} h_x(y)$. Then, the following statements hold:

i) g is continuous.

ii) Assume that $\mathbb{S}_2$ is convex and, for all $x \in \mathbb{S}_1$, h_x is convex. Then, g is convex.

iii) Assume that $\mathbb{S}_2$ is open and, for all $(x,y) \in \mathbb{S}_1 \times \mathbb{S}_2$, $\partial f/\partial x$ exists and is continuous on $\mathbb{S}_1 \times \mathbb{S}_2$. Then, for all $z \in \mathbb{F}^n$,

$$\mathrm{D}g(y;z) = \max_{x\in\mathbb{S}_1} \frac{\partial f}{\partial x}(x,y)z.$$

Remark: See [1267, p. 20]. A related result is given in [962, p. 208]. **Remark:** This is *Danskin's theorem.*

Fact 12.13.9. Define $f\colon\ \mathbb{R}^2 \mapsto \mathbb{R}$ by

$$f(x,y) \triangleq \begin{cases} \min\{x/y, y/x\}, & x>0 \text{ and } y>0,\\ 0, & \text{otherwise}. \end{cases}$$

Furthermore, for all $x \in \mathbb{R}$, define $g_x\colon\ \mathbb{R} \mapsto \mathbb{R}$ by $g_x(y) \triangleq f(x,y)$, and, for all $y \in \mathbb{R}$, define $h_y\colon\ \mathbb{R} \mapsto \mathbb{R}$ by $h_y(x) \triangleq f(x,y)$. Then, the following statements hold:

i) For all $x \in \mathbb{R}$, g_x is continuous.

ii) For all $y \in \mathbb{R}$, h_y is continuous.

iii) f is not continuous at $(0,0)$.

Fact 12.13.10. Let $l_1, \dots, l_n$ and $m_1, \dots, m_n$ be positive integers, and define $f\colon \mathbb{R}^n \mapsto \mathbb{R}$ by

$$f(x) = \begin{cases} \dfrac{\prod_{i=1}^n x_{(i)}^{l_i}}{\sum_{i=1}^n x_{(i)}^{2m_i}}, & x \neq 0, \\ 0, & x = 0. \end{cases}$$

Then, the following statements are equivalent:

i) f is continuous.

ii) $\sum_{i=1}^n \frac{l_i}{m_i} > 2$.

Now, assume that $n \ge 2$ and $m_1 \le \cdots \le m_n$. Then, the following conditions are equivalent:

iii) For every hyperplane $\mathcal{S} \subset \mathbb{R}^n$, f is continuous on $\mathcal{S}$.

iv) For all $k \in \{2, \dots, n\}$, $2 + \frac{l_k}{m_k} - \frac{l_k}{m_{k-1}} < \sum_{i=1}^n \frac{l_i}{m_i}$.

Example: $f\colon \mathbb{R}^2 \mapsto \mathbb{R}$ defined by

$$f(x,y) = \begin{cases} \dfrac{xy^2}{x^2 + y^4}, & x^2 + y^2 > 0, \\ 0, & x = y = 0, \end{cases}$$

is not continuous at $x = y = 0$ but is continuous on every line that contains zero. Likewise, $f\colon \mathbb{R}^3 \mapsto \mathbb{R}$ defined by

$$f(x,y,z) = \begin{cases} \dfrac{xyz^2}{x^2 + y^4 + z^8}, & x^2 + y^2 + z^2 > 0, \\ 0, & x = y = z = 0, \end{cases}$$

is not continuous at $x = y = z = 0$ but is continuous on every plane that contains zero. **Source:** [746].

Fact 12.13.11. Let $\mathcal{S} \subseteq \mathbb{F}^n$, assume that $\mathcal{S}$ is pathwise connected, let $f\colon \mathcal{S} \mapsto \mathbb{F}^n$, and assume that f is continuous. Then, $f(\mathcal{S})$ is pathwise connected. **Source:** [2588, p. 65].

Fact 12.13.12. Let $f\colon [0,\infty) \to \mathbb{R}$, assume that f is continuous, and define

$$g(t) \triangleq \int_0^t f(\tau)\,\mathrm{d}\tau, \quad h(t) \triangleq \frac{1}{t}\int_0^t f(\tau)\,\mathrm{d}\tau$$

Then, the following statements hold:

i) Let $\mathcal{I}$ be a bounded subset of $[0,\infty)$. Then, f, g, and h are bounded on $\mathcal{I}$.

ii) If f is bounded, then h is bounded.

iii) If $\lim_{t\to\infty} f(t) = 0$, then either $\lim_{t\to\infty} g(t)$ exists, $\lim_{t\to\infty} g(t) = \infty$, or $\lim_{t\to\infty} g(t) = -\infty$.

iv) Assume that $\lim_{t\to\infty} f(t)$ exists. If $\lim_{t\to\infty} f(t) > 0$, then $\lim_{t\to\infty} g(t) = \infty$. If $\lim_{t\to\infty} f(t) < 0$, then $\lim_{t\to\infty} g(t) = -\infty$.

v) If $\lim_{t\to\infty} f(t) = \infty$, then $\lim_{t\to\infty} g(t) = \infty$. If $\lim_{t\to\infty} f(t) = -\infty$, then $\lim_{t\to\infty} g(t) = -\infty$.

vi) If $\lim_{t\to\infty} f(t) = \infty$, then $\lim_{t\to\infty} h(t) = \infty$. If $\lim_{t\to\infty} f(t) = -\infty$, then $\lim_{t\to\infty} h(t) = -\infty$.

vii) If $\lim_{t\to\infty} f(t)$ exists, then $\lim_{t\to\infty} h(t) = \lim_{t\to\infty} f(t)$.

Fact 12.13.13. Let $f\colon [0,1] \to \mathbb{R}$, and assume that f is continuous. Then,

$$\lim_{\alpha\downarrow 0} \int_0^1 \alpha t^{\alpha-1} f(t)\,\mathrm{d}t = f(0).$$

Source: [116, p. 439].

Fact 12.13.14. Let $f\colon\ \mathbb{R} \to \mathbb{R}$, assume that f is continuous, assume that, for all $x \in \mathbb{R}$, $f(x+1) = f(x)$, and let a be an irrational real number. Then,

$$\lim_{n\to\infty} \frac{1}{n} \sum_{i=1}^{n} f(ia) = \int_0^1 f(x)\,\mathrm{d}x.$$

Source: [1158, pp. 166, 167].

Fact 12.13.15. Let a and b be real numbers such that $a < b$, let $f\colon\ [a,b] \to \mathbb{R}$, assume that f is continuous, let $g\colon\ [a,b] \to \mathbb{R}$, and assume that g is integrable. If, for all $x \in [a,b]$, $g(x) \ge 0$, then there exists $c \in (a,b)$ such that

$$\int_a^b f(x)g(x)\,\mathrm{d}x = f(c)\int_a^b g(x)\,\mathrm{d}x.$$

In particular, there exists $c \in (a,b)$ such that

$$\int_a^b f(x)\,\mathrm{d}x = f(c)(b-a).$$

If f is nondecreasing on $[a,b]$, then there exists $c \in (a,b)$ such that

$$\int_a^b f(x)g(x)\,\mathrm{d}x = f(a)\int_a^c g(x)\,\mathrm{d}x + f(b)\int_c^b g(x)\,\mathrm{d}x.$$

Source: [2687]. **Remark:** The second statement is the *mean-value theorem.*

Fact 12.13.16. Let a be a positive number, let $f\colon\ [0,a] \to \mathbb{R}$, assume that $f(0) = 0$, assume that f is continuous and increasing, and let $b \in [0, f(a)]$. Then,

$$ab \le \int_0^a f(x)\,\mathrm{d}x + \int_0^b f^{\mathsf{Inv}}(x)\,\mathrm{d}x.$$

Equality holds if and only if $b = f(a)$. **Source:** [3002]. **Remark:** This is *Young's inequality.* **Related:** Fact 2.2.39.

Fact 12.13.17. Let $a > 0$, let $p > 1$, let $f\colon\ [0,a] \to \mathbb{R}$, and assume that f^p is integrable. Then,

$$\int_0^a \left(\frac{1}{x}\int_0^x f(t)\,\mathrm{d}t\right)^p \mathrm{d}x \le \left(\frac{p}{p-1}\right)^p \int_0^a f^p(x)\,\mathrm{d}x.$$

In particular, if f^2 is integrable, then

$$\int_0^a \left(\frac{1}{x}\int_0^x f(t)\,\mathrm{d}t\right)^2 \mathrm{d}x \le 4\int_0^a f^2(x)\,\mathrm{d}x.$$

Source: [2527, p. 166]. **Remark:** This is the *Hardy inequality.* **Remark:** 4 cannot be replaced by a smaller value. **Related:** Fact 2.11.132.

Fact 12.13.18. Let $a, b \in \mathbb{R}$, where $a < b$, let $f\colon\ [a,b] \to \mathbb{R}$, and assume that f is convex. Then,

$$f[\tfrac{1}{2}(a+b)] \le \frac{1}{b-a}\int_a^b f(x)\,\mathrm{d}x \le \tfrac{1}{2}[f(a) + f(b)].$$

Source: [943]. **Remark:** This is the *Hermite-Hadamard inequality.*

Fact 12.13.19. Let $\mathcal{I} \subseteq \mathbb{R}$ be either a finite or infinite interval, let $f\colon\ \mathcal{I} \to \mathbb{R}$, assume that f is continuous, and assume that, for all $x, y \in \mathcal{I}$, it follows that $f[\frac{1}{2}(x+y)] \le \frac{1}{2}f(x+y)$. Then, f is convex. **Source:** [2128, p. 10]. **Credit:** J. Jensen. **Related:** Fact 1.21.7.

Fact 12.13.20. Let $\mathcal{I} \subseteq \mathbb{R}$ be either a finite or infinite interval, let $f, g\colon\ \mathcal{I} \to \mathbb{R}$, assume that f and g are nonincreasing. Then, $f + g$ is nonincreasing. Now, assume that $f, g\colon\ \mathcal{I} \to [0, \infty)$. Then, fg is nonincreasing.

Fact 12.13.21. Let $A_0 \in \mathbb{F}^{n\times n}$, let $\|\cdot\|$ be a norm on $\mathbb{F}^{n\times n}$, and let $\varepsilon > 0$. Then, there exists $\delta > 0$ such that, if $A \in \mathbb{F}^{n\times n}$ and $\|A - A_0\| < \delta$, then

$$\operatorname{dist}[\operatorname{mspec}(A) - \operatorname{mspec}(A_0)] < \varepsilon,$$

where

$$\operatorname{dist}[\operatorname{mspec}(A) - \operatorname{mspec}(A_0)] \triangleq \min_{\sigma} \max_{i\in\{1,\ldots,n\}} |\lambda_{\sigma(i)}(A) - \lambda_i(A_0)|$$

and the minimum is taken over all permutations σ of $(1, \ldots, n)$. **Source:** [1399, p. 399].

Fact 12.13.22. Let $\mathcal{I} \subseteq \mathbb{R}$ be an interval, let $A\colon\ \mathcal{I} \mapsto \mathbb{F}^{n\times n}$, and assume that A is continuous. Then, for $i \in \{1, \ldots, n\}$, there exist continuous functions $\lambda_i\colon \mathcal{I} \mapsto \mathbb{C}$ such that, for all $t \in \mathcal{I}$, $\operatorname{mspec}(A(t)) = \{\lambda_1(t), \ldots, \lambda_n(t)\}_{\mathrm{ms}}$. **Source:** [1399, p. 399]. **Remark:** If A is continuously parameterized by either two or more variables, then the spectrum of A is not necessarily continuously parameterizable. See [1399, p. 399].

Fact 12.13.23. Let a be a positive number, let $f\colon\ [0, a) \mapsto \mathbb{R}$, assume that $f(0) \ge 0$, assume that, for all $x \in (0, a)$, f is differentiable at x, assume that f' is convex on $(0, a)$, and define $g\colon\ (0, a) \mapsto \mathbb{R}$ by $g(x) \triangleq f(x)/x$. Then, g is convex. **Source:** [1378].

Fact 12.13.24. Let $\mathcal{S} \subseteq \mathbb{F}^n$ be convex, let $f\colon\ \mathcal{S} \mapsto \mathbb{R}$, assume that f is convex, and let $\alpha \in \mathbb{R}$. Then, $f^{\mathsf{inv}}[(-\infty, \alpha)]$ and $f^{\mathsf{inv}}[(-\infty, \alpha]]$ are convex sets. **Remark:** $f^{\mathsf{inv}}[(-\infty, \alpha)]$ and $f^{\mathsf{inv}}[(-\infty, \alpha]]$ are *sublevel sets*. **Remark:** The converse is false. Let $f(x) = x^3$.

Fact 12.13.25. Let $\mathcal{S} \subseteq \mathbb{R}^n$, assume that $\mathcal{S}$ is convex, let $f\colon\ \mathcal{S} \mapsto \mathbb{R}$, and define

$$\operatorname{epi}(f) \triangleq \{\left[\begin{smallmatrix} x \\ y \end{smallmatrix}\right]\colon\ x \in \mathcal{S}, y \ge f(x)\}.$$

Then, f is convex if and only if $\operatorname{epi}(f)$ is convex. **Remark:** $\operatorname{epi}(f)$ is the *epigraph* of f.

Fact 12.13.26. Let $A \in \mathbb{F}^{n\times m}$, and let $\|\cdot\|$ be a norm on $\mathbb{F}^{n\times m}$. Then, the following statements are equivalent:

i) $\operatorname{rank} A = \min\{n, m\}$.

ii) There exists $\varepsilon > 0$ such that, for all $B \in \mathbb{F}^{n\times m}$ such that $\|B\| < \varepsilon$, $\operatorname{rank}(A + B) = \operatorname{rank} A$.

Furthermore, the following statements are equivalent:

iii) $\operatorname{rank} A < \min\{n, m\}$.

iv) For all $\varepsilon > 0$, there exists $B \in \mathbb{F}^{n\times m}$ such that $\|B\| < \varepsilon$ and $\operatorname{rank}(A + B) > \operatorname{rank} A$.

Now, let $(A_i)_{i=1}^{\infty} \subset \mathbb{F}^{n\times m}$, and assume that $A_\infty \triangleq \lim_{i\to\infty} A_i$ exists. Then,

$$\operatorname{rank} A_\infty \le \liminf_{i\to\infty} \operatorname{rank} A_i.$$

Source: To prove *iii)* $\Longrightarrow$*iv)*, let $r \triangleq \operatorname{rank} A$, and let $A_0 \in \mathbb{F}^{r\times r}$ be a nonsingular submatrix of A. By continuity of the determinant, there exists $\varepsilon > 0$ such that, for all $C \in \mathbb{F}^{r\times r}$ such that $\|C\|_{\mathrm{F}} < \varepsilon$, $\det(A_0 + C) \ne 0$. Now, let $B \in \mathbb{F}^{n\times m}$ satisfy $\|B\|_{\mathrm{F}} < \varepsilon$, and let B_0 be the submatrix of B whose rows and columns correspond to those of A_0 as a submatrix of A. Then, $\|B_0\|_{\mathrm{F}} < \varepsilon$, and thus $\det(A_0 + B_0) \ne 0$, and therefore $\operatorname{rank}(A + B) \ge r$. By norm equivalence, $\|\cdot\|_{\mathrm{F}}$ can be replaced by $\|\cdot\|$. **Remark:** This result shows that the rank function is lower semicontinuous. See [850, 857].

12.14 Facts on Functions of a Complex Variable

Fact 12.14.1. Define $f\colon\ \mathbb{C}\backslash\{0\} \mapsto \mathbb{C}$ by $f(z) \triangleq e^{-1/z^2}$, and define $g\colon\ \mathbb{C} \mapsto \mathbb{C}$ by $g(z) \triangleq f(z)$ for $z \ne 0$ and $g(0) \triangleq 0$. Then, the following statements hold:

i) f is analytic.

ii) 0 is an essential singularity of f.

iii) g is not continuous at 0.

iv) The restriction of g to $\mathbb{R}$ is C^∞.

Fact 12.14.2. Let $f \in \mathbb{F}(s)$, assume that f has no pole at zero, and let $\sum_{i=0}^\infty \beta_i z^i$ be the Taylor series of f at 0. Then, the following statements are equivalent:

i) f is analytic in the CIUD.

ii) CIUD is a subset of the domain of convergence of the Taylor series of f at 0.

iii) $(\beta_i)_{i=1}^\infty \in \ell_1$.

iv) For all $p \in [1, \infty)$, $(\beta_i)_{i=1}^\infty \in \ell_p$.

v) There exists $p \in [1, \infty)$ such that $(\beta_i)_{i=1}^\infty \in \ell_p$.

Fact 12.14.3. Let G be a proper rational function, let $z_0 \in \mathbb{C}$, consider the Laurent series of G given by (12.6.9), and define $R \triangleq \max\{|\lambda|: \lambda \text{ is a pole of } G\}$. Then, the following statements hold:

i) $R = \limsup_{i\to\infty} |\beta_i|^{1/i}$.

ii) $R \in [0, 1)$ if and only if $\lim_{i\to\infty} \beta_i = 0$.

iii) $R = 1$ and G has no repeated poles in $\mathbb{S}_1(z_0)$ if and only if $0 < \limsup_{i\to\infty} |\beta_i| < \infty$.

iv) Either $R \in (1, \infty)$ or both $R = 1$ and G has at least one repeated pole in $\mathbb{S}_1(z_0)$ if and only if $\limsup_{i\to\infty} |\beta_i| = \infty$.

Credit: S. Dai.

Fact 12.14.4. Let f: OIUD $\mapsto$ OIUD, and assume that f is analytic in OIUD, and $f(0) = 0$. Then, $|f'(0)| \le 1$, and, for all $z \in$ OIUD, $|f(z)| \le |z|$. Furthermore, if either $|f'(0)| = 1$ or there exists nonzero $z \in$ OIUD such that $|f(z)| = |z|$, then there exists $\theta \in \mathbb{R}$ such that, for all $z \in$ OIUD, $f(z) = e^{\theta J} z$. **Remark:** f is a rotation. **Remark:** This is the *Schwarz lemma*. See [1697, p. 78].

Fact 12.14.5. Let f: OIUD $\mapsto$ OIUD, assume that f is analytic in OIUD, and assume that there exist $a, b \in$ OIUD such that $f(a) = b$. Then, $|f'(a)| \le \frac{1-|b|^2}{1-|a|^2}$. Furthermore, let $a_1, a_2 \in$ OIUD. Then,

$$\left|\frac{f(a_2) - f(a_1)}{1 - \overline{f(a_1)} f(a_2)}\right| \le \left|\frac{a_2 - a_1}{1 - \overline{a_1} a_2}\right|.$$

Finally, if equality holds in either of these equalities with $a_1 \ne a_2$ in the second inequality, then f is one-to-one and $f(\text{OIUD}) = \text{OIUD}$. **Remark:** f is a conformal mapping. **Remark:** This is the *Schwarz-Pick lemma*. See [1697, p. 78].

Fact 12.14.6. Let f: OIUD $\mapsto$ OIUD and assume that f is analytic in OIUD. Then, f is a conformal mapping if and only if there exist $\theta \in \mathbb{R}$ and $a \in$ OIUD such that, for all $z \in$ OIUD,

$$f(z) = e^{\theta J} \frac{z - a}{1 - \overline{a} z}.$$

Remark: Every conformal mapping on the unit disk is the product of a rotation and a *Möbius transformation*

Fact 12.14.7. Let $\mathcal{D} \subseteq \mathbb{C}$, assume that $\mathcal{D}$ is open and connected, let $f: \mathcal{D} \mapsto \mathbb{C}$, and assume that f is analytic in $\mathcal{D}$. Then, $f(\mathcal{D})$ is either a single point or an open set. **Remark:** This is the *open mapping theorem*. See [1697, p. 73].

Fact 12.14.8. Let $\mathcal{D} \subseteq \mathbb{C}$, assume that $\mathcal{D}$ is open and connected, let $f: \mathcal{D} \mapsto \mathbb{C}$, and assume that f is analytic in $\mathcal{D}$. Then, the following statements hold:

i) If there exists $z_0 \in \mathcal{D}$ such that, for all $z \in \mathcal{D}$, $|f(z)| \le |f(z_0)|$, then f is a constant function.

ii) If $\mathcal{D} = \mathbb{C}$ and f is bounded, then f is a constant function.

iii) If there exists $\varepsilon > 0$ and $z_0 \in \mathcal{D}$ such that, for all $z \in \mathbb{B}_\varepsilon(z_0)$, $|f(z)| \le |f(z_0)|$, then f is a constant function.

iv) If there exists a connected compact set $\mathcal{D}_0 \subset \mathcal{D}$ and $z_0 \in \mathcal{D}_0$ such that, for all $z \in \mathcal{D}_0$, $|f(z)| \le |f(z_0)|$, then $z_0 \in \operatorname{bd} \mathcal{D}_0$.

v) If, for all $z \in \mathcal{D}$, $f(z) \neq 0$ and there exists $z_0 \in \mathcal{D}$ such that, for all $z \in \mathcal{D}$, $|f(z_0)| \le |f(z)|$, then f is a constant function.

vi) If there exists a connected compact set $\mathcal{D}_0 \subset \mathcal{D}$ and $z_0 \in \mathcal{D}_0$ such that, for all $z \in \mathcal{D}_0$, $|f(z_0)| \le |f(z)|$, then $z_0 \in \operatorname{bd} \mathcal{D}_0$.

vii) Assume that $\mathcal{D}$ is bounded, let $g\colon \operatorname{cl} \mathcal{D} \mapsto \mathbb{C}$, assume that g is analytic on $\mathcal{D}$ and continuous on $\operatorname{cl} \mathcal{D}$, and assume that, for all $z \in \operatorname{cl} \mathcal{D}$, $g(z) \neq 0$. Then, there exists $z_0 \in \operatorname{bd} \mathcal{D}$ such that, for all $z \in \operatorname{cl} \mathcal{D}$, $|g(z_0)| \le |g(z)|$.

Remark: See [1697, pp. 31, 32, 76, 77]. **Remark:** These are versions of the *maximum modulus principle*. *ii*) is *Liouville's theorem*. Note that *i*) does not imply *ii*). **Remark:** The absolute value of $f\colon$ OIUD $\mapsto \mathbb{C}$ defined by $f(z) = z^2$ is minimized by $z = 0$. Therefore, *v*) is false if the assumption that f is nonzero on $\mathcal{D}$ is removed.

Fact 12.14.9. Let $f\colon \mathcal{D} \mapsto \mathbb{C}$, and assume that f is analytic and not constant in $\mathbb{C}$. Then, $\operatorname{card}(\mathbb{C}\backslash f(\mathbb{C})) \le 1$. **Source:** [2964]. **Remark:** Let $f(z) = e^z$. Then, $f(\mathbb{C}) = \mathbb{C}\backslash\{0\}$. **Remark:** This is *Picard's theorem*.

Fact 12.14.10. Let $\mathcal{D} \subset \mathbb{C}$, assume that $\mathcal{D}$ is open and simply connected. Then, there exists a conformal mapping $f\colon \mathcal{D} \mapsto$ OIUD. **Remark:** This is the *Riemann mapping theorem*. See [1697, p. 86]. **Remark:** $\mathcal{D}$ and OIUD are *homeomorphic*, and ϕ is a *homeomorphism*. **Remark:** Note that $\mathcal{D}$ is assumed to be a proper subset of $\mathbb{C}$.

Fact 12.14.11. Let $\mathcal{D} \subseteq \mathbb{C}$, assume that $\mathcal{D}$ is open and connected, let $f\colon \mathcal{D} \mapsto \mathbb{C}$, let $g\colon \mathcal{D} \mapsto \mathbb{C}$, assume that f and g are analytic in $\mathcal{D}$, and assume that there exists a nonempty, open set $\mathcal{D}_0 \subseteq \mathcal{D}$ such that, for all $z \in \mathcal{D}_0$, $f(z) = g(z)$. Then, for all $z \in \mathcal{D}$, $f(z) = g(z)$. **Remark:** This result shows that every analytic function has at most one analytic extension. See [1697, p. 123]. The process of constructing the analytic extension is called *analytic continuation*. See [1697, pp. 123–135] and [2249, Chapter 13].

Fact 12.14.12. Let $f\colon$ OIUD $\mapsto \mathbb{C}$, and assume that f is one-to-one, $f(0) = 0$, f is analytic in OIUD, and $f'(0) = 1$. Then, the following statements hold:

i) There exist $\beta_2, \beta_3, \ldots \in \mathbb{C}$ such that, for all $z \in$ OIUD, $f(z) = z + \sum_{i=2}^{\infty} \beta_i z^i$.

ii) For all $i \ge 2$, $|\beta_i| \le i$.

iii) For all $i \ge 2$, $|\beta_i| = i$ if and only if there exists $\theta \in [0, 2\pi)$ such that, for all $z \in$ OIUD,

$$f(z) = \frac{z}{(1 + e^{\theta \jmath} z)^2}.$$

Remark: f is a *Schlicht function*. f given by *iii*) is a *Köbe function*. See [1697, pp. 149, 150]. **Remark:** *ii*) is *de Branges's theorem*. See [1671].

Fact 12.14.13. Let $\mathcal{D} \subseteq \mathbb{C}$, assume that $\mathcal{D}$ is open and contains OIUD, let $f\colon \mathcal{D} \mapsto \mathbb{C}$, assume that f is analytic in $\mathcal{D}$, for all $i \ge 0$, define $\beta_i \triangleq f^{(i)}(z_0)/i!$, and assume that the radius of convergence of $\sum_{i=0}^{\infty} \beta_i z^i$ is at least 1. Then, for all $z \in$ OIUD,

$$\frac{f(z)}{1 - z} = \sum_{i=0}^{\infty} \left(\sum_{j=0}^{i} \beta_j \right) z^i, \quad \frac{f(z^2)}{1 - z} = \sum_{i=0}^{\infty} \left(\sum_{j=0}^{i} \beta_j z^j \right) z^i.$$

Source: [2880, p. 39].

Fact 12.14.14. Let $(\beta_i)_{i=0}^{\infty} \subset \mathbb{C}$, let $z_0 \in \mathbb{C}$, define the power series f with domain of convergence $\mathcal{D}$ by (12.3.1), let ρ be the radius of convergence of (12.3.1), assume that $\rho \ge 1$, and assume that, for all $z \in$ OIUD, $|f(z)| < 1$. Then, for all $z \in \mathbb{C}$ such that $|z| < \frac{1}{3}$, $\sum_{i=0}^{\infty} |\beta_i z^i| < 1$. **Remark:** $1/3$ is the *Bohr radius*. See [22].

Fact 12.14.15. Let $f \in \mathbb{F}(s)$, assume that f is proper, let $\rho > 0$ be such that f has no poles in ρCOUD, and, for all $z \in \rho$COUD, let $g(z) = \sum_{i=0}^{\infty} \alpha_i z^{-i}$. Then, the following statements are equivalent:

i) f has no poles in COUD.

ii) For all $p \in [1, \infty)$, $(\sum_{i=0}^{\infty} |\alpha_i|^p)^{1/p}$ exists.

iii) There exists $p \in [1, \infty)$ such that $(\sum_{i=0}^{\infty} |\alpha_i|^p)^{1/p}$ exists.

iv) $\lim_{i\to\infty} \alpha_i = 0$.

Source: [430].

12.15 Facts on Functions of a Matrix

Fact 12.15.1. Let $A \in \mathbb{C}^{n\times n}$, and assume that A is group invertible and has no eigenvalues in $(-\infty, 0)$. Then,

$$A^{1/2} = \frac{2}{\pi} A \int_0^\infty (t^2 I + A)^{-1}\, \mathrm{d}t.$$

Source: [1391, p. 133].

Fact 12.15.2. Let $A \in \mathbb{C}^{n\times n}$, and assume that A has no eigenvalues on the imaginary axis. Then, the following statements hold:

i) $\mathrm{Sign}(A)$ is involutory.

ii) $A = \mathrm{Sign}(A)$ if and only if A is involutory.

iii) $[A, \mathrm{Sign}(A)] = 0$.

iv) $\mathrm{Sign}(A) = \mathrm{Sign}(A^{-1})$.

v) If A is real, then $\mathrm{Sign}(A)$ is real.

vi) $\mathrm{Sign}(A) = A(A^2)^{-1/2} = A^{-1}(A^2)^{1/2}$.

vii) $\mathrm{Sign}(A)$ is given by

$$\mathrm{Sign}(A) = \frac{2}{\pi} A \int_0^\infty (t^2 I + A^2)^{-1}\, \mathrm{d}t.$$

Source: [1391, pp. 39, 40 and Chapter 5] and [1602]. **Remark:** The square root in *vi)* is the principal square root.

Fact 12.15.3. Let $A, B \in \mathbb{C}^{n\times n}$, assume that AB has no eigenvalues in $(-\infty, 0]$, and define $C \triangleq A(BA)^{-1/2}$. Then,

$$\mathrm{Sign}\left(\begin{bmatrix} 0 & A \\ B & 0 \end{bmatrix}\right) = \begin{bmatrix} 0 & C \\ C^{-1} & 0 \end{bmatrix}.$$

If, in addition, A has no eigenvalues in $(-\infty, 0]$, then

$$\mathrm{Sign}\left(\begin{bmatrix} 0 & A \\ I & 0 \end{bmatrix}\right) = \begin{bmatrix} 0 & A^{1/2} \\ A^{-1/2} & 0 \end{bmatrix}.$$

Source: *vi)* of Fact 12.15.2 and [1391, p. 108]. **Remark:** The square root is the principal square root.

Fact 12.15.4. Let $A, B \in \mathbb{C}^{n\times n}$, and assume that A and B are positive definite. Then,

$$\mathrm{Sign}\left(\begin{bmatrix} 0 & B \\ A^{-1} & 0 \end{bmatrix}\right) = \begin{bmatrix} 0 & A\#B \\ (A\#B)^{-1} & 0 \end{bmatrix}.$$

Source: [1391, p. 131]. **Remark:** The geometric mean is defined in Fact 10.11.68.

Fact 12.15.5. Let $A \in \mathbb{F}^{n\times n}$, $B \in \mathbb{F}^{m\times m}$, and $C \in \mathbb{F}^{n\times m}$, and assume that $\rho_{\max}(B) < \rho_{\min}(A)$. Then, the unique solution $X \in \mathbb{F}^{n\times m}$ of $AX + XB + C = 0$ is given by

$$X = \sum_{i=0}^{\infty}(-1)^{i+1}A^{-i-1}CB^{i}.$$

Source: [2979, p. 42]. Fact 9.4.18 implies that $\rho_{\max}(B^{\mathsf{T}} \otimes A^{-1}) = \rho_{\max}(B)\rho_{\max}(A^{-1}) = \rho_{\max}(B)/\rho_{\min}(A) < 1$. Using Proposition 11.3.10, it follows that

$$X = -\operatorname{vec}^{-1}(I + B^{\mathsf{T}}\otimes A^{-1})^{-1}\operatorname{vec}(A^{-1}C) = -\operatorname{vec}^{-1}\sum_{i=0}^{\infty}(-B^{\mathsf{T}}\otimes A^{-1})^{i}\operatorname{vec}(A^{-1}C)$$

$$= -\operatorname{vec}^{-1}\sum_{i=0}^{\infty}(-1)^{i}(B^{i\mathsf{T}}\otimes A^{-i})\operatorname{vec}(A^{-1}C) = \sum_{i=0}^{\infty}(-1)^{i+1}A^{-i-1}CB^{i}.$$

12.16 Facts on Derivatives

Fact 12.16.1. Let $(\beta_i)_{i=0}^{\infty} \subset \mathbb{C}$, let $z_0 \in \mathbb{C}$, define the power series f with domain of convergence $\mathcal{D}$ by (12.3.1), let ρ be the radius of convergence of (12.3.1), and assume that $\rho > 0$. Then, f is analytic in $\mathcal{D}$, f' is given by

$$f'(z) \triangleq \sum_{i=1}^{\infty}\beta_i i(z - z_0)^{i-1},$$

and the radius of convergence of f' is ρ. **Source:** [152].

Fact 12.16.2. Let $p \in \mathbb{R}[s]$, assume that $\operatorname{roots}(p) \subset \mathbb{R}$, and define $n \triangleq \deg p$. Then, for all $x \in \mathbb{R}$,

$$p(x)p''(x) \le \frac{n-1}{n}[p'(x)]^2 \le [p'(x)]^2.$$

Source: [108, pp. 420, 421].

Fact 12.16.3. Let $p \in \mathbb{C}[s]$, define $n \triangleq \deg p$. Then,

$$\max_{z\in \mathrm{UC}}|p'(z)| \le n \max_{z\in \mathrm{UC}}|p(z)|.$$

If, in addition, $\rho_{\min}(p) \ge 1$, then

$$\max_{z\in \mathrm{UC}}|p'(z)| \le \left\{\begin{array}{c}\dfrac{n}{2}\left(\max\limits_{z\in \mathrm{UC}}|p(z)| - \min\limits_{z\in \mathrm{UC}}|p(z)|\right)\\[2ex] \dfrac{n}{1+\rho_{\min}(p)}\max\limits_{z\in \mathrm{UC}}|p(z)|\end{array}\right\} \le \frac{n}{2}\max_{z\in \mathrm{UC}}|p(z)|.$$

Source: [2295, pp. 508–510] and [3011]. **Remark:** The first inequality is *Bernstein's theorem.* **Related:** Fact 12.13.7.

Fact 12.16.4. Let $p \in \mathbb{C}[s]$, define $n \triangleq \deg p$, assume that $\operatorname{roots}(p) \subset \mathrm{UC}$, and define $q \in \mathbb{C}[s]$ by $q(z) \triangleq 2zp'(z) - np(z)$. Then, $\operatorname{roots}(q) \subset \mathrm{UC}$. **Source:** [108, p. 426].

Fact 12.16.5. Let $p \in \mathbb{C}[s]$. Then, $\operatorname{roots}(p') \subseteq \operatorname{conv}\operatorname{roots}(p)$. Furthermore, let $\operatorname{mroots}(p) = \{\lambda_1, \dots, \lambda_n\}_{\mathrm{ms}}$ and $\operatorname{mroots}(p') = \{\mu_1, \dots, \mu_{n-1}\}_{\mathrm{ms}}$. Then, $\frac{1}{n}\sum_{i=1}^{n}\lambda_i = \frac{1}{n-1}\sum_{i=1}^{n-1}\mu_i$. If, in addition, the roots of p are distinct and not contained in a single line, then $\operatorname{roots}(p') \cap \operatorname{bd}\operatorname{conv}\operatorname{roots}(p) = \varnothing$. **Source:** [970, p. 488], [1574], and [2295, pp. 71–74]. **Remark:** This is the *Gauss-Lucas theorem.* **Related:** Fact 5.5.3 and Fact 15.18.1.

Fact 12.16.6. Let $p, q \in \mathbb{C}[s]$, assume that $n \triangleq \deg q \ge 3$, assume that q has n distinct roots

$\lambda_1,\dots,\lambda_n \in \mathbb{C}$, and assume that $\deg p \le n-2$. Then,

$$\sum_{i=1}^{n} \frac{p(\lambda_i)}{q'(\lambda_i)} = 0.$$

Source: [2979, pp. 225, 226]. **Remark:** This is *Abel's theorem.* **Related:** Fact 2.11.4.

Fact 12.16.7. Let $f\colon\ \mathbb{R}^2 \mapsto \mathbb{R}$, $g\colon\ \mathbb{R} \mapsto \mathbb{R}$, and $h\colon\ \mathbb{R} \mapsto \mathbb{R}$, and assume that g and h are differentiable. Then, assuming each of the following integrals exists,

$$\frac{\mathrm{d}}{\mathrm{d}\alpha}\int_{g(\alpha)}^{h(\alpha)} f(t,\alpha)\,\mathrm{d}t = f(h(\alpha),\alpha)h'(\alpha) - f(g(\alpha),\alpha)g'(\alpha) + \int_{g(\alpha)}^{h(\alpha)} \frac{\partial}{\partial\alpha} f(t,\alpha)\,\mathrm{d}t.$$

Remark: This is *Leibniz's rule*.

Fact 12.16.8. Let $\mathcal{D} \subset \mathbb{R}$, let $f\colon\ \mathcal{D} \mapsto \mathbb{R}$ and $g\colon\ \mathcal{D} \mapsto \mathbb{R}$, let $x \in \mathcal{D}$, assume that $1/x \in \mathcal{D}$ and $x \ne 0$, let $n \ge 1$, and assume that f and g are n times differentiable at x. Then,

$$\frac{\mathrm{d}^n}{\mathrm{d}x^n}[g(x)f(1/x)] = \sum_{i=0}^{n}(-1)^i\binom{n}{i}\frac{1}{x^i}f^{(i)}(1/x)\frac{\mathrm{d}^{n-i}}{\mathrm{d}x^{n-i}}\frac{g(x)}{x^i}.$$

Therefore,

$$\frac{\mathrm{d}^n}{\mathrm{d}x^n}[x^{n-1}f(1/x)] = \frac{(-1)^n}{x^{n+1}}f^{(n)}(1/x).$$

In particular,

$$\frac{\mathrm{d}^n}{\mathrm{d}x^n}(x^{n-1}\log x) = \frac{(n-1)!}{x}, \quad \frac{\mathrm{d}^n}{\mathrm{d}x^n}(x^n\log x) = n!(H_n + \log x), \quad \frac{\mathrm{d}^n}{\mathrm{d}x^n}(x^{n-1}e^{1/x}) = \frac{(-1)^n}{x^{n+1}}e^{1/x}.$$

Source: [771, p. 161].

Fact 12.16.9. Let $n \ge 1$, let $0 \le m \le n$, and define $f\colon\ [0,\infty) \mapsto \mathbb{R}$ by $f(x) \triangleq \binom{x+n}{m}$. Then,

$$f'(0) = \binom{n}{m}(H_n - H_{n-m}), \quad f''(0) = \binom{n}{m}[(H_n - H_{n-m})^2 - H_{n,2} + H_{n-m,2}],$$

$$f'''(0) = \binom{n}{m}[(H_n - H_{n-m})^3 - 3(H_n - H_{n-m})(H_{n,2} - H_{n-m,2}) + 2(H_{n,3} - H_{n-m,3})].$$

Source: [2830].

Fact 12.16.10. Let $\mathcal{D} \subseteq \mathbb{R}^m$, assume that $\mathcal{D}$ is an open, convex set, let $f\colon\ \mathcal{D} \mapsto \mathbb{R}$, and assume that f is C^1 on $\mathcal{D}$. Then, the following statements hold:

i) f is convex if and only if, for all $x,y \in \mathcal{D}$, $f(x) + (y-x)^{\mathsf{T}}f'(x) \le f(y)$.

ii) f is strictly convex if and only if, for all distinct $x,y \in \mathcal{D}$, $f(x) + (y-x)^{\mathsf{T}}f'(x) < f(y)$.

Remark: If f is not differentiable, then these inequalities can be stated in terms of either directional differentials of f or the *subdifferential* of f. See [2128, pp. 29–31, 128–145].

Fact 12.16.11. Let $f\colon\ \mathcal{D} \subseteq \mathbb{F}^m \mapsto \mathbb{F}^n$, and assume that $\mathrm{D}_+f(0;\xi)$ exists. Then, for all $\beta > 0$,

$$\mathrm{D}_+f(0;\beta\xi) = \beta\mathrm{D}_+f(0;\xi).$$

Fact 12.16.12. Let $f\colon\ \mathbb{R} \mapsto \mathbb{R}$, assume that f is C^2, and assume that $a \triangleq \sup_{x\in\mathbb{R}}|f(x)|$ and $b \triangleq \sup_{x\in\mathbb{R}}|f''(x)|$ are finite. Then,

$$\sup_{x\in\mathbb{R}}|f'(x)| \le \sqrt{2ab}.$$

Source: [2294, p. 239]. **Remark:** This is *Landau's inequality*. Extensions to higher order derivatives are given in [2294, pp. 240–242].

Fact 12.16.13. Let $f\colon\ [0,\infty)\mapsto\mathbb{R}$, assume that f is C^2, and assume that $a\triangleq\int_0^\infty|f(x)|^2\,\mathrm{d}x$ and $b\triangleq\int_0^\infty|f''(x)|^2\,\mathrm{d}x$ are finite. Then,

$$\int_0^\infty|f'(x)|^2\,\mathrm{d}x\le 2\sqrt{ab}.$$

Source: [2294, p. 239]. **Remark:** This is *Hardy and Littlewood's inequality.*

Fact 12.16.14. Define $f\colon\ \mathbb{R}\mapsto\mathbb{R}$ by $f(x)\triangleq|x|$. Then, for all $\xi\in\mathbb{R}$,

$$\mathrm{D}_+f(0;\xi)=|\xi|.$$

Now, define $f\colon\ \mathbb{R}^n\mapsto\mathbb{R}^n$ by $f(x)\triangleq\sqrt{x^{\mathsf{T}}x}$. Then, for all $\xi\in\mathbb{R}^n$,

$$\mathrm{D}_+f(0;\xi)=\sqrt{\xi^{\mathsf{T}}\xi}.$$

Fact 12.16.15. For all $k\ge0$, let $f_k\colon\ [0,T]\mapsto\mathbb{F}^n$ be continuous, let $c_k\ge0$, and assume that $\max_{t\in[0,T]}\|f_k(t)\|_2\le c_k$ and $\sum_{k=0}^\infty c_k$ exists. Then, $\sum_{k=0}^\infty f_k$ converges absolutely and uniformly. **Source:** [2271, p. 217]. **Remark:** This is the *Weierstrass M-test.*

Fact 12.16.16. For all $k\ge0$, let $f_k\colon\ [0,T]\mapsto\mathbb{F}^n$ be continuous, and assume that $(f_k)_{k=1}^\infty$ converges uniformly. Then, $f\triangleq\lim_{k\to\infty}f_k$ is continuous. Now, assume, in addition, that, for all $k\ge0$, f_k is differentiable on $[0,T]$ and that $(\dot f_k)_{k=1}^\infty$ converges uniformly. Then, f is differentiable on $[0,T]$, and, for all $t\in[0,T]$, $\dot f(t)=\lim_{k\to\infty}\dot f_k(t)$. **Source:** [2271, pp. 213, 219, 220].

Fact 12.16.17. Let $A,B\in\mathbb{F}^{n\times n}$. Then, for all $s\in\mathbb{F}$,

$$\frac{\mathrm{d}}{\mathrm{d}s}(A+sB)^2=AB+BA+2sB^2.$$

Hence,

$$\frac{\mathrm{d}}{\mathrm{d}s}(A+sB)^2\bigg|_{s=0}=AB+BA.$$

Furthermore, for all $k\ge1$,

$$\frac{\mathrm{d}}{\mathrm{d}s}(A+sB)^k\bigg|_{s=0}=\sum_{i=0}^{k-1}A^iBA^{k-1-i}.$$

Fact 12.16.18. Let $A,B\in\mathbb{F}^{n\times n}$, and let $\mathcal{D}\triangleq\{s\in\mathbb{F}\colon\ \det(A+sB)\ne0\}$. Then, for all $s\in\mathcal{D}$,

$$\frac{\mathrm{d}}{\mathrm{d}s}(A+sB)^{-1}=-(A+sB)^{-1}B(A+sB)^{-1}.$$

Hence, if A is nonsingular, then

$$\frac{\mathrm{d}}{\mathrm{d}s}(A+sB)^{-1}\bigg|_{s=0}=-A^{-1}BA^{-1}.$$

Fact 12.16.19. Let $A,B\in\mathbb{F}^{n\times n}$. Then, for all $s\in\mathbb{F}$,

$$\frac{\mathrm{d}}{\mathrm{d}s}\det(A+sB)=\operatorname{tr}B(A+sB)^{\mathsf{A}}.$$

Hence,

$$\frac{\mathrm{d}}{\mathrm{d}s}\det(A+sB)\bigg|_{s=0}=\operatorname{tr}BA^{\mathsf{A}}=\sum_{i=1}^n\det[A\overset{i}{\leftarrow}\operatorname{col}_i(B)].$$

Source: Fact 3.19.8 and Fact 12.16.22. **Related:** This result generalizes Lemma 6.4.8.

Fact 12.16.20. Let $A,B\in\mathbb{F}^{n\times n}$, and let $\mathcal{D}\triangleq\{s\in\mathbb{F}\colon\ \det(A+sB)\ne0\}$. Then, for all $s\in\mathcal{D}$,

$$\frac{\mathrm{d}}{\mathrm{d}s}(A+sB)^{\mathsf{A}}=\frac{\mathrm{d}}{\mathrm{d}s}[\det(A+sB)](A+sB)^{-1}$$

$$= [\mathrm{tr}\; B(A+sB)^{\mathsf{A}}](A+sB)^{-1} - [\det(A+sB)](A+sB)^{-1}B(A+sB)^{-1}$$

Hence, if A is nonsingular, then

$$\frac{\mathrm{d}}{\mathrm{d}s}(A+sB)^{\mathsf{A}}\bigg|_{s=0} = (\mathrm{tr}\; BA^{\mathsf{A}})A^{-1} - (\det A)A^{-1}BA^{-1}.$$

Fact 12.16.21. Let $A, B \in \mathbb{F}^{n\times n}$, and let $\mathcal{D} \triangleq \{s \in \mathbb{F}\colon\ \det(A+sB) \neq 0\}$. Then, for all $s \in \mathcal{D}$,

$$\begin{aligned}\frac{\mathrm{d}^2}{\mathrm{d}s^2}\det(A+sB) &= \frac{\mathrm{d}}{\mathrm{d}s}\,\mathrm{tr}\; B(A+sB)^{\mathsf{A}}\\ &= \frac{\mathrm{d}}{\mathrm{d}s}([\det(A+sB)]\;\mathrm{tr}\; B(A+sB)^{-1})\\ &= [\mathrm{tr}\; B(A+sB)^{\mathsf{A}}]\;\mathrm{tr}\; B(A+sB)^{-1} - [\det(A+sB)]\;\mathrm{tr}\,[B(A+sB)^{-1}]^2.\end{aligned}$$

Hence, if A is nonsingular, then

$$\frac{\mathrm{d}^2}{\mathrm{d}s^2}\det(A+sB)\bigg|_{s=0} = (\mathrm{tr}\; BA^{\mathsf{A}})\;\mathrm{tr}\; BA^{-1} - (\det A)\;\mathrm{tr}\,(BA^{-1})^2.$$

Fact 12.16.22. Let $\mathcal{D} \subseteq \mathbb{F}$, let $A\colon\ \mathcal{D} \mapsto \mathbb{F}^{n\times n}$, and assume that A is differentiable. Then,

$$\frac{\mathrm{d}}{\mathrm{d}s}\det A(s) = \mathrm{tr}\, A^{\mathsf{A}}(s)A'(s) = \frac{1}{n-1}\,\mathrm{tr}\left(A(s)\frac{\mathrm{d}}{\mathrm{d}s}A^{\mathsf{A}}(s)\right) = \sum_{i=1}^{n}\det A_i(s),$$

where $A_i(s)$ is obtained by differentiating the entries of the ith row of $A(s)$. If $A(s)$ is nonsingular for all $s \in \mathcal{D}$, then

$$\frac{\mathrm{d}}{\mathrm{d}s}\log\det A(s) = \mathrm{tr}\, A^{-1}(s)A'(s).$$

If $A(s)$ is positive definite for all $s \in \mathcal{D}$, then

$$\frac{\mathrm{d}}{\mathrm{d}s}\det A^{1/n}(s) = \frac{1}{n}[\det A^{1/n}(s)]\;\mathrm{tr}\, A^{-1}(s)A'(s).$$

Finally, $A(s)$ is nonsingular and has no negative eigenvalues for all $s \in \mathcal{D}$, then

$$\frac{\mathrm{d}}{\mathrm{d}s}\log A(s) = \int_0^1 [(A(s)-I)t+I]^{-1}A'(s)[(A(s)-I)t+I]^{-1}\,\mathrm{d}t,$$

$$\frac{\mathrm{d}}{\mathrm{d}s}[\log A(s)]^2 = 2[\mathrm{tr}\log A(s)]A^{-1}(s)A'(s).$$

Source: [799, p. 267], [1190, 2065], [2263, pp. 199, 212], [2314, p. 430], and [2417]. **Related:** Fact 16.20.6.

Fact 12.16.23. Let $\mathcal{D} \subseteq \mathbb{F}$, let $A\colon\ \mathcal{D} \mapsto \mathbb{F}^{n\times n}$, let $k \geq 1$, and assume that A is k-times differentiable. Then,

$$\frac{\mathrm{d}^k}{\mathrm{d}s^k}\det A(s) = \sum \binom{k}{i_1,\ldots,i_n}\det\begin{bmatrix}\frac{\mathrm{d}^{i_1}}{\mathrm{d}s^{i_1}}\mathrm{row}_1[A(s)]\\ \vdots\\ \frac{\mathrm{d}^{i_n}}{\mathrm{d}s^{i_n}}\mathrm{row}_n[A(s)]\end{bmatrix},$$

where the sum is taken over all n-tuples $(i_1,\ldots,i_n)$ of nonnegative integers whose components sum to k. **Source:** [578]. **Remark:** An alternative expression is given in [465].

Fact 12.16.24. Let $\mathcal{D} \subseteq \mathbb{F}$, let $A\colon\ \mathcal{D} \mapsto \mathbb{F}^{n\times n}$, assume that A is differentiable, and assume that $A(s)$ is nonsingular for all $s \in \mathcal{D}$. Then,

$$\frac{\mathrm{d}}{\mathrm{d}s}A^{-1}(s) = -A^{-1}(s)A'(s)A^{-1}(s), \quad \mathrm{tr}\, A^{-1}(s)A'(s) = -\,\mathrm{tr}\left(A(s)\frac{\mathrm{d}}{\mathrm{d}s}A^{-1}(s)\right).$$

Source: [1450, p. 491] and [2263, pp. 198, 212].

Fact 12.16.25. Let $A \in \mathbb{R}^{n\times n}$, assume that A is symmetric, let $X \in \mathbb{R}^{m\times n}$, and assume that XAX^T is nonsingular. Then,

$$\frac{\mathrm{d}}{\mathrm{d}X}\det XAX^\mathsf{T} = 2(\det XAX^\mathsf{T})A^\mathsf{T}X^\mathsf{T}(XAX^\mathsf{T})^{-1}.$$

Source: [786].

Fact 12.16.26. Let $\mathcal{D} \subseteq \mathbb{R}$ be an open interval, and, for all $i \in \{1,\dots,n\}$, let $f_i\colon \mathcal{D} \mapsto \mathbb{R}$ and assume that f_i is $n-1$ times differentiable. Furthermore, for all $x \in \mathcal{D}$, define

$$A(x) \triangleq \begin{bmatrix} f_1(x) & \cdots & f_n(x) \\ f_1'(x) & \cdots & f_n'(x) \\ \vdots & \therefore & \vdots \\ f_1^{(n-1)}(x) & \cdots & f_n^{(n-1)}(x) \end{bmatrix},$$

and assume that, for all $x \in \mathcal{D}$, $\det A(x) = 0$. Then, there exist an open interval $\mathcal{D}' \subseteq \mathcal{D}$ and $\alpha_1,\dots,\alpha_n \in \mathbb{R}$ such that $\alpha_1,\dots,\alpha_n$ are not all zero and such that, for all $x \in \mathcal{D}'$, $\sum_{i=1}^n \alpha_i f_i(x) = 0$. **Source:** [1706]. **Remark:** $\det A(x)$ is the *Wronskian of* $f_1,\dots,f_n$. **Example:** The open interval $\mathcal{D}'$ may need to be a proper subset of $\mathcal{D}$. For example, let $\mathcal{D} = \mathbb{R}$, $f_1(x) = x^2$, and $f_2(x) = x|x|$, so that $\det A(x) = 0$ for all $x \in \mathbb{R}$. Then, f_1 and f_2 are linearly dependent on every open interval contained in $(-\infty, 0]$ and every open interval contained in $[0,\infty)$. However, on every open interval that includes zero, f_1 and f_2 are linearly independent. See [1706]. **Remark:** Extensions to analytic functions are considered in [530].

Fact 12.16.27. Let $\phi\colon \mathbb{R}^3 \mapsto \mathbb{R}$, let $\psi\colon \mathbb{R}^3 \mapsto \mathbb{R}$, and define

$$\nabla \triangleq \begin{bmatrix} \dfrac{\partial}{\partial x} \\ \dfrac{\partial}{\partial y} \\ \dfrac{\partial}{\partial z} \end{bmatrix}, \quad \nabla^2 \triangleq \nabla^\mathsf{T}\nabla = \frac{\partial^2}{\partial x^2} + \frac{\partial^2}{\partial y^2} + \frac{\partial^2}{\partial z^2}.$$

Then,

$$\nabla\phi \triangleq \begin{bmatrix} \dfrac{\partial\phi}{\partial x} \\ \dfrac{\partial\phi}{\partial y} \\ \dfrac{\partial\phi}{\partial z} \end{bmatrix}, \quad \nabla^\mathsf{T}\phi = \frac{\partial\phi}{\partial x} + \frac{\partial\phi}{\partial y} + \frac{\partial\phi}{\partial z}, \quad \nabla^2\phi = \frac{\partial^2\phi}{\partial x^2} + \frac{\partial^2\phi}{\partial y^2} + \frac{\partial^2\phi}{\partial z^2}, \quad \nabla\times\nabla\phi = 0,$$

$$\nabla(\phi+\psi) = \nabla\phi + \nabla\psi, \quad \nabla(\phi\psi) = \psi\nabla\phi + \phi\nabla\psi, \quad \nabla^2(\phi\psi) = \phi\nabla^2\psi + \psi\nabla^2\phi + 2(\nabla\phi)^\mathsf{T}\nabla\psi.$$

Now, let $f = [f_1\ \ f_2\ \ f_3]^\mathsf{T}\colon \mathbb{R}^3 \mapsto \mathbb{R}^3$. Then,

$$\nabla\times f = K(\nabla)f = \begin{bmatrix} 0 & -\dfrac{\partial}{\partial z} & \dfrac{\partial}{\partial y} \\ \dfrac{\partial}{\partial z} & 0 & -\dfrac{\partial}{\partial x} \\ -\dfrac{\partial}{\partial y} & \dfrac{\partial}{\partial x} & 0 \end{bmatrix} f = \begin{bmatrix} \dfrac{\partial f_3}{\partial y} - \dfrac{\partial f_2}{\partial z} \\ \dfrac{\partial f_1}{\partial z} - \dfrac{\partial f_3}{\partial x} \\ \dfrac{\partial f_2}{\partial x} - \dfrac{\partial f_1}{\partial y} \end{bmatrix},$$

$$\nabla^{\mathsf{T}}(\nabla\times f)=0,\quad K(\nabla\times f)=\tfrac{1}{2}[(\nabla f^{\mathsf{T}})^{\mathsf{T}}-\nabla f^{\mathsf{T}}],\quad \nabla\times(\nabla\times f)=\nabla(\nabla^{\mathsf{T}}f)-\nabla^2 f,$$
$$\nabla^{\mathsf{T}}(\phi f)=\phi\nabla^{\mathsf{T}}f+f^{\mathsf{T}}\nabla\phi,\quad \nabla\times(\phi f)=\phi\nabla\times f+(\nabla\phi)\times f,$$
$$\nabla(f^{\mathsf{T}}f)=2(f^{\mathsf{T}}\nabla)f-2(\nabla\times f)\times f,\quad \nabla^2 f=\nabla(\nabla^{\mathsf{T}}f)-\nabla\times(\nabla\times f),$$
$$\nabla^2(\nabla^{\mathsf{T}}f)=\nabla^{\mathsf{T}}(\nabla^2 f),\quad \nabla^2(f^{\mathsf{T}}f)=2\nabla^{\mathsf{T}}(f^{\mathsf{T}}\nabla)f-\nabla^{\mathsf{T}}[(\nabla\times f)\times f],$$
$$\operatorname{tr}\nabla f^{\mathsf{T}}=\nabla^{\mathsf{T}}f,\quad \operatorname{tr}(\nabla f^{\mathsf{T}})^2=\nabla^{\mathsf{T}}(f^{\mathsf{T}}\nabla)f-(f^{\mathsf{T}}\nabla)(\nabla^{\mathsf{T}}f).$$

Now, let $g=[g_1\ \ g_2\ \ g_3]^{\mathsf{T}}\colon\mathbb{R}^3\mapsto\mathbb{R}^3$. Then,

$$\nabla(f^{\mathsf{T}}g)=f\times(\nabla\times g)+g\times(\nabla\times f)+(f^{\mathsf{T}}\nabla)g+(g^{\mathsf{T}}\nabla)f,$$
$$\nabla^{\mathsf{T}}(f\times g)=g^{\mathsf{T}}(\nabla\times f)-f^{\mathsf{T}}(\nabla\times g),$$
$$\nabla\times(f\times g)=(g^{\mathsf{T}}\nabla)f-(f^{\mathsf{T}}\nabla)g+f(\nabla^{\mathsf{T}}g)-g(\nabla^{\mathsf{T}}f).$$

Remark: ∇, ∇^{T}, and $K(\nabla)$ denote *gradient*, *divergence*, and *curl*, respectively. **Remark:** For $x\in\mathbb{R}^3$, $K(x)$ is defined in Fact 4.12.1. **Remark:** $\phi'(x)=[\nabla\phi(x)]^{\mathsf{T}}$, and $f'(x)=\begin{bmatrix}[\nabla f_1(x)]^{\mathsf{T}}\\ [\nabla f_2(x)]^{\mathsf{T}}\\ [\nabla f_3(x)]^{\mathsf{T}}\end{bmatrix}$.

12.17 Facts on Limits of Functions

Fact 12.17.1. Let α be a positive number. Then,

$$\lim_{x\downarrow 0}\alpha^x=\lim_{x\uparrow 0}\alpha^x=\lim_{x\to 0}\alpha^x=\lim_{x\to\infty}\alpha^{1/x}=\lim_{x\to-\infty}\alpha^{1/x}=1.$$

Furthermore,

$$\lim_{x\downarrow 0}x^x=\lim_{x\to\infty}x^{1/x}=1.$$

Remark: $\lim_{x\downarrow 0}x^x=1$ motivates the convention $0^0=1$.

Fact 12.17.2. Let α be a real number. Then,

$$\lim_{x\to-\infty}\alpha^x=\begin{cases}\infty, & 0<\alpha<1,\\ 0, & \alpha>1,\end{cases}\qquad \lim_{x\to\infty}\alpha^x=\begin{cases}0, & 0<\alpha<1,\\ \infty, & \alpha>1,\end{cases}$$

$$\lim_{x\downarrow 0}x^\alpha=\begin{cases}\infty, & \alpha<0,\\ 0, & \alpha>0,\end{cases}\qquad \lim_{x\to\infty}x^\alpha=\begin{cases}0, & \alpha<0,\\ \infty, & \alpha>0.\end{cases}$$

If $\alpha<0$, then

$$\lim_{x\to-\infty}|x^\alpha|=0,\quad \lim_{x\uparrow 0}|x^\alpha|=\infty.$$

If $\alpha>0$, then

$$\lim_{x\uparrow 0}|x^\alpha|=0.$$

Source: [2294, p. 116]. **Remark:** In the last three limits, x^α may be complex.

Fact 12.17.3. Let $\alpha>0$, and assume that $\alpha\neq 1$. Then,

$$\lim_{x\to\infty}\left[\frac{\alpha^x-1}{(\alpha-1)x}\right]^{1/x}=\begin{cases}1, & 0<\alpha<1,\\ \alpha, & \alpha>1.\end{cases}$$

Source: [2294, p. 134].

Fact 12.17.4. Let a be a real number. Then,

$$\lim_{x\to\infty}x\tan\frac{a}{x}=\lim_{x\to-\infty}x\tan\frac{a}{x}=a.$$

Fact 12.17.5. Define $S\colon (0,\infty) \mapsto \mathbb{R}$ by

$$S(x) \triangleq \begin{cases} \dfrac{x^{1/(x-1)}}{e\log x^{1/(x-1)}}, & x \neq 1,\\ 1, & x = 1. \end{cases}$$

Then, the following statements hold:

i) S is continuous and strictly convex.

ii) For all $x \in (0,\infty)$, $S(x) \geq 1$.

iii) $S(x) = 1$ if and only if $x = 1$.

iv) If $x > 0$, then $S(x) = S(1/x)$.

v) $\lim_{x\downarrow 0} S(x) = \lim_{x\to\infty} S(x) = \infty$.

Remark: S is *Specht's ratio*. **Related:** Fact 2.2.54, Fact 2.2.66, Fact 2.11.95, Fact 11.10.7, and Fact 15.15.23.

Fact 12.17.6.

$$\lim_{x\to\infty}\left(1+\frac{1}{x}\right)^x = \lim_{x\to-\infty}\left(1+\frac{1}{x}\right)^x = \lim_{x\to\infty}\left(\frac{x+1}{x-1}\right)^{x/2} = e, \quad \lim_{x\to\infty}\left(\frac{x-1}{x}\right)^{x-1} = \frac{1}{e},$$

$$\lim_{x\to\infty}\left(\frac{x-2}{x}\right)^{x-1} = \frac{1}{e^2}, \quad \lim_{x\to\infty}\left[\frac{(x+1)^{x+1}}{x^x} - \frac{x^x}{(x-1)^{x-1}}\right] = e.$$

Now, let $z \in \mathbb{C}$. Then,

$$\lim_{x\to\infty}\left(1+\frac{z}{x}\right)^x = e^z.$$

Source: The first two limits are given in [2294, p. 123], the third limit is given in [2363], and the sixth limit is given in [1648].

Fact 12.17.7. Let $a > 0$. As $x \to 0$,

$$a^x = 1 + (\log a)x + \tfrac{1}{2}(\log^2 a)x^2 + \tfrac{1}{6}(\log^3 a)x^3 + O(x^4).$$

Therefore,

$$\lim_{x\to 0}\frac{a^x - 1}{x} = \log a, \quad \lim_{x\to 0}\frac{a^x - 1 - (\log a)x}{x^2} = \tfrac{1}{2}\log^2 a.$$

Fact 12.17.8. Let $\alpha > 0$ and $p > 1$. Then,

$$\lim_{x\to\infty}\frac{x^\alpha}{p^x} = 0.$$

Fact 12.17.9. Let $\alpha > 0$. Then,

$$\lim_{x\to\infty}\frac{\log x}{x^\alpha} = \lim_{x\downarrow 0} x^\alpha \log x = 0.$$

Fact 12.17.10. Let x, y, p, q be positive numbers, and assume that $1 \leq p \leq q$. Then,

$$\lim_{r\to 0}\left[\tfrac{1}{2}(x^r + y^r)\right]^{1/r} = \sqrt{xy} \leq \tfrac{1}{2}(x+y) \leq \max\{x,y\} = \lim_{r\to\infty}\left\|\left[\begin{smallmatrix} x\\ y\end{smallmatrix}\right]\right\|_r = \left\|\left[\begin{smallmatrix} x\\ y\end{smallmatrix}\right]\right\|_\infty$$

$$\leq \left\|\left[\begin{smallmatrix} x\\ y\end{smallmatrix}\right]\right\|_q \leq \left\|\left[\begin{smallmatrix} x\\ y\end{smallmatrix}\right]\right\|_p \leq \left\|\left[\begin{smallmatrix} x\\ y\end{smallmatrix}\right]\right\|_1 = \lim_{r\to 1}\left\|\left[\begin{smallmatrix} x\\ y\end{smallmatrix}\right]\right\|_r = x + y.$$

Source: [1566, p. 49]. **Related:** Proposition 11.1.5 and Fact 12.17.11.

Fact 12.17.11. Let $x \in \mathbb{R}^n$, assume that $x \geq\geq 0$, let p, q be positive numbers, and assume that $1 \le p \le q$. Then,

$$\lim_{r\to 0}\left(\frac{1}{n}\sum_{i=1}^n x_{(i)}^r\right)^{1/r} = \sqrt[n]{\prod_{i=1}^n x_{(i)}} \le \frac{1}{n}\sum_{i=1}^n x_{(i)} \le \max\{x_{(1)},\ldots x_{(n)}\} = \lim_{r\to\infty}\|x\|_r$$

$$= \|x\|_\infty \le \|x\|_q \le \|x\|_p \le \|x\|_1 = \lim_{r\to\infty}\|x\|_1 = \sum_{i=1}^n x_{(i)}.$$

Related: Proposition 11.1.5 and Fact 12.17.10.

Fact 12.17.12. Let $f\colon [0,\infty) \to \mathbb{R}$. Then, the following statements hold:

i) Let $f(x) = \sqrt{x}$. Then, $\lim_{x\to\infty} f(x) = \infty$ and $\lim_{x\to\infty} f'(x) = 0$.

ii) Let $f(x) = e^{-x}\sin e^{x^2}$. Then, $\lim_{x\to\infty} f(x) = 0$, $\lim_{x\to\infty} f'(x)$ does not exist, $\liminf_{x\to\infty} f'(x) = -\infty$, and $\limsup_{x\to\infty} f'(x) = \infty$.

iii) Let $f(x) = \sin\sqrt{x}$. Then, $\lim_{x\to\infty} f(x)$ does not exist, $\liminf_{x\to\infty} f(x) = -1$, $\limsup_{x\to\infty} f(x) = 1$, and $\lim_{x\to\infty} f'(x) = 0$.

Fact 12.17.13. Let $\mathcal{S} \subset \mathbb{R}$, let $f\colon \mathcal{S} \mapsto \mathbb{R}$ and $g\colon \mathcal{S} \mapsto \mathbb{R}$, assume that f and g are differentiable, let $a \in \mathbb{R}$, and let $\varepsilon > 0$. Then, the following statements hold:

i) Assume that $\mathcal{I} \triangleq (a-\varepsilon, a) \cup (a, a+\varepsilon) \subseteq \mathcal{S}$, $0 \notin g(\mathcal{I}) \cup g'(\mathcal{I})$, either $\lim_{x\to a} f(x) = \lim_{x\to a} g(x) = 0$ or both $\lim_{x\to a} f(x) = \pm\infty$ and $\lim_{x\to a} g(x) = \pm\infty$, and either $\lim_{x\to a}\frac{f'(x)}{g'(x)}$ exists or $\lim_{x\to a}\frac{f'(x)}{g'(x)} = \pm\infty$. Then,

$$\lim_{x\to a}\frac{f(x)}{g(x)} = \lim_{x\to a}\frac{f'(x)}{g'(x)}.$$

ii) Assume that $\mathcal{I} \triangleq (a, a+\varepsilon) \subseteq \mathcal{S}$, $0 \notin g(\mathcal{I}) \cup g'(\mathcal{I})$, either $\lim_{x\downarrow a} f(x) = \lim_{x\downarrow a} g(x) = 0$ or both $\lim_{x\downarrow a} f(x) = \pm\infty$ and $\lim_{x\downarrow a} g(x) = \pm\infty$, and either $\lim_{x\downarrow a}\frac{f'(x)}{g'(x)}$ exists or $\lim_{x\downarrow a}\frac{f'(x)}{g'(x)} = \pm\infty$. Then,

$$\lim_{x\downarrow a}\frac{f(x)}{g(x)} = \lim_{x\downarrow a}\frac{f'(x)}{g'(x)}.$$

iii) Assume that $\mathcal{I} \triangleq (a-\varepsilon, a) \subseteq \mathcal{S}$, $0 \notin g(\mathcal{I}) \cup g'(\mathcal{I})$, either $\lim_{x\uparrow a} f(x) = \lim_{x\uparrow a} g(x) = 0$ or both $\lim_{x\uparrow a} f(x) = \pm\infty$ and $\lim_{x\uparrow a} g(x) = \pm\infty$, and either $\lim_{x\uparrow a}\frac{f'(x)}{g'(x)}$ exists or $\lim_{x\uparrow a}\frac{f'(x)}{g'(x)} = \pm\infty$. Then,

$$\lim_{x\uparrow a}\frac{f(x)}{g(x)} = \lim_{x\uparrow a}\frac{f'(x)}{g'(x)}.$$

iv) Assume that $\mathcal{I} \triangleq (-\infty, a) \subseteq \mathcal{S}$, $0 \notin g(\mathcal{I}) \cup g'(\mathcal{I})$, either $\lim_{x\to-\infty} f(x) = \lim_{x\to-\infty} g(x) = 0$ or both $\lim_{x\to-\infty} f(x) = \pm\infty$ and $\lim_{x\to-\infty} g(x) = \pm\infty$, and either $\lim_{x\to-\infty}\frac{f'(x)}{g'(x)}$ exists or $\lim_{x\to-\infty}\frac{f'(x)}{g'(x)} = \pm\infty$. Then,

$$\lim_{x\to-\infty}\frac{f(x)}{g(x)} = \lim_{x\to-\infty}\frac{f'(x)}{g'(x)}.$$

v) Assume that $\mathcal{I} \triangleq (a, \infty) \subseteq \mathcal{S}$, $0 \notin g(\mathcal{I}) \cup g'(\mathcal{I})$, either $\lim_{x\to\infty} f(x) = \lim_{x\to\infty} g(x) = 0$ or both $\lim_{x\to\infty} f(x) = \pm\infty$ and $\lim_{x\to\infty} g(x) = \pm\infty$, and either $\lim_{x\to\infty}\frac{f'(x)}{g'(x)}$ exists or $\lim_{x\to\infty}\frac{f'(x)}{g'(x)} = \pm\infty$. Then,

$$\lim_{x\to\infty}\frac{f(x)}{g(x)} = \lim_{x\to\infty}\frac{f'(x)}{g'(x)}.$$

Source: [2406, p. 200]. **Remark:** This is *l'Hôpital's rule*. **Example:** $\lim_{x\to 0}\frac{\sin x}{x} = \lim_{x\to 0}\cos x = 1$ and $\lim_{x\to\infty}\frac{e^x}{\sqrt{x}} = \lim_{x\to\infty} 2\sqrt{x}e^x = \infty$.

Fact 12.17.14. Let $f\colon [0,\infty) \mapsto \mathbb{R}$, assume that f is uniformly continuous, and assume that $\lim_{t\to\infty} \int_0^t f(x)\,\mathrm{d}x$ exists. Then, $\lim_{t\to\infty} f(t) = 0$. **Source:** [781, 1021]. **Remark:** This is *Barbalat's lemma*. An equivalent form is given by Fact 12.17.17.

Fact 12.17.15. Let $f\colon [a,b] \mapsto \mathbb{R}$, and assume that f is integrable. Then,

$$\lim_{n\to\infty} \int_a^b f(x)\sin nx\,\mathrm{d}x = \lim_{n\to\infty} \int_a^b f(x)\cos nx\,\mathrm{d}x = 0.$$

Source: [1103, p. 261].

Fact 12.17.16. Let $f\colon [0,\infty) \mapsto \mathbb{R}$, and assume that f is C^1, f' is uniformly continuous, and $\lim_{x\to\infty} f(x)$ exists. Then, $\lim_{x\to\infty} f'(x) = 0$. **Remark:** This result follows from Fact 12.17.14.

Fact 12.17.17. Let $f\colon [0,\infty) \mapsto \mathbb{R}$, and assume that f is C^2, f'' is bounded, and $\lim_{x\to\infty} f(x)$ exists. Then, $\lim_{x\to\infty} f'(x) = 0$. **Source:** This result follows from Fact 12.17.14. See [2527, p. 118].

Fact 12.17.18. Let $\mathcal{D} \subseteq \mathbb{R}^2$, assume that $\mathcal{D}$ is open and convex, let $z \in \mathcal{D}$, let $\mathcal{I}_1$ and $\mathcal{I}_2$ be open intervals in $\mathbb{R}$ such that $z \in \mathcal{I}_1 \times \mathcal{I}_2 \subseteq \mathcal{D}$, let $f\colon \mathcal{D} \mapsto \mathbb{R}$, and assume that, for all convex, C^1 functions $g_1\colon \mathcal{I}_1 \mapsto \mathbb{R}$ and $g_2\colon \mathcal{I}_2 \mapsto \mathbb{R}$ such that $g_1(z_{(2)}) = z_{(1)}$ and $g_2(z_{(1)}) = z_{(2)}$, it follows that $\lim_{x\to z_{(2)}} f(g_1(x), x) = \lim_{x\to z_{(1)}} f(x, g_2(x)) = f(z)$. Then, $\lim_{\xi\to z} f(\xi) = f(z)$. **Source:** [2111]. **Credit:** A. Rosenthal.

12.18 Facts on Limits of Sequences and Series

Fact 12.18.1. The following statements hold:

i) Let $X = (x_i)_{i=1}^\infty \subset \mathbb{F}$, assume that $\sum_{i=1}^\infty x_i$ converges absolutely, and let $\hat{X} = (\hat{x}_i)_{i=1}^\infty$ be a rearrangement of X. Then, $\sum_{i=1}^\infty \hat{x}_i$ converges absolutely, and $\sum_{i=1}^\infty \hat{x}_i = \sum_{i=1}^\infty x_i$.

ii) $\sum_{i=1}^\infty (-1)^{i+1} H_i = \log 2$.

iii) Let $(x_i)_{i=1}^\infty = (1, -H_2, -H_4, H_3, -H_6, -H_8, H_5, -H_{10}, -H_{12}, \ldots)$. Then, $\sum_{i=1}^\infty x_i = \frac{1}{2}\log 2$.

iv) Let $(x_i)_{i=1}^\infty = (1, H_3, -H_2, H_5, H_7, -H_4, H_9, H_{11}, -H_6, \ldots)$. Then, $\sum_{i=1}^\infty x_i = \frac{3}{2}\log 2$.

Source: [1350, p. 102]. **Remark:** If $X = (x_i)_{i=1}^\infty \subset \mathbb{F}$ converges, then Proposition 12.2.8 implies that every rearrangement of X converges to the same limit as X. Assuming that $\sum_{i=1}^\infty x_i$ converges but not absolutely, this example shows that a rearrangement of X may yield a series $(\sum_{i=1}^k \hat{x}_i)_{k=1}^\infty$ with a limit that is different from the limit of the series based on X; the series is *conditionally convergent*. Let $\mathbb{F} = \mathbb{R}$. If the series $(\sum_{i=1}^k x_i)_{k=1}^\infty$ is conditionally convergent, then, for all $\alpha \in \mathbb{R}$, the sequence X can be rearranged so that $\sum_{i=1}^\infty \hat{x}_i = \alpha$. For a power series with a finite radius of convergence, conditional convergence can occur only on the boundary of the domain of convergence. **Credit:** B. Riemann.

Fact 12.18.2. The following statements hold:

i) If $X = (-1,-2,-3,-4,\ldots)$, then $-\infty = \inf X = \liminf X = \limsup X < \sup X = -1$.

ii) If $X = (-1,0,-2,0,-3,0,\ldots)$, then $-\infty = \inf X = \liminf X < \limsup X = \sup X = 0$.

iii) If $X = (-1,1,-2,0,-3,0,\ldots)$, then $-\infty = \inf X = \liminf X < \limsup X = 0 < \sup X = 1$.

iv) If $X = (-1,1,-2,2,-3,3,\ldots)$, then $-\infty = \inf X = \liminf X < \limsup X = \sup X = \infty$.

v) If $X = (1,1,1,1,\ldots)$, then $\inf X = \liminf X = \limsup X = \sup X = 1$.

vi) If $X = (0,1,1,1,1,\ldots)$, then $0 = \inf X < \liminf X = \limsup X = \sup X = 1$.

vii) If $X = (1,2,1,2,1,2,\ldots)$, then $1 = \inf X = \liminf X < \limsup X = \sup X = 2$.

viii) If $X = (2,1,1,1,1,\ldots)$, then $1 = \inf X = \liminf X = \limsup X = 1 < \sup X = 2$.

ix) If $X = (0,1,2,1,2,1,2,\ldots)$, then $0 = \inf X < \liminf X = 1 < \limsup X = \sup X = 2$.

x) If $X = (0,2,1,1,1,1,\ldots)$, then $0 = \inf X < \liminf X = \limsup X = 1 < \sup X = 2$.

xi) If $X = (2,4,1,3,1,3,1,3,\dots)$, then $1 = \inf X = \liminf X < \limsup X = 3 < \sup X = 4$.

xii) If $X = (0,3,1,2,1,2,1,2,\dots)$, then $0 = \inf X < \liminf X < \limsup X = 2 < \sup X = 3$.

xiii) If $X = (1,2,1,3,1,4,\dots)$, then $1 = \inf X = \liminf X < \limsup X = \sup X = \infty$.

xiv) If $X = (0,1,2,1,3,1,4,\dots)$, then $0 = \inf X < \liminf X = 1 < \limsup X = \sup X = \infty$.

xv) If $X = (1,2,3,4,\dots)$, then $1 = \inf X < \liminf X = \limsup X = \sup X = \infty$.

Remark: These 15 cases exhaust all possible forms of *i)* in Proposition 12.2.12.

Fact 12.18.3. Let $\mathcal{S} \triangleq (S_i)_{i=1}^\infty$ be a sequence of subsets of $\mathbb{F}^n$, and define

$$\operatorname{liminf}(\mathcal{S}) \triangleq \{x \in \mathbb{F}^n\colon \text{there exists } (x_i)_{i=1}^\infty \text{ such that, for all } i \ge 1, x_i \in S_i, \text{ and } x = \lim_{i\to\infty} x_i\},$$

$$\operatorname{limsup}(\mathcal{S}) = \{x \in \mathbb{F}^n\colon \text{there exists a subsequence } (S_{i_j})_{j=1}^\infty \text{ of } \mathcal{S} \text{ and a sequence } (x_{i_j})_{j=1}^\infty \text{ such that, for all } j \ge 1, x_i \in S_{i_j}, \text{ such that } x = \lim_{j\to\infty} x_{i_j}\}.$$

Furthermore, if $\operatorname{liminf}(\mathcal{S}) = \operatorname{limsup}(\mathcal{S})$, then define $\lim(\mathcal{S}) \triangleq \operatorname{liminf}(\mathcal{S}) = \operatorname{limsup}(\mathcal{S})$. Then,

$$\operatorname{glb}(\mathcal{S}) \subseteq \operatorname{essglb}(\mathcal{S}) \subseteq \begin{cases} \operatorname{esslub}(\mathcal{S}) \subseteq \begin{cases} \operatorname{lub}(\mathcal{S}) \\ \operatorname{limsup}(\mathcal{S}) \end{cases} \\ \operatorname{liminf}(\mathcal{S}) \subseteq \operatorname{limsup}(\mathcal{S}). \end{cases}$$

Furthermore, the following statements hold:

i) If $\lim(\mathcal{S})$ and $\operatorname{esslim}(\mathcal{S})$ exist, then $\operatorname{esslim}(\mathcal{S}) \subseteq \lim(\mathcal{S})$.

ii) $\operatorname{liminf}(\mathcal{S})$ and $\operatorname{limsup}(\mathcal{S})$ are closed.

iii) If $\lim(\mathcal{S})$ exists, then $\lim(\mathcal{S})$ is closed.

iv) Let $A \in \mathbb{F}^{m\times n}$, and assume that there exists a bounded set $S_0 \subset \mathbb{F}^n$ such that, for all $i \ge 1$, $S_i \subset S_0$. If $\lim(\mathcal{S})$ exists, then $\lim(A\mathcal{S}) = A\lim(\mathcal{S})$.

Source: [1185, pp. 97–101].

Fact 12.18.4. Let $(x_i)_{i=1}^\infty \subset \mathbb{F}^n$. Then, $\lim_{i\to\infty} x_i = x$ if and only if, for all $j \in \{1,\dots,n\}$, $\lim_{i\to\infty} x_{i(j)} = x_{(j)}$.

Fact 12.18.5. Let $(x_i)_{i=1}^\infty \subset \mathbb{R}$. Then, the following statements are equivalent:

i) $\lim_{i\to\infty} x_i = 0$.

ii) $\lim_{i\to\infty} |x_i| = 0$.

iii) $\liminf_{i\to\infty} -|x_i| = 0$.

iv) $\limsup_{i\to\infty} |x_i| = 0$.

Fact 12.18.6. Let $(x_i)_{i=1}^\infty \subset (0,\infty)$. Then,

$$\liminf_{i\to\infty} \frac{x_{i+1}}{x_i} \le \liminf_{i\to\infty} \sqrt[i]{x_i} \le \limsup_{i\to\infty} \sqrt[i]{x_i} \le \limsup_{i\to\infty} \frac{x_{i+1}}{x_i}.$$

Source: [2294, p. 6].

Fact 12.18.7. Let $X = (x_i)_{i=1}^\infty \subset \mathbb{F}^n$ and $Y = (y_i)_{i=1}^\infty \subset \mathbb{F}^n$, and assume that X and Y converge. Then, $X + Y$ and $X \odot Y$ converge, and

$$\lim(X+Y) = \lim X + \lim Y, \quad \lim(X \odot Y) = (\lim X) \odot (\lim Y).$$

Fact 12.18.8. Let $X = (x_i)_{i=1}^\infty \subseteq \mathbb{R}$ and $Y = (y_i)_{i-1}^\infty \subseteq \mathbb{R}$. Then, the following statements hold:

i) If X is a rearrangement of Y, then $\liminf X = \liminf Y$ and $\limsup X = \limsup Y$.

ii) If X is either nonincreasing or nondecreasing, then $\liminf X = \limsup X = \lim X$.

iii) $\liminf X + \liminf Y \le \liminf(X+Y) \le \liminf X + \limsup Y \le \limsup(X+Y) \le \limsup X + \limsup Y$.

iv) If $X \subset [0,\infty)$ and $Y \subset [0,\infty)$ are bounded, then $(\liminf X)\liminf Y \le \liminf(XY)$ and $\limsup(XY) \le (\limsup X)\limsup Y$.

Source: [1566, p. 44]. **Remark:** $XY \triangleq (x_iy_i)_{i=1}^\infty$.

Fact 12.18.9. Let $(x_i)_{i=1}^\infty \subset \mathbb{C}$, and assume that at least one of the following statements holds:

i) $\limsup_{i\to\infty} |x_i|^{1/i} < 1$.

ii) $0 \notin (x_i)_{i=1}^\infty$ and $\lim_{i\to\infty} |x_{i+1}/x_i| < 1$.

Then, $\sum_{i=1}^\infty x_i$ converges. **Source:** Under *i*) and *ii*), the Cauchy-Hadamard formula and ratio test of Proposition 12.3.7 imply that the unit circle in $\mathbb{C}$ is contained in the domain of convergence of $\sum_{i=1}^\infty x_iz^i$.

Fact 12.18.10. Let $(x_i)_{i=1}^\infty \subset (0,\infty)$, assume that $(x_i)_{i=1}^\infty$ is nonincreasing, and assume that $\lim_{i\to\infty} x_{2i}/x_i$ exists. Then, the following statements hold:

i) If $\lim_{i\to\infty} x_{2i}/x_i < 1/2$, then $\sum_{i=1}^\infty x_i$ converges.

ii) If $\lim_{i\to\infty} x_{2i}/x_i > 1/2$, then $\sum_{i=1}^\infty x_i = \infty$.

Source: [703].

Fact 12.18.11. Let $(x_i)_{i=1}^\infty \subset (0,\infty)$, and assume that $(x_i)_{i=1}^\infty$ is nonincreasing and $\lim_{i\to\infty} x_i = 0$. Then, $\sum_{i=1}^\infty (-1)^i x_i$ converges. **Remark:** This is *Leibniz's test*.

Fact 12.18.12. Let $(x_i)_{i=1}^\infty \subset \mathbb{R}$, assume that $(\sum_{i=1}^n x_i)_{n=1}^\infty$ is bounded, let $(y_i)_{i=1}^\infty \subset [0,\infty)$, assume that $(y_i)_{i=1}^\infty$ is nonincreasing, and assume that $\lim_{i\to\infty} y_i = 0$. Then, $\sum_{i=1}^\infty x_iy_i$ converges. **Source:** [2294, p. 63]. **Remark:** This is *Dirichlet's test*. **Example:** $1 - \frac{1}{2} + \frac{1}{3} - \frac{1}{4} + \cdots$ converges. See [1024].

Fact 12.18.13. Let $(x_i)_{i=1}^\infty \subset \mathbb{R}$, assume that $\sum_{i=1}^\infty x_i$ converges, let $(y_i)_{i=1}^\infty \subset \mathbb{R}$, and assume that $(y_i)_{i=1}^\infty$ is bounded and either nonincreasing or nondecreasing. Then, $\sum_{i=1}^\infty x_iy_i$ converges. **Source:** [2294, p. 63]. **Remark:** This is *Abel's test*.

Fact 12.18.14. Let $(x_i)_{i=1}^\infty \subset (0,\infty)$. Then, the following statements are equivalent:

i) $\sum_{i=1}^\infty x_i = \infty$.

ii) For every sequence $(y_i)_{i=1}^\infty \subset \mathbb{R}$ with a limit in $\mathbb{R} \cup \{-\infty,\infty\}$, it follows that

$$\lim_{n\to\infty} \frac{\sum_{i=1}^n x_iy_i}{\sum_{i=1}^n x_i} = \lim_{i\to\infty} y_i.$$

In particular, if $(y_i)_{i=1}^\infty \subset \mathbb{R}$ has a limit in $\mathbb{R} \cup \{-\infty,\infty\}$, then

$$\lim_{n\to\infty} \frac{1}{n}\sum_{i=1}^n y_i = \lim_{i\to\infty} y_i.$$

Source: [2294, p. 89]. **Remark:** This is *Cesaro's lemma*.

Fact 12.18.15. Let $(x_i)_{i=1}^\infty \subset \mathbb{C}$, assume that $\sum_{i=1}^\infty x_i$ exists, let $(a_i)_{i=1}^\infty \subset (0,\infty)$, and assume that $(a_i)_{i=1}^\infty$ is nondecreasing and $\lim_{i\to\infty} a_i = \infty$. Then, $\lim_{n\to\infty} \frac{1}{a_n}\sum_{i=1}^n a_ix_i = 0$. **Source:** [75]. **Remark:** This is *Kronecker's lemma*. **Related:** Fact 10.11.76.

Fact 12.18.16. Let $(x_i)_{i=1}^\infty \subset \mathbb{C}$, and assume that $\sum_{i=1}^\infty \frac{x_i}{i}$ exists. Then, $\lim_{n\to\infty} \frac{1}{n}\sum_{i=1}^n x_i = 0$. **Source:** Fact 12.18.15 and [2527, p. 177].

Fact 12.18.17. Let $(a_i)_{i=1}^\infty \subset \mathbb{R}$, and assume that $(\frac{1}{i}\sum_{j=1}^i \sum_{l=1}^j a_l)_{i=1}^\infty$ converges. Then,

$$\lim_{x\uparrow 1} \sum_{i=1}^\infty a_ix^i = \lim_{i\to\infty} \frac{1}{i}\sum_{j=1}^i \sum_{l=1}^j a_l.$$

Source: [116, pp. 599, 600]. **Remark:** This result shows that *Cesaro summability* implies *Abel summability*. **Example:** For all $i \geq 1$, let $a_i = (-1)^{i+1}$. Then, both limits exist and equal $\frac{1}{2}$. For all $i \geq 1$, let $a_i = (-1)^{i+1}i$. Then, the first limit equals $\frac{1}{4}$, but the second limit does not exist.

Fact 12.18.18. Let $(a_i)_{i=1}^{\infty} \subset \mathbb{C}$ and $(b_i)_{i=1}^{\infty} \subset \mathbb{C}$, assume that $\sum_{i=1}^{\infty} a_i$ and $\sum_{i=1}^{\infty} b_i$ converge, for all $i \geq 1$, define $c_i \triangleq \sum_{j=1}^{i} a_j b_{i+1-j}$, and assume that $\sum_{i=1}^{\infty} c_i$ converges. Then,

$$\sum_{i=1}^{\infty} c_i = \left(\sum_{i=1}^{\infty} a_i\right)\sum_{i=1}^{\infty} b_i.$$

Source: [116, p. 600].

Fact 12.18.19. Let $(a_i)_{i=1}^{\infty} \subset \mathbb{R}$ and $(b_i)_{i=1}^{\infty} \subset \mathbb{R}$. Then, the following statements hold:

i) Assume that one of the following statements holds:

a) $(b_i)_{i=1}^{\infty} \subset \mathbb{R}$ is decreasing, and $\lim_{i\to\infty} a_i = \lim_{i\to\infty} b_i = 0$.

b) $(b_i)_{i=1}^{\infty} \subset \mathbb{R}$ is increasing, and $\lim_{i\to\infty} b_i = \infty$.

If

$$\alpha \triangleq \lim_{i\to\infty} \frac{a_{i+1} - a_i}{b_{i+1} - b_i} \in \mathbb{R} \cup \{-\infty, \infty\},$$

then

$$\lim_{i\to\infty} \frac{a_i}{b_i} = \alpha.$$

ii) Assume that $(b_i)_{i=1}^{\infty} \subset \mathbb{R}$ is increasing, $\lim_{i\to\infty} b_i = \infty$, $\alpha \triangleq \lim_{i\to\infty} a_i/b_i$ exists, $\beta \triangleq \lim_{i\to\infty} b_i/b_{i+1}$ exists, and $\beta \neq 1$. Then,

$$\lim_{i\to\infty} \frac{a_{i+1} - a_i}{b_{i+1} - b_i} = \alpha.$$

Source: [1103, pp. 263–266] and [2294, p. 7]. **Remark:** *i)* is the *Stolz-Cesaro lemma*, which is a discrete analogue of l'Hôpital's rule given by Fact 12.17.13. *ii)* is a converse. **Remark:** In *ii)*, α and β are real numbers.

Fact 12.18.20.

$$\liminf_{i\to\infty} \cos i = -1, \quad \limsup_{i\to\infty} \cos i = 1.$$

Source: [968, p. 287].

Fact 12.18.21. Let ϕ denote Euler's totient function, and let $\sigma(n)$ denote the sum of the divisors of $n \geq 1$. If $\alpha \geq 0$, then

$$\lim_{n\to\infty} \frac{1}{n^{\alpha+1}} \sum_{i=1}^{n} i^{\alpha-1}\phi(i) = \frac{6}{\pi^2(1+\alpha)}, \quad \lim_{n\to\infty} \sum_{i=1}^{n} \frac{\sigma(i)}{n^2 + \alpha i^2} = \frac{\pi^2}{12\alpha}\log(1+\alpha).$$

In particular,

$$\lim_{n\to\infty} \frac{1}{n^2} \sum_{i=1}^{n} \phi(i) = \frac{3}{\pi^2}, \quad \lim_{n\to\infty} \frac{1}{n} \sum_{i=1}^{n} \frac{\phi(i)}{i} = \frac{6}{\pi^2}.$$

Furthermore,

$$\lim_{n\to\infty} \frac{1}{n^2} \sum_{i=1}^{n} \sigma(i) = \frac{\pi^2}{12}.$$

Source: [109, pp. 382, 383]. **Related:** Fact 1.20.3 and Fact 1.20.4.

Fact 12.18.22. Let α and p be real numbers. Then, the following statements hold:

i) If $\alpha > 0$ and $p > 0$, then $\lim_{n\to\infty} \alpha^{1/n^p} = 1$.

ii) If $p > 0$, then $\lim_{n\to\infty} n^{1/n^p} = 1$.

iii) If $p > 1$, then $\lim_{n\to\infty}(n!)^{1/n^p} = 1$.

Fact 12.18.23. Let $a, b, c, d, f, g \in \mathbb{R}$, and assume that $ac > 0$. Then,

$$\lim_{n\to\infty}\left(\frac{an+b}{cn+d}\right)^{fn+g} = \begin{cases} 0, & f(|c|-|a|) > 0,\\ \infty, & f(|c|-|a|) < 0,\\ \left(\frac{a}{c}\right)^g, & f = 0,\\ e^{f(b-d)/a}, & a = c \text{ and } f \neq 0. \end{cases}$$

Fact 12.18.24. Let a be a real number. Then,

$$\lim_{n\to\infty}\frac{e^{an}}{\left(1+\frac{a}{n}\right)^{n^2}} = e^{a^2/2}, \quad \lim_{n\to\infty}\frac{e^{an^2}}{\left(1+\frac{a}{n}+\frac{a^2}{2n^2}\right)^{n^3}} = e^{a^3/6}, \quad \lim_{n\to\infty}\frac{e^{an^3}}{\left(1+\frac{a}{n}+\frac{a^2}{2n^2}+\frac{a^3}{6n^3}\right)^{n^4}} = e^{a^4/24}.$$

Source: [1563].

Fact 12.18.25. Let $f\colon (0,1] \mapsto (0,\infty)$, assume that f is differentiable, assume that $f'(1) > 0$, and assume that f'/f is nonincreasing. Furthermore, let $a \in (0,\infty)$ and $b \in [0,\infty)$, and, for all $n \geq 1$, define

$$x_n \triangleq \sum_{i=1}^{n}[f(\tfrac{i}{n})]^{ai+b}.$$

Then,

$$\lim_{n\to\infty} x_n = \begin{cases} 0, & f(1) < 1,\\ \dfrac{e^{af'(1)}}{e^{af'(1)}-1}, & f(1) = 1,\\ \infty, & f(1) > 1. \end{cases}$$

Now, let z be a complex number. Then,

$$\lim_{n\to\infty}\sum_{i=1}^{n}\left(\frac{z+i}{n}\right)^{ai+b} = \lim_{n\to\infty}\sum_{i=1}^{n}\left(\frac{z+i}{n}\right)^{an+b} = \frac{e^{a(z+1)}}{e^a-1}.$$

In particular,

$$\lim_{n\to\infty}\sum_{i=1}^{n}\left(\frac{i}{n}\right)^{ai} = \lim_{n\to\infty}\sum_{i=1}^{n}\left(\frac{i}{n}\right)^{an} = \frac{e^a}{e^a-1}.$$

Furthermore,

$$\lim_{n\to\infty}\frac{1}{e^n}\sum_{i=0}^{n}\frac{n^i}{i!} = \frac{1}{2}.$$

Source: [1103, pp. 2, 3, 32], [1564, 1733], [2294, p. 56], and [2546]. **Credit:** The first result is due to O. Furdui. **Related:** Fact 12.18.26.

Fact 12.18.26. Let a and b be real numbers such that $0 \leq a < b$. Then,

$$\lim_{n\to\infty}\sum_{i=1}^{n}\left(\frac{bi-a}{bn}\right)^{n} = \frac{e^{1-a/b}}{e-1}.$$

In particular,

$$\lim_{n\to\infty}\sum_{i=1}^{n}\left(\frac{i}{n}\right)^{n} = \frac{e}{e-1}.$$

Source: [1565]. **Related:** Fact 12.18.25.

Fact 12.18.27. Let $(\alpha_i)_{i=1}^{\infty} \subset \mathbb{R}$. Then, the following statements are equivalent:

i) For all $a, b \in [0,1)$ such that $a < b$, $\lim_{n\to\infty} \frac{1}{n}\operatorname{card}(\{i \le n\colon \alpha_i - \lfloor \alpha_i \rfloor \in [a,b]\}) = b - a$.

ii) For all $k \ge 1$, $\lim_{n\to\infty} \frac{1}{n}\sum_{i=1}^n e^{2\pi k\alpha_i J} = 0$.

If $(\alpha_i)_{i=1}^{\infty} \subset [0,1)$, then the following statement is equivalent to *i)* and *ii)*:

iii) For all $k \ge 1$, $\lim_{i\to\infty} \frac{1}{i}\sum_{j=1}^i \alpha_j^k = \frac{1}{k+1}$.

Furthermore, the following sequences satisfy *i)* and *ii)*:

iv) $(ie)_{i=1}^{\infty}$.

v) $(i^p)_{i=1}^{\infty}$, where $p \in (0,1)$.

vi) $(\log^p i)_{i=1}^{\infty}$, where $p \in (1,\infty)$.

vii) $(i\log^p i)_{i=1}^{\infty}$, where $p \in (-\infty, 0)$.

Source: [968, pp. 269–275, 287]. **Remark:** The equivalence of *i)* and *ii)* is *Weyl's criterion.*

Fact 12.18.28. Let $z \in \mathbb{C}$, and assume that $|z| < 1$. Then,

$$\lim_{n\to\infty} \frac{1}{n}\sum_{k=0}^{n-1} \sin^2 \arg(z + e^{2\pi Jk/n}) = \lim_{n\to\infty} \frac{1}{n}\sum_{k=0}^{n-1} \cos^2 \arg(z + e^{2\pi Jk/n}) = \frac{1}{2},$$

$$\lim_{n\to\infty} \frac{1}{n}\sum_{k=0}^{n-1} [\sin \arg(z + e^{2\pi Jk/n})] \cos \arg(z + e^{2\pi Jk/n}) = 0.$$

Credit: N. Crasta.

Fact 12.18.29. Let $k \ge 1$. Then,

$$\lim_{n\to\infty} \frac{1}{n^{k+1}}\sum_{i=1}^n i^k = \frac{1}{k+1}.$$

Source: [141] and Fact 1.12.1.

Fact 12.18.30. Let x be a positive number. Then,

$$\lim_{n\to\infty} \sum_{i=1}^n (x^{i/n^2} - 1) = \tfrac{1}{2}\log x.$$

Source: [1566, p. 61].

Fact 12.18.31. Let p be a nonzero real number, and let q be a positive number. Then,

$$\lim_{n\to\infty} \sum_{i=1}^n \left(\sqrt[p]{1 + \frac{i^{q-1}}{n^q}} - 1 \right) = \frac{1}{pq}.$$

In particular,

$$\lim_{n\to\infty} \sum_{i=1}^n \left(\sqrt{1 + \frac{i}{n^2}} - 1 \right) = \frac{1}{4}, \quad \lim_{n\to\infty} \sum_{i=1}^n \left(\sqrt[3]{1 + \frac{i^2}{n^3}} - 1 \right) = \frac{1}{9}.$$

Source: [1566, pp. 32, 61] and [2294, p. 15].

Fact 12.18.32. Let $n \ge 1$. Then,

$$\frac{1}{2n + (2\log 2 - 1)/(1 - \log 2)} \le \left| \sum_{i=1}^n (-1)^{i+1}\frac{1}{i} - \log 2 \right| < \frac{1}{2n+1},$$

$$\frac{1}{4n^2 + 2} < \left| \sum_{i=1}^n (-1)^{i+1}\frac{1}{i} - \log 2 - (-1)^{n+1}\frac{1}{2n} \right| \le \frac{1}{4n^2 + (6 - 8\log 2)/(2\log 2 - 1)}.$$

Source: [2463, 2690].

Fact 12.18.33. If $n \ge 1$, then

$$H_n = \gamma + \log n + \frac{1}{2n} - \sum_{i=1}^{\infty} \frac{B_{2i}}{2in^{2i}}.$$

For all $k \ge 1$, as $n \to \infty$,

$$H_n \sim \gamma + \log n - \sum_{i=1}^{k} \frac{B_i}{in^i}.$$

In particular,

$$H_n \sim \gamma + \log n + \frac{1}{2n}, \quad H_n \sim \gamma + \log n + \frac{1}{2n} - \frac{1}{12n^2}.$$

Furthermore,

$$\lim_{n\to\infty} H_n = \sum_{i=1}^{\infty} \frac{1}{i} = \infty, \quad \lim_{n\to\infty} \frac{H_n}{\log n} = \left(\lim_{n\to\infty} \frac{H_n}{\gamma + \log n + \frac{1}{2n}}\right) \lim_{n\to\infty} \frac{\gamma + \log n + \frac{1}{2n}}{\log n} = 1,$$

$$\begin{aligned}
\gamma &\triangleq \lim_{n\to\infty} [H_n - \log(n+1)] = \lim_{n\to\infty} (H_n - \log n) \\
&= \sum_{i=1}^{\infty} \left[\frac{1}{i} - \log\left(1 + \frac{1}{i}\right)\right] = 1 + \sum_{i=2}^{\infty} \left[\frac{1}{i} + \log\left(1 - \frac{1}{i}\right)\right] \\
&= \sum_{i=2}^{\infty} (-1)^i \frac{\zeta(i)}{i} = \sum_{i=1}^{\infty} (-1)^i \frac{1}{i} \left\lfloor \frac{\log i}{\log 2} \right\rfloor \\
&\approx 0.5772156649015328606065120900824024310421593359399235.
\end{aligned}$$

Furthermore, let $k \ge 2$. Then,

$$\gamma = \frac{1}{k-1} \lim_{n\to\infty} (kH_n - H_{n^k}).$$

In particular,

$$\gamma = \lim_{n\to\infty} (2H_n - H_{n^2}) = \tfrac{1}{2} \lim_{n\to\infty} (3H_n - H_{n^3}).$$

Source: The first equality is given in [721]. The third expression for γ is given in [1566, p. 69]. The fourth expression for γ is given in [516, p. 327]. The fifth expression for γ is given in [2294, p. 112]. The sixth expression for γ is given in [506, pp. 140, 141]. The second to last expression for γ is given in [1920]. **Remark:** γ is the *Euler constant*. See [880, 1350]. **Remark:** $\gamma \approx 323007/559595 \approx 0.5772156649005$, $\gamma \approx \pi/(2.002e) \approx 0.577286$, and $\frac{\pi}{2e} \approx 0.5778$. **Related:** Fact 13.5.47.

Fact 12.18.34. Let $n \ge 2$. Then,

$$H_n - \log(n+1) < \gamma < H_n - \log n \le 1,$$

where the (first, third) term is (an increasing, a decreasing) function of n. Furthermore,

$$-\frac{1}{48n^2} < H_n - \log\left(n + \frac{1}{2} + \frac{1}{24n}\right) - \gamma < -\frac{1}{48(n+1)^3},$$

$$\begin{aligned}
\frac{1}{2n+2} &< \frac{1}{2n + \frac{2}{5}} < \frac{1}{2n - 2 + \frac{1}{1-\gamma}} < \frac{1}{2n} - \frac{1}{8n^2} \\
&< \frac{1}{2n + \frac{1}{3} + \frac{1}{18n}} < H_n - \log n - \gamma < \frac{1}{2n + \frac{1}{3} + \frac{1}{32n}} < \frac{1}{2n},
\end{aligned}$$

$$\frac{1}{2n} - \frac{1}{12n^2 + 2(7-12\gamma)/(2\gamma-1)} \le H_n - \log n - \gamma < \frac{1}{2n} - \frac{1}{12n^2 + 6/5},$$

$$\frac{1}{24(n+1)^2} < \frac{1}{24\left(n - 1 + 1/\sqrt{24(1-\gamma-\log\frac{3}{2})}\right)^2} < \frac{1}{24n(n+1) + \frac{51}{5} - \frac{1}{n(n+1)}}$$

$$< H_n - \log(n+\tfrac{1}{2}) - \gamma < \frac{1}{24n(n+1) + \frac{51}{5} - \frac{3}{2n(n+1)}} < \frac{1}{24(n+\frac{1}{2})^2},$$

$$\frac{1}{6n(n+1) + \frac{6}{5} - \frac{1}{6n(n+1)}} < H_n - \tfrac{1}{2}\log(n^2+n) - \gamma < \frac{1}{6n(n+1) + \frac{6}{5} - \frac{1}{4n(n+1)}},$$

$$\tfrac{1}{n} + \log n \le \tfrac{1}{2n} + \tfrac{1}{2} + \log n < \log(n+\tfrac{1}{2}) + \gamma < H_n \le \log(n + e^{1-\gamma} - 1) + \gamma,$$

$$1 + \log(\sqrt{e} - 1) - \log(e^{1/(n+1)} - 1) \le H_n < \gamma - \log(e^{1/(n+1)} - 1),$$

$$\frac{1}{2} - \frac{1}{4n+2} + \frac{1}{2}\log(2n+1) < \sum_{i=1}^{n} \frac{1}{2i-1} < 1 + \frac{1}{2}\log n,$$

$$\frac{1}{2n - 2 + 1/(1-\log 2)} \le \left|\sum_{i=n+1}^{\infty} (-1)^{i+1}\frac{1}{i}\right| < \frac{1}{2n+1},$$

$$\tfrac{1}{2}H_{n+1,2} - \tfrac{1}{2} + \log(n+1) \le H_n \le \tfrac{1}{2}H_{n,2} + \log(n+1).$$

Finally, as $n \to \infty$,

$$\gamma = H_n - \log n - \frac{1}{2n} + \frac{1}{12n^2} - \frac{1}{120n^4} + \frac{1}{252n^6} - \frac{1}{240n^8} + \frac{1}{132n^{10}} + O(n^{-12}),$$

and thus $H_n \sim \gamma + \log n$. **Source:** [689, 721, 1271, 1408, 2462, 2519]. **Remark:** $1/(2n) - 1/(8n^2) < H_n - \log n - \gamma < 1/(2n)$ is *Franel's inequality*. **Related:** Fact 1.12.46.

Fact 12.18.35. If $n \ge 1$, then

$$\frac{1}{n + \frac{\gamma}{1-\gamma}} \le 2H_n - H_{n^2} - \gamma < \frac{1}{n + \frac{2}{3}}, \quad \frac{1}{3n^2 + \frac{2}{\gamma} - 3} \le \gamma - (2H_n - H_{n^2} - \frac{1}{n}) < \frac{2}{3n^2 + \frac{43}{200}}.$$

If $n \ge 2$, then

$$\lim_{i\to\infty} i[nH_i - H_{i^n} - (n-1)\gamma] = \frac{n}{2}.$$

In particular,

$$\lim_{i\to\infty} i(2H_i - H_{i^2} - \gamma) = 1.$$

Furthermore, if $n, m \ge 1$, then

$$\gamma < H_n + H_m - H_{nm} < \gamma + \tfrac{1}{2}[1 - (1 - \tfrac{1}{n})(1 - \tfrac{1}{m})].$$

Finally,

$$\lim_{i,j\to\infty} (H_i + H_j - H_{ij}) = \gamma.$$

Source: [2464, 2519].

Fact 12.18.36. Let $k \ge 1$. As $n \to \infty$,

$$\sum_{i=n+1}^{\infty} \frac{1}{i^{2k+1}} = \zeta(2k+1) - H_{n,2k+1} = \frac{1}{2^k k n^k (n+1)^k} + O(1/n^{2k+2}).$$

In particular, as $n \to \infty$,

$$\sum_{i=n+1}^{\infty} \frac{1}{i^3} = \frac{1}{2n(n+1)} + O(1/n^4), \quad \sum_{i=n+1}^{\infty} \frac{1}{i^3} = \frac{1}{8n^2(n+1)^2} + O(1/n^6).$$

Finally, as $n \to \infty$,

$$H_n = \gamma + \frac{1}{2}\log n(n+1) + \frac{1}{6n(n+1)} + O(1/n^4).$$

Source: [1409].

Fact 12.18.37. $\sum_{i=1}^{\infty} \frac{1}{iH_i} = \infty$. **Source:** [506, pp. 125, 126].

Fact 12.18.38. Let $x_1, \ldots, x_n$ be positive numbers. Then,

$$\lim_{k\to\infty} \sqrt[k]{\frac{1}{n}\sum_{i=1}^{n} x_i^k} = \max\{x_1, \ldots, x_n\}.$$

Fact 12.18.39. Let $(x_i)_{i=1}^{\infty} \subset \mathbb{R}$, and assume that $x \triangleq \lim_{i\to\infty} \frac{1}{i}\sum_{j=1}^{i} x_j$ exists. Then,

$$\lim_{n\to\infty} \frac{1}{\log n}\sum_{i=1}^{n} \frac{x_i}{i} = x.$$

Source: [1566, p. 39].

Fact 12.18.40. Let $(x_i)_{i=1}^{\infty} \subset \mathbb{C}$ be a convergent sequence, and let $x \triangleq \lim_{i\to\infty} x_i$. Then,

$$\lim_{n\to\infty} \frac{1}{n}\sum_{i=1}^{n} x_i = \lim_{n\to\infty} \frac{2}{n^2}\sum_{i=1}^{n}(n-i+1)x_i = \lim_{n\to\infty} \frac{1}{\log n}\sum_{i=1}^{n} \frac{x_i}{i} = x.$$

Now, assume that $(x_i)_{i=1}^{\infty} \subset (0,\infty)$. Then,

$$\lim_{n\to\infty} \sqrt[n]{\prod_{i=1}^{n} x_i} = x.$$

Furthermore, if $\alpha \triangleq \lim_{i\to\infty} x_{i+1}/x_i$ exists, then $\lim_{n\to\infty} \sqrt[n]{x_n} = \alpha$. **Source:** [1566, pp. 35, 36, 39].

Fact 12.18.41. Let $(p_i)_{i=1}^{\infty}$ denote the prime numbers, and, for all $x > 0$, let $\pi(x)$ denote the number of prime numbers that are either less than or equal to x. Then, the following statements hold:

i) $\lim_{i\to\infty} p_i = \infty$.

ii) If $n \ge 10$, then $\frac{p_n}{2} - \frac{9n}{4} < \frac{1}{n}\sum_{i=1}^{n} p_i < \frac{p_n}{2} - \frac{n}{12}$. Thus, $\lim_{n\to\infty} \frac{\frac{1}{n}\sum_{i=1}^{n} p_i}{p_n} = \frac{1}{2}$.

iii) $\log\log(n+1) \le \sum_{i=1}^{n} \frac{1}{p_i} + \log\frac{\pi^2}{6}$, and thus $\sum_{i=1}^{\infty} \frac{1}{p_i} = \infty$.

iv) If $n > 2$, then there exists a prime between n and $2n$.

v) If $n \ge 25$, then there exists a prime between n and $6n/5$.

vi) If $n \ge 26$, then $p_n < (1.2)^n$.

vii) If $n \ge 6$, then $p_n < n\log n + \log\log n$. If $n \ge 7022$, then $p_n \le n\log n + n[\log\log n - 0.9385]$.

viii) For all $m \ge 1$, there exists $n \ge 2$ such that $n/\pi(n) = m$.

ix) If $n \ge 2$, then

$$\pi(n) < \frac{n}{\log n}\left(1 + \frac{3}{2\log n}\right).$$

x) If $n \geq 67$, then

$$\frac{n}{\log n - \frac{1}{2}} < \pi(n) < \frac{n}{\log n - \frac{3}{2}}, \quad \log n - \tfrac{3}{2} < \frac{n}{\pi(n)} < \log n - \tfrac{1}{2}.$$

xi) There exists an increasing sequence $(m_i)_{i=1}^\infty \subset \mathbb{P}$ such that, for all $i \geq 1$,

$$\pi(m_i) = \frac{m_i}{\lfloor \log m_i - \frac{1}{2} \rfloor}, \quad \frac{m_i}{\pi(m_i)} = \lfloor \log m_i - \tfrac{1}{2} \rfloor.$$

xii) As $x \to \infty$, $\pi(x) \sim \frac{x}{\log x}$.

xiii) As $x \to \infty$, $\pi(x) = \frac{x}{\log x} + O\Big(\frac{x}{\log^2 x}\Big)$.

xiv) As $x \to \infty$, $\pi(x) \sim \int_2^x \frac{1}{\log t}\,\mathrm{d}t$.

xv) If $x \geq 3$, then $\log\log x < \sum_{i=1}^{\pi(x)} \frac{1}{p_i} < 1 + \log\log x$.

xvi) $\lim_{x\to\infty} \frac{\log \sum_{i=1}^{\lfloor x \rfloor} \frac{1}{i}}{\sum_{i=1}^{\pi(x)} \frac{1}{p_i}} = 1$.

xvii) For all $n \geq 2$, there exists $i > n$ such that $p_{i+1} - p_i \leq 246$.

xviii) If $n \geq 3$ and $p_{n+1} - p_n = 2$, then there exists $k \geq 1$ such that $p_n = 6k - 1$ and $p_{n+1} = 6k + 1$.

xix) Let $n \geq 1$. Then, $p_{n+1} - p_n = 2$ if and only if $4[(p_n - 1)! + 1] \overset{p_n p_{n+1}}{\equiv} p_n(p_{n+1} - 1)$.

xx) $\limsup_{n\to\infty} \frac{p_{n+1} - p_n}{\log n} = \infty$.

Source: *ii*) is given in [1235, 1346]; *iii*) is proved in [506, p. 84]; *iv*)–*vii*) are given in [1522]; *ix*) is given in [972]; *x*)–*xii*) are given in [1131]; *xiii*) is given in [202]; *xv*) and *xvi*) are given in [332]; *xvii*) is given in [139]; *xviii*) and *xix*) are given in [143]. **Remark:** $\lim_{i\to\infty} p_i = \infty$ in *i*) is due to Euclid. *xiii*) implies *xii*). *xii*) and *xiv*) are equivalent versions of the *prime number theorem.* See [1350, pp. 181, 182]. This result was conjectured by Gauss and proved by Hadamard and de la Vallée-Poussin. See [1515]. A stronger version is given by the Riemann hypothesis. See Fact 13.3.1. **Remark:** *xvii*) can be written as $\liminf_{i\to\infty}(p_{i+1} - p_i) \leq 246$. The *twin prime conjecture* states that $\liminf_{i\to\infty}(p_{i+1} - p_i) = 2$.

Fact 12.18.42. As $n \to \infty$,

$$\sum_{i=1}^n \frac{1}{n+i} = \log 2 - \frac{1}{4n} + \frac{1}{16n^2} + O(n^{-4}).$$

Hence,

$$\lim_{n\to\infty} \sum_{i=1}^n \frac{1}{n+i} = \log 2.$$

Source: [1566, p. 230].

Fact 12.18.43. Let $x, y \in (0, \infty)$. Then,

$$\lim_{n\to\infty} \sum_{i=1}^n \frac{1}{nx + iy} = \frac{1}{y} \log \frac{x+y}{x}.$$

Fact 12.18.44. As $n \to \infty$,

$$\sum_{i=1}^{n^2} \frac{n}{n^2 + i^2} = \frac{\pi}{2} - \frac{3}{2n} + \frac{5}{6n^3} + O(n^{-5}).$$

Hence,

$$\lim_{n\to\infty}\sum_{i=1}^{n^2}\frac{n}{n^2+i^2}=\frac{\pi}{2}.$$

Source: [2294, p. 54].

Fact 12.18.45.

$$\lim_{n\to\infty}\sum_{i=1}^{n-1}\frac{n^2}{i^2(n-i)^2}=\frac{\pi^2}{3}.$$

Source: [2549].

Fact 12.18.46.

$$\lim_{n\to\infty}\sum_{i=1}^{n}\frac{1}{\sqrt{n^2+i}}=1,\quad \lim_{n\to\infty}\sum_{i=1}^{n}\frac{1}{\sqrt{n^2+i^2}}=\log(1+\sqrt{2}).$$

Source: [1564].

Fact 12.18.47. Let $k\ge 2$. Then, as $n\to\infty$,

$$\binom{kn}{n}\sim\sqrt{\frac{k}{2\pi(k-1)n}}\left[\frac{k^k}{(k-1)^{k-1}}\right]^n,\quad \lim_{n\to\infty}\sqrt[n]{\binom{kn}{n}}=\frac{k^k}{(k-1)^{k-1}},\quad \binom{2n}{n}\sim\frac{4^n}{\sqrt{n\pi}}.$$

Source: The second result is given in [1566, pp. 38, 185]. The third result is given in [571, p. 619]. **Remark:** The second result does not imply the first, nor does it imply $\binom{2n}{n}\sim 4^n$, which is false.

Fact 12.18.48. As $n\to\infty$,

$$\binom{2n}{n}\sim\frac{4^n}{\sqrt{n\pi}}\left(1-\frac{1}{8n}+\frac{1}{128n^2}+\frac{5}{1024n^3}-\frac{21}{32768n^4}-\frac{399}{262144n^5}\right),$$

$$\binom{2n}{n}^{-1}\sim\frac{\sqrt{n\pi}}{4^n}\left(1-\frac{1}{8n}+\frac{1}{128n^2}+\frac{5}{1024n^3}-\frac{21}{32768n^4}-\frac{399}{262144n^5}\right),$$

$$\sum_{i=0}^{n}\binom{2i}{i}\sim\frac{4^{n+1}}{3\sqrt{n\pi}}\left(1+\frac{1}{24n}+\frac{59}{384n^2}+\frac{2425}{9216n^3}+\frac{576793}{884736n^4}+\frac{5000317}{2359296n^5}+\frac{953111599}{113246208n^6}\right),$$

$$\sum_{i=0}^{n}C_n\sim\frac{4^{n+1}}{3n\sqrt{n\pi}}\left(1-\frac{5}{8n}+\frac{475}{384n^2}+\frac{1225}{9216n^3}+\frac{395857}{98304n^4}+\frac{27786605}{2359296n^5}+\frac{6798801295}{113246208n^6}\right).$$

Source: [984, 985]. **Remark:** An asymptotic approximation of $\prod_{i=0}^{n}\binom{n}{i}$ is given in [1410].

Fact 12.18.49. Let $\alpha\in(0,1)$, let $f\colon\mathbb{P}\mapsto[0,\infty)$, assume that $\lim_{n\to\infty}f(n)=0$, and assume that, for all $n\ge 1$, $[\alpha+f(n)]n$ is a positive integer. Then,

$$\lim_{n\to\infty}\frac{1}{n}\log\binom{n}{[\alpha+f(n)]n}=\alpha\log\frac{1}{\alpha}+(1-\alpha)\log\frac{1}{1-\alpha}.$$

For example,

$$\lim_{n\to\infty}\frac{1}{n}\log\binom{2n}{n}=2\log 2.$$

Source: Fact 12.18.47 and [2586, p. 38]. **Remark:** The expression for the limit is the *entropy*.

Fact 12.18.50. Let $k\ge 1$. Then, as $n\to\infty$,

$$\sum_{i=0}^{n}\binom{n}{i}^k\sim\left(2^n\sqrt{\frac{2}{\pi n}}\right)^k\sqrt{\frac{\pi n}{2k}},\quad \sum_{i=0}^{n}\binom{n}{i}^2\sim\frac{4^n}{\sqrt{n\pi}},\quad \sum_{i=0}^{n}\binom{n}{i}^3\sim\frac{2\sqrt{3}8^n}{3\pi n}.$$

Source: [771, p. 90] and [1022]. **Remark:** The first asymptotic approximation is an equality for $k = 1$. See Fact 1.16.10. An exact expression for $k = 2$ is given by Fact 1.16.13. No analytical expression exists for $k = 3$. See [1022] and [2228, p. 222].

Fact 12.18.51. Define $\alpha \triangleq \frac{1}{2}(1+\sqrt{5})$. Then, as $n \to \infty$,

$$\lim_{n\to\infty} \frac{n}{\alpha^{5n+5/2}} \sum_{i=0}^{n} \binom{n}{i}^2 \binom{n+i}{n} = \frac{5^{3/4}}{10\pi}.$$

Source: [1411]

Fact 12.18.52.

$$\lim_{n\to\infty} \left[\prod_{i=1}^{n} \binom{n}{i}\right]^{1/n^2} = \sqrt{e}, \quad \lim_{n\to\infty} \frac{\sqrt{n}\sqrt[n]{\prod_{i=1}^{n}\binom{n}{i}}}{e^{n/2}} = \frac{e}{\sqrt{2\pi}}.$$

Source: [1103, pp. 2, 27, 28] and [1158, pp. 154, 155].

Fact 12.18.53. Let $n \ge 1$. Then,

$$\lim_{k\to\infty} \frac{1}{k^2} \sum_{i=0}^{k} \log\binom{nk}{ni} = \frac{n}{2}, \quad \lim_{k\to\infty} \frac{1}{k^2} \sum_{i=0}^{k} \log\binom{k}{i} = \frac{1}{2}.$$

Source: [771, p. 295].

Fact 12.18.54. Let $n \ge 1$ and $k \ge 1$. As $n \to \infty$,

$$\begin{bmatrix} n+1 \\ k+1 \end{bmatrix} \sim \frac{n!}{k!} \log^k n, \quad \begin{Bmatrix} n \\ k \end{Bmatrix} \sim \frac{k^n}{k!}.$$

Source: [771, p. 293].

Fact 12.18.55. Let $a \ge b > 0$, and define the sequences $(a_i)_{i=1}^\infty$ and $(b_i)_{i=1}^\infty$ by $a_1 \triangleq a$, $b_1 = b$, and, for all $i \ge 1$, $a_{i+1} \triangleq \frac{1}{2}(a_i + b_i)$ and $b_{i+1} = \sqrt{a_i b_i}$. Then, $(a_i)_{i=1}^\infty$ is decreasing, $(b_i)_{i=1}^\infty$ is increasing, and thus $\alpha \triangleq \lim_{i\to\infty} a_i$ and $\beta \triangleq \lim_{i\to\infty} b_i$ exist. Furthermore,

$$\alpha = \beta = \frac{\pi(a+b)}{4K(\gamma)},$$

where

$$K(\gamma) \triangleq \int_0^{\pi/2} \frac{1}{\sqrt{1-\gamma^2 \sin^2 x}}\,dx, \quad \gamma \triangleq \frac{a-b}{a+b}.$$

Equivalently,

$$\alpha = \beta = \frac{\pi}{2I(a,b)},$$

where

$$I(a,b) \triangleq \int_0^{\pi/2} \frac{1}{\sqrt{a^2\cos^2 x + b^2 \sin^2 x}}\,dx.$$

Source: [1566, p. 24]. **Remark:** Computational techniques based on iteration are discussed in [518, 519, 634] and [2013, pp. 68–70]. **Remark:** The limit is the *arithmetic-geometric mean* of a and b. $K(\gamma)$ is a *complete elliptic integral of the first kind.* See [116, pp. 132–135] and [511, p. 88]. **Remark:** Note that

$$I(a,b) - \frac{1}{a}K\left(\frac{\sqrt{a^2-b^2}}{a}\right) = \frac{2}{a+b}K\left(\frac{a-b}{a+b}\right).$$

Related: Fact 14.3.14.

Fact 12.18.56. Let $a > 0$ and $b > 0$, and define the sequences $(a_i)_{i=1}^\infty$ and $(b_i)_{i=1}^\infty$ by $a_1 \triangleq \frac{1}{2}(a+b)$, $b_1 \triangleq \sqrt{ab}$, and, for all $i \geq 1$, $a_{i+1} \triangleq \frac{1}{2}(a_i + b_i)$ and $b_{i+1} = \sqrt{a_{i+1}b_i}$. Then,

$$\lim_{i\to\infty} a_i = \lim_{i\to\infty} b_i = \begin{cases} \dfrac{a-b}{\log a - \log b}, & a \neq b, \\ a, & a = b. \end{cases}$$

Source: [635]. **Related:** Fact 2.2.63.

Fact 12.18.57. Let $a_2 = \frac{1}{2}$ and $b_2 = \frac{1}{4}$, and define the sequences $(a_i)_{i=2}^\infty$ and $(b_i)_{i=2}^\infty$ by $a_{i+1} = \sqrt{a_i b_i}$ and $b_{i+1} = \frac{1}{2}(a_{i+1} + b_i)$. Then, the following statements hold:

i) For all $i \geq 3$, $1/a_i$ is the area of a regular 2^i-sided polygon inscribed in a circle of radius 1.

ii) For all $i \geq 3$, $1/b_i$ is the area of a regular 2^i-sided polygon that circumscribes a circle of radius 1.

iii) $\lim_{i\to\infty} a_i = \lim_{i\to\infty} b_i = \pi$.

Remark: As discussed in [518], Archimedes calculated the areas of regular 96-sided inscribed and circumscribed polygons to obtain

$$\frac{25344}{8069} \approx 3.1409 < \pi < \frac{29376}{9347} \approx 3.1428.$$

Fact 12.18.58. As $n \to \infty$, $n! \sim \sqrt{2n\pi}(n/e)^n$,

$$n! \sim \sqrt{2\pi}\left(\frac{n+1/2}{e}\right)^{n+1/2} = \sqrt{2n\pi}\left(\frac{n}{e}\right)^n \frac{1}{\sqrt{e}}\left(1 + \frac{1}{2n}\right)^{n+1/2}$$
$$= \sqrt{2n\pi}\left(\frac{n}{e}\right)^n \frac{1}{\sqrt{e}}\left(1 + \frac{1}{2n}\right)^n \left(1 + \frac{1}{4n} - \frac{1}{32n^2} + \frac{1}{128n^3} - \frac{5}{2048n^4} + \cdots\right).$$

$$n! \sim \sqrt{2n\pi}\left(\frac{n}{e}\right)^n\left(1 + \frac{1}{12n^2}\right)^n = \sqrt{2n\pi}\left(\frac{n}{e}\right)^n\left(1 + \frac{1}{12n} + \frac{n-1}{288n^3} + \cdots + \frac{1}{12^{n-1}n^{2n-1}} + \frac{1}{12^n n^{2n}}\right),$$

$$n! \sim \sqrt{2n\pi}\left(\frac{n}{e}\right)^n \sqrt{\frac{n}{n-1/6}} = \sqrt{2n\pi}\left(\frac{n}{e}\right)^n\left(1 + \frac{1}{12n} + \frac{1}{96n^2} + \frac{5}{3456n^3} + \frac{35}{165888n^4} + \cdots\right).$$

For all $k \geq 1$, as $n \to \infty$,

$$n! \sim \sqrt{2n\pi}\left(\frac{n}{e}\right)^n \sqrt[k]{\sum_{i=0}^\infty \frac{P_i}{n^i}},$$

where $P_0 \triangleq 1$ and, for all $i \geq 1$,

$$P_i \triangleq \frac{k}{i}\sum_{j=1}^{\lfloor (i+1)/2 \rfloor} \frac{B_{2j}}{2j} P_{i-2j+1}.$$

Therefore, for all $k \geq 1$, as $n \to \infty$,

$$n! \sim \sqrt{2n\pi}\left(\frac{n}{e}\right)^n\left[1 + \frac{k}{2^2 3n} + \frac{k^2}{2^5 3^2 n^2} + \left(\frac{k^3}{2^7 3^4} - \frac{k}{2^3 3^2 5}\right)\frac{1}{n^3} + \left(\frac{k^4}{2^{11}3^5} - \frac{k^2}{2^5 3^3 5}\right)\frac{1}{n^4}\right]^{1/k}.$$

In particular, as $n \to \infty$,

$$n! \sim \sqrt{2n\pi}\left(\frac{n}{e}\right)^n\left(1 + \frac{1}{12n} + \frac{1}{288n^2} - \frac{139}{51840n^3} - \frac{571}{2488320n^4}\right),$$

$$n! \sim \sqrt{2n\pi}\left(\frac{n}{e}\right)^n \sqrt{1 + \frac{1}{6n} + \frac{1}{72n^2} - \frac{31}{6480n^3} - \frac{139}{155520n^4}},$$

$$n! \sim \sqrt{2n\pi}\left(\frac{n}{e}\right)^n \sqrt[3]{1 + \frac{1}{4n} + \frac{1}{32n^2} - \frac{11}{1920n^3} - \frac{59}{30720n^4}},$$

$$n! \sim \sqrt{2n\pi}\left(\frac{n}{e}\right)^n \sqrt[4]{1 + \frac{1}{3n} + \frac{1}{18n^2} - \frac{2}{405n^3} - \frac{31}{9720n^4}},$$

$$n! \sim \sqrt{2n\pi}\left(\frac{n}{e}\right)^n \sqrt[6]{1 + \frac{1}{2n} + \frac{1}{8n^2} + \frac{1}{240n^3} - \frac{11}{1920n^4}},$$

$$n! \sim \sqrt{2n\pi}\left(\frac{n}{e}\right)^n \sqrt[24]{1 + \frac{2}{n} + \frac{2}{n^2} + \frac{19}{15n^3} + \frac{8}{15n^4} + \frac{16}{105n^5}}.$$

As $n \to \infty$,

$$n! \sim \sqrt{2n\pi}\left(\frac{n}{e}\right)^n e^{1/(12n)} \sqrt[n]{\sum_{i=0}^{\infty} \frac{Q_{2i}}{n^{2i}}},$$

where $Q_0 \triangleq 1$ and, for all $i \ge 1$,

$$Q_{2i} \triangleq \frac{1}{2i}\sum_{j=1}^{i} \frac{jB_{2j+2}}{(j+1)(2j+1)} Q_{2i-2j}.$$

Hence,

$$n! \sim \sqrt{2n\pi}\left(\frac{n}{e}\right)^n e^{1/(12n)} \sqrt[n]{1 - \frac{1}{360n^2} + \frac{1447}{1814400n^4} - \frac{1170727}{1959552000n^6} + \cdots}.$$

For all $k \ge 1$, as $n \to \infty$,

$$n! \sim \sqrt{2\pi}\left(\frac{n+\frac{1}{2}}{e}\right)^{n+\frac{1}{2}} \sqrt[k]{\sum_{i=0}^{\infty} \frac{R_i}{(n+\frac{1}{2})^i}},$$

where $R_0 \triangleq 1$ and, for all $i \ge 1$,

$$R_i \triangleq \frac{k}{i}\sum_{j=1}^{\lfloor (i+1)/2 \rfloor} \frac{(2^{1-2j}-1)B_{2j}}{2j} R_{i-2j+1}.$$

Hence, as $n \to \infty$,

$$n! \sim \sqrt{2\pi}\left(\frac{n+\frac{1}{2}}{e}\right)^n \sqrt[6]{1 - \frac{1}{4(n+\frac{1}{2})} + \frac{1}{32(n+\frac{1}{2})^2} + \frac{23}{1920(n+\frac{1}{2})^3} + \cdots},$$

$$n! \sim \sqrt{2\pi}\left(\frac{n+\frac{1}{2}}{e}\right)^n \sqrt[12]{1 - \frac{1}{2(n+\frac{1}{2})} + \frac{1}{8(n+\frac{1}{2})^2} + \frac{1}{120(n+\frac{1}{2})^3} + \cdots},$$

$$n! \sim \sqrt{2\pi}\left(\frac{n+\frac{1}{2}}{e}\right)^n \sqrt[24]{1 - \frac{1}{n+\frac{1}{2}} + \frac{1}{2(n+\frac{1}{2})^2} - \frac{13}{120(n+\frac{1}{2})^3} + \cdots}.$$

For all $k \ge 1$, as $n \to \infty$,

$$\log \frac{n!}{\sqrt{2n\pi}}\left(\frac{e}{n}\right)^n \sim \sum_{i=2}^{2k} \frac{B_i}{i(i-1)n^{i-1}}.$$

Hence, as $n \to \infty$,

$$n! \sim \sqrt{2n\pi}\left(\frac{n}{e}\right)^n \exp\left(\sum_{i=2}^{2k} \frac{B_i}{i(i-1)n^{i-1}}\right)$$

$$= \sqrt{2n\pi}\left(\frac{n}{e}\right)^n \exp\left(\frac{B_2}{2n} + \frac{B_4}{12n^3} + \frac{B_6}{30n^5} + \cdots + \frac{B_{2k}}{2k(2k-1)n^{2k-1}}\right)$$

$$= \sqrt{2n\pi}\left(\frac{n}{e}\right)^n \exp\left(\frac{1}{12n} - \frac{1}{360n^3} + \frac{1}{1260n^5} + \cdots + \frac{B_{2k}}{2k(2k-1)n^{2k-1}}\right)$$

$$= \sqrt{2n\pi}\left(\frac{n}{e}\right)^n \left[1 + \frac{1}{12n} - \frac{1}{360n^3} + \frac{1}{1260n^5} + \cdots + \frac{B_{2k}}{2k(2k-1)n^{2k-1}} + \frac{1}{2}\left(\frac{1}{12n} - \frac{1}{360n^3} + \frac{1}{1260n^5} + \cdots + \frac{B_{2k}}{2k(2k-1)n^{2k-1}}\right)^2 + \frac{1}{6}\left(\frac{1}{12n} - \frac{1}{360n^3} + \frac{1}{1260n^5} + \cdots + \frac{B_{2k}}{2k(2k-1)n^{2k-1}}\right)^3 + \frac{1}{24}\left(\frac{1}{12n} - \frac{1}{360n^3} + \frac{1}{1260n^5} + \cdots + \frac{B_{2k}}{2k(2k-1)n^{2k-1}}\right)^4 + \frac{1}{120}\left(\frac{1}{12n} - \frac{1}{360n^3} + \frac{1}{1260n^5} + \cdots + \frac{B_{2k}}{2k(2k-1)n^{2k-1}}\right)^5 + \cdots\right]$$

$$= \sqrt{2n\pi}\left(\frac{n}{e}\right)^n \left(1 + \frac{1}{12n} + \frac{1}{288n^2} - \frac{139}{51840n^3} - \frac{571}{2488320n^4} + \frac{163879}{209018880n^5} + \cdots\right).$$

Finally,

$$n! = \lim_{i\to\infty} \frac{i!i^n}{(n+1)^{\bar{i}}}, \quad \lim_{n\to\infty} \frac{1}{\sqrt[n]{n!}} = 0, \quad \lim_{n\to\infty} \frac{n}{\sqrt[n]{n!}} = e, \quad \lim_{n\to\infty} \frac{\frac{1}{n}\sum_{i=1}^n i}{\sqrt[n]{n!}} = \frac{e}{2},$$

$$\lim_{n\to\infty}\left(\sqrt[n+1]{(n+1)!} - \sqrt[n]{n!}\right) = \frac{1}{e}, \quad \lim_{n\to\infty} \frac{\log n!}{\log n^n} = 1, \quad \lim_{n\to\infty} \frac{4^n(n!)^2}{(2n)!\sqrt{n}} = \sqrt{\pi}.$$

Source: [968, pp. 60–66], [985], [2502, Chapter 1], and Fact 1.13.14. The asymptotic approximation involving the Bernoulli numbers is given in [2013, pp. 125–127] and is based on the Euler summation formula given by Fact 13.1.6 with $f(n) = \log n!$. **Remark:** The first asymptotic approximation for $n!$ is *Stirling's formula*. The second asymptotic approximation for $n!$ is *Burnside's formula*. See [2085]. The next three asymptotic approximations are given in [2086]. **Remark:** For all $n \geq 1$, the series $\sum_{i=2}^{\infty} \frac{B_i}{i(i-1)n^{i-1}}$ diverges. For all $k \geq 1$, the last series (which depends on k) converges for all $n \geq 1$ and provides an asymptotic approximation for $n!$ as $n \to \infty$. This series is an *asymptotic expansion*. See [116, p. 611]. Consequently, for all $k \geq 1$, every truncation of this series provides an asymptotic approximation for $n!$ as $n \to \infty$. Stirling's formula is the first term of this series. However, for all $n \geq 1$, this series diverges as $k \to \infty$. The coefficients of this series are discussed in [563], [771, p. 267], and [2117]. The approximation involving P_i is also an asymptotic expansion. See [985]. **Remark:** Refinements are discussed in [846]. **Related:** Fact 1.13.14.

Fact 12.18.59.

$$\lim_{n\to\infty} \frac{(2n-1)!!(2n+1)!!}{[(2n)!!]^2} = \frac{2}{\pi} \approx 0.6367788676.$$

Source: [644, p. 106]. **Related:** Fact 1.13.15 and Fact 13.10.3.

Fact 12.18.60.

$$\lim_{n\to\infty}\left(\sqrt[n+1]{(n+1)!} - \sqrt[n]{n!}\right) = \frac{1}{e}.$$

Source: [1158, pp. 106, 107] and [2294, p. 9]. **Remark:** A generalization is given in [1103, p. 13]. **Credit:** T. Lalescu.

Fact 12.18.61. Let $(a_i)_{i=1}^\infty \subset [0,\infty)$, and, for all $i \ge 1$, define

$$\beta_i \triangleq \sqrt{a_1 + \sqrt{a_2 + \sqrt{a_3 + \cdots + \sqrt{a_i}}}}.$$

Then, $\lim_{i\to\infty} \beta_i$ exists if and only if $(a_i^{1/2^i})_{i=1}^\infty$ is bounded. **Source:** [1910].

Fact 12.18.62.

$$\sqrt{1+\sqrt{1+\sqrt{1+\sqrt{1+\cdots}}}} = \tfrac{1}{2}(1+\sqrt{5}), \qquad \sqrt{1+2\sqrt{1+3\sqrt{1+4\sqrt{1+\cdots}}}} = 3.$$

Source: The first equality is given in [1035, p. 24]. The second equality, which is due to S. Ramanujan, is given in [977].

Fact 12.18.63. If $a > 1$, then

$$\sqrt{a(a-1)+\sqrt{a(a-1)+\sqrt{a(a-1)+\sqrt{a(a-1)+\cdots}}}} = a.$$

Equivalently, if $a > 0$, then

$$\sqrt{a+\sqrt{a+\sqrt{a+\sqrt{a+\cdots}}}} = \frac{1}{2}(1+\sqrt{4a+1}).$$

In particular,

$$\sqrt{2+\sqrt{2+\sqrt{2+\sqrt{2+\cdots}}}} = 2, \qquad \sqrt{6+\sqrt{6+\sqrt{6+\sqrt{6+\cdots}}}} = 3,$$

$$\sqrt{12+\sqrt{12+\sqrt{12+\sqrt{12+\cdots}}}} = 4, \qquad \sqrt{20+\sqrt{20+\sqrt{20+\sqrt{20+\cdots}}}} = 5.$$

Source: [3009].

Fact 12.18.64. Let $n \ge 2$ and $1 \le m \le n-1$. Then,

$$\sqrt{n(n-m)+m\sqrt{n(n-m)+m\sqrt{n(n-m)+\sqrt{n(n-m)+\cdots}}}} = n.$$

In particular,

$$\sqrt{3+2\sqrt{3+2\sqrt{3+2\sqrt{3+\cdots}}}} = 3, \qquad \sqrt{5+4\sqrt{5+4\sqrt{5+4\sqrt{5+\cdots}}}} = 5,$$

$$\sqrt{4+3\sqrt{4+3\sqrt{4+3\sqrt{4+\cdots}}}} = \sqrt{8+2\sqrt{8+2\sqrt{8+2\sqrt{8+\cdots}}}} = 4,$$

$$\sqrt{15+2\sqrt{15+2\sqrt{15+2\sqrt{15+\cdots}}}} = \sqrt{10+3\sqrt{10+3\sqrt{10+3\sqrt{10+\cdots}}}} = 5.$$

Source: [3009].

Fact 12.18.65. Let $a > 0$. Then,

$$\sqrt{a-\sqrt{a-\sqrt{a-\sqrt{a-\cdots}}}} = \frac{1}{2}(\sqrt{4a+1}-1).$$

Source: [2219].

Fact 12.18.66. Let x be a positive number, let $b \in (0, \frac{1}{2}(\sqrt{5}+1)x)$, and define $a \triangleq x^2 + bx$. Then,

$$\sqrt{a-b\sqrt{a-b\sqrt{a-b\sqrt{a-\cdots}}}} = x.$$

In particular,

$$\sqrt{2-\sqrt{2-\sqrt{2-\sqrt{2-\cdots}}}} = 1, \quad \sqrt{6-\sqrt{6-\sqrt{6-\sqrt{6-\cdots}}}} = 2,$$

$$\sqrt{8-2\sqrt{8-2\sqrt{8-2\sqrt{8-\cdots}}}} = \sqrt{10-3\sqrt{10-3\sqrt{10-3\sqrt{10-\cdots}}}} = 2.$$

Source: [3009].

Fact 12.18.67. Define

$$\alpha \triangleq \frac{2}{\sqrt{\frac{4}{3}\left(2+\sqrt[3]{\frac{25}{2}-\frac{3}{2}\sqrt{69}}+\sqrt[3]{\frac{25}{2}+\frac{3}{2}\sqrt{69}}\right)-3}-1} \approx 1.9903,$$

let x be a positive number, let $b \in (0, \alpha x)$, and define $a \triangleq x^2 + bx + b^2$. Then,

$$\sqrt{a-b\sqrt{a+b\sqrt{a-b\sqrt{a+\cdots}}}} = x, \quad \sqrt{a+b\sqrt{a-b\sqrt{a+b\sqrt{a-\cdots}}}} = x+b.$$

In particular,

$$\sqrt{3-\sqrt{3+\sqrt{3-\sqrt{3+\cdots}}}} = 1, \quad \sqrt{3+\sqrt{3-\sqrt{3+\sqrt{3-\cdots}}}} = 2,$$

$$\sqrt{7-\sqrt{7+\sqrt{7-\sqrt{7+\cdots}}}} = 2, \quad \sqrt{7+\sqrt{7-\sqrt{7+\sqrt{7-\cdots}}}} = 3,$$

$$\sqrt{12-2\sqrt{12+2\sqrt{12-2\sqrt{12+\cdots}}}} = 2, \quad \sqrt{12+2\sqrt{12-2\sqrt{12+2\sqrt{12-\cdots}}}} = 4,$$

$$\sqrt{19-3\sqrt{19+3\sqrt{19-3\sqrt{19+\cdots}}}} = 2, \quad \sqrt{19+3\sqrt{19-3\sqrt{19+3\sqrt{19-\cdots}}}} = 5,$$

$$\sqrt{13-\sqrt{13+\sqrt{13-\sqrt{13+\cdots}}}} = 3, \quad \sqrt{13+\sqrt{13-\sqrt{13+\sqrt{13-\cdots}}}} = 4.$$

Source: [3009].

Fact 12.18.68.

$$\sqrt{6+\sqrt[3]{-7-\sqrt[4]{3-\sqrt{6+\sqrt[3]{-7-\sqrt[4]{3-\cdots}}}}}} = \cdots \sqrt[4]{14+\sqrt[3]{10-\sqrt{6-\sqrt[4]{14+\sqrt[3]{10-\sqrt{6}}}}}} = 2.$$

Source: [1786].

Fact 12.18.69.

$$\sqrt{2-\sqrt{2+\sqrt{2+\sqrt{2-\cdots}}}} = 2\sin\frac{\pi}{18},$$

$$\lim_{i\to\infty} 2^i \underbrace{\sqrt{2-\sqrt{2+\cdots+\sqrt{2+\sqrt{2}}}}}_{i \text{ square roots}} = \pi, \qquad \prod_{i=1}^{\infty}\left(\frac{1}{2}\underbrace{\sqrt{2+\sqrt{2+\cdots+\sqrt{2+\sqrt{2}}}}}_{i \text{ square roots}}\right) = \frac{2}{\pi}.$$

Source: [977] and [2425].

Fact 12.18.70.

$$\sqrt{1+F_2\sqrt{1+F_4\sqrt{1+F_6\sqrt{1+\cdots}}}} = \sqrt{F_2^2+\sqrt{F_4^2+\sqrt{F_8^2+\sqrt{1+\cdots}}}} = 3.$$

Source: [2158, 2147]. **Remark:** F_n is the nth Fibonacci number. **Remark:** It is shown in [1676] that $\sqrt{F_1+\sqrt{F_2+\sqrt{F_3+\sqrt{F_4+\cdots}}}}$ exists.

12.19 Notes

In more standard terminology, f is *holomorphic* on $\mathcal{D}$ if $f'(z)$ exists on $\mathcal{D}$, and f is *analytic* on $\mathcal{D}$ if f has infinitely many derivatives on $\mathcal{D}$ and is equal to its power series expansion in a neighborhood of every point in $\mathcal{D}$. Then, f is holomorphic on $\mathcal{D}$ if and only if $\mathcal{D}$ is analytic on $\mathcal{D}$.

The convergence of a power series on the boundary of its domain of convergence is discussed in [733, pp. 151, 152].

Differentials of functions of a complex matrix are studied in [1424].

Generating functions are often studied as formal power series [166, pp. 10–12], [133], where the variable is not viewed as a complex number and convergence is not considered. The relationship between power series and formal power series is discussed in [1054, pp. 223, 224].

Chapter Thirteen
Infinite Series, Infinite Products, and Special Functions

13.1 Facts on Series for Subset, Eulerian, Partition, Bell, Ordered Bell, Bernoulli, Genocchi, Euler, and Up/Down Numbers

Fact 13.1.1. Let $n \geq 1$, let $\mathcal{D} \subseteq \mathbb{C}$, let $f\colon \mathcal{D} \mapsto \mathbb{C}$, assume that there exists $r > n$ such that f is analytic on $\mathcal{D}_0 \triangleq \{z \in \mathbb{C}\colon |z| < r\} \subseteq \mathcal{D}$, let $(\beta_i)_{i=0}^{\infty} \subset \mathbb{C}$, and assume that, for all $z \in \mathcal{D}_0$, $f(z) = \sum_{i=1}^{\infty} \beta_i z^i$. Then,

$$\sum_{i=0}^{n} (-1)^i \binom{n}{i} f(i) = (-1)^n n! \sum_{i=n}^{\infty} \beta_i \left\{ {i \atop n} \right\}.$$

Source: [555].

Fact 13.1.2. Let $n \geq 1$. Then, for all $k \geq 1$ and $z \in \text{OIUD}$,

$$\frac{z^k}{(1-z)^{n+1}} = \sum_{i=k}^{\infty} \binom{i+n-k}{n} z^i.$$

Now, define $A_n \in \mathbb{F}[s]$ by

$$A_n(s) \triangleq \sum_{i=1}^{n} \left\langle {n \atop i-1} \right\rangle s^i.$$

Then, for all $z \in \text{OIUD}$,

$$\frac{A_n(z)}{(1-z)^{n+1}} = \sum_{i=1}^{\infty} i^n z^i.$$

Source: In [2225], $\left\langle {n \atop i-1} \right\rangle$ is denoted by $A_{n,i}$. **Remark:** $\left\langle {n \atop i} \right\rangle$ is an Eulerian number. See Fact 1.19.5.

Fact 13.1.3. For all $n \geq 1$, let p_n denote the nth partition number. Then,

$$p_n - \frac{1}{\sqrt{2}\pi} \sum_{i=1}^{\infty} \alpha_{n,i} \sqrt{i} \, \frac{\mathrm{d}}{\mathrm{d}x} \left(\frac{\sinh \frac{\sqrt{6}\pi}{3i} \sqrt{x - \frac{1}{24}}}{\sqrt{x - \frac{1}{24}}} \right) \Bigg|_{x=n},$$

where, for all $i \geq 1$,

$$\alpha_{n,i} \triangleq \sum_{j=1}^{i} \operatorname{truth}(\gcd\{i,j\} = 1) e^{(\beta_{i,j} - 2nj/i)\pi \jmath}$$

and, for all $i, j \geq 1$ such that $j \leq i$,

$$\beta_{i,j} \triangleq \sum_{l=1}^{i-1} \left(\frac{l}{i} - \left\lfloor \frac{l}{i} \right\rfloor - \frac{1}{2} \right) \left(\frac{jl}{i} - \left\lfloor \frac{jl}{i} \right\rfloor - \frac{1}{2} \right).$$

Furthermore, as $n \to \infty$,

$$p_n \sim \frac{\sqrt{3}}{12n} e^{\sqrt{\frac{2n}{3}}\pi},$$

Furthermore, $\lim_{n\to\infty} \sqrt[n]{p_n} = 1$. **Source:** [114, p. 70], [118, p. 63], and [747]. **Remark:** The infinite series for p_n is *Rademacher's formula.* **Remark:** The asymptotic expression corrects a misprint in [505, p. 67]. **Related:** Fact 1.20.1.

Fact 13.1.4. For all $n \geq 0$, let $\mathcal{B}_n$ denote the nth Bell number. Then, for all $n \geq 0$,

$$\mathcal{B}_n = \left.\frac{\mathrm{d}^n}{\mathrm{d}z^n} e^{e^z - 1}\right|_{z=0}.$$

Therefore, for all $z \in \mathbb{C}$,

$$e^{e^z-1} = \sum_{i=0}^{\infty} \frac{\mathcal{B}_i}{i!} z^i.$$

Consequently,

$$e^{e-1} = \sum_{i=0}^{\infty} \frac{\mathcal{B}_i}{i!}, \quad e^{\sqrt{e}-1} = \sum_{i=0}^{\infty} \frac{\mathcal{B}_i}{2^i i!}, \quad e^{\sqrt[3]{e}-1} = \sum_{i=0}^{\infty} \frac{\mathcal{B}_i}{3^i i!}.$$

Furthermore, for all $n \geq 1$,

$$\mathcal{B}_n = \frac{1}{e} \sum_{i=0}^{\infty} \frac{i^n}{i!}.$$

In particular,

$$\sum_{i=1}^{\infty} \frac{i}{i!} = e, \quad \sum_{i=1}^{\infty} \frac{i^2}{i!} = 2e, \quad \sum_{i=1}^{\infty} \frac{i^3}{i!} = 5e, \quad \sum_{i=1}^{\infty} \frac{i^4}{i!} = 15e,$$

$$\sum_{i=1}^{\infty} \frac{i^5}{i!} = 52e, \quad \sum_{i=1}^{\infty} \frac{i^6}{i!} = 203e, \quad \sum_{i=1}^{\infty} \frac{i^7}{i!} = 877e, \quad \sum_{i=1}^{\infty} \frac{i^8}{i!} = 4140e.$$

Source: [34, pp. 159, 160] and [571, p. 623]. **Remark:** The series for $\mathcal{B}_n$ is *Dobinski's formula.* **Related:** Fact 1.19.6.

Fact 13.1.5. For all $n \geq 0$, let $\mathcal{O}_n$ denote the nth ordered Bell number. Then, for all $n \geq 0$,

$$\mathcal{O}_n = \left.\frac{\mathrm{d}^n}{\mathrm{d}z^n} \frac{1}{2 - e^z}\right|_{z=0}.$$

Therefore, for all $z \in \mathbb{C}$ such that $|z| < \log 2$,

$$\frac{1}{2 - e^z} = \sum_{i=0}^{\infty} \frac{\mathcal{O}_i}{i!} z^i.$$

Consequently,

$$\frac{1}{2 - \sqrt{e}} = \sum_{i=0}^{\infty} \frac{\mathcal{O}_i}{2^i i!}, \quad \frac{1}{2 - \sqrt[3]{e}} = \sum_{i=0}^{\infty} \frac{\mathcal{O}_i}{3^i i!}.$$

Finally, as $n \to \infty$,

$$\mathcal{O}_n \sim \frac{n!}{2(\log 2)^{n+1}}, \quad \lim_{n\to\infty} \frac{n\mathcal{O}_{n-1}}{\mathcal{O}_n} = \log 2.$$

Source: [899, 2100]. **Related:** Fact 1.19.7.

Fact 13.1.6. Define $f\colon \mathbb{R} \mapsto \mathbb{R}$ by

$$f(x) \triangleq \begin{cases} \dfrac{x}{e^x - 1}, & x \neq 0, \\ 1, & x = 0, \end{cases}$$

and define the sequence $(B_i)_{i=0}^{\infty} \subset \mathbb{R}$ by $B_i \triangleq f^{(i)}(0)$. Then, the following statements hold:

i) For all $z \in \mathbb{C}$ such that $|z| < 2\pi$,

$$\frac{z}{e^z - 1} = \sum_{i=0}^{\infty} \frac{B_i}{i!} z^i.$$

ii) $(B_i)_{i=0}^{21} = (1, -\frac{1}{2}, \frac{1}{6}, 0, -\frac{1}{30}, 0, \frac{1}{42}, 0, -\frac{1}{30}, 0, \frac{5}{66}, 0, \frac{-691}{2730}, 0, \frac{7}{6}, 0, \frac{-3617}{510}, 0, \frac{43867}{798}, 0, \frac{-174611}{330}, 0)$.

iii) For all $n \geq 1$, $B_{2n+1} = 0$ and $(-1)^{n+1} B_{2n} > 0$.

iv) For all $n \geq 1$,

$$B_{2n} + \sum_{p-1|2n} \frac{1}{p}$$

is an integer, where the sum is taken over all primes p such that $p - 1$ divides $2n$.

v) For all $z \in \mathbb{C}$ such that $|z| < 2\pi$,

$$\frac{z}{e^z - 1} = 1 - \frac{z}{2} + \sum_{i=1}^{\infty} \frac{B_{2i}}{(2i)!} z^{2i}.$$

Therefore,

$$\frac{1}{e - 1} = \frac{1}{2} + \sum_{i=1}^{\infty} \frac{B_{2i}}{(2i)!}, \quad \frac{1}{\sqrt{e} - 1} = \frac{3}{2} + \sum_{i=1}^{\infty} \frac{B_{2i}}{2^{2i-1}(2i)!}, \quad \frac{1}{\sqrt[3]{e} - 1} = \frac{5}{2} + \sum_{i=1}^{\infty} \frac{B_{2i}}{3^{2i-1}(2i)!}.$$

vi) For all $z \in \mathbb{C}$ such that $|z| < \pi$,

$$1 - \frac{z}{2} \cot \frac{z}{2} = \sum_{i=1}^{\infty} \frac{|B_{2i}|}{(2i)!} z^{2i}.$$

vii) For all $n \geq 1$,

$$\sum_{i=0}^{n} \binom{n+1}{i} B_i = \sum_{i=1}^{n} \binom{n}{i} B_{n-i} = \sum_{i=0}^{n} \binom{n}{i} (n+i) B_{n-1+i} = 0, \quad B_n = (-1)^n \sum_{i=0}^{n} \binom{n}{i} B_i,$$

$$\sum_{i=0}^{n} \binom{2n+1}{2i} (1 - 2^{2i-1}) B_{2i} = 0, \quad \sum_{i=1}^{n} 2^{2i-1} \binom{2n+1}{2i} B_{2i} = n, \quad \sum_{i=1}^{n} \binom{2n+1}{2i} B_{2i} = n - \tfrac{1}{2},$$

$$B_{n+1} = (n+1) \sum_{i=0}^{n} \binom{n}{i} \frac{B_i}{n+2-i} = -\sum_{i=0}^{n} \binom{n+1}{i} \frac{B_i}{n+2-i} = -\frac{1}{n+2} \sum_{i=0}^{n} \binom{n+2}{i} B_i,$$

$$B_{2n} = -\frac{1}{(n+1)(2n+1)} \sum_{i=0}^{n-1} \binom{n+1}{i} (n+1+i) B_{n+i},$$

$$B_{2n} = \frac{1}{2} - \frac{1}{2n+1} \sum_{i=0}^{n-1} \binom{2n+1}{2i} B_{2i} = \frac{1}{2n+1} - \frac{1}{(n+1)(2n+1)} \sum_{i=0}^{n-1} \binom{2n+2}{2i} B_{2i}.$$

viii) For all $n \geq 0$,

$$B_n = \sum_{i=0}^{n} \frac{1}{i+1} \sum_{j=0}^{i} (-1)^j \binom{i}{j} j^n = \sum_{i=1}^{n} (-1)^i \frac{i!}{i+1} \left\{ {n \atop i} \right\}.$$

For all $n \geq 1$,

$$B_n = \frac{1}{n+1}\sum_{i=1}^{n}\sum_{j=1}^{i}(-1)^j j^n \frac{\binom{n+1}{i-j}}{\binom{n}{i}} = -\sum_{i=1}^{n+1}(-1)^i \frac{1}{i}\binom{n+1}{i}\sum_{j=1}^{i} j^n,$$

$$B_n = (-1)^n \frac{n}{2^n-1}\sum_{i=1}^{n}\frac{1}{2^i}\sum_{j=0}^{i-1}(-1)^j\binom{k-1}{j}(j+1)^{n-1},$$

$$B_n = \sum_{i=0}^{n}(-1)^i \frac{n!}{(n+i)!}\binom{n+1}{i+1}\sum_{j=0}^{i}(-1)^{i-j}\binom{i}{j}j^{n+i}, \quad B_{2n} = \sum_{i=2}^{2n+1}(-1)^{i-1}\frac{1}{i}\binom{2n+1}{i}\sum_{j=1}^{i-1} j^{2n}.$$

ix) For all $n, m \geq 0$,

$$(-1)^m \sum_{i=0}^{m}\binom{m}{i}B_{n+i} = (-1)^n \sum_{i=0}^{n}\binom{n}{i}B_{m+i}.$$

x) For all $n \geq 1$,

$$\sum_{i=0}^{n}\begin{bmatrix} n \\ i \end{bmatrix} B_i = -\frac{(n-1)!}{n+1}.$$

xi) For all $n \geq 2$,

$$B_{2n} = -\sum_{i=1}^{n-1}\frac{2^{2i}-1}{2^{2n}-1}\binom{2n}{2i}B_{2i}B_{2n-2i} = -\frac{1}{2n+1}\sum_{i=1}^{n-1}\binom{2n}{2i}B_{2i}B_{2n-2i} = (-1)^{n-1}\frac{(2n)!\zeta(2n)}{2^{2n-1}\pi^{2n}}.$$

Therefore, as $n \to \infty$,

$$B_{2n} \sim (-1)^{n+1}\frac{2(2n)!}{(2\pi)^{2n}}.$$

xii) As $n \to \infty$,

$$|B_n| \sim 4\sqrt{n\pi}\left(\frac{n}{\pi e}\right)^{2n}, \quad B_{2n} \sim \frac{(-1)^{n+1}4n^{2n}\sqrt{n\pi}}{(\pi e)^{2n}}.$$

xiii) For all $n \geq 0$,

$$\sum_{i=0}^{n}\frac{4^i B_{2i}}{(2i)!(2n+1-2i)!} = \frac{1}{(2n)!}, \quad \sum_{i=0}^{2n+1}(-1)^i \frac{B_{4n-2i+2}B_{2i}}{(2i)!(4n-2i+2)!} = 0,$$

$$\sum_{i=0}^{n}(-4)^i\left[\frac{B_{4n-4i+2}B_{4i}}{(4i)!(4n-4i+2)!} + 2\frac{B_{4n-4i}B_{4i+2}}{(4n-4i)!(4i+2)!}\right] = 0.$$

xiv) For all $n \geq 1$,

$$\zeta(-n) = -\frac{B_{n+1}}{n+1}.$$

Furthermore, $\zeta(0) = B_1 = -\frac{1}{2}$.

xv) For all $n \geq 4$,

$$\sum_{i=2}^{n-2}\left[1-\binom{n}{i}\right]\frac{B_i B_{n-i}}{i(n-i)} = \frac{2H_n B_n}{n}, \quad \sum_{i=2}^{n-2}\left[n+2-2\binom{n+2}{i}\right]B_i B_{n-i} = n(n+1)B_n.$$

xvi) For all $n \ge 1$,

$$B_n = n!\det\begin{bmatrix} 1 & 0 & \cdots & 0 & 1\\ \frac{1}{2!} & 1 & \ddots & 0 & 0\\ \vdots & \ddots & \ddots & \ddots & \vdots\\ \frac{1}{n!} & \frac{1}{(n-1)!} & \ddots & 1 & 0\\ \frac{1}{(n+1)!} & \frac{1}{n!} & \cdots & \frac{1}{2!} & 0 \end{bmatrix} = (-1)^n n!\det\begin{bmatrix} \frac{1}{2!} & 1 & \ddots & 0 & 0\\ \frac{1}{3!} & \frac{1}{2!} & \ddots & 0 & 0\\ \vdots & \ddots & \ddots & \ddots & \vdots\\ \frac{1}{n!} & \frac{1}{(n-1)!} & \ddots & \frac{1}{2!} & 1\\ \frac{1}{(n+1)!} & \frac{1}{n!} & \cdots & \frac{1}{3!} & \frac{1}{2!} \end{bmatrix},$$

$$B_{2n} = -\frac{(2n)!}{4^n-2}\det\begin{bmatrix} 1 & 0 & 0 & \cdots & 0 & 1\\ \frac{1}{3!} & 1 & 0 & \cdots & 0 & 0\\ \frac{1}{5!} & \frac{1}{3!} & 1 & \ddots & 0 & 0\\ \vdots & \ddots & \ddots & \ddots & \ddots & \vdots\\ \frac{1}{(2n-1)!} & \frac{1}{(2n-3)!} & \frac{1}{(2n-5)!} & \cdots & 1 & 0\\ \frac{1}{(2n+1)!} & \frac{1}{(2n-1)!} & \frac{1}{(2n-3)!} & \cdots & \frac{1}{3!} & 0 \end{bmatrix}$$

$$= (-1)^n\frac{(2n)!}{4^n-2}\det\begin{bmatrix} \frac{1}{3!} & 1 & 0 & \cdots & 0 & 0\\ \frac{1}{5!} & \frac{1}{3!} & 1 & \ddots & 0 & 0\\ \frac{1}{7!} & \frac{1}{5!} & \frac{1}{3!} & \ddots & 0 & 0\\ \vdots & \ddots & \ddots & \ddots & \ddots & \vdots\\ \frac{1}{(2n-1)!} & \frac{1}{(2n-3)!} & \frac{1}{(2n-5)!} & \cdots & \frac{1}{3!} & 1\\ \frac{1}{(2n+1)!} & \frac{1}{(2n-1)!} & \frac{1}{(2n-3)!} & \cdots & \frac{1}{5!} & \frac{1}{3!} \end{bmatrix}.$$

xvii) For all $n \ge 1$,

$$B_{2n} = (-1)^{n+1}2^{2n+1}\int_{-\infty}^{\infty}\left(\frac{\mathrm{d}^{n-1}}{\mathrm{d}x^{n-1}}\operatorname{sech}^2 x\right)^2 \mathrm{d}x.$$

xviii) For all $n \ge 1$,

$$B_n = n!\sum(-1)^{\sum_{i=1}^n k_i}\binom{\sum_{i=1}^n k_i}{k_1,\dots,k_n}\prod_{i=2}^{n+1}\left(\frac{1}{i!}\right)^{k_{i-1}} = \frac{(2n)!}{4^n-2}\sum(-1)^{\sum_{i=1}^n k_i}\binom{\sum_{i=1}^n k_i}{k_1,\dots,k_n}\prod_{i=1}^{n}\left(\frac{1}{(2i+1)!}\right)^{k_i},$$

where both sums are taken over all $(k_1,\dots,k_n)$ such that $\sum_{i=1}^n ik_i = n$.

xix) For all $n \ge 1$,

$$\sum_{i=1}^n \frac{4^i(4^i-1)d_{n-1,i-1}|B_{2i}|}{i} = 2(2n-1)!,$$

where $d_{n,n} \triangleq 1$ and, for all $0 \le i < n$, $d_{n,i} \triangleq 4\sum\prod_{j=1}^{n-i} j_i^2$, where the sum is taken over all subsets $\{j_1,\dots,j_{n-i}\}$ of $\{1,\dots,n\}$.

xx) Let $n, m \ge 1$, and assume that m is odd. Then,

$$\sum_{i=0}^{n+m}\binom{n+m}{i}\binom{n+m+i}{m}B_{n+i} = 0.$$

Source: [968, Chapter 11]. *iv*) is given in [155, p. 275]; *vi*) is given in [2504, pp. 142, 143]; *vii*) is given in [155, pp. 265, 275], [348], [511, p. 100], [2468], and [2513, p. 129]; *viii*) is given in [155,

p. 275], [555], [968, p. 308], and [1209]; *ix*) is given in [2267]; *x*) is given in [2880, pp. 137, 138]; *xi*) is given in [155, p. 267], [511, p. 131], and [2513, p. 403]; *xii*) is given in [155, p. 267] and [2504, pp. 142, 143]; *xiii*) is given in [155, p. 275] and [2792]; *xiv*) is given in [155, p. 266]; *xv*) is given in [2572]; *xvi*) is given in [702, 1930, 2016]; *xvii*) is given in [2068, p. 384]; *xviii*) is given in [2016]; *xix*) is given in [1149]; *xx*) is given in [348]. **Remark:** B_n is the nth *Bernoulli number*. See [1219, pp. 283–290, 367]. **Related:** Fact 1.12.1, Fact 3.16.29, and Fact 13.2.1.

Fact 13.1.7. Define $f\colon \mathbb{R} \mapsto \mathbb{R}$ by

$$f(x) \triangleq \frac{2x}{e^x + 1},$$

and define the sequence $(G_i)_{i=0}^{\infty} \subset \mathbb{R}$ by $G_i \triangleq f^{(i)}(0)$. Then, the following statements hold:

i) For all $z \in \mathbb{C}$ such that $|z| < \pi$,

$$\frac{2z}{e^z + 1} = \sum_{i=0}^{\infty} \frac{G_i}{i!} z^i.$$

ii) $(G_i)_{i=0}^{21} = (0, 1, -1, 0, 1, 0, -3, 0, 17, 0, -155, 0, 2073, 0, -38227, 0, 929569, 0, -28820619)$.

iii) For all $n \ge 0$, $G_n = 2(1 - 2^n)B_n$, where B_n is the nth Bernoulli number.

iv) If $n \ge 0$, then $G_{2n+1} = 0$ and $(-1)^n G_{2n}$ is an odd positive integer.

v) If $n \ge 2$, then

$$G_{2n} = -n - \frac{1}{2}\sum_{i=1}^{n-1}\binom{2n}{2i}G_{2i}, \quad G_{2n} = -1 - \sum_{i=1}^{n-1}\binom{2n}{2i-1}\frac{G_{2i}}{2i}.$$

Source: [2513, pp. 123, 124]. **Related:** Fact 14.6.45.

Fact 13.1.8. Define $f\colon \mathbb{R} \mapsto \mathbb{R}$ by

$$f(x) \triangleq \frac{2e^x}{e^{2x} + 1} = \frac{1}{\cosh x} = \operatorname{sech} x,$$

and define the sequence $(E_i)_{i=0}^{\infty} \subset \mathbb{R}$ by $E_i \triangleq f^{(i)}(0)$. Then, the following statements hold:

i) For all $z \in \mathbb{C}$ such that $|z| < \pi$,

$$\frac{2e^z}{e^{2z} + 1} = \sum_{i=0}^{\infty} \frac{E_i}{i!} z^i.$$

ii) $(E_i)_{i=0}^{15} = (1, 0, -1, 0, 5, 0, -61, 0, 1385, 0, -50521, 0, 2702765, 0, -199360981, 0)$.

iii) For all $n \ge 0$, $E_{2n+1} = 0$. For all $n \ge 1$, $E_{4n} \overset{60}{\equiv} 5$ and $E_{4n+2} \overset{60}{\equiv} -1$.

iv) For all $n \ge 1$,

$$\sum_{i=0}^{n} \operatorname{rem}_2(n - i + 1)\binom{n}{i}E_i = \sum_{i=0}^{n}\binom{2n}{2i}E_{2i} = 0.$$

v) For all $n \ge 1$,

$$E_{2n} = (2n)!\det\begin{bmatrix} 1 & 0 & 0 & \cdots & 0 & 1 \\ \frac{1}{2!} & 1 & 0 & \cdots & 0 & 0 \\ \frac{1}{4!} & \frac{1}{2!} & 1 & \ddots & 0 & 0 \\ \vdots & \ddots & \ddots & \ddots & \ddots & \vdots \\ \frac{1}{(2n-2)!} & \frac{1}{(2n-4)!} & \frac{1}{(2n-5)!} & \cdots & 1 & 0 \\ \frac{1}{(2n)!} & \frac{1}{(2n-2)!} & \frac{1}{(2n-4)!} & \cdots & \frac{1}{2!} & 0 \end{bmatrix}.$$

vi) For all $n \geq 1$, let $A_n \in \mathbb{R}^{2n\times 2n}$, where, for all $i, j \in \{1,\dots,2n\}$, $A_{(i,j)} \triangleq \binom{i}{j-1}\cos\frac{1}{2}(i-j+1)\pi$. Then, for all $n \geq 1$, $E_{2n} = (-1)^n \det A_n$.

vii) For all $n \geq 1$,

$$E_{2n} = (2n+1)\sum_{i=1}^{2n}(-1)^i \frac{1}{2^i(i+1)}\binom{2n}{i}\sum_{j=0}^{i}\binom{i}{j}(2j-i)^{2n} = \sum_{i=1}^{2n}(-1)^i\frac{1}{2^i}\sum_{j=0}^{2i}(-1)^j\binom{2i}{j}(i-j)^{2n}.$$

viii) For all $n \geq 1$,

$$E_n = 1+\sum_{i=1}^{n}\frac{(i+1)!}{2^i}\left\{ {n \atop i} \right\}\sum_{j=1}^{i}(-1)^j\frac{2^j}{j+1}\binom{j+1}{i-j} = 1+\sum_{i=1}^{n}\frac{(-1)^i}{i+1}\sum_{j=0}^{n-i}\frac{(i+j+1)!}{2^j}\binom{i+1}{j}\left\{ {n \atop i+j} \right\}.$$

ix) For all $n \geq 1$,

$$\sum_{i=0}^{n} e_{n,i}|E_{2i}| = (2n)!,$$

where $e_{n,n} \triangleq 1$ and, for all $i \in [0, n-1]$, $e_{n,i} \triangleq \sum\prod_{k=1}^{n-i}(2j_k-1)^2$, where the sum is taken over all subsets $\{j_1,\dots,j_{n-i}\}$ of $\{1,\dots,n\}$.

Source: [771, pp. 48, 49], [1149], [1219, p. 559], [1412] [1524, p. 40], and [1930, 2849]. **Remark:** E_n is the nth *Euler number*. **Related:** Fact 13.2.3.

Fact 13.1.9. Define $f\colon \{x \in \mathbb{R}\colon |x| < \frac{\pi}{2}\} \mapsto \mathbb{R}$ by

$$f(x) \triangleq \sec x + \tan x,$$

and define the sequence $(U_i)_{i=0}^{\infty} \subset \mathbb{R}$ by $U_i \triangleq f^{(i)}(0)$. Then, the following statements hold:

i) For all $z \in \mathbb{C}$ such that $|z| < \frac{\pi}{2}$,

$$\sec z + \tan z = \sum_{i=0}^{\infty}\frac{U_i}{i!}z^i.$$

ii) $(U_i)_{i=0}^{13} = (1, 1, 1, 2, 5, 16, 61, 272, 1385, 7936, 50521, 353792, 2702765, 22368256)$.

iii) For all $i \geq 1$, U_i is the number of permutations $(i_1,\dots,i_n)$ of $(1,\dots,n)$ such that $i_1 < i_2 < i_3 < i_4 \cdots$.

Source: [1368]. **Remark:** U_n is the nth *up/down number*. **Example:** $(1,3,2)$ and $(2,3,1)$ are up/down permutations of $(1,2,3)$.

13.2 Facts on Bernoulli, Euler, Chebyshev, Legendre, Laguerre, Hermite, Bell, Ordered Bell, Harmonic, Fibonacci, and Lucas Polynomials

Fact 13.2.1. For all $n \geq 0$, define $B_{\mathrm{p},n} \in \mathbb{R}[s]$ by

$$B_{\mathrm{p},n}(x) \triangleq \left.\frac{\mathrm{d}^n}{\mathrm{d}z^n}\frac{ze^{xz}}{e^z-1}\right|_{z=0},$$

where $\left.\frac{ze^{xz}}{e^z-1}\right|_{z=0} \triangleq 1$. Then, the following statements hold:

i) For all $x, z \in \mathbb{C}$ such that $|z| < 2\pi$,

$$\frac{ze^{xz}}{e^z-1} = \sum_{i=0}^{\infty}\frac{B_{\mathrm{p},i}(x)}{i!}z^i.$$

ii) For all $n \geq 0$ and $z \in \mathbb{C}$,

$$B_{\mathrm{p},n}(z) \triangleq \sum_{i=0}^{n} B_{n-i}\binom{n}{i}z^i = \sum_{i=0}^{n} B_i\binom{n}{i}z^{n-i} = \sum_{i=0}^{n}\frac{1}{i+1}\sum_{j=0}^{i}(-1)^j\binom{i}{j}(z+j)^n.$$

In particular,

$$B_{\mathrm{p},0}(z) = 1, \quad B_{\mathrm{p},1}(z) = z - \tfrac{1}{2}, \quad B_{\mathrm{p},2}(z) = z^2 - z + \tfrac{1}{6}, \quad B_{\mathrm{p},3}(z) = z^3 - \tfrac{3}{2}z^2 + \tfrac{1}{2}z,$$
$$B_{\mathrm{p},4}(z) = z^4 - 2z^3 + z^2 - \tfrac{1}{30}, \quad B_{\mathrm{p},5}(z) = z^5 - \tfrac{5}{2}z^4 + \tfrac{5}{3}z^3 - \tfrac{1}{6}z, \quad B_{\mathrm{p},6}(z) = z^6 - 3z^5 + \tfrac{5}{2}z^4 - \tfrac{1}{2}z^2 + \tfrac{1}{42}.$$

iii) For all $n \geq 1$, $B_{\mathrm{p},n}(0) = B_n$, $B_{\mathrm{p},2n}(\frac{1}{2}) = (2^{1-2n} - 1)B_n$, and $B_{\mathrm{p},2n+1}(\frac{1}{2}) = 0$. For all $n \geq 2$, $B_{\mathrm{p},n}(1) = B_{\mathrm{p},n}(0) = B_n$.

iv) For all $n \geq 1$ and $z \in \mathbb{C}$,

$$z^n = \sum_{i=1}^{n}\binom{n}{i}\frac{1}{n-i+1}B_{\mathrm{p},i}(z) = \frac{1}{n+1}\sum_{i=0}^{n}\binom{n+1}{i}B_{\mathrm{p},i}(z),$$

$$B_{\mathrm{p},n}(1-z) = (-1)^n B_{\mathrm{p},n}(z), \quad B_{\mathrm{p},n}(z+1) = B_{\mathrm{p},n}(z) + nz^{n-1}, \quad B'_{\mathrm{p},n}(z) = nB_{\mathrm{p},n-1}(z).$$

v) Let $n \geq 0$ and $a, b \in \mathbb{R}$. Then,

$$\int_a^b B_{\mathrm{p},n}(x)\,\mathrm{d}x = \frac{B_{\mathrm{p},n+1}(b) - B_{\mathrm{p},n+1}(a)}{n+1}, \qquad \int_a^{a+1} B_{\mathrm{p},n}(x)\,\mathrm{d}x = a^n.$$

vi) For all $n, m \geq 1$,

$$\int_0^1 B_{\mathrm{p},n}(x)B_{\mathrm{p},m}(x)\,\mathrm{d}x = (-1)^{n+m-1}\frac{B_{n+m}}{\binom{n+m}{n}}.$$

vii) Let $n, m \geq 1$ and $z \in \mathbb{C}$. Then,

$$B_{\mathrm{p},n}(mz) = m^{n-1}\sum_{i=0}^{m-1} B_{\mathrm{p},n}(z + \tfrac{i}{m}).$$

In particular,

$$B_{\mathrm{p},n}(2z) = 2^{n-1}[B_{\mathrm{p},n}(z) + B_{\mathrm{p},n}(z + \tfrac{1}{2})].$$

viii) Let $n \geq 0$ and $x, y \in \mathbb{C}$. Then,

$$B_{\mathrm{p},n}(x+y) = \sum_{i=0}^{n}\binom{n}{i}B_{\mathrm{p},i}(x)y^{n-i}.$$

ix) For all $n \geq 0$ and $z \in \mathbb{C}$,

$$B_{\mathrm{p},n+1}(z) = B_{n+1} + \sum_{i=0}^{n}\frac{n+1}{i+1}\begin{Bmatrix} n \\ i \end{Bmatrix} z^{\underline{i+1}}, \qquad z^{\underline{n+1}} = \sum_{i=0}^{n}\frac{n+1}{i+1}\begin{bmatrix} n \\ i \end{bmatrix}[B_{\mathrm{p},i+1}(z) - B_{i+1}].$$

x) Let $n, k \geq 1$. Then,

$$\sum_{i=1}^{n} i^k = \frac{B_{\mathrm{p},k+1}(n+1) - B_{\mathrm{p},k+1}(0)}{k+1}.$$

xi) Let k, l be integers, where $k < l$, let $f\colon [k, l] \mapsto \mathbb{R}$, let m be a positive integer, and assume that f is C^m. Then,

$$\sum_{i=k}^{l-1} f(i) = \int_k^l f(x)\,\mathrm{d}x + \sum_{i=1}^{m}\frac{B_i}{i!}[f^{(i-1)}(l) - f^{(i-1)}(k)] + R_m,$$

where

$$R_m \triangleq (-1)^{m+1}\int_k^l \frac{B_{\mathrm{p},m}(x-\lfloor x\rfloor)}{m!}f^{(m)}(x)\,\mathrm{d}x.$$

Source: *ii*) is given in [645, p. 100]; *iii*) is given in [155, pp. 265, 274], [771, p. 48], and [2068, p. 361]; *iv*) is given in [155, p. 274] and [771, pp. 48, 49, 165]; *vi*) is given in [2068, p. 369]; *vii*) and *viii*) are given in [155, p. 275]; *xi*) is given in [1219, pp. 469, 470]. **Remark:** $B_{\mathrm{p},n}$ is the nth *Bernoulli polynomial.* See [1219, pp. 283–290, 367] and [2264, pp. 112–117]. B_n denotes the nth Bernoulli number. *x*) is the *Euler summation formula.* See [645, pp. 246–248] and [1219, pp. 469, 470]. **Related:** Fact 1.12.1 and Fact 13.1.6.

Fact 13.2.2. For all $n \ge 1$, define $R_n\colon (0,1) \mapsto \mathbb{R}$ by

$$R_n(x) \triangleq \left.\frac{\mathrm{d}^{n-1}}{\mathrm{d}z^{n-1}}\frac{1}{e^z+1}\right|_{z=\log(1/x-1)}.$$

Then, for all $n \ge 1$, R_n is a polynomial on $[0,1]$ with real coefficients. Hence, let $R_n \in \mathbb{R}[s]$ denote the extension of R_n to $\mathbb{C}$. Then, the following statements hold:

i) For all $n \ge 0$ and $z \in \mathbb{C}$,

$$R_{n+1}(z) = \sum_{i=1}^{n+1} a_{n,i}z^i,$$

where

$$a_{n,i} \triangleq (-1)^n\sum_{j=0}^{i-1}(-1)^j\binom{i-1}{j}(j+1)^n = (-1)^{n+k-1}(k-1)!\left\{\begin{matrix} n+1\\ k\end{matrix}\right\}.$$

ii) For all $z \in \mathbb{C}$ and $n \ge 1$,

$$R_{n+1}(z) = (z^2-z)R_n'(z).$$

In particular,

$$R_1(z) = z,\quad R_2(z) = z^2-z,\quad R_3(z) = 2z^3-3z^2+z,\quad R_4(z) = 6z^4-12z^3+7z^2-z,$$
$$R_5(z) = 24z^5-60z^4+50z^3-15z^2+z,\quad R_6(z) = 120z^6-360z^5+390z^4-180z^3+31z^2-z.$$

iii) If $n \ge 2$ and $z \in [0,1]$, then $R_n(z) = (-1)^nR_n(1-z)$. Hence, if $n \ge 3$ is odd, then $R_n(\tfrac{1}{2}) = 0$.

iv) If $n \ge 2$, then R_n has n nonrepeated roots, all of which are contained in $[0,1]$.

v) Let $x \in (0,1)$, and define $\alpha \triangleq \sqrt{\pi^2+\log^2\frac{x}{1-x}}$. Then, for all $z \in \mathbb{C}$ such that $|z| < \alpha$,

$$\frac{x}{x+(1-x)e^z} = \sum_{i=0}^{\infty}\frac{R_{i+1}(x)}{i!}z^i.$$

vi) For all $n \ge 1$,

$$\int_0^1 R_n(x)\,\mathrm{d}x = -B_n.$$

vii) For all $x \in \mathbb{R}$, define $f(x) \triangleq 1/(e^x+1)$. Then, for all $n \ge k \ge 1$,

$$\int_0^1 \frac{R_{n-k+1}(x)R_{k+1}(x)}{x^2-x}\,\mathrm{d}x = (-1)^kB_n,\qquad \int_{-\infty}^{\infty} f^{(n-k)}(x)f^{(k)}(x)\,\mathrm{d}x = (-1)^{k+1}B_n.$$

Source: [2354, 2355].

Fact 13.2.3. For all $n \ge 0$, define $E_{\mathrm{p},n} \in \mathbb{R}[s]$ by

$$E_{\mathrm{p},n}(x) \triangleq \left.\frac{\mathrm{d}^n}{\mathrm{d}z^n}\frac{2e^{xz}}{e^z+1}\right|_{z=0}.$$

Then, the following statements hold:

i) For all $x, z \in \mathbb{C}$ such that $|z| < \pi$,

$$\frac{2e^{xz}}{e^z + 1} = \sum_{i=0}^{\infty} \frac{E_{\mathrm{p},i}(x)}{i!} z^i.$$

ii) For all $n \geq 0$ and $z \in \mathbb{C}$,

$$E_{\mathrm{p},n}(z) \triangleq \sum_{i=0}^{n} \binom{n}{i} \frac{E_n}{2^i} (z - \tfrac{1}{2})^{n-i} = \sum_{i=0}^{n} \frac{1}{2^i} \sum_{j=0}^{i} (-1)^j \binom{i}{j} (z + j)^n.$$

In particular,

$$E_{\mathrm{p},0}(z) = 1, \quad E_{\mathrm{p},1}(z) = z - \tfrac{1}{2}, \quad E_{\mathrm{p},2}(z) = z^2 - z, \quad E_{\mathrm{p},3}(z) = z^3 - \tfrac{3}{2}z^2 + \tfrac{1}{4},$$

$$E_{\mathrm{p},4}(z) = z^4 - 2z^3 + z, \quad E_{\mathrm{p},5}(z) = z^5 - \tfrac{5}{2}z^4 + \tfrac{5}{2}z^2 - \tfrac{1}{2}, \quad E_{\mathrm{p},6}(z) = z^6 - 3z^5 + 5z^3 - 3z.$$

iii) For all $n \geq 1$, $E_n = 2^n E_{\mathrm{p},n}(\tfrac{1}{2})$.

iv) For all $n \geq 1$ and $z \in \mathbb{C}$,

$$z^n = E_{\mathrm{p},n}(z) + \frac{1}{2} \sum_{i=0}^{n-1} \binom{n}{i} E_{\mathrm{p},i}(z), \quad E_{\mathrm{p},n}(1 - z) = (-1)^n E_{\mathrm{p},n}(z),$$

$$E_{\mathrm{p},n}(z + 1) = E_{\mathrm{p},n}(z) + 2z^n, \quad E'_{\mathrm{p},n}(z) = nE_{\mathrm{p},n-1}(z).$$

v) Let $n \geq 0$ and $x, y \in \mathbb{C}$. Then,

$$E_{\mathrm{p},n}(x + y) = \sum_{i=0}^{n} \binom{n}{i} E_{\mathrm{p},i}(x) y^{n-i}.$$

vi) Let $n \geq 0$ and $a, b \in \mathbb{R}$. Then,

$$\int_a^b E_{\mathrm{p},n}(x)\, \mathrm{d}x = \frac{E_{\mathrm{p},n+1}(b) - E_{\mathrm{p},n+1}(a)}{n + 1}.$$

vii) Let $n, m \geq 1$ and $x \in \mathbb{C}$. If n is odd, then

$$E_{\mathrm{p},n}(mx) = m^n \sum_{i=1}^{m-1} (-1)^i E_{\mathrm{p},n}(x + \tfrac{i}{m}).$$

If n is even, then

$$E_{\mathrm{p},n}(mx) = \frac{-2m^n}{n + 1} \sum_{i=1}^{m-1} (-1)^i B_{\mathrm{p},i+1}(x + \tfrac{i}{m}).$$

Remark: $E_{\mathrm{p},n}$ is the nth *Euler polynomial*. See [771, pp. 48, 49], [1219, p. 559], and [1524, p. 40]. E_n denotes the nth Euler number. **Related:** Fact 13.1.8.

Fact 13.2.4. For all $n \geq 0$, define $\mathcal{B}_n \in \mathbb{R}[s]$ by

$$\mathcal{B}_n(x) \triangleq \left. \frac{\mathrm{d}^n}{\mathrm{d}z^n} e^{x(e^z - 1)} \right|_{z=0}.$$

Then, the following statements hold:

i) For all $x, z \in \mathbb{C}$,

$$e^{x(e^z - 1)} = \sum_{i=0}^{\infty} \frac{\mathcal{B}_i(x)}{i!} z^i.$$

ii) For all $n \geq 0$ and $z \in \mathbb{C}$,

$$\mathcal{B}_n(z) = \sum_{i=0}^{n} \left\{ {n \atop i} \right\} z^i.$$

In particular,

$$\mathcal{B}_0(z) = 1, \quad \mathcal{B}_1(z) = z, \quad \mathcal{B}_2(z) = z^2 + z,$$
$$\mathcal{B}_3(z) = z^3 + 3z^2 + z, \quad \mathcal{B}_4(z) = z^4 + 6z^3 + 7z^2 + z,$$
$$\mathcal{B}_5(z) = z^5 + 10z^4 + 25z^3 + 15z^2 + z, \quad \mathcal{B}_6(z) = z^6 + 15z^5 + 65z^4 + 90z^3 + 31z^2 + z.$$

iii) For all $n \geq 0$, $\mathcal{B}_n = \mathcal{B}_n(1)$.

iv) For all $n \geq 0$ and $z \in \mathbb{C}$,

$$\mathcal{B}_{n+1}(z) = z \sum_{i=0}^{n} \binom{n}{i} \mathcal{B}_i(z) = z[\mathcal{B}_n(z) + \mathcal{B}_n'(z)], \quad \sum_{i=0}^{n-1} (-1)^i \binom{n}{i} \mathcal{B}_i(z) = \sum_{i=1}^{n} (-1)^{i-1} \binom{n}{i} \mathcal{B}_i'(z).$$

Remark: $\mathcal{B}_n$ is the nth *Bell polynomial*. See [898].

Fact 13.2.5. For all $n \geq 0$, define $\mathcal{O}_n \in \mathbb{R}[s]$ by

$$\mathcal{O}_n(x) \triangleq \left. \frac{\mathrm{d}^n}{\mathrm{d}z^n} \frac{1}{1 - x(e^z - 1)} \right|_{z=0}.$$

Then, the following statements hold:

i) For all $x, z \in \mathbb{C}$ such that $|z| < \log(1 + 1/|x|)$,

$$\frac{1}{1 - x(e^z - 1)} = \sum_{i=0}^{\infty} \frac{\mathcal{O}_i(x)}{i!} z^i.$$

ii) For all $n \geq 1$ and $z \in \mathbb{C}$,

$$\mathcal{O}_n(z) = \sum_{i=1}^{n} i! \left\{ {n \atop i} \right\} z^i.$$

In particular,

$$\mathcal{O}_0(z) = 1, \quad \mathcal{O}_1(z) = z, \quad \mathcal{O}_2(z) = 2z^2 + z, \quad \mathcal{O}_3(z) = 6z^3 + 6z^2 + z,$$
$$\mathcal{O}_4(z) = 24z^4 + 36z^3 + 14z^2 + z, \quad \mathcal{O}_5(z) = 120z^5 + 240z^4 + 150z^3 + 30z^2 + z,$$
$$\mathcal{O}_6(z) = 720z^6 + 1800z^5 + 1560z^4 + 540z^3 + 62z^2 + z.$$

iii) For all $n \geq 0$ and $z \in \mathbb{C}$,

$$\mathcal{O}_n = \mathcal{O}_n(1), \quad \mathcal{O}_{n+1}(z) = z\mathcal{O}_n(z) + z(z+1)\mathcal{O}_n'(z) = z \sum_{i=0}^{n} \binom{n+1}{i} \mathcal{O}_i(z),$$
$$\sum_{i=0}^{n} \binom{n}{i} z\mathcal{O}_i'(z) = \sum_{i=1}^{n} \binom{n}{i-1} \mathcal{O}_i(z), \quad \mathcal{O}_n(z) = \int_0^{\infty} e^{-t} \mathcal{B}_n(zt)\, \mathrm{d}t.$$

Source: [899, 2100]. **Remark:** $\mathcal{O}_n$ is the nth *ordered Bell polynomial*. **Related:** Fact 1.19.7 and Fact 13.1.5.

Fact 13.2.6. For all $n \geq 0$, define $T_n \in \mathbb{R}[s]$ by

$$T_n(x) \triangleq \left. \frac{1}{n!} \frac{\mathrm{d}^n}{\mathrm{d}z^n} \frac{1 - xz}{z^2 - 2xz + 1} \right|_{z=0}.$$

Then, the following statements hold:

i) For all $x, z \in \mathbb{C}$ such that $|z| < \rho_{\max}(p)$, where $p(s) \triangleq s^2 - 2xs + 1$,

$$\frac{1 - xz}{z^2 - 2xz + 1} = \sum_{i=0}^{\infty} T_i(x)z^i.$$

ii) For all $n \geq 0$, $T_{n+2}(z) = 2zT_{n+1}(z) - T_n(z)$. Furthermore,

$$T_0(z) = 1, \quad T_1(z) = z, \quad T_2(z) = 2z^2 - 1, \quad T_3(z) = 4z^3 - 3z,$$
$$T_4(z) = 8z^4 - 8z^2 + 1 \quad T_5(z) = 16z^5 - 20z^3 + 5z \quad T_6(z) = 32z^6 - 48z^4 - 18z^2 - 1.$$

iii) For all $n \geq 0$ and $z \in \mathbb{C}$,

$$T_n(z) = \tfrac{1}{2}[(z - \sqrt{z^2 - 1})^n + (z + \sqrt{z^2 - 1})^n],$$
$$T_n(\cos z) = \cos nz, \quad T_{n+1}^2(z) = T_n(z)T_{n+2}(z) + 1 - z^2,$$
$$T_n(z) = \sum_{i=0}^{\lfloor n/2 \rfloor} \binom{n}{2i}(z^2 - 1)^i z^{n-2i} = 2^{n-1}n \sum_{i=0}^{\lfloor n/2 \rfloor} (-1)^i \frac{(n - i - 1)!}{4^i i!(n - 2i)!} z^{n-2i},$$
$$(1 - z^2)T_n''(z) - zT_n'(z) + n^2 T_n(z) = 0, \quad (n + 1)zT_{n+1}(z) + (1 - z^2)T_{n+1}'(z) = (n + 1)T_n(z).$$

iv) For all $n \geq 0$,

$$2T_n^2 = 1 + T_{2n}, \quad T_n(1) = 1, \quad T_n(-1) = (-1)^n, \quad T_{2n}(0) = (-1)^n, \quad T_{2n+1}(0) = 0.$$

v) For all $n, m \geq 0$,

$$T_{nm} = T_n \circ T_m = T_m \circ T_n.$$

vi) For all $n, m \geq 0$ and $z \in \mathbb{C}$,

$$(z^2 - 1)[T_{n+m}(z) - T_n(z)T_m(z)] = [T_{m+1}(z) - zT_m(z)][T_{n+1}(z) - zT_n(z)].$$

vii) For all $n \geq m \geq 0$,

$$T_{n+m} + T_{n-m} = 2T_m T_n.$$

viii) For all $n \geq 0$,

$$\frac{1}{\pi}\int_{-1}^{1} \frac{1}{\sqrt{1 - x^2}} T_0(x)T_n(x)\,\mathrm{d}x = \delta_{0,n}.$$

ix) For all $n, m \geq 1$,

$$\frac{2}{\pi}\int_{-1}^{1} \frac{1}{\sqrt{1 - x^2}} T_n(x)T_m(x)\,\mathrm{d}x = \delta_{m,n}.$$

x) For all $n \geq 1$,

$$T_{2n}(\sqrt{5}/2) = \tfrac{1}{2}(F_{2n-1} + F_{2n+1}) = \frac{F_{4n}}{2F_{2n}}, \quad T_{2n+1}(\sqrt{5}/2) = \frac{\sqrt{5}}{2}F_{2n+1}.$$

xi) let $x \in \mathbb{C}$, let $z \in \mathbb{C}$, and assume that $|xz|$ is sufficiently small. Then,

$$\operatorname{atan} \frac{2xz}{1 - z^2} = 2\sum_{i=0}^{\infty} (-1)^i \frac{T_{2i+1}(x)z^{2i+1}}{2i + 1}.$$

Remark: T_n is the nth *Chebyshev polynomial of the first kind*. See [352, pp. 187–196], [666], [1158, p. 58], and [1976]. *x*) and *xi*) are given in [649].

Fact 13.2.7. For all $n \geq 0$, define $U_n \in \mathbb{R}[s]$ by

$$U_n(x) \triangleq \frac{1}{n!}\frac{\mathrm{d}^n}{\mathrm{d}z^n} \frac{1}{z^2 - 2xz + 1}\bigg|_{z=0}.$$

Then, the following statements hold:

i) For all $x, z \in \mathbb{C}$ such that $|z| < \rho_{\max}(p)$, where $p(s) \triangleq s^2 - 2xs + 1$,

$$\frac{1}{z^2 - 2xz + 1} = \sum_{i=0}^{\infty} U_i(x)z^i.$$

ii) For all $n \geq 0$ and $z \in \mathbb{C}$, $U_{n+2}(z) = 2zU_{n+1}(z) - U_n(z)$. Furthermore,

$$U_0(z) = 1, \quad U_1(z) \triangleq 2z, \quad U_2(z) = 4z^2 - 1, \quad U_3(z) = 8z^3 - 4z,$$
$$U_4(z) = 16z^4 - 12z^2 + 1, \quad U_5(z) = 32z^5 - 32z^3 + 6z, \quad U_6(z) = 64z^6 - 80z^4 + 24z^2 - 1.$$

iii) For all $n \geq 0$ and $z \in \mathbb{C}$,

$$U_n(z) = \det(2zI_n + N_n + N_n^{\mathsf{T}}),$$
$$U_n(z) = \frac{(z + \sqrt{z^2 - 1})^{n+1} - (z - \sqrt{z^2 - 1})^{n+1}}{2\sqrt{z^2 - 1}},$$
$$(1 - z^2)U_n''(z) - 3zU_n'(z) + n(n+2)U_n(z) = 0,$$
$$U_n(z) = \sum_{i=0}^{\lfloor n/2 \rfloor} \binom{n+1}{2i+1}(z^2 - 1)^i z^{n-2i} = 2^n \prod_{i=1}^{n} \left(x - \cos\frac{i\pi}{n+1}\right),$$
$$U_n(\cos z) = \frac{\sin{(n+1)z}}{\sin z}, \quad (1 - z^2)U_n'(z) = zU_n(z) - (n+1)T_{n+1}(z),$$
$$T_{n+1}(z) = U_{n+1}(z) - zU_n(z), \quad T_{n+2}(z) = zT_{n+1}(z) - (1 - z^2)U_n(z).$$

iv) For all $n \geq 0$,

$$U_{n+1}^2 = U_nU_{n+2} + 1, \quad T_{n+1}' = (n+1)U_n, \quad T_{n+2} = \tfrac{1}{2}(U_{n+2} - U_n), \quad U_n(-\jmath/2) = (-\jmath)^n F_n.$$

v) For all $n, m \geq 0$,

$$\frac{2}{\pi}\int_{-1}^{1} \sqrt{1 - x^2}U_n(x)U_m(x)\,\mathrm{d}x = \delta_{m,n}.$$

vi) For all $n, m \geq 0$,

$$U_{m+n+2} = U_{m+1}U_{n+1} - U_mU_n.$$

vii) For all $n \geq 1$,

$$U_{2n}(\sqrt{5}/2) = F_{2n} + F_{2n+2} = \frac{F_{4n+2}}{2F_{2n+1}}, \quad U_{2n-1}(\sqrt{5}/2) = \sqrt{5}F_{2n}.$$

Source: *vi)* is given in [666]; *vii)* is given in [649]. **Remark:** U_n is the nth *Chebyshev polynomial of the second kind.* **Remark:** F_n is the nth Fibonacci number. See Fact 1.17.1.

Fact 13.2.8. For all $n \geq 0$, define $P_n \in \mathbb{R}[s]$ by

$$P_n(x) \triangleq \frac{1}{n!}\frac{\mathrm{d}^n}{\mathrm{d}z^n}\frac{1}{\sqrt{z^2 - 2xz + 1}}\bigg|_{z=0}.$$

Then, the following statements hold:

i) For all $x, z \in \mathbb{C}$ such that $|z| < \rho_{\max}(p)$, where $p(s) \triangleq s^2 - 2xs + 1$,

$$\frac{1}{\sqrt{z^2 - 2xz + 1}} = \sum_{i=0}^{\infty} P_i(x)z^i.$$

ii) $\frac{1}{\sqrt{z^2+1}} = \sum_{i=0}^{\infty} P_i(0)z^i.$

iii) For all $n \geq 0$ and $z \in \mathbb{C}$, $P_{n+2}(z) = \frac{1}{n+2}[(2n+3)zP_{n+1}(z) - (n+1)P_n(z)]$. Furthermore,

$$P_0(z) = 1, \quad P_1(z) = z, \quad P_2(z) = \tfrac{1}{2}(3z^2 - 1), \quad P_3(z) = \tfrac{1}{2}(5z^3 - 3z),$$
$$P_4(z) = \tfrac{1}{8}(35z^4 - 30z^2 + 3), \quad P_5(z) = \tfrac{1}{8}(63z^5 - 70z^3 + 15z),$$
$$P_6(z) = \tfrac{1}{16}(231z^6 - 315z^4 + 105z^2 - 5).$$

iv) For all $n \geq 0$ and $z \in \mathbb{C}$,

$$P_n(z) = 2^n \sum_{i=0}^{n} \binom{n}{i} \binom{\frac{1}{2}(n+i-1)}{n} z^i = \frac{1}{2^n} \sum_{i=0}^{\lfloor n/2 \rfloor} (-1)^i \binom{n}{i} \binom{2n-2i}{n} z^{n-2i},$$

$$P_n(z) = \frac{1}{2^n} \sum_{i=0}^{n} \binom{n}{i}^2 (z-1)^i (z+1)^{n-i} = \frac{1}{2^n n!} \frac{\mathrm{d}^n}{\mathrm{d}z^n} (z^2-1)^n = \frac{1}{\pi} \int_0^\pi (z + \sqrt{z^2-1}\cos\theta)^n \,\mathrm{d}\theta,$$

$$(1-z^2)P_n''(z) - 2zP_n'(z) + n(n+1)P_n(z) = 0,$$

$$P_{n+1}(z) = \frac{1}{n+1}[zP_{n+1}'(z) - P_n'(z)] = \frac{1}{n+2}[P_{n+2}'(z) - zP_{n+1}'(z)],$$

$$(1-z)\sum_{i=0}^{n}(2i+1)P_i(z) = (n+1)[P_n(z) - P_{n+1}(z)], \quad z^n = \frac{n!}{2^n} \sum_{i=0}^{\lfloor n/2 \rfloor} \frac{2n-4i+1}{i!(\frac{3}{2})^{\overline{n-i}}} P_{n-2i}(z).$$

v) For all $n \geq 0$,

$$P_{n+1} = \frac{1}{2n+1}[P_{n+2}' - P_n'], \quad \sum_{i=0}^{n}(2i+1)P_i = P_{n+1}' + P_n',$$

$$P_n(1) = 1, \quad P_n(-1) = (-1)^n, \quad P_n'(1) = \tfrac{1}{2}n(n+1), \quad P_n'(-1) = (-1)^{n-1}\tfrac{1}{2}n(n+1),$$

$$P_{2n}(0) = (-1)^n \frac{(2n)!}{4^n (n!)^2}, \quad P_{2n+1}(0) = 0.$$

vi) For all $n \geq 1$,

$$\int_{-1}^{1} xP_n(x)P_{n-1}(x)\,\mathrm{d}x = \frac{2n}{4n^2-1}.$$

vii) For all $n \geq 1$ and $k \geq 0$,

$$\int_{-1}^{1} P_n(x)x^{n+2k}\,\mathrm{d}x = \frac{(2k+1)^{\overline{n}}}{2^n(k+\frac{1}{2})^{\overline{n+1}}}.$$

viii) For all $n, m \geq 0$,

$$\frac{2n+1}{2}\int_{-1}^{1} P_n(x)P_m(x)\,\mathrm{d}x = \frac{2n+1}{2n(n+1)}\int_{-1}^{1}(1-x)^2 P_n'(x)P_m'(x)\,\mathrm{d}x = \delta_{n,m}.$$

ix) For all $n \geq 1$ and $z \in \mathbb{C}$,

$$P_n(z) = \frac{2}{\sqrt{\pi}n!}\int_0^\infty x^n e^{-x^2} H_n(zx)\,\mathrm{d}x.$$

x) Let $n \geq 1$ and $\theta \in (0, \pi)$. Then,

$$P_n(\cos\theta) = \frac{\sqrt{2}}{\pi}\int_0^\theta \frac{\cos(n+\frac{1}{2})x}{\sqrt{\cos x - \cos\theta}}\,\mathrm{d}x = \frac{\sqrt{2}}{\pi}\int_\theta^\pi \frac{\sin(n+\frac{1}{2})x}{\sqrt{\cos\theta - \cos x}}\,\mathrm{d}x.$$

xi) Let $x \in (-1, 1)$ and $n \geq 1$. Then,

$$P_{n-1}(x)P_{n+1}(x) \leq P_n^2(x).$$

xii) For all $n \geq 1$ and $z \in \mathbb{C}$,

$$(1-z)^n P_n\left(\frac{1+z}{1-z}\right) = \sum_{i=0}^{n} \binom{n}{i}^2 z^i.$$

xiii) Let $n \geq 0$ and $z \in \mathbb{C}$. Then,

$$P_n(z) = 1 + (z-1)\sum_{i=0}^{n-1}(2i+1)(H_n - H_i)P_i(z).$$

xiv) Let $n \geq 0$ and $0 \leq m \leq n-1$. Then,

$$\int_{-1}^{1} \frac{1-P_n(x)}{1-x} P_m(x)\,\mathrm{d}x = 2(H_n - H_m).$$

In particular,

$$\int_{-1}^{1} \frac{1-P_n(x)}{1-x}\,\mathrm{d}x = 2H_n.$$

xv) Let $n \geq 1$. Then,

$$\int_{0}^{1} P_n(2x-1)\log x\,\mathrm{d}x = \frac{(-1)^{n+1}}{n(n+1)}.$$

xvi) Let $n \geq 1$ and $x \in \mathbb{R}$. Then,

$$\int_{x}^{1} P_n(t)\,\mathrm{d}t = \frac{(1-x^2)P_n'(x)}{n(n+1)}.$$

xvii) Let $n \geq 1$. Then,

$$\frac{\mathrm{d}^n}{\mathrm{d}x^n}\frac{(1-x^2)^n}{n!}P_n\left(\frac{1+x}{1-x}\right)\bigg|_{x=0} = \sum_{i=0}^{n}\binom{n}{i}^3, \quad \frac{\mathrm{d}^n}{\mathrm{d}x^n}\frac{(1-x)^{2n}}{n!}P_n^2\left(\frac{1+x}{1-x}\right)\bigg|_{x=0} = \sum_{i=0}^{n}\binom{n}{i}^4.$$

xviii) Let $n \geq 1$ and $x \in (-1, 1]$. Then,

$$\sum_{i=0}^{n} \frac{3^{\overline{n-i}}}{(n-i)!}(i + \tfrac{1}{2})P_i(x) > 0.$$

Remark: P_n is the nth *Legendre polynomial*. See [116, pp. 313, 342, 343], [178, p. 4], [352, pp. 42–55, 90, 91, 164], [771, p. 165], and [2068, pp. 169, 170, 388, 393, 396–400]. **Remark:** In *ix*), H_n is the nth Hermite polynomial. See Fact 13.2.7. In *xiii*) and *xiv*), H_n is the nth harmonic number. **Remark:** *xi*) is *Turan's inequality*. See [116, p. 342]. **Related:** Fact 1.16.13.

Fact 13.2.9. For all $n \geq 0$, define $L_n \in \mathbb{R}[s]$ by

$$L_n(x) \triangleq \frac{1}{n!}\frac{\mathrm{d}^n}{\mathrm{d}z^n}\frac{e^{xz/(z-1)}}{1-z}\bigg|_{z=0}.$$

Then, the following statements hold:

i) For all $x, z \in \mathbb{C}$ such that $|z| < 1$,

$$\frac{e^{xz/(z-1)}}{1-z} = \sum_{i=0}^{\infty} L_i(x)z^i.$$

ii) For all $n \geq 0$ and $z \in \mathbb{C}$, $L_{n+2}(z) = \frac{1}{n+2}[(2n+3-z)L_{n+1}(z) - (n+1)L_n(z)]$. Therefore,

$$L_0(z) = 1, \quad L_1(z) = -z+1, \quad L_2(z) = \tfrac{1}{2}(z^2 - 4z + 2),$$
$$L_3(z) = \tfrac{1}{6}(-z^3 + 9z^2 - 18z + 6), \quad L_4(z) = \tfrac{1}{24}(z^4 - 16z^3 + 72z^2 - 96z + 24),$$

$$L_5(z) = \tfrac{1}{120}(-z^5 + 25z^4 - 200z^3 + 600z^2 - 600z + 120),$$
$$L_6(z) = \tfrac{1}{720}(z^6 - 36z^5 + 450z^4 - 2400z^3 + 5400z^2 - 4320z + 720).$$

iii) For all $n \geq 0$ and $z \in \mathbb{C}$,

$$L_n(z) = \sum_{i=0}^{n}(-1)^i\frac{1}{i!}\binom{n}{i}z^i = \frac{e^z}{n!}\frac{\mathrm{d}^n}{\mathrm{d}z^n}(e^{-z}z^n),$$

$$zL_n''(z) + (1-z)L_n'(z) + nL_n(z) = 0, \quad zL_{n+1}'(z) = nL_{n+1}(z) - nL_n(z), \quad L_n^{(n)}(z) = (-1)^n.$$

iv) For all $n, m \geq 0$ and $z \in \mathbb{C}$,

$$zL_{n+m}^{(m+2)}(z) + (m+1-z)L_{n+m}^{(m+1)}(z) + nL_{n+m}^{(m)}(z) = 0.$$

v) For all $n \geq 0$,

$$\sum_{i=0}^{n} L_i = -L_{n+1}', \quad L_n(0) = 1, \quad L_n'(0) = -n, \quad L_n''(0) = \tfrac{1}{2}n(n-1),$$

$$\int_0^\infty e^{-x}L_n^3(x)\,\mathrm{d}x = (-1)^n\sum_{i=0}^{n}\binom{n}{i}^3.$$

vi) For all $n, m \geq 0$

$$\int_0^\infty e^{-x}L_n(x)L_m(x)\,\mathrm{d}x = \delta_{n,m}.$$

vii) For all $n \geq m \geq 0$,

$$\int_0^\infty x^m e^{-x}L_n(x)\,\mathrm{d}x = \begin{cases} 0, & m < n, \\ (-1)^n n!, & m = n. \end{cases}$$

Remark: L_n is the nth *Laguerre polynomial*. See [352, pp. 168–185]. **Related:** Fact 1.18.3.

Fact 13.2.10. For all $n \geq 1$, define $H_n \in \mathbb{R}[s]$ by

$$H_n(x) \triangleq \left.\frac{\mathrm{d}^n}{\mathrm{d}z^n}e^{2xz-z^2}\right|_{z=0}.$$

Then, the following statements hold:

i) For all $x, z \in \mathbb{C}$,

$$e^{2xz-z^2} = \sum_{i=0}^{\infty}\frac{H_i(x)}{i!}z^i.$$

ii) For all $n \geq 0$ and $z \in \mathbb{C}$, $H_{n+2}(z) = 2zH_{n+1}(z) - 2(n+1)H_n(z)$. Furthermore,

$$H_0(z) = 1, \quad H_1(z) = 2z, \quad H_2(z) = 4z^2 - 2, \quad H_3(z) = 8z^3 - 12z,$$
$$H_4(z) = 16z^4 - 48z^2 + 12, \quad H_5(z) = 32z^5 - 160z^3 + 120z,$$
$$H_6(z) = 64z^6 - 480z^4 + 720z^2 - 120.$$

iii) For all $n \geq 1$ and $z \in \mathbb{C}$,

$$H_n(z) = n!\sum_{i=0}^{\lfloor n/2\rfloor}(-1)^i\frac{1}{i!(n-2i)!}(2z)^{n-2i} = (-1)^n e^{z^2}\frac{\mathrm{d}^n}{\mathrm{d}z^n}e^{-z^2} = \frac{(-2\jmath)^n e^{z^2}}{\sqrt{\pi}}\int_{\infty}^{\infty} e^{-x^2}x^n e^{2zx\jmath}\,\mathrm{d}x,$$

$$H_n''(z) - 2zH_n'(z) + 2nH_n(z) = 0, \quad H_{n+1}(z) = 2zH_n(z) - 2nH_{n-1}(z), \quad H_n'(z) = 2nH_{n-1}(z),$$

$$z^{2n} = \frac{(2n)!}{4^n}\sum_{i=0}^{n}\frac{1}{(2i)!(n-i)!}H_{2i}(z), \quad z^{2n+1} = \frac{(2n+1)!}{2^{2n+1}}\sum_{i=0}^{n}\frac{1}{(2i+1)!(n-i)!}H_{2i+1}(z).$$

iv) For all $n \ge 0$,

$$H_n(0) = \begin{cases} 0, & n\,\text{odd},\\ (-1)^{n/2}2^{n/2}(n-1)!!, & n\,\text{even}.\end{cases}$$

v) For all $n \ge 0$,

$$\int_{-\infty}^{\infty} e^{-2x^2}H_n^2(x)\,\mathrm{d}x = 2^{n-1}\sqrt{2}\Gamma(n+\tfrac{1}{2}), \quad \int_{-\infty}^{\infty} x^2e^{-x^2}H_n^2(x)\,\mathrm{d}x = 2^n n!(n+\tfrac{1}{2})\sqrt{\pi}.$$

vi) For all $n, m \ge 0$,

$$\frac{1}{2^n n!\sqrt{\pi}}\int_{-\infty}^{\infty} e^{-x^2}H_n(x)H_m(x)\,\mathrm{d}x = \delta_{n,m},$$

$$\int_{-\infty}^{\infty} xe^{-x^2}H_n(x)H_m(x)\,\mathrm{d}x = 2^{n-1}n!\sqrt{\pi}\delta_{n-1,m} + 2^n(n+1)!\sqrt{\pi}\delta_{n+1,m}.$$

vii) Let $0 \le l \le m \le n$, and assume that $l+m+n$ is even and $n \le l+m$. Then,

$$\int_{-\infty}^{\infty} e^{-x^2}H_l(x)H_m(x)H_n(x)\,\mathrm{d}x = \frac{2^{(l+m+n)/2}l!m!n!\sqrt{\pi}}{[\frac{1}{2}(l+m-n)]![\frac{1}{2}(m+n-l)]![\frac{1}{2}(n+l-m)]!}.$$

viii) For all $n, m \ge 0$ and $z \in \mathbb{C}$,

$$H_m(z)H_n(z) = \sum_{i=0}^{\min\{m,n\}} 2^i i!\binom{m}{i}\binom{n}{i}H_{m+n-2i}(z).$$

ix) For all $n \ge m \ge 0$,

$$H_n^{(m)} = \frac{2^m n!}{(n-m)!}H_{n-m}, \quad H_n^{(n)} = 2^n n!.$$

x) Let $x \in \mathbb{R}$, and assume that $x \ne 0$. Then,

$$\sum_{i=0}^{\infty}(-1)^i\frac{1}{4^i(2i+1)i!}H_{2i+1}(x) = \operatorname{sign}(x)\sqrt{\pi}.$$

xi) Let $x, y \in \mathbb{C}$ and $n \ge 0$. Then,

$$H_n(x+y) = \sum_{i=0}^{n}\binom{n}{i}H_i(x)(xy)^{n-i}, \quad L_n(x^2+y^2) = \frac{(-1)^n}{4^n}\sum_{i=0}^{n}\frac{H_{2i}(x)H_{2n-2i}(y)}{i!(n-i)!}.$$

xii) Let $n \ge 1$. Then, the n roots $r_1, \ldots, r_n$ of H_n are real. Now, define $r \triangleq [r_1 \ \cdots \ r_n]^\mathsf{T}$ and $f\colon \mathbb{R}^n \mapsto \mathbb{R}$ by

$$f(x) = \exp\left(-\frac{1}{2}\sum_{i=1}^{n}x_{(i)}^2\right)\prod |x_{(i)} - x_{(j)}|,$$

where the product is taken over all $i, j \in \{1, \ldots, n\}$ such that $i < j$. Then, for all $x \in \mathbb{R}^n$ such that $x \ne r$, $f(x) < f(r)$.

xiii) Let $m \ge 1$ and $z_1, \ldots, z_m \in \mathbb{C}$. Then,

$$H_n\left(\frac{1}{\sqrt{m}}\sum_{i=1}^{m}z_i\right) = \frac{n!}{m^{n/2}}\sum\prod_{j=1}^{m}\frac{H_{i_j}(z_j)}{i_j!},$$

where the sum is taken over all m-tuples $(i_1,\dots,i_m)$ of nonnegative integers such that $\sum_{j=1}^m i_j = n$.

Remark: H_n is the nth *Hermite polynomial*. See [116, pp. 278, 279, 318, 328, 339, 340–342, 418, 419], [352, pp. 156–167], and [771, p. 165]. *xiii*) is given in [2799]. **Remark:** L_n is the nth Laguerre polynomial. See Fact 13.2.9.

Fact 13.2.11. For all $n \ge 0$, define $H_{\mathrm{p},n} \in \mathbb{R}[s]$ by

$$H_{\mathrm{p},n}(x) \triangleq \sum_{i=1}^{n} \frac{1}{i} x^i.$$

Then, for all $n \ge 1$, $H_n = H_{\mathrm{p},n}(1) = \sum_{i=1}^n \frac{1}{i}$. If $n \ge 1$ and $x \in \mathbb{C}$, where $x \ne 1$, then

$$\sum_{i=1}^{n} H_i x^i = \frac{H_n x^{n+1} - H_{\mathrm{p},n}(x)}{x-1}.$$

If $n \ge 1$ and $x \in \mathbb{C}$, where $x \notin \{0,1\}$, then

$$\sum_{i=1}^{n} H_i x^{n-i} = \frac{H_{\mathrm{p},n}(1/x)x^{n+1} - H_n}{x-1}.$$

Furthermore, if $n \ge 1$, then

$$\sum_{i=1}^{n} (-1)^{n-i} H_i = \frac{1}{2}[H_n + (-1)^n H_{\mathrm{p},n}(-1)].$$

Source: [899]. **Remark:** $H_{\mathrm{p},n}$ is the nth *harmonic polynomial*, and H_n is the nth harmonic number.

Fact 13.2.12. For all $n \ge 0$, define $F_{\mathrm{p},n} \in \mathbb{R}[s]$ by

$$F_{\mathrm{p},n}(x) \triangleq \frac{1}{n!} \frac{\mathrm{d}^n}{\mathrm{d}z^n} \frac{-z}{z^2 + xz - 1}\bigg|_{z=0}.$$

Then, the following statements hold:

i) For all $x, z \in \mathbb{C}$ such that $|z| < \rho_{\max}(p)$, where $p(s) \triangleq s^2 + xs - 1$,

$$\frac{-z}{z^2 + xz - 1} = \sum_{i=0}^{\infty} F_{\mathrm{p},i}(x) z^i.$$

ii) For all $n \ge 2$, $F_{\mathrm{p},n}(z) = zF_{\mathrm{p},n-1}(z) + F_{\mathrm{p},n-2}(z)$. In particular,

$$F_{\mathrm{p},0}(z) = 0, \quad F_{\mathrm{p},1}(z) = 1, \quad F_{\mathrm{p},2}(z) = z, \quad F_{\mathrm{p},3}(z) = z^2 + 1,$$
$$F_{\mathrm{p},4}(z) = z^3 + 2z, \quad F_{\mathrm{p},5}(z) = z^4 + 3z^2 + 1, \quad F_{\mathrm{p},6}(z) = z^5 + 4z^3 + 3z.$$

iii) For all $n \ge 1$, $F_n = F_{\mathrm{p},n}(1)$.

iv) For all $x \in \mathbb{R}$, define $\alpha(x) \triangleq \frac{1}{2}(x + \sqrt{x^2+4})$ and $\beta(x) \triangleq \frac{1}{2}(x - \sqrt{x^2+4})$. Then, for all $n \ge 0$ and $x \in \mathbb{R}$,

$$F_{\mathrm{p},n}(x) = \frac{\alpha^n(x) - \beta^n(x)}{\alpha(x) - \beta(x)}.$$

v) For all $n \ge 1$ and $z \in \mathbb{C}$,

$$F_{\mathrm{p},n}(z) = \sum_{i=0}^{\lfloor (n-1)/2 \rfloor} \binom{n-i-1}{i} z^{n-2i-1}.$$

vi) For all $n \ge 2$, $\operatorname{roots}(F_{\mathrm{p},n}) = \{2(\cos \frac{i\pi}{n})\jmath\colon\ 1 \le i \le n-1\}$.

vii) For all $n \geq 0$,

$$x^n = \sum_{i=0}^{n} (-1)^i \binom{n}{i} F_{\mathrm{p},n+1-2i}(x).$$

viii) If $n \geq 4$ is even, then

$$\int_{-\infty}^{\infty} \frac{1}{F_{\mathrm{p},n}(x)} \, \mathrm{d}x = \frac{\pi}{n}\left(1 + \sec\frac{\pi}{n}\right).$$

ix) If $n \geq 3$ is odd, then

$$\int_{-\infty}^{\infty} \frac{x}{F_{\mathrm{p},n}(x)} \, \mathrm{d}x = \frac{\pi}{n}\left(\tan\frac{\pi}{2n} + \tan\frac{3\pi}{2n}\right).$$

Source: [595], [2068, pp. 124, 125], and [2683]. **Remark:** $F_{\mathrm{p},n}$ is the nth *Fibonacci polynomial.*

Fact 13.2.13. For all $n \geq 0$, define $L_{\mathrm{p},n} \in \mathbb{R}[s]$ by

$$L_{\mathrm{p},n}(x) \triangleq \frac{1}{n!} \frac{\mathrm{d}^n}{\mathrm{d}z^n} \frac{xz - 2}{z^2 + xz - 1}\bigg|_{z=0}.$$

Then, the following statements hold:

i) For all $x, z \in \mathbb{C}$ such that $|z| < \rho_{\max}(p)$, where $p(s) \triangleq s^2 + xs - 1$,

$$\frac{xz - 2}{z^2 + xz - 1} = \sum_{i=0}^{\infty} L_{\mathrm{p},i}(x) z^i.$$

ii) For all $n \geq 2$, $L_{\mathrm{p},n}(z) = zL_{\mathrm{p},n-1}(z) + F_{\mathrm{p},n-2}(z)$. In particular,

$$L_{\mathrm{p},0}(z) = 2, \quad L_{\mathrm{p},1}(z) = z, \quad L_{\mathrm{p},2}(z) = z^2 + 2, \quad L_{\mathrm{p},3}(z) = z^3 + 3z,$$
$$L_{\mathrm{p},4}(z) = z^4 + 4z^2 + 2, \quad L_{\mathrm{p},5}(z) = z^5 + 5z^3 + 5z, \quad L_{\mathrm{p},6}(z) = z^6 + 6z^4 + 9z^2 + 2.$$

iii) For all $n \geq 1$, $L_{\mathrm{p},n}(z) = \frac{1}{2^n}[(z - \sqrt{z^2+4})^n + (z + \sqrt{z^2+4})^n]$.

iv) For all $n \geq 1$, $L_n = L_{\mathrm{p},n}(1)$ and $L_{\mathrm{p},n}(0) = 1 + (-1)^n$.

v) For all $x \in \mathbb{R}$, define $\alpha(x) \triangleq \frac{1}{2}(x + \sqrt{x^2+4})$ and $\beta(x) \triangleq \frac{1}{2}(x - \sqrt{x^2+4})$. Then, for all $n \geq 0$ and $x \in \mathbb{R}$, $L_{\mathrm{p},n}(x) = \alpha^n(x) + \beta^n(x)$.

vi) For all $n \geq 1$ and $z \in \mathbb{C}$,

$$F_{\mathrm{p},n}(z) = \sum_{i=0}^{\lfloor n/2 \rfloor} \frac{n}{n-i}\binom{n-i}{i} z^{n-2i}.$$

vii) For all $n \geq 1$, $L'_{\mathrm{p},n}(z) = \frac{n}{z^2+4}[zL_{\mathrm{p},n}(z) + 2L_{\mathrm{p},n-1}(z)]$.

viii) For all $n \geq 1$, $\mathrm{roots}(L_{\mathrm{p},n}) = \{2(\sin\frac{i\pi}{n})J\colon 1 \leq i \leq n-1\}$.

ix) $L_{\mathrm{p},m}$ divides $L_{\mathrm{p},n}$ if and only if n/m is an odd integer.

x) If n is prime, then $\mathrm{roots}(L_{\mathrm{p},n}) \subset \mathrm{IA}$.

Source: [2683]. **Remark:** $L_{\mathrm{p},n}$ is the nth *Lucas polynomial.*

Fact 13.2.14. Let $n \geq 0$. Then, there exists $Q_n \in \mathbb{R}[s]$ such that

$$\frac{\mathrm{d}^n}{\mathrm{d}x^n} \sec x = (\sec x) Q_n(\tan x) = (\sec x) \sum_{i=0}^{n} E_{n,i} \tan^i x.$$

Furthermore,

$$E_{n,0} = Q_n(0) = \frac{\mathrm{d}^n}{\mathrm{d}x^n} \sec x\bigg|_{x=0} = \begin{cases} (-1)^{n/2} E_n, & n \text{ even}, \\ 0, & n \text{ odd}. \end{cases}$$

$$E_{n,n} = n!, \quad E_{n+1,0} = E_{n,1}, \quad E_{n+1,n} = nE_{n,n-1}, \quad E_{n+1,n+1} = (n+1)E_{n,n}.$$

If, in addition, $1 \le m \le n-1$, then

$$E_{n+1,m} = mE_{n,m-1} + (m+1)E_{n,m+1}.$$

For all $x \in \mathbb{R}$,

$$Q_{n+1}(x) = xQ_n(x) + (x^2+1)Q_n'(x).$$

In particular, $Q_0(x) = 1$, $Q_1(x) = x$, $Q_2(x) = 2x^2+1$, and $Q_3(x) = 6x^3+5x$. For all $t, x \in \mathbb{R}$ such that $\cos t - x\sin t \ne 0$,

$$\frac{1}{\cos t - x\sin t} = \sum_{i=0}^{\infty} \frac{Q_i(x)}{n!}t^i.$$

Furthermore, there exists $P_n \in \mathbb{R}[s]$ such that

$$\frac{\mathrm{d}^n}{\mathrm{d}x^n}\tan x = P_n(\tan x) = \sum_{i=0}^{n} D_{n,i}\tan^i x.$$

For all $t, x \in \mathbb{R}$ such that $\cos t - x\sin t \ne 0$,

$$\frac{\sin t + x\cos t}{\cos t - x\sin t} = \sum_{i=0}^{\infty} \frac{P_i(x)}{n!}t^i.$$

Source: [630, 1430]. **Remark:** Q_n and P_n are *derivative polynomials*. E_n is the nth Euler number. See Fact 13.1.8. **Related:** Fact 14.8.19.

Fact 13.2.15. Let $A_n \in \mathbb{C}[s]^{(n+1)\times(n+1)}$ be the Hankel matrix whose (i,j) entry is $p_{i+j-2}(x)$, where, for all $k \in \{0,\dots,2n\}$, $p_k(x) \triangleq C_0x^k + \cdots + C_{k-1}x + C_k$, and define $H_n(x) \triangleq \det A_n(x)$. Then, for all $x \in \mathbb{C}$,

$$H_n(x) = \sum_{i=0}^{n}(-1)^i\binom{n+i}{n-i}x^i.$$

In particular, $H(0) = 1$. Alternatively, for all $x \in \mathbb{C}$ and $k \in \{0,\dots,2n\}$, define

$$p_k(x) \triangleq \sum_{i=1}^{k}\binom{2k-2i}{k-i}x^i.$$

Then,

$$H_n(x) = 2^n\sum_{i=0}^{n}(-1)^i\binom{n+i}{n-i}x^i.$$

In both cases, for all $x \in \mathbb{C}$, $H_n(x)$ satisfies

$$x(x-4)H_n''(x) + 2(x-1)H_n'(x) - n(n+1)H_n(x) = 0.$$

Source: [979]. **Remark:** $(H_i(x))_{i=0}^{\infty}$ is the *Hankel transform* of $(p_i(x))_{i=0}^{\infty}$.

13.3 Facts on the Zeta, Gamma, Digamma, Generalized Harmonic, Dilogarithm, and Dirichlet L Functions

Fact 13.3.1. For all $z \in \mathbb{C}$ such that $\operatorname{Re} z > 1$, define

$$\zeta(z) \triangleq \sum_{i=1}^{\infty}\frac{1}{i^z}.$$

Then, the following statements hold:

i) The series $\sum_{i=1}^{\infty}\frac{1}{i^z}$ converges for all $z\in\mathcal{D}_0\triangleq\{z\in\mathbb{C}\colon\ \operatorname{Re}z>1\}$. Furthermore, $\zeta\colon\mathcal{D}_0\mapsto\mathbb{C}$ is given by

$$\zeta(z)=\frac{1}{1-2^{-z}}\sum_{i=1}^{\infty}\frac{1}{(2i-1)^z}=\frac{1}{\Gamma(z)}\int_0^\infty\frac{x^{z-1}}{e^x-1}\,\mathrm{d}x=\int_0^1\frac{\log^{z-1}1/x}{1-x}\,\mathrm{d}x.$$

ii) ζ has an analytic continuation to $\mathcal{D}_1\triangleq\text{ORHP}\backslash\{1\}$. Furthermore, $\zeta\colon\mathcal{D}_1\mapsto\mathbb{C}$ is given by

$$\zeta(z)=\frac{1}{1-2^{1-z}}\sum_{i=1}^{\infty}(-1)^{i-1}\frac{1}{i^z}=\frac{1}{(1-2^{1-z})\Gamma(z)}\int_0^\infty\frac{x^{z-1}}{e^x+1}\,\mathrm{d}x=\frac{z}{z-1}+z\int_1^\infty\frac{\lfloor x\rfloor-x}{x^{z+1}}\,\mathrm{d}x.$$

iii) ζ has an analytic continuation to $\mathcal{D}\triangleq\mathbb{C}\backslash\{1\}$. Furthermore, $\zeta\colon\mathcal{D}\mapsto\mathbb{C}$ is given by

$$\zeta(z)=\frac{2^{z-1}}{z-1}-2^z\int_0^\infty\frac{\sin(z\,\operatorname{atan}x)}{(1+x^2)^{z/2}(e^{\pi x}+1)}\mathrm{d}x=\frac{1}{1-2^{1-z}}\sum_{i=0}^{\infty}\frac{\sum_{j=0}^{i}(-1)^j\binom{i}{j}(j+1)^{-z}}{2^{i+1}}.$$

iv) For all $z\in\mathcal{D}$,

$$\zeta(z)=\frac{1}{z-1}\sum_{i=0}^{\infty}\frac{1}{i+1}\sum_{j=0}^{i}(-1)^j\binom{i}{j}\frac{1}{(j+1)^{z-1}}=\frac{1}{z-1}+\sum_{i=0}^{\infty}(-1)^i\frac{\gamma_i}{i!}(z-1)^i,$$

where $\gamma_0\triangleq\gamma$ and, for all $i\ge 0$,

$$\gamma_i\triangleq\lim_{j\to\infty}\left(\sum_{k=1}^{j}\frac{\log^i k}{k}-\frac{\log^{i+1}j}{i+1}\right).$$

Hence, $\lim_{z\to1}(z-1)\zeta(z)=1$ and $\lim_{z\to1}\left(\zeta(z)-\frac{1}{z-1}\right)=\gamma$.

v) For all $z\in\mathcal{D}$,

$$\zeta(z)=\frac{\pi^{z/2}}{\Gamma(\frac{z}{2})}\left(\frac{1}{z(z-1)}+\int_1^\infty\tfrac{1}{2}(x^{\frac{z}{2}-1}+x^{-\frac{z}{2}-1})\psi(x)\,\mathrm{d}x\right),$$

where, for all $x>0$,

$$\psi(x)\triangleq\sum_{i=1}^{\infty}e^{-i^2\pi x}.$$

vi) Define $f\colon\mathbb{C}\mapsto\mathbb{C}$ and $g\colon\mathbb{C}\mapsto\mathbb{C}$ by

$$f(z)\triangleq\begin{cases}(z-1)\zeta(z), & z\ne1,\\ 1, & z=1,\end{cases}\qquad g(z)\triangleq\begin{cases}\zeta(z)-\frac{1}{z-1}, & z\ne1,\\ \gamma, & z=1.\end{cases}$$

Then, f and g are analytic.

vii) Let $n\ge1$. Then, $\zeta(1-2n)=-\frac{1}{2n}B_{2n}$, $\zeta(-2n)=0$, and

$$\zeta'(-2n)=(-1)^n\frac{(2n)!}{2(2\pi)^{2n}}\zeta(2n+1).$$

viii) $\zeta(0)=-\frac{1}{2}$, $\zeta'(0)=-\frac{1}{2}\log2\pi$, and $\zeta''(0)=\gamma_1+\frac{\gamma^2}{2}-\frac{\pi^2}{24}-\frac{1}{2}\log^2 2\pi$.

ix) For all $z\in\mathbb{C}\backslash\mathbb{N}$,

$$\zeta(z)=2(2\pi)^{z-1}\left(\sin\frac{\pi z}{2}\right)\Gamma(1-z)\zeta(1-z).$$

x) For all $z \in \mathbb{C}\backslash(-\mathbb{P} \cup \{0, 1\})$,

$$\zeta(1-z) = \frac{2}{(2\pi)^z}\left(\cos\frac{\pi z}{2}\right)\Gamma(z)\zeta(z).$$

xi) For all $n \geq 1$,

$$\zeta(2n) = \frac{2^{2n-1}\pi^{2n}}{(2n)!}|B_{2n}|, \quad \lim_{z\to -n}\Gamma(z)\zeta(2z) = (-1)^n\frac{2}{n!}\zeta'(-2n),$$

$$\lim_{z\to -n}\Gamma(\tfrac{1}{2}-z)\zeta(1-2z) = (-1)^n\frac{2\pi^{2n+1/2}}{n!}\zeta'(-2n) = \frac{\sqrt{\pi}(2n-1)!!}{2^n}\zeta(2n+1).$$

xii) The function $\xi(z) \triangleq \frac{1}{2}z(z-1)\pi^{-z/2}\Gamma(\frac{z}{2})\zeta(z)$ is analytic on $\mathbb{C}$ and, for all $z \in \mathbb{C}$, satisfies $\xi(z) = \xi(1-z)$.

xiii) If $z \in \mathbb{C}$ and $\operatorname{Re} z = \frac{1}{2}$, then $\pi^{-z/2}\Gamma(\frac{z}{2})\zeta(z)$ is real.

xiv) For all $\alpha > 1$, $\lim_{n\to\infty} H_{n,\alpha} = \zeta(\alpha)$.

xv) For all $n \geq 2$, $(n + \frac{1}{2})\zeta(2n) = \sum_{i=1}^{n-1}\zeta(2i)\zeta(2n-2i)$.

xvi) For all $x > 1$, $\zeta^2(x) = \sum_{i=1}^{\infty}\frac{d(i)}{n^x}$, where $d(i)$ is the number of positive divisors of i. (Example: $d(12) = 6$.)

xvii) For all $x > 2$, $\zeta(x)\zeta(1-x) = \sum_{i=1}^{\infty}\frac{\sigma(i)}{n^x}$, where $\sigma(i)$ is the sum of the positive divisors of i. (Example: $\sigma(12) = 28$.)

xviii) For all $x > 2$, $\frac{\zeta(x-1)}{\zeta(x)} = \sum_{i=1}^{\infty}\frac{\phi(i)}{n^x}$, where $\phi(i)$ is the number of positive integers $j \leq i$ such that $\{i, j\}$ is coprime. (Example: $\phi(12) = 4$.)

xix) If $z \in -2\mathbb{P}$, then $\zeta(z) = 0$.

xx) If $z \in \mathcal{D}$, $z \notin -2\mathbb{P}$, and $\zeta(z) = 0$, then $0 \leq \operatorname{Re} z < 1$, $\zeta(\overline{z}) = 0$, and $\zeta(1-z) = 0$.

xxi) $\operatorname{card}\{z \in \mathbb{C}\colon \operatorname{Re} z = \frac{1}{2} \text{ and } \zeta(z) = 0\} = \infty$.

xxii) For all $i \geq 1$, let p_i denote the ith prime number, where $p_1 = 2$. Then, for all $z \in \mathbb{C}$ such that $\operatorname{Re} z > 1$,

$$\zeta(z) = \prod_{i=1}^{\infty}\frac{1}{1-p_i^{-z}}.$$

In particular,

$$\prod_{i=1}^{\infty}\frac{1}{1-p_i^{-2}} = \frac{\pi^2}{6}.$$

xxiii) The probability that k random positive integers are coprime is $1/\zeta(k)$.

In addition, the following statements are equivalent:

xxiv) If $z \notin -2\mathbb{P}$ and $\zeta(z) = 0$, then $\operatorname{Re} z = \frac{1}{2}$.

xxv) For all $\varepsilon > 0$ there exists $c > 0$ such that, for all $x \geq 2$, $|\pi(x) - \int_2^x \frac{1}{\log t}\,\mathrm{d}t| < cx^{1/2+\varepsilon}$.

xxvi) If $a, b \in \mathbb{R}$ and $\sum_{i=1}^{\infty}(-1)^i\frac{1}{i^a}\sin(b\log i) = \sum_{i=1}^{\infty}(-1)^i\frac{1}{i^a}\cos(b\log i) = 0$, then either $a = 1/2$ or $a = 1$.

xxvii) For all $n \geq 1$, let $\omega(n)$ denote the number of prime divisors of n. (Example: $\omega(1) = 0$ and $\omega(24) = 4$.) Then, for all $\varepsilon > 0$, $\lim_{n\to\infty}\frac{1}{n^{1/2+\varepsilon}}\sum_{i=1}^{n}(-1)^{\omega(i)} = 0$.

Source: [116, p. 16], [525, pp. 6, 13, 14], [723], [968, pp. 276–288], [1350, pp. 68, 207], [2106, pp. 352–360], [2486], [2513, pp. 164, 167, 226], and [2528, Chapter 7]. **Remark:** ζ is the *zeta function.* **Remark:** Analytic continuation of ζ is discussed in [1004, 2491]. **Remark:** *xxi)* is *Hardy's theorem.* See [525, pp. 24–47]. **Remark:** *xxii)* is *Euler's product formula.* **Remark:** *xxiv)*–*xxvii)* are equivalent statements of the *Riemann hypothesis.* See [2493]. An additional equivalent statement is given by Fact 1.11.47. Further additional statements are given in [525, Chapter 5]. **Remark:** *iv*) is the Laurent series of ζ at $z = 1$. **Remark:** *xxv*) is a strengthened version of the prime number theorem given by Fact 12.18.41, where $\pi(x)$ denotes the number of primes that are either less than or equal to x. **Remark:** ϕ is the Euler totient function. See Fact 1.20.4. **Remark:** The definitions of $d(i)$ and $\sigma(i)$ are different from the definitions in Fact 1.20.3, where $\sqrt{i}$ is counted twice in the case where $\sqrt{i}$ is an integer. **Remark:** γ_n is the nth *Stieltjes constant*. **Remark:** $\psi(x) = \frac{1}{2}(\theta(x) - 1)$, where $\theta(x) \triangleq \sum_{i=-\infty}^{\infty} e^{-i^2\pi x}$ is the *Jacobi theta function*. For all $x > 0$, $x^{1/2}\theta(x) = \theta(\frac{1}{x})$. See [525, p. 13]. **Remark:** On the line $\{z \in \mathbb{C}\colon \operatorname{Re} z = \frac{1}{2}\}$, $\xi(z)$ is real, and this can be used to compute the zeros of ζ whose real part is $\frac{1}{2}$. Note that $\zeta(z)$ is not necessarily real for all z such that $\operatorname{Re} z = \frac{1}{2}$. See [975, pp. 119–135]. **Remark:** The zeros of ζ are related to the eigenvalues of random matrices. See [525, p. 40]. **Remark:** Historical and background material on Riemann's contribution to the zeta function is given in [525, 975].

Fact 13.3.2. For all $z \in \text{ORHP}$, define

$$\Gamma(z) \triangleq \int_0^\infty x^{z-1}e^{-x}\,\mathrm{d}x = \int_0^1 \left(\log\frac{1}{x}\right)^{z-1}\mathrm{d}x.$$

Then, the following statements hold:

i) Both integrals exist for all $z \in \text{ORHP}$.

ii) If $z \in \mathbb{C}$ and $\operatorname{Re} z > 1$, then

$$\Gamma(z) = \frac{1}{\zeta(z)}\int_0^\infty \frac{x^{z-1}}{e^x - 1}\,\mathrm{d}x = \frac{1}{(1-2^{1-z})\zeta(z)}\int_0^\infty \frac{x^{z-1}}{e^x + 1}\,\mathrm{d}x.$$

If $x \in \mathbb{R}\backslash\{1, 0, -1, -2, \ldots\}$, then

$$\Gamma(x) = \frac{1}{\zeta(x)}\int_0^\infty \frac{t^{x-1}}{e^t - 1}\,\mathrm{d}t.$$

iii) Γ has an analytic continuation to $\mathcal{D} \triangleq \mathbb{C}\backslash(-\mathbb{N})$. In particular, for all $z \in \mathcal{D}$ and $n \geq 1$,

$$\Gamma(z) = \frac{\Gamma(z+n)}{z^{\overline{n}}}.$$

iv) If $z \in \mathcal{D}$, then $\overline{z} \in \mathcal{D}$ and $\Gamma(\overline{z}) = \overline{\Gamma(z)}$.

v) If $n \geq 0$, then

$$\Gamma(n + \tfrac{1}{2}) = \frac{(2n)!}{4^n n!}\sqrt{\pi} = (\tfrac{1}{2})^{\overline{n}}\sqrt{\pi} = \frac{\sqrt{\pi}}{2^n}(2n-1)!!.$$

In particular, $\Gamma(\frac{1}{2}) = \sqrt{\pi}$, $\Gamma(\frac{3}{2}) = \frac{\sqrt{\pi}}{2}$, and $\Gamma(\frac{5}{2}) = \frac{3\sqrt{\pi}}{4}$.

vi) If $n \geq 0$, then

$$\Gamma(\tfrac{1}{2} - n) = \frac{(-4)^n n!}{(2n)!}\sqrt{\pi} = \frac{(-1)^n}{(\frac{1}{2})^{\overline{n}}}\sqrt{\pi} = (-1)^n\frac{2^n\sqrt{\pi}}{(2n-1)!!}.$$

In particular, $\Gamma(-\frac{1}{2}) = -2\sqrt{\pi}$, $\Gamma(-\frac{3}{2}) = \frac{4\sqrt{\pi}}{3}$, and $\Gamma(-\frac{5}{2}) = -\frac{8\sqrt{\pi}}{15}$.

vii) If $n \geq 1$, then $\Gamma(n) = (n-1)!$. In particular, $\Gamma(1) = \Gamma(2) = 1$.

viii) If $n \geq 0$, then $\lim_{x\downarrow -n}\Gamma(x) = (-1)^n\infty$ and $\lim_{x\uparrow -n}\Gamma(x) = (-1)^{n+1}\infty$.

ix) If $z \in \mathcal{D}$, then

$$\Gamma(z) = \lim_{i\to\infty}\frac{i!i^z}{z^{\overline{i+1}}} = \lim_{i\to\infty}\frac{i!i^{z-1}}{z^{\overline{i}}} = \frac{1}{z}\prod_{i=1}^{\infty}\frac{(1+\frac{1}{i})^z}{1+\frac{z}{i}}.$$

x) If $z \in \mathcal{D}$, then

$$\Gamma(z) = \frac{1}{ze^{\gamma z}} \prod_{i=1}^{\infty} \frac{e^{z/i}}{1 + \frac{z}{i}}.$$

xi) $\lim_{z \to 0} z\Gamma(z) = 1$.

xii) If $z \in \mathcal{D}$ and $1 - z \in \mathcal{D}$, then

$$\Gamma(z)\Gamma(1 - z) = \frac{\pi}{\sin \pi z}.$$

xiii) If $z \in \mathcal{D}$ and $z + 1 \in \mathcal{D}$, then

$$\Gamma(z + 1) = z\Gamma(z).$$

In particular,

$$\Gamma(-\tfrac{1}{3}) = -3\Gamma(\tfrac{2}{3}), \quad \Gamma(-\tfrac{1}{4}) = -4\Gamma(\tfrac{3}{4}), \quad \Gamma(\tfrac{1}{4}) = 4\Gamma(\tfrac{5}{4}), \quad \Gamma(\tfrac{1}{3}) = 3\Gamma(\tfrac{4}{3}),$$
$$\Gamma(\tfrac{1}{2}) = 2\Gamma(\tfrac{3}{2}) = -\tfrac{1}{2}\Gamma(-\tfrac{1}{2}) = \sqrt{\pi}.$$

xiv) If $z \in \mathcal{D}$ and $-z \in \mathcal{D}$, then

$$\Gamma(z)\Gamma(-z) = \frac{-\pi}{z \sin \pi z}.$$

xv) If $\frac{1}{2} + z \in \mathcal{D}$ and $\frac{1}{2} - z \in \mathcal{D}$, then

$$\Gamma(\tfrac{1}{2} + z)\Gamma(\tfrac{1}{2} - z) = \frac{\pi}{\cos \pi z}.$$

In particular,

$$\Gamma(\tfrac{1}{6})\Gamma(\tfrac{5}{6}) = 2\pi, \quad \Gamma(\tfrac{1}{4})\Gamma(\tfrac{3}{4}) = \sqrt{2}\pi, \quad \Gamma(\tfrac{1}{3})\Gamma(\tfrac{2}{3}) = \frac{2\sqrt{3}\pi}{3}.$$

xvi) If $x \geq 0$, then

$$\lfloor x \rfloor! x^{x - \lfloor x \rfloor} \leq \Gamma(x + 1) \leq \lfloor x \rfloor!(\lfloor x \rfloor + 1)^{x - \lfloor x \rfloor}.$$

xvii)

$$\lim_{x \to \infty} \frac{\Gamma(x + 1)e^x}{x^{x + 1/2}} = \sqrt{2\pi}.$$

xviii) If $a \in (0, 1)$, then

$$\Gamma(a)\Gamma(1 - a) = \frac{\pi}{\sin \pi a} = \int_0^\infty \frac{x^{a-1}}{x + 1}\, \mathrm{d}x.$$

xix) Let $n \geq 1$. Then,

$$\prod_{i=1}^{n-1} \Gamma\left(\tfrac{i}{n}\right) = \sqrt{\frac{(2\pi)^{n-1}}{n}}.$$

xx) If $z \in \mathcal{D}$, $z + \frac{1}{2} \in \mathcal{D}$, and $2z \in \mathcal{D}$, then

$$\Gamma(z)\Gamma(z + \tfrac{1}{2}) = 2^{1-2z}\sqrt{\pi}\Gamma(2z).$$

xxi) If $n \geq 1$, then

$$\binom{2n}{n} = \frac{\Gamma(2n + 1)}{\Gamma^2(n + 1)} = \frac{4^n\Gamma(n + \frac{1}{2})}{\sqrt{\pi}\Gamma(n + 1)},$$

$$C_n = \frac{1}{n + 1}\binom{2n}{n} = \frac{\Gamma(2n + 1)}{\Gamma(n + 1)\Gamma(n + 2)} = \frac{4^n\Gamma(n + \frac{1}{2})}{\sqrt{\pi}\Gamma(n + 2)}.$$

xxii)

$$\sum_{i=1}^{\infty} \frac{\Gamma(i+\frac{1}{2})}{i^2\Gamma(i)} = \sqrt{\pi}\log 4.$$

xxiii) If $a > 0$, then

$$\int_0^{\infty} x^{a-1}e^{-x}\log x \, \mathrm{d}x = \Gamma'(a).$$

In particular, $\Gamma'(1) = -\gamma$.

xxiv) If $z \in \mathcal{D}$, then

$$\frac{\mathrm{d}^2}{\mathrm{d}z^2}\log\Gamma(z) = \sum_{i=0}^{\infty} \frac{1}{(z+i)^2}.$$

xxv) Let $a, b \in \mathbb{C}$, and assume that $\operatorname{Re} a > 0$ and $\operatorname{Re} b > 0$. Then,

$$\int_0^1 x^{a-1}(1-x)^{b-1}\,\mathrm{d}x = \frac{\Gamma(a)\Gamma(b)}{\Gamma(a+b)}.$$

xxvi) If $z \in$ ORHP, then

$$\begin{aligned}\log\Gamma(z) &= \int_0^{\infty} \frac{1}{x}\left(\frac{z-1}{e^x} - \frac{(x+1)^{-1}-(x+1)^{-z}}{\log(x+1)}\right)\mathrm{d}x = \int_0^{\infty} \frac{1}{x}\left(\frac{z-1}{e^x} - \frac{e^{-x}-e^{-zx}}{1-e^{-x}}\right)\mathrm{d}x \\ &= (z-\tfrac{1}{2})\log z - z + \tfrac{1}{2}\log 2\pi + \int_0^{\infty} \frac{e^{-zx}}{x}\left(\frac{1}{2} - \frac{1}{x} + \frac{1}{e^x-1}\right)\mathrm{d}x \\ &= (z-\tfrac{1}{2})\log z - z + \tfrac{1}{2}\log 2\pi + 2\int_0^{\infty} \frac{\operatorname{atan}\frac{x}{z}}{e^{2\pi x}-1}\,\mathrm{d}x.\end{aligned}$$

xxvii) If $x \in (0, 1)$, then

$$\log\Gamma(x) = \tfrac{1}{2}\log 2\pi - \tfrac{1}{2}\log(2\sin\pi x) + \tfrac{1}{2}(\gamma + \log 2\pi)(1-2x) + \frac{1}{\pi}\sum_{i=1}^{\infty}\frac{\log i}{i}\sin 2\pi i x.$$

xxviii) For all $i \in \{1, \ldots, n\}$, let $0 < \alpha_i \le \beta_i \le 1$. Then,

$$1 \le \frac{\Gamma(1+\sum_{i=1}^n \alpha_i)}{\prod_{i=1}^n \Gamma(1+\frac{1}{\alpha_i})} \le \frac{\Gamma(1+\sum_{i=1}^n \beta_i)}{\prod_{i=1}^n \Gamma(1+\frac{1}{\beta_i})} \le n!.$$

xxix) Let $x > 0$. Then,

$$\frac{\Gamma(x+1)}{\Gamma(x+\frac{1}{2})} < \frac{2x+1}{\sqrt{4x+3}}, \quad [\Gamma'(x)]^2 < \Gamma(x)\Gamma''(x).$$

xxx) Let $n \ge 2$. Then,

$$\Gamma\left(\frac{1}{2^{n+1}-2}\right)\prod_{i=1}^{n-1}\Gamma\left(\frac{2^i+2^n-1}{2^{n+1}-2}\right) = 2^{n-1}\pi^{n/2}.$$

In particular,

$$\Gamma(\tfrac{1}{6})\Gamma(\tfrac{5}{6}) = 2\pi, \quad \Gamma(\tfrac{1}{14})\Gamma(\tfrac{9}{14})\Gamma(\tfrac{11}{14}) = 4\pi^{3/2},$$
$$\Gamma(\tfrac{1}{30})\Gamma(\tfrac{17}{30})\Gamma(\tfrac{19}{30})\Gamma(\tfrac{23}{30}) = 8\pi^2, \quad \Gamma(\tfrac{1}{62})\Gamma(\tfrac{33}{62})\Gamma(\tfrac{35}{62})\Gamma(\tfrac{39}{62})\Gamma(\tfrac{47}{62}) = 16\pi^{5/2}.$$

xxxi) Let $n \ge 2$, define $\Phi(n) \triangleq \{i \in \{1, \ldots, n\} \colon \gcd\{n, i\} = 1\}$, and define $\phi(n) \triangleq \operatorname{card}\Phi(n)$. Then,

$$\prod_{i\in\Phi(n)}\Gamma(\tfrac{i}{n}) = (2\pi)^{\phi(n)/2}.$$

In particular,

$$\Gamma(\tfrac{1}{14})\Gamma(\tfrac{3}{14})\Gamma(\tfrac{5}{14})\Gamma(\tfrac{9}{14})\Gamma(\tfrac{11}{14})\Gamma(\tfrac{13}{14}) = (2\pi)^3.$$

xxxii) If $z \in \mathbb{C}/(-\mathbb{P})$, then

$$\Gamma(z+1) = \lim_{n\to\infty} \frac{n!(n+1)^z}{z^{\overline{n}}}.$$

xxxiii)

$$\lim_{n\to\infty}\left(\frac{\pi^{n/2}}{\Gamma(n/2+1)}\right)^{1/\log n} = 0,\ \lim_{n\to\infty}\left(\frac{\pi^{n/2}}{\Gamma(n/2+1)}\right)^{1/(n\log n)} = \frac{1}{\sqrt{e}},\ \lim_{n\to\infty}\left(\frac{\pi^{n/2}}{\Gamma(n/2+1)}\right)^{1/(n^2\log n)} = 1.$$

xxxiv) $\lim_{n\to\infty}[n - \Gamma(\frac{1}{n})] = \gamma$ and $\lim_{n\to\infty} \frac{1}{n}\sqrt[n]{\prod_{i=1}^n \Gamma(\frac{1}{i})} = \frac{1}{e}$.

xxxv) Let x be a nonzero real number. Then,

$$|\Gamma(\jmath x)|^2 = \frac{\pi}{x\sinh \pi x}, \quad |\Gamma(1+\jmath x)|^2 = \frac{\pi x}{\sinh \pi x}.$$

xxxvi) Let x be a real number. Then,

$$|\Gamma(\tfrac{1}{2}+\jmath x)|^2 = \frac{\pi}{\cosh \pi x}.$$

xxxvii) Let x and y be real numbers such that $1+x+\jmath y, 1+x-\jmath y, 1-x+\jmath y, 1-x-\jmath y \in \mathcal{D}$. Then,

$$\Gamma(1+x+\jmath y)\Gamma(1+x-\jmath y)\Gamma(1-x+\jmath y)\Gamma(1-x-\jmath y) = \frac{2\pi^2(x^2+y^2)}{\cosh 2\pi x - \cos 2\pi x}.$$

xxxviii) Let x be a positive number, and let $n \geq 1$. Then,

$$\Gamma(nx) = \frac{n^{nx-1/2}}{(2\pi)^{(n-1)/2}}\prod_{i=0}^{n-1}\Gamma(x+\tfrac{i}{n}).$$

In particular,

$$\Gamma(2x) = \frac{2^{2x-1}}{\sqrt{\pi}}\Gamma(x)\Gamma(x+\tfrac{1}{2}), \quad \Gamma(3x) = \frac{3^{3x-1/2}}{2\pi}\Gamma(x)\Gamma(x+\tfrac{1}{3})\Gamma(x+\tfrac{2}{3}).$$

xxxix)

$$\prod_{i=1}^{8}\Gamma(\tfrac{i}{3}) = \frac{640\sqrt{3}\pi^3}{6561}.$$

Source: [116, Chapter 1], [134], [208], [516, pp. 202, 206], [352, p. 41], [969], [1217, pp. 892–898], [1350, p. 60], [1568, pp. 35, 41, 42], [1890], and [2500, Chapter 43]. *xiv*) is given in [1145], see Fact 5.5.14; *xxx*) is given in [2135]; *xxxi*) is given in [667]; *xxxiii*) is given in [78]; *xxxiv*) is given in [1103, p. 33]; *xxxvii*) is given in [1217, p. 896]. **Remark:** *xxvii*) is *Kummer's expansion*. See [116, pp. 29–32]. **Remark:** Combining *xxx*) and *xxxi*) yields $\Gamma(\frac{3}{14})\Gamma(\frac{5}{14})\Gamma(\frac{13}{14}) = 2\pi^{3/2}$. **Remark:** Γ is the *Gamma function*. **Related:** For *xxix*), see Fact 1.13.15.

Fact 13.3.3. Let $\mathcal{D} \triangleq \mathbb{C}\backslash -\mathbb{N}$, and, for all $z \in \mathcal{D}$, define

$$\psi(z) = \frac{\mathrm{d}}{\mathrm{d}z}\log\Gamma(z) = \frac{\Gamma'(z)}{\Gamma(z)}.$$

Then, the following statements hold:

i) If $z \in \mathcal{D}\backslash\mathbb{Z}$, then $\psi(1-z) = \psi(z) + \pi\cot\pi z$.

ii) Let $z \in \mathcal{D}$ and $n \geq 1$. Then, $\psi(z+n) = \psi(z) + \sum_{i=0}^{n-1}\frac{1}{z+i}$. In particular, $\psi(z+1) = \psi(z) + \frac{1}{z}$.

iii) Let $z \in \mathcal{D}$ and $n \geq 2$. Then, $\psi(nz) = \frac{1}{n}\sum_{i=0}^{n-1}\psi\left(z+\frac{i}{n}\right) + \log n$.

iv) If $z \in \mathbb{C}$, then $\lim_{n\to\infty}[\psi(z+n) - \log n] = 0$.

v) If $z \in \mathcal{D}$, then

$$\psi(z) = -\gamma + \sum_{i=0}^{\infty} \frac{z-1}{(i+1)(i+z)} = -\gamma + \sum_{i=0}^{\infty}\left(\frac{1}{i+1} - \frac{1}{i+z}\right).$$

vi) If $z \in$ OIUD, then $\psi(z+1) = -\gamma - \sum_{i=1}^{\infty}(-1)^i\zeta(i+1)z^i$.

vii) If $z \in$ ORHP, then

$$\psi(z) = \int_0^{\infty}\left(\frac{1}{xe^x} + \frac{e^{(1-z)x}}{1-e^x}\right)\mathrm{d}x = \int_0^1 \frac{1-x^{z-1}}{1-x}\,\mathrm{d}x - \gamma = H_{z-1} - \gamma.$$

viii) If $n \ge 1$, then $\psi(n) = H_{n-1} - \gamma$. In particular, $\psi(1) = -\gamma$, $\psi(2) = 1-\gamma$, and $\psi(3) = \frac{3}{2} - \gamma$.

ix) If $k \ge 0$ and $1 \le m < n$, then

$$\psi(\tfrac{m}{n} - k) = \sum_{i=0}^{k-1} \frac{n}{n(i+1)-m} + 2\sum_{i=1}^{\lfloor (n-1)/2\rfloor}\left(\cos\frac{2mi\pi}{n}\right)\log\sin\frac{i\pi}{n} - \frac{\pi}{2}\cot\frac{m\pi}{n} - \log 2n - \gamma.$$

x) If $k \ge 0$ and $1 \le m < n$, then

$$\psi(\tfrac{m}{n} + k) = \sum_{i=0}^{k-1} \frac{n}{ni+m} + 2\sum_{i=1}^{\lfloor (n-1)/2\rfloor}\left(\cos\frac{2mi\pi}{n}\right)\log\sin\frac{i\pi}{n} - \frac{\pi}{2}\cot\frac{m\pi}{n} - \log 2n - \gamma.$$

xi) If $n \in \mathbb{Z}$, then $\psi(n + \frac{1}{2}) = -2\log 2 + \sum_{i=1}^{|n|}\frac{2}{2i-1} - \gamma$. In particular, $\psi(\frac{1}{2}) = -2\log 2 - \gamma$, $\psi(\frac{3}{2}) = \psi(-\frac{1}{2}) = 2 - 2\log 2 - \gamma$, and $\psi(\frac{5}{2}) = \psi(-\frac{3}{2}) = \frac{8}{3} - 2\log 2 - \gamma$.

xii) If $n \ge 0$, then $\psi(n+\frac{1}{3}) = \sum_{i=0}^{n-1}\frac{3}{3i+1} - \frac{3\log 3}{2} - \frac{\sqrt{3}\pi}{6} - \gamma$ and $\psi(\frac{1}{3}-n) = \sum_{i=0}^{n-1}\frac{3}{3i+2} - \frac{3\log 3}{2} - \frac{\sqrt{3}\pi}{6} - \gamma$

xiii) If $n \ge 0$, then $\psi(n+\frac{2}{3}) = \sum_{i=0}^{n-1}\frac{3}{3i+2} + \frac{\sqrt{3}\pi}{6} - \frac{3\log 3}{2} - \gamma$ and $\psi(\frac{2}{3}-n) = \sum_{i=0}^{n-1}\frac{3}{3i+1} + \frac{\sqrt{3}\pi}{6} - \frac{3\log 3}{2} - \gamma$.

xiv) If $n \ge 0$, then $\psi(n+\frac{1}{4}) = \sum_{i=0}^{n-1}\frac{4}{4i+1} - \frac{\pi}{2} - 3\log 2 - \gamma$ and $\psi(\frac{1}{4}-n) = \sum_{i=0}^{n-1}\frac{4}{4i+3} - \frac{\pi}{2} - 3\log 2 - \gamma$.

xv) If $n \ge 0$, then $\psi(n+\frac{3}{4}) = \sum_{i=0}^{n-1}\frac{4}{4i+3} + \frac{\pi}{2} - 3\log 2 - \gamma$ and $\psi(\frac{3}{4}-n) = \sum_{i=0}^{n-1}\frac{4}{4i+1} + \frac{\pi}{2} - 3\log 2 - \gamma$.

xvi) If $n \ge 1$, then $\psi'(n) = \frac{\pi^2}{6} - H_{n-1,2}$. In particular, $\psi'(1) = \frac{\pi^2}{6}$, $\psi'(2) = \frac{\pi^2}{6} - 1$, and $\psi'(3) = \frac{\pi^2}{6} - \frac{5}{4}$.

xvii) If $n \ge 0$, then $\psi'(n + \frac{1}{2}) = \frac{\pi^2}{2} - \sum_{i=1}^{n}\frac{4}{(2i-1)^2}$. In particular, $\psi'(\frac{1}{2}) = \frac{\pi^2}{2}$, and $\psi'(\frac{3}{2}) = \frac{\pi^2}{2} - 4$.

xviii) If $n \ge 0$, then $\psi'(\frac{1}{2} - n) = \frac{\pi^2}{2} + \sum_{i=1}^{n}\frac{4}{(2i-1)^2}$. In particular, $\psi'(-\frac{1}{2}) = \frac{\pi^2}{2} + 4$, and $\psi'(-\frac{3}{2}) = \frac{\pi^2}{2} + \frac{40}{9}$.

xix) If $n \ge 0$, then $\psi'(n - \frac{1}{4}) = \pi^2 - 8G + 16 - \sum_{i=0}^{n-1}\frac{16}{(4i-1)^2}$ and $\psi'(\frac{1}{4} - n) = \pi^2 + 8G + \sum_{i=1}^{n}\frac{16}{(4i-1)^2}$.

xx) If $n \ge 0$, then $\psi'(n + \frac{1}{4}) = \pi^2 + 8G - 16 - \sum_{i=1}^{n-1}\frac{16}{(4i+1)^2}$ and $\psi'(\frac{3}{4} - n) = \pi^2 - 8G + \sum_{i=0}^{n-1}\frac{16}{(4i+1)^2}$.

Source: [1217, pp. 902–905], [138], and [2513, pp. 24–37]. **Remark:** ψ is the *digamma function*. **Remark:** $H_{z,p}$ is the generalized harmonic function. See Fact 13.3.4. **Related:** Fact 13.5.71.

Fact 13.3.4. For all $z \in \mathbb{C}$ such that $\operatorname{Re} z > -1$, define

$$H_z \triangleq \int_0^1 \frac{1-x^z}{1-x}\,\mathrm{d}x.$$

Then, the following statements hold:

i) If $n \ge 1$, then $H_n = \sum_{i=1}^{n}\frac{1}{i}$.

ii) If $z \in \mathbb{C}$ and $\operatorname{Re} z > -1$, then $H_z = \psi(z+1) + \gamma$.

iii) If $p > 0$, then $H_p = H_{p-1} + \frac{1}{p}$.

iv) If $p > -1$ and p is not an integer, then $H_p - H_{1-p} = \frac{1}{p} - \frac{1}{1-p} - \pi\cot p\pi$.

v) If $p > -\frac{1}{2}$, then $H_{2p} = \frac{1}{2}(H_p + H_{p-\frac{1}{2}}) + \log 2$.

vi) If $p > -\frac{2}{3}$, then $H_{3p} = \frac{1}{3}(H_p + H_{p-\frac{1}{3}} + H_{p-\frac{2}{3}}) + \log 3$.

vii) If $p > 0$, then

$$H_p = p\sum_{i=1}^{\infty}\frac{1}{i(p+i)}.$$

viii) If $p > 0$ and $n \ge 1$, then

$$\int_0^n H_x\,\mathrm{d}x = n\gamma + \log n!.$$

Next, define $H_{p,1} \triangleq H_p$, and, for all $p > -1$ and $n \ge 1$, define

$$H_{p,n+1} \triangleq \zeta(n+1) - \frac{1}{n}\frac{\mathrm{d}}{\mathrm{d}p}H_{p,n}.$$

Then, the following statements hold:

ix) If $n \ge 1$ and $k \ge 1$, then $H_{n,k} = \sum_{i=1}^{n}\frac{1}{i^k}$.

x) For all $n \ge 1$,

$$H_{p,n+1} = \zeta(n+1) + (-1)^{n+1}\frac{1}{n!}\int_0^1 \frac{x^p \log^n x}{1-x}\,\mathrm{d}x.$$

In particular,

$$H_{p,2} = \frac{\pi^2}{6} + \int_0^1 \frac{x^p\log x}{1-x}\,\mathrm{d}x, \quad H_{p,3} = \zeta(3) - \frac{1}{2}\int_0^1 \frac{x^p\log^2 x}{1-x}\,\mathrm{d}x,$$

$$H_{p,4} = \frac{\pi^4}{90} + \frac{1}{6}\int_0^1 \frac{x^p\log^3 x}{1-x}\,\mathrm{d}x, \quad H_{p,5} = \zeta(5) - \frac{1}{24}\int_0^1 \frac{x^p\log^4 x}{1-x}\,\mathrm{d}x.$$

xi) Let $n \ge 1$ and $z \in \mathbb{C}$, and assume that $\operatorname{Re} z > -1$. Then,

$$H_{z,n+1} = \zeta(n+1) + (-1)^n\frac{1}{n!}\psi^{(n)}(z+1).$$

xii) If $z \in \mathbb{C}$ and $\operatorname{Re} z > -1$, then $H_{z,2} = \frac{\pi^2}{6} - \psi'(z+1)$. In particular, $H_{\frac{1}{4},2} = 16 - 8G - \frac{5\pi^2}{6}$, $H_{\frac{1}{2},2} = 4 - \frac{\pi^2}{3}$, and $H_{\frac{3}{4},2} = 8G + \frac{16}{9} - \frac{5\pi^2}{6}$.

xiii) If $z \in \mathbb{C}$ and $\operatorname{Re} z > -1$, then $H_{z,3} = \zeta(3) + \frac{1}{2}\psi''(z+1)$. In particular, $H_{\frac{1}{4},3} = 64 - \pi^2 - 27\zeta(3)$, $H_{\frac{1}{2},3} = 8 - 6\zeta(3)$, and $H_{\frac{3}{4},3} = \frac{64}{27} + \pi^3 - 27\zeta(3)$.

xiv) If $z \in \mathbb{C}$ and $\operatorname{Re} z > -1$, then $H_{z,4} = \zeta(4) + \frac{1}{2}\psi'''(z+1)$. In particular, $H_{\frac{1}{2},4} = 16 - \frac{7\pi^4}{45}$.

Remark: $H_{z,p}$ is the *generalized harmonic function*. **Related:** Fact 13.3.3 and Fact 14.2.7.

Fact 13.3.5. For all $z \in$ CIUD, define

$$\mathrm{Li}_2(z) \triangleq \int_0^1 \frac{z\log x}{zx-1}\,\mathrm{d}x.$$

Then, the following statements hold:

i) The integral exists for all $z \in$ CIUD.

ii) Let $z \in$ CIUD. Then,

$$\mathrm{Li}_2(z) = \sum_{i=1}^{\infty}\frac{z^i}{i^2},$$

and the series converges absolutely.

iii) Li_2 has an analytic continuation on $\mathcal{D} \triangleq \mathbb{C}\backslash[1,\infty)$. In particular, for all $z \in \mathbb{C}\backslash[1,\infty)$,

$$\mathrm{Li}_2(z) = -\int_0^z \frac{\log(1-t)}{t}\,\mathrm{d}t.$$

where the integral is along an arbitrary path from 0 to z and contained in $\mathbb{C}\backslash[1,\infty)$. If the path is the line segment connecting 0 to z, then

$$\mathrm{Li}_2(z) = -\int_0^1 \frac{\log(1-tz)}{t}\,\mathrm{d}t.$$

iv) Let $x \in [0,1]$. Then,

$$\mathrm{Li}_2(x) = -\int_0^x \frac{\log(1-t)}{t}\,\mathrm{d}t = -(\log x)\log(1-x) + \int_0^x \frac{\log t}{t-1}\,\mathrm{d}t.$$

v) Let $x \in (0,1)$. Then,

$$\mathrm{Li}_2(x) + \mathrm{Li}_2(-x) = \tfrac{1}{2}\mathrm{Li}_2(x^2), \quad \mathrm{Li}_2(x) + \mathrm{Li}_2(1-x) = \frac{\pi^2}{6} - (\log x)\log(1-x),$$

$$\mathrm{Li}_2(x) + \mathrm{Li}_2\left(\frac{x}{x-1}\right) = -\tfrac{1}{2}\log^2(1-x).$$

vi) $\mathrm{Li}_2(-1) = -\frac{\pi^2}{12}, \quad \mathrm{Li}_2(0) = 0, \quad \mathrm{Li}_2(\frac{1}{2}) = \frac{\pi^2}{12} - \frac{1}{2}\log^2 2, \quad \mathrm{Li}_2(1) = \frac{\pi^2}{6}.$

vii) Let $\alpha \triangleq \frac{1}{2}(\sqrt{5}+1)$ and $\beta \triangleq 1/\alpha \triangleq \frac{1}{2}(\sqrt{5}-1)$. Then,

$$\mathrm{Li}_2(-\alpha) = -\frac{\pi^2}{10} - \log^2\alpha, \quad \mathrm{Li}_2(\beta) = \frac{\pi^2}{10} - \log^2\beta,$$

$$\mathrm{Li}_2(-\beta) = -\frac{\pi^2}{15} + \frac{1}{2}\log^2\beta, \quad \mathrm{Li}_2(1-\beta) = \frac{\pi^2}{15} - \log^2\beta = \frac{\pi^2}{15} - \frac{1}{4}\log^2(1-\beta).$$

viii) Let $x \in [-1,1]$. Then,

$$\int_0^1 \frac{\mathrm{Li}_2(t)}{(1-xt)^2}\,\mathrm{d}t = \frac{\pi^2}{6(1-x)} - \frac{\mathrm{Li}_2(x)}{x} - \frac{\log^2(1-x)}{2x}.$$

In particular,

$$\int_0^1 \frac{\mathrm{Li}_2(t)}{(1+t)^2}\,\mathrm{d}t = \frac{1}{2}\log^2 2, \quad \int_0^1 \frac{\mathrm{Li}_2(t)}{(2-t)^2}\,\mathrm{d}t = \frac{\pi^2}{24}.$$

Source: [116, pp. 102–106], [324], [1158, p. 179], and [2513, pp. 175–179]. **Remark:** Li_2 is the *dilogarithm function*. **Related:** Fact 13.5.35.

Fact 13.3.6. For all $x \in (0,1)$, the series

$$L(x) \triangleq \sum_{i=0}^{\infty}(-1)^i\frac{1}{(2i+1)^x}$$

converges. In addition, L can be extended to $\mathbb{C}$ by analytic continuation. Furthermore, for all $z \in \mathbb{C}$,

$$L(1-z) = \left(\frac{2}{\pi}\right)^z\left(\sin\frac{\pi z}{2}\right)\Gamma(z)L(z).$$

Finally,

$$L(0) = \frac{1}{2}, \quad L(1) = \frac{\pi}{4}, \quad L'(0) = \log\Gamma(\tfrac{1}{4}) - \log\Gamma(\tfrac{3}{4}) - \log 2, \quad L'(1) = \frac{\pi\gamma}{4} + \frac{\pi}{2}\log\frac{\sqrt{2\pi}\Gamma(\frac{3}{4})}{\Gamma(\frac{1}{4})}.$$

Source: [2783]. **Remark:** L is the *Dirichlet L function*. **Related:** Fact 14.6.23.

13.4 Facts on Power Series, Laurent Series, and Partial Fraction Expansions

Fact 13.4.1. Consider the power series

$$f(z) \triangleq \sum_{i=1}^{\infty} z^i, \quad g(z) \triangleq \sum_{i=1}^{\infty} \frac{z^i}{i}, \quad h(z) \triangleq \sum_{i=1}^{\infty} \frac{z^i}{i^2}.$$

Then, the following statements hold:

i) The radius of convergence of f, g, and h is 1.

ii) The domain of convergence of f is $\{z \in \mathbb{C}\colon |z| < 1\}$.

iii) The domain of convergence of g is $\{z \in \mathbb{C}\colon |z| \le 1\} \backslash \{1\}$.

iv) The domain of convergence of h is $\{z \in \mathbb{C}\colon |z| \le 1\}$.

Source: [2113, p. 113].

Fact 13.4.2. Let $z \in \mathbb{C}$. Then, the following statements hold:

i) Let $\alpha \in \mathbb{C}$, and assume that one of the following conditions is satisfied:

a) $|z| < 1$.

b) $z = -1$ and $\alpha \in \text{ORHP} \cup \{0\}$.

c) $z \in \text{UC} \backslash \{-1\}$ and $\operatorname{Re} \alpha > -1$.

d) $|z| > 1$ and $\alpha \in \mathbb{N}$.

Then,

$$(1+z)^\alpha = \sum_{i=0}^{\infty} \frac{\alpha^{\underline{i}}}{i!} z^i.$$

Furthermore, as $z \to 0$,

$$\begin{aligned}(1+z)^\alpha &= \binom{\alpha}{0} + \binom{\alpha}{1} z + \binom{\alpha}{2} z^2 + \binom{\alpha}{3} z^3 + \binom{\alpha}{4} z^4 + O(z^5) \\ &= 1 + \alpha z + \frac{\alpha(\alpha-1)}{2!} z^2 + \frac{\alpha(\alpha-1)(\alpha-2)}{3!} z^3 + \frac{\alpha(\alpha-1)(\alpha-2)(\alpha-3)}{4!} z^4 + O(z^5).\end{aligned}$$

ii) Let $\alpha \in \mathbb{C}$. For all $|z| < 1$ and, as $z \to 0$,

$$\begin{aligned}\frac{1}{(1-z)^\alpha} &= \sum_{i=0}^{\infty} \frac{\alpha^{\overline{i}}}{i!} z^i = \binom{\alpha-1}{0} + \binom{\alpha}{1} z + \binom{\alpha+1}{2} z^2 + \binom{\alpha+2}{3} z^3 + \binom{\alpha+3}{4} z^4 + O(z^5) \\ &= 1 + \alpha z + \frac{(\alpha+1)\alpha}{2!} z^2 + \frac{(\alpha+2)(\alpha+1)\alpha}{3!} z^3 + \frac{(\alpha+3)(\alpha+2)(\alpha+1)\alpha}{4!} z^4 + O(z^5).\end{aligned}$$

iii) For all $|z| < 1$ and, as $z \to 0$,

$$\frac{1}{1-z} = \sum_{i=0}^{\infty} z^i = 1+z+z^2+z^3+z^4+O(z^5), \quad \frac{1}{1+z} = \sum_{i=0}^{\infty} (-1)^i z^i = 1-z+z^2-z^3+z^4-O(z^5).$$

iv) For all $|z| < 1$ and, as $z \to 0$,

$$\frac{1}{(1-z)^2} = \sum_{i=0}^{\infty} (i+1) z^i = 1 + 2z + 3z^2 + 4z^3 + 5z^4 + O(z^5),$$

$$\frac{1}{(1+z)^2} = \sum_{i=0}^{\infty} (-1)^i (i+1) z^i = 1 - 2z + 3z^2 - 4z^3 + 5z^4 - O(z^5).$$

v) Let $k \geq 1$. Then, for all $|z| < 1$,

$$\frac{1}{(1-z)^k} = \sum_{i=0}^{\infty} \binom{i+k-1}{i} z^i, \quad \frac{1}{(1+z)^k} = \sum_{i=0}^{\infty} (-1)^i \binom{i+k-1}{i} z^i,$$

$$\frac{1}{(1-z)^{k+1}} = \sum_{i=0}^{\infty} \binom{k+i}{i} z^i, \quad \frac{z^k}{(1-z)^{k+1}} = \sum_{i=k}^{\infty} \binom{i}{k} z^i, \quad \frac{(k+1)z}{(1-z)^{k+2}} = \sum_{i=1}^{\infty} i \binom{k+i}{i} z^i.$$

vi) For all $|z| < 1$ and, as $z \to 0$,

$$\frac{z}{1-z} = \sum_{i=1}^{\infty} z^i = z + z^2 + z^3 + z^4 + z^5 + O(z^6),$$

$$\frac{z}{1+z} = \sum_{i=1}^{\infty} (-1)^{i+1} z^i = z - z^2 + z^3 - z^4 + z^5 - O(z^6),$$

$$\frac{z}{(1-z)^2} = \sum_{i=1}^{\infty} i z^i = z + 2z^2 + 3z^3 + 4z^4 + 5z^5 + O(z^6),$$

$$\frac{z}{(1+z)^2} = \sum_{i=1}^{\infty} (-1)^{i+1} i z^i = z - 2z^2 + 3z^3 - 4z^4 + 5z^5 - O(z^6),$$

$$\frac{z}{(1-z)^3} = \sum_{i=1}^{\infty} \tfrac{1}{2} i(i+1) z^i = z + 3z^2 + 6z^3 + 10z^4 + 15z^5 + O(z^6),$$

$$\frac{z}{(1+z)^3} = \sum_{i=1}^{\infty} (-1)^{i+1} \tfrac{1}{2} i(i+1) z^i = z - 3z^2 + 6z^3 - 10z^4 + 15z^5 - O(z^6),$$

$$\frac{z(z+1)}{(1-z)^2} = \sum_{i=i}^{\infty} i^2 z^i = z + 4z^2 + 9z^3 + 16z^4 + O(z^5).$$

vii) For all $|z| > 1$ and, as $z \to \infty$,

$$\frac{z}{z-1} = \sum_{i=0}^{\infty} \frac{1}{z^i} = 1 + \frac{1}{z} + \frac{1}{z^2} + \frac{1}{z^3} + \frac{1}{z^4} + O(z^{-5}),$$

$$\frac{z}{z+1} = \sum_{i=0}^{\infty} (-1)^i \frac{1}{z^i} = 1 - \frac{1}{z} + \frac{1}{z^2} - \frac{1}{z^3} + \frac{1}{z^4} - O(z^{-5}).$$

viii) For all $|z| > 1$ and, as $z \to \infty$,

$$\frac{z^2}{(z-1)^2} = \sum_{i=0}^{\infty} \frac{i+1}{z^i} = 1 + \frac{2}{z} + \frac{3}{z^2} + \frac{4}{z^3} + \frac{5}{z^4} + O(z^{-5}),$$

$$\frac{z^2}{(z+1)^2} = \sum_{i=0}^{\infty} (-1)^i \frac{i+1}{z^i} = 1 - \frac{2}{z} + \frac{3}{z^2} - \frac{4}{z^3} + \frac{5}{z^4} - O(z^{-5}).$$

ix) For all $|z| > 1$ and, as $z \to \infty$,

$$\frac{z}{(z-1)^2} = \sum_{i=1}^{\infty} \frac{i}{z^i} = \frac{1}{z} + \frac{2}{z^2} + \frac{3}{z^3} + \frac{4}{z^4} + \frac{5}{z^5} + O(z^{-6}),$$

$$\frac{z}{(z+1)^2} = \sum_{i=1}^{\infty} (-1)^{i+1} \frac{i}{z^i} = \frac{1}{z} - \frac{2}{z^2} + \frac{3}{z^3} - \frac{4}{z^4} + \frac{5}{z^5} - O(z^{-6}).$$

x) Let $a, b \in \mathbb{C}$. Then, for all $|z| < \min\{1/|a|, 1/|b|\}$ and, as $z \to 0$,

$$\frac{1+bz}{1+az} = 1 + \sum_{i=1}^{\infty}(-1)^i a^{i-1}(a-b)z^i = 1 + (b-a)z + a(a-b)z^2 + a^2(b-a)z^3 + O(z^4).$$

xi) Let $n \geq 1$. Then, as $z \to 0$,

$$\left(\frac{1+bz}{1+az}\right)^n = 1 + n(b-a)z + \tfrac{1}{2}n(a-b)[(n+1)a-(n-1)b]z^2$$
$$+ [na^2(b-a) + \tfrac{1}{6}n(n-1)(n-2)(b-a)^3 - n(n-1)a(a-b)^2]z^3 + O(z^4).$$

In particular, as $z \to 0$,

$$\left(\frac{1+bz}{1+az}\right)^2 = 1 + 2(b-a)z + (a-b)(3a-b)z^2 + 2a(b-a)(2a-b)z^3 + O(z^4).$$

xii) For all $|z| < 1$ and, as $z \to 0$,

$$\left(\frac{1+z}{1-z}\right)^3 = 1 + \sum_{i=1}^{\infty}(4i^2+2)z^i = 1 + 6z + 18z^2 + 38z^3 + 66z^4 + 102z^5 + O(z^6).$$

xiii) For all $|z| < 1$ and, as $z \to 0$,

$$\frac{z^3}{(1-z)^2(1+z)} = \frac{z^3+z^4}{(1-z^2)^2} = \sum_{i=3}^{\infty}\left\lfloor\frac{i-1}{2}\right\rfloor z^i = z^3 + z^4 + 2z^5 + 2z^6 + O(z^7).$$

xiv) For all $|z| < 1$,

$$\sqrt{z+1} = 1 + \sum_{i=1}^{\infty}(-1)^{i+1}\frac{(2i)!}{4^i(i!)^2(2i-1)}z^i, \qquad \sqrt{1-z} = 1 - \sum_{i=1}^{\infty}\frac{(2i)!}{4^i(i!)^2(2i-1)}z^i,$$

$$\frac{1}{\sqrt{1-z}} = \sum_{i=0}^{\infty}\binom{-\frac{1}{2}}{i}(-z)^i = \sum_{i=0}^{\infty}\frac{(2i-1)!!}{(2i)!!}z^i, \qquad \frac{1}{\sqrt{1-z^2}} = \sum_{i=0}^{\infty}\frac{1}{4^i}\binom{2i}{i}z^{2i}.$$

xv) For all $|z| < \frac{1}{4}$,

$$\frac{1}{\sqrt{1-4z}} = \sum_{i=0}^{\infty}\binom{2i}{i}z^i.$$

xvi) For all $|z| < \frac{1}{4}$,

$$\frac{2z}{(1-4z)^{3/2}} = \sum_{i=1}^{\infty}i\binom{2i}{i}z^i.$$

xvii) For all $|z| < \frac{1}{4}$,

$$\frac{2z(2z+1)}{(1-4z)^{5/2}} = \sum_{i=1}^{\infty}i^2\binom{2i}{i}z^i.$$

xviii) For all $|z| < 1$,

$$\frac{z\operatorname{asin} z}{(1-z^2)^{3/2}} + \frac{z^2}{1-z^2} = \sum_{i=1}^{\infty}\frac{4^i}{\binom{2i}{i}}z^{2i}.$$

xix) For all $|z| < \frac{1}{4}$,

$$2\log\frac{1-\sqrt{1-4z}}{2z} = \sum_{i=1}^{\infty}\frac{1}{i}\binom{2i}{i}z^i.$$

xx) For all $|z| < \frac{1}{4}$,

$$\frac{1-\sqrt{1-4z}}{2z} = \frac{2}{1+\sqrt{1-4z}} = \sum_{i=0}^{\infty} \frac{1}{i+1}\binom{2i}{i} z^i = \sum_{i=0}^{\infty} C_i z^i.$$

xxi) For all $|z| < 4$,

$$\frac{10z-z^2}{(z-4)^2} + 24\sqrt{\frac{z}{(4-z)^5}}\,\mathrm{asin}\,\frac{\sqrt{z}}{2} = \frac{10z-z^2}{(z-4)^2} + 24\sqrt{\frac{z}{(4-z)^5}}\,\mathrm{atan}\,\sqrt{\frac{z}{4-z}}$$
$$= \sum_{i=1}^{\infty} \frac{i+1}{\binom{2i}{i}} z^i = \sum_{i=1}^{\infty} \frac{z^i}{C_i}.$$

xxii) For all $|z| < \frac{1}{4}$,

$$\frac{2}{\sqrt{1-4z}} \log \frac{1+\sqrt{1-4z}}{2\sqrt{1-4z}} = \sum_{i=0}^{\infty} \binom{2i}{i} H_i z^i.$$

xxiii) For all $|z| < \frac{1}{4}$,

$$\frac{2}{\sqrt{1+4z}} \log \frac{2\sqrt{1+4z}}{1+\sqrt{1+4z}} = \sum_{i=0}^{\infty} (-1)^{i+1}\binom{2i}{i} H_i z^i.$$

xxiv) For all $|z| < \frac{1}{4}$,

$$\sqrt{1-4z}\log(2\sqrt{1-4z}) - (1+\sqrt{1-4z})\log(1+\sqrt{1-4z}) + \log 2 = \sum_{i=1}^{\infty} C_i H_i z^{i+1}.$$

xxv) Let $k \ge 1$. Then, for all $|z| < \frac{1}{4}$,

$$\frac{1}{\sqrt{1-4z}} \sum_{i=0}^{k} i!\binom{2i}{i}\left\{ k \atop i \right\}\left(\frac{z}{1-4z}\right)^i = \sum_{i=0}^{\infty} i^k \binom{2i}{i} z^i.$$

In particular,

$$\frac{2z}{(1-4z)\sqrt{1-4z}} = \sum_{i=0}^{\infty} i\binom{2i}{i} z^i, \quad \frac{2z(2z+1)}{(1-4z)^2\sqrt{1-4z}} = \sum_{i=0}^{\infty} i^2\binom{2i}{i} z^i,$$
$$\frac{2z(4z^2+10z+1)}{(1-4z)^3\sqrt{1-4z}} = \sum_{i=0}^{\infty} i^3\binom{2i}{i} z^i.$$

xxvi) For all $|z| < \frac{1}{4}$,

$$\frac{1}{\sqrt{1+4z}} \sum_{i=1}^{\infty} \frac{1}{i}\binom{2i}{i}\left(\frac{z}{1+4z}\right)^i = \sum_{i=0}^{\infty} (-1)^{i+1} H_i \binom{2i}{i} z^i.$$

xxvii) For all $|z| < \frac{1}{3}$,

$$\frac{1}{\sqrt{1-2z-3z^2}} = \sum_{i=0}^{\infty} \sum_{j=0}^{\lfloor i/2 \rfloor} \binom{i}{2j}\binom{2j}{j} z^i.$$

xxviii) Let $a, b \in \mathbb{C}$, where $a \ne b$. Then, for all $|z-b| < |a-b|$,

$$\frac{1}{z-a} = \sum_{i=0}^{\infty} \frac{(-1)^i}{(b-a)^{i+1}} (z-b)^i.$$

xxix) Let $b \in \mathbb{R}$, and define $\phi \triangleq \arg(\jmath - b)$. Then, for all $x \in \mathbb{R}$ such that $|x - b| < \sqrt{1+b^2}$,

$$\frac{1}{x^2+1} = \sum_{i=0}^{\infty} \frac{\sin[(i+1)\phi]}{(\sqrt{1+b^2})^{i+1}}(x-b)^i.$$

xxx) Let $k \geq 0$. Then, for all $|z| < 1$,

$$\sum_{i=1}^{\infty} \frac{i^k z^i}{1-z^i} = \sum_{i=1}^{\infty} s_{i,k} z^i,$$

where $s_{i,k}$ is the sum of the kth powers of the distinct divisors of i. In particular, $s_{i,0}$ is the number of distinct divisors of i. (Example: $s_{12,0} = 6$ and $s_{12,1} = 28$.) Hence, as $z \to 0$,

$$\sum_{i=1}^{\infty} \frac{z^i}{1-z^i} = z+2z^2+2z^3+3z^4+2z^5+O(z^6), \quad \sum_{i=1}^{\infty} \frac{iz^i}{1-z^i} = z+3z^2+4z^3+7z^4+6z^5+O(z^6),$$

$$\sum_{i=1}^{\infty} \frac{i^2 z^i}{1-z^i} = z+5z^2+10z^3+21z^4+26z^5+O(z^6), \quad \sum_{i=1}^{\infty} \frac{i^3 z^i}{1-z^i} = z+9z^2+28z^3+73z^4+126z^5+O(z^6).$$

xxxi) For all $|z| < 1$ and, as $z \to 0$,

$$\frac{z^3}{(1-z^2)(1-z^3)(1-z^4)} = \sum_{i=0}^{\infty} \left(\left\lfloor \frac{i^2+6}{12} \right\rfloor - \left\lfloor \frac{i}{4} \right\rfloor \left\lfloor \frac{i+2}{4} \right\rfloor \right) z^i = z^3 + z^5 + z^6 + 2z^7 + z^8 + O(z^9).$$

xxxii) For all $|z| < 1$ and, as $z \to 0$,

$$\frac{2z\log(1-z)}{(z-2)^3} + \frac{z(4-3z)}{(z-1)^2(z-2)^2} = \sum_{i=1}^{\infty} \left[\sum_{j=1}^{i} j\binom{n}{j}^{-1} \right] z^i = z + \frac{5}{2}z^2 + 4z^3 + \frac{16}{3}z^4 + O(z^5).$$

Source: *xii)* is given in [1757, p. 191]; *xiii)* is given in [1165]; *xiv)* is given in [1206] and [2880, p. 134]; *xv)* is given in [1217, p. 11] and [511, p. 66]; *xvi)* and *xvii)* are given in [1781]; *xviii)* is given in [1675, p. 86]; *xix)* is given in [1675, p. 87]; *xx)* is given in [1158, pp. 299, 300]; *xxi)* is given in [343]; *xxii)–xxvi)* are given in [556]; *xxvii)* is given in [2068, p. 100]; *xxviii)* and *xxix)* are given in [2113, pp. 76, 77]; *xxx)* is given in [900, 2175, 2800]; *xxxi)* is given in [335]; *xxxii)* is given in [1944]. **Remark:** In *xii)–xvi)* the square root function is the principal branch, whose values have imaginary part in $(-\frac{\pi}{2}, \frac{\pi}{2}]$. **Remark:** *i)* is the *binomial series*. Convergence for real z and α is discussed in [1790]. **Remark:** $\binom{i+k-1}{i}$ in *v)* is the binomial coefficient with repetition $\left(\binom{i}{k}\right)_{\rm r}$. See [1974, pp. 16, 49] and Fact 1.12.1. **Remark:** *xx)* is the generating function for the Catalan numbers. See Fact 1.18.4. **Remark:** *xxvii)* is the generating function for the *central trinomial coefficients*. See Fact 2.1.10. **Remark:** $s_{n,1}$ in *xxx)* is written as s_n in Fact 1.20.2. The generating function for the sequence of divisor sums $(s_i)_{i=1}^{\infty}$ given by $\sum_{i=1}^{\infty} \frac{iz^i}{1-z^i} = \sum_{i=1}^{\infty} s_i z^i$.

Fact 13.4.3. Let $z_1, \ldots, z_n \in \mathbb{C}$ be nonzero, and define $p \in \mathbb{C}[s]$ by

$$p(z) = \prod_{i=1}^{n} (1 - z_i z) = \sum_{i=1}^{n} \alpha_i(z_1, \ldots, z_n) z^i.$$

Furthermore, for all $z \in \mathcal{D} \triangleq \{z \in \mathbb{C} \colon |z| < \min\{1/|z_1|, \ldots, 1/|z_n|\}$, consider the power series

$$\frac{1}{\prod_{i=1}^{n}(1 - z_i z)} = \sum_{i=0}^{\infty} \beta_i(z_1, \ldots, z_n) z^i.$$

Then, for all $z \in \mathcal{D}$ and $k \geq 1$,

$$\sum_{i=1}^{k} i\alpha_i(z_1, \ldots, z_n)\beta_{k-i}(z_1, \ldots, z_n) = \sum_{i=1}^{n} z_i^k.$$

Source: [1862].

Fact 13.4.4. The following statements hold:

i) Let $n \geq 1$ and $|z| < 1$. Then,

$$\frac{\log^n(1+z)}{n!} = \sum_{i=n}^{\infty} (-1)^{i-n} \frac{1}{i!} \begin{bmatrix} n \\ i \end{bmatrix} z^i.$$

ii) Let $n \geq 1$ and $z \in \mathbb{C}$. Then,

$$\frac{(e^z - 1)^n}{n!} = \sum_{i=n}^{\infty} \frac{1}{i!} \begin{Bmatrix} i \\ n \end{Bmatrix} z^i.$$

iii) Let $n \geq 1$ and $|z| < 1/n$. Then,

$$\prod_{i=1}^{n} \frac{z}{1 - iz} = \sum_{i=n}^{\infty} \begin{Bmatrix} i \\ n \end{Bmatrix} z^i.$$

In particular, as $z \to 0$,

$$\frac{z}{1-z} = \sum_{i=1}^{\infty} \begin{Bmatrix} i \\ 1 \end{Bmatrix} z^i = z + z^2 + z^3 + z^4 + z^5 + O(z^6),$$

$$\frac{z^2}{(1-z)(1-2z)} = \sum_{i=2}^{\infty} \begin{Bmatrix} i \\ 2 \end{Bmatrix} z^i = z^2 + 3z^3 + 7z^4 + 15z^5 + 31z^6 + O(z^7),$$

$$\frac{z^3}{(1-z)(1-2z)(1-3z)} = \sum_{i=3}^{\infty} \begin{Bmatrix} i \\ 3 \end{Bmatrix} z^i = z^3 + 6z^4 + 25z^5 + 90z^6 + 301z^7 + O(z^8).$$

iv) Let $n \geq 1$ and $|z| > 1$. Then,

$$\frac{1}{\prod_{i=1}^{n}(z+i)} = \sum_{i=n}^{\infty} (-1)^{i-n} \begin{Bmatrix} i \\ n \end{Bmatrix} \frac{1}{z^i}.$$

v) Let $x, y \in \mathbb{C}$. Then,

$$\exp[y(e^x - 1)] = \sum_{i,j=0}^{\infty} \frac{1}{i!} \begin{Bmatrix} i \\ j \end{Bmatrix} x^i y^j.$$

vi) Let $n \geq 1$ and $|z| < 1$. Then,

$$\frac{1}{(1-z)^{n+1}} \sum_{i=0}^{n} \left\langle \begin{matrix} n \\ i \end{matrix} \right\rangle z^{i+1} = \sum_{i=1}^{\infty} i^n z^i.$$

Source: *i*) is given in [571, p. 624], [2100], and [3024, p. 139]. *ii*) is given in [34, p. 158], [555], and [571, p. 625]; *iii*) is given in [349] and [411, p. 104]; *iv*) follows from *iii*) and is given in [2235]; *v*) is given in [571, p. 625]. **Related:** Fact 1.19.3.

Fact 13.4.5. For all $n \geq 1$, let H_n denote the nth harmonic number, and let $x \in \mathbb{C}$, where $|x| < 1$. Then,

$$\frac{x^2[3 - 2\log(1-x)]}{(1-x)^3} = \sum_{i=2}^{\infty} i(i-1)H_i x^i, \quad \frac{x^2[6 + 5x - (4+2x)\log(1-x)]}{(1-x)^4} = \sum_{i=2}^{\infty} i^2(i-1)H_i x^i,$$

$$\frac{x^3[11 - 6\log(1 - x)]}{(1 - x)^4} = \sum_{i=3}^{\infty} i(i - 1)(i - 2)H_i x^i,$$

$$\frac{x^3[33 + 17x - (18 + 6x)\log(1 - x)]}{(1 - x)^5} = \sum_{i=3}^{\infty} i^2(i - 1)(i - 2)H_i x^i.$$

Source: [899].

Fact 13.4.6. For all $n \geq 1$, let $T_n \triangleq \frac{1}{2}n(n + 1)$ denote the nth triangular number. Then, the following statements hold:

i) For all $z \in \mathbb{C}$ such that $|z| < 1$,

$$\frac{z}{(1 - z)^3} = \sum_{i=1}^{\infty} T_i z^i.$$

ii) For all $z \in \mathbb{C}$,

$$(1 + 2z + \tfrac{1}{2}z^2)e^z = \sum_{i=1}^{\infty} \frac{T_i}{i!} z^i.$$

iii) $$\sum_{i=1}^{\infty} \frac{1}{T_i} = 2, \quad \sum_{i=1}^{\infty} \frac{1}{T_i^2} = \frac{4}{3}(\pi^2 - 9), \quad \sum_{i=1}^{\infty} \frac{1}{T_i^3} = 8(10 - \pi^2), \quad \sum_{i=1}^{\infty} \frac{1}{T_i^4} = \frac{16}{45}(\pi^4 + 150\pi^2 - 1575),$$

$$\sum_{i=1}^{\infty} \frac{1}{T_i^5} = \frac{32}{9}(1134 - 105\pi^2 - \pi^4), \quad \sum_{i=1}^{\infty} \frac{1}{T_i^6} = \frac{64}{945}(2\pi^6 + 441\pi^4 + 39690\pi^2 - 436590).$$

Related: Fact 1.12.3.

Fact 13.4.7. For all $n \geq 1$, let $P_n \triangleq \frac{1}{2}n(3n - 1)$ denote the nth pentagonal number, let $P'_n \triangleq \frac{1}{2}n(3n + 1)$ denote the nth dual pentagonal number, and let g_i denote the nth generalized pentagonal number. Then, the following statements hold:

i) For all $z \in \mathbb{C}$ such that $|z| < 1$,

$$\frac{z(2z + 1)}{(1 - z)^3} = \sum_{i=1}^{\infty} P_i z^i, \quad \frac{2z^2 + 1}{(1 - z^2)^3} = \sum_{i=0}^{\infty} P_{i+1} z^{2i}.$$

ii) For all $z \in \mathbb{C}$ such that $|z| < 1$,

$$\frac{z(z + 2)}{(1 - z)^3} = \sum_{i=1}^{\infty} P'_i z^i, \quad \frac{z(z^2 + 2)}{(1 - z^2)^3} = \sum_{i=1}^{\infty} P'_i z^{2i-1}.$$

iii) For all $z \in \mathbb{C}$ such that $|z| \neq 1$,

$$\frac{z^2 + z + 1}{(1 + z)^2(1 - z)^3} = \frac{2z^2 + 1}{(1 - z^2)^3} + \frac{z(z^2 + 2)}{(1 - z^2)^3}.$$

iv) For all $z \in \mathbb{C}$ such that $|z| < 1$,

$$\frac{z^2 + z + 1}{(1 + z)^2(1 - z)^3} = \sum_{i=1}^{\infty} g_i z^i.$$

v) For all $z \in \mathbb{C}$ such that $|z| > 1$,

$$\frac{z^2 + z + 1}{(1 + z)^2(1 - z)^3} = -\sum_{i=1}^{\infty} g_i \frac{1}{z^i}.$$

vi) $$\sum_{i=1}^{\infty} \frac{1}{P_i} = 3\log 3 - \frac{\sqrt{3}\pi}{3}, \quad \sum_{i=1}^{\infty} \frac{1}{P'_i} = 6 - 3\log 3 - \frac{\sqrt{3}\pi}{3},$$

$$\sum_{i=1}^{\infty} \frac{1}{P_i + P'_i} = \frac{\pi^2}{18}, \quad \sum_{i=1}^{\infty} \frac{1}{(P_i + P'_i)^2} = \frac{\pi^4}{810},$$

$$\sum_{i=1}^{\infty} \frac{1}{P_i P'_i} = 18 - 2\sqrt{3} - \frac{2\pi^2}{3}, \quad \sum_{i=1}^{\infty} \frac{1}{(P_i P'_i)^2} = 180\sqrt{3}\pi + 96\pi^2 + \frac{8\pi^4}{45} - 1944.$$

Remark: *i*) and *ii*) give generating functions for the pentagonal and dual pentagonal numbers. *iv*) shows that $g(z) = \frac{z^2+z+1}{(1+z)^2(1-z)^3}$ is a generating function for the generalized pentagonal numbers. See [2242, p. A.141]. *v*) follows from $g(1/z) = -z^3 g(z)$. **Related:** Fact 1.12.5 and Fact 13.10.27.

Fact 13.4.8. The following statements hold:

i) For all $z \in \mathbb{C}$ and, as $z \to 0$,

$$\sin z = z - \frac{1}{3!}z^3 + \frac{1}{5!}z^5 - \frac{1}{7!}z^7 + O(z^9), \quad \cos z = 1 - \frac{1}{2!}z^2 + \frac{1}{4!}z^4 - \frac{1}{6!}z^6 + O(z^8).$$

ii) For all $z \in \mathbb{C}$ such that $|z| < \frac{\pi}{2}$ and, as $z \to 0$,

$$\tan z = \sum_{i=1}^{\infty} (-1)^{i+1} \frac{4^i(4^i - 1)B_{2i}}{(2i)!} z^{2i-1} = z + \frac{1}{3}z^3 + \frac{2}{15}z^5 + \frac{17}{315}z^7 + \frac{62}{2835}z^9 + O(z^{11}).$$

iii) For all $z \in \mathbb{C}$ such that $0 < |z| < \pi$ and, as $z \to 0$,

$$\csc z = \frac{1}{z} + \sum_{i=1}^{\infty} \frac{2(2^{2i-1} - 1)|B_{2i}|}{(2i)!} z^{2i-1} = \frac{1}{z} + \frac{1}{6}z + \frac{7}{360}z^3 + \frac{31}{15120}z^5 + O(z^7).$$

iv) For all $z \in \mathbb{C}$ such that $|z| < \frac{\pi}{2}$ and, as $z \to 0$,

$$\sec z = 1 + \sum_{i=1}^{\infty} \frac{|E_{2i}|}{(2i)!} z^{2i} = 1 + \frac{1}{2}z^2 + \frac{5}{24}z^4 + \frac{61}{720}z^6 + O(z^8).$$

v) For all $z \in \mathbb{C}$ such that $0 < |z| < \pi$ and, as $z \to 0$,

$$\cot z = \frac{1}{z} - \sum_{i=1}^{\infty} \frac{4^i |B_{2i}|}{(2i)!} z^{2i-1} = \frac{1}{z} - \frac{1}{3}z - \frac{1}{45}z^3 - \frac{2}{945}z^5 - O(z^7).$$

vi) For all $z \in \mathbb{C}$ such that $|z| < \pi$ and, as $z \to 0$,

$$z \cot z = -2 \sum_{i=0}^{\infty} \frac{\zeta(2i)}{\pi^{2i}} z^{2i} = 1 - \frac{1}{3}z^2 - \frac{1}{45}z^4 - \frac{2}{945}z^6 - O(z^8).$$

vii) For all $z \in \mathbb{C}$ such that $|z| < \pi$ and, as $z \to 0$,

$$\log \frac{\sin z}{z} = -\sum_{i=1}^{\infty} \frac{2^{2i-1}|B_{2i}|}{i(2i)!} z^{2i} = -\frac{1}{6}z^2 - \frac{1}{180}z^4 - \frac{1}{2835}z^6 - O(z^8).$$

viii) For all $z \in \mathbb{C}$ such that $|z| < \frac{\pi}{2}$ and, as $z \to 0$,

$$\log \cos z = -\sum_{i=1}^{\infty} \frac{2^{2i-1}(2^{2i} - 1)|B_{2i}|}{i(2i)!} z^{2i} = -\frac{1}{2}z^2 - \frac{1}{12}z^4 - \frac{1}{45}z^6 - O(z^8).$$

ix) For all $z \in \mathbb{C}$ such that $|z| < \frac{\pi}{2}$ and, as $z \to 0$,

$$\log \frac{\tan z}{z} = \sum_{i=1}^{\infty} \frac{4^i(2^{2i-1} - 1)|B_{2i}|}{i(2i)!} z^{2i} = \frac{1}{3}z^2 + \frac{7}{90}z^4 + \frac{62}{2835}z^6 + O(z^8).$$

Remark: Each series is either a power series or a Laurent series.

Fact 13.4.9. The following statements hold:

i) For all $z \in \mathbb{C}$ such that $|z| \le 1$ and, as $z \to 0$,

$$\operatorname{asin} z = \sum_{i=0}^{\infty} \frac{\binom{2i}{i}}{4^i(2i+1)} z^{2i+1} = \sum_{i=0}^{\infty} \frac{(2i)!}{4^i(i!)^2(2i+1)} z^{2i+1} = \sum_{i=0}^{\infty} \frac{(2i-1)!!}{(2i)!!(2i+1)} z^{2i+1}$$
$$= z + \frac{1}{6}z^3 + \frac{3}{40}z^5 + \frac{5}{112}z^7 + \frac{35}{1152}z^3 + \frac{63}{2816}z^{11} + \frac{231}{13312}z^{13} + O(z^{15}).$$

ii) For all $z \in \mathbb{C}$ such that $|z| \le 1$ and, as $z \to 0$,

$$\operatorname{acos} z = \frac{\pi}{2} - \operatorname{asin} z = \frac{\pi}{2} - \sum_{i=0}^{\infty} \frac{\binom{2i}{i}}{4^i(2i+1)} z^{2i+1} = \frac{\pi}{2} - \sum_{i=0}^{\infty} \frac{(2i)!}{4^i(i!)^2(2i+1)} z^{2i+1}$$
$$= \frac{\pi}{2} - \sum_{i=0}^{\infty} \frac{(2i-1)!!}{(2i)!!(2i+1)} z^{2i+1} = \frac{\pi}{2} - \left(z + \frac{1}{6}z^3 + \frac{3}{40}z^5 + \frac{5}{112}z^7 + O(z^9)\right).$$

iii) For all $z \in \mathbb{C}$ such that $|z| \le 1$ and $z \notin \{-\jmath, \jmath\}$ and, as $z \to 0$,

$$\operatorname{atan} z = \sum_{i=0}^{\infty} (-1)^i \frac{1}{2i+1} z^{2i+1} = z - \frac{1}{3}z^3 + \frac{1}{5}z^5 - \frac{1}{7}z^7 + \frac{1}{9}z^9 - \frac{1}{11}z^{11} + O(z^{13}).$$

iv) For all $z \in \mathbb{C}$ such that $|z| \ge 1$ and $z \notin \{-\jmath, \jmath\}$ and, as $z \to \infty$,

$$\operatorname{atan} z = (\operatorname{sign} \operatorname{Re} z)\frac{\pi}{2} - \operatorname{atan} \frac{1}{z} = (\operatorname{sign} \operatorname{Re} z)\frac{\pi}{2} + \sum_{i=0}^{\infty} (-1)^{i+1} \frac{1}{2i+1} \frac{1}{z^{2i+1}}$$
$$= (\operatorname{sign} \operatorname{Re} z)\frac{\pi}{2} - \frac{1}{z} + \frac{1}{3}\frac{1}{z^3} - \frac{1}{5}\frac{1}{z^5} + \frac{1}{7}\frac{1}{z^7} - \frac{1}{9}\frac{1}{z^9} + \frac{1}{11}\frac{1}{z^{11}} - O(z^{-13}).$$

v) For all $z \in \mathbb{C}$ such that $|z| \ge 1$ and, as $z \to \infty$,

$$\operatorname{acsc} z = \operatorname{asin} \frac{1}{z} = \sum_{i=0}^{\infty} \frac{\binom{2i}{i}}{4^i(2i+1)} \frac{1}{z^{2i+1}} = \sum_{i=0}^{\infty} \frac{(2i)!}{4^i(i!)^2(2i+1)} \frac{1}{z^{2i+1}}$$
$$= \sum_{i=0}^{\infty} \frac{(2i-1)!!}{(2i)!!(2i+1)} \frac{1}{z^{2i+1}} = \frac{1}{z} + \frac{1}{6}\frac{1}{z^3} + \frac{3}{40}\frac{1}{z^5} + \frac{5}{112}\frac{1}{z^7} + \frac{35}{1152}\frac{1}{z^9} + O(z^{-11}).$$

vi) For all $z \in \mathbb{C}$ such that $|z| \ge 1$ and, as $z \to \infty$,

$$\operatorname{asec} z = \operatorname{acos} \frac{1}{z} = \frac{\pi}{2} - \operatorname{asin} \frac{1}{z} = \frac{\pi}{2} - \sum_{i=0}^{\infty} \frac{(2i)!}{4^i(i!)^2(2i+1)} \frac{1}{z^{2i+1}}$$
$$= \frac{\pi}{2} - \sum_{i=0}^{\infty} \frac{\binom{2i}{i}}{4^i(2i+1)} \frac{1}{z^{2i+1}} = \frac{\pi}{2} - \sum_{i=0}^{\infty} \frac{(2i-1)!!}{(2i)!!(2i+1)} \frac{1}{z^{2i+1}}$$
$$= \frac{\pi}{2} - \left(\frac{1}{z} + \frac{1}{6}\frac{1}{z^3} + \frac{3}{40}\frac{1}{z^5} + \frac{5}{112}\frac{1}{z^7} + \frac{35}{1152}\frac{1}{z^9} + O(z^{-11})\right).$$

vii) For all $z \in \mathbb{C}$ such that $|z| \le 1$ and $z \notin \{-\jmath, \jmath\}$ and, as $z \to 0$,

$$\operatorname{acot} z = \frac{\pi}{2} - \operatorname{atan} z = \frac{\pi}{2} - \sum_{i=0}^{\infty} (-1)^i \frac{1}{2i+1} z^{2i+1} = \frac{\pi}{2} - z + \frac{1}{3}z^3 - \frac{1}{5}z^5 + \frac{1}{7}z^7 - O(z^9).$$

viii) For all $z \in \mathbb{C}$ such that $|z| \geq 1$ and $z \notin \{-\jmath, \jmath\}$ and, as $z \to \infty$,

$$\operatorname{acot} z = \frac{\pi}{2} - \operatorname{atan} z = \frac{\pi}{2} - (\operatorname{sign}\operatorname{Re} z)\frac{\pi}{2} + \operatorname{atan}\frac{1}{z} = (1 - \operatorname{sign}\operatorname{Re} z)\frac{\pi}{2} + \sum_{i=0}^{\infty}(-1)^i \frac{1}{2i+1}\frac{1}{z^{2i+1}}$$

$$= (1 - \operatorname{sign}\operatorname{Re} z)\frac{\pi}{2} + \frac{1}{z} - \frac{1}{3}\frac{1}{z^3} + \frac{1}{5}\frac{1}{z^5} - \frac{1}{7}\frac{1}{z^7} + O(z^{-9}).$$

ix) For all $z \in \mathbb{C}$ such that $|z| \leq \frac{1}{2}$, and, as $z \to 0$,

$$\frac{\operatorname{asin} 2z}{2z} = \sum_{i=0}^{\infty} \frac{\binom{2i}{i}}{2i+1} z^{2i} = 1 + \frac{2}{3}z^2 + \frac{6}{5}z^4 + \frac{20}{7}z^6 + \frac{70}{9}z^8 + \frac{252}{11}z^{10} + O(z^{12}).$$

x) For all $z \in \mathbb{C}$ such that $|z| < 1$ and, as $z \to 0$,

$$\operatorname{asin}^2 z = \sum_{i=1}^{\infty} \frac{4^i}{2i^2\binom{2i}{i}} z^{2i} = z^2 + \frac{1}{3}z^4 + \frac{8}{45}z^6 + \frac{4}{35}z^8 + \frac{128}{1575}z^{10} + \frac{128}{2079}z^{12} + O(z^{14}).$$

xi) For all $z \in \mathbb{C}$ such that $|z| \leq 2$ and, as $z \to 0$,

$$2\operatorname{asin}^2\frac{z}{2} + 2\pi\operatorname{asin}\frac{z}{2} = \sum_{i=1}^{\infty} \frac{\Gamma^2(\frac{i}{2})}{i!} z^i = \pi z + \frac{1}{2}z^2 + \frac{\pi}{24}z^3 + \frac{1}{24}z^4 + \frac{3\pi}{640}z^5 + \frac{1}{180}z^6 + O(z^7).$$

xii) For all $z \in \mathbb{C}$ such that $|z| < 1$ and, as $z \to 0$,

$$\sqrt{\frac{z}{1-z}}\operatorname{atan}\sqrt{\frac{z}{1-z}} = \sum_{i=1}^{\infty} \frac{4^i}{2i\binom{2i}{i}} z^i = \int_0^1 \frac{2zx}{1-4zx(1-x)}\,\mathrm{d}x$$

$$= z + \frac{2}{3}z^2 + \frac{8}{15}z^3 + \frac{16}{35}z^4 + \frac{128}{315}z^5 + \frac{256}{693}z^6 + \frac{1024}{3003}z^7 + O(z^8).$$

xiii) For all $z \in \mathbb{C}$ such that $|z| < 1$ and, as $z \to 0$,

$$\frac{z\operatorname{asin} z}{\sqrt{1-z^2}} = \frac{z}{\sqrt{1-z^2}}\operatorname{atan}\frac{z}{\sqrt{1-z^2}} = \sum_{i=1}^{\infty} \frac{4^i}{2i\binom{2i}{i}} z^{2i} = \int_0^1 \frac{2z^2x}{1-4z^2x(1-x)}\,\mathrm{d}x$$

$$= z^2 + \frac{2}{3}z^4 + \frac{8}{15}z^6 + \frac{16}{35}z^8 + \frac{128}{315}z^{10} + \frac{256}{693}z^{12} + \frac{1024}{3003}z^{14} + O(z^{16}).$$

xiv) For all $z \in \mathbb{C}$ such that $|z| < 2$ and, as $z \to 0$,

$$\frac{z^2\sqrt{4-z^2} + 4z\operatorname{asin}(z/2)}{(4-z^2)\sqrt{4-z^2}} = \sum_{i=1}^{\infty} \frac{1}{\binom{2i}{i}} z^{2i} = \frac{1}{2}z^2 + \frac{1}{6}z^4 + \frac{1}{20}z^6 + \frac{1}{70}z^8 + \frac{1}{252}z^{10} + O(z^{12}).$$

xv) For all $z \in \mathbb{C}$ and, as $z \to 0$,

$$e^z = \sum_{i=0}^{\infty} \frac{1}{i!} z^i = 1 + z + \frac{1}{2}z^2 + \frac{1}{6}z^3 + \frac{1}{24}z^4 + \frac{1}{120}z^5 + \frac{1}{720}z^6 + \frac{1}{40320}z^8 + O(z^9).$$

xvi) For all z such that $0 < |z| < 2\pi$ and, as $z \to 0$,

$$\frac{1}{e^z - 1} = \frac{1}{z} - \frac{1}{2} + \sum_{i=1}^{\infty} \frac{B_{i+1}}{(i+1)!} z^i = \frac{1}{z} - \frac{1}{2} + \frac{1}{12}z - \frac{1}{720}z^3 + \frac{1}{30240}z^5 - O(z^6).$$

xvii) For all z such that $|z| < 1$ and, as $z \to 0$,

$$\frac{e^z}{1-z} = \sum_{i=1}^{\infty} \frac{a_i}{i!} z^i = 1 + 2z + \frac{5}{2}z^2 + \frac{8}{3}z^3 + \frac{65}{24}z^4 + \frac{163}{60}z^5 + \frac{1957}{720}z^6 + \frac{685}{252}z^7 + O(z^8).$$

xviii) For all z such that $|z| < 1$ and, as $z \to 0$,

$$\frac{1}{e^z(1-z)} = \sum_{i=1}^{\infty} \frac{d_i}{i!} z^i = 1 + \frac{1}{2}z^2 + \frac{1}{3}z^3 + \frac{3}{8}z^4 + \frac{11}{30}z^5 + \frac{53}{144}z^6 + \frac{103}{280}z^7 + \frac{2119}{5760}z^8 + O(z^9).$$

xix) For all $z \neq 0$ and, as $z \to \infty$,

$$e^{1/z} = \sum_{i=0}^{\infty} \frac{1}{i! z^i} = 1 + \frac{1}{z} + \frac{1}{2}\frac{1}{z^2} + \frac{1}{6}\frac{1}{z^3} + \frac{1}{24}\frac{1}{z^4} + \frac{1}{120}\frac{1}{z^5} + \frac{1}{720}\frac{1}{z^6} + O(z^{-7}).$$

xx) For all $z \neq 0$ and, as $z \to \infty$,

$$e^{-1/z^2} = \sum_{i=0}^{\infty} (-1)^i \frac{1}{i! z^{2i}} = 1 - \frac{1}{z^2} + \frac{1}{2}\frac{1}{z^4} - \frac{1}{6}\frac{1}{z^6} + \frac{1}{24}\frac{1}{z^8} - \frac{1}{120}\frac{1}{z^{10}} + \frac{1}{720}\frac{1}{z^{12}} - O(z^{-14}).$$

xxi) For all $z \neq 0$ and, as $z \to 0$,

$$e^{ze^z} = 1 + \sum_{i=1}^{\infty} \frac{1}{i!} \sum_{j=1}^{i} \binom{i}{j} j^{i-j} z^i = 1 + z + \frac{3}{2}z^2 + \frac{5}{3}z^3 + \frac{41}{24}z^4 + \frac{49}{30}z^5 + \frac{1057}{720}z^6 + O(z^7).$$

xxii) For all $z \in \mathbb{C}$ and, as $z \to 0$,

$$e^z + 2e^{-z/2} \cos \frac{\sqrt{3}z}{2} = \sum_{i=0}^{\infty} \frac{3}{(3i)!} z^{3i} = 3 + \frac{1}{2}z^3 + \frac{1}{240}z^6 + \frac{1}{120960}z^9 + O(z^{12}).$$

xxiii) For all $z \in \mathbb{C}$ and, as $z \to 0$,

$$e^z \sin z = \sum_{i=1}^{\infty} \frac{2^{i/2} \sin i\pi/4}{i!} z^i = z + z^2 + \frac{1}{3}z^3 - \frac{1}{30}z^5 - \frac{1}{90}z^6 - \frac{1}{630}z^7 + \frac{1}{22680}z^9 + O(z^{10}).$$

Source: *xi*) is given in [323]; *xxii*) is given in [2880, pp. 53, 54]; *xxiii*) is given in [2880, p. 57]. **Remark:** *iv*)–*vi*), *viii*), *xvi*), *xix*), and *xx*) are Laurent series at $z = 0$; *xvi*) has a regular singularity at 0, whereas the remaining Laurent series have essential singularities. **Remark:** a_i in *xvii*) is the ith arrangement number. See Fact 1.18.1. d_i in *xviii*) is the ith derangement number. See Fact 1.18.2. $\sum_{j=1}^{i} \binom{i}{j} j^{i-j}$ of z^i in *xxi*) is the ith *idempotent number*. See [771, pp. 91, 135]. **Remark:** $f\colon \mathbb{R} \mapsto \mathbb{R}$ defined by $f(x) = e^{1/x}$ for nonzero x and $f(0) = 0$ is not differentiable at 0. In contrast, the real-valued function $f\colon \mathbb{R} \mapsto \mathbb{R}$ defined by $f(x) = e^{-1/x^2}$ for nonzero x and $f(0) = 0$ is infinitely differentiable at 0 and satisfies $f^{(i)}(0) = 0$ for all $i \in \mathbb{P}$. However, the complex-valued function $g : \mathbb{C} \mapsto \mathbb{C}$ defined by $g(z) = e^{-1/z^2}$ for nonzero z and $g(0) = 0$ is not differentiable at $z = 0$. **Remark:** Setting $z = x^2/(x^2+1)$ in *xii*), it follows that, for all $x \in \mathbb{R}$,

$$\operatorname{atan} x = \sum_{i=0}^{\infty} \frac{4^i (i!)^2}{(2i+1)!} \frac{x^{2i+1}}{(x^2+1)^{i+1}} = \frac{x}{x^2+1} + \frac{2}{3}\frac{x^3}{(x^2+1)^2} + \frac{8}{15}\frac{x^5}{(x^2+1)^3} + \cdots,$$

which is a power series for $x \operatorname{atan} x$ in the variable $x^2/(x^2+1)$. Likewise, setting $z = (x^2+1)/x^2$ in *xii*), it follows that, for all $x \in \mathbb{R}$,

$$\operatorname{atan} \frac{1}{x} = \sum_{i=0}^{\infty} \frac{4^i (i!)^2}{(2i+1)!} \frac{x}{(x^2+1)^{i+1}} = \frac{x}{x^2+1} + \frac{2}{3}\frac{x}{(x^2+1)^2} + \frac{8}{15}\frac{x}{(x^2+1)^3} + \cdots,$$

which is a Laurent series for $(\operatorname{atan} 1/x)/x$ in the variable $x^2 + 1$. For details, see [649].

Fact 13.4.10. The following statements hold:

i) For all $z \in \mathbb{C}$ such that, for all $i \in \mathbb{N}$, $z^2 \neq (i+\frac{1}{2})^2\pi^2$,

$$\tan z = -2\sum_{i=0}^{\infty}\frac{z}{z^2-(i+\frac{1}{2})^2\pi^2}.$$

ii) For all $z \in \mathbb{C}$ such that, for all $i \in \mathbb{N}$, $z^2 \neq i^2\pi^2$,

$$\csc z = \frac{1}{z} + 2\sum_{i=1}^{\infty}(-1)^i\frac{z}{z^2-i^2\pi^2}.$$

iii) For all $z \in \mathbb{C}$ such that, for all $i \in \mathbb{N}$, $z^2 \neq (i+\frac{1}{2})^2\pi^2$,

$$\sec z = 2\pi\sum_{i=0}^{\infty}(-1)^{i+1}\frac{i+\frac{1}{2}}{z^2-(i+\frac{1}{2})^2\pi^2} = \sum_{i=-\infty}^{\infty}(-1)^i\frac{1}{z-(i+\frac{1}{2})\pi}.$$

iv) For all $z \in \mathbb{C}$ such that, for all $i \in \mathbb{N}$, $z^2 \neq i^2\pi^2$,

$$\cot z = \frac{1}{z} + 2\sum_{i=1}^{\infty}\frac{z}{z^2-i^2\pi^2}.$$

v) For all $z \in \mathbb{C}$ such that, for all $i \in \mathbb{P}$, $z^2 \neq i^2$,

$$\sum_{i=1}^{\infty}\frac{1}{i^2-z^2} = \frac{1-\pi z\cot \pi z}{2z^2}.$$

vi) For all $z \in \mathbb{C}$ such that $|z| < 1$,

$$\sum_{i=0}^{\infty}\zeta(2i+2)z^{2i} = \sum_{i=1}^{\infty}\frac{1}{i^2-z^2} = \frac{1-\pi z\cot \pi z}{2z^2}.$$

vii) For all $z \in \mathbb{C}$ such that, for all $i \in \mathbb{Z}$, $z^3 \neq i^3$,

$$\sum_{i=-\infty}^{\infty}\frac{1}{z^3-i^3} = \frac{\pi}{3z^2}[\cot \pi z + e^{2\pi \jmath/3}\cot e^{2\pi \jmath/3}\pi z + e^{4\pi \jmath/3}\cot e^{4\pi \jmath/3}\pi z].$$

viii) For all $z \in \mathbb{C}$ such that, for all $i \in \mathbb{Z}$, $z \neq i\pi$,

$$\csc^2 z = \sum_{i=-\infty}^{\infty}\frac{1}{(z-i\pi)^2}.$$

ix) For all $z \in \mathbb{C}$ such that, for all $i \in \mathbb{Z}$, $z \neq (i+\frac{1}{2})\pi$,

$$\sec^2 z = \sum_{i=-\infty}^{\infty}\frac{1}{[z-(i+\frac{1}{2})\pi]^2}.$$

x) For all $z \in \mathbb{C}$ such that, for all $i \in \mathbb{N}$, $z \neq i2\pi \jmath$,

$$\frac{1}{e^z-1} = \frac{1}{z} - \frac{1}{2} + 2\sum_{i=1}^{\infty}\frac{z}{z^2+4\pi^2 i^2}.$$

Source: [516, pp. 252, 253], [645, pp. 96, 98], [1136, pp. 352, 357], and [1219, p. 286]. **Remark:** These are *partial fraction expansions*.

Fact 13.4.11. The following statements hold:

i) For all $z \in \mathbb{C}$ and, as $z \to 0$,

$$\sinh z = \sin \jmath z = \sum_{i=0}^{\infty} \frac{1}{(2i+1)!} z^{2i+1} = z + \frac{1}{6}z^3 + \frac{1}{120}z^5 + \frac{1}{5040}z^7 + O(z^9).$$

ii) For all $z \in \mathbb{C}$ and, as $z \to 0$,

$$\cosh z = \cos \jmath z = \sum_{i=0}^{\infty} \frac{1}{(2i)!} z^{2i} = 1 + \frac{1}{2}z^2 + \frac{1}{24}z^4 + \frac{1}{720}z^6 + O(z^8).$$

iii) For all $z \in \mathbb{C}$ such that $|z| < \frac{\pi}{2}$ and, as $z \to 0$,

$$\tanh z = \tan \jmath z = \sum_{i=1}^{\infty} \frac{2^{2i}(2^{2i}-1)B_{2i}}{(2i)!} z^{2i-1} = z - \frac{1}{3}z^3 + \frac{2}{15}z^5 - \frac{17}{315}z^7 + \frac{62}{2835}z^9 - O(z^{11}).$$

iv) For all $z \in \mathbb{C}$ such that $0 < |z| < \pi$ and, as $z \to 0$,

$$\operatorname{csch} z = \frac{1}{z} + \sum_{i=1}^{\infty} \frac{(2-2^{2i})B_{2i}}{(2i)!} z^{2i-1} = \frac{1}{z} - \frac{1}{6}z + \frac{7}{360}z^3 - \frac{31}{15120}z^5 + O(z^7).$$

v) For all $z \in \mathbb{C}$ such that $|z| < \frac{\pi}{2}$ and, as $z \to 0$,

$$\operatorname{sech} z = 1 + \sum_{i=1}^{\infty} \frac{E_{2i}}{(2i)!} z^{2i} = 1 - \frac{1}{2}z^2 + \frac{5}{24}z^4 - \frac{61}{720}z^6 + \frac{277}{8064}z^8 - O(z^{10}).$$

vi) For all $z \in \mathbb{C}$ such that $0 < |z| < \pi$ and, as $z \to 0$,

$$\coth z = \frac{1}{z} + \sum_{i=1}^{\infty} \frac{2^{2i}B_{2i}}{(2i)!} z^{2i-1} = \frac{1}{z} + \frac{1}{3}z - \frac{1}{45}z^3 + \frac{2}{945}z^5 - \frac{1}{4725}z^7 + O(z^9).$$

vii) For all $z \in \mathbb{C}$ and, as $z \to 0$,

$$\frac{z}{\sinh z} = \sum_{i=0}^{\infty} \frac{(2-4^i)B_{2i}}{(2i)!} z^{2i} = 1 - \frac{1}{6}z^2 + \frac{7}{360}z^4 - \frac{31}{15120}z^6 + O(z^8).$$

Fact 13.4.12. The following statements hold:

i) For all $z \in \mathbb{C}$ such that, for all $i \in \mathbb{N}$, $z^2 \neq -(i+\frac{1}{2})^2\pi^2$,

$$\tanh z = 2\sum_{i=0}^{\infty} \frac{z}{z^2 + (i+\frac{1}{2})^2\pi^2}.$$

ii) For all $z \in \mathbb{C}$ such that, for all $i \in \mathbb{N}$, $z^2 \neq -i^2\pi^2$,

$$\operatorname{csch} z = \frac{1}{z} + 2\sum_{i=1}^{\infty} (-1)^i \frac{z}{z^2 + i^2\pi^2}.$$

iii) For all $z \in \mathbb{C}$ such that, for all $i \in \mathbb{N}$, $z^2 \neq -(i+\frac{1}{2})^2\pi^2$,

$$\operatorname{sech} z = 2\pi \sum_{i=0}^{\infty} (-1)^i \frac{i+\frac{1}{2}}{z^2 + (i+\frac{1}{2})^2\pi^2}.$$

iv) For all $z \in \mathbb{C}$ such that, for all $i \in \mathbb{N}$, $z^2 \neq i^2\pi^2$,

$$\coth z = \frac{1}{z} + 2\sum_{i=1}^{\infty} \frac{z}{z^2 + i^2\pi^2}.$$

v) For all $z \in \mathbb{C}$ such that, for all $i \in \mathbb{P}$, $z^2 \neq -i^2$,

$$\sum_{i=1}^{\infty} \frac{1}{i^2 + z^2} = \frac{\pi z \coth \pi z - 1}{2z^2}.$$

vi) For all $z \in \mathbb{C}$ such that, for all $i \in \mathbb{Z}$, $z \neq i\pi$,

$$-\operatorname{csch}^2 z = \sum_{i=-\infty}^{\infty} \frac{1}{(zJ - i\pi)^2}.$$

vii) For all $z \in \mathbb{C}$ such that, for all $i \in \mathbb{Z}$, $z \neq (i + \frac{1}{2})\pi$,

$$\operatorname{sech}^2 z = \sum_{i=-\infty}^{\infty} \frac{1}{[zJ - (i + \frac{1}{2})\pi]^2}.$$

Source: [516, pp. 252, 253], [645, pp. 96, 98], [1136, pp. 352, 357], and [1219, p. 286]. **Remark:** These are partial fraction expansions.

Fact 13.4.13. The following statements hold:

i) For all $z \in \mathbb{C}$ such that $|z| < 1$ and, as $z \to 0$,

$$\operatorname{asinh} z = \sum_{i=0}^{\infty} \frac{(-1)^i (2i)!}{4^i (2i+1)(i!)^2} z^{2i+1} = z - \frac{1}{6}z^3 + \frac{3}{40}z^5 - \frac{5}{112}z^7 + \frac{35}{1152}z^9 - O(z^{11}).$$

ii) For all $z \in \mathbb{C}$ such that $|z| > 1$ and, as $z \to \infty$,

$$\operatorname{acosh} z = \log 2z - \sum_{i=1}^{\infty} \frac{(2i-1)!}{4^i (i!)^2} \frac{1}{z^{2i}} = \log 2z - \left(\frac{1}{4}\frac{1}{z^2} + \frac{3}{32}\frac{1}{z^4} + \frac{5}{96}\frac{1}{z^6} + \frac{35}{1024}\frac{1}{z^8} + O(z^{-10}) \right).$$

iii) For all $z \in \mathbb{C}$ such that $|z| < 1$ and, as $z \to 0$,

$$\operatorname{atanh} z = \frac{1}{2} \log \frac{1+z}{1-z} = \sum_{i=0}^{\infty} \frac{1}{2i+1} z^{2i+1} = z + \frac{1}{3}z^3 + \frac{1}{5}z^5 + \frac{1}{7}z^7 + \frac{1}{9}z^9 + O(z^{11}).$$

iv) For all $z \in \mathbb{C}$ such that $|z| > 1$ and, as $z \to \infty$,

$$\operatorname{acsch} z = \operatorname{asinh} \frac{1}{z} = \sum_{i=0}^{\infty} \frac{(-1)^i (2i)!}{4^i (2i+1)(i!)^2} \frac{1}{z^{2i+1}} = \frac{1}{z} - \frac{1}{6}\frac{1}{z^3} + \frac{3}{40}\frac{1}{z^5} - \frac{5}{112}\frac{1}{z^7} + O(z^{-9}).$$

v) For all $z \in \mathbb{C}$ such that $0 < |z| < 1$ and, as $z \to 0$,

$$\begin{aligned} \operatorname{asech} z = \operatorname{acosh} \frac{1}{z} &= \log \frac{2}{z} - \sum_{i=1}^{\infty} \frac{(2i-1)!}{4^i (i!)^2} z^{2i} + 2\pi J \left\lfloor \frac{\arg(z) + \pi}{2\pi} \right\rfloor \\ &= \log \frac{2}{z} - \left(\frac{1}{4}z^2 + \frac{3}{32}z^4 + \frac{5}{96}z^6 + \frac{35}{1024}z^6 + O(z^8) \right) + 2\pi J \left\lfloor \frac{\arg(z) + \pi}{2\pi} \right\rfloor. \end{aligned}$$

vi) For all $z \in \mathbb{C}$ such that $|z| > 1$ and, as $z \to \infty$,

$$\operatorname{acoth} z = \operatorname{atanh} \frac{1}{z} = \sum_{i=0}^{\infty} \frac{1}{2i+1} \frac{1}{z^{2i+1}} = \frac{1}{z} + \frac{1}{3}\frac{1}{z^3} + \frac{1}{5}\frac{1}{z^5} + \frac{1}{7}\frac{1}{z^7} + \frac{1}{9}\frac{1}{z^9} + O(z^{-11}).$$

Fact 13.4.14. The following statements hold:

i) For all $z \in \mathbb{C}$ such that $|z - 1| \le 1$ and $z \neq 0$ and, as $z \to 1$,

$$\log z = \sum_{i=1}^{\infty} \frac{1}{i}(1-z)^i = -\left[1 - z + \frac{1}{2}(1-z)^2 + \frac{1}{3}(1-z)^3 + \frac{1}{4}(1-z)^4 + O[(1-z)^5] \right].$$

ii) For all $z \in \mathbb{C}$ such that $z \in \text{CIUD}\backslash\{1\}$ and, as $z \to 0$,

$$\log(1-z) = -\sum_{i=1}^{\infty} \frac{1}{i} z^i = -\left(z + \frac{1}{2}z^2 + \frac{1}{3}z^3 + \frac{1}{4}z^4 + O(z^5)\right).$$

iii) For all $z \in \mathbb{C}$ such that $z \in \text{CIUD}\backslash\{-1\}$ and, as $z \to 0$,

$$\log(1+z) = \sum_{i=1}^{\infty} (-1)^{i+1} \frac{1}{i} z^i = z - \frac{1}{2}z^2 + \frac{1}{3}z^3 - \frac{1}{4}z^4 + \frac{1}{5}z^5 - \frac{1}{6}z^6 + O(z^7).$$

iv) For all $z \in \mathbb{C}$ such that $|z| > 1$ and, as $z \to 0$,

$$\log \frac{z}{z-1} = \sum_{i=1}^{\infty} \frac{1}{iz^i} = z + \frac{1}{2}\frac{1}{z^2} + \frac{1}{3}\frac{1}{z^3} + \frac{1}{4}\frac{1}{z^4} + \frac{1}{5}\frac{1}{z^5} + \frac{1}{6}\frac{1}{z^6} + O(z^{-7}).$$

v) Let $a, b \in \mathbb{C}$. For all $z \in \mathbb{C}$ such that $|z| \le \min\{1/|a|, 1/|b|\}$ and $z \notin \{-1/a, -1/b\}$ and, as $z \to 0$,

$$\log \frac{1+bz}{1+az} = \sum_{i=1}^{\infty} (-1)^i \frac{a^i - b^i}{i} z^i = (b-a)z + \frac{1}{2}(a^2 - b^2)z^2 + \frac{1}{3}(b^3 - a^3)z^3 + O(z^4).$$

In particular, if $z \in \text{CIUD}\backslash\{-1, 1\}$ and, as $z \to 0$,

$$\log \frac{1+z}{1-z} = \sum_{i=0}^{\infty} \frac{2}{2i+1} z^{2i+1} = 2z + \frac{2}{3}z^3 + \frac{2}{5}z^5 + \frac{2}{7}z^7 + \frac{2}{9}z^9 + O(z^{11}).$$

vi) For all $z \in \mathbb{C}$ such that $\operatorname{Re} z > 0$,

$$\log z = \sum_{i=0}^{\infty} \frac{2}{2i+1} \left(\frac{z-1}{z+1}\right)^{2i+1}.$$

vii) For all $z \in \mathbb{C}$ such that $|z| < 1$ and, as $z \to 0$,

$$\log \frac{1}{1-z} = \sum_{i=1}^{\infty} \frac{1}{i} z^i = z + \frac{1}{2}z^2 + \frac{1}{3}z^3 + \frac{1}{4}z^4 + \frac{1}{5}z^5 + \frac{1}{6}z^6 + O(z^7).$$

viii) For all $z \in \mathbb{C}$ such that $|z| < 1$ and, as $z \to 0$,

$$\frac{\log(1-z)}{z-1} = \sum_{i=1}^{\infty} H_i z^i = z + \frac{3}{2}z^2 + \frac{11}{6}z^3 + \frac{25}{12}z^4 + \frac{137}{60}z^5 + \frac{49}{20}z^6 + O(z^7).$$

ix) For all $z \in \mathbb{C}$ such that $|z| < 1$ and, as $z \to 0$,

$$\log^2(1-z) = 2\sum_{i=1}^{\infty} \frac{H_i}{i+1} z^{i+1} = z^2 + z^3 + \frac{11}{12}z^4 + \frac{5}{6}z^5 + \frac{136}{180}z^6 + \frac{7}{10}z^7 + O(z^8).$$

x) For all $z \in \mathbb{C}$ such that $|z| < 1$ and, as $z \to 0$,

$$\log^3(1-z) = -6\sum_{i=1}^{\infty} \frac{1}{i+2} \left(\sum_{j=1}^{i} \frac{H_j}{j+1}\right) z^{i+2} = -z^3 - \frac{3}{2}z^4 - \frac{7}{4}z^5 - \frac{15}{8}z^6 - \frac{29}{15}z^7 - O(z^8).$$

xi) For all $z \in \mathbb{C}$ such that $|z| < \frac{1}{4}$ and, as $z \to 0$,

$$\log\left(\frac{1 - \sqrt{1-4z}}{2z}\right) = \log\left(\frac{2}{1 + \sqrt{1-4z}}\right) = \frac{1}{2}\sum_{i=1}^{\infty} \frac{1}{i}\binom{2i}{i} z^i$$

$$= z + \frac{3}{2}z^2 + \frac{10}{3}z^3 + \frac{35}{4}z^4 + \frac{126}{5}z^5 + 77z^6 + \frac{1716}{7}z^7 + O(z^8).$$

xii) For all $z \in \mathbb{C}$ such that $|z| < 1$ and, as $z \to 0$,

$$\frac{z}{\log(1+z)} = \sum_{i=0}^{\infty} \frac{1}{i!} \sum_{j=0}^{i} (-1)^{i-j} \frac{1}{j+1} \begin{bmatrix} i \\ j \end{bmatrix} z^i = 1 + \frac{1}{2}z - \frac{1}{12}z^2 + \frac{1}{24}z^3 - \frac{19}{720}z^4 + \frac{3}{160}z^5 - O(z^6).$$

xiii) For all $z \in \mathbb{C}$ such that $|z| < 1$ and, as $z \to 0$,

$$[\log(1+z)]\log(1-z) = \sum_{i=1}^{\infty} \frac{1}{i}\left(H_i - H_{2i} - \frac{1}{2i}\right) z^{2i} = -z^2 - \frac{5}{12}z^4 - \frac{47}{180}z^6 - \frac{319}{1680}z^8 - O(z^{10}).$$

Source: [1524, pp. 11, 12]. For $x \in \mathbb{R}$ such that $|x| < 1$, it follows that

$$\frac{\mathrm{d}}{\mathrm{d}x}\log(1-x) = \frac{-1}{1-x} = -[1 + x + x^2 + x^3 + x^4 + O(x^5)].$$

Integrating yields

$$\log(1-x) = -\left(x + \frac{1}{2}x^2 + \frac{1}{3}x^3 + O(x^4)\right),$$

which yields *ii*). *xii*) is given in [2276]. **Remark:** *iii*) is *Mercator's series*, while *iv*) and *v*) are equivalent forms of *Gregory's series*. See [1391, p. 273]. **Remark:** *viii*) is the generating function for the harmonic numbers. **Related:** Setting $a = \beta$ and $b = \alpha$ in *v*) yields a series involving the Fibonacci numbers. See Fact 13.9.3.

Fact 13.4.15. The following statements hold:

i) For all $A \in \mathbb{F}^{n\times n}$,

$$\sin A = A - \tfrac{1}{3!}A^3 + \tfrac{1}{5!}A^5 - \tfrac{1}{7!}A^7 + \cdots.$$

ii) For all $A \in \mathbb{F}^{n\times n}$,

$$\cos A = I - \tfrac{1}{2!}A^2 + \tfrac{1}{4!}A^4 - \tfrac{1}{6!}A^6 + \cdots.$$

iii) For all $A \in \mathbb{F}^{n\times n}$ such that $\rho_{\max}(A) < \frac{\pi}{2}$,

$$\tan A = A + \tfrac{1}{3}A^3 + \tfrac{2}{15}A^5 + \tfrac{17}{315}A^7 + \tfrac{62}{2835}A^9 + \cdots.$$

iv) For all $A \in \mathbb{F}^{n\times n}$, then

$$e^A = I + A + \tfrac{1}{2!}A^2 + \tfrac{1}{3!}A^3 + \tfrac{1}{4!}A^4 + \cdots.$$

v) For all $A \in \mathbb{F}^{n\times n}$ such that $\rho_{\max}(A - I) < 1$,

$$\log A = -[I - A + \tfrac{1}{2}(I-A)^2 + \tfrac{1}{3}(I-A)^3 + \tfrac{1}{4}(I-A)^4 + \cdots].$$

vi) For all $A \in \mathbb{F}^{n\times n}$ such that $\rho_{\max}(A) < 1$,

$$\log(I - A) = -(A + \tfrac{1}{2}A^2 + \tfrac{1}{3}A^3 + \tfrac{1}{4}A^4 + \cdots).$$

vii) For all $A \in \mathbb{F}^{n\times n}$ such that $\rho_{\max}(A) < 1$,

$$\log(I + A) = A - \tfrac{1}{2}A^2 + \tfrac{1}{3}A^3 - \tfrac{1}{4}A^4 + \cdots.$$

viii) For all $A \in \mathbb{F}^{n\times n}$ such that $\mathrm{spec}(A) \subset \mathrm{ORHP}$,

$$\log A = \sum_{i=0}^{\infty} \frac{2}{2i+1}[(A - I)(A + I)^{-1}]^{2i+1}.$$

ix) For all $A \in \mathbb{F}^{n\times n}$,

$$\sinh A = \sin \jmath A = A + \tfrac{1}{3!}A^3 + \tfrac{1}{5!}A^5 + \tfrac{1}{7!}A^7 + \cdots .$$

x) For all $A \in \mathbb{F}^{n\times n}$,

$$\cosh A = \cos \jmath A = I + \tfrac{1}{2!}A^2 + \tfrac{1}{4!}A^4 + \tfrac{1}{6!}A^6 + \cdots .$$

xi) For all $A \in \mathbb{F}^{n\times n}$ such that $\rho_{\max}(A) < \frac{\pi}{2}$,

$$\tanh A = \tan \jmath A = A - \tfrac{1}{3}A^3 + \tfrac{2}{15}A^5 - \tfrac{17}{315}A^7 + \tfrac{62}{2835}A^9 - \cdots .$$

xii) Let $\alpha \in \mathbb{R}$. For all $A \in \mathbb{F}^{n\times n}$ such that $\rho_{\max}(A) < 1$,

$$\begin{aligned}(I + A)^\alpha &= I + \alpha A + \tfrac{\alpha(\alpha-1)}{2!}A^2 + \tfrac{\alpha(\alpha-1)(\alpha-2)}{3!}A^3 + \tfrac{1}{4}A^4 + \cdots \\ &= I + \tbinom{\alpha}{1}A + \tbinom{\alpha}{2}A^2 + \tbinom{\alpha}{3}A^3 + \tbinom{\alpha}{4}A^4 + \cdots .\end{aligned}$$

xiii) For all $A \in \mathbb{F}^{n\times n}$ such that $\rho_{\max}(A) < 1$,

$$(I - A)^{-1} = I + A + A^2 + A^3 + A^4 + \cdots .$$

Source: Fact 13.4.8.

Fact 13.4.16. Define

$$A \triangleq \begin{bmatrix} 63 & 104 & -68 \\ 64 & 104 & -67 \\ 80 & 131 & -85 \end{bmatrix}.$$

Then, $\det A = -1$, and thus, for all $n \in \mathbb{Z}$, every entry of A^n is an integer. In particular,

$$A^{-1} = \begin{bmatrix} 63 & 68 & -104 \\ -80 & -85 & 131 \\ -64 & -67 & 104 \end{bmatrix}.$$

Next, for all $n \in \mathbb{Z}$, define

$$\begin{bmatrix} a_n \\ b_n \\ c_n \end{bmatrix} \triangleq A^n \begin{bmatrix} 1 \\ 2 \\ 2 \end{bmatrix}.$$

Then, for all $z \in \mathbb{C}$ such that $|z| < \frac{1}{2}(83 - 9\sqrt{85}) \approx 0.01205$,

$$\sum_{i=0}^\infty a_i z^i = \frac{1 + 53z + 9z^2}{1 - 82z - 82z^2 + z^3}, \quad \sum_{i=0}^\infty b_i z^i = \frac{2 - 26z - 12z^2}{1 - 82z - 82z^2 + z^3},$$

$$\sum_{i=0}^\infty c_i z^i = \frac{2 + 8z - 10z^2}{1 - 82z - 82z^2 + z^3}, \quad a_n^3 + b_n^3 = c_n^3 + (-1)^n.$$

Remark: Extensions are given in [2000]. **Credit:** S. A. Ramanujan. See [1308, 1309]. **Example:** $a_2 = 11161$, $b_2 = 11468$, and $c_2 = 14258$, and thus $11161^3 + 11468^3 - 14258^3 = 1$.

Fact 13.4.17. Let $n \geq 1$ and $z_1, \ldots, z_n \in \mathrm{OIUD}$. Then,

$$\frac{\prod_{i=1}^n z_i}{\left(1 - \prod_{i=1}^n z_i\right)\prod_{i=1}^n (1 - z_i)} = \sum \min\{i_1, \ldots, i_n\} \prod_{j=1}^n z_{i_j},$$

where the sum is taken over all n-tuples $(i_1, \ldots, i_n)$ of positive integers. **Source:** [771, p. 87].

Fact 13.4.18. Let $z \in \mathbb{C}$, and assume that $|z| < 1/e$. Then,

$$\exp\left(\sum_{i=1}^{\infty} \frac{i^{i-1}}{i!} z^i\right) = \sum_{i=0}^{\infty} \frac{(i+1)^{i-1}}{i!} z^i.$$

Source: [771, p. 174].

Fact 13.4.19. Let $x, y \in \mathbb{C}$. Then,

$$\sum_{i=0}^{\infty} \sum_{j=i}^{2i} \frac{1}{2^{j-i}(j-i)!(2i-j)!} x^j y^i = e^{y(x+x^2/2)}.$$

Source: [2880, p. 206].

13.5 Facts on Series of Rational Functions

Fact 13.5.1. Let $z \in \mathbb{C}$. Then, the following statements hold:

i) If $|z| < 1$, then

$$\frac{z}{1-z} = \sum_{i=0}^{\infty} \frac{2^i z^{2^i}}{1+z^{2^i}}, \quad \frac{z}{1-z} = \sum_{i=0}^{\infty} \frac{z^{2^i}}{1-z^{2^{i+1}}}.$$

ii) If $|z| > 1$, then

$$\frac{1}{1-z} = \sum_{i=0}^{\infty} \frac{z^{2^i}}{1-z^{2^{i+1}}}.$$

Source: [1647, p. 267].

Fact 13.5.2. Let $n \geq 1$, let $z \in \mathbb{C}$, and assume that $|z| < 1$. Then,

$$\sum_{i=1}^{\infty} \frac{4^i z^{n2^i}}{(1+z^{n2^i})^2} = \frac{4z^{2n}}{(1-z^{2n})^2}.$$

Source: [2150]. **Related:** Fact 2.1.14.

Fact 13.5.3. Let $n \geq 1$ and $z \in \mathrm{OIUD}$. Then,

$$\frac{z^n}{(1-z)^n} = \sum_{i=n}^{\infty} \alpha_{n,i} \frac{z^i}{1-z^i},$$

where

$$\alpha_{n,m} \triangleq \operatorname{card}\{(k_1, \ldots, k_m) \in \mathbb{P}^n \colon k_1 \leq \cdots \leq k_m, \ \gcd\{k_1, \ldots, k_m\} = 1, \text{ and } \textstyle\sum_{i=1}^m k_i = n\}.$$

Source: [771, p. 123]. **Remark:** $\alpha_{3,2} = 1$, $\alpha_{5,3} = 2$, $\alpha_{7,3} = 3$, and $\alpha_{10,2} = 2$.

Fact 13.5.4. Let $z \in \mathrm{OIUD}$. Then,

$$\sum_{i=1}^{\infty} \frac{z^i}{1+z^{2i}} = \sum_{i=0}^{\infty} (-1)^i \frac{z^{2i+1}}{1-z^{2i+1}}.$$

Source: [2013, p. 130].

Fact 13.5.5. Let $z \in \mathrm{OIUD}$. Then,

$$\left(\sum_{i=-\infty}^{\infty} (-1)^i z^{i^2}\right)^4 = 1 + 8 \sum_{i=1}^{\infty} \frac{z^i}{[z^i + (-1)^i]^2}.$$

Source: [117, 2970]. **Related:** Fact 1.11.25 and Fact 2.1.17.

Fact 13.5.6. For all $n \geq 1$ and $z \in \mathbb{C}$, define $g_n(z) \triangleq \prod_{i=1}^{n}(1 - z^i)$. Then, for all $z \in$ OIUD,

$$\sum_{i=0}^{\infty} \frac{z^{i^2}}{g_i(z)} = \prod_{i=1}^{\infty} \frac{1}{(1 - z^{5i-4})(1 - z^{5i-1})}, \quad \sum_{i=0}^{\infty} \frac{z^{i^2+i}}{g_i(z)} = \prod_{i=1}^{\infty} \frac{1}{(1 - z^{5i-3})(1 - z^{5i-2})}.$$

Source: [2970]. **Remark:** These are the *Rogers-Ramanujan identities.* **Related:** Fact 2.1.19 and Fact 13.10.27.

Fact 13.5.7. Let $z \in \mathbb{C}$, and assume that $z \notin \{0, -1, -1/2, -1/3, \ldots\}$. Then,

$$\sum_{i=1}^{\infty} \frac{i}{\prod_{j=1}^{i}(jz + 1)} = \frac{1}{z}.$$

In particular,

$$\sum_{i=1}^{\infty} \frac{i}{(i + 1)!} = 1.$$

Source: [1647, p. 267].

Fact 13.5.8. Let $(x_i)_{i=1}^{\infty} \subset \mathbb{R}$ and $(y_i)_{i=1}^{\infty} \subset \mathbb{R}$, assume that $\sum_{i=1}^{\infty} x_i$ and $\sum_{i=1}^{\infty} y_i$ are convergent, assume that at least one of these series is absolutely convergent, and define $(z_i)_{i=1}^{\infty}$, where, for all $i \geq 1$, $z_i \triangleq \sum_{j=1}^{i} x_j y_{i-j+1}$. Then, $\sum_{i=1}^{\infty} z_i$ converges. Furthermore,

$$\sum_{i=1}^{\infty} z_i = \left(\sum_{i=1}^{\infty} x_i\right) \sum_{i=1}^{\infty} y_i.$$

Source: [1566, p. 102]. **Remark:** This is *Mertens's theorem.* The series $\sum_{i=1}^{\infty} z_i$ is the *Cauchy product* of $\sum_{i=1}^{\infty} x_i$ and $\sum_{i=1}^{\infty} y_i$.

Fact 13.5.9. Let $(x_i)_{i=1}^{\infty} \subset (0, \infty)$. Then,

$$\sum_{i=1}^{\infty} \frac{i}{\sum_{j=1}^{i} x_j} \leq 4 \sum_{i=1}^{\infty} \frac{1}{x_i}.$$

Source: [506, pp. 177, 178]. **Related:** Fact 2.11.45.

Fact 13.5.10. Let $(x_i)_{i=1}^{\infty} \subset (0, \infty)$. If, for all $i \in \mathbb{P}$, $x_{i+1} < x_i$, then

$$\sum_{i=1}^{\infty} \frac{1}{x_i}(x_i - x_{i+1}) = \infty.$$

Alternatively, if, for all $i \in \mathbb{P}$, $x_i < x_{i+1}$, then

$$\sum_{i=1}^{\infty} \frac{1}{x_{i+1}}(x_{i+1} - x_i) = \infty.$$

Source: [2801].

Fact 13.5.11. Let $(x_i)_{i=1}^{\infty} \subset \mathbb{R}$. Then, the following statements hold:

i) Assume that $\prod_{i=1}^{\infty}(1 + x_i)$ converges to $\alpha \in (0, \infty)$. Then,

$$\sum_{i=1}^{\infty} \frac{x_i}{\prod_{j=1}^{i}(1 + x_j)} = 1 - \frac{1}{\alpha}.$$

ii) Assume that $\lim_{n\to\infty} \prod_{i=1}^{n}(1 + x_i) = \pm\infty$. Then,

$$\sum_{i=1}^{\infty} \frac{x_i}{\prod_{j=1}^{i}(1 + x_j)} = 1.$$

Source: [1566, p. 65].

Fact 13.5.12. Let $(x_i)_{i=1}^\infty \subset (0,\infty)$, and assume that $\sum_{i=1}^\infty \frac{1}{x_i}$ converges. Then,

$$\sum_{i=1}^\infty \frac{1}{\sum_{j=1}^i x_j} \le 2\sum_{i=1}^\infty \frac{1}{x_i}.$$

Source: [2294, p. 93].

Fact 13.5.13. Let $x > 2$. Then,

$$\sum_{i=1}^\infty \frac{i!}{\prod_{j=0}^{n-i}(x+j)} = \frac{1}{x-2}.$$

Source: [1566, p. 65].

Fact 13.5.14.

$$\sum_{i=0}^\infty \frac{1}{i!} = e,\quad \sum_{i=1}^\infty \frac{2^i i^2}{i!} = 6e^2,\quad \sum_{i=0}^\infty (-1)^i\frac{1}{i!} = \frac{1}{e},\quad \sum_{i=0}^\infty \frac{9i^2+1}{(3i)!} = e,\quad \sum_{i=1}^\infty \frac{i}{(i+1)!} = 1,$$

$$\sum_{i=0}^\infty (-1)^i\frac{4^i}{i!} = \frac{1}{e^4},\quad \sum_{i=1}^\infty \frac{i^2}{(i-1)!} = 5e,\quad \sum_{i=1}^\infty \frac{i}{(2i+1)!} = \frac{1}{2e},\quad \sum_{i=0}^\infty (-1)^i\frac{1}{i!4^i} = \frac{1}{\sqrt[4]{e}},$$

$$\sum_{i=0}^\infty \frac{2}{(i^4+i^2+1)i!} = e,\quad \sum_{i=0}^\infty \frac{1}{(2i)!} = \frac{1}{2}\left(e+\frac{1}{e}\right),\quad \sum_{i=1}^\infty \frac{i}{(2i)!} = \frac{1}{4}\left(e-\frac{1}{e}\right),$$

$$\sum_{i=0}^\infty \frac{4i+3}{4^i(2i+1)!} = 2\sqrt{e},\quad \sum_{i=0}^\infty (-1)^{i+1}\frac{(i+1)^3}{i!} = \frac{1}{e},\quad \sum_{i=0}^\infty (-1)^i\frac{9i^2-6i+1}{(3i)!} = \frac{1}{e},$$

$$\sum_{i=0}^\infty \frac{1}{(2i+1)!} = \frac{1}{2}\left(e-\frac{1}{e}\right),\quad \sum_{i=1}^\infty \frac{i^2}{(2i+1)!} = \frac{1}{8}\left(e-\frac{3}{e}\right),\quad \sum_{i=1}^\infty \frac{1}{i(i+1)(i+1)!} = 3-e.$$

Source: [506, pp. 112, 129, 130, 140], [579], [1217, p. 13], and [1566, pp. 49, 64].

Fact 13.5.15. Let $0 \le m < n$. Then,

$$\sum_{i=1}^\infty \frac{1}{(i+m)(i+n)} = \frac{1}{n-m}\sum_{i=m+1}^n \frac{1}{i},\quad \sum_{i=1}^\infty \frac{1}{i(i+n)} = \frac{H_n}{n}.$$

For example,

$$\sum_{i=1}^\infty \frac{1}{i(i+1)} = 1,\quad \sum_{i=1}^\infty \frac{1}{i(i+2)} = \frac{3}{4},\quad \sum_{i=1}^\infty \frac{1}{i(i+3)} = \frac{11}{18},$$

$$\sum_{i=1}^\infty \frac{1}{i(i+4)} = \frac{25}{48},\quad \sum_{i=1}^\infty \frac{1}{i(i+5)} = \frac{137}{300},\quad \sum_{i=1}^\infty \frac{1}{i(i+6)} = \frac{49}{120}.$$

Source: [1217, p. 12]. **Remark:** The second equality follows from the penultimate equality in Fact 13.7.8 with $n = 1$. **Remark:** $\sum_{i=1}^{n-1} \frac{1}{i(i+1)} = \frac{n-1}{n}$ implies $\sum_{i=n}^\infty \frac{1}{i(i+1)} = \frac{1}{n}$.

Fact 13.5.16.

$$\sum_{i=1}^\infty \frac{H_{i+1}}{i(i+1)} = 2,\quad \sum_{i=1}^\infty \frac{H_i}{i(i+1)} = \frac{\pi^2}{6},\quad \sum_{i=1}^\infty \frac{H_i}{(i+1)(i+2)} = 1,\quad \sum_{i=1}^\infty \frac{H_i}{i(i+2)} = \frac{\pi^2}{12}+\frac{1}{2},$$

$$\sum_{i=1}^\infty \frac{H_i}{i(i+3)} = \frac{\pi^2}{18}+\frac{7}{12},\quad \sum_{i=1}^\infty \frac{H_i}{i(i+4)} = \frac{\pi^2}{24}+\frac{85}{144},\quad \sum_{i=1}^\infty \frac{H_i}{i(i+5)} = \frac{\pi^2}{30}+\frac{83}{144},$$

$$\frac{1}{2}\sum_{i=1}^{\infty}\frac{H_i}{i^2} = \sum_{i=1}^{\infty}\frac{H_i}{(i+1)^2} = \sum_{i=1}^{\infty}\frac{(2i+1)H_i}{i^2(i+1)^2} = \zeta(3), \quad \sum_{i=1}^{\infty}\frac{H_i}{(i+2)^2} = \zeta(3) + \frac{\pi^2}{6} - 2,$$

$$\sum_{i=1}^{\infty}\frac{H_i}{(i+3)^2} = \zeta(3) + \frac{\pi^2}{4} - 3, \quad \sum_{i=1}^{\infty}\frac{H_i}{(i+4)^2} = \zeta(3) + \frac{11\pi^2}{36} - \frac{395}{108},$$

$$\sum_{i=1}^{\infty}\left(\frac{H_i}{i}\right)^2 = \frac{17\pi^4}{360}, \quad \sum_{i=1}^{\infty}\left(\frac{H_i}{i+1}\right)^2 = \frac{11\pi^4}{360}, \quad \sum_{i=1}^{\infty}\left(\frac{H_i}{i+2}\right)^2 = 2\zeta(3) - 3 + \frac{11\pi^4}{360},$$

$$\sum_{i=1}^{\infty}\frac{H_i^2}{i(i+1)} = 3\zeta(3), \quad \sum_{i=1}^{\infty}\frac{H_i^2}{i(i+2)} = \frac{3}{2}\zeta(3) + \frac{\pi^2}{12} + \frac{1}{2}, \quad \sum_{i=1}^{\infty}\frac{H_i^2}{(i+1)(i+2)} = \frac{\pi^2}{6} + 1,$$

$$\sum_{i=1}^{\infty}\frac{H_iH_{i+1}}{(i+1)^2} = \frac{\pi^4}{30}, \quad \sum_{i=1}^{\infty}\frac{H_iH_{i+1}}{i(i+1)} = 2\zeta(3) + \frac{\pi^2}{6}, \quad \sum_{i=1}^{\infty}\frac{H_iH_{i+2}}{i(i+2)} = \zeta(3) + \frac{\pi^2}{8} + \frac{5}{4},$$

$$\sum_{i=1}^{\infty}\frac{H_{2i}}{i(2i-1)} = \frac{\pi^2}{6} - \log^2 2 + 2\log 2, \quad \sum_{i=1}^{\infty}\frac{H_{2i}\zeta(2i)}{i(2i+1)} = \frac{\pi^2}{6},$$

$$\sum_{i=1}^{\infty}\frac{H_i}{i^3} = \frac{\pi^4}{72}, \quad \sum_{i=1}^{\infty}\frac{H_i}{i^4} = 3\zeta(5) - \frac{\pi^2}{6}\zeta(3), \quad \sum_{i=1}^{\infty}\frac{H_i}{i^5} = \frac{\pi^6}{540} - \frac{1}{2}\zeta^2(3),$$

$$\sum_{i=1}^{\infty}\frac{H_i^2}{i^3} = \frac{7}{2}\zeta(5) - \frac{\pi^2}{6}\zeta(3), \quad \sum_{i=1}^{\infty}\frac{H_i^2}{i^4} = \frac{97\pi^6}{22680} - 2\zeta^2(3),$$

$$\sum_{i=1}^{\infty}\frac{H_{i,2}}{i^2} = \frac{7\pi^4}{360}, \quad \sum_{i=1}^{\infty}\frac{H_{i,2}}{(i+1)^2} = \frac{\pi^4}{120}, \quad \sum_{i=1}^{\infty}\frac{H_{i,2}}{(i+2)^2} = 3 - \frac{\pi^2}{3} + \frac{\pi^4}{120},$$

$$\sum_{i=1}^{\infty}\frac{H_{i,2}}{(i+3)^2} = \frac{59}{16} - \frac{5\pi^2}{12} + \frac{\pi^4}{120}, \quad \sum_{i=1}^{\infty}\frac{H_{i,2}}{(i+4)^2} = \frac{1717}{432} - \frac{49\pi^2}{108} + \frac{\pi^4}{120},$$

$$\sum_{i=1}^{\infty}\frac{H_{i,2}}{i(i+1)} = \zeta(3), \quad \sum_{i=1}^{\infty}\frac{H_{i,2}}{i(i+2)} = \frac{1}{2}\zeta(3) + \frac{\pi^2}{12} - \frac{1}{2}, \quad \sum_{i=1}^{\infty}\frac{H_{i,2}}{(i+1)(i+2)} = \frac{\pi^2}{6} - 1,$$

$$\sum_{i=1}^{\infty}\frac{(\frac{1}{2})^{\underline{i+1}}H_i}{(i+1)(i+1)!} = \frac{\pi^2}{6} + 2\log^2 2, \quad \sum_{i=1}^{\infty}\frac{(\frac{1}{2})^{\underline{i+1}}H_i^2}{(i+1)(i+1)!} = \pi^2\log 2 + \frac{7}{2}\zeta(3) + \frac{4}{3}\log^3 2,$$

$$\sum_{i=1}^{\infty}\frac{(\frac{1}{2})^{\underline{i+1}}H_{i,2}}{(i+1)(i+1)!} = \frac{\pi^2}{3}\log 2 - \frac{1}{2}\zeta(3) - \frac{4}{3}\log^3 2.$$

Let $n \ge 2$. Then,

$$\sum_{i=1}^{\infty}\frac{H_i}{i^n} = \frac{1}{2}\left((n+2)\zeta(n+1) + \sum_{i=1}^{n-2}\zeta(n-i)\zeta(i+1)\right), \quad \sum_{i=1}^{\infty}\frac{H_{i,n}}{i^n} = \tfrac{1}{2}[\zeta^2(n) + \zeta(2n)],$$

$$\sum_{i=1}^{\infty}\frac{H_i}{(i+1)^{\overline{n}}} = \frac{1}{(n-1)!(n-1)^2}.$$

Let $p > 1$. Then,

$$\sum_{i=1}^{\infty}\left(\frac{H_i}{i}\right)^p \le \frac{p}{p-1}\zeta(p).$$

Source: [506, pp. 126, 134, 135], [511, p. 78], [515, 650, 1101, 2220, 2483], [2513, p. 354], and [2859]. **Related:** Fact 13.5.41.

Fact 13.5.17. Let $n \geq 2$. Then,

$$\sum_{i=1}^{\infty} \frac{H_i}{n^i} = \frac{n}{n-1} \log \frac{n}{n-1}, \quad \sum_{i=1}^{\infty} \frac{H_i}{2^i} = 2 \log 2, \quad \sum_{i=1}^{\infty} \frac{H_i}{3^i} = \frac{3}{2} \log \frac{3}{2}.$$

Furthermore,

$$\sum_{i=1}^{\infty} \frac{H_i}{2^i i} = \frac{\pi^2}{12}, \quad \sum_{i=1}^{\infty} \frac{H_{i+1}}{2^i i} = \frac{\pi^2}{12} + 1 - \log 2, \quad \sum_{i=1}^{\infty} \frac{H_{i+2}}{2^i i} = \frac{\pi^2}{12} + \frac{27}{12} - \frac{5}{2} \log 2,$$

$$\sum_{i=1}^{\infty} \frac{H_i}{2^i (i+1)} = \log^2 2, \quad \sum_{i=1}^{\infty} \frac{H_i}{2^i (i+2)} = 2 \log^2 2 + 2 \log 2 - 2, \quad \sum_{i=1}^{\infty} \frac{H_i^2}{2^i} = \log^2 2 + \frac{\pi^2}{6},$$

$$\sum_{i=1}^{\infty} \frac{H_i}{2^i i^2} = \zeta(3) - \frac{\pi^2 \log 2}{12}, \quad \sum_{i=1}^{\infty} \frac{H_{i,2}}{i^4} = \zeta^2(3) - \frac{\pi^6}{2835}, \quad \sum_{i=1}^{\infty} \frac{H_{i,2}}{2^i i} = \frac{5}{8}\zeta(3).$$

Source: [2483, 2859].

Fact 13.5.18.

$$\sum_{i=1}^{\infty} \frac{H_i}{4^i i}\binom{2i}{i} = \frac{\pi^2}{3}, \quad \sum_{i=0}^{\infty} \frac{H_i}{8^i}\binom{2i}{i} = 2\sqrt{2} \log \frac{1+\sqrt{2}}{2}, \quad \sum_{i=0}^{\infty} (-1)^{i+1} \frac{H_i}{4^i}\binom{2i}{i} = \sqrt{2} \log \frac{2\sqrt{2}}{1+\sqrt{2}},$$

$$\sum_{i=1}^{\infty} \frac{H_{i+1}}{4^i (i+1)}\binom{2i}{i} = 3, \quad \sum_{i=1}^{\infty} \left(\frac{H_{2i} - \frac{1}{2}H_i}{i}\right)^2 = \frac{\pi^4}{32}, \quad \sum_{i=1}^{\infty} \frac{H_{2i} - \frac{1}{2}H_i}{4^i i}\binom{2i}{i} = \frac{\pi^2}{4},$$

$$\sum_{i=1}^{\infty} \frac{H_i}{4^i (2i-1)}\binom{2i}{i} = 2, \quad \sum_{i=1}^{\infty} \frac{i(H_{2i} - \frac{1}{2}H_i)}{4^i (2i-1)^2}\binom{2i}{i} = \frac{\pi \log 2}{2},$$

$$\sum_{i=1}^{\infty} \frac{H_i}{i}\left[\left(\sum_{j=1}^{i} \frac{1}{4j-1}\right)\prod_{j=1}^{i}\left(1 - \frac{1}{4j}\right) - \left(\sum_{j=1}^{i} \frac{1}{4j-3}\right)\prod_{j=1}^{i}\left(1 - \frac{3}{4j}\right)\right] = \pi^3,$$

$$\sum_{i=1}^{\infty} \frac{H_i}{4^i i}\binom{2i}{i}\left[\left(\sum_{j=1}^{i} \frac{1}{2j-1}\right)^2 - \sum_{j=1}^{i} \frac{1}{(2j-1)^2}\right] = \frac{\pi^4}{4},$$

$$\sum_{i=1}^{\infty} (-1)^{i+1} \frac{H_{i+1}}{4^i (i+1)}\binom{2i}{i} = 5 + 4\sqrt{2}\left(\log \frac{2\sqrt{2}}{1+\sqrt{2}} - 1\right),$$

$$\sum_{i=0}^{\infty} \frac{C_i H_i}{4^i} = 4 \log 2, \quad \sum_{i=0}^{\infty} (-1)^{i+1} \frac{C_i H_i}{4^{i+1}} = (1 + 3\sqrt{2}/2) \log 2 - (1+\sqrt{2}) \log(1+\sqrt{2}).$$

Source: [62].

Fact 13.5.19. Let $n \geq 1$. Then,

$$\lim_{i\to\infty}(H_{ni} - H_i) = \log n.$$

In particular,

$$\log 2 = \lim_{i\to\infty}(H_{2i} - H_i) = \tfrac{1}{1} - \tfrac{1}{2} + \tfrac{1}{3} - \tfrac{1}{4} + \tfrac{1}{5} - \tfrac{1}{6} + \cdots,$$

$$\log 3 = \lim_{i\to\infty}(H_{3i} - H_i) = \tfrac{1}{1} + \tfrac{1}{2} - \tfrac{2}{3} + \tfrac{1}{4} + \tfrac{1}{5} - \tfrac{2}{6} + \cdots.$$

Source: [506, p. 125], [1158, pp. 153, 154], and [1566, p. 48].

Fact 13.5.20. Let $n \geq 2$. Then,

$$\sum_{i=1}^{\infty}\left(\log n + H_i - H_{ni} - \frac{n-1}{2in}\right) = \frac{n-1}{2n} - \frac{1}{2}\log n - \frac{\pi}{2n^2}\sum_{i=1}^{n-1} i\cot\frac{i\pi}{n}.$$

In particular,

$$\sum_{i=1}^{\infty}\left(\log 2 + H_i - H_{2i} - \frac{1}{4i}\right) = \frac{1}{4} - \frac{1}{2}\log 2, \quad \sum_{i=1}^{\infty}\left(\log 3 + H_i - H_{3i} - \frac{1}{3i}\right) = \frac{1}{3} - \frac{1}{2}\log 3 + \frac{\sqrt{3}\pi}{54}.$$

Source: [1103, pp. 150, 214–216].

Fact 13.5.21. Let $n \geq 2$. Then,

$$\sum_{i=1}^{\infty}\frac{1}{i}(\log n + H_i - H_{ni}) = \frac{(n-1)(n+2)}{24n}\pi^2 - \frac{1}{2}\log^2 n - \frac{1}{2}\sum_{i=1}^{n-1}\log^2\left(2\sin\frac{i\pi}{n}\right).$$

In particular,

$$\sum_{i=1}^{\infty}\frac{1}{i}(\log 2 + H_i - H_{2i}) = \frac{\pi^2}{12} - \log^2 2, \quad \sum_{i=1}^{\infty}\frac{1}{i}(\log 3 + H_i - H_{3i}) = \frac{5\pi^2}{36} - \frac{3}{4}\log^2 3,$$

$$\sum_{i=1}^{\infty}\frac{1}{i}(\log 4 + H_i - H_{4i}) = \frac{3\pi^2}{16} - \frac{11}{4}\log^2 2, \quad \sum_{i=1}^{\infty}\frac{1}{i}(\log 5 + H_i - H_{5i}) = \frac{7\pi^2}{30} - \frac{5}{8}\log^2 5 - \frac{1}{2}\log^2\frac{1+\sqrt{5}}{2}.$$

Source: [1682].

Fact 13.5.22. Let $n \geq 2$. Then,

$$\begin{aligned}\sum_{i=1}^{\infty}(-1)^{i+1}(\log n + H_i - H_{ni}) &= \frac{n-1}{2n}\log 2 + \frac{1}{2}\log n - \frac{\pi}{4n}\sum_{i=1}^{n}\cot\frac{i\pi}{2n}\\ &= \frac{n-1}{2n}\log 2 + \frac{1}{2}\log n - \frac{\pi}{2n^2}\sum_{i=1}^{\lfloor n/2\rfloor}(n+1-2i)\cot\frac{(2i-1)\pi}{2n}.\end{aligned}$$

In particular,

$$\sum_{i=1}^{\infty}(-1)^{i-1}(\log 2 + H_i - H_{2i}) = \frac{3}{4}\log 2 - \frac{\pi}{8}, \quad \sum_{i=1}^{\infty}(-1)^{i-1}(\log 3 + H_i - H_{3i}) = \frac{1}{3}\log 2 + \frac{1}{2}\log 3 - \frac{\sqrt{3}\pi}{9},$$

$$\sum_{i=1}^{\infty}(-1)^{i-1}(\log 4 + H_i - H_{4i}) = \frac{11}{8}\log 2 - (1 + 2\sqrt{2})\frac{\pi}{16}.$$

Furthermore,

$$\sum_{i=1}^{\infty}(-1)^{i-1}(\log 2 + H_i - H_{2i})^2 = \frac{\pi^2}{48} - \frac{\pi}{8}\log 2 - \frac{7}{8}\log^2 2 + \frac{G}{2}.$$

Source: [1682, 1683]. The last equality is given in [1103, pp. 146, 196–199].

Fact 13.5.23.

$$\sum_{i=1}^{\infty}(\log 2 + H_i - H_{2i})^2 = \frac{\pi^2}{48} + \frac{1}{2}\log 2 - \log^2 2.$$

Source: [1103, pp. 143, 184–186].

Fact 13.5.24.

$$\sum_{i=0}^{\infty}\left(\sum_{j=1}^{\infty}(-1)^{j+1}\frac{1}{i+j}\right)^2 = \log 2, \quad \sum_{i=0}^{\infty}(-1)^i\left(\sum_{j=1}^{\infty}(-1)^{j+1}\frac{1}{i+j}\right)^2 = \frac{\pi^2}{24}.$$

Source: [1103, pp. 143, 146, 186, 199, 200].

Fact 13.5.25. Let $n \ge 2$. Then,

$$\sum_{i=1}^{\infty}(-1)^{i+1}(H_{ni} - H_{n(i-1)}) = \int_0^1 \frac{x^n - 1}{(x^n+1)(x-1)}\,dx = \frac{\log 2}{n} - \frac{\pi}{n^2}\sum_{i=1}^{\lfloor n/2\rfloor}(n+1-2i)\cot\frac{(2i-1)\pi}{2n}.$$

In particular,

$$\sum_{i=1}^{\infty}(-1)^{i+1}(H_i - H_{i-1}) = \int_0^1 \frac{1}{x+1}\,dx = \log 2,$$

$$\sum_{i=1}^{\infty}(-1)^{i+1}(H_{2i} - H_{2(i-1)}) = \int_0^1 \frac{x^2-1}{(x^2+1)(x-1)}\,dx = \tfrac{1}{2}\log 2 + \frac{1}{4}\pi,$$

$$\sum_{i=1}^{\infty}(-1)^{i+1}(H_{3i} - H_{3(i-1)}) = \int_0^1 \frac{x^3-1}{(x^3+1)(x-1)}\,dx = \frac{1}{3}\log 2 + \frac{2\sqrt{3}}{9}\pi,$$

$$\sum_{i=1}^{\infty}(-1)^{i+1}(H_{4i} - H_{4(i-1)}) = \int_0^1 \frac{x^4-1}{(x^4+1)(x-1)}\,dx = \frac{1}{4}\log 2 + \frac{1+2\sqrt{2}}{8}\pi.$$

Finally,

$$\sum_{i=1}^{\infty}(-1)^{i+1}(H_{ni} - H_{n(i-1)}) = \log\frac{2n}{\pi} + \gamma + \frac{1}{n}\log 2 - \frac{\pi^2}{72n^2} + O\left(\frac{1}{n^4}\right).$$

Source: [1202, 1683].

Fact 13.5.26. Let $n, m \ge 1$. Then,

$$\sum_{i=m}^{\infty}\frac{1}{\prod_{j=0}^{n}(i+j)} = \frac{1}{n\prod_{j=0}^{n-1}(m+j)}.$$

In particular,

$$\sum_{i=1}^{\infty}\frac{1}{\prod_{j=0}^{n}(i+j)} = \frac{1}{nn!}.$$

Hence,

$$\sum_{i=1}^{\infty}\frac{1}{i(i+1)} = 1, \quad \sum_{i=1}^{\infty}\frac{1}{i(i+1)(i+2)} = \frac{1}{4}, \quad \sum_{i=1}^{\infty}\frac{1}{i(i+1)(i+2)(i+3)} = \frac{1}{18}.$$

Source: [113], [506, pp. 123, 124], and [2577]. **Related:** Fact 13.5.27 and Fact 14.2.11.

Fact 13.5.27. Let $m, n \ge 1$ and $l \ge 0$. Then,

$$\sum_{i=1}^{\infty}\frac{1}{\prod_{j=0}^{l+1}[n+m(i+j-1)]} = \frac{1}{(l+1)m\prod_{j=0}^{l}(n+mj)}.$$

In particular,

$$\sum_{i=1}^{\infty}\frac{1}{[n+m(i-1)](n+mi)} = \frac{1}{mn},$$

$$\sum_{i=1}^{\infty} \frac{1}{[n+m(i-1)](n+mi)[n+m(i+1)]} = \frac{1}{2mn(n+m)},$$

$$\sum_{i=1}^{\infty} \frac{1}{\prod_{j=0}^{l+1}(i+j)} = \frac{1}{(l+1)\prod_{j=0}^{l}(1+j)}.$$

Furthermore,

$$\sum_{i=1}^{\infty} \frac{1}{i(i+1)} = 1, \quad \sum_{i=1}^{\infty} \frac{1}{4i^2-1} = \frac{1}{2}, \quad \sum_{i=1}^{\infty} \frac{1}{(i+1)(i+2)} = \frac{1}{2},$$

$$\sum_{i=1}^{\infty} \frac{1}{(i+2)(i+3)} = \frac{1}{3}, \quad \sum_{i=1}^{n} \frac{1}{(3i+1)(3i-2)} = \frac{1}{3}, \quad \sum_{i=1}^{\infty} \frac{1}{i(i+1)(i+2)} = \frac{1}{4}.$$

Source: [1217, p. 12]. **Related:** Fact 1.12.25, Fact 1.12.27, Fact 13.5.26, and Fact 13.5.67.

Fact 13.5.28.

$$\sum_{i=1}^{\infty} \left(\frac{1}{i} - \log\frac{i+1}{i}\right) = \gamma, \quad \sum_{i=1}^{\infty} (-1)^{i+1}\left(\frac{1}{i} - \log\frac{i+1}{i}\right) = \log\frac{4}{\pi},$$

$$\sum_{i=1}^{\infty} (-1)^{i+1}\frac{\log i}{i} = \frac{\log^2 2}{2} - \gamma\log 2, \quad \sum_{i=1}^{\infty} \left(H_i - \log i - \gamma - \frac{1}{2i}\right) = \tfrac{1}{2}(\gamma + 1 - \log 2\pi),$$

$$\lim_{x\downarrow 1} \sum_{i=1}^{\infty} \left(\frac{1}{i^x} - \frac{1}{x^i}\right) = \gamma, \quad \lim_{x\to\infty} [x - \Gamma(\tfrac{1}{x})] = \gamma.$$

Source: [1103, pp. 144, 192–195]. The last two equalities are given in [1350, p. 109].

Fact 13.5.29. Let n be a positive integer. Then,

$$\sum_{i=n+1}^{\infty} \frac{1}{i^2-n^2} = \frac{3}{4n^2} + \sum_{i=1}^{n-1} \frac{1}{n^2-i^2}.$$

In particular,

$$\sum_{i=2}^{\infty} \frac{1}{i^2-1} = \frac{3}{4}, \quad \sum_{i=3}^{\infty} \frac{1}{i^2-4} = \frac{25}{48}, \quad \sum_{i=4}^{\infty} \frac{1}{i^2-9} = \frac{49}{120}.$$

Source: [1217, p. 10]. **Related:** Fact 1.12.27 and Fact 13.5.59.

Fact 13.5.30. Let n be an even positive integer. Then,

$$\sum_{i=n+1}^{\infty} (-1)^i \frac{1}{i^2-n^2} = \frac{3}{4n^2} + \sum_{i=1}^{n-1} (-1)^i \frac{1}{n^2-i^2}.$$

In particular,

$$\sum_{i=3}^{\infty} (-1)^i \frac{1}{i^2-4} = -\frac{7}{48}, \quad \sum_{i=5}^{\infty} (-1)^i \frac{1}{i^2-16} = -\frac{533}{6720}.$$

Source: [1217, p. 10]. **Related:** Fact 13.5.61.

Fact 13.5.31. Let n be an odd positive integer. Then,

$$\sum_{i=n+1}^{\infty} (-1)^i \frac{1}{i^2-n^2} = \frac{1}{4n^2} + \sum_{i=1}^{n-1} (-1)^i \frac{1}{n^2-i^2}.$$

In particular,

$$\sum_{i=2}^{\infty}(-1)^i\frac{1}{i^2-1}=\frac{1}{4},\quad \sum_{i=4}^{\infty}(-1)^i\frac{1}{i^2-9}=\frac{37}{360}.$$

Fact 13.5.32. Let n be an even positive integer. Then,

$$\sum_{i=0}^{\infty}\frac{1}{4i^2+4i+1-n^2}=0.$$

Source: Fact 13.5.29 and Fact 13.5.30.

Fact 13.5.33. Let n be an odd positive integer. Then,

$$\sum_{i=(n+1)/2}^{\infty}\frac{1}{(2i+1)^2-n^2}=\frac{1}{4n^2}+\sum_{i=0}^{(n-3)/2}\frac{1}{n^2-(2i+1)^2}.$$

In particular,

$$\sum_{k=1}^{\infty}\frac{1}{(2i+1)^2-1}=\frac{1}{4},\quad \sum_{k=2}^{\infty}\frac{1}{(2i+1)^2-9}=\frac{11}{72},\quad \sum_{k=3}^{\infty}\frac{1}{(2i+1)^2-25}=\frac{137}{1200}.$$

Source: Fact 13.5.29 and Fact 13.5.31.

Fact 13.5.34. Let z be a complex number, and assume that $|z|<1$. Then,

$$\sum_{i=0}^{\infty}z^i=\frac{1}{1-z},\quad \sum_{i=1}^{\infty}iz^i=\frac{z}{(1-z)^2},\quad \sum_{i=1}^{\infty}i^2z^i=\frac{z^2+z}{(1-z)^3},\quad \sum_{i=1}^{\infty}i^3z^i=\frac{z^3+4z^2+z}{(1-z)^4}.$$

In particular,

$$\sum_{i=1}^{\infty}\frac{1}{2^i}=1,\quad \sum_{i=1}^{\infty}\frac{i}{2^i}=2,\quad \sum_{i=1}^{\infty}\frac{i^2}{2^i}=6,\quad \sum_{i=1}^{\infty}\frac{i^3}{2^i}=26,$$

$$\sum_{i=1}^{\infty}\frac{1}{3^i}=\frac{1}{2},\quad \sum_{i=1}^{\infty}\frac{i}{3^i}=\frac{3}{4},\quad \sum_{i=1}^{\infty}\frac{i^2}{3^i}=\frac{3}{2},\quad \sum_{i=1}^{\infty}\frac{i^3}{3^i}=\frac{33}{8}.$$

Furthermore,

$$\sum_{i=1}^{\infty}\frac{i^4}{2^i}=150,\quad \sum_{i=1}^{\infty}\frac{i^5}{2^i}=1082,\quad \sum_{i=1}^{\infty}\frac{i^6}{2^i}=9366,$$

$$\sum_{i=1}^{\infty}\frac{i^4}{3^i}=15,\quad \sum_{i=1}^{\infty}\frac{i^5}{3^i}=\frac{273}{4},\quad \sum_{i=1}^{\infty}\frac{i^6}{3^i}=\frac{1491}{4}.$$

Source: Fact 1.19.3. **Remark:** $\sum_{i=1}^{\infty}i^2z^i=1/(1-z)-3/(1-z)^2+2/(1-z)^3$.

Fact 13.5.35.

$$\sum_{i=1}^{\infty}\frac{1}{2^i}=1,\quad \sum_{i=1}^{\infty}\frac{1}{2^ii}=\log 2,\quad \sum_{i=1}^{\infty}\frac{1}{2^ii^2}=\frac{\pi^2}{12}-\frac{1}{2}\log^2 2,\quad \sum_{i=1}^{\infty}\frac{1}{2^ii^3}=\frac{7\zeta(3)}{8}+\frac{\log^3 2}{6}-\frac{\pi^2\log 2}{12},$$

$$\sum_{i=1}^{\infty}\frac{1}{2^i(i+1)}=2\log 2-1,\quad \sum_{i=1}^{\infty}\frac{1}{2^ii(i+1)}=1-\log 2,\quad \sum_{i=1}^{\infty}\frac{1}{2^ii^2(i+1)}=\frac{\pi^2}{12}+\log 2-\frac{1}{2}\log^2 2-1,$$

$$\sum_{i=1}^{\infty}\frac{1}{3^i}=\frac{1}{2},\quad \sum_{i=1}^{\infty}\frac{1}{3^ii}=\log\frac{3}{2},\quad \sum_{i=1}^{\infty}\frac{1}{3^ii^2}=\mathrm{Li}_2(\tfrac{1}{3})\approx 0.366213,$$

$$\sum_{i=1}^{\infty} \frac{1}{3^i(i+1)} = \log \frac{27}{8} - 1, \quad \sum_{i=1}^{\infty} \frac{1}{3^i i(i+1)} = 1 - \log \frac{9}{4}, \quad \sum_{i=1}^{\infty} \frac{1}{3^i i^2(i+1)} = \mathrm{Li}_2(\tfrac{1}{3}) + \log \frac{9}{4} - 1.$$

Remark: Li_2 is the dilogarithm function defined by Fact 13.3.5.

Fact 13.5.36. Let $n \geq 1$. Then, the following statements hold:

i) $\zeta(2n) = \sum_{i=1}^{\infty} \frac{1}{i^{2n}} = \frac{2^{2n-1}\pi^{2n}}{(2n)!}|B_{2n}|.$

ii) $\sum_{i=1}^{\infty} (-1)^{i+1} \frac{1}{i^{2n}} = \frac{(2^{2n-1}-1)\pi^{2n}}{(2n)!}|B_{2n}|.$

iii) $\sum_{i=1}^{\infty} \frac{1}{(2i-1)^{2n}} = \frac{(4^n-1)\pi^{2n}}{2(2n)!}|B_{2n}|.$

In particular,

$$\sum_{i=1}^{\infty} \frac{1}{i^2} = \frac{\pi^2}{6}, \quad \sum_{i=1}^{\infty} \frac{1}{i^4} = \frac{\pi^4}{90}, \quad \sum_{i=1}^{\infty} \frac{1}{i^6} = \frac{\pi^6}{945}, \quad \sum_{i=1}^{\infty} \frac{1}{i^8} = \frac{\pi^8}{9450}, \quad \sum_{i=1}^{\infty} \frac{1}{i^{10}} = \frac{\pi^{10}}{93555},$$

$$\sum_{i=1}^{\infty} (-1)^{i+1} \frac{1}{i^2} = \frac{\pi^2}{12}, \quad \sum_{i=1}^{\infty} (-1)^{i+1} \frac{1}{i^4} = \frac{7\pi^4}{720}, \quad \sum_{i=1}^{\infty} (-1)^{i+1} \frac{1}{i^6} = \frac{31\pi^6}{30240}, \quad \sum_{i=1}^{\infty} (-1)^{i+1} \frac{1}{i^8} = \frac{127\pi^8}{1209600},$$

$$\sum_{i=1}^{\infty} \frac{1}{(2i-1)^2} = \frac{\pi^2}{8}, \quad \sum_{i=1}^{\infty} \frac{1}{(2i-1)^4} = \frac{\pi^4}{96}, \quad \sum_{i=1}^{\infty} \frac{1}{(2i-1)^6} = \frac{\pi^6}{960}, \quad \sum_{i=1}^{\infty} \frac{1}{(2i-1)^8} = \frac{17\pi^8}{161280}.$$

Finally, if $n \geq 2$, then

$$\zeta(2n) = \frac{2}{2n+1} \sum_{i=1}^{n-1} \zeta(2i)\zeta(2n-2i).$$

Source: [968, p. 297] and [2513, p. 167]. **Remark:**

$$\frac{\pi^2}{6} = \sum_{i=1}^{\infty} \frac{1}{i^2} = \sum_{i=1}^{\infty} \frac{1}{(2i)^2} + \sum_{i=1}^{\infty} \frac{1}{(2i-1)^2} = \frac{1}{4} \sum_{i=1}^{\infty} \frac{1}{i^2} + \sum_{i=1}^{\infty} \frac{1}{(2i-1)^2} = \frac{\pi^2}{24} + \frac{\pi^2}{8} = \frac{\pi^2}{6}.$$

Fact 13.5.37. Let $p > 1$. Then,

$$\sum \frac{1}{i^p} = \frac{\zeta(p)}{\zeta(2p)},$$

where the sum is taken over all positive integers i that are not divisible by a square that is greater than 1. In particular, with the same summation index,

$$\sum \frac{1}{i^2} = \frac{15}{\pi^2}, \quad \sum \frac{1}{i^4} = \frac{105}{\pi^4}, \quad \sum \frac{1}{i^6} = \frac{675675}{691\pi^6}, \quad \sum \frac{1}{i^6} = \frac{34459425}{3617\pi^8}.$$

Furthermore, with the same summation index,

$$\sum (-1)^{i+1} \frac{1}{i^2} = \frac{9}{\pi^2}.$$

Source: [1406, 2414].

Fact 13.5.38. Let $n \geq 2$. Then,

$$\sum \frac{1}{i^n} = \frac{1}{2}\zeta(n) - \frac{\zeta(2n)}{2\zeta(n)},$$

where the sum is taken over all positive integers that have an odd number of prime divisors. Furthermore,

$$\sum \frac{1}{i^n} = \frac{\zeta(n)}{\zeta(2n)} - 1,$$

where the sum is taken over all positive integers that have distinct prime divisors. In addition,

$$\sum \frac{1}{i^n} = \frac{\zeta(n)}{2\zeta(2n)} - \frac{1}{2\zeta(2n)},$$

where the sum is taken over all positive integers that have an odd number of distinct prime divisors. Finally,

$$\sum \frac{1}{i^n} = \zeta(n) - \frac{\zeta(n)}{\zeta(2n)},$$

where the sum is taken over all positive integers that have at least two equal prime divisors. **Source:** [1317, pp. 20, 21]. **Related:** Fact 13.10.24.

Fact 13.5.39. For $x \in (1, \infty)$, define $\zeta(x) \triangleq \sum_{i=1}^{\infty} 1/i^x$. Then,

$$\frac{1}{x-1} \le \zeta(x) \le \frac{x}{x-1}.$$

Furthermore,

$$\lim_{x\to\infty} \zeta(x) = 1, \quad \lim_{x\downarrow 1} \zeta(x) = \infty, \quad \lim_{x\downarrow 1} (x-1)\zeta(x) = 1.$$

Source: [2294, p. 67].

Fact 13.5.40. Let $n \ge 1$. Then,

$$\zeta(n) = \sum_{i=n-1}^{\infty} \frac{1}{i(i!)} \begin{bmatrix} i \\ n-1 \end{bmatrix}.$$

In particular,

$$\frac{\pi^2}{6} = \sum_{i=1}^{\infty} \frac{1}{i(i!)} \begin{bmatrix} i \\ 1 \end{bmatrix} = \sum_{i=1}^{\infty} \frac{1}{i^2}, \quad \zeta(3) = \sum_{i=2}^{\infty} \frac{1}{i(i!)} \begin{bmatrix} i \\ 2 \end{bmatrix} = \sum_{i=2}^{\infty} \frac{H_{i-1}}{i^2},$$

$$\frac{\pi^4}{90} = \sum_{i=3}^{\infty} \frac{1}{i(i!)} \begin{bmatrix} i \\ 3 \end{bmatrix} = \frac{1}{2} \sum_{i=3}^{\infty} \frac{H_{i-1}^2 - H_{i-1,2}}{i^2},$$

$$\zeta(5) = \sum_{i=4}^{\infty} \frac{1}{i(i!)} \begin{bmatrix} i \\ 4 \end{bmatrix} = \frac{1}{6} \sum_{i=4}^{\infty} \frac{H_{i-1}^3 - 3H_{i-1}H_{i-1,2} + 2H_{i-1,3}}{i^2}.$$

Source: [610]. **Related:** Fact 1.19.1 and Fact 13.5.16.

Fact 13.5.41.

$$\begin{aligned}
\zeta(3) &= \sum_{i=1}^{\infty} \frac{1}{i^3} = \frac{1}{2} \sum_{i=1}^{\infty} \frac{H_i}{i^2} = \sum_{i=1}^{\infty} \frac{H_i}{(i+1)^2} = \sum_{i=2}^{\infty} \frac{1}{i(i!)} \begin{bmatrix} i \\ 2 \end{bmatrix} = \frac{8}{7} \sum_{i=0}^{\infty} \frac{1}{(2i+1)^3} = \frac{4}{7} \sum_{i=1}^{\infty} \frac{H_{2i} - \frac{1}{2}H_i}{i^2} \\
&= \frac{4}{3} \sum_{i=0}^{\infty} (-1)^i \frac{1}{(i+1)^3} = \frac{5}{2} \sum_{i=1}^{\infty} (-1)^{i+1} \frac{(i!)^2}{i^3 (2i)!} = \frac{7\pi^3}{180} - 2 \sum_{i=1}^{\infty} \frac{1}{i^3(e^{2i\pi} - 1)} = -4\pi^2 \zeta'(-2) \\
&= \frac{1}{7} \sum_{i=1}^{\infty} \frac{1}{4^i i} \binom{2i}{i} (H_{2i} - \tfrac{1}{2}H_i) = \frac{4}{7} \left(\pi G + \sum_{i=1}^{\infty} (-1)^i \frac{H_{2i} - \frac{1}{2}H_i}{i^2} \right) \\
&= -\frac{8\pi^2}{9} \sum_{i=0}^{\infty} \frac{\zeta(2i)}{(2i+1)(2i+3)4^i} = -\frac{8\pi^2}{5} \sum_{i=0}^{\infty} \frac{\zeta(2i)}{(2i+1)(2i+2)(2i+3)4^i} = -\frac{2\pi^2}{7} \sum_{i=0}^{\infty} \frac{\zeta(2i)}{(i+1)(2i+1)4^i}
\end{aligned}$$

$$= -\frac{6\pi^2}{23}\sum_{i=0}^{\infty}\frac{(98i+121)\zeta(2i)}{(2i+1)(2i+2)(2i+3)(2i+4)(2i+5)4^i}$$

$$= \frac{\pi^2}{6}\left(\frac{3}{4} - \frac{1}{2}\log\frac{\pi}{3} + \sum_{i=1}^{\infty}\frac{\zeta(2i)}{i(2i+1)(2i+2)36^i}\right) = \frac{4}{7} + \frac{1}{14}\sum_{i=1}^{\infty}\frac{16^i(16i^3+12i^2+6i+1)(i!)^4}{i^3(2i+1)^3[(2i)!]^2}$$

$$= \frac{\pi^2}{2}\left(\frac{11}{18} - \frac{1}{3}\log\pi + \sum_{i=1}^{\infty}\frac{\zeta(2i)}{i(i+1)(2i+1)(2i+3)4^i}\right)$$

$$= -\frac{120\pi^2}{1573}\sum_{i=0}^{\infty}\frac{(8576i^2+24286i+17283)\zeta(2i)}{(2i+1)(2i+2)(2i+3)(2i+4)(2i+5)(2i+6)(2i+7)4^i}$$

$$= \frac{1}{4}\sum_{i=1}^{\infty}(-1)^{i+1}\frac{56i^2-32i+5}{(2i-1)^2i^3\binom{3i}{i}\binom{2i}{i}} = \frac{1}{72}\sum_{i=0}^{\infty}(-1)^i\frac{5265i^4+13878i^3+13761i^2+6120i+1040}{(4i+1)(4i+3)(i+1)(3i+1)^2(3i+2)^2\binom{3i}{i}\binom{4i}{i}}$$

$$= \int_0^{\infty}\frac{\frac{\pi}{4}\operatorname{sech}^2\pi x - x\tanh\pi x}{(x^2+\frac{1}{4})^2}\,dx = \frac{37\pi^3}{900} - \frac{2}{5}\sum_{i=1}^{\infty}\frac{1}{i^3}\left(\frac{4}{e^{i\pi}-1} + \frac{1}{e^{4i\pi}-1}\right)$$

$$= \frac{2}{7}\sum_{i=1}^{\infty}\frac{1}{4^i i}\binom{2i}{i}\left[\left(\sum_{j=1}^{i}\frac{1}{2j-1}\right)^2 - \sum_{j=1}^{i}\frac{1}{(2j-1)^2}\right] = \frac{2\pi^2}{9}\left(\log 2 + 2\sum_{i=0}^{\infty}\frac{\zeta(2i)}{(2i+3)4^i}\right)$$

$$= \frac{9}{52}\sum_{i=1}^{\infty}\frac{1}{i}\sum_{k=1}^{2}\left(\prod_{j=1}^{i}\frac{3j-k}{3j}\right)\left[\left(\sum_{j=1}^{i}\frac{1}{3j-k}\right)^2 - \sum_{j=1}^{i}\frac{1}{(3j-k)^2}\right]$$

$$\approx 1.2020569031595942853997381615114499907649862923404988817922715553418 38.$$

Furthermore, $\zeta(3)$ is an irrational number. **Source:** [62, 66, 323], [511, pp. 78, 236], [645, p. 99], [1352, Chapter 5], [2513, pp. 246, 405, 431, 432, 443], and [2792]. **Remark:** $\zeta(3)$ is *Apery's constant*. See [2389]. **Related:** Fact 13.5.16.

Fact 13.5.42.

$$\zeta(5) = \frac{\pi^2}{13}\zeta(3) - \frac{8\pi^4}{13}\sum_{i=0}^{\infty}\frac{\zeta(2i)}{(2i+3)(2i+4)(2i+5)4^i} = \frac{3\pi^2}{31}\zeta(3) + \frac{4\pi^4}{93}\sum_{i=0}^{\infty}\frac{\zeta(2i)}{(2i+3)(2i+4)4^i}$$

$$= \frac{\pi^2}{11}\zeta(3) + \frac{\pi^4}{33}\sum_{i=0}^{\infty}\frac{\zeta(2i)}{(2i+4)(2i+5)4^i} = \frac{7\pi^2}{75}\zeta(3) + \frac{8\pi^4}{255}\sum_{i=0}^{\infty}\frac{\zeta(2i)}{(2i+3)(2i+5)4^i}$$

$$= \frac{4\pi^2}{31}\zeta(3) + \frac{8\pi^4}{13}\sum_{i=0}^{\infty}\frac{\zeta(2i)}{(2i+1)(2i+2)(2i+3)(2i+4)4^i}$$

$$= \frac{2\pi^2}{27}\zeta(3) - \frac{4\pi^4}{9}\sum_{i=0}^{\infty}\frac{\zeta(2i)}{(2i+2)(2i+3)(2i+4)(2i+5)4^i}$$

$$= \frac{2\pi^4}{45}\left(\log\pi - \frac{47}{60} - 30\sum_{i=1}^{\infty}\frac{\zeta(2i)}{i(i+2)(2i+3)(2i+5)4^i}\right).$$

Source: [2513, pp. 429–431]. **Remark:** At least one of the numbers $\zeta(5), \zeta(7), \zeta(9), \zeta(11)$ is irrational. See [1350, p. 42].

Fact 13.5.43. Let $n \geq 2$. Then,

$$\sum_{i=1}^{\infty}\frac{H_i}{i^n} = \frac{n+2}{2}\zeta(n+1) - \frac{1}{2}\sum_{i=1}^{n-2}\zeta(i+1)\zeta(n-i), \qquad \sum_{i=1}^{\infty}\frac{H_i}{(i+1)^n} = \frac{n}{2}\zeta(n+1) - \frac{1}{2}\sum_{i=1}^{n-2}\zeta(i+1)\zeta(n-i).$$

In particular,

$$\sum_{i=1}^{\infty}\frac{H_i}{i^2}=2\zeta(3),\quad \sum_{i=1}^{\infty}\frac{H_i}{i^3}=\frac{5}{2}\zeta(4)-\frac{1}{2}\zeta^2(2)=\frac{\pi^4}{72},\quad \sum_{i=1}^{\infty}\frac{H_i}{i^4}=3\zeta(5)-\frac{\pi^2}{6}\zeta(3),$$

$$\sum_{i=1}^{\infty}\frac{H_i}{(i+1)^2}=\zeta(3),\quad \sum_{i=1}^{\infty}\frac{H_i}{(i+1)^3}=\frac{\pi^4}{360},\quad \sum_{i=1}^{\infty}\frac{H_i}{(i+1)^4}=2\zeta(5)-\frac{\pi^2}{6}\zeta(3).$$

Now, let $n\ge 1$. Then,

$$\sum_{i=1}^{\infty}\frac{H_i}{i(i+n)}=\frac{1}{2n}[2\zeta(2)+H_{n-1}^2+H_{n-1,2}],\quad \sum_{i=1}^{\infty}\frac{H_i}{(i+n)^2}=\zeta(3)+\zeta(2)H_{n-1}-H_{n-1}H_{n-1,2}-H_{n-1,3},$$

$$\sum_{i=1}^{\infty}\frac{H_i}{i^{2n+1}}=\frac{1}{2}\sum_{i=2}^{2n}(-1)^i\zeta(i)\zeta(2n-i+2),$$

$$\sum_{i=1}^{\infty}\frac{H_{i,2}}{i^{2n+1}}=\zeta(2)\zeta(2n+1)-\frac{1}{2}(n+2)(2n+1)\zeta(2n+3)+2\sum_{i=2}^{n+1}(i-1)\zeta(2i-1)\zeta(2n+4-2i).$$

In particular,

$$\sum_{i=1}^{\infty}\frac{H_i}{i(i+1)}=\frac{\pi^2}{6},\quad \sum_{i=1}^{\infty}\frac{H_i}{i(i+2)}=\frac{1}{2}+\frac{\pi^2}{12},\quad \sum_{i=1}^{\infty}\frac{H_i}{i(i+3)}=\frac{7}{12}+\frac{\pi^2}{18},\quad \sum_{i=1}^{\infty}\frac{H_i}{i(i+4)}=\frac{85}{144}+\frac{\pi^2}{24},$$

$$\sum_{i=1}^{\infty}\frac{H_i}{(i+2)^2}=\zeta(3)+\frac{\pi^2}{6}-2,\quad \sum_{i=1}^{\infty}\frac{H_i}{(i+3)^2}=\zeta(3)+\frac{\pi^2}{4}-3,\quad \sum_{i=1}^{\infty}\frac{H_i}{(i+4)^2}=\zeta(3)+\frac{11\pi^2}{36}-\frac{395}{108}.$$

Source: [2485] and [2513, pp. 228, 229]. **Remark:** $\sum_{i=1}^{\infty}\frac{H_i}{i}=\infty$. **Credit:** L. Euler.

Fact 13.5.44. If $n\ge 1$, then

$$\sum_{i=1}^{\infty}\frac{H_{i+n}}{\overline{i^{n+1}}}=\frac{1}{n(n!)}\left(\frac{1}{n}+H_n\right).$$

In particular,

$$\sum_{i=1}^{\infty}\frac{H_{i+1}}{i(i+1)}=2,\quad \sum_{i=1}^{\infty}\frac{H_{i+2}}{i(i+1)(i+2)}=\frac{1}{2},\quad \sum_{i=1}^{\infty}\frac{H_{i+3}}{i(i+1)(i+2)(i+3)}=\frac{13}{108}.$$

If $n\ge 2$, then

$$\sum_{i=1}^{\infty}\frac{H_i}{\overline{i^{n+1}}}=\frac{1}{n!}\left(\frac{\pi^2}{6}-H_{n-1,2}\right).$$

In particular,

$$\sum_{i=1}^{\infty}\frac{H_i}{i(i+1)(i+2)}=\frac{\pi^2}{12}-\frac{1}{2},\quad \sum_{i=1}^{\infty}\frac{H_i}{i(i+1)(i+2)(i+3)}=\frac{\pi^2}{36}-\frac{5}{24}.$$

Furthermore,

$$\sum_{i=1}^{\infty}\frac{H_{i+2}}{i(i+1)}=\frac{9}{4},\quad \sum_{i=1}^{\infty}\frac{H_{i+3}}{i(i+1)}=\frac{22}{9},\quad \sum_{i=1}^{\infty}\frac{H_{i+4}}{i(i+1)}=\frac{125}{48},\quad \sum_{i=1}^{\infty}\frac{H_{i+5}}{i(i+1)}=\frac{137}{50}.$$

Finally,

$$\sum_{i=1}^{\infty}\frac{H_i}{i}\sum_{j=i+1}^{\infty}\frac{1}{j^2}=\frac{7}{4}\zeta(4),\quad \sum_{i=1}^{\infty}\frac{1}{i}(\log 3+H_i-H_{3i})=\frac{5\pi^2}{36}-\frac{3\log^2 3}{4}.$$

Source: [1099] and [1103, pp. 149, 208–214].

Fact 13.5.45. Let $n \geq 1$. Then,

$$\sum_{i=1}^{\infty}(-1)^{i-1}\frac{H_{ni}}{i}=\frac{(n^2+1)\pi^2}{24n}-\frac{1}{2}\sum_{i=0}^{n-1}\log^2\left(2\sin\frac{(2i+1)\pi}{2n}\right).$$

In particular,

$$\sum_{i=1}^{\infty}(-1)^{i-1}\frac{H_i}{i}=\frac{\pi^2}{12}-\frac{1}{2}\log^2 2,\quad \sum_{i=1}^{\infty}(-1)^{i-1}\frac{H_{2i}}{i}=\frac{5\pi^2}{48}-\frac{1}{4}\log^2 2,$$

$$\sum_{i=1}^{\infty}(-1)^{i-1}\frac{H_{3i}}{i}=\frac{5\pi^2}{36}-\frac{1}{2}\log^2 2,\quad \sum_{i=1}^{\infty}(-1)^{i-1}\frac{H_{4i}}{i}=\frac{17\pi^2}{96}-\frac{1}{8}\log^2 2-\frac{1}{2}\log^2(1+\sqrt{2}).$$

Furthermore,

$$\sum_{i=1}^{\infty}(-1)^{i+1}\frac{H_i}{i^2}=\frac{5}{8}\zeta(3),\ \sum_{i=1}^{\infty}(-1)^{i+1}\frac{H_{i+1}}{i^2}=\frac{5}{8}\zeta(3)+\frac{\pi^2}{12}-2\log 2+1,\ \sum_{i=1}^{\infty}(-1)^{i}\frac{H_i}{i(i+1)}=\log^2 2-\frac{\pi^2}{12},$$

$$\sum_{i=1}^{\infty}(-1)^{i}\frac{H_i}{i(i+1)(i+2)}=\frac{1}{2}-\frac{\pi^2}{24}-\log 2+\log^2 2,\quad \sum_{i=1}^{\infty}(-1)^{i+1}\frac{H_i}{i^2(i+1)^2}=\frac{3}{4}\zeta(3)+\frac{1}{3}\log^3 2-\frac{\pi^2}{12}\log 2,$$

$$\sum_{i=1}^{\infty}(-1)^{i+1}\frac{H_i^2}{i}=\frac{3}{4}\zeta(3)+\frac{1}{3}\log^3 2-\frac{\pi^2}{12}\log 2,\quad \sum_{i=1}^{\infty}(-1)^{i+1}\frac{H_i^2}{i(i+1)}=\frac{\pi^2}{6}\log 2-\frac{2}{3}\log^3 2-\zeta(3).$$

Source: [1682] and [2513, pp. 357, 358]. **Remark:** The last expression corrects (41) given in [2513, p. 358].

Fact 13.5.46. Let x be a real number. If $|x| > 1$, then

$$\sum_{i=1}^{\infty}\frac{\zeta(2i)-1}{x^{2i}}=\frac{1}{2}+\frac{1}{1-x^2}-\frac{\pi}{2x}\cot\frac{\pi}{x}.$$

In particular,

$$\sum_{i=1}^{\infty}\frac{\zeta(2i)-1}{4^i}=\frac{1}{6},\quad \sum_{i=1}^{\infty}\frac{\zeta(2i)-1}{16^i}=\frac{13}{30}-\frac{\pi}{8},\quad \sum_{i=1}^{\infty}\frac{\zeta(2i)-1}{64^i}=\frac{61}{126}-\frac{(1+\sqrt{2})\pi}{16}.$$

If $|x| < 1$, then

$$\sum_{i=2}^{\infty}\zeta(i)x^{i-1}=-\psi(1-x)-\gamma,\quad \sum_{i=1}^{\infty}\zeta(2i)x^{2i-1}=\tfrac{1}{2}[\psi(1+x)-\psi(1-x)]=\frac{1}{2x}-\frac{\pi}{2}\cot\pi x,$$

$$\sum_{i=1}^{\infty}\zeta(2i)x^{2i}=\psi(1+x)+\psi(1-x)=\log\frac{\pi x}{\sin\pi x},\quad \sum_{i=1}^{\infty}\zeta(2i+1)x^{2i}=-\tfrac{1}{2}[\psi(1+x)+\psi(1-x)]-\gamma,$$

$$\sum_{i=2}^{\infty}\frac{\zeta(i)}{i}x^i=\log\Gamma(1-x)-\gamma x,\quad \sum_{i=1}^{\infty}\frac{\zeta(2i)}{i}x^{2i}=\log\Gamma(1+x)+\log\Gamma(1-x)=\log\frac{\pi x}{\sin\pi x},$$

$$\sum_{i=1}^{\infty}\frac{\zeta(2i+1)}{2i+1}x^{2i+1}=\tfrac{1}{2}[\log\Gamma(1-x)-\log\Gamma(1-x)]-\gamma x,\quad \sum_{i=1}^{\infty}\frac{\zeta(2i)}{i^2}x^{2i}=\log(\pi x\csc\pi x),$$

$$\sum_{i=1}^{\infty}(-1)^{i+1}\zeta(2i)x^{2i-1} = \frac{\pi}{2}\coth\pi x - \frac{1}{2x}.$$

If $0 < |x| \leq 1$, then

$$\sum_{i=1}^{\infty}(-1)^{i+1}[\zeta(2i)-1]x^{2i-1} = \frac{\pi}{2}\coth\pi x - \frac{3x^2+1}{2x(x^2+1)}, \quad \sum_{i=1}^{\infty}(-1)^{i+1}\frac{\zeta(2i)}{i}x^{2i} = \log\frac{\sinh\pi x}{\pi x}.$$

In particular,

$$\sum_{i=1}^{\infty}(-1)^{i+1}\frac{\zeta(2i)}{4^i} = \frac{\pi}{4}\coth\pi - \frac{1}{2}, \quad \sum_{i=1}^{\infty}(-1)^{i+1}[\zeta(2i)-1] = \frac{\pi}{2}\coth\pi - 1,$$

$$\sum_{i=1}^{\infty}(-1)^{i+1}\frac{\zeta(2i)}{i} = \log\frac{\sinh\pi}{\pi}, \quad \sum_{i=1}^{\infty}(-1)^{i+1}\frac{\zeta(2i)}{4^i i} = \log\frac{2\sinh\frac{\pi}{2}}{\pi}.$$

If $z \in \mathbb{C}$ and $|z| < 2$, then

$$\sum_{i=0}^{\infty}(-1)^{i+1}\frac{\zeta(2i)}{4^i}z^{2i} = \frac{\pi z}{4} + \frac{\pi z}{2(e^{\pi z}-1)}.$$

In particular,

$$\sum_{i=0}^{\infty}(-1)^{i+1}\frac{\zeta(2i)}{4^i} = \frac{\pi}{4} + \frac{\pi}{2(e^{\pi}-1)} = \frac{\pi e^{\pi z}}{2(e^{\pi z}-1)} - \frac{\pi}{4}.$$

Source: [2513, pp. 268, 270, 271, 366, 423]. **Remark:** ψ is the digamma function.

Fact 13.5.47.

$$\sum_{i=2}^{\infty}\frac{\zeta(i)}{2^i} = \log 2, \quad \sum_{i=2}^{\infty}\frac{\zeta(i)}{3^i} = \frac{1}{2}\log 3 - \frac{\sqrt{3}\pi}{18}, \quad \sum_{i=2}^{\infty}\frac{\zeta(i)}{4^i} = \frac{3}{4}\log 2 - \frac{\pi}{8},$$

$$\sum_{i=2}^{\infty}(-1)^i\frac{\zeta(i)}{2^i} = 1 - \log 2, \quad \sum_{i=2}^{\infty}(-1)^i\frac{\zeta(i)}{3^i} = 1 - \frac{1}{2}\log 3 - \frac{\sqrt{3}\pi}{18}, \quad \sum_{i=2}^{\infty}(-1)^i\frac{\zeta(i)}{4^i} = 1 - \frac{\pi}{8} - \frac{3}{4}\log 2,$$

$$\sum_{i=1}^{\infty}\frac{\zeta(2i)}{4^i} = \frac{1}{2}, \quad \sum_{i=1}^{\infty}\frac{\zeta(2i+1)}{4^i} = 2\log 2 - 1, \quad \sum_{i=1}^{\infty}\frac{\zeta(2i)}{9^i} = \frac{1}{2} - \frac{\sqrt{3}\pi}{18}, \quad \sum_{i=1}^{\infty}\frac{\zeta(2i+1)}{9^i} = \frac{3}{2}(\log 3 - 1),$$

$$\sum_{i=1}^{\infty}[\zeta(2i)-1] = \frac{3}{4}, \quad \sum_{i=1}^{\infty}[\zeta(2i+1)-1] = \frac{1}{4}, \quad \sum_{i=2}^{\infty}[\zeta(i)-1] = 1, \quad \sum_{i=1}^{\infty}[\zeta(4i)-1] = \frac{7}{8} - \frac{\pi}{4}\coth\pi,$$

$$\sum_{i=1}^{\infty}\frac{i\zeta(2i+1)}{4^i} = \frac{1}{2}, \quad \sum_{i=1}^{\infty}\frac{(2i+1)\zeta(2i+1)}{4^i} = 2\log 2,$$

$$\sum_{i=2}^{\infty}\frac{\zeta(i)-1}{i} = 1 - \gamma, \quad \sum_{i=1}^{\infty}\frac{\zeta(2i)-1}{i} = \log 2, \quad \sum_{i=1}^{\infty}\frac{\zeta(3i)-1}{i} = \log\left(3\pi\operatorname{sech}\frac{3\pi}{2}\right),$$

$$\sum_{i=2}^{\infty}(-1)^i\frac{\zeta(i)-1}{i} = \gamma + \log 2 - 1, \quad \sum_{i=2}^{\infty}\frac{\zeta(i)-1}{i+1} = \frac{3}{2} - \frac{\gamma}{2} - \frac{1}{2}\log 2\pi, \quad \sum_{i=2}^{\infty}\frac{\zeta(2i)-1}{i+1} = \frac{3}{2} - 2\log\pi,$$

$$\sum_{i=2}^{\infty}\frac{\zeta(i)-1}{2^i} = \log 2 - \frac{1}{2}, \quad \sum_{i=2}^{\infty}\frac{\zeta(i)-1}{3^i} = \frac{1}{2}\log 3 - \frac{1}{6} - \frac{\sqrt{3}\pi}{18}, \quad \sum_{i=2}^{\infty}\frac{\zeta(i)-1}{4^i} = \frac{3}{4}\log 2 - \frac{1}{12} - \frac{\pi}{8},$$

$$\sum_{i=1}^{\infty}\frac{\zeta(4i)-1}{i} = \log(4\pi\operatorname{csch}\pi), \quad \sum_{i=2}^{\infty}(-1)^i\frac{\zeta(i)}{i} = \gamma, \quad \sum_{i=1}^{\infty}(-1)^i\frac{\zeta(2i)}{i} = \log(\pi\operatorname{csch}\pi),$$

$$\sum_{i=1}^{\infty}(-1)^i\frac{\zeta(3i)}{i} = \log\left(\pi\,\text{sech}\,\frac{\sqrt{3}\pi}{2}\right), \quad \sum_{i=2}^{\infty}\frac{\zeta(i)-1}{2^i i} = \frac{1}{2} - \frac{\gamma}{2} + \log\frac{\sqrt{\pi}}{2},$$

$$\sum_{i=1}^{\infty}\frac{\zeta(2i)}{4^i i} = \log\frac{\pi}{2}, \quad \sum_{i=1}^{\infty}\frac{\zeta(2i)}{9^i i} = \log\frac{2\sqrt{3}\pi}{9}, \quad \sum_{i=1}^{\infty}\frac{\zeta(2i)}{16^i i} = \log\frac{2\pi}{4},$$

$$\sum_{i=2}^{\infty}\frac{[2(\frac{3}{2})^i - 3][\zeta(i)-1]}{i} = \log\pi, \quad \sum_{i=2}^{\infty}\frac{(\text{Im}[(1+\jmath)^n - 1 - \jmath^n])[\zeta(i)-1]}{i} = \frac{\pi}{4},$$

$$\sum_{i=1}^{\infty}\frac{\zeta(2i)}{(i+1)(2i+1)} = \frac{1}{2}, \quad \sum_{i=1}^{\infty}\frac{\zeta(2i)}{i(i+1)} = \log 2\pi - \frac{1}{2}, \quad \sum_{i=1}^{\infty}\frac{\zeta(2i)}{i(2i+1)} = \log 2\pi - 1,$$

$$\sum_{i=1}^{\infty}\frac{\zeta(2i)}{4^i i(2i+1)} = \log\pi - 1, \quad \sum_{i=1}^{\infty}\frac{\zeta(2i)}{4^i(i+1)} = \frac{1}{2} - \log 2 + \frac{7\zeta(3)}{2\pi^2}, \quad \sum_{i=1}^{\infty}\frac{\zeta(2i)}{4^i(2i+1)} = \frac{1}{2}(1-\log 2),$$

$$\sum_{i=1}^{\infty}\frac{\zeta(2i)}{4^i(i+1)(2i+1)} = \frac{1}{2} - \frac{7\zeta(3)}{2\pi^2}, \quad \sum_{i=1}^{\infty}\frac{\zeta(2i)}{4^i(2i+3)} = \frac{9\zeta(3)}{4\pi^2} - \frac{1}{2}\log 2 + \frac{1}{6},$$

$$\sum_{i=1}^{\infty}\frac{\zeta(2i)}{4^i(i+1)(2i+3)} = 1 - \frac{\zeta(3)}{2\pi^2}, \quad \sum_{i=1}^{\infty}(-1)^i\frac{\zeta(i+1)}{2^i(i+1)} = \log\frac{4}{\pi} - \gamma,$$

$$\sum_{i=1}^{\infty}\frac{\zeta(2i+1)-1}{i+1} = \log 2 - \gamma, \quad \sum_{i=1}^{\infty}\frac{\zeta(2i+1)-1}{2i+1} = 1 - \frac{1}{2}\log 2 - \gamma,$$

$$\sum_{i=1}^{\infty}\frac{\zeta(2i+1)}{(2i+1)4^i} = \log 2 - \gamma, \quad \sum_{i=1}^{\infty}\frac{\zeta(2i+1)-1}{(2i+1)4^i} = 1 - \log\frac{3}{2} - \gamma,$$

$$\sum_{i=1}^{\infty}\frac{\zeta(2i+1)}{(i+1)(2i+1)} = 1 - \gamma, \quad \sum_{i=1}^{\infty}\frac{\zeta(2i+1)-1}{(i+1)(2i+1)} = 2 - 2\log 2 - \gamma,$$

$$\sum_{i=1}^{\infty}\left(\frac{3}{2}\right)^{2i}\frac{\zeta(2i+1)-1}{(2i+1)} = 1 + \frac{1}{3}\log\frac{8}{15} - \gamma, \quad \sum_{i=2}^{\infty}(-1)^i\frac{2^i[\zeta(i)-1]}{i} = 2\gamma + \log 6 - 2.$$

Now, let $a \in \mathbb{R}$, and assume that $|a| > 1$. Then,

$$\sum_{i=1}^{\infty}\frac{\zeta(2i)-1}{a^{2i}} = \frac{1}{2} + \frac{1}{1-a^2} - \frac{\pi}{2a}\cot\frac{\pi}{a}.$$

In particular,

$$\sum_{i=1}^{\infty}\frac{\zeta(2i)-1}{4^i} = \frac{1}{6}, \quad \sum_{i=1}^{\infty}\frac{\zeta(2i)-1}{16^i} = \frac{13}{30} - \frac{\pi}{8}, \quad \sum_{i=1}^{\infty}\frac{\zeta(2i)-1}{64^i} = \frac{61}{126} - \frac{\pi}{16}(1+\sqrt{2}).$$

Source: [511, pp. 131, 248], [559, 650, 700], and [2513, pp. 272–314, 365, 556, 557]. **Related:** Fact 12.18.33.

Fact 13.5.48. Let $n \ge 1$, and define

$$S_n \triangleq \sum \frac{1}{\prod_{j=1}^{n} i_j^2},$$

where the sum is taken over all n-tuples $(i_1,\dots,i_n)$ such that $i_1\le\cdots\le i_n$. Then, $S_n=(2-4^{1-n})\zeta(2n)$. Furthermore,

$$\sum_{i=0}^{n}(-1)^i\frac{1}{\pi^{2i}(2n-2i+1)!}S_i=\sum_{i=0}^{n}(-1)^i\frac{2-4^{1-i}}{\pi^{2i}(2n-2i+1)!}\zeta(2i)=0.$$

Source: [2468].

Fact 13.5.49. Let $n\ge 1$. Then,

$$\sum_{i=1}^{\infty}\frac{2}{i^{4n-1}(e^{2i\pi}-1)}=-\zeta(4n-1)+\frac{(2\pi)^{4n-1}}{(4n)!}\sum_{i=0}^{2n}(-1)^i\binom{4n}{2i}B_{2i}B_{4n-2i}.$$

Furthermore,

$$\zeta(3)=\frac{37\pi^3}{900}-\frac{2}{5}\sum_{i=1}^{\infty}\frac{1}{i^3}\left(\frac{4}{e^{i\pi}-1}+\frac{1}{e^{4i\pi}-1}\right),\quad \zeta(5)=\frac{\pi^5}{294}-\frac{72}{35}\sum_{i=1}^{\infty}\frac{1}{i^5(e^{2i\pi}-1)}-\frac{2}{35}\sum_{i=1}^{\infty}\frac{1}{i^5(e^{2i\pi}+1)},$$

$$\zeta(7)=\frac{409\pi^7}{94500}-\frac{2}{5}\sum_{i=1}^{\infty}\frac{1}{i^7}\left(\frac{4}{e^{i\pi}-1}+\frac{1}{e^{4i\pi}-1}\right),\quad \frac{\pi^3}{180}=\sum_{i=1}^{\infty}\frac{1}{i^3}\left(\frac{4}{e^{i\pi}-1}-\frac{5}{e^{2i\pi}-1}+\frac{1}{e^{4i\pi}-1}\right).$$

Source: [2792].

Fact 13.5.50. Let $n\ge 1$. Then,

$$\sum_{i=1}^{\infty}\frac{i^{4n+1}}{e^{2i\pi}-1}=\frac{B_{4n+2}}{8n+4},\quad \sum_{i=1}^{\infty}\frac{(2i-1)^{4n+1}}{e^{(2i-1)\pi}+1}=\frac{(2^{4n+1}-1)B_{4n+2}}{8n+4}.$$

In particular,

$$\sum_{i=1}^{\infty}\frac{i^5}{e^{2i\pi}-1}=\frac{1}{504},\quad \sum_{i=1}^{\infty}\frac{i^{13}}{e^{2i\pi}-1}=\frac{1}{24},\quad \sum_{i=1}^{\infty}\frac{(2i-1)^5}{e^{(2i-1)\pi}+1}=\frac{31}{504},\quad \sum_{i=1}^{\infty}\frac{(2i-1)^9}{e^{(2i-1)\pi}+1}=\frac{511}{264}.$$

Source: [1317, p. xxvi] and [2792].

Fact 13.5.51. Let $n\ge 1$. Then,

$$\sum_{i=1}^{\infty}\frac{1}{(4i^2-1)^n}=(-1)^n\sum_{i=1}^{\lfloor n/2\rfloor}\frac{4^{i-n}(4^i-1)|B_{2i}|\pi^{2i}}{(2i)!}\binom{2n-2i-1}{n-1}+(-1)^{n+1}\frac{1}{2}.$$

In particular,

$$\sum_{i=1}^{\infty}\frac{1}{4i^2-1}=\frac{1}{2},\quad \sum_{i=1}^{\infty}\frac{1}{(4i^2-1)^2}=\frac{\pi^2}{16}-\frac{1}{2},$$

$$\sum_{i=1}^{\infty}\frac{1}{(4i^2-1)^3}=\frac{1}{2}-\frac{3\pi^2}{64},\quad \sum_{i=1}^{\infty}\frac{1}{(4i^2-1)^4}=\frac{\pi^4}{768}+\frac{5\pi^2}{128}-\frac{1}{2}.$$

Source: [1888].

Fact 13.5.52. Let $n\ge 1$. Then,

$$\sum_{i=1}^{\infty}\frac{1}{i^n(i+1)^n}=(-1)^n\sum_{i=1}^{\lfloor n/2\rfloor}\frac{4^i|B_{2i}|\pi^{2i}}{(2i)!}\binom{2n-2i-1}{n-1}+(-1)^{n+1}\binom{2n-1}{n}.$$

In particular,

$$\sum_{i=1}^{\infty}\frac{1}{i(i+1)}=1,\quad \sum_{i=1}^{\infty}\frac{1}{i^2(i+1)^2}=\frac{\pi^2}{3}-3,\quad \sum_{i=1}^{\infty}\frac{1}{i^3(i+1)^3}=10-\pi^2,$$

$$\sum_{i=1}^{\infty}\frac{1}{i^4(i+1)^4}=\frac{\pi^4}{45}+\frac{10\pi^2}{3}-35,\quad \sum_{i=1}^{\infty}\frac{1}{i^5(i+1)^5}=126-\frac{\pi^4}{9}-\frac{35\pi^2}{3},$$

$$\sum_{i=1}^{\infty}\frac{1}{i^6(i+1)^6}=\frac{2\pi^6}{945}+\frac{7\pi^4}{15}+42\pi^2-462,\quad \sum_{i=1}^{\infty}\frac{1}{i^7(i+1)^7}=1716-\frac{2\pi^6}{135}-\frac{28\pi^4}{15}-154\pi^2.$$

Source: [1888].

Fact 13.5.53. Let x be a nonzero real number. Then,

$$\sum_{i=1}^{\infty}\frac{1}{i^2+x^2}=\frac{\pi\coth\pi x}{2x}-\frac{1}{2x^2}.$$

In particular,

$$\sum_{i=1}^{\infty}\frac{1}{i^2+1}=\frac{\pi\coth\pi}{2}-\frac{1}{2}=\frac{1+\pi+e^{2\pi}(\pi-1)}{2(e^{2\pi}-1)}.$$

Source: Fact 13.4.12. **Remark:** $\lim_{x\to 0}\left(\frac{\pi\coth\pi x}{2x}-\frac{1}{2x^2}\right)=\frac{\pi^2}{6}$.

Fact 13.5.54. Let x be a nonzero real number. Then,

$$\sum_{i=1}^{\infty}\frac{1}{(i^2+x^2)^2}=\frac{\pi^2x^2\operatorname{csch}^2\pi x+\pi x\coth\pi x-2}{4x^4}.$$

Remark: $\lim_{x\to 0}\frac{\pi^2x^2\operatorname{csch}^2\pi x+\pi x\coth\pi x-2}{4x^4}=\frac{\pi^4}{90}$.

Fact 13.5.55. Let x be a nonzero real number. Then,

$$\sum_{i=1}^{\infty}\frac{(-1)^i}{i^2+x^2}=\frac{\pi\operatorname{csch}\pi x}{2x}-\frac{1}{2x^2}.$$

In particular,

$$\sum_{i=1}^{\infty}\frac{(-1)^i}{i^2+1}=\frac{\pi\operatorname{csch}\pi}{2}-\frac{1}{2}.$$

Remark: $\lim_{x\to 0}\left(\frac{1}{2x^2}-\frac{\pi\operatorname{csch}\pi x}{2x}\right)=\frac{\pi^2}{12}$.

Fact 13.5.56. Let x be a nonzero real number. Then,

$$\sum_{i=1}^{\infty}\frac{(-1)^{i+1}}{(i^2+x^2)^2}=\frac{1}{2x^4}-\frac{\pi(\pi x\coth\pi x+1)\operatorname{csch}\pi x}{4x^3}.$$

Remark: $\lim_{x\to 0}\left(\frac{1}{2x^4}-\frac{\pi(\pi x\coth\pi x+1)\operatorname{csch}\pi x}{4x^3}\right)=\frac{7\pi^4}{720}$.

Fact 13.5.57. Let x be a nonzero real number. Then,

$$\sum_{i=1}^{\infty}\frac{1}{(2i-1)^2+x^2}=\frac{\pi\tanh\frac{\pi x}{2}}{4x}.$$

Source: [1167, p. 301]. **Remark:** $\lim_{x\to 0}\frac{\pi\tanh\frac{\pi x}{2}}{4x}=\frac{\pi^2}{8}$.

Fact 13.5.58. Let x be a nonzero real number. Then,

$$\sum_{i=1}^{\infty}\frac{1}{[(2i-1)^2+x^2]^2}=\frac{\pi}{8x^3}\left(\tanh\frac{\pi x}{2}-\frac{\pi x}{2}\operatorname{sech}^2\frac{\pi x}{2}\right)=\frac{\pi(\sinh\pi x-\pi x)}{8x^3(\cosh\pi x+1)}.$$

Source: [1167, p. 301]. **Remark:** $\lim_{x\to 0}\frac{\pi}{8x^3}\left(\tanh\frac{\pi x}{2}-\frac{\pi x}{2}\operatorname{sech}^2\frac{\pi x}{2}\right)=\frac{\pi^4}{96}$.

Fact 13.5.59. Let x be a complex number, and assume that x is not an integer. Then,

$$\sum_{i=1}^{\infty}\frac{1}{i^2-x^2}=\frac{1}{2x^2}-\frac{\pi\cot\pi x}{2x}\qquad\sum_{i=-\infty}^{\infty}\frac{1}{x-i}=\pi\cot\pi x.$$

In particular,

$$\sum_{i=1}^{\infty}\frac{1}{4i^2-1}=\frac{1}{2},\quad\sum_{i=1}^{\infty}\frac{1}{16i^2-1}=\frac{1}{2}-\frac{\pi}{8},\quad\sum_{i=1}^{\infty}\frac{1}{64i^2-1}=\frac{1}{2}-\frac{(1+\sqrt{2})\pi}{16}.$$

Now, let $n\ge 2$, and assume that $\sqrt{n}$ is not an integer. Then,

$$\sum_{i=1}^{\infty}\frac{1}{i^2-n}=\frac{1-\sqrt{n}\pi\cot\sqrt{n}\pi}{2n}.$$

In particular,

$$\sum_{i=1}^{\infty}\frac{1}{i^2-2}=\frac{1-\sqrt{2}\pi\cot\sqrt{2}\pi}{4},\quad\sum_{i=1}^{\infty}\frac{1}{i^2-3}=\frac{1-\sqrt{3}\pi\cot\sqrt{3}\pi}{6}.$$

Source: [116, p. 11]. **Remark:** $\lim_{x\to 0}\left(\frac{1}{2x^2}-\frac{\pi\cot\pi x}{2x}\right)=\frac{\pi^2}{6}$. **Related:** Fact 13.5.29.

Fact 13.5.60. Let x be a real number, and assume that x is not an integer. Then,

$$\sum_{i=1}^{\infty}\frac{1}{(i^2-x^2)^2}=\frac{\pi^2x^2\csc^2\pi x+\pi x\cot\pi x-2}{4x^4}.$$

In particular,

$$\sum_{i=1}^{\infty}\frac{1}{(4i^2-1)^2}=\frac{\pi^2}{16}-\frac{1}{2},\quad\sum_{i=1}^{\infty}\frac{1}{(16i^2-1)^2}=\frac{\pi^2+2\pi}{32}-\frac{1}{2}.$$

Remark: $\lim_{x\to 0}\frac{\pi^2x^2\csc^2\pi x+\pi x\cot\pi x-2}{4x^4}=\frac{\pi^4}{90}$.

Fact 13.5.61. Let $x\in\mathbb{C}$, and assume that x is not an integer. Then,

$$\sum_{i=1}^{\infty}(-1)^i\frac{1}{i^2-x^2}=\frac{\pi}{2x\sin\pi x}-\frac{1}{2x^2},\quad\sum_{i=-\infty}^{\infty}(-1)^i\frac{1}{x-i}=\pi\sin\pi x,$$

$$\sum_{i=-\infty}^{\infty}(-1)^i\frac{1}{i+x+\frac{1}{2}}=\frac{\pi}{\cos\pi x},\quad\sum_{i=-\infty}^{\infty}\frac{1}{i-x+\frac{1}{2}}=\pi\tan\pi x.$$

In particular,

$$\sum_{i=1}^{\infty}(-1)^i\frac{1}{4i^2-1}=\frac{1}{2}-\frac{\pi}{4},\quad\sum_{i=1}^{\infty}(-1)^i\frac{1}{16i^2-1}=\frac{1}{2}-\frac{\sqrt{2}\pi}{8}.$$

Source: [116, p. 11] and [968, p. 236]. **Related:** Fact 13.5.30.

Fact 13.5.62. Let x be a real number, and assume that x is not an integer. Then,

$$\sum_{i=1}^{\infty}\frac{(-1)^{i+1}}{(i^2-x^2)^2}=\frac{1-\cos 2\pi x-\pi x\sin\pi x-\pi^2x^2\cos\pi x}{4(\sin^2\pi x)x^4}.$$

In particular,

$$\sum_{i=1}^{\infty}\frac{(-1)^{i+1}}{(4i^2-1)^2}=\frac{1}{2}-\frac{\pi}{8},\quad\sum_{i=1}^{\infty}\frac{(-1)^{i+1}}{(16i^2-1)^2}=\frac{1}{2}-\frac{\sqrt{2}(\pi^2+4\pi)}{64}.$$

Fact 13.5.63. Let x be a nonzero real number. Then,

$$\sum_{i=1}^{\infty}\frac{1}{i^4+x^4}=\frac{\sqrt{2}\pi}{4x^3}\frac{\sinh\sqrt{2}\pi x+\sin\sqrt{2}\pi x}{\cosh\sqrt{2}\pi x-\cos\sqrt{2}\pi x}-\frac{1}{2x^4}.$$

In particular,

$$\sum_{i=1}^{\infty}\frac{1}{i^4+4}=\frac{\pi\coth\pi-1}{8}.$$

Source: [645, p. 97]. **Remark:** As $y\to 0$, $\frac{y^4}{y\frac{\sinh y+\sin y}{\cosh y-\cos y}-2}=90+\frac{3}{14}y^4+O(y^6)$. Hence, $\lim_{x\to 0}\left(\frac{\sqrt{2}\pi}{4x^3}\frac{\sinh\sqrt{2}\pi x+\sin\sqrt{2}\pi x}{\cosh\sqrt{2}\pi x-\cos\sqrt{2}\pi x}-\frac{1}{2x^4}\right)=\frac{\pi^4}{90}$.

Fact 13.5.64. Let x be a real number, and assume that x is not an integer. Then,

$$\sum_{i=1}^{\infty}\frac{1}{i^4-x^4}=\frac{1}{2x^4}-\frac{\pi\cot\pi x+\pi\coth\pi x}{4x^3}.$$

In particular,

$$\sum_{i=1}^{\infty}\frac{1}{16i^4-1}=\frac{1}{2}-\frac{\pi\coth\frac{\pi}{2}}{8}.$$

Remark: $\lim_{x\to 0}\left(\frac{1}{2x^4}-\frac{\pi\cot\pi x+\pi\coth\pi x}{4x^3}\right)=\frac{\pi^4}{90}$.

Fact 13.5.65. Let x be a positive number. Then,

$$\sum_{i=1}^{\infty}(-1)^{i+1}\frac{i}{(i+1)(i+2)x^i}=2(x+\tfrac{1}{2})\log\left(1+\frac{1}{x}\right).$$

In particular,

$$\sum_{i=1}^{\infty}(-1)^{i+1}\frac{i}{(i+1)(i+2)}=2\log 2-2,\quad \sum_{i=1}^{\infty}(-1)^{i+1}\frac{i}{(i+1)(i+2)(3/2)^i}=4\log\frac{5}{3}-2.$$

Source: [2085].

Fact 13.5.66. Let x and y be complex numbers such that division by zero does not occur in the expressions below. Then,

$$\sum_{i=0}^{\infty}\frac{1}{(xi+y)(xi+x+y)}=\frac{1}{xy},\quad \sum_{i=0}^{\infty}\frac{1}{(xi+y)(xi+2x+y)}=\frac{x+2y}{2xy(x+y)},$$

$$\sum_{i=0}^{\infty}\frac{1}{(xi+y)(xi+3x+y)}=\frac{2x^2+6xy+3y^2}{3xy(x+y)(2x+y)},$$

$$\sum_{i=0}^{\infty}\frac{1}{(xi+y)(xi+4x+y)}=\frac{3x^3+11x^2y+9xy^2+2y^3}{3xy(x+y)(2x+y)(3x+y)},$$

$$\sum_{i=0}^{\infty}\frac{1}{(xi+y)(xi-x+y)}=\frac{1}{xy-x^2},\quad \sum_{i=0}^{\infty}\frac{1}{(xi+y)(xi-2x+y)}=\frac{3x-2y}{2x(2x-y)(y-x)},$$

$$\sum_{i=0}^{\infty}\frac{1}{(xi+y)(xi-3x+y)}=\frac{11x^2-12xy+3y^2}{3x(3x-y)(2x-y)(y-x)},$$

$$\sum_{i=0}^{\infty}\frac{1}{(xi+y)(xi-4x+y)}=\frac{25x^3-35x^2y+15xy^2-2y^3}{2x(4x-y)(3x-y)(2x-y)(y-x)}.$$

Fact 13.5.67. Let $n, k \geq 1$. Then,

$$\sum_{i=0}^{\infty} \frac{1}{(ki+n)(ki+n+k)} = \frac{1}{kn}.$$

In particular,

$$\sum_{i=0}^{\infty} \frac{1}{(ki+1)(ki+1+k)} = \frac{1}{k}, \quad \sum_{i=0}^{\infty} \frac{1}{(ki+2)(ki+2+k)} = \frac{1}{2k}, \quad \sum_{i=0}^{\infty} \frac{1}{(ki+3)(ki+3+k)} = \frac{1}{3k},$$

$$\sum_{i=0}^{\infty} \frac{1}{(i+n)(i+n+1)} = \frac{1}{n}, \quad \sum_{i=0}^{\infty} \frac{1}{(2i+n)(2i+n+2)} = \frac{1}{2n}, \quad \sum_{i=0}^{\infty} \frac{1}{(3i+n)(3i+n+3)} = \frac{1}{3n}.$$

Source: Fact 1.12.25, Fact 13.5.27, and Fact 13.5.66.

Fact 13.5.68. Let $n, k \geq 1$. Then,

$$\sum_{i=0}^{\infty} \frac{1}{(i+n)(i+n+k)} = \frac{H_{n+k-1} - H_{n-1}}{k}.$$

In particular,

$$\sum_{i=0}^{\infty} \frac{1}{(i+1)(i+1+k)} = \frac{H_k}{k}, \quad \sum_{i=0}^{\infty} \frac{1}{(i+2)(i+2+k)} = \frac{H_{k+1} - 1}{k},$$

$$\sum_{i=0}^{\infty} \frac{1}{(i+3)(i+3+k)} = \frac{2H_{k+2} - 3}{2k}, \quad \sum_{i=0}^{\infty} \frac{1}{(i+4)(i+4+k)} = \frac{6H_{k+3} - 11}{6k}.$$

Hence,

$$\sum_{i=0}^{\infty} \frac{1}{(i+1)(i+2)} = 1, \quad \sum_{i=0}^{\infty} \frac{1}{(i+1)(i+3)} = \frac{3}{4}, \quad \sum_{i=0}^{\infty} \frac{1}{(i+1)(i+4)} = \frac{11}{18},$$

$$\sum_{i=0}^{\infty} \frac{1}{(i+2)(i+3)} = \frac{1}{2}, \quad \sum_{i=0}^{\infty} \frac{1}{(i+2)(i+4)} = \frac{5}{12}, \quad \sum_{i=0}^{\infty} \frac{1}{(i+2)(i+5)} = \frac{13}{36},$$

$$\sum_{i=0}^{\infty} \frac{1}{(i+3)(i+4)} = \frac{1}{3}, \quad \sum_{i=0}^{\infty} \frac{1}{(i+3)(i+5)} = \frac{7}{24}, \quad \sum_{i=0}^{\infty} \frac{1}{(i+3)(i+6)} = \frac{47}{180}.$$

Furthermore,

$$\sum_{i=0}^{\infty} \frac{1}{(i+n)(i+n+1)} = \frac{1}{n},$$

$$\sum_{i=0}^{\infty} \frac{1}{(i+n)(i+n+2)} = \frac{2n+1}{2n(n+1)},$$

$$\sum_{i=0}^{\infty} \frac{1}{(i+n)(i+n+3)} = \frac{3n^2+6n+2}{3n(n+1)(n+2)}.$$

Hence,

$$\sum_{i=0}^{\infty} \frac{1}{(i+2)(i+3)} = \frac{1}{2}, \quad \sum_{i=0}^{\infty} \frac{1}{(i+3)(i+4)} = \frac{1}{3}, \quad \sum_{i=0}^{\infty} \frac{1}{(i+4)(i+5)} = \frac{1}{4},$$

$$\sum_{i=0}^{\infty} \frac{1}{(i+2)(i+4)} = \frac{5}{12}, \quad \sum_{i=0}^{\infty} \frac{1}{(i+3)(i+5)} = \frac{7}{24}, \quad \sum_{i=0}^{\infty} \frac{1}{(i+4)(i+6)} = \frac{9}{40},$$

$$\sum_{i=0}^{\infty} \frac{1}{(i+2)(i+5)} = \frac{13}{36}, \quad \sum_{i=0}^{\infty} \frac{1}{(i+3)(i+6)} = \frac{47}{180}, \quad \sum_{i=0}^{\infty} \frac{1}{(i+4)(i+7)} = \frac{37}{180}.$$

Related: Fact 1.12.26.

Fact 13.5.69. Let $n \ge 0$. Then,

$$\sum_{i=0}^{\infty} \frac{2n-1}{(i+1)(2i+2n+1)} = \sum_{i=0}^{n-1} \frac{2}{2i+1} - \log 4.$$

In particular,

$$\sum_{i=0}^{\infty} \frac{1}{(i+1)(2i+1)} = \log 4, \quad \sum_{i=0}^{\infty} \frac{1}{(i+1)(2i+3)} = 2 - \log 4,$$

$$\sum_{i=0}^{\infty} \frac{1}{(i+1)(2i+5)} = \frac{8}{9} - \frac{\log 4}{3}, \quad \sum_{i=0}^{\infty} \frac{1}{(i+1)(2i+7)} = \frac{46}{75} - \frac{\log 4}{5}.$$

Fact 13.5.70. Let x and y be complex numbers such that division by zero does not occur in the expressions below, and define $\alpha \triangleq \cot \pi y/x$. Then,

$$\sum_{i=0}^{\infty} \frac{1}{(xi-y)(xi+x+y)} = -\frac{\alpha\pi}{x(x+2y)}, \quad \sum_{i=0}^{\infty} \frac{1}{(xi-y)(xi+2x+y)} = \frac{x-\alpha\pi(x+y)}{2x(x+y)^2},$$

$$\sum_{i=0}^{\infty} \frac{1}{(xi-y)(xi+3x+y)} = \frac{3x^2+2xy-\alpha\pi(2x^2+3xy+y^2)}{x(x+y)(2x+y)(3x+2y)},$$

$$\sum_{i=0}^{\infty} \frac{1}{(xi-y)(xi+4x+y)} = \frac{11x^3+12x^2y+3xy^2+y^2-\alpha\pi(6x^3+11x^2y+6xy^2+y^3)}{x(x+y)(2x+y)(3x+2y)},$$

$$\sum_{i=0}^{\infty} \frac{1}{(xi-y)(xi-x+y)} = \frac{x^2-2xy+\alpha\pi(xy-y^2)}{xy(x-2y)(x-y)},$$

$$\sum_{i=0}^{\infty} \frac{1}{(xi-y)(xi-2x+y)} = \frac{2x^3-6x^2y+3xy^2+\alpha\pi(2x^2y-3xy^2+y^3)}{2xy(2x-y)(x-y)^2},$$

$$\sum_{i=0}^{\infty} \frac{1}{(xi-y)(xi-3x+y)} = \frac{6x^4-22x^3y+18x^2y^2-4xy^3+\alpha\pi(6x^3y-11x^2y^2+6xy^3-y^4)}{xy(3x-2y)(x-y)(2x-y)(3x-y)}.$$

In particular,

$$\sum_{i=0}^{\infty} \frac{1}{(3i+1)(3i+2)} = \frac{\sqrt{3}\pi}{9}, \quad \sum_{i=0}^{\infty} \frac{1}{(3i+1)(3i+5)} = \frac{1}{8} + \frac{\sqrt{3}\pi}{36}, \quad \sum_{i=0}^{\infty} \frac{1}{(4i+1)(4i+3)} = \frac{\pi}{8},$$

$$\sum_{i=0}^{\infty} \frac{1}{(5i+1)(5i+4)} = \frac{\pi}{15}\sqrt{1+\frac{2\sqrt{5}}{5}}, \quad \sum_{i=0}^{\infty} \frac{1}{(5i+2)(5i+3)} = \frac{\pi}{5}\sqrt{1-\frac{2\sqrt{5}}{5}},$$

$$\sum_{i=0}^{\infty} \frac{1}{(6i+1)(6i+5)} = \frac{\sqrt{3}\pi}{24}, \quad \sum_{i=0}^{\infty} \frac{1}{(6i+5)(6i+7)} = \frac{1}{2} - \frac{\sqrt{3}\pi}{12},$$

$$\sum_{i=0}^{\infty} \frac{1}{(8i+1)(8i+7)} = \frac{(1+\sqrt{2})\pi}{48}, \quad \sum_{i=0}^{\infty} \frac{1}{(8i+3)(8i+5)} = \frac{(\sqrt{2}-1)\pi}{16},$$

$$\sum_{i=0}^{\infty} \frac{1}{(8i+7)(8i+9)} = \frac{1}{2} - \frac{(1+\sqrt{2})\pi}{16}, \quad \sum_{i=0}^{\infty} \frac{1}{(8i+11)(8i+13)} = \frac{(\sqrt{2}-1)\pi}{16} - \frac{1}{15}.$$

Fact 13.5.71. Let $a, b \in (0, \infty)$, and assume that $a \neq b$. Then,

$$\sum_{i=1}^{\infty} \frac{1}{(i+a)(i+b)} = \frac{\psi(a+1) - \psi(b+1)}{a-b}.$$

In particular,

$$\sum_{i=0}^{\infty} \frac{1}{(i+1)(3i+1)} = \frac{\sqrt{3}\pi}{12} + \frac{3\log 3}{4}, \quad \sum_{i=0}^{\infty} \frac{1}{(i+1)(3i+2)} = \frac{3\log 3}{2} - \frac{\sqrt{3}\pi}{6},$$

$$\sum_{i=0}^{\infty} \frac{1}{(i+1)(4i+1)} = \frac{\pi}{6} + \log 2, \quad \sum_{i=0}^{\infty} \frac{1}{(i+1)(4i+3)} = 3\log 2 - \frac{\pi}{2},$$

$$\sum_{i=0}^{\infty} \frac{1}{(i+1)(4i+5)} = 4 - 3\log 2 - \frac{\pi}{2}, \quad \sum_{i=0}^{\infty} \frac{1}{(i+1)(4i+7)} = \frac{4}{9} + \frac{\pi}{6} - \log 2,$$

$$\sum_{i=0}^{\infty} \frac{1}{(2i+1)(3i+1)} = \frac{\sqrt{3}\pi}{6} + \frac{3\log 3}{2} - 2\log 2, \quad \sum_{i=0}^{\infty} \frac{1}{(2i+1)(3i+2)} = \frac{\sqrt{3}\pi}{6} - \frac{3\log 3}{2} + 12\log 2,$$

$$\sum_{i=0}^{\infty} \frac{1}{(2i+1)(4i+1)} = \frac{\pi}{4} + \frac{\log 2}{2}, \quad \sum_{i=0}^{\infty} \frac{1}{(2i+1)(4i+3)} = \frac{\pi}{4} - \frac{\log 2}{2},$$

$$\sum_{i=0}^{\infty} \frac{1}{(2i+1)(6i+1)} = \frac{\sqrt{3}\pi + 3\log 3}{8}, \quad \sum_{i=0}^{\infty} \frac{1}{(2i+1)(6i+5)} = \frac{\sqrt{3}\pi - 3\log 3}{8},$$

$$\sum_{i=0}^{\infty} \frac{1}{(2i+1)(8i+1)} = \frac{(\sqrt{2}+1)\pi}{12} + \frac{\sqrt{2}\log(\sqrt{2}+1)}{6} + \frac{\log 2}{3},$$

$$\sum_{i=0}^{\infty} \frac{1}{(2i+1)(8i+3)} = \frac{(\sqrt{2}-1)\pi}{4} - \frac{\sqrt{2}}{2} - \log(\sqrt{2}+1) + \log 2,$$

$$\sum_{i=0}^{\infty} \frac{1}{(2i+1)(8i+5)} = \frac{(\sqrt{2}-1)\pi}{4} + \frac{\sqrt{2}\log(\sqrt{2}+1)}{2} - \log 2,$$

$$\sum_{i=0}^{\infty} \frac{1}{(3i+1)(3i+2)(3i+3)} = \frac{1}{12}(\sqrt{3}\pi - 3\log 3),$$

$$\sum_{i=0}^{\infty} \frac{1}{(4i+1)(2i+3)} = \frac{1}{5} + \frac{\pi}{20} + \frac{\log 2}{10}, \quad \sum_{i=0}^{\infty} \frac{1}{(4i+1)(2i+5)} = \frac{4}{27} + \frac{\pi}{36} + \frac{\log 2}{18},$$

$$\sum_{i=0}^{\infty} \frac{1}{(6i+1)(6i+4)} = \frac{\sqrt{3}\pi}{27} + \frac{\log 2}{9}, \quad \sum_{i=0}^{\infty} \frac{1}{(6i+2)(6i+5)} = \frac{\sqrt{3}\pi}{27} - \frac{\log 2}{9},$$

$$\sum_{i=0}^{\infty} \frac{1}{(6i+3)(6i+5)} = \frac{\sqrt{3}\pi}{24} - \frac{\log 3}{8}, \quad \sum_{i=0}^{\infty} \frac{1}{(6i+4)(6i+5)} = \frac{\sqrt{3}\pi}{18} - \frac{\log 2}{3},$$

$$\sum_{i=0}^{\infty} \frac{1}{(8i+1)(8i+3)} = \frac{\pi + \sqrt{2}\log(3+2\sqrt{2})}{16}, \quad \sum_{i=0}^{\infty} \frac{1}{(8i+1)(8i+5)} = \frac{\sqrt{2}\pi + \sqrt{2}\log(3+2\sqrt{2})}{32},$$

$$\sum_{i=0}^{\infty} \frac{1}{(8i+3)(8i+7)} = \frac{\sqrt{2}\pi - 2\sqrt{2}\log(1+\sqrt{2})}{32}, \quad \sum_{i=0}^{\infty} \frac{1}{(8i+5)(8i+7)} = \frac{\pi - 2\sqrt{2}\log(1+\sqrt{2})}{16},$$

$$\sum_{i=0}^{\infty}\frac{1}{(5i+2)(5i+4)}=\frac{\sqrt{5}[\log(5-\sqrt{5})-\log(5+\sqrt{5})]}{20}.$$

Source: [116, pp. 58, 59], [593], and [1311, p. 17]. **Remark:** ψ is the digamma function. See Fact 13.3.3.

Fact 13.5.72.

$$\sum_{i=0}^{\infty}\frac{1}{(i+1)^2}=\frac{\pi^2}{6},\quad \sum_{i=0}^{\infty}\frac{1}{(2i+1)^2}=\frac{\pi^2}{8},\quad \sum_{i=0}^{\infty}\frac{1}{(3i+1)^2}=\frac{1}{9}\frac{\mathrm{d}^2}{\mathrm{d}z^2}\Gamma(z)\bigg|_{z=1/3},$$

$$\sum_{i=0}^{\infty}\frac{1}{(4i+1)^2}=\frac{\pi^2}{16}+\frac{G}{2},\quad \sum_{i=0}^{\infty}\frac{1}{(4i+3)^2}=\frac{\pi^2}{16}-\frac{G}{2},$$

$$\sum_{i=0}^{\infty}\frac{1}{(2i+1)^3}=\frac{7\zeta(3)}{8},\quad \sum_{i=0}^{\infty}\frac{1}{(2i+1)^4}=\frac{\pi^4}{96},\quad \sum_{i=0}^{\infty}\frac{1}{(2i+1)^5}=\frac{31\zeta(5)}{32},$$

$$\sum_{i=0}^{\infty}\frac{1}{(3i+1)^3}=\frac{13\zeta(3)}{27}+\frac{2\sqrt{3}\pi^3}{243},\quad \sum_{i=0}^{\infty}\frac{1}{(3i+2)^3}=\frac{13\zeta(3)}{27}-\frac{2\sqrt{3}\pi^3}{243},$$

$$\sum_{i=0}^{\infty}\frac{1}{(3i+1)^5}=\frac{121\zeta(5)}{243}+\frac{2\sqrt{3}\pi^5}{2187},\quad \sum_{i=0}^{\infty}\frac{1}{(3i+2)^5}=\frac{121\zeta(5)}{243}-\frac{2\sqrt{3}\pi^5}{2187},$$

$$\sum_{i=0}^{\infty}\frac{1}{(4i+1)^3}=\frac{7\zeta(3)}{16}+\frac{\pi^3}{64},\quad \sum_{i=0}^{\infty}\frac{1}{(4i+3)^3}=\frac{7\zeta(3)}{16}-\frac{\pi^3}{64},$$

$$\sum_{i=0}^{\infty}\frac{1}{(4i+1)^5}=\frac{31\zeta(5)}{64}+\frac{5\pi^5}{3072},\quad \sum_{i=0}^{\infty}\frac{1}{(4i+3)^5}=\frac{31\zeta(5)}{64}-\frac{5\pi^5}{3072},$$

$$\sum_{i=0}^{\infty}\frac{1}{(6i+1)^3}=\frac{91\zeta(3)+2\sqrt{3}\pi^3}{216},\qquad \sum_{i=0}^{\infty}\frac{1}{(6i+1)^5}=\frac{11(1023\zeta(5)+2\sqrt{3}\pi^5)}{23328}.$$

Source: [2545]. **Related:** Fact 13.3.2, Fact 13.5.86, and Fact 13.5.95.

Fact 13.5.73. Let a, b, c be real numbers. Then,

$$\sum_{i=0}^{\infty}\left(\frac{a}{4i+1}+\frac{b}{4i+2}+\frac{c}{4i+3}-\frac{a+b+c}{4i+4}\right)=(3a+2b+3c)\frac{\log 2}{4}+(a-c)\frac{\pi}{8}.$$

In particular,

$$\sum_{i=0}^{\infty}\frac{8i+5}{(4i+1)(4i+3)(2i+2)}=\sum_{i=0}^{\infty}\left(\frac{1}{4i+1}+\frac{1}{4i+3}-\frac{2}{4i+4}\right)=\frac{3}{2}\log 2,$$

$$\sum_{i=0}^{\infty}\frac{1}{(4i+1)(4i+2)(4i+3)(4i+4)}=\sum_{i=0}^{\infty}\left(\frac{1/6}{4i+1}-\frac{1/2}{4i+2}+\frac{1/2}{4i+3}-\frac{1/6}{4i+4}\right)=\frac{\log 2}{4}-\frac{\pi}{24}.$$

Source: [506, pp. 128, 129], [1647, p. 268], and [1686].

Fact 13.5.74.

$$\sum_{i=0}^{\infty}\frac{1}{(2i+1)(2i+3)(2i+5)}=\frac{1}{12},\quad \sum_{i=1}^{\infty}\frac{i}{(2i+1)(2i+3)(2i+5)}=\frac{1}{24},$$

$$\sum_{i=0}^{\infty}\frac{1}{(2i+1)(2i+3)(2i+7)}=\frac{11}{180},\quad \sum_{i=1}^{\infty}\frac{i}{(2i+1)(2i+3)(2i+5)}=\frac{13}{360},$$

$$\sum_{i=0}^{\infty}\frac{1}{(2i+1)(2i+5)(2i+7)}=\frac{7}{180},\quad \sum_{i=1}^{\infty}\frac{i}{(2i+1)(2i+5)(2i+7)}=\frac{11}{360},$$

$$\sum_{i=0}^{\infty}\frac{1}{(2i+3)(2i+5)(2i+7)}=\frac{1}{60},\quad \sum_{i=1}^{\infty}\frac{i}{(2i+3)(2i+5)(2i+7)}=\frac{1}{40}.$$

Fact 13.5.75.

$$\sum_{i=0}^{\infty}\frac{1}{(4i+1)(4i+3)(4i+5)}=\frac{\pi}{16}-\frac{1}{8},\quad \sum_{i=1}^{\infty}\frac{i}{(4i+1)(4i+3)(4i+5)}=\frac{5}{32}-\frac{3\pi}{64},$$

$$\sum_{i=0}^{\infty}\frac{1}{(4i+1)(4i+3)(4i+7)}=\frac{\pi}{48}-\frac{1}{72},\quad \sum_{i=1}^{\infty}\frac{i}{(4i+1)(4i+3)(4i+5)}=\frac{7}{288}-\frac{3\pi}{192},$$

$$\sum_{i=0}^{\infty}\frac{1}{(4i+1)(4i+5)(4i+7)}=\frac{7}{72}-\frac{\pi}{48},\quad \sum_{i=1}^{\infty}\frac{i}{(4i+1)(4i+5)(4i+7)}=\frac{7\pi}{192}-\frac{31}{288},$$

$$\sum_{i=0}^{\infty}\frac{1}{(4i+3)(4i+5)(4i+7)}=\frac{5}{24}-\frac{\pi}{16},\quad \sum_{i=1}^{\infty}\frac{i}{(4i+3)(4i+5)(4i+7)}=\frac{5\pi}{64}-\frac{23}{96}.$$

Fact 13.5.76. Define $\alpha \triangleq \frac{1}{2}(1+\sqrt{5})$ and $\beta \triangleq -1/\alpha$. Then,

$$\sum_{i=0}^{\infty}\left(\frac{1}{5i+1}-\frac{1}{5i+2}-\frac{1}{5i+3}+\frac{1}{5i+4}\right)=\sum_{i=0}^{\infty}\frac{10(2i+1)}{(5i+1)(5i+2)(5i+3)(5i+4)}=\frac{2\sqrt{5}}{5}\log\alpha,$$

$$\sum_{i=0}^{\infty}(-1)^i\beta^{5i}\left(\frac{1}{5i+1}-\frac{\beta^2}{5i+3}-\frac{\beta^4}{5i+4}-\frac{\beta^5}{5i+5}\right)=\pi\sqrt[4]{\frac{\alpha^6}{5^5}},$$

$$\sum_{i=0}^{\infty}(-1)^i\beta^{5i}\left(\frac{1}{5i+1}+\frac{1}{5i+2}-\frac{\beta^2}{5i+4}-\frac{\beta^4}{5i+5}\right)=\pi\sqrt[4]{\frac{\alpha^{10}}{5^5}},$$

$$\sum_{i=0}^{\infty}(-1)^i\beta^{5i}\left(\frac{\alpha^3}{5i+2}+\frac{\alpha}{5i+3}-\frac{1}{5i+4}-\frac{1}{5i+5}\right)=\pi\sqrt[4]{\frac{\alpha^{14}}{5^5}},$$

$$\sum_{i=0}^{\infty}(-1)^i\beta^{5i}\left(\frac{1}{5i+1}+\frac{\beta^2}{5i+2}+\frac{\beta^3}{5i+3}+\frac{\beta^3}{5i+4}\right)=2\pi\sqrt[4]{\frac{\alpha^2}{5^5}},$$

$$\sum_{i=0}^{\infty}(-1)^i\beta^{5i}\left(\frac{1}{5i+1}+\frac{2}{5i+2}+\frac{\beta^2}{5i+3}+\frac{\beta}{5i+4}-\frac{\beta^2}{5i+5}\right)=\pi\sqrt[4]{\frac{\alpha^6}{5^3}},$$

$$\sum_{i=0}^{\infty}\frac{1}{16^i}\left(\frac{4}{8i+1}-\frac{2}{8i+4}-\frac{1}{8i+5}-\frac{1}{8i+6}\right)=\pi,$$

$$\sum_{i=0}^{\infty}\frac{1}{\alpha^{5i}}\left[\frac{\alpha^2}{(5i+1)^2}-\frac{\alpha}{(5i+2)^2}-\frac{\alpha^2}{(5i+3)^2}+\frac{\alpha^5}{(5i+4)^2}+\frac{2\alpha^5}{(5i+5)^2}\right]=\frac{\pi^2}{6},$$

$$\sum_{i=0}^{\infty}\frac{1}{16^i}\left[\frac{16}{(8i+1)^2}-\frac{16}{(8i+2)^2}-\frac{8}{(8i+3)^2}-\frac{16}{(8i+4)^2}-\frac{4}{(8i+5)^2}-\frac{4}{(8i+6)^2}+\frac{2}{(8i+7)^2}\right]=\pi^2,$$

$$\sum_{i=0}^{\infty}\left(\frac{1}{7i+1}+\frac{1}{7i+2}-\frac{1}{7i+3}+\frac{1}{7i+4}-\frac{1}{7i+5}-\frac{1}{7i+6}\right)$$

$$= \sum_{i=0}^{\infty} \frac{7(2401i^4 + 4802i^3 + 3437i^2 + 1036i + 108)}{(7i+1)(7i+2)(7i+3)(7i+4)(7i+5)(7i+6)} = \frac{\sqrt{7}\pi}{7}.$$

Source: [666, 1229, 1230, 1682].

Fact 13.5.77.

$$\sum_{i=1}^{\infty} \frac{1}{i(4i^2-1)} = 2\log 2 - 1, \quad \sum_{i=1}^{\infty} \frac{1}{i(9i^2-1)} = \frac{3}{2}[\log 3 - 1],$$

$$\sum_{i=1}^{\infty} \frac{1}{i(36i^2-1)} = \frac{3}{2}\log 3 + 2\log 2 - 3, \quad \sum_{i=1}^{\infty} \frac{1}{i(2i+1)^2} = 4 - \frac{\pi^2}{4} - 2\log 2,$$

$$\sum_{i=1}^{\infty} \frac{1}{(4i^2-1)^2} = \frac{\pi^2}{16} - \frac{1}{2}, \quad \sum_{i=1}^{\infty} \frac{i}{(4i^2-1)^2} = \frac{1}{8}, \quad \sum_{i=1}^{\infty} \frac{i^2}{(4i^2-1)^2} = \frac{\pi^2}{64},$$

$$\sum_{i=1}^{\infty} \frac{1}{i(4i^2-1)^2} = \frac{3}{2} - 2\log 2, \quad \sum_{i=1}^{\infty} \frac{12i^2-1}{i(4i^2-1)^2} = 2\log 2,$$

$$\sum_{i=1}^{\infty} \frac{1}{(4i^2-1)^3} = \frac{1}{2} - \frac{3\pi^2}{64}, \quad \sum_{i=1}^{\infty} \frac{i}{(4i^2-1)^3} = \frac{7\zeta(3)-6}{64},$$

$$\sum_{i=1}^{\infty} \frac{i^2}{(4i^2-1)^3} = \frac{\pi^2}{256}, \quad \sum_{i=1}^{\infty} \frac{i^3}{(4i^2-1)^3} = \frac{7\zeta(3)+2}{256},$$

$$\sum_{i=1}^{\infty} \frac{1}{i(i+1)(2i+1)} = 3 - 4\log 2, \quad \sum_{i=1}^{\infty} \frac{2i+1}{i^2(i+1)^2} = \sum_{i=1}^{\infty} \frac{3i^2+3i+1}{i^3(i+1)^3} = 1,$$

$$\sum_{i=0}^{\infty} \left[\frac{1}{(4i+1)(4i+2)(4i+3)} - \frac{1}{(4i+3)(4i+4)(4i+5)}\right] = \frac{1}{2} - \frac{\log 2}{2},$$

$$\sum_{i=0}^{\infty} \left[\frac{1}{(4i+2)(4i+3)(4i+4)} - \frac{1}{(4i+4)(4i+5)(4i+6)}\right] = \frac{\pi}{4} - \frac{3}{4}.$$

Source: Fact 1.12.27, [1217, pp. 9–12], [1524, p. 75], and [1787]. The first equality is given in [506, p. 119]. The third equality is given in [1757, pp. 235, 236]. The first expression in the third to last equality is given in [1566, p. 63]. The last two equalities are given in [1647, p. 269]. **Related:** Fact 1.12.26 and Fact 1.12.27.

Fact 13.5.78.

$$\sum_{i=2}^{\infty} \frac{3i^2-1}{(i^3-i)^2} = \frac{3}{8}, \quad \sum_{i=0}^{\infty} \frac{1}{[4(2i+1)^2-1]^2} = \frac{\pi(\pi-2)}{32}, \quad \sum_{i=1}^{\infty} \frac{i}{i^4+i^2+1} = \frac{1}{2},$$

$$\sum_{i=0}^{\infty} \frac{6i+3}{4i^4+8i^3+8i^2+4i+3} = \frac{3}{2}, \quad \sum_{i=1}^{\infty} \frac{i^2+3i+3}{i^4+2i^3-3i^2-4i+2} = -\frac{1}{2},$$

$$\sum_{i=1}^{\infty} \frac{1}{2i(2i+1)(2i+2)} = \frac{3}{4} - \log 2, \quad \sum_{i=0}^{\infty} \frac{1}{(2i+1)(2i+2)(2i+3)} = \log 2 - \frac{1}{2},$$

$$\sum_{i=0}^{\infty} \frac{1}{(3i+1)(3i+2)(3i+3)} = \frac{\sqrt{3}\pi}{12} - \frac{\log 3}{4}, \quad \sum_{i=0}^{\infty} \frac{9i+5}{(3i+1)(3i+2)(3i+3)} = \log 3,$$

$$\sum_{i=0}^{\infty} \frac{1}{(i+1)(i+2)(i+3)(i+4)} = \frac{1}{18}, \quad \sum_{i=0}^{\infty} \frac{1}{(i+1)(i+2)(i+3)(i+5)} = \frac{13}{288},$$

$$\sum_{i=0}^{\infty}\frac{1}{(i+1)(i+2)(i+4)(i+5)}=\frac{5}{144},\quad \sum_{i=0}^{\infty}\frac{1}{(i+1)(i+2)(i+4)(i+6)}=\frac{211}{7200},$$

$$\sum_{i=1}^{\infty}\frac{1}{(2i+1)(2i+2)(2i+4)(2i+5)}=\frac{1}{360},\quad \sum_{i=1}^{\infty}\frac{i^2}{(i+1)(i+2)(i+3)(i+4)}=\frac{5}{36},$$

$$\sum_{i=0}^{\infty}\frac{1}{(3i+1)(3i+2)(3i+3)(3i+4)}=\frac{1}{6}+\frac{\sqrt{3}\pi}{36}-\frac{\log 3}{4},$$

$$\sum_{i=0}^{\infty}\frac{1}{(i+1)^3(i+2)^3}=10-\pi^2,\quad \sum_{i=0}^{\infty}\frac{1}{(i+1)^3(i+2)^3(i+3)^3}=\frac{29}{32}-\frac{3\zeta(3)}{4}.$$

Source: The first equality is given in [109, pp. 272, 273]. The second equality is given in [645, p. 99]. The third equality is given in [2294, p. 70]. The sixth and seventh equalities are given in [1568, p. 199]. The eighth equality is given in [1647, p. 268]. The ninth equality is given in [506, p. 125], and is given by Fact 13.5.80. The 14th equality is given in [1647, p. 268]. The 15th equality is given in [1566, p. 64]. The 16th equality is given in [1217, p. 10]. The 17th equality is given in [1647, p. 272].

Fact 13.5.79. Let x be a real number. Then,

$$\sum_{i=1}^{\infty}\frac{i(i^4+4-x^2)}{i^8+2(x^2+4)i^4+16x^2i^2+(x^2+4)^2}=\frac{3(x^2+2)}{4(x^2+1)(x^2+4)}.$$

In particular,

$$\sum_{i=1}^{\infty}\frac{i}{i^4+4}=\frac{3}{8},\quad \sum_{i=1}^{\infty}\frac{i(i^4+3)}{i^8+10i^4+16i^2+25}=\frac{9}{40},$$

$$\sum_{i=1}^{\infty}\frac{i^5}{i^8+16i^4+64i^2+64}=\frac{9}{80},\quad \sum_{i=1}^{\infty}\frac{i(i^4-5)}{i^8+26i^4+144i^2+169}=\frac{33}{520}.$$

Source: [512].

Fact 13.5.80. Let $k\ge 2$. Then,

$$\sum_{i=0}^{\infty}\left(\sum_{j=1}^{k-1}\frac{1}{ki+j}-\frac{k-1}{ki+k}\right)=\log k.$$

In particular,

$$\sum_{i=0}^{\infty}\left(\frac{1}{2i+1}-\frac{1}{2i+2}\right)=\log 2,\quad \sum_{i=0}^{\infty}\left(\frac{1}{3i+1}+\frac{1}{3i+2}-\frac{2}{3i+3}\right)=\log 3.$$

Source: [1787].

Fact 13.5.81. Let m and n be positive integers. Then,

$$\sum_{i=1}^{\infty}\left(\frac{1}{(n+m)i-n}-\frac{1}{(n+m)i-m}\right)=\frac{\pi}{n+m}\cot\frac{m\pi}{n+m}.$$

In particular,

$$\sum_{i=1}^{\infty}\left(\frac{1}{7i-5}-\frac{1}{7i-2}\right)=\frac{\pi}{7}\cot\frac{2\pi}{7},\quad \sum_{i=1}^{\infty}\left(\frac{1}{11i-8}-\frac{1}{11i-3}\right)=\frac{\pi}{11}\cot\frac{3\pi}{11}.$$

Source: [1787].

Fact 13.5.82.

$$\sum_{i=0}^{\infty}\frac{(2i)!}{(2i+2)!}=\log 2,\quad \sum_{i=0}^{\infty}\frac{(2i)!}{(2i+3)!}=\log 2-\frac{1}{2},$$

$$\sum_{i=0}^{\infty}\frac{(2i)!}{(2i+4)!}=\frac{2\log 2}{3}-\frac{5}{12},\quad \sum_{i=0}^{\infty}\frac{(2i)!}{(2i+5)!}=\frac{\log 2}{3}-\frac{2}{9},\quad \sum_{i=0}^{\infty}\frac{(2i)!}{(2i+6)!}=\frac{2\log 2}{15}-\frac{131}{1440},$$

$$\sum_{i=0}^{\infty}\frac{(i!)^2}{(2i+1)!}=\frac{2\sqrt{3}\pi}{9},\quad \sum_{i=0}^{\infty}\frac{2^i(i!)^2}{(2i+1)!}=\frac{\pi}{2},\quad \sum_{i=0}^{\infty}\frac{3^i(i!)^2}{(2i+1)!}=\frac{4\sqrt{3}\pi}{9},$$

$$\sum_{i=0}^{\infty}\frac{(i!)^2}{(2i+2)!}=\frac{\pi^2}{18},\quad \sum_{i=0}^{\infty}\frac{2^i(i!)^2}{(2i+2)!}=\frac{\pi^2}{16},\quad \sum_{i=0}^{\infty}\frac{3^i(i!)^2}{(2i+2)!}=\frac{2\pi^2}{27},\quad \sum_{i=0}^{\infty}\frac{4^i(i!)^2}{(2i+2)!}=\frac{\pi^2}{8},$$

$$\sum_{i=0}^{\infty}\frac{(i!)^2}{(2i+3)!}=\frac{\pi^2}{18}+\frac{2\sqrt{3}\pi}{3}-4,\quad \sum_{i=0}^{\infty}\frac{2^i(i!)^2}{(2i+3)!}=\frac{\pi^2}{16}+\frac{\pi}{2}-2,\quad \sum_{i=0}^{\infty}\frac{3^i(i!)^2}{(2i+3)!}=\frac{2\pi^2}{27}+\frac{4\sqrt{3}\pi}{27}-\frac{4}{3},$$

$$\sum_{i=0}^{\infty}\frac{4^i(i!)^2}{(2i+3)!}=\frac{\pi^2}{8}-1,\quad \sum_{i=0}^{\infty}\frac{(i!)^2}{(2i+4)!}=\frac{\pi^2}{12}+\frac{\sqrt{3}\pi}{2}-\frac{42}{12},\quad \sum_{i=0}^{\infty}\frac{2^i(i!)^2}{(2i+4)!}=\frac{\pi^2}{16}+\frac{3\pi}{8}-\frac{7}{4},$$

$$\sum_{i=0}^{\infty}\frac{3^i(i!)^2}{(2i+4)!}=\frac{5\pi^2}{81}+\frac{\sqrt{3}\pi}{9}-\frac{7}{6},\quad \sum_{i=0}^{\infty}\frac{4^i(i!)^2}{(2i+4)!}=\frac{3\pi^2}{32}-\frac{7}{8},\quad \sum_{i=0}^{\infty}\left[\frac{i!}{(i+1)!}\right]^2=\frac{\pi^2}{6},$$

$$\sum_{i=0}^{\infty}\left[\frac{i!}{(i+2)!}\right]^2=\frac{\pi^2}{3}-3,\quad \sum_{i=0}^{\infty}\left[\frac{i!}{(i+3)!}\right]^2=\frac{\pi^2}{4}-\frac{39}{16},\quad \sum_{i=0}^{\infty}\left[\frac{i!}{(i+4)!}\right]^2=\frac{5\pi^2}{54}-\frac{197}{216},$$

$$\sum_{i=1}^{\infty}\frac{(2i)!}{4^i(i!)^2(2i+1)}=\frac{\pi}{2}-1,\quad \sum_{i=1}^{\infty}(-1)^i\frac{(2i)!}{4^i(i!)^2}=\frac{\sqrt{2}}{2}-1,\quad \sum_{i=1}^{\infty}(-1)^i\frac{(2i)!}{4^i(i!)^2(2i-1)}=1-\sqrt{2},$$

$$\sum_{i=1}^{\infty}\frac{16^i(i!)^4}{i^3[(2i)!]^2}=8\pi G-14\zeta(3),\quad \sum_{i=1}^{\infty}\frac{16^i(i!)^4}{(2i+1)^3[(2i)!]^2}=\frac{7}{2}\zeta(3)-\pi G.$$

Source: [323], [968, p. 103], and [1647, p. 269].

Fact 13.5.83.

$$\sum_{i=1}^{\infty}\frac{(i+1)!}{(2i+1)!!}=\frac{\pi}{2},\quad \sum_{i=0}^{\infty}\frac{(2i-1)!!}{(2i)!!(2i+1)}=\frac{\pi}{2},\quad \sum_{i=0}^{\infty}\frac{(2i-1)!!}{(2i)!!(3i+1)}=\frac{\sqrt{\pi}\Gamma(\frac{4}{3})}{\Gamma(\frac{5}{6})},$$

$$\sum_{i=0}^{\infty}\frac{(2i+1)!!}{(2i+2)!!(i+1)}=\log 4,\quad \sum_{i=0}^{\infty}\frac{[(2i+1)!!]^2}{[(2i+2)!!]^2(i+1)}=4\log 2-\frac{8G}{\pi},\quad \sum_{i=0}^{\infty}\frac{[(2i+1)!!]^2}{[(2i+2)!!]^2(i+2)}=\frac{4}{\pi},$$

$$\sum_{i=0}^{\infty}\frac{(2i-1)!!}{(2i)!!(2i+1)^2}=\frac{\pi}{2}\log 2,\quad \sum_{i=0}^{\infty}\frac{(2i-1)!!}{(2i)!!(2i+1)^3}=\frac{\pi}{4}\log^2 2+\frac{\pi^3}{48},$$

$$\sum_{i=0}^{\infty}\frac{(2i-1)!!}{(2i)!!(2i+1)^4}=\frac{\pi}{12}\log^3 2+\frac{\pi^3}{48}\log 2+\frac{\pi\zeta(3)}{8},$$

$$\sum_{i=0}^{\infty}\frac{(2i+1)!!}{(2i+2)!!(i+1)^2}=\frac{\pi^2}{6}-2\log^2 2,\quad \sum_{i=0}^{\infty}\frac{(2i+1)!!}{(2i+2)!!(i+1)^3}=\frac{4\log^3 2}{3}-\frac{\pi^2\log 2}{3}+2\zeta(3),$$

$$\sum_{i=0}^{\infty}\frac{(2i-1)!!}{(2i)!!(4i+1)}=\frac{\Gamma^2(\frac{1}{4})}{4\sqrt{2\pi}},\quad \sum_{i=0}^{\infty}\frac{(2i-1)!!}{(2i)!!(4i+1)^2}=\frac{\pi}{4}\frac{\Gamma^2(\frac{1}{4})}{4\sqrt{2\pi}},$$

$$\sum_{i=0}^{\infty}\frac{(2i-1)!!}{(2i)!!(4i+1)^3}=\left(\frac{\pi^2}{32}+\frac{G}{2}\right)\frac{\Gamma^2(\frac{1}{4})}{4\sqrt{2\pi}},\quad \sum_{i=0}^{\infty}\frac{(2i-1)!!}{(2i)!!(4i+1)^4}=\left(\frac{5\pi^3}{384}+\frac{\pi G}{8}\right)\frac{\Gamma^2(\frac{1}{4})}{4\sqrt{2\pi}},$$

$$\sum_{i=0}^{\infty}(-1)^i(4i+1)\left[\frac{(2i-1)!!}{(2i)!!}\right]^3=\frac{2}{\Gamma^2(\frac{1}{2})}=\frac{2}{\pi},\quad \sum_{i=0}^{\infty}(-1)^i(4i+1)\left[\frac{(2i-1)!!}{(2i)!!}\right]^5=\frac{2}{\Gamma^4(\frac{3}{4})},$$

$$\sum_{i=0}^{\infty}(-1)^i\left[\frac{(2i-1)!!}{(2i)!!}\right]^3=\left[\frac{\Gamma(9/8)}{\Gamma(5/4)\Gamma(7/8)}\right]^2.$$

Source: [116, p. 182], [650, 1206], [1317, pp. 16, 17], [1647, pp. 268, 269], and [1675, pp. 51, 58, 59].

Fact 13.5.84. Let $x \in (1,\infty)$. Then,

$$\sum_{i=1}^{\infty}\frac{1}{x^i i}=\log\frac{x}{x-1}.$$

In particular,

$$\sum_{i=1}^{\infty}\frac{1}{(\frac{3}{2})^i i}=\log 3,\quad \sum_{i=1}^{\infty}\frac{1}{2^i i}=\log 2,\quad \sum_{i=1}^{\infty}\frac{1}{10^i i}=\log\frac{10}{9}.$$

Source: [516, p. 129].

Fact 13.5.85.

$$\sum_{i=0}^{\infty}\frac{2i-1}{2^i i!}=0,\quad \sum_{i=0}^{\infty}\frac{2^i(i!)^2}{(2i+1)!}=\frac{\pi}{2},\quad \sum_{i=1}^{\infty}\frac{1}{2^i i^2}=\frac{\pi^2}{12}-\frac{1}{2}\log^2 2,\quad \sum_{i=0}^{\infty}\frac{1}{4^i(2i+1)}=8\log 3,$$

$$\sum_{i=0}^{\infty}\frac{1}{9^i(2i+1)}=\frac{3}{2}\log 2,\quad \sum_{i=0}^{\infty}\frac{1}{16^i(i+1)}=8\log\frac{256}{225},\quad \sum_{i=0}^{\infty}\frac{1}{16^i(2i+1)}=2\log\frac{5}{3},$$

$$\sum_{i=0}^{\infty}\frac{1}{16^i(4i+1)}=\frac{1}{2}\log 3+\operatorname{atan}\frac{1}{2},\quad \sum_{i=0}^{\infty}\frac{1}{16^i(4i+3)}=2\log 3-4\operatorname{atan}\frac{1}{2},$$

$$\sum_{i=1}^{\infty}\frac{i}{(i+1)!}=1,\quad \sum_{i=0}^{\infty}\frac{i^2+i-1}{(i+2)!}=0,\quad \sum_{i=0}^{\infty}\frac{i^2+3i+1}{(i+2)!}=2,\quad \sum_{i=0}^{\infty}\frac{(4i+1)i!}{(2i+1)!}=2,$$

$$\sum_{i=0}^{\infty}\frac{i^3+6i^2+11i+5}{(i+3)!}=\frac{5}{2},\quad \sum_{i=1}^{\infty}\frac{(i+1)(i+2)(i+3)}{(2i+4)!}=\frac{e^2-4e+5}{16e},$$

$$\sum_{i=0}^{\infty}\frac{(i+1)(i+2)(2i+3)}{(2i+4)!}=\frac{e^2-1}{8e^2},\quad \sum_{i=1}^{\infty}\frac{i}{\prod_{j=1}^{i}(2j+1)}=\frac{1}{2},\quad \sum_{i=2}^{\infty}\frac{1}{i^2\prod_{j=2}^{i}\left(1-\frac{1}{j^2}\right)}=\frac{1}{2},$$

$$\sum_{i=0}^{\infty}\frac{(4i)!(1103+26390i)}{(i!)^4 396^{4i}}=\frac{9801\sqrt{2}}{4\pi},\quad \sum_{i=0}^{\infty}(-1)^i\frac{(6i)!(13591409+545140134i)}{(3i!)(i!)^3 640320^{3i+3/2}}=\frac{1}{12\pi},$$

$$1+\sum_{i=1}^{\infty}(-1)^k(4i+1)\left[\frac{\prod_{j=0}^{i-1}(2j+1)}{2^i i!}\right]^3=\frac{2}{\pi},\quad 1+\sum_{i=1}^{\infty}(8i+1)\left[\frac{\prod_{j=0}^{i-1}(4j+1)}{4^i i!}\right]^4=\frac{2\sqrt{2}}{\sqrt{\pi}\Gamma^2(\frac{3}{4})}.$$

Source: [213], [516, p. 108], [1566, pp. 28, 64, 65], and [1581]. The last two equalities are given in [1317, pp. xxv, xxvi].

Fact 13.5.86. Define

$$G \triangleq \sum_{i=0}^{\infty}(-1)^i\frac{1}{(2i+1)^2} \approx 0.915965594177219015054603514932384110774149374281672 1.$$

Then,

$$\sum_{i=0}^{\infty}\frac{1}{(4i+1)^2} = \frac{\pi^2}{16}+\frac{G}{2}, \quad \sum_{i=0}^{\infty}\frac{1}{(4i+3)^2} = \frac{\pi^2}{16}-\frac{G}{2},$$

$$\sum_{i=0}^{\infty}\frac{2k+1}{(4i+1)^2(4k+3)^2} = \frac{G}{8}, \quad \sum_{i=1}^{\infty}\frac{16^i(12i^3+12i^2+6i+1)(i!)^4}{i^3(2i+1)^3[(2i)!]^2} = 4\pi G - 4.$$

Remark: G is *Catalan's constant*. See Fact 14.9.1. **Source:** [323]. **Related:** Fact 13.5.72.

Fact 13.5.87.

$$\sum_{i=1}^{\infty}\frac{1}{i}\prod_{j=1}^{i}\left(1-\frac{1}{4j}\right) = 3\log 2+\frac{\pi}{2}, \quad \sum_{i=1}^{\infty}\frac{1}{i}\prod_{j=1}^{i}\left(1-\frac{3}{4j}\right) = 3\log 2-\frac{\pi}{2},$$

$$\sum_{i=1}^{\infty}\frac{1}{i}\left(\prod_{j=1}^{i}\frac{4j-1}{4j}+\prod_{j=1}^{i}\frac{4j-3}{4j}\right) = 6\log 2, \quad \sum_{i=1}^{\infty}\frac{1}{i}\left(\prod_{j=1}^{i}\frac{4j-1}{4j}-\prod_{j=1}^{i}\frac{4j-3}{4j}\right) = \pi,$$

$$\sum_{i=1}^{\infty}\frac{1}{i^2}\left(\prod_{j=1}^{i}\frac{4j}{4j-1}+\prod_{j=1}^{i}\frac{4j}{4j-3}\right) = 2\pi^2, \quad \sum_{i=1}^{\infty}\frac{1}{i^2}\left(\prod_{j=1}^{i}\frac{4j}{4j-1}-\prod_{j=1}^{i}\frac{4j}{4j-3}\right) = 16G,$$

$$\sum_{i=1}^{\infty}\frac{1}{i}\left[\left(\sum_{j=1}^{i}\frac{1}{4j-1}\right)\prod_{j=1}^{i}\left(1-\frac{1}{4j}\right)+\left(\sum_{j=1}^{i}\frac{1}{4j-3}\right)\prod_{j=1}^{i}\left(1-\frac{3}{4j}\right)\right] = \frac{\pi^2}{2},$$

$$\sum_{i=1}^{\infty}\frac{1}{i}\left[\left(\sum_{j=1}^{i}\frac{1}{4j-1}\right)\prod_{j=1}^{i}\left(1-\frac{1}{4j}\right)-\left(\sum_{j=1}^{i}\frac{1}{4j-3}\right)\prod_{j=1}^{i}\left(1-\frac{3}{4j}\right)\right] = 4G,$$

$$\sum_{i=1}^{\infty}\frac{1}{i}\left[\prod_{j=1}^{i}\left(1-\frac{1}{3j}\right)+\prod_{j=1}^{i}\left(1-\frac{2}{3j}\right)\right] = 3\log 3, \quad \sum_{i=1}^{\infty}\frac{1}{i}\left[\prod_{j=1}^{i}\left(1-\frac{1}{3j}\right)-\prod_{j=1}^{i}\left(1-\frac{2}{3j}\right)\right] = \frac{\sqrt{3}\pi}{3},$$

$$\sum_{i=1}^{\infty}\frac{1}{i}\left[\left(\prod_{j=1}^{i}\frac{3j-1}{3j}\right)\sum_{j=1}^{i}\frac{1}{3j-1}+\left(\prod_{j=1}^{i}\frac{3j-2}{3j}\right)\sum_{j=1}^{i}\frac{1}{3j-2}\right] = \frac{4\pi^2}{9},$$

$$\sum_{i=1}^{\infty}\frac{1}{i}\sum_{k=1}^{2}(-1)^k\left(\prod_{j=1}^{i}\frac{3j-k}{3j}\right)\left[\sum_{j=1}^{i}\frac{1}{(3j-k)^2}-\left(\sum_{j=1}^{i}\frac{1}{3j-k}\right)^2\right] = \frac{8\sqrt{3}\pi^3}{81},$$

$$\sum_{i=2}^{\infty}\frac{H_i}{i}\sum_{k=1,3}\left(\prod_{j=1}^{i}\frac{4j-k}{4j}\right)\left[\left(\sum_{j=1}^{i}\frac{1}{4j-k}\right)^2-\sum_{j=1}^{i}\frac{1}{(4j-k)^2}\right] = \pi^4.$$

If, in addition, $x \in (0,1)$, then

$$\sum_{i=1}^{\infty}\frac{(1-x)^{\overline{i}}-x^{\overline{i}}}{i(i!)} = \pi\cot\pi x.$$

Source: [62].

Fact 13.5.88.

$$\sum_{i=1}^{\infty}\frac{1}{2i+1}\prod_{j=1}^{2i}\frac{2j-1}{2j} = \sqrt{2}-1, \quad \sum_{i=1}^{\infty}\frac{1}{2i+1}\prod_{j=1}^{2i}\frac{3j-2}{3j} = \frac{3\sqrt[3]{4}}{4}-1.$$

Source: [2461].

Fact 13.5.89. Let $n \geq 1$. Then,

$$\sum_{i=1}^{\infty}(-1)^i \frac{1}{\prod_{j=i}^{i+n} j} = -\frac{2^n \log 2}{n!} + \frac{1}{n!}\sum_{i=1}^{n}\frac{2^{n-i}}{i}.$$

In particular,

$$\sum_{i=1}^{\infty}(-1)^i \frac{1}{i(i+1)} = 1 - 2\log 2, \quad \sum_{i=1}^{\infty}(-1)^i \frac{1}{i(i+1)(i+2)} = \frac{5}{4} - 2\log 2,$$

$$\sum_{i=1}^{\infty}(-1)^i \frac{1}{i(i+1)(i+2)(i+3)} = \frac{8}{9} - \frac{4\log 2}{3}, \quad \sum_{i=1}^{\infty}(-1)^i \frac{1}{i(i+1)(i+2)(i+3)(i+4)} = \frac{131}{288} - \frac{2\log 2}{3}.$$

Source: [107, pp. 37, 210, 211].

Fact 13.5.90. Let $n \geq 1$, let $m \in \mathbb{Z}$, and assume that $n + m \geq 1$. Then,

$$\sum_{i=1}^{\infty}(-1)^{i+1}\frac{1}{in+m} = \int_0^1 \frac{x^{n+m-1}}{x^n+1}\,\mathrm{d}x.$$

Therefore, for all $n \geq 1$,

$$\frac{1}{n}\sum_{i=1}^{\infty}(-1)^{i+1}\frac{1}{i} = \int_0^1 \frac{x^{n-1}}{x^n+1}\,\mathrm{d}x = \frac{\log 2}{n}.$$

In particular,

$$\sum_{i=1}^{\infty}(-1)^{i+1}\frac{1}{i+1} = 1 - \log 2, \quad \sum_{i=1}^{\infty}(-1)^{i+1}\frac{1}{i+2} = \log 2 - \frac{1}{2}, \quad \sum_{i=1}^{\infty}(-1)^{i+1}\frac{1}{i+3} = \frac{5}{6} - \log 2,$$

$$\sum_{i=1}^{\infty}(-1)^{i+1}\frac{1}{i+4} = \log 2 - \frac{7}{12}, \quad \sum_{i=1}^{\infty}(-1)^{i+1}\frac{1}{i+5} = \frac{47}{60} - \log 2, \quad \sum_{i=1}^{\infty}(-1)^{i+1}\frac{1}{i+6} = \log 2 - \frac{37}{60},$$

$$\sum_{i=1}^{\infty}(-1)^{i+1}\frac{1}{2i-1} = \frac{1}{2}\log 2, \quad \sum_{i=1}^{\infty}(-1)^{i+1}\frac{1}{2i+1} = 1 - \frac{\pi}{4}, \quad \sum_{i=1}^{\infty}(-1)^{i+1}\frac{1}{2i+3} = \frac{\pi}{4} - \frac{2}{3},$$

$$\sum_{i=1}^{\infty}(-1)^{i+1}\frac{1}{2i+5} = \frac{13}{15} - \frac{\pi}{4}, \quad \sum_{i=1}^{\infty}(-1)^{i+1}\frac{1}{2i+7} = \frac{\pi}{4} - \frac{76}{105}, \quad \sum_{i=1}^{\infty}(-1)^{i+1}\frac{1}{2i+9} = \frac{263}{315} - \frac{\pi}{4},$$

$$\sum_{i=1}^{\infty}(-1)^{i+1}\frac{1}{3i-2} = \frac{\sqrt{3}\pi}{9} + \frac{\log 2}{3}, \quad \sum_{i=1}^{\infty}(-1)^{i+1}\frac{1}{3i-1} = \frac{\sqrt{3}\pi}{9} - \frac{\log 2}{3},$$

$$\sum_{i=1}^{\infty}(-1)^{i+1}\frac{1}{3i+1} = 1 - \frac{\sqrt{3}\pi}{9} - \frac{\log 2}{3}, \quad \sum_{i=1}^{\infty}(-1)^{i+1}\frac{1}{3i+2} = \frac{1}{2} - \frac{\sqrt{3}\pi}{9} + \frac{\log 2}{3},$$

$$\sum_{i=1}^{\infty}(-1)^{i+1}\frac{1}{3i+4} = \frac{\sqrt{3}\pi}{9} + \frac{\log 2}{3} - \frac{1}{4}, \quad \sum_{i=1}^{\infty}(-1)^{i+1}\frac{1}{3i+5} = \frac{\sqrt{3}\pi}{9} - \frac{\log 2}{3} - \frac{3}{10},$$

$$\sum_{i=1}^{\infty}(-1)^{i+1}\frac{1}{4i-1} = \frac{\sqrt{2}\pi}{8} + \frac{\sqrt{2}\log(3-2\sqrt{2})}{8}, \quad \sum_{i=1}^{\infty}(-1)^{i+1}\frac{1}{4i-3} = \frac{\sqrt{2}\pi}{8} + \frac{\sqrt{2}\log(1+\sqrt{2})}{4},$$

$$\sum_{i=1}^{\infty}(-1)^{i+1}\frac{1}{4i+1} = 1 + \frac{\sqrt{2}\log(3-2\sqrt{2}) - \sqrt{2}\pi}{8}, \quad \sum_{i=1}^{\infty}(-1)^{i+1}\frac{1}{4i+3} = \frac{1}{3} - \frac{\sqrt{2}\log(3-2\sqrt{2}) - \sqrt{2}\pi}{8},$$

$$\sum_{i=1}^{\infty}(-1)^{i+1}\frac{1}{4i+5} = \frac{\sqrt{2}\pi - \sqrt{2}\log(3-2\sqrt{2})}{8} - \frac{4}{5},$$

$$\sum_{i=1}^{\infty}(-1)^{i+1}\frac{1}{4i+7} = \frac{\sqrt{2}\log(3-2\sqrt{2}) + \sqrt{2}\pi}{8} - \frac{4}{21}.$$

Source: [712].

Fact 13.5.91. Let $n \ge 2$ and $1 \le m \le n-1$. Then,

$$\sum_{i=1}^{\infty}(-1)^{i+1}\frac{1}{in+m} = \int_0^1 \frac{x^{n+m-1}}{x^n+1}\,dx = \frac{1}{m} - \frac{\pi}{2n}\csc\frac{m\pi}{n} + \frac{2}{n}\sum_{i=0}^{\lfloor (n-2)/2\rfloor}\left(\cos\frac{(2i+1)m\pi}{n}\right)\log\sin\frac{(2i+1)\pi}{2}.$$

In particular,

$$\sum_{i=1}^{\infty}(-1)^{i+1}\frac{1}{2i+1} = \int_0^1 \frac{x^2}{x^2+1}\,dx = 1 - \frac{\pi}{4},$$

$$\sum_{i=1}^{\infty}(-1)^{i+1}\frac{1}{3i+1} = \int_0^1 \frac{x^3}{x^3+1}\,dx = 1 - \frac{\sqrt{3}\pi}{9} - \frac{\log 2}{3},$$

$$\sum_{i=1}^{\infty}(-1)^{i+1}\frac{1}{3i+2} = \int_0^1 \frac{x^4}{x^3+1}\,dx = \frac{1}{2} - \frac{\sqrt{3}\pi}{9} + \frac{\log 2}{3}.$$

Source: [628].

Fact 13.5.92.

$$\sum_{i=0}^{\infty}(-1)^i\frac{1}{i+1} = \sum_{i=0}^{\infty}\frac{1}{(2i+1)(2i+2)} = \log 2, \quad \sum_{i=0}^{\infty}(-1)^i\frac{1}{2i+1} = \sum_{i=0}^{\infty}\frac{2}{(4i+1)(4i+3)} = \frac{\pi}{4},$$

$$\sum_{i=0}^{\infty}(-1)^i\frac{1}{3i+1} = \sum_{i=0}^{\infty}\frac{3}{(6i+1)(6i+4)} = \frac{\sqrt{3}\pi}{9} + \frac{\log 2}{3},$$

$$\sum_{i=0}^{\infty}(-1)^i\frac{1}{4i+1} = \sum_{i=0}^{\infty}\frac{4}{(8i+1)(8i+5)} = \frac{\sqrt{2}\pi}{8} + \frac{\sqrt{2}\log(1+\sqrt{2})}{4} = \frac{\sqrt{2}\pi}{8} + \frac{\sqrt{2}\operatorname{asinh} 1}{4}.$$

Fact 13.5.93. If $n \ge 2$, then

$$\sum_{i=1}^{\infty}(-1)^{i+1}\frac{1}{i^n} = \frac{2^n-2}{2^n}\sum_{i=1}^{\infty}\frac{1}{i^n}.$$

If $n \ge 1$, then

$$\sum_{i=1}^{\infty}(-1)^{i+1}\frac{1}{i^{2n}} = \frac{(2^{2n-1}-1)\pi^{2n}}{(2n)!}|B_{2n}|.$$

If $n \ge 0$, then

$$\sum_{i=1}^{\infty}(-1)^{i+1}\frac{1}{(2i-1)^{2n+1}} = \frac{\pi^{2n+1}}{2^{2n+2}(2n)!}|E_{2n}| = \frac{4^n\pi^{2n+1}}{(2n+1)!}|B_{2n+1}(1/4)|.$$

In particular,

$$\sum_{i=1}^{\infty}(-1)^{i+1}\frac{1}{i^2} = \frac{\pi^2}{12}, \quad \sum_{i=1}^{\infty}(-1)^{i+1}\frac{1}{i^4} = \frac{7\pi^4}{720}, \quad \sum_{i=1}^{\infty}(-1)^{i+1}\frac{1}{i^6} = \frac{31\pi^6}{30240},$$

$$\sum_{i=1}^{\infty}(-1)^{i+1}\frac{1}{2i-1}=\frac{\pi}{4},\quad \sum_{i=1}^{\infty}(-1)^{i+1}\frac{1}{(2i-1)^3}=\frac{\pi^3}{32},$$

$$\sum_{i=1}^{\infty}(-1)^{i+1}\frac{1}{(2i-1)^5}=\frac{5\pi^5}{1536},\quad \sum_{i=1}^{\infty}(-1)^{i+1}\frac{1}{(2i-1)^7}=\frac{61\pi^7}{184320}.$$

Source: The expression involving the Bernoulli polynomial is given in [116, p. 56] and [705].
Remark: Each alternating series can be rewritten as a nonalternating series. For example,

$$\sum_{i=1}^{\infty}(-1)^{i+1}\frac{1}{i^2}=\sum_{i=0}^{\infty}\frac{4i+3}{(2i+1)^2(2i+2)^2}=\frac{\pi^2}{12}.$$

Related: Fact 14.6.15.

Fact 13.5.94.

$$\sum_{i=1}^{\infty}(-1)^{i+1}\frac{1}{i(i+1)}=2\log 2-1,\quad \sum_{i=1}^{\infty}(-1)^{i+1}\frac{1}{i^2(i+1)}=\frac{\pi^2}{12}-2\log 2+1,$$

$$\sum_{i=1}^{\infty}(-1)^{i+1}\frac{1}{i^3(i+1)}=2\log 2-\frac{\pi^2}{12}+\frac{3}{4}\zeta(3)-1,$$

$$\sum_{i=1}^{\infty}(-1)^{i+1}\frac{1}{i(i+2)}=\frac{1}{4},\ \sum_{i=1}^{\infty}(-1)^{i+1}\frac{1}{i^2(i+2)}=\frac{\pi^2}{24}-\frac{1}{8},\ \sum_{i=1}^{\infty}(-1)^{i+1}\frac{1}{i^3(i+2)}=\frac{1}{16}-\frac{\pi^2}{48}+\frac{3}{8}\zeta(3),$$

$$\sum_{i=1}^{\infty}(-1)^{i+1}\frac{1}{i(i+3)}=\frac{2}{3}\log 2-\frac{5}{18},\quad \sum_{i=1}^{\infty}(-1)^{i+1}\frac{1}{i^2(i+3)}=\frac{\pi^2}{36}-\frac{2}{9}\log 2+\frac{5}{54},$$

$$\sum_{i=1}^{\infty}(-1)^{i+1}\frac{1}{i^3(i+3)}=\frac{2}{27}\log 2+\frac{1}{4}\zeta(3)-\frac{\pi^2}{108}-\frac{5}{162},$$

$$\sum_{i=1}^{\infty}(-1)^{i+1}\frac{1}{i(i+1)^2}=\frac{\pi^2}{12}+2\log 2-2,\quad \sum_{i=1}^{\infty}(-1)^{i+1}\frac{1}{i^2(i+1)^2}=3-4\log 2,$$

$$\sum_{i=1}^{\infty}(-1)^{i+1}\frac{1}{i^3(i+1)^2}=6\log 2+\frac{3}{4}\zeta(3)-\frac{\pi^2}{12}-4,$$

$$\sum_{i=1}^{\infty}(-1)^{i+1}\frac{1}{i(i+2)^2}=\frac{1}{2}-\frac{\pi^2}{24},\quad \sum_{i=1}^{\infty}(-1)^{i+1}\frac{1}{i^2(i+2)^2}=\frac{\pi^2}{24}-\frac{5}{16},$$

$$\sum_{i=1}^{\infty}(-1)^{i+1}\frac{1}{i^3(i+2)^2}=\frac{3}{16}-\frac{\pi^2}{32}+\frac{3}{16}\zeta(3),$$

$$\sum_{i=1}^{\infty}(-1)^{i+1}\frac{1}{i(i+3)^2}=\frac{\pi^2}{36}+\frac{2}{9}\log 2-\frac{41}{108},\quad \sum_{i=1}^{\infty}(-1)^{i+1}\frac{1}{i^2(i+3)^2}=\frac{17}{108}-\frac{4}{27}\log 2,$$

$$\sum_{i=1}^{\infty}(-1)^{i+1}\frac{1}{i^3(i+3)^2}=\frac{2}{27}\log 2+\frac{1}{12}\zeta(3)-\frac{61}{972}-\frac{\pi^2}{324},$$

$$\sum_{i=1}^{\infty}(-1)^{i+1}\frac{1}{i(i+1)^3}=\frac{\pi^2}{12}+2\log 2+\frac{3}{4}\zeta(3)-3,\quad \sum_{i=1}^{\infty}(-1)^{i+1}\frac{1}{i^2(i+1)^3}=6-\frac{\pi^2}{12}-6\log 2-\frac{3}{4}\zeta(3),$$

$$\sum_{i=1}^{\infty}(-1)^{i+1}\frac{1}{i^3(i+1)^3}=12\log 2+\frac{3}{2}\zeta(3)-10,$$

$$\sum_{i=1}^{\infty}(-1)^{i+1}\frac{1}{i(i+2)^3}=\frac{11}{16}-\frac{\pi^2}{48}-\frac{3}{8}\zeta(3),\quad \sum_{i=1}^{\infty}(-1)^{i+1}\frac{1}{i^2(i+2)^3}=\frac{\pi^2}{32}+\frac{3}{16}\zeta(3)-\frac{1}{2},$$

$$\sum_{i=1}^{\infty}(-1)^{i+1}\frac{1}{i^3(i+2)^3}=\frac{11}{32}-\frac{\pi^2}{32},\quad \sum_{i=1}^{\infty}(-1)^{i+1}\frac{1}{i(i+3)^3}=\frac{\pi^2}{108}+\frac{2}{27}\log 2+\frac{1}{4}\zeta(3)-\frac{31}{72},$$

$$\sum_{i=1}^{\infty}(-1)^{i+1}\frac{1}{i^2(i+3)^3}=\frac{127}{648}-\frac{\pi^2}{324}-\frac{2}{27}\log 2-\frac{\zeta(3)}{12},\ \sum_{i=1}^{\infty}(-1)^{i+1}\frac{1}{i^3(i+3)^3}=\frac{4}{81}\log 2+\frac{\zeta(3)}{18}-\frac{503}{5832}.$$

Fact 13.5.95. Let x be a real number, and assume that x is not an integer. Then,

$$\sum_{i=-\infty}^{\infty}\frac{1}{(i+x)^2}=\frac{\pi^2}{\sin^2\pi x},\quad \sum_{i=-\infty}^{\infty}\frac{1}{\left(\frac{i}{x}+1\right)^2}=\left(\frac{\pi x}{\sin\pi x}\right)^2.$$

In particular,

$$\sum_{i=-\infty}^{\infty}\frac{1}{(2i+1)^2}=\frac{\pi^2}{4},\quad \sum_{i=-\infty}^{\infty}\frac{1}{(3i+1)^2}=\frac{4\pi^2}{27},\quad \sum_{i=-\infty}^{\infty}\frac{1}{(4i+1)^2}=\frac{\pi^2}{8}.$$

Source: Fact 13.4.10. **Related:** Fact 13.5.72.

Fact 13.5.96. Let $n\ge 1$, and define $Q_n\in\mathbb{R}[s]$ and $P_n\in\mathbb{R}[s]$ as in Fact 13.2.14. If n is odd, then

$$\sum_{i=1}^{\infty}\frac{1}{i^{n+1}}=\frac{\pi^{n+1}P_n(0)}{2(2^{n+1}-1)n!},\quad \sum_{i=0}^{\infty}(-1)^{\lfloor (i+1)/2\rfloor}\frac{1}{(2i+1)^{n+1}}=\frac{\sqrt{2}\pi^{n+1}Q_n(1)}{4^{n+1}n!}.$$

If n is even, then

$$\sum_{i=0}^{\infty}(-1)^{i}\frac{1}{(2i+1)^{n+1}}=\frac{\pi^{n+1}Q_n(0)}{2^{n+2}n!},\quad \sum_{i=0}^{\infty}(-1)^{\lfloor i/2\rfloor}\frac{1}{(2i+1)^{n+1}}=\frac{\sqrt{2}\pi^{n+1}Q_n(1)}{4^{n+1}n!}.$$

In particular,

$$\sum_{i=0}^{\infty}(-1)^{i}\frac{1}{2i+1}=\frac{\pi}{4},\quad \sum_{i=0}^{\infty}(-1)^{i}\frac{1}{(2i+1)^3}=\frac{\pi^3}{32},\quad \sum_{i=0}^{\infty}(-1)^{i}\frac{1}{(2i+1)^5}=\frac{5\pi^5}{1536},$$

$$\sum_{i=0}^{\infty}(-1)^{\lfloor (i+1)/2\rfloor}\frac{1}{(2i+1)^2}=\frac{\sqrt{2}\pi^2}{16},\quad \sum_{i=0}^{\infty}(-1)^{\lfloor (i+1)/2\rfloor}\frac{1}{(2i+1)^4}=\frac{11\sqrt{2}\pi^4}{6144},$$

$$\sum_{i=0}^{\infty}(-1)^{\lfloor (i+1)/2\rfloor}\frac{1}{(2i+1)^6}=\frac{361\sqrt{2}\pi^6}{491520},\quad \sum_{i=0}^{\infty}(-1)^{\lfloor i/2\rfloor}\frac{1}{2i+1}=\frac{\sqrt{2}\pi}{4},$$

$$\sum_{i=0}^{\infty}(-1)^{\lfloor i/2\rfloor}\frac{1}{(2i+1)^3}=\frac{3\sqrt{2}\pi^3}{128},\quad \sum_{i=0}^{\infty}(-1)^{\lfloor i/2\rfloor}\frac{1}{(2i+1)^5}=\frac{19\sqrt{2}\pi^5}{8192}.$$

Source: [1430].

Fact 13.5.97.

$$\sum_{i=0}^{\infty}(-1)^{i}\frac{1}{i!}=\frac{1}{e},\quad \sum_{i=1}^{\infty}(-1)^{i+1}\frac{1}{i}=\log 2,\quad \sum_{i=0}^{\infty}(-1)^{i}\frac{1}{(2i)!}=\cos 1,\quad \sum_{i=1}^{\infty}(-1)^{i+1}\frac{2i+1}{i(i+1)}=1,$$

$$\sum_{i=1}^{\infty}(-1)^{i}\frac{H_i}{i}=\frac{\log^2 2}{2}-\frac{\pi^2}{12},\quad \sum_{i=1}^{\infty}(-1)^{i+1}\frac{H_i}{i+1}=\frac{\log^2 2}{2},\quad \sum_{i=0}^{\infty}(-1)^{i+1}\frac{1}{(2i-1)!}=\sin 1,$$

$$\sum_{i=1}^{\infty}(-1)^{\lfloor (i+5)/3\rfloor}\frac{1}{2i-1}=\frac{5\pi}{12},\quad \sum_{i=1}^{\infty}(-1)^{\lfloor (i+3)/2\rfloor}\frac{1}{2i-1}=\frac{\sqrt{2}\pi}{4},\quad \sum_{i=1}^{\infty}(-1)^{i+1}\frac{1}{i(4i^2-1)}=1-\log 2,$$

$$\sum_{i=1}^{\infty}(-1)^{i+1}\frac{1}{(2i+1)(2i+3)}=\frac{5}{6}-\frac{\pi}{4},\quad \sum_{i=1}^{\infty}(-1)^{i+1}\frac{1}{(2i+1)(2i+3)(2i+5)}=\frac{2}{5}-\frac{\pi}{8},$$

$$\sum_{i=1}^{\infty}(-1)^{i+1}\frac{1}{(2i+1)(2i+3)(2i+5)(2i+7)}=\frac{83}{630}-\frac{\pi}{24},\quad \sum_{i=1}^{\infty}(-1)^{i+1}\frac{1}{(3i+1)(3i+2)}=\frac{1}{2}-\frac{2\log 2}{3},$$

$$\sum_{i=1}^{\infty}(-1)^{i+1}\frac{1}{(3i+1)(3i+2)(3i+3)}=\frac{1}{6}+\frac{\sqrt{3}\pi}{18}-\frac{2\log 2}{3},\quad \sum_{i=0}^{\infty}(-1)^{i}\frac{2i+1}{(4i+1)(4i+3)}=\frac{\sqrt{2}\pi}{16},$$

$$\sum_{i=1}^{\infty}(-1)^{i+1}\frac{i}{(i+1)^2}=\frac{\pi^2}{12}-\log 2,\quad \sum_{i=1}^{\infty}(-1)^{i+1}\frac{1}{3i-1}=\frac{\sqrt{3}\pi}{9}-\frac{\log 2}{3},$$

$$\sum_{i=0}^{\infty}(-1)^{i}\frac{1}{3i+1}=\sum_{i=1}^{\infty}(-1)^{i+1}\frac{1}{3i-2}=\sum_{i=0}^{\infty}\frac{3}{(6i+1)(6i+4)}=\frac{\sqrt{3}\pi}{9}+\frac{\log 2}{3},$$

$$\sum_{i=1}^{\infty}(-1)^{i+1}\frac{1}{i(i+1)}=\log 4-1,\quad \sum_{i=1}^{\infty}(-1)^{i+1}\frac{1}{i(i+1)(i+2)}=\log 4-\frac{5}{4},$$

$$\sum_{i=1}^{\infty}(-1)^{i+1}\frac{1}{i(i+1)(i+2)(i+3)}=\frac{4}{3}\log 2-\frac{8}{9},$$

$$\sum_{i=1}^{\infty}(-1)^{i+1}\frac{1}{i(i+1)(i+2)(i+3)(i+4)}=\frac{2}{3}\log 2-\frac{131}{288},$$

$$\sum_{i=1}^{\infty}(-1)^{i+1}\frac{1}{i(i+2)}=\frac{1}{4},\quad \sum_{i=1}^{\infty}(-1)^{i+1}\frac{1}{i(i+2)(i+4)}=\frac{5}{96},$$

$$\sum_{i=1}^{\infty}(-1)^{i+1}\frac{1}{i(i+2)(i+4)(i+6)}=\frac{11}{1440},\quad \sum_{i=1}^{\infty}(-1)^{i+1}\frac{1}{i(i+2)(i+4)(i+6)(i+8)}=\frac{31}{35840},$$

$$\sum_{i=1}^{\infty}(-1)^{i+1}\frac{1}{i(i+3)}=\frac{2\log 2}{3}-\frac{5}{18},\quad \sum_{i=1}^{\infty}(-1)^{i+1}\frac{1}{i(i+3)(i+6)}=\frac{2\log 2}{9}-\frac{137}{1080},$$

$$\sum_{i=1}^{\infty}(-1)^{i+1}\frac{1}{i(2i+1)(3i-1)}=\frac{1}{5}[4-8\log 2+(\sqrt{3}-1)\pi],$$

$$\sum_{i=1}^{\infty}(-1)^{i}\frac{1}{i(i+1)(2i+1)}=3-\pi,\quad \sum_{i=2}^{\infty}(-1)^{i}\frac{1}{i(i-1)(2i-1)}=\pi-3,$$

$$\sum_{i=0}^{\infty}(-1)^{i}\frac{1}{(2i+1)(2i+2)}=\frac{\pi}{4}-\frac{\log 2}{2},\quad \sum_{i=0}^{\infty}(-1)^{i}\frac{1}{(2i+1)(2i+2)(2i+3)}=\frac{1}{2}-\frac{\log 2}{2},$$

$$\sum_{i=0}^{\infty}(-1)^{i}\frac{1}{\prod_{j=1}^{4}(2i+j)}=\frac{5-\pi}{12}-\frac{\log 2}{6},\quad \sum_{i=0}^{\infty}(-1)^{i}\frac{1}{\prod_{j=1}^{5}(2i+j)}=\frac{5}{36}-\frac{\pi}{24},$$

$$\sum_{i=0}^{\infty}(-1)^{i}\frac{1}{\prod_{j=1}^{6}(2i+j)}=\frac{23}{140}-\frac{3\pi}{380}+\frac{\log 2}{60},\quad \sum_{i=0}^{\infty}(-1)^{i}\frac{1}{\prod_{j=1}^{7}(2i+j)}=\frac{\log 2}{180}-\frac{79}{21600},$$

$$\sum_{i=0}^{\infty}(-1)^i\frac{1}{\prod_{j=1}^{8}(2i+j)} = \frac{\pi}{2520} - \frac{\log 2}{1260} - \frac{67}{37800}, \quad \sum_{i=0}^{\infty}(-1)^i\frac{1}{\prod_{j=1}^{9}(2i+j)} = \frac{1}{2520}\left(\frac{\pi}{4} - \frac{109}{140}\right),$$

$$\sum_{i=0}^{\infty}(-1)^i\frac{1}{(2i+1)(2i+2)(2i+4)(2i+5)} = \frac{5}{36} - \frac{\log 2}{6},$$

$$\sum_{i=0}^{\infty}(-1)^i\frac{1}{(2i+1)(2i+2)(2i+6)(2i+7)} = \frac{\pi}{60} - \frac{149}{3600},$$

$$\sum_{i=1}^{\infty}(-1)^{i+1}\frac{i^2}{i^3+1} = \frac{1}{3}\left(1 - \log 2 + \pi\,\mathrm{sech}\,\frac{\sqrt{3}\pi}{2}\right),$$

$$\sum_{i=1}^{\infty}(-1)^{\lfloor (i+3)/2\rfloor}\frac{1}{i} = \sum_{i=0}^{\infty}\frac{64i^2+80i+22}{(4i+1)(4i+2)(4i+3)(4i+4)} = \frac{\pi}{4} + \frac{\log 2}{2},$$

$$\sum_{i=0}^{\infty}(-1)^i\frac{(2i+1)^3}{(2i+1)^4+4} = 0, \quad \sum_{i=0}^{\infty}(-1)^i\frac{1}{(2i+1)[(2i+1)^4+4]} = \frac{\pi}{16},$$

$$\sum_{i=1}^{\infty}(-1)^i\frac{1}{i(4i^4+1)} = \frac{1}{2} - \log 2, \quad \sum_{i=0}^{\infty}(-1)^i\left[\frac{1}{(6i+5)3^{3i+2}} + \frac{1}{(6i+7)3^{3i+3}}\right] = 1 - \frac{\log 7}{2}.$$

Source: [116, p. 60], [506, p. 25], [712], [1217, pp. 10–12], [1566, pp. 69, 104], [1568, p. 199], [1647, pp. 268, 269], [1757, p. 181], [2481], and [2577].

Fact 13.5.98. Let $a \ge 1$. Then,

$$\sum_{i=1}^{\infty}(-1)^{i+1}\frac{1}{a^i i} = \log\frac{a+1}{a}, \quad \sum_{i=0}^{\infty}(-1)^i\frac{1}{a^i(i+1)} = a\log\frac{a+1}{a}.$$

Fact 13.5.99. Let $a \ge 1$. Then

$$\sum_{i=0}^{\infty}(-1)^i\frac{1}{a^i(2i+1)} = \sqrt{a}\,\mathrm{acot}\,\sqrt{a}.$$

In particular,

$$\sum_{i=0}^{\infty}(-1)^i\frac{1}{2i+1} = \frac{\pi}{4}, \quad \sum_{i=0}^{\infty}(-1)^i\frac{1}{2^i(2i+1)} = \sqrt{2}\,\mathrm{acot}\,\sqrt{2}, \quad \sum_{i=0}^{\infty}(-1)^i\frac{1}{3^i(2i+1)} = \frac{\sqrt{3}\pi}{6}.$$

Fact 13.5.100.

$$\sum_{i=1}^{\infty}\frac{i-\sqrt{i^2-1}}{\sqrt{i(i+1)}} = 1, \quad \sum_{i=1}^{\infty}\frac{1}{(\sqrt{i}+\sqrt{i+1})\sqrt{i(i+1)}} = 1.$$

Source: [1566, p. 63].

Fact 13.5.101.

$$\sum_{i=1}^{\infty}\log\frac{i^2+2i+1}{i^2+2i} = \log 2, \quad \sum_{i=1}^{\infty}\log\frac{i(2i+1)}{(i+1)(2i-1)} = \log 2,$$

$$\sum_{i=1}^{\infty}\log\frac{(i+1)(3i+1)}{i(3i+4)} = \log\frac{4}{3}, \quad \sum_{i=1}^{\infty}\left(i\log\frac{2i+1}{2i-1} - 1\right) = \frac{1}{2}(1-\log 2),$$

$$\sum_{i=1}^{\infty}(-1)^i\frac{\log i}{i} = \gamma\log 2 - \frac{1}{2}\log^2 2, \quad \sum_{i=1}^{\infty}(-1)^{i+1}\log\left(1+\frac{1}{i}\right) = \log\frac{\pi}{2},$$

$$\sum_{i=1}^{\infty}(-1)^i \log\left(1-\frac{1}{(i+1)^2}\right)=\log\frac{\pi^2}{8},\quad \lim_{n\to\infty}\sum_{i=0}^{n}\left(\frac{1}{2i+1}-\frac{1}{2}\log n\right)=\frac{\gamma}{2}+\log 2.$$

Furthermore, for all $n\ge 1$,

$$\sum_{i=1}^{\infty}(-1)^{i-1}\frac{\log^n i}{i}=\frac{\log^{n+1}2}{n+1}-\sum_{i=0}^{n-1}\binom{n}{i}\gamma_i\log^{n-i}2.$$

Source: [968, p. 68] and [1566, pp. 64, 69, 70]. **Remark:** γ_i is defined in Fact 13.3.1.

Fact 13.5.102. Let x be a real number such that $x>1$. Then,

$$\sum_{i=1}^{\infty}\frac{x^i}{\prod_{j=1}^{i}(1+x^j)}=1.$$

Source: Fact 13.10.1 and [1566, p. 117].

Fact 13.5.103. Let $n\ge 1$. Then,

$$\sum_{i=1}^{\infty}\frac{1}{\sum_{j=1}^{i}j^3}=\frac{4\pi^2}{3}-12,\quad \sum_{i=1}^{\infty}(-1)^{i+1}\frac{1}{\sum_{j=1}^{i}j^3}=12-16\log 2.$$

Source: [108, pp. 34, 203–205]. **Related:** Fact 1.12.28.

13.6 Facts on Series of Trigonometric and Hyperbolic Functions

Fact 13.6.1. Let $x\in\mathbb{C}$ and $n\ge 0$. Then,

$$\sin nx=n\sin x+\sum_{i=1}^{\infty}(-1)^i\frac{n\prod_{j=1}^{i}[n^2-(2j-1)^2]}{(2i+1)!}\sin^{2i+1}x.$$

Source: [1745, 1756]. **Remark:** If n is odd, then the series is finite.

Fact 13.6.2.

$$\sum_{i=1}^{\infty}\frac{\sin\frac{i\pi}{2}}{i^2}=-\int_0^{\pi/2}\log\left(2\sin\frac{x}{2}\right)\mathrm{d}x=G.$$

If $x\in(0,2\pi)$, then

$$\sum_{i=1}^{\infty}\frac{\sin ix}{i}=\frac{\pi-x}{2},\quad \sum_{i=1}^{\infty}\frac{\cos ix}{i}=-\log\left(2\sin\frac{x}{2}\right),$$

$$\sum_{i=1}^{\infty}\frac{\sin ix}{i^2}=-\int_0^{x}\log\left(2\sin\frac{t}{2}\right)\mathrm{d}t,\quad \sum_{i=1}^{\infty}\frac{\cos ix}{i^2}=\frac{1}{4}x^2-\frac{\pi}{2}x+\frac{\pi^2}{6},$$

$$\sum_{i=1}^{\infty}\frac{\sin ix}{i^3}=\frac{1}{12}x^3-\frac{\pi}{4}x^2+\frac{\pi^2}{6}x,\quad \sum_{i=1}^{\infty}\frac{\cos ix}{i^4}=-\frac{1}{48}x^4+\frac{\pi}{12}x^3-\frac{\pi^2}{12}x^2+\frac{\pi^4}{90},$$

$$\sum_{i=1}^{\infty}\frac{\sin ix}{i^5}=-\frac{1}{240}x^5+\frac{\pi}{48}x^4-\frac{\pi^2}{36}x^3+\frac{\pi^4}{90}x.$$

If $x\in(0,2\pi)$ and $k\ge 1$, then

$$\sum_{i=1}^{\infty}\frac{\sin ix}{i^{2k+1}}=(-1)^{k+1}\frac{4^k\pi^{2k+1}}{(2k+1)!}B_{2k+1}[x/(2\pi)]=(-1)^k\frac{\pi}{2(2k)!}x^{2k}+\sum_{i=0}^{k}(-1)^i\frac{\zeta(2k-2i)}{(2i+1)!}x^{2i+1},$$

$$\sum_{i=1}^{\infty}\frac{\cos ix}{i^{2k}}=(-1)^{k+1}\frac{4^k\pi^{2k}}{2(2k)!}B_{2k}[x/(2\pi)]=(-1)^k\frac{\pi}{2(2k-1)!}x^{2k-1}+\sum_{i=0}^{k}(-1)^i\frac{\zeta(2k-2i)}{(2i)!}x^{2i}.$$

If $x \in (-\pi, \pi)$, then

$$\sum_{i=1}^{\infty}(-1)^i\frac{\sin ix}{i} = -\frac{x}{2}, \quad \sum_{i=1}^{\infty}(-1)^i\frac{\cos ix}{i} = -\log\left(2\cos\frac{x}{2}\right),$$

$$\sum_{i=1}^{\infty}(-1)^i\frac{\sin ix}{i^2} = \int_0^{\pi-x}\log\left(2\sin\frac{t}{2}\right)\mathrm{d}t, \quad \sum_{i=1}^{\infty}(-1)^i\frac{\cos ix}{i^2} = \frac{1}{4}x^2 - \frac{\pi^2}{12},$$

$$\sum_{i=1}^{\infty}(-1)^i\frac{\sin ix}{i^3} = \frac{1}{12}x^3 - \frac{\pi^2}{12}x.$$

If $x \in (0, \pi)$, then

$$\sum_{i=1}^{\infty}\frac{\sin(2i-1)x}{2i-1} = \frac{\pi}{4}, \quad \sum_{i=1}^{\infty}\frac{\sin(2i-1)x}{(2i-1)^2} = -\frac{1}{2}\int_0^{x}\log\left(2\sin\frac{t}{2}\right)\mathrm{d}t - \frac{1}{2}\int_0^{\pi-x}\log\left(2\sin\frac{t}{2}\right)\mathrm{d}t,$$

$$\sum_{i=1}^{\infty}\frac{\sin(2i-1)x}{(2i-1)^3} = -\frac{\pi}{8}x^2 + \frac{\pi^2}{8}x,$$

$$\sum_{i=1}^{\infty}\frac{\cos(2i-1)x}{2i-1} = \frac{1}{2}\log\cot\frac{x}{2}, \quad \sum_{i=1}^{\infty}\frac{\cos(2i-1)x}{(2i-1)^2} = -\frac{\pi}{4}x + \frac{\pi^2}{8}.$$

If $n \ge 0$ and $x \in (n\pi, (n+1)\pi)$, then

$$\sum_{i=1}^{\infty}\frac{\sin(2i-1)x}{2i-1} = (-1)^n\frac{\pi}{4}.$$

If $x \in [-\frac{\pi}{2}, \frac{\pi}{2}]$, then

$$\sum_{i=1}^{\infty}(-1)^{i+1}\frac{\cos(2i-1)x}{2i-1} = \frac{\pi}{4}, \quad \sum_{i=1}^{\infty}(-1)^{i+1}\frac{\sin(2i-1)x}{(2i-1)^2} = \frac{\pi}{4}x,$$

$$\sum_{i=1}^{\infty}(-1)^{i+1}\frac{\cos(2i-1)x}{(2i-1)^3} = -\frac{\pi}{8}x^2 + \frac{\pi^3}{32}.$$

If $x \in [-\pi, \pi]$ and $\alpha \in \mathbb{R}\backslash\mathbb{Z}$, then

$$\sum_{i=1}^{\infty}(-1)^i\frac{\cos ix}{\alpha^2 - i^2} = \frac{\pi\cos\alpha x}{2\alpha\sin\alpha\pi} - \frac{1}{2\alpha^2}.$$

Source: [116, p. 371], [154, p. 338], [672], [968, p. 210], and [2163, pp. 324–326]. **Remark:** The indefinite integral is *Clausen's integral.* Expressions for Catalan's constant G are given in Fact 14.9.1.

Fact 13.6.3.

$$\sum_{i=1}^{\infty}\frac{\sin i}{i} = \sum_{i=1}^{\infty}\frac{\sin^2 i}{i^2} = \frac{\pi-1}{2}, \quad \sum_{i=1}^{\infty}\frac{\sin 2i}{i} = \frac{\pi-2}{2}, \quad \sum_{i=1}^{\infty}\frac{\sin 3i}{i} = \frac{\pi-3}{2},$$

$$\sum_{i=1}^{\infty}\frac{\sin^2 i}{i^4} = \frac{(\pi-1)^2}{6}, \quad \sum_{i=1}^{\infty}\frac{\sin^2 2i}{i^2} = \pi - 2, \quad \sum_{i=1}^{\infty}(-1)^{i+1}\frac{\sin i}{i} = \sum_{i=1}^{\infty}(-1)^{i+1}\frac{\sin^2 i}{i^2} = \frac{1}{2},$$

$$\sum_{i=1}^{\infty}\frac{\sin^3 i}{i} = \sum_{i=1}^{\infty}\frac{\sin^4 i}{i^2} = \frac{\pi}{4}, \quad \sum_{i=1}^{\infty}\frac{\sin^5 i}{i} = \sum_{i=1}^{\infty}\frac{\sin^6 i}{i^2} = \frac{3\pi}{16}, \quad \sum_{i=1}^{\infty}\frac{\sin^2 2i}{16i^4} = \frac{(\pi-2)^2}{24},$$

$$\sum_{i=1}^{\infty} \frac{\sin(2i-1)}{2i-1} = \sum_{i=1}^{\infty} \frac{\sin^2(2i-1)}{(2i-1)^2} = \sum_{i=1}^{\infty} (-1)^{i-1} \frac{\sin(2i-1)}{(2i-1)^2} = \sum_{i=1}^{\infty} (-1)^{i-1} \frac{1}{2i-1} = \frac{\pi}{4},$$

$$\sum_{i=1}^{\infty} \frac{\sin^2(2i-1)}{(2i-1)^4} = \frac{3\pi^2 - 4\pi}{24}, \quad \sum_{i=0}^{\infty} \frac{1}{2^i} \tan \frac{\pi}{2^{i+2}} = \frac{4}{\pi}.$$

If a, b are real numbers such that $b < a < (2\pi - b)/3$, then

$$\sum_{i=1}^{\infty} \frac{(\sin^3 ai) \cos bi}{i} = \frac{\pi}{4}, \quad \sum_{i=1}^{\infty} \frac{(\sin^3 ai) \cos bi}{i^3} = \frac{3a^2\pi - b^2\pi - 4a^3}{8}.$$

If $n \in \{0, 1, 2, 3\}$, then

$$\sum_{i=1}^{\infty} \frac{(\sin^n i) \sin 3i}{i^{n+1}} = \frac{\pi - 3}{2}.$$

If $n \in \{1, 2, 3\}$, then

$$\sum_{i=1}^{\infty} \frac{(\sin^n i) \cos i}{i^n} = \frac{\pi - 2}{4}.$$

If $x \in [0, 1]$, then

$$\sum_{i=1}^{\infty} \frac{(\sin i) \sin ix}{i^2} = \frac{(\pi - 1)x}{2}.$$

If $x \in [1, 2\pi - 1)$, then

$$\sum_{i=1}^{\infty} \frac{(\sin i) \sin ix}{i^2} = \frac{\pi - x}{2}.$$

If $x \in (1, \pi - 1)$, then

$$\sum_{i=1}^{\infty} \frac{\sin 2ix}{2i} = \sum_{i=1}^{\infty} \frac{(\sin 2i) \sin 2ix}{4i^2} = \frac{\pi - 2x}{4},$$

$$\sum_{i=1}^{\infty} \frac{\sin (2i-1)x}{2i-1} = \sum_{i=1}^{\infty} \frac{[\sin(2i-1)] \sin (2i-1)x}{(2i-1)^2} = \frac{\pi}{4},$$

$$\sum_{i=1}^{\infty} (-1)^{i+1} \frac{\sin ix}{i} = \sum_{i=1}^{\infty} (-1)^{i+1} \frac{(\sin i) \sin ix}{i^2} = \frac{x}{2}.$$

If $x \in (0, \frac{2\pi}{3})$, then

$$\sum_{i=1}^{\infty} \frac{\sin^3 ix}{i} = \frac{\pi}{4}.$$

If $x \in [0, \frac{\pi}{2}]$, then

$$\sum_{i=1}^{\infty} \frac{\sin^4 ix}{i^2} = \frac{\pi x}{4}.$$

If $x \in (0, \frac{2\pi}{5})$, then

$$\sum_{i=1}^{\infty} \frac{\sin^5 ix}{i} = \frac{3\pi}{16}.$$

If $x \in [0, \frac{\pi}{3}]$, then

$$\sum_{i=1}^{\infty} \frac{\sin^6 ix}{i^2} = \frac{3\pi x}{16}.$$

If $x \in (-\pi, \pi)$, then

$$\sum_{i=1}^{\infty}(-1)^{i+1}\frac{\sin ix}{i} = \frac{x}{2}, \quad \sum_{i=1}^{\infty}(-1)^{i+1}\frac{\cos ix}{i^2} = \frac{\pi^2 - 3x^2}{12}.$$

Source: [216], [516, p. 316], [968, p. 240], and [1647, p. 268]. **Remark:** These equalities are Fourier series.

Fact 13.6.4.

$$\sum_{i=0}^{\infty}(-1)^{\lfloor i/2\rfloor}\frac{1}{2i+1} = \frac{\sqrt{2}\pi}{4}, \quad \sum_{i=0}^{\infty}\frac{1}{(2i+1)^2} = \frac{\pi^2}{8}.$$

Source: [216].

Fact 13.6.5. Let $x \in \mathbb{R}$. Then,

$$\sum_{i=1}^{\infty} 3^{i-1}\sin^3\frac{x}{3^i} = \frac{1}{4}(x - \sin x), \quad \sum_{i=0}^{\infty}\frac{1}{3^i}\sin^3\frac{x}{3^{1-i}} = \frac{3}{4}\sin\frac{x}{3}, \quad \sum_{i=0}^{\infty}(-1)^i\frac{\cos^3 3^i x}{3^i} = \frac{3}{4}\cos x,$$

$$\sum_{i=1}^{\infty} 4^i\sin^4\frac{x}{2^i} = x^2 - \sin^2 x, \quad |\sin x| = \frac{2}{\pi} - \frac{4}{\pi}\sum_{i=1}^{\infty}\frac{\cos 2ix}{4i^2-1} = \frac{8}{\pi}\sum_{i=1}^{\infty}\frac{\sin^2 ix}{4i^2-1}.$$

If $2x/\pi$ is not an integer, then

$$\sum_{i=0}^{\infty}\frac{1}{2^i}\tan\frac{x}{2^i} = \frac{1}{x} - 2\cot 2x = \frac{1}{x} + \tan x - \cot x, \quad \sum_{i=1}^{\infty}\frac{1}{2^i}\tan\frac{x}{2^i} = \frac{1}{x} - \cot x.$$

If $|x| < \pi$, then

$$\sum_{i=1}^{\infty} 2^i\left(1 - \cos\frac{x}{2^i}\right)^2\csc\frac{x}{2^{i-1}} = \tan\frac{x}{2} - \frac{x}{2}, \quad \sum_{i=1}^{\infty}\left(1 - \cos\frac{x}{2^i}\right)\csc\frac{x}{2^{i-1}} = \frac{1}{2}\tan\frac{x}{2}.$$

Source: [311], [516, p. 213], [968, p. 236], [1566, p. 66], and [1757, p. 173].

Fact 13.6.6. Let $x \in (0,1)$ and $\theta \in (0, 2\pi)$. Then,

$$\sum_{i=1}^{\infty}\frac{x^i\sin i\theta}{i} = \operatorname{atan}\frac{x\sin\theta}{1 - x\cos\theta}, \quad \sum_{i=1}^{\infty}\frac{x^i\cos i\theta}{i} = -\log\sqrt{x^2 - 2(\cos\theta)x + 1}.$$

Source: [1757, p. 189].

Fact 13.6.7. Let $a \in (-1,1)$ and $x \in \mathbb{R}$. Then,

$$\sum_{i=-\infty}^{\infty} a^{|i|}e^{ix\jmath} = 1 + 2\sum_{i=1}^{\infty} a^i\cos ix = \frac{1-a^2}{1+a^2-2a\cos x} = \operatorname{Re}\frac{1+ae^{x\jmath}}{1-ae^{x\jmath}} \ge 0,$$

$$\sum_{i=1}^{\infty} a^i\cos ix = \frac{a\cos x - a^2}{1+a^2-2a\cos x}.$$

Source: [116, pp. 243, 601]. **Related:** Fact 14.4.54.

Fact 13.6.8. If x be a real number. Then,

$$\sum_{i=1}^{\infty}\operatorname{atan}\frac{x}{i^2+i+x^2} = \operatorname{atan} x, \quad \sum_{i=1}^{\infty}\operatorname{atan}\frac{4ix}{i^4+4+x^2} = \operatorname{atan} x + \operatorname{atan}\frac{x}{2}.$$

In particular,

$$\sum_{i=1}^{\infty}\operatorname{atan}\frac{1}{i^2+i+1} = \frac{\pi}{4}, \quad \sum_{i=1}^{\infty}\operatorname{atan}\frac{\sqrt{3}}{i^2+i+3} = \frac{\pi}{3},$$

$$\sum_{i=1}^{\infty} \operatorname{atan} \frac{4i}{i^2+5} = \frac{\pi}{4} + \operatorname{atan} \frac{1}{2}, \quad \sum_{i=1}^{\infty} \operatorname{atan} \frac{4\sqrt{2}i}{i^2+6} = \frac{\pi}{2}.$$

Source: [512, 2180]. **Related:** Fact 2.16.9.

Fact 13.6.9. Let a, b be real numbers, and assume that $a \geq 0$ and $a + b \geq 0$. Then,

$$\sum_{i=1}^{\infty} \operatorname{atan} \frac{a}{a^2 i^2 + a(a+2b)i + 1 + ab + b^2} = \frac{\pi}{2} - \operatorname{atan}(a+b).$$

In particular,

$$\sum_{i=1}^{\infty} \operatorname{atan} \frac{1}{i^2+i+1} = \frac{\pi}{4}, \quad \sum_{i=1}^{\infty} \operatorname{atan} \frac{2}{(2i+1)^2} = \frac{\pi}{2}, \quad \sum_{i=1}^{\infty} \operatorname{atan} \frac{1}{i^2-i+1} = \frac{\pi}{2}.$$

Source: [512], [1566, p. 68], and [2180].

Fact 13.6.10. Let a, b, c be real numbers, and assume that $b^2 \leq 4ac$. Then,

$$\sum_{i=1}^{\infty} \operatorname{atan} \frac{2ai + a + b}{a^2 i^4 + 2a(a+b)i^3 + (a^2 + 3ab + b^2 + 2ac)i^2 + (ab + b^2 + 2ac + 2bc)i + 1 + ac + bc + c^2}$$
$$= \frac{\pi}{2} - \operatorname{atan}(a+b+c).$$

In particular,

$$\sum_{i=1}^{\infty} \operatorname{atan} \frac{8ai}{4a^2 i^4 + a^2 + 4} = \frac{\pi}{2} - \operatorname{atan} \frac{a}{2}, \quad \sum_{i=1}^{\infty} \operatorname{atan} \frac{2ai}{a^2 i^4 - a^2 i^2 + 1} = \frac{\pi}{2},$$

$$\sum_{i=1}^{\infty} \operatorname{atan} \frac{8i}{4i^4+5} = \frac{\pi}{2} - \operatorname{atan} \frac{1}{2}, \quad \sum_{i=1}^{\infty} \operatorname{atan} \frac{8i}{i^4 - 2i^2 + 5} = \frac{\pi}{2} + \operatorname{atan} 2.$$

Source: [512], [1566, p. 68], and [2068, p. 349].

Fact 13.6.11. Let $x \in \mathbb{R}$. Then,

$$\sum_{i=0}^{\infty} (-1)^i \operatorname{atan} \frac{x}{2i+1} = \frac{\pi}{4} - \operatorname{atan} e^{-\frac{1}{2}\pi x} = \frac{1}{2} \operatorname{atan} \sinh \frac{\pi x}{2}.$$

If $n \geq 1$, then

$$\sum_{i=0}^{\infty} (-1)^i (2i+1) \operatorname{atan} \frac{8n^2}{(2i+1)^2} = (-1)^{n-1} 4n \operatorname{atan} e^{-\pi n}.$$

Furthermore,

$$\sum_{i=0}^{\infty} (-1)^i \operatorname{atan} \frac{2}{i^2} = -\frac{\pi}{4}.$$

Source: [1311, pp. 276, 277] and [1317, p. 42].

Fact 13.6.12. Let $x, y \in \mathbb{R}$, and assume that $x \neq 0$ and, for all $n \geq 1$, $n^2 + y^2 \neq x^2$. Then,

$$\sum_{i=1}^{\infty} \operatorname{atan} \frac{2xy}{i^2 + y^2 - x^2} = \operatorname{atan} \frac{y}{x} - \operatorname{atan} \frac{\tanh \pi y}{\tan \pi x}.$$

In particular,

$$\sum_{i=1}^{\infty} \operatorname{atan} \frac{2x^2}{i^2} = \frac{\pi}{4} - \operatorname{atan} \frac{\tanh \pi x}{\tan \pi x}, \quad \sum_{i=1}^{\infty} \operatorname{atan} \frac{1}{2i^2} = \frac{\pi}{4}, \quad \sum_{i=1}^{\infty} \operatorname{atan} \frac{2}{i^2} = \frac{3\pi}{4},$$

$$\sum_{i=1}^{\infty} \operatorname{atan} \frac{2}{(i+1)^2} = \frac{\pi}{4} + \operatorname{atan} \frac{1}{2}, \quad \sum_{i=1}^{\infty} \operatorname{atan} \frac{1}{i^2} = \frac{\pi}{4} - \operatorname{atan} \frac{\tanh \frac{\sqrt{2}\pi}{2}}{\tan \frac{\sqrt{2}\pi}{2}} = \operatorname{atan} \frac{\tan \frac{\sqrt{2}\pi}{2} - \tanh \frac{\sqrt{2}\pi}{2}}{\tan \frac{\sqrt{2}\pi}{2} + \tanh \frac{\sqrt{2}\pi}{2}}.$$

Furthermore,

$$\sum_{i=1}^{\infty} \frac{1}{2^i} \operatorname{atan} \frac{\sinh 2^i x}{\sin 2^i x} = \operatorname{atan} \frac{\tanh x}{\tan x}, \quad \sum_{i=1}^{\infty} (-1)^{i+1} \operatorname{atan} \frac{2x^2}{i^2} = -\frac{\pi}{4} + \operatorname{atan} \frac{\sinh \pi x}{\sin \pi x},$$

$$\sum_{i=1}^{\infty} \frac{i^2}{i^4 + 4x^4} = \frac{\pi(\sin 2\pi x - \sinh 2\pi x)}{4x(\cos 2\pi x - \cosh 2\pi x)}, \quad \sum_{i=1}^{\infty} \frac{i^2}{i^4 + 4} = \frac{\pi \coth \pi}{4},$$

$$\sum_{i=1}^{\infty} \frac{i^2}{i^4 + i^2 x^2 + x^4} = \frac{\pi(\sinh \sqrt{3}\pi x - \sqrt{3} \sin \pi x)}{2\sqrt{3}x(\cosh \sqrt{3}\pi x - \cos \pi x)}.$$

Source: [506, pp. 105–107], [512], [1566, p. 68], and [2068, p. 349]. **Remark:** Note that $\lim_{x\uparrow 1} \operatorname{atan} \frac{\tanh \pi x}{\tan \pi x} = -\frac{\pi}{2}$.

Fact 13.6.13.

$$\sum_{i=1}^{\infty} \operatorname{asin} \frac{2(i^2 - i + 1)}{(i^2+1)(i^2 - 2i + 2)} = \pi, \quad \sum_{i=1}^{\infty} \operatorname{csch} 2^i \pi = \coth \pi - 1,$$

$$\sum_{i=1}^{\infty} \operatorname{acos} \frac{1 + \sqrt{4i^2 - 1}}{2\sqrt{i^2 + i}} = \frac{\pi}{4}, \quad \sum_{i=1}^{\infty} \operatorname{acos} \frac{1 + \sqrt{(i^2 + 2i)(i^2 + 4i + 3)}}{(i+1)(i+2)} = \frac{\pi}{6}.$$

Source: [512, 2145, 2180, 2573].

Fact 13.6.14. If $x \in (-1, 1)$,

$$\sum_{i=1}^{\infty} \operatorname{atanh} \frac{x}{i^2 + i - x^2} = \operatorname{atanh} x.$$

Source: [2180]. **Related:** Fact 2.16.9.

Fact 13.6.15. Let $n \ge 1$. Then,

$$\sum_{i=1}^{\infty} \frac{\coth i\pi}{i^{2n+1}} = 4^n \pi^{2n+1} \sum_{i=0}^{n+1} (-1)^{i+1} \frac{B_{2i} B_{2n-2i+2}}{(2i)!(2n-2i+2)!}.$$

Consequently,

$$\sum_{i=1}^{\infty} \frac{\coth i\pi}{i^{4n-1}} = \frac{1}{2}(2\pi)^{4n-1} \sum_{i=0}^{2n} (-1)^{i+1} \frac{B_{2i} B_{4n-2i}}{(2i)!(4n-2i)!}, \quad \sum_{i=1}^{\infty} \frac{\coth i\pi}{i^{4n+1}} = 0.$$

In particular,

$$\sum_{i=1}^{\infty} \frac{\coth i\pi}{i^3} = \frac{7\pi^3}{180}, \quad \sum_{i=1}^{\infty} \frac{\coth i\pi}{i^5} = 0, \quad \sum_{i=1}^{\infty} \frac{\coth i\pi}{i^7} = \frac{19\pi^7}{56700},$$

$$\sum_{i=1}^{\infty} \frac{\coth i\pi}{i^9} = 0, \quad \sum_{i=1}^{\infty} \frac{\coth i\pi}{i^{11}} = \frac{1453\pi^{11}}{425675250}.$$

Source: [645, p. 99], [1311, p. 285], and [1317, p. xxvi]. **Related:** Fact 13.1.6.

Fact 13.6.16. If $x \ne 0$, then

$$\sum_{i=1}^{\infty} \operatorname{csch} 2^i x = \frac{2}{e^{2x} - 1}.$$

In particular,

$$\sum_{i=0}^{\infty} \frac{1}{\sinh 2^i} = \coth 1 - 1.$$

Source: [2159].

Fact 13.6.17.

$$\sum_{i=0}^{\infty} (-1)^i \frac{1}{(2i+1)^5 \cosh \frac{(2i+1)\pi}{2}} = \frac{\pi^5}{768}.$$

Now, let n be an odd integer. Then,

$$\sum_{i=0}^{\infty} (-1)^i \frac{1}{(2i+1)[\cos \frac{(2i+1)\pi}{2n} + \cosh \frac{(2i+1)\pi}{2n}]} = \frac{\pi}{8}.$$

Source: [1317, pp. xxvi, xxviii].

Fact 13.6.18. Let $z \in \mathbb{C}$, and assume that $z \neq 0$. Then,

$$\sum_{i=0}^{\infty} \left(\frac{z}{\sinh \frac{z}{2^i}} - 2^i \right) = 1 - \frac{z}{\tanh z}.$$

Furthermore,

$$\sum_{i=0}^{\infty} \frac{2^i - \coth \frac{1}{2^i}}{2^i \sinh \frac{1}{2^i}} = \frac{1 + 4e^2 - e^4}{1 - 2e^2 + e^4}, \quad \sum_{i=0}^{\infty} \frac{2 - 4^i + \operatorname{csch}^2 \frac{1}{2^i} - \operatorname{sech}^2 \frac{1}{2^i}}{4^i \sinh 2^{1-i}} = \frac{e^{12} - 17e^8 - 17e^4 + 1}{e^{12} - 3e^8 + 3e^4 - 1}.$$

Source: [512].

Fact 13.6.19. Let $z \in \mathbb{C}$, and assume that $\frac{z}{2\pi}$ is not an integer. Then,

$$\lim_{n\to\infty} \frac{1}{n} \sum_{j=1}^{n} \sum_{i=1}^{j} \cos iz = -\frac{1}{2}.$$

Source: Use Fact 2.16.9.

13.7 Facts on Series of Binomial Coefficients

Fact 13.7.1. Let p be a real number, and assume that $|p| > 4$. Then,

$$\sum_{i=0}^{\infty} \frac{\binom{2i}{i}}{p^i} = \sqrt{\frac{p}{p-4}}, \quad \sum_{i=0}^{\infty} \frac{\binom{4i}{2i}}{p^{2i}} = \frac{1}{2}\left(\sqrt{\frac{p}{p-4}} + \sqrt{\frac{p}{p+4}} \right).$$

In particular,

$$\sum_{i=0}^{\infty} \frac{\binom{2i}{i}}{8^i} = \sqrt{2}, \quad \sum_{i=0}^{\infty} (-1)^i \frac{\binom{2i}{i}}{8^i} = \sqrt{\frac{2}{3}}, \quad \sum_{i=0}^{\infty} \frac{\binom{4i}{2i}}{8^{2i}} = \frac{3\sqrt{2} + \sqrt{6}}{6}.$$

Source: Use the power series for $1/\sqrt{1-4z}$ given by Fact 13.4.2. See [1781].

Fact 13.7.2. Let p be a real number, and assume that $|p| \geq 4$. Then,

$$\sum_{i=1}^{\infty} \frac{\binom{2i}{i}}{ip^i} = 2\log\left[\frac{p}{2}\left(1 - \sqrt{\frac{p-4}{p}} \right) \right], \quad \sum_{i=1}^{\infty} \frac{\binom{4i}{2i}}{ip^{2i}} = 2\log\left[\frac{p^2}{4}\left(1 - \sqrt{\frac{p-4}{p}} \right)\left(\sqrt{\frac{p+4}{p}} - 1 \right) \right].$$

In particular,

$$\sum_{i=1}^{\infty} \frac{\binom{2i}{i}}{i4^i} = \log 4, \quad \sum_{i=1}^{\infty} (-1)^i \frac{\binom{2i}{i}}{i4^i} = 2\log 2(\sqrt{2} - 1), \quad \sum_{i=1}^{\infty} \frac{\binom{4i}{2i}}{i4^{2i}} = 2\log 4(\sqrt{2} - 1).$$

Source: Use the power series for $2\log\left(\frac{1-\sqrt{1-4z}}{2z}\right)$ given by Fact 13.4.2. See [1781].

Fact 13.7.3. Let p be a real number, and assume that $|p| \ge 4$. Then,

$$\sum_{i=0}^{\infty} \frac{\binom{2i}{i}}{(i+1)p^i} = \frac{p}{2}\left(1 - \sqrt{\frac{p-4}{p}}\right), \quad \sum_{i=0}^{\infty} \frac{\binom{4i}{2i}}{(2i+1)p^{2i}} = \frac{p}{4}\left(\sqrt{\frac{p+4}{p}} - \sqrt{\frac{p-4}{p}}\right).$$

In particular,

$$\sum_{i=0}^{\infty} \frac{\binom{2i}{i}}{(i+1)7^i} = \frac{7-\sqrt{21}}{2}, \quad \sum_{i=0}^{\infty} (-1)^i \frac{\binom{2i}{i}}{(i+1)7^i} = \frac{\sqrt{77}-7}{2}, \quad \sum_{i=0}^{\infty} \frac{\binom{4i}{2i}}{(2i+1)7^{2i}} = \frac{\sqrt{77}-\sqrt{21}}{4}.$$

Source: Use the power series for $(1-\sqrt{1-4z})/(2z)$ given by Fact 13.4.2. See [1781].

Fact 13.7.4. Let p be a real number, and assume that $|p| \ge 1$. Then,

$$\sum_{i=0}^{\infty} \frac{\binom{2i}{i}}{4^i(2i+1)p^{2i}} = p \operatorname{asin} \frac{1}{p}.$$

In particular,

$$\sum_{i=0}^{\infty} \frac{\binom{2i}{i}}{4^i(2i+1)} = \frac{\pi}{2}, \quad \sum_{i=0}^{\infty} \frac{3^i\binom{2i}{i}}{16^i(2i+1)} = \frac{2\sqrt{3}\pi}{9}, \quad \sum_{i=0}^{\infty} \frac{\binom{2i}{i}}{16^i(2i+1)} = \frac{\pi}{3}.$$

Source: Use the power series for $\operatorname{asin} z$ given by Fact 13.4.9.

Fact 13.7.5. Let p be a real number, and assume that $|p| > 1$. Then,

$$\sum_{i=1}^{\infty} \frac{4^i}{i\binom{2i}{i}p^{2i}} = \frac{2\operatorname{asin}\frac{1}{|p|}}{\sqrt{p^2-1}}.$$

In particular,

$$\sum_{i=1}^{\infty} \frac{1}{i\binom{2i}{i}} = \frac{\sqrt{3}\pi}{9}, \quad \sum_{i=1}^{\infty} \frac{3^i}{i\binom{2i}{i}} = \frac{2\sqrt{3}\pi}{3}, \quad \sum_{i=1}^{\infty} \frac{2^i}{i\binom{2i}{i}} = \frac{\pi}{2}, \quad \sum_{i=1}^{\infty} \frac{1}{i\binom{2i}{i}} = \frac{\sqrt{3}\pi}{9}.$$

Source: Use the power series for $(z\operatorname{asin} z)/\sqrt{1-z^2}$ given by Fact 13.4.9.

Fact 13.7.6. Let p be a real number, and assume that $|p| > \frac{1}{2}$. Then,

$$\sum_{i=1}^{\infty} \frac{1}{\binom{2i}{i}p^{2i}} = \frac{\sqrt{4p^2-1} + 4p^2\operatorname{asin}\frac{1}{2|p|}}{(4p^2-1)^{3/2}}.$$

In particular,

$$\sum_{i=1}^{\infty} \frac{1}{\binom{2i}{i}} = \frac{1}{3} + \frac{2\sqrt{3}\pi}{27}.$$

Source: Use the power series for $\frac{z^2\sqrt{4-z^2}+4z\operatorname{asin}(z/2)}{(4-z^2)\sqrt{4-z^2}}$ given by Fact 13.4.9.

Fact 13.7.7. Let $k \ge 0$, and define

$$f_k(z) \triangleq \sum_{i=1}^{\infty} \frac{i^k}{\binom{2i}{i}} z^i.$$

Then, the radius of convergence of f is 4. Furthermore,

$$\sum_{i=1}^{\infty} \frac{1}{\binom{2i}{i}} = \frac{1}{3} + \frac{2\sqrt{3}\pi}{27}, \quad \sum_{i=1}^{\infty} \frac{i}{\binom{2i}{i}} = \frac{2}{3} + \frac{2\sqrt{3}\pi}{27}, \quad \sum_{i=1}^{\infty} \frac{i^2}{\binom{2i}{i}} = \frac{4}{3} + \frac{10\sqrt{3}\pi}{81}, \quad \sum_{i=1}^{\infty} \frac{i^3}{\binom{2i}{i}} = \frac{10}{3} + \frac{74\sqrt{3}\pi}{243},$$

$$\sum_{i=1}^{\infty} \frac{2^i}{\binom{2i}{i}} = 1 + \frac{\pi}{2}, \quad \sum_{i=1}^{\infty} \frac{2^i i}{\binom{2i}{i}} = 3 + \pi, \quad \sum_{i=1}^{\infty} \frac{2^i i^2}{\binom{2i}{i}} = 11 + \frac{7\pi}{2}, \quad \sum_{i=1}^{\infty} \frac{2^i i^3}{\binom{2i}{i}} = 55 + \frac{35\pi}{2},$$

$$\sum_{i=1}^{\infty} \frac{3^i}{\binom{2i}{i}} = 3 + \frac{4\sqrt{3}\pi}{3}, \quad \sum_{i=1}^{\infty} \frac{3^i i}{\binom{2i}{i}} = 18 + \frac{20\sqrt{3}\pi}{3},$$

$$\sum_{i=1}^{\infty} \frac{3^i i^2}{\binom{2i}{i}} = 156 + \frac{172\sqrt{3}\pi}{3}, \quad \sum_{i=1}^{\infty} \frac{3^i i^3}{\binom{2i}{i}} = 1890 + \frac{2084\sqrt{3}\pi}{3}.$$

Source: [971, 1781], **Remark:** A closed-form expression for $f_k(z)$ is given in [971].

Fact 13.7.8.

$$\sum_{i=1}^{\infty} \frac{1}{i\binom{2i}{i}} = \frac{\sqrt{3}\pi}{9}, \quad \sum_{i=1}^{\infty} \frac{1}{i^2\binom{2i}{i}} = \frac{\pi^2}{18}, \quad \sum_{i=1}^{\infty} \frac{1}{i^4\binom{2i}{i}} = \frac{17\pi^4}{3240}, \quad \sum_{i=0}^{\infty} \frac{2^i}{\binom{2i}{i}} = \frac{\pi}{2} + 2,$$

$$\sum_{i=1}^{\infty} \frac{2^i}{i\binom{2i}{i}} = \frac{\pi}{2}, \quad \sum_{i=1}^{\infty} \frac{2^i}{i^2\binom{2i}{i}} = \frac{\pi^2}{8}, \quad \sum_{i=1}^{\infty} \frac{2^i}{i^3\binom{2i}{i}} = \pi G - \frac{35\zeta(3)}{16} + \frac{\pi^2 \log 2}{8},$$

$$\sum_{i=2}^{\infty} \frac{1}{(i-1)\binom{2i}{i}} = \frac{1}{2} - \frac{\sqrt{3}\pi}{18}, \quad \sum_{i=2}^{\infty} \frac{1}{(i-1)^2\binom{2i}{i}} = \frac{\sqrt{3}\pi}{6} + \frac{\pi^2}{36} - 1, \quad \sum_{i=0}^{\infty} \frac{1}{(i+1)\binom{2i}{i}} = \frac{4\sqrt{3}\pi}{9} - \frac{\pi^2}{9},$$

$$\sum_{i=0}^{\infty} \frac{1}{(2i+1)\binom{2i}{i}} = \frac{2\sqrt{3}\pi}{9}, \quad \sum_{i=0}^{\infty} \frac{1}{(2i+1)^2\binom{2i}{i}} = \frac{8}{3}G - \frac{\pi}{3}\log(2+\sqrt{3}),$$

$$\sum_{i=1}^{\infty} \frac{3^i}{\binom{2i}{i}} = 3 + \frac{4\sqrt{3}\pi}{3}, \quad \sum_{i=1}^{\infty} \frac{3^i i}{\binom{2i}{i}} = 18 + \frac{20\sqrt{3}\pi}{3}, \quad \sum_{i=1}^{\infty} \frac{3^i i^2}{\binom{2i}{i}} = 156 + \frac{172\sqrt{3}\pi}{3},$$

$$\sum_{i=1}^{\infty} \frac{3^i}{i\binom{2i}{i}} = \frac{2\sqrt{3}\pi}{3}, \quad \sum_{i=1}^{\infty} \frac{3^i}{i^2\binom{2i}{i}} = \frac{2\pi^2}{9}, \quad \sum_{i=1}^{\infty} \frac{(5-\sqrt{5})^i}{2^i i\binom{2i}{i}} = \frac{2\pi\sqrt{5-2\sqrt{5}}}{5}, \quad \sum_{i=1}^{\infty} \frac{\binom{3i}{i}}{8^i} = \frac{3\sqrt{5}}{5},$$

$$\sum_{i=1}^{\infty} \frac{\binom{3i}{i}}{9^i} = 2\cos\frac{\pi}{9} - 1, \quad \sum_{i=1}^{\infty} \frac{\binom{3i}{i}}{27^i} = \frac{2\sqrt{3}\cos\frac{\pi}{18}}{3} - 1, \quad \sum_{i=1}^{\infty} \frac{1}{2^i\binom{3i}{i}} = \frac{2}{25} - \frac{6\log 2}{125} + \frac{11\pi}{250},$$

$$\sum_{i=0}^{\infty} \frac{i}{2^i\binom{3i}{i}} = \frac{81}{625} - \frac{18\log 2}{3125} + \frac{79\pi}{3125}, \quad \sum_{i=0}^{\infty} \frac{50i-6}{2^i\binom{3i}{i}} = \pi + 6, \quad \sum_{i=1}^{\infty} \frac{1}{2^i i\binom{3i}{i}} = \frac{\pi}{10} - \frac{\log 2}{5},$$

$$\sum_{i=1}^{\infty} \frac{1}{2^i i^2\binom{3i}{i}} = \frac{\pi^2}{24} - \frac{\log^2 2}{2}, \quad \sum_{i=0}^{\infty} \frac{1}{(2i+1)^2\binom{2i}{i}} = \frac{8G}{3} - \frac{\pi}{3}\log(2+\sqrt{3}), \quad \sum_{i=4}^{\infty}\sum_{j=2}^{i-2} \frac{1}{\binom{i}{j}} = \frac{3}{2},$$

$$\sum_{i=1}^{\infty} \frac{1}{\binom{3i}{i}} = \left(\frac{36\sqrt{23}\tau}{529(\tau^2-\tau+1)} + \frac{18\sqrt{3}\tau(\tau^2+1)}{23(\tau^2-\tau+1)^2}\right)\operatorname{atan}\frac{\sqrt{3}}{2\tau-1}$$
$$+ \left(\frac{9\tau(t^4-2\tau^3-2\tau+1)}{23(\tau^3+1)^2} + \frac{6\sqrt{69}\tau(\tau-1)}{529(\tau^3+1)}\right)\log\frac{\tau^3+1}{(\tau+1)^3} + \frac{108\tau^3}{23(\tau^3+1)},$$

where $\tau \triangleq \left(\frac{25}{2} + \frac{3}{2}\sqrt{69}\right)^{1/3}$,

$$\sum_{i=1}^{\infty} \frac{1}{i^2\binom{3i}{i}} = 6\left(\operatorname{atan} \frac{\sqrt{3}}{1-(100+12\sqrt{69})^{1/3}}\right)^2 - \frac{1}{2}\log^2 \frac{12(9+\sqrt{69})}{[2+(100+12\sqrt{69})^{1/3}]^3},$$

$$\sum_{i=1}^{\infty} \frac{6^i}{\binom{3i}{i}} = 2[240+96(2^{1/3})+75(2^{2/3})]^{1/2} \operatorname{atan} \frac{\sqrt{3}}{2^{4/3}-1} + 2^{1/3}[4(2^{1/3})-5]\log(2^{1/3}-1) + 8,$$

$$\sum_{i=1}^{\infty} \frac{6^i}{i\binom{3i}{i}} = 2^{4/3}(1+2^{1/3})\sqrt{3}\operatorname{atan} \frac{\sqrt{3}}{2^{4/3}-1} - 2^{1/3}(1-2^{1/3})\log(2^{1/3}-1),$$

$$\sum_{i=1}^{\infty} \frac{6^i}{i^2\binom{3i}{i}} = 6\operatorname{atan}^2 \frac{\sqrt{3}}{2^{4/3}-1} - \frac{1}{2}\log^2(2^{1/3}-1), \quad \sum_{i=1}^{\infty} \frac{27^i}{4^i i^2\binom{3i}{i}} = \frac{2\pi^2}{3} - 2\log^2 2,$$

$$\sum_{i=0}^{\infty} \frac{1}{4^i(2i+1)^2}\binom{2i}{i} = \frac{\pi \log 2}{2}, \quad \sum_{i=0}^{\infty} \frac{1}{16^i(2i+1)^3}\binom{2i}{i} = \frac{7\pi^3}{216}, \quad \sum_{i=1}^{\infty} \frac{1}{4^i(i+1)}\binom{2i}{i} = 1,$$

$$\sum_{i=1}^{\infty} \frac{1}{4^i i}\binom{2i}{i} = \log 4, \quad \sum_{i=1}^{\infty} \frac{1}{4^i(2i+1)}\binom{2i}{i} = \frac{\pi}{2} - 1, \quad \sum_{i=1}^{\infty} \frac{4^i}{i^2\binom{2i}{i}} = \frac{\pi^2}{2}, \quad \sum_{i=1}^{\infty} \frac{4^i}{i^3\binom{2i}{i}} = \pi^2 \log 2 - \frac{7\zeta(3)}{2},$$

$$\sum_{i=2}^{\infty} \frac{4^i}{(i-1)^2\binom{2i}{i}} = \pi^2 - 4, \quad \sum_{i=2}^{\infty} \frac{1}{i^2(i^2-1)\binom{2i}{i}} = \frac{11}{8} - \frac{\sqrt{3}\pi}{4}, \quad \sum_{i=2}^{\infty} \frac{4^i}{i^2(i^2-1)\binom{2i}{i}} = 4 - \frac{3\pi^2}{8},$$

$$\sum_{i=1}^{\infty} \frac{H_{2i+2} - \frac{1}{2}H_{i+1}}{4^i(2i+1)}\binom{2i}{i} = \pi \log 2 - 1, \quad \sum_{i=0}^{\infty} \frac{1}{\binom{4i}{2i}} = \frac{16}{15} + \frac{\sqrt{3}\pi}{27} - \frac{2\sqrt{5}}{25}\log \tfrac{1}{2}(1+\sqrt{5}),$$

$$\sum_{i=0}^{\infty}(-1)^i \frac{1}{\binom{4i}{2i}} = \frac{16}{17} + \frac{4\sqrt{34}(\sqrt{17}-2)}{289\sqrt{\sqrt{17}-1}} \operatorname{atan} \frac{\sqrt{2}}{\sqrt{\sqrt{17}-1}} + \frac{2\sqrt{34}(\sqrt{17}+2)}{289\sqrt{\sqrt{17}+1}} \log \frac{\sqrt{\sqrt{17}+1} - \sqrt{2}}{\sqrt{\sqrt{17}+1} + \sqrt{2}},$$

$$\sum_{i=1}^{\infty}(-1)^{i+1} \frac{1}{4^i i}\binom{2i}{i} = \log \tfrac{1}{2}(1+\sqrt{2}), \quad \sum_{i=1}^{\infty}(-1)^i \frac{4^i}{i\binom{2i}{i}} = \sqrt{2}\log(\sqrt{2}-1), \quad \sum_{i=0}^{\infty}(-1)^i \frac{1}{8^i(2i+1)^2}\binom{2i}{i} = \frac{\pi^2}{10},$$

$$\sum_{i=1}^{\infty}(-1)^i \frac{4^i}{i^2\binom{2i}{i}} = \sqrt{2}\log^2(\sqrt{2}-1), \quad \sum_{i=0}^{\infty}(-1)^i \frac{4^i}{(2i+1)\binom{2i}{i}} = -\frac{\sqrt{2}}{2}\log(\sqrt{2}-1),$$

$$\sum_{i=0}^{\infty}(-1)^i \frac{4^i}{(i+1)\binom{2i}{i}} = -\sqrt{2}\log(\sqrt{2}-1) - \log^2(\sqrt{2}-1), \quad \sum_{i=2}^{\infty}(-1)^i \frac{4^i}{(i-1)\binom{2i}{i}} = -3\sqrt{2}\log(\sqrt{2}-1) - 2,$$

$$\sum_{i=2}^{\infty}(-1)^i \frac{4^{i-1}}{(i-1)^2\binom{2i}{i}} = \sqrt{2}\log(\sqrt{2}-1) + \log^2(\sqrt{2}-1) + 1,$$

$$\sum_{i=1}^{\infty}(-1)^{i+1} \frac{1}{\binom{2i}{i}} = \frac{4\sqrt{5}\log \frac{1}{2}(1+\sqrt{5})}{25} + \frac{1}{5}, \quad \sum_{i=1}^{\infty}(-1)^{i+1} \frac{i}{\binom{2i}{i}} = \frac{4\sqrt{5}\log \frac{1}{2}(1+\sqrt{5})}{125} + \frac{6}{25},$$

$$\sum_{i=1}^{\infty}(-1)^{i+1} \frac{i^2}{\binom{2i}{i}} = \frac{4}{25} - \frac{4\sqrt{5}\log \frac{1}{2}(1+\sqrt{5})}{125}, \quad \sum_{i=1}^{\infty}(-1)^{i+1} \frac{i^3}{\binom{2i}{i}} = \frac{56\sqrt{5}\log \frac{1}{2}(1+\sqrt{5})}{625} + \frac{2}{125},$$

$$\sum_{i=1}^{\infty}(-1)^{i+1} \frac{1}{i\binom{2i}{i}} = \frac{2\sqrt{5}\log \frac{1}{2}(1+\sqrt{5})}{5}, \quad \sum_{i=1}^{\infty}(-1)^{i+1} \frac{1}{i^2\binom{2i}{i}} = 2\log^2 \tfrac{1}{2}(1+\sqrt{5}) = 2\operatorname{acsch}^2 2,$$

$$\sum_{i=1}^{\infty}(-1)^{i+1}\frac{1}{i^3\binom{2i}{i}}=\frac{2\zeta(3)}{5},\quad \sum_{i=0}^{\infty}(-1)^{i}\frac{1}{(i+1)\binom{2i}{i}}=\frac{8\sqrt{5}}{5}\log\tfrac{1}{2}(1+\sqrt{5})-4\log^2\tfrac{1}{2}(1+\sqrt{5}),$$

$$\sum_{i=0}^{\infty}(-1)^{i}\frac{1}{(2i+1)\binom{2i}{i}}=\frac{4\sqrt{5}}{5}\log\tfrac{1}{2}(1+\sqrt{5}),\quad \sum_{i=2}^{\infty}(-1)^{i}\frac{1}{(i-1)\binom{2i}{i}}=\frac{3\sqrt{5}}{5}\log\tfrac{1}{2}(1+\sqrt{5})-\frac{1}{2},$$

$$\sum_{i=2}^{\infty}(-1)^{i}\frac{1}{(i-1)^2\binom{2i}{i}}=1-\sqrt{5}\log\tfrac{1}{2}(1+\sqrt{5})+\log^2\tfrac{1}{2}(1+\sqrt{5}),$$

$$\sum_{i=2}^{\infty}(-1)^{i}\frac{1}{i^2(i^2-1)\binom{2i}{i}}=4\log^2\tfrac{1}{2}(1+\sqrt{5})-\frac{\sqrt{5}}{2}\log\tfrac{1}{2}(1+\sqrt{5})-\frac{3}{8},$$

$$\sum_{i=2}^{\infty}(-1)^{i}\frac{4^i}{i^2(i^2-1)\binom{2i}{i}}=\frac{5}{2}\log^2(\sqrt{2}-1)-\sqrt{2}\log(\sqrt{2}-1)-3,$$

$$\sum_{i=0}^{\infty}(-1)^{i}\frac{1}{16^i(2i+1)^2}\binom{2i}{i}=\frac{\pi^2}{10},\quad \sum_{i=0}^{\infty}(-1)^{i}\frac{1}{32^i(2i+1)^2}\binom{2i}{i}=\sqrt{2}\left(\frac{\pi^2}{12}-\frac{\log^2 2}{4}\right),$$

$$\sum_{i=1}^{\infty}(-1)^{i+1}\frac{2^i}{\binom{2i}{i}}=\frac{3+\sqrt{3}\log(2+\sqrt{3})}{9},\quad \sum_{i=1}^{\infty}(-1)^{i+1}\frac{2^i}{i\binom{2i}{i}}=\frac{\sqrt{3}\log(2+\sqrt{3})}{3},$$

$$\sum_{i=1}^{\infty}(-1)^{i+1}\frac{2^i i^2}{\binom{2i}{i}}=\frac{3-\sqrt{3}\log(2+\sqrt{3})}{27},\quad \sum_{i=1}^{\infty}(-1)^{i+1}\frac{2^i i^3}{\binom{2i}{i}}=\frac{15+\sqrt{3}\log(2+\sqrt{3})}{81},$$

$$\sum_{i=1}^{\infty}(-1)^{i+1}\frac{1}{i^3\binom{2i}{i}}=\frac{2}{5}\zeta(3)=\frac{1}{5}\sum_{i=1}^{\infty}\frac{H_i}{i^2}=\frac{2}{5}\sum_{i=1}^{\infty}\frac{H_i}{(i+1)^2},$$

$$\sum_{i=1}^{\infty}(-1)^{i+1}\frac{2^i i}{\binom{2i}{i}}=\frac{1}{3},\quad \sum_{i=1}^{\infty}(-1)^{i+1}\frac{205i^2-160i+32}{i^5\binom{2i}{i}^5}=2\sum_{i=1}^{\infty}\frac{1}{i^3},$$

$$\sum_{i=1}^{\infty}(-1)^{i}\frac{1}{4^i\binom{3i}{i}}=\frac{39\sqrt{7}}{784}\operatorname{acot}(2\sqrt{3}+\sqrt{7})-\frac{1}{28}-\frac{3}{32}\log 2,$$

$$\sum_{i=1}^{\infty}(-1)^{i}\frac{1}{4^i i^2\binom{3i}{i}}=6[\operatorname{acot}(2\sqrt{3}+\sqrt{7})]^2-\frac{1}{2}\log^2 2,$$

$$\sum_{i=0}^{\infty}\frac{\binom{i+3}{i}}{\binom{3i+5}{3i}}=\frac{100\sqrt{3}\pi}{243}-\frac{10}{9},\quad \sum_{i=0}^{\infty}(-1)^{i}\frac{\binom{i+3}{i}}{\binom{3i+5}{3i}}=\frac{10}{9}+\frac{40\log 2}{27}-\frac{160\sqrt{3}\pi}{729},$$

$$\sum_{i=0}^{\infty}\frac{\binom{i+\frac{7}{2}}{i}}{\binom{2i+9}{2i}}=\frac{9\pi}{2}-\frac{456}{35},\quad \sum_{i=0}^{\infty}(-1)^{i}\frac{\binom{i+\frac{7}{2}}{i}}{\binom{2i+9}{2i}}=9\log(1+\sqrt{2})+\frac{2559\sqrt{2}}{35}-\frac{552}{5},$$

$$\sum_{i=0}^{\infty}(-1)^{i}\frac{(4i+1)}{2^{6i}}\binom{2i}{i}^3=\frac{2}{\pi},\quad \sum_{i=0}^{\infty}(-1)^{i}\frac{(20i+3)}{2^{10i}}\binom{2i}{i}^2\binom{4i}{2i}=\frac{8}{\pi},\quad \sum_{i=0}^{\infty}\frac{(42i+5)}{2^{12i}}\binom{2i}{i}^3=\frac{16}{\pi},$$

$$\sum_{i=0}^{\infty}\frac{\binom{6i}{3i}\binom{3i}{i}}{2(2i+1)\binom{2i}{i}108^i}=\frac{3\sqrt{3}}{8},\quad \sum_{i=0}^{\infty}\frac{\binom{6i}{3i}\binom{3i}{i}}{2(2i+1)(2i+3)\binom{2i}{i}108^i}=\frac{3\sqrt{3}}{8},$$

$$\lim_{n\to\infty}\sum_{i=0}^{n}\frac{1}{\binom{n}{i}}=\lim_{n\to\infty}\frac{n+1}{2^n}\sum_{i=0}^{n}\frac{2^i}{i+1}=2,\quad \lim_{n\to\infty}\sum_{i=0}^{n}(-1)^i\sqrt{\binom{n}{i}}=\lim_{n\to\infty}\sum_{i=0}^{n}(-1)^i\sqrt[3]{\binom{n}{i}}=0,$$

$$2\sum_{i=1}^{\infty}(-1)^{i+1}\frac{1}{i^5\binom{2i}{i}}-\frac{5}{2}\sum_{i=2}^{\infty}(-1)^{i+1}\frac{H_{i-1,2}}{i^3\binom{2i}{i}}=\zeta(5),\quad \frac{5}{2}\sum_{i=1}^{\infty}(-1)^{i+1}\frac{1}{i^7\binom{2i}{i}}+\frac{25}{2}\sum_{i=2}^{\infty}(-1)^{i+1}\frac{H_{i-1,4}}{i^3\binom{2i}{i}}=\zeta(7).$$

Let $n > m \geq 1$. Then,

$$\sum_{i=0}^{\infty}\frac{\binom{i+m-1}{i}}{\binom{i+n}{i}}=\frac{n}{n-m}.$$

In particular,

$$\sum_{i=0}^{\infty}\frac{1}{\binom{i+n}{i}}=\frac{n}{n-1},\quad \sum_{i=0}^{\infty}\frac{\binom{i+n-2}{i}}{\binom{i+n}{i}}=n.$$

Let $n \geq 2$. Then,

$$\sum_{i=1}^{\infty}\frac{1}{\binom{n+i}{i}}=\frac{1}{n-1},\quad \sum_{i=n}^{\infty}\frac{1}{\binom{i}{n}}=\frac{n}{n-1},\quad \sum_{i=1}^{\infty}\frac{H_i}{\binom{n+i}{i}}=\frac{n}{(n-1)^2}.$$

Let $n \geq 1$. Then,

$$\sum_{i=1}^{\infty}\frac{1}{i\binom{n+i}{i}}=\frac{1}{n},\quad \sum_{i=1}^{\infty}\frac{1}{i^2\binom{n+i}{i}}=\frac{\pi^2}{6}-H_{n,2},\quad \sum_{i=0}^{\infty}2^i\frac{\binom{2n+i}{n+i+1}}{\binom{2n+2i+1}{n+i+1}}=\frac{(2n-1)!!\pi}{(2i-2)!!4},\quad \sum_{i=1}^{\infty}\frac{1}{\binom{i+n}{n+1}}=\frac{n+1}{n}.$$

Furthermore,

$$\sum_{i=1}^{\infty}\frac{1}{i\binom{i+1}{2}}=\frac{\pi^2}{3}-\frac{1}{2},\quad \sum_{i=1}^{\infty}\frac{1}{i\binom{i+2}{3}}=\frac{\pi^2}{2}-\frac{15}{4},\quad \sum_{i=1}^{\infty}\frac{1}{i\binom{i+3}{4}}=\frac{2\pi^2}{3}-\frac{49}{9},\quad \sum_{i=1}^{\infty}\frac{1}{i\binom{i+4}{5}}=\frac{5\pi^2}{6}-\frac{1025}{144},$$

$$\sum_{i=1}^{\infty}\frac{1}{i^2\binom{i+1}{2}}=2\zeta(3)+2-\frac{\pi^2}{3},\quad \sum_{i=1}^{\infty}\frac{1}{i^2\binom{i+2}{3}}=3\zeta(3)+\frac{39}{8}-\frac{3\pi^2}{4},\quad \sum_{i=1}^{\infty}\frac{1}{i^2\binom{i+3}{4}}=4\zeta(3)+\frac{449}{54}-\frac{11\pi^2}{9},$$

$$\sum_{i=1}^{\infty}\frac{1}{i^2\binom{i+4}{5}}=5\zeta(3)+\frac{21035}{1728}-\frac{125\pi^2}{72},\quad \sum_{i=1}^{\infty}\frac{1}{i^3\binom{i+2}{3}}=\frac{7\pi^2}{8}+\frac{\pi^4}{30}-\frac{9}{2}\zeta(3)-\frac{87}{16},$$

$$\sum_{i=1}^{\infty}\frac{1}{i^3\binom{1+i}{i}}=\zeta(3)+1-\frac{\pi^2}{6},\quad \sum_{i=1}^{\infty}\frac{1}{i^3\binom{2+i}{i}}=\zeta(3)+\frac{13}{8}-\frac{\pi^2}{4},\quad \sum_{i=1}^{\infty}\frac{1}{i^3\binom{3+i}{i}}=\zeta(3)+\frac{449}{216}-\frac{11\pi^2}{36},$$

$$\sum_{i=1}^{\infty}\frac{1}{i^4\binom{1+i}{i}}=\frac{\pi^2}{6}+\frac{\pi^4}{90}-1-\zeta(3),\quad \sum_{i=1}^{\infty}\frac{1}{i^4\binom{2+i}{i}}=\frac{7\pi^2}{24}+\frac{\pi^4}{90}-\frac{29}{16}-\frac{3\zeta(3)}{2},$$

$$\sum_{i=1}^{\infty}\frac{1}{i^4\binom{4i+2}{2}^3}=\zeta(4)+864\zeta(3)+32\pi^3+840\zeta(2)+1024(\pi+G)+5216\log 2-11184,$$

$$\sum_{i=1}^{\infty}\frac{1}{i^5\binom{1+i}{i}}=\zeta(3)+\zeta(5)+1-\frac{\pi^2}{6}-\frac{\pi^4}{90},\quad \sum_{i=1}^{\infty}\frac{1}{i^5\binom{2+i}{i}}=\frac{7}{4}\zeta(3)+\zeta(5)+\frac{61}{32}-\frac{5\pi^2}{16}-\frac{\pi^4}{60},$$

$$\sum_{i=1}^{\infty}\frac{H_i}{i^4\binom{4i+3}{3}}=3\zeta(5)-\zeta(3)\zeta(2)-\frac{55}{6}\zeta(4)+\frac{680}{9}\zeta(3)+\frac{1312}{9}\zeta(2)-\frac{2560}{9}\pi\log 2-\frac{2480}{3}\log^2 2+\frac{20480}{27}G,$$

$$\sum_{i=1}^{\infty}\frac{H_i}{i^3\binom{4i+3}{3}^2}=\frac{\pi^4}{72}-\frac{3602}{3}\zeta(3)-\frac{52}{3}\pi^3-\frac{8}{3}\pi^2+(130\pi^2-32\pi+312\log 2)\log 2+\frac{64}{3}(7\pi+4+39\log 2)G,$$

$$\sum_{i=1}^{\infty}\frac{H_i}{i^4\binom{i+5}{5}}=3\zeta(5)-\zeta(3)\zeta(2)-\frac{137}{48}\zeta(4)+\frac{12019}{1800}\zeta(3)-\frac{874853}{216000}\zeta(2)+\frac{131891}{172800},$$

$$\sum_{i=1}^{\infty}\frac{H_i}{i^4\binom{i+5}{5}^2}=3\zeta(5)-\zeta(3)\zeta(2)-\frac{137}{24}\zeta(4)+\frac{311383}{3600}\zeta(3)+\frac{2296919}{108000}\zeta(2)-\frac{642641}{4800},$$

$$\sum_{i=1}^{\infty}\frac{H_i}{i^4\binom{i+5}{5}^3}=3\zeta(5)-\zeta(3)\zeta(2)-\frac{3209}{120}\zeta(4)+\frac{269701}{900}\zeta(3)+\frac{11312047}{216000}\zeta(2)-\frac{41259977}{34560},$$

$$\sum_{i=1}^{\infty}\frac{H_i}{\binom{i+2}{2}^3}=120-56\zeta(3)-\frac{16\pi^2}{3},\quad \sum_{i=1}^{\infty}\frac{iH_i}{\binom{i+2}{2}^3}=88\zeta(3)+\frac{28\pi^2}{3}+\frac{\pi^4}{45}-200,$$

$$\sum_{i=1}^{\infty}\frac{i^2H_i}{\binom{i+2}{2}^3}=328-136\zeta(3)-16\pi^2-\frac{\pi^4}{15},\quad \sum_{i=1}^{\infty}\frac{i^3H_i}{\binom{i+2}{2}^3}=208\zeta(3)-528+\frac{80\pi^2}{3}+\frac{7\pi^4}{45}.$$

If $n\in\{0,1,2\}$, then

$$\sum_{i=1}^{\infty}\frac{i^nH_i}{\binom{i+2}{2}^3}=(-2)^n[\tfrac{1}{4}n(n-5)\zeta(4)-(n^2-13n+56)\zeta(3)+(4n-32)\zeta(2)+(n^2-21n+120)].$$

Let $n\ge 1$. Then,

$$\sum_{i=0}^{\infty}(-1)^i\frac{1}{\binom{n+i}{i}}=n2^{n-1}(\log 2-H_{n-1})-n\sum_{i=1}^{n-1}(-1)^i\binom{n-1}{i}\frac{2^{n-1-i}}{i}.$$

In particular,

$$\sum_{i=0}^{\infty}(-1)^i\frac{1}{i+1}=\log 2,\quad \sum_{i=0}^{\infty}(-1)^i\frac{1}{(i+1)(i+2)}=2\log 2-1.$$

Let $n\ge 1$. Then,

$$\sum_{i=0}^{\infty}\frac{1}{2^i\binom{n+i}{i}}=2n(-1)^{n-1}\left(\log 2+\sum_{i=1}^{n-1}\frac{1+(-2)^i\binom{n-1}{i}}{i}\right).$$

In particular,

$$\sum_{i=0}^{\infty}\frac{1}{2^i(i+1)}=2\log 2,\quad \sum_{i=0}^{\infty}\frac{1}{2^{i-1}(i+1)(i+2)}=4-4\log 2.$$

Let $n\ge 1$. Then,

$$\sum_{i=0}^{\infty}(-1)^i\frac{1}{\prod_{j=1}^{n+1}(i+j)}=\frac{2^n\log 2}{n!}+\frac{2^n}{n!}\sum_{i=1}^{n}(-1)^i\frac{2^i-1}{2^ii}\binom{n}{i}.$$

In particular,

$$\sum_{i=0}^{\infty}(-1)^i\frac{1}{(i+1)(i+2)}=2\log 2-1.$$

Let $n \geq 2$, and define $\omega \triangleq e^{(2\pi/n)J}$. Then,

$$\sum_{i=0}^{\infty} \frac{1}{\binom{ni}{i}} = \sum_{i=1}^{n-1} -\omega^i(1-\omega^i)^{n-1}\log(1-\omega^{-i}).$$

Let $n \geq 1$. Then,

$$\sum_{i=n}^{\infty} \frac{i}{2^{i+2}}\binom{n}{i} = n + \frac{1}{2}, \quad \sum_{i=n}^{\infty} \frac{i^2+i}{2^{i+3}}\binom{n}{i} = (n+1)^2.$$

Furthermore,

$$\lim_{n\to\infty} \sum_{i=1}^{n} (-4)^i \frac{\binom{2n-i}{n}}{\binom{2n}{n}} = \frac{1}{3}, \quad \sum_{i=0}^{\infty} \frac{1054i+233}{480^i}\binom{2i}{i}\sum_{j=0}^{i}(-1)^j 8^{2j-i}\binom{i}{j}^2\binom{2j}{i} = \frac{520}{\pi}.$$

Source: [42, 69], [217, pp. 163, 164, 167], [324, 408], [506, pp. 127, 136, 137], [511, p. 56], [516, p. 239], [700], [517, pp. 20, 25, 26, 126, 136–139], [520, 521], [570, pp. 166, 167, 175], [1207, 1563], [1566, p. 56], [1675, pp. 72, 84, 86, 87], [1781], [2013, p. 178], [2325, 2481, 2484, 2485, 2486, 2509, 2571, 2577, 3021]. **Credit:** The equality involving $(4i+1)\binom{2i}{i}^3$ and the next two equalities are due to S. A. Ramanujan.

Fact 13.7.9. Let $n \geq 1$, and let α_n be the positive root of $x^n - x^{n-1} - \cdots - x - 1 = 0$. Then,

$$\alpha_n = 2 - 2\sum_{i=1}^{\infty} \frac{1}{2^{i(n+1)}i}\binom{i(n+1)-2}{i-1}, \quad \frac{1}{\alpha_n} = \frac{1}{2} + \frac{1}{2}\sum_{i=1}^{\infty} \frac{1}{2^{i(n+1)}i}\binom{i(n+1)}{i-1},$$

$$\frac{1}{2-\alpha_n} = 2^n - \frac{n}{2} - \frac{1}{2}\sum_{i=1}^{\infty} \frac{1}{2^{i(n+1)}i}\binom{i(n+1)}{i-1}.$$

In particular, $\alpha_2 = \frac{1}{2}(1+\sqrt{5})$,

$$\alpha_3 = \frac{1}{3}\left(1 + \sqrt[3]{19+3\sqrt{33}} + \sqrt[3]{19-3\sqrt{33}}\right) \approx 1.83929,$$

$$\alpha_4 = \frac{1}{4}\left(1 + a + \sqrt{11 - a^2 + \frac{26}{a}}\right) \approx 1.92756,$$

where

$$a \triangleq \frac{\sqrt{3}}{3}\sqrt{11 + 2\sqrt[3]{12\sqrt{1689}-260} - 2\sqrt[3]{12\sqrt{1689}+260}}.$$

Therefore,

$$\sum_{i=1}^{\infty} \frac{1}{8^i i}\binom{3i-2}{i-1} = \frac{1}{4}(3-\sqrt{5}), \quad \sum_{i=1}^{\infty} \frac{1}{8^i i}\binom{3i}{i-1} = \sqrt{5}-2, \quad \sum_{i=1}^{\infty} \frac{1}{8^i i}\binom{3i}{i+1} = 3-\sqrt{5},$$

$$\sum_{i=1}^{\infty} \frac{1}{16^i i}\binom{4i-2}{i-1} = \frac{1}{6}\left(5 - \sqrt[3]{19+3\sqrt{33}} - \sqrt[3]{19-3\sqrt{33}}\right).$$

Source: [1320, 1854].

Fact 13.7.10. Let $z, w \in \mathbb{C}$, and assume that $z \notin -\mathbb{N}$ and $\mathrm{Re}(z-w) > 1$. Then,

$$\sum_{i=1}^{\infty} \frac{\binom{w+i}{i}}{\binom{z+i}{i}} \sum_{j=1}^{i} \frac{1}{z+j} = \frac{w+1}{(z-w-1)^2}, \quad \sum_{i=1}^{\infty} \frac{\binom{w+i}{i}}{\binom{z+i}{i}} \sum_{1\leq j\leq l\leq i} \frac{1}{z^2+(j+l)z+jl} = \frac{w+1}{(z-w-1)^3}.$$

Source: [2516].

Fact 13.7.11. Let x be a real number, assume that $0 < |x| < 27/4$, and let y be the real root of $(y-1)^2 y = 1/x$; that is,

$$y = \frac{1}{3}\left(\frac{27 - 2x + 3\sqrt{81 - 12x}}{2x}\right)^{1/3} + \frac{1}{3}\left(\frac{27 - 2x + 3\sqrt{81 - 12x}}{2x}\right)^{-1/3} + \frac{2}{3}.$$

Then,

$$\sum_{i=0}^{\infty} \frac{x^i}{(3i+1)\binom{3i}{i}} = \frac{y(y-1)}{3y-1}\left(\frac{3}{2}\log\left|\frac{y}{y-1}\right| + \frac{3y-2}{\sqrt{3y^2-4y}}\left[\operatorname{atan}\frac{y}{\sqrt{3y^2-4y}} + \operatorname{atan}\frac{2-y}{\sqrt{3y^2-4y}}\right]\right).$$

Source: [324, 326].

13.8 Facts on Double-Summation Series

Fact 13.8.1.

$$\lim_{n\to\infty} \frac{1}{n}\sum_{i=2}^{n}\sum_{j=2}^{i}(-1)^j \log j = \frac{1}{2}\log\frac{\pi}{2}.$$

Source: [1675, p. 55].

Fact 13.8.2.

$$\sum_{i=1}^{\infty}\sum_{j=1}^{i} \frac{1}{j(j+1)(i-j+1)!} = e - 1.$$

Source: [1566, p. 103].

Fact 13.8.3.

$$\sum_{i=1}^{\infty}\sum_{j=1}^{2i-1}(-1)^{n+i}\frac{1}{i^2 j} = \pi G - \frac{27}{16}\zeta(3), \quad \sum_{i=1}^{\infty}\sum_{j=1}^{2i-1}(-1)^{i-1}\frac{1}{i^2 j} = \pi G - \frac{29}{16}\zeta(3).$$

Source: [2513, p. 235].

Fact 13.8.4. Let z be a complex number, and assume that $|z| < 1$. Then,

$$\sum_{i,j=1}^{\infty}(-1)^{i+j}\frac{z^{i+j}}{i+j} = \log(z+1) + \frac{1}{z+1} - 1.$$

Source: [1757, p. 182].

Fact 13.8.5.

$$\sum_{i,j=-\infty}^{\infty}(-1)^{i+j}\frac{1}{i^2 + (3j+1)^2} = \frac{2\pi}{9}\log(2\sqrt{3} - 2).$$

Source: [3022].

Fact 13.8.6.

$$\sum_{i,j=1}^{\infty}(-1)^{i+j}\frac{1}{ij(i+j)} = \frac{1}{4}\zeta(3), \quad \sum_{i,j=1}^{\infty}(-1)^{i}\frac{1}{ij(i+j)} = \frac{5}{8}\zeta(3).$$

Source: [2513, p. 228].

Fact 13.8.7.

$$\sum_{i,j=0}^{\infty}\frac{1}{i^6(i^2+j^2)} = \frac{13\pi^8}{28350}, \quad \sum_{i,j=0}^{\infty}\frac{1}{i^2(i^2 - ij + j^2)} = \frac{\sqrt{3}\pi^4}{30},$$

where $i = j = 0$ is excluded from both summations. **Source:** [2513, p. 229].

Fact 13.8.8. Let n be a positive integer. Then,

$$\sum \frac{1}{(\prod_{j=1}^n i_j)(1+\sum_{j=1}^n i_j)} = n!, \quad \sum \frac{1}{(\prod_{j=1}^n i_j)\sum_{j=1}^n i_j} = n!\zeta(n+1),$$

where both sums are taken over all n-tuples $(i_1,\ldots,i_n)$ of positive integers. In particular,

$$\sum_{i=1}^\infty \frac{1}{i(i+1)} = 1, \quad \sum_{i,j=1}^\infty \frac{1}{ij(i+j+1)} = 2, \quad \sum_{i,j,k=1}^\infty \frac{1}{ijk(i+j+k+1)} = 6,$$

$$\sum_{i=1}^\infty \frac{1}{i^2} = \frac{\pi^2}{6}, \quad \sum_{i,j=1}^\infty \frac{1}{ij(i+j)} = 2\zeta(3), \quad \sum_{i,j,k=1}^\infty \frac{1}{ijk(i+j+k)} = \frac{\pi^4}{15}.$$

Furthermore,

$$\sum \frac{1}{(\prod_{j=1}^n i_j)\sum_{j=1}^n i_j} = n!,$$

where the sum is taken over all n-tuples $(i_1,\ldots,i_n)$ of positive integers such that $\gcd\{i_1,\ldots,i_n\} = 1$.
Source: [1103, pp. 158, 239] and [2124].

Fact 13.8.9. Let $n \ge 1$ and $k \ge 1$. Then,

$$\sum \frac{1}{(\prod_{j=1}^n i_j)(k+\sum_{j=1}^n i_j)} = n!\sum_{i=0}^{k-1}(-1)^i\frac{1}{(i+1)^{n+1}}\binom{k-1}{i},$$

where the sum is taken over all n-tuples $(i_1,\ldots,i_n)$ of positive integers. In particular,

$$\sum_{i,j=1}^\infty \frac{1}{ij(i+j+1)} = 2, \quad \sum_{i,j=1}^\infty \frac{1}{ij(i+j+2)} = \frac{7}{4}, \quad \sum_{i,j=1}^\infty \frac{1}{ij(i+j+3)} = \frac{85}{54},$$

$$\sum_{i,j=1}^\infty \frac{1}{ij(i+j+4)} = \frac{415}{288}, \quad \sum_{i,j=1}^\infty \frac{1}{ij(i+j+5)} = \frac{12019}{9000}, \quad \sum_{i,j=1}^\infty \frac{1}{ij(i+j+6)} = \frac{13489}{10800}.$$

Source: [1103, pp. 158, 239] and [1757, p. 172].

Fact 13.8.10.

$$\sum_{i=2,j=1}^\infty \frac{1}{(2j)^i} = \log 2, \quad \sum_{i=2,j=1}^\infty \frac{1}{(3j)^i} = \frac{\log 3}{2} - \frac{\sqrt{3}\pi}{18}, \quad \sum_{i=2,j=1}^\infty \frac{1}{(4j)^i} = \frac{3\log 2}{4} - \frac{\pi}{8},$$

$$\sum_{i,j=1}^\infty \frac{1}{(4i-1)^{2j}} = \frac{\log 2}{4}, \quad \sum_{i,j=1}^\infty \frac{1}{(6i-1)^{2j}} = \frac{\log 3}{8} - \frac{\sqrt{3}\pi}{72}, \quad \sum_{i,j=1}^\infty \frac{1}{(8i-1)^{2j}} = \frac{3\log 2}{16} - \frac{\pi}{32},$$

$$\sum_{i,j=1}^\infty \frac{i!j!}{(i+j+1)!} = 1, \quad \sum_{i,j=0}^\infty \frac{1}{i!j!(i+j+1)} = \frac{1}{2}(e^2-1), \quad \sum_{i,j=0}^\infty \frac{i!j!}{(i+j+2)!} = \frac{\pi^2}{6},$$

$$\sum_{i,j=1}^\infty \frac{1}{(i+j)!} = 1, \quad \sum_{i,j=1}^\infty \frac{i^2}{(i+j)!} = \frac{5}{6}e, \quad \sum_{i,j=1}^\infty \frac{ij}{(i+j)!} = \frac{2}{3}e,$$

$$\sum_{i,j,k=1}^\infty \frac{ij}{(i+j+k)!} = \frac{5}{24}e, \quad \sum_{i,j,k=1}^\infty \frac{ijk}{(i+j+k)!} = \frac{31}{120}e.$$

If $a > -1$, then

$$\sum_{i,j=2}^\infty \frac{1}{(a+i)^j} = \frac{1}{a+1}.$$

Source: [107, pp. 47, 259], [1103, pp. 158, 159, 163, 241, 242], and [1566, pp. 110, 111].

Fact 13.8.11. Let $a > 1$. Then,

$$\sum_{i,j=1}^{\infty} \frac{i^2 j}{a^i(ja^i + ia^j)} = \frac{a^2}{2(a-1)^4}.$$

Source: [387].

Fact 13.8.12. Let x be a real number. Then,

$$\sum_{i,j=1}^{\infty} \frac{1}{(i+j)!} x^{i+j} = (x-1)e^x + 1, \quad \sum_{i,j=1}^{\infty} \frac{ij}{(i+j)!} x^{i+j} = x^2 e^x \left(\frac{x}{6} + \frac{1}{2}\right).$$

Now, let y be a real number such that $y \neq x$. Then,

$$\sum_{i,j=1}^{\infty} \frac{1}{(i+j)!} x^i y^j = \frac{x(e^y - e^x)}{y - x} + 1 - e^x.$$

Source: [1103, pp. 158, 241].

Fact 13.8.13.

$$\sum_{i=1}^{\infty} (-1)^i \frac{H_i}{i} = \frac{\log^2 2}{2} - \frac{\pi^2}{12}, \quad \sum_{i=1}^{\infty} \sum_{j=1}^{\infty} (-1)^{i+j} \frac{H_{i+j}}{i+j} = \frac{\pi^2}{12} - \frac{\log 2}{2} - \frac{\log^2 2}{2},$$

$$\sum_{i=1}^{\infty} \sum_{j=1}^{\infty} \sum_{k=1}^{\infty} (-1)^{i+j+k} \frac{H_{i+j+k}}{i+j+k} = -\frac{\pi^2}{12} + \frac{7\log 2}{8} + \frac{\log^2 2}{2} - \frac{1}{8},$$

$$\sum_{i=1}^{\infty} \sum_{j=1}^{\infty} (-1)^{i+j} \frac{H_{i+j}}{ij} = \frac{1}{6}\pi^2 \log 2 - \frac{2}{3}\log^3 2 - \frac{1}{4}\zeta(3),$$

$$\sum_{i=1}^{\infty} \sum_{j=1}^{\infty} (-1)^{i+j} \frac{H_i H_j}{i+j} = \frac{1}{4}\zeta(3) + \frac{1}{2}\log^2 2 - \frac{1}{3}\log^3 2 + \log 2 - 1,$$

$$\sum_{i=1}^{\infty} \sum_{j=1}^{\infty} (-1)^{i+j} \frac{H_i H_j}{i+j+1} = 1 - \log 2 - \frac{1}{2}\log^2 2.$$

Source: [1100, 1102] and [1103, pp. 156, 231, 232].

Fact 13.8.14.

$$\sum_{i=1}^{\infty} \sum_{j=1}^{\infty} \frac{H_i(H_{j+1} - 1)}{ij(i+j)(j+1)} = 2\zeta(5) + 4\zeta(2)\zeta(3) - 2\zeta(3) - 4\zeta(2).$$

Source: [2215].

Fact 13.8.15. Let $n \geq 2$. Then,

$$\sum_{i=1}^{\infty} \frac{n!}{i^{\overline{n+1}}}(H_{i+n} - H_n) = \sum_{j=0}^{\infty} \sum_{i=0}^{j} (-1)^i \frac{1}{(i+n+1)^2}\binom{j}{i} = \frac{1}{n^2}.$$

Source: [739].

Fact 13.8.16.

$$\sum_{i=0}^{\infty} \sum_{j=1}^{\infty} (-1)^{i+j} \frac{\log(i+j)}{i+j} = \frac{1}{2}\log\frac{\pi}{2}, \quad \sum_{i=1}^{\infty} \sum_{j=1}^{\infty} (-1)^{i+j} \frac{\log(i+j)}{i+j} = \frac{1}{2}\log\frac{\pi}{2} + \frac{1}{2}\log^2 2 - \gamma \log 2.$$

Source: [976] and [1103, pp. 156, 231]. **Remark:** A generalization is given in [1100].

Fact 13.8.17. Let $m \geq 1$. Then,

$$\lim_{n\to\infty} \frac{1}{n^{m+2}} \sum_{i=1}^{n} \sum_{j=1}^{i} j^m = \frac{1}{(m+1)(m+2)}.$$

In particular,

$$\lim_{n\to\infty} \frac{1}{n^3} \sum_{i=1}^{n} \sum_{j=1}^{i} j = \frac{1}{6}, \quad \lim_{n\to\infty} \frac{1}{n^4} \sum_{i=1}^{n} \sum_{j=1}^{i} j^2 = \frac{1}{12}.$$

Source: [1566, p. 28].

Fact 13.8.18.

$$\sum_{i=1}^{\infty} \sum_{j=i+1}^{\infty} \frac{1}{(ij)^2} = \frac{\pi^4}{120}, \quad \sum_{i=1}^{\infty} \sum_{j=i+1}^{\infty} \frac{1}{ij^2} = \zeta(3), \quad \sum_{i=1}^{\infty} \sum_{j=i+1}^{\infty} \frac{1}{ij^3} = \frac{\pi^4}{360},$$

$$\sum_{i=1}^{\infty} \sum_{j=i+1}^{\infty} \frac{1}{(ij)^3} = \frac{\zeta^2(3)}{2} - \frac{\pi^6}{1890}, \quad \sum_{i=1}^{\infty} \sum_{j=i+1}^{\infty} \frac{1}{(ij)^4} = \frac{\pi^8}{113400}, \quad \sum_{i=1}^{\infty} \sum_{j=i+1}^{\infty} \frac{1}{(ij)^5} = \frac{\zeta^2(5)}{2} - \frac{\pi^{10}}{187100}.$$

Source: [2174].

Fact 13.8.19.

$$\sum_{i=1}^{\infty} \sum_{j=i}^{\infty} \frac{1}{i^2 j} = 2\zeta(3), \quad \sum_{i=1}^{\infty} \sum_{j=i}^{\infty} \frac{1}{i^3 j} = \frac{5}{4}\zeta(4).$$

Source: [2068, p. 268].

Fact 13.8.20. Let n be a positive integer. Then,

$$\sum_{i=1}^{\infty} \frac{1}{i^2} \sum \frac{1}{j} = 2, \quad \sum_{i=1}^{\infty} \frac{1}{i^3} \sum \frac{1}{j} = \frac{5}{4},$$

where the second summation in both equalities is taken over all positive integers $j < i$ such that $\gcd\{i, j\} = 1$. **Source:** [506, pp. 127, 128] and [2068, pp. 267, 268].

Fact 13.8.21. Define $\mathcal{S} \triangleq \{n^m \colon n, m \geq 2\}$. Then,

$$\sum_{i\in\mathcal{S}} \frac{1}{i-1} = 1.$$

Source: [723], [1566, p. 112], and [1647, p. 273]. **Remark:** The summation is over distinct integers. Note that

$$1 = \sum_{i=2}^{\infty} \frac{1}{i(i-1)} = \sum_{i,j=2}^{\infty} \frac{1}{i^j} < \sum_{i,j=2}^{\infty} \frac{1}{i^j - 1} \approx 1.128.$$

Credit: C. Goldbach.

13.9 Facts on Miscellaneous Series

Fact 13.9.1. For all $n \geq 0$, let C_n denote the nth Catalan number. Then, the following statements hold:

$$\lim_{n\to\infty} \frac{C_{n+1}}{C_n} = 4, \quad \sum_{i=0}^{\infty} \frac{1}{C_i} = 2 + \frac{4\sqrt{3}\pi}{27}, \quad \sum_{i=1}^{\infty} \frac{i}{C_i} = 2 + \frac{16\sqrt{3}\pi}{81},$$

$$\sum_{i=1}^{\infty} \frac{i(i+1)}{C_{i+1}} = \frac{8}{3} + \frac{56\sqrt{3}\pi}{243}, \quad \sum_{i=0}^{\infty} (-1)^i \frac{1}{C_i} = \frac{14}{25} - \frac{24\sqrt{5}}{125} \log \tfrac{1}{2}(1+\sqrt{5}),$$

$$\sum_{i=0}^{\infty}(-1)^i\frac{2^i}{C_i} = \frac{1}{3} - \frac{\sqrt{3}}{9}\log(2+\sqrt{3}), \quad \sum_{i=0}^{\infty}(-1)^i\frac{3^i}{C_i} = \frac{10}{49} - \frac{12\sqrt{21}}{343}\log\tfrac{1}{2}(5+\sqrt{21}).$$

As $n \to \infty$,

$$C_n \sim \frac{4^n}{n^{3/2}\sqrt{\pi}}.$$

Now, let x be a real number such that $|x| < \frac{\pi}{2}$. Then,

$$\sum_{i=1}^{\infty}\frac{1}{4^i}C_{i-1}\sin^{2i}x = \tfrac{1}{2}(1-\cos x), \quad \sum_{i=1}^{\infty}\frac{C_{i-1}}{2^{2n-1}}\sin^{2i+1}x = \sin x - \tfrac{1}{2}\sin 2x,$$

$$\sum_{i=1}^{\infty}\frac{1}{4^i}(8C_{i-1}-C_i)\sin^{2i+3}x = \tfrac{1}{2}\sin 4x - 2\sin x + 5\sin^3 x.$$

In particular,

$$\sum_{i=1}^{\infty}\frac{1}{4^i}C_{i-1} = \frac{1}{2}, \quad \sum_{i=1}^{\infty}\left(\frac{3}{16}\right)^i C_{i-1} = \frac{1}{4}, \quad \sum_{i=1}^{\infty}\frac{1}{8^i}C_{i-1} = \frac{1}{2}\left(1-\frac{\sqrt{2}}{2}\right), \quad \sum_{i=1}^{\infty}\frac{1}{16^i}C_{i-1} = \frac{1}{2}\left(1-\frac{\sqrt{3}}{2}\right),$$

$$\sum_{i=1}^{\infty}\frac{8C_{i-1}-C_i}{4^i} = 3, \quad \sum_{i=1}^{\infty}\left(\frac{3}{16}\right)^i(8C_{i-1}-C_i) = \frac{5}{3}, \quad \sum_{i=1}^{\infty}\frac{8C_{i-1}-C_i}{8^i} = 1, \quad \sum_{i=1}^{\infty}\frac{18C_{i-1}-C_i}{16^i} = 2\sqrt{3}-3.$$

Source: [1678, 1744, 1746, 1748, 1756]. **Related:** Fact 1.18.4. The generating functions for $(C_i)_{i=1}^{\infty}$ and $(1/C_i)_{i=1}^{\infty}$ are given by Fact 13.4.2.

Fact 13.9.2. For all $n \ge 0$, let F_n and L_n denote the nth Fibonacci and Lucas number, respectively, and define $A \triangleq \left[\begin{smallmatrix}1&1\\1&0\end{smallmatrix}\right]$. Then, $\chi_A(s) = s^2 - s - 1$ and $\operatorname{spec}(A) = \{\alpha,\beta\}$, where $\alpha \triangleq \frac{1}{2}(1+\sqrt{5}) \approx 1.61803$ and $\beta \triangleq \frac{1}{2}(1-\sqrt{5}) \approx -0.61803$. Furthermore,

$$\alpha^2 = \alpha+1, \quad \alpha = 1+1/\alpha, \quad 2\alpha^2 = \alpha^3+1, \quad \sqrt{5}\alpha = \alpha+2 = \alpha^2+1,$$

$$\alpha^3 = 2\alpha+1, \quad \alpha^6 = 4\alpha^3+1, \quad \sqrt{5}\alpha = \tfrac{1}{2}(\alpha^3+1), \quad \sqrt{5} = 2\alpha-1, \quad \alpha = e^{(\pi/5)\jmath} + e^{-(\pi/5)\jmath},$$

$$\beta^2 = \beta+1, \quad \beta-1 = 1/\beta, \quad \alpha\beta = -1, \quad 1/\alpha = -\beta, \quad 1/\beta = -\alpha,$$

$$\alpha+\beta = 1, \quad \alpha-\beta = \sqrt{5}, \quad \alpha^2+\beta^2 = 3, \quad \alpha^2-\beta^2 = \sqrt{5}, \quad \alpha^3+\beta^3 = 4, \quad \alpha^3-\beta^3 = 2\sqrt{5},$$

$$\alpha^4+\beta^4 = 7, \quad \alpha\sqrt{3-\alpha} = \sqrt{\alpha+2}, \quad \sqrt{3-\beta} = \frac{1}{2}\sqrt{10+2\sqrt{5}}, \quad (\beta+2)^2 = 5\beta^2,$$

$$\frac{\beta}{\beta+2} = \frac{1}{\beta-\alpha}, \quad \alpha = 2\cos\frac{\pi}{5} = \cot\frac{\operatorname{atan}2}{2}, \quad \beta = 2\cos\frac{3\pi}{5}, \quad \alpha = \max_{x\in[1,\infty)}\min\left\{x, \frac{1}{x-1}\right\},$$

$$\operatorname{atan}\alpha = \operatorname{atan}1 + \frac{1}{2}\operatorname{atan}\frac{1}{2}, \quad \operatorname{atan}\alpha^3 = 2\operatorname{atan}1 - \frac{1}{2}\operatorname{atan}\frac{1}{2}.$$

For all $k \ge 0$,

$$\operatorname{atan}\alpha^{4k-1} = 3\operatorname{atan}1 - \frac{1}{2}\operatorname{atan}\frac{1}{2} - \operatorname{atan}\frac{F_{2k-1}}{F_{2k}}, \quad \operatorname{atan}\alpha^{4k-3} = \operatorname{atan}1 + \frac{1}{2}\operatorname{atan}\frac{1}{2} + \operatorname{atan}\frac{F_{2k-2}}{F_{2k-1}},$$

$$\operatorname{atan}\frac{1}{\alpha^{2k}} = \operatorname{atan}\frac{2}{\sqrt{5}F_{2k}}.$$

For all $k \ge 1$,

$$\sum_{i=1}^{k}\operatorname{atan}\frac{1}{\sqrt{5}F_{2k+1}} = \operatorname{atan}\frac{1}{\alpha^2} - \operatorname{atan}\frac{1}{\alpha^{2k+2}}.$$

For all $k \geq 3$, $\alpha^{k-2} < F_k$. Furthermore, $\left[\begin{smallmatrix}\alpha\\1\end{smallmatrix}\right]$ is an eigenvector of A associated with α. Now, for all $k \geq 0$, consider the difference equation $x_{k+1} = Ax_k$. Then, for all $k \geq 0$,

$$x_{k+2(1)} = x_{k+1(1)} + x_{k(1)}, \quad x_k = A^k x_0,$$

where, for all $k \geq 2$,

$$A^k = \begin{bmatrix} F_{k+1} & F_k \\ F_k & F_{k-1} \end{bmatrix} = F_k A + F_{k-1} I_n.$$

If $x_0 >> 0$, then

$$\lim_{k\to\infty} \frac{x_{k(1)}}{x_{k(2)}} = \alpha.$$

In particular, if $x_0 \triangleq \left[\begin{smallmatrix}1\\1\end{smallmatrix}\right]$, then, for all $k \geq 0$,

$$x_k = \begin{bmatrix} F_{k+2} \\ F_{k+1} \end{bmatrix},$$

where $F_1 = F_2 = 1$ and, for all $k \geq 1$, F_k satisfies

$$F_{k+2} = F_{k+1} + F_k, \quad F_k = \frac{\sqrt{5}}{5}(\alpha^k - \beta^k) = \frac{\alpha^k - \beta^k}{\alpha - \beta} = \frac{2^k\sqrt{5}}{5}\left(\cos^k \frac{\pi}{5} - \cos^k \frac{3\pi}{5}\right),$$

$$\alpha^k - \beta^k = F_k\sqrt{5}, \quad \alpha^{k+1}F_k = \alpha^k F_{k+1} + (-1)^{k+1}, \quad \alpha^{k+1} = \alpha F_{k+1} + F_k, \quad \sum_{i=0}^{k}\binom{k}{i}\alpha^{3i-2k} = 2^k,$$

$$F_k + F_{k+2} = \sqrt{5}F_{k+1} + 2\beta^{k+1}, \quad (1+\alpha)^k + (1-\alpha)^k = 2\sum_{i=0}^{\lfloor n/2\rfloor}\binom{n}{2i}\alpha^{2i} = \alpha(F_{2k} - F_k) + F_{2k-1} + F_{k+1}.$$

For all $k \geq 2$,

$$\sqrt[k]{\alpha F_k + F_{k-1}} + (-1)^{k+1}\sqrt[k]{F_{k+1} - \alpha F_k} = 1.$$

For all $k \in \mathbb{Z}$,

$$\alpha^{k+2} = \alpha^{k+1} + \alpha^k, \quad (\alpha^3+1)^k = 2^k\alpha^{2k} = 2^k(\alpha F_{2k} + F_{2k-1}), \quad (\sqrt{5}\alpha)^k = 5^{k/2}(\alpha F_k + F_{k-1}),$$
$$\alpha^{k+1} = \alpha F_{k+1} + F_k, \quad \alpha^{-k} = (-1)^k(F_{k+1} - \alpha F_k), \quad (1+\alpha^2)^k + (1-\alpha^2)^k = 5^{k/2}\alpha^k + (-\alpha)^k,$$
$$(1+\alpha^3)^k + (1-\alpha^3)^k = 2^k[\alpha^k + (-1)^k\alpha^k], \quad (1+\alpha^3)^k - (1-\alpha^3)^k = 2^k[\alpha^{2k} - (-1)^k\alpha^k],$$
$$\frac{1}{F_kF_{k+1}} = \frac{1}{\alpha^kF_k} + \frac{1}{\alpha^{k+1}F_{k+1}}, \quad \frac{1}{F_{2k}} = \frac{1}{\alpha^kF_k} + (-1)^{k+1}\frac{1}{\alpha^{2k}F_{2k}}.$$

For all $k, l \in \mathbb{Z}$,

$$\alpha^l(\alpha^3+1)^k = 2^k\alpha^{2k+l}.$$

For all $k > l \geq 1$,

$$\alpha F_{k-l+1} + F_{k-l} = \alpha^{k-l+1}.$$

Alternatively, if $x_0 \triangleq \left[\begin{smallmatrix}3\\1\end{smallmatrix}\right]$, then, for all $k \geq 0$,

$$x_k = \begin{bmatrix} L_{k+2} \\ L_{k+1} \end{bmatrix},$$

where $L_1 = 1$, $L_2 = 3$ and, for all $k \geq 1$, L_k satisfies

$$L_{k+2} = L_{k+1} + L_k, \quad \sqrt{5}\alpha^{k+1} = \alpha L_{k+1} + L_k, \quad L_k = \alpha^k + \beta^k.$$

Consequently,

$$\lim_{k\to\infty}\frac{F_{k+1}}{F_k} = \lim_{k\to\infty}\frac{L_{k+1}}{L_k} = \alpha.$$

Source: [649, 1240, 1239], [1241, pp. 59, 74], and [2156, 2480]. The limit follows from Fact 6.11.5. **Remark:** α is the *golden ratio*. The expressions for F_k and L_k involving powers of α and β are *Binet's formulas*. **Remark:** $\alpha = e^{\operatorname{acsch} 2}$. **Related:** Fact 1.17.1.

Fact 13.9.3. For all $n \ge 0$, let F_n, L_n, and P_n denote the nth Fibonacci, Lucas number, and Pell number, respectively, and define $\alpha \triangleq \frac{1}{2}(1+\sqrt{5})$. Then,

$$\lim_{i\to\infty} \sqrt[i]{F_i} = \lim_{k\to\infty} \frac{F_{k+1}}{F_k} = \alpha, \quad \sum_{i=1}^{\infty} \frac{(-1)^i}{F_iF_{i+1}} = \beta.$$

If $|z| < 1/\alpha$, then

$$\frac{-z}{z^2+z-1} = \sum_{i=1}^{\infty} F_i z^i, \quad \frac{z-2}{z^2+z-1} = \sum_{i=0}^{\infty} L_i z^i,$$

$$\frac{-z}{z^2+2z-1} = \sum_{i=0}^{\infty} P_i z^i, \quad \log \frac{1}{1-z-z^2} = \sum_{i=1}^{\infty} \frac{L_i}{i} z^i.$$

Consequently, if $k \ge 2$, then

$$\sum_{i=1}^{\infty} \frac{F_i}{k^i} = \frac{k}{k^2-k-1}.$$

If $|z| < \frac{1}{2}(3-\sqrt{5})$, then

$$\frac{z}{1-3z+z^2} = \sum_{i=1}^{\infty} F_{2i} z^i, \quad \frac{1-z}{1-3z+z^2} = \sum_{i=0}^{\infty} F_{2i+1} z^i, \quad \frac{2-3z}{1-3z+z^2} = \sum_{i=0}^{\infty} L_{2i} z^i,$$

$$\frac{z(1-z)}{(1+z)(1-3z+z^2)} = \sum_{i=1}^{\infty} F_i^2 z^i.$$

If $|z| < \sqrt{5}-2$, then

$$\frac{2z}{1-4z-z^2} = \sum_{i=1}^{\infty} F_{3i} z^i, \quad \frac{z(1-2z-z^2)}{(1+z-z^2)(1-4z-z^2)} = \sum_{i=1}^{\infty} F_i^3 z^i.$$

For all $|z|$ sufficiently small,

$$\frac{z(1-4z-4z^2+z^3)}{(1-z)(1-7z+z^2)(1-3z+z^2)} = \sum_{i=1}^{\infty} F_i^4 z^i,$$

$$\frac{2-z}{1-z-z^2} = \sum_{i=0}^{\infty} L_i z^i, \quad \frac{4-7z-z^2}{(1+z)(1-3z+z^2)} = \sum_{i=0}^{\infty} L_i^2 z^i,$$

$$\frac{8-13z-24z^2+z^3}{(1+z-z^2)(1-4z-z^2)} = \sum_{i=0}^{\infty} L_i^3 z^i, \quad \frac{16-79z-164z^2+76z^3+z^4}{(1-z)(1+3z+z^2)(1-7z+z^2)} = \sum_{i=0}^{\infty} L_i^4 z^i,$$

$$\frac{z(1-z)}{(1+z)(1-6z+z^2)} = \sum_{i=0}^{\infty} P_i^2 z^i, \quad \frac{z(1-4z-z^2)}{(1+2z-z^2)(1-14z-z^2)} = \sum_{i=0}^{\infty} P_i^3 z^i,$$

$$\frac{z(1+z)(1-14z+z^2)}{(1-z)(1+6z+z^2)(1-34z-z^2} = \sum_{i=0}^{\infty} P_i^4 z^i.$$

If $|z| < 1/\alpha$, then

$$\log \frac{1+\alpha z}{1+\beta z} = \sqrt{5} \sum_{i=1}^{\infty} (-1)^{i+1} \frac{F_i}{i} z^i.$$

For all $k \geq 1$ and all $|z|$ sufficiently small,

$$\frac{F_k z}{1 - L_k z + (-1)^k z^2} = \sum_{i=1}^{\infty} F_{ki} z^i, \qquad \frac{2 - L_k z}{1 - L_k z + (-1)^k z^2} = \sum_{i=0}^{\infty} L_{ki} z^i.$$

Furthermore,

$$\sum_{i=1}^{\infty} \frac{F_i}{2^{i+1}} = 1, \quad \sum_{i=1}^{\infty} \frac{iF_i}{2^{i+1}} = 5, \quad \sum_{i=1}^{\infty} \frac{i^2 F_i}{2^{i+1}} = 47, \quad \sum_{i=1}^{\infty} \frac{i^3 F_i}{2^{i+1}} = 665, \quad \sum_{i=1}^{\infty} \frac{i^4 F_i}{2^{i+1}} = 12551,$$

$$\sum_{i=1}^{\infty} \frac{F_i}{3^{i+1}} = \frac{1}{5}, \quad \sum_{i=1}^{\infty} \frac{iF_i}{3^{i+1}} = \frac{2}{5}, \quad \sum_{i=1}^{\infty} \frac{i^2 F_i}{3^{i+1}} = \frac{32}{25}, \quad \sum_{i=1}^{\infty} \frac{i^3 F_i}{3^{i+1}} = \frac{154}{25}, \quad \sum_{i=1}^{\infty} \frac{i^4 F_i}{3^{i+1}} = \frac{4984}{125},$$

$$\sum_{i=1}^{\infty} \frac{F_i}{4^{i+1}} = \frac{1}{11}, \quad \sum_{i=1}^{\infty} \frac{iF_i}{4^{i+1}} = \frac{17}{121}, \quad \sum_{i=1}^{\infty} \frac{i^2 F_i}{4^{i+1}} = \frac{413}{1331}, \quad \sum_{i=1}^{\infty} \frac{i^3 F_i}{4^{i+1}} = \frac{14705}{14641}, \quad \sum_{i=1}^{\infty} \frac{i^4 F_i}{4^{i+1}} = \frac{710981}{161051},$$

$$\sum_{i=0}^{\infty} (-1)^i \frac{4^i F_{2i+1}}{(2i+1)(3+\sqrt{5})^{2i+1}} = \frac{\sqrt{5}\pi}{40}, \quad \sum_{i=0}^{\infty} (-1)^i \frac{F_{2i+1}^2}{(2i+1)(3+\sqrt{10})^{2i+1}} = \frac{\pi}{20},$$

$$\sum_{i=0}^{\infty} \frac{1}{1 + F_{2i+1}} = \frac{\sqrt{5}}{2}, \quad \sum_{i=0}^{\infty} \frac{1}{3\sqrt{5} + 5F_{2i+1}} = \frac{1}{5}, \quad \sum_{i=0}^{\infty} \frac{1}{F_{2^i}} = 4 - \alpha = \frac{7 - \sqrt{5}}{2}, \quad \sum_{i=1}^{\infty} \frac{1}{F_i F_{i+2}} = 1,$$

$$\sum_{i=1}^{\infty} \frac{F_{i+1}}{F_i F_{i+2}} = 2, \quad \sum_{i=1}^{\infty} (-1)^{i+1} \frac{1}{F_i F_{i+2}} = \sqrt{5} - 2, \quad \sum_{i=1}^{\infty} (-1)^{i+1} \frac{P_{6i+3}}{P_{3i}^2 P_{3i+3}^2} = \frac{1}{125},$$

$$\sum_{i=1}^{\infty} (-1)^{i-1} \frac{1}{F_i F_{i+1}} = \sum_{i=1}^{\infty} \frac{1}{F_{2i-1} F_{2i+1}} = \frac{1}{\alpha}, \quad \sum_{i=0}^{\infty} (-1)^i \frac{1}{L_{i+1} L_i} = \sum_{i=0}^{\infty} \frac{1}{L_{2i+2} L_{2i}} = \frac{\sqrt{5}}{10},$$

$$\sum_{i=1}^{\infty} \frac{L_{2^{i+1}}}{F_{3(2^i)}} = \frac{5}{4}, \quad \sum_{i=1}^{\infty} \frac{F_{2^{i-1}}^2}{L_{2^i}^2 - 1} = \sum_{i=1}^{\infty} \frac{2^i F_{2^i}^2}{L_{3(2^i)}} = \frac{3}{20}, \quad \sum_{i=1}^{\infty} \frac{2^i F_{2^i}^3}{L_{2^{i+1}} L_{3(2^i)}} = \frac{7}{300},$$

$$\sum_{i=1}^{\infty} (-1)^{i+1} \frac{1}{\sum_{j=1}^{i} F_j^2} = \frac{1}{\alpha}, \quad \prod_{i=2}^{\infty} \left(1 + (-1)^i \frac{1}{F_i^2}\right) = \alpha, \quad \prod_{i=3}^{\infty} \frac{F_{2i-1}^2 + F_{2i-1} - 2}{F_{2i-1}^2 - F_{2i-1} - 2} = 2,$$

$$\sum_{i=3}^{\infty} \sum_{j=1}^{\infty} \frac{1}{F_i^{4j-2}} = \frac{4}{9}, \quad \sum_{i=1}^{\infty} \operatorname{atan} \frac{1}{F_{2i+1}} = \frac{\pi}{4}, \quad \sum_{i=1}^{\infty} \operatorname{atan} \frac{1}{\sqrt{5} F_{2i+1}} = \operatorname{atan} \frac{1}{\alpha^2},$$

$$\prod_{i=1}^{\infty} \left(1 + \frac{2}{L_{2^i}}\right) = \sqrt{5}, \quad \prod_{i=1}^{\infty} \left(1 - \frac{2}{L_{2^i}}\right) = \frac{\sqrt{5}}{4}.$$

If $n \geq 1$, then

$$\sum_{i=1}^{\infty} (-1)^{n(i-1)} \frac{1}{F_{n(i-1)+1} F_{ni+1}} = \frac{1}{\alpha F_n}, \quad \sum_{i=1}^{\infty} \frac{(F_i F_{i+2})^n - F_{i+1}^{2n}}{F_i F_{i+1}} = \alpha^n - 1,$$

$$\sum_{i=0}^{\infty} (-1)^{ni} \frac{1}{L_{n(i+1)} L_{ni}} = \frac{\sqrt{5}}{10 F_n}, \quad \sum_{i=1}^{\infty} \frac{4^i}{L_{n2^i}^2} = \frac{4}{5 F_{2n}^2},$$

$$\sum_{i=n}^{\infty} \operatorname{atan} \frac{1}{F_{2i+1}} = \operatorname{atan} \frac{1}{F_{2n}}, \quad \sum_{i=1}^{\infty} \operatorname{atan} \frac{1}{L_{2i}} = \operatorname{atan} \frac{1}{\alpha}, \quad \sum_{i=0}^{\infty} \operatorname{atan} \frac{\sqrt{5}}{L_{4i+2}} = \frac{\pi}{4},$$

$$F_n = \sum_{i=-\infty}^{\infty} (-1)^i \binom{n-1}{\lfloor (n-1-5i)/2 \rfloor} = \sum_{i=-\infty}^{\infty} (-1)^i \binom{n}{\lfloor (n-1-5i)/2 \rfloor}.$$

If $n \geq 2$, then

$$\left\lfloor \left(\sum_{i=2n}^{\infty} \frac{1}{F_i} \right)^{-1} \right\rfloor = F_{2n-2}, \quad \left\lfloor \left(\sum_{i=2n+1}^{\infty} \frac{1}{F_i} \right)^{-1} \right\rfloor = F_{2n-1} - 1.$$

If $n \geq 1$, then

$$\left\lfloor \left(\sum_{i=2n}^{\infty} \frac{1}{F_i^2} \right)^{-1} \right\rfloor = F_{2n}F_{2n-1} - 1, \quad \left\lfloor \left(\sum_{i=2n-1}^{\infty} \frac{1}{F_i^2} \right)^{-1} \right\rfloor = F_{2n-1}F_{2n-2},$$

$$\left\lfloor \left(\sum_{i=n}^{\infty} \frac{1}{F_{2i}^2} \right)^{-1} \right\rfloor = F_{4n-2} - 1, \quad \left\lfloor \left(\sum_{i=n}^{\infty} \frac{1}{F_{2i-1}^2} \right)^{-1} \right\rfloor = F_{4n-4}.$$

If $n, m \geq 1$, then

$$\prod_{i=1}^{\infty} \left(1 + \frac{F_m}{F_{2^i n+m}} \right) = \frac{1 - (-1)^m \alpha^{-2(n+m)}}{1 - \alpha^{-2n}}, \quad \prod_{i=1}^{\infty} \left(1 + \frac{L_m}{L_{2^i n+m}} \right) = \frac{1 + (-1)^m \alpha^{-2(n+m)}}{1 - \alpha^{-2n}}.$$

In particular,

$$\prod_{i=1}^{\infty} \left(1 + \frac{1}{F_{2^i+1}} \right) = \frac{3}{\alpha}, \quad \prod_{i=1}^{\infty} \left(1 + \frac{1}{L_{2^i+1}} \right) = 3 - \alpha.$$

Source: [12], [112, p. 40], [411, p. 141], [804], [993, p. 206], [1165, 1236], [1566, p. 67], [1757, p. 175], [1926, 1927, 2027], [2068, pp. 131, 350], [2100, 2150, 2151, 2152, 2153, 2154, 2160, 2180, 2489], [2764, p. 180], and [2810, 2811]. **Remark:** $\frac{-z}{z^2+z-1}$ is the generating function for the Fibonacci numbers, $\frac{z-2}{z^2+z-1}$ is the generating function for the Lucas numbers, and $\frac{-z}{z^2+2z-1}$ is the generating function for the Pell numbers. See [1612, 2841]. **Remark:** If $|z + z^2| < 1$, then

$$\frac{z}{1 - z - z^2} = z \sum_{i=0}^{\infty} (z + z^2)^i.$$

This series converges for $z = -1$. However, -1 is not in the domain of convergence of $\frac{-z}{z^2+z-1} = \sum_{i=0}^{\infty} F_i z^i$, and thus $\sum_{i=0}^{\infty} (-1)^i F_i = -1$ is false. **Related:** Fact 1.17.1.

Fact 13.9.4. Let $n \geq 1$. Then,

$$\left\lfloor \left(\sum_{i=n}^{\infty} \frac{1}{i^2} \right)^{-1} \right\rfloor = n - 1, \quad \left\lfloor \left(\sum_{i=n}^{\infty} \frac{1}{i^3} \right)^{-1} \right\rfloor = 2n(n - 1),$$

$$\left\lfloor \left(\sum_{i=2n}^{\infty} \frac{1}{i^4} \right)^{-1} \right\rfloor = 24n^3 - 18n^2 + \left\lfloor \frac{15n - 3}{2} \right\rfloor, \quad \left\lfloor \left(\sum_{i=2n-1}^{\infty} \frac{1}{i^4} \right)^{-1} \right\rfloor = 24n^3 - 54n^2 + \left\lfloor \frac{174n - 51}{4} \right\rfloor.$$

Source: [2931].

Fact 13.9.5. Let a and b be positive numbers, where $b < a$, let $\mathcal{E}$ be the ellipse defined by

$$\mathcal{E} \triangleq \left\{ x \in \mathbb{R}^2 \colon \left(\frac{x_{(1)}}{a} \right)^2 + \left(\frac{x_{(2)}}{b} \right)^2 = 1 \right\},$$

let $\varepsilon \triangleq \sqrt{a^2 - b^2}/a$ denote the eccentricity of $\mathcal{E}$, and let L denote the arc length of $\mathcal{E}$. Then,

$$\begin{aligned} L &= 2a \int_0^{\pi} \sqrt{1 - \varepsilon^2 \sin^2 x} \, dx = 4a \int_0^1 \frac{\sqrt{1 - \varepsilon^2 x^2}}{\sqrt{1 - x^2}} \, dx \\ &= 2\pi a \sum_{i=0}^{\infty} \left[\frac{(2i)!}{4^i (i!)^2} \right]^2 \frac{\varepsilon^{2i}}{1 - 2i} = 2\pi a \left[1 - \sum_{i=1}^{\infty} \left(\frac{(2i-1)!!}{(2i)!!} \right)^2 \frac{\varepsilon^{2i}}{2i - 1} \right]. \end{aligned}$$

Furthermore,

$$\pi(a+b) \le 2\pi\left(\frac{a^{3/2}+b^{3/2}}{2}\right)^{2/3} \le L \le \pi\sqrt{2(a^2+b^2)}.$$

In addition, let $A = \frac{1}{2}(a+b)$, $G \triangleq \sqrt{ab}$, and $H \triangleq 2ab/(a+b)$. If $a > b$, then,

$$8A - (16-5\pi)G - (3\pi-8)H \le L \le \frac{\pi}{8}[21A - 2G - 3H].$$

Finally, for $a - b \to 0$,

$$L \sim \pi[3(a+b) - \sqrt{(3a+b)(a+3b)}], \quad L \sim \pi(a+b)\left[1 + \frac{3\left(\frac{a-b}{a+b}\right)^2}{10 + \sqrt{4 - 3\left(\frac{a-b}{a+b}\right)^2}}\right].$$

Source: [41, 130, 670, 2828] and [1568, p. 56]. **Related:** Fact 5.5.8 for the area of an ellipse.

13.10 Facts on Infinite Products

Fact 13.10.1. Let $(x_i)_{i=1}^\infty \subset \mathbb{R}$. Then, the following statements hold:

i) If two of the quantities $\sum_{i=1}^\infty x_i$, $\sum_{i=1}^\infty x_i^2$, $\prod_{i=1}^\infty (1+x_i)$, and $\prod_{i=1}^\infty (1-x_i)$ converge, then all four quantities converge.

ii) If $\prod_{i=1}^\infty (1+x_i)$ converges, then $\lim_{i\to\infty} x_i = 0$.

iii) If $\sum_{i=1}^\infty x_i^2 = \infty$ and $\sum_{i=1}^\infty x_i$ converges, then $\prod_{i=1}^\infty (1+x_i) = 0$.

iv) If $\sum_{i=1}^\infty x_i^2 = \infty$ and $\prod_{i=1}^\infty (1+x_i)$ converges, then $\sum_{i=1}^\infty x_i = \infty$.

v) Assume that $(x_i)_{i=1}^\infty \subset (-1,\infty)$, and assume that $\sum_{i=1}^\infty x_i$ converges. Then, $\prod_{i=1}^\infty (1+x_i)$ converges. Furthermore, $\sum_{i=1}^\infty x_i^2 = \infty$ if and only if $\prod_{i=1}^\infty (1+x_i) = 0$.

vi) Assume that $(x_i)_{i=1}^\infty \subset [0,\infty)$. Then, $\prod_{i=1}^\infty (1+x_i)$ converges if and only if $\sum_{i=1}^\infty x_i$ converges.

vii) Assume that $(x_i)_{i=1}^\infty \subset [0,1) \cup (1,\infty)$. Then, $\prod_{i=1}^\infty (1-x_i)$ converges if and only if $\sum_{i=1}^\infty x_i$ converges.

viii) Assume that $\lim_{i\to\infty} x_i = 0$ and, for all $i \ge 1$, $x_i \ge x_{i+1} \ge 0$. Then, $\prod_{i=1}^\infty [1 + (-1)^i x_i]$ converges if and only if $\sum_{i=1}^\infty x_i^2$ converges.

ix) If there exist distinct nonzero real numbers α_1 and α_2 such that $\prod_{i=1}^\infty (1+\alpha_1 x_i)$ and $\prod_{i=1}^\infty (1+\alpha_2 x_i)$ converge, then $\prod_{i=1}^\infty (1+\alpha x_i)$ converges for all $\alpha \in \mathbb{R}$.

x) If $\sum_{i=1}^\infty x_i^2$ converges and $\sum_{i=1}^\infty x_i$ does not converge, then

$$\lim_{n\to\infty} \frac{\prod_{i=1}^n (1+x_i)}{e^{\sum_{i=1}^n x_i}}$$

exists.

xi) Assume that $(x_i)_{i=1}^\infty \subset (0,\infty)$, and assume that $\sum_{i=1}^\infty x_i$ converges. Then,

$$x_1 \prod_{i=2}^\infty \left(1 + \frac{x_i}{\sum_{j=1}^{i-1} x_j}\right) = \sum_{i=1}^\infty x_i.$$

xii) Assume that $(x_i)_{i=1}^\infty \subset (-1,\infty)$, and assume that $\prod_{i=1}^\infty (1+x_i)$ converges. Then,

$$\sum_{i=1}^\infty \frac{x_i}{\prod_{j=1}^i (1+x_j)} = 1 - \frac{1}{\prod_{i=1}^\infty (1+x_i)}.$$

xiii) Assume that $(x_i)_{i=1}^\infty \subset (0,\infty)$, and assume that $\prod_{i=1}^\infty (1+x_i)$ does not converge. Then,

$$\sum_{i=1}^\infty \frac{x_i}{\prod_{j=1}^i (1+x_j)} = 1.$$

xiv) $\prod_{i=1}^\infty (1+x_i)$ converges if and only if there exists $k \ge 1$ such that $\lim_{n\to\infty} \prod_{i=k}^n (1+x_i)$ exists and is nonzero. If these conditions hold, then $\operatorname{card}(\{i \ge 1\colon\ x_i = -1\})$ is finite.

xv) Assume that $(x_i)_{i=1}^\infty \subset (0,1)$. Then, $\prod_{i=1}^\infty (1-x_i)$ converges if and only if $\sum_{i=1}^\infty x_i$ converges.

xvi) Assume that $(x_i)_{i=1}^\infty \subset [0,\infty)$. Then, $\prod_{i=1}^\infty (1+x_i) \le e^{\sum_{i=1}^\infty x_i}$.

Source: [217, pp. 241–249], [1136, p. 356], [1566, pp. 113–117], and [2294, pp. 63–65, 86].

Fact 13.10.2. Let $(z_i)_{i=1}^\infty \subset \text{OIUD}\backslash\{0\}$, and assume that $\prod_{i=1}^\infty |z_i|$ exists. Then,

$$\sum_{i=1}^\infty \frac{1-|z_i|^2}{|z_i|} \le \sum_{i=1}^\infty \left(\frac{1}{|z_i|} - |z_i|\right) = -2\sum_{i=1}^\infty \sinh\log|z_i| \le -2\sinh\sum_{i=1}^\infty \log|z_i| = \frac{1-\prod_{i=1}^\infty |z_i|^2}{\prod_{i=1}^\infty |z_i|}.$$

Source: [2089].

Fact 13.10.3. The following statements hold:

i) If $n \ge 1$, then

$$\prod_{i=n+1}^\infty \frac{i^2}{i^2-n^2} = \binom{2n}{n}.$$

In particular,

$$\prod_{i=2}^\infty \frac{i^2}{i^2-1} = 2,\quad \prod_{i=3}^\infty \frac{i^2}{i^2-4} = 6,\quad \prod_{i=4}^\infty \frac{i^2}{i^2-9} = 20,\quad \prod_{i=5}^\infty \frac{i^2}{i^2-16} = 70.$$

ii) If $n \ge 2$, then

$$\prod_{i=1,i\ne n}^\infty \frac{i^2}{i^2-n^2} = (-1)^{n+1}2.$$

iii) Let $n \ge 1$. Then,

$$\prod_{i=n+1}^\infty \frac{i^2-1}{i^2-n^2} = \binom{2n}{n-1}.$$

In particular,

$$\prod_{i=3}^\infty \frac{i^2-1}{i^2-4} = 4,\quad \prod_{i=4}^\infty \frac{i^2-1}{i^2-9} = 15,\quad \prod_{i=5}^\infty \frac{i^2-1}{i^2-16} = 56,\quad \prod_{i=6}^\infty \frac{i^2-1}{i^2-25} = 210.$$

iv) Let $n \ge 2$. Then,

$$\prod_{i=1}^\infty \frac{n^2i^2}{n^2i^2-1} = \frac{\pi}{n\sin\pi/n}.$$

In particular,

$$\prod_{i=1}^\infty \frac{4i^2}{4i^2-1} = \frac{\pi}{2},\quad \prod_{i=1}^\infty \frac{9i^2}{9i^2-1} = \frac{2\sqrt{3}\pi}{9},\quad \prod_{i=1}^\infty \frac{16i^2}{16i^2-1} = \frac{\sqrt{2}\pi}{4},$$

$$\prod_{i=1}^\infty \frac{25i^2}{25i^2-1} = \frac{\pi}{10}\sqrt{\frac{5-\sqrt{5}}{2}},\quad \prod_{i=1}^\infty \frac{36i^2}{36i^2-1} = \frac{\pi}{3},\quad \prod_{i=1}^\infty \frac{64i^2}{64i^2-1} = \frac{\pi}{16}\sqrt{2-\sqrt{2}}.$$

v) Let $n \geq 1$. Then,

$$\prod_{i=1}^{\infty}\left(1+\frac{4n^4}{i^4}\right)=\frac{\sinh^2 n\pi}{2n^2\pi^2}.$$

vi) Let z be a complex number, and assume that $|z|<1$. Then,

$$\prod_{i=0}^{\infty}(1+z^{2^i})=\sum_{i=0}^{\infty}z^i=\frac{1}{1-z},\quad \prod_{i=1}^{\infty}(1+z^{2^i})=\frac{1}{1+z}\sum_{i=0}^{\infty}z^i=\frac{1}{1-z^2}=\sum_{i=0}^{\infty}z^{2i}.$$

Consequently, if p is a real number and $|p|>1$, then

$$\prod_{i=0}^{\infty}\left(1+\frac{1}{p^{2^i}}\right)=\frac{p}{p-1}.$$

In particular,

$$\prod_{i=0}^{\infty}\left(1+\frac{1}{2^{2^i}}\right)=2,\quad \prod_{i=0}^{\infty}\left(1+\frac{1}{3^{2^i}}\right)=\frac{3}{2}.$$

vii) Let α be a real number. Then, $\prod_{i=1}^{\infty}(1+1/i^\alpha)$ converges if and only if $\alpha>1$. In particular,

$$\prod_{i=1}^{\infty}\left(1+\frac{1}{i}\right)=\infty,\quad \prod_{i=1}^{\infty}\left(1+\frac{1}{i^2}\right)=\frac{\sinh\pi}{\pi},\quad \prod_{i=1}^{\infty}\left(1+\frac{1}{i^3}\right)=\frac{\cosh\frac{\sqrt{3}\pi}{2}}{\pi},$$

$$\prod_{i=1}^{\infty}\left(1+\frac{1}{i^4}\right)=\frac{1}{\pi^2}\left(\sin^2\frac{\sqrt{2}\pi}{2}+\sinh^2\frac{\sqrt{2}\pi}{2}\right)=\frac{\cosh\sqrt{2}\pi-\cos\sqrt{2}\pi}{2\pi^2},$$

$$\prod_{i=1}^{\infty}\left(1+\frac{1}{i^6}\right)=\frac{(\sinh\pi)(\cosh\pi-\cos\sqrt{3}\pi)}{2\pi^3}.$$

viii) Let p_i denote the ith prime. Then,

$$\prod_{i=1}^{\infty}\left(1-\frac{1}{p_i}\right)=0,\quad \sum_{i=1}^{\infty}\prod_{j=1}^{i}\frac{p_j-1}{p_{j+1}}=1,$$

$$\prod_{i=1}^{\infty}\left(1+\frac{\sin p_i\pi/2}{p_i}\right)=\prod_{i=2}^{\infty}\left(1+\frac{(-1)^{(p_i-1)/2}}{p_i}\right)=\frac{2}{\pi},$$

$$\lim_{i\to\infty}\frac{1}{\log i}\prod_{p_j\leq i}\left(1+\frac{1}{p_j}\right)^{-1}=\frac{6e^\gamma}{\pi^2},\quad \lim_{i\to\infty}\frac{1}{\log i}\prod_{p_j\leq i}\left(1-\frac{1}{p_j}\right)^{-1}=e^\gamma.$$

Source: *i*) and *ii*) are given in [968, p. 141]. *iii*) for $n=2$ is given in [1136, p. 356]. To prove *iv*) for $n=2$, Stirling's formula given by Fact 12.18.58 implies that

$$\begin{aligned}\sqrt{\pi}&=\lim_{n\to\infty}\frac{e^n n!}{n^n\sqrt{2n}}=\lim_{n\to\infty}\frac{2^{2n}(n!)^2}{(2n)!\sqrt{n}}=\lim_{n\to\infty}\frac{2^{2n}(n!)2^{-n}(2n)!!}{(2n)!!(2n-1)!!\sqrt{n}}\\&=\lim_{n\to\infty}\frac{(2n)!!}{(2n-1)!!\sqrt{n}}=\lim_{n\to\infty}\frac{(2n)!!}{(2n-1)!!\sqrt{\frac{2n+1}{2}}}=\sqrt{2}\lim_{n\to\infty}\frac{(2n)!!}{(2n-1)!!\sqrt{2n+1}}\\&=\sqrt{2}\sqrt{\frac{2}{1}\cdot\frac{2}{3}\cdot\frac{4}{3}\cdot\frac{4}{5}\cdot\frac{6}{5}\cdot\frac{6}{7}\cdots}=\sqrt{2}\sqrt{\prod_{i=1}^{\infty}\left(\frac{2i}{2i-1}\right)\left(\frac{2i}{2i+1}\right)}.\end{aligned}$$

See [1693]. *iv*) for $n = 3$ is given in [1566, p. 113]; *v*) is given in [2013, p. 102]; *vi*) follows from [1757, p. 169]; *vii*) is given in [1136, p. 354]; the cases $\alpha = 2$ and $\alpha = 4$ are given in [2013, p. 102]; *viii*) is given in [1350, p. 109] and [2067, 2861]. **Remark:** *iv*) for $n = 2$ is due to J. Wallis. This can be written as

$$\frac{\pi}{2} = \lim_{i\to\infty} \frac{[(2i)!!]^2}{(2i-1)!!(2i+1)!!} = \lim_{i\to\infty} \frac{[(2i)!!]^2}{[(2i-1)!!]^2(2i+1)}.$$

See [1206], Fact 1.13.15, and Fact 12.18.59. **Remark:** It is noted in [69] that

$$\prod_{i=1}^{n} \frac{4i^2}{4i^2-1} = \frac{2^{4n}}{(n+1)\binom{2n}{n}\binom{2n+1}{n}}, \quad \lim_{n\to\infty} \frac{2^{4n}}{(n+1)\binom{2n}{n}\binom{2n+1}{n}} = \frac{\pi}{2}.$$

Credit: *iv*) for $n = 4$ is due to E. Catalan.

Fact 13.10.4. Let z be a complex number. Then,

$$\sin z = z\prod_{i=1}^{\infty}\left(1 - \frac{z^2}{i^2\pi^2}\right), \quad \cos z = \prod_{i=1}^{\infty}\left(1 - \frac{4z^2}{(2i-1)^2\pi^2}\right),$$

$$\sinh z = z\prod_{i=1}^{\infty}\left(1 + \frac{z^2}{i^2\pi^2}\right), \quad \cosh z = \prod_{i=1}^{\infty}\left(1 + \frac{4z^2}{(2i-1)^2\pi^2}\right),$$

$$\cosh 2z - \cos 2z = 4z^2\prod_{i=1}^{\infty}\left(1 + \frac{4z^2}{i^4\pi^4}\right), \quad \cosh 2z + \cos 2z = 2\prod_{i=1}^{\infty}\left(1 + \frac{16z^4}{(2i-1)^4\pi^4}\right),$$

$$e^z - 1 = ze^{z/2}\prod_{i=1}^{\infty}\left(1 + \frac{z^2}{4\pi^2 i^2}\right).$$

If $|z| < 1$, then

$$\frac{1}{1-z} = \prod_{i=0}^{\infty}(1 + z^{2^i}).$$

If $z \neq 0$, then

$$\log z = (z-1)\prod_{i=1}^{\infty}\frac{2}{1 + z^{1/2^i}}.$$

Source: [217, p. 255], [635], [1136, pp. 356, 357], [1167, pp. 291–293], and [2173].

Fact 13.10.5. Let $z \in \mathbb{C}\backslash\mathbb{N}$ and $n \geq 2$. Then,

$$\prod_{i=1}^{\infty}\left[1 - \left(\frac{z}{i}\right)^n\right] = -\frac{1}{z^n}\prod_{i=1}^{n}\frac{1}{\Gamma(-ze^{(2i\pi/n)J})}.$$

Now, let x be a real number. Then,

$$\sin x = x\prod_{i=1}^{\infty}\left(1 - \frac{x^2}{i^2\pi^2}\right) = x\prod_{i=1}^{\infty}\cos\frac{x}{2^i} = \frac{x}{3}\prod_{i=1}^{\infty}\left(4\cos^2\frac{x}{3^i} - 1\right),$$

$$\frac{\sin \pi x}{\pi x} = \prod_{i=1}^{\infty}\left(1 - \frac{x^2}{i^2}\right), \quad \cos x = \prod_{i=1}^{\infty}\left(1 - \frac{4x^2}{(2i-1)^2\pi^2}\right),$$

$$\sinh x = x\prod_{i=1}^{\infty}\left(1 + \frac{x^2}{i^2\pi^2}\right) = x\prod_{i=1}^{\infty}\cosh\frac{x}{2^i}, \quad \frac{\sinh \pi x}{\pi x} = \prod_{i=1}^{\infty}\left(1 + \frac{x^2}{i^2}\right),$$

$$\cosh x = \prod_{i=1}^{\infty}\left(1 + \frac{4x^2}{(2i-1)^2\pi^2}\right), \quad \frac{(\sin \pi x)\sinh \pi x}{\pi^2 x^2} = \prod_{i=1}^{\infty}\left(1 - \frac{x^4}{i^4}\right),$$

$$\frac{1}{\pi^2 x^2}\left(\sin^2\frac{\pi x}{\sqrt{2}} + \sinh^2\frac{\pi x}{\sqrt{2}}\right) = \prod_{i=1}^{\infty}\left(1 + \frac{x^4}{i^4}\right).$$

If $|x| < 1$, then

$$\prod_{i=1}^{\infty}(1 + x^i) = \prod_{i=1}^{\infty}\frac{1}{1 - x^{2i-1}}.$$

If $x > -1$, then

$$e^x = \lim_{n\to\infty}\prod_{i=1}^{n}\left(1 + \frac{i^2 x}{n^3}\right)^3.$$

If $x \ge 0$, then

$$e^x = \prod_{n=1}^{\infty}\sqrt[n]{\prod_{i=1}^{n}(ix+1)^{(-1)^{i+1}\binom{n}{i}}}.$$

If y is a real number, then

$$\sin^2\pi x - \sin^2\pi y = (\sin^2\pi x)\prod_{i=-\infty}^{\infty}\left[1 - \frac{y^2}{(i+x)^2}\right].$$

Source: [1266], [1566, pp. 56, 113, 117], and [2013, pp. 99–104]. The first equality is given in [1217, p. 896].

Fact 13.10.6. Let $n > k \ge 1$. Then,

$$\prod_{i=0}^{\infty}\frac{n^2(i+1)^2}{(ni+n-k)(ni+n+k)} = \frac{k\pi}{n}\csc\frac{k\pi}{n}.$$

In particular,

$$\prod_{i=0}^{\infty}\frac{4(i+1)^2}{(2i+1)(2i+3)} = \frac{\pi}{2},\quad \prod_{i=0}^{\infty}\frac{9(i+1)^2}{(3i+2)(3i+4)} = \frac{2\sqrt{3}\pi}{9},\quad \prod_{i=0}^{\infty}\frac{9(i+1)^2}{(3i+1)(3i+5)} = \frac{4\sqrt{3}\pi}{9},$$

$$\prod_{i=0}^{\infty}\frac{16(i+1)^2}{(4i+3)(4i+5)} = \frac{\sqrt{2}\pi}{4},\quad \prod_{i=0}^{\infty}\frac{16(i+1)^2}{(4i+1)(4i+7)} = \frac{3\sqrt{2}\pi}{4},$$

$$\prod_{i=0}^{\infty}\frac{25(i+1)^2}{(5i+4)(5i+6)} = \frac{\sqrt{10}}{25}\sqrt{5+\sqrt{5}}\pi,\quad \prod_{i=0}^{\infty}\frac{25(i+1)^2}{(5i+3)(5i+7)} = \frac{2\sqrt{10}}{25}\sqrt{5-\sqrt{5}}\pi,$$

$$\prod_{i=0}^{\infty}\frac{25(i+1)^2}{(5i+2)(5i+8)} = \frac{3\sqrt{10}}{25}\sqrt{5-\sqrt{5}}\pi,\quad \prod_{i=0}^{\infty}\frac{25(i+1)^2}{(5i+1)(5i+9)} = \frac{4\sqrt{10}}{25}\sqrt{5+\sqrt{5}}\pi,$$

$$\prod_{i=0}^{\infty}\frac{36(i+1)^2}{(6i+5)(6i+7)} = \frac{\pi}{3},\quad \prod_{i=0}^{\infty}\frac{36(i+1)^2}{(6i+1)(6i+11)} = \frac{5\pi}{3}.$$

Source: [357]. **Remark:** The second equality is *Wallis's equality*. Related equalities are given in [2495]. **Related:** Fact 13.10.4 and Fact 13.10.5.

Fact 13.10.7. Let $n > k \ge 1$. Then,

$$\prod_{i=0}^{\infty}\frac{n^2(2i+1)^2}{(2ni+n-k)(2ni+n+k)} = \sec\frac{k\pi}{2n}.$$

In particular,

$$\prod_{i=0}^{\infty}\frac{4(2i+1)^2}{(4i+1)(4i+3)} = \sqrt{2}, \quad \prod_{i=0}^{\infty}\frac{9(2i+1)^2}{4(3i+1)(3i+2)} = \frac{2\sqrt{3}}{3}, \quad \prod_{i=0}^{\infty}\frac{9(2i+1)^2}{(6i+1)(6i+5)} = 2,$$

$$\prod_{i=0}^{\infty}\frac{16(2i+1)^2}{(8i+3)(8i+5)} = \sqrt{4-2\sqrt{2}}, \quad \prod_{i=0}^{\infty}\frac{16(2i+1)^2}{(8i+1)(8i+7)} = \sqrt{4+2\sqrt{2}}.$$

Source: [357]. **Related:** Fact 13.10.4 and Fact 13.10.5.

Fact 13.10.8. Let $n \geq 1$. Then,

$$\prod_{i=1}^{\infty}\prod_{j=1}^{n}\frac{4n^2(2i-1)^2-4(j-1)^2}{4n^2(2i-1)^2-(2j-1)^2} = \sqrt{2n},$$

$$\prod_{i=1}^{\infty}\prod_{j=1}^{n}\frac{(2n+1)^2(2i-1)^2-(2j-1)^2}{(2n+1)^2(2i-1)^2-4j^2} = \sqrt{2n+1}.$$

Source: [2137].

Fact 13.10.9. Let $n \geq 1$. Then,

$$\prod_{i=1}^{\infty}\frac{(i+n)^2}{i(i+2n)} = \binom{2n}{n}.$$

In particular,

$$\prod_{i=1}^{\infty}\frac{(i+1)^2}{i(i+2)} = 2, \quad \prod_{i=1}^{\infty}\frac{(i+2)^2}{i(i+4)} = 6, \quad \prod_{i=1}^{\infty}\frac{(i+3)^2}{i(i+6)} = 20, \quad \prod_{i=1}^{\infty}\frac{(i+4)^2}{i(i+8)} = 70.$$

Furthermore,

$$\prod_{i=1}^{\infty}\frac{(i+n+\frac{1}{2})^2}{i(i+2n+1)} = \frac{\Gamma(2n+2)}{\Gamma^2(n+\frac{3}{2})}.$$

In particular,

$$\prod_{i=1}^{\infty}\frac{(i+\frac{3}{2})^2}{i(i+3)} = \frac{32}{3\pi}, \quad \prod_{i=1}^{\infty}\frac{(i+\frac{5}{2})^2}{i(i+5)} = \frac{512}{15\pi}, \quad \prod_{i=1}^{\infty}\frac{(i+\frac{7}{2})^2}{i(i+7)} = \frac{4096}{35\pi}, \quad \prod_{i=1}^{\infty}\frac{(i+\frac{9}{2})^2}{i(i+9)} = \frac{131072}{315\pi}.$$

Finally,

$$\prod_{i=1}^{\infty}\frac{(i+n+1)^2}{(i+n)(i+n+2)} = \frac{n+2}{n+1}.$$

Fact 13.10.10.

$$\prod_{i=0}^{\infty}\frac{(24i+5)(24i+7)}{(24i+1)(24i+11)} = \sqrt{6+3\sqrt{3}}, \quad \prod_{i=0}^{\infty}\frac{(4i+3)(12i+5)}{8(2i+1)(3i+2)} = \frac{\sqrt{1+\sqrt{3}}}{2^{1/4}3^{3/8}},$$

$$\prod_{i=0}^{\infty}\frac{(10i+3)(30i+19)}{20(3i+1)(5i+3)} = \frac{\sqrt{\sqrt{15}+\sqrt{5+2\sqrt{5}}}}{2^{19/20}3^{1/20}5^{1/3}}, \quad \prod_{i=0}^{\infty}\frac{(6i+5)^2}{3(2i+1)(6i+1)}\left(\frac{3i+1}{3i+2}\right)^3 = 1,$$

$$\prod_{i=0}^{\infty}\frac{(14i+1)(14i+9)(14i+11)}{243(2i+1)^3} = \frac{1}{4}, \quad \prod_{i=1}^{\infty}\frac{i(i+3)(i+5)(i+6)}{(i+1)(i+2)(i+4)(i+7)} = \frac{7}{15},$$

$$\prod_{i=0}^{\infty}\prod_{k=1}^{3}\frac{(18i+9-2k)(18i+9+2k)}{81(2i+1)^2}=\frac{1}{8}.$$

Furthermore, let $k \geq 1$. Then,

$$\prod_{i=1}^{\infty}\prod_{j=1}^{k}\left(\frac{(2k+1)i+2j-1}{(2k+1)i+2j}\right)^{(-1)^i}=\frac{4^k}{\sqrt{2k+1}\binom{2k}{k}}.$$

Source: [357, 667].

Fact 13.10.11.

$$\prod_{i=2}^{\infty}\left(1-\frac{1}{i}\right)=0,\quad \prod_{i=2}^{\infty}\left(1-\frac{1}{i^2}\right)=\frac{1}{2},$$

$$\prod_{i=2}^{\infty}\left(1-\frac{1}{i^3}\right)=\frac{\cosh\frac{\sqrt{3}\pi}{2}}{3\pi},\quad \prod_{i=2}^{\infty}\left(1-\frac{1}{i^4}\right)=\frac{\sinh\pi}{4\pi},\quad \prod_{i=2}^{\infty}\left(1-\frac{1}{i^6}\right)=\frac{1+\cosh\sqrt{3}\pi}{12\pi^2},$$

$$\prod_{i=3}^{\infty}\left(1-\frac{8}{i^3}\right)=\frac{\sqrt{3}\sinh\sqrt{3}\pi}{126\pi},\quad \prod_{i=3}^{\infty}\left(1-\frac{16}{i^4}\right)=\frac{\sinh 2\pi}{120\pi},\quad \prod_{i=3}^{\infty}\frac{i^2-1}{i^3-4}=4,$$

$$\prod_{i=2}^{\infty}\frac{i^2-1}{i^2+1}=\frac{1}{2}\prod_{i=2}^{\infty}\frac{i^2}{i^2+1}=\frac{\pi}{\sinh\pi}\approx 0.2720,\quad \prod_{i=2}^{\infty}\frac{i^3-1}{i^3+1}=\frac{2}{3},\quad \prod_{i=2}^{\infty}e\left(1-\frac{1}{i^2}\right)^{i^2}=\frac{\pi}{e^{3/2}},$$

$$\prod_{i=2}^{\infty}\frac{i^4-1}{i^4+1}=\frac{\pi\sinh\pi}{\cosh\sqrt{2}\pi-\cos\sqrt{2}\pi}\approx 0.8480,\quad \prod_{i=1}^{\infty}\frac{(i^2+i+1)^2}{i^4+2i^3+3i^2}=\frac{3\sqrt{2}\cosh^2\frac{\sqrt{3}\pi}{2}}{\pi\sinh\sqrt{2}\pi},$$

$$\prod_{i=1}^{\infty}\left[1+\frac{1}{i(i+2)}\right]=2,\quad \prod_{i=2}^{\infty}\left[1-\frac{2}{i(i+1)}\right]=\frac{1}{3},\quad \prod_{i=2}^{\infty}\left[1-\frac{1}{i(i-1)}\right]=-\frac{1}{\pi}\cos\frac{\sqrt{5}\pi}{2},$$

$$\prod_{i=1}^{\infty}\left(1+(-1)^{i+1}\frac{1}{2i+1}\right)=\frac{\sqrt{2}\pi}{4},\quad \prod_{i=1}^{\infty}\left[1+(-1)^{i+1}\frac{1}{(2i+1)^3}\right]=\frac{\pi}{12}+\frac{\sqrt{2}\pi}{12}\cosh\frac{\sqrt{3}\pi}{4},$$

$$\prod_{i=1}^{\infty}\frac{1}{e}\left(1+\frac{1}{i}\right)^{i+1/2}=\frac{e}{\sqrt{2\pi}},\quad \prod_{i=1}^{\infty}\frac{(2i-1)^2}{(4i-1)(4i-3)}=\frac{\sqrt{2}}{4},$$

$$\prod_{i=0}^{\infty}\frac{100i(i+1)+25}{100i(i+1)+9}=1+\sqrt{5},\quad \prod_{i=0}^{\infty}\frac{100i(i+1)+49}{100i(i+1)+25}=\cosh\frac{\sqrt{6}\pi}{5},$$

$$\prod_{i=1}^{\infty}\frac{1}{1-\tan^2 2^{-i}}=\tan 1,\quad \prod_{i=1}^{\infty}\frac{\sqrt[i]{e}}{1+\frac{1}{i}}=e^{\gamma},\quad \prod_{i=1}^{\infty}\left(1+(-1)^i\frac{1}{F_i^2}\right)=\frac{\sqrt{5}+1}{2},$$

$$\prod_{i=0}^{\infty}\left(\frac{i+1}{i+2}\right)^{2i+2}\left(\frac{2i+3}{2i+1}\right)^{2i+1}=\lim_{n\to\infty}\prod_{i=1}^{2n}\left(1+\frac{2}{i}\right)^{(-1)^{i+1}i}=\frac{\pi}{2e}\approx 0.577863674895,$$

$$\prod_{i=1}^{\infty}\left(\frac{i+1}{i}\right)^{2i}\left(\frac{2i+1}{2i+3}\right)^{2i+1}=\lim_{n\to\infty}\prod_{i=2}^{2n+1}\left(1+\frac{2}{i}\right)^{(-1)^i i}=\frac{6}{\pi e}\approx 0.70259797829182,$$

$$\prod_{i=1}^{\infty}\left(\frac{4i-1}{4i-3}\right)^{4i-2}\left(\frac{4i-1}{4i+1}\right)^{4i}=e^{4G/\pi-1},\quad \prod_{i=1}^{\infty}\frac{1}{e^{1/i}}\left(1+\frac{1}{i}+\frac{1}{2i^2}\right)=\frac{e^{\pi/2}+e^{-\pi/2}}{\pi e^{\gamma}},$$

$$\prod_{i=0}^{\infty}\left(1+\frac{1}{e^{(2i+1)\pi}}\right)=\frac{\sqrt[4]{2}}{e^{\pi/24}},\quad \prod_{i=1}^{\infty}(1+2e^{-i\sqrt{3}\pi}\cosh i\sqrt{3}\pi/3)=\frac{e^{\sqrt{3}\pi/18}}{\sqrt[4]{3}},$$

$$\lim_{n\to\infty}\prod_{i=1}^{n}\frac{\cosh(i^2+i+\frac{1}{2})+[\sinh(i+\frac{1}{2})]J}{\cosh(i^2+i+\frac{1}{2})-[\sinh(i+\frac{1}{2})]J}=\frac{e^2-1+2eJ}{e^2+1},$$

$$\prod_{i=1}^{\infty}\frac{(4i-1)^2[(4i+1)^2-1]}{(4i+1)^2[(4i-1)^2-1]}=\frac{1}{16\pi^2}\Gamma^4(\tfrac{1}{4}).$$

Source: [109, pp. 288, 289], [112, pp. 39], [517, pp. 4–6], [522, 667], [1158, pp. 124, 125], [1217, p. 897], [1410, 1433, 1434, 1563], [1566, pp. 112, 113], [2012], [2013, pp. 99–104], [2079], [2249, p. 421], and [2517, 2531, 2861]. **Related:** Fact 1.13.14 and Fact 13.5.86.

Fact 13.10.12.

$$\lim_{n\to\infty}\prod_{i=0}^{2n-1}\frac{2n+1+3i}{2n+3i}=\sqrt[3]{4},\quad \lim_{n\to\infty}\prod_{i=1}^{n-1}\frac{n+1+7i}{n+7i}=\sqrt[7]{8}.$$

Source: [2461].

Fact 13.10.13. Let $x>1$. Then,

$$\prod_{i=1}^{\infty}\left(1+(-1)^{i+1}\frac{1}{ix}\right)=\frac{\sqrt{\pi}}{\Gamma(1-\frac{1}{2x})\Gamma(\frac{x+1}{2x})}.$$

In particular,

$$\prod_{i=1}^{\infty}\left(1+(-1)^{i+1}\frac{1}{i}\right)=1,\quad \prod_{i=1}^{\infty}\left(1+(-1)^{i+1}\frac{1}{2i}\right)=\frac{\sqrt{\pi}}{\Gamma^2(\frac{3}{4})},\quad \prod_{i=1}^{\infty}\left(1+(-1)^{i+1}\frac{1}{3i}\right)=\frac{\sqrt{\pi}}{\Gamma(\frac{2}{3})\Gamma(\frac{5}{6})}.$$

Fact 13.10.14.

$$\prod_{i=1}^{\infty}\left(1+\frac{1}{i^2}\right)^{(-1)^{i-1}}=\frac{\pi}{2}\tanh\frac{\pi}{2},\quad \prod_{i=1}^{\infty}\left(\frac{i^2+1}{i^2-1}\right)^{(-1)^{i-1}}=\frac{2}{\pi}\tanh\frac{\pi}{2},\quad \prod_{i=2}^{\infty}\left(1-\frac{1}{i^2}\right)^{(-1)^{i-1}}=\frac{\pi^2}{8},$$

$$\prod_{i=1}^{\infty}\left(\sqrt{\frac{\pi}{2}}\frac{(2i-1)!!\,\sqrt{2i+1}}{2^i i!}\right)^{(-1)^{i-1}}=\frac{\sqrt[4]{8}}{\sqrt{\pi}}.$$

If $n\ge 2$, then

$$\prod_{i=1}^{\infty}\left(1+\frac{1}{\lfloor\sqrt[n]{i}\rfloor}\right)^{(-1)^{i-1}}=\prod_{i=2^n}^{\infty}\left(1-\frac{1}{\lfloor\sqrt[n]{i}\rfloor}\right)^{(-1)^{i-1}}=\prod_{i=1}^{\infty}\left(1+\frac{1}{i}\right)^{(-1)^{i-1}}=\frac{\pi}{2}.$$

If x is a real number such that $x\notin\pi\mathbb{Z}$, then

$$\prod_{i=1}^{\infty}\left(\frac{x}{\sin x}\prod_{j=1}^{i}\left(1-\frac{x^2}{(j\pi)^2}\right)\right)^{(-1)^{i-1}}=\frac{2}{x\sin\frac{x}{2}}.$$

If $z\in\mathrm{ORHP}$, then

$$\prod_{i=1}^{\infty}\left(1+\frac{z}{i}\right)^{(-1)^{i-1}}=\frac{2\Gamma^2(z/2)2^{z-2}}{\Gamma(z)}.$$

In particular, if $n\ge 1$, then

$$\prod_{i=1}^{\infty}\left(1+\frac{2n}{i}\right)^{(-1)^{i-1}}=\prod_{i=1}^{\infty}\left(\frac{i}{n+i}\right)\left(\frac{2n+2i-1}{2i-1}\right)=\frac{4^n}{\binom{2n}{n}}.$$

Source: [1103, pp. 146, 147, 201–204].

Fact 13.10.15. If $x \in [-1, 1)$, then

$$\frac{\sqrt{1-x^2}}{\operatorname{acos} x} = \prod_{i=1}^{\infty} \underbrace{\sqrt{\frac{1}{2} + \frac{1}{2}\sqrt{\frac{1}{2} + \frac{1}{2}\sqrt{\frac{1}{2} + \cdots + \frac{1}{2}\sqrt{\frac{1}{2} + \frac{x}{2}}}}}}_{i \text{ roots}}.$$

If $x > 1$, then

$$\frac{\sqrt{x^2-1}}{\log(x + \sqrt{x^2-1})} = \prod_{i=1}^{\infty} \underbrace{\sqrt{\frac{1}{2} + \frac{1}{2}\sqrt{\frac{1}{2} + \frac{1}{2}\sqrt{\frac{1}{2} + \cdots + \frac{1}{2}\sqrt{\frac{1}{2} + \frac{x}{2}}}}}}_{i \text{ roots}}.$$

In particular,

$$\prod_{i=1}^{\infty} \frac{1}{2} \underbrace{\sqrt{2 + \sqrt{2 + \sqrt{2 + \cdots + \sqrt{2}}}}}_{i \text{ roots}} = \prod_{i=1}^{\infty} \underbrace{\sqrt{\frac{1}{2} + \frac{1}{2}\sqrt{\frac{1}{2} + \frac{1}{2}\sqrt{\frac{1}{2} + \cdots + \frac{1}{2}\sqrt{\frac{1}{2}}}}}}_{i \text{ roots}} = \prod_{i=2}^{\infty} \cos\frac{\pi}{2^i} = \frac{2}{\pi},$$

$$\prod_{i=1}^{\infty} \underbrace{\sqrt{\frac{1}{2} + \frac{1}{2}\sqrt{\frac{1}{2} + \cdots + \frac{1}{2}\sqrt{\frac{3}{4}}}}}_{i \text{ roots}} = \frac{3\sqrt{3}}{2\pi}, \quad \prod_{i=1}^{\infty} \underbrace{\sqrt{\frac{1}{2} + \frac{1}{2}\sqrt{\frac{1}{2} + \cdots + \frac{1}{2}\sqrt{\frac{2+\sqrt{3}}{4}}}}}_{i \text{ roots}} = \frac{3}{\pi}.$$

Source: [1410] and [2107, p. 64]. **Remark:** The fifth equality is due to F. Viète. The equalities of Wallis (see Fact 13.10.6) and Viète are unified and generalized in [2172].

Fact 13.10.16. If $x \in [-1, 1)$, then

$$\frac{2x+1}{3} = \prod_{i=1}^{\infty} \left(2 \underbrace{\sqrt{\frac{1}{2} + \frac{1}{2}\sqrt{\frac{1}{2} + \frac{1}{2}\sqrt{\frac{1}{2} + \cdots + \frac{1}{2}\sqrt{\frac{1}{2} + \frac{x}{2}}}}}}_{i \text{ roots}} - 1 \right).$$

If $x > 2$, then

$$\frac{x+1}{3} = \prod_{i=1}^{\infty} \left(\underbrace{\sqrt{2 + \sqrt{2 + \sqrt{2 + \cdots + \sqrt{2 + x}}}}}_{i \text{ roots}} - 1 \right).$$

Source: [1410].

Fact 13.10.17. Let $n, m \geq 1$. Then,

$$\prod_{i=1}^{n} [(2i-1)^2 - 4m^2] = (-1)^m \frac{4^n}{\pi} \Gamma(n-m+\tfrac{1}{2})\Gamma(n+m+\tfrac{1}{2}), \quad \prod_{i=1}^{m} \frac{2n+2i-1}{2n-2i+1} = \frac{\Gamma(n-m+\frac{1}{2})\Gamma(n+m+\frac{1}{2})}{\Gamma^2(n+\frac{1}{2})}.$$

Source: [1745].

Fact 13.10.18. Let $a, b \in \mathbb{R}$. Then,

$$\prod_{i=1}^{\infty} \left[1 + \left(\frac{a+b}{i+a} \right)^3 \right] = \frac{\Gamma^3(1+a)}{\Gamma(2a+b+1)\Gamma[1 + \frac{1}{2}(a-b+(a+b)\sqrt{3}\jmath)]\Gamma[1 + \frac{1}{2}(a-b-(a+b)\sqrt{3}\jmath)]}.$$

In particular,

$$\prod_{i=1}^{\infty}\left[1+8\left(\frac{a}{i+a}\right)^3\right]=\frac{\Gamma^3(1+a)}{\Gamma(3a+1)\Gamma(1+a\sqrt{3}\jmath)\Gamma(1-a\sqrt{3}\jmath)}=\frac{a^2\Gamma^3(a)\sinh a\sqrt{3}\pi}{\Gamma(3a+1)\sqrt{3}\pi},$$

$$\prod_{i=1}^{\infty}\left[1+\frac{8}{(i+1)^3}\right]=\frac{\sinh\sqrt{3}\pi}{6\sqrt{3}\pi},\quad \prod_{i=1}^{\infty}\left[1+\frac{64}{(i+2)^3}\right]=\frac{\sinh 2\sqrt{3}\pi}{180\sqrt{3}\pi}.$$

Source: [667].

Fact 13.10.19. Let $n\ge 1$, let $a_1,\dots,a_n,b_1,\dots,b_n\in\mathbb{C}\backslash(-\mathbb{N})$, and assume that $\sum_{i=1}^n a_i=\sum_{i=1}^n b_i$. Then,

$$\prod_{j=0}^{\infty}\prod_{i=1}^{n}\frac{j+a_i}{j+b_i}=\prod_{i=1}^{n}\frac{\Gamma(b_i)}{\Gamma(a_i)}.$$

In particular,

$$\prod_{i=0}^{\infty}\frac{(i+9)(i+10)(i+12)(i+15)}{(i+8)(i+11)(i+13)(i+14)}=\frac{15}{14},\quad \prod_{i=0}^{\infty}\frac{(i+3/14)(i+5/14)(i+13/14)}{(i+1/14)(i+9/14)(i+11/14)}=2,$$

$$\prod_{i=0}^{\infty}\frac{(i+15/62)(i+23/62)(i+27/62)(i+29/62)(i+61/62)}{(i+1/62)(i+33/62)(i+35/62)(i+39/62)(i+47/62)}=8.$$

Source: [667]. **Related:** Fact 13.3.2.

Fact 13.10.20. Let $a\in\mathbb{R}$ and $p\in(1,\infty)$. Then,

$$\lim_{n\to\infty}\prod_{i=1}^{n}\frac{n^p+(a-1)i^{p-1}}{n^p-i^{p-1}}=e^{a/p}.$$

In particular,

$$\lim_{n\to\infty}\prod_{i=1}^{n}\frac{n^2+i}{n^2-i}=e,\quad \lim_{n\to\infty}\prod_{i=1}^{n}\frac{n^2}{n^2-i}=\lim_{n\to\infty}\prod_{i=1}^{n}\left(1+\frac{i}{n^2}\right)=\sqrt{e}.$$

Source: [1103, pp. 3, 35], [1566, pp. 48, 61], and [2294, p. 230].

Fact 13.10.21.

$$\lim_{n\to\infty}\prod_{i=1}^{n}\left(1+\frac{i}{n}\right)^{1/i}=e^{\pi^2/12},\quad \lim_{n\to\infty}\prod_{i=1}^{n}\left(1+\frac{i}{n}\right)^{n/i^3}=e^{\pi^2/6},$$

$$\lim_{n\to\infty}\frac{n^{3n/2}}{n!}\prod_{i=1}^{n}\sin\frac{i}{n^{3/2}}=e^{-1/18},\quad \lim_{n\to\infty}\frac{1}{n^4}\prod_{i=1}^{2n}\sqrt[n]{n^2+i^2}=25e^{2\,\mathrm{atan}\,2-4}.$$

Let $x>0$. Then,

$$\lim_{n\to\infty}\prod_{i=1}^{n}\left(1+\frac{i^2x}{n^3}\right)=e^{x/3}.$$

In particular,

$$\lim_{n\to\infty}\prod_{i=1}^{n}\left(1+\frac{i^2}{n^3}\right)=\sqrt[3]{e}.$$

Let $x\in(0,1)$. Then,

$$\lim_{n\to\infty}\prod_{i=1}^{n}\left(1-\frac{i^2x}{n^3}\right)=e^{-x/3}.$$

In particular,

$$\lim_{n\to\infty}\prod_{i=1}^{n}\left(1-\frac{i^2}{2n^3}\right)=e^{-1/6}.$$

Source: [108, pp. 33, 199, 200], [1566, pp. 48, 56, 61, 231], [1757, pp. 230, 231], and [2294, p. 55].

Fact 13.10.22. Let $n \ge 1$. Then,

$$\int_0^1 \frac{1}{\sqrt{1-x^n}}\,\mathrm{d}x = \frac{n+2}{n}\prod_{i=1}^{\infty}\frac{[2(ni+1)+n]i}{(2i+1)(ni+1)}.$$

In particular,

$$\int_0^1 \frac{1}{\sqrt{1-x^2}}\,\mathrm{d}x = 2\prod_{i=1}^{\infty}\frac{4(i+1)i}{(2i+1)^2}=\frac{\pi}{2}.$$

Source: [1492]. **Related:** Fact 14.3.10.

Fact 13.10.23. Let $n \ge 1$. Then,

$$\prod_{i=1}^{\infty}\frac{\left(1+\frac{1}{i}\right)^n}{1+\frac{n}{i}} = n!.$$

Source: [2086]. **Credit:** L. Euler.

Fact 13.10.24. Let $n \ge 2$. Then,

$$\prod\left(1+\frac{1}{i^n}\right)=\frac{\zeta(n)}{\zeta(2n)},$$

where the product is taken over all primes. In particular,

$$\prod\frac{1}{i^2}=\frac{15}{\pi^2},\quad \prod\frac{1}{i^4}=\frac{105}{\pi^4}.$$

Source: [1317, pp. 20, 21]. **Related:** Fact 13.5.38.

Fact 13.10.25. For all $n \ge 0$, let p_n denote the nth partition number, where $p_0 \triangleq 1$, and let $z \in \mathbb{C}$, where $|z| < 1$. Then,

$$\prod_{i=1}^{\infty}\frac{1}{1-z^i}=\sum_{i=0}^{\infty}p_iz^i.$$

Furthermore,

$$\prod_{i=1}^{\infty}\frac{1}{1-z^i}=1+\sum_{i=1}^{\infty}\frac{z^{i^2}}{\prod_{j=1}^{i}(1-z^j)^2}.$$

Remark: See [621, p. 210]. $\prod_{i=1}^{\infty}\frac{1}{1-z^i}$ is the generating function for the partition numbers. The last equality is the *Durfee square identity*. See [771, p. 119]. **Related:** Fact 1.20.1 and Fact 13.1.3.

Fact 13.10.26. Let a and z be complex numbers, and assume that $a \ne 0$ and $|z| < 1$. Then,

$$\prod_{i=1}^{\infty}(1-z^{2i})(1+az^{2i-1})(1-\tfrac{1}{a}z^{2i-1})=\sum_{i=-\infty}^{\infty}a^iz^{i^2}.$$

Equivalently,

$$\prod_{i=1}^{\infty}(1-z^{i})(1+az^{i})(1+\tfrac{1}{a}z^{i-1})=\sum_{i=-\infty}^{\infty}a^iz^{i(i+1)/2}.$$

Source: [118, pp. 80, 81], [155, pp. 319, 320], and [621, pp. 216, 217]. **Remark:** This is *Jacobi's triple product identity.*

Fact 13.10.27. Let $z \in \text{OIUD}$, and define $(e_i)_{i=1}^{\infty}$ as in Fact 1.20.1. Then,

$$\prod_{i=1}^{\infty}(1-z^i) = \sum_{i=-\infty}^{\infty}(-1)^i z^{i(3i-1)/2} = 1 + \sum_{i=1}^{\infty}(-1)^i[z^{i(3i-1)/2} + z^{i(3i+1)/2}] = \sum_{i=0}^{\infty} e_i z^i,$$

$$\prod_{i=1}^{\infty}\frac{(1-z^{2i})(1+z^i)^2}{(1+z^{2i})^2} = \prod_{i=1}^{\infty}\frac{[1-(-z)^i]^2}{1-z^{2i}} = \sum_{i=-\infty}^{\infty} z^{i^2}, \quad \prod_{i=1}^{\infty}\frac{1-z^i}{1+z^i} = \sum_{i=-\infty}^{\infty}(-1)^i z^{i^2},$$

$$\prod_{i=1}^{\infty}(1-z^{2i})(1+z^{i-1}) = \sum_{i=-\infty}^{\infty} z^{i(i+1)/2} = 2\sum_{i=0}^{\infty} z^{i(i+1)/2},$$

$$\sum_{i=0}^{\infty} z^{i(i+1)/2} = \prod_{i=1}^{\infty}\frac{1-z^{2i}}{1-z^{2i-1}} = \prod_{i=1}^{\infty}\frac{(1-z^{2i})^2}{1-z^i} = 1 + z + \sum_{i=1}^{\infty}\left(\prod_{j=1}^{i}\frac{1-z^{2j}}{1-z^{2j+1}}\right)z^{2i+1},$$

$$\sum_{i=2}^{\infty} z^{i(i+1)/2} = \sum_{i=1}^{\infty}\left(\prod_{j=1}^{i}\frac{1-z^{2j}}{1-z^{2j+1}}\right)z^{2i+1}, \quad \prod_{i=1}^{\infty}(1-z^{2i})^3 = \sum_{i=-\infty}^{\infty}(-1)^i(2i-1)z^{i(i+1)},$$

$$\prod_{i=1}^{\infty}(1-z^i)^3 = \sum_{i=-\infty}^{\infty}(-1)^i i z^{i(i+1)/2} = 1 + \sum_{i=1}^{\infty}(-1)^i(2i+1)z^{i(i+1)/2}.$$

Source: These equalities are special cases of Fact 13.10.26. See [118, p. 81], [155, p. 321], [621, p. 215], and [2374, 2885]. **Remark:** The first equality, which is due to L. Euler, is the *pentagonal number theorem.* This infinite product has the form

$$\prod_{i=1}^{\infty}(1-z^i) = 1 - z - z^2 + z^5 + z^7 - z^{12} - z^{15} + z^{22} + z^{26} - z^{35} - z^{40} + \cdots,$$

where the exponents $(g_0, g_1, g_2, \ldots)$ are the generalized pentagonal numbers defined by Fact 1.12.5. The last equality is the *Gauss identity.* See [1083, p. 54]. **Related:** Fact 13.4.7 and Fact 13.5.6.

Fact 13.10.28. Let $z \in \text{OIUD}$. Then,

$$\prod_{i=1}^{\infty}(1+z^{2i-1})^8 = \prod_{i=1}^{\infty}(1-z^{2i-1})^8 + 16z\prod_{i=1}^{\infty}(1+z^{2i})^8,$$

$$\prod_{i=1}^{\infty}(1-z^{2i})^6(1-z^{6i})^6 = \prod_{i=1}^{\infty}(1-z^i)^4(1-z^{3i})^4(1-z^{4i})^2(1-z^{12i})^2 + 4z\prod_{i=1}^{\infty}(1-z^i)^2(1-z^{3i})^2(1-z^{4i})^4(1-z^{12i})^4,$$

$$\prod_{i=1}^{\infty}(1-z^{2i})^4(1-z^{3i})^9 = \prod_{i=1}^{\infty}(1-z^i)^8(1-z^{3i})(1-z^{6i})^4 + 8z\prod_{i=1}^{\infty}(1-z^i)^3(1-z^{2i})(1-z^{6i})^9,$$

$$\prod_{i=1}^{\infty}(1-z^{2i})^9(1-z^{3i})(1-z^{12i})^2 + 2\prod_{i=1}^{\infty}(1-z^i)^3(1-z^{4i})^6(1-z^{6i})^3$$
$$= 3\prod_{i=1}^{\infty}(1-z^i)^2(1-z^{2i})^2(1-z^{3i})^3(1-z^{4i})^3(1-z^{6i})(1-z^{12i}).$$

Now, let $n \geq 1$. Then,

$$\sum_{j=0}^{n}\binom{n}{j}2^j z^j\prod_{i=1}^{\infty}(1-z^i)^{2n+j}(1-z^{3i})^{2n-3j}(1-z^{4i})^{3n-j}(1-z^{12i})^{3j} = \left[\prod_{i=1}^{\infty}(1-z^{2i})^7(1+z^{6i})\right]^n.$$

Source: [27]. **Remark:** The first equality is *Jacobi's abstruse identity.* **Remark:** Additional equalities are given in [28].

Fact 13.10.29. Let $a \in (0, 1)$, let $z \in \mathbb{C}$, and assume that $a < |z| < 1/a$. Then,

$$\prod_{i=0}^{\infty} \frac{(1 + a^{2i+2})(1 + a^{2i+1}z)(z + a^{2i+1})}{(1 - a^{2i+2})(1 - a^{2i+1}z)(z - a^{2i+1})} = \sum_{i=-\infty}^{\infty} \frac{1}{\cosh i \log a} z^i.$$

Source: [2399]. **Remark:** The infinite product is analytic on $\mathbb{C}\backslash(\{0\} \cup \{a^{2i+1} : i \in \mathbb{Z}\})$.

13.11 Notes

An introductory treatment of limits and continuity is given in [2112]. An essential closure point is traditionally called either a *limit point* or an *accumulation point.* *i)* $\Longrightarrow$ *iii)* in Proposition 12.2.8 is given in [1647, p. 70].

A power series is also called a *Taylor series.*

Generating functions are presented within the context of hypergeometric functions in [2514].

The derivative and the directional differential are typically called the Fréchet derivative and the Gâteaux differential, respectively [1060]. Differentiation of matrix functions is considered in [1343, 1924, 1981, 2247, 2324, 2416]. An extensive treatment of matrix functions is given in Chapter 6 of [1450]; see also [1456]. The identity theorem is discussed in [1498]. A chain rule for matrix functions is considered in [1924, 1989]. Differentiation with respect to complex matrices is discussed in [1554]. Extensive tables of derivatives of matrix functions are given in [835, pp. 586–593].

Chapter Fourteen
Integrals

14.1 Facts on Indefinite Integrals

Fact 14.1.1. Let $x > 0$ and $a \in \mathbb{R}$. Then, the following statements hold:

i) $\int_0^x t^a \,\mathrm{d}t$ exists if and only if $a > -1$.

ii) $\int_x^\infty t^a \,\mathrm{d}t$ exists if and only if $a < -1$.

Fact 14.1.2. Let $x \in \mathbb{R}$ and $a > -1$. Then

$$\int_0^x t^a \,\mathrm{d}t = \frac{x^{a+1}}{a+1}.$$

In particular,

$$\int_{-1}^0 \frac{1}{\sqrt{t}} \,\mathrm{d}t = -2\jmath, \quad \int_0^1 \frac{1}{\sqrt{t}} \,\mathrm{d}t = 2, \quad \int_{-(3/2)^{2/3}}^0 \sqrt{t} \,\mathrm{d}t = \jmath,$$

$$\int_{-1}^0 \sqrt{t} \,\mathrm{d}t = \frac{2\jmath}{3}, \quad \int_0^1 \sqrt{t} \,\mathrm{d}t = \frac{2}{3}, \quad \int_0^{(3/2)^{2/3}} \sqrt{t} \,\mathrm{d}t = 1.$$

Fact 14.1.3. Let $x > 0$. If $a < -1$, then

$$\int_x^\infty t^a \,\mathrm{d}t = -\frac{x^{a+1}}{a+1}.$$

If $a > 1$, then

$$\int_x^\infty \frac{1}{t^a} \,\mathrm{d}t = \frac{x^{1-a}}{a-1}, \quad \int_1^\infty \frac{1}{t^a} \,\mathrm{d}t = \frac{1}{a-1}.$$

Fact 14.1.4. Let $x \in \mathbb{R}$ and $a > 0$. Then,

$$\int_0^x a^t \,\mathrm{d}t = \frac{a^x - 1}{\log a}.$$

Fact 14.1.5. Let a, b, c, x be real numbers. Then,

$$\int_0^x \frac{1}{at^2 + bt + c} \,\mathrm{d}t = \begin{cases} \dfrac{2}{\sqrt{4ac - b^2}}\left(\operatorname{atan} \dfrac{2ax + b}{\sqrt{4ac - b^2}} - \operatorname{atan} \dfrac{b}{\sqrt{4ac - b^2}}\right), & b^2 < 4ac, \\ \dfrac{4ax}{2abx + b^2}, & b^2 = 4ac \neq 0, \\ \dfrac{-2}{\sqrt{b^2 - 4ac}}\left(\operatorname{atanh} \dfrac{2ax + b}{\sqrt{b^2 - 4ac}} - \operatorname{atanh} \dfrac{b}{\sqrt{b^2 - 4ac}}\right), & 4ac < b^2. \end{cases}$$

In particular,

$$\int_0^x \frac{1}{t^2 + t + 1} \,\mathrm{d}t = \frac{2\sqrt{3}}{3}\left(\operatorname{atan} \frac{\sqrt{3}(2x + 1)}{3} - \frac{\pi}{6}\right),$$

$$\int_0^1 \frac{1}{t^2 + 3t + 1} \,\mathrm{d}t = \frac{2\sqrt{5}}{5}\left(\operatorname{atanh} \frac{3\sqrt{5}}{5} - \operatorname{atanh} \sqrt{5}\right) = \frac{2\sqrt{5}\operatorname{acoth}\sqrt{5}}{5}.$$

Fact 14.1.6. Let x be a real number. Then,

$$\int_0^x \frac{1}{t^3+t^2+t+1}\,\mathrm{d}t = \frac{1}{4}\log\frac{(x+1)^2}{x^2+1} + \frac{1}{2}\operatorname{atan} x.$$

Related: Fact 14.2.31.

Fact 14.1.7. Let $a > 0$ and $x > -a$. Then,

$$\int_0^x \frac{1}{t^3+a^3}\,\mathrm{d}t = \frac{1}{6a^2}\log\frac{(x+a)^2}{x^2-ax+a^2} + \frac{\sqrt{3}}{3a^2}\left(\operatorname{atan}\frac{\sqrt{3}(2x-a)}{3a} + \frac{\pi}{6}\right).$$

In particular, if $x > -1$, then

$$\int_0^x \frac{1}{t^3+1}\,\mathrm{d}t = \frac{1}{6}\log\frac{(x+1)^2}{x^2-x+1} + \frac{\sqrt{3}}{3}\left(\operatorname{atan}\frac{\sqrt{3}(2x-1)}{3} + \frac{\pi}{6}\right).$$

Remark: A similar formula can be obtained for $\int_0^x \frac{1}{t^5+a}\,\mathrm{d}t$.

Fact 14.1.8. Let $x \in \mathbb{R}$. Then,

$$\int_0^x \frac{t^4(t-1)^4}{t^2+1}\,\mathrm{d}t = \frac{1}{7}x^7 - \frac{2}{3}x^6 + x^5 - \frac{4}{3}x^3 + 4x - 4\operatorname{atan} x.$$

In particular,

$$\int_0^1 \frac{t^4(t-1)^4}{t^2+1}\,\mathrm{d}t = \frac{22}{7} - \pi.$$

Source: [522]. **Remark:** $\pi < 22/7$.

Fact 14.1.9. Let $a > 0$ and $x \in [0, a)$. Then,

$$\int_0^x \frac{t^2}{a^2-t^2}\,\mathrm{d}t = \frac{a}{2}\log\frac{a+x}{a-x} - x = a\operatorname{atanh}\frac{x}{a} - x,$$

$$\int_0^x \frac{t^2}{a^4-t^4}\,\mathrm{d}t = \frac{1}{2a}\operatorname{atan}\frac{a}{x} + \frac{1}{4a}\log\frac{a+x}{a-x} - \frac{\pi}{4a} = \frac{1}{2a}\operatorname{atan}\frac{a}{x} + \frac{1}{2a}\operatorname{atanh}\frac{x}{a} - \frac{\pi}{4a},$$

$$\int_0^x \frac{t^2}{a^6-t^6}\,\mathrm{d}t = \frac{1}{6a^3}\log\frac{a^3+x^3}{a^3-x^3}.$$

Source: [1563].

Fact 14.1.10. Let $a > 0$, let $n \ge 2$, assume that n is even, and let $x \in [0, \sqrt[n+1]{a})$. Then,

$$\int_0^x \frac{t^n}{t^{n+1}-a}\,\mathrm{d}t = \frac{1}{n+1}\log\frac{a-x^{n+1}}{a}.$$

In particular, if $x \in [0, \sqrt[3]{a})$, then

$$\int_0^x \frac{t^2}{t^3-a}\,\mathrm{d}t = \frac{1}{3}\log\frac{a-x^3}{a}.$$

Source: [1563].

Fact 14.1.11. Let $a, c > 0$, $b \ge 0$, and $x \in [0, \infty)$. Then,

$$\int_0^x \sqrt{at^2+c}\,\mathrm{d}t = \tfrac{1}{2}x\sqrt{ax^2+c} + \frac{\sqrt{ac}}{2a}\log\frac{ax+\sqrt{a^2x^2+ac}}{\sqrt{ac}},$$

$$\int_0^x \sqrt{at^2+bt}\,\mathrm{d}t = \frac{ax(2a^2x^2+3abx+b^2) - b^2\sqrt{a^2x^2+abx}\log(\sqrt{a^2x+ab}+a\sqrt{x})}{4a^2\sqrt{ax^2+bx}} + \frac{b^2}{8a^{3/2}}\log ab.$$

Fact 14.1.12. Let $a > 0$ and $x \in [0, a)$. Then,

$$\int_0^x \sqrt{a-t}\,\mathrm{d}t = \frac{2}{3}[a^{3/2} - (a-x)^{3/2}], \quad \int_0^x t\sqrt{a-t}\,\mathrm{d}t = \frac{2}{15}[2a^{5/2} - (2a+3x)(a-x)^{3/2}],$$

$$\int_0^x t^2\sqrt{a-t}\,\mathrm{d}t = \frac{2}{105}[8a^{7/2} - (8a^3 + 4a^2x + 3ax^2 - 15x^3)\sqrt{a-x}],$$

$$\int_0^x \sqrt{a^2-t^2}\,\mathrm{d}t = \frac{1}{2}\left(x\sqrt{a^2-x^2} + a^2 \operatorname{atan}\frac{x}{\sqrt{a^2-x^2}}\right), \quad \int_0^x t\sqrt{a^2-t^2}\,\mathrm{d}t = \frac{1}{3}[a^3 - (a^2-x^2)^{3/2}],$$

$$\int_0^x t^2\sqrt{a^2-t^2}\,\mathrm{d}t = \frac{1}{8}\left(x(2x^2-a^2)\sqrt{a^2-x^2} + a^4 \operatorname{asin}\frac{x}{a}\right).$$

Fact 14.1.13. Let $a > 0$ and $x \in [0, a)$. Then,

$$\int_0^x \frac{1}{\sqrt{a-t}}\,\mathrm{d}t = 2(\sqrt{a} - \sqrt{a-x}), \quad \int_0^x \frac{t}{\sqrt{a-t}}\,\mathrm{d}t = \frac{2}{3}[2a^{3/2} - (x+2a)\sqrt{a-x}],$$

$$\int_0^x \frac{t^2}{\sqrt{a-t}}\,\mathrm{d}t = \frac{2}{15}[8a^{5/2} - (3x^2 + 4ax + 8a^2)\sqrt{a-x}],$$

$$\int_0^x \frac{1}{\sqrt{a^2-t^2}}\,\mathrm{d}t = \operatorname{asin}\frac{x}{a} = -\operatorname{atan}\frac{t\sqrt{a^2-t^2}}{t^2-a^2}, \quad \int_0^x \frac{t}{\sqrt{a^2-t^2}}\,\mathrm{d}t = a - \sqrt{a^2-x^2},$$

$$\int_0^x \frac{t^2}{\sqrt{a^2-t^2}}\,\mathrm{d}t = \frac{1}{2}\left(a^2\operatorname{asin}\frac{x}{a} - x\sqrt{a^2-x^2}\right), \quad \int_0^x \frac{t^3}{\sqrt{a^2-t^2}}\,\mathrm{d}t = \frac{1}{3}[2a^3 - (x^2+2a^2)\sqrt{a^2-x^2}].$$

Fact 14.1.14. Let $x \in [1, \infty)$. Then,

$$\int_1^x \sqrt{\frac{t-1}{t+1}}\,\mathrm{d}t = \sqrt{x^2-1} - \log(x + \sqrt{x^2-1}).$$

Fact 14.1.15. Let x and a be real numbers, and assume that $0 < x < a$. Then,

$$\int_0^x \frac{t+a}{t-a}\,\mathrm{d}x = 2a\log\frac{a-x}{a} + x.$$

Fact 14.1.16. Let x be a real number and a be a positive number. Then,

$$\int_0^x \frac{\sin^2 t}{\sin^2 t + a}\,\mathrm{d}t = x - \sqrt{\frac{a}{a+1}}\operatorname{atan}\left(\sqrt{\frac{a+1}{a}}\tan x\right).$$

Source: Fact 14.4.47.

Fact 14.1.17. Let $x \in (-\frac{\pi}{2}, \frac{\pi}{2})$. Then,

$$\int_0^x \sec t\,\mathrm{d}t = \log(\sec x + \tan x) = \log\tan(\tfrac{x}{2} + \tfrac{\pi}{4}).$$

Source: [2309]. **Remark:** This integral has applications to cartography. See [49, p. 175], [1027, p. 64], and [1945, p. 176].

Fact 14.1.18. Let $x \in [0, \frac{\pi}{2}]$. Then,

$$\int_0^x \left(\frac{1}{t} - \cot t\right)\mathrm{d}x = \log\frac{x}{\sin x}.$$

In particular,

$$\int_0^{\pi/6} \left(\frac{1}{t} - \cot t\right)\mathrm{d}t = \log\frac{\pi}{3}, \quad \int_0^{\pi/4} \left(\frac{1}{t} - \cot t\right)\mathrm{d}t = \log\pi - \frac{3}{2}\log 2,$$

$$\int_0^{\pi/3}\left(\frac{1}{t}-\cot t\right)\mathrm{d}t=\log\frac{2\sqrt{3}\pi}{9},\qquad \int_0^{\pi/2}\left(\frac{1}{t}-\cot t\right)\mathrm{d}t=\log\frac{\pi}{2}.$$

Fact 14.1.19. Let $x\in[-1,1]$. Then,

$$\int_0^x\frac{\operatorname{asin} t}{\sqrt{1-t^2}}\,\mathrm{d}t=\frac{1}{2}\operatorname{asin}^2 x.$$

Fact 14.1.20. Let $x\ge 0$, $a>0$, and $b>1$. Then,

$$\int_0^x\frac{1}{be^{at}-1}\,\mathrm{d}t=\frac{1}{a}\log\frac{b-e^{-at}}{b-1}.$$

14.2 Facts on Definite Integrals of Rational Functions

Fact 14.2.1. The following statements hold:

i) Let $n\ge 1$. Then,

$$\int_0^1(1-x^2)^n\,\mathrm{d}x=\frac{(2n)!!}{(2n+1)!!}=\frac{4^n(n!)^2}{(2n+1)!}.$$

ii) Let $n\ge 1$, assume that n is odd, and let $a>0$. Then,

$$\int_0^a(a^2-x^2)^{n/2}\,\mathrm{d}x=\frac{n!!}{(n+1)!!}\frac{a^{n+1}\pi}{2}.$$

iii) Let a,b,c be real numbers, and assume that $a>0$, $b>-1$, and $c>-2$. Then,

$$\int_0^a x^b(a^2-x^2)^{c/2}\,\mathrm{d}x=\frac{a^{b+c+1}\Gamma\left(\frac{b+1}{2}\right)\Gamma\left(\frac{c+2}{2}\right)}{2\Gamma\left(\frac{b+c+3}{2}\right)}.$$

iv) Let $n\ge 1$ and $m\ge 0$. Then,

$$\int_0^1(1-x^{1/n})^m\,\mathrm{d}x=\frac{1}{\binom{n+m}{m}}.$$

v) Let $a,b\in(-1,\infty)$. Then,

$$\int_0^1 x^a(1-x)^b\,\mathrm{d}x=\int_0^\infty\frac{x^a}{(x+1)^{a+b+2}}\,\mathrm{d}x=\frac{\Gamma(a+1)\Gamma(b+1)}{\Gamma(a+b+2)}.$$

Now, let $n,m\ge 0$. Then,

$$\int_0^1 x^n(1-x)^m\,\mathrm{d}x=\int_0^\infty\frac{x^n}{(x+1)^{n+m+2}}\,\mathrm{d}x=\frac{n!m!}{(n+m+1)!}.$$

If, in particular, $n\le m$, then

$$\int_0^1 x^n(1-x)^{m-n}\,\mathrm{d}x=\frac{1}{(m+1)\binom{m}{n}}.$$

vi) Let $n\ge 0$ and $\alpha>-1$. Then,

$$\int_0^1 x^n(1-x)^\alpha\,\mathrm{d}x=\frac{n!}{(\alpha+n+1)^{\underline{n+1}}}.$$

vii) Let $p, q \in (-1, \infty)$ and $a, b \in \mathbb{R}$, where $a < b$. Then,

$$\int_a^b (x-a)^p(b-x)^q\,\mathrm{d}x = (b-a)^{p+q+1}\frac{\Gamma(p+1)\Gamma(q+1)}{\Gamma(p+q+2)}.$$

Now, let $n, m \geq 0$. Then,

$$\int_a^b (x-a)^n(b-x)^m\,\mathrm{d}x = (b-a)^{n+m+1}\frac{n!m!}{(n+m+1)!}.$$

In particular,

$$\int_a^b (x-a)^n(b-x)^n\,\mathrm{d}x = (b-a)^{2n+1}\frac{(n!)^2}{(2n+1)!} = \frac{(2n)!!}{(2n+1)!!}\left(\frac{b-a}{2}\right)^{2n+1}.$$

Source: *i*) is given in [819], [1568, p. 145], and [1675, p. 56]; *ii*) and *iii*) are given in [3024, p. 330]; *iv*) is given in [1610]; *v*) is given in [1885]; *vi*) is given in [1568, p. 165]. **Remark:** Recursions involving the integrals in *vii*) and their connection with series involving binomial coefficients are given in [2947].

Fact 14.2.2.

$$\int_0^1 x^x\,\mathrm{d}x = \sum_{i=1}^\infty (-1)^{i+1}\frac{1}{i^i} \approx 0.78343051071, \quad \int_0^1 x^{x+1}\,\mathrm{d}x = \sum_{i=1}^\infty (-1)^{i+1}\frac{1}{(i+1)^i} \approx 0.40303444442,$$

$$\int_0^1 \frac{1}{x^x}\,\mathrm{d}x = \sum_{i=1}^\infty \frac{1}{i^i} \approx 1.2912859971, \quad \int_0^1 \frac{x}{x^x}\,\mathrm{d}x = \sum_{i=1}^\infty \frac{1}{(i+1)^i} \approx 0.6284737129,$$

$$\int_0^1 \frac{1-x^x}{x}\,\mathrm{d}x = \sum_{i=1}^\infty (-1)^{i+1}\frac{1}{i^{i+1}} \approx 0.88642971056, \quad \int_0^1 \frac{x^{-x}-1}{x}\,\mathrm{d}x = \sum_{i=1}^\infty \frac{1}{i^{i+1}} \approx 1.138389995.$$

Source: [506, pp. 103, 104], [517, pp. 4, 44], and [1179]. **Related:** Fact 14.8.7.

Fact 14.2.3. The following statements hold:

i) Let $n \geq 0$. Then,

$$\int_0^1 \frac{x^n}{x+1}\,\mathrm{d}x = (-1)^n(H_{\lfloor n/2\rfloor} - H_n + \log 2).$$

ii) Let $n \geq 0$. Then,

$$\int_0^1 \frac{x^{2n}}{x+1}\,\mathrm{d}x = H_n - H_{2n} + \log 2 = \sum_{i=2n+1}^\infty (-1)^{i+1}\frac{1}{i}.$$

In particular,

$$\int_0^1 \frac{1}{x+1}\,\mathrm{d}x = \log 2, \quad \int_0^1 \frac{x^2}{x+1}\,\mathrm{d}x = \log 2 - \tfrac{1}{2}.$$

iii) Let $n \geq 0$. Then,

$$\int_0^1 \frac{x^{2n+1}}{x+1}\,\mathrm{d}x = H_{2n+1} - H_n - \log 2.$$

In particular,

$$\int_0^1 \frac{x}{x+1}\,\mathrm{d}x = 1 - \log 2, \quad \int_0^1 \frac{x^3}{x+1}\,\mathrm{d}x = \frac{5}{6} - \log 2.$$

iv) Let $n \geq 0$. Then,

$$\int_0^1 \frac{x^{2n}}{x^2+1}\,\mathrm{d}x = (-1)^n\frac{\pi}{4} + \sum_{i=1}^n (-1)^{i+1}\frac{1}{2n-2i+1}.$$

In particular,

$$\int_0^1 \frac{1}{x^2+1}\,\mathrm{d}x = \frac{\pi}{4}, \quad \int_0^1 \frac{x^2}{x^2+1}\,\mathrm{d}x = 1-\frac{\pi}{4}, \quad \int_0^1 \frac{x^4}{x^2+1}\,\mathrm{d}x = \frac{\pi}{4}-\frac{2}{3},$$

$$\int_0^1 \frac{x^6}{x^2+1}\,\mathrm{d}x = \frac{13}{15}-\frac{\pi}{4}, \quad \int_0^1 \frac{x^8}{x^2+1}\,\mathrm{d}x = \frac{\pi}{4}-\frac{76}{105}, \quad \int_0^1 \frac{x^{10}}{x^2+1}\,\mathrm{d}x = \frac{263}{315}-\frac{\pi}{4}.$$

v) Let $n \ge 0$. Then,

$$\int_0^1 \frac{x^{4n+1}}{x^2+1}\,\mathrm{d}x = \frac{1}{2}(H_n - H_{2n} + \log 2).$$

In particular,

$$\int_0^1 \frac{x}{x^2+1}\,\mathrm{d}x = \frac{\log 2}{2}, \quad \int_0^1 \frac{x^5}{x^2+1}\,\mathrm{d}x = \frac{\log 4 - 1}{4}, \quad \int_0^1 \frac{x^9}{x^2+1}\,\mathrm{d}x = \frac{12\log 2 - 7}{24}.$$

vi) Let $n \ge 0$. Then,

$$\int_0^1 \frac{x^{4n+3}}{x^2+1}\,\mathrm{d}x = \frac{1}{2}(H_{2n+1} - H_n - \log 2).$$

In particular,

$$\int_0^1 \frac{x^3}{x^2+1}\,\mathrm{d}x = \frac{1-\log 2}{2}, \quad \int_0^1 \frac{x^7}{x^2+1}\,\mathrm{d}x = \frac{5-6\log 2}{12}, \quad \int_0^1 \frac{x^{11}}{x^2+1}\,\mathrm{d}x = \frac{47-60\log 2}{120}.$$

vii) Let $n \ge 1$. Then,

$$\int_0^1 \frac{x^{n-1}}{(x+1)^n}\,\mathrm{d}x = \log 2 - \sum_{i=1}^{n-1} \frac{1}{i2^i}.$$

Source: *ii*) is given in [2013, p. 178]. *vii*) is given in [1100].

Fact 14.2.4.

$$\int_0^1 \frac{1}{x^3+1}\,\mathrm{d}x = \frac{\sqrt{3}\pi}{9} + \frac{\log 2}{3}, \quad \int_0^1 \frac{x}{x^3+1}\,\mathrm{d}x = \frac{\sqrt{3}\pi}{9} - \frac{\log 2}{3}, \quad \int_0^1 \frac{x^2}{x^3+1}\,\mathrm{d}x = \frac{\log 2}{3},$$

$$\int_0^1 \frac{1}{x^4+1}\,\mathrm{d}x = \frac{\sqrt{2}\pi}{8} + \frac{\sqrt{2}\log(3+2\sqrt{2})}{8}, \quad \int_0^1 \frac{x}{x^4+1}\,\mathrm{d}x = \frac{\pi}{8},$$

$$\int_0^1 \frac{x^2}{x^4+1}\,\mathrm{d}x = \frac{\sqrt{2}\pi}{8} - \frac{\sqrt{2}\log(3+2\sqrt{2})}{8}, \quad \int_0^1 \frac{x^3}{x^4+1}\,\mathrm{d}x = \frac{\log 2}{4},$$

$$\int_0^1 \frac{x^2+1}{x^4+1}\,\mathrm{d}x = \frac{\sqrt{2}\pi}{4}, \quad \int_0^1 \frac{x^6}{(x^2+1)(x^4+1)}\,\mathrm{d}x = 1 - \frac{(1+\sqrt{2})\pi}{8},$$

$$\int_0^1 \frac{1}{x^5+1}\,\mathrm{d}x = \frac{1}{10}\left(\sqrt{\frac{10+2\sqrt{5}}{5}}\pi + \sqrt{5}\,\mathrm{atanh}\,\frac{\sqrt{5}}{3} + 2\log 2\right).$$

Fact 14.2.5. Let a and b be positive numbers. Then,

$$\int_0^1 \frac{x}{(a^2x^2+1)(bx+1)}\,\mathrm{d}x = \frac{1}{a^2+b^2}\left(\log\frac{\sqrt{a^2+1}}{b+1} + \frac{b}{a}\,\mathrm{atan}\,a\right),$$

$$\int_0^1 \frac{x^2}{(a^2x^2+1)(bx+1)}\,\mathrm{d}x = \frac{1}{b(a^2+b^2)}\left[\frac{b^2}{2a^2}\log(a^2+1) + \log(b+1) - \frac{b}{a}\,\mathrm{atan}\,a\right].$$

Source: [560].

Fact 14.2.6. The following statements hold:

i) Let $n \geq 1$. Then,

$$\int_0^1 \frac{1-x^n}{1-x}\,\mathrm{d}x = \int_0^1 \frac{1-(1-x)^n}{x}\,\mathrm{d}x = H_n.$$

ii) Let $m, n \geq 1$. Then,

$$\int_0^1 \frac{1-x^m}{1-x^n}\,\mathrm{d}x = \sum_{i=0}^{\infty} \frac{m}{(ni+1)(ni+m+1)}.$$

In particular,

$$\int_0^1 \frac{1-x}{1-x^2}\,\mathrm{d}x = \log 2, \quad \int_0^1 \frac{1-x}{1-x^3}\,\mathrm{d}x = \frac{\sqrt{3}\pi}{9}, \quad \int_0^1 \frac{1-x^2}{1-x^3}\,\mathrm{d}x = \frac{\sqrt{3}\pi}{18} + \frac{\log 3}{2},$$

$$\int_0^1 \frac{1-x}{1-x^4}\,\mathrm{d}x = \frac{\pi}{8} + \frac{\log 2}{4}, \quad \int_0^1 \frac{1-x^2}{1-x^4}\,\mathrm{d}x = \frac{\pi}{4}, \quad \int_0^1 \frac{1-x^3}{1-x^4}\,\mathrm{d}x = \frac{\pi}{8} + \frac{3\log 2}{4}.$$

iii) Let $n \geq 1$. Then,

$$\int_0^1 \frac{1-(1-x)^n}{x} \log \frac{1}{1-x}\,\mathrm{d}x = H_{n,2}.$$

Source: *i)* follows from $\frac{1-x^n}{1-x} = 1 + x + \cdots + x^{n-1}$ and [511, p. 176]; *ii)* is given in [2455]; *iii)* is given in [2013, p. 178]. **Related:** Fact 13.5.68 gives expressions for the infinite series in *ii)*.

Fact 14.2.7. Let $p > -1$. Then,

$$\int_0^1 \frac{1-x^p}{1-x}\,\mathrm{d}x = H_p = \psi(p+1) + \gamma.$$

In particular,

$$\int_0^1 \frac{1-x^{\ 1/2}}{1-x}\,\mathrm{d}x = -2\log 2, \quad \int_0^1 \frac{1-x^{-1/3}}{1-x}\,\mathrm{d}x = \frac{\sqrt{3}\pi}{6} - \frac{3\log 3}{2},$$

$$\int_0^1 \frac{1-x^{-1/4}}{1-x}\,\mathrm{d}x = \frac{\pi}{2} - 3\log 2, \quad \int_0^1 \frac{1-x^{1/6}}{1-x}\,\mathrm{d}x = 6 - \frac{\sqrt{3}\pi}{2} - \frac{3\log 3}{2} - 2\log 2,$$

$$\int_0^1 \frac{1-x^{1/4}}{1-x}\,\mathrm{d}x = 4 - \frac{\pi}{2} - 3\log 2, \quad \int_0^1 \frac{1-x^{1/3}}{1-x}\,\mathrm{d}x = 3 - \frac{\sqrt{3}\pi}{6} - \frac{3\log 3}{2},$$

$$\int_0^1 \frac{1-x^{1/2}}{1-x}\,\mathrm{d}x = 2 - 2\log 2, \quad \int_0^1 \frac{1-x^{2/3}}{1-x}\,\mathrm{d}x = \frac{3}{2} + \frac{\sqrt{3}\pi}{6} - \frac{3\log 3}{2},$$

$$\int_0^1 \frac{1-x^{3/4}}{1-x}\,\mathrm{d}x = \frac{4}{3} + \frac{\pi}{2} - 3\log 2, \quad \int_0^1 \frac{1-x^{3/2}}{1-x}\,\mathrm{d}x = \frac{8}{3} - 2\log 2.$$

$$\int_0^1 \frac{1-x^{5/2}}{1-x}\,\mathrm{d}x = \frac{46}{15} - 2\log 2, \quad \int_0^1 \frac{1-x^{7/2}}{1-x}\,\mathrm{d}x = \frac{352}{105} - 2\log 2.$$

If, in addition, m and n are positive integers such that $m < n$, then

$$\int_0^1 \frac{1-x^{m/n}}{1-x}\,\mathrm{d}x = \frac{n}{m} + 2\sum_{i=1}^{\lfloor (n-1)/2 \rfloor} \left(\cos \frac{2mi\pi}{n}\right) \log \sin \frac{i\pi}{n} - \frac{\pi}{2} \cot \frac{m\pi}{n} - \log 2n.$$

In particular,

$$\int_0^1 \frac{1-x^{1/7}}{1-x}\,\mathrm{d}x = 7 - \log 14 - \frac{\pi}{2}\cot\frac{\pi}{7} + 2(\sin \tfrac{3\pi}{14}) \log \sin \tfrac{\pi}{7} - 2(\sin \tfrac{\pi}{14}) \log \cos \tfrac{3\pi}{14} - 2(\cos \tfrac{\pi}{7}) \log \cos \tfrac{\pi}{14}.$$

Related: Fact 13.3.4.

Fact 14.2.8. Let $n \geq 1$ and $k \geq 2$, and define $p \in \mathbb{R}[s]$ by $p(s) = \sum_{i=0}^{k-1} s^i$. Then,

$$\int_0^1 \frac{x^{nk} p'(x)}{p(x)} \, \mathrm{d}x = \log k + H_n - H_{kn}.$$

Source: [1682].

Fact 14.2.9. Let $a \in (-1, 1)$. Then,

$$\int_0^1 \frac{x^a}{(1-x)^a} \, \mathrm{d}x = \frac{a\pi}{\sin a\pi}.$$

Fact 14.2.10. Let $a \in (0, 1)$. Then,

$$\int_0^1 \frac{1}{(1-x)^{1-a} x^a} \, \mathrm{d}x = \int_0^1 \frac{1}{(1-x)^a x^{1-a}} \, \mathrm{d}x = \frac{\pi}{\sin a\pi}.$$

Fact 14.2.11. Let a and b be positive numbers, and let $n \geq 1$. Then,

$$\int_0^1 \frac{x^{b-1}(1-x)^n}{1-x^a} \, \mathrm{d}x = \sum_{i=0}^{\infty} \frac{n!}{\prod_{j=0}^{n}(ia + b + j)}.$$

In particular,

$$\int_0^1 \frac{(1-x)^n}{1-x^a} \, \mathrm{d}x = \sum_{i=0}^{\infty} \frac{n!}{\prod_{j=1}^{n+1}(ia + j)}, \quad \int_0^1 (1-x)^{n-1} \, \mathrm{d}x = \frac{1}{n} = \sum_{i=0}^{\infty} \frac{n!}{\prod_{j=1}^{n+1}(i + j)}.$$

Source: [113]. **Remark:** The last equality is given by Fact 13.5.26.

Fact 14.2.12. Let $a > 0$ and $n \geq 1$. Then,

$$\int_0^1 \frac{(1-x)^n}{1-x^a} \, \mathrm{d}x = \sum_{i=0}^{\infty} \frac{n!}{\prod_{j=1}^{n+1}(ai + j)} = \frac{1}{(n+1)!} \sum_{i=0}^{\infty} \frac{1}{\binom{ai+n+1}{ai}},$$

$$\int_0^1 \frac{(1-x)^n}{1+x^a} \, \mathrm{d}x = \sum_{i=0}^{\infty} (-1)^i \frac{n!}{\prod_{j=1}^{n+1}(ai + j)} = \frac{1}{(n+1)!} \sum_{i=0}^{\infty} (-1)^i \frac{1}{\binom{ai+n+1}{ai}}.$$

Now, assume that $1/a = 2k$, where $k \geq 1$. Then,

$$\int_0^1 \frac{(1-x)^n}{1-x^{1/(2k)}} \, \mathrm{d}x = \sum_{i=0}^{2k-1} \frac{(n-1)!}{\prod_{j=1}^{n}[i/(2k) + j]}, \quad \int_0^1 \frac{(1-x)^n}{1+x^{1/(2k)}} \, \mathrm{d}x = \sum_{i=0}^{2k-1} (-1)^i \frac{(n-1)!}{\prod_{j=1}^{n}[i/(2k) + j]}.$$

In particular,

$$\int_0^1 \frac{1-x}{1-x^{1/2}} \, \mathrm{d}x = \frac{5}{3}, \quad \int_0^1 \frac{(1-x)^2}{1-x^{1/2}} \, \mathrm{d}x = \frac{23}{30}, \quad \int_0^1 \frac{1-x}{1-x^{1/4}} \, \mathrm{d}x = \frac{319}{105}, \quad \int_0^1 \frac{(1-x)^2}{1-x^{1/4}} \, \mathrm{d}x = \frac{9217}{6930},$$

$$\int_0^1 \frac{1-x}{1+x^{1/2}} \, \mathrm{d}x = \frac{1}{3}, \quad \int_0^1 \frac{(1-x)^2}{1+x^{1/2}} \, \mathrm{d}x = \frac{7}{30}, \quad \int_0^1 \frac{1-x}{1+x^{1/4}} \, \mathrm{d}x = \frac{31}{105}, \quad \int_0^1 \frac{(1-x)^2}{1+x^{1/4}} \, \mathrm{d}x = \frac{1409}{6930}.$$

Source: [2481].

Fact 14.2.13.

$$\int_0^1 \frac{1}{1-x+x^2} \, \mathrm{d}x = \frac{2\sqrt{3}\pi}{9}, \quad \int_0^1 \frac{x}{1-x+x^2} \, \mathrm{d}x = \frac{\sqrt{3}\pi}{9},$$

$$\int_0^1 \frac{x^2}{1-x+x^2} \, \mathrm{d}x = 1 - \frac{\sqrt{3}\pi}{9}, \quad \int_0^1 \frac{1}{(1-x+x^2)^2} \, \mathrm{d}x = \frac{2}{3} + \frac{4\sqrt{3}\pi}{27},$$

$$\int_0^1 \frac{x}{(1-x+x^2)^2}\,\mathrm{d}x = \frac{1}{3} + \frac{2\sqrt{3}\pi}{27}, \quad \int_0^1 \frac{x^2}{(1-x+x^2)^2}\,\mathrm{d}x = \frac{4\sqrt{3}\pi}{27} - \frac{1}{3},$$

$$\int_0^1 \frac{x^3}{(1-x+x^2)^2}\,\mathrm{d}x = \frac{5\sqrt{3}\pi}{27} - \frac{2}{3}, \quad \int_0^1 \frac{x^4}{(1-x+x^2)^2}\,\mathrm{d}x = \frac{2}{3} - \frac{2\sqrt{3}\pi}{27},$$

$$\int_0^1 \frac{1}{1+x-x^2}\,\mathrm{d}x = \frac{4\sqrt{5}}{5}\log\tfrac{1}{2}(1+\sqrt{5}), \quad \int_0^1 \frac{x}{1+x-x^2}\,\mathrm{d}x = \frac{\sqrt{5}}{5}\log\tfrac{1}{2}(3+\sqrt{5}),$$

$$\int_0^1 \frac{x^2}{1+x-x^2}\,\mathrm{d}x = \frac{\sqrt{5}}{5}\log(9+4\sqrt{5}) - 1, \quad \int_0^1 \frac{1}{(1+x-x^2)^2}\,\mathrm{d}x = \frac{2}{5} + \frac{8\sqrt{5}}{25}\log\tfrac{1}{2}(1+\sqrt{5}),$$

$$\int_0^1 \frac{x}{(1+x-x^2)^2}\,\mathrm{d}x = \frac{1}{5} + \frac{4\sqrt{5}}{25}\log\tfrac{1}{2}(1+\sqrt{5}), \quad \int_0^1 \frac{x^2}{(1+x-x^2)^2}\,\mathrm{d}x = \frac{3}{5} - \frac{8\sqrt{5}}{25}\log\tfrac{1}{2}(1+\sqrt{5}),$$

$$\int_0^1 \frac{x^3}{(1+x-x^2)^2}\,\mathrm{d}x = \frac{4}{5} - \frac{7\sqrt{5}}{25}\log\tfrac{1}{2}(3+\sqrt{5}),$$

$$\int_0^1 \frac{x^4}{(1+x-x^2)^2}\,\mathrm{d}x = \frac{12}{5} - \frac{26\sqrt{5}}{25}\log\tfrac{1}{2}(3+\sqrt{5}),$$

$$\int_0^1 \frac{1}{1+x+x^2}\,\mathrm{d}x = \frac{\sqrt{3}\pi}{9}, \quad \int_0^1 \frac{x}{1+x+x^2}\,\mathrm{d}x = \frac{\log 3}{2} - \frac{\sqrt{3}\pi}{18},$$

$$\int_0^1 \frac{x^2}{1+x+x^2}\,\mathrm{d}x = 1 - \frac{\log 3}{2} - \frac{\sqrt{3}\pi}{18}, \quad \int_0^1 \frac{1}{(1+x+x^2)^2}\,\mathrm{d}x = \frac{2\sqrt{3}\pi}{27},$$

$$\int_0^1 \frac{x}{(1+x+x^2)^2}\,\mathrm{d}x = \frac{1}{3} - \frac{\sqrt{3}\pi}{27}, \quad \int_0^1 \frac{x^2}{(1+x+x^2)^2}\,\mathrm{d}x = \frac{2\sqrt{3}\pi}{27} - \frac{1}{3},$$

$$\int_0^1 \frac{x^3}{(1+x+x^2)^2}\,\mathrm{d}x = \frac{\log 3}{2} - \frac{5\sqrt{3}\pi}{54}, \quad \int_0^1 \frac{x^4}{(1+x+x^2)^2}\,\mathrm{d}x = \frac{4}{3} - \frac{\sqrt{3}\pi}{27} - \log 3.$$

Now, let $n \ge 1$. Then,

$$\int_0^1 \frac{1}{(1+x+x^2)^n}\,\mathrm{d}x = \frac{1-3^{n-2}}{3^{n-1}(n-1)} + \frac{(2n-3)!!}{2(n-1)!}\sum_{i=2}^{n-3}(3^{1-n}-3^{-i})\frac{2^i(n-i-1)!}{(2n-2i-1)!!} + \frac{2^{n-1}(2n-3)!!\sqrt{3}\pi}{(3n+1)(n-1)!},$$

$$\int_0^1 \frac{x}{(1+x+x^2)^n}\,\mathrm{d}x = \frac{1-3^{n-1}}{2(1-n)3^{n-1}} - \frac{1}{2}\int_0^1 \frac{1}{(1+x+x^2)^n}\,\mathrm{d}x.$$

Source: [817].

Fact 14.2.14.

$$\int_0^1 \frac{1-x}{x^2+x+1}\,\mathrm{d}x = \sum_{i=0}^\infty \frac{2(3i)!}{(3i+3)!} = \sum_{i=0}^\infty \frac{2}{(3i+1)(3i+2)(3i+3)} = \frac{\sqrt{3}\pi - 3\log 3}{6}.$$

Source: [2577].

Fact 14.2.15. Let $n > m \ge 1$. Then,

$$\int_0^1 \frac{(n-1)x^m(1-x)^{n-m}+1}{[x^m(1-x)^{n-m}-1]^2}\,\mathrm{d}x = \sum_{i=0}^\infty \frac{1}{\binom{ni}{mi}}.$$

In particular,

$$\int_0^1 \frac{mx^m(1-x)+1}{[x^m(1-x)-1]^2}\,\mathrm{d}x = \sum_{i=0}^\infty \frac{1}{\binom{(m+1)i}{mi}}, \quad \int_0^1 \frac{(m+1)x^m(1-x)^2+1}{[x^m(1-x)^2-1]^2}\,\mathrm{d}x = \sum_{i=0}^\infty \frac{1}{\binom{(m+2)i}{mi}},$$

$$\int_0^1 \frac{x(1-x)+1}{[x(1-x)-1]^2}\,\mathrm{d}x = \sum_{i=0}^{\infty} \frac{1}{\binom{2i}{i}} = \frac{4}{3} + \frac{2\sqrt{3}\pi}{27},$$

$$\int_0^1 \frac{3x^2(1-x)^2+1}{[x^2(1-x)^2-1]^2}\,\mathrm{d}x = \sum_{i=0}^{\infty} \frac{1}{\binom{4i}{2i}} = \frac{16}{15} + \frac{\sqrt{3}\pi}{27} - \frac{2\sqrt{5}}{25}\log \tfrac{1}{2}(1+\sqrt{5}).$$

Source: [2509, 2728].

Fact 14.2.16. Let $x \in (-4, 4)$. Then,

$$\int_0^1 \frac{2x(1-t)}{[1-xt(1-t)]^3}\,\mathrm{d}t = \frac{10x-x^2}{(x-4)^2} + 24\sqrt{\frac{x}{(4-x)^5}}\,\mathrm{asin}\,\frac{\sqrt{x}}{2} = \sum_{i=1}^{\infty} \frac{x^i}{C_i}.$$

Source: [343]. **Related:** Fact 13.4.2 and Fact 13.9.1.

Fact 14.2.17. Let $a > 0$ and $b > 0$, and assume that $a \neq b$. Then,

$$\int_0^1 a^x b^{1-x}\,\mathrm{d}x = \prod_{i=1}^{\infty} \tfrac{1}{2}(a^{1/2^i} + b^{1/2^i}) = \frac{a-b}{\log a - \log b}, \quad \int_0^1 \frac{1}{ax+b(1-x)}\,\mathrm{d}x = \frac{\log a - \log b}{a-b}.$$

Source: [635]. **Related:** Fact 2.2.63.

Fact 14.2.18. Let $a > 0$ and $b > 0$. Then,

$$\int_0^1 \frac{1}{[ax+b(1-x)]^2}\,\mathrm{d}x = \frac{1}{ab}.$$

Source: [2106, p. 18].

Fact 14.2.19. Let $a > 0$ and $b > 0$. Then,

$$\int_0^{\infty} \left(\frac{1}{x+b} - \frac{1}{x+a}\right)\mathrm{d}x = \log\frac{a}{b}, \quad \int_0^{\infty} \frac{1}{(ax+b)(x/b+1/a)}\,\mathrm{d}x = 1, \quad \int_0^{\infty} \frac{1}{(ax+b)^2}\,\mathrm{d}x = \frac{1}{ab}.$$

If $a \neq b$, then

$$\int_0^{\infty} \frac{1}{(ax+b)(x+1)}\,\mathrm{d}x = \int_0^{\infty} \frac{1}{(x+a)(x+b)}\,\mathrm{d}x = \frac{\log a - \log b}{a-b}, \quad \int_0^1 a^x b^{1-x}\,\mathrm{d}x = \frac{a-b}{\log a - \log b}.$$

Now, let $c > 0$ and $d > 0$, and assume that $ad \neq bc$. Then,

$$\int_0^{\infty} \frac{1}{(ax+b)(cx+d)}\,\mathrm{d}x = \frac{\log ad - \log bc}{ad-bc}.$$

Related: Fact 2.2.63, Fact 10.10.49, and Fact 10.11.69.

Fact 14.2.20. Let a, b, c, d, e, f be positive numbers, and assume that $ad \neq bc$, $be \neq af$, and $cf \neq de$. Then,

$$\int_0^{\infty} \frac{1}{(ax+b)(cx+d)(ex+f)}\,\mathrm{d}x = \frac{a\log\frac{b}{a}}{(ad-bc)(be-af)} + \frac{c\log\frac{d}{c}}{(ad-bc)(cf-de)} + \frac{e\log\frac{f}{e}}{(be-af)(cf-de)}.$$

Fact 14.2.21. The following statements hold:

i) Let $n \geq 1$ and $a > 0$. Then,

$$\int_0^{\infty} \frac{1}{(x^2+a^2)^n}\,\mathrm{d}x = \frac{\pi}{(2a)^{2n-1}}\binom{2n-2}{n-1} = \frac{(2n-3)!!\pi}{2^n(n-1)!a^{2n-1}}.$$

In particular,

$$\int_0^{\infty} \frac{1}{x^2+a^2}\,\mathrm{d}x = \frac{\pi}{2a}, \quad \int_0^{\infty} \frac{1}{(x^2+a^2)^2}\,\mathrm{d}x = \frac{\pi}{4a^3}.$$

ii) Let $n \geq 1$. Then,

$$\int_0^\infty \frac{1}{(x^2+1)^n}\,\mathrm{d}x = \frac{\pi}{2^{2n-1}}\binom{2n-2}{n-1} = \frac{(2n-3)!!\pi}{2^n(n-1)!} = \frac{(2n-2)!\pi}{2^{2n-1}[(n-1)!]^2}.$$

iii) Let $n \geq 2$ and $a > 0$. Then,

$$\int_0^\infty \frac{x}{(x^2+a^2)^n}\,\mathrm{d}x = \frac{1}{(2n-2)a^{2n-2}}, \qquad \int_0^\infty \frac{x^2}{(x^2+a^2)^n}\,\mathrm{d}x = \frac{(2n-5)!!\pi}{2^n(n-1)!a^{2n-3}}.$$

iv) Let $n \geq 3$ and $a > 0$. Then,

$$\int_0^\infty \frac{x^3}{(x^2+a^2)^n}\,\mathrm{d}x = \frac{1}{(2n^2-6n+4)a^{2n-4}}, \qquad \int_0^\infty \frac{x^4}{(x^2+a^2)^n}\,\mathrm{d}x = \frac{3(2n-7)!!\pi}{2^n(n-1)!a^{2n-5}}.$$

v) Let $a > 0$. Then,

$$\int_0^\infty \frac{x}{(x^2+a^2)^2}\,\mathrm{d}x = \frac{1}{2a^2}, \qquad \int_0^\infty \frac{x^2}{(x^2+a^2)^2}\,\mathrm{d}x = \frac{\pi}{4a},$$

$$\int_0^\infty \frac{1}{(x^2+a^2)^3}\,\mathrm{d}x = \frac{3\pi}{16a^5}, \qquad \int_0^\infty \frac{x}{(x^2+a^2)^3}\,\mathrm{d}x = \frac{1}{4a^4}, \qquad \int_0^\infty \frac{x^2}{(x^2+a^2)^3}\,\mathrm{d}x = \frac{\pi}{16a^3},$$

$$\int_0^\infty \frac{x^3}{(x^2+a^2)^3}\,\mathrm{d}x = \frac{1}{4a^2}, \qquad \int_0^\infty \frac{x^4}{(x^2+a^2)^3}\,\mathrm{d}x = \frac{3\pi}{16a}.$$

vi) Let $a > 0$ and $b \in (-1,1) \cup (1,3)$. Then,

$$\int_0^\infty \frac{x^b}{(x^2+a^2)^2}\,\mathrm{d}x = \frac{(1-b)\pi}{4a^{3-b}} \sec \frac{b\pi}{2}.$$

In particular,

$$\int_0^\infty \frac{\sqrt{x}}{(x^2+a^2)^2}\,\mathrm{d}x = \frac{\sqrt{2}\pi}{8a^{5/2}}, \qquad \int_0^\infty \frac{x^{3/2}}{(x^2+a^2)^2}\,\mathrm{d}x = \frac{\sqrt{2}\pi}{8a^{3/2}}.$$

vii) Let $a > 0$, $k \geq 0$, and $n \geq 2$, and assume that $k+2 \leq n$. Then,

$$\int_0^\infty \frac{x^k}{x^n+a^2}\,\mathrm{d}x = \frac{\pi}{na^{2(n-k-1)/n}\sin\frac{(k+1)\pi}{n}}.$$

Hence,

$$\int_0^\infty \frac{x^k}{x^{2k+2}+a^2}\,\mathrm{d}x = \frac{\pi}{(2k+2)a^{(2k+2)/n}}, \qquad \int_0^\infty \frac{x^k}{x^{3k+3}+a^2}\,\mathrm{d}x = \frac{2\sqrt{3}\pi}{(9k+9)a^{(4k+4)/n}},$$

$$\int_0^\infty \frac{x^k}{x^{4k+4}+a^2}\,\mathrm{d}x = \frac{\sqrt{2}\pi}{(2k+2)a^{(6k+6)/n}}.$$

If, in addition, $n \geq 3$, then

$$\int_0^\infty \frac{x}{x^n+a^2}\,\mathrm{d}x = \frac{\pi}{na^{2(n-2)/n}\sin\frac{2\pi}{n}},$$

$$\int_0^\infty \frac{x}{x^n+1}\,\mathrm{d}x = \int_0^\infty \frac{x-1}{x^n-1}\,\mathrm{d}x = \int_0^\infty \frac{1}{\sum_{i=0}^{n-1} x^i}\,\mathrm{d}x = \frac{\pi}{n}\csc\frac{2\pi}{n}.$$

viii) Let $n \geq 2$ and $a > 0$. Then,

$$\int_0^\infty \frac{1}{x^n+a^2}\,\mathrm{d}x = \frac{\pi}{na^{2(n-1)/n}\sin\frac{\pi}{n}}.$$

ix) Let $n \ge 3$ and $a > 0$. Then,

$$\int_0^\infty \frac{x}{x^n + a^2}\,\mathrm{d}x = \frac{\pi}{na^{2(n-2)/n}\sin\frac{2\pi}{n}}.$$

x) Let $n \ge 4$ and $a > 0$. Then,

$$\int_0^\infty \frac{x^2}{x^n + a^2}\,\mathrm{d}x = \frac{\pi}{na^{2(n-3)/n}\sin\frac{3\pi}{n}}.$$

xi) Let $a > 0$ and $b \in (-1, 1)$. Then,

$$\int_0^\infty \frac{x^b}{x^2 + a^2}\,\mathrm{d}x = \frac{1}{2}a^{b-1}\pi\sec\frac{b\pi}{2}.$$

xii) Let $n \ge 1$. Then,

$$\int_0^\infty \prod_{i=1}^n \frac{1}{x^2 + i^2}\,\mathrm{d}x = \frac{\pi}{2(2n-1)(n-1)!n!},$$

$$\int_0^\infty \prod_{i=1}^n \frac{1}{x^2 + (2i-1)^2}\,\mathrm{d}x = \frac{\pi}{2^{2n-1}(2n-1)[(n-1)!]^2}.$$

xiii) Let $n \ge 1$. Then,

$$\int_0^\infty \frac{x^2}{(x^2+1)^{n+2}}\,\mathrm{d}x = \int_0^1 \frac{8x^2(1-x^2)^{2n}}{(x^2+1)^{2n+3}}\,\mathrm{d}x = \frac{\sqrt{\pi}\Gamma(n+\frac{1}{2})}{4(n+1)!}$$
$$= \frac{(2n)!\pi}{4^{n+1}n!(n+1)!} = \frac{\pi}{4^{n+1}(n+1)}\binom{2n}{n} = \frac{\pi C_n}{4^{n+1}}.$$

Source: *i*) is given in [511, p. 114]; *ii*) is given in [69], [1568, p. 166], and [2068, pp. 247, 248]; *v*) is given in [1136, p. 202]; *vi*) is given in [2106, p. 340]; *vii*) is given in [561], [645, p. 82], and [2447]; *viii*) and *x*) are given in [1136, p. 202]; *ix*) and *xii*) are given in [69]; *xi*) is given in [953]; *xiii*) is given in [822]. **Remark:** *ii*) is *Wallis's formula*. See [2068, p. 247].

Fact 14.2.22. Let $b > 0$ and $c > 0$. If $a \in (-1, b-1)$, then

$$\int_0^\infty \frac{x^a}{x^b + c}\,\mathrm{d}x = \frac{c^{(a+1-b)/b}\pi}{b\sin\frac{(a+1)\pi}{b}}.$$

If $a \in (-1, 2b-1)$, then

$$\int_0^\infty \frac{x^a}{(x^b + c)^2}\,\mathrm{d}x = \frac{(b-a-1)c^{(a+1-2b)/b}\pi}{b^2\sin\frac{(a+1)\pi}{b}}.$$

In particular,

$$\int_0^\infty \frac{\sqrt{x}}{x^2+1}\,\mathrm{d}x = \frac{\sqrt{2}\pi}{2}, \quad \int_0^\infty \frac{x}{x^3+1}\,\mathrm{d}x = \frac{2\sqrt{3}\pi}{9}, \quad \int_0^\infty \frac{x}{x^4+1}\,\mathrm{d}x = \frac{\pi}{4},$$

$$\int_0^\infty \frac{\sqrt{x}}{(x^2+1)^2}\,\mathrm{d}x = \frac{\sqrt{2}\pi}{8}, \quad \int_0^\infty \frac{x}{(x^3+1)^2}\,\mathrm{d}x = \frac{2\sqrt{3}\pi}{27}, \quad \int_0^\infty \frac{x}{(x^4+1)^2}\,\mathrm{d}x = \frac{\pi}{8}.$$

Furthermore, if $d > 1$, then

$$\int_0^\infty \frac{x^{d-1}}{x^{2d}+1}\,\mathrm{d}x = \frac{\pi}{2d}.$$

Fact 14.2.23. Let $a \in (-1, 1)$. Then,

$$\int_0^\infty \frac{x^a}{x^2+x+1}\,\mathrm{d}x = \frac{2\sqrt{3}\pi}{3+6\cos\frac{2a\pi}{3}}, \quad \int_0^\infty \frac{x^a}{x^2-x+1}\,\mathrm{d}x = \frac{4\sqrt{3}\pi\cos\frac{a\pi}{3}}{3+6\cos\frac{2a\pi}{3}}.$$

In particular,

$$\int_0^\infty \frac{\sqrt{x}}{x^2+x+1}\,\mathrm{d}x = \frac{\sqrt{3}\pi}{3}, \quad \int_0^\infty \frac{\sqrt{x}}{x^2-x+1}\,\mathrm{d}x = \pi.$$

Source: [150, p. 13].

Fact 14.2.24. Let a and b be real numbers such that $0 < a < b$. Then,

$$\int_0^\infty \frac{1}{(x^{1/a}+1)^b}\,\mathrm{d}x = \frac{\Gamma(a+1)\Gamma(b-a)}{\Gamma(b)}.$$

In particular,

$$\int_0^\infty \frac{1}{(x^\alpha+1)^\alpha}\,\mathrm{d}x = 1.$$

Source: [2574]. Setting $b = \frac{1}{2}(1+\sqrt{5})$ and $a = 1/b = b-1 = \frac{1}{2}(\sqrt{5}-1)$ yields the special case.

Fact 14.2.25.

$$\int_0^\infty \frac{1}{x^2+x+2}\,\mathrm{d}x = \frac{2\sqrt{7}}{7}\left(\frac{\pi}{2} - \operatorname{atan}\frac{\sqrt{7}}{7}\right), \quad \int_0^\infty \frac{1}{x^3+x+2}\,\mathrm{d}x = \frac{\log 2}{8} + \frac{3\sqrt{7}(\pi - \operatorname{atan}\sqrt{7})}{28}.$$

Source: [645, p. 82].

Fact 14.2.26. The following statements hold:

i) Let a and b be positive numbers. Then,

$$\int_0^\infty \frac{1}{(x^2+a^2)(x+b)}\,\mathrm{d}x = \frac{1}{a^2+b^2}\left(\frac{b\pi}{2a} + \log\frac{a}{b}\right).$$

ii) Let a and b be positive numbers. Then,

$$\int_0^\infty \frac{x}{(x^2+a^2)(x+b)}\,\mathrm{d}x = \frac{1}{2(a^2+b^2)}\left(a\pi + 2b\log\frac{b}{a}\right).$$

iii) Let a and b be positive numbers. If $a \neq b$ and $c \in (0, 4)$, then

$$\int_0^\infty \frac{x^{c-1}}{(x^2+a^2)(x^2+b^2)}\,\mathrm{d}x = \begin{cases} \dfrac{\pi(b^{c-2}-a^{c-2})}{2(a^2-b^2)\sin\frac{c\pi}{2}}, & a \neq b \text{ and } c \neq 2,\\ \dfrac{\pi(2-c)}{4a^{4-c}\sin\frac{c\pi}{2}}, & a = b \text{ and } c \neq 2,\\ \dfrac{1}{a^2-b^2}\log\dfrac{a}{b}, & a \neq b \text{ and } c = 2,\\ \dfrac{1}{2a^2}, & a = b \text{ and } c = 2. \end{cases}$$

In particular,

$$\int_0^\infty \frac{1}{(x^2+a^2)(x^2+b^2)}\,\mathrm{d}x = \frac{\pi}{2ab(a+b)}, \quad \int_0^\infty \frac{x^2}{(x^2+a^2)(x^2+b^2)}\,\mathrm{d}x = \frac{\pi}{2(a+b)}.$$

iv) Let a, b, c be positive numbers. Then,

$$\int_0^\infty \frac{1}{(x^2+a^2)(x^2+b^2)(x^2+c^2)}\,\mathrm{d}x = \frac{(a+b+c)\pi}{2abc(a+b)(b+c)(c+a)},$$

$$\int_0^\infty \frac{x^2}{(x^2+a^2)(x^2+b^2)(x^2+c^2)}\,\mathrm{d}x = \frac{\pi}{2(a+b)(b+c)(c+a)},$$

$$\int_0^\infty \frac{x^4}{(x^2+a^2)(x^2+b^2)(x^2+c^2)}\,\mathrm{d}x = \frac{(ab+bc+ca)\pi}{2(a+b)(b+c)(c+a)}.$$

If, in addition, a, b, c are distinct, then

$$\int_0^\infty \frac{1}{(x^2+a^2)(x^2+b^2)(x^2+c^2)}\,\mathrm{d}x = -\frac{[ab(a^2-b^2)+bc(b^2-c^2)+ca(c^2-a^2)]\pi}{2abc(a^2-b^2)(b^2-c^2)(c^2-a^2)},$$

$$\int_0^\infty \frac{x}{(x^2+a^2)(x^2+b^2)(x^2+c^2)}\,\mathrm{d}x = \frac{(a^2-b^2)\log c+(b^2-c^2)\log a+(c^2-a^2)\log b}{(a^2-b^2)(b^2-c^2)(c^2-a^2)},$$

$$\int_0^\infty \frac{x^3}{(x^2+a^2)(x^2+b^2)(x^2+c^2)}\,\mathrm{d}x = -\frac{(a^2-b^2)c^2\log c+(b^2-c^2)a^2\log a+(c^2-a^2)b^2\log b}{(a^2-b^2)(b^2-c^2)(c^2-a^2)}.$$

v) Let a,b,c,d be positive numbers. Then,

$$\int_0^\infty \frac{1}{(x^2+a^2)(x^2+b^2)(x^2+c^2)(x^2+d^2)}\,\mathrm{d}x = \frac{[(a+b+c+d)^3-(a^3+b^3+c^3+d^3)]\pi}{6abcd(a+b)(a+c)(a+d)(b+c)(b+d)(c+d)}$$
$$= \frac{[a^2(b+c+d)+a(b+c+d)^2+(b+c)(b+d)(c+d)]\pi}{2abcd(a+b)(a+c)(a+d)(b+c)(b+d)(c+d)},$$

$$\int_0^\infty \frac{x^2}{(x^2+a^2)(x^2+b^2)(x^2+c^2)(x^2+d^2)}\,\mathrm{d}x = \frac{(a+b+c+d)\pi}{2(a+b)(a+c)(a+d)(b+c)(b+d)(c+d)},$$

$$\int_0^\infty \frac{x^4}{(x^2+a^2)(x^2+b^2)(x^2+c^2)(x^2+d^2)}\,\mathrm{d}x = \frac{(abc+bcd+cda+dab)\pi}{2(a+b)(a+c)(a+d)(b+c)(b+d)(c+d)},$$

$$\int_0^\infty \frac{x^6}{(x^2+a^2)(x^2+b^2)(x^2+c^2)(x^2+d^2)}\,\mathrm{d}x$$
$$= \frac{[a^2(b+c)(b+d)(c+d)+a(bc+bd+cd)^2+bcd(bc+bd+cd)]\pi}{2(a+b)(a+c)(a+d)(b+c)(b+d)(c+d)}.$$

vi) Let $z \in \mathbb{C}$, assume that $\operatorname{Re} z \neq 0$. Then,

$$\int_0^\infty \frac{1}{x^2+z^2}\,\mathrm{d}x = \frac{(\operatorname{sign}\operatorname{Re} z)\pi}{2z}.$$

In particular, if a is a nonzero real number, then

$$\int_0^\infty \frac{a}{x^2+a^2}\,\mathrm{d}x = \frac{(\operatorname{sign} a)\pi}{2}.$$

vii) Let $z_1,\ldots,z_n \in \mathbb{C}$, assume that $z_1,\ldots,z_n$ are distinct, and assume that, for all $i \in \{1,\ldots,n\}$, $\operatorname{Re} z_i \neq 0$. Then,

$$\int_0^\infty \prod_{i=1}^n \frac{1}{x^2+z_i^2}\,\mathrm{d}x = \sum_{i=1}^n \frac{(\operatorname{sign}\operatorname{Re} z_i)\pi}{2z_i} \prod_{\substack{j=1\\ j\neq i}}^n \frac{1}{z_j^2-z_i^2}.$$

Source: *i)* is given in [1963]; *ii)* is given in [1217, p. 330]; *vii)* is given in [2106, p. 56]. **Related:** Fact 14.10.1.

Fact 14.2.27. Let a be a complex number such that $a \notin \pi\mathbb{Z} + \frac{1}{2}\pi$. Then,

$$\int_0^\infty \frac{1}{x^4+2x^2\cos 2a+1}\,\mathrm{d}x = \int_0^\infty \frac{x^2}{x^4+2x^2\cos 2a+1}\,\mathrm{d}x = \frac{\pi}{4\cos a}.$$

In particular,

$$\int_0^\infty \frac{1}{x^4 - x^2 + 1}\,\mathrm{d}x = \frac{\pi}{2}, \quad \int_0^\infty \frac{1}{x^4 + 1}\,\mathrm{d}x = \frac{\sqrt{2}\pi}{4},$$

$$\int_0^\infty \frac{1}{x^4 + x^2 + 1}\,\mathrm{d}x = \frac{\sqrt{3}\pi}{6}, \quad \int_0^\infty \frac{1}{x^4 + 2x^2 + 1}\,\mathrm{d}x = \frac{\pi}{4},$$

$$\int_0^\infty \frac{1}{x^4 - \sqrt{2}x^2 + 1}\,\mathrm{d}x = \frac{\pi}{4}\sqrt{4 + 2\sqrt{2}}, \quad \int_0^\infty \frac{1}{x^4 + \sqrt{2}x^2 + 1}\,\mathrm{d}x = \frac{\pi}{4}\sqrt{4 - 2\sqrt{2}}.$$

Source: [2106, pp. 56–60]. Replacing x by $1/x$ changes the numerator from 1 to x^2.

Fact 14.2.28. Let $a \in (-\frac{\pi}{2}, \frac{\pi}{2})$. Then,

$$\int_0^\infty \frac{1}{x^2 + 2x\cos 2a + 1}\,\mathrm{d}x = \int_0^\infty \frac{2x}{x^4 + 2x^2\cos 2a + 1}\,\mathrm{d}x = \frac{2a}{\sin 2a}.$$

In particular,

$$\int_0^\infty \frac{1}{x^2 - x + 1}\,\mathrm{d}x = \frac{4\sqrt{3}\pi}{9}, \quad \int_0^\infty \frac{1}{x^2 + 1}\,\mathrm{d}x = \frac{\pi}{2},$$

$$\int_0^\infty \frac{1}{x^2 + x + 1}\,\mathrm{d}x = \frac{2\sqrt{3}\pi}{9}, \quad \int_0^\infty \frac{1}{x^2 + 2x + 1}\,\mathrm{d}x = 1,$$

$$\int_0^\infty \frac{1}{x^2 - \sqrt{2}x + 1}\,\mathrm{d}x = \frac{3\sqrt{2}\pi}{4}, \quad \int_0^\infty \frac{1}{x^2 + \sqrt{2}x + 1}\,\mathrm{d}x = \frac{\sqrt{2}\pi}{4}.$$

Fact 14.2.29. Let a be a complex number such that $\jmath a \notin \pi\mathbb{Z} + \frac{1}{2}\pi$. Then,

$$\int_0^\infty \frac{1}{x^4 + 2x^2\cosh 2a + 1}\,\mathrm{d}x = \int_0^\infty \frac{x^2}{x^4 + 2x^2\cosh 2a + 1}\,\mathrm{d}x = \frac{\pi}{4\cosh a}.$$

In particular,

$$\int_0^\infty \frac{1}{x^4 + \frac{17}{4}x^2 + 1}\,\mathrm{d}x = \frac{\pi}{5}, \quad \int_0^\infty \frac{1}{x^4 + \frac{82}{9}x^2 + 1}\,\mathrm{d}x = \frac{3\pi}{20}, \quad \int_0^\infty \frac{1}{x^4 + \frac{257}{16}x^2 + 1}\,\mathrm{d}x = \frac{2\pi}{17}.$$

Source: Replacing x by $1/x$ changes the numerator from 1 to x^2.

Fact 14.2.30. Let a be a real number such that $a \notin (\pi\jmath/2)\mathbb{Z}$. Then,

$$\int_0^\infty \frac{1}{x^2 + 2x\cosh 2a + 1}\,\mathrm{d}x = \int_0^\infty \frac{2x}{x^4 + 2x^2\cosh 2a + 1}\,\mathrm{d}x = \frac{2a}{\sinh 2a}.$$

In particular,

$$\int_0^\infty \frac{1}{x^2 + \frac{17}{4}x + 1}\,\mathrm{d}x = \frac{16\log 2}{15}, \int_0^\infty \frac{1}{x^2 + \frac{82}{9}x + 1}\,\mathrm{d}x = \frac{9\log 3}{20}, \int_0^\infty \frac{1}{x^2 + \frac{257}{16}x + 1}\,\mathrm{d}x = \frac{128\log 2}{255}.$$

Fact 14.2.31. The following statements hold:

i) Let a be a complex number, let b be a positive number, and assume that $0 < \operatorname{Re} a < b$. Then,

$$\int_0^\infty \frac{x^{a-1}}{x^b + 1}\,\mathrm{d}x = \frac{\pi}{b\sin\frac{a\pi}{b}}.$$

In particular,

$$\int_0^\infty \frac{1}{\sqrt{x}(x+1)}\,\mathrm{d}x = \pi, \quad \int_0^\infty \frac{x^{a-1}}{x+1}\,\mathrm{d}x = \frac{\pi}{\sin a\pi}, \quad \int_0^\infty \frac{1}{x^4 + 1}\,\mathrm{d}x = \frac{\sqrt{2}\pi}{4}.$$

ii) Let $a \in (0,1)$ and $b \in (0,\infty)$. Then,

$$\int_0^\infty \frac{1}{x^a(x+b)}\,\mathrm{d}x = \frac{\pi}{b^a \sin a\pi}.$$

In particular,

$$\int_0^\infty \frac{1}{\sqrt{x}(x+b)}\,\mathrm{d}x = \frac{\pi}{\sqrt{b}}, \quad \int_0^\infty \frac{1}{\sqrt[3]{x}(x+b)}\,\mathrm{d}x = \frac{2\sqrt{3}\pi}{3\sqrt[3]{b}}, \quad \int_0^\infty \frac{1}{\sqrt[4]{x}(x+b)}\,\mathrm{d}x = \frac{\sqrt{2}\pi}{\sqrt[4]{b}}.$$

iii) Let $a \in (-1,1)$ and $b \in (0,\infty)$. Then,

$$\int_0^\infty \frac{1}{x^a(x+b)^2}\,\mathrm{d}x = \frac{a\pi}{b^{a+1}\sin a\pi}.$$

iv) Let $a \in (-2,1)$ and $b \in (0,\infty)$. Then,

$$\int_0^\infty \frac{1}{x^a(x+b)^3}\,\mathrm{d}x = \frac{a(a+1)\pi}{2b^{a+2}\sin a\pi}.$$

v) Let $a \in (-3,1)$ and $b \in (0,\infty)$. Then,

$$\int_0^\infty \frac{1}{x^a(x^2+b^2)^2}\,\mathrm{d}x = \frac{(a+1)\pi}{4b^{a+3}\cos\frac{a\pi}{2}}.$$

In particular, if $c \in (-2,2)$, then

$$\int_0^\infty \frac{x^{c+1}}{(x^2+b^2)^2}\,\mathrm{d}x = \frac{c\pi}{4b^{2-c}\sin\frac{c\pi}{2}}.$$

vi) Let $a,b \in (0,1)$. Then,

$$\int_0^\infty \frac{x^{a-1}-x^{b-1}}{1-x}\,\mathrm{d}x = \pi(\cot\pi a - \cot\pi b).$$

vii) Let $n \ge 0$ and $m \ge n+2$. Then,

$$\int_0^\infty \frac{x^n}{(x+1)^m}\,\mathrm{d}x = \frac{n!(m-n-2)!}{(m-1)!} = \frac{1}{(m-1-n)\binom{m-1}{n}}.$$

viii) Let a,b,c,d be real numbers such that $a > -1$, $d > 0$, $c > (a+1)/d$, and $b > 0$. Then,

$$\int_0^\infty \frac{x^a}{(x^d+b)^c}\,\mathrm{d}x = \frac{b^{(a-dc+1)/d}\Gamma\left(\frac{a+1}{d}\right)\Gamma\left(c-\frac{a+1}{d}\right)}{d\Gamma(c)}.$$

ix) Let a,b,c be real numbers such that $b > 0$, $a < 1$, and $1-a < c$. Then,

$$\int_0^\infty \frac{1}{x^a(x+b)^c}\,\mathrm{d}x = \frac{\Gamma(1-a)\Gamma(a+c-1)}{b^{a+c-1}\Gamma(c)}.$$

In particular,

$$\int_0^\infty \frac{x}{(x+b)^3}\,\mathrm{d}x = \frac{1}{2b}, \quad \int_0^\infty \frac{x^2}{(x+b)^4}\,\mathrm{d}x = \frac{1}{3b}, \quad \int_0^\infty \frac{1}{\sqrt{x}(x+b)}\,\mathrm{d}x = \frac{\pi}{\sqrt{b}}.$$

Source: *i*) is given in [1136, p. 208]; *ii*) is given in [1146, p. 69], [1568, p. 40], and [3024, p. 330, formula 589]; *iii*) is given in [1136, p. 206]; *iv*) follows from *iii*); *v*) is given in [1136, p. 208] and [1217, p. 325]; *vi*) is given in [3024, p. 330]; *vii*) is given in [1568, p. 40]. **Related:** Fact 14.1.6.

Fact 14.2.32. Let a and b be positive numbers, and let $m > n \geq 0$. Then,

$$\int_0^\infty \frac{x^n}{(ax+b)^{m+1}}\,\mathrm{d}x = \frac{1}{a^{n+1}b^{m-n}(m-n)\binom{m}{n}}.$$

Remark: See [511, pp. 48–52]. **Related:** Fact 1.16.11.

Fact 14.2.33. Let a be a real number, and let b be a positive number. Then,

$$\int_0^\infty \frac{1}{x^2+2ax+b}\,\mathrm{d}x = \begin{cases} \dfrac{1}{\sqrt{b-a^2}}\left[\dfrac{\pi}{2} - \operatorname{atan}\left(\dfrac{a}{\sqrt{b-a^2}}\right)\right], & a^2 < b, \\ \dfrac{1}{2\sqrt{a^2-b}}\log\left(\dfrac{a+\sqrt{a^2-b}}{a-\sqrt{a^2-b}}\right), & b < a^2 \text{ and } a > 0, \\ \dfrac{1}{a}, & b = a^2 \text{ and } a > 0. \end{cases}$$

Source: [511, p. 32]. **Remark:** Extensions are given in [1217, pp. 325, 326]. **Remark:** If $a > 0$ and $b \leq a^2$, then allowing cancellation of singularities yields

$$\int_{-\infty}^\infty \frac{1}{x^2+2ax+b}\,\mathrm{d}x = 0.$$

See [2106, p. 301].

Fact 14.2.34. Let a, b, c be real numbers, assume that $b \geq 0$ and $b^2 < ac$, and let $n \geq 0$. Then,

$$\int_0^\infty \frac{1}{(ax^2+2bx+c)^{n+1}}\,\mathrm{d}x = \frac{a^n\binom{2n}{n}}{4^n(ac-b^2)^{n+1/2}}\left[\operatorname{acot}\left(\frac{b}{\sqrt{ac-b^2}}\right) - b\sum_{i=1}^n \frac{2^{2i-1}(ac-b^2)^{(i-1)/2}}{i(ac)^i\binom{2i}{i}}\right].$$

In particular,

$$\int_0^\infty \frac{1}{ax^2+2bx+c}\,\mathrm{d}x = \frac{1}{\sqrt{ac-b^2}}\operatorname{acot}\frac{b}{\sqrt{ac-b^2}}.$$

Source: [2068, pp. 233, 234]. **Example:**

$$\int_0^\infty \frac{1}{x^2+4x+5}\,\mathrm{d}x = \frac{\pi}{2} - \operatorname{atan} 2, \qquad \int_0^\infty \frac{1}{(x^2+4x+5)^2}\,\mathrm{d}x = -\frac{1}{5} + \frac{1}{2}\operatorname{atan}\frac{1}{2}.$$

Fact 14.2.35. Let a, b, c be real numbers, assume that $a > 0$ and $b^2 < 4ac$, and let $n \geq 1$. Then,

$$\int_{-\infty}^\infty \frac{1}{(ax^2+bx+c)^n}\,\mathrm{d}x = \frac{2^n(2n-3)!!a^{n-1}\pi}{(n-1)!(4ac-b^2)^{n-1/2}}.$$

In particular,

$$\int_{-\infty}^\infty \frac{1}{ax^2+bx+c}\,\mathrm{d}x = \frac{2\pi}{\sqrt{4ac-b^2}}, \qquad \int_{-\infty}^\infty \frac{1}{(ax^2+bx+c)^2}\,\mathrm{d}x = \frac{4a\pi}{(4ac-b^2)^{3/2}}.$$

Now, assume that $n \geq 2$. Then,

$$\int_{-\infty}^\infty \frac{x}{(ax^2+bx+c)^n}\,\mathrm{d}x = -\frac{2^{n-1}(2n-3)!!a^{n-2}b\pi}{(n-1)!(4ac-b^2)^{n-1/2}}.$$

In particular,

$$\int_{-\infty}^\infty \frac{x}{(ax^2+bx+c)^2}\,\mathrm{d}x = -\frac{2b\pi}{(4ac-b^2)^{3/2}}, \qquad \int_{-\infty}^\infty \frac{x}{(ax^2+bx+c)^3}\,\mathrm{d}x = -\frac{6ab\pi}{(4ac-b^2)^{5/2}}.$$

Source: [645, p. 82], [1217, pp. 325, 326], and [1524, p. 267].

Fact 14.2.36. Let $a \in (-1, \infty)$ and $n \geq 1$. Then,

$$\int_0^\infty \frac{1}{(x^4 + 2ax^2 + 1)^{n+1}}\,\mathrm{d}x = \frac{\pi}{2^{3n+3/2}(a+1)^{n+1/2}} \sum_{i=0}^{n} 2^i \binom{2n-2i}{n-i}\binom{n+i}{n}(a+1)^i.$$

Source: [509].

Fact 14.2.37. Let a, b, r be real numbers such that $a > -1$ and $r > \frac{1}{2}$. Then,

$$\begin{aligned}\int_0^\infty \left(\frac{x^2}{x^4+2ax^2+1}\right)^r \mathrm{d}x &= \int_0^\infty \left(\frac{x^2}{x^4+2ax^2+1}\right)^r \frac{1}{x^2}\,\mathrm{d}x = \frac{1}{2}\int_0^\infty \left(\frac{x^2}{x^4+2ax^2+1}\right)^r \frac{x^2+1}{x^2}\,\mathrm{d}x\\ &= \int_0^\infty \left(\frac{x^2}{x^4+2ax^2+1}\right)^r \frac{x^2+1}{x^2(x^b+1)}\,\mathrm{d}x = \frac{(a+1)^{1/2-r}}{2^{r+1/2}}\int_0^1 \frac{x^{r-3/2}}{\sqrt{1-x}}\,\mathrm{d}x\\ &= \frac{(a+1)^{1/2-r}\sqrt{\pi}\,\Gamma(r-\frac{1}{2})}{2^{r+1/2}\Gamma(r)}.\end{aligned}$$

Now, let c and d be real numbers such that $c \geq 0$, $d > 0$, and $-a < \sqrt{cd}$. Then,

$$\int_0^\infty \left(\frac{x^2}{dx^4+2ax^2+c}\right)^r \mathrm{d}x = \frac{(a+\sqrt{cd})^{1/2-r}\sqrt{\pi}\,\Gamma(r-\frac{1}{2})}{2^{r+1/2}\sqrt{d}\,\Gamma(r)}.$$

Now, assume that r is a positive integer. Then,

$$\int_0^\infty \left(\frac{x^2}{x^4+2ax^2+1}\right)^r \mathrm{d}x = \frac{(a+1)^{1/2-r}(\frac{1}{2})^{\overline{r-1}}\pi}{2^{r+1/2}(r-1)!},$$

$$\int_0^\infty \left(\frac{x^2}{dx^4+2ax^2+c}\right)^r \mathrm{d}x = \frac{(a+\sqrt{cd})^{1/2-r}(\frac{1}{2})^{\overline{r-1}}\pi}{2^{r+1/2}\sqrt{d}(r-1)!}.$$

In particular,

$$\int_0^\infty \frac{x^4}{(x^4+x^2+1)^3}\,\mathrm{d}x = \frac{\sqrt{3}\pi}{144}, \quad \int_0^\infty \frac{x^3}{(x^4+7x^2+1)^{5/2}}\,\mathrm{d}x = \frac{2}{243},$$

$$\int_0^\infty \frac{\sqrt{x}}{(x^4+14x^2+1)^{5/4}}\,\mathrm{d}x = \frac{\Gamma^2(3/4)}{4\sqrt{2\pi}}, \quad \int_0^\infty \frac{1}{\sqrt{x}(x^4+x^2+1)^{3/4}}\,\mathrm{d}x = \frac{\pi^{3/2}}{\Gamma^2(3/4)\sqrt[4]{12}}.$$

Source: [510]. **Remark:** The value of the fourth integral is independent of b.

Fact 14.2.38. Let $a \in (1, \infty)$. Then,

$$\int_0^\infty \frac{x^a - 2x + 1}{x^{2a} - 1}\,\mathrm{d}x = 0.$$

Furthermore, if $n \geq 2$, then

$$\int_0^\infty \frac{x^n - 2x + 1}{x^{2n} - 1}\,\mathrm{d}x = 0.$$

Source: [156, 2447].

Fact 14.2.39. Let $a \in (0, 1)$. Then,

$$\int_0^\infty \frac{1}{\prod_{i=0}^\infty (1 + a^{2i}x^2)}\,\mathrm{d}x = \frac{\pi}{2\sum_{i=0}^\infty a^{i(i+1)/2}}.$$

Source: [1317, p. xxv].

14.3 Facts on Definite Integrals of Radicals

Fact 14.3.1. Let $a > 0$. Then,

$$\int_0^1 (1 - x^a)^{1/a}\,dx = \frac{\Gamma^2(\frac{1}{a})}{2a\Gamma(\frac{2}{a})}.$$

In particular,

$$\int_0^1 (1 - \sqrt{x})^2\,dx = \frac{1}{6}, \quad \int_0^1 (1 - \sqrt[3]{x})^3\,dx = \frac{1}{20}, \quad \int_0^1 (1 - \sqrt[4]{x})^4\,dx = \frac{1}{70},$$

$$\int_0^1 \sqrt{1 - x^2}\,dx = \frac{\pi}{4}, \quad \int_0^1 \sqrt[3]{1 - x^3}\,dx = \frac{\Gamma^2(\frac{1}{3})}{6\Gamma(\frac{2}{3})} = \frac{\sqrt{\pi}\Gamma(\frac{1}{3})}{3\sqrt[3]{4}\Gamma(\frac{5}{6})}, \quad \int_0^1 \sqrt[4]{1 - x^4}\,dx = \frac{\Gamma^2(\frac{1}{4})}{8\sqrt{\pi}}.$$

Remark: $\Gamma(\frac{1}{3})\Gamma(\frac{5}{6}) = \sqrt[3]{2}\sqrt{\pi}\Gamma(\frac{2}{3})$.

Fact 14.3.2. Let $a > 0$, $b > -1$, and $k \ge 0$. Then,

$$\int_0^1 x^k(1 - \sqrt[a]{x})^b\,dx = \frac{\Gamma[(k+1)a+1]\Gamma(b+1)}{(k+1)\Gamma[(k+1)a+b+1]}.$$

In particular,

$$\int_0^1 \sqrt{1 - x}\,dx = \frac{2}{3}, \quad \int_0^1 \sqrt{1 - x^2}\,dx = \frac{\pi}{4}, \quad \int_0^1 \sqrt{1 - x^3}\,dx = \frac{\sqrt{\pi}\Gamma(\frac{1}{3})}{6\Gamma(\frac{11}{6})} = \frac{\sqrt{3}\Gamma^3(\frac{1}{3})}{10\sqrt[3]{2}\pi},$$

$$\int_0^1 \sqrt{1 - x^4}\,dx = \frac{\sqrt{\pi}\Gamma(\frac{1}{4})}{8\Gamma(\frac{7}{4})} = \frac{\sqrt{2\pi}\Gamma^2(\frac{1}{4})}{12\pi}, \quad \int_0^1 \sqrt{1 - x^5}\,dx = \frac{5\sqrt{\pi}\Gamma(\frac{6}{5})}{7\Gamma(\frac{7}{10})},$$

$$\int_0^1 \sqrt[3]{1 - x}\,dx = \frac{3}{4}, \quad \int_0^1 \sqrt[3]{1 - x^2}\,dx = \frac{\sqrt{\pi}\Gamma(\frac{1}{3})}{5\Gamma(\frac{5}{6})}, \quad \int_0^1 \sqrt[3]{1 - x^4}\,dx = \frac{4\Gamma(\frac{1}{3})\Gamma(\frac{5}{4})}{7\Gamma(\frac{7}{12})},$$

$$\int_0^1 \sqrt[4]{1 - x}\,dx = \frac{4}{5}, \quad \int_0^1 \sqrt[4]{1 - x^2}\,dx = \frac{\sqrt{\pi}\Gamma(\frac{1}{4})}{6\Gamma(\frac{3}{4})}, \quad \int_0^1 \sqrt[4]{1 - x^3}\,dx = \frac{\Gamma(\frac{1}{4})\Gamma(\frac{1}{3})}{7\Gamma(\frac{7}{12})},$$

$$\int_0^1 \frac{1}{\sqrt{1 - x}}\,dx = 2, \quad \int_0^1 \frac{1}{\sqrt{1 - x^2}}\,dx = \frac{\pi}{2},$$

$$\int_0^1 \frac{1}{\sqrt{1 - x^3}}\,dx = \frac{\sqrt{\pi}\Gamma(\frac{4}{3})}{\Gamma(\frac{5}{6})}, \quad \int_0^1 \frac{1}{\sqrt{1 - x^4}}\,dx = \frac{\sqrt{\pi}\Gamma(\frac{5}{4})}{\Gamma(\frac{3}{4})} = \frac{\sqrt{2}\Gamma^2(\frac{1}{4})}{8\sqrt{\pi}},$$

$$\int_0^1 \frac{1}{\sqrt[3]{1 - x}}\,dx = \frac{3}{2}, \quad \int_0^1 \frac{1}{\sqrt[3]{1 - x^2}}\,dx = \frac{3\sqrt{\pi}\Gamma(\frac{2}{3})}{\Gamma(\frac{1}{6})}, \quad \int_0^1 \frac{1}{\sqrt[3]{1 - x^3}}\,dx = \frac{2\sqrt{3}\pi}{9},$$

$$\int_0^1 \frac{1}{\sqrt[4]{1 - x}}\,dx = \frac{4}{3}, \quad \int_0^1 \frac{1}{\sqrt[4]{1 - x^2}}\,dx = \frac{2\sqrt{\pi}\Gamma(\frac{3}{4})}{\Gamma(\frac{1}{4})}, \quad \int_0^1 \frac{1}{\sqrt[4]{1 - x^3}}\,dx = \frac{4\Gamma(\frac{3}{4})\Gamma(\frac{1}{3})}{\Gamma(\frac{1}{12})}.$$

Furthermore, if $n \ge 1$, then

$$\int_0^1 \frac{1}{(1 - \sqrt{x})^{n/2}}\,dx = \frac{8}{n^2 - 6n + 8}, \quad \int_0^1 (1 - \sqrt{x})^n\,dx = \frac{2}{n^2 + 3n + 2},$$

$$\int_0^1 (1 - \sqrt[3]{x})^n\,dx = \frac{6}{(n+3)(n+2)(n+1)}, \quad \int_0^1 (1 - \sqrt[4]{x})^n\,dx = \frac{24}{(n+4)(n+3)(n+2)(n+1)}.$$

Fact 14.3.3. Let $n \geq 0$ and $a > 0$. Then,

$$\int_0^a x^n \sqrt{a-x}\,\mathrm{d}x = \frac{4^{n+1}n!(n+1)!a^{3/2+n}}{(2n+3)!}.$$

In particular,

$$\int_0^a \sqrt{a-x}\,\mathrm{d}x = \frac{2a^{3/2}}{3}, \quad \int_0^a x\sqrt{a-x}\,\mathrm{d}x = \frac{4a^{5/2}}{15}, \quad \int_0^a x^2\sqrt{a-x}\,\mathrm{d}x = \frac{16a^{7/2}}{105}.$$

Source: [815].

Fact 14.3.4. Let $n \geq 0$ and $a > 0$. Then,

$$\int_0^a x^{2n}\sqrt{a^2-x^2}\,\mathrm{d}x = \left(\frac{a}{2}\right)^{2n+2}\pi C_n, \quad \int_0^a x^{2n+1}\sqrt{a^2-x^2}\,\mathrm{d}x = \frac{2^{2n+1}a^{2n+3}}{(2n+3)(2n+2)(2n+1)C_n}.$$

In particular,

$$\int_0^a \sqrt{a^2-x^2}\,\mathrm{d}x = \frac{a^2\pi}{4}, \quad \int_0^a x\sqrt{a^2-x^2}\,\mathrm{d}x = \frac{a^3}{3}, \quad \int_0^a x^2\sqrt{a^2-x^2}\,\mathrm{d}x = \frac{a^4\pi}{16}.$$

Source: [816].

Fact 14.3.5. Let $a \in (0,1]$. Then,

$$\int_0^1 \sqrt{x-ax^2}\,\mathrm{d}x = \frac{\sqrt{a-a^2}(2a-1)+\operatorname{asin}\sqrt{a}}{4a^{3/2}}.$$

In particular,

$$\int_0^1 \sqrt{x-\tfrac{1}{2}x^2}\,\mathrm{d}x = \frac{\sqrt{2}\pi}{8}, \quad \int_0^1 \sqrt{x-\tfrac{3}{4}x^2}\,\mathrm{d}x = \frac{1}{12}+\frac{2\sqrt{3}\pi}{27}, \quad \int_0^1 \sqrt{x-x^2}\,\mathrm{d}x = \frac{\pi}{8}.$$

Fact 14.3.6. Let $a \geq 0$ and $b > 0$. Then,

$$\int_0^1 \frac{x^a}{\sqrt{1-x^b}}\,\mathrm{d}x = \frac{\sqrt{\pi}\Gamma\left(\frac{a+1}{b}\right)}{b\Gamma\left(\frac{2a+b+2}{2b}\right)}.$$

In particular, if $n \geq 0$, then

$$\int_0^1 \frac{1}{\sqrt{1-x}}\,\mathrm{d}x = 2, \quad \int_0^1 \frac{1}{\sqrt{1-x^2}}\,\mathrm{d}x = \frac{\pi}{2}, \quad \int_0^1 \frac{1}{\sqrt{1-x^3}}\,\mathrm{d}x = \frac{\sqrt{\pi}\Gamma\left(\frac{4}{3}\right)}{\Gamma\left(\frac{5}{6}\right)},$$

$$\int_0^1 \frac{1}{\sqrt{1-x^4}}\,\mathrm{d}x = \frac{\sqrt{\pi}\Gamma\left(\frac{1}{4}\right)}{4\Gamma\left(\frac{3}{4}\right)} = \frac{\sqrt{\pi}\Gamma\left(\frac{5}{4}\right)}{\Gamma\left(\frac{3}{4}\right)} = \frac{\sqrt{2}\Gamma^2\left(\frac{1}{4}\right)}{8\sqrt{\pi}},$$

$$\int_0^1 \frac{x^n}{\sqrt{1-x}}\,\mathrm{d}x = \frac{\sqrt{\pi}\Gamma(n+1)}{\Gamma(n+3/2)}, \quad \int_0^1 \frac{x^n}{\sqrt{1-x^2}}\,\mathrm{d}x = \frac{\sqrt{\pi}\Gamma[(n+1)/2]}{2\Gamma(1+n/2)},$$

$$\int_0^1 \frac{x^{2n+1}}{\sqrt{1-x^2}}\,\mathrm{d}x = \frac{(2n)!!}{(2n+1)!!}, \quad \int_0^1 \frac{x^n}{\sqrt{1-x^3}}\,\mathrm{d}x = \frac{\sqrt{\pi}\Gamma[(n+1)/3]}{3\Gamma[(2n+5)/6]}.$$

Source: [2960].

Fact 14.3.7. Let $n \geq 0$ and $a > 0$. Then,

$$\int_{-a}^a x^{2n}\sqrt{\frac{a+x}{a-x}}\,\mathrm{d}x = \frac{a^{2n+1}(2n)!\pi}{4^n(n!)^2}, \quad \int_{-a}^a x^{2n+1}\sqrt{\frac{a+x}{a-x}}\,\mathrm{d}x = \frac{a^{2n+2}(2n+2)!\pi}{4^{n+1}[(n+1)!]^2}.$$

In particular,

$$\int_{-a}^{a} \sqrt{\frac{a+x}{a-x}}\,\mathrm{d}x = a\pi, \quad \int_{-a}^{a} x\sqrt{\frac{a+x}{a-x}}\,\mathrm{d}x = \frac{a^2\pi}{2}, \quad \int_{-a}^{a} x^2\sqrt{\frac{a+x}{a-x}}\,\mathrm{d}x = \frac{a^3\pi}{2}, \quad \int_{-a}^{a} x^3\sqrt{\frac{a+x}{a-x}}\,\mathrm{d}x = \frac{3a^4\pi}{8}.$$

Source: [823].

Fact 14.3.8. Let $n \ge 0$. Then,

$$\int_0^1 \sqrt{\frac{1-x}{1+x}}\,\mathrm{d}x = \frac{\pi}{2} - 1, \quad \int_0^1 x\sqrt{\frac{1-x}{1+x}}\,\mathrm{d}x = 1 - \frac{\pi}{4}, \quad \int_0^1 x^2\sqrt{\frac{1-x}{1+x}}\,\mathrm{d}x = \frac{\pi}{4} - \frac{2}{3}, \quad \int_0^1 x^3\sqrt{\frac{1-x}{1+x}}\,\mathrm{d}x = \frac{2}{3} - \frac{3\pi}{16}.$$

Source: [820].

Fact 14.3.9. Let $n \ge 0$. Then,

$$\begin{aligned} C_n &= \frac{1}{2\pi}\int_0^4 x^n \sqrt{\frac{4-x}{x}}\,\mathrm{d}x = \frac{1}{\pi}\int_0^2 x^{2n}\sqrt{4-x^2}\,\mathrm{d}x \\ &= \frac{2^{2n+1}}{\pi}\int_0^1 x^{n-1/2}(1-x)^{1/2}\,\mathrm{d}x = \frac{4^{n+1}}{\pi}\int_0^{\pi/2} (\sin^2 x)\cos^{2n} x\,\mathrm{d}x. \end{aligned}$$

Source: [816, 1753, 2217].

Fact 14.3.10. Let $b \ge 0$ and $a > b + 1$. Then,

$$\int_0^1 \frac{x^b}{(1-x^a)^{(b+1)/a}}\,\mathrm{d}x = \frac{\pi}{a\sin\frac{(b+1)\pi}{a}}.$$

In particular,

$$\int_0^1 \frac{1}{\sqrt[a]{1-x^a}}\,\mathrm{d}x = \frac{\pi}{a\sin\frac{\pi}{a}}, \quad \int_0^1 \frac{1}{\sqrt{1-x^2}}\,\mathrm{d}x = \frac{\pi}{2}, \quad \int_0^1 \frac{1}{\sqrt[3]{1-x^3}}\,\mathrm{d}x = \frac{2\sqrt{3}\pi}{9}.$$

Source: [352, p. 41]. **Related:** Fact 13.10.22.

Fact 14.3.11. If $a > 0$, then

$$\int_0^1 \frac{\sqrt{1-x^2}}{x^2+a^2}\,\mathrm{d}x = \frac{\pi}{2a(a+\sqrt{a^2+1})}.$$

If $a > 1$, then

$$\int_0^1 \frac{\sqrt{1-x^2}}{a^2-x^2}\,\mathrm{d}x = \frac{a\pi}{2}(a-\sqrt{a^2-1}).$$

Source: [150, p. 14].

Fact 14.3.12. Let $a \in (-1,0) \cup (0,1)$. Then,

$$\int_0^1 \left(\frac{1}{x} - 1\right)^{1/a}\,\mathrm{d}x = \frac{a\pi}{\sin a\pi}.$$

In particular,

$$\int_0^1 \sqrt{\frac{1}{x} - 1}\,\mathrm{d}x = \int_0^1 \frac{1}{\sqrt{\frac{1}{x} - 1}}\,\mathrm{d}x = \frac{\pi}{2}, \quad \int_0^1 \sqrt[3]{\frac{1}{x} - 1}\,\mathrm{d}x = \int_0^1 \frac{1}{\sqrt[3]{\frac{1}{x} - 1}}\,\mathrm{d}x = \frac{2\sqrt{3}\pi}{9},$$

$$\int_0^1 \sqrt[4]{\frac{1}{x} - 1}\,\mathrm{d}x = \int_0^1 \frac{1}{\sqrt[4]{\frac{1}{x} - 1}}\,\mathrm{d}x = \frac{\sqrt{2}\pi}{4}, \quad \int_0^1 \sqrt[5]{\frac{1}{x} - 1}\,\mathrm{d}x = \int_0^1 \frac{1}{\sqrt[5]{\frac{1}{x} - 1}}\,\mathrm{d}x = \frac{\sqrt{2}\pi}{5}\sqrt{1+\frac{\sqrt{5}}{5}}.$$

Source: [352, p. 41].

Fact 14.3.13. Let $a > 0$ and $b > 0$. Then,

$$\int_0^1 \frac{1}{\sqrt{x(1-x)}[ax+b(1-x)]}\,\mathrm{d}x = \frac{\pi}{\sqrt{ab}}.$$

Source: [1381, p. 223]. **Related:** Fact 10.11.68.

Fact 14.3.14. Let $a > 0$ and $b > 0$. Then,

$$\begin{aligned}\int_0^\infty \frac{1}{\sqrt{(x^2+a^2)(x^2+b^2)}}\,\mathrm{d}x &= \int_0^\infty \frac{1}{\sqrt{(x^2+[(\frac{1}{2}(a+b)]^2)(x^2+ab)}}\,\mathrm{d}x\\ &= \int_0^{\pi/2} \frac{1}{\sqrt{a^2\cos^2 x + b^2\sin^2 x}}\,\mathrm{d}x\\ &= \int_0^{\pi/2} \frac{2}{\sqrt{(a+b)^2-(a-b)^2\sin^2 x}}\,\mathrm{d}x.\end{aligned}$$

In particular,

$$\int_0^\infty \frac{1}{\sqrt{(x^2+1)(x^2+2)}} = \int_0^1 \frac{1}{\sqrt{1-x^2}}\,\mathrm{d}x = \frac{\sqrt{\pi}\Gamma\left(\frac{1}{4}\right)}{4\Gamma\left(\frac{3}{4}\right)}.$$

Source: [2960]. **Related:** Fact 12.18.55.

Fact 14.3.15. Let $a > 0$ and $b > 1$. Then,

$$\int_0^\infty (\sqrt{x^2+a^2}-x)^b\,\mathrm{d}x = \frac{ba^{b+1}}{b^2-1}, \quad \int_0^\infty \frac{1}{(\sqrt{x^2+a^2}+x)^b}\,\mathrm{d}x = \frac{b}{a^{b-1}(b^2-1)}.$$

Source: [823] and [1524, p. 267].

Fact 14.3.16. Let a be a positive number. Then,

$$\int_1^\infty \frac{1}{(x+a)\sqrt{x-1}}\,\mathrm{d}x = \frac{\pi}{\sqrt{a+1}}.$$

Source: [2106, pp. 43–45].

14.4 Facts on Definite Integrals of Trigonometric Functions

Fact 14.4.1. Let $a > -1$. Then,

$$\int_0^{\pi/2} \sin^a x\,\mathrm{d}x = \int_0^{\pi/2} \cos^a x\,\mathrm{d}x = \frac{\sqrt{\pi}\Gamma\left(\frac{a+1}{2}\right)}{2\Gamma\left(\frac{a}{2}+1\right)}.$$

In particular,

$$\int_0^{\pi/2} \frac{1}{\sqrt[4]{\sin x}}\,\mathrm{d}x = \frac{\sqrt{\pi}\Gamma(\frac{3}{8})}{2\Gamma(\frac{7}{8})}, \quad \int_0^{\pi/2} \frac{1}{\sqrt{\sin x}}\,\mathrm{d}x = \frac{\Gamma(\frac{1}{4})}{2\sqrt{2\pi}},$$

$$\int_0^{\pi/2} \sqrt[4]{\sin x}\,\mathrm{d}x = \frac{\sqrt{\pi}\Gamma\left(\frac{5}{8}\right)}{2\Gamma\left(\frac{9}{8}\right)} = \frac{4\sqrt{\pi}\Gamma(\frac{5}{8})}{\Gamma(\frac{1}{8})}, \quad \int_0^{\pi/2} \sqrt[3]{\sin x}\,\mathrm{d}x = \frac{\sqrt{\pi}\Gamma\left(\frac{2}{3}\right)}{2\Gamma\left(\frac{7}{6}\right)} = \frac{3\sqrt{\pi}\Gamma(\frac{2}{3})}{\Gamma(\frac{1}{6})},$$

$$\int_0^{\pi/2} \sqrt{\sin x}\,\mathrm{d}x = \frac{\sqrt{\pi}\Gamma\left(\frac{3}{4}\right)}{2\Gamma\left(\frac{5}{4}\right)} = \sqrt{\frac{2}{\pi}}\Gamma^2(\tfrac{3}{4}), \quad \int_0^{\pi/2} \sin^{3/2} x\,\mathrm{d}x = \frac{\sqrt{\pi}\Gamma\left(\frac{5}{4}\right)}{2\Gamma\left(\frac{7}{4}\right)} = \frac{\Gamma^2(\frac{1}{4})}{6\sqrt{2\pi}}.$$

Furthermore, if $n \ge 0$, then

$$\int_0^{\pi/2} \sin^{2n} x\,\mathrm{d}x = \int_0^{\pi/2} \cos^{2n} x\,\mathrm{d}x = \int_0^1 \frac{2(1-x^2)^{2n}}{(1+x^2)^{2n+1}}\,\mathrm{d}x = \frac{(2n-1)!!\pi}{2(2n)!!} = \frac{(2n)!\pi}{2^{2n+1}(n!)^2} = \frac{\pi}{2^{2n+1}}\binom{2n}{n},$$

$$\int_0^{\pi/2} \sin^{2n+1} x\,\mathrm{d}x = \int_0^{\pi/2} \cos^{2n+1} x\,\mathrm{d}x = \int_0^1 \frac{2(1-x^2)^{2n+1}}{(1+x^2)^{2n+2}}\,\mathrm{d}x = \frac{(2n)!!}{(2n+1)!!} = \frac{4^n(n!)^2}{(2n+1)!} = \frac{4^n}{(2n+1)\binom{2n}{n}}.$$

Remark: See [511, p. 113], [822], [1136, p. 205], and [1158, p. 153]. **Related:** Fact 14.4.60.

Fact 14.4.2. Let $a > -1$ and $b > -1$. Then,

$$\int_0^{\pi/2} (\sin^a x)\cos^b x\,\mathrm{d}x = \frac{\Gamma(\frac{a+1}{2})\Gamma(\frac{b+1}{2})}{2\Gamma(\frac{a+b+2}{2})}.$$

In particular,

$$\int_0^{\pi/2} \frac{1}{\sqrt{(\sin x)\cos x}}\,\mathrm{d}x = \frac{\Gamma^2(\frac{1}{4})}{2\sqrt{\pi}}, \quad \int_0^{\pi/2} \sqrt{(\sin x)\cos x}\,\mathrm{d}x = \frac{2\pi^{3/2}}{\Gamma^2(\frac{1}{4})} = \frac{\Gamma^2(\frac{3}{4})}{\sqrt{\pi}}.$$

Fact 14.4.3. Let a be a nonzero real number. Then,

$$\int_0^{\pi} \sin^2 ax\,\mathrm{d}x = \frac{\pi}{2} - \frac{\sin 2a\pi}{4a}, \quad \int_0^{\pi} \cos^2 ax\,\mathrm{d}x = \frac{\pi}{2} + \frac{\sin 2a\pi}{4a}, \quad \int_0^{\pi} (\sin ax)\cos ax\,\mathrm{d}x = \frac{\sin^2 a\pi}{2a}.$$

If $2a$ is an integer, then

$$\int_0^{\pi} \sin^2 ax\,\mathrm{d}x = \int_0^{\pi} \cos^2 ax\,\mathrm{d}x = \frac{\pi}{2}.$$

If a is an integer, then

$$\int_0^{\pi} \sin^2 ax\,\mathrm{d}x = \int_0^{\pi} \cos^2 ax\,\mathrm{d}x = \frac{\pi}{2}, \quad \int_0^{\pi} (\sin ax)\cos ax\,\mathrm{d}x = 0.$$

If $2a$ is an integer but a is not an integer, then

$$\int_0^{\pi} (\sin ax)\cos ax\,\mathrm{d}x = \frac{1}{2a}.$$

Fact 14.4.4. Let a be a nonzero real number. Then,

$$\int_0^{2\pi} \sin^2 ax\,\mathrm{d}x = \pi - \frac{\sin 4a\pi}{4a}, \quad \int_0^{2\pi} \cos^2 ax\,\mathrm{d}x = \pi + \frac{\sin 4a\pi}{4a}, \quad \int_0^{2\pi} (\sin ax)\cos ax\,\mathrm{d}x = \frac{\sin^2 2a\pi}{2a}.$$

If $4a$ is an integer, then

$$\int_0^{2\pi} \sin^2 ax\,\mathrm{d}x = \int_0^{2\pi} \cos^2 ax\,\mathrm{d}x = \pi.$$

If $2a$ is an integer, then

$$\int_0^{2\pi} \sin^2 ax\,\mathrm{d}x = \int_0^{2\pi} \cos^2 ax\,\mathrm{d}x = \pi, \quad \int_0^{2\pi} (\sin ax)\cos ax\,\mathrm{d}x = 0.$$

If $4a$ is an integer but $2a$ is not an integer, then

$$\int_0^{2\pi} (\sin ax)\cos ax\,\mathrm{d}x = \frac{1}{2a}.$$

Fact 14.4.5. Let $n \ge 0$. Then,

$$\int_0^{2\pi} \cos^{2n} x\,\mathrm{d}x = \frac{2\pi(2n)!}{4^n(n!)^2}.$$

Source: [2106, p. 328].

Fact 14.4.6. Let a and b be real numbers, and assume that $a^2 \neq b^2$. Then,

$$\int_0^\pi (\sin ax)\sin bx\,\mathrm{d}x = \frac{b(\sin a\pi)\cos b\pi - a(\cos a\pi)\sin b\pi}{a^2-b^2},$$

$$\int_0^\pi (\cos ax)\cos bx\,\mathrm{d}x = \frac{a(\sin a\pi)\cos b\pi - b(\cos a\pi)\sin b\pi}{a^2-b^2},$$

$$\int_0^\pi (\sin ax)\cos bx\,\mathrm{d}x = \frac{a - a(\cos a\pi)\cos b\pi - b(\sin a\pi)\sin b\pi}{a^2-b^2}.$$

Now, let n and m be integers. If $n \neq m$, then

$$\int_0^\pi (\sin nx)\sin mx\,\mathrm{d}x = \int_0^\pi (\cos nx)\cos mx\,\mathrm{d}x = 0.$$

If $n-m$ is even, then

$$\int_0^\pi (\sin nx)\cos mx\,\mathrm{d}x = 0.$$

If $n-m$ is odd, then

$$\int_0^\pi (\sin nx)\cos mx\,\mathrm{d}x = \frac{2n}{n^2-m^2}.$$

Fact 14.4.7. Let $n \ge 0$. Then,

$$\int_0^{\pi/2} (\sin^{2n} x)\sin 2nx\,\mathrm{d}x = (-1)^{n+1}\frac{1}{2^{2n+1}}\sum_{i=1}^{2n}\frac{2^i}{i}, \quad \int_0^{\pi/2} (\sin^{2n+1} x)\sin(2n+1)x\,\mathrm{d}x = (-1)^n\frac{\pi}{4^{n+1}},$$

$$\int_0^{\pi/2} (\cos^n x)\cos nx\,\mathrm{d}x = \frac{\pi}{2^{n+1}}, \quad \int_0^{\pi/2} (\cos^n x)\sin nx\,\mathrm{d}x = \frac{1}{2^{n+1}}\sum_{i=1}^{n}\frac{2^i}{i},$$

$$\int_0^{\pi/2} (\sin^{2n} x)\cos 2nx\,\mathrm{d}x = (-1)^n\frac{\pi}{2^{2n+1}}, \quad \int_0^{\pi/2} (\sin^{2n+1} x)\cos(2n+1)x\,\mathrm{d}x = (-1)^n\frac{1}{4^{n+1}}\sum_{i=1}^{2n+1}\frac{2^i}{i},$$

$$\int_0^{\pi} (\sin^{2n} x)\sin 2nx\,\mathrm{d}x = 0, \quad \int_0^{\pi} (\sin^{2n+1} x)\sin(2n+1)x\,\mathrm{d}x = (-1)^n\frac{\pi}{2^{2n+1}},$$

$$\int_0^{\pi} (\cos^n x)\cos nx\,\mathrm{d}x = \frac{\pi}{2^n}, \quad \int_0^{\pi} (\cos^n x)\sin nx\,\mathrm{d}x = 0,$$

$$\int_0^{\pi} (\sin^{2n} x)\cos 2nx\,\mathrm{d}x = (-1)^n\frac{\pi}{4^n}, \quad \int_0^{\pi} (\sin^{2n+1} x)\cos(2n+1)x\,\mathrm{d}x = 0.$$

Fact 14.4.8. Let a and b be real numbers, and assume that $a^2 \neq b^2$. Then,

$$\int_0^{2\pi} (\sin ax)\sin bx\,\mathrm{d}x = \frac{b(\sin 2a\pi)\cos 2b\pi - a(\cos 2a\pi)\sin 2b\pi}{a^2-b^2},$$

$$\int_0^{2\pi} (\cos ax)\cos bx\,\mathrm{d}x = \frac{a(\sin 2a\pi)\cos 2b\pi - b(\cos 2a\pi)\sin 2b\pi}{a^2-b^2},$$

$$\int_0^{2\pi} (\sin ax)\cos bx\,\mathrm{d}x = \frac{a - a(\cos 2a\pi)\cos 2b\pi - b(\sin 2a\pi)\sin 2b\pi}{a^2-b^2}.$$

Now, let n and m be integers. Then,

$$\int_0^{2\pi} (\sin nx)\sin mx\,\mathrm{d}x = \int_0^{2\pi} (\cos nx)\cos mx\,\mathrm{d}x = \int_0^{2\pi} (\sin nx)\cos mx\,\mathrm{d}x = 0.$$

Fact 14.4.9. Let $n \ge 0$. Then,

$$\int_0^{\pi/2} x^2 \cos^{2n+2} x \, dx = \frac{\pi}{4} \frac{(2n+1)!!}{(2n+2)!!} \left(\frac{\pi^2}{6} - H_{n+1,2} \right).$$

In particular,

$$\int_0^{\pi/2} x^2 \cos^2 x \, dx = \frac{\pi}{48}(\pi^2 - 6), \quad \int_0^{\pi/2} x^2 \cos^4 x \, dx = \frac{3\pi}{64}\left(\frac{\pi^2}{3} - \frac{5}{2} \right),$$

$$\int_0^{\pi/2} x^2 \cos^6 x \, dx = \frac{15\pi}{1152}\left(\pi^2 - \frac{49}{6} \right).$$

Source: [826]. **Related:** Fact 13.5.44.

Fact 14.4.10.

$$\int_0^{\pi/6} \frac{x}{\sin x} \, dx = \frac{4G}{3} - \frac{\pi}{6} \operatorname{acosh} 2,$$

$$\int_0^{\pi/4} x \sin x \, dx = \frac{(4-\pi)\sqrt{2}}{8}, \quad \int_0^{\pi/4} x \cos x \, dx = \frac{\sqrt{2}}{2} + \frac{\sqrt{2}\pi}{8} - 1,$$

$$\int_0^{\pi/4} \frac{x^2}{\sin^2 x} \, dx = G - \frac{\pi^2}{16} + \frac{\pi}{4} \log 2, \quad \int_0^{\pi/4} \frac{x^3}{\sin^2 x} \, dx = \frac{3G\pi}{4} - \frac{\pi^3}{64} + \frac{3\pi^2}{32} - \frac{105}{64}\zeta(3),$$

$$\int_0^{\pi/4} \frac{1}{\cos^2 x} \, dx = 1, \quad \int_0^{\pi/4} \frac{x}{\cos^2 x} \, dx = \frac{\pi}{4} - \frac{1}{2} \log 2, \quad \int_0^{\pi/4} \frac{x^2}{\cos^2 x} \, dx = \frac{\pi^2}{16} + \frac{\pi}{4} \log 2 - G,$$

$$\int_0^{\pi/4} \frac{x^3}{\cos^2 x} \, dx = \frac{\pi^3}{64} + \frac{3\pi^2}{32} \log 2 + \frac{63}{64}\zeta(3) - \frac{3G\pi}{4},$$

$$\int_0^{\pi/4} \tan x \, dx = \frac{1}{2} \log 2, \quad \int_0^{\pi/4} x \tan x \, dx = \frac{G}{2} - \frac{\pi}{8} \log 2, \quad \int_0^{\pi/4} x^2 \tan x \, dx = \frac{G\pi}{4} - \frac{\pi^2}{32} \log 2 - \frac{21}{64}\zeta(3),$$

$$\int_0^{\pi/4} \tan^2 x \, dx = 1 - \frac{\pi}{4}, \quad \int_0^{\pi/4} x \tan^2 x \, dx = \frac{\pi}{4} - \frac{\pi^2}{32} - \frac{1}{2} \log 2,$$

$$\int_0^{\pi/4} x^2 \tan^2 x \, dx = \frac{\pi^2}{16} - \frac{\pi^3}{192} + \frac{\pi}{4} \log 2 - G, \quad \int_0^{\pi/4} \tan^3 x \, dx = \frac{1}{2} - \frac{1}{2} \log 2,$$

$$\int_0^{\pi/2} \left(\frac{\pi}{2} - x \right) \tan x \, dx = \int_{\pi/2}^{\pi} \left(\frac{\pi}{2} - x \right) \tan x \, dx = \frac{1}{2} \int_0^{\pi} \left(\frac{\pi}{2} - x \right) \tan x \, dx = \frac{\pi}{2} \log 2,$$

$$\int_0^{\pi/4} x \tan^3 x \, dx = \frac{\pi}{4} + \frac{\pi}{8} \log 2 - \frac{1}{2} - \frac{G}{2},$$

$$\int_0^{\pi/4} x^2 \tan^3 x \, dx = \frac{1}{2} \log 2 + \frac{\pi^2}{16} + \frac{\pi^2}{32} \log 2 + \frac{21}{64}\zeta(3) - \frac{\pi}{4} - \frac{G\pi}{4},$$

$$\int_0^{\pi/4} x \cot x \, dx = \frac{G}{2} + \frac{\pi}{8} \log 2, \quad \int_0^{\pi/4} x^2 \cot x \, dx = \frac{G\pi}{4} + \frac{\pi^2}{32} \log 2 - \frac{35}{64}\zeta(3),$$

$$\int_0^{\pi/4} x^2 \cot^2 x \, dx = G - \frac{\pi^2}{16} - \frac{\pi^3}{192} + \frac{\pi}{4} \log 2,$$

$$\int_0^{\pi/4} x^3 \cot^2 x \, dx = \frac{3G\pi}{4} - \frac{\pi^3}{64} - \frac{\pi^4}{1024} + \frac{3\pi^2}{32} \log 2 - \frac{105}{64}\zeta(3),$$

$$\int_0^{\pi/4} \frac{x^2 \tan x}{\cos^2 x} \, dx = \frac{\pi^2}{16} + \frac{1}{2} \log 2 - \frac{\pi}{4}, \quad \int_0^{\pi/3} \frac{x^2 \tan x}{\cos^2 x} \, dx = \frac{2\pi^2}{9} + \log 2 - \frac{\sqrt{3}\pi}{3},$$

$$\int_0^{\pi/4} \frac{x^2\tan^2 x}{\cos^2 x}\,\mathrm{d}x = \frac{1}{3} + \frac{\pi^2}{48} + \frac{G}{3} - \frac{\pi}{12}\log 2 - \frac{\pi}{6},$$

$$\int_0^{\pi/2} x\cot x\,\mathrm{d}x = \frac{\pi}{2}\log 2, \qquad \int_0^{\pi/2} x^2\cot x\,\mathrm{d}x = \frac{\pi^2}{4}\log 2 - \frac{7}{8}\zeta(3),$$

$$\int_0^{\pi/2} x^3\cot x\,\mathrm{d}x = \frac{\pi^3}{8}\log 2 - \frac{9\pi}{16}\zeta(3), \qquad \int_0^{\pi/2} x^4\cot x\,\mathrm{d}x = \frac{\pi^4}{16}\log 2 - \frac{9\pi^2}{16}\zeta(3) + \frac{93}{32}\zeta(5),$$

$$\int_0^{\pi/2} \frac{x}{\sin x}\,\mathrm{d}x = 2G, \qquad \int_0^{\pi/2} \frac{x^2}{\sin x}\,\mathrm{d}x = 2G\pi - \frac{7}{2}\zeta(3),$$

$$\int_0^{\pi/2} \frac{x^2}{\sin^2 x}\,\mathrm{d}x = \pi\log 2, \qquad \int_0^{\pi/2} \frac{x^3}{\sin^2 x}\,\mathrm{d}x = \frac{3\pi^2}{4}\log 2 - 7\zeta(3),$$

$$\int_0^{\pi/2} \frac{x}{\tan x}\,\mathrm{d}x = \frac{\pi}{2}\log 2, \qquad \int_0^{\pi/2} \frac{x^2}{\tan x}\,\mathrm{d}x = \frac{\pi^2}{4}\log 2 - \frac{7}{8}\zeta(3),$$

$$\int_0^{\pi/2} \frac{x^2\cos x}{\sin^2 x}\,\mathrm{d}x = 4G - \frac{\pi^2}{4}, \qquad \int_0^{\pi/2} \frac{x^3\cos x}{\sin^2 x}\,\mathrm{d}x = 6G\pi - \frac{\pi^2}{8} - \frac{21}{2}\zeta(3),$$

$$\int_0^{\pi/2} \frac{x^3\cos x}{\sin^3 x}\,\mathrm{d}x = \frac{3\pi}{2}\log 2 - \frac{\pi^3}{16}, \qquad \int_0^{\pi/2} \frac{x^4\cos x}{\sin^3 x}\,\mathrm{d}x = \frac{3\pi^2}{2}\log 2 - \frac{\pi^4}{32} - \frac{21}{4}\zeta(3).$$

Fact 14.4.11. Let $n \ge 1$. Then,

$$\int_0^{\pi/2} \frac{\sin 2nx}{\sin x}\,\mathrm{d}x = \sum_{i=1}^n (-1)^{i+1}\frac{2}{2i-1}, \qquad \int_0^{\pi/2} \frac{\sin(2n-1)x}{\sin x}\,\mathrm{d}x = \frac{\pi}{2},$$

$$\int_0^{\pi/2} \frac{\sin 2nx}{\cos x}\,\mathrm{d}x = \frac{1}{2}\int_0^{\pi} \frac{\sin 2nx}{\cos x}\,\mathrm{d}x = (-1)^{n-1}\sum_{i=1}^n (-1)^{i+1}\frac{2}{2i-1},$$

$$\int_0^{\pi} \frac{\sin 2nx}{\sin x}\,\mathrm{d}x = 0, \qquad \int_0^{\pi} \frac{\sin(2n-1)x}{\sin x}\,\mathrm{d}x = \pi, \qquad \int_0^{\pi} \frac{x\sin 2nx}{\sin x}\,\mathrm{d}x = -\sum_{i=1}^n \frac{4}{(2i-1)^2}.$$

In particular,

$$\int_0^{\pi/2} \frac{\sin 2x}{\sin x}\,\mathrm{d}x = 2, \qquad \int_0^{\pi/2} \frac{\sin 4x}{\sin x}\,\mathrm{d}x = \frac{4}{3}, \qquad \int_0^{\pi/2} \frac{\sin 6x}{\sin x}\,\mathrm{d}x = \frac{26}{15}, \qquad \int_0^{\pi/2} \frac{\sin 8x}{\sin x}\,\mathrm{d}x = \frac{152}{105}.$$

Source: [815] and [1217, p. 471].

Fact 14.4.12.

$$\int_0^{\pi/4} \left(\frac{1}{x} - \frac{1}{\sin x}\right)\mathrm{d}x = \log[(1+\sqrt{2})\pi] - 3\log 2, \qquad \int_0^{\pi/4} \left(\frac{1}{x^2} - \frac{1}{\sin^2 x}\right)\mathrm{d}x = 1 - \frac{4}{\pi},$$

$$\int_0^{\pi/4} \left(\frac{1}{2x} + \frac{1}{x^3} - \frac{1}{\sin^3 x}\right)\mathrm{d}x = \frac{1}{12} + \frac{\sqrt{2}}{2} - \frac{8}{\pi^2} - \frac{3\pi}{2} + \frac{1}{2}\log[(1+\sqrt{2})\pi],$$

$$\int_0^{\pi/3} \left(\frac{1}{x} - \frac{1}{\sin x}\right)\mathrm{d}x = \log\frac{\pi}{2} - \frac{1}{2}\log 3, \qquad \int_0^{\pi/3} \left(\frac{1}{x^2} - \frac{1}{\sin^2 x}\right)\mathrm{d}x = \frac{\sqrt{3}}{3} - \frac{3}{\pi},$$

$$\int_0^{\pi/3} \left(\frac{1}{2x} + \frac{1}{x^3} - \frac{1}{\sin^3 x}\right)\mathrm{d}x = \frac{5}{12} - \frac{9}{2\pi^2} + \frac{1}{2}\log\frac{\pi}{2} - \frac{1}{4}\log 3,$$

$$\int_0^{\pi/2} \left(\frac{1}{x} - \frac{1}{\sin x}\right)\mathrm{d}x = \log\frac{\pi}{4}, \qquad \int_0^{\pi/2} \left(\frac{1}{x^2} - \frac{1}{\sin^2 x}\right)\mathrm{d}x = -\frac{2}{\pi},$$

$$\int_0^{\pi/2}\left(\frac{1}{2x}+\frac{1}{x^3}-\frac{1}{\sin^3 x}\right)\mathrm{d}x = \frac{1}{12}-\frac{2}{\pi^2}+\frac{1}{2}\log\pi-\log 2,$$

Fact 14.4.13. Let a and b be positive numbers. Then,

$$\int_0^\infty \frac{(\sin ax)\cos bx}{x}\,\mathrm{d}x = \begin{cases}\frac{\pi}{2}, & b<a,\\ \frac{\pi}{4}, & b=a,\\ 0, & a<b,\end{cases}\qquad \int_0^\infty \frac{(\sin ax)\sin bx}{x^2}\,\mathrm{d}x = \frac{\pi}{4}\min\{a,b\}.$$

If, in addition, $a\neq b$, then

$$\int_0^\infty \frac{(\sin ax)\sin bx}{x}\,\mathrm{d}x = \frac{1}{2}\log\frac{a+b}{|a-b|}.$$

Fact 14.4.14. Let $n\geq 1$ and $a>0$. Then,

$$\int_0^\infty \frac{\sin^{2n+1} ax}{x}\,\mathrm{d}x = \int_0^\infty \frac{(2n+2)(\sin^{2n+1} ax)\cos ax}{x}\,\mathrm{d}x = \frac{\pi}{2^{2n+1}}\binom{2n}{n},$$

$$\int_0^\infty \frac{(\sin^{2n+1} ax)\cos^{2n-1} ax}{x}\,\mathrm{d}x = \frac{\pi}{2^{4n}}\binom{2n-1}{n-1},\qquad \int_0^\infty \frac{\sin^{2n} ax}{x^2}\,\mathrm{d}x = \frac{a\pi}{2^{2n-1}}\binom{2n-2}{n-1},$$

$$\int_0^\infty \frac{(\cos^n x)\sin nx}{x}\,\mathrm{d}x = \frac{\pi}{2}\left(1-\frac{1}{2^n}\right),\qquad \int_0^\infty \frac{\sin^n ax}{x^n}\,\mathrm{d}x = \frac{a^{n-1}\pi}{2^n(n-1)!}\sum_{i=0}^{\lfloor n/2\rfloor}(-1)^i(n-2i)^{n-1}\binom{n}{i},$$

$$\int_0^\infty \frac{(\cos^{2n} x)\tan x}{x}\,\mathrm{d}x = \frac{\pi}{2^{2n+1}}\binom{2n}{n}.$$

In particular,

$$\int_0^\infty \frac{\sin ax}{x}\,\mathrm{d}x = \frac{\pi}{2},\qquad \int_0^\infty \frac{\sin^3 ax}{x}\,\mathrm{d}x = \frac{\pi}{4},\qquad \int_0^\infty \frac{\sin^5 ax}{x}\,\mathrm{d}x = \frac{3\pi}{16},$$

$$\int_0^\infty \frac{\sin^2 ax}{x^2}\,\mathrm{d}x = \frac{a\pi}{2},\qquad \int_0^\infty \frac{\sin^3 ax}{x^3}\,\mathrm{d}x = \frac{3a^2\pi}{8},$$

$$\int_0^\infty \frac{\sin^4 ax}{x^4}\,\mathrm{d}x = \frac{a^3\pi}{3},\qquad \int_0^\infty \frac{\sin^5 ax}{x^5}\,\mathrm{d}x = \frac{115a^4\pi}{384},\qquad \int_0^\infty \frac{\sin^6 ax}{x^6}\,\mathrm{d}x = \frac{11a^5\pi}{40},$$

$$\int_0^\infty \frac{\sin^7 ax}{x^7}\,\mathrm{d}x = \frac{5887a^6\pi}{23040},\qquad \int_0^\infty \frac{\sin^8 ax}{x^8}\,\mathrm{d}x = \frac{a^7\pi}{630}.$$

Source: [513, 758], [1217, pp. 462, 463, 472], [1444], [1568, p. 39], and [1889, 2050, 2691].
Remark: $\int_0^\infty \frac{\sin ax}{x}\,\mathrm{d}x$ is the *Dirichlet integral*. See [1945, p. 132].

Fact 14.4.15. Let $n\geq 1$ and $m\geq 0$. Then, the following statements hold:

i) If $1\leq m\leq n$, then

$$\int_0^\infty \frac{\sin^{2n} x}{x^{2m}}\,\mathrm{d}x = \frac{\pi}{4^n}\sum_{i=1}^{n}(-1)^{i+m}\frac{(2i)^{2m-1}}{(2m-1)!}\binom{2n}{n-i}.$$

ii) If $0\leq m\leq n$, then

$$\int_0^\infty \frac{\sin^{2n+1} x}{x^{2m+1}}\,\mathrm{d}x = \frac{\pi}{2^{2n+1}}\sum_{i=0}^{n}(-1)^{i+m}\frac{(2i+1)^{2m}}{(2m)!}\binom{2n+1}{n-i}.$$

iii) If $2 \le m \le n$, then

$$\int_0^\infty \frac{\sin^{2n} x}{x^{2m-1}}\,\mathrm{d}x = \frac{1}{2^{2n-1}}\sum_{i=1}^{n}(-1)^{i+m}\frac{(2i)^{2m-2}}{(2m-2)!}\binom{2n}{n-i}\log i.$$

iv) If $1 \le m \le n$, then

$$\int_0^\infty \frac{\sin^{2n+1} x}{x^{2m}}\,\mathrm{d}x = \frac{1}{4^n}\sum_{i=0}^{n}(-1)^{i+m}\frac{(2i+1)^{2m-1}}{(2m-1)!}\binom{2n+1}{n-i}\log(2i+1).$$

In particular,

$$\int_0^\infty \frac{\sin^3 ax}{x^2}\,\mathrm{d}x = \frac{3a}{4}\log 3, \quad \int_0^\infty \frac{\sin^5 ax}{x^2}\,\mathrm{d}x = \frac{5a}{16}(3\log 3 - \log 5),$$

$$\int_0^\infty \frac{\sin^7 ax}{x^2}\,\mathrm{d}x = \frac{7a}{64}(9\log 3 - 5\log 5 + \log 7), \quad \int_0^\infty \frac{\sin^4 ax}{x^2}\,\mathrm{d}x = \frac{a\pi}{4}, \quad \int_0^\infty \frac{\sin^6 ax}{x^2}\,\mathrm{d}x = \frac{3a\pi}{16},$$

$$\int_0^\infty \frac{\sin^8 ax}{x^2}\,\mathrm{d}x = \frac{5a\pi}{32}, \quad \int_0^\infty \frac{\sin^4 ax}{x^3}\,\mathrm{d}x = a^2\log 2, \quad \int_0^\infty \frac{\sin^5 ax}{x^3}\,\mathrm{d}x = \frac{5a^2\pi}{32},$$

$$\int_0^\infty \frac{\sin^6 ax}{x^3}\,\mathrm{d}x = \frac{3a^2}{16}(8\log 2 - 3\log 3), \quad \int_0^\infty \frac{\sin^5 ax}{x^4}\,\mathrm{d}x = \frac{5a^3}{96}(25\log 5 - 27\log 3),$$

$$\int_0^\infty \frac{\sin^6 ax}{x^4}\,\mathrm{d}x = \frac{\pi a^3}{8}, \quad \int_0^\infty \frac{\sin^7 ax}{x^4}\,\mathrm{d}x = \frac{7a^3}{384}(125\log 5 - 81\log 3 - 49\log 7),$$

$$\int_0^\infty \frac{\sin^6 ax}{x^5}\,\mathrm{d}x = a^4\left(\frac{27}{16}\log 3 - 2\log 2\right), \quad \int_0^\infty \frac{\sin^7 ax}{x^5}\,\mathrm{d}x = \frac{77\pi a^3}{768}.$$

Source: [1680].

Fact 14.4.16. Let $a > 0$ and $b > 1$. Then,

$$\int_0^\infty \sin ax^b\,\mathrm{d}x = \frac{\Gamma(\frac{1}{b})}{ba^{1/b}}\sin\frac{\pi}{2b}, \quad \int_0^\infty \cos ax^b\,\mathrm{d}x = \frac{\Gamma(\frac{1}{b})}{ba^{1/b}}\cos\frac{\pi}{2b}.$$

In particular,

$$\int_0^\infty \sin ax^2\,\mathrm{d}x = \int_0^\infty \cos ax^2\,\mathrm{d}x = \sqrt{\frac{\pi}{8a}}.$$

Fact 14.4.17. Let $a > 0$ and $b \ge 0$. Then,

$$\int_0^\infty (\sin ax^2)\cos 2bx\,\mathrm{d}x = \frac{\pi}{8a}\left(\cos\frac{b^2}{a} - \sin\frac{b^2}{a}\right), \quad \int_0^\infty (\cos ax^2)\cos 2bx\,\mathrm{d}x = \frac{\pi}{8a}\left(\cos\frac{b^2}{a} + \sin\frac{b^2}{a}\right).$$

Source: [1524, p. 269].

Fact 14.4.18. Let a and b be real numbers, and assume that $b \ne 0$. If $a \in (0,2)$, then

$$\int_0^\infty \frac{\sin bx}{x^a}\,\mathrm{d}x = (\operatorname{sign} b)|b|^{a-1}\Gamma(1-a)\cos\frac{a\pi}{2} = \frac{(\operatorname{sign} b)|b|^{a-1}\pi}{2\Gamma(a)}\csc\frac{a\pi}{2}.$$

If $a \in (0,1)$, then

$$\int_0^\infty \frac{\cos bx}{x^a}\,\mathrm{d}x = |b|^{a-1}\Gamma(1-a)\sin\frac{a\pi}{2} = \frac{|b|^{a-1}\pi}{2\Gamma(a)}\sec\frac{a\pi}{2}.$$

In particular,

$$\int_0^\infty \frac{\sin bx}{\sqrt{x}}\,\mathrm{d}x = (\operatorname{sign} b)\sqrt{\frac{\pi}{2|b|}}, \quad \int_0^\infty \frac{\cos bx}{\sqrt{x}}\,\mathrm{d}x = \sqrt{\frac{\pi}{2|b|}}, \quad \int_0^\infty \frac{\sin bx}{x^{3/2}}\,\mathrm{d}x = (\operatorname{sign} b)\sqrt{2\pi|b|}.$$

Source: [116, p. 50], [968, pp. 261, 262], and [1136, p. 211].

Fact 14.4.19. Let a and b be positive numbers. Then,

$$\int_0^\infty \frac{\sin ax^b}{x}\,\mathrm{d}x = \frac{\pi}{2b}.$$

Source: [1217, p. 483].

Fact 14.4.20. Let $a, b \in \mathbb{R}$, and assume that $c > 0$ and either $a > 1$ and $b > 1$ or $a \in (-\infty, 1)$ and $a + b > 1$. Then,

$$\int_0^\infty \frac{\sin cx^b}{x^a}\,\mathrm{d}x = \frac{c^{(a-1)/b}}{b}\Gamma\left(\frac{1-a}{b}\right)\cos\frac{(a+b-1)\pi}{2b}.$$

In particular, if either $a > 1$ or $a \in (1/2, 1)$, then

$$\int_0^\infty \frac{\sin cx^a}{x^a}\,\mathrm{d}x = \frac{c^{(a-1)/b}}{a-1}\Gamma\left(\frac{1}{a}\right)\cos\frac{\pi}{2a} = -\frac{c^{(a-1)/b}}{a}\Gamma\left(\frac{1-a}{a}\right)\cos\frac{\pi}{2a}.$$

Hence,

$$\int_0^\infty \frac{\sin cx^2}{x^2}\,\mathrm{d}x = \frac{1}{2}\sqrt{c\pi}.$$

Source: [2106, p. 135].

Fact 14.4.21. Let $a > 0$. Then,

$$\int_0^\infty \frac{\sin^2 ax^2}{x^2}\,\mathrm{d}x = \frac{1}{2}\sqrt{a\pi}, \quad \int_0^\infty \frac{\sin^2 ax^2}{x^3}\,\mathrm{d}x = \frac{a\pi}{4}, \quad \int_0^\infty \frac{\sin^2 ax^2}{x^4}\,\mathrm{d}x = \frac{2}{3}\sqrt{a^3\pi},$$

$$\int_0^\infty \frac{\sin^3 ax^2}{x^2}\,\mathrm{d}x = \frac{3-\sqrt{3}}{4}\sqrt{\frac{a\pi}{2}}, \quad \int_0^\infty \frac{\sin^3 ax^2}{x^3}\,\mathrm{d}x = \frac{3a}{8}\log 3,$$

$$\int_0^\infty \frac{\sin^3 ax^2}{x^4}\,\mathrm{d}x = \frac{\sqrt{3}-1}{2}\sqrt{\frac{a^3\pi}{2}}, \quad \int_0^\infty \frac{\sin^3 ax^2}{x^5}\,\mathrm{d}x = \frac{3a^2\pi}{16},$$

$$\int_0^\infty \frac{\sin^3 ax^2}{x^6}\,\mathrm{d}x = \frac{3\sqrt{3}-1}{5}\sqrt{\frac{a^5\pi}{2}}, \quad \int_0^\infty \frac{\sin ax^3}{x}\,\mathrm{d}x = \frac{\pi}{6},$$

$$\int_0^\infty \frac{\sin ax^3}{x^2}\,\mathrm{d}x = \frac{\sqrt[3]{a}}{2}\Gamma(\tfrac{2}{3}), \quad \int_0^\infty \frac{\sin ax^3}{x^3}\,\mathrm{d}x = \frac{\sqrt{3}a^{2/3}}{4}\Gamma(\tfrac{1}{3}),$$

$$\int_0^\infty \frac{\sin ax^4}{x^2}\,\mathrm{d}x = \frac{1}{2}\sqrt{2-\sqrt{2}}a^{1/4}\Gamma(\tfrac{3}{4}), \quad \int_0^\infty \frac{\sin ax^4}{x^3}\,\mathrm{d}x = \frac{1}{4}\sqrt{2a\pi},$$

$$\int_0^\infty \frac{\sin ax^4}{x^4}\,\mathrm{d}x = \frac{1}{6}\sqrt{2+\sqrt{2}}a^{3/4}\Gamma(\tfrac{1}{4}), \quad \int_0^\infty \frac{\sin ax^5}{x^4}\,\mathrm{d}x = \frac{1}{12}(1+\sqrt{5})a^{3/5}\Gamma(\tfrac{2}{5}).$$

Fact 14.4.22. If a is a real number, then

$$\int_0^\infty \frac{\sin ax - ax\cos ax}{x^3}\,\mathrm{d}x = (\operatorname{sign} a)\frac{\pi a^2}{4}.$$

If a is a positive number, then

$$\int_0^\infty \frac{\cos ax + x\sin ax}{x^2+1}\,\mathrm{d}x = \frac{\pi}{e^a}.$$

Source: [1217, p. 447].

Fact 14.4.23.

$$\int_0^\infty \frac{x^2-\sin^2 x}{x^4}\,\mathrm{d}x = \frac{\pi}{3}, \quad \int_0^\infty \frac{x^3-\sin^3 x}{x^5}\,\mathrm{d}x = \frac{13\pi}{32}, \quad \int_0^\infty \frac{x^4-\sin^4 x}{x^6}\,\mathrm{d}x = \frac{7\pi}{15}.$$

Fact 14.4.24. Let a and b be nonnegative numbers. Then,

$$\int_0^\infty \frac{\cos ax - \cos bx}{x^2}\,\mathrm{d}x = \frac{\pi}{2}(b-a).$$

In particular,

$$\int_0^\infty \frac{1-\cos ax}{x^2}\,\mathrm{d}x = \frac{a\pi}{2}.$$

Source: [1136, p. 211].

Fact 14.4.25. Let a and b be positive numbers. Then,

$$\int_0^\infty \frac{\cos ax - \cos bx}{x}\,\mathrm{d}x = \log\frac{b}{a}, \quad \int_0^\infty \frac{b\sin ax - a\sin bx}{x^2}\,\mathrm{d}x = ab\log\frac{b}{a},$$

$$\int_0^\infty \frac{\sin^2 ax - \sin^2 bx}{x}\,\mathrm{d}x = \frac{1}{2}\log\frac{a}{b}.$$

Now, let $c \in (0,1)$. Then,

$$\int_0^\infty \frac{\cos ax - \cos bx}{x^{c+1}}\,\mathrm{d}x = \left(\cos\frac{c\pi}{2}\right)\Gamma(-c)(a^c - b^c).$$

In particular,

$$\int_0^\infty \frac{\cos ax - \cos bx}{x^{5/4}}\,\mathrm{d}x = \frac{1}{2}\sqrt{2+\sqrt{2}}\,\Gamma(-\tfrac{1}{4})(\sqrt[4]{a} - \sqrt[4]{b}),$$

$$\int_0^\infty \frac{\cos ax - \cos bx}{x^{4/3}}\,\mathrm{d}x = \frac{\sqrt{3}}{2}\Gamma(-\tfrac{1}{3})(\sqrt[3]{a} - \sqrt[3]{b}), \quad \int_0^\infty \frac{\cos ax - \cos bx}{x^{3/2}}\,\mathrm{d}x = \sqrt{2\pi}(\sqrt{b} - \sqrt{a}).$$

Source: [1217, pp. 447, 448]. **Remark:** $\lim_{c\to 1}\cos(c\pi/2)\Gamma(-c) = -\frac{\pi}{2}$.

Fact 14.4.26. Let a and b be positive numbers. Then,

$$\int_0^\infty \frac{\cos ax}{x^2+b^2}\,\mathrm{d}x = \frac{\pi}{2be^{ab}}, \quad \int_0^\infty \frac{\cos ax}{(x^2+b^2)^2}\,\mathrm{d}x = \frac{\pi(ab+1)}{4b^3e^{ab}}.$$

Source: [1136, pp. 201, 203].

Fact 14.4.27. Let a, b, and c be positive numbers. Then,

$$\int_{-\infty}^\infty \frac{\sin ax}{(x+b)^2+c^2}\,\mathrm{d}x = -\frac{\pi\sin ab}{ce^{ac}}, \quad \int_{-\infty}^\infty \frac{\cos ax}{(x+b)^2+c^2}\,\mathrm{d}x = \frac{\pi\cos ab}{ce^{ac}},$$

$$\int_{-\infty}^\infty \frac{x\sin ax}{(x+b)^2+c^2}\,\mathrm{d}x = \frac{\pi(b\sin ab + c\cos ab)}{ce^{ac}}, \quad \int_{-\infty}^\infty \frac{x\cos ax}{(x+b)^2+c^2}\,\mathrm{d}x = \frac{\pi(c\sin ab - b\cos ab)}{ce^{ac}}.$$

Source: [2106, p. 340].

Fact 14.4.28. Let a, b, c be positive numbers, and assume that $b \ne c$. Then,

$$\int_0^\infty \frac{x\sin ax}{(x^2+b^2)(x^2+c^2)}\,\mathrm{d}x = \frac{(e^{-ac}-e^{-ab})\pi}{2(b^2-c^2)}, \quad \int_0^\infty \frac{\cos ax}{(x^2+b^2)(x^2+c^2)}\,\mathrm{d}x = \frac{(be^{-ac}-ce^{-ab})\pi}{2bc(b^2-c^2)}.$$

Fact 14.4.29. Let a and b be positive numbers. Then,

$$\int_0^\infty \frac{x\sin ax}{x^2+b^2}\,\mathrm{d}x = \frac{\pi e^{-ab}}{2}, \quad \int_0^\infty \frac{x\sin ax}{(x^2+b^2)^2}\,\mathrm{d}x = \frac{a\pi e^{-ab}}{4b}, \quad \int_0^\infty \frac{\sin ax}{x(x^2+b^2)}\,\mathrm{d}x = \frac{\pi(1-e^{-ab})}{2b^2}.$$

Source: [1136, p. 211].

Fact 14.4.30. Let a and b be distinct positive numbers. Then,

$$\int_0^\infty \frac{x\sin x}{(x^2+a^2)(x^2+b^2)}\,\mathrm{d}x = \frac{(\sinh a - \cosh a - \sinh b + \cosh b)\pi}{2(a^2+b^2)}.$$

In particular,

$$\int_0^\infty \frac{x\sin x}{(x^2+1)(x^2+4)}\,\mathrm{d}x = \frac{(e-1)\pi}{6e^2}.$$

Fact 14.4.31. Let a and b be positive numbers. Then,

$$\int_0^\infty \frac{\cos ax}{x^4+b^2}\,\mathrm{d}x = \frac{\sqrt{2}\pi}{4b^{3/2}e^{a\sqrt{2b}/2}}\left(\sin\frac{a\sqrt{2b}}{2} + \cos\frac{a\sqrt{2b}}{2}\right).$$

Source: [1136, p. 202].

Fact 14.4.32. Let a and b be positive numbers. Then,

$$\int_0^\infty \frac{\sin^2 ax}{x^2+b^2}\,\mathrm{d}x = \frac{\pi}{4b}(1-e^{-2ab}), \qquad \int_0^\infty \frac{\cos^2 ax}{x^2+b^2}\,\mathrm{d}x = \frac{\pi}{4b}(1+e^{-2ab}),$$

$$\int_0^\infty \frac{(\sin^2 ax)\cos^2 ax}{x^2+b^2}\,\mathrm{d}x = \frac{\pi}{16b}(1-e^{-4ab}).$$

Source: [1136, p. 203] and [1524, p. 268].

Fact 14.4.33. Let a, b, and c be positive numbers. Then,

$$\int_0^\infty \frac{(\sin^2 ax)\cos^2 bx}{x^2+c^2}\,\mathrm{d}x = \frac{\pi}{8c}\left(1 + e^{-2bc} - e^{-2ac} - \frac{1}{2}e^{-2(a+b)c} - \frac{1}{2}e^{2(b-a)c}\right).$$

Source: [1217, p. 461].

Fact 14.4.34. Let a, b, and c be positive numbers, and assume that $b \neq c$. Then,

$$\int_0^\infty \frac{\sin^2 ax}{(x^2+b^2)(x^2+c^2)}\,\mathrm{d}x = \frac{(b-c+ce^{-2ab}-be^{-2ac})\pi}{4bc(b^2-c^2)},$$

$$\int_0^\infty \frac{\cos^2 ax}{(x^2+b^2)(x^2+c^2)}\,\mathrm{d}x = \frac{(b-c-ce^{-2ab}+be^{-2ac})\pi}{4bc(b^2-c^2)},$$

$$\int_0^\infty \frac{\sin^2 ax}{x^2(x^2+b^2)}\,\mathrm{d}x = \frac{(2ab-1+e^{-2ab})\pi}{4b^3},$$

$$\int_0^\infty \frac{\sin^2 ax}{(x^2+b^2)^2}\,\mathrm{d}x = \frac{[1-(2ab+1)e^{-2ab}]\pi}{8b^3}, \qquad \int_0^\infty \frac{\cos^2 ax}{(x^2+b^2)^2}\,\mathrm{d}x = \frac{[1+(2ab+1)e^{-2ab}]\pi}{8b^3}.$$

Source: [1217, p. 462].

Fact 14.4.35. Let $a \in (0,\infty)$ and $m,n \geq 1$. If m is odd, then

$$\int_0^\infty \frac{\sin m\pi x}{(\sin \pi x)\Gamma(a+x)\Gamma(a-x)}\,\mathrm{d}x = \frac{2^{2a-3}}{\Gamma(2a-1)}, \qquad \int_0^\infty \frac{\sin m\pi x}{x\prod_{i=1}^n(1-\frac{x^2}{i^2})}\,\mathrm{d}x = \frac{4^n(n!)^2\pi}{2(2n)!}.$$

If m is even, then both integrals are 0. **Source:** [1317, p. 216]. **Remark:** These integrals depend on the parity of m but not its value.

Fact 14.4.36. If $x \in (-\frac{\pi}{4}, \frac{3\pi}{4})$, then

$$\int_0^x \frac{1}{\sin t + \cos t}\,\mathrm{d}t = \sqrt{2}\,\mathrm{atanh}\left[\frac{\sqrt{2}}{2}\left(\tan\frac{x}{2} - 1\right)\right] + \sqrt{2}\,\mathrm{atanh}\,\frac{\sqrt{2}}{2}$$

$$= \frac{\sqrt{2}}{2}\log\frac{\sin(x-\frac{\pi}{4})+1}{(\sqrt{2}-1)\cos(x-\frac{\pi}{4})} = \frac{\sqrt{2}}{2}\log\frac{\tan(\frac{x}{2}+\frac{\pi}{8})}{\tan\frac{\pi}{8}}.$$

In particular,

$$\int_0^{\pi/6}\frac{1}{\sin x+\cos x}\,\mathrm{d}x = \sqrt{2}\,\mathrm{atanh}\sqrt{\frac{2-\sqrt{3}}{3}} = \frac{\sqrt{2}}{2}\log[\tfrac{1}{2}(\sqrt{6}-2)(4+3\sqrt{2})] \approx 0.435953,$$

$$\int_0^{\pi/4}\frac{1}{\sin x+\cos x}\,\mathrm{d}x = \frac{\sqrt{2}}{2}\,\mathrm{atanh}\,\frac{\sqrt{2}}{2} = \frac{\sqrt{2}}{2}\log(1+\sqrt{2}) \approx 0.623225,$$

$$\int_0^{\pi/3}\frac{1}{\sin x+\cos x}\,\mathrm{d}x = \sqrt{2}\,\mathrm{atanh}\sqrt{2-\sqrt{3}} = \frac{\sqrt{2}}{2}\log(\sqrt{2}+\sqrt{3}) \approx 0.810497,$$

$$\int_0^{\pi/2}\frac{1}{\sin x+\cos x}\,\mathrm{d}x = \sqrt{2}\,\mathrm{atanh}\,\frac{\sqrt{2}}{2} = \sqrt{2}\log(1+\sqrt{2}) \approx 1.24645.$$

Furthermore,

$$\int_0^{\pi/2}\frac{\sin^2 x}{\sin x+\cos x}\,\mathrm{d}x = \int_0^{\pi/2}\frac{\cos^2 x}{\sin x+\cos x}\,\mathrm{d}x = \frac{\sqrt{2}}{2}\log(1+\sqrt{2}).$$

Source: The last integral is given in [2106, p. 52]. **Remark:** $\mathrm{atanh}\,\frac{\sqrt{2}}{2} = 2\,\mathrm{atanh}(\sqrt{2}-1) = \log(1+\sqrt{2}) = \frac{1}{2}\log(3+2\sqrt{2})$.

Fact 14.4.37.

$$\int_0^{\pi/4}\frac{1}{\sin x+1}\,\mathrm{d}x = 2-\sqrt{2}, \quad \int_0^{\pi/4}\frac{x}{\sin x+1}\,\mathrm{d}x = \frac{(1-\sqrt{2})\pi}{4} + \frac{1}{4}\log\left(\frac{17}{4}+3\sqrt{2}\right),$$

$$\int_0^{\pi/4}\frac{1}{\cos x+1}\,\mathrm{d}x = \sqrt{2}-1, \quad \int_0^{\pi/4}\frac{x}{\cos x+1}\,\mathrm{d}x = \log(2+\sqrt{2}) - 2\log 2 + \frac{\pi}{4}(\sqrt{2}-1),$$

$$\int_0^{\pi/3}\frac{1}{\sin x+1}\,\mathrm{d}x = \sqrt{3}-1, \quad \int_0^{\pi/3}\frac{x}{\sin x+1}\,\mathrm{d}x = \frac{\pi}{3}(\sqrt{3}-2) - 2\log(\sqrt{3}-1),$$

$$\int_0^{\pi/3}\frac{1}{\cos x+1}\,\mathrm{d}x = \frac{\sqrt{3}}{3}, \quad \int_0^{\pi/3}\frac{x}{\cos x+1}\,\mathrm{d}x = \frac{\sqrt{3}\pi}{9} - \log\frac{4}{3},$$

$$\int_0^{\pi/2}\frac{1}{\sin x+1}\,\mathrm{d}x = 1, \quad \int_0^{\pi/2}\frac{x}{\sin x+1}\,\mathrm{d}x = \log 2, \quad \int_0^{\pi/2}\frac{x^2}{\sin x+1}\,\mathrm{d}x = 2\pi\log 2 - 4G,$$

$$\int_0^{\pi/2}\frac{x^3}{\sin x+1}\,\mathrm{d}x = \frac{3\pi^2}{2}\log 2 - \frac{63}{8}\zeta(3), \quad \int_0^{\pi/2}\frac{1}{\cos x+1}\,\mathrm{d}x = 1, \quad \int_0^{\pi/2}\frac{x}{\cos x+1}\,\mathrm{d}x = \frac{\pi}{2} - \log 2,$$

$$\int_0^{\pi/2}\frac{x^2}{\cos x+1}\,\mathrm{d}x = \frac{\pi^2}{4} + \pi\log 2 - 4G, \quad \int_0^{\pi/2}\frac{x^3}{\cos x+1}\,\mathrm{d}x = \frac{\pi^3}{8} + \frac{3\pi^2}{4}\log 2 + \frac{63}{8}\zeta(3) - 6\pi G,$$

$$\int_0^{\pi}\frac{1}{\sin x+1}\,\mathrm{d}x = 2, \quad \int_0^{\pi}\frac{x}{\sin x+1}\,\mathrm{d}x = \pi, \quad \int_0^{\pi}\frac{x^2}{\sin x+1}\,\mathrm{d}x = \pi^2 + 2\pi\log 2 - 8G,$$

$$\int_0^{\pi}\frac{x^3}{\sin x+1}\,\mathrm{d}x = \pi^3 + 3\pi^2\log 2 - 12\pi G, \quad \int_{-\pi/2}^{\pi/2}\frac{x^2}{\cos x+1}\,\mathrm{d}x = \frac{\pi^2}{2} + 2\pi\log 2 - 8G,$$

$$\int_0^{\pi/4}\frac{\sin x}{\sin x+1}\,\mathrm{d}x = \sqrt{2} - 2 + \frac{\pi}{4}, \quad \int_0^{\pi/4}\frac{\cos x}{\sin x+1}\,\mathrm{d}x = \log\left(1+\frac{\sqrt{2}}{2}\right),$$

$$\int_0^{\pi/4}\frac{\sin x}{\cos x+1}\,\mathrm{d}x = 2\log 2 - \log(2+\sqrt{2}), \quad \int_0^{\pi/4}\frac{\cos x}{\cos x+1}\,\mathrm{d}x = 1 - \sqrt{2} + \frac{\pi}{4},$$

$$\int_0^{\pi/3} \frac{\sin x}{\sin x + 1}\, dx = 1 - \sqrt{3} + \frac{\pi}{3}, \quad \int_0^{\pi/3} \frac{\cos x}{\sin x + 1}\, dx = \log\left(1 + \frac{\sqrt{3}}{2}\right),$$

$$\int_0^{\pi/3} \frac{\sin x}{\cos x + 1}\, dx = \log\frac{4}{3}, \quad \int_0^{\pi/3} \frac{\cos x}{\cos x + 1}\, dx = \frac{\pi}{3} - \frac{\sqrt{3}}{3},$$

$$\int_0^{\pi/2} \frac{\sin x}{\sin x + 1}\, dx = \int_0^{\pi/2} \frac{\cos x}{\cos x + 1}\, dx = \frac{\pi}{2} - 1, \quad \int_0^{\pi/2} \frac{\cos x}{\sin x + 1}\, dx = \int_0^{\pi/2} \frac{\sin x}{\cos x + 1}\, dx = \log 2,$$

$$\int_0^{\pi/2} \frac{x\sin x}{\sin x + 1}\, dx = \frac{\pi^2}{8} - \log 2, \quad \int_0^{\pi/2} \frac{x\cos x}{\sin x + 1}\, dx = \pi\log 2 - 2G,$$

$$\int_0^{\pi/2} \frac{x\sin x}{\cos x + 1}\, dx = 2G - \frac{\pi}{2}\log 2, \quad \int_0^{\pi/2} \frac{x\cos x}{\cos x + 1}\, dx = \frac{\pi^2}{8} - \frac{\pi}{2} + \log 2,$$

$$\int_0^{\pi} \frac{\sin x}{\sin x + 1}\, dx = \pi - 2, \quad \int_0^{\pi} \frac{\cos x}{\sin x + 1}\, dx = 0, \quad \int_0^{\pi} \frac{(x-\pi)^2}{\cos x + 1}\, dx = 4\pi\log 2,$$

$$\int_0^{\pi} \frac{x\sin x}{\sin x + 1}\, dx = \frac{\pi^2}{2} - \pi, \quad \int_0^{\pi} \frac{x\cos x}{\sin x + 1}\, dx = \pi\log 2 - 4G,$$

$$\int_0^{\pi} \frac{(x-\pi)^2\sin x}{\cos x + 1}\, dx = 2\pi^2\log 2 - 7\zeta(3), \quad \int_0^{\pi} \frac{(x-\pi)^2\cos x}{\cos x + 1}\, dx = \frac{\pi^2}{3} - 4\pi\log 2.$$

Fact 14.4.38.

$$\int_0^{\pi/2} \frac{(x-\frac{\pi}{2})^2}{1-\sin x}\, dx = \int_0^{\pi/2} \frac{x^2}{1-\cos x}\, dx = 4G + \pi\log 2 - \frac{\pi^2}{4},$$

$$\int_0^{\pi/2} \frac{x(x-\frac{\pi}{2})^2}{1-\sin x}\, dx = \frac{105}{8}\zeta(3) - 4\pi G - \frac{\pi^2}{4}\log 2,$$

$$\int_0^{\pi/2} \frac{x^3}{1-\cos x}\, dx = 6\pi G - \frac{\pi^3}{8} + \frac{3\pi^2}{4}\log 2 - \frac{105}{8}\zeta(3),$$

$$\int_0^{\pi} \frac{(x-\frac{\pi}{2})^2}{1-\sin x}\, dx = 8G + 2\pi\log 2 - \frac{\pi^2}{2}, \quad \int_0^{\pi} \frac{x(x-\frac{\pi}{2})^2}{1-\sin x}\, dx = 4\pi G + \pi^2\log 2 - \frac{\pi^3}{4},$$

$$\int_0^{\pi/2} \frac{(\frac{\pi}{2}-x)\cos x}{1-\sin x}\, dx = 2G + \frac{\pi}{2}\log 2, \quad \int_0^{\pi} \frac{(\frac{\pi}{2}-x)\cos x}{1-\sin x}\, dx = 4G + \pi\log 2,$$

$$\int_0^{\pi} \frac{x^2}{1-\cos x}\, dx = 4\pi\log 2, \quad \int_0^{\pi} \frac{x^3}{1-\cos x}\, dx = 6\pi^2\log 2 - 21\zeta(3),$$

$$\int_0^{\pi/2} \frac{x\sin x}{1-\cos x}\, dx = 2G + \frac{\pi}{2}\log 2, \quad \int_0^{\pi/2} \frac{x^2\sin x}{1-\cos x}\, dx = 2\pi G + \frac{\pi^2}{4}\log 2 - \frac{35}{8}\zeta(3),$$

$$\int_0^{\pi} \frac{x\sin x}{1-\cos x}\, dx = 2\pi\log 2, \quad \int_0^{\pi} \frac{x^2\sin x}{1-\cos x}\, dx = 2\pi^2\log 2 - 7\zeta(3),$$

$$\int_0^{\pi/2} \frac{x-\sin x}{1-\cos x}\, dx = 2 - \frac{\pi}{2}, \quad \int_0^{\pi/2} \frac{(x-\sin x)^2}{1-\cos x}\, dx = 1 + \frac{\pi}{2} - \frac{\pi^2}{4},$$

$$\int_0^{\pi/2} \frac{(x-\sin x)^3}{1-\cos x}\, dx = \frac{1}{8}(\pi^2 - 12)(3-\pi),$$

$$\int_0^{\pi} \frac{x-\sin x}{1-\cos x}\, dx = 2, \quad \int_0^{\pi} \frac{(x-\sin x)^2}{1-\cos x}\, dx = \pi, \quad \int_0^{\pi} \frac{(x-\sin x)^3}{1-\cos x}\, dx = \frac{3\pi^2}{2} - 8.$$

Fact 14.4.39. Let a and b be real numbers, and assume that $b \neq 0$ and $|a| < b$. Then,

$$\int_0^{\pi/2} \frac{1}{a\sin x + b}\,\mathrm{d}x = \frac{\operatorname{acos}\frac{a}{b}}{\sqrt{b^2 - a^2}}, \quad \int_0^{\pi} \frac{1}{a\sin x + b}\,\mathrm{d}x = \frac{2\operatorname{acos}\frac{a}{b}}{\sqrt{b^2 - a^2}},$$

$$\int_0^{3\pi/2} \frac{1}{a\sin x + b}\,\mathrm{d}x = \frac{\pi + \operatorname{acos}\frac{a}{b}}{\sqrt{b^2 - a^2}}, \quad \int_0^{2\pi} \frac{1}{a\sin x + b}\,\mathrm{d}x = \frac{2\pi}{\sqrt{b^2 - a^2}},$$

$$\int_0^{\pi/2} \frac{1}{a\cos x + b}\,\mathrm{d}x = \frac{\operatorname{acos}\frac{a}{b}}{\sqrt{b^2 - a^2}}, \quad \int_0^{\pi} \frac{1}{a\cos x + b}\,\mathrm{d}x = \frac{\pi}{\sqrt{b^2 - a^2}},$$

$$\int_0^{3\pi/2} \frac{1}{a\cos x + b}\,\mathrm{d}x = \frac{2\pi - \operatorname{acos}\frac{a}{b}}{\sqrt{b^2 - a^2}}, \quad \int_0^{2\pi} \frac{1}{a\cos x + b}\,\mathrm{d}x = \frac{2\pi}{\sqrt{b^2 - a^2}}.$$

Fact 14.4.40. Let a and b be real numbers, and assume that $0 < |a| < b$. Then,

$$\int_0^{\pi} \frac{x\sin x}{a\cos x + b}\,\mathrm{d}x = \frac{\pi}{a}\log\frac{b + \sqrt{b^2 - a^2}}{2(b - a)}, \quad \int_0^{2\pi} \frac{x\sin x}{a\cos x + b}\,\mathrm{d}x = \frac{2\pi}{a}\log\frac{b + \sqrt{b^2 - a^2}}{2(a + b)}.$$

Fact 14.4.41. Let $a > 1$ and $n \geq 0$. Then,

$$\int_0^{\pi} \frac{\cos nx}{\cos x + a}\,\mathrm{d}x = \frac{\pi(\sqrt{a^2 - 1} - a)^n}{\sqrt{a^2 - 1}}.$$

Source: [1136, p. 205] and [2249, p. 326].

Fact 14.4.42. Let a and b be real numbers, and assume that $a \geq b > 0$. Then,

$$\int_0^{\pi} \frac{\sin^2 x}{b\cos x + a}\,\mathrm{d}x = \frac{\pi}{b^2}(a - \sqrt{a^2 - b^2}).$$

Source: [1136, p. 205] and [1472, p. 161].

Fact 14.4.43. Let $a > -1$, and assume that $a \neq 0$. Then,

$$\int_0^{\pi/2} \frac{x\sin 2x}{a\sin^2 x + 1}\,\mathrm{d}x = \frac{\pi}{a}\log(a + 1 - \sqrt{a + 1}), \quad \int_0^{\pi/2} \frac{x\sin 2x}{a\cos^2 x + 1}\,\mathrm{d}x = \frac{\pi}{a}\log\frac{1}{2}(1 + \sqrt{a + 1}).$$

Source: [1217, p. 453].

Fact 14.4.44. Let a be a real number. If $|a| > 1$, then

$$\int_0^{\pi} \frac{x}{a^2 - \cos^2 x}\,\mathrm{d}x = \frac{\pi^2}{2a\sqrt{a^2 - 1}}, \quad \int_0^{\pi} \frac{x\sin 2x}{a^2 - \cos^2 x}\,\mathrm{d}x = 2\pi\log[2(1 - a^2 + a\sqrt{a^2 - 1})].$$

If $a \in (0, 1)$, then

$$\int_0^{\pi} \frac{x\sin x}{a^2 - \cos^2 x}\,\mathrm{d}x = \frac{\pi}{2a}\log\left|\frac{1 + a}{1 - a}\right|.$$

Source: [1217, pp. 453, 454].

Fact 14.4.45. Let a and b be real numbers. If $|a| < |b|$, then

$$\int_0^{\infty} \frac{\sin x}{(a\cos 2x + b)x}\,\mathrm{d}x = \frac{\pi}{2\sqrt{b^2 - a^2}}.$$

Fact 14.4.46. Let $a \in (-1, 1)$. Then,

$$\int_0^{\pi} \frac{1}{(a\cos x + 1)^2}\,\mathrm{d}x = \frac{\pi}{(1 - a^2)^{3/2}}.$$

Source: [2106, p. 331].

Fact 14.4.47. Let a be a positive number. Then,

$$\int_0^\pi \frac{1}{\sin^2 x + a}\,\mathrm{d}x = \frac{\pi}{\sqrt{a^2+a}}, \qquad \int_0^\pi \frac{\sin x}{\sin^2 x + a}\,\mathrm{d}x = \frac{2}{\sqrt{a+1}}\operatorname{acoth}\sqrt{a+1},$$

$$\int_0^\pi \frac{\sin^2 x}{\sin^2 x + a}\,\mathrm{d}x = \pi\left(1-\sqrt{\frac{a}{a+1}}\right), \qquad \int_0^\pi \frac{\sin^3 x}{\sin^2 x + a}\,\mathrm{d}x = 2 - \frac{a}{\sqrt{a+1}}\operatorname{atanh}\frac{2\sqrt{a+1}}{a+2},$$

$$\int_0^\pi \frac{\sin^4 x}{\sin^2 x + a}\,\mathrm{d}x = \pi\left(\frac{a^{3/2}}{\sqrt{a+1}}+\frac{1}{2}-a\right), \qquad \int_0^\pi \frac{\sin^5 x}{\sin^2 x + a}\,\mathrm{d}x = \frac{4}{3} - 2a + \frac{a^2}{\sqrt{a+1}}\operatorname{atanh}\frac{2\sqrt{a+1}}{a+2}.$$

Source: [1136, p. 205]. **Remark:** $\operatorname{acoth}\sqrt{a+1} = \frac{1}{2}\log\frac{a+2+2\sqrt{a+1}}{a}$. **Related:** Fact 14.1.16.

Fact 14.4.48. Let a be a positive number. If $n \ge 1$ is odd, then

$$\int_0^\pi \frac{\cos^n x}{\cos^2 x + a}\,\mathrm{d}x = \int_0^\pi \frac{\cos^n x}{\sin^2 x + a}\,\mathrm{d}x = 0.$$

Furthermore,

$$\int_0^\pi \frac{1}{\cos^2 x + a}\,\mathrm{d}x = \frac{\pi}{\sqrt{a^2+a}}, \qquad \int_0^\pi \frac{\cos^2 x}{\cos^2 x + a}\,\mathrm{d}x = \pi\left(1-\sqrt{\frac{a}{a+1}}\right),$$

$$\int_0^\pi \frac{\cos^4 x}{\cos^2 x + a}\,\mathrm{d}x = \pi\left(\frac{a^{3/2}}{\sqrt{a+1}}+\frac{1}{2}-a\right), \qquad \int_0^\pi \frac{\cos^6 x}{\cos^2 x + a}\,\mathrm{d}x = \pi\left(a^2-\frac{a}{2}+\frac{3}{8}-\frac{a^{5/2}}{\sqrt{a+1}}\right).$$

$$\int_0^\pi \frac{\cos^2 x}{\sin^2 x + a}\,\mathrm{d}x = \int_0^\pi \frac{\sin^2 x}{\cos^2 x + a}\,\mathrm{d}x = \pi\left(\sqrt{\frac{a+1}{a}}-1\right),$$

$$\int_0^\pi \frac{\sin^3 x}{\cos^2 x + a}\,\mathrm{d}x = \frac{1}{2\sqrt{a}}\left[(a+1)\pi - 4\sqrt{a} - 2(a+1)\operatorname{atan}\frac{a-1}{2\sqrt{a}}\right].$$

Fact 14.4.49. Let $a \ge b > 0$. Then,

$$\int_0^\pi \frac{x\sin x}{a+b\sin^2 x}\,\mathrm{d}x = \frac{\pi}{\sqrt{(a+b)b}}\operatorname{atanh}\sqrt{\frac{b}{a+b}}, \qquad \int_0^\pi \frac{x\sin x}{a+b\cos^2 x}\,\mathrm{d}x = \frac{\pi}{\sqrt{ab}}\operatorname{atan}\sqrt{\frac{b}{a}}.$$

In particular,

$$\int_0^\pi \frac{x\sin x}{\sin^2 x+1}\,\mathrm{d}x = \frac{\sqrt{2}\pi}{2}\operatorname{atanh}\frac{\sqrt{2}}{2} = \frac{\sqrt{2}\pi}{2}\log(\sqrt{2}+1), \qquad \int_0^\pi \frac{x\sin x}{\cos^2 x+1}\,\mathrm{d}x = \frac{\pi^2}{4}.$$

Source: [1158, p. 150] and [2106, p. 153].

Fact 14.4.50. If $n \ge 0$, then

$$\int_0^{\pi/2} \frac{\sin(2n+1)x}{\sin x}\,\mathrm{d}x = \frac{\pi}{2}.$$

If $n \ge 1$, then

$$\int_0^{\pi/2} \frac{\sin 2nx}{\sin x}\,\mathrm{d}x = 2\sum_{i=0}^{n-1}(-1)^i\frac{1}{2i+1}.$$

If $n \ge 0$, then

$$\int_0^\pi \frac{\cos nx - 1}{\cos x - 1}\,\mathrm{d}x = n\pi.$$

Source: [1158, p. 153].

Fact 14.4.51. Let $a > 0$ and $n \geq 1$. Then,

$$\int_0^\pi \frac{\cos nx - \cos na}{\cos x - \cos a}\,dx = \frac{\pi \sin na}{\sin a}.$$

Source: [2106, p. 62] and [2294, p. 413].

Fact 14.4.52.

$$\int_0^{\pi/4} \frac{\cos^{3/2} 2x}{\cos^3 x}\,dx = \frac{(4\sqrt{2} - 5)\pi}{4}.$$

Source: [1502].

Fact 14.4.53.

$$\int_0^{\pi/2} \frac{x(\pi - x)}{\sin x}\,dx = \tfrac{7}{2}\zeta(3), \qquad \int_0^{\pi/4} \frac{x^2}{\sin^2 x}\,dx = \frac{\pi \log 2}{4} - \frac{\pi^2}{16} + G.$$

Source: [516, p. 63] and [2278].

Fact 14.4.54. Let a be a real number. If $|a| < 1$, then

$$\int_0^{\pi/2} \frac{\sin^2 x}{1 + a^2 - 2a\cos 2x}\,dx = \frac{\pi}{4(1+a)}, \qquad \int_0^{\pi/2} \frac{\cos^2 x}{1 + a^2 - 2a\cos 2x}\,dx = \frac{\pi}{4(1-a)},$$

$$\int_0^{\pi} \frac{1 - a^2}{1 + a^2 - 2a\cos 2x}\,dx = \pi, \qquad \int_{-\pi}^{\pi} \frac{1}{1 + a^2 - 2a\cos 2x}\,dx = \frac{2\pi}{1 - a^2}.$$

If $|a| > 1$, then

$$\int_0^{\pi/2} \frac{\cos^2 x}{1 + a^2 - 2a\cos 2x}\,dx = \frac{\pi}{4a(a-1)}.$$

Now, let $n \geq 1$. Then,

$$\int_0^{\pi} \frac{\cos 2nx}{1 + a^2 - 2a\cos 2x}\,dx = \begin{cases} \dfrac{a^n\pi}{1 - a^2}, & |a| < 1, \\ \dfrac{\pi}{(a^2 - 1)a^n}, & |a| > 1. \end{cases}$$

Related: Fact 13.6.7.

Fact 14.4.55. Let a be a real number. Then,

$$\int_0^{\pi} \frac{x\sin x}{1 + a^2 - 2a\cos x}\,dx = \begin{cases} \dfrac{\pi}{a}\log(1 + a), & a \in (-1, 0) \cup (0, 1], \\ \dfrac{\pi}{a}\log\left(1 + \dfrac{1}{a}\right), & |a| > 1, \end{cases}$$

$$\int_0^{2\pi} \frac{x\sin x}{1 + a^2 - 2a\cos x}\,dx = \begin{cases} \dfrac{2\pi}{a}\log(1 - a), & a \in [-1, 0) \cup (0, 1), \\ \dfrac{2\pi}{a}\log\left(1 - \dfrac{1}{a}\right), & |a| > 1. \end{cases}$$

Source: [1217, p. 449].

Fact 14.4.56. Let $a \in \mathbb{R}$ and $n \geq 0$. Then,

$$\int_0^{\pi} \frac{\cos nx}{1 + a^2 - 2a\cos x}\,dx = \begin{cases} \dfrac{a^n\pi}{1 - a^2}, & |a| < 1, \\ \dfrac{\pi}{(a^2 - 1)a^n}, & |a| > 1. \end{cases}$$

Fact 14.4.57. Let a be a real number, and assume that $a \neq 0$ and $a \neq 1$. Then,

$$\int_0^\infty \frac{\sin x}{(1 + a^2 - 2a\cos x)x}\,\mathrm{d}x = \frac{\pi}{4a}\left(\left|\frac{a+1}{a-1}\right| - 1\right).$$

Source: [1217, p. 450].

Fact 14.4.58. Let $a, b \in \mathbb{R}$ and $n \geq 0$. Then,

$$\int_0^{2\pi} (a\sin x + b\cos x)^{2n+1} = 0, \qquad \int_0^{2\pi} (a\sin x + b\cos x)^{2n} = \frac{(2n-1)!!}{(2n)!!}2\pi(a^2+b^2)^n.$$

Source: [1217, p. 405].

Fact 14.4.59. Let $a, b > 0$ and $n \geq 1$. Then,

$$\int_0^{\pi/2} \frac{1}{(a\sin^2 x + b\cos^2 x)^n}\,\mathrm{d}x = \frac{\pi}{2^{2n-1}\sqrt{ab}}\sum_{i=0}^{n-1}\binom{2i}{i}\binom{2n-2i-2}{n-i-1}\frac{1}{a^i b^{n-i-1}}.$$

In particular,

$$\int_0^{\pi/2} \frac{1}{a\sin^2 x + b\cos^2 x}\,\mathrm{d}x = \frac{\pi}{2\sqrt{ab}}, \qquad \int_0^{\pi/2} \frac{1}{(a\sin^2 x + b\cos^2 x)^2}\,\mathrm{d}x = \frac{(a+b)\pi}{4(ab)^{3/2}},$$

$$\int_0^{\pi/2} \frac{1}{(a\sin^2 x + b\cos^2 x)^3}\,\mathrm{d}x = \frac{\pi}{16\sqrt{ab}}\left(\frac{3}{a^2} + \frac{3}{b^2} + \frac{2}{ab}\right).$$

Source: [2013, p. 182] and [2106, p. 100].

Fact 14.4.60.

$$\int_0^{\pi/2} \sqrt{\sin x}\,\mathrm{d}x = \int_0^{\pi/2} \sqrt{\cos x}\,\mathrm{d}x = \sqrt{\frac{2}{\pi}}\Gamma^2(\tfrac{3}{4}), \qquad \int_0^{\pi/2} \sqrt{\csc x}\,\mathrm{d}x = \int_0^{\pi/2} \sqrt{\sec x}\,\mathrm{d}x = \frac{1}{2\sqrt{2\pi}}\Gamma^2(\tfrac{1}{4}),$$

$$\int_0^{\pi/2} \sqrt{\tan x}\,\mathrm{d}x = \int_0^{\pi/2} \sqrt{\cot x}\,\mathrm{d}x = \frac{\sqrt{2}\pi}{2}, \qquad \int_0^{\pi/2} \frac{\sqrt{\sin x}}{\sqrt{\sin x} + \sqrt{\cos x}}\,\mathrm{d}x = \frac{\pi}{4}.$$

Source: [2106, p. 49]. **Related:** Fact 14.4.1.

Fact 14.4.61.

$$\int_0^{\pi} \frac{1}{\sqrt{3 - \cos x}}\,\mathrm{d}x = \frac{\Gamma^2(\frac{1}{4})}{4\sqrt{\pi}}, \qquad \int_0^{\pi/2} \frac{1}{\sqrt{2\cos^2 x + \sin^2 x}}\,\mathrm{d}x = \frac{\left(\frac{\pi}{2}\right)^{3/2}}{\Gamma^2(\frac{3}{4})}.$$

Fact 14.4.62. Let $n \geq 0$. Then,

$$\int_0^{\pi/4} \tan^{2n+1} x\,\mathrm{d}x = (-1)^n\frac{\log 2}{2} + \sum_{i=1}^{n}(-1)^{i-1}\frac{1}{2n-2i-2},$$

$$\int_0^{\pi/4} \tan^{2n} x\,\mathrm{d}x = (-1)^n\frac{\pi}{4} + \sum_{i=1}^{n}(-1)^{i-1}\frac{1}{2n-2i+1}.$$

In particular,

$$\int_0^{\pi/4} \tan x\,\mathrm{d}x = \frac{\log 2}{2}, \qquad \int_0^{\pi/4} \tan^2 x\,\mathrm{d}x = 1 - \frac{\pi}{4},$$

$$\int_0^{\pi/4} \tan^3 x\,\mathrm{d}x = \frac{1}{2} - \frac{\log 2}{2}, \qquad \int_0^{\pi/4} \tan^4 x\,\mathrm{d}x = \frac{\pi}{4} - \frac{2}{3}.$$

Source: [1158, p. 152] and [1217, p. 395].

Fact 14.4.63. Let $a \in (-1, 1)$. Then,

$$\int_0^{\pi/2} \tan^a x \, dx = \frac{\pi}{2\cos\frac{\pi a}{2}}.$$

Source: [1568, p. 42].

Fact 14.4.64. Let $a \ge 0$. Then,

$$\int_0^\infty \frac{1}{(x^2+1)(x^a+1)} \, dx = \frac{\pi}{4}, \quad \int_0^{\pi/2} \frac{1}{1+\tan^a x} \, dx = \frac{\pi}{4}.$$

Source: A proof and generalization of the first integral are given in [630]. The second integral can be obtained from the first integral by defining $x = \tan\theta$. See [511, p. 254]. The second integral is given in [1757, pp. 32, 33] for the case $a = \sqrt{2}$, and in [3024, p. 451] for the case where a is an integer. **Remark:** Both integrals are independent of a.

Fact 14.4.65. Let $n \ge 1$. Then,

$$\zeta(2n+1) = (-1)^{n+1} \frac{(2\pi)^{2n+1}}{2(2n+1)!} \int_0^1 B_{2n+1}(x) \cot \pi x \, dx,$$

where B_n is the nth Bernoulli polynomial. **Source:** [2513, p. 167]. **Related:** Fact 13.2.1.

Fact 14.4.66.

$$\int_0^{\pi/2} \frac{1}{1+8\sin^2 \tan x} \, dx = \frac{1}{2} \int_{-\infty}^\infty \frac{1}{(1+8\sin^2 x)(1+x^2)} \, dx = \frac{\pi(2e^2+1)}{6(2e^2-1)}.$$

Source: [1684].

14.5 Facts on Definite Integrals of Inverse Trigonometric Functions

Fact 14.5.1. Let $n \ge 1$. Then,

$$\int_0^1 x^{2n} \operatorname{asin} x \, dx = \frac{1}{2n+1}\left(\frac{\pi}{2} - \frac{4^n (n!)^2}{(2n+1)!}\right), \quad \int_0^1 x^{2n+1} \operatorname{asin} x \, dx = \frac{\pi}{2n+2}\left(\frac{1}{2} - \frac{(2n+2)!}{2^{2n+3}(n+1)!}\right).$$

Source: [822].

Fact 14.5.2.

$$\int_0^1 \operatorname{asin} x \, dx = \frac{\pi}{2} - 1, \quad \int_0^1 x \operatorname{asin} x \, dx = \frac{\pi}{8}, \quad \int_0^1 x^2 \operatorname{asin} x \, dx = \frac{\pi}{6} - \frac{2}{9},$$

$$\int_0^1 \frac{\operatorname{asin} x}{x} \, dx = \frac{\pi}{2} \log 2, \quad \int_0^{\sqrt{2}/2} \frac{\operatorname{asin} x}{x} \, dx = \frac{G}{2} + \frac{\pi}{8} \log 2, \quad \int_0^1 \frac{\operatorname{asin} \sqrt{x}}{x^2 - x + 1} \, dx = \frac{\sqrt{3}\pi^2}{18},$$

$$\int_0^1 \frac{\operatorname{acos} x}{1-x} \, dx = \frac{\pi}{2} \log 2 + G, \quad \int_0^1 \frac{\operatorname{acos} x}{1+x} \, dx = -\frac{\pi}{2} \log 2 + G,$$

$$\int_0^{\pi/2} \operatorname{acos} \frac{1}{1+2\cos x} \, dx = \frac{\pi^2}{6}, \quad \int_0^{\pi/2} \operatorname{acos} \frac{\cos x}{1+2\cos x} \, dx = \frac{5\pi^2}{24},$$

$$\int_0^1 \frac{\operatorname{atan} x}{\sqrt{x}} \, dx = \sqrt{2} \operatorname{acoth} \sqrt{2} + \frac{1}{2}(1-\sqrt{2})\pi, \quad \int_0^1 \frac{\operatorname{atan} x}{x} \, dx = G,$$

$$\int_0^1 \frac{\operatorname{atan} x}{x+1} \, dx = \frac{\pi}{8} \log 2, \quad \int_0^1 \frac{\operatorname{atan} x}{x^2+1} \, dx = \frac{\pi^2}{32}, \quad \int_0^1 \frac{x \operatorname{atan} x}{x^2+1} \, dx = \frac{G}{2} - \frac{\pi}{8} \log 2,$$

$$\int_0^1 \frac{x^2 \operatorname{atan} x}{x^2+1} \, dx = \frac{\pi}{4} - \frac{\pi^2}{32} - \frac{1}{2} \log 2, \quad \int_0^1 \frac{x \operatorname{atan} x}{(x^2+1)(x+1)} \, dx = \frac{G}{4} - \frac{\pi^2}{16} - \frac{\pi}{8} \log 2,$$

$$\int_0^1 \frac{\operatorname{atan} x}{x^3+x}\,\mathrm{d}x = \frac{\pi}{8}\log 2 + G, \qquad \int_0^1 \frac{\operatorname{acot} x}{x^2+x}\,\mathrm{d}x = -\frac{\pi}{8}\log 2 + G,$$

$$\int_0^1 \frac{\operatorname{atan}^2 x}{x}\,\mathrm{d}x = \frac{\pi G}{2} - \frac{7}{8}\zeta(3), \qquad \int_0^1 \frac{\operatorname{atan}^2 x}{x^2}\,\mathrm{d}x = G - \frac{\pi^2}{16} + \frac{\pi}{4}\log 2,$$

$$\int_0^1 \frac{\operatorname{atan}^2 x}{x^2+1}\,\mathrm{d}x = \frac{\pi^3}{192}, \qquad \int_0^1 \frac{\operatorname{atan}\sqrt{x^2+2}}{(x^2+1)\sqrt{x^2+2}}\,\mathrm{d}x = \frac{5\pi^2}{96}.$$

Source: [388], [1107], [1217, pp. 600–602], and [2106, pp. 190–201, 220].

Fact 14.5.3.

$$\int_0^1 (\operatorname{asin} x)\log x\,\mathrm{d}x = 1 - \log 2 - \frac{\pi}{2}, \qquad \int_0^1 (\operatorname{acos} x)\log x\,\mathrm{d}x = \log 2 - 2,$$

$$\int_0^1 (\operatorname{atan} x)\log x\,\mathrm{d}x = \frac{1}{2}\log 2 - \frac{\pi}{4} + \frac{\pi^2}{48}, \qquad \int_0^1 (\operatorname{acot} x)\log x\,\mathrm{d}x = -\frac{1}{2}\log 2 - \frac{\pi}{4} - \frac{\pi^2}{48}.$$

Source: [1217, p. 607].

Fact 14.5.4. Let $a > 0$. Then,

$$\int_0^1 \frac{\operatorname{atan} ax}{a^2x+1}\,\mathrm{d}x = \frac{1}{2a^2}(\operatorname{atan} a)\log(a^2+1), \qquad \int_0^1 \frac{\operatorname{acot} ax}{a^2x+1}\,\mathrm{d}x = \frac{1}{a^2}\left(\frac{\pi}{4} + \frac{1}{2}\operatorname{acot} a\right)\log(a^2+1).$$

Source: [1217, p. 603].

Fact 14.5.5. Let $n \ge 0$. Then,

$$\int_0^1 x^{2n}\operatorname{asin} x\,\mathrm{d}x = \frac{1}{2n+1}\left(\frac{\pi}{2} - \frac{2^n n!}{(2n+1)!!}\right), \qquad \int_0^1 x^{2n+1}\operatorname{asin} x\,\mathrm{d}x = \frac{\pi}{4n+4}\left(1 - \frac{(2n+1)!!}{2^{n+1}(n+1)!}\right),$$

$$\int_0^1 x^{2n}\operatorname{acos} x\,\mathrm{d}x = \frac{2^n n!}{(2n+1)(2n+1)!!}, \qquad \int_0^1 x^{2n+1}\operatorname{acos} x\,\mathrm{d}x = \frac{\pi(2n+1)!!}{2^{n+3}(n+1)(n+1)!}.$$

Source: [1217, p. 601].

Fact 14.5.6. Let $a \in (-\infty, -1) \cup (0, \infty)$. Then,

$$\int_{1/(a+1)}^{1/a} \frac{\operatorname{atan} ax}{x}\,\mathrm{d}x = \int_{1/(a+1)}^{1/a} \frac{\operatorname{acot}(a+1)x}{x}\,\mathrm{d}x.$$

Source: [109, pp. 225, 226].

Fact 14.5.7. Let $a \ge 0$. Then,

$$\int_0^\infty \frac{\operatorname{atan} ax}{x^3+x}\,\mathrm{d}x = \frac{\pi}{2}\log(a+1), \qquad \int_0^\infty \frac{\operatorname{atan} ax}{x-x^3}\,\mathrm{d}x = \frac{\pi}{4}\log(a^2+1),$$

$$\int_0^\infty \frac{\operatorname{atan} ax}{x\sqrt{1-x^2}}\,\mathrm{d}x = \frac{\pi}{2}\log(a + \sqrt{a^2+1}).$$

Fact 14.5.8. If $a \in [1, 2]$, then

$$\int_0^\infty \frac{\operatorname{atan} x}{x^a}\,\mathrm{d}x = \frac{\pi}{2(a-1)}\csc\frac{a\pi}{2}.$$

If $a \in [0, 1]$, then

$$\int_0^\infty \frac{\operatorname{acot} x}{x^a}\,\mathrm{d}x = \frac{\pi}{2(1-a)}\csc\frac{a\pi}{2}.$$

Source: [1217, p. 602].

Fact 14.5.9. Let $a > 0$ and $b > 0$. Then,

$$\int_0^\infty \frac{\operatorname{atan} bx}{(x+a)^2}\,\mathrm{d}x = \frac{b}{a^2b^2+1}\left(\frac{ab\pi}{2} - \log ab\right), \quad \int_0^\infty \frac{\operatorname{acot} bx}{(x+a)^2}\,\mathrm{d}x = \frac{b}{a^2b^2+1}\left(\frac{\pi}{2ab} + \log ab\right),$$

$$\int_0^\infty \frac{x \operatorname{acot} bx}{x^2+a^2}\,\mathrm{d}x = \frac{\pi}{2}\log\frac{ab+1}{ab}, \quad \int_0^\infty \frac{x \operatorname{atan} bx}{x(x^2+a^2)}\,\mathrm{d}x = \frac{\pi}{2a^2}\log(ab+1),$$

$$\int_0^\infty \frac{x \operatorname{atan} bx}{(x^2+a^2)^2}\,\mathrm{d}x = \frac{b\pi}{4a(ab+1)}, \quad \int_0^\infty \frac{x \operatorname{acot} bx}{(x^2+a^2)^2}\,\mathrm{d}x = \frac{\pi}{4a^2(ab+1)}.$$

Source: [1217, p. 603].

Fact 14.5.10. Let $a > 0$ and $b > 0$. Then,

$$\int_0^\infty \frac{\operatorname{atan} ax - \operatorname{atan} bx}{x}\,\mathrm{d}x = \frac{\pi}{2}\log\frac{a}{b}, \quad \int_0^\infty \frac{(\operatorname{atan} ax)\operatorname{atan} bx}{x^2}\,\mathrm{d}x = \frac{\pi}{2}\log\frac{(a+b)^{a+b}}{a^a b^b}.$$

Source: [701] and [1217, p. 604].

Fact 14.5.11. Let $a > 0$ and $b > 0$. Then,

$$\int_0^\infty (\operatorname{acot} ax)\operatorname{acot} bx\,\mathrm{d}x = \frac{\pi}{2}\left[\frac{1}{a}\log\left(1+\frac{a}{b}\right) + \frac{1}{b}\log\left(1+\frac{b}{a}\right)\right].$$

Source: [1217, p. 599].

Fact 14.5.12.

$$\int_0^\infty \frac{\operatorname{atan} x}{x^3+x}\,\mathrm{d}x = \frac{\pi}{2}\log 2, \quad \int_0^\infty \frac{x\operatorname{atan} x}{x^4+1}\,\mathrm{d}x = \frac{\pi^2}{16},$$

$$\int_0^\infty \frac{\operatorname{atan} x}{x^4+x}\,\mathrm{d}x = \frac{\pi}{12}\log 2 + \frac{2\pi}{9}\log(2+\sqrt{3}) - \frac{G}{9},$$

$$\int_0^\infty \frac{\operatorname{atan} x}{x^2+x}\,\mathrm{d}x = \int_0^\infty \frac{\operatorname{acot} x}{x+1}\,\mathrm{d}x = \frac{\pi}{4}\log 2 + G,$$

$$\int_0^\infty \frac{\operatorname{acot} x}{\sqrt{x^2+1}}\,\mathrm{d}x = \int_0^\infty \frac{\operatorname{atan} x}{x\sqrt{x^2+1}}\,\mathrm{d}x = 2G.$$

Source: [1217, p. 601, 602].

14.6 Facts on Definite Integrals of Logarithmic Functions

Fact 14.6.1. Let $z \in \mathbb{C}$, and assume that $\operatorname{Re} z > 1$. Then,

$$\int_0^\infty \frac{e^{-x} - e^{-xz}}{x}\,\mathrm{d}x = \int_0^1 \frac{x^{z-1}-1}{\log x}\,\mathrm{d}x = \log z.$$

Source: [2513, p. 18].

Fact 14.6.2. The following statements hold:

i) Let $a \in (-1,\infty)$ and $n \ge 0$. Then,

$$\int_0^1 x^a \log^n x\,\mathrm{d}x = (-1)^n \frac{n!}{(a+1)^{n+1}}.$$

In particular,

$$\int_0^1 \log^n x\,\mathrm{d}x = (-1)^n n!, \quad \int_0^1 x\log^n x\,\mathrm{d}x = (-1)^n \frac{n!}{2^{n+1}}.$$

ii) Let $a, b \in (-1,\infty)$. Then,

$$\int_0^1 x^a \log^b \frac{1}{x}\,\mathrm{d}x = \frac{\Gamma(b+1)}{(a+1)^{b+1}}.$$

In particular,

$$\int_0^1 \frac{1}{\sqrt{\log \frac{1}{x}}}\,\mathrm{d}x = \sqrt{\pi}, \quad \int_0^1 \sqrt{\log \frac{1}{x}}\,\mathrm{d}x = \frac{\sqrt{\pi}}{2}, \quad \int_0^1 \log^{3/2} \frac{1}{x}\,\mathrm{d}x = \frac{3\sqrt{\pi}}{4},$$

$$\int_0^1 \frac{1}{\sqrt{x \log \frac{1}{x}}}\,\mathrm{d}x = \sqrt{2\pi}, \quad \int_0^1 \sqrt{x \log \frac{1}{x}}\,\mathrm{d}x = \frac{\sqrt{6\pi}}{9}.$$

iii) Let $n \geq 0$. Then,

$$\int_1^e \log^n x\,\mathrm{d}x = \left(\sum_{i=0}^{n} (-1)^{n-i} \frac{n!}{i!}\right) e - (-1)^n n!.$$

In particular,

$$\int_1^e \log x\,\mathrm{d}x = 1, \quad \int_1^e \log^2 x\,\mathrm{d}x = e - 2, \quad \int_1^e \log^3 x\,\mathrm{d}x = 6 - 2e, \quad \int_1^e \log^4 x\,\mathrm{d}x = 9e - 24.$$

Source: [511, p. 97] and [815, 818].

Fact 14.6.3. If $n \geq 0$, then

$$\int_0^1 x^n \log(1 - x)\,\mathrm{d}x = -\frac{H_{n+1}}{n+1}.$$

In particular,

$$\int_0^1 \log(1 - x)\,\mathrm{d}x = -1, \quad \int_0^1 x \log(1 - x)\,\mathrm{d}x = -\frac{3}{4}, \quad \int_0^1 x^2 \log(1 - x)\,\mathrm{d}x = -\frac{11}{18}.$$

If $n \geq 0$, then

$$\int_0^1 x^n \log^2(1 \quad x)\,\mathrm{d}x = \frac{H_{n+1}^2 + H_{n+1,2}}{n+1}.$$

In particular,

$$\int_0^1 \log^2(1 - x)\,\mathrm{d}x = 2, \quad \int_0^1 x \log^2(1 - x)\,\mathrm{d}x = \frac{7}{4}, \quad \int_0^1 x^2 \log^2(1 - x)\,\mathrm{d}x = \frac{85}{54}.$$

If $n \geq 0$, then

$$\int_0^1 x^n \log(x + 1)\,\mathrm{d}x = \frac{H_{n/2} - H_{(n-1)/2}}{2n + 2} + \frac{\log 2}{n + 1} - \frac{1}{(n + 1)^2},$$

$$\int_0^1 x^{2n} \log(x + 1)\,\mathrm{d}x = \frac{1}{2n + 1}\left(\log 4 + \sum_{i=1}^{2n+1} (-1)^i \frac{1}{i}\right),$$

$$\int_0^1 x^{2n+1} \log(x + 1)\,\mathrm{d}x = \frac{1}{2n + 2} \sum_{i=1}^{2n+2} (-1)^{i+1} \frac{1}{i},$$

$$\int_0^1 x^{n+1/2} \log(x + 1)\,\mathrm{d}x = \frac{2}{2n + 3} \log 2 + (-1)^{n+1} \frac{4}{2n + 3}\left(\frac{\pi}{4} - \sum_{i=0}^{n+1} (-1)^i \frac{1}{2i + 1}\right).$$

In particular,

$$\int_0^1 \log(x + 1)\,\mathrm{d}x = 2 \log 2 - 1, \quad \int_0^1 x \log(x + 1)\,\mathrm{d}x = \frac{1}{4}, \quad \int_0^1 x^2 \log(x + 1)\,\mathrm{d}x = \frac{2 \log 2}{3} - \frac{5}{18},$$

$$\int_0^1 x^3\log(x+1)\,\mathrm{d}x = \frac{7}{48}, \quad \int_0^1 x^4\log(x+1)\,\mathrm{d}x = \frac{2\log 2}{5} - \frac{47}{300}.$$

Furthermore,

$$\int_0^1 \log^2(x+1)\,\mathrm{d}x = 2(\log 2 - 1)^2, \quad \int_0^1 x\log^2(x+1)\,\mathrm{d}x = 2\log 2 - \frac{5}{4},$$

$$\int_0^1 x^2\log^2(x+1)\,\mathrm{d}x = \frac{(6\log 2 - 11)(6\log 2 - 5)}{54},$$

$$\int_0^1 x^3\log^2(x+1)\,\mathrm{d}x = \frac{4\log 2}{3} - \frac{241}{288}, \quad \int_0^1 x^4\log^2(x+1)\,\mathrm{d}x = \frac{2\log^2 2}{5} - \frac{92\log 2}{75} + \frac{6589}{9000}.$$

Source: [1101, 2766]. **Remark:** For noninteger p, H_p is defined in Fact 13.3.4.

Fact 14.6.4. If $a > 0$, then

$$\int_0^1 \frac{\log(x^a+1)}{x}\,\mathrm{d}x = \frac{\pi^2}{12a}, \quad \int_0^1 \frac{\log(1-x^a)}{x}\,\mathrm{d}x = -\frac{\pi^2}{6a}.$$

In particular,

$$\int_0^1 \frac{\log(x+1)}{x}\,\mathrm{d}x = \frac{\pi^2}{12}, \quad \int_0^1 \frac{\log(1-x)}{x}\,\mathrm{d}x = \int_0^1 \frac{\log x}{1-x}\,\mathrm{d}x = -\frac{\pi^2}{6}.$$

Fact 14.6.5. The following statements hold:

i) Let $a > 0$. Then,

$$\int_0^a \frac{\log(ax+1)}{x^2+1}\,\mathrm{d}x = \frac{1}{2}(\operatorname{atan} a)\log(1+a^2), \quad \int_0^a \frac{\log(x+a)}{x^2+a^2}\,\mathrm{d}x = \frac{\pi}{8a}\log 2a^2.$$

In particular,

$$\int_0^1 \frac{\log(1+x)}{1+x^2}\,\mathrm{d}x = \frac{\pi}{8}\log 2.$$

ii) Let $n \ge 1$. Then,

$$\int_0^1 \frac{\log^{2n} x}{1+x^2}\,\mathrm{d}x = \frac{\pi^{2n+1}|E_{2n}|}{4^{n+1}}.$$

In particular,

$$\int_0^1 \frac{\log^2 x}{x^2+1}\,\mathrm{d}x = \frac{\pi^3}{16}, \quad \int_0^1 \frac{\log^4 x}{x^2+1}\,\mathrm{d}x = \frac{5\pi^5}{64}, \quad \int_0^1 \frac{\log^6 x}{x^2+1}\,\mathrm{d}x = \frac{61\pi^7}{256}.$$

iii) Let $n \ge 1$. Then,

$$\int_0^1 \frac{\log^{2n-1} x}{1-x^2}\,\mathrm{d}x = (-1)^n\frac{(2^{2n}-1)\pi^{2n}B_{2n}}{4n}.$$

In particular,

$$\int_0^1 \frac{\log x}{1-x^2}\,\mathrm{d}x = -\frac{\pi^2}{8}, \quad \int_0^1 \frac{\log^3 x}{1-x^2}\,\mathrm{d}x = -\frac{\pi^4}{16}, \quad \int_0^1 \frac{\log^5 x}{1-x^2}\,\mathrm{d}x = -\frac{\pi^6}{8}.$$

iv) Let $a, b > -1$. Then,

$$\int_0^1 \frac{x^a - x^b}{\log x}\,\mathrm{d}x = \log\frac{a+1}{b+1}.$$

Source: *i)* is given in [511, p. 243], [2013, p. 182], and [2106, p. 55]; *ii)* is given in [1217, p. 550]; *iii)* is given in [511, p. 241]; *iv)* is given in [3024, p. 335].

Fact 14.6.6. Let $a \in (0,1)$. Then,

$$\int_0^1 \frac{\log x}{(1-x)^a}\,dx = \int_0^1 \frac{\log(1-x)}{x^a}\,dx = \frac{H_{1-a}}{a-1}.$$

In particular,

$$\int_0^1 \frac{\log x}{\sqrt[4]{1-x}}\,dx = 4\log 2 - \frac{16}{9} - \frac{2\pi}{3}, \quad \int_0^1 \frac{\log x}{\sqrt[3]{1-x}}\,dx = \frac{9\log 3 - 9 - \sqrt{3}}{4},$$

$$\int_0^1 \frac{\log x}{\sqrt{1-x}}\,dx = 4\log 2 - 4, \quad \int_0^1 \frac{\log x}{(1-x)^{2/3}}\,dx = \frac{\sqrt{3}\pi + 9\log 3}{2} - 9.$$

Furthermore,

$$\int_0^1 \frac{\log^2 x}{\sqrt{1-x}}\,dx = 16+8\log^2 2-16\log 2-\frac{2\pi^2}{3}, \quad \int_0^1 \frac{\sqrt[4]{x}\log^2 x}{\sqrt{1-x}}\,dx = \frac{\sqrt{\pi}\Gamma(\frac{5}{4})}{9\Gamma(\frac{7}{4})}(9\pi^2-48\pi+144G-64),$$

$$\int_0^1 \frac{\sqrt{x}\log^2 x}{\sqrt{1-x}}\,dx = \frac{\pi^3}{6}-\pi+2\pi\log^2 2-2\pi\log 2, \quad \int_0^1 \frac{x\log^2 x}{\sqrt{1-x}}\,dx = \frac{224}{27}+12\log^2 2-\frac{80}{9}\log 2-\frac{4\pi^2}{9},$$

$$\int_0^1 \frac{\log x}{\sqrt{x(1-x^2)}}\,dx = -\frac{\sqrt{2\pi}}{8}\Gamma^2(\tfrac{1}{4}), \quad \int_0^1 \frac{\log^2 x}{\sqrt{x(1-x^2)}}\,dx = \frac{16G+\pi^2}{8\sqrt{2\pi}}\Gamma^2(\tfrac{1}{4}),$$

$$\int_0^1 \frac{\log^3 x}{\sqrt{x(1-x^2)}}\,dx = -\frac{48G+5\pi^2}{16}\sqrt{\frac{\pi}{2}}\Gamma^2(\tfrac{1}{4}), \quad \int_0^1 \frac{x\log x}{\sqrt{x(1-x^2)}}\,dx = \frac{\sqrt{2}(\pi-4)\pi^{3/2}}{\Gamma^2(\frac{1}{4})}.$$

Source: [1217, p. 538].

Fact 14.6.7. Let $n \ge 0$. Then,

$$\int_0^1 \frac{x^{2n}\log x}{\sqrt{1-x^2}}\,dx = \frac{(2n-1)!!}{(2n)!!}\frac{\pi}{2}\left(\sum_{i=1}^{2n}(-1)^{i-1}\frac{1}{i} - \log 2\right),$$

$$\int_0^1 \frac{x^{2n+1}\log x}{\sqrt{1-x^2}}\,dx = \frac{(2n)!!}{(2n+1)!!}\frac{\pi}{2}\left(\sum_{i=1}^{2n+1}(-1)^{i-1}\frac{1}{i} + \log 2\right),$$

$$\int_0^1 x^{2n}(\log x)\sqrt{1-x^2}\,dx = \frac{(2n-1)!!}{(2n+2)!!}\frac{\pi}{2}\left(\sum_{i=1}^{2n}(-1)^{i-1}\frac{1}{i} - \log 2 - \frac{1}{2n+2}\right),$$

$$\int_0^1 x^{2n+1}(\log x)\sqrt{1-x^2}\,dx = \frac{(2n)!!}{(2n+3)!!}\frac{\pi}{2}\left(\sum_{i=1}^{2n+1}(-1)^{i-1}\frac{1}{i} + \log 2 - \frac{1}{2n+3}\right).$$

In particular,

$$\int_0^1 \frac{\log x}{\sqrt{1-x^2}}\,dx = -\frac{\pi}{4}\log 2, \quad \int_0^1 \frac{x\log x}{\sqrt{1-x^2}}\,dx = \log 2 - 1,$$

$$\int_0^1 (\log x)\sqrt{1-x^2}\,dx = -\frac{\pi}{8} - \frac{\pi}{4}\log 2, \quad \int_0^1 x(\log x)\sqrt{1-x^2}\,dx = \frac{1}{3}\log 2 - \frac{4}{9}.$$

Furthermore,

$$\int_0^1 \frac{\log^2 x}{\sqrt{1-x^2}}\,dx = \frac{\pi^3}{24} + \frac{\pi}{2}\log^2 2, \quad \int_0^1 \frac{x\log^2 x}{\sqrt{1-x^2}}\,dx = 2 - \frac{\pi^2}{12} + \log^2 2 - 2\log 2,$$

$$\int_0^1 \frac{\log^3 x}{\sqrt{1-x^2}}\,dx = -\frac{\pi^3}{8}\log 2 - \frac{\pi}{2}\log^3 2 - \frac{3\pi}{4}\zeta(3).$$

Source: [1217, p. 538].

Fact 14.6.8. Let a be a real number. If $|a| \le 1$, then

$$\int_0^1 \frac{\log(ax+1)}{x\sqrt{1-x^2}}\,\mathrm{d}x = \frac{\pi^2}{8} - \frac{1}{2}\operatorname{acos}^2 a.$$

If $|a| \le 1$ and $a \ne 0$, then

$$\int_0^1 \frac{x\log(ax+1)}{\sqrt{1-x^2}}\,\mathrm{d}x = \frac{\sqrt{1-a^2}}{a}\operatorname{asin} a + \frac{(1-\sqrt{1-a^2})\pi}{2a} - 1.$$

If $|a| \ge 1$, then

$$\int_0^1 \frac{x\log(ax+1)}{\sqrt{1-x^2}}\,\mathrm{d}x = \frac{\sqrt{a^2-1}}{a}\log(1+\sqrt{a^2-1}) + \frac{\pi}{2a} - 1.$$

Furthermore,

$$\int_0^1 \frac{\log(1+x)}{\sqrt{1-x^2}}\,\mathrm{d}x = -\frac{\pi}{2}\log 2 + 2G, \quad \int_0^1 \frac{x\log(1+x)}{\sqrt{1-x^2}}\,\mathrm{d}x = \frac{\pi}{2} - 1,$$

$$\int_0^1 \frac{\log(1-x)}{\sqrt{1-x^2}}\,\mathrm{d}x = -\frac{\pi}{2}\log 2 - 2G, \quad \int_0^1 \frac{x\log(1-x)}{\sqrt{1-x^2}}\,\mathrm{d}x = -\frac{\pi}{2} - 1.$$

Source: [1217, p. 558].

Fact 14.6.9. Let $a \in [-1, 1)$. Then,

$$\int_0^1 \frac{(a-x)\log(1-x)}{x^2-2ax+1}\,\mathrm{d}x = \frac{\pi^2}{12} - \frac{(\pi - \operatorname{acos} a)^2}{8} - \frac{\log^2(2-2a)}{8}.$$

In particular,

$$\int_0^1 \frac{(x+1)\log(1-x)}{x^2+2x+1}\,\mathrm{d}x = \frac{\log^2 2}{4} - \frac{\pi^2}{12}, \quad \int_0^1 \frac{(2x+1)\log(1-x)}{x^2+x+1}\,\mathrm{d}x = \frac{1}{4}\log^2 3 - \frac{5\pi^2}{36},$$

$$\int_0^1 \frac{x\log(1-x)}{x^2+1}\,\mathrm{d}x = \frac{1}{8}\log^2 2 - \frac{5\pi^2}{96}, \quad \int_0^1 \frac{(1-2x)\log(1-x)}{x^2-x+1}\,\mathrm{d}x = \frac{\pi^2}{18}.$$

Source: [1682].

Fact 14.6.10. Let $p > -1$. Then,

$$\int_0^1 \frac{x^p \log x}{1-x}\,\mathrm{d}x = H_{p,2} - \frac{\pi^2}{6} = -\psi'(p+1).$$

In particular,

$$\int_0^1 \frac{\log x}{x^{3/4}(1-x)}\,\mathrm{d}x = -\pi^2 - 8G, \quad \int_0^1 \frac{\log x}{x^{1/2}(1-x)}\,\mathrm{d}x = -\frac{\pi^2}{2}, \quad \int_0^1 \frac{\log x}{x^{1/4}(1-x)}\,\mathrm{d}x = 8G - \pi^2,$$

$$\int_0^1 \frac{\log x}{1-x}\,\mathrm{d}x = -\frac{\pi^2}{6}, \quad \int_0^1 \frac{x^{1/4}\log x}{1-x}\,\mathrm{d}x = -\pi^2 - 8G + 16, \quad \int_0^1 \frac{x^{1/2}\log x}{1-x}\,\mathrm{d}x = 4 - \frac{\pi^2}{2},$$

$$\int_0^1 \frac{x^{3/4}\log x}{1-x}\,\mathrm{d}x = -\pi^2 + 8G + \frac{16}{9}, \quad \int_0^1 \frac{x\log x}{1-x}\,\mathrm{d}x = 1 - \frac{\pi^2}{6}, \quad \int_0^1 \frac{x^{5/4}\log x}{1-x}\,\mathrm{d}x = -\pi^2 - 8G + \frac{416}{25},$$

$$\int_0^1 \frac{x^{3/2}\log x}{1-x}\,\mathrm{d}x = \frac{40}{9} - \frac{\pi^2}{2}, \quad \int_0^1 \frac{x^{7/4}\log x}{1-x}\,\mathrm{d}x = -\pi^2 + 8G + \frac{928}{441}, \quad \int_0^1 \frac{x^2\log x}{1-x}\,\mathrm{d}x = \frac{5}{4} - \frac{\pi^2}{6},$$

$$\int_0^1 \frac{x^3 \log x}{1-x}\,\mathrm{d}x = \frac{49}{36} - \frac{\pi^2}{6}, \quad \int_0^1 \frac{x^4 \log x}{1-x}\,\mathrm{d}x = \frac{205}{144} - \frac{\pi^2}{6}, \quad \int_0^1 \frac{x^5 \log x}{1-x}\,\mathrm{d}x = \frac{5269}{3600} - \frac{\pi^2}{6}.$$

Remark: ψ is the digamma function. See Fact 13.3.3. **Related:** Fact 13.3.4 and Fact 14.2.7.

Fact 14.6.11. Let $p > -1$ and $n \geq 1$. Then,

$$\int_0^1 \frac{x^p \log^n x}{1-x}\,\mathrm{d}x = (-1)^n n![\zeta(n+1) - H_{p,n+1}] = -\psi^{(n)}(p+1).$$

In particular,

$$\int_0^1 \frac{\log^{2n-1} x}{1-x}\,\mathrm{d}x = -\frac{4^{n-1}}{n}\pi^{2n}|B_{2n}|,$$

$$\int_0^1 \frac{\log^2 x}{x^{3/4}(1-x)}\,\mathrm{d}x = 2\pi^3 + 56\zeta(3), \quad \int_0^1 \frac{\log^2 x}{x^{1/2}(1-x)}\,\mathrm{d}x = 14\zeta(3),$$

$$\int_0^1 \frac{\log^2 x}{x^{1/4}(1-x)}\,\mathrm{d}x = 56\zeta(3) - 2\pi^3, \quad \int_0^1 \frac{x^{1/2}\log^2 x}{1-x}\,\mathrm{d}x = 14\zeta(3) - 16,$$

$$\int_0^1 \frac{x^{5/2}\log^2 x}{1-x}\,\mathrm{d}x = 14\zeta(3) - \frac{56432}{3375}, \quad \int_0^1 \frac{x^{3/2}\log^2 x}{1-x}\,\mathrm{d}x = 14\zeta(3) - \frac{448}{27}.$$

Related: Fact 13.3.4 and Fact 14.6.12.

Fact 14.6.12. Let $m \geq 0$ and $n \geq 1$. Then,

$$\int_0^1 \frac{x^m \log^n x}{1-x}\,\mathrm{d}x = (-1)^n n![\zeta(n+1) - H_{m,n+1}].$$

In particular,

$$\int_0^1 \frac{\log^2 x}{1-x}\,\mathrm{d}x = 2\zeta(3), \quad \int_0^1 \frac{x\log^2 x}{1-x}\,\mathrm{d}x = 2\zeta(3) - 2, \quad \int_0^1 \frac{x^2\log^2 x}{1-x}\,\mathrm{d}x = 2\zeta(3) - \frac{9}{4},$$

$$\int_0^1 \frac{\log^3 x}{1-x}\,\mathrm{d}x = -\frac{\pi^4}{15}, \quad \int_0^1 \frac{x\log^3 x}{1-x}\,\mathrm{d}x = 6 - \frac{\pi^4}{15}, \quad \int_0^1 \frac{x^2\log^3 x}{1-x}\,\mathrm{d}x = \frac{51}{8} - \frac{\pi^4}{15}.$$

Furthermore,

$$\int_0^1 \frac{\log^{2n} x}{1-x^2}\,\mathrm{d}x = \frac{2^{2n+1}-1}{2^{2n+1}}(2n)!\zeta(2n+1), \quad \int_0^1 \frac{x^{2n}\log^2 x}{1-x^2}\,\mathrm{d}x = \frac{7}{4}\zeta(3) - \sum_{i=1}^n \frac{2}{(2i-1)^3},$$

$$\int_0^1 \frac{\log^{2n+1} x}{1-x^2}\,\mathrm{d}x = \frac{(1-4^{n+1})\pi^{2n+2}}{4n+4}|B_{2n+2}|, \quad \int_0^1 \frac{x\log^{2n+1} x}{1-x^2}\,\mathrm{d}x = -\frac{\pi^{2n+2}}{4n+4}|B_{2n+2}|,$$

$$\int_0^1 \frac{(1+x^2)\log^{2n} x}{(1-x^2)^2}\,\mathrm{d}x = \frac{(4^n-1)}{2}\pi^{2n}|B_{2n}|, \quad \int_0^1 \frac{\log^{2n} x}{1+x^2}\,\mathrm{d}x = \frac{\pi^{2n+1}}{4^{n+1}}|E_{2n}|.$$

Related: Fact 13.3.4, Fact 14.6.11, and Fact 14.8.15.

Fact 14.6.13. Let $a > 0$. Then,

$$\int_0^1 \frac{(-\log x)^a}{1-x}\,\mathrm{d}x = \Gamma(a+1)\zeta(a+1).$$

In particular,

$$\int_0^1 \frac{(-\log x)^{1/2}}{1-x}\,\mathrm{d}x = \frac{\sqrt{\pi}}{2}\zeta(\tfrac{3}{2}), \quad \int_0^1 \frac{(-\log x)^{3/2}}{1-x}\,\mathrm{d}x = \frac{3\sqrt{\pi}}{4}\zeta(\tfrac{5}{2}), \quad \int_0^1 \frac{(-\log x)^{5/2}}{1-x}\,\mathrm{d}x = \frac{15\sqrt{\pi}}{8}\zeta(\tfrac{7}{2}).$$

Fact 14.6.14. Let $n \ge 0$. Then,

$$\int_0^1 \frac{x^{2n}\log x}{x+1}\,\mathrm{d}x = -\frac{\pi^2}{12} + \sum_{i=1}^{2n}(-1)^{i+1}\frac{1}{i^2}, \quad \int_0^1 \frac{x^{2n+1}\log x}{x+1}\,\mathrm{d}x = \frac{\pi^2}{12} + \sum_{i=1}^{2n+1}(-1)^{i}\frac{1}{i^2},$$

$$\int_0^1 \frac{x^n\log^2 x}{x+1}\,\mathrm{d}x = (-1)^n\left(\frac{3}{2}\zeta(3) + 2\sum_{i=1}^{n}(-1)^i\frac{1}{i^3}\right),$$

$$\int_0^1 \frac{\log^{2n} x}{x+1}\,\mathrm{d}x = \frac{(4^n-1)(2n)!}{4^n}\zeta(2n+1), \quad \int_0^1 \frac{\log^{2n+1} x}{x+1}\,\mathrm{d}x = \frac{(1-2^{2n+1})\pi^{2n+2}}{2n+2}|B_{2n+2}|,$$

Source: [1217, pp. 540, 550].

Fact 14.6.15.

$$\int_0^1 \left(\frac{1}{1-x} + \frac{x\log x}{(1-x)^2}\right)\mathrm{d}x = \frac{\pi^2}{6} - 1,$$

$$\int_0^1 \frac{\log^2 x}{(1-x)^2}\,\mathrm{d}x = \frac{\pi^2}{3}, \quad \int_0^1 \frac{\log^3 x}{(1-x)^2}\,\mathrm{d}x = -6\zeta(3), \quad \int_0^1 \frac{\log^4 x}{(1-x)^2}\,\mathrm{d}x = \frac{4\pi^2}{15},$$

$$\int_0^1 \frac{\log^5 x}{(1-x)^2}\,\mathrm{d}x = -120\zeta(5), \quad \int_0^1 \frac{\log^3 x}{(1-x)^3}\,\mathrm{d}x = -3\zeta(3) - \frac{\pi^2}{2}, \quad \int_0^1 \frac{\log^4 x}{(1-x)^3}\,\mathrm{d}x = 12\zeta(3) + \frac{2\pi^4}{15},$$

$$\int_0^1 \frac{\log x}{1-x^2}\,\mathrm{d}x = -\frac{\pi^2}{8}, \quad \int_0^1 \frac{\log^2 x}{1-x^2}\,\mathrm{d}x = \frac{7}{4}\zeta(3), \quad \int_0^1 \frac{\log^3 x}{1-x^2}\,\mathrm{d}x = -\frac{\pi^4}{16},$$

$$\int_0^1 \frac{x\log x}{1-x^2}\,\mathrm{d}x = -\frac{\pi^2}{24}, \quad \int_0^1 \frac{x\log^2 x}{1-x^2}\,\mathrm{d}x = \frac{1}{4}\zeta(3), \quad \int_0^1 \frac{x\log^3 x}{1-x^2}\,\mathrm{d}x = -\frac{\pi^4}{240},$$

$$\int_0^1 \frac{\log x}{x+1}\,\mathrm{d}x = -\frac{\pi^2}{12}, \quad \int_0^1 \frac{\log^2 x}{x+1}\,\mathrm{d}x = \frac{3}{2}\zeta(3), \quad \int_0^1 \frac{\log^3 x}{x+1}\,\mathrm{d}x = -\frac{7\pi^4}{120},$$

$$\int_0^1 \frac{x\log x}{x+1}\,\mathrm{d}x = \frac{\pi^2}{12} - 1, \quad \int_0^1 \frac{x\log^2 x}{x+1}\,\mathrm{d}x = \frac{3}{4} - \frac{\pi^2}{12}, \quad \int_0^1 \frac{x\log^3 x}{x+1}\,\mathrm{d}x = \frac{\pi^2}{12} - \frac{31}{36},$$

$$\int_0^1 \frac{\log x}{(x+1)^2}\,\mathrm{d}x = -\log 2, \quad \int_0^1 \frac{\log^2 x}{(x+1)^2}\,\mathrm{d}x = \frac{\pi^2}{6}, \quad \int_0^1 \frac{\log^3 x}{(x+1)^2}\,\mathrm{d}x = -\frac{9}{2}\zeta(3),$$

$$\int_0^1 \frac{\log x}{(x+1)^3}\,\mathrm{d}x = -\frac{1}{4} - \frac{\log 2}{2}, \quad \int_0^1 \frac{\log^2 x}{(x+1)^3}\,\mathrm{d}x = \frac{\pi^2}{12} + \log 2, \quad \int_0^1 \frac{\log^3 x}{(x+1)^3}\,\mathrm{d}x = -\frac{9\zeta(3)+\pi^2}{4},$$

$$\int_0^1 \frac{\log x}{(x+1)^4}\,\mathrm{d}x = -\frac{7}{24} - \frac{\log 2}{3}, \quad \int_0^1 \frac{\log^2 x}{(x+1)^4}\,\mathrm{d}x = \frac{1}{6} + \frac{\pi^2}{18} + \log 2,$$

$$\int_0^1 \frac{\log^3 x}{(x+1)^4}\,\mathrm{d}x = -\frac{3}{2}\zeta(3) - \frac{\pi^2}{4} - \log 2, \quad \int_0^1 \frac{\log^4 x}{(x+1)^4}\,\mathrm{d}x = 9\zeta(3) + \frac{\pi^2}{3} + \frac{7\pi^4}{90},$$

$$\int_0^1 \frac{x-1}{\log x}\,\mathrm{d}x = \log 2, \quad \int_0^1 \frac{\log x}{\sqrt[4]{1-x^2}}\,\mathrm{d}x = \frac{(-8+\pi+2\log 2)\sqrt{\pi}\Gamma\left(\frac{3}{4}\right)}{2\Gamma\left(\frac{1}{4}\right)},$$

$$\int_0^1 \frac{\log x}{\sqrt[3]{1-x^2}}\,\mathrm{d}x = \frac{(-12+\sqrt{3}\pi+3\log 3)\sqrt{\pi}\Gamma\left(\frac{2}{3}\right)}{8\Gamma\left(\frac{7}{6}\right)}, \quad \int_0^1 \frac{\log x}{\sqrt{1-x^2}}\,\mathrm{d}x = -\frac{\pi\log 2}{2},$$

$$\int_0^1 \frac{\log(1-x)}{1+x}\,\mathrm{d}x = \frac{\log^2 2}{2} - \frac{\pi^2}{12}, \quad \int_0^1 \log\frac{1+x}{1-x}\,\mathrm{d}x = 2\log 2, \quad \int_0^1 \log\frac{1+x^2}{1-x^2}\,\mathrm{d}x = \frac{\pi}{2} - \log 2,$$

$$\int_0^1 \log\frac{1+x^3}{1-x^3}\,\mathrm{d}x = \frac{\sqrt{3}\pi}{6} - \frac{3}{2}\log 3 + 2\log 2, \quad \int_0^1 \log\frac{1+x^3}{1-x^2}\,\mathrm{d}x = \frac{\sqrt{3}\pi}{3} - 1,$$

$$\int_0^1 \frac{1}{x}\log\frac{1+x}{1-x}\,dx = \int_0^\infty \log\frac{e^x+1}{e^x-1}\,dx = \frac{\pi^2}{4}, \quad \int_0^1 \frac{1}{x}\log\frac{1+x^2}{1-x^2}\,dx = \frac{\pi^2}{8},$$

$$\int_0^1 \frac{(\log x)\log(2-x)}{1-x}\,dx = -\frac{5}{8}\zeta(3), \quad \int_0^1 \frac{(\log x)\log(2+x)}{1+x}\,dx = -\frac{13}{24}\zeta(3),$$

$$\int_0^1 (\log x)\log(1+x)\,dx = 2 - 2\log 2 - \frac{\pi^2}{12}, \quad \int_0^1 (\log x)\log(1-x)\,dx = 2 - \frac{\pi^2}{6},$$

$$\int_0^1 \frac{(\log x)\log(1-x)}{(1+x)^2}\,dx = \frac{\pi^2}{24} - \frac{\log^2 2}{2}, \quad \int_0^1 \frac{\log(x+1)}{x^2+1}\,dx = \frac{\pi}{8}\log 2,$$

$$\int_0^1 \frac{\log^2(1-x)}{x}\,dx = 2\zeta(3), \quad \int_0^1 \frac{\log^3(1-x)}{x}\,dx = -\frac{\pi^4}{15}, \quad \int_0^1 \frac{\log^4(1-x)}{x}\,dx = 24\zeta(5),$$

$$\int_0^1 \frac{\log^2(1-x)}{x^2}\,dx = \frac{\pi^2}{3}, \quad \int_0^1 \frac{\log^4(1-x)}{x^4}\,dx = 12\zeta(3) + \frac{2\pi^2}{3} + \frac{4\pi^4}{45},$$

$$\int_0^1 \frac{\log(x+1)}{x}\,dx = \frac{\pi^2}{12}, \quad \int_0^1 \frac{\log^2(x+1)}{x}\,dx = \frac{1}{4}\zeta(3), \quad \int_0^1 \frac{\log^2(x+1)}{x^2}\,dx = \frac{\pi^2}{6} - 2\log^2 2,$$

$$\int_0^1 \frac{\log^3(x+1)}{x^2}\,dx = \frac{3}{4}\zeta(3) - 2\log^3 2, \quad \int_0^1 \frac{\log^3(x+1)}{x^3}\,dx = \frac{\pi^2}{4} - 3\log^2 2 - \frac{3}{8}\zeta(3),$$

$$\int_0^1 \frac{\log^3(1-x)}{x^2}\,dx = -6\zeta(3), \quad \int_0^1 \frac{\log^3(x+1)}{x^2}\,dx = \frac{3\zeta(4)}{6} - 2\log^3 2,$$

$$\int_0^1 \frac{\log^3(1-x)}{x^3}\,dx = -3\zeta(3) - \frac{\pi^2}{2}, \quad \int_0^1 \frac{\log^3(x+1)}{x^3}\,dx = \frac{\pi^2}{4} - \frac{3\zeta(3)}{8} - 3\log^2 2,$$

$$\int_0^1 \frac{(\log x)\log^2(1-x)}{x}\,dx = -\frac{\pi^4}{180}, \quad \int_0^1 \frac{x^2\log x}{(x^2-1)(x^4+1)}\,dx = \frac{(2-\sqrt{2})\pi^2}{32},$$

$$\int_0^1 \log(1-x+x^2)\,dx = \frac{\sqrt{3}\pi}{3} - 2, \quad \int_0^1 \log(1+x-x^2)\,dx = 2\sqrt{5}\log\frac{1+\sqrt{5}}{2} - 2,$$

$$\int_0^1 \frac{\log(x^2+1)}{x}\,dx = \frac{\pi^2}{24}, \quad \int_0^1 \frac{\log(1-x^2)}{x}\,dx = -\frac{\pi^2}{12},$$

$$\int_0^1 \frac{\log(x^2+1)}{x^2}\,dx = \frac{\pi}{2} - \log 2, \quad \int_0^1 \frac{\log(1-x^2)}{x^2}\,dx = -2\log 2,$$

$$\int_0^1 \frac{\log(x^3+1)}{x^3}\,dx = -\frac{\sqrt{3}\pi}{12} - \frac{3\log 3}{4}, \quad \int_0^1 \frac{\log(1-x^3)}{x^3}\,dx = \frac{\sqrt{3}\pi}{6},$$

$$\int_0^1 \frac{\log(x+1)}{x+1}\,dx = \frac{1}{2}\log^2 2, \quad \int_0^1 \frac{x\log(x+1)}{x+1}\,dx = 2\log 2 - \frac{1}{2}\log^2 2 - 1,$$

$$\int_0^1 \frac{x^2\log(x+1)}{x+1}\,dx = \frac{1}{2}\log^2 2 + \frac{5}{4} - 2\log 2, \quad \int_0^1 \frac{\log(x+1)}{(x+1)^2}\,dx = \frac{1}{2}(1-\log 2),$$

$$\int_0^1 \frac{x\log(x+1)}{(x+1)^2}\,dx = \frac{1}{2}(\log 2 + \log^2 2 - 1), \quad \int_0^1 \frac{x^2\log(x+1)}{(x+1)^2}\,dx = \frac{1}{2}(3\log 2 - 2\log^2 2 - 1),$$

$$\int_0^1 \frac{\log(x+1)}{(x+1)^3}\,dx = \frac{3}{16} - \frac{1}{8}\log 2, \quad \int_0^1 \frac{x\log(x+1)}{(x+1)^3}\,dx = \frac{5}{16} - \frac{3}{8}\log 2,$$

$$\int_0^1 \frac{x^2\log(x+1)}{(x+1)^3}\,dx = \frac{7}{8}\log 2 + \frac{1}{2}\log^2 2 - \frac{13}{16},$$

$$\int_0^1 \frac{\log(x^2+1)}{x+1}\,\mathrm{d}x = \frac{3\log^2 2}{4} - \frac{\pi^2}{48}, \quad \int_0^1 \frac{\log(x^2+1)}{x^2+1}\,\mathrm{d}x = \frac{\pi\log 2}{2} - G,$$

$$\int_0^1 \log(x^3+1)\,\mathrm{d}x = \frac{\sqrt{3}\pi}{3} + 2\log 2 - 3, \quad \int_0^1 \frac{\log(x^3+1)}{x}\,\mathrm{d}x = \frac{\pi^2}{36},$$

$$\int_0^1 \frac{\log(x^3+1)}{x^2}\,\mathrm{d}x = \frac{\sqrt{3}\pi}{3} - 2\log 2, \quad \int_0^1 \frac{\log(x^3+1)}{x^3}\,\mathrm{d}x = \frac{\sqrt{3}\pi}{6},$$

$$\int_0^1 \frac{\log x}{x^2+1}\,\mathrm{d}x = -G, \quad \int_0^1 \frac{\log x}{(x^2+1)^2}\,\mathrm{d}x = -\frac{\pi}{8} - \frac{G}{2}, \quad \int_0^1 \frac{\log x}{(x^2+1)^3}\,\mathrm{d}x = -\frac{\pi}{8} - \frac{3G}{8} - \frac{1}{16},$$

$$\int_0^1 \frac{x\log x}{x^2+1}\,\mathrm{d}x = -\frac{\pi^2}{48}, \quad \int_0^1 \frac{x\log x}{(x^2+1)^2}\,\mathrm{d}x = -\frac{1}{4}\log 2, \quad \int_0^1 \frac{x\log x}{(x^2+1)^3}\,\mathrm{d}x = -\frac{1}{8}\log 2 - \frac{1}{16},$$

$$\int_0^1 \frac{\log^2 x}{x^2+1}\,\mathrm{d}x = \frac{\pi^3}{16}, \quad \int_0^1 \frac{\log^2 x}{(x^2+1)^2}\,\mathrm{d}x = G + \frac{\pi^3}{32}, \quad \int_0^1 \frac{\log^2 x}{(x^2+1)^3}\,\mathrm{d}x = G + \frac{\pi}{16} + \frac{3\pi^2}{128},$$

$$\int_0^1 \frac{\log(x+1)}{x^2+1}\,\mathrm{d}x = \frac{\pi}{8}\log 2, \quad \int_0^1 \frac{\log(x+1)}{(x^2+1)^2}\,\mathrm{d}x = \frac{\pi}{16}(\log 2 - 1) + \frac{3}{8}\log 2,$$

$$\int_0^1 \frac{\log(x+1)}{(x^2+1)^3}\,\mathrm{d}x = \frac{3\pi}{64}(\log 2 - 1) + \frac{3}{8}\log 2 - \frac{1}{16}, \quad \int_0^1 \frac{x\log(x+1)}{x^2+1}\,\mathrm{d}x = \frac{\pi^2}{96} + \frac{1}{8}\log^2 2,$$

$$\int_0^1 \frac{x\log(x+1)}{(x^2+1)^2}\,\mathrm{d}x = \frac{\pi}{16} - \frac{1}{8}\log 2, \quad \int_0^1 \frac{x\log(x+1)}{(x^2+1)^3}\,\mathrm{d}x = \frac{\pi}{32} - \frac{1}{32}\log 2,$$

$$\int_0^1 \frac{x^2\log(x+1)}{x^2+1}\,\mathrm{d}x = 2\log 2 - \frac{\pi}{8}\log 2 - 1, \quad \int_0^1 \frac{x^2\log(x+1)}{(x^2+1)^2}\,\mathrm{d}x = \frac{\pi}{16} - \frac{3}{8} + \frac{\pi}{16}\log 2,$$

$$\int_0^1 \frac{x^2\log(x+1)}{(x^2+1)^3}\,\mathrm{d}x = \frac{1}{16} - \frac{\pi}{64} + \frac{\pi}{64}\log 2,$$

$$\int_0^1 \frac{\log(1-x)}{x^2+1}\,\mathrm{d}x = \frac{\pi}{8}\log 2 - G, \quad \int_0^1 \frac{\log(1-x)}{(x^2+1)^2}\,\mathrm{d}x = \frac{\pi}{16}(\log 2 - 1) + \frac{1}{8}\log 2 - \frac{1}{2}G,$$

$$\int_0^1 \frac{\log(1-x)}{(x^2+1)^3}\,\mathrm{d}x = \frac{3\pi}{64}(\log 2 - 1) + \frac{1}{8}\log 2 - \frac{3G}{8}, \quad \int_0^1 \frac{x\log(1-x)}{x^2+1}\,\mathrm{d}x = \frac{1}{4}\log^2 2 - \frac{5\pi^2}{96},$$

$$\int_0^1 \frac{x\log(1-x)}{(x^2+1)^2}\,\mathrm{d}x = -\frac{\pi}{16} - \frac{1}{8}\log 2, \quad \int_0^1 \frac{x\log(1-x)}{(x^2+1)^3}\,\mathrm{d}x = -\frac{\pi}{32} - \frac{1}{32}\log 2 - \frac{1}{16},$$

$$\int_0^1 \frac{x^2\log(1-x)}{x^2+1}\,\mathrm{d}x = \frac{\pi}{8}\log 2 + G - 1, \quad \int_0^1 \frac{x^2\log(1-x)}{(x^2+1)^2}\,\mathrm{d}x = \frac{\pi}{16}(\log 2 + 1) - \frac{1}{8}\log 2 - \frac{G}{2},$$

$$\int_0^1 \frac{x^2\log(1-x)}{(x^2+1)^3}\,\mathrm{d}x = \frac{\pi}{64}(\log 2 - 1) - \frac{G}{8}, \quad \int_0^1 \frac{\log(x+1)}{x^2+x}\,\mathrm{d}x = \frac{\pi^2}{12} - \frac{1}{2}\log^2 2,$$

$$\int_0^1 \frac{\log^2(x+1)}{x^2+x}\,\mathrm{d}x = 1 + \frac{\pi^2}{6} - \log 2 - \frac{5}{2}\log^2 2 + \frac{2}{3}\log^3 2 - \frac{1}{2}\zeta(3),$$

$$\int_0^1 \frac{\log(1-x^2)}{x}\,\mathrm{d}x = -\frac{\pi^2}{12}, \quad \int_0^1 \frac{\log(1-x^2)}{x^2}\,\mathrm{d}x = -2\log 2, \quad \int_0^1 \frac{\log^2(1-x^2)}{x}\,\mathrm{d}x = \zeta(3),$$

$$\int_0^1 \frac{\log(1-x^2)}{x^2+1}\,\mathrm{d}x = \frac{\pi}{4}\log 2 - G, \quad \int_0^1 \frac{\log(1-x^2)}{(x^2+1)^2}\,\mathrm{d}x = \frac{\pi}{8}\log 2 + \frac{1}{2}\log 2 - \frac{G}{2} - \frac{\pi}{8},$$

$$\int_0^1 \frac{\log(x^2+1)}{x^2+1}\,\mathrm{d}x = \frac{\pi}{2}\log 2 - G, \quad \int_0^1 \frac{\log(x^2+1)}{(x^2+1)^2}\,\mathrm{d}x = \frac{1}{4} - \frac{\pi}{8} + \frac{\pi+1}{4}\log 2 - \frac{G}{2},$$

$$\int_0^1 \frac{x\log(x^2+1)}{x^2+1}\,dx = \frac{\log^2 2}{4}, \quad \int_0^1 \frac{\log(x^2+1)}{(x^2+1)^2}\,dx = \frac{\log 2+1}{4} + \frac{(2\log 2-1)\pi}{8} - \frac{G}{2},$$

$$\int_0^1 \frac{x\log(x^2+1)}{(x^2+1)^2}\,dx = \frac{1-\log 2}{4}, \quad \int_0^1 \frac{\log(x^2+1)}{x^3+x}\,dx = \frac{\pi^2}{24} - \frac{1}{4}\log^2 2,$$

$$\int_0^1 \frac{\log(x^2+1)}{(x^2+1)(x+1)}\,dx = \frac{\log^2 2}{4} - \frac{\pi^2}{96} + \frac{\pi\log 2}{4} - \frac{G}{2}, \quad \int_0^1 \frac{[\log(1-x)]\log^2(1+x)}{x}\,dx = -\frac{\pi^4}{240},$$

$$\int_0^1 \frac{[\log(1-x)]\log(1+x)}{1+x}\,dx = \frac{1}{8}\zeta(3) + \frac{1}{3}\log^3 2 - \frac{\pi^2}{12}\log 2,$$

$$\int_0^1 \frac{[\log(1-x)]\log(1+x)}{(1+x)^2}\,dx = \frac{\pi^2}{24} - \frac{1}{2}(\log 2)(1+\log 2),$$

$$\int_0^1 \frac{[\log(1-x)]\log(1+x)}{x}\,dx = -\frac{5}{8}\zeta(3), \quad \int_0^1 \frac{[\log(1-x)]\log(1+x)}{x^2}\,dx = -\frac{\pi^2}{12} - \log^2 2,$$

$$\int_0^1 \frac{\log(1-x)[\log(2-x)-1-\log 2]}{(2-x)^2}\,dx = \frac{\log^2 2}{2} + \frac{\pi^2}{12}, \quad \int_0^1 \frac{\log(x+1/x)}{x^2+1}\,dx = \frac{\pi}{2}\log 2.$$

Source: [107, pp. 37, 206–209], [511, p. 97], [514], [516, p. 63], [1107], [1158, pp. 160, 161], [1217, pp. 240, 241], [1341], [1568, pp. 197, 198], [2013, p. 198], [2106, pp. 54, 55], [2483, 2765, 2946].

Fact 14.6.16. Let $a \in [-1, 1]$. Then,

$$\int_0^1 \frac{\log(1-a^2x^2)}{\sqrt{1-x^2}}\,dx = \pi\log\tfrac{1}{2}(1+\sqrt{1-a^2}), \quad \int_0^1 \frac{\log(1-a^2x^2)}{x\sqrt{1-x^2}}\,dx = -\left(\frac{\pi}{2} - \operatorname{acos}|a|\right)^2,$$

$$\int_0^1 \frac{\log(1+a^2x^2)}{\sqrt{1-x^2}}\,dx = \pi\log\tfrac{1}{2}(1+\sqrt{1+a^2}), \quad \int_0^1 \frac{1}{x\sqrt{1-x^2}}\log\frac{1+ax}{1-ax}\,dx = \pi\operatorname{asin} a.$$

In particular,

$$\int_0^1 \frac{\log(1-x^2)}{\sqrt{1-x^2}}\,dx = -\pi\log 2, \quad \int_0^1 \frac{\log(1-a^2x^2)}{x\sqrt{1-x^2}}\,dx = -\frac{\pi^2}{4},$$

$$\int_0^1 \frac{\log(1+x^2)}{\sqrt{1-x^2}}\,dx = \pi\log\tfrac{1}{2}(1+\sqrt{2}), \quad \int_0^1 \frac{1}{x\sqrt{1-x^2}}\log\frac{1+x}{1-x}\,dx = \frac{\pi^2}{2}.$$

Source: [1217, pp. 562, 563].

Fact 14.6.17. Let $a > 0$. Then,

$$\int_0^1 \frac{\log(a+x)}{x^2+a}\,dx = \frac{1}{2\sqrt{a}}(\operatorname{acot}\sqrt{a})\log(a^2+a),$$

$$\int_0^1 \frac{\log(ax+1)}{ax^2+1}\,dx = \frac{1}{2\sqrt{a}}(\operatorname{atan}\sqrt{a})\log(a+1), \quad \int_0^1 \frac{1}{1-x^2}\log\frac{a^2x^2+1}{a^2+1}\,dx = -\operatorname{atan}^2 a,$$

$$\int_0^1 \sqrt{1-x^2}\log(ax^2+1)\,dx = \frac{\pi}{2}\log\frac{1+\sqrt{1+a}}{2} + \frac{\pi}{4}\frac{1-\sqrt{1+a}}{1+\sqrt{1+a}}.$$

Source: [1217, pp. 556, 557, 560, 562].

Fact 14.6.18. Let a and b be distinct positive numbers. Then,

$$\int_0^1 \frac{\log(x+1)}{(ax+b)^2}\,dx = \frac{1}{a(a-b)}\log\frac{a+b}{b} + \frac{2\log 2}{b^2-a^2},$$

$$\int_0^1 \frac{\log(ax+b)}{(x+1)^2}\,\mathrm{d}x = \frac{1}{a-b}[\tfrac{1}{2}(a+b)\log(a+b) - b\log b - a\log 2].$$

Source: [1217, pp. 556, 557].

Fact 14.6.19. Let $p \in \mathbb{R}[s]$, assume that p has no roots on the unit circle, and let $\operatorname{mroots}(p) = \{\lambda_1, \ldots, \lambda_n\}_{\mathrm{ms}}$. Then,

$$\int_0^1 \frac{p'(x)\log(1-x)}{p(x)}\,\mathrm{d}x = \frac{1}{2}\sum_{i=1}^n [\log^2|1-\lambda_i| + \arg^2(1-\lambda_i)] - \frac{n\pi^2}{12}.$$

Source: [1682].

Fact 14.6.20. Let $n \ge 0$. Then,

$$\int_0^1 \frac{\log \sum_{i=0}^n x^i}{x}\,\mathrm{d}x = \frac{n\pi^2}{6(n+1)}.$$

Source: [2465].

Fact 14.6.21. Let $n \ge 0$ and $a \in [0, 2\pi]$. Then,

$$\int_0^1 \frac{(\log^{2n} x)\log(x^2 - 2x\cos a + 1)}{x}\,\mathrm{d}x = (-1)^{n+1}\frac{(2\pi)^{2n+2}}{(2n+2)(2n+1)}B_{2n+2}(\tfrac{a}{2\pi}).$$

In particular,

$$\int_0^1 \frac{\log(x^2 - 2x\cos a + 1)}{x}\,\mathrm{d}x = \frac{\pi^2}{6} - \frac{1}{2}(a-\pi)^2,$$

$$\int_0^1 \frac{(\log^{2n} x)\log(x^2+1)}{x}\,\mathrm{d}x = (-1)^{n+1}\frac{(2\pi)^{2n+2}}{(2n+2)(2n+1)}B_{2n+2}(\tfrac{1}{4}),$$

$$\int_0^1 \frac{(\log^n x)\log(1+x)}{x}\,\mathrm{d}x = (-1)^n n!\left(1 - \frac{1}{2^{n+1}}\right)\zeta(n+2),$$

$$\int_0^1 \frac{(\log^n x)\log(1-x)}{x}\,\mathrm{d}x = (-1)^{n+1} n!\zeta(n+2),$$

$$\int_0^1 \frac{(\log^n x)\log(1-x^2)}{x}\,\mathrm{d}x = (-1)^{n+1}\frac{n!}{2^{n+1}}\zeta(n+2),$$

$$\int_0^1 \frac{\log^n x}{x}\log\frac{1+x}{1-x}\,\mathrm{d}x = (-1)^n n!\left(2 - \frac{1}{2^{n+1}}\right)\zeta(n+2),$$

$$\int_0^1 \frac{(\log^2 x)\log(1-x)}{x}\,\mathrm{d}x = -\frac{\pi^4}{45}, \quad \int_0^1 \frac{(\log^2 x)\log(1+x)}{x}\,\mathrm{d}x = \frac{7\pi^4}{360},$$

$$\int_0^1 \frac{(\log^4 x)\log(1-x)}{x}\,\mathrm{d}x = -\frac{8\pi^6}{315}, \quad \int_0^1 \frac{(\log^4 x)\log(1+x)}{x}\,\mathrm{d}x = \frac{31\pi^6}{1260},$$

$$\int_0^1 \frac{(\log^2 x)\log(1+x^2)}{x}\,\mathrm{d}x = \frac{7\pi^4}{2880}, \quad \int_0^1 \frac{(\log^4 x)\log(1+x^2)}{x}\,\mathrm{d}x = \frac{31\pi^6}{40320}.$$

Furthermore,

$$\int_0^1 \frac{\log^{2n} x}{x^2 - 2x\cos a + 1}\,\mathrm{d}x = -\frac{4^n\pi^{2n+1}}{(2n+1)\sin a}B_{2n+1}(\tfrac{a}{2\pi}).$$

In particular,

$$\int_0^1 \frac{\log x}{x^2+1}\,\mathrm{d}x = -G, \quad \int_0^1 \frac{\log^2 x}{x^2+1}\,\mathrm{d}x = \frac{\pi^3}{16}, \quad \int_0^1 \frac{\log^4 x}{x^2+1}\,\mathrm{d}x = \frac{5\pi^5}{64},$$

$$\int_0^1 \frac{\log^2 x}{x^2-x+1}\,\mathrm{d}x = \frac{10\sqrt{3}\pi^3}{243}, \quad \int_0^1 \frac{\log^2 x}{(1-x)^2}\,\mathrm{d}x = \frac{\pi^2}{3}.$$

If $n \ge 1$, then

$$\int_0^1 \frac{x\log^{2n} x}{(x^2+1)^2}\,\mathrm{d}x = 4^{n-1}\pi^{2n}B_{2n}(\tfrac{1}{4}).$$

Finally,

$$\int_0^4 \frac{\log x}{\sqrt{4x-x^2}}\,\mathrm{d}x = \int_0^2 \frac{\log 2x}{\sqrt{2x-x^2}}\,\mathrm{d}x = 0.$$

Source: [705] and [2106, p. 70]. **Related:** Fact 13.5.93.

Fact 14.6.22.

$$\int_0^1 \frac{\log[\frac{1}{2}(1+\sqrt{4x+1})]}{x}\,\mathrm{d}x = \frac{\pi^2}{15}, \quad \int_0^1 \frac{\log[\frac{1}{2}(1+\sqrt{8x+1})]}{x}\,\mathrm{d}x = \frac{\pi^2}{12} + \frac{1}{2}\log^2 2.$$

Fact 14.6.23.

$$\int_0^1 \log\log\frac{1}{x}\,\mathrm{d}x = -\gamma, \quad \int_0^1 \frac{\log\log 1/x}{\sqrt{\log 1/x}}\,\mathrm{d}x = -\sqrt{\pi}(2\log 2 + \gamma).$$

Furthermore,

$$\int_0^1 \frac{\log\log 1/x}{x^2+1}\,\mathrm{d}x = \int_{\pi/4}^{\pi/2} \log\log\tan x\,\mathrm{d}x = \frac{\mathrm{d}}{\mathrm{d}x}\Gamma(x)L(x)\bigg|_{x=1} = -\frac{\pi\gamma}{4} + L'(1)$$
$$= \frac{\pi}{2}\log\frac{\sqrt{2\pi}\Gamma(\frac{3}{4})}{\Gamma(\frac{1}{4})} = \frac{\pi}{4}\log\frac{4\pi^3}{\Gamma^4(\frac{1}{4})},$$

where L is the Dirichlet L function defined in Fact 13.3.6. Furthermore,

$$\int_0^1 \frac{\log\log 1/x}{x+1}\,\mathrm{d}x = -\frac{1}{2}\log^2 2, \quad \int_0^1 \frac{\log\log 1/x}{(x+1)^2}\,\mathrm{d}x = \frac{1}{2}\log\frac{\pi}{2} - \frac{\gamma}{2},$$

$$\int_0^1 \frac{\log\log 1/x}{x^2+x+1}\,\mathrm{d}x = \frac{\sqrt{3}\pi}{3}\log\frac{\sqrt[3]{2\pi}\Gamma(\frac{2}{3})}{\Gamma(\frac{1}{3})}, \quad \int_0^1 \frac{\log\log 1/x}{x^2-x+1}\,\mathrm{d}x = \frac{2\sqrt{3}\pi}{3}\left(\frac{5}{6}\log 2\pi - \log\Gamma(\tfrac{1}{6})\right).$$

If $a > 0$, then

$$\int_0^1 x^{a-1}\log\log 1/x\,\mathrm{d}x = -\frac{\gamma}{a} - \frac{\log a}{a}, \quad \int_0^1 \frac{x^{a-1}}{x^a+1}\log\log 1/x\,\mathrm{d}x = -\frac{\log 2}{2a}\log 2a^2.$$

Finally, let $a \in (2, 6.1\cdot 10^{13})$, and define $\phi \triangleq \frac{1}{2}(\sqrt{a+2} - \sqrt{a-2})$. Then,

$$\int_0^1 \frac{\log\log 1/x}{x^2+ax+1}\,\mathrm{d}x = \frac{2\log 2\pi}{\sqrt{k^2-1}}\log\phi + \frac{2\pi}{\sqrt{k^2-1}}\arg\Gamma\left(\frac{1}{2} + \frac{J}{\pi}\log\phi\right).$$

Source: [1217, pp. 534, 535, 570] and [1688, 2783].

Fact 14.6.24.

$$\int_0^1 \frac{1-x}{(1+x)\log x}\,\mathrm{d}x = \log\frac{2}{\pi}, \quad \int_0^1 \frac{1-x}{(1+x)(1+x^2)\log x}\,\mathrm{d}x = -\frac{1}{2}\log 2,$$

$$\int_0^1 \frac{x(1-x)}{(1+x)(1+x^2)\log x}\,\mathrm{d}x = \log 2 + \frac{3}{2}\log\pi - 2\log\Gamma(\tfrac{1}{4}),$$

$$\int_0^1 \frac{x^2(1-x)}{(1+x)(1+x^2)\log x}\,\mathrm{d}x = \log\frac{2\sqrt{2}}{\pi}, \quad \int_0^1 \frac{(1-x)^2}{(1+x^2)\log x}\,\mathrm{d}x = \log\frac{\pi}{4},$$

$$\int_0^1 \frac{x(1-x)^2}{(1+x^2)\log x}\,\mathrm{d}x = \log 16\pi^2 - 4\log\Gamma(\tfrac{1}{4}), \qquad \int_0^1 \frac{1-x^2}{(1+x^2)\log x}\,\mathrm{d}x = \log 8\pi^2 - 4\log\Gamma(\tfrac{1}{4}).$$

Source: [1217, p. 545].

Fact 14.6.25. Let $r \ge 0$, let $a \ge 0$, and define

$$f(a,r) \triangleq \int_1^\infty \frac{\log^r x}{x^a}\,\mathrm{d}x.$$

If $a \in [0,1]$, then $f(a,r) = \infty$. If $a > 1$, then $f(a,r) \in \mathbb{R}$.

Fact 14.6.26. Let $n \ge 1$. Then,

$$\int_0^\infty \frac{\log^{2n} x}{x^2+1}\,\mathrm{d}x = \left(\frac{\pi}{2}\right)^{2n+1}|E_{2n}|, \quad \int_0^\infty \frac{\log^{2n+1} x}{x^2+1}\,\mathrm{d}x = 0, \quad \int_0^\infty \frac{\log^{2n-1} x}{1-x^2}\,\mathrm{d}x = \frac{(1-4^n)\pi^{2n}}{2n}|B_{2n}|.$$

In particular,

$$\int_0^\infty \frac{\log^2 x}{x^2+1}\,\mathrm{d}x = \frac{\pi^3}{8}, \quad \int_0^\infty \frac{\log^4 x}{x^2+1}\,\mathrm{d}x = \frac{5\pi^5}{32}, \quad \int_0^\infty \frac{\log^6 x}{x^2+1}\,\mathrm{d}x = \frac{61\pi^7}{128}.$$

Source: [630, 650] and [1217, p. 550]. **Remark:** $\int_0^\infty \frac{\log^{2n} x}{x^2+1}\,\mathrm{d}x = 2\int_0^1 \frac{\log^{2n} x}{x^2+1}\,\mathrm{d}x$. See Fact 14.6.21.

Fact 14.6.27. Let $a > 0$, $k \ge 0$, and $n \ge 2$, and assume that $k+2 \le n$. Then,

$$\int_0^\infty \frac{x^k \log x}{x^n + a^2}\,\mathrm{d}x = \frac{a^{(2k+2)/n}\pi}{a^2 n^2 \sin\frac{(k+1)\pi}{n}}\left(2\log a - \pi\cot\tfrac{(k+1)\pi}{n}\right).$$

In particular,

$$\int_0^\infty \frac{x^k \log x}{x^{2k+2}+a^2}\,\mathrm{d}x = \frac{\pi\log a}{2a(k+1)^2}, \quad \int_0^\infty \frac{x^k\log x}{x^{3k+3}+a^2}\,\mathrm{d}x = \frac{2\pi}{27a^{4/3}(k+1)^2}(2\sqrt{3}\log a - \pi),$$

$$\int_0^\infty \frac{x^k \log x}{x^{4k+4}+a^2}\,\mathrm{d}x = \frac{\sqrt{2}\pi}{16a^{3/2}(k+1)^2}(2\log a - \pi).$$

Source: [561] and use Fact 14.2.21.

Fact 14.6.28. Let $n \ge 1$ and $k \ge 2$, and define $Q_n \in \mathbb{R}[s]$ as in Fact 13.2.14. Then,

$$\int_0^\infty \frac{\log^n x}{x^k+1}\,\mathrm{d}x = (-1)^n\left(\frac{\pi}{k}\right)^{n+1}\sec\left[\frac{\pi(k-2)}{2k}\right]Q_n\left(\tan\left[\frac{\pi(k-2)}{2k}\right]\right).$$

In particular,

$$\int_0^\infty \frac{\log x}{x^2+1}\,\mathrm{d}x = 0, \quad \int_0^\infty \frac{\log^2 x}{x^2+1}\,\mathrm{d}x = \frac{\pi^3}{8}, \quad \int_0^\infty \frac{\log^3 x}{x^2+1}\,\mathrm{d}x = 0,$$

$$\int_0^\infty \frac{\log x}{x^3+1}\,\mathrm{d}x = -\frac{2\pi^2}{27}, \quad \int_0^\infty \frac{\log^2 x}{x^3+1}\,\mathrm{d}x = \frac{10\sqrt{3}\pi^3}{243}, \quad \int_0^\infty \frac{\log^3 x}{x^3+1}\,\mathrm{d}x = -\frac{14\pi^4}{243},$$

$$\int_0^\infty \frac{\log x}{x^4+1}\,\mathrm{d}x = -\frac{\sqrt{2}\pi^2}{16}, \quad \int_0^\infty \frac{\log^2 x}{x^4+1}\,\mathrm{d}x = \frac{3\sqrt{2}\pi^3}{64}, \quad \int_0^\infty \frac{\log^3 x}{x^4+1}\,\mathrm{d}x = -\frac{11\sqrt{2}\pi^4}{256},$$

$$\int_0^\infty \frac{\log^n x}{x^3+1}\,\mathrm{d}x = (-1)^n\frac{2\sqrt{3}}{3}\left(\frac{\pi}{3}\right)^{n+1}\sum_{i=0}^n E_{n,i}\left(\frac{\sqrt{3}}{3}\right)^i, \quad \int_0^\infty \frac{\log^n x}{x^4+1}\,\mathrm{d}x = (-1)^n\sqrt{2}\left(\frac{\pi}{4}\right)^{n+1}\sum_{i=0}^n E_{n,i}.$$

Source: [630, 650].

Fact 14.6.29. Let $a > 0$. Then,

$$\int_0^\infty \frac{\log x}{x^2+a^2}\,\mathrm{d}x = \frac{\pi\log a}{2a}, \quad \int_0^\infty \frac{\log^2 x}{x^2+a^2}\,\mathrm{d}x = \frac{\pi\log^2 a}{2} + \frac{\pi^3}{8a},$$

$$\int_0^\infty \frac{\log x}{(x+a)^2}\,\mathrm{d}x = \frac{\log a}{a}, \quad \int_0^\infty \frac{\log^2 x}{(x+a)^2}\,\mathrm{d}x = \frac{\log^2 a}{a} + \frac{\pi^2}{3a},$$

$$\int_0^\infty \frac{\log x}{(x^2+a^2)^2}\,\mathrm{d}x = \frac{\pi(\log a - 1)}{4a^3}, \quad \int_0^\infty \frac{\log^2 x}{(x^2+a^2)^2}\,\mathrm{d}x = \frac{\pi(\log a - 2)\log a}{4a^3} + \frac{\pi^3}{16a^3},$$

$$\int_0^\infty \left(\frac{\log x}{x+1} - \frac{\log x}{x+a}\right)\mathrm{d}x = \frac{1}{2}\log^2 a, \quad \int_0^\infty \left(\frac{\log^2 x}{x+1} - \frac{\log^2 x}{x+a}\right)\mathrm{d}x = \frac{1}{3}\log^3 a + \frac{\pi^2}{3}\log a.$$

Source: [1158, p. 151], [1568, p. 166], and [2013, pp. 197, 198].

Fact 14.6.30.

$$\int_0^\infty \frac{\log x}{(x+1)^2}\,\mathrm{d}x = 0, \quad \int_0^\infty \frac{\log^2 x}{(x+1)^2}\,\mathrm{d}x = \frac{\pi^2}{3}, \quad \int_0^\infty \frac{\log^3 x}{(x+1)^2}\,\mathrm{d}x = 0,$$

$$\int_0^\infty \frac{\log x}{(x^2+1)^2}\,\mathrm{d}x = -\frac{\pi}{4}, \quad \int_0^\infty \frac{\log^2 x}{(x^2+1)^2}\,\mathrm{d}x = \frac{\pi^3}{16}, \quad \int_0^\infty \frac{\log^3 x}{(x^2+1)^2}\,\mathrm{d}x = -\frac{3\pi^3}{16},$$

$$\int_0^\infty \frac{x\log x}{(x^2+1)^2}\,\mathrm{d}x = 0, \quad \int_0^\infty \frac{x\log^2 x}{(x^2+1)^2}\,\mathrm{d}x = \frac{\pi^2}{24}, \quad \int_0^\infty \frac{x\log^3 x}{(x^2+1)^2}\,\mathrm{d}x = 0,$$

$$\int_0^\infty \frac{x^2\log x}{(x^2+1)^2}\,\mathrm{d}x = \frac{\pi}{4}, \quad \int_0^\infty \frac{x^2\log^2 x}{(x^2+1)^2}\,\mathrm{d}x = \frac{\pi^3}{16}, \quad \int_0^\infty \frac{x^2\log^3 x}{(x^2+1)^2}\,\mathrm{d}x = \frac{3\pi^3}{16},$$

$$\int_0^\infty \frac{\log x}{x^2-1}\,\mathrm{d}x = \frac{\pi^2}{4}, \quad \int_0^\infty \frac{\log^2 x}{x^2-1}\,\mathrm{d}x = 0, \quad \int_0^\infty \frac{\log^3 x}{x^2-1}\,\mathrm{d}x = \frac{\pi^4}{8},$$

$$\int_0^\infty \frac{\log x}{x^3-1}\,\mathrm{d}x = \frac{4\pi^2}{27}, \quad \int_0^\infty \frac{\log^2 x}{x^3-1}\,\mathrm{d}x = -\frac{8\sqrt{3}\pi^3}{243}, \quad \int_0^\infty \frac{\log^3 x}{x^3-1}\,\mathrm{d}x = \frac{16\pi^4}{243},$$

$$\int_0^\infty \frac{x\log x}{x^3-1}\,\mathrm{d}x = 0, \quad \int_0^\infty \frac{x\log^2 x}{x^3-1}\,\mathrm{d}x = \frac{8\sqrt{3}\pi^3}{243}, \quad \int_0^\infty \frac{x\log^3 x}{x^3-1}\,\mathrm{d}x = \frac{16\pi^4}{243},$$

$$\int_0^\infty \frac{\log x}{(x+1)^3}\,\mathrm{d}x = -\frac{1}{2}, \quad \int_0^\infty \frac{\log^2 x}{(x+1)^3}\,\mathrm{d}x = \frac{\pi^2}{6}, \quad \int_0^\infty \frac{\log^3 x}{(x+1)^3}\,\mathrm{d}x = -\frac{\pi^2}{2},$$

$$\int_0^\infty \frac{x\log x}{(x+1)^3}\,\mathrm{d}x = \frac{1}{2}, \quad \int_0^\infty \frac{x\log^2 x}{(x+1)^3}\,\mathrm{d}x = \frac{\pi^2}{6}, \quad \int_0^\infty \frac{x\log^3 x}{(x+1)^3}\,\mathrm{d}x = \frac{\pi^2}{2},$$

$$\int_0^\infty \frac{\log x}{(x^2+1)^3}\,\mathrm{d}x = -\frac{\pi}{4}, \quad \int_0^\infty \frac{\log^2 x}{(x^2+1)^3}\,\mathrm{d}x = \frac{3\pi^3}{64} + \frac{\pi}{8}, \quad \int_0^\infty \frac{\log^3 x}{(x^2+1)^3}\,\mathrm{d}x = -\frac{3\pi^3}{16},$$

$$\int_0^\infty \frac{x\log x}{(x^2+1)^3}\,\mathrm{d}x = -\frac{1}{8}, \quad \int_0^\infty \frac{x\log^2 x}{(x^2+1)^3}\,\mathrm{d}x = \frac{\pi^2}{48}, \quad \int_0^\infty \frac{x\log^3 x}{(x^2+1)^3}\,\mathrm{d}x = -\frac{\pi^2}{32},$$

$$\int_0^\infty \frac{x\log x}{(x^2+1)^3}\,\mathrm{d}x = 0, \quad \int_0^\infty \frac{x\log^2 x}{(x^2+1)^3}\,\mathrm{d}x = \frac{\pi^3}{64} - \frac{\pi}{8}, \quad \int_0^\infty \frac{x\log^3 x}{(x^2+1)^3}\,\mathrm{d}x = 0.$$

Fact 14.6.31. The following statements hold:

i) Let $a > 0$. Then,

$$\int_0^\infty \frac{\log x}{(x+a)(x-1)}\,\mathrm{d}x = \frac{\pi^2 + \log^2 a}{2(a+1)}.$$

ii) Let $a \ge 0$. Then,

$$\int_0^\infty \frac{\log x}{x^2+ax+1}\,\mathrm{d}x = 0.$$

iii) Let $a \in (0,1)$. Then,

$$\int_0^\infty \frac{\log x}{x^a(x+1)}\,\mathrm{d}x = \frac{\pi^2\cos a\pi}{\sin^2 a\pi}, \quad \int_0^\infty \frac{\log x}{x^a(x-1)}\,\mathrm{d}x = \frac{2\pi^2}{1-\cos 2a\pi}.$$

iv) Let $a \in (-1,1)$. Then,

$$\int_0^\infty \frac{\log x}{x^a(x+1)^2}\,\mathrm{d}x = \frac{a\pi^2\cos a\pi - \pi\sin a\pi}{\sin^2 a\pi}.$$

Source: [1136, pp. 208, 211], [1217, p. 536], and [2106, p. 68].

Fact 14.6.32. If $n \ge 0$, then

$$\int_0^\infty \frac{\log^{2n} x}{\sqrt{x}(x+1)}\,\mathrm{d}x = |E_{2n}|\pi^{2n+1}, \quad \int_0^\infty \frac{\log^{2n+1} x}{\sqrt{x}(x+1)}\,\mathrm{d}x = 0.$$

In particular,

$$\int_0^\infty \frac{\log^2 x}{\sqrt{x}(x+1)}\,\mathrm{d}x = \pi^3, \quad \int_0^\infty \frac{\log^4 x}{\sqrt{x}(x+1)}\,\mathrm{d}x = 5\pi^5, \quad \int_0^\infty \frac{\log^6 x}{\sqrt{x}(x+1)}\,\mathrm{d}x = 61\pi^7,$$

$$\int_0^\infty \frac{\log^8 x}{\sqrt{x}(x+1)}\,\mathrm{d}x = 1385\pi^9, \quad \int_0^\infty \frac{\log^{10} x}{\sqrt{x}(x+1)}\,\mathrm{d}x = 50521\pi^{11}.$$

Source: [2513, pp. 374, 375].

Fact 14.6.33. Let $a \in (1,2)$. Then,

$$\int_0^\infty \frac{\log(x+1)}{x^a} = \frac{\pi}{(1-a)\sin a\pi}.$$

In particular,

$$\int_0^\infty \frac{\log(x+1)}{x^{9/8}}\,\mathrm{d}x = 8\sqrt{4+2\sqrt{2}}\pi, \quad \int_0^\infty \frac{\log(x+1)}{x^{5/4}}\,\mathrm{d}x = 4\sqrt{2}\pi, \quad \int_0^\infty \frac{\log(x+1)}{x^{4/3}}\,\mathrm{d}x = 2\sqrt{3}\pi,$$

$$\int_0^\infty \frac{\log(x+1)}{x^{11/8}}\,\mathrm{d}x = \frac{8}{3}\sqrt{4-2\sqrt{2}}\pi, \quad \int_0^\infty \frac{\log(x+1)}{x^{3/2}}\,\mathrm{d}x = 2\pi, \quad \int_0^\infty \frac{\log(x+1)}{x^{13/8}}\,\mathrm{d}x = \frac{8}{5}\sqrt{4-2\sqrt{2}}\pi,$$

$$\int_0^\infty \frac{\log(x+1)}{x^{5/3}}\,\mathrm{d}x = \sqrt{3}\pi, \quad \int_0^\infty \frac{\log(x+1)}{x^{7/4}}\,\mathrm{d}x = \frac{4\sqrt{2}}{3}\pi, \quad \int_0^\infty \frac{\log(x+1)}{x^{15/8}}\,\mathrm{d}x = \frac{8}{7}\sqrt{4+2\sqrt{2}}\pi.$$

Fact 14.6.34. Let $a \ge 0$ and $b \in (1,3)$. Then,

$$\int_0^\infty \frac{\log(a^2x^2+1)}{x^b}\,\mathrm{d}x = \frac{a^{b-1}\pi}{1-b}\sec\frac{b\pi}{2}.$$

In particular,

$$\int_0^\infty \frac{\log(a^2x^2+1)}{x^{3/2}}\,\mathrm{d}x = 2\sqrt{2a}\pi, \quad \int_0^\infty \frac{\log(a^2x^2+1)}{x^2}\,\mathrm{d}x = a\pi, \quad \int_0^\infty \frac{\log(a^2x^2+1)}{x^{5/2}}\,\mathrm{d}x = \frac{2}{3}\sqrt{2a^3}.$$

Fact 14.6.35. Let $a \ge 0$ and $b > 1$. Then,

$$\int_0^\infty \log\left(1+\frac{a^b}{x^b}\right)\mathrm{d}x = a\pi\csc\frac{\pi}{b}.$$

In particular,

$$\int_0^\infty \log\left(1+\frac{a^2}{x^2}\right)\mathrm{d}x = a\pi,\quad \int_0^\infty \log\left(1+\frac{a^3}{x^3}\right)\mathrm{d}x = \frac{2\sqrt{3}a\pi}{3},\quad \int_0^\infty \log\left(1+\frac{a^4}{x^4}\right)\mathrm{d}x = \sqrt{2}a\pi.$$

Fact 14.6.36. Let $a > 0$. Then,

$$\int_0^\infty \frac{\log(x^2+1)}{(x+a)^2}\,\mathrm{d}x = \frac{\pi}{a^2+1} + \frac{2a}{a^2+1}\log a,\qquad \int_0^\infty (\log x)\log\left(1+\frac{a^2}{x^2}\right)\mathrm{d}x = a\pi(\log a - 1).$$

Source: [1217, pp. 530, 560].

Fact 14.6.37. Let $a > 0$ and $b > 0$. Then,

$$\int_0^\infty \frac{\log ax}{x^2+b^2}\,\mathrm{d}x = \frac{\pi}{2b}\log ab,\qquad \int_0^\infty \frac{\log x}{a^2+b^2x^2}\,\mathrm{d}x = \frac{\pi}{2ab}\log\frac{a}{b},\qquad \int_0^\infty \frac{\log x}{a^2-b^2x^2}\,\mathrm{d}x = \frac{\pi^2}{4ab},$$

$$\int_0^\infty \log\frac{x^2+a^2}{x^2+b^2}\,\mathrm{d}x = (a-b)\pi,\qquad \int_0^\infty (\log x)\log\frac{x^2+a^2}{x^2+b^2}\,\mathrm{d}x = \pi(b-a) + \pi\log\frac{a^a}{b^b},$$

$$\int_0^\infty [\log(a^2+x^2)]\log\left(1+\frac{b^2}{x^2}\right)\mathrm{d}x = 2\pi[(a+b)\log(a+b) - a\log a - b],$$

$$\int_0^\infty \left[\log\left(1+\frac{a^2}{x^2}\right)\right]\log\left(1+\frac{b^2}{x^2}\right)\mathrm{d}x = 2\pi[(a+b)\log(a+b) - a\log a - b\log b],$$

$$\int_0^\infty [\log(a^2x^2+1)]\log\left(1+\frac{b^2}{x^2}\right)\mathrm{d}x = 2\pi\left(\frac{ab+1}{a}\log(ab+1) - b\right),$$

$$\int_0^\infty \left[\log\left(a^2+\frac{1}{x^2}\right)\right]\log\left(1+\frac{b^2}{x^2}\right)\mathrm{d}x = 2\pi\left(\frac{ab+1}{a}\log(ab+1) - b\log b\right),$$

$$\int_0^\infty \frac{x^{a-1}-x^{b-1}}{\log x}\,\mathrm{d}x = \log\frac{a}{b},\qquad \int_0^\infty \frac{x^{a-1}-x^{b-1}}{(x+1)\log x}\,\mathrm{d}x = \log\frac{\Gamma(\frac{b}{2})\Gamma(\frac{a+1}{2})}{\Gamma(\frac{a}{2})\Gamma(\frac{b+1}{2})},$$

$$\int_0^\infty \frac{\log(a^2x^2+1) - \log(b^2x^2+1)}{x^2}\,\mathrm{d}x = \pi(a-b).$$

Source: [1217, pp. 530, 545, 560].

Fact 14.6.38. Let a and b be distinct positive numbers. Then,

$$\int_0^\infty \frac{\log x}{(x+a)(x+b)}\,\mathrm{d}x = \frac{\log^2 a - \log^2 b}{2(a-b)},\qquad \int_0^\infty \frac{\log x}{(x^2+a^2)(x^2+b^2)}\,\mathrm{d}x = \frac{\pi}{2ab(a^2-b^2)}\log\frac{b^a}{a^b},$$

$$\int_0^\infty \frac{x\log x}{(x^2+a^2)(x^2+b^2)}\,\mathrm{d}x = \frac{\log^2 a - \log^2 b}{2(a^2-b^2)},\qquad \int_0^\infty \frac{x^2\log x}{(x^2+a^2)(x^2+b^2)}\,\mathrm{d}x = \frac{\pi}{2(a^2-b^2)}\log\frac{a^a}{b^b},$$

$$\int_0^\infty \frac{\log(x+1)}{(ax+b)^2}\,\mathrm{d}x = \frac{1}{a(a-b)}\log\frac{a}{b},\qquad \int_0^\infty \frac{\log(x+a)}{(x+b)^2}\,\mathrm{d}x = \frac{1}{b(a-b)}\log\frac{a^a}{b^b},$$

$$\int_0^\infty \frac{\log(x+1)}{(ax+b)^2}\,\mathrm{d}x = \frac{1}{a(a-b)}\log\frac{a}{b},\qquad \int_0^\infty \frac{\log(ax+b)}{(x+1)^2}\,\mathrm{d}x = \frac{1}{a-b}\log\frac{a^a}{b^b}.$$

Source: [1136, p. 208] and [1217, pp. 537, 556, 557].

Fact 14.6.39. Let a, b, c, d be positive numbers. Then,

$$\int_0^\infty \frac{\log(a^2x^2+b^2)}{c^2x^2+d^2}\,\mathrm{d}x = \frac{\pi}{cd}\log\left(\frac{ad}{c}+b\right).$$

Source: [1217, p. 560].

Fact 14.6.40. Let a, b, c, d be positive numbers. If $ad \neq bc$, then

$$\int_0^\infty \frac{\log(ax+b)}{(cx+d)^2}\,\mathrm{d}x = \frac{a(\log a - \log c + \log d)}{c(ad-bc)} - \frac{b\log b}{d(ad-bc)}.$$

Furthermore,

$$\int_0^\infty \frac{x\log(ax+b)}{(cx+d)^2}\,\mathrm{d}x = \frac{a^2(2\log a - \log c + \log d)}{4c(a^2d+b^2c)} + \frac{b(a\sqrt{cd}\pi + 2bc\log b)}{4cd(a^2d+b^2c)}.$$

Fact 14.6.41. Let $a \geq 0$ and $b > 0$. Then,

$$\int_0^\infty \frac{\log(a^2x^2+1)}{x^2+b^2}\,\mathrm{d}x = \frac{\pi\log(ab+1)}{b}, \quad \int_0^\infty \frac{\log(a^2x^2+1)}{(x^2+b^2)^2}\,\mathrm{d}x = \frac{[\log(ab+1) - ab/(ab+1)]\pi}{2b^3},$$

$$\int_0^\infty \frac{\log(a^2x^2+1)}{(x^2+b^2)^3}\,\mathrm{d}x = \frac{[3(ab+1)^2\log(ab+1) - ab(4ab+3)]\pi}{8b^5(ab+1)^2},$$

$$\int_0^\infty \frac{\log x}{(x+a)^2+b^2}\,\mathrm{d}x = \frac{1}{2b}\left(\operatorname{atan}\frac{b}{a}\right)\log(a^2+b^2),$$

$$\int_0^\infty \frac{\log^2 x}{(x+a)^2+b^2}\,\mathrm{d}x = \frac{1}{b}\left(\operatorname{atan}\frac{b}{a}\right)\left(\frac{\pi^2}{3} - \frac{1}{3}\operatorname{atan}^2\frac{b}{a} + \frac{1}{4}\log^2(a^2+b^2)\right).$$

Source: [511, pp. 259–261] and [2106, pp. 333–339].

Fact 14.6.42.

$$\int_0^\infty \frac{\log(1-x^2)^2}{x^2}\,\mathrm{d}x = \int_0^\infty \frac{\log(1-x^2)^4}{x^2}\,\mathrm{d}x = \int_1^\infty \frac{\log(1-x^2)^2}{x^3}\,\mathrm{d}x = 0,$$

$$\int_0^\infty \frac{\log(x+1)}{x^2+1}\,\mathrm{d}x = G + \frac{\pi}{4}\log 2, \quad \int_0^\infty \frac{\log(x+1)}{x^2+x}\,\mathrm{d}x = \frac{\pi^2}{6}, \quad \int_0^\infty \frac{\log(x+1)}{x^3+x}\,\mathrm{d}x = \frac{5\pi^2}{48},$$

$$\int_0^\infty \frac{\log(x^2+1)}{x^2}\,\mathrm{d}x = \pi, \quad \int_0^\infty \frac{\log^2(x^2+1)}{x^3}\,\mathrm{d}x = \frac{\pi^2}{6}, \quad \int_0^\infty \frac{\log^3(x^2+1)}{x^3}\,\mathrm{d}x = 3\zeta(3),$$

$$\int_0^\infty \frac{\log(x^2+1)}{x^2+1}\,\mathrm{d}x = \pi\log 2, \quad \int_0^\infty \frac{\log(x^2+1)}{x^2+x}\,\mathrm{d}x = \frac{5\pi^2}{24}, \quad \int_0^\infty \frac{\log(x^2+1)}{x^3+x}\,\mathrm{d}x = \frac{\pi^2}{12},$$

$$\int_0^\infty \frac{\log(x^2+1)}{(x^2+1)^2}\,\mathrm{d}x = \frac{(2\log 2 - 1)\pi}{4}, \quad \int_0^\infty \frac{x\log(x^2+1)}{(x^2+1)^2}\,\mathrm{d}x = \frac{1}{2},$$

$$\int_0^\infty \frac{x^2\log(x^2+1)}{(x^2+1)^2}\,\mathrm{d}x = \frac{(2\log 2 + 1)\pi}{4}, \quad \int_0^\infty \frac{\log(x^2+1)}{(x^2+1)^3}\,\mathrm{d}x = \frac{(12\log 2 - 7)\pi}{32},$$

$$\int_0^\infty \frac{x\log(x^2+1)}{(x^2+1)^3}\,\mathrm{d}x = \frac{1}{8}, \quad \int_0^\infty \frac{x^2\log(x^2+1)}{(x^2+1)^3}\,\mathrm{d}x = \frac{(4\log 2 - 1)\pi}{32},$$

$$\int_0^\infty \frac{x^3\log(x^2+1)}{(x^2+1)^3}\,\mathrm{d}x = \frac{3}{8}, \quad \int_0^\infty \frac{x^4\log(x^2+1)}{(x^2+1)^3}\,\mathrm{d}x = \frac{3(4\log 2 + 3)\pi}{32},$$

$$\int_0^\infty \frac{\log(x^2+1)}{(x^2+1)^4}\,\mathrm{d}x = \frac{(60\log 2 - 37)\pi}{192}, \quad \int_0^\infty \frac{x\log(x^2+1)}{(x^2+1)^4}\,\mathrm{d}x = \frac{1}{18},$$

$$\int_0^\infty \frac{x^2\log(x^2+1)}{(x^2+1)^4}\,\mathrm{d}x = \frac{(12\log 2 - 5)\pi}{192}, \quad \int_0^\infty \frac{x^3\log(x^2+1)}{(x^2+1)^4}\,\mathrm{d}x = \frac{5}{72},$$

$$\int_0^\infty \frac{x^4\log(x^2+1)}{(x^2+1)^4}\,\mathrm{d}x = \frac{(12\log 2 - 1)\pi}{192}, \quad \int_0^\infty \frac{x^5\log(x^2+1)}{(x^2+1)^4}\,\mathrm{d}x = \frac{11}{36},$$

$$\int_0^\infty \frac{x^6\log(x^2+1)}{(x^2+1)^4}\,\mathrm{d}x = \frac{5(12\log 2+11)\pi}{192},\quad \int_0^\infty \frac{\log^2(x^2+1)}{(x^2+1)^2}\,\mathrm{d}x = \left(\frac{\pi^2}{12}-\frac{1}{2}+\log^2 2-\log 2\right)\pi,$$

$$\int_0^\infty \frac{1}{(x^2+1)^2}\log\frac{x}{x^2+1}\,\mathrm{d}x = -\frac{\pi\log 2}{2},\quad \int_0^\infty \frac{1}{(x^2+1)^2}\log^2\frac{x}{x^2+1}\,\mathrm{d}x = \frac{\pi^3}{48}+\pi\log^2 2.$$

If $n\ge 1$, then

$$\int_0^\infty \frac{1}{x^2}\left(\frac{x}{x^2+1}\right)^{2n+1}\log\frac{x}{x^2+1}\,\mathrm{d}x = \frac{H_n-H_{2n}-1/(2n)}{2n\binom{2n}{n}}.$$

Source: [510] and [511, pp. 259–261].

Fact 14.6.43. Let $a>0$. Then,

$$\int_0^\infty \frac{1}{(x+a)(\log^2 x+\pi^2)}\,\mathrm{d}x = \begin{cases}\dfrac{1}{1-a}+\dfrac{1}{\log a}, & a\ne 1,\\[2mm] \dfrac{1}{2}, & a=1.\end{cases}$$

Source: [2106, p. 184]. **Remark:** $\lim_{a\to 1}[1/(1-a)+1/\log a] = 1/2$.

Fact 14.6.44. If $a\in(1,2)$, then

$$\int_0^\infty \frac{\log|x-1|}{x^a}\,\mathrm{d}x = \frac{\pi}{a-1}\cot a\pi,\quad \int_0^\infty \frac{\log(x+1)}{x^a}\,\mathrm{d}x = \frac{\pi}{1-a}\csc a\pi.$$

Source: [1217, p. 558].

Fact 14.6.45. Let $n\ge 1$. Then,

$$\int_0^\infty \log^{2n}\frac{|x-1|}{x+1}\,\mathrm{d}x = 2|(1-4^n)B_{2n}|\pi^{2n}.$$

In particular,

$$\int_0^\infty \log^2\frac{|x-1|}{x+1}\,\mathrm{d}x = \pi^2,\quad \int_0^\infty \log^4\frac{|x-1|}{x+1}\,\mathrm{d}x = \pi^4,\quad \int_0^\infty \log^6\frac{|x-1|}{x+1}\,\mathrm{d}x = 3\pi^6,$$

$$\int_0^\infty \log^8\frac{|x-1|}{x+1}\,\mathrm{d}x = 17\pi^8,\quad \int_0^\infty \log^{10}\frac{|x-1|}{x+1}\,\mathrm{d}x = 155\pi^{10},\quad \int_0^\infty \log^{12}\frac{|x-1|}{x+1}\,\mathrm{d}x = 2073\pi^{12}.$$

Furthermore,

$$\int_0^\infty \log^3\frac{|x-1|}{x+1}\,\mathrm{d}x = -21\zeta(3),\quad \int_0^\infty \log^5\frac{|x-1|}{x+1}\,\mathrm{d}x = -465\zeta(5),$$

$$\int_0^\infty \log^7\frac{|x-1|}{x+1}\,\mathrm{d}x = -\frac{1}{2}40005\zeta(7),\quad \int_0^\infty \log^9\frac{|x-1|}{x+1}\,\mathrm{d}x = -1448685\zeta(9).$$

Source: [2106, p. 275]. **Remark:** B_{2n} is a Bernoulli number, and $2|(1-4^n)B_{2n}|$ is the nth *unsigned Genocchi number of even index*. **Related:** Fact 13.1.7.

Fact 14.6.46. If $a>0$, then

$$\int_a^{a+1}\log\Gamma(x+a)\,\mathrm{d}x = \int_0^1\log\Gamma(x+a)\,\mathrm{d}x = \frac{1}{2}\log 2\pi + a\log a - a.$$

Furthermore,

$$\int_0^1\log\Gamma(x)\,\mathrm{d}x = \int_0^1\log\Gamma(1-x)\,\mathrm{d}x = \frac{1}{2}\log 2\pi.$$

If $n \geq 1$, then

$$\int_0^1 [\log\Gamma(x)]\sin 2\pi nx\,dx = \frac{1}{2\pi n}(\log 2\pi n + G), \quad \int_0^1 [\log\Gamma(x)]\cos 2\pi nx\,dx = \frac{1}{4n}.$$

If $n \geq 0$, then

$$\int_0^1 [\log\Gamma(x)]\sin(2n+1)\pi x\,dx = \frac{1}{(2n+1)\pi}\left(\log\frac{\pi}{2} + \sum_{i=1}^n \frac{2}{2i-1} + \frac{1}{2n+1}\right).$$

Source: [1103, p. 27] and [1217, p. 656].

14.7 Facts on Definite Integrals of Logarithmic, Trigonometric, and Hyperbolic Functions

Fact 14.7.1. For all $n \geq 0$,

$$\int_0^{\pi/4} \log^{2n}\tan x\,dx = \frac{\pi^{2n+1}}{4^{n+1}}|E_{2n}|.$$

In particular,

$$\int_0^{\pi/4} \log^2\tan x\,dx = \frac{\pi^3}{16}, \quad \int_0^{\pi/4} \log^4\tan x\,dx = \frac{5\pi^5}{64}, \quad \int_0^{\pi/4} \log^6\tan x\,dx = \frac{61\pi^7}{256}.$$

Furthermore,

$$\int_0^{\pi/4} \log\sin x\,dx = -\frac{G}{2} - \frac{\pi}{4}\log 2, \quad \int_0^{\pi/4} \log\cos x\,dx = \frac{G}{2} - \frac{\pi}{4}\log 2, \quad \int_0^{\pi/4} \log\tan x\,dx = -G,$$

$$\int_0^{\pi/4} x\log\sin x\,dx = -\frac{G\pi}{8} - \frac{\pi^2}{32}\log 2 + \frac{35}{128}\zeta(3), \quad \int_0^{\pi/4} x\log\cos x\,dx = \frac{G\pi}{8} - \frac{\pi^2}{32}\log 2 - \frac{21}{128}\zeta(3),$$

$$\int_0^{\pi/4} x\log\tan x\,dx = 7\zeta(3) - \frac{G\pi}{4},$$

$$\int_0^{\pi/4} \log(1+\tan x)\,dx = \frac{\pi}{8}\log 2, \quad \int_0^{\pi/4} \log(1-\tan x)\,dx = \frac{\pi}{8}\log 2 - G,$$

$$\int_0^{\pi/4} x\log(1+\tan x)\,dx = \frac{\pi^2}{64}\log 2 - \frac{G\pi}{8} + \frac{21}{64}\zeta(3),$$

$$\int_0^{\pi/4} x\log(1-\tan x)\,dx = \frac{\pi^2}{64}\log 2 - \frac{G\pi}{8} - \frac{7}{64}\zeta(3),$$

$$\int_0^{\pi/4} \log(1+\tan^2 x)\,dx = \frac{\pi}{2}\log 2 - G, \quad \int_0^{\pi/4} \log(1-\tan^2 x)\,dx = \frac{\pi}{4}\log 2 - G,$$

$$\int_0^{\pi/4} x\log(1+\tan^2 x)\,dx = \frac{\pi^2}{16}\log 2 - \frac{G\pi}{4} + \frac{21}{64}\zeta(3),$$

$$\int_0^{\pi/4} x\log(1-\tan^2 x)\,dx = \frac{\pi^2}{32}\log 2 - \frac{G\pi}{4} + \frac{7}{32}\zeta(3),$$

$$\int_0^{\pi/4} \log(\cot x + 1)\,dx = \frac{\pi}{8}\log 2 + G, \quad \int_0^{\pi/4} \log(\cot x - 1)\,dx = \frac{\pi}{8}\log 2,$$

$$\int_0^{\pi/4} x\log(\cot x+1)\,dx = \frac{\pi^2}{64}\log 2 - \frac{G\pi}{8} - \frac{7}{64}\zeta(3), \quad \int_0^{\pi/4} x\log(\cot x-1)\,dx = \frac{\pi^2}{64}\log 2 + \frac{G\pi}{8} - \frac{35}{64}\zeta(3),$$

$$\int_0^{\pi/4} \log(\cos x + \sin x)\,\mathrm{d}x = \frac{G}{2} - \frac{\pi}{8}\log 2, \quad \int_0^{\pi/4} \log(\cos x - \sin x)\,\mathrm{d}x = -\frac{G}{2} - \frac{\pi}{8}\log 2,$$

$$\int_0^{\pi/4} x\log(\cos x+\sin x)\,\mathrm{d}x = \frac{21}{128}\zeta(3) - \frac{\pi^2}{64}\log 2, \quad \int_0^{\pi/4} x\log(\cos x-\sin x)\,\mathrm{d}x = -\frac{35}{128}\zeta(3) - \frac{\pi^2}{64}\log 2,$$

$$\int_0^{\pi/4} \log(\tan x + \cot x)\,\mathrm{d}x = \frac{\pi}{2}\log 2, \quad \int_0^{\pi/4} x\log(\tan x + \cot x)\,\mathrm{d}x = \frac{\pi^2}{16}\log 2 - \frac{7}{64}\zeta(3),$$

$$\int_0^{\pi/4} \log(\cot x - \tan x)\,\mathrm{d}x = \frac{\pi}{4}\log 2, \quad \int_0^{\pi/4} x\log(\cot x - \tan x)\,\mathrm{d}x = \frac{\pi^2}{32}\log 2 - \frac{7}{32}\zeta(3),$$

$$\int_0^{\pi/4} \log(\sqrt{\tan x} + \sqrt{\cot x})\,\mathrm{d}x = \frac{\pi}{8}\log 2 + \frac{G}{2}, \quad \int_0^{\pi/4} \log(\sqrt{\tan x} + \sqrt{\cot x})\,\mathrm{d}x = \frac{\pi}{8}\log 2 + \frac{G}{2},$$

$$\int_0^{\pi/4} \log(\sqrt{\tan x} - \sqrt{\cot x})^2\,\mathrm{d}x = \frac{\pi}{4}\log 2 - G, \quad \int_0^{\pi/4} x\log(\sqrt{\tan x} - \sqrt{\cot x})^2\,\mathrm{d}x = \frac{\pi^2}{32}\log 2 - \frac{21}{32}\zeta(3).$$

Fact 14.7.2.

$$\int_0^{\pi/3} \log^2\left(2\sin\frac{x}{2}\right)\mathrm{d}x = \frac{7\pi^3}{108}, \quad \int_0^{\pi/3} x\log^2\left(2\sin\frac{x}{2}\right)\mathrm{d}x = \frac{17\pi^4}{6480},$$

$$\int_0^{\pi/3} \left(x - \frac{\pi}{3}\right)\log\left(2\sin\frac{x}{2}\right)\mathrm{d}x = \frac{2}{3}\zeta(3).$$

Source: [2513, p. 581] and [3022].

Fact 14.7.3.

$$\int_0^{\pi/2} \log\sin x\,\mathrm{d}x = \int_0^{\pi/2} \log\cos x\,\mathrm{d}x = -\frac{\pi}{2}\log 2, \quad \int_0^{\pi/2} \log\tan x\,\mathrm{d}x = 0,$$

$$\int_0^{\pi/2} x\log\sin x\,\mathrm{d}x = \frac{7}{16}\zeta(3) - \frac{\pi^2}{8}\log 2, \quad \int_0^{\pi/2} x^2\log\sin x\,\mathrm{d}x = \frac{3\pi}{16}\zeta(3) - \frac{\pi^3}{24}\log 2,$$

$$\int_0^{\pi/2} (\sin x)\log\sin x\,\mathrm{d}x = \log 2 - 1,$$

$$\int_0^{\pi/2} x\log\cos x\,\mathrm{d}x = -\frac{7}{16}\zeta(3) - \frac{\pi^2}{8}\log 2, \quad \int_0^{\pi/2} x^2\log\cos x\,\mathrm{d}x = -\frac{\pi}{4}\zeta(3) - \frac{\pi^3}{24}\log 2,$$

$$\int_0^{\pi/2} x\log\tan x\,\mathrm{d}x = \frac{7}{8}\zeta(3), \quad \int_0^{\pi/2} x^2\log\tan x\,\mathrm{d}x = \frac{7\pi}{16}\zeta(3), \quad \int_0^{\pi/2} \frac{\log\sec x}{\tan x}\,\mathrm{d}x = \frac{\pi^2}{24},$$

$$\int_0^{\pi/2} \frac{\log^2\sec x}{\tan x}\,\mathrm{d}x = \frac{1}{4}\zeta(3), \quad \int_0^{\pi/2} \frac{\log^3\sec x}{\tan x}\,\mathrm{d}x = \frac{\pi^4}{240}, \quad \int_0^{\pi/2} \frac{\log^4\sec x}{\tan x}\,\mathrm{d}x = \frac{3}{4}\zeta(5),$$

$$\int_0^{\pi/2} \log(1 + \sin x)\,\mathrm{d}x = 2G - \frac{\pi}{2}\log 2, \quad \int_0^{\pi/2} \log(1 - \sin x)\,\mathrm{d}x = -2G - \frac{\pi}{2}\log 2,$$

$$\int_0^{\pi/2} \log(1 + \tan x)\,\mathrm{d}x = \frac{\pi}{4}\log 2 + G, \quad \int_0^{\pi/2} x\log(1 + \tan x)\,\mathrm{d}x = \frac{\pi^2}{16}\log 2 + \frac{G\pi}{4} + \frac{7}{16}\zeta(3),$$

$$\int_0^{\pi/2} \log(\cos x + \sin x)\,\mathrm{d}x = G - \frac{\pi}{4}\log 2, \quad \int_0^{\pi/2} x\log(\cos x + \sin x)\,\mathrm{d}x = \frac{G\pi}{4} - \frac{\pi^2}{16}\log 2,$$

$$\int_0^{\pi/2} \log(\tan x + \cot x)\,\mathrm{d}x = \pi\log 2, \quad \int_0^{\pi/2} x\log(\tan x + \cot x)\,\mathrm{d}x = \frac{\pi^2}{4}\log 2,$$

$$\int_0^{\pi/2} x^2 \log(\tan x + \cot x)\,\mathrm{d}x = \frac{\pi^3}{12}\log 2 + \frac{\pi}{16}\zeta(3),$$

$$\int_0^{\pi/2} x^3 \log(\tan x + \cot x)\,\mathrm{d}x = \frac{\pi^4}{32}\log 2 + \frac{3\pi}{64}\zeta(3), \quad \int_0^{\pi/2} \log\frac{1-\cos x}{1+\cos x}\,\mathrm{d}x = -4G,$$

$$\int_0^{\pi/2} \log\,(\tan x - \cot x)^2\,\mathrm{d}x = \pi\log 2, \quad \int_0^{\pi/2} x\log\,(\tan x - \cot x)^2\,\mathrm{d}x = \frac{\pi^2}{4}\log 2,$$

$$\int_0^{\pi/2} \log(\sqrt{\tan x} + \sqrt{\cot x})\,\mathrm{d}x = \frac{\pi}{4}\log 2 + G, \quad \int_0^{\pi/2} \log(\sqrt{\tan x} + \sqrt{\cot x})\,\mathrm{d}x = \frac{\pi^2}{16}\log 2 + \frac{G\pi}{4},$$

$$\int_0^{\pi/2} \log\,(\sqrt{\tan x} - \sqrt{\cot x})^2\,\mathrm{d}x = \frac{\pi}{2}\log 2 - 2G, \quad \int_0^{\pi/2} x\log\,(\sqrt{\tan x} - \sqrt{\cot x})^2\,\mathrm{d}x = \frac{\pi^2}{8}\log 2 - \frac{G\pi}{4}.$$

Furthermore, the following statements hold:

i) Let $n \ge 0$. Then,

$$\int_0^{\pi/2} \log^{2n+1}\tan x\,\mathrm{d}x = 0.$$

ii) Let $a > 0$. Then,

$$\int_0^{\pi/2} \log(1 + a\sin^2 x)\,\mathrm{d}x = \int_0^{\pi/2} \log(1 + a\cos^2 x)\,\mathrm{d}x = \pi\log\tfrac{1}{2}(\sqrt{a+1} + 1).$$

iii) Let $a > 0$. Then,

$$\int_0^{\pi/2} \log^2(a\sin x)\,\mathrm{d}x = \int_0^{\pi/2} \log^2(a\cos x)\,\mathrm{d}x = \frac{\pi^3}{24} + \frac{\pi}{2}\log^2\frac{2}{a},$$

$$\int_0^{\pi/2} [\log(a\sin x)]\log(a\cos x)\,\mathrm{d}x = \frac{\pi}{2}\log^2\frac{2}{a} - \frac{\pi^3}{48}.$$

iv) Let $a \in [-1, 1]$. Then,

$$\log(a^2 - \sin^2 x)^2\,\mathrm{d}x = -2\pi\log 2.$$

v) Let a and b be real numbers such that $0 < b < a$. Then,

$$\int_0^{\pi/2} (\csc x)\log\frac{a + b\sin x}{a - b\sin x}\,\mathrm{d}x = \frac{\pi}{2}\operatorname{asin}\frac{b}{a}.$$

vi) Let a and b be positive numbers. Then,

$$\int_0^{\pi/2} (a^2\cos^2 x + b^2\sin^2 x)\,\mathrm{d}x = \pi\log\tfrac{1}{2}(a+b).$$

vii) Let a and b be positive numbers. Then,

$$\int_0^{\pi/2} (a^2 + b^2\tan^2 x)\,\mathrm{d}x = \pi\log(a+b).$$

Source: [441], [511, pp. 242, 260], [1890], [2513, pp. 570, 581], and [3022]. *i)* is given in [1217, p. 533]; *ii)* given in [1217, p. 532] and [2013, pp. 70, 71]; *iii)* is given in [2106, p. 236]; *iv)* is given in [1217, p. 532]; *v)* is given in [701]; *vi)* is given in [1217, p. 532]; *vii)* is given in [1217, p. 534].

Fact 14.7.4.

$$\int_0^{\pi} \log\sin\frac{x}{2}\,\mathrm{d}x = \int_0^{\pi} \log\sin x\,\mathrm{d}x = -\pi\log 2, \quad \int_0^{\pi} x\log\sin x\,\mathrm{d}x = -\frac{\pi^2}{2}\log 2,$$

$$\int_0^\pi \log^2 \sin x\,\mathrm{d}x = \frac{\pi^3}{12} + \pi\log^2 2, \quad \int_0^\pi \log^3 \sin x\,\mathrm{d}x = -\frac{3\pi}{2}\zeta(3) - \pi\log^3 2 - \frac{\pi^3}{4}\log 2,$$

$$\int_0^\pi \log^4 \sin x\,\mathrm{d}x = 6\pi\zeta(3)\log 2 + \pi\log^4 2 + \frac{\pi^3}{2}\log^2 2 + \frac{19\pi^5}{240},$$

$$\int_0^\pi \log\left(2\sin\frac{x}{2}\right)\mathrm{d}x = 0, \quad \int_0^\pi \log^2\left(2\sin\frac{x}{2}\right)\mathrm{d}x = \frac{\pi^3}{12},$$

$$\int_0^\pi \log^3\left(2\sin\frac{x}{2}\right)\mathrm{d}x = -\frac{3\pi}{2}\zeta(3), \quad \int_0^\pi \log^4\left(2\sin\frac{x}{2}\right)\mathrm{d}x = \frac{19\pi^5}{240},$$

$$\int_0^\pi \log(1+\sin x)\,\mathrm{d}x = 4G - \pi\log 2, \quad \int_0^\pi \log(1-\sin x)\,\mathrm{d}x = -4G - \pi\log 2,$$

$$\int_0^\pi x\log(1+\sin x)\,\mathrm{d}x = 2\pi G - \frac{\pi^2}{2}\log 2, \quad \int_0^\pi x\log(1-\sin x)\,\mathrm{d}x = -2\pi G - \frac{\pi^2}{2}\log 2,$$

$$\int_0^\pi \log(1+\cos x)\,\mathrm{d}x = \int_0^\pi \log(1-\cos x)\,\mathrm{d}x = -\pi\log 2,$$

$$\int_0^\pi x\log(1+\cos x)\,\mathrm{d}x = -\frac{7}{2}\zeta(3) - \frac{\pi^2}{2}\log 2, \quad = \int_0^\pi x\log(1-\cos x)\,\mathrm{d}x = \frac{7}{2}\zeta(3) - \frac{\pi^2}{2}\log 2,$$

$$\int_0^\pi x\log\cos^2 x\,\mathrm{d}x = -\pi^2\log 2, \quad \int_0^\pi x\log\tan^2 x\,\mathrm{d}x = 0,$$

$$\int_0^\pi x^2\log^2\left(2\cos\frac{x}{2}\right)\mathrm{d}x = \int_0^\pi (\pi - x)^2\log^2\left(2\sin\frac{x}{2}\right)\mathrm{d}x = \frac{11\pi^5}{180},$$

$$\int_{-\pi/2}^{\pi/2} \log\cos x\,\mathrm{d}x = -\pi\log 2, \quad \int_{-\pi/2}^{\pi/2} \log(1-\cos x)\,\mathrm{d}x = -\pi\log 2 - 4G.$$

Furthermore, the following statements hold:

i) Let $a \in \mathbb{R}$. If $|a| \le 1$, then

$$\int_0^\pi \log(a^2 + 1 - 2a\cos x)\,\mathrm{d}x = 0.$$

If $|a| \ge 1$, then

$$\int_0^\pi \log(a^2 + 1 - 2a\cos x)\,\mathrm{d}x = 2\pi\log|a|.$$

ii) Let a, b be real numbers such that $0 < |b| \le a$. Then,

$$\int_0^\pi \log(a + b\cos x)\,\mathrm{d}x = \pi\log\tfrac{1}{2}(a + \sqrt{a^2 - b^2}).$$

iii) Let a, b be positive numbers. Then,

$$\int_0^\pi \log(a^2 + b^2 - 2ab\cos x)\,\mathrm{d}x = 2\pi\log\max\{a, b\}.$$

iv) Let a be a real number, and assume that $|a| \ge 1$. Then,

$$\int_0^\pi \log|1 + a\cos x|\,\mathrm{d}x = \frac{\pi}{2}\log\frac{a^2}{4}.$$

In particular,

$$\int_0^\pi \log|1 \pm 2\cos x|\,\mathrm{d}x = 0, \quad \int_0^\pi \log|1 \pm 3\cos x|\,\mathrm{d}x = \frac{\pi}{2}\log\frac{9}{4}.$$

Furthermore,

$$\int_0^{2\pi/3} \log(1+2\cos x)\,\mathrm{d}x = -\int_{2\pi/3}^{\pi} \log|1+2\cos x|\,\mathrm{d}x \approx 0.676628.$$

v) Let $a \in [1,\infty)$. Then,

$$\int_0^{\pi} (1+\cos x)\log(a+\cos x)\,\mathrm{d}x = [a-\sqrt{a^2-1}+\log\tfrac{1}{2}(a+\sqrt{a^2-1})]\pi.$$

vi) Let $a \in (-1,1)$. Then,

$$\int_0^{\pi} (\sec x)\log(1+a\cos x)\,\mathrm{d}x = \pi \operatorname{asin} a.$$

vii) Let $a \in [-1,1]$. Then,

$$\int_0^{\pi} \log(1+a\cos x)^2\,\mathrm{d}x = 2\pi\log\tfrac{1}{2}(1+\sqrt{1-a^2}).$$

viii) Let $a \geq -1$. Then,

$$\int_0^{\pi} \log(1+a\sin^2 x)\,\mathrm{d}x = \int_0^{\pi} \log(1+a\cos^2 x)\,\mathrm{d}x = 2\pi\log\tfrac{1}{2}(1+\sqrt{a+1}),$$

$$\int_0^{\pi} x\log(1+a\sin^2 x)\,\mathrm{d}x = \int_0^{\pi} x\log(1+a\cos^2 x)\,\mathrm{d}x = \pi^2\log\tfrac{1}{2}(1+\sqrt{a+1}).$$

ix) Let a and b be positive numbers. Then,

$$\int_0^{\pi} (a^2\cos^2 x + b^2\sin^2 x)\,\mathrm{d}x = 2\pi\log\tfrac{1}{2}(a+b).$$

x) Let a and b be positive numbers. Then,

$$\int_0^{\pi} (a^2 + b^2\tan^2 x)\,\mathrm{d}x = 2\pi\log(a+b).$$

Source: [441], [516, p. 58], [1101], [1568, p. 167], [2013, pp. 185, 186], and [2513, p. 193]. *i)–iii)* are given in [1158, pp. 155, 542, 543] and [1217, p. 531]. *iv)* corrects a misprint in [1217, p. 531] and [1524, p. 274]; *v)* is given in [2962]; *vi)* is given in [2106, p. 97]; *vii)–x)* are given in [1217, pp. 531–534].

Fact 14.7.5.

$$\int_0^{\infty} \frac{\log\cos^2 x}{x^2}\,\mathrm{d}x = -\pi, \qquad \int_0^{\infty} \frac{\log\cos^2 x}{1+e^{2x}}\,\mathrm{d}x = -\frac{1}{2}\log^2 2.$$

Source: [1217, p. 569] and [1685].

Fact 14.7.6.

$$\int_0^{\infty} \frac{1}{x}\log\left(\frac{1+\sin x}{1-\sin x}\right)^2\mathrm{d}x = \pi^2, \qquad \int_0^{\infty} \frac{1}{x}\log\left(\frac{1+\tan x}{1-\tan x}\right)^2\mathrm{d}x = \frac{\pi^2}{2}.$$

Let $a > 0$ and $b > 0$. Then,

$$\int_0^{\infty} \frac{\log\sin^2 ax}{x^2+b^2}\,\mathrm{d}x = \frac{\pi}{b}\log\frac{1-e^{-2ab}}{2}, \qquad \int_0^{\infty} \frac{\log\cos^2 ax}{x^2+b^2}\,\mathrm{d}x = \frac{\pi}{b}\log\frac{1+e^{-2ab}}{2},$$

$$\int_0^{\infty} \frac{\log\tan^2 ax}{x^2+b^2}\,\mathrm{d}x = \frac{\pi}{b}\tanh ab, \qquad \int_0^{\infty} \frac{x\sin x}{(\cosh 2a-\cos x)(x^2+b^2)}\,\mathrm{d}x = \frac{\pi}{e^{2a+b}-1},$$

$$\int_0^\infty \frac{1}{(\sin^2 x + \sinh^2 a)(x^2+b^2)}\,\mathrm{d}x = \frac{\pi(e^{2a}+e^{-2b})}{b(\sinh 2a)(e^{2a}-e^{-2b})}.$$

Source: [1217, pp. 568, 569] and [1684].

Fact 14.7.7. Let $a > 0$. Then,

$$\int_0^\infty \frac{(\log x)\sin ax}{x}\,\mathrm{d}x = -\frac{\pi}{2}(\gamma + \log a), \quad \int_0^\infty \frac{(\log x)\sin^2 ax}{x^2}\,\mathrm{d}x = -\frac{a\pi}{2}(\gamma + \log 2a - 1),$$

$$\int_0^\infty \frac{(\log^2 x)\sin ax}{x}\,\mathrm{d}x = \frac{\pi\gamma^2}{2} + \frac{\pi^3}{24} + \pi\gamma\log a + \frac{\pi}{2}\log^2 a.$$

Source: [1217, p. 594].

Fact 14.7.8. Let $a > 0$ and $b > 0$. Then,

$$\int_0^\infty \frac{(\log x)(\cos ax - \cos bx)}{x}\,\mathrm{d}x = (\gamma + \tfrac{1}{2}\log ab)\log\frac{a}{b},$$

$$\int_0^\infty \frac{(\log x)(\cos ax - \cos bx)}{x^2}\,\mathrm{d}x = \frac{\pi}{2}[(a-b)(\gamma-1) + a\log a - b\log b].$$

Source: [1217, p. 594].

Fact 14.7.9. Let $a > 0$ and $b > 0$. Then,

$$\int_0^\infty \frac{(\log x)(\sin bx)}{e^{ax}}\,\mathrm{d}x = \frac{1}{a^2+b^2}\left(a\,\mathrm{atan}\,\frac{b}{a} - b\gamma - \frac{b}{2}\log(a^2+b^2)\right),$$

$$\int_0^\infty \frac{(\log x)(\cos bx)}{e^{ax}}\,\mathrm{d}x = -\frac{1}{a^2+b^2}\left(b\,\mathrm{atan}\,\frac{b}{a} + a\gamma + \frac{a}{2}\log(a^2+b^2)\right).$$

Source: [1217, p. 594].

Fact 14.7.10. Let $a > 0$. Then,

$$\int_0^\infty (\log x)\sin ax^2\,\mathrm{d}x = -\frac{1}{8}\sqrt{\frac{2\pi}{a}}\left(\log 4a + \gamma - \frac{\pi}{2}\right),$$

$$\int_0^\infty (\log x)\cos ax^2\,\mathrm{d}x = -\frac{1}{8}\sqrt{\frac{2\pi}{a}}\left(\log 4a + \gamma + \frac{\pi}{2}\right),$$

$$\int_0^\infty (\log x)\sin ax^3\,\mathrm{d}x = -\frac{1}{108\sqrt[3]{a}}(6\gamma - 2\sqrt{3}\pi + 9\log 3 + 6\log a)\Gamma(\tfrac{1}{3}),$$

$$\int_0^\infty (\log x)\cos ax^3\,\mathrm{d}x = -\frac{1}{36\sqrt[3]{a}}[2\pi + 2\sqrt{3}(a+\gamma) + 3\sqrt{3}\log 3]\Gamma(\tfrac{1}{3}),$$

$$\int_0^\infty (\log x)\sin x^4\,\mathrm{d}x = \frac{1}{32}\left(\sqrt{\frac{2-\sqrt{2}}{2}}\pi - \sqrt{2-\sqrt{2}}(\gamma + 3\log 2)\right)\Gamma(\tfrac{1}{4}),$$

$$\int_0^\infty (\log x)\cos x^4\,\mathrm{d}x = -\frac{1}{64}\left(\sqrt{2-\sqrt{2}}\pi + \sqrt{2+\sqrt{2}}(2\gamma + \pi + 6\log 2)\right)\Gamma(\tfrac{1}{4}).$$

Source: [1217, pp. 593]. The second equality corrects a misprint.

Fact 14.7.11. Let $n \ge 1$. Then,

$$\int_0^\infty \log^n \tanh x\,\mathrm{d}x = (-1)^n n!(1 - 2^{-n-1})\zeta(n+1).$$

In particular,

$$\int_0^\infty \log\tanh x\,\mathrm{d}x = -\frac{\pi^2}{8},\quad \int_0^\infty \log^2\tanh x\,\mathrm{d}x = \frac{7}{4}\zeta(3),\quad \int_0^\infty \log^3\tanh x\,\mathrm{d}x = -\frac{\pi^4}{16}.$$

Fact 14.7.12.

$$\int_0^\infty \frac{\log x}{\cosh x}\mathrm{d}x = \frac{\pi}{2}\log\frac{4\pi^3}{\Gamma^4(\frac{1}{4})},\quad \int_0^\infty \frac{\log x}{\cosh^2 x}\mathrm{d}x = \log\frac{\pi}{4} - \gamma,\quad \int_0^\infty \frac{\log x}{\cosh^3 x}\mathrm{d}x = \frac{\pi}{4}\log\frac{4\pi^3}{\Gamma^4(\frac{1}{4})} - \frac{2G}{\pi}.$$

Fact 14.7.13. Let $p, q \in \mathbb{R}[s]$, where $p(s) = b_n s^n + \cdots + b_1 s + b_0$ and $q(s) = a_n s^n + \cdots + a_1 s + a_0$, assume that $\deg p \le \deg q = n$, assume that $a_n + b_n \ne 0$ and $p + q$ is asymptotically stable, and define $\{\lambda_1, \ldots, \lambda_r\}_{\mathrm{ms}} \triangleq \operatorname{mroots}(q) \cap \mathrm{ORHP}$. Then,

$$\int_0^\infty \log\left|\frac{q(\omega j)}{q(\omega j) + p(\omega j)}\right| \mathrm{d}\omega = \pi\left(\frac{b_n a_{n-1} - a_n b_{n-1}}{2a_n(a_n + b_n)} + \sum_{i=1}^r \lambda_i\right).$$

In particular, if $\deg p \le (\deg q) - 1$, then

$$\int_0^\infty \log\left|\frac{q(\omega j)}{q(\omega j) + p(\omega j)}\right| \mathrm{d}\omega = \pi\left(\frac{-b_{n-1}}{2a_n} + \sum_{i=1}^r \lambda_i\right),$$

and, if $\deg p \le (\deg q) - 2$, then

$$\int_0^\infty \log\left|\frac{q(\omega j)}{q(\omega j) + p(\omega j)}\right| \mathrm{d}\omega = \pi\sum_{i=1}^r \lambda_i.$$

Source: [931, pp. 99–101], [1076], and [2422, pp. 54–56]. The cases $\deg q - \deg p \in \{0, 1\}$ are considered in [2422, pp. 54–56]. **Example:** For all $\alpha > 0$,

$$\int_0^\infty \log\left|\frac{(\omega j + 1)^2}{(\omega j + 1)^2 + \alpha}\right| \mathrm{d}\omega = 0,\quad \int_0^\infty \log\left|\frac{(\omega j + 2)(\omega j - 1)}{(\omega j + 2)(\omega j - 1) + 2 + \alpha}\right| \mathrm{d}\omega = \pi,$$

$$\int_0^\infty \log\left|\frac{(\omega j + 3)(\omega j - 1)^2}{(\omega j + 3)(\omega j - 1)^2 + (8 + \alpha)\omega j}\right| \mathrm{d}\omega = 2\pi.$$

Remark: This integral is the *Bode sensitivity integral*, which shows that the presence of open-right-half-plane poles in the loop transfer function $L \triangleq p/q \in \mathbb{R}(s)$ of an asymptotically stable servo loop with reference-to-error sensitivity transfer function $S \triangleq 1/(1 + L) = q/(q + p) \in \mathbb{R}(s)$ limits the achievable control-system performance.

Fact 14.7.14. Let $L \in \mathbb{R}(s)$, assume that L is strictly proper, and let $\omega_0 \in (0, \infty)$. Then,

$$\begin{aligned}\arg L(\omega_0 j) &= \frac{2\omega_0}{\pi}\int_0^\infty \frac{\log|L(\omega j)| - \log|L(\omega_0 j)|}{\omega^2 - \omega_0^2}\mathrm{d}\omega \\ &= \frac{2}{\pi}\int_0^\infty \frac{\log|L(\omega_0 e^x j)| - \log|L(\omega_0 j)|}{\sinh x}\mathrm{d}x \\ &= \frac{2}{\pi}\int_0^\infty \left(\frac{\mathrm{d}}{\mathrm{d}x}\log|L(\omega_0 e^x j)|\right)\log\coth\frac{x}{2}\,\mathrm{d}x.\end{aligned}$$

Source: [931, pp. 109–113]. **Example:** Let $L(s) = 1/s$ and $\omega_0 > 0$. Then,

$$\begin{aligned}\arg\frac{1}{\omega_0 j} &= \frac{2\omega_0}{\pi}\int_0^\infty \frac{\log\frac{\omega_0}{\omega}}{\omega^2 - \omega_0^2}\mathrm{d}\omega = -\frac{2}{\pi}\int_0^\infty \frac{\log x}{x^2 - 1}\mathrm{d}x = -\frac{2}{\pi}\int_0^\infty \frac{x}{\sinh x}\mathrm{d}x \\ &= -\frac{2}{\pi}\int_0^\infty \left(\frac{\mathrm{d}}{\mathrm{d}x}\log\omega_0 e^x\right)\log\coth\frac{x}{2}\,\mathrm{d}x = -\frac{2}{\pi}\int_0^\infty \log\coth\frac{x}{2}\,\mathrm{d}x = -\frac{\pi}{2}.\end{aligned}$$

Remark: This is the *Bode phase integral*, which relates the phase angle of the loop transfer function of a servo loop to the slope of the magnitude of the loop transfer function.

14.8 Facts on Definite Integrals of Exponential Functions

Fact 14.8.1. Let $n \geq 0$. Then,

$$\int_0^1 x^n e^{-x}\,\mathrm{d}x = \sum_{i=0}^{\infty}(-1)^i \frac{1}{(i+n+1)i!} = n! - \frac{a_n}{e}.$$

In particular,

$$\int_0^1 e^{-x}\,\mathrm{d}x = \sum_{i=0}^{\infty}(-1)^i \frac{1}{(i+1)i!} = 1 - \frac{1}{e}, \quad \int_0^1 xe^{-x}\,\mathrm{d}x = \sum_{i=0}^{\infty}(-1)^i \frac{1}{(i+2)i!} = 1 - \frac{2}{e},$$

$$\int_0^1 x^2e^{-x}\,\mathrm{d}x = \sum_{i=0}^{\infty}(-1)^i \frac{1}{(i+3)i!} = 2 - \frac{5}{e}, \quad \int_0^1 x^3e^{-x}\,\mathrm{d}x = \sum_{i=0}^{\infty}(-1)^i \frac{1}{(i+4)i!} = 6 - \frac{16}{e},$$

$$\int_0^1 x^4e^{-x}\,\mathrm{d}x = \sum_{i=0}^{\infty}(-1)^i \frac{1}{(i+5)i!} = 24 - \frac{65}{e}, \quad \int_0^1 x^5e^{-x}\,\mathrm{d}x = \sum_{i=0}^{\infty}(-1)^i \frac{1}{(i+5)i!} = 120 - \frac{326}{e}.$$

Source: [116, p. 7]. **Remark:** a_n is the nth arrangement number. See Fact 1.18.1. **Related:** Fact 14.8.9.

Fact 14.8.2. Let $n \geq 1$. Then,

$$\int_0^\infty [1-(1-e^{-x})^n]\,\mathrm{d}x = H_n.$$

Fact 14.8.3. Let $n \geq 0$. Then,

$$\int_0^\pi e^x \sin nx\,\mathrm{d}x = \frac{n}{n^2+1}[(-1)^{n+1}e^\pi + 1], \quad \int_0^\pi e^x \cos nx\,\mathrm{d}x = \frac{1}{n^2+1}[(-1)^n e^\pi - 1].$$

Source: [821].

Fact 14.8.4. Let $a \in \mathbb{R}$ and $n \geq 0$. Then,

$$\int_0^\pi e^{ax}\sin^{2n} x\,\mathrm{d}x = \frac{(2n)!(e^{a\pi}-1)}{a\prod_{i=0}^{n-1}[a^2+4(n-i)^2]},$$

$$\int_0^\pi e^{ax}\sin^{2n+1} x\,\mathrm{d}x = \frac{(2n+1)!(e^{a\pi}+1)}{(1+a^2)\prod_{i=0}^{n-1}[a^2+(2n-2i+1)^2]}.$$

In particular,

$$\int_0^\pi e^{ax}\,\mathrm{d}x = \frac{e^{a\pi}-1}{a}, \quad \int_0^\pi e^{ax}\sin x\,\mathrm{d}x = \frac{e^{a\pi}+1}{a^2+1},$$

$$\int_0^\pi e^{ax}\sin^2 x\,\mathrm{d}x = \frac{2(e^{a\pi}-1)}{a(a^2+4)}, \quad \int_0^\pi e^{ax}\sin^3 x\,\mathrm{d}x = \frac{6(e^{a\pi}+1)}{a^4+10a^2+9}.$$

Source: [821].

Fact 14.8.5. Let $x \in \mathbb{C}^n$ and $y \in \mathbb{C}^n$. Then,

$$\int_{-\pi}^\pi e^{-tJ}\|x + e^{tJ}y\|_2^2\,\mathrm{d}t = 2\pi x^*y.$$

Remark: This is the integral analogue of x) of Fact 11.8.3.

Fact 14.8.6.

$$\sum_{i=1}^{\infty}\frac{1}{\Gamma(i+\frac{1}{2})}=\frac{2e}{\sqrt{\pi}}\int_0^1 e^{-x^2}\,\mathrm{d}x,\quad \sum_{i=0}^{\infty}(-1)^i\frac{1}{\Gamma(i+\frac{3}{2})}=\frac{2}{\sqrt{\pi}e}\int_0^1 e^{x^2}\,\mathrm{d}x.$$

Source: [1904].

Fact 14.8.7.

$$\int_0^1 \sin(x\log x)\,\mathrm{d}x=\sum_{i=1}^{\infty}\frac{(-1)^i}{(2i)^{2i}}\approx -0.246115,\quad \int_0^1 \cos(x\log x)\,\mathrm{d}x=\sum_{i=1}^{\infty}\frac{(-1)^{i+1}}{(2i-1)^{2i-1}}\approx 0.963281,$$

$$\int_0^1 \sinh(x\log x)\,\mathrm{d}x=-\sum_{i=1}^{\infty}\frac{1}{(2i)^{2i}}\approx -0.253927,\quad \int_0^1 \cosh(x\log x)\,\mathrm{d}x=\sum_{i=1}^{\infty}\frac{1}{(2i-1)^{2i-1}}\approx 1.03735.$$

Source: [1179]. **Related:** Fact 14.2.2.

Fact 14.8.8. Let $n\ge 0$ and $a>0$. Then,

$$\int_0^\infty \frac{x^n}{a^x}\,\mathrm{d}x=\frac{n!}{\log^{n+1}a},\quad \int_0^\infty \frac{x^n}{e^{ax}}\,\mathrm{d}x=\frac{n!}{a^{n+1}}.$$

In particular,

$$\int_0^\infty \frac{1}{a^x}\,\mathrm{d}x=\frac{1}{\log a},\quad \int_0^\infty \frac{x}{a^x}\,\mathrm{d}x=\frac{1}{\log^2 a},\quad \int_0^\infty \frac{x^2}{a^x}\,\mathrm{d}x=\frac{2}{\log^3 a},$$

$$\int_0^\infty \frac{1}{e^{ax}}\,\mathrm{d}x=\frac{1}{a},\quad \int_0^\infty \frac{x}{e^{ax}}\,\mathrm{d}x=\frac{1}{a^2},\quad \int_0^\infty \frac{x^2}{e^{ax}}\,\mathrm{d}x=\frac{2}{a^3}.$$

Fact 14.8.9. Let $n\ge 1$. Then,

$$\int_0^\infty x^n e^{-x}\,\mathrm{d}x=n!,\quad \int_1^\infty x^n e^{-x}\,\mathrm{d}x=\frac{\lfloor en!\rfloor}{e}.$$

Source: [1345]. **Related:** Fact 14.8.1.

Fact 14.8.10. Let $a\in(0,\infty)$ and $b\in(-1,\infty)$. Then,

$$\int_0^\infty x^b e^{-ax}\,\mathrm{d}x=\frac{\Gamma(b+1)}{a^{b+1}}.$$

In particular,

$$\int_0^\infty \frac{e^{-ax}}{\sqrt{x}}\,\mathrm{d}x=\sqrt{\frac{\pi}{a}},\quad \int_0^\infty \sqrt{x}e^{-ax}\,\mathrm{d}x=\frac{1}{2}\sqrt{\frac{\pi}{a^3}}.$$

Fact 14.8.11. Let $a>0$. Then,

$$\int_0^\infty \frac{1}{\sqrt{e^{ax}-1}}\,\mathrm{d}x=\frac{\pi}{a},\quad \int_0^\infty \frac{x}{\sqrt{e^{ax}-1}}\,\mathrm{d}x=\frac{2\pi\log 2}{a^2},$$

$$\int_0^\infty \frac{x^2}{\sqrt{e^{ax}-1}}\,\mathrm{d}x=\frac{\pi^3+12\pi\log^2 2}{3a^3},\quad \int_0^\infty \frac{x^3}{\sqrt{e^{ax}-1}}\,\mathrm{d}x=\frac{12\pi\zeta(3)+8\pi\log^3 2+2\pi^3\log 2}{a^4}.$$

Fact 14.8.12. Let $n\ge 0$. Then,

$$\int_0^\infty \frac{xe^{-nx}}{\sqrt{e^x-1}}\,\mathrm{d}x=\frac{(2n)!\pi}{4^n(n!)^2}\left(H_n+2\log 2-\sum_{i=1}^{n}\frac{2}{2i-1}\right).$$

In particular,

$$\int_0^\infty \frac{xe^{-x}}{\sqrt{e^x-1}}\,dx = \pi\log 2 - \frac{\pi}{2}, \quad \int_0^\infty \frac{xe^{-2x}}{\sqrt{e^x-1}}\,dx = \frac{3\pi}{4}\log 2 - \frac{7\pi}{16}.$$

In addition,

$$\int_0^\infty \frac{x^2e^{-x}}{\sqrt{e^x-1}}\,dx = \frac{\pi^3}{6} - \pi + 2\pi\log^2 2 - 2\pi\log 2,$$

$$\int_0^\infty \frac{x^3e^{-x}}{\sqrt{e^x-1}}\,dx = \pi[6\zeta(3) + 4\log^3 2 - 6\log^2 2 - 6\log 2 - 3] + \pi^3(\log 2 - \tfrac{1}{2}).$$

Fact 14.8.13. Let $a > 0$ and $b \ge -1$. Then,

$$\int_0^\infty \frac{x^b}{e^{ax}+1}\,dx = \frac{(2^b-1)\Gamma(b+1)\zeta(b+1)}{2^b a^{b+1}}.$$

In particular,

$$\int_0^\infty \frac{1}{e^{ax}+1}\,dx = \frac{\log 2}{a}, \quad \int_0^\infty \frac{x}{e^{ax}+1}\,dx = \frac{\pi^2}{12a^2},$$

$$\int_0^\infty \frac{x^2}{e^{ax}+1}\,dx = \frac{3\zeta(3)}{2a^3}, \quad \int_0^\infty \frac{x^3}{e^{ax}+1}\,dx = \frac{7\pi^4}{120a^4},$$

$$\int_0^\infty \frac{x^4}{e^{ax}+1}\,dx = \frac{45\zeta(5)}{2a^5}, \quad \int_0^\infty \frac{x^5}{e^{ax}+1}\,dx = \frac{31\pi^6}{252a^6},$$

$$\int_0^\infty \frac{x^6}{e^{ax}+1}\,dx = \frac{2835\zeta(7)}{4a^7}, \quad \int_0^\infty \frac{x^7}{e^{ax}+1}\,dx = \frac{127\pi^8}{240a^8},$$

$$\int_0^\infty \frac{x^8}{e^{ax}+1}\,dx = \frac{80325\zeta(9)}{2a^9}, \quad \int_0^\infty \frac{x^9}{e^{ax}+1}\,dx = \frac{511\pi^{10}}{132a^{10}}.$$

Source: [511, p. 98]. **Remark:** $\lim_{b\to 0}(2^b-1)\Gamma(b+1)\zeta(b+1) = \log 2$.

Fact 14.8.14. Let $n \ge 0$. Then,

$$\int_0^\infty \frac{xe^{-nx}}{e^x+1}\,dx = (-1)^n\left(\frac{\pi^2}{12} + \sum_{i=1}^n (-1)^i\frac{1}{i^2}\right).$$

Furthermore, if $n \ge 1$, then

$$\int_0^\infty \frac{xe^{-x}}{e^{x/(2n)}+1}\,dx = n^2\left(\frac{\pi^2}{3} + H_{n,2} - \sum_{i=1}^n \frac{4}{(2i-1)^2}\right),$$

$$\int_0^\infty \frac{xe^{-x}}{e^{x/(2n-1)}+1}\,dx = (n-\tfrac{1}{2})^2\left(\sum_{i=1}^n \frac{4}{(2i-1)^2} - H_{n-1,2} - \frac{\pi^2}{3}\right).$$

In particular,

$$\int_0^\infty \frac{xe^{-x}}{e^x+1}\,dx = 1 - \frac{\pi^2}{12}, \quad \int_0^\infty \frac{xe^{-x}}{e^{x/2}+1}\,dx = \frac{\pi^2}{3} - 3, \quad \int_0^\infty \frac{xe^{-x}}{e^{x/3}+1}\,dx = \frac{31}{4} - \frac{3\pi^2}{4},$$

$$\int_0^\infty \frac{xe^{-x}}{e^{x/4}+1}\,dx = \frac{4\pi^2}{3} - \frac{115}{9}, \quad \int_0^\infty \frac{xe^{-x}}{e^{x/5}+1}\,dx = \frac{3019}{144} - \frac{25\pi^2}{12}, \quad \int_0^\infty \frac{xe^{-x}}{e^{x/6}+1}\,dx = 3\pi^2 - \frac{2919}{100}.$$

Furthermore,

$$\int_0^\infty \frac{xe^{-x}}{e^{2x}+1}\,dx = 1 - G.$$

Source: [1217, p. 354].

Fact 14.8.15. Let $a > 0$ and $b > 0$. Then,

$$\int_0^\infty \frac{x^b}{e^{ax}-1}\,\mathrm{d}x = \frac{\Gamma(b+1)\zeta(b+1)}{a^{b+1}}.$$

In particular,

$$\int_0^\infty \frac{x}{e^{ax}-1}\,\mathrm{d}x = \frac{\pi^2}{6a^2}, \quad \int_0^\infty \frac{x^2}{e^{ax}-1}\,\mathrm{d}x = \frac{2\zeta(3)}{a^3},$$

$$\int_0^\infty \frac{x^3}{e^{ax}-1}\,\mathrm{d}x = \frac{\pi^4}{15a^4}, \quad \int_0^\infty \frac{x^4}{e^{ax}-1}\,\mathrm{d}x = \frac{24\zeta(5)}{a^5}.$$

Hence, if $n \ge 1$, then

$$\int_0^\infty \frac{x^{2n-1}}{e^{ax}-1}\,\mathrm{d}x = (-1)^{n-1}\left(\frac{2\pi}{a}\right)^{2n}\frac{B_{2n}}{4n}.$$

Furthermore, if $m \ge 1$ and $n \ge 0$, then

$$\int_0^\infty \frac{x^m e^{-nx}}{e^x-1}\,\mathrm{d}x = m![\zeta(m+1) - H_{n,m+1}].$$

In particular,

$$\int_0^\infty \frac{x}{e^x-1}\,\mathrm{d}x = \frac{\pi^2}{6}, \quad \int_0^\infty \frac{xe^{-x}}{e^x-1}\,\mathrm{d}x = \frac{\pi^2}{6} - 1, \quad \int_0^\infty \frac{xe^{-2x}}{e^x-1}\,\mathrm{d}x = \frac{\pi^2}{6} - \frac{5}{4},$$

$$\int_0^\infty \frac{x^2}{e^x-1}\,\mathrm{d}x = 2\zeta(3), \quad \int_0^\infty \frac{x^2e^{-x}}{e^x-1}\,\mathrm{d}x = 2\zeta(3) - 2, \quad \int_0^\infty \frac{x^2e^{-2x}}{e^x-1}\,\mathrm{d}x = 2\zeta(3) - \frac{9}{4},$$

$$\int_0^\infty \frac{x^3}{e^x-1}\,\mathrm{d}x = \frac{\pi^4}{15}, \quad \int_0^\infty \frac{x^3e^{-x}}{e^x-1}\,\mathrm{d}x = \frac{\pi^4}{15} - 6, \quad \int_0^\infty \frac{x^3e^{-2x}}{e^x-1}\,\mathrm{d}x = \frac{\pi^4}{15} - \frac{51}{8}.$$

If $a > 0$, then

$$\int_0^\infty \frac{x}{2e^{ax}-1}\,\mathrm{d}x = \frac{\pi^2 - 6\log^2 2}{12a^2}, \quad \int_0^\infty \frac{x^2}{2e^{ax}-1}\,\mathrm{d}x = \frac{21\zeta(3) + 4\log^3 2 - \pi^2\log 4}{12a^3}.$$

Furthermore,

$$\int_0^\infty \frac{xe^{-x}}{e^{2x}-1}\,\mathrm{d}x = \frac{\pi^2}{8} - 1, \quad \int_0^\infty \frac{xe^{-x}}{e^{4x}-1}\,\mathrm{d}x = \frac{\pi^2}{16} + \frac{G}{2} - 1, \quad \int_0^\infty \frac{x^2e^{-x}}{e^{2x}-1}\,\mathrm{d}x = \frac{7}{4}\zeta(3) - 2,$$

$$\int_0^\infty \frac{x^2e^{-x}}{e^{3x}-1}\,\mathrm{d}x = \frac{4\sqrt{3}\pi^3}{243} + \frac{26}{27}\zeta(3) - 2, \quad \int_0^\infty \frac{x^2e^{-x}}{e^{4x}-1}\,\mathrm{d}x = \frac{\pi^3}{32} + \frac{7}{8}\zeta(3) - 2,$$

$$\int_0^\infty \frac{x^3e^{-x}}{e^{2x}-1}\,\mathrm{d}x = \frac{\pi^4}{16} - 6, \quad \int_0^\infty \frac{x^4e^{-x}}{e^x-1}\,\mathrm{d}x = 24\zeta(5) - 24,$$

$$\int_0^\infty \frac{x^4e^{-x}}{e^{2x}-1}\,\mathrm{d}x = \frac{93}{4}\zeta(5) - 24, \quad \int_0^\infty \frac{x^4e^{-x}}{e^{3x}-1}\,\mathrm{d}x = \frac{968}{81}\zeta(5) + \frac{16\sqrt{3}}{729}\pi^5 - 24,$$

$$\int_0^\infty \frac{x^5e^{-x}}{e^x-1}\,\mathrm{d}x = \frac{8\pi^6}{63} - 120, \quad \int_0^\infty \frac{x^5e^{-x}}{e^{2x}-1}\,\mathrm{d}x = \frac{\pi^6}{8} - 120.$$

Source: [511, p. 232]. **Remark:** $\int_0^\infty \frac{x^3}{e^x-1}\,\mathrm{d}x = \frac{\pi^4}{15}$ is *Planck's integral*. **Related:** Fact 14.6.11.

Fact 14.8.16. Let $n \ge 1$. Then,

$$\int_0^\infty \frac{x^{2n-1}}{e^{2\pi x}-1}\,\mathrm{d}x = \frac{\Gamma(2n)\zeta(2n)}{(2\pi)^{2n}} = \frac{(-1)^{n+1}B_{2n}}{4n}.$$

Source: [116, p. 29] and [645, p. 99].

Fact 14.8.17. Let $a, b > 0$. Then,

$$\int_0^\infty \frac{e^{-ax} - e^{-bx}}{x}\,\mathrm{d}x = \log\frac{b}{a}, \quad \int_0^\infty \frac{(e^{-ax} - e^{-bx})^2}{x^2}\,\mathrm{d}x = 2b\log\frac{b}{a} + 2(a+b)\log\frac{2a}{a+b},$$

$$\int_0^\infty \frac{(e^{-ax} - e^{-bx})^3}{x^3}\,\mathrm{d}x$$
$$= \frac{3}{2}[3b^2\log b - 3a^2\log 3a + 4a(a+b)\log(2a+b) + b^2\log[27(2a+b)] - (a+2b)^2\log(a+2b),$$

$$\int_0^\infty \frac{(1 - e^{-bx})^2}{x^2}\,\mathrm{d}x = 2b\log 2, \quad \int_0^\infty \frac{(1 - e^{-bx})^3}{x^3}\,\mathrm{d}x = \frac{3}{2}b^2\log\frac{27}{16}.$$

Source: [1158, p. 179].

Fact 14.8.18. Let a and b be real numbers. If either $0 < a < b$ or $b < a < 0$, then

$$\int_{-\infty}^\infty \frac{e^{-ax}}{1 + e^{-bx}}\,\mathrm{d}x = \frac{\pi}{b}\csc\frac{a\pi}{b}.$$

If $0 < a < 1$ and $b > 0$, then

$$\int_{-\infty}^\infty \frac{e^{-ax}}{b + e^{-x}}\,\mathrm{d}x = \frac{\pi}{b^{1-a}}\csc a\pi, \quad \int_{-\infty}^\infty \frac{e^{-ax}}{b - e^{-x}}\,\mathrm{d}x = \frac{\pi}{b^{1-a}}\cot a\pi.$$

Source: [1524, p. 270].

Fact 14.8.19. Let $a \in (0, 1)$. Then,

$$\int_{-\infty}^\infty \frac{e^{ax} - e^{(1-a)x}}{e^x - 1}\,\mathrm{d}x = -2\pi\cot a\pi.$$

Next, for all $n \ge 0$, define $Q_n \in \mathbb{R}[s]$ in Fact 13.2.14. If $n \ge 0$, then

$$\int_{-\infty}^\infty \frac{x^n e^{ax}}{e^x + 1}\,\mathrm{d}x = \pi^{n+1}(\csc a\pi)Q_n(-\cot a\pi).$$

In particular,

$$\int_{-\infty}^\infty \frac{e^{ax}}{e^x + 1}\,\mathrm{d}x = \pi\csc a\pi, \quad \int_{-\infty}^\infty \frac{xe^{ax}}{e^x + 1}\,\mathrm{d}x = -\pi^2(\csc a\pi)\cot a\pi,$$

$$\int_{-\infty}^\infty \frac{x^2 e^{ax}}{e^x + 1}\,\mathrm{d}x = \pi^3(\csc a\pi)(2\cot^2 a\pi + 1) = \tfrac{1}{2}\pi^3(\cos 2a\pi + 3)\csc^3 a\pi.$$

Source: [1430]. **Remark:** Ignoring removable singularities, it follows that, for all $n \ge 0$,

$$\int_{-\infty}^\infty \frac{x^n e^{ax}}{e^x - 1}\,\mathrm{d}x = \pi^{n+1}P_n(-\cot a\pi),$$

where $P_n \in \mathbb{R}[s]$ is defined in Fact 13.2.14. See [1430] and [2106, p. 305].

Fact 14.8.20.

$$\int_0^\infty \frac{xe^x}{(x+1)^2}\,\mathrm{d}x = \frac{e}{2} - 1.$$

Source: [1217, p. 341] and [1524, p. 270].

Fact 14.8.21. Let $a > 0$ and $b > -1$. Then,

$$\int_0^\infty x^b e^{-ax^2}\,\mathrm{d}x = \frac{\Gamma\left(\frac{b+1}{2}\right)}{2a^{(b+1)/2}}.$$

Consequently, if $n \ge 0$, then

$$\int_0^\infty x^{2n}e^{-ax^2}\,\mathrm{d}x = \frac{\Gamma\left(n+\frac{1}{2}\right)}{2a^{n+\frac{1}{2}}} = \frac{(2n)!\sqrt{\pi}}{2^{2n+1}n!a^{n+\frac{1}{2}}} = \frac{(2n-1)!!}{2^{n+1}a^n}\sqrt{\frac{\pi}{a}}, \quad \int_0^\infty x^{2n+1}e^{-ax^2}\,\mathrm{d}x = \frac{n!}{2a^{n+1}}.$$

In particular,

$$\int_0^\infty e^{-ax^2}\,\mathrm{d}x = \frac{1}{2}\sqrt{\frac{\pi}{a}}, \quad \int_0^\infty xe^{-ax^2}\,\mathrm{d}x = \frac{1}{2a}, \quad \int_0^\infty x^2e^{-ax^2}\,\mathrm{d}x = \frac{\sqrt{\pi}}{4a^{3/2}}.$$

Fact 14.8.22. Let $a \ge 0$ and $b \ge 0$. Then,

$$\int_0^\infty \frac{e^{-ax^2}-e^{-bx^2}}{x^2}\,\mathrm{d}x = \sqrt{b\pi}-\sqrt{a\pi}.$$

Source: [2106, p. 90].

Fact 14.8.23. Let $a > 0$, $b > -1$, and $c > 0$. Then,

$$\int_0^\infty x^be^{-ax^c}\,\mathrm{d}x = \frac{\Gamma\left(\frac{b+1}{c}\right)}{ca^{(b+1)/c}}, \quad \int_0^\infty x^be^{-ax}\,\mathrm{d}x = \frac{\Gamma(b+1)}{a^{b+1}}, \quad \int_0^\infty e^{-ax^c}\,\mathrm{d}x = \frac{\Gamma\left(\frac{c+1}{c}\right)}{a^{1/c}}.$$

In particular,

$$\int_0^\infty e^{-x^3}\,\mathrm{d}x = \Gamma(\tfrac{4}{3}), \quad \int_0^\infty \sqrt{x}e^{-x}\,\mathrm{d}x = \frac{\pi}{2}, \quad \int_0^\infty x^{3/2}e^{-x^4}\,\mathrm{d}x = \frac{1}{4}\Gamma(\tfrac{5}{8}).$$

Furthermore, if $m \ge 1$ and $n \ge 0$, then

$$\int_0^\infty x^ne^{-ax^m}\,\mathrm{d}x = \frac{\Gamma\left(\frac{n+1}{m}\right)}{ma^{(n+1)/m}}, \quad \int_0^\infty x^ne^{-ax}\,\mathrm{d}x = \frac{n!}{a^{n+1}}.$$

Furthermore,

$$\int_0^\infty \frac{e^{-ax}}{\sqrt{x}}\,\mathrm{d}x = \sqrt{\frac{\pi}{a}}, \quad \int_0^\infty e^{-ax}\,\mathrm{d}x = \frac{1}{a}, \quad \int_0^\infty \sqrt{x}e^{-ax}\,\mathrm{d}x = \frac{1}{2a}\sqrt{\frac{\pi}{a}}, \quad \int_0^\infty xe^{-ax}\,\mathrm{d}x = \frac{1}{a^2}.$$

Source: [352, p. 41], [1158, p. 178], and [2106, pp. 119, 125].

Fact 14.8.24. Let z be a complex number. Then,

$$e^{-z^2} = \frac{1}{\sqrt{\pi}}\int_{-\infty}^\infty e^{-x^2}e^{2zx\jmath}\,\mathrm{d}x.$$

Source: [116, p. 278].

Fact 14.8.25. Let $a > 0$ and $b \ge 0$. Then,

$$\int_0^\infty e^{-ax^2-b/x^2}\,\mathrm{d}x = \frac{1}{2}\sqrt{\frac{\pi}{a}}e^{-2\sqrt{ab}}, \quad \int_0^\infty \frac{e^{-ax-b/x}}{\sqrt{x}}\,\mathrm{d}x = \sqrt{\frac{\pi}{a}}e^{-2\sqrt{ab}}.$$

In particular,

$$\int_0^\infty e^{-ax^2}\,\mathrm{d}x = \frac{1}{2}\sqrt{\frac{\pi}{a}}, \quad \int_0^\infty \frac{e^{-ax}}{\sqrt{x}}\,\mathrm{d}x = \sqrt{\frac{\pi}{a}}.$$

Source: [2013, p. 200] and [2504, p. 113].

Fact 14.8.26. Let a and b be real numbers, and assume that $a > 0$. Then,

$$\int_{-\infty}^\infty e^{-ax^2+bx}\,\mathrm{d}x = \sqrt{\frac{\pi}{a}}\exp\left(\frac{b^2}{4a}\right).$$

Furthermore,

$$\int_{-\infty}^{\infty} xe^{-x^2-x}\,\mathrm{d}x = -\frac{1}{2}\sqrt{\pi\sqrt{e}}, \quad \int_{-\infty}^{\infty} x^2e^{-x^2-x}\,\mathrm{d}x = \frac{3}{4}\sqrt{\pi\sqrt{e}}.$$

Source: [1524, p. 271] and [2106, pp. 115, 116].

Fact 14.8.27. Let a, b, c be real numbers, and assume that $a > 0$. Then,

$$\int_0^{\infty} e^{-ax}\sin(bx+c)\,\mathrm{d}x = \frac{b\cos c + a\sin c}{a^2+b^2}, \quad \int_0^{\infty} e^{-ax}\cos(bx+c)\,\mathrm{d}x = \frac{a\cos c - b\sin c}{a^2+b^2}.$$

In particular,

$$\int_0^{\infty} e^{-ax}\sin bx\,\mathrm{d}x = \frac{b}{a^2+b^2}, \quad \int_0^{\infty} e^{-ax}\cos bx\,\mathrm{d}x = \frac{a}{a^2+b^2}.$$

Furthermore,

$$\int_0^{\infty} xe^{-ax}\sin bx\,\mathrm{d}x = \frac{2ab}{(a^2+b^2)^2}, \quad \int_0^{\infty} xe^{-ax}\cos bx\,\mathrm{d}x = \frac{a^2-b^2}{(a^2+b^2)^2}.$$

Fact 14.8.28. Let a and b be real numbers, and assume that $a > 0$. Then,

$$\int_0^{\infty} e^{-ax^2}\cos bx\,\mathrm{d}x = \frac{1}{2}\sqrt{\frac{\pi}{a}}\exp\left(-\frac{b^2}{4a}\right), \quad \int_0^{\infty} xe^{-ax^2}\sin bx\,\mathrm{d}x = \sqrt{\frac{\pi}{a^3}}\frac{b}{4}\exp\left(-\frac{b^2}{4a}\right),$$

$$\int_0^{\infty} x^2e^{-ax^2}\cos bx\,\mathrm{d}x = \sqrt{\frac{\pi}{a^5}}\frac{2a-b^2}{8}\exp\left(-\frac{b^2}{4a}\right),$$

$$\int_0^{\infty} x^3e^{-ax^2}\sin bx\,\mathrm{d}x = \sqrt{\frac{\pi}{a^7}}\frac{3(6ab-b^3)}{16}\exp\left(-\frac{b^2}{4a}\right).$$

Fact 14.8.29. Let $a > 0$ and $b > 0$. Then,

$$\int_0^{\infty} \frac{e^{-ax}\sin bx}{x}\,\mathrm{d}x = \operatorname{atan}\frac{b}{a},$$

$$\int_0^{\infty} \frac{e^{-ax}\sin bx}{\sqrt{x}}\,\mathrm{d}x = \sqrt{\frac{(\sqrt{a^2+b^2}-a)\pi}{2(a^2+b^2)}}, \quad \int_0^{\infty} \frac{e^{-ax}\cos bx}{\sqrt{x}}\,\mathrm{d}x = \sqrt{\frac{(\sqrt{a^2+b^2}+a)\pi}{2(a^2+b^2)}}.$$

Source: [1158, p. 178] and [1890].

Fact 14.8.30. Let a, b, c be real numbers, and assume that $a \ge 0$ and $a^2 + b^2 > 0$. Then,

$$\int_0^{\infty} e^{-ax^2}(\sin bx^2)\cos cx\,\mathrm{d}x = \frac{\sqrt{\pi}}{2\sqrt[4]{a^2+b^2}}e^{-c^2a/[4(a^2+b^2)]}\sin\left(\frac{1}{2}\operatorname{atan}\frac{b}{a} - \frac{c^2b}{4(a^2+b^2)}\right),$$

$$\int_0^{\infty} e^{-ax^2}(\cos bx^2)\cos cx\,\mathrm{d}x = \frac{\sqrt{\pi}}{2\sqrt[4]{a^2+b^2}}e^{-c^2a/[4(a^2+b^2)]}\cos\left(\frac{1}{2}\operatorname{atan}\frac{b}{a} - \frac{c^2b}{4(a^2+b^2)}\right).$$

In particular,

$$\int_0^{\infty} \sin|b|x^2\,\mathrm{d}x = \int_0^{\infty} \cos bx^2\,\mathrm{d}x = \frac{1}{2}\sqrt{\frac{\pi}{2|b|}}, \quad \int_0^{\infty} e^{-x^2}\cos cx\,\mathrm{d}x = \frac{\sqrt{\pi}}{2}e^{-c^2/4}.$$

Source: [352, p. 166] and [2013, p. 184].

Fact 14.8.31. Let $a \ge 1$ and $b > 0$. Then,

$$\int_0^{\infty} \frac{e^{-bx}-\cos x}{\sqrt[a]{x}}\,\mathrm{d}x = \frac{1}{b}\left(\sqrt[a]{b} - b\sin\frac{\pi}{2a}\right)\Gamma\left(\frac{a-1}{a}\right).$$

In particular,

$$\int_0^\infty \frac{e^{-x}-\cos x}{\sqrt[6]{x}}\,\mathrm{d}x = \frac{1}{4}(4+\sqrt{2}-\sqrt{6})\Gamma(\tfrac{5}{6}), \quad \int_0^\infty \frac{e^{-x}-\cos x}{\sqrt[5]{x}}\,\mathrm{d}x = \frac{1}{4}(5-\sqrt{5})\Gamma(\tfrac{4}{5}),$$

$$\int_0^\infty \frac{e^{-x}-\cos x}{\sqrt[4]{x}}\,\mathrm{d}x = \frac{1}{2}\left(2-\sqrt{2-\sqrt{2}}\right)\Gamma(\tfrac{3}{4}), \quad \int_0^\infty \frac{e^{-x}-\cos x}{\sqrt[3]{x}}\,\mathrm{d}x = \frac{1}{2}\Gamma(\tfrac{2}{3}),$$

$$\int_0^\infty \frac{e^{-x}-\cos x}{\sqrt{x}}\,\mathrm{d}x = \frac{1}{2}(2-\sqrt{2})\sqrt{\pi}, \quad \int_0^\infty \frac{e^{-x}-\cos x}{x}\,\mathrm{d}x = 0.$$

Fact 14.8.32. Let $a \ge 0$, $b \ge 0$, and $c, d \in \mathbb{R}$, and assume that $a^2+c^2 > 0$ and $b^2+d^2 > 0$. Then,

$$\int_0^\infty \frac{e^{-ax}\cos cx - e^{-bx}\cos dx}{x}\,\mathrm{d}x = \frac{1}{2}\log\frac{b^2+d^2}{a^2+c^2}.$$

Fact 14.8.33. Let a be a positive number. Then,

$$\int_0^\infty \frac{x}{e^{ax}-e^{-ax}}\,\mathrm{d}x = \frac{1}{2}\int_0^\infty \frac{x}{\sinh ax}\,\mathrm{d}x = \int_0^\infty \log\coth ax\,\mathrm{d}x = \frac{\pi^2}{8a},$$

$$\int_0^\infty \frac{x}{e^{ax}+e^{-ax}}\,\mathrm{d}x = \frac{1}{2}\int_0^\infty \frac{x}{\cosh ax}\,\mathrm{d}x = \frac{\pi}{2a}.$$

Fact 14.8.34. Let $a > 0$, and define

$$\phi(a) \triangleq \int_0^\infty \frac{\cos ax}{e^{2\pi\sqrt{x}}-1}\,\mathrm{d}x.$$

Then,

$$\int_0^\infty \frac{\sin ax}{e^{2\pi\sqrt{x}}-1}\,\mathrm{d}x = \phi(a) + \sqrt{\frac{2\pi^3}{a^3}}\phi\left(\frac{\pi^2}{a}\right) - \frac{1}{2a}, \quad \lim_{a\to\infty}\phi(a) = 0.$$

In particular,

$$\int_0^\infty \frac{1}{e^{2\pi\sqrt{x}}-1}\,\mathrm{d}x = \frac{1}{12}, \quad \int_0^\infty \frac{\cos\frac{\pi}{5}x}{e^{2\pi\sqrt{x}}-1}\,\mathrm{d}x = \frac{3}{2} + \frac{\sqrt{5}}{4} - \frac{5\sqrt{10}}{8}, \quad \int_0^\infty \frac{\cos\frac{2\pi}{5}x}{e^{2\pi\sqrt{x}}-1}\,\mathrm{d}x = \frac{1}{2} - \frac{3\sqrt{5}}{16},$$

$$\int_0^\infty \frac{\cos\frac{\pi}{2}x}{e^{2\pi\sqrt{x}}-1}\,\mathrm{d}x = \frac{1}{4\pi}, \quad \int_0^\infty \frac{\cos\frac{2\pi}{3}x}{e^{2\pi\sqrt{x}}-1}\,\mathrm{d}x = \frac{1}{3} + \frac{\sqrt{3}}{8\pi} - \frac{3\sqrt{3}}{16}, \quad \int_0^\infty \frac{\cos\pi x}{e^{2\pi\sqrt{x}}-1}\,\mathrm{d}x = \frac{1}{4} - \frac{\sqrt{2}}{8},$$

$$\int_0^\infty \frac{\cos 2\pi x}{e^{2\pi\sqrt{x}}-1}\,\mathrm{d}x = \frac{1}{16}, \quad \int_0^\infty \frac{\cos 4\pi x}{e^{2\pi\sqrt{x}}-1}\,\mathrm{d}x = \frac{3-\sqrt{2}}{32}, \quad \int_0^\infty \frac{\cos 6\pi x}{e^{2\pi\sqrt{x}}-1}\,\mathrm{d}x = \frac{13-4\sqrt{3}}{144},$$

$$\int_0^\infty \frac{\sin\frac{\pi}{2}x}{e^{2\pi\sqrt{x}}-1}\,\mathrm{d}x = \frac{1}{4} - \frac{3}{4\pi}, \quad \int_0^\infty \frac{\sin\pi x}{e^{2\pi\sqrt{x}}-1}\,\mathrm{d}x = \frac{\sqrt{2}}{8} - \frac{1}{2\pi}, \quad \int_0^\infty \frac{\sin 2\pi x}{e^{2\pi\sqrt{x}}-1}\,\mathrm{d}x = \frac{1}{16} - \frac{1}{8\pi}.$$

Furthermore,

$$\int_0^\infty \frac{x\cos\frac{\pi}{2}x}{e^{2\pi\sqrt{x}}-1}\,\mathrm{d}x = \frac{13}{8\pi^2} - \frac{1}{2\pi}, \quad \int_0^\infty \frac{x\cos 2\pi x}{e^{2\pi\sqrt{x}}-1}\,\mathrm{d}x = \frac{1}{64}\left(\frac{1}{2} - \frac{3}{\pi} + \frac{5}{\pi^2}\right),$$

$$\int_0^\infty \frac{x^2\cos 2\pi x}{e^{2\pi\sqrt{x}}-1}\,\mathrm{d}x = \frac{1}{256}\left(1 - \frac{5}{\pi} + \frac{5}{\pi^2}\right).$$

Source: [1317, pp. xxv, 66, 67].

Fact 14.8.35. Let $a, b \in \mathbb{R}$. If a is nonzero and b is positive, then

$$\int_0^\infty \frac{\sin ax}{e^{bx}-1}\,\mathrm{d}x = \frac{\pi}{2b}\coth\frac{a\pi}{b} - \frac{1}{2a}.$$

If a and b are positive, then

$$\int_0^\infty \frac{\sin ax}{e^{bx}+1}\,\mathrm{d}x = \frac{1}{2a} - \frac{\pi}{2b}\operatorname{csch}\frac{a\pi}{b}.$$

If a is positive, then

$$\int_0^\infty \frac{e^{x/2}\sin ax}{e^x-1}\,\mathrm{d}x = \frac{\pi}{2}\tanh a\pi.$$

Source: [1217, p. 489] and [1317, p. 56].

Fact 14.8.36. Let $a > 0$ and $b > 0$. Then,

$$\int_0^\infty \frac{(1-e^{-ax})\sin bx}{x}\,\mathrm{d}x = \frac{\pi}{2} - \operatorname{atan}\frac{b}{a}, \quad \int_0^\infty \frac{(1-e^{-ax})\sin bx}{x^2}\,\mathrm{d}x = \frac{b}{2}\log\left(\frac{a^2}{b^2}+1\right) + \frac{a\pi}{2} - a\operatorname{atan}\frac{a}{b},$$

$$\int_0^\infty \frac{(1-e^{-ax})\cos bx}{x}\,\mathrm{d}x = \frac{1}{2}\log\left(\frac{a^2}{b^2}+1\right).$$

Source: [1217, p. 501].

Fact 14.8.37. Let $a \in (0,\infty)$. Then,

$$\int_0^\infty \frac{\operatorname{atan}\frac{x}{a}}{e^{2\pi x}-1}\,\mathrm{d}x = \frac{1}{2}\log\Gamma(a) + \frac{a}{2} + \left(\frac{1}{4} - \frac{a}{2}\right)\log a - \frac{1}{4}\log 2\pi.$$

Furthermore,

$$\int_0^\infty \frac{\operatorname{atan} x^3}{e^{2\pi x}-1}\,\mathrm{d}x = \frac{1}{4}\log 2\pi - \frac{\sqrt{3}\pi}{12} - \frac{1}{2}\log(1+e^{-\sqrt{3}\pi}),$$

$$\int_0^\infty \frac{\operatorname{atan} x^3}{e^{4\pi x}-1}\,\mathrm{d}x = \frac{1}{8}\log 12\pi - \frac{\sqrt{3}\pi}{12} - \frac{1}{4}\log(1-e^{-2\sqrt{3}\pi}).$$

Source: [1317, p. 51].

Fact 14.8.38. Let a and b be real numbers, and assume that $|b| < a$. Then,

$$\int_0^\infty e^{-ax}\sinh bx\,\mathrm{d}x = \frac{b}{a^2-b^2}, \quad \int_0^\infty e^{-ax}\cosh bx\,\mathrm{d}x = \frac{a}{a^2-b^2}.$$

Fact 14.8.39. Let $z \in \mathbb{C}$, and assume that $\operatorname{Im} z \in (-\frac{1}{2}, \frac{1}{2})$. Then,

$$\int_{-\infty}^\infty \frac{e^{-2\pi z t \jmath}}{\cosh \pi t}\,\mathrm{d}t = \frac{1}{\cosh \pi z}.$$

Now, let $a \in (-1,1)$. Then,

$$\int_{-\infty}^\infty \frac{e^{-ax}}{\cosh x}\,\mathrm{d}x = \frac{\pi}{\cos a\pi/2}.$$

In particular,

$$\int_{-\infty}^\infty \frac{e^{-x/2}}{\cosh x}\,\mathrm{d}x = \sqrt{2}\pi, \quad \int_{-\infty}^\infty \frac{e^{-x/3}}{\cosh x}\,\mathrm{d}x = \frac{2\sqrt{3}\pi}{3}, \quad \int_{-\infty}^\infty \frac{e^{-x/4}}{\cosh x}\,\mathrm{d}x = \sqrt{4-2\sqrt{2}}\pi.$$

Furthermore, if $n \ge 0$, then

$$\int_0^\infty \frac{x^{2n}}{\cosh x}\,\mathrm{d}x = \left(\frac{\pi}{2}\right)^{2n+1}|E_{2n}|.$$

In particular,

$$\int_0^\infty \frac{1}{\cosh x}\,\mathrm{d}x = \frac{\pi}{2}, \quad \int_0^\infty \frac{x^2}{\cosh x}\,\mathrm{d}x = \frac{\pi^3}{8}, \quad \int_0^\infty \frac{x^4}{\cosh x}\,\mathrm{d}x = \frac{5\pi^5}{32}.$$

Finally,

$$\int_0^\infty \frac{x}{\cosh x}\,\mathrm{d}x = 2G.$$

Source: [629] and [2513, p. 115].

Fact 14.8.40. Let $a > 0$ and $b > 0$. Then,

$$\int_0^\infty \frac{x^a}{\sinh bx}\,\mathrm{d}x = \frac{2^{a+1}-1}{2^a b^{a+1}}\Gamma(a+1)\zeta(a+1).$$

Furthermore, if $n \ge 1$, then

$$\int_0^\infty \frac{x^n}{\sinh bx}\,\mathrm{d}x = \frac{(2^{n+1}-1)n!}{2^n b^{n+1}}\zeta(n+1).$$

In particular,

$$\int_0^\infty \frac{x}{\sinh bx}\,\mathrm{d}x = \frac{\pi^2}{4b^2}, \quad \int_0^\infty \frac{x^2}{\sinh bx}\,\mathrm{d}x = \frac{7}{2b^3}\zeta(3),$$

$$\int_0^\infty \frac{x^3}{\sinh bx}\,\mathrm{d}x = \frac{\pi^4}{8b^4}, \quad \int_0^\infty \frac{x^4}{\sinh bx}\,\mathrm{d}x = \frac{93}{2b^5}\zeta(5).$$

Fact 14.8.41. Let a and b be real numbers, and assume that $b > 0$. Then,

$$\int_0^\infty \frac{\sin ax}{\sinh bx}\,\mathrm{d}x = \frac{\pi}{2b}\tanh\frac{a\pi}{2b}, \quad \int_0^\infty \frac{\cos ax}{\cosh bx}\,\mathrm{d}x = \frac{\pi}{2b}\operatorname{sech}\frac{a\pi}{2b},$$

$$\int_0^\infty \frac{\cos ax}{\cosh^2 bx}\,\mathrm{d}x = \frac{a\pi}{2b^2}\operatorname{csch}\frac{a\pi}{2b}, \quad \int_0^\infty \frac{\cos ax}{\cosh^3 bx}\,\mathrm{d}x = \frac{(a^2+b^2)\pi}{4b^3}\operatorname{sech}\frac{a\pi}{2b}.$$

If, in addition, $|a| < b$, then

$$\int_0^\infty \frac{\sinh ax}{\sinh bx}\,\mathrm{d}x = \frac{\pi}{2b}\tan\frac{a\pi}{2b}, \quad \int_0^\infty \frac{\cosh ax}{\cosh bx}\,\mathrm{d}x = \frac{\pi}{2b}\sec\frac{a\pi}{2b}.$$

Source: [1524, p. 273] and [1217, pp. 509, 510].

Fact 14.8.42. Let $a \in \mathbb{R}$. Then,

$$\int_0^\infty \frac{\cos 2ax}{1+2\cosh\frac{2\pi}{3}x}\,\mathrm{d}x = \frac{\sqrt{3}}{2(1+2\cosh 2a)},$$

$$\int_0^\infty \frac{\cos ax}{(x^2+1)\cosh\frac{\pi}{2}x}\,\mathrm{d}x = (\cosh a)\log(2\cosh a) - a\sinh a,$$

$$\int_0^\infty \frac{\cos 2ax}{(x^2+1)\cosh\pi x}\,\mathrm{d}x = 2\cosh a - e^{2a}\operatorname{atan} e^{-a} - e^{-2a}\operatorname{atan} e^{a},$$

$$\int_0^\infty \frac{(\sin\pi x^2)\cos 2\pi ax}{\cosh\pi x}\,\mathrm{d}x = \frac{-1+\sqrt{2}\sin\pi a^2}{2\sqrt{2}\cosh\pi a}, \quad \int_0^\infty \frac{(\cos\pi x^2)\cos 2\pi ax}{\cosh\pi x}\,\mathrm{d}x = \frac{1+\sqrt{2}\sin\pi a^2}{2\sqrt{2}\cosh\pi a},$$

$$\int_0^\infty \frac{(\sin\pi x^2)\sin 2\pi ax}{\sinh\pi x}\,\mathrm{d}x = \frac{\sin\pi a^2}{2\sinh\pi a}, \quad \int_0^\infty \frac{(\cos\pi x^2)\sin 2\pi ax}{\sinh\pi x}\,\mathrm{d}x = \frac{\cosh\pi a - \cos\pi a^2}{2\sinh\pi a},$$

$$\int_0^\infty \frac{(\sin\pi x^2)\sin 2\pi ax}{\tanh\pi x}\,\mathrm{d}x = \tfrac{1}{2}(\tanh\pi a)\sin(\tfrac{\pi}{4}+\pi a^2),$$

$$\int_0^\infty \frac{(\cos\pi x^2)\sin 2\pi ax}{\tanh\pi x}\,\mathrm{d}x = \tfrac{1}{2}(\tanh\pi a)[1-\cos(\tfrac{\pi}{4}+\pi a^2)],$$

$$\int_0^\infty \frac{(\sin \pi x^2)\cos \pi a x}{1 + 2\cosh \frac{2\sqrt{3}\pi}{3}x}\,\mathrm{d}x = \frac{2\cos\frac{\pi - 3\pi a^2}{12} - \sqrt{3}}{8\cosh\frac{\sqrt{3}\pi}{3}a - 4}, \quad \int_0^\infty \frac{(\cos \pi x^2)\cos \pi a x}{1 + 2\cosh \frac{2\sqrt{3}\pi}{3}x}\,\mathrm{d}x = \frac{1 - 2\sin\frac{\pi - 3\pi a^2}{12}}{8\cosh\frac{\sqrt{3}\pi}{3}a - 4}.$$

Furthermore,

$$\int_0^\infty \frac{x}{(x^2+1)\sinh \pi x}\,\mathrm{d}x = \log 2 - \frac{1}{2}, \quad \int_0^\infty \frac{x}{(x^2+1)\sinh \frac{\pi}{2}x}\,\mathrm{d}x = \frac{\pi}{2} - 1,$$

$$\int_0^\infty \frac{x}{(x^2+1)\sinh \frac{\pi}{4}x}\,\mathrm{d}x = \frac{\sqrt{2}\pi}{2} + \sqrt{2}\log(1+\sqrt{2}) - 2,$$

$$\int_0^\infty \frac{x}{(x^2+1)\cosh \frac{\pi}{4}x}\,\mathrm{d}x = \frac{\sqrt{2}\pi}{2} - \sqrt{2}\log(1+\sqrt{2}).$$

Source: [1217, pp. 375, 376] and [1317, pp. 55, 61–66].

Fact 14.8.43.

$$\int_0^\infty \frac{x\sin \pi x^2}{\sinh \pi x}\,\mathrm{d}x = \frac{1}{4\pi}, \quad \int_0^\infty \frac{x\cos \pi x^2}{\sinh \pi x}\,\mathrm{d}x = \frac{1}{8},$$

$$\int_0^\infty \frac{x^2\sin \pi x^2}{\cosh \pi x}\,\mathrm{d}x = \frac{1}{8} - \frac{\sqrt{2}}{16}, \quad \int_0^\infty \frac{x^2\cos \pi x^2}{\cosh \pi x}\,\mathrm{d}x = \frac{\sqrt{2}}{16} - \frac{1}{4\pi},$$

$$\int_0^\infty \frac{x^3\sin \pi x^2}{\sinh \pi x}\,\mathrm{d}x = \frac{1}{16\pi}, \quad \int_0^\infty \frac{x^3\cos \pi x^2}{\sinh \pi x}\,\mathrm{d}x = \frac{1}{64} - \frac{3}{16\pi^2}.$$

Source: [1317, pp. 63, 65].

Fact 14.8.44. If $a > 0$ and $b > 1$, then

$$\int_0^\infty \frac{x^b}{\sinh^2 ax}\,\mathrm{d}x = \frac{4}{(2a)^{b+1}}\Gamma(b+1)\zeta(b).$$

If $a > 0$ and $n \ge 1$, then

$$\int_0^\infty \frac{x^{2n}}{\sinh^2 ax}\,\mathrm{d}x = \frac{\pi^{2n}}{a^{2n+1}}|B_{2n}|, \quad \int_0^\infty \frac{x^{2n}\cosh ax}{\sinh^2 ax}\,\mathrm{d}x = \frac{(4^n-1)}{a}\left(\frac{\pi}{a}\right)^{2n}|B_{2n}|.$$

If $a \ne 0$ and $n \ge 1$, then

$$\int_0^\infty \frac{x^{2n+1}\cosh ax}{\sinh^2 ax}\,\mathrm{d}x = \frac{(2^{2n+1}-1)(2n+1)!}{4^n a^{2n+2}}\zeta(2n+1).$$

Source: [1217, pp. 379, 380].

Fact 14.8.45. If $a > 0$, $b > -1$, and $b \ne 1$, then

$$\int_0^\infty \frac{x^b}{\cosh^2 ax}\,\mathrm{d}x = \frac{4(1-2^{1-b})}{(2a)^{b+1}}\Gamma(b+1)\zeta(b).$$

In particular,

$$\int_0^\infty \frac{1}{\sqrt{x}\cosh^2 ax}\,\mathrm{d}x = 2(\sqrt{2}-4)\sqrt{\pi}\zeta(-\tfrac{1}{2}), \quad \int_0^\infty \frac{\sqrt{x}}{\cosh^2 ax}\,\mathrm{d}x = \left(\frac{\sqrt{2}}{2} - 1\right)\sqrt{\pi}\zeta(\tfrac{1}{2}).$$

If $a \ne 0$, then

$$\int_0^\infty \frac{1}{\cosh^2 ax}\,\mathrm{d}x = \frac{1}{a}, \quad \int_0^\infty \frac{x}{\cosh^2 ax}\,\mathrm{d}x = \frac{\log 2}{a^2}.$$

If $a > 0$ and $n \ge 1$, then

$$\int_0^\infty \frac{x^{2n}}{\cosh^2 ax}\,\mathrm{d}x = \frac{(4^n-2)\pi^{2n}}{4^n a^{2n+1}}|B_{2n}|.$$

If $a > 0$ and $n \geq 0$, then

$$\int_0^\infty \frac{x^{2n+1}\sinh ax}{\cosh^2 ax}\,\mathrm{d}x = \frac{2n+1}{a}\left(\frac{\pi}{2a}\right)^{2n+1}|E_{2n}|.$$

If $a > 0$ and $b > 0$, then

$$\int_0^\infty \frac{x\sinh ax}{\cosh^{2b+1} ax}\,\mathrm{d}x = \frac{\sqrt{\pi}\Gamma(b)}{4a^2 b\Gamma(b+\frac{1}{2})}.$$

Source: [1217, pp. 379, 380].

Fact 14.8.46. Let a, b, c be real numbers, and assume that $|b| < c$ and $a^2 + b^2 > 0$. Then,

$$\int_0^\infty \frac{(\cos ax)\sinh bx}{e^{cx}+1}\,\mathrm{d}x = \frac{\pi(\sin\frac{b\pi}{c})\cosh\frac{a\pi}{c}}{c(\cosh\frac{2a\pi}{c} - \cos\frac{2b\pi}{c})} - \frac{b}{2(a^2+b^2)},$$

$$\int_0^\infty \frac{(\cos ax)\sinh bx}{e^{cx}-1}\,\mathrm{d}x = \frac{b}{2(a^2+b^2)} - \frac{\pi(\sin\frac{2b\pi}{c})}{2c(\cosh\frac{2a\pi}{c} - \cos\frac{2b\pi}{c})}.$$

In particular, if $0 < |b| < c$, then

$$\int_0^\infty \frac{\sinh bx}{e^{cx}+1}\,\mathrm{d}x = \frac{\pi}{2c}\csc\frac{a\pi}{c} - \frac{1}{2b}, \quad \int_0^\infty \frac{\sinh bx}{e^{cx}-1}\,\mathrm{d}x = \frac{1}{2b} - \frac{\pi}{2c}\cot\frac{b\pi}{c}.$$

Source: [1217, p. 523] and [3024, p. 336].

Fact 14.8.47. Let a and b be nonzero real numbers such that division by zero does not occur in the expressions below. Then,

$$\int_0^\infty \frac{1}{a+b\sinh x}\,\mathrm{d}x = \frac{1}{\sqrt{a^2+b^2}}\log\frac{a+b+\sqrt{a^2+b^2}}{a+b-\sqrt{a^2+b^2}},$$

$$\int_0^\infty \frac{1}{a+b\cosh x}\,\mathrm{d}x = \begin{cases} \dfrac{1}{\sqrt{a^2+b^2}}\log\dfrac{a+b+\sqrt{a^2+b^2}}{a+b-\sqrt{a^2+b^2}}, & |b| < |a|, \\ \dfrac{2}{\sqrt{b^2-a^2}}\operatorname{atan}\dfrac{\sqrt{b^2-a^2}}{a+b}, & |a| < |b|, \end{cases}$$

$$\int_0^\infty \frac{1}{a\sinh x + b\cosh x}\,\mathrm{d}x = \begin{cases} \dfrac{1}{\sqrt{a^2-b^2}}\log\dfrac{a+b+\sqrt{a^2-b^2}}{a+b-\sqrt{a^2-b^2}}, & |b| < |a|, \\ \dfrac{2}{\sqrt{b^2-a^2}}\operatorname{atan}\dfrac{\sqrt{b^2-a^2}}{a+b}, & |a| < |b|. \end{cases}$$

Source: [1524, p. 273].

Fact 14.8.48. Let $a > 0$. Then,

$$\int_0^\infty \frac{\log x}{e^{ax}+1}\,\mathrm{d}x = -\frac{(\log 2)(2\log a + \log 2)}{2a}, \quad \int_0^\infty \frac{x^2\log x}{e^{ax^2}}\,\mathrm{d}x = \frac{1}{8}\sqrt{\frac{\pi}{a^3}}(2 - \log 4a - \gamma).$$

Source: [1217, p. 574].

Fact 14.8.49. Let a and b be positive numbers. Then,

$$\int_0^\infty \frac{\log bx}{e^{ax}}\,\mathrm{d}x = \frac{1}{a}\left(\log\frac{b}{a} - \gamma\right), \quad \int_0^\infty \frac{\log^2 bx}{e^{ax}}\,\mathrm{d}x = \frac{\pi^2}{6a} + \frac{\gamma^2}{a} + \frac{1}{a}\left(\log\frac{a}{b}\right)\left(2\gamma + \log\frac{a}{b}\right),$$

$$\int_0^\infty \frac{\log^3 bx}{e^{ax}}\,\mathrm{d}x = -\frac{1}{2a}\left[\pi^2 + 2\gamma^2 + 2\left(\log\frac{a}{b}\right)\left(2\gamma + \log\frac{a}{b}\right)\right]\left(\gamma + \log\frac{a}{b}\right) - \frac{2}{a}\zeta(3),$$

$$\int_0^\infty \frac{\log bx}{e^{ax^2}}\,\mathrm{d}x = \frac{1}{4}\sqrt{\frac{\pi}{a}}(2\log b - \gamma - \log 4a),$$

$$\int_0^\infty \frac{\log^2 bx}{e^{ax^2}}\,\mathrm{d}x = \frac{1}{16}\sqrt{\frac{\pi}{a}}[\pi^2 + 2(\gamma + \log 4a - 2\log b)^2],$$

$$\int_0^\infty \frac{\log bx}{e^{ax^3}}\,\mathrm{d}x = -\frac{1}{\sqrt[3]{a}}\Gamma(\tfrac{4}{3})\left(\frac{\sqrt{3}\pi}{18} + \frac{\gamma}{3} + \frac{1}{2}\log 3 + \frac{1}{3}\log a - \log b\right),$$

$$\int_0^\infty \frac{\log bx}{e^{ax^4}}\,\mathrm{d}x = -\frac{1}{\sqrt[4]{a}}\Gamma(\tfrac{5}{4})\left(\frac{\pi}{8} + \frac{\gamma}{4} + \frac{3}{4}\log 2 + \frac{1}{4}\log a - \log b\right).$$

In particular,

$$\int_0^\infty \frac{\log x}{e^x}\,\mathrm{d}x = -\gamma, \quad \int_0^\infty \frac{\log^2 x}{e^x}\,\mathrm{d}x = \frac{\pi^2}{6} + \gamma^2, \quad \int_0^\infty \frac{\log^3 x}{e^x}\,\mathrm{d}x = -\gamma^3 - \frac{1}{2}\gamma\pi^2 - 2\zeta(3),$$

$$\int_0^\infty \frac{\log^4 x}{e^x}\,\mathrm{d}x = \frac{3\pi^4}{20} + \gamma^4 + \pi^2\gamma^2 + 8\gamma\zeta(3),$$

$$\int_0^\infty \frac{\log x}{e^{x^2}}\,\mathrm{d}x = -\frac{\sqrt{\pi}}{4}(\gamma + 2\log 2), \qquad \int_0^\infty \frac{\log^2 x}{e^{x^2}}\,\mathrm{d}x = \frac{\sqrt{\pi}}{16}[\pi^2 + 2(\gamma + 2\log 2)^2].$$

Fact 14.8.50. Let $a > 0$ and $n \ge 0$. Then,

$$\int_0^\infty \frac{x^n \log x}{e^{ax}}\,\mathrm{d}x = \frac{n!}{a^{n+1}}(H_n - \log a - \gamma),$$

$$\int_0^\infty \frac{x^{n+1/2}\log x}{e^{ax}}\,\mathrm{d}x - \frac{\sqrt{\pi}(2n+1)!!}{2^{n+1}a^{n+3/2}}\left(\sum_{i=1}^{2n+1}\frac{2}{2i+1} - \log 4a - \gamma\right).$$

Source: [1217, p. 573].

Fact 14.8.51. Let $a > 0$. Then,

$$\int_0^\infty \log(1 + e^{-ax})\,\mathrm{d}x = \frac{\pi^2}{12a}, \quad \int_0^\infty \log(1 + e^{-ax})\,\mathrm{d}x = -\frac{\pi^2}{6a}, \quad \int_0^\infty \log^2(1 + e^{-ax})\,\mathrm{d}x = \frac{1}{4a}\zeta(3).$$

If $n \ge 1$, then,

$$\int_0^\infty \log^n(1 - e^{-ax})\,\mathrm{d}x = (-1)^n\frac{n!}{a}\zeta(n+1).$$

In particular,

$$\int_0^\infty \log^2(1-e^{-ax})\,\mathrm{d}x = \frac{2}{a}\zeta(3), \quad \int_0^\infty \log^3(1-e^{-ax})\,\mathrm{d}x = -\frac{\pi^4}{15a}, \quad \int_0^\infty \log^4(1-e^{-ax})\,\mathrm{d}x = \frac{24}{a}\zeta(5).$$

Source: [1217, p. 530].

14.9 Facts on Integral Representations of G and γ

Fact 14.9.1.

$$G = \frac{1}{2}\int_0^{\pi/2}\frac{x}{\sin x}\,\mathrm{d}x = \int_0^1 \frac{\operatorname{atan} x}{x}\,\mathrm{d}x = -\int_0^1 \frac{\log x}{x^2+1}\,\mathrm{d}x = \frac{\pi^2}{4}\int_0^1 \frac{\frac{1}{2} - x}{\cos \pi x}\,\mathrm{d}x$$

$$= 2\int_0^{\pi/4}\log 2\cos x\,\mathrm{d}x = -2\int_0^{\pi/4}\log 2\sin x\,\mathrm{d}x = -\int_0^{\pi/2}\log 2\sin\tfrac{x}{2}\,\mathrm{d}x$$

$$= \int_0^{\pi/4} \log\cot x\,\mathrm{d}x = \frac{1}{2}\int_0^\infty \frac{x}{\cosh x}\,\mathrm{d}x = \frac{3}{4}\int_0^{\pi/2} \log|1+2\cos x|\,\mathrm{d}x$$

$$= \frac{1}{4}\int_0^\infty \frac{x^2\sinh x}{\cosh^2 x}\,\mathrm{d}x = \frac{\pi^2}{16} - \frac{\pi}{4}\log 2 + \int_0^{\pi/4} \frac{x^2}{\sin^2 x}\,\mathrm{d}x = \lim_{n\to\infty}\sum_{i=1}^n \frac{1}{i}\operatorname{atan}\frac{i}{n}.$$

Furthermore,

$$0.91528 \approx \frac{\pi}{2} - \frac{\pi}{2}\log\frac{\pi}{2} + \frac{\pi^3}{576} < G < \frac{\pi}{2} - \frac{\pi}{4}\log 2 - \frac{\pi^3}{288} \approx 0.91874.$$

Source: [11], [1217, p. 380], and [2411]. The inequalities are given in [2952]. **Related:** Fact 13.5.97 and Fact 13.6.2.

Fact 14.9.2. Let $a > 0$ and $b > 0$. Then,

$$\gamma = -\int_0^1 \log\log\frac{1}{x}\,\mathrm{d}x = \int_0^1 \left(\frac{1}{1-x} + \frac{1}{\log x}\right)\mathrm{d}x = \int_0^1 \frac{1}{x}(1 - e^{-x} - e^{-1/x})\,\mathrm{d}x$$

$$= \int_0^1 \frac{1}{\log x} - \frac{1}{x(\log x)(1-\log x)}\,\mathrm{d}x = -\int_0^\infty \frac{\log x}{e^x}\,\mathrm{d}x = \int_0^\infty \frac{1}{x}\left(\frac{1}{2} - \cos x\right)\mathrm{d}x$$

$$= \int_0^\infty \left(\frac{1}{e^x - 1} - \frac{1}{e^x x}\right)\mathrm{d}x = \int_0^\infty \frac{2}{x}\left(\frac{1}{e^{x^2}} - \frac{1}{e^x}\right)\mathrm{d}x = \int_0^\infty \frac{2}{x}(\cos x^2 - \cos x)\,\mathrm{d}x$$

$$= \int_0^\infty \left(1 - \frac{1}{e^{e^x}} - \frac{1}{e^{e^{-x}}}\right)\mathrm{d}x = \int_0^\infty \frac{b}{x}\left(\frac{1}{x^a+1} - \frac{1}{e^{x^b}}\right)\mathrm{d}x = \int_0^\infty \frac{b}{x}\left(\frac{1}{x^a+1} - \cos x^b\right)\mathrm{d}x$$

$$= \log\frac{1}{a} + \int_0^\infty \frac{1}{x}\left(\frac{2}{\pi}\operatorname{acot} x - e^{-ax}\right)\mathrm{d}x = \log\frac{1}{a} - \frac{2}{\pi}\int_0^\infty \frac{1}{x}(\sin ax)\log x\,\mathrm{d}x$$

$$= 1 - \log 2a - \frac{2}{a\pi}\int_0^\infty \frac{1}{x^2}(\sin^2 ax)\log x\,\mathrm{d}x = \frac{1}{2} + 2\int_0^\infty \frac{x}{(x^2+1)(e^{2\pi x}-1)}\,\mathrm{d}x$$

$$= 1 + \int_0^\infty \frac{a}{x}\left(\frac{1}{x^b+1} + \frac{e^{-x^a}-1}{x^a}\right)\mathrm{d}x = 1 + \int_0^\infty \frac{a}{x}\left(\frac{1}{x^b+1} - \frac{\sin x^a}{x^a}\right)\mathrm{d}x$$

$$= 1 + \int_0^\infty \frac{\log x}{x}\left(\cos x - \frac{\sin x}{x}\right)\mathrm{d}x = \frac{3}{2} + \int_0^\infty \frac{2a}{x}\left(\frac{1}{2(x^b+1)} + \frac{\cos x^a - 1}{x^{2a}}\right)\mathrm{d}x.$$

If $a \ne b$, then

$$\gamma = \frac{ab}{b-a}\int_0^\infty \frac{e^{-x^b} - e^{-x^a}}{x}\,\mathrm{d}x = \frac{ab}{b-a}\int_0^\infty \frac{\cos x^b - \cos x^a}{x}\,\mathrm{d}x$$

$$= \frac{ab}{b-a}\int_0^\infty \frac{\cos x^a - e^{-x^a}}{x}\,\mathrm{d}x = 1 + \frac{ab}{b-a}\int_0^\infty \frac{1}{x}\left(\frac{\sin x^b}{x^b} - \frac{\sin x^a}{x^a}\right)\mathrm{d}x$$

$$= 1 + \frac{1}{a-b}\log\frac{b^b}{a^a} + \frac{2}{\pi(a-b)}\int_0^\infty \frac{1}{x^2}(\cos ax - \cos bx)\log x\,\mathrm{d}x.$$

Furthermore,

$$\int_0^\infty \frac{1}{x}\left(\frac{1}{x^a+1} - \frac{1}{x^b+1}\right)\mathrm{d}x = \int_0^\infty \frac{x^{b-1} - x^{a-1}}{(x^a+1)(x^b+1)}\,\mathrm{d}x = 0.$$

Source: [208], [511, pp. 176–179], [1217, p. 906], [2013, p. 198], [2256], [2504, p. 113], and [2513, pp. 15–22].

14.10 Facts on Definite Integrals of the Gamma Function

Fact 14.10.1. Let $a, b \in (0, \infty)$. Then,

$$\int_0^\infty |\Gamma(a + x\jmath)|^2 \cos 2bx \, \mathrm{d}x = \tfrac{1}{2}\sqrt{\pi}\Gamma(a)\Gamma(a + \tfrac{1}{2}) \operatorname{sech}^{2a} b,$$

$$\int_0^\infty (\operatorname{sech}^{2a} x) \cos 2bx \, \mathrm{d}x = \frac{\sqrt{\pi}|\Gamma(a + b\jmath)|^2}{2\Gamma(a)\Gamma(a + \frac{1}{2})}.$$

If $a + \frac{1}{2} < b$, then

$$\int_0^\infty \left|\frac{\Gamma(a + x\jmath)}{\Gamma(b + x\jmath)}\right|^2 \mathrm{d}x = \frac{\sqrt{\pi}\Gamma(a)\Gamma(a + \frac{1}{2})\Gamma(b - a - \frac{1}{2})}{2\Gamma(b)\Gamma(b - \frac{1}{2})\Gamma(b - a)},$$

$$\int_0^\infty \prod_{i=0}^\infty \frac{1 + \frac{x^2}{(b+i)^2}}{1 + \frac{x^2}{(a+i)^2}} \mathrm{d}x = \frac{\sqrt{\pi}\Gamma(b)\Gamma(a + \frac{1}{2})\Gamma(b - a - \frac{1}{2})}{2\Gamma(a)\Gamma(b - \frac{1}{2})\Gamma(b - a)}.$$

Remark: If $b - a$ is an integer, then the infinite product in the last integral is a finite product. For example, setting $a = 11/10$ and $b = 61/10$ yields

$$\int_0^\infty \frac{1}{(x^2 + 11^2)(x^2 + 21^2)(x^2 + 31^2)(x^2 + 41^2)(x^2 + 51^2)} \mathrm{d}x$$
$$= \frac{5\pi}{12 \cdot 13 \cdot 16 \cdot 17 \cdot 18 \cdot 22 \cdot 23 \cdot 24 \cdot 31 \cdot 32 \cdot 41} = \frac{5\pi}{377244828499968}.$$

Source: [1317, pp. 20, 21, 54]. **Related:** Fact 14.2.26.

Fact 14.10.2. Let $a, b \in (0, \infty)$. If $a > \frac{1}{2}$, then

$$\int_0^\infty \frac{\cos \pi x}{\Gamma^2(a + x)\Gamma^2(a - x)} \mathrm{d}x = \frac{1}{4\Gamma(2a - 1)\Gamma^2(a)}.$$

If $a + b > \frac{3}{2}$, then

$$\int_0^\infty \frac{1}{\Gamma^2(a + x)\Gamma^2(b - x)} \mathrm{d}x = \frac{\Gamma(2a + 2b - 3)}{\Gamma^4(a + b - 1)},$$

$$\int_0^\infty \frac{1}{\Gamma(a + x)\Gamma(a - x)\Gamma(b + x)\Gamma(b - x)} \mathrm{d}x = \frac{\Gamma(2a + 2b - 3)}{2\Gamma(2a - 1)\Gamma(2b - 1)\Gamma^2(a + b - 1)},$$

$$\int_0^\infty \frac{\cos \pi x}{\Gamma(a + x)\Gamma(a - x)\Gamma(b + 2x)\Gamma(b - 2x)} \mathrm{d}x = \frac{4^{a+b-3}\Gamma(a + b - \frac{3}{2})}{\sqrt{\pi}\Gamma(a)\Gamma(2b - 1)\Gamma(2a + b - 2)}.$$

Source: [1317, pp. 226–229].

14.11 Facts on Integral Inequalities

Fact 14.11.1. The following statements hold:

i) If $n \geq 1$, then

$$\frac{2^{n+1} - 1}{n + 1} \leq \int_0^{\pi/2} (\cos x + 1)^n \, \mathrm{d}x.$$

ii) If $x \in (0, \infty)$, then

$$\int_x^\infty e^{-t^2/2} \, \mathrm{d}t \leq \frac{1}{x} e^{-x^2/2}.$$

iii) If $x \in (0, 1)$, then

$$\int_x^1 \frac{\log(t + 1)}{t} \mathrm{d}t < (2 \log 2)\frac{1 - x}{1 + x}.$$

Source: [2527, pp. 117, 118].

14.12 Facts on the Gaussian Density

Fact 14.12.1. Let $A \in \mathbb{R}^{n\times n}$, assume that A is positive definite, let $b \in \mathbb{R}^n$, and let $a \in \mathbb{R}$. Then,

$$\int_{\mathbb{R}^n} e^{-x^{\mathsf{T}}Ax+2b^{\mathsf{T}}x+a}\,\mathrm{d}x = \frac{\pi^{n/2}}{\sqrt{\det A}}e^{b^{\mathsf{T}}A^{-1}b+a}, \quad \int_{\mathbb{R}^n} e^{-x^{\mathsf{T}}Ax+2b^{\mathsf{T}}xj+a}\,\mathrm{d}x = \frac{\pi^{n/2}}{\sqrt{\det A}}e^{-b^{\mathsf{T}}A^{-1}b+a}.$$

In particular,

$$\int_{\mathbb{R}^n} e^{-x^{\mathsf{T}}Ax}\,\mathrm{d}x = \frac{\pi^{n/2}}{\sqrt{\det A}}.$$

Source: [2013, pp. 103, 104]. **Related:** Fact 15.14.12.

Fact 14.12.2. Let $A \in \mathbb{R}^{n\times n}$, assume that A is positive definite, and define $f\colon \mathbb{R}^n \mapsto \mathbb{R}$ by

$$f(x) = \frac{e^{-\frac{1}{2}x^{\mathsf{T}}A^{-1}x}}{(2\pi)^{n/2}\sqrt{\det A}}.$$

Then,

$$\int_{\mathbb{R}^n} f(x)\,\mathrm{d}x = 1, \quad \int_{\mathbb{R}^n} f(x)xx^{\mathsf{T}}\,\mathrm{d}x = A, \quad \int_{\mathbb{R}^n} f(x)\log f(x)\,\mathrm{d}x = -\tfrac{1}{2}\log[(2\pi e)^n\det A].$$

Source: [788] and Fact 14.12.5. **Remark:** f is the multivariate Gaussian density. The *entropy* of f is $-\int_{\mathbb{R}^n} f(x)\log f(x)\,\mathrm{d}x$.

Fact 14.12.3. Let $A, B \in \mathbb{R}^{n\times n}$, assume that A and B are positive definite, and, for all $k \in \{0, 1, 2, 3\}$, define

$$\mathcal{I}_k \triangleq \frac{1}{(2\pi)^{n/2}\sqrt{\det A}}\int_{\mathbb{R}^n} (x^{\mathsf{T}}Bx)^k e^{-\frac{1}{2}x^{\mathsf{T}}A^{-1}x}\,\mathrm{d}x.$$

Then,

$$\mathcal{I}_0 = 1, \quad \mathcal{I}_1 = \operatorname{tr} AB, \quad \mathcal{I}_2 = (\operatorname{tr} AB)^2 + 2\operatorname{tr}(AB)^2,$$

$$\mathcal{I}_3 = (\operatorname{tr} AB)^3 + 6(\operatorname{tr} AB)\operatorname{tr}(AB)^2 + 8\operatorname{tr}(AB)^3.$$

$$\mathcal{I}_4 = (\operatorname{tr} AB)^4 + 12(\operatorname{tr} AB)^2\operatorname{tr}(AB)^2 + 32(\operatorname{tr} AB)\operatorname{tr}(AB)^3 + 12[\operatorname{tr}(AB)^2]^2 + 48\operatorname{tr}(AB)^4.$$

Source: [2043, p. 80], which uses a moment generating function. A Kronecker product approach is used in [2402]. **Remark:** These are *Lancaster's formulas.*

Fact 14.12.4. Let $A, B, C, D \in \mathbb{R}^{n\times n}$, assume that A is positive definite, assume that B, C, and D are symmetric, and let $\mu \in \mathbb{R}^n$. Then,

$$\frac{1}{(2\pi)^{n/2}\sqrt{\det A}}\int_{\mathbb{R}^n} x^{\mathsf{T}}Bxe^{-\frac{1}{2}(x-\mu)^{\mathsf{T}}A^{-1}(x-\mu)}\,\mathrm{d}x = \operatorname{tr} AB + \mu^{\mathsf{T}}B\mu,$$

$$\begin{aligned}\frac{1}{(2\pi)^{n/2}\sqrt{\det A}}\int_{\mathbb{R}^n} x^{\mathsf{T}}Bxx^{\mathsf{T}}Cxe^{-\frac{1}{2}(x-\mu)^{\mathsf{T}}A^{-1}(x-\mu)}\,\mathrm{d}x &= (\operatorname{tr} AB)\operatorname{tr} AC + 2\operatorname{tr} ACAB + \mu^{\mathsf{T}}C\mu\operatorname{tr} AB\\ &\quad + 4\mu^{\mathsf{T}}BAC\mu + \mu^{\mathsf{T}}B\mu\operatorname{tr} CA + \mu^{\mathsf{T}}B\mu\mu^{\mathsf{T}}C\mu,\end{aligned}$$

$$\begin{aligned}&\frac{1}{(2\pi)^{n/2}\sqrt{\det A}}\int_{\mathbb{R}^n} x^{\mathsf{T}}Bxx^{\mathsf{T}}Cxx^{\mathsf{T}}Dxe^{-\frac{1}{2}x^{\mathsf{T}}A^{-1}x}\,\mathrm{d}x\\ &\quad = (\operatorname{tr} AB)(\operatorname{tr} AC)\operatorname{tr} AD + 2[(\operatorname{tr} AB)\operatorname{tr} ACAD + (\operatorname{tr} AC)\operatorname{tr} ADAB + (\operatorname{tr} AD)\operatorname{tr} ABAC]\\ &\qquad + 4(\operatorname{tr} ABACAD + \operatorname{tr} ABADAC).\end{aligned}$$

Source: [2043, p. 83] and [2403, pp. 414–419]. **Remark:** These are the mean and covariance of quadratic forms in multivariate Gaussian random variables. Setting $\mu = 0$ and $C = B$ yields $\mathcal{I}_2$ of Fact 14.12.3.

Fact 14.12.5. Let $A \in \mathbb{R}^{n\times n}$, assume that A is positive definite, let $B \in \mathbb{R}^{n\times n}$, let $a, b \in \mathbb{R}^n$, and let $\alpha, \beta \in \mathbb{R}$. Then,

$$\int_{\mathbb{R}^n} (x^\mathsf{T}Bx + b^\mathsf{T}x + \beta)e^{-(x^\mathsf{T}Ax + a^\mathsf{T}x + \alpha)}\,\mathrm{d}x = \frac{\pi^{n/2}}{2\sqrt{\det A}}(2\beta + \operatorname{tr} A^{-1}B - b^\mathsf{T}A^{-1}a + \tfrac{1}{2}a^\mathsf{T}A^{-1}BA^{-1}a)e^{\frac{1}{4}a^\mathsf{T}A^{-1}a-\alpha}.$$

Source: [1343, p. 322].

Fact 14.12.6. Let $A_1, A_2 \in \mathbb{R}^{n\times n}$, assume that A_1 and A_2 are positive definite, and let $\mu_1, \mu_2 \in \mathbb{R}^n$. Then,

$$e^{\frac{1}{2}(x-\mu_1)^\mathsf{T}A_1^{-1}(x-\mu_1)}e^{\frac{1}{2}(x-\mu_2)^\mathsf{T}A_2^{-1}(x-\mu_2)} = \alpha e^{\frac{1}{2}(x-\mu_3)^\mathsf{T}A_3^{-1}(x-\mu_3)},$$

where

$$A_3 \triangleq (A_1^{-1} + A_2^{-1})^{-1}, \quad \mu_3 \triangleq A_3(A_1^{-1}\mu_1 + A_2^{-1}\mu_2), \quad \alpha \triangleq e^{\frac{1}{2}(\mu_1^\mathsf{T}A_1^{-1}\mu_1 + \mu_2^\mathsf{T}A_2^{-1}\mu_2 - \mu_3^\mathsf{T}A_3^{-1}\mu_3)}.$$

Remark: A product of Gaussian densities is a weighted Gaussian density.

14.13 Facts on Multiple Integrals

Fact 14.13.1. Let $n \ge 1$. Then,

$$\int_0^1\int_0^1 |x-y|^n\,\mathrm{d}x\,\mathrm{d}y = \frac{2}{(n+1)(n+2)}.$$

Fact 14.13.2.

$$\int_0^1\int_0^1 \frac{1}{1-xy}\,\mathrm{d}x\,\mathrm{d}y = \frac{\pi^2}{6}, \quad \int_0^1\int_0^1 \frac{1}{1+xy}\,\mathrm{d}x\,\mathrm{d}y = \frac{\pi^2}{12},$$

$$\int_0^1\int_0^1 \frac{1}{1-xy^2}\,\mathrm{d}x\,\mathrm{d}y = 2\log 2, \quad \int_0^1\int_0^1 \frac{1}{1+xy^2}\,\mathrm{d}x\,\mathrm{d}y = \frac{\pi}{2} - \log 2,$$

$$\int_0^1\int_0^1 \frac{1}{1-xy^3}\,\mathrm{d}x\,\mathrm{d}y = \frac{\sqrt{3}\pi}{12} + \frac{3}{4}\log 3, \quad \int_0^1\int_0^1 \frac{1}{1+xy^3}\,\mathrm{d}x\,\mathrm{d}y = \frac{\sqrt{3}\pi}{6},$$

$$\int_0^1\int_0^1 \frac{1}{1-x^2y^2}\,\mathrm{d}x\,\mathrm{d}y = \frac{\pi^2}{8}, \quad \int_0^1\int_0^1 \frac{1}{1+x^2y^2}\,\mathrm{d}x\,\mathrm{d}y = \frac{1}{2}\int_0^{\pi/2}\int_0^{\pi/2} \frac{1}{1+(\cos x)\cos y}\,\mathrm{d}x\,\mathrm{d}y = G,$$

$$\int_0^1\int_0^1 \frac{1}{1-x^2y^3}\,\mathrm{d}x\,\mathrm{d}y = \int_0^1 \frac{\operatorname{atanh} x^{3/2}}{x^{3/2}}\,\mathrm{d}x = \frac{\sqrt{3}\pi}{6} + \frac{3}{2}\log 3 - 2\log 2,$$

$$\int_0^1\int_0^1 \frac{xy}{1-xy}\,\mathrm{d}x\,\mathrm{d}y = \frac{\pi^2}{6} - 1, \quad \int_0^1\int_0^1 \frac{xy}{1+xy}\,\mathrm{d}x\,\mathrm{d}y = 1 - \frac{\pi^2}{12},$$

$$\int_0^1\int_0^1 \frac{xy}{1-xy^2}\,\mathrm{d}x\,\mathrm{d}y = \frac{1}{2}, \quad \int_0^1\int_0^1 \frac{xy}{1+xy^2}\,\mathrm{d}x\,\mathrm{d}y = \log 2 - \frac{1}{2},$$

$$\int_0^1\int_0^1 \frac{xy}{1-xy^3}\,\mathrm{d}x\,\mathrm{d}y = \frac{1}{4} - \frac{\sqrt{3}\pi}{24} + \frac{3}{8}\log 3, \quad \int_0^1\int_0^1 \frac{xy}{1+xy^3}\,\mathrm{d}x\,\mathrm{d}y = \frac{\sqrt{3}\pi}{12} - \frac{1}{4},$$

$$\int_0^1\int_0^1 \frac{xy}{1-x^2y^2}\,\mathrm{d}x\,\mathrm{d}y = \frac{\pi^2}{24}, \quad \int_0^1\int_0^1 \frac{xy}{1+x^2y^2}\,\mathrm{d}x\,\mathrm{d}y = \frac{\pi^2}{48},$$

$$\int_0^1\int_0^1 \frac{1}{1+x^2y^3}\,\mathrm{d}x\,\mathrm{d}y = \left(\frac{\sqrt{3}}{3} - \frac{1}{2}\right)\pi + \log 2, \quad \int_0^1\int_0^1 \frac{xy}{1+x^2y^3}\,\mathrm{d}x\,\mathrm{d}y = \frac{\sqrt{3}}{6} - \log 2,$$

$$\int_0^1\int_0^1 \frac{x}{1-x^3y^3}\,\mathrm{d}x\,\mathrm{d}y = \frac{\sqrt{3}\pi}{9},\quad \int_0^1\int_0^1 \frac{x^2y[x^2y(1-x)(1-y)+1]}{[1-x^2y(1-x)(1-y)]^3}\,\mathrm{d}x\,\mathrm{d}y = \sum_{i=0}^{\infty}\frac{(i!)^3}{(3i)!} \approx 1.178403258,$$

$$\int_0^1\int_0^1 \frac{1}{2-xy}\,\mathrm{d}x\,\mathrm{d}y = \frac{\pi^2}{12} - \frac{1}{2}\log^2 2,\quad \int_0^1\int_0^1 \frac{x}{2-xy}\,\mathrm{d}x\,\mathrm{d}y = 1 - \log 2,$$

$$\int_0^1\int_0^1 \frac{xy}{2-xy}\,\mathrm{d}x\,\mathrm{d}y = \frac{\pi^2}{6} - 1 - \log^2 2,\quad \int_0^1\int_0^1 \frac{xy}{2-xy^2}\,\mathrm{d}x\,\mathrm{d}y = \frac{1}{2}(1-\log 2),$$

$$\int_0^1\int_0^1 \frac{(1-y)[2-xy^2(1-x)+x^2y^4(1-x)^2]}{[1-xy^2(1-x)]^3}\,\mathrm{d}x\,\mathrm{d}y = \sum_{i=0}^{\infty}\frac{(i!)^2}{(2i+1)(2i+1)!} = \frac{8G}{3} - \frac{\pi}{3}\log(2+\sqrt{3}),$$

$$\int_0^1\int_0^1 \frac{x^2y^2}{(1+x^2y^2)(1+x)(1+y)}\,\mathrm{d}x\,\mathrm{d}y = \sum_{i=1}^{\infty}(-1)^{i+1}(H_{2i}-H_i-\log 2)^2 = \frac{\pi\log 2}{8} + \frac{7\log^2 2}{8} - \frac{13\pi^2}{192} - \frac{G}{2},$$

$$\int_0^1\int_0^1 \frac{1}{1-x^2y^2\tan^2\frac{\pi}{8}}\,\mathrm{d}x\,\mathrm{d}y = \frac{\pi^2}{16\tan\frac{\pi}{8}} - \frac{\log^2\tan\frac{\pi}{8}}{4\tan\frac{\pi}{8}}.$$

Source: [511, pp. 226, 234] and [1107, 1266, 1491, 2483].

Fact 14.13.3. The following statements hold:

i) If $\alpha > -1$ and $\beta > -1$, where $\alpha \neq \beta$, then

$$\int_0^1\int_0^1 \frac{x^\alpha y^\beta}{\log xy}\,\mathrm{d}x\,\mathrm{d}y = \frac{1}{\beta-\alpha}\log\frac{1+\alpha}{1+\beta}.$$

In particular,

$$\int_0^1\int_0^1 \frac{\sqrt{x}}{\log xy}\,\mathrm{d}x\,\mathrm{d}y = 2\log\frac{2}{3},\quad \int_0^1\int_0^1 \frac{x}{\log xy}\,\mathrm{d}x\,\mathrm{d}y = -\log 2,$$

$$\int_0^1\int_0^1 \frac{\sqrt{x}y}{\log xy}\,\mathrm{d}x\,\mathrm{d}y = 2\log\frac{3}{4},\quad \int_0^1\int_0^1 \frac{xy^2}{\log xy}\,\mathrm{d}x\,\mathrm{d}y = \log\frac{2}{3}.$$

ii) If $\alpha > -1$, then

$$\int_0^1\int_0^1 \frac{(xy)^\alpha}{\log xy}\,\mathrm{d}x\,\mathrm{d}y = -\frac{1}{1+\alpha}.$$

In particular,

$$\int_0^1\int_0^1 \frac{1}{\log xy}\,\mathrm{d}x\,\mathrm{d}y = -1,\quad \int_0^1\int_0^1 \frac{1}{\sqrt{xy}\log xy}\,\mathrm{d}x\,\mathrm{d}y = -2,$$

$$\int_0^1\int_0^1 \frac{\sqrt{xy}}{\log xy}\,\mathrm{d}x\,\mathrm{d}y = -\frac{2}{3},\quad \int_0^1\int_0^1 \frac{xy}{\log xy}\,\mathrm{d}x\,\mathrm{d}y = -\frac{1}{2}.$$

iii) If $\alpha > -1$, then

$$\int_0^1\int_0^1 \frac{1}{(1+\alpha xy)\log xy}\,\mathrm{d}x\,\mathrm{d}y = -\frac{\log(1+\alpha)}{\alpha}.$$

In particular,

$$\int_0^1\int_0^1 \frac{1}{(2-xy)\log xy}\,\mathrm{d}x\,\mathrm{d}y = -\log 2,\quad \int_0^1\int_0^1 \frac{1}{(2+xy)\log xy}\,\mathrm{d}x\,\mathrm{d}y = \log\frac{2}{3},$$

$$\int_0^1\int_0^1 \frac{1}{(1+xy)\log xy}\,\mathrm{d}x\,\mathrm{d}y = -\log 2,\quad \int_0^1\int_0^1 \frac{1}{(1+2xy)\log xy}\,\mathrm{d}x\,\mathrm{d}y = -\frac{1}{2}\log 3.$$

iv) If $k \ge 1$, then

$$\int_0^1\int_0^1 \frac{(1-y)^k y\log(1-x)}{(1-xy)^2}\,\mathrm{d}x\,\mathrm{d}y = -\frac{1}{k^2}.$$

Source: [1266].

Fact 14.13.4. Let $n \ge 2$. Then,

$$\int_0^1\int_0^1 \frac{\log^{n-2} xy}{1-xy}\,\mathrm{d}x\,\mathrm{d}y = (-1)^n(n-1)!\zeta(n).$$

In particular,

$$\int_0^1\int_0^1 \frac{1}{1-xy}\,\mathrm{d}x\,\mathrm{d}y = \frac{\pi^2}{6},\quad \int_0^1\int_0^1 \frac{\log xy}{1-xy}\,\mathrm{d}x\,\mathrm{d}y = -2\zeta(3).$$

Source: [2106, p. 164].

Fact 14.13.5.

$$\int_0^1\int_0^1 \frac{\log x}{1+xy}\,\mathrm{d}x\,\mathrm{d}y = -\frac{3}{4}\zeta(3),\quad \int_0^1\int_0^1 \frac{\log xy}{1-xy}\,\mathrm{d}x\,\mathrm{d}y = -2\zeta(3),$$

$$\int_0^1\int_0^1 \frac{\log(1-x)}{1-xy}\,\mathrm{d}x\,\mathrm{d}y = 2\int_0^1\int_0^1 \frac{\log(1-xy)}{1-xy}\,\mathrm{d}x\,\mathrm{d}y = -2\zeta(3),$$

$$\int_0^1\int_0^1 \frac{x-1}{(1-xy)\log xy}\,\mathrm{d}x\,\mathrm{d}y = \gamma,\quad \int_0^1\int_0^1 \frac{\log(2-xy)}{1-xy}\,\mathrm{d}x\,\mathrm{d}y = \int_0^1\int_0^1 \frac{\log(1+x)}{1-xy}\,\mathrm{d}x\,\mathrm{d}y = \frac{5}{8}\zeta(3),$$

$$\int_0^1\int_0^1 \frac{\log(2-x)}{1-xy}\,\mathrm{d}x\,\mathrm{d}y = \int_0^1\int_0^1 \frac{\log(1+xy)}{1-xy}\,\mathrm{d}x\,\mathrm{d}y = \frac{\pi^2\log 2}{4} - \zeta(3),$$

$$\int_0^1\int_0^1 \frac{1}{(2-xy)\log xy}\,\mathrm{d}x\,\mathrm{d}y = -\log 2,\quad \int_0^1\int_0^1 \frac{\log xy}{2-xy}\,\mathrm{d}x\,\mathrm{d}y = \frac{\pi^2}{6}\log 2 - \frac{1}{3}\log^3 2 - \frac{7}{4}\zeta(3),$$

$$\int_0^1\int_0^1 \frac{x}{(2-xy)\log xy}\,\mathrm{d}x\,\mathrm{d}y = \int_0^1 \frac{x-1}{(x-2)\log x}\,\mathrm{d}x = \sum_{i=1}^\infty \frac{1}{2^i}\log\frac{1}{i} = \sum_{i=1}^\infty \frac{1}{2^i}\log\frac{i}{i+1} \approx -0.507834,$$

$$\int_0^1\int_0^1 \frac{\log xy}{1+x^2y^2}\,\mathrm{d}x\,\mathrm{d}y = -\frac{\pi^3}{16},\quad \int_0^1\int_0^1 \frac{1}{(1+x^2y^2)\log xy}\,\mathrm{d}x\,\mathrm{d}y = -\frac{\pi}{4},$$

$$\int_0^1\int_0^1 \frac{\log x}{1+x^2y^2}\,\mathrm{d}x\,\mathrm{d}y = -\frac{\pi^3}{32},\quad \int_0^1\int_0^1 \frac{x\log xy}{1+x^2y^2}\,\mathrm{d}x\,\mathrm{d}y = \frac{\pi^2}{48} - G,\quad \int_0^1\int_0^1 \frac{x\log xy}{1-x^2y^2}\,\mathrm{d}x\,\mathrm{d}y = -\frac{\pi^2}{12},$$

$$\int_0^1\int_0^1 \frac{1+x}{(1+xy)\log xy}\,\mathrm{d}x\,\mathrm{d}y = -\log\pi,\quad \int_0^1\int_0^1 \frac{1-x}{(1+xy)\log xy}\,\mathrm{d}x\,\mathrm{d}y = \log\frac{\pi}{4},$$

$$\int_0^1\int_0^1 \frac{x}{(1+xy)\log xy}\,\mathrm{d}x\,\mathrm{d}y = \log\frac{2}{\pi},\quad \int_0^1\int_0^1 \frac{x}{(1+x^2y^2)\log xy}\,\mathrm{d}x\,\mathrm{d}y = \log\frac{\Gamma^2(3/4)}{\sqrt{2\pi}},$$

$$\int_0^1\int_0^1 \frac{(1+\frac{1}{2}xy)[\log(1-x)]\log(1-y)}{(1-\frac{1}{2}xy)^3}\,\mathrm{d}x\,\mathrm{d}y = 2\log^2 2 + \frac{\pi^2}{3},\quad \int_0^1\int_0^1 \frac{1}{(-\log xy)^{3/2}}\,\mathrm{d}x\,\mathrm{d}y = \sqrt{\pi},$$

$$\int_0^1\int_0^1 \frac{x}{(-\log xy)^{3/2}}\,\mathrm{d}x\,\mathrm{d}y = 2(\sqrt{2}-1)\sqrt{\pi},\quad \int_0^1\int_0^1 \frac{x}{(-\log xy)^{5/4}}\,\mathrm{d}x\,\mathrm{d}y = \Gamma(\tfrac{3}{4}),$$

$$\int_0^1\int_0^1 \frac{1-x}{(-\log xy)^{5/2}}\,\mathrm{d}x\,\mathrm{d}y = \frac{\sqrt{\pi}}{3}(8\sqrt{2}-10),\quad \int_0^1\int_0^1 \frac{1-x^2}{(1+x^2y^2)\log^2 xy}\,\mathrm{d}x\,\mathrm{d}y = \log\frac{\Gamma(\frac{1}{4})}{2\Gamma(\frac{3}{4})} + \frac{2G}{\pi} - \frac{1}{2},$$

$$\int_0^1\int_0^1 \frac{\log^4 xy}{(1+xy)^2}\,\mathrm{d}x\,\mathrm{d}y = \frac{225}{2}\zeta(5),\quad \int_0^1\int_0^1 \frac{x\log^2 xy}{(1+x^2y^2)^2}\,\mathrm{d}x\,\mathrm{d}y = G - \frac{\pi^2}{48} + \frac{\pi^3}{32},$$

$$\int_0^1\int_0^1 \frac{1}{(1+x^2y^2)^2\log xy}\,\mathrm{d}x\,\mathrm{d}y = -\frac{1}{8}(\pi+2), \quad \int_0^1\int_0^1 \frac{x^2}{(1+x^2y^2)^2\log xy}\,\mathrm{d}x\,\mathrm{d}y = -\frac{G}{\pi},$$

$$\int_0^1\int_0^1 \frac{1}{(1+x)(1+y)(1+xy)}\,\mathrm{d}x\,\mathrm{d}y = \sum_{i=0}^\infty (-1)^i \int_0^1\int_0^1 \frac{x^iy^i}{(1+x)(1+y)}\,\mathrm{d}x\,\mathrm{d}y$$

$$= \sum_{i=0}^\infty (-1)^i\left(\int_0^1 \frac{x^i}{x+1}\,\mathrm{d}x\right)^2 = \sum_{i=0}^\infty (-1)^i\left(\sum_{j=1}^\infty (-1)^{j-1}\frac{1}{i+j}\right)^2 = \frac{\pi^2}{24}.$$

If $z \in \mathbb{C}\backslash[1,\infty)$, then

$$\int_0^1\int_0^1 \frac{1}{(1-xyz)\log xy}\,\mathrm{d}x\,\mathrm{d}y = \frac{\log(1-z)}{z}.$$

If $n \ge 1$, then

$$\int_0^1\int_0^1 \frac{(xy)^n\log xy}{1-xy}\,\mathrm{d}x\,\mathrm{d}y = \sum_{i=1}^n \frac{2}{i^3} - 2\zeta(3).$$

If $n \ge 0$, then

$$\int_0^1\int_0^1 \frac{(\log x)(\log y)\log^n xy}{1-xy}\,\mathrm{d}x\,\mathrm{d}y = (-1)^n\frac{(n+3)!\zeta(n+4)}{6}.$$

If $m \ge 2$ and $n \ge \max\{0, m-3\}$, then

$$\int_0^1\int_0^1 \frac{(\log x)(\log y)\log^n xy}{(1-xy)^m}\,\mathrm{d}x\,\mathrm{d}y = (-1)^n\frac{(n+3)!}{6(m-1)!}\sum_{i=1}^{m-1}\begin{bmatrix} m-1 \\ i \end{bmatrix}\zeta(n+4-i).$$

If $n \ge 1$, then

$$\int_0^1\int_0^1 \frac{[(x(1-x)y(1-y)]^n}{(1-xy)\log xy}\,\mathrm{d}x\,\mathrm{d}y = -\sum_{i=n+1}^\infty \int_i^\infty \left(\frac{n!}{\overline{x^{n+1}}}\right)^2 \mathrm{d}x.$$

Source: [116, p. 392] and [1105, 1106, 1266, 2483, 2484, 2492]. **Remark:** $\left[\begin{smallmatrix} m-1 \\ i \end{smallmatrix}\right]$ is a Stirling number of the first kind.

Fact 14.13.6. Let $\alpha = \frac{1}{2}(1+\sqrt{5})$. Then,

$$\int_0^1\int_0^1 \frac{1}{\alpha-xy}\,\mathrm{d}x\,\mathrm{d}y = \frac{\pi^2}{10} - \log^2\alpha, \quad \int_0^1\int_0^1 \frac{1}{\alpha^2-xy}\,\mathrm{d}x\,\mathrm{d}y = \frac{\pi^2}{15} - \log^2\alpha,$$

$$\int_0^1\int_0^1 \frac{1}{1+\alpha xy}\,\mathrm{d}x\,\mathrm{d}y = \frac{\pi^2}{10\alpha} + \frac{1}{\alpha}\log^2\alpha, \quad \int_0^1\int_0^1 \frac{1}{\alpha+xy}\,\mathrm{d}x\,\mathrm{d}y = \frac{\pi^2}{15} - \frac{1}{2}\log^2\alpha,$$

$$\int_0^1\int_0^1 \frac{1}{(\alpha-xy)\log xy}\,\mathrm{d}x\,\mathrm{d}y = -2\log\alpha, \quad \int_0^1\int_0^1 \frac{1}{(\alpha^2-xy)\log xy}\,\mathrm{d}x\,\mathrm{d}y = -\log\alpha,$$

$$\int_0^1\int_0^1 \frac{1}{(1+\alpha xy)\log xy}\,\mathrm{d}x\,\mathrm{d}y = -\frac{2}{\alpha}\log\alpha, \quad \int_0^1\int_0^1 \frac{1}{(\alpha+xy)\log xy}\,\mathrm{d}x\,\mathrm{d}y = -\log\alpha,$$

$$\int_0^1\int_0^1 \frac{\log xy}{\alpha^2-xy}\,\mathrm{d}x\,\mathrm{d}y = \frac{4\pi^2}{15}\log\alpha - \frac{4}{3}\log^3\alpha - \frac{8}{5}\zeta(3),$$

$$\int_0^1\int_0^1 \frac{1}{\alpha^6-x^2y^2}\,\mathrm{d}x\,\mathrm{d}y = \frac{\pi^2}{24\alpha^3} - \frac{3}{4\alpha^3}\log^2\alpha.$$

Source: [1266].

Fact 14.13.7.

$$\int_0^1\int_0^1\int_0^1 \frac{1}{1-(\cos x)(\cos y)\cos z}\,\mathrm{d}x\,\mathrm{d}y\,\mathrm{d}z = \frac{1}{4}\Gamma^2\left(\frac{1}{4}\right).$$

Source: [2106, p. 212].

Fact 14.13.8. Let $n \ge 2$. Then,

$$\int_0^1\cdots\int_0^1 \frac{1}{1-x_1\cdots x_n}\,\mathrm{d}x_1\cdots\mathrm{d}x_n = \zeta(n).$$

Source: [2513, p. 402].

Fact 14.13.9. Let $n \ge 2$. Then,

$$\int_0^1\cdots\int_0^1 \frac{1}{\sum_{i=1}^n x_i}\,\mathrm{d}x_1\cdots\mathrm{d}x_n = \int_0^\infty \left(\frac{1-e^{-x}}{x}\right)^n \mathrm{d}x = \frac{n}{(n-1)!}\sum_{i=2}^n (-1)^{n-i} i^{n-2}\binom{n-1}{i-1}\log i.$$

Source: [1103, pp. 48–50] and [1104].

Fact 14.13.10. Let $n \ge 1$. Then,

$$\int_0^1\cdots\int_0^1 \frac{1}{\left(1+\sum_{i=1}^n x_i^2\right)^{(n+1)/2}}\,\mathrm{d}x_1\cdots\mathrm{d}x_n = \frac{\pi^{(n+1)/2}}{2^n(n+1)\Gamma\left(\frac{n+1}{2}\right)}.$$

In particular,

$$\int_0^1 \frac{1}{1+x^2}\,\mathrm{d}x = \frac{\pi}{4}, \qquad \int_0^1\int_0^1 \frac{1}{(1+x^2+y^2)^{3/2}}\,\mathrm{d}x\,\mathrm{d}y = \frac{\pi}{6}.$$

Fact 14.13.11. Let $n \ge 1$, $\mathcal{S} \triangleq \{x \in (0,\infty)^n\colon \|x\|_1 \le 1\}$, and $\alpha_1,\ldots,\alpha_n \in \mathrm{ORHP}$. Then,

$$\int\cdots\int_{\mathcal{S}} \prod_{i=1}^n x_i^{\alpha_i-1}\,\mathrm{d}x_1\cdots\mathrm{d}x_n = \frac{\prod_{i=1}^n \Gamma(\alpha_i)}{\Gamma(1+\sum_{i=1}^n \alpha_i)}.$$

In particular, if k and l are nonnegative integers, then

$$\int_0^1\int_0^{1-y} x^k y^l\,\mathrm{d}x\,\mathrm{d}y = \frac{1}{k+1}\int_0^1 y^l(1-y)^{k+1}\,\mathrm{d}y = \frac{k!l!}{(k+l+2)!}.$$

Source: [116, pp. 32, 33].

Fact 14.13.12. Let $n \ge 1$, $\mathcal{S} \triangleq \{x \in (0,\infty)^n\colon \|x\|_1 = 1\}$, and $\alpha_1,\ldots,\alpha_n \in \mathrm{ORHP}$. Then,

$$\int\cdots\int_{\mathcal{S}} \prod_{i=1}^n x_i^{\alpha_i-1}\,\mathrm{d}x_1\cdots\mathrm{d}x_{n-1} = \frac{\prod_{i=1}^n \Gamma(\alpha_i)}{\Gamma(\sum_{i=1}^n \alpha_i)}.$$

Source: [116, p. 35].

Fact 14.13.13. Let $n \ge 1$, for all $i \in \{1,\ldots,n\}$, let $\alpha_i \in \mathrm{ORHP}$, $a_i > 0$, and $p_i > 0$, and let $\mathcal{S} \triangleq \{x \in (0,\infty)^n\colon \sum_{i=1}^n (x_{(i)}/a_i)^{p_i} \le 1\}$. Then,

$$\int\cdots\int_{\mathcal{S}} \prod_{i=1}^n x_i^{\alpha_i-1}\,\mathrm{d}x_1\cdots\mathrm{d}x_n = \frac{\prod_{i=1}^n (a_i^{\alpha_i}/p_i)\Gamma(\alpha_i/p_i)}{\Gamma(1+\sum_{i=1}^n \alpha_i/p_i)}.$$

In particular, the volume of the hyperellipsoidal solid $\mathcal{E} \triangleq \{x \in \mathbb{R}^n\colon \sum_{i=1}^n (x_{(i)}/a_i)^2 \le 1\}$ is

$$\mathrm{vol}(\mathcal{E}) = \frac{\pi^{n/2}\prod_{i=1}^n a_i}{\Gamma\left(\frac{n}{2}+1\right)}.$$

Source: [116, p. 33]. **Related:** Fact 5.5.14 and Fact 5.5.15.

Fact 14.13.14. Let $n \ge 2$ and $a < -1/n$. Then,

$$\int_{-\pi}^{\pi} \cdots \int_{-\pi}^{\pi} \prod |e^{\jmath x_i} - e^{\jmath x_j}|^{2a}\, \mathrm{d}x_1 \cdots \mathrm{d}x_n = \frac{(2\pi)^n \Gamma(na+1)}{\Gamma^n(a+1)},$$

where the product is taken over all $i, j \in \{1, \ldots, n\}$ such that $i < j$. In particular,

$$\int_{-\pi}^{\pi} \int_{-\pi}^{\pi} |e^{\jmath x} - e^{\jmath y}|^2\, \mathrm{d}x\, \mathrm{d}y = 2\pi^2, \quad \int_{-\pi}^{\pi} \int_{-\pi}^{\pi} \int_{-\pi}^{\pi} |(e^{\jmath x} - e^{\jmath y})(e^{\jmath x} - e^{\jmath z})(e^{\jmath y} - e^{\jmath z})|^2\, \mathrm{d}x\, \mathrm{d}y\, \mathrm{d}z = 48\pi^3.$$

Source: [116, p. 426] and [179, p. 144]. **Remark:** This is *Dyson's integral.*

Fact 14.13.15. Let $n \ge 2$ and $a > -1/n$. Then,

$$\int_{-\infty}^{\infty} \cdots \int_{-\infty}^{\infty} \exp\left(-\frac{1}{2}\sum_{i=1}^{n} x_i^2\right) \prod |x_i - x_j|^{2a}\, \mathrm{d}x_1 \cdots \mathrm{d}x_n = (2\pi)^{n/2} \prod_{i=1}^{n} \frac{\Gamma(ia+1)}{\Gamma(a+1)},$$

where the product under the integral is taken over all $i, j \in \{1, \ldots, n\}$ such that $i < j$. **Source:** [116, p. 407] and [179, p. 143].

14.14 Notes

Historical notes on the accuracy of integral tables are given in [2585].

Proofs of many of the formulas in [1217] are given in [2069, 2070].

An extensive collection of formulas is given in [2268].

Integrals with matrix arguments are considered in [2253, pp. 230–237].

Chapter Fifteen
The Matrix Exponential and Stability Theory

The matrix exponential function is fundamental to the study of linear ordinary differential equations. This chapter focuses on the properties of the matrix exponential as well as on stability theory.

15.1 Definition of the Matrix Exponential

The scalar differential equation

$$\dot{x}(t) = ax(t), \tag{15.1.1}$$

$$x(0) = x_0, \tag{15.1.2}$$

where $t \in [0, \infty)$, $a \in \mathbb{R}$, and $x(t) \in \mathbb{R}$, has the solution

$$x(t) = e^{at}x_0, \tag{15.1.3}$$

where $t \in [0, \infty)$. We are interested in systems of linear differential equations of the form

$$\dot{x}(t) = Ax(t), \tag{15.1.4}$$

$$x(0) = x_0, \tag{15.1.5}$$

where $t \in [0, \infty)$, $x(t) \in \mathbb{R}^n$, and $A \in \mathbb{R}^{n\times n}$. Here $\dot{x}(t)$ denotes $\frac{\mathrm{d}x(t)}{\mathrm{d}t}$, where the derivative is one-sided for $t = 0$ and two-sided for $t > 0$. The solution of (15.1.4), (15.1.5) is given by

$$x(t) = e^{tA}x_0, \tag{15.1.6}$$

where $t \in [0, \infty)$ and e^{tA} is the *matrix exponential.* The following definition is based on (12.8.2).

Definition 15.1.1. Let $A \in \mathbb{F}^{n\times n}$. Then, the *matrix exponential*, which is written as either $e^A \in \mathbb{F}^{n\times n}$ or $\exp(A) \in \mathbb{F}^{n\times n}$, is the matrix

$$e^A \triangleq \sum_{i=0}^{\infty} \frac{1}{i!}A^i. \tag{15.1.7}$$

Note that $0! \triangleq 1$, $A^0 = I_n$, and $e^{0_{n\times n}} = I_n$.

Proposition 15.1.2. Let $A \in \mathbb{F}^{n\times n}$. Then, the following statements hold:

i) The series (15.1.7) converges absolutely.

ii) The series (15.1.7) converges to e^A.

iii) Let $\|\cdot\|$ be a normalized submultiplicative norm on $\mathbb{F}^{n\times n}$. Then,

$$e^{-\|A\|} \le \|e^A\| \le e^{\|A\|}. \tag{15.1.8}$$

Proof. To prove *i)*, let $\|\cdot\|$ be a normalized submultiplicative norm on $\mathbb{F}^{n\times n}$. Then, for all $k \ge 1$,

$$\sum_{i=0}^{k} \tfrac{1}{i!}\|A^i\| \le \sum_{i=0}^{k} \tfrac{1}{i!}\|A\|^i \le e^{\|A\|}.$$

Since the sequence $\left(\sum_{i=0}^{k}\frac{1}{i!}\|A^i\|\right)_{k=0}^{\infty}$ of partial sums is increasing and bounded, there exists $\alpha > 0$ such that the series $\sum_{i=0}^{\infty}\frac{1}{i!}\|A^i\|$ converges to α. Hence, the series $\sum_{i=0}^{\infty}\frac{1}{i!}A^i$ converges absolutely.

ii) follows from *i*) using Proposition 12.3.4. Next, to prove the second inequality in (15.1.8), note that

$$\|e^A\| = \left\|\sum_{i=0}^{\infty}\frac{1}{i!}A^i\right\| \le \sum_{i=0}^{\infty}\frac{1}{i!}\|A^i\| \le \sum_{i=0}^{\infty}\frac{1}{i!}\|A\|^i = e^{\|A\|}.$$

Finally, $1 \le \|e^A\|\|e^{-A}\| \le \|e^A\|e^{\|A\|}$, and thus $e^{-\|A\|} \le \|e^A\|$. □

Proposition 15.1.3. Let $A \in \mathbb{F}^{n\times n}$. Then,

$$e^A = \lim_{k\to\infty}\left(I + \tfrac{1}{k}A\right)^k. \tag{15.1.9}$$

Proof. It follows from the binomial theorem that

$$\left(I + \tfrac{1}{k}A\right)^k = \sum_{i=0}^{k}\alpha_i(k)A^i,$$

where

$$\alpha_i(k) \triangleq \frac{1}{k^i}\binom{k}{i} = \frac{1}{k^i}\frac{k!}{i!(k-i)!}.$$

For all $i \in \mathbb{P}$, it follows that $\alpha_i(k) \to 1/i!$ as $k \to \infty$. Hence,

$$\lim_{k\to\infty}\left(I + \tfrac{1}{k}A\right)^k = \lim_{k\to\infty}\sum_{i=0}^{k}\alpha_i(k)A^i = \sum_{i=0}^{\infty}\frac{1}{i!}A^i = e^A.$$ □

Proposition 15.1.4. Let $A \in \mathbb{F}^{n\times n}$. Then, for all $t \in \mathbb{R}$,

$$e^{tA} - I = \int_0^t Ae^{\tau A}\,\mathrm{d}\tau, \tag{15.1.10}$$

$$\frac{\mathrm{d}}{\mathrm{d}t}e^{tA} = Ae^{tA}. \tag{15.1.11}$$

Proof. Note that

$$\int_0^t Ae^{\tau A}\,\mathrm{d}\tau = \int_0^t\sum_{k=0}^{\infty}\frac{1}{k!}\tau^kA^{k+1}\,\mathrm{d}\tau = \sum_{k=0}^{\infty}\frac{1}{k!}\frac{t^{k+1}}{k+1}A^{k+1} = e^{tA} - I,$$

which yields (15.1.10), while differentiating (15.1.10) with respect to t yields (15.1.11). □

Proposition 15.1.5. Let $A, B \in \mathbb{F}^{n\times n}$. Then, $AB = BA$ if and only if, for all $t \in [0,\infty)$,

$$e^{tA}e^{tB} = e^{t(A+B)}. \tag{15.1.12}$$

Proof. Suppose that $AB = BA$. By expanding e^{tA}, e^{tB}, and $e^{t(A+B)}$, it can be seen that the expansions of $e^{tA}e^{tB}$ and $e^{t(A+B)}$ are identical. Conversely, differentiating (15.1.12) twice with respect to t and setting $t = 0$ yields $AB = BA$. □

Corollary 15.1.6. Let $A, B \in \mathbb{F}^{n\times n}$, and assume that $AB = BA$. Then,

$$e^Ae^B = e^Be^A = e^{A+B}. \tag{15.1.13}$$

The converse of Corollary 15.1.6 is false. Let $A \triangleq \begin{bmatrix} 0 & \pi \\ -\pi & 0 \end{bmatrix}$ and $B \triangleq \begin{bmatrix} 0 & (7+4\sqrt{3})\pi \\ (-7+4\sqrt{3})\pi & 0 \end{bmatrix}$. Then, $e^A = e^B = -I$ and $e^{A+B} = e^Ae^B = e^Be^A = I$. However, $AB \ne BA$. A partial converse is given by Fact 15.15.1.

Proposition 15.1.7. Let $A \in \mathbb{F}^{n\times n}$ and $B \in \mathbb{F}^{m\times m}$. Then,

$$e^{A\otimes I_m} = e^A \otimes I_m, \tag{15.1.14}$$

$$e^{I_n\otimes B} = I_n \otimes e^B, \tag{15.1.15}$$

$$e^{A\oplus B} = e^A \otimes e^B. \tag{15.1.16}$$

Proof. Note that

$$\begin{aligned} e^{A\otimes I_m} &= I_{nm} + A\otimes I_m + \tfrac{1}{2!}(A\otimes I_m)^2 + \cdots = I_n\otimes I_m + A\otimes I_m + \tfrac{1}{2!}(A^2\otimes I_m) + \cdots \\ &= (I_n + A + \tfrac{1}{2!}A^2 + \cdots)\otimes I_m = e^A\otimes I_m \end{aligned}$$

and similarly for (15.1.15). To prove (15.1.16), note that $(A\otimes I_m)(I_n\otimes B) = A\otimes B$ and $(I_n\otimes B)(A\otimes I_m) = A\otimes B$, which shows that $A\otimes I_m$ and $I_n\otimes B$ commute. Thus, by Corollary 15.1.6,

$$e^{A\oplus B} = e^{A\otimes I_m + I_n\otimes B} = e^{A\otimes I_m}e^{I_n\otimes B} = (e^A\otimes I_m)(I_n\otimes e^B) = e^A\otimes e^B. \qquad \square$$

15.2 Structure of the Matrix Exponential

To elucidate the structure of the matrix exponential, recall that, by Theorem 6.6.1, each term A^k in (15.1.7), where $k \ge r \triangleq \deg \mu_A$, can be expressed as a linear combination of $I, A, \dots, A^{r-1}$. The following result provides an expression for e^{tA} in terms of $I, A, \dots, A^{r-1}$.

Proposition 15.2.1. Let $A \in \mathbb{F}^{n\times n}$. Then, for all $t \in \mathbb{R}$,

$$e^{tA} = \frac{1}{2\pi j}\oint_{\mathcal{C}} (zI - A)^{-1} e^{tz}\,\mathrm{d}z = \sum_{i=0}^{n-1} \psi_i(t) A^i, \tag{15.2.1}$$

where, for all $i \in \{0, \dots, n-1\}$, $\psi_i(t)$ is given by

$$\psi_i(t) \triangleq \frac{1}{2\pi j}\oint_{\mathcal{C}} \frac{\chi_A^{[i+1]}(z)}{\chi_A(z)} e^{tz}\,\mathrm{d}z, \tag{15.2.2}$$

where $\mathcal{C}$ is a simple, closed contour in the complex plane enclosing $\operatorname{spec}(A)$,

$$\chi_A(s) = s^n + \beta_{n-1}s^{n-1} + \cdots + \beta_1 s + \beta_0, \tag{15.2.3}$$

and the polynomials $\chi_A^{[1]}, \dots, \chi_A^{[n]}$ are defined by the recursion

$$s\chi_A^{[i+1]}(s) = \chi_A^{[i]}(s) - \beta_i, \quad i = 0, \dots, n-1,$$

where $\chi_A^{[0]} \triangleq \chi_A$ and $\chi_A^{[n]}(s) = 1$. Furthermore, for all $i \in \{0, \dots, n-1\}$ and $t \ge 0$, $\psi_i(t)$ satisfies

$$\psi_i^{(n)}(t) + \beta_{n-1}\psi_i^{(n-1)}(t) + \cdots + \beta_1\psi_i'(t) + \beta_0\psi_i(t) = 0, \tag{15.2.4}$$

where, for all $i, j \in \{0, \dots, n-1\}$,

$$\psi_i^{(j)}(0) = \delta_{i,j}. \tag{15.2.5}$$

Proof. See [1785, 1881], [2912, p. 31], and Fact 6.9.7. $\square$

The coefficient $\psi_i(t)$ of A^i in (15.2.1) can be further characterized in terms of the Laplace transform. Define

$$\hat{x}(s) \triangleq \mathcal{L}\{x(t)\} \triangleq \int_0^\infty e^{-st}x(t)\,\mathrm{d}t. \tag{15.2.6}$$

Then,

$$\mathcal{L}\{\dot{x}(t)\} = s\hat{x}(s) - x(0), \tag{15.2.7}$$

$$\mathcal{L}\{\ddot{x}(t)\} = s^2\hat{x}(s) - sx(0) - \dot{x}(0). \tag{15.2.8}$$

The following result shows that the resolvent of A is the Laplace transform of the exponential of A. See (6.4.24).

Proposition 15.2.2. Let $A \in \mathbb{F}^{n\times n}$, and define $\psi_0, \ldots, \psi_{n-1}$ as in Proposition 15.2.1. Then, for all $s \in \mathbb{C}\backslash\mathrm{spec}(A)$,

$$\mathcal{L}\{e^{tA}\} = \int_0^\infty e^{-st}e^{tA}\,\mathrm{d}t = (sI - A)^{-1}. \tag{15.2.9}$$

Furthermore, for all $i \in \{0, \ldots, n-1\}$, the Laplace transform $\hat{\psi}_i(s)$ of $\psi_i(t)$ is given by

$$\hat{\psi}_i(s) = \frac{\chi_A^{[i+1]}(s)}{\chi_A(s)}, \tag{15.2.10}$$

$$(sI - A)^{-1} = \sum_{i=0}^{n-1} \hat{\psi}_i(s)A^i. \tag{15.2.11}$$

Proof. Let $s \in \mathbb{C}$ satisfy $\mathrm{Re}\, s > \alpha_{\max}(A)$ so that $A - sI$ is asymptotically stable. It thus follows from Lemma 15.10.2 that

$$\mathcal{L}\{e^{tA}\} = \int_0^\infty e^{-st}e^{tA}\,\mathrm{d}t = \int_0^\infty e^{t(A-sI)}\,\mathrm{d}t = (sI - A)^{-1}.$$

By analytic continuation, the expression $\mathcal{L}\{e^{tA}\}$ is given by (15.2.9) for all $s \in \mathbb{C}\backslash\,\mathrm{spec}(A)$. □

Comparing (15.2.11) with the expression for $(sI - A)^{-1}$ given by (6.4.24) shows that there exist $B_0, \ldots, B_{n-2} \in \mathbb{F}^{n\times n}$ such that

$$\sum_{i=0}^{n-1} \hat{\psi}_i(s)A^i = \frac{s^{n-1}}{\chi_A(s)}I + \frac{s^{n-2}}{\chi_A(s)}B_{n-2} + \cdots + \frac{s}{\chi_A(s)}B_1 + \frac{1}{\chi_A(s)}B_0. \tag{15.2.12}$$

To further illustrate the structure of e^{tA}, where $A \in \mathbb{F}^{n\times n}$, let $A = SBS^{-1}$, where $B = \mathrm{diag}(B_1, \ldots, B_k)$ is the Jordan form of A. Hence, by Proposition 15.2.8,

$$e^{tA} = Se^{tB}S^{-1}, \tag{15.2.13}$$

where

$$e^{tB} = \mathrm{diag}(e^{tB_1}, \ldots, e^{tB_k}). \tag{15.2.14}$$

The structure of e^{tB} can thus be determined by considering the block $B_i \in \mathbb{F}^{\alpha_i\times\alpha_i}$, which, for all $i \in \{1, \ldots, k\}$ has the form

$$B_i = \lambda_i I_{\alpha_i} + N_{\alpha_i}. \tag{15.2.15}$$

Since $\lambda_i I_{\alpha_i}$ and N_{α_i} commute, it follows from Proposition 15.1.5 that

$$e^{tB_i} = e^{t(\lambda_i I_{\alpha_i} + N_{\alpha_i})} = e^{\lambda_i t I_{\alpha_i}}e^{tN_{\alpha_i}} = e^{\lambda_i t}e^{tN_{\alpha_i}}. \tag{15.2.16}$$

Since $N_{\alpha_i}^{\alpha_i} = 0$, it follows that $e^{tN_{\alpha_i}}$ is a finite sum of powers of tN_{α_i}. Specifically,

$$e^{tN_{\alpha_i}} = I_{\alpha_i} + tN_{\alpha_i} + \tfrac{1}{2}t^2N_{\alpha_i}^2 + \cdots + \tfrac{1}{(\alpha_i-1)!}t^{\alpha_i-1}N_{\alpha_i}^{\alpha_i-1}, \tag{15.2.17}$$

and thus

$$e^{tN_{\alpha_i}} = \begin{bmatrix} 1 & t & \frac{t^2}{2} & \cdots & \frac{t^{\alpha_i-2}}{(\alpha_i-2)!} & \frac{t^{\alpha_i-1}}{(\alpha_i-1)!} \\ 0 & 1 & t & \ddots & \frac{t^{\alpha_i-3}}{(\alpha_i-3)!} & \frac{t^{\alpha_i-2}}{(\alpha_i-2)!} \\ 0 & 0 & 1 & \ddots & \frac{t^{\alpha_i-4}}{(\alpha_i-4)!} & \frac{t^{\alpha_i-3}}{(\alpha_i-3)!} \\ \vdots & \vdots & \ddots & \ddots & \ddots & \vdots \\ 0 & 0 & 0 & \ddots & 1 & t \\ 0 & 0 & 0 & \cdots & 0 & 1 \end{bmatrix}, \tag{15.2.18}$$

which is upper triangular and Toeplitz (see Fact 15.14.1). Alternatively, (15.2.18) follows from (12.8.5) with $f(s) = e^{st}$.

Note that (15.2.16) follows from (12.8.5) with $f(\lambda) = e^{\lambda t}$. Furthermore, every entry of e^{tB_i} is of the form $\frac{1}{r!}t^r e^{\lambda_i t}$, where $r \in \{0, \alpha_i - 1\}$ and λ_i is an eigenvalue of A. Reconstructing A by means of $A = SBS^{-1}$ shows that every entry of A is a linear combination of the entries of the blocks e^{tB_i}. If A is real, then e^{tA} is also real. Thus, the term $e^{\lambda_i t}$ for complex $\lambda_i = \nu_i + \omega_i \jmath \in \operatorname{spec}(A)$, where ν_i and ω_i are real, yields terms of the form $e^{\nu_i t}\cos\omega_i t$ and $e^{\nu_i t}\sin\omega_i t$.

The following result follows from either (15.2.18) or Corollary 12.8.4.

Proposition 15.2.3. Let $A \in \mathbb{F}^{n\times n}$. Then,

$$\operatorname{mspec}(e^A) = \{e^\lambda\colon\ \lambda \in \operatorname{mspec}(A)\}_{\rm ms}. \tag{15.2.19}$$

Proof. Note that every diagonal entry of the Jordan form of e^A is of the form e^λ, where $\lambda \in \operatorname{spec}(A)$. □

Corollary 15.2.4. Let $A \in \mathbb{F}^{n\times n}$. Then,

$$\det e^A = e^{\operatorname{tr} A}. \tag{15.2.20}$$

Corollary 15.2.5. Let $A \in \mathbb{F}^{n\times n}$, and assume that $\operatorname{tr} A = 0$. Then, $\det e^A = 1$.

Corollary 15.2.6. Let $A \in \mathbb{F}^{n\times n}$. Then, the following statements hold:

i) If e^A is unitary, then $\operatorname{spec}(A) \subset \jmath\mathbb{R}$.

ii) $\operatorname{spec}(e^A) \subset \mathbb{R}$ if and only if $\operatorname{Im}\operatorname{spec}(A) \subset \pi\mathbb{Z}$.

Proposition 15.2.7. Let $A \in \mathbb{F}^{n\times n}$. Then, the following statements hold:

i) A and e^A have the same number of Jordan blocks of corresponding sizes.

ii) e^A is semisimple if and only if A is semisimple.

iii) If $\mu \in \operatorname{spec}(e^A)$, then

$$\operatorname{amult}_{\exp(A)}(\mu) = \sum_{\{\lambda\in\operatorname{spec}(A)\colon\ e^\lambda=\mu\}} \operatorname{amult}_A(\lambda), \tag{15.2.21}$$

$$\operatorname{gmult}_{\exp(A)}(\mu) = \sum_{\{\lambda\in\operatorname{spec}(A)\colon\ e^\lambda=\mu\}} \operatorname{gmult}_A(\lambda). \tag{15.2.22}$$

iv) If e^A is simple, then A is simple.

v) If e^A is cyclic, then A is cyclic.

vi) e^A is a scalar multiple of the identity matrix if and only if A is semisimple and every pair of eigenvalues of A differs by an integer multiple of $2\pi\jmath$.

vii) e^A is a real scalar multiple of the identity matrix if and only if A is semisimple, every pair of eigenvalues of A differs by an integer multiple of $2\pi\jmath$, and the imaginary part of every eigenvalue of A is an integer multiple of $\pi\jmath$.

Proof. To prove *i)*, note that, for all $t \neq 0$, $\operatorname{def}(e^{tN_{\alpha_i}} - I_{\alpha_i}) = 1$, and thus the geometric multiplicity of (15.2.18) is 1. Since (15.2.18) has one distinct eigenvalue, it follows that (15.2.18) is cyclic. Hence, by Proposition 7.7.15, (15.2.18) is similar to a single Jordan block. Now, *i)* follows by setting $t = 1$ and applying this argument to each Jordan block of A. *ii)*–*v)* follow by similar arguments.

To prove *vi)*, note that, for all $\lambda_i, \lambda_j \in \operatorname{spec}(A)$, it follows that $e^{\lambda_i} = e^{\lambda_j}$. Furthermore, since A is semisimple, it follows from *ii)* that e^A is also semisimple. Since all of the eigenvalues of e^A are equal, it follows that e^A is a scalar multiple of the identity matrix. Finally, *vii)* is an immediate consequence of *vii)*. □

Proposition 15.2.8. Let $A \in \mathbb{F}^{n\times n}$. Then, the following statements hold:

i) $(e^A)^{\mathsf{T}} = e^{A^{\mathsf{T}}}$.

ii) $\overline{(e^A)} = e^{\overline{A}}$.

iii) $(e^A)^* = e^{A^*}$.

iv) e^A is nonsingular, and $(e^A)^{-1} = e^{-A}$.

v) If $S \in \mathbb{F}^{n\times n}$ is nonsingular, then $e^{SAS^{-1}} = Se^AS^{-1}$.

vi) If $A = \operatorname{diag}(A_1, \ldots, A_k)$, where $A_i \in \mathbb{F}^{n_i\times n_i}$ for all $i \in \{1, \ldots, k\}$, then $e^A = \operatorname{diag}(e^{A_1}, \ldots, e^{A_k})$.

vii) If A is Hermitian, then e^A is positive definite.

Furthermore, the following statements are equivalent:

viii) A is normal.

ix) $\operatorname{tr} e^{A^*}e^A = \operatorname{tr} e^{A^*+A}$.

x) $e^{A^*}e^A = e^{A^*+A}$.

xi) $e^Ae^{A^*} = e^{A^*}e^A = e^{A^*+A}$.

Finally, the following statements hold:

xii) If A is normal, then e^A is normal.

xiii) e^A is normal if and only if A is unitarily similar to a block-diagonal matrix $\operatorname{diag}(A_1, \ldots, A_k)$ such that, for all $i \in \{1, \ldots, k\}$, A_i is semisimple and each pair of eigenvalues of A_i differ by an integer multiple of $2\pi\jmath$, and, for all distinct $i, j \in \{1, \ldots, k\}$, $\operatorname{spec}(e^{A_i}) \neq \operatorname{spec}(e^{A_j})$.

xiv) If e^A is normal and no pair of eigenvalues of A differ by an integer multiple of $2\pi\jmath$, then A is normal.

xv) A is skew Hermitian if and only if A is normal and e^A is unitary.

xvi) If $\mathbb{F} = \mathbb{R}$ and A is skew symmetric, then e^A is orthogonal and $\det e^A = 1$.

xvii) If e^A is unitary, then either A is skew Hermitian or at least two eigenvalues of A differ by a nonzero integer multiple of $2\pi\jmath$.

xviii) e^A is unitary if and only if A is unitarily similar to a block-diagonal matrix $\operatorname{diag}(A_1, \ldots, A_k)$ such that, for all $i \in \{1, \ldots, k\}$, A_i is semisimple, every eigenvalue of A_i is imaginary, and each pair of eigenvalues of A_i differ by an integer multiple of $2\pi\jmath$, and, for all distinct $i, j \in \{1, \ldots, k\}$, $\operatorname{spec}(e^{A_i}) \neq \operatorname{spec}(e^{A_j})$.

xix) e^A is Hermitian if and only if A is unitarily similar to a block-diagonal matrix $\operatorname{diag}(A_1, \ldots, A_k)$ such that, for all $i \in \{1, \ldots, k\}$, A_i is semisimple, the imaginary part of every eigenvalue of A_i is an integer multiple of $\pi\jmath$, and each pair of eigenvalues of A_i differ by an integer multiple of $2\pi\jmath$, and, for all distinct $i, j \in \{1, \ldots, k\}$, $\operatorname{spec}(e^{A_i}) \neq \operatorname{spec}(e^{A_j})$.

xx) e^A is positive definite if and only if A is unitarily similar to a block-diagonal matrix $\operatorname{diag}(A_1,$

$\ldots, A_k)$ such that, for all $i \in \{1,\ldots,k\}$, A_i is semisimple, the imaginary part of every eigenvalue of A_i is an integer multiple of $2\pi\jmath$, and each pair of eigenvalues of A_i differ by an integer multiple of $2\pi\jmath$, and, for all distinct $i, j \in \{1,\ldots,k\}$, $\operatorname{spec}(e^{A_i}) \neq \operatorname{spec}(e^{A_j})$.

Proof. The equivalence of *viii)* and *ix)* is given in [987, 2476], while the equivalence of *viii)* and *xi)* is given in [2395]. *xi)* $\Longrightarrow$ *x)* $\Longrightarrow$ *ix)* is immediate. *xii)* follows from the fact that *viii)* $\Longrightarrow$ *xi)*. *xiii)* is given in [2934]. *xiv)* is a consequence of *xiii)*. To prove sufficiency in *xv)*, note that $e^{A+A^*} = e^A e^{A^*} = e^A(e^A)^* = I = e^0$. Since $A + A^*$ is Hermitian, it follows from *iii)* of Proposition 15.4.3 that $A + A^* = 0$. The converse is immediate. To prove *xvi)*, note that $e^A(e^A)^{\mathrm{T}} = e^A e^{A^{\mathrm{T}}} = e^A e^{-A} = e^A(e^A)^{-1} = I$, and, using Corollary 15.2.5, $\det e^A = e^{\operatorname{tr} A} = e^0 = 1$. To prove *xvii)*, note that it follows from *xiii)* that, if every block A_i is scalar, then A is skew Hermitian, while, if at least one block A_i is not scalar, then A has at least two eigenvalues that differ by an integer multiple of $2\pi\jmath$. Finally, *xviii)*–*xx)* are analogous to *xiii)*. □

The converse of *xii)* is false. Let $A \triangleq \begin{bmatrix} -2\pi & 4\pi \\ -2\pi & 2\pi \end{bmatrix}$, which satisfies $e^A = I$ but is not normal. Likewise, $A = \begin{bmatrix} \pi\jmath & 1 \\ 0 & -\pi\jmath \end{bmatrix}$ satisfies $e^A = -I$ but is not normal. For both matrices, $e^{A^*}e^A = e^A e^{A^*} = I$, but $e^{A^*}e^A \neq e^{A^*+A}$, which confirms that *xi)* does not hold. Both matrices have eigenvalues $\pm\pi\jmath$.

Proposition 15.2.9. The following statements hold:

i) If $A, B \in \mathbb{F}^{n\times n}$ are similar, then e^A and e^B are similar.

ii) If $A, B \in \mathbb{F}^{n\times n}$ are unitarily similar, then e^A and e^B are unitarily similar.

The converse of *i)* is false. Let $A \triangleq \begin{bmatrix} 0 & 0 \\ 0 & 0 \end{bmatrix}$ and $B \triangleq \begin{bmatrix} 0 & 2\pi \\ -2\pi & 0 \end{bmatrix}$, which satisfy $e^A = e^B = I$ but are not similar.

15.3 Explicit Expressions

This section presents explicit expressions for the exponential of 2×2 matrices. These expressions are given in terms of either the entries of A or the eigenvalues of A.

Lemma 15.3.1. Let $A \triangleq \begin{bmatrix} a & b \\ 0 & d \end{bmatrix} \in \mathbb{C}^{2\times 2}$. Then,

$$e^A = \begin{cases} e^a \begin{bmatrix} 1 & b \\ 0 & 1 \end{bmatrix}, & a = d, \\ \begin{bmatrix} e^a & b\frac{e^a - e^d}{a-d} \\ 0 & e^d \end{bmatrix}, & a \neq d. \end{cases} \tag{15.3.1}$$

The following result gives an expression for e^A in terms of the eigenvalues of A.

Proposition 15.3.2. Let $A \in \mathbb{C}^{2\times 2}$ and $\operatorname{mspec}(A) = \{\lambda, \mu\}_{\mathrm{ms}}$. Then,

$$e^A = \begin{cases} e^\lambda[(1-\lambda)I + A], & \lambda = \mu, \\ \frac{\mu e^\lambda - \lambda e^\mu}{\mu - \lambda} I + \frac{e^\mu - e^\lambda}{\mu - \lambda} A, & \lambda \neq \mu. \end{cases} \tag{15.3.2}$$

Proof. This result follows from Theorem 12.8.2. Alternatively, suppose that $\lambda = \mu$. Then, there exists a nonsingular matrix $S \in \mathbb{C}^{2\times 2}$ such that $A = S\begin{bmatrix} \lambda & \alpha \\ 0 & \lambda \end{bmatrix}S^{-1}$, where $\alpha \in \mathbb{C}$. Hence, $e^A = e^\lambda S\begin{bmatrix} 1 & \alpha \\ 0 & 1 \end{bmatrix}S^{-1} = e^\lambda[(1-\lambda)I + A]$. Now, suppose that $\lambda \neq \mu$. Then, there exists a nonsingular matrix $S \in \mathbb{C}^{2\times 2}$ such that $A = S\begin{bmatrix} \lambda & 0 \\ 0 & \mu \end{bmatrix}S^{-1}$. Hence, $e^A = S\begin{bmatrix} e^\lambda & 0 \\ 0 & e^\mu \end{bmatrix}S^{-1}$. Then, the equality $\begin{bmatrix} e^\lambda & 0 \\ 0 & e^\mu \end{bmatrix} = \frac{\mu e^\lambda - \lambda e^\mu}{\mu - \lambda} I + \frac{e^\mu - e^\lambda}{\mu - \lambda}\begin{bmatrix} \lambda & 0 \\ 0 & \mu \end{bmatrix}$ yields the desired result. □

Next, we give an expression for e^A in terms of the entries of $A \in \mathbb{R}^{2\times 2}$.

Corollary 15.3.3. Let $A \triangleq \begin{bmatrix} a & b \\ c & d \end{bmatrix} \in \mathbb{R}^{2\times 2}$, and define $\gamma \triangleq (a-d)^2 + 4bc$ and $\delta \triangleq \frac{1}{2}|\gamma|^{1/2}$. Then,

$$e^A = \begin{cases} e^{\frac{a+d}{2}}\begin{bmatrix} \cos\delta + \frac{a-d}{2\delta}\sin\delta & \frac{b}{\delta}\sin\delta \\ \frac{c}{\delta}\sin\delta & \cos\delta - \frac{a-d}{2\delta}\sin\delta \end{bmatrix}, & \gamma < 0, \\ e^{\frac{a+d}{2}}\begin{bmatrix} 1 + \frac{a-d}{2} & b \\ c & 1 - \frac{a-d}{2} \end{bmatrix}, & \gamma = 0, \\ e^{\frac{a+d}{2}}\begin{bmatrix} \cosh\delta + \frac{a-d}{2\delta}\sinh\delta & \frac{b}{\delta}\sinh\delta \\ \frac{c}{\delta}\sinh\delta & \cosh\delta - \frac{a-d}{2\delta}\sinh\delta \end{bmatrix}, & \gamma > 0. \end{cases} \tag{15.3.3}$$

Proof. The eigenvalues of A are $\lambda \triangleq \frac{1}{2}(a+d-\sqrt{\gamma})$ and $\mu \triangleq \frac{1}{2}(a+d+\sqrt{\gamma})$. Hence, $\lambda = \mu$ if and only if $\gamma = 0$. The result now follows from Proposition 15.3.2. □

Example 15.3.4. Let $A \triangleq \begin{bmatrix} \nu & \omega \\ -\omega & \nu \end{bmatrix} \in \mathbb{R}^{2\times 2}$. Then,

$$e^{tA} = e^{\nu t}\begin{bmatrix} \cos\omega t & \sin\omega t \\ -\sin\omega t & \cos\omega t \end{bmatrix}. \tag{15.3.4}$$

On the other hand, if $A \triangleq \begin{bmatrix} \nu & \omega \\ \omega & -\nu \end{bmatrix}$, then

$$e^{tA} = \begin{bmatrix} \cosh\delta t + \frac{\nu}{\delta}\sinh\delta t & \frac{\omega}{\delta}\sinh\delta t \\ \frac{\omega}{\delta}\sinh\delta t & \cosh\delta t - \frac{\nu}{\delta}\sinh\delta t \end{bmatrix}, \tag{15.3.5}$$

where $\delta \triangleq \sqrt{\omega^2 + \nu^2}$. ◇

Example 15.3.5. Let $\alpha \in \mathbb{F}$, and define $A \triangleq \begin{bmatrix} 0 & 1 \\ 0 & \alpha \end{bmatrix}$. Then,

$$e^{tA} = \begin{cases} \begin{bmatrix} 1 & \alpha^{-1}(e^{\alpha t} - 1) \\ 0 & e^{\alpha t} \end{bmatrix}, & \alpha \neq 0, \\ \begin{bmatrix} 1 & t \\ 0 & 1 \end{bmatrix}, & \alpha = 0. \end{cases}$$

◇

Example 15.3.6. Let $\theta \in \mathbb{R}$, and define $A \triangleq \begin{bmatrix} 0 & \theta \\ -\theta & 0 \end{bmatrix}$. Then,

$$e^A = \begin{bmatrix} \cos\theta & \sin\theta \\ -\sin\theta & \cos\theta \end{bmatrix}.$$

Furthermore, define $B \triangleq \begin{bmatrix} 0 & \frac{\pi}{2} - \theta \\ \frac{-\pi}{2} + \theta & 0 \end{bmatrix}$. Then,

$$e^B = \begin{bmatrix} \sin\theta & \cos\theta \\ -\cos\theta & \sin\theta \end{bmatrix}.$$

◇

Example 15.3.7. Consider the second-order mechanical vibration equation

$$m\ddot{q} + c\dot{q} + kq = 0, \tag{15.3.6}$$

where m is positive and c and k are nonnegative. Here m, c, and k denote mass, damping, and stiffness parameters, respectively. Equation (15.3.6) can be written in companion form as

$$\dot{x} = Ax, \tag{15.3.7}$$

where

$$x \triangleq \begin{bmatrix} q \\ \dot{q} \end{bmatrix}, \quad A \triangleq \begin{bmatrix} 0 & 1 \\ -k/m & -c/m \end{bmatrix}. \tag{15.3.8}$$

In the inelastic case, $k = 0$, and thus A is upper triangular. In this case,

$$e^{tA} = \begin{cases} \begin{bmatrix} 1 & t \\ 0 & 1 \end{bmatrix}, & k = c = 0, \\ \begin{bmatrix} 1 & \frac{m}{c}(1 - e^{-ct/m}) \\ 0 & e^{-ct/m} \end{bmatrix}, & k = 0,\ c > 0, \end{cases} \tag{15.3.9}$$

where $c = 0$ and $c > 0$ correspond to a rigid body and a damped rigid body, respectively.

Next, we consider the elastic case where $c \ge 0$ and $k > 0$. In this case, we define

$$\omega_{\rm n} \triangleq \sqrt{\frac{k}{m}}, \quad \zeta \triangleq \frac{c}{2\sqrt{mk}}, \tag{15.3.10}$$

where $\omega_{\rm n} > 0$ denotes the (undamped) *natural frequency* of vibration and $\zeta \ge 0$ denotes the *damping ratio*. Now, A can be written as

$$A = \begin{bmatrix} 0 & 1 \\ -\omega_{\rm n}^2 & -2\zeta\omega_{\rm n} \end{bmatrix}, \tag{15.3.11}$$

and Corollary 15.3.3 yields

$$e^{tA} = \begin{cases} \begin{bmatrix} \cos\omega_{\rm n}t & \frac{1}{\omega_{\rm n}}\sin\omega_{\rm n}t \\ -\omega_{\rm n}\sin\omega_{\rm n}t & \cos\omega_{\rm n}t \end{bmatrix}, & \zeta = 0, \\ e^{-\zeta\omega_{\rm n}t}\begin{bmatrix} \cos\omega_{\rm d}t + \frac{\zeta}{\sqrt{1-\zeta^2}}\sin\omega_{\rm d}t & \frac{1}{\omega_{\rm d}}\sin\omega_{\rm d}t \\ \frac{-\omega_{\rm d}}{1-\zeta^2}\sin\omega_{\rm d}t & \cos\omega_{\rm d}t - \frac{\zeta}{\sqrt{1-\zeta^2}}\sin\omega_{\rm d}t \end{bmatrix}, & 0 < \zeta < 1, \\ e^{-\omega_{\rm n}t}\begin{bmatrix} 1 + \omega_{\rm n}t & t \\ -\omega_{\rm n}^2 t & 1 - \omega_{\rm n}t \end{bmatrix}, & \zeta = 1, \\ e^{-\zeta\omega_{\rm n}t}\begin{bmatrix} \cosh\omega_{\rm d}t + \frac{\zeta}{\sqrt{\zeta^2-1}}\sinh\omega_{\rm d}t & \frac{1}{\omega_{\rm d}}\sinh\omega_{\rm d}t \\ \frac{-\omega_{\rm d}}{\zeta^2-1}\sinh\omega_{\rm d}t & \cosh\omega_{\rm d}t - \frac{\zeta}{\sqrt{\zeta^2-1}}\sinh\omega_{\rm d}t \end{bmatrix}, & \zeta > 1, \end{cases}$$

where $\zeta = 0$, $0 < \zeta < 1$, $\zeta = 1$, and $\zeta > 1$ correspond to *undamped, underdamped, critically damped*, and *overdamped oscillators*, and where the *damped natural frequency* $\omega_{\rm d}$ is the positive number

$$\omega_{\rm d} \triangleq \begin{cases} \omega_{\rm n}\sqrt{1-\zeta^2}, & 0 < \zeta < 1, \\ \omega_{\rm n}\sqrt{\zeta^2 - 1}, & \zeta > 1. \end{cases} \tag{15.3.12}$$

15.4 Matrix Logarithms

Definition 15.4.1. Let $A \in \mathbb{C}^{n\times n}$. Then, $B \in \mathbb{C}^{n\times n}$ is a *logarithm* of A if $e^B = A$.

The following result shows that every complex, nonsingular matrix has a complex logarithm.

Proposition 15.4.2. Let $A \in \mathbb{C}^{n\times n}$. Then, there exists $B \in \mathbb{C}^{n\times n}$ such that $A = e^B$ if and only if A is nonsingular.

Proof. See [1299, pp. 35, 60] and [1450, p. 474]. □

Proposition 15.4.3. The following statements hold:

i) $B \in \mathbb{C}^{n\times n}$ is positive definite if and only if there exists a Hermitian matrix $A \in \mathbb{F}^{n\times n}$ such that $e^A = B$. If these conditions hold, then A is unique.

ii) $B \in \mathbb{C}^{n\times n}$ is Hermitian and nonsingular if and only if there exists a normal matrix $A \in \mathbb{C}^{n\times n}$ such that, for all $\lambda \in \operatorname{spec}(A)$, $\operatorname{Im}\lambda$ is an integer multiple of πJ and $e^A = B$.

iii) $B \in \mathbb{C}^{n\times n}$ is normal and nonsingular if and only if there exists a normal matrix $A \in \mathbb{C}^{n\times n}$ such that $e^A = B$.

iv) $B \in \mathbb{C}^{n\times n}$ is unitary if and only if there exists a skew-Hermitian matrix $A \in \mathbb{C}^{n\times n}$ such that $e^A = B$.

Proof. To prove *i)*, let $B = S\operatorname{diag}(b_1, \dots, b_n)S^{-1}$, where $S \in \mathbb{F}^{n\times n}$ is unitary and $b_1, \dots, b_n$ are positive. Then, define $A \triangleq S\operatorname{diag}(\log b_1, \dots, \log b_n)S^{-1}$. Next, to prove uniqueness, let A and $\hat{A}$ be Hermitian matrices such that $B = e^A = e^{\hat{A}}$. Then, for all $t \ge 0$, it follows that $e^{tA} = (e^A)^t = (e^{\hat{A}})^t = e^{t\hat{A}}$. Differentiating yields $Ae^{tA} = \hat{A}e^{t\hat{A}}$, while setting $t = 0$ implies that $A = \hat{A}$. As another proof, *x)* of Fact 15.15.1 implies that A and $\hat{A}$ commute. Therefore, $I = e^Ae^{-\hat{A}} = e^{A-\hat{A}}$, which, since $A - \hat{A}$ is Hermitian, implies that $A - \hat{A} = 0$. The result also follows from *xiii)* of Fact 15.15.1. *iv)* is given by *v)* of Proposition 15.6.7. □

Corollary 15.4.4. Let $A, B \in \mathbb{C}^{n\times n}$, assume that A and B are normal, and assume that $e^A = e^B$. Then, $A + A^* = B + B^*$.

Proof. Note that $e^{A+A^*} = e^{B+B^*}$, which, by *vii)* of Proposition 15.2.8, is positive definite. The result now follows from *i)* of Proposition 15.4.3. □

Although the real number -1 does not have a real logarithm, the real matrix $B = \begin{bmatrix} 0 & \pi \\ -\pi & 0 \end{bmatrix}$ satisfies $e^B = \begin{bmatrix} -1 & 0 \\ 0 & -1 \end{bmatrix}$. The following result characterizes the real matrices that have a real logarithm.

Proposition 15.4.5. Let $A \in \mathbb{R}^{n\times n}$. Then, there exists $B \in \mathbb{R}^{n\times n}$ such that $A = e^B$ if and only if A is nonsingular and, for every negative eigenvalue λ of A and for every positive integer k, the Jordan form of A has an even number of $k \times k$ blocks associated with λ.

Proof. See [1450, p. 475]. □

Corollary 15.4.6. Let $A \in \mathbb{R}^{n\times n}$ be nonsingular. Then, there exists $C \in \mathbb{R}^{n\times n}$ such that $A = C^2$ if and only if there exists $B \in \mathbb{R}^{n\times n}$ such that $A = e^B$.

Proof. To prove sufficiency, let $C = e^{\frac{1}{2}B}$. To prove necessity, let $\lambda \in \operatorname{spec}(A)$, and assume that $\lambda < 0$. Hence, there exists $\mu \in \operatorname{spec}(C) \cap \mathrm{IA}$ such that $\mu^2 = \lambda$. Thus, $\overline{\mu} \in \operatorname{spec}(C)$, $\overline{\mu}^2 = \lambda$, and $\operatorname{amult}_C(\mu) = \operatorname{amult}_C(\mu)$. Therefore, for every positive integer k, the Jordan form of A has an even number of $k \times k$ blocks associated with λ. □

Corollary 15.4.7. Let $A \in \mathbb{R}^{n\times n}$ be nonsingular, and let $k \ge 1$. Then, there exists $B \in \mathbb{R}^{n\times n}$ such that $A^{2k} = e^B$.

Proposition 15.4.8. Let $B \in \mathbb{R}^{n\times n}$. Then, the following statements hold:

i) B is positive definite if and only if there exists a symmetric matrix $A \in \mathbb{R}^{n\times n}$ such that $e^A = B$. If these conditions hold, then A is unique.

ii) B is symmetric and $I \le B$ if and only if there exists a positive-semidefinite matrix $A \in \mathbb{R}^{n\times n}$ such that $e^A = B$.

iii) B is symmetric and $I < B$ if and only if there exists a positive-definite matrix $A \in \mathbb{R}^{n\times n}$ such that $e^A = B$.

iv) B is normal and nonsingular and, for all $\lambda \in \operatorname{spec}(B)$, $\operatorname{amult}_B(\lambda)$ is even if and only if there exists a normal matrix $A \in \mathbb{R}^{n\times n}$ such that $e^A = B$.

v) B is normal, nonsingular, and, for all $\lambda \in \operatorname{spec}(B)$, $\operatorname{amult}_B(\lambda)$ is even and $|\log\lambda| = 1$ if and only if there exists an orthogonal matrix $A \in \mathbb{R}^{n\times n}$ such that $e^A = B$.

vi) B is orthogonal and $\det B = 1$ if and only if there exists a skew-symmetric matrix $A \in \mathbb{R}^{n\times n}$ such that $e^A = B$.

Proof. See [2442]. □

Replacing A and B in Proposition 15.4.5 by e^A and A, respectively, yields the following result.

Corollary 15.4.9. Let $A \in \mathbb{R}^{n\times n}$. Then, for every negative eigenvalue λ of e^A and for every positive integer k, the Jordan form of e^A has an even number of $k \times k$ blocks associated with λ.

Since the matrix $A \triangleq \begin{bmatrix} -2\pi & 4\pi \\ -2\pi & 2\pi \end{bmatrix}$ satisfies $e^A = I$, it follows that a positive-definite matrix can have a logarithm that is not normal. However, the following result shows that every positive-definite matrix has exactly one Hermitian logarithm.

Proposition 15.4.10. The function exp: $\mathbf{H}^n \mapsto \mathbf{P}^n$ is one-to-one and onto.

Proof. The result follows from *vii*) of Proposition 15.2.8 and *iii*) of Proposition 15.4.3. □

Let $A \in \mathbb{R}^{n\times n}$. If there exists $B \in \mathbb{R}^{n\times n}$ such that $A = e^B$, then Corollary 15.2.4 implies that $\det A = \det e^B = e^{\operatorname{tr} B} > 0$. However, the converse is not true. Consider, for example, $A \triangleq \begin{bmatrix} -1 & 0 \\ 0 & -2 \end{bmatrix}$, which satisfies $\det A > 0$. However, Proposition 15.4.5 implies that there does not exist a matrix $B \in \mathbb{R}^{2\times 2}$ such that $A = e^B$. On the other hand, note that $A = e^B e^C$, where $B \triangleq \begin{bmatrix} 0 & \pi \\ -\pi & 0 \end{bmatrix}$ and $C \triangleq \begin{bmatrix} 0 & 0 \\ 0 & \log 2 \end{bmatrix}$. While the product of two exponentials of real matrices has positive determinant, the following result shows that the converse is also true.

Proposition 15.4.11. Let $A \in \mathbb{R}^{n\times n}$. Then, there exist $B, C \in \mathbb{R}^{n\times n}$ such that $A = e^B e^C$ if and only if $\det A > 0$.

Proof. Suppose that there exist $B, C \in \mathbb{R}^{n\times n}$ such that $A = e^B e^C$. Then, it follows from Corollary 15.2.4 that $\det A = (\det e^B)(\det e^C) = e^{\operatorname{tr} B} e^{\operatorname{tr} C} > 0$. Conversely, suppose that $\det A > 0$. In the case where A has no negative eigenvalues, Proposition 15.4.5 implies that there exists $B \in \mathbb{R}^{n\times n}$ such that $A = e^B$. Hence, $A = e^B e^{0_{n\times n}}$. Now, suppose that A has at least one negative eigenvalue. Then, Theorem 7.4.6 on the real Jordan form implies that there exist a nonsingular matrix $S \in \mathbb{R}^{n\times n}$ and matrices $A_1 \in \mathbb{R}^{n_1\times n_1}$ and $A_2 \in \mathbb{R}^{n_2\times n_2}$ such that $A = S\begin{bmatrix} A_1 & 0 \\ 0 & A_2 \end{bmatrix}S^{-1}$, where every eigenvalue of A_1 is negative and none of the eigenvalues of A_2 are negative. Since $\det A$ and $\det A_2$ are positive, it follows that $\det A_1$ is positive, and thus n_1 is even. Now, write $A = S\begin{bmatrix} -I_{n_1} & 0 \\ 0 & I_{n_2} \end{bmatrix}\begin{bmatrix} -A_1 & 0 \\ 0 & A_2 \end{bmatrix}S^{-1}$, and note that $\begin{bmatrix} -I_{n_1} & 0 \\ 0 & I_{n_2} \end{bmatrix} = e^{\hat{B}}$, where

$$\hat{B} \triangleq \begin{bmatrix} I_{n_1/2} \otimes \begin{bmatrix} 0 & -\pi \\ \pi & 0 \end{bmatrix} & 0_{n_1\times n_2} \\ 0_{n_2\times n_1} & 0_{n_2\times n_2} \end{bmatrix}.$$

Furthermore, since $\begin{bmatrix} -A_1 & 0 \\ 0 & A_2 \end{bmatrix}$ has no negative eigenvalues, it follows from Proposition 15.4.5 that there exists $\hat{C} \in \mathbb{R}^{n\times n}$ such that $\begin{bmatrix} -A_1 & 0 \\ 0 & A_2 \end{bmatrix} = e^{\hat{C}}$. Hence, $e^A = Se^{\hat{B}}e^{\hat{C}}S^{-1} = e^{S\hat{B}S^{-1}}e^{S\hat{C}S^{-1}}$. See [2091]. □

Although $e^A e^B$ may be different from e^{A+B}, the following result, known as the *Baker-Campbell-Hausdorff series*, provides an expansion of a matrix function $C(t)$ that satisfies $e^{C(t)} = e^{tA}e^{tB}$.

Proposition 15.4.12. Let $A_1, \ldots, A_l \in \mathbb{F}^{n\times n}$. Then, there exists $\varepsilon > 0$ such that, for all $t \in (-\varepsilon, \varepsilon)$,

$$e^{tA_1} \cdots e^{tA_l} = e^{C(t)}, \tag{15.4.1}$$

where, as $t \to 0$,

$$C(t) \triangleq \sum_{i=1}^{l} tA_i + \sum_{1\le i<j\le l} \tfrac{1}{2}t^2[A_i, A_j] + O(t^3). \tag{15.4.2}$$

Proof. See [1299, Chapter 3], [2379, p. 35], and [2781, p. 97]. □

To illustrate (15.4.1), let $l = 2$, $A = A_1$, and $B = A_2$. Then, the first few terms of the series are

$$e^{tA}e^{tB} = e^{tA+tB+(t^2/2)[A,B]+(t^3/12)[[B,A],A+B]+\cdots}. \tag{15.4.3}$$

The radius of convergence of this series is discussed in [490, 508, 842, 994, 2126, 2304].

The following result is the *Lie-Trotter product formula*.

Corollary 15.4.13. Let $A, B \in \mathbb{F}^{n\times n}$. Then,

$$e^{A+B} = \lim_{p\to\infty}\left(e^{\frac{1}{p}A}e^{\frac{1}{p}B}\right)^p. \tag{15.4.4}$$

Proof. Setting $l = 2$ and $t = 1/p$ in (15.4.1) yields, as $p \to \infty$,

$$\left(e^{\frac{1}{p}A}e^{\frac{1}{p}B}\right)^p = \left(e^{\frac{1}{p}(A+B)+O(1/p^2)}\right)^p = e^{A+B+O(1/p)} \to e^{A+B}. \qquad \square$$

15.5 Principal Logarithm

Let $A \in \mathbb{F}^{n\times n}$ be positive definite so that $A = SBS^* \in \mathbb{F}^{n\times n}$, where $S \in \mathbb{F}^{n\times n}$ is unitary and $B \in \mathbb{R}^{n\times n}$ is diagonal with positive diagonal entries. In Section 8.5, $\log A$ is defined as $\log A = S(\log B)S^* \in \mathbf{H}^n$, where $(\log B)_{(i,i)} \triangleq \log B_{(i,i)}$. Since $\log A$ satisfies $A = e^{\log A}$, it follows that $\log A$ is a logarithm of A. The following result extends the definition of $\log A$ to a larger class of matrices $A \in \mathbb{C}^{n\times n}$. A logarithm function defined on the set of nonsingular matrices can be based on the principal branch of the log function given by Fact 2.21.29 along with Definition 12.8.1; however, this function is not continuous, and thus an alternative approach to defining the principal logarithm is taken.

Definition 15.5.1. Let $A \in \mathbb{C}^{n\times n}$, and assume that A has no eigenvalues in $(-\infty, 0]$. Then, the *principal logarithm* of A is the unique logarithm of A whose eigenvalues are elements of $\{z \in \mathbb{C}\colon |\mathrm{Im}\, z| < \pi\}$. The notation $\log A$ denotes the principal logarithm of A.

Theorem 15.5.2. Let $A \in \mathbb{C}^{n\times n}$. Then, the following statements hold:

i) If A is nonsingular, then $\log A$ is a logarithm of A; that is, $e^{\log A} = A$.

ii) $\log e^A = A$ if and only if, for all $\lambda \in \mathrm{spec}(A)$, it follows that $|\mathrm{Im}\,\lambda| < \pi$.

iii) If A is nonsingular and $\rho_{\max}(A - I) < 1$, then $\log A$ is given by the series

$$\log A = \sum_{i=1}^{\infty} \frac{(-1)^{i+1}}{i}(A - I)^i, \tag{15.5.1}$$

which converges absolutely for every submultiplicative norm $\|\cdot\|$ such that $\|A - I\| < 1$.

iv) If $\mathrm{spec}(A) \subset \mathrm{ORHP}$, then $\log A$ is given by the series

$$\log A = \sum_{i=0}^{\infty} \frac{2}{2i+1}[(A - I)(A + I)^{-1}]^{2i+1}. \tag{15.5.2}$$

v) If A has no eigenvalues in $(-\infty, 0]$, then

$$\log A = \int_0^1 (A - I)[t(A - I) + I]^{-1}\, \mathrm{d}t. \tag{15.5.3}$$

vi) If A has no eigenvalues in $(-\infty, 0]$ and $\alpha \in [-1, 1]$, then

$$\log A^\alpha = \alpha \log A. \tag{15.5.4}$$

In particular,

$$\log A^{-1} = -\log A, \quad \log A^{1/2} = \tfrac{1}{2}\log A. \tag{15.5.5}$$

vii) If A is real and $\mathrm{spec}(A) \subset \mathrm{ORHP}$, then $\log A$ is real.

viii) If A is real and nonsingular, then A has a real logarithm if and only if A is nonsingular and, for every negative eigenvalue λ of A and for every positive integer k, the Jordan form of A has an even number of $k \times k$ blocks associated with λ.

Now, let $\|\cdot\|$ be a submultiplicative norm on $\mathbb{C}^{n\times n}$. Then, the following statements hold:

ix) The function log is continuous on $\{X \in \mathbb{C}^{n\times n}\colon \|X - I\| < 1\}$.

x) If $B \in \mathbb{C}^{n\times n}$ and $\|B\| < \log 2$, then $\|e^B - I\| < 1$ and $\log e^B = B$.

xi) $\exp\colon \ \mathbb{B}_{\log 2}(0) \mapsto \mathbb{F}^{n\times n}$ is one-to-one.

xii) If $\|A - I\| < 1$, then

$$\|\log A\| \le -\log(1 - \|A - I\|) \le \frac{\|A - I\|}{1 - \|A - I\|}. \tag{15.5.6}$$

xiii) If $\|A - I\| < 2/3$, then

$$\|A - I\|\left(1 - \frac{\|A - I\|}{2(1 - \|A - I\|)}\right) \le \|\log A\|. \tag{15.5.7}$$

xiv) Assume that A is nonsingular, and let $\operatorname{mspec}(A) = \{\lambda_1, \ldots, \lambda_n\}_{\rm ms}$. Then,

$$\operatorname{mspec}(\log A) = \{\log \lambda_1, \ldots, \log \lambda_n\}_{\rm ms}. \tag{15.5.8}$$

Hence,

$$\log \det A = \operatorname{tr} \log A. \tag{15.5.9}$$

Proof. *i*) follows from the discussion in [1450, p. 420]; *ii*) is given in [1391, p. 32]; *iii*) and *iv*) are given by Fact 13.4.15, see [1299, pp. 34–35] and [1391, p. 273]; *v*) is given in [1391, p. 269]; *vi*) is given in [1391, p. 270]; *vii*) is immediate; *viii*) follows from Proposition 15.4.5 and the discussion in [1450, pp. 474–475]. *ix*) and *x*) are proved in [1299, pp. 34–35]. To prove the inequality in *x*), let $\|B\| < 2$, so that $e^{\|B\|} < 2$, and thus

$$\|e^B - I\| \le \sum_{i=1}^{\infty} (i!)^{-1}\|B\|^i = e^{\|B\|} - 1 < 1.$$

To prove *xi*), let $B_1, B_2 \in \mathbb{B}_{\log 2}(0)$, and assume that $e^{B_1} = e^{B_2}$. Then, *ii*) implies that $B_1 = \log e^{B_1} = \log e^{B_2} = B_2$. Finally, to prove *xii*), let $\alpha \triangleq \|A - I\| < 1$. Then, it follows from (15.5.1) and *iv*) of Fact 2.21.29 that $\|\log A\| \le \sum_{i=1}^{\infty} \alpha^i/i = -\log(1 - \alpha)$. For *xiii*), see [1391, p. 647]. □

15.6 Lie Groups

Definition 15.6.1. Let $\mathcal{S} \subset \mathbb{F}^{n\times n}$, and assume that $\mathcal{S}$ is a group. Then, $\mathcal{S}$ is a *Lie group* if $\mathcal{S}$ is closed relative to $\mathrm{GL}_{\mathbb{F}}(n)$.

Proposition 15.6.2. Let $\mathcal{S} \subset \mathbb{F}^{n\times n}$, and assume that $\mathcal{S}$ is a multiplication group. Then, $\mathcal{S}$ is a Lie group if and only if the limit of every convergent sequence in $\mathcal{S}$ is either an element of $\mathcal{S}$ or is singular.

The multiplication groups $\mathrm{SL}_{\mathbb{F}}(n)$, $\mathrm{U}(n)$, $\mathrm{O}(n)$, $\mathrm{SU}(n)$, $\mathrm{SO}(n)$, $\mathrm{U}(n,m)$, $\mathrm{O}(n,m)$, $\mathrm{SU}(n,m)$, $\mathrm{SO}(n,m)$, $\mathrm{Symp}_{\mathbb{F}}(2n)$, $\mathrm{Aff}_{\mathbb{F}}(n)$, $\mathrm{SE}_{\mathbb{F}}(n)$, and $\mathrm{Trans}_{\mathbb{F}}(n)$ defined in Proposition 4.6.6 are closed sets, and thus are Lie groups. Although the multiplication groups $\mathrm{GL}_{\mathbb{F}}(n)$, $\mathrm{PL}_{\mathbb{F}}(n)$, and $\mathrm{UT}(n)$ (see Fact 4.31.11) are not closed sets, they are closed relative to $\mathrm{GL}_{\mathbb{F}}(n)$, and thus they are Lie groups. Finally, the multiplication group $\mathcal{S} \subset \mathbb{C}^{2\times 2}$ defined by

$$\mathcal{S} \triangleq \left\{ \begin{bmatrix} e^{\jmath t} & 0 \\ 0 & e^{\pi \jmath t} \end{bmatrix} : t \in \mathbb{R} \right\} \tag{15.6.1}$$

is not closed relative to $\mathrm{GL}_{\mathbb{C}}(2)$, and thus is not a Lie group. For details, see [1299, p. 4].

Proposition 15.6.3. Let $\mathcal{S} \subset \mathbb{F}^{n\times n}$, and assume that $\mathcal{S}$ is a Lie group. Furthermore, define

$$\mathcal{S}_0 \triangleq \left\{ A \in \mathbb{F}^{n\times n} \colon\ e^{tA} \in \mathcal{S} \text{ for all } t \in \mathbb{R} \right\}. \tag{15.6.2}$$

Then, $\mathcal{S}_0$ is a Lie algebra.

Proof. See [1299, pp. 39, 43, 44]. □

The Lie algebra $\mathcal{S}_0$ defined by (15.6.2) is *the Lie algebra of* $\mathcal{S}$.

Proposition 15.6.4. Let $\mathcal{S} \subset \mathbb{F}^{n\times n}$, assume that $\mathcal{S}$ is a Lie group, and let $\mathcal{S}_0 \subseteq \mathbb{F}^{n\times n}$ be the Lie algebra of $\mathcal{S}$. Furthermore, let $S \in \mathcal{S}$ and $A \in \mathcal{S}_0$. Then, $SAS^{-1} \in \mathcal{S}_0$.

Proof. For all $t \in \mathbb{R}$, $e^{tA} \in \mathcal{S}$, and thus $e^{tSAS^{-1}} = Se^{tA}S^{-1} \in \mathcal{S}$. Hence, $SAS^{-1} \in \mathcal{S}_0$. □

Proposition 15.6.5. The following statements hold:

i) $\mathrm{gl}_{\mathbb{F}}(n)$ is the Lie algebra of $\mathrm{GL}_{\mathbb{F}}(n)$.

ii) $\mathrm{gl}_{\mathbb{R}}(n) = \mathrm{pl}_{\mathbb{R}}(n)$ is the Lie algebra of $\mathrm{PL}_{\mathbb{R}}(n)$.

iii) $\mathrm{pl}_{\mathbb{C}}(n)$ is the Lie algebra of $\mathrm{PL}_{\mathbb{C}}(n)$.

iv) $\mathrm{sl}_{\mathbb{F}}(n)$ is the Lie algebra of $\mathrm{SL}_{\mathbb{F}}(n)$.

v) $\mathrm{u}(n)$ is the Lie algebra of $\mathrm{U}(n)$.

vi) $\mathrm{so}(n)$ is the Lie algebra of $\mathrm{O}(n)$.

vii) $\mathrm{su}(n)$ is the Lie algebra of $\mathrm{SU}(n)$.

viii) $\mathrm{so}(n)$ is the Lie algebra of $\mathrm{SO}(n)$.

ix) $\mathrm{su}(n,m)$ is the Lie algebra of $\mathrm{U}(n,m)$.

x) $\mathrm{so}(n,m)$ is the Lie algebra of $\mathrm{O}(n,m)$.

xi) $\mathrm{su}(n,m)$ is the Lie algebra of $\mathrm{SU}(n,m)$.

xii) $\mathrm{so}(n,m)$ is the Lie algebra of $\mathrm{SO}(n,m)$.

xiii) $\mathrm{symp}_{\mathbb{F}}(2n)$ is the Lie algebra of $\mathrm{Symp}_{\mathbb{F}}(2n)$.

xiv) $\mathrm{osymp}_{\mathbb{F}}(2n)$ is the Lie algebra of $\mathrm{OSymp}_{\mathbb{F}}(2n)$.

xv) $\mathrm{aff}_{\mathbb{F}}(n)$ is the Lie algebra of $\mathrm{Aff}_{\mathbb{F}}(n)$.

xvi) $\mathrm{se}_{\mathbb{C}}(n)$ is the Lie algebra of $\mathrm{SE}_{\mathbb{C}}(n)$.

xvii) $\mathrm{se}_{\mathbb{R}}(n)$ is the Lie algebra of $\mathrm{SE}_{\mathbb{R}}(n)$.

xviii) $\mathrm{trans}_{\mathbb{F}}(n)$ is the Lie algebra of $\mathrm{Trans}_{\mathbb{F}}(n)$.

Proof. See [1299, pp. 38–41]. □

Proposition 15.6.6. Let $\mathcal{S} \subset \mathbb{F}^{n\times n}$, assume that $\mathcal{S}$ is a Lie group, and let $\mathcal{S}_0 \subseteq \mathbb{F}^{n\times n}$ be the Lie algebra of $\mathcal{S}$. Then, $\exp\colon\ \mathcal{S}_0 \mapsto \mathcal{S}$. Furthermore, if exp is onto, then $\mathcal{S}$ is pathwise connected.

Proof. Let $A \in \mathcal{S}_0$ so that $e^{tA} \in \mathcal{S}$ for all $t \in \mathbb{R}$. Hence, setting $t = 1$ implies that $\exp\colon\ \mathcal{S}_0 \mapsto \mathcal{S}$. Now, suppose that exp is onto, let $B \in \mathcal{S}$, and let $A \in \mathcal{S}_0$ satisfy $e^A = B$. Then, $f(t) \triangleq e^{tA}$ satisfies $f(0) = I$ and $f(1) = B$, which implies that $\mathcal{S}$ is pathwise connected. □

A Lie group can consist of multiple pathwise-connected components.

Proposition 15.6.7. Let $n \geq 1$. Then, the following functions are onto:

i) $\exp\colon\ \mathrm{gl}_{\mathbb{C}}(n) \mapsto \mathrm{GL}_{\mathbb{C}}(n)$.

ii) $\exp\colon\ \mathrm{gl}_{\mathbb{R}}(1) \mapsto \mathrm{PL}_{\mathbb{R}}(1)$.

iii) $\exp\colon\ \mathrm{pl}_{\mathbb{C}}(n) \mapsto \mathrm{PL}_{\mathbb{C}}(n)$.

iv) $\exp\colon\ \mathrm{sl}_{\mathbb{C}}(n) \mapsto \mathrm{SL}_{\mathbb{C}}(n)$.

v) $\exp\colon\ \mathrm{u}(n) \mapsto \mathrm{U}(n)$.

vi) $\exp\colon\ \mathrm{su}(n) \mapsto \mathrm{SU}(n)$.

vii) $\exp\colon\ \mathrm{so}(n) \mapsto \mathrm{SO}(n)$.

Furthermore, the following functions are not onto:

viii) $\exp\colon\ \mathrm{gl}_{\mathbb{R}}(n) \mapsto \mathrm{PL}_{\mathbb{R}}(n)$, where $n \geq 2$.

ix) $\exp\colon\ \mathrm{sl}_{\mathbb{R}}(n) \mapsto \mathrm{SL}_{\mathbb{R}}(n)$.

x) $\exp\colon\ \mathrm{so}(n) \mapsto \mathrm{O}(n)$.

xi) exp: $\mathrm{symp}_{\mathbb{R}}(2n) \mapsto \mathrm{Symp}_{\mathbb{R}}(2n)$.

Proof. *i)* follows from Proposition 15.4.2; *ii)* is immediate; *iii)*–*vii)* can be verified by construction; see [2263, pp. 199, 212] for the proof of *v)* and *vii)*. The example $A \triangleq \begin{bmatrix} -1 & 0 \\ 0 & -2 \end{bmatrix}$ and Proposition 15.4.5 show that *viii)* is not onto. For $\lambda < 0$, where $\lambda \neq -1$, Proposition 15.4.5 and the example $\begin{bmatrix} \lambda & 0 \\ 0 & 1/\lambda \end{bmatrix}$ given in [2379, p. 39] show that *ix)* is not onto; see also [218, pp. 84, 85]. *viii)* shows that *x)* is not onto. For *xi)*, see [896]. □

Proposition 15.6.8. The Lie groups $\mathrm{GL}_{\mathbb{C}}(n), \mathrm{SL}_{\mathbb{F}}(n), \mathrm{U}(n), \mathrm{SU}(n)$, and $\mathrm{SO}(n)$ are pathwise connected. The Lie groups $\mathrm{GL}_{\mathbb{R}}(n), \mathrm{O}(n), \mathrm{O}(n,1)$, and $\mathrm{SO}(n,1)$ are not pathwise connected.

Proof. See [1299, p. 15]. □

Proposition 15.6.8 and *ix)* of Proposition 15.6.7 show that the converse of Proposition 15.6.6 does not hold; that is, pathwise connectedness does not imply that exp is onto. See [2379, p. 39].

15.7 Linear Time-Varying Differential Equations

Let $A\colon [0,\infty) \mapsto \mathbb{F}^{n\times n}$ be continuous, let $t \in [0,\infty)$, let $x_0 \in \mathbb{F}^n$, and consider

$$\dot{x}(t) = A(t)x(t), \tag{15.7.1}$$

$$x(0) = x_0. \tag{15.7.2}$$

Theorem 15.7.1. There exists a unique C^1 solution $x\colon [0,\infty) \mapsto \mathbb{F}^n$ of (15.7.1) and (15.7.2). Furthermore, for all $t \in [0,\infty)$, $x(t)$ is given by the series

$$x(t) = x_0 + \sum_{n=1}^{\infty} \int_0^t \cdots \int_0^{t_{n-1}} A(t_1)\cdots A(t_n)\,\mathrm{d}t_n \cdots \mathrm{d}t_1 x_0, \tag{15.7.3}$$

which, for all $T > 0$, converges absolutely and uniformly on $[0,T]$.

Proof. To prove existence, let $T > 0$ and define $c \triangleq \max_{t\in[0,T]} \sigma_{\max}[A(t)]$. For all $t \in [0,T]$, define $\psi_0 : [0,T] \mapsto \mathbb{F}^n$ by $\psi_0(t) \triangleq x_0$, and, for all $k \geq 1$ and $t \in [0,T]$, define $\psi_k\colon [0,T] \mapsto \mathbb{F}^n$ by

$$\psi_k(t) \triangleq \int_0^t \cdots \int_0^{t_{k-1}} A(t_1)\cdots A(t_k)x_0\,\mathrm{d}t_k \cdots \mathrm{d}t_1. \tag{15.7.4}$$

Then, for all $k \geq 1$,

$$\begin{aligned} \max_{t\in[0,T]} \|\psi_k(t)\|_2 &\leq \int_0^t \cdots \int_0^{t_{k-1}} \sigma_{\max}[A(t_1)]\cdots\sigma_{\max}[A(t_k)]\,\mathrm{d}t_k \cdots \mathrm{d}t_1 \|x_0\|_2 \\ &= \frac{1}{k!}\int_0^t \cdots \int_0^t \sigma_{\max}[A(t_1)]\cdots\sigma_{\max}[A(t_k)]\|\,\mathrm{d}t_k \cdots \mathrm{d}t_1 \|x_0\|_2 \\ &\leq \frac{c^kT^k}{k!}\|x_0\|_2. \end{aligned} \tag{15.7.5}$$

Since

$$\sum_{k=0}^{\infty} \frac{c^kT^k}{k!}\|x_0\|_2 = \exp(cT)\|x_0\|_2, \tag{15.7.6}$$

Fact 12.16.15 implies that $\sum_{k=0}^{\infty} \psi_k(t)$ converges absolutely and uniformly on $[0,T]$. Furthermore, Fact 12.16.16 implies that $x\colon [0,T] \mapsto \mathbb{F}^n$ defined by

$$x(t) \triangleq \sum_{k=0}^{\infty} \psi_k(t) \tag{15.7.7}$$

is continuous. Note that, for all $t \in [0,T]$,

$$x(t) = \lim_{n\to\infty} \phi_n(t), \tag{15.7.8}$$

where, for all $n \ge 0$ and all $t \in [0,T]$, $\phi_n : [0,T] \mapsto \mathbb{F}^n$ is the partial sum

$$\phi_n(t) \triangleq \sum_{k=0}^{n} \psi_k(t). \tag{15.7.9}$$

Next, for all $n \ge 0$, ϕ_n is differentiable and, furthermore, for all $n \ge 2$ and $t \in [0,T]$,

$$\begin{aligned}\dot\phi_n(t) &= A(t)x_0 + \sum_{k=2}^{n} A(t)\int_0^t \cdots \int_0^{t_{k-1}} A(t_2)\cdots A(t_k)\mathrm{d}t_k \cdots \mathrm{d}t_2 x_0\\ &= A(t)\left(x_0 + \sum_{k=1}^{n-1}\int_0^t \cdots \int_0^{t_{k-1}} A(t_1)\cdots A(t_k)\mathrm{d}t_k \cdots \mathrm{d}t_1 x_0\right) = A(t)\phi_{n-1}(t). \end{aligned}\tag{15.7.10}$$

Since A is continuous on $[0,T]$ and $(\phi_n)_{n=1}^\infty$ converges to x on $[0,T]$, it follows from (15.7.10) that, for all $t \in [0,T]$,

$$\lim_{n\to\infty} \dot\phi_n(t) = \lim_{n\to\infty} A(t)\phi_{n-1}(t) = A(t)x(t). \tag{15.7.11}$$

Since $(\phi_n)_{n=1}^\infty$ converges uniformly to x on $[0,T]$, it follows from (15.7.11) that $(\dot\phi_n)_{n=1}^\infty$ converges uniformly on $[0,T]$. Since, for all $n \ge 1$, ϕ_n is continuous on $[0,T]$ and $(\phi_n)_{n=1}^\infty$ converges uniformly on $[0,T]$, it follows from Fact 12.16.16 that x is differentiable on $[0,T]$ and satisfies

$$\dot x(t) = \lim_{n\to\infty} \dot\phi_n(t) = \sum_{k=0}^{\infty} \dot\psi_k(t) = A(t)x(t). \tag{15.7.12}$$

Hence, x satisfies (15.7.1) and (15.7.2).

To prove uniqueness of the solution (15.7.3), let x satisfy (15.7.1), and let y satisfy

$$\dot y(t) = A(t)y(t), \tag{15.7.13}$$

$$y(0) = x_0. \tag{15.7.14}$$

Then, for all $t \ge 0$, $z(t) \triangleq x(t) - y(t)$ satisfies

$$\dot z(t) = A(t)z(t), \tag{15.7.15}$$

$$z(0) = 0. \tag{15.7.16}$$

Integrating (15.7.15) yields

$$z(t) = \int_0^t A(s)z(s)\,\mathrm{d}s. \tag{15.7.17}$$

Hence, for all $t \ge 0$,

$$\|z(t)\|_2 \le f(t), \tag{15.7.18}$$

where

$$f(t) \triangleq \int_0^t \sigma_{\max}[A(s)]\|z(s)\|_2\,\mathrm{d}s. \tag{15.7.19}$$

Differentiating (15.7.19) with respect to t, and using (15.7.18), yields

$$\dot f(t) = \sigma_{\max}[A(t)]\|z(t)\|_2 \le \sigma_{\max}[A(t)]f(t). \tag{15.7.20}$$

Since, for all $t \geq 0$, $\sigma_{\max}[A(t)] \geq 0$, it follows from (15.7.20) that, for all $t \geq 0$,

$$0 \leq f(t)\exp^{-\int_0^t \sigma_{\max}[A(s)]\,\mathrm{d}s} = \int_0^t \frac{\mathrm{d}}{\mathrm{d}q}[f(q)\exp^{-\int_0^q \sigma_{\max}[A(s)]\mathrm{d}s}]\,\mathrm{d}q$$

$$= \int_0^t [\dot{f}(q) - \sigma_{\max}[A(q)]f(q)]\exp^{-\int_0^q \sigma_{\max}[A(s)]\mathrm{d}s}\,\mathrm{d}q \leq 0.$$

Hence, for all $t \geq 0$, $f(t) = 0$, and thus (15.7.18) implies that, for all $t \geq 0$, $z(t) = 0$. Therefore, (15.7.3) is the unique solution of (15.7.1), (15.7.2). □

15.8 Lyapunov Stability Theory

Let $\mathcal{D} \subseteq \mathbb{R}^n$, assume that $\mathcal{D}$ is open, let $f\colon\ \mathcal{D} \to \mathbb{R}^n$, and assume that f is locally Lipschitz. Consider the ordinary differential equation

$$\dot{x}(t) = f[x(t)], \tag{15.8.1}$$

where $t \geq 0$. Throughout this section, let $\|\cdot\|$ be a norm on $\mathbb{R}^n$. We first consider the case $n = 1$.

Theorem 15.8.1. Assume that $n = 1$. Then, for all $x_0 \in \mathcal{D}$, exactly one of the following statements holds:

i) There exists $T_{x_0} > 0$ and a unique C^1 solution $x\colon\ [0, T_{x_0}) \mapsto \mathcal{D}$ satisfying (15.8.1) and $x(0) = x_0$ and such that either $\lim_{t\to T_{x_0}} x(t) \in \operatorname{bd}\mathcal{D}$ or $\lim_{t\to T_{x_0}} |x(t)| = \infty$.

ii) There exists a unique C^1 solution $x\colon\ [0, \infty) \mapsto \mathcal{D}$ satisfying (15.8.1) and $x(0) = x_0$.

Furthermore, the following statements hold:

iii) If *i)* holds and f is bounded, then $\lim_{t\to T_{x_0}} x(t) \in \operatorname{bd}\mathcal{D}$.

iv) If $\mathcal{D} = \mathbb{R}$ and f is either bounded or globally Lipschitz, then *ii)* holds.

Now, we consider the more general case where $n \geq 1$.

Theorem 15.8.2. For all $x_0 \in \mathcal{D}$, exactly one of the following statements holds:

i) There exists $T_{x_0} > 0$ and a unique C^1 solution $x\colon\ [0, T_{x_0}) \mapsto \mathcal{D}$ satisfying (15.8.1) and $x(0) = x_0$ and such that, for every compact set $\mathcal{D}' \subset \mathcal{D}$, there exists $T \in [0, T_{x_0})$ such that, for all $t \in (T, T_{x_0})$, $x(t) \notin \mathcal{D}'$.

ii) There exists a unique C^1 solution $x\colon\ [0, \infty) \mapsto \mathcal{D}$ satisfying (15.8.1) and $x(0) = x_0$.

Furthermore, the following statements hold:

iii) If *i)* holds and f is bounded, then $\lim_{t\to T_{x_0}} x(t) \in \operatorname{bd}\mathcal{D}$.

iv) If $\mathcal{D} = \mathbb{R}^n$ and f is either bounded or globally Lipschitz, then *ii)* holds.

Proof. The result follows from [1297, Theorem 2.1, Theorem 3.1]. □

Consider the example

$$\dot{r} = \frac{1}{1-r}, \qquad \dot{\theta} = \frac{1}{(1-r)^2}, \tag{15.8.2}$$

which represents (15.8.1) with $n = 2$ and $\mathcal{D} = \mathbb{B}_1(0)\backslash\{0\}$, where x is expressed in polar coordinates. For all $r_0 \in (0, 1)$, it follows that

$$r(t) = 1 - \sqrt{(1-r_0)^2 - 2t}, \tag{15.8.3}$$

where $r(0) = r_0$. Therefore, $r(t) \to 1$ as $t \to T_{x_0} \triangleq \frac{1}{2}(1-r_0)^2$. Furthermore, $\theta(t) \to \infty$ as $t \to T_{x_0}$. Since $r(t)$ and $\theta(t)$ are increasing, it follows that $x(t)$ spirals counterclockwise toward the unit circle with increasing speed, encircling the origin an infinite number of times in a finite interval of time. Consequently, $x(t)$ does not converge as t approaches T_{x_0}. Since, for all $\varepsilon \in (0, 1)$, $x(t)$ leaves $\mathbb{B}_\varepsilon(0)$ in finite time, it follows that *i)* of Theorem 15.8.2 holds. The fact that $x(t)$ does not converge as t

approaches T_{x_0} is consistent with the fact that f is not bounded on $\mathcal{D}$.

If $x_e \in \mathcal{D}$ satisfies $f(x_e) = 0$, then $x(t) \equiv x_e$ is an *equilibrium* of (15.8.1). The following definition concerns the stability of an equilibrium of (15.8.1).

Definition 15.8.3. Let $x_e \in \mathcal{D}$ be an equilibrium of (15.8.1).

i) x_e is *Lyapunov stable* if, for all $\varepsilon > 0$ such that $\operatorname{cl} \mathbb{B}_\varepsilon(x_e) \subset \mathcal{D}$, there exists $\delta \in (0, \varepsilon)$ such that, for all $x_0 \in \mathbb{B}_\delta(x_e)$, there exists a unique C^1 solution $x\colon [0,\infty) \mapsto \mathbb{B}_\varepsilon(x_e)$ satisfying (15.8.1) and $x(0) = x_0$.

ii) x_e is *asymptotically stable* if it is Lyapunov stable and there exists $\delta > 0$ such that $\mathbb{B}_\delta(x_e) \subseteq \mathcal{D}$ and such that, for all $x_0 \in \mathbb{B}_\delta(x_e)$, there exists a unique C^1 solution $x\colon [0,\infty) \mapsto \mathbb{B}_\delta(x_e)$ satisfying (15.8.1), $x(0) = x_0$, and $\lim_{t\to\infty} x(t) = x_e$.

iii) x_e is *globally asymptotically stable* if it is Lyapunov stable, $\mathcal{D} = \mathbb{R}^n$, and, for all $x_0 \in \mathbb{R}^n$, there exists a unique C^1 solution $x\colon [0,\infty) \mapsto \mathbb{R}^n$ satisfying (15.8.1), $x(0) = x_0$, and $\lim_{t\to\infty} x(t) = x_e$.

iv) x_e is *unstable* if it is not Lyapunov stable.

Note that, if $x_e \in \mathbb{R}^n$ is a globally asymptotically stable equilibrium of (15.8.1), then x_e is the unique equilibrium of (15.8.1).

Lyapunov's direct method gives sufficient conditions for Lyapunov stability and asymptotic stability of an equilibrium of (15.8.1). The following definition is needed.

Definition 15.8.4. The function $V\colon \mathbb{R}^n \mapsto \mathbb{R}$ is *radially unbounded* if, for every sequence $(x_i)_{i=1}^\infty \subset \mathbb{R}^n$ such that $\lim_{i\to\infty} \|x_i\| = \infty$, it follows that $\lim_{i\to\infty} V(x_i) = \infty$.

Theorem 15.8.5. Let $x_e \in \mathcal{D}$ be an equilibrium of (15.8.1), and let $V\colon \mathcal{D} \mapsto \mathbb{R}$ be a C^1 function such that $V(x_e) = 0$ and such that, for all $x \in \mathcal{D}\backslash\{x_e\}$, $V(x) > 0$. Then, the following statements hold:

i) Assume that, for all $x \in \mathcal{D}$,

$$V'(x)f(x) \le 0. \tag{15.8.4}$$

Then, x_e is Lyapunov stable.

ii) Assume that, for all $x \in \mathcal{D}\backslash\{x_e\}$,

$$V'(x)f(x) < 0. \tag{15.8.5}$$

Then, x_e is asymptotically stable. If, in addition, $\mathcal{D} = \mathbb{R}^n$ and V is radially unbounded, then x_e is globally asymptotically stable.

Proof. To prove *i)*, let $\varepsilon > 0$ satisfy $\operatorname{cl} \mathbb{B}_\varepsilon(x_e) \subset \mathcal{D}$. Since $\mathbb{S}_\varepsilon(x_e)$ is a compact subset of $\mathcal{D}$ and V is continuous on $\mathbb{S}_\varepsilon(x_e)$, Corollary 12.4.12 implies that there exists $x_1 \in \mathbb{S}_\varepsilon(x_e)$ such that, for all $x \in \mathbb{S}_\varepsilon(x_e)$, $V(x_1) \le V(x)$. Since $x_1 \in \mathbb{S}_\varepsilon(x_e)$ and $V(x) > 0$ for all $x \in \mathbb{S}_\varepsilon(x_e)$, it follows that $V(x_1) > 0$.

Next, since V is continuous, it follows from Theorem 12.4.9 that $\mathcal{D}_1 \triangleq \{x \in \mathbb{B}_\varepsilon(x_e)\colon V(x) < V(x_1)\}$ is open. Since $\mathcal{D}_1$ is open and contains x_e, it follows that there exists $\delta \in (0,\varepsilon)$ such that $\mathbb{B}_\delta(x_e) \subseteq \mathcal{D}_1$. Therefore, for all $x \in \mathbb{B}_\delta(x_e)$, $V(x) < V(x_1)$.

Now, let $x_0 \in \mathbb{B}_\delta(x_e)$, and suppose that *i)* of Theorem 15.8.2 holds. Then, there exists $T_{x_0} > 0$ and a unique C^1 solution $x\colon\ [0, T_{x_0}) \mapsto \mathcal{D}$ satisfying (15.8.1) and $x(0) = x_0$. It therefore follows from (15.8.4) that, for all $t \in [0, T_{x_0})$,

$$V[x(t)] - V[x(0)] = \int_0^t V'[x(\tau)]f[x(\tau)]\,d\tau \le 0. \tag{15.8.6}$$

Since $x(0) \in \mathbb{B}_\delta(x_e)$, it follows that, for all $t \in [0, T_{x_0})$, $V[x(t)] \le V[x(0)] < V(x_1)$. Since, for all $x \in \mathbb{S}_\varepsilon(x_e)$, $V(x) \ge V(x_1)$, it follows that, for all $t \in [0, T_{x_0})$, $\|x(t) - x_e\| \ne \varepsilon$. Next, let $t_1 \in [0, T_{x_0})$ satisfy $\|x(t_1) - x_e\| > \varepsilon$. Then, since $\|x_0 - x_e\| < \varepsilon$, it follows from the intermediate value theorem that there exists $t_2 \in (0, t_1)$ such that $\|x(t_2) - x_e\| = \varepsilon$, which is a contradiction. Therefore, *i)* of

Theorem 15.8.2 does not hold.

Since *ii*) of Theorem 15.8.2 holds, it follows that there exists a unique C^1 solution $x\colon [0,\infty) \mapsto \mathcal{D}$ satisfying (15.8.1) and $x(0) = x_0$. It therefore follows from (15.8.4) that, for all $t \in [0,\infty)$, (15.8.6) holds. Since $x(0) \in \mathbb{B}_\delta(x_\mathrm{e})$, it follows that, for all $t \in [0,\infty)$, $V[x(t)] \le V[x(0)] < V(x_1)$. Since, for all $x \in \mathbb{S}_\varepsilon(x_\mathrm{e})$, $V(x) \ge V(x_1)$, it follows that, for all $t \in [0,\infty)$, $\|x(t) - x_\mathrm{e}\| \ne \varepsilon$. Next, suppose there exists $t_1 > 0$ such that $\|x(t_1) - x_\mathrm{e}\| > \varepsilon$. Then, since $\|x_0 - x_\mathrm{e}\| < \varepsilon$, it follows from the intermediate value theorem that there exists $t_2 \in (0, t_1)$ such that $\|x(t_2) - x_\mathrm{e}\| = \varepsilon$, which is a contradiction. Hence, for all $t \in [0,\infty)$, $x(t) \in \mathbb{B}_\varepsilon(x_\mathrm{e})$, and thus x_e is Lyapunov stable.

To prove *ii*), note that, since (15.8.5) holds for all $x \in \mathcal{D}\backslash\{x_\mathrm{e}\}$ and since $f(x_\mathrm{e}) = 0$, it follows that (15.8.4) holds for all $x \in \mathcal{D}$. Therefore, $x_\mathrm{e} = 0$ is Lyapunov stable. Let $\varepsilon > 0$ satisfy $\mathrm{cl}\,\mathbb{B}_\varepsilon(x_\mathrm{e}) \subseteq \mathcal{D}$, and let $\delta \in (0,\varepsilon)$ be such that $\mathbb{B}_\delta(x_\mathrm{e}) \subseteq \mathcal{D}$ and such that, for all $x_0 \in \mathbb{B}_\delta(x_\mathrm{e})$, there exists a unique solution $x\colon [0,\infty) \mapsto \mathbb{B}_\varepsilon(x_\mathrm{e})$ satisfying (15.8.1) and $x(0) = x_0$.

Let $x_0 \in \mathbb{B}_\delta(x_\mathrm{e})$. Then, for all $t \ge 0$, $\frac{\mathrm{d}}{\mathrm{d}t}V[x(t)] = V'[x(t)]f[x(t)] < 0$, and thus $V[x(t)]$ is a decreasing function on $[0,\infty)$ that is bounded from below by zero. Now, suppose that $V[x(t)]$ does not converge to zero. Thus, there exists $L > 0$ such that, for all $t \ge 0$, $V[x(t)] \ge L$. Since $\mathcal{D}_1 \triangleq \{x \in \mathbb{B}_\delta(x_\mathrm{e})\colon V(x) < L\}$ is open and contains x_e, there exists $\eta > 0$ such that $\mathbb{B}_\eta(x_\mathrm{e}) \subseteq \mathcal{D}_1$. Therefore, for all $t \ge 0$, $\|x(t) - x_\mathrm{e}\| \ge \eta$. Next, define $\gamma < 0$ by $\gamma \triangleq \max_{x\in\{x\in\mathbb{R}^n\colon\, \eta\le\|x-x_\mathrm{e}\|\le\varepsilon\}} V'(x)f(x)$. Since, for all $t \ge 0$, $\eta \le \|x(t) - x_\mathrm{e}\| < \varepsilon$, it follows that

$$V[x(t)] - V[x(0)] = \int_0^t V'[x(\tau)]f[x(\tau)]\,\mathrm{d}\tau \le \gamma t.$$

Therefore, for all $t \ge 0$, $V(x(t)) \le V[x(0)] + \gamma t$. However, for all $t > -V[x(0)]/\gamma$, $V[x(t)] < 0$, which is a contradiction. Therefore, $\lim_{t\to\infty} V[x(t)] = 0$.

Now, suppose that either $\lim_{t\to\infty} x(t)$ does not exist or $\lim_{t\to\infty} x(t)$ exists but is not x_e. In both cases, there exists $\rho > 0$ such that, for every positive integer i, there exists $t_i > i$ such that $\rho < \|x(t_i) - x_\mathrm{e}\| < \varepsilon$. Let $(t_{i_j})_{j=1}^\infty$ be an increasing subsequence of $(t_i)_{i=1}^\infty$. Since, for all $j \ge 1$, $\|x(t_{i_j}) - x_\mathrm{e}\| > \rho$ and since $\{V(x_{i_j})\}_{j=1}^\infty$ is decreasing and nonnegative, it follows that $\lim_{j\to\infty} V[x(t_{i_j})] \ge \min_{x\in\{x\in\mathbb{R}\colon\, \rho\le\|x-x_\mathrm{e}\|\le\varepsilon} V(x) > 0$. However, $\lim_{j\to\infty} V[x(t_{i_j})] = 0$, which is a contradiction. Hence, $\lim_{t\to\infty} x(t) = x_\mathrm{e}$, which proves that x_e is asymptotically stable.

To prove the last statement, let $x_0 \in \mathbb{R}^n$. Suppose that *i*) of Theorem 15.8.2 holds. Let $(t_i)_{i=1}^\infty \subset (0, T_{x_0})$ satisfy $\lim_{i\to\infty} t_i = T_{x_0}$ and $\lim_{i\to\infty} \|x(t_i)\| = \infty$. Since V is radially unbounded, it follows that $\lim_{i\to\infty} V[x(t_i)] = \infty$. However, it follows from (15.8.4) that, for all $t \in [0, T_{x_0})$,

$$V[x(t)] - V[x(0)] = \int_0^t V'[x(\tau)]f[x(\tau)]\,\mathrm{d}\tau \le 0,$$

and thus, for all $t \in [0, T_{x_0})$, $V[x(t)] \le V[x(0)]$. Therefore, $\lim_{i\to\infty} V[x(t_i)] \le V(x_0)$, which is a contradiction. Hence, *ii*) of Theorem 15.8.2 holds.

Since *ii*) of Theorem 15.8.2 holds, there exists a unique C^1 solution $x\colon [0,\infty) \mapsto \mathbb{R}^n$ satisfying (15.8.1) and $x(0) = x_0$. Since $V[x(t)]$ is decreasing and nonnegative on $[0,\infty)$, it follows that $\eta \triangleq \lim_{t\to\infty} V[x(t)]$ exists and is nonnegative. Suppose that $\eta > 0$. Therefore, for all $t \ge 0$, $x(t) \in \mathcal{D}_1 \triangleq \{x \in \mathbb{R}^n\colon \eta \le V(x) \le V(x_0)\}$. Since V is radially unbounded, it follows that $\mathcal{D}_1$ is compact. Therefore, define $\gamma \triangleq \max_{x\in\mathcal{D}_1} V'(x)f(x)$. Since $\eta > 0$, it follows that $x_\mathrm{e} \notin \mathcal{D}_1$, and thus $\gamma < 0$. Since, for all $t \ge 0$, $x(t) \in \mathcal{D}_1$, it follows that

$$V[x(t)] - V[x(0)] = \int_0^t V'[x(\tau)]f[x(\tau)]\,\mathrm{d}\tau \le \gamma t.$$

Therefore, for all $t \ge 0$, $V(x(t)) \le V[x(0)] + \gamma t$. However, for all $t > -V[x(0)]/\gamma$ it follows that $V[x(t)] < 0$, which is a contradiction. Therefore, $\eta = 0$, and thus $\lim_{t\to\infty} V[x(t)] = 0$. As in the proof of asymptotic stability, it can be shown that $\lim_{t\to\infty} x(t) = x_\mathrm{e}$. $\square$

15.9 Linear Stability Theory

We now specialize Definition 15.8.3 to the linear system

$$\dot{x}(t) = Ax(t), \tag{15.9.1}$$

where $t \geq 0$, $x(t) \in \mathbb{R}^n$, and $A \in \mathbb{R}^{n \times n}$. Note that $x_e = 0$ is an equilibrium of (15.9.1), and that $x_e \in \mathbb{R}^n$ is an equilibrium of (15.9.1) if and only if $x_e \in \mathcal{N}(A)$. Hence, if x_e is the globally asymptotically stable equilibrium of (15.9.1), then A is nonsingular and $x_e = 0$.

We consider three types of stability for the linear system (15.9.1). Unlike Definition 15.8.3, these definitions are stated in terms of the dynamics matrix rather than the equilibrium.

Definition 15.9.1. For $A \in \mathbb{C}^{n \times n}$, define the following classes of matrices:

i) A is *Lyapunov stable* if $\operatorname{spec}(A) \subset \text{CLHP}$ and, if $\lambda \in \operatorname{spec}(A) \cap \text{IA}$, then λ is semisimple.

ii) A is *semistable* if $\operatorname{spec}(A) \subset \text{OLHP} \cup \{0\}$ and, if $0 \in \operatorname{spec}(A)$, then 0 is semisimple.

iii) A is *asymptotically stable* if $\operatorname{spec}(A) \subset \text{OLHP}$.

iv) A is *unstable* if A is not Lyapunov stable.

The following result concerns Lyapunov stability, semistability, and asymptotic stability.

Proposition 15.9.2. Let $A \in \mathbb{R}^{n \times n}$. Then, the following statements are equivalent:

i) $x_e = 0$ is a Lyapunov-stable equilibrium of (15.9.1).

ii) At least one equilibrium of (15.9.1) is Lyapunov stable.

iii) Every equilibrium of (15.9.1) is Lyapunov stable.

iv) A is Lyapunov stable.

v) For every initial condition $x(0) \in \mathbb{R}^n$, $\{x(t) \colon t \geq 0\}$ is bounded.

vi) $\{e^{tA} \colon t \geq 0\}$ is bounded.

vii) For every initial condition $x(0) \in \mathbb{R}^n$, $\{e^{tA}x(0) \colon t \geq 0\}$ is bounded.

The following statements are equivalent:

viii) A is semistable.

ix) $\lim_{t\to\infty} e^{tA}$ exists.

x) For every initial condition $x(0)$, $\lim_{t\to\infty} x(t)$ exists.

If these conditions hold, then

$$\lim_{t\to\infty} e^{tA} = I - AA^{\#}. \tag{15.9.2}$$

The following statements are equivalent:

xi) $x_e = 0$ is an asymptotically stable equilibrium of (15.9.1).

xii) A is asymptotically stable.

xiii) $\alpha_{\max}(A) < 0$.

xiv) For every initial condition $x(0) \in \mathbb{R}^n$, $\lim_{t\to\infty} x(t) = 0$.

xv) For every initial condition $x(0) \in \mathbb{R}^n$, $e^{tA}x(0) \to 0$ as $t \to \infty$.

xvi) $e^{tA} \to 0$ as $t \to \infty$.

The following definition concerns the stability of a polynomial.

Definition 15.9.3. Let $p \in \mathbb{R}[s]$. Then, define the following terminology:

i) p is *Lyapunov stable* if $\operatorname{roots}(p) \subset \text{CLHP}$ and, for all $\lambda \in \operatorname{roots}(p) \cap \text{IA}$, $\operatorname{mult}_p(\lambda) = 1$.

ii) p is *semistable* if $\operatorname{roots}(p) \subset \text{OLHP} \cup \{0\}$ and, if $0 \in \operatorname{roots}(p)$, then $\operatorname{mult}_p(0) = 1$.

iii) p is *asymptotically stable* if $\operatorname{roots}(p) \subset \text{OLHP}$.

iv) p is *unstable* if p is not Lyapunov stable.

For the following result, recall Definition 15.9.1.

Proposition 15.9.4. Let $A \in \mathbb{R}^{n\times n}$. Then, the following statements hold:

i) A is Lyapunov stable if and only if μ_A is Lyapunov stable.

ii) A is semistable if and only if μ_A is semistable.

Furthermore, the following statements are equivalent:

iii) A is asymptotically stable

iv) μ_A is asymptotically stable.

v) χ_A is asymptotically stable.

Next, consider the factorization of the minimal polynomial μ_A of A given by

$$\mu_A = \mu_A^{\mathrm{s}}\mu_A^{\mathrm{u}}, \tag{15.9.3}$$

where μ_A^{s} and μ_A^{u} are monic polynomials such that

$$\mathrm{roots}(\mu_A^{\mathrm{s}}) \subset \mathrm{OLHP}, \quad \mathrm{roots}(\mu_A^{\mathrm{u}}) \subset \mathrm{CRHP}. \tag{15.9.4}$$

Proposition 15.9.5. Let $A \in \mathbb{R}^{n\times n}$, and let $S \in \mathbb{R}^{n\times n}$ be a nonsingular matrix such that

$$A = S\begin{bmatrix} A_1 & A_{12} \\ 0 & A_2 \end{bmatrix}S^{-1}, \tag{15.9.5}$$

where $A_1 \in \mathbb{R}^{r\times r}$ is asymptotically stable, $A_{12} \in \mathbb{R}^{r\times(n-r)}$, and $A_2 \in \mathbb{R}^{(n-r)\times(n-r)}$ satisfies $\mathrm{spec}(A_2) \subset$ CRHP. Then,

$$\mu_A^{\mathrm{s}}(A) = S\begin{bmatrix} 0 & C_{12\mathrm{s}} \\ 0 & \mu_A^{\mathrm{s}}(A_2) \end{bmatrix}S^{-1}, \tag{15.9.6}$$

where $C_{12\mathrm{s}} \in \mathbb{R}^{r\times(n-r)}$ and $\mu_A^{\mathrm{s}}(A_2)$ is nonsingular, and

$$\mu_A^{\mathrm{u}}(A) = S\begin{bmatrix} \mu_A^{\mathrm{u}}(A_1) & C_{12\mathrm{u}} \\ 0 & 0 \end{bmatrix}S^{-1}, \tag{15.9.7}$$

where $C_{12\mathrm{u}} \in \mathbb{R}^{r\times(n-r)}$ and $\mu_A^{\mathrm{u}}(A_1)$ is nonsingular. Consequently,

$$\mathcal{N}[\mu_A^{\mathrm{s}}(A)] = \mathcal{R}[\mu_A^{\mathrm{u}}(A)] = \mathcal{R}\left(S\begin{bmatrix} I_r \\ 0 \end{bmatrix}\right). \tag{15.9.8}$$

If, in addition, $A_{12} = 0$, then

$$\mu_A^{\mathrm{s}}(A) = S\begin{bmatrix} 0 & 0 \\ 0 & \mu_A^{\mathrm{s}}(A_2) \end{bmatrix}S^{-1}, \quad \mu_A^{\mathrm{u}}(A) = S\begin{bmatrix} \mu_A^{\mathrm{u}}(A_1) & 0 \\ 0 & 0 \end{bmatrix}S^{-1}. \tag{15.9.9}$$

Consequently,

$$\mathcal{R}[\mu_A^{\mathrm{s}}(A)] = \mathcal{N}[\mu_A^{\mathrm{u}}(A)] = \mathcal{R}\left(S\begin{bmatrix} 0 \\ I_{n-r} \end{bmatrix}\right). \tag{15.9.10}$$

Corollary 15.9.6. Let $A \in \mathbb{R}^{n\times n}$. Then,

$$\mathcal{N}[\mu_A^{\mathrm{s}}(A)] = \mathcal{R}[\mu_A^{\mathrm{u}}(A)], \tag{15.9.11}$$

$$\mathcal{N}[\mu_A^{\mathrm{u}}(A)] = \mathcal{R}[\mu_A^{\mathrm{s}}(A)]. \tag{15.9.12}$$

Proof. Theorem 7.4.6 implies that there exists a nonsingular matrix $S \in \mathbb{R}^{n\times n}$ such that (15.9.5) holds, where $A_1 \in \mathbb{R}^{r\times r}$ is asymptotically stable, $A_{12} = 0$, and $A_2 \in \mathbb{R}^{(n-r)\times(n-r)}$ satisfies $\mathrm{spec}(A_2) \subset$ CRHP. The result now follows from Proposition 15.9.5. □

In view of Corollary 15.9.6, we define the *asymptotically stable subspace* $\mathcal{S}_{\mathrm{s}}(A)$ of A by

$$\mathcal{S}_{\rm s}(A) \triangleq \mathcal{N}[\mu_A^{\rm s}(A)] = \mathcal{R}[\mu_A^{\rm u}(A)] \tag{15.9.13}$$

and the *unstable subspace* $\mathcal{S}_{\rm u}(A)$ of A by

$$\mathcal{S}_{\rm u}(A) \triangleq \mathcal{N}[\mu_A^{\rm u}(A)] = \mathcal{R}[\mu_A^{\rm s}(A)]. \tag{15.9.14}$$

Note that

$$\dim \mathcal{S}_{\rm s}(A) = \operatorname{def} \mu_A^{\rm s}(A) = \operatorname{rank} \mu_A^{\rm u}(A) = \sum_{\substack{\lambda\in\operatorname{spec}(A)\\ \operatorname{Re}\lambda<0}} \operatorname{amult}_A(\lambda), \tag{15.9.15}$$

$$\dim \mathcal{S}_{\rm u}(A) = \operatorname{def} \mu_A^{\rm u}(A) = \operatorname{rank} \mu_A^{\rm s}(A) = \sum_{\substack{\lambda\in\operatorname{spec}(A)\\ \operatorname{Re}\lambda\geq 0}} \operatorname{amult}_A(\lambda). \tag{15.9.16}$$

Lemma 15.9.7. Let $A \in \mathbb{R}^{n\times n}$, assume that $\operatorname{spec}(A) \subset$ CRHP, let $x \in \mathbb{R}^n$, and assume that $\lim_{t\to\infty} e^{tA}x = 0$. Then, $x = 0$.

For the following result, note Proposition 4.8.3, Fact 4.15.5, Proposition 8.1.8, Proposition 15.9.2, and Fact 15.19.4.

Proposition 15.9.8. Let $A \in \mathbb{R}^{n\times n}$. Then, the following statements hold:

i) $\mathcal{S}_{\rm s}(A) = \{x \in \mathbb{R}^n\colon\ \lim_{t\to\infty} e^{tA}x = 0\}$.

ii) $\mu_A^{\rm s}(A)$ and $\mu_A^{\rm u}(A)$ are group invertible.

iii) $P_{\rm s} \triangleq I - \mu_A^{\rm s}(A)[\mu_A^{\rm s}(A)]^\#$ and $P_{\rm u} \triangleq I - \mu_A^{\rm u}(A)[\mu_A^{\rm u}(A)]^\#$ are idempotent.

iv) $P_{\rm s} + P_{\rm u} = I$.

v) $P_{{\rm s}\perp} = P_{\rm u}$ and $P_{{\rm u}\perp} = P_{\rm s}$.

vi) $\mathcal{S}_{\rm s}(A) = \mathcal{R}(P_{\rm s}) = \mathcal{N}(P_{\rm u})$.

vii) $\mathcal{S}_{\rm u}(A) = \mathcal{R}(P_{\rm u}) = \mathcal{N}(P_{\rm s})$.

viii) $\mathcal{S}_{\rm s}(A)$ and $\mathcal{S}_{\rm u}(A)$ are invariant subspaces of A.

ix) $\mathcal{S}_{\rm s}(A)$ and $\mathcal{S}_{\rm u}(A)$ are complementary subspaces.

x) $P_{\rm s}$ is the idempotent matrix onto $\mathcal{S}_{\rm s}(A)$ along $\mathcal{S}_{\rm u}(A)$.

xi) $P_{\rm u}$ is the idempotent matrix onto $\mathcal{S}_{\rm u}(A)$ along $\mathcal{S}_{\rm s}(A)$.

Proof. To prove *i*), let $S \in \mathbb{R}^{n\times n}$ be a nonsingular matrix such that

$$A = S\begin{bmatrix} A_1 & 0 \\ 0 & A_2 \end{bmatrix} S^{-1},$$

where $A_1 \in \mathbb{R}^{r\times r}$ is asymptotically stable and $\operatorname{spec}(A_2) \subset$ CRHP. It then follows from Proposition 15.9.5 that

$$\mathcal{S}_{\rm s}(A) = \mathcal{N}[\mu_A^{\rm s}(A)] = \mathcal{R}\left(S\begin{bmatrix} I_r \\ 0 \end{bmatrix}\right).$$

Furthermore,

$$e^{tA} = S\begin{bmatrix} e^{tA_1} & 0 \\ 0 & e^{tA_2} \end{bmatrix} S^{-1}.$$

To prove $\mathcal{S}_{\rm s}(A) \subseteq \{z \in \mathbb{R}^n\colon\ \lim_{t\to\infty} e^{tA}z = 0\}$, let $x \triangleq S\left[\begin{smallmatrix} x_1 \\ 0 \end{smallmatrix}\right] \in \mathcal{S}_{\rm s}(A)$, where $x_1 \in \mathbb{R}^r$. Then, $e^{tA}x = S\left[\begin{smallmatrix} e^{tA_1}x_1 \\ 0 \end{smallmatrix}\right] \to 0$ as $t \to \infty$. Hence, $x \in \{z \in \mathbb{R}^n\colon\ \lim_{t\to\infty} e^{tA}z = 0\}$. Conversely, to prove $\{z \in \mathbb{R}^n\colon\ \lim_{t\to\infty} e^{tA}z = 0\} \subseteq \mathcal{S}_{\rm s}(A)$, let $x \triangleq S\left[\begin{smallmatrix} x_1 \\ x_2 \end{smallmatrix}\right] \in \mathbb{R}^n$ satisfy $\lim_{t\to\infty} e^{tA}x = 0$. Hence, $e^{tA_2}x_2 \to 0$ as $t \to \infty$. Since $\operatorname{spec}(A_2) \subset$ CRHP, it follows from Lemma 15.9.7 that $x_2 = 0$. Hence, $x \in \mathcal{R}\left(S\left[\begin{smallmatrix} I_r \\ 0 \end{smallmatrix}\right]\right) = \mathcal{S}_{\rm s}(A)$. The remaining statements follow from Proposition 15.9.5. $\square$

15.10 The Lyapunov Equation

In this section we specialize Theorem 15.8.5 to the linear system (15.9.1).

Corollary 15.10.1. Let $A \in \mathbb{R}^{n\times n}$, and assume that there exist a positive-semidefinite matrix $R \in \mathbb{R}^{n\times n}$ and a positive-definite matrix $P \in \mathbb{R}^{n\times n}$ satisfying

$$A^\mathsf{T}P + PA + R = 0. \tag{15.10.1}$$

Then, A is Lyapunov stable. If, in addition, for all nonzero $\omega \in \mathbb{R}$,

$$\operatorname{rank}\begin{bmatrix} \omega jI - A \\ R \end{bmatrix} = n, \tag{15.10.2}$$

then A is semistable. Finally, if R is positive definite, then A is asymptotically stable.

Proof. Define $V(x) \triangleq x^\mathsf{T}Px$, which satisfies $V(x_\mathrm{e}) = 0$ with $x_\mathrm{e} = 0$ and satisfies $V(x) > 0$ for all nonzero $x \in \mathcal{D} = \mathbb{R}^n$. Furthermore, Theorem 15.8.5 with $f(x) \triangleq Ax$ implies that $V'(x)f(x) = 2x^\mathsf{T}PAx = x^\mathsf{T}(A^\mathsf{T}P + PA)x = -x^\mathsf{T}Rx$, which satisfies (15.8.4) for all $x \in \mathbb{R}^n$. Thus, Theorem 15.8.5 implies that A is Lyapunov stable. If, in addition, R is positive definite, then, for all $x \neq 0$, (15.8.5) holds, and thus A is asymptotically stable.

Alternatively, we now prove the first and third statements without using Theorem 15.8.5. Letting $\lambda \in \operatorname{spec}(A)$, and letting $x \in \mathbb{C}^n$ be an associated eigenvector, it follows that $0 \geq -x^*Rx = x^*(A^\mathsf{T}P + PA)x = (\overline{\lambda} + \lambda)x^*Px$. Therefore, $\operatorname{spec}(A) \subset \mathrm{CLHP}$. Now, suppose that $\omega j \in \operatorname{spec}(A)$, where $\omega \in \mathbb{R}$, and let $x \in \mathcal{N}[(\omega jI - A)^2]$. Defining $y \triangleq (\omega jI - A)x$, it follows that $(\omega jI - A)y = 0$, and thus $Ay = \omega jy$. Therefore, $-y^*Ry = y^*(A^\mathsf{T}P + PA)y = -\omega jy^*Py + \omega jy^*Py = 0$, and thus $Ry = 0$. Hence, $0 = x^*Ry = -x^*(A^\mathsf{T}P + PA)y = -x^*(A^\mathsf{T} + \omega jI)Py = y^*Py$. Since P is positive definite, it follows that $y = 0$, and thus $(\omega jI - A)x = 0$. Therefore, $x \in \mathcal{N}(\omega jI - A)$. Now, Proposition 7.7.8 implies that ωj is semisimple. Therefore, A is Lyapunov stable.

Next, to prove that A is asymptotically stable, let $\lambda \in \operatorname{spec}(A)$, and let $x \in \mathbb{C}^n$ be an associated eigenvector. Thus, $0 > -x^*Rx = (\overline{\lambda} + \lambda)x^*Px$, which implies that A is asymptotically stable.

Finally, to prove that A is semistable, let $\omega j \in \operatorname{spec}(A)$, where $\omega \in \mathbb{R}$ is nonzero, and let $x \in \mathbb{C}^n$ be an associated eigenvector. Then,

$$-x^*Rx = x^*(A^\mathsf{T}P + PA)x = -x^*[(\omega jI - A)^*P + P(\omega jI - A]x = 0.$$

Therefore, $Rx = 0$, and thus

$$\begin{bmatrix} \omega jI - A \\ R \end{bmatrix} x = 0,$$

which implies that $x = 0$, which contradicts $x \neq 0$. Consequently, $\omega j \notin \operatorname{spec}(A)$ for all nonzero $\omega \in \mathbb{R}$, and thus A is semistable. □

Equation (15.10.1) is a *Lyapunov equation.* Converse results for Corollary 15.10.1 are given by Corollary 15.10.4, Proposition 15.10.6, Proposition 15.10.5, and Proposition 15.10.6. The following lemma is useful for analyzing (15.10.1).

Lemma 15.10.2. Assume that $A \in \mathbb{F}^{n\times n}$ is asymptotically stable. Then,

$$\int_0^\infty e^{tA}\,\mathrm{d}t = -A^{-1}. \tag{15.10.3}$$

Proof. Proposition 15.1.4 implies $\int_0^t e^{\tau A}\,\mathrm{d}\tau = A^{-1}(e^{tA} - I)$. Letting $t \to \infty$ yields (15.10.3). □

The following result concerns Sylvester's equation. See Fact 7.11.26 and Proposition 9.2.5.

Proposition 15.10.3. Let $A, B, C \in \mathbb{R}^{n\times n}$. Then, there exists a unique matrix $X \in \mathbb{R}^{n\times n}$ satisfying

$$AX + XB + C = 0 \tag{15.10.4}$$

if and only if $B^{\mathsf{T}} \oplus A$ is nonsingular. If these conditions hold, then X is given by

$$X = -\operatorname{vec}^{-1}[(B^{\mathsf{T}} \oplus A)^{-1} \operatorname{vec} C]. \tag{15.10.5}$$

If, in addition, $B^{\mathsf{T}} \oplus A$ is asymptotically stable, then X is given by

$$X = \int_0^\infty e^{tA} C e^{tB} \, \mathrm{d}t. \tag{15.10.6}$$

Proof. The first two statements follow from Proposition 9.2.5. Now, assume that $B^{\mathsf{T}} \oplus A$ is asymptotically stable. Then, it follows from (15.10.5), Lemma 15.10.2, and Proposition 15.1.7 that

$$\begin{aligned} X &= \int_0^\infty \operatorname{vec}^{-1}\left(e^{t(B^{\mathsf{T}} \oplus A)} \operatorname{vec} C\right) \mathrm{d}t = \int_0^\infty \operatorname{vec}^{-1}\left(e^{tB^{\mathsf{T}}} \otimes e^{tA}\right) \operatorname{vec} C \, \mathrm{d}t \\ &= \int_0^\infty \operatorname{vec}^{-1} \operatorname{vec}\left(e^{tA} C e^{tB}\right) \mathrm{d}t = \int_0^\infty e^{tA} C e^{tB} \, \mathrm{d}t. \end{aligned}$$

□

The following result provides a converse to Corollary 15.10.1 for the case of asymptotic stability.

Corollary 15.10.4. Let $A \in \mathbb{R}^{n \times n}$ and $R \in \mathbb{R}^{n \times n}$. Then, there exists a unique matrix $P \in \mathbb{R}^{n \times n}$ satisfying (15.10.1) if and only if $A \oplus A$ is nonsingular. If these conditions hold and R is symmetric, then P is symmetric. Now, assume that A is asymptotically stable. Then, P is given by

$$P = \int_0^\infty e^{tA^{\mathsf{T}}} R e^{tA} \, \mathrm{d}t. \tag{15.10.7}$$

Finally, if R is (positive semidefinite, positive definite), then so is P.

Proof. First note that $A \oplus A$ is nonsingular if and only if $(A \oplus A)^{\mathsf{T}} = A^{\mathsf{T}} \oplus A^{\mathsf{T}}$ is nonsingular. Now, the first statement follows from Proposition 15.10.3. To prove the second statement, note that $A^{\mathsf{T}}(P - P^{\mathsf{T}}) + (P - P^{\mathsf{T}})A = 0$, which implies that P is symmetric. Now, suppose that A is asymptotically stable. Then, Fact 15.19.33 implies that $A \oplus A$ is asymptotically stable. Consequently, (15.10.7) follows from (15.10.6). □

The following results include converse statements. We first consider asymptotic stability.

Proposition 15.10.5. Let $A \in \mathbb{R}^{n \times n}$. Then, the following statements are equivalent:

i) A is asymptotically stable.

ii) For every positive-definite matrix $R \in \mathbb{R}^{n \times n}$ there exists a positive-definite matrix $P \in \mathbb{R}^{n \times n}$ that satisfies (15.10.1).

iii) There exist a positive-definite matrix $R \in \mathbb{R}^{n \times n}$ and a positive-definite matrix $P \in \mathbb{R}^{n \times n}$ that satisfy (15.10.1).

Proof. It follows from Corollary 15.10.4 that *i)* implies *ii)*. *ii)* immediately implies *iii)*. Finally, it follows from Corollary 15.10.1 that *iii)* implies *i)*. □

Next, we consider the case of Lyapunov stability.

Proposition 15.10.6. Let $A \in \mathbb{R}^{n \times n}$. Then, the following statements hold:

i) If A is Lyapunov stable, then there exist a positive-definite matrix $P \in \mathbb{R}^{n \times n}$ and a positive-semidefinite matrix $R \in \mathbb{R}^{n \times n}$ such that $\operatorname{rank} R = \nu_-(A)$ and such that (15.10.1) is satisfied.

ii) If there exist a positive-definite matrix $P \in \mathbb{R}^{n \times n}$ and a positive-semidefinite matrix $R \in \mathbb{R}^{n \times n}$ that satisfy (15.10.1), then A is Lyapunov stable.

Proof. To prove *i)*, suppose that A is Lyapunov stable. Then, it follows from Theorem 7.4.6 and Definition 15.9.1 that there exists a nonsingular matrix $S \in \mathbb{R}^{n \times n}$ such that $A = S\left[\begin{smallmatrix} A_1 & 0 \\ 0 & A_2 \end{smallmatrix}\right]S^{-1}$ is in real Jordan form, where $A_1 \in \mathbb{R}^{n_1 \times n_1}$, $A_2 \in \mathbb{R}^{n_2 \times n_2}$, $\operatorname{spec}(A_1) \subset \mathrm{IA}$, A_1 is semisimple, and $\operatorname{spec}(A_2) \subset \mathrm{OLHP}$. Next, Fact 7.10.5 implies that there exists a nonsingular matrix $S_1 \in \mathbb{R}^{n_1 \times n_1}$ such that $A_1 = S_1 J_1 S_1^{-1}$, where $J_1 \in \mathbb{R}^{n_1 \times n_1}$ is skew symmetric. Then, it follows that $A_1^{\mathsf{T}} P_1 + P_1 A_1 =$

$S_1^{-\mathsf{T}}(J_1 + J_1^{\mathsf{T}})S_1^{-1} = 0$, where $P_1 \triangleq S_1^{-\mathsf{T}}S_1^{-1}$ is positive definite. Next, let $R_2 \in \mathbb{R}^{n_2 \times n_2}$ be positive definite, and let $P_2 \in \mathbb{R}^{n_2 \times n_2}$ be the positive-definite solution of $A_2^{\mathsf{T}}P_2 + P_2A_2 + R_2 = 0$. Hence, (15.10.1) is satisfied with $P \triangleq S^{-\mathsf{T}}\begin{bmatrix} P_1 & 0 \\ 0 & P_2 \end{bmatrix}S^{-1}$ and $R \triangleq S^{-\mathsf{T}}\begin{bmatrix} 0 & 0 \\ 0 & R_2 \end{bmatrix}S^{-1}$. Finally, Corollary 15.10.1 implies *ii*). □

Corollary 15.10.7. Let $A \in \mathbb{R}^{n \times n}$. Then, the following statements hold:

i) A is Lyapunov stable if and only if there exists a positive-definite matrix $P \in \mathbb{R}^{n \times n}$ such that $A^{\mathsf{T}}P + PA$ is negative semidefinite.

ii) A is asymptotically stable if and only if there exists a positive-definite matrix $P \in \mathbb{R}^{n \times n}$ such that $A^{\mathsf{T}}P + PA$ is negative definite.

15.11 Discrete-Time Stability Theory

The theory of difference equations is concerned with solutions of discrete-time dynamical systems of the form

$$x(k+1) = f[x(k)], \tag{15.11.1}$$

where $f\colon \mathbb{R}^n \to \mathbb{R}^n$, $k \in \mathbb{N}$, $x(k) \in \mathbb{R}^n$, and $x(0) = x_0$ is the initial condition. The solution $x(k) \equiv x_{\mathrm{e}}$ is an equilibrium of (15.11.1) if $x_{\mathrm{e}} = f(x_{\mathrm{e}})$. A linear discrete-time system has the form

$$x(k+1) = Ax(k), \tag{15.11.2}$$

where $A \in \mathbb{R}^{n \times n}$. For $k \in \mathbb{N}$, $x(k)$ is given by

$$x(k) = A^k x_0. \tag{15.11.3}$$

The asymptotic behavior of the sequence $(x(k))_{k=0}^{\infty}$ is determined by the stability of A.

Definition 15.11.1. For $A \in \mathbb{C}^{n \times n}$, define the following classes of matrices:

i) A is *discrete-time Lyapunov stable* if $\operatorname{spec}(A) \subset \mathrm{CIUD}$ and, if $\lambda \in \operatorname{spec}(A)$ and $|\lambda| = 1$, then λ is semisimple.

ii) A is *discrete-time semistable* if $\operatorname{spec}(A) \subset \mathrm{OIUD} \cup \{1\}$ and, if $1 \in \operatorname{spec}(A)$, then 1 is semisimple.

iii) A is *discrete-time asymptotically stable* if $\operatorname{spec}(A) \subset \mathrm{OIUD}$.

iv) A is *discrete-time unstable* if A is not discrete-time Lyapunov stable.

Proposition 15.11.2. Let $A \in \mathbb{R}^{n \times n}$ and consider the linear discrete-time system (15.11.2). Then, the following statements are equivalent:

i) A is discrete-time Lyapunov stable.

ii) For every initial condition $x_0 \in \mathbb{R}^n$, $(x(k))_{k=1}^{\infty}$ is bounded.

iii) For every initial condition $x_0 \in \mathbb{R}^n$, $(A^k x_0)_{k=1}^{\infty}$ is bounded.

iv) $(A^k)_{k=1}^{\infty}$ is bounded.

The following statements are equivalent:

v) A is discrete-time semistable.

vi) $\lim_{k\to\infty} A^k$ exists. In fact, $\lim_{k\to\infty} A^k = I - (I - A)(I - A)^{\#}$.

vii) For every initial condition $x_0 \in \mathbb{R}^n$, $\lim_{k\to\infty} x(k)$ exists.

The following statements are equivalent:

viii) A is discrete-time asymptotically stable.

ix) $\rho_{\max}(A) < 1$.

x) For every initial condition $x_0 \in \mathbb{R}^n$, $\lim_{k\to\infty} x(k) = 0$.

xi) For every initial condition $x_0 \in \mathbb{R}^n$, $A^k x_0 \to 0$ as $k \to \infty$.

xii) $A^k \to 0$ as $k \to \infty$.

The following definition concerns discrete-time stability of a polynomial.

Definition 15.11.3. For $p \in \mathbb{R}[s]$, define the following terminology:

i) p is *discrete-time Lyapunov stable* if $\operatorname{roots}(p) \subset \mathrm{CIUD}$ and, if λ is an imaginary root of p, then $\operatorname{mult}_p(\lambda) = 1$.

ii) p is *discrete-time semistable* if $\operatorname{roots}(p) \subset \mathrm{OIUD} \cup \{1\}$ and, if $1 \in \operatorname{roots}(p)$, then $\operatorname{mult}_p(1) = 1$.

iii) p is *discrete-time asymptotically stable* if $\operatorname{roots}(p) \subset \mathrm{OIUD}$.

iv) p is *discrete-time unstable* if p is not discrete-time Lyapunov stable.

Proposition 15.11.4. Let $A \in \mathbb{R}^{n\times n}$. Then, the following statements hold:

i) A is discrete-time Lyapunov stable if and only if μ_A is discrete-time Lyapunov stable.

ii) A is discrete-time semistable if and only if μ_A is discrete-time semistable.

Furthermore, the following statements are equivalent:

iii) A is discrete-time asymptotically stable.

iv) μ_A is discrete-time asymptotically stable.

v) χ_A is discrete-time asymptotically stable.

We now consider the *discrete-time Lyapunov equation*

$$P = A^{\mathsf{T}}PA + R. \tag{15.11.4}$$

Proposition 15.11.5. Let $A \in \mathbb{R}^{n\times n}$. Then, the following statements are equivalent:

i) A is discrete-time asymptotically stable.

ii) For every positive-definite matrix $R \in \mathbb{R}^{n\times n}$ there exists a positive-definite matrix $P \in \mathbb{R}^{n\times n}$ that satisfies (15.11.4).

iii) There exist a positive-definite matrix $R \in \mathbb{R}^{n\times n}$ and a positive-definite matrix $P \in \mathbb{R}^{n\times n}$ that satisfy (15.11.4).

Proposition 15.11.6. Let $A \in \mathbb{R}^{n\times n}$. Then, A is discrete-time Lyapunov-stable if and only if there exist a positive-definite matrix $P \in \mathbb{R}^{n\times n}$ and a positive-semidefinite matrix $R \in \mathbb{R}^{n\times n}$ that satisfy (15.11.4).

15.12 Facts on Matrix Exponential Formulas

Fact 15.12.1. Let $A \in \mathbb{F}^{n\times n}$ and $t \in \mathbb{R}$. Then, the following statements hold:

i) If $A^2 = 0$, then $e^{tA} = I + tA$.

ii) If $A^2 = I$, then $e^{tA} = (\cosh t)I + (\sinh t)A$.

iii) If $A^2 = -I$, then $e^{tA} = (\cos t)I + (\sin t)A$.

iv) If $A^2 = A$, then $e^{tA} = I + (e^t - 1)A$.

v) If $A^2 = -A$, then $e^{tA} = I + (1 - e^{-t})A$.

vi) If $\operatorname{rank} A = 1$ and $\operatorname{tr} A = 0$, then $e^{tA} = I + tA$.

vii) If $\operatorname{rank} A = 1$ and $\operatorname{tr} A \neq 0$, then $e^{tA} = I + \frac{e^{(\operatorname{tr} A)t} - 1}{\operatorname{tr} A}A$.

viii) If there exists nonzero $r \in \mathbb{F}$ such that $A^2 = r^2 I$, then

$$e^{tA} = (\cosh rt)I + \frac{\sinh rt}{r}A.$$

ix) If there exists nonzero $r \in \mathbb{F}$ such that $A^3 = r^2 A$, then

$$e^{tA} = I + \frac{\sinh rt}{r}A + \frac{(\cosh rt) - 1}{r^2}A^2.$$

x) If there exist $k \geq 1$ and nonzero $r \in \mathbb{F}$ such that $A^{k+1} = rA^k$, then

$$e^{tA} = \sum_{i=0}^{k-1} \frac{1}{i!} t^i A^i + \frac{e^{rt} - \sum_{i=0}^{k-1} \frac{1}{i!} r^i t^i}{r^k} A^k.$$

xi) If $\rho_{\max}(A) < 1$, then

$$e^{\sum_{i=1}^{\infty} \frac{1}{i} \operatorname{tr} A^i} = \frac{1}{\det(I - A)}.$$

xii) $e^{1_{n\times n}t} = I + \frac{e^{nt}-1}{n} 1_{n\times n}$.

Source: *viii*)–*x*) are given in [2913]. *xi*) follows from the power series for $\log(I - A)$ given by Fact 13.4.15 along with Corollary 15.2.4. *xii*) is given in [904]. **Remark:** Explicit expressions for the cases where A satisfies $A^{k+2} = r^2A^k$ and $A^{k+3} = r^3A^k$ are given in [2913]. See also [2234]. **Related:** Setting $r = \theta_J$ in *ix*) yields Fact 15.12.6.

Fact 15.12.2. Let $A \triangleq \begin{bmatrix} 0 & I_n \\ I_n & 0 \end{bmatrix}$. Then,

$$e^{tA} = (\cosh t)I_{2n} + (\sinh t)A, \quad e^{tJ_{2n}} = (\cos t)I_{2n} + (\sin t)J_{2n}.$$

Fact 15.12.3. Let $A \in \mathbb{R}^{n\times n}$, and assume that A is skew symmetric. Then, $\{e^{\theta A}\colon\ \theta \in \mathbb{R}\} \subseteq \mathrm{SO}(n)$ is a multiplication group. If, in addition, $n = 2$, then

$$\{e^{\theta J_2}\colon\ \theta \in \mathbb{R}\} = \mathrm{SO}(2).$$

Remark: Note that $e^{\theta J_2} = \begin{bmatrix} \cos\theta & \sin\theta \\ -\sin\theta & \cos\theta \end{bmatrix}$. See Fact 4.14.1.

Fact 15.12.4. Let $n \geq 2$ and $A \in \mathbb{R}^{n\times n}$, where

$$A \triangleq \begin{bmatrix} 0 & 1 & 0 & 0 & \cdots & 0 \\ 0 & 0 & 2 & 0 & \cdots & 0 \\ 0 & 0 & 0 & 3 & \cdots & 0 \\ \vdots & \vdots & \vdots & \ddots & \ddots & \vdots \\ 0 & 0 & 0 & 0 & \ddots & n-1 \\ 0 & 0 & 0 & 0 & \cdots & 0 \end{bmatrix}.$$

Then,

$$e^A = \begin{bmatrix} \binom{0}{0} & \binom{1}{0} & \binom{2}{0} & \binom{3}{0} & \cdots & \binom{n-1}{0} \\ 0 & \binom{1}{1} & \binom{2}{1} & \binom{3}{1} & \cdots & \binom{n-1}{1} \\ 0 & 0 & \binom{2}{2} & \binom{3}{2} & \cdots & \binom{n-1}{2} \\ \vdots & \vdots & \vdots & \ddots & \ddots & \vdots \\ 0 & 0 & 0 & 0 & \ddots & \binom{n-1}{n-2} \\ 0 & 0 & 0 & 0 & \cdots & \binom{n-1}{n-1} \end{bmatrix}.$$

Furthermore, if $k \geq n$, then

$$\sum_{i=1}^{k} i^{n-1} = [1^{n-1} \ \ 2^{n-1} \ \ \cdots \ \ n^{n-1}]e^{-A} \begin{bmatrix} \binom{k}{1} \\ \vdots \\ \binom{k}{n} \end{bmatrix}.$$

Source: [168]. **Remark:** A is called the *creation matrix* in [9]. **Related:** Fact 7.18.5.

Fact 15.12.5. Let $A \in \mathbb{F}^{3\times 3}$. If $\operatorname{spec}(A) = \{\lambda\}$, then

$$e^{tA} = e^{\lambda t}\big[I + t(A - \lambda I) + \tfrac{1}{2}t^2(A - \lambda I)^2\big].$$

If $\operatorname{mspec}(A) = \{\lambda, \lambda, \mu\}_{\mathrm{ms}}$, where $\mu \neq \lambda$, then

$$e^{tA} = e^{\lambda t}[I + t(A - \lambda I)] + \left[\frac{e^{\mu t} - e^{\lambda t}}{(\mu - \lambda)^2} - \frac{te^{\lambda t}}{\mu - \lambda}\right](A - \lambda I)^2.$$

If $\operatorname{spec}(A) = \{\lambda, \mu, \nu\}$, then

$$e^{tA} = \frac{e^{\lambda t}}{(\lambda - \mu)(\lambda - \nu)}(A - \mu I)(A - \nu I) + \frac{e^{\mu t}}{(\mu - \lambda)(\mu - \nu)}(A - \lambda I)(A - \nu I) + \frac{e^{\nu t}}{(\nu - \lambda)(\nu - \mu)}(A - \lambda I)(A - \mu I).$$

Source: [153]. **Remark:** Additional expressions are given in [4, 361, 436, 717, 1326, 2234, 2246].

Fact 15.12.6. Let $x \in \mathbb{R}^3$, assume that x is nonzero, and define $\theta \triangleq \|x\|_2$. Then,

$$\begin{aligned} e^{K(x)} &= I + \frac{\sin\theta}{\theta}K(x) + \frac{1 - \cos\theta}{\theta^2}K^2(x) \\ &= I + \frac{\sin\theta}{\theta}K(x) + \frac{2\sin^2\frac{\theta}{2}}{\theta^2}K^2(x) \\ &= (\cos\theta)I + \frac{\sin\theta}{\theta}K(x) + \frac{1 - \cos\theta}{\theta^2}xx^{\mathsf{T}}, \end{aligned}$$

$$e^{K(x)}x = x, \quad \operatorname{spec}[e^{K(x)}] = \{1, e^{\|x\|_2 J}, e^{-\|x\|_2 J}\}, \quad \operatorname{tr} e^{K(x)} = 1 + 2\cos\theta = 1 + 2\cos\|x\|_2.$$

Source: The Cayley-Hamilton theorem and Fact 4.12.1 imply that $K^3(x) + \theta^2 K(x) = 0$. Then, every term $K^k(x)$ in the expansion of $e^{K(x)}$ can be expressed in terms of either $K(x)$ or $K^2(x)$. Finally, Fact 4.12.1 implies that $\theta^2 I + K^2(x) = xx^{\mathsf{T}}$. **Remark:** Fact 15.12.7 shows that, for all $z \in \mathbb{R}^3$, $e^{K(x)}z$ is the counterclockwise (right-hand-rule) rotation of z about the vector x through the angle θ, which is given by the Euclidean norm of x. In Fact 4.14.9, the cross product is used to construct the pivot vector x for a given pair of vectors having the same length. **Remark:** $K(x)y = x \times y$. See Fact 4.12.1. **Related:** Fact 15.12.1.

Fact 15.12.7. Let $x, y \in \mathbb{R}^3$, and assume that x and y are nonzero. Then, there exists a skew-symmetric matrix $A \in \mathbb{R}^{3\times 3}$ such that $y = e^A x$ if and only if $x^{\mathsf{T}}x = y^{\mathsf{T}}y$. If $x \neq -y$, then one such matrix is $A = \theta K(z)$, where

$$z \triangleq \frac{1}{\|x \times y\|_2}x \times y, \quad \theta \triangleq \operatorname{acos}\frac{x^{\mathsf{T}}y}{\|x\|_2\|y\|_2}.$$

If $x = -y$, then one such matrix is $A = \pi K(z)$, where $z \triangleq \|y\|_2^{-1}\nu \times y$ and $\nu \in \{y\}^{\perp}$ satisfies $\nu^{\mathsf{T}}\nu = 1$. **Source:** This result follows from Fact 4.14.9 and Fact 15.12.6, which provide equivalent expressions for an orthogonal matrix that transforms a given vector into another given vector having the same length. This result thus provides a geometric interpretation for Fact 15.12.6. **Remark:** Note that z is the unit vector perpendicular to the plane containing x and y, where the direction of z is determined by the right-hand rule. An intuitive proof is to let x be the initial condition $w(0) = x$ to the differential equation $\dot{w}(t) = K(z)w(t)$, where $t \in [0, \theta]$. Then, the derivative $\dot{w}(t)$ lies in the x, y plane and is perpendicular to $w(t)$ for all $t \in [0, \theta]$. Hence, $y = w(\theta)$. **Remark:** Since $\det e^A = e^{\operatorname{tr} A} = 1$, it follows that every pair of vectors in $\mathbb{R}^3$ having the same Euclidean length are related by a *proper rotation*. See Fact 4.11.5 and Fact 4.19.4. This is a linear interpolation problem. See Fact 4.11.5, Fact 4.14.9, and [1549]. **Remark:** Parameterizations of SO(3) are considered in [2448, 2562]. **Related:** Fact 4.14.9. **Problem:** Extend this result to $\mathbb{R}^n$. See [285, 2383].

Fact 15.12.8. Let $A \in \mathrm{SO}(3)$, let $z \in \mathbb{R}^3$ be an eigenvector of A corresponding to the eigenvalue 1, assume that $\|z\|_2 = 1$, assume that $\operatorname{tr} A > -1$, and let $\theta \in (-\pi, \pi)$ satisfy $\operatorname{tr} A = 1 + 2\cos\theta$. Then,

$$A = e^{\theta K(z)}.$$

Related: Fact 7.12.2.

Fact 15.12.9. Let $x, y \in \mathbb{R}^3$, and assume that x and y are nonzero. Then, $x^\mathsf{T} x = y^\mathsf{T} y$ if and only if

$$y = e^{\frac{\theta}{\|x\times y\|_2}(yx^\mathsf{T} - xy^\mathsf{T})}x,$$

where

$$\theta \triangleq \operatorname{acos} \frac{x^\mathsf{T} y}{\|x\|_2\|y\|_2}.$$

Source: Fact 15.12.7. **Remark:** Note that $K(x \times y) = yx^\mathsf{T} - xy^\mathsf{T}$.

Fact 15.12.10. Let $A \in \mathbb{R}^{3\times 3}$, assume that $A \in \mathrm{SO}(3)$ and $\operatorname{tr} A > -1$, and let $\theta \in (-\pi, \pi)$ satisfy $\operatorname{tr} A = 1 + 2\cos\theta$. Then,

$$\log A = \begin{cases} 0, & \theta = 0, \\ \frac{\theta}{2\sin\theta}(A - A^\mathsf{T}), & \theta \neq 0. \end{cases}$$

Source: [1506, p. 364] and [2064]. **Related:** Fact 15.16.11.

Fact 15.12.11. Let $x \in \mathbb{R}^3$, assume that x is nonzero, and define $\theta \triangleq \|x\|_2$. Then,

$$K(x) = \frac{\theta}{2\sin\theta}[e^{K(x)} - e^{-K(x)}].$$

Source: Fact 15.12.10. **Related:** Fact 4.12.1.

Fact 15.12.12. Let $A \in \mathrm{SO}(3)$, let $x, y \in \mathbb{R}^3$, and assume that $x^\mathsf{T} x = y^\mathsf{T} y$. Then, $Ax = y$ if and only if, for all $t \in \mathbb{R}$,

$$Ae^{tK(x)}A^{-1} = e^{tK(y)}.$$

Source: [1784].

Fact 15.12.13. Let $x, y, z \in \mathbb{R}^3$. Then, the following statements are equivalent:

i) For every $A \in \mathrm{SO}(3)$, there exist $\alpha, \beta, \gamma \in \mathbb{R}$ such that $A = e^{\alpha K(x)}e^{\beta K(y)}e^{\gamma K(z)}$.

ii) $y^\mathsf{T} x = 0$ and $y^\mathsf{T} z = 0$.

Source: [1784]. **Credit:** P. Davenport. **Problem:** Given $A \in \mathrm{SO}(3)$, determine α, β, γ.

Fact 15.12.14. Let $x, y, z \in \mathbb{R}^3$, assume that $x^\mathsf{T}(y \times z) \neq 0$, assume that $\|x\|_2 = \|y\|_2 = \|z\|_2 = 1$, and let $\alpha, \beta, \gamma \in [0, 2\pi)$ satisfy

$$\tan\tfrac{\alpha}{2} = \frac{z^\mathsf{T}(x\times y)}{(x\times y)^\mathsf{T}(x\times z)}, \quad \tan\tfrac{\beta}{2} = \frac{x^\mathsf{T}(y\times z)}{(y\times z)^\mathsf{T}(y\times x)}, \quad \tan\tfrac{\gamma}{2} = \frac{y^\mathsf{T}(z\times x)}{(z\times x)^\mathsf{T}(z\times y)}.$$

Then,

$$e^{\alpha K(x)}e^{\beta K(y)}e^{\gamma K(z)} = I.$$

Source: [2077, p. 5]. **Remark:** This is the *Rodrigues-Hamilton theorem*, which concerns a *closed rotation sequence*. As stated in [929], "Successive rotations about three concurrent lines fixed in space, through twice the dihedral angles of the planes formed by them, restore a body to its original position." An equivalent result is given by *Donkin's theorem*. See [446] and [2077, p. 6]. **Remark:** A converse result given in [446] shows that, if a sequence of three nonzero rotations about linearly independent rotation axes is closed, then the rotation angles are uniquely determined. **Remark:** See [2334, 2902].

Fact 15.12.15. Let $A \in \mathbb{R}^{4\times 4}$, and assume that A is skew symmetric with $\operatorname{mspec}(A) = \{\omega J, -\omega J, \mu J, -\mu J\}_{\mathrm{ms}}$. If $\omega \neq \mu$, then $e^A = a_3A^3 + a_2A^2 + a_1A + a_0I$, where

$$a_3 = (\omega^2 - \mu^2)^{-1}\left(\tfrac{1}{\mu}\sin\mu - \tfrac{1}{\omega}\sin\omega\right), \quad a_2 = (\omega^2 - \mu^2)^{-1}(\cos\mu - \cos\omega),$$

$$a_1 = (\omega^2 - \mu^2)^{-1}\left(\tfrac{\omega^2}{\mu}\sin\mu - \tfrac{\mu^2}{\omega}\sin\omega\right), \quad a_0 = (\omega^2 - \mu^2)^{-1}(\omega^2\cos\mu - \mu^2\cos\omega).$$

If $\omega = \mu$, then

$$e^A = (\cos\omega)I + \frac{\sin\omega}{\omega}A.$$

Source: [1277, p. 18] and [2246]. **Remark:** There are misprints in [1277, p. 18] and [2246]. **Related:** Fact 6.9.19 and Fact 6.10.8.

Fact 15.12.16. Let $a, b, c \in \mathbb{R}$, define the skew-symmetric matrix $A \in \mathbb{R}^{4\times 4}$ by either

$$A \triangleq \begin{bmatrix} 0 & a & b & c \\ -a & 0 & -c & b \\ -b & c & 0 & -a \\ -c & -b & a & 0 \end{bmatrix}, \quad A \triangleq \begin{bmatrix} 0 & a & b & c \\ -a & 0 & c & -b \\ -b & -c & 0 & a \\ -c & b & -a & 0 \end{bmatrix},$$

and define $\theta \triangleq \sqrt{a^2 + b^2 + c^2}$. Then,

$$\operatorname{mspec}(A) = \{\theta J, -\theta J, \theta J, -\theta J\}_{\mathrm{ms}}, \quad e^A = (\cos\theta)I + \frac{\sin\theta}{\theta}A,$$

$$A^k = \begin{cases} (-1)^{k/2}\theta^k I, & k \text{ even}, \\ (-1)^{(k-1)/2}\theta^{k-1}A, & k \text{ odd}. \end{cases}$$

Source: [2771]. **Remark:** $(\sin 0)/0 = 1$. **Remark:** The skew-symmetric matrix A arises in the kinematic relationship between the angular velocity vector and quaternion (Euler-parameter) rates. See [308, p. 385]. **Remark:** The two matrices A are similar. To show this, note that Fact 7.10.10 implies that A and $-A$ are similar. Then, apply the similarity transformation $S = \operatorname{diag}(-1, 1, 1, 1)$. **Related:** Fact 6.9.19 and Fact 6.10.8.

Fact 15.12.17. Let $x \in \mathbb{R}^3$, and define the skew-symmetric matrix $A \in \mathbb{R}^{4\times 4}$ by

$$A = \begin{bmatrix} 0 & -x^{\mathsf{T}} \\ x & -K(x) \end{bmatrix}.$$

Then, for all $t \in \mathbb{R}$,

$$e^{\frac{1}{2}tA} = (\cos\tfrac{1}{2}\|x\|t)I_4 + \frac{\sin\frac{1}{2}\|x\|t}{\|x\|}A.$$

Source: [1486, p. 34]. **Remark:** $\frac{1}{2}A$ expresses quaternion rates in terms of the angular velocity vector.

Fact 15.12.18. Let $a, b \in \mathbb{R}^3$, define the skew-symmetric matrix $A \in \mathbb{R}^{4\times 4}$ by

$$A = \begin{bmatrix} K(a) & b \\ -b^{\mathsf{T}} & 0 \end{bmatrix},$$

assume that $a^{\mathsf{T}}b = 0$, and define $\alpha \triangleq \sqrt{a^{\mathsf{T}}a + b^{\mathsf{T}}b}$. Then,

$$e^A = I_4 + \frac{\sin\alpha}{\alpha}A + \frac{1 - \cos\alpha}{\alpha^2}A^2.$$

Source: [2723]. **Related:** Fact 6.9.19 and Fact 6.10.8.

Fact 15.12.19. Let $a, b \in \mathbb{R}^{n-1}$, define $A \in \mathbb{R}^{n\times n}$ by

$$A \triangleq \begin{bmatrix} 0 & a^\mathsf{T} \\ b & 0_{(n-1)\times(n-1)} \end{bmatrix},$$

and define $\alpha \triangleq \sqrt{|a^\mathsf{T}b|}$. Then, the following statements hold:

i) If $a^\mathsf{T}b < 0$, then

$$e^{\alpha A} = I + \frac{\sin\alpha}{\alpha}A + \frac{2\sin^2\frac{\alpha}{2}}{\alpha^2}A^2.$$

ii) If $a^\mathsf{T}b = 0$, then, for all $t \in \mathbb{R}$,

$$e^{tA} = I + A + \tfrac{1}{2}A^2.$$

iii) If $a^\mathsf{T}b > 0$, then

$$e^{\alpha A} = I + \frac{\sinh\alpha}{\alpha}A + \frac{2\sinh^2\frac{\alpha}{2}}{\alpha^2}A^2.$$

Source: [2967].

15.13 Facts on the Matrix Sine and Cosine

Fact 15.13.1. Let $A \in \mathbb{C}^{n\times n}$, and define

$$\sin A \triangleq \tfrac{1}{2\jmath}(e^{\jmath A} - e^{-\jmath A}), \quad \cos A \triangleq \tfrac{1}{2}(e^{\jmath A} + e^{-\jmath A}).$$

Then, the following statements hold:

i) $\sin^2 A + \cos^2 A = I$, $\sin 2A = 2(\sin A)\cos A$, $\cos 2A = 2(\cos^2 A) - I$.

ii) If A is real, then $\sin A = \operatorname{Im} e^{\jmath A}$ and $\cos A = \operatorname{Re} e^{\jmath A}$.

iii) $\sin(A \oplus B) = (\sin A) \otimes \cos B + (\cos A) \otimes \sin B$.

iv) $\cos(A \oplus B) = (\cos A) \otimes \cos B - (\sin A) \otimes \sin B$.

v) If A is involutory and k is an integer, then $\cos k\pi A = (-1)^k I$.

Furthermore, the following statements are equivalent:

vi) For all $t \in \mathbb{R}$, $\sin t(A + B) = (\sin tA)\cos tB + (\cos tA)\sin tB$.

vii) For all $t \in \mathbb{R}$, $\cos t(A + B) = (\cos tA)\cos tB - (\sin tA)\sin tB$.

viii) $AB = BA$.

Source: [1391, pp. 287, 288, 300].

15.14 Facts on the Matrix Exponential for One Matrix

Fact 15.14.1. Let $A \in \mathbb{F}^{n\times n}$, and assume that A is (lower triangular, upper triangular). Then, so is e^A. If, in addition, A is Toeplitz, then so is e^A. **Related:** Fact 4.23.7.

Fact 15.14.2. Let $A \in \mathbb{F}^{n\times n}$. Then, $\rho_{\max}(e^A) = e^{\alpha_{\max}(A)}$.

Fact 15.14.3. Let $S\colon\ [t_0, t_1] \to \mathbb{R}^{n\times n}$ be differentiable. Then, for all $t \in [t_0, t_1]$,

$$\frac{\mathrm{d}}{\mathrm{d}t}S^2(t) = \dot{S}(t)S(t) + S(t)\dot{S}(t).$$

Let $S_1\colon\ [t_0, t_1] \to \mathbb{R}^{n\times m}$ and $S_2\colon\ [t_0, t_1] \to \mathbb{R}^{m\times l}$ be differentiable. Then, for all $t \in [t_0, t_1]$,

$$\frac{\mathrm{d}}{\mathrm{d}t}S_1(t)S_2(t) = \dot{S}_1(t)S_2(t) + S_1(t)\dot{S}_2(t).$$

Fact 15.14.4. Let $A \in \mathbb{F}^{n\times n}$, and define $A_\mathrm{H} \triangleq \frac{1}{2}(A + A^*)$ and $A_\mathrm{S} \triangleq \frac{1}{2}(A - A^*)$. Then, $[A_\mathrm{H}, A_\mathrm{S}] = \frac{1}{2}[A^*, A]$. Hence, $A_\mathrm{H}A_\mathrm{S} = A_\mathrm{S}A_\mathrm{H}$ if and only if A is normal. In this case, $e^{A_\mathrm{H}}e^{A_\mathrm{S}}$ is the polar decomposition of e^A. **Related:** Fact 7.20.2 and Fact 7.20.3. **Problem:** Find a decomposition of nonnormal

A that yields the polar decomposition of e^A.

Fact 15.14.5. Let $A \in \mathbb{F}^{n\times n}$, and assume that either $\operatorname{spec}(A) \subset \text{OLHP}$ or $\operatorname{spec}(A) \subset \text{OLHP}$. Then,

$$A^{-1} = \frac{2}{\pi}\int_0^\infty (x^2 I + A^2)^{-1}\,\mathrm{d}x.$$

Source: [953].

Fact 15.14.6. Let $A \in \mathbb{F}^{n\times m}$, and assume that $\operatorname{rank} A = m$. Then,

$$A^+ = \int_0^\infty e^{-tA^*A}A^*\,\mathrm{d}t.$$

Fact 15.14.7. Let $A \in \mathbb{F}^{n\times n}$, and assume that A is nonsingular. Then,

$$A^{-1} = \int_0^\infty e^{-tA^*A}\,\mathrm{d}t\,A^*.$$

Fact 15.14.8. Let $A \in \mathbb{F}^{n\times n}$, and define $k \triangleq \operatorname{ind} A$. Then,

$$A^{\mathsf{D}} = \int_0^\infty e^{-tA^kA^{(2k+1)*}A^{k+1}}\,\mathrm{d}t\,A^kA^{(2k+1)*}A^k.$$

Source: [1198].

Fact 15.14.9. Let $A \in \mathbb{F}^{n\times n}$, and assume that $\operatorname{ind} A = 1$. Then,

$$A^{\#} = \int_0^\infty e^{-tAA^{3*}A^2}\,\mathrm{d}t\,AA^{3*}A.$$

Source: Fact 15.14.8.

Fact 15.14.10. Let $A \in \mathbb{F}^{n\times n}$, and define $k \triangleq \operatorname{ind} A$. Then,

$$\int_0^t e^{\tau A}\,\mathrm{d}\tau = A^{\mathsf{D}}(e^{tA} - I) + (I - AA^{\mathsf{D}})\left(tI + \tfrac{1}{2!}t^2A + \cdots + \tfrac{1}{k!}t^kA^{k-1}\right).$$

If, in particular, A is group invertible, then

$$\int_0^t e^{\tau A}\,\mathrm{d}\tau = A^{\#}(e^{tA} - I) + t(I - AA^{\#}).$$

Fact 15.14.11. Let $A \in \mathbb{F}^{n\times n}$, let $\operatorname{mspec}(A) = \{\lambda_1,\ldots,\lambda_r,0,\ldots,0\}_{\mathrm{ms}}$, where $\lambda_1,\ldots,\lambda_r$ are nonzero, and let $t > 0$. Then,

$$\det\int_0^t e^{\tau A}\,\mathrm{d}\tau = t^{n-r}\prod_{i=1}^r \lambda_i^{-1}(e^{\lambda_i t} - 1).$$

Hence, $\det\int_0^t e^{\tau A}\,\mathrm{d}\tau \neq 0$ if and only if, for every nonzero integer k, $2k\pi\jmath/t \notin \operatorname{spec}(A)$. Finally, $\det(e^{tA} - I) \neq 0$ if and only if $\det A \neq 0$ and $\det\int_0^t e^{\tau A}\,\mathrm{d}\tau \neq 0$. **Remark:** $e^{tA} - I = A\int_0^t e^{\tau A}\,\mathrm{d}\tau$.

Fact 15.14.12. Let $A \in \mathbb{R}^{n\times n}$ and $a \in (0,\infty)$. Then,

$$\int_{-\infty}^\infty e^{-\tau^2/a}e^{\tau A}\,\mathrm{d}\tau = \sqrt{\sigma\pi}e^{(a/4)A^2}.$$

Credit: S. Bhat. **Related:** Fact 14.12.1.

Fact 15.14.13. Let $A \in \mathbb{F}^{n\times n}$, and assume that there exists $\alpha \in \mathbb{R}$ such that $\operatorname{spec}(A) \subset \{z \in \mathbb{C}\colon \alpha \le \operatorname{Im} z < 2\pi + \alpha\}$. Then, e^A is (diagonal, upper triangular, lower triangular) if and only if A is.
Source: [1887].

Fact 15.14.14. Let $A \in \mathbb{F}^{n\times n}$. Then, the following statements hold:

i) If A is unipotent, then (15.5.1) is a finite series, $\log A$ exists and is nilpotent, and $e^{\log A} = A$.

ii) If A is nilpotent, then e^A is unipotent and $\log e^A = A$.

Source: [1299, p. 60]. **Remark:** A is unipotent if and only if $A - I$ is nilpotent.

Fact 15.14.15. Let $A \in \mathbb{R}^{n\times n}$, assume that $\det A > 0$, and assume that A is either symmetric, skew symmetric, or orthogonal. Then, there exists $B \in \mathbb{R}^{n\times n}$ such that $e^A = e^{\frac{1}{2}(B^{\mathsf{T}}-B)}e^B$. **Source:** [576]. **Remark:** A has a *twisted logarithm.*

Fact 15.14.16. Let $A \in \mathbb{C}^{n\times n}$, let $\lambda \in \operatorname{spec}(A)$, assume that λ is simple, assume that, for all $\mu \in \operatorname{spec}(A)\backslash\{\lambda\}$, $\operatorname{Re}\mu < \operatorname{Re}\lambda$, and let $x, y \in \mathbb{C}^n$ be nonzero vectors that satisfy $Ax = \lambda x$ and $A^{\mathsf{T}}y = \lambda y$. Then, $y^{\mathsf{T}}x \neq 0$ and

$$\lim_{t\to\infty} \frac{1}{e^{\lambda t}}e^{tA} = \frac{1}{y^{\mathsf{T}}x}xy^{\mathsf{T}}.$$

Source: [1742]. **Remark:** λ is a *CT-dominant eigenvalue.* **Related:** Fact 6.11.5 and Fact 15.22.24.

15.15 Facts on the Matrix Exponential for Two or More Matrices

Fact 15.15.1. Let $A, B \in \mathbb{F}^{n\times n}$, and consider the following statements:

i) $A = B$.

ii) $e^A = e^B$.

iii) $AB = BA$.

iv) $Ae^B = e^BA$.

v) $e^Ae^B = e^Be^A$.

vi) $e^Ae^B = e^{A+B}$.

vii) $e^Ae^B = e^Be^A = e^{A+B}$.

viii) For all $i, j \in \mathbb{Z}$, $e^{iA+jB} = e^{iA}e^{jB}$.

Then, the following statements hold:

ix) *iii)* $\Longrightarrow$ *iv)* $\Longrightarrow$ *v)*.

x) *iii)* $\Longrightarrow$ *vii)* $\Longrightarrow$ *vi)*.

xi) If $\operatorname{spec}(A)$ is $2\pi\jmath$ congruence free, then *ii)* $\Longrightarrow$ *iii)* $\Longrightarrow$ *iv)* $\Longleftrightarrow$ *v)*.

xii) If $\operatorname{spec}(A)$ and $\operatorname{spec}(B)$ are $2\pi\jmath$ congruence free, then *ii)* $\Longrightarrow$ *iii)* $\Longleftrightarrow$ *iv)* $\Longleftrightarrow$ *v)*.

xiii) If $\operatorname{spec}(A + B)$ is $2\pi\jmath$ congruence free, then *iii)* $\Longleftrightarrow$ *vii)*.

xiv) If, for all $\lambda \in \operatorname{spec}(A)$ and $\mu \in \operatorname{spec}(B)$, it follows that $(\lambda - \mu)/(2\pi\jmath)$ is not a nonzero integer, then *ii)* $\Longrightarrow$ *i)*.

xv) *iii)* $\Longleftrightarrow$ *viii)*.

xvi) If A and B are Hermitian, then *i)* $\Longleftrightarrow$ *ii)* $\Longrightarrow$ *iii)* $\Longleftrightarrow$ *iv)* $\Longleftrightarrow$ *v)* $\Longleftrightarrow$ *vi)*.

Remark: The set $\mathcal{S} \subset \mathbb{C}$ is $2\pi\jmath$ *congruence free* if no two elements of $\mathcal{S}$ differ by a nonzero integer multiple of $2\pi\jmath$. **Source:** [856, 1073], [1304, pp. 88, 89, 270–272], and [2190, 2392, 2393, 2394, 2476, 2866, 2867]. The assumption of normality in operator versions of several of these statements in [2190, 2394] is not needed in the matrix case. *xiii)* and the first implication in *x)* are given in [1391, p. 32]. **Remark:** The matrices $A = \begin{bmatrix} 0 & 1 \\ 0 & 2\pi\jmath \end{bmatrix}$ and $B = \begin{bmatrix} 2\pi\jmath & 0 \\ 0 & -2\pi\jmath \end{bmatrix}$ satisfy $e^A = e^B = e^{A+B} = I$, but $AB \neq BA$. Therefore, *vii)* $\Longrightarrow$ *iii)* does not hold. Furthermore, since *vii)* holds and *iii)* does not hold, it follows from *xii)* that $\operatorname{spec}(A + B)$ is not $2\pi\jmath$ congruence free. In fact, $\operatorname{spec}(A + B) = \{0, 2\pi\jmath\}$. The same observation holds for the real matrices

$$A = 2\pi\begin{bmatrix} 0 & 0 & \sqrt{3}/2 \\ 0 & 0 & -1/2 \\ -\sqrt{3}/2 & 1/2 & 0 \end{bmatrix}, \quad B = 2\pi\begin{bmatrix} 0 & 0 & -\sqrt{3}/2 \\ 0 & 0 & -1/2 \\ \sqrt{3}/2 & 1/2 & 0 \end{bmatrix}.$$

Remark: The matrices $A = \begin{bmatrix} \log 2+\pi\jmath & 0 \\ 0 & 0 \end{bmatrix}$ and $B = \begin{bmatrix} 2\log 2+2\pi\jmath & 1 \\ 0 & 0 \end{bmatrix}$ satisfy $e^Ae^B = e^{A+B} \neq e^Be^A$. Therefore, *vi*) $\Longrightarrow$ *vii*) does not hold. Furthermore, $\operatorname{spec}(A+B) = \{3\log 2 + 3\pi\jmath, 0\}$, which is $2\pi\jmath$ congruence free. Therefore, since *vii*) does not hold, it follows from *xii*) that *iii*) does not hold; that is, $AB \neq BA$. In fact, $AB - BA = \left[\begin{smallmatrix} 0 & \log 2+\pi\jmath \\ 0 & 0 \end{smallmatrix}\right]$. Consequently, under the additional statement that $\operatorname{spec}(A+B)$ is $2\pi\jmath$ congruence free, *vi*) $\Longrightarrow$ *iii*) does not hold. However, *xiv*) implies that *vi*) $\Longrightarrow$ *iii*) holds under the additional assumption that A and B are Hermitian, in which case $\operatorname{spec}(A+B)$ is $2\pi\jmath$ congruence free. This example is due to G. Bourgeois.

Fact 15.15.2. Let $A \in \mathbb{F}^{n\times n}$, $B \in \mathbb{F}^{n\times m}$, and $C \in \mathbb{F}^{m\times m}$. Then,

$$e^{t\left[\begin{smallmatrix} A & B \\ 0 & C \end{smallmatrix}\right]} = \begin{bmatrix} e^{tA} & \int_0^t e^{(t-\tau)A}Be^{\tau C}\,\mathrm{d}\tau \\ 0 & e^{tC} \end{bmatrix}, \quad \int_0^t e^{\tau A}\,\mathrm{d}\tau = [I \ \ 0]e^{t\left[\begin{smallmatrix} A & I \\ 0 & 0 \end{smallmatrix}\right]}\begin{bmatrix} 0 \\ I \end{bmatrix}.$$

Remark: This result can be extended to arbitrary upper block-triangular matrices. See [2774]. For an application to sampled-data control, see [2171].

Fact 15.15.3. Let $\mathcal{I}$ be an interval, let $A\colon \mathcal{I} \mapsto \mathbb{F}^{n\times n}$, and assume that A is differentiable. Then, for all $t \in \mathcal{I}$,

$$\frac{\mathrm{d}}{\mathrm{d}t}e^{A(t)} = \int_0^1 e^{\tau A(t)}\left(\frac{\mathrm{d}}{\mathrm{d}t}A(t)\right)e^{(1-\tau)A(t)}\,\mathrm{d}\tau.$$

Source: [2586, pp. 256, 257]. **Remark:** This is the *Duhamel formula*. **Related:** Fact 15.15.4.

Fact 15.15.4. Let $A, B \in \mathbb{F}^{n\times n}$. Then,

$$\frac{\mathrm{d}}{\mathrm{d}t}e^{A+tB} = \int_0^1 e^{\tau(A+tB)}Be^{(1-\tau)(A+tB)}\,\mathrm{d}\tau.$$

Hence,

$$\operatorname{Dexp}(A;B) = \left.\frac{\mathrm{d}}{\mathrm{d}t}e^{A+tB}\right|_{t=0} = \int_0^1 e^{\tau A}Be^{(1-\tau)A}\,\mathrm{d}\tau.$$

Furthermore,

$$\frac{\mathrm{d}}{\mathrm{d}t}\operatorname{tr} e^{A+tB} = \operatorname{tr} e^{A+tB}B.$$

Hence,

$$\left.\frac{\mathrm{d}}{\mathrm{d}t}\operatorname{tr} e^{A+tB}\right|_{t=0} = \operatorname{tr} e^{A}B.$$

Source: [353, p. 175], [962, p. 371], and [1774, 1986, 2108].

Fact 15.15.5. Let $A, B \in \mathbb{F}^{n\times n}$. Then,

$$\left.\frac{\mathrm{d}}{\mathrm{d}t}e^{A+tB}\right|_{t=0} = \left(\frac{e^{\mathrm{ad}_A} - I}{\mathrm{ad}_A}\right)(B)e^A = e^A\left(\frac{I - e^{-\mathrm{ad}_A}}{\mathrm{ad}_A}\right)(B) = \sum_{k=0}^\infty \tfrac{1}{(k+1)!}\mathrm{ad}_A^k(B)e^A.$$

Source: The second and fourth expressions are given in [218, p. 49] and [1506, p. 248], while the third expression appears in [2754]. See also [2781, pp. 107–110]. **Related:** Fact 3.23.5.

Fact 15.15.6. Let $A, B \in \mathbb{F}^{n\times n}$, and assume that $e^A = e^B$. Then, the following statements hold:

i) If $|\lambda| < \pi$ for all $\lambda \in \operatorname{spec}(A) \cup \operatorname{spec}(B)$, then $A = B$.

ii) If $\lambda - \mu \neq 2k\pi\jmath$ for all $\lambda \in \operatorname{spec}(A)$, $\mu \in \operatorname{spec}(B)$, and $k \in \mathbb{Z}$, then $[A, B] = 0$.

iii) If A is normal and $\sigma_{\max}(A) < \pi$, then $[A, B] = 0$.

iv) If A is normal and $\sigma_{\max}(A) = \pi$, then $[A^2, B] = 0$.

Source: [2396, 2476] and [2781, p. 111]. **Remark:** If $[A, B] = 0$, then $[A^2, B] = 0$.

Fact 15.15.7. Let $A, B \in \mathbb{F}^{n\times n}$, and assume that A and B are skew Hermitian. Then, $e^{tA}e^{tB}$ is unitary, and there exists a skew-Hermitian matrix $C(t)$ such that $e^{tA}e^{tB} = e^{C(t)}$. **Remark:** Convergence of (15.4.1) is discussed in [490, 508, 842, 994, 2126, 2304], and closed-form expressions are given in [2770, 1068].

Fact 15.15.8. Let $A, B \in \mathbb{F}^{n\times n}$, and assume that A and B are Hermitian. Then,

$$\lim_{p\to 0}\left(e^{\frac{p}{2}A}e^{pB}e^{\frac{p}{2}A}\right)^{1/p} = e^{A+B}.$$

Source: [101]. **Remark:** This result is related to the Lie-Trotter formula given by Corollary 15.4.13. For extensions, see [19, 1127].

Fact 15.15.9. Let $A, B \in \mathbb{F}^{n\times n}$, and assume that A and B are Hermitian. Then,

$$\lim_{p\to\infty}\left[\tfrac{1}{2}\left(e^{pA}+e^{pB}\right)\right]^{1/p} = e^{\frac{1}{2}(A+B)}.$$

Source: [443].

Fact 15.15.10. Let $A, B \in \mathbb{F}^{n\times n}$. Then,

$$\lim_{k\to\infty}\left[e^{\frac{1}{k}A}e^{\frac{1}{k}B}e^{-\frac{1}{k}A}e^{-\frac{1}{k}B}\right]^{k^2} = e^{[A,B]}.$$

Fact 15.15.11. Let $A \in \mathbb{F}^{n\times m}$, $X \in \mathbb{F}^{m\times l}$, and $B \in \mathbb{F}^{l\times n}$. Then,

$$\frac{\mathrm{d}}{\mathrm{d}X}\operatorname{tr} e^{AXB} = Be^{AXB}A.$$

Fact 15.15.12. Let $A, B \in \mathbb{F}^{n\times n}$. Then,

$$\frac{\mathrm{d}}{\mathrm{d}t}e^{tA}e^{tB}e^{-tA}e^{-tB}\bigg|_{t=0} = 0, \quad \frac{\mathrm{d}}{\mathrm{d}t}e^{\sqrt{t}A}e^{\sqrt{t}B}e^{-\sqrt{t}A}e^{-\sqrt{t}B}\bigg|_{t=0} = AB - BA.$$

Fact 15.15.13. Let $A, B, C \in \mathbb{F}^{n\times n}$, assume there exists $\beta \in \mathbb{F}$ such that $[A, B] = \beta B + C$, and assume that $[A, C] = [B, C] = 0$. Then,

$$e^{A+B} = e^{A}e^{\phi(\beta)B}e^{\psi(\beta)C},$$

where

$$\phi(\beta) \triangleq \begin{cases}\frac{1}{\beta}(1-e^{-\beta}), & \beta \neq 0,\\ 1, & \beta = 0,\end{cases} \qquad \psi(\beta) \triangleq \begin{cases}\frac{1}{\beta^2}(1-\beta-e^{-\beta}), & \beta \neq 0,\\ -\frac{1}{2}, & \beta = 0.\end{cases}$$

Source: [1181, 2599].

Fact 15.15.14. Let $A, B \in \mathbb{F}^{n\times n}$, and assume that there exist $\alpha, \beta \in \mathbb{F}$ such that $[A, B] = \alpha A + \beta B$. Then,

$$e^{t(A+B)} = e^{\phi(t)A}e^{\psi(t)B},$$

where

$$\phi(t) \triangleq \begin{cases} t, & \alpha = \beta = 0,\\ \alpha^{-1}\log(1+\alpha t), & \alpha = \beta \neq 0,\ 1+\alpha t > 0,\\ \int_0^t \frac{\alpha-\beta}{\alpha e^{(\alpha-\beta)\tau}-\beta}\,\mathrm{d}\tau, & \alpha \neq \beta,\end{cases} \qquad \psi(t) \triangleq \int_0^t e^{-\beta\phi(\tau)}\,\mathrm{d}\tau.$$

Source: [2600].

Fact 15.15.15. Let $A, B \in \mathbb{F}^{n\times n}$, and assume that there exists nonzero $\beta \in \mathbb{F}$ such that $[A, B] = \alpha B$. Then, for all $t > 0$,

$$e^{t(A+B)} = e^{tA}e^{[(1-e^{-\alpha t})/\alpha]B}.$$

Source: Apply Fact 15.15.13 with $[tA, tB] = \alpha t(tB)$ and $\beta = \alpha t$.

Fact 15.15.16. Let $A, B \in \mathbb{F}^{n\times n}$, and assume that $[[A,B],A] = 0$ and $[[A,B],B] = 0$. Then, for all $t \in \mathbb{R}$,

$$e^{tA}e^{tB} = e^{tA+tB+(t^2/2)[A,B]}.$$

In particular,

$$e^Ae^B = e^{A+B+\frac{1}{2}[A,B]} = e^{A+B}e^{\frac{1}{2}[A,B]} = e^{\frac{1}{2}[A,B]}e^{A+B}, \quad e^Be^{2A}e^B = e^{2A+2B}.$$

Source: [1299, pp. 64–66] and [2878].

Fact 15.15.17. Let $A, B \in \mathbb{F}^{n\times n}$, and assume that $[A, B] = B^2$. Then,

$$e^{A+B} = e^A(I + B).$$

Fact 15.15.18. Let $A, B \in \mathbb{F}^{n\times n}$. Then, for all $t \in [0, \infty)$,

$$e^{t(A+B)} = e^{tA}e^{tB} + \sum_{k=2}^{\infty} C_k t^k,$$

where, for all $k \in \mathbb{N}$,

$$C_{k+1} \triangleq \tfrac{1}{k+1}([A+B]C_k + [B, D_k]), \qquad C_0 \triangleq 0,$$
$$D_{k+1} \triangleq \tfrac{1}{k+1}(AD_k + D_kB), \qquad D_0 \triangleq I.$$

Source: [2308].

Fact 15.15.19. Let $A, B \in \mathbb{F}^{n\times n}$. Then, for all $t \in [0, \infty)$,

$$e^{t(A+B)} = e^{tA}e^{tB}e^{tC_2}e^{tC_3}\cdots,$$

where

$$C_2 \triangleq -\tfrac{1}{2}[A,B], \quad C_3 \triangleq \tfrac{1}{3}[B,[A,B]] + \tfrac{1}{6}[A,[A,B]].$$

Remark: This is the *Zassenhaus product formula*. See [1391, p. 236] and [2401]. **Remark:** Higher order terms are given in [2401]. **Remark:** Conditions for convergence appear to be unknown.

Fact 15.15.20. Let $A \in \mathbb{R}^{2n\times 2n}$, and assume that A is symplectic and discrete-time Lyapunov stable. Then, $\operatorname{spec}(A) \subset \mathrm{UC}$, $\operatorname{amult}_A(1)$ and $\operatorname{amult}_A(-1)$ are even, A is semisimple, and there exists a Hamiltonian matrix $B \in \mathbb{R}^{2n\times 2n}$ such that $A = e^B$. **Source:** Since A is symplectic and discrete-time Lyapunov stable, it follows that the spectrum of A is a subset of the unit circle and A is semisimple. Therefore, the only negative eigenvalue that A can have is -1. Since all nonreal eigenvalues appear in complex conjugate pairs and A has even size, and since, by Fact 4.28.10, $\det A = 1$, it follows that the eigenvalues -1 and 1 (if present) have even algebraic multiplicity. Theorem 2.6 of [896] implies that A has a Hamiltonian logarithm. **Related:** *xiii*) of Proposition 15.6.5.

Fact 15.15.21. Let $A, B \in \mathbb{F}^{n\times n}$, assume that A is positive definite, and assume that B is positive semidefinite. Then,

$$A + B \le A^{1/2}e^{A^{-1/2}BA^{-1/2}}A^{1/2}.$$

Hence,

$$\frac{\det(A+B)}{\det A} \le e^{\operatorname{tr} A^{-1}B}.$$

Furthermore, for each inequality, equality holds if and only if $B = 0$. **Source:** If C is positive semidefinite, then $I + C \le e^C$.

Fact 15.15.22. Let $A, B \in \mathbb{F}^{n\times n}$, and assume that A and B are Hermitian. Then,

$$I \odot (A + B) \le \log(e^A \odot e^B).$$

Source: [88, 2977]. **Related:** Fact 10.25.58.

Fact 15.15.23. Let $A, B \in \mathbb{F}^{n\times n}$, assume that A and B are Hermitian, assume that $A \le B$, let $\alpha, \beta \in \mathbb{R}$, assume that either $\alpha I \le A \le \beta I$ or $\alpha I \le B \le \beta I$, and let $t > 0$. Then,

$$e^{tA} \le S(t, e^{\beta-\alpha})e^{tB},$$

where, for all $t > 0$ and $h > 0$,

$$S(t,h) \triangleq \begin{cases} \dfrac{(h^t-1)h^{t/(h^t-1)}}{et\log h}, & h \ne 1, \\ 1, & h = 1. \end{cases}$$

Source: [1094]. **Related:** $S(t,h)$ is Specht's ratio. See Fact 2.2.54 and Fact 2.11.95.

Fact 15.15.24. Let $A, B \in \mathbb{F}^{n\times n}$, assume that A and B are Hermitian, let $\alpha, \beta \in \mathbb{R}$, assume that $\alpha I \le A \le \beta I$ and $\alpha I \le B \le \beta I$, let $t > 0$, and define $S(t,h)$ as in Fact 15.15.23. Then,

$$\frac{1}{S(1,e^{\beta-\alpha})S^{1/t}(t,e^{\beta-\alpha})}[\alpha e^{tA} + (1-\alpha)e^{tB}]^{1/t} \le e^{\alpha A+(1-\alpha)B} \le S(1,e^{\beta-\alpha})[\alpha e^{tA} + (1-\alpha)e^{tB}]^{1/t}.$$

Source: [1094].

Fact 15.15.25. Let $A, B \in \mathbb{F}^{n\times n}$, and assume that A and B are positive definite. Then,

$$\log\det A = \operatorname{tr}\log A, \quad \log\det AB = \operatorname{tr}(\log A + \log B).$$

Fact 15.15.26. Let $\mathcal{S} \subset \mathbb{R}$, assume that $\mathcal{S}$ is an open set, let $f\colon \mathcal{S} \mapsto \mathbb{R}$, assume that f is convex and differentiable, let $A, B \in \mathbf{H}^n$, and assume that all of the eigenvalues of A and B are elements of $\mathcal{S}$. Then,

$$\operatorname{tr}[(A-B)f'(B)] \le \operatorname{tr}[f(A) - f(B)].$$

Source: [454, pp. 115, 116].

Fact 15.15.27. Let $A, B \in \mathbb{F}^{n\times n}$, and assume that A and B are positive definite. Then,

$$\operatorname{tr}(A-B) \le \operatorname{tr} A(\log A - \log B),$$
$$(\log\operatorname{tr} A - \log\operatorname{tr} B)\operatorname{tr} A \le \operatorname{tr} A(\log A - \log B).$$

Source: The first inequality follows from Fact 15.15.26 with $\mathcal{S} = (0,\infty)$ and $f(x) = x\log x$. See [337] and [449, p. 281]. **Remark:** The first inequality is *Klein's inequality*. See [454, p. 118]. **Remark:** $\operatorname{tr} A(\log A - \log B)$ is the *relative entropy of Umegaki*. **Related:** The second inequality is equivalent to the thermodynamic inequality. See Fact 15.15.35.

Fact 15.15.28. Let $A, B \in \mathbb{F}^{n\times n}$, assume that A and B are positive definite, and define

$$\mu(A,B) \triangleq e^{\log A + \log B}.$$

Then, the following statements hold:

i) $\mu(A, A^{-1}) = I$.

ii) $\mu(A,B) = \mu(B,A)$.

iii) If $AB = BA$, then $\mu(A,B) = AB$.

Source: [169]. **Remark:** With multiplication defined by μ, $\mathbf{P}^n$ is a commutative Lie group.

Fact 15.15.29. Let $A, B \in \mathbb{F}^{n\times n}$, assume that A and B are positive definite, and let $p > 0$. Then,

$$\tfrac{1}{p}\operatorname{tr} A\log(B^{p/2}A^pB^{p/2}) \le \operatorname{tr} A(\log A + \log B) \le \tfrac{1}{p}\operatorname{tr} A\log(A^{p/2}B^pA^{p/2}).$$

Furthermore,

$$\lim_{p\downarrow 0}\tfrac{1}{p}\operatorname{tr} A\log(B^{p/2}A^pB^{p/2}) = \operatorname{tr} A(\log A + \log B) = \lim_{p\downarrow 0}\tfrac{1}{p}\operatorname{tr} A\log(A^{p/2}B^pA^{p/2}).$$

Source: [101, 338, 1127, 1380]. **Remark:** This result arises in quantum information theory.

Fact 15.15.30. Let $A, B \in \mathbb{F}^{n\times n}$, assume that A and B are Hermitian, let $q \geq p > 0$, and define $h \triangleq \lambda_{\max}(e^A)/\lambda_{\min}(e^B)$ and

$$S(1,h) \triangleq \frac{(h-1)h^{1/(h-1)}}{e\log h}.$$

Then, there exist unitary matrices $U, V \in \mathbb{F}^{n\times n}$ such that

$$\tfrac{1}{S(1,h)}Ue^{A+B}U^* \leq e^{\frac{1}{2}A}e^Be^{\frac{1}{2}A} \leq S(1,h)Ve^{A+B}V^*.$$

Furthermore,

$$\operatorname{tr} e^{A+B} \leq \operatorname{tr} e^Ae^B \leq S(1,h)\operatorname{tr} e^{A+B},$$

$$\operatorname{tr}(e^{pA}\#e^{pB})^{2/p} \leq \operatorname{tr} e^{A+B} \leq \operatorname{tr}(e^{\frac{p}{2}B}e^{pA}e^{\frac{p}{2}B})^{1/p} \leq \operatorname{tr}(e^{\frac{q}{2}B}e^{qA}e^{\frac{q}{2}B})^{1/q},$$

$$\operatorname{tr} e^{A+B} = \lim_{p\downarrow 0}\operatorname{tr}(e^{\frac{p}{2}B}e^{pA}e^{\frac{p}{2}B})^{1/p}, \quad e^{A+B} = \lim_{p\downarrow 0}(e^{pA}\#e^{pB})^{2/p}.$$

Moreover, $\operatorname{tr} e^{A+B} = \operatorname{tr} e^Ae^B$ if and only if $AB = BA$. Furthermore, for all $i \in \{1,\ldots,n\}$,

$$\tfrac{1}{S(1,h)}\lambda_i(e^{A+B}) \leq \lambda_i(e^Ae^B) \leq S(1,h)\lambda_i(e^{A+B}).$$

Finally, let $\alpha \in [0,1]$. Then,

$$\lim_{p\downarrow 0}(e^{pA}\#_\alpha e^{pB})^{1/p} = e^{(1-\alpha)A+\alpha B}, \quad \operatorname{tr}(e^{pA}\#_\alpha e^{pB})^{1/p} \leq \operatorname{tr} e^{(1-\alpha)A+\alpha B}.$$

Source: [550] and [2586, pp. 254–256]. **Remark:** The left-hand inequality in the second string of inequalities is the *Golden-Thompson inequality*. See Fact 15.17.4. **Remark:** Since $S(1,h) > 1$ for all $h > 1$, the left-hand inequality in the first string of inequalities does not imply the Golden-Thompson inequality. **Remark:** For $i=1$, the stronger eigenvalue inequality $\lambda_{\max}(e^{A+B}) \leq \lambda_{\max}(e^Ae^B)$ holds. See Fact 15.17.4. **Remark:** $S(1,h)$ is Specht's ratio given by Fact 15.15.23. **Remark:** The generalized geometric mean is defined in Fact 10.11.72. **Related:** Fact 15.15.31.

Fact 15.15.31. Let $A, B \in \mathbb{F}^{n\times n}$, and assume that A and B are normal. Then,

$$\sigma(e^{A+B}) \overset{\text{slog}}{\prec} \lambda(\langle e^A\rangle\langle e^B\rangle), \quad \lambda(e^{A+B}) \overset{\text{slog}}{\prec} \lambda(\langle e^A\rangle\langle e^B\rangle), \quad |\operatorname{tr} e^{A+B}| \leq \operatorname{tr}\langle e^A\rangle\langle e^B\rangle.$$

Furthermore, if $\|\cdot\|$ is a unitarily invariant norm on $\mathbb{F}^{n\times n}$, then

$$\|\langle e^{A+B}\rangle\| \leq \|\langle\langle e^A\rangle^{1/2}\langle e^B\rangle\langle e^A\rangle^{1/2}\rangle\|.$$

Source: [1810]. **Related:** Fact 15.15.30.

Fact 15.15.32. Let $A, B \in \mathbb{F}^{n\times n}$, and assume that A and B are Hermitian. Then,

$$(\operatorname{tr} e^A)e^{(\operatorname{tr} e^AB)/\operatorname{tr} e^A} \leq \operatorname{tr} e^{A+B}.$$

Source: [337]. **Remark:** This is the *Peierls-Bogoliubov inequality*. **Related:** Fact 15.15.35.

Fact 15.15.33. Let $A, B, C \in \mathbb{F}^{n\times n}$, and assume that A, B, and C are positive definite. Then,

$$\operatorname{tr} e^{\log A - \log B + \log C} \leq \operatorname{tr}\int_0^\infty A(B+xI)^{-1}C(B+xI)^{-1}\,\mathrm{d}x.$$

Source: [1817, 1891]. **Remark:** $-\log B$ is correct. **Remark:** $\operatorname{tr} e^{A+B+C} \leq |\operatorname{tr} e^Ae^Be^C|$ is false.

Fact 15.15.34. Let $A, B, C \in \mathbb{F}^{n\times n}$, and assume that A, B, and C are Hermitian. Then,

$$\operatorname{tr} e^{A+B+C} \leq \int_0^\infty \operatorname{tr} e^A(xI+e^{-C})^{-1}e^B(xI+e^{-C})^{-1}\,\mathrm{d}x.$$

Source: [1381, pp. 180, 181].

Fact 15.15.35. Let $A, B \in \mathbb{F}^{n\times n}$, and assume that A is positive definite, $\operatorname{tr} A = 1$, and B is Hermitian. Then,

$$\operatorname{tr} AB \le \operatorname{tr} A \log A + \log \operatorname{tr} e^B.$$

Furthermore, equality holds if and only if $(\operatorname{tr} e^B)A = e^B$. **Source:** [337]. **Remark:** This is the *thermodynamic inequality*. Equivalent forms are given by Fact 15.15.27 and Fact 15.15.32.

Fact 15.15.36. Let $A, B \in \mathbb{F}^{n\times n}$, and assume that A and B are Hermitian. Then,

$$\|A - B\|_{\rm F} \le \|\log(e^{-\frac{1}{2}A} e^B e^{\frac{1}{2}A})\|_{\rm F}.$$

Source: [454, p. 203]. **Remark:** This result has a distance interpretation in terms of geodesics. See [454, p. 203] and [464, 2064, 2065].

Fact 15.15.37. Let $A, B \in \mathbb{F}^{n\times n}$, and assume that A and B are skew Hermitian. Then, there exist unitary matrices $S_1, S_2 \in \mathbb{F}^{n\times n}$ such that

$$e^A e^B = e^{S_1 A S_1^{-1} + S_2 B S_2^{-1}}.$$

Source: [2478, 2614, 2615]. **Related:** Fact 10.11.70.

Fact 15.15.38. Let $A, B \in \mathbb{F}^{n\times n}$, and assume that A and B are Hermitian. Then, there exist unitary matrices $S_1, S_2 \in \mathbb{F}^{n\times n}$ such that

$$e^{\frac{1}{2}A} e^B e^{\frac{1}{2}A} = e^{S_1 A S_1^{-1} + S_2 B S_2^{-1}}.$$

Source: [2477, 2478, 2614, 2615]. **Problem:** Relate this result to Fact 15.15.37.

Fact 15.15.39. Let $A, B \in \mathbb{F}^{n\times n}$, assume that A and B are positive semidefinite, and assume that $B \le A$. Furthermore, let $p, q, r, t \in \mathbb{R}$, and assume that $r \ge t \ge 0$, $p \ge 0$, $p+q \ge 0$, and $p+q+r > 0$. Then,

$$\left[e^{\frac{r}{2}A} e^{qA+pB} e^{\frac{r}{2}A}\right]^{t/(p+q+r)} \le e^{tA}.$$

Source: [2757].

Fact 15.15.40. Let $A \in \mathbb{F}^{n\times n}$ and $B \in \mathbb{F}^{m\times m}$. Then,

$$\operatorname{tr} e^{A\oplus B} = (\operatorname{tr} e^A) \operatorname{tr} e^B.$$

Fact 15.15.41. Let $A \in \mathbb{F}^{n\times n}$, $B \in \mathbb{F}^{m\times m}$, and $C \in \mathbb{F}^{l\times l}$. Then,

$$e^{A\oplus B\oplus C} = e^A \otimes e^B \otimes e^C.$$

Fact 15.15.42. Let $A \in \mathbb{F}^{n\times n}$, $B \in \mathbb{F}^{m\times m}$, $C \in \mathbb{F}^{k\times k}$, and $D \in \mathbb{F}^{l\times l}$. Then,

$$\operatorname{tr} e^{A\otimes I\otimes B\otimes I + I\otimes C\otimes I\otimes D} = (\operatorname{tr} e^{A\otimes B}) \operatorname{tr} e^{C\otimes D}.$$

Source: By Fact 9.4.38, a similarity transformation involving the Kronecker permutation matrix can be used to reorder the inner two terms. See [2525].

Fact 15.15.43. Let $A, B \in \mathbb{R}^{n\times n}$, and assume that A and B are positive definite. Then, $A\#B$ is the unique positive-definite solution X of the matrix equation

$$\log A^{-1}X + \log B^{-1}X = 0.$$

Source: [2065].

15.16 Facts on the Matrix Exponential and Eigenvalues, Singular Values, and Norms for One Matrix

Fact 15.16.1. Let $A \in \mathbb{F}^{n\times n}$, assume that e^A is positive definite, and assume that $\sigma_{\max}(A) < 2\pi$. Then, A is Hermitian. **Source:** [1712, 2395].

Fact 15.16.2. Let $A \in \mathbb{F}^{n\times n}$, and define $f\colon [0,\infty) \mapsto (0,\infty)$ by $f(t) \triangleq \sigma_{\max}(e^{At})$. Then,

$$f'(0) = \tfrac{1}{2}\lambda_{\max}(A + A^*).$$

Hence, there exists $\varepsilon > 0$ such that $f(t) \triangleq \sigma_{\max}(e^{tA})$ is decreasing on $[0,\varepsilon)$ if and only if A is dissipative. **Source:** *iii*) of Fact 15.16.7 and [2834]. **Remark:** The derivative is one-sided.

Fact 15.16.3. Let $A \in \mathbb{F}^{n\times n}$. Then, for all $t \ge 0$,

$$\frac{\mathrm{d}}{\mathrm{d}t}\|e^{tA}\|_{\mathrm{F}}^2 = \operatorname{tr} e^{tA}(A + A^*)e^{tA^*}.$$

Hence, if A is dissipative, then $f(t) \triangleq \|e^{tA}\|_{\mathrm{F}}$ is decreasing on $[0,\infty)$. **Source:** [2834].

Fact 15.16.4. Let $A \in \mathbb{F}^{n\times n}$. Then,

$$|\operatorname{tr} e^{2A}| \le \operatorname{tr} e^{A}e^{A^*} \le \operatorname{tr} e^{A+A^*} \le [n \operatorname{tr} e^{2(A+A^*)}]^{1/2} \le \tfrac{n}{2} + \tfrac{1}{2}\operatorname{tr} e^{2(A+A^*)}.$$

In addition, $\operatorname{tr} e^{A}e^{A^*} = \operatorname{tr} e^{A+A^*}$ if and only if A is normal. **Source:** [426], [1450, p. 515], and [2476]. **Remark:** $\operatorname{tr} e^{A}e^{A^*} \le \operatorname{tr} e^{A+A^*}$ is *Bernstein's inequality*. See [93]. **Related:** Fact 4.10.12.

Fact 15.16.5. Let $A \in \mathbb{F}^{n\times n}$. Then,

$$\lambda^{\odot 1/2}(e^{A^*}e^{A}) = \lambda(\langle e^{A}\rangle) = \sigma(e^{A}) \overset{\mathrm{slog}}{\prec} \sigma(e^{\frac{1}{2}(A+A^*)}) = \lambda(e^{\frac{1}{2}(A+A^*)}) = \lambda^{\odot 1/2}(e^{A+A^*}).$$

That is, for all $k \in \{1,\ldots,n\}$,

$$\prod_{i=1}^{k}\sigma_i(e^{A}) \le \prod_{i=1}^{k}\lambda_i\Big(e^{\frac{1}{2}(A+A^*)}\Big) = \prod_{i=1}^{k}e^{\lambda_i[\frac{1}{2}(A+A^*)]} \le \prod_{i=1}^{k}e^{\sigma_i(A)}$$

with equality for $k = n$. Furthermore, for all $k \in \{1,\ldots,n\}$,

$$\sum_{i=1}^{k}\sigma_i(e^{A}) \le \sum_{i=1}^{k}\lambda_i\Big(e^{\frac{1}{2}(A+A^*)}\Big) = \sum_{i=1}^{k}e^{\lambda_i[\frac{1}{2}(A+A^*)]} \le \sum_{i=1}^{k}e^{\sigma_i(A)}.$$

In particular,

$$\sigma_{\max}(e^{A}) \le \lambda_{\max}\Big(e^{\frac{1}{2}(A+A^*)}\Big) = e^{\frac{1}{2}\lambda_{\max}(A+A^*)} \le e^{\sigma_{\max}(A)}.$$

Equivalently,

$$\lambda_{\max}(e^{A}e^{A^*}) \le \lambda_{\max}(e^{A+A^*}) = e^{\lambda_{\max}(A+A^*)} \le e^{2\sigma_{\max}(A)}.$$

Furthermore,

$$|\det e^{A}| = |e^{\operatorname{tr} A}| \le e^{|\operatorname{tr} A|} \le e^{\operatorname{tr}\langle A\rangle}, \quad \operatorname{tr}\langle e^{A}\rangle \le \sum_{i=1}^{n}e^{\sigma_i(A)}.$$

Source: [767, 1878, 2479], and use Fact 3.25.15, Fact 10.21.9, and Fact 10.21.10.

Fact 15.16.6. Let $A \in \mathbb{F}^{n\times n}$, and let $\|\cdot\|$ be a unitarily invariant norm on $\mathbb{F}^{n\times n}$. Then,

$$\|e^{A}e^{A^*}\| \le \|e^{A+A^*}\|.$$

In particular,

$$\lambda_{\max}(e^{A}e^{A^*}) \le \lambda_{\max}(e^{A+A^*}), \quad \operatorname{tr} e^{A}e^{A^*} \le \operatorname{tr} e^{A+A^*}.$$

Source: [767].

Fact 15.16.7. Let $A, B \in \mathbb{F}^{n\times n}$, let $\|\cdot\|$ be the norm on $\mathbb{F}^{n\times n}$ induced by the norm $\|\cdot\|'$ on $\mathbb{F}^n$, let $\operatorname{mspec}(A) = \{\lambda_1,\ldots,\lambda_n\}_{\mathrm{ms}}$, and define

$$\mu(A) \triangleq \lim_{\varepsilon\downarrow 0}\frac{\|I + \varepsilon A\| - 1}{\varepsilon}.$$

Then, the following statements hold:

i) $\mu(A) = \mathrm{D}_+ f(A; I)$, where $f\colon \mathbb{F}^{n\times n} \mapsto \mathbb{R}$ is defined by $f(A) \triangleq \|A\|$.

ii) $\mu(A) = \lim_{t\downarrow 0} t^{-1}\log\|e^{tA}\| = \sup_{t>0} t^{-1}\log\|e^{tA}\|$.

iii) $\mu(A) = \frac{\mathrm{d}^+}{\mathrm{d}t}\|e^{tA}\|\big|_{t=0} = \frac{\mathrm{d}^+}{\mathrm{d}t}\log\|e^{tA}\|\big|_{t=0}$.

iv) $\mu(I) = 1$, $\mu(-I) = -1$, and $\mu(0) = 0$.

v) $\alpha_{\max}(A) = \lim_{t\to\infty} t^{-1}\log\|e^{tA}\| = \inf_{t>0} t^{-1}\log\|e^{tA}\|$.

vi) For all $i \in \{1,\ldots,n\}$, $-\|A\| \le -\mu(-A) \le \mathrm{Re}\,\lambda_i \le \alpha_{\max}(A) \le \mu(A) \le \|A\|$.

vii) For all $\alpha \in \mathbb{R}$, $\mu(\alpha A) = |\alpha|\mu[(\mathrm{sign}\,\alpha)A]$.

viii) For all $\alpha \in \mathbb{F}$, $\mu(A + \alpha I) = \mu(A) + \mathrm{Re}\,\alpha$.

ix) $\max\{\mu(A) - \mu(-B), -\mu(-A) + \mu(B)\} \le \mu(A + B) \le \mu(A) + \mu(B)$.

x) $\mu\colon \mathbb{F}^{n\times n} \mapsto \mathbb{R}$ is convex.

xi) $|\mu(A) - \mu(B)| \le \max\{|\mu(A - B)|, |\mu(B - A)|\} \le \|A - B\|$.

xii) For all $x \in \mathbb{F}^n$, $\max\{-\mu(-A), -\mu(A)\}\|x\|' \le \|Ax\|'$.

xiii) If A is nonsingular, then $\max\{-\mu(-A), -\mu(A)\} \le 1/\|A^{-1}\|$.

xiv) For all $t \ge 0$ and $i \in \{1,\ldots,n\}$,

$$e^{-\|A\|t} \le e^{-\mu(-A)t} \le e^{(\mathrm{Re}\,\lambda_i)t} \le e^{\alpha_{\max}(A)t} \le \|e^{tA}\| \le e^{\mu(A)t} \le e^{\|A\|t}.$$

xv) $\mu(A) = \min\{\beta \in \mathbb{R}\colon \|e^{tA}\| \le e^{\beta t} \text{ for all } t \ge 0\}$.

xvi) If $\|\cdot\|' = \|\cdot\|_1$, and thus $\|\cdot\| = \|\cdot\|_{\mathrm{col}}$, then

$$\mu(A) = \max_{j\in\{1,\ldots,n\}}\left(\mathrm{Re}\,A_{(j,j)} + \sum_{\substack{i=1\\ i\ne j}}^{n} |A_{(i,j)}|\right).$$

xvii) If $\|\cdot\|' = \|\cdot\|_2$ and thus $\|\cdot\| = \sigma_{\max}(\cdot)$, then $\mu(A) = \lambda_{\max}[\frac{1}{2}(A + A^*)]$.

xviii) If $\|\cdot\|' = \|\cdot\|_\infty$, and thus $\|\cdot\| = \|\cdot\|_{\mathrm{row}}$, then

$$\mu(A) = \max_{i\in\{1,\ldots,n\}}\left(\mathrm{Re}\,A_{(i,i)} + \sum_{\substack{j=1\\ j\ne i}}^{n} |A_{(i,j)}|\right).$$

Source: [879, 883, 2194, 2561], [1399, pp. 653–655], and [2695, p. 150]. **Remark:** μ is called either the *matrix measure*, *logarithmic derivative*, or *initial growth rate*. For applications, see [1399, 2798]. See Fact 15.19.11 for the logarithmic derivative of an asymptotically stable matrix. **Remark:** The directional differential $\mathrm{D}_+ f(A; I)$ is defined in (12.5.2). **Remark:** *vi)* and *xvii)* yield Fact 7.12.27. **Remark:** Higher order logarithmic derivatives are studied in [460].

Fact 15.16.8. Let $A \in \mathbb{F}^{n\times n}$, let $\beta > \alpha_{\max}(A)$, let $\gamma \ge 1$, and let $\|\cdot\|$ be a normalized, submultiplicative norm on $\mathbb{F}^{n\times n}$. Then, for all $t \ge 0$, $\|e^{tA}\| \le \gamma e^{\beta t}$ if and only if, for all $k \ge 1$ and $\alpha > \beta$,

$$\|(\alpha I - A)^{-k}\| \le \frac{\gamma}{(\alpha - \beta)^k}.$$

Remark: This is a consequence of the *Hille-Yosida theorem.* See [802, p. 26] and [1399, p. 672]. **Related:** Fact 15.16.9.

Fact 15.16.9. Let $A \in \mathbb{F}^{n\times n}$, let $\beta > \alpha_{\max}(A)$, and let $\|\cdot\|$ be a submultiplicative norm on $\mathbb{F}^{n\times n}$. Then, there exists $\gamma \ge 1$ such that, for all $t \ge 0$, $\|e^{tA}\| \le \gamma e^{\beta t}$. **Source:** [1184, pp. 201–206], [1399, pp. 649–651], and [1572]. Note that $\gamma \ge \sup_{t\ge 0}\|e^{(A-\beta I)t}\|$. **Remark:** If A is asymptotically stable, then β can be chosen such that $\alpha_{\max}(A) < \beta < 0$ so that the bound converges to zero. **Related:** Fact

11.13.11 and Fact 15.16.8.

Fact 15.16.10. Let $A \in \mathbb{R}^{n\times n}$, let $\beta \in \mathbb{R}$, and assume that there exists a positive-definite matrix $P \in \mathbb{R}^{n\times n}$ such that

$$A^{\mathsf{T}}P + PA \le 2\beta P.$$

Then, for all $t \ge 0$,

$$\sigma_{\max}(e^{tA}) \le \sqrt{\sigma_{\max}(P)/\sigma_{\min}(P)}e^{\beta t}.$$

Remark: See [1399, p. 665]. **Related:** Fact 15.19.9.

Fact 15.16.11. Let $A \in \mathrm{SO}(3)$, and define $\theta \triangleq 2\,\mathrm{acos}(\frac{1}{2}\sqrt{1+\mathrm{tr}\,A})$. Then,

$$\theta = \sigma_{\max}(\log A) = \frac{1}{\sqrt{2}}\|\log A\|_{\mathrm{F}}.$$

Remark: θ is a Riemannian metric giving the length of the shortest geodesic curve on SO(3) between A and I. See [2064]. **Related:** Fact 4.14.6 and Fact 15.12.10.

Fact 15.16.12. Let $A \in \mathbb{C}^{n\times n}$, for all $i \in \{1,\ldots,n\}$, define $f_i\colon\ [0,\infty) \mapsto \mathbb{R}$ by $f_i(t) \triangleq \log\sigma_i(e^{tA})$, and, for all $x \in \mathbb{C}^n$, define $g_x(t)\colon\ \mathbb{R} \mapsto \mathbb{R}$ by $g_x(t) \triangleq \log\sigma_{\max}(e^{tA}x)$. Then, the following statements are equivalent:

i) A is normal.

ii) For all $i \in \{1,\ldots,n\}$, f_i is convex.

iii) For all $x \in \mathbb{C}^n$, g_x is convex

Source: [199, 987, 1077]. **Remark:** The statement in [199, 987] that convexity holds on $\mathbb{R}$ is erroneous. A counterexample is $A \triangleq \begin{bmatrix} 1 & 0 \\ 0 & -1 \end{bmatrix}$ for which $\log\sigma_1(e^{tA}) = |t|$ and $\log\sigma_2(e^{tA}) = -|t|$. **Credit:** *iii)* is due to S. Friedland.

15.17 Facts on the Matrix Exponential and Eigenvalues, Singular Values, and Norms for Two or More Matrices

Fact 15.17.1. Let $A, B \in \mathbb{F}^{n\times n}$. Then,

$$|\,\mathrm{tr}\,e^{A+B}| \le \mathrm{tr}\,e^{\frac{1}{2}(A+B)}e^{\frac{1}{2}(A+B)^*} \le \mathrm{tr}\,e^{\frac{1}{2}(A+A^*+B+B^*)} \le \mathrm{tr}\,e^{\frac{1}{2}(A+A^*)}e^{\frac{1}{2}(B+B^*)}$$

$$\le (\mathrm{tr}\,e^{A+A^*})^{1/2}(\mathrm{tr}\,e^{B+B^*})^{1/2} \le \tfrac{1}{2}\mathrm{tr}(e^{A+A^*} + e^{B+B^*}),$$

$$\left.\begin{matrix} \mathrm{tr}\,e^Ae^B \\ \frac{1}{2}\mathrm{tr}(e^{2A}+e^{2B}) \end{matrix}\right\} \le \tfrac{1}{2}\mathrm{tr}(e^Ae^{A^*} + e^Be^{B^*}) \le \tfrac{1}{2}\mathrm{tr}(e^{A+A^*} + e^{B+B^*}).$$

Now, assume that A and B are Hermitian. Then,

$$\mathrm{tr}\,e^{A+B} \le \mathrm{tr}\,e^Ae^B \le (\mathrm{tr}\,e^{2A})^{1/2}(\mathrm{tr}\,e^{2B})^{1/2} \le \tfrac{1}{2}\mathrm{tr}(e^{2A}+e^{2B}).$$

Source: [426, 768], [1450, p. 514], and [2206, 2523, 2524]. **Remark:** The following conjecture is given in [2523]. Assume that A and B are Hermitian, let $p, q \in (1,\infty)$, and assume that $1/p+1/q = 1$. Then,

$$(\mathrm{tr}\,e^{pA})^{1/p}(\mathrm{tr}\,e^{qB})^{1/q} \le \tfrac{1}{2}\mathrm{tr}(e^{2A}+e^{2B}).$$

Fact 15.17.2. Let $A, B \in \mathbb{F}^{n\times n}$ and $p \in (0,\infty)$. Then,

$$\sigma_{\max}\!\left[e^{A+B} - \left(e^{\frac{1}{p}A}e^{\frac{1}{p}B}\right)^p\right] \le \tfrac{1}{2p}\sigma_{\max}([A,B])e^{\sigma_{\max}(A)+\sigma_{\max}(B)}.$$

Source: [1391, p. 237] and [2066]. **Related:** Corollary 12.11.15 and Fact 15.17.3.

Fact 15.17.3. Let $A \in \mathbb{F}^{n\times n}$, define $A_{\rm H} \triangleq \frac{1}{2}(A+A^*)$ and $A_{\rm S} \triangleq \frac{1}{2}(A-A^*)$, and let $p\in(0,\infty)$. Then,

$$\sigma_{\max}\left[e^A - \left(e^{\frac{1}{p}A_{\rm H}}e^{\frac{1}{p}A_{\rm S}}\right)^p\right] \le \tfrac{1}{4p}\sigma_{\max}([A^*,A])e^{\frac{1}{2}\lambda_{\max}(A+A^*)}.$$

Source: [2066]. **Related:** Fact 12.11.15.

Fact 15.17.4. Let $A,B \in \mathbb{F}^{n\times n}$, assume that A and B are Hermitian, and let $\|\cdot\|$ be a unitarily invariant norm on $\mathbb{F}^{n\times n}$. Then,

$$\|e^{A+B}\| \le \left\|e^{\frac{1}{2}A}e^Be^{\frac{1}{2}A}\right\| \le \|e^Ae^B\|.$$

If, in addition, $p>0$, then

$$\|e^{A+B}\| \le \left\|e^{\frac{p}{2}A}e^Be^{\frac{p}{2}A}\right\|^{1/p}, \quad \|e^{A+B}\| = \lim_{p\downarrow 0}\left\|e^{\frac{p}{2}A}e^Be^{\frac{p}{2}A}\right\|^{1/p}.$$

Furthermore, for all $k\in\{1,\ldots,n\}$,

$$\prod_{i=1}^k \lambda_i(e^A)\lambda_{n-i+1}(e^B) \le \prod_{i=1}^k \lambda_i(e^{A+B}) \le \prod_{i=1}^k \lambda_i(e^Ae^B) \le \begin{cases}\displaystyle\prod_{i=1}^k \lambda_i(e^A)\lambda_i(e^B)\\ \displaystyle\prod_{i=1}^k \sigma_i(e^Ae^B)\end{cases}$$

with equality for $k=n$; that is,

$$\prod_{i=1}^n \lambda_i(e^A)\lambda_{n-i+1}(e^B) = \prod_{i=1}^n \lambda_i(e^{A+B}) = \prod_{i=1}^n \lambda_i(e^Ae^B) = \prod_{i=1}^n \sigma_i(e^Ae^B) = \det e^Ae^B.$$

Hence, $\lambda(e^{A+B}) \overset{\rm slog}{\prec} \lambda(e^Ae^B)$. In fact,

$$\begin{aligned}\det e^{A+B} &= \prod_{i=1}^n \lambda_i(e^{A+B}) = \prod_{i=1}^n e^{\lambda_i(A+B)} = e^{{\rm tr}(A+B)} = e^{({\rm tr}\,A)+({\rm tr}\,B)}\\ &= e^{{\rm tr}\,A}e^{{\rm tr}\,B} = (\det e^A)\det e^B = \det e^Ae^B = \prod_{i=1}^n \sigma_i(e^Ae^B).\end{aligned}$$

Furthermore, for all $k\in\{1,\ldots,n\}$,

$$\sum_{i=1}^k \lambda_i(e^{A+B}) \le \sum_{i=1}^k \lambda_i(e^Ae^B) \le \sum_{i=1}^k \sigma_i(e^Ae^B).$$

In particular,

$$\lambda_{\max}(e^{A+B}) \le \lambda_{\max}(e^Ae^B) \le \sigma_{\max}(e^Ae^B), \quad {\rm tr}\,e^{A+B} \le {\rm tr}\,e^Ae^B \le {\rm tr}\,\langle e^Ae^B\rangle,$$

and, for all $p>0$,

$${\rm tr}\,e^{A+B} \le {\rm tr}\,e^{\frac{p}{2}A}e^Be^{\frac{p}{2}A}.$$

Finally, ${\rm tr}\,e^{A+B} = {\rm tr}\,e^Ae^B$ if and only if A and B commute. **Source:** [101], [449, p. 261], [2607, 2750], Fact 3.25.15, Fact 7.12.32, and Fact 10.13.2. For the last statement, see [2476]. **Remark:** Note that $\det e^{A+B} = (\det e^A)\det e^B$ despite the fact that e^{A+B} and e^Ae^B may not be equal. See [1391, p. 265] and [1450, p. 442]. **Remark:** ${\rm tr}\,e^{A+B} \le {\rm tr}\,e^Ae^B$ is the Golden-Thompson inequality. See Fact 15.15.30. **Remark:** $\|e^{A+B}\| \le \|e^{\frac{1}{2}A}e^Be^{\frac{1}{2}A}\|$ is *Segal's inequality*. See [93]. **Problem:** Compare the upper bound ${\rm tr}\,\langle e^Ae^B\rangle$ for ${\rm tr}\,e^Ae^B$ with the upper bound $S(1,h)\,{\rm tr}\,e^{A+B}$ given by Fact 15.15.30.

Fact 15.17.5. Let $A, B \in \mathbb{F}^{n\times n}$. Then,

$$\lambda(\langle e^{A+B}\rangle) \overset{\text{slog}}{\prec} \lambda(e^{\frac{1}{2}(A+A^*+B+B^*)}) \overset{\text{wlog}}{\prec} \lambda(e^{\frac{1}{2}(A+A^*)}e^{\frac{1}{2}(B+B^*)}).$$

If, in addition, A and B are normal, then

$$\lambda(\langle e^{A+B}\rangle) \overset{\text{slog}}{\prec} \lambda(\langle e^{A}\rangle\langle e^{B}\rangle).$$

Source: [1878].

Fact 15.17.6. Let $A, B \in \mathbb{F}^{n\times n}$. Then,

$$\begin{bmatrix} \sigma[e^{\frac{1}{2}(A+B)}] \\ \sigma[e^{\frac{1}{2}(A+B)}] \end{bmatrix} \overset{\text{slog}}{\prec} \begin{bmatrix} \sigma(e^{A}) \\ \sigma(e^{B}) \end{bmatrix}.$$

Source: [2750]. **Related:** Fact 10.22.4.

Fact 15.17.7. Let $A, B \in \mathbb{F}^{n\times n}$, assume that A and B are Hermitian, let $q, p > 0$, where $q \le p$, and let $\|\cdot\|$ be a unitarily invariant norm on $\mathbb{F}^{n\times n}$. Then,

$$\left\|\left(e^{\frac{q}{2}A}e^{qB}e^{\frac{q}{2}A}\right)^{1/q}\right\| \le \left\|\left(e^{\frac{p}{2}A}e^{pB}e^{\frac{p}{2}A}\right)^{1/p}\right\|.$$

Source: [101].

Fact 15.17.8. Let $A, B \in \mathbb{F}^{n\times n}$, and assume that A and B are positive semidefinite. Then,

$$e^{\sigma_{\max}^{1/2}(AB)} - 1 \le \sigma_{\max}^{1/2}[(e^{A} - I)(e^{B} - I)], \quad e^{\sigma_{\max}^{1/3}(BAB)} - 1 \le \sigma_{\max}^{1/3}[(e^{B} - I)(e^{A} - I)(e^{B} - I)].$$

Source: [2756]. **Related:** Fact 10.22.42.

Fact 15.17.9. Let $A, B \in \mathbb{F}^{n\times n}$ and $t \ge 0$. Then,

$$e^{t(A+B)} = e^{tA} + \int_0^t e^{(t-\tau)A}Be^{\tau(A+B)}\,\mathrm{d}\tau.$$

Source: [1391, p. 238].

Fact 15.17.10. Let $A, B \in \mathbb{F}^{n\times n}$, and let $\|\cdot\|$ be a submultiplicative norm on $\mathbb{F}^{n\times n}$. Then, for all $t \ge 0$,

$$\|e^{tA} - e^{tB}\| \le e^{\|A\|t}(e^{\|A-B\|t} - 1).$$

Fact 15.17.11. Let $A, B \in \mathbb{F}^{n\times n}$, let $\|\cdot\|$ be a normalized submultiplicative norm on $\mathbb{F}^{n\times n}$, and let $t \ge 0$. Then,

$$\|e^{tA} - e^{tB}\| \le t\|A - B\|e^{t\max\{\|A\|,\|B\|\}}.$$

Source: [1391, p. 265].

Fact 15.17.12. Let $A, B \in \mathbb{R}^{n\times n}$, and assume that A is normal. Then, for all $t \ge 0$,

$$\sigma_{\max}(e^{tA} - e^{tB}) \le \sigma_{\max}(e^{tA})[e^{\sigma_{\max}(A-B)t} - 1].$$

Source: [2866].

Fact 15.17.13. Let $A \in \mathbb{F}^{n\times n}$, let $\|\cdot\|$ be an induced norm on $\mathbb{F}^{n\times n}$, and let $\alpha > 0$ and $\beta \in \mathbb{R}$ be such that, for all $t \ge 0$,

$$\|e^{tA}\| \le \alpha e^{\beta t}.$$

Then, for all $B \in \mathbb{F}^{n\times n}$ and $t \ge 0$,

$$\|e^{t(A+B)}\| \le \alpha e^{(\beta+\alpha\|B\|)t}.$$

Source: [1399, p. 406].

Fact 15.17.14. Let $A, B \in \mathbb{C}^{n\times n}$, assume that A and B are idempotent, assume that $A \neq B$, and let $\|\cdot\|$ be a norm on $\mathbb{C}^{n\times n}$. Then,

$$\|e^{JA} - e^{JB}\| = |e^J - 1|\|A - B\| < \|A - B\|.$$

Source: [2109]. **Remark:** $|e^J - 1| \approx 0.96$.

Fact 15.17.15. Let $A, B \in \mathbb{C}^{n\times n}$, assume that A and B are Hermitian, let $X \in \mathbb{C}^{n\times n}$, and let $\|\cdot\|$ be a unitarily invariant norm on $\mathbb{C}^{n\times n}$. Then,

$$\|e^{JA}X - Xe^{JB}\| \le \|AX - XB\|.$$

Source: [2109]. **Remark:** This is a matrix version of *xi*) of Fact 2.21.27.

Fact 15.17.16. Let $A, B \in \mathbf{H}^n$, let $k \ge 1$, and define $f\colon (0,\infty) \mapsto \mathbb{R}$ by $f(t) \triangleq \sum_{i=1}^k \frac{1}{t}\log\lambda_i(e^{tA}e^{tB})$. Then, f is nondecreasing. Furthermore,

$$\lim_{t\downarrow 0} f(t) = \sum_{i=1}^k \lambda_i(A + B) \le \sum_{i=1}^k \log\lambda_i(e^A e^B).$$

Source: [1079, pp. 378–380]. **Remark:** Setting $k = n$ yields the Golden-Thompson inequality.

Fact 15.17.17. Let $A, B \in \mathbf{H}^n$, for all $t \in (0,\infty)$, define $C(t) \triangleq \frac{1}{t}\log\lambda_i(e^{\frac{1}{2}tA}e^{tB}e^{\frac{1}{2}tA})$, let $k \in \{1,\dots,n\}$, and define $f\colon (0,\infty) \mapsto (0,\infty)$ by $f(t) \triangleq \sum_{i=1}^k \lambda_i[C(t)]$. Then, f is nondecreasing. Furthermore, for all $t \in (0,\infty)$, $\lambda[C(t)] \overset{\mathrm{w}}{\prec} \lambda(A) + \lambda(B)$. Finally, $C_0 \triangleq \lim_{t\to\infty} C(t)$ exists, $C_0 \in \mathbf{H}^n$, and $[A, C_0] = 0$. **Source:** [1079, pp. 380–387].

Fact 15.17.18. Let $A, B \in \mathbb{F}^{n\times n}$, assume that A and B are positive semidefinite, and let $\|\cdot\|$ be a unitarily invariant norm on $\mathbb{F}^{n\times n}$. Then,

$$\|e^{\langle A-B\rangle} - I\| \le \|e^A - e^B\|, \quad \|e^A + e^B\| \le \|e^{A+B} + I\|.$$

Source: [106] and [449, p. 294]. **Related:** Fact 11.10.86.

Fact 15.17.19. Let $A, X, B \in \mathbb{F}^{n\times n}$, assume that A and B are Hermitian, and let $\|\cdot\|$ be a unitarily invariant norm on $\mathbb{F}^{n\times n}$. Then,

$$\|AX - XB\| \le \|e^{\frac{1}{2}A}Xe^{-\frac{1}{2}B} - e^{-\frac{1}{2}B}Xe^{\frac{1}{2}A}\|.$$

Source: [477]. **Related:** Fact 11.10.87.

Fact 15.17.20. Let $A \in \mathbb{F}^{n\times n}$, $B \in \mathbb{F}^{n\times m}$, $C \in \mathbb{F}^{m\times n}$, and $D \in \mathbb{F}^{m\times m}$, let $(M_i)_{i=1}^\infty \subset \mathbb{F}^{n\times n}$, and assume that $\lim_{i\to\infty}\lambda_{\min}(M_i + M_i^*) = \infty$. Then,

$$\lim_{i\to\infty}\exp\left(\begin{bmatrix} A - M_i & B \\ C & D\end{bmatrix}\right) = \begin{bmatrix} 0 & 0 \\ 0 & e^D\end{bmatrix}.$$

Source: [2729].

15.18 Facts on Stable Polynomials

Fact 15.18.1. Let $p \in \mathbb{C}[s]$, and assume that p is (asymptotically stable, discrete-time asymptotically stable). Then, so is p'. **Related:** Fact 12.16.5.

Fact 15.18.2. Let $a_1,\dots,a_n$ be real numbers, let $\Delta \triangleq \{i \in \{1,\dots,n-1\}\colon \frac{a_{i+1}}{a_i} < 0\}$, let $b_1,\dots,b_n$ be real numbers such that $b_1 < \cdots < b_n$, define $f\colon (0,\infty) \mapsto \mathbb{R}$ by $f(x) = a_nx^{b_n} + \cdots + a_1x^{b_1}$, and define $\mathcal{S} \triangleq \{x \in (0,\infty)\colon f(x) = 0\}$. Furthermore, for all $x \in \mathcal{S}$, define the multiplicity of x to be the positive integer m such that $f(x) = f'(x) = \cdots = f^{(m-1)} = 0$ and $f^{(m)}(x) \neq 0$, and let $\mathcal{S}'$ denote the multiset consisting of all elements of $\mathcal{S}$ counting multiplicity as roots of f. Then, $\operatorname{card}(\mathcal{S}') \le \operatorname{card}(\Delta)$. If, in addition, $b_1,\dots,b_n$ are nonnegative integers, then $\operatorname{card}(\Delta) - \operatorname{card}(\mathcal{S}')$ is even. **Source:** [414, 1666, 2831]. **Remark:** This is the *Descartes rule of signs*.

Fact 15.18.3. Let $p \in \mathbb{R}[s]$, where $p(s) = s^n + a_{n-1}s^{n-1} + \cdots + a_0$. If p is asymptotically stable, then $a_0, \ldots, a_{n-1}$ are positive. Now, assume that $a_0, \ldots, a_{n-1}$ are positive. Then, the following statements hold:

i) If either $n = 1$ or $n = 2$, then p is asymptotically stable.

ii) If $n = 3$, then p is asymptotically stable if and only if $a_0 < a_1a_2$.

iii) If $n = 4$, then p is asymptotically stable if and only if $a_1^2 + a_0a_3^2 < a_1a_2a_3$.

iv) If $n = 5$, then p is asymptotically stable if and only if

$$a_2 < a_3a_4, \quad a_2^2 + a_1a_4^2 < a_0a_4 + a_2a_3a_4,$$
$$a_0^2 + a_1a_2^2 + a_1^2a_4^2 + a_0a_3^2a_4 < a_0a_2a_3 + 2a_0a_1a_4 + a_1a_2a_3a_4.$$

Remark: These are special cases of the *Routh criterion*, which provides necessary and sufficient conditions for asymptotic stability of a polynomial. See [676]. **Remark:** The Jury criterion for discrete-time asymptotic stability of a polynomial is given by Fact 15.21.2. **Related:** Fact 10.21.2 and Fact 15.18.15.

Fact 15.18.4. Let $p \in \mathbb{R}[s]$, where $p(s) = s^3 + a_2s^2 + a_1s + a_0$, assume that $a_1 > 0$, and define $\alpha \triangleq \min\{-a_0/a_1, -a_2\}$ and $\beta \triangleq \max\{-a_0/a_1, -a_2\}$. Then, there exists $\lambda \in \operatorname{roots}(p)$ such that $\lambda \in (\alpha, \beta)$. **Source:** [290]. **Related:** Fact 2.21.2, Fact 15.21.27, and Fact 15.21.29.

Fact 15.18.5. Let $\varepsilon \in [0, 1]$, let $n \in \{2, 3, 4\}$, let $p_\varepsilon \in \mathbb{R}[s]$, where $p_\varepsilon(s) = s^n + a_{n-1}s^{n-1} + \cdots + a_1s + \varepsilon a_0$, and assume that p_1 is asymptotically stable. Then, for all $\varepsilon \in (0, 1]$, p_ε is asymptotically stable. Furthermore, $p_0(s)/s$ is asymptotically stable. **Remark:** This result does not hold for $n = 5$. A counterexample is $p(s) = s^5 + 2s^4 + 3s^3 + 5s^2 + 2s + 2.5\varepsilon$, which is asymptotically stable if and only if $\varepsilon \in (4/5, 1]$. This is another instance of the quartic barrier. See [787], Fact 10.17.8, and Fact 10.19.24.

Fact 15.18.6. Let $p \in \mathbb{R}[s]$ be monic, and define $q(s) \triangleq s^n p(1/s)$, where $n \triangleq \deg p$. Then, p is asymptotically stable if and only if q is asymptotically stable. **Related:** Fact 6.8.3 and Fact 15.18.7.

Fact 15.18.7. Let $p \in \mathbb{R}[s]$ be monic, and assume that p is semistable. Then, $q(s) \triangleq p(s)/s$ and $\hat{q}(s) \triangleq s^n p(1/s)$ are asymptotically stable. **Related:** Fact 6.8.3 and Fact 15.18.6.

Fact 15.18.8. Let $p, q \in \mathbb{R}[s]$, and assume that p is even, q is odd, and every coefficient of $p+q$ is positive. Then, $p + q$ is asymptotically stable if and only if every root of p and every root of q is imaginary, and the roots of p and the roots of q are interlaced on the imaginary axis. **Source:** [482, 676, 1438]. **Remark:** This is called either the *Hermite-Biehler* or *interlacing theorem.* **Example:** $s^2 + 2s + 5 = (s^2 + 5) + 2s$.

Fact 15.18.9. Let $p \in \mathbb{R}[s]$ be asymptotically stable, and let $p(s) = \beta_n s^n + \beta_{n-1}s^{n-1} + \cdots + \beta_1 s + \beta_0$, where $\beta_n > 0$. Then, for all $i \in \{1, \ldots, n-2\}$, $\beta_{i-1}\beta_{i+2} < \beta_i\beta_{i+1}$. **Remark:** This is a necessary condition for asymptotic stability. **Remark:** See [482, p. 68]. **Credit:** X. Xie. See [2949].

Fact 15.18.10. Let $n \ge 2$ be even, let $m \triangleq n/2$, let $p \in \mathbb{R}[s]$, where $p(s) = \beta_n s^n + \beta_{n-1}s^{n-1} + \cdots + \beta_1 s + \beta_0$ and $\beta_n > 0$, and assume that p is asymptotically stable. Then, for all $i \in \{1, \ldots, m-1\}$,

$$\binom{m}{i}\beta_0^{(m-i)/m}\beta_n^{i/m} \le \beta_{2i}.$$

Remark: This is a necessary condition for asymptotic stability. See [2949, 2950] for extensions to polynomials of odd degree, and [328] for inequalities involving either four or more terms. **Credit:** A. Borobia and S. Dormido.

Fact 15.18.11. Let $p \in \mathbb{R}[s]$, where $p(s) = \alpha_n s^n + \alpha_{n-1}s^{n-1} + \cdots + \alpha_1 s + \alpha_0$, assume that $\alpha_0, \ldots, \alpha_n$ are positive, let $r \in [1, \infty)$, and define $p^{\odot r}(s) \triangleq \alpha_n^r s^n + \alpha_{n-1}^r s^{n-1} + \cdots + \alpha_1^r s + \alpha_0^r$. Then, the following statements hold:

i) Assume that $n = 3$ and all of the roots of p are real. Then, all of the roots of $p^{\odot r}$ are real.

ii) Assume that n is even, all of the roots of p are real, and $r > 1/[\log_2(n+2) - \log_2 n]$. Then, all of the roots of $p^{\odot r}$ are real.

iii) Assume that n is odd, all of the roots of p are real, and $r > 2/[\log_2(n+3) - \log_2(n-1)]$. Then, all of the roots of $p^{\odot r}$ are real.

iv) If p is asymptotically stable, then so is $p^{\odot r}$.

Source: [2832]. **Remark:** $p^{\odot r}$ is a *Schur power* of p. **Remark:** *i)* is false for $n = 4$.

Fact 15.18.12. Let $p, q \in \mathbb{R}[s]$, where $p(s) = \alpha_n s^n + \alpha_{n-1}s^{n-1} + \cdots + \alpha_1 s + \alpha_0$ and $q(s) = \beta_m s^m + \beta_{m-1}s^{m-1} + \cdots + \beta_1 s + \beta_0$, and define $(p \odot q)(s) \triangleq \alpha_l\beta_l s^l + \alpha_{l-1}\beta_{l-1}s^{l-1} + \cdots + \alpha_1\beta_1 s + \alpha_0\beta_0$, where $l \triangleq \min\{m, n\}$. If p and q are (Lyapunov stable, asymptotically stable), then so is $p \odot q$. **Source:** [1147]. **Remark:** $p \odot q$ is the *Schur product* of p and q. See [184, 1539].

Fact 15.18.13. Let $p, q \in \mathbb{R}[s]$, where $p(s) = \alpha_n s^n + \alpha_{n-1}s^{n-1} + \cdots + \alpha_1 s + \alpha_0$ and $q(s) = \beta_m s^m + \beta_{m-1}s^{m-1} + \cdots + \beta_1 s + \beta_0$, assume that α_n and β_n are nonzero, define $(p \odot q)(s) \triangleq \alpha_n\beta_n s^n + \cdots + \alpha_1\beta_1 s + \alpha_0\beta_0$, assume that all of the roots of p and q are real, and assume that all of the roots of either p or q have the same sign. Then, all of the roots of $p \odot q$ are real. **Source:** [2832].

Fact 15.18.14. Let $A \in \mathbb{R}^{n\times n}$, and assume that A is diagonalizable over $\mathbb{R}$. Then, χ_A has all positive coefficients if and only if A is asymptotically stable. **Source:** Sufficiency follows from Fact 15.18.3. For necessity, note that all of the roots of χ_A are real and that $\chi_A(\lambda) > 0$ for all $\lambda \geq 0$. Hence, $\operatorname{roots}(\chi_A) \subset (-\infty, 0)$.

Fact 15.18.15. Let $A \in \mathbb{R}^{n\times n}$. Then, the following statements are equivalent:

i) $\chi_{A\oplus A}$ has all positive coefficients.

ii) $\chi_{A\oplus A}$ is asymptotically stable.

iii) $A \oplus A$ is asymptotically stable.

iv) A is asymptotically stable.

Source: If A is not asymptotically stable, then Fact 15.19.32 implies that $A \oplus A$ has a nonnegative eigenvalue λ. Since $\chi_{A\oplus A}(\lambda) = 0$, it follows that $\chi_{A\oplus A}$ cannot have all positive coefficients. See [1097, Theorem 5]. **Remark:** A similar proof is used in Proposition 10.2.8. **Related:** Fact 15.18.3.

Fact 15.18.16. Let $A \in \mathbb{R}^{n\times n}$. Then, the following statements are equivalent:

i) χ_A and $\chi_{A^{(2,1)}}$ have all positive coefficients.

ii) A is asymptotically stable.

Source: [2559]. **Remark:** The additive compound $A^{(2,1)}$ is defined in Fact 9.5.18.

Fact 15.18.17. For all $i \in \{1, \ldots, n-1\}$, let $a_i, b_i \in \mathbb{R}$ satisfy $0 < a_i \leq b_i$, define $\phi_1, \phi_2, \psi_1, \psi_2 \in \mathbb{R}[s]$ by

$$\phi_1(s) = b_n s^n + a_{n-2}s^{n-2} + b_{n-4}s^{n-4} + \cdots, \qquad \phi_2(s) = a_n s^n + b_{n-2}s^{n-2} + a_{n-4}s^{n-4} + \cdots,$$

$$\psi_1(s) = b_{n-1}s^{n-1} + a_{n-3}s^{n-3} + b_{n-5}s^{n-5} + \cdots, \qquad \psi_2(s) = a_{n-1}s^{n-1} + b_{n-3}s^{n-3} + a_{n-5}s^{n-5} + \cdots,$$

assume that $\phi_1 + \psi_1$, $\phi_1 + \psi_2$, $\phi_2 + \psi_1$, and $\phi_2 + \psi_2$ are asymptotically stable, let $p \in \mathbb{R}[s]$, where $p(s) = \beta_n s^n + \beta_{n-1}s^{n-1} + \cdots + \beta_1 s + \beta_0$, and assume that, for all $i \in \{1, \ldots, n\}$, $a_i \leq \beta_i \leq b_i$. Then, p is asymptotically stable. **Source:** [970, pp. 466, 467]. **Remark:** This is *Kharitonov's theorem*.

Fact 15.18.18. Let $p \in \mathbb{R}[s]$, where $p(s) = \beta_n s^n + \beta_{n-1}s^{n-1} + \cdots + \beta_1 s + \beta_0$, and assume that $\beta_0, \ldots, \beta_{n-1} \geq 0$. Then, the following statements hold:

i) $\alpha_{\max}(p) \in \operatorname{roots}(p)$.

ii) If $p(1) \leq 2$, then $\alpha_{\max}(p) \leq [p(1) - 1]^{1/n}$.

iii) If $p(1) \geq 2$, then $\alpha_{\max}(p) \leq p(1) - 1$.

iv) $\alpha_{\max}(p) \leq 2\max\{\beta_{n-1}, \sqrt[2]{\beta_{n-2}}, \ldots, \sqrt[n]{\beta_0}\}$.

Source: [1172, pp. 6, 7].

Fact 15.18.19. Let $p \in \mathbb{R}[s]$, where $p(s) = \beta_n s^n + \beta_{n-1}s^{n-1} + \cdots + \beta_1 s + \beta_0$, and assume that, for all $i \in \{0, 1, \ldots, n-1\}$, $\beta_i \in [0, 1]$. Then, $\alpha_{\max}(p) < \frac{1}{2}(1 + \sqrt{5})$. **Source:** [107, pp. 53, 288].

15.19 Facts on Stable Matrices

Fact 15.19.1. Let $A \in \mathbb{F}^{n\times n}$. Then, the following statements are equivalent:

i) A is semistable.

ii) For all nonzero $\omega \in \mathbb{R}$, $\operatorname{rank}(\omega \jmath I - A) = n$, and there exist a positive-semidefinite matrix $R \in \mathbb{F}^{n\times n}$ and a positive-definite matrix $P \in \mathbb{F}^{n\times n}$ such that

$$A^{\mathsf{T}}P + PA + R = 0. \tag{15.19.1}$$

iii) There exist a positive-semidefinite matrix $R \in \mathbb{F}^{n\times n}$ and a positive-definite matrix $P \in \mathbb{F}^{n\times n}$ such that, for all nonzero $\omega \in \mathbb{R}$, $\operatorname{rank}\begin{bmatrix}\omega \jmath I - A\\ R\end{bmatrix} = n$ and (15.19.1) holds.

iv) There exist a positive-semidefinite matrix $R \in \mathbb{F}^{n\times n}$ and a positive-definite matrix $P \in \mathbb{F}^{n\times n}$ satisfying $\bigcap_{k=1}^{n} \mathcal{N}(RA^{k-1}) = \mathcal{N}(A)$ and (15.19.1).

v) For each positive-semidefinite matrix $R \in \mathbb{F}^{n\times n}$ satisfying $\bigcap_{k=1}^{n} \mathcal{N}(RA^{k-1}) = \mathcal{N}(A)$, there exists a positive-definite matrix $P \in \mathbb{F}^{n\times n}$ satisfying (15.19.1).

vi) For each positive-semidefinite matrix $R \in \mathbb{F}^{n\times n}$ such that $k \triangleq \min\{l \geq 0\colon \bigcap_{i=1}^{n} \mathcal{N}(RA^{i+l-1}) = \mathcal{N}(A)\}$ exists, there exists a positive-semidefinite matrix $P \in \mathbb{F}^{n\times n}$ satisfying

$$A^{k\mathsf{T}}(A^{\mathsf{T}}P + PA + R)A^k = 0.$$

vii) For each positive-semidefinite matrix $R \in \mathbb{F}^{n\times n}$ such that $k \triangleq \min\{l \geq 0\colon \bigcap_{i=1}^{n} \mathcal{N}(RA^{i+l-1}) = \mathcal{N}(A)\}$ exists, there exists a positive-definite matrix $P \in \mathbb{F}^{n\times n}$ such that

$$A^{\mathsf{T}}P + PA + A^{k\mathsf{T}}RA^k = 0.$$

Credit: Q. Hui.

Fact 15.19.2. Let $A \in \mathbb{F}^{n\times n}$, and assume that A is semistable. Then, A is Lyapunov stable.

Fact 15.19.3. Let $A \in \mathbb{F}^{n\times n}$, and assume that A is Lyapunov stable. Then, A is group invertible.

Fact 15.19.4. Let $A \in \mathbb{F}^{n\times n}$, and assume that A is semistable. Then, A is group invertible.

Fact 15.19.5. Let $A, B \in \mathbb{F}^{n\times n}$, and assume that A and B are similar. Then, A is (Lyapunov stable, semistable, asymptotically stable, discrete-time Lyapunov stable, discrete-time semistable, discrete-time asymptotically stable) if and only if B is.

Fact 15.19.6. Let $A \in \mathbb{R}^{n\times n}$, and assume that A is Lyapunov stable. Then,

$$\lim_{t\to\infty} \frac{1}{t}\int_0^t e^{\tau A}\,\mathrm{d}\tau = I - AA^{\#}.$$

Related: Fact 15.19.3.

Fact 15.19.7. Let $A \in \mathbb{F}^{n\times n}$, and assume that A is semistable. Then,

$$\lim_{t\to\infty} e^{tA} = I - AA^{\#},$$

and thus

$$\lim_{t\to\infty} \frac{1}{t}\int_0^t e^{\tau A}\,\mathrm{d}\tau = I - AA^{\#}.$$

Related: Fact 12.13.12, Fact 15.19.2, and Fact 15.19.3.

Fact 15.19.8. Let $A, B \in \mathbb{F}^{n\times n}$. Then, $\lim_{\alpha\to\infty} e^{A+\alpha B}$ exists if and only if B is semistable. If these conditions hold, then

$$\lim_{\alpha\to\infty} e^{A+\alpha B} = e^{(I-BB^{\#})A}(I - BB^{\#}) = (I - BB^{\#})e^{A(I-BB^{\#})}.$$

Source: [625].

Fact 15.19.9. Let $A \in \mathbb{R}^{n\times n}$, assume that A is asymptotically stable, let $\beta \in (\alpha_{\max}(A), 0)$, let $P \in \mathbb{R}^{n\times n}$ be positive definite and satisfy

$$A^{\mathsf{T}}P + PA \le 2\beta P,$$

and let $\|\cdot\|$ be a normalized, submultiplicative norm on $\mathbb{R}^{n\times n}$. Then, for all $t \ge 0$,

$$\|e^{tA}\| \le \sqrt{\|P\|\|P^{-1}\|}e^{\beta t}.$$

Source: [1398] and Fact 15.16.10.

Fact 15.19.10. Let $A \in \mathbb{F}^{n\times n}$, assume that A is asymptotically stable, let $R \in \mathbb{F}^{n\times n}$, assume that R is positive definite, and let $P \in \mathbb{F}^{n\times n}$ be the positive-definite solution of $A^*P + PA + R = 0$. Then, for all $t \ge 0$,

$$\sigma_{\max}(e^{tA}) \le \sqrt{\frac{\sigma_{\max}(P)}{\sigma_{\min}(P)}}e^{-t\lambda_{\min}(RP^{-1})/2}, \quad \|e^{tA}\|_{\mathrm{F}} \le \sqrt{\|P\|_{\mathrm{F}}\|P^{-1}\|_{\mathrm{F}}}e^{-t\lambda_{\min}(RP^{-1})/2}.$$

If, in addition, $A + A^*$ is negative definite, then, for all $t \ge 0$,

$$\|e^{tA}\|_{\mathrm{F}} \le e^{-t\lambda_{\min}(-A-A^*)/2}.$$

Source: [1936].

Fact 15.19.11. Let $A \in \mathbb{R}^{n\times n}$, assume that A is asymptotically stable, let $R \in \mathbb{R}^{n\times n}$, assume that R is positive definite, and let $P \in \mathbb{R}^{n\times n}$ be the positive-definite solution of $A^{\mathsf{T}}P + PA + R = 0$. Furthermore, define the vector norm $\|x\|' \triangleq \sqrt{x^{\mathsf{T}}Px}$ on $\mathbb{R}^n$, let $\|\cdot\|$ denote the induced norm on $\mathbb{R}^{n\times n}$, and let $\mu(\cdot)$ denote the corresponding logarithmic derivative. Then,

$$\mu(A) = -\lambda_{\min}(RP^{-1})/2.$$

Consequently, for all $t \ge 0$,

$$\|e^{tA}\| \le e^{-t\lambda_{\min}(RP^{-1})/2}.$$

Source: [1475] and *xiv*) of Fact 15.16.7. **Related:** Fact 15.16.7 for the logarithmic derivative.

Fact 15.19.12. Let $A \in \mathbb{F}^{n\times n}$. Then, A is similar to a skew-Hermitian matrix if and only if there exists a positive-definite matrix $P \in \mathbb{F}^{n\times n}$ such that $A^*P + PA = 0$. **Related:** Fact 7.10.5.

Fact 15.19.13. Let $A \in \mathbb{R}^{n\times n}$. Then, A and A^2 are asymptotically stable if and only if, for all $\lambda \in \mathrm{spec}(A)$, there exist $r > 0$ and $\theta \in \left(\frac{\pi}{2}, \frac{3\pi}{4}\right) \cup \left(\frac{5\pi}{4}, \frac{3\pi}{2}\right)$ such that $\lambda = re^{j\theta}$.

Fact 15.19.14. Let $A \in \mathbb{R}^{n\times n}$. Then, A is group invertible and $2k\pi j \notin \mathrm{spec}(A)$ for all $k \ge 1$ if and only if

$$AA^{\#} = (e^A - I)(e^A - I)^{\#}.$$

In particular, if A is semistable, then this equality holds. **Source:** *ii*) of Fact 15.22.13 and *ix*) of Proposition 15.9.2.

Fact 15.19.15. Let $A \in \mathbb{F}^{n\times n}$. Then, A is asymptotically stable if and only if A^{-1} is asymptotically stable. Hence, $e^{tA} \to 0$ as $t \to \infty$ if and only if $e^{tA^{-1}} \to 0$ as $t \to \infty$.

Fact 15.19.16. Let $A, B \in \mathbb{R}^{n\times n}$, assume that A is asymptotically stable, and assume that $\sigma_{\max}(B \oplus B) < \sigma_{\min}(A \oplus A)$. Then, $A + B$ is asymptotically stable. **Source:** Since $A \oplus A$ is nonsingular, Fact 11.16.22 implies that $A \oplus A + \alpha(B \oplus B) = (A + \alpha B) \oplus (A + \alpha B)$ is nonsingular for all $0 \le \alpha \le 1$. Now, suppose that $A + B$ is not asymptotically stable. Then, there exists $\alpha_0 \in (0, 1]$ such that $A + \alpha_0 B$ has an imaginary eigenvalue, and thus $(A + \alpha_0 B) \oplus (A + \alpha_0 B) = A \oplus A + \alpha_0(B \oplus B)$ is singular, which is a contradiction. **Remark:** This is a suboptimal solution of a nearness problem. See [1387, Sect. 7] and Fact 11.16.22.

Fact 15.19.17. Let $A \in \mathbb{C}^{n\times n}$, assume that A is asymptotically stable, let $\|\cdot\|$ denote either $\sigma_{\max}(\cdot)$ or $\|\cdot\|_{\mathrm{F}}$, and define

$$\beta(A) \triangleq \{\|B\|\colon\ B \in \mathbb{C}^{n\times n} \text{ and } A + B \text{ is not asymptotically stable}\}.$$

Then,

$$\tfrac{1}{2}\sigma_{\min}(A \otimes A) \le \beta(A) = \min_{\gamma\in\mathbb{R}} \sigma_{\min}(A + \gamma \jmath I) \le \min\{\alpha_{\max}(A), \sigma_{\min}(A), \tfrac{1}{2}\sigma_{\max}(A + A^*)\}.$$

Furthermore, let $R \in \mathbb{F}^{n\times n}$, assume that R is positive definite, and let $P \in \mathbb{F}^{n\times n}$ be the positive-definite solution of $A^*P + PA + R = 0$. Then,

$$\tfrac{1}{2}\sigma_{\min}(R)/\|P\| \le \beta(A).$$

If, in addition, $A + A^*$ is negative definite, then

$$-\tfrac{1}{2}\lambda_{\min}(A + A^*) \le \beta(A).$$

Source: [1387, 2775]. **Remark:** Real matrices and real perturbations are discussed in [2283].

Fact 15.19.18. Let $A \in \mathbb{F}^{n\times n}$, assume that A is asymptotically stable, let $V \in \mathbb{F}^{n\times n}$, assume that V is positive definite, and let $Q \in \mathbb{F}^{n\times n}$ be the positive-definite solution of $AQ + QA^* + V = 0$. Then, for all $t \ge 0$,

$$\|e^{tA}\|_{\mathrm{F}}^2 = \operatorname{tr} e^{tA}e^{tA^*} \le \kappa(Q)\operatorname{tr} e^{-tS^{-1}VS^{-*}} \le \kappa(Q)\operatorname{tr} e^{-[t/\sigma_{\max}(Q)]V},$$

where $S \in \mathbb{F}^{n\times n}$ satisfies $Q = SS^*$ and $\kappa(Q) \triangleq \sigma_{\max}(Q)/\sigma_{\min}(Q)$. If, in particular, $AQ + QA^* + I = 0$, then

$$\|e^{tA}\|_{\mathrm{F}}^2 \le n\kappa(Q)e^{-t/\sigma_{\max}(Q)}.$$

Source: [2934]. **Remark:** Fact 15.16.4 yields $e^{tA}e^{tA^*} \le e^{t(A+A^*)}$. However, this bound is poor in the case where $A + A^*$ is not asymptotically stable. See [427]. **Related:** Fact 15.19.19.

Fact 15.19.19. Let $A \in \mathbb{F}^{n\times n}$, assume that A is asymptotically stable, let $Q \in \mathbb{F}^{n\times n}$ be the positive-definite solution of $AQ + QA^* + I = 0$, and define $\kappa(Q) \triangleq \sigma_{\max}(Q)/\sigma_{\min}(Q)$. Then, for all $t \ge 0$,

$$\sigma_{\max}^2(e^{tA}) \le \kappa(Q)e^{-t/\sigma_{\max}(Q)}.$$

Source: [2795, 2796] and references therein. **Remark:** Since $\|e^{tA}\|_{\mathrm{F}} \le \sqrt{n}\sigma_{\max}(e^{tA})$, it follows that this inequality implies the last inequality in Fact 15.19.18.

Fact 15.19.20. Let $A \in \mathbb{R}^{n\times n}$, and assume that $A >> 0$. Then, A is unstable. **Source:** Fact 6.11.5.

Fact 15.19.21. Let $A \in \mathbb{R}^{n\times n}$. Then, A is asymptotically stable if and only if there exist $B, C \in \mathbb{R}^{n\times n}$ such that B is positive definite, C is dissipative, and $A = BC$. **Source:** $A = P^{-1}(-A^{\mathsf{T}}P - R)$. **Remark:** To reverse the ordering of the factors, consider A^{T}.

Fact 15.19.22. Let $A \in \mathbb{F}^{n\times n}$. Then, the following statements hold:

i) All of the real eigenvalues of A are positive if and only if A is the product of two dissipative matrices.

ii) A is nonsingular and $A \ne \alpha I$ for all $\alpha < 0$ if and only if A is the product of two asymptotically stable matrices.

iii) A is nonsingular if and only if A is the product of three or fewer asymptotically stable matrices.

Source: [272, 2919].

Fact 15.19.23. Let $p \in \mathbb{R}[s]$, where $p(s) = s^n + \beta_{n-1}s^{n-1} + \cdots + \beta_1 s + \beta_0$ and $\beta_0, \ldots, \beta_n > 0$.

Furthermore, define $A \in \mathbb{R}^{n\times n}$ by

$$A \triangleq \begin{bmatrix} \beta_{n-1} & \beta_{n-3} & \beta_{n-5} & \beta_{n-7} & \cdots & \cdots & 0 \\ 1 & \beta_{n-2} & \beta_{n-4} & \beta_{n-6} & \cdots & \cdots & 0 \\ 0 & \beta_{n-1} & \beta_{n-3} & \beta_{n-5} & \cdots & \cdots & 0 \\ 0 & 1 & \beta_{n-2} & \beta_{n-4} & \cdots & \cdots & 0 \\ \vdots & \vdots & \vdots & \vdots & \ddots & \vdots & \vdots \\ 0 & 0 & 0 & \cdots & \cdots & \beta_1 & 0 \\ 0 & 0 & 0 & \cdots & \cdots & \beta_2 & \beta_0 \end{bmatrix}.$$

If p is Lyapunov stable, then every subdeterminant of A is nonnegative. **Source:** [184]. **Remark:** A is totally nonnegative. Furthermore, p is asymptotically stable if and only if every leading principal subdeterminant of A is positive. **Remark:** The diagonal entries of A are $\beta_{n-1},\dots,\beta_0$. **Credit:** The second statement is due to A. Hurwitz. **Problem:** Show that this stability condition is equivalent to the condition given in [1040, p. 183].

Fact 15.19.24. Let $A \in \mathbb{R}^{n\times n}$, assume that A is tridiagonal, assume that $A_{(i,i)} > 0$ for all $i \in \{1,\dots,n\}$, and assume that $A_{(i,i+1)}A_{(i+1,i)} > 0$ for all $i \in \{1,\dots,n-1\}$. Then, A is asymptotically stable. **Source:** [636]. **Credit:** S. Barnett and C. Storey.

Fact 15.19.25. Let $A \in \mathbb{R}^{n\times n}$, and assume that A is cyclic. Then, there exists a nonsingular matrix $S \in \mathbb{R}^{n\times n}$ such that $A_S \triangleq SAS^{-1}$ is given by the tridiagonal matrix

$$A_S = \begin{bmatrix} 0 & 1 & 0 & \cdots & 0 & 0 \\ -\alpha_n & 0 & 1 & \cdots & 0 & 0 \\ 0 & -\alpha_{n-1} & 0 & \ddots & 0 & 0 \\ \vdots & \vdots & \ddots & \ddots & \ddots & \vdots \\ 0 & 0 & 0 & \ddots & 0 & 1 \\ 0 & 0 & 0 & \cdots & -\alpha_2 & -\alpha_1 \end{bmatrix},$$

where $\alpha_1,\dots,\alpha_n$ are real numbers. If $\alpha_1\alpha_2\cdots\alpha_n \neq 0$, then the number of eigenvalues of A in the OLHP is equal to the number of positive elements in $\{\alpha_1,\alpha_1\alpha_2,\dots,\alpha_1\alpha_2\cdots\alpha_n\}_{\rm ms}$. Furthermore, $A_S^{\rm T}P + PA_S + R = 0$, where

$$P \triangleq \operatorname{diag}(\alpha_1\alpha_2\cdots\alpha_n, \alpha_1\alpha_2\cdots\alpha_{n-1},\dots,\alpha_1\alpha_2,\alpha_1), \quad R \triangleq \operatorname{diag}(0,\dots,0,2\alpha_1^2).$$

Finally, A_S is asymptotically stable if and only if $\alpha_1,\dots,\alpha_n$ are positive. **Source:** [302, pp. 52, 95]. **Remark:** A_S is a *Schwarz matrix*. See Fact 15.19.26 and Fact 15.19.27.

Fact 15.19.26. Let $\alpha_1,\dots,\alpha_n$ be real numbers, and define $A \in \mathbb{R}^{n\times n}$ by

$$A = \begin{bmatrix} 0 & 1 & 0 & \cdots & 0 & 0 \\ -\alpha_n & 0 & 1 & \cdots & 0 & 0 \\ 0 & -\alpha_{n-1} & 0 & \ddots & 0 & 0 \\ \vdots & \vdots & \ddots & \ddots & \ddots & \vdots \\ 0 & 0 & 0 & \ddots & 0 & 1 \\ 0 & 0 & 0 & \cdots & -\alpha_2 & \alpha_1 \end{bmatrix}.$$

Then, $\operatorname{spec}(A) \subset$ ORHP if and only if $\alpha_1,\dots,\alpha_n$ are positive. **Source:** [1450, p. 111]. **Remark:** Note the absence of the minus sign in the (n,n) entry compared to the matrix in Fact 15.19.25. This minus sign changes the signs of all of the eigenvalues of A.

Fact 15.19.27. Let $\alpha_1, \alpha_2, \alpha_3 > 0$, and define $A_{\rm R}, P, R \in \mathbb{R}^{3\times 3}$ by the tridiagonal matrix

$$A_{\rm R} \triangleq \begin{bmatrix} -\alpha_1 & \alpha_2^{1/2} & 0 \\ -\alpha_2^{1/2} & 0 & \alpha_3^{1/2} \\ 0 & -\alpha_3^{1/2} & 0 \end{bmatrix}$$

and the diagonal matrices

$$P \triangleq I, \quad R \triangleq \operatorname{diag}(2\alpha_1, 0, 0).$$

Then, $A_{\rm R}^{\rm T}P + PA_{\rm R} + R = 0$. **Remark:** The matrix $A_{\rm R}$ is a *Routh matrix*. The Routh matrix $A_{\rm R}$ and the Schwarz matrix $A_{\rm S}$ are related by $A_{\rm R} = S_{\rm RS}A_{\rm S}S_{\rm RS}^{-1}$, where

$$S_{\rm RS} \triangleq \begin{bmatrix} 0 & 0 & \alpha_1^{1/2} \\ 0 & -(\alpha_1\alpha_2)^{1/2} & 0 \\ (\alpha_1\alpha_2\alpha_3)^{1/2} & 0 & 0 \end{bmatrix}.$$

Related: Fact 15.19.25.

Fact 15.19.28. Let $\alpha_1, \alpha_2, \alpha_3 > 0$, and define $A_{\rm C}, P, R \in \mathbb{R}^{3\times 3}$ by the tridiagonal matrix

$$A_{\rm C} \triangleq \begin{bmatrix} 0 & 1/a_3 & 0 \\ -1/a_2 & 0 & 1/a_2 \\ 0 & -1/a_1 & -1/a_1 \end{bmatrix}$$

and the diagonal matrices

$$P \triangleq \operatorname{diag}(a_3, a_2, a_1), \quad R \triangleq \operatorname{diag}(0, 0, 2),$$

where $a_1 \triangleq 1/\alpha_1$, $a_2 \triangleq \alpha_1/\alpha_2$, and $a_3 \triangleq \alpha_2/(\alpha_1\alpha_3)$. Then, $A_{\rm C}^{\rm T}P + PA_{\rm C} + R = 0$. **Source:** [696, p. 346]. **Remark:** The matrix $A_{\rm C}$ is a *Chen matrix*. The Schwarz matrix $A_{\rm S}$ and the Chen matrix $A_{\rm C}$ are related by $A_{\rm S} = S_{\rm SC}A_{\rm C}S_{\rm SC}^{-1}$, where

$$S_{\rm SC} \triangleq \begin{bmatrix} 1/(\alpha_1\alpha_3) & 0 & 0 \\ 0 & 1/\alpha_2 & 0 \\ 0 & 0 & 1/\alpha_1 \end{bmatrix}.$$

Remark: Schwarz, Routh, and Chen matrices are used to prove the Routh criterion. See [74, 590, 696, 2204]. **Remark:** A circuit interpretation of the Chen matrix is given in [1955].

Fact 15.19.29. Let $\alpha_1, \ldots, \alpha_n > 0$ and $\beta_1, \ldots, \beta_n > 0$, and define $A \in \mathbb{R}^{n\times n}$ by

$$A \triangleq \begin{bmatrix} -\alpha_1 & 0 & \cdots & 0 & -\beta_1 \\ \beta_2 & -\alpha_2 & \cdots & 0 & 0 \\ \vdots & \ddots & \ddots & \vdots & \vdots \\ 0 & 0 & \ddots & -\alpha_{n-1} & 0 \\ 0 & 0 & \cdots & \beta_n & -\alpha_n \end{bmatrix}.$$

Then,

$$\chi_A(s) = \prod_{i=1}^n (s + \alpha_i) + \prod_{i=1}^n \beta_i.$$

Furthermore, if

$$\cos^n \frac{\pi}{n} < \prod_{i=1}^{n} \frac{\alpha_i}{\beta_i},$$

then A is asymptotically stable. **Source:** [2497]. **Remark:** For $n = 2$, A is asymptotically stable for all positive $\alpha_1, \beta_1, \alpha_2, \beta_2$. **Remark:** This is the *secant condition*.

Fact 15.19.30. Let $A \in \mathbb{F}^{n\times n}$. Then, the following statements are equivalent:

i) A is asymptotically stable.

ii) There exist a negative-definite matrix $B \in \mathbb{F}^{n\times n}$, a skew-Hermitian matrix $C \in \mathbb{F}^{n\times n}$, and a nonsingular matrix $S \in \mathbb{F}^{n\times n}$ such that $A = B + SCS^{-1}$.

iii) There exist a negative-definite matrix $B \in \mathbb{F}^{n\times n}$, a skew-Hermitian matrix $C \in \mathbb{F}^{n\times n}$, and a nonsingular matrix $S \in \mathbb{F}^{n\times n}$ such that $A = S(B + C)S^{-1}$.

Source: [831].

Fact 15.19.31. Let $A \in \mathbb{R}^{n\times n}$ and $k \ge 2$. Then, there exist asymptotically stable matrices $A_1, \dots, A_k \in \mathbb{R}^{n\times n}$ such that $A = \sum_{i=1}^{k} A_i$ if and only if $\operatorname{tr} A < 0$. **Source:** [1508].

Fact 15.19.32. Let $A \in \mathbb{C}^{n\times n}$. Then, A is (Lyapunov stable, semistable, asymptotically stable) if and only if $A \oplus A$ is. **Source:** Use Fact 9.5.8 and $\operatorname{vec}(e^{tA}Ve^{tA^*}) = e^{t(A\oplus\overline{A})}\operatorname{vec} V$.

Fact 15.19.33. Let $A \in \mathbb{R}^{n\times n}$ and $B \in \mathbb{R}^{m\times m}$. Then, the following statements hold:

i) If A and B are (Lyapunov stable, semistable, asymptotically stable), then so is $A \oplus B$.

ii) If $A \oplus B$ is (Lyapunov stable, semistable, asymptotically stable), then so is either A or B.

Source: Fact 9.5.8.

Fact 15.19.34. Let $A \in \mathbb{R}^{n\times n}$, and assume that A is asymptotically stable. Then,

$$\int_{-\infty}^{\infty} (\jmath\omega I - A)^{-1}\, d\omega = \pi I, \qquad \int_{-\infty}^{\infty} (\omega^2 I + A^2)^{-1}\, d\omega = -\pi A^{-1},$$

$$(A \oplus A)^{-1} = \frac{-1}{2\pi}\int_{-\infty}^{\infty} (-\jmath\omega I - A)^{-1} \otimes (\jmath\omega I - A)^{-1}\, d\omega.$$

Source: Use $(\jmath\omega I - A)^{-1} + (-\jmath\omega I - A)^{-1} = -2A(\omega^2 I + A^2)^{-1}$ and Corollary 16.11.5.

Fact 15.19.35. Let $A \in \mathbb{R}^{2\times 2}$. Then, A is asymptotically stable if and only if $\operatorname{tr} A < 0$ and $\det A > 0$.

Fact 15.19.36. Let $A \in \mathbb{C}^{n\times n}$. Then, there exists a unique asymptotically stable matrix $B \in \mathbb{C}^{n\times n}$ such that $B^2 = -A$. **Source:** [2540]. **Remark:** The uniqueness of the square root of a complex matrix that has no eigenvalues in $(-\infty, 0]$ is implicitly assumed in [2541]. **Related:** Fact 7.17.21.

Fact 15.19.37. Let $A \in \mathbb{R}^{n\times n}$. Then, the following statements hold:

i) If A is semidissipative, then A is Lyapunov stable.

ii) If A is dissipative, then A is asymptotically stable.

iii) If A is Lyapunov stable and normal, then A is semidissipative.

iv) If A is asymptotically stable and normal, then A is dissipative.

v) If A is discrete-time Lyapunov stable and normal, then A is semicontractive.

Fact 15.19.38. Let $M \in \mathbb{R}^{r\times r}$, assume that M is positive definite, let $C, K \in \mathbb{R}^{r\times r}$, assume that C and K are positive semidefinite, and consider the equation

$$M\ddot{q} + C\dot{q} + Kq = 0.$$

Furthermore, define

$$A \triangleq \begin{bmatrix} 0 & I \\ -M^{-1}K & -M^{-1}C \end{bmatrix}.$$

Then, the following statements hold:

i) A is Lyapunov stable if and only if $C + K$ is positive definite.

ii) A is Lyapunov stable if and only if $\operatorname{rank}\left[\begin{smallmatrix} C \\ K \end{smallmatrix}\right] = r$.

iii) A is semistable if and only if $(M^{-1}K, C)$ is observable.

iv) A is asymptotically stable if and only if A is semistable and K is positive definite.

Source: [431]. **Related:** Fact 7.13.31.

15.20 Facts on Almost Nonnegative Matrices

Fact 15.20.1. Let $A \in \mathbb{R}^{n \times n}$. Then, e^{tA} is nonnegative for all $t \geq 0$ if and only if A is almost nonnegative. **Source:** Let $\alpha > 0$ satisfy $\alpha I + A \geq\geq 0$, and consider $e^{t(\alpha I + A)}$. See [422, p. 74], [423, p. 146], [434, 812], and [2451, p. 37].

Fact 15.20.2. Let $A \in \mathbb{R}^{n \times n}$, and assume that A is almost nonnegative. Then, e^{tA} is positive for all $t > 0$ if and only if A is irreducible. **Source:** [2418, p. 208].

Fact 15.20.3. Let $A \in \mathbb{R}^{n \times n}$, where $n \geq 2$, and assume that A is almost nonnegative. Then, the following statements are equivalent:

i) There exist $\alpha \in (0, \infty)$ and $B \in \mathbb{R}^{n \times n}$ such that $A = B - \alpha I$, $B \geq\geq 0$, and $\rho_{\max}(B) \leq \alpha$.

ii) $\operatorname{spec}(A) \subset \text{OLHP} \cup \{0\}$.

iii) $\operatorname{spec}(A) \subset \text{CLHP}$.

iv) If $\lambda \in \operatorname{spec}(A)$ is real, then $\lambda \leq 0$.

v) Every principal subdeterminant of $-A$ is nonnegative.

vi) For every diagonal, positive-definite matrix $B \in \mathbb{R}^{n \times n}$, it follows that $A - B$ is nonsingular.

Remark: A is an *N-matrix* if A is almost nonnegative and *i)*–*vi)* hold. **Related:** This result follows from Fact 6.11.13. **Example:** $A = \left[\begin{smallmatrix} 0 & 1 \\ 0 & 0 \end{smallmatrix}\right]$.

Fact 15.20.4. Let $A \in \mathbb{R}^{n \times n}$, where $n \geq 2$, and assume that A is almost nonnegative. Then, the following statements are equivalent:

i) A is a group-invertible N-matrix.

ii) A is a Lyapunov-stable N-matrix.

iii) A is a semistable N-matrix.

iv) A is Lyapunov stable.

v) A is semistable.

vi) A is an N-matrix, and there exist $\alpha \in (0, \infty)$ and $B \in \mathbb{R}^{n \times n}$ such that $A = B - \alpha I$, $B \geq\geq 0$, and $\alpha^{-1}B$ is discrete-time semistable.

vii) There exists a positive-definite matrix $P \in \mathbb{R}^{n \times n}$ such that $A^{\mathsf{T}}P + PA$ is negative semidefinite.

Furthermore, consider the following statements:

viii) There exists a positive vector $p \in \mathbb{R}^n$ such that $-Ap$ is nonnegative.

ix) There exists a nonzero nonnegative vector $p \in \mathbb{R}^n$ such that $-Ap$ is nonnegative.

Then, *viii)* $\Longrightarrow$ [*i)*–*vii)*] $\Longrightarrow$ *ix)*. **Source:** [423, pp. 152–155] and [424]. The statement [*i)*–*vii)*] $\Longrightarrow$ *ix)* is given by Fact 6.11.15. **Remark:** The converse of *viii)* $\Longrightarrow$ [*i)*–*vii)*] is false. For example, $A = \left[\begin{smallmatrix} 0 & 1 \\ 0 & -1 \end{smallmatrix}\right]$ is almost negative and semistable, but there does not exist a positive vector $p \in \mathbb{R}^2$ such that $-Ap$ is nonnegative. However, note that *viii)* holds for A^{T}, but not for either $\operatorname{diag}(A, A^{\mathsf{T}})$ or its transpose. **Remark:** A discrete-time semistable matrix is called *semiconvergent* in [423, p. 152]. **Remark:** The last statement follows from the fact that the function $V(x) = p^{\mathsf{T}}x$ is a Lyapunov function for the system $\dot{x} = -Ax$ for $x \in [0, \infty)^n$ with Lyapunov derivative $\dot{V}(x) = -A^{\mathsf{T}}p$. See [683, 1286].

Fact 15.20.5. Let $A \in \mathbb{R}^{n\times n}$, where $n \ge 2$, and assume that A is almost nonnegative. Then, the following statements are equivalent:

i) A is a nonsingular N-matrix.

ii) A is asymptotically stable.

iii) A is an asymptotically stable N-matrix.

iv) There exist $\alpha \in (0,\infty)$ and $B \in \mathbb{R}^{n\times n}$ such that $A = B - \alpha I$, $B \geq\geq 0$, and $\rho_{\max}(B) < \alpha$.

v) If $\lambda \in \operatorname{spec}(A)$ is real, then $\lambda < 0$.

vi) If $B \in \mathbb{R}^{n\times n}$ is nonnegative and diagonal, then $A - B$ is nonsingular.

vii) Every principal subdeterminant of $-A$ is positive.

viii) Every leading principal subdeterminant of $-A$ is positive.

ix) For all $i \in \{1,\dots,n\}$, the sign of the ith leading principal subdeterminant of A is $(-1)^i$.

x) For all $k \in \{1,\dots,n\}$, the sum of all $k\times k$ principal subdeterminants of $-A$ is positive.

xi) There exists a positive-definite matrix $P \in \mathbb{R}^{n\times n}$ such that $A^{\mathsf{T}}P + PA$ is negative definite.

xii) There exists a positive vector $p \in \mathbb{R}^n$ such that $-Ap$ is positive.

xiii) There exists a nonnegative vector $p \in \mathbb{R}^n$ such that $-Ap$ is positive.

xiv) If $p \in \mathbb{R}^n$ and $-Ap$ is nonnegative, then $p \geq\geq 0$ is nonnegative.

xv) For every nonnegative vector $y \in \mathbb{R}^n$, there exists a unique nonnegative vector $x \in \mathbb{R}^n$ such that $Ax = -y$.

xvi) A is nonsingular and $-A^{-1}$ is nonnegative.

Source: [422, pp. 134–140] and [1450, pp. 114–116]. **Remark:** $-A$ is a nonsingular M-matrix. See Fact 6.11.13.

Fact 15.20.6. For all $i,j \in \{1,\dots,n\}$, let $\sigma_{ij} \in [0,\infty)$, and define $A \in \mathbb{R}^{n\times n}$ by $A_{(i,j)} \triangleq \sigma_{ij}$ for all $i \ne j$ and $A_{(i,i)} \triangleq -\sum_{j=1}^n \sigma_{ij}$. Then, the following statements hold:

i) A is almost nonnegative.

ii) $-A1_{n\times 1} = [\sigma_{11} \ \cdots \ \sigma_{nn}]^{\mathsf{T}}$ is nonnegative.

iii) $\operatorname{spec}(A) \subset \mathrm{OLHP} \cup \{0\}$.

iv) A is an N-matrix.

v) A is a group-invertible N-matrix.

vi) A is a Lyapunov-stable N-matrix.

vii) A is a semistable N-matrix.

If, in addition, $\sigma_{11},\dots,\sigma_{nn}$ are positive, then A is a nonsingular N-matrix. **Source:** It follows from the Gershgorin circle theorem given by Fact 6.10.22 that every eigenvalue λ of A is an element of a disk in $\mathbb{C}$ centered at $-\sum_{j=1}^n \sigma_{ij} \le 0$ and with radius $\sum_{j=1,j\ne i}^n \sigma_{ij}$. Hence, if $\sigma_{ii} = 0$, then either $\lambda = 0$ or $\operatorname{Re}\lambda < 0$, whereas, if $\sigma_{ii} > 0$, then $\operatorname{Re}\lambda \le \sigma_{ii} < 0$. Thus, *iii)* holds. *iv)*–*vii)* follow from *ii)* and Fact 15.20.4. The last statement follows from the Gershgorin circle theorem. **Remark:** A^{T} is a *compartmental matrix*. See [434, 1288, 2809]. **Problem:** Determine necessary and sufficient conditions on the parameters σ_{ij} such that A is a nonsingular N-matrix.

Fact 15.20.7. Let $\mathcal{G} = (\mathcal{X},\mathcal{R})$ be a directed graph, where $\mathcal{X} = \{x_1,\dots,x_n\}$, and let $L \in \mathbb{R}^{n\times n}$ denote either the inbound Laplacian or the outbound Laplacian of $\mathcal{G}$. Then, the following statements hold:

i) $-L$ is semistable.

ii) $\lim_{t\to\infty} e^{-Lt}$ exists.

Remark: Fact 15.20.6. **Remark:** The spectrum of the Laplacian is discussed in [16].

Fact 15.20.8. Let $A \in \mathbb{R}^{n\times n}$, and assume that A is asymptotically stable. Then, at least one of the following statements holds:

i) All of the diagonal entries of A are negative.

ii) At least one diagonal entry of A is negative and at least one off-diagonal entry of A is negative.

Source: [1080]. **Remark:** *sign stability* is discussed in [1525].

15.21 Facts on Discrete-Time-Stable Polynomials

Fact 15.21.1. Let $p \in \mathbb{R}[s]$, where $p(s) = s^n + a_{n-1}s^{n-1} + \cdots + a_0$, and assume that p is discrete-time asymptotically stable. Then, the following statements hold:

i) $p(1) > 0$.

ii) If n is even, then $p(-1) > 0$.

iii) If n is odd, then $p(-1) < 0$.

Fact 15.21.2. Let $p \in \mathbb{R}[s]$, where $p(s) = s^n + a_{n-1}s^{n-1} + \cdots + a_0$. Then, the following statements hold:

i) If $n = 1$, then p is discrete-time asymptotically stable if and only if $|a_0| < 1$.

ii) If $n = 2$, then p is discrete-time asymptotically stable if and only if $|a_0| < 1$ and $|a_1| < 1 + a_0$.

iii) If $n = 3$, then the following conditions are equivalent:

a) p is discrete-time asymptotically stable.

b) $|a_0 + a_2| < |1 + a_1|$, and $|a_1 - a_0a_2| < 1 - a_0^2$.

c) $|a_0 + a_2| < 1 + a_1$, and $|a_1 - a_0a_2| < 1 - a_0^2$.

d) $|a_0 + a_2| < 1 + a_1$, $|3a_0 - a_2| < 3 - a_1$, and $a_0^2 + a_1 - a_0a_2 < 1$.

iv) If $n = 4$, then the following conditions are equivalent:

a) p is discrete-time asymptotically stable.

b) $|a_0| < 1$, $|a_1 + a_3| < 1 + a_0 + a_2$, $(a_0a_3 - a_1)(a_1 + a_3) + (1 - a_0^2)(1 + a_0 + a_2) > 0$, and $(a_0a_3 - a_1)(a_1 - a_3) + (1 - a_0)^2(1 + a_0 - a_2) > 0$.

c) $|a_1 - a_0a_3| < 1 - a_0^2$, $[(a_0a_3 - a_1)(a_1 + a_3) + (1 - a_0^2)(1 + a_0 + a_2)][(a_0a_3 - a_1)(a_1 - a_3) + (1 - a_0)^2(1 + a_0 - a_2)] > 0$, $|(a_3 - a_0a_1)(1 + a_0) - a_2(a_1 - a_0a_3)| < |(a_0a_3 - a_1)(a_1 + a_3) + (1 - a_0^2)(1 + a_0 + a_2)|$.

Source: [288, p. 185], [622, p. 6], [1399, p. 355], and [1562, pp. 34, 35]. **Remark:** These conditions are the *Schur-Cohn criterion* and *Jury criterion*. **Remark:** The Routh criterion for the asymptotic stability of a polynomial is given by Fact 15.18.3. **Problem:** Show that the inequalities *b*), *c*), and *d*) for $n = 3$ equivalent and that the inequalities *b*) and *c*) for $n = 4$ equivalent.

Fact 15.21.3. Let $p \in \mathbb{C}[s]$, where $p(s) = s^n + a_{n-1}s^{n-1} + \cdots + a_0$, and define $\hat{p} \in \mathbb{C}[s]$ by

$$\hat{p}(s) \triangleq s^{n-1} + \frac{a_{n-1} - a_0\overline{a}_1}{1 - |a_0|^2}s^{n-1} + \frac{a_{n-2} - a_0\overline{a}_2}{1 - |a_0|^2}s^{n-2} + \cdots + \frac{a_1 - a_0\overline{a}_{n-1}}{1 - |a_0|^2}.$$

Then, p is discrete-time asymptotically stable if and only if $|a_0| < 1$ and $\hat{p}$ is discrete-time asymptotically stable. **Source:** [1399, p. 354].

Fact 15.21.4. Let $p \in \mathbb{R}[s]$, where $p(s) = s^n + a_{n-1}s^{n-1} + \cdots + a_0$. Then, the following statements hold:

i) If $a_0 \le \cdots \le a_{n-1} \le 1$, then $\operatorname{roots}(p) \subset \{z \in \mathbb{C}\colon |z| \le 1 + |a_0| - a_0\}$.

ii) If $0 < a_0 \le \cdots \le a_{n-1} \le 1$, then $\operatorname{roots}(p) \subset \mathrm{CIUD}$.

iii) If $0 < a_0 < \cdots < a_{n-1} < 1$, then p is discrete-time asymptotically stable.

Source: For *i*), see [2427]. For *ii*), see [2046, p. 272]. For *iii*), use Fact 15.21.3. See [1399, p.

355]. **Remark:** If there exists $r > 0$ such that $0 < ra_0 < \cdots < r^{n-1}a_{n-1} < r^n$, then $\operatorname{roots}(p) \subset \{z \in \mathbb{C}\colon |z| \le r\}$. **Remark:** *ii)* is the *Enestrom-Kakeya theorem.*

Fact 15.21.5. Let $p \in \mathbb{C}[s]$, where $p(s) = s^n + a_{n-1}s^{n-1} + \cdots + a_0$, and assume that $a_0, \ldots, a_{n-1}$ are nonzero. Then,

$$\rho_{\max}(p) \le \max\{2|a_{n-1}|, 2|a_{n-2}/a_{n-1}|, \ldots, 2|a_1/a_2|, |a_0/a_1|\}.$$

Credit: N. Bourbaki. See [2047]. This is *Kojima's bound.* See [2979, pp. 222, 223].

Fact 15.21.6. Let $p \in \mathbb{R}[s]$, where $p(s) = a_n s^n + a_{n-1}s^{n-1} + \cdots + a_0$, assume that $a_0, \ldots, a_n$ are positive, and let $\lambda \in \operatorname{roots}(p)$. Then,

$$\min_{1\le i\le n} \frac{a_{i-1}}{a_i} \le |\lambda| \le \max_{1\le i\le n} \frac{a_{i-1}}{a_i}.$$

Source: [2979, p. 225]. **Related:** This result implies *ii)* of Fact 15.21.4.

Fact 15.21.7. Let $p \in \mathbb{R}[s]$, where $p(s) = a_n s^n + a_{n-1}s^{n-1} + \cdots + a_0$, assume that $a_0, \ldots, a_n$ are nonzero, and let $\lambda \in \operatorname{roots}(p)$. Then,

$$\frac{3}{2}\min_{1\le i\le n}\left(\frac{2^n\binom{n}{i}F_i}{F_{4i}}\left|\frac{a_0}{a_i}\right|\right)^{1/i} \le |\lambda| \le \frac{2}{3}\max_{1\le i\le n}\left(\frac{F_{4i}}{2^n\binom{n}{i}F_i}\left|\frac{a_{n-i}}{a_n}\right|\right)^{1/i}.$$

Source: [885]. **Remark:** Extensions are given in [485]. **Remark:** F_i is a Fibonacci number.

Fact 15.21.8. Let $p \in \mathbb{C}[s]$, where $p(s) = s^n + a_{n-1}s^{n-1} + \cdots + a_0$, assume that $a_0, \ldots, a_{n-1}$ are nonzero, and let $\lambda \in \operatorname{roots}(p)$. Then,

$$|\lambda| \le \sum_{i=1}^{n-1} |a_i|^{1/(n-i)}, \qquad |\lambda + \tfrac{1}{2}a_{n-1}| \le \tfrac{1}{2}|a_{n-1}| + \sum_{i=0}^{n-2} |a_i|^{1/(n-i)}.$$

Credit: J. L. Walsh. See [2047].

Fact 15.21.9. Let $p \in \mathbb{C}[s]$, where $p(s) = s^n + a_{n-1}s^{n-1} + \cdots + a_0$, and let $\lambda \in \operatorname{roots}(p)$. Then,

$$\frac{|a_0|}{|a_0| + \max\{|a_1|, \ldots, |a_{n-1}|, 1\}} < |\lambda| \le \max\{|a_0|, 1 + |a_1|, \ldots, 1 + |a_{n-1}|\}.$$

Source: The lower bound is given in [2047], while the upper bound is given in [882] and [2979, p. 222]. **Remark:** The upper bound is *Cauchy's estimate.* **Remark:** The weaker upper bound $|\lambda| < 1 + \max_{i=0,\ldots,n-1}|a_i|$ is given in [288, p. 184] and [2047].

Fact 15.21.10. Let $p \in \mathbb{C}[s]$, where $p(s) = s^n + a_{n-1}s^{n-1} + \cdots + a_0$, and define $\alpha \triangleq \max_{i=0,\ldots,n-1}|a_i|$. Then,

$$\rho_{\max}(p) \le \begin{cases} \sqrt[n]{\dfrac{\alpha(1-n\alpha)}{1-(n\alpha)^{1/n}}}, & \alpha \le 1/n,\\[2ex] \min\left\{(1+\alpha)\left(1 - \dfrac{\alpha}{(1+\alpha)^{n+1} - n\alpha}\right), 1 + \dfrac{2(n\alpha - 1)}{n+1}\right\}, & 1/n \le \alpha. \end{cases}$$

Source: [500]. **Remark:** Extensions are given in [1513].

Fact 15.21.11. Let $p \in \mathbb{C}[s]$, where $p(s) = s^n + a_{n-1}s^{n-1} + \cdots + a_0$. Then,

$$\rho_{\max}(p) \le \sqrt{2 + \max_{i=0,\ldots,n-2}|a_i|^2 + |a_{n-1}|^2},$$

$$\rho_{\max}(p) \le \tfrac{1}{2}(1 + |a_{n-1}|) + \sqrt{\max_{i=0,\ldots,n-2}|a_i| + \tfrac{1}{4}(1 - |a_{n-1}|)^2},$$

$$\rho_{\max}(p) \le \max\{2, |a_0| + |a_{n-1}|, |a_1| + |a_{n-1}|, \ldots, |a_{n-2}| + |a_{n-1}|\}.$$

Source: [882]. **Credit:** The second inequality is due to A. Joyal, G. Labelle, and Q. I. Rahman. See [2047].

Fact 15.21.12. Let $p \in \mathbb{C}[s]$, where $p(s) = s^n + a_{n-1}s^{n-1} + \cdots + a_0$, assume that $a_0, \ldots, a_{n-1}$ are nonzero, define

$$\alpha \triangleq \max\left\{\left|\frac{a_0}{a_1}\right|, \left|\frac{a_1}{a_2}\right|, \ldots, \left|\frac{a_{n-2}}{a_{n-1}}\right|\right\}, \quad \beta \triangleq \max\left\{\left|\frac{a_1}{a_2}\right|, \left|\frac{a_2}{a_3}\right|, \ldots, \left|\frac{a_{n-2}}{a_{n-1}}\right|\right\},$$

and let $\lambda \in \operatorname{roots}(p)$. Then,

$$|\lambda| \le \tfrac{1}{2}(\beta + |a_{n-1}|) + \sqrt{\alpha|a_{n-1}| + \tfrac{1}{4}(\beta - |a_{n-1}|)^2}, \quad |\lambda| \le |a_{n-1}| + \alpha,$$

$$|\lambda| \le \max\left\{\left|\frac{a_0}{a_1}\right|, 2\beta, 2|a_{n-1}|\right\}, \quad |\lambda| \le 2 \max_{i=0,\ldots,n-1} |a_i|^{1/(n-i)}, \quad |\lambda| \le \sqrt{2|a_{n-1}|^2 + \alpha^2 + \beta^2}.$$

Source: [882, 1857]. **Remark:** The third inequality is *Kojima's bound*, while the fourth inequality is *Fujiwara's bound*.

Fact 15.21.13. Let $p \in \mathbb{C}[s]$, where $p(s) = s^n + a_{n-1}s^{n-1} + \cdots + a_0$, define $\alpha \triangleq 1 + \sum_{i=0}^{n-1} |a_i|^2$, and let $\lambda \in \operatorname{roots}(p)$. Then,

$$|\lambda| \le \frac{1}{n}|a_{n-1}| + \sqrt{\frac{n}{n-1}\left(n - 1 + \sum_{i=0}^{n-1} |a_i|^2 - \frac{1}{n}|a_{n-1}|^2\right)},$$

$$|\lambda| \le \frac{1}{2}\left(|a_{n-1}| + 1 + \sqrt{(|a_{n-1}| - 1)^2 + 4\sqrt{\sum_{i=0}^{n-2} |a_i|^2}}\right),$$

$$|\lambda| \le \frac{1}{2}\left(|a_{n-1}| + \cos\frac{\pi}{n} + \sqrt{\left(|a_{n-1}| - \cos\frac{\pi}{n}\right)^2 + (|a_{n-2}| + 1)^2 + \sum_{i=0}^{n-3} |a_i|^2}\right),$$

$$|\lambda| \le \cos\frac{\pi}{n+1} + \frac{1}{2}\left(|a_{n-1}| + \sqrt{\sum_{i=0}^{n-1} |a_i|^2}\right),$$

$$\sqrt{\tfrac{1}{2}\left(\alpha - \sqrt{\alpha^2 - 4|a_0|^2}\right)} \le |\lambda| \le \sqrt{\tfrac{1}{2}\left(\alpha + \sqrt{\alpha^2 - 4|a_0|^2}\right)},$$

$$|\operatorname{Re}\lambda| \le \frac{1}{2}\left(|\operatorname{Re} a_{n-1}| + \cos\frac{\pi}{n} + \sqrt{\left(|\operatorname{Re} a_{n-1}| - \cos\frac{\pi}{n}\right)^2 + (|a_{n-2}| - 1)^2 + \sum_{i=0}^{n-3} |a_i|^2}\right),$$

$$|\operatorname{Im}\lambda| \le \frac{1}{2}\left(|\operatorname{Im} a_{n-1}| + \cos\frac{\pi}{n} + \sqrt{\left(|\operatorname{Im} a_{n-1}| - \cos\frac{\pi}{n}\right)^2 + (|a_{n-2}| + 1)^2 + \sum_{i=0}^{n-3} |a_i|^2}\right).$$

Source: [1090, 1633, 1638, 1857]. **Remark:** The Parodi bound refines the Carmichael-Mason bound. See Fact 15.21.14. **Credit:** The first bound is due to H. Linden (see [1638]), the fourth bound is due to M. Fujii and F. Kubo, and the upper bound in the fifth string, which follows from Fact 7.12.24 and Fact 7.12.34, is due to M. Parodi, see [2047]. See also [1601, 1628].

Fact 15.21.14. Let $p \in \mathbb{C}[s]$, where $p(s) = s^n + a_{n-1}s^{n-1} + \cdots + a_0$, let $r, q \in (1, \infty)$, assume that $1/r + 1/q = 1$, and define $\alpha \triangleq (\sum_{i=0}^{n-1} |a_i|^r)^{1/r}$. Then,

$$\rho_{\max}(p) \le (1 + \alpha^q)^{1/q}.$$

In particular, if $r = q = 2$, then

$$\rho_{\max}(p) \le \sqrt{1 + |a_{n-1}|^2 + \cdots + |a_0|^2}.$$

Furthermore, for $r \to 1$ and $q \to \infty$,

$$\rho_{\max}(p) \le \max\{1, |a_0| + \cdots + |a_{n-1}|\}.$$

Source: [1857, 2047]. For the last inequality, see [2979, p. 222]. **Credit:** The case $r = q = 2$ is due to R. D. Carmichael and T. E. Mason. The last inequality is *Montel's bound.*

Fact 15.21.15. Let $p \in \mathbb{C}[s]$, where $p(s) = s^n + a_{n-1}s^{n-1} + \cdots + a_0$, let $m \in \{1, \ldots, n\}$, and define $\mathcal{S}_m \triangleq \{z \in \mathbb{C}\colon |z| < 1 + \max_{i \in \{1,\ldots,m-1\}} |a_i|^{1/(n-m+1)}\}$. Then, at least m roots of p are contained in $\mathcal{S}_m$. In particular,

$$\rho_{\max}(p) < 1 + \max\{|a_0|, \ldots, |a_{n-1}|\}.$$

Source: [1954, pp. 123, 151]. **Related:** Fact 15.21.14.

Fact 15.21.16. Let $p \in \mathbb{C}[s]$, where $p(s) = s^n + a_{n-1}s^{n-1} + \cdots + a_0$, let $m \in \{1, \ldots, n\}$, let $r \ge 1$, and define $\mathcal{S}_{m,r} \triangleq \{z \in \mathbb{C}\colon |z| \le 1 + (\sum_{i=0}^{m-1} |a_i|^r)^{1/r}\}$. Then, at least m roots of p are contained in $\mathcal{S}_{m,r}$. In particular,

$$\rho_{\max}(p) \le 1 + \sqrt{|a_{n-1}|^2 + \cdots + |a_0|^2}.$$

Source: [1954, p. 151] and [1978]. **Related:** Fact 15.21.14.

Fact 15.21.17. Let $p \in \mathbb{C}[s]$, where $p(s) = s^n + a_{n-1}s^{n-1} + \cdots + a_0$. Then,

$$\rho_{\max}(p) \le \sqrt{1 + |1 - a_{n-1}|^2 + |a_{n-1} - a_{n-2}|^2 + \cdots + |a_1 - a_0|^2 + |a_0|^2}.$$

Source: [1978, 2965]. **Credit:** K. P. Williams.

Fact 15.21.18. Let $p \in \mathbb{C}[s]$, where $p(s) = s^n + a_{n-1}s^{n-1} + \cdots + a_0$, let $\operatorname{mroots}(p) = \{\lambda_1, \ldots, \lambda_n\}_{\mathrm{ms}}$, and let $r > 0$ be the unique positive root of $\hat{p}(s) \triangleq s^n - |a_{n-1}|s^{n-1} - \cdots - |a_0|$. Then,

$$(\sqrt[n]{2} - 1)r \le \max_{i \in \{1,\ldots,n\}} |\lambda_i| \le r, \quad (\sqrt[n]{2} - 1)r \le \frac{1}{n}\sum_{i=1}^{n} |\lambda_i| < r.$$

Finally, the third inequality is an equality if and only if $\lambda_1 = \cdots = \lambda_n$. **Credit:** The first inequality is due to A. Cohn, the second inequality is due to A. L. Cauchy, and the third and fourth inequalities are due to L. Berwald. See [2046, p. 245] and [2047].

Fact 15.21.19. Let $p \in \mathbb{C}[s]$, where $p(s) = s^n + a_{n-1}s^{n-1} + \cdots + a_0$, define $\alpha \triangleq 1 + \sum_{i=0}^{n-1} |a_i|^2$, and let $\lambda \in \operatorname{roots}(p)$. Then,

$$\sqrt{\tfrac{1}{2}\left(\alpha - \sqrt{\alpha^2 - 4|a_0|^2}\right)} \le |\lambda| \le \sqrt{\tfrac{1}{2}\left(\alpha + \sqrt{\alpha^2 - 4|a_0|^2}\right)}.$$

Source: Fact 7.12.33, Fact 7.12.34, and [1634].

Fact 15.21.20. Let $p \in \mathbb{C}[s]$, where $p(s) = a_n s^n + a_{n-1}s^{n-1} + \cdots + a_0$, and assume that $a_1, \ldots, a_n$ are nonzero. Then, for all $i \in \{1, \ldots, n\}$,

$$\left\{z \in \operatorname{roots}(p)\colon |z| \le \left[\binom{n}{i}\left|\frac{a_0}{a_i}\right|\right]^{1/i}\right\} \ne \emptyset.$$

Source: [205].

Fact 15.21.21. Let $n \ge 2$, let $p \in \mathbb{C}[s]$, where $p(s) = a_n s^n + a_1 s + a_0$, and assume that a_1 and a_n are nonzero. Then,

$$\left\{z \in \operatorname{roots}(p)\colon \left|z + \frac{a_0}{a_1}\right| \le \left|\frac{a_0}{a_1}\right|\right\} \ne \emptyset, \quad \left\{z \in \operatorname{roots}(p)\colon \left|\frac{a_0}{a_1}\right| \le \left|z + \frac{a_0}{a_1}\right|\right\} \ne \emptyset.$$

Consequently, $\{z \in \operatorname{roots}(p)\colon |z| \le 2|a_0/a_1|\} \ne \varnothing$. **Source:** [205]. **Remark:** p is a trinomial. **Credit:** L. Fejér.

Fact 15.21.22. Let $n \ge 1$, $0 \le m < n$, $p \in \mathbb{C}[s]$, where $p(s) = a_n s^n + a_m s^m + a_1 s + a_0$, and assume that a_1, a_m, and a_n are nonzero. Then,

$$\left\{z \in \operatorname{roots}(p)\colon |z| \le \frac{17}{3}\left|\frac{a_0}{a_1}\right|\right\} \ne \varnothing.$$

Source: [205]. **Remark:** p is a quadrinomial.

Fact 15.21.23. Let $n \ge 3$, $2 \le m < n$, $p \in \mathbb{C}[s]$, where $p(s) = a_n s^n + a_m s^m + a_1 s + a_0$, and assume that a_1, a_m, and a_n are nonzero. Then,

$$\left\{z \in \operatorname{roots}(p)\colon |z| \le \frac{2n}{n-1}\left|\frac{a_0}{a_1}\right|\right\} \ne \varnothing.$$

Source: [205]. **Remark:** $2n/(n-1) \le 3$.

Fact 15.21.24. Let $n \ge 3$, $1 \le m \le n-2$, $p \in \mathbb{C}[s]$, where $p(s) = s^n + a_{n-1}s^{n-1} + a_m s^m + a_0$, and assume that a_0, a_m, and a_{n-1} are nonzero. Then,

$$\left\{z \in \operatorname{roots}(p)\colon \frac{n-1}{2n}|a_{n-1}| \le |z|\right\} \ne \varnothing.$$

Source: [205].

Fact 15.21.25. Let $n \ge 3$, let $p \in \mathbb{C}[s]$, where $p(s) = a_n s^n + a_2 s^2 + a_1 s + a_0$, and assume that a_2 and a_n are nonzero. Then,

$$\left\{z \in \operatorname{roots}(p)\colon |z| \le \left(\frac{n}{n-2}\left|\frac{a_0}{a_2}\right|\right)^{1/2}\right\} \ne \varnothing, \quad \left\{z \in \operatorname{roots}(p)\colon \left|\frac{a_1}{2a_2}\right| \le \left|z + \frac{a_1}{2a_2}\right|\right\} \ne \varnothing.$$

Source: [205].

Fact 15.21.26. Let $p \in \mathbb{R}[s]$, where $p(s) = s^n + a_{n-1}s^{n-1} + \cdots + a_0$, assume that $a_0, \ldots, a_{n-1}$ are nonnegative, and let $x_1, \ldots, x_m \in [0, \infty)$. Then,

$$p(\sqrt[m]{x_1 \cdots x_m}) \le \sqrt[m]{p(x_1) \cdots p(x_m)}.$$

Source: [2129]. **Remark:** This result extends a result of C. Huygens for the case $p(x) = x + 1$. **Credit:** D. Mihet.

Fact 15.21.27. Let $p \in \mathbb{R}[s]$, where $p(s) = s^n + a_{n-1}s^{n-1} + \cdots + a_0$, and assume that $\operatorname{mroots}(p) = \{\lambda_1, \ldots, \lambda_n\} \subset \mathbb{R}$, where $\lambda_n \le \cdots \le \lambda_1$. Then, $2na_{n-2} \le (n-1)a_{n-1}^2$, and

$$\frac{2(n-1)a_{n-1}^2 - 4na_{n-2}}{n(n-1)} \le (\lambda_n - \lambda_1)^2 \le \frac{2(n-1)a_{n-1}^2 - 4na_{n-2}}{n}.$$

Source: [400]. **Related:** Fact 15.21.28.

Fact 15.21.28. Let $p \in \mathbb{R}[s]$, where $p(s) = s^n + a_{n-1}s^{n-1} + \cdots + a_0$, and assume that at least one of the following statements holds:

i) $(n-1)a_{n-1}^2 < 2na_{n-2}$.

ii) $a_{n-1}^2 < 2a_{n-2} + na_0^{2/n}$.

iii) $a_0 \ne 0$ and $(n-1)a_1^2 < 2na_2a_0$.

iv) $a_0 \ne 0$ and $a_1^2 < 2a_2a_0 + na_0^{(2n-2)/n}$.

Then, p has at least two roots that are not real. **Source:** [2680]. **Related:** Fact 15.18.4, Fact 15.21.27, and Fact 15.21.29.

Fact 15.21.29. Let $p \in \mathbb{R}[s]$, where $p(s) = as^3 + bs^2 + cs + d$, and assume that $a \neq 0$, and assume that at least one of the following conditions is satisfied:

i) $b^2 < 3ac$.

ii) $b^2 < 2ac + 3\sqrt[3]{a^4d^2}$.

iii) $d \neq 0$ and $c^2 < 3bd$.

iv) $d \neq 0$ and $c^2 < 2bd + 3\sqrt[3]{a^2d^4}$.

v) $d \neq 0$ and $b^2c^2 + 18abcd < 4b^3d + 4ac^3 + 27a^2d^2$.

Then, p has exactly one real root. **Source:** [2680]. **Related:** Fact 2.21.2, Fact 6.10.7, Fact 15.18.4, and Fact 15.21.28.

Fact 15.21.30. Let $p \in \mathbb{C}[s]$, where $p(s) = s^n + a_{n-1}s^{n-1} + \cdots + a_0$, and let α and β be the largest nonnegative roots, respectively, of the polynomials

$$q(s) \triangleq s^n - |a_{n-1}|s^{n-1} - |a_{n-2}|s^{n-2} - \cdots - |a_1|s - |a_0|,$$
$$r(s) \triangleq s^n + |a_{n-1}|s^{n-1} - |a_{n-2}|s^{n-2} - \cdots - |a_1|s - |a_0|.$$

Then,

$$\operatorname{roots}(p) \subset \mathbb{B}_\alpha(0) \cap \mathbb{B}_{\beta+|a_{n-1}|}(-a_{n-1}).$$

Source: [2009]. **Remark:** This result extends a result due to A. L. Cauchy.

Fact 15.21.31. Let $p \in \mathbb{C}[s]$, where $p(s) = a_n s^n + a_{n-1}s^{n-1} + \cdots + a_0$, and assume that there exists $m \in \{0, \ldots, n-1\}$ such that $|a_0|\binom{n}{m} < |a_{n-m}|$. Then, there exists $\lambda \in \operatorname{roots}(p)$ such that $|\lambda| < 1$. **Source:** [108, pp. 49, 280].

Fact 15.21.32. Let $p, q \in \mathbb{C}[s]$, assume that p and q are monic, $\deg p = \deg q = n$, $\rho_{\max}(p) \le 1$, and $\rho_{\max}(q) \le 1$, and let $\alpha \in [0, 1]$. Then, $\rho_{\max}[\alpha p + (1-\alpha)q] \le \cot \frac{\pi}{2n}$. **Source:** [1029].

Fact 15.21.33. Let $p, q \in \mathbb{R}[s]$, assume that the leading coefficients of p and q are positive, assume that p and q have no roots in ORHP, and define $\theta \triangleq \pi/\max\{2, \deg p, \deg q\}$. Then, $p + q$ has no roots in $\{z \in \mathbb{C}\colon |\arg z| < \theta\}$. Now, let $n \ge 2$, let $p_1, \ldots, p_n \in \mathbb{R}[s]$, assume that, for all $i \in \{1, \ldots, n\}$, the leading coefficient of p_i is positive, p_i has no roots in ORHP, and $\deg p_i \le n$. Then, $\sum_{i=1}^n p_i$ has no roots in $\{z \in \mathbb{C}\colon |\arg z| < \pi/n\}$. **Source:** [1440].

15.22 Facts on Discrete-Time-Stable Matrices

Fact 15.22.1. Let $A \in \mathbb{R}^{2\times 2}$. Then, the following statements are equivalent:

i) A is discrete-time asymptotically stable.

ii) $|\operatorname{tr} A| < 1 + \det A$ and $|\det A| < 1$.

Fact 15.22.2. Let $A \in \mathbb{R}^{n\times n}$, and assume that A is nonnegative. Then, the following statements are equivalent:

i) A is discrete-time asymptotically stable.

ii) Every principal subdeterminant of $I - A$ is positive.

Source: [353, p. 303].

Fact 15.22.3. Let $A \in \mathbb{F}^{n\times n}$. Then, A is discrete-time (Lyapunov stable, semistable, asymptotically stable) if and only if A^2 is.

Fact 15.22.4. Let $A \in \mathbb{R}^{n\times n}$, and let $\chi_A(s) = s^n + a_{n-1}s^{n-1} + \cdots + a_1 s + a_0$. Then, for all $k \ge 0$,

$$A^k = x_1(k)I + x_2(k)A + \cdots + x_n(k)A^{n-1},$$

where, for all $i \in \{1, \ldots, n\}$ and $k \ge 0$, $x_i\colon\ \mathbb{N} \mapsto \mathbb{R}$ satisfies

$$x_i(k+n) + a_{n-1}x_i(k+n-1) + \cdots + a_1x_i(k+1) + a_0x_i(k) = 0,$$

with, for all $i, j \in \{1, \ldots, n\}$, the initial conditions $x_i(j-1) = \delta_{i,j}$. **Source:** [1715].

Fact 15.22.5. Let $A \in \mathbb{R}^{n\times n}$. Then, the following statements hold:

i) If A is semicontractive, then A is discrete-time Lyapunov stable.

ii) If A is contractive, then A is discrete-time asymptotically stable.

iii) If A is discrete-time Lyapunov stable and normal, then A is semicontractive.

iv) If A is discrete-time asymptotically stable and normal, then A is contractive.

Problem: Prove these results using Fact 15.16.6.

Fact 15.22.6. Let $x \in \mathbb{F}^n$, let $A \in \mathbb{F}^{n\times n}$, and assume that A is discrete-time semistable. Then, $\sum_{i=0}^{\infty} A^i x$ exists if and only if $x \in \mathcal{R}(A - I)$. If these conditions hold, then

$$\sum_{i=0}^{\infty} A^i x = -(A - I)^{\#} x.$$

Source: [1487].

Fact 15.22.7. Let $A \in \mathbb{F}^{n\times n}$. Then, A is discrete-time (Lyapunov stable, semistable, asymptotically stable) if and only if $A \otimes A$ is. **Source:** Fact 9.4.21. **Remark:** See [1487].

Fact 15.22.8. Let $A \in \mathbb{R}^{n\times n}$ and $B \in \mathbb{R}^{m\times m}$. Then, the following statements hold:

i) If A and B are discrete-time (Lyapunov stable, semistable, asymptotically stable), then $A \otimes B$ is discrete-time (Lyapunov stable, semistable, asymptotically stable).

ii) If $A \otimes B$ is discrete-time (Lyapunov stable, semistable, asymptotically stable), then either A or B is discrete-time (Lyapunov stable, semistable, asymptotically stable).

Source: Fact 9.4.21.

Fact 15.22.9. Let $A \in \mathbb{R}^{n\times n}$, and assume that A is (Lyapunov stable, semistable, asymptotically stable). Then, e^A is discrete-time (Lyapunov stable, semistable, asymptotically stable). **Problem:** If $B \in \mathbb{R}^{n\times n}$ is discrete-time (Lyapunov stable, semistable, asymptotically stable), under what conditions does there exist a (Lyapunov-stable, semistable, asymptotically stable) matrix $A \in \mathbb{R}^{n\times n}$ such that $B = e^A$? See Proposition 15.4.5.

Fact 15.22.10. The following statements hold:

i) If $A \in \mathbb{R}^{n\times n}$ is discrete-time asymptotically stable, then $B \triangleq (A+I)^{-1}(A-I)$ is asymptotically stable.

ii) If $B \in \mathbb{R}^{n\times n}$ is asymptotically stable, then $A \triangleq (I+B)(I-B)^{-1}$ is discrete-time asymptotically stable.

iii) If $A \in \mathbb{R}^{n\times n}$ is discrete-time asymptotically stable, then there exists a unique asymptotically stable matrix $B \in \mathbb{R}^{n\times n}$ such that $A = (I + B)(I - B)^{-1}$. In fact, $B = (A + I)^{-1}(A - I)$.

iv) If $B \in \mathbb{R}^{n\times n}$ is asymptotically stable, then there exists a unique discrete-time asymptotically stable matrix $A \in \mathbb{R}^{n\times n}$ such that $B = (A + I)^{-1}(A - I)$. In fact, $A = (I + B)(I - B)^{-1}$.

Source: [1354]. **Related:** Results on the Cayley transform are given in Fact 4.13.24, Fact 4.13.25, Fact 4.13.26, Fact 4.28.12, and Fact 10.10.35. **Problem:** Obtain analogous results for Lyapunov-stable and semistable matrices.

Fact 15.22.11. Let $\begin{bmatrix} P_1 & P_{12} \\ P_{12}^{\mathsf{T}} & P_2 \end{bmatrix} \in \mathbb{R}^{2n\times 2n}$ be positive definite, where $P_1, P_{12}, P_2 \in \mathbb{R}^{n\times n}$. If $P_1 \geq P_2$, then $A \triangleq P_1^{-1}P_{12}^{\mathsf{T}}$ is discrete-time asymptotically stable, while, if $P_2 \geq P_1$, then $A \triangleq P_2^{-1}P_{12}$ is discrete-time asymptotically stable. **Source:** If $P_1 \geq P_2$, then $P_1 - P_{12}P_1^{-1}P_1P_1^{-1}P_{12}^{\mathsf{T}} \geq P_1 - P_{12}P_2^{-2}P_{12}^{\mathsf{T}} > 0$. See [744].

Fact 15.22.12. Let $A \in \mathbb{R}^{n\times n}$, where $n \geq 2$, and assume that A is row stochastic. Then, the following statements hold:

i) A is discrete-time Lyapunov stable.

ii) If A is primitive, then A is discrete-time semistable.

Source: For all $k \geq 1$, $A^k 1_{n\times 1} = 1_{n\times 1}$. Since A^k is nonnegative, it follows that every entry of A is bounded. If A is primitive, then the result follows from Fact 6.11.11, which implies that $\rho_{\max}(A) = 1$, and *viii*) and *xv*) of Fact 6.11.5, which imply that 1 is a simple eigenvalue of A as well as the unique eigenvalue of A on the unit circle.

Fact 15.22.13. Let $A \in \mathbb{R}^{n\times n}$, and let $\|\cdot\|$ be a norm on $\mathbb{R}^{n\times n}$. Then, the following statements hold:

i) A is discrete-time Lyapunov stable if and only if $\{\|A^k\|\}_{k=0}^{\infty}$ is bounded.

ii) A is discrete-time semistable if and only if $A_\infty \triangleq \lim_{k\to\infty} A^k$ exists.

iii) Assume that A is discrete-time semistable. Then, $A_\infty \triangleq I - (A - I)(A - I)^{\#}$ is idempotent and $\operatorname{rank} A_\infty = \operatorname{amult}_A(1)$. If, in addition, $\operatorname{rank} A = 1$, then, for each eigenvector x of A associated with the eigenvalue 1, there exists $y \in \mathbb{F}^n$ such that $y^* x = 1$ and $A_\infty = xy^*$.

iv) A is discrete-time asymptotically stable if and only if $\lim_{k\to\infty} A^k = 0$.

Source: For *ii*), see [2036, p. 640]. **Related:** Fact 15.22.18.

Fact 15.22.14. Let $A \in \mathbb{F}^{n\times n}$. Then, A is discrete-time Lyapunov stable if and only if

$$A_\infty \triangleq \lim_{k\to\infty} \frac{1}{k}\sum_{i=0}^{k-1} A^i$$

exists. If these conditions hold, then $A_\infty = I - (A - I)(A - I)^{\#}$. **Source:** [2036, p. 633]. **Remark:** A is *Cesaro summable*. **Related:** Fact 8.3.37.

Fact 15.22.15. Let $A \in \mathbb{F}^{n\times n}$. Then, A is discrete-time asymptotically stable if and only if $\lim_{k\to\infty} A^k = 0$. If these conditions hold, then

$$(I - A)^{-1} = \sum_{i-1}^{\infty} A^i,$$

where the series converges absolutely.

Fact 15.22.16. Let $A \in \mathbb{F}^{n\times n}$, assume that A is discrete-time asymptotically stable, let $\|\cdot\|$ be a norm on $\mathbb{F}^n$, let $x_0 \in \mathbb{C}^n$, and, for all $k \geq 0$, let $x_k \in \mathbb{C}^n$ satisfy $x_{k+1} = Ax_k$. Then, $\sum_{k=1}^{\infty} \|x_k\|$ is finite. **Source:** Fact 11.13.11.

Fact 15.22.17. Let $A \in \mathbb{F}^{n\times n}$, where A is unitary. Then, A is discrete-time Lyapunov stable.

Fact 15.22.18. Let $A, B \in \mathbb{R}^{n\times n}$, assume that A is discrete-time semistable, and let $A_\infty \triangleq \lim_{k\to\infty} A^k$. Then,

$$\lim_{k\to\infty} (A + \tfrac{1}{k}B)^k = A_\infty e^{A_\infty B A_\infty}.$$

Source: [499, 2876]. **Remark:** If A is idempotent, then $A_\infty = A$. The existence of A_∞ is guaranteed by Fact 15.22.13, which also implies that A_∞ is idempotent.

Fact 15.22.19. Let $A \in \mathbb{R}^{n\times n}$. Then, thc following statements hold:

i) A is discrete-time Lyapunov stable if and only if there exists a positive-definite matrix $P \in \mathbb{R}^{n\times n}$ such that $P - A^{\mathsf{T}}PA$ is positive semidefinite.

ii) A is discrete-time asymptotically stable if and only if there exists a positive-definite matrix $P \in \mathbb{R}^{n\times n}$ such that $P - A^{\mathsf{T}}PA$ is positive definite.

Remark: The *discrete-time Lyapunov equation* (also called the *Stein equation*) is $P = A^{\mathsf{T}}PA + R$.

Fact 15.22.20. Let $A, B, C \in \mathbb{F}^{n\times n}$, and assume that $\rho_{\max}(A\otimes B) < 1$. Then, there exists a unique matrix $X \in \mathbb{F}^{n\times n}$ such that $X = AXB + C$. In particular,

$$X = \sum_{i=0}^{\infty} A^i C B^i.$$

Source: [2991, p. 324].

Fact 15.22.21. Let $A \in \mathbb{R}^{n\times n}$, assume that A is discrete-time asymptotically stable, let $P \in \mathbb{R}^{n\times n}$ be positive definite, and assume that P satisfies $P = A^\mathsf{T} PA + I$. Then, P is given by

$$P = \frac{1}{2\pi}\int_{-\pi}^{\pi} (A^\mathsf{T} - e^{-\theta J} I)^{-1}(A - e^{\theta J} I)\, \mathrm{d}\theta.$$

Furthermore,

$$1 \le \lambda_n(P) \le \lambda_1(P) \le \frac{[\sigma_{\max}(A) + 1]^{2n-2}}{[1 - \rho_{\max}(A)]^{2n}}.$$

Source: [1328, pp. 167–169].

Fact 15.22.22. Let $(A_k)_{k=0}^{\infty} \subset \mathbb{R}^{n\times n}$ and, for $k \in \mathbb{N}$, consider the discrete-time, time-varying system

$$x(k + 1) = A_k x(k).$$

Furthermore, assume that there exist real numbers $\beta \in (0, 1)$, $\gamma > 0$, and $\varepsilon > 0$ such that $\rho(\beta^2 + \rho\varepsilon^2)^2 < 1$, where $\rho \triangleq (\gamma + 1)^{2n-2}/(1 - \beta)^{2n}$ and such that, for all $k \in \mathbb{N}$,

$$\rho_{\max}(A_k) < \beta, \quad \|A_k\| < \gamma, \quad \|A_{k+1} - A_k\| < \varepsilon,$$

where $\|\cdot\|$ is an induced norm on $\mathbb{R}^{n\times n}$. Then, $x(k) \to 0$ as $k \to \infty$. **Source:** [1328, pp. 170–173]. **Remark:** This result arises from the theory of *infinite matrix products*. See [171, 494, 495, 836, 1278, 1437, 1725].

Fact 15.22.23. Let $A \in \mathbb{F}^{n\times n}$, and define

$$r(A) \triangleq \sup_{\{z\in\mathbb{C}:\, |z|>1\}} \frac{|z| - 1}{\sigma_{\min}(zI - A)}.$$

Then,

$$r(A) \le \sup_{k\ge 0} \sigma_{\max}(A^k) \le ner(A).$$

Hence, if A is discrete-time Lyapunov stable, then $r(A)$ is finite. **Source:** [2848]. **Remark:** This is the *Kreiss matrix theorem*. **Remark:** The constant ne is the best possible. See [2848].

Fact 15.22.24. Let $A \in \mathbb{C}^{n\times n}$, let $\lambda \in \operatorname{spec}(A)$, assume that λ is simple, assume that, for all $\mu \in \operatorname{spec}(A)\backslash\{\lambda\}$, $|\mu| < |\lambda|$, and let $x, y \in \mathbb{C}^n$ be nonzero vectors such that $Ax = \lambda x$ and $A^\mathsf{T} y = \lambda y$. Then, $y^\mathsf{T} x \ne 0$ and

$$\lim_{i\to\infty} \frac{1}{\lambda^i} A^i = \frac{1}{y^\mathsf{T} x} xy^\mathsf{T}.$$

Source: [1742]. **Remark:** λ is a *DT-dominant eigenvalue*. **Related:** Fact 6.11.5 and Fact 15.14.16.

Fact 15.22.25. Let $p \in \mathbb{R}[s]$, and assume that p is discrete-time semistable. Then, $C(p)$ is discrete-time semistable, and there exists $v \in \mathbb{R}^n$ such that

$$\lim_{k\to\infty} C^k(p) = 1_{n\times 1} v^\mathsf{T}.$$

Source: Since $C(p)$ is a companion form matrix, it follows from Proposition 15.11.4 that its minimal polynomial is p. Hence, $C(p)$ is discrete-time semistable. Now, it follows from Proposition 15.11.2 that $\lim_{k\to\infty} C^k(p)$ exists, and thus the state $x(k)$ of the difference equation $x(k + 1) = C(p)x(k)$ converges for all initial conditions $x(0) = x_0$. The structure of $C(p)$ shows that all components of $x(k)$ converge to the same value as $k \to \infty$. Hence, all rows of $\lim_{k\to\infty} C^k(p)$ are equal.

15.23 Facts on Lie Groups

Fact 15.23.1. The groups $\mathrm{UT}(n), \mathrm{UT}_{+}(n), \mathrm{UT}_{\pm 1}(n), \mathrm{SUT}(n)$, and $\{I_n\}$ are Lie groups. Furthermore, $\mathrm{ut}(n)$ is the Lie algebra of $\mathrm{UT}(n)$, $\mathrm{sut}(n)$ is the Lie algebra of $\mathrm{SUT}(n)$, and $\{0_{n\times n}\}$ is the Lie algebra of $\{I_n\}$. **Related:** Fact 4.31.10 and Fact 4.31.11. **Problem:** Determine the Lie algebras of $\mathrm{UT}_{+}(n)$ and $\mathrm{UT}_{\pm 1}(n)$.

15.24 Facts on Subspace Decomposition

Fact 15.24.1. Let $A \in \mathbb{R}^{n\times n}$, and let $S \in \mathbb{R}^{n\times n}$ be a nonsingular matrix such that

$$A = S\begin{bmatrix} A_1 & A_{12} \\ 0 & A_2 \end{bmatrix} S^{-1},$$

where $A_1 \in \mathbb{R}^{r\times r}$ is asymptotically stable, $A_{12} \in \mathbb{R}^{r\times(n-r)}$, and $A_2 \in \mathbb{R}^{(n-r)\times(n-r)}$. Then,

$$\mu_A^{\mathrm{s}}(A) = S\begin{bmatrix} 0 & B_{12\mathrm{s}} \\ 0 & \mu_A^{\mathrm{s}}(A_2) \end{bmatrix} S^{-1},$$

where $B_{12\mathrm{s}} \in \mathbb{R}^{r\times(n-r)}$, and

$$\mu_A^{\mathrm{u}}(A) = S\begin{bmatrix} \mu_A^{\mathrm{u}}(A_1) & B_{12\mathrm{u}} \\ 0 & \mu_A^{\mathrm{u}}(A_2) \end{bmatrix} S^{-1},$$

where $B_{12\mathrm{u}} \in \mathbb{R}^{r\times(n-r)}$ and $\mu_A^{\mathrm{u}}(A_1)$ is nonsingular. Consequently,

$$\mathcal{R}\left(S\begin{bmatrix} I_r \\ 0 \end{bmatrix}\right) \subseteq \mathcal{S}_{\mathrm{s}}(A).$$

If, in addition, $A_{12} = 0$, then

$$\mu_A^{\mathrm{s}}(A) = S\begin{bmatrix} 0 & 0 \\ 0 & \mu_A^{\mathrm{s}}(A_2) \end{bmatrix} S^{-1}, \quad \mu_A^{\mathrm{u}}(A) = S\begin{bmatrix} \mu_A^{\mathrm{u}}(A_1) & 0 \\ 0 & \mu_A^{\mathrm{u}}(A_2) \end{bmatrix} S^{-1}, \quad \mathcal{S}_{\mathrm{u}}(A) \subseteq \mathcal{R}\left(S\begin{bmatrix} 0 \\ I_{n-r} \end{bmatrix}\right).$$

Source: Fact 6.10.17.

Fact 15.24.2. Let $A \in \mathbb{R}^{n\times n}$, and let $S \in \mathbb{R}^{n\times n}$ be a nonsingular matrix such that

$$A = S\begin{bmatrix} A_1 & A_{12} \\ 0 & A_2 \end{bmatrix} S^{-1},$$

where $A_1 \in \mathbb{R}^{r\times r}$, $A_{12} \in \mathbb{R}^{r\times(n-r)}$, and $A_2 \in \mathbb{R}^{(n-r)\times(n-r)}$ satisfies $\mathrm{spec}(A_2) \subset \mathrm{CRHP}$. Then,

$$\mu_A^{\mathrm{s}}(A) = S\begin{bmatrix} \mu_A^{\mathrm{s}}(A_1) & C_{12\mathrm{s}} \\ 0 & \mu_A^{\mathrm{s}}(A_2) \end{bmatrix} S^{-1},$$

where $C_{12\mathrm{s}} \in \mathbb{R}^{r\times(n-r)}$ and $\mu_A^{\mathrm{s}}(A_2)$ is nonsingular, and

$$\mu_A^{\mathrm{u}}(A) = S\begin{bmatrix} \mu_A^{\mathrm{u}}(A_1) & C_{12\mathrm{u}} \\ 0 & 0 \end{bmatrix} S^{-1},$$

where $C_{12\mathrm{u}} \in \mathbb{R}^{r\times(n-r)}$. Consequently,

$$\mathcal{S}_{\mathrm{s}}(A) \subseteq \mathcal{R}\left(S\begin{bmatrix} I_r \\ 0 \end{bmatrix}\right).$$

If, in addition, $A_{12} = 0$, then

$$\mu_A^{\mathrm{s}}(A) = S\begin{bmatrix} \mu_A^{\mathrm{s}}(A_1) & 0 \\ 0 & \mu_A^{\mathrm{s}}(A_2) \end{bmatrix} S^{-1}, \quad \mu_A^{\mathrm{u}}(A) = S\begin{bmatrix} \mu_A^{\mathrm{u}}(A_1) & 0 \\ 0 & 0 \end{bmatrix} S^{-1}, \quad \mathcal{R}\left(S\begin{bmatrix} 0 \\ I_{n-r} \end{bmatrix}\right) \subseteq \mathcal{S}_{\mathrm{u}}(A).$$

Fact 15.24.3. Let $A \in \mathbb{R}^{n\times n}$, and let $S \in \mathbb{R}^{n\times n}$ be a nonsingular matrix such that

$$A = S\begin{bmatrix} A_1 & A_{12} \\ 0 & A_2 \end{bmatrix}S^{-1},$$

where $A_1 \in \mathbb{R}^{r\times r}$ satisfies $\operatorname{spec}(A_1) \subset \mathrm{CRHP}$, $A_{12} \in \mathbb{R}^{r\times(n-r)}$, and $A_2 \in \mathbb{R}^{(n-r)\times(n-r)}$. Then,

$$\mu_A^{\mathrm{s}}(A) = S\begin{bmatrix} \mu_A^{\mathrm{s}}(A_1) & B_{12\mathrm{s}} \\ 0 & \mu_A^{\mathrm{s}}(A_2) \end{bmatrix}S^{-1},$$

where $\mu_A^{\mathrm{s}}(A_1)$ is nonsingular and $B_{12\mathrm{s}} \in \mathbb{R}^{r\times(n-r)}$, and

$$\mu_A^{\mathrm{u}}(A) = S\begin{bmatrix} 0 & B_{12\mathrm{u}} \\ 0 & \mu_A^{\mathrm{u}}(A_2) \end{bmatrix}S^{-1},$$

where $B_{12\mathrm{u}} \in \mathbb{R}^{r\times(n-r)}$. Consequently,

$$\mathcal{R}\left(S\begin{bmatrix} I_r \\ 0 \end{bmatrix}\right) \subseteq \mathcal{S}_{\mathrm{u}}(A).$$

If, in addition, $A_{12} = 0$, then

$$\mu_A^{\mathrm{s}}(A) = S\begin{bmatrix} \mu_A^{\mathrm{s}}(A_1) & 0 \\ 0 & \mu_A^{\mathrm{s}}(A_2) \end{bmatrix}S^{-1}, \quad \mu_A^{\mathrm{u}}(A) = S\begin{bmatrix} 0 & 0 \\ 0 & \mu_A^{\mathrm{u}}(A_2) \end{bmatrix}S^{-1}, \quad \mathcal{S}_{\mathrm{s}}(A) \subseteq \mathcal{R}\left(S\begin{bmatrix} 0 \\ I_{n-r} \end{bmatrix}\right).$$

Fact 15.24.4. Let $A \in \mathbb{R}^{n\times n}$, and let $S \in \mathbb{R}^{n\times n}$ be a nonsingular matrix such that

$$A = S\begin{bmatrix} A_1 & A_{12} \\ 0 & A_2 \end{bmatrix}S^{-1},$$

where $A_1 \in \mathbb{R}^{r\times r}$, $A_{12} \in \mathbb{R}^{r\times(n-r)}$, and $A_2 \in \mathbb{R}^{(n-r)\times(n-r)}$ is asymptotically stable. Then,

$$\mu_A^{\mathrm{s}}(A) = S\begin{bmatrix} \mu_A^{\mathrm{s}}(A_1) & C_{12\mathrm{s}} \\ 0 & 0 \end{bmatrix}S^{-1},$$

where $C_{12\mathrm{s}} \in \mathbb{R}^{r\times(n-r)}$, and

$$\mu_A^{\mathrm{u}}(A) = S\begin{bmatrix} \mu_A^{\mathrm{u}}(A_1) & C_{12\mathrm{u}} \\ 0 & \mu_A^{\mathrm{u}}(A_2) \end{bmatrix}S^{-1},$$

where $\mu_A^{\mathrm{u}}(A_2)$ is nonsingular and $C_{12\mathrm{u}} \in \mathbb{R}^{r\times(n-r)}$. Consequently,

$$\mathcal{S}_{\mathrm{u}}(A) \subseteq \mathcal{R}\left(S\begin{bmatrix} I_r \\ 0 \end{bmatrix}\right).$$

If, in addition, $A_{12} = 0$, then

$$\mu_A^{\mathrm{s}}(A) = S\begin{bmatrix} \mu_A^{\mathrm{s}}(A_1) & 0 \\ 0 & 0 \end{bmatrix}S^{-1}, \quad \mu_A^{\mathrm{u}}(A) = S\begin{bmatrix} \mu_A^{\mathrm{u}}(A_1) & 0 \\ 0 & \mu_A^{\mathrm{u}}(A_2) \end{bmatrix}S^{-1}, \quad \mathcal{R}\left(S\begin{bmatrix} 0 \\ I_{n-r} \end{bmatrix}\right) \subseteq \mathcal{S}_{\mathrm{s}}(A).$$

Fact 15.24.5. Let $A \in \mathbb{R}^{n\times n}$, and let $S \in \mathbb{R}^{n\times n}$ be a nonsingular matrix such that

$$A = S\begin{bmatrix} A_1 & A_{12} \\ 0 & A_2 \end{bmatrix}S^{-1},$$

where $A_1 \in \mathbb{R}^{r\times r}$ satisfies $\operatorname{spec}(A_1) \subset \mathrm{CRHP}$, $A_{12} \in \mathbb{R}^{r\times(n-r)}$, and $A_2 \in \mathbb{R}^{(n-r)\times(n-r)}$ is asymptotically stable. Then,

$$\mu_A^{\mathrm{s}}(A) = S\begin{bmatrix} \mu_A^{\mathrm{s}}(A_1) & C_{12\mathrm{s}} \\ 0 & 0 \end{bmatrix}S^{-1},$$

where $C_{12\mathrm{s}} \in \mathbb{R}^{r\times(n-r)}$ and $\mu_A^{\mathrm{s}}(A_1)$ is nonsingular, and

$$\mu_A^{\mathrm{u}}(A) = S\begin{bmatrix} 0 & C_{12\mathrm{u}} \\ 0 & \mu_A^{\mathrm{u}}(A_2) \end{bmatrix}S^{-1},$$

where $C_{12\mathrm{u}} \in \mathbb{R}^{r\times(n-r)}$ and $\mu_A^{\mathrm{u}}(A_2)$ is nonsingular. Consequently,

$$\mathcal{S}_{\mathrm{u}}(A) = \mathcal{R}\left(S\begin{bmatrix} I_r \\ 0 \end{bmatrix}\right).$$

If, in addition, $A_{12} = 0$, then

$$\mu_A^{\mathrm{s}}(A) = S\begin{bmatrix} \mu_A^{\mathrm{s}}(A_1) & 0 \\ 0 & 0 \end{bmatrix}S^{-1}, \quad \mu_A^{\mathrm{u}}(A) = S\begin{bmatrix} 0 & 0 \\ 0 & \mu_A^{\mathrm{u}}(A_2) \end{bmatrix}S^{-1}, \quad \mathcal{S}_{\mathrm{s}}(A) = \mathcal{R}\left(S\begin{bmatrix} 0 \\ I_{n-r} \end{bmatrix}\right).$$

Fact 15.24.6. Let $A \in \mathbb{R}^{n\times n}$, and let $S \in \mathbb{R}^{n\times n}$ be a nonsingular matrix such that

$$A = S\begin{bmatrix} A_1 & 0 \\ A_{21} & A_2 \end{bmatrix}S^{-1},$$

where $A_1 \in \mathbb{R}^{r\times r}$ is asymptotically stable, $A_{21} \in \mathbb{R}^{(n-r)\times r}$, and $A_2 \in \mathbb{R}^{(n-r)\times(n-r)}$. Then,

$$\mu_A^{\mathrm{s}}(A) = S\begin{bmatrix} 0 & 0 \\ B_{21\mathrm{s}} & \mu_A^{\mathrm{s}}(A_2) \end{bmatrix}S^{-1},$$

where $B_{21\mathrm{s}} \in \mathbb{R}^{(n-r)\times r}$, and

$$\mu_A^{\mathrm{u}}(A) = S\begin{bmatrix} \mu_A^{\mathrm{u}}(A_1) & 0 \\ B_{21\mathrm{u}} & \mu_A^{\mathrm{u}}(A_2) \end{bmatrix}S^{-1},$$

where $B_{21\mathrm{u}} \in \mathbb{R}^{(n-r)\times r}$ and $\mu_A^{\mathrm{u}}(A_1)$ is nonsingular. Consequently,

$$\mathcal{S}_{\mathrm{u}}(A) \subseteq \mathcal{R}\left(S\begin{bmatrix} 0 \\ I_{n-r} \end{bmatrix}\right).$$

If, in addition, $A_{21} = 0$, then

$$\mu_A^{\mathrm{s}}(A) = S\begin{bmatrix} 0 & 0 \\ 0 & \mu_A^{\mathrm{s}}(A_2) \end{bmatrix}S^{-1}, \quad \mu_A^{\mathrm{u}}(A) = S\begin{bmatrix} \mu_A^{\mathrm{u}}(A_1) & 0 \\ 0 & \mu_A^{\mathrm{u}}(A_2) \end{bmatrix}S^{-1}, \quad \mathcal{R}\left(S\begin{bmatrix} I_r \\ 0 \end{bmatrix}\right) \subseteq \mathcal{S}_{\mathrm{s}}(A).$$

Fact 15.24.7. Let $A \in \mathbb{R}^{n\times n}$, and let $S \in \mathbb{R}^{n\times n}$ be a nonsingular matrix such that

$$A = S\begin{bmatrix} A_1 & 0 \\ A_{21} & A_2 \end{bmatrix}S^{-1},$$

where $A_1 \in \mathbb{R}^{r\times r}$, $A_{21} \in \mathbb{R}^{(n-r)\times r}$, and $A_2 \in \mathbb{R}^{(n-r)\times(n-r)}$ satisfies $\operatorname{spec}(A_2) \subset \mathrm{CRHP}$. Then,

$$\mu_A^{\mathrm{s}}(A) = S\begin{bmatrix} \mu_A^{\mathrm{s}}(A_1) & 0 \\ C_{21\mathrm{s}} & \mu_A^{\mathrm{s}}(A_2) \end{bmatrix}S^{-1},$$

where $C_{21\mathrm{s}} \in \mathbb{R}^{(n-r)\times r}$ and $\mu_A^{\mathrm{s}}(A_2)$ is nonsingular, and

$$\mu_A^{\mathrm{u}}(A) = S\begin{bmatrix} \mu_A^{\mathrm{u}}(A_1) & 0 \\ C_{21\mathrm{u}} & 0 \end{bmatrix}S^{-1},$$

where $C_{21\mathrm{u}} \in \mathbb{R}^{(n-r)\times r}$. Consequently,

$$\mathcal{R}\left(S\begin{bmatrix} 0 \\ I_{n-r} \end{bmatrix}\right) \subseteq \mathcal{S}_{\mathrm{u}}(A).$$

If, in addition, $A_{21} = 0$, then

$$\mu_A^{\mathrm{s}}(A) = S\begin{bmatrix} \mu_A^{\mathrm{s}}(A_1) & 0 \\ 0 & \mu_A^{\mathrm{s}}(A_2) \end{bmatrix}S^{-1}, \quad \mu_A^{\mathrm{u}}(A) = S\begin{bmatrix} \mu_A^{\mathrm{u}}(A_1) & 0 \\ 0 & 0 \end{bmatrix}S^{-1}, \quad \mathcal{S}_{\mathrm{s}}(A) \subseteq \mathcal{R}\left(S\begin{bmatrix} I_r \\ 0 \end{bmatrix}\right).$$

Fact 15.24.8. Let $A \in \mathbb{R}^{n\times n}$, and let $S \in \mathbb{R}^{n\times n}$ be a nonsingular matrix such that

$$A = S\begin{bmatrix} A_1 & 0 \\ A_{21} & A_2 \end{bmatrix}S^{-1},$$

where $A_1 \in \mathbb{R}^{r\times r}$ is asymptotically stable, $A_{21} \in \mathbb{R}^{(n-r)\times r}$, and $A_2 \in \mathbb{R}^{(n-r)\times(n-r)}$ satisfies $\operatorname{spec}(A_2) \subset$ CRHP. Then,

$$\mu_A^{\mathrm{s}}(A) = S\begin{bmatrix} 0 & 0 \\ C_{21\mathrm{s}} & \mu_A^{\mathrm{s}}(A_2) \end{bmatrix}S^{-1},$$

where $C_{21\mathrm{s}} \in \mathbb{R}^{(n-r)\times r}$ and $\mu_A^{\mathrm{s}}(A_2)$ is nonsingular, and

$$\mu_A^{\mathrm{u}}(A) = S\begin{bmatrix} \mu_A^{\mathrm{u}}(A_1) & 0 \\ C_{21\mathrm{u}} & 0 \end{bmatrix}S^{-1},$$

where $C_{21\mathrm{u}} \in \mathbb{R}^{(n-r)\times r}$ and $\mu_A^{\mathrm{u}}(A_1)$ is nonsingular. Consequently,

$$\mathcal{S}_{\mathrm{u}}(A) = \mathcal{R}\left(S\begin{bmatrix} 0 \\ I_{n-r} \end{bmatrix}\right).$$

If, in addition, $A_{21} = 0$, then

$$\mu_A^{\mathrm{s}}(A) = S\begin{bmatrix} 0 & 0 \\ 0 & \mu_A^{\mathrm{s}}(A_2) \end{bmatrix}S^{-1}, \quad \mu_A^{\mathrm{u}}(A) = S\begin{bmatrix} \mu_A^{\mathrm{u}}(A_1) & 0 \\ 0 & 0 \end{bmatrix}S^{-1}, \quad \mathcal{S}_{\mathrm{s}}(A) = \mathcal{R}\left(S\begin{bmatrix} I_r \\ 0 \end{bmatrix}\right).$$

Fact 15.24.9. Let $A \in \mathbb{R}^{n\times n}$, and let $S \in \mathbb{R}^{n\times n}$ be a nonsingular matrix such that

$$A = S\begin{bmatrix} A_1 & 0 \\ A_{21} & A_2 \end{bmatrix}S^{-1},$$

where $A_1 \in \mathbb{R}^{r\times r}$, $A_{21} \in \mathbb{R}^{(n-r)\times r}$, and $A_2 \in \mathbb{R}^{(n-r)\times(n-r)}$ is asymptotically stable. Then,

$$\mu_A^{\mathrm{s}}(A) = S\begin{bmatrix} \mu_A^{\mathrm{s}}(A_1) & 0 \\ B_{21\mathrm{s}} & 0 \end{bmatrix}S^{-1},$$

where $B_{21\mathrm{s}} \in \mathbb{R}^{(n-r)\times r}$, and

$$\mu_A^{\mathrm{u}}(A) = S\begin{bmatrix} \mu_A^{\mathrm{u}}(A_1) & 0 \\ B_{21\mathrm{u}} & \mu_A^{\mathrm{u}}(A_2) \end{bmatrix}S^{-1},$$

where $B_{21\mathrm{u}} \in \mathbb{R}^{(n-r)\times r}$ and $\mu_A^{\mathrm{u}}(A_2)$ is nonsingular. Consequently,

$$\mathcal{R}\left(S\begin{bmatrix} 0 \\ I_{n-r} \end{bmatrix}\right) \subseteq \mathcal{S}(A).$$

If, in addition, $A_{21} = 0$, then

$$\mu_A^{\mathrm{s}}(A) = S\begin{bmatrix} \mu_A^{\mathrm{s}}(A_1) & 0 \\ 0 & 0 \end{bmatrix}S^{-1}, \quad \mu_A^{\mathrm{u}}(A) = S\begin{bmatrix} \mu_A^{\mathrm{u}}(A_1) & 0 \\ 0 & \mu_A^{\mathrm{u}}(A_2) \end{bmatrix}S^{-1}, \quad \mathcal{S}_{\mathrm{u}}(A) \subseteq \mathcal{R}\left(S\begin{bmatrix} I_r \\ 0 \end{bmatrix}\right).$$

Fact 15.24.10. Let $A \in \mathbb{R}^{n\times n}$, and let $S \in \mathbb{R}^{n\times n}$ be a nonsingular matrix such that

$$A = S\begin{bmatrix} A_1 & 0 \\ A_{21} & A_2 \end{bmatrix}S^{-1},$$

where $A_1 \in \mathbb{R}^{r\times r}$ satisfies $\operatorname{spec}(A_1) \subset \mathrm{CRHP}$, $A_{21} \in \mathbb{R}^{(n-r)\times r}$, and $A_2 \in \mathbb{R}^{(n-r)\times(n-r)}$. Then,

$$\mu_A^{\mathrm{s}}(A) = S\begin{bmatrix} \mu_A^{\mathrm{s}}(A_1) & 0 \\ C_{12\mathrm{s}} & \mu_A^{\mathrm{s}}(A_2) \end{bmatrix} S^{-1},$$

where $C_{21\mathrm{s}} \in \mathbb{R}^{(n-r)\times r}$ and $\mu_A^{\mathrm{s}}(A_1)$ is nonsingular, and

$$\mu_A^{\mathrm{u}}(A) = S\begin{bmatrix} 0 & 0 \\ C_{21\mathrm{u}} & \mu_A^{\mathrm{u}}(A_2) \end{bmatrix} S^{-1},$$

where $C_{21\mathrm{u}} \in \mathbb{R}^{(n-r)\times r}$. Consequently,

$$\mathcal{S}_{\mathrm{s}}(A) \subseteq \mathcal{R}\left(S\begin{bmatrix} 0 \\ I_{n-r} \end{bmatrix}\right).$$

If, in addition, $A_{21} = 0$, then

$$\mu_A^{\mathrm{s}}(A) = S\begin{bmatrix} \mu_A^{\mathrm{s}}(A_1) & 0 \\ 0 & \mu_A^{\mathrm{s}}(A_2) \end{bmatrix} S^{-1}, \quad \mu_A^{\mathrm{u}}(A) = S\begin{bmatrix} 0 & 0 \\ 0 & \mu_A^{\mathrm{u}}(A_2) \end{bmatrix} S^{-1}, \quad \mathcal{R}\left(S\begin{bmatrix} I_r \\ 0 \end{bmatrix}\right) \subseteq \mathcal{S}_{\mathrm{u}}(A).$$

Fact 15.24.11. Let $A \in \mathbb{R}^{n\times n}$, and let $S \in \mathbb{R}^{n\times n}$ be a nonsingular matrix such that

$$A = S\begin{bmatrix} A_1 & 0 \\ A_{21} & A_2 \end{bmatrix} S^{-1},$$

where $A_1 \in \mathbb{R}^{r\times r}$ satisfies $\operatorname{spec}(A_1) \subset \mathrm{CRHP}$, $A_{21} \in \mathbb{R}^{(n-r)\times r}$, and $A_2 \in \mathbb{R}^{(n-r)\times(n-r)}$ is asymptotically stable. Then,

$$\mu_A^{\mathrm{s}}(A) = S\begin{bmatrix} \mu_A^{\mathrm{s}}(A_1) & 0 \\ C_{21\mathrm{s}} & 0 \end{bmatrix} S^{-1},$$

where $C_{21\mathrm{s}} \in \mathbb{R}^{(n-r)\times r}$ and $\mu_A^{\mathrm{s}}(A_1)$ is nonsingular, and

$$\mu_A^{\mathrm{u}}(A) = S\begin{bmatrix} 0 & 0 \\ C_{21\mathrm{u}} & \mu_A^{\mathrm{u}}(A_2) \end{bmatrix} S^{-1},$$

where $C_{21\mathrm{u}} \in \mathbb{R}^{(n-r)\times r}$ and $\mu_A^{\mathrm{u}}(A_2)$ is nonsingular. Consequently,

$$\mathcal{S}_{\mathrm{s}}(A) = \mathcal{R}\left(S\begin{bmatrix} 0 \\ I_{n-r} \end{bmatrix}\right).$$

If, in addition, $A_{21} = 0$, then

$$\mu_A^{\mathrm{s}}(A) = S\begin{bmatrix} \mu_A^{\mathrm{s}}(A_1) & 0 \\ 0 & 0 \end{bmatrix} S^{-1}, \quad \mu_A^{\mathrm{u}}(A) = S\begin{bmatrix} 0 & 0 \\ 0 & \mu_A^{\mathrm{u}}(A_2) \end{bmatrix} S^{-1}, \quad \mathcal{S}_{\mathrm{u}}(A) = \mathcal{R}\left(S\begin{bmatrix} I_r \\ 0 \end{bmatrix}\right).$$

15.25 Notes

The Laplace transform (15.2.10) is given in [2469, p. 34]. Computational methods are discussed in [1391, 2066]. An arithmetic-mean–geometric-mean iteration for computing the matrix exponential and matrix logarithm is given in [2541].

The exponential function plays a central role in the theory of Lie groups, see [350, 648, 1299, 1471, 1495, 2379, 2781]. Applications to robotics and kinematics are given in [1997, 2104, 2200]. Additional applications are discussed in [647].

The real logarithm is discussed in [800, 1364, 2143, 2272]. The multiplicity and properties of logarithms are discussed in [1002].

An asymptotically stable polynomial is traditionally called *Hurwitz*. Semistability is defined in [624] and developed in [431, 444]. Stability theory is treated in [1292, 1782, 2257] and [1140,

Chapter XV]. Solutions of the Lyapunov equation under weak conditions are considered in [2475]. Structured solutions of the Lyapunov equation are discussed in [1586].

Linear and nonlinear difference equations are studied in [17, 622, 1710].

Chapter Sixteen
Linear Systems and Control Theory

This chapter considers linear state space systems with inputs and outputs. These systems are considered in both the time domain and frequency (Laplace) domain. Some basic results in control theory are also presented.

16.1 State Space Models

Let $A \in \mathbb{R}^{n\times n}$ and $B \in \mathbb{R}^{n\times m}$, and, for all $t \geq t_0$, consider the *state equation*

$$\dot{x}(t) = Ax(t) + Bu(t), \tag{16.1.1}$$

with the *initial condition*

$$x(t_0) = x_0. \tag{16.1.2}$$

In (16.1.1), $x\colon\ [0,\infty) \mapsto \mathbb{R}^n$ is the *state*, and $u\colon\ [0,\infty) \mapsto \mathbb{R}^m$ is the *input*. The function x is the *solution* of (16.1.1).

The following result gives the solution of (16.1.1) and is known as the *variation of constants formula*.

Proposition 16.1.1. For $t \geq t_0$ the state $x(t)$ of the dynamical equation (16.1.1) with initial condition (16.1.2) is given by

$$x(t) = e^{(t-t_0)A}x_0 + \int_{t_0}^{t} e^{(t-\tau)A}Bu(\tau)\,\mathrm{d}\tau. \tag{16.1.3}$$

Proof. Multiplying (16.1.1) by e^{-tA} yields

$$e^{-tA}[\dot{x}(t) - Ax(t)] = e^{-tA}Bu(t),$$

which is equivalent to

$$\frac{\mathrm{d}}{\mathrm{d}t}[e^{-tA}x(t)] = e^{-tA}Bu(t).$$

Integrating over $[t_0, t]$ yields

$$e^{-tA}x(t) = e^{-t_0A}x(t_0) + \int_{t_0}^{t} e^{-\tau A}Bu(\tau)\,\mathrm{d}\tau.$$

Now, multiplying by e^{tA} yields (16.1.3).

Alternatively, let $x(t)$ be given by (16.1.3). Then, it follows from Leibniz's rule Fact 12.16.7 that

$$\dot{x}(t) = \frac{\mathrm{d}}{\mathrm{d}t}e^{(t-t_0)A}x_0 + \frac{\mathrm{d}}{\mathrm{d}t}\int_{t_0}^{t} e^{(t-\tau)A}Bu(\tau)\,\mathrm{d}\tau$$

$$= Ae^{(t-t_0)A}x_0 + \int_{t_0}^{t} Ae^{(t-\tau)A}Bu(\tau)\,\mathrm{d}\tau + Bu(t)$$

$$= Ax(t) + Bu(t).$$

□

For convenience, we can reset the clock by replacing t_0 by 0, and therefore assume without loss of generality that $t_0 = 0$. In this case, $x(t)$ for all $t \ge 0$ is given by

$$x(t) = e^{tA}x_0 + \int_0^t e^{(t-\tau)A}Bu(\tau)\,\mathrm{d}\tau. \tag{16.1.4}$$

If $u(t) = 0$ for all $t \ge 0$, then, for all $t \ge 0$, $x(t)$ is given by

$$x(t) = e^{tA}x_0. \tag{16.1.5}$$

Now, let $u(t) = \delta(t)v$, where $\delta(t)$ is the *unit impulse* occurring at $t = 0$ and $v \in \mathbb{R}^m$. Loosely speaking, the unit impulse occurring at $t = 0$ is zero for all $t \ne 0$ and is infinite at $t = 0$. More precisely, let $a < b$. Then,

$$\int_a^b \delta(\tau)\,\mathrm{d}\tau = \begin{cases} 0, & a > 0 \text{ or } b \le 0, \\ 1, & a \le 0 < b. \end{cases} \tag{16.1.6}$$

More generally, if $g\colon [a,b] \to \mathbb{R}^n$ and g is continuous at $t_0 \in [a,b]$, then

$$\int_a^b \delta(\tau - t_0)g(\tau)\,\mathrm{d}\tau = \begin{cases} 0, & a > t_0 \text{ or } b \le t_0, \\ g(t_0), & a \le t_0 < b. \end{cases} \tag{16.1.7}$$

Consequently, for all $t \ge 0$, $x(t)$ is given by

$$x(t) = e^{tA}x_0 + e^{tA}Bv. \tag{16.1.8}$$

The unit impulse has the physical dimensions of 1/time. This convention makes the integral in (16.1.6) dimensionless.

Alternatively, let the input $u(t)$ be a *step function*; that is, $u(t) = 0$ for all $t < 0$ and $u(t) = v$ for all $t \ge 0$, where $v \in \mathbb{R}^m$. Then, by replacing $t - \tau$ by τ in the integral in (16.1.4), it follows that, for all $t \ge 0$,

$$x(t) = e^{tA}x_0 + \int_0^t e^{\tau A}\,\mathrm{d}\tau Bv. \tag{16.1.9}$$

Using Fact 15.14.10, (16.1.9) can be written for all $t \ge 0$ as

$$x(t) = e^{tA}x_0 + \left(A^{\mathrm{D}}(e^{tA} - I) + (I - AA^{\mathrm{D}})\sum_{i=1}^{\operatorname{ind} A} \frac{t^i}{i!}A^{i-1}\right)Bv. \tag{16.1.10}$$

If A is group invertible, then, for all $t \ge 0$, (16.1.10) becomes

$$x(t) = e^{tA}x_0 + [A^{\#}(e^{tA} - I) + t(I - AA^{\#})]Bv. \tag{16.1.11}$$

If, in addition, A is nonsingular, then, for all $t \ge 0$, (16.1.11) becomes

$$x(t) = e^{tA}x_0 + A^{-1}(e^{tA} - I)Bv. \tag{16.1.12}$$

Next, consider the *output equation*

$$y(t) = Cx(t) + Du(t), \tag{16.1.13}$$

where $t \geq 0$, $y(t) \in \mathbb{R}^l$ is the *output*, $C \in \mathbb{R}^{l\times n}$, and $D \in \mathbb{R}^{l\times m}$. Then, for all $t \geq 0$, the *total response* of (16.1.1), (16.1.13) is

$$y(t) = Ce^{tA}x_0 + \int_0^t Ce^{(t-\tau)A}Bu(\tau)\,\mathrm{d}\tau + Du(t). \tag{16.1.14}$$

If $u(t) = 0$ for all $t \geq 0$, then the *free response* is given by

$$y(t) = Ce^{tA}x_0, \tag{16.1.15}$$

while, if $x_0 = 0$, then the *forced response* is given by

$$y(t) = \int_0^t Ce^{(t-\tau)A}Bu(\tau)\,\mathrm{d}\tau + Du(t). \tag{16.1.16}$$

Setting $u(t) = \delta(t)v$, where $v \in \mathbb{R}^m$, yields, for all $t > 0$, the total response

$$y(t) = Ce^{tA}x_0 + H(t)v, \tag{16.1.17}$$

where, for all $t \geq 0$, the *impulse response function* $H(t)$ is defined by

$$H(t) \triangleq Ce^{tA}B + \delta(t)D. \tag{16.1.18}$$

The corresponding forced response is the *impulse response*

$$y(t) = H(t)v = Ce^{tA}Bv + \delta(t)Dv. \tag{16.1.19}$$

Alternatively, if $u(t) = v \in \mathbb{R}^m$ for all $t \geq 0$, then the total response is

$$y(t) = Ce^{tA}x_0 + \int_0^t Ce^{\tau A}\,\mathrm{d}\tau Bv + Dv, \tag{16.1.20}$$

and the forced response is the *step response*

$$y(t) = \int_0^t H(\tau)\,\mathrm{d}\tau v = \int_0^t Ce^{\tau A}\,\mathrm{d}\tau Bv + Dv. \tag{16.1.21}$$

For an arbitrary input u, the forced response can be written as the *convolution integral*

$$y(t) = \int_0^t H(t-\tau)u(\tau)\,\mathrm{d}\tau. \tag{16.1.22}$$

Setting $u(t) = \delta(t)v$ in (16.1.22) yields (16.1.19) by noting that

$$\int_0^t \delta(t-\tau)\delta(\tau)\,\mathrm{d}\tau = \delta(t). \tag{16.1.23}$$

Proposition 16.1.2. Let $D = 0$ and $m = 1$, and assume that $x_0 = Bv$. Then, the free response and the impulse response are equal and given by

$$y(t) = Ce^{tA}x_0 = Ce^{tA}Bv. \tag{16.1.24}$$

16.2 Laplace Transform Analysis and Transfer Functions

Now, consider the linear system

$$\dot{x}(t) = Ax(t) + Bu(t), \tag{16.2.1}$$

$$y(t) = Cx(t) + Du(t), \tag{16.2.2}$$

with state $x(t) \in \mathbb{R}^n$, input $u(t) \in \mathbb{R}^m$, and *output* $y(t) \in \mathbb{R}^l$, where $t \geq 0$ and $x(0) = x_0$. Taking Laplace transforms yields

$$s\hat{x}(s) - x_0 = A\hat{x}(s) + B\hat{u}(s), \tag{16.2.3}$$

$$\hat{y}(s) = C\hat{x}(s) + D\hat{u}(s), \tag{16.2.4}$$

where

$$\hat{x}(s) \triangleq \mathcal{L}\{x(t)\} = \int_0^\infty e^{-st}x(t)\,\mathrm{d}t, \tag{16.2.5}$$

$$\hat{u}(s) \triangleq \mathcal{L}\{u(t)\}, \quad \hat{y}(s) \triangleq \mathcal{L}\{y(t)\}. \tag{16.2.6}$$

Hence,

$$\hat{x}(s) = (sI - A)^{-1}x_0 + (sI - A)^{-1}B\hat{u}(s), \tag{16.2.7}$$

and thus

$$\hat{y}(s) = C(sI - A)^{-1}x_0 + [C(sI - A)^{-1}B + D]\hat{u}(s). \tag{16.2.8}$$

We can also obtain (16.2.8) from the time-domain expression for $y(t)$ given by (16.1.14). Using Proposition 15.2.2, it follows from (16.1.14) that

$$\begin{aligned}\hat{y}(s) &= \mathcal{L}\{Ce^{tA}x_0\} + \mathcal{L}\left\{\int_0^t Ce^{(t-\tau)A}Bu(\tau)\,\mathrm{d}\tau\right\} + D\hat{u}(s)\\ &= C\mathcal{L}\{e^{tA}\}x_0 + C\mathcal{L}\{e^{tA}\}B\hat{u}(s) + D\hat{u}(s)\\ &= C(sI - A)^{-1}x_0 + [C(sI - A)^{-1}B + D]\hat{u}(s),\end{aligned} \tag{16.2.9}$$

which coincides with (16.2.8). We define

$$G(s) \triangleq C(sI - A)^{-1}B + D. \tag{16.2.10}$$

Note that $G \in \mathbb{R}^{l\times m}(s)$, and thus, by Definition 6.7.2, G is a proper rational transfer function. Since $\mathcal{L}\{\delta(t)\} = 1$, it follows from the definition (16.1.18) of the impulse response function H that

$$G(s) = \mathcal{L}\{H(t)\}. \tag{16.2.11}$$

Using (6.7.2), G can be written as

$$G(s) = \frac{1}{\chi_A(s)}C(sI - A)^{\mathsf{A}}B + D. \tag{16.2.12}$$

It follows from (6.7.3) that G is a proper rational transfer function. Furthermore, G is a strictly proper rational transfer function if and only if $D = 0$, whereas G is an exactly proper rational transfer function if and only if $D \neq 0$. Finally, if A is nonsingular, then

$$G(0) = -CA^{-1}B + D. \tag{16.2.13}$$

Let $A \in \mathbb{R}^{n\times n}$. If $|s| > \rho_{\max}(A)$, then Proposition 11.3.10 implies that

$$(sI - A)^{-1} = \frac{1}{s}\left(I - \frac{1}{s}A\right)^{-1} = \sum_{k=0}^{\infty} \frac{1}{s^{k+1}}A^k, \tag{16.2.14}$$

where the Laurent series is absolutely convergent, and thus

$$G(s) = D + \frac{1}{s}CB + \frac{1}{s^2}CAB + \cdots = \sum_{k=0}^{\infty} \frac{1}{s^k}H_k, \tag{16.2.15}$$

where, for all $k \ge 0$, the *Markov parameter* $H_k \in \mathbb{R}^{l\times m}$ is defined by

$$H_k \triangleq \begin{cases} D, & k = 0, \\ CA^{k-1}B, & k \ge 1. \end{cases} \tag{16.2.16}$$

It follows from (16.2.14) that $\lim_{s\to\infty}(sI - A)^{-1} = 0$ and from (16.2.15) that

$$H_0 = D = \lim_{s\to\infty} G(s), \tag{16.2.17}$$

$$H_1 = CB = \lim_{s\to\infty} s[G(s) - H_0], \tag{16.2.18}$$

$$H_2 = CAB = \lim_{s\to\infty} (s^2[G(s) - H_0] - sH_1). \tag{16.2.19}$$

Furthermore, Definition 6.7.3 implies that

$$\operatorname{reldeg} G = \min\{k \ge 0\colon\ H_k \ne 0\}. \tag{16.2.20}$$

16.3 The Unobservable Subspace and Observability

Let $A \in \mathbb{R}^{n\times n}$ and $C \in \mathbb{R}^{l\times n}$, and, for all $t \ge 0$, consider the linear system

$$\dot{x}(t) = Ax(t), \tag{16.3.1}$$

$$x(0) = x_0, \tag{16.3.2}$$

$$y(t) = Cx(t). \tag{16.3.3}$$

Definition 16.3.1. The *unobservable subspace* $\mathcal{U}_{t_{\rm f}}(A, C)$ of (A, C) at time $t_{\rm f} > 0$ is the subspace

$$\mathcal{U}_{t_{\rm f}}(A, C) \triangleq \{x_0 \in \mathbb{R}^n\colon\ y(t) = 0 \text{ for all } t \in [0, t_{\rm f}]\}. \tag{16.3.4}$$

Let $t_{\rm f} > 0$. Then, Definition 16.3.1 states that $x_0 \in \mathcal{U}_{t_{\rm f}}(A, C)$ if and only if $y(t) = 0$ for all $t \in [0, t_{\rm f}]$. Since $y(t) = 0$ for all $t \in [0, t_{\rm f}]$ is the free response corresponding to $x_0 = 0$, it follows that $0 \in \mathcal{U}_{t_{\rm f}}(A, C)$. Now, suppose there exists a nonzero vector $x_0 \in \mathcal{U}_{t_{\rm f}}(A, C)$. Then, with $x(0) = x_0$, the free response is given by $y(t) = 0$ for all $t \in [0, t_{\rm f}]$, and thus x_0 cannot be determined from knowledge of $y(t)$ for all $t \in [0, t_{\rm f}]$.

The following result provides explicit expressions for $\mathcal{U}_{t_{\rm f}}(A, C)$.

Lemma 16.3.2. Let $t_{\rm f} > 0$. Then, the following subspaces are equal:

i) $\mathcal{U}_{t_{\rm f}}(A, C)$.

ii) $\bigcap_{t\in[0,t_{\rm f}]} \mathcal{N}(Ce^{tA})$.

iii) $\bigcap_{i=0}^{n-1} \mathcal{N}(CA^i)$.

iv) $\mathcal{N}\left(\begin{bmatrix} C \\ CA \\ \vdots \\ CA^{n-1} \end{bmatrix}\right)$.

v) $\mathcal{N}\left(\int_0^{t_{\rm f}} e^{tA^{\rm T}}C^{\rm T}Ce^{tA}\,{\rm d}t\right)$.

If, in addition, $\lim_{t_f\to\infty}\int_0^{t_f} e^{tA^\mathsf{T}}C^\mathsf{T}Ce^{tA}\mathrm{d}t$ exists, then the following subspace is equal to *i)*–*v)*:

vi) $\mathcal{N}\left(\int_0^{\infty} e^{tA^\mathsf{T}}C^\mathsf{T}Ce^{tA}\mathrm{d}t\right)$.

Proof. See the proof of Lemma 16.6.2. □

The expressions given by *iii)* and *iv)* of Lemma 16.3.2 show that $\mathcal{U}_{t_f}(A,C)$ is independent of t_f. We thus write $\mathcal{U}(A,C)$ for $\mathcal{U}_{t_f}(A,C)$, and call $\mathcal{U}(A,C)$ the *unobservable subspace* of (A,C). (A,C) is *observable* if $\mathcal{U}(A,C)=\{0\}$. For convenience, define the $nl\times n$ *observability matrix*

$$\mathcal{O}(A,C) \triangleq \begin{bmatrix} C \\ CA \\ \vdots \\ CA^{n-1} \end{bmatrix} \tag{16.3.5}$$

so that

$$\mathcal{U}(A,C) = \mathcal{N}[\mathcal{O}(A,C)]. \tag{16.3.6}$$

Define

$$p \triangleq n - \dim \mathcal{U}(A,C) = n - \operatorname{def} \mathcal{O}(A,C). \tag{16.3.7}$$

Corollary 16.3.3. For all $t_f > 0$,

$$p = \dim \mathcal{U}(A,C)^\perp = \operatorname{rank} \mathcal{O}(A,C) = \operatorname{rank} \int_0^{t_f} e^{tA^\mathsf{T}}C^\mathsf{T}Ce^{tA}\mathrm{d}t. \tag{16.3.8}$$

If, in addition, $\lim_{t_f\to\infty}\int_0^{t_f} e^{tA^\mathsf{T}}C^\mathsf{T}Ce^{tA}\mathrm{d}t$ exists, then

$$p = \operatorname{rank} \int_0^{\infty} e^{tA^\mathsf{T}}C^\mathsf{T}Ce^{tA}\mathrm{d}t. \tag{16.3.9}$$

Corollary 16.3.4. $\mathcal{U}(A,C)$ is an invariant subspace of A.

The following result shows that the unobservable subspace $\mathcal{U}(A,C)$ is unchanged by output injection

$$\dot{x}(t) = Ax(t) + Fy(t). \tag{16.3.10}$$

Proposition 16.3.5. Let $F\in\mathbb{R}^{n\times l}$. Then,

$$\mathcal{U}(A+FC,C) = \mathcal{U}(A,C). \tag{16.3.11}$$

In particular, (A,C) is observable if and only if $(A+FC,C)$ is observable.

Proof. See the proof of Proposition 16.6.5. □

Let $\tilde{\mathcal{U}}(A,C)\subseteq\mathbb{R}^n$ be a subspace that is complementary to $\mathcal{U}(A,C)$. Then, $\tilde{\mathcal{U}}(A,C)$ is an *observable subspace* in the sense that, if $x_0 = x_0' + x_0''$, where $x_0'\in\tilde{\mathcal{U}}(A,C)$ is nonzero and $x_0''\in\mathcal{U}(A,C)$, then it is possible to determine x_0' from knowledge of $y(t)$ for all $t\in[0,t_f]$. Using Proposition 4.8.3, let $\mathcal{P}\in\mathbb{R}^{n\times n}$ be the unique idempotent matrix such that $\mathcal{R}(\mathcal{P})=\tilde{\mathcal{U}}(A,C)$ and $\mathcal{N}(\mathcal{P})=\mathcal{U}(A,C)$. Then, $x_0'=\mathcal{P}x_0$. The following result constructs $\mathcal{P}$ and provides an expression for x_0' in terms of $y(t)$ for $\tilde{\mathcal{U}}(A,C)=\mathcal{U}(A,C)^\perp$. In this case, $\mathcal{P}$ is a projector.

Lemma 16.3.6. Let $t_f>0$, and define $\mathcal{P}\in\mathbb{R}^{n\times n}$ by

$$\mathcal{P} \triangleq \left(\int_0^{t_f} e^{tA^\mathsf{T}}C^\mathsf{T}Ce^{tA}\,\mathrm{d}t\right)^+ \int_0^{t_f} e^{tA^\mathsf{T}}C^\mathsf{T}Ce^{tA}\,\mathrm{d}t. \tag{16.3.12}$$

Then, $\mathcal{P}$ is the projector onto $\mathcal{U}(A,C)^{\perp}$, and $\mathcal{P}_{\perp}$ is the projector onto $\mathcal{U}(A,C)$. Hence,

$$\mathcal{R}(\mathcal{P}) = \mathcal{N}(\mathcal{P}_{\perp}) = \mathcal{U}(A,C)^{\perp}, \tag{16.3.13}$$

$$\mathcal{N}(\mathcal{P}) = \mathcal{R}(\mathcal{P}_{\perp}) = \mathcal{U}(A,C), \tag{16.3.14}$$

$$\operatorname{rank}\mathcal{P} = \operatorname{def}\mathcal{P}_{\perp} = \dim\mathcal{U}(A,C)^{\perp} = p, \tag{16.3.15}$$

$$\operatorname{def}\mathcal{P} = \operatorname{rank}\mathcal{P}_{\perp} = \dim\mathcal{U}(A,C) = n-p. \tag{16.3.16}$$

If $x_0 = x_0' + x_0''$, where $x_0' \in \mathcal{U}(A,C)^{\perp}$ and $x_0'' \in \mathcal{U}(A,C)$, then

$$x_0' = \mathcal{P}x_0 = \left(\int_0^{t_{\mathrm{f}}} e^{tA^{\mathsf{T}}}C^{\mathsf{T}}Ce^{tA}\,\mathrm{d}t\right)^{+}\int_0^{t_{\mathrm{f}}} e^{tA^{\mathsf{T}}}C^{\mathsf{T}}y(t)\,\mathrm{d}t. \tag{16.3.17}$$

Finally, (A,C) is observable if and only if $\mathcal{P} = I_n$. If these conditions hold, then, for all $x_0 \in \mathbb{R}^n$,

$$x_0 = \left(\int_0^{t_{\mathrm{f}}} e^{tA^{\mathsf{T}}}C^{\mathsf{T}}Ce^{tA}\,\mathrm{d}t\right)^{-1}\int_0^{t_{\mathrm{f}}} e^{tA^{\mathsf{T}}}C^{\mathsf{T}}y(t)\,\mathrm{d}t. \tag{16.3.18}$$

Lemma 16.3.7. Let $\alpha \in \mathbb{R}$. Then,

$$\mathcal{U}(A+\alpha I, C) = \mathcal{U}(A,C). \tag{16.3.19}$$

The following result uses a coordinate transformation to characterize the observable dynamics of a system.

Theorem 16.3.8. There exists an orthogonal matrix $S \in \mathbb{R}^{n\times n}$ such that

$$A = S\begin{bmatrix} A_1 & 0 \\ A_{21} & A_2 \end{bmatrix}S^{-1}, \quad C = [C_1 \ \ 0]S^{-1}, \tag{16.3.20}$$

where $A_1 \in \mathbb{R}^{p\times p}$, $C_1 \in \mathbb{R}^{l\times p}$, and (A_1, C_1) is observable.

Proof. See the proof of Theorem 16.6.8. □

Proposition 16.3.9. Let $S \in \mathbb{R}^{n\times n}$, and assume that S is orthogonal. Then, the following statements are equivalent:

i) A and C have the form (16.3.20), where $A_1 \in \mathbb{R}^{p\times p}$, $C_1 \in \mathbb{R}^{l\times p}$, and (A_1, C_1) is observable.

ii) $\mathcal{U}(A,C) = \mathcal{R}\left(S\begin{bmatrix} 0 \\ I_{n-p}\end{bmatrix}\right)$.

iii) $\mathcal{U}(A,C)^{\perp} = \mathcal{R}\left(S\begin{bmatrix} I_p \\ 0\end{bmatrix}\right)$.

iv) $\mathcal{P} = S\begin{bmatrix} I_p & 0 \\ 0 & 0\end{bmatrix}S^{\mathsf{T}}$.

Proposition 16.3.10. Let $S \in \mathbb{R}^{n\times n}$, and assume that S is nonsingular. Then, the following statements are equivalent:

i) A and C have the form (16.3.20), where $A_1 \in \mathbb{R}^{p\times p}$, $C_1 \in \mathbb{R}^{l\times p}$, and (A_1, C_1) is observable.

ii) $\mathcal{U}(A,C) = \mathcal{R}\left(S\begin{bmatrix} 0 \\ I_{n-p}\end{bmatrix}\right)$ and $\tilde{\mathcal{U}}(A,C) = \mathcal{R}\left(S\begin{bmatrix} I_p \\ 0\end{bmatrix}\right)$.

iii) $\mathcal{P} = S\begin{bmatrix} I_p & 0 \\ 0 & 0\end{bmatrix}S^{-1}$.

Definition 16.3.11. Let $S \in \mathbb{R}^{n\times n}$, assume that S is nonsingular, and let A and C have the form (16.3.20), where $A_1 \in \mathbb{R}^{p\times p}$, $C_1 \in \mathbb{R}^{l\times p}$, and (A_1, C_1) is observable. Then, $\lambda \in \mathbb{C}$ is an *unobservable eigenvalue* of (A,C) if $\lambda \in \operatorname{spec}(A_2)$, the *unobservable spectrum* of (A,C) is $\operatorname{spec}(A_2)$, and the *unobservable multispectrum* of (A,C) is $\operatorname{mspec}(A_2)$. Finally, $\lambda \in \operatorname{spec}(A)$ is an *observable eigenvalue* of (A,C) if λ is not an unobservable eigenvalue of (A,C).

Definition 16.3.12. The *observability pencil* $\mathcal{O}_{A,C}(s)$ is the pencil

$$\mathcal{O}_{A,C} = P_{\left[\begin{smallmatrix} A \\ -C \end{smallmatrix}\right],\left[\begin{smallmatrix} I \\ 0 \end{smallmatrix}\right]}. \tag{16.3.21}$$

Equivalently,

$$\mathcal{O}_{A,C}(s) = \begin{bmatrix} sI - A \\ C \end{bmatrix}. \tag{16.3.22}$$

Proposition 16.3.13. Let $\lambda \in \operatorname{spec}(A)$. Then, λ is an unobservable eigenvalue of (A, C) if and only if

$$\operatorname{rank}\begin{bmatrix} \lambda I - A \\ C \end{bmatrix} < n. \tag{16.3.23}$$

Proof. See the proof of Proposition 16.6.13. □

Proposition 16.3.14. Let $\lambda \in \operatorname{mspec}(A)$ and $F \in \mathbb{R}^{n\times l}$. Then, λ is an unobservable eigenvalue of (A, C) if and only if λ is an unobservable eigenvalue of $(A + FC, C)$.

Proof. See the proof of Proposition 16.6.14. □

Proposition 16.3.15. Assume that (A, C) is observable. Then, the Smith form of $\mathcal{O}_{A,C}$ is $\left[\begin{smallmatrix} I_n \\ 0_{l\times n} \end{smallmatrix}\right]$.

Proof. See the proof of Proposition 16.6.15. □

Proposition 16.3.16. $S \in \mathbb{R}^{n\times n}$, assume that S is nonsingular, and let A and C have the form (16.3.20), where $A_1 \in \mathbb{R}^{p\times p}$, $C_1 \in \mathbb{R}^{l\times p}$, and (A_1, C_1) is observable. Furthermore, let $p_1, \ldots, p_{n-p}$ be the similarity invariants of A_2, where, for all $i \in \{1, \ldots, n-p-1\}$, p_i divides p_{i+1}. Then, there exist unimodular matrices $S_1 \in \mathbb{R}[s]^{(n+l)\times(n+l)}$ and $S_2 \in \mathbb{R}[s]^{n\times n}$ such that, for all $s \in \mathbb{C}$,

$$\begin{bmatrix} sI - A \\ C \end{bmatrix} = S_1(s)\begin{bmatrix} I_p & & & \\ & p_1(s) & & \\ & & \ddots & \\ & & & p_{n-p}(s) \\ & 0_{l\times n} & & \end{bmatrix} S_2(s). \tag{16.3.24}$$

Consequently,

$$\operatorname{Szeros}(\mathcal{O}_{A,C}) = \bigcup_{i=1}^{n-p} \operatorname{roots}(p_i) = \operatorname{roots}(\chi_{A_2}) = \operatorname{spec}(A_2), \tag{16.3.25}$$

$$\operatorname{mSzeros}(\mathcal{O}_{A,C}) = \bigcup_{i=1}^{n-p} \operatorname{mroots}(p_i) = \operatorname{mroots}(\chi_{A_2}) = \operatorname{mspec}(A_2). \tag{16.3.26}$$

Proof. See the proof of Proposition 16.6.16. □

Proposition 16.3.17. Let $s \in \mathbb{C}$. Then,

$$\mathcal{O}(A, C) \subseteq \operatorname{Re}\mathcal{R}\left(\begin{bmatrix} sI - A \\ C \end{bmatrix}\right). \tag{16.3.27}$$

Proof. See the proof of Proposition 16.6.17. □

The next result characterizes observability in several equivalent ways.

Theorem 16.3.18. The following statements are equivalent:

i) (A, C) is observable.

ii) There exists $t > 0$ such that $\int_0^t e^{\tau A^\mathsf{T}} C^\mathsf{T} C e^{\tau A}\, d\tau$ is positive definite.

iii) For all $t > 0$, $\int_0^t e^{\tau A^\mathsf{T}} C^\mathsf{T} C e^{\tau A}\, d\tau$ is positive definite.

iv) rank $\mathcal{O}(A, C) = n$.

v) Every eigenvalue of (A, C) is observable.

vi) If $x \in \mathbb{C}^n$ is an eigenvector of A, then $B^{\mathsf{T}}x \neq 0$.

vii) There exists $F \in \mathbb{R}^{n\times l}$ such that $(A + FC, C)$ is observable.

viii) For all $F \in \mathbb{R}^{n\times l}$, $(A + FC, C)$ is observable.

If, in addition, $\lim_{t\to\infty} \int_0^t e^{\tau A^{\mathsf{T}}} C^{\mathsf{T}} C e^{\tau A}\, \mathrm{d}\tau$ exists, then the following statement is equivalent to *i*)–*viii*):

ix) $\int_0^\infty e^{tA^{\mathsf{T}}} C^{\mathsf{T}} C e^{tA}\, \mathrm{d}t$ is positive definite.

Proof. See the proof of Theorem 16.6.18. □

The following result, which is a restatement of the equivalence of *i*) and *v*) of Theorem 16.3.18, is the *PBH test for observability*.

Corollary 16.3.19. The following statements are equivalent:

i) (A, C) is observable.

ii) For all $s \in \mathbb{C}$,

$$\operatorname{rank}\begin{bmatrix} sI - A \\ C \end{bmatrix} = n. \tag{16.3.28}$$

The following result implies that arbitrary eigenvalue placement is possible for (16.3.10) if (A, C) is observable.

Proposition 16.3.20. The pair (A, C) is observable if and only if, for every polynomial $p \in \mathbb{R}[s]$ such that $\deg p = n$, there exists $F \in \mathbb{R}^{n\times l}$ such that $\operatorname{mspec}(A + FC) = \operatorname{mroots}(p)$.

Proof. See the proof of Proposition 16.6.20. □

16.4 Observable Asymptotic Stability

Let $A \in \mathbb{R}^{n\times n}$ and $C \in \mathbb{R}^{l\times n}$, and define $p \triangleq n - \dim \mathcal{U}(A, C)$.

Definition 16.4.1. (A, C) is *observably asymptotically stable* if

$$\mathcal{S}_{\mathrm{u}}(A) \subseteq \mathcal{U}(A, C). \tag{16.4.1}$$

Proposition 16.4.2. Let $F \in \mathbb{R}^{n\times l}$. Then, (A, C) is observably asymptotically stable if and only if $(A + FC, C)$ is observably asymptotically stable.

Proposition 16.4.3. The following statements are equivalent:

i) (A, C) is observably asymptotically stable.

ii) There exists an orthogonal matrix $S \in \mathbb{R}^{n\times n}$ such that (16.3.20) holds, where $A_1 \in \mathbb{R}^{p\times p}$ is asymptotically stable and $C_1 \in \mathbb{R}^{l\times p}$.

iii) There exists a nonsingular matrix $S \in \mathbb{R}^{n\times n}$ such that (16.3.20) holds, where $A_1 \in \mathbb{R}^{p\times p}$ is asymptotically stable and $C_1 \in \mathbb{R}^{l\times p}$.

iv) $\lim_{t\to\infty} Ce^{tA} = 0$.

v) The positive-semidefinite matrix $P \in \mathbb{R}^{n\times n}$ defined by

$$P \triangleq \int_0^\infty e^{tA^{\mathsf{T}}} C^{\mathsf{T}} C e^{tA}\, \mathrm{d}t \tag{16.4.2}$$

exists.

vi) There exists a positive-semidefinite matrix $P \in \mathbb{R}^{n\times n}$ satisfying

$$A^{\mathsf{T}}P + PA + C^{\mathsf{T}}C = 0. \tag{16.4.3}$$

If these conditions hold, then the positive-semidefinite matrix $P \in \mathbb{R}^{n\times n}$ defined by (16.4.2) satisfies (16.4.3).

Proof. See the proof of Proposition 16.7.3. □

The matrix P defined by (16.4.2) is the *observability Gramian*, and (16.4.3) is the *observation Lyapunov equation.*

Proposition 16.4.4. Assume that (A, C) is observably asymptotically stable, let $P \in \mathbb{R}^{n\times n}$ be the positive-semidefinite matrix defined by (16.4.2), and define $\mathcal{P} \in \mathbb{R}^{n\times n}$ by (16.3.12). Then, the following statements hold:

i) $PP^+ = \mathcal{P}$.

ii) $\mathcal{R}(P) = \mathcal{R}(\mathcal{P}) = \mathcal{U}(A, C)^\perp$.

iii) $\mathcal{N}(P) = \mathcal{N}(\mathcal{P}) = \mathcal{U}(A, C)$.

iv) $\operatorname{rank} P = \operatorname{rank} \mathcal{P} = p$.

v) P is the unique positive-semidefinite solution of (16.4.3) whose rank is p.

Proof. See the proof of Proposition 16.7.4. □

Proposition 16.4.5. Assume that (A, C) is observably asymptotically stable, let $P \in \mathbb{R}^{n\times n}$ be the positive-semidefinite matrix defined by (16.4.2), and let $\hat{P} \in \mathbb{R}^{n\times n}$. Then, the following statements are equivalent:

i) $\hat{P}$ is positive semidefinite and satisfies (16.4.3).

ii) There exists a positive-semidefinite matrix $P_0 \in \mathbb{R}^{n\times n}$ such that $\hat{P} = P + P_0$ and $A^\mathsf{T} P_0 + P_0 A = 0$.

If these conditions hold, then

$$\operatorname{rank} \hat{P} = p + \operatorname{rank} P_0, \tag{16.4.4}$$

$$\operatorname{rank} P_0 \le \sum_{\substack{\lambda \in \operatorname{spec}(A) \\ \lambda \in \mathrm{I}\mathbb{A}}} \operatorname{gmult}_A(\lambda). \tag{16.4.5}$$

Proof. See the proof of Proposition 16.7.5. □

Proposition 16.4.6. The following statements are equivalent:

i) (A, C) is observably asymptotically stable, every imaginary eigenvalue of A is semisimple, and A has no ORHP eigenvalues.

ii) (16.4.3) has a positive-definite solution $P \in \mathbb{R}^{n\times n}$.

Proof. See the proof of Proposition 16.7.6. □

Proposition 16.4.7. The following statements are equivalent:

i) (A, C) is observably asymptotically stable, and A has no imaginary eigenvalues.

ii) (16.4.3) has exactly one positive-semidefinite solution $P \in \mathbb{R}^{n\times n}$.

If these conditions hold, then $P \in \mathbb{R}^{n\times n}$ is given by (16.4.2) and satisfies $\operatorname{rank} P = p$.

Proof. See the proof of Proposition 16.7.7. □

Corollary 16.4.8. Assume that A is asymptotically stable. Then, the positive-semidefinite matrix $P \in \mathbb{R}^{n\times n}$ defined by (16.4.2) is the unique solution of (16.4.3) and satisfies $\operatorname{rank} P = p$.

Proof. See the proof of Corollary 16.7.8. □

Proposition 16.4.9. The following statements are equivalent:

i) (A, C) is observable, and A is asymptotically stable.

ii) (16.4.3) has exactly one positive-semidefinite solution $P \in \mathbb{R}^{n\times n}$, and P is positive definite.

If these conditions hold, then $P \in \mathbb{R}^{n\times n}$ is given by (16.4.2).

Proof. See the proof of Proposition 16.7.9. □

Corollary 16.4.10. Assume that A is asymptotically stable. Then, the positive-semidefinite matrix $P \in \mathbb{R}^{n\times n}$ defined by (16.4.2) exists. Furthermore, P is positive definite if and only if (A, C) is observable.

16.5 Detectability

Let $A \in \mathbb{R}^{n\times n}$ and $C \in \mathbb{R}^{l\times n}$, and define $p \triangleq n - \dim \mathcal{U}(A, C)$.

Definition 16.5.1. The *undetectable subspace* $\mathcal{U}_{\rm u}(A, C)$ of (A, C) is

$$\mathcal{U}_{\rm u}(A, C) \triangleq \mathcal{U}(A, C) \cap \mathcal{S}_{\rm u}(A). \tag{16.5.1}$$

Furthermore, (A, C) is *detectable* if $\mathcal{U}_{\rm u}(A, C) = \{0\}$.

Proposition 16.5.2. (A, C) is detectable if and only if

$$\mathcal{U}(A, C) \subseteq \mathcal{S}_{\rm s}(A). \tag{16.5.2}$$

Proposition 16.5.3. Let $F \in \mathbb{R}^{n\times l}$. Then, (A, C) is detectable if and only if $(A + FC, C)$ is detectable.

Proposition 16.5.4. The following statements are equivalent:

i) (A, C) is detectable.

ii) There exists an orthogonal matrix $S \in \mathbb{R}^{n\times n}$ such that (16.3.20) holds, where $A_1 \in \mathbb{R}^{p\times p}$, $C_1 \in \mathbb{R}^{l\times p}$, (A_1, C_1) is observable, and $A_2 \in \mathbb{R}^{(n-p)\times(n-p)}$ is asymptotically stable.

iii) There exists a nonsingular matrix $S \in \mathbb{R}^{n\times n}$ such that (16.3.20) holds, where $A_1 \in \mathbb{R}^{p\times p}$, $C_1 \in \mathbb{R}^{l\times p}$, (A_1, C_1) is observable, and $A_2 \in \mathbb{R}^{(n-p)\times(n-p)}$ is asymptotically stable.

iv) Every CRHP eigenvalue of (A, C) is observable.

Proof. See the proof of Proposition 16.8.4. □

The following result, which is a restatement of the equivalence of *i*) and *iv*) of Proposition 16.5.4, is the *PBH test for detectability*.

Corollary 16.5.5. The following statements are equivalent:

i) (A, C) is detectable.

ii) For all $s \in$ CRHP,

$$\operatorname{rank}\begin{bmatrix} sI - A \\ C \end{bmatrix} = n. \tag{16.5.3}$$

Proposition 16.5.6. The following statements are equivalent:

i) A is asymptotically stable.

ii) (A, C) is observably asymptotically stable and detectable.

Proof. See the proof of Proposition 16.8.6. □

Corollary 16.5.7. The following statements are equivalent:

i) There exists a positive-semidefinite matrix $P \in \mathbb{R}^{n\times n}$ satisfying (16.4.3), and (A, C) is detectable.

ii) A is asymptotically stable.

Proof. See the proof of Proposition 16.8.7. □

16.6 The Controllable Subspace and Controllability

Let $A \in \mathbb{R}^{n\times n}$ and $B \in \mathbb{R}^{n\times m}$, and, for $t \ge 0$, consider the linear system

$$\dot{x}(t) = Ax(t) + Bu(t), \tag{16.6.1}$$

$$x(0) = 0. \tag{16.6.2}$$

Definition 16.6.1. The *controllable subspace* $\mathcal{C}_{t_f}(A,B)$ of (A,B) at time $t_f > 0$ is the subspace

$$\mathcal{C}_{t_f}(A,B) \triangleq \{x_f \in \mathbb{R}^n\colon \text{ there exists a continuous control } u\colon [0,t_f] \mapsto \mathbb{R}^m \text{ such that the solution } x(\cdot) \text{ of (16.6.1), (16.6.2) satisfies } x(t_f) = x_f\}. \tag{16.6.3}$$

Let $t_f > 0$. Then, Definition 16.6.1 states that $x_f \in \mathcal{C}_{t_f}(A,B)$ if and only if there exists a continuous control $u\colon [0,t_f] \mapsto \mathbb{R}^m$ such that

$$x_f = \int_0^{t_f} e^{(t_f - t)A}Bu(t)\,dt. \tag{16.6.4}$$

The following result provides expressions for $\mathcal{C}_{t_f}(A,B)$.

Lemma 16.6.2. Let $t_f > 0$. Then, the following subspaces are equal:

i) $\mathcal{C}_{t_f}(A,B)$.

ii) $\left[\bigcap_{t\in[0,t_f]} \mathcal{N}(B^{\mathsf{T}}e^{tA^{\mathsf{T}}})\right]^{\perp}$.

iii) $\left[\bigcap_{i=0}^{n-1} \mathcal{N}(B^{\mathsf{T}}A^{i\mathsf{T}})\right]^{\perp}$.

iv) $\mathcal{R}([B\ \ AB\ \ \cdots\ \ A^{n-1}B])$.

v) $\mathcal{R}\left(\int_0^{t_f} e^{tA}BB^{\mathsf{T}}e^{tA^{\mathsf{T}}}\,dt\right)$.

If, in addition, $\lim_{t_f\to\infty} \int_0^{t_f} e^{tA}BB^{\mathsf{T}}e^{tA^{\mathsf{T}}}\,dt$ exists, then the following subspace is equal to *i)*–*v)*:

vi) $\mathcal{R}\left(\int_0^{\infty} e^{tA}BB^{\mathsf{T}}e^{tA^{\mathsf{T}}}\,dt\right)$.

Proof. To prove *i)* $\subseteq$ *ii)*, let $\eta \in \bigcap_{t\in[0,t_f]} \mathcal{N}(B^{\mathsf{T}}e^{tA^{\mathsf{T}}})$ so that $\eta^{\mathsf{T}}e^{tA}B = 0$ for all $t \in [0,t_f]$. Now, let $u\colon [0,t_f] \mapsto \mathbb{R}^m$ be continuous. Then, $\eta^{\mathsf{T}}\int_0^{t_f} e^{(t_f-t)A}Bu(t)\,dt = 0$, which implies that $\eta \in \mathcal{C}_{t_f}(A,B)^{\perp}$.

To prove *ii)* $\subseteq$ *iii)*, let $\eta \in \bigcap_{i=0}^{n-1} \mathcal{N}(B^{\mathsf{T}}A^{i\mathsf{T}})$ so that $\eta^{\mathsf{T}}A^iB = 0$ for all $i \in \{0,1,\ldots,n-1\}$. It follows from the Cayley-Hamilton theorem given by Theorem 6.4.7 that $\eta^{\mathsf{T}}A^iB = 0$ for all $i \geq 0$. Now, let $t \in [0,t_f]$. Then, $\eta^{\mathsf{T}}e^{tA}B = \sum_{i=0}^{\infty} t^i(i!)^{-1}\eta^{\mathsf{T}}A^iB = 0$, and thus $\eta \in \mathcal{N}(B^{\mathsf{T}}e^{tA^{\mathsf{T}}})$.

To show that *iii)* $\subseteq$ *iv)*, let $\eta \in \mathcal{R}([B\ \ AB\ \ \cdots\ \ A^{n-1}B])^{\perp}$. It thus follows that $\eta \in \mathcal{N}([B\ \ AB\ \ \cdots\ \ A^{n-1}B]^{\mathsf{T}})$, which implies that $\eta^{\mathsf{T}}A^iB = 0$ for all $i \in \{0,1,\ldots,n-1\}$.

To prove *iv)* $\subseteq$ *v)*, let $\eta \in \mathcal{N}\left(\int_0^{t_f} e^{tA}BB^{\mathsf{T}}e^{tA^{\mathsf{T}}}\,dt\right)$. Therefore, $\eta^{\mathsf{T}}\int_0^{t_f} e^{tA}BB^{\mathsf{T}}e^{tA^{\mathsf{T}}}\,dt\eta = 0$, which implies that $\eta^{\mathsf{T}}e^{tA}B = 0$ for all $t \in [0,t_f]$. Differentiating with respect to t and setting $t = 0$ implies that $\eta^{\mathsf{T}}A^iB = 0$ for all $i \in \{0,1,\ldots,n-1\}$. Hence, $\eta \in \mathcal{R}([B\ \ AB\ \ \cdots\ \ A^{n-1}B])^{\perp}$.

To prove *v)* $\subseteq$ *i)*, let $\eta \in \mathcal{C}_{t_f}(A,B)^{\perp}$. Then, $\eta^{\mathsf{T}}\int_0^{t_f} e^{(t_f-t)A}Bu(t)\,dt = 0$ for all continuous $u\colon [0,t_f] \mapsto \mathbb{R}^m$. Letting $u(t) = B^{\mathsf{T}}e^{(t_f-t)A^{\mathsf{T}}}\eta$ implies $\eta^{\mathsf{T}}\int_0^{t_f} e^{tA}BB^{\mathsf{T}}e^{tA^{\mathsf{T}}}\,dt\eta = 0$, and thus $\eta \in \mathcal{N}\left(\int_0^{t_f} e^{tA}BB^{\mathsf{T}}e^{tA^{\mathsf{T}}}\,dt\right)$. □

The expressions given by *iii)* and *iv)* of Lemma 16.6.2 show that $\mathcal{C}_{t_f}(A,B)$ is independent of t_f. We thus write $\mathcal{C}(A,B)$ for $\mathcal{C}_{t_f}(A,B)$, and call $\mathcal{C}(A,B)$ the *controllable subspace* of (A,B). (A,B) is *controllable* if $\mathcal{C}(A,B) = \mathbb{R}^n$. For convenience, define the $n \times nm$ *controllability matrix*

$$\mathcal{K}(A,B) \triangleq [B\ \ AB\ \ \cdots\ \ A^{n-1}B] \tag{16.6.5}$$

so that

$$\mathcal{C}(A,B) = \mathcal{R}[\mathcal{K}(A,B)]. \tag{16.6.6}$$

Define

$$q \triangleq \dim \mathcal{C}(A,B) = \operatorname{rank} \mathcal{K}(A,B). \tag{16.6.7}$$

Corollary 16.6.3. For all $t_\mathrm{f} > 0$,

$$q = \dim \mathcal{C}(A,B) = \operatorname{rank} \mathcal{K}(A,B) = \operatorname{rank} \int_0^{t_\mathrm{f}} e^{tA}BB^\mathsf{T}e^{tA^\mathsf{T}}\,\mathrm{d}t. \tag{16.6.8}$$

If, in addition, $\lim_{t_\mathrm{f}\to\infty} \int_0^{t_\mathrm{f}} e^{tA}BB^\mathsf{T}e^{tA^\mathsf{T}}\,\mathrm{d}t$ exists, then

$$q = \operatorname{rank} \int_0^{\infty} e^{tA}BB^\mathsf{T}e^{tA^\mathsf{T}}\,\mathrm{d}t. \tag{16.6.9}$$

Corollary 16.6.4. $\mathcal{C}(A,B)$ is an invariant subspace of A.

The following result shows that the controllable subspace $\mathcal{C}(A,B)$ is unchanged by full-state feedback $u(t) = Kx(t) + v(t)$.

Proposition 16.6.5. Let $K \in \mathbb{R}^{m\times n}$. Then,

$$\mathcal{C}(A+BK,B) = \mathcal{C}(A,B). \tag{16.6.10}$$

In particular, (A,B) is controllable if and only if $(A+BK,B)$ is controllable.

Proof. Note that

$$\begin{aligned}\mathcal{C}(A+BK,B) &= \mathcal{R}\left[\mathcal{K}(A+BK,B)\right] = \mathcal{R}([B \quad AB+BKB \quad A^2B+ABKB+BKAB+BKBKB \quad \cdots]) \\ &= \mathcal{R}[\mathcal{K}(A,B)] = \mathcal{C}(A,B).\end{aligned}$$

□

Let $\tilde{\mathcal{C}}(A,B) \subseteq \mathbb{R}^n$ be a subspace that is complementary to $\mathcal{C}(A,B)$. Then, $\tilde{\mathcal{C}}(A,B)$ is an *uncontrollable subspace* in the sense that, if $x_\mathrm{f} = x_\mathrm{f}' + x_\mathrm{f}'' \in \mathbb{R}^n$, where $x_\mathrm{f}' \in \mathcal{C}(A,B)$ and $x_\mathrm{f}'' \in \tilde{\mathcal{C}}(A,B)$ is nonzero, then there exists a continuous control $u\colon\ [0,t_\mathrm{f}] \to \mathbb{R}^m$ such that $x(t_\mathrm{f}) = x_\mathrm{f}'$, but there exists no continuous control such that $x(t_\mathrm{f}) = x_\mathrm{f}$. Using Proposition 4.8.3, let $\mathcal{Q} \in \mathbb{R}^{n\times n}$ be the unique idempotent matrix such that $\mathcal{R}(\mathcal{Q}) = \mathcal{C}(A,B)$ and $\mathcal{N}(\mathcal{Q}) = \tilde{\mathcal{C}}(A,B)$. Then, $x_\mathrm{f}' = \mathcal{Q}x_\mathrm{f}$. The following result constructs $\mathcal{Q}$ and a continuous control $u(\cdot)$ that yields $x(t_\mathrm{f}) = x_\mathrm{f}'$ for $\tilde{\mathcal{C}}(A,B) \triangleq \mathcal{C}(A,B)^\perp$. In this case, $\mathcal{Q}$ is a projector.

Lemma 16.6.6. Let $t_\mathrm{f} > 0$, and define $\mathcal{Q} \in \mathbb{R}^{n\times n}$ by

$$\mathcal{Q} \triangleq \left(\int_0^{t_\mathrm{f}} e^{tA}BB^\mathsf{T}e^{tA^\mathsf{T}}\,\mathrm{d}t\right)^{+} \int_0^{t_\mathrm{f}} e^{tA}BB^\mathsf{T}e^{tA^\mathsf{T}}\,\mathrm{d}t. \tag{16.6.11}$$

Then, $\mathcal{Q}$ is the projector onto $\mathcal{C}(A,B)$, and $\mathcal{Q}_\perp$ is the projector onto $\mathcal{C}(A,B)^\perp$. Hence,

$$\mathcal{R}(\mathcal{Q}) = \mathcal{N}(\mathcal{Q}_\perp) = \mathcal{C}(A,B), \tag{16.6.12}$$

$$\mathcal{N}(\mathcal{Q}) = \mathcal{R}(\mathcal{Q}) = \mathcal{C}(A,B)^\perp, \tag{16.6.13}$$

$$\operatorname{rank} \mathcal{Q} = \operatorname{def} \mathcal{Q}_\perp = \dim \mathcal{C}(A,B) = q, \tag{16.6.14}$$

$$\operatorname{def} \mathcal{Q} = \operatorname{rank} \mathcal{Q}_\perp = \dim \mathcal{C}(A,B)^\perp = n - q. \tag{16.6.15}$$

Now, define $u\colon\ [0,t_\mathrm{f}] \mapsto \mathbb{R}^m$ by

$$u(t) \triangleq B^\mathsf{T}e^{(t_\mathrm{f}-t)A^\mathsf{T}} \left(\int_0^{t_\mathrm{f}} e^{\tau A}BB^\mathsf{T}e^{\tau A^\mathsf{T}}\,\mathrm{d}\tau\right)^{+} x_\mathrm{f}. \tag{16.6.16}$$

If $x_{\rm f} = x'_{\rm f} + x''_{\rm f}$, where $x'_{\rm f} \in \mathcal{C}(A, B)$ and $x''_{\rm f} \in \mathcal{C}(A, B)^\perp$, then

$$x'_{\rm f} = \mathcal{Q}x_{\rm f} = \int_0^{t_{\rm f}} e^{(t_{\rm f}-t)A}Bu(t)\,{\rm d}t. \tag{16.6.17}$$

Finally, (A, B) is controllable if and only if $\mathcal{Q} = I_n$. If these conditions hold, then, for all $x_{\rm f} \in \mathbb{R}^n$,

$$x_{\rm f} = \int_0^{t_{\rm f}} e^{(t_{\rm f}-t)A}Bu(t)\,{\rm d}t, \tag{16.6.18}$$

where $u\colon\ [0, t_{\rm f}] \mapsto \mathbb{R}^m$ is given by

$$u(t) = B^{\mathsf T}e^{(t_{\rm f}-t)A^{\mathsf T}}\left(\int_0^{t_{\rm f}} e^{\tau A}BB^{\mathsf T}e^{\tau A^{\mathsf T}}\,{\rm d}\tau\right)^{-1} x_{\rm f}. \tag{16.6.19}$$

Lemma 16.6.7. Let $\alpha \in \mathbb{R}$. Then,

$$\mathcal{C}(A + \alpha I, B) = \mathcal{C}(A, B). \tag{16.6.20}$$

The following result uses a coordinate transformation to characterize the controllable dynamics of (16.6.1).

Theorem 16.6.8. There exists an orthogonal matrix $S \in \mathbb{R}^{n\times n}$ such that

$$A = S\begin{bmatrix} A_1 & A_{12} \\ 0 & A_2 \end{bmatrix}S^{-1}, \quad B = S\begin{bmatrix} B_1 \\ 0 \end{bmatrix}, \tag{16.6.21}$$

where $A_1 \in \mathbb{R}^{q\times q}$, $B_1 \in \mathbb{R}^{q\times m}$, and (A_1, B_1) is controllable.

Proof. Let $\alpha < 0$ be such that $A_\alpha \triangleq A + \alpha I$ is asymptotically stable, and let $Q \in \mathbb{R}^{n\times n}$ be the positive-semidefinite solution of

$$A_\alpha Q + QA_\alpha^{\mathsf T} + BB^{\mathsf T} = 0 \tag{16.6.22}$$

given by

$$Q = \int_0^\infty e^{tA_\alpha}BB^{\mathsf T}e^{tA_\alpha^{\mathsf T}}\,{\rm d}t.$$

It now follows from Lemma 16.6.2 and Lemma 16.6.7 that

$$\mathcal{R}(Q) = \mathcal{R}[\mathcal{C}(A_\alpha, B)] = \mathcal{R}[\mathcal{C}(A, B)].$$

Hence,

$$\operatorname{rank} Q = \dim \mathcal{C}(A_\alpha, B) = \dim \mathcal{C}(A, B) = q.$$

Next, let $S \in \mathbb{R}^{n\times n}$ be an orthogonal matrix such that $Q = S\left[\begin{smallmatrix} Q_1 & 0 \\ 0 & 0 \end{smallmatrix}\right]S^{\mathsf T}$, where $Q_1 \in \mathbb{R}^{q\times q}$ is positive definite. Writing $A_\alpha = S\left[\begin{smallmatrix} \hat{A}_1 & \hat{A}_{12} \\ \hat{A}_{21} & \hat{A}_2 \end{smallmatrix}\right]S^{-1}$ and $B = S\left[\begin{smallmatrix} B_1 \\ B_2 \end{smallmatrix}\right]$, where $\hat{A}_1 \in \mathbb{R}^{q\times q}$ and $B_1 \in \mathbb{R}^{q\times m}$, it follows from (16.6.22) that

$$\hat{A}_1Q_1 + Q_1\hat{A}_1^{\mathsf T} + B_1B_1^{\mathsf T} = 0,$$
$$\hat{A}_{21}Q_1 + B_2B_1^{\mathsf T} = 0,$$
$$B_2B_2^{\mathsf T} = 0.$$

Therefore, $B_2 = 0$ and $\hat{A}_{21} = 0$, and thus

$$A_\alpha = S\begin{bmatrix} \hat{A}_1 & \hat{A}_{12} \\ 0 & \hat{A}_2 \end{bmatrix}S^{-1}, \quad B = S\begin{bmatrix} B_1 \\ 0 \end{bmatrix}.$$

Furthermore,

$$A = S\begin{bmatrix} \hat{A}_1 & \hat{A}_{12} \\ 0 & \hat{A}_2 \end{bmatrix}S^{-1} - \alpha I = S\left(\begin{bmatrix} \hat{A}_1 & \hat{A}_{12} \\ 0 & \hat{A}_2 \end{bmatrix} - \alpha I\right)S^{-1}.$$

Hence,

$$A = S\begin{bmatrix} A_1 & A_{12} \\ 0 & A_2 \end{bmatrix}S^{-1},$$

where $A_1 \triangleq \hat{A}_1 - \alpha I_q$, $A_{12} \triangleq \hat{A}_{12}$, and $A_2 \triangleq \hat{A}_2 - \alpha I_{n-q}$. □

Proposition 16.6.9. Let $S \in \mathbb{R}^{n\times n}$, and assume that S is orthogonal. Then, the following statements are equivalent:

i) A and B have the form (16.6.21), where $A_1 \in \mathbb{R}^{q\times q}$, $B_1 \in \mathbb{R}^{q\times m}$, and (A_1, B_1) is controllable.

ii) $\mathcal{C}(A,B) = \mathcal{R}\left(S\begin{bmatrix} I_q \\ 0 \end{bmatrix}\right)$.

iii) $\mathcal{C}(A,B)^\perp = \mathcal{R}\left(S\begin{bmatrix} 0 \\ I_{n-q} \end{bmatrix}\right)$.

iv) $\mathcal{Q} = S\begin{bmatrix} I_q & 0 \\ 0 & 0 \end{bmatrix}S^{\mathsf{T}}$.

Proposition 16.6.10. Let $S \in \mathbb{R}^{n\times n}$, and assume that S is nonsingular. Then, the following statements are equivalent:

i) A and B have the form (16.6.21), where $A_1 \in \mathbb{R}^{q\times q}$, $B_1 \in \mathbb{R}^{q\times m}$, and (A_1, B_1) is controllable.

ii) $\mathcal{C}(A,B) = \mathcal{R}\left(S\begin{bmatrix} I_q \\ 0 \end{bmatrix}\right)$ and $\tilde{\mathcal{C}}(A,B) = \mathcal{R}\left(S\begin{bmatrix} 0 \\ I_{n-q} \end{bmatrix}\right)$.

iii) $\mathcal{Q} = S\begin{bmatrix} I_q & 0 \\ 0 & 0 \end{bmatrix}S^{-1}$.

Definition 16.6.11. Let $S \in \mathbb{R}^{n\times n}$, assume that S is nonsingular, and let A and B have the form (16.6.21), where $A_1 \in \mathbb{R}^{q\times q}$, $B_1 \in \mathbb{R}^{q\times m}$, and (A_1, B_1) is controllable. Then, $\lambda \in \mathbb{C}$ is an *uncontrollable eigenvalue* of (A,B) if $\lambda \in \operatorname{spec}(A_2)$, the *uncontrollable spectrum* of (A,B) is $\operatorname{spec}(A_2)$, and the *uncontrollable multispectrum* of (A,B) is $\operatorname{mspec}(A_2)$. Finally, $\lambda \in \operatorname{spec}(A)$ is a *controllable eigenvalue* of (A,B) if λ is not an uncontrollable eigenvalue of (A,B).

Definition 16.6.12. The *controllability pencil* $\mathcal{C}_{A,B}(s)$ is the pencil

$$\mathcal{C}_{A,B} = P_{[A\ -B],[I\ 0]}. \tag{16.6.23}$$

Equivalently,

$$\mathcal{C}_{A,B}(s) = [sI - A\ \ B]. \tag{16.6.24}$$

Proposition 16.6.13. Let $\lambda \in \operatorname{spec}(A)$. Then, λ is an uncontrollable eigenvalue of (A,B) if and only if

$$\operatorname{rank}\,[\lambda I - A\ \ B] < n. \tag{16.6.25}$$

Proof. Since (A_1, B_1) is controllable, it follows from (16.6.21) that

$$\begin{aligned}\operatorname{rank}\,[\lambda I - A\ \ B] &= \operatorname{rank}\begin{bmatrix} \lambda I - A_1 & A_{12} & B_1 \\ 0 & \lambda I - A_2 & 0 \end{bmatrix} \\ &= \operatorname{rank}\,[\lambda I - A_1\ \ B_1] + \operatorname{rank}(\lambda I - A_2) \\ &= q + \operatorname{rank}(\lambda I - A_2).\end{aligned}$$

Hence, $\operatorname{rank}\,[\lambda I - A\ \ B] < n$ if and only if $\operatorname{rank}(\lambda I - A_2) < n - q$; that is, if and only if $\lambda \in \operatorname{spec}(A_2)$. □

Proposition 16.6.14. Let $\lambda \in \operatorname{mspec}(A)$ and $K \in \mathbb{R}^{m\times n}$. Then, λ is an uncontrollable eigenvalue of (A, B) if and only if λ is an uncontrollable eigenvalue of $(A + BK, B)$.

Proof. Theorem 16.6.8 implies that there exists an orthogonal matrix $S \in \mathbb{R}^{n\times n}$ such that (16.6.21) holds, where $A_1 \in \mathbb{R}^{q\times q}$, $B_1 \in \mathbb{R}^{q\times m}$, and (A_1, B_1) is controllable. Partitioning $KS = [K_1\ \ K_2]$, where $K_1 \in \mathbb{R}^{m\times q}$ and $K_2 \in \mathbb{R}^{m\times(n-q)}$, it follows that

$$A + BK = S\begin{bmatrix} A_1 + B_1K_1 & A_{12} + B_1K_2 \\ 0 & A_2 \end{bmatrix}S^{-1}.$$

Therefore, the uncontrollable eigenvalues of $A + BK$ are the elements of $\operatorname{spec}(A_2)$. □

Proposition 16.6.15. Assume that (A, B) is controllable. Then, the Smith form of $\mathcal{C}_{A,B}$ is $[I_n\ \ 0_{n\times m}]$.

Proof. First, note that, if $\lambda \in \mathbb{C}$ is not an eigenvalue of A, then $n = \operatorname{rank}(\lambda I - A) = \operatorname{rank}[\lambda I - A\ \ B] = \operatorname{rank}\mathcal{C}_{A,B}(\lambda)$. Therefore, $\operatorname{rank}\mathcal{C}_{A,B} = n$, and thus $\mathcal{C}_{A,B}$ has n Smith polynomials. Furthermore, since (A, B) is controllable, it follows that (A, B) has no uncontrollable eigenvalues. Therefore, it follows from Proposition 16.6.13 that, for all $\lambda \in \operatorname{spec}(A)$, $\operatorname{rank}[\lambda I - A\ \ B] = n$. Consequently, $\operatorname{rank}\mathcal{C}_{A,B}(\lambda) = n$ for all $\lambda \in \mathbb{C}$. Thus, every Smith polynomial $\mathcal{C}_{A,B}$ is 1. □

Proposition 16.6.16. Let $S \in \mathbb{R}^{n\times n}$, assume that S is nonsingular, and let A and B have the form (16.6.21), where $A_1 \in \mathbb{R}^{q\times q}$, $B_1 \in \mathbb{R}^{q\times m}$, and (A_1, B_1) is controllable. Furthermore, let $p_1, \ldots, p_{n-q}$ be the similarity invariants of A_2, where, for all $i \in \{1, \ldots, n-q-1\}$, p_i divides p_{i+1}. Then, there exist unimodular matrices $S_1 \in \mathbb{R}[s]^{n\times n}$ and $S_2 \in \mathbb{R}[s]^{(n+m)\times(n+m)}$ such that, for all $s \in \mathbb{C}$,

$$[sI - A\ \ B] = S_1(s)\begin{bmatrix} I_q & & & \\ & p_1(s) & & 0_{n\times m} \\ & & \ddots & \\ & & & p_{n-q}(s) \end{bmatrix}S_2(s). \tag{16.6.26}$$

Consequently,

$$\operatorname{Szeros}(\mathcal{C}_{A,B}) = \bigcup_{i=1}^{n-q}\operatorname{roots}(p_i) = \operatorname{roots}(\chi_{A_2}) = \operatorname{spec}(A_2), \tag{16.6.27}$$

$$\operatorname{mSzeros}(\mathcal{C}_{A,B}) = \bigcup_{i=1}^{n-q}\operatorname{mroots}(p_i) = \operatorname{mroots}(\chi_{A_2}) = \operatorname{mspec}(A_2). \tag{16.6.28}$$

Proof. Define $S \in \mathbb{R}^{n\times n}$ as in Theorem 16.6.8, let $\hat{S}_1 \in \mathbb{R}[s]^{q\times q}$ and $\hat{S}_2 \in \mathbb{R}[s]^{(q+m)\times(q+m)}$ be unimodular matrices such that

$$\hat{S}_1(s)[sI_q - A_1\ \ B_1]\hat{S}_2(s) = [I_q\ \ 0_{q\times m}],$$

and let $\hat{S}_3, \hat{S}_4 \in \mathbb{R}^{(n-q)\times(n-q)}$ be unimodular matrices such that

$$\hat{S}_3(s)(sI - A_2)\hat{S}_4(s) = \hat{P}(s),$$

where $\hat{P} \triangleq \operatorname{diag}(p_1, \ldots, p_{n-q})$. Then,

$$[sI - A\ \ B] = S\begin{bmatrix} \hat{S}_1^{-1}(s) & 0 \\ 0 & \hat{S}_3^{-1}(s) \end{bmatrix}\begin{bmatrix} I_q & 0 & 0_{q\times m} \\ 0 & \hat{P}(s) & 0 \end{bmatrix}$$

$$\times\begin{bmatrix} I_q & 0 & -\hat{S}_1(s)A_{12} \\ 0 & 0 & \hat{S}_4^{-1}(s) \\ 0 & I_m & 0 \end{bmatrix}\begin{bmatrix} \hat{S}_2^{-1}(s) & 0 \\ 0 & I_{n-q} \end{bmatrix}\begin{bmatrix} I_q & 0 & 0_{q\times m} \\ 0 & 0 & I_m \\ 0 & I_{n-q} & 0 \end{bmatrix}\begin{bmatrix} S^{-1} & 0 \\ 0 & I_m \end{bmatrix}.$$ □

Proposition 16.6.17. Let $s \in \mathbb{C}$. Then,

$$\mathcal{C}(A,B) \subseteq \operatorname{Re}\mathcal{R}([sI - A \;\; B]). \tag{16.6.29}$$

Proof. Using Proposition 16.6.9 and the notation in the proof of Proposition 16.6.16, it follows that, for all $s \in \mathbb{C}$,

$$\mathcal{C}(A,B) = \mathcal{R}\left(S\begin{bmatrix} I_q \\ 0 \end{bmatrix}\right) \subseteq \mathcal{R}\left(S\begin{bmatrix} \hat{S}_1^{-1}(s) & 0 \\ 0 & \hat{S}_3^{-1}(s)\hat{P}(s) \end{bmatrix}\right) = \mathcal{R}([sI - A \;\; B]).$$

Finally, (16.6.29) follows from the fact that $\mathcal{C}(A,B)$ is a real subspace. □

The next result characterizes controllability in several equivalent ways.

Theorem 16.6.18. The following statements are equivalent:

i) (A,B) is controllable.

ii) There exists $t > 0$ such that $\int_0^t e^{\tau A}BB^{\mathsf{T}}e^{\tau A^{\mathsf{T}}}\,\mathrm{d}\tau$ is positive definite.

iii) For all $t > 0$, $\int_0^t e^{\tau A}BB^{\mathsf{T}}e^{\tau A^{\mathsf{T}}}\,\mathrm{d}\tau$ is positive definite.

iv) $\operatorname{rank}\mathcal{K}(A,B) = n$.

v) Every eigenvalue of (A,B) is controllable.

vi) If $x \in \mathbb{C}^n$ is an eigenvector of A, then $Cx \neq 0$.

vii) There exists $K \in \mathbb{R}^{m\times n}$ such that $(A+BK, B)$ is controllable.

viii) For all $K \in \mathbb{R}^{m\times n}$, $(A+BK, B)$ is controllable.

If, in addition, $\lim_{t\to\infty}\int_0^t e^{\tau A}BB^{\mathsf{T}}e^{\tau A^{\mathsf{T}}}\,\mathrm{d}\tau$ exists, then the following statement is equivalent to *i)*–*viii)*:

ix) $\int_0^\infty e^{tA}BB^{\mathsf{T}}e^{tA^{\mathsf{T}}}\,\mathrm{d}t$ is positive definite.

Proof. The equivalence of *i)*–*iv)* follows from Lemma 16.6.2.

To prove *iv)* $\Longrightarrow$ *v)*, suppose that *v)* does not hold; that is, there exist $\lambda \in \operatorname{spec}(A)$ and a nonzero vector $x \in \mathbb{C}^n$ such that $x^*A = \lambda x^*$ and $x^*B = 0$. It thus follows that $x^*AB = \lambda x^*B = 0$. Similarly, $x^*A^iB = 0$ for all $i \in \{0,1,\ldots,n-1\}$. Hence, $(\operatorname{Re}x)^{\mathsf{T}}\mathcal{K}(A,B) = 0$ and $(\operatorname{Im}x)^{\mathsf{T}}\mathcal{K}(A,B) = 0$. Since $\operatorname{Re}x$ and $\operatorname{Im}x$ are not both zero, it follows that $\dim\mathcal{C}(A,B) < n$.

Conversely, to show that *v)* implies *iv)*, suppose that $\operatorname{rank}\mathcal{K}(A,B) < n$. Then, there exists a nonzero vector $x \in \mathbb{R}^n$ such that $x^{\mathsf{T}}A^iB = 0$ for all $i \in \{0,\ldots,n-1\}$. Now, let $p \in \mathbb{R}[s]$ be a nonzero polynomial of minimal degree such that $x^{\mathsf{T}}p(A) = 0$. Note that p is not a constant polynomial and that $x^{\mathsf{T}}\mu_A(A) = 0$. Thus, $1 \le \deg p \le \deg\mu_A$. Now, let $\lambda \in \mathbb{C}$ satisfy $p(\lambda) = 0$, and let $q \in \mathbb{C}[s]$ satisfy $p(s) = q(s)(s-\lambda)$. Since $\deg q < \deg p$, it follows that $x^{\mathsf{T}}q(A) \neq 0$. Therefore, $\eta \triangleq q(A)^*x$ is nonzero. Furthermore, $\eta^*(A - \lambda I) = x^{\mathsf{T}}p(A) = 0$. Since, for all $i \in \{0,\ldots,n-1\}$, $x^{\mathsf{T}}A^iB = 0$, it follows that $\eta^*B = x^{\mathsf{T}}q(A)B = 0$. Consequently, *v)* does not hold. □

The following result, which is a restatement of the equivalence of *i)* and *v)* of Theorem 16.6.18, is the *PBH test for controllability.*

Corollary 16.6.19. The following statements are equivalent:

i) (A,B) is controllable.

ii) For all $s \in \mathbb{C}$,

$$\operatorname{rank}\,[sI - A \;\; B] = n. \tag{16.6.30}$$

The following result implies that arbitrary eigenvalue placement is possible for (16.6.1) if (A,B) is controllable.

Proposition 16.6.20. The pair (A,B) is controllable if and only if, for every polynomial $p \in \mathbb{R}[s]$ such that $\deg p = n$, there exists a matrix $K \in \mathbb{R}^{m\times n}$ such that $\operatorname{mspec}(A+BK) = \operatorname{mroots}(p)$.

Proof. For the case $m = 1$, define $A_{\mathrm{c}} \triangleq C(\chi_A)$ and $B_{\mathrm{c}} \triangleq e_n$ as in (16.9.6). Then, Proposition

16.9.3 implies that $\mathcal{K}(A_c, B_c)$ is nonsingular, while Corollary 16.9.21 implies that $A_c = S^{-1}AS$ and $B_c = S^{-1}B$. Now, let $\text{mroots}(p) = \{\lambda_1, \ldots, \lambda_n\}_{\text{ms}} \subset \mathbb{C}$. Letting $K \triangleq e_n^{\mathsf{T}}[C(p) - A_c]S^{-1}$ it follows that

$$A + BK = S(A_c + B_c KS)S^{-1} = S(A_c + E_{n,n}[C(p) - A_c])S^{-1} = SC(p)S^{-1}.$$

The case $m \geq 2$ requires the multivariable controllable canonical form. See [2349, p. 248]. □

16.7 Controllable Asymptotic Stability

Let $A \in \mathbb{R}^{n\times n}$ and $B \in \mathbb{R}^{n\times m}$, and define $q \triangleq \dim \mathcal{C}(A, C)$.

Definition 16.7.1. (A, B) is *controllably asymptotically stable* if

$$\mathcal{C}(A, B) \subseteq \mathcal{S}_{\mathrm{s}}(A). \tag{16.7.1}$$

Proposition 16.7.2. Let $K \in \mathbb{R}^{m\times n}$. Then, (A, B) is controllably asymptotically stable if and only if $(A + BK, B)$ is controllably asymptotically stable.

Proposition 16.7.3. The following statements are equivalent:

i) (A, B) is controllably asymptotically stable.

ii) There exists an orthogonal matrix $S \in \mathbb{R}^{n\times n}$ such that (16.6.21) holds, where $A_1 \in \mathbb{R}^{q\times q}$ is asymptotically stable and $B_1 \in \mathbb{R}^{q\times m}$.

iii) There exists a nonsingular matrix $S \in \mathbb{R}^{n\times n}$ such that (16.6.21) holds, where $A_1 \in \mathbb{R}^{q\times q}$ is asymptotically stable and $B_1 \in \mathbb{R}^{q\times m}$.

iv) $\lim_{t\to\infty} e^{tA}B = 0$.

v) The positive-semidefinite matrix $Q \in \mathbb{R}^{n\times n}$ defined by

$$Q \triangleq \int_0^\infty e^{tA}BB^{\mathsf{T}}e^{tA^{\mathsf{T}}}\,\mathrm{d}t \tag{16.7.2}$$

exists.

vi) There exists a positive-semidefinite matrix $Q \in \mathbb{R}^{n\times n}$ satisfying

$$AQ + QA^{\mathsf{T}} + BB^{\mathsf{T}} = 0. \tag{16.7.3}$$

If these conditions hold, then the positive-semidefinite matrix $Q \in \mathbb{R}^{n\times n}$ defined by (16.7.2) satisfies (16.7.3).

Proof. To prove *i)* $\Longrightarrow$ *ii)*, note that, since (A, B) is controllably asymptotically stable, it follows that $\mathcal{C}(A, B) \subseteq \mathcal{S}_{\mathrm{s}}(A) = \mathcal{N}[\mu_A^{\mathrm{s}}(A)] = \mathcal{R}[\mu_A^{\mathrm{u}}(A)]$. Then, Theorem 16.6.8 implies that there exists an orthogonal matrix $S \in \mathbb{R}^{n\times n}$ such that (16.6.21) holds, where $A_1 \in \mathbb{R}^{q\times q}$ and (A_1, B_1) is controllable. Thus, $\mathcal{R}\big(S\big[\begin{smallmatrix} I_q \\ 0 \end{smallmatrix}\big]\big) = \mathcal{C}(A, B) \subseteq \mathcal{R}[\mu_A^{\mathrm{s}}(A)]$. Next, note that

$$\mu_A^{\mathrm{s}}(A) = S\begin{bmatrix} \mu_A^{\mathrm{s}}(A_1) & B_{12\mathrm{s}} \\ 0 & \mu_A^{\mathrm{s}}(A_2) \end{bmatrix}S^{-1},$$

where $B_{12\mathrm{s}} \in \mathbb{R}^{q\times(n-q)}$, and suppose that A_1 is not asymptotically stable with CRHP eigenvalue λ. Then, $\lambda \notin \text{roots}(\mu_A^{\mathrm{s}})$, and thus $\mu_A^{\mathrm{s}}(A_1) \neq 0$. Let $x_1 \in \mathbb{R}^{n-q}$ satisfy $\mu_A^{\mathrm{s}}(A_1)x_1 \neq 0$. Then,

$$\begin{bmatrix} x_1 \\ 0 \end{bmatrix} \in \mathcal{R}\left(S\begin{bmatrix} I_q \\ 0 \end{bmatrix}\right) = \mathcal{C}(A, B), \quad \mu_A^{\mathrm{s}}(A)S\begin{bmatrix} x_1 \\ 0 \end{bmatrix} = S\begin{bmatrix} \mu_A^{\mathrm{s}}(A_1)x_1 \\ 0 \end{bmatrix},$$

and thus $\big[\begin{smallmatrix} x_1 \\ 0 \end{smallmatrix}\big] \notin \mathcal{N}[\mu_A^{\mathrm{s}}(A)] = \mathcal{S}_{\mathrm{s}}(A)$, which implies that $\mathcal{C}(A, B)$ is not contained in $\mathcal{S}_{\mathrm{s}}(A)$. Therefore, A_1 is asymptotically stable.

To prove *iii)* $\Longrightarrow$ *iv)*, note that $e^{tA}B = S\big[\begin{smallmatrix} e^{tA_1}B_1 \\ 0 \end{smallmatrix}\big] \to 0$ as $t \to \infty$.

Next, to prove *iv)* $\Longrightarrow$ *v)*, note that, since $e^{tA}B \to 0$ as $t \to \infty$, it follows that every entry of $e^{tA}B$ involves exponentials of t, where the coefficients of t have negative real part. Hence, so does every entry of $e^{tA}BB^{\mathsf{T}}e^{tA^{\mathsf{T}}}$, which implies that $\int_0^\infty e^{tA}BB^{\mathsf{T}}e^{tA^{\mathsf{T}}}\,\mathrm{d}t$ exists.

To prove *v)* $\Longrightarrow$ *vi)*, note that, since $Q = \int_0^\infty e^{tA}BB^{\mathsf{T}}e^{tA^{\mathsf{T}}}\,\mathrm{d}t$ exists, it follows that $e^{tA}BB^{\mathsf{T}}e^{tA^{\mathsf{T}}} \to 0$ as $t \to \infty$. Thus,

$$AQ + QA^{\mathsf{T}} = \int_0^\infty \Big[Ae^{tA}BB^{\mathsf{T}}e^{tA^{\mathsf{T}}} + e^{tA}BB^{\mathsf{T}}e^{tA^{\mathsf{T}}}A^{\mathsf{T}}\Big]\,\mathrm{d}t = \int_0^\infty \tfrac{\mathrm{d}}{\mathrm{d}t}e^{tA}BB^{\mathsf{T}}e^{tA^{\mathsf{T}}}\,\mathrm{d}t = \lim_{t\to\infty} e^{tA}BB^{\mathsf{T}}e^{tA^{\mathsf{T}}} - BB^{\mathsf{T}} = -BB^{\mathsf{T}},$$

which shows that Q satisfies (16.7.3).

To prove *vi)* $\Longrightarrow$ *i)*, note that, for all $t \ge 0$,

$$\int_0^t e^{\tau A}BB^{\mathsf{T}}e^{\tau A^{\mathsf{T}}}\,\mathrm{d}\tau = -\int_0^t e^{\tau A}(AQ + QA^{\mathsf{T}})e^{\tau A^{\mathsf{T}}}\,\mathrm{d}\tau = -\int_0^t \frac{\mathrm{d}}{\mathrm{d}\tau}e^{\tau A}Qe^{\tau A^{\mathsf{T}}}\,\mathrm{d}\tau = Q - e^{tA}Qe^{tA^{\mathsf{T}}} \le Q.$$

Next, Theorem 16.6.8 implies that there exists an orthogonal matrix $S \in \mathbb{R}^{n\times n}$ such that (16.6.21) holds, where $A_1 \in \mathbb{R}^{q\times q}$, $B_1 \in \mathbb{R}^{q\times m}$, and (A_1, B_1) is controllable. Consequently,

$$\int_0^t e^{\tau A_1}B_1B_1^{\mathsf{T}}e^{\tau A_1^{\mathsf{T}}}\,\mathrm{d}\tau = [I\ \ 0]S^{\mathsf{T}}\int_0^t e^{\tau A}BB^{\mathsf{T}}e^{\tau A^{\mathsf{T}}}\,\mathrm{d}\tau S\begin{bmatrix} I \\ 0 \end{bmatrix} \le [I\ \ 0]S^{\mathsf{T}}QS\begin{bmatrix} I \\ 0 \end{bmatrix}.$$

Thus, Proposition 10.6.3 implies that $Q_1 \triangleq \int_0^\infty e^{tA_1}B_1B_1^{\mathsf{T}}e^{tA_1^{\mathsf{T}}}\,\mathrm{d}t$ exists. Since (A_1, B_1) is controllable, it follows from *vii)* of Theorem 16.6.18 that Q_1 is positive definite.

Now, let λ be an eigenvalue of A_1^{T}, and let $x_1 \in \mathbb{C}^n$ be an associated eigenvector. Consequently, $\alpha \triangleq x_1^*Q_1x_1$ is positive, and

$$\alpha = x_1^*\int_0^\infty e^{\overline{\lambda}t}BB_1^{\mathsf{T}}e^{\lambda t}\,\mathrm{d}t x_1 = x_1^*B_1B_1^{\mathsf{T}}x_1\int_0^\infty e^{2(\mathrm{Re}\,\lambda)t}\,\mathrm{d}t.$$

Hence, $\int_0^\infty e^{2(\mathrm{Re}\,\lambda)t}\,\mathrm{d}t = \alpha/(x_1^*B_1B_1^{\mathsf{T}}x_1)$ exists, and thus $\mathrm{Re}\,\lambda < 0$. Consequently, A_1 is asymptotically stable, and thus $\mathcal{C}(A, B) \subseteq \mathcal{S}_{\mathrm{s}}(A)$; that is, (A, B) is controllably asymptotically stable. □

The matrix $Q \in \mathbb{R}^{n\times n}$ defined by (16.7.2) is the *controllability Gramian*, and (16.7.3) is the *control Lyapunov equation*.

Proposition 16.7.4. Assume that (A, B) is controllably asymptotically stable, let $Q \in \mathbb{R}^{n\times n}$ be the positive-semidefinite matrix defined by (16.7.2), and define $\mathcal{Q} \in \mathbb{R}^{n\times n}$ by (16.6.11). Then, the following statements hold:

i) $QQ^+ = \mathcal{Q}$.

ii) $\mathcal{R}(Q) = \mathcal{R}(\mathcal{Q}) = \mathcal{C}(A, B)$.

iii) $\mathcal{N}(Q) = \mathcal{N}(\mathcal{Q}) = \mathcal{C}(A, B)^\perp$.

iv) $\operatorname{rank} Q = \operatorname{rank} \mathcal{Q} = q$.

v) Q is the unique positive-semidefinite solution of (16.7.3) whose rank is q.

Proof. See [2475] for the proof of *v)*. □

Proposition 16.7.5. Assume that (A, B) is controllably asymptotically stable, let $Q \in \mathbb{R}^{n\times n}$ be the positive-semidefinite matrix defined by (16.7.2), and let $\hat{Q} \in \mathbb{R}^{n\times n}$. Then, the following statements are equivalent:

i) $\hat{Q}$ is positive semidefinite and satisfies (16.7.3).

ii) There exists a positive-semidefinite matrix $Q_0 \in \mathbb{R}^{n\times n}$ such that $\hat{Q} = Q + Q_0$ and $AQ_0 + Q_0A^\mathsf{T} = 0$.

If these conditions hold, then

$$\operatorname{rank} \hat{Q} = q + \operatorname{rank} Q_0, \tag{16.7.4}$$

$$\operatorname{rank} Q_0 \le \sum_{\substack{\lambda\in\operatorname{spec}(A)\\ \lambda\in\mathrm{IA}}} \operatorname{gmult}_A(\lambda). \tag{16.7.5}$$

Proof. See [2475]. □

Proposition 16.7.6. The following statements are equivalent:

i) (A, B) is controllably asymptotically stable, every imaginary eigenvalue of A is semisimple, and A has no ORHP eigenvalues.

ii) (16.7.3) has a positive-definite solution $Q \in \mathbb{R}^{n\times n}$.

Proof. See [2475]. □

Proposition 16.7.7. The following statements are equivalent:

i) (A, B) is controllably asymptotically stable, and A has no imaginary eigenvalues.

ii) (16.7.3) has exactly one positive-semidefinite solution $Q \in \mathbb{R}^{n\times n}$.

If these conditions hold, then $Q \in \mathbb{R}^{n\times n}$ is given by (16.7.2) and satisfies $\operatorname{rank} Q = q$.

Proof. See [2475]. □

Corollary 16.7.8. Assume that A is asymptotically stable. Then, the positive-semidefinite matrix $Q \in \mathbb{R}^{n\times n}$ defined by (16.7.2) is the unique solution of (16.7.3) and satisfies $\operatorname{rank} Q = q$.

Proof. See [2475]. □

Proposition 16.7.9. The following statements are equivalent:

i) (A, B) is controllable, and A is asymptotically stable.

ii) (16.7.3) has exactly one positive-semidefinite solution $Q \in \mathbb{R}^{n\times n}$, and Q is positive definite.

If these conditions hold, then $Q \in \mathbb{R}^{n\times n}$ is given by (16.7.2).

Proof. See [2475]. □

Corollary 16.7.10. Assume that A is asymptotically stable. Then, the positive-semidefinite matrix $Q \in \mathbb{R}^{n\times n}$ defined by (16.7.2) exists. Furthermore, Q is positive definite if and only if (A, B) is controllable.

16.8 Stabilizability

Let $A \in \mathbb{R}^{n\times n}$ and $B \in \mathbb{R}^{n\times m}$, and define $q \triangleq \dim \mathcal{C}(A, B)$.

Definition 16.8.1. The *stabilizable subspace* $\mathcal{C}_\mathrm{s}(A, B)$ of (A, B) is

$$\mathcal{C}_\mathrm{s}(A, B) \triangleq \mathcal{C}(A, B) + \mathcal{S}_\mathrm{s}(A). \tag{16.8.1}$$

Furthermore, (A, B) is *stabilizable* if $\mathcal{C}_\mathrm{s}(A, B) = \mathbb{R}^n$.

Proposition 16.8.2. (A, B) is stabilizable if and only if

$$\mathcal{S}_\mathrm{u}(A) \subseteq \mathcal{C}(A, B). \tag{16.8.2}$$

Proposition 16.8.3. Let $K \in \mathbb{R}^{m\times n}$. Then, (A, B) is stabilizable if and only if $(A + BK, B)$ is stabilizable.

Proposition 16.8.4. The following statements are equivalent:

i) (A, B) is stabilizable.

ii) There exists an orthogonal matrix $S \in \mathbb{R}^{n\times n}$ such that (16.6.21) holds, where $A_1 \in \mathbb{R}^{q\times q}$, $B_1 \in \mathbb{R}^{q\times m}$, (A_1, B_1) is controllable, and $A_2 \in \mathbb{R}^{(n-q)\times(n-q)}$ is asymptotically stable.

iii) There exists a nonsingular matrix $S \in \mathbb{R}^{n\times n}$ such that (16.6.21) holds, where $A_1 \in \mathbb{R}^{q\times q}$, $B_1 \in \mathbb{R}^{q\times m}$, (A_1, B_1) is controllable, and $A_2 \in \mathbb{R}^{(n-q)\times(n-q)}$ is asymptotically stable.

iv) Every CRHP eigenvalue of (A, B) is controllable.

Proof. To prove *i*) $\Longrightarrow$ *ii*), note that, since (A, B) is stabilizable, it follows that $\mathcal{S}_{\mathrm{u}}(A) = \mathcal{N}[\mu_A^{\mathrm{u}}(A)] = \mathcal{R}[\mu_A^{\mathrm{s}}(A)] \subseteq \mathcal{C}(A, B)$. Using Theorem 16.6.8, it follows that there exists an orthogonal matrix $S \in \mathbb{R}^{n\times n}$ such that (16.6.21) holds, where $A_1 \in \mathbb{R}^{q\times q}$ and (A_1, B_1) is controllable. Thus, $\mathcal{R}[\mu_A^{\mathrm{s}}(A)] \subseteq \mathcal{C}(A, B) = \mathcal{R}\left(S\left[\begin{smallmatrix} I_q \\ 0 \end{smallmatrix}\right]\right)$.

Next, note that

$$\mu_A^{\mathrm{s}}(A) = S\begin{bmatrix} \mu_A^{\mathrm{s}}(A_1) & B_{12\mathrm{s}} \\ 0 & \mu_A^{\mathrm{s}}(A_2) \end{bmatrix} S^{-1},$$

where $B_{12\mathrm{s}} \in \mathbb{R}^{q\times(n-q)}$, and suppose that A_2 is not asymptotically stable with CRHP eigenvalue λ. Then, $\lambda \notin \mathrm{roots}(\mu_A^{\mathrm{s}})$, and thus $\mu_A^{\mathrm{s}}(A_2) \neq 0$. Let $x_2 \in \mathbb{R}^{n-q}$ satisfy $\mu_A^{\mathrm{s}}(A_2)x_2 \neq 0$. Then,

$$\mu_A^{\mathrm{s}}(A)S\begin{bmatrix} 0 \\ x_2 \end{bmatrix} = S\begin{bmatrix} B_{12\mathrm{s}}x_2 \\ \mu_A^{\mathrm{s}}(A_2)x_2 \end{bmatrix} \notin \mathcal{R}\left(S\begin{bmatrix} I_q \\ 0 \end{bmatrix}\right) = \mathcal{C}(A, B),$$

which implies that $\mathcal{S}_{\mathrm{u}}(A)$ is not contained in $\mathcal{C}(A, B)$. Hence, A_2 is asymptotically stable.

The statement *ii*) $\Longrightarrow$ *iii*) is immediate. To prove *iii*) $\Longrightarrow$ *iv*), let $\lambda \in \mathrm{spec}(A)$ be a CRHP eigenvalue of A. Since A_2 is asymptotically stable, it follows that $\lambda \notin \mathrm{spec}(A_2)$. Consequently, Proposition 16.6.13 implies that λ is not an uncontrollable eigenvalue of (A, B), and thus λ is a controllable eigenvalue of (A, B).

To prove *iv*) $\Longrightarrow$ *i*), let $S \in \mathbb{R}^{n\times n}$ be nonsingular and such that A and B have the form (16.6.21), where $A_1 \in \mathbb{R}^{q\times q}$, $B_1 \in \mathbb{R}^{q\times m}$, and (A_1, B_1) is controllable. Since every CRHP eigenvalue of (A, B) is controllable, it follows from Proposition 16.6.13 that A_2 is asymptotically stable. From Fact 15.24.4 it follows that $\mathcal{S}_{\mathrm{u}}(A) \subseteq \mathcal{R}\left(S\left[\begin{smallmatrix} I_q \\ 0 \end{smallmatrix}\right]\right) = \mathcal{C}(A, B)$, which implies that (A, B) is stabilizable. $\square$

The following result, which is a restatement of the equivalence of *i*) and *iv*) of Proposition 16.8.4, is the *PBH test for stabilizability*.

Corollary 16.8.5. The following statements are equivalent:

i) (A, B) is stabilizable.

ii) For all $s \in$ CRHP,

$$\mathrm{rank}\,[sI - A \;\; B] = n. \tag{16.8.3}$$

Proposition 16.8.6. The following statements are equivalent:

i) A is asymptotically stable.

ii) (A, B) is controllably asymptotically stable and stabilizable.

Proof. To prove *i*) $\Longrightarrow$ *ii*), note that, since A is asymptotically stable, it follows that $\mathcal{S}_{\mathrm{u}}(A) = \{0\}$, and $\mathcal{S}_{\mathrm{s}}(A) = \mathbb{R}^n$. Thus, $\mathcal{S}_{\mathrm{u}}(A) \subseteq \mathcal{C}(A, B)$, and $\mathcal{C}(A, B) \subseteq \mathcal{S}_{\mathrm{s}}(A)$.

To prove *ii*) $\Longrightarrow$ *i*), note that, since (A, B) is stabilizable and controllably asymptotically stable, it follows that $\mathcal{S}_{\mathrm{u}}(A) \subseteq \mathcal{C}(A, B) \subseteq \mathcal{S}_{\mathrm{s}}(A)$, and thus $\mathcal{S}_{\mathrm{u}}(A) = \{0\}$.

As an alternative proof that *ii*) $\Longrightarrow$ *i*), note that, since (A, B) is stabilizable, it follows from Proposition 16.8.4 that there exists a nonsingular matrix $S \in \mathbb{R}^{n\times n}$ such that (16.6.21) holds, where $A_1 \in \mathbb{R}^{q\times q}$, $B_1 \in \mathbb{R}^{q\times m}$, (A_1, B_1) is controllable, and $A_2 \in \mathbb{R}^{(n-q)\times(n-q)}$ is asymptotically stable. Then,

$$\int_0^\infty e^{tA}BB^{\mathsf{T}}e^{tA^{\mathsf{T}}}\,\mathrm{d}t = S\begin{bmatrix} \int_0^\infty e^{tA_1}B_1B_1^{\mathsf{T}}e^{tA_1^{\mathsf{T}}}\,\mathrm{d}t & 0 \\ 0 & 0 \end{bmatrix} S^{-1}.$$

Since the integral on the left-hand side exists by assumption, the integral on the right-hand side also exists. Since (A_1, B_1) is controllable, it follows from *vii*) of Theorem 16.6.18 that $Q_1 \triangleq \int_0^\infty e^{tA_1}B_1B_1^{\mathsf{T}}e^{tA_1^{\mathsf{T}}}\,\mathrm{d}t$ is positive definite.

Now, let λ be an eigenvalue of A_1^{T}, and let $x_1 \in \mathbb{C}^q$ be an associated eigenvector. Consequently, $\alpha \triangleq x_1^*Q_1x_1$ is positive, and

$$\alpha = x_1^*\int_0^\infty e^{\bar{\lambda}t}B_1B_1^{\mathsf{T}}e^{\lambda t}\,\mathrm{d}t x_1 = x_1^*B_1B_1^{\mathsf{T}}x_1\int_0^\infty e^{2(\mathrm{Re}\,\lambda)t}\,\mathrm{d}t.$$

Hence, $\int_0^\infty e^{2(\mathrm{Re}\,\lambda)t}\,\mathrm{d}t$ exists, and thus $\mathrm{Re}\,\lambda < 0$. Consequently, A_1 is asymptotically stable, and thus A is asymptotically stable. □

Corollary 16.8.7. The following statements are equivalent:

i) There exists a positive-semidefinite matrix $Q \in \mathbb{R}^{n\times n}$ satisfying (16.7.3), and (A, B) is stabilizable.

ii) A is asymptotically stable.

Proof. This result follows from Proposition 16.7.3 and Proposition 16.8.6. □

16.9 Realization Theory

Given a proper rational transfer function G, we wish to determine (A, B, C, D) such that (16.2.10) holds. The following terminology is convenient.

Definition 16.9.1. Let $G \in \mathbb{R}^{l\times m}(s)$. If $l = m = 1$, then G is a *single-input/single-output (SISO)* rational transfer function; if $l = 1$ and $m > 1$, then G is a *multiple-input/single-output (MISO)* rational transfer function; if $l > 1$ and $m = 1$, then G is a *single-input/multiple-output (SIMO)* rational transfer function; and, if $l > 1$ and $m > 1$, then G is a *multiple-input/multiple-output (MIMO)* rational transfer function.

Definition 16.9.2. Let $G \in \mathbb{R}(s)^{l\times m}_{\mathrm{prop}}$, and assume that $A \in \mathbb{R}^{n\times n}$, $B \in \mathbb{R}^{n\times m}$, $C \in \mathbb{R}^{l\times n}$, and $D \in \mathbb{R}^{l\times m}$ satisfy $G(s) = C(sI - A)^{-1}B + D$. Then, $\left[\begin{array}{c|c} A & B \\ \hline C & D \end{array}\right]$ is a *realization* of G, which is written as

$$G \sim \left[\begin{array}{c|c} A & B \\ \hline C & D \end{array}\right]. \tag{16.9.1}$$

The *order* of the realization (16.9.1) is the size of A. Finally, (A, B, C) is *controllable and observable* if (A, B) is controllable and (A, C) is observable.

Let $n = 0$. Then, A, B, and C are empty matrices, and $G \in \mathbb{R}(s)^{l\times m}_{\mathrm{prop}}$ is given by

$$G(s) = 0_{l\times 0}(sI_{0\times 0} - 0_{0\times 0})^{-1}0_{0\times m} + D = 0_{l\times m} + D = D. \tag{16.9.2}$$

Therefore, the order of the realization $\left[\begin{array}{c|c} 0_{0\times 0} & 0_{0\times m} \\ \hline 0_{l\times 0} & D \end{array}\right]$ is zero.

Although the realization (16.9.1) is not unique, the matrix D is unique and is given by

$$D = G(\infty) \triangleq \lim_{s\to\infty} G(s). \tag{16.9.3}$$

Furthermore, note that $G \sim \left[\begin{array}{c|c} A & B \\ \hline C & D \end{array}\right]$ if and only if $G - D \sim \left[\begin{array}{c|c} A & B \\ \hline C & 0 \end{array}\right]$.

The following result shows that every proper, SISO rational transfer function G has a realization. In fact, two realizations are the *controllable canonical form* $G \sim \left[\begin{array}{c|c} A_{\mathrm{c}} & B_{\mathrm{c}} \\ \hline C_{\mathrm{c}} & D_{\mathrm{c}} \end{array}\right]$ and the *observable canonical form* $G \sim \left[\begin{array}{c|c} A_{\mathrm{o}} & B_{\mathrm{o}} \\ \hline C_{\mathrm{o}} & D_{\mathrm{o}} \end{array}\right]$.

Proposition 16.9.3. Let $G \in \mathbb{R}(s)_{\text{prop}}$ be the SISO proper rational transfer function

$$G(s) = \frac{\alpha_n s^n + \alpha_{n-1}s^{n-1} + \cdots + \alpha_1 s + \alpha_0}{s^n + \beta_{n-1}s^{n-1} + \cdots + \beta_1 s + \beta_0}. \tag{16.9.4}$$

Then, $G \sim \left[\begin{array}{c|c} A_\mathrm{c} & B_\mathrm{c} \\ \hline C_\mathrm{c} & D_\mathrm{c} \end{array}\right]$, where $A_\mathrm{c}, B_\mathrm{c}, C_\mathrm{c}, D_\mathrm{c}$ are defined by

$$A_\mathrm{c} \triangleq \begin{bmatrix} 0 & 1 & 0 & \cdots & 0 \\ 0 & 0 & 1 & \cdots & 0 \\ \vdots & \vdots & \vdots & \ddots & \vdots \\ 0 & 0 & 0 & \cdots & 1 \\ -\beta_0 & -\beta_1 & -\beta_2 & \cdots & -\beta_{n-1} \end{bmatrix}, \quad B_\mathrm{c} \triangleq \begin{bmatrix} 0 \\ 0 \\ \vdots \\ 0 \\ 1 \end{bmatrix}, \tag{16.9.5}$$

$$C_\mathrm{c} \triangleq [\alpha_0 - \beta_0\alpha_n \;\; \alpha_1 - \beta_1\alpha_n \;\; \cdots \;\; \alpha_{n-1} - \beta_{n-1}\alpha_n], \quad D_\mathrm{c} \triangleq \alpha_n, \tag{16.9.6}$$

and $G \sim \left[\begin{array}{c|c} A_\mathrm{o} & B_\mathrm{o} \\ \hline C_\mathrm{o} & D_\mathrm{o} \end{array}\right]$, where $A_\mathrm{o}, B_\mathrm{o}, C_\mathrm{o}, D_\mathrm{o}$ are defined by

$$A_\mathrm{o} \triangleq \begin{bmatrix} 0 & 0 & \cdots & 0 & -\beta_0 \\ 1 & 0 & \cdots & 0 & -\beta_1 \\ 0 & 1 & \cdots & 0 & -\beta_2 \\ \vdots & \vdots & \ddots & & \vdots \\ 0 & 0 & \cdots & 1 & -\beta_{n-1} \end{bmatrix}, \quad B_\mathrm{o} \triangleq \begin{bmatrix} \alpha_0 - \beta_0\alpha_n \\ \alpha_1 - \beta_1\alpha_n \\ \alpha_2 - \beta_2\alpha_n \\ \vdots \\ \alpha_{n-1} - \beta_{n-1}\alpha_n \end{bmatrix}, \tag{16.9.7}$$

$$C_\mathrm{o} \triangleq [0 \;\; 0 \;\; \cdots \;\; 0 \;\; 1], \quad D_\mathrm{o} \triangleq \alpha_n. \tag{16.9.8}$$

Furthermore,

$$\mathcal{K}(A_\mathrm{c}, B_\mathrm{c}) = \mathcal{O}(A_\mathrm{o}, C_\mathrm{o}) = \begin{bmatrix} 0 & 0 & 0 & \cdots & 0 & 1 \\ 0 & 0 & 0 & .\cdot^{\cdot} & 1 & -\beta_{n-1} \\ \vdots & \vdots & .\cdot^{\cdot} & .\cdot^{\cdot} & .\cdot^{\cdot} & \vdots \\ 0 & 0 & 1 & .\cdot^{\cdot} & \circ & \circ \\ 0 & 1 & -\beta_{n-1} & .\cdot^{\cdot} & \circ & \circ \\ 1 & -\beta_{n-1} & \circ & \cdots & \circ & \circ \end{bmatrix}, \tag{16.9.9}$$

where each "$\circ$" denotes a possibly nonzero entry, and

$$\mathcal{K}(A_\mathrm{c}, B_\mathrm{c})^{-1} = \mathcal{O}(A_\mathrm{o}, C_\mathrm{o})^{-1} = \begin{bmatrix} \beta_1 & \beta_2 & \beta_3 & \cdots & \beta_{n-1} & 1 \\ \beta_2 & \beta_3 & \beta_4 & .\cdot^{\cdot} & 1 & 0 \\ \vdots & \vdots & .\cdot^{\cdot} & .\cdot^{\cdot} & .\cdot^{\cdot} & \vdots \\ \beta_{n-2} & \beta_{n-1} & 1 & .\cdot^{\cdot} & 0 & 0 \\ \beta_{n-1} & 1 & 0 & .\cdot^{\cdot} & 0 & 0 \\ 1 & 0 & 0 & \cdots & 0 & 0 \end{bmatrix}. \tag{16.9.10}$$

Hence, $(A_\mathrm{c}, B_\mathrm{c})$ is controllable, and $(A_\mathrm{o}, C_\mathrm{o})$ is observable.

Proof. The realizations can be verified by forming the corresponding transfer functions. It follows from Fact 3.16.15 that $\det \mathcal{K}(A_\mathrm{c}, B_\mathrm{c}) = \det \mathcal{O}(A_\mathrm{o}, C_\mathrm{o}) = (-1)^{\lfloor n/2 \rfloor}$, which implies that $(A_\mathrm{c}, B_\mathrm{c})$ is controllable and $(A_\mathrm{o}, C_\mathrm{o})$ is observable. $\square$

The following result shows that every proper rational transfer function has a realization.

Theorem 16.9.4. Let $G \in \mathbb{R}(s)^{l\times m}_{\text{prop}}$. Then, there exist $A \in \mathbb{R}^{n\times n}$, $B \in \mathbb{R}^{n\times m}$, $C \in \mathbb{R}^{l\times n}$, and $D \in \mathbb{R}^{l\times m}$ such that $G \sim \left[\begin{array}{c|c} A & B \\ \hline C & D \end{array}\right]$.

Proof. By Proposition 16.9.3, every entry $G_{(i,j)}$ of G has a realization $G_{(i,j)} \sim \left[\begin{array}{c|c} A_{ij} & B_{ij} \\ \hline C_{ij} & D_{ij} \end{array}\right]$. Combining these realizations yields a realization of G. □

Proposition 16.9.5. Let $G \in \mathbb{R}(s)^{l\times m}_{\text{prop}}$ have the nth-order realization $\left[\begin{array}{c|c} A & B \\ \hline C & D \end{array}\right]$, let $S \in \mathbb{R}^{n\times n}$, and assume that S is nonsingular. Then,

$$G \sim \left[\begin{array}{c|c} SAS^{-1} & SB \\ \hline CS^{-1} & D \end{array}\right]. \tag{16.9.11}$$

If, in addition, (A, B, C) is (controllable, observable), then so is (SAS^{-1}, SB, CS^{-1}).

Definition 16.9.6. Let $G \in \mathbb{R}(s)^{l\times m}_{\text{prop}}$, and let $\left[\begin{array}{c|c} A & B \\ \hline C & D \end{array}\right]$ and $\left[\begin{array}{c|c} \hat{A} & \hat{B} \\ \hline \hat{C} & D \end{array}\right]$ be nth-order realizations of G. Then, $\left[\begin{array}{c|c} A & B \\ \hline C & D \end{array}\right]$ and $\left[\begin{array}{c|c} \hat{A} & \hat{B} \\ \hline \hat{C} & D \end{array}\right]$ are *equivalent* if there exists a nonsingular matrix $S \in \mathbb{R}^{n\times n}$ such that $\hat{A} = SAS^{-1}$, $\hat{B} = SB$, and $\hat{C} = CS^{-1}$.

The following result shows that the Markov parameters of a proper rational transfer function are independent of the realization.

Lemma 16.9.7. Let $G \in \mathbb{R}(s)^{l\times m}_{\text{prop}}$, and assume that $G \sim \left[\begin{array}{c|c} A & B \\ \hline C & D \end{array}\right]$, where $A \in \mathbb{R}^{n\times n}$, and $G \sim \left[\begin{array}{c|c} \hat{A} & \hat{B} \\ \hline \hat{C} & \hat{D} \end{array}\right]$, where $\hat{A} \in \mathbb{R}^{\hat{n}\times\hat{n}}$. Then, $D = \hat{D}$, and, for all $k \geq 0$, $CA^kB = \hat{C}\hat{A}^k\hat{B}$.

Proof. Since $D = G(\infty)$ and $\hat{D} = G(\infty)$, it follows that $D = \hat{D}$. Next, it follows from (16.2.11) and the definition (16.1.18) of the impulse response function H that, for all $t \geq 0$, $H(t) = \mathcal{L}^{-1}\{G(s)\} = Ce^{tA}B + \delta(t)D = \hat{C}e^{t\hat{A}}\hat{B} + \delta(t)D$. Hence, for all $t \geq 0$, $Ce^{tA}B = \hat{C}e^{t\hat{A}}\hat{B}$. Setting $t = 0$ yields $CB = \hat{C}\hat{B}$. Furthermore, for all $t \geq 0$ and $k \geq 1$,

$$CA^ke^{tA}B = \frac{\mathrm{d}^k}{\mathrm{d}t^k}Ce^{tA}B = \frac{\mathrm{d}^k}{\mathrm{d}t^k}\hat{C}e^{t\hat{A}}\hat{B} = \hat{C}\hat{A}^ke^{t\hat{A}}\hat{B}.$$

Setting $t = 0$ implies that, for all $k \geq 1$, $CA^kB = \hat{C}\hat{A}^k\hat{B}$. □

The following result shows that, if two controllable and observable realizations of a proper rational transfer function have the same order, then they are equivalent.

Proposition 16.9.8. Let $G \in \mathbb{R}(s)^{l\times m}_{\text{prop}}$, and assume that G has the nth-order controllable and observable realizations $\left[\begin{array}{c|c} A_1 & B_1 \\ \hline C_1 & D \end{array}\right]$ and $\left[\begin{array}{c|c} A_2 & B_2 \\ \hline C_2 & D \end{array}\right]$. Then, there exists a unique nonsingular matrix $S \in \mathbb{R}^{n\times n}$ such that $A_2 = SA_1S^{-1}$, $B_2 = SB_1$, and $C_2 = C_1S^{-1}$. In fact, S and S^{-1} are given by

$$S = (\mathcal{O}_2^\mathsf{T}\mathcal{O}_2)^{-1}\mathcal{O}_2^\mathsf{T}\mathcal{O}_1 = \mathcal{K}_2\mathcal{K}_2^\mathsf{T}(\mathcal{K}_1\mathcal{K}_2^\mathsf{T})^{-1}, \tag{16.9.12}$$

$$S^{-1} = \mathcal{K}_1\mathcal{K}_2^\mathsf{T}(\mathcal{K}_2\mathcal{K}_2^\mathsf{T})^{-1} = (\mathcal{O}_2^\mathsf{T}\mathcal{O}_1)^{-1}\mathcal{O}_2^\mathsf{T}\mathcal{O}_2, \tag{16.9.13}$$

where $\mathcal{K}_1 \triangleq \mathcal{K}(A_1, B_1)$, $\mathcal{K}_2 \triangleq \mathcal{K}(A_2, B_2)$, $\mathcal{O}_1 \triangleq \mathcal{O}(A_1, C_1)$, and $\mathcal{O}_2 \triangleq \mathcal{O}(A_2, C_2)$. If, in addition, $m = l = 1$, then S and S^{-1} are given by

$$S = \mathcal{K}_2\mathcal{K}_1^{-1} = \mathcal{O}_2^{-1}\mathcal{O}_1, \quad S^{-1} = \mathcal{K}_1\mathcal{K}_2^{-1} = \mathcal{O}_1^{-1}\mathcal{O}_2. \tag{16.9.14}$$

Proof. By Lemma 16.9.7, the realizations $\left[\begin{array}{c|c} A_1 & B_1 \\ \hline C_1 & D \end{array}\right]$ and $\left[\begin{array}{c|c} A_2 & B_2 \\ \hline C_2 & D \end{array}\right]$ generate the same Markov parameters. Hence, $\mathcal{O}_1\mathcal{K}_1 = \mathcal{O}_2\mathcal{K}_2$, and thus $\mathcal{O}_2^\mathsf{T}\mathcal{O}_1\mathcal{K}_1\mathcal{K}_2^\mathsf{T} = \mathcal{O}_2^\mathsf{T}\mathcal{O}_2\mathcal{K}_2\mathcal{K}_2^\mathsf{T}$. Since (A_2, B_2, C_2) is controllable and observable, it follows that the $n \times n$ matrices $\mathcal{K}_2\mathcal{K}_2^{\mathsf{T}}$ and $\mathcal{O}_2^{\mathsf{T}}\mathcal{O}_2$ are nonsingular. Therefore, $(\mathcal{O}_2^{\mathsf{T}}\mathcal{O}_2)^{-1}\mathcal{O}_2^{\mathsf{T}}\mathcal{O}_1\mathcal{K}_1\mathcal{K}_2^{\mathsf{T}}(\mathcal{K}_2\mathcal{K}_2^{\mathsf{T}})^{-1} = I$. Defining $S \in \mathbb{R}^{n\times n}$ as in (16.9.12), it follows that S^{-1} is given by (16.9.13). Since $\mathcal{O}_2A_2\mathcal{K}_2 = \mathcal{O}_1A_1\mathcal{K}_1$, it follows that $\mathcal{O}_2^{\mathsf{T}}\mathcal{O}_2A_2\mathcal{K}_2\mathcal{K}_2^{\mathsf{T}} = \mathcal{O}_2^{\mathsf{T}}\mathcal{O}_1A_1\mathcal{K}_1\mathcal{K}_2^{\mathsf{T}}$, and thus $A_2 = SA_1S^{-1}$. Likewise, it follows from $\mathcal{O}_1B_1 = \mathcal{O}_2B_2$ and $C_1\mathcal{K}_1 = C_2\mathcal{K}_2$ that $B_2 = SB_1$ and $C_2 = C_1S^{-1}$.

To prove uniqueness, suppose that there exists a nonsingular matrix $\hat{S} \in \mathbb{R}^{n\times n}$ such that $A_2 = \hat{S}A_1\hat{S}^{-1}$, $B_2 = \hat{S}B_1$, and $C_2 = C_1\hat{S}^{-1}$. Then, it follows that $\mathcal{O}_1 = \mathcal{O}_2\hat{S}$. Since $\mathcal{O}_1 = \mathcal{O}_2S$, it follows that $\mathcal{O}_2(S - \hat{S}) = 0$. Consequently, $S = \hat{S}$. □

Proposition 16.9.9. Let $A_1 \in \mathbb{R}^{n\times n}$ and $B_1 \in \mathbb{R}^{n\times m}$, assume that (A_1, B_1) is controllable, let $S \in \mathbb{R}^{n\times n}$, assume that S is nonsingular, and define $A_2 \triangleq SA_1S^{-1}$ and $B_2 \triangleq SB_1$. Then, (A_2, B_2) is controllable, $S = \mathcal{K}_2\mathcal{K}_2^{\mathsf{T}}(\mathcal{K}_1\mathcal{K}_2^{\mathsf{T}})^{-1}$, and $S^{-1} = \mathcal{K}_1\mathcal{K}_2^{\mathsf{T}}(\mathcal{K}_2\mathcal{K}_2^{\mathsf{T}})^{-1}$, where $\mathcal{K}_1 \triangleq \mathcal{K}(A_1, B_1)$ and $\mathcal{K}_2 \triangleq \mathcal{K}(A_2, B_2)$. If, in addition, $m = 1$, then $S = \mathcal{K}_2\mathcal{K}_1^{-1}$ and $S^{-1} = \mathcal{K}_1\mathcal{K}_2^{-1}$.

Proof. Note that $\mathcal{K}_2 = S\mathcal{K}_1$, and thus $\mathcal{K}_2\mathcal{K}_2^{\mathsf{T}} = S\mathcal{K}_1\mathcal{K}_2^{\mathsf{T}}$. □

Proposition 16.9.10. Let $A_1 \in \mathbb{R}^{n\times n}$ and $C_1 \in \mathbb{R}^{p\times n}$, assume that (A_1, C_1) is observable, let $S \in \mathbb{R}^{n\times n}$, assume that S is nonsingular, and define $A_2 \triangleq SA_1S^{-1}$ and $C_2 \triangleq CS^{-1}$. Then, (A_2, C_2) is observable, $S = (\mathcal{O}_2^{\mathsf{T}}\mathcal{O}_2)^{-1}\mathcal{O}_2^{\mathsf{T}}\mathcal{O}_1$, and $S^{-1} = (\mathcal{O}_2^{\mathsf{T}}\mathcal{O}_1)^{-1}\mathcal{O}_2^{\mathsf{T}}\mathcal{O}_2$, where $\mathcal{O}_1 \triangleq \mathcal{O}(A_1, C_1)$ and $\mathcal{O}_2 \triangleq \mathcal{O}(A_2, C_2)$. If, in addition, $m = 1$, then $S = \mathcal{O}_1\mathcal{O}_2^{-1}$ and $S^{-1} = \mathcal{O}_2\mathcal{O}_1^{-1}$.

Proof. Note that $\mathcal{O}_2 = \mathcal{O}_1S^{-1}$, and thus $\mathcal{O}_2^{\mathsf{T}}\mathcal{O}_2 = \mathcal{O}_2^{\mathsf{T}}\mathcal{O}_1S^{-1}$. □

Proposition 16.9.11. Let $G \in \mathbb{R}(s)_{\mathrm{prop}}^{l\times m}$, assume that G has the nth-order controllable and observable realization $\left[\begin{array}{c|c} A_1 & B_1 \\ \hline C_1 & D \end{array}\right]$, let $S \in \mathbb{R}^{n\times n}$, assume that S is nonsingular, and define $A_2 \triangleq SA_1S^{-1}$, $B_2 \triangleq SB_1$, and $C_2 \triangleq C_1S^{-1}$. Then, $\left[\begin{array}{c|c} A_2 & B_2 \\ \hline C_2 & D \end{array}\right]$ is an nth-order controllable and observable realization of G, and S and S^{-1} are given by (16.9.12) and (16.9.13), where $\mathcal{K}_1 \triangleq \mathcal{K}(A_1, B_1)$, $\mathcal{K}_2 \triangleq \mathcal{K}(A_2, B_2)$, $\mathcal{O}_1 \triangleq \mathcal{O}(A_1, C_1)$, and $\mathcal{O}_2 \triangleq \mathcal{O}(A_2, C_2)$. If, in addition, $m = l = 1$, then S and S^{-1} are given by (16.9.14).

The following result, known as the *Kalman decomposition*, is useful for constructing controllable and observable realizations.

Proposition 16.9.12. Let $G \in \mathbb{R}(s)_{\mathrm{prop}}^{l\times m}$, where $G \sim \left[\begin{array}{c|c} A & B \\ \hline C & D \end{array}\right]$. Then, there exists a nonsingular matrix $S \in \mathbb{R}^{n\times n}$ such that

$$A = S\begin{bmatrix} A_1 & 0 & A_{13} & 0 \\ A_{21} & A_2 & A_{23} & A_{24} \\ 0 & 0 & A_3 & 0 \\ 0 & 0 & A_{43} & A_4 \end{bmatrix}S^{-1}, \quad B = S\begin{bmatrix} B_1 \\ B_2 \\ 0 \\ 0 \end{bmatrix}, \quad C = [C_1 \ \ 0 \ \ C_3 \ \ 0]S^{-1}, \tag{16.9.15}$$

where, for all $i \in \{1,\dots,4\}$, $A_i \in \mathbb{R}^{n_i\times n_i}$, $\left(\begin{bmatrix} A_1 & 0 \\ A_{21} & A_2 \end{bmatrix}, \begin{bmatrix} B_1 \\ B_2 \end{bmatrix}\right)$ is controllable, and $\left(\begin{bmatrix} A_1 & A_{13} \\ 0 & A_3 \end{bmatrix}, [C_1 \ \ C_3]\right)$ is observable. Furthermore, the following statements hold:

i) (A, B) is stabilizable if and only if A_3 and A_4 are asymptotically stable.

ii) (A, B) is controllable if and only if A_3 and A_4 are empty.

iii) (A, C) is detectable if and only if A_2 and A_4 are asymptotically stable.

iv) (A, C) is observable if and only if A_2 and A_4 are empty.

v) $G \sim \left[\begin{array}{c|c} A_1 & B_1 \\ \hline C_1 & D \end{array}\right]$.

vi) (A_1, B_1, C_1) is controllable and observable.

Proof. Let $\alpha \le 0$ be such that $A + \alpha I$ is asymptotically stable, and let $Q \in \mathbb{R}^{n\times n}$ and $P \in \mathbb{R}^{n\times n}$ denote the controllability and observability Gramians of the system $(A + \alpha I, B, C)$. Then, Theorem

10.3.5 implies that there exists a nonsingular matrix $S \in \mathbb{R}^{n\times n}$ such that

$$Q = S\begin{bmatrix} Q_1 & 0 & 0 & 0\\ 0 & Q_2 & 0 & 0\\ 0 & 0 & 0 & 0\\ 0 & 0 & 0 & 0 \end{bmatrix}S^{\mathrm{T}}, \quad P = S^{-\mathrm{T}}\begin{bmatrix} P_1 & 0 & 0 & 0\\ 0 & 0 & 0 & 0\\ 0 & 0 & P_2 & 0\\ 0 & 0 & 0 & 0 \end{bmatrix}S^{-1},$$

where Q_1 and P_1 are the same size, and where Q_1, Q_2, P_1, and P_2 are positive definite and diagonal. The form of SAS^{-1}, SB, and CS^{-1} given by (16.9.15) now follows from (16.7.3) and (16.4.3) with A replaced by $A+\alpha I$, where, as in the proof of Theorem 16.6.8, $SAS^{-1} = S(A+\alpha I)S^{-1} - \alpha I$. Finally, statements *i)*–*v)* are immediate, while it can be verified that $\left[\begin{array}{c|c} A_1 & B_1\\ \hline C_1 & D_1 \end{array}\right]$ is a realization of G. □

Note that the uncontrollable multispectrum of (A,B) is given by $\mathrm{mspec}(A_3)\cup\mathrm{mspec}(A_4)$, while the unobservable multispectrum of (A,C) is given by $\mathrm{mspec}(A_2)\cup\mathrm{mspec}(A_4)$. Likewise, the *uncontrollable-unobservable multispectrum* of (A,B,C) is given by $\mathrm{mspec}(A_4)$.

Let $G \sim \left[\begin{array}{c|c} A & B\\ \hline C & D \end{array}\right]$. Then, define the *i-step observability matrix* $\mathcal{O}_i(A,C) \in \mathbb{R}^{il\times n}$ by

$$\mathcal{O}_i(A,C) \triangleq \begin{bmatrix} C\\ CA\\ \vdots\\ CA^{i-1} \end{bmatrix} \tag{16.9.16}$$

and the *j-step controllability matrix* $\mathcal{K}_j(A,B) \in \mathbb{R}^{n\times jm}$ by

$$\mathcal{K}_j(A,B) \triangleq [B\ \ AB\ \ \cdots\ \ A^{j-1}B]. \tag{16.9.17}$$

Note that $\mathcal{O}(A,C) = \mathcal{O}_n(A,C)$ and $\mathcal{K}(A,B) = \mathcal{K}_n(A,B)$. Furthermore, define the *Markov block-Hankel matrix* $\mathcal{H}_{i,j,k}(G) \in \mathbb{R}^{il\times jm}$ of G by

$$\mathcal{H}_{i,j,k}(G) \triangleq \mathcal{O}_i(A,C)A^k\mathcal{K}_j(A,B). \tag{16.9.18}$$

Note that $\mathcal{H}_{i,j,k}(G)$ is the block-Hankel matrix of Markov parameters given by

$$\mathcal{H}_{i,j,k}(G) = \begin{bmatrix} CA^kB & CA^{k+1}B & CA^{k+2}B & \cdots & CA^{k+j-1}B\\ CA^{k+1}B & CA^{k+2}B & \iddots & \iddots & \iddots\\ CA^{k+2}B & \iddots & \iddots & \iddots & \iddots\\ \vdots & \iddots & \iddots & \iddots & \iddots\\ CA^{k+i-1}B & \iddots & \iddots & \iddots & CA^{k+j+i-2}B \end{bmatrix}$$

$$= \begin{bmatrix} H_{k+1} & H_{k+2} & H_{k+3} & \cdots & H_{k+j}\\ H_{k+2} & H_{k+3} & \iddots & \iddots & \iddots\\ H_{k+3} & \iddots & \iddots & \iddots & \iddots\\ \vdots & \iddots & \iddots & \iddots & \iddots\\ H_{k+i} & \iddots & \iddots & \iddots & H_{k+j+i-1} \end{bmatrix}. \tag{16.9.19}$$

Note that

$$\mathcal{H}_{i,j,0}(G) = \mathcal{O}_i(A,C)\mathcal{K}_j(A,B), \tag{16.9.20}$$

$$\mathcal{H}_{i,j,1}(G) = \mathcal{O}_i(A,C)A\mathcal{K}_j(A,B). \tag{16.9.21}$$

Furthermore, define

$$\mathcal{H}(G) \triangleq \mathcal{H}_{n,n,0}(G) = \mathcal{O}(A,C)\mathcal{K}(A,B). \tag{16.9.22}$$

The following result provides a MIMO extension of Fact 6.8.10.

Proposition 16.9.13. Let $G \sim \left[\begin{array}{c|c} A & B \\ \hline C & D \end{array}\right]$, where $A \in \mathbb{R}^{n\times n}$. Then, the following statements are equivalent:

i) (A,B,C) is controllable and observable.

ii) $\operatorname{rank}\mathcal{H}(G) = n$.

iii) For all $i,j \geq n$, $\operatorname{rank}\mathcal{H}_{i,j,0}(G) = n$.

iv) There exist $i,j \geq n$ such that $\operatorname{rank}\mathcal{H}_{i,j,0}(G) = n$.

Proof. The equivalence of *ii)*, *iii)*, and *iv)* follows from Fact 3.14.13. To prove *i)* $\Longrightarrow$ *ii)*, note that, since the $n\times n$ matrices $\mathcal{O}(A,C)^{\mathsf{T}}\mathcal{O}(A,C)$ and $\mathcal{K}(A,B)\mathcal{K}(A,B)^{\mathsf{T}}$ are positive definite, it follows that

$$n = \operatorname{rank}\mathcal{O}(A,C)^{\mathsf{T}}\mathcal{O}(A,C)\mathcal{K}(A,B)\mathcal{K}(A,B)^{\mathsf{T}} \leq \operatorname{rank}\mathcal{H}(G) \leq n.$$

Conversely, $n = \operatorname{rank}\mathcal{H}(G) \leq \min\{\operatorname{rank}\mathcal{O}(A,C), \operatorname{rank}\mathcal{K}(A,B)\} \leq n$. □

Proposition 16.9.14. Let $G \sim \left[\begin{array}{c|c} A & B \\ \hline C & D \end{array}\right]$, where $A \in \mathbb{R}^{n\times n}$, assume that (A,B,C) is controllable and observable, and let $i,j \geq 1$ satisfy $\operatorname{rank}\mathcal{O}_i(A,C) = \operatorname{rank}\mathcal{K}_j(A,B) = n$. Then, $G \sim \left[\begin{array}{c|c} \hat{A} & \hat{B} \\ \hline \hat{C} & D \end{array}\right]$, where

$$\hat{A} \triangleq \mathcal{O}_i^{+}(A,C)\mathcal{H}_{i,j,1}(G)\mathcal{K}_j^{+}(A,B), \tag{16.9.23}$$

$$\hat{B} \triangleq \mathcal{K}_j(A,B)\begin{bmatrix} I_m \\ 0_{(j-1)n\times m} \end{bmatrix}, \tag{16.9.24}$$

$$\hat{C} \triangleq [I_l \;\; 0_{l\times(i-1)l}]\mathcal{O}_i(A,C). \tag{16.9.25}$$

Proposition 16.9.15. Let $G \in \mathbb{R}(s)^{l\times m}_{\text{prop}}$, let $i,j \geq 1$, define $n \triangleq \operatorname{rank}\mathcal{H}_{i,j,0}(G)$, and let $L \in \mathbb{R}^{il\times n}$ and $R \in \mathbb{R}^{n\times jm}$ satisfy $\mathcal{H}_{i,j,0}(G) = LR$. Then, the realization

$$G \sim \left[\begin{array}{c|c} L^{+}\mathcal{H}_{i,j,1}(G)R^{+} & R\begin{bmatrix} I_m \\ 0_{(j-1)n\times m} \end{bmatrix} \\ \hline [I_l \;\; 0_{l\times(i-1)l}]L & D \end{array}\right] \tag{16.9.26}$$

is controllable and observable.

A rational transfer function $G \in \mathbb{R}(s)^{l\times m}_{\text{prop}}$ can have realizations of different orders. For example, letting

$$A = 1, \quad B = 1, \quad C = 1, \quad D = 0,$$

$$\hat{A} = \begin{bmatrix} 1 & 0 \\ 0 & 1 \end{bmatrix}, \quad \hat{B} = \begin{bmatrix} 1 \\ 0 \end{bmatrix}, \quad \hat{C} = [1 \;\; 0], \quad \hat{D} = 0,$$

it follows that

$$G(s) = C(sI-A)^{-1}B + D = \hat{C}(sI-\hat{A})^{-1}\hat{B} + \hat{D} = \frac{1}{s-1}.$$

It is usually desirable to find realizations whose order is as small as possible.

Definition 16.9.16. Let $G \in \mathbb{R}(s)_{\text{prop}}^{l\times m}$, and assume that $G \sim \left[\begin{array}{c|c} A & B \\ \hline C & D \end{array}\right]$. Then, $\left[\begin{array}{c|c} A & B \\ \hline C & D \end{array}\right]$ is a *minimal realization* of G if its order is either less than or equal to the order of every realization of G. In this case, we write

$$G \overset{\min}{\sim} \left[\begin{array}{c|c} A & B \\ \hline C & D \end{array}\right]. \tag{16.9.27}$$

Note that the minimality of a realization is independent of D.

The following result shows that every proper transfer function has a minimal realization.

Proposition 16.9.17. Let $G \in \mathbb{R}(s)_{\text{prop}}^{l\times m}$. Then, G has a minimal realization.

Proof. Theorem 16.9.4 implies that G has a realization $G \sim \left[\begin{array}{c|c} A & B \\ \hline C & D \end{array}\right]$. Let n be the order of this realization. For all $i \in \{1,\ldots,n\}$, let $\mathcal{R}_i$ denote the set of realizations of G whose order is i. Let i_0 denotes the smallest $i \in \{1,\ldots,n\}$ such that $\mathcal{R}_i$ is nonempty. Then, each realization in $\mathcal{R}_{i_0}$ is a minimal realization. Hence, G has a minimal realization. □

The following result shows that a realization is controllable and observable if and only if it is minimal.

Proposition 16.9.18. Let $G \in \mathbb{R}^{l\times m}(s)$, and assume that $G \sim \left[\begin{array}{c|c} A & B \\ \hline C & D \end{array}\right]$. Then, $\left[\begin{array}{c|c} A & B \\ \hline C & D \end{array}\right]$ is minimal if and only if (A,B,C) is controllable and observable.

Proof. To prove necessity, suppose that (A,B,C) is either not controllable or not observable. Then, Proposition 16.9.12 implies that G has a realization of order less than n. Hence, $\left[\begin{array}{c|c} A & B \\ \hline C & D \end{array}\right]$ is not minimal.

To prove sufficiency, let $A \in \mathbb{R}^{n\times n}$, where (A,B,C) is controllable and observable, but suppose that $\left[\begin{array}{c|c} A & B \\ \hline C & D \end{array}\right]$ is not a minimal realization of G. Hence, Proposition 16.9.12 implies that G has a minimal realization $\left[\begin{array}{c|c} \hat{A} & \hat{B} \\ \hline \hat{C} & D \end{array}\right]$ of order $\hat{n} < n$. Since the Markov parameters of G are independent of the realization, it follows from Proposition 16.9.13 that $\operatorname{rank}\mathcal{H}(G) = \hat{n} < n$. However, since (A,B,C) is controllable and observable, it follows from Proposition 16.9.13 that $\operatorname{rank}\mathcal{H}(G) = n$, which is a contradiction. □

Proposition 16.9.3 implies that the realizations $\left[\begin{array}{c|c} A_{\text{c}} & B_{\text{c}} \\ \hline C_{\text{c}} & D_{\text{c}} \end{array}\right]$ and $\left[\begin{array}{c|c} A_{\text{o}} & B_{\text{o}} \\ \hline C_{\text{o}} & D_{\text{o}} \end{array}\right]$ of G are such that $(A_{\text{c}}, B_{\text{c}})$ is controllable and $(A_{\text{o}}, C_{\text{o}})$ is observable. However, Proposition 16.9.3 does not imply that $(A_{\text{c}}, C_{\text{c}})$ is observable and $(A_{\text{o}}, B_{\text{o}})$ is controllable, and thus $\left[\begin{array}{c|c} A_{\text{c}} & B_{\text{c}} \\ \hline C_{\text{c}} & D_{\text{c}} \end{array}\right]$ and $\left[\begin{array}{c|c} A_{\text{o}} & B_{\text{o}} \\ \hline C_{\text{o}} & D_{\text{o}} \end{array}\right]$ are not necessarily minimal. It turns out that minimality of these realizations depends on whether or not the numerator and denominator of G given by (16.9.4) are coprime.

Proposition 16.9.19. Let $G \in \mathbb{R}(s)_{\text{prop}}$, assume that $G = p/q$, where $p, q \in \mathbb{R}[s]$, assume that $G \sim \left[\begin{array}{c|c} A & B \\ \hline C & D \end{array}\right]$, where $A \in \mathbb{R}^{n\times n}$, and consider the following statements:

i) $n = \deg q$.

ii) p and q are coprime.

iii) $\left[\begin{array}{c|c} A & B \\ \hline C & D \end{array}\right]$ is a minimal realization of G.

Then, if two of the above statements are satisfied, then the remaining statement is also satisfied.

Proof. To prove [*i*), *ii*)] $\Longrightarrow$ *iii*), suppose that the realization $\left[\begin{array}{c|c} A & B \\ \hline C & D \end{array}\right]$ of G is not minimal. Therefore, Proposition 16.9.17 implies that G has a realization $\left[\begin{array}{c|c} \hat{A} & \hat{B} \\ \hline \hat{C} & D \end{array}\right]$, where $\hat{A} \in \mathbb{R}^{\hat{n}\times\hat{n}}$ and $\hat{n} < n$. Therefore, $G = \hat{p}/\hat{q}$, where $\hat{p}(s) \triangleq \hat{C}(sI - \hat{A})^{\mathsf{A}}\hat{B} + D\hat{q}(s)$ and $\hat{q} \triangleq \chi_{\hat{A}}$. Since $G = p/q = \hat{p}/\hat{q}$ and

$\deg \hat{q} = \hat{n} < n = \deg q$, it follows that p and q are not coprime, which is a contradiction.

To prove $[ii), iii)] \Longrightarrow i)$, note that $G = p/q = \hat{p}/\hat{q}$, where $\hat{p}(s) \triangleq C(sI - A)^{\mathsf{A}}B + D\hat{q}(s)$ and $\hat{q} \triangleq \chi_A$. Since p and q are coprime, it follows that $\deg q \le \deg \hat{q} = n$. Assume that $\deg q < n$. Then, Proposition 16.9.3 implies that $G \sim \left[\begin{array}{c|c} A_c & B_c \\ \hline C_c & D \end{array}\right]$, where $A_c \in \mathbb{R}^{(\deg q)\times(\deg q)}$. Hence, $\left[\begin{array}{c|c} A & B \\ \hline C & D \end{array}\right]$ is not minimal, which is a contradiction.

To prove $[iii), i)] \Longrightarrow ii)$, suppose that p and q are not coprime. Hence, let $\hat{p}, \hat{q} \in \mathbb{R}[s]$ satisfy $G = \hat{p}/\hat{q}$, where $\hat{n} \triangleq \deg \hat{q} < n$. Next, Proposition 16.9.3 implies that $G \sim \left[\begin{array}{c|c} A_c & B_c \\ \hline C_c & D \end{array}\right]$, where $A_c \in \mathbb{R}^{\hat{n}\times\hat{n}}$. Since $\left[\begin{array}{c|c} A & B \\ \hline C & D \end{array}\right]$ is an nth-order realization of G and $\hat{n} < n$, it follows that $\left[\begin{array}{c|c} A & B \\ \hline C & D \end{array}\right]$ is not minimal, which is a contradiction. □

Proposition 16.9.20. Let $G \in \mathbb{R}(s)_{\text{prop}}$ be the SISO proper rational transfer function defined in (16.9.4), define (A_c, B_c, C_c) by (16.9.5) and (16.9.6), and define (A_o, B_o, C_o) by (16.9.7) and (16.9.8). Then, the following statements are equivalent:

i) The numerator and denominator of G given in (16.9.4) are coprime.

ii) (A_c, C_c) is observable.

iii) (A_c, B_c, C_c) is controllable and observable.

iv) (A_o, B_o) is controllable.

v) (A_o, B_o, C_o) is controllable and observable.

Proof. Let $p, q \in \mathbb{R}[s]$ denote the numerator and denominator of G in (16.9.4). To prove $i) \Longrightarrow iii)$, note that, since p and q are coprime, $\deg q = n$, and $\left[\begin{array}{c|c} A_c & B_c \\ \hline C_c & D_c \end{array}\right]$ is a realization of G, Proposition 16.9.19 implies that $\left[\begin{array}{c|c} A_c & B_c \\ \hline C_c & D_c \end{array}\right]$ is a minimal realization of G. Proposition 16.9.18 thus implies that (A_c, B_c, C_c) is controllable and observable. Next, $iii) \Longrightarrow ii)$ is immediate. To prove $ii) \Longrightarrow i)$, note that Proposition 16.9.3 implies that (A_c, B_c) is controllable. Since (A_c, B_c, C_c) is controllable and observable, Proposition 16.9.18 implies that $\left[\begin{array}{c|c} A_c & B_c \\ \hline C_c & D_c \end{array}\right]$ is a minimal realization of G. Since $A_c \in \mathbb{R}^{n\times n}$ and $\deg q = n$, Proposition 16.9.19 implies that p and q are coprime. Next, note that $(A_o^{\mathsf{T}}, C_o^{\mathsf{T}}, B_o^{\mathsf{T}}) = (A_c, B_c, C_c)$. Since $i) \Longleftrightarrow ii) \Longleftrightarrow iii)$, it follows that $i) \Longleftrightarrow [(A_o^{\mathsf{T}}, B_o^{\mathsf{T}})$ is observable$]$ $\Longleftrightarrow [(A_o^{\mathsf{T}}, C_o^{\mathsf{T}}, B_o^{\mathsf{T}})$ is controllable and observable$]$. Equivalently, $i) \Longleftrightarrow iv) \Longleftrightarrow v)$. □

Note that

$$\mathcal{K}(A_c, B_c) = \mathcal{K}(A_o^{\mathsf{T}}, C_o^{\mathsf{T}}) = \mathcal{O}(A_o, C_o)^{\mathsf{T}}, \quad \mathcal{O}(A_c, C_c) = \mathcal{O}(A_o^{\mathsf{T}}, B_o^{\mathsf{T}}) = \mathcal{K}(A_o, B_o)^{\mathsf{T}}. \tag{16.9.28}$$

The following results are consequences of Proposition 16.9.9, Proposition 16.9.10, and Proposition 16.9.11.

Corollary 16.9.21. Let $G \in \mathbb{R}(s)_{\text{prop}}$ be given by (16.9.4), assume that G has the nth-order realization $\left[\begin{array}{c|c} A & B \\ \hline C & D \end{array}\right]$, where (A, B) is controllable, and define A_c, B_c, C_c, D_c by (16.9.5) and (16.9.6). Furthermore, define $S_c \triangleq \mathcal{K}(A_c, B_c)\mathcal{K}(A, B)^{-1}$. Then,

$$\left[\begin{array}{c|c} A_c & B_c \\ \hline C_c & D_c \end{array}\right] = \left[\begin{array}{c|c} S_cAS_c^{-1} & S_cB \\ \hline CS_c^{-1} & D \end{array}\right]. \tag{16.9.29}$$

Corollary 16.9.22. Let $G \in \mathbb{R}(s)_{\text{prop}}$ be given by (16.9.4), assume that G has the nth-order realization $\left[\begin{array}{c|c} A & B \\ \hline C & D \end{array}\right]$, where (A, C) is observable, and define A_o, B_o, C_o, D_o by (16.9.7) and (16.9.8).

Furthermore, define $S_o \triangleq \mathcal{O}(A_o, C_o)^{-1}\mathcal{O}(A, C)$. Then,

$$\left[\begin{array}{c|c} A_o & B_o \\ \hline C_o & D_o \end{array}\right] = \left[\begin{array}{c|c} S_oAS_o^{-1} & S_oB \\ \hline CS_o^{-1} & D \end{array}\right]. \tag{16.9.30}$$

Corollary 16.9.23. Let $G \in \mathbb{R}(s)_{\rm prop}$ be given by (16.9.4), assume that G has the nth-order realization $\left[\begin{array}{c|c} A & B \\ \hline C & D \end{array}\right]$, where (A, B, C) is controllable and observable, and define A_c, B_c, C_c, D_c by (16.9.5) and (16.9.6) and A_o, B_o, C_o, D_o by (16.9.7) and (16.9.8). Furthermore, define $S_c \triangleq \mathcal{K}(A_c, B_c)\mathcal{K}(A, B)^{-1}$ and $S_o \triangleq \mathcal{O}(A_o, C_o)^{-1}\mathcal{O}(A, C)$. Then, (16.9.29) and (16.9.30) hold.

Theorem 16.9.24. Let $G \in \mathbb{R}(s)^{l\times m}_{\rm prop}$, where $G \sim \left[\begin{array}{c|c} A & B \\ \hline C & D \end{array}\right]$ and $A \in \mathbb{R}^{n\times n}$. Then,

$$\text{poles}(G) \subseteq \text{spec}(A), \tag{16.9.31}$$

$$\text{mpoles}(G) \subseteq \text{mspec}(A). \tag{16.9.32}$$

Furthermore, the following statements are equivalent:

i) $G \overset{\rm min}{\sim} \left[\begin{array}{c|c} A & B \\ \hline C & D \end{array}\right]$.

ii) $\text{Mcdeg}(G) = n$.

iii) $\text{mpoles}(G) = \text{mspec}(A)$.

Proof. See [2349, p. 319]. □

Definition 16.9.25. Let $G \in \mathbb{R}(s)^{l\times m}_{\rm prop}$, where $G \overset{\rm min}{\sim} \left[\begin{array}{c|c} A & B \\ \hline C & D \end{array}\right]$. Then, G is (asymptotically stable, semistable, Lyapunov stable) if A is.

Proposition 16.9.26. Let $G = p/q \in \mathbb{R}(s)_{\rm prop}$, where $p, q \in \mathbb{R}[s]$, and assume that p and q are coprime. Then, G is (asymptotically stable, semistable, Lyapunov stable) if and only if q is.

Proposition 16.9.27. Let $G \in \mathbb{R}(s)^{l\times m}_{\rm prop}$. Then, G is (asymptotically stable, semistable, Lyapunov stable) if and only if every entry of G is.

Definition 16.9.28. Let $G \in \mathbb{R}(s)^{l\times m}_{\rm prop}$, where $G \overset{\rm min}{\sim} \left[\begin{array}{c|c} A & B \\ \hline C & D \end{array}\right]$ and A is asymptotically stable. Then, the realization $\left[\begin{array}{c|c} A & B \\ \hline C & D \end{array}\right]$ is *balanced* if the controllability and observability Gramians (16.7.2) and (16.4.2) are diagonal and equal.

Proposition 16.9.29. Let $G \in \mathbb{R}(s)^{l\times m}_{\rm prop}$, where $G \overset{\rm min}{\sim} \left[\begin{array}{c|c} A & B \\ \hline C & D \end{array}\right]$ and A is asymptotically stable. Then, there exists a nonsingular matrix $S \in \mathbb{R}^{n\times n}$ such that the realization $G \sim \left[\begin{array}{c|c} SAS^{-1} & SB \\ \hline CS^{-1} & D \end{array}\right]$ is balanced.

Proof. It follows from Corollary 10.3.4 that there exists a nonsingular matrix $S \in \mathbb{R}^{n\times n}$ such that $SQS^{\rm T}$ and $S^{-\rm T}PS^{-1}$ are diagonal, where Q and P are the positive-definite controllability and observability Gramians (16.7.2) and (16.4.2). Hence, the realization $\left[\begin{array}{c|c} SAS^{-1} & SB \\ \hline CS^{-1} & D \end{array}\right]$ is balanced. □

16.10 Zeros

In Section 4.7 the Smith-McMillan decomposition is used to define transmission zeros and blocking zeros of a transfer function $G(s)$. We now define the invariant zeros of a realization of $G(s)$ and relate these zeros to the transmission zeros. These zeros are related to the Smith zeros of a polynomial matrix as well as the spectrum of a pencil.

Definition 16.10.1. Let $G \in \mathbb{R}(s)^{l\times m}_{\rm prop}$, where $G \sim \left[\begin{array}{c|c} A & B \\ \hline C & D \end{array}\right]$. Then, the *Rosenbrock system* *matrix* $\mathcal{Z} \in \mathbb{R}[s]^{(n+l)\times(n+m)}$ is the polynomial matrix

$$\mathcal{Z}(s) \triangleq \begin{bmatrix} sI - A & B \\ C & -D \end{bmatrix}. \tag{16.10.1}$$

Furthermore, $z \in \mathbb{C}$ is an *invariant zero* of the realization $\left[\begin{array}{c|c} A & B \\ \hline C & D \end{array}\right]$ if

$$\operatorname{rank} \mathcal{Z}(z) < \operatorname{rank} \mathcal{Z}. \tag{16.10.2}$$

Note that $\mathcal{Z}$ is square if and only if G is square.

Let $G \in \mathbb{R}(s)^{l\times m}_{\mathrm{prop}}$, where $G \sim \left[\begin{array}{c|c} A & B \\ \hline C & D \end{array}\right]$ and $A \in \mathbb{R}^{n\times n}$, and note that $\mathcal{Z}$ is the pencil

$$\mathcal{Z}(s) = P_{\left[\begin{smallmatrix} A & -B \\ -C & D \end{smallmatrix}\right],\left[\begin{smallmatrix} I_n & 0 \\ 0 & 0 \end{smallmatrix}\right]}(s) = s\begin{bmatrix} I_n & 0 \\ 0 & 0 \end{bmatrix} - \begin{bmatrix} A & -B \\ -C & D \end{bmatrix}. \tag{16.10.3}$$

Thus,

$$\operatorname{Szeros}(\mathcal{Z}) = \operatorname{spec}\left(\left[\begin{smallmatrix} A & -B \\ -C & D \end{smallmatrix}\right],\left[\begin{smallmatrix} I_n & 0 \\ 0 & 0 \end{smallmatrix}\right]\right), \tag{16.10.4}$$

$$\operatorname{mSzeros}(\mathcal{Z}) = \operatorname{mspec}\left(\left[\begin{smallmatrix} A & -B \\ -C & D \end{smallmatrix}\right],\left[\begin{smallmatrix} I_n & 0 \\ 0 & 0 \end{smallmatrix}\right]\right). \tag{16.10.5}$$

Hence, we define the set of invariant zeros of $\left[\begin{array}{c|c} A & B \\ \hline C & D \end{array}\right]$ by

$$\operatorname{izeros}\left(\left[\begin{array}{c|c} A & B \\ \hline C & D \end{array}\right]\right) \triangleq \operatorname{Szeros}(\mathcal{Z})$$

and the multiset of invariant zeros of $\left[\begin{array}{c|c} A & B \\ \hline C & D \end{array}\right]$ by

$$\operatorname{mizeros}\left(\left[\begin{array}{c|c} A & B \\ \hline C & D \end{array}\right]\right) \triangleq \operatorname{mSzeros}(\mathcal{Z}).$$

Note that $P_{\left[\begin{smallmatrix} A & -B \\ -C & D \end{smallmatrix}\right],\left[\begin{smallmatrix} I_n & 0 \\ 0 & 0 \end{smallmatrix}\right]}$ is regular if and only if $\operatorname{rank} \mathcal{Z} = n + \min\{l, m\}$.

The following result shows that a strictly proper transfer function with either full-state observation or full-state actuation has no invariant zeros.

Proposition 16.10.2. Let $G \in \mathbb{R}(s)^{l\times m}_{\mathrm{prop}}$, where $G \sim \left[\begin{array}{c|c} A & B \\ \hline C & 0 \end{array}\right]$ and $A \in \mathbb{R}^{n\times n}$. Then, the following statements hold:

i) If $m = n$ and B is nonsingular, then $\operatorname{rank} \mathcal{Z} = n + \operatorname{rank} C$ and $\left[\begin{array}{c|c} A & B \\ \hline C & 0 \end{array}\right]$ has no invariant zeros.

ii) If $l = n$ and C is nonsingular, then $\operatorname{rank} \mathcal{Z} = n + \operatorname{rank} B$ and $\left[\begin{array}{c|c} A & B \\ \hline C & 0 \end{array}\right]$ has no invariant zeros.

iii) If $m = n$ and B is nonsingular, then $P_{\left[\begin{smallmatrix} I_n & 0 \\ 0 & 0 \end{smallmatrix}\right],\left[\begin{smallmatrix} A & -B \\ -C & 0 \end{smallmatrix}\right]}$ is regular if and only if $\operatorname{rank} C = \min\{l, n\}$.

iv) If $l = n$ and C is nonsingular, then $P_{\left[\begin{smallmatrix} I_n & 0 \\ 0 & 0 \end{smallmatrix}\right],\left[\begin{smallmatrix} A & -B \\ -C & 0 \end{smallmatrix}\right]}$ is regular if and only if $\operatorname{rank} B = \min\{m, n\}$.

It is useful to note that, for all $s \notin \operatorname{spec}(A)$,

$$\mathcal{Z}(s) = \begin{bmatrix} I & 0 \\ C(sI - A)^{-1} & I \end{bmatrix}\begin{bmatrix} sI - A & B \\ 0 & -G(s) \end{bmatrix} \tag{16.10.6}$$

$$= \begin{bmatrix} sI - A & 0 \\ C & -G(s) \end{bmatrix}\begin{bmatrix} I & (sI - A)^{-1}B \\ 0 & I \end{bmatrix}. \tag{16.10.7}$$

Proposition 16.10.3. Let $G \in \mathbb{R}(s)^{l\times m}_{\mathrm{prop}}$, where $G \sim \left[\begin{array}{c|c} A & B \\ \hline C & D \end{array}\right]$ and $A \in \mathbb{R}^{n\times n}$. If $s \notin \mathrm{spec}(A)$, then

$$\operatorname{rank} \mathcal{Z}(s) = n + \operatorname{rank} G(s). \tag{16.10.8}$$

Furthermore,

$$\operatorname{rank} \mathcal{Z} = n + \operatorname{rank} G. \tag{16.10.9}$$

Proof. Let $s \notin \mathrm{spec}(A)$. Then, it follows from Theorem 16.9.24 that $s \notin \mathrm{poles}(G)$. Hence, (16.10.6) implies (16.10.8). Furthermore, Proposition 6.3.7 and Proposition 6.7.8 imply that

$$\operatorname{rank} \mathcal{Z} = \max_{s\in\mathbb{C}} \operatorname{rank} \mathcal{Z}(s) = \max_{s\in\mathbb{C}\setminus\mathrm{spec}(A)} \operatorname{rank} \mathcal{Z}(s) = n + \max_{s\in\mathbb{C}\setminus\mathrm{spec}(A)} \operatorname{rank} G(s) = n + \operatorname{rank} G. \qquad \square$$

Proposition 16.10.3 implies that $P_{\left[\begin{smallmatrix} A & -B \\ -C & D \end{smallmatrix}\right],\left[\begin{smallmatrix} I_n & 0 \\ 0 & 0 \end{smallmatrix}\right]}$ is (regular, singular) for one realization of G if and only if it is (regular, singular) for every realization of G. The following result shows that $P_{\left[\begin{smallmatrix} A & -B \\ -C & D \end{smallmatrix}\right],\left[\begin{smallmatrix} I_n & 0 \\ 0 & 0 \end{smallmatrix}\right]}$ is regular if and only if G has full rank.

Corollary 16.10.4. Let $G \in \mathbb{R}(s)^{l\times m}_{\mathrm{prop}}$, where $G \sim \left[\begin{array}{c|c} A & B \\ \hline C & D \end{array}\right]$. Then, $P_{\left[\begin{smallmatrix} A & -B \\ -C & D \end{smallmatrix}\right],\left[\begin{smallmatrix} I_n & 0 \\ 0 & 0 \end{smallmatrix}\right]}$ is regular if and only if $\operatorname{rank} G = \min\{l, m\}$.

Let G be SISO. Then, it follows from (16.10.6) and (16.10.7) that, for all $s \in \mathbb{C}\setminus\mathrm{spec}(A)$,

$$\det \mathcal{Z}(s) = -[\det(sI - A)]G(s). \tag{16.10.10}$$

Consequently, for all $s \in \mathbb{C}$,

$$\det \mathcal{Z}(s) = -C(sI - A)^{\mathsf{A}}B - [\det(sI - A)]D. \tag{16.10.11}$$

The equality (16.10.11) also follows from Fact 3.17.2. If $s \in \mathrm{spec}(A)$, then (16.10.11) implies that

$$\det \mathcal{Z}(s) = -C(sI - A)^{\mathsf{A}}B. \tag{16.10.12}$$

If, in addition, $n \ge 2$ and $\operatorname{rank}(sI - A) \le n - 2$, then Fact 3.19.3 implies that $(sI - A)^{\mathsf{A}} = 0$, and thus

$$\det \mathcal{Z}(s) = 0. \tag{16.10.13}$$

In the case $n = 1$, it follows that, for all $s \in \mathbb{C}$, $(sI - A)^{\mathsf{A}} = 1$, and thus, for all $s \in \mathbb{C}$,

$$\det \mathcal{Z}(s) = -CB - (sI - A)D. \tag{16.10.14}$$

Finally, let $s \in \mathbb{C}\setminus\mathrm{spec}(A)$. Then, it follows from (16.10.10) and (16.10.11) that

$$G(s) = \frac{C(sI - A)^{\mathsf{A}}B + [\det(sI - A)]D}{\det(sI - A)} = \frac{-\det \mathcal{Z}(s)}{\det(sI - A)}. \tag{16.10.15}$$

Consequently, $G(s) \ne 0$ if and only if $\det \mathcal{Z}(s) \ne 0$. These observations are summarized by the following result for scalar transfer functions.

Corollary 16.10.5. Let $G \in \mathbb{R}(s)_{\mathrm{prop}}$, where $G \sim \left[\begin{array}{c|c} A & B \\ \hline C & D \end{array}\right]$. Then, the following statements are equivalent:

i) $P_{\left[\begin{smallmatrix} A & -B \\ -C & D \end{smallmatrix}\right],\left[\begin{smallmatrix} I_n & 0 \\ 0 & 0 \end{smallmatrix}\right]}$ is regular.

ii) $G \ne 0$.

iii) $\operatorname{rank} G = 1$.

iv) $\det \mathcal{Z} \ne 0$.

v) $\operatorname{rank} \mathcal{Z} = n + 1$.

vi) $C(sI - A)^{\mathsf{A}}B + [\det(sI - A)]D$ is not the zero polynomial.

If these conditions hold, then

$$\operatorname{mizeros}\left(\left[\begin{array}{c|c} A & B \\ \hline C & D \end{array}\right]\right) = \operatorname{mroots}(\det \mathcal{Z}), \tag{16.10.16}$$

$$\operatorname{mizeros}\left(\left[\begin{array}{c|c} A & B \\ \hline C & D \end{array}\right]\right) = \operatorname{mtzeros}(G) \cup [\operatorname{mspec}(A) \backslash \operatorname{mpoles}(G)]. \tag{16.10.17}$$

If, in addition, $G \overset{\min}{\sim} \left[\begin{array}{c|c} A & B \\ \hline C & D \end{array}\right]$, then

$$\operatorname{mizeros}\left(\left[\begin{array}{c|c} A & B \\ \hline C & D \end{array}\right]\right) = \operatorname{mtzeros}(G). \tag{16.10.18}$$

Now, suppose that G is square; that is, $l = m$. Then, it follows from (16.10.6) and (16.10.7) that, for all $s \in \mathbb{C} \backslash \operatorname{spec}(A)$,

$$\det \mathcal{Z}(s) = (-1)^l [\det(sI - A)] \det G(s), \tag{16.10.19}$$

and thus

$$\det G(s) = \frac{(-1)^l \det \mathcal{Z}(s)}{\det(sI - A)}. \tag{16.10.20}$$

Furthermore, for all $s \in \mathbb{C}$,

$$[\det(sI - A)]^{l-1} \det \mathcal{Z}(s) = (-1)^l \det[C(sI - A)^{\mathsf{A}} B + [\det(sI - A)] D]. \tag{16.10.21}$$

Hence, if $l \geq 2$, then, for all $s \in \operatorname{spec}(A)$,

$$\det C(sI - A)^{\mathsf{A}} B = 0. \tag{16.10.22}$$

We thus have the following result for square transfer functions G that satisfy $\det G \neq 0$.

Corollary 16.10.6. Let $G \in \mathbb{R}(s)^{l \times l}_{\text{prop}}$, where $G \sim \left[\begin{array}{c|c} A & B \\ \hline C & D \end{array}\right]$. Then, the following statements are equivalent:

i) $P_{\left[\begin{smallmatrix} A & -B \\ -C & D \end{smallmatrix}\right], \left[\begin{smallmatrix} I_n & 0 \\ 0 & 0 \end{smallmatrix}\right]}$ is regular.

ii) $\det G \neq 0$.

iii) $\operatorname{rank} G = l$.

iv) $\det \mathcal{Z} \neq 0$.

v) $\operatorname{rank} \mathcal{Z} = n + l$.

vi) $\det[C(sI - A)^{\mathsf{A}} B + [\det(sI - A)] D]$ is not the zero polynomial.

If these conditions hold, then

$$\operatorname{mizeros}\left(\left[\begin{array}{c|c} A & B \\ \hline C & D \end{array}\right]\right) = \operatorname{mroots}(\det \mathcal{Z}), \tag{16.10.23}$$

$$\operatorname{mizeros}\left(\left[\begin{array}{c|c} A & B \\ \hline C & D \end{array}\right]\right) = \operatorname{mtzeros}(G) \cup [\operatorname{mspec}(A) \backslash \operatorname{mpoles}(G)], \tag{16.10.24}$$

$$\operatorname{izeros}\left(\left[\begin{array}{c|c} A & B \\ \hline C & D \end{array}\right]\right) = \operatorname{tzeros}(G) \cup [\operatorname{spec}(A) \backslash \operatorname{poles}(G)]. \tag{16.10.25}$$

If, in addition, $G \overset{\min}{\sim} \left[\begin{array}{c|c} A & B \\ \hline C & D \end{array}\right]$, then

$$\operatorname{mizeros}\left(\left[\begin{array}{c|c} A & B \\ \hline C & D \end{array}\right]\right) = \operatorname{mtzeros}(G). \tag{16.10.26}$$

Example 16.10.7. Consider $G \in \mathbb{R}^{2\times 2}(s)$ defined by

$$G(s) \triangleq \begin{bmatrix} \frac{s-1}{s+1} & 0 \\ 0 & \frac{s+1}{s-1} \end{bmatrix}. \tag{16.10.27}$$

Then, the Smith-McMillan form of G is given by

$$G(s) \triangleq S_1(s)\begin{bmatrix} \frac{1}{s^2-1} & 0 \\ 0 & s^2-1 \end{bmatrix} S_2(s), \tag{16.10.28}$$

where $S_1, S_2 \in \mathbb{R}[s]^{2\times 2}$ are the unimodular matrices

$$S_1(s) \triangleq \begin{bmatrix} (s-1)^2 & -1 \\ -\frac{1}{4}(s+1)^2(s-2) & \frac{1}{4}(s+2) \end{bmatrix}, \quad S_2(s) \triangleq \begin{bmatrix} \frac{1}{4}(s-1)^2(s+2) & (s+1)^2 \\ \frac{1}{4}(s-2) & 1 \end{bmatrix}. \tag{16.10.29}$$

A minimal realization of G is given by

$$G \overset{\min}{\sim} \left[\begin{array}{cc|cc} -1 & 0 & 1 & 0 \\ 0 & 1 & 0 & 1 \\ \hline -2 & 0 & 1 & 0 \\ 0 & 2 & 0 & 1 \end{array}\right]. \tag{16.10.30}$$

Thus, $\operatorname{mtzeros}(G) = \operatorname{mpoles}(G) = \operatorname{mspec}(A) = \{1,-1\}$. Finally, $\det \mathcal{Z}(s) = (-1)^2[\det(sI-A)]\det G = s^2-1$, which confirms (16.10.26). ◊

Theorem 16.10.8. Let $G \in \mathbb{R}(s)^{l\times m}_{\text{prop}}$, where $G \sim \left[\begin{array}{c|c} A & B \\ \hline C & D \end{array}\right]$. Then,

$$\operatorname{izeros}\left(\left[\begin{array}{c|c} A & B \\ \hline C & D \end{array}\right]\right)\backslash\operatorname{spec}(A) \subseteq \operatorname{tzeros}(G), \tag{16.10.31}$$

$$\operatorname{tzeros}(G)\backslash\operatorname{poles}(G) \subseteq \operatorname{izeros}\left(\left[\begin{array}{c|c} A & B \\ \hline C & D \end{array}\right]\right). \tag{16.10.32}$$

If, in addition, $G \overset{\min}{\sim} \left[\begin{array}{c|c} A & B \\ \hline C & D \end{array}\right]$, then

$$\operatorname{izeros}\left(\left[\begin{array}{c|c} A & B \\ \hline C & D \end{array}\right]\right)\backslash\operatorname{poles}(G) = \operatorname{tzeros}(G)\backslash\operatorname{poles}(G). \tag{16.10.33}$$

Proof. To prove (16.10.31), let $z \in \operatorname{izeros}\left(\left[\begin{array}{c|c} A & B \\ \hline C & D \end{array}\right]\right)\backslash\operatorname{spec}(A)$. Since $z \notin \operatorname{spec}(A)$, Theorem 16.9.24 implies that $z \notin \operatorname{poles}(G)$. It now follows from Proposition 16.10.3 that $n + \operatorname{rank} G(z) = \operatorname{rank} \mathcal{Z}(z) < \operatorname{rank} \mathcal{Z} = n + \operatorname{rank} G$, which implies that $\operatorname{rank} G(z) < \operatorname{rank} G$. Thus, $z \in \operatorname{tzeros}(G)$.

To prove (16.10.32), let $z \in \operatorname{tzeros}(G)\backslash\operatorname{poles}(G)$. Then, Proposition 16.10.3 implies that $\operatorname{rank} \mathcal{Z}(z) = n + \operatorname{rank} G(z) < n + \operatorname{rank} G = \operatorname{rank} \mathcal{Z}$, which implies that $z \in \operatorname{izeros}\left(\left[\begin{array}{c|c} A & B \\ \hline C & D \end{array}\right]\right)$. The last statement follows from (16.10.31), (16.10.32), and Theorem 16.9.24. □

The following result is a stronger form of Theorem 16.10.8.

Theorem 16.10.9. Let $G \in \mathbb{R}(s)^{l\times m}_{\text{prop}}$, where $G \sim \left[\begin{array}{c|c} A & B \\ \hline C & D \end{array}\right]$, let $S \in \mathbb{R}^{n\times n}$, assume that S is nonsingular, and let A, B, and C have the form (16.9.15), where $\left(\begin{bmatrix} A_1 & 0 \\ A_{21} & A_2 \end{bmatrix}, \begin{bmatrix} B_1 \\ B_2 \end{bmatrix}\right)$ is controllable and $\left(\begin{bmatrix} A_1 & A_{13} \\ 0 & A_3 \end{bmatrix}, [C_1 \ C_3]\right)$ is observable. Then,

$$\operatorname{mtzeros}(G) = \operatorname{mizeros}\left(\left[\begin{array}{c|c} A_1 & B_1 \\ \hline C_1 & D \end{array}\right]\right), \tag{16.10.34}$$

$$\operatorname{mizeros}\left(\left[\begin{array}{c|c} A & B \\ \hline C & D \end{array}\right]\right) = \operatorname{mspec}(A_2) \cup \operatorname{mspec}(A_3) \cup \operatorname{mspec}(A_4) \cup \operatorname{mtzeros}(G). \tag{16.10.35}$$

Proof. Defining $\mathcal{Z}$ by (16.10.1) and using the notation of Proposition 16.9.12, $\mathcal{Z}$ has the same Smith form as

$$\tilde{\mathcal{Z}} \triangleq \begin{bmatrix} sI - A_4 & -A_{43} & 0 & 0 & 0 \\ 0 & sI - A_3 & 0 & 0 & 0 \\ -A_{24} & -A_{23} & sI - A_2 & -A_{21} & B_2 \\ 0 & -A_{13} & 0 & sI - A_1 & B_1 \\ 0 & C_3 & 0 & C_1 & -D \end{bmatrix}.$$

Hence, Proposition 16.10.3 implies that $\operatorname{rank} \mathcal{Z} = \operatorname{rank} \tilde{\mathcal{Z}} = n + r$, where $r \triangleq \operatorname{rank} G$. Let $\tilde{p}_1, \ldots, \tilde{p}_{n+r}$ be the Smith polynomials of $\tilde{\mathcal{Z}}$. Then, since $\tilde{p}_{n+r}$ is the monic greatest common divisor of all $(n + r) \times (n + r)$ subdeterminants of $\tilde{\mathcal{Z}}$, it follows that $\tilde{p}_{n+r} = \chi_{A_1}\chi_{A_2}\chi_{A_3}p_r$, where p_r is the rth Smith polynomial of $\left[\begin{smallmatrix} sI-A_1 & B_1 \\ C_1 & -D \end{smallmatrix}\right]$. Therefore,

$$\operatorname{mSzeros}(\mathcal{Z}) = \operatorname{mspec}(A_2) \cup \operatorname{mspec}(A_3) \cup \operatorname{mspec}(A_4) \cup \operatorname{mSzeros}\left(\left[\begin{smallmatrix} sI-A_1 & B_1 \\ C_1 & -D \end{smallmatrix}\right]\right).$$

Next, using the Smith-McMillan decomposition Theorem 6.7.5, there exist unimodular matrices $S_1 \in \mathbb{R}[s]^{l\times l}$ and $S_2 \in \mathbb{R}[s]^{m\times m}$ such that $G = S_1 D_0^{-1} N_0 S_2$, where

$$D_0 \triangleq \begin{bmatrix} q_1 & & & 0 \\ & \ddots & & \\ & & q_r & \\ 0 & & & I_{l-r} \end{bmatrix}, \quad N_0 \triangleq \begin{bmatrix} p_1 & & & 0 \\ & \ddots & & \\ & & p_r & \\ 0 & & & 0_{(l-r)\times(m-r)} \end{bmatrix}.$$

Now, define the polynomial matrix $\hat{\mathcal{Z}} \in \mathbb{R}[s]^{(n+l)\times(n+m)}$ by

$$\hat{\mathcal{Z}} \triangleq \begin{bmatrix} I_{n-l} & 0_{(n-l)\times l} & 0_{(n-l)\times m} \\ 0_{l\times(n-l)} & D_0 & N_0 S_2 \\ 0_{l\times(n-l)} & S_1 & 0_{l\times m} \end{bmatrix}.$$

Since S_1 is unimodular, it follows that the Smith form $\mathcal{S}$ of $\hat{\mathcal{Z}}$ is given by

$$\mathcal{S} = \begin{bmatrix} I_n & 0_{n\times m} \\ 0_{l\times n} & N_0 \end{bmatrix}.$$

Consequently, $\operatorname{mSzeros}(\hat{\mathcal{Z}}) = \operatorname{mSzeros}(\mathcal{S}) = \operatorname{mtzeros}(G)$.

Next, note that

$$\operatorname{rank} \begin{bmatrix} I_{n-l} & 0_{(n-l)\times l} & 0_{(n-l)\times m} \\ 0_{l\times(n-l)} & D_0 & N_0 S_2 \end{bmatrix} = \operatorname{rank} \begin{bmatrix} I_{n-l} & 0_{(n-l)\times l} \\ 0_{l\times(n-l)} & D_0 \\ 0_{l\times(n-l)} & S_1 \end{bmatrix} = n,$$

$$G = \begin{bmatrix} 0_{l\times(n-l)} & S_1 & 0_{l\times m} \end{bmatrix} \begin{bmatrix} I_{n-l} & 0_{(n-l)\times l} \\ 0_{l\times(n-l)} & D_0 \end{bmatrix}^{-1} \begin{bmatrix} 0_{(n-l)\times m} \\ N_0 S_2 \end{bmatrix}.$$

Furthermore, $G \overset{\min}{\sim} \left[\begin{array}{c|c} A_1 & B_1 \\ \hline C_1 & D \end{array}\right]$. Consequently, $\hat{\mathcal{Z}}$ and $\left[\begin{smallmatrix} sI-A_1 & B_1 \\ C_1 & D \end{smallmatrix}\right]$ have no decoupling zeros [2339, pp. 64–70], and it thus follows from Theorem 3.1 of [2339, p. 106] that $\hat{\mathcal{Z}}$ and $\left[\begin{smallmatrix} sI-A_1 & B_1 \\ C_1 & D \end{smallmatrix}\right]$ have the same Smith form. Thus,

$$\operatorname{mSzeros}\left(\begin{bmatrix} sI - A_1 & B_1 \\ C_1 & -D \end{bmatrix}\right) = \operatorname{mSzeros}(\hat{\mathcal{Z}}) = \operatorname{mtzeros}(G).$$

Consequently,

$$\text{mizeros}\left(\left[\begin{array}{c|c} A_1 & B_1 \\ \hline C_1 & D \end{array}\right]\right) = \text{mSzeros}\left(\left[\begin{smallmatrix} sI-A_1 & B_1 \\ C_1 & -D \end{smallmatrix}\right]\right) = \text{mtzeros}(G),$$

which proves (16.10.34).

Finally, to prove (16.10.31) note that

$$\begin{aligned}\text{mizeros}\left(\left[\begin{array}{c|c} A & B \\ \hline C & D \end{array}\right]\right) &= \text{mSzeros}(\mathcal{Z}) \\ &= \text{mspec}(A_2) \cup \text{mspec}(A_3) \cup \text{mspec}(A_4) \cup \text{mSzeros}\left(\left[\begin{smallmatrix} sI-A_1 & B_1 \\ -C_1 & -D \end{smallmatrix}\right]\right) \\ &= \text{mspec}(A_2) \cup \text{mspec}(A_3) \cup \text{mspec}(A_4) \cup \text{mtzeros}(G). \end{aligned}$$

□

Proposition 16.10.10. Equivalent realizations have the same invariant zeros. Furthermore, invariant zeros are not changed by output feedback.

Proof. Letting $G \sim \left[\begin{array}{c|c} A & B \\ \hline C & D \end{array}\right]$ and $u = Ky + v$, where $\det(I - KD) \neq 0$, the closed-loop transfer function from v to y has the realization

$$\tilde{G} \sim \left[\begin{array}{c|c} A + BK_0C & B(I-KD)^{-1} \\ \hline C + DK_0C & D(I-KD)^{-1} \end{array}\right],$$

where $K_0 \triangleq (I - KD)^{-1}K$. Since

$$\begin{bmatrix} zI - (A + BK_0C) & B(I-KD)^{-1} \\ C + DK_0C & -D(I-KD)^{-1} \end{bmatrix} = \begin{bmatrix} zI - A & B \\ C & -D \end{bmatrix}\begin{bmatrix} I & 0 \\ -K_0C & (I-KD)^{-1} \end{bmatrix},$$

it follows that $\left[\begin{array}{c|c} A & B \\ \hline C & D \end{array}\right]$ and $\left[\begin{array}{c|c} A + BK_0C & B(I-KD)^{-1} \\ \hline C + DK_0C & D(I-KD)^{-1} \end{array}\right]$ have the same invariant zeros. □

The special case $C = I$ is full-state feedback. Fact 16.24.15 also concerns invariant zeros under output feedback. The following result provides an interpretation of *i*) of Theorem 16.17.9.

Proposition 16.10.11. Let $G \in \mathbb{R}(s)^{l \times m}_{\text{prop}}$, where $G \sim \left[\begin{array}{c|c} A & B \\ \hline C & D \end{array}\right]$, and assume that $R \triangleq D^{\mathsf{T}}D$ is positive definite. Then, the following statements hold:

i) $\operatorname{rank} \mathcal{Z} = n + m$.

ii) $z \in \mathbb{C}$ is an invariant zero of $\left[\begin{array}{c|c} A & B \\ \hline C & D \end{array}\right]$ if and only if z is an unobservable eigenvalue of $(A - BR^{-1}D^{\mathsf{T}}C, [I - DR^{-1}D^{\mathsf{T}}]C)$.

Proof. To prove *i*), suppose that $\operatorname{rank} \mathcal{Z} < n + m$. Then, for all $s \in \mathbb{C}$, there exists a nonzero vector $\left[\begin{smallmatrix} x \\ y \end{smallmatrix}\right] \in \mathcal{N}[\mathcal{Z}(s)]$; that is,

$$\begin{bmatrix} sI - A & B \\ C & -D \end{bmatrix}\begin{bmatrix} x \\ y \end{bmatrix} = 0.$$

Consequently, $Cx - Dy = 0$, which implies that $D^{\mathsf{T}}Cx - Ry = 0$, and thus $y = R^{-1}D^{\mathsf{T}}Cx$. Furthermore, since $(sI - A + BR^{-1}D^{\mathsf{T}}C)x = 0$, choosing $s \notin \text{spec}(A - BR^{-1}D^{\mathsf{T}}C)$ implies $x = 0$, and thus $y = 0$, which is a contradiction.

To prove *ii*), note that it follows from *i*) that z is an invariant zero of $\left[\begin{array}{c|c} A & B \\ \hline C & D \end{array}\right]$ if and only if $\operatorname{rank} \mathcal{Z}(z) < n+m$, which holds if and only if there exists a nonzero vector $\left[\begin{smallmatrix} x \\ y \end{smallmatrix}\right] \in \mathcal{N}[\mathcal{Z}(z)]$. Therefore,

$$\begin{bmatrix} zI - A + BR^{-1}D^{\mathsf{T}}C \\ (I - DR^{-1}D^{\mathsf{T}})C \end{bmatrix} x = 0,$$

where $x \neq 0$, which holds if and only if z is an unobservable eigenvalue of $(A - BR^{-1}D^{\mathsf{T}}C, [I - DR^{-1}D^{\mathsf{T}}]C)$. □

Corollary 16.10.12. Assume that $R \triangleq D^{\mathsf{T}}D$ is positive definite, and assume that $(A - BR^{-1}D^{\mathsf{T}}C,$ $[I - DR^{-1}D^{\mathsf{T}}]C)$ is observable. Then, $\left[\begin{array}{c|c} A & B \\ \hline C & D \end{array}\right]$ has no invariant zeros.

16.11 H_2 System Norm

Consider the system

$$\dot{x}(t) = Ax(t) + Bu(t), \tag{16.11.1}$$

$$y(t) = Cx(t), \tag{16.11.2}$$

where $A \in \mathbb{R}^{n\times n}$ is asymptotically stable, $B \in \mathbb{R}^{n\times m}$, and $C \in \mathbb{R}^{l\times n}$. Then, for all $t \geq 0$, the impulse response function defined by (16.1.18) is given by

$$H(t) = Ce^{tA}B. \tag{16.11.3}$$

The L_2 *norm* of H is defined by

$$\|H\|_{\mathrm{L}_2} \triangleq \left(\int_0^\infty \|H(t)\|_{\mathrm{F}}^2 \,\mathrm{d}t \right)^{1/2}. \tag{16.11.4}$$

The following result provides expressions for $\|H\|_{\mathrm{L}_2}$ in terms of the controllability and observability Gramians.

Theorem 16.11.1. Assume that A is asymptotically stable. Then, the L_2 norm of H is given by

$$\|H\|_{\mathrm{L}_2}^2 = \operatorname{tr} CQC^{\mathsf{T}} = \operatorname{tr} B^{\mathsf{T}}PB, \tag{16.11.5}$$

where $Q, P \in \mathbb{R}^{n\times n}$ satisfy

$$AQ + QA^{\mathsf{T}} + BB^{\mathsf{T}} = 0, \tag{16.11.6}$$

$$A^{\mathsf{T}}P + PA + C^{\mathsf{T}}C = 0. \tag{16.11.7}$$

Proof. Note that

$$\|H\|_{\mathrm{L}_2}^2 = \int_0^\infty \operatorname{tr} Ce^{tA}BB^{\mathsf{T}}e^{tA^{\mathsf{T}}}C^{\mathsf{T}}\,\mathrm{d}t = \operatorname{tr} CQC^{\mathsf{T}},$$

where Q satisfies (16.11.6). The dual expression (16.11.7) follows either in a similar manner or by noting that

$$\operatorname{tr} CQC^{\mathsf{T}} = \operatorname{tr} C^{\mathsf{T}}CQ = -\operatorname{tr}(A^{\mathsf{T}}P + PA)Q = -\operatorname{tr}(AQ + QA^{\mathsf{T}})P = \operatorname{tr} BB^{\mathsf{T}}P = \operatorname{tr} B^{\mathsf{T}}PB. \qquad \square$$

For the following definition, note that, for all $s \in \mathbb{C}\backslash\mathrm{poles}(G)$,

$$\|G(s)\|_{\mathrm{F}} = [\operatorname{tr} G(s)G^*(s)]^{1/2}. \tag{16.11.8}$$

Definition 16.11.2. Let $G \in \mathbb{R}^{l\times m}(s)$, and assume that G is asymptotically stable. Then, the H_2 *norm* of $G \in \mathbb{R}^{l\times m}(s)$ is given by

$$\|G\|_{\mathrm{H}_2} \triangleq \left(\frac{1}{2\pi} \int_{-\infty}^\infty \|G(\omega \jmath)\|_{\mathrm{F}}^2 \,\mathrm{d}\omega \right)^{1/2}. \tag{16.11.9}$$

The following result is *Parseval's theorem*, which relates the L_2 norm of the impulse response function to the H_2 norm of its Fourier transform.

Theorem 16.11.3. Let $G \in \mathbb{R}(s)^{l\times m}_{\text{prop}}$, where $G \sim \left[\begin{array}{c|c} A & B \\ \hline C & 0 \end{array}\right]$, define H by (16.11.3), and assume that $A \in \mathbb{R}^{n\times n}$ is asymptotically stable. Then,

$$\int_0^\infty H(t)H^{\mathsf{T}}(t)\,\mathrm{d}t = \frac{1}{2\pi}\int_{-\infty}^{\infty} G(\omega\jmath)G^*(\omega\jmath)\,\mathrm{d}\omega. \tag{16.11.10}$$

Therefore,

$$\|H\|_{\mathrm{L}_2} = \|G\|_{\mathrm{H}_2}. \tag{16.11.11}$$

Proof. First note that

$$G(s) = \mathcal{L}\{H(t)\} = \int_0^\infty H(t)e^{-st}\,\mathrm{d}t$$

and that

$$H(t) = \frac{1}{2\pi}\int_{-\infty}^{\infty} G(\omega\jmath)e^{\omega\jmath t}\,\mathrm{d}\omega.$$

Hence,

$$\begin{aligned}\int_0^\infty H(t)H^{\mathsf{T}}(t)e^{-st}\,\mathrm{d}t &= \int_0^\infty \left(\frac{1}{2\pi}\int_{-\infty}^{\infty} G(\omega\jmath)e^{\omega\jmath t}\,\mathrm{d}\omega\right)H^{\mathsf{T}}(t)e^{-st}\,\mathrm{d}t\\ &= \frac{1}{2\pi}\int_{-\infty}^{\infty} G(\omega\jmath)\left(\int_0^\infty H^{\mathsf{T}}(t)e^{-(s-\omega\jmath)t}\,\mathrm{d}t\right)\mathrm{d}\omega\\ &= \frac{1}{2\pi}\int_{-\infty}^{\infty} G(\omega\jmath)G^{\mathsf{T}}(s-\omega\jmath)\,\mathrm{d}\omega.\end{aligned}$$

Setting $s = 0$ yields (16.11.7), while taking the trace of (16.11.10) yields (16.11.11). □

Corollary 16.11.4. Let $G \in \mathbb{R}(s)^{l\times m}_{\text{prop}}$, where $G \sim \left[\begin{array}{c|c} A & B \\ \hline C & 0 \end{array}\right]$, and assume that $A \in \mathbb{R}^{n\times n}$ is asymptotically stable. Then,

$$\|G\|^2_{\mathrm{H}_2} = \|H\|^2_{\mathrm{L}_2} = \operatorname{tr} CQC^{\mathsf{T}} = \operatorname{tr} B^{\mathsf{T}}PB, \tag{16.11.12}$$

where $Q, P \in \mathbb{R}^{n\times n}$ satisfy (16.11.6) and (16.11.7), respectively.

The following corollary of Theorem 16.11.3 provides a frequency-domain expression for the solution of the Lyapunov equation.

Corollary 16.11.5. Let $A \in \mathbb{R}^{n\times n}$, assume that A is asymptotically stable, let $B \in \mathbb{R}^{n\times m}$, and define $Q \in \mathbb{R}^{n\times n}$ by

$$Q = \frac{1}{2\pi}\int_{-\infty}^{\infty} (\omega\jmath I - A)^{-1}BB^{\mathsf{T}}(\omega\jmath I - A)^{-*}\,\mathrm{d}\omega. \tag{16.11.13}$$

Then, Q satisfies

$$AQ + QA^{\mathsf{T}} + BB^{\mathsf{T}} = 0. \tag{16.11.14}$$

Proof. The result follows from (16.11.10) with $H(t) = e^{tA}B$ and $G(s) = (sI-A)^{-1}B$. Alternatively, to show that the solution Q of (16.11.14) is given by (16.11.13), note that

$$\int_{-\infty}^{\infty}(\omega\jmath I - A)^{-1}\,\mathrm{d}\omega Q + Q\int_{-\infty}^{\infty}(\omega\jmath I - A)^{-*}\,\mathrm{d}\omega = \int_{-\infty}^{\infty}(\omega\jmath I - A)^{-1}BB^{\mathsf{T}}(\omega\jmath I - A)^{-*}\,\mathrm{d}\omega.$$

Assuming that A is diagonalizable with eigenvalues $\lambda_i = -\sigma_i + \omega_i\jmath$, it follows that

$$\int_{-\infty}^{\infty}\frac{\mathrm{d}\omega}{\omega\jmath - \lambda_i} = \int_{-\infty}^{\infty}\frac{\sigma_i - \omega\jmath}{\sigma_i^2 + \omega^2}\,\mathrm{d}\omega = \frac{\sigma_i\pi}{|\sigma_i|} - \jmath\lim_{r\to\infty}\int_{-r}^{r}\frac{\omega}{\sigma_i^2 + \omega^2}\,\mathrm{d}\omega = \pi,$$

which implies that

$$\int_{-\infty}^{\infty}(\omega\jmath I - A)^{-1}\,\mathrm{d}\omega = \pi I_n,$$

which yields (16.11.13). See [685] for the case where A is not diagonalizable. □

Proposition 16.11.6. Let $G_1, G_2 \in \mathbb{R}_{\mathrm{prop}}^{l\times m}(s)$ be asymptotically stable rational transfer functions. Then,

$$\|G_1 + G_2\|_{\mathrm{H}_2} \le \|G_1\|_{\mathrm{H}_2} + \|G_2\|_{\mathrm{H}_2}. \tag{16.11.15}$$

Proof. Let $G_1 \overset{\min}{\sim} \left[\begin{array}{c|c} A_1 & B_1 \\ \hline C_1 & 0 \end{array}\right]$ and $G_2 \overset{\min}{\sim} \left[\begin{array}{c|c} A_2 & B_2 \\ \hline C_2 & 0 \end{array}\right]$, where $A_1 \in \mathbb{R}^{n_1\times n_1}$ and $A_2 \in \mathbb{R}^{n_2\times n_2}$. It follows from Proposition 16.13.2 that

$$G_1 + G_2 \sim \left[\begin{array}{cc|c} A_1 & 0 & B_1 \\ 0 & A_2 & B_2 \\ \hline C_1 & C_2 & 0 \end{array}\right].$$

Now, Corollary 16.11.4 implies that $\|G_1\|_{\mathrm{H}_2} = \sqrt{\operatorname{tr} C_1Q_1C_1^{\mathsf{T}}}$ and $\|G_2\|_{\mathrm{H}_2} = \sqrt{\operatorname{tr} C_2Q_2C_2^{\mathsf{T}}}$, where $Q_1 \in \mathbb{R}^{n_1\times n_1}$ and $Q_2 \in \mathbb{R}^{n_2\times n_2}$ are the unique positive-definite matrices satisfying $A_1Q_1 + Q_1A_1^{\mathsf{T}} + B_1B_1^{\mathsf{T}} = 0$ and $A_2Q_2 + Q_2A_2^{\mathsf{T}} + B_2B_2^{\mathsf{T}} = 0$. Furthermore,

$$\|G_1 + G_2\|_{\mathrm{H}_2}^2 = \operatorname{tr}\,[C_1\ \ C_2]Q\begin{bmatrix} C_1^{\mathsf{T}} \\ C_2^{\mathsf{T}} \end{bmatrix},$$

where $Q \in \mathbb{R}^{(n_1+n_2)\times(n_1+n_2)}$ is the unique, positive-semidefinite matrix satisfying

$$\begin{bmatrix} A_1 & 0 \\ 0 & A_2 \end{bmatrix}Q + Q\begin{bmatrix} A_1 & 0 \\ 0 & A_2 \end{bmatrix}^{\mathsf{T}} + \begin{bmatrix} B_1 \\ B_2 \end{bmatrix}\begin{bmatrix} B_1 \\ B_2 \end{bmatrix}^{\mathsf{T}} = 0.$$

It can be seen that $Q = \begin{bmatrix} Q_1 & Q_{12} \\ Q_{12}^{\mathsf{T}} & Q_2 \end{bmatrix}$, where Q_1 and Q_2 are as given above and where Q_{12} satisfies $A_1Q_{12} + Q_{12}A_2^{\mathsf{T}} + B_1B_2^{\mathsf{T}} = 0$. Now, using the Cauchy-Schwarz inequality (11.3.17) and *iii*) of Proposition 10.2.5, it follows that

$$\begin{aligned}
\|G_1 + G_2\|_{\mathrm{H}_2}^2 &= \operatorname{tr}(C_1Q_1C_1^{\mathsf{T}} + C_2Q_2C_2^{\mathsf{T}} + C_2Q_{12}^{\mathsf{T}}C_1^{\mathsf{T}} + C_1Q_{12}C_2^{\mathsf{T}}) \\
&= \|G_1\|_{\mathrm{H}_2}^2 + \|G_2\|_{\mathrm{H}_2}^2 + 2\operatorname{tr} C_1Q_{12}Q_2^{-1/2}Q_2^{1/2}C_2^{\mathsf{T}} \\
&\le \|G_1\|_{\mathrm{H}_2}^2 + \|G_2\|_{\mathrm{H}_2}^2 + 2[\operatorname{tr}(C_1Q_{12}Q_2^{-1}Q_{12}^{\mathsf{T}}C_1^{\mathsf{T}})\operatorname{tr}(C_2Q_2C_2^{\mathsf{T}})]^{1/2} \\
&\le \|G_1\|_{\mathrm{H}_2}^2 + \|G_2\|_{\mathrm{H}_2}^2 + 2[\operatorname{tr}(C_1Q_1C_1^{\mathsf{T}})\operatorname{tr}(C_2Q_2C_2^{\mathsf{T}})]^{1/2}
\end{aligned}$$

$$= (\|G_1\|_{\mathrm{H}_2} + \|G_2\|_{\mathrm{H}_2})^2. \qquad \square$$

16.12 Harmonic Steady-State Response

The following result concerns the response of a linear system to a harmonic input.

Theorem 16.12.1. For $t \ge 0$, consider the linear system

$$\dot{x}(t) = Ax(t) + Bu(t), \tag{16.12.1}$$

with harmonic input

$$u(t) = \mathrm{Re}\, u_0 e^{\omega_0 \jmath t}, \tag{16.12.2}$$

where $u_0 \in \mathbb{C}^m$ and $\omega_0 \in \mathbb{R}$ is such that $\omega_0 \jmath \notin \mathrm{spec}(A)$. Then, $x(t)$ is given by

$$x(t) = e^{tA}(x(0) - \mathrm{Re}[(\omega_0 \jmath I - A)^{-1}Bu_0]) + \mathrm{Re}[(\omega_0 \jmath I - A)^{-1}Bu_0 e^{\omega_0 \jmath t}]. \tag{16.12.3}$$

Proof. We have

$$x(t) = e^{tA}x(0) + \int_0^t e^{(t-\tau)A}B\mathrm{Re}(u_0 e^{\omega_0 \jmath \tau})\mathrm{d}\tau = e^{tA}x(0) + e^{tA}\mathrm{Re}\int_0^t e^{\tau(\omega_0 \jmath I - A)}\,\mathrm{d}\tau Bu_0$$

$$= e^{tA}x(0) + e^{tA}\mathrm{Re}[(\omega_0 \jmath I - A)^{-1}(e^{t(\omega_0 \jmath I - A)} - I)Bu_0]$$

$$= e^{tA}x(0) + \mathrm{Re}[(\omega_0 \jmath I - A)^{-1}(e^{\omega_0 \jmath t I} - e^{tA})Bu_0]$$

$$= e^{tA}x(0) + \mathrm{Re}[(\omega_0 \jmath I - A)^{-1}(-e^{tA})Bu_0] + \mathrm{Re}[(\omega_0 \jmath I - A)^{-1}e^{\omega_0 \jmath t}Bu_0]$$

$$= e^{tA}(x(0) - \mathrm{Re}[(\omega_0 \jmath I - A)^{-1}Bu_0]) + \mathrm{Re}[(\omega_0 \jmath I - A)^{-1}Bu_0 e^{\omega_0 \jmath t}]. \qquad \square$$

Theorem 16.12.1 shows that the total response $y(t)$ of the linear system $G \sim \left[\begin{array}{c|c} A & B \\ \hline C & 0 \end{array}\right]$ to a harmonic input can be written as $y(t) = y_{\mathrm{trans}}(t) + y_{\mathrm{hss}}(t)$, where the transient component

$$y_{\mathrm{trans}}(t) \triangleq Ce^{tA}(x(0) - \mathrm{Re}[(\omega_0 \jmath I - A)^{-1}Bu_0]) \tag{16.12.4}$$

depends on the initial condition and the input, and the harmonic steady-state component

$$y_{\mathrm{hss}}(t) = \mathrm{Re}[G(\omega_0 \jmath)u_0 e^{\omega_0 \jmath t}] \tag{16.12.5}$$

depends only on the input.

If A is asymptotically stable, then $\lim_{t\to\infty} y_{\mathrm{trans}}(t) = 0$, and thus $y(t)$ approaches its harmonic steady-state component $y_{\mathrm{hss}}(t)$ for large t. Since the harmonic steady-state component is sinusoidal, it follows that $y(t)$ does not converge in the usual sense.

Finally, if A is semistable, then it follows from Proposition 15.9.2 that

$$\lim_{t\to\infty} y_{\mathrm{trans}}(t) = C(I - AA^{\#})(x(0) - \mathrm{Re}[(\omega_0 \jmath I - A)^{-1}Bu_0]), \tag{16.12.6}$$

which represents a constant offset to the harmonic steady-state component.

In the SISO case, let $u_0 \triangleq a_0(\sin\phi_0 - \cos\phi_0 \jmath)$, where $a_0, \phi_0 \in \mathbb{R}$, and consider the input

$$u(t) = a_0 \sin(\omega_0 t + \phi_0) = \mathrm{Re}\, u_0 e^{\omega_0 \jmath t}. \tag{16.12.7}$$

Then, writing $G(\omega_0 \jmath) = Me^{\theta \jmath}$, where $M \in \mathbb{R}$, it follows that

$$y_{\mathrm{hss}}(t) = a_0 M \sin(\omega_0 t + \phi_0 + \theta). \tag{16.12.8}$$

16.13 System Interconnections

Let $G \in \mathbb{R}(s)^{l\times m}_{\rm prop}$. We define the *parahermitian conjugate* $G^\sim$ of G by

$$G^\sim(s) \triangleq G^{\mathsf{T}}(-s). \tag{16.13.1}$$

The following result provides realizations for G^{T}, $G^\sim$, and G^{-1}.

Proposition 16.13.1. Let $G \in \mathbb{R}(s)^{l\times m}_{\rm prop}$, and assume that $G \sim \left[\begin{array}{c|c} A & B \\ \hline C & D \end{array}\right]$. Then,

$$G^{\mathsf{T}} \sim \left[\begin{array}{c|c} A^{\mathsf{T}} & C^{\mathsf{T}} \\ \hline B^{\mathsf{T}} & D^{\mathsf{T}} \end{array}\right], \quad G^\sim \sim \left[\begin{array}{c|c} -A^{\mathsf{T}} & -C^{\mathsf{T}} \\ \hline B^{\mathsf{T}} & D^{\mathsf{T}} \end{array}\right]. \tag{16.13.2}$$

Furthermore, if G is square and D is nonsingular, then

$$G^{-1} \sim \left[\begin{array}{c|c} A - BD^{-1}C & BD^{-1} \\ \hline -D^{-1}C & D^{-1} \end{array}\right]. \tag{16.13.3}$$

Proof. Since $y = Gu$, it follows that G^{-1} satisfies $u = G^{-1}y$. Since $\dot{x} = Ax + Bu$ and $y = Cx + Du$, it follows that $u = -D^{-1}Cx + D^{-1}y$, and thus $\dot{x} = Ax + B(-D^{-1}Cx + D^{-1}y) = (A - BD^{-1}C)x + BD^{-1}y$. □

Note that, if $G \in \mathbb{R}(s)_{\rm prop}$ and $G \sim \left[\begin{array}{c|c} A & B \\ \hline C & D \end{array}\right]$, then $G \sim \left[\begin{array}{c|c} A^{\mathsf{T}} & C^{\mathsf{T}} \\ \hline B^{\mathsf{T}} & D \end{array}\right]$.

Let $G_1 \in \mathbb{R}(s)^{l_1\times m_1}_{\rm prop}$ and $G_2 \in \mathbb{R}(s)^{l_2\times m_2}_{\rm prop}$. If $m_2 = l_2$, then the *cascade interconnection* of G_1 and G_2 shown in Figure 16.13.1 is the product G_2G_1, while, if $m_1 = m_2$ and $l_1 = l_2$, then the *parallel interconnection* shown in Figure 16.13.2 is the sum $G_1 + G_2$.

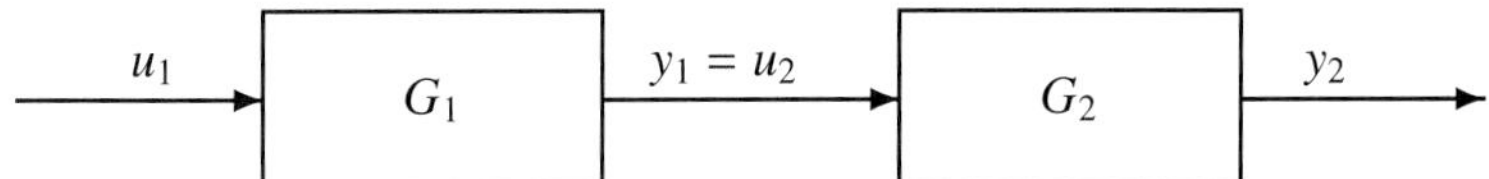

Figure 16.13.1
Cascade Interconnection of Linear Systems

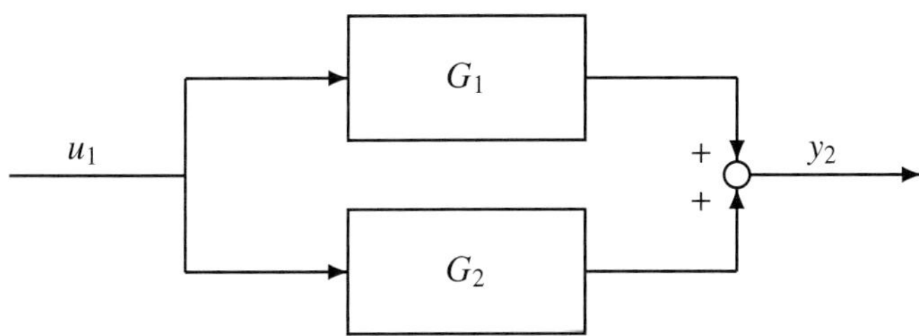

Figure 16.13.2
Parallel Interconnection of Linear Systems

Proposition 16.13.2. For $i \in \{1, 2\}$, let $G_i \in \mathbb{R}(s)^{l_i\times m_i}_{\rm prop}$, where $G_i \sim \left[\begin{array}{c|c} A_i & B_i \\ \hline C_i & D_i \end{array}\right]$. If $l_1 = m_2$, then

$$G_2G_1 \sim \left[\begin{array}{cc|c} A_1 & 0 & B_1 \\ B_2C_1 & A_2 & B_2D_1 \\ \hline D_2C_1 & C_2 & D_2D_1 \end{array}\right]. \tag{16.13.4}$$

If $l_1 = l_2$ and $m_1 = m_2$, then

$$G_1 + G_2 \sim \left[\begin{array}{cc|c} A_1 & 0 & B_1 \\ 0 & A_2 & B_2 \\ \hline C_1 & C_2 & D_1 + D_2 \end{array}\right]. \tag{16.13.5}$$

Proof. Consider the state space equations

$$\dot{x}_1 = A_1 x_1 + B_1 u_1, \quad \dot{x}_2 = A_2 x_2 + B_2 u_2,$$
$$y_1 = C_1 x_1 + D_1 u_1, \quad y_2 = C_2 x_2 + D_2 u_2.$$

Since $u_2 = y_1$, it follows that

$$\dot{x}_2 = A_2 x_2 + B_2 C_1 x_1 + B_2 D_1 u_1,$$
$$y_2 = C_2 x_2 + D_2 C_1 x_1 + D_2 D_1 u_1,$$

and thus

$$\begin{bmatrix} \dot{x}_1 \\ \dot{x}_2 \end{bmatrix} = \begin{bmatrix} A_1 & 0 \\ B_2 C_1 & A_2 \end{bmatrix} \begin{bmatrix} x_1 \\ x_2 \end{bmatrix} + \begin{bmatrix} B_1 \\ B_2 D_1 \end{bmatrix} u_1,$$

$$y_2 = [D_2 C_1 \ \ C_2] \begin{bmatrix} x_1 \\ x_2 \end{bmatrix} + D_2 D_1 u_1,$$

which yields the realization of $G_2 G_1$. The realization of $G_1 + G_2$ can be obtained similarly. □

It is often useful to combine transfer functions by concatenating them into either row, column, or block-diagonal transfer functions.

Proposition 16.13.3. Let $G_1 \sim \left[\begin{array}{c|c} A_1 & B_1 \\ \hline C_1 & D_1 \end{array}\right]$ and $G_2 \sim \left[\begin{array}{c|c} A_2 & B_2 \\ \hline C_2 & D_2 \end{array}\right]$. Then,

$$[G_1 \ \ G_2] \sim \left[\begin{array}{cc|cc} A_1 & 0 & B_1 & 0 \\ 0 & A_2 & 0 & B_2 \\ \hline C_1 & C_2 & D_1 & D_2 \end{array}\right], \quad \begin{bmatrix} G_1 \\ G_2 \end{bmatrix} \sim \left[\begin{array}{cc|c} A_1 & 0 & B_1 \\ 0 & A_2 & B_2 \\ \hline C_1 & 0 & D_1 \\ 0 & C_2 & D_2 \end{array}\right], \tag{16.13.6}$$

$$\begin{bmatrix} G_1 & 0 \\ 0 & G_2 \end{bmatrix} \sim \left[\begin{array}{cc|cc} A_1 & 0 & B_1 & 0 \\ 0 & A_2 & 0 & B_2 \\ \hline C_1 & 0 & D_1 & 0 \\ 0 & C_2 & 0 & D_2 \end{array}\right]. \tag{16.13.7}$$

Next, we interconnect a pair of systems G_1, G_2 by means of feedback as shown in Figure 16.13.3. It can be seen that u and y are related by

$$\hat{y} = (I + G_1 G_2)^{-1} G_1 \hat{u}, \tag{16.13.8}$$
$$\hat{y} = G_1 (I + G_2 G_1)^{-1} \hat{u}. \tag{16.13.9}$$

The equivalence of (16.13.8) and (16.13.9) follows from the push-through identity given by Fact 3.20.6,

$$(I + G_1 G_2)^{-1} G_1 = G_1 (I + G_2 G_1)^{-1}. \tag{16.13.10}$$

A realization of this rational transfer function is given by the following result.

Proposition 16.13.4. Let $G_1 \sim \left[\begin{array}{c|c} A_1 & B_1 \\ \hline C_1 & D_1 \end{array}\right]$ and $G_2 \sim \left[\begin{array}{c|c} A_2 & B_2 \\ \hline C_2 & D_2 \end{array}\right]$, and assume that $\det(I +$

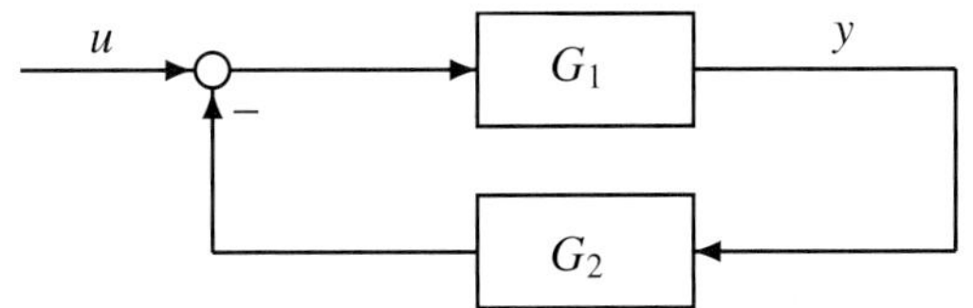

Figure 16.13.3
Feedback Interconnection of Linear Systems

$D_1D_2) \neq 0$. Then,

$$(I+G_1G_2)^{-1}G_1 \sim \left[\begin{array}{cc|c} A_1 - B_1(I+D_2D_1)^{-1}D_2C_1 & -B_1(I+D_2D_1)^{-1}C_2 & B_1(I+D_2D_1)^{-1} \\ B_2(I+D_1D_2)^{-1}C_1 & A_2 - B_2(I+D_1D_2)^{-1}D_1C_2 & B_2(I+D_1D_2)^{-1}D_1 \\ \hline (I+D_1D_2)^{-1}C_1 & -(I+D_1D_2)^{-1}D_1C_2 & (I+D_1D_2)^{-1}D_1 \end{array}\right]. \tag{16.13.11}$$

16.14 Standard Control Problem

The standard control problem shown in Figure 16.14.1 involves four distinct signals, namely, an *exogenous input* w, a *control input* u, a *performance variable* z, and a *feedback signal* y. This system can be written as

$$\begin{bmatrix} \hat{z}(s) \\ \hat{y}(s) \end{bmatrix} = \mathcal{G}(s) \begin{bmatrix} \hat{w}(s) \\ \hat{u}(s) \end{bmatrix}, \tag{16.14.1}$$

where $\mathcal{G}(s)$ is partitioned as

$$\mathcal{G} \triangleq \begin{bmatrix} G_{11} & G_{12} \\ G_{21} & G_{22} \end{bmatrix} \tag{16.14.2}$$

with the realization

$$\mathcal{G} \sim \left[\begin{array}{c|cc} A & D_1 & B \\ \hline E_1 & E_0 & E_2 \\ C & D_2 & D \end{array}\right], \tag{16.14.3}$$

which represents the equations

$$\dot{x} = Ax + D_1w + Bu, \tag{16.14.4}$$

$$z = E_1x + E_0w + E_2u, \tag{16.14.5}$$

$$y = Cx + D_2w + Du. \tag{16.14.6}$$

Consequently,

$$\mathcal{G}(s) = \begin{bmatrix} E_1(sI-A)^{-1}D_1 + E_0 & E_1(sI-A)^{-1}B + E_2 \\ C(sI-A)^{-1}D_1 + D_2 & C(sI-A)^{-1}B + D \end{bmatrix}, \tag{16.14.7}$$

which shows that G_{11}, G_{12}, G_{21}, and G_{22} have the realizations

$$G_{11} \sim \left[\begin{array}{c|c} A & D_1 \\ \hline E_1 & E_0 \end{array}\right], \quad G_{12} \sim \left[\begin{array}{c|c} A & B \\ \hline E_1 & E_2 \end{array}\right], \quad G_{21} \sim \left[\begin{array}{c|c} A & D_1 \\ \hline C & D_2 \end{array}\right], \quad G_{22} \sim \left[\begin{array}{c|c} A & B \\ \hline C & D \end{array}\right]. \tag{16.14.8}$$

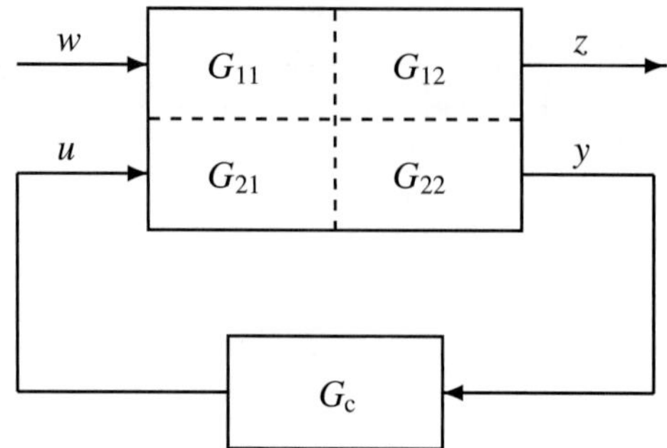

Figure 16.14.1
Standard Control Problem

Letting G_c denote a feedback controller with realization

$$G_c \sim \left[\begin{array}{c|c} A_c & B_c \\ \hline C_c & D_c \end{array}\right], \tag{16.14.9}$$

we interconnect G and G_c according to

$$\hat{u}(s) = G_c(s)\hat{y}(s). \tag{16.14.10}$$

The resulting rational transfer function $\tilde{\mathcal{G}}$ satisfying $\hat{z}(s) = \tilde{\mathcal{G}}(s)\hat{w}(s)$ is thus given by

$$\tilde{\mathcal{G}} = G_{11} + G_{12}G_c(I - G_{22}G_c)^{-1}G_{21}, \tag{16.14.11}$$

$$= G_{11} + G_{12}(I - G_cG_{22})^{-1}G_cG_{21}. \tag{16.14.12}$$

A realization of $\tilde{\mathcal{G}}$ is given by the following result.

Proposition 16.14.1. Let $\mathcal{G}$ and G_c have the realizations (16.14.3) and (16.14.9), define $\Omega \triangleq I - DD_c$, and assume that Ω is nonsingular. Then,

$$\tilde{\mathcal{G}} \sim \left[\begin{array}{cc|c} A + BD_c\Omega^{-1}C & B(I + D_c\Omega^{-1}D)C_c & D_1 + BD_c\Omega^{-1}D_2 \\ B_c\Omega^{-1}C & A_c + B_c\Omega^{-1}DC_c & B_c\Omega^{-1}D_2 \\ \hline E_1 + E_2D_c\Omega^{-1}C & E_2(I + D_c\Omega^{-1}D)C_c & E_0 + E_2D_c\Omega^{-1}D_2 \end{array}\right]. \tag{16.14.13}$$

The realization (16.14.13) can be simplified if $DD_c = 0$. For example, if $D = 0$, then

$$\tilde{\mathcal{G}} \sim \left[\begin{array}{cc|c} A + BD_cC & BC_c & D_1 + BD_cD_2 \\ B_cC & A_c & B_cD_2 \\ \hline E_1 + E_2D_cC & E_2C_c & E_0 + E_2D_cD_2 \end{array}\right], \tag{16.14.14}$$

whereas, if $D_c = 0$, then

$$\tilde{\mathcal{G}} \sim \left[\begin{array}{cc|c} A & BC_c & D_1 \\ B_cC & A_c + B_cDC_c & B_cD_2 \\ \hline E_1 & E_2C_c & E_0 \end{array}\right]. \tag{16.14.15}$$

Finally, if both $D = 0$ and $D_c = 0$, then

$$\tilde{\mathcal{G}} \sim \left[\begin{array}{cc|c} A & BC_c & D_1 \\ B_cC & A_c & B_cD_2 \\ \hline E_1 & E_2C_c & E_0 \end{array}\right]. \tag{16.14.16}$$

The feedback interconnection shown in Figure 16.14.1 forms the basis for the *standard control problem* in feedback control. For this problem the signal w is an exogenous signal representing either a command or a disturbance, while the signal z is the *performance variable*, which is the variable whose behavior reflects the performance of the closed-loop system. The performance variable may or may not be physically measured. The *controlled input* (also called the *control*) u is the output of the feedback controller G_c, while the *measurement* signal y serves as the input to the *feedback controller* G_c. The standard control problem is the following: Given knowledge of w, determine G_c that minimizes a performance criterion $J(G_c)$.

16.15 Linear-Quadratic Control

Let $A \in \mathbb{R}^{n\times n}$ and $B \in \mathbb{R}^{n\times m}$, and consider the system

$$\dot{x}(t) = Ax(t) + Bu(t), \tag{16.15.1}$$

$$x(0) = x_0, \tag{16.15.2}$$

where $t \geq 0$. Furthermore, let $K \in \mathbb{R}^{m\times n}$, and consider the full-state-feedback control law

$$u(t) = Kx(t). \tag{16.15.3}$$

The objective of the *linear-quadratic control problem* is to minimize the *quadratic performance measure*

$$J(K, x_0) = \int_0^\infty [x^{\mathrm{T}}(t)R_1x(t) + 2x^{\mathrm{T}}(t)R_{12}u(t) + u^{\mathrm{T}}(t)R_2u(t)]\,\mathrm{d}t, \tag{16.15.4}$$

with respect to the feedback gain $K \in \mathbb{R}^{m\times n}$, where $R_1 \in \mathbb{R}^{n\times n}$, $R_{12} \in \mathbb{R}^{n\times m}$, and $R_2 \in \mathbb{R}^{m\times m}$. We assume that $\left[\begin{smallmatrix} R_1 & R_{12} \\ R_{12}^{\mathrm{T}} & R_2 \end{smallmatrix}\right]$ is positive semidefinite and R_2 is positive definite.

The performance measure (16.15.4) indicates the desire to maintain the state vector $x(t)$ close to the zero equilibrium without an excessive expenditure of control effort. Specifically, the term $x^{\mathrm{T}}(t)R_1x(t)$ is a measure of the deviation of the state $x(t)$ from the zero state, where the $n\times n$ positive-semidefinite matrix R_1 determines how much weighting is associated with each component of the state. Likewise, the $m \times m$ positive-definite matrix R_2 weights the magnitude of the control input. Finally, the cross-weighting term R_{12} arises due to the use of filters to shape the system response.

Using (16.15.1) and (16.15.3), the closed-loop dynamic system can be written as

$$\dot{x}(t) = (A + BK)x(t) \tag{16.15.5}$$

so that

$$x(t) = e^{t\tilde{A}}x_0, \tag{16.15.6}$$

where $\tilde{A} \triangleq A + BK$. Thus, the quadratic performance measure (16.15.4) becomes

$$J(K, x_0) = \int_0^\infty x^{\mathrm{T}}(t)\tilde{R}x(t)\,\mathrm{d}t = \int_0^\infty x_0^{\mathrm{T}}e^{t\tilde{A}^{\mathrm{T}}}\tilde{R}e^{t\tilde{A}}x_0\,\mathrm{d}t = \operatorname{tr} x_0^{\mathrm{T}}\int_0^\infty e^{t\tilde{A}^{\mathrm{T}}}\tilde{R}e^{t\tilde{A}}\,\mathrm{d}t x_0 = \operatorname{tr}\int_0^\infty e^{t\tilde{A}^{\mathrm{T}}}\tilde{R}e^{t\tilde{A}}\,\mathrm{d}t x_0x_0^{\mathrm{T}}, \tag{16.15.7}$$

where

$$\tilde{R} \triangleq R_1 + R_{12}K + K^{\mathsf{T}}R_{12}^{\mathsf{T}} + K^{\mathsf{T}}R_2K. \tag{16.15.8}$$

Now, consider the standard control problem with plant

$$\mathcal{G} \sim \left[\begin{array}{c|cc} A & D_1 & B \\ \hline E_1 & 0 & E_2 \\ I_n & 0 & 0 \end{array}\right] \tag{16.15.9}$$

and full-state feedback $u = Kx$. Then, the closed-loop transfer function is given by

$$\tilde{\mathcal{G}} \sim \left[\begin{array}{c|c} A + BK & D_1 \\ \hline E_1 + E_2K & 0 \end{array}\right]. \tag{16.15.10}$$

The following result shows that the quadratic performance measure (16.15.4) is equal to the H_2 norm of a transfer function.

Proposition 16.15.1. Assume that $D_1 = x_0$ and

$$\begin{bmatrix} R_1 & R_{12} \\ R_{12}^{\mathsf{T}} & R_2 \end{bmatrix} = \begin{bmatrix} E_1^{\mathsf{T}} \\ E_2^{\mathsf{T}} \end{bmatrix} [E_1 \ \ E_2], \tag{16.15.11}$$

and let $\tilde{\mathcal{G}}$ be given by (16.15.10). Then,

$$J(K, x_0) = \|\tilde{\mathcal{G}}\|_{\mathrm{H}_2}^2. \tag{16.15.12}$$

Proof. This result follows from Proposition 16.1.2. □

For the following development, we assume that (16.15.11) holds so that R_1, R_{12}, and R_2 are given by

$$R_1 = E_1^{\mathsf{T}}E_1, \quad R_{12} = E_1^{\mathsf{T}}E_2, \quad R_2 = E_2^{\mathsf{T}}E_2. \tag{16.15.13}$$

To develop necessary conditions for the linear-quadratic control problem, we restrict K to the set of stabilizing gains

$$\mathcal{S} \triangleq \{K \in \mathbb{R}^{m\times n}\colon\ A + BK \text{ is asymptotically stable}\}. \tag{16.15.14}$$

Obviously, $\mathcal{S}$ is nonempty if and only if (A, B) is stabilizable. The following result gives necessary conditions that characterize a stabilizing solution K of the linear-quadratic control problem.

Theorem 16.15.2. Assume that (A, B) is stabilizable, assume that $K \in \mathcal{S}$ solves the linear-quadratic control problem, and assume that $(A + BK, D_1)$ is controllable. Then, there exists an $n \times n$ positive-semidefinite matrix P such that K is given by

$$K = -R_2^{-1}(B^{\mathsf{T}}P + R_{12}^{\mathsf{T}}) \tag{16.15.15}$$

and such that P satisfies

$$\hat{A}_{\mathrm{R}}^{\mathsf{T}}P + P\hat{A}_{\mathrm{R}} + \hat{R}_1 - PBR_2^{-1}B^{\mathsf{T}}P = 0, \tag{16.15.16}$$

where

$$\hat{A}_{\mathrm{R}} \triangleq A - BR_2^{-1}R_{12}^{\mathsf{T}}, \quad \hat{R}_1 \triangleq R_1 - R_{12}R_2^{-1}R_{12}^{\mathsf{T}}. \tag{16.15.17}$$

Furthermore, the minimal cost is given by

$$J(K) = \operatorname{tr} PV, \tag{16.15.18}$$

where $V \triangleq D_1D_1^{\mathsf{T}}$.

Proof. Since $K \in \mathcal{S}$, it follows that $\tilde{A}$ is asymptotically stable. It then follows that $J(K)$ is given by (16.15.18), where $P \triangleq \int_0^\infty e^{t\tilde{A}^\mathsf{T}}\tilde{R}e^{t\tilde{A}}\,\mathrm{d}t$ is positive semidefinite and satisfies the Lyapunov equation

$$\tilde{A}^\mathsf{T}P + P\tilde{A} + \tilde{R} = 0. \tag{16.15.19}$$

Note that (16.15.19) can be written as

$$(A + BK)^\mathsf{T}P + P(A + BK) + R_1 + R_{12}K + K^\mathsf{T}R_{12}^\mathsf{T} + K^\mathsf{T}R_2K = 0. \tag{16.15.20}$$

To optimize (16.15.18) subject to the constraint (16.15.19) over the open set $\mathcal{S}$, form the Lagrangian

$$\mathcal{L}(K, P, Q, \lambda_0) \triangleq \operatorname{tr}[\lambda_0 PV + Q(\tilde{A}^\mathsf{T}P + P\tilde{A} + \tilde{R})], \tag{16.15.21}$$

where the Lagrange multipliers $\lambda_0 \ge 0$ and $Q \in \mathbb{R}^{n\times n}$ are not both zero. Note that the $n\times n$ Lagrange multiplier Q accounts for the $n \times n$ constraint equation (16.15.19).

The necessary condition $\partial\mathcal{L}/\partial P = 0$ implies

$$\tilde{A}Q + Q\tilde{A}^\mathsf{T} + \lambda_0 V = 0. \tag{16.15.22}$$

Since $\tilde{A}$ is asymptotically stable, it follows from Proposition 15.10.3 that, for all $\lambda_0 \ge 0$, (16.15.22) has a unique solution Q and, furthermore, Q is positive semidefinite. In particular, if $\lambda_0 = 0$, then $Q = 0$. Since λ_0 and Q are not both zero, we can set $\lambda_0 = 1$ so that (16.15.22) becomes

$$\tilde{A}Q + Q\tilde{A}^\mathsf{T} + V = 0. \tag{16.15.23}$$

Since $(\tilde{A}, D_1)$ is controllable, it follows from Corollary 16.7.10 that Q is positive definite.

Next, evaluating $\partial\mathcal{L}/\partial K = 0$ yields

$$R_2KQ + (B^\mathsf{T}P + R_{12}^\mathsf{T})Q = 0. \tag{16.15.24}$$

Since Q is positive definite, it follows from (16.15.24) that (16.15.15) holds. Furthermore, using (16.15.15), it follows that (16.15.19) is equivalent to (16.15.16). □

With K given by (16.15.15), the closed-loop dynamics matrix $\tilde{A} = A + BK$ is given by

$$\tilde{A} = A - BR_2^{-1}(B^\mathsf{T}P + R_{12}^\mathsf{T}), \tag{16.15.25}$$

where P is the solution of the *Riccati equation* (16.15.16).

16.16 Solutions of the Riccati Equation

For convenience in the following development, we assume that $R_{12} = 0$. With this assumption, the gain K given by (16.15.15) becomes

$$K = -R_2^{-1}B^\mathsf{T}P. \tag{16.16.1}$$

Defining

$$\Sigma \triangleq BR_2^{-1}B^\mathsf{T}, \tag{16.16.2}$$

(16.15.25) becomes

$$\tilde{A} = A - \Sigma P, \tag{16.16.3}$$

while the Riccati equation (16.15.16) can be written as

$$A^\mathsf{T}P + PA + R_1 - P\Sigma P = 0. \tag{16.16.4}$$

Note that (16.16.4) has the alternative representation

$$(A - \Sigma P)^\mathsf{T}P + P(A - \Sigma P) + R_1 + P\Sigma P = 0, \tag{16.16.5}$$

which is equivalent to the Lyapunov equation

$$\tilde{A}^{\mathsf{T}}P + P\tilde{A} + \tilde{R} = 0, \tag{16.16.6}$$

where

$$\tilde{R} \triangleq R_1 + P\Sigma P. \tag{16.16.7}$$

By comparing (16.15.16) and (16.16.4), it can be seen that the linear-quadratic control problems with (A, B, R_1, R_{12}, R_2) and $(\hat{A}_{\mathrm{R}}, B, \hat{R}_1, 0, R_2)$ are equivalent. Hence, there is no loss of generality in assuming that $R_{12} = 0$ in the following development, where A and R_1 take the place of $\hat{A}_{\mathrm{R}}$ and $\hat{R}_1$, respectively.

To motivate the subsequent development, the following examples demonstrate the existence of solutions under various assumptions on (A, B, E_1). In the following four examples, (A, B) is not stabilizable.

Example 16.16.1. Let $n = 1$, $A = 1$, $B = 0$, $E_1 = 0$, and $R_2 > 0$. Hence, (A, B, E_1) has an ORHP eigenvalue that is uncontrollable and unobservable. In this case, (16.16.4) has the unique solution $P = 0$. Furthermore, since $B = 0$, it follows that $\tilde{A} = A$. ◇

Example 16.16.2. Let $n = 1$, $A = 1$, $B = 0$, $E_1 = 1$, and $R_2 > 0$. Hence, (A, B, E_1) has an ORHP eigenvalue that is uncontrollable and observable. In this case, (16.16.4) has the unique solution $P = -1/2 < 0$. Furthermore, since $B = 0$, it follows that $\tilde{A} = A$. ◇

Example 16.16.3. Let $n = 1$, $A = 0$, $B = 0$, $E_1 = 0$, and $R_2 > 0$. Hence, (A, B, E_1) has an imaginary eigenvalue that is uncontrollable and unobservable. In this case, (16.16.4) has infinitely many solutions $P \in \mathbb{R}$. Hence, (16.16.4) has no maximal solution. Furthermore, since $B = 0$, it follows that, for every solution P, $\tilde{A} = A$. ◇

Example 16.16.4. Let $n = 1$, $A = 0$, $B = 0$, $E_1 = 1$, and $R_2 > 0$. Hence, (A, B, E_1) has an imaginary eigenvalue that is uncontrollable and observable. In this case, (16.16.4) becomes $R_1 = 0$. Thus, (16.16.4) has no solution. ◇

In the remaining examples, (A, B) is controllable.

Example 16.16.5. Let $n = 1$, $A = 1$, $B = 1$, $E_1 = 0$, and $R_2 > 0$. Hence, (A, B, E_1) has an ORHP eigenvalue that is controllable and unobservable. In this case, (16.16.4) has the solutions $P = 0$ and $P = 2R_2 > 0$. The corresponding closed-loop dynamics matrices are $\tilde{A} = 1 > 0$ and $\tilde{A} = -1 < 0$. Hence, the solution $P = 2R_2$ is stabilizing, and the closed-loop eigenvalue -1, which does not depend on R_2, is the reflection of the open-loop eigenvalue 1 across the imaginary axis. ◇

Example 16.16.6. Let $n = 1$, $A = 1$, $B = 1$, $E_1 = 1$, and $R_2 > 0$. Hence, (A, B, E_1) has an ORHP eigenvalue that is controllable and observable. In this case, (16.16.4) has the solutions $P = R_2 - \sqrt{R_2^2 + R_2} < 0$ and $P = R_2 + \sqrt{R_2^2 + R_2} > 0$. The corresponding closed-loop dynamics matrices are $\tilde{A} = \sqrt{1 + 1/R_2} > 0$ and $\tilde{A} = -\sqrt{1 + 1/R_2} < 0$. Hence, the positive-definite solution $P = R_2 + \sqrt{R_2^2 + R_2}$ is stabilizing. ◇

Example 16.16.7. Let $n = 1$, $A = 0$, $B = 1$, $E_1 = 0$, and $R_2 > 0$. Hence, (A, B, E_1) has an imaginary eigenvalue that is controllable and unobservable. In this case, (16.16.4) has the unique solution $P = 0$, which is not stabilizing. ◇

Example 16.16.8. Let $n = 1$, $A = 0$, $B = 1$, $E_1 = 1$, and $R_2 > 0$. Hence, (A, B, E_1) has an imaginary eigenvalue that is controllable and observable. In this case, (16.16.4) has the solutions $P = -\sqrt{R_2} < 0$ and $P = \sqrt{R_2} > 0$. The corresponding closed-loop dynamics matrices are $\tilde{A} = 1/\sqrt{R_2} > 0$ and $\tilde{A} = -1/\sqrt{R_2} < 0$. Hence, the positive-definite solution $P = \sqrt{R_2}$ is stabilizing. ◇

Example 16.16.9. Let $n = 2$, $A = \begin{bmatrix} 0 & 1 \\ -1 & 0 \end{bmatrix}$, $B = I_2$, $E_1 = 0$, and $R_2 = 1$. Hence, as in Example 16.16.7, both eigenvalues of (A, B, E_1) are imaginary, controllable, and unobservable. Taking the trace of (16.16.4) yields $\operatorname{tr} P^2 = 0$. Thus, the unique symmetric matrix P that satisfies (16.16.4) is $P = 0$, which implies that $\tilde{A} = A$. Consequently, the open-loop eigenvalues $\pm \jmath$ are unaffected by the feedback gain (16.15.15) even though (A, B) is controllable. ◇

Example 16.16.10. Let $n = 2$, $A = 0$, $B = I_2$, $E_1 = I_2$, and $R_2 = I$. Hence, as in Example 16.16.8, both eigenvalues of (A, B, E_1) are imaginary, controllable, and observable. Furthermore, (16.16.4) becomes $P^2 = I$. Requiring that P be symmetric, it follows that P is a reflector. Hence, $P = I$ is the unique positive-semidefinite solution. In fact, P is positive definite and stabilizing since $\tilde{A} = -I$. ◇

Example 16.16.11. Let $A = \left[\begin{smallmatrix} 1 & 0 \\ 0 & 2 \end{smallmatrix}\right]$, $B = \left[\begin{smallmatrix} 1 \\ 1 \end{smallmatrix}\right]$, $E_1 = 0$, and $R_2 = 1$ so that (A, B) is controllable, although neither of the states is weighted. In this case, (16.16.4) has four positive-semidefinite solutions, which are given by

$$P_1 = \begin{bmatrix} 18 & -24 \\ -24 & 36 \end{bmatrix}, \quad P_2 = \begin{bmatrix} 2 & 0 \\ 0 & 0 \end{bmatrix}, \quad P_3 = \begin{bmatrix} 0 & 0 \\ 0 & 4 \end{bmatrix}, \quad P_4 = \begin{bmatrix} 0 & 0 \\ 0 & 0 \end{bmatrix}.$$

The corresponding feedback matrices are $K_1 = [6 \ \ -12]$, $K_2 = [-2 \ \ 0]$, $K_3 = [0 \ \ -4]$, and $K_4 = [0 \ \ 0]$. Letting $\tilde{A}_i = A - \Sigma P_i$, it follows that $\operatorname{spec}(\tilde{A}_1) = \{-1, -2\}$, $\operatorname{spec}(\tilde{A}_2) = \{-1, 2\}$, $\operatorname{spec}(\tilde{A}_3) = \{1, -2\}$, and $\operatorname{spec}(\tilde{A}_4) = \{1, 2\}$. Thus, P_1 is the unique solution that stabilizes the closed-loop system, while the solutions P_2 and P_3 partially stabilize the closed-loop system. Note also that the closed-loop poles that differ from those of the open-loop system are mirror images of the open-loop poles as reflected across the imaginary axis. Finally, note that these solutions satisfy the partial orderings $P_1 \ge P_2 \ge P_4$ and $P_1 \ge P_3 \ge P_4$, and that "larger" solutions are more stabilizing than "smaller" solutions. Moreover, letting $J(K_i) = \operatorname{tr} P_i V$, it can be seen that larger solutions incur a greater closed-loop cost, with the greatest cost incurred by the stabilizing solution P_1. However, the cost expression $J(K) = \operatorname{tr} PV$ does not follow from (16.15.4) if $A + BK$ is not asymptotically stable. ◇

The following definition concerns solutions of the Riccati equation.

Definition 16.16.12. A matrix $P \in \mathbb{R}^{n \times n}$ is a *solution* of the Riccati equation (16.16.4) if P is symmetric and satisfies (16.16.4). Furthermore, P is the *stabilizing solution* of (16.16.4) if $\tilde{A} = A - \Sigma P$ is asymptotically stable. Finally, a solution $P_{\max}$ of (16.16.4) is the *maximal solution* to (16.16.4) if $P \le P_{\max}$ for every solution P to (16.16.4).

Since the ordering "$\le$" is antisymmetric, it follows that (16.16.4) has at most one maximal solution. The uniqueness of the stabilizing solution is shown in the following section.

Next, define the $2n \times 2n$ *Hamiltonian*

$$\mathcal{H} \triangleq \begin{bmatrix} A & \Sigma \\ R_1 & -A^{\mathsf{T}} \end{bmatrix}. \tag{16.16.8}$$

Proposition 16.16.13. The following statements hold:

i) The Hamiltonian $\mathcal{H}$ is a Hamiltonian matrix.

ii) $\chi_{\mathcal{H}}$ has a spectral factorization; that is, there exists a monic polynomial $p \in \mathbb{R}[s]$ such that, for all $s \in \mathbb{C}$, $\chi_{\mathcal{H}}(s) = p(s)p(-s)$.

iii) $\chi_{\mathcal{H}}(\omega \jmath) \ge 0$ for all $\omega \in \mathbb{R}$.

iv) If either $R_1 = 0$ or $\Sigma = 0$, then $\operatorname{mspec}(\mathcal{H}) = \operatorname{mspec}(A) \cup \operatorname{mspec}(-A)$.

v) $\chi_{\mathcal{H}}$ is even.

vi) $\lambda \in \operatorname{spec}(\mathcal{H})$ if and only if $-\lambda \in \operatorname{spec}(\mathcal{H})$.

vii) If $\lambda \in \operatorname{spec}(\mathcal{H})$, then $\operatorname{amult}_{\mathcal{H}}(\lambda) = \operatorname{amult}_{\mathcal{H}}(-\lambda)$.

viii) Every imaginary root of $\chi_{\mathcal{H}}$ has even multiplicity.

ix) Every imaginary eigenvalue of $\mathcal{H}$ has even algebraic multiplicity.

Proof. This result follows from Proposition 6.1.1 and Fact 6.9.33. □

It is helpful to keep in mind that spectral factorizations are not unique. For example, if $\chi_{\mathcal{H}}(s) = (s+1)(s+2)(-s+1)(-s+2)$, then $\chi_{\mathcal{H}}(s) = p(s)p(-s) = \hat{p}(s)\hat{p}(-s)$, where $p(s) = (s+1)(s+2)$ and $\hat{p}(s) = (s+1)(s-2)$. Thus, the spectral factors $p(s)$ and $p(-s)$ can "trade" roots. These roots are the eigenvalues of $\mathcal{H}$. The following result shows that the Hamiltonian $\mathcal{H}$ is closely linked to the Riccati equation (16.16.4).

Proposition 16.16.14. Let $P \in \mathbb{R}^{n\times n}$ be symmetric. Then, the following statements are equivalent:

i) P is a solution of (16.16.4).

ii) P satisfies

$$[P\ \ I]\mathcal{H}\begin{bmatrix} I \\ -P \end{bmatrix} = 0. \tag{16.16.9}$$

iii) P satisfies

$$\mathcal{H}\begin{bmatrix} I \\ -P \end{bmatrix} = \begin{bmatrix} I \\ -P \end{bmatrix}(A - \Sigma P). \tag{16.16.10}$$

iv) P satisfies

$$\mathcal{H} = \begin{bmatrix} I & 0 \\ -P & I \end{bmatrix}\begin{bmatrix} A - \Sigma P & \Sigma \\ 0 & -(A-\Sigma P)^{\mathsf{T}} \end{bmatrix}\begin{bmatrix} I & 0 \\ P & I \end{bmatrix}. \tag{16.16.11}$$

If these conditions hold, then the following statements hold:

v) $\operatorname{mspec}(\mathcal{H}) = \operatorname{mspec}(A - \Sigma P) \cup \operatorname{mspec}[-(A - \Sigma P)]$.

vi) $\chi_{\mathcal{H}}(s) = (-1)^n \chi_{A-\Sigma P}(s)\chi_{A-\Sigma P}(-s)$.

vii) $\mathcal{R}(\left[\begin{smallmatrix} I \\ -P \end{smallmatrix}\right])$ is an invariant subspace of $\mathcal{H}$.

Corollary 16.16.15. Assume that (16.16.4) has a stabilizing solution. Then, $\mathcal{H}$ has no imaginary eigenvalues.

For the next two results, P is not necessarily a solution of (16.16.4).

Lemma 16.16.16. Assume that $\lambda \in \operatorname{spec}(A)$ is an observable eigenvalue of (A, R_1), and let $P \in \mathbb{R}^{n\times n}$ be symmetric. Then, $\lambda \in \operatorname{spec}(A)$ is an observable eigenvalue of $(\tilde{A}, \tilde{R})$.

Proof. Suppose that $\operatorname{rank}\left[\begin{smallmatrix} \lambda I - \tilde{A} \\ \tilde{R} \end{smallmatrix}\right] < n$. Then, there exists a nonzero vector $v \in \mathbb{C}^n$ such that $\tilde{A}v = \lambda v$ and $\tilde{R}v = 0$. Hence, $v^* R_1 v = -v^* P \Sigma P v \le 0$, which implies that $R_1 v = 0$ and $P\Sigma P v = 0$. Hence, $\Sigma P v = 0$, and thus $Av = \lambda v$. Therefore, $\operatorname{rank}\left[\begin{smallmatrix} \lambda I - A \\ R_1 \end{smallmatrix}\right] < n$. □

Lemma 16.16.17. Assume that (A, R_1) is (observable, detectable), and let $P \in \mathbb{R}^{n\times n}$ be symmetric. Then, $(\tilde{A}, \tilde{R})$ is (observable, detectable).

Lemma 16.16.18. Assume that (A, E_1) is observable, and assume that (16.16.4) has a solution P. Then, the following statements hold:

i) $\nu_-(\tilde{A}) = \nu_+(P)$.

ii) $\nu_0(\tilde{A}) = \nu_0(P) = 0$.

iii) $\nu_+(\tilde{A}) = \nu_-(P)$.

Proof. Since (A, R_1) is observable, it follows from Lemma 16.16.17 that $(\tilde{A}, \tilde{R})$ is observable. By writing (16.16.4) as the Lyapunov equation (16.16.6), the result now follows from Fact 16.22.1. □

16.17 The Stabilizing Solution of the Riccati Equation

Proposition 16.17.1. The following statements hold:

i) (16.16.4) has at most one stabilizing solution.

ii) If P is the stabilizing solution of (16.16.4), then P is positive semidefinite.

iii) If P is the stabilizing solution of (16.16.4), then

$$\operatorname{rank} P = \operatorname{rank} \mathcal{O}(\tilde{A}, \tilde{R}). \tag{16.17.1}$$

Proof. To prove *i)*, suppose that (16.16.4) has stabilizing solutions P_1 and P_2. Then,

$$A^\mathsf{T}P_1 + P_1A + R_1 - P_1\Sigma P_1 = 0,$$
$$A^\mathsf{T}P_2 + P_2A + R_1 - P_2\Sigma P_2 = 0.$$

Subtracting these equations and rearranging yields

$$(A - \Sigma P_1)^\mathsf{T}(P_1 - P_2) + (P_1 - P_2)(A - \Sigma P_2) = 0.$$

Since $A - \Sigma P_1$ and $A - \Sigma P_2$ are asymptotically stable, it follows from Proposition 15.10.3 and Fact 15.19.33 that $P_1 - P_2 = 0$. Hence, (16.16.4) has at most one stabilizing solution.

Next, to prove *ii)*, let P be a stabilizing solution of (16.16.4). Then, it follows from (16.16.4) that

$$P = \int_0^\infty e^{t(A-\Sigma P)^\mathsf{T}}(R_1 + P\Sigma P)e^{t(A-\Sigma P)}\,\mathrm{d}t,$$

which shows that P is positive semidefinite. Finally, *iii)* follows from Corollary 16.3.3. □

Theorem 16.17.2. Assume that (16.16.4) has a positive-semidefinite solution P, and assume that (A, E_1) is detectable. Then, P is the stabilizing solution of (16.16.4) and the unique positive-semidefinite solution of (16.16.4). If, in addition, (A, E_1) is observable, then P is positive definite.

Proof. Since (A, R_1) is detectable, Lemma 16.16.17 implies that $(\tilde{A}, \tilde{R})$ is detectable. Next, since (16.16.4) has a positive-semidefinite solution P, it follows from Corollary 16.5.7 that $\tilde{A}$ is asymptotically stable. Hence, by Proposition 16.17.1, P is the stabilizing solution of (16.16.4) and thus the unique positive-semidefinite solution of (16.16.4). The last statement follows from Lemma 16.16.18. □

Corollary 16.17.3. Assume that (A, E_1) is detectable. Then, (16.16.4) has at most one positive-semidefinite solution.

Lemma 16.17.4. Let $\lambda \in \mathbb{C}$, and assume that λ is either an uncontrollable eigenvalue of (A, B) or an unobservable eigenvalue of (A, E_1). Then, $\lambda \in \operatorname{spec}(\mathcal{H})$.

Proof. Note that

$$\lambda I - \mathcal{H} = \begin{bmatrix} \lambda I - A & -\Sigma \\ -R_1 & \lambda I + A^\mathsf{T} \end{bmatrix}.$$

In the case where λ is an uncontrollable eigenvalue of (A, B), the first n rows of $\lambda I - \mathcal{H}$ are linearly dependent, and thus $\lambda \in \operatorname{spec}(\mathcal{H})$. On the other hand, in the case where λ is an unobservable eigenvalue of (A, E_1), the first n columns of $\lambda I - \mathcal{H}$ are linearly dependent, and thus $\lambda \in \operatorname{spec}(\mathcal{H})$. □

The following result is a consequence of Lemma 16.17.4.

Proposition 16.17.5. Let $S \in \mathbb{R}^{n\times n}$ be a nonsingular matrix such that

$$A = S\begin{bmatrix} A_1 & 0 & A_{13} & 0 \\ A_{21} & A_2 & A_{23} & A_{24} \\ 0 & 0 & A_3 & 0 \\ 0 & 0 & A_{43} & A_4 \end{bmatrix}S^{-1}, \quad B = S\begin{bmatrix} B_1 \\ B_2 \\ 0 \\ 0 \end{bmatrix}, \quad E_1 = [E_{11}\ \ 0\ \ E_{13}\ \ 0]S^{-1}, \tag{16.17.2}$$

where $\left(\begin{bmatrix} A_1 & 0 \\ A_{21} & A_2 \end{bmatrix}, \begin{bmatrix} B_1 \\ B_2 \end{bmatrix}\right)$ is controllable and $\left(\begin{bmatrix} A_1 & A_{13} \\ 0 & A_3 \end{bmatrix}, [E_{11}\ \ E_{13}]\right)$ is observable. Then,

$$\operatorname{mspec}(A_2) \cup \operatorname{mspec}(-A_2) \subseteq \operatorname{mspec}(\mathcal{H}), \tag{16.17.3}$$

$$\operatorname{mspec}(A_3) \cup \operatorname{mspec}(-A_3) \subseteq \operatorname{mspec}(\mathcal{H}), \tag{16.17.4}$$

$$\operatorname{mspec}(A_4) \cup \operatorname{mspec}(-A_4) \subseteq \operatorname{mspec}(\mathcal{H}). \tag{16.17.5}$$

Next, we present a partial converse of Lemma 16.17.4.

Lemma 16.17.6. Let $\lambda \in \operatorname{spec}(\mathcal{H})$, and assume that $\operatorname{Re}\lambda = 0$. Then, λ is either an uncontrollable eigenvalue of (A, B) or an unobservable eigenvalue of (A, E_1).

Proof. Let $\lambda = \omega\jmath$ be an eigenvalue of $\mathcal{H}$, where $\omega \in \mathbb{R}$. Then, there exist $x, y \in \mathbb{C}^n$ such that $\left[\begin{smallmatrix} x \\ y \end{smallmatrix}\right] \neq 0$ and $\mathcal{H}\left[\begin{smallmatrix} x \\ y \end{smallmatrix}\right] = \omega\jmath\left[\begin{smallmatrix} x \\ y \end{smallmatrix}\right]$. Consequently,

$$Ax + \Sigma y = \omega\jmath x, \quad R_1 x - A^{\mathsf{T}} y = \omega\jmath y.$$

Rewriting these equalities as

$$(A - \omega\jmath I)x = -\Sigma y, \quad (A - \omega\jmath I)^* y = R_1 x$$

yields

$$y^*(A - \omega\jmath I)x = -y^*\Sigma y, \quad x^*(A - \omega\jmath I)^* y = x^* R_1 x.$$

Since $x^*(A - \omega\jmath I)^* y$ is real, it follows that $x^*R_1x = -y^*\Sigma y$. Since $0 \le x^*R_1x = -y^*\Sigma y \le 0$, it follows that $x^*R_1x = y^*\Sigma y = 0$, and thus $B^{\mathsf{T}}y = 0$ and $E_1x = 0$. Therefore,

$$(A - \omega\jmath I)x = 0, \quad (A - \omega\jmath I)^* y = 0,$$

and hence

$$\begin{bmatrix} A - \omega\jmath I \\ E_1 \end{bmatrix} x = 0, \quad y^*[A - \omega\jmath I \;\; B] = 0.$$

Since $\left[\begin{smallmatrix} x \\ y \end{smallmatrix}\right] \neq 0$, it follows that either $x \neq 0$ or $y \neq 0$. Hence, either $\operatorname{rank}\left[\begin{smallmatrix} A-\omega\jmath I \\ E_1 \end{smallmatrix}\right] < n$ or $\operatorname{rank}\,[A - \omega\jmath I \;\; B] < n$. □

The following result is a restatement of Lemma 16.17.6.

Proposition 16.17.7. Let $S \in \mathbb{R}^{n\times n}$ be a nonsingular matrix such that (16.17.2) holds, where $\left(\left[\begin{smallmatrix} A_1 & 0 \\ A_{21} & A_2 \end{smallmatrix}\right], \left[\begin{smallmatrix} B_1 \\ B_2 \end{smallmatrix}\right]\right)$ is controllable and $\left(\left[\begin{smallmatrix} A_1 & A_{13} \\ 0 & A_3 \end{smallmatrix}\right], [E_{11} \;\; E_{13}]\right)$ is observable. Then,

$$\operatorname{mspec}(\mathcal{H}) \cap \mathrm{IA} \subseteq \operatorname{mspec}(A_2) \cup \operatorname{mspec}(-A_2) \cup \operatorname{mspec}(A_3) \cup \operatorname{mspec}(-A_3) \cup \operatorname{mspec}(A_4) \cup \operatorname{mspec}(-A_4). \tag{16.17.6}$$

Combining Lemma 16.17.4 and Lemma 16.17.6 yields the following result.

Proposition 16.17.8. Let $\lambda \in \mathbb{C}$, assume that $\operatorname{Re}\lambda = 0$, and let $S \in \mathbb{R}^{n\times n}$ be a nonsingular matrix such that (16.17.2) holds, where (A_1, B_1, E_{11}) is controllable and observable, (A_2, B_2) is controllable, and (A_3, E_{13}) is observable. Then, the following statements are equivalent:

i) λ is either an uncontrollable eigenvalue of (A, B) or an unobservable eigenvalue of (A, E_1).

ii) $\lambda \in \operatorname{mspec}(A_2) \cup \operatorname{mspec}(A_3) \cup \operatorname{mspec}(A_4)$.

iii) λ is an eigenvalue of $\mathcal{H}$.

The next result gives necessary and sufficient conditions under which (16.16.4) has a stabilizing solution. This result also provides a constructive characterization of the stabilizing solution. *ii)* of Proposition 16.10.11 shows that the statement in *i)* that every imaginary eigenvalue of (A, E_1) is observable is equivalent to the statement that $\left[\begin{array}{c|c} A & B \\ \hline E_1 & E_2 \end{array}\right]$ has no imaginary invariant zeros.

Theorem 16.17.9. The following statements are equivalent:

i) (A, B) is stabilizable, and every imaginary eigenvalue of (A, E_1) is observable.

ii) There exists a nonsingular matrix $S \in \mathbb{R}^{n\times n}$ such that (16.17.2) holds, where $\left(\begin{bmatrix} A_1 & 0 \\ A_{21} & A_2 \end{bmatrix}, \begin{bmatrix} B_1 \\ B_2 \end{bmatrix}\right)$ is controllable, $\left(\begin{bmatrix} A_1 & A_{13} \\ 0 & A_3 \end{bmatrix}, [E_{11}\ \ E_{13}]\right)$ is observable, $\nu_0(A_2) = 0$, and A_3 and A_4 are asymptotically stable.

iii) (16.16.4) has a stabilizing solution.

Now, assume that these conditions hold, and let

$$M = \begin{bmatrix} M_1 & M_{12} \\ M_{21} & M_2 \end{bmatrix} \in \mathbb{R}^{2n\times 2n} \tag{16.17.7}$$

be a nonsingular matrix such that $\mathcal{H} = MZM^{-1}$, where

$$Z = \begin{bmatrix} Z_1 & Z_{12} \\ 0 & Z_2 \end{bmatrix} \in \mathbb{R}^{2n\times 2n} \tag{16.17.8}$$

and $Z_1 \in \mathbb{R}^{n\times n}$ is asymptotically stable. Then, M_1 is nonsingular, and

$$P \triangleq -M_{21}M_1^{-1} \tag{16.17.9}$$

is the stabilizing solution of (16.16.4).

Proof. The equivalence of *i*) and *ii*) is immediate. To prove *i*) $\Longrightarrow$ *iii*), first note that Lemma 16.17.6 implies that $\mathcal{H}$ has no imaginary eigenvalues. Hence, since $\mathcal{H}$ is Hamiltonian, it follows that there exists $M \in \mathbb{R}^{2n\times 2n}$ of the form (16.17.7) such that M is nonsingular and $\mathcal{H} = MZM^{-1}$, where $Z \in \mathbb{R}^{2n\times 2n}$ is of the form (16.17.8) and $Z_1 \in \mathbb{R}^{n\times n}$ is asymptotically stable.

Next, note that $\mathcal{H}M = MZ$ implies that

$$\mathcal{H}\begin{bmatrix} M_1 \\ M_{21} \end{bmatrix} = M\begin{bmatrix} Z_1 \\ 0 \end{bmatrix} = \begin{bmatrix} M_1 \\ M_{21} \end{bmatrix} Z_1.$$

Therefore,

$$\begin{bmatrix} M_1 \\ M_{21} \end{bmatrix}^{\mathsf{T}} J_n\mathcal{H}\begin{bmatrix} M_1 \\ M_{21} \end{bmatrix} = \begin{bmatrix} M_1 \\ M_{21} \end{bmatrix}^{\mathsf{T}} J_n \begin{bmatrix} M_1 \\ M_{21} \end{bmatrix} Z_1 = \begin{bmatrix} M_1^{\mathsf{T}} & M_{21}^{\mathsf{T}} \end{bmatrix} \begin{bmatrix} M_{21} \\ -M_1 \end{bmatrix} Z_1 = LZ_1,$$

where $L \triangleq M_1^{\mathsf{T}}M_{21} - M_{21}^{\mathsf{T}}M_1$. Since $J_n\mathcal{H} = (J_n\mathcal{H})^{\mathsf{T}}$, it follows that LZ_1 is symmetric; that is, $LZ_1 = Z_1^{\mathsf{T}}L^{\mathsf{T}}$. Since, in addition, L is skew symmetric, it follows that $0 = Z_1^{\mathsf{T}}L + LZ_1$. Now, since Z_1 is asymptotically stable, it follows that $L = 0$. Hence, $M_1^{\mathsf{T}}M_{21} = M_{21}^{\mathsf{T}}M_1$, which shows that $M_{21}^{\mathsf{T}}M_1$ is symmetric.

To show that M_1 is nonsingular, note that

$$[I\ \ 0]\mathcal{H}\begin{bmatrix} M_1 \\ M_{21} \end{bmatrix} = [I\ \ 0]\begin{bmatrix} M_1 \\ M_{21} \end{bmatrix} Z_1$$

implies that

$$AM_1 + \Sigma M_{21} = M_1Z_1.$$

Now, let $x \in \mathbb{R}^n$ satisfy $M_1x = 0$. Then,

$$x^{\mathsf{T}}M_{21}^{\mathsf{T}}\Sigma M_{21}x = x^{\mathsf{T}}M_{21}^{\mathsf{T}}(AM_1 + \Sigma M_{21})x = x^{\mathsf{T}}M_{21}^{\mathsf{T}}M_1Z_1x = x^{\mathsf{T}}M_1^{\mathsf{T}}M_{21}Z_1x = 0,$$

which implies that $B^{\mathsf{T}}M_{21}x = 0$. Hence, $M_1Z_1x = (AM_1 + \Sigma M_{21})x = 0$. Thus, $Z_1\mathcal{N}(M_1) \subseteq \mathcal{N}(M_1)$.

Now, suppose that M_1 is singular. Since $Z_1\mathcal{N}(M_1) \subset \mathcal{N}(M_1)$, it follows that there exist $\lambda \in \operatorname{spec}(Z_1)$ and $x \in \mathbb{C}^n$ such that $Z_1x = \lambda x$ and $M_1x = 0$. Noting

$$[0\ \ I]\mathcal{H}\begin{bmatrix} M_1 \\ M_{21} \end{bmatrix} x = [0\ \ I]\begin{bmatrix} M_1 \\ M_{21} \end{bmatrix} Z_1x$$

yields $-A^{\mathsf{T}}M_{21}x = M_{21}\lambda x$, and thus $(\lambda I + A^{\mathsf{T}})M_{21}x = 0$. Since, in addition, as shown above, $B^{\mathsf{T}}M_{21}x = 0$, it follows that $x^*M_{21}^{\mathsf{T}}[-\overline{\lambda}I - A \;\; B] = 0$. Since $\lambda \in \mathrm{spec}(Z_1)$, it follows that $\mathrm{Re}(-\overline{\lambda}) > 0$. Furthermore, since, by assumption, (A, B) is stabilizable, it follows that $\mathrm{rank}\,[\overline{\lambda}I - A \;\; B] = n$. Therefore, $M_{21}x = 0$. Combining this fact with $M_1x = 0$ yields $\left[\begin{smallmatrix} M_1 \\ M_{21} \end{smallmatrix}\right]x = 0$. Since x is nonzero, it follows that M is singular, which is a contradiction. Consequently, M_1 is nonsingular. Next, define $P \triangleq -M_{21}M_1^{-1}$ and note that, since $M_1^{\mathsf{T}}M_{21}$ is symmetric, it follows that $P = -M_1^{-\mathsf{T}}(M_1^{\mathsf{T}}M_{21})M_1^{-1}$ is also symmetric.

Since $\mathcal{H}\left[\begin{smallmatrix} M_1 \\ M_{21} \end{smallmatrix}\right] = \left[\begin{smallmatrix} M_1 \\ M_{21} \end{smallmatrix}\right]Z_1$, it follows that

$$\mathcal{H}\begin{bmatrix} I \\ M_{21}M_1^{-1} \end{bmatrix} = \begin{bmatrix} I \\ M_{21}M_1^{-1} \end{bmatrix} M_1Z_1M_1^{-1},$$

and thus

$$\mathcal{H}\begin{bmatrix} I \\ -P \end{bmatrix} = \begin{bmatrix} I \\ -P \end{bmatrix} M_1Z_1M_1^{-1}.$$

Multiplying on the left by $[P \;\; I]$ yields

$$0 = [P \;\; I]\mathcal{H}\begin{bmatrix} I \\ -P \end{bmatrix} = A^{\mathsf{T}}P + PA + R_1 - P\Sigma P,$$

which shows that P is a solution of (16.16.4). Similarly, multiplying on the left by $[I \;\; 0]$ yields $A - \Sigma P = M_1Z_1M_1^{-1}$. Since Z_1 is asymptotically stable, it follows that $A - \Sigma P$ is also asymptotically stable.

To prove *iii)* $\Longrightarrow$ *i)*, note that the existence of a stabilizing solution P implies that (A, B) is stabilizable, and that (16.16.11) implies that $\mathcal{H}$ has no imaginary eigenvalues. □

Corollary 16.17.10. Assume that (A, B) is stabilizable and (A, E_1) is detectable. Then, (16.16.4) has a stabilizing solution.

16.18 The Maximal Solution of the Riccati Equation

In this section we consider the existence of the maximal solution of (16.16.4). Example 16.16.3 shows that (16.16.4) may not have a maximal solution.

Theorem 16.18.1. The following statements are equivalent:

i) (A, B) is stabilizable.

ii) (16.16.4) has a solution $P_{\max}$ that is positive semidefinite, maximal, and satisfies

$$\mathrm{spec}(A - \Sigma P_{\max}) \subset \mathrm{CLHP}. \tag{16.18.1}$$

Proof. *i)* $\Longrightarrow$ *ii)* is given by Theorem 2.1 and Theorem 2.2 of [1188]. See also (*i*) of Theorem 13.11 of [2999]. The converse follows from Corollary 3 of [2386]. □

Proposition 16.18.2. Assume that (16.16.4) has a maximal solution $P_{\max}$, let P be a solution of (16.16.4), and assume that $\mathrm{spec}(A - \Sigma P_{\max}) \subset \mathrm{CLHP}$ and $\mathrm{spec}(A - \Sigma P) \subset \mathrm{CLHP}$. Then, $P = P_{\max}$.

Proof. It follows from *i)* of Proposition 16.16.14 that $\mathrm{spec}(A - \Sigma P) = \mathrm{spec}(A - \Sigma P_{\max})$. Since $P_{\max}$ is the maximal solution of (16.16.4), it follows that $P \le P_{\max}$. Consequently, it follows from the contrapositive form of the second statement of Theorem 10.4.9 that $P = P_{\max}$. □

Proposition 16.18.3. Assume that (16.16.4) has a solution P such that $\mathrm{spec}(A - \Sigma P) \subset \mathrm{CLHP}$. Then, P is stabilizing if and only if $\mathcal{H}$ has no imaginary eigenvalues.

It follows from Proposition 16.18.2 that (16.16.4) has at most one positive-semidefinite solution P such that $\mathrm{spec}(A - \Sigma P) \subset \mathrm{CLHP}$. Consequently, (16.16.4) has at most one positive-semidefinite stabilizing solution.

Theorem 16.18.4. The following statements hold:

i) (16.16.4) has at most one stabilizing solution.

ii) If P is the stabilizing solution of (16.16.4), then P is positive semidefinite.

iii) If P is the stabilizing solution of (16.16.4), then P is maximal.

Proof. To prove *i*), suppose that (16.16.4) has stabilizing solutions P_1 and P_2. Then, (A, B) is stabilizable, and Theorem 16.18.1 implies that (16.16.4) has a maximal solution $P_{\max}$ such that $\operatorname{spec}(A - \Sigma P_{\max}) \subset \text{CLHP}$. Now, Proposition 16.18.2 implies that $P_1 = P_{\max}$ and $P_2 = P_{\max}$. Hence, $P_1 = P_2$. Alternatively, suppose that (16.16.4) has the stabilizing solutions P_1 and P_2. Then,

$$A^{\mathsf{T}}P_1 + P_1A + R_1 - P_1\Sigma P_1 = 0,$$
$$A^{\mathsf{T}}P_2 + P_2A + R_1 - P_2\Sigma P_2 = 0.$$

Subtracting these equations and rearranging yields

$$(A - \Sigma P_1)^{\mathsf{T}}(P_1 - P_2) + (P_1 - P_2)(A - \Sigma P_2) = 0.$$

Since $A - \Sigma P_1$ and $A - \Sigma P_2$ are asymptotically stable, it follows from Proposition 15.10.3 and Fact 15.19.33 that $P_1 - P_2 = 0$. Hence, (16.16.4) has at most one stabilizing solution.

Next, to prove *ii*), suppose that P is a stabilizing solution of (16.16.4). Then, it follows from (16.16.4) that

$$P = \int_0^\infty e^{t(A-\Sigma P)^{\mathsf{T}}}(R_1 + P\Sigma P)e^{t(A-\Sigma P)}\,\mathrm{d}t,$$

which shows that P is positive semidefinite. To prove *iii*), let P' be a solution of (16.16.4). Then,

$$(A - \Sigma P)^{\mathsf{T}}(P - P') + (P - P')(A - \Sigma P) + (P - P')\Sigma(P - P') = 0,$$

which implies that $P' \le P$. Thus, P is also the maximal solution of (16.16.4). □

The following result concerns the monotonicity of solutions of the Riccati equation (16.16.4).

Proposition 16.18.5. Assume that (A, B) is stabilizable, and let $P_{\max}$ denote the maximal solution of (16.16.4). Furthermore, let $\hat{R}_1 \in \mathbb{R}^{n\times n}$ be positive semidefinite, let $\hat{R}_2 \in \mathbb{R}^{m\times m}$ be positive definite, let $\hat{A} \in \mathbb{R}^{n\times n}$, let $\hat{B} \in \mathbb{R}^{n\times m}$, define $\hat{\Sigma} \triangleq \hat{B}\hat{R}_2^{-1}B^{\mathsf{T}}$, assume that

$$\begin{bmatrix} \hat{R}_1 & \hat{A}^{\mathsf{T}} \\ \hat{A} & -\hat{\Sigma} \end{bmatrix} \le \begin{bmatrix} R_1 & A^{\mathsf{T}} \\ A & -\Sigma \end{bmatrix},$$

and let $\hat{P}$ be a solution of

$$\hat{A}^{\mathsf{T}}\hat{P} + \hat{P}\hat{A} + \hat{R}_1 - \hat{P}\hat{\Sigma}\hat{P} = 0. \tag{16.18.2}$$

Then,

$$\hat{P} \le P_{\max}. \tag{16.18.3}$$

Proof. See Theorem 1 of [2895]. □

Corollary 16.18.6. Assume that (A, B) is stabilizable, let $\hat{R}_1 \in \mathbb{R}^{n\times n}$ be positive semidefinite, assume that $\hat{R}_1 \le R_1$, and let $P_{\max}$ and $\hat{P}_{\max}$ denote, respectively, the maximal solutions of (16.16.4) and

$$A^{\mathsf{T}}P + PA + \hat{R}_1 - P\Sigma P = 0. \tag{16.18.4}$$

Then,

$$\hat{P}_{\max} \le P_{\max}. \tag{16.18.5}$$

Proof. This result follows from either Proposition 16.18.5 or Theorem 2.3 of [1188]. □

The following result shows that, if $R_1 = 0$, then the closed-loop eigenvalues of the closed-loop dynamics obtained from the maximal solution consist of the CLHP open-loop eigenvalues and reflections of the ORHP open-loop eigenvalues.

Proposition 16.18.7. Assume that (A, B) is stabilizable, assume that $R_1 = 0$, and let $P \in \mathbb{R}^{n\times n}$ be a positive-semidefinite solution of (16.16.4). Then, P is the maximal solution of (16.16.4) if and only if

$$\operatorname{mspec}(A - \Sigma P) = [\operatorname{mspec}(A) \cap \mathrm{CLHP}] \cup [\operatorname{mspec}(-A) \cap \mathrm{OLHP}]. \tag{16.18.6}$$

Proof. Sufficiency follows from Proposition 16.18.2. To prove necessity, note that it follows from the definition (16.16.8) of $\mathcal{H}$ with $R_1 = 0$ and from *iv*) of Proposition 16.16.14 that

$$\operatorname{mspec}(A) \cup \operatorname{mspec}(-A) = \operatorname{mspec}(A - \Sigma P) \cup \operatorname{mspec}[-(A - \Sigma P)].$$

Now, Theorem 16.18.1 implies that $\operatorname{mspec}(A-\Sigma P) \subseteq \mathrm{CLHP}$, which implies that (16.18.6) holds. □

Corollary 16.18.8. Let $R_1 = 0$, and assume that $\operatorname{spec}(A) \subset \mathrm{CLHP}$. Then, $P = 0$ is the unique positive-semidefinite solution of (16.16.4).

16.19 Positive-Semidefinite and Positive-Definite Solutions of the Riccati Equation

The following result gives sufficient conditions under which (16.16.4) has a positive-semidefinite solution.

Proposition 16.19.1. Assume that there exists a nonsingular matrix $S \in \mathbb{R}^{n\times n}$ such that (16.17.2) holds, where $\left(\left[\begin{smallmatrix} A_1 & 0 \\ A_{21} & A_2 \end{smallmatrix}\right], \left[\begin{smallmatrix} B_1 \\ B_2 \end{smallmatrix}\right]\right)$ is controllable, $\left(\left[\begin{smallmatrix} A_1 & A_{13} \\ 0 & A_3 \end{smallmatrix}\right], [E_{11} \ E_{13}]\right)$ is observable, and A_3 is asymptotically stable. Then, (16.16.4) has a positive-semidefinite solution.

Proof. First, rewrite (16.17.2) as

$$A = S\begin{bmatrix} A_1 & A_{13} & 0 & 0 \\ 0 & A_3 & 0 & 0 \\ A_{21} & A_{23} & A_2 & A_{24} \\ 0 & A_{43} & 0 & A_4 \end{bmatrix} S^{-1}, \quad B = S\begin{bmatrix} B_1 \\ 0 \\ B_2 \\ 0 \end{bmatrix},$$

$$E_1 = \begin{bmatrix} E_{11} & E_{13} & 0 & 0 \end{bmatrix} S^{-1},$$

where $\left(\left[\begin{smallmatrix} A_1 & 0 \\ A_{21} & A_2 \end{smallmatrix}\right], \left[\begin{smallmatrix} B_1 \\ B_2 \end{smallmatrix}\right]\right)$ is controllable, $\left(\left[\begin{smallmatrix} A_1 & A_{13} \\ 0 & A_3 \end{smallmatrix}\right], [E_{11} \ E_{13}]\right)$ is observable, and A_3 is asymptotically stable. Since $\left(\left[\begin{smallmatrix} A_1 & A_{13} \\ 0 & A_3 \end{smallmatrix}\right], \left[\begin{smallmatrix} B_1 \\ 0 \end{smallmatrix}\right]\right)$ is stabilizable, it follows from Theorem 16.18.1 that there exists a positive-semidefinite matrix $\hat{P}_1$ that satisfies

$$\begin{bmatrix} A_1 & A_{13} \\ 0 & A_3 \end{bmatrix}^{\mathrm{T}} \hat{P}_1 + \hat{P}_1 \begin{bmatrix} A_1 & A_{13} \\ 0 & A_3 \end{bmatrix} + \begin{bmatrix} E_{11}^{\mathrm{T}} E_{11} & E_{11}^{\mathrm{T}} E_{13} \\ E_{13}^{\mathrm{T}} E_{11} & E_{13}^{\mathrm{T}} E_{13} \end{bmatrix} - \hat{P}_1 \begin{bmatrix} B_1 R_2^{-1} B_1^{\mathrm{T}} & 0 \\ 0 & 0 \end{bmatrix} \hat{P}_1 = 0.$$

Consequently, $P \triangleq S^{\mathrm{T}} \operatorname{diag}(\hat{P}_1, 0, 0) S$ is a positive-semidefinite solution of (16.16.4). □

Corollary 16.19.2. Assume that (A, B) is stabilizable. Then, (16.16.4) has a positive-semidefinite solution P. If, in addition, (A, E_1) is detectable, then P is the stabilizing solution of (16.16.4), and thus P is the unique positive-semidefinite solution of (16.16.4). Finally, if (A, E_1) is observable, then P is positive definite.

Proof. The first statement is given by Theorem 16.18.1. Next, since (A, E_1) is detectable, Theorem 16.17.2 implies that P is a stabilizing solution of (16.16.4), which is the unique positive-semidefinite solution of (16.16.4). Finally, using Theorem 16.17.2, (A, E_1) observable implies that P is positive definite. □

The next result gives necessary and sufficient conditions under which (16.16.4) has a positive-definite solution.

Proposition 16.19.3. The following statements are equivalent:

i) (16.16.4) has a positive-definite solution.

ii) There exists a nonsingular matrix $S \in \mathbb{R}^{n\times n}$ such that (16.17.2) holds, where $\left(\left[\begin{smallmatrix} A_1 & 0 \\ A_{21} & A_2 \end{smallmatrix}\right], \left[\begin{smallmatrix} B_1 \\ B_2 \end{smallmatrix}\right]\right)$ is controllable, $\left(\left[\begin{smallmatrix} A_1 & A_{13} \\ 0 & A_3 \end{smallmatrix}\right], [E_{11}\ E_{13}]\right)$ is observable, A_3 is asymptotically stable, $-A_2$ is asymptotically stable, $\operatorname{spec}(A_4) \subset \mathrm{IA}$, and A_4 is semisimple.

If these conditions hold, then (16.16.4) has exactly one positive-definite solution if and only if A_4 is empty, and infinitely many positive-definite solutions if and only if A_4 is nonempty.

Proof. See [2306]. □

Proposition 16.19.4. Assume that (16.16.4) has a stabilizing solution P, and let $S \in \mathbb{R}^{n\times n}$ be a nonsingular matrix such that (16.17.2) holds, where (A_1, B_1, E_{11}) is controllable and observable, (A_2, B_2) is controllable, (A_3, E_{13}) is observable, $\nu_0(A_2) = 0$, and A_3 and A_4 are asymptotically stable. Then,

$$\operatorname{def} P = \nu_-(A_2). \tag{16.19.1}$$

Hence, P is positive definite if and only if $\operatorname{spec}(A_2) \subset \mathrm{ORHP}$.

16.20 Facts on Linear Differential Equations

Fact 16.20.1. Let $A \in \mathbb{F}^{n\times n}$ and $b \in \mathbb{F}^n$, assume that A is nonsingular, define $x_{\mathrm{s}} \triangleq -A^{-1}b$, and let $x\colon [0,\infty) \mapsto \mathbb{F}^n$ satisfy $\dot{x}(t) = Ax(t) + b$. Then, for all $t \ge 0$,

$$x(t) = e^{tA}x(0) + (I - e^{tA})x_{\mathrm{s}} = x_{\mathrm{s}} + e^{tA}[x(0) - x_{\mathrm{s}}].$$

If, in addition, A is asymptotically stable, then

$$\lim_{t\to\infty} x(t) = x_{\mathrm{s}} = \int_0^\infty e^{\tau A}\,\mathrm{d}\tau b.$$

Related: Fact 16.20.2.

Fact 16.20.2. Let $A_1 \in \mathbb{F}^{n\times n}$ and $A_2 \in \mathbb{F}^{m\times m}$, assume that $A_1 \oplus A_2$ is nonsingular, let $R \in \mathbb{F}^{n\times m}$, define $P_{\mathrm{s}} \triangleq -\operatorname{vec}^{-1}[(A_2^{\mathsf{T}} \otimes A_1)^{-1}\operatorname{vec} R] \in \mathbb{F}^{n\times m}$, and let $P\colon [0,\infty) \mapsto \mathbb{F}^{n\times m}$ satisfy

$$\dot{P}(t) = A_1P(t) + P(t)A_2 + R.$$

Then, for all $t \ge 0$,

$$P(t) = e^{tA_1}P(0)e^{tA_2} + P_{\mathrm{s}} - e^{tA_1}P_{\mathrm{s}}e^{tA_2} = P_{\mathrm{s}} + e^{tA_1}[P(0) - P_{\mathrm{s}}]e^{tA_2}.$$

If, in addition, $A_2^{\mathsf{T}} \oplus A_1$ is asymptotically stable, then

$$\lim_{t\to\infty} P(t) = P_{\mathrm{s}} = \int_0^\infty e^{\tau A_1}Re^{\tau A_2}\,\mathrm{d}\tau.$$

Related: Fact 16.20.1.

Fact 16.20.3. Let $A \in \mathbb{F}^{n\times n}$ and $X_0 \in \mathbb{F}^{n\times n}$. Then, the matrix differential equation

$$\dot{X}(t) = AX(t), \quad X(0) = X_0,$$

where $t \ge 0$, has the unique solution $X(t) = e^{tA}X_0$.

Fact 16.20.4. Let $A \in \mathbb{C}^{n\times n}$, let $\lambda \in \operatorname{spec}(A)$, and let $v \in \mathbb{C}^n$ be an eigenvector of A associated with λ. Then, for all $t \ge 0$, $x(t) \triangleq \operatorname{Re}(e^{\lambda t}v)$ satisfies $\dot{x}(t) = Ax(t)$. **Remark:** $x(t)$ is an *eigensolution*.

Fact 16.20.5. Let $A \in \mathbb{R}^{n\times n}$, let $\lambda \in \operatorname{spec}(A)$, let $(v_1, \ldots, v_k) \in \bigtimes_{i=1}^k \mathbb{C}^n$ be a Jordan chain of A

associated with λ, let $i \in \{1, \ldots, k\}$, and define

$$x(t) \triangleq \operatorname{Re}\left(e^{\lambda t}\left[\tfrac{1}{(i-1)!}t^{i-1}v_1 + \cdots + tv_{i-1} + v_i\right]\right).$$

Then, for all $t \geq 0$, $x(t)$ satisfies $\dot{x}(t) = Ax(t)$. **Related:** Fact 7.15.12 for the definition of a Jordan chain. **Remark:** $x(t)$ is a *generalized eigensolution*. For $i = 1$, $x(t)$ is an eigensolution. See Fact 16.20.4. **Example:** Let $A = \left[\begin{smallmatrix} 0 & 1 \\ 0 & 0 \end{smallmatrix}\right]$, $\lambda = 0$, $v_1 = \left[\begin{smallmatrix} \beta \\ 0 \end{smallmatrix}\right]$, $v_2 = \left[\begin{smallmatrix} 0 \\ \beta \end{smallmatrix}\right]$. If $i = 2$, then $x(t) = tv_1 + v_2 = \left[\begin{smallmatrix} \beta t \\ \beta \end{smallmatrix}\right]$ is a generalized eigensolution. Alternatively, $i = 1$ yields the eigensolution $x(t) = v_1 = \left[\begin{smallmatrix} \beta \\ 0 \end{smallmatrix}\right]$. Note that β represents velocity for the generalized eigensolution, and position for the eigensolution. See [2187].

Fact 16.20.6. Let $A\colon\ [0, T] \mapsto \mathbb{R}^{n \times n}$, assume that A is continuous, and let $X_0 \in \mathbb{R}^{n \times n}$. Then, the matrix differential equation

$$\dot{X}(t) = A(t)X(t), \quad X(0) = X_0$$

has a unique solution $X\colon\ [0, T] \mapsto \mathbb{R}^{n \times n}$. For all $t \in [0, T]$,

$$\det X(t) = e^{\int_0^t \operatorname{tr} A(\tau)\, \mathrm{d}\tau} \det X_0.$$

If X_0 is nonsingular, then, for all $t \in [0, T]$, $X(t)$ is nonsingular. If, for all $t_1, t_2 \in [0, T]$,

$$A(t_2)\int_{t_1}^{t_2} A(\tau)\, \mathrm{d}\tau = \int_{t_1}^{t_2} A(\tau)\, \mathrm{d}\tau A(t_2),$$

then, for all $t \in [0, T]$,

$$X(t) = e^{\int_0^t A(\tau)\, \mathrm{d}\tau} X_0.$$

Source: Fact 12.16.22 implies that $(\mathrm{d}/\mathrm{d}t) \det X = \operatorname{tr} X^{\mathsf{A}}\dot{X} = \operatorname{tr} X^{\mathsf{A}}AX = \operatorname{tr} XX^{\mathsf{A}}A = (\det X)\operatorname{tr} A$. See See [1190], [1450, pp. 507, 508], and [2349, pp. 64–66]. **Remark:** The first result is *Jacobi's identity*. **Remark:** If the commutativity assumption does not hold, then the solution is given by the *Peano-Baker series*. See [2349, Chapter 3]. Alternative expressions for $X(t)$ are given by the Magnus, Fer, Baker-Campbell-Hausdorff-Dynkin, Wei-Norman, Goldberg, and Zassenhaus expansions. See [491, 963], [1293, pp. 118–120], and [1505, 1506, 1654, 1925, 2178, 2560, 2616, 2851, 2852, 2857]. **Related:** Fact 16.20.6.

Fact 16.20.7. Let $A\colon\ [t_0, t_1] \mapsto \mathbb{R}^{n \times n}$, assume that A is continuous, let $B\colon\ [t_0, t_1] \mapsto \mathbb{R}^{n \times m}$, assume that B is continuous, let $X\colon\ [t_0, t_1] \mapsto \mathbb{R}^{n \times n}$ be the unique solution of the matrix differential equation

$$\dot{X}(t) = A(t)X(t), \quad X(t_0) = I,$$

define $\Phi(t, \tau) \triangleq X(t)X^{-1}(\tau)$, let $u\colon\ [t_0, t_1] \mapsto \mathbb{R}^m$, and assume that u is continuous. Then, the unique solution of the vector differential equation

$$\dot{x}(t) = A(t)x(t) + B(t)u(t), \quad x(t_0) = x_0$$

is

$$x(t) = \Phi(t, t_0)x_0 + \int_{t_0}^t \Phi(t, \tau)B(\tau)u(\tau)\, \mathrm{d}\tau.$$

If, in addition, $x_1 = x(t_1)$, then, for all $t \in [t_0, t_1]$,

$$x(t) = \Phi(t, t_1)x_1 + \int_{t_1}^t \Phi(t, \tau)B(\tau)u(\tau)\, \mathrm{d}\tau.$$

Remark: Fact 16.20.6 implies that, for all $t \in [t_0, t_1]$, $X(t)$ is nonsingular. **Remark:** $\Phi(t, \tau)$ is the *state transition matrix*.

Fact 16.20.8. Let $A\colon [0,\infty) \mapsto \mathbb{C}^{n\times n}$, assume that A is continuous, assume that, for all $t \ge 0$, $A(t+T) = A(t)$, let $X_0 \in \mathbb{C}^{n\times n}$, and let $X\colon [0,\infty) \mapsto \mathbb{C}^{n\times n}$ be the unique solution of the matrix differential equation

$$\dot{X}(t) = A(t)X(t), \quad X(0) = X_0.$$

Then, there exist $P\colon [0,\infty) \mapsto \mathbb{C}^{n\times n}$ and $B \in \mathbb{C}^{n\times n}$ such that $P(0) = X_0$ and, for all $t \ge 0$, $P(t+T) = P(t)$ and

$$X(t) = P(t)e^{Bt}.$$

Now, assume that X_0 is nonsingular. Then, for all $t \ge 0$, $X(t+T)X^{-1}(t) = e^{BT}$. **Source:** [1297, pp. 118, 119]. **Remark:** e^{BT} is a *monodromy matrix*, the eigenvalues of e^{BT} are *characteristic multipliers*, and the eigenvalues of B are *characteristic exponents*. See [1297, pp. 118, 119] and [2936, pp. 70–109].

Fact 16.20.9. Let $A\colon [0,\infty) \mapsto \mathbb{R}^{n\times n}$, assume that A is continuous, assume that, for all $t \ge 0$, $A(t+T) = A(t)$, let $X_0 \in \mathbb{R}^{n\times n}$, and let $X\colon [0,\infty) \mapsto \mathbb{R}^{n\times n}$ be the unique solution of the matrix differential equation

$$\dot{X}(t) = A(t)X(t), \quad X(0) = X_0.$$

Then, there exist $P\colon [0,\infty) \mapsto \mathbb{R}^{n\times n}$ and $B \in \mathbb{R}^{n\times n}$ such that $P(0) = X_0$ and such that, for all $t \ge 0$, $P(t+2T) = P(t)$ and

$$X(t) = P(t)e^{Bt}.$$

Now, assume that X_0 is nonsingular. Then, for all $t \ge 0$, $X(t+T)X^{-1}(t) = e^{BT}$. **Source:** [1297, pp. 118, 119].

16.21 Facts on Stability, Observability, and Controllability

Fact 16.21.1. Let $A \in \mathbb{R}^{n\times n}$, let $x\colon [0,\infty) \mapsto \mathbb{R}^n$ satisfy $\dot{x} = Ax$, let $C \in \mathbb{R}^{p\times n}$, and define $y\colon [0,\infty) \mapsto \mathbb{R}^p$ by $y = Cx$. Then, for all $t \ge 0$, $\chi_A(\frac{\mathrm{d}}{\mathrm{d}t})y(t) = 0$. **Remark:** $(\mathrm{d}/\mathrm{d}t)^k \triangleq \mathrm{d}^k/\mathrm{d}t^k$.

Fact 16.21.2. Let $A \in \mathbb{R}^{n\times n}$, $B \in \mathbb{R}^{n\times m}$, and $C \in \mathbb{R}^{p\times n}$, and assume that (A,B) is controllable and (A,C) is observable. Then, for all $v \in \mathbb{R}^m$, the step response

$$y(t) = \int_0^t Ce^{tA}\,\mathrm{d}\tau Bv + Dv$$

is bounded on $[0,\infty)$ if and only if A is Lyapunov stable and nonsingular.

Fact 16.21.3. Let $A \in \mathbb{R}^{n\times n}$ and $C \in \mathbb{R}^{p\times n}$, assume that (A,C) is detectable, and let $x(t)$ and $y(t)$ satisfy $\dot{x}(t) = Ax(t)$ and $y(t) = Cx(t)$ for all $t \in [0,\infty)$. Then, the following statements hold:

i) y is bounded if and only if x is bounded.

ii) $\lim_{t\to\infty} y(t)$ exists if and only if $\lim_{t\to\infty} x(t)$ exists.

iii) $y(t) \to 0$ as $t \to \infty$ if and only if $x(t) \to 0$ as $t \to \infty$.

Fact 16.21.4. Let $x(0) = x_0$, and let $x_\mathrm{f} - e^{t_\mathrm{f}A}x_0 \in \mathcal{C}(A,B)$. Then, for all $t \in [0,t_\mathrm{f}]$, the control $u\colon [0,t_\mathrm{f}] \mapsto \mathbb{R}^m$ defined by

$$u(t) \triangleq B^\mathrm{T}e^{(t_\mathrm{f}-t)A^\mathrm{T}}\left(\int_0^{t_\mathrm{f}} e^{\tau A}BB^\mathrm{T}e^{\tau A^\mathrm{T}}\,\mathrm{d}\tau\right)^{+}(x_\mathrm{f} - e^{t_\mathrm{f}A}x_0)$$

yields $x(t_\mathrm{f}) = x_\mathrm{f}$.

Fact 16.21.5. Let $x(0) = x_0$, let $x_\mathrm{f} \in \mathbb{R}^n$, and assume that (A,B) is controllable. Then, for all

$t \in [0, t_f]$, the control $u\colon\ [0, t_f] \mapsto \mathbb{R}^m$ defined by

$$u(t) \triangleq B^{\mathrm{T}} e^{(t_f - t)A^{\mathrm{T}}} \left(\int_0^{t_f} e^{\tau A} B B^{\mathrm{T}} e^{\tau A^{\mathrm{T}}} \mathrm{d}\tau \right)^{-1} (x_f - e^{t_f A} x_0)$$

yields $x(t_f) = x_f$.

Fact 16.21.6. Let $A \in \mathbb{R}^{n\times n}$, let $B \in \mathbb{R}^{n\times m}$, assume that A is skew symmetric, assume that (A, B) is controllable, and let $\alpha > 0$. Then, $A - \alpha BB^{\mathrm{T}}$ is asymptotically stable.

Fact 16.21.7. Let $A \in \mathbb{R}^{n\times n}$ and $B \in \mathbb{R}^{n\times m}$. Then, (A, B) is (controllable, stabilizable) if and only if (A, BB^{T}) is (controllable, stabilizable). Now, assume that B is positive semidefinite. Then, (A, B) is (controllable, stabilizable) if and only if $(A, B^{1/2})$ is (controllable, stabilizable).

Fact 16.21.8. Let $A \in \mathbb{R}^{n\times n}$, $B \in \mathbb{R}^{n\times m}$, and $\hat{B} \in \mathbb{R}^{n\times \hat{m}}$, and assume that (A, B) is (controllable, stabilizable) and $\mathcal{R}(B) \subseteq \mathcal{R}(\hat{B})$. Then, $(A, \hat{B})$ is also (controllable, stabilizable).

Fact 16.21.9. Let $A \in \mathbb{R}^{n\times n}$, $B \in \mathbb{R}^{n\times m}$, and $\hat{B} \in \mathbb{R}^{n\times \hat{m}}$, and assume that (A, B) is (controllable, stabilizable) and $BB^{\mathrm{T}} \le \hat{B}\hat{B}^{\mathrm{T}}$. Then, $(A, \hat{B})$ is also (controllable, stabilizable). **Source:** Lemma 10.6.1 and Fact 16.21.8.

Fact 16.21.10. Let $A \in \mathbb{R}^{n\times n}$, $B \in \mathbb{R}^{n\times m}$, $\hat{B} \in \mathbb{R}^{n\times \hat{m}}$, and $\hat{C} \in \mathbb{R}^{\hat{m}\times n}$, and assume that (A, B) is (controllable, stabilizable). Then,

$$(A + \hat{B}\hat{C}, [BB^{\mathrm{T}} + \hat{B}\hat{B}^{\mathrm{T}}]^{1/2})$$

is also (controllable, stabilizable). **Source:** [2912, p. 79].

Fact 16.21.11. Let $A \in \mathbb{R}^{n\times n}$ and $B \in \mathbb{R}^{n\times m}$. Then, the following statements are equivalent:

i) (A, B) is controllable.

ii) There exists $\alpha \in \mathbb{R}$ such that $(A + \alpha I, B)$ is controllable.

iii) $(A + \alpha I, B)$ is controllable for all $\alpha \in \mathbb{R}$.

Fact 16.21.12. Let $A \in \mathbb{R}^{n\times n}$ and $B \in \mathbb{R}^{n\times m}$. Then, the following statements are equivalent:

i) (A, B) is stabilizable.

ii) There exists $\alpha \le \max\{0, -\alpha_{\max}(A)\}$ such that $(A + \alpha I, B)$ is stabilizable.

iii) $(A + \alpha I, B)$ is stabilizable for all $\alpha \le \max\{0, -\alpha_{\max}(A)\}$.

Fact 16.21.13. Let $A \in \mathbb{R}^{n\times n}$, assume that A is diagonal, and let $B \in \mathbb{R}^{n\times 1}$. Then, (A, B) is controllable if and only if the diagonal entries of A are distinct and every entry of B is nonzero. **Source:** Note that

$$\det \mathcal{K}(A, B) = \det \begin{bmatrix} b_1 & & 0 \\ & \ddots & \\ 0 & & b_n \end{bmatrix} \begin{bmatrix} 1 & a_1 & \cdots & a_1^{n-1} \\ \vdots & \vdots & .\because. & \vdots \\ 1 & a_n & \cdots & a_n^{n-1} \end{bmatrix} = \left(\prod_{i=1}^n b_i \right) \prod_{i<j} (a_i - a_j).$$

Fact 16.21.14. Let $A \in \mathbb{R}^{n\times n}$ and $B \in \mathbb{R}^{n\times 1}$, and assume that (A, B) is controllable. Then, A is cyclic. **Source:** Fact 7.15.10. **Related:** Fact 16.21.15.

Fact 16.21.15. Let $G \in \mathbb{R}(s)_{\mathrm{prop}}$, where $G \sim \left[\begin{array}{c|c} A & B \\ \hline C & D \end{array}\right]$ and $A \in \mathbb{R}^{n\times n}$. Then, the following statements are equivalent:

i) $G \overset{\min}{\sim} \left[\begin{array}{c|c} A & B \\ \hline C & D \end{array}\right]$.

ii) There exist $p, q \in \mathbb{R}[s]$ such that p and q are coprime and $\deg q = n$.

iii) $\mathrm{mpoles}(G) = \mathrm{mspec}(A)$.

Now, assume that *i)*–*iii)* hold. Then, the following statements hold:

iv) A has exactly one Jordan block associated with each pole of G.

v) Let $\lambda \in \operatorname{spec}(A)$. Then, λ is semisimple if and only if λ is a nonrepeated pole of G.

vi) Let $\lambda \in \operatorname{spec}(A)$. Then, λ is defective if and only if λ is a repeated pole of G.

Related: Theorem 16.9.24 and Fact 16.21.14.

Fact 16.21.16. Let $A \in \mathbb{R}^{n\times n}$ and $B \in \mathbb{R}^{n\times m}$, and assume that (A, B) is controllable. Then,

$$\max_{\lambda\in\operatorname{spec}(A)} \operatorname{gmult}_A(\lambda) \le m.$$

Fact 16.21.17. Let $A \in \mathbb{R}^{n\times n}$ and $B \in \mathbb{R}^{n\times m}$. Then, the following statements are equivalent:

i) (A, B) is (controllable, stabilizable) and A is nonsingular.

ii) (A, AB) is (controllable, stabilizable).

Fact 16.21.18. Let $A \in \mathbb{R}^{n\times n}$ and $B \in \mathbb{R}^{n\times m}$, and assume that (A, B) is controllable. Then, $(A, B^\mathsf{T}S^{-\mathsf{T}})$ is observable, where $S \in \mathbb{R}^{n\times n}$ is a nonsingular matrix that satisfies $A^\mathsf{T} = S^{-1}AS$.

Fact 16.21.19. Let (A, B) be controllable, let $t_1 > 0$, and define

$$P = \left(\int_0^{t_1} e^{-tA}BB^\mathsf{T}e^{-tA^\mathsf{T}}\,\mathrm{d}t\right)^{-1}.$$

Then, $A - BB^\mathsf{T}P$ is asymptotically stable. **Source:** P satisfies

$$(A - BB^\mathsf{T}P)^\mathsf{T}P + P(A - BB^\mathsf{T}P) + P(BB^\mathsf{T} + e^{t_1A}BB^\mathsf{T}e^{t_1A^\mathsf{T}})P = 0.$$

Since $(A - BB^\mathsf{T}P, BB^\mathsf{T} + e^{t_1A}BB^\mathsf{T}e^{t_1A^\mathsf{T}})$ is observable and P is positive definite, it follows from Proposition 15.10.5 that $A - BB^\mathsf{T}P$ is asymptotically stable. **Credit:** D. L. Lukes and D. Kleinman. See [2351, pp. 113, 114].

Fact 16.21.20. Let $A \in \mathbb{R}^{n\times n}$ and $B \in \mathbb{R}^{n\times m}$, and assume that A is asymptotically stable. Furthermore, let $u\colon [0,\infty) \mapsto \mathbb{R}^m$, assume that u is continuous and bounded, let $x\colon [0,\infty) \mapsto \mathbb{R}^n$, and assume that x satisfies $\dot{x} = Ax + Bu$. Then, x is bounded. If, in addition, $u(t) \to 0$ as $t \to \infty$, then $x(t) \to 0$ as $t \to \infty$. **Source:** [2496, p. 330]. **Remark:** These are consequences of *input-to-state stability*.

Fact 16.21.21. Let $A \in \mathbb{R}^{n\times n}$ and $C \in \mathbb{R}^{p\times n}$, assume that (A, C) is observable, and let $k \ge n$. Then,

$$A = \begin{bmatrix} 0_{p\times n} \\ \mathcal{O}_k(A, C) \end{bmatrix}^+ \mathcal{O}_{k+1}(A, C).$$

Credit: H. Palanthandalam-Madapusi.

Fact 16.21.22. Let $G_1, G_2 \in \mathbb{R}(s)_{\rm prop}$, where $G_2 \overset{\min}{\sim} \left[\begin{array}{c|c} A_1 & B_1 \\ \hline C_1 & 0 \end{array}\right]$, $G_2 \overset{\min}{\sim} \left[\begin{array}{c|c} A_2 & B_2 \\ \hline C_2 & 0 \end{array}\right]$, let $\lambda \in \operatorname{spec}(A_1)$, assumc that $G_2(\lambda) = 0$, and define $\tilde{A} \triangleq \left[\begin{smallmatrix} A_1 & B_1C_2 \\ B_2C_1 & A_2 \end{smallmatrix}\right]$. Then, $\lambda \in \operatorname{spec}(\tilde{A})$. **Remark:** This result implies that CRHP pole-zero cancellation cannot occur in an asymptotically stable feedback loop.

Fact 16.21.23. Let $\beta_0, \ldots, \beta_{n-1} \in \mathbb{F}$, define $A \triangleq C_{\rm b}(p)$ as in Fact 7.18.3, and let $b, c \in \mathbb{F}^n$. Then, $\mathcal{K}(A, b)c = \mathcal{K}(A, c)b$. **Source:** [600].

16.22 Facts on the Lyapunov Equation and Inertia

Fact 16.22.1. Let $A, P \in \mathbb{F}^{n\times n}$, assume that P is Hermitian, let $C \in \mathbb{F}^{l\times n}$, and assume that $A^*P + PA + C^*C = 0$. Then, the following statements hold:

i) $|\nu_-(A) - \nu_+(P)| \le n - \operatorname{rank} \mathcal{O}(A, C)$.

ii) $|\nu_+(A) - \nu_-(P)| \le n - \operatorname{rank} \mathcal{O}(A, C)$.

iii) If $\nu_0(A) = 0$, then $|\nu_-(A) - \nu_+(P)| + |\nu_+(A) - \nu_-(P)| \le n - \operatorname{rank} \mathcal{O}(A, C)$.

If, in addition, (A, C) is observable, then the following statements hold:

iv) $\nu_-(A) = \nu_+(P)$.

v) $\nu_0(A) = \nu_0(P) = 0$.

vi) $\nu_+(A) = \nu_-(P)$.

vii) If P is positive definite, then A is asymptotically stable.

Source: [147, 695], [1738, p. 448], and [1883, 2891]. **Remark:** *v*) does not follow from *i*)–*iii*). **Remark:** For related results, see [638, 833, 1883, 2176].

Fact 16.22.2. Let $A, P \in \mathbb{F}^{n\times n}$, assume that P is nonsingular and Hermitian, and assume that $A^*P + PA$ is negative semidefinite. Then, the following statements hold:

i) $\nu_-(A) \le \nu_+(P)$.

ii) $\nu_+(A) \le \nu_-(P)$.

iii) If P is positive definite, then $\operatorname{spec}(A) \subset \mathrm{CLHP}$.

Source: [1738, p. 447]. **Remark:** If P is positive definite, then A is Lyapunov stable, although this result does not follow from *i*) and *ii*).

Fact 16.22.3. Let $A, P \in \mathbb{F}^{n\times n}$, and assume that $\nu_0(A) = 0$, P is Hermitian, and $A^*P + PA$ is negative semidefinite. Then, the following statements hold:

i) $\nu_-(P) \le \nu_+(A)$.

ii) $\nu_+(P) \le \nu_-(A)$.

iii) If P is nonsingular, then $\nu_-(P) = \nu_+(A)$ and $\nu_+(P) = \nu_-(A)$.

iv) If P is positive definite, then A is asymptotically stable.

Source: [1738, p. 447].

Fact 16.22.4. Let $A, P \in \mathbb{F}^{n\times n}$, and assume that $\nu_0(A) = 0$, P is nonsingular and Hermitian, and $A^*P + PA$ is negative semidefinite. Then, the following statements hold:

i) $\nu_-(A) = \nu_+(P)$.

ii) $\nu_+(A) = \nu_-(P)$.

Source: Fact 16.22.2 and Fact 16.22.3. See [1738, p. 448]. **Credit:** D. Carlson and H. Schneider.

Fact 16.22.5. Let $A, P \in \mathbb{F}^{n\times n}$, assume that P is Hermitian, and assume that $A^*P + PA$ is negative definite. Then, the following statements hold:

i) $\nu_-(A) = \nu_+(P)$.

ii) $\nu_0(A) = 0$.

iii) $\nu_+(A) = \nu_-(P)$.

iv) P is nonsingular.

v) If P is positive definite, then A is asymptotically stable.

Source: [970, pp. 441, 442], [1738, p. 445], and [2176]. This result follows from Fact 16.22.1 with positive-definite $C = -(A^*P + PA)^{1/2}$. **Remark:** These statements are the *classical constraints*. An analogous result holds for the discrete-time Lyapunov equation, where the analogous definition of inertia counts the numbers of eigenvalues inside the open unit disk, outside the open unit disk, and on the unit circle. See [613, 862]. **Credit:** M. G. Krein, A. Ostrowski, and H. Schneider.

Fact 16.22.6. Let $A \in \mathbb{F}^{n\times n}$. Then, the following statements are equivalent:

i) $\nu_0(A) = 0$.

ii) There exists a nonsingular Hermitian matrix $P \in \mathbb{F}^{n\times n}$ such that A^*P+PA is negative definite.

iii) There exists a Hermitian matrix $P \in \mathbb{F}^{n\times n}$ such that $A^*P + PA$ is negative definite.

If these conditions hold, then the following statements hold for P given by *ii*) and *iii*):

iv) $\nu_-(A) = \nu_+(P)$.

v) $\nu_0(A) = \nu_0(P) = 0$.

vi) $\nu_+(A) = \nu_-(P)$.

vii) P is nonsingular.

viii) If P is positive definite, then A is asymptotically stable.

Source: For *i)* $\Longrightarrow$ *ii)*, see [1738, p. 445]. *iii)* $\Longrightarrow$ *i)* follows from Fact 16.22.5. See [97, 613, 640].

Fact 16.22.7. Let $A \in \mathbb{F}^{n\times n}$. Then, the following statements are equivalent:

i) A is Lyapunov stable.

ii) There exists a positive-definite matrix $P \in \mathbb{F}^{n\times n}$ such that $A^*P + PA$ is negative semidefinite.

Furthermore, the following statements are equivalent:

iii) A is asymptotically stable.

iv) There exists a positive-definite matrix $P \in \mathbb{F}^{n\times n}$ such that $A^*P + PA$ is negative definite.

v) For every positive-definite matrix $R \in \mathbb{F}^{n\times n}$, there exists a positive-definite matrix $P \in \mathbb{F}^{n\times n}$ such that $A^*P + PA$ is negative definite.

Remark: See Proposition 15.10.5 and Proposition 15.10.6.

Fact 16.22.8. Let $A, P \in \mathbb{F}^{n\times n}$, and assume that P is Hermitian. Then, the following statements hold:

i) $\nu_+(A^*P + PA) \le \operatorname{rank} P$.

ii) $\nu_-(A^*P + PA) \le \operatorname{rank} P$.

If, in addition, A is asymptotically stable, then the following statement holds:

iii) $1 \le \nu_-(A^*P + PA) \le \operatorname{rank} P$.

Source: [266, 862].

Fact 16.22.9. Let $A, P \in \mathbb{R}^{n\times n}$, assume that $\nu_0(A) = n$, and assume that P is positive semidefinite. Then, exactly one of the following statements holds:

i) $A^{\mathsf{T}}P + PA = 0$.

ii) $\nu_-(A^{\mathsf{T}}P + PA) \ge 1$ and $\nu_+(A^{\mathsf{T}}P + PA) \ge 1$.

Source: [2755].

Fact 16.22.10. Let $R \in \mathbb{F}^{n\times n}$, and assume that R is Hermitian and $\nu_+(R) \ge 1$. Then, there exist an asymptotically stable matrix $A \in \mathbb{F}^{n\times n}$ and a positive-definite matrix $P \in \mathbb{F}^{n\times n}$ such that $A^*P + PA + R = 0$. **Source:** [266].

Fact 16.22.11. Let $A \in \mathbb{F}^{n\times n}$, assume that A is cyclic, and let a, b, c, d, e be nonnegative integers such that $a + b = c + d + e = n$, $c \ge 1$, and $e \ge 1$. Then, there exists a nonsingular, Hermitian matrix $P \in \mathbb{F}^{n\times n}$ such that

$$\operatorname{In} P = \begin{bmatrix} a \\ 0 \\ b \end{bmatrix}, \quad \operatorname{In}(A^*P + PA) = \begin{bmatrix} c \\ d \\ e \end{bmatrix}.$$

Source: [2452, 2453].

Fact 16.22.12. Let $P, R \in \mathbb{F}^{n\times n}$, and assume that P is positive and R is Hermitian. Then, the following statements are equivalent:

i) $\operatorname{tr} RP^{-1} > 0$.

ii) There exists an asymptotically stable matrix $A \in \mathbb{F}^{n\times n}$ such that $A^*P + PA + R = 0$.

Source: [266].

Fact 16.22.13. Let $A_1 \in \mathbb{R}^{n_1\times n_1}$, $A_2 \in \mathbb{R}^{n_2\times n_2}$, $B \in \mathbb{R}^{n_1\times m}$, and $C \in \mathbb{R}^{m\times n_2}$, assume that $A_1 \oplus A_2$ is nonsingular, and assume that $\operatorname{rank} B = \operatorname{rank} C = m$. Furthermore, let $X \in \mathbb{R}^{n_1\times n_2}$ be the unique solution of

$$A_1X + XA_2 + BC = 0.$$

Then,

$$\operatorname{rank} X \le \min\{\operatorname{rank} \mathcal{K}(A_1, B), \operatorname{rank} \mathcal{O}(A_2, C)\}.$$

Furthermore, if $m = 1$, then equality holds. **Source:** [858]. **Remark:** Related results are given in [2891, 2897].

Fact 16.22.14. Let $A_1, A_2 \in \mathbb{R}^{n\times n}$, $B \in \mathbb{R}^n$, $C \in \mathbb{R}^{1\times n}$, assume that $A_1 \oplus A_2$ is nonsingular, let $X \in \mathbb{R}^{n\times n}$ satisfy

$$A_1X + XA_2 + BC = 0,$$

and assume that (A_1, B) is controllable and (A_2, C) is observable. Then, X is nonsingular. **Source:** Fact 16.22.13 and [2897].

Fact 16.22.15. Let $A, P, R \in \mathbb{R}^{n\times n}$, and assume that P and R are positive semidefinite, $A^{\mathsf{T}}P + PA + R = 0$, and $\mathcal{N}[\mathcal{O}(A, R)] = \mathcal{N}(A)$. Then, A is semistable. **Source:** [444].

Fact 16.22.16. Let $A, V \in \mathbb{R}^{n\times n}$, assume that A is asymptotically stable, assume that V is positive semidefinite, and let $Q \in \mathbb{R}^{n\times n}$ be the unique, positive-definite solution to $AQ + QA^{\mathsf{T}} + V = 0$. Furthermore, let $C \in \mathbb{R}^{l\times n}$, and assume that CVC^{T} is positive definite. Then, CQC^{T} is positive definite.

Fact 16.22.17. Let $A, R \in \mathbb{R}^{n\times n}$, assume that A is asymptotically stable, assume that $R \in \mathbb{R}^{n\times n}$ is positive semidefinite, and let $P \in \mathbb{R}^{n\times n}$ satisfy $A^{\mathsf{T}}P + PA + R = 0$. Then, for all $i, j \in \{1, \dots, n\}$, there exist $\alpha_{ij} \in \mathbb{R}$ such that

$$P = \sum_{i,j=1}^{n} \alpha_{ij} A^{(i-1)\mathsf{T}} R A^{j-1}.$$

In particular, for all $i, j \in \{1, \dots, n\}$, $\alpha_{ij} = \hat{P}_{(i,j)}$, where $\hat{P} \in \mathbb{R}^{n\times n}$ satisfies $\hat{A}^{\mathsf{T}}\hat{P} + \hat{P}\hat{A} + \hat{R} = 0$, where $\hat{A} = C(\chi_A)$ and $\hat{R} = E_{1,1}$. **Source:** [2472]. **Remark:** This is *Smith's method.* See [859, 911, 1330, 1911] for finite-sum solutions of linear matrix equations.

Fact 16.22.18. Let $\lambda_1, \dots, \lambda_n \in \mathbb{C}$, assume that, for all $i \in \{1, \dots, n\}$, $\operatorname{Re} \lambda_i < 0$, define $\Lambda \triangleq \operatorname{diag}(\lambda_1, \dots, \lambda_n)$, let k be a nonnegative integer, and, for all $i, j \in \{1, \dots, n\}$, define $P \in \mathbb{C}^{n\times n}$ by

$$P \triangleq \frac{1}{k!} \int_0^\infty t^k e^{\overline{\Lambda} t} e^{\Lambda t} \, \mathrm{d}t.$$

Then, P is positive definite, P satisfies the Lyapunov equation $\overline{\Lambda}P + P\Lambda + I = 0$, and, for all $i, j \in \{1, \dots, n\}$,

$$P_{(i,j)} = \left(\frac{-1}{\overline{\lambda_i} + \lambda_j} \right)^{k+1}.$$

Source: For all nonzero $x \in \mathbb{C}^n$,

$$x^*Px = \int_0^\infty t^k \|e^{\Lambda t}x\|_2^2 \, \mathrm{d}t$$

is positive. Hence, P is positive definite. Furthermore, note that

$$P_{(i,j)} = \int_0^\infty t^k e^{\overline{\lambda_i} t} e^{\lambda_j t} \, \mathrm{d}t = \frac{(-1)^{k+1} k!}{(\overline{\lambda_i} + \lambda_j)^{k+1}}.$$

Remark: See [575] and [1450, p. 348]. **Related:** Fact 10.9.20 and Fact 16.22.19.

Fact 16.22.19. Let $\lambda_1, \dots, \lambda_n \in \mathbb{C}$, assume that, for all $i \in \{1, \dots, n\}$, $\operatorname{Re} \lambda_i < 0$, define $\Lambda \triangleq \operatorname{diag}(\lambda_1, \dots, \lambda_n)$, let k be a nonnegative integer, let $R \in \mathbb{C}^{n\times n}$, assume that R is positive semidefinite, and, for all $i, j \in \{1, \dots, n\}$, define $P \in \mathbb{C}^{n\times n}$ by

$$P \triangleq \frac{1}{k!}\int_0^\infty t^k e^{\overline{\Lambda}t} R e^{\Lambda t}\,\mathrm{d}t.$$

Then, P is positive semidefinite, P satisfies the Lyapunov equation $\overline{\Lambda}P + P\Lambda + R = 0$, and, for all $i, j \in \{1, \dots, n\}$,

$$P_{(i,j)} = R_{(i,j)}\left(\frac{-1}{\overline{\lambda_i} + \lambda_j}\right)^{k+1}.$$

If, in addition, $I \odot R$ is positive definite, then P is positive definite. **Source:** Fact 10.25.16 and Fact 16.22.18. **Related:** Fact 10.9.20 and Fact 16.22.18. Note that $P = \hat{P} \odot R$, where $\hat{P}$ is the solution to the Lyapunov equation with $R = I$.

Fact 16.22.20. Let $A, R \in \mathbb{R}^{n\times n}$, assume that $R \in \mathbb{R}^{n\times n}$ is positive semidefinite, let $q, r \in \mathbb{R}$, where $r > 0$, and assume that there exists a positive-definite matrix $P \in \mathbb{R}^{n\times n}$ that satisfies

$$[A - (q+r)I]^{\mathsf{T}}P + P[A - (q+r)I] + \tfrac{1}{r}A^{\mathsf{T}}PA + R = 0.$$

Then, the spectrum of A is contained in a disk centered at $q + \jmath 0$ with radius r. **Remark:** The disk is an *eigenvalue inclusion region*. See [297, 1285, 2833] for related results concerning elliptical, parabolic, hyperbolic, sector, and vertical strip regions.

16.23 Facts on the Discrete-Time Lyapunov Equation

Fact 16.23.1. Let $A, R \in \mathbb{F}^{n\times n}$. Then, the following statements are equivalent:

i) There exists a unique matrix $Q \in \mathbb{F}^{n\times n}$ that satisfies $Q = AQA^* + R$.

ii) For all $\lambda, \mu \in \operatorname{spec}(A)$, $\lambda\overline{\mu} \neq 1$.

iii) $I - A \otimes \overline{A}$ is nonsingular.

Now, assume that these conditions hold.Then, the following statements hold:

iv) $Q = \operatorname{vec}^{-1}[(I - A\otimes\overline{A})^{-1}\operatorname{vec} R]$.

v) If R is Hermitian, then Q is Hermitian.

vi) If R is positive semidefinite and A is discrete-time asymptotically stable, then A is positive semidefinite. In addition, (A, R) is controllable if and only if Q is positive definite.

vii) If A is in companion form and $R = E_{n,n}$, then Q is positive definite and Toeplitz.

16.24 Facts on Realizations and the H_2 System Norm

Fact 16.24.1. Let $x\colon [0,\infty) \mapsto \mathbb{R}^n$ and $y\colon [0,\infty) \mapsto \mathbb{R}^n$, assume that the integrals $\int_0^\infty x^{\mathsf{T}}(t)x(t)\,\mathrm{d}t$ and $\int_0^\infty y^{\mathsf{T}}(t)y(t)\,\mathrm{d}t$ exist, and let $\hat{x}\colon \mathrm{I}\mathbb{A} \mapsto \mathbb{C}^n$ and $\hat{y}\colon \mathrm{I}\mathbb{A} \mapsto \mathbb{C}^n$ denote the Fourier transforms of x and y, respectively. Then,

$$\int_0^\infty x^{\mathsf{T}}(t)x(t)\,\mathrm{d}t = \int_{-\infty}^\infty \hat{x}^*(\omega\jmath)\hat{x}(\omega\jmath)\,\mathrm{d}\omega, \qquad \int_0^\infty x^{\mathsf{T}}(t)y(t)\,\mathrm{d}t = \operatorname{Re}\int_{-\infty}^\infty \hat{x}^*(\omega\jmath)\hat{y}(\omega\jmath)\,\mathrm{d}\omega.$$

Remark: These equalities are equivalent versions of Parseval's theorem. The second equality follows from the first equality by replacing x with $x + y$.

Fact 16.24.2. Let $G \in \mathbb{R}(s)^{l\times m}_{\mathrm{prop}}$, where $G \overset{\min}{\sim} \left[\begin{array}{c|c} A & B \\ \hline C & D \end{array}\right]$, and assume that, for all $i \in \{1, \dots, l\}$ and $j \in \{1, \dots, m\}$, $G_{(i,j)} = p_{i,j}/q_{i,j}$, where $p_{i,j}, q_{i,j} \in \mathbb{R}[s]$ are coprime. Then,

$$\operatorname{spec}(A) = \bigcup_{i,j=1}^{l,m} \operatorname{roots}(p_{i,j}).$$

Fact 16.24.3. Let $G \sim \left[\begin{array}{c|c} A & B \\ \hline C & D \end{array}\right]$, let $a, b \in \mathbb{R}$, where $a \neq 0$, and define $H(s) \triangleq G(as+b)$. Then,

$$H \sim \left[\begin{array}{c|c} a^{-1}(A - bI) & B \\ \hline a^{-1}C & D \end{array}\right].$$

Fact 16.24.4. Let $G \sim \left[\begin{array}{c|c} A & B \\ \hline C & D \end{array}\right]$, where A is nonsingular, and define $H(s) \triangleq G(1/s)$. Then,

$$H \sim \left[\begin{array}{c|c} A^{-1} & -A^{-1}B \\ \hline CA^{-1} & D - CA^{-1}B \end{array}\right].$$

Fact 16.24.5. Let $G(s) = C(sI - A)^{-1}B$. Then,

$$G(\omega J) = -CA(\omega^2 I + A^2)^{-1}B - \omega J C(\omega^2 I + A^2)^{-1}B.$$

Fact 16.24.6. Let $G(s) = C(sI - A)^{-1}B$, and define $H(s) = sG(s)$. Then,

$$H \sim \left[\begin{array}{c|c} A & B \\ \hline CA & CB \end{array}\right].$$

Consequently,

$$sC(sI - A)^{-1}B = CA(sI - A)^{-1}B + CB.$$

Fact 16.24.7. Let $G(s) = C(sI - A)^{-1}B$, and assume that A is nonsingular. Then,

$$G(s) = sCA^{-1}(sI - A)^{-1}B + G(0).$$

Fact 16.24.8. Let $G = \begin{bmatrix} G_{11} & G_{12} \\ G_{21} & G_{22} \end{bmatrix}$, where $G_{ij} \sim \left[\begin{array}{c|c} A_{ij} & B_{ij} \\ \hline C_{ij} & D_{ij} \end{array}\right]$ for all $i, j = 1, 2$. Then,

$$\begin{bmatrix} G_{11} & G_{12} \\ G_{21} & G_{22} \end{bmatrix} \sim \left[\begin{array}{cccc|cc} A_{11} & 0 & 0 & 0 & B_{11} & 0 \\ 0 & A_{12} & 0 & 0 & 0 & B_{12} \\ 0 & 0 & A_{21} & 0 & B_{21} & 0 \\ 0 & 0 & 0 & A_{22} & 0 & B_{22} \\ \hline C_{11} & C_{12} & 0 & 0 & D_{11} & D_{12} \\ 0 & 0 & C_{21} & C_{22} & D_{21} & D_{22} \end{array}\right].$$

Fact 16.24.9. Let $G \sim \left[\begin{array}{c|c} A & B \\ \hline C & 0 \end{array}\right]$, where $G \in \mathbb{R}^{l \times m}(s)$, and let $M \in \mathbb{R}^{m \times l}$. Then,

$$(I + GM)^{-1} \sim \left[\begin{array}{c|c} A - BMC & BM \\ \hline -C & I \end{array}\right], \quad (I + GM)^{-1}G \sim \left[\begin{array}{c|c} A - BMC & B \\ \hline C & 0 \end{array}\right].$$

Fact 16.24.10. Let $G \sim \left[\begin{array}{c|c} A & B \\ \hline C & D \end{array}\right]$, where $G \in \mathbb{R}^{l \times m}(s)$. If D has a left inverse $D^{\mathsf{L}} \in \mathbb{R}^{m \times l}$, then

$$G^{\mathsf{L}} \sim \left[\begin{array}{c|c} A - BD^{\mathsf{L}}C & BD^{\mathsf{L}} \\ \hline -D^{\mathsf{L}}C & D^{\mathsf{L}} \end{array}\right]$$

satisfies $G^{\mathsf{L}}G = I$. If D has a right inverse $D^{\mathsf{R}} \in \mathbb{R}^{m \times l}$, then

$$G^{\mathsf{R}} \sim \left[\begin{array}{c|c} A - BD^{\mathsf{R}}C & BD^{\mathsf{R}} \\ \hline -D^{\mathsf{R}}C & D^{\mathsf{R}} \end{array}\right]$$

satisfies $GG^{\mathsf{R}} = I$.

Fact 16.24.11. Let $G \sim \left[\begin{array}{c|c} A & B \\ \hline C & 0 \end{array}\right]$ be a SISO rational transfer function, and let $\lambda \in \mathbb{C}$. Then, there exists a rational function H such that

$$G(s) = \frac{1}{(s+\lambda)^r}H(s)$$

and such that λ is neither a pole nor a zero of H if and only if the Jordan form of A has exactly one block associated with λ, which is of size r.

Fact 16.24.12. Let $G \sim \left[\begin{array}{c|c} A & B \\ \hline C & D \end{array}\right]$. Then, $G(s)$ is given by

$$G(s) = (A - sI)\left|\left[\begin{matrix} A - sI & B \\ C & D \end{matrix}\right]\right..$$

Remark: See [307]. **Remark:** The vertical bar denotes the Schur complement.

Fact 16.24.13. Let $G \in \mathbb{F}(s)^{n\times m}$, where $G \overset{\min}{\sim} \left[\begin{array}{c|c} A & B \\ \hline C & D \end{array}\right]$, and, for all $i \in \{1,\dots,n\}$ and $j \in \{1,\dots,m\}$, let $G_{(i,j)} = p_{ij}/q_{ij}$, where $p_{ij}, q_{ij} \in \mathbb{F}[s]$ are coprime. Then,

$$\bigcup_{i,j=1}^{n,m} \operatorname{roots}(q_{ij}) = \operatorname{spec}(A).$$

Fact 16.24.14. Let $A \in \mathbb{R}^{n\times n}$, $B \in \mathbb{R}^{n\times m}$, and $C \in \mathbb{R}^{m\times n}$. If $s \in \mathbb{C}\backslash\operatorname{spec}(A)$, then

$$\det[sI - (A + BC)] = \det[I - C(sI - A)^{-1}B]\det(sI - A).$$

If, in addition, $n = m = 1$ and $s \in \mathbb{C}$, then

$$\det[sI - (A + BC)] = \det(sI - A) - C(sI - A)^{\mathsf{A}}B.$$

Source: Note that

$$\det[I - C(sI - A)^{-1}B]\det(sI - A) = \det\begin{bmatrix} sI - A & B \\ C & I \end{bmatrix} = \det\begin{bmatrix} sI - A & B \\ C & I \end{bmatrix}\begin{bmatrix} I & 0 \\ -C & I \end{bmatrix}$$
$$= \det\begin{bmatrix} sI - A - BC & B \\ 0 & I \end{bmatrix} = \det(sI - A - BC).$$

Remark: The last expression is used in [2054] to compute the frequency response of a transfer function.

Fact 16.24.15. Let $A \in \mathbb{R}^{n\times n}$, $B \in \mathbb{R}^{n\times m}$, $C \in \mathbb{R}^{p\times n}$, and $K \in \mathbb{R}^{m\times p}$, assume that either $m = 1$ or $p = 1$, and let $s \in \mathbb{C}$. Then,

$$C(sI - A)^{\mathsf{A}}B = C[sI - (A + BKC)]^{\mathsf{A}}B.$$

Source: Fact 3.19.11. **Remark:** This result shows that the invariant zeros of SISO, SIMO, and MISO transfer function are invariant under feedback. **Related:** Proposition 16.10.10. **Problem:** Consider the case of MIMO transfer functions.

Fact 16.24.16. Let $A \in \mathbb{R}^{n\times n}$, $B \in \mathbb{R}^{n\times m}$, $C \in \mathbb{R}^{m\times n}$, and $K \in \mathbb{R}^{m\times n}$, and assume that $A + BK$ is nonsingular. Then,

$$\det\begin{bmatrix} A & B \\ C & 0 \end{bmatrix} = (-1)^m\det(A + BK)\det[C(A + BK)^{-1}B].$$

Hence, $\left[\begin{smallmatrix} A & B \\ C & 0 \end{smallmatrix}\right]$ is nonsingular if and only if $C(A + BK)^{-1}B$ is nonsingular. **Source:** Note that

$$\det\begin{bmatrix} A & B \\ C & 0 \end{bmatrix} = \det\begin{bmatrix} A & B \\ C & 0 \end{bmatrix}\begin{bmatrix} I & 0 \\ K & I \end{bmatrix} = \det\begin{bmatrix} A + BK & B \\ C & 0 \end{bmatrix} = \det(A + BK)\det[-C(A + BK)^{-1}B].$$

Fact 16.24.17. Let $A_1 \in \mathbb{R}^{n\times n}$, $C_1 \in \mathbb{R}^{1\times n}$, $A_2 \in \mathbb{R}^{m\times m}$, and $B_2 \in \mathbb{R}^{m\times 1}$, let $\lambda \in \mathbb{C}$, assume that λ is an observable eigenvalue of (A_1, C_1) and a controllable eigenvalue of (A_2, B_2), and define the dynamics matrix $\mathcal{A}$ of the cascaded system by

$$\mathcal{A} \triangleq \begin{bmatrix} A_1 & 0 \\ B_2C_1 & A_2 \end{bmatrix}.$$

Then, $\operatorname{amult}_{\mathcal{A}}(\lambda) = \operatorname{amult}_{A_1}(\lambda) + \operatorname{amult}_{A_2}(\lambda)$ and $\operatorname{gmult}_{\mathcal{A}}(\lambda) = 1$. **Remark:** The eigenvalue λ is a cyclic eigenvalue of both subsystems as well as the cascaded system. In other words, λ, which occurs in a single Jordan block of each subsystem, occurs in a single Jordan block in the cascaded system. The Jordan blocks of the subsystems corresponding to λ are merged.

Fact 16.24.18. Let $G_1 \in \mathbb{R}^{l_1\times m}(s)$ and $G_2 \in \mathbb{R}^{l_2\times m}(s)$ be strictly proper. Then,

$$\left\|\begin{bmatrix} G_1 \\ G_2 \end{bmatrix}\right\|_{\mathrm{H}_2}^2 = \|G_1\|_{\mathrm{H}_2}^2 + \|G_2\|_{\mathrm{H}_2}^2.$$

Fact 16.24.19. Let $G_1 \in \mathbb{R}^{l\times m_1}(s)$ and $G_2 \in \mathbb{R}^{l\times m_2}(s)$ be strictly proper. Then,

$$\|[G_1 \;\; G_2]\|_{\mathrm{H}_2} = \|G_1\|_{\mathrm{H}_2}^2 + \|G_2\|_{\mathrm{H}_2}^2.$$

Fact 16.24.20. Let $G_1, G_2 \in \mathbb{R}^{m\times m}(s)$ be strictly proper. Then,

$$\left\|\begin{bmatrix} G_1 \\ G_2 \end{bmatrix}\right\|_{\mathrm{H}_2} = \|[G_1 \;\; G_2]\|_{\mathrm{H}_2}.$$

Fact 16.24.21. Let $G(s) \triangleq \frac{\alpha}{s+\beta}$, where $\beta > 0$. Then,

$$\|G\|_{\mathrm{H}_2} = \frac{|\alpha|}{\sqrt{2\beta}}.$$

Fact 16.24.22. Let $G(s) \triangleq \frac{\alpha_1 s+\alpha_0}{s^2+\beta_1 s+\beta_0}$, where $\beta_0, \beta_1 > 0$. Then,

$$\|G\|_{\mathrm{H}_2} = \sqrt{\frac{\alpha_0^2}{2\beta_0\beta_1} + \frac{\alpha_1^2}{2\beta_1}}.$$

Fact 16.24.23. Let $G_1(s) = \frac{\alpha_1}{s+\beta_1}$ and $G_2(s) = \frac{\alpha_2}{s+\beta_2}$, where $\beta_1, \beta_2 > 0$. Then,

$$\|G_1G_2\|_{\mathrm{H}_2} \le \|G_1\|_{\mathrm{H}_2}\|G_2\|_{\mathrm{H}_2}$$

if and only if $\beta_1 + \beta_2 \ge 2$. **Remark:** The H_2 norm is not submultiplicative.

16.25 Facts on the Riccati Equation

Fact 16.25.1. Assume that (A, B) is stabilizable, and assume that $\mathcal{H}$ defined by (16.16.8) has an imaginary eigenvalue λ. Then, every Jordan block of $\mathcal{H}$ associated with λ has even size. **Source:** Let P be a solution of (16.16.4), and let $\mathcal{J}$ denote the Jordan form of $A - \Sigma P$. Then, there exists a nonsingular $2n \times 2n$ block-diagonal matrix $\mathcal{S}$ such that $\hat{\mathcal{H}} \triangleq \mathcal{S}^{-1}\mathcal{H}\mathcal{S} = \left[\begin{smallmatrix} \mathcal{J} & \hat{\Sigma} \\ 0 & -\mathcal{J}^{\mathsf{T}} \end{smallmatrix}\right]$, where $\hat{\Sigma}$ is positive semidefinite. Next, let $\mathcal{J}_\lambda \triangleq \lambda I_r + N_r$ be a Jordan block of $\mathcal{J}$ associated with λ, and consider the submatrix of $\lambda I - \hat{\mathcal{H}}$ consisting of the rows and columns of $\lambda I - \mathcal{J}_\lambda$ and $\lambda I + \mathcal{J}_\lambda^{\mathsf{T}}$. Since (A, B) is stabilizable, it follows that the rank of this submatrix is $2r - 1$. Hence, every Jordan block of $\mathcal{H}$ associated with λ has even size. **Remark:** Canonical forms for symplectic and Hamiltonian matrices are discussed in [1763].

Fact 16.25.2. Let $A, B \in \mathbb{C}^{n\times n}$, assume that A and B are positive definite, let $S \in \mathbb{C}^{n\times n}$ satisfy $A = S^*S$, and define

$$X \triangleq S^{-1}(SBS^*)^{1/2}S^{-*}.$$

Then, X satisfies $XAX = B$. **Source:** [1391, p. 52].

Fact 16.25.3. Let $A, B \in \mathbb{C}^{n\times n}$, assume that either A or B is nonsingular. Then, there exist $X, Y \in \mathbb{C}^{n\times n}$ such that $XAY - YBX = I$. **Source:** [107, pp. 37, 204–206].

Fact 16.25.4. Let $A, B \in \mathbb{C}^{n\times n}$, and assume that the $2n \times 2n$ matrix

$$\begin{bmatrix} A & -2I \\ 2B - \frac{1}{2}A^2 & A \end{bmatrix}$$

is simple. Then, there exists $X \in \mathbb{C}^{n\times n}$ that satisfies $X^2 + AX + B = 0$. **Source:** [2727].

Fact 16.25.5. Let $P, K, X \in \mathbb{R}^{n\times n}$, and assume that P is positive definite and K is skew symmetric, and assume that $XP - PX^T = 2K$. Then, there exists a symmetric matrix $S \in \mathbb{R}^{n\times n}$ such that $X = (K + S)P^{-1}$. Furthermore, X is orthogonal if and only if there exists a symmetric matrix $S \in \mathbb{R}^{n\times n}$ such that $X = (K + S)P^{-1}$ and $S^2 + SK + K^{\mathrm{T}}S = P^2 + K^2$. **Source:** [631]. **Remark:** $XP - PX^T = 2K$ is the *Moser-Veselov equation*.

Fact 16.25.6. Let $A, B \in \mathbb{F}^{n\times n}$, and assume that A and B are positive semidefinite. Then, the following statements hold:

i) If A is positive definite, then $X = A\#B$ is the unique positive-definite solution of

$$XA^{-1}X - B = 0.$$

ii) If A is positive definite, then $X = \frac{1}{2}[-A+A\#(A+4B)]$ is the unique positive-definite solution of

$$XA^{-1}X + X - B = 0.$$

iii) If A is positive definite, then $X = \frac{1}{2}[A+A\#(A+4B)]$ is the unique positive-definite solution of

$$XA^{-1}X - X - B = 0.$$

iv) If B is positive definite, then $X = A\#B$ is the unique positive-definite solution of

$$XB^{-1}X = A.$$

v) If A is positive definite, then $X = \frac{1}{2}[A + A\#(A + 4BA^{-1}B)]$ is the unique positive-definite solution of

$$BX^{-1}B - X + A = 0.$$

vi) If A is positive definite, then $X = \frac{1}{2}[-A + A\#(A + 4BA^{-1}B)]$ is the unique positive-definite solution of

$$BX^{-1}B - X - A = 0.$$

vii) If $0 < A \leq B$, then $X = \frac{1}{2}[A + A\#(4B - 3A)]$ is the unique positive-definite solution of

$$XA^{-1}X - X - (B - A) = 0.$$

viii) If $0 < A \leq B$, then $X = \frac{1}{2}[-A + A\#(4B - 3A)]$ is the unique positive-definite solution of

$$XA^{-1}X + X - (B - A) = 0.$$

ix) If $0 < A < B$, $X(0)$ is positive definite, and $X(t)$ satisfies

$$\dot{X} = -XA^{-1}X + X + B - A,$$

then

$$\lim_{t\to\infty} X(t) = \frac{1}{2}[A + A\#(4B - 3A)].$$

x) If $0 < A < B$, $X(0)$ is positive definite, and $X(t)$ satisfies

$$\dot{X} = -XA^{-1}X - X + B - A,$$

then

$$\lim_{t\to\infty} X(t) = \tfrac{1}{2}[A + A\#(4B - 3A)].$$

xi) If $0 < A < B$, $X(0)$ and $Y(0)$ are positive definite, and $X(t)$ and $Y(t)$ satisfy

$$\dot{X} = -XA^{-1}X + X + B - A,$$
$$\dot{Y} = -YA^{-1}Y - Y + B - A,$$

then

$$\lim_{t\to\infty} X(t)\#Y(t) = A\#(B - A).$$

Source: [1824]. **Remark:** $A\#B$ is the geometric mean of A and B. See Fact 10.11.68. **Remark:** The solution X given by *vii)* is the *golden mean* of A and B. In the scalar case with $A = 1$ and $B = 2$, the solution X of $X^2 - X - 1 = 0$ is the *golden ratio* $\frac{1}{2}(1 + \sqrt{5})$. See Fact 1.17.1.

Fact 16.25.7. Let $P_0 \in \mathbb{R}^{n\times n}$, assume that P_0 is positive definite, let $V \in \mathbb{R}^{n\times n}$ be positive semidefinite, and, for all $t \ge 0$, let $P(t) \in \mathbb{R}^{n\times n}$ satisfy

$$\dot{P}(t) = A^{\mathsf{T}}P(t) + P(t)A + P(t)VP(t), \quad P(0) = P_0.$$

Then, for all $t \ge 0$,

$$P(t) = e^{tA^{\mathsf{T}}}\left[P_0^{-1} - \int_0^t e^{\tau A}Ve^{\tau A^{\mathsf{T}}}\,\mathrm{d}\tau\right]^{-1} e^{tA}.$$

Remark: $P(t)$ satisfies a Riccati differential equation.

Fact 16.25.8. Let $G_{\mathrm{c}} \sim \left[\begin{array}{c|c} A_{\mathrm{c}} & B_{\mathrm{c}} \\ \hline C_{\mathrm{c}} & 0 \end{array}\right]$ denote an nth-order dynamic controller for the standard control problem. If G_{c} minimizes $\|\tilde{\mathcal{G}}\|_2$, then G_{c} is given by

$$A_{\mathrm{c}} \triangleq A + BC_{\mathrm{c}} - B_{\mathrm{c}}C - B_{\mathrm{c}}DC_{\mathrm{c}},$$
$$B_{\mathrm{c}} \triangleq (QC^{\mathsf{T}} + V_{12})V_2^{-1},$$
$$C_{\mathrm{c}} \triangleq -R_2^{-1}(B^{\mathsf{T}}P + R_{12}^{\mathsf{T}}),$$

where P and Q are positive-semidefinite solutions to the algebraic Riccati equations

$$\hat{A}_{\mathrm{R}}^{\mathsf{T}}P + P\hat{A}_{\mathrm{R}} - PBR_2^{-1}B^{\mathsf{T}}P + \hat{R}_1 = 0,$$
$$\hat{A}_{\mathrm{E}}Q + Q\hat{A}_{\mathrm{E}}^{\mathsf{T}} - QC^{\mathsf{T}}V_2^{-1}CQ + \hat{V}_1 = 0,$$

where $\hat{A}_{\mathrm{R}}$ and $\hat{R}_1$ are defined by

$$\hat{A}_{\mathrm{R}} \triangleq A - BR_2^{-1}R_{12}^{\mathsf{T}}, \quad \hat{R}_1 \triangleq R_1 - R_{12}R_2^{-1}R_{12}^{\mathsf{T}},$$

and $\hat{A}_{\mathrm{E}}$ and $\hat{V}_1$ are defined by

$$\hat{A}_{\mathrm{E}} \triangleq A - V_{12}V_2^{-1}C, \quad \hat{V}_1 \triangleq V_1 - V_{12}V_2^{-1}V_{12}^{\mathsf{T}}.$$

Furthermore, the eigenvalues of the closed-loop system are given by

$$\operatorname{mspec}\left(\begin{bmatrix} A & BC_{\mathrm{c}} \\ B_{\mathrm{c}}C & A_{\mathrm{c}} + B_{\mathrm{c}}DC_{\mathrm{c}} \end{bmatrix}\right) = \operatorname{mspec}(A + BC_{\mathrm{c}}) \cup \operatorname{mspec}(A - B_{\mathrm{c}}C).$$

Fact 16.25.9. Let $G_{\mathrm{c}} \sim \left[\begin{array}{c|c} A_{\mathrm{c}} & B_{\mathrm{c}} \\ \hline C_{\mathrm{c}} & 0 \end{array}\right]$ denote an nth-order dynamic controller for the discrete-time standard control problem. If G_{c} minimizes $\|\tilde{\mathcal{G}}\|_2$, then G_{c} is given by

$$A_{\mathrm{c}} \triangleq A + BC_{\mathrm{c}} - B_{\mathrm{c}}C - B_{\mathrm{c}}DC_{\mathrm{c}},$$
$$B_{\mathrm{c}} \triangleq (AQC^{\mathsf{T}} + V_{12})(V_2 + CQC^{\mathsf{T}})^{-1},$$

$$C_c \triangleq -(R_2 + B^{\mathsf{T}}PB)^{-1}(R_{12}^{\mathsf{T}} + B^{\mathsf{T}}PA),$$

and the eigenvalues of the closed-loop system are given by

$$\operatorname{mspec}\left(\begin{bmatrix} A & BC_c \\ B_cC & A_c + B_cDC_c \end{bmatrix}\right) = \operatorname{mspec}(A + BC_c) \cup \operatorname{mspec}(A - B_cC).$$

Now, assume that $D = 0$ and $G_c \sim \left[\begin{array}{c|c} A_c & B_c \\ \hline C_c & D_c \end{array}\right]$. Then,

$$\begin{aligned}
A_c &\triangleq A + BC_c - B_cC - BD_cC,\\
B_c &\triangleq (AQC^{\mathsf{T}} + V_{12})(V_2 + CQC^{\mathsf{T}})^{-1} + BD_c,\\
C_c &\triangleq -(R_2 + B^{\mathsf{T}}PB)^{-1}(R_{12}^{\mathsf{T}} + B^{\mathsf{T}}PA) - D_cC,\\
D_c &\triangleq (R_2 + B^{\mathsf{T}}PB)^{-1}[B^{\mathsf{T}}PAQC^{\mathsf{T}} + R_{12}^{\mathsf{T}}QC^{\mathsf{T}} + B^{\mathsf{T}}PV_{12}](V_2 + CQC^{\mathsf{T}})^{-1},
\end{aligned}$$

and the eigenvalues of the closed-loop system are given by

$$\operatorname{mspec}\left(\begin{bmatrix} A + BD_cC & BC_c \\ B_cC & A_c \end{bmatrix}\right) = \operatorname{mspec}(A + BC_c) \cup \operatorname{mspec}(A - B_cC).$$

In both cases, P and Q are positive-semidefinite solutions to the discrete-time algebraic Riccati equations

$$\begin{aligned}
P &= \hat{A}_{\mathrm{R}}^{\mathsf{T}}P\hat{A}_{\mathrm{R}} - \hat{A}_{\mathrm{R}}^{\mathsf{T}}PB(R_2 + B^{\mathsf{T}}PB)^{-1}B^{\mathsf{T}}P\hat{A}_{\mathrm{R}} + \hat{R}_1,\\
Q &= \hat{A}_{\mathrm{E}}Q\hat{A}_{\mathrm{E}}^{\mathsf{T}} - \hat{A}_{\mathrm{E}}QC^{\mathsf{T}}(V_2 + CQC^{\mathsf{T}})^{-1}CQ\hat{A}_{\mathrm{E}}^{\mathsf{T}} + \hat{V}_1,
\end{aligned}$$

where $\hat{A}_{\mathrm{R}}$ and $\hat{R}_1$ are defined by

$$\hat{A}_{\mathrm{R}} \triangleq A - BR_2^{-1}R_{12}^{\mathsf{T}}, \quad \hat{R}_1 \triangleq R_1 - R_{12}R_2^{-1}R_{12}^{\mathsf{T}},$$

and $\hat{A}_{\mathrm{E}}$ and $\hat{V}_1$ are defined by

$$\hat{A}_{\mathrm{E}} \triangleq A - V_{12}V_2^{-1}C, \quad \hat{V}_1 \triangleq V_1 - V_{12}V_2^{-1}V_{12}^{\mathsf{T}}.$$

Source: [1289].

16.26 Notes

Linear system theory is treated in [574, 2349, 2726, 2907]. Time-varying linear systems are considered in [827, 2349], while discrete-time systems are emphasized in [1360]. The PBH test is given in [1349]. Spectral factorization results are given in [760]. Stabilization aspects are discussed in [934]. Observable asymptotic stability and controllable asymptotic stability were introduced and used to analyze Lyapunov equations in [2475]. Zeros are treated in [39, 1037, 1573, 1583, 1915, 2205, 2359, 2404]. Matrix-based methods for linear system identification are developed in [2778], while stochastic theory is considered in [1310].

Solutions of the LQR problem under weak conditions are given in [1156]. Solutions of the Riccati equation are considered in [901, 1189, 1701, 1707, 1735, 1736, 1973, 2306, 2883, 2895, 2900]. Proposition 16.16.16 is based on Theorem 3.6 of [2912, p. 79]. A variation of Theorem 16.18.1 is given without proof by Theorem 7.2.1 of [1511, p. 125].

There are numerous extensions to the results given in this chapter relating to various generalizations of (16.16.4). These generalizations include the case where R_1 is indefinite [1188, 2892, 2894] as well as the case where Σ is indefinite [2386]. The latter case is relevant to H_∞ optimal control theory [432]. Additional extensions include the Riccati inequality $A^{\mathsf{T}}P + PA + R_1 - P\Sigma P \geq 0$ [2297, 2385, 2386, 2387], the discrete-time Riccati equation [18, 1360, 1503, 1735, 2297, 2899], and fixed-order control [1493].

Bibliography

[1] A. Abdessemed and E. B. Davies, "Some Commutator Estimates in the Schatten Classes," *J. London Math. Soc.*, Vol. s2-39, pp. 299–308, 1989. (Cited on p. 885.)

[2] U. Abel, "A Generalization of the Leibniz Rule," *Amer. Math. Monthly*, Vol. 120, pp. 924–928, 2013. (Cited on pp. 68 and 86.)

[3] U. Abel, "On an Identity Involving Powers of Binomial Coefficients," *Coll. Math. J.*, Vol. 46, p. 138, 2015. (Cited on p. 76.)

[4] R. Ablamowicz, "Matrix Exponential via Clifford Algebras," *J. Nonlin. Math. Phys.*, Vol. 5, pp. 294–313, 1998. (Cited on p. 1206.)

[5] H. Abou-Kandil, G. Freiling, V. Ionescu, and G. Jank, *Matrix Riccati Equations in Control and Systems Theory*. Basel: Birkhauser, 2003. (Cited on p. xxi.)

[6] S. Abramovich, J. Baric, M. Matic, and J. Pecaric, "On Van De Lune-Alzer's Inequality," *J. Math. Ineq.*, Vol. 1, pp. 563–587, 2007. (Cited on p. 47.)

[7] S. Abramovich, J. Baric, and J. E. Pecaric, "A New Proof of an Inequality of Bohr for Hilbert Space Operators," *Lin. Alg. Appl.*, Vol. 430, pp. 1432–1435, 2009. (Cited on p. 747.)

[8] A. Abu-Omar and F. Kittaneh, "Numerical Radius Inequalities for $n \times n$ Operator Matrices," *Lin. Alg. Appl.*, Vol. 468, pp. 18–26, 2015. (Cited on p. 866.)

[9] L. Aceto and D. Trigiante, "The Matrices of Pascal and Other Greats," *Amer. Math. Monthly*, Vol. 108, pp. 232–245, 2001. (Cited on pp. 613, 619, 727, and 1205.)

[10] F. S. Acton, "Summation of Tangents," *Amer. Math. Monthly*, Vol. 59, pp. 337–338, 1952. (Cited on p. 242.)

[11] V. Adamchik, "33 Representations for Catalan's Constant," http://www-2.cs.cmu.edu/~adamchik/articles/catalan/catalan.htm. (Cited on p. 1170.)

[12] K. Adegoke, "The Golden Ratio, Fibonacci Numbers and BBP-Type Formulas," *Fibonacci Quart.*, Vol. 52, pp. 129–138, 2014. (Cited on pp. 103 and 1079.)

[13] C. Aebi and G. Cairns, "Morley's Other Miracle," *Math. Mag.*, Vol. 85, pp. 205–211, 2012. (Cited on p. 55.)

[14] C. Aebi and G. Cairns, "Generalizations of Wilson's Theorem for Double-, Hyper-, Sub- and Superfactorials," *Amer. Math. Monthly*, Vol. 122, pp. 433–443, 2015. (Cited on pp. 31, 49, 104, and 613.)

[15] S. Afriat, "Orthogonal and Oblique Projectors and the Characteristics of Pairs of Vector Spaces," *Proc. Cambridge Phil. Soc.*, Vol. 53, pp. 800–816, 1957. (Cited on pp. 415 and 595.)

[16] R. Agaev and P. Chebotarev, "On the Spectra of Nonsymmetric Laplacian Matrices," *Lin. Alg. Appl.*, Vol. 399, pp. 157–168, 2005. (Cited on p. 1233.)

[17] R. P. Agarwal, *Difference Equations and Inequalities: Theory, Methods, and Applications*, 2nd ed. New York: Marcel Dekker, 2000. (Cited on p. 1248.)

[18] C. D. Ahlbrandt and A. C. Peterson, *Discrete Hamiltonian Systems: Difference Equations, Continued Fractions, and Riccati Equations*. Dordrecht: Kluwer, 1996. (Cited on p. 1319.)

[19] E. Ahn, S. Kim, and Y. Lim, "An Extended Lie-Trotter Formula and Its Application," *Lin. Alg. Appl.*, Vol. 427, pp. 190–196, 2007. (Cited on p. 1213.)

[20] M. Aigner and G. M. Ziegler, *Proofs from THE BOOK*, 4th ed. Berlin: Springer, 2010. (Cited on pp. 35 and 52.)

[21] A. C. Aitken, *Determinants and Matrices*, 9th ed. Edinburgh: Oliver and Boyd, 1956. (Cited on p. xxii.)

[22] L. Aizenberg, "Generalization of Results about the Bohr Radius for Power Series," *Studia Mathematica*, Vol. 180, pp. 161–168, 2007. (Cited on p. 948.)

[23] A. O. Ajibade and M. A. Rashid, "A Strange Property of the Determinant of Minors," *Int. J. Math. Educ. Sci. Tech.*, Vol. 38, pp. 852–858, 2007. (Cited on p. 341.)

[24] M. Al-Ahmar, "An Identity of Jacobi," *Amer. Math. Monthly*, Vol. 103, pp. 78–79, 1996. (Cited on p. 389.)

[25] A. M. Al-Rashed, "Norm Inequalities and Characterizations of Inner Product Spaces," *J. Math. Anal. Appl.*, Vol. 176, pp. 587–593, 1993. (Cited on p. 852.)

[26] Z. Al Zhour and A. Kilicman, "Matrix Equalities and Inequalities Involving Khatri-Rao and Tracy-Singh Sums," *J. Ineq. Pure Appl. Math.*, Vol. 7, pp. 1–8, 2006, article 34. (Cited on pp. 700, 821, 827, and 873.)

[27] A. Alaca, S. Alaca, and K. S. Williams, "An Infinite Class of Identities," *Bull. Austral. Math. Soc.*, Vol. 75, pp. 239–246, 2007. (Cited on pp. 119 and 1092.)

[28] A. Alaca, S. Alaca, and K. S. Williams, "Some Infinite Products of Ramanujan Type," *Canad. Math. Bull.*, Vol. 52, pp. 481–492, 2009. (Cited on p. 1092.)

[29] A. A. Albert and B. Muckenhoupt, "On Matrices of Trace Zero," *Michigan Math. J.*, Vol. 4, pp. 1–3, 1957. (Cited on p. 571.)

[30] A. E. Albert, "Conditions for Positive and Nonnegative Definiteness in Terms of Pseudoinverses," *SIAM J. Appl. Math.*, Vol. 17, pp. 434–440, 1969. (Cited on p. 831.)

[31] A. E. Albert, *Regression and the Moore-Penrose Pseudoinverse*. New York: Academic Press, 1972. (Cited on pp. 633, 637, 638, 665, and 911.)

[32] J. M. Aldaz, "A Refinement of the Inequality between Arithmetic and Geometric Means," *J. Math. Ineq.*, Vol. 2, pp. 473–477, 2007. (Cited on pp. 210 and 855.)

[33] J. M. Aldaz, "Strengthened Cauchy-Schwarz Inequalities and Hölder Inequalities," *J. Ineq. Pure. Appl. Math.*, Vol. 10, no. 4/116, pp. 1–6, 2009. (Cited on p. 855.)

[34] R. Aldrovandi, *Special Matrices of Mathematical Physics: Stochastic, Circulant and Bell Matrices*. Singapore: World Scientific, 2001. (Cited on pp. xxi, 108, 112, 539, 615, 976, and 1009.)

[35] M. Aleksiejczuk and A. Smoktunowicz, "On Properties of Quadratic Matrices," *Mathematica Pannonica*, Vol. 112, pp. 239–248, 2000. (Cited on p. 667.)

[36] A. Y. Alfakih, "On the Nullspace, the Rangespace and the Characteristic Polynomial of Euclidean Distance Matrices," *Lin. Alg. Appl.*, Vol. 416, pp. 348–354, 2006. (Cited on p. 861.)

[37] M. Alic, P. S. Bullen, J. E. Pecaric, and V. Volenec, "On the Geometric-Arithmetic Mean Inequality for Matrices," *Mathematical Communications*, Vol. 2, pp. 125–128, 1997. (Cited on p. 745.)

[38] M. Alic, B. Mond, J. E. Pecaric, and V. Volenec, "Bounds for the Differences of Matrix Means," *SIAM J. Matrix Anal. Appl.*, Vol. 18, pp. 119–123, 1997. (Cited on p. 744.)

[39] H. Aling and J. M. Schumacher, "A Nine-Fold Decomposition for Linear Systems," *Int. J. Contr.*, Vol. 39, pp. 779–805, 1984. (Cited on p. 1319.)

[40] K. Aljanaideh and D. S. Bernstein, "An Adjugate Identity," *IMAGE*, Vol. 51, p. 35, 2013. (Cited on p. 346.)

[41] G. Almkvist and B. Berndt, "Gauss, Landen, Ramanujan, the Arithmetic-Geometric Mean, Ellipses, π, and the *Ladies Diary*," *Amer. Math. Monthly*, Vol. 95, pp. 585–608, 1988. (Cited on p. 1080.)

[42] G. Almkvist and A. Granville, "Borwein and Bradley's Apery-Like Formulae for $\zeta(4n+3)$," *Exp. Math.*, Vol. 8, pp. 197–203, 1999. (Cited on pp. 44, 45, and 1070.)

[43] G. Alpargu and G. P. H. Styan, "Some Remarks and a Bibliography on the Kantorovich Inequality," in *Multidimensional Statistical Analysis and Theory of Random Matrices*, A. K. Gupta and V. L. Girko, Eds. Utrecht: VSP, 1996, pp. 1–13. (Cited on p. 792.)

[44] R. C. Alperin, "The Matrix of a Rotation," *Coll. Math. J.*, Vol. 20, p. 230, 1989. (Cited on p. 396.)

[45] C. Alsina and R. Ger, "On Some Inequalities and Stability Results Related to the Exponential Function," *J. Ineq. Appl.*, Vol. 2, pp. 373–380, 1998. (Cited on p. 141.)

[46] C. Alsina and R. B. Nelsen, "On Candido's Identity," *Math. Mag.*, Vol. 80, pp. 226–228, 2007. (Cited on p. 135.)

[47] C. Alsina and R. B. Nelsen, *When Less Is More: Visualizing Basic Inequalities*. Washington, DC: Mathematical Association of America, 2009. (Cited on pp. 133, 157, 168, 177, 212, 215, 222, 225, 250, 252, 446, 452, 456, 466, 472, 480, 484, 485, 486, 489, 491, and 495.)

[48] C. Alsina and R. B. Nelsen, *Charming Proofs: A Journey into Elegant Mathematics*. Washington, DC: Mathematical Association of America, 2010. (Cited on p. 149.)

[49] C. Alsina and R. B. Nelsen, *A Mathematical Space Odyssey*. Washington, DC: Mathematical Association of America, 2015. (Cited on p. 1095.)

[50] C. Alsina and R. B. Nelsen, "The Maclaurin Inequality and Rectangular Boxes," *Math. Mag.*, Vol. 90, pp. 228–230, 2017. (Cited on p. 161.)

[51] C. Alsina and R. B. Nelson, "Proof without Words: Inequalities for Two Numbers Whose Sum Is One," *Math. Mag.*, Vol. 84, p. 228, 2011. (Cited on p. 126.)

[52] A. Alt, "An Acute Triangle Inequality," *Coll. Math. J.*, Vol. 38, pp. 392–393, 2007. (Cited on p. 484.)

[53] A. Alt, "Several Symmetric Inequalities of Exponential Kind," *Crux Math.*, Vol. 35, pp. 457–462, 2009. (Cited on p. 186.)

[54] A. Alt, "A Medians and Radii Inequality," *Math. Mag.*, Vol. 87, pp. 233–234, 2014. (Cited on p. 480.)

[55] A. Alt, "An Inequality between Symmetric Polynomials," *Math. Mag.*, Vol. 89, p. 297, 2016. (Cited on p. 177.)

[56] A. Alt, "Sum of Medians of a Triangle," *Amer. Math. Monthly*, Vol. 124, p. 186, 2017. (Cited on p. 480.)

[57] S. L. Altmann, *Rotations, Quaternions, and Double Groups*. New York: Oxford University Press, 1986. (Cited on pp. xxi, 438, and 439.)

[58] S. L. Altmann, "Hamilton, Rodrigues, and the Quaternion Scandal," *Math. Mag.*, Vol. 62, pp. 291–308, 1989. (Cited on p. 394.)

[59] N. Altshiller-Court, *College Geometry: An Introduction to the Modern Geometry of the Triangle and the Circle*. Mineola: Dover, 2007. (Cited on pp. 472, 480, and 491.)

[60] H. Alzer, "A Lower Bound for the Difference between the Arithmetic and Geometric Means," *Nieuw. Arch. Wisk.*, Vol. 8, pp. 195–197, 1990. (Cited on p. 203.)

[61] H. Alzer, "A Trigonometric Double Inequality," *Elem. Math.*, Vol. 65, pp. 45–48, 2010. (Cited on p. 250.)

[62] H. Alzer, D. Karayannakis, and H. M. Srivastava, "Series Representations for Some Mathematical Constants," *J. Math. Anal. Appl.*, Vol. 320, pp. 145–162, 2006. (Cited on pp. 1025, 1032, and 1050.)

[63] H. Alzer and S. Koumandos, "Sharp Inequalities for Trigonometric Sums," *Math. Proc. Cambridge Phil. Soc.*, Vol. 134, pp. 139–152, 2003. (Cited on p. 254.)

[64] H. Alzer and M. K. Kwong, "Extension of a Trigonometric Inequality of Turan," *Acta Sci. Math (Szeged)*, Vol. 80, pp. 21–26, 2014. (Cited on p. 254.)

[65] H. Alzer and M. K. Kwong, "Rogosinski-Szego Type Inequalities for Trigonometric Sums," *J. Approx. Theory*, Vol. 190, pp. 62–72, 2015. (Cited on p. 254.)

[66] T. Amdeberhan, "Faster and Faster Convergent Series for $\zeta(3)$," *Elec. J. Combinatorics*, Vol. 3, pp. 1–2, 1996. (Cited on p. 1032.)

[67] T. Amdeberhan, V. de Angelis, M. Lin, V. H. Moll, and B. Sury, "A Pretty Binomial Identity," *Elem. Math.*, Vol. 67, pp. 18–25, 2012. (Cited on p. 86.)

[68] T. Amdeberhan and S. B. Ekhad, "Play Time with Determinants," arXiv.math/1010.3881v3. (Cited on p. 334.)

[69] T. Amdeberhan, O. Espinosa, V. H. Moll, and A. Straub, "Wallis-Ramanujan-Schur-Feynman," *Amer. Math. Monthly*, Vol. 117, pp. 618–632, 2010. (Cited on pp. 86, 1070, 1083, and 1104.)

[70] S. Amghibech, "A Matrix Inequality," *Amer. Math. Monthly*, Vol. 115, p. 671, 2008. (Cited on p. 776.)

[71] M. Andelic and C. M. da Fonseca, "Sufficient Conditions for Positive Definiteness of Tridiagonal Matrices Revisited," *Positivity*, Vol. 15, pp. 155–159, 2011. (Cited on p. 424.)

[72] B. D. O. Anderson, "Orthogonal Decomposition Defined by a Pair of Skew-Symmetric Forms," *Lin. Alg. Appl.*, Vol. 8, pp. 91–93, 1974. (Cited on pp. xxi and 605.)

[73] B. D. O. Anderson, "Weighted Hankel-Norm Approximation: Calculation of Bounds," *Sys. Contr. Lett.*, Vol. 7, pp. 247–255, 1986. (Cited on p. 793.)

[74] B. D. O. Anderson, E. I. Jury, and M. Mansour, "Schwarz Matrix Properties for Continuous and Discrete Time Systems," *Int. J. Contr.*, Vol. 23, pp. 1–16, 1976. (Cited on p. 1230.)

[75] B. D. O. Anderson and J. B. Moore, "A Matrix Kronecker Lemma," *Lin. Alg. Appl.*, Vol. 15, pp. 227–234, 1976. (Cited on pp. 745 and 959.)

[76] E. C. Anderson, "Morrie's Law and Experimental Mathematics," *J. Recreat. Math.*, Vol. 29, pp. 85–88, 1998. (Cited on p. 245.)

[77] G. Anderson, M. Vamanamurthy, and M. Vuorinen, "Monotonicity Rules in Calculus," *Amer. Math. Monthly*, Vol. 113, pp. 805–816, 2006. (Cited on pp. 123, 124, 227, and 249.)

[78] G. D. Anderson and S. L. Qiu, "A Monotonicity Property of the Gamma Function," *Proc. Amer. Math. Soc.*, Vol. 125, pp. 3355–3362, 1997. (Cited on p. 1000.)

[79] M. Anderson and T. Feil, *A First Course in Abstract Algebra*, 3rd ed. Boca Raton: CRC Press, 2015. (Cited on pp. 28 and 370.)

[80] T. W. Anderson and I. Olkin, "An Extremal Problem for Positive Definite Matrices," *Lin. Multilin. Alg.*, Vol. 6, pp. 257–262, 1978. (Cited on p. 740.)

[81] W. N. Anderson, "Shorted Operators," *SIAM J. Appl. Math.*, Vol. 20, pp. 520–525, 1971. (Cited on p. 819.)

[82] W. N. Anderson and R. J. Duffin, "Series and Parallel Addition of Matrices," *J. Math. Anal. Appl.*, Vol. 26, pp. 576–594, 1969. (Cited on p. 819.)

[83] W. N. Anderson, E. J. Harner, and G. E. Trapp, "Eigenvalues of the Difference and Product of Projections," *Lin. Multilin. Alg.*, Vol. 17, pp. 295–299, 1985. (Cited on p. 590.)

[84] W. N. Anderson and M. Schreiber, "On the Infimum of Two Projections," *Acta Sci. Math.*, Vol. 33, pp. 165–168, 1972. (Cited on pp. 657 and 658.)

[85] W. N. Anderson and G. E. Trapp, "Shorted Operators II," *SIAM J. Appl. Math.*, Vol. 28, pp. 60–71, 1975. (Cited on pp. 819 and 831.)

[86] W. N. Anderson and G. E. Trapp, "Inverse Problems for Means of Matrices," *SIAM J. Alg. Disc. Meth.*, Vol. 7, pp. 188–192, 1986. (Cited on p. 744.)

[87] W. N. Anderson and G. E. Trapp, "Symmetric Positive Definite Matrices," *Amer. Math. Monthly*, Vol. 95, pp. 261–262, 1988. (Cited on p. 760.)

[88] T. Ando, "Concavity of Certain Maps on Positive Definite Matrices and Applications to Hadamard Products," *Lin. Alg. Appl.*, Vol. 26, pp. 203–241, 1979. (Cited on pp. 719, 825, 827, 830, 831, and 1214.)

[89] T. Ando, "Inequalities for M-Matrices," *Lin. Multilin. Alg.*, Vol. 8, pp. 291–316, 1980. (Cited on p. 824.)

[90] T. Ando, "On the Arithmetic-Geometric-Harmonic-Mean Inequalities for Positive Definite Matrices," *Lin. Alg. Appl.*, Vol. 52/53, pp. 31–37, 1983. (Cited on p. 744.)

[91] T. Ando, "On Some Operator Inequalities," *Math. Ann.*, Vol. 279, pp. 157–159, 1987. (Cited on pp. 744 and 745.)

[92] T. Ando, "Majorization, Doubly Stochastic Matrices, and Comparison of Eigenvalues," *Lin. Alg. App.*, Vol. 118, pp. 163–248, 1989. (Cited on p. 741.)

[93] T. Ando, "Majorizations and Inequalities in Matrix Theory," *Lin. Alg. Appl.*, Vol. 199, pp. 17–67, 1994. (Cited on pp. 738, 757, 779, 780, 871, 893, 904, 1218, and 1221.)

[94] T. Ando, "Majorization Relations for Hadamard Products," *Lin. Alg. Appl.*, Vol. 223–224, pp. 57–64, 1995. (Cited on p. 825.)

[95] T. Ando, "Matrix Young Inequalities," *Oper. Theory Adv. Appl.*, Vol. 75, pp. 33–38, 1995. (Cited on pp. 738 and 904.)

[96] T. Ando, "Problem of Infimum in the Positive Cone," in *Analytic and Geometric Inequalities and Applications*, T. M. Rassias and H. M. Srivastava, Eds. Dordrecht: Kluwer, 1999, pp. 1–12. (Cited on pp. 740 and 819.)

[97] T. Ando, "Löwner Inequality of Indefinite Type," *Lin. Alg. Appl.*, Vol. 385, pp. 73–80, 2004. (Cited on pp. 738 and 1311.)

[98] T. Ando, "Some Homogeneous Cyclic Inequalities of Three Variables of Degree Three and Four," *Austr. J. Math. Anal. Appl.*, Vol. 7, pp. 1–14, 2010. (Cited on pp. 151, 152, 154, and 180.)

[99] T. Ando, "Cubic and Quartic Cyclic Homogeneous Inequalities of Three Variables," *Math. Ineq. Appl.*, Vol. 16, pp. 127–142, 2013. (Cited on pp. 151 and 177.)

[100] T. Ando and R. Bhatia, "Eigenvalue Inequalities Associated with the Cartesian Decomposition," *Lin. Multilin. Alg.*, Vol. 22, pp. 133–147, 1987. (Cited on pp. 777, 806, and 807.)

[101] T. Ando and F. Hiai, "Log-Majorization and Complementary Golden-Thompson Type Inequalities," *Lin. Alg. Appl.*, Vol. 197/198, pp. 113–131, 1994. (Cited on pp. 746, 1213, 1215, 1221, and 1222.)

[102] T. Ando and F. Hiai, "Hölder Type Inequalities for Matrices," *Math. Ineq. Appl.*, Vol. 1, pp. 1–30, 1998. (Cited on pp. 721, 741, and 742.)

[103] T. Ando, F. Hiai, and K. Okubo, "Trace Inequalities for Multiple Products of Two Matrices," *Math. Ineq. Appl.*, Vol. 3, pp. 307–318, 2000. (Cited on pp. 592 and 766.)

[104] T. Ando, R. A. Horn, and C. R. Johnson, "The Singular Values of a Hadamard Product: A Basic Inequality," *Lin. Multilin. Alg.*, Vol. 21, pp. 345–365, 1987. (Cited on p. 907.)

[105] T. Ando, C.-K. Li, and R. Mathias, "Geometric Means," *Lin. Alg. Appl.*, Vol. 385, pp. 305–334, 2004. (Cited on p. 744.)

[106] T. Ando and X. Zhan, "Norm Inequalities Related to Operator Monotone Functions," *Math. Ann.*, Vol. 315, pp. 771–780, 1999. (Cited on pp. 806, 870, 883, and 1223.)

[107] T. Andreescu, *Mathematical Reflections: The First Two Years*. Plano: XYZ Press, 2011. (Cited on pp. 31, 42, 46, 92, 100, 103, 136, 138, 151, 155, 156, 160, 161, 164, 166, 169, 173, 177, 183, 215, 238, 242, 243, 269, 456, 466, 472, 480, 484, 1051, 1073, 1141, 1226, and 1317.)

[108] T. Andreescu, *Mathematical Reflections: The Next Two Years (2008–2009)*. Plano: XYZ Press, 2012. (Cited on pp. 31, 39, 42, 45, 55, 119, 137, 148, 151, 160, 164, 167, 169, 173, 177, 180, 183, 185, 186, 191, 207, 232, 242, 243, 456, 466, 472, 480, 484, 489, 492, 494, 949, 1057, 1090, and 1239.)

[109] T. Andreescu, *Mathematical Reflections: Two More Years (2010–2011)*. Plano: XYZ Press, 2014. (Cited on pp. 32, 50, 100, 103, 133, 144, 151, 169, 173, 177, 183, 197, 201, 242, 269, 452, 466, 472, 480, 484, 489, 849, 960, 1047, 1087, and 1131.)

[110] T. Andreescu and D. Andrica, *Complex Numbers from A to Z*. Boston: Birkhauser, 2006. (Cited on pp. 266, 271, 272, 273, 442, 443, 452, 472, 491, 493, and 496.)

[111] T. Andreescu and Z. Feng, *103 Trigonometry Problems*. Boston: Birkhauser, 2004. (Cited on p. 491.)

[112] T. Andreescu and R. Gelca, *Mathematical Olympiad Challenges*. Boston: Birkhauser, 2002. (Cited on pp. 45, 50, 52, 124, 135, 150, 155, 180, 190, 239, 244, 258, 329, 428, 493, 1079, and 1087.)

[113] M. Andreoli, "An Interesting and Well-Known Series," *Coll. Math. J.*, Vol. 39, pp. 65–67, 2007. (Cited on pp. 1027 and 1100.)

[114] G. E. Andrews, *The Theory of Partitions*. Reading: Addison-Wesley, 1976. (Cited on pp. 114 and 976.)

[115] G. E. Andrews, "EΥPHKA! num = $\Delta + \Delta + \Delta$," *J. Number Theory*, Vol. 23, pp. 285–293, 1986. (Cited on p. 40.)

[116] G. E. Andrews, R. Askey, and R. Roy, *Special Functions*. Cambridge: Cambridge University Press, 1999. (Cited on pp. 35, 86, 92, 114, 203, 208, 210, 238, 239, 254, 930, 943, 960, 968, 971, 989, 992, 996, 1000, 1003, 1039, 1044, 1049, 1053, 1056, 1058, 1060, 1121, 1157, 1161, 1162, 1176, 1177, and 1178.)

[117] G. E. Andrews, S. B. Ekhad, and D. Zeilberger, "A Short Proof of Jacobi's Formula for the Number of Representations of an Integer as a Sum of Four Squares," *Amer. Math. Monthly*, Vol. 100, pp. 273–276, 1993. (Cited on pp. 33, 122, and 1021.)

[118] G. E. Andrews and K. Eriksson, *Integer Partitions*. Cambridge: Cambridge University Press, 2004. (Cited on pp. 114, 616, 976, and 1091.)

[119] D. Andrica and C. Barbu, "A Geometric Proof of Blundon's Inequalities," *Math. Ineq. Appl.*, Vol. 15, pp. 361–370, 2012. (Cited on p. 452.)

[120] S. P. Andriopoulos, "Sines and Exponentials," *Coll. Math. J.*, Vol. 47, pp. 139–140, 2016. (Cited on p. 466.)

[121] E. Andruchow, G. Corach, and D. Stojanoff, "Geometric Operator Inequalities," *Lin. Alg. Appl.*, Vol. 258, pp. 295–310, 1997. (Cited on pp. 865 and 882.)

[122] E. Angel, *Interactive Computer Graphics*, 3rd ed. Reading: Addison-Wesley, 2002. (Cited on pp. xxi and 396.)

[123] N. Anghel, "Square Roots of Real 2×2 Matrices," *Gazeta Matematica Serie B*, Vol. 118, pp. 489–491, 2013. (Cited on p. 329.)

[124] Anonymous, "Catalan number," http://en.wikipedia.org/wiki/Catalannumber. (Cited on p. 105.)

[125] Anonymous, "Computing Minimal Equal Sums of Like Powers," http://euler.free.fr/identities.htm#ref 01. (Cited on p. 75.)

[126] Anonymous, "Cyclotomic polynomial," https://en.wikipedia.org/wiki/Cyclotomic_polynomial. (Cited on p. 269.)

[127] Anonymous, "Degen's eight-square identity," https://en.wikipedia.org/wiki/Degen%27s_eight-square_identity. (Cited on p. 188.)

[128] Anonymous, "Discriminant," http://en.wikipedia.org/wiki/Discriminant. (Cited on p. 523.)

[129] Anonymous, "Dodgson condensation," http://en.wikipedia.org/wiki/Dodgson_condensation. (Cited on p. 341.)

[130] Anonymous, "Ellipse," http://en.wikipedia.org/wiki/Ellipse. (Cited on p. 1080.)

[131] Anonymous, "Erdős-Straus Conjecture," https://en.wikipedia.org/wiki/ErdosStraus_conjecture. (Cited on p. 36.)

[132] Anonymous, "Fascinating Triangular Numbers," http://www.shyamsundergupta.com/triangle.htm. (Cited on p. 39.)

[133] Anonymous, "Formal Power Series," https://en.wikipedia.org/wiki/Formal_power_series. (Cited on p. 974.)

[134] Anonymous, "Gamma function," http://en.wikipedia.org/wiki/Gammafunction. (Cited on p. 1000.)

[135] Anonymous, "Like the Carlson's inequality," http://www.mathlinks.ro/viewtopic. php?t=151210. (Cited on p. 173.)

[136] Anonymous, "Pentagonal number theorem," http://en.wikipedia.org/wiki/Pentagonalnumbertheorem. (Cited on p. 40.)

[137] Anonymous, "Poincaré Metric," http://en.wikipedia.org/wiki/Poincar%c3%a9_metric. (Cited on p. 276.)

[138] Anonymous, "PolyGamma," http://functions.wolfram.com/GammaBetaErf/PolyGamma/03/01/. (Cited on p. 1001.)

[139] Anonymous, "Prime gap," http://en.wikipedia.org/wiki/Prime_gap. (Cited on p. 966.)

[140] Anonymous, "Schur's inequality," http://en.wikipedia.org/wiki/Schur%27s_inequality. (Cited on p. 154.)

[141] Anonymous, "Stolz-Cesaro Theorem," http://topologicalmusings.wordpress.com/2008/05/08/stolz-cesaro-theorem/. (Cited on p. 962.)

[142] Anonymous, "Triangle," http://en.wikipedia.org/wiki/Triangle. (Cited on p. 466.)

[143] Anonymous, "Twin primes," http://en.wikipedia.org/wiki/Twin_prime. (Cited on p. 966.)

[144] Anonymous, "Weyr Canonical Form," https://en.wikipedia.org/wiki/Weyr_canonical_form. (Cited on p. 619.)

[145] Anonymous, "Wolstenholme's theorem," http://en.wikipedia.org/wiki/Wolstenholme%27s_theorem. (Cited on p. 49.)

[146] Anonymous, "75th Annual William Lowell Putnam Mathematical Competition," *Math. Mag.*, Vol. 88, pp. 81–88, 2015. (Cited on p. 334.)

[147] A. C. Antoulas and D. C. Sorensen, "Lyapunov, Lanczos, and Inertia," *Lin. Alg. Appl.*, Vol. 326, pp. 137–150, 2001. (Cited on p. 1310.)

[148] M. Apagodu and D. Zeilberger, "Some nice sums are almost as nice if you turn them upside down," *J. Combinatorics Number Theory*, Vol. 2, pp. 73–78, 2010. (Cited on p. 68.)

[149] M. Apagodu and D. Zeilberger, "Using the "Freshman's Dream" to Prove Combinatorial Congruences," *Amer. Math. Monthly*, Vol. 124, pp. 597–608, 2017. (Cited on p. 32.)

[150] A. Apelblat, *Tables of Integrals and Series*. Frankfurt am Main: Harri Deutsch, 1996. (Cited on pp. 1105 and 1113.)

[151] J. D. Aplevich, *Implicit Linear Systems*. Berlin: Springer, 1991. (Cited on pp. xxi and 563.)

[152] T. M. Apostol, "Term-Wise Differentiation of Power Series," *Amer. Math. Monthly*, Vol. 59, pp. 323–326, 1952. (Cited on p. 949.)

[153] T. M. Apostol, "Some Explicit Formulas for the Exponential Matrix," *Amer. Math. Monthly*, Vol. 76, pp. 289–292, 1969. (Cited on p. 1206.)

[154] T. M. Apostol, *Mathematical Analysis*, 2nd ed. Reading: Addison-Wesley, 1974. (Cited on pp. xl, 27, 215, 226, 935, 937, 939, and 1058.)

[155] T. M. Apostol, *Introduction to Analytic Number Theory*. New York: Springer, 1998. (Cited on pp. 75, 116, 727, 979, 980, 983, and 1091.)

[156] G. Apostolopoulos, "A Sequence of Null Integrals," *Math. Mag.*, Vol. 87, p. 64, 2014. (Cited on p. 1110.)

[157] G. Apostolopoulos, "An Erdős-Mordell Type Inequality," *Math. Mag.*, Vol. 87, p. 295, 2014. (Cited on p. 472.)

[158] G. Apostolopoulos, "Inequalities Related to Circumradius and Incenter," *Coll. Math. J.*, Vol. 45, pp. 396–397, 2014. (Cited on p. 466.)

[159] G. Apostolopoulos, "A Product and Sums of Sines and Cosines," *Coll. Math. J.*, Vol. 47, pp. 303–304, 2016. (Cited on p. 466.)

[160] G. Apostolopoulos, "An Inequality for the Medians of a Triangle," *Coll. Math. J.*, Vol. 47, pp. 66–66, 2016. (Cited on p. 480.)

[161] G. Apostolopoulos, "A Half-angle Identity for Triangles," *Coll. Math. J.*, Vol. 48, p. 59, 2017. (Cited on p. 456.)

[162] G. Apostolopoulos, "A Symmetric Bound," *Amer. Math. Monthly*, Vol. 124, pp. 662–663, 2017. (Cited on p. 160.)

[163] H. Araki, "On an Inequality of Lieb and Thirring," *Lett. Math. Phys.*, Vol. 19, pp. 167–170, 1990. (Cited on pp. 767 and 875.)

[164] H. Araki and S. Yamagami, "An Inequality for Hilbert-Schmidt Norm," *Comm. Math. Phys.*, Vol. 81, pp. 89–96, 1981. (Cited on p. 871.)

[165] J. Araujo and J. D. Mitchell, "An Elementary Proof that Every Singular Matrix Is a Product of Idempotent Matrices," *Amer. Math. Monthly*, Vol. 112, pp. 641–645, 2005. (Cited on p. 609.)

[166] F. Ardila, "Algebraic and Geometric Methods in Enumerative Combinatorics," in *Handbook of Enumerative Combinatorics*, M. Bona, Ed. Boca Raton: CRC Press, 2015, pp. 3–172. (Cited on pp. 334 and 974.)

[167] A. Arimoto, "A Simple Proof of the Classification of Normal Toeplitz Matrices," *Elec. J. Lin. Alg.*, Vol. 9, pp. 108–111, 2002. (Cited on p. 615.)

[168] T. Arponen, "A Matrix Approach to Polynomials," *Lin. Alg. Appl.*, Vol. 359, pp. 181–196, 2003. (Cited on p. 1205.)

[169] V. Arsigny, P. Fillard, X. Pennec, and N. Ayache, "Geometric Means in a Novel Vector Space Structure on Symmetric Positive-Definite Matrices," *SIAM J. Matrix Anal. Appl.*, Vol. 29, pp. 328–347, 2007. (Cited on p. 1215.)

[170] M. Artin, *Algebra*. Englewood Cliffs: Prentice-Hall, 1991. (Cited on pp. 436 and 440.)

[171] M. Artzrouni, "A Theorem on Products of Matrices," *Lin. Alg. Appl.*, Vol. 49, pp. 153–159, 1983. (Cited on p. 1242.)

[172] A. Arvanitoyeorgos, *An Introduction to Lie Groups and the Geometry of Homogeneous Spaces*. Providence: American Mathematical Society, 2003. (Cited on p. 440.)

[173] A. Ash and R. Gross, *Fearless Symmetry: Exposing the Hidden Patterns of Numbers*. Princeton: Princeton University Press, 2006. (Cited on p. 436.)

[174] P. Ash, "de Gua's Theorem," *Coll. Math. J.*, Vol. 43, p. 343, 2012. (Cited on p. 149.)

[175] M. Ashkartizabi, M. Aminghafari, and A. Mohammadpour, "Square Roots of 3×3 Matrices," *Calcolo*, Vol. 54, pp. 643–656, 2017. (Cited on p. 608.)

[176] I. Asiba and M. Nihei, "On a Unified Method of Solving Some Inequalities," *Math. Gaz.*, Vol. 84, pp. 292–298, 2004. (Cited on pp. 137, 164, and 191.)

[177] M. A. Asiru, "A Generalization of the Formula for the Triangular Number of the Sum and Product of Natural Numbers," *Int. J. Math. Educ. Sci. Tech.*, Vol. 39, pp. 979–985, 2008. (Cited on p. 39.)

[178] R. Askey, *Orthogonal Polynomials and Special Functions*. Philadelphia: SIAM, 1975. (Cited on pp. 239, 254, and 989.)

[179] R. A. Askey and R. Roy, "Gamma Function," in *NIST Handbook of Mathematical Functions*, F. W. J. Olver, D. W. Lozier, R. F. Boisvert, and C. W. Clark, Eds. Cambridge: Cambridge University Press, 2010, pp. 135–147. (Cited on p. 1178.)

[180] H. Aslaksen, "SO(2) Invariants of a Set of 2×2 Matrices," *Math. Scand.*, Vol. 65, pp. 59–66, 1989. (Cited on p. 528.)

[181] H. Aslaksen, "Laws of Trigonometry on SU(3)," *Trans. Amer. Math. Soc.*, Vol. 317, pp. 127–142, 1990. (Cited on pp. 357, 382, and 528.)

[182] H. Aslaksen, "Quaternionic Determinants," *Mathematical Intelligencer*, Vol. 18, no. 3, pp. 57–65, 1996. (Cited on pp. 438 and 440.)

[183] H. Aslaksen, "Defining Relations of Invariants of Two 3×3 Matrices," *J. Algebra*, Vol. 298, pp. 41–57, 2006. (Cited on p. 529.)

[184] B. A. Asner, "On the Total Nonnegativity of the Hurwitz Matrix," *SIAM J. Appl. Math.*, Vol. 18, pp. 407–414, 1970. (Cited on pp. 1225 and 1229.)

[185] Y.-H. Au-Yeung, "A Note on Some Theorems on Simultaneous Diagonalization of Two Hermitian Matrices," *Proc. Cambridge Phil. Soc.*, Vol. 70, pp. 383–386, 1971. (Cited on pp. 797 and 799.)

[186] Y.-H. Au-Yeung, "Some Inequalities for the Rational Power of a Nonnegative Definite Matrix," *Lin. Alg. Appl.*, Vol. 7, pp. 347–350, 1973. (Cited on p. 738.)

[187] Y.-H. Au-Yeung, "On the Semi-Definiteness of the Real Pencil of Two Hermitian Matrices," *Lin. Alg. Appl.*, Vol. 10, pp. 71–76, 1975. (Cited on p. 619.)

[188] K. M. R. Audenaert, "Trace Inequalities for Completely Monotone Functions and Bernstein Functions," http://arxiv.org/abs/1109.3057. (Cited on p. 770.)

[189] K. M. R. Audenaert, "A Norm Compression Inequality for Block Partitioned Positive Semidefinite Matrices," *Lin. Alg. Appl.*, Vol. 413, pp. 155–176, 2006. (Cited on p. 890.)

[190] K. M. R. Audenaert, "A Singular Value Inequality for Heinz Means," *Lin. Alg. Appl.*, Vol. 422, pp. 279–283, 2007. (Cited on p. 882.)

[191] K. M. R. Audenaert, "On a Norm Compression Inequality for $2 \times N$ Partitioned Block Matrices," *Lin. Alg. Appl.*, Vol. 428, pp. 781–795, 2008. (Cited on p. 890.)

[192] K. M. R. Audenaert, "On the Araki-Lieb-Thirring Inequality," *Int. J. Inform. Sys. Sci.*, Vol. 4, pp. 78–83, 2008. (Cited on p. 767.)

[193] K. M. R. Audenaert, "Spectral Radius of Hadamard Product Versus Conventional Product for Nonnegative Matrices," *Lin. Alg. Appl.*, Vol. 432, pp. 366–368, 2010. (Cited on p. 699.)

[194] K. M. R. Audenaert, "Trace Inequalities for Completely Monotone Functions and Bernstein Functions," *Lin. Alg. Appl.*, Vol. 437, pp. 601–611, 2012. (Cited on pp. 770 and 890.)

[195] J. S. Aujla, "Some Norm Inequalities for Completely Monotone Functions," *SIAM J. Matrix Anal. Appl.*, Vol. 22, pp. 569–573, 2000. (Cited on p. 806.)

[196] J. S. Aujla and J.-C. Bourin, "Eigenvalue Inequalities for Convex and Log-Convex Functions," *Lin. Alg. Appl.*, Vol. 424, pp. 25–35, 2007. (Cited on pp. 739, 813, 827, 870, and 876.)

[197] J. S. Aujla and F. C. Silva, "Weak Majorization Inequalities and Convex Functions," *Lin. Alg. Appl.*, Vol. 369, pp. 217–233, 2003. (Cited on p. 806.)

[198] J. S. Aujla and H. L. Vasudeva, "Inequalities Involving Hadamard Product and Operator Means," *Math. Japonica*, Vol. 42, pp. 265–272, 1995. (Cited on pp. 822, 828, and 830.)

[199] B. Aupetit and J. Zemanek, "A Characterization of Normal Matrices by Their Exponentials," *Lin. Alg. Appl.*, Vol. 132, pp. 119–121, 1990. (Cited on p. 1220.)

[200] M. Avriel, *Nonlinear Programming: Analysis and Methods*. Englewood Cliffs: Prentice-Hall, 1976, reprinted by Dover, Mineola, 2003. (Cited on p. 798.)

[201] O. Axelsson, *Iterative Solution Methods*. Cambridge: Cambridge University Press, 1994. (Cited on p. xxi.)

[202] C. Axler and T. Lessmann, "On the First k-Ramanujan Prime," *Amer. Math. Monthly*, Vol. 124, pp. 642–646, 2017. (Cited on pp. 35 and 966.)

[203] F. Ayres, *Schaum's Outline of Theory and Problems of Plane and Spherical Trigonometry*. New York: McGraw-Hill, 1954. (Cited on p. 236.)

[204] L. E. Azar, "On Some Extensions of Hardy-Hilbert's Inequality and Applications," *J. Ineq. Appl.*, pp. 1–14, 2008, article ID 546829. (Cited on p. 220.)

[205] A. Aziz and N. A. Rather, "Location of the Zeros of Trinomials and Quadrinomials," *Math. Ineq. Appl.*, Vol. 17, pp. 823–829, 2014. (Cited on pp. 1237 and 1238.)

[206] J. C. Baez, "Symplectic, Quaternionic, Fermionic," http://math.ucr.edu/home /baez/symplectic.html. (Cited on pp. 434 and 438.)

[207] J. C. Baez, "The Octonions," *Bull. Amer. Math. Soc.*, Vol. 39, pp. 145–205, 2001. (Cited on p. 438.)

[208] R. Bagby, "A Simple Proof that $\Gamma'(1) = -\gamma$," *Amer. Math. Monthly*, Vol. 117, pp. 83–85, 2010. (Cited on pp. 1000 and 1170.)

[209] O. Bagdasar, *Inequalities and Applications*. Cluj-Napoca: Babes-Bolyai University, 2006, Bachelor's Degree Thesis, http://rgmia.vu.edu.au/mono-graphs/index.html. (Cited on pp. 117, 142, 143, 165, and 208.)

[210] O. Bagdasar, "Inequalities for Chains of Normalized Symmetric Sums," *J. Ineq. Pure Appl. Math.*, Vol. 9, no. 1, pp. 1–7, 2008, Article 24. (Cited on p. 199.)

[211] Z. Bai and G. H. Golub, "Bounds for the Trace of the Inverse and the Determinant of Symmetric Positive Definite Matrices," *Ann. Numer. Math.*, Vol. 4, pp. 29–38, 1997. (Cited on p. 762.)

[212] Z.-J. Bai and Z.-Z. Bai, "On Nonsingularity of Block Two-by-two Matrices," *Lin. Alg. Appl.*, Vol. 439, pp. 2388–2404, 2013. (Cited on pp. 340 and 661.)

[213] D. H. Bailey, J. M. Borwein, P. B. Borwein, and S. Plouffe, "The Quest for Pi," *Math. Intelligencer*, Vol. 19, no. 1, pp. 50–57, 1997. (Cited on p. 1049.)

[214] D. W. Bailey and D. E. Crabtree, "Bounds for Determinants," *Lin. Alg. Appl.*, Vol. 2, pp. 303–309, 1969. (Cited on p. 535.)

[215] H. Bailey and Y. A. Rubinstein, "A Variety of Triangle Inequalities," *Coll. Math. J.*, Vol. 31, pp. 350–355, 2000. (Cited on pp. 446 and 452.)

[216] R. Baillie, "Fun with Fourier Series," http://arxiv.org/abs/0806.0150. (Cited on p. 1060.)

[217] J. Bak and D. J. Newman, *Complex Analysis*, 3rd ed. New York: Springer, 2010. (Cited on pp. 94, 276, 1070, 1081, and 1083.)

[218] A. Baker, *Matrix Groups: An Introduction to Lie Group Theory*. New York: Springer, 2001. (Cited on pp. 382, 428, 438, 440, 607, 1193, and 1212.)

[219] M. Bakke, B. Wu, and B. Suceava, "An Inequality," *Amer. Math. Monthly*, Vol. 121, p. 745, 2014. (Cited on pp. 200 and 201.)

[220] J. K. Baksalary and O. M. Baksalary, "Idempotency of Linear Combinations of Two Idempotent Matrices," *Lin. Alg. Appl.*, Vol. 321, pp. 3–7, 2000. (Cited on p. 415.)

[221] J. K. Baksalary and O. M. Baksalary, "Nonsingularity of Linear Combinations of Idempotent Matrices," *Lin. Alg. Appl.*, Vol. 388, pp. 25–29, 2004. (Cited on pp. 400, 402, and 406.)

[222] J. K. Baksalary and O. M. Baksalary, "Particular Formulae for the Moore-Penrose Inverse of a Columnwise Partitioned Matrix," *Lin. Alg. Appl.*, Vol. 421, pp. 16–23, 2007. (Cited on p. 667.)

[223] J. K. Baksalary, O. M. Baksalary, and X. Liu, "Further Relationships between Certain Partial Orders of Matrices and Their Squares," *Lin. Alg. Appl.*, Vol. 375, pp. 171–180, 2003. (Cited on p. 431.)

[224] J. K. Baksalary, O. M. Baksalary, and X. Liu, "Further Properties of Generalized and Hypergeneralized Projectors," *Lin. Alg. Appl.*, Vol. 389, pp. 295–303, 2004. (Cited on p. 674.)

[225] J. K. Baksalary, O. M. Baksalary, X. Liu, and G. Trenkler, "Further Results on Generalized and Hypergeneralized Projectors," *Lin. Alg. Appl.*, Vol. 429, pp. 1038–1050, 2008. (Cited on pp. 673 and 674.)

[226] J. K. Baksalary, O. M. Baksalary, and T. Szulc, "Properties of Schur Complements in Partitioned Idempotent Matrices," *Lin. Alg. Appl.*, Vol. 397, pp. 303–318, 2004. (Cited on p. 662.)

[227] J. K. Baksalary, O. M. Baksalary, and G. Trenkler, "A Revisitation of Formulae for the Moore-Penrose Inverse of Modified Matrices," *Lin. Alg. Appl.*, Vol. 372, pp. 207–224, 2003. (Cited on p. 635.)

[228] J. K. Baksalary, K. Nordstrom, and G. P. H. Styan, "Löwner-Ordering Antitonicity of Generalized Inverses of Hermitian Matrices," *Lin. Alg. Appl.*, Vol. 127, pp. 171–182, 1990. (Cited on pp. 568, 738, and 817.)

[229] J. K. Baksalary and F. Pukelsheim, "On the Löwner, Minus, and Star Partial Orderings of Nonnegative Definite Matrices," *Lin. Alg. Appl.*, Vol. 151, pp. 135–141, 1990. (Cited on pp. 740, 814, and 816.)

[230] J. K. Baksalary, F. Pukelsheim, and G. P. H. Styan, "Some Properties of Matrix Partial Orderings," *Lin. Alg. Appl.*, Vol. 119, pp. 57–85, 1989. (Cited on pp. 431, 817, and 825.)

[231] J. K. Baksalary, P. Semrl, and G. P. H. Styan, "A Note on Rank Additivity and Range Additivity," *Lin. Alg. Appl.*, Vol. 237/238, pp. 489–498, 1996. (Cited on p. 320.)

[232] J. K. Baksalary and G. P. H. Styan, "Generalized Inverses of Bordered Matrices," *Lin. Alg. Appl.*, Vol. 354, pp. 41–47, 2002. (Cited on p. 667.)

[233] O. M. Baksalary, "Revisitation of Generalized and Hypergeneralized Projectors," in *Statistical Inference, Econometric Analysis and Matrix Algebra*, B. Schipp and W. Kramer, Eds. Heidelberg: Springer, 2009, pp. 317–324. (Cited on pp. 641, 642, 673, and 674.)

[234] O. M. Baksalary, D. S. Bernstein, and G. Trenkler, "On the Equality between Rank and Trace of an Idempotent Matrix," *App. Math. Comput.*, Vol. 217, pp. 4076–4080, 2010. (Cited on pp. 408, 410, 411, 573, 647, 650, 651, and 652.)

[235] O. M. Baksalary and P. Kik, "On Commutativity of Projectors," *Lin. Alg. Appl.*, Vol. 417, pp. 31–41, 2006. (Cited on pp. 411, 638, and 654.)

[236] O. M. Baksalary, G. P. H. Styan, and G. Trenkler, "On a Matrix Decomposition of Hartwig and Spindelböck," *Lin. Alg. Appl.*, Vol. 430, pp. 2798–2812, 2009. (Cited on p. 573.)

[237] O. M. Baksalary and G. Trenkler, "Rank of a Generalized Projector," *IMAGE*, Vol. 39, pp. 25–27, 2007. (Cited on p. 673.)

[238] O. M. Baksalary and G. Trenkler, "Rank of a Nonnegative Definite Matrix," *IMAGE*, Vol. 39, pp. 27–28, 2007. (Cited on p. 815.)

[239] O. M. Baksalary and G. Trenkler, "An Inequality Involving a Product of Two Orthogonal Projectors," *IMAGE*, Vol. 41, p. 44, 2008. (Cited on p. 772.)

[240] O. M. Baksalary and G. Trenkler, "Characterizations of EP, Normal, and Hermitian Matrices," *Lin. Multilin. Alg.*, Vol. 56, pp. 299–304, 2008. (Cited on pp. 379, 380, 573, 630, 644, 650, and 672.)

[241] O. M. Baksalary and G. Trenkler, "An Inequality with Positive-Definite Matrices," *Coll. Math. J.*, Vol. 40, p. 296, 2009. (Cited on p. 796.)

[242] O. M. Baksalary and G. Trenkler, "Column Space Equalities for Orthogonal Projectors," *Appl. Math. Comp.*, Vol. 212, pp. 519–529, 2009. (Cited on pp. 411, 413, and 657.)

[243] O. M. Baksalary and G. Trenkler, "Eigenvalues of Functions of Orthogonal Projectors," *Lin. Alg. Appl.*, Vol. 431, pp. 2172–2186, 2009. (Cited on p. 590.)

[244] O. M. Baksalary and G. Trenkler, "On Angles and Distances between Subspaces," *Lin. Alg. Appl.*, Vol. 431, pp. 2243–2260, 2009. (Cited on pp. 412 and 657.)

[245] O. M. Baksalary and G. Trenkler, "Revisitation of the Product of Two Orthogonal Projectors," *Lin. Alg. Appl.*, Vol. 430, pp. 2813–2833, 2009. (Cited on pp. 589, 591, 652, 654, and 677.)

[246] O. M. Baksalary and G. Trenkler, "Capturing Eigenvalues in an Interval," *Amer. Math. Monthly*, Vol. 117, p. 746, 2010. (Cited on pp. 590 and 736.)

[247] O. M. Baksalary and G. Trenkler, "Core Inverse of Matrices," *Lin. Multilin. Alg.*, Vol. 58, pp. 681–697, 2010. (Cited on pp. 573, 644, 648, and 652.)

[248] O. M. Baksalary and G. Trenkler, "Functions of Orthogonal Projectors Involving the Moore-Penrose Inverse," *Computers Math. Appl.*, Vol. 59, pp. 764–778, 2010. (Cited on pp. 402, 415, and 656.)

[249] O. M. Baksalary and G. Trenkler, "On a Subspace Metric Based on Matrix Rank," *Lin. Alg. Appl.*, Vol. 432, pp. 1475–1491, 2010. (Cited on pp. 314, 595, and 653.)

[250] O. M. Baksalary and G. Trenkler, "On Disjoint Range Matrices," *Lin. Alg. Appl.*, Vol. 435, pp. 1222–1240, 2011. (Cited on pp. 376, 644, and 647.)

[251] O. M. Baksalary and G. Trenkler, "On the Projectors FF^+ and F^+F," *Appl. Math. Comp.*, Vol. 217, pp. 10 213–10 223, 2011. (Cited on pp. 415, 590, 644, and 647.)

[252] O. M. Baksalary and G. Trenkler, "Rank Formulae from the Perspective of Orthogonal Projectors," *Lin. Multilin. Alg.*, Vol. 59, pp. 607–625, 2011. (Cited on pp. 320, 410, 411, 415, 633, and 654.)

[253] O. M. Baksalary and G. Trenkler, "A Simple Metric for Finite Dimensional Vector Spaces," *Acta Univ. Apulensis*, Vol. 31, pp. 327–330, 2012. (Cited on p. 314.)

[254] O. M. Baksalary and G. Trenkler, "Bounding the Antidiagonal of an Orthogonal Projector," *Math. Mag.*, Vol. 86, p. 292, 2013. (Cited on p. 408.)

[255] O. M. Baksalary and G. Trenkler, "On a Pair of Vector Spaces," *Appl. Math. Comp.*, Vol. 219, pp. 9572–9580, 2013. (Cited on pp. 313, 314, and 409.)

[256] O. M. Baksalary and G. Trenkler, "On Column and Null Spaces of Functions of a Pair of Oblique Projectors," *Lin. Multilin. Alg.*, Vol. 61, pp. 1116–1129, 2013. (Cited on pp. 404, 405, and 413.)

[257] O. M. Baksalary and G. Trenkler, "On K-Potent Matrices," *Elec. J. Lin. Alg.*, Vol. 26, pp. 446–470, 2013. (Cited on pp. 317, 318, 319, 320, 377, 397, 407, 418, 573, 631, 642, 643, and 673.)

[258] O. M. Baksalary and G. Trenkler, "Range-Hermitianness of Certain Functions of Projectors," *IMAGE*, Vol. 50, p. 44, 2013. (Cited on pp. 413 and 651.)

[259] O. M. Baksalary and G. Trenkler, "On a Craig-Sakamoto Theorem for Orthogonal Projectors," in *Empirical Economic and Financial Research*, Y. Feng and H. Hebbel, Eds. Berlin: Springer, 2015, pp. 479–484. (Cited on pp. 331 and 412.)

[260] O. M. Baksalary and G. Trenkler, "On the Entries of Orthogonal Projection Matrices," in *Combinatorial Matrix Theory and Generalized Inverses of Matrices*, R. B. Bapat, S. J. Kirkland, K. M. Prasad, and S. Puntanen, Eds. New Delhi: Springer, 2015, pp. 101–118. (Cited on p. 733.)

[261] O. M. Baksalary and G. Trenkler, "Further Characterizations of Functions of a Pair of Orthogonal Projectors," 2017. (Cited on pp. 411, 412, and 654.)

[262] K. Ball, E. Carlen, and E. Lieb, "Sharp Uniform Convexity and Smoothness Inequalities for Trace Norms," *Invent. Math.*, Vol. 115, pp. 463–482, 1994. (Cited on pp. 859 and 879.)

[263] C. S. Ballantine, "Products of Positive Semidefinite Matrices," *Pac. J. Math.*, Vol. 23, pp. 427–433, 1967. (Cited on pp. 608 and 831.)

[264] C. S. Ballantine, "Products of Positive Definite Matrices II," *Pac. J. Math.*, Vol. 24, pp. 7–17, 1968. (Cited on p. 831.)

[265] C. S. Ballantine, "Products of Positive Definite Matrices III," *J. Algebra*, Vol. 10, pp. 174–182, 1968. (Cited on p. 831.)

[266] C. S. Ballantine, "A Note on the Matrix Equation $H = AP + PA^*$," *Lin. Alg. Appl.*, Vol. 2, pp. 37–47, 1969. (Cited on pp. 571 and 1311.)

[267] C. S. Ballantine, "Products of Positive Definite Matrices IV," *Lin. Alg. Appl.*, Vol. 3, pp. 79–114, 1970. (Cited on p. 831.)

[268] C. S. Ballantine, "Products of EP Matrices," *Lin. Alg. Appl.*, Vol. 12, pp. 257–267, 1975. (Cited on p. 377.)

[269] C. S. Ballantine, "Some Involutory Similarities," *Lin. Multilin. Alg.*, Vol. 3, pp. 19–23, 1975. (Cited on p. 609.)

[270] C. S. Ballantine, "Products of Involutory Matrices I," *Lin. Multilin. Alg.*, Vol. 5, pp. 53–62, 1977. (Cited on p. 609.)

[271] C. S. Ballantine, "Products of Idempotent Matrices," *Lin. Alg. Appl.*, Vol. 19, pp. 81–86, 1978. (Cited on p. 609.)

[272] C. S. Ballantine and C. R. Johnson, "Accretive Matrix Products," *Lin. Multilin. Alg.*, Vol. 3, pp. 169–185, 1975. (Cited on p. 1228.)

[273] A. Banerjee, "Polynomials Satisfied by Square Matrices: A Converse to the Cayley-Hamilton Theorem," *Resonance*, Vol. 7, pp. 47–58, 2002. (Cited on p. 528.)

[274] S. Banerjee, "Revisiting Spherical Trigonometry with Orthogonal Projectors," *Coll. Math. J.*, Vol. 35, pp. 375–381, 2004. (Cited on p. 497.)

[275] J. Bang-Jensen and G. Gutin, *Digraphs: Theory, Algorithms and Applications*, 2nd ed. New York: Springer, 2009. (Cited on p. xxi.)

[276] J. Bang-Jensen and G. Z. Gutin, *Digraphs, Theory, Algorithms and Applications*, 2nd ed. London: Springer, 2010. (Cited on p. 26.)

[277] W. Bani-Domi and F. Kittaneh, "Norm Equalities and Inequalities for Operator Matrices," *Lin. Alg. Appl.*, Vol. 429, pp. 57–67, 2008. (Cited on pp. 887, 888, and 889.)

[278] R. Bapat and A. Dey, "An Identity Involving the Inverse of LCL^{T}," *IMAGE*, Vol. 55, pp. 35–36, 2015. (Cited on p. 723.)

[279] R. B. Bapat, "Two Inequalities for the Perron Root," *Lin. Alg. Appl.*, Vol. 85, pp. 241–248, 1987. (Cited on pp. 225 and 544.)

[280] R. B. Bapat, "On Generalized Inverses of Banded Matrices," *Elec. J. Lin. Alg.*, Vol. 16, pp. 284–290, 2007. (Cited on p. 423.)

[281] R. B. Bapat, *Graphs and Matrices*. London: Springer, 2010. (Cited on pp. 329, 426, 628, and 910.)

[282] R. B. Bapat and M. K. Kwong, "A Generalization of $A \odot A^{-1} \geq I$," *Lin. Alg. Appl.*, Vol. 93, pp. 107–112, 1987. (Cited on p. 821.)

[283] R. B. Bapat and T. E. S. Raghavan, *Nonnegative Matrices and Applications*. Cambridge: Cambridge University Press, 1997. (Cited on p. 539.)

[284] R. B. Bapat and B. Zheng, "Generalized Inverses of Bordered Matrices," *Elec. J. Lin. Alg.*, Vol. 10, pp. 16–30, 2003. (Cited on pp. 320 and 667.)

[285] I. Y. Bar-Itzhack, D. Hershkowitz, and L. Rodman, "Pointing in Real Euclidean Space," *J. Guid. Contr. Dyn.*, Vol. 20, pp. 916–922, 1997. (Cited on pp. 396 and 1206.)

[286] R. Barbara, "A Triangle Formula," *Math. Gaz.*, Vol. 90, pp. 298–299, 2006. (Cited on p. 472.)

[287] R. Barbara, "A Useful Inequality," *Crux Math.*, Vol. 34, pp. 38–43, 2008. (Cited on pp. 164, 167, 169, 173, and 489.)

[288] E. J. Barbeau, *Polynomials*. New York: Springer, 1989. (Cited on pp. 133, 148, 149, 177, 179, 184, 466, 1234, and 1235.)

[289] E. J. Barbeau, M. S. Klamkin, and W. O. J. Moser, *Five Hundred Mathematical Challenges*. Washington, DC: The Mathematical Association of America, 1995. (Cited on pp. 29, 31, 32, 36, 41, 43, 50, 56, 119, 124, 131, 134, 146, 148, 149, 173, 178, 180, 183, 480, and 492.)

[290] C. Barboianu, "Where Are the Zeros?" *Amer. Math. Monthly*, Vol. 120, p. 181, 2013. (Cited on p. 1224.)

[291] E. R. Barnes and A. J. Hoffman, "On Bounds for Eigenvalues of Real Symmetric Matrices," *Lin. Alg. Appl.*, Vol. 40, pp. 217–223, 1981. (Cited on p. 536.)

[292] J. K. R. Barnett, "A Systematization of Some Trigonometric Identities," *Math. Gaz.*, Vol. 87, pp. 539–545, 2006. (Cited on pp. 456 and 491.)

[293] J. K. R. Barnett, "Generalizing a 'Trigonometric' Identity," *Math. Gaz.*, Vol. 90, pp. 110–112, 2006. (Cited on p. 149.)

[294] S. Barnett, "A Note on the Bezoutian Matrix," *SIAM J. Appl. Math.*, Vol. 22, pp. 84–86, 1972. (Cited on p. 520.)

[295] S. Barnett, "Congenial Matrices," *Lin. Alg. Appl.*, Vol. 41, pp. 277–298, 1981. (Cited on pp. 611 and 614.)

[296] S. Barnett, "Inversion of Partitioned Matrices with Patterned Blocks," *Int. J. Sys. Sci.*, Vol. 14, pp. 235–237, 1983. (Cited on p. 420.)

[297] S. Barnett, *Polynomials and Linear Control Systems*. New York: Marcel Dekker, 1983. (Cited on pp. 611 and 1313.)

[298] S. Barnett, *Matrices in Control Theory*. Malabar: Krieger, 1984. (Cited on p. xxi.)

[299] S. Barnett, "Leverrier's Algorithm: A New Proof and Extensions," *SIAM J. Matrix Anal. Appl.*, Vol. 10, pp. 551–556, 1989. (Cited on p. 544.)

[300] S. Barnett, *Matrices: Methods and Applications*. Oxford: Clarendon, 1990. (Cited on pp. 350, 610, 611, 612, 632, 669, 734, and 911.)

[301] S. Barnett and P. Lancaster, "Some Properties of the Bezoutian for Polynomial Matrices," *Lin. Multilin. Alg.*, Vol. 9, pp. 99–110, 1980. (Cited on p. 520.)

[302] S. Barnett and C. Storey, *Matrix Methods in Stability Theory*. New York: Barnes and Noble, 1970. (Cited on pp. xxi, 337, 619, 730, and 1229.)

[303] W. Barrett, "A Theorem on Inverses of Tridiagonal Matrices," *Lin. Alg. Appl.*, Vol. 27, pp. 211–217, 1979. (Cited on p. 424.)

[304] W. Barrett, "Hermitian and Positive Definite Matrices," in *Handbook of Linear Algebra*, L. Hogben, Ed. Boca Raton: Chapman & Hall/CRC, 2007, pp. 8–1–8–12. (Cited on p. 822.)

[305] W. Barrett, C. R. Johnson, and P. Tarazaga, "The Real Positive Definite Completion Problem for a Simple Cycle," *Lin. Alg. Appl.*, Vol. 192, pp. 3–31, 1993. (Cited on p. 729.)

[306] J. Barria and P. R. Halmos, "Vector Bases for Two Commuting Matrices," *Lin. Multilin. Alg.*, Vol. 27, pp. 147–157, 1990. (Cited on p. 577.)

[307] H. Bart, I. Gohberg, M. A. Kaashoek, and A. C. M. Ran, "Schur Complements and State Space Realizations," *Lin. Alg. Appl.*, Vol. 399, pp. 203–224, 2005. (Cited on p. 1315.)

[308] H. Baruh, *Analytical Dynamics*. Boston: McGraw-Hill, 1999. (Cited on pp. 393, 438, and 1208.)

[309] A. Barvinok, *A Course in Convexity*. Providence: American Mathematical Society, 2002. (Cited on pp. 178, 179, 180, 307, 310, 788, 789, 803, 937, 938, and 939.)

[310] S. Barza, J. Pecaric, and L.-E. Persson, "Carlson Type Inequalities," *J. Ineq. Appl.*, Vol. 2, pp. 121–135, 1998. (Cited on p. 213.)

[311] E. L. Basor, S. N. Evans, and K. E. Morrison, "Hidden Telescope," *Amer. Math. Monthly*, Vol. 118, p. 943, 2011. (Cited on p. 1060.)

[312] M. Bataille, "Conditions for rank(A) = rank(B)," *Coll. Math. J.*, Vol. 42, pp. 237–238, 2011. (Cited on p. 398.)

[313] M. Bataille, "Extrema," *Amer. Math. Monthly*, Vol. 118, pp. 278–279, 2011. (Cited on p. 489.)

[314] M. Bataille, "A Binomial Coefficient Identity," *Coll. Math. J.*, Vol. 43, pp. 260, 261, 2012. (Cited on p. 92.)

[315] M. Bataille, "A Limit with the Harmonic Numbers," *Coll. Math. J.*, Vol. 44, pp. 235–237, 2013. (Cited on p. 48.)

[316] M. Bataille, "Largest Constant for an Equality of a Triangle," *Coll. Math. J.*, Vol. 44, p. 326, 2013. (Cited on p. 452.)

[317] D. M. Batinetu-Giurgiu, M. Basarab, and N. Stanciu, "Correction to 1018," *Coll. Math. J.*, Vol. 45, p. 222, 2014. (Cited on p. 480.)

[318] D. M. Batinetu-Giurgiu and O. T. Pop, "A Generalization of Radon's Inequality," *Creative Math. Inf.*, Vol. 19, pp. 116–121, 2010. (Cited on pp. 177 and 218.)

[319] D. M. Batinetu-Giurgiu and N. Stanciu, "Inequalities with the Since and Cosine Functions," *Coll. Math. J.*, Vol. 44, pp. 438–439, 2013. (Cited on p. 484.)

[320] D. M. Batinetu-Giurgiu and N. Stanciu, "Nesbitt Type Inequality with Fibonacci Numbers," *Fibonacci Quart.*, Vol. 51, pp. 278–279, 2013. (Cited on p. 183.)

[321] D. M. Batinetu-Giurgiu and N. Stanciu, "Triangle Inequalities and the Tangent Function," *Coll. Math. J.*, Vol. 45, pp. 397–398, 2014. (Cited on p. 466.)

[322] D. M. Batinetu-Giurgiu, N. Stanciu, M. Bataille, and O. Kouba, "An Inequality with the Angle Bisectors of a Triangle," *Coll. Math. J.*, Vol. 45, p. 225, 2014. (Cited on p. 480.)

[323] N. Batir, "Integral Representations of Some Series Involving $\binom{2k}{k}^{-1} k^{-n}$ and Some Related Series," *Appl. Math. Comp.*, Vol. 218, pp. 645–667, 2004. (Cited on pp. 1014, 1032, 1048, and 1050.)

[324] N. Batir, "On the Series $\sum_{k=1}^{\infty} \binom{3k}{k}^{-1} k^{-n} x^k$," *Proc. Indian Acad. Sci.*, Vol. 115, pp. 371–381, 2005. (Cited on pp. 1003, 1070, and 1071.)

[325] N. Batir, "Very Accurate Approximations for the Factorial Function," *J. Math. Ineq.*, Vol. 4, pp. 335–344, 2010. (Cited on p. 51.)

[326] N. Batir, "On Certain Series Involving Reciprocals of Binomial Coefficients," *J. Classical Anal.*, Vol. 2, pp. 1–8, 2013. (Cited on p. 1071.)

[327] N. Batir and M. Cancan, "Sharp Inequalities Involving the Constant e and the Sequence $(1 + 1/n)^n$," *Int. J. Math. Educ. Sci. Tech.*, Vol. 40, pp. 1101–1109, 2009. (Cited on pp. 128 and 216.)

[328] P. Batra, "A Family of Necessary Stability Inequalities via Quadratic Forms," *Math. Ineq. Appl.*, Vol. 14, pp. 313–321, 2011. (Cited on p. 1224.)

[329] F. L. Bauer, J. Stoer, and C. Witzgall, "Absolute and Monotonic Norms," *Numer. Math.*, Vol. 3, pp. 257–264, 1961. (Cited on p. 912.)

[330] D. S. Bayard, "An Optimization Result with Application to Optimal Spacecraft Reaction Wheel Orientation Design," in *Proc. Amer. Contr. Conf.*, Arlington, VA, June 2001, pp. 1473–1478. (Cited on p. 593.)

[331] M. Bayat and H. Teimoori, "A Generalization of Mason's Theorem for Four Polynomials," *Elem. Math.*, Vol. 59, pp. 23–28, 2004. (Cited on p. 523.)

[332] J. Bayless and D. Klyve, "Reciprocal Sums as a Knowledge Metric: Theory, Computation, and Perfect Numbers," *Amer. Math. Monthly*, Vol. 120, pp. 822–831, 2013. (Cited on p. 966.)

[333] M. S. Bazaraa, H. D. Sherali, and C. M. Shetty, *Nonlinear Programming*, 2nd ed. New York: Wiley, 1993. (Cited on pp. 543, 922, and 924.)

[334] A. F. Beardon, "Sums of Powers of Integers," *Amer. Math. Monthly*, Vol. 103, pp. 201–213, 1996. (Cited on pp. 38 and 56.)

[335] R. A. Beauregard and V. A. Dobrushkin, "Finite Sums of the Alcuin Numbers," *Math. Mag.*, Vol. 86, pp. 280–287, 2013. (Cited on pp. 41 and 1008.)

[336] N. Bebiano and J. da Providencia, "On the Determinant of Certain Strictly Dissipative Matrices," *Lin. Alg. Appl.*, Vol. 83, pp. 117–128, 1986. (Cited on p. 777.)

[337] N. Bebiano, J. da Providencia, and R. Lemos, "Matrix Inequalities in Statistical Mechanics," *Lin. Alg. Appl.*, Vol. 376, pp. 265–273, 2004. (Cited on pp. xxi, 1215, 1216, and 1217.)

[338] N. Bebiano, R. Lemos, and J. da Providencia, "Inequalities for Quantum Relative Entropy," *Lin. Alg. Appl.*, Vol. 401, pp. 159–172, 2005. (Cited on pp. xxi and 1215.)

[339] N. Bebiano and M. E. Miranda, "On a Recent Determinantal Inequality," *Lin. Alg. Appl.*, Vol. 201, pp. 99–102, 1994. (Cited on p. 594.)

[340] E. F. Beckenbach and R. Bellman, *Inequalities*. Berlin: Springer, 1965. (Cited on pp. 276 and 834.)

[341] R. I. Becker, "Necessary and Sufficient Conditions for the Simultaneous Diagonability of Two Quadratic Forms," *Lin. Alg. Appl.*, Vol. 30, pp. 129–139, 1980. (Cited on pp. 619 and 797.)

[342] W. Beckner, "Inequalities in Fourier Analysis," *Ann. Math.*, Vol. 102, p. 159182, 1975. (Cited on p. 879.)

[343] D. Beckwith, "Reciprocal Catalan Sums," *Amer. Math. Monthly*, Vol. 123, pp. 405–406, 2016. (Cited on pp. 1008 and 1102.)

[344] G. Beer, *Topologies on Closed and Closed Convex Sets*. Dordrecht: Kluwer, 1993. (Cited on p. 940.)

[345] L. W. Beineke and R. J. Wilson, Eds., *Topics in Algebraic Graph Theory*. Cambridge: Cambridge University Press, 2005. (Cited on p. xxi.)

[346] T. N. Bekjan, "On Joint Convexity of Trace Functions," *Lin. Alg. Appl.*, Vol. 390, pp. 321–327, 2004. (Cited on p. 790.)

[347] P. A. Bekker, "The Positive Semidefiniteness of Partitioned Matrices," *Lin. Alg. Appl.*, Vol. 111, pp. 261–278, 1988. (Cited on pp. 751, 772, and 831.)

[348] H. Belbachir, M. Rahmani, and I. M. Gessel, "On Gessel-Kaneko's Identity for Bernoulli Numbers," *Applic. Anal. Disc. Math.*, Vol. 7, pp. 1–10, 2013. (Cited on pp. 979 and 980.)

[349] H. Belbachir, M. Rahmani, and B. Sury, "Alternating Sums of the Reciprocals of Binomial Coefficients," *J. Integer Sequences*, Vol. 15, pp. 1–16, 2012, article 12.2.8. (Cited on pp. 112 and 1009.)

[350] J. G. Belinfante, B. Kolman, and H. A. Smith, "An Introduction to Lie Groups and Lie Algebras with Applications," *SIAM Rev.*, Vol. 8, pp. 11–46, 1966. (Cited on p. 1247.)

[351] G. R. Belitskii and Y. I. Lyubich, *Matrix Norms and Their Applications*. Basel: Birkhauser, 1988. (Cited on p. 912.)

[352] W. W. Bell, *Special Functions for Scientists and Engineers*. Mineola: Dover, 2004. (Cited on pp. 986, 989, 990, 992, 1000, 1113, 1162, and 1163.)

[353] R. Bellman, *Introduction to Matrix Analysis*, 2nd ed. New York: McGraw-Hill, 1960, reprinted by SIAM, Philadelphia, 1997. (Cited on pp. 350, 734, 1212, and 1239.)

[354] R. Bellman, "Some Inequalities for the Square Root of a Positive Definite Matrix," *Lin. Alg. Appl.*, Vol. 1, pp. 321–324, 1968. (Cited on p. 741.)

[355] E.-V. Belmega, S. Lasaulce, and M. Debbah, "A Trace Inequality for Positive Definite Matrices," *J. Ineq. Pure Appl. Math.*, Vol. 10, pp. 1–4, 2009, article 5. (Cited on pp. 181 and 773.)

[356] I. Ben-Ari and K. Conrad, "Maclaurin's Inequality and a Generalized Bernoulli Inequality," *Math. Mag.*, Vol. 87, pp. 14–24, 2014. (Cited on pp. 124 and 195.)

[357] I. Ben-Ari, D. Hay, and A. Roitershtein, "On Wallis-type Products and Polya's Urn Schemes," *Amer. Math. Monthly*, Vol. 121, pp. 422–432, 2014. (Cited on pp. 245, 1084, 1085, and 1086.)

[358] A. Ben-Israel, "A Note on Partitioned Matrices and Equations," *SIAM Rev.*, Vol. 11, pp. 247–250, 1969. (Cited on p. 667.)

[359] A. Ben-Israel, "The Moore of the Moore-Penrose Inverse," *Elect. J. Lin. Alg.*, Vol. 9, pp. 150–157, 2002. (Cited on p. 679.)

[360] A. Ben-Israel and T. N. E. Greville, *Generalized Inverses: Theory and Applications*, 2nd ed. New York: Springer, 2003. (Cited on pp. 628, 631, 642, 667, 670, 674, 679, and 799.)

[361] B. Ben Taher and M. Rachidi, "Some Explicit Formulas for the Polynomial Decomposition of the Matrix Exponential and Applications," *Lin. Alg. Appl.*, Vol. 350, pp. 171–184, 2002. (Cited on p. 1206.)

[362] A. Ben-Tal and A. Nemirovski, *Lectures on Modern Convex Optimization*. Philadelphia: SIAM, 2001. (Cited on pp. 308 and 362.)

[363] M. Bencze, "Identities and Inequalities for Determinants," *Octogon*, Vol. 13, pp. 955–966, 2005. (Cited on p. 529.)

[364] M. Bencze, "Inequalities for the Angles of a Triangle," *Octogon Math. Mag.*, Vol. 15, pp. 267–273, 2005. (Cited on p. 466.)

[365] M. Bencze, "Joszef Wildt International Mathematical Competition," *Octogon*, Vol. 13, pp. 413–417, 2005. (Cited on pp. 271 and 529.)

[366] M. Bencze, "About One Identity," *Octogon*, Vol. 14, pp. 138–150, 2006. (Cited on pp. 41, 58, and 103.)

[367] M. Bencze, "About One Trigonometrical Identity," *Octogon*, Vol. 14, pp. 250–253, 2006. (Cited on pp. 237, 452, and 466.)

[368] M. Bencze, "New Cauchy-Type Inequalities," *Octogon*, Vol. 16, pp. 137–140, 2006. (Cited on p. 223.)

[369] M. Bencze, "New Inequalities for the Number *e*," *Octogon*, Vol. 14, pp. 476–480, 2006. (Cited on p. 128.)

[370] M. Bencze, "Some New Inequalities for the Number "e''," *Octogon*, Vol. 14, pp. 559–562, 2006. (Cited on pp. 49 and 103.)

[371] M. Bencze, "About a Chain of Inequalities," *Octogon Math. Mag.*, Vol. 15, pp. 185–189, 2007. (Cited on p. 173.)

[372] M. Bencze, "About Some Identities and Inequalities for Complex Numbers, Vectors and Determinants," *Octogon*, Vol. 15, pp. 790–794, 2007. (Cited on p. 529.)

[373] M. Bencze, "Some New Trigonometric Inequalities," *Octogon*, Vol. 15, pp. 818–822, 2007. (Cited on pp. 452 and 466.)

[374] M. Bencze, "A New Inequality for the Function $\ln x$ (1)," *Octogon*, Vol. 16, pp. 965–980, 2008. (Cited on pp. 143 and 228.)

[375] M. Bencze, "A New Inequality for the Function $\ln x$ and Its Applications (2)," *Octogon*, Vol. 16, pp. 981–983, 2008. (Cited on pp. 143 and 228.)

[376] M. Bencze, "About One Inequality and Its Corollaries," *Octogon*, Vol. 16, pp. 303–305, 2008. (Cited on p. 144.)

[377] M. Bencze, "About a Partition Inequality," *Octogon*, Vol. 17, pp. 272–274, 2009. (Cited on pp. 137, 157, 180, and 183.)

[378] M. Bencze, "About Dumitru Acu's Inequality," *Octogon*, Vol. 17, pp. 257–264, 2009. (Cited on p. 125.)

[379] M. Bencze, "About Some Elementary Inequalities," *Octogon*, Vol. 17, pp. 707–713, 2009. (Cited on pp. 125, 133, 134, 214, and 250.)

[380] M. Bencze, "New Identities and Inequalities in Triangle," *Octogon*, Vol. 17, pp. 653–659, 2009. (Cited on p. 149.)

[381] M. Bencze, "The Integral Method in Inequalities Theory," *Octogon*, Vol. 17, pp. 19–52, 2009. (Cited on p. 177.)

[382] M. Bencze, "An Inequality with a Finite Sum of Products," *Coll. Math. J.*, Vol. 41, p. 412, 2010. (Cited on p. 211.)

[383] M. Bencze, "Inequalities: One Out of Two," *Amer. Math. Monthly*, Vol. 118, p. 851, 2011. (Cited on p. 194.)

[384] M. Bencze, "Square Root Equality," *Crux Math.*, Vol. 37, p. 13, 2011. (Cited on p. 45.)

[385] M. Bencze, "Problem 2004," *Math. Mag.*, Vol. 89, p. 294, 2016. (Cited on p. 68.)

[386] M. Bencze, "Bounds for a Sum," *Amer. Math. Monthly*, Vol. 124, p. 376, 2017. (Cited on p. 229.)

[387] M. Bencze, "Problem 1999," *Math. Mag.*, Vol. 90, p. 303, 2017. (Cited on p. 1073.)

[388] M. Bencze and D. M. Batinetu-Giurgiu, "About Integrals," *Octogon*, Vol. 14, pp. 227–232, 2006. (Cited on p. 1131.)

[389] M. Bencze and Z. Changjian, "About New Trigonometric Identities," *Octogon*, Vol. 14, pp. 197–203, 2006. (Cited on p. 252.)

[390] M. Bencze and Z. Changjian, "About Marcel Chirita's Inequalities," *Octogon Math. Mag.*, Vol. 15, pp. 189–191, 2007. (Cited on p. 225.)

[391] M. Bencze and Z. Changjian, "New Trigonometric Identities and Inequalities," *Octogon*, Vol. 15, pp. 557–566, 2007. (Cited on pp. 237 and 252.)

[392] M. Bencze and Z. Changjian, "About Euler's Inequality," *Octogon*, Vol. 16, pp. 269–270, 2008. (Cited on p. 452.)

[393] M. Bencze and Z. Changjian, "A Refinement of Jensen's Inequality," *Octogon*, Vol. 17, pp. 209–214, 2009. (Cited on pp. 125 and 138.)

[394] M. Bencze and J. L. Diaz-Barrero, "About Sums," *Octogon*, Vol. 16, pp. 280–283, 2008. (Cited on pp. 40 and 189.)

[395] M. Bencze and M. Dinca, "Refinements for Euler's $R \geq 2r$ Inequalities," *Octogon*, Vol. 13, pp. 263–307, 2005. (Cited on pp. 452 and 480.)

[396] M. Bencze, N. Minculette, and O. T. Pop, "Certain Inequalities for the Triangle," *Gazeta Matematica Serie A*, Vol. 30, pp. 86–93, 2012. (Cited on pp. 472 and 480.)

[397] M. Bencze and O. T. Pop, "Generalizations and Refinements for Nesbitt's Inequality," *J. Math. Ineq.*, Vol. 5, pp. 13–20, 2011. (Cited on pp. 165, 200, and 452.)

[398] M. Bencze and J. Sandor, "About Some Identities," *Octogon*, Vol. 15, pp. 265–266, 2007. (Cited on pp. 46 and 120.)

[399] M. Bencze and J. Sandor, "About Some Special Identities," *Octogon*, Vol. 15, pp. 250–253, 2007. (Cited on pp. 48, 100, 121, and 132.)

[400] M. Bencze and L. Smarandache, "A Property of Polynomials," *Octogon*, Vol. 17, pp. 740–741, 2009. (Cited on p. 1238.)

[401] M. Bencze and S.-H. Wu, "Ratio-Type Inequalities for Bisectors, Medians, Altitudes, and Sides of a Triangle," *Crux Math.*, Vol. 36, pp. 304–308, 2010. (Cited on p. 480.)

[402] J. Benitez, "A New Decomposition for Square Matrices," *Elec. J. Lin. Alg.*, Vol. 20, pp. 207–225, 2010. (Cited on pp. 575, 669, and 678.)

[403] J. Benitez and X. Liu, "On the Continuity of the Group Inverse," *Oper. Matrices*, Vol. 6, pp. 859–868, 2012. (Cited on pp. 632 and 670.)

[404] J. Benitez and X. Liu, "Expressions for Generalized Inverses of Square Matrices," *Lin. Multilin. Alg.*, Vol. 61, pp. 1536–1554, 2013. (Cited on p. 575.)

[405] J. Benitez, X. Liu, and J. Zhong, "Some Results on Matrix Partial Orderings and Reverse Order Law," *Elec. J. Lin. Alg.*, Vol. 20, pp. 254–273, 2010. (Cited on pp. 440 and 649.)

[406] J. Benitez and V. Rakocevic, "Applications of CS Decomposition in Linear Combinations of Two Orthogonal Projectors," *Appl. Math. Comp.*, Vol. 203, pp. 761–769, 2008. (Cited on pp. 413, 414, and 655.)

[407] A. T. Benjamin and N. T. Cameron, "Counting on Determinants," *Amer. Math. Monthly*, Vol. 112, pp. 481–492, 2005. (Cited on p. 425.)

[408] A. T. Benjamin, B. Chen, and K. Kindred, "Sums of Evenly Spaced Binomial Coefficients," *Math. Mag.*, Vol. 83, no. 3, pp. 370–373, 2010. (Cited on pp. 76 and 1070.)

[409] A. T. Benjamin and F. Juhnke, "Another Way of Counting N^N," *SIAM J. Disc. Math.*, Vol. 5, pp. 377–379, 1992. (Cited on p. 94.)

[410] A. T. Benjamin and M. E. Orrison, "Two Quick Combinatorial Proofs of $\sum_{k=1}^{n} = \binom{n+1}{2}^2$," *Coll. Math. J.*, Vol. 33, no. 5, pp. 406–408, 2002. (Cited on p. 37.)

[411] A. T. Benjamin and J. J. Quinn, *Proofs That Really Count: The Art of Combinatorial Proof.* Washington, DC: Mathematical Association of America, 2003. (Cited on pp. 23, 32, 37, 48, 50, 58, 67, 68, 75, 86, 92, 100, 103, 106, 107, 108, 111, 114, 115, 1009, and 1079.)

[412] A. T. Benjamin and J. N. Scott, "Third and Fourth Binomial Coefficients," *Fibonacci Quart.*, Vol. 49, pp. 99–101, 2011. (Cited on p. 77.)

[413] G. Bennett, "Meaningful Inequalities," *J. Math. Ineq.*, Vol. 1, pp. 449–471, 2007. (Cited on pp. 47, 51, and 141.)

[414] G. Bennett, "p-free ℓ^p Inequalities," *Amer. Math. Monthly*, Vol. 117, pp. 334–351, 2010. (Cited on pp. 186 and 1223.)

[415] G. Bercu, "The Natural Approach of Trigonometric Inequalities—Padé Approximant," *J. Math. Ineq.*, Vol. 11, pp. 181–191, 2017. (Cited on p. 249.)

[416] A. Berele and S. Catoiu, "Rationalizing Denominators," *Math. Math.*, Vol. 88, pp. 121–136, 2015. (Cited on p. 53.)

[417] L. Berg, "Three Results in Connection with Inverse Matrices," *Lin. Alg. Appl.*, Vol. 84, pp. 63–77, 1986. (Cited on p. 330.)

[418] C. Berge, *The Theory of Graphs.* London: Methuen, 1962, reprinted by Dover, Mineola, 2001. (Cited on p. xxi.)

[419] V. Berinde, *Exploring, Investigating and Discovering in Mathematics.* Basel: Birkhauser, 2004. (Cited on p. 138.)

[420] V. Berinde, "On a Homogeneous Inequality Given at the 2004 J.B.M.O." *Gazeta Matematica Seria B*, Vol. 115, pp. 340–343, 2010. (Cited on p. 135.)

[421] L. D. Berkovitz, *Convexity and Optimization in $\mathbb{R}^n$.* New York: Wiley, 2002. (Cited on pp. 362 and 940.)

[422] A. Berman, M. Neumann, and R. J. Stern, *Nonnegative Matrices in Dynamic Systems.* New York: Wiley, 1989. (Cited on pp. 539, 1232, and 1233.)

[423] A. Berman and R. J. Plemmons, *Nonnegative Matrices in the Mathematical Sciences.* New York: Academic Press, 1979, reprinted by SIAM, Philadelphia, 1994. (Cited on pp. 375, 440, 542, 544, and 1232.)

[424] A. Berman, R. S. Varga, and R. C. Ward, "ALPS: Matrices with Nonpositive Off-Diagonal Entries," *Lin. Alg. Appl.*, Vol. 21, pp. 233–244, 1978. (Cited on p. 1232.)

[425] B. C. Berndt and B. P. Yeap, "Explicit evaluations and reciprocity theorems for finite trigonometric sums," *Adv. Appl. Math.*, Vol. 29, pp. 358–385, 2002. (Cited on p. 242.)

[426] D. S. Bernstein, "Inequalities for the Trace of Matrix Exponentials," *SIAM J. Matrix Anal. Appl.*, Vol. 9, pp. 156–158, 1988. (Cited on pp. 1218 and 1220.)

[427] D. S. Bernstein, "Some Open Problems in Matrix Theory Arising in Linear Systems and Control," *Lin. Alg. Appl.*, Vol. 162–164, pp. 409–432, 1992. (Cited on p. 1228.)

[428] D. S. Bernstein, *Matrix Mathematics: Theory, Fact, and Formulas with Application to Linear Systems Theory.* Princeton: Princeton University Press, 2005. (Not cited.)

[429] D. S. Bernstein, *Matrix Mathematics: Theory, Fact, and Formulas*, 2nd ed. Princeton: Princeton University Press, 2009. (Not cited.)

[430] D. S. Bernstein, K. Aljanaideh, and A. E. Frazho, "Laurent Series and ℓ^p Sequences," *Amer. Math. Monthly*, Vol. 123, p. 398, 2016. (Cited on p. 948.)

[431] D. S. Bernstein and S. P. Bhat, "Lyapunov Stability, Semistability, and Asymptotic Stability of Matrix Second-Order Systems," *ASME Trans. J. Vibr. Acoustics*, Vol. 117, pp. 145–153, 1995. (Cited on pp. 597, 1232, and 1247.)

[432] D. S. Bernstein and W. M. Haddad, "LQG Control with an H_∞ Performance Bound: A Riccati Equation Approach," *IEEE Trans. Autom. Contr.*, Vol. 34, pp. 293–305, 1989. (Cited on p. 1319.)

[433] D. S. Bernstein, W. M. Haddad, D. C. Hyland, and F. Tyan, "Maximum Entropy-Type Lyapunov Functions for Robust Stability and Performance Analysis," *Sys. Contr. Lett.*, Vol. 21, pp. 73–87, 1993. (Cited on p. 604.)

[434] D. S. Bernstein and D. C. Hyland, "Compartmental Modeling and Second-Moment Analysis of State Space Systems," *SIAM J. Matrix Anal. Appl.*, Vol. 14, pp. 880–901, 1993. (Cited on pp. 440, 1232, and 1233.)

[435] D. S. Bernstein and M. Lin, "A Norm Identity," *IMAGE*, Vol. 52, pp. 42–43, 2014. (Cited on p. 858.)

[436] D. S. Bernstein and W. So, "Some Explicit Formulas for the Matrix Exponential," *IEEE Trans. Autom. Contr.*, Vol. 38, pp. 1228–1232, 1993. (Cited on p. 1206.)

[437] D. S. Bernstein and C. F. Van Loan, "Rational Matrix Functions and Rank-1 Updates," *SIAM J. Matrix Anal. Appl.*, Vol. 22, pp. 145–154, 2000. (Cited on p. 327.)

[438] H. J. Bernstein, "An Inequality for Selfadjoint Operators on a Hilbert Space," *Proc. Amer. Math. Soc.*, Vol. 100, pp. 319–321, 1987. (Cited on p. 892.)

[439] A. Besenyei, "Trace Inequality for Positive Block Matrices," *IMAGE*, Vol. 50, p. 44, 2013. (Cited on p. 751.)

[440] A. Besenyei and D. Petz, "Partial Subadditivity of Entropies," *Lin. Alg. Appl.*, Vol. 439, pp. 3297–3305, 2013. (Cited on p. 220.)

[441] M. G. Beumer, "Some Special Integrals," *Amer. Math. Monthly*, Vol. 68, pp. 645–647, 1961. (Cited on pp. 1152 and 1154.)

[442] W. A. Beyer, J. D. Louck, and D. Zeilberger, "Math Bite: A Generalization of a Curiosity that Feynman Remembered All His Life," *Math. Mag.*, Vol. 69, pp. 43–44, 1996. (Cited on p. 245.)

[443] K. V. Bhagwat and R. Subramanian, "Inequalities between Means of Positive Operators," *Math. Proc. Camb. Phil. Soc.*, Vol. 83, pp. 393–401, 1978. (Cited on pp. 716, 741, 742, and 1213.)

[444] S. P. Bhat and D. S. Bernstein, "Nontangency-Based Lyapunov Tests for Convergence and Stability in Systems Having a Continuum of Equilibria," *SIAM J. Contr. Optim.*, Vol. 42, pp. 1745–1775, 2003. (Cited on pp. 1247 and 1312.)

[445] S. P. Bhat and D. S. Bernstein, "Average-Preserving Symmetries and Energy Equipartition in Linear Hamiltonian Systems," *Math. Contr. Sig. Sys.*, Vol. 21, pp. 127–146, 2009. (Cited on p. 439.)

[446] S. P. Bhat and N. Crasta, "Closed Rotation Sequences," *Disc.Comput. Geom.*, Vol. 53, pp. 366–396, 2015. (Cited on p. 1207.)

[447] R. Bhatia, *Perturbation Bounds for Matrix Eigenvalues*. Essex: Longman Scientific and Technical, 1987. (Cited on p. 893.)

[448] R. Bhatia, "Perturbation Inequalities for the Absolute Value Map in Norm Ideas of Operators," *J. Operator Theory*, Vol. 19, pp. 129–136, 1988. (Cited on p. 871.)

[449] R. Bhatia, *Matrix Analysis*. New York: Springer, 1997. (Cited on pp. 361, 362, 384, 583, 584, 716, 718, 719, 721, 722, 766, 767, 802, 805, 812, 831, 847, 853, 866, 867, 871, 872, 874, 875, 876, 881, 883, 885, 886, 893, 898, 1215, 1221, and 1223.)

[450] R. Bhatia, "Linear Algebra to Quantum Cohomology: The Story of Alfred Horn's Inequalities," *Amer. Math. Monthly*, Vol. 108, pp. 289–318, 2001. (Cited on pp. 803, 805, and 831.)

[451] R. Bhatia, "Infinitely Divisible Matrices," *Amer. Math. Monthly*, Vol. 113, pp. 221–235, 2006. (Cited on pp. 427, 726, 727, 728, and 820.)

[452] R. Bhatia, "Mean Matrices and Infinite Divisibility," *Lin. Alg. Appl.*, Vol. 424, pp. 36–54, 2007. (Cited on p. 726.)

[453] R. Bhatia, *Perturbation Bounds for Matrix Eigenvalues*. Philadelphia: SIAM, 2007. (Cited on pp. 537, 809, 862, 893, and 894.)

[454] R. Bhatia, *Positive Definite Matrices*. Princeton: Princeton University Press, 2007. (Cited on pp. 328, 721, 725, 726, 727, 728, 729, 737, 744, 903, 1215, and 1217.)

[455] R. Bhatia, "Min Matrices and Mean Matrices," *Math. Intelligencer*, Vol. 33, pp. 22–28, 2011. (Cited on p. 726.)

[456] R. Bhatia, "Trace Inequalities for Products of Positive Definite Matrices," *J. Math. Physics*, Vol. 55, pp. 013 509–1–3, 2014. (Cited on pp. 767 and 873.)

[457] R. Bhatia and C. Davis, "More Matrix Forms of the Arithmetic-Geometric Mean Inequality," *SIAM J. Matrix Anal. Appl.*, Vol. 14, pp. 132–136, 1993. (Cited on p. 882.)

[458] R. Bhatia and C. Davis, "A Cauchy-Schwarz Inequality for Operators with Applications," *Lin. Alg. Appl.*, Vol. 223/224, pp. 119–129, 1995. (Cited on pp. 875 and 882.)

[459] R. Bhatia and D. Drissi, "Generalized Lyapunov Equations and Positive Definite Functions," *SIAM J. Matrix Anal. Appl.*, Vol. 27, pp. 103–114, 2005. (Cited on p. 728.)

[460] R. Bhatia and L. Elsner, "Higher Order Logarithmic Derivatives of Matrices in the Spectral Norm," *SIAM J. Matrix Anal. Appl.*, Vol. 25, pp. 662–668, 2003. (Cited on p. 1219.)

[461] R. Bhatia, L. Elsner, and G. Krause, "Bounds for the Variation of the Roots of a Polynomial and the Eigenvalues of a Matrix," *Lin. Alg. Appl.*, Vol. 142, pp. 195–209, 1990. (Cited on pp. 893, 941, and 942.)

[462] R. Bhatia and P. Grover, "Norm Inequalities Related to the Matrix Geometric Mean," *Lin. Alg. Appl.*, Vol. 437, pp. 726–733, 2012. (Cited on p. 742.)

[463] R. Bhatia and J. Holbrook, "Frechet Derivatives of the Power Function," *Indiana Univ. Math. J.*, Vol. 49, pp. 1155–1173, 2000. (Cited on pp. 126 and 807.)

[464] R. Bhatia and J. Holbrook, "Riemannian Geometry and Matrix Geometric Means," *Lin. Alg. Appl.*, Vol. 413, pp. 594–618, 2006. (Cited on p. 1217.)

[465] R. Bhatia and T. Jain, "Higher Order Derivatives and Perturbation Bounds," *Lin. Alg. Appl.*, Vol. 431, pp. 2102–2108, 2009. (Cited on p. 952.)

[466] R. Bhatia and F. Kittaneh, "On Some Perturbation Inequalities for Operators," *Lin. Alg. Appl.*, Vol. 106, pp. 271–279, 1988. (Cited on p. 871.)

[467] R. Bhatia and F. Kittaneh, "Norm Inequalities for Partitioned Operators and an Application," *Math. Ann.*, Vol. 287, pp. 719–726, 1990. (Cited on p. 887.)

[468] R. Bhatia and F. Kittaneh, "On the Singular Values of a Product of Operators," *SIAM J. Matrix Anal. Appl.*, Vol. 11, pp. 272–277, 1990. (Cited on pp. 739, 882, and 905.)

[469] R. Bhatia and F. Kittaneh, "Norm Inequalities for Positive Operators," *Lett. Math. Phys.*, Vol. 43, pp. 225–231, 1998. (Cited on pp. 870 and 872.)

[470] R. Bhatia and F. Kittaneh, "Cartesian Decompositions and Schatten Norms," *Lin. Alg. Appl.*, Vol. 318, pp. 109–116, 2000. (Cited on p. 879.)

[471] R. Bhatia and F. Kittaneh, "Notes on Matrix Arithmetic-Geometric Mean Inequalities," *Lin. Alg. Appl.*, Vol. 308, pp. 203–211, 2000. (Cited on pp. 809, 874, and 876.)

[472] R. Bhatia and F. Kittaneh, "Clarkson Inequalities with Several Operators," *Bull. London Math. Soc.*, Vol. 36, pp. 820–832, 2004. (Cited on p. 876.)

[473] R. Bhatia and F. Kittaneh, "Commutators, Pinchings, and Spectral Variation," *Oper. Matrices*, Vol. 2, pp. 143–151, 2008. (Cited on p. 885.)

[474] R. Bhatia and F. Kittaneh, "The Matrix Arithmetic-Geometric Mean Inequality Revisited," *Lin. Alg. Appl.*, Vol. 428, pp. 2177–2191, 2008. (Cited on pp. 760 and 872.)

[475] R. Bhatia and F. Kittaneh, "The Singular Values of $A + B$ and $A + iB$," *Lin. Alg. Appl.*, Vol. 431, pp. 1502–1508, 2009. (Cited on p. 880.)

[476] R. Bhatia, Y. Lim, and T. Yamazaki, "Some Norm Inequalities for Matrix Means," *Lin. Alg. Appl.*, Vol. 501, pp. 112–122, 2016. (Cited on p. 771.)

[477] R. Bhatia and K. R. Parthasarathy, "Positive Definite Functions and Operator Inequalities," *Bull. London Math. Soc.*, Vol. 32, pp. 214–228, 2000. (Cited on pp. 727, 882, 883, and 1223.)

[478] R. Bhatia and P. Rosenthal, "How and Why to Solve the Operator Equation $AX - XB = Y$," *Bull. London Math. Soc.*, Vol. 29, pp. 1–21, 1997. (Cited on pp. 578 and 579.)

[479] R. Bhatia and P. Semrl, "Orthogonality of Matrices and Some Distance Problems," *Lin. Alg. Appl.*, Vol. 287, pp. 77–85, 1999. (Cited on p. 853.)

[480] R. Bhatia and X. Zhan, "Norm Inequalities for Operators with Positive Real Part," *J. Operator Theory*, Vol. 50, pp. 67–76, 2003. (Cited on pp. 777, 879, and 880.)

[481] R. Bhattacharya and K. Mukherjea, "On Unitary Similarity of Matrices," *Lin. Alg. Appl.*, Vol. 126, pp. 95–105, 1989. (Cited on p. 576.)

[482] S. P. Bhattacharyya, H. Chapellat, and L. Keel, *Robust Control: The Parametric Approach*. Englewood Cliffs: Prentice-Hall, 1995. (Cited on p. 1224.)

[483] K. Bibak and M. H. S. Haghighi, "Some Trigonometric Identities Involving Fibonacci and Lucas Numbers," *J. Integer Sequences*, Vol. 12, pp. 1–5, 2009, article 09.8.4. (Cited on pp. 100 and 103.)

[484] M. R. Bicknell, "The Lambda Number of a Matrix: The Sum of Its n^2 Cofactors," *Amer. Math. Monthly*, Vol. 72, pp. 260–264, 1965. (Cited on p. 346.)

[485] M. Bidkham, A. Zireh, and H. A. S. Mezerji, "Bound for the Zeros of Polynomials," *J. Classical Anal.*, Vol. 3, pp. 149–155, 2013. (Cited on p. 1235.)

[486] N. Biggs, *Algebraic Graph Theory*, 2nd ed. Cambridge: Cambridge University Press, 2000. (Cited on pp. xxi and 596.)

[487] P. Binding, B. Najman, and Q. Ye, "A Variational Principle for Eigenvalues of Pencils of Hermitian Matrices," *Integ. Equ. Oper. Theory*, Vol. 35, pp. 398–422, 1999. (Cited on p. 619.)

[488] K. Binmore, *Fun and Games: A Text on Game Theory*. Lexington: D. C. Heath and Co., 1992. (Cited on p. xxi.)

[489] A. Bjorck, *Numerical Methods for Least Squares Problems*. Philadelphia: SIAM, 1996. (Cited on pp. 679 and 911.)

[490] S. Blanes and F. Casas, "On the Convergence and Optimization of the Baker-Campbell-Hausdorff Formula," *Lin. Alg. Appl.*, Vol. 378, pp. 135–158, 2004. (Cited on pp. 1189 and 1213.)

[491] S. Blanes, F. Casas, J. A. Oteo, and J. Ros, "Magnus and Fer Expansions for Matrix Differential Equations: The Convergence Problem," *J. Phys. A: Math. Gen.*, Vol. 31, pp. 259–268, 1998. (Cited on p. 1306.)

[492] W. D. Blizard, "Multiset Theory," *Notre Dame J. Formal Logic*, Vol. 30, pp. 36–66, 1989. (Cited on p. 118.)

[493] E. D. Bloch, *Proofs and Fundamentals: A First Course in Abstract Mathematics*. Boston: Birkhauser, 2000. (Cited on pp. 22 and 118.)

[494] V. Blondel and J. N. Tsitsiklis, "When Is a Pair of Matrices Mortal?" *Inform. Proc. Lett.*, Vol. 63, pp. 283–286, 1997. (Cited on p. 1242.)

[495] V. Blondel and J. N. Tsitsiklis, "The Boundedness of All Products of a Pair of Matrices Is Undecidable," *Sys. Contr. Lett.*, Vol. 41, pp. 135–140, 2000. (Cited on p. 1242.)

[496] L. M. Blumenthal, *Theory and Applications of Distance Geometry*. Oxford: Oxford University Press, 1953. (Cited on p. 495.)

[497] W. J. Blundon, "Inequalities Associated with the Triangle," *Can. Math. Bull.*, Vol. 8, pp. 615–626, 1965. (Cited on p. 452.)

[498] H. P. Boas, "Mocposite Functions," *Amer. Math. Monthly*, Vol. 123, pp. 427–438, 2016. (Cited on p. 118.)

[499] W. Boehm, "An Operator Limit," *SIAM Rev.*, Vol. 36, p. 659, 1994. (Cited on p. 1241.)

[500] F. G. Boese and W. J. Luther, "A Note on a Classical Bound for the Moduli of All Zeros of a Polynomial," *IEEE Trans. Autom. Contr.*, Vol. AC-34, pp. 998–1001, 1989. (Cited on p. 1235.)

[501] B. Bogosel and C. Lupu, "Carlson's Inequality," *Amer. Math. Monthly*, Vol. 122, p. 77, 2015. (Cited on p. 214.)

[502] A. W. Bojanczyk and A. Lutoborski, "Computation of the Euler Angles of a Symmetric 3×3 Matrix," *SIAM J. Matrix Anal. Appl.*, Vol. 12, pp. 41–48, 1991. (Cited on p. 532.)

[503] B. Bollobas, *Modern Graph Theory*. New York: Springer, 1998. (Cited on p. xxi.)

[504] M. Bona, *A Walk Through Combinatorics: An Introduction to Enumeration and Graph Theory*, 2nd ed. New Jersey: World Scientific, 2006. (Cited on pp. 68, 86, and 94.)

[505] M. Bona, *Introduction to Enumerative and Analytic Combinatorics*, 2nd ed. Boca Raton: CRC Press, 2016. (Cited on p. 976.)

[506] D. D. Bonar and M. J. Khoury, *Real Infinite Series*. Washington, DC: The Mathematical Association of America, 2006. (Cited on pp. 45, 68, 197, 217, 963, 965, 966, 1022, 1023, 1024, 1025, 1027, 1044, 1046, 1047, 1056, 1062, 1070, 1074, and 1097.)

[507] J. V. Bondar, "Comments on and Complements to *Inequalities: Theory of Majorization and Its Applications*," *Lin. Alg. Appl.*, Vol. 199, pp. 115–129, 1994. (Cited on pp. 216 and 489.)

[508] A. Bonfuglioli and R. Fulci, *Topics in Noncommutative Algebra. The Theorem of Campbell, Baker, Hausdorff and Dynkin*. Springer, 2012. (Cited on pp. 1189 and 1213.)

[509] G. Boros and V. Moll, "An Integral Hidden in Gradshteyn and Ryzhik," *J. Comp. Appl. Math.*, Vol. 106, pp. 361–368, 1999. (Cited on p. 1110.)

[510] G. Boros and V. H. Moll, "An Integral with Three Parameters," *SIAM Rev.*, Vol. 40, pp. 972–980, 1998. (Cited on pp. 1110 and 1149.)

[511] G. Boros and V. H. Moll, *Irresistible Integrals*. Cambridge: Cambridge University Press, 2004. (Cited on pp. 67, 68, 75, 86, 92, 968, 979, 980, 1008, 1024, 1032, 1036, 1070, 1099, 1104, 1109, 1115, 1130, 1133, 1134, 1141, 1148, 1149, 1152, 1159, 1160, 1170, and 1174.)

[512] G. Boros and V. H. Moll, "Sums of Arctangents and Some Formulae of Ramanujan," *Scientia*, Vol. 11, pp. 13–24, 2005. (Cited on pp. 239, 262, 1047, 1061, 1062, and 1063.)

[513] D. Borwein, J. M. Borwein, and I. E. Leonard, "L_p Norms and the Sinc Function," *Amer. Math. Monthly*, Vol. 117, pp. 528–538, 2010. (Cited on p. 1119.)

[514] D. Borwein, J. M. Borwein, and J. Wan, "Integrals Arising from a Plane Random Walk," *Amer. Math. Monthly*, Vol. 121, pp. 553–555, 2014. (Cited on p. 1141.)

[515] J. M. Borwein, "Hilbert's Inequality and Witten's Zeta-Function," *Amer. Math. Monthly*, Vol. 115, pp. 125–137, 2008. (Cited on pp. 219 and 1024.)

[516] J. M. Borwein and D. H. Bailey, *Mathematics by Experiment: Plausible Reasoning in the 21st Century*, 2nd ed. Natick: A K Peters, 2008. (Cited on pp. 51, 54, 242, 963, 1000, 1015, 1017, 1049, 1060, 1070, 1128, 1141, and 1154.)

[517] J. M. Borwein, D. H. Bailey, and R. Girgensohn, *Experimentation in Mathematics: Computational Paths to Discovery*. Natick: A K Peters, 2004. (Cited on pp. 50, 68, 94, 117, 210, 495, 540, 1070, 1087, and 1097.)

[518] J. M. Borwein and P. B. Borwein, "The Arithmetic-Geometric Mean and Fast Computation of Elementary Functions," *SIAM Rev.*, Vol. 26, pp. 351–366, 1984. (Cited on pp. 968 and 969.)

[519] J. M. Borwein and P. B. Borwein, *Pi and the AGM*. New York: Wiley, 1987. (Cited on p. 968.)

[520] J. M. Borwein and P. B. Borwein, "On the Complexity of Familiar Functions and Numbers," *SIAM Rev.*, Vol. 30, pp. 589–601, 1988. (Cited on p. 1070.)

[521] J. M. Borwein and D. Bradley, "Empirically Determined Apery-Like Formulae for $\zeta(4n+3)$," *Exp. Math.*, Vol. 6, pp. 181–194, 1997. (Cited on p. 1070.)

[522] J. M. Borwein and S. T. Chapman, "I Prefer Pi: A Brief History and Anthology of Articles in the American Mathematical Monthly," *Amer. Math. Monthly*, Vol. 122, pp. 195–216, 2015. (Cited on pp. 1087 and 1094.)

[523] J. M. Borwein and A. S. Lewis, *Convex Analysis and Nonlinear Optimization*. New York: Springer, 2000. (Cited on pp. 228, 229, 231, 362, 543, 592, 593, 721, 741, 787, 935, 938, and 939.)

[524] P. Borwein, S. Choi, B. Rooney, and A. Weirathmueller, *The Riemann Hypothesis: A Resource for the Afficionado and Virtuoso Alike*. New York: Springer, 2007. (Cited on p. 36.)

[525] P. Borwein, S. Choi, B. Rooney, and A. Weirathmueller, *The Riemann Hypothesis*. New York: Springer, 2008. (Cited on pp. 996 and 997.)

[526] P. Borwein and T. Erdelyi, *Polynomials and Polynomial Inequalities*. New York: Springer, 1995. (Cited on p. 239.)

[527] A. J. Bosch, "The Factorization of a Square Matrix into Two Symmetric Matrices," *Amer. Math. Monthly*, Vol. 93, pp. 462–464, 1986. (Cited on pp. 608 and 619.)

[528] A. J. Bosch, "Note on the Factorization of a Square Matrix into Two Hermitian or Symmetric Matrices," *SIAM Rev.*, Vol. 29, pp. 463–468, 1987. (Cited on pp. 608 and 619.)

[529] A. Bostan and T. Combot, "A Binomial-like Matrix Equation," *Amer. Math. Monthly*, Vol. 119, pp. 593–597, 2012. (Cited on p. 529.)

[530] A. Bostan and P. Dumas, "Wronskians and Linear Independence," *Amer. Math. Monthly*, Vol. 117, pp. 722–727, 2010. (Cited on p. 953.)

[531] A. Bottcher and I. M. Spitkovsky, "A Gentle Guide to the Basics of Two Projections Theory," *Lin. Alg. Appl.*, Vol. 432, pp. 1412–1459, 2010. (Cited on pp. 595 and 737.)

[532] A. Bottcher and D. Wenzel, "How Big Can the Commutator of Two Matrices Be and How Big Is it Typically?" *Lin. Alg. Appl.*, Vol. 403, pp. 216–228, 2005. (Cited on p. 884.)

[533] A. Bottcher and D. Wenzel, "The Frobenius Norm and the Commutator," *Lin. Alg. Appl.*, Vol. 429, pp. 1864–1885, 2008. (Cited on p. 884.)

[534] O. Bottema, *Topics in Elementary Geometry*, 2nd ed. New York: Springer, 2008. (Cited on p. 446.)

[535] O. Bottema, R. Z. Djordjevic, R. R. Janic, D. S. Mitrinovic, and P. M. Vasic, *Geometric Inequalities*. Groningen: Wolters-Noordhoff, 1969. (Cited on pp. 452, 466, and 484.)

[536] T. L. Boullion and P. L. Odell, *Generalized Inverse Matrices*. New York: Wiley, 1971. (Cited on pp. 638, 666, and 679.)

[537] J.-C. Bourin, "Some Inequalities for Norms on Matrices and Operators," *Lin. Alg. Appl.*, Vol. 292, pp. 139–154, 1999. (Cited on pp. 768, 769, 865, and 900.)

[538] J.-C. Bourin, "Convexity or Concavity Inequalities for Hermitian Operators," *Math. Ineq. Appl.*, Vol. 7, pp. 607–620, 2004. (Cited on p. 771.)

[539] J.-C. Bourin, "Hermitian Operators and Convex Functions," *J. Ineq. Pure Appl. Math.*, Vol. 6, no. 5, pp. 1–6, 2005, Article 139. (Cited on pp. 771 and 772.)

[540] J.-C. Bourin, "Reverse Inequality to Araki's Inequality Comparison of $A^pZ^pZ^p$ and $(AZA)^p$," *Math. Ineq. Appl.*, Vol. 232, pp. 373–378, 2005. (Cited on p. 746.)

[541] J.-C. Bourin, "Matrix Versions of Some Classical Inequalities," *Lin. Alg. Appl.*, Vol. 407, pp. 890–907, 2006. (Cited on pp. 593 and 773.)

[542] J.-C. Bourin, "Reverse Rearrangement Inequalities via Matrix Technics," *J. Ineq. Pure Appl. Math.*, Vol. 7, no. 2, pp. 1–6, 2006, Article 43. (Cited on pp. 216 and 739.)

[543] J.-C. Bourin, "A Matrix Subadditivity Inequality for Symmetric Norms," *Proc. Amer. Math. Soc.*, Vol. 138, pp. 495–504, 2010. (Cited on p. 870.)

[544] J.-C. Bourin and F. Hiai, "Norm and Anti-norm Inequalities for Positive Semi-definite Matrices," *Int. J. Math.*, Vol. 22, pp. 1121–1138, 2011. (Cited on pp. 759 and 869.)

[545] J.-C. Bourin and F. Hiai, "Norm and Anti-norm Inequalities for Positive Semi-definite Matrices," *Int. J. Math.*, Vol. 22, pp. 1121–1138, 2011. (Cited on p. 876.)

[546] J.-C. Bourin and E.-Y. Lee, "Decomposition and Partial Trace of Positive Matrices with Hermitian Blocks," *Int. J. Math.*, Vol. 24, pp. 1 350 010–1–1 350 010–13, 2013. (Cited on pp. 759, 771, 779, and 869.)

[547] J.-C. Bourin and E.-Y. Lee, "Direct Sums of Positive Semi-definite Matrices," *Lin. Alg. Appl.*, Vol. 463, pp. 273–281, 2014. (Cited on p. 805.)

[548] J.-C. Bourin, E.-Y. Lee, and M. Lin, "On a Decomposition Lemma for Positive Semi-definite Block-matrices," *Lin. Alg. Appl.*, Vol. 437, pp. 1906–1912, 2012. (Cited on pp. 759 and 886.)

[549] J.-C. Bourin, E.-Y. Lee, and M. Lin, "Positive Matrices Partitioned into a Small Number of Hermitian Blocks," *Lin. Alg. Appl.*, Vol. 438, pp. 2591–2598, 2013. (Cited on pp. 759 and 888.)

[550] J.-C. Bourin and Y. Seo, "Reverse Inequality to Golden-Thompson Type Inequalities: Comparison of e^{A+B} and e^Ae^B," *Lin. Alg. Appl.*, Vol. 426, pp. 312–316, 2007. (Cited on p. 1216.)

[551] J.-C. Bourin and M. Uchiyama, "A Matrix Subadditivity Inequality for $f(A+B)$ and $f(A)+f(B)$," *Lin. Alg. Appl.*, Vol. 423, pp. 512–518, 2007. (Cited on p. 870.)

[552] K. Bourque and S. Ligh, "Matrices Associated with Classes of Arithmetical Functions," *J. Number Theory*, Vol. 45, pp. 367–376, 1993. (Cited on p. 727.)

[553] R. M. Bowen and C.-C. Wang, *Introduction to Vectors & Tensors*, 2nd ed. Mineola: Dover, 2008. (Cited on p. 691.)

[554] K. N. Boyadzhiev, "Harmonic Number Identities Via Euler's Transform," *J. Integer Sequences*, Vol. 12, pp. 1–8, 2009, article 09.6.1. (Cited on pp. 57, 68, and 75.)

[555] K. N. Boyadzhiev, "Close Encounters with the Stirling Numbers of the Second Kind," *Math. Mag.*, Vol. 85, pp. 252–266, 2012. (Cited on pp. 111, 975, 980, and 1009.)

[556] K. N. Boyadzhiev, "Series with Central Binomial Coefficients, Catalan Numbers, and Harmonic Numbers," *J. Integer Sequences*, Vol. 15, pp. 1–11, 2012, article 12.1.7. (Cited on pp. 75 and 1008.)

[557] S. Boyd, "Entropy and Random Feedback," in *Open Problems in Mathematical Systems and Control Theory*, V. D. Blondel, E. D. Sontag, M. Vidyasagar, and J. C. Willems, Eds. New York: Springer, 1998, pp. 71–74. (Cited on p. 866.)

[558] S. Boyd and L. Vandenberghe, *Convex Optimization*. Cambridge: Cambridge University Press, 2004. (Cited on pp. xxi, 362, and 756.)

[559] P. Bracken, "A Riemann (Zeta) Sum," *Amer. Math. Monthly*, Vol. 117, pp. 933–934, 2010. (Cited on p. 1036.)

[560] B. Bradie, "Two Closely Related Definite Integrals," *Coll. Math. J.*, Vol. 40, pp. 297–299, 2009. (Cited on p. 1098.)

[561] B. Bradie, "Yet More Ways to Skin a Definite Integral," *Math. Mag.*, Vol. 47, pp. 11–18, 2016. (Cited on pp. 1104 and 1144.)

[562] C. J. Bradley, "Cyclic Quadrilaterals," *Math. Gaz.*, Vol. 88, pp. 417–431, 2004. (Cited on p. 492.)

[563] S. Brassesco and M. A. Mendez, "The Asymptotic Expansion for $n!$ and the Lagrange Inversion Formula," *Ramanujan J.*, Vol. 24, pp. 219–234, 2011. (Cited on p. 971.)

[564] A. Bremner and J.-J. Delorme, "Equal Sums of Four Seventh Powers," *Math. Comp.*, Vol. 79, pp. 603–612, 2010. (Cited on p. 34.)

[565] M. Brennan, "A Note on the Converse to Lagrange's Theorem," *Math. Gaz.*, Vol. 82, pp. 286–288, 1998. (Cited on p. 436.)

[566] M. Brennan and D. Machale, "Variations on a Theme: $A4$ Definitely Has No Subgroup of Order Six!" *Math. Mag.*, Vol. 73, pp. 36–40, 2000. (Cited on p. 436.)

[567] J. L. Brenner, "A Bound for a Determinant with Dominant Main Diagonal," *Proc. Amer. Math. Soc.*, Vol. 5, pp. 631–634, 1954. (Cited on p. 535.)

[568] J. L. Brenner, "Expanded Matrices from Matrices with Complex Elements," *SIAM Rev.*, Vol. 3, pp. 165–166, 1961. (Cited on pp. 359 and 440.)

[569] J. L. Brenner and J. S. Lim, "The Matrix Equations $A = XYZ$ and $B = ZYX$ and Related Ones," *Canad. Math. Bull.*, Vol. 17, pp. 179–183, 1974. (Cited on p. 606.)

[570] D. M. Bressoud, *Proofs and Confirmations, The Story of the Alternating Sign Matrix Conjecture*. Cambridge: Cambridge University Press, 1999. (Cited on pp. 67, 121, 617, and 1070.)

[571] M. Bressoud, "Combinatorial Analysis," in *NIST Handbook of Mathematical Functions*, F. W. J. Olver, D. W. Lozier, R. F. Boisvert, and C. W. Clark, Eds. Cambridge: Cambridge University Press, 2010, pp. 617–636. (Cited on pp. 112, 114, 967, 976, and 1009.)

[572] J. W. Brewer, "Kronecker Products and Matrix Calculus in System Theory," *IEEE Trans. Circ. Sys.*, Vol. CAS-25, pp. 772–781, 1978, correction: CAS-26, p. 360, 1979. (Cited on p. 701.)

[573] L. Brickman, "On the Field of Values of a Matrix," *Proc. Amer. Math. Soc.*, Vol. 12, pp. 61–66, 1961. (Cited on p. 789.)

[574] R. Brockett, *Finite Dimensional Linear Systems*. New York: Wiley, 1970. (Citcd on p. 1319.)

[575] R. W. Brockett, "Using Feedback to Improve System Identification," in *Control of Uncertain Systems*, B. A. Francis, M. C. Smith, and J. C. Willems, Eds. Berlin: Springer, 2006, pp. 45–65. (Cited on pp. 803 and 1312.)

[576] R. W. Brockett, "Nonholonomic Trajectory Optimization and the Existence of Twisted Matrix Logarithms," in *Analysis and Design of Nonlinear Control Systems*, A. Astolfi and L. Marconi, Eds. Berlin: Springer, 2008, pp. 65–76. (Cited on p. 1211.)

[577] G. Brookfield, "The Coefficients of Cyclotomic Polynomials," *Math. Mag.*, Vol. 89, pp. 179–188, 2016. (Cited on p. 269.)

[578] B. P. Brooks, "The Coefficients of the Characteristic Polynomial in Terms of the Eigenvalues and the Elements of an $n \times n$ Matrix," *Appl. Math. Lett.*, Vol. 19, pp. 511–515, 2006. (Cited on p. 952.)

[579] H. Brothers, "Improving the Convergence of Newton's Series Approximation for e," *Coll. Math. J.*, Vol. 35, pp. 34–39, 2004. (Cited on p. 1023.)

[580] K. A. Broughan, "Characterizing the Sum of Two Cubes," *J. Integer Sequences*, Vol. 6, pp. 1–7, 2003, article 03.4.6. (Cited on p. 35.)

[581] G. Brown, "Convexity and Minkowski's Inequality," *Amer. Math. Monthly*, Vol. 112, pp. 740–742, 2005. (Cited on p. 224.)

[582] J. W. Brown and R. V. Churchill, *Complex Variables and Applications*, 8th ed. Boston: McGraw-Hill, 2009. (Cited on pp. 271, 276, 927, 928, and 929.)

[583] E. T. Browne, *Introduction to the Theory of Determinants and Matrices*. Chapel Hill: University of North Carolina Press, 1958. (Cited on p. 831.)

[584] R. Bru, J. J. Climent, and M. Neumann, "On the Index of Block Upper Triangular Matrices," *SIAM J. Matrix Anal. Appl.*, Vol. 16, pp. 436–447, 1995. (Cited on p. 604.)

[585] R. Bru and N. Thome, "Group Inverse and Group Involutory Matrices," *Lin. Multilin. Alg.*, Vol. 45, pp. 207–218, 1998. (Cited on p. 418.)

[586] R. A. Brualdi, *Combinatorial Matrix Classes*. Cambridge: Cambridge University Press, 2006. (Cited on p. xxi.)

[587] R. A. Brualdi, "Combinatorial Matrix Theory," in *Handbook of Linear Algebra*, L. Hogben, Ed. Boca Raton: Chapman & Hall/CRC, 2007, pp. 27–1–27–12. (Cited on p. 540.)

[588] R. A. Brualdi and D. Cvetkovic, *A Combinatorial Approach to Matrix Theory and Its Applications*. Boca Raton: CRC Press, 2009. (Cited on pp. xxi, 362, and 527.)

[589] R. A. Brualdi and J. Q. Massey, "Some Applications of Elementary Linear Algebra in Combinatorics," *Coll. Math. J.*, Vol. 24, pp. 10–19, 1993. (Cited on p. 332.)

[590] R. A. Brualdi and S. Mellendorf, "Regions in the Complex Plane Containing the Eigenvalues of a Matrix," *Amer. Math. Monthly*, Vol. 101, pp. 975–985, 1994. (Cited on pp. 535 and 1230.)

[591] R. A. Brualdi and H. J. Ryser, *Combinatorial Matrix Theory*. Cambridge: Cambridge University Press, 1991. (Cited on pp. xxi, 326, 332, and 792.)

[592] R. A. Brualdi and H. Schneider, "Determinantal Identities: Gauss, Schur, Cauchy, Sylvester, Kronecker, Jacobi, Binet, Laplace, Muir, and Cayley," *Lin. Alg. Appl.*, Vol. 52/53, pp. 769–791, 1983. (Cited on p. 330.)

[593] P. Bruckman, J. B. Dence, T. P. Dence, and J. Young, "Series of Triangular Numbers," *Coll. Math. J.*, Vol. 44, pp. 177–184, 2013. (Cited on p. 1044.)

[594] P. S. Bruckman, "A Proof of the Strong Goldbach Conjecture," *Int. J. Math. Educ. Sci. Tech.*, Vol. 39, pp. 1102–1109, 2008. (Cited on pp. 36 and 94.)

[595] P. S. Bruckman, "An "Inverse" Relation," *Fibonacci Quart.*, Vol. 49, pp. 369–370, 2011. (Cited on p. 993.)

[596] P. S. Bruckman, ""Snake Oiled" Identity!" *Fibonacci Quart.*, Vol. 50, pp. 275–276, 2012. (Cited on p. 75.)

[597] P. S. Bruckman, "The "Snake Oil Method" Again!" *Fibonacci Quart.*, Vol. 50, pp. 276–277, 2012. (Cited on p. 68.)

[598] P. S. Bruckman, "A Sum Yielding Pell Numbers," *Fibonacci Quart.*, Vol. 51, pp. 92–93, 2013. (Cited on p. 68.)

[599] P. S. Bruckman, "On a Power Series with Binomial Coefficients," *Fibonacci Quart.*, Vol. 51, pp. 91–92, 2015. (Cited on p. 86.)

[600] A. Brzezinski, E. Wu, and D. S. Bernstein, "Curiously Commuting Vectors," *IMAGE*, Vol. 45, pp. 42–43, 2010. (Cited on p. 1309.)

[601] D. Buckholtz, "Inverting the Difference of Hilbert Space Projections," *Amer. Math. Monthly*, Vol. 104, pp. 60–61, 1997. (Cited on p. 415.)

[602] D. Buckholtz, "Hilbert Space Idempotents and Involutions," *Proc. Amer. Math. Soc.*, Vol. 128, pp. 1415–1418, 1999. (Cited on p. 595.)

[603] P. S. Bullen, *A Dictionary of Inequalities*. Essex: Longman, 1998. (Cited on pp. 173, 204, 205, 222, 249, 252, 264, 276, 489, 835, and 860.)

[604] P. S. Bullen, *Handbook of Means and Their Inequalities*. Dordrecht: Kluwer, 2003. (Cited on pp. 123, 127, 227, 228, 249, and 276.)

[605] P. S. Bullen, D. S. Mitrinovic, and P. M. Vasic, *Means and Their Inequalities*. Dordrecht: Reidel, 1988. (Cited on pp. 144, 203, 276, and 850.)

[606] F. Bullo, J. Cortes, and S. Martinez, *Distributed Control of Robotic Networks*. Princeton: Princeton University Press, 2009. (Cited on pp. xxi and 541.)

[607] A. Bultheel and M. Van Barel, *Linear Algebra, Rational Approximation and Orthogonal Polynomials*. Amsterdam: Elsevier, 1997. (Cited on p. 544.)

[608] F. Burns, D. Carlson, E. V. Haynsworth, and T. L. Markham, "Generalized Inverse Formulas Using the Schur-Complement," *SIAM J. Appl. Math*, Vol. 26, pp. 254–259, 1974. (Cited on p. 667.)

[609] P. J. Bushell and G. B. Trustrum, "Trace Inequalities for Positive Definite Matrix Power Products," *Lin. Alg. Appl.*, Vol. 132, pp. 173–178, 1990. (Cited on p. 810.)

[610] P. L. Butzer, C. Markett, and M. Schmidt, "Stirling Numbers, Central Factorial Numbers, and Representations of the Riemann Zeta Function," *Results in Mathematics*, Vol. 19, pp. 257–274, 1991. (Cited on p. 1031.)

[611] N. D. Cahill, J. R. D'Errico, D. A. Narayan, and J. Y. Narayan, "Fibonacci Determinants," *Coll. Math. J.*, Vol. 33, pp. 221–225, 2002. (Cited on p. 425.)

[612] L. Cai, W. Xu, and W. Li, "Additive and Multiplicative Perturbation Bounds for the Moore-Penrose Inverse," *Lin. Alg. Appl.*, Vol. 434, pp. 480–489, 2011. (Cited on p. 881.)

[613] B. E. Cain, "Inertia Theory," *Lin. Alg. Appl.*, Vol. 30, pp. 211–240, 1980. (Cited on pp. 567, 1310, and 1311.)

[614] N. J. Calkin, "A Curious Binomial Identity," *Disc.Math.*, Vol. 131, pp. 335–337, 1994. (Cited on p. 86.)

[615] D. Callan, "When Is 'Rank' Additive?" *Coll. Math. J.*, Vol. 29, pp. 145–147, 1998. (Cited on p. 321.)

[616] D. Callan, "A Combinatorial Interpretation for an Identity of Barrucand," *J. Integer Sequences*, Vol. 11, pp. 1–4, 2008, article 08.3.4. (Cited on p. 86.)

[617] D. Callan, "Divisibility of a Central Binomial Sum," *Amer. Math. Monthly*, Vol. 116, pp. 468–470, 2009. (Cited on p. 55.)

[618] D. Callan, "Flexagons Lead to a Catalan Number Identity," *Amer. Math. Monthly*, Vol. 119, pp. 415–418, 2012. (Cited on p. 86.)

[619] D. K. Callebaut, "Generalization of the Cauchy-Schwarz Inequality," *J. Math. Anal. Appl.*, Vol. 12, pp. 491–494, 1965. (Cited on p. 221.)

[620] F. M. Callier, J. Winkin, and J. L. Willems, "Convergence of the Time-Invariant Riccati Differential Equation and LQ-Problem: Mechanisms of Attraction," *Int. J. Contr.*, Vol. 59, pp. 983–1000, 1994. (Cited on p. 632.)

[621] P. J. Cameron, *Combinatorics: Topics, Techniques, Algorithms.* Cambridge: Cambridge University Press, 1994. (Cited on pp. 49, 104, 1090, and 1091.)

[622] E. Camouzis and G. Ladas, *Dynamics of Third-Order Rational Difference Equations with Open Problems and Conjectures.* Boca Raton: CRC Press, 2007. (Cited on pp. 1234 and 1248.)

[623] S. L. Campbell, *Singular Systems.* London: Pitman, 1980. (Cited on p. 679.)

[624] S. L. Campbell and C. D. Meyer, *Generalized Inverses of Linear Transformations.* London: Pitman, 1979, reprinted by Dover, Mineola, 1991. (Cited on pp. 632, 639, 667, 671, 675, 679, and 1247.)

[625] S. L. Campbell and N. J. Rose, "Singular Perturbation of Autonomous Linear Systems," *SIAM J. Math. Anal.*, Vol. 10, pp. 542–551, 1979. (Cited on p. 1227.)

[626] M. Can, "An Inequality Involving Sines," *Coll. Math. J.*, Vol. 47, pp. 304–305, 2016. (Cited on p. 466.)

[627] L. Cao and H. M. Schwartz, "A Decomposition Method for Positive Semidefinite Matrices and Its Application to Recursive Parameter Estimation," *SIAM J. Matrix Anal. Appl.*, Vol. 22, pp. 1095–1111, 2001. (Cited on pp. 704 and 737.)

[628] E. Capelas de Oliveira, "On a Particular Case of Series," *Int. J. Math. Educ. Sci. Tech.*, Vol. 43, pp. 674–702, 2012. (Cited on p. 1052.)

[629] E. Capelas de Oliveira and A. O. Chiacchio, "A Real Integral By Means of Complex Integration," *Int. J. Math. Educ. Sci. Tech.*, Vol. 35, pp. 596–601, 2004. (Cited on p. 1166.)

[630] E. Capelas de Oliveira, M. A. F. Rosa, and J. Vaz, "On a Class of Real Integrals," *Int. J. Math. Educ. Sci. Tech.*, Vol. 40, pp. 541–550, 2009. (Cited on pp. 994, 1130, and 1144.)

[631] J. R. Cardoso and F. S. Leite, "The Moser-Veselov Equation," *Lin. Alg. Appl.*, Vol. 360, pp. 237–248, 2003. (Cited on p. 1317.)

[632] E. A. Carlen, R. L. Frank, and E. H. Lieb, "Some Operator and Trace Function Convexity Theorems," *Lin. Alg. Appl.*, Vol. 490, pp. 174–185, 2016. (Cited on pp. 790 and 791.)

[633] E. A. Carlen and E. H. Lieb, "A Minkowski Type Trace Inequality and Strong Subadditivity of Quantum Entropy," *Amer. Math. Soc. Transl.*, Vol. 189, pp. 59–62, 1999. (Cited on p. 721.)

[634] B. C. Carlson, "Algorithms Involving Arithmetic and Geometric Means," *Amer. Math. Monthly*, Vol. 78, pp. 496–505, 1971. (Cited on p. 968.)

[635] B. C. Carlson, "The Logarithmic Mean," *Amer. Math. Monthly*, Vol. 79, pp. 615–618, 1972. (Cited on pp. 143, 969, 1083, and 1102.)

[636] D. Carlson, "Controllability, Inertia, and Stability for Tridiagonal Matrices," *Lin. Alg. Appl.*, Vol. 56, pp. 207–220, 1984. (Cited on p. 1229.)

[637] D. Carlson, "What Are Schur Complements Anyway?" *Lin. Alg. Appl.*, Vol. 74, pp. 257–275, 1986. (Cited on p. 831.)

[638] D. Carlson, "On the Controllability of Matrix Pairs (A, K) with K Positive Semidefinite, II," *SIAM J. Matrix Anal. Appl.*, Vol. 15, pp. 129–133, 1994. (Cited on p. 1310.)

[639] D. Carlson, E. V. Haynsworth, and T. L. Markham, "A Generalization of the Schur Complement by Means of the Moore-Penrose Inverse," *SIAM J. Appl. Math.*, Vol. 26, pp. 169–175, 1974. (Cited on pp. 661, 721, and 831.)

[640] D. Carlson and R. Hill, "Generalized Controllability and Inertia Theory," *Lin. Alg. Appl.*, Vol. 15, pp. 177–187, 1976. (Cited on p. 1311.)

[641] D. Carlson, C. R. Johnson, D. C. Lay, and A. D. Porter, Eds., *Linear Algebra Gems: Assets for Undergraduate Mathematics.* Washington, DC: Mathematical Association of America, 2002. (Cited on p. 362.)

[642] D. Carlson, C. R. Johnson, D. C. Lay, A. D. Porter, A. E. Watkins, and W. Watkins, Eds., *Resources for Teaching Linear Algebra.* Washington, DC: Mathematical Association of America, 1997. (Cited on p. 362.)

[643] N. L. Carothers, *Real Analysis.* Cambridge: Cambridge University Press, 2000. (Cited on p. 27.)

[644] G. S. Carr, *Formulas and Theorems in Pure Mathematics.* New York: Chelsea Publishing Company, 1970. (Cited on pp. 144, 161, 212, 238, 456, and 971.)

[645] G. F. Carrier, M. Krook, and C. E. Pearson, *Functions of a Complex Variable.* New York: McGraw-Hill, 1966. (Cited on pp. 276, 983, 1015, 1017, 1032, 1040, 1047, 1062, 1104, 1105, 1109, and 1161.)

[646] N. Carter, *Visual Group Theory*. Washington, DC: Mathematical Association of America, 2009. (Cited on p. 440.)

[647] P. Cartier, "Mathemagics, A Tribute to L. Euler and R. Feynman," in *Noise, Oscillators and Algebraic Randomness*, M. Planat, Ed. New York: Springer, 2000, pp. 6–67. (Cited on p. 1247.)

[648] D. I. Cartwright and M. J. Field, "A Refinement of the Arithmetic Mean-Geometric Mean Inequality," *Proc. Amer. Math. Soc.*, Vol. 71, pp. 36–38, 1978. (Cited on pp. 203 and 1247.)

[649] D. Castellanos, "Rapidly Converging Expansions with Fibonacci Coefficients," *Fibonacci Quart.*, Vol. 24, pp. 70–82, 1986. (Cited on pp. 259, 986, 987, 1014, and 1077.)

[650] D. Castellanos, "The Ubiquitous π, Part 1," *Math. Mag.*, Vol. 61, pp. 67–98, 1988. (Cited on pp. 51, 52, 54, 257, 1024, 1036, 1049, and 1144.)

[651] D. Castellanos, "The Ubiquitous π, Part 2," *Math. Mag.*, Vol. 61, pp. 148–163, 1988. (Cited on p. 34.)

[652] N. Castro-Gonzalez and E. Dopazo, "Representations of the Drazin Inverse for a Class of Block Matrices," *Lin. Alg. Appl.*, Vol. 400, pp. 253–269, 2005. (Cited on p. 667.)

[653] N. Castro-Gonzalez and M. F. Martinez-Serrano, "Drazin Inverse of Partitioned Matrices in Terms of Banachiewicz-Schur Forms," *Lin. Alg. Appl.*, Vol. 432, pp. 1691–1702, 2010. (Cited on pp. 667, 675, 676, and 678.)

[654] N. Castro-Gonzalez, M. F. Martinez-Serrano, and J. Robles, "Expressions for the Moore-Penrose Inverse of Block Matrices Involving the Schur Complement," *Lin. Alg. Appl.*, Vol. 471, pp. 353–368, 2015. (Cited on p. 667.)

[655] H. Caswell, *Matrix Population Models: Construction, Analysis, and Interpretation*, 2nd ed. Sunderland: Sinauer Associates, Inc., 2006. (Cited on pp. xxi and 610.)

[656] F. S. Cater, "Products of Central Collineations," *Lin. Alg. Appl.*, Vol. 19, pp. 251–274, 1978. (Cited on p. 607.)

[657] M. Catral, D. D. Olesky, and P. van den Driessche, "Block Representations of the Drazin Inverse of a Bipartite Matrix," *Elec. J. Lin. Alg.*, Vol. 18, pp. 98–107, 2009. (Cited on pp. 675 and 678.)

[658] T. Ceausu, "About the Equivalence of Some Classical Inequalities," *Acta Univ. Apulensis*, Vol. 41, pp. 221–241, 2015. (Cited on pp. 124, 138, 195, 209, and 225.)

[659] A. Cerny, "Characterization of the Oblique Projector $U(VU)^+V$ with Application to Constrained Least Squares," *Lin. Alg. Appl.*, Vol. 431, pp. 1564–1570, 2009. (Cited on pp. 396 and 651.)

[660] P. Cerone and S. S. Dragomir, *Mathematical Inequalities: A Perspective*. Boca Raton: CRC Press, 2011. (Cited on p. 222.)

[661] Y. Chabrillac and J.-P. Crouzeix, "Definiteness and Semidefiniteness of Quadratic Forms Revisited," *Lin. Alg. Appl.*, Vol. 63, pp. 283–292, 1984. (Cited on p. 798.)

[662] G. D. Chakerian, "A Distorted View of Geometry," in *Mathematical Plums*, R. Honsberger, Ed. Washington, DC: Mathematical Association of America, 1979, pp. 130–150. (Cited on p. 496.)

[663] M. Chamberland, "Using Integer Relations Algorithms for Finding Relationships Among Functions," in *Contemporary Mathematics, Vol. 457* , T. Amdeberhan and V. Moll, Eds. Berlin: Springer, 2008, pp. 127–133. (Cited on pp. 100 and 148.)

[664] M. Chamberland, "Ramanujan's 6-8-10 Equation and Beyond," *Amer. Math. Monthly*, Vol. 82, pp. 135–140, 2009. (Cited on pp. 132 and 179.)

[665] M. Chamberland, "Finite Trigonometric Product and Sum Identities," *Fibonacci Quart.*, Vol. 50, pp. 217–221, 2012. (Cited on pp. 100, 103, 130, 244, and 245.)

[666] M. Chamberland, "Factored Matrices Can Generate Combinatorial Identities," *Lin. Alg. Appl.*, Vol. 438, pp. 1667–1677, 2013. (Cited on pp. 86, 100, 105, 122, 986, 987, and 1046.)

[667] M. Chamberland and A. Straub, "On gamma quotients and infinite products," *Adv. Appl. Math.*, Vol. 51, pp. 546–562, 2013. (Cited on pp. 1000, 1086, 1087, and 1089.)

[668] M. Chamberland and D. Zeilberger, "A Short Proof of McDougall's Circle Theorem," *Amer. Math. Monthly*, Vol. 121, pp. 263–265, 2014. (Cited on p. 491.)

[669] N. N. Chan and M. K. Kwong, "Hermitian Matrix Inequalities and a Conjecture," *Amer. Math. Monthly*, Vol. 92, pp. 533–541, 1985. (Cited on p. 738.)

[670] T. R. Chandrupatla, "The Perimeter of an Ellipse," *Math. Scientist*, Vol. 35, pp. 122–131, 2010. (Cited on p. 1080.)

[671] G. Chang and C. Xu, "Generalization and Probabilistic Proof of a Combinatorial Identity," *Amer. Math. Monthly*, Vol. 118, pp. 175–177, 2011. (Cited on p. 86.)

[672] X.-Q. Chang, "A Two-Parameter Trigonometric Series," *Coll. Math. J.*, Vol. 36, pp. 408–412, 2005. (Cited on pp. 268 and 1058.)

[673] X. Q. Chang, "Eigenvalues, Trace, and Determinant," *Amer. Math. Monthly*, Vol. 118, pp. 749–750, 2011. (Cited on p. 775.)

[674] X.-Q. Chang, "Matrix Inequalities," *Coll. Math. J.*, Vol. 44, pp. 70–72, 2013. (Cited on p. 801.)

[675] Z. Changjian, C.-J. Chen, and W.-S. Cheung, "On Precaric-Rajic-Dragomir-Type Inequalities in Normed Linear Spaces," *J. Ineq. Appl.*, pp. 1–7, 2009, article Number 137301. (Cited on p. 852.)

[676] H. Chapellat, M. Mansour, and S. P. Bhattacharyya, "Elementary Proofs of Some Classical Stability Criteria," *IEEE Trans. Educ.*, Vol. 33, pp. 232–239, 1990. (Cited on p. 1224.)

[677] G. Chartrand, *Graphs and Digraphs*, 4th ed. Boca Raton: Chapman & Hall, 2004. (Cited on p. xxi.)

[678] G. Chartrand and L. Lesniak, *Graphs and Digraphs*, 4th ed. Boca Raton: Chapman & Hall/CRC, 2004. (Cited on p. xxi.)

[679] F. Chatelin, *Eigenvalues of Matrices*. New York: Wiley, 1993. (Cited on p. xxi.)

[680] J.-J. Chattot, *Computational Aerodynamics and Fluid Dynamics*. Berlin: Springer, 2002. (Cited on p. xxi.)

[681] N. A. Chaturvedi, N. H. McClamroch, and D. S. Bernstein, "Asymptotic Smooth Stabilization of the Inverted 3D Pendulum," *IEEE Trans. Autom. Contr.*, Vol. 54, pp. 1204–1215, 2009. (Cited on p. 387.)

[682] J.-P. Chehab and M. Raydan, "Geometrical Properties of the Frobenius Condition Number for Positive Definite Matrices," *Lin. Alg. Appl.*, Vol. 429, pp. 2089–2097, 2008. (Cited on p. 762.)

[683] V. Chellaboina, S. P. Bhat, W. M. Haddad, and D. S. Bernstein, "Modeling and Analysis of Mass-Action Kinetics: Nonnegativity, Realizability, Reducibility, and Semistability," *IEEE Contr. Sys. Mag.*, Vol. 29, no. August, pp. 60–78, 2009. (Cited on p. 1232.)

[684] V.-S. Chellaboina and W. M. Haddad, "Is the Frobenius Matrix Norm Induced?" *IEEE Trans. Autom. Contr.*, Vol. 40, pp. 2137–2139, 1995. (Cited on p. 868.)

[685] V.-S. Chellaboina and W. M. Haddad, "Solution to 'Some Matrix Integral Identities'," *SIAM Rev.*, Vol. 39, pp. 763–765, 1997. (Cited on p. 1287.)

[686] V.-S. Chellaboina, W. M. Haddad, D. S. Bernstein, and D. A. Wilson, "Induced Convolution Operator Norms of Linear Dynamical Systems," *Math. Contr. Sig. Sys.*, Vol. 13, pp. 216–239, 2000. (Cited on p. 912.)

[687] A. Chen and H. Chen, "Identities for the Fibonacci Powers," *Int. J. Math. Educ. Sci. Tech.*, Vol. 39, pp. 534–541, 2008. (Cited on pp. 100 and 103.)

[688] B. M. Chen, Z. Lin, and Y. Shamash, *Linear Systems Theory: A Structural Decomposition Approach*. Boston: Birkhauser, 2004. (Cited on pp. xxi, 311, 312, 315, 345, 582, 592, and 619.)

[689] C.-P. Chen, "Inequalities for the Euler-Mascheroni Constant," *Appl. Math. Lett.*, Vol. 23, pp. 161–164, 2010. (Cited on p. 964.)

[690] C.-P. Chen and W.-S. Cheung, "Sharp Cusa and Becker-Stark Inequalities," *J. Ineq. Appl.*, pp. 1–6, 2011, article ID 136. (Cited on p. 249.)

[691] C.-P. Chen and W.-S. Cheung, "Sharpness of Wilker and Huygens Type Inequalities," *J. Ineq. Appl.*, pp. 1–11, 2012, article 72. (Cited on p. 249.)

[692] C.-P. Chen and F. Qi, "The Best Bounds in Wallis' Inequality," *Proc. Amer. Math. Soc.*, Vol. 133, pp. 397–401, 2005. (Cited on p. 52.)

[693] C.-P. Chen and J. Sandor, "Inequality Chains for Wilker, Huygens and Lazarevic Type Inequalities," *J. Math. Ineq.*, Vol. 8, pp. 55–67, 2014. (Cited on p. 249.)

[694] C.-P. Chen and J. Sandor, "Sharp Inequalities for Trigonometric and Hyperbolic Functions," *J. Math. Ineq.*, Vol. 9, pp. 203–217, 2015. (Cited on p. 249.)

[695] C. T. Chen, "A Generalization of the Inertia Theorem," *SIAM J. Appl. Math.*, Vol. 25, pp. 158–161, 1973. (Cited on p. 1310.)

[696] C.-T. Chen, *Linear System Theory and Design*. New York: Holt, Rinehart, Winston, 1984. (Cited on pp. xxi and 1230.)

[697] D. Chen and Y. Zhang, "On the Spectral Radius of Hadamard Products," *Banach J. Math. Anal.*, Vol. 9, pp. 127–13, 2015. (Cited on p. 699.)

[698] H. Chen, "A Unified Elementary Approach to Some Classical Inequalities," *Int. J. Math. Educ. Sci. Tech.*, Vol. 31, pp. 289–292, 2000. (Cited on pp. 203 and 217.)

[699] H. Chen, "On a New Trigonometric Identity," *Int. J. Math. Educ. Sci. Tech.*, Vol. 33, pp. 306–309, 2002. (Cited on p. 243.)

[700] H. Chen, "A Power Series Expansion and Its Applications," *Int. J. Math. Educ. Sci. Tech.*, Vol. 37, pp. 362–368, 2006. (Cited on pp. 86, 1036, and 1070.)

[701] H. Chen, "Parametric Differentiation and Integration," *Int. J. Math. Educ. Sci. Tech.*, Vol. 40, pp. 559–570, 2009. (Cited on pp. 1132 and 1152.)

[702] H. Chen, "Bernoulli Numbers via Determinants," *Int. J. Math. Educ. Sci. Tech.*, Vol. 34, pp. 291–297, 2010. (Cited on p. 980.)

[703] H. Chen, "A New Ratio Test for Positive Monotone Series," *Coll. Math. J.*, Vol. 44, pp. 139–141, 2013. (Cited on p. 959.)

[704] H. Chen and M. Fulford, "Trigonometric Integrals via Partial Fractions," *Int. J. Math. Educ. Sci. Tech.*, Vol. 36, pp. 559–565, 2005. (Cited on p. 123.)

[705] H. Chen and P. Khalili, "On a Class of Logarithmic Integrals," *Int. J. Math. Educ. Sci. Tech.*, Vol. 37, pp. 119–125, 2006. (Cited on pp. 1053 and 1143.)

[706] J. Chen, "On Bounds of Matrix Eigenvalues," *Math. Ineq. Appl.*, Vol. 10, pp. 723–726, 2007. (Cited on p. 891.)

[707] J.-Q. Chen, J. Ping, and F. Wang, *Group Representation for Physicists*, 2nd ed. Singapore: World Scientific, 2002. (Cited on p. 436.)

[708] K.-W. Chen, "Extensions of an Amazing Identity of Ramanujan," *Fibonacci Quart.*, Vol. 50, pp. 227–230, 2012. (Cited on p. 179.)

[709] L. Chen and C. S. Wong, "Inequalities for Singular Values and Traces," *Lin. Alg. Appl.*, Vol. 171, pp. 109–120, 1992. (Cited on pp. 765 and 905.)

[710] S. Chen, "Inequalities for M-Matrices and Inverse M-Matrices," *Lin. Alg. Appl.*, Vol. 426, pp. 610–618, 2007. (Cited on p. 824.)

[711] W. Y. C. Chen, Q.-H. Hou, and Y.-P. Mu, "A Telescoping Method for Double Summations," *J. Comp. Appl. Math.*, Vol. 196, pp. 553–566, 2006. (Cited on p. 86.)

[712] Z. Chen, S. Wei, and X. Xiao, "Finding Sums for an Infinite Class of Alternating Series," *Int. J. Math. Educ. Sci. Tech.*, Vol. 43, pp. 674–702, 2012. (Cited on pp. 1052 and 1056.)

[713] C.-M. Cheng, "Cases of Equality for a Singular Value Inequality for the Hadamard Product," *Lin. Alg. Appl.*, Vol. 177, pp. 209–231, 1992. (Cited on p. 907.)

[714] C.-M. Cheng, R. A. Horn, and C.-K. Li, "Inequalities and Equalities for the Cartesian Decomposition of Complex Matrices," *Lin. Alg. Appl.*, Vol. 341, pp. 219–237, 2002. (Cited on pp. 776, 777, and 807.)

[715] D. Cheng, "Semi-tensor Product of Matrices and Its Applications—A Survey," in *Proc. 4th Int. Conf. Chinese Mathematicians*, 2007, pp. 641–668. (Cited on p. 701.)

[716] D. Cheng, H. Qi, and Y. Zhao, *An Introduction to Semi-Tensor Product of Matrices and Its Applications*. Singapore: World Scientific, 2012. (Cited on p. 701.)

[717] H.-W. Cheng and S. S.-T. Yau, "More Explicit Formulas for the Matrix Exponential," *Lin. Alg. Appl.*, Vol. 262, pp. 131–163, 1997. (Cited on p. 1206.)

[718] S. Cheng and Y. Tian, "Moore-Penrose Inverses of Products and Differences of Orthogonal Projectors," *Acta Scientiarum Math.*, Vol. 69, pp. 533–542, 2003. (Cited on pp. 651, 653, 654, and 660.)

[719] S. Cheng and Y. Tian, "Two Sets of New Characterizations for Normal and EP Matrices," *Lin. Alg. Appl.*, Vol. 375, pp. 181–195, 2003. (Cited on pp. 377, 379, 641, 650, and 672.)

[720] M.-T. Chien and M. Neumann, "Positive Definiteness of Tridiagonal Matrices via the Numerical Range," *Elec. J. Lin. Alg.*, Vol. 3, pp. 93–102, 1998. (Cited on p. 424.)

[721] E. Chlebus, "A Recursive Scheme for Improving the Original Rate of Convergence to the Euler-Mascheroni Constant," *Amer. Math. Monthly*, Vol. 118, pp. 268–273, 2011. (Cited on pp. 963 and 964.)

[722] D. Choi, "A Generalization of the Cauchy-Schwarz Inequality," *J. Math. Ineq.*, Vol. 10, pp. 1009–1012, 2016. (Cited on p. 856.)

[723] J. Choi and H. M. Srivastava, "Series Involving the Zeta Functions and a Family of Generalized Goldbach-Euler Series," *Amer. Math. Monthly*, Vol. 121, pp. 229–236, 2014. (Cited on pp. 996 and 1074.)

[724] M.-D. Choi, "Tricks or Treats with the Hilbert Matrix," *Amer. Math. Monthly*, Vol. 90, pp. 301–312, 1983. (Cited on p. 420.)

[725] M. D. Choi, T. Y. Lam, and B. Reznick, "Sums of Squares of Real Polynomials," in *K Theory and Algebraic Geometry*, B. Jacob and A. Rosenberg, Eds. Providence: American Mathematical Society, 1995, pp. 103–126. (Cited on p. 173.)

[726] M.-D. Choi and P. Y. Wu, "Convex Combinations of Projections," *Lin. Alg. Appl.*, Vol. 136, pp. 25–42, 1990. (Cited on pp. 413 and 619.)

[727] J. Chollet, "Some Inequalities for Principal Submatrices," *Amer. Math. Monthly*, Vol. 104, pp. 609–617, 1997. (Cited on p. 758.)

[728] A. Choudhry, "On Equal Sums of Fifth Powers," *Indian J. Pure Appl. Math.*, Vol. 28, pp. 1443–1450, 1997. (Cited on p. 148.)

[729] A. Choudhry, "On Equal Sums of Cubes," *Rocky Mountain J. Math.*, Vol. 28, pp. 1251–1257, 1998. (Cited on p. 148.)

[730] A. Choudhry, "Extraction of nth Roots of 2×2 Matrices," *Lin. Alg. Appl.*, Vol. 387, pp. 183–192, 2004. (Cited on p. 608.)

[731] A. Choudhry, "Equal Sums of Like Powers, Both Positive and Negative," *Rocky Mountain J. Math.*, Vol. 41, pp. 737–763, 2011. (Cited on p. 34.)

[732] G. Chrystal, *Algebra, Vol. 1*, 5th ed. New York: Chelsea Publishing Company, 1965. (Cited on pp. 120, 130, 131, 148, 149, 178, and 179.)

[733] G. Chrystal, *Algebra, Vol. 2*. New York: Forgotten Books, 2015. (Cited on p. 974.)

[734] M. T. Chu, R. E. Funderlic, and G. H. Golub, "A Rank-One Reduction Formula and Its Application to Matrix Factorizations," *SIAM Rev.*, Vol. 37, pp. 512–530, 1995. (Cited on pp. 635, 662, and 663.)

[735] M. T. Chu, R. E. Funderlic, and G. H. Golub, "Rank Modifications of Semidefinite Matrices Associated with a Secant Update Formula," *SIAM J. Matrix Anal. Appl.*, Vol. 20, pp. 428–436, 1998. (Cited on p. 737.)

[736] W. Chu, "Partial-Fraction Decompositions and Harmonic Number Identities," arXiv:math/0502537v2. (Cited on pp. 86 and 92.)

[737] W. Chu, "Harmonic Number Identities and HermitePade Approximations to the Logarithm Function," *J. Approximation Theory*, Vol. 137, pp. 42–56, 2005. (Cited on pp. 75, 86, and 92.)

[738] W. Chu, "Elementary Proofs for Convolution Identities of Abel and Hagen-Rothe," *Elec. J. Combinatorics*, Vol. 17, pp. 1–5, 2010. (Cited on p. 86.)

[739] W. Chu, "Summation Formulae Involving Harmonic Numbers," *Filomat*, Vol. 26, pp. 143–152, 2012. (Cited on pp. 68, 75, 76, 92, and 1073.)

[740] X.-G. Chu, C.-E. Zhang, and F. Qi, "Two New Algebraic Inequalities with $2n$ Variables," *J. Ineq. Pure Appl. Math.*, Vol. 8, no. 4, pp. 1–6, 2007, Article 102. (Cited on pp. 58, 67, 68, and 94.)

[741] Y.-M. Chu, M.-K. Wang, and Z.-K. Wang, "Best Possible Inequalities Among Harmonic, Geometric, Logarithmic, and Seiffert Means," *Math. Ineq. Appl.*, Vol. 15, pp. 415–422, 2012. (Cited on p. 143.)

[742] Y.-M. Chu and W.-F. Xia, "Two Sharp Inequalities for Power Mean, Geometric Mean, and Harmonic Mean," *J. Ineq. Appl.*, pp. 1–6, 2009, article 741923. (Cited on p. 143.)

[743] J. Chuai and Y. Tian, "Rank Equalities and Inequalities for Kronecker Products of Matrices with Applications," *Appl. Math. Comp.*, Vol. 150, pp. 129–137, 2004. (Cited on pp. 688 and 692.)

[744] N. L. C. Chui and J. M. Maciejowski, "Realization of Stable Models with Subspace Methods," *Automatica*, Vol. 32, pp. 1587–1595, 1996. (Cited on p. 1240.)

[745] F. R. K. Chung, *Spectral Graph Theory*. Providence: American Mathematical Society, 2000. (Cited on p. xxi.)

[746] K. C. Ciesielski, "On a Genocchi-Peano Example," *Coll. Math. J.*, Vol. 48, pp. 205–213, 2017. (Cited on p. 943.)

[747] J. Cilleruelo and F. Luca, "On the Largest Prime Factor of the Partition Function of n," *Acta Arithmetica*, Vol. 156, pp. 29–38, 2012. (Cited on p. 976.)

[748] V. Cirtoaje, *Algebraic Inequalities: Old and New Methods*. Zalau: Gil Publishing House, 2006. (Cited on pp. 150, 151, 152, 153, 154, 155, 156, 157, 160, 164, 167, 169, 173, 177, 180, 183, 184, 186, 193, 194, 202, 206, 218, and 489.)

[749] V. Cirtoaje, "Problem 244," *Gazeta Matematica Seria A*, Vol. 26, pp. 250–252, 2008. (Cited on p. 173.)

[750] V. Cirtoaje, "On the Cyclic Homogeneous Polynomial Inequalities of Degree Four," *J. Ineq. Pure. Appl. Math.*, Vol. 10, pp. 1–10, 2009, article 67. (Cited on pp. 148, 151, and 152.)

[751] V. Cirtoaje, "Necessary and Sufficient Conditions for Symmetric Homogeneous Polynomial Inequalities in Nonnegative Real Variables," *Math. Ineq. Appl.*, Vol. 16, pp. 413–438, 2013. (Cited on pp. 151, 152, 168, 177, and 452.)

[752] V. Cirtoaje, V. Q. B. Can, and T. Q. Anh, *Inequalities with Beautiful Solutions*. Zalau: Gil Publishing House, 2009. (Cited on pp. 134, 139, 145, 151, 153, 154, 155, 157, 160, 161, 164, 167, 169, 173, 177, 180, 183, 184, 185, 187, 193, 196, 197, 198, 201, 202, 206, 217, 220, and 489.)

[753] A. Cizmesija and J. Pecaric, "Classical Hardy's and Carleman's Inequalities and Mixed Means," in *Survey on Classical Inequalities*, T. M. Rassias, Ed. Dordrecht: Kluwer, 2000, pp. 27–65. (Cited on pp. 212 and 213.)

[754] J. Clark and D. A. Holton, *A First Look at Graph Theory*. Singapore: World Scientific, 1991. (Cited on p. 26.)

[755] G. Clarke, "Quadratic Identities," *Fibonacci Quart.*, Vol. 51, pp. 370–371, 2013. (Cited on p. 100.)

[756] R. J. Clarke, "A Different Aspect of Wallis's Formula," *Math. Gaz.*, Vol. 82, pp. 289–290, 1998. (Cited on p. 75.)

[757] R. J. Clarke, "Geometrical Interpretation of $\tan \frac{B-C}{2} = \frac{b-c}{b+c} \cot \frac{A}{2}$," *Math. Gaz.*, Vol. 82, pp. 8–18, 1998. (Cited on p. 446.)

[758] R. J. Clarke, "Some Integrals of the Form $\int_0^\infty \frac{\sin^m x \cos^n x}{x^p} dx$," *Math. Gaz.*, Vol. 82, pp. 290–293, 1998. (Cited on p. 1119.)

[759] R. J. Clarke, "Some Rules of the Triangle," *Math. Gaz.*, Vol. 92, pp. 319–320, 2008. (Cited on p. 446.)

[760] D. J. Clements, B. D. O. Anderson, A. J. Laub, and J. B. Matson, "Spectral Factorization with Imaginary-Axis Zeros," *Lin. Alg. Appl.*, Vol. 250, pp. 225–252, 1997. (Cited on p. 1319.)

[761] R. E. Cline, "Representations for the Generalized Inverse of a Partitioned Matrix," *SIAM J. Appl. Math.*, Vol. 12, pp. 588–600, 1964. (Cited on p. 666.)

[762] R. E. Cline and R. E. Funderlic, "The Rank of a Difference of Matrices and Associated Generalized Inverses," *Lin. Alg. Appl.*, Vol. 24, pp. 185–215, 1979. (Cited on pp. 319, 321, 430, 638, and 639.)

[763] A. H. J. J. Cloot and J. H. Meyer, "An Interesting Family of Identities," *Int. J. Math. Educ. Sci. Tech.*, Vol. 37, pp. 247–252, 2006. (Cited on p. 189.)

[764] M. J. Cloud and B. C. Drachman, *Inequalities with Applications to Engineering*. New York: Springer, 1998. (Cited on p. 276.)

[765] E. S. Coakley, F. M. Dopico, and C. R. Johnson, "Matrices for Orthogonal Groups Admitting Only Determinant One," *Lin. Alg. Appl.*, Vol. 428, pp. 796–813, 2008. (Cited on p. 434.)

[766] J. Cofman, "Catalan Numbers for the Classroom?" *Elem. Math.*, Vol. 52, pp. 108–117, 1997. (Cited on p. 105.)

[767] J. E. Cohen, "Spectral Inequalities for Matrix Exponentials," *Lin. Alg. Appl.*, Vol. 111, pp. 25–28, 1988. (Cited on p. 1218.)

[768] J. E. Cohen, S. Friedland, T. Kato, and F. P. Kelly, "Eigenvalue Inequalities for Products of Matrix Exponentials," *Lin. Alg. Appl.*, Vol. 45, pp. 55–95, 1982. (Cited on p. 1220.)

[769] P. M. Cohn, *Further Algebra and Applications*. London: Springer, 2003. (Cited on p. 328.)

[770] D. K. Cohoon, "Sufficient Conditions for the Zero Matrix," *Amer. Math. Monthly*, Vol. 96, pp. 448–449, 1989. (Cited on p. 327.)

[771] L. Comtet, *Advanced Combinatorics*. Dordrecht: Reidel, 1974. (Cited on pp. 37, 42, 43, 52, 55, 67, 68, 75, 86, 94, 103, 106, 108, 114, 130, 131, 274, 424, 441, 493, 495, 613, 950, 968, 971, 981, 983, 984, 989, 992, 1014, 1020, 1021, and 1090.)

[772] D. Constales, "A Closed Formula for the Moore-Penrose Generalized Inverse of a Complex Matrix of Given Rank," *Acta Math. Hung.*, Vol. 80, pp. 83–88, 1998. (Cited on p. 630.)

[773] J. C. Conway and D. A. Smith, *On Quaternions and Octonions: Their Geometry, Arithmetic, and Symmetry*. Natick: A. K. Peters, 2003. (Cited on pp. 178, 187, 436, and 438.)

[774] W. J. Cook, "An n-dimensional Pythagorean Theorem," *Coll. Math. J.*, Vol. 44, pp. 98–101, 2013. (Cited on p. 387.)

[775] I. D. Coope and P. F. Renaud, "Trace Inequalities with Applications to Orthogonal Regression and Matrix Nearness Problems," *J. Ineq. Pure Appl. Math.*, Vol. 10, pp. 1–7, 2009, article 92. (Cited on pp. 593, 594, and 911.)

[776] C. Cooper, "Identities in the Spirit of Ramanujan's Amazing Identity," http://www.math-cs.ucmo.edu/ curtisc/talks/ramanujan/hungary12.pdf. (Cited on p. 130.)

[777] B. Cooperstein, *Advanced Linear Algebra*, 2nd ed. Boca Raton: CRC, 2015. (Cited on p. 701.)

[778] G. Corach and A. Maestripieri, "Polar Decomposition of Oblique Projections," *Lin. Alg. Appl.*, Vol. 433, pp. 511–519, 2010. (Cited on pp. 399, 650, 651, and 656.)

[779] G. Corach and A. Maestripieri, "Products of Orthogonal Projections and Polar Decomposition," *Lin. Alg. Appl.*, Vol. 434, pp. 1594–1609, 2011. (Cited on p. 595.)

[780] G. Corach, H. Porta, and L. Recht, "An Operator Inequality," *Lin. Alg. Appl.*, Vol. 142, pp. 153–158, 1990. (Cited on p. 881.)

[781] C. Corduneanu, "Barbalat and his lemma," *Gazeta Matematica Serie A*, Vol. 29, pp. 97–102, 2011. (Cited on p. 957.)

[782] M. J. Corless and A. E. Frazho, *Linear Systems and Control: An Operator Perspective*. New York: Marcel Dekker, 2003. (Cited on p. xxi.)

[783] A. Coronel and F. Huancas, "The Proof of Three Power-Exponential Inequalities," *J. Ineq. Appl.*, pp. 1–10, 2014, article 509. (Cited on pp. 145, 161, and 209.)

[784] E. B. Corrachano and G. Sobczyk, Eds., *Geometric Algebra with Applications in Science and Engineering*. Boston: Birkhauser, 2001. (Cited on pp. 387, 438, and 691.)

[785] G. E. Cossali, "A Common Generating Function for Catalan Numbers and Other Integer Sequences," *J. Integer Sequences*, Vol. 6, pp. 1–8, 2003, article 03.1.8. (Cited on p. 86.)

[786] P. J. Costa and S. Rabinowitz, "Matrix Differentiation Identities," *SIAM Rev.*, Vol. 36, pp. 657–659, 1994. (Cited on p. 953.)

[787] R. W. Cottle, "Quartic Barriers," *Comp. Optim. Appl.*, Vol. 12, pp. 81–105, 1999. (Cited on pp. 788, 798, and 1224.)

[788] T. M. Cover and J. A. Thomas, "Determinant Inequalities via Information Theory," *SIAM J. Matrix Anal. Appl.*, Vol. 9, pp. 384–392, 1988. (Cited on pp. 721, 776, and 1172.)

[789] T. M. Cover and J. A. Thomas, *Elements of Information Theory*, 2nd ed. New York: Wiley, 2006. (Cited on pp. xxi, 231, 721, 776, and 782.)

[790] D. A. Cox and U. Thieu, "Avoiding Voids," *Amer. Math. Monthly*, Vol. 124, pp. 473–474, 2017. (Cited on p. 92.)

[791] D. E. Crabtree, "The Characteristic Vector of the Adjoint Matrix," *Amer. Math. Monthly*, Vol. 75, pp. 1127–1128, 1968. (Cited on p. 601.)

[792] L. Craciun and I. Totolici, "Demonstrarea Unor Inegalitati cu Ajutorul Generalizarii Inegalitatii lui Cauchy la Matrici," *Gazeta Matematica Seria A*, Vol. 26, pp. 235–244, 2008. (Cited on p. 200.)

[793] C. R. Crawford and Y. S. Moon, "Finding a Positive Definite Linear Combination of Two Hermitian Matrices," *Lin. Alg. Appl.*, Vol. 51, pp. 37–48, 1983. (Cited on p. 798.)

[794] T. Crilly, "Cayley's Anticipation of a Generalised Cayley-Hamilton Theorem," *Historia Mathematica*, Vol. 5, pp. 211–219, 1978. (Cited on p. 529.)

[795] G. W. Cross and P. Lancaster, "Square Roots of Complex Matrices," *Lin. Multilin. Alg.*, Vol. 1, pp. 289–293, 1974. (Cited on p. 608.)

[796] M. D. Crossley, *Essential Topology*. New York: Springer, 2005. (Cited on p. 392.)

[797] J. Cui, C.-K. Li, and N.-S. Sze, "Products of Positive Semi-definite Matrices," *Lin. Alg. Appl.*, Vol. 528, pp. 17–24, 2017. (Cited on pp. 608 and 609.)

[798] C. G. Cullen, "A Note on Normal Matrices," *Amer. Math. Monthly*, Vol. 72, pp. 643–644, 1965. (Cited on p. 604.)

[799] C. G. Cullen, *Matrices and Linear Transformations*, 2nd ed. Reading: Addison-Wesley, 1972, reprinted by Dover, Mineola, 1990. (Cited on pp. 932 and 952.)

[800] W. J. Culver, "On the Existence and Uniqueness of the Real Logarithm of a Matrix," *Proc. Amer. Math. Soc.*, Vol. 17, pp. 1146–1151, 1966. (Cited on p. 1247.)

[801] B. Curgus, "A Matrix with a Geometric Sequence of Eigenvalues," *Math. Mag.*, Vol. 89, p. 152, 2016. (Cited on p. 614.)

[802] R. F. Curtain and H. J. Zwart, *An Introduction to Infinite-Dimensional Linear Systems Theory*. New York: Springer, 1995. (Cited on p. 1219.)

[803] M. L. Curtis, *Matrix Groups*, 2nd ed. New York: Springer, 1984. (Cited on pp. 394, 438, and 440.)

[804] A. Cusamano, "Infinite Sums Defined by the Fibonacci Sequence," *Coll. Math. J.*, Vol. 37, pp. 309–310, 2006. (Cited on p. 1079.)

[805] D. Cvetkovic, P. Rowlinson, and S. Simic, *Eigenspaces of Graphs*. Cambridge: Cambridge University Press, 1997. (Cited on p. xxi.)

[806] Z. Cvetkovski, *Inequalities: Theorems, Techniques and Selected Problems*. Heidelberg: Springer, 2012. (Cited on pp. 47, 48, 125, 134, 137, 144, 145, 148, 151, 153, 154, 155, 156, 157, 160, 161, 162, 164, 165, 167, 169, 173, 177, 183, 185, 187, 190, 193, 195, 199, 201, 224, 225, 227, 237, 249, 452, 456, 466, 472, 480, 484, 485, 489, and 492.)

[807] C. M. da Fonseca and V. Kowalenko, "On a Finite Sum with Powers of Cosines," *Applic. Anal. Disc. Math.*, Vol. 7, pp. 354–377, 2013. (Cited on p. 243.)
[808] P. J. Daboul and R. Delbourgo, "Matrix Representations of Octonions and Generalizations," *J. Math. Phys.*, Vol. 40, pp. 4134–4150, 1999. (Cited on p. 438.)
[809] S. Daboul, J. Mangaldan, M. Z. Spivey, and P. J. Taylor, "The Lah Numbers and the nth Derivative of $e^{1/x}$," *Math. Mag.*, Vol. 86, pp. 39–47, 2013. (Cited on p. 111.)
[810] F. Dadipour, M. Fujii, and M. S. Moslehian, "Dunkl-Williams Inequality for Operators Associated with p-Angular Distance," *Nihonkai. Math. J.*, Vol. 21, pp. 11–20, 2010. (Cited on p. 748.)
[811] F. Dadipour and M. S. Moslehian, "A Characterization of Inner Product Spaces Related to the p-Angular Distance," *J. Math. Anal. Appl.*, Vol. 371, pp. 677–681, 2010. (Cited on pp. 852 and 856.)
[812] G. Dahlquist, "On Matrix Majorants and Minorants, with Application to Differential Equations," *Lin. Alg. Appl.*, Vol. 52/53, pp. 199–216, 1983. (Cited on p. 1232.)
[813] P. P. Dalyay, "A Twice Told Problem," *Amer. Math. Monthly*, Vol. 118, pp. 852–853, 2011. (Cited on p. 480.)
[814] P. P. Dalyay, "A Four-Number Symmetric Inequality," *Amer. Math. Monthly*, Vol. 119, p. 706, 2012. (Cited on p. 183.)
[815] T. Dana-Picard, "Explicit Closed Forms for Parametric Integrals," *Int. J. Math. Educ. Sci. Tech.*, Vol. 35, pp. 456–467, 2004. (Cited on pp. 1112, 1118, and 1133.)
[816] T. Dana-Picard, "Parametric Integrals and Catalan Numbers," *Int. J. Math. Educ. Sci. Tech.*, Vol. 36, pp. 410–414, 2005. (Cited on pp. 1112 and 1113.)
[817] T. Dana-Picard, "Sequences of Definite Integrals, Factorials and Double Factorials," *J. Integer Seq.*, Vol. 8, pp. 1–10, 2005. (Cited on p. 1101.)
[818] T. Dana-Picard, "Sequences of Definite Integrals, Factorials and Double Factorials," *J. Integer Sequences*, Vol. 8, pp. 1–10, 2005, article 05.4.6. (Cited on p. 1133.)
[819] T. Dana-Picard, "Sequences of Definite Integrals," *Int. J. Math. Educ. Sci. Tech.*, Vol. 38, pp. 393–401, 2007. (Cited on p. 1097.)
[820] T. Dana-Picard, "Two Related Parametric Integrals," *Int. J. Math. Educ. Sci. Tech.*, Vol. 38, pp. 1106–1113, 2007. (Cited on p. 1113.)
[821] T. Dana-Picard, "Closed Forms for 4-Parameter Families of Integrals," *Int. J. Math. Educ. Sci. Tech.*, Vol. 40, pp. 828–837, 2009. (Cited on p. 1157.)
[822] T. Dana-Picard, "Integral Presentations of Catalan Numbers and Wallis Formula," *Int. J. Math. Educ. Sci. Tech.*, Vol. 42, pp. 122–129, 2011. (Cited on pp. 1104, 1115, and 1130.)
[823] T. Dana-Picard, "Parametric Improper Integrals, Wallis Formula and Catalan Numbers," *Int. J. Math. Educ. Sci. Tech.*, Vol. 43, pp. 515–520, 2012. (Cited on pp. 1113 and 1114.)
[824] J. Dancis, "Poincare's Inequalities and Hermitian Completions of Certain Partial Matrices," *Lin. Alg. Appl.*, Vol. 167, pp. 219–225, 1992. (Cited on p. 569.)
[825] R. D'Andrea, "Extension of Parrott's Theorem to Nondefinite Scalings," *IEEE Trans. Autom. Contr.*, Vol. 45, pp. 937–940, 2000. (Cited on p. 902.)
[826] D. Daners, "A Short Elementary Proof of $\sum_{k=1}^{n} 1/k^2 = \pi^2/6$," *Math. Mag.*, Vol. 85, pp. 361–364, 2012. (Cited on p. 1117.)
[827] H. D'Angelo, *Linear Time-Varying Systems: Analysis and Synthesis*. Boston: Allyn and Bacon, 1970. (Cited on p. 1319.)
[828] J. P. D'Angelo, *Inequalities from Complex Analysis*. Washington, DC: The Mathematical Association of America, 2002. (Cited on pp. 271 and 855.)
[829] F. M. Dannan, "Matrix and Operator Inequalities," *J. Ineq. Pure. Appl. Math.*, Vol. 2, no. 3/34, pp. 1–7, 2001. (Cited on pp. 768 and 779.)
[830] F. M. Dannan, "Convexity of $f(A) = (\det A)^m$," *Math. Ineq. Appl.*, Vol. 14, pp. 455–458, 2011. (Cited on pp. 214 and 778.)
[831] R. Datko and V. Seshadri, "A Characterization and a Canonical Decomposition of Hurwitzian Matrices," *Amer. Math. Monthly*, Vol. 77, pp. 732–733, 1970. (Cited on p. 1231.)
[832] B. N. Datta, *Numerical Linear Algebra and Applications*. Pacific Grove: Brooks/Cole, 1995. (Cited on p. xxi.)
[833] B. N. Datta, "Stability and Inertia," *Lin. Alg. Appl.*, Vol. 302–303, pp. 563–600, 1999. (Cited on p. 1310.)

[834] B. N. Datta, *Numerical Methods for Linear Control Systems*. San Diego: Elsevier Academic Press, 2003. (Cited on p. xxi.)

[835] J. Dattorro, *Convex Optimization and Euclidean Distance Geometry*. Palo Alto: Meboo Publishing, 2005. (Cited on pp. xxi, 318, 326, 581, and 1092.)

[836] I. Daubechies and J. C. Lagarias, "Sets of Matrices all Infinite Products of Which Converge," *Lin. Alg. Appl.*, Vol. 162, pp. 227–263, 1992. (Cited on p. 1242.)

[837] K. R. Davidson and A. P. Donsig, *Real Analysis with Real Applications*. Upper Saddle River: Prentice Hall, 2002. (Cited on pp. 922, 923, and 924.)

[838] P. J. Davis, *The Mathematics of Matrices*, 2nd ed. Boston: Ginn, 1965. (Cited on p. xxii.)

[839] P. J. Davis, *Circulant Matrices*, 2nd ed. New York: Chelsea, 1994. (Cited on pp. 615 and 616.)

[840] P. J. Davis, "The Rise, Fall, and Possible Transfiguration of Triangle Geometry: A Mini-History," *Amer. Math. Monthly*, Vol. 102, pp. 204–214, 1995. (Cited on p. 446.)

[841] R. J. H. Dawlings, "Products of Idempotents in the Semigroup of Singular Endomorphisms of a Finite-Dimensional Vector Space," *Proc. Royal Soc. Edinburgh*, Vol. 91A, pp. 123–133, 1981. (Cited on p. 609.)

[842] J. Day, W. So, and R. C. Thompson, "Some Properties of the Campbell-Baker-Hausdorff Formula," *Lin. Multilin. Alg.*, Vol. 29, pp. 207–224, 1991. (Cited on pp. 1189 and 1213.)

[843] J. Day, W. So, and R. C. Thompson, "The Spectrum of a Hermitian Matrix Sum," *Lin. Alg. Appl.*, Vol. 280, pp. 289–332, 1998. (Cited on p. 805.)

[844] P. W. Day, "Rearrangement Inequalities," *Canad. J. Math.*, Vol. 24, pp. 930–943, 1972. (Cited on p. 216.)

[845] J. de Andrade Bezerra, "EP Product of EP Matrices," *IMAGE*, Vol. 55, 2015. (Cited on p. 377.)

[846] V. De Angelis, "Stirling's Series Revisited," *Amer. Math. Monthly*, Vol. 116, pp. 839–843, 2009. (Cited on p. 971.)

[847] C. de Boor, "An Empty Exercise," *SIGNUM*, Vol. 25, pp. 2–6, 1990. (Cited on p. 362.)

[848] P. P. N. de Groen, "A Counterexample on Vector Norms and the Subordinate Matrix Norms," *Amer. Math. Monthly*, Vol. 97, pp. 406–407, 1990. (Cited on p. 868.)

[849] F. R. de Hoog, R. P. Speed, and E. R. Williams, "On a Matrix Identity Associated with Generalized Least Squares," *Lin. Alg. Appl.*, Vol. 127, pp. 449–456, 1990. (Cited on p. 636.)

[850] I. De Hoyos, "Points of Continuity of the Kronecker Canonical Form," *SIAM J. Matrix Anal. Appl.*, Vol. 11, pp. 278–300, 1990. (Cited on p. 945.)

[851] W. de Launey and J. Seberry, "The Strong Kronecker Product," *J. Combinatorial Thy., Series A*, Vol. 66, pp. 192–213, 1994. (Cited on p. 700.)

[852] J. de Pillis, "Transformations on Partitioned Matrices," *Duke Math. J.*, Vol. 36, pp. 511–515, 1969. (Cited on pp. 760 and 785.)

[853] J. de Pillis, "Inequalities for Partitioned Semidefinite Matrices," *Lin. Alg. Appl.*, Vol. 4, pp. 79–94, 1971. (Cited on p. 785.)

[854] J. de Pillis, "Linear Operators and Their Partitioned Matrices," *J. Arch. Math.*, Vol. 22, pp. 79–84, 1971. (Cited on p. 785.)

[855] L. G. De Pillis, "Determinants and Polynomial Root Structure," *Int. J. Math. Educ. Sci. Technol.*, Vol. 36, pp. 469–481, 2005. (Cited on p. 519.)

[856] C. de Seguins Pazzis, "A Condition on the Powers of exp(A) and exp(B) that implies $AB = BA$," arXiv:1012.4420v4. (Cited on p. 1211.)

[857] V. De Silva, "Tensor Rank and the Ill-Posedness of the Best Low-Rank Approximation Problem," *SIAM J. Matrix Anal. Appl.*, Vol. 30, pp. 1084–1127, 2008. (Cited on p. 945.)

[858] E. de Souza and S. P. Bhattacharyya, "Controllability, Observability and the Solution of $AX - XB = C$," *Lin. Alg. Appl.*, Vol. 39, pp. 167–188, 1981. (Cited on p. 1312.)

[859] E. de Souza and S. P. Bhattarcharyya, "Controllability, Observability, and the Solution of $AX - XB = -C$," *Lin. Alg. Appl.*, Vol. 39, pp. 167–188, 1981. (Cited on p. 1312.)

[860] P. N. De Souza and J.-N. Silva, *Berkeley Problems in Mathematics*, 3rd ed. New York: Springer, 2004. (Cited on pp. 314, 318, 355, 356, and 738.)

[861] L. M. DeAlba, "Determinants and Eigenvalues," in *Handbook of Linear Algebra*, L. Hogben, Ed. Boca Raton: Chapman & Hall/CRC, 2007, pp. 4–1–4–11. (Cited on p. 695.)

[862] L. M. DeAlba and C. R. Johnson, "Possible Inertia Combinations in the Stein and Lyapunov Equations," *Lin. Alg. Appl.*, Vol. 222, pp. 227–240, 1995. (Cited on pp. 711, 1310, and 1311.)

[863] H. P. Decell, "An Application of the Cayley-Hamilton Theorem to Generalized Matrix Inversion," *SIAM Rev.*, Vol. 7, pp. 526–528, 1965. (Cited on p. 631.)

[864] A. Defant and K. Floret, *Tensor Norms and Operator Ideals*. Amsterdam: North-Holland, 1993. (Cited on pp. 843 and 864.)

[865] N. Del Buono, L. Lopez, and T. Politi, "Computation of Functions of Hamiltonian and Skew-Symmetric Matrices," *Math. Computers Simulation*, Vol. 79, pp. 1284–1297, 2008. (Cited on p. 428.)

[866] J. DeMaio, "Counting Triangles to Sum Squares," *Coll. Math. J.*, Vol. 43, pp. 297–303, 2012. (Cited on p. 56.)

[867] J. DeMaio, "Proof without Words: An Alternating Sum of Squares," *Coll. Math. J.*, Vol. 44, p. 170, 2013. (Cited on p. 44.)

[868] A. Dembo, T. M. Cover, and J. A. Thomas, "Information Theoretic Inequalities," *IEEE Trans. Inform. Theory*, Vol. 37, pp. 1501–1518, 1991. (Cited on p. 782.)

[869] J. W. Demmel, *Applied Numerical Linear Algebra*. Philadelphia: SIAM, 1997. (Cited on p. xxi.)

[870] C. Deng and Y. Wei, "Characterizations and Representations of the Drazin Inverse Involving Idempotents," *Lin. Alg. Appl.*, Vol. 431, pp. 1526–1538, 2009. (Cited on p. 675.)

[871] C. Y. Deng, "The Drazin Inverses of Products and Differences of Orthogonal Projections," *J. Math. Anal. Appl.*, Vol. 335, pp. 64–71, 2007. (Cited on p. 676.)

[872] C. Y. Deng, "The Drazin Inverses of Sum and Difference of Idempotents," *Lin. Alg. Appl.*, Vol. 430, pp. 1282–1291, 2009. (Cited on pp. 675 and 676.)

[873] C. Y. Deng, "Characterizations and Representations of the Group Inverse Involving Idempotents," *Lin. Alg. Appl.*, Vol. 434, pp. 1067–1079, 2011. (Cited on pp. 402, 407, 674, 677, and 678.)

[874] Z. Denkowski, S. Migorski, and N. S. Papageorgiou, *An Introduction to Nonlinear Analysis: Theory*. Boston: Kluwer, 2003. (Cited on pp. 922 and 923.)

[875] E. D. Denman and A. N. Beavers, "The Matrix Sign Function and Computations in Systems," *Appl. Math. Computation*, Vol. 2, pp. 63–94, 1976. (Cited on p. 608.)

[876] C. R. DePrima and C. R. Johnson, "The Range of $A^{-1}A^*$ in $\mathrm{GL}(n, C)$," *Lin. Alg. Appl.*, Vol. 9, pp. 209–222, 1974. (Cited on p. 605.)

[877] D. Desbrow, "A Trigonometric Identity," *Math. Gaz.*, Vol. 85, pp. 477–479, 2001. (Cited on p. 238.)

[878] M. N. Deshpande, "Another Division by 5," *Fibonacci Quart.*, Vol. 51, p. 339, 2013. (Cited on p. 103.)

[879] C. A. Desoer and H. Haneda, "The Measure of a Matrix as a Tool to Analyze Computer Algorithms for Circuit Analysis," *IEEE Trans. Circ. Thy.*, Vol. 19, pp. 480–486, 1972. (Cited on p. 1219.)

[880] D. W. DeTemple, "A Geometric Look at Sequences that Converge to Euler's Constant," *Coll. Math. J.*, Vol. 37, pp. 128–131, 2006. (Cited on p. 963.)

[881] E. Deutsch, "Matricial Norms," *Numer. Math.*, Vol. 16, pp. 73–84, 1970. (Cited on p. 887.)

[882] E. Deutsch, "Matricial Norms and the Zeros of Polynomials," *Lin. Alg. Appl.*, Vol. 3, pp. 483–489, 1970. (Cited on pp. 1235 and 1236.)

[883] E. Deutsch and M. Mlynarski, "Matricial Logarithmic Derivatives," *Lin. Alg. Appl.*, Vol. 19, pp. 17–31, 1978. (Cited on p. 1219.)

[884] P. J. Dhrymes, *Mathematics for Econometrics*, 3rd ed. New York: Springer, 2000. (Cited on p. xxi.)

[885] J. L. Diaz-Barrero, "An Annulus for the Zeros of Polynomials," *J. Math. Anal. Appl.*, Vol. 273, pp. 349–352, 2002. (Cited on pp. 100 and 1235.)

[886] J. L. Diaz-Barrero, "Endpoint-approximations of a Definite Integral," *Coll. Math. J.*, Vol. 37, p. 145, 2006. (Cited on p. 252.)

[887] J. L. Diaz-Barrero, "A Recurrent Identity," *Amer. Math. Monthly*, Vol. 114, pp. 364–365, 2007. (Cited on pp. 76 and 92.)

[888] J. L. Diaz-Barrero, "A Consequence of the Arithmetic-Geometric Inequality," *Coll. Math. J.*, Vol. 40, pp. 59–60, 2009. (Cited on p. 180.)

[889] J. L. Diaz-Barrero, "An Inequality of a Triangle with Its Circumradius and Inradius," *Coll. Math. J.*, Vol. 42, pp. 66–67, 2011. (Cited on p. 452.)

[890] J. L. Diaz-Barrero, "It Is Two!" *Fibonacci Quart.*, Vol. 50, p. 182, 2012. (Cited on p. 100.)

[891] J. L. Diaz-Barrero, "Applications of Radon and Bergstrom Inequalities," *Coll. Math. J.*, Vol. 44, pp. 441–442, 2013. (Cited on p. 202.)

[892] J. L. Diaz-Barrero, "The Fifth and Seven Powers of Fibonacci Numbers," *Fibonacci Quart.*, Vol. 51, pp. 277–278, 2013. (Cited on p. 100.)

[893] J. L. Diaz-Barrero and J. J. Egozcue, "Inequalities for the Zeros of a Complex Polynomial," *Octogon*, Vol. 15, pp. 106–109, 2007. (Cited on p. 103.)

[894] J. L. Diaz-Barrero and J. Gibergans-Baguena, "Inequalities Derived Using Telescopic Sums," *Gazeta Matematica Seria A*, Vol. 27, pp. 295–301, 2009. (Cited on pp. 46, 94, and 197.)

[895] J. L. Diaz-Barrero, J. Gibergans-Baguena, and P. G. Popescu, "Some Identities Involving Rational Sums," *Applic. Anal. Disc. Math.*, Vol. 1, pp. 397–402, 2007. (Cited on p. 92.)

[896] L. Dieci, "Real Hamiltonian Logarithm of a Symplectic Matrix," *Lin. Alg. Appl.*, Vol. 281, pp. 227–246, 1998. (Cited on pp. 1193 and 1214.)

[897] R. Diestel, *Graph Theory*, 3rd ed. Berlin: Springer, 2006. (Cited on p. xxi.)

[898] A. Dil and V. Kurt, "Investigating Geometric and Exponential Polynomials with Euler-Seidel Matrices," *J. Integer Sequences*, Vol. 14, pp. 1–12, 2011, article 11.4.6. (Cited on pp. 113 and 985.)

[899] A. Dil and V. Kurt, "Polynomials Related to Harmonic Numbers and Evaluation of Harmonic Number Series II," *Applic. Anal. Disc.Math.*, Vol. 5, pp. 212–229, 2011. (Cited on pp. 113, 976, 985, 992, and 1010.)

[900] K. Dilcher, "Some q-series Identities Related to Divisor Functions," *Disc.Math.*, Vol. 145, pp. 83–93, 1995. (Cited on pp. 75 and 1008.)

[901] A. S. A. Dilip and H. Pillai, "Yet Another Characterization of Solutions of the Algebraic Riccati Equation," *Lin. Alg. Appl.*, Vol. 481, pp. 1–35, 2015. (Cited on p. 1319.)

[902] C. R. Diminnie, E. Z. Andalafte, and R. W. Freese, "Angles in Normed Linear Spaces and a Characterization of Real Inner Product Spaces," *Math. Nachr.*, Vol. 129, pp. 197–204, 1986. (Cited on p. 856.)

[903] L. L. Dines, "On the Mapping of Quadratic Forms," *Bull. Amer. Math. Soc.*, Vol. 47, pp. 494–498, 1941. (Cited on p. 789.)

[904] F. Ding, "Computation of Matrix Exponentials of Special Matrices," *Appl. Math. Comp.*, Vol. 223, pp. 311–326, 2013. (Cited on p. 1205.)

[905] J. Ding, "Perturbation of Systems of Linear Algebraic Equations," *Lin. Multilin. Alg.*, Vol. 47, pp. 119–127, 2000. (Cited on p. 909.)

[906] J. Ding, "Lower and Upper Bounds for the Perturbation of General Linear Algebraic Equations," *Appl. Math. Lett.*, Vol. 14, pp. 49–52, 2001. (Cited on p. 909.)

[907] J. Ding and W. C. Pye, "On the Spectrum and Pseudoinverse of a Special Bordered Matrix," *Lin. Alg. Appl.*, Vol. 331, pp. 11–20, 2001. (Cited on p. 527.)

[908] A. Dittmer, "Cross Product Identities in Arbitrary Dimension," *Amer. Math. Monthly*, Vol. 101, pp. 887–891, 1994. (Cited on pp. 386, 387, and 691.)

[909] G. M. Dixon, *Division Algebras: Octonions, Quaternions, Complex Numbers and the Algebraic Design of Physics*. Dordrecht: Kluwer, 1994. (Cited on p. 438.)

[910] J. D. Dixon, "How Good Is Hadamard's Inequality for Determinants?" *Canad. Math. Bull.*, Vol. 27, pp. 260–264, 1984. (Cited on p. 775.)

[911] T. E. Djaferis and S. K. Mitter, "Algebraic Methods for the Study of Some Linear Matrix Equations," *Lin. Alg. Appl.*, Vol. 44, pp. 125–142, 1982. (Cited on p. 1312.)

[912] D. Z. Djokovic, "Product of Two Involutions," *Arch. Math.*, Vol. 18, pp. 582–584, 1967. (Cited on p. 609.)

[913] D. Z. Djokovic, "On Some Representations of Matrices," *Lin. Multilin. Alg.*, Vol. 4, pp. 33–40, 1976. (Cited on pp. 355 and 575.)

[914] D. Z. Djokovic and O. P. Lossers, "A Determinant Inequality," *Amer. Math. Monthly*, Vol. 83, pp. 483–484, 1976. (Cited on p. 776.)

[915] D. Z. Djokovic and J. Malzan, "Products of Reflections in the Unitary Group," *Proc. Amer. Math. Soc.*, Vol. 73, pp. 157–160, 1979. (Cited on p. 607.)

[916] D. S. Djordjevic, "Products of EP Operators on Hilbert Spaces," *Proc. Amer. Math. Soc.*, Vol. 129, pp. 1727–1731, 2001. (Cited on p. 377.)

[917] R. Z. Djordjevic, "Some Inequalities for Triangle: Old and New Results," in *Analytic and Geometric Inequalities and Applications*, T. M. Rassias and H. M. Srivastava, Eds. Dordrecht: Kluwer, 1999, pp. 69–92. (Cited on p. 452.)

[918] D. Z. Dokovic, "On the Product of Two Alternating Matrices," *Amer. Math. Monthly*, Vol. 98, pp. 935–936, 1991. (Cited on p. 605.)

[919] D. Z. Dokovic, "Unitary Similarity of Projectors," *Aequationes Mathematicae*, Vol. 42, pp. 220–224, 1991. (Cited on pp. 572 and 577.)

[920] D. Z. Dokovic, F. Szechtman, and K. Zhao, "An Algorithm that Carries a Square Matrix into Its Transpose by an Involutory Congruence Transformation," *Elec. J. Lin. Alg.*, Vol. 10, pp. 320–340, 2003. (Cited on p. 570.)

[921] V. Dolotin and A. Morozov, *Introduction to Non-Linear Algebra*. Singapore: World Scientific, 2007. (Cited on p. 701.)

[922] W. F. Donoghue, *Monotone Matrix Functions and Analytic Continuation*. New York: Springer, 1974. (Cited on pp. 726, 727, 822, and 831.)

[923] F. M. Dopico and C. R. Johnson, "Complementary Bases in Symplectic Matrices and a Proof That Their Determinant Is One," *Lin. Alg. Appl.*, Vol. 419, pp. 772–778, 2006. (Cited on p. 428.)

[924] F. M. Dopico, C. R. Johnson, and J. M. Molera, "Multiple LU Factorizations of a Singular Matrix," *Lin. Alg. Appl.*, Vol. 419, pp. 24–36, 2006. (Cited on p. 606.)

[925] C. Doran and A. Lasenby, *Geometric Algebra for Physicists*. Cambridge: Cambridge University Press, 2005. (Cited on pp. 387, 438, and 691.)

[926] L. Dorst, D. Fontijne, and S. Mann, *Geometric Algebra for Computer Science: An Object-Oriented Approach to Geometry*. Amsterdam: Elsevier, 2007. (Cited on pp. 438 and 858.)

[927] G. Dospinescu, M. Lascu, C. Pohoata, and M. Tetiva, "An Elementary Proof of Blundon's Inequality," *J. Ineq. Pure Appl. Math.*, Vol. 9, pp. 1–3, 2008, article 100. (Cited on p. 452.)

[928] Y.-N. Dou, H. Du, and Y. Li, "Concavity of the Function $f(A) = \det(I - A)$ for Density Operator A," *Math. Ineq. Appl.*, Vol. 18, pp. 581–588, 2015. (Cited on pp. 225 and 762.)

[929] S. P. Doughty and E. F. Infante, "Matrix Proof of the Theorem of Rodrigues and Hamilton," *Amer. J. Physics*, Vol. 32, pp. 712–713, 1964. (Cited on p. 1207.)

[930] R. G. Douglas, "On Majorization, Factorization, and Range Inclusion of Operators on Hilbert Space," *Proc. Amer. Math. Soc.*, Vol. 17, pp. 413–415, 1966. (Cited on p. 714.)

[931] J. C. Doyle, B. A. Francis, and A. R. Tannenbaum, *Feedback Control Theory*. New York: McMillan Publishing Co., 1992. (Cited on p. 1156.)

[932] P. G. Doyle and J. L. Snell, *Random Walks and Electric Networks*. Washington, DC: Mathematical Association of America, 1984. (Cited on p. xxi.)

[933] M. Dragan and M. Bencze, "The Finsler-Hadwiger Reverse in an Acute Triangle," *Scientia Magna*, Vol. 7, pp. 92–95, 2011. (Cited on p. 452.)

[934] V. Dragan and A. Halanay, *Stabilization of Linear Systems*. Boston: Birkhauser, 1999. (Cited on p. 1319.)

[935] S. S. Dragomir, "A Survey on Cauchy-Bunyakovsky-Schwarz Type Discrete Inequalities," *J. Ineq. Pure Appl. Math.*, Vol. 4, no. 3, pp. 1–142, 2003, Article 3. (Cited on pp. 140, 212, 217, 220, 221, 223, 224, and 271.)

[936] S. S. Dragomir, *Discrete Inequalities of the Cauchy-Bunyakovsky-Schwarz Type*. Hauppauge: Nova Science Publishers, 2004. (Cited on pp. 140, 212, 217, 220, 221, 223, 224, and 271.)

[937] S. S. Dragomir, "Refinements of Buzano's and Kurepa's Inequalities in Inner Product Spaces," *Facta Universitatis*, Vol. 20, pp. 65–73, 2005. (Cited on p. 856.)

[938] S. S. Dragomir, "Some Reverses of the Generalised Triangle Inequality in Complex Inner Product Spaces," *Lin. Alg. Appl.*, Vol. 402, pp. 245–254, 2005. (Cited on p. 274.)

[939] S. S. Dragomir, "Bounds for the Normalised Jensen Functional," *Bull. Austral. Math. Soc.*, Vol. 74, pp. 471–478, 2006. (Cited on p. 852.)

[940] S. S. Dragomir, "Inequalities for Normal Operators in Hilbert Spaces," *Applic. Anal. Disc. Math.*, Vol. 1, pp. 92–110, 2007. (Cited on pp. 584 and 855.)

[941] S. S. Dragomir, "Generalization of the Pecaric-Rajic Inequality in Normed Linear Spaces," *Math. Ineq. Appl.*, Vol. 12, pp. 53–65, 2009. (Cited on p. 852.)

[942] S. S. Dragomir, "Some Lipschitz Type Inequalities for Complex Functions," *Appl. Math. Comp.*, Vol. 230, pp. 516–529, 2014. (Cited on pp. 271, 272, and 275.)

[943] S. S. Dragomir, "Symmetrized Convexity and Hermite-Hadamard Type Inequalities," *J. Math. Ineq.*, Vol. 10, pp. 901–918, 2016. (Cited on p. 944.)

[944] S. S. Dragomir and C. J. Goh, "A Counterpart of Jensen's Discrete Inequality for Differentiable Convex Mappings and Applications in Information Theory," *Math. Comput. Modelling*, Vol. 24, pp. 1–11, 1996. (Cited on p. 230.)

[945] S. S. Dragomir and C. J. Goh, "Some Bounds on Entropy Measures in Information Theory," *Appl. Math. Lett.*, Vol. 10, pp. 23–28, 1997. (Cited on p. 230.)

[946] S. S. Dragomir, C. E. Pearce, and J. Pecaric, "Some New Inequalities for the Logarithmic Map, with Applications to Entropy and Mutual Information," *Kyungpook Math. J.*, Vol. 41, pp. 115–125, 2001. (Cited on pp. 223, 226, 228, and 230.)

[947] V. Dragovic, "Polynomial Dynamics and a Proof of the Fermat Little Theorem," *Amer. Math. Monthly*, Vol. 120, pp. 171–173, 2013. (Cited on p. 32.)

[948] M. P. Drazin, "A Note on Skew-Symmetric Matrices," *Math. Gaz.*, Vol. 36, pp. 253–255, 1952. (Cited on p. 522.)

[949] D. Drissi, "Sharp Inequalities for Some Operator Means," *SIAM J. Matrix Anal. Appl.*, Vol. 28, pp. 822–828, 2006. (Cited on p. 144.)

[950] D. Drivaliaris, S. Karanasios, and D. Pappas, "Factorizations of EP Operators," *Lin. Alg. Appl.*, Vol. 429, pp. 1555–1567, 2008. (Cited on p. 643.)

[951] R. Drnovsek, H. Radjavi, and P. Rosenthal, "A Characterization of Commutators of Idempotents," *Lin. Alg. Appl.*, Vol. 347, pp. 91–99, 2002. (Cited on pp. 419 and 420.)

[952] M. Drukker, "A Trigonometric Proof of a Trigonometric Inequality," *Math. Gaz.*, Vol. 83, pp. 503–504, 1999. (Cited on p. 252.)

[953] S. Drury, "Principal Powers of Matrices with Positive Definite Real Part," *Lin. Multilin. Alg.*, Vol. 63, pp. 296–301, 2015. (Cited on pp. 1104 and 1210.)

[954] S. W. Drury, "On a Question of Bhatia and Kittaneh," *Lin. Alg. Appl.*, Vol. 437, pp. 1955–1960, 2012. (Cited on p. 812.)

[955] S. W. Drury, "Positive Semidefiniteness of a 3×3 Matrix Related to Partitioning," *Lin. Alg. Appl.*, Vol. 446, pp. 369–376, 2014. (Cited on pp. 729 and 888.)

[956] S. W. Drury and G. P. H. Styan, "Normal Matrix and a Commutator," *IMAGE*, Vol. 31, p. 24, 2003. (Cited on p. 429.)

[957] H. Du, C. Deng, and Q. Li, "On the Infimum Problem of Hilbert Space Effects," *Science in China: Series A Math.*, Vol. 49, pp. 545–556, 2006. (Cited on p. 740.)

[958] K. Du, "Norm Inequalities Involving Hadamard Products," *Lin. Multilin. Alg.*, Vol. 58, pp. 235–238, 2009. (Cited on p. 908.)

[959] F. Dubeau, "Linear Algebra and the Sums of Powers of Integers," *Elec. J. Lin. Alg.*, Vol. 17, pp. 577–596, 2008. (Cited on p. 37.)

[960] P. H. Duc, "Problem 3488," *Crux Math.*, Vol. 36, pp. 553–554, 2010. (Cited on p. 160.)

[961] P. H. Duc, "Problem 3507," *Crux Math.*, Vol. 37, pp. 58–59, 2011. (Cited on p. 173.)

[962] J. J. Duistermaat and J. A. C. Kolk, *Multidimensional Real Analysis I: Differentiation*. Cambridge: Cambridge University Press, 2004. (Cited on pp. 938, 942, and 1212.)

[963] I. Duleba, "On a Computationally Simple Form of the Generalized Campbell-Baker-Hausdorff-Dynkin Formula," *Sys. Contr. Lett.*, Vol. 34, pp. 191–202, 1998. (Cited on p. 1306.)

[964] G. E. Dullerud and F. Paganini, *A Course in Robust Control Theory: A Convex Approach*, 2nd ed. New York: Springer, 1999. (Cited on p. xxi.)

[965] D. S. Dummit, "Solving Solvable Quintics," *Math. Comp.*, Vol. 57, pp. 387–401, 1991. (Cited on p. 436.)

[966] D. S. Dummit and R. M. Foote, *Abstract Algebra*, 3rd ed. New York: Wiley, 2004. (Cited on pp. 370, 371, 434, 436, 616, and 619.)

[967] C. F. Dunkl and K. S. Williams, "A Simple Inequality," *Amer. Math. Monthly*, Vol. 71, pp. 53–54, 1964. (Cited on p. 855.)

[968] P. Duren, *Invitation to Classical Analysis*. Providence: American Mathematical Society, 2012. (Cited on pp. 27, 127, 271, 386, 960, 962, 971, 979, 980, 996, 1030, 1039, 1048, 1057, 1058, 1060, 1082, and 1121.)

[969] D. E. Dutkay, C. P. Niculescu, and F. Popovici, "Stirling's Formula and Its Extension for the Gamma Function," *Amer. Math. Monthly*, Vol. 120, pp. 737–740, 2013. (Cited on p. 1000.)

[970] H. Dym, *Linear Algebra in Action*. Providence: American Mathematical Society, 2006. (Cited on pp. 275, 330, 333, 335, 494, 544, 568, 594, 595, 667, 714, 738, 799, 803, 816, 883, 902, 937, 940, 949, 1225, and 1310.)

[971] F. J. Dyson, N. E. Frankel, and M. L. Glasser, "Lehmer's Interesting Series," *Amer. Math. Monthly*, Vol. 120, pp. 116–130, 2013. (Cited on p. 1065.)

[972] E. Early, P. Kim, and M. Proulx, "Goldbach's Pigeonhole," *Coll. Math. J.*, Vol. 46, pp. 131–135, 2015. (Cited on p. 966.)

[973] A. Edelman and G. Strang, "Pascal Matrices," *Amer. Math. Monthly*, Vol. 111, pp. 189–197, 2004. (Cited on p. 727.)

[974] C. H. Edwards, *Advanced Calculus of Several Variables*. New York: Academic Press, 1973. (Cited on p. 926.)

[975] H. M. Edwards, *Riemann's Zeta Function*. Mineola: Dover, 2001. (Cited on p. 997.)

[976] C. Efthimiou, "A Weakly Convergent Series of Logs," *Math. Mag.*, Vol. 84, pp. 65–67, 2011. (Cited on p. 1074.)

[977] C. J. Efthimiou, "A Class of Periodic Continued Radicals," *Amer. Math. Monthly*, Vol. 119, pp. 52–58, 2012. (Cited on pp. 232, 972, and 974.)

[978] O. Egecioglu, "Parallelogram-Law-Type Identities," *Lin. Alg. Appl.*, Vol. 225, pp. 1–12, 1995. (Cited on pp. 271 and 272.)

[979] O. Egecioglu, "A Catalan-Hankel Determinant," *Congressus Numerantium*, Vol. 195, pp. 49–63, 2009. (Cited on p. 994.)

[980] H. G. Eggleston, *Convexity*. Cambridge: Cambridge University Press, 1958. (Cited on p. 362.)

[981] R. L. Ekl, "Equal Sums of Four Seventh Powers," *Math. Comp.*, Vol. 66, pp. 1755–1756, 1996. (Cited on p. 34.)

[982] R. L. Ekl, "New Results in Equal Sums of Like Powers," *Math. Comp.*, Vol. 67, pp. 1309–1315, 1998. (Cited on p. 35.)

[983] M. El-Haddad and F. Kittaneh, "Numerical Radius Inequalities for Hilbert Space Operators II," *Studia Mathematica*, Vol. 182, pp. 133–140, 2007. (Cited on p. 866.)

[984] N. Elezovic, "Asymptotic Expansions of Central Binomial Coefficients and Catalan Numbers," *J. Integer Sequences*, Vol. 17, pp. 1–14, 2014, article 14.2.1. (Cited on p. 967.)

[985] N. Elezovic, "Asymptotic Expansions of Gamma and Related Functions," *J. Math. Ineq.*, Vol. 9, pp. 177–196, 2015. (Cited on pp. 967 and 971.)

[986] L. Elsner, D. Hershkowitz, and H. Schneider, "Bounds on Norms of Compound Matrices and on Products of Eigenvalues," *Bull. London Math. Soc.*, Vol. 32, pp. 15–24, 2000. (Cited on pp. 94 and 694.)

[987] L. Elsner and K. D. Ikramov, "Normal Matrices: An Update," *Lin. Alg. Appl.*, Vol. 285, pp. 291–303, 1998. (Cited on pp. 379, 584, 733, 763, 1185, and 1220.)

[988] L. Elsner, C. R. Johnson, and J. A. D. Da Silva, "The Perron Root of a Weighted Geometric Mean of Nonnegative Matrices," *Lin. Multilin. Alg.*, Vol. 24, pp. 1–13, 1988. (Cited on p. 699.)

[989] L. Elsner and V. Monov, "The Bialternate Matrix Product Revisited," *Lin. Alg. Appl.*, Vol. 434, pp. 1058–1066, 2011. (Cited on pp. 694, 695, and 697.)

[990] L. Elsner and M. H. C. Paardekooper, "On Measures of Nonnormality of Matrices," *Lin. Alg. Appl.*, Vol. 92, pp. 107–124, 1987. (Cited on p. 379.)

[991] L. Elsner and P. Rozsa, "On Eigenvectors and Adjoints of Modified Matrices," *Lin. Multilin. Alg.*, Vol. 10, pp. 235–247, 1981. (Cited on p. 352.)

[992] L. Elsner and T. Szulc, "Convex Sets of Schur Stable and Stable Matrices," *Lin. Multilin. Alg.*, Vol. 48, pp. 1–19, 2000. (Cited on p. 536.)

[993] A. Engel, *Problem-Solving Strategies*. New York: Springer, 1998. (Cited on pp. 31, 32, 67, 68, 100, 111, 121, 130, 133, 134, 137, 144, 150, 153, 164, 167, 168, 173, 177, 183, 193, 194, 198, 200, 206, 213, 216, 271, 272, 452, 472, 480, 484, 486, 489, 491, and 1079.)

[994] K. Engo, "On the BCH Formula in so(3)," *Numer. Math. BIT*, Vol. 41, pp. 629–632, 2001. (Cited on pp. 1189 and 1213.)

[995] M. Epstein and H. Flanders, "On the Reduction of a Matrix to Diagonal Form," *Amer. Math. Monthly*, Vol. 62, pp. 168–171, 1955. (Cited on p. 599.)

[996] M. P. Epstein, "On a Class of Determinants Associated with a Matrix," *Amer. Math. Monthly*, Vol. 63, pp. 160–162, 1956. (Cited on p. 599.)

[997] P. Erdős and J. L. Selfridge, "The Product of Consecutive Integers Is Never a Power," *Illinois J. Math.*, Vol. 19, pp. 292–301, 1975. (Cited on p. 28.)

[998] K. Erdmann and M. J. Wildon, *Introduction to Lie Algebras*. New York: Springer, 2006. (Cited on p. 440.)

[999] J. A. Erdos, "On Products of Idempotent Matrices," *Glasgow Math. J.*, Vol. 8, pp. 118–122, 1967. (Cited on p. 609.)

[1000] R. Eriksson, "On the Measure of Solid Angles," *Math. Mag.*, Vol. 63, pp. 184–187, 1990. (Cited on p. 497.)

[1001] C. A. Eschenbach and Z. Li, "How Many Negative Entries Can A^2 Have?" *Lin. Alg. Appl.*, Vol. 254, pp. 99–117, 1997. (Cited on p. 540.)

[1002] J.-C. Evard and F. Uhlig, "On the Matrix Equation $f(X) = A$," *Lin. Alg. Appl.*, Vol. 162–164, pp. 447–519, 1992. (Cited on p. 1247.)

[1003] S. Even and J. Gillis, "Derangements and Laguerre Polynomials," *Math. Proc. Cambridge Phil. Soc.*, Vol. 79, pp. 135–143, 1976. (Cited on p. 104.)

[1004] G. Everest, C. Rottger, and T. Ward, "The Continuing Story of Zeta," *Math. Intelligencer*, Vol. 31, no. 3, pp. 13–17, 2009. (Cited on p. 996.)

[1005] W. N. Everitt, "A Note on Positive Definite Matrices," *Proc. Glasgow Math. Assoc.*, Vol. 3, pp. 173–175, 1958. (Cited on p. 783.)

[1006] J. Fabrykowski and S. R. Dunbar, "41st USA Mathematical Olympiad, 3rd USA Junior Mathematical Olympiad," *Math. Mag.*, Vol. 85, pp. 304–312, 2012. (Cited on p. 169.)

[1007] J. Fabrykowski and T. Smotzer, "Q1035," *Math. Mag.*, Vol. 86, pp. 382, 387, 2013. (Cited on p. 486.)

[1008] S. Fairgrieve and H. W. Gould, "Product Difference Fibonacci Identities of Simson, Gelin-Cesaro, Tagiuri and Generalizations," *Fibonacci Quart.*, Vol. 43, pp. 137–141, 2005. (Cited on p. 100.)

[1009] S. Fallat and M. J. Tsatsomeros, "On the Cayley Transform of Positivity Classes of Matrices," *Elec. J. Lin. Alg.*, Vol. 9, pp. 190–196, 2002. (Cited on p. 733.)

[1010] K. Fan, "Generalized Cayley Transforms and Strictly Dissipative Matrices," *Lin. Alg. Appl.*, Vol. 5, pp. 155–172, 1972. (Cited on p. 734.)

[1011] K. Fan, "On Real Matrices with Positive Definite Symmetric Component," *Lin. Multilin. Alg.*, Vol. 1, pp. 1–4, 1973. (Cited on pp. 734 and 774.)

[1012] K. Fan, "On Strictly Dissipative Matrices," *Lin. Alg. Appl.*, Vol. 9, pp. 223–241, 1974. (Cited on pp. 734 and 774.)

[1013] M. Fang, "Bounds on Eigenvalues of the Hadamard Product and the Fan Product of Matrices," *Lin. Alg. Appl.*, Vol. 425, pp. 7–15, 2007. (Cited on p. 699.)

[1014] Y. Fang, K. A. Loparo, and X. Feng, "Inequalities for the Trace of Matrix Product," *IEEE Trans. Autom. Contr.*, Vol. 39, pp. 2489–2490, 1994. (Cited on p. 593.)

[1015] R. W. Farebrother, "A Class of Square Roots of Involutory Matrices," *IMAGE*, Vol. 28, pp. 26–28, 2002. (Cited on p. 674.)

[1016] R. W. Farebrother, "Matrix Proof of Dickson's Equality," *IMAGE*, Vol. 55, 2015. (Cited on pp. 178 and 184.)

[1017] R. W. Farebrother, J. Gross, and S.-O. Troschke, "Matrix Representation of Quaternions," *Lin. Alg. Appl.*, Vol. 362, pp. 251–255, 2003. (Cited on p. 440.)

[1018] R. W. Farebrother and I. Wrobel, "Regular and Reflected Rotation Matrices," *IMAGE*, Vol. 29, pp. 24–25, 2002. (Cited on p. 607.)

[1019] D. R. Farenick, M. Krupnik, N. Krupnik, and W. Y. Lee, "Normal Toeplitz Matrices," *SIAM J. Matrix Anal. Appl.*, Vol. 17, pp. 1037–1043, 1996. (Cited on p. 615.)

[1020] B. Farhi, "An Identity Involving the Least Common Multiple of Binomial Coefficients and Its Application," *Amer. Math. Monthly*, Vol. 116, pp. 836–839, 2009. (Cited on p. 55.)

[1021] B. Farkas and S.-A. Wegner, "Variations on Barbalat's Lemma," *Amer. Math. Monthly*, Vol. 123, pp. 825–830, 2016. (Cited on p. 957.)

[1022] J. D. Farmer and S. C. Leth, "An Asymptotic Formula for Powers of Binomial Coefficients," *Math. Gaz.*, Vol. 89, pp. 385–391, 2005. (Cited on p. 968.)

[1023] A. Fassler and E. Stiefel, *Group Theoretical Methods and Their Applications*. Boston: Birkhauser, 1992. (Cited on p. 440.)

[1024] T. H. Fay and G. L. Walls, "Summing Sequences Having Mixed Signs," *Int. J. Math. Educ. Sci. Tech.*, Vol. 34, pp. 470–477, 2003. (Cited on pp. 239 and 959.)

[1025] O. Faynshteyn, "An Inequality for the Excircles of a Triangle," *Math. Mag.*, Vol. 83, pp. 307–308, 2010. (Cited on p. 173.)

[1026] O. Faynshteyn, "Some Triangle Inequalities," *Amer. Math. Monthly*, Vol. 117, pp. 372–374, 2010. (Cited on pp. 446, 456, 472, and 480.)

[1027] T. G. Feeman, *Portraits of the Earth: A Mathematician Looks at Maps*. Providence: American Mathematical Society, 2002. (Cited on p. 1095.)

[1028] A. E. Fekete, *Real Linear Algebra*. New York: Marcel Dekker, 1985. (Cited on pp. 386 and 398.)

[1029] H. J. Fell, "On the Zeros of Convex Combinations of Polynomials," *Pacific J. Math.*, Vol. 89, pp. 43–50, 1980. (Cited on p. 1239.)

[1030] B. Q. Feng, "Equivalence Constants for Certain Matrix Norms," *Lin. Alg. Appl.*, Vol. 374, pp. 247–253, 2003. (Cited on pp. 863, 864, and 875.)

[1031] B. Q. Feng and A. Tonge, "Equivalence Constants for Certain Matrix Norms, II," *Lin. Alg. Appl.*, Vol. 420, pp. 388–399, 2007. (Cited on pp. 863, 864, and 875.)

[1032] B. Q. Feng and A. Tonge, "A Cauchy-Khinchin Integral Inequality," *Lin. Alg. Appl.*, Vol. 433, pp. 1024–1030, 2010. (Cited on p. 329.)

[1033] Z. Feng and Y. Sun, "53rd International Mathematical Olympiad," *Math. Mag.*, Vol. 85, pp. 312–317, 2012. (Cited on pp. 125 and 137.)

[1034] D. Fengming, H. W. Kim, and L. T. Yeong, "A Family of Identities via Arbitrary Polynomials," *Coll. Math. J.*, Vol. 44, pp. 43–47, 2013. (Cited on p. 75.)

[1035] R. Fenn, *Geometry*. New York: Springer, 2001. (Cited on pp. 387, 393, 438, 497, and 972.)

[1036] A. Fernandez and A. I. Stan, "An Inequality About Pairs of Conjugate Hölder Numbers," *J. Math. Ineq.*, Vol. 10, pp. 1093–1104, 2016. (Cited on p. 140.)

[1037] P. G. Ferreira and S. P. Bhattacharyya, "On Blocking Zeros," *IEEE Trans. Autom. Contr.*, Vol. AC-22, pp. 258–259, 1977. (Cited on p. 1319.)

[1038] J. H. Ferziger and M. Peric, *Computational Methods for Fluid Dynamics*, 3rd ed. Berlin: Springer, 2002. (Cited on p. xxi.)

[1039] M. Fiedler, "A Note on the Hadamard Product of Matrices," *Lin. Alg. Appl.*, Vol. 49, pp. 233–235, 1983. (Cited on pp. 821 and 825.)

[1040] M. Fiedler, *Special Matrices and Their Applications in Numerical Mathematics*. Dordrecht: Martinus Nijhoff, 1986, reprinted by Dover, Mineola, 2008. (Cited on pp. xxi, 519, 520, 539, 586, 614, 665, 694, and 1229.)

[1041] M. Fiedler, "Some Applications of Matrices and Graphs in Euclidean Geometry," in *Handbook of Linear Algebra*, L. Hogben, Ed. Boca Raton: Chapman & Hall/CRC, 2007, pp. 66–1–66–15. (Cited on pp. 495 and 821.)

[1042] M. Fiedler, "Notes on Hilbert and Cauchy Matrices," *Lin. Alg. Appl.*, Vol. 432, pp. 351–356, 2010. (Cited on pp. 188, 427, 728, and 821.)

[1043] M. Fiedler and T. L. Markham, "A Characterization of the Moore-Penrose Inverse," *Lin. Alg. Appl.*, Vol. 179, pp. 129–133, 1993. (Cited on pp. 354 and 632.)

[1044] M. Fiedler and T. L. Markham, "An Observation on the Hadamard Product of Hermitian Matrices," *Lin. Alg. Appl.*, Vol. 215, pp. 179–182, 1995. (Cited on p. 823.)

[1045] M. Fiedler and T. L. Markham, "Some Results on the Bergstrom and Minkowski Inequalities," *Lin. Alg. Appl.*, Vol. 232, pp. 199–211, 1996. (Cited on p. 750.)

[1046] M. Fiedler and V. Ptak, "A New Positive Definite Geometric Mean of Two Positive Definite Matrices," *Lin. Alg. Appl.*, Vol. 251, pp. 1–20, 1997. (Cited on p. 744.)

[1047] J. A. Fill and D. E. Fishkind, "The Moore-Penrose Generalized Inverse for Sums of Matrices," *SIAM J. Matrix Anal. Appl.*, Vol. 21, pp. 629–635, 1999. (Cited on p. 636.)

[1048] P. A. Fillmore, "On Similarity and the Diagonal of a Matrix," *Amer. Math. Monthly*, Vol. 76, pp. 167–169, 1969. (Cited on p. 571.)

[1049] P. A. Fillmore, "On Sums of Projections," *J. Funct. Anal.*, Vol. 4, pp. 146–152, 1969. (Cited on p. 619.)

[1050] P. A. Fillmore and J. P. Williams, "On Operator Ranges," *Adv. Math.*, Vol. 7, pp. 254–281, 1971. (Cited on p. 714.)

[1051] S. R. Finch, *Mathematical Constants*. Cambridge: Cambridge University Press, 2003. (Cited on p. 219.)

[1052] A. M. Fink, "Shapiro's Inequality," in *Recent Progress in Inequalities*, G. V. Milovanovic, Ed. Dordrecht: Kluwer, 1998, pp. 241–248. (Cited on p. 198.)

[1053] C. H. Fitzgerald and R. A. Horn, "On Fractional Hadamard Powers of Positive Definite Matrices," *J. Math. Anal. Appl.*, Vol. 61, pp. 633–642, 1977. (Cited on p. 820.)

[1054] P. Flajolet and R. Sedgewick, *Analytic Combinatorics*. Cambridge: Cambridge University Press, 2010. (Cited on p. 974.)

[1055] H. Flanders, "Methods of Proof in Linear Algebra," *Amer. Math. Monthly*, Vol. 63, pp. 1–15, 1956. (Cited on p. 603.)

[1056] H. Flanders, "On the Norm and Spectral Radius," *Lin. Multilin. Alg.*, Vol. 2, pp. 239–240, 1974. (Cited on p. 897.)

[1057] H. Flanders, "An Extremal Problem in the Space of Positive Definite Matrices," *Lin. Multilin. Alg.*, Vol. 3, pp. 33–39, 1975. (Cited on p. 740.)

[1058] C. Fleischhack and S. Friedland, "Asymptotic Positivity of Hurwitz Product Traces: Two Proofs," *Lin. Alg. Appl.*, Vol. 432, pp. 1363–1383, 2010. (Cited on p. 771.)

[1059] W. Fleming, *Functions of Several Variables*, 2nd ed. New York: Springer, 1987. (Cited on p. 785.)

[1060] T. M. Flett, *Differential Analysis*. Cambridge: Cambridge University Press, 1980. (Cited on p. 1092.)

[1061] S. Foldes, *Fundamental Structures of Algebra and Discrete Mathematics*. New York: Wiley, 1994. (Cited on p. 434.)

[1062] J. Foley, A. van Dam, S. Feiner, and J. Hughes, *Computer Graphics Principles and Practice*, 2nd ed. Reading: Addison-Wesley, 1990. (Cited on pp. xxi and 396.)

[1063] A. Fonda, "On a Geometrical Formula Involving Medians and Bimedians," *Math. Mag.*, Vol. 86, pp. 351–357, 2013. (Cited on p. 494.)

[1064] C. K. Fong, "Norm Estimates Related to Self-Commutators," *Lin. Alg. Appl.*, Vol. 74, pp. 151–156, 1986. (Cited on p. 897.)

[1065] E. Formanek, "Polynomial Identities and the Cayley-Hamilton Theorem," *Mathematical Intelligencer*, Vol. 11, pp. 37–39, 1989. (Cited on pp. 420, 429, and 528.)

[1066] E. Formanek, *The Polynomial Identities and Invariants of $n \times n$ Matrices*. Providence: American Mathematical Society, 1991. (Cited on pp. 420, 429, and 528.)

[1067] L. R. Foulds, *Graph Theory Applications*. New York: Springer, 1992. (Cited on p. xxi.)

[1068] D. L. Foulis, "The Algebra of Complex 2×2 Matrices and a General Closed Baker-Campbell-Hausdorff Formula," *J. Phys. A: Math. Theor.*, Vol. 48, pp. 1–19, 2017. (Cited on p. 1213.)

[1069] J. B. Fraleigh, *A First Course in Abstract Algebra*, 6th ed. Reading, MA: Addison-Wesley Longman, 1999. (Cited on p. 436.)

[1070] B. A. Francis, *A Course in H_∞ Control Theory*. New York: Springer, 1987. (Cited on p. xxi.)

[1071] J. Franklin, *Matrix Theory*. Englewood Cliffs: Prentice-Hall, 1968. (Cited on p. xxii.)

[1072] M. Frazier, *An Introduction to Wavelets through Linear Algebra*. New York: Springer, 1999. (Cited on p. xxi.)

[1073] M. Frechet, "Les Solutions Non Commutables de L'Equation," *Rendiconti del Circolo Matematico di Palermo*, Vol. 1, pp. 11–27, 1952. (Cited on p. 1211.)

[1074] W. Freedman, "Minimizing Areas and Volumes and a Generalized AM-GM Inequality," *Math. Mag.*, Vol. 85, pp. 116–123, 2012. (Cited on p. 204.)

[1075] R. W. Freese, C. R. Diminnie, and E. Z. Andalafte, "Angle Bisectors in Normed Linear Spaces," *Math. Nachr.*, Vol. 131, pp. 167–173, 1987. (Cited on p. 856.)

[1076] J. S. Freudenberg and D. P. Looze, "Right Half Plane Poles and Zeros and Design Tradeoffs in Feedback Systems," *IEEE Trans. Autom. Contr.*, Vol. AC-30, pp. 555–565, 1985. (Cited on p. 1156.)

[1077] S. Friedland, "A Characterization of Normal Matrices," *Israel J. Math.*, Vol. 42, pp. 235–240, 1982. (Cited on p. 1220.)

[1078] S. Friedland, "A Note on a Determinantal Inequality," *Lin. Alg. Appl.*, Vol. 141, pp. 221–222, 1990. (Cited on p. 895.)

[1079] S. Friedland, *Matrices: Algebra, Analysis and Applications*. Hackensack: World Scientific, 2016. (Cited on p. 1223.)

[1080] S. Friedland, D. Hershkowitz, and S. M. Rump, "Positive Entries of Stable Matrices," *Elec. J. Lin. Alg.*, Vol. 12, pp. 17–24, 2005. (Cited on p. 1234.)

[1081] S. Friedland and A. Torokhti, "Generalized Rank-Constrained Matrix Approximation," *SIAM J. Matrix Anal. Appl.*, Vol. 29, pp. 656–659, 2007. (Cited on pp. 906 and 911.)

[1082] M. Frumosu and A. Teodorescu-Frumosu, "A Remarkable Combinatorial Identity," *Math. Mag.*, Vol. 85, pp. 201–205, 2012. (Cited on pp. 50, 108, and 111.)

[1083] D. Fuchs and S. Tabachnikov, *Mathematical Omnibus*. Providence: American Mathematical Society, 2007. (Cited on pp. 39, 55, 58, 100, 236, 523, 532, and 1091.)

[1084] P. A. Fuhrmann, *A Polynomial Approach to Linear Algebra*. New York: Springer, 1996. (Cited on pp. 390, 519, 520, 521, 544, 568, and 724.)

[1085] J. I. Fujii and M. Fujii, "Kolmogorov's Complexity for Positive Definite Matrices," *Lin. Alg. Appl.*, Vol. 341, pp. 171–180, 2002. (Cited on p. 794.)

[1086] J. I. Fujii, M. Fujii, T. Furuta, and R. Nakamoto, "Norm Inequalities Equivalent to Heinz Inequality," *Proc. Amer. Math. Soc.*, Vol. 118, pp. 827–830, 1993. (Cited on p. 882.)

[1087] J.-I. Fujii, S. Izumino, and Y. Seo, "Determinant for Positive Operators and Specht's Theorem," *Scientiae Mathematicae*, Vol. 1, pp. 307–310, 1998. (Cited on p. 206.)

[1088] M. Fujii, E. Kamei, and R. Nakamoto, "On a Question of Furuta on Chaotic Order," *Lin. Alg. Appl.*, Vol. 341, pp. 119–129, 2002. (Cited on pp. 742, 747, and 814.)

[1089] M. Fujii, E. Kamei, and R. Nakamoto, "On a Question of Furuta on Chaotic Order, II," *Math. J. Okayama University*, Vol. 45, pp. 123–131, 2003. (Cited on p. 747.)

[1090] M. Fujii and F. Kubo, "Buzano's Inequality and Bounds for Roots of Algebraic Equations," *Proc. Amer. Math. Soc.*, Vol. 117, pp. 359–361, 1993. (Cited on pp. 226, 856, and 1236.)

[1091] M. Fujii, S. H. Lee, Y. Seo, and D. Jung, "Reverse Inequalities on Chaotically Geometric Mean via Specht Ratio," *Math. Ineq. Appl.*, Vol. 6, pp. 509–519, 2003. (Cited on p. 141.)

[1092] M. Fujii, S. H. Lee, Y. Seo, and D. Jung, "Reverse Inequalities on Chaotically Geometric Mean via Specht Ratio, II," *J. Ineq. Pure Appl. Math.*, Vol. 4/40, pp. 1–8, 2003. (Cited on p. 206.)

[1093] M. Fujii and R. Nakamoto, "Rota's Theorem and Heinz Inequalities," *Lin. Alg. Appl.*, Vol. 214, pp. 271–275, 1995. (Cited on p. 881.)

[1094] M. Fujii, Y. Seo, and M. Tominaga, "Golden-Thompson Type Inequalities Related to a Geometric Mean via Specht's Ratio," *Math. Ineq. Appl.*, Vol. 5, pp. 573–582, 2002. (Cited on p. 1215.)

[1095] M. Fujii and H. Zuo, "Matrix Order in Bohr Inequality for Operators," *Banach J. Math. Anal.*, Vol. 4, pp. 21–27, 2010. (Cited on pp. 271, 273, 747, 748, 855, and 856.)

[1096] H. Fuks, "Adam Adamandy Kochanski's Approximation of π: Reconstruction of the Algorithm," *Math. Intelligencer*, Vol. 34, pp. 40–45, 2012. (Cited on pp. 54 and 185.)

[1097] A. T. Fuller, "Conditions for a Matrix to Have Only Characteristic Roots with Negative Real Parts," *J. Math. Anal. Appl.*, Vol. 23, pp. 71–98, 1968. (Cited on pp. 687, 695, and 1225.)

[1098] W. Fulton and J. Harris, *Representation Theory*. New York: Springer, 2004. (Cited on p. 436.)

[1099] O. Furdui, "An Equality Related to the Harmonic Series," *Coll. Math. J.*, Vol. 39, pp. 243–245, 2008. (Cited on p. 1034.)

[1100] O. Furdui, "Closed Form Evaluation of a Multiple Harmonic Series," *Automation Computers Appl. Math.*, Vol. 20, pp. 19–24, 2011. (Cited on pp. 1073, 1074, and 1098.)

[1101] O. Furdui, "Series Involving Products of Two Harmonic Numbers," *Math. Mag.*, Vol. 84, pp. 371–377, 2011. (Cited on pp. 1024, 1134, and 1154.)

[1102] O. Furdui, "A Harmonious Sum," *Amer. Math. Monthly*, Vol. 119, pp. 431–433, 2012. (Cited on p. 1073.)

[1103] O. Furdui, *Limits, Series, and Fractional Part Integrals, Problems in Mathematical Analysis*. New York: Springer, 2013. (Cited on pp. 957, 960, 961, 968, 971, 1000, 1026, 1027, 1028, 1034, 1072, 1073, 1074, 1087, 1089, 1150, and 1177.)

[1104] O. Furdui, "The Limit of a Multiple Sum," *Fibonacci Quart.*, Vol. 51, pp. 94–96, 2014. (Cited on pp. 75 and 1177.)

[1105] O. Furdui, "A Stirling Integral," *Amer. Math. Monthly*, Vol. 122, pp. 290–291, 2015. (Cited on p. 1176.)

[1106] O. Furdui, "An Alternating Sum of Squares of Alternating Sums," *Amer. Math. Monthly*, Vol. 122, pp. 78–79, 2015. (Cited on p. 1176.)

[1107] O. Furdui and H. Qin, "Catalan's Constant, π and $\ln 2$," *Fibonacci Quart.*, Vol. 50, pp. 90–92, 2012. (Cited on pp. 1131, 1141, and 1174.)

[1108] S. Furuichi, "Matrix Trace Inequalities on the Tsallis Entropies," *J. Ineq. Pure Appl. Math.*, Vol. 9, no. 1, pp. 1–7, 2008, Article 1. (Cited on p. 767.)

[1109] S. Furuichi, "On Refined Young Inequalities and Reverse Inequalities," *J. Math. Ineq.*, Vol. 5, pp. 21–31, 2011. (Cited on pp. 140 and 141.)

[1110] S. Furuichi, "Refined Young Inequalities with Specht's Ratio," *J. Egyptian Math. Soc.*, Vol. 20, pp. 46–49, 2012. (Cited on p. 144.)

[1111] S. Furuichi, K. Kuriyama, and K. Yanagi, "Trace Inequalities for Products of Matrices," *Lin. Alg. Appl.*, Vol. 430, pp. 2271–2276, 2009. (Cited on p. 766.)

[1112] S. Furuichi and M. Lin, "A Matrix Trace Inequality and Its Application," *Lin. Alg. Appl.*, Vol. 433, pp. 1324–1328, 2010. (Cited on p. 770.)

[1113] S. Furuichi and M. Lin, "Refinements of the Trace Inequality of Belmega, Lasaulce and Debbah," *Austr. J. Math. Anal. Appl.*, Vol. 7, pp. 1–4, 2011, article 23. (Cited on p. 773.)

[1114] S. Furuichi and N. Minculete, "Alternative Reverse Inequalities for Young's Inequality," *J. Math. Ineq.*, Vol. 5, pp. 595–600, 2011. (Cited on p. 141.)

[1115] S. Furuichi, N. Minculete, and F.-C. Mitroi, "Some Inequalities on Generalized Entropies," *J. Ineq. Appl.*, pp. 1–16, 2012, article 226. (Cited on p. 188.)

[1116] T. Furuta, "$A \geq B \geq 0$ Assures $(B^rA^pB^r)^{1/q} \geq B^{(p+2r)/q}$ for $r \geq 0,\ p \geq 0,\ q \geq 1$ with $(1+2r)q \geq p+2r$," *Proc. Amer. Math. Soc.*, Vol. 101, pp. 85–88, 1987. (Cited on p. 715.)

[1117] T. Furuta, "Norm Inequalities Equivalent to Löwner-Heinz Theorem," *Rev. Math. Phys.*, Vol. 1, pp. 135–137, 1989. (Cited on pp. 812 and 875.)

[1118] T. Furuta, "Two Operator Functions with Monotone Property," *Proc. Amer. Math. Soc.*, Vol. 111, pp. 511–516, 1991. (Cited on p. 743.)

[1119] T. Furuta, "A Note on the Arithmetic-Geometric-Mean Inequality for Every Unitarily Invariant Matrix Norm," *Lin. Alg. Appl.*, Vol. 208/209, pp. 223–228, 1994. (Cited on p. 882.)

[1120] T. Furuta, "Extension of the Furuta Inequality and Ando-Hiai Log-Majorization," *Lin. Alg. Appl.*, Vol. 219, pp. 139–155, 1995. (Cited on pp. 742, 743, and 747.)

[1121] T. Furuta, "Generalizations of Kosaki Trace Inequalities and Related Trace Inequalities on Chaotic Order," *Lin. Alg. Appl.*, Vol. 235, pp. 153–161, 1996. (Cited on p. 768.)

[1122] T. Furuta, "Operator Inequalities Associated with Hölder-McCarthy and Kantorovich Inequalities," *J. Ineq. Appl.*, Vol. 2, pp. 137–148, 1998. (Cited on p. 746.)

[1123] T. Furuta, "Simple Proof of the Concavity of Operator Entropy $f(A) = -A\log A$," *Math. Ineq. Appl.*, Vol. 3, pp. 305–306, 2000. (Cited on p. 721.)

[1124] T. Furuta, *Invitation to Linear Operators: From Matrices to Bounded Linear Operators on a Hilbert Space*. London: Taylor and Francis, 2001. (Cited on pp. 126, 411, 412, 715, 716, 717, 735, 743, 745, 746, 747, 794, 803, 814, and 849.)

[1125] T. Furuta, "Spectral Order $A \succ B$ if and only if $A^{2p-r} \geq \left(A^{-r/2}B^pA^{-r/2}\right)^{(2p-r)/(p-r)}$ for all $p > r \geq 0$ and Its Application," *Math. Ineq. Appl.*, Vol. 4, pp. 619–624, 2001. (Cited on pp. 747 and 814.)

[1126] T. Furuta, "The Hölder-McCarthy and the Young Inequalities Are Equivalent for Hilbert Space Operators," *Amer. Math. Monthly*, Vol. 108, pp. 68–69, 2001. (Cited on p. 794.)

[1127] T. Furuta, "Convergence of Logarithmic Trace Inequalities via Generalized Lie-Trotter Formulae," *Lin. Alg. Appl.*, Vol. 396, pp. 353–372, 2005. (Cited on pp. 1213 and 1215.)

[1128] T. Furuta, "Concrete Examples of Operator Monotone Functions Obtained by an Elementary Method without Appealing to Löwner Integral Representation," *Lin. Alg. Appl.*, Vol. 429, pp. 972–980, 2008. (Cited on p. 749.)

[1129] T. Furuta, "Precise Lower Bound of $f(A) - f(B)$ for $A > B > 0$ and non-constant Operator Monotone Function f on $[0, \infty)$," *J. Math. Ineq.*, Vol. 9, pp. 47–52, 2015. (Cited on p. 742.)

[1130] F. Gaines, "A Note on Matrices with Zero Trace," *Amer. Math. Monthly*, Vol. 73, pp. 630–631, 1966. (Cited on pp. 429 and 571.)

[1131] K. Gaitanas, "An Explicit Formula for the Prime Counting Function Which Is Valid Infinitely Often," *Amer. Math. Monthly*, Vol. 122, p. 283, 2015. (Cited on p. 966.)

[1132] K. Gaitanas, "Euler's Favorite Proof Meets a Theorem of Vantieghem," *Math. Mag.*, Vol. 90, pp. 70–72, 2017. (Cited on p. 32.)

[1133] A. Galantai, *Projectors and Projection Methods*. Dordrecht: Kluwer, 2004. (Cited on pp. 404, 440, 573, 574, 585, 590, 595, 736, 860, and 940.)

[1134] A. Galantai, "Subspaces, Angles and Pairs of Orthogonal Projections," *Lin. Multilin. Alg.*, Vol. 56, pp. 227–260, 2008. (Cited on pp. 314, 412, 415, 572, 574, 585, 586, 591, 595, and 651.)

[1135] J. Gallier and D. Xu, "Computing Exponentials of Skew-Symmetric Matrices and Logarithms of Orthogonal Matrices," *Int. J. Robotics Automation*, Vol. 17, pp. 1–11, 2002. (Cited on p. 394.)

[1136] T. W. Gamelin, *Complex Analysis*. New York: Springer, 2001. (Cited on pp. 276, 928, 930, 1015, 1017, 1081, 1082, 1083, 1104, 1108, 1115, 1121, 1122, 1123, 1126, 1127, 1146, and 1147.)

[1137] O. Ganea and C. Lupu, "A Skew-Symmetric Determinant Inequality," *Amer. Math. Monthly*, Vol. 122, p. 1017, 2015. (Cited on p. 383.)

[1138] L. Gangsong and Z. Guobiao, "Inverse Forms of Hadamard Inequality," *SIAM J. Matrix Anal. Appl.*, Vol. 23, pp. 990–997, 2002. (Cited on p. 803.)

[1139] F. R. Gantmacher, *The Theory of Matrices*. New York: Chelsea, 1959, Vol. I. (Cited on pp. xxii and 831.)

[1140] F. R. Gantmacher, *The Theory of Matrices*. New York: Chelsea, 1959, Vol. II. (Cited on pp. xxii, 563, 576, and 1248.)

[1141] F. Gao, "Volumes of Generalized Balls," *Amer. Math. Monthly*, Vol. 120, p. 130, 2013. (Cited on p. 498.)

[1142] X. Gao, G. Wang, X. Zhang, and J. Tan, "Several Matrix Trace Inequalities on Hermitian and Skew-Hermitian Matrices," *J. Ineq. Appl.*, pp. 1–8, 2014, article Number 358. (Cited on pp. 763 and 793.)

[1143] S. R. Garcia, D. Sherman, and G. Weiss, "On the Similarity of AB and BA for Normal and Other Matrices," *Lin. Alg. Appl.*, Vol. 508, pp. 14–21, 2016. (Cited on p. 606.)

[1144] S. R. Garcia and J. E. Tener, "Unitary Equivalence of a Matrix to Its Transpose," *J. Operator Theory*, Vol. 68, pp. 179–203, 2012. (Cited on p. 570.)

[1145] E. M. Garcia-Caballero and S. G. Moreno, "Yet Another Generalization of a Celebrated Inequality of the Gamma Function," *Amer. Math. Monthly*, Vol. 120, p. 821, 2013. (Cited on p. 1000.)

[1146] D. J. H. Garling, *Inequalities: A Journey into Linear Analysis*. Cambridge: Cambridge University Press, 2007. (Cited on pp. 135, 136, 206, 207, 213, 219, 361, 852, 864, and 1108.)

[1147] J. Garloff and D. G. Wagner, "Hadamard Products of Stable Polynomials Are Stable," *J. Math. Anal. Appl.*, Vol. 202, pp. 797–809, 1996. (Cited on p. 1225.)

[1148] N. Garnier and O. Ramare, "Fibonacci Numbers and Trigonometric Identities," *Fibonacci Quart.*, Vol. 46/47, pp. 56–61, 2009. (Cited on p. 100.)

[1149] N. Gauthier, "Two Identities for the Bernoulli-Euler Numbers," *Int. J. Math. Educ. Sci. Tech.*, Vol. 39, pp. 937–944, 2008. (Cited on pp. 246, 980, and 981.)

[1150] N. Gauthier, "Exact Closed Forms for Determinants of Binomial Coefficients," *Math. Gaz.*, Vol. 93, pp. 98–104, 2009. (Cited on p. 334.)

[1151] N. Gauthier, "Families of Linear Recurrences for Catalan Numbers," *Int. J. Math. Educ. Sci. Tech.*, Vol. 42, pp. 131–139, 2011. (Cited on p. 105.)

[1152] N. Gauthier, "Binomial Coefficients and Fibonacci and Lucas Numbers," *Fibonacci Quart.*, Vol. 52, pp. 379–381, 2012. (Cited on p. 68.)

[1153] N. Gauthier, "Convolutions with Catalan Numbers," *Fibonacci Quart.*, Vol. 51, pp. 287–288, 2013. (Cited on p. 105.)

[1154] N. Gauthier, "Evaluating a Sum Involving Binomial Coefficients," *Fibonacci Quart.*, Vol. 51, pp. 284–285, 2013. (Cited on p. 86.)

[1155] N. Gauthier, "Identities Involving Fibonacci, Lucas and Pell Numbers," *Fibonacci Quart.*, Vol. 52, pp. 286–288, 2014. (Cited on p. 103.)

[1156] T. Geerts, "A Necessary and Sufficient Condition for Solvability of the Linear-Quadratic Control Problem without Stability," *Sys. Contr. Lett.*, Vol. 11, pp. 47–51, 1988. (Cited on p. 1319.)

[1157] E. Gekeler, "On the Pointwise Matrix Product and the Mean Value Theorem," *Lin. Alg. Appl.*, Vol. 35, pp. 183–191, 1981. (Cited on p. 907.)

[1158] R. Gelca and T. Andreescu, *Putnam and Beyond*. New York: Springer, 2007. (Cited on pp. 29, 31, 32, 46, 47, 49, 50, 51, 55, 68, 75, 86, 100, 105, 122, 124, 132, 144, 149, 150, 151, 152, 153, 160, 161, 164, 167, 168, 180, 183, 197, 210, 211, 216, 225, 239, 242, 244, 250, 252, 327, 329, 355, 380, 386, 429, 472, 484, 486, 491, 495, 736, 944, 968, 971, 986, 1003, 1008, 1025, 1087, 1115, 1127, 1129, 1141, 1145, 1154, 1161, 1162, and 1163.)

[1159] I. M. Gelfand, M. M. Kapranov, and A. V. Zelevinsky, *Discriminants, Resultants, and Multidimensional Determinants*. Boston: Birkhauser, 1994. (Cited on p. 701.)

[1160] M. G. Genton, "Classes of Kernels for Machine Learning: A Statistics Perspective," *J. Machine Learning Res.*, Vol. 2, pp. 299–312, 2001. (Cited on p. 725.)

[1161] A. George and K. D. Ikramov, "Common Invariant Subspaces of Two Matrices," *Lin. Alg. Appl.*, Vol. 287, pp. 171–179, 1999. (Cited on pp. 604 and 617.)

[1162] P. Gerdes, *Adventures in the World of Matrices*. Hauppauge: Nova Publishers, 2007. (Cited on p. 615.)

[1163] A. Gerrard and J. M. Burch, *Introduction to Matrix Methods in Optics*. New York: Wiley, 1975. (Cited on p. xxi.)

[1164] I. M. Gessel and P. J. Larcombe, "The Sum $16^n \sum_{k=0}^{2n} 4^k \binom{\frac{1}{2}}{k}\binom{-\frac{1}{2}}{k}\binom{-2k}{2n-k}$: A Third Proof of Its Closed Form," *Utilitas Math.*, Vol. 80, pp. 59–63, 2009. (Cited on p. 86.)

[1165] I. M. Gessel and J. Li, "Compositions and Fibonacci Identities," *J. Integer Sequences*, Vol. 16, pp. 1–16, 2013, article 12.4.5. (Cited on pp. 1008 and 1079.)

[1166] B. Ghalayini, "An Identity Involving Middle Binomial Coefficients," *Fibonacci Quart.*, Vol. 52, pp. 282–283, 2014. (Cited on p. 86.)

[1167] M. S. Ghausi and J. J. Kelly, *Introduction to Distributed-Parameter Networks with Application to Integrated Circuits*. New York: Hot, Rinehart and Winston, Inc., 1968. (Cited on pp. 1038 and 1083.)

[1168] A. Gheondea, "When Are the Products of Normal Operators Normal?" *Bull. Math. Soc. Sci. Math. Roumanie*, Vol. 52, pp. 129–150, 2009. (Cited on p. 604.)

[1169] A. Gheondea, S. Gudder, and P. Jonas, "On the Infimum of Quantum Effects," *J. Math. Phys.*, Vol. 46, pp. 1–11, 2005, paper 062102. (Cited on p. 740.)

[1170] F. A. A. Ghouraba and M. A. Seoud, "Set Matrices," *Int. J. Math. Educ. Sci. Technol.*, Vol. 30, pp. 651–659, 1999. (Cited on p. 440.)

[1171] M. Gidea and C. P. Niculescu, "A Brief Account on Lagrange's Algebraic Identity," *Math. Intelligencer*, Vol. 34, pp. 55–61, 2012. (Cited on pp. 178, 217, 858, and 910.)

[1172] M. Gil', *Operator Functions and Localization of Spectra*. Berlin: Springer, 2003. (Cited on pp. 206, 804, 891, 897, 898, and 1225.)

[1173] M. I. Gil', "On Inequalities for Eigenvalues of Matrices," *Lin. Alg. Appl.*, Vol. 184, pp. 201–206, 1993. (Cited on pp. 582 and 583.)

[1174] D. Gilat, "Gauss's Lemma and the Irrationality of Roots, Revisited," *Math. Mag.*, Vol. 85, pp. 114–16, 2012. (Cited on p. 517.)

[1175] R. A. Gilbert, "A Fine Rediscovery," *Amer. Math. Monthly*, Vol. 122, pp. 322–331, 2015. (Cited on p. 115.)

[1176] R. Gilmore, *Lie Groups, Lie Algebras, and Some of Their Applications*. New York: Wiley, 1974, reprinted by Dover, Mineola, 2005. (Cited on p. 440.)

[1177] R. Gilmore, *Lie Groups, Physics, and Geometry: An Introduction for Physicists, Engineers, and Chemists*. Cambridge: Cambridge University Press, 2008. (Cited on p. 440.)

[1178] P. R. Girard, *Quaternions, Clifford Algebras and Relativistic Physics*. Boston: Birkhauser, 2007. (Cited on pp. 387, 393, 394, 437, and 691.)

[1179] P. Glaister, "Taylor's Theorem and Integral Expansions—A Class of Infinite Series," *Int. J. Math. Educ. Sci. Tech.*, Vol. 33, pp. 910–926, 2002. (Cited on pp. 1097 and 1158.)

[1180] P. Glaister, "On Integer Solutions to an Arctan Problem," *Int. J. Math. Educ. Sci. Tech.*, Vol. 39, pp. 829–833, 2008. (Cited on p. 258.)

[1181] M. L. Glasser, "Exponentials of Certain Hilbert Space Operators," *SIAM Rev.*, Vol. 34, pp. 498–500, 1992. (Cited on p. 1213.)

[1182] M. S. Gockenbach, *Finite-Dimensional Linear Algebra*. Boca Raton: CRC Press, 2010. (Cited on pp. 538 and 619.)

[1183] C. Godsil and G. Royle, *Algebraic Graph Theory*. New York: Springer, 2001. (Cited on p. xxi.)

[1184] S. K. Godunov, *Modern Aspects of Linear Algebra*. Providence: American Mathematical Society, 1998. (Cited on pp. xxii, 416, and 1219.)

[1185] R. Goebel, R. G. Sanfelice, and A. R. Teel, *Hybrid Dynamical Systems: Modeling, Stability, and Robustness*. Princeton: Princeton University Press, 2012. (Cited on p. 958.)

[1186] I. Gohberg, P. Lancaster, and L. Rodman, *Matrix Polynomials*. New York: Academic Press, 1982. (Cited on pp. 502, 544, and 619.)

[1187] I. Gohberg, P. Lancaster, and L. Rodman, *Invariant Subspaces of Matrices with Applications*. New York: Wiley, 1986. (Cited on pp. 595, 619, and 941.)

[1188] I. Gohberg, P. Lancaster, and L. Rodman, "On Hermitian Solutions of the Symmetric Algebraic Riccati Equation," *SIAM J. Contr. Optim.*, Vol. 24, pp. 1323–1334, 1986. (Cited on pp. 1302, 1303, and 1319.)

[1189] I. Gohberg, P. Lancaster, and L. Rodman, *Indefinite Linear Algebra and Applications*. Boston: Birkhauser, 2005. (Cited on pp. 619 and 1319.)

[1190] M. A. Golberg, "The Derivative of a Determinant," *Amer. Math. Monthly*, Vol. 79, pp. 1124–1126, 1972. (Cited on pp. 952 and 1306.)

[1191] M. Goldberg, "Mixed Multiplicativity and l_p Norms for Matrices," *Lin. Alg. Appl.*, Vol. 73, pp. 123–131, 1986. (Cited on p. 875.)

[1192] M. Goldberg, "Equivalence Constants for l_p Norms of Matrices," *Lin. Multilin. Alg.*, Vol. 21, pp. 173–179, 1987. (Cited on p. 875.)

[1193] M. Goldberg, "Multiplicativity Factors and Mixed Multiplicativity," *Lin. Alg. Appl.*, Vol. 97, pp. 45–56, 1987. (Cited on p. 875.)

[1194] M. Goldberg and G. Zwas, "On Matrices Having Equal Spectral Radius and Spectral Norm," *Lin. Alg. Appl.*, Vol. 8, pp. 427–434, 1974. (Cited on p. 897.)

[1195] H. Goller, "Shorted Operators and Rank Decomposition Matrices," *Lin. Alg. Appl.*, Vol. 81, pp. 207–236, 1986. (Cited on p. 819.)

[1196] G. H. Golub and C. F. Van Loan, *Matrix Computations*, 4th ed. Baltimore: Johns Hopkins University Press, 2013. (Cited on pp. xxi, 906, and 911.)

[1197] L. Gonzalez, "A New Approach for Proving or Generating Combinatorial Identities," *Int. J. Math. Educ. Sci. Tech.*, Vol. 41, pp. 359–372, 2010. (Cited on pp. 39, 67, 68, 75, 86, and 111.)

[1198] N. C. Gonzalez, J. J. Koliha, and Y. Wei, "Integral Representation of the Drazin Inverse," *Elect. J. Lin. Alg.*, Vol. 9, pp. 129–131, 2002. (Cited on p. 1210.)

[1199] F. M. Goodman, *Algebra: Abstract and Concrete*, 2nd ed. Englewood Cliffs: Prentice-Hall, 2003. (Cited on p. 436.)

[1200] L. E. Goodman and W. H. Warner, *Statics*. Belmont: Wadsworth Publishing Company, 1964, reprinted by Dover, Mineola, 2001. (Cited on p. 386.)

[1201] G. R. Goodson, "On Commutators in Matrix Theory," *Operators and Matrices*, Vol. 4, pp. 283–292, 2010. (Cited on pp. 603 and 604.)

[1202] G. Gordon and P. McGrath, "Convergent Sign Patterns of the Harmonic Series," *Math. Mag.*, Vol. 86, pp. 68–70, 2013. (Cited on p. 1027.)

[1203] N. Gordon and D. Salmond, "Bayesian Pattern Matching Technique for Target Acquisition," *J. Guid. Contr. Dyn.*, Vol. 22, pp. 68–77, 1999. (Cited on p. 340.)

[1204] P. Gorkin and E. Skubak, "Polynomials, Ellipses, and Matrices: Two Questions, One Answer," *Amer. Math. Monthly*, Vol. 118, pp. 522–533, 2011. (Cited on p. 496.)

[1205] A. Goroncy and T. Rychlik, "How Deviant Can You Be? The Complete Solution," *Math. Ineq. Appl.*, Vol. 9, pp. 633–647, 2006. (Cited on p. 195.)

[1206] H. Gould and J. Quaintance, "Double Fun with Double Factorials," *Math. Mag.*, Vol. 85, pp. 177–192, 2012. (Cited on pp. 58, 1008, 1049, and 1083.)

[1207] H. W. Gould, *Combinatorial Identities*. Morgantown, WV: Morgantown Printing and Binding, 1972. (Cited on pp. 67 and 1070.)

[1208] H. W. Gould, *Combinatorial Identities*. Morgantown, WV: Morgantown Printing and Binding Co., 1972. (Cited on p. 276.)

[1209] H. W. Gould, "Formulas for Bernoulli Numbers," *Amer. Math. Monthly*, Vol. 79, pp. 44–51, 1972. (Cited on p. 980.)

[1210] H. W. Gould, "A Curious Identity Which Is Not So Curious," *Math. Gaz.*, Vol. 88, p. 87, 2004. (Cited on p. 86.)

[1211] H. W. Gould and J. Quaintance, "Generalizations of Vosmansky's Identity," *Fibonacci Quart.*, Vol. 48, pp. 56–61, 2010. (Cited on p. 92.)

[1212] W. Govaerts and J. D. Pryce, "A Singular Value Inequality for Block Matrices," *Lin. Alg. Appl.*, Vol. 125, pp. 141–145, 1989. (Cited on pp. 901 and 902.)

[1213] W. Govaerts and B. Sijnave, "Matrix Manifolds and the Jordan Structure of the Bialternate Matrix Product," *Lin. Alg. Appl.*, Vol. 292, pp. 245–266, 1999. (Cited on p. 695.)

[1214] E. Gover, "Monotonicity of Sequences Approximating e^x," *Math. Mag.*, Vol. 83, pp. 380–384, 2010. (Cited on p. 128.)

[1215] R. Gow, "The Equivalence of an Invertible Matrix to Its Transpose," *Lin. Alg. Appl.*, Vol. 8, pp. 329–336, 1980. (Cited on pp. 570 and 609.)

[1216] R. Gow and T. J. Laffey, "Pairs of Alternating Forms and Products of Two Skew-Symmetric Matrices," *Lin. Alg. Appl.*, Vol. 63, pp. 119–132, 1984. (Cited on pp. 605 and 610.)

[1217] I. S. Gradshteyn and I. M. Ryzhik, *Table of Integrals, Series, and Products*, 7th ed. San Diego: Academic Press, 2007, edited by A. Jeffrey and D. Zwillinger. (Cited on pp. 37, 43, 44, 45, 67, 75, 92, 132, 215, 239, 242, 243, 244, 245, 257, 259, 1000, 1001, 1008, 1023, 1028, 1046, 1047, 1056, 1084, 1087, 1106, 1108, 1109, 1118, 1119, 1121, 1122, 1123, 1126, 1128, 1129, 1131, 1132, 1134, 1135, 1136, 1138, 1141, 1142, 1143, 1144, 1146, 1147, 1149, 1150, 1152, 1154, 1155, 1160, 1161, 1165, 1166, 1167, 1168, 1169, 1170, and 1178.)

[1218] A. Graham, *Kronecker Products and Matrix Calculus with Applications*. Chichester: Ellis Horwood, 1981. (Cited on p. 701.)

[1219] R. L. Graham, D. E. Knuth, and O. Patashnik, *Concrete Mathematics*, 2nd ed. Reading: Addison-Wesley, 1994. (Cited on pp. 37, 48, 58, 67, 68, 75, 86, 92, 106, 108, 111, 112, 118, 188, 217, 980, 981, 983, 984, 1015, and 1017.)

[1220] I. Grattan-Guinness, "Numbers as Moments of Multisets: A New-Old Formulation of Arithmetic," *Math. Intelligencer*, Vol. 33, no. 4, pp. 19–29, 2011. (Cited on p. 118.)

[1221] F. A. Graybill, *Matrices with Applications in Statistics*, 2nd ed. Belmont: Wadsworth, 1983. (Cited on p. xxi.)

[1222] J. Grcar, "Linear Algebra People," http://seesar.lbl.gov/ccse/people/grcar/index. html. (Cited on p. 362.)

[1223] J. F. Grcar, "A Matrix Lower Bound," *Lin. Alg. Appl.*, Vol. 433, pp. 203–220, 2010. (Cited on pp. 846 and 847.)

[1224] W. L. Green and T. D. Morley, "Operator Means and Matrix Functions," *Lin. Alg. Appl.*, Vol. 137/138, pp. 453–465, 1990. (Cited on pp. 744 and 819.)

[1225] D. H. Greene and D. E. Knuth, *Mathematics for the Analysis of Algorithms*, 3rd ed. Boston: Birkhauser, 1990. (Cited on pp. 67, 86, and 92.)

[1226] J. Gregg, "A Survey of Methods for Euler's Sum of Powers Conjecture," http://www.hillsdalesites.org/personal/dmurphy/GreggThesis.pdf. (Cited on p. 148.)

[1227] W. H. Greub, *Multilinear Algebra*. New York: Springer, 1967. (Cited on p. 701.)

[1228] W. H. Greub, *Linear Algebra*. New York: Springer, 1981. (Cited on p. xxii.)

[1229] G. C. Greubel, "The Bilateral Binomial Theorem and Fibonacci Numbers," *Fibonacci Quart.*, Vol. 48, pp. 185–187, 2010. (Cited on p. 1046.)

[1230] G. C. Greubel, "Series with Powers of the Golden Section," *Fibonacci Quart.*, Vol. 49, pp. 375–378, 2011. (Cited on p. 1046.)

[1231] T. N. E. Greville, "Notes on the Generalized Inverse of a Matrix Product," *SIAM Rev.*, Vol. 8, pp. 518–521, 1966. (Cited on p. 637.)

[1232] T. N. E. Greville, "Solutions of the Matrix Equation $XAX = X$ and Relations between Oblique and Orthogonal projectors," *SIAM J. Appl. Math*, Vol. 26, pp. 828–832, 1974. (Cited on pp. 415, 655, and 656.)

[1233] D. Gries, "Characterizations of Certain Classes of Norms," *Numer. Math.*, Vol. 10, pp. 30–41, 1967. (Cited on p. 853.)

[1234] H. B. Griffiths and A. J. Oldknow, "A General Approach to Three-Variable Inequalities," *Math. Gaz.*, Vol. 82, pp. 8–18, 1998. (Cited on p. 466.)

[1235] M. Griffiths, "Some Obvious Facts about Primes?" *Math. Gaz.*, Vol. 89, pp. 39–41, 2005. (Cited on p. 966.)

[1236] M. Griffiths, "Families of Fibonacci and Lucas Sums via the Moments of a Random Variable," *Fibonacci Quart.*, Vol. 49, pp. 76–81, 2011. (Cited on pp. 100 and 1079.)

[1237] M. Griffiths, "Extending the Domains of Definition of Some Fibonacci Identities," *Fibonacci Quart.*, Vol. 50, pp. 352–359, 2012. (Cited on p. 100.)

[1238] M. Griffiths, "From Golden-Ratio Equalities to Fibonacci and Lucas Identities," *Math. Gaz.*, Vol. 97, pp. 234–241, 2013. (Cited on p. 100.)

[1239] M. Griffiths, "From Golden-Ratio Equalities to Fibonacci and Lucas Identities," *Math. Gaz.*, Vol. 97, pp. 234–241, 2013. (Cited on pp. 100, 103, and 1077.)

[1240] M. Griffiths, "Fibonacci Identities via Some Simple Expansions," *Math. Spectrum*, Vol. 46, no. 2, pp. 50–55, 2013/14. (Cited on pp. 100 and 1077.)

[1241] R. P. Grimaldi, *Fibonacci and Catalan Numbers: An Introduction*. Hoboken: Wiley, 2012. (Cited on pp. 28, 58, 94, 100, 103, and 1077.)

[1242] R. Grone, C. R. Johnson, E. M. Sa, and H. Wolkowicz, "Normal Matrices," *Lin. Alg. Appl.*, Vol. 87, pp. 213–225, 1987. (Cited on pp. 379, 388, 584, 600, 650, and 733.)

[1243] J. Gross, "A Note on a Partial Ordering in the Set of Hermitian Matrices," *SIAM J. Matrix Anal. Appl.*, Vol. 18, pp. 887–892, 1997. (Cited on pp. 814 and 816.)

[1244] J. Gross, "Some Remarks on Partial Orderings of Hermitian Matrices," *Lin. Multilin. Alg.*, Vol. 42, pp. 53–60, 1997. (Cited on pp. 814 and 816.)

[1245] J. Gross, "More on Concavity of a Matrix Function," *SIAM J. Matrix Anal. Appl.*, Vol. 19, pp. 365–368, 1998. (Cited on p. 819.)

[1246] J. Gross, "On Oblique Projection, Rank Additivity and the Moore-Penrose Inverse of the Sum of Two Matrices," *Lin. Multilin. Alg.*, Vol. 46, pp. 265–275, 1999. (Cited on p. 656.)

[1247] J. Gross, "On the Product of Orthogonal Projectors," *Lin. Alg. Appl.*, Vol. 289, pp. 141–150, 1999. (Cited on pp. 413 and 651.)

[1248] J. Gross, "The Moore-Penrose Inverse of a Partitioned Nonnegative Definite Matrix," *Lin. Alg. Appl.*, Vol. 321, pp. 113–121, 2000. (Cited on p. 667.)

[1249] J. Gross and G. Trenkler, "Generalized and Hypergeneralized Projectors," *Lin. Alg. Appl.*, Vol. 264, pp. 463–474, 1997. (Cited on pp. 673 and 674.)

[1250] J. Gross and G. Trenkler, "On the Product of Oblique Projectors," *Lin. Multilin. Alg.*, Vol. 44, pp. 247–259, 1998. (Cited on p. 404.)

[1251] J. Gross and G. Trenkler, "Nonsingularity of the Difference of Two Oblique Projectors," *SIAM J. Matrix Anal. Appl.*, Vol. 21, pp. 390–395, 1999. (Cited on pp. 400, 402, 403, 406, 407, and 415.)

[1252] J. Gross and G. Trenkler, "Product and Sum of Projections," *Amer. Math. Monthly*, Vol. 111, pp. 261–262, 2004. (Cited on p. 591.)

[1253] J. Gross, G. Trenkler, and S.-O. Troschke, "The Vector Cross Product in $\mathbb{C}$," *Int. J. Math. Educ. Sci. Tech.*, Vol. 30, pp. 549–555, 1999. (Cited on p. 387.)

[1254] J. Gross, G. Trenkler, and S.-O. Troschke, "Quaternions: Further Contributions to a Matrix Oriented Approach," *Lin. Alg. Appl.*, Vol. 326, pp. 205–213, 2001. (Cited on p. 440.)

[1255] J. Gross and S.-O. Troschke, "Some Remarks on Partial Orderings of Nonnegative Definite Matrices," *Lin. Alg. Appl.*, Vol. 264, pp. 457–461, 1997. (Cited on pp. 815 and 816.)

[1256] J. Gross and J. Yellen, *Graph Theory and Its Applications*, 2nd ed. Boca Raton: Chapman & Hall/CRC, 2005. (Cited on p. xxi.)

[1257] I. Grossman and W. Magnus, *Groups and Their Graphs*. Washington, DC: The Mathematical Association of America, 1992. (Cited on p. 436.)

[1258] E. Grosswald, *Representations of Integers as Sums of Squares*. New York: Springer, 1985. (Cited on pp. 35, 39, and 178.)

[1259] L. C. Grove and C. T. Benson, *Finite Reflection Groups*, 2nd ed. New York: Springer, 1996. (Cited on p. 436.)

[1260] P. M. Gruber, *Convex and Discrete Geometry*. Berlin: Springer, 2007. (Cited on pp. 33, 116, 142, 307, 935, 936, and 938.)

[1261] B. Grunbaum, *Convex Polytopes*, 2nd ed. New York: Springer, 2003. (Cited on p. 495.)

[1262] B. Grunbaum, "An Enduring Error," *Elem. Math.*, Vol. 64, pp. 89–101, 2009. (Cited on p. 495.)

[1263] K. Guan and H. Zhu, "The Generalized Heronian Mean and Its Inequalities," *Publikacije Elektrotehnickog Fakulteta–Serija: Matematika*, Vol. 17, pp. 60–75, 2006. (Cited on p. 144.)

[1264] H. W. Guggenheimer, A. S. Edelman, and C. R. Johnson, "A Simple Estimate of the Condition Number of a Linear System," *Coll. Math. J.*, Vol. 26, pp. 2–5, 1995. (Cited on p. 896.)

[1265] V. Guillemin and A. Pollack, *Differential Topology*. Englewood Cliffs: Prentice Hall, 1974. (Cited on p. 691.)

[1266] J. Guillera and J. Sondow, "Double Integrals and Infinite Products for Some Classical Constants via Analytic Continuations of Lerch's Transcendent," *Ramanujan J.*, Vol. 16, pp. 247–270, 2008. (Cited on pp. 1084, 1174, 1175, and 1176.)

[1267] O. Guler, *Foundations of Optimization*. New York: Springer, 2010. (Cited on pp. 212, 278, 306, 307, 311, 312, 937, 940, and 942.)

[1268] S. Gull, A. Lasenby, and C. Doran, "Imaginary Numbers Are Not Real: The Geometric Algebra of Spacetime," *Found. Phys.*, Vol. 23, pp. 1175–1201, 1993. (Cited on pp. 387, 438, 691, and 858.)

[1269] M. Gunther and L. Klotz, "Schur's Theorem for a Block Hadamard Product," *Lin. Alg. Appl.*, Vol. 437, pp. 948–956, 2012. (Cited on p. 700.)

[1270] B.-N. Guo and F. Qi, "Alternative Proofs for Inequalities of Some Trigonometric Functions," *Int. J. Math. Educ. Sci. Tech.*, Vol. 39, pp. 384–389, 2008. (Cited on pp. 249 and 250.)

[1271] B.-N. Guo and F. Qi, "Sharp Bounds for Harmonic Numbers," *Appl. Math. Comp.*, Vol. 218, pp. 991–995, 2011. (Cited on p. 964.)

[1272] Q.-P. Guo, H.-B. Li, and M.-Y. Song, "New Inequalities on Eigenvalues of the Hadamard Product and the Fan Product of Matrices," *J. Ineq. Appl.*, pp. 1–11, 2013, article ID 433. (Cited on p. 699.)

[1273] R.-N. Guo and F. Qi, "An Inductive Proof for an Identity Involving $\binom{n}{k}$ and the Partial Sums of Some Series," *Int. J. Math. Educ. Sci. Tech.*, Vol. 33, pp. 249–317, 2002. (Cited on p. 76.)

[1274] V. J. W. Guo, "On Jensen's and Related Combinatorial Identities," *Applic. Anal. Disc. Math.*, Vol. 5, pp. 201–211, 2011. (Cited on pp. 68 and 86.)

[1275] W. Guorong, W. Yimin, and Q. Sanzheng, *Generalized Inverses: Theory and Computations.* Beijing/New York: Science Press, 2004. (Cited on pp. 400, 401, 404, 609, 632, 639, 667, 670, and 910.)

[1276] A. K. Gupta and D. K. Nagar, *Matrix Variate Distributions.* Boca Raton: CRC, 1999. (Cited on p. xxi.)

[1277] K. Gurlebeck and W. Sprossig, *Quaternionic and Clifford Calculus for Physicists and Engineers.* New York: Chichester, 1997. (Cited on pp. 438, 439, and 1208.)

[1278] L. Gurvits, "Stability of Discrete Linear Inclusion," *Lin. Alg. Appl.*, Vol. 231, pp. 47–85, 1995. (Cited on p. 1242.)

[1279] K. E. Gustafson, "Matrix Trigonometry," *Lin. Alg. Appl.*, Vol. 217, pp. 117–140, 1995. (Cited on p. 866.)

[1280] K. E. Gustafson and D. K. M. Rao, *Numerical Range.* New York: Springer, 1997. (Cited on pp. 788 and 866.)

[1281] W. H. Gustafson, P. R. Halmos, and H. Radjavi, "Products of Involutions," *Lin. Alg. Appl.*, Vol. 13, pp. 157–162, 1976. (Cited on p. 609.)

[1282] R. K. Guy, "Every Number Is Expressible as the Sum of How Many Polygonal Numbers?" *Amer. Math. Monthly*, Vol. 101, pp. 169–172, 1994. (Cited on p. 40.)

[1283] P. W. Gwanyama, "The HM-GM-QM-QM Inequalities," *Coll. Math. J.*, Vol. 35, pp. 47–50, 2004. (Cited on p. 205.)

[1284] W. M. Haddad and D. S. Bernstein, "Robust Stabilization with Positive Real Uncertainty: Beyond the Small Gain Theorem," *Sys. Contr. Lett.*, Vol. 17, pp. 191–208, 1991. (Cited on p. 779.)

[1285] W. M. Haddad and D. S. Bernstein, "Controller Design with Regional Pole Constraints," *IEEE Trans. Autom. Contr.*, Vol. 37, pp. 54–69, 1992. (Cited on p. 1313.)

[1286] W. M. Haddad and V. Chellaboina, "Stability and Dissipativity Theory for Nonnegative Dynamical Systems: A Unified Analysis Framework for Biological and Physiological Systems," *Nonlinear Anal. Real World Appl.*, Vol. 6, pp. 35–65, 2005. (Cited on p. 1232.)

[1287] W. M. Haddad and V. Chellaboina, *Nonlinear Dynamical Systems and Control.* Princeton: Princeton University Press, 2008. (Cited on p. xxi.)

[1288] W. M. Haddad, V. Chellaboina, and S. G. Nersesov, *Thermodynamics: A Dynamical Systems Approach.* Princeton: Princeton University Press, 2005. (Cited on pp. 440 and 1233.)

[1289] W. M. Haddad, V. Kapila, and E. G. Collins, "Optimality Conditions for Reduced-Order Modeling, Estimation, and Control for Discrete-Time Linear Periodic Plants," *J. Math. Sys. Est. Contr.*, Vol. 6, pp. 437–460, 1996. (Cited on p. 1319.)

[1290] C. R. Hadlock, *Field Theory and Its Classical Problems.* Washington, DC: The Mathematical Association of America, 2000. (Cited on p. 436.)

[1291] W. W. Hager, "Updating the Inverse of a Matrix," *SIAM Rev.*, Vol. 31, pp. 221–239, 1989. (Cited on p. 362.)

[1292] W. Hahn, *Stability of Motion.* Berlin: Springer, 1967. (Cited on p. 1247.)

[1293] E. Hairer, C. Lubich, and G. Wanner, *Geometric Numerical Integration: Structure-Preserving Algorithms for Ordinary Differential Equations.* Berlin: Springer, 2002. (Cited on p. 1306.)

[1294] M. Hajja, "A Trigonometric Identity that G. S. Carr Missed and a Trigonometric Proof of a Theorem of Bang," *Math. Gaz.*, Vol. 85, pp. 293–296, 2001. (Cited on p. 238.)

[1295] M. Hajja, "A Method for Establishing Certain Trigonometric Inequalities," *J. Ineq. Pure Appl. Math.*, Vol. 8, no. 1, pp. 1–11, 2007, Article 29. (Cited on p. 472.)

[1296] M. Hajja and P. Walker, "The Measure of Solid Angles in n-dimensional Euclidean Space," *Int. J. Math. Educ. Sci. Tech.*, Vol. 33, pp. 725–729, 2002. (Cited on p. 497.)

[1297] J. K. Hale, *Ordinary Differential Equations.* New York: Wiley, 1969. (Cited on pp. 1195 and 1307.)

[1298] A. Hall, "Conditions for a Matrix Kronecker Lemma," *Lin. Alg. Appl.*, Vol. 76, pp. 271–277, 1986. (Cited on p. 745.)

[1299] B. C. Hall, *Lie Groups, Lie Algebras, and Representations: An Elementary Introduction.* New York: Springer, 2003. (Cited on pp. 428, 1187, 1189, 1191, 1192, 1193, 1211, 1214, and 1247.)

[1300] G. T. Halliwell and P. R. Mercer, "A Refinement of an Inequality from Information Theory," *J. Ineq. Pure Appl. Math.*, Vol. 5, no. 1, pp. 1–3, 2004, Article 3. (Cited on pp. 228, 231, and 264.)

[1301] P. R. Halmos, *Finite-Dimensional Vector Spaces*. Princeton: Van Nostrand, 1958, reprinted by Springer, New York, 1974. (Cited on pp. 571, 599, and 632.)

[1302] P. R. Halmos, *A Hilbert Space Problem Book*, 2nd ed. New York: Springer, 1982. (Cited on pp. 576, 604, 657, and 658.)

[1303] P. R. Halmos, "Bad Products of Good Matrices," *Lin. Multilin. Alg.*, Vol. 29, pp. 1–20, 1991. (Cited on p. 608.)

[1304] P. R. Halmos, *Problems for Mathematicians Young and Old*. Washington, DC: Mathematical Association of America, 1991. (Cited on pp. 315, 543, 731, and 1211.)

[1305] P. R. Halmos, *Linear Algebra Problem Book*. Washington, DC: Mathematical Association of America, 1995. (Cited on pp. 279, 400, and 604.)

[1306] M. Hamermesh, *Group Theory and Its Application to Physical Problems*. Reading: Addison-Wesley, 1962, reprinted by Dover, Mineola, 1989. (Cited on p. 436.)

[1307] M. Hampton, "Cosines and Cayley, Triangles and Tetrahedra," *Amer. Math. Monthly*, Vol. 121, pp. 937–941, 2014. (Cited on pp. 456 and 495.)

[1308] J. H. Han and M. D. Hirschhorn, "An Amazing Identity of Ramanujan," *Math. Mag.*, Vol. 68, pp. 199–201, 1995. (Cited on pp. 130 and 1020.)

[1309] J. H. Han and M. D. Hirschhorn, "Another Look at an Amazing Identity of Ramanujan," *Math. Mag.*, Vol. 79, pp. 302–304, 2006. (Cited on pp. 130 and 1020.)

[1310] E. J. Hannan and M. Deistler, *The Statistical Theory of Linear Systems*. New York: Wiley, 1988. (Cited on p. 1319.)

[1311] E. R. Hansen, *A Table of Series and Products*. Englewood Cliffs: Prentice-Hall, 1975. (Cited on pp. 258, 1044, 1061, and 1062.)

[1312] F. Hansen, "Extrema for Concave Operator Mappings," *Math. Japonica*, Vol. 40, pp. 331–338, 1994. (Cited on p. 772.)

[1313] F. Hansen, "Operator Inequalities Associated with Jensen's Inequality," in *Survey on Classical Inequalities*, T. M. Rassias, Ed. Dordrecht: Kluwer, 2000, pp. 67–98. (Cited on p. 772.)

[1314] F. Hansen, "Some Operator Monotone Functions," *Lin. Alg. Appl.*, Vol. 430, pp. 795–799, 2009. (Cited on p. 142.)

[1315] A. J. Hanson, *Visualizing Quaternions*. Amsterdam: Elsevier, 2006. (Cited on pp. 438 and 439.)

[1316] G. Hardy, J. E. Littlewood, and G. Polya, *Inequalities*. Cambridge: Cambridge University Press, 1988. (Cited on pp. 215 and 276.)

[1317] G. H. Hardy, P. V. Seshu Aiyar, and B. M. Wilson, *Collected Papers of Srinivasa Ramanujan*. Providence: AMS Chelsea, 1962. (Cited on pp. 54, 1031, 1037, 1049, 1061, 1062, 1063, 1090, 1110, 1123, 1164, 1165, 1167, and 1171.)

[1318] G. H. Hardy and E. M. Wright, *An Introduction to the Theory of Numbers*, 6th ed. Oxford: Oxford University Press, 2008. (Cited on p. 130.)

[1319] M. Hardy, "On Tangents and Secants of Infinite Sums," *Amer. Math. Monthly*, Vol. 123, pp. 701–703, 2016. (Cited on p. 238.)

[1320] K. Hare, H. Prodinger, and J. Shallit, "Three Series for the Generalized Golden Mean," *Fibonacci Quart.*, Vol. 52, pp. 307–313, 2014. (Cited on p. 1070.)

[1321] J. D. Harper, "Ramanujan, Quadratic Forms, and the Sum of Three Cubes," *Math. Mag.*, Vol. 86, pp. 275–279, 2013. (Cited on pp. 34 and 130.)

[1322] J. Harries, "Area of a Quadrilateral," *Math. Gaz.*, Vol. 86, pp. 310–311, 2002. (Cited on p. 491.)

[1323] J. M. Harris, J. L. Hirst, and M. J. Missinghoff, *Combinatorics and Graph Theory*, 2nd ed. New York: Springer, 2008. (Cited on p. xxi.)

[1324] L. A. Harris, "Factorizations of Operator Matrices," *Lin. Alg. Appl.*, Vol. 225, pp. 37–41, 1995. (Cited on pp. 724 and 758.)

[1325] L. A. Harris, "The Inverse of a Block Matrix," *Amer. Math. Monthly*, Vol. 102, pp. 656–657, 1995. (Cited on p. 354.)

[1326] W. A. Harris, J. P. Fillmore, and D. R. Smith, "Matrix Exponentials–Another Approach," *SIAM Rev.*, Vol. 43, pp. 694–706, 2001. (Cited on p. 1206.)

[1327] G. W. Hart, *Multidimensional Analysis: Algebras and Systems for Science and Engineering*. New York: Springer, 1995. (Cited on pp. xxi and 440.)

[1328] D. J. Hartfiel, *Nonhomogeneous Matrix Products*. Singapore: World Scientific, 2002. (Cited on pp. xxi and 1242.)

[1329] R. Hartwig, X. Li, and Y. Wei, "Representations for the Drazin Inverse of a 2×2 Block Matrix," *SIAM J. Matrix Anal. Appl.*, Vol. 27, pp. 757–771, 2006. (Cited on p. 667.)

[1330] R. E. Hartwig, "Resultants and the Solutions of $AX - XB = -C$," *SIAM J. Appl. Math.*, Vol. 23, pp. 104–117, 1972. (Cited on p. 1312.)

[1331] R. E. Hartwig, "Block Generalized Inverses," *Arch. Rat. Mech. Anal.*, Vol. 61, pp. 197–251, 1976. (Cited on p. 667.)

[1332] R. E. Hartwig, "A Note on the Partial Ordering of Positive Semi-Definite Matrices," *Lin. Multilin. Alg.*, Vol. 6, pp. 223–226, 1978. (Cited on pp. 815 and 817.)

[1333] R. E. Hartwig, "How to Partially Order Regular Elements," *Math. Japonica*, Vol. 25, pp. 1–13, 1980. (Cited on p. 430.)

[1334] R. E. Hartwig, "A Note on Rank-additivity," *Lin. Multilin. Alg.*, Vol. 10, pp. 59–61, 1981. (Cited on p. 320.)

[1335] R. E. Hartwig and I. J. Katz, "On Products of EP Matrices," *Lin. Alg. Appl.*, Vol. 252, pp. 339–345, 1997. (Cited on p. 643.)

[1336] R. E. Hartwig and M. S. Putcha, "When Is a Matrix a Difference of Two Idempotents?" *Lin. Multilin. Alg.*, Vol. 26, pp. 267–277, 1990. (Cited on pp. 580 and 596.)

[1337] R. E. Hartwig and M. S. Putcha, "When Is a Matrix a Sum of Idempotents?" *Lin. Multilin. Alg.*, Vol. 26, pp. 279–286, 1990. (Cited on p. 619.)

[1338] R. E. Hartwig and K. Spindelbock, "Matrices for which A^* and A^+ Commute," *Lin. Multilin. Alg.*, Vol. 14, pp. 241–256, 1984. (Cited on pp. 573, 630, 631, and 644.)

[1339] R. E. Hartwig and G. P. H. Styan, "On Some Characterizations of the "Star" Partial Ordering for Matrices and Rank Subtractivity," *Lin. Alg. Appl.*, Vol. 82, pp. 145–161, 1989. (Cited on pp. 430, 431, 640, and 816.)

[1340] R. E. Hartwig, G. Wang, and Y. Wei, "Some Additive Results on Drazin Inverse," *Lin. Alg. Appl.*, Vol. 322, pp. 207–217, 2001. (Cited on p. 675.)

[1341] H. Haruki and S. Haruki, "Euler's Integrals," *Amer. Math. Monthly*, Vol. 90, pp. 464–466, 1983. (Cited on p. 1141.)

[1342] N. Harvey, "A Generalization of the Cauchy-Schwarz Inequality Involving Four Vectors," *J. Math. Ineq.*, Vol. 9, pp. 489–491, 2015. (Cited on p. 856.)

[1343] D. A. Harville, *Matrix Algebra from a Statistician's Perspective*. New York: Springer, 1997. (Cited on pp. xxi, 397, 400, 629, 634, 636, 638, 777, 1092, and 1173.)

[1344] M. A. Hasan, "Generalized Wielandt and Cauchy-Schwarz Inequalities," in *Proc. Amer. Contr. Conf.*, Boston, June 2041, pp. 2142–2147. (Cited on pp. 794, 795, 796, 800, and 801.)

[1345] M. Hassani, "Derangements and Applications," *J. Integer Sequences*, Vol. 6, pp. 1–8, 2003, article 03.1.2. (Cited on pp. 49, 104, and 1158.)

[1346] M. Hassani, "On the Ratio of the Arithmetic and Geometric Means of the Prime Numbers and the Number e," *Int. J. Number Theory*, Vol. 9, pp. 1593–1603, 2013. (Cited on p. 966.)

[1347] J. Hauke and A. Markiewicz, "On Partial Orderings on the Set of Rectangular Matrices," *Lin. Alg. Appl.*, Vol. 219, pp. 187–193, 1995. (Cited on p. 815.)

[1348] P. Haukkanen, M. Mattila, J. K. Merikoski, and A. Kovacec, "Bounds for Sine and Cosine via Eigenvalue Estimation," *Special Matrices*, Vol. 2, pp. 19–29, 2014. (Cited on p. 249.)

[1349] M. L. J. Hautus, "Controllability and Observability Conditions of Linear Autonomous Systems," *Proc. Koniklijke Akademic Van Wetenshappen*, Vol. 72, pp. 443–448, 1969. (Cited on p. 1319.)

[1350] J. Havil, *Gamma: Exploring Euler's Constant*. Princeton: Princeton University Press, 2003. (Cited on pp. 35, 47, 48, 54, 332, 957, 963, 966, 996, 1000, 1028, 1032, and 1083.)

[1351] J. Havil, *Impossible? Surprising Solutions to Counterintuitive Conundrums*. Princeton: Princeton University Press, 2011. (Cited on p. 275.)

[1352] J. Havil, *The Irrationals*. Princeton: Princeton University Press, 2012. (Cited on pp. 52, 54, 517, and 1032.)

[1353] S. Hayajneh and F. Kittaneh, "Lieb-Thirring Trace Inequalities and a Question of Bourin," *J. Math. Physics*, Vol. 54, pp. 033 504–1–8, 2013. (Cited on pp. 767, 873, and 875.)

[1354] T. Haynes, "Stable Matrices, the Cayley Transform, and Convergent Matrices," *Int. J. Math. Math. Sci.*, Vol. 14, pp. 77–81, 1991. (Cited on p. 1240.)

[1355] E. V. Haynsworth, "Applications of an Inequality for the Schur Complement," *Proc. Amer. Math. Soc.*, Vol. 24, pp. 512–516, 1970. (Cited on pp. 721, 758, and 831.)

[1356] C. He and L. Zou, "Some Inequalities Involving Unitarily Invariant Norms," *Math. Ineq. Appl.*, Vol. 12, pp. 767–776, 2012. (Cited on pp. 140 and 768.)

[1357] G. He, "Equivalence of Some Matrix Inequalities," *J. Math. Ineq.*, Vol. 8, pp. 153–158, 2014. (Cited on p. 739.)

[1358] D. J. Heath, "Straightedge and Compass Constructions in Spherical Geometry," *Math. Mag.*, Vol. 87, pp. 350–359, 2014. (Cited on p. 232.)

[1359] E. Hecht, *Optics*, 4th ed. Reading: Addison-Wesley, 2002. (Cited on p. xxi.)

[1360] C. Heij, A. Ran, and F. van Schagen, *Introduction to Mathematical Systems Theory: Linear Systems, Identification and Control*. Basel: Birkhauser, 2007. (Cited on p. 1319.)

[1361] G. Heinig, "Matrix Representations of Bezoutians," *Lin. Alg. Appl.*, Vol. 223/224, pp. 337–354, 1995. (Cited on p. 520.)

[1362] U. Helmke and P. A. Fuhrmann, "Bezoutians," *Lin. Alg. Appl.*, Vol. 122–124, pp. 1039–1097, 1989. (Cited on p. 520.)

[1363] U. Helmke and J. Rosenthal, "Eigenvalue Inequalities and Schubert Calculus," *Math. Nachr.*, Vol. 171, pp. 207–225, 1995. (Cited on p. 805.)

[1364] B. W. Helton, "Logarithms of Matrices," *Proc. Amer. Math. Soc.*, Vol. 19, pp. 733–738, 1968. (Cited on p. 1247.)

[1365] H. V. Henderson, F. Pukelsheim, and S. R. Searle, "On the History of the Kronecker Product," *Lin. Multilin. Alg.*, Vol. 14, pp. 113–120, 1983. (Cited on p. 701.)

[1366] H. V. Henderson and S. R. Searle, "On Deriving the Inverse of a Sum of Matrices," *SIAM Rev.*, Vol. 23, pp. 53–60, 1981. (Cited on pp. 350 and 362.)

[1367] H. V. Henderson and S. R. Searle, "The Vec-Permutation Matrix, The Vec Operator and Kronecker Products: A Review," *Lin. Multilin. Alg.*, Vol. 9, pp. 271–288, 1981. (Cited on p. 701.)

[1368] F. Heneghan and K. Petersen, "Power Series for Up-Down Min-Max Permutations," *Coll. Math. J.*, Vol. 45, pp. 83–91, 2014. (Cited on p. 981.)

[1369] P. Henrici, *Applied and Computational Complex Analysis: Vol. 1. Power Series–Integration–Conformal Mapping–Location of Zeros*. New York: John Wiley & Sons, 1974. (Cited on pp. 92 and 276.)

[1370] E. A. Herman, "AGM Inequality for Determinants," *IMAGE*, Vol. 54, p. 38, 2015. (Cited on p. 784.)

[1371] J. Herman, R. Kucera, and J. Simsa, *Equations and Inequalities*. New York: Springer, 2000. (Cited on pp. 37, 48, 50, 58, 67, 75, 86, 94, 123, 124, 125, 126, 134, 137, 143, 150, 151, 155, 161, 164, 167, 168, 173, 177, 180, 183, 191, 193, 197, 215, 484, and 485.)

[1372] J. Herman, R. Kucera, and J. Simsa, *Counting and Configurations: Problems in Combinatorics, Arithmetic, and Geometry*. New York: Springer, 2003. (Cited on pp. 23, 67, 68, 86, and 92.)

[1373] D. Hershkowitz, "Positive Semidefinite Pattern Decompositions," *SIAM J. Matrix Anal. Appl.*, Vol. 11, pp. 612–619, 1990. (Cited on p. 759.)

[1374] D. Hestenes, *Space-Time Algebra*. New York: Gordon and Breach, 1966. (Cited on p. 438.)

[1375] D. Hestenes, *New Foundations for Classical Mechanics*. Dordrecht: Kluwer, 1986. (Cited on pp. 387, 438, and 691.)

[1376] D. Hestenes and G. Sobczyk, *Clifford Algebra to Geometric Calculus: A Unified Language for Mathematics and Physics*. Dordrecht: D. Riedel, 1984. (Cited on pp. 387, 438, and 691.)

[1377] M. R. Hestenes, "Relative Hermitian Matrices," *Pacific J. Math.*, Vol. 11, pp. 225–245, 1961. (Cited on p. 319.)

[1378] F. Hiai, "Monotonicity for Entrywise Functions of Matrices," *Lin. Alg. Appl.*, Vol. 431, pp. 1125–1146, 2009. (Cited on pp. 830 and 945.)

[1379] F. Hiai and H. Kosaki, *Means of Hilbert Space Operators*. Berlin: Springer, 2003. (Cited on pp. 264 and 725.)

[1380] F. Hiai and D. Petz, "The Golden-Thompson Trace Inequality Is Complemented," *Lin. Alg. Appl.*, Vol. 181, pp. 153–185, 1993. (Cited on pp. 743 and 1215.)

[1381] F. Hiai and D. Petz, *Introduction to Matrix Analysis and Applications*. Heidelberg: Springer, 2014. (Cited on pp. 657, 721, 735, 744, 749, 770, 871, 1114, and 1216.)

[1382] F. Hiai and X. Zhan, "Submultiplicativity vs Subadditivity for Unitarily Invariant Norms," *Lin. Alg. Appl.*, Vol. 377, pp. 155–164, 2004. (Cited on p. 873.)

[1383] F. Hiai and X. Zhan, "Inequalities Involving Unitarily Invariant Norms and Operator Monotone Functions," *Lin. Alg. Appl.*, Vol. 436, pp. 3354–3361, 2012. (Cited on pp. 865, 870, 882, and 883.)

[1384] D. J. Higham and N. J. Higham, *Matlab Guide*, 2nd ed. Philadelphia: SIAM, 2005. (Cited on p. 362.)

[1385] N. J. Higham, "Newton's Method for the Matrix Square Root," *Math. Computation*, Vol. 46, pp. 537–549, 1986. (Cited on p. 608.)

[1386] N. J. Higham, "Computing Real Square Roots of a Real Matrix," *Lin. Alg. Appl.*, Vol. 88/89, pp. 405–430, 1987. (Cited on p. 608.)

[1387] N. J. Higham, "Matrix Nearness Problems and Applications," in *Applications of Matrix Theory*, M. J. C. Gover and S. Barnett, Eds. Oxford: Oxford University Press, 1989, pp. 1–27. (Cited on pp. 900, 1227, and 1228.)

[1388] N. J. Higham, "Estimating the Matrix p-Norm," *Numer. Math.*, Vol. 62, pp. 539–555, 1992. (Cited on pp. 856 and 863.)

[1389] N. J. Higham, *Accuracy and Stability of Numerical Algorithms*, 2nd ed. Philadelphia: SIAM, 2002. (Cited on pp. xxi, 420, 427, 584, 585, 587, 856, 862, 863, 864, 866, 897, 900, 907, and 912.)

[1390] N. J. Higham, "Cayley, Sylvester, and Early Matrix Theory," *Lin. Alg. Appl.*, Vol. 428, pp. 39–43, 2008. (Cited on p. 529.)

[1391] N. J. Higham, *Functions of Matrices: Theory and Computation*. Philadelphia: SIAM, 2008. (Cited on pp. xxi, 275, 529, 608, 723, 724, 871, 932, 948, 1019, 1191, 1209, 1211, 1214, 1220, 1221, 1222, 1247, and 1317.)

[1392] G. N. Hile and P. Lounesto, "Matrix Representations of Clifford Algebras," *Lin. Alg. Appl.*, Vol. 128, pp. 51–63, 1990. (Cited on p. 438.)

[1393] R. D. Hill, R. G. Bates, and S. R. Waters, "On Centrohermitian Matrices," *SIAM J. Matrix Anal. Appl.*, Vol. 11, pp. 128–133, 1990. (Cited on p. 427.)

[1394] R. D. Hill and E. E. Underwood, "On the Matrix Adjoint (Adjugate)," *SIAM J. Alg. Disc. Meth.*, Vol. 6, pp. 731–737, 1985. (Cited on pp. 345 and 352.)

[1395] C.-J. Hillar, "Advances on the Bessis-Moussa-Villani Trace Conjecture," *Lin. Alg. Appl.*, Vol. 426, pp. 130–142, 2007. (Cited on p. 771.)

[1396] L. O. Hilliard, "The Case of Equality in Hopf's Inequality," *SIAM J. Alg. Disc. Meth.*, Vol. 8, pp. 691–709, 1987. (Cited on p. 544.)

[1397] H. J. Hindin, "Trigonometric Identity to Hyperbolic Identity to Fibonacci-Lucas Identity," *Math. Gaz.*, Vol. 93, pp. 485–488, 2009. (Cited on pp. 103, 236, and 262.)

[1398] D. Hinrichsen, E. Plischke, and F. Wirth, "Robustness of Transient Behavior," in *Unsolved Problems in Mathematical Systems and Control Theory*, V. D. Blondel and A. Megretski, Eds. Princeton: Princeton University Press, 2004, pp. 197–202. (Cited on p. 1227.)

[1399] D. Hinrichsen and A. J. Pritchard, *Mathematical Systems Theory I: Modelling, State Space Analysis, Stability and Robustness*. Berlin: Springer, 2005. (Cited on pp. 577, 582, 584, 843, 884, 945, 1219, 1220, 1222, 1234, and 1235.)

[1400] J.-B. Hiriart-Urruty, "Potpourri of Conjectures and Open Questions in Nonlinear Analysis and Optimization," *SIAM J. Matrix Anal. Appl.*, Vol. 49, pp. 255–273, 2007. (Cited on pp. 792 and 797.)

[1401] M. W. Hirsch and S. Smale, *Differential Equations, Dynamical Systems, and Linear Algebra*. San Diego: Academic Press, 1974. (Cited on pp. xxi and 570.)

[1402] M. W. Hirsch, S. Smale, and R. L. Devaney, *Differential Equations, Dynamical Systems and an Introduction to Chaos*, 2nd ed. New York: Elsevier, 2003. (Cited on p. xxi.)

[1403] M. Hirschhorn, "Two or Three Identities of Ramanujan," *Amer. Math. Monthly*, Vol. 105, pp. 52–55, 1998. (Cited on p. 179.)

[1404] M. D. Hirschhorn, "Ramanujan's Partition Congruences," *Disc.Math.*, Vol. 131, pp. 351–355, 1994. (Cited on pp. 114 and 391.)

[1405] M. D. Hirschhorn, "Some Binomial Coefficient Identities," *Math. Gaz.*, Vol. 87, pp. 288–291, 2003. (Cited on p. 86.)

[1406] M. D. Hirschhorn, "Sums Involving Square-free Integers," *Math. Gaz.*, Vol. 87, pp. 527–528, 2003. (Cited on p. 1030.)

[1407] M. D. Hirschhorn, "An Identity Involving Binomial Coefficients," *Math. Gaz.*, Vol. 89, pp. 256–258, 2005. (Cited on p. 56.)

[1408] M. D. Hirschhorn, "Approximating Euler's Constant," *Fibonacci Quart.*, Vol. 49, pp. 243–248, 2011. (Cited on p. 964.)

[1409] M. D. Hirschhorn, "Ramanujan's Enigmatic Formula for the Harmonic Series," *Ramanujan J.*, Vol. 27, pp. 343–347, 2012. (Cited on p. 965.)

[1410] M. D. Hirschhorn, "The Asymptotic Behavior of $\prod_{k=0}^{n}\binom{n}{k}$," *Fibonacci Quart.*, Vol. 51, pp. 163–173, 2013. (Cited on pp. 56, 967, 1087, and 1088.)

[1411] M. D. Hirschhorn, "A Connection between π and ϕ," *Fibonacci Quart.*, Vol. 53, pp. 42–47, 2015. (Cited on p. 968.)

[1412] M. D. Hirschhorn, "Binomial Identities and Congruences for Euler Numbers," *Fibonacci Quart.*, Vol. 53, pp. 319–322, 2015. (Cited on p. 981.)

[1413] M. D. Hirschhorn, "Wallis's Product and the Central Binomial Coefficient," *Amer. Math. Monthly*, Vol. 122, p. 689, 2015. (Cited on p. 94.)

[1414] M. D. Hirschhorn and V. Sinescu, "Elementary Algebra in Ramanujan's Lost Notebook," *Fibonacci Quart.*, Vol. 51, pp. 123–129, 2013. (Cited on p. 53.)

[1415] M. D. Hirschhorn and M. B. Villarino, "A Refinement of Ramanujan's Factorial Approximation," http://arxiv.org/pdf/1212.1428.pdf. (Cited on pp. 51 and 228.)

[1416] O. Hirzallah, "Non-commutative Operator Bohr Inequality," *J. Math. Anal. Appl.*, Vol. 282, pp. 578–583, 2003. (Cited on p. 747.)

[1417] O. Hirzallah, "Inequalities for Sums and Products of Operators," *Lin. Alg. Appl.*, Vol. 407, pp. 32–42, 2005. (Cited on p. 905.)

[1418] O. Hirzallah and F. Kittaneh, "Matrix Young Inequalities for the Hilbert-Schmidt Norm," *Lin. Alg. Appl.*, Vol. 308, pp. 77–84, 2000. (Cited on pp. 140, 738, and 904.)

[1419] O. Hirzallah and F. Kittaneh, "Commutator Inequalities for Hilbert-Schmidt Norm," *J. Math. Anal. Appl.*, Vol. 268, pp. 67–73, 2002. (Cited on pp. 145, 271, and 855.)

[1420] O. Hirzallah and F. Kittaneh, "Non-Commutative Clarkson Inequalities for Unitarily Invariant Norms," *Pacific J. Math.*, Vol. 202, pp. 363–369, 2002. (Cited on p. 876.)

[1421] O. Hirzallah and F. Kittaneh, "Norm Inequalities for Weighted Power Means of Operators," *Lin. Alg. Appl.*, Vol. 341, pp. 181–193, 2002. (Cited on p. 794.)

[1422] O. Hirzallah and F. Kittaneh, "Inequalities for Sums and Direct Sums of Hilbert Space Operators," *Lin. Alg. Appl.*, Vol. 424, pp. 71–82, 2007. (Cited on pp. 904 and 905.)

[1423] O. Hirzallah, F. Kittaneh, and M. S. Moslehian, "Schatten p-Norm Inequalities Related to a Characterization of Inner Product Spaces," *Math. Ineq. Appl.*, Vol. 13, pp. 235–241, 2010. (Cited on pp. 142, 211, and 877.)

[1424] A. Hjorungnes, *Complex-Valued Matrix Derivatives with Application in Signal Processing and Communications*. Cambridge: Cambridge University Press, 2011. (Cited on p. 974.)

[1425] A. Hmamed, "A Matrix Inequality," *Int. J. Contr.*, Vol. 49, pp. 363–365, 1989. (Cited on p. 831.)

[1426] D. T. Hoa, H. Osaka, and H. M. Toan, "On Generalized Powers-Stormer's Inequality," *Lin. Alg. Appl.*, Vol. 438, pp. 242–249, 2013. (Cited on pp. 139 and 768.)

[1427] J. B. Hoagg, J. Chandrasekar, and D. S. Bernstein, "On the Zeros, Initial Undershoot, and Relative Degree of Lumped-Parameter Structures," *ASME J. Dyn. Sys. Meas. Contr.*, Vol. 129, pp. 493–502, 2007. (Cited on p. 424.)

[1428] A. J. Hoffman and C. W. Wu, "A Simple Proof of a Generalized Cauchy-Binet Theorem," *Amer. Math. Monthly*, Vol. 123, pp. 928–930, 2016. (Cited on p. 530.)

[1429] K. Hoffman and R. Kunze, *Linear Algebra*, 2nd ed. Englewood Cliffs: Prentice-Hall, 1971. (Cited on p. xxii.)

[1430] M. E. Hoffman, "Derivative Polynomials for Tangent and Secant," *Amer. Math. Monthly*, Vol. 102, pp. 23–30, 1995. (Cited on pp. 994, 1054, and 1161.)

[1431] L. Hogben, Ed., *Handbook of Linear Algebra*. Boca Raton: Chapman & Hall/CRC, 2007. (Cited on p. xxi.)

[1432] J. Holbrook and K. C. O'Meara, "Some Thoughts on Gerstenhaber's Theorem," *Lin. Alg. Appl.*, Vol. 466, pp. 267–295, 2015. (Cited on p. 577.)

[1433] S. R. Holcombe, "A Product Representation for π," http://arxiv.org/abs/1204.2451. (Cited on p. 1087.)

[1434] S. R. Holcombe, "A Product Representation of π," *Amer. Math. Monthly*, Vol. 120, p. 705, 2013. (Cited on p. 1087.)

[1435] F. Holland and S. Wills, "Schur and Definite," *Amer. Math. Monthly*, Vol. 118, p. 89, 2011. (Cited on p. 820.)

[1436] R. R. Holmes and T. Y. Tam, "Group Representations," in *Handbook of Linear Algebra*, L. Hogben, Ed. Boca Raton: Chapman & Hall/CRC, 2007, pp. 68–1–68–11. (Cited on p. 436.)

[1437] O. Holtz, "On Convergence of Infinite Matrix Products," *Elec. J. Lin. Alg.*, Vol. 7, pp. 178–181, 2000. (Cited on p. 1242.)

[1438] O. Holtz, "Hermite-Biehler, Routh-Hurwitz, and Total Positivity," *Lin. Alg. Appl.*, Vol. 372, pp. 105–110, 2003. (Cited on p. 1224.)

[1439] O. Holtz and M. Karow, "Real and Complex Operator Norms," arXiv:math/0512608v1. (Cited on p. 843.)

[1440] M. Holzel, "An Inequality for the Zeros of a Sum of Stable Polynomials," 2015, unpublished. (Cited on p. 1239.)

[1441] Y. Hong and R. A. Horn, "The Jordan Canonical Form of a Product of a Hermitian and a Positive Semidefinite Matrix," *Lin. Alg. Appl.*, Vol. 147, pp. 373–386, 1991. (Cited on p. 609.)

[1442] Y. P. Hong and C.-T. Pan, "A Lower Bound for the Smallest Singular Value," *Lin. Alg. Appl.*, Vol. 172, pp. 27–32, 1992. (Cited on p. 899.)

[1443] R. Honsberger, *Episodes in Nineteenth and Twentieth Century Euclidean Geometry*. Washington, DC: The Mathematical Association of America, 1995. (Cited on p. 446.)

[1444] D. Hopkins, "Will My Numbers Add Up Correctly If I Round Them?" *Math. Gaz.*, Vol. 100, pp. 396–409, 2016. (Cited on p. 1119.)

[1445] K. J. Horadam, *Hadamard Matrices and Their Applications*. Princeton: Princeton University Press, 2007. (Cited on p. 617.)

[1446] A. Horn, "Doubly Stochastic Matrices and the Diagonal of a Rotation Matrix," *Amer. J. Math.*, Vol. 76, pp. 620–630, 1954. (Cited on p. 803.)

[1447] A. Horn, "Eigenvalues of Sums of Hermitian Matrices," *Pacific J. Math.*, Vol. 16, pp. 225–241, 1962. (Cited on p. 805.)

[1448] R. A. Horn and C. R. Johnson, *Matrix Analysis*. Cambridge: Cambridge University Press, 1985. (Cited on pp. 341, 384, 387, 518, 536, 538, 539, 544, 553, 568, 578, 603, 605, 606, 617, 694, 714, 724, 727, 758, 778, 782, 792, 794, 798, 799, 802, 803, 804, 821, 822, 827, 831, 835, 838, 839, 841, 860, 862, 866, 872, 894, and 895.)

[1449] R. A. Horn and C. R. Johnson, "Hadamard and Conventional Submultiplicativity for Unitarily Invariant Norms on Matrices," *Lin. Multilin. Alg.*, Vol. 20, pp. 91–106, 1987. (Cited on pp. 880 and 882.)

[1450] R. A. Horn and C. R. Johnson, *Topics in Matrix Analysis*. Cambridge: Cambridge University Press, 1991. (Cited on pp. 361, 390, 542, 543, 582, 584, 603, 608, 685, 688, 700, 714, 722, 753, 788, 802, 847, 848, 849, 866, 867, 870, 897, 898, 899, 903, 904, 907, 908, 953, 1092, 1187, 1188, 1191, 1218, 1220, 1221, 1229, 1233, 1306, and 1312.)

[1451] R. A. Horn and C. R. Johnson, *Matrix Analysis*, 2nd ed. Cambridge: Cambridge University Press, 2013. (Cited on pp. 325, 330, 346, 347, 538, 546, 570, 572, 576, 663, 695, 711, 778, 782, 792, 793, 794, 797, 838, 839, 868, 894, 895, and 901.)

[1452] R. A. Horn and R. Mathias, "An Analog of the Cauchy-Schwarz Inequality for Hadamard Products and Unitarily Invariant Norms," *SIAM J. Matrix Anal. Appl.*, Vol. 11, pp. 481–498, 1990. (Cited on pp. 874, 884, 897, and 908.)

[1453] R. A. Horn and R. Mathias, "Cauchy-Schwarz Inequalities Associated with Positive Semidefinite Matrices," *Lin. Alg. Appl.*, Vol. 142, pp. 63–82, 1990. (Cited on pp. 828, 866, and 873.)

[1454] R. A. Horn and R. Mathias, "Block-Matrix Generalizations of Schur's Basic Theorems on Hadamard Products," *Lin. Alg. Appl.*, Vol. 172, pp. 337–346, 1992. (Cited on p. 700.)

[1455] R. A. Horn and I. Olkin, "When Does $A^*A = B^*B$ and Why Does One Want to Know?" *Amer. Math. Monthly*, Vol. 103, pp. 470–482, 1996. (Cited on p. 578.)

[1456] R. A. Horn and G. G. Piepmeyer, "Two Applications of the Theory of Primary Matrix Functions," *Lin. Alg. Appl.*, Vol. 361, pp. 99–106, 2003. (Cited on p. 1092.)

[1457] R. A. Horn and X. Zhan, "Inequalities for C-S Seminorms and Lieb Functions," *Lin. Alg. Appl.*, Vol. 291, pp. 103–113, 1999. (Cited on p. 870.)

[1458] R. A. Horn and F. Zhang, "Basic Properties of the Schur Complement," in *The Schur Complement and Its Applications*, F. Zhang, Ed. New York: Springer, 2004, pp. 17–46. (Cited on pp. 591, 669, 742, 749, 781, and 784.)

[1459] R. A. Horn and F. Zhang, "Bounds on the Spectral Radius of a Hadamard Product of Nonnegative or Positive Semidefinite Matrices," *Elec. J. Lin. Alg.*, Vol. 20, pp. 90–94, 2010. (Cited on pp. 699 and 823.)

[1460] R. A. Horn and F. Zhang, "Matrix Diagonal Entries," *IMAGE*, Vol. 50, pp. 38–40, 2013. (Cited on p. 729.)

[1461] B. G. Horne, "Lower Bounds for the Spectral Radius of a Matrix," *Lin. Alg. Appl.*, Vol. 263, pp. 261–273, 1997. (Cited on p. 897.)

[1462] A. Horwitz, "Ellipses of Maximal Area and of Minimal Eccentricity Inscribed in a Convex Quadrilateral," *Australian J. Math. Anal. Appl.*, Vol. 2, pp. 1–12, 2005. (Cited on p. 496.)

[1463] A. Horwitz, "An Area Inequality for Ellipses Inscribed in Quadrilaterals," *J. Math. Ineq.*, Vol. 4, pp. 431–443, 2010. (Cited on p. 496.)

[1464] A. Horwitz, "Ellipses Inscribed in Parallelograms," *Australian J. Math. Anal. Appl.*, Vol. 9, pp. 1–12, 2012. (Cited on p. 496.)

[1465] H.-C. Hou and H.-K. Du, "Norm Inequalities of Positive Operator Matrices," *Integ. Eqns. Operator Theory*, Vol. 22, pp. 281–294, 1995. (Cited on pp. 751, 761, and 902.)

[1466] S.-H. Hou, "A Simple Proof of the Leverrier-Faddeev Characteristic Polynomial Algorithm," *SIAM Rev.*, Vol. 40, pp. 706–709, 1998. (Cited on p. 544.)

[1467] A. S. Householder, *The Theory of Matrices in Numerical Analysis*. New York: Blaisdell, 1964, reprinted by Dover, New York, 1975. (Cited on pp. xxi, 635, 903, and 937.)

[1468] A. S. Householder, "Bezoutiants, Elimination and Localization," *SIAM Rev.*, Vol. 12, pp. 73–78, 1970. (Cited on p. 520.)

[1469] A. S. Householder and J. A. Carpenter, "The Singular Values of Involutory and of Idempotent Matrices," *Numer. Math.*, Vol. 5, pp. 234–237, 1963. (Cited on p. 585.)

[1470] J. Howard, "Problem M419," *Crux Math.*, Vol. 36, p. 368, 2010. (Cited on p. 489.)

[1471] R. Howe, "Very Basic Lie Theory," *Amer. Math. Monthly*, Vol. 90, pp. 600–623, 1983. (Cited on pp. 440 and 1247.)

[1472] J. M. Howie, *Complex Analysis*. New York: Springer, 2003. (Cited on pp. 276 and 1126.)

[1473] R. A. Howland, *Intermediate Dynamics: A Linear Algebraic Approach*. New York: Springer, 2006. (Cited on p. xxi.)

[1474] P.-F. Hsieh and Y. Sibuya, *Basic Theory of Ordinary Differential Equations*. New York: Springer, 1999. (Cited on pp. xxi and 570.)

[1475] G.-D. Hu and G.-H. Hu, "A Relation between the Weighted Logarithmic Norm of a Matrix and the Lyapunov Equation," *Numer. Math. BIT*, Vol. 40, pp. 606–610, 2000. (Cited on p. 1227.)

[1476] X. Hu, "Some Inequalities for Unitarily Invariant Norms," *J. Math. Ineq.*, Vol. 6, pp. 615–623, 2012. (Cited on p. 861.)

[1477] S. Hu-yun, "Estimation of the Eigenvalues of AB for $A > 0$ and $B > 0$," *Lin. Alg. Appl.*, Vol. 73, pp. 147–150, 1986. (Cited on pp. 327 and 811.)

[1478] Y. Hua, "Refinements and Sharpness of Some New Huygens Type Inequalities," *J. Math. Ineq.*, Vol. 6, pp. 493–500, 2012. (Cited on pp. 249 and 264.)

[1479] J. Huang, Y. Peng, and L. Zou, "A Note on Norm Inequalities for Positive Operators," *J. Ineq. Appl.*, pp. 1–4, 2014, article Number 171. (Cited on pp. 869 and 870.)

[1480] R. Huang, "Some Inequalities for Hadamard Products of Matrices," *Lin. Multilin. Alg.*, Vol. 56, pp. 543–554, 2008. (Cited on pp. 897, 908, and 909.)

[1481] R. Huang, "Some Inequalities for the Hadamard Product and the Fan Product of Matrices," *Lin. Alg. Appl.*, Vol. 428, pp. 1551–1559, 2008. (Cited on pp. 699 and 700.)

[1482] T.-Z. Huang and L. Wang, "Improving Bounds for Eigenvalues of Complex Matrices Using Traces," *Lin. Alg. Appl.*, Vol. 426, pp. 841–854, 2007. (Cited on pp. 195 and 891.)

[1483] X. Huang, W. Lin, and B. Yang, "Global Finite-Time Stabilization of a Class of Uncertain Nonlinear Systems," *Automatica*, Vol. 41, pp. 881–888, 2005. (Cited on p. 145.)

[1484] Z. Huang, "On the Spectral Radius and the Spectral Norm of Hadamard Products of Nonnegative Matrices," *Lin. Alg. Appl.*, Vol. 434, pp. 457–462, 2011. (Cited on p. 699.)

[1485] J. H. Hubbard and B. B. Hubbard, *Vector Calculus, Linear Algebra, and Differential Forms*, 2nd ed. Upper Saddle River: Prentice Hall, 2002. (Cited on pp. 267 and 861.)

[1486] P. C. Hughes, *Spacecraft Attitude Dynamics*. New York: Wiley, 1986. (Cited on pp. xxi and 1208.)

[1487] Q. Hui and W. M. Haddad, "$\mathcal{H}_2$ Optimal Semistable Stabilisation for Linear Discrete-Time Dynamical Systems with Applications to Network Consensus," *Int. J. Contr.*, Vol. 82, pp. 456–469, 2009. (Cited on p. 1240.)

[1488] S. Humphries and C. Krattenthaler, "Trace Identities from Identities for Determinants," *Lin. Alg. Appl.*, Vol. 411, pp. 328–342, 2005. (Cited on p. 333.)

[1489] C. H. Hung and T. L. Markham, "The Moore-Penrose Inverse of a Partitioned Matrix," *Lin. Alg. Appl.*, Vol. 11, pp. 73–86, 1975. (Cited on p. 667.)

[1490] J. J. Hunter, "Generalized Inverses and Their Application to Applied Probability Problems," *Lin. Alg. Appl.*, Vol. 45, pp. 157–198, 1982. (Cited on p. 679.)

[1491] D. Huylebrouck, "Similarities in Irrationality Proofs for π, $\ln 2$, $\zeta(2)$, and $\zeta(3)$," *Amer. Math. Monthly*, Vol. 108, pp. 222–231, 2001. (Cited on p. 1174.)

[1492] T. Hyde, "A Wallis Product on Clovers," *Amer. Math. Monthly*, Vol. 121, pp. 237–243, 2014. (Cited on p. 1090.)

[1493] D. C. Hyland and D. S. Bernstein, "The Optimal Projection Equations for Fixed-Order Dynamic Compensation," *IEEE Trans. Autom. Contr.*, Vol. AC-29, pp. 1034–1037, 1984. (Cited on p. 1319.)

[1494] D. C. Hyland and E. G. Collins, "Block Kronecker Products and Block Norm Matrices in Large-Scale Systems Analysis," *SIAM J. Matrix Anal. Appl.*, Vol. 10, pp. 18–29, 1989. (Cited on p. 700.)

[1495] N. H. Ibragimov, *Elementary Lie Group Analysis and Ordinary Differential Equations*. Chichester: Wiley, 1999. (Cited on p. 1247.)

[1496] A. Ibrahim, S. S. Dragomir, and M. Darus, "Power Series Inequalities via Young's Inequality with Applications," *J. Ineq. Appl.*, pp. 1–13, 2013, article 314. (Cited on p. 275.)

[1497] H. Ibstedt, "Integral Powers of an Irrational Number," *J. Recreat. Math.*, Vol. 38, p. 62, 2014. (Cited on p. 39.)

[1498] Y. Ikebe and T. Inagaki, "An Elementary Approach to the Functional Calculus for Matrices," *Amer. Math. Monthly*, Vol. 93, pp. 390–392, 1986. (Cited on p. 1092.)

[1499] K. D. Ikramov, "A Simple Proof of the Generalized Schur Inequality," *Lin. Alg. Appl.*, Vol. 199, pp. 143–149, 1994. (Cited on pp. 761 and 891.)

[1500] K. D. Ikramov, "Determinantal Inequalities for Accretive-Dissipative Matrices," *J. Math. Sci.*, Vol. 121, pp. 2458–2464, 2004. (Cited on p. 783.)

[1501] K. D. Ikramov, "The Inertia of the Components in the Cartesian Decomposition of a Square Matrix," *Dokl. Math.*, Vol. 78, pp. 738–740, 2008. (Cited on p. 618.)

[1502] E. J. Ionascu, "Problem 376," *Gazeta Matematica Serie A*, Vol. 31, pp. 60–61, 2013. (Cited on p. 1128.)

[1503] V. Ionescu, C. Oar, and M. Weiss, *Generalized Riccati Theory and Robust Control*. Chichester: Wiley, 1999. (Cited on pp. xxi and 1319.)

[1504] I. Ipsen and C. Meyer, "The Angle between Complementary Subspaces," *Amer. Math. Monthly*, Vol. 102, pp. 904–911, 1995. (Cited on pp. 314, 415, 585, 586, 595, and 656.)

[1505] A. Iserles, "Solving Linear Ordinary Differential Equations by Exponentials of Iterated Commutators," *Numer. Math.*, Vol. 45, pp. 183–199, 1984. (Cited on p. 1306.)

[1506] A. Iserles, H. Z. Munthe-Kaas, S. P. Norsett, and A. Zanna, "Lie-Group Methods," *Acta Numerica*, Vol. 9, pp. 215–365, 2000. (Cited on pp. 386, 1207, 1212, and 1306.)

[1507] T. Ito, "Mixed Arithmetic and Geometric Means and Related Inequalities," *J. Ineq. Pure Appl. Math.*, Vol. 9, no. 3, pp. 1–21, 2008, Article 65. (Cited on pp. 161 and 173.)

[1508] Y. Ito, S. Hattori, and H. Maeda, "On the Decomposition of a Matrix into the Sum of Stable Matrices," *Lin. Alg. Appl.*, Vol. 297, pp. 177–182, 1999. (Cited on p. 1231.)

[1509] N. V. Ivanov, "Arnol'd, the Jacob Identity, and Orthocenters," *Amer. Math. Monthly*, Vol. 118, pp. 41–65, 2011. (Cited on p. 386.)

[1510] S. Izumino, H. Mori, and Y. Seo, "On Ozeki's Inequality," *J. Ineq. Appl.*, Vol. 2, pp. 235–253, 1998. (Cited on pp. 223 and 793.)

[1511] D. H. Jacobson, D. H. Martin, M. Pachter, and T. Geveci, *Extensions of Linear Quadratic Control Theory*. Berlin: Springer, 1980. (Cited on p. 1319.)

[1512] V. K. Jain, "Visser's Inequality and Its Sharp Refinement," *Bull. Math. Soc. Sci. Math. Roumanie*, Vol. 49, pp. 171–175, 2006. (Cited on p. 942.)

[1513] V. K. Jain, "On the Location of Zeros of a Polynomial," *Bull. Math. Soc. Sci. Math. Roumanie*, Vol. 54, pp. 337–352, 2011. (Cited on p. 1235.)

[1514] G. Jameson and N. Jameson, "Three Answers to an Integral," *Math. Gaz.*, Vol. 90, pp. 457–459, 2006. (Cited on p. 257.)

[1515] G. J. O. Jameson, *The Prime Number Theorem*. Cambridge: Cambridge University Press, 2003. (Cited on p. 966.)

[1516] G. J. O. Jameson, "Inequalities for Gamma Function Ratios," *Amer. Math. Monthly*, Vol. 120, pp. 936–940, 2013. (Cited on pp. 52 and 145.)

[1517] G. J. O. Jameson, "Convexity of $\Gamma(x)\Gamma(1/x)$," *Math. Ineq. Appl.*, Vol. 19, pp. 949–953, 2016. (Cited on p. 117.)

[1518] D. Janezic, A. Milicevic, S. Nikolic, and N. Trinajstic, *Graph-Theoretical Matrices in Chemistry*. Boca Raton: CRC, 2015. (Cited on p. 440.)

[1519] M. Janjic, "Hessenberg Matrices and Integer Sequences," *J. Integer Sequences*, Vol. 13, pp. 1–10, 2010, article 10.7.8. (Cited on pp. 92, 425, 525, and 527.)

[1520] M. Janjic, "Determinants and Recurrent Sequences," *J. Integer Sequences*, Vol. 15, pp. 1–21, 2012, article 12.3.5. (Cited on pp. 425 and 526.)

[1521] W. Janous, "A Short Note on the Erdős-Debrunner Inequality," *Elem. Math.*, Vol. 61, pp. 32–35, 2006. (Cited on p. 472.)

[1522] J. H. Jaroma, "An Upper Bound on the *n*th Prime," *Coll. Math. J.*, Vol. 36, pp. 158–159, 2005. (Cited on p. 966.)

[1523] R. N. Jazar, *Theory of Applied Robotics: Kinematics, Dynamics, and Control*. New York: Springer, 2007. (Cited on p. 237.)

[1524] A. Jeffrey and H.-H. Dai, *Handbook of Mathematical Formulas and Integrals*, 4th ed. San Diego: Academic Press, 1995. (Cited on pp. 43, 67, 75, 76, 236, 237, 262, 981, 984, 1019, 1046, 1109, 1114, 1120, 1123, 1154, 1161, 1163, 1166, and 1168.)

[1525] C. Jeffries, V. Klee, and P. van den Driessche, "Quality Stability of Linear Systems," *Lin. Alg. Appl.*, Vol. 87, pp. 1–48, 1987. (Cited on p. 1234.)

[1526] A. Jennings and J. J. McKeown, *Matrix Computation*, 2nd ed. New York: Wiley, 1992. (Cited on p. xxi.)

[1527] G. A. Jennings, *Modern Geometry with Applications*. New York: Springer, 1997. (Cited on p. 497.)

[1528] S. T. Jensen and G. P. H. Styan, "Some Comments and a Bibliography on the Laguerre-Samuelson Inequality with Extensions and Applications in Statistics and Matrix Theory," in *Analytic and Geometric Inequalities and Applications*, T. M. Rassias and H. M. Srivastava, Eds. Dordrecht: Kluwer, 1999, pp. 151–181. (Cited on pp. 195 and 734.)

[1529] J. Ji, "Explicit Expressions of the Generalized Inverses and Condensed Cramer Rules," *Lin. Alg. Appl.*, Vol. 404, pp. 183–192, 2005. (Cited on p. 330.)

[1530] G. Jia and J. Cao, "A New Upper Bound of the Logarithmic Mean," *J. Ineq. Pure Appl. Math.*, Vol. 4, no. 4, pp. 1–4, 2003, Article 80. (Cited on pp. 144 and 228.)

[1531] C.-C. Jiang, "On Products of Two Hermitian Operators," *Lin. Alg. Appl.*, Vol. 430, pp. 553–557, 2009. (Cited on p. 571.)

[1532] W.-D. Jiang, "A Triangle Inequality," *Amer. Math. Monthly*, Vol. 118, p. 467, 2011. (Cited on p. 466.)

[1533] W.-D. Jiang and M. Bencze, "Some Geometric Inequalities Involving Angle Bisectors and Medians of a Triangle," *J. Math. Ineq.*, Vol. 5, pp. 363–369, 2011. (Cited on pp. 446, 452, and 480.)

[1534] Y. Jiang, W. W. Hager, and J. Li, "The Geometric Mean Decomposition," *Lin. Alg. Appl.*, Vol. 396, pp. 373–384, 2005. (Cited on p. 575.)

[1535] D. Jocic and F. Kittaneh, "Some Perturbation Inequalities for Self-Adjoint Operators," *J. Operator Theory*, Vol. 31, pp. 3–10, 1994. (Cited on p. 872.)

[1536] C. R. Johnson, "An Inequality for Matrices Whose Symmetric Part Is Positive Definite," *Lin. Alg. Appl.*, Vol. 6, pp. 13–18, 1973. (Cited on p. 774.)

[1537] C. R. Johnson, "Inequalities for a Complex Matrix whose Real Part Is Positive Definite," *Trans. Amer. Math. Soc.*, Vol. 212, pp. 149–154, 1975. (Cited on p. 774.)

[1538] C. R. Johnson, "The Inertia of a Product of Two Hermitian Matrices," *J. Math. Anal. Appl.*, Vol. 57, pp. 85–90, 1977. (Cited on p. 567.)

[1539] C. R. Johnson, "Closure Properties of Certain Positivity Classes of Matrices under Various Algebraic Operations," *Lin. Alg. Appl.*, Vol. 97, pp. 243–247, 1987. (Cited on p. 1225.)

[1540] C. R. Johnson, "A Gersgorin-Type Lower Bound for the Smallest Singular Value," *Lin. Alg. Appl.*, Vol. 112, pp. 1–7, 1989. (Cited on p. 899.)

[1541] C. R. Johnson and L. Elsner, "The Relationship between Hadamard and Conventional Multiplication for Positive Definite Matrices," *Lin. Alg. Appl.*, Vol. 92, pp. 231–240, 1987. (Cited on p. 823.)

[1542] C. R. Johnson, M. Neumann, and M. J. Tsatsomeros, "Conditions for the Positivity of Determinants," *Lin. Multilin. Alg.*, Vol. 40, pp. 241–248, 1996. (Cited on p. 424.)

[1543] C. R. Johnson and M. Newman, "A Surprising Determinantal Inequality for Real Matrices," *Math. Ann.*, Vol. 247, pp. 179–186, 1980. (Cited on p. 333.)

[1544] C. R. Johnson and P. Nylen, "Monotonicity Properties of Norms," *Lin. Alg. Appl.*, Vol. 148, pp. 43–58, 1991. (Cited on p. 912.)

[1545] C. R. Johnson, K. Okubo, and R. Beams, "Uniqueness of Matrix Square Roots," *Lin. Alg. Appl.*, Vol. 323, pp. 51–60, 2001. (Cited on p. 608.)

[1546] C. R. Johnson and L. Rodman, "Convex Sets of Hermitian Matrices with Constant Inertia," *SIAM J. Alg. Disc. Meth.*, Vol. 6, pp. 351–359, 1985. (Cited on pp. 566 and 569.)

[1547] C. R. Johnson and R. Schreiner, "The Relationship between AB and BA," *Amer. Math. Monthly*, Vol. 103, pp. 578–582, 1996. (Cited on p. 606.)

[1548] C. R. Johnson and H. M. Shapiro, "Mathematical Aspects of the Relative Gain Array $A \odot A^{-\mathrm{T}}$," *SIAM J. Alg. Disc. Meth.*, Vol. 7, pp. 627–644, 1986. (Cited on pp. 698 and 821.)

[1549] C. R. Johnson and R. L. Smith, "Linear Interpolation Problems for Matrix Classes and a Transformational Characterization of M-Matrices," *Lin. Alg. Appl.*, Vol. 330, pp. 43–48, 2001. (Cited on pp. 383, 396, and 1206.)

[1550] C. R. Johnson and T. Szulc, "Further Lower Bounds for the Smallest Singular Value," *Lin. Alg. Appl.*, Vol. 272, pp. 169–179, 1998. (Cited on p. 899.)

[1551] C. R. Johnson and F. Zhang, "An Operator Inequality and Matrix Normality," *Lin. Alg. Appl.*, Vol. 240, pp. 105–110, 1996. (Cited on p. 794.)

[1552] W. P. Johnson, "The Curious History of Faà di Bruno's Formula," *Amer. Math. Monthly*, Vol. 109, pp. 217–234, 2002. (Cited on p. 108.)

[1553] W. P. Johnson, "Trigonometric Identities à la Hermite," *Amer. Math. Monthly*, Vol. 117, pp. 310–327, 2010. (Cited on pp. 245 and 246.)

[1554] M. Jolly, "On the Calculus of Complex Matrices," *Int. J. Contr.*, Vol. 61, pp. 749–755, 1995. (Cited on p. 1092.)

[1555] T. F. Jordan, *Quantum Mechanics in Simple Matrix Form*. New York: Wiley, 1986, reprinted by Dover, Mineola, 2005. (Cited on p. 439.)

[1556] E. A. Jorswieck and H. Boche, "Majorization and Matrix Monotone Functions in Wireless Communications," *Found. Trends Comm. Inform. Theory*, Vol. 3, pp. 553–701, 2006. (Cited on p. xxi.)

[1557] E. A. Jorswieck and H. Boche, "Performance Analysis of MIMO Systems in Spatially Correlated Fading Using Matrix-Monotone Functions," *IEICE Trans. Fund.*, Vol. E89-A, pp. 1454–1472, 2006. (Cited on pp. xxi and 360.)

[1558] A. Joseph, A. Melnikov, and R. Rentschler, Eds., *Studies in Memory of Issai Schur*. Boston: Birkhauser, 2002. (Cited on p. 831.)

[1559] M. Jovanovic and V. Jovanovic, "Convexity of Elementary Functions with Applications to Inequalities," *Elem. Math.*, Vol. 66, pp. 10–18, 2011. (Cited on pp. 209, 210, 252, 264, 446, and 472.)

[1560] D. Joyner, *Adventures in Group Theory*. Baltimore: Johns Hopkins University Press, 2002. (Cited on pp. 329, 370, 417, 436, 437, and 440.)

[1561] S. Jukna, *Extremal Combinators with Applications to Computer Science*, 2nd ed. New York: Springer, 2011. (Cited on p. 24.)

[1562] E. I. Jury, *Inners and Stability of Dynamic Systems*, 2nd ed. Malabar: Krieger, 1982. (Cited on pp. 695 and 1234.)

[1563] Y. Kachi and P. Tzermias, "Infinite Products Involving $\zeta(3)$ and Catalan's Constant," *J. Integer Sequences*, Vol. 15, pp. 1–16, 2012, article 12.9.4. (Cited on pp. 961, 1070, 1087, and 1094.)

[1564] S. Kaczkowski, "The Limiting Value of a Series with Exponential Terms," *Math. Mag.*, Vol. 89, pp. 282–287, 2016. (Cited on pp. 961 and 967.)

[1565] S. Kaczkowski, "A Sum with an Exponential Limit," *Coll. Math. J.*, Vol. 48, pp. 62–63, 2017. (Cited on p. 961.)

[1566] W. J. Kaczor and M. T. Nowak, *Problems in Mathematical Analysis I: Real Numbers, Sequences and Series*. American Mathematical Society, 2000. (Cited on pp. 24, 47, 48, 49, 51, 94, 115, 125, 156, 191, 197, 198, 208, 210, 211, 212, 232, 253, 859, 955, 959, 962, 963, 965, 966, 967, 968, 1022, 1023, 1025, 1046, 1047, 1049, 1056, 1057, 1060, 1061, 1062, 1070, 1071, 1073, 1074, 1079, 1081, 1083, 1084, 1087, 1089, and 1090.)

[1567] W. J. Kaczor and M. T. Nowak, *Problems in Mathematical Analysis II: Continuity and Differentiation*. American Mathematical Society, 2001. (Cited on pp. 125, 144, 225, 227, 229, 249, 250, 261, and 264.)

[1568] W. J. Kaczor and M. T. Nowak, *Problems in Mathematical Analysis III: Integration*. American Mathematical Society, 2003. (Cited on pp. 1000, 1047, 1056, 1080, 1097, 1104, 1108, 1119, 1130, 1141, 1145, and 1154.)

[1569] Z. Kadelburg, D. Dukic, M. Lukic, and I. Matic, "Inequalities of Karamata, Schur and Muirhead, and Some Applications," *Teaching Math.*, Vol. VIII, pp. 31–45, 2005. (Cited on pp. 116, 139, 164, 167, 177, 201, 253, 361, and 362.)

[1570] R. V. Kadison, "Order Properties of Bounded Self-Adjoint Operators," *Proc. Amer. Math. Soc.*, Vol. 2, pp. 505–510, 1951. (Cited on p. 740.)

[1571] A. Kagan and P. J. Smith, "A Stronger Version of Matrix Convexity as Applied to Functions of Hermitian Matrices," *J. Ineq. Appl.*, Vol. 3, pp. 143–152, 1999. (Cited on p. 772.)

[1572] J. B. Kagstrom, "Bounds and Perturbation Bounds for the Matrix Exponential," *Numer. Math. BIT*, Vol. 17, pp. 39–57, 1977. (Cited on p. 1219.)

[1573] T. Kailath, *Linear Systems*. Englewood Cliffs: Prentice-Hall, 1980. (Cited on pp. 504, 563, 612, 619, and 1319.)

[1574] B. Kalantari, "A Geometric Modulus Principle for Polynomials," *Amer. Math. Monthly*, Vol. 118, pp. 931–935, 2011. (Cited on p. 949.)

[1575] D. Kalman, "A Matrix Proof of Newton's Identities," *Math. Mag.*, Vol. 73, pp. 313–315, 2000. (Cited on p. 518.)

[1576] D. Kalman, "An Elementary Proof of Marden's Theorem," *Amer. Math. Monthly*, Vol. 115, pp. 330–338, 2008. (Cited on p. 496.)

[1577] D. Kalman and J. E. White, "Polynomial Equations and Circulant Matrices," *Amer. Math. Monthly*, Vol. 108, pp. 821–840, 2001. (Cited on p. 615.)

[1578] T. R. Kane, P. W. Likins, and D. A. Levinson, *Spacecraft Dynamics*. New York: McGraw-Hill, 1983. (Cited on p. xxi.)

[1579] R. K. Kanwar, "A Generalization of the Cayley-Hamilton Theorem," *Adv. Pure Math.*, Vol. 3, pp. 109–115, 2013. (Cited on p. 529.)

[1580] I. Kaplansky, *Linear Algebra and Geometry: A Second Course*. New York: Chelsea, 1974, reprinted by Dover, Mineola, 2003. (Cited on pp. xxii and 576.)

[1581] N. Karagiannis, "Two Inequalities for Real Numbers," *Octogon*, Vol. 17, pp. 742–745, 2009. (Cited on p. 1049.)

[1582] N. S. Karamzadeh, "Schur's Inequality for the Dimension of Commuting Families of Matrices," *Math. Ineq. Appl.*, Vol. 13, pp. 625–628, 2010. (Cited on p. 577.)

[1583] N. Karcanias, "Matrix Pencil Approach to Geometric System Theory," *Proc. IEE*, Vol. 126, pp. 585–590, 1979. (Cited on pp. 619 and 1319.)

[1584] S. Karlin and F. Ost, "Some Monotonicity Properties of Schur Powers of Matrices and Related Inequalities," *Lin. Alg. Appl.*, Vol. 68, pp. 47–65, 1985. (Cited on p. 699.)

[1585] N. N. Kasturiwale and M. N. Deshpande, "A Surprising Identity Involving Fibonacci Numbers," *Math. Gaz.*, Vol. 87, pp. 277–279, 2003. (Cited on p. 100.)

[1586] E. Kaszkurewicz and A. Bhaya, *Matrix Diagonal Stability in Systems and Computation*. Boston: Birkhauser, 2000. (Cited on p. 1248.)

[1587] K. K. Kataria, "Some Probabilistic Interpretations of the Multinomial Theorem," *Math. Mag.*, Vol. 90, pp. 221–224, 2017. (Cited on pp. 86 and 189.)

[1588] M. Kato, K.-S. Saito, and T. Tamura, "Sharp Triangle Inequality and Its Reverse in Banach Space," *Math. Ineq. Appl.*, Vol. 10, pp. 451–460, 2007. (Cited on pp. 851 and 852.)

[1589] T. Kato, "Spectral Order and a Matrix Limit Theorem," *Lin. Multilin. Alg.*, Vol. 8, pp. 15–19, 1979. (Cited on p. 741.)

[1590] T. Kato, *Perturbation Theory for Linear Operators*, 2nd ed. Berlin: Springer, 1980. (Cited on pp. 595, 651, 893, 912, and 941.)

[1591] H. Katsuura, "Generalizations of the Arithmetic-Geometric Mean Inequality and a Three Dimensional Puzzle," *Coll. Math. J.*, Vol. 34, pp. 280–282, 2003. (Cited on pp. 183 and 206.)

[1592] H. Katsuura, "Summations Involving Binomial Coefficients," *Coll. Math. J.*, Vol. 40, pp. 275–278, 2009. (Cited on p. 75.)

[1593] M. Kauderer, *Symplectic Matrices: First Order Systems and Special Relativity*. Singapore: World Scientific, 1994. (Cited on p. xxi.)

[1594] M. Kauers and S.-L. Ko, "A Stirling Sum," *Amer. Math. Monthly*, Vol. 119, pp. 885–886, 2012. (Cited on p. 107.)

[1595] S. Kayalar and H. L. Weinert, "Oblique Projections: Formulas, Algorithms, and Error Bounds," *Math. Contr. Sig. Sys.*, Vol. 2, pp. 33–45, 1989. (Cited on p. 574.)

[1596] J. Y. Kazakia, "Orthogonal Transformation of a Trace Free Symmetric Matrix into One with Zero Diagonal Elements," *Int. J. Eng. Sci.*, Vol. 26, pp. 903–906, 1988. (Cited on p. 571.)

[1597] N. D. Kazarinoff, *Analytic Inequalities*. New York: Holt, Rinehart, and Winston, 1961, reprinted by Dover, Mineola, 2003. (Cited on pp. 141 and 835.)

[1598] D. Keedwell, "Square-Triangular Numbers," *Math. Gaz.*, Vol. 84, pp. 292–294, 2000. (Cited on p. 39.)

[1599] D. G. Kendall and R. Osborn, "Two Simple Lower Bounds for Euler's Function," *Texas J. Sci.*, Vol. 17, 1965. (Cited on p. 115.)

[1600] M. G. Kendall, *A Course in the Geometry of n Dimensions*. London: Griffin, 1961, reprinted by Dover, Mineola, 2004. (Cited on p. 498.)

[1601] C. Kenney and A. J. Laub, "Controllability and Stability Radii for Companion Form Systems," *Math. Contr. Sig. Sys.*, Vol. 1, pp. 239–256, 1988. (Cited on pp. 584 and 1236.)

[1602] C. Kenney and A. J. Laub, "Rational Iteration Methods for the Matrix Sign Function," *SIAM J. Matrix Anal. Appl.*, Vol. 12, pp. 273–291, 1991. (Cited on p. 948.)

[1603] R. Keskin and B. Demirturk, "Some New Fibonacci and Lucas Identities by Matrix Methods," *Int. J. Math. Educ. Sci. Tech.*, Vol. 41, pp. 379–387, 2010. (Cited on pp. 100 and 103.)

[1604] H. Kestelman, "Eigenvectors of a Cross-Diagonal Matrix," *Amer. Math. Monthly*, Vol. 93, p. 566, 1986. (Cited on p. 599.)

[1605] N. Keyfitz, *Introduction to the Mathematics of Population*. Reading: Addison-Wesley, 1968. (Cited on pp. xxi and 540.)

[1606] W. Khalil and E. Dombre, *Modeling, Identification, and Control of Robots*. New York: Taylor and Francis, 2002. (Cited on p. xxi.)

[1607] C. G. Khatri, "A Note on Idempotent Matrices," *Lin. Alg. Appl.*, Vol. 70, pp. 185–195, 1985. (Cited on p. 651.)

[1608] C. G. Khatri and S. K. Mitra, "Hermitian and Nonnegative Definite Solutions of Linear Matrix Equations," *SIAM J. Appl. Math.*, Vol. 31, pp. 579–585, 1976. (Cited on p. 639.)

[1609] S. K. Khattri, "Three Proofs of the Inequality $e < \left(1 + \frac{1}{n}\right)^{n+0.5}$," *Amer. Math. Monthly*, Vol. 117, pp. 273–277, 2010. (Cited on p. 128.)

[1610] S. Khrushchev, "Two Great Theorems of Lord Brouncker and His Formula $b(s-1)b(s+1) = s^2$," *Math. Intelligencer*, Vol. 32, no. 1, pp. 19–31, 2010. (Cited on p. 1097.)

[1611] D. Khuyrana, T. Y. Lam, and N. Shomron, "A Quantum-Trace Determinantal Formula for Matrix Commutators, and Applications," *Lin. Alg. Appl.*, Vol. 436, pp. 2380–2397, 2012. (Cited on pp. 327, 429, and 528.)

[1612] E. Kilic and H. Prodinger, "Some double binomial sums related with the fibonacci, pell and generalized order-k fibonacci numbers," *Rocky Mountain J. Math.*, Vol. 43, pp. 975–987, 2013. (Cited on pp. 100, 103, and 1079.)

[1613] H. Kim and Y. Lim, "An Extended Matrix Exponential Formula," *J. Math. Ineq.*, Vol. 1, pp. 443–447, 2007. (Cited on p. 744.)

[1614] S. Kim and Y. Lim, "A Converse Inequality of Higher Order Weighted Arithmetic and Geometric Means of Positive Definite Operators," *Lin. Alg. Appl.*, Vol. 426, pp. 490–496, 2007. (Cited on p. 744.)

[1615] C. Kimberling and P. Moses, "Inequalities Involving Six Numbers, with Applications to Triangle Geometry," *Math. Mag.*, Vol. 85, pp. 221–227, 2012. (Cited on pp. 151 and 184.)

[1616] C. King, "Inequalities for Trace Norms of 2×2 Block Matrices," *Comm. Math. Phys.*, Vol. 242, pp. 531–545, 2003. (Cited on p. 889.)

[1617] C. King and M. Nathanson, "New Trace Norm Inequalities for 2×2 Blocks of Diagonal Matrices," *Lin. Alg. Appl.*, Vol. 389, pp. 77–93, 2004. (Cited on p. 879.)

[1618] R. B. King, *Beyond the Quartic Equation*. Boston: Birkhauser, 1996. (Cited on p. 436.)

[1619] M. K. Kinyon, "Revisited: arctan 1 + arctan 2 + arctan 3 = π," *Coll. Math. J.*, Vol. 37, pp. 218–219, 2006. (Cited on p. 257.)

[1620] W. A. Kirk and M. F. Smiley, "Another Characterization of Inner Product Spaces," *Amer. Math. Monthly*, Vol. 71, pp. 890–891, 1964. (Cited on p. 850.)

[1621] U. S. Kirmaci, M. K. Bakula, M. E. Ozdemir, and J. Pecaric, "On Some Inequalities for p-norms," *J. Ineq. Pure. Appl. Math.*, Vol. 4, no. 1/27, pp. 1–8, 2008. (Cited on p. 205.)

[1622] A. P. Kisielewicz, "Some Binomial Identities Arising from a Partition of an n-Dimensional Cube," *Fibonacci Quart.*, Vol. 52, pp. 325–330, 2014. (Cited on p. 68.)

[1623] F. Kittaneh, "Inequalities for the Schatten p-Norm III," *Commun. Math. Phys.*, Vol. 104, pp. 307–310, 1986. (Cited on pp. 869 and 871.)

[1624] F. Kittaneh, "Inequalities for the Schatten p-Norm. IV," *Commun. Math. Phys.*, Vol. 106, pp. 581–585, 1986. (Cited on p. 871.)

[1625] F. Kittaneh, "On Zero-Trace Matrices," *Lin. Alg. Appl.*, Vol. 151, pp. 119–124, 1991. (Cited on p. 571.)

[1626] F. Kittaneh, "A Note on the Arithmetic-Geometric-Mean Inequality for Matrices," *Lin. Alg. Appl.*, Vol. 171, pp. 1–8, 1992. (Cited on p. 882.)

[1627] F. Kittaneh, "Norm Inequalities for Fractional Powers of Positive Operators," *Lett. Math. Phys.*, Vol. 27, pp. 279–285, 1993. (Cited on pp. 807 and 905.)

[1628] F. Kittaneh, "Singular Values of Companion Matrices and Bounds on Zeros of Polynomials," *SIAM J. Matrix Anal. Appl.*, Vol. 16, pp. 333–340, 1995. (Cited on pp. 584 and 1236.)

[1629] F. Kittaneh, "Norm Inequalities for Certain Operator Sums," *J. Funct. Anal.*, Vol. 143, pp. 337–348, 1997. (Cited on pp. 142, 211, 808, 812, 877, and 881.)

[1630] F. Kittaneh, "Some Norm Inequalities for Operators," *Canad. Math. Bull.*, Vol. 42, pp. 87–96, 1999. (Cited on p. 882.)

[1631] F. Kittaneh, "Commutator Inequalities Associated with the Polar Decomposition," *Proc. Amer. Math. Soc.*, Vol. 130, pp. 1279–1283, 2001. (Cited on pp. 871, 873, 884, and 897.)

[1632] F. Kittaneh, "Norm Inequalities for Sums of Positive Operators," *J. Operator Theory*, Vol. 48, pp. 95–103, 2002. (Cited on pp. 808, 881, 902, and 903.)

[1633] F. Kittaneh, "Bounds for the Zeros of Polynomials from Matrix Inequalities," *Arch. Math.*, Vol. 81, pp. 601–608, 2003. (Cited on pp. 588, 887, and 1236.)

[1634] F. Kittaneh, "A Numerical Radius Inequality and an Estimate for the Numerical Radius of the Frobenius Companion Matrix," *Studia Mathematica*, Vol. 158, pp. 11–17, 2003. (Cited on pp. 866 and 1237.)

[1635] F. Kittaneh, "Norm Inequalities for Sums and Differences of Positive Operators," *Lin. Alg. Appl.*, Vol. 383, pp. 85–91, 2004. (Cited on pp. 808, 812, and 897.)

[1636] F. Kittaneh, "Numerical Radius Inequalities for Hilbert Space Operators," *Studia Mathematica*, Vol. 168, pp. 73–80, 2005. (Cited on pp. 866 and 903.)

[1637] F. Kittaneh, "Norm Inequalities for Sums of Positive Operators II," *Positivity*, Vol. 10, pp. 251–260, 2006. (Cited on pp. 870 and 881.)

[1638] F. Kittaneh, "Bounds for the Zeros of Polynomials from Matrix Inequalities–II," *Lin. Multilin. Alg.*, Vol. 55, pp. 147–158, 2007. (Cited on p. 1236.)

[1639] F. Kittaneh and H. Kosaki, "Inequalities for the Schatten p-Norm. V," *Publications of RIMS Kyoto University*, Vol. 23, pp. 433–443, 1987. (Cited on pp. 871 and 872.)

[1640] F. Kittaneh and Y. Manasrah, "Improved Young and Heinz Inequalities for Matrices," *J. Math. Anal. Appl.*, Vol. 361, pp. 262–269, 2010. (Cited on pp. 140, 141, 768, 778, and 882.)

[1641] M. S. Klamkin and K. R. S. Sastry, "Solution to Problem 3024," *Crux Math.*, Vol. 32, pp. 181–183, 2006. (Cited on p. 333.)

[1642] A.-L. Klaus and C.-K. Li, "Isometries for the Vector (p,q) Norm and the Induced (p,q) Norm," *Lin. Multilin. Alg.*, Vol. 38, pp. 315–332, 1995. (Cited on p. 875.)

[1643] L. F. Klosinski, G. L. Alexanderson, and M. Krusemeyer, "The Seventy-Sixth William Lowell Putnam Mathematical Competition," *Amer. Math. Monthly*, Vol. 123, pp. 739–752, 2016. (Cited on p. 530.)

[1644] A. A. Klyachko, "Stable Bundles, Representation Theory and Hermitian Operators," *Selecta Math.*, Vol. 4, pp. 419–445, 1998. (Cited on p. 805.)

[1645] M. P. Knapp, "Sines and Cosines of Angles in Arithmetic Progression," *Math. Mag.*, Vol. 82, pp. 371–372, 2009. (Cited on p. 242.)

[1646] O. Knill, "Cauchy-Binet for Pseudo-determinants," *Lin. Alg. Appl.*, Vol. 459, pp. 522–547, 2014. (Cited on p. 530.)

[1647] K. Knopp, *Theory and Application of Infinite Series*. New York: Dover, 1990. (Cited on pp. 1021, 1022, 1044, 1046, 1047, 1048, 1049, 1056, 1060, 1074, and 1092.)

[1648] J. A. Knox and H. J. Brothers, "Novel Series-Based Approximations to e," *Coll. Math. J.*, Vol. 30, pp. 269–275, 1999. (Cited on pp. 127 and 955.)

[1649] D. E. Knuth, "Two Notes on Notation," *Amer. Math. Monthly*, Vol. 99, pp. 403–422, 1992. (Cited on pp. 108 and 118.)

[1650] D. E. Knuth, *The Art of Computer Programming, Vol. 1, Fundamental Algorithms*, 3rd ed. Reading: Addison-Wesley, 1997. (Cited on p. 188.)

[1651] D. E. Knuth, *The Art of Computer Programming, Vol. 2: Seminumerical Algorithms*, 3rd ed. Upper Saddle River: Addison-Wesley, 1998. (Cited on p. 119.)

[1652] D. E. Knuth, *The Art of Computer Programming, Vol. 3: Sorting and Searching*, 3rd ed. Upper Saddle River: Addison-Wesley, 1998. (Cited on p. 112.)

[1653] A. Knutson and T. Tao, "Honeycombs and Sums of Hermitian Matrices," *Notices Amer. Math. Soc.*, Vol. 48, pp. 175–186, 2001. (Cited on p. 805.)

[1654] C. T. Koch and J. C. H. Spence, "A Useful Expansion of the Exponential of the Sum of Two Non-Commuting Matrices, One of Which Is Diagonal," *J. Phys. A: Math. Gen.*, Vol. 36, pp. 803–816, 2003. (Cited on p. 1306.)

[1655] M. Koeber and U. Schafer, "The Unique Square Root of a Positive Semidefinite Matrix," *Int. J. Math. Educ. Sci. Tech.*, Vol. 37, pp. 990–992, 2006. (Cited on p. 714.)

[1656] D. Koks, *Explorations in Mathematical Physics*. New York: Springer, 2006. (Cited on pp. 438 and 570.)

[1657] J. J. Koliha, "A Generalized Drazin Inverse," *Glasgow Math. J.*, Vol. 38, pp. 367–381, 1996. (Cited on p. 669.)

[1658] J. J. Koliha, "A Simple Proof of the Product Theorem for EP Matrices," *Lin. Alg. Appl.*, Vol. 294, pp. 213–215, 1999. (Cited on p. 643.)

[1659] J. J. Koliha and V. Rakocevic, "Invertibility of the Sum of Idempotents," *Lin. Multilin. Alg.*, Vol. 50, pp. 285–292, 2002. (Cited on p. 400.)

[1660] J. J. Koliha and V. Rakocevic, "Invertibility of the Difference of Idempotents," *Lin. Multilin. Alg.*, Vol. 51, pp. 97–110, 2003. (Cited on p. 402.)

[1661] J. J. Koliha and V. Rakocevic, "The Nullity and Rank of Linear Combinations of Idempotent Matrices," *Lin. Alg. Appl.*, Vol. 418, pp. 11–14, 2006. (Cited on p. 400.)

[1662] J. J. Koliha, V. Rakocevic, and I. Straskraba, "The Difference and Sum of Projectors," *Lin. Alg. Appl.*, Vol. 388, pp. 279–288, 2004. (Cited on pp. 400, 402, 405, and 406.)

[1663] N. Komaroff, "Bounds on Eigenvalues of Matrix Products with an Applicaton to the Algebraic Riccati Equation," *IEEE Trans. Autom. Contr.*, Vol. 35, pp. 348–350, 1990. (Cited on p. 592.)

[1664] N. Komaroff, "Rearrangement and Matrix Product Inequalities," *Lin. Alg. Appl.*, Vol. 140, pp. 155–161, 1990. (Cited on pp. 216 and 809.)

[1665] N. Komaroff, "Enhancements to the von Neumann Trace Inequality," *Lin. Alg. Appl.*, Vol. 428, pp. 738–741, 2008. (Cited on p. 593.)

[1666] V. Komornik, "Another Short Proof of Descartes's Rule of Signs," *Amer. Math. Monthly*, Vol. 113, pp. 829–830, 2006. (Cited on p. 1223.)

[1667] H. Konig, *Eigenvalue Distribution of Compact Operators*. Birkhauser, 1986. (Cited on p. 892.)

[1668] R. H. Koning, H. Neudecker, and T. Wansbeek, "Block Kronecker Products and the vecb Operator," *Lin. Alg. Appl.*, Vol. 149, pp. 165–184, 1991. (Cited on p. 700.)

[1669] J. Konvalina, "A Combinatorial Formula for Powers of 2×2 Matrices," *Math. Mag.*, Vol. 88, pp. 280–284, 2015. (Cited on p. 531.)

[1670] T. H. Koornwinder and M. J. Schlosser, "On an Identity by Chaundy and Bullard, I," *Indag. Math.*, Vol. 19, pp. 239–261, 2008. (Cited on p. 68.)

[1671] J. Korevaar, "A Conjecture and Its Proof by Louis de Branges," *Amer. Math. Monthly*, Vol. 93, pp. 505–514, 1986. (Cited on p. 947.)

[1672] P. Korus, "Some Applications of Retkes' Identity," *Arch. Math.*, Vol. 104, pp. 217–222, 2015. (Cited on pp. 50, 76, and 189.)

[1673] T. Koshy, "A Generalization of a Curious Sum," *Math. Gaz.*, Vol. 83, pp. 97–101, 1999. (Cited on p. 42.)

[1674] T. Koshy, *Fibonacci and Lucas Numbers with Applications*. New York: Wiley, 2001. (Cited on pp. xxi, 100, 103, and 423.)

[1675] T. Koshy, *Catalan Numbers with Applications*. Oxford: Oxford University Press, 2009. (Cited on pp. 52, 55, 58, 68, 75, 76, 86, 92, 94, 100, 103, 105, 1008, 1049, 1070, 1071, and 1097.)

[1676] T. Koshy, "Nested Fibonacci and Lucas Radical Sums," *Math. Gaz.*, Vol. 93, pp. 83–87, 2009. (Cited on p. 974.)

[1677] T. Koshy, "Fibonacci, Lucas, and Pell Numbers, and Pascal's Triangle," *Math. Spectrum*, Vol. 43, pp. 125–132, 2011. (Cited on pp. 68, 103, and 131.)

[1678] T. Koshy and Z. Gao, "Convergence of a Catalan Series," *Coll. Math. J.*, Vol. 43, pp. 141–146, 2012. (Cited on p. 1075.)

[1679] O. Kouba, "An Optimal Inequality for the Tangent Function," http://arxiv.org/abs/1205.0377. (Cited on p. 249.)

[1680] O. Kouba, "Elementary Evaluation of $\int_0^\infty \frac{\sin^p t}{t^q}\, dt$," http://arxiv.org/math/0401099. (Cited on p. 1120.)

[1681] O. Kouba, "A Nonobtuse Altitude Inequality," *Amer. Math. Monthly*, Vol. 118, pp. 752–753, 2011. (Cited on pp. 466, 472, and 480.)

[1682] O. Kouba, "The Sum of Certain Series Related to Harmonic Numbers," *Octogon Math. Mag.*, Vol. 19, pp. 3–18, 2011. (Cited on pp. 1026, 1034, 1046, 1100, 1136, and 1142.)

[1683] O. Kouba, "A Series with Harmonic Numbers," *Amer. Math. Monthly*, Vol. 119, pp. 253–254, 2012. (Cited on pp. 1026 and 1027.)

[1684] O. Kouba, "Exact Evaluation of Some Highly Oscillatory Integrals," *J. Classical Anal.*, Vol. 3, pp. 45–57, 2013. (Cited on pp. 1130 and 1155.)

[1685] O. Kouba, "A Mixed Parseval-Plancherel Formula," *J. Classical Anal.*, Vol. 4, pp. 159–165, 2014. (Cited on p. 1154.)

[1686] O. Kouba, "Adding a Linear Combination of Harmonic Series," *Math. Mag.*, Vol. 87, p. 235, 2014. (Cited on p. 1044.)

[1687] O. Kouba, "Bounds for the Ratios of Differences of Power Means in Two Arguments," *Math. Ineq. Appl.*, Vol. 17, pp. 913–928, 2014. (Cited on p. 143.)

[1688] O. Kouba, "Generalized Stieltjes Constants and Integrals Involving the Log-log Function: Kummer's Theorem in Action," *J. Classical Anal.*, Vol. 9, pp. 79–88, 2016. (Cited on p. 1143.)

[1689] O. Kouba, "Inequalities for Finite Trigonometric Sums. An Interplay: With Some Series Related to Harmonic Numbers," *J. Ineq. Appl.*, pp. 1–15, 2016, article 173. (Cited on p. 242.)

[1690] O. Kouba, "On Certain New Refinements of Finsler-Hadwiger Inequalities," *J. Ineq. Appl.*, pp. 1–13, 2017, article 80. (Cited on p. 452.)

[1691] S. Koumandos, "Some Inequalities for Cosine Sums," *Math. Ineq. Appl.*, Vol. 4, pp. 267–279, 2001. (Cited on p. 254.)

[1692] J. Kovac-Striko and K. Veselic, "Trace Minimization and Definiteness of Symmetric Pencils," *Lin. Alg. Appl.*, Vol. 216, pp. 139–158, 1995. (Cited on p. 619.)

[1693] M. Kovalyov, "Removing Magic from the Normal Distribution and Stirling and Wallis Formulas," *Math. Intelligencer*, Vol. 33, no. 4, pp. 32–36, 2011. (Cited on p. 1083.)

[1694] V. Kowalenko, "On a Finite Sum Involving Inverse Powers of Cosines," *Acta Appl. Math.*, Vol. 115, pp. 139–151, 2011. (Cited on p. 242.)

[1695] T. Kowalski, "The Sine of a Single Degree," *Coll. Math. J.*, Vol. 47, pp. 322–332, 2016. (Cited on p. 232.)

[1696] O. Krafft, "An Arithmetic-Harmonic-Mean Inequality for Nonnegative Definite Matrices," *Lin. Alg. Appl.*, Vol. 268, pp. 243–246, 1998. (Cited on p. 819.)

[1697] S. G. Krantz, *Handbook of Complex Variables*. Boston: Birkhauser, 1999. (Cited on pp. 276, 946, and 947.)

[1698] S. G. Krantz, "A Matter of Gravity," *Amer. Math. Monthly*, Vol. 110, pp. 465–481, 2003. (Cited on p. 472.)

[1699] C. Krattenthaler, "Advanced Determinant Calculus," http://arxiv.math/9902004v3. (Cited on pp. 334 and 341.)

[1700] C. Krattenthaler, "Advanced Determinant Calculus: A Complement," *Lin. Alg. Appl.*, Vol. 411, pp. 68–166, 2005. (Cited on p. xxi.)

[1701] W. Kratz, *Quadratic Functionals in Variational Analysis and Control Theory*. New York: Wiley, 1995. (Cited on p. 1319.)

[1702] M. Krebs and N. C. Martinez, "The Combinatorial Trace Method in Action," *Coll. Math. J.*, Vol. 44, pp. 32–36, 2013. (Cited on p. 76.)

[1703] E. Kreindler and A. Jameson, "Conditions for Nonnegativeness of Partitioned Matrices," *IEEE Trans. Autom. Contr.*, Vol. AC-17, pp. 147–148, 1972. (Cited on p. 831.)

[1704] R. Kress, H. L. de Vries, and R. Wegmann, "On Nonnormal Matrices," *Lin. Alg. Appl.*, Vol. 8, pp. 109–120, 1974. (Cited on p. 890.)

[1705] D. Kressner, *Numerical Methods for General and Structured Eigenvalue Problems.* New York: Springer, 2005. (Cited on p. 597.)

[1706] M. Krusemeyer, "Why Does the Wronskian Work?" *Amer. Math. Monthly*, Vol. 95, pp. 46–49, 1988. (Cited on p. 953.)

[1707] V. Kucera, "On Nonnegative Definite Solutions to Matrix Quadratic Equations," *Automatica*, Vol. 8, pp. 413–423, 1972. (Cited on p. 1319.)

[1708] A. Kufner, L. Maligranda, and L.-E. Persson, "The Prehistory of the Hardy Inequality," *Amer. Math. Monthly*, Vol. 113, pp. 715–732, 2006. (Cited on pp. 212 and 219.)

[1709] J. B. Kuipers, *Quaternions and Rotation Sequences: A Primer with Applications to Orbits, Aerospace, and Virtual Reality.* Princeton: Princeton University Press, 1999. (Cited on pp. xxi, 438, and 439.)

[1710] M. R. S. Kulenovic and G. Ladas, *Dynamics of Second Order Rational Difference Equations with Open Problems and Conjectures.* Boca Raton: Chapman & Hall/CRC Press, 2002. (Cited on p. 1248.)

[1711] M. U. Kumar, V. Chellaboina, S. P. Bhat, S. Prasad, and A. Bhatia, "Discrete-Time Optimal Hedging for Multi-Asset Path-Dependent European Contingent Claims," in *Proc. Conf. Dec. Contr.*, Shanghai, China, December 2009, pp. 3679–3684. (Cited on p. 697.)

[1712] S. Kurepa, "A Note on Logarithms of Normal Operators," *Proc. Amer. Math. Soc.*, Vol. 13, pp. 307–311, 1962. (Cited on p. 1217.)

[1713] H. Kurzweil and B. Stellmacher, *The Theory of Finite Groups: An Introduction.* New York: Springer, 2004. (Cited on p. 370.)

[1714] K. Kwakernaak and R. Sivan, *Linear Optimal Control Systems.* New York: Wiley, 1972. (Cited on p. xxi.)

[1715] M. Kwapisz, "The Power of a Matrix," *SIAM Rev.*, Vol. 40, pp. 703–705, 1998. (Cited on p. 1239.)

[1716] M. K. Kwong, "Some Results on Matrix Monotone Functions," *Lin. Alg. Appl.*, Vol. 118, pp. 129–153, 1989. (Cited on pp. 427 and 741.)

[1717] I. I. Kyrchei, "Analogs of the Adjoint Matrix for Generalized Inverses and Corresponding Cramer Rules," *Lin. Multilin. Alg.*, Vol. 56, pp. 453–469, 2008. (Cited on p. 330.)

[1718] K. R. Laberteaux, "Hermitian Matrices," *Amer. Math. Monthly*, Vol. 104, p. 277, 1997. (Cited on p. 380.)

[1719] T. J. Laffey, "Simultaneous Triangularization of a Pair of Matrices–Low Rank Cases and the Nonderogatory Case," *Lin. Multilin. Alg.*, Vol. 6, pp. 269–306, 1978. (Cited on p. 617.)

[1720] T. J. Laffey, "Products of Skew-Symmetric Matrices," *Lin. Alg. Appl.*, Vol. 68, pp. 249–251, 1985. (Cited on p. 610.)

[1721] T. J. Laffey, "Products of Matrices," in *Generators and Relations in Groups and Geometries*, A. Barlotti, E. W. Ellers, P. Plaumann, and K. Strambach, Eds. Dordrecht: Kluwer, 1991, pp. 95–123. (Cited on p. 610.)

[1722] T. J. Laffey and S. Lazarus, "Two-Generated Commutative Matrix Subalgebras," *Lin. Alg. Appl.*, Vol. 147, pp. 249–273, 1991. (Cited on p. 577.)

[1723] T. J. Laffey and E. Meehan, "A Refinement of an Inequality of Johnson, Loewy and London on Nonnegative Matrices and Some Applications," *Elec. J. Lin. Alg.*, Vol. 3, pp. 119–128, 1998. (Cited on p. 544.)

[1724] J. C. Lagarias, "An Elementary Problem Equivalent to the Riemann Hypothesis," *Amer. Math. Monthly*, Vol. 109, no. 6, pp. 534–543, 2002. (Cited on p. 36.)

[1725] J. C. Lagarias and Y. Wang, "The Finiteness Conjecture for the Generalized Spectral Radius of a Set of Matrices," *Lin. Alg. Appl.*, Vol. 214, pp. 17–42, 1995. (Cited on p. 1242.)

[1726] S. Lakshminarayanan, S. L. Shah, and K. Nandakumar, "Cramer's Rule for Non-Square Matrices," *Amer. Math. Monthly*, Vol. 106, p. 865, 1999. (Cited on p. 330.)

[1727] M. S. Lambrou and W. E. Longstaff, "On the Lengths of Pairs of Complex Matrices of Size Six," *Bull. Austral. Math. Soc.*, Vol. 80, pp. 177–201, 2009. (Cited on p. 578.)

[1728] E. Lampakis, "A Trigonometric Inequality for Triangles," *Math. Mag.*, Vol. 85, pp. 231–232, 2012. (Cited on p. 466.)

[1729] E. Lampakis, "A Lower Bound on a Sum," *Coll. Math. J.*, Vol. 44, p. 144, 2013. (Cited on p. 202.)

[1730] E. Lampakis, "An Inequality with Semiperimeter, Inradius, and Circumradius," *Coll. Math. J.*, Vol. 46, pp. 65–66, 2015. (Cited on p. 452.)

[1731] E. Lampakis, "Another Algebraic Inequality," *Coll. Math. J.*, Vol. 46, p. 221, 2015. (Cited on p. 169.)
[1732] E. Lampakis, "An Inequality for the Sides of a Triangle," *Coll. Math. J.*, Vol. 48, p. 145, 2017. (Cited on p. 452.)
[1733] V. Lampret, "A Sharp Double Inequality for Sums of Powers," *J. Ineq. Appl.*, pp. 1–7, 2011, article 721827. (Cited on p. 961.)
[1734] P. Lancaster, *Lambda-Matrices and Vibrating Systems*. Oxford: Pergamon, 1966, reprinted by Dover, Mineola, 2002. (Cited on p. xxi.)
[1735] P. Lancaster and L. Rodman, "Solutions of the Continuous and Discrete Time Algebraic Riccati Equations: A Review," in *The Riccati Equation*, S. Bittanti, J. C. Willems, and A. Laub, Eds. New York: Springer, 1991, pp. 11–51. (Cited on p. 1319.)
[1736] P. Lancaster and L. Rodman, *Algebraic Riccati Equations*. Oxford: Clarendon, 1995. (Cited on pp. xxi and 1319.)
[1737] P. Lancaster and L. Rodman, "Canonical Forms for Hermitian Matrix Pairs under Strict Equivalence and Congruence," *SIAM Rev.*, Vol. 47, pp. 407–443, 2005. (Cited on pp. 563 and 797.)
[1738] P. Lancaster and M. Tismenetsky, *The Theory of Matrices*, 2nd ed. Orlando: Academic Press, 1985. (Cited on pp. 578, 599, 687, 847, 868, 912, 1310, and 1311.)
[1739] S. Landau, "How to Tangle with a Nested Radical," *Math. Intelligencer*, Vol. 16, no. 3, pp. 49–55, 1994. (Cited on p. 53.)
[1740] L. J. Lander, T. R. Parkin, and J. L. Selfridge, "A Survey of Equal Sums of Like Powers," *Math. Comp.*, Vol. 21, pp. 446–459, 1967. (Cited on p. 35.)
[1741] C. L. Lang and M. L. Lang, "Fibonacci Numbers and Identities," *Fibonacci Quart.*, Vol. 51, pp. 330–338, 2013. (Cited on p. 100.)
[1742] H. Langer, "Matrices with a Strictly Dominant Eigenvalue," *Elem. Math.*, Vol. 56, pp. 55–61, 2001. (Cited on pp. 1211 and 1242.)
[1743] P. J. Larcombe, "A Forgotten Convolution Type Identity of Catalan," *Utilitas Math.*, Vol. 57, pp. 65–72, 2000. (Cited on p. 105.)
[1744] P. J. Larcombe, "On Catalan Numbers and Expanding the Sine Function," *Bull. Inst. Combinatorics Appl.*, Vol. 28, pp. 39–47, 2000. (Cited on p. 1075.)
[1745] P. J. Larcombe, "On Certain Series Expansions of the Sine Function Containing Embedded Catalan Numbers: A Complete Analytic Formulation," *J. Combinatorial Math. Combinatorial Computing*, Vol. 59, pp. 3–16, 2006. (Cited on pp. 1057 and 1088.)
[1746] P. J. Larcombe, "Closed Form Evaluations of Some Series Involving Catalan Numbers," *Bull. Inst. Combinatorics Appl.*, Vol. 71, p. 117119, 2014. (Cited on p. 1075.)
[1747] P. J. Larcombe, "A Note on the Invariance of the General 2×2 Matrix Anti-Diagonals Ratio with Increasing Matrix Power: Four Proofs," *Fibonacci Quart.*, Vol. 53, pp. 360–364, 2015. (Cited on p. 328.)
[1748] P. J. Larcombe, "Closed Form Evaluations of Some Series Comprising Sums of Exponentiated Multiples of Two-Term and Three-Term Catalan Number Linear Combinations," *Fibonacci Quart.*, Vol. 53, pp. 253–260, 2015. (Cited on p. 1075.)
[1749] P. J. Larcombe and E. J. Fennessey, "Applying Integer Programming to Enumerate Equilibrium States of a Multi-Link Inverted Pendulum: A Strange Binomial Coefficient Identity and Its Proof," *Bull. Inst. Combinatorics Appl.*, Vol. 64, pp. 83–108, 2012. (Cited on p. 67.)
[1750] P. J. Larcombe and E. J. Fennessey, "Some Properties of the Sum $\sum_{i=0}^{n} i^p \binom{n+i}{i}$," *Congressus Numerantium*, Vol. 214, pp. 49–64, 2012. (Cited on p. 68.)
[1751] P. J. Larcombe and E. J. Fennessey, "A Condition for Anti-Diagonals Product Invariance Across Powers of 2×2 Matrix Sets Characterizing a Particular Class of Polynomial Families," *Fibonacci Quart.*, Vol. 53, pp. 175–179, 2015. (Cited on p. 328.)
[1752] P. J. Larcombe, E. J. Fennessey, and W. A. Koepf, "Integral Proofs of Two Alternating Sign Binomial Coefficient Identities," *Utilitas Math.*, Vol. 66, pp. 93–103, 2004. (Cited on p. 75.)
[1753] P. J. Larcombe and D. R. French, "A New Proof of the Integral Form for the General Catalan Number Using a Trigonometric Identity of Bullard," *Bull. Inst. Combinatorics Appl.*, Vol. 36, pp. 37–45, 2002. (Cited on p. 1113.)
[1754] P. J. Larcombe, D. R. French, and I. M. Gessel, "On the Identity of von Szily: Original Derivation and a New Proof," *Utilitas Math.*, Vol. 64, pp. 167–181, 2003. (Cited on p. 92.)

[1755] P. J. Larcombe and I. M. Gessel, "A Forgotten Convolution Type Identity of Catalan: Two Hypergeometric Proofs," *Utilitas Math.*, Vol. 59, pp. 97–109, 2001. (Cited on p. 105.)

[1756] P. J. Larcombe, S. T. O'Neill, and E. J. Fennessey, "On Certain Series Expansions of the Sine Function: Catalan Numbers and Convergence," *Fibonacci Quart.*, Vol. 52, pp. 236–242, 2014. (Cited on pp. 1057 and 1075.)

[1757] L. Larson, *Problem Solving Through Problems.* New York: Springer, 1983. (Cited on pp. 28, 29, 31, 35, 43, 45, 47, 48, 50, 75, 86, 94, 100, 116, 127, 128, 132, 138, 145, 164, 167, 173, 192, 198, 207, 236, 239, 244, 249, 250, 252, 261, 331, 433, 443, 472, 485, 489, 493, 1008, 1046, 1056, 1060, 1071, 1072, 1079, 1083, 1090, and 1130.)

[1758] L. Larsson, L. Maligranda, J. Pecaric, and L.-E. Persson, *Multiplicative Inequalities of Carlson Type and Interpolation.* Singapore: World Scientific, 2006. (Cited on p. 213.)

[1759] J. Lasenby, A. N. Lasenby, and C. J. L. Doran, "A Unified Mathematical Language for Physics and Engineering in the 21st Century," *Phil. Trans. Math. Phys. Eng. Sci.*, Vol. 358, pp. 21–39, 2000. (Cited on pp. 387, 438, 691, and 858.)

[1760] J. B. Lasserre, "A Trace Inequality for Matrix Product," *IEEE Trans. Autom. Contr.*, Vol. 40, pp. 1500–1501, 1995. (Cited on p. 592.)

[1761] J. Latulippe and J. Switkes, "Even with That Step Size?" *Math. Mag.*, Vol. 84, pp. 377–385, 2011. (Cited on p. 92.)

[1762] A. J. Laub, *Matrix Analysis for Scientists and Engineers.* Philadelphia: SIAM, 2005. (Cited on pp. xxii, 310, 563, and 565.)

[1763] A. J. Laub and K. Meyer, "Canonical Forms for Symplectic and Hamiltonian Matrices," *Celestial Mechanics*, Vol. 9, pp. 213–238, 1974. (Cited on p. 1316.)

[1764] C. Laurie, B. Mathes, and H. Radjavi, "Sums of Idempotents," *Lin. Alg. Appl.*, Vol. 208/209, pp. 175–197, 1994. (Cited on p. 619.)

[1765] J. L. Lavoie, "A Determinantal Inequality Involving the Moore-Penrose Inverse," *Lin. Alg. Appl.*, Vol. 31, pp. 77–80, 1980. (Cited on p. 668.)

[1766] C. L. Lawson, *Solving Least Squares Problems.* Englewood Cliffs: Prentice-Hall, 1974, reprinted by SIAM, Philadelphia, 1995. (Cited on p. 679.)

[1767] J. D. Lawson and Y. Lim, "The Geometric Mean, Matrices, Metrics, and More," *Amer. Math. Monthly*, Vol. 108, pp. 797–812, 2001. (Cited on pp. 714 and 744.)

[1768] P. D. Lax, *Linear Algebra.* New York: Wiley, 1997. (Cited on pp. 494, 522, 737, and 893.)

[1769] S. R. Lay, *Convex Sets and Their Applications.* New York: Wiley, 1982. (Cited on pp. 307, 362, and 940.)

[1770] T. Le, "A Generalization of Nesbitt's Inequality," *Math. Mag.*, Vol. 84, pp. 69–70, 2011. (Cited on p. 165.)

[1771] T. Le, "A Symmetric Inequality," *Amer. Math. Monthly*, Vol. 118, p. 850, 2011. (Cited on p. 160.)

[1772] T. Le, "On the Lower Bounds of a Symmetric Inequality Involving Roots," *Math. Scientist*, Vol. 36, pp. 10–18, 2011. (Cited on pp. 164, 177, and 186.)

[1773] B. Leclerc, "On Identities Satisfied by Minors of a Matrix," *Adv. Math.*, Vol. 100, pp. 101–132, 1993. (Cited on p. 330.)

[1774] K. J. LeCouteur, "Representation of the Function Tr(exp(A-λB)) as a Laplace Transform with Positive Weight and Some Matrix Inequalities," *J. Phys. A: Math. Gen.*, Vol. 13, pp. 3147–3159, 1980. (Cited on pp. 810 and 1212.)

[1775] A. Lee, "On S-Symmetric, S-Skewsymmetric, and S-Orthogonal Matrices," *Periodica Math. Hungar.*, Vol. 7, pp. 71–76, 1976. (Cited on p. 571.)

[1776] A. Lee, "Centrohermitian and Skew-Centrohermitian Matrices," *Lin. Alg. Appl.*, Vol. 29, pp. 205–210, 1980. (Cited on pp. 427 and 632.)

[1777] C. Lee and V. Peterson, "The Rank of Recurrence Matrices," *Coll. Math. J.*, Vol. 45, pp. 207–215, 2014. (Cited on p. 320.)

[1778] E.-Y. Lee, "How to Compare the Absolute Values of Operator Sums and the Sums of Absolute Values," *Oper. Matrices*, Vol. 6, pp. 613–619, 2012. (Cited on p. 870.)

[1779] E.-Y. Lee, "On the Fan and Lidskii Majorisations of Positive Semidefinite Matrices," *Math. Ineq. Appl.*, Vol. 18, pp. 569–580, 2015. (Cited on pp. 764, 859, 869, 882, and 893.)

[1780] J. M. Lee and D. A. Weinberg, "A Note on Canonical Forms for Matrix Congruence," *Lin. Alg. Appl.*, Vol. 249, pp. 207–215, 1996. (Cited on p. 619.)

[1781] D. H. Lehmer, "Interesting Series Involving the Central Binomial Coefficient," *Amer. Math. Monthly*, Vol. 92, pp. 449–457, 1985. (Cited on pp. 1008, 1063, 1064, 1065, and 1070.)

[1782] S. H. Lehnigk, *Stability Theorems for Linear Motions*. Englewood Cliffs: Prentice-Hall, 1966. (Cited on p. 1247.)

[1783] T.-G. Lei, C.-W. Woo, and F. Zhang, "Matrix Inequalities by Means of Embedding," *Elect. J. Lin. Alg.*, Vol. 11, pp. 66–77, 2004. (Cited on pp. 733, 761, 772, 783, 784, 792, 794, 798, and 800.)

[1784] F. S. Leite, "Bounds on the Order of Generation of so(n, r) by One-Parameter Subgroups," *Rocky Mountain J. Math.*, Vol. 21, pp. 879–911, 1183–1188, 1991. (Cited on pp. 607 and 1207.)

[1785] E. Leonard, "The Matrix Exponential," *SIAM Rev.*, Vol. 38, pp. 507–512, 1996. (Cited on p. 1181.)

[1786] D. A. Lesher and C. D. Lynd, "Convergence Results for the Class of Periodic Left Nested Radicals," *Math. Mag.*, Vol. 89, pp. 319–335, 2016. (Cited on p. 974.)

[1787] J. Lesko, "Sums of Harmonic-Type Series," *Coll. Math. J.*, Vol. 35, pp. 171–182, 2004. (Cited on pp. 1046 and 1047.)

[1788] G. Letac, "A Matrix and Its Matrix of Reciprocals Both Positive Semi-Definite," *Amer. Math. Monthly*, Vol. 82, pp. 80–81, 1975. (Cited on p. 820.)

[1789] G. Letac, "Complex Hermitian Matrix," *Amer. Math. Monthly*, Vol. 117, pp. 464–465, 2010. (Cited on p. 725.)

[1790] P. Levrie, "The Asumptotic Behavior of the Binomial Coefficients," *Math. Mag.*, Vol. 90, pp. 119–123, 2017. (Cited on p. 1008.)

[1791] J. S. Lew, "The Cayley Hamilton Theorem in n Dimensions," *Z. Angew. Math. Phys.*, Vol. 17, pp. 650–653, 1966. (Cited on p. 528.)

[1792] A. S. Lewis, "Convex Analysis on the Hermitian Matrices," *SIAM J. Optim.*, Vol. 6, pp. 164–177, 1996. (Cited on pp. 592 and 593.)

[1793] B. Lewis, "Trigonometric and Fibonacci Numbers," *Math. Gaz.*, Vol. 91, pp. 216–226, 2007. (Cited on pp. 103, 239, and 245.)

[1794] D. C. Lewis, "A Qualitative Analysis of S-Systems: Hopf Bifurcations," in *Canonical Nonlinear Modeling*, E. O. Voit, Ed. New York: Van Nostrand Reinhold, 1991, pp. 304–344. (Cited on p. 698.)

[1795] C.-K. Li, "A Simple Proof of the Craig-Sakamoto Theorem," *Lin. Alg. Appl.*, Vol. 321, pp. 281–283, 2000. (Cited on p. 331.)

[1796] C.-K. Li, "Positive Semidefinite Matrices and Tensor Product," *IMAGE*, Vol. 41, p. 44, 2008. (Cited on p. 827.)

[1797] C.-K. Li and R.-C. Li, "A Note on Eigenvalues of Perturbed Hermitian Matrices," *Lin. Alg. Appl.*, Vol. 395, pp. 183–190, 2005. (Cited on p. 537.)

[1798] C.-K. Li and R. Mathias, "The Determinant of the Sum of Two Matrices," *Bull. Austral. Math. Soc.*, Vol. 52, pp. 425–429, 1995. (Cited on pp. 903 and 904.)

[1799] C.-K. Li and R. Mathias, "The Lidskii-Mirsky-Wielandt Theorem–Additive and Multiplicative Versions," *Numer. Math.*, Vol. 81, pp. 377–413, 1999. (Cited on p. 893.)

[1800] C.-K. Li and R. Mathias, "Extremal Characterizations of the Schur Complement and Resulting Inequalities," *SIAM Rev.*, Vol. 42, pp. 233–246, 2000. (Cited on pp. 717, 721, 822, and 831.)

[1801] C.-K. Li and R. Mathias, "Inequalities on Singular Values of Block Triangular Matrices," *SIAM J. Matrix Anal. Appl.*, Vol. 24, pp. 126–131, 2002. (Cited on p. 585.)

[1802] C.-K. Li and S. Nataraj, "Some Matrix Techniques in Game Theory," *Math. Ineq. Appl.*, Vol. 3, pp. 133–141, 2000. (Cited on p. xxi.)

[1803] C.-K. Li and E. Poon, "Additive Decomposition of Real Matrices," *Lin. Multilin. Alg.*, Vol. 50, pp. 321–326, 2002. (Cited on p. 618.)

[1804] C.-K. Li and Y.-T. Poon, "Sum of Hermitian Matrices with Given Eigenvalues: Inertia, Rank, and Multiple Eigenvalues," *Canadian J. Math.*, Vol. 62, pp. 109–132, 2010. (Cited on p. 805.)

[1805] C.-K. Li and L. Rodman, "Some Extremal Problems for Positive Definite Matrices and Operators," *Lin. Alg. Appl.*, Vol. 140, pp. 139–154, 1990. (Cited on pp. 740 and 741.)

[1806] C.-K. Li and H. Schneider, "Orthogonality of Matrices," *Lin. Alg. Appl.*, Vol. 347, pp. 115–122, 2002. (Cited on p. 853.)

[1807] C.-K. Li and N. Sze, "Determinantal and Eigenvalue Inequalities for Matrices with Numerical Ranges in a Sector," *J. Math. Anal. Appl.*, Vol. 410, pp. 487–491, 2014. (Cited on p. 783.)

[1808] C.-K. Li and F. Zhang, "Positivity of Partitioned Hermitian Matrices with Unitarily Invariant Norms," *Positivity*, Vol. 19, pp. 439–444, 2015. (Cited on p. 888.)

[1809] H. Li, Z. Gao, and D. Zhao, "A Note on the Frobenius Conditional Number with Positive Definite Matrices," *J. Ineq. Appl.*, pp. 1–9, 2011, article ID 120. (Cited on pp. 193, 214, 861, and 874.)

[1810] H. Li and D. Zhao, "An Extension of the Golden-Thompson Theorem," *J. Ineq. Appl.*, pp. 1–6, 2014, article Number 14. (Cited on pp. 898, 900, and 1216.)

[1811] H.-B. Li, T.-Z. Huang, and H. Li, "Some New Results on Determinantal Inequalities and Applications," *J. Ineq. Appl.*, pp. 1–16, 2010, article ID 847357. (Cited on p. 896.)

[1812] J.-L. Li and Y.-L. Li, "On the Strengthened Jordan's Inequality," *J. Ineq. Appl.*, pp. 1–8, 2007, article ID 74328. (Cited on pp. 249 and 250.)

[1813] Q. Li, "Commutators of Orthogonal Projections," *Nihonkai. Math. J.*, Vol. 15, pp. 93–99, 2004. (Cited on p. 884.)

[1814] X. Li and Y. Wei, "A Note on the Representations for the Drazin Inverse of 2×2 Block Matrices," *Lin. Alg. Appl.*, Vol. 423, pp. 332–338, 2007. (Cited on p. 667.)

[1815] J. Liao, H. Liu, M. Shao, and X. Xu, "A Matrix Identity and Its Applications," *Lin. Alg. Appl.*, Vol. 471, pp. 346–352, 2015. (Cited on p. 612.)

[1816] B. V. Lidskii, "Spectral Polyhedron of a Sum of Hermitian Matrices," *Functional Analysis Appl.*, Vol. 10, pp. 139–140, 1982. (Cited on p. 805.)

[1817] E. H. Lieb, "Convex Trace Functions and the Wigner-Yanase-Dyson Conjecture," *Adv. Math.*, Vol. 11, pp. 267–288, 1973. (Cited on pp. 721, 791, 831, and 1216.)

[1818] E. H. Lieb and M. Loss, *Analysis*. Providence: American Mathematical Society, 2001. (Cited on p. 859.)

[1819] E. H. Lieb and M. B. Ruskai, "Some Operator Inequalities of the Schwarz Type," *Adv. Math.*, Vol. 12, pp. 269–273, 1974. (Cited on pp. 758 and 831.)

[1820] E. H. Lieb and R. Seiringer, "Equivalent Forms of the Bessis-Moussa-Villani Conjecture," *J. Stat. Phys.*, Vol. 115, pp. 185–190, 2004. (Cited on p. 771.)

[1821] E. H. Lieb and W. E. Thirring, "Inequalities for the Moments of the Eigenvalues of the Schrodinger Hamiltonian and Their Relation to Sobolev Inequalities," in *Studies in Mathematical Physics*, E. H. Lieb, B. Simon, and A. Wightman, Eds. Princeton: Princeton University Press, 1976, pp. 269–303. (Cited on p. 810.)

[1822] P. Ligouras, "An Inequality for Triangles," *Amer. Math. Monthly*, Vol. 118, pp. 561–562, 2011. (Cited on p. 452.)

[1823] P. Ligouras, "Hexagon Inequality," *Coll. Math. J.*, Vol. 44, pp. 69–70, 2013. (Cited on p. 489.)

[1824] Y. Lim, "The Matrix Golden Mean and Its Applications to Riccati Matrix Equations," *SIAM J. Matrix Anal. Appl.*, Vol. 29, pp. 54–66, 2006. (Cited on p. 1318.)

[1825] C.-S. Lin, "On Halmos's Sharpening of Reid's Inequality," *Math. Reports Acad. Sci. Canada*, Vol. 20, pp. 62–64, 1998. (Cited on p. 795.)

[1826] C.-S. Lin, "Inequalities of Reid Type and Furuta," *Proc. Amer. Math. Soc.*, Vol. 129, pp. 855–859, 2000. (Cited on p. 795.)

[1827] C.-S. Lin, "Heinz-Kato-Furuta Inequalities with Bounds and Equality Conditions," *Math. Ineq. Appl.*, Vol. 5, pp. 735–743, 2002. (Cited on p. 855.)

[1828] C.-S. Lin, "On Operator Order and Chaotic Operator Order for Two Operators," *Lin. Alg. Appl.*, Vol. 425, pp. 1–6, 2007. (Cited on pp. 746 and 814.)

[1829] M. Lin, "Remarks on Krein's Inequality," *Math. Intelligencer*, Vol. 34, pp. 3–4, 2012. (Cited on pp. 252, 326, and 857.)

[1830] M. Lin, "Reversed Determinantal Inequalities for Accretive-Dissipative Matrices," *Math. Ineq. Appl.*, Vol. 15, pp. 955–958, 2012. (Cited on pp. 750 and 783.)

[1831] M. Lin, "A Lewent Type Determinantal Inequality," *Taiwanese J. Math.*, Vol. 17, pp. 1303–1309, 2013. (Cited on p. 779.)

[1832] M. Lin, "Fischer Type Determinantal Inequalities for Accretive-Dissipative Matrices," *Lin. Alg. Appl.*, Vol. 438, pp. 2808–2812, 2013. (Cited on pp. 777 and 783.)

[1833] M. Lin, "Notes on an Anderson-Taylor Type Inequality," *Elec. J. Lin. Alg.*, Vol. 26, pp. 63–70, 2013. (Cited on pp. 744, 745, and 798.)

[1834] M. Lin, "A Completely PPT Map," *Lin. Alg. Appl.*, Vol. 459, pp. 404–410, 2014. (Cited on pp. 751 and 755.)

[1835] M. Lin, "A Determinantal Inequality for Positive Definite Matrices," *Elec. J. Lin. Alg.*, Vol. 27, pp. 821–826, 2014. (Cited on pp. 226, 360, 778, and 781.)

[1836] M. Lin, "A Hadamard-like Inequality," *IMAGE*, Vol. 53, p. 45, 2014. (Cited on p. 775.)

[1837] M. Lin, "A Positive Semidefinite Matrix," *IMAGE*, Vol. 54, pp. 40–41, 2015. (Cited on p. 755.)

[1838] M. Lin, "Determinantal Inequalities for Block Triangular Matrices," *Math. Ineq. Appl.*, Vol. 18, pp. 1079–1086, 2015. (Cited on pp. 776 and 784.)

[1839] M. Lin, "Extension of a Result of Haynsworth and Hartfiel," *Arch. Math.*, Vol. 104, pp. 93–100, 2015. (Cited on p. 778.)

[1840] M. Lin, "Inequalities Related to 2×2 Block PPT Matrices," *Operators Matrices*, Vol. 9, pp. 917–924, 2015. (Cited on pp. 751, 755, 756, 757, 872, 885, and 905.)

[1841] M. Lin, "On an Inequality of Marcus," *Math. Notes*, Vol. 97, pp. 644–646, 2015. (Cited on p. 782.)

[1842] M. Lin, "Orthogonal Sets of Normal or Conjugate-Normal Matrices," *Lin. Alg. Appl.*, Vol. 483, pp. 227–235, 2015. (Cited on p. 764.)

[1843] M. Lin, "A Determinant Inequality," *IMAGE*, Vol. 57, p. 30, 2016. (Cited on p. 778.)

[1844] M. Lin, "The Hua Matrix and Inequalities Related to Contractive Matrices," *Lin. Alg. Appl.*, Vol. 511, pp. 22–30, 2016. (Cited on p. 757.)

[1845] M. Lin, "Comparison of the Trace of Two Inverse Matrices," *IMAGE*, 2017. (Cited on pp. 359 and 770.)

[1846] M. Lin, "On a Determinantal Inequality Arising from Diffusion Tensor Imaging," *Comm. Contemp. Math.*, Vol. 19, pp. 1 650 044–1–1 650 044–6, 2017. (Cited on pp. 776 and 778.)

[1847] M. Lin and G. Sinnamon, "A Condition for Convexity of a Product of Positive Definite Quadratic Forms," *SIAM J. Matrix Anal. Appl.*, Vol. 32, pp. 457–462, 2011. (Cited on pp. 787 and 792.)

[1848] M. Lin and S. Sra, "A Proof of Thompson's Determinantal Inequality," *Math. Notes*, Vol. 99, pp. 164–165, 2016. (Cited on p. 785.)

[1849] M. Lin and P. van den Driessche, "Positive Semidefinite 3×3 Block Matrices," *Elec. J. Lin. Alg.*, Vol. 27, pp. 827–836, 2014. (Cited on pp. 783 and 888.)

[1850] M. Lin and H. Wolkowicz, "An Eigenvalue Majorization Inequality for Positive Semidefinite Block Matrices," *Lin. Multilin. Alg.*, Vol. 60, pp. 1365–1368, 2012. (Cited on pp. 752, 770, 805, and 806.)

[1851] M. Lin and H. Wolkowicz, "A General Hua-Type Matrix Inequality and Its Applications," 2013, unpublished. (Cited on p. 350.)

[1852] M. Lin and H. Wolkowicz, "Hiroshima's Theorem and Matrix Norm Inequalities," *Acta Sci. Math (Szeged)*, Vol. 81, pp. 45–53, 2015. (Cited on pp. 870 and 872.)

[1853] M. Lin and P. Zhang, "Unifying a Result of Thompson and a Result of Fiedler and Markham on Block Positive Definite Matrices," *Lin. Alg. Appl.*, Vol. 533, pp. 380–385, 2017. (Cited on pp. 760 and 784.)

[1854] P. Y. Lin, "De Moivre-type Identities for the Tetrabonacci Numbers," in *Applications of Fibonacci Numbers, Vol. 4*, G. E. Bergum, A. N. Philippou, and A. F. Horadam, Eds. New York: Springer, 1991, pp. 215–218. (Cited on p. 1070.)

[1855] T.-P. Lin, "The Power Mean and the Logarithmic Mean," *Amer. Math. Monthly*, Vol. 81, pp. 879–883, 1974. (Cited on pp. 141 and 143.)

[1856] W.-W. Lin, "The Computation of the Kronecker Canonical Form of an Arbitrary Symmetric Pencil," *Lin. Alg. Appl.*, Vol. 103, pp. 41–71, 1988. (Cited on p. 619.)

[1857] H. Linden, "Numerical Radii of Some Companion Matrices and Bounds for the Zeros of Polynomials," in *Analytic and Geometric Inequalities and Applications*, T. M. Rassias and H. M. Srivastava, Eds. Dordrecht: Kluwer, 1999, pp. 205–229. (Cited on pp. 1236 and 1237.)

[1858] R. A. Lippert and G. Strang, "The Jordan Forms of AB and BA," *Elec. J. Lin. Alg.*, Vol. 18, pp. 281–288, 2009. (Cited on p. 606.)

[1859] L. Lipsky, *Queuing Theory: A Linear Algebraic Approach*, 2nd ed. New York: Springer, 2008. (Cited on p. xxi.)

[1860] A. Liu and B. Shawyer, Eds., *Problems from Murray Klamkin: The Canadian Collection*. Washington: Mathematical Association of America, 2009. (Cited on pp. 148, 151, 152, 153, 173, 177, 200, 202, 249, 333, 386, 452, 456, 466, 484, 485, 489, 491, and 492.)

[1861] B. Liu and H.-J. Lai, *Matrices in Combinatorics and Graph Theory*. New York: Springer, 2000. (Cited on p. xxi.)

[1862] G. Liu, "Identities with Symmetric Polynomials," *Fibonacci Quart.*, Vol. 49, p. 190, 2011. (Cited on p. 1009.)

[1863] H. Liu and L. Zhu, "New Strengthened Carleman's Inequality and Hardy's Inequality," *J. Ineq. Appl.*, pp. 1–7, 2007, article ID 84104. (Cited on p. 128.)

[1864] J. Liu, "On a Geometric Inequality of Oppenheim," *J. Science Arts*, Vol. 12, pp. 5–12, 2012. (Cited on p. 472.)

[1865] J. Liu, "A Geometric Inequality with One Parameter for a Point in the Plane of a Triangle," *J. Math. Ineq.*, Vol. 8, pp. 91–106, 2014. (Cited on p. 472.)

[1866] J. Liu, "On a Geometric Inequality of Klamkin," 2014, http://ajmaa.org/RGMIA/papers/v17/v17a08.pdf. (Cited on p. 472.)

[1867] J. Liu, "A Geometric Inequality with Applications," *J. Math. Ineq.*, Vol. 10, pp. 641–648, 2016. (Cited on p. 472.)

[1868] J. Liu and R. Huang, "Generalized Schur Complements of Matrices and Compound Matrices," *Elec. J. Lin. Alg.*, Vol. 21, pp. 12–24, 2010. (Cited on pp. 694 and 830.)

[1869] J. Liu and J. Wang, "Some Inequalities for Schur Complements," *Lin. Alg. Appl.*, Vol. 293, pp. 233–241, 1999. (Cited on pp. 721 and 831.)

[1870] J. Liu and Q. Xie, "Inequalities Involving Khatri-Rao Products of Positive Semidefinite Hermitian Matrices," *Int. J. Inform. Sys. Sci.*, Vol. 4, pp. 30–40–135, 2008. (Cited on p. 700.)

[1871] Q. Liu and G. Chen, "On Two Inequalities for the Hadamard Product and the Fan Product of Matrices," *Lin. Alg. Appl.*, Vol. 431, pp. 974–984, 2009. (Cited on p. 699.)

[1872] R.-W. Liu and R. J. Leake, "Exhaustive Equivalence Classes of Optimal Systems with Separable Controls," *SIAM Rev.*, Vol. 4, pp. 678–685, 1966. (Cited on p. 383.)

[1873] S. Liu, "Matrix Results on the Khatri-Rao and Tracy-Singh Products," *Lin. Alg. Appl.*, Vol. 289, pp. 267–277, 1999. (Cited on pp. 700, 822, 825, and 828.)

[1874] S. Liu, "An Inequality Involving a Special Hadamard Product," *IMAGE*, Vol. 24, p. 9, 2000. (Cited on p. 821.)

[1875] S. Liu, "Several Inequalities Involving Khatri-Rao Products of Positive Semidefinite Matrices," *Lin. Alg. Appl.*, Vol. 354, pp. 175–186, 2002. (Cited on p. 700.)

[1876] S. Liu and H. Neudecker, "Several Matrix Kantorovich-Type Inequalities," *Math. Anal. Appl.*, Vol. 197, pp. 23–26, 1996. (Cited on pp. 212 and 792.)

[1877] S. Liu and G. Trenkler, "Hadamard, Khatri-Rao, Kronecker and Other Matrix Products," *Int. J. Inform. Sys. Sci.*, Vol. 4, pp. 160–177, 2008. (Cited on p. 700.)

[1878] X. Liu, "Generalization of Golden-Thompson Type Inequalities for Normal Matrices," *Math. Ineq. Appl.*, Vol. 19, pp. 1097–1106, 2016. (Cited on pp. 1218 and 1222.)

[1879] X. Liu, L. Wu, and J. Benitez, "On the Group Inverse of Linear Combinations of Two Group Invertible Matrices," *Elec. J. Lin. Alg.*, Vol. 22, pp. 490–503, 2011. (Cited on p. 677.)

[1880] X. Liu and H. Yang, "Some Results on the Partial Orderings of Block Matrices," *J. Ineq. Appl.*, pp. 1–7, 2011, article ID 54. (Cited on pp. 430, 431, 640, and 666.)

[1881] E. Liz, "A Note on the Matrix Exponential," *SIAM Rev.*, Vol. 40, pp. 700–702, 1998. (Cited on p. 1181.)

[1882] N. Loehr, *Advanced Linear Algebra*. Boca Raton: CRC Press, 2014. (Cited on p. 330.)

[1883] R. Loewy, "An Inertia Theorem for Lyapunov's Equation and the Dimension of a Controllability Space," *Lin. Alg. Appl.*, Vol. 260, pp. 1–7, 1997. (Cited on p. 1310.)

[1884] D. O. Logofet, *Matrices and Graphs: Stability Problems in Mathematical Ecology*. Boca Raton: CRC Press, 1993. (Cited on p. xxi.)

[1885] M. Lopez and J. Marengo, "An Upper Bound for the Expected Difference between Order Statistics," *Math. Mag.*, Vol. 84, pp. 365–369, 2011. (Cited on pp. 94 and 1097.)

[1886] D. Lopez-Aguayo, "Non-Integrality of Binomial Sums and Fermat's Little Theorem," *Math. Mag.*, Vol. 88, pp. 231–234, 2015. (Cited on p. 68.)

[1887] R. Lopez-Valcarce and S. Dasgupta, "Some Properties of the Matrix Exponential," *IEEE Trans. Circ. Sys. Analog. Dig. Sig. Proc.*, Vol. 48, pp. 213–215, 2001. (Cited on p. 1210.)

[1888] N. Lord, "The Closed Form Evaluation of $\sum_{n=1}^{\infty} \frac{1}{n^k(n+1)^k}$ and Related Eulerian Sums," *Math. Gaz.*, Vol. 87, pp. 527–528, 2003. (Cited on pp. 1037 and 1038.)

[1889] N. Lord, "An Amusing Sequence of Trigonometrical Integrals," *Math. Gaz.*, Vol. 91, pp. 281–285, 2007. (Cited on pp. 239 and 1119.)

[1890] N. Lord, "Intriguing Integrals: An Euler-Inspired Odyssey," *Math. Gaz.*, Vol. 91, pp. 415–427, 2007. (Cited on pp. 1000, 1152, and 1163.)

[1891] M. Loss and M. B. Ruskai, Eds., *Inequalities: Selecta of Elliott H. Lieb*. New York: Springer, 2002. (Cited on pp. 791 and 1216.)

[1892] P. Lounesto, *Clifford Algebras and Spinors*, 2nd ed. Cambridge: Cambridge University Press, 2001. (Cited on pp. 387, 438, and 691.)

[1893] Z. Lu, "An Optimal Inequality," *Math. Gaz.*, Vol. 91, pp. 521–523, 2007. (Cited on p. 161.)

[1894] D. G. Luenberger, *Optimization by Vector Space Methods*. New York: Wiley, 1969. (Cited on p. xxi.)

[1895] D. G. Luenberger, *Introduction to Linear and Nonlinear Programming*, 2nd ed. Reading: Addison-Wesley, 1984. (Cited on pp. 756 and 798.)

[1896] M. Lundquist and W. Barrett, "Rank Inequalities for Positive Semidefinite Matrices," *Lin. Alg. Appl.*, Vol. 248, pp. 91–100, 1996. (Cited on pp. 324, 723, and 782.)

[1897] C. Lupu, "Another Trace Inequality for Unitary Matrices," *Gazeta Matematica Serie A*, Vol. 28, pp. 137–141, 2010. (Cited on pp. 391 and 856.)

[1898] C. Lupu, "An Adjugate Problem," *Coll. Math. J.*, Vol. 42, p. 411, 2011. (Cited on p. 346.)

[1899] C. Lupu, "Matrix Adjugates and Additive Commutators," *Gazeta Matematica Serie A*, Vol. 29, pp. 90–96, 2011. (Cited on p. 420.)

[1900] C. Lupu, "A Triangle Inequality," *Amer. Math. Monthly*, Vol. 119, pp. 523–524, 2012. (Cited on p. 484.)

[1901] C. Lupu, "Equation $x_1 + x_2 + x_3 = x_1x_2x_3$ Is a Disguised Triangle," *Amer. Math. Monthly*, Vol. 121, pp. 84–85, 2014. (Cited on p. 466.)

[1902] C. Lupu and T. Lupu, "An Application of Popoviciu's Inequality," *Amer. Math. Monthly*, Vol. 116, pp. 752–753, 2009. (Cited on p. 484.)

[1903] C. Lupu and T. Lupu, "A Symmetric Inequality in Three Variables," *Coll. Math. J.*, Vol. 42, pp. 408–409, 2011. (Cited on p. 173.)

[1904] C. Lupu and T. Lupu, "Problem 303," *Gazeta Matematica Serie A*, Vol. 29, pp. 61–62, 2011. (Cited on p. 1158.)

[1905] C. Lupu, C. Mateescu, V. Matei, and M. Opincariu, "A Refinement of the Finsler-Hadwiger Reverse Inequality," *Gazeta Matematica Serie A*, Vol. 28, pp. 49–53, 2010. (Cited on p. 452.)

[1906] C. Lupu and V. Nicula, "Improved Finsler-Hadwiger Inequality," *Gazeta Matematica Serie A*, Vol. 28, pp. 130–133, 2010. (Cited on p. 452.)

[1907] C. Lupu and C. Pohoata, "Sharpening the Hadwiger-Finsler Inequality," *Crux Math.*, Vol. 34, pp. 97–101, 2008. (Cited on p. 452.)

[1908] C. Lupu and D. Schwarz, "Another Look at Some New Cauchy-Schwarz Type Inner Product Inequalities," *Appl. Math. Comp.*, Vol. 231, pp. 463–477, 2014. (Cited on pp. 226 and 856.)

[1909] H. Lutkepohl, *Handbook of Matrices*. Chichester: Wiley, 1996. (Cited on p. xxii.)

[1910] C. D. Lynd, "Using Difference Equations to Generalize Results for Periodic Nested Radicals," *Amer. Math. Monthly*, Vol. 121, pp. 45–59, 2014. (Cited on p. 972.)

[1911] E.-C. Ma, "A Finite Series Solution of the Matrix Equation $AX - XB = C$," *SIAM J. Appl. Math.*, Vol. 14, pp. 490–495, 1966. (Cited on p. 1312.)

[1912] Y. Ma, S. Soatto, J. Kosecka, and S. S. Sastry, *An Invitation to 3-D Vision*. New York: Springer, 2004. (Cited on p. xxi.)

[1913] R. Mabry, "Crosscut Convex Quadrilaterals," *Math. Mag.*, Vol. 84, pp. 16–25, 2011. (Cited on p. 492.)

[1914] C. C. MacDuffee, *The Theory of Matrices*. New York: Chelsea, 1956. (Cited on pp. 687, 692, and 695.)

[1915] A. G. J. MacFarlane and N. Karcanias, "Poles and Zeros of Linear Multivariable Systems: A Survey of the Algebraic, Geometric, and Complex Variable Theory," *Int. J. Contr.*, Vol. 24, pp. 33 74, 1976. (Cited on p. 1319.)

[1916] D. S. Mackey, N. Mackey, and F. Tisseur, "Structured Tools for Structured Matrices," *Elec. J. Lin. Alg.*, Vol. 10, pp. 106–145, 2003. (Cited on p. 440.)

[1917] G. Mackiw, "A Combinatorial Approach to Sums of Integer Powers," *Math. Mag.*, Vol. 73, pp. 44–46, 2000. (Cited on p. 37.)

[1918] K. MacMillan and J. Sondow, "Proofs of Power Sum and Binomial Coefficient Congruences via Pascal's Identity," *Amer. Math. Monthly*, Vol. 118, pp. 549–551, 2011. (Cited on pp. 32 and 68.)

[1919] K. MacMillan and J. Sondow, "Divisibility of Power Sums and the Generalized Erdős-Moser Equation," *Elem. Math.*, Vol. 67, pp. 182–186, 2012. (Cited on p. 37.)

[1920] J. J. Macys, "A New Problem," *Amer. Math. Monthly*, Vol. 119, p. 82, 2011. (Cited on p. 963.)

[1921] J. H. Maddocks, "Restricted Quadratic Forms, Inertia Theorems, and the Schur Complement," *Lin. Alg. Appl.*, Vol. 108, pp. 1–36, 1988. (Cited on pp. 568 and 798.)

[1922] J. R. Magnus, "A Representation Theorem for $(\operatorname{tr} A^p)^{1/p}$," *Lin. Alg. Appl.*, Vol. 95, pp. 127–134, 1987. (Cited on pp. 763, 764, and 769.)

[1923] J. R. Magnus, *Linear Structures*. London: Griffin, 1988. (Cited on pp. xxi and 701.)

[1924] J. R. Magnus and H. Neudecker, *Matrix Differential Calculus with Applications in Statistics and Econometrics*. Chichester: Wiley, 1988, revised edition. (Cited on pp. xxi, 630, 639, 660, 667, 679, 701, 763, 764, 796, and 1092.)

[1925] W. Magnus, "On the Exponential Solution of Differential Equations for a Linear Operator," *Commun. Pure Appl. Math.*, Vol. VII, pp. 649–673, 1954. (Cited on p. 1306.)

[1926] B. J. Mahon, "A Sum of Pell Numbers," *Fibonacci Quart.*, Vol. 49, pp. 87–88, 2011. (Cited on pp. 103 and 1079.)

[1927] B. J. Mahon, "It Is Two, Too!" *Fibonacci Quart.*, Vol. 50, p. 183, 2012. (Cited on p. 1079.)

[1928] B. J. M. Mahon, "Much Ado about F_n," *Fibonacci Quart.*, Vol. 50, p. 86, 2012. (Cited on p. 100.)

[1929] K. N. Majindar, "On Simultaneous Hermitian Congruence Transformations of Matrices," *Amer. Math. Monthly*, Vol. 70, pp. 842–844, 1963. (Cited on p. 799.)

[1930] J. Malenfant, "Finite, Closed-form Expressions for the Partition Function and for Euler, Bernoulli, and Stirling Numbers," http://arxiv.org/pdf/1103.1585.pdf. (Cited on pp. 106, 108, 980, and 981.)

[1931] L. Maligranda, "Simple Norm Inequalities," *Amer. Math. Monthly*, Vol. 113, pp. 256–260, 2006. (Cited on pp. 851 and 852.)

[1932] L. Maligranda, "Some Remarks on the Triangle Inequality for Norms," *Banach J. Math. Anal.*, Vol. 2, pp. 31–41, 2008. (Cited on pp. 138, 850, and 851.)

[1933] L. Maligranda and N. Sabourova, "Real and Complex Operator Norms between Quasi-Banach $L^p - L_q$ Spaces," *Math. Ineq. Appl.*, Vol. 14, pp. 247–270, 2011. (Cited on p. 843.)

[1934] S. B. Malik, "Some More Properties of Core Partial Order," *Appl. Math. Comp.*, Vol. 221, pp. 192–201, 2013. (Cited on pp. 648 and 649.)

[1935] A. Malter, D. Schleicher, and D. Zagier, "New Looks at Old Number Theory," *Amer. Math. Monthly*, Vol. 120, pp. 243–264, 2013. (Cited on p. 119.)

[1936] A. N. Malyshev and M. Sadkane, "On the Stability of Large Matrices," *J. Comp. Appl. Math.*, Vol. 102, pp. 303–313, 1999. (Cited on p. 1227.)

[1937] Y.-K. Man, "On Computing Closed Forms for Indefinite Summations," *J. Symb. Comp.*, Vol. 16, pp. 355–376, 1993. (Cited on p. 43.)

[1938] R. B. Manfrino, J. A. G. Ortega, and R. V. Delgado, *Inequalities: A Mathematical Olympiad Approach*. Basel: Birkhauser, 2009. (Cited on pp. 133, 134, 137, 138, 139, 150, 151, 155, 160, 162, 164, 165, 167, 168, 169, 173, 177, 183, 184, 193, 196, 198, 199, 216, 218, 224, 226, 252, 361, 446, 452, 466, 472, 480, 485, 489, and 492.)

[1939] O. Mangasarian, *Nonlinear Programming*. New York: McGraw-Hill, 1969, reprinted by Krieger, Malabar, 1982. (Cited on p. xxi.)

[1940] S. M. Manjegani, "Hölder and Young Inequalities for the Trace of Operators," *Positivity*, Vol. 11, pp. 239–250, 2007. (Cited on p. 764.)

[1941] S. M. Manjegani and A. Norouzi, "Matrix Form of the Inverse Young Inequalities," *Lin. Alg. Appl.*, Vol. 486, pp. 484–493, 2015. (Cited on p. 140.)

[1942] H. B. Mann, "Quadratic Forms with Linear Constraints," *Amer. Math. Monthly*, Vol. 50, pp. 430–433, 1943. (Cited on p. 798.)

[1943] L. E. Mansfield, *Linear Algebra with Geometric Application*. New York: Marcel-Dekker, 1976. (Cited on p. xxii.)

[1944] T. Mansour, "Combinatorial identities and inverse binomial coefficients," *Adv. Appl. Math.*, Vol. 28, pp. 196–202, 2002. (Cited on pp. 68 and 1008.)

[1945] E. Maor, *Trigonometric Delights*. Princeton: Princeton University Press, 1998. (Cited on pp. 1095 and 1119.)

[1946] M. Marcus, "An Eigenvalue Inequality for the Product of Normal Matrices," *Amer. Math. Monthly*, Vol. 63, pp. 173–174, 1956. (Cited on p. 592.)

[1947] M. Marcus, *Finite Dimensional Multilinear Algebra, Part I*. Marcel Dekker, 1973. (Cited on pp. 694 and 701.)

[1948] M. Marcus, *Finite Dimensional Multilinear Algebra, Part II*. Marcel Dekker, 1975. (Cited on p. 701.)

[1949] M. Marcus, "Two Determinant Condensation Formulas," *Lin. Multilin. Alg.*, Vol. 22, pp. 95–102, 1987. (Cited on pp. 337 and 338.)

[1950] M. Marcus and S. M. Katz, "Matrices of Schur Functions," *Duke Math. J.*, Vol. 36, pp. 343–352, 1969. (Cited on p. 785.)

[1951] M. Marcus and N. A. Khan, "A Note on the Hadamard Product," *Canad. Math. J.*, Vol. 2, pp. 81–83, 1959. (Cited on pp. 685, 701, and 824.)

[1952] M. Marcus and H. Minc, *A Survey of Matrix Theory and Matrix Inequalities*. Boston: Prindle, Weber, and Schmidt, 1964, reprinted by Dover, New York, 1992. (Cited on pp. xxii, 204, 276, 440, 582, 834, and 891.)

[1953] M. Marcus and W. Watkins, "Partitioned Hermitian Matrices," *Duke Math. J.*, Vol. 38, pp. 237–249, 1971. (Cited on pp. 730, 751, 760, and 785.)

[1954] M. Marden, *Geometry of Polynomials*, 2nd ed. Providence: American Mathematical Society, 1965. (Cited on p. 1237.)

[1955] M. Margaliot and G. Langholz, "The Routh-Hurwitz Array and Realization of Characteristic Polynomials," *IEEE Trans. Autom. Contr.*, Vol. 45, pp. 2424–2427, 2000. (Cited on p. 1230.)

[1956] D. S. Marinescu, M. Monea, M. Opincariu, and M. Stroe, "Note on Hadwiger-Finsler's Inequalities," *J. Math. Ineq.*, Vol. 6, pp. 57–64, 2012. (Cited on pp. 157, 160, 452, and 486.)

[1957] D. S. Marinescu, M. Monea, M. Opincariu, and M. Stroe, "A Unitary Approach to Some Classical Inequalities," *J. Math. Ineq.*, Vol. 7, pp. 151–159, 2013. (Cited on pp. 218, 220, 224, and 225.)

[1958] C. Marion, "Proof Without Words: Triangular Sums and Perfect Quartics," *Math. Mag.*, Vol. 90, p. 258, 2017. (Cited on p. 39.)

[1959] T. L. Markham, "An Application of Theorems of Schur and Albert," *Proc. Amer. Math. Soc.*, Vol. 59, pp. 205–210, 1976. (Cited on p. 830.)

[1960] T. L. Markham, "Oppenheim's Inequality for Positive Definite Matrices," *Amer. Math. Monthly*, Vol. 93, pp. 642–644, 1986. (Cited on p. 824.)

[1961] F. L. Markley and J. L. Crassidis, *Fundamentals of Spacecraft Attitude Determination and Control*. New York: Springer, 2014. (Cited on p. 386.)

[1962] J.-B. R. Marquez, "An Inequality Involving Cubes," *Coll. Math. J.*, Vol. 38, pp. 317–318, 2007. (Cited on p. 164.)

[1963] J.-B. R. Marquez, "Two Improper Integrals and L'Hôpital's Rule," *Coll. Math. J.*, Vol. 38, pp. 155–156, 2007. (Cited on p. 1106.)

[1964] J.-B. R. Marquez, "Inequalities with Inradius and Circumradius," *Coll. Math. J.*, Vol. 39, pp. 155–157, 2008. (Cited on p. 452.)

[1965] J.-B. R. Marquez, "Problem Totten-04," *Crux Math.*, Vol. 36, pp. 324–325, 2010. (Cited on p. 229.)

[1966] G. Marsaglia and G. P. H. Styan, "Equalities and Inequalities for Ranks of Matrices," *Lin. Multilin. Alg.*, Vol. 2, pp. 269–292, 1974. (Cited on pp. 321, 362, 661, 663, and 664.)

[1967] J. E. Marsden and R. S. Ratiu, *Introduction to Mechanics and Symmetry: A Basic Exposition of Classical Mechanical Systems*, 2nd ed. New York: Springer, 1994. (Cited on p. 393.)

[1968] J. E. Marsden and T. S. Ratiu, *Introduction to Mechanics and Symmetry*. New York: Springer, 1994. (Cited on p. xxi.)

[1969] A. W. Marshall and I. Olkin, *Inequalities: Theory of Majorization and Its Applications*. New York: Academic Press, 1979. (Cited on pp. 202, 216, 276, 360, 361, 362, 384, 466, 489, 582, 583, 593, 594, 694, 721, 722, 752, 802, 805, 831, 904, 911, and 912.)

[1970] A. W. Marshall and I. Olkin, "Matrix Versions of the Cauchy and Kantorovich Inequalities," *Aequationes Math.*, Vol. 40, pp. 89–93, 1990. (Cited on p. 735.)

[1971] A. W. Marshall, I. Olkin, and B. Arnold, *Inequalities: Theory of Majorization and Its Applications*, 2nd ed. New York: Springer, 2009. (Cited on pp. 202, 216, 276, 360, 361, 362, 384, 466, 489, 582, 583, 593, 594, 694, 721, 722, 752, 805, 831, 904, 911, and 912.)

[1972] R. Marsli and F. J. Hall, "Geometric Multiplicities and Gersgorin Discs," *Amer. Math. Monthly*, Vol. 120, pp. 452–455, 2013. (Cited on p. 535.)

[1973] K. Martensson, "On the Matrix Riccati Equation," *Inform. Sci.*, Vol. 3, pp. 17–49, 1971. (Cited on p. 1319.)

[1974] G. E. Martin, *Counting: The Art of Enumerative Combinatorics*. New York: Springer, 2001. (Cited on p. 1008.)

[1975] V. Mascioni, "Inequalities for the Sum of Two Sines or Cosines," 2011, unpublished. (Cited on pp. 250, 252, and 261.)

[1976] J. C. Mason and D. C. Handscomb, *Chebyshev Polynomials*. Boca Raton: Chapman & Hall CRC, 2002. (Cited on p. 986.)

[1977] W. S. Massey, "Cross Products of Vectors in Higher Dimensional Euclidean Spaces," *Amer. Math. Monthly*, Vol. 90, pp. 697–701, 1983. (Cited on pp. 387 and 691.)

[1978] L. Matejicka, "Further Development and Refining Some Inequalities for Polynomial Zeros," *Math. Ineq. Appl.*, Vol. 6, pp. 205–214, 2012. (Cited on p. 1237.)

[1979] L. Matejicka, "On the Cirtoaje's Conjecture," *J. Ineq. Appl.*, pp. 1–6, 2016, article 152. (Cited on p. 145.)

[1980] L. Matejicka, "On the Cirtoaje's Conjecture," *J. Ineq. Appl.*, pp. 1–11, 2016, article 269. (Cited on p. 145.)

[1981] A. M. Mathai, *Jacobians of Matrix Transformations and Functions of Matrix Argument*. Singapore: World Scientific, 1997. (Cited on p. 1092.)

[1982] J. S. Matharu and J. S. Aujla, "Hadamard Product Versions of the Chebyshev and Kantorovich Inequalities," *J. Ineq. Pure Appl. Math.*, Vol. 10, pp. 1–6, 2009, article 51. (Cited on pp. 128 and 830.)

[1983] J. S. Matharu and J. S. Aujla, "Some Inequalities for Unitarily Invariant Norms," *Lin. Alg. Appl.*, Vol. 436, pp. 1623–1631, 2012. (Cited on p. 880.)

[1984] J. S. Matharu, M. S. Moslehian, and J. S. Aujla, "Eigenvalue Extensions of Bohr's Inequality," *Lin. Alg. Appl.*, Vol. 435, pp. 270–276, 2011. (Cited on pp. 271, 274, 855, and 877.)

[1985] R. Mathias, "The Spectral Norm of a Nonnegative Matrix," *Lin. Alg. Appl.*, Vol. 131, pp. 269–284, 1990. (Cited on p. 907.)

[1986] R. Mathias, "Evaluating the Frechet Derivative of the Matrix Exponential," *Numer. Math.*, Vol. 63, pp. 213–226, 1992. (Cited on p. 1212.)

[1987] R. Mathias, "Matrices with Positive Definite Hermitian Part: Inequalities and Linear Systems," *SIAM J. Matrix Anal. Appl.*, Vol. 13, pp. 640–654, 1992. (Cited on pp. 720 and 804.)

[1988] R. Mathias, "An Arithmetic-Geometric-Harmonic Mean Inequality Involving Hadamard Products," *Lin. Alg. Appl.*, Vol. 184, pp. 71–78, 1993. (Cited on p. 728.)

[1989] R. Mathias, "A Chain Rule for Matrix Functions and Applications," *SIAM J. Matrix Anal. Appl.*, Vol. 17, pp. 610–620, 1996. (Cited on p. 1092.)

[1990] R. Mathias, "Singular Values and Singular Value Inequalities," in *Handbook of Linear Algebra*, L. Hogben, Ed. Boca Raton: Chapman & Hall/CRC, 2007, pp. 17–1–17–16. (Cited on p. 589.)

[1991] M. Matic, C. E. M. Pearce, and J. Pecaric, "Shannon's and Related Inequalities in Information Theory," in *Survey on Classical Inequalities*, T. M. Rassias, Ed. Dordrecht: Kluwer, 2000, pp. 127–164. (Cited on pp. 223, 230, and 231.)

[1992] J. Matousek, *Thirty-three Miniatures*. Providence: American Mathematical Society, 2010. (Cited on pp. 24, 25, and 860.)

[1993] T. Matsuda, "An Inductive Proof of a Mixed Arithmetic-Geometric Mean Inequality," *Amer. Math. Monthly*, Vol. 102, pp. 634–637, 1995. (Cited on p. 213.)

[1994] D. Mavlo, "Beauty Is Truth: Geometric Inequalities," *Math. Gaz.*, Vol. 84, pp. 423–432, 2000. (Cited on pp. 472 and 480.)

[1995] V. Mazorchuk and S. Rabanovich, "Multicommutators and Multianticommutators of Orthogonal Projectors," *Lin. Multilin. Alg.*, Vol. 56, pp. 639–646, 2008. (Cited on p. 884.)

[1996] J. E. McCarthy, "Pick's Theorem—What's the Big Deal?" *Amer. Math. Monthly*, Vol. 110, pp. 36–45, 2003. (Cited on p. 729.)

[1997] J. M. McCarthy, *Geometric Design of Linkages*. New York: Springer, 2000. (Cited on p. 1247.)

[1998] J. P. McCloskey, "Characterizations of r-Potent Matrices," *Math. Proc. Camb. Phil. Soc.*, Vol. 96, pp. 213–222, 1984. (Cited on p. 418.)

[1999] J. McLaughlin, "Combinatorial Identities Deriving from the n-th Power of a 2×2 Matrix," *Elec. J. Combinatorial Number Theory*, Vol. 4, pp. 1–15, 2004, article A19. (Cited on p. 531.)

[2000] J. Mclaughlin, "An Identity Motivated by an Amazing Identity of Ramanujan," *Fibonacci Quart.*, Vol. 48, pp. 34–38, 2010. (Cited on pp. 34, 148, and 1020.)

[2001] A. R. Meenakshi and C. Rajian, "On a Product of Positive Semidefinite Matrices," *Lin. Alg. Appl.*, Vol. 295, pp. 3–6, 1999. (Cited on p. 820.)

[2002] M. H. Mehrabi, "Some Inequalities for Triangles," *Amer. Math. Monthly*, Vol. 120, pp. 175–176, 2013. (Cited on pp. 164 and 452.)

[2003] C. L. Mehta, "Some Inequalities Involving Traces of Operators," *J. Math. Phys.*, Vol. 9, pp. 693–697, 1968. (Cited on p. 791.)
[2004] R. S. Melham, "Families of Identities Involving Sums of Powers of the Fibonacci and Lucas Numbers," *Fibonacci Quart.*, Vol. 37, pp. 315–319, 1999. (Cited on pp. 100 and 103.)
[2005] R. S. Melham, "A Fibonacci Identity in the Spirit of Simson and Gelin-Cesaro," *Fibonacci Quart.*, Vol. 41, pp. 142–143, 2003. (Cited on p. 100.)
[2006] R. S. Melham, "More on Combinations of Higher Powers of Fibonacci Numbers," *Fibonacci Quart.*, Vol. 48, pp. 307–311, 2010. (Cited on pp. 92 and 100.)
[2007] R. S. Melham, "On Certain Combinations of Higher Powers of Fibonacci Numbers," *Fibonacci Quart.*, Vol. 48, pp. 256–259, 2010. (Cited on p. 103.)
[2008] R. S. Melham, "Some Conjectures Concerning Sums of Odd Powers of Fibonacci and Lucas Numbers," *Fibonacci Quart.*, Vol. 48, pp. 1–4, 2010. (Cited on p. 100.)
[2009] A. Melman, "The Twin of a Theorem by Cauchy," *Amer. Math. Monthly*, Vol. 120, pp. 164–168, 2013. (Cited on p. 1239.)
[2010] Y. A. Melnikov, *Influence Functions and Matrices*. New York: Marcel Dekker, 1998. (Cited on p. xxi.)
[2011] J. Melville, "Some Simple Geometric Inequalities," *Math. Gaz.*, Vol. 88, pp. 136–137, 2004. (Cited on p. 480.)
[2012] Z. A. Melzak, "Infinite Products for πe and π/e," *Amer. Math. Monthly*, Vol. 68, pp. 39–41, 1961. (Cited on p. 1087.)
[2013] Z. A. Melzak, *Companion to Concrete Mathematics*. Mineola: Dover, 2007. (Cited on pp. 68, 259, 968, 971, 1021, 1070, 1083, 1084, 1087, 1098, 1099, 1129, 1134, 1141, 1145, 1152, 1154, 1162, 1163, 1170, and 1172.)
[2014] L. Menikhes and V. Karachik, "Looking for a Closed-Form Solution," *Math. Mag.*, Vol. 87, pp. 234–235, 2014. (Cited on p. 92.)
[2015] M. Merca, "A Note on Cosine Power Sums," *J. Integer Sequences*, Vol. 15, pp. 1–7, 2012, article 12.5.3. (Cited on pp. 67, 75, and 242.)
[2016] M. Merca, "A Note on the Determinant of a Toeplitz-Hessenberg Matrix," *Special Matrices*, Vol. 1, pp. 10–16, 2013. (Cited on pp. 86, 426, and 980.)
[2017] M. Merca, "On Some Power Sums of Sine or Cosine," *Amer. Math. Monthly*, Vol. 121, pp. 244–248, 2014. (Cited on p. 243.)
[2018] M. Merca, "An Alternative to Faulhaber's Formula," *Amer. Math. Monthly*, Vol. 122, pp. 599–601, 2015. (Cited on p. 111.)
[2019] M. Merca, "A Sum of Cosine Powers," *Coll. Math. J.*, Vol. 48, pp. 140–141, 2017. (Cited on p. 252.)
[2020] M. Merca, "On the Arithmetic Mean of the Square Roots of the First n Positive Integers," *Coll. Math. J.*, Vol. 48, pp. 129–133, 2017. (Cited on p. 46.)
[2021] A. Mercer, U. Abel, and D. Caccia, "A Sharpening of Jordan's Inequality," *Amer. Math. Monthly*, Vol. 93, pp. 568–569, 1986. (Cited on p. 249.)
[2022] P. R. Mercer, "A Refined Cauchy-Schwarz Inequality," *Int. J. Math. Educ. Sci. Tech.*, Vol. 38, pp. 839–843, 2007. (Cited on p. 855.)
[2023] P. R. Mercer, "The Dunkl-Williams Inequality in an Inner Product Space," *Math. Ineq. Appl.*, Vol. 10, pp. 447–450, 2007. (Cited on p. 851.)
[2024] J. K. Merikoski and X. Liu, "On Rank Subtractivity between Normal Matrices," *J. Ineq. Pure Appl. Math.*, Vol. 9, pp. 1–8, 2008, article 4. (Cited on pp. 431 and 648.)
[2025] J. K. Merikoski, H. Sarria, and P. Tarazaga, "Bounds for Singular Values Using Traces," *Lin. Alg. Appl.*, Vol. 210, pp. 227–254, 1994. (Cited on p. 897.)
[2026] D. I. Merino, "The Sum of Orthogonal Matrices," *Lin. Alg. Appl.*, Vol. 436, pp. 1960–1968, 2012. (Cited on p. 618.)
[2027] D. Merlini and R. Sprugnoli, "Playing with Some Identities of Andrews," *J. Integer Sequences*, Vol. 10, pp. 1–16, 2007, article 07.9.5. (Cited on p. 1079.)
[2028] R. Merris, "Laplacian Matrices of Graphs: A Survey," *Lin. Alg. Appl.*, Vol. 198, pp. 143–176, 1994. (Cited on pp. 541 and 792.)
[2029] R. Merris, *Multilinear Algebra*. Amsterdam: Gordon and Breach, 1997. (Cited on pp. 701 and 825.)
[2030] R. Merris and S. Pierce, "Monotonicity of Positive Semidefinite Hermitian Matrices," *Proc. Amer. Math. Soc.*, Vol. 31, pp. 437–440, 1972. (Cited on pp. 742 and 753.)

[2031] B. G. Mertzios and M. A. Christodoulou, "On the Generalized Cayley-Hamilton Theorem," *IEEE Trans. Autom. Contr.*, Vol. AC-31, pp. 156–157, 1986. (Cited on p. 529.)

[2032] M. Mesbahi and M. Egerstedt, *Graph Theoretic Methods in Multiagent Networks*. Princeton: Princeton University Press, 2010. (Cited on p. 541.)

[2033] G. Meurant, "A Review on the Inverse of Symmetric Tridiagonal and Block Tridiagonal Matrices," *SIAM J. Matrix Anal. Appl.*, Vol. 13, pp. 707–728, 1992. (Cited on p. 424.)

[2034] C. D. Meyer, "The Moore-Penrose Inverse of a Bordered Matrix," *Lin. Alg. Appl.*, Vol. 5, pp. 375–382, 1972. (Cited on p. 667.)

[2035] C. D. Meyer, "Generalized Inverses and Ranks of Block Matrices," *SIAM J. Appl. Math*, Vol. 25, pp. 597–602, 1973. (Cited on pp. 661 and 667.)

[2036] C. D. Meyer, *Matrix Analysis and Applied Linear Algebra*. Philadelphia: SIAM, 2000. (Cited on pp. 374, 375, 376, 417, 494, 508, 543, 615, 651, and 1241.)

[2037] C. D. Meyer and N. J. Rose, "The Index and the Drazin Inverse of Block Triangular Matrices," *SIAM J. Appl. Math.*, Vol. 33, pp. 1–7, 1977. (Cited on pp. 604, 667, 671, and 678.)

[2038] J.-M. Miao, "General Expressions for the Moore-Penrose Inverse of a 2×2 Block Matrix," *Lin. Alg. Appl.*, Vol. 151, pp. 1–15, 1991. (Cited on p. 667.)

[2039] Y. Miao, L.-M. Liu, and F. Qi, "Refinements of Inequalities between the Sum of Squares and the Exponential of Sum of a Nonnegative Sequence," *J. Ineq. Pure. Appl. Math.*, Vol. 9, pp. 1–5, 2008, article 53. (Cited on pp. 127 and 214.)

[2040] C. Michels, "Estimating Eigenvalues of Matrices by Induced Norms," *Oper. Matrices*, Vol. 5, pp. 389–407, 2011. (Cited on pp. 843, 864, 884, and 892.)

[2041] L. Mihalyffy, "An Alternative Representation of the Generalized Inverse of Partitioned Matrices," *Lin. Alg. Appl.*, Vol. 4, pp. 95–100, 1971. (Cited on p. 667.)

[2042] Z. Mijalkovic and M. Mijalkovic, "Sharpening of Cauchy Inequality," in *Recent Progress in Inequalities*, G. V. Milovanovic, Ed. Dordrecht: Kluwer, 1998, pp. 485–487. (Cited on p. 207.)

[2043] K. S. Miller, *Some Eclectic Matrix Theory*. Malabar: Krieger, 1987. (Cited on pp. 518, 692, 1172, and 1173.)

[2044] G. A. Milliken and F. Akdeniz, "A Theorem on the Difference of the Generalized Inverses of Two Nonnegative Matrices," *Commun. Statist. Theory Methods*, Vol. 6, pp. 73–79, 1977. (Cited on p. 817.)

[2045] G. V. Milovanovic and I. Z. Milovanovic, "Discrete Inequalities of Wirtinger's Type," in *Recent Progress in Inequalities*, G. V. Milovanovic, Ed. Dordrecht: Kluwer, 1998, pp. 289–308. (Cited on p. 192.)

[2046] G. V. Milovanovic, D. S. Mitrinovic, and T. M. Rassias, *Topics in Polynomials: Extremal Problems, Inequalities, Zeros*. Singapore: World Scientific, 1994. (Cited on pp. 1234 and 1237.)

[2047] G. V. Milovanovic and T. M. Rassias, "Inequalities for Polynomial Zeros," in *Survey on Classical Inequalities*, T. M. Rassias, Ed. Dordrecht: Kluwer, 2000, pp. 165–202. (Cited on pp. 941, 1235, 1236, and 1237.)

[2048] N. Minamide, "An Extension of the Matrix Inversion Lemma," *SIAM J. Alg. Disc. Meth.*, Vol. 6, pp. 371–377, 1985. (Cited on pp. 635 and 636.)

[2049] D. Minda and S. Phelps, "Triangles, Ellipses, and Cubic Polynomials," *Amer. Math. Monthly*, Vol. 115, pp. 679–689, 2008. (Cited on p. 496.)

[2050] G. Minton, "A Lobachevsky Integral," *Amer. Math. Monthly*, Vol. 117, pp. 655–656, 2010. (Cited on p. 1119.)

[2051] H. Miranda and R. C. Thompson, "A Trace Inequality with a Subtracted Term," *Lin. Alg. Appl.*, Vol. 185, pp. 165–172, 1993. (Cited on p. 831.)

[2052] L. Mirsky, "The Norms of Adjugate and Inverse Matrices," *Arch. Math.*, Vol. VII, pp. 276–277, 1956. (Cited on p. 861.)

[2053] L. Mirsky, *An Introduction to Linear Algebra*. Oxford: Clarendon, 1972, reprinted by Dover, Mineola, 1990. (Cited on pp. xxii and 395.)

[2054] P. Misra and R. V. Patel, "A Determinant Identity and Its Application in Evaluating Frequency Response Matrices," *SIAM J. Matrix Anal. Appl.*, Vol. 9, pp. 248–255, 1988. (Cited on p. 1315.)

[2055] K.-I. Mitani and K.-S. Saito, "On Sharp Triangle Inequalities in Banach Spaces II," *J. Ineq. Appl.*, pp. 1–17, 2010, article ID 323609. (Cited on p. 852.)

[2056] D. W. Mitchell, "Heron-Type Formula in Terms of Sines," *Math. Gaz.*, Vol. 93, pp. 108–109, 2009. (Cited on p. 446.)

[2057] D. W. Mitchell, "The Area of a Quadrilateral," *Math. Gaz.*, Vol. 93, pp. 306–309, 2009. (Cited on pp. 491 and 492.)

[2058] S. K. Mitra, "On Group Inverses and the Sharp Order," *Lin. Alg. Appl.*, Vol. 92, pp. 17–37, 1987. (Cited on pp. 431, 640, and 648.)

[2059] S. K. Mitra, P. Bhimasankaram, and S. B. Malik, *Matrix Partial Orders, Shorted Operators and Applications*. New Jersey: World Scientific, 2010. (Cited on pp. 431, 640, and 648.)

[2060] D. S. Mitrinovic, *Analytic Inequalities*. Berlin: Springer, 1970. (Cited on pp. 205, 210, 215, 217, and 271.)

[2061] D. S. Mitrinovic, J. E. Pecaric, and A. M. Fink, *Classical and New Inequalities in Analysis*. Dordrecht: Kluwer, 1993. (Cited on pp. 123, 222, 229, 271, 276, 536, 850, 851, 855, 856, and 858.)

[2062] D. S. Mitrinovic, J. E. Pecaric, and V. Volenec, *Recent Advances in Geometric Inequalities*. Dordrecht: Kluwer, 1989. (Cited on pp. 446, 452, 456, 466, 472, 480, 484, and 489.)

[2063] B. Mityagin, "An Inequality in Linear Algebra," *SIAM Rev.*, Vol. 33, pp. 125–127, 1991. (Cited on p. 740.)

[2064] M. Moakher, "Means and Averaging in the Group of Rotations," *SIAM J. Matrix Anal. Appl.*, Vol. 24, pp. 1–16, 2002. (Cited on pp. 389, 724, 1207, 1217, and 1220.)

[2065] M. Moakher, "A Differential Geometric Approach to the Geometric Mean of Symmetric Positive-Definite Matrices," *SIAM J. Matrix Anal. Appl.*, Vol. 26, pp. 735–747, 2005. (Cited on pp. 744, 952, and 1217.)

[2066] C. Moler and C. F. Van Loan, "Nineteen Dubious Ways to Compute the Exponential of a Matrix, Twenty-Five Years Later," *SIAM Rev.*, Vol. 45, pp. 3–49, 2000. (Cited on pp. 1220, 1221, and 1247.)

[2067] G. Molinari, "Elementary Prime Identities," *Amer. Math. Monthly*, Vol. 122, pp. 151–154, 2015. (Cited on p. 1083.)

[2068] V. H. Moll, *Numbers and Functions*. Providence: American Mathematical Society, 2012. (Cited on pp. 31, 32, 35, 49, 50, 86, 100, 103, 105, 121, 980, 983, 989, 993, 1008, 1061, 1062, 1074, 1079, 1104, and 1109.)

[2069] V. H. Moll, *Special Integrals of Gradshteyn and Ryzhik: The Proofs—Vol. I*. Boca Raton: CRC Press, 2014. (Cited on p. 1178.)

[2070] V. H. Moll, *Special Integrals of Gradshteyn and Ryzhik: The Proofs—Vol. II*. Boca Raton: CRC Press, 2015. (Cited on p. 1178.)

[2071] B. Mond, "Generalized Inverse Extensions of Matrix Inequalities," *Lin. Alg. Appl.*, Vol. 2, pp. 393–399, 1969. (Cited on p. 793.)

[2072] B. Mond and J. E. Pecaric, "Reverse Forms of a Convex Matrix Inequality," *Lin. Alg. Appl.*, Vol. 220, pp. 359–364, 1995. (Cited on p. 737.)

[2073] B. Mond and J. E. Pecaric, "Inequalities for the Hadamard Product of Matrices," *SIAM J. Matrix Anal. Appl.*, Vol. 19, pp. 66–70, 1998. (Cited on p. 827.)

[2074] B. Mond and J. E. Pecaric, "On Inequalities Involving the Hadamard Product of Matrices," *Elec. J. Lin. Alg.*, Vol. 6, pp. 56–61, 2000. (Cited on pp. 826 and 828.)

[2075] V. V. Monov, "On the Spectrum of Convex Sets of Matrices," *IEEE Trans. Autom. Contr.*, Vol. 44, pp. 1009–1012, 1992. (Cited on p. 536.)

[2076] J. P. Morais, S. Georgiev, and W. Sprossig, *Real Quaternionic Calculus Handbook*. Basel: Birkhauser, 2014. (Cited on p. 438.)

[2077] A. Morawiec, *Orientations and Rotations: Computations in Crystallographic Textures*. New York: Springer, 2003. (Cited on pp. 438 and 1207.)

[2078] T. Moreland and S. Gudder, "Infima of Hilbert Space Effects," *Lin. Alg. Appl.*, Vol. 286, pp. 1–17, 1999. (Cited on p. 740.)

[2079] S. G. Moreno and E. M. Garcia-Caballero, "An Infinite Product for the Golden Ratio," *Amer. Math. Monthly*, Vol. 119, p. 645, 2012. (Cited on p. 1087.)

[2080] S. G. Moreno and E. M. Garcia-Caballero, "A Geometric Proof of a Morrie-Type Formula," *Math. Mag.*, Vol. 89, pp. 214–215, 2016. (Cited on p. 245.)

[2081] F. Morgan, "Eigenvalues of the Sum of Two Orthogonal Projection Matrices," *Coll. Math. J.*, Vol. 40, pp. 56–57, 2009. (Cited on p. 590.)

[2082] T. Mori, "Comments on 'A Matrix Inequality Associated with Bounds on Solutions of Algebraic Riccati and Lyapunov Equation'," *IEEE Trans. Autom. Contr.*, Vol. 33, p. 1088, 1988. (Cited on p. 831.)

[2083] H. C. Morris, "A Large Power of n! That Divides (mn)!," *Math. Mag.*, Vol. 88, pp. 288–289, 2015. (Cited on p. 52.)

[2084] D. Mortari, "Attitude Analysis in Flatland: The Plane Truth," *J. Astron. Sci.*, Vol. 52, pp. 195–209, 2004. (Cited on p. 416.)

[2085] C. Mortici, "Product Approximations via Asymptotic Integration," *Amer. Math. Monthly*, Vol. 117, pp. 434–441, 2010. (Cited on pp. 971 and 1040.)

[2086] C. Mortici, "A Substantial Improvement of the Stirling Formula," *Appl. Math. Lett.*, Vol. 24, pp. 1351–1354, 2011. (Cited on pp. 971 and 1090.)

[2087] C. Mortici, "The Natural Approach of Wilker-Cusa-Huygens Inequalities," *Math. Ineq. Appl.*, Vol. 14, pp. 535–541, 2011. (Cited on p. 249.)

[2088] C. Mortici, "A Power Series Approach to Some Inequalities," *Amer. Math. Monthly*, Vol. 119, pp. 147–151, 2012. (Cited on pp. 135, 151, 161, and 162.)

[2089] R. Mortini and J. Noel, "A Sum and Product Inequality," *Amer. Math. Monthly*, Vol. 120, pp. 661–662, 2013. (Cited on p. 1081.)

[2090] R. Mortini and J. Noel, "A Probabilistically Expected Inequality," *Math. Mag.*, Vol. 88, p. 288, 2015. (Cited on p. 252.)

[2091] R. Mortini and R. Rupp, "Logarithms and Exponentials in Banach Algebras," *Banach J. Math. Anal.*, Vol. 9, pp. 164–172, 2015. (Cited on p. 1189.)

[2092] Y. Moschovakis, *Notes on Set Theory*, 2nd ed. New York: Springer, 2005. (Cited on p. 28.)

[2093] M. S. Moslehian, "An Operator Extension of the Parallelogram Law and Related Norm Inequalities," *Math. Ineq. Appl.*, Vol. 14, pp. 717–725, 2011. (Cited on pp. 275, 748, 858, and 878.)

[2094] M. S. Moslehian, F. Dadipour, R. Rajic, and A. Maric, "A Glimpse at the Dunkl-Williams Inequality," *Banach J. Math. Anal.*, Vol. 5, pp. 138–151, 2011. (Cited on pp. 850, 851, and 852.)

[2095] M. S. Moslehian and H. Najafi, "An Extension of the Löwner-Heinz Inequality," *Lin. Alg. Appl.*, Vol. 437, pp. 2359–2365, 2012. (Cited on p. 742.)

[2096] M. S. Moslehian, M. Tominaga, and K.-S. Saito, "Schatten p-norm Inequalities Related to an Extended Operator Parallelogram Law," *Lin. Alg. App.*, Vol. 435, pp. 823–829, 2011. (Cited on p. 878.)

[2097] T. Muir, *A Treatise on the Theory of Determinants*. New York: Dover, 1960, revised and enlarged by W. H. Metzler. (Cited on p. 362.)

[2098] T. Muir, *The Theory of Determinants in the Historical Order of Development*. New York: Dover, 1966. (Cited on p. 362.)

[2099] W. W. Muir, "Inequalities Concerning the Inverses of Positive Definite Matrices," *Proc. Edinburgh Math. Soc.*, Vol. 19, pp. 109–113, 1974–75. (Cited on pp. 721 and 831.)

[2100] E. Munarini, "Riordan Matrices and Sums of Harmonic Numbers," *Applic. Anal. Disc.Math.*, Vol. 5, pp. 176–200, 2011. (Cited on pp. 75, 86, 103, 113, 976, 985, 1009, and 1079.)

[2101] J. R. Munkres, *Analysis on Manifolds*. Boulder: Westview, 1991. (Cited on p. 691.)

[2102] W. D. Munn, "Ellipses Circumscribing Convex Quadrilaterals," *Math. Gaz.*, Vol. 92, pp. 566–568, 2008. (Cited on p. 496.)

[2103] I. S. Murphy, "A Note on the Product of Complementary Principal Minors of a Positive Definite Matrix," *Lin. Alg. Appl.*, Vol. 44, pp. 169–172, 1982. (Cited on p. 782.)

[2104] R. M. Murray, Z. Li, and S. S. Sastry, *A Mathematical Introduction to Robotic Manipulation*. Boca Raton: CRC, 1994. (Cited on pp. xxi and 1247.)

[2105] M. R. Murty and C. Weatherby, "A Generalization of Euler's Theorem for $\zeta(2k)$," *Amer. Math. Monthly*, Vol. 123, pp. 53–65, 2016. (Cited on p. 108.)

[2106] P. Nahin, *Inside Interesting Integrals*. Princeton: Princeton University Press, 2015. (Cited on pp. 996, 1102, 1104, 1106, 1107, 1109, 1114, 1116, 1121, 1122, 1124, 1126, 1127, 1128, 1129, 1131, 1134, 1141, 1143, 1146, 1148, 1149, 1152, 1154, 1161, 1162, 1163, 1175, and 1177.)

[2107] P. J. Nahin, *An Imaginary Tale: The Story of* $\sqrt{-1}$. Princeton: Princeton University Press, 1998. (Cited on pp. 21, 35, 178, and 1088.)

[2108] I. Najfeld and T. F. Havel, "Derivatives of the Matrix Exponential and Their Computation," *Adv. Appl. Math.*, Vol. 16, pp. 321–375, 1995. (Cited on p. 1212.)

[2109] R. Nakamoto, "A Norm Inequality for Hermitian Operators," *Amer. Math. Monthly*, Vol. 110, pp. 238–239, 2003. (Cited on p. 1223.)

[2110] Y. Nakamura, "Any Hermitian Matrix Is a Linear Combination of Four Projections," *Lin. Alg. Appl.*, Vol. 61, pp. 133–139, 1984. (Cited on p. 619.)

[2111] O. Nanyes, "Continuously Differentiable Curves Detect Limits of Functions of Two Variables," *Coll. Math. J.*, Vol. 45, pp. 54–56, 2014. (Cited on p. 957.)

[2112] A. W. Naylor and G. R. Sell, *Linear Operator Theory in Engineering and Science*. New York: Springer, 1986. (Cited on pp. 118, 271, 849, 855, 917, 922, and 1092.)

[2113] T. Needham, *Visual Complex Analysis*. Oxford: Oxford University Press, 1997. (Cited on pp. 276, 920, 1004, and 1008.)

[2114] R. B. Nelsen, "Proof without Words: Squares Modulo 3," *Coll. Math. J.*, Vol. 44, p. 183, 2013. (Cited on p. 30.)

[2115] R. B. Nelsen, "Proof without Words: A Trigonometric Proof of the Arithmetic Mean–Geometric Mean Inequality," *Coll. Math. J.*, Vol. 46, p. 42, 2015. (Cited on p. 249.)

[2116] R. B. Nelsen, "Proof without Words: Sums of Odd and Even Squares," *Math. Mag.*, Vol. 88, pp. 278–279, 2015. (Cited on pp. 33 and 178.)

[2117] G. Nemes, "On the Coefficients of the Asymptotic Expansion of $n!$," *J. Integer Sequences*, Vol. 13, pp. 1–5, 2010, article 10.6.6. (Cited on p. 971.)

[2118] C. N. Nett and W. M. Haddad, "A System-Theoretic Appropriate Realization of the Empty Matrix Concept," *IEEE Trans. Autom. Contr.*, Vol. 38, pp. 771–775, 1993. (Cited on p. 362.)

[2119] M. G. Neubauer, "An Inequality for Positive Definite Matrices with Applications to Combinatorial Matrices," *Lin. Alg. Appl.*, Vol. 267, pp. 163–174, 1997. (Cited on p. 775.)

[2120] E. Neuman, "Inequalities Involving Inverse Circular and Inverse Hyperbolic Function II," *J. Math. Ineq.*, Vol. 4, pp. 11–14, 2010. (Cited on p. 266.)

[2121] E. Neuman, "On Wilker and Huygens Type Inequalities," *Math. Ineq. Appl.*, Vol. 15, pp. 271–279, 2012. (Cited on pp. 249 and 264.)

[2122] M. F. Neuts, *Matrix-Geometric Solutions in Stochastic Models*. Baltimore: Johns Hopkins University Press, 1981, reprinted by Dover, Mineola, 1994. (Cited on p. xxi.)

[2123] R. W. Newcomb, "On the Simultaneous Diagonalization of Two Semi-Definite Matrices," *Quart. Appl. Math.*, Vol. 19, pp. 144–146, 1961. (Cited on p. 831.)

[2124] D. J. Newman, "A Sum over k-tuples," *Amer. Math. Monthly*, Vol. 73, pp. 1025–1026, 1966. (Cited on p. 1072.)

[2125] M. Newman, "Lyapunov Revisited," *Lin. Alg. Appl.*, Vol. 52/53, pp. 525–528, 1983. (Cited on p. 567.)

[2126] M. Newman, W. So, and R. C. Thompson, "Convergence Domains for the Campbell-Baker-Hausdorff Formula," *Lin. Multilin. Alg.*, Vol. 24, pp. 301–310, 1989. (Cited on pp. 1189 and 1213.)

[2127] D. W. Nicholson, "Eigenvalue Bounds for of $AB + BA$ for A, B Positive Definite Matrices," *Lin. Alg. Appl.*, Vol. 24, pp. 173–183, 1979. (Cited on p. 811.)

[2128] C. Niculescu and L.-E. Persson, *Convex Functions and Their Applications: A Contemporary Approach*. New York: Springer, 2006. (Cited on pp. 117, 143, 173, 204, 208, 209, 219, 252, 333, 772, 856, 944, and 950.)

[2129] C. P. Niculescu, "Convexity According to the Geometric Mean," *Math. Ineq. Appl.*, Vol. 3, pp. 155–167, 2000. (Cited on pp. 161, 208, 466, and 1238.)

[2130] C. P. Niculescu and F. Popovici, "A Refinement of Popoviciu's Inequality," *Bull. Math. Soc. Sci. Math. Roumanie*, Vol. 49(97), pp. 285–290, 2006. (Cited on p. 117.)

[2131] M. A. Nielsen and I. L. Chuang, *Quantum Computation and Quantum Information*. Cambridge: Cambridge University Press, 2000. (Cited on p. xxi.)

[2132] M. Niezgoda, "Laguerre-Samuelson Type Inequalities," *Lin. Alg. Appl.*, Vol. 422, pp. 574–581, 2007. (Cited on pp. 195 and 734.)

[2133] M. Nihei, "Two Proofs of a Trigonometric Identity," *Math. Gaz.*, Vol. 86, pp. 298–299, 2002. (Cited on p. 244.)

[2134] G. C. Nihous, "A Combinatorial Proof of a Novel Binomial Identity," *Math. Scientist*, Vol. 35, pp. 54–56, 2010. (Cited on p. 86.)

[2135] A. Nijenhuis, "Short Gamma Products with Simple Values," *Amer. Math. Monthly*, Vol. 117, pp. 733–737, 2010. (Cited on p. 1000.)

[2136] N. Nikolov and S. Hristova, "On an Inequality from the IMO 2008," *Crux Math.*, Vol. 36, pp. 42–43, 2010. (Cited on pp. 139 and 214.)

[2137] A. S. Nimbran, "An Infinite Product for the Square Root of an Integer," *Amer. Math. Monthly*, Vol. 124, pp. 647–650, 2017. (Cited on pp. 244 and 1085.)

[2138] K. Nishio, "The Structure of a Real Linear Combination of Two Projections," *Lin. Alg. Appl.*, Vol. 66, pp. 169–176, 1985. (Cited on p. 405.)

[2139] B. Noble and J. W. Daniel, *Applied Linear Algebra*, 3rd ed. Englewood Cliffs: Prentice-Hall, 1988. (Cited on pp. xxii and 362.)

[2140] K. Nomakuchi, "On the Characterization of Generalized Inverses by Bordered Matrices," *Lin. Alg. Appl.*, Vol. 33, pp. 1–8, 1980. (Cited on p. 667.)

[2141] K. Nordstrom, "Some Further Aspects of the Löwner-Ordering Antitonicity of the Moore-Penrose Inverse," *Commun. Statist. Theory Meth.*, Vol. 18, pp. 4471–4489, 1989. (Cited on pp. 738 and 817.)

[2142] K. Nordstrom, "Convexity of the Inverse and Moore-Penrose Inverse," *Lin. Alg. Appl.*, Vol. 434, pp. 1489–1512, 2011. (Cited on pp. 349, 797, and 818.)

[2143] J. Nunemacher, "Which Real Matrices Have Real Logarithms?" *Math. Mag.*, Vol. 62, pp. 132–135, 1989. (Cited on p. 1247.)

[2144] M. Nyblom, "More Nested Square Roots of 2," *Amer. Math. Monthly*, Vol. 112, pp. 822–825, 2005. (Cited on p. 53.)

[2145] M. Nyblom, "Can You See the Telescope?" *Amer. Math. Monthly*, Vol. 117, p. 284, 2010. (Cited on p. 1062.)

[2146] H. Ogawa, "An Operator Pseudo-Inversion Lemma," *SIAM J. Appl. Math.*, Vol. 48, pp. 1527–1531, 1988. (Cited on p. 636.)

[2147] H. Ohtsuka, "A Sequence of Nested Radicals," *Fibonacci Quart.*, Vol. 49, p. 182, 2011. (Cited on p. 974.)

[2148] H. Ohtsuka, "Higher Powers Equalities," *Fibonacci Quart.*, Vol. 49, p. 275, 2011. (Cited on p. 100.)

[2149] H. Ohtsuka, "One of Many!" *Fibonacci Quart.*, Vol. 49, pp. 370–371, 2011. (Cited on p. 103.)

[2150] H. Ohtsuka, "Sums of Reciprocals of Squares of Lucas Numbers," *Fibonacci Quart.*, Vol. 50, pp. 378–379, 2012. (Cited on pp. 122, 1021, and 1079.)

[2151] H. Ohtsuka, "Convergent Series Involving Fibonacci and Lucas Numbers with Rational Sums," *Fibonacci Quart.*, Vol. 51, pp. 283–284, 2013. (Cited on p. 1079.)

[2152] H. Ohtsuka, "Evaluating a Double Sum with Inverse Fibonacci Numbers," *Fibonacci Quart.*, Vol. 51, pp. 180–181, 2013. (Cited on p. 1079.)

[2153] H. Ohtsuka, "Problem B-1155," *Fibonacci Quart.*, Vol. 52, p. 275, 2014. (Cited on p. 1079.)

[2154] H. Ohtsuka, "Problem B-1156," *Fibonacci Quart.*, Vol. 52, p. 367, 2014. (Cited on p. 1079.)

[2155] H. Ohtsuka, "Sum of Cubes and the Square of Sum of Cubes," *Fibonacci Quart.*, Vol. 52, p. 369, 2014. (Cited on p. 100.)

[2156] H. Ohtsuka, "Sums of Sums Reciprocals of Fibonacci Numbers," *Fibonacci Quart.*, Vol. 52, p. 191, 2014. (Cited on p. 1077.)

[2157] H. Ohtsuka, "Problem H766," *Fibonacci Quart.*, Vol. 53, p. 88, 2015. (Cited on p. 100.)

[2158] H. Ohtsuka, "Problem H767," *Fibonacci Quart.*, Vol. 53, p. 88, 2015. (Cited on p. 974.)

[2159] H. Ohtsuka, "A Hyperbolic Sine Series," *Amer. Math. Monthly*, Vol. 124, p. 469, 2017. (Cited on p. 1063.)

[2160] H. Ohtsuka and S. Nakamura, "On the Sum of Reciprocal Fibonacci Numbers," *Fibonacci Quart.*, Vol. 46/47, pp. 153–159, 2008/2009. (Cited on p. 1079.)

[2161] I. Ojeda, "Kronecker Square Roots and the Block Vec Matrix," *Amer. Math. Monthly*, Vol. 122, pp. 60–64, 2015. (Cited on p. 700.)

[2162] C. Olah and M. Bencze, "About Problem PP. 6752," *Octogon*, Vol. 14, pp. 707–708, 2006. (Cited on pp. 456 and 466.)

[2163] K. B. Oldham, J. Myland, and J. Spanier, *An Atlas of Functions*, 2nd ed. New York: Springer, 2009. (Cited on p. 1058.)

[2164] I. Olkin, "An Inequality for a Sum of Forms," *Lin. Alg. Appl.*, Vol. 52–53, pp. 529–532, 1983. (Cited on p. 737.)

[2165] R. L. Ollerton, "Fibonacci Cubes," *Int. J. Math. Educ. Sci. Tech.*, Vol. 37, pp. 754–756, 2006. (Cited on p. 100.)

[2166] K. C. O'Meara, J. Clark, and C. I. Vinsonhaler, *Advanced Topics in Linear Algebra: Weaving Matrix Problems Through the Weyr Form*. New York: Oxford University Press, 2011. (Cited on pp. 576, 577, and 619.)

[2167] M. E. Omidvar, M. S. Moslehian, and A. Niknam, "Unitarily Invariant Norm Inequalities for Operators," *J. Egyptian Math. Soc.*, Vol. 20, pp. 38–42, 2012. (Cited on pp. 870 and 882.)

[2168] A. Oppenheim, "Inequalities for Non-obtuse Triangles," *Amer. Math. Monthly*, Vol. 85, pp. 596–597, 1978. (Cited on p. 484.)
[2169] J. M. Ortega, *Matrix Theory, A Second Course*. New York: Plenum, 1987. (Cited on p. xxii.)
[2170] B. Ortner and A. R. Krauter, "Lower Bounds for the Determinant and the Trace of a Class of Hermitian Matrices," *Lin. Alg. Appl.*, Vol. 236, pp. 147–180, 1996. (Cited on p. 762.)
[2171] S. L. Osburn and D. S. Bernstein, "An Exact Treatment of the Achievable Closed-Loop H_2 Performance of Sampled-Data Controllers: From Continuous-Time to Open-Loop," *Automatica*, Vol. 31, pp. 617–620, 1995. (Cited on p. 1212.)
[2172] T. J. Osler, "The Union of Vieta's and Wallis's Product for π," *Amer. Math. Monthly*, Vol. 106, pp. 774–776, 1999. (Cited on p. 1088.)
[2173] T. J. Osler, "Interesting Finite and Infinite Products from Simple Algebraic Identities," *Math. Gaz.*, Vol. 90, pp. 90–93, 2006. (Cited on pp. 130 and 1083.)
[2174] T. J. Osler, "Novel Method for Finding $\zeta(2p)$ from a Product of Sines," *Int. J. Math. Educ. Sci. Tech.*, Vol. 37, pp. 231–235, 2006. (Cited on p. 1074.)
[2175] T. J. Osler, A. Hassen, and T. R. Chandrupatla, "Connections between Partitions and Divisors," *Coll. Math. J.*, Vol. 38, pp. 278–287, 2007. (Cited on pp. 114, 115, and 1008.)
[2176] A. Ostrowski and H. Schneider, "Some Theorems on the Inertia of General Matrices," *J. Math. Anal. Appl.*, Vol. 4, pp. 72–84, 1962. (Cited on p. 1310.)
[2177] A. M. Ostrowski, "On Some Metrical Properties of Operator Matrices and Matrices Partitioned into Blocks," *J. Math. Anal. Appl.*, Vol. 2, pp. 161–209, 1961. (Cited on p. 887.)
[2178] J. A. Oteo, "The Baker-Campbell-Hausdorff Formula and Nested Commutator Identities," *J. Math. Phys.*, Vol. 32, pp. 419–424, 1991. (Cited on p. 1306.)
[2179] D. V. Ouellette, "Schur Complements and Statistics," *Lin. Alg. Appl.*, Vol. 36, pp. 187–295, 1981. (Cited on pp. 749, 750, 778, and 831.)
[2180] S. Ovchinnikov, "Series with Inverse Function Terms," *Coll. Math. J.*, Vol. 42, pp. 283–288, 2011. (Cited on pp. 258, 1061, 1062, and 1079.)
[2181] D. A. Overdijk, "Skew-Symmetric Matrices in Classical Mechanics," Eindhoven University, Memorandum COSOR 89-23, 1989. (Cited on p. 386.)
[2182] E. Packard and M. Reitenbach, "A Generalization of the Identity $4\cos\frac{\pi}{3} = \frac{1}{2}$," *Math. Mag.*, Vol. 81, pp. 3–15, 2008. (Cited on p. 242.)
[2183] C. C. Paige and M. Saunders, "Towards a Generalized Singular Value Decomposition," *SIAM J. Numer. Anal.*, Vol. 18, pp. 398–405, 1981. (Cited on p. 574.)
[2184] C. C. Paige, G. P. H. Styan, B. Y. Wang, and F. Zhang, "Hua's Matrix Equality and Schur Complements," *Int. J. Inform. Sys. Sci.*, Vol. 4, pp. 124–135, 2008. (Cited on pp. 350, 568, 757, 780, and 796.)
[2185] C. C. Paige and M. Wei, "History and Generality of the CS Decomposition," *Lin. Alg. Appl.*, Vol. 208/209, pp. 303–326, 1994. (Cited on p. 574.)
[2186] V. E. Paksoy, R. Turkmen, and F. Zhang, "Inequalities of Generalized Matrix Functions via Tensor Products," *Elec. J. Lin. Alg.*, Vol. 27, pp. 332–341, 2014. (Cited on p. 781.)
[2187] H. Palanthandalam-Madapusi, D. S. Bernstein, and R. Venugopal, "Dimensional Analysis of Matrices: State-Space Models and Dimensionless Units," *IEEE Contr. Sys. Mag.*, Vol. 27, no. December, pp. 100–109, 2007. (Cited on pp. 440 and 1306.)
[2188] H. Palanthandalam-Madapusi, T. Van Pelt, and D. S. Bernstein, "Existence Conditions for Quadratic Programming with a Sign-Indefinite Quadratic Equality Constraint," *J. Global Optimization*, Vol. 45, pp. 533–549, 2009. (Cited on p. 789.)
[2189] H. Palanthandalam-Madapusi, T. Van Pelt, and D. S. Bernstein, "Parameter Consistency and Quadratically Constrained Errors-in-Variables Least-Squares Identification," *Int. J. Contr.*, Vol. 83, pp. 862–877, 2010. (Cited on p. 789.)
[2190] F. C. Paliogiannis, "On Commuting Operator Exponentials," *Proc. Amer. Math. Soc.*, Vol. 131, pp. 3777–3781, 2003. (Cited on p. 1211.)
[2191] B. P. Palka, *An Introduction to Complex Function Theory*. New York: Springer, 1991. (Cited on p. 276.)
[2192] W. Pan and L. Zhu, "Generalizations of Shafer-Fink-Type Inequalities for the Arc Sine Function," *J. Ineq. Appl.*, pp. 1–6, 2009, article ID 705317. (Cited on p. 261.)
[2193] K. P. Pandeyend, "Characteristics of Triangular Numbers," *Octogon*, Vol. 17, pp. 282–284, 2009. (Cited on p. 39.)

[2194] C. V. Pao, "Logarithmic Derivatives of a Square Matrix," *Lin. Alg. Appl.*, Vol. 6, pp. 159–164, 1973. (Cited on p. 1219.)

[2195] J. G. Papastavridis, *Tensor Calculus and Analytical Dynamics*. Boca Raton: CRC, 1998. (Cited on p. xxi.)

[2196] J. G. Papastavridis, *Analytical Mechanics*. Oxford: Oxford University Press, 2002. (Cited on pp. xxi and 731.)

[2197] C. J. Pappacena, "An Upper Bound for the Length of a Finite-Dimensional Algebra," *J. Algebra*, Vol. 197, pp. 535–545, 1997. (Cited on p. 578.)

[2198] R. B. Paris and P. J. Larcombe, "On the Asymptotic Expansion of a Binomial Sum Involving Powers of the Summation Index," *J. Classical Anal.*, Vol. 1, pp. 113–123, 2012. (Cited on p. 68.)

[2199] J. L. Parish, "On the Derivative of a Vertex Polynomial," *Forum Geometricorum*, Vol. 6, pp. 285–288, 2006. (Cited on p. 496.)

[2200] F. C. Park, "Computational Aspects of the Product-of-Exponentials Formula for Robot Kinematics," *IEEE Trans. Autom. Contr.*, Vol. 39, pp. 643–647, 1994. (Cited on p. 1247.)

[2201] P. Park, "On the Trace Bound of a Matrix Product," *IEEE Trans. Autom. Contr.*, Vol. 41, pp. 1799–1802, 1996. (Cited on p. 593.)

[2202] P.-S. Park, "Some Logarithmic Approximations for π and e," *Math. Mag.*, Vol. 87, p. 43, 2014. (Cited on p. 54.)

[2203] D. F. Parker, *Fields, Flow and Waves: An Introduction to Continuum Models*. London: Springer, 2003. (Cited on p. 619.)

[2204] P. C. Parks, "A New Proof of the Routh-Hurwitz Stability Criterion Using the Second Method of Liapunov," *Proc. Camb. Phil. Soc.*, Vol. 58, pp. 694–702, 1962. (Cited on p. 1230.)

[2205] R. V. Patel, "On Blocking Zeros in Linear Multivariable Systems," *IEEE Trans. Autom. Contr.*, Vol. AC-31, pp. 239–241, 1986. (Cited on p. 1319.)

[2206] R. V. Patel and M. Toda, "Trace Inequalities Involving Hermitian Matrices," *Lin. Alg. Appl.*, Vol. 23, pp. 13–20, 1979. (Cited on pp. 751, 760, and 1220.)

[2207] G. N. Patruno, "Sums of Irrational Square Roots Are Irrational," *Math. Mag.*, Vol. 61, pp. 44–234, 1988. (Cited on p. 52.)

[2208] A. Paz, "An Application of the Cayley-Hamilton Theorem to Matrix Polynomials in Several Variables," *Lin. Multilin. Alg.*, Vol. 15, pp. 161–170, 1984. (Cited on p. 578.)

[2209] C. Pearcy, "A Complete Set of Unitary Invariants for Operators Generating Finite W^*-Algebras of Type I," *Pacific J. Math.*, Vol. 12, pp. 1405–1414, 1962. (Cited on p. 576.)

[2210] M. C. Pease, *Methods of Matrix Algebra*. New York: Academic Press, 1965. (Cited on p. 440.)

[2211] J. Pecaric and R. Rajic, "The Dunkl-Williams Inequality with n Elements in Normed Linear Spaces," *Math. Ineq. Appl.*, Vol. 10, pp. 461–470, 2007. (Cited on pp. 851 and 852.)

[2212] J. Pecaric and R. Rajic, "Inequalities of the Dunkl-Williams Type for Absolute Value Operators," *J. Math. Ineq.*, Vol. 4, pp. 1–10, 2010. (Cited on pp. 748 and 850.)

[2213] J. E. Pecaric, F. Proschan, and Y. L. Tong, *Finite-Dimensional Linear Algebra*. San Diego: Academic Press, 1992. (Cited on pp. 118 and 362.)

[2214] J. E. Pecaric, S. Puntanen, and G. P. H. Styan, "Some Further Matrix Extensions of the Cauchy-Schwarz and Kantorovich Inequalities, with Some Statistical Applications," *Lin. Alg. Appl.*, Vol. 237/238, pp. 455–476, 1996. (Cited on p. 793.)

[2215] R. Pemantle and C. Schneider, "When Is 0.999... Equal to 1?" *Amer. Math. Monthly*, Vol. 114, pp. 344–350, 2007. (Cited on p. 1073.)

[2216] L. Pennisi, *Elements of Complex Variables*. New York: Holt, Rhinehart and Winston, 1963. (Cited on p. 21.)

[2217] K. A. Penson and J.-M. Sixdeniers, "Integral Representations of Catalan and Related Numbers," *J. Integer Sequences*, Vol. 4, pp. 1–6, 2001, article 01.2.6. (Cited on pp. 105 and 1113.)

[2218] A. Peperko, "Bounds on the Generalized and the Joint Spectral Radius of Hadamard Products of Bounded Sets of Positive Operators on Sequence Spaces," *Lin. Alg. Appl.*, Vol. 437, pp. 189–201, 2012. (Cited on p. 699.)

[2219] J. Peralta, "On Some Infinite Sums of Iterated Radicals," *Int. J. Math. Educ. Sci. Tech.*, Vol. 40, pp. 290–300, 2009. (Cited on p. 973.)

[2220] P. Perfetti, "Problem Q1009," *Math. Mag.*, Vol. 84, pp. 151, 156, 2011. (Cited on p. 1024.)

[2221] S. Perlis, *Theory of Matrices*. Reading: Addison-Wesley, 1952, reprinted by Dover, New York, 1991. (Cited on pp. 362, 501, 502, 504, and 619.)

[2222] D. Perrin, *Cours d'Algèbre*. France: Ellipses, 1996. (Cited on p. 607.)

[2223] T. Peter, "Maximizing the Area of a Quadrilateral," *Coll. Math. J.*, Vol. 34, pp. 315–316, 2003. (Cited on p. 492.)

[2224] I. R. Petersen and C. V. Hollot, "A Riccati Equation Approach to the Stabilization of Uncertain Systems," *Automatica*, Vol. 22, pp. 397–411, 1986. (Cited on p. 796.)

[2225] K. Petersen, "Two-Sided Eulerian Numbers via Balls in Boxes," *Math. Mag.*, Vol. 86, pp. 159–176, 2013. (Cited on pp. 112 and 975.)

[2226] J. Peterson, "A Probabilistic Proof of a Binomial Identity," *Amer. Math. Monthly*, Vol. 120, pp. 558–562, 2013. (Cited on p. 75.)

[2227] L. D. Petkovic and M. S. Petkovic, "Inequalities in Circular Arithmetic: A Survey," in *Recent Progress in Inequalities*, G. V. Milovanovic, Ed. Dordrecht: Kluwer, 1998, pp. 325–340. (Cited on pp. 128 and 210.)

[2228] M. Petkovsek, H. S. Wilf, and D. Zeilberger, $A = B$. Wellesley: A K Peters, 1996. (Cited on pp. 43, 49, 67, 75, 76, 86, 92, 187, and 968.)

[2229] D. Petz and R. Temesi, "Means of Positive Numbers and Matrices," *SIAM J. Matrix Anal. Appl.*, Vol. 27, pp. 712–720, 2005. (Cited on pp. 744 and 749.)

[2230] D. Phillips, "Improving Spectral Variation Bounds with Chebyshev Polynomials," *Lin. Alg. Appl.*, Vol. 133, pp. 165–173, 1990. (Cited on p. 893.)

[2231] G. Piazza and T. Politi, "An Upper Bound for the Condition Number of a Matrix in Spectral Norm," *J. Comp. Appl. Math.*, Vol. 143, pp. 141–144, 2002. (Cited on p. 896.)

[2232] T. Piezas, "A Collection of Algebraic Identities," https://sites.google.com/site/tpiezas/Home. (Cited on pp. 34, 178, and 189.)

[2233] A. Pinkus, "Interpolation by Matrices," *Elec. J. Lin. Alg.*, Vol. 11, pp. 281–291, 2004. (Cited on p. 383.)

[2234] L. A. Pipes, "Applications of Laplace Transforms of Matrix Functions," *J. Franklin Inst.*, Vol. 285, pp. 436–451, 1968. (Cited on pp. 1205 and 1206.)

[2235] N. Pippenger, "The Hypercube of Resistors, Asymptotic Expansions, and Preferential Arrangements," *Math. Mag.*, Vol. 83, pp. 331–346, 2010. (Cited on pp. 68 and 1009.)

[2236] E. Pite, "Orthogonality of Matrices under Additivity of Traces of Powers," *Amer. Math. Monthly*, Vol. 119, pp. 166–167, 2012. (Cited on p. 763.)

[2237] A. O. Pittenger and M. H. Rubin, "Convexity and Separation Problem of Quantum Mechanical Density Matrices," *Lin. Alg. Appl.*, Vol. 346, pp. 47–71, 2002. (Cited on p. xxi.)

[2238] R. Piziak and P. L. Odell, *Matrix Theory: From Generalized Inverses to Jordan Form*. Boca Raton: Chapman & Hall/CRC, 2007. (Cited on pp. 226, 271, 272, 312, 314, 315, 319, 325, 326, 343, 379, 397, 407, 408, 411, 530, 579, 591, 623, 628, 629, 633, 634, 636, 637, 638, 639, 640, 651, 657, 658, 661, 666, 669, 670, 674, 675, 736, 855, 856, 857, and 910.)

[2239] R. Piziak, P. L. Odell, and R. Hahn, "Constructing Projections on Sums and Intersections," *Computers Math. Appl.*, Vol. 37, pp. 67–74, 1999. (Cited on pp. 319, 413, and 656.)

[2240] A. Plaza, "Proof without Words: Partial Column Sums in Pascal's Triangle," *Math. Mag.*, Vol. 90, pp. 117–118, 2017. (Cited on p. 67.)

[2241] A. Plaza and S. Falcon, "A Combinatorial Sum," *Coll. Math. J.*, Vol. 45, pp. 227–228, 2014. (Cited on p. 75.)

[2242] S. Plouffe, "Approximations of Generating Functions and a Few Conjectures," http://arxiv.org/abs/0911.4975. (Cited on p. 1011.)

[2243] M. Poghosyan, "A Combinatorial Identity from Arcsin," *Amer. Math. Monthly*, Vol. 117, pp. 183–184, 2010. (Cited on p. 86.)

[2244] C. Pohoata and D. Grinberg, "More Triangle Inequalities," *Amer. Math. Monthly*, Vol. 121, pp. 656–657, 2014. (Cited on p. 452.)

[2245] I. Polik and T. Terlaky, "A Survey of the S-Lemma," *SIAM Rev.*, Vol. 49, pp. 371–418, 2007. (Cited on p. 789.)

[2246] T. Politi, "A Formula for the Exponential of a Real Skew-Symmetric Matrix of Order 4," *Numer. Math. BIT*, Vol. 41, pp. 842–845, 2001. (Cited on pp. 1206 and 1208.)

[2247] D. S. G. Pollock, "Tensor Products and Matrix Differential Calculus," *Lin. Alg. Appl.*, Vol. 67, pp. 169–193, 1985. (Cited on p. 1092.)

[2248] B. T. Polyak, "Convexity of Quadratic Transformations and Its Use in Control and Optimization," *J. Optim. Theory Appl.*, Vol. 99, pp. 553–583, 1998. (Cited on pp. 788 and 789.)

[2249] S. Ponnusamy and H. Silverman, *Complex Variables with Applications*. Boston: Birkhauser, 2006. (Cited on pp. xl, 21, 132, 244, 947, 1087, and 1126.)

[2250] B. Poonen, "A Unique $(2k + 1)$-th Root of a Matrix," *Amer. Math. Monthly*, Vol. 98, p. 763, 1991. (Cited on p. 608.)

[2251] B. Poonen, "Positive Deformations of the Cauchy Matrix," *Amer. Math. Monthly*, Vol. 102, pp. 842–843, 1995. (Cited on p. 728.)

[2252] O. T. Pop, "About Bergstrom's Inequality," *J. Math. Ineq.*, Vol. 3, pp. 237–242, 2009. (Cited on pp. 217, 218, 271, and 273.)

[2253] V. Pop and O. Furdui, *Square Matrices of Order 2: Theory, Applications, and Problems*. New York: Springer, 2017. (Cited on pp. 362 and 1178.)

[2254] D. Popa and I. Rasa, "Inequalities Involving the Inner Product," *J. Ineq. Pure Appl. Math.*, Vol. 8, no. 3, pp. 1–4, 2007, Article 86. (Cited on p. 856.)

[2255] I. Popescu, "Problem 401," *Gazeta Matematica Serie A*, Vol. 31, p. 37, 2013. (Cited on p. 86.)

[2256] A. Popescu-Zorica, "La Constante d'Euler Exprimeé par des Intégrales," *Gazeta Matematica Seria A*, Vol. 26, pp. 220–231, 2008. (Cited on p. 1170.)

[2257] V. M. Popov, *Hyperstability of Control Systems*. Berlin: Springer, 1973. (Cited on pp. xxi and 1247.)

[2258] G. J. Porter, "Linear Algebra and Affine Planar Transformations," *Coll. Math. J.*, Vol. 24, pp. 47–51, 1993. (Cited on p. 396.)

[2259] A. S. Posamentier and I. Lehmann, *Mathematical Curiosities*. Amherst: Prometheus Books, 2014. (Cited on p. 34.)

[2260] B. H. Pourciau, "Modern Multiplier Rules," *Amer. Math. Monthly*, Vol. 87, pp. 433–452, 1980. (Cited on p. 940.)

[2261] A. Prach and D. S. Bernstein, "An Additive Decomposition Involving Rank-One Matrices," *IMAGE*, Vol. 53, pp. 50–51, 2014. (Cited on p. 382.)

[2262] C. R. Pranesachar, "Ratio Circum-to-In-Radius," *Amer. Math. Monthly*, Vol. 114, p. 648, 2007. (Cited on p. 452.)

[2263] V. V. Prasolov, *Problems and Theorems in Linear Algebra*. Providence: American Mathematical Society, 1994. (Cited on pp. xxii, 346, 347, 355, 390, 407, 419, 429, 438, 519, 520, 522, 533, 536, 539, 571, 578, 585, 590, 593, 599, 600, 604, 605, 608, 609, 618, 669, 694, 721, 758, 777, 779, 783, 799, 822, 861, 881, 893, 894, 900, 952, 953, and 1193.)

[2264] V. V. Prasolov, *Polynomials*. Berlin: Springer, 2004. (Cited on pp. 134, 361, 436, and 983.)

[2265] U. Prells, M. I. Friswell, and S. D. Garvey, "Use of Geometric Algebra: Compound Matrices and the Determinant of the Sum of Two Matrices," *Proc. Royal Soc. Lond. A*, Vol. 459, pp. 273–285, 2003. (Cited on pp. 387, 691, 694, and 695.)

[2266] C. Pritchard, "Varignon's Theorem," in *The Changing Shape of Geometry*, C. Pritchard, Ed. Cambridge: Cambridge University Press, 2003, pp. 175–176. (Cited on p. 492.)

[2267] H. Prodinger, "A Short Proof of Carlitz's Bernoulli Number Identity," *J. Integer Sequences*, Vol. 17, pp. 1–2, 2014, article 14.4.1. (Cited on p. 980.)

[2268] A. P. Prudnikov, Y. A. Brychkov, and O. I. Marichev, *Integrals and Series—Vol. 1: Elementary Functions, Vol. 2: Special Functions*. Boca Raton: CRC Press, 1998. (Cited on p. 1178.)

[2269] J. S. Przemieniecki, *Theory of Matrix Structural Analysis*. New York: McGraw-Hill, 1968. (Cited on p. xxi.)

[2270] P. J. Psarrakos, "On the *m*th Roots of a Complex Matrix," *Elec. J. Lin. Alg.*, Vol. 9, pp. 32–41, 2002. (Cited on p. 608.)

[2271] C. C. Pugh, *Real Mathematical Analysis*, 2nd ed. New York: Springer, 2015. (Cited on p. 951.)

[2272] N. J. Pullman, *Matrix Theory and Its Applications: Selected Topics*. New York: Marcel Dekker, 1976. (Cited on p. 1247.)

[2273] S. Puntanen and G. P. H. Styan, "Historical Introduction: Issai Schur and the Early Development of the Schur Complement," in *The Schur Complement and Its Applications*, F. Zhang, Ed. New York: Springer, 2004, pp. 1–16. (Cited on p. 662.)

[2274] F. Qi, "Inequalities and Monotonicity of the Ratio of the Geometric Means of a Positive Arithmetic Sequence with Unit Difference," *Int. J. Math. Educ. Sci. Tech.*, Vol. 34, pp. 601–607, 2003. (Cited on p. 47.)

[2275] F. Qi, "Inequalities between the Sum of Squares and the Exponential of Sum of a Nonnegative Sequence," *J. Ineq. Pure Appl. Math.*, Vol. 8, no. 3, pp. 1–5, 2007, Article 78. (Cited on p. 214.)

[2276] F. Qi, "Explicit Formulas for Computing Bernoulli Numbers of the Second Kind and Stirling Numbers of the First Kind," *Filomat*, Vol. 28, pp. 319–327, 2014. (Cited on p. 1019.)

[2277] F. Qi and L. Debnath, "Inequalities for Power-Exponential Functions," *J. Ineq. Pure. Appl. Math.*, Vol. 1, no. 2/15, pp. 1–5, 2000. (Cited on p. 145.)

[2278] F. Qi, D.-W. Niu, and B.-N. Guo, "Refinements, Generalizations, and Applications of Jordan's Inequality and Related Problems," *J. Ineq. Appl.*, pp. 1–52, 2009, article ID 271923. (Cited on pp. 249, 250, 264, 466, 484, and 1128.)

[2279] F. Qi, D.-W. Niu, and B.-N. Guo, "Refinements, Generalizations, and Applications of Jordan's Inequality and Related Problems," *J. Ineq. Appl.*, pp. 1–52, 2009, article 271923. (Cited on p. 250.)

[2280] F. Qi, S.-Q. Zhang, and B.-N. Guo, "Sharpening and Generalizations of Shafer's Inequality for the Arc Tangent Function," *J. Ineq. Appl.*, pp. 1–6, 2009, article ID 930294. (Cited on p. 261.)

[2281] C. Qian and J. Li, "Global Finite-Time Stabilization by Output Feedback for Planar Systems without Observation Linearization," *IEEE Trans. Autom. Contr.*, Vol. 50, pp. 885–890, 2005. (Cited on p. 145.)

[2282] R. X. Qian and C. L. DeMarco, "An Approach to Robust Stability of Matrix Polytopes through Copositive Homogeneous Polynomials," *IEEE Trans. Autom. Contr.*, Vol. 37, pp. 848–852, 1992. (Cited on p. 536.)

[2283] L. Qiu, B. Bernhardsson, A. Rantzer, E. J. Davison, P. M. Young, and J. C. Doyle, "A Formula for Computation of the Real Stability Radius," *Automatica*, Vol. 31, pp. 879–890, 1995. (Cited on p. 1228.)

[2284] L. Qiu and X. Zhan, "On the Span of Hadamard Products of Vectors," *Lin. Alg. Appl.*, Vol. 422, pp. 304–307, 2007. (Cited on p. 698.)

[2285] J. F. Queiro and A. L. Duarte, "On the Cartesian Decomposition of a Matrix," *Lin. Multilin. Alg.*, Vol. 18, pp. 77–85, 1985. (Cited on p. 777.)

[2286] J. F. Queiro and A. Kovacec, "A Bound for the Determinant of the Sum of Two Normal Matrices," *Lin. Multilin. Alg.*, Vol. 33, pp. 171–173, 1993. (Cited on p. 594.)

[2287] V. Rabanovich, "Every Matrix Is a Linear Combination of Three Idempotents," *Lin. Alg. Appl.*, Vol. 390, pp. 137–143, 2004. (Cited on p. 619.)

[2288] S. Rabinowitz, "The Tangent Function and $\sqrt{11}$," *Coll. Math. J.*, Vol. 38, pp. 310–311, 2007. (Cited on p. 232.)

[2289] H. Radjavi, "Decomposition of Matrices into Simple Involutions," *Lin. Alg. Appl.*, Vol. 12, pp. 247–255, 1975. (Cited on p. 607.)

[2290] H. Radjavi and P. Rosenthal, *Simultaneous Triangularization*. New York: Springer, 2000. (Cited on p. 617.)

[2291] H. Radjavi and J. P. Williams, "Products of Self-Adjoint Operators," *Michigan Math. J.*, Vol. 16, pp. 177–185, 1969. (Cited on pp. 577, 607, and 655.)

[2292] S. Radulescu, M. Dragan, and I. V. Maftei, "Some Consequences of W. J. Blundon's Inequality," *Gazeta Matematica Serie B*, Vol. 116, pp. 3–9, 2011. (Cited on p. 452.)

[2293] S. Radulescu, M. Radulescu, J. L. Diaz-Barrero, and P. Alexandrescu, "Two Families of Cyclic Inequalities," *Math. Ineq. Appl.*, Vol. 15, pp. 199–209, 2012. (Cited on p. 153.)

[2294] T.-L. T. Radulescu, V. D. Radulescu, and T. Andreescu, *Problems in Real Analysis: Advanced Calculus on the Real Axis*. Dordrecht: Springer, 2009. (Cited on pp. 25, 47, 48, 50, 144, 160, 161, 194, 198, 207, 208, 209, 213, 217, 225, 227, 228, 229, 249, 250, 252, 261, 271, 472, 855, 934, 950, 951, 954, 955, 958, 959, 960, 961, 962, 963, 967, 971, 1023, 1031, 1047, 1081, 1089, 1090, and 1128.)

[2295] Q. I. Rahman and G. Schmeisser, *Analytic Theory of Polynomials: Critical Points, Zeros and Extremal Properties*. London: Oxford University Press, 2002. (Cited on pp. 891 and 949.)

[2296] V. Rakocevic, "On the Norm of Idempotent Operators in a Hilbert Space," *Amer. Math. Monthly*, Vol. 107, pp. 748–750, 2000. (Cited on pp. 415 and 595.)

[2297] A. C. M. Ran and R. Vreugdenhil, "Existence and Comparison Theorems for Algebraic Riccati Equations for Continuous- and Discrete-Time Systems," *Lin. Alg. Appl.*, Vol. 99, pp. 63–83, 1988. (Cited on p. 1319.)

[2298] A. Rantzer, "On the Kalman-Yakubovich-Popov Lemma," *Sys. Contr. Lett.*, Vol. 28, pp. 7–10, 1996. (Cited on p. 578.)

[2299] C. R. Rao and S. K. Mitra, *Generalized Inverse of Matrices and Its Applications*. New York: Wiley, 1971. (Cited on pp. 679, 819, and 831.)

[2300] C. R. Rao and M. B. Rao, *Matrix Algebra and Its Applications to Statistics and Econometrics*. Singapore: World Scientific, 1998. (Cited on pp. xxi, 425, and 700.)

[2301] J. V. Rao, "Some More Representations for the Generalized Inverse of a Partitioned Matrix," *SIAM J. Appl. Math.*, Vol. 24, pp. 272–276, 1973. (Cited on p. 667.)

[2302] U. A. Rauhala, "Array Algebra Expansion of Matrix and Tensor Calculus: Part 1," *SIAM J. Matrix Anal. Appl.*, Vol. 24, pp. 490–508, 2003. (Cited on p. 701.)

[2303] P. A. Regalia and S. K. Mitra, "Kronecker Products, Unitary Matrices and Signal Processing Applications," *SIAM Rev.*, Vol. 31, pp. 586–613, 1989. (Cited on p. 701.)

[2304] M. W. Reinsch, "A Simple Expression for the Terms in the Baker-Campbell-Hausdorff Series," *J. Math. Physics*, Vol. 41, pp. 2434–2442, 2000. (Cited on pp. 1189 and 1213.)

[2305] P. Ressel, "The Hornick-Hlawka Inequality and Bernstein Functions," *J. Math. Ineq.*, Vol. 9, pp. 883–888, 2015. (Cited on pp. 153 and 856.)

[2306] T. J. Richardson and R. H. Kwong, "On Positive Definite Solutions to the Algebraic Riccati Equation," *Sys. Contr. Lett.*, Vol. 7, pp. 99–104, 1986. (Cited on pp. 1305 and 1319.)

[2307] D. S. Richeson, *Euler's Gem: The Polyhedron Formula and the Birth of Topology*. Princeton: Princeton University Press, 2008. (Cited on pp. 26 and 495.)

[2308] A. N. Richmond, "Expansions for the Exponential of a Sum of Matrices," in *Applications of Matrix Theory*, M. J. C. Gover and S. Barnett, Eds. Oxford: Oxford University Press, 1989, pp. 283–289. (Cited on p. 1214.)

[2309] V. F. Rickey and P. M. Tuchinsky, "An Application of Geography to Mathematics: History of the Integral of the Secant," *Math. Math.*, Vol. 53, pp. 162–166, 1980. (Cited on p. 1095.)

[2310] K. S. Riedel, "A Sherman-Morrison-Woodbury Identity for Rank Augmenting Matrices with Application to Centering," *SIAM J. Matrix Anal. Appl.*, Vol. 13, pp. 659–662, 1992. (Cited on p. 636.)

[2311] J. R. Ringrose, *Compact Non-Self-Adjoint Operators*. New York: Van Nostrand Reinhold, 1971. (Cited on p. 912.)

[2312] J. Riordan, *Combinatorial Identities*. New York: Wiley, 1968. (Cited on pp. 92 and 111.)

[2313] R. S. Rivlin, "Further Remarks on the Stress Deformation Relations for Isotropic Materials," *J. Rational Mech. Anal.*, Vol. 4, pp. 681–702, 1955. (Cited on p. 528.)

[2314] J. W. Robbin, *Matrix Algebra Using MINImal MATlab*. Wellesley: A. K. Peters, 1995. (Cited on pp. 118, 315, 362, 416, 440, 544, 545, 578, 606, and 952.)

[2315] J. W. Robbin and D. A. Salamon, "The Exponential Vandermonde Matrix," *Lin. Alg. Appl.*, Vol. 317, pp. 225–226, 2000. (Cited on p. 698.)

[2316] J. Roberts, *Lure of the Integers*. Washington, DC: The Mathematical Association of America, 1992. (Cited on pp. 33, 35, 178, and 188.)

[2317] D. W. Robinson, "Nullities of Submatrices of the Moore-Penrose Inverse," *Lin. Alg. Appl.*, Vol. 94, pp. 127–132, 1987. (Cited on p. 423.)

[2318] P. Robinson and A. J. Wathen, "Variational Bounds on the Entries of the Inverse of a Matrix," *IMA J. Numer. Anal.*, Vol. 12, pp. 463–486, 1992. (Cited on p. 762.)

[2319] R. T. Rockafellar, *Convex Analysis*. Princeton: Princeton University Press, 1990. (Cited on pp. 362, 922, 935, and 940.)

[2320] R. T. Rockafellar and R. J. B. Wets, *Variational Analysis*. Berlin: Springer, 1998. (Cited on p. 935.)

[2321] A. M. Rockett, "Sums of the Inverses of Binomial Coefficients," *Fibonacci Quart.*, Vol. 19, pp. 433–437, 1981. (Cited on p. 68.)

[2322] L. Rodman, "Products of Symmetric and Skew Symmetric Matrices," *Lin. Multilin. Alg.*, Vol. 43, pp. 19–34, 1997. (Cited on p. 610.)

[2323] L. Rodman, *Topics in Quaternion Linear Algebra*. Princeton: Princeton University Press, 2014. (Cited on p. 438.)

[2324] G. S. Rogers, *Matrix Derivatives*. New York: Marcel Dekker, 1980. (Cited on p. 1092.)

[2325] M. D. Rogers and A. Straub, "A Solution of Sun's \$520 Challenge Concerning $\frac{520}{\pi}$," *Int. J. Number Theory*, Vol. 9, pp. 1273–1288, 2013. (Cited on p. 1070.)

[2326] R.-A. Rohan, "The Exponential Map and the Euclidean Isometries," *Acta Univ. Apulensis*, Vol. 31, pp. 199–204, 2012. (Cited on p. 394.)

[2327] C. A. Rohde, "Generalized Inverses of Partitioned Matrices," *SIAM J. Appl. Math.*, Vol. 13, pp. 1033–1035, 1965. (Cited on p. 667.)

[2328] J. Rohn, "Computing the Norm $\|A\|_{\infty,1}$ Is NP-Hard," *Lin. Multilin. Alg.*, Vol. 47, pp. 195–204, 2000. (Cited on p. 864.)

[2329] O. Rojo, "Further Bounds for the Smallest Singular Value and the Spectral Condition Number," *Computers Math. Appl.*, Vol. 38, pp. 215–228, 1999. (Cited on p. 898.)

[2330] O. Rojo, "Inequalities Involving the Mean and the Standard Deviation of Nonnegative Real Numbers," *J. Ineq. Appl.*, Vol. 2006, pp. 1–15, 2006, article ID 43465. (Cited on pp. 195 and 196.)

[2331] J. Rooin, "On Ky Fan's Inequality and Its Additive Analogues," *Math. Ineq. Appl.*, Vol. 6, pp. 595–604, 2003. (Cited on p. 209.)

[2332] J. Rooin, "An Approach to Ky Fan Type Inequalities from Binomial Expansions," *Math. Ineq. Appl.*, Vol. 11, pp. 679–688, 2008. (Cited on p. 124.)

[2333] J. Rooin and A. Alikhani, "A Geometric Proof of the Beppo-Levi Inequality," *Amer. Math. Monthly*, Vol. 122, p. 566, 2015. (Cited on p. 855.)

[2334] T. G. Room, "The Composition of Rotations in Euclidean Three-Space," *Amer. Math. Monthly*, Vol. 59, pp. 688–692, 1952. (Cited on p. 1207.)

[2335] D. J. Rose, "Matrix Identities of the Fast Fourier Transform," *Lin. Alg. Appl.*, Vol. 29, pp. 423–443, 1980. (Cited on p. 615.)

[2336] H. E. Rose, *A Course on Finite Groups*. London: Springer, 2009. (Cited on pp. 36, 370, 432, 433, 434, 436, 437, and 575.)

[2337] J. S. Rose, *A Course on Group Theory*. Mineola: Dover, 2012. (Cited on p. 437.)

[2338] K. H. Rosen, Ed., *Handbook of Discrete and Combinatorial Mathematics*. Boca Raton: CRC, 2000. (Cited on pp. xxi and xxii.)

[2339] H. H. Rosenbrock, *State-Space and Multivariable Theory*. New York: Wiley, 1970. (Cited on p. 1283.)

[2340] M. Rosenfeld, "A Sufficient Condition for Nilpotence," *Amer. Math. Monthly*, Vol. 103, pp. 907–909, 1996. (Cited on pp. xxi and 420.)

[2341] A. Rosoiu, "Triangle Radii Inequality," *Amer. Math. Monthly*, Vol. 114, p. 640, 2007. (Cited on p. 452.)

[2342] J. B. Rosser and L. Schoenfeld, "Approximate Formulas for Some Functions of Prime Numbers," *Ill. J. Math.*, Vol. 6, pp. 64–94, 1962. (Cited on p. 35.)

[2343] W. Rossmann, *Lie Groups: An Introduction Through Linear Groups*. Oxford: Oxford University Press, 2002. (Cited on p. 440.)

[2344] U. G. Rothblum, "Nonnegative Matrices and Stochastic Matrices," in *Handbook of Linear Algebra*, L. Hogben, Ed. Boca Raton: Chapman & Hall/CRC, 2007, pp. 9–1–9–25. (Cited on pp. 538, 539, 544, and 675.)

[2345] J. J. Rotman, *An Introduction to the Theory of Groups*, 4th ed. New York: Springer, 1999. (Cited on pp. 432, 436, and 616.)

[2346] R. Roy and F. W. J. Olver, "Elementary Functions," in *NIST Handbook of Mathematical Functions*, F. W. J. Olver, D. W. Lozier, R. F. Boisvert, and C. W. Clark, Eds. Cambridge: Cambridge University Press, 2010, pp. 103–134. (Cited on pp. 245, 262, 264, and 269.)

[2347] M. Rozhkova, "Problem 3504," *Crux Math.*, Vol. 37, pp. 54–56, 2011. (Cited on p. 484.)

[2348] J. Rubio-Massegu and J. L. Diaz-Barrero, "Zero and Coefficient Inequalities for Stable Polynomials," *Applic. Anal. Disc. Math.*, Vol. 3, pp. 69–77, 2009. (Cited on p. 127.)

[2349] W. J. Rugh, *Linear System Theory*, 2nd ed. Upper Saddle River: Prentice Hall, 1996. (Cited on pp. 516, 517, 1266, 1278, 1306, and 1319.)

[2350] A. M. Russell and C. J. F. Upton, "A Class of Positive Semidefinite Matrices," *Lin. Alg. Appl.*, Vol. 93, pp. 121–126, 1987. (Cited on p. 729.)

[2351] D. L. Russell, *Mathematics of Finite-Dimensional Control Systems*. New York: Marcel Dekker, 1979. (Cited on p. 1309.)

[2352] G. Rzadkowski, "A Product of Sines Once Again," *Math. Gaz.*, Vol. 84, pp. 306–307, 2000. (Cited on p. 244.)

[2353] G. Rzadkowski, "On Identities Following from Trigonometric Formulae of Multiple Angles," *Math. Gaz.*, Vol. 85, pp. 289–293, 2001. (Cited on pp. 86, 236, 242, and 244.)

[2354] G. Rzadkowski, "On a Family of Polynomials," *Math. Gaz.*, Vol. 90, pp. 283–286, 2006. (Cited on p. 983.)

[2355] G. Rzadkowski, "Some Integral Representations for Bernoulli Numbers," *Math. Gaz.*, Vol. 93, pp. 112–114, 2009. (Cited on p. 983.)

[2356] T. L. Saaty, *The Analytic Hierarchy Process*. New York: McGraw-Hill, 1980. (Cited on p. 540.)

[2357] A. Saberi, P. Sannuti, and B. M. Chen, *H_2 Optimal Control*. Englewood Cliffs: Prentice-Hall, 1995. (Cited on p. xxi.)

[2358] N. Sabourova, *Real and Complex Operator Norms*. Lulea, Sweden: Lulea University of Technology, 2007, Ph.D. Dissertation, http://epubl.ltu.se/1402-1757/2007/09/LTU-LIC-0709-SE.pdf. (Cited on p. 843.)

[2359] M. K. Sain and C. B. Schrader, "The Role of Zeros in the Performance of Multiinput, Multioutput Feedback Systems," *IEEE Trans. Educ.*, Vol. 33, pp. 244–257, 1990. (Cited on p. 1319.)

[2360] K.-S. Saito and M. Tominaga, "A Dunkl-Williams Type Inequality for Absolute Value Operators," *Lin. Alg. App.*, Vol. 432, pp. 3258–3264, 2010. (Cited on p. 748.)

[2361] J. Sandor, "On Certain Inequalities for Means, II," *J. Math. Anal. Appl.*, Vol. 199, pp. 629–635, 1996. (Cited on p. 143.)

[2362] J. Sandor, "Inequalities for Generalized Convex Functions with Applications I," *Studia University Babes-Bolyai, Mathematica*, Vol. XLIV, pp. 93–107, 1999. (Cited on p. 117.)

[2363] J. Sandor, "On Certain Limits Related to the Number *e*," *Libertas Mathematica*, Vol. XX, pp. 155–159, 2000. (Cited on p. 955.)

[2364] J. Sandor, "Inequalities for Generalized Convex Functions with Applications II," *Studia University Babes-Bolyai, Mathematica*, Vol. XLVI, pp. 79–91, 2001. (Cited on p. 117.)

[2365] J. Sandor, "On an Inequality of Ky Fan," *Int. J. Math. Educ. Sci. Tech.*, Vol. 32, pp. 133–160, 2001. (Cited on p. 209.)

[2366] J. Sandor, *Geometric Theorems, Diophantine Equations, and Arithmetic Functions*. Rehoboth: American Research Press, 2002, http://www.gallup.unm.edu/ smarandache/JozsefSandor2.pdf. (Cited on pp. 452 and 466.)

[2367] J. Sandor, "On Child's Inequality," *Octogon Math. Mag.*, Vol. 13, pp. 392–393, 2005. (Cited on p. 466.)

[2368] J. Sandor, "A Note on Inequalities for the Logarithmic Function," *Octogon*, Vol. 17, pp. 299–301, 2009. (Cited on p. 143.)

[2369] J. Sandor, "Two Sharp Inequalities for Trigonometric and Hyperbolic Functions," *Math. Ineq. Appl.*, Vol. 15, pp. 409–413, 2012. (Cited on pp. 250 and 264.)

[2370] J. Sandor, "On Certain Inequalities for Hyperbolic and Trigonometric Functions," *J. Math. Ineq.*, Vol. 7, pp. 421–425, 2013. (Cited on pp. 261 and 266.)

[2371] J. Sandor, "On D'Aurizio's Trigonometric Inequality," *J. Math. Ineq.*, Vol. 10, pp. 885–888, 2016. (Cited on p. 249.)

[2372] J. Sandor and M. Bencze, "On Two Inequalities for the Arctangent Function," *Octogon*, Vol. 15, pp. 202–206, 2007. (Cited on p. 261.)

[2373] J. Sandor and L. Debnath, "On Certain Inequalities Involving the Constant *e* and Their Applications," *J. Math. Anal. Appl.*, Vol. 249, pp. 569–582, 2000. (Cited on pp. 51, 128, and 225.)

[2374] J. P. O. Santos and R. Da Silva, "On a Combinatorial Proof for an Identity Involving the Triangular Numbers," *Bull. Austral. Math. Soc.*, Vol. 83, pp. 46–49, 2011. (Cited on pp. 114 and 1091.)

[2375] R. Satnoianu and I. Sofair, "More Triangle Inequalities," *Amer. Math. Monthly*, Vol. 111, pp. 67–69, 2004. (Cited on p. 472.)

[2376] R. A. Satnoianu, "A General Method for Establishing Geometric Inequalities in a Triangle," *Amer. Math. Monthly*, Vol. 108, pp. 360–364, 2001. (Cited on p. 480.)

[2377] R. A. Satnoianu, "General Power Inequalities between the Sides and the Circumscribed and Inscribed Radii Related to the Fundamental Triangle Inequality," *Math. Ineq. Appl.*, Vol. 5, pp. 745–751, 2002. (Cited on p. 452.)

[2378] R. A. Satnoianu, "Improved GA-Convexity Inequalities," *J. Ineq. Pure Appl. Math.*, Vol. 3, pp. 1–6, 2002, article 82. (Cited on p. 203.)

[2379] D. H. Sattinger and O. L. Weaver, *Lie Groups and Algebras with Applications to Physics, Geometry, and Mechanics*. New York: Springer, 1986. (Cited on pp. 355, 372, 1189, 1193, and 1247.)

[2380] S. Savchev and T. Andreescu, *Mathematical Miniatures*. Washington, DC: Mathematical Association of America, 2003. (Cited on pp. 28, 29, 33, 35, 36, 47, 49, 52, 137, 151, 155, 173, 178, 202, and 226.)

[2381] M. Sawhney, "Problem 1100," *Coll. Math. J.*, Vol. 48, p. 139, 2017. (Cited on p. 164.)

[2382] A. H. Sayed, *Fundamentals of Adaptive Filtering*. New York: Wiley, 2003. (Cited on p. xxi.)

[2383] H. Schaub, P. Tsiotras, and J. L. Junkins, "Principal Rotation Representations of Proper $N \times N$ Orthogonal Matrices," *Int. J. Eng. Sci.*, Vol. 33, pp. 2277–2295, 1995. (Cited on p. 1206.)

[2384] A. R. Schep, "Bounds on the Spectral Radius of Hadamard Products of Positive Operators on ℓ_p-Spaces," *Elec. J. Lin. Alg.*, Vol. 22, pp. 443–447, 2011. (Cited on p. 699.)

[2385] C. W. Scherer, "The Solution Set of the Algebraic Riccati Equation and the Algebraic Riccati Inequality," *Lin. Alg. Appl.*, Vol. 153, pp. 99–122, 1991. (Cited on p. 1319.)

[2386] C. W. Scherer, "The Algebraic Riccati Equation and Inequality for Systems with Uncontrollable Modes on the Imaginary Axis," *SIAM J. Matrix Anal. Appl.*, Vol. 16, pp. 1308–1327, 1995. (Cited on pp. 1302 and 1319.)

[2387] C. W. Scherer, "The General Nonstrict Algebraic Riccati Inequality," *Lin. Alg. Appl.*, Vol. 219, pp. 1–33, 1995. (Cited on p. 1319.)

[2388] P. Scherk, "On the Decomposition of Orthogonalities into Symmetries," *Proc. Amer. Math. Soc.*, Vol. 1, pp. 481–491, 1950. (Cited on p. 607.)

[2389] E. E. Scheufens, "From Fourier Series to Rapidly Convergent Series for Zeta(3)," *Math. Mag.*, Vol. 84, pp. 26–32, 2011. (Cited on p. 1032.)

[2390] C. Schiebold, "Cauchy-type Determinants and Integrable Systems," *Lin. Alg. Appl.*, Vol. 433, pp. 447–475, 2010. (Cited on pp. 427 and 692.)

[2391] R. L. Schilling, R. Song, and Z. Vondracek, *Bernstein Functions*. Berlin: de Gruyter, 2010. (Cited on p. 153.)

[2392] C. Schmoeger, "Remarks on Commuting Exponentials in Banach Algebras," *Proc. Amer. Math. Soc.*, Vol. 127, pp. 1337–1338, 1999. (Cited on p. 1211.)

[2393] C. Schmoeger, "Remarks on Commuting Exponentials in Banach Algebras, II," *Proc. Amer. Math. Soc.*, Vol. 128, pp. 3405–3409, 2000. (Cited on p. 1211.)

[2394] C. Schmoeger, "On Normal Operator Exponentials," *Proc. Amer. Math. Soc.*, Vol. 130, pp. 697–702, 2001. (Cited on p. 1211.)

[2395] C. Schmoeger, "On Logarithms of Linear Operators on Hilbert Spaces," *Demonstratio Math.*, Vol. XXXV, pp. 375–384, 2002. (Cited on pp. 1185 and 1217.)

[2396] C. Schmoeger, "On the Operator Equation $e^A = e^B$," *Lin. Alg. Appl.*, Vol. 359, pp. 169–179, 2003. (Cited on p. 1212.)

[2397] H. Schneider, "Olga Taussky-Todd's Influence on Matrix Theory and Matrix Theorists," *Lin. Multilin. Alg.*, Vol. 5, pp. 197–224, 1977. (Cited on p. 535.)

[2398] R. Schneider, *Convex Bodies: The Brunn-Minkowski Theory*. Cambridge: Cambridge University Press, 2008. (Cited on p. 311.)

[2399] I. J. Schoenberg, "A Direct Derivation of a Jacobian Identity from Elliptic Functions," *Amer. Math. Monthly*, Vol. 88, pp. 616–618, 1981. (Cited on p. 1092.)

[2400] B. Scholkopf and A. J. Smola, *Learning with Kernels: Support Vector Machines, Regularization, Optimization, and Beyond*. Cambridge: MIT Press, 2001. (Cited on p. 725.)

[2401] D. Scholz and M. Weyrauch, "A Note on the Zassenhaus Product Formula," *J. Math. Phys.*, Vol. 47, pp. 033 505/1–7, 2006. (Cited on p. 1214.)

[2402] J. R. Schott, "Kronecker Product Permutation Matrices and Their Application to Moment Matrices of the Normal Distribution," *J. Multivariate Analysis*, Vol. 87, pp. 177–190, 2003. (Cited on pp. 86 and 1172.)

[2403] J. R. Schott, *Matrix Analysis for Statistics*, 2nd ed. New York: Wiley, 2005. (Cited on pp. xxi, 313, 416, 422, 617, 632, 635, 636, 637, 638, 666, 689, 711, 722, 776, 804, 805, 809, 810, 811, 821, 822, 824, 825, 839, 892, 910, 937, and 1173.)

[2404] C. B. Schrader and M. K. Sain, "Research on System Zeros: A Survey," *Int. J. Contr.*, Vol. 50, pp. 1407–1433, 1989. (Cited on p. 1319.)

[2405] B. S. W. Schroder, *Ordered Sets: An Introduction*. Boston: Birkhauser, 2003. (Cited on pp. 25 and 118.)

[2406] B. S. W. Schroder, *Mathematical Analysis: A Concise Introduction*. Hoboken: Wiley, 2008. (Cited on p. 956.)

[2407] A. J. Schwenk, "Tight Bounds on the Spectral Radius of Asymmetric Nonnegative Matrices," *Lin. Alg. Appl.*, Vol. 75, pp. 257–265, 1986. (Cited on p. 698.)

[2408] H. Schwerdtfeger, *Introduction to Linear Algebra and The Theory of Matrices*. Groningen: P. Noordhoff, 1950. (Cited on p. 440.)

[2409] J. A. Scott, "Examples of the Use of Areal Coordinates in Triangle Geometry," *Math. Gaz.*, Vol. 83, pp. 472–477, 1999. (Cited on p. 472.)

[2410] J. A. Scott, "A Product of Sines Revisited," *Math. Gaz.*, Vol. 84, pp. 304–306, 2000. (Cited on p. 244.)

[2411] J. A. Scott, "In Praise of the Catalan Constant," *Math. Gaz.*, Vol. 86, pp. 102–103, 2002. (Cited on p. 1170.)

[2412] J. A. Scott, "Yet another look at $n^{1/n}$," *Math. Gaz.*, Vol. 87, p. 311, 2003. (Cited on p. 127.)

[2413] J. A. Scott, "Another Cotangent Identity for the Triangle," *Math. Gaz.*, Vol. 88, pp. 114–115, 2004. (Cited on p. 456.)

[2414] J. A. Scott, "Square-Freedom Revisited," *Math. Gaz.*, Vol. 90, pp. 112–113, 2006. (Cited on p. 1030.)

[2415] S. R. Searle, *Matrix Algebra Useful for Statistics*. New York: Wiley, 1982. (Cited on p. xxi.)

[2416] P. Sebastian, "On the Derivatives of Matrix Powers," *SIAM J. Matrix Anal. Appl.*, Vol. 17, pp. 640–648, 1996. (Cited on p. 1092.)

[2417] P. Sebastiani, "On the Derivatives of Matrix Powers," *SIAM J. Matrix Anal. Appl.*, Vol. 17, pp. 640–648, 1996. (Cited on p. 952.)

[2418] G. A. F. Seber, *A Matrix Handbook for Statisticians*. New York: Wiley, 2008. (Cited on pp. 217, 225, 231, 312, 319, 341, 382, 398, 407, 410, 424, 440, 442, 443, 534, 536, 616, 636, 651, 669, 736, 778, 795, 816, 824, 899, and 1232.)

[2419] H.-J. Seiffert, "Some Identities Involving Pell Numbers and Binomial Coefficients," *Fibonacci Quart.*, Vol. 48, p. 93, 2010. (Cited on p. 103.)

[2420] E. Seiler and B. Simon, "An Inequality Among Determinants," *Proc. Nat. Acad. Sci.*, Vol. 72, pp. 3277–3278, 1975. (Cited on pp. 774 and 779.)

[2421] J. M. Selig, *Geometric Fundamentals of Robotics*, 2nd ed. New York: Springer, 2005. (Cited on pp. xxi, 393, 438, and 440.)

[2422] M. M. Seron, J. H. Braslavsky, and G. C. Goodwin, *Fundamental Limitations in Filtering and Control*. New York: Springer, 1997. (Cited on p. 1156.)

[2423] D. Serre, *Matrices: Theory and Applications*. New York: Springer, 2002. (Cited on pp. 372, 428, 528, and 665.)

[2424] J.-P. Serre and L. L. Scott, *Linear Representations of Finite Groups*. New York: Springer, 1996. (Cited on p. 436.)

[2425] L. D. Servi, "Nested Square Roots of 2," *Amer. Math. Monthly*, Vol. 110, pp. 326–330, 2003. (Cited on pp. 232 and 974.)

[2426] C. Shafroth, "A Generalization of the Formula for Computing the Inverse of a Matrix," *Amer. Math. Monthly*, Vol. 88, pp. 614–616, 1981. (Cited on pp. 341 and 389.)

[2427] W. M. Shah and A. Liman, "On the Zeros of a Class of Polynomials," *Math. Ineq. Appl.*, Vol. 10, pp. 733–744, 2007. (Cited on p. 1234.)

[2428] L. Shaikhet, *Lyapunov Functionals and Stability of Stochastic Difference Equations*. New York: Springer, 2011. (Cited on p. 142.)

[2429] H. Shapiro, "A Survey of Canonical Forms and Invariants for Unitary Similarity," *Lin. Alg. Appl.*, Vol. 147, pp. 101–167, 1991. (Cited on p. 576.)

[2430] H. Shapiro, "Notes from Math 223: Olga Taussky Todd's Matrix Theory Course, 1976–1977," *Mathematical Intelligencer*, Vol. 19, no. 1, pp. 21–27, 1997. (Cited on p. 609.)

[2431] H. Shapiro, "The Weyr Characteristic," *Amer. Math. Monthly*, Vol. 106, pp. 919–929, 1999. (Cited on p. 619.)

[2432] H. Shapiro, *Linear Algebra and Matrices: Topics for a Second Course*. Providence: American Mathematical Society, 2015. (Cited on pp. 538, 617, 619, 775, and 788.)

[2433] R. Sharma, M. Gupta, and G. Kapoor, "Some Better Bounds on the Variance with Applications," *J. Math. Ineq.*, Vol. 4, pp. 355–363, 2010. (Cited on pp. 196 and 588.)

[2434] M. Shattuck and T. Waldhauser, "Proofs of Some Binomial Identities Using the Method of Last Squares," *Fibonacci Quart.*, Vol. 48, pp. 290–297, 2010. (Cited on p. 86.)

[2435] R. Shaw and F. I. Yeadon, "On $(a \times b) \times c$," *Amer. Math. Monthly*, Vol. 96, pp. 623–629, 1989. (Cited on p. 386.)

[2436] K. Shebrawi and H. Albadawi, "Operator Norm Inequalities of Minkowski Type," *J. Ineq. Pure. Appl. Math.*, Vol. 9, pp. 1–11, 2008, article 26. (Cited on pp. 865, 870, 877, 878, and 879.)

[2437] K. Shebrawi and H. Albadawi, "Numerical Radius and Operator Norm Inequalities," *J. Ineq. Appl.*, pp. 1–11, 2009, article 492154. (Cited on pp. 866 and 903.)

[2438] K. Shebrawi and H. Albadawi, "Norm Inequalities for the Absolute Value of Hilbert Space Operators," *Lin. Multilin. Alg.*, Vol. 58, pp. 453–463, 2010. (Cited on pp. 865, 871, and 876.)

[2439] K. Shebrawi and H. Albadawi, "Trace Inequalities for Matrices," *Bull. Aust. Math. Soc.*, Vol. 87, pp. 139–148, 2013. (Cited on pp. 765, 766, 772, and 903.)
[2440] D. Shemesh, "A Simultaneous Triangularization Result," *Lin. Alg. Appl.*, Vol. 498, pp. 394–398, 2016. (Cited on p. 617.)
[2441] S.-Q. Shen and T.-Z. Huang, "Several Inequalities for the Largest Singular Value and the Spectral Radius of Matrices," *Math. Ineq. Appl.*, Vol. 10, pp. 713–722, 2007. (Cited on pp. 539, 700, and 908.)
[2442] N. Sherif and E. Morsy, "Computing Real Logarithm of a Real Matrix," *Int. J. Algebra*, Vol. 127, pp. 131–142, 2008. (Cited on p. 1188.)
[2443] H.-N. Shi, "Generalizations of Bernoulli's Inequality with Applications," *J. Math. Ineq.*, Vol. 2, pp. 101–107, 2008. (Cited on p. 190.)
[2444] S.-C. Shi and Y.-D. Wu, "The Best Constant in a Geometric Inequality Relating Medians, Inradius and Circumradius in a Triangle," *J. Math. Ineq.*, Vol. 7, pp. 183–194, 2013. (Cited on p. 480.)
[2445] G. E. Shilov, *Linear Algebra*. Englewood Cliffs: Prentice-Hall, 1971, reprinted by Dover, New York, 1977. (Cited on p. xxii.)
[2446] S. A. Shirali, "The Bhaskara-Aryabhata Approximation to the Sine Function," *Math. Mag.*, Vol. 84, pp. 98–107, 2011. (Cited on p. 249.)
[2447] P. Shiu, "The Equality of Two Integrals," *Math. Gaz.*, Vol. 90, pp. 459–460, 2006. (Cited on pp. 1104 and 1110.)
[2448] M. D. Shuster, "A Survey of Attitude Representations," *J. Astron. Sci.*, Vol. 41, pp. 439–517, 1993. (Cited on pp. 438 and 1206.)
[2449] M. D. Shuster, "Attitude Analysis in Flatland: The Plane Truth," *J. Astron. Sci.*, Vol. 52, pp. 195–209, 2004. (Cited on p. 394.)
[2450] W. Sierpinski, *Elementary Theory of Numbers*. Amsterdam: North-Holland, 1988. (Cited on p. 35.)
[2451] D. D. Siljak, *Large-Scale Dynamic Systems: Stability and Structure*. New York: North-Holland, 1978. (Cited on pp. xxi, 440, and 1232.)
[2452] F. C. Silva and R. Simoes, "On the Lyapunov and Stein Equations," *Lin. Alg. Appl.*, Vol. 420, pp. 329–338, 2007. (Cited on p. 1311.)
[2453] F. C. Silva and R. Simoes, "On the Lyapunov and Stein Equations, II," *Lin. Alg. Appl.*, Vol. 426, pp. 305–311, 2007. (Cited on p. 1311.)
[2454] C. S. Simons and M. Wright, "Fibonacci Imposters," *Int. J. Math. Educ. Sci. Tech.*, Vol. 38, pp. 677–682, 2007. (Cited on p. 100.)
[2455] S. Simons, "How Do We Tackle $\left(\int_0^1 \frac{1-x^m}{1-x^n}\right)dx$," *Math. Gaz.*, Vol. 82, pp. 290–293, 1998. (Cited on p. 1099.)
[2456] S. Simons, "A Curious Identity," *Math. Gaz.*, Vol. 85, pp. 296–298, 2001. (Cited on p. 121.)
[2457] S. Simons, "Some Results concerning Angle Bisectors," *Math. Gaz.*, Vol. 93, pp. 115–116, 2009. (Cited on p. 472.)
[2458] A. Simoson, "A Constant Function with Arcsin," *Coll. Math. J.*, Vol. 43, pp. 411–412, 2012. (Cited on p. 257.)
[2459] S. F. Singer, *Symmetry in Mechanics: A Gentle, Modern Introduction*. Boston: Birkhauser, 2001. (Cited on p. xxi.)
[2460] D. Singh and J. N. Singh, "Some Combinatorics of Multisets," *Int. J. Math. Educ. Sci. Tech.*, Vol. 34, pp. 489–499, 2003. (Cited on pp. 24 and 118.)
[2461] A. Sintamarian, "Some Convergent Sequences and Series," *Int. J. Pure Appl. Math.*, Vol. 57, pp. 885–902, 2009. (Cited on pp. 1051 and 1087.)
[2462] A. Sintamarian, "Euler's Constant, Sequences and Some Estimates," *Surveys Math. Appl.*, Vol. 8, pp. 103–114, 2013. (Cited on p. 964.)
[2463] A. Sintamarian, "Sharp Estimates Regarding the Remainder of the Alternating Harmonic Series," *Math. Ineq. Appl.*, Vol. 18, pp. 347–352, 2015. (Cited on p. 963.)
[2464] A. Sintamarian, "Estimating a Special Difference of Harmonic Numbers," *Creative Mathematics Informatics*, Vol. 25, pp. 99–106, 2016. (Cited on p. 964.)
[2465] A. Sintamarian and O. Furdui, "Q1055," *Math. Mag.*, Vol. 88, pp. 378, 383, 384, 2015. (Cited on p. 1142.)
[2466] C. C. Siong, "A Simple Proof of Ljunggren's Binomial Congruence," *Amer. Math. Monthly*, Vol. 121, pp. 162–163, 2014. (Cited on p. 55.)

[2467] D. Sitaru, "Problem 11984," *Amer. Math. Monthly*, Vol. 124, p. 466, 2017. (Cited on p. 452.)

[2468] B. D. Sittinger, "Computing $\zeta(2m)$ by Using Telescomping Sums," *Amer. Math. Monthly*, Vol. 123, pp. 710–715, 2016. (Cited on pp. 979 and 1037.)

[2469] R. E. Skelton, T. Iwasaki, and K. Grigoriadis, *A Unified Algebraic Approach to Linear Control Design*. London: Taylor and Francis, 1998. (Cited on pp. xxi and 1247.)

[2470] D. M. Smiley and M. F. Smiley, "The Polygonal Inequalities," *Amer. Math. Monthly*, Vol. 71, pp. 755–760, 1964. (Cited on pp. 274 and 857.)

[2471] O. K. Smith, "Eigenvalues of a Symmetric 3×3 Matrix," *Comm. ACM*, Vol. 4, p. 168, 1961. (Cited on p. 532.)

[2472] R. A. Smith, "Matrix Calculations for Lyapunov Quadratic Forms," *J. Diff. Eqns.*, Vol. 2, pp. 208–217, 1966. (Cited on p. 1312.)

[2473] A. Smoktunowicz, "Block Matrices and Symmetric Perturbations," *Lin. Alg. Appl.*, Vol. 429, pp. 2628–2635, 2008. (Cited on pp. 383 and 887.)

[2474] R. Snieder, *A Guided Tour of Mathematical Methods for the Physical Sciences*, 2nd ed. Cambridge: Cambridge University Press, 2004. (Cited on p. 630.)

[2475] J. Snyders and M. Zakai, "On Nonnegative Solutions of the Equation $AD + DA' = C$," *SIAM J. Appl. Math.*, Vol. 18, pp. 704–714, 1970. (Cited on pp. 1248, 1267, 1268, and 1319.)

[2476] W. So, "Equality Cases in Matrix Exponential Inequalities," *SIAM J. Matrix Anal. Appl.*, Vol. 13, pp. 1154–1158, 1992. (Cited on pp. 379, 763, 861, 1185, 1211, 1212, 1218, and 1221.)

[2477] W. So, "The High Road to an Exponential Formula," *Lin. Alg. Appl.*, Vol. 379, pp. 69–75, 2004. (Cited on p. 1217.)

[2478] W. So and R. C. Thompson, "Product of Exponentials of Hermitian and Complex Symmetric Matrices," *Lin. Multilin. Alg.*, Vol. 29, pp. 225–233, 1991. (Cited on p. 1217.)

[2479] W. So and R. C. Thompson, "Singular Values of Matrix Exponentials," *Lin. Multilin. Alg.*, Vol. 47, pp. 249–258, 2000. (Cited on pp. 582, 801, and 1218.)

[2480] A. Sofo, "Generalization of a Radical Identity," *Math. Gaz.*, Vol. 89, pp. 39–41, 2005. (Cited on pp. 53 and 1077.)

[2481] A. Sofo, "General Properties Involving Reciprocals of Binomial Coefficients," *J. Integer Sequences*, Vol. 9, pp. 1–11, 2006, article 06.4.5. (Cited on pp. 1056, 1070, and 1100.)

[2482] A. Sofo, "Some More Identities Involving Rational Sums," *Applic. Anal. Disc. Math.*, Vol. 2, pp. 56–66, 2008. (Cited on pp. 48 and 92.)

[2483] A. Sofo, "Harmonic Numbers and Double Binomial Coefficients," *Integral Transforms and Special Functions*, Vol. 20, pp. 847–857, 2009. (Cited on pp. 1024, 1025, 1141, 1174, and 1176.)

[2484] A. Sofo, "Triple Integral Identities and Zeta Functions," *Applic. Anal. Disc. Math.*, Vol. 4, pp. 347–360, 2010. (Cited on pp. 1070 and 1176.)

[2485] A. Sofo and D. Cvijovic, "Extensions of Euler Harmonic Sums," *Applic. Anal. Disc. Math.*, Vol. 6, pp. 317–328, 2012. (Cited on pp. 1033 and 1070.)

[2486] A. Sofo and H. M. Srivastava, "Identities for the Harmonic Numbers and Binomial Coefficients," *Ramanujan J.*, Vol. 25, pp. 93–113, 2011. (Cited on pp. 996 and 1070.)

[2487] V. Soltan, *Lectures on Convex Sets*. New Jersey: World Scientific, 2015. (Cited on pp. 278, 306, 308, 309, 311, 312, 314, 315, 855, 935, 936, 937, and 938.)

[2488] A. M. Sommariva, "The Generating Identity of Cauchy-Schwarz-Bunyakovsky Inequality," *Elem. Math.*, Vol. 63, pp. 1–5, 2008. (Cited on p. 855.)

[2489] J. Sondow, "Evaluation of Tachiya's Qlgebraic Infinite Products Involving Fibonacci and Lucas Numbers," http://arxiv.org/abs/1106.4246v1. (Cited on p. 1079.)

[2490] J. Sondow, "An Identity Involving Harmonic Numbers," *Amer. Math. Monthly*, Vol. 112, pp. 367–369, 1966. (Cited on p. 68.)

[2491] J. Sondow, "Analytic Continuation fo Riemann's Zeta Function and Values at Negative Integers via Euler's Transformation of Series," *Proc. Amer. Math. Soc.*, Vol. 120, pp. 421–424, 1994. (Cited on p. 996.)

[2492] J. Sondow, "Criteria for Irrationality of Euler's Constant," *Proc. Amer. Math. Soc.*, Vol. 131, pp. 3335–3344, 2003. (Cited on pp. 76 and 1176.)

[2493] J. Sondow, "A Simple Counterexample to Havil's "Reformulation" of the Riemann Hypothesis," *Elem. Math.*, Vol. 67, pp. 61–67, 2012. (Cited on p. 996.)

[2494] J. Sondow and K. MacMillan, "Primary Pseudoperfect Numbers, Arithmetic Progressions, and the Erdos-Moser Equation," *Amer. Math. Monthly*, Vol. 124, pp. 232–240, 2017. (Cited on p. 36.)

[2495] J. Sondow and H. Yi, "New Wallis- and Catalan-Type Infinite Products for π, e, and $\sqrt{2+\sqrt{2}}$," *Amer. Math. Monthly*, Vol. 117, pp. 912–917, 2010. (Cited on p. 1084.)

[2496] E. D. Sontag, *Mathematical Control Theory: Deterministic Finite-Dimensional Systems*, 2nd ed. New York: Springer, 1998. (Cited on pp. xxi and 1309.)

[2497] E. D. Sontag, "Passivity Gains and the 'Secant Condition' for Stability," *Sys. Contr. Lett.*, Vol. 55, pp. 177–183, 2006. (Cited on p. 1231.)

[2498] A. R. Sourour, "A Factorization Theorem for Matrices," *Lin. Multilin. Alg.*, Vol. 19, pp. 141–147, 1986. (Cited on p. 609.)

[2499] A. R. Sourour, "Nilpotent Factorization of Matrices," *Lin. Multilin. Alg.*, Vol. 31, pp. 303–308, 1992. (Cited on p. 609.)

[2500] J. Spanier and K. B. Oldham, *An Atlas of Functions*. New York: Taylor and Francis, 1987. (Cited on p. 1000.)

[2501] W. Specht, "Zur Theorie der Elementaren Mittel," *Math. Z.*, Vol. 74, pp. 91–98, 1960. (Cited on p. 206.)

[2502] J. Spencer and L. Florescu, *Asymptopia*. Providence: American Mathematical Society, 2014. (Cited on p. 971.)

[2503] E. Spiegel, "Sums of Projections," *Lin. Alg. Appl.*, Vol. 187, pp. 239–249, 1993. (Cited on p. 619.)

[2504] M. R. Spiegel, S. Lipschutz, and J. Liu, *Mathematical Handbook of Formulas and Tables*, 3rd ed. New York: McGraw-Hill, 2009. (Cited on pp. 37, 130, 131, 979, 980, 1162, and 1170.)

[2505] M. Spivak, *A Comprehensive Introduction to Differential Geometry, Vol. One*, 3rd ed. Houston: Publish or Perish, 1999. (Cited on pp. 438, 691, and 923.)

[2506] M. Z. Spivey, "A Generalized Recurrence for Bell Numbers," *J. Integer Sequences*, Vol. 11, pp. 1–3, 2008, article 08.2.5. (Cited on p. 112.)

[2507] M. Z. Spivey, "A Combinatorial Proof for the Alternating Convolution of the Central Binomial Coefficients," *Amer. Math. Monthly*, Vol. 121, pp. 537–540, 2014. (Cited on p. 92.)

[2508] M. Z. Spivey, "Probabilistic Proofs of a Binomial Identity, Its Inverse, and Generalizations," *Amer. Math. Monthly*, Vol. 123, pp. 175–180, 2016. (Cited on p. 75.)

[2509] R. Sprugnoli, "Sums of Reciprocals of the Central Binomial Coefficients," *Elec. J. Combinatorial Number Theory*, Vol. 6, pp. 1–18, 2006, article A27. (Cited on pp. 68, 75, 76, 86, 1070, and 1102.)

[2510] R. Sprugnoli, "Moments of Reciprocals of Binomial Coefficients," *J. Integer Sequences*, Vol. 14, pp. 1–14, 2011, article 11.7.8. (Cited on p. 68.)

[2511] R. Sprugnoli, "Alternating Weighted Sums of Inverses of Binomial Coefficients," *J. Integer Sequences*, Vol. 15, pp. 1–12, 2012, article 12.6.3. (Cited on p. 75.)

[2512] S. Sra, "The Determinant Kernel," *IMAGE*, Vol. 53, pp. 45–46, 2014. (Cited on p. 727.)

[2513] H. M. Srivastava and J. Choi, *Zeta and q-Zeta Functions and Associated Series and Integrals*. Amsterdam: Elsevier, 2012. (Cited on pp. 75, 76, 979, 980, 996, 1001, 1003, 1024, 1030, 1032, 1033, 1034, 1035, 1036, 1071, 1130, 1132, 1146, 1151, 1152, 1154, 1166, 1170, and 1177.)

[2514] H. M. Srivastava and H. L. Manocha, *A Treatise on Generating Functions*. New York: Wiley, 1984. (Cited on p. 1092.)

[2515] R. Srivastava, "Some Combinatorial Series Identities and Rational Sums," *Integral Transforms and Special Functions*, Vol. 20, pp. 83–91, 2009. (Cited on p. 92.)

[2516] R. Srivastava, "Some Families Series of Combinatorial and Other Identities and Their Applications," *Appl. Math. Comp.*, Vol. 218, pp. 1077–1083, 2011. (Cited on pp. 92 and 1070.)

[2517] A. Stadler, "Dedekind η Function Disguised," *Amer. Math. Monthly*, Vol. 121, pp. 951–952, 2014. (Cited on p. 1087.)

[2518] W. Stanford, "A Combinatorial Identity," *Amer. Math. Monthly*, Vol. 119, pp. 429–430, 2012. (Cited on p. 68.)

[2519] M. S. Stankovic, B. M. Dankovic, and S. B. Trickovic, "Some Inequalities Involving Harmonic Numbers," in *Recent Progress in Inequalities*, G. V. Milovanovic, Ed. Dordrecht: Kluwer, 1998, pp. 493–498. (Cited on p. 964.)

[2520] R. P. Stanley, *Enumerative Combinatorics, Vol. 1*, 2nd ed. Cambridge: Cambridge University Press, 2000. (Cited on p. 23.)

[2521] R. P. Stanley, *Catalan Numbers*. Cambridge: Cambridge University Press, 2015. (Cited on p. 105.)

[2522] W.-H. Steeb, *Matrix Calculus and Kronecker Product with Applications and C++ Programs*. Singapore: World Scientific, 2001. (Cited on p. 701.)

[2523] W. H. Steeb and C. M. Villet, "A Conjectured Matrix Inequality," *SIAM Rev.*, Vol. 31, p. 495, 1989. (Cited on p. 1220.)

[2524] W. H. Steeb and C. M. Villet, "On Lower Bounds for the Free Energy of the Hubbard Model," *Z. Naturforsch*, Vol. 47a, pp. 469–470, 1992. (Cited on p. 1220.)

[2525] W.-H. Steeb and F. Wilhelm, "Exponential Functions of Kronecker Products and Trace Calculation," *Lin. Multilin. Alg.*, Vol. 9, pp. 345–346, 1981. (Cited on p. 1217.)

[2526] J. M. Steele, "Variations on the Monotone Subsequence Theme of Erdős and Szekeres," in *Discrete Probability and Algorithms*, D. Aldous and et al, Eds. New York: Springer-Verlag, 1995, pp. 111–131. (Cited on p. 25.)

[2527] J. M. Steele, *The Cauchy-Schwarz Master Class*. Washington, DC: Mathematical Association of America, 2004. (Cited on pp. 124, 125, 126, 138, 160, 161, 164, 169, 177, 178, 180, 199, 200, 201, 203, 207, 208, 211, 212, 213, 215, 217, 221, 225, 226, 273, 274, 446, 856, 857, 944, 957, 959, and 1172.)

[2528] E. M. Stein and R. Shakarchi, *Complex Analysis*. Princeton: Princeton University Press, 2003. (Cited on p. 996.)

[2529] E. M. Stein and R. Shakarchi, *Fourier Analysis: An Introduction*. Princeton: Princeton University Press, 2003. (Cited on p. 31.)

[2530] R. F. Stengel, *Flight Dynamics*. Princeton: Princeton University Press, 2004. (Cited on p. xxi.)

[2531] A. Stenger, "Experimental Math for Math Monthly Problems," *Amer. Math. Monthly*, Vol. 124, pp. 116–131, 2017. (Cited on p. 1087.)

[2532] C. Stepniak, "Ordering of Nonnegative Definite Matrices with Application to Comparison of Linear Models," *Lin. Alg. Appl.*, Vol. 70, pp. 67–71, 1985. (Cited on p. 816.)

[2533] H. J. Stetter, *Numerical Polynomial Algebra*. Philadelphia: SIAM, 2004. (Cited on p. xxi.)

[2534] G. W. Stewart, *Introduction to Matrix Computations*. New York: Academic Press, 1973. (Cited on p. xxi.)

[2535] G. W. Stewart, "On the Perturbation of Pseudo-Inverses, Projections and Linear Least Squares Problems," *SIAM Rev.*, Vol. 19, pp. 634–662, 1977. (Cited on p. 911.)

[2536] G. W. Stewart, *Matrix Algorithms Volume I: Basic Decompositions*. Philadelphia: SIAM, 1998. (Cited on p. xxi.)

[2537] G. W. Stewart, "On the Adjugate Matrix," *Lin. Alg. Appl.*, Vol. 283, pp. 151–164, 1998. (Cited on pp. xli and 362.)

[2538] G. W. Stewart, *Matrix Algorithms Volume II: Eigensystems*. Philadelphia: SIAM, 2001. (Cited on p. xxi.)

[2539] G. W. Stewart and J. Sun, *Matrix Perturbation Theory*. Boston: Academic Press, 1990. (Cited on pp. 563, 565, 574, 595, 597, 617, 798, 831, 860, 866, 867, 893, 906, 907, 912, 939, 940, and 941.)

[2540] E. U. Stickel, "Fast Computation of Matrix Exponential and Logarithm," *Analysis*, Vol. 5, pp. 163–173, 1985. (Cited on p. 1231.)

[2541] E. U. Stickel, "An Algorithm for Fast High Precision Computation of Matrix Exponential and Logarithm," *Analysis*, Vol. 10, pp. 85–95, 1990. (Cited on pp. 1231 and 1247.)

[2542] L. Stiller, "Multilinear Algebra and Chess Endgames," in *Games of No Chance*, R. Nowakowski, Ed. Berkeley: Mathematical Sciences Research Institute, 1996, pp. 151–192. (Cited on p. xxi.)

[2543] J. Stoer, "On the Characterization of Least Upper Bound Norms in Matrix Space," *Numer. Math*, Vol. 6, pp. 302–314, 1964. (Cited on p. 868.)

[2544] J. Stoer and C. Witzgall, *Convexity and Optimization in Finite Dimensions I*. Berlin: Springer, 1970. (Cited on pp. 362 and 940.)

[2545] M. Stofka, "An Infinite Sum Introduces a Zeta," *Amer. Math. Monthly*, Vol. 122, pp. 608–609, 2015. (Cited on p. 1044.)

[2546] G. Stoica, "Problem 372," *Gazeta Matematica Serie A*, Vol. 31, pp. 52–53, 2013. (Cited on p. 961.)

[2547] G. Stoica, "Problem 11947," *Amer. Math. Monthly*, Vol. 123, p. 1051, 2016. (Cited on p. 272.)

[2548] G. Stoica, "A Complex Inequality," *Amer. Math. Monthly*, Vol. 124, p. 279, 2017. (Cited on p. 273.)

[2549] G. Stoica, "A Limit of Sums of Reciprocals of Quartics," *Math. Mag.*, Vol. 90, pp. 232–233, 2017. (Cited on p. 967.)

[2550] G. Stoica, "An Optimal Hölder Exponent," *Amer. Math. Monthly*, Vol. 124, pp. 664–665, 2017. (Cited on p. 229.)

[2551] K. B. Stolarsky, "Generalizations of the Logarithmic Mean," *J. Math. Anal. Appl.*, Vol. 48, pp. 87–92, 1975. (Cited on p. 143.)

[2552] M. G. Stone, "A Mnemonic for Areas of Polygons," *Amer. Math. Monthly*, Vol. 93, pp. 479–480, 1986. (Cited on p. 493.)

[2553] G. Strang, *Linear Algebra and Its Applications*, 3rd ed. San Diego: Harcourt, Brace, Jovanovich, 1988. (Cited on pp. xxi and xxii.)

[2554] G. Strang, "The Fundamental Theorem of Linear Algebra," *Amer. Math. Monthly*, Vol. 100, pp. 848–855, 1993. (Cited on p. 362.)

[2555] G. Strang, "Every Unit Matrix Is a LULU," *Lin. Alg. Appl.*, Vol. 265, pp. 165–172, 1997. (Cited on p. 606.)

[2556] G. Strang and K. Borre, *Linear Algebra, Geodesy, and GPS*. Wellesley-Cambridge Press, 1997. (Cited on p. xxi.)

[2557] G. Strang and T. Nguyen, "The Interplay of Ranks of Submatrices," *SIAM Rev.*, Vol. 46, pp. 637–646, 2004. (Cited on pp. 325 and 423.)

[2558] V. Strehl, "Binomial Identities–Combinatorial and Algorithmic Aspects," *Disc.Math.*, Vol. 136, pp. 309–346, 1994. (Cited on p. 86.)

[2559] S. Strelitz, "On the Routh-Hurwitz Problem," *Amer. Math. Monthly*, Vol. 84, pp. 542–544, 1977. (Cited on p. 1225.)

[2560] R. S. Strichartz, "The Campbell-Baker-Hausdorff-Dynkin Formula and Solutions of Differential Equations," *J. Funct. Anal.*, Vol. 72, pp. 320–345, 1987. (Cited on p. 1306.)

[2561] T. Strom, "On Logarithmic Norms," *SIAM J. Numer. Anal.*, Vol. 12, pp. 741–753, 1975. (Cited on p. 1219.)

[2562] J. Stuelpnagel, "On the Parametrization of the Three-Dimensional Rotation Group," *SIAM Rev.*, Vol. 6, pp. 422–430, 1964. (Cited on p. 1206.)

[2563] G. P. H. Styan, "On Lavoie's Determinantal Inequality," *Lin. Alg. Appl.*, Vol. 37, pp. 77–80, 1981. (Cited on p. 668.)

[2564] J. Su, X. Miao, and H. Li, "Generalization of Hua's Inequalities and an Application," *J. Math. Ineq.*, Vol. 9, pp. 27–45, 2015. (Cited on p. 780.)

[2565] T. Sugimoto, "On an Inverse Formula of a Tridiagonal Matrix," *Operators and Matrices*, Vol. 6, pp. 465–480, 2012. (Cited on p. 424.)

[2566] P. Sullivan, "The FA Cup Draw and Pairing Up Probabilities," *Coll. Math. J.*, Vol. 47, pp. 282–291, 2016. (Cited on p. 94.)

[2567] R. P. Sullivan, "Products of Nilpotent Matrices," *Lin. Multilin. Alg.*, Vol. 56, pp. 311–317, 2008. (Cited on p. 609.)

[2568] H. Sun, T. Zhao, Y. Chu, and B. Liu, "A Note on the Neuman-Sandor Mean," *J. Math. Ineq.*, Vol. 8, pp. 287–297, 2014. (Cited on p. 143.)

[2569] X.-J. Sun, "On Sums of Three Pentagonal Numbers," *Amer. Math. Monthly*, Vol. 123, pp. 189–191, 2016. (Cited on pp. 39, 40, and 148.)

[2570] Z.-W. Sun, "On Sums Involving Products of Three Binomial Coefficients," *Acta Arithmetica*, Vol. 156, pp. 123–141, 2012. (Cited on p. 86.)

[2571] Z.-W. Sun, "Products and Sums Divisible by Central Binomial Coefficients," *Elec. J. Combinatorics*, Vol. 20, pp. 1–14, 2013. (Cited on pp. 55 and 1070.)

[2572] Z.-W. Sun and H. Pan, "Identities Concerning Bernoulli and Euler Polynomials," *Acta Arithmetica*, Vol. 125, pp. 21–39, 2006. (Cited on p. 980.)

[2573] W. F. Sung, "A Sum of Angles with Radical Cosines," *Math. Mag.*, Vol. 89, p. 61, 2016. (Cited on p. 1062.)

[2574] W. F. Sung, "Problem 1059," *Math. Mag.*, Vol. 89, pp. 148, 154, 2016. (Cited on p. 1105.)

[2575] B. Sury, "A Polynomial Parent to a Fibonacci-Lucas Relation," *Amer. Math. Monthly*, Vol. 121, p. 236, 2014. (Cited on p. 103.)

[2576] B. Sury, "A New Fibonacci-Lucas Relation," *Amer. Math. Monthly*, Vol. 122, p. 683, 2015. (Cited on p. 103.)

[2577] B. Sury, T. Wang, and F.-Z. Zhao, "Identities Involving Reciprocals of Binomial Coefficients," *J. Integer Sequences*, Vol. 7, pp. 1–12, 2004, article 04.2.8. (Cited on pp. 67, 68, 75, 1027, 1056, 1070, and 1101.)

[2578] M. A. Sustik and I. S. Dhillon, "On a Zero-Finding Problem Involving the Matrix Exponential," *SIAM J. Matrix Anal. Appl.*, Vol. 33, pp. 1237–1249, 2012. (Cited on p. 791.)

[2579] K. N. Swamy, "On Sylvester's Criterion for Positive-Semidefinite Matrices," *IEEE Trans. Autom. Contr.*, Vol. AC-18, p. 306, 1973. (Cited on p. 831.)

[2580] P. Szekeres, *A Course in Modern Mathematical Physics: Groups, Hilbert Space and Differential Geometry*. Cambridge: Cambridge University Press, 2004. (Cited on p. 438.)

[2581] G. Szep, "Simultaneous Triangularization of Projector Matrices," *Acta Math. Hung.*, Vol. 48, pp. 285–288, 1986. (Cited on p. 617.)

[2582] T. Szirtes, *Applied Dimensional Analysis and Modeling*. New York: McGraw-Hill, 1998. (Cited on p. xxi.)

[2583] H. Takagi, T. Miura, T. Kanzo, and S.-E. Takahasi, "A Reconsideration of Hua's Inequality," *J. Ineq. Appl.*, Vol. 2005, pp. 15–23, 2005. (Cited on pp. 196 and 859.)

[2584] Y. Takahashi, S.-E. Takahasi, and S. Wada, "A General Hlawka Inequality and Its Reverse Inequality," *Math. Ineq. Appl.*, Vol. 12, pp. 1–10, 2009. (Cited on p. 858.)

[2585] E. Talvila, "Some Divergent Trigonometric Integrals," *Amer. Math. Monthly*, Vol. 108, pp. 432–436, 2001. (Cited on p. 1178.)

[2586] T. Tao, *Topics in Random Matrix Theory*. Providence: American Mathematical Society, 2012. (Cited on pp. 51, 717, 899, 967, 1212, and 1216.)

[2587] Y. Tao, "More Results on Singular Value Inequalities of Matrices," *Lin. Alg. Appl.*, Vol. 416, pp. 724–729, 2006. (Cited on pp. 760 and 807.)

[2588] K. Tapp, *Matrix Groups for Undergraduates*. Providence: American Mathematical Society, 2005. (Cited on pp. 394, 434, 436, 438, and 943.)

[2589] K. Tapp, *Symmetry: A Mathematical Exploration*. New York: Springer, 2012. (Cited on p. 436.)

[2590] R. Tauraso, "A Rational Function Identity," *Amer. Math. Monthly*, Vol. 124, pp. 86–87, 2017. (Cited on p. 122.)

[2591] O. Taussky, "Positive-Definite Matrices and Their Role in the Study of the Characteristic Roots of General Matrices," *Adv. Math.*, Vol. 2, pp. 175–186, 1968. (Cited on p. 619.)

[2592] O. Taussky, "The Role of Symmetric Matrices in the Study of General Matrices," *Lin. Alg. Appl.*, Vol. 5, pp. 147–154, 1972. (Cited on p. 619.)

[2593] O. Taussky, "How I Became a Torchbearer for Matrix Theory," *Amer. Math. Monthly*, Vol. 95, pp. 801–812, 1988. (Cited on p. 362.)

[2594] O. Taussky and J. Todd, "Another Look at a Matrix of Mark Kac," *Lin. Alg. Appl.*, Vol. 150, pp. 341–360, 1991. (Cited on p. 587.)

[2595] O. Taussky and H. Zassenhaus, "On the Similarity Transformation between a Matrix and Its Transpose," *Pac. J. Math.*, Vol. 9, pp. 893–896, 1959. (Cited on p. 619.)

[2596] W. Tempelman, "The Linear Algebra of Cross Product Operations," *J. Astron. Sci.*, Vol. 36, pp. 447–461, 1988. (Cited on p. 386.)

[2597] G. ten Have, "Structure of the *n*th Roots of a Matrix," *Lin. Alg. Appl.*, Vol. 187, pp. 59–66, 1993. (Cited on p. 608.)

[2598] D. Terrana and H. Chen, "Generating Functions for the Powers of Fibonacci Sequences," *Int. J. Math. Educ. Sci. Tech.*, Vol. 38, pp. 531–537, 2007. (Cited on pp. 100 and 103.)

[2599] R. E. Terrell, "Solution to 'Exponentials of Certain Hilbert Space Operators'," *SIAM Rev.*, Vol. 34, pp. 498–500, 1992. (Cited on p. 1213.)

[2600] R. E. Terrell, "Matrix Exponentials," *SIAM Rev.*, Vol. 38, pp. 313–314, 1996. (Cited on p. 1213.)

[2601] M. Tetiva, "Problem 245," *Gazeta Matematica Seria A*, Vol. 26, pp. 252–254, 2008. (Cited on pp. 127 and 177.)

[2602] M. Tetiva, "A Bisector Inequality," *Amer. Math. Monthly*, Vol. 116, p. 749, 2009. (Cited on p. 480.)

[2603] M. Tetiva, "A New Lower Bound for the Sum of the Altitudes," *Amer. Math. Monthly*, Vol. 116, pp. 748–749, 2009. (Cited on p. 480.)

[2604] M. Tetiva, "Blundon's Inequality Improved," *Amer. Math. Monthly*, Vol. 117, pp. 839–840, 2010. (Cited on p. 466.)

[2605] M. Tetiva, "A Geometric Inequality for the Secants of a Triangle," *Math. Mag.*, Vol. 84, p. 154, 2011. (Cited on p. 466.)

[2606] A. C. Thompson, *Minkowski Geometry*. Cambridge: Cambridge University Press, 1996. (Cited on p. 306.)

[2607] C. J. Thompson, "Inequalities and Partial Orders on Matrix Spaces," *Indiana Univ. Math. J.*, Vol. 21, pp. 469–480, 1971. (Cited on p. 1221.)

[2608] R. C. Thompson, "On Matrix Commutators," *J. Washington Acad. Sci.*, Vol. 48, pp. 306–307, 1958. (Cited on p. 429.)

[2609] R. C. Thompson, "A Determinantal Inequality for Positive Definite Matrices," *Canad. Math. Bull.*, Vol. 4, pp. 57–62, 1961. (Cited on p. 785.)

[2610] R. C. Thompson, "Some Matrix Factorization Theorems," *Pac. J. Math.*, Vol. 33, pp. 763–810, 1970. (Cited on p. 609.)

[2611] R. C. Thompson, "Dissipative Matrices and Related Results," *Lin. Alg. Appl.*, Vol. 11, pp. 155–169, 1975. (Cited on pp. 774 and 776.)

[2612] R. C. Thompson, "A Matrix Inequality," *Comment. Math. Univ. Carolinae*, Vol. 17, pp. 393–397, 1976. (Cited on p. 779.)

[2613] R. C. Thompson, "Matrix Type Metric Inequalities," *Lin. Multilin. Alg.*, Vol. 5, pp. 303–319, 1978. (Cited on pp. 268, 271, and 738.)

[2614] R. C. Thompson, "Proof of a Conjectured Exponential Formula," *Lin. Multilin. Alg.*, Vol. 19, pp. 187–197, 1986. (Cited on p. 1217.)

[2615] R. C. Thompson, "Special Cases of a Matrix Exponential Formula," *Lin. Alg. Appl.*, Vol. 107, pp. 283–292, 1988. (Cited on p. 1217.)

[2616] R. C. Thompson, "Convergence Proof for Goldberg's Exponential Series," *Lin. Alg. Appl.*, Vol. 121, pp. 3–7, 1989. (Cited on p. 1306.)

[2617] R. C. Thompson, "Pencils of Complex and Real Symmetric and Skew Matrices," *Lin. Alg. Appl.*, Vol. 147, pp. 323–371, 1991. (Cited on p. 619.)

[2618] R. C. Thompson, "High, Low, and Quantitative Roads in Linear Algebra," *Lin. Alg. Appl.*, Vol. 162–164, pp. 23–64, 1992. (Cited on p. 362.)

[2619] P. V. Thuan, "Problem 3511," *Crux Math.*, Vol. 37, p. 62, 2011. (Cited on p. 183.)

[2620] P. V. Thuan, "Problem 3602," *Crux Math.*, Vol. 38, pp. 30–31, 2012. (Cited on pp. 94 and 183.)

[2621] P. V. Thuan and L. Vi, "A Useful Inequality Revisited," *Crux Math.*, Vol. 35, pp. 164–171, 2009. (Cited on p. 183.)

[2622] J.-F. Tian, "Reversed Version of a Generalized Aczel's Inequality and Its Application," *J. Ineq. Appl.*, pp. 1–8, 2012, article 202. (Cited on p. 222.)

[2623] Y. Tian, "Expressions for the Greatest Lower Bound and Least Upper Bound of a Pair of Hermitian Matrices in the Löwner Partial Ordering," unpublished note. (Cited on p. 741.)

[2624] Y. Tian, "The Moore-Penrose Inverses of $m \times n$ Block Matrices and Their Applications," *Lin. Alg. Appl.*, Vol. 283, pp. 35–60, 1998. (Cited on pp. 667 and 668.)

[2625] Y. Tian, "Equality of Two Nonnegative Definite Matrices," *IMAGE*, Vol. 30, pp. 28–30, 2000. (Cited on p. 816.)

[2626] Y. Tian, "Matrix Representations of Octonions and Their Applications," *Adv. Appl. Clifford Algebras*, Vol. 10, pp. 61–90, 2000. (Cited on p. 438.)

[2627] Y. Tian, "Commutativity of EP Matrices," *IMAGE*, Vol. 27, pp. 25–27, 2001. (Cited on p. 643.)

[2628] Y. Tian, "How to Characterize Equalities for the Moore-Penrose Inverse of a Matrix," *Kyungpook Math. J.*, Vol. 41, pp. 1–15, 2001. (Cited on pp. 357, 401, 631, 641, and 651.)

[2629] Y. Tian, "Some Equalities for Generalized Inverses of Matrix Sums and Block Circulant Matrices," *Archivum Mathematicum (BRNO)*, Vol. 37, pp. 301–306, 2001. (Cited on pp. 354, 358, 668, and 679.)

[2630] Y. Tian, "Some Inequalities for Sums of Matrices," *Scientiae Mathematicae Japonicae*, Vol. 54, pp. 355–361, 2001. (Cited on p. 745.)

[2631] Y. Tian, "A Range Equality for Moore-Penrose Inverses," *IMAGE*, Vol. 29, pp. 28–29, 2002. (Cited on p. 667.)

[2632] Y. Tian, "How to Express a Parallel Sum of k Matrices," *J. Math. Anal. Appl.*, Vol. 266, pp. 333–341, 2002. (Cited on p. 819.)

[2633] Y. Tian, "The Minimal Rank of the Matrix Expression $A - BX - YC$," *Missouri J. Math. Sci.*, Vol. 14, pp. 40–48, 2002. (Cited on p. 662.)

[2634] Y. Tian, "Upper and Lower Bounds for Ranks of Matrix Expressions Using Generalized Inverses," *Lin. Alg. Appl.*, Vol. 355, pp. 187–214, 2002. (Cited on p. 661.)

[2635] Y. Tian, "A range equality for idempotent matrix," *IMAGE*, Vol. 30, pp. 26–27, 2003. (Cited on p. 397.)

[2636] Y. Tian, "A range equality for the difference of orthogonal projectors," *IMAGE*, Vol. 30, p. 36, 2003. (Cited on pp. 653 and 654.)

[2637] Y. Tian, "On Mixed-Type Reverse Order Laws for the Moore-Penrose Inverse of a Matrix Product," *Int. J. Math. Math. Sci.*, Vol. 58, pp. 3103–3116, 2004. (Cited on pp. 637, 652, and 666.)

[2638] Y. Tian, "Rank Equalities for Block Matrices and Their Moore-Penrose Inverses," *Houston J. Math.*, Vol. 30, pp. 483–510, 2004. (Cited on pp. 325 and 661.)

[2639] Y. Tian, "Using Rank Formulas to Characterize Equalities for Moore-Penrose Inverses of Matrix Products," *Appl. Math. Comp.*, Vol. 147, pp. 581–600, 2004. (Cited on p. 637.)

[2640] Y. Tian, "A Range Equality for the Commutator with Two Involutory Matrices," *IMAGE*, Vol. 35, pp. 32–33, 2005. (Cited on pp. 404 and 417.)

[2641] Y. Tian, "A Range Equality for the Kronecker Product of Matrices," *IMAGE*, Vol. 34, pp. 26–27, 2005. (Cited on p. 686.)

[2642] Y. Tian, "Rank Inequalities for Idempotent Matrices," *Coll. Math. J.*, Vol. 36, p. 162, 2005. (Cited on p. 406.)

[2643] Y. Tian, "Similarity of Two Block Matrices," *IMAGE*, Vol. 34, pp. 27–32, 2005. (Cited on p. 579.)

[2644] Y. Tian, "Some Rank Equalities and Inequalities for Kronecker Products of Matrices," *Lin. Multilin. Alg.*, Vol. 53, pp. 445–454, 2005. (Cited on p. 688.)

[2645] Y. Tian, "Special Forms of Generalized Inverses of Row Block Matrices," *Elec. J. Lin. Alg.*, Vol. 13, pp. 249–261, 2005. (Cited on p. 665.)

[2646] Y. Tian, "The Schur Complement in an Orthogonal Projector," *IMAGE*, Vol. 35, pp. 34–36, 2005. (Cited on p. 662.)

[2647] Y. Tian, "Two Characterizations of an EP Matrix," *IMAGE*, Vol. 34, p. 32, 2005. (Cited on p. 641.)

[2648] Y. Tian, "An Orthogonal Projector Consisting of Nonnegative Matrices," *Coll. Math. J.*, Vol. 37, pp. 233–234, 2006. (Cited on p. 754.)

[2649] Y. Tian, "Comparing Ranges of Keonecker Products of Matrices," *Coll. Math. J.*, Vol. 37, p. 397, 2006. (Cited on p. 686.)

[2650] Y. Tian, "Inequalities Involving Rank and Kronecker Product," *Amer. Math. Monthly*, Vol. 113, p. 851, 2006. (Cited on p. 688.)

[2651] Y. Tian, "On the Inverse of a Product of Orthogonal Projectors," *Amer. Math. Monthly*, Vol. 113, pp. 467–468, 2006. (Cited on p. 659.)

[2652] Y. Tian, "The Equivalence between $(AB)^+ = B^+A^+$ and Other Mixed-type Reverse-order Laws," *Int. J. Math. Educ. Sci. Tech.*, Vol. 37, pp. 331–339, 2006. (Cited on pp. 636 and 637.)

[2653] Y. Tian, "Two Commutativity Equalities for the Regularized Tikhonov Inverse," *IMAGE*, Vol. 41, pp. 37–39, 2008. (Cited on p. 735.)

[2654] Y. Tian, "Two Equalities for the Moore-Penrose Inverse of a Row Block Matrix," *IMAGE*, Vol. 41, pp. 39–42, 2008. (Cited on p. 667.)

[2655] Y. Tian, "Eight Expressions for Generalized Inverses of a Bordered Matrix," *Lin. Multilin. Alg.*, Vol. 58, pp. 203–220, 2009. (Cited on p. 667.)

[2656] Y. Tian, "Equalities and Inequalities for Inertias of Hermitian Matrices with Applications," *Lin. Alg. Appl.*, Vol. 433, pp. 263–296, 2010. (Cited on pp. 569, 661, and 662.)

[2657] Y. Tian, "Equalities for Orthogonal Projectors and Their Operations," *Cent. Eur. J. Math.*, Vol. 8, pp. 855–870, 2010. (Cited on pp. 400, 401, 405, 406, 411, 412, 414, 630, 631, 637, 638, 652, 654, 657, and 658.)

[2658] Y. Tian, "Characterizations of EP Matrices and Weighted-EP Matrices," *Lin. Alg. Appl.*, Vol. 434, pp. 1295–1318, 2011. (Cited on pp. 377 and 570.)

[2659] Y. Tian, "Expansion Formulas for the Inertias of Hermitian Matrix Polynomials and Matrix Pencils of Orthogonal Projectors," *J. Math. Anal. Appl.*, Vol. 376, pp. 162–186, 2011. (Cited on pp. 317, 414, and 567.)

[2660] Y. Tian, "Expansion Formulas for the Inertias of Hermitian Matrix Polynomials and Matrix Pencils of Orthogonal Projectors," *J. Math. Anal. Appl.*, Vol. 376, pp. 162–186, 2011. (Cited on p. 414.)

[2661] Y. Tian, "Eigenvalues of Sums and Differences of Idempotent Matrices," *Amer. Math. Monthly*, Vol. 119, pp. 162–163, 2012. (Cited on p. 568.)

[2662] Y. Tian, "Inertia Formula for a Partitioned Hermitian Matrix," *IMAGE*, Vol. 55, pp. 38–41, 2015. (Cited on p. 569.)

[2663] Y. Tian, "Equalities and Inequalities for Ranks of Products of Generalized Inverses of Two Matrices and Their Applications," *J. Ineq. Appl.*, pp. 1–51, 2016, article 182. (Cited on p. 636.)

[2664] Y. Tian, "Upper and Lower Bounds of Ranks of Matrices," *IMAGE*, Vol. 58, p. 44, 2017. (Cited on p. 325.)

[2665] Y. Tian and S. Cheng, "Some Identities for Moore-Penrose Inverses of Matrix Products," *Lin. Multilin. Alg.*, Vol. 52, pp. 405–420, 2004. (Cited on p. 636.)

[2666] Y. Tian and S. Cheng, "The Moore-Penrose Inverse for Sums of Matrices under Rank Additivity Conditions," *Lin. Multilin. Alg.*, Vol. 53, pp. 45–65, 2005. (Cited on pp. 321, 638, 668, and 820.)

[2667] Y. Tian and Y. Li, "Distributions of Eigenvalues and Inertias of Some Block Hermitian Matrices Consisting of Orthogonal Projectors," *Lin. Multilin. Alg.*, Vol. 60, pp. 1027–1069, 2012. (Cited on pp. 354, 415, and 569.)

[2668] Y. Tian and Y. Liu, "Extremal Ranks of Some Symmetric Matrix Expressions with Applications," *SIAM J. Matrix Anal. Appl.*, Vol. 28, pp. 890–905, 2006. (Cited on p. 662.)

[2669] Y. Tian and C. Meyer, "Injective Properties of Complex Matrices," *Amer. Math. Monthly*, Vol. 111, p. 728, 2004. (Cited on p. 667.)

[2670] Y. Tian and G. P. H. Styan, "Universal Factorization Inequalities for Quaternion Matrices and Their Applications," *Math. J. Okayama University*, Vol. 41, pp. 45–62, 1999. (Cited on pp. 439 and 440.)

[2671] Y. Tian and G. P. H. Styan, "How to Establish Universal Block-Matrix Factorizations," *Elec. J. Lin. Alg.*, Vol. 8, pp. 115–127, 2001. (Cited on pp. 342, 439, and 440.)

[2672] Y. Tian and G. P. H. Styan, "Rank Equalities for Idempotent and Involutory Matrices," *Lin. Alg. Appl.*, Vol. 335, pp. 101–117, 2001. (Cited on pp. 398, 400, 401, 403, 405, 406, 407, 414, 417, and 630.)

[2673] Y. Tian and G. P. H. Styan, "Rank Equalities for Idempotent and Involutory Matrices," *Lin. Alg. Appl.*, Vol. 335, pp. 101–117, 2001. (Cited on pp. 400, 402, 414, and 641.)

[2674] Y. Tian and G. P. H. Styan, "A New Rank Formula for Idempotent Matrices with Applications," *Comment. Math. University Carolinae*, Vol. 43, pp. 379–384, 2002. (Cited on pp. 324 and 398.)

[2675] Y. Tian and G. P. H. Styan, "When Does $\operatorname{rank}(ABC) = \operatorname{rank}(AB) + \operatorname{rank}(BC) - \operatorname{rank}(B)$ Hold?" *Int. J. Math. Educ. Sci. Tech.*, Vol. 33, pp. 127–137, 2002. (Cited on pp. 324, 401, 409, 418, 663, and 664.)

[2676] Y. Tian and G. P. H. Styan, "Rank Equalities for Idempotent Matrices with Applications," *J. Comp. Appl. Math.*, Vol. 191, pp. 77–97, 2006. (Cited on pp. 400, 414, and 416.)

[2677] Y. Tian and Y. Takane, "Schur Complements and Banachiewicz-Schur Forms," *Elec. J. Lin. Alg.*, Vol. 13, pp. 405–418, 2005. (Cited on p. 667.)

[2678] Y. Tian and Y. Takane, "The Inverse of Any Two-by-two Nonsingular Partitioned Matrx and Three Matrix Inverse Completion Problems," *Comp. Math. Appl.*, Vol. 57, pp. 1294–1304, 2009. (Cited on p. 667.)

[2679] Y. Tian and Y. Wang, "Expansion Formulas for Orthogonal Projectors onto Ranges of Row Block Matrices," *J. Math. Res. Appl.*, Vol. 34, pp. 147–154, 2014. (Cited on p. 666.)

[2680] L. Tie, "On the Existence of Complex Roots of Real Polynomials," *Lin. Multilin. Alg.*, Vol. 64, pp. 980–990, 2016. (Cited on pp. 1238 and 1239.)

[2681] L. Tie, K.-Y. Cai, and Y. Lin, "Rearrangement Inequalities for Hermitian Matrices," *Lin. Alg. Appl.*, Vol. 434, pp. 443–456, 2011. (Cited on pp. 773, 774, 781, 782, 811, and 826.)

[2682] C. Tigaeru, "An Inequality in the Complex Domain," *J. Math. Ineq.*, Vol. 6, pp. 167–173, 2012. (Cited on p. 271.)

[2683] W. Tingting and Z. Wenpeng, "Some Identities Involving Fibonacci, Lucas Polynomials and Their Applications," *Bull. Math. Soc. Sci. Math. Roumanie*, Vol. 55, pp. 95–103, 2012. (Cited on p. 993.)

[2684] T. Toffoli and J. Quick, "Three-Dimensional Rotations by Three Shears," *Graph. Models Image Process.*, Vol. 59, pp. 89–96, 1997. (Cited on p. 606.)

[2685] F. A. F. Tojo, "A Hyperbolic Analog of the Phasor Addition Formula," *Amer. Math. Monthly*, Vol. 123, pp. 574–582, 2016. (Cited on pp. 237 and 262.)

[2686] M. Tominaga, "A Generalized Reverse Inequality of the Cordes Inequality," *Math. Ineq. Appl.*, Vol. 11, pp. 221–227, 2008. (Cited on p. 812.)

[2687] J. Tong, "Mean Value Theorem for Integrals Generalised to Involve Two Functions," *Math. Gaz.*, Vol. 90, pp. 126–127, 2006. (Cited on p. 944.)

[2688] A. Tonge, "Equivalence Constants for Matrix Norms: A Problem of Goldberg," *Lin. Alg. Appl.*, Vol. 306, pp. 1–13, 2000. (Cited on p. 875.)

[2689] M. Torabi-Dashti, "Faulhaber's Triangle," *Coll. Math. J.*, Vol. 42, no. 2, pp. 96–97, 2011. (Cited on p. 37.)

[2690] L. Toth and J. Bukor, "On the Alternating Series $1 - \frac{1}{2} + \frac{1}{3} - \frac{1}{4} \cdots$," *J. Math. Anal. Appl.*, Vol. 282, pp. 21–25, 2003. (Cited on p. 963.)

[2691] J. Trainin, "Integrating Expressions of the Form $\frac{\sin^n x}{x^m}$ and Others," *Math. Gaz.*, Vol. 94, pp. 216–233, 2010. (Cited on p. 1119.)

[2692] G. E. Trapp, "Hermitian Semidefinite Matrix Means and Related Matrix Inequalities—An Introduction," *Lin. Multilin. Alg.*, Vol. 16, pp. 113–123, 1984. (Cited on p. 744.)

[2693] M. V. Travaglia, "On an Inequality Involving Power and Contraction of Matrices with and without Trace," *J. Ineq. Pure Appl. Math.*, Vol. 7, pp. 1–8, 2006, article 65. (Cited on p. 768.)

[2694] L. N. Trefethen and D. Bau, *Numerical Linear Algebra*. Philadelphia: SIAM, 1997. (Cited on p. xxi.)

[2695] L. N. Trefethen and M. Embree, *Spectra and Pseudospectra*. Princeton: Princeton University Press, 2005. (Cited on p. 1219.)

[2696] D. Trenkler, G. Trenkler, C.-K. Li, and H. J. Werner, "Square Roots and Additivity," *IMAGE*, Vol. 29, p. 30, 2002. (Cited on p. 740.)

[2697] G. Trenkler, "A Trace Inequality," *Amer. Math. Monthly*, Vol. 102, pp. 362–363, 1995. (Cited on p. 764.)

[2698] G. Trenkler, "Four Square Roots of the Vector Cross Product," *Math. Gaz.*, Vol. 82, pp. 100–102, 1998. (Cited on p. 386.)

[2699] G. Trenkler, "The Vector Cross Product from an Algebraic Point of View," *Discuss. Math.*, Vol. 21, pp. 67–82, 2001. (Cited on p. 665.)

[2700] G. Trenkler, "An Application for the Trace Operator," *Coll. Math. J.*, Vol. 33, pp. 326–327, 2002. (Cited on p. 408.)

[2701] G. Trenkler, "The Moore-Penrose Inverse and the Vector Product," *Int. J. Math. Educ. Sci. Tech.*, Vol. 33, pp. 431–436, 2002. (Cited on pp. 386 and 664.)

[2702] G. Trenkler, "A Matrix Related to an Idempotent Matrix," *IMAGE*, Vol. 31, pp. 39–40, 2003. (Cited on p. 651.)

[2703] G. Trenkler, "An Extension of Lagrange's Identity to Matrices," *Int. J. Math. Educ. Sci. Technol.*, Vol. 35, pp. 245–249, 2004. (Cited on pp. 217, 329, 391, and 628.)

[2704] G. Trenkler, "Matrices Which Take a Given Vector into a Given Vector: Revisited," *Amer. Math. Monthly*, Vol. 111, pp. 50–52, 2004. (Cited on p. 383.)

[2705] G. Trenkler, "On the Product of Orthogonal Projectors," *IMAGE*, Vol. 32, pp. 30–31, 2004. (Cited on pp. 407, 591, and 654.)

[2706] G. Trenkler, "On the Product of Orthogonal Projectors," *IMAGE*, Vol. 32, pp. 30–31, 2004. (Cited on pp. 411, 591, and 654.)

[2707] G. Trenkler, "A Property for the Sum of a Matrix A and Its Moore-Penrose Inverse A^+," *IMAGE*, Vol. 35, pp. 38–40, 2005. (Cited on p. 642.)

[2708] G. Trenkler, "Factorization of a Projector," *IMAGE*, Vol. 34, pp. 33–35, 2005. (Cited on p. 609.)

[2709] G. Trenkler, "On the Product of Orthogonal Projectors," *IMAGE*, Vol. 35, pp. 42–43, 2005. (Cited on p. 739.)

[2710] G. Trenkler, "Projectors and Similarity," *IMAGE*, Vol. 34, pp. 35–36, 2005. (Cited on pp. 651 and 667.)

[2711] G. Trenkler, "Property of the Cross Product," *IMAGE*, Vol. 34, pp. 36–37, 2005. (Cited on p. 386.)

[2712] G. Trenkler, "A Range Equality for Idempotent Hermitian Matrices," *IMAGE*, Vol. 36, pp. 34–35, 2006. (Cited on p. 413.)

[2713] G. Trenkler, "Necessary and Sufficient Conditions for $A + A^*$ to be a Nonnegative Definite Matrix," *IMAGE*, Vol. 37, pp. 28–30, 2006. (Cited on p. 815.)

[2714] G. Trenkler, "Necessary and Sufficient Conditions for $A^2 = -I$," *Coll. Math. J.*, Vol. 37, pp. 67–68, 2006. (Cited on p. 579.)

[2715] G. Trenkler, "Necessary and Sufficient Conditions for $A^2 = -A$," *Coll. Math. J.*, Vol. 37, pp. 67, 68, 2006. (Cited on p. 397.)

[2716] G. Trenkler, "On the Rotations Taking One Vector Into Another," *Int. J. Math. Educ. Sci. Tech.*, Vol. 37, pp. 614–618, 2006. (Cited on p. 393.)

[2717] G. Trenkler, "Another Property for the Sum of a Matrix A and Its Moore-Penrose Inverse A^+," *IMAGE*, Vol. 39, pp. 23–25, 2007. (Cited on p. 642.)

[2718] G. Trenkler, "Characterization of EP-ness," *IMAGE*, Vol. 39, pp. 30–31, 2007. (Cited on p. 641.)

[2719] G. Trenkler, "Column Space of a Square Matrix and Its Conjugate Transpose," *Coll. Math. J.*, Vol. 38, pp. 152–153, 2007. (Cited on p. 377.)

[2720] G. Trenkler and S. Puntanen, "A Multivariate Version of Samuelson's Inequality," *Lin. Alg. Appl.*, Vol. 410, pp. 143–149, 2005. (Cited on pp. 195 and 734.)

[2721] G. Trenkler and D. Trenkler, "The Sherman-Morrison Formula and Eigenvalues of a Special Bordered Matrix," *Acta Math. University Comenianae*, Vol. 74, pp. 255–258, 2005. (Cited on p. 527.)

[2722] G. Trenkler and D. Trenkler, "On the Product of Rotations," *Int. J. Math. Educ. Sci. Tech.*, Vol. 39, pp. 94–104, 2008. (Cited on pp. 393 and 394.)

[2723] G. Trenkler and D. Trenkler, "The Vector Cross Product and 4×4 Skew-Symmetric Matrices," in *Recent Advances in Linear Models and Related Areas*, Shalabh and C. Heumann, Eds. Berlin: Springer, 2008, pp. 95–104. (Cited on pp. 386, 528, 665, and 1208.)

[2724] G. Trenkler and D. Trenkler, "Simplifying a Matrix Expression," *Coll. Math. J.*, Vol. 46, p. 63, 2015. (Cited on p. 386.)

[2725] G. Trenkler and H. J. Werner, "Partial Isometry and Idempotent Matrices," *IMAGE*, Vol. 29, pp. 30–32, 2002. (Cited on p. 407.)

[2726] H. L. Trentelman, A. A. Stoorvogel, and M. L. J. Hautus, *Control Theory for Linear Systems*. New York: Springer, 2001. (Cited on pp. xxi, 345, and 1319.)

[2727] P. Treuenfels, "The Matrix Equation $X^2 - 2AX + B = 0$," *Amer. Math. Monthly*, Vol. 66, pp. 145–146, 1959. (Cited on p. 1317.)

[2728] T. Trif, "Combinatorial Sums and Series Involving Inverses of Binomial Coefficients," *Fibonacci Quart.*, Vol. 38, pp. 79–84, 2000. (Cited on pp. 68, 75, and 1102.)

[2729] S. Trimpe and R. D'Andrea, "A Limiting Property of the Matrix Exponential with Application to Multi-loop Control," in *Proc. Conf. Dec. Contr.*, Shanghai, China, December 2009, pp. 6419–6425. (Cited on p. 1223.)

[2730] N. Trinajstic, *Chemical Graph Theory*, 2nd ed. Boca Raton: CRC, 1992. (Cited on p. 440.)

[2731] B. Tromborg and S. Waldenstrom, "Bounds on the Diagonal Elements of a Unitary Matrix," *Lin. Alg. Appl.*, Vol. 20, pp. 189–195, 1978. (Cited on p. 389.)

[2732] J. A. Tropp, A. C. Gilbert, and M. J. Strauss, "Algorithms for Simultaneous Sparse Approximation, Part I: Greedy Pursuit," *Sig. Proc.*, Vol. 86, pp. 572–588, 2006. (Cited on p. 868.)

[2733] T. Trotter, "Some Identities for the Triangular Numbers," *J. Recreat. Math.*, Vol. 6, pp. 127–134, 1973. (Cited on p. 39.)

[2734] W. T. Trotter, *Combinatorics and Partially Ordered Sets: Dimension Theory*. Baltimore: The Johns Hopkins University Press, 1992. (Cited on pp. 25 and 118.)

[2735] R. J. Trudeau, *Introduction to Graph Theory*. New York: Dover, 1993. (Cited on p. 26.)

[2736] T. Trudgian, "Updating the Error Term in the Prime Number Theorem," *Ramanujan J.*, Vol. 39, pp. 225–234, 2016. (Cited on p. 35.)

[2737] M. Tsatsomeros, J. S. Maybee, D. D. Olesky, and P. van den Driessche, "Nullspaces of Matrices and Their Compounds," *Lin. Multilin. Alg.*, Vol. 34, pp. 291–300, 1993. (Cited on p. 598.)

[2738] N.-K. Tsing and F. Uhlig, "Inertia, Numerical Range, and Zeros of Quadratic Forms for Matrix Pencils," *SIAM J. Matrix Anal. Appl.*, Vol. 12, pp. 146–159, 1991. (Cited on pp. 619 and 797.)

[2739] G. A. Tsintsifas, "Weitzenbock Generalized," *Amer. Math. Monthly*, Vol. 95, pp. 658–659, 1988. (Cited on p. 452.)

[2740] P. Tsiotras, "Further Passivity Results for the Attitude Control Problem," *IEEE Trans. Autom. Contr.*, Vol. 43, pp. 1597–1600, 1998. (Cited on p. 386.)

[2741] P. Tsiotras, J. L. Junkins, and H. Schaub, "Higher-Order Cayley Transforms with Applications to Attitude Representations," *J. Guid. Contr. Dyn.*, Vol. 20, pp. 528–534, 1997. (Cited on p. 396.)

[2742] L. W. Tu, *An Introduction to Manifolds*, 2nd ed. New York: Springer, 2011. (Cited on p. 691.)

[2743] N. M. Tuan and L. Q. Thuong, "On an Extension of Shapiro's Cyclic Inequality," *J. Ineq. Appl.*, pp. 1–5, 2009, article 491576. (Cited on p. 166.)

[2744] H. J. H. Tuenter, "A Double Binomial Sum: Putnam 35-A4 Revisited," *Amer. Math. Monthly*, Vol. 122, pp. 685–689, 2015. (Cited on pp. 68 and 86.)

[2745] Y. Tuncbilek, “A Straight-forward Inequality,” *Coll. Math. J.*, Vol. 46, p. 146, 2015. (Cited on p. 145.)

[2746] S. H. Tung, “On Lower and Upper Bounds of the Difference between the Arithmetic and the Geometric Mean,” *Math. Comput.*, Vol. 29, pp. 834–836, 1975. (Cited on p. 203.)

[2747] D. A. Turkington, *Matrix Calculus and Zero-One Matrices*. Cambridge: Cambridge University Press, 2002. (Cited on p. 701.)

[2748] D. A. Turkington, *Generalized Vectorization, Cross-Products, and Matrix Calculus*. Cambridge: Cambridge University Press, 2013. (Cited on p. 701.)

[2749] R. Turkmen and D. Bozkurt, “A Note for Bounds of Norms of Hadamard Product of Matrices,” *Math. Ineq. Appl.*, Vol. 9, pp. 211–217, 2006. (Cited on p. 907.)

[2750] R. Turkmen, V. E. Paksoy, and F. Zhang, “Some Inequalities of Majorization Type,” *Lin. Alg. Appl.*, Vol. 437, pp. 1305–1316, 2012. (Cited on pp. 193, 360, 361, 752, 770, 771, 805, 806, 819, 886, 903, 906, 1221, and 1222.)

[2751] R. Turkmen and Z. Ulukok, “On the Frobenius Condition Number of Positive Definite Matrices,” *J. Ineq. Appl.*, pp. 1–9, 2010, article ID 897279. (Cited on pp. 762, 861, and 874.)

[2752] H. W. Turnbull, *The Theory of Determinants, Matrices and Invariants*. London: Blackie, 1950. (Cited on p. 362.)

[2753] J. D. Turner, “Quaternion Analysis Tools for Engineering and Scientific Applications,” *AIAA J. Guid. Contr. Dyn.*, Vol. 32, pp. 686–693, 2009. (Cited on p. 438.)

[2754] G. M. Tuynman, “The Derivation of the Exponential Map of Matrices,” *Amer. Math. Monthly*, Vol. 102, pp. 818–820, 1995. (Cited on p. 1212.)

[2755] F. Tyan and D. S. Bernstein, “Global Stabilization of Systems Containing a Double Integrator Using a Saturated Linear Controller,” *Int. J. Robust Nonlin. Contr.*, Vol. 9, pp. 1143–1156, 1999. (Cited on p. 1311.)

[2756] M. Uchiyama, “Norms and Determinants of Products of Logarithmic Functions of Positive Semi-Definite Operators,” *Math. Ineq. Appl.*, Vol. 1, pp. 279–284, 1998. (Cited on pp. 813 and 1222.)

[2757] M. Uchiyama, “Some Exponential Operator Inequalities,” *Math. Ineq. Appl.*, Vol. 2, pp. 469–471, 1999. (Cited on p. 1217.)

[2758] F. E. Udwadia and R. E. Kalaba, *Analytical Dynamics: A New Approach*. Cambridge: Cambridge University Press, 1996. (Cited on p. xxi.)

[2759] F. Uhlig, “A Recurring Theorem about Pairs of Quadratic Forms and Extensions: A Survey,” *Lin. Alg. Appl.*, Vol. 25, pp. 219–237, 1979. (Cited on pp. 619, 797, and 799.)

[2760] F. Uhlig, “On the Matrix Equation $AX = B$ with Applications to the Generators of a Controllability Matrix,” *Lin. Alg. Appl.*, Vol. 85, pp. 203–209, 1987. (Cited on p. 317.)

[2761] F. Uhlig, “Constructive Ways for Generating (Generalized) Real Orthogonal Matrices as Products of (Generalized) Symmetries,” *Lin. Alg. Appl.*, Vol. 332–334, pp. 459–467, 2001. (Cited on p. 607.)

[2762] Z. Ulukok and R. Turkmen, “On Some Matrix Trace Inequalities,” *J. Ineq. Appl.*, pp. 1–8, 2010, article ID 201486. (Cited on p. 884.)

[2763] D. Underwood, *Elementary Number Theory*, 2nd ed. Mineola: Dover, 1978. (Cited on pp. 30, 31, and 32.)

[2764] S. Vajda, *Fibonacci and Lucas Numbers and the Golden Section: Theory and Applications*. Mineola: Dover, 2007. (Cited on pp. 100, 103, 493, and 1079.)

[2765] C. I. Valean, “Problem 11993,” *Amer. Math. Monthly*, Vol. 124, p. 659, 2017. (Cited on p. 1141.)

[2766] C. I. Valean and O. Furdui, “Reviving the Quadratic Series of Au-Yeung,” *J. Classical Anal.*, Vol. 6, pp. 113–118, 2015. (Cited on p. 1134.)

[2767] F. A. Valentine, *Convex Sets*. New York: McGraw-Hill, 1964. (Cited on p. 362.)

[2768] M. Van Barel, V. Ptak, and Z. Vavrin, “Bezout and Hankel Matrices Associated with Row Reduced Matrix Polynomials, Barnett-Type Formulas,” *Lin. Alg. Appl.*, Vol. 332–334, pp. 583–606, 2001. (Cited on p. 520.)

[2769] G. Van Brummelen, *Heavenly Mathematics: The Forgotten Art of Spherical Trigonometry*. Princeton: Princeton University Press, 2013. (Cited on p. 497.)

[2770] A. Van-Brunt and M. Visser, “Special-Case Closed Form of the Baker-Campbell-Hausdorff Formula,” *J. Phys. A: Math. Theor.*, Vol. 48, pp. 1–10, 2015. (Cited on p. 1213.)

[2771] R. van der Merwe, *Sigma-Point Kalman Filters for Probabilistic Inference in Dynamic State-Space Models*. Portland: Oregon Health and Science University, 2004, Ph.D. Dissertation. (Cited on p. 1208.)

[2772] P. Van Dooren, "The Computation of Kronecker's Canonical Form of a Singular Pencil," *Lin. Alg. Appl.*, Vol. 27, pp. 103–140, 1979. (Cited on p. 619.)
[2773] S. Van Huffel and J. Vandewalle, *The Total Least Squares Problem: Computational Aspects and Analysis*. Philadelphia: SIAM, 1987. (Cited on p. 912.)
[2774] C. F. Van Loan, "Computing Integrals Involving the Matrix Exponential," *IEEE Trans. Autom. Contr.*, Vol. AC-23, pp. 395–404, 1978. (Cited on p. 1212.)
[2775] C. F. Van Loan, "How Near Is a Stable Matrix to an Unstable Matrix," *Contemporary Math.*, Vol. 47, pp. 465–478, 1985. (Cited on p. 1228.)
[2776] C. F. Van Loan, *Computational Frameworks for the Fast Fourier Transform*. Philadelphia: SIAM, 1992. (Cited on pp. xxi and 615.)
[2777] C. F. Van Loan, "The Ubiquitous Kronecker Product," *J. Comp. Appl. Math.*, Vol. 123, pp. 85–100, 2000. (Cited on p. 701.)
[2778] P. Van Overschee and B. De Moor, *Subspace Identification for Linear Systems: Theory, Implementation, Applications*. Dordrecht: Kluwer, 1996. (Cited on p. 1319.)
[2779] R. Vandebril and M. Van Barel, "A Note on the Nullity Theorem," *J. Comp. Appl. Math.*, Vol. 189, pp. 179–190, 2006. (Cited on p. 325.)
[2780] R. Vandebril, M. Van Barel, and N. Mastronardi, *Matrix Computations and Semiseparable Matrices*. Baltimore: Johns Hopkins University Press, 2008. (Cited on pp. 325 and 341.)
[2781] V. S. Varadarajan, *Lie Groups, Lie Algebras, and Their Representations*. New York: Springer, 1984. (Cited on pp. 440, 1189, 1212, and 1247.)
[2782] J. M. Varah, "A Lower Bound for the Smallest Singular Value of a Matrix," *Lin. Alg. Appl.*, Vol. 11, pp. 3–5, 1975. (Cited on p. 899.)
[2783] I. Vardi, "Integrals, an Introduction to Analytic Number Theory," *Amer. Math. Monthly*, Vol. 95, pp. 308–315, 1988. (Cited on pp. 1003 and 1143.)
[2784] A. I. G. Vardulakis, *Linear Multivariable Control: Algebraic Analysis and Synthesis Methods*. Chichester: Wiley, 1991. (Cited on pp. xxi and 544.)
[2785] R. S. Varga, *Matrix Iterative Analysis*. Englewood Cliffs: Prentice-Hall, 1962. (Cited on pp. xxi and 539.)
[2786] R. S. Varga, *Gersgorin and His Circles*. New York: Springer, 2004. (Cited on p. 535.)
[2787] S. Varosanec, "A Generalized Beckenback-Dresher Inequality and Related Results," *Banach J. Math. Anal.*, Vol. 4, pp. 13–20, 2010. (Cited on p. 224.)
[2788] P. R. Vein, "A Short Survey of Some Recent Applications of Determinants," *Lin. Alg. Appl.*, Vol. 42, pp. 287–297, 1982. (Cited on p. 362.)
[2789] R. Vein and P. Dale, *Determinants and Their Applications in Mathematical Physics*. New York: Springer, 1999. (Cited on pp. xxi and 362.)
[2790] D. Veljan, "The Sine Theorem and Inequalities for Volumes of Simplices and Determinants," *Lin. Alg. Appl.*, Vol. 219, pp. 79–91, 1995. (Cited on p. 495.)
[2791] D. Veljan and S. Wu, "Parametrized Klamkin's Inequality and Improved Euler's Inequality," *Math. Ineq. Appl.*, Vol. 11, pp. 729–737, 2008. (Cited on pp. 173, 177, 452, and 466.)
[2792] L. Vepstas, "On Plouffe's Ramanujan Identities," *Ramanujan J.*, Vol. 27, pp. 387–408, 2012. (Cited on pp. 980, 1032, and 1037.)
[2793] J. Vermeer, "Orthogonal Similarity of a Real Matrix and Its Transpose," *Lin. Alg. Appl.*, Vol. 428, pp. 382–392, 2008. (Cited on p. 605.)
[2794] K. Veselic, "On Real Eigenvalues of Real Tridiagonal Matrices," *Lin. Alg. Appl.*, Vol. 27, pp. 167–171, 1979. (Cited on p. 586.)
[2795] K. Veselic, "Estimating the Operator Exponential," *Lin. Alg. Appl.*, Vol. 280, pp. 241–244, 1998. (Cited on p. 1228.)
[2796] K. Veselic, "Bounds for Exponentially Stable Semigroups," *Lin. Alg. Appl.*, Vol. 358, pp. 309–333, 2003. (Cited on p. 1228.)
[2797] W. J. Vetter, "Matrix Calculus Operations and Taylor Expansions," *SIAM Rev.*, Vol. 15, pp. 352–369, 1973. (Cited on p. 701.)
[2798] M. Vidyasagar, "On Matrix Measures and Convex Liapunov Functions," *J. Math. Anal. Appl.*, Vol. 62, pp. 90–103, 1978. (Cited on p. 1219.)
[2799] C. Vignat and V. H. Moll, "A Probabilistic Approach to Some Binomial Identities," *Elem. Math.*, Vol. 69, pp. 1–12, 2014. (Cited on pp. 86 and 992.)

[2800] J. L. C. Villacorta, "Certain Inequalities Associated with the Divisor Function," *Amer. Math. Monthly*, Vol. 120, pp. 832–837, 2013. (Cited on p. 1008.)

[2801] A. Villani, "Measurable Functions with a Given Set of Integrability Exponents," *Amer. Math. Monthly*, Vol. 118, pp. 77–82, 2011. (Cited on p. 1022.)

[2802] G. Visick, "A Weak Majorization Involving the Matrices $A \odot B$ and AB," *Lin. Alg. Appl.*, Vol. 223/224, pp. 731–744, 1995. (Cited on p. 825.)

[2803] G. Visick, "Majorizations of Hadamard Products of Matrix Powers," *Lin. Alg. Appl.*, Vol. 269, pp. 233–240, 1998. (Cited on pp. 825 and 830.)

[2804] G. Visick, "A Quantitative Version of the Observation that the Hadamard Product Is a Principal Submatrix of the Kronecker Product," *Lin. Alg. Appl.*, Vol. 304, pp. 45–68, 2000. (Cited on pp. 821, 827, and 828.)

[2805] G. Visick, "Another Inequality for Hadamard Products," *IMAGE*, Vol. 29, pp. 32–33, 2002. (Cited on p. 821.)

[2806] N. G. Voll, "The Cassini Identity and Its Relatives," *Fibonacci Quart.*, Vol. 48, pp. 197–201, 2010. (Cited on p. 103.)

[2807] S.-W. Vong and X.-Q. Jin, "Proof of Bottcher and Wenzel's Conjecture," *Oper. Matrices*, Vol. 2, pp. 435–442, 2008. (Cited on p. 884.)

[2808] S. Wada, "On Some Refinement of the Cauchy-Schwarz Inequality," *Lin. Alg. Appl.*, Vol. 420, pp. 433–440, 2007. (Cited on pp. 220 and 221.)

[2809] G. G. Walter and M. Contreras, *Compartmental Modeling with Networks*. Boston: Birkhauser, 1999. (Cited on pp. xxi and 1233.)

[2810] A. Y. Z. Wang and P. Wen, "On the Partial Finite Sums of the Reciprocals of the Fibonacci Numbers," *J. Ineq. Appl.*, pp. 1–13, 2015, article 73. (Cited on pp. 100 and 1079.)

[2811] A. Y. Z. Wang and F. Zhang, "The Reciprocal Sums of Even and Odd Terms in the Fibonacci Sequence," *J. Ineq. Appl.*, pp. 1–13, 2016, article 376. (Cited on p. 1079.)

[2812] B. Wang and F. Zhang, "Some Inequalities for the Eigenvalues of the Product of Positive Semidefinite Hermitian Matrices," *Lin. Alg. Appl.*, Vol. 160, pp. 113–118, 1992. (Cited on pp. 810, 811, and 905.)

[2813] B.-Y. Wang and M.-P. Gong, "Some Eigenvalue Inequalities for Positive Semidefinite Matrix Power Products," *Lin. Alg. Appl.*, Vol. 184, pp. 249–260, 1993. (Cited on p. 810.)

[2814] B.-Y. Wang and B.-Y. Xi, "Some Inequalities for Singular Values of Matrix Products," *Lin. Alg. Appl.*, Vol. 264, pp. 109–115, 1997. (Cited on p. 905.)

[2815] B.-Y. Wang, B.-Y. Xi, and F. Zhang, "Some Inequalities for Sum and Product of Positive Semidefinite Matrices," *Lin. Alg. Appl.*, Vol. 293, pp. 39–49, 1999. (Cited on pp. 818 and 907.)

[2816] B.-Y. Wang and F. Zhang, "A Trace Inequality for Unitary Matrices," *Amer. Math. Monthly*, Vol. 101, pp. 453–455, 1994. (Cited on p. 390.)

[2817] B.-Y. Wang and F. Zhang, "Trace and Eigenvalue Inequalities for Ordinary and Hadamard Products of Positive Semidefinite Hermitian Matrices," *SIAM J. Matrix Anal. Appl.*, Vol. 16, pp. 1173–1183, 1995. (Cited on pp. 810 and 823.)

[2818] B.-Y. Wang and F. Zhang, "Schur Complements and Matrix Inequalities of Hadamard Products," *Lin. Multilin. Alg.*, Vol. 43, pp. 315–326, 1997. (Cited on p. 830.)

[2819] C.-L. Wang, "On Development of Inverses of the Cauchy and Hölder Inequalities," *SIAM Rev.*, Vol. 21, pp. 550–557, 1979. (Cited on pp. 212 and 224.)

[2820] D. Wang, "The Polar Decomposition and a Matrix Inequality," *Amer. Math. Monthly*, Vol. 96, pp. 517–519, 1989. (Cited on p. 715.)

[2821] G. Wang, Y. Wei, and S. Qiao, *Generalized Inverses: Theory and Computations*. Beijing/New York: Science Press, 2004. (Cited on pp. 330, 398, 635, and 679.)

[2822] H. Wang and M. Wei, "New Representations for the Moore-Penrose Inverse," *Elec. J. Lin. Alg.*, Vol. 22, pp. 720–728, 2011. (Cited on p. 630.)

[2823] J. Wang and Y. Zhang, "A Proof of an Inequality Conjectured by J. Chen," *Math. Gaz.*, Vol. 93, pp. 105–108, 2009. (Cited on p. 252.)

[2824] J.-H. Wang, "The Length Problem for a Sum of Idempotents," *Lin. Alg. Appl.*, Vol. 215, pp. 135–159, 1995. (Cited on p. 619.)

[2825] L. Wang, M.-Z. Xu, and T.-Z. Huang, "Some Lower Bounds for the Spectral Radius of Matrices Using Traces," *Lin. Alg. Appl.*, Vol. 432, pp. 1007–1016, 2010. (Cited on p. 588.)

[2826] L. Wang and H. Yang, "Several Matrix Euclidean Norm Inequalities Involving Kantorovich Inequality," *J. Ineq. Appl.*, pp. 1–9, 2009, article ID 291984. (Cited on pp. 739 and 874.)

[2827] L.-C. Wang and C.-L. Li, "On Some New Mean Value Inequalities," *J. Ineq. Pure Appl. Math.*, Vol. 8, no. 3, pp. 1–8, 2007, Article 87. (Cited on p. 212.)

[2828] M.-K. Wang, Y.-M. Chu, Y.-P. Jiang, and S.-L. Qiu, "Bounds of the Perimeter of an Ellipse Using Arithmetic, Geometric and Harmonic Means," *Math. Ineq. Appl.*, Vol. 17, pp. 101–111, 2014. (Cited on p. 1080.)

[2829] Q.-G. Wang, "Necessary and Sufficient Conditions for Stability of a Matrix Polytope with Normal Vertex Matrices," *Automatica*, Vol. 27, pp. 887–888, 1991. (Cited on p. 536.)

[2830] W. Wang and C. Jia, "Harmonic Number Identities via the Newton-Andrews Method," *Ramanujan J.*, Vol. 35, pp. 263–285, 2014. (Cited on pp. 76, 86, 92, and 950.)

[2831] X. Wang, "A Simple Proof of Descartes's Rule of Signs," *Amer. Math. Monthly*, Vol. 111, pp. 525–526, 2004. (Cited on p. 1223.)

[2832] Y. Wang and B. Zhang, "Hadamard Powers of Chebyshev Polynomials with Only Real Zeros," *Lin. Alg. Appl.*, Vol. 439, pp. 3173–3176, 1990. (Cited on p. 1225.)

[2833] Y. W. Wang and D. S. Bernstein, "Controller Design with Regional Pole Constraints: Hyperbolic and Horizontal Strip Regions," *AIAA J. Guid. Contr. Dyn.*, Vol. 16, pp. 784–787, 1993. (Cited on p. 1313.)

[2834] Y. W. Wang and D. S. Bernstein, "L_2 Controller Synthesis with L_∞-Bounded Closed-Loop Impulse Response," *Int. J. Contr.*, Vol. 60, pp. 1295–1306, 1994. (Cited on p. 1218.)

[2835] A. J. B. Ward and F. Gerrish, "A Constructive Proof by Elementary Transformations of Roth's Removal Theorems in the Theory of Matrix Equations," *Int. J. Math. Educ. Sci. Tech.*, Vol. 31, pp. 425–429, 2000. (Cited on p. 578.)

[2836] J. Warga, *Optimal Control of Differential and Functional Equations*. New York: Academic Press, 1972. (Cited on pp. 924 and 925.)

[2837] R. H. Wasserman, *Tensors and Manifolds with Applications to Mechanics and Relativity*. New York: Oxford University Press, 1992. (Cited on p. 691.)

[2838] W. C. Waterhouse, "Bhattacharyya's Matrix Theorem," *Lin. Alg. Appl.*, Vol. 438, pp. 7–9, 2013. (Cited on p. 618.)

[2839] W. E. Waterhouse, "A Determinant Identity with Matrix Entries," *Amer. Math. Monthly*, Vol. 97, pp. 249–250, 1990. (Cited on pp. 337 and 338.)

[2840] W. Watkins, "Convex Matrix Functions," *Proc. Amer. Math. Soc.*, Vol. 44, pp. 31–34, 1974. (Cited on pp. 778 and 826.)

[2841] W. Watkins, "Generating Functions," *Coll. Math. J.*, Vol. 18, pp. 195–211, 1987. (Cited on p. 1079.)

[2842] W. Watkins, "A Determinantal Inequality for Correlation Matrices," *Lin. Alg. Appl.*, Vol. 104, pp. 59–63, 1988. (Cited on p. 822.)

[2843] G. S. Watson, G. Alpargu, and G. P. H. Styan, "Some Comments on Six Inequalities Associated with the Inefficiency of Ordinary Least Squares with One Regressor," *Lin. Alg. Appl.*, Vol. 264, pp. 13–54, 1997. (Cited on pp. 212 and 220.)

[2844] J. R. Weaver, "Centrosymmetric (Cross-Symmetric) Matrices, Their Basic Properties, Eigenvalues, and Eigenvectors," *Amer. Math. Monthly*, Vol. 92, pp. 711–717, 1985. (Cited on pp. 427 and 572.)

[2845] J. H. Webb, "An Inequality in Algebra and Geometry," *Coll. Math. J.*, Vol. 36, p. 164, 2005. (Cited on p. 489.)

[2846] R. Webster, *Convexity*. Oxford: Oxford University Press, 1994. (Cited on p. 362.)

[2847] P. A. Wedin, "On Angles between Subspaces of a Finite Dimensional Inner Product Space," in *Matrix Pencils*, B. Kagstrom and A. Ruhe, Eds. New York: Springer, 1983, pp. 263–285. (Cited on p. 574.)

[2848] E. Wegert and L. N. Trefethen, "From the Buffon Needle Problem to the Kreiss Matrix Theorem," *Amer. Math. Monthly*, Vol. 101, pp. 132–139, 1994. (Cited on p. 1242.)

[2849] C.-F. Wei and F. Qi, "Several Closed Expressions for the Euler Numbers," *J. Ineq. Appl.*, pp. 1–8, 2016, article 219. (Cited on p. 981.)

[2850] F. Wei and S. Wu, "Generalizations and Analogues of the Nesbitt's Inequality," *Octogon Math. Mag.*, Vol. 17, pp. 215–220, 2009. (Cited on pp. 160, 166, and 199.)

[2851] J. Wei and E. Norman, "Lie Algebraic Solution of Linear Differential Equations," *J. Math. Phys.*, Vol. 4, pp. 575–581, 1963. (Cited on p. 1306.)

[2852] J. Wei and E. Norman, "On Global Representations of the Solutions of Linear Differential Equations as a Product of Exponentials," *Proc. Amer. Math. Soc.*, Vol. 15, pp. 327–334, 1964. (Cited on p. 1306.)

[2853] M. Wei, "Reverse Order Laws for Generalized Inverses of Multiple Matrix Products," *Lin. Alg. Appl.*, Vol. 293, pp. 273–288, 1999. (Cited on p. 637.)

[2854] Y. Wei, "A Characterization and Representation of the Drazin Inverse," *SIAM J. Matrix Anal. Appl.*, Vol. 17, pp. 744–747, 1996. (Cited on p. 670.)

[2855] Y. Wei, "Expressions for the Drazin Inverse of a 2×2 Block Matrix," *Lin. Multilin. Alg.*, Vol. 45, pp. 131–146, 1998. (Cited on p. 667.)

[2856] S. H. Weintraub, *Differential Forms: A Complement to Vector Calculus*. San Diego: Academic Press, 1997. (Cited on p. 691.)

[2857] G. H. Weiss and A. A. Maradudin, "The Baker-Hausdorff Formula and a Problem in Crystal Physics," *J. Math. Phys.*, Vol. 3, pp. 771–777, 1962. (Cited on p. 1306.)

[2858] E. W. Weisstein, "Barycentric Coordinates," http://mathworld.wolfram.com/Homogeneous BarycentricCoordinates.html. (Cited on p. 472.)

[2859] E. W. Weisstein, "Harmonic Numbers," http://mathworld.wolfram.com/HarmonicNumber.html. (Cited on pp. 68, 1024, and 1025.)

[2860] E. W. Weisstein, "Homogeneous Barycentric Coordinates," http://mathworld.wolfram.com/Barycentric Coordinates.html. (Cited on p. 472.)

[2861] E. W. Weisstein, "Infinite Product," http://mathworld.wolfram.com/InfiniteProduct.html. (Cited on pp. 1083 and 1087.)

[2862] E. W. Weisstein, "Pentagonal Number," http://mathworld.wolfram.com/PentagonalNumber.html. (Cited on p. 39.)

[2863] E. W. Weisstein, "Polygonal Number," http://mathworld.wolfram.com/PolygonalNumber.html. (Cited on p. 40.)

[2864] R. M. Welukar and M. N. Deshpande, "A Result on Fibonacci Numbers," *Math. Gaz.*, Vol. 89, pp. 455–456, 2005. (Cited on p. 100.)

[2865] M. Werman and D. Zeilberger, "A Bijective Proof of Cassini's Fibonacci Identity," *Disc. Math.*, Vol. 58, p. 109, 1986. (Cited on p. 100.)

[2866] E. M. E. Wermuth, "Two Remarks on Matrix Exponentials," *Lin. Alg. Appl.*, Vol. 117, pp. 127–132, 1989. (Cited on pp. 1211 and 1222.)

[2867] E. M. E. Wermuth, "A Remark on Commuting Operator Exponentials," *Proc. Amer. Math. Soc.*, Vol. 125, pp. 1685–1688, 1997. (Cited on p. 1211.)

[2868] H.-J. Werner, "On the Matrix Monotonicity of Generalized Inversion," *Lin. Alg. Appl.*, Vol. 27, pp. 141–145, 1979. (Cited on p. 817.)

[2869] H. J. Werner, "On the Product of Orthogonal Projectors," *IMAGE*, Vol. 32, pp. 30–37, 2004. (Cited on pp. 411, 413, 651, and 652.)

[2870] H. J. Werner, "Solution 34-4.2 to 'A Range Equality for the Commutator with Two Involutory Matrices'," *IMAGE*, Vol. 35, pp. 32–33, 2005. (Cited on p. 404.)

[2871] J. R. Wertz and W. J. Larson, Eds., *Space Mission Analysis and Design*. Dordrecht: Kluwer, 1999. (Cited on p. 497.)

[2872] P. Wesseling, *Principles of Computational Fluid Dynamics*. Berlin: Springer, 2001. (Cited on p. xxi.)

[2873] D. B. West, *Introduction to Graph Theory*, 2nd ed. Delhi: Prentice Hall of India, 2000. (Cited on p. 118.)

[2874] J. R. Westlake, *A Handbook of Numerical Matrix Inversion and Solution of Linear Equations*. New York: Wiley, 1968. (Cited on p. xxi.)

[2875] N. A. Wiegmann, "Normal Products of Matrices," *Duke Math. J.*, Vol. 15, pp. 633–638, 1948. (Cited on p. 604.)

[2876] Z. Wiener, "An Interesting Matrix Exponent Formula," *Lin. Alg. Appl.*, Vol. 257, pp. 307–310, 1997. (Cited on p. 1241.)

[2877] E. P. Wigner and M. M. Yanase, "On the Positive Semidefinite Nature of a Certain Matrix Expression," *Canad. J. Math.*, Vol. 16, pp. 397–406, 1964. (Cited on p. 737.)

[2878] R. M. Wilcox, "Exponential Operators and Parameter Differentiation in Quantum Physics," *J. Math. Phys.*, Vol. 8, pp. 962–982, 1967. (Cited on pp. xxi and 1214.)

[2879] N. J. Wildberger, "A New Look at Multisets," 2003, http://web.maths.unsw.edu.au/ norman/papers /NewMultisets5.pdf. (Cited on pp. 24 and 118.)

[2880] H. S. Wilf, *Generatingfunctionology*, 3rd ed. Wellesley: CRC Press, 2006. (Cited on pp. 68, 75, 86, 232, 947, 980, 1008, 1014, and 1021.)

[2881] J. B. Wilker, J. S. Sumner, A. A. Jagers, M. Vowe, and J. Anglesio, "Inequalities Involving Trigonometric Functions," *Amer. Math. Monthly*, Vol. 98, pp. 264–266, 1991. (Cited on p. 249.)

[2882] J. H. Wilkinson, *The Algebraic Eigenvalue Problem*. London: Oxford University Press, 1965. (Cited on p. xxi.)

[2883] J. C. Willems, "Least Squares Stationary Optimal Control and the Algebraic Riccati Equation," *IEEE Trans. Autom. Contr.*, Vol. AC-16, pp. 621–634, 1971. (Cited on p. 1319.)

[2884] K. S. Williams, "The nth Power of a 2×2 Matrix," *Math. Mag.*, Vol. 65, p. 336, 1992. (Cited on p. 531.)

[2885] K. S. Williams, "The Parents of Jacobi's Four Squares Theorem Are Unique," *Amer. Math. Monthly*, Vol. 120, pp. 329–345, 2013. (Cited on p. 1091.)

[2886] K. S. Williams, "A "Four Integers" Theorem and a "Five Integers" Theorem," *Amer. Math. Monthly*, Vol. 122, pp. 528–536, 2015. (Cited on p. 33.)

[2887] K. S. Williams, "Bounds for the Representations of Integers by Positive Quadratic Forms," *Math. Mag.*, Vol. 89, pp. 122–131, 2016. (Cited on p. 795.)

[2888] D. A. Wilson, "Convolution and Hankel Operator Norms for Linear Systems," *IEEE Trans. Autom. Contr.*, Vol. AC-34, pp. 94–97, 1989. (Cited on p. 912.)

[2889] P. M. H. Wilson, *Curved Spaces: From Classical Geometries to Elementary Differential Geometry*. Cambridge: Cambridge University Press, 2008. (Cited on pp. 394 and 497.)

[2890] R. A. Wilson, *The Finite Simple Groups*. New York: Springer, 2009. (Cited on p. 437.)

[2891] H. K. Wimmer, "Inertia Theorems for Matrices, Controllability, and Linear Vibrations," *Lin. Alg. Appl.*, Vol. 8, pp. 337–343, 1974. (Cited on pp. 1310 and 1312.)

[2892] H. K. Wimmer, "The Algebraic Riccati Equation without Complete Controllability," *SIAM J. Alg. Disc. Math.*, Vol. 3, pp. 1–12, 1982. (Cited on p. 1319.)

[2893] H. K. Wimmer, "On Ostrowski's Generalization of Sylvester's Law of Inertia," *Lin. Alg. Appl.*, Vol. 52/53, pp. 739–741, 1983. (Cited on p. 568.)

[2894] H. K. Wimmer, "The Algebraic Riccati Equation: Conditions for the Existence and Uniqueness of Solutions," *Lin. Alg. Appl.*, Vol. 58, pp. 441–452, 1984. (Cited on p. 1319.)

[2895] H. K. Wimmer, "Monotonicity of Maximal Solutions of Algebraic Riccati Equations," *Sys. Contr. Lett.*, Vol. 5, pp. 317–319, 1985. (Cited on pp. 1303 and 1319.)

[2896] H. K. Wimmer, "Extremal Problems for Hölder Norms of Matrices and Realizations of Linear Systems," *SIAM J. Matrix Anal. Appl.*, Vol. 9, pp. 314–322, 1988. (Cited on pp. 740 and 876.)

[2897] H. K. Wimmer, "Linear Matrix Equations, Controllability and Observability, and the Rank of Solutions," *SIAM J. Matrix Anal. Appl.*, Vol. 9, pp. 570–578, 1988. (Cited on p. 1312.)

[2898] H. K. Wimmer, "On the History of the Bezoutian and the Resultant Matrix," *Lin. Alg. Appl.*, Vol. 128, pp. 27–34, 1990. (Cited on p. 520.)

[2899] H. K. Wimmer, "Normal Forms of Symplectic Pencils and the Discrete-Time Algebraic Riccati Equation," *Lin. Alg. Appl.*, Vol. 147, pp. 411–440, 1991. (Cited on p. 1319.)

[2900] H. K. Wimmer, "Lattice Properties of Sets of Semidefinite Solutions of Continuous-Time Algebraic Riccati Equations," *Automatica*, Vol. 31, pp. 173–182, 1995. (Cited on p. 1319.)

[2901] A. Witkowski, "A New Proof of the Monotonicity of Power Means," *J. Ineq. Pure Appl. Math.*, Vol. 5, no. 1, pp. 1–2, 2004, Article 6. (Cited on pp. 123 and 227.)

[2902] J. Wittenburg and L. Lilov, "Decomposition of a Finite Rotation into Three Rotations about Given Axes," *Multibody Sys. Dyn.*, Vol. 9, pp. 353–375, 2003. (Cited on p. 1207.)

[2903] R. Witula, "Ramanujan Type Trigonometric Formulas: The General Form for the Argument $\frac{2\pi}{7}$," *J. Integer Sequences*, Vol. 12, pp. 1–23, 2009, article 09.8.5. (Cited on pp. 53 and 232.)

[2904] R. Witula, E. Hetmaniok, and D. Slota, "Factoring a Homogeneous Polynomial," *Math. Mag.*, Vol. 85, pp. 296–297, 2012. (Cited on p. 131.)

[2905] H. Wolkowicz and G. P. H. Styan, "Bounds for Eigenvalues Using Traces," *Lin. Alg. Appl.*, Vol. 29, pp. 471–506, 1980. (Cited on pp. 196, 580, and 588.)

[2906] H. Wolkowicz and G. P. H. Styan, "More Bounds for Eigenvalues Using Traces," *Lin. Alg. Appl.*, Vol. 31, pp. 1–17, 1980. (Cited on p. 588.)

[2907] W. A. Wolovich, *Linear Multivariable Systems*. New York: Springer, 1974. (Cited on p. 1319.)

[2908] M. J. Wonenburger, "A Decomposition of Orthogonal Transformations," *Canad. Math. Bull.*, Vol. 7, pp. 379–383, 1964. (Cited on p. 609.)

[2909] M. J. Wonenburger, "Transformations Which Are Products of Two Involutions," *J. Math. Mech.*, Vol. 16, pp. 327–338, 1966. (Cited on p. 609.)

[2910] C. S. Wong, "Characterizations of Products of Symmetric Matrices," *Lin. Alg. Appl.*, Vol. 42, pp. 243–251, 1982. (Cited on p. 609.)

[2911] E. Wong, *Stochastic Processes in Information and Dynamical Systems*. New York: McGraw-Hill, 1971. (Cited on p. 725.)

[2912] W. M. Wonham, *Linear Multivariable Control: A Geometric Approach*, 2nd ed. New York: Springer, 1979. (Cited on pp. xxi, 525, 1181, 1308, and 1319.)

[2913] B. Wu, "Explicit Formulas for the Exponentials of Some Special Matrices," *Appl. Math. Lett.*, Vol. 24, pp. 642–647, 2011. (Cited on p. 1205.)

[2914] C.-F. Wu, "On Some Ordering Properties of the Generalized Inverses of Nonnegative Definite Matrices," *Lin. Alg. Appl.*, Vol. 32, pp. 49–60, 1980. (Cited on p. 817.)

[2915] D. W. Wu, "Bernoulli Numbers and Sums of Powers," *Int. J. Math. Educ. Sci. Tech.*, Vol. 32, pp. 440–443, 2001. (Cited on p. 37.)

[2916] J. Wu, P. Zhang, and W. Liao, "Upper Bounds for the Spread of a Matrix," *Lin. Alg. Appl.*, Vol. 437, pp. 2813–2822, 2012. (Cited on p. 892.)

[2917] P. Y. Wu, "Products of Nilpotent Matrices," *Lin. Alg. Appl.*, Vol. 96, pp. 227–232, 1987. (Cited on p. 609.)

[2918] P. Y. Wu, "Products of Positive Semidefinite Matrices," *Lin. Alg. Appl.*, Vol. 111, pp. 53–61, 1988. (Cited on pp. 608, 609, and 831.)

[2919] P. Y. Wu, "The Operator Factorization Problems," *Lin. Alg. Appl.*, Vol. 117, pp. 35–63, 1989. (Cited on pp. 608, 609, 610, and 1228.)

[2920] P. Y. Wu, "Sums of Idempotent Matrices," *Lin. Alg. Appl.*, Vol. 142, pp. 43–54, 1990. (Cited on p. 619.)

[2921] S. Wu, "Improvement of Aczel's Inequality and Popoviciu's Inequality," *J. Ineq. Appl.*, pp. 1–9, 2007, article ID 72173. (Cited on p. 222.)

[2922] S. Wu, "A Sharpened Version of the Fundamental Triangle Inequality," *Math. Ineq. Appl.*, Vol. 11, pp. 477–482, 2008. (Cited on pp. 452 and 472.)

[2923] S. Wu, "Note on a Conjecture of R. A. Satnoianu," *Math. Ineq. Appl.*, Vol. 12, pp. 147–151, 2009. (Cited on p. 214.)

[2924] S.-H. Wu and Z.-H. Zhang, "A Class of Inequalities Related to the Angle Bisectors and the Sides of a Triangle," *J. Ineq. Pure Appl. Math.*, Vol. 7, no. 3, pp. 1–7, 2006, Article 108. (Cited on p. 480.)

[2925] Y.-D. Wu, Z.-H. Zhang, and X.-G. Chu, "On a Geometric Inequality by J. Sandor," *J. Ineq. Pure Appl. Math.*, Vol. 10, pp. 1–8, 2009, article 118. (Cited on pp. 161, 452, and 472.)

[2926] Y.-D. Wu, Z.-H. Zhang, and V. Lokesha, "Sharpening on Mircea's Inequality," *J. Ineq. Pure Appl. Math.*, Vol. 8, no. 4, pp. 1–6, 2007, Article 116. (Cited on p. 452.)

[2927] Y.-L. Wu, "The Refinement and Reverse of a Geometric Inequality," *J. Ineq. Pure Appl. Math.*, Vol. 10, pp. 1–7, 2009, article 82. (Cited on pp. 148 and 452.)

[2928] Y.-L. Wu, "The Refinement and Reverse of a Geometric Inequality," *J. Ineq. Pure Appl. Math.*, Vol. 10, no. 3, pp. 1–7, 2009, Article 82. (Cited on p. 466.)

[2929] J. X. Xiang, "A Note on the Cauchy-Schwarz Inequality," *Amer. Math. Monthly*, Vol. 120, pp. 456–459, 2013. (Cited on p. 221.)

[2930] Z.-G. Xiao and Z.-H. Zhang, "The Inequalities $G \leq L \leq I \leq A$ in n Variables," *J. Ineq. Pure. Appl. Math.*, Vol. 4, no. 2/39, pp. 1–6, 2003. (Cited on p. 208.)

[2931] L. Xin and L. Xiaoxue, "A Reciprocal Sum Related to the Riemann ζ-Function," *J. Math. Ineq.*, Vol. 11, pp. 209–215, 2017. (Cited on p. 1079.)

[2932] C. Xu, "Bellman's Inequality," *Lin. Alg. Appl.*, Vol. 229, pp. 9–14, 1995. (Cited on p. 766.)

[2933] C. Xu, Z. Xu, and F. Zhang, "Revisiting Hua-Marcus-Bellman-Ando Inequalities on Contractive Matrices," *Lin. Alg. Appl.*, Vol. 430, pp. 1499–1508, 2009. (Cited on pp. 271, 350, 757, 780, and 855.)

[2934] H. Xu, "Two Results about the Matrix Exponential," *Lin. Alg. Appl.*, Vol. 262, pp. 99–109, 1997. (Cited on pp. 1185 and 1228.)

[2935] X. Xu and C. He, "Inequalities for Eigenvalues of Matrices," *J. Ineq. Appl.*, pp. 1–6, 2013, article ID 6. (Cited on pp. 809 and 874.)

[2936] V. A. Yakubovich and V. M. Starzhinskii, *Linear Differential Equations with Periodic Coefficients, Vol. 1*. New York: John Wiley and Sons, 1975. (Cited on p. 1307.)

[2937] Y. Yamamoto and M. Lin, "On a Matrix Equality," *J. East China Normal Univ. (Natural Science)*, Vol. 1, pp. 39–43, 2010. (Cited on p. 328.)

[2938] T. Yamazaki, "An Extension of Specht's Theorem via Kantorovich Inequality and Related Results," *Math. Ineq. Appl.*, Vol. 3, pp. 89–96, 2000. (Cited on p. 206.)
[2939] T. Yamazaki, "Further Characterizations of Chaotic Order via Specht's Ratio," *Math. Ineq. Appl.*, Vol. 3, pp. 259–268, 2000. (Cited on p. 814.)
[2940] K. Yan, "A Direct and Simple Proof of Jacobi Identities for Determinants," arXiv:0712.1932v1. (Cited on pp. 341 and 342.)
[2941] Z. Yan, "New Representations of the Moore-Penrose Inverse of 2×2 Block Matrices," *Lin. Alg. Appl.*, Vol. 456, pp. 3–15, 2014. (Cited on p. 661.)
[2942] H. Yanai, K. Takeuchi, and Y. Takane, *Projection Matrices, Generalized Inverse Matrices, and Singular Value Decomposition*. New York: Springer, 2011. (Cited on p. 410.)
[2943] A. L. Yandl and C. Swenson, "A Class of Matrices with Zero Determinant," *Math. Mag.*, Vol. 85, pp. 126–130, 2012. (Cited on p. 613.)
[2944] B. Yang, "On a New Inequality Similar to Hardy-Hilbert's Inequality," *Math. Ineq. Appl.*, Vol. 6, pp. 37–44, 2003. (Cited on p. 220.)
[2945] B. Yang and T. M. Rassias, "On the Way of Weight Coefficient and Research for the Hilbert-Type Inequalities," *Math. Ineq. Appl.*, Vol. 6, pp. 625–658, 2003. (Cited on p. 220.)
[2946] J.-H. Yang and F.-Z. Zhao, "Sums Involving the Inverses of Binomial Coefficients," *J. Integer Sequences*, Vol. 9, pp. 1–11, 2006, article 06.4.2. (Cited on p. 1141.)
[2947] J.-H. Yang and F.-Z. Zhao, "Certain Sums Involving Inverses of Binomial Coefficients and Some Integrals," *J. Integer Sequences*, Vol. 10, pp. 1–11, 2007, article 07.8.7. (Cited on p. 1097.)
[2948] X. Yang, "On Carleman's Inequality," *J. Math. Anal. Appl.*, Vol. 253, pp. 691–694, 2001. (Cited on p. 128.)
[2949] X. Yang, "Necessary Conditions of Hurwitz Polynomials," *Lin. Alg. Appl.*, Vol. 359, pp. 21–27, 2003. (Cited on p. 1224.)
[2950] X. Yang, "Some Necessary Conditions for Hurwitz Stability," *Automatica*, Vol. 40, pp. 527–529, 2004. (Cited on p. 1224.)
[2951] X. M. Yang, "A Matrix Trace Inequality," *J. Math. Anal. Appl.*, Vol. 263, pp. 327–331, 2001. (Cited on p. 766.)
[2952] Z.-H. Yang, "Refinements of a Two-sided Inequality for Trigonometric Functions," *J. Math. Ineq.*, Vol. 7, pp. 601–615, 2013. (Cited on pp. 249 and 1170.)
[2953] Z.-H. Yang and Y.-M. Chu, "Inequalities for Certain Means in Two Arguments," *J. Ineq. Appl.*, pp. 1–11, 2016, article 299. (Cited on pp. 143 and 266.)
[2954] Z. P. Yang and X. X. Feng, "A Note on the Trace Inequality for Products of Hermitian Matrix Power," *J. Ineq. Pure. Appl. Math.*, Vol. 3, no. 5/78, pp. 1–12, 2002. (Cited on pp. 592 and 766.)
[2955] S. F. Yau and Y. Bresler, "A Generalization of Bergstrom's Inequality and Some Applications," *Lin. Alg. Appl.*, Vol. 161, pp. 135–151, 1992. (Cited on pp. 824 and 828.)
[2956] K. Ye and L.-H. Lim, "Every Matrix Is a Product of Toeplitz Matrices," *Found. Comput. Math.*, Vol. 16, pp. 577–598, 2017. (Cited on p. 610.)
[2957] S. Yin, "A Note on Generalized Cauchy-Schwarz Inequality," *J. Math. Ineq.*, Vol. 11, pp. 891–895, 2017. (Cited on p. 869.)
[2958] D. M. Young, *Iterative Solution of Large Linear Systems*. New York: Academic Press, 1971, reprinted by Dover, Mineola, 2003. (Cited on p. xxi.)
[2959] N. J. Young, "Matrices Which Maximise Any Analytic Function," *Acta Math. Acad. Scient. Hung.*, Vol. 34, pp. 239–243, 1979. (Cited on p. 901.)
[2960] R. M. Young, "On the Area Enclosed by the Curve $x^4 + y^4 = 1$," *Math. Gaz.*, Vol. 93, pp. 295–299, 2009. (Cited on pp. 1112 and 1114.)
[2961] J. Zacharias, "Problem 1072," *Math. Mag.*, Vol. 90, pp. 232, 237, 238, 2017. (Cited on p. 239.)
[2962] K. Zacharias, "Some Intriguing Definite Integrals," 1999, Weierstrass-Institut Preprint No. 493. (Cited on p. 1154.)
[2963] D. Zagier, "A One-sentence Proof that Every Prime $p \equiv 1 \pmod 4$ Is a Sum of Two Squares," *Amer. Math. Monthly*, Vol. 97, p. 144, 1990. (Cited on p. 35.)
[2964] L. Zalcman, "A Tale of Three Theorems," *Amer. Math. Monthly*, Vol. 123, pp. 643–656, 2016. (Cited on p. 947.)
[2965] R. Zamfir, "Refining Some Inequalities," *J. Ineq. Pure Appl. Math.*, Vol. 9, no. 3, pp. 1–5, 2008, Article 77. (Cited on p. 1237.)

[2966] R. Zamfir, "Bounds for Polynomial Roots Using Powers of the Generalized Frobenius Companion Matrix," *Gazeta Matematica Serie A*, Vol. 31, pp. 1–9, 2013. (Cited on p. 610.)

[2967] A. Zanna and H. Z. Munthe-Kaas, "Generalized Polar Decompositions for the Approximation of the Matrix Exponential," *SIAM J. Matrix Anal. Appl.*, Vol. 23, pp. 840–862, 2002. (Cited on p. 1209.)

[2968] D. Zeilberger, "$\binom{5}{2}$ Proofs that $\binom{n}{k} \leq \binom{n}{k+1}$," http://www.math.rutgers.edu/zeilberg/mamarim/ mamarimhtml/trivial.html. (Cited on p. 94.)

[2969] D. Zeilberger, "A Combinatorial Approach to Matrix Algebra," *Disc.Math.*, Vol. 56, pp. 61–72, 1985. (Cited on p. 362.)

[2970] D. Zeilberger, "Theorems for a Price: Tomorrow's Semi-Rigorous Mathematical Culture," *Math. Intelligencer*, Vol. 16, no. 4, pp. 11–18, 1994. (Cited on pp. 33, 122, 1021, and 1022.)

[2971] D. Zeilberger, "Proof of the Alternating Sign Matrix Conjecture," *Elec. J. Combinatorics*, Vol. 3, pp. 1–84, 1996. (Cited on p. 617.)

[2972] S. Zhan, "On the Determinantal Inequalities," *J. Ineq. Pure Appl. Math.*, Vol. 6, pp. 1–7, 2005, article 105. (Cited on p. 594.)

[2973] X. Zhan, "Inequalities for the Singular Values of Hadamard Products," *SIAM J. Matrix Anal. Appl.*, Vol. 18, pp. 1093–1095, 1997. (Cited on p. 907.)

[2974] X. Zhan, "Inequalities for Unitarily Invariant Norms," *SIAM J. Matrix Anal. Appl.*, Vol. 20, pp. 466–470, 1998. (Cited on p. 728.)

[2975] X. Zhan, "Singular Values of Differences of Positive Semidefinite Matrices," *SIAM J. Matrix Anal. Appl.*, Vol. 22, pp. 819–823, 2000. (Cited on pp. 807 and 808.)

[2976] X. Zhan, "Span of the Orthogonal Orbit of Real Matrices," *Lin. Multilin. Alg.*, Vol. 49, pp. 337–346, 2001. (Cited on p. 618.)

[2977] X. Zhan, *Matrix Inequalities*. New York: Springer, 2002. (Cited on pp. 362, 584, 716, 721, 728, 738, 802, 803, 812, 819, 820, 825, 827, 828, 830, 861, 874, 904, and 1214.)

[2978] X. Zhan, "On Some Matrix Inequalities," *Lin. Alg. Appl.*, Vol. 376, pp. 299–303, 2004. (Cited on p. 808.)

[2979] X. Zhan, *Matrix Theory*. Providence: American Mathematical Society, 2013. (Cited on pp. 326, 350, 443, 495, 535, 539, 540, 541, 542, 543, 544, 576, 578, 579, 583, 596, 687, 694, 711, 777, 788, 805, 807, 839, 847, 849, 867, 873, 882, 892, 893, 894, 895, 899, 901, 905, 949, 950, 1235, and 1237.)

[2980] F. Zhang, *Linear Algebra: Challenging Problems for Students*. Baltimore: Johns Hopkins University Press, 1996. (Cited on pp. 331, 346, 347, 429, and 540.)

[2981] F. Zhang, "Quaternions and Matrices of Quaternions," *Lin. Alg. Appl.*, Vol. 251, pp. 21–57, 1997. (Cited on pp. 438 and 440.)

[2982] F. Zhang, "A Compound Matrix with Positive Determinant," *Amer. Math. Monthly*, Vol. 105, p. 958, 1998. (Cited on p. 359.)

[2983] F. Zhang, *Matrix Theory: Basic Results and Techniques*. New York: Springer, 1999. (Cited on pp. 326, 327, 421, 571, 603, 615, 737, 745, 755, 765, 780, 830, and 855.)

[2984] F. Zhang, "Schur Complements and Matrix Inequalities in the Lowner Ordering," *Lin. Alg. Appl.*, Vol. 321, pp. 399–410, 2000. (Cited on p. 780.)

[2985] F. Zhang, "Matrix Inequalities by Means of Block Matrices," *Math. Ineq. Appl.*, Vol. 4, pp. 481–490, 2001. (Cited on pp. 732, 751, 752, 783, 801, 815, 822, 823, 828, and 829.)

[2986] F. Zhang, "Inequalities Involving Square Roots," *IMAGE*, Vol. 29, pp. 33–34, 2002. (Cited on p. 738.)

[2987] F. Zhang, "Block Matrix Techniques," in *The Schur Complement and Its Applications*, F. Zhang, Ed. New York: Springer, 2004, pp. 83–110. (Cited on pp. 733, 761, 772, 783, 784, 792, 794, 798, and 800.)

[2988] F. Zhang, "A Matrix Identity on the Schur Complement," *Lin. Multilin. Alg.*, Vol. 52, pp. 367–373, 2004. (Cited on p. 669.)

[2989] F. Zhang, "On the Bohr Inequality of Operator," *J. Math. Anal. Appl.*, Vol. 333, pp. 1264–1271, 2007. (Cited on pp. 273, 748, and 856.)

[2990] F. Zhang, *Linear Algebra: Challenging Problems for Students*, 2nd ed. Baltimore: Johns Hopkins University Press, 2009. (Cited on p. 736.)

[2991] F. Zhang, *Matrix Theory: Basic Results and Techniques*, 2nd ed. New York: Springer, 2011. (Cited on pp. 126, 193, 197, 216, 230, 314, 315, 316, 319, 320, 326, 327, 329, 330, 336, 338, 340, 350, 359, 360, 361, 362, 379, 380, 389, 390, 397, 407, 427, 428, 429, 441, 531, 537, 544, 566, 571, 576, 577, 579, 580, 582, 583, 584, 592, 593, 597, 600, 603, 604, 605, 610, 613, 615, 616, 617, 662, 686, 691,

694, 699, 700, 713, 714, 722, 723, 724, 725, 730, 731, 733, 735, 736, 737, 738, 740, 741, 742, 745, 750, 751, 752, 753, 754, 755, 756, 757, 760, 761, 762, 763, 764, 766, 771, 772, 773, 774, 776, 777, 778, 779, 780, 782, 785, 788, 792, 793, 794, 796, 797, 798, 799, 801, 802, 803, 804, 805, 807, 808, 809, 811, 812, 813, 821, 822, 823, 824, 825, 827, 828, 829, 830, 831, 852, 853, 861, 865, 866, 867, 870, 871, 872, 874, 881, 886, 889, 891, 892, 894, 895, 896, 898, 901, 903, 904, 905, 906, 907, 908, and 1242.)

[2992] F. Zhang, "An Inequality for a Contraction Matrix," *IMAGE*, Vol. 55, pp. 41–42, 2015. (Cited on p. 861.)

[2993] L. Zhang, "A Characterization of the Drazin Inverse," *Lin. Alg. Appl.*, Vol. 335, pp. 183–188, 2001. (Cited on p. 670.)

[2994] R. Zhang, "Some Optimal Elementary Inequalities on the Unit Disc," *Rocky Mountain J. Math.*, Vol. 41, pp. 1011–1023, 2011. (Cited on p. 269.)

[2995] Z. Zhang, Q. Song, and Z.-S. Wang, "Some Strengthened Results on Euler's Inequality," *Octogon*, Vol. 13, pp. 1010–1013, 2005. (Cited on pp. 452 and 466.)

[2996] Q.-C. Zhong, "*J*-Spectral Factorization of Regular Para-Hermitian Transfer Matrices," *Automatica*, Vol. 41, pp. 1289–1293, 2005. (Cited on p. 399.)

[2997] Q.-C. Zhong, *Robust Control of Time-delay Systems*. London: Springer, 2006. (Cited on p. 399.)

[2998] H. Zhou, "On Some Trace Inequalities for Positive Definite Hermitian Matrices," *J. Ineq. Appl.*, pp. 1–6, 2014, article ID 64. (Cited on p. 771.)

[2999] K. Zhou, *Robust and Optimal Control*. Upper Saddle River: Prentice-Hall, 1996. (Cited on pp. xxi, 619, 899, 902, and 1302.)

[3000] X. Zhou, "Some Generalizations of Aczel, Bellman's Inequalities and Related Power Sums," *J. Ineq. Appl.*, pp. 1–11, 2012, article 130. (Cited on p. 222.)

[3001] K. Zhu, H. Zhang, and K. Weng, "A Class of New Trigonometric Inequalities and Their Sharpenings," *J. Math. Ineq.*, Vol. 2, pp. 429–436, 2007. (Cited on pp. 466 and 484.)

[3002] L. Zhu, "On Young's Inequality," *Int. J. Math. Educ. Sci. Tech.*, Vol. 35, pp. 601–603, 2004. (Cited on p. 944.)

[3003] L. Zhu, "A New Simple Proof of Wilker's Inequality," *Math. Ineq. Appl.*, Vol. 8, pp. 749–750, 2005. (Cited on pp. 249 and 250.)

[3004] L. Zhu, "On Wilker-Type Inequalities," *Math. Ineq. Appl.*, Vol. 10, pp. 727–731, 2005. (Cited on pp. 249, 250, and 264.)

[3005] L. Zhu, "New Inequalities of Shafer-Fink Type for Arc Hyperbolic Sine," *J. Ineq. Appl.*, pp. 1–5, 2008, article 368275. (Cited on p. 261.)

[3006] L. Zhu, "Generalized Lazarevic's Inequality and Its Applications—Part II," *J. Ineq. Appl.*, pp. 1–4, 2009, article ID 379142. (Cited on p. 264.)

[3007] Y. Zi-zong, "Schur Complements and Determinant Inequalities," *J. Math. Ineq.*, Vol. 3, pp. 161–167, 2009. (Cited on pp. 742, 759, and 781.)

[3008] G. Zielke, "Some Remarks on Matrix Norms, Condition Numbers, and Error Estimates for Linear Equations," *Lin. Alg. Appl.*, Vol. 110, pp. 29–41, 1988. (Cited on pp. 862 and 909.)

[3009] S. Zimmerman and C. Ho, "On Infinitely Nested Radicals," *Math. Mag.*, Vol. 81, pp. 3–15, 2008. (Cited on pp. 972 and 973.)

[3010] J. Zinn, "A Sum Inequality," *Amer. Math. Monthly*, Vol. 113, p. 766, 2006. (Cited on p. 197.)

[3011] A. Zireh, "Inequalities for a Polynomial and Its Derivative," *J. Ineq. Appl.*, pp. 1–12, 2012, article Number 210. (Cited on pp. 942 and 949.)

[3012] X. Zitian and Z. Yibing, "A Best Approximation for the Constant *e* and an Improvement to Hardy's Inequality," *J. Math. Anal. Appl.*, Vol. 252, pp. 994–998, 2000. (Cited on p. 128.)

[3013] M. R. Zizovic and M. R. Stevanovic, "Some Inequalities for Altitudes and Other Elements of Triangle," in *Recent Progress in Inequalities*, G. V. Milovanovic, Ed. Dordrecht: Kluwer, 1998, pp. 505–510. (Cited on p. 480.)

[3014] S. Zlobec, "An Explicit Form of the Moore-Penrose Inverse of an Arbitrary Complex Matrix," *SIAM Rev.*, Vol. 12, pp. 132–134, 1970. (Cited on p. 633.)

[3015] L. Zou, "A Lower Bound for the Smallest Singular Value," *J. Math. Ineq.*, Vol. 6, pp. 625–629, 2012. (Cited on p. 896.)

[3016] L. Zou, "On Some Matrix Norm Inequalities," pp. 343–349, 2016. (Cited on p. 870.)

[3017] L. Zou and C. He, "On Some Inequalities for Unitarily Invariant Norms and Singular Values," *Lin. Alg. Appl.*, Vol. 436, pp. 3354–3361, 2012. (Cited on p. 875.)

[3018] L. Zou, C. He, and S. Qaisar, "Inequalities for Absolute Value Operators," *Lin. Alg. Appl.*, Vol. 438, pp. 436–442, 2013. (Cited on p. 747.)

[3019] L. Zou and Y. Jiang, "Improved Arithmetic-Geometric Mean Inequality and Its Application," *J. Math. Ineq.*, Vol. 9, pp. 107–111, 2015. (Cited on pp. 144 and 744.)

[3020] L. Zou and Y. Peng, "Some Trace Inequalities for Matrix Means," *J. Ineq. Appl.*, pp. 1–5, 2016, article 283. (Cited on p. 771.)

[3021] I. J. Zucker, "On the Series $\sum_{k=1}^{\infty}\binom{2k}{k}k^{-n}$ and Related Sums," *J. Number Theory*, Vol. 20, pp. 92–102, 1985. (Cited on p. 1070.)

[3022] J. Zucker and R. McPhedran, "More than Meets the Eye," *Amer. Math. Monthly*, Vol. 118, pp. 937–940, 2011. (Cited on pp. 1071, 1151, and 1152.)

[3023] K. Zuo, "Nonsingularity of the Difference and the Sum of Two Idempotent Matrices," *Lin. Alg. Appl.*, Vol. 433, pp. 476–482, 2010. (Cited on pp. 400 and 402.)

[3024] D. Zwillinger, *Standard Mathematical Tables and Formulae*, 32nd ed. Boca Raton: Chapman & Hall/CRC, 2012. (Cited on pp. 106, 107, 108, 114, 446, 491, 497, 519, 616, 1009, 1097, 1108, 1130, 1134, and 1168.)

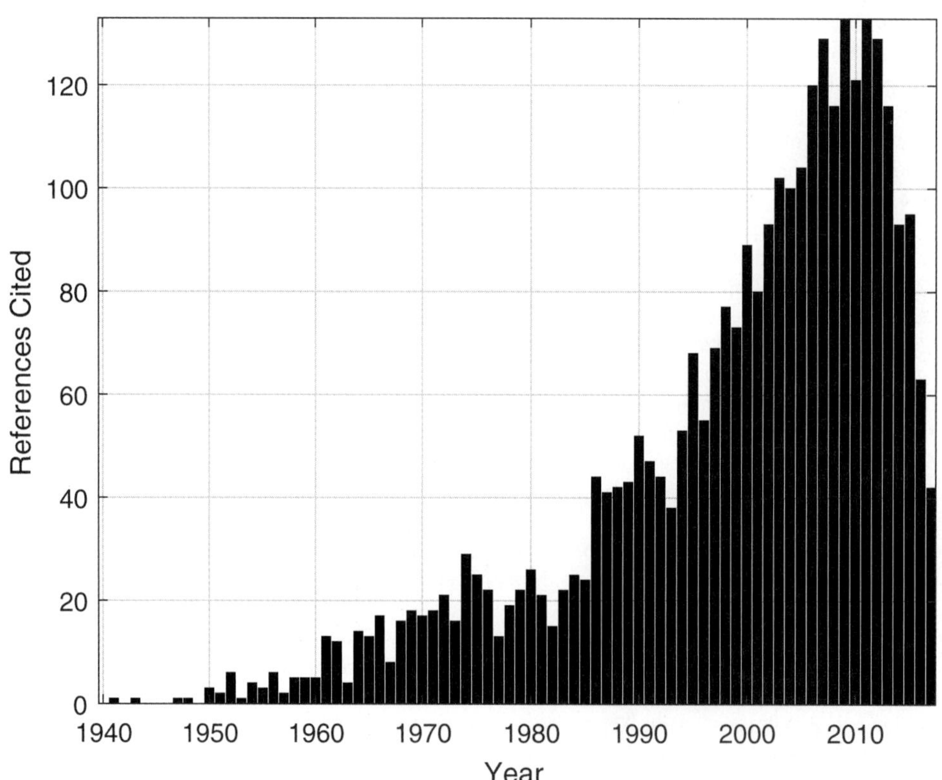

Author Index

Index

Symbols

B

C

D

E

F

G

H

I

J

K

L

M

N

P

Q

R

S

T

U

V

W